CHILTON®

ASIAN
SERVICE MANUAL
2008 EDITION
VOLUME II
Hyundai
Infiniti
Kia
Nissan

CENGAGE
Learning™

Australia • Brazil • Japan • Korea • Mexico • Singapore • Spain • United Kingdom • United States

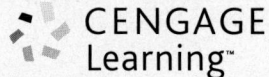
CENGAGE Learning™

CHILTON®
Asian Service Manual
2008 Edition
Volume II
Hyundai, Infiniti, Kia, Nissan

**Vice President,
Technology & Trades Professional
Business Unit:**
Gregory L. Clayton

**Publisher,
Technology & Trades Professional
Business Unit:**
David Koontz

Director of Marketing:
Beth A. Lutz

Marketing Manager:
Jennifer Stall

Marketing Assistant:
Rachael Conover

Production Director:
Carolyn Miller

Editorial Assistant:
Jason Yager

Production Manager:
Andrew Crouth

Publishing Coordinator:
Paula Baillie

Sr. Content Project Manager:
Elizabeth C. Hough

Managing Editor:
Terry L. Blomquist

Editors:
Dennis Bailey

Jim Bailey

Ken Burdette

Sherry Burdette

Jacques Gordon

Eugene F. Hannon, Jr.

John Howard

Doug Lee

David G. Olson

Christine Sheeky

Lance Williams

Graphical Designer:
Melinda Possinger

For more information contact:
Cengage Learning
Executive Woods
5 Maxwell Drive, PO Box 8007,
Clifton Park, NY 12065-8007
Visit us at **www.chilton.cengage.com**
Visit our corporate website at **www.cengage.com**
For permission to use material from
the text or product, contact us by
Tel. (800) 730-2214
Fax (800) 730-2215
www.cengage.com/permissions

Cengage Learning products are represented in Canada by Nelson Education, Ltd.

ISBN 10: 1-4283-2216-7
ISBN 13: 978-1-4283-2216-5
ISSN: 1939-621X

NOTICE TO THE READER

Publisher does not warrant or guarantee any of the products described herein or perform any independent analysis in connection with any of the product information contained herein. Publisher does not assume, and expressly disclaims, any obligation to obtain and include information other than that provided to it by the manufacturer.

The reader is expressly warned to consider and adopt all safety precautions that might be indicated by the activities herein and to avoid all potential hazards. By following the instructions contained herein, the reader willingly assumes all risks in connection with such instructions.

The publisher makes no representation or warranties of any kind, including but not limited to, the warranties of fitness for particular purpose or merchantability, nor are any such representations implied with respect to the material set forth herein, and the publisher takes no responsibility with respect to such material. The publisher shall not be liable for any special, consequential, or exemplary damages resulting, in whole or part, from the readers' use of, or reliance upon, this material.

Printed in the United States of America
1 2 3 4 5 xx 13 12 11 10 09 08

Table of Contents

Model Index

USING THIS INFORMATION

Organization

To find where a particular model section or procedure is located, look in the Table of Contents. Main topics are listed with the page number on which they may be found. Following the main topics is an alphabetical listing of all of the procedures within the section and their page numbers.

Manufacturer and Model Coverage

This product covers 2007–2008 Asian models that are produced in sufficient quantities to warrant coverage, and which have technical content available from the vehicle manufacturers before our publication date. Although this information is as complete as possible at the time of publication, some manufacturers may make changes which cannot be included here. While striving for total accuracy, the publisher cannot assume responsibility for any errors, changes, or omissions that may occur in the compilation of this data.

Part Numbers & Special Tools

Part numbers and special tools are recommended by the publisher and vehicle manufacturer to perform specific jobs. Before substituting any part or tool for the one recommended, you must be completely satisfied that neither your personal safety, nor the performance of the vehicle will be endangered.

ACKNOWLEDGEMENT

The publisher would like to express appreciation to the following vehicle manufacturers for their assistance in producing this manual. No further reproduction or distribution of the material in this manual is allowed without the expressed written permission of the vehicle manufacturers and the publisher. Hyundai Group, including Hyundai and Kia Motor, Nissan North America including Infiniti and Nissan divisions.

PRECAUTIONS

Before servicing any vehicle, please be sure to read all of the following precautions, which deal with personal safety, prevention of component damage, and important points to take into consideration when servicing a motor vehicle:

- Always wear safety glasses or goggles when drilling, cutting, grinding or prying.
- Steel-toed work shoes should be worn when working with heavy parts. Pockets should not be used for carrying tools. A slip or fall can drive a screwdriver into your body.
- Work surfaces, including tools and the floor should be kept clean of grease, oil or other slippery material.
- When working around moving parts, don't wear loose clothing. Long hair should be tied back under a hat or cap, or in a hair net.
- Always use tools only for the purpose for which they were designed. Never pry with a screwdriver.
- Keep a fire extinguisher and first aid kit handy.
- Always properly support the vehicle with approved stands or lift.
- Always have adequate ventilation when working with chemicals or hazardous material.
- Carbon monoxide is colorless, odorless and dangerous. If it is necessary to operate the engine with vehicle in a closed area such as a garage, always use an exhaust collector to vent the exhaust gases outside the closed area.
- When draining coolant, keep in mind that small children and some pets are attracted by ethylene glycol antifreeze, and are quite likely to drink any left in an open container, or in puddles on the ground. This will prove fatal in sufficient quantity. Always drain the coolant into a sealable container.

- To avoid personal injury, do not remove the coolant pressure relief cap while the engine is operating or hot. The cooling system is under pressure; steam and hot liquid can come out forcefully when the cap is loosened slightly. Failure to follow these instructions may result in personal injury. The coolant must be recovered in a suitable, clean container for reuse. If the coolant is contaminated it must be recycled or disposed of correctly.

- When carrying out maintenance on the starting system be aware that heavy gauge leads are connected directly to the battery. Make sure the protective caps are in place when maintenance is completed. Failure to follow these instructions may result in personal injury.

- Do not remove any part of the engine emission control system. Operating the engine without the engine emission control system will reduce fuel economy and engine ventilation. This will weaken engine performance and shorten engine life. It is also a violation of Federal law.

- Due to environmental concerns, when the air conditioning system is drained, the refrigerant must be collected using refrigerant recovery/recycling equipment. Federal law requires that refrigerant be recovered into appropriate recovery equipment and the process be conducted by qualified technicians who have been certified by an approved organization, such as MACS, ASI, etc. Use of a recovery machine dedicated to the appropriate refrigerant is necessary to reduce the possibility of oil and refrigerant incompatibility concerns. Refer to the instructions provided by the equipment manufacturer when removing refrigerant from or charging the air conditioning system.

- Always disconnect the battery ground when working on or around the electrical system.

- Batteries contain sulfuric acid. Avoid contact with skin, eyes, or clothing. Also, shield your eyes when working near batteries to protect against possible splashing of the acid solution. In case of acid contact with skin or eyes, flush immediately with water for a minimum of 15 minutes and get prompt medical attention. If acid is swallowed, call a physician immediately. Failure to follow these instructions may result in personal injury.

- Batteries normally produce explosive gases. Therefore, do not allow flames, sparks or lighted substances to come near the battery. When charging or working near a battery, always shield your face and protect your eyes. Always provide ventilation. Failure to follow these instructions may result in personal injury.

- When lifting a battery, excessive pressure on the end walls could cause acid to spew through the vent caps, resulting in personal injury, damage to the vehicle or battery. Lift with a battery carrier or with your hands on opposite corners. Failure to follow these instructions may result in personal injury.

- Observe all applicable safety precautions when working around fuel. Whenever servicing the fuel system, always work in a well-ventilated area. Do not allow fuel spray or vapors to come in contact with a spark, open flame, or excessive heat (a hot drop light, for example). Keep a dry chemical fire extinguisher near the work area. Always keep fuel in a container specifically designed for fuel storage; also, always properly seal fuel containers to avoid the possibility of fire or explosion. Do not smoke or carry lighted tobacco or open flame of any type when working on or near any fuel-related components.

• Fuel injection systems often remain pressurized, even after the engine has been turned OFF. The fuel system pressure must be relieved before disconnecting any fuel lines. Failure to do so may result in fire and/or personal injury.

• The evaporative emissions system contains fuel vapor and condensed fuel vapor. Although not present in large quantities, it still presents the danger of explosion or fire. Disconnect the battery ground cable from the battery to minimize the possibility of an electrical spark occurring, possibly causing a fire or explosion if fuel vapor or liquid fuel is present in the area. Failure to follow these instructions can result in personal injury.

• The EPA warns that prolonged contact with used engine oil may cause a number of skin disorders, including cancer! You should make every effort to minimize your exposure to used engine oil. Protective gloves should be worn when changing oil. Wash your hands and any other exposed skin areas as soon as possible after exposure to used engine oil. Soap and water, or waterless hand cleaner should be used.

• Some vehicles are equipped with an air bag system, often referred to as a Supple-

mental Restraint System (SRS) or Supplemental Inflatable Restraint (SIR) system. The system must be disabled before performing service on or around system components, steering column, instrument panel components, wiring and sensors. Failure to follow safety and disabling procedures could result in accidental air bag deployment, possible personal injury and unnecessary system repairs.

• Always wear safety goggles when working with, or around, the air bag system. When carrying a non-deployed air bag, be sure the bag and trim cover are pointed away from your body. When placing a non-deployed air bag on a work surface, always face the bag and trim cover upward, away from the surface. This will reduce the motion of the module if it is accidentally deployed.

• Electronic modules are sensitive to electrical charges. The ABS module can be damaged if exposed to these charges.

• Brake pads and shoes may contain asbestos, which has been determined to be a cancer-causing agent. Never clean brake surfaces with compressed air. Avoid inhaling brake dust. Clean all brake surfaces with a commercially available brake cleaning fluid.

• When replacing brake pads, shoes, discs or drums, replace them as complete axle sets.

• When servicing drum brakes, disassemble and assemble one side at a time, leaving the remaining side intact for reference.

• Brake fluid often contains polyglycol ethers and polyglycols. Avoid contact with the eyes and wash your hands thoroughly after handling brake fluid. If you do get brake fluid in your eyes, flush your eyes with clean, running water for 15 minutes. If eye irritation

persists, or if you have taken brake fluid internally, immediately seek medical assistance.

• Clean, high quality brake fluid from a sealed container is essential to the safe and proper operation of the brake system. You should always buy the correct type of brake fluid for your vehicle. If the brake fluid becomes contaminated, completely flush the system with new fluid. Never reuse any brake fluid. Any brake fluid that is removed from the system should be discarded. Also, do not allow any brake fluid to come in contact with a painted or plastic surface; it will damage the paint.

• Never operate the engine without the proper amount and type of engine oil; doing so will result in severe engine damage.

• Timing belt maintenance is extremely important! Many models utilize an interference-type, non-freewheeling engine. If the timing belt breaks, the valves in the cylinder head may strike the pistons, causing potentially serious (also time-consuming and expensive) engine damage.

• Disconnecting the negative battery cable on some vehicles may interfere with the functions of the on-board computer system (s) and may require the computer to undergo a relearning process once the negative battery cable is reconnected.

• Steering and suspension fasteners are critical parts because they affect performance of vital components and systems and their failure can result in major service expense. They must be replaced with the same grade or part number or an equivalent part if replacement is necessary. Do not use a replacement part of lesser quality or substitute design. Torque values must be used as specified during reassembly to ensure proper retention of these parts.

HYUNDAI

Accent • Azera • Elantra • Sonata • Tiburon

SPECIFICATIONS AND MAINTENANCE CHARTS

ENGINE AND VEHICLE IDENTIFICATION

	Engine							Model Year	
Code ①	Liters (cc)	Cu. In.	Cyl.	Fuel Sys.	Engine Type	Eng. Mfg.		Code ②	Year
C	1.6 (1599)	97.57	I4	MPFI	DOHC	Hyundai		6	2006
D	2.0 (1975)	120.52	I4	MPFI	DOHC	Hyundai		7	2007
C	2.4 (2359)	143.90	I4	MPFI	DOHC	Hyundai		8	2008
F	2.7 (2656)	164.30	V6	MPFI	DOHC	Hyundai			
D	3.3 (3342)	203.86	V6	MPFI	DOHC	Hyundai			
F	3.3 (3342)	203.86	V6	MPFI	DOHC	Hyundai			
F	3.8 (3778)	230.55	V6	MPFI	DOHC	Hyundai			

MPFI: Multi-Point Fuel Injection

DOHC: Double Overhead Camshafts

① 8th digit of VIN

② 10th digit of VIN

22140_HYUN_C0001

GENERAL ENGINE SPECIFICATIONS

All measurements are given in inches.

Year	Model	Engine Displacement Liters	Engine Series VIN	Net Horsepower @ rpm	Net Torque @ rpm (ft. lbs.)	Bore x Stroke (in.)	Com- pression Ratio	Oil Pressure @ rpm
2006	Accent	1.6	C	110@6000	106@4500	3.012 x 3.425	10.0:1	15.6@Idle
	Azera	3.8	F	263@6000	255@4500	3.780 x 3.743	10.4:1	18.8@1000
	Elantra	2.0	D	138@6000	136@4500	3.228 x 3.681	10.1:1	35.5@1500
	Sonata	2.4	C	162@5800	164@4250	3.464 x 3.819	10.5:1	NA
	Sonata	3.3	F	235@6000	226@3500	3.622 x 3.299	10.4:1	18.8@1000
	Tiburon	2.0	D	138@6000	136@4500	3.228 x 3.681	10.1:1	35.5@1500
	Tiburon	2.7	F	172@6000	181@3800	3.413 x 2.953	10.0:1	7.3@Idle
2007	Accent	1.6	C	110@6000	106@4500	3.012 x 3.425	10.0:1	15.6@Idle
	Azera	3.3	D	247@6000	238@3500	3.622 x 3.299	10.4:1	18.8@1000
	Azera	3.8	F	263@6000	255@4500	3.780 x 3.743	10.4:1	18.8@1000
	Elantra	2.0	D	138@6000	136@4500	3.228 x 3.681	10.1:1	35.5@1500
	Sonata	2.4	C	162@5800	164@4250	3.464 x 3.819	10.5:1	NA
	Sonata	3.3	F	235@6000	226@3500	3.622 x 3.299	10.4:1	18.8@1000
	Tiburon	2.0	D	138@6000	136@4500	3.228 x 3.681	10.1:1	35.5@1500
	Tiburon	2.7	F	172@6000	181@3800	3.413 x 2.953	10.0:1	7.3@Idle
2008	Accent	1.6	C	110@6000	106@4500	3.012 x 3.425	10.0:1	15.6@Idle
	Azera	3.3	D	247@6000	238@3500	3.622 x 3.299	10.4:1	18.8@1000
	Azera	3.8	F	263@6000	255@4500	3.780 x 3.743	10.4:1	18.8@1000
	Elantra	2.0	D	138@6000	136@4500	3.228 x 3.681	10.1:1	35.5@1500
	Sonata	2.4	C	162@5800	164@4250	3.464 x 3.819	10.5:1	NA
	Sonata	3.3	F	235@6000	226@3500	3.622 x 3.299	10.4:1	18.8@1000
	Tiburon	2.0	D	138@6000	136@4500	3.228 x 3.681	10.1:1	35.5@1500
	Tiburon	2.7	F	172@6000	181@3800	3.413 x 2.953	10.0:1	7.3@Idle

NA: Not Available

22140_HYUN_C0002

GASOLINE ENGINE TUNE-UP SPECIFICATIONS

Year	Engine Displacement Liters	Engine VIN	Spark Plug Gap (in.)	Ignition Timing (deg.) MT	AT	Fuel Pump (psi)	Idle Speed (rpm) MT	AT	Valve Clearance In.	Ex.
2006	1.6	C	0.039-0.043	①	①	49.8	②	②	HYD	HYD
	2.0	D	0.039-0.043	①	①	49.0-50.5	②	②	HYD	HYD
	2.4	C	0.039-0.043	①	①	49.8	②	②	HYD	HYD
	2.7	F	0.039-0.043	①	①	49.8	②	②	HYD	HYD
	3.3	F	0.039-0.043	①	①	54.3-55.8	②	②	HYD	HYD
	3.8	F	0.039-0.043	①	①	54.3-55.8	②	②	HYD	HYD
2007	1.6	C	0.039-0.043	①	①	49.8	②	②	HYD	HYD
	2.0	D	0.039-0.043	①	①	49.0-50.5	②	②	HYD	HYD
	2.4	C	0.039-0.043	①	①	49.8	②	②	HYD	HYD
	2.7	F	0.039-0.043	①	①	49.8	②	②	HYD	HYD
	3.3	D	0.039-0.043	①	①	54.3-55.8	②	②	HYD	HYD
	3.3	F	0.039-0.043	①	①	54.3-55.8	②	②	HYD	HYD
	3.8	F	0.039-0.043	①	①	54.3-55.8	②	②	HYD	HYD
2008	1.6	C	0.039-0.043	①	①	49.8	②	②	HYD	HYD
	2.0	D	0.039-0.043	①	①	49.0-50.5	②	②	HYD	HYD
	2.4	C	0.039-0.043	①	①	49.8	②	②	HYD	HYD
	2.7	F	0.039-0.043	①	①	49.8	②	②	HYD	HYD
	3.3	D	0.039-0.043	①	①	54.3-55.8	②	②	HYD	HYD
	3.3	F	0.039-0.043	①	①	54.3-55.8	②	②	HYD	HYD
	3.8	F	0.039-0.043	①	①	54.3-55.8	②	②	HYD	HYD

NOTE: The Vehicle Emission Control Information label reflects specification changes made during production.

Follow the figures on the label if they differ from those in this chart.

HYD: Hydraulic

① Ignition timing is preset and cannot be adjusted

② Idle speed is maintained by the Electronic Control Module (ECM)

22140_HYUN_C0003

CAPACITIES

Year	Model	Engine Displacement Liters	Engine VIN	Engine Oil with Filter (qts.)	Transmission (pts.)		Fuel Tank (gal.)	Cooling System (qts.)
					Manual	Auto. ①		
2006	Accent	1.6	C	3.5	4.2	12.9	11.9	5.8-6.1
	Azera	3.8	F	6.8	—	11.5	19.8	9.1
	Elantra	2.0	D	4.3	4.2	13.8	14.0	6.9-7.0
	Sonata	2.4	C	4.2	4.6	16.4	18.5	8.2
	Sonata	3.3	F	6.3	—	11.5	17.7	9.4
	Tiburon	2.0	D	4.2	4.6	16.4	14.5	7.7
	Tiburon	2.7	F	4.8	4.6	16.4	14.5	9.1
2007	Accent	1.6	C	3.5	4.2	12.9	11.9	5.8-6.1
	Azera	3.3	D	6.8	—	11.5	19.8	9.1
	Azera	3.8	F	6.8	—	11.5	19.8	9.1
	Elantra	2.0	D	4.3	4.2	13.8	14.0	6.9-7.0
	Sonata	2.4	C	4.2	4.6	16.4	18.5	8.2
	Sonata	3.3	F	6.3	—	11.5	17.7	9.4
	Tiburon	2.0	D	4.2	4.6	16.4	14.5	7.7
	Tiburon	2.7	F	4.8	4.6	16.4	14.5	9.1
2008	Accent	1.6	C	3.5	4.2	12.9	11.9	5.8-6.1
	Azera	3.3	D	6.8	—	11.5	19.8	9.1
	Azera	3.8	F	6.8	—	11.5	19.8	9.1
	Elantra	2.0	D	4.3	4.2	13.8	14.0	6.9-7.0
	Sonata	2.4	C	4.2	4.6	16.4	18.5	8.2
	Sonata	3.3	F	6.3	—	11.5	17.7	9.4
	Tiburon	2.0	D	4.2	4.6	16.4	14.5	7.7
	Tiburon	2.7	F	4.8	4.6	16.4	14.5	9.1

NOTE: All capacities are approximate. Add fluid gradually and check to be sure a proper fluid level is obtained.

① Drain and refill

22140_HYUN_C0004

FLUID SPECIFICATIONS

Year	Model	Engine Displacement Liters	Engine ID/VIN	Engine Oil	Auto. Trans.	Manual Trans.	Power Steering Fluid	Brake Master Cylinder
2006	Accent	1.6	C	5W-20	①	②	PSF-3	③
	Azera	3.8	F	5W-20	①	—	PSF-3	③
	Elantra	2.0	D	5W-20	①	②	PSF-3	③
	Sonata	2.4	C	5W-20	①	②	PSF-3	③
	Sonata	3.3	F	5W-20	①	②	PSF-3	③
	Tiburon	2.0	D	5W-20	①	②	PSF-3	③
	Tiburon	2.7	F	5W-20	①	②	PSF-3	③
2007	Accent	1.6	C	5W-20	①	②	PSF-3	③
	Azera	3.3	D	5W-20	①	—	PSF-3	③
	Azera	3.8	F	5W-20	①	—	PSF-3	③
	Elantra	2.0	D	5W-20	①	②	PSF-3	③
	Sonata	2.4	C	5W-20	①	②	PSF-3	③
	Sonata	3.3	F	5W-20	①	②	PSF-3	③
	Tiburon	2.0	D	5W-20	①	②	PSF-3	③
	Tiburon	2.7	F	5W-20	①	②	PSF-3	③
2008	Accent	1.6	C	5W-20	①	②	PSF-3	③
	Azera	3.3	D	5W-20	①	—	PSF-3	③
	Azera	3.8	F	5W-20	①	—	PSF-3	③
	Elantra	2.0	D	5W-20	①	②	PSF-3	③
	Sonata	2.4	C	5W-20	①	②	PSF-3	③
	Sonata	3.3	F	5W-20	①	②	PSF-3	③
	Tiburon	2.0	D	5W-20	①	②	PSF-3	③
	Tiburon	2.7	F	5W-20	①	②	PSF-3	③

DOT: Department Of Transportation

① DIAMOND ATF SP-III, SK ATF SP-III

② GENUINE PART MTF 75W/85 (API GL - 4)

③ DOT 3, DOT 4, or equivalent

22140_HYUN_C0005

VALVE SPECIFICATIONS

Year	Engine Displacement Liters	Engine VIN	Seat Angle (deg.)	Face Angle (deg.)	Spring Test Pressure (lbs. @ in.)	Spring Installed Height (in.)	Stem-to-Guide Clearance (in.)		Stem Diameter (in.)	
							Intake	Exhaust	Intake	Exhaust
2006	1.6	C	45	45	94.5-104.3 @1.071	NA	0.0008-0.0020	0.0014-0.0026	0.2344-0.2350	0.2348-0.2354
	2.0	D	45	NA	87.1-93.7 @1.201	1.535	0.0008-0.0019	0.0014-0.0026	0.2348-0.2354	0.2343-0.2348
	2.4	C	44.75-45.10	45.25-45.75	85.1-90.4 @1.024	NA	0.0008-0.0019	0.0012-0.0021	0.2151-0.2157	0.2149-0.2153
	2.7	F	NA	45	48.4 @1.378	NA	0.0008-0.0020	0.0012-0.0026	0.2350-0.2354	0.2304-0.2350
	3.3	F	44.75-45.20	45.25-45.75	90.4-96.2 @0.953	NA	0.0008-0.0019	0.0012-0.0021	0.2151-0.2157	0.2149-0.2153
	3.8	F	44.75-45.20	45.25-45.75	90.4-96.2 @0.953	NA	0.0008-0.0019	0.0012-0.0021	0.2151-0.2157	0.2149-0.2153
2007	1.6	C	45	45	94.5-104.3 @1.071	NA	0.0008-0.0020	0.0014-0.0026	0.2344-0.2350	0.2348-0.2354
	2.0	D	45	NA	87.1-93.7 @1.201	1.535	0.0008-0.0019	0.0014-0.0026	0.2348-0.2354	0.2343-0.2348
	2.4	C	44.75-45.10	45.25-45.75	85.1-90.4 @1.024	NA	0.0008-0.0019	0.0012-0.0021	0.2151-0.2157	0.2149-0.2153
	2.7	F	NA	45	48.4 @1.378	NA	0.0008-0.0020	0.0012-0.0026	0.2350-0.2354	0.2304-0.2350
	3.3	D	44.75-45.20	45.25-45.75	90.4-96.2 @0.953	NA	0.0008-0.0019	0.0012-0.0021	0.2151-0.2157	0.2149-0.2153
	3.3	F	44.75-45.20	45.25-45.75	90.4-96.2 @0.953	NA	0.0008-0.0019	0.0012-0.0021	0.2151-0.2157	0.2149-0.2153
	3.8	F	44.75-45.20	45.25-45.75	90.4-96.2 @0.953	NA	0.0008-0.0019	0.0012-0.0021	0.2151-0.2157	0.2149-0.2153
2008	1.6	C	45	45	94.5-104.3 @1.071	NA	0.0008-0.0020	0.0014-0.0026	0.2344-0.2350	0.2348-0.2354
	2.0	D	45	NA	87.1-93.7 @1.201	1.535	0.0008-0.0019	0.0014-0.0026	0.2348-0.2354	0.2343-0.2348
	2.4	C	44.75-45.10	45.25-45.75	85.1-90.4 @1.024	NA	0.0008-0.0019	0.0012-0.0021	0.2151-0.2157	0.2149-0.2153
	2.7	F	NA	45	48.4 @1.378	NA	0.0008-0.0020	0.0012-0.0026	0.2350-0.2354	0.2304-0.2350
	3.3	D	44.75-45.20	45.25-45.75	90.4-96.2 @0.953	NA	0.0008-0.0019	0.0012-0.0021	0.2151-0.2157	0.2149-0.2153
	3.3	F	44.75-45.20	45.25-45.75	90.4-96.2 @0.953	NA	0.0008-0.0019	0.0012-0.0021	0.2151-0.2157	0.2149-0.2153
	3.8	F	44.75-45.20	45.25-45.75	90.4-96.2 @0.953	NA	0.0008-0.0019	0.0012-0.0021	0.2151-0.2157	0.2149-0.2153

NA: Not Available

22140_HYUN_C0006

CAMSHAFT AND BEARING SPECIFICATIONS CHART
All measurements are given in inches.

Year	Engine Displ. Liters	Engine ID/VIN	Journal Dia.	Brg. Oil Clearance	Shaft End-play	Runout	Journal Bore	Lobe Height Intake	Lobe Height Exhaust
2006	1.6	C	1.0616-1.0622	0.0008-0.0024	0.0039-0.0079	NA	NA	1.7224-1.7303	1.7382-1.7460
	2.0	D	1.1023	0.0008-0.0024	0.0040-0.0079	NA	NA	1.7527-1.7566	1.7487-1.7527
	2.4	C	①	②	0.0039-0.0086	NA	NA	1.7244	1.7716
	2.7	F	1.0222-1.0228	0.0007-0.0024	0.0039-0.0059	NA	NA	1.7303-1.7382	1.7303-1.7382
	3.3	F	③	④	0.0022-0.0025	NA	NA	1.8228	1.8031
	3.8	F	③	④	0.0008-0.0071	NA	NA	1.8425	1.8031
2007	1.6	C	1.0616-1.0622	0.0008-0.0024	0.0039-0.0079	NA	NA	1.7224-1.7303	1.7382-1.7460
	2.0	D	1.1023	0.0008-0.0024	0.0040-0.0079	NA	NA	1.7527-1.7566	1.7487-1.7527
	2.4	C	①	②	0.0039-0.0086	NA	NA	1.7244	1.7716
	2.7	F	1.0222-1.0228	0.0007-0.0024	0.0039-0.0059	NA	NA	1.7303-1.7382	1.7303-1.7382
	3.3	D	③	④	0.0008-0.0071	NA	NA	1.8228	1.8031
	3.3	F	③	④	0.0022-0.0025	NA	NA	1.8228	1.8031
	3.8	F	③	④	0.0008-0.0071	NA	NA	1.8425	1.8031
2008	1.6	C	1.0616-1.0622	0.0008-0.0024	0.0039-0.0079	NA	NA	1.7224-1.7303	1.7382-1.7460
	2.0	D	1.1023	0.0008-0.0024	0.0040-0.0079	NA	NA	1.7527-1.7566	1.7487-1.7527
	2.4	C	①	②	0.0039-0.0086	NA	NA	1.7244	1.7716
	2.7	F	1.0222-1.0228	0.0007-0.0024	0.0039-0.0059	NA	NA	1.7303-1.7382	1.7303-1.7382
	3.3	D	③	④	0.0008-0.0071	NA	NA	1.8228	1.8031
	3.3	F	③	④	0.0022-0.0025	NA	NA	1.8228	1.8031
	3.8	F	③	④	0.0008-0.0071	NA	NA	1.8425	1.8031

NA: Not Available

① Intake No. 1 is 1.1811 inch
　Intake No. 2, 3, 4, 5 are 0.9449 inch
　Exhaust No.1 is 1.5748 inch
　Exhaust No. 2, 3, 4, 5 are 0.9449 inch
② Intake No. 1 is 0.0008-0.0022 inch
　Intake No. 2, 3, 4, 5 are 0.0018-0.0032 inch
　Exhaust No. 1, 2, 3, 4, 5 are 0.0018-0.0032 inch

③ Intake No. 1 is 1.1009-1.1016 inch
　Intake No. 2, 3, 4 are 0.9430-0.9437 inch
　Exhaust No.1 is 1.1009-1.1016 inch
　Exhaust No. 2, 3, 4 are 0.9430-0.9437 inch
④ Intake No. 1 is 0.0008-0.0022 inch
　Intake No. 2, 3, 4 are 0.0012-0.0026 inch
　Exhaust No.1 is 0.0008-0.0022 inch
　Exhaust No. 2. 3. 4 are 0.0012-0.0026 inch

CRANKSHAFT AND CONNECTING ROD SPECIFICATIONS

All measurements are given in inches.

Year	Engine Displacement Liters	Engine VIN	Crankshaft				Connecting Rod		
			Main Brg. Journal Dia.	Main Brg. Oil Clearance	Shaft End-play	Thrust on No.	Journal Diameter	Oil Clearance	Side Clearance
2006	1.6	C	1.9665-1.9672	①	0.0020-0.0069	3	1.8898-1.8905	0.0007-0.0014	0.0039-0.0098
	2.0	D	2.2418-2.2426	0.0011-0.0018	0.0023-0.0100	3	1.7695-1.7703	0.0009-0.0017	0.0039-0.0100
	2.4	C	2.0449-2.0456	0.0010-0.0019	0.0027-0.0098	3	1.8879-1.8886	0.0011-0.0018	0.0039-0.0100
	2.7	F	2.4402-2.4409	0.0002-0.0009	0.0028-0.0098	3	1.8891-1.8898	0.0007-0.0014	0.0039-0.0098
	3.3	F	2.7142-2.7149	0.0008-0.0016	0.0039-0.0110	3	2.1635-2.1642	0.0012-0.0019	0.0039-0.0098
	3.8	F	2.7142-2.7149	0.0008-0.0016	0.0039-0.0110	3	2.1635-2.1642	0.0015-0.0022	0.0039-0.0098
2007	1.6	C	1.9665-1.9672	①	0.0020-0.0069	3	1.8898-1.8905	0.0007-0.0014	0.0039-0.0098
	2.0	D	2.2418-2.2426	0.0011-0.0018	0.0023-0.0100	3	1.7695-1.7703	0.0009-0.0017	0.0039-0.0100
	2.4	C	2.0449-2.0456	0.0010-0.0019	0.0027-0.0098	3	1.8879-1.8886	0.0011-0.0018	0.0039-0.0100
	2.7	F	2.4402-2.4409	0.0002-0.0009	0.0028-0.0098	3	1.8891-1.8898	0.0007-0.0014	0.0039-0.0098
	3.3	D	2.7142-2.7149	0.0008-0.0016	0.0039-0.0110	3	2.1635-2.1642	0.0015-0.0022	0.0039-0.0098
	3.3	F	2.7142-2.7149	0.0008-0.0016	0.0039-0.0110	3	2.1635-2.1642	0.0012-0.0019	0.0039-0.0098
	3.8	F	2.7142-2.7149	0.0008-0.0016	0.0039-0.0110	3	2.1635-2.1642	0.0015-0.0022	0.0039-0.0098
2008	1.6	C	1.9665-1.9672	①	0.0020-0.0069	3	1.8898-1.8905	0.0007-0.0014	0.0039-0.0098
	2.0	D	2.2418-2.2426	0.0011-0.0018	0.0023-0.0100	3	1.7695-1.7703	0.0009-0.0017	0.0039-0.0100
	2.4	C	2.0449-2.0456	0.0010-0.0019	0.0027-0.0098	3	1.8879-1.8886	0.0011-0.0018	0.0039-0.0100
	2.7	F	2.4402-2.4409	0.0002-0.0009	0.0028-0.0098	3	1.8891-1.8898	0.0007-0.0014	0.0039-0.0098
	3.3	D	2.7142-2.7149	0.0008-0.0016	0.0039-0.0110	3	2.1635-2.1642	0.0015-0.0022	0.0039-0.0098
	3.3	F	2.7142-2.7149	0.0008-0.0016	0.0039-0.0110	3	2.1635-2.1642	0.0012-0.0019	0.0039-0.0098
	3.8	F	2.7142-2.7149	0.0008-0.0016	0.0039-0.0110	3	2.1635-2.1642	0.0015-0.0022	0.0039-0.0098

① No. 1, 2, 4, 5 are 0.0009-0.0016 inch

No. 3 is 0.0011-0.0018 inch

22140_HYUN_C0008

PISTON AND RING SPECIFICATIONS

All measurements are given in inches.

Year	Engine Displ. Liters	Engine VIN	Piston Clearance	Ring Gap			Ring Side Clearance		
				Top Compression	Bottom Compression	Oil Control	Top Compression	Bottom Compression	Oil Control
2006	1.6	C	0.0008-0.0016	0.0059-0.0118	0.0138-0.0197	0.0079-0.0276	0.0016-0.0033	0.0016-0.0033	0.0031-0.0069
	2.0	D	0.0008-0.0016	0.0090-0.0149	0.0177-0.0236	0.0078-0.0236	0.0015-0.0031	0.0012-0.0027	NA
	2.4	C	0.0008-0.0016	0.0059-0.0118	0.0118-0.0177	0.0078-0.0275	0.0012-0.0027	0.0012-0.0027	0.0024-0.0059
	2.7	F	0.0008-0.0016	0.0059-0.0118	0.0118-0.0177	0.0079-0.0276	0.0016-0.0031	0.0012-0.0028	NA
	3.3	F	0.0008-0.0016	0.0067-0.0126	0.0126-0.0185	0.0078-0.0275	0.0016-0.0031	0.0012-0.0027	0.0024-0.0059
	3.8	F	0.0012-0.0020	0.0067-0.0126	0.0126-0.0185	0.0078-0.0275	0.0012-0.0027	0.0012-0.0027	0.0024-0.0059
2007	1.6	C	0.0008-0.0016	0.0059-0.0118	0.0138-0.0197	0.0079-0.0276	0.0016-0.0033	0.0016-0.0033	0.0031-0.0069
	2.0	D	0.0008-0.0016	0.0090-0.0149	0.0177-0.0236	0.0078-0.0236	0.0015-0.0031	0.0012-0.0027	NA
	2.4	C	0.0008-0.0016	0.0059-0.0118	0.0118-0.0177	0.0078-0.0275	0.0012-0.0027	0.0012-0.0027	0.0024-0.0059
	2.7	F	0.0008-0.0016	0.0059-0.0118	0.0118-0.0177	0.0079-0.0276	0.0016-0.0031	0.0012-0.0028	NA
	3.3	D	0.0008-0.0016	0.0067-0.0126	0.0126-0.0185	0.0078-0.0275	0.0016-0.0031	0.0012-0.0027	0.0024-0.0059
	3.3	F	0.0008-0.0016	0.0067-0.0126	0.0126-0.0185	0.0078-0.0275	0.0016-0.0031	0.0012-0.0027	0.0024-0.0059
	3.8	F	0.0012-0.0020	0.0067-0.0126	0.0126-0.0185	0.0078-0.0275	0.0012-0.0027	0.0012-0.0027	0.0024-0.0059
2008	1.6	C	0.0008-0.0016	0.0059-0.0118	0.0138-0.0197	0.0079-0.0276	0.0016-0.0033	0.0016-0.0033	0.0031-0.0069
	2.0	D	0.0008-0.0016	0.0090-0.0149	0.0177-0.0236	0.0078-0.0236	0.0015-0.0031	0.0012-0.0027	NA
	2.4	C	0.0008-0.0016	0.0059-0.0118	0.0118-0.0177	0.0078-0.0275	0.0012-0.0027	0.0012-0.0027	0.0024-0.0059
	2.7	F	0.0008-0.0016	0.0059-0.0118	0.0118-0.0177	0.0079-0.0276	0.0016-0.0031	0.0012-0.0028	NA
	3.3	D	0.0008-0.0016	0.0067-0.0126	0.0126-0.0185	0.0078-0.0275	0.0016-0.0031	0.0012-0.0027	0.0024-0.0059
	3.3	F	0.0008-0.0016	0.0067-0.0126	0.0126-0.0185	0.0078-0.0275	0.0016-0.0031	0.0012-0.0027	0.0024-0.0059
	3.8	F	0.0012-0.0020	0.0067-0.0126	0.0126-0.0185	0.0078-0.0275	0.0012-0.0027	0.0012-0.0027	0.0024-0.0059

NA: Not Available

22140_HYUN_C0009

TORQUE SPECIFICATIONS
All readings in ft. lbs.

Year	Engine Displacement Liters	Engine VIN	Cylinder Head Bolts	Main Bearing Bolts	Rod Bearing Bolts	Crankshaft Damper Bolts	Flywheel Bolts	Manifold		Spark Plugs	Oil Pan Drain Plug
								Intake	Exhaust		
2006	1.6	C	①	40-43	23-25	101-109	87-94	11-15	22-25	15-22	29-33
	2.0	D	①	40-43	36-38	116-123	87-94	13-18	31-40	15-22	29-33
	2.4	C	②	③	④	123-130	87-94	14-20	⑤	15-22	29-33
	2.7	F	⑥	⑦	⑧	123-130	53-56	14-17	⑨	15-22	25-33
	3.3	F	⑩	⑪	④	210-224	53-56	14-17	29-33	15-22	25-33
	3.8	F	⑩	⑪	④	210-224	53-56	14-17	29-33	15-22	25-33
2007	1.6	C	①	40-43	23-25	101-109	87-94	11-15	22-25	15-22	29-33
	2.0	D	①	40-43	36-38	116-123	87-94	13-18	31-40	15-22	29-33
	2.4	C	②	③	④	123-130	87-94	14-20	⑤	15-22	29-33
	2.7	F	⑥	⑦	⑧	123-130	53-56	14-17	⑨	15-22	25-33
	3.3	D	⑩	⑪	④	210-224	53-56	14-17	29-33	15-22	25-33
	3.3	F	⑩	⑪	④	210-224	53-56	14-17	29-33	15-22	25-33
	3.8	F	⑩	⑪	④	210-224	53-56	14-17	29-33	15-22	25-33
2008	1.6	C	①	40-43	23-25	101-109	87-94	11-15	22-25	15-22	29-33
	2.0	D	①	40-43	36-38	116-123	87-94	13-18	31-40	15-22	29-33
	2.4	C	②	③	④	123-130	87-94	14-20	⑤	15-22	29-33
	2.7	F	⑥	⑦	⑧	123-130	53-56	14-17	⑨	15-22	25-33
	3.3	D	⑩	⑪	④	210-224	53-56	14-17	29-33	15-22	25-33
	3.3	F	⑩	⑪	④	210-224	53-56	14-17	29-33	15-22	25-33
	3.8	F	⑩	⑪	④	210-224	53-56	14-17	29-33	15-22	25-33

① Step 1: 22 ft. lbs.
Step 2: Plus 90 degrees
Step 3: Loosen to 0 ft. lbs.
Step 4: 22 ft. lbs.
Step 5: Plus 90 degrees

② Step 1: 25 ft. lbs.
Step 2: Plus 90 degrees
Step 3: Plus 90 degrees

③ Step 1: 20 ft. lbs.
Step 2: Plus 45 degrees

④ Step 1: 15 ft. lbs.
Step 2: Plus 90 degrees

⑤ Exhaust manifold heat protector bolt and stay bolt (M8): 14-20 ft. lbs.
Exhaust manifold nut: 29-33 ft. lbs.
Exhaust manifold stay bolt (M10): 38-43 ft. lbs.

⑥ Step 1: 18 ft. lbs.
Step 2: Plus 58-62 degrees
Step 3: Plus 43-47 degrees

⑦ M10 bolts Step 1: 20-24 ft. lbs.
M10 bolts Step 2: Plus 90-95 degrees
M8 bolts Step 1: 9-14 ft. lbs.
M8 bolts Step 2: Plus 90-95 degrees

⑧ Step 1: 12-15 ft. lbs.
Step 2: Plus 90-94 degrees

⑨ Heat protector exhaust manifold: 12-16 ft. lbs.
Exhaust manifold to cylinder head (Self-locking nut): 18-22 ft. lbs.

⑩ Step 1: 29 ft. lbs.
Step 2: Plus 120 degrees
Step 3: Plus 90 degrees

⑪ M11 bolts (inner) Step 1: 36 ft. lbs.
M11 bolts (inner) Step 2: Plus 90 degrees
M8 bolts (outer) Step 1: 15 ft. lbs.
M8 bolts (outer) Step 2: Plus 120 degrees
M8 bolts (side): 22-23 ft. lbs.

22140_HYUN_C0010

WHEEL ALIGNMENT

Year	Model		Caster		Camber		Toe-in
			Range (+/-Deg.)	Preferred Setting (Deg.)	Range (+/-Deg.)	Preferred Setting (Deg.)	(Deg.)
2006	Accent	Front	0.50	+1.80	0.50	0.00	0 +/- 0.12
		Rear	—	—	0.50	-0.68	0.12 +/- 0.08
	Azera	Front	1.00	+3.25	0.50	0.00	0 +/- 0.08
		Rear	—	—	0.50	-0.50	0.08 +/- 0.08
	Elantra	Front	0.50	+2.81	0.50	0.00	0 +/- 0.08
		Rear	—	—	0.50	-0.91	0.10 +/- 0.04
	Sonata	Front	1.00	+3.25	0.50	0.00	0 +/- 0.08
		Rear	—	—	0.50	-0.50	0.08 +/- 0.08
	Tiburon	Front	0.50	+3.38	0.50	-0.20	0.08 +/- 0.08
		Rear	—	—	0.50	-1.18	0.16 +0.12 -0.04
2007	Accent	Front	0.50	+1.80	0.50	0.00	0 +/- 0.12
		Rear	—	—	0.50	-0.68	0.12 +/- 0.08
	Azera	Front	1.00	+3.25	0.50	0.00	0 +/- 0.08
		Rear	—	—	0.50	-0.50	0.08 +/- 0.08
	Elantra	Front	0.50	+2.81	0.50	0.00	0 +/- 0.08
		Rear	—	—	0.50	-0.91	0.10 +/- 0.04
	Sonata	Front	1.00	+3.25	0.50	0.00	0 +/- 0.08
		Rear	—	—	0.50	-0.50	0.08 +/- 0.08
	Tiburon	Front	0.50	+3.38	0.50	-0.20	0.08 +/- 0.08
		Rear	—	—	0.50	-1.18	0.16 +0.12 -0.04
2008	Accent	Front	0.50	+1.80	0.50	0.00	0 +/- 0.12
		Rear	—	—	0.50	-0.68	0.12 +/- 0.08
	Azera	Front	1.00	+3.25	0.50	0.00	0 +/- 0.08
		Rear	—	—	0.50	-0.50	0.08 +/- 0.08
	Elantra	Front	0.50	+2.81	0.50	0.00	0 +/- 0.08
		Rear	—	—	0.50	-0.91	0.10 +/- 0.04
	Sonata	Front	1.00	+3.25	0.50	0.00	0 +/- 0.08
		Rear	—	—	0.50	-0.50	0.08 +/- 0.08
	Tiburon	Front	0.50	+3.38	0.50	-0.20	0.08 +/- 0.08
		Rear	—	—	0.50	-1.18	0.16 +0.12 -0.04

22140_HYUN_C0011

TIRE, WHEEL AND BALL JOINT SPECIFICATIONS

| Year | Model | OEM Tires | | Tire Pressures (psi) | | Wheel Size | Ball Joint Inspection | Lug Nut Torque (ft. lbs.) |
		Standard	Optional	Front	Rear			
2006	Accent	P175/70R14	P185/65R14 P195/55R15	①	①	5/5.5J x 14 5.5J x 15	②	65-79
	Azera	P225/60R16	P235/55R17	①	①	6.5J x 16 7.0J x 17	②	65-80
	Elantra	P195/65R15	P205/55R16	①	①	6.0J x 15 6.5J x 16/17	②	65-80
	Sonata	P215/60R16	P225/50R17	①	①	6.5J×16 6.5J×17	②	65-80
	Tiburon	P205/55R16	P215/45R17	①	①	6.5J×16 7.0J×17	②	67-82
2007	Accent	P175/70R14	P185/65R14 P195/55R15	①	①	5/5.5J x 14 5.5J x 15	②	65-79
	Azera	P225/60R16	P235/55R17	①	①	6.5J x 16 7.0J x 17	②	65-80
	Elantra	P195/65R15	P205/55R16	①	①	6.0J x 15 6.5J x 16/17	②	65-80
	Sonata	P215/60R16	P225/50R17	①	①	6.5J×16 6.5J×17	②	65-80
	Tiburon	P205/55R16	P215/45R17	①	①	6.5J×16 7.0J×17	②	67-82
2008	Accent	P175/70R14	P185/65R14 P195/55R15	①	①	5/5.5J x 14 5.5J x 15	②	65-79
	Azera	P225/60R16	P235/55R17	①	①	6.5J x 16 7.0J x 17	②	65-80
	Elantra	P195/65R15	P205/55R16	①	①	6.0J x 15 6.5J x 16/17	②	65-80
	Sonata	P215/60R16	P225/50R17	①	①	6.5J×16 6.5J×17	②	65-80
	Tiburon	P205/55R16	P215/45R17	①	①	6.5J×16 7.0J×17	②	67-82

① Refer to placard on vehicle for proper inflation pressure.

② Replace if any measurable movement is found.

22140_HYUN_C0012

BRAKE SPECIFICATIONS
All measurements in inches unless noted

Year	Model		Brake Disc			Brake Drum Diameter			Minimum Lining Thickness	Brake Caliper	
			Original Thickness	Minimum Thickness	Maximum Runout	Original Inside Diameter	Max. Wear Limit	Maximum Machine Diameter		Bracket Bolts (ft. lbs.)	Mounting Bolts (ft. lbs.)
2006	Accent	F	0.870	0.790	0.001	—	—	—	0.079	62-69	16-23
		R	—	—	—	8.000	①	①	0.039	—	—
	Azera	F	1.100	1.040	0.002	—	—	—	0.079	58-72	16-23
		R	0.390	0.310	0.002	—	—	—	0.080	58-72	16-23
	Elantra	F	1.020	0.940	0.002	—	—	—	0.079	58-72	16-23
		R	0.390	0.330	0.002	—	—	—	0.079	36-43	16-23
	Elantra	F	1.020	0.940	0.002	—	—	—	0.079	58-72	16-23
		R	—	—	—	8.000	①	①	0.039	—	—
	Sonata (2.4L)	F	1.024	0.961	0.002	—	—	—	0.120-0.160	59-74	18-22
		R	0.390	0.330	0.002	—	—	—	0.120	59-74	18-22
	Sonata (3.3L)	F	1.100	1.040	0.002	—	—	—	0.120-0.160	59-74	18-22
		R	0.390	0.330	0.002	—	—	—	0.120	59-74	18-22
	Tiburon	F	1.024	0.961	0.003	—	—	—	0.079	48-55	16-24
		R	0.400	0.330	0.002	—	—	—	0.080	48-55	16-24
2007	Accent	F	0.870	0.790	0.001	—	—	—	0.079	62-69	16-23
		R	—	—	—	8.000	①	①	0.039	—	—
	Azera	F	1.100	1.040	0.002	—	—	—	0.079	58-72	16-23
		R	0.390	0.310	0.002	—	—	—	0.080	58-72	16-23
	Elantra	F	1.020	0.940	0.002	—	—	—	0.079	58-72	16-23
		R	0.390	0.330	0.002	—	—	—	0.079	36-43	16-23
	Elantra	F	1.020	0.940	0.002	—	—	—	0.079	58-72	16-23
		R	—	—	—	8.000	①	①	0.039	—	—
	Sonata (2.4L)	F	1.024	0.961	0.002	—	—	—	0.120-0.160	59-74	18-22
		R	0.390	0.330	0.002	—	—	—	0.120	59-74	18-22
	Sonata (3.3L)	F	1.100	1.040	0.002	—	—	—	0.120-0.160	59-74	18-22
		R	0.390	0.330	0.002	—	—	—	0.120	59-74	18-22
	Tiburon	F	1.024	0.961	0.003	—	—	—	0.079	48-55	16-24
		R	0.400	0.330	0.002	—	—	—	0.080	48-55	16-24
2008	Accent	F	0.870	0.790	0.001	—	—	—	0.079	62-69	16-23
		R	—	—	—	8.000	①	①	0.039	—	—
	Azera	F	1.100	1.040	0.002	—	—	—	0.079	58-72	16-23
		R	0.390	0.310	0.002	—	—	—	0.080	58-72	16-23
	Elantra	F	1.020	0.940	0.002	—	—	—	0.079	58-72	16-23
		R	0.390	0.330	0.002	—	—	—	0.079	36-43	16-23
	Elantra	F	1.020	0.940	0.002	—	—	—	0.079	58-72	16-23
		R	—	—	—	8.000	①	①	0.039	—	—
	Sonata (2.4L)	F	1.024	0.961	0.002	—	—	—	0.120-0.160	59-74	18-22
		R	0.390	0.330	0.002	—	—	—	0.120	59-74	18-22
	Sonata (3.3L)	F	1.100	1.040	0.002	—	—	—	0.120-0.160	59-74	18-22
		R	0.390	0.330	0.002	—	—	—	0.120	59-74	18-22
	Tiburon	F	1.024	0.961	0.003	—	—	—	0.079	48-55	16-24
		R	0.400	0.330	0.002	—	—	—	0.080	48-55	16-24

① Drum roundness Service Limit: 0.00236 inch

22140_HYUN_C0013

SCHEDULED MAINTENANCE INTERVALS
HYUNDAI—ACCENT, AZERA, ELANTRA, SONATA, & TIBURON

TO BE SERVICED	TYPE OF SERVICE	VEHICLE MILEAGE INTERVAL (x1000)												
		7.5	15	22.5	30	37.5	45	52.5	60	67.5	75	82.5	90	97.5
Engine oil & filter	R	✓	✓	✓	✓	✓	✓	✓	✓	✓	✓	✓	✓	✓
Automatic transaxle fluid	S/I		✓		✓		✓		✓		✓		✓	
Brake pads, calipers & rotors	S/I		✓		✓		✓		✓		✓		✓	
Driveshafts & boots	S/I		✓		✓		✓		✓		✓		✓	
Wheel bearing grease	S/I				✓				✓				✓	
Air cleaner filter	R				✓				✓				✓	
Automatic transaxle fluid & filter	R				✓				✓				✓	
Brake fluid	R				✓				✓				✓	
Engine coolant	R				✓				✓				✓	
Fuel hose, vapor hose & fuel filter cap	S/I							✓						
Spark plugs	R				✓				✓				✓	
Spark plugs (Platinum coated)	R								✓					
Spark plugs (Iridium coated) 100,000 mile replacement	R													
Bolts & nuts on chassis & body (Accent)	S/I				✓				✓				✓	
Drive belts	S/I				✓				✓				✓	
Exhaust pipe connections, muffler & suspension bolts	S/I				✓				✓				✓	
Manual transaxle oil	S/I				✓				✓				✓	
Rear brake drums, linings & parking brake	S/I				✓				✓				✓	
Steering gear rack, linkage & boots	S/I				✓				✓				✓	
Suspension ball joints & dust covers (Accent)	S/I				✓				✓				✓	
Timing belt (Accent & Elantra)	S/I				✓				✓				✓	
Timing belt (except Accent & Elantra)	R								✓					
Fuel filter	R							✓						
Fuel lines & connections	S/I								✓					
Vacuum & crankcase ventilation hoses	S/I								✓					

R: Replace   S/I: Service or Inspect

FREQUENT OPERATION MAINTENANCE (SEVERE SERVICE)

If a vehicle is operated under any of the following conditions it is considered severe service:

- Extremely dusty areas.

- 50% or more of the vehicle operation is in 90°F (32°C) or higher temperatures, or constant operation in temperatures below 32°F (0°C).

- Prolonged idling (vehicle operation in stop and go traffic).

- Frequent short running periods (engine does not warm to normal operating temperatures).

- Police, taxi, delivery usage or trailer towing usage.

Oil & oil filter: change every 3,000 miles.

Brake pads, calipers & rotors: service or inspect every 7,500 miles.

Driveshaft boots: service or inspect every 7,500 miles

Steering gear rack, linkage & boots: service or inspect every 7,500 miles.

Air cleaner filter: service or inspect every 15,000 miles.

Automatic transaxle fluid & filter: replace every 15,000 miles.

Rear brake drums & linings: service or inspect every 15,000 miles.

Spark plugs: service or inspect every 24,000 miles.

PRECAUTIONS

Before servicing any vehicle, please be sure to read all of the following precautions, which deal with personal safety, prevention of component damage, and important points to take into consideration when servicing a motor vehicle:

• Never open, service or drain the radiator or cooling system when the engine is hot; serious burns can occur from the steam and hot coolant.

• Observe all applicable safety precautions when working around fuel. Whenever servicing the fuel system, always work in a well-ventilated area. Do not allow fuel spray or vapors to come in contact with a spark, open flame, or excessive heat (a hot drop light, for example). Keep a dry chemical fire extinguisher near the work area. Always keep fuel in a container specifically designed for fuel storage; also, always properly seal fuel containers to avoid the possibility of fire or explosion. Refer to the additional fuel system precautions later in this section.

• Fuel injection systems often remain pressurized, even after the engine has been turned **OFF**. The fuel system pressure must be relieved before disconnecting any fuel lines. Failure to do so may result in fire and/or personal injury.

• Brake fluid often contains polyglycol ethers and polyglycols. Avoid contact with the eyes and wash your hands thoroughly after handling brake fluid. If you do get brake fluid in your eyes, flush your eyes with clean, running water for 15 minutes. If eye irritation persists, or if you have taken

brake fluid internally, IMMEDIATELY seek medical assistance.

• The EPA warns that prolonged contact with used engine oil may cause a number of skin disorders, including cancer. You should make every effort to minimize your exposure to used engine oil. Protective gloves should be worn when changing oil. Wash your hands and any other exposed skin areas as soon as possible after exposure to used engine oil. Soap and water, or waterless hand cleaner should be used.

• All new vehicles are now equipped with an air bag system, often referred to as a Supplemental Restraint System (SRS) or Supplemental Inflatable Restraint (SIR) system. The system must be disabled before performing service on or around system components, steering column, instrument panel components, wiring and sensors. Failure to follow safety and disabling procedures could result in accidental air bag deployment, possible personal injury and unnecessary system repairs.

• Always wear safety goggles when working with, or around, the air bag system. When carrying a non-deployed air bag, be sure the bag and trim cover are pointed away from your body. When placing a non-deployed air bag on a work surface, always face the bag and trim cover upward, away from the surface. This will reduce the motion of the module if it is accidentally deployed. Refer to the additional air bag system precautions later in this section.

• Clean, high quality brake fluid from a sealed container is essential to the safe and

proper operation of the brake system. You should always buy the correct type of brake fluid for your vehicle. If the brake fluid becomes contaminated, completely flush the system with new fluid. Never reuse any brake fluid. Any brake fluid that is removed from the system should be discarded. Also, do not allow any brake fluid to come in contact with a painted surface; it will damage the paint.

• Never operate the engine without the proper amount and type of engine oil; doing so WILL result in severe engine damage.

• Timing belt maintenance is extremely important. Many models utilize an interference-type, non-freewheeling engine. If the timing belt breaks, the valves in the cylinder head may strike the pistons, causing potentially serious (also time-consuming and expensive) engine damage. Refer to the maintenance interval charts for the recommended replacement interval for the timing belt, and to the timing belt section for belt replacement and inspection.

• Disconnecting the negative battery cable on some vehicles may interfere with the functions of the on-board computer system(s) and may require the computer to undergo a relearning process once the negative battery cable is reconnected.

• When servicing drum brakes, only disassemble and assemble one side at a time, leaving the remaining side intact for reference.

• Only an MVAC-trained, EPA-certified automotive technician should service the air conditioning system or its components.

BRAKES

ANTI-LOCK BRAKE SYSTEM (ABS)

GENERAL INFORMATION

See Figure 1.

The Anti-Lock Brake System (ABS) controls the hydraulic brake pressure of all four wheels during sudden braking and braking on hazardous road surfaces, preventing the wheels from locking. The ABS provides the following benefits: (1) Enables steering around obstacles with a greater degree of certainty during panic braking. (2) Enables stopping during panic braking while allowing stability and control, even on curves. If a malfunction occurs in the ABS, the system will operate as a normal brake (fail safe mode). A diagnostic function and a fail-safe system have been included for serviceability.

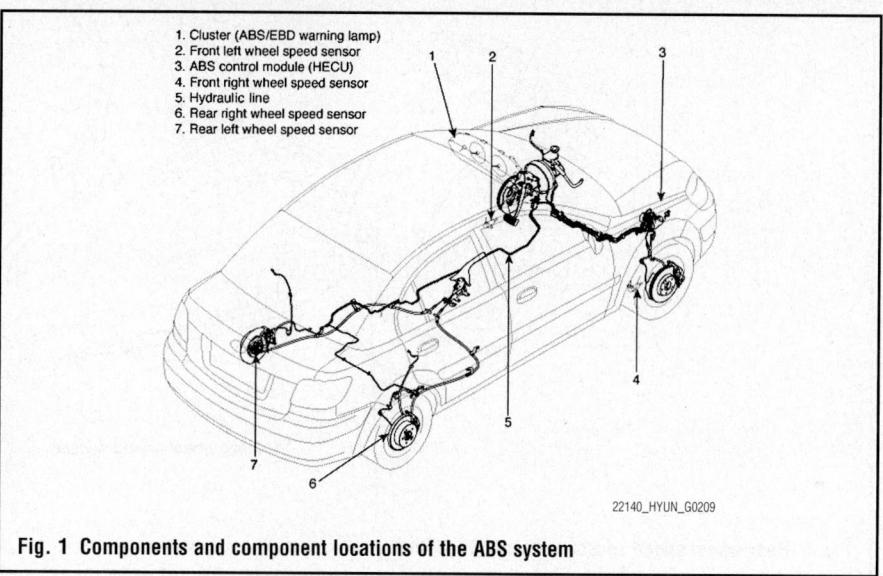

1. Cluster (ABS/EBD warning lamp)
2. Front left wheel speed sensor
3. ABS control module (HECU)
4. Front right wheel speed sensor
5. Hydraulic line
6. Rear right wheel speed sensor
7. Rear left wheel speed sensor

22140_HYUN_G0209

Fig. 1 Components and component locations of the ABS system

WHEEL SPEED SENSORS

REMOVAL & INSTALLATION

Front

See Figure 2.

1. Remove the front wheel speed sensor mounting bolt.
2. Remove the front wheel guard.
3. Disconnect the wheel speed sensor connector.

4. Remove the front wheel speed sensor.
5. Installation is the reverse of the removal procedure.

Rear

See Figure 3.

1. Remove the rear wheel speed sensor mounting bolt.
2. Remove the rear seat side pad then disconnect the rear wheel speed sensor connector.

3. Installation is the reverse of the removal procedure.

WHEEL SPEED SENSOR RINGS (TOOTHED RINGS)

REMOVAL & INSTALLATION

See Figure 4.

1. Before servicing the vehicle, refer to the Precautions Section.
2. Remove the halfshaft(s). Refer to Half-shafts, removal & installation.
3. Using the special tool (09432-11000), remove the wheel speed sensor ring (also called tone wheel).

To install:

4. Installation is the reverse of the removal.
5. Replace with wheel speed sensor ring of the same number of teeth.

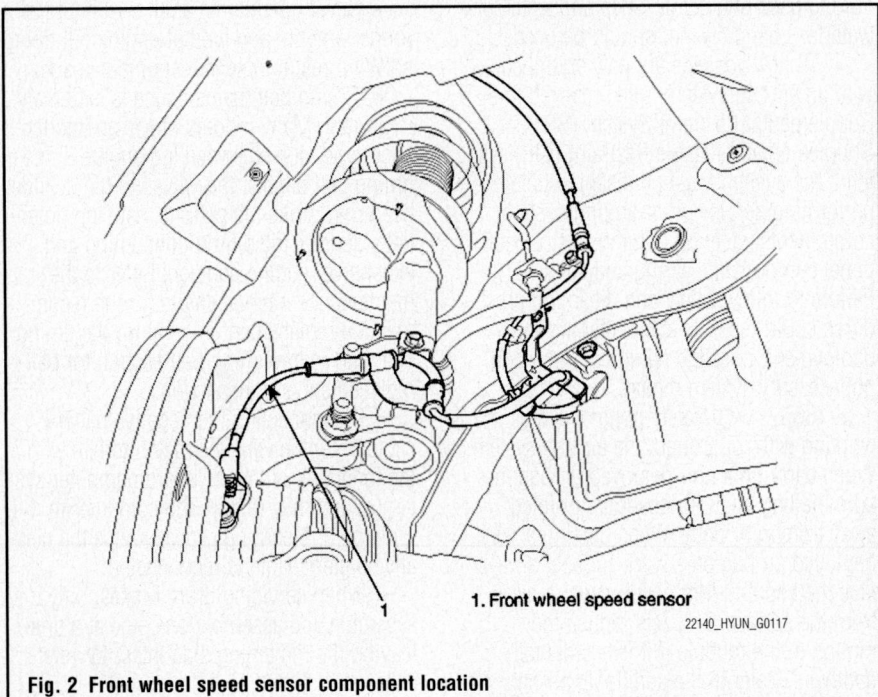

1. Front wheel speed sensor

22140_HYUN_G0117

Fig. 2 Front wheel speed sensor component location

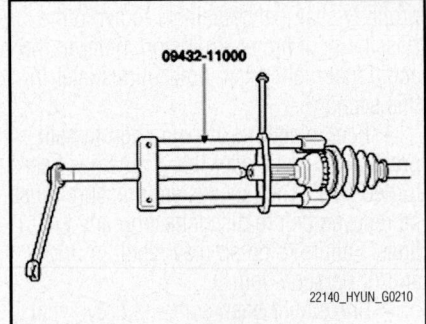

09432-11000

22140_HYUN_G0210

Fig. 4 Using the special tool (09432-11000) to remove the wheel speed sensor ring

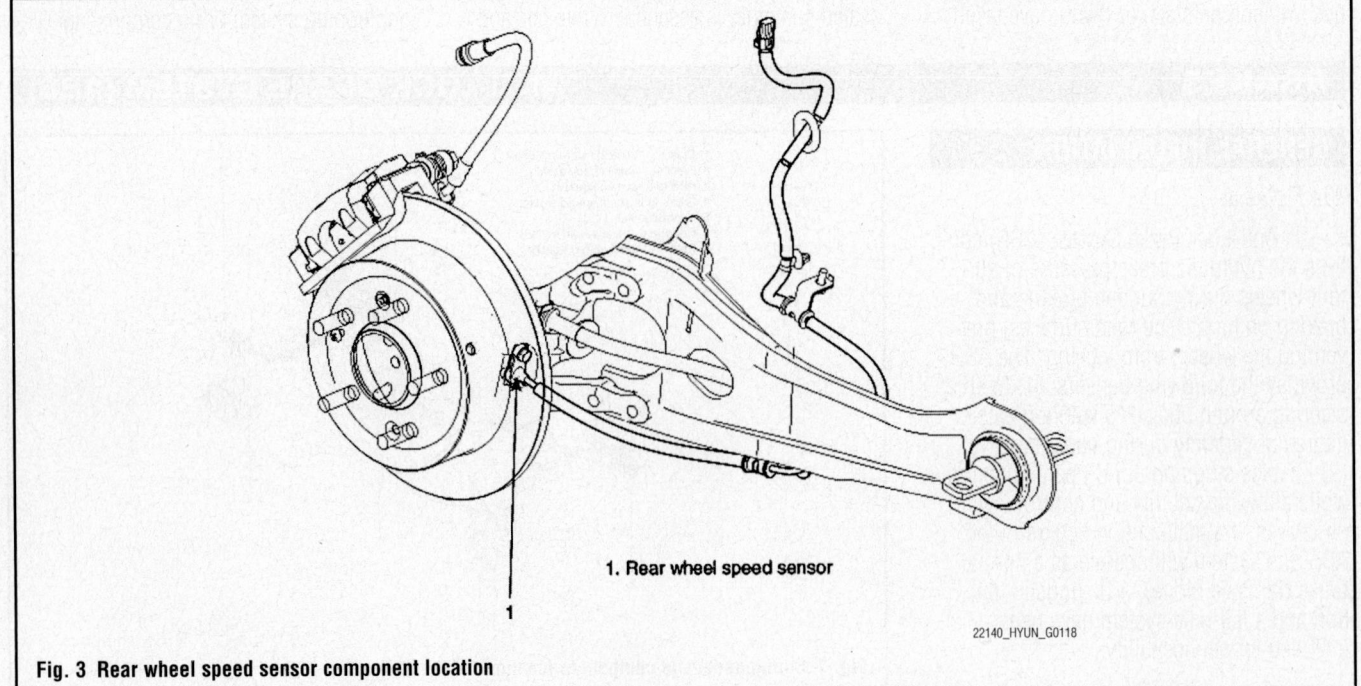

1. Rear wheel speed sensor

22140_HYUN_G0118

Fig. 3 Rear wheel speed sensor component location

BLEEDING THE BRAKE SYSTEM

BLEEDING PROCEDURE

These vehicles come standard with a 4-wheel Anti-Lock Braking System (ABS). Please refer to the Bleeding the ABS System procedure.

BLEEDING THE ABS SYSTEM

This procedure should be followed to ensure adequate bleeding of air and the filling of the ABS unit, the brake lines, and the master cylinder with brake fluid.

1. Before servicing the vehicle, refer to the Precautions Section.
2. Remove the reservoir cap and fill the brake reservoir with brake fluid.

✳ WARNING

If there is any brake fluid on any painted surface, wash it off immediately.

➡**When pressure bleeding, do not depress the brake pedal. Recommended brake fluid: DOT3 or DOT4.**

3. Connect a clear plastic tube to the wheel cylinder bleeder plug and insert the other end of the tube into a clear plastic bottle that is half filled with clean brake fluid.
4. Connect the Hi-Scan Pro® to the data link connector located underneath the dash panel.
5. Select and operate according to the instructions on the Hi-Scan Pro® screen.

✳ CAUTION

You must obey the maximum operating time of the ABS motor with the Hi-Scan Pro® to prevent the motor pump from burning.

6. Select Hyundai vehicle diagnosis.
7. Select vehicle name.
8. Select Anti-Lock Brake system.
9. Select air bleeding mode.
10. Press "YES" to operate motor pump and solenoid valve.

✳ WARNING

Wait 60 seconds before operating the air bleeding or damage to the motor may occur.

11. Wait 60 seconds before operating the air bleeding.
12. Pump the brake pedal several times, and then loosen the bleeder screw until fluid starts to run out without bubbles. Then, close the bleeder screw.
13. Repeat until there are no more bubbles in the fluid for each wheel.

BRAKES

✳ CAUTION

Dust and dirt accumulating on brake parts during normal use may contain asbestos fibers from production or aftermarket brake linings. Breathing excessive concentrations of asbestos fibers can cause serious bodily harm. Exercise care when servicing brake parts. Do not sand or grind brake lining unless equipment used is designed to contain the dust residue. Do not clean brake parts with compressed air or by dry brushing. Cleaning should be done by dampening the brake components with a fine mist of water, then wiping the brake components clean with a dampened cloth. Dispose of cloth and all residue containing asbestos fibers in an impermeable container with the appropriate label. Follow practices prescribed by the Occupational Safety and Health Administration (OSHA) and the Environmental Protection Agency (EPA) for the handling, processing, and

disposing of dust or debris that may contain asbestos fibers.

BRAKE CALIPER

REMOVAL & INSTALLATION

1. Before servicing the vehicle, refer to the Precautions Section.
2. Remove or disconnect the following:
   - Wheel
   - Brake hose from the caliper
   - Caliper mounting bolts
   - Caliper

To install:
3. Install or connect the following:
   - Caliper
   - Caliper mounting bolts and tighten to 58–73 ft. lbs. (80–100 Nm) on the front caliper, or 37–44 ft. lbs. (50–60 Nm) on the rear caliper.
   - Brake hose to the caliper and tighten the fitting to 18–22 ft. lbs. (25–30 Nm)
   - Wheel
4. Bleed the brake system.

FRONT DISC BRAKES

DISC BRAKE PADS

REMOVAL & INSTALLATION

1. Before servicing the vehicle, refer to the Precautions Section.
2. Remove or disconnect the following:
   - Wheel
   - Caliper mounting bolts
   - Caliper and support to one side with a wire.

✳ WARNING

Do not let the caliper hang by the hose.

   - Pads and shims

To install:
3. Install the pads, clips, and shims.
4. Bottom the caliper piston using tool 09581-11000 or a C-clamp.
5. Install the caliper mounting bolts and tighten to 16–24 ft. lbs. (22–32 Nm).
6. Install the wheel.

BRAKES

✳✳ CAUTION

Dust and dirt accumulating on brake parts during normal use may contain asbestos fibers from production or aftermarket brake linings. Breathing excessive concentrations of asbestos fibers can cause serious bodily harm. Exercise care when servicing brake parts. Do not sand or grind brake lining unless equipment used is designed to contain the dust residue. Do not clean brake parts with compressed air or by dry brushing. Cleaning should be done by dampening the brake components with a fine mist of water, then wiping the brake components clean with a dampened cloth. Dispose of cloth and all residue containing asbestos fibers in an impermeable container with the appropriate label. Follow practices prescribed by the Occupational Safety and Health Administration (OSHA) and the Environmental Protection Agency (EPA) for the handling, processing, and disposing of dust or debris that may contain asbestos fibers.

BRAKE CALIPER

REMOVAL & INSTALLATION

1. Before servicing the vehicle, refer to the Precautions Section.
2. Release the parking brake.
3. Remove or disconnect the following:
   • Wheel
   • Brake line at the caliper
   • Caliper mounting bolts
   • Caliper

To install:

4. Install the caliper onto its mounting.
5. Tighten the caliper mounting bolts:
   a. Azera and Sonata: 59–74 ft. lbs. (80–100 Nm).
   b. Elantra: 36–43 ft. lbs. (49–58 Nm).
   c. Tiburon: 48–55 ft. lbs. (65–75 Nm).
6. Install the brake line to the caliper with 2 new metal gaskets. Torque the

REAR DISC BRAKES

brake line union bolt to 18–22 ft. lbs. (24–30 Nm).
7. Bleed the system.
8. Install the wheel.

DISC BRAKE PADS

REMOVAL & INSTALLATION

1. Before servicing the vehicle, refer to the Precautions Section.
2. Remove or disconnect the following:
   • Rear wheels
   • Lower caliper mounting bolt and rotate the caliper upward
   • Pads from the caliper support
   • Pad retainers, if necessary

To install:

3. Install or connect the following:
   • Pad retainers, if removed
   • Pads onto the pad retainers
4. Compress the caliper piston using a C-clamp. Rotate the caliper downward and install the mounting bolt. Tighten to: 16–24 ft. lbs. (22–32 Nm).
5. Install the wheel.

BRAKES

✳✳ CAUTION

Dust and dirt accumulating on brake parts during normal use may contain asbestos fibers from production or aftermarket brake linings. Breathing excessive concentrations of asbestos fibers can cause serious bodily harm. Exercise care when servicing brake parts. Do not sand or grind brake lining unless equipment used is designed to contain the dust residue. Do not clean brake parts with compressed air or by dry brushing. Cleaning should be done by dampening the brake components with a fine mist of water, then wiping the brake components clean with a dampened cloth. Dispose of cloth and all residue containing asbestos fibers in an impermeable container with the appropriate label. Follow practices prescribed by the Occupational Safety and Health Administration (OSHA) and the Environmental Protection Agency (EPA) for the handling, processing, and disposing of dust or debris that may contain asbestos fibers.

BRAKE DRUM

REMOVAL & INSTALLATION
See Figure 5.

✳✳ CAUTION

Frequent inhalation of brake pad dust, regardless of material composi-

REAR DRUM BRAKES

tion, could be hazardous to your health. Avoid breathing dust particles. Never use an air hose or brush to clean brake assemblies.

1. Before servicing the vehicle, refer to the Precautions Section.
2. Check to ensure that the park brake is fully released.

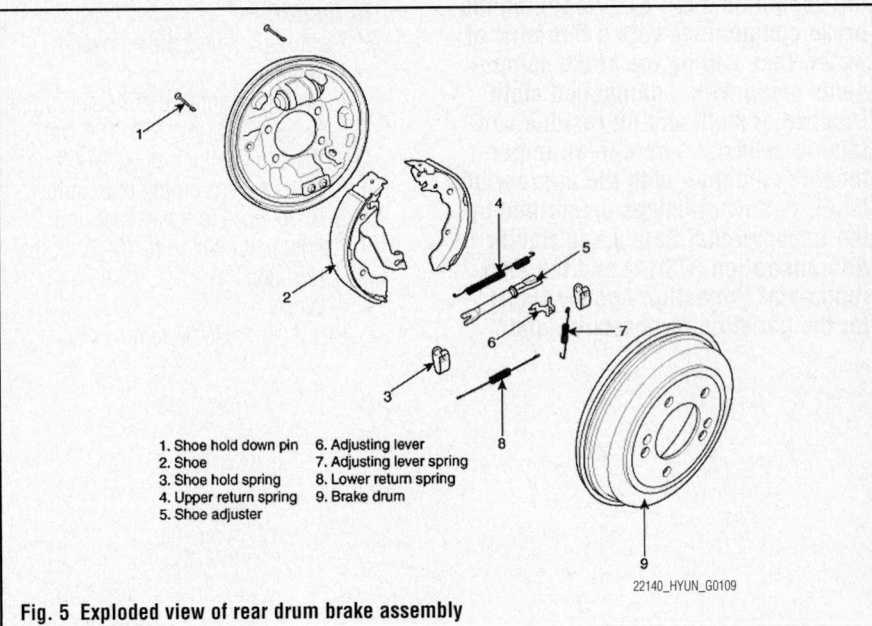

1. Shoe hold down pin
2. Shoe
3. Shoe hold spring
4. Upper return spring
5. Shoe adjuster
6. Adjusting lever
7. Adjusting lever spring
8. Lower return spring
9. Brake drum

22140_HYUN_G0109

Fig. 5 Exploded view of rear drum brake assembly

3. Raise and safely support the vehicle.
4. Remove the tire and wheel assembly.
5. Remove the brake drum.

To install:

6. If installing a new brake drum, use denatured alcohol or an equivalent approved brake cleaner and a clean shop towel to remove the protective coating from the friction surface of the drum.
7. Install the brake drum.
8. Install the tire and wheel assembly.
9. Apply the brakes approximately 3 times in order to seat and center the brake shoes within the drum.
10. Lower the vehicle.

BRAKE SHOES

REMOVAL & INSTALLATION

See Figure 6 through 8.

✳✳ CAUTION

Frequent inhalation of brake pad dust, regardless of material composition, could be hazardous to your health. Avoid breathing dust particles. Never use an air hose or brush to clean brake assemblies.

1. Before servicing the vehicle, refer to the Precautions Section.
2. Check to ensure that the park brake is fully released.
3. Raise and safely support the vehicle.
4. Remove the tire and wheel assembly.
5. Remove the brake drum.
6. Remove the shoe hold spring and shoe hold pin (B).
7. Remove the upper return spring (A).
8. Lower the brake shoe assembly (A), and remove the lower return spring (B). Make sure not to damage the dust cover on the wheel cylinder.
9. Remove the parking brake cable (A) from the brake assembly.
10. Remove the brake shoe assembly.

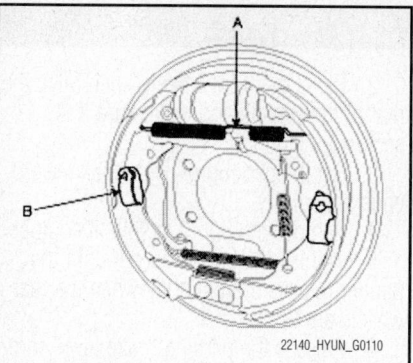

Fig. 6 Remove the shoe hold spring and shoe hold pin (B). Remove the upper return spring (A)

To install:

11. Connect the parking brake cable (A) to the brake assembly.
12. Clean the threaded portions of adjuster sleeve and push rod female. Coat the threads of the adjuster assembly with grease. To shorten the clevises, turn the adjuster bolt.
13. Hook the shoe adjuster lever, then install it to the brake shoe.
14. Install the adjuster assembly and upper return spring.

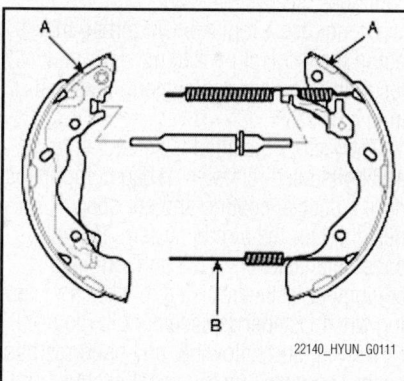

Fig. 7 Lower the brake shoe assembly (A), and remove the lower return spring (B)

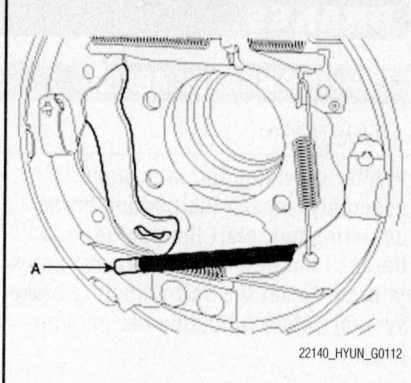

Fig. 8 Remove the parking brake cable (A) from the brake assembly

✳✳ WARNING

Be careful not to damage the wheel cylinder dust covers.

15. Install the lower return spring.
16. Apply brake cylinder grease, or equivalent rubber grease, to the sliding surfaces and brake shoe ends and opposite edges of the shoes.

➡ **Be careful not to get grease on the brake linings.**

17. Install the brake shoes onto the backing plate.
18. Install the shoe hole down pins and the shoe hole down springs.
19. Install the rear brake drum.
20. Install the tire and wheel assembly.
21. Depress the brake pedal several times to set the self-adjusting brake.
22. Adjust the parking brake, as necessary.

ADJUSTMENT

These vehicles have a self-adjusting mechanism in the rear drum brake assembly.

BRAKES

PARKING BRAKE CABLES

ADJUSTMENT

→**After servicing the rear brake assembly, loosen the parking brake adjusting nut, start the engine, and depress the brake pedal several times in order to set the self-adjusting brake system before adjusting the parking brake.**

1. Before servicing the vehicle, refer to the Precautions Section.

2. Block the front wheels, then raise the rear of the vehicle and make sure it is securely supported.

3. Set the parking brake 1 click toward engagement.

4. Remove the floor console, if equipped.

5. Tighten the adjusting nut until the parking brakes drag slightly when the rear wheels are turned.

6. Release the parking brake lever completely.

7. Check if the parking brakes drag when the rear wheels are turned. Readjust if

PARKING BRAKE

necessary until there is no drag from the parking brakes.

8. Check the proper operation of the parking brakes by fully applying the parking brakes.

9. Reinstall the floor console, if equipped.

PARKING BRAKE SHOES

REMOVAL & INSTALLATION

The rear drum brake shoes serve as the parking brakes. Refer to the procedures under Rear Drum Brakes, removal & installation.

CHASSIS ELECTRICAL

AIR BAG (SUPPLEMENTAL RESTRAINT SYSTEM)

GENERAL INFORMATION

✴ CAUTION

Some vehicles are equipped with an air bag system. The system must be disarmed before performing service on, or around, system components, the steering column, instrument panel components, wiring, and sensors. Failure to follow the safety precautions and the disarming procedure could result in accidental air bag deployment, possible injury, and unnecessary system repairs.

SERVICE PRECAUTIONS

Disconnect and isolate the battery negative cable before beginning any airbag system component diagnosis, testing, removal, or installation procedures. Allow system capacitor to discharge for 3 minutes before beginning any component service. This will disable the airbag system. Failure to disable the airbag system may result in accidental airbag deployment, personal injury, or death.

Do not place an intact undeployed airbag face down on a solid surface. The airbag will propel into the air if accidentally deployed and may result in personal injury or death.

When carrying or handling an undeployed airbag, the trim side (face) of the airbag should be pointing towards the body to minimize possibility of injury if accidental deployment occurs. Failure to do this may result in personal injury or death.

Replace airbag system components with OEM replacement parts. Substitute parts may appear interchangeable, but internal differences may result in inferior occupant

protection. Failure to do so may result in occupant personal injury or death.

Wear safety glasses, rubber gloves, and long sleeved clothing when cleaning powder residue from vehicle after an airbag deployment. Powder residue emitted from a deployed airbag can cause skin irritation. Flush affected area with cool water if irritation is experienced. If nasal or throat irritation is experienced, exit the vehicle for fresh air until the irritation ceases. If irritation continues, see a physician.

Do not use a replacement airbag that is not in the original packaging. This may result in improper deployment, personal injury, or death.

The factory installed fasteners, screws and bolts used to fasten airbag components have a special coating and are specifically designed for the airbag system. Do not use substitute fasteners. Use only original equipment fasteners listed in the parts catalog when fastener replacement is required.

During, and following, any child restraint anchor service, due to impact event or vehicle repair, carefully inspect all mounting hardware, tether straps, and anchors for proper installation, operation, or damage. If a child restraint anchor is found damaged in any way, the anchor must be replaced. Failure to do this may result in personal injury or death.

Deployed and non-deployed airbags may or may not have live pyrotechnic material within the airbag inflator.

Do not dispose of driver/passenger/curtain airbags or seat belt tensioners unless you are sure of complete deployment. Refer to the Hazardous Substance Control System for proper disposal.

Dispose of deployed airbags and tensioners consistent with state, provincial, local, and federal regulations.

After any airbag component testing or service, do not connect the battery negative cable. Personal injury or death may result if the system test is not performed first.

If the vehicle is equipped with the Occupant Classification System (OCS), do not connect the battery negative cable before performing the OCS Verification Test using the scan tool and the appropriate diagnostic information. Personal injury or death may result if the system test is not performed properly.

Never replace both the Occupant Restraint Controller (ORC) and the Occupant Classification Module (OCM) at the same time. If both require replacement, replace one, then perform the Airbag System test before replacing the other.

Both the ORC and the OCM store Occupant Classification System (OCS) calibration data, which they transfer to one another when one of them is replaced. If both are replaced at the same time, an irreversible fault will be set in both modules and the OCS may malfunction and cause personal injury or death.

If equipped with OCS, the Seat Weight Sensor is a sensitive, calibrated unit and must be handled carefully. Do not drop or handle roughly. If dropped or damaged, replace with another sensor. Failure to do so may result in occupant injury or death.

If equipped with OCS, the front passenger seat must be handled carefully as well. When removing the seat, be careful when it setting on the floor not to drop it. If dropped, the sensor may be inoperative, could result in occupant injury, or possibly death.

If equipped with OCS, when the passenger front seat is on the floor, no one should sit in the front passenger seat. This uneven force may damage the sensing ability of the seat weight sensors. If sat on and damaged, the sensor may be inoperative, could result in occupant injury, or possibly death.

Several precautions must be observed when handling the inflator module to avoid accidental deployment and possible personal injury.

• Never carry the inflator module by the wires or connector on the underside of the module

• When carrying a live inflator module, hold it securely with both hands, and ensure that the bag and trim cover are pointed away

• Place the inflator module on a bench or other surface with the bag and trim cover facing up

• With the inflator module on the bench, never place anything on or close to the module which may be thrown in the event of an accidental deployment

Before servicing the vehicle, make sure to refer to the precautions in the beginning of this section as well.

DISARMING THE SYSTEM

1. Disconnect and isolate the negative battery cable.
2. Wait 3 minutes for the system capacitor to discharge before performing any service.

ARMING THE SYSTEM

1. Reconnect the negative battery cable.
2. Turn the ignition switch to the **RUN** position.
3. Verify the SIR indicator light flashes 7–9 times, if not, inspect the system for malfunction.

CLOCKSPRING CENTERING

See Figures 9 through 11.

1. Disconnect and isolate the battery negative cable. Allow the system capacitor to discharge for 3 minutes before beginning any component service. This will disable the airbag system.

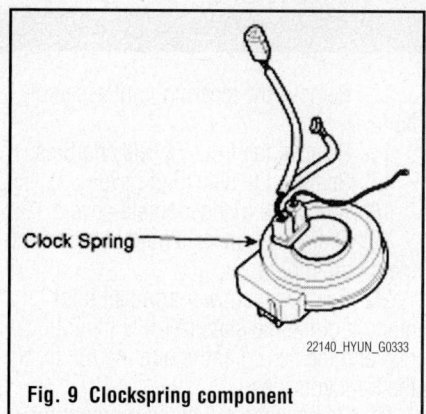

Fig. 9 Clockspring component

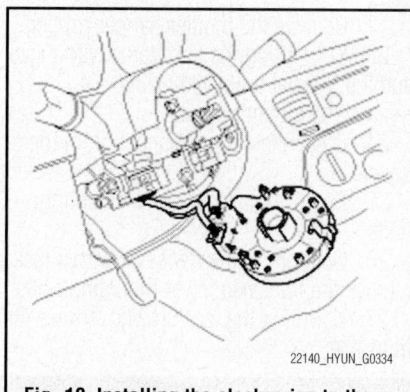

Fig. 10 Installing the clockspring to the steering column

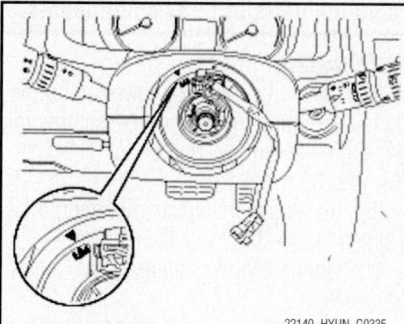

Fig. 11 Aligning the marks to center the clockspring

2. Remove the ignition key from the vehicle.
3. Connect the clockspring harness connector and horn harness connector to the clockspring.
4. Set the clockspring on neutral position and after turning the front wheels to the straight-ahead position, install the clockspring.

a. Check connectors and protective tube for damage, and terminals for deformities.

b. If even one abnormal point is discovered, replace the clockspring with a new one.

5. Connect the clockspring harness connector and the steering switch harness connector to the clockspring.
6. Set the center position by getting the marks between the clockspring and the cover into line.
7. Turn the clockspring clockwise to the stop and then 2.4 revolutions counterclockwise. See the illustration of marks in line.
8. Install the steering wheel column cover and the steering wheel.
9. Connect the Driver Airbag (DAB) module connector and the horn connector, then install the DAB module on the steering wheel.
10. Secure the DAB with the new mounting bolts and tighten to 70–96 inch lbs. (8–11 Nm).
11. Connect the battery negative cable.
12. After installing the airbag, confirm proper system operation:

a. Turn the ignition switch ON; the SRS indicator light should be turned on for about 6 seconds and then go off.

b. Make sure the horn works.

DRIVE TRAIN

AUTOMATIC TRANSAXLE ASSEMBLY

REMOVAL & INSTALLATION

1.6L, 2.0L & 2.4L Engines

See Figure 12.

Use fender covers to avoid damaging painted surfaces. To avoid damage, unplug the wiring connectors carefully while holding the connector portion. Mark all wiring and hoses to avoid misconnection.

1. Before servicing the vehicle, refer to the Precautions Section.
2. Remove the engine cover.
3. Remove the battery after removing the battery terminal.
4. Remove the air duct assembly.
5. Remove the air cleaner assembly by disconnecting the Air Flow Sensor (AFS) connector, the clamp, and the ECM connector.
6. Remove the battery tray.
7. Remove the ground cable from the transaxle.
8. Disconnect the inhibitor switch connector, solenoid valve connector, and the input shaft speed sensor connector.
9. Disconnect the output shaft speed sensor connector.
10. Remove the control cable assembly.
11. Remove the oil cooler hoses.
12. Install the special tools (09200-38001), the engine support fixture and the adapter on the engine assembly.
13. Remove the transaxle upper mounting bolts and the starter motor mounting bolts.
14. Remove the 4 bolts and take off the transaxle support bracket.

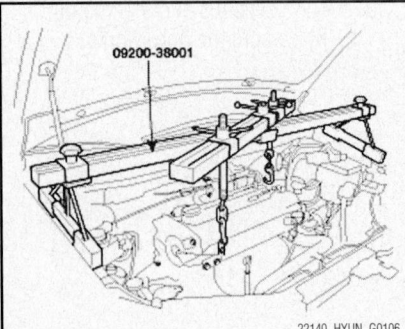

Fig. 12 Install the special tools (09200-38001), the engine support fixture and the adapter on the engine assembly—1.6L, 2.0L and 2.4L engines

15. Remove the steering joint assembly bolt.
16. Remove the front wheels and tires.
17. Remove the side mud cover.
18. Remove the under shield cover.
19. Drain the transaxle fluid by removing the oil drain plug.
20. Remove the lower arm ball joint mounting nut, the stabilizer link mounting nut, and the tie rod end mounting nut from the front knuckles.
21. Remove the roll stopper mounting bolts.
22. Remove the muffler hanger rubber.
23. Supporting the sub frame with a jack and the Special tool (09624-38000), remove the mounting bolts.
24. Disconnect the halfshafts from the transaxle.
25. Remove the drive plate mounting bolts.
26. Supporting the transaxle with a jack, remove the transaxle lower mounting bolts.
27. Lowering the jack slowly, remove the transaxle.

✳✳ WARNING

When removing the transaxle assembly, be careful not to damage any surrounding parts or body components.

To install:

28. Install the transaxle lower mounting bolts after fitting the transaxle assembly into the engine assembly. Tighten to: 31–40 ft. lbs. (43–55 Nm).
29. Install the drive plate mounting bolts. Tighten to: 33–38 ft. lbs. (46–53 Nm).
30. Connect the halfshafts to the transaxle.
31. Supporting the sub frame with a jack and the Special tool(09624-38000), install the mounting bolts. Tighten the bolts to: 101–118 ft. lbs. (140–160 Nm).
32. Install the muffler hanger rubber.
33. Install the roll stopper mounting bolts. Tighten to: 36–47 ft. lbs. (50–65 Nm).
34. Install the lower arm ball joint mounting nut, the stabilizer link mounting nut, and the tie rod end mounting nut to the front knuckles.
35. Install the under shield cover.
36. Install the side mud cover.
37. Install the front wheels and tires.
38. Install the steering joint assembly bolt.
39. Install the transaxle support bracket bolts. Tighten to: 43–58 ft. lbs. (60–80 Nm).
40. Install the transaxle upper mounting bolts. Tighten to: 43–58 ft. lbs. (60–80 Nm).

41. Install the starter motor mounting bolts. Tighten to: 28–43 ft. lbs. (39–60 Nm).
42. Remove the special tool (09200-38001).
43. Connect the transaxle oil cooler hoses to the tubes by fastening the clamps.
44. Install the control cable assembly.
45. Install the output speed sensor connector.
46. Connect the inhibitor switch connector, solenoid valve connector and the input shaft speed sensor connector.
47. Install the ground cable to transaxle.
48. Install the battery tray.
49. Install the air cleaner assembly by connecting the AFS connector, the clamp, and the ECM connector.
50. Install the air duct assembly.
51. Install the battery and the battery terminal.
52. Install the engine cover.
53. After completing the installation perform the following procedure:
   a. Adjust the shift cable.
   b. Refill the transaxle fluid.
   c. Clean the battery posts and cable terminals with sandpaper and grease them to prevent corrosion before installing.
   d. When replacing the automatic transaxle, reset the automatic transaxle's values by using the High-Scan Pro:
   • Connect the Hi-Scan Pro connector to the data link connector under the crash pad and power cable to the cigar jack under the center fascia
   • Turn the ignition switch on and power on the Hi-Scan Pro
   • Select the vehicle name
   • Select AUTOMATIC TRANSAXLE
   • Select RESETTING AUTO T/A VALUES and perform the procedure
   • Perform the procedure by pressing F1 (REST)
54. Install the transaxle control cable and adjust as follows:
   a. Move the shift lever and the transaxle range switch to the "N" Position, and install the control cable.
   b. When connecting the control cable to the transaxle mounting bracket, install the clip until it contacts the control cable.
   c. Remove any free-play in the control cable by adjusting the nut and then check to see that the select lever moves smoothly.
   d. Check to see that the control cable has been adjusted correctly.

55. Fill the transaxle fluid to the proper level.

56. Start the vehicle, check for leaks and repair if necessary.

2.7L, 3.3L & 3.8L Engines

See Figures 13 and 14.

Use fender covers to avoid damaging painted surfaces. To avoid damage, unplug the wiring connectors carefully while holding the connector portion. Mark all wiring and hoses to avoid misconnection.

1. Before servicing the vehicle, refer to the Precautions Section.

2. Disconnect the negative terminal from the battery.

3. Remove the engine cover.

4. Remove the intake air hose and the air cleaner assembly.

   a. Disconnect the AFS connector.

   b. Disconnect the breather hose from air cleaner hose.

   c. Disconnect the PCM connectors.

   d. Remove the intake air hose and air cleaner.

5. Disconnect the positive terminal from the battery and remove the battery.

6. Remove the transaxle oil cooler hoses.

7. Remove engine wiring.

   a. Disconnect the RH rear oxygen sensor connector.

   b. Disconnect the LH rear oxygen sensor connector and the CPS connector.

8. Disconnect the transaxle wire harness connector and remove transaxle control cable.

   a. Remove the wiring brackets.

   b. After removing a transaxle bracket, remove the inhibiter switch connector and shift cable.

   c. Remove the solenoid valve connector.

   d. Remove the input speed sensor, output speed sensor and vehicle speed sensor connector.

   e. Disconnect the ground wire.

9. Disconnect the power steering pressure sensor connector.

10. Remove the power steering hose mounting bolts.

11. Remove the front wheels.

12. Disconnect the EPS connector around the left hand side front wheel.

13. Remove the transaxle mounting bolts.

14. Using the SST (09200-38001), hold the engine and transaxle assembly safely.

15. Remove the transaxle insulator mounting bolt.

16. Raise and safely support the vehicle.

17. Remove the undercover.

18. Drain the transaxle oil.

19. Disconnect the power steering pump hose.

20. Disconnect the lower arm assembly from the knuckle.

21. Disconnect the tie rod end ball joint from the knuckle after removing the split pin.

22. Disconnect the stabilizer bar link.

23. Remove the front roll stopper mounting bolt.

24. Remove the front exhaust pipe.

25. Remove the rear roll stopper mounting bolt.

26. Using the SST (09624-38000) and holding the cross member with a jack, remove the steering bolt.

27. Remove the cross member.

28. Remove the halfshaft from transaxle.

29. Install a jack for supporting the transaxle assembly.

30. Remove the transaxle under mounting bolts and the drive plate bolts.

31. Lifting the vehicle up or lowering the jack slowly, remove the transaxle assembly.

To install:

32. Lowering the vehicle or lifting up a jack, install the transaxle assembly.

33. Tighten the transaxle under mounting bolts. Tighten to:

   a. Bolts (A): 25–30 ft. lbs. (34–41.2 Nm).

   b. Bolt (B): 33–38 ft. lbs. (45–52 Nm).

34. Remove the jack and insert the halfshafts.

35. Supporting the cross member with the SST (09624-38000), tighten the steering column bolt and the cross member mounting bolts.

36. Tighten the rear roll stopper mounting bolt to: 36–47 ft. lbs. (49–64 Nm).

37. Install the front exhaust pipe.

38. Tighten the front roll stopper mounting bolt to: 36–47 ft. lbs. (49–64 Nm).

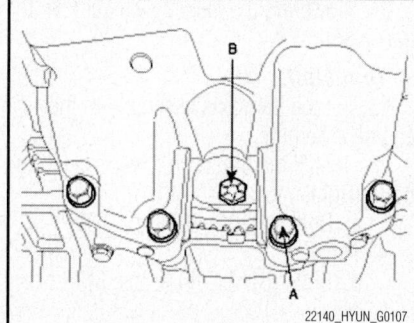

Fig. 13 Tighten the transaxle under mounting bolts (A, B)—2.7L, 3.3L, and 3.8L engines

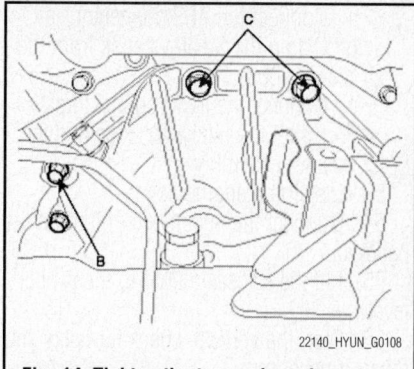

Fig. 14 Tighten the transaxle under mounting bolts (B, C)—2.7L, 3.3L, and 3.8L engines

39. Install the steering bar tie rod, the stabilizer bar link, and the lower arm assembly.

40. Clamp the power steering pump hose.

41. Install the undercover.

42. After lowering the vehicle, tighten the transaxle insulator mounting bolt (A). Tighten to: 47–62 ft. lbs. (63–83 Nm).

43. Tighten the transaxle mounting bolts (B, C).

   a. Bolts (B): 24–36 ft. lbs. (32–49 Nm).

   b. Bolts (C): 47–62 ft. lbs. (64–83 Nm).

44. Remove the SST (09200-38001) holding the engine and transaxle assembly.

45. Connect the EPS connector and install the front wheels and tires.

46. Install the power steering hose mounting bolts.

47. Connect the power steering pressure sensor connector.

48. Connect the transaxle wire harness connector and the control cable.

   a. Install the wiring brackets.

   b. Connect the inhibitor switch connector and the shift cable and install the transaxle bracket.

   c. Connect the solenoid valve connector.

   d. Connect the input/output speed sensor connectors and vehicle speed sensor connector.

   e. Connect the ground wire.

49. Connect the engine wiring.

   a. Connect the RH rear oxygen sensor connector.

   b. Connect the LH rear oxygen sensor connector and the CPS connector.

50. Clamp the transaxle oil cooler hoses(A).

51. After disconnecting the positive terminal from the battery, remove the battery.

52. Install the intake air hose and the air cleaner assembly.

a. Connect the AFS connector.

b. Clamp the breather hose from the air cleaner hose.

c. Connect the PCM connectors.

d. Install the intake air hose and the air cleaner assembly.

53. Install the engine cover.

54. Connect the negative terminal on the battery.

55. Fill the transaxle fluid to the proper level.

56. Start the vehicle, check for leaks and repair if necessary.

MANUAL TRANSAXLE ASSEMBLY

REMOVAL & INSTALLATION

1.6L, 2.0L & 2.4L Engines

See Figures 15 through 17.

Use fender covers to avoid damaging painted surfaces. To avoid damage, unplug the wiring connectors carefully while holding the connector portion. Mark all wiring and hoses to avoid misconnection.

1. Before servicing the vehicle, refer to the Precautions Section.

2. Remove the engine cover.

3. Remove the battery after removing the battery terminal connections.

4. Remove the air duct assembly.

5. Remove the air cleaner assembly by disconnecting the clamp and the ECM connector.

6. Remove the ground cable from the transaxle.

7. Disconnect the vehicle speed sensor and the back lamp switch integrated connector.

8. Remove the control cable assembly by removing the snap pins and clips.

9. Remove the control cable bracket.

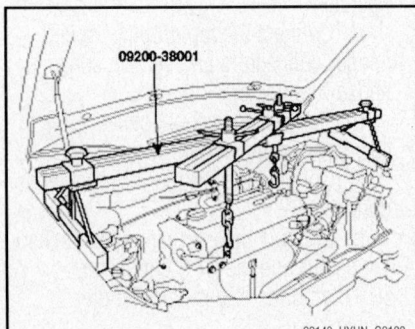

Fig. 15 Using the special tool (09200-38001) to support the engine assembly—1.6L, 2.0L and 2.4L engines

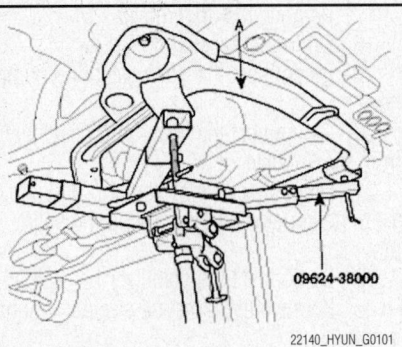

Fig. 16 Support the sub frame (A) with a jack and the Special tool (09624-38000) and remove the mounting bolts—1.6L, 2.0L and 2.4L engines

10. Using the special tool (09200-38001), support the engine assembly safely.

11. Remove the transaxle upper mounting bolts and the starter motor mounting bolts.

12. After removing the bolts, take the transaxle insulator mounting bracket off.

13. Remove the steering joint assembly bolt.

14. Raise and safely support the vehicle.

15. Remove the front wheels and tires.

16. Remove the lower arm ball joint mounting nut, the stabilizer link mounting nut, and the tie rod end mounting nut from the front knuckles.

17. Remove the under shield cover.

18. Remove the roll stopper mounting bolts.

19. Disconnect the muffler hanger rubber.

20. Supporting the sub frame (A) with a jack and the Special tool (09624-38000), remove the mounting bolts.

21. Disconnect the halfshafts from the transaxle.

22. Remove the clutch release cylinder assembly.

23. Supporting the transaxle with a jack, remove the transaxle lower mounting bolts.

24. Lowering the jack slowly, remove the transaxle.

To install:

25. Fit the transaxle assembly to the engine assembly.

26. Install the transaxle lower mounting bolts. Tighten to:

    a. Bolts (A): 22–30 ft. lbs. (30–42 Nm).

    b. Bolts (B): 31–40 ft. lbs. (43–55 Nm).

27. Install the clutch release cylinder assembly.

28. Connect the halfshafts to the transaxle.

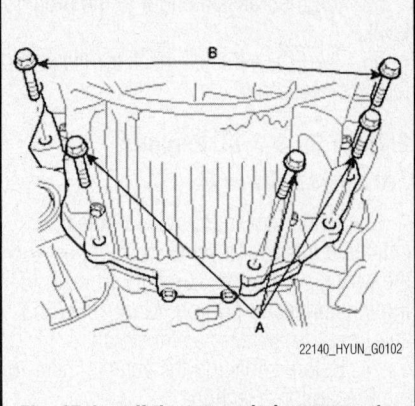

Fig. 17 Install the transaxle lower mounting bolts—1.6L, 2.0L and 2.4L engines

29. Supporting the sub frame with a jack and the Special tool(09624-38000), install the mounting bolts. Tighten to: 101–118 ft. lbs. (140–160 Nm).

30. Install the muffler hanger rubber.

31. Install the roll stopper bracket bolts. Tighten to: 36–47 ft. lbs. (50–65 Nm).

32. Install the under shield cover.

33. Install the lower arm ball joint mounting nut, the stabilizer link mounting nut, and the tie rod end mounting nut to the front knuckles.

34. Install the front wheels and tires.

35. Install the steering joint assembly bolt. Tighten to: 13–18 ft. lbs. (18–25 Nm).

36. Install the transaxle insulator mounting bracket bolts. Tighten to: 43–58 ft. lbs. (60–80 Nm).

37. Remove the special tool (09200-38001).

38. Install the transaxle upper mounting bolts. Tighten to: 43–60 ft. lbs. (60–80 Nm).

39. Install the starter motor mounting bolts. Tighten to: 28–43 ft. lbs. (39–60 Nm).

40. Install the control cable bracket. Tighten to: 11–16 ft. lbs. (15–22 Nm).

41. Connect the vehicle speed sensor and the back lamp switch integrated connector.

42. Install the control cable assembly by installing the clips and pins.

43. Install the ground cable to the transaxle.

44. Install the air cleaner assembly and the ECM connector.

45. Install the air duct assembly.

46. Install the battery and connect the battery terminals.

47. Install the engine cover.

2.7L Engine

Hyundai recommends that the engine and manual transaxle be removed as a single unit. Refer to Engine Assembly, removal & installation.

CLUTCH DRIVEN DISC & PRESSURE PLATE

REMOVAL & INSTALLATION

See Figures 18 through 23.

1. Before servicing the vehicle, refer to the Precautions Section.

2. Remove the transaxle assembly. Refer to Manual Transaxle Assembly, removal & installation.

3. Insert the special tool (09411-11000) in the clutch disc to prevent the disc from falling.

Fig. 18 Special tool placement diagram

4. Loosen the bolts which attach the clutch cover to the flywheel in a star pattern. Loosen the bolts in succession, 1 or 2 turns at a time, to avoid bending the cover flange.

➡**Do not clean the clutch disc or the release bearing with cleaning solvent.**

5. Remove the release fork shaft and bushing.

6. Remove clutch cover assembly and then driven disc and pressure plate.

To install:

❉❉ CAUTION

When installing the clutch, apply grease to each part, but be careful not to apply excessive grease. It can cause clutch slippage and shudder.

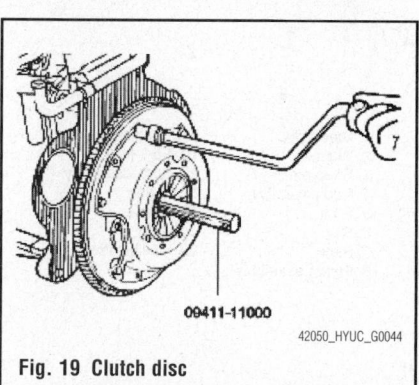

Fig. 19 Clutch disc

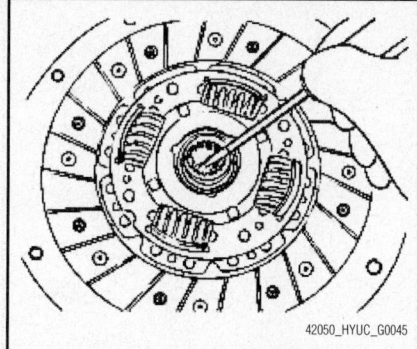

Fig. 20 Clutch grease application

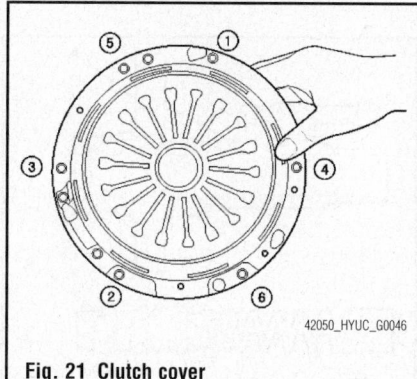

Fig. 21 Clutch cover

7. Install the driven disc and pressure plate in to the clutch cover assembly.

8. Install the clutch disc assembly to the flywheel using the special tool (09411-11000).

9. Install the clutch cover assembly to the flywheel and temporarily tighten the bolts 1 or 2 steps at a time in a star pattern to a final torque of 11–16 ft. lbs. (15–22 Nm).

10. Align the bearing (A) to the release fork (B) and then install it to the sleeve of the housing.

❉❉ CAUTION

Apply multipurpose grease (CAS-MOLY L9508) to the bearing sleeve, contact point of the release fork (B) and the bushing inner surface (C).

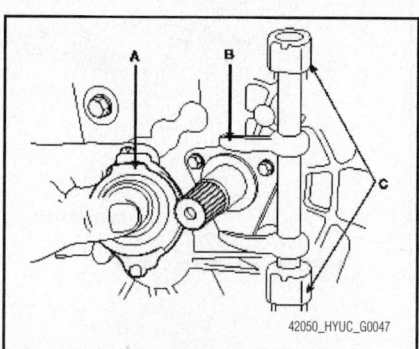

Fig. 22 Release fork grease application

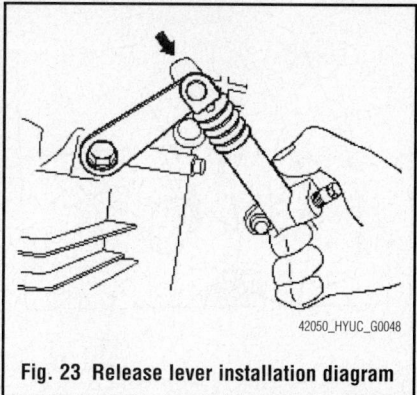

Fig. 23 Release lever installation diagram

11. Install the release lever to the release fork.

❉❉ CAUTION

If the transaxle assembly is installed to the engine without performing this step, the release bearing can be separated, as the release fork rotates freely.

12. Install the transaxle assembly to the engine.

ADJUSTMENTS

The clutch system is hydraulic and requires no adjustment.

CLUTCH MASTER CYLINDER

REMOVAL & INSTALLATION

See Figures 24 and 25.

❉❉ WARNING

Do not spill brake fluid on the vehicle; it may damage the paint; if brake fluid does contact the paint, wash it off immediately with water.

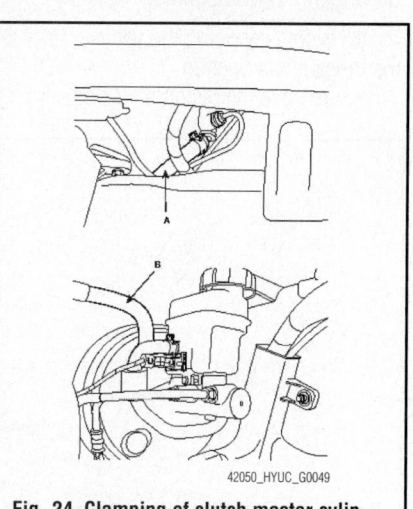

Fig. 24 Clamping of clutch master cylinder hose diagram

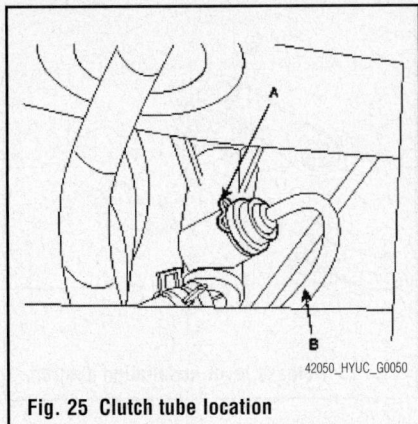

Fig. 25 Clutch tube location

1. Before servicing the vehicle, refer to the Precautions Section.

2. Remove the brake fluid from the clutch master cylinder reservoir with a syringe.

3. Clamp the clutch master cylinder hose (A). If there is not enough room for clamping, you can also clamp hose (B) from the brake master cylinder side.

4. Disconnect the hose from the cylinder by releasing the clutch master cylinder clamp.

5. Remove the clip (A) and disconnect the clutch tube (B).

6. Remove the pin and washer which connect the clutch pedal with the clutch master cylinder.

7. After loosening the clutch master cylinder assembly mounting bolts under the driver's seat, remove the clutch master cylinder. It can be helpful to do this step after removing the clutch pedal mounting bracket.

To install:

8. Installation is the reverse of removal.

OVERHAUL

See Figures 26 through 29.

1. Before servicing the vehicle, refer to the Precautions Section.

2. Remove the piston stop ring.

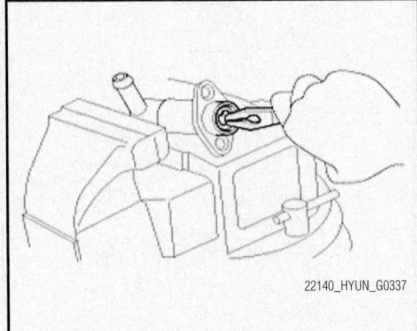

Fig. 26 Removing the piston stop ring of the clutch master cylinder

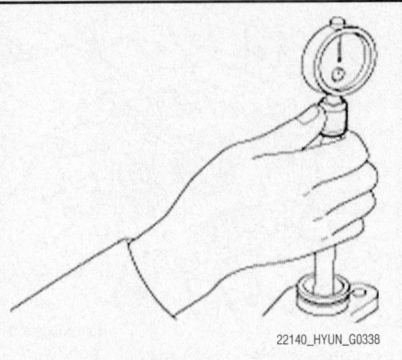

Fig. 27 Using a cylinder gauge to measure the master cylinder inside diameter

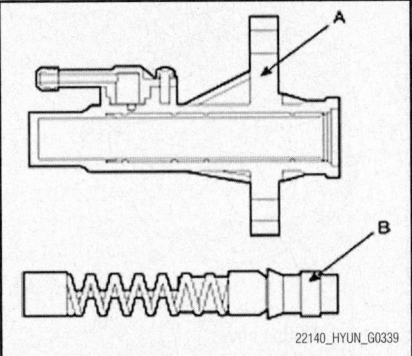

Fig. 28 Apply brake fluid to the inner surface of the master cylinder body (A) and to the entire periphery of the piston assembly (B)

3. Pull out the push rod and piston assembly.

4. Remove the reserve tank band, reserve tank cap, and reserve tank.

    a. Use care not to damage the master cylinder body and piston assembly.

    b. Do not disassemble the piston assembly.

5. Check the inside of the cylinder body for rust, pitting, or scoring.

6. Check the piston cup for wear or distortion.

7. Check the piston for rust, pitting, or scoring.

8. Check to make sure the clutch line tube is not clogged or restricted in any way.

9. Measure the master cylinder inside diameter and the piston outside diameter with a cylinder gauge micrometer.

➡**Measure the inside diameter of the master cylinder at 3 places (bottom, middle, and top) in a perpendicular direction.**

10. If the master cylinder-to-piston clearance exceeds the limit, replace the master cylinder and/or piston assembly. Limit: 0.006 inch (0.15mm).

11. Apply the specified fluid to the inner surface of the master cylinder body (A) and to the entire periphery of the piston assembly (B). Specified fluid: Brake fluid DOT 3 or DOT 4.

12. Install the piston assembly.

13. Install the piston snap ring.

14. Install the push rod assembly.

CLUTCH SLAVE CYLINDER

REMOVAL & INSTALLATION

See Figures 30 and 31.

1. Before servicing the vehicle, refer to the Precautions Section.

2. Disconnect the clutch tube.

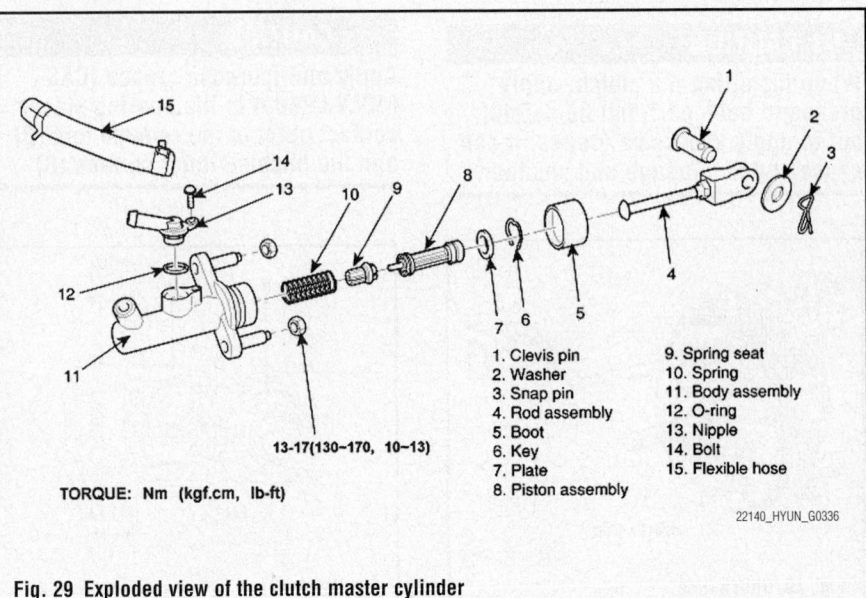

1. Clevis pin
2. Washer
3. Snap pin
4. Rod assembly
5. Boot
6. Key
7. Plate
8. Piston assembly
9. Spring seat
10. Spring
11. Body assembly
12. O-ring
13. Nipple
14. Bolt
15. Flexible hose

13-17(130~170, 10~13)

TORQUE: Nm (kgf.cm, lb-ft)

Fig. 29 Exploded view of the clutch master cylinder

3. Remove the 2 clutch slave cylinder mounting bolts (A).

4. Remove the clutch slave cylinder (also called the clutch release cylinder).

To install:

5. Check the clutch release cylinder for fluid leakage.

6. Check the clutch release cylinder boots for damage.

7. Coat the clutch clevis push rod with specified grease: CASMOLY® L9508.

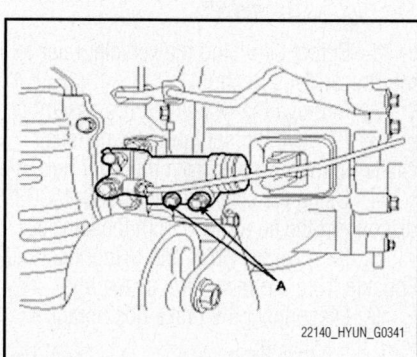

22140_HYUN_G0341

Fig. 30 Disconnect the clutch tube and remove the 2 clutch release slave cylinder mounting bolts (A)

8. Install the release cylinder (A) to the transaxle. Tighten the bolts to: 11–16 ft. lbs. (15–22 Nm).

9. Install the clutch tube.

CLUTCH HYDRAULIC SYSTEM BLEEDING

See Figure 32.

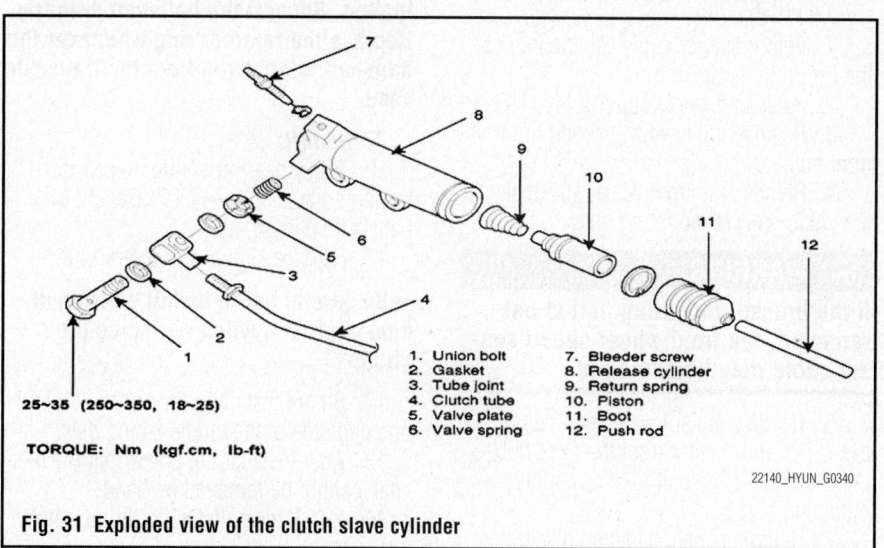

25~35 (250~350, 18~25)

TORQUE: Nm (kgf.cm, lb-ft)

1. Union bolt
2. Gasket
3. Tube joint
4. Clutch tube
5. Valve plate
6. Valve spring
7. Bleeder screw
8. Release cylinder
9. Return spring
10. Piston
11. Boot
12. Push rod

22140_HYUN_G0340

Fig. 31 Exploded view of the clutch slave cylinder

1. Before servicing the vehicle, refer to the Precautions Section.

2. Connect a hose to the bleeder screw and place the other end of hose into a container of clean brake fluid. Open the bleeder screw.

3. Have an assistant pump the clutch pedal slowly until no air bubbles are present at the bleeder screw.

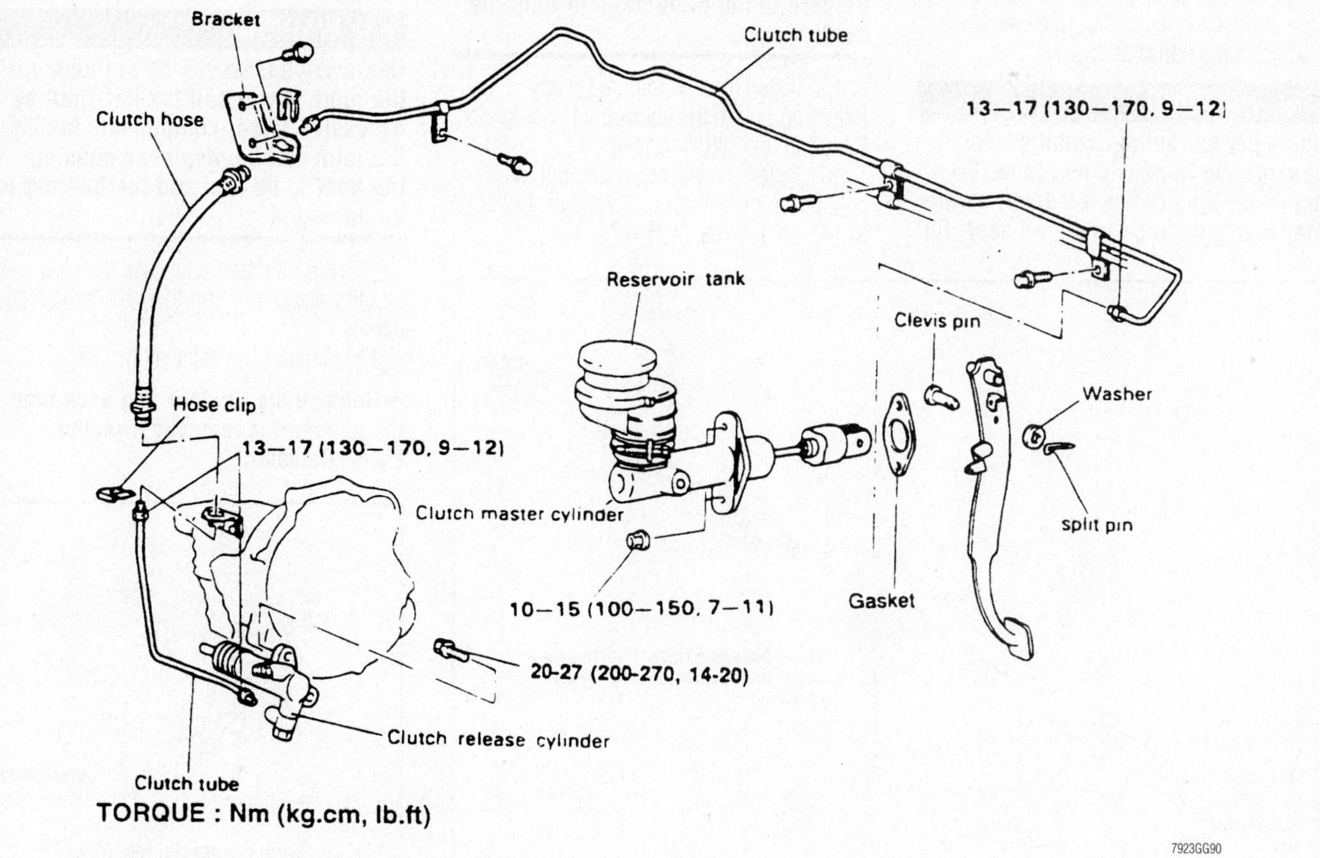

Bracket

Clutch tube

Clutch hose

13—17 (130—170, 9—12)

Hose clip

13—17 (130—170, 9—12)

Reservoir tank

Clevis pin

Washer

split pin

Clutch master cylinder

10—15 (100—150, 7—11)

Gasket

20-27 (200-270, 14-20)

Clutch release cylinder

Clutch tube

TORQUE : Nm (kg.cm, lb.ft)

7923GG90

Fig. 32 Exploded view of the clutch hydraulic system

4. Close the bleeder screw.
5. Fill the clutch master cylinder.
6. Check the clutch operation.

FRONT HALFSHAFTS

REMOVAL & INSTALLATION

1.6L, 2.0L, 2.4L & 2.7L Engines

See Figure 33.

1. Before servicing the vehicle, refer to the Precautions Section.
2. Raise and safely support the vehicle.
3. Remove the front wheel and tire from front hub.
4. Remove the front wheel speed sensor cable bracket mounting bolt.

❊❊ WARNING

If the bracket mounting bolt is not removed, the front wheel speed sensor cable may be damaged.

5. Remove the split pin, then remove castle nut and washer from the front hub.
6. Remove the ball joint assembly mounting bolt from the knuckle.
7. Using a plastic hammer, disconnect halfshaft from the axle hub.
8. Insert a pry bar between the transaxle case and joint case, and separate the halfshaft from the transaxle case.

❊❊ WARNING

Use a pry bar being careful not to damage the transaxle and joint. Do not insert the pry bar too deep, as this may cause damage to the oil seal. Do not pull the halfshaft by excessive force it may cause components inside the joint to dislodge resulting in a torn boot or a damaged bearing.

9. Pull out the halfshaft from the transaxle case.

➡ **Plug the hole of the transaxle case with the oil seal cap to prevent contamination. Support the halfshaft properly. Replace the retainer ring whenever the halfshaft is removed from the transaxle case.**

To install:

10. Apply gear oil on the oil seal contacting surface of transaxle case and the halfshaft splines.
11. Replace circlip with a new one.

➡ **Be careful not to install a different kind of circlip, when replacing the circlip.**

12. Before installing the halfshaft, set the opening side of the circlip facing downward.
13. After installation, check that the halfshaft cannot be removed by hand.
14. Install the halfshaft to the axle hub.

❊❊ WARNING

Be careful not to damage or dent the boots.

15. Install the ball joint assembly mounting bolt to the knuckle and tighten to 72–87 ft. lbs. (98–118 Nm).
16. Install the washer, castle nut, and split pin to the front hub assembly. Tighten to 145–203 ft. lbs. (196–275 Nm).

➡ **The washer should be assembled with convex surface outward when installing the castle nut and split pin.**

17. Install the front wheel speed sensor cable bracket mounting bolt and tighten to 61–96 inch lbs. (7–11 Nm).
18. Install the wheel and the tire to the front hub. Tighten the lug nuts to 65–80 ft. lbs. (88–108 Nm).

3.3L & 3.8L Engines

See Figures 34 and 35.

1. Before servicing the vehicle, refer to the Precautions Section.
2. Remove the wheel and tire assembly.
3. Remove the split pin and halfshaft castle nut and washer from the front hub.
4. Using the special tool (09568-4A000), disconnect the tie rod end from the knuckle.
5. Remove the bolts and disconnect the knuckle from the lower arm assembly.
6. Disconnect the brake hose bracket from the knuckle.
7. Disconnect the wheel speed sensor bracket from the knuckle.
8. Using a plastic hammer, disconnect the halfshaft from the axle assembly.
9. Remove the left-hand halfshaft from the transaxle using a pry bar.

❊❊ WARNING

Use a pry bar so you do not damage the joint. If you pull the halfshaft by excessive force, components inside the joint can be displaced causing the boot to be torn and the bearing to be damaged.

10. Plug the transaxle case opening with an oil seal cap in order to avoid contamination.
11. Support the halfshaft properly.

➡ **Replace the retainer ring each time the halfshaft is removed from the transaxle case.**

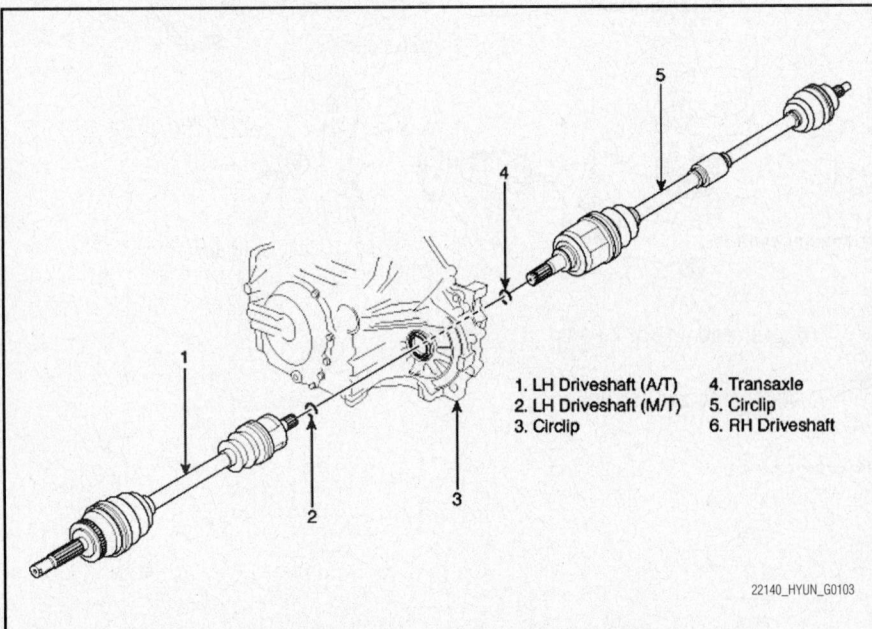

1. LH Driveshaft (A/T)	4. Transaxle
2. LH Driveshaft (M/T)	5. Circlip
3. Circlip	6. RH Driveshaft

22140_HYUN_G0103

Fig. 33 Exploded view of halfshaft components—1.6L, 2.0L, 2.4L, and 2.7L engines

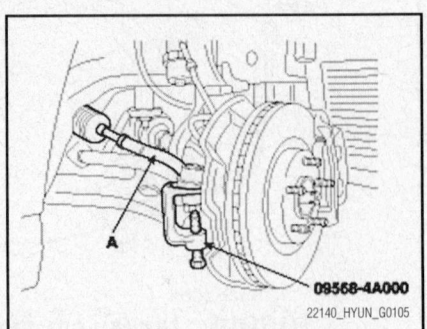

09568-4A000

22140_HYUN_G0105

Fig. 34 Using the special tool (09568-4A000), disconnect the tie rod end from the knuckle—3.3L & 3.8L engines

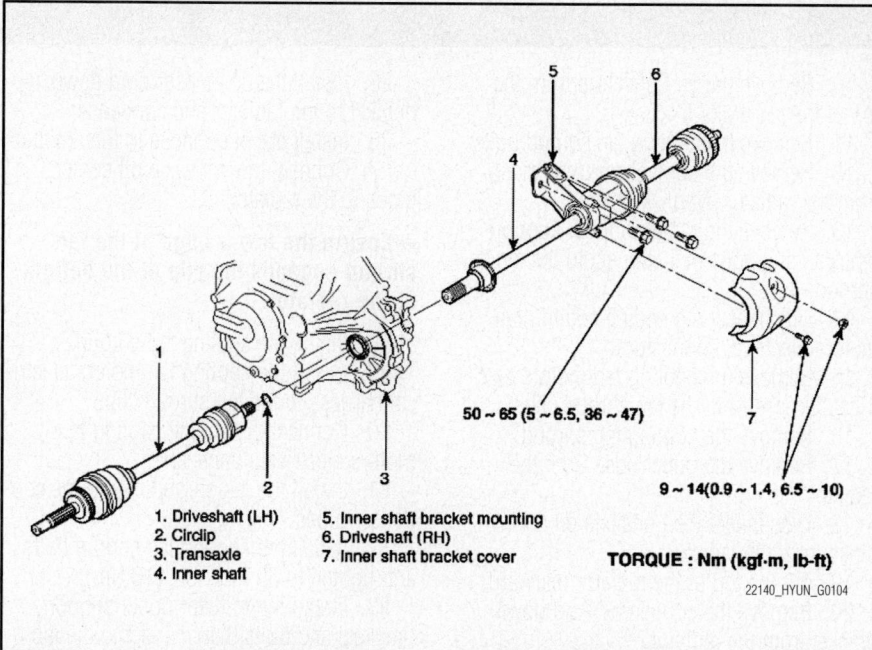

1. Driveshaft (LH)
2. Circlip
3. Transaxle
4. Inner shaft
5. Inner shaft bracket mounting
6. Driveshaft (RH)
7. Inner shaft bracket cover

50 ~ 65 (5 ~ 6.5, 36 ~ 47)

9 ~ 14(0.9 ~ 1.4, 6.5 ~ 10)

TORQUE : Nm (kgf·m, lb-ft)

22140_HYUN_G0104

Fig. 35 Exploded view of halfshaft components—3.3L & 3.8L engines

✳✳ WARNING

While loosening the halfshaft nut, do not allow vehicle weight to be concentrated on the wheel bearing. If the vehicle moves, hold the wheel bearing using the special tool.

12. Remove the right-hand halfshaft.
   a. Remove the stabilizer link from the fork.
   b. Remove the fork from the front lower arm.

✳✳ WARNING

Be careful not to damage to the aluminum lower arm.

   c. Remove the fork from the front strut assembly.
   d. Remove the inner shaft cover from the inner shaft bracket.
   e. Remove the inner shaft bracket mounting bolts.
   f. Remove the front halfshaft assembly with the inner shaft from the transaxle.

✳✳ WARNING

Do not try to disconnect the inner shaft from the halfshaft. Because they cannot be disconnected once assembled. Do not reuse the halfshaft which is disassembled from the inner shaft.

13. Use the special tool (09432-11000), to remove the tone wheel.

To install:

14. Replace the circlips with new ones after removal.
15. Apply gear oil on the halfshaft splines and the contacting surface of differential case oil seal.
16. After installation, check that the halfshaft cannot be removed.
17. Install the right-hand halfshaft.
   a. Install the inner shaft bracket mounting bolt and tighten to 36–47 ft. lbs. (50–65 Nm).
   b. Install the inner shaft cover by installing the cover mounting bolts. Tighten to 84–120 inch lbs. (9–14 Nm).

   c. Install the fork to the front strut assembly. Tighten to 44–59 ft. lbs. (60–80 Nm).
   d. Install the connecting bolt between the fork and the lower arm and tighten to 101–118 ft. lbs. (140–160 Nm).
   e. Install the stabilizer link to the fork and tighten to 74–88 ft. lbs. (100–120 Nm).
18. Install the halfshaft into the front axle assembly.

✳✳ WARNING

Be careful not to damage the boot.

19. Install the knuckle in the lower arm assembly and tighten the bolts. Tightening Torque: 74–88 ft. lbs. (100–120 Nm).
20. Install the tie rod end in the knuckle. Tighten to 18–25 ft. lbs. (24–34 Nm).
21. Install the wheel speed sensor in the knuckle.
22. Install the brake hose bracket to the front knuckle.
23. After installing the washer with convex surface outward, install the lock nut and the split pin. Tighten the lock nut to: 148–207 ft. lbs. (200–280 Nm).
24. Install the wheel & tire assembly.

CV-BOOTS INSPECTION

1. Check the Double Offset Joint (DOJ) outer race, inner race, cage and balls for rust or damage.
2. Check splines for wear.
3. Check for tears, water, foreign matter, or rust in the Birfield Joint (BJ) boot.
4. Check the DOJ boot for tears, cuts or holes.

✳✳ CAUTION

When the Birfield Joint assembly is to be reused, do not wipe away the grease. Check that there are no foreign substances in the grease. If necessary, clean the Birfield Joint assembly and replace grease.

ENGINE COOLING

ENGINE FAN

REMOVAL & INSTALLATION

1. Before servicing the vehicle, refer to the Precautions Section.
2. Disconnect the ground cable from the battery cable.
3. Disconnect the connectors from the fan motor and the harness from the shroud.
4. For vehicles with automatic transaxles, remove the oil cooler hose from the shroud.
5. Remove the 4 bolts holding the shroud.
6. Remove the shroud with the fan motor.
7. Remove the fan mounting clip and detach the fan from the fan motor.
8. Remove the 3 screws and detach the fan motor.

To install:

9. Reattach the fan motor to the fan shroud and insert the 3 screws to secure.
10. Reattach the fan to the fan motor and install the fan mounting clip to secure.
11. Install the fan shroud with attached fan and fan motor to the radiator.
12. Install and tighten the 4 bolts holding the fan shroud in place to the radiator.
13. Install the oil cooler hose back on the shroud, if previously removed.
14. Reconnect the connectors from the fan motor and the harness from the shroud.
15. Reconnect the battery ground or negative cable.

RADIATOR

REMOVAL & INSTALLATION

1. Before servicing the vehicle, refer to the Precautions Section.
2. Remove the air duct.
3. Disconnect the negative battery cable.
4. Remove the air cleaner assembly.
5. Drain the cooling system. Remove the radiator cap to speed draining.
6. Remove the inlet hose from the radiator.
7. Remove the transmission oil cooler lines from the retainer clip at the bottom of the cooling fan shroud.
8. Remove the fan shroud clip from the condenser tubes.
9. Remove the bolt that connects the fan shroud to the condenser hold down bracket.

10. Remove the air deflectors from the top of the radiator.
11. Remove the cooling fan shroud bolts.
12. Remove the coolant reservoir hose from the radiator overflow neck.
13. Remove the radiator upper support brackets and bolts that connect to the fan shroud.
14. Disconnect the engine cooling fan motors electrical connectors.
15. Remove the cooling fan motors electrical harness from the fan shroud clips.
16. Remove the cooling fan shroud.
17. Remove the outlet hose from the radiator.
18. Disconnect the transaxle oil cooler pipes from the radiator.
19. Tilt the top of the radiator rearward.
20. Remove the condenser hold down bracket from the radiator.
21. Lift the condenser from the mounting tabs on the radiator and position the condenser aside.
22. Remove the radiator.

To install:

23. Install the radiator to the lower mounts.

➡**Verify that the condenser is fully seated in the radiator mounting tabs.**

24. Install the condenser to the mounting tabs on the radiator.

25. Install the condenser hold down bracket to the radiator and condenser.
26. Install the outlet hose to the radiator.
27. Connect the transaxle oil cooler pipes to the radiator.

➡**Ensure the lower edge of the fan shroud engages the clip at the bottom of the radiator.**

28. Install the cooling fan shroud.
29. Install the cooling fan motors electrical harness to the fan shroud clips.
30. Connect the engine cooling fan motors electrical connectors.
31. Install the fan shroud clip to the condenser tubes.
32. Install the cooling fan shroud bolts and tighten to 89 inch lbs. (10 Nm).
33. Install the radiator upper support brackets and bolts that connect to the fan shroud.
34. Install the air deflectors to the top of the radiator.
35. Install the bolt that connects the fan shroud to the condenser hold down bracket.
36. Install the inlet hose to the radiator.
37. Install the air cleaner assembly.
38. Install the coolant reservoir hose to the radiator overflow neck.
39. Install the transmission oil cooler lines to the retainer clip at the bottom of the cooling fan shroud.

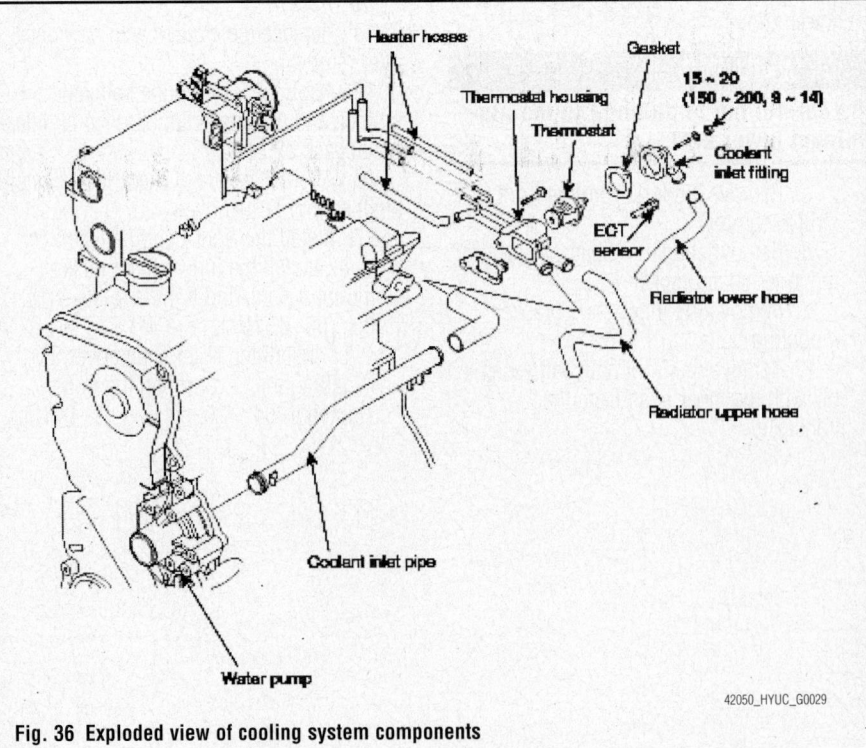

Fig. 36 Exploded view of cooling system components

40. Fill the cooling system.
41. Connect the negative battery cable.
42. Check for leaks.

THERMOSTAT

REMOVAL & INSTALLATION

See Figure 36.

❋❋ CAUTION

Make sure the engine and radiator are cool to the touch prior to beginning this procedure in order to prevent scalding or burns.

1. Before servicing the vehicle, refer to the Precautions Section.
2. Drain the coolant down to thermostat level or below.
3. Remove the coolant outlet fitting and gasket.
4. Remove the thermostat.
5. Installation is the reverse of the removal procedure.

WATER PUMP

REMOVAL & INSTALLATION

1.6L Engine

See Figures 37 and 38.

❋❋ CAUTION

The system is under high pressure when the engine is hot. To avoid danger of releasing scalding engine coolant, remove the cap only when the engine is cool.

1. Before servicing the vehicle, refer to the Precautions Section.
2. Drain the engine coolant.
3. Loosen the water pump pulley bolts.
4. Remove or disconnect the following:

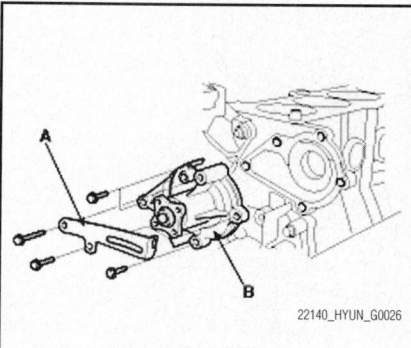

Fig. 37 Water pump removal showing brace—1.6L engine

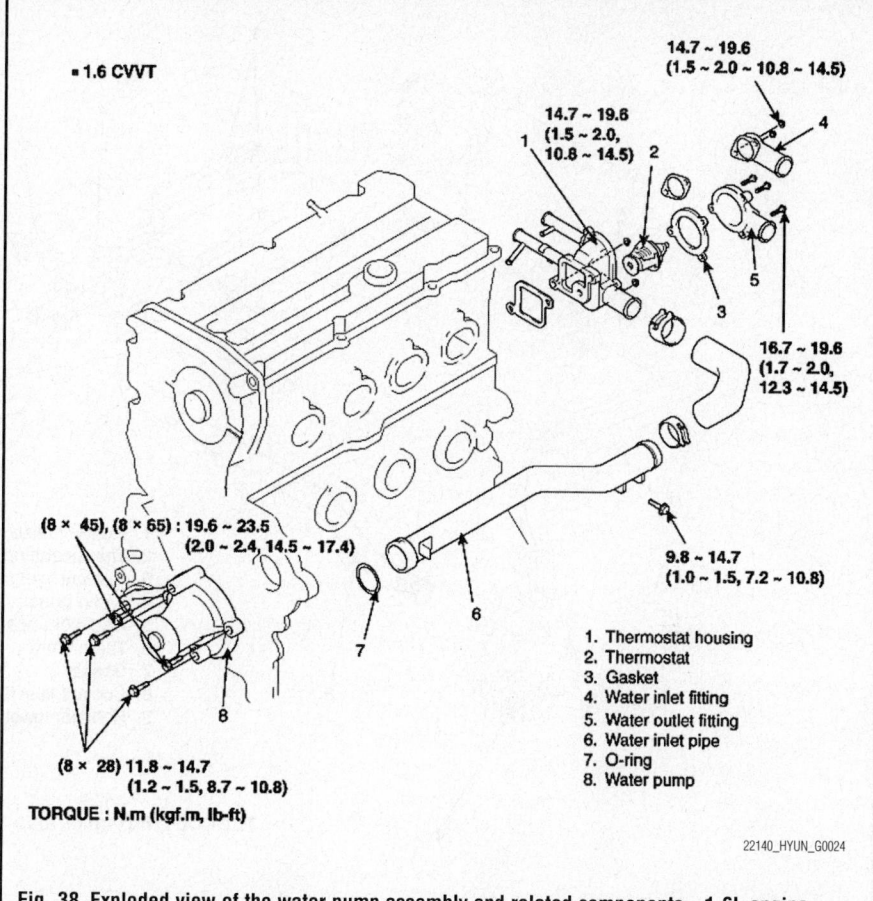

• 1.6 CVVT

14.7 ~ 19.6
(1.5 ~ 2.0 ~ 10.8 ~ 14.5)

14.7 ~ 19.6
(1.5 ~ 2.0,
10.8 ~ 14.5)

16.7 ~ 19.6
(1.7 ~ 2.0,
12.3 ~ 14.5)

(8 × 45), (8 × 65) : 19.6 ~ 23.5
(2.0 ~ 2.4, 14.5 ~ 17.4)

9.8 ~ 14.7
(1.0 ~ 1.5, 7.2 ~ 10.8)

(8 × 28) 11.8 ~ 14.7
(1.2 ~ 1.5, 8.7 ~ 10.8)

TORQUE : N.m (kgf.m, lb-ft)

1. Thermostat housing
2. Thermostat
3. Gasket
4. Water inlet fitting
5. Water outlet fitting
6. Water inlet pipe
7. O-ring
8. Water pump

Fig. 38 Exploded view of the water pump assembly and related components—1.6L engine

• The accessory drive belt
• The water pump pulley
• The timing belt
• The timing belt idler
5. Remove the water pump:
a. Remove the 2 bolts and alternator brace (A).
b. Remove the 3 bolts and remove the water pump (B) and gasket.

To install:

6. Install the water pump and a new gasket with the 3 bolts. Tightening torque: 9–11 ft. lbs. (12–15 Nm).
7. Install the alternator brace with the 2 bolts. Tightening torque: 15–17 ft. lbs. (20–24 Nm).
8. Install or connect the following:
• The timing belt idler
• The timing belt
• The water pump pulley
• The accessory drive belt
9. Tighten the water pump pulley bolts. Tightening torque: 72–84 inch lbs. (8–10 Nm).
10. Fill with engine coolant.
11. Start engine and check for leaks.
12. Recheck engine coolant level.

2.0L Engine

See Figures 39 and 40.

❋❋ CAUTION

The system is under high pressure when the engine is hot. To avoid danger of releasing scalding engine coolant, remove the cap only when the engine is cool.

1. Before servicing the vehicle, refer to the Precautions Section.

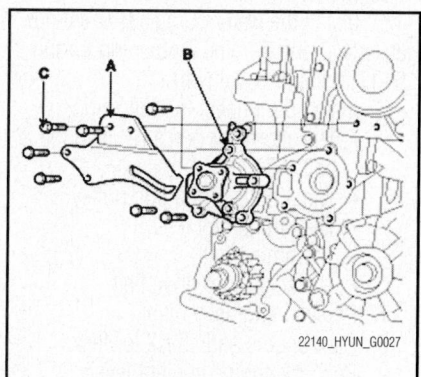

Fig. 39 Water pump removal showing brace—2.0L engine

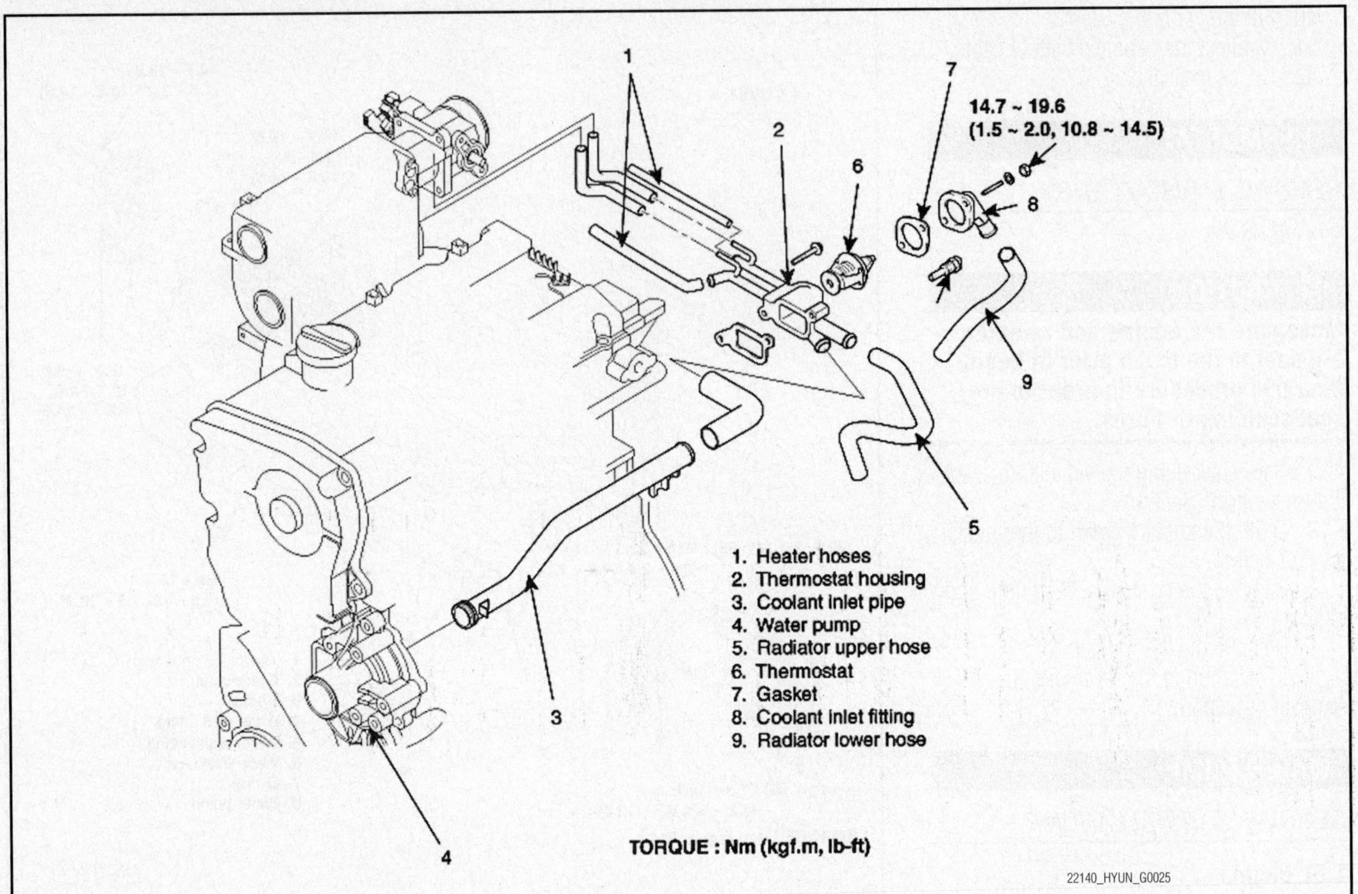

14.7 ~ 19.6
(1.5 ~ 2.0, 10.8 ~ 14.5)

1. Heater hoses
2. Thermostat housing
3. Coolant inlet pipe
4. Water pump
5. Radiator upper hose
6. Thermostat
7. Gasket
8. Coolant inlet fitting
9. Radiator lower hose

TORQUE : Nm (kgf.m, lb-ft)

22140_HYUN_G0025

Fig. 40 Exploded view of the water pump assembly and related components—2.0L engine

2. Drain the engine coolant.
3. Remove or disconnect the following:
   • The accessory drive belt
   • The timing belt
   • The timing belt idler
4. Remove the water pump:
   a. Remove the 4 bolts and pump pulley.
   b. Remove the 2 bolts (C), then
   remove the alternator brace (A).
   c. Remove the water pump (B) and
   gasket.

To install:

5. Install the water pump (B) and a new
gasket with the 3 bolts. Tightening torque:
15–17 ft. lbs. (20–24 Nm).
6. Install or connect the following:
   • The alternator brace (A) with the 2
     bolts (C)
   • The 4 bolts and pump pulley
   • The timing belt idler
   • The timing belt
   • The accessory drive belt
7. Fill with engine coolant.
8. Start engine and check for leaks.
9. Recheck engine coolant level.

2.4L Engine
See Figure 41.

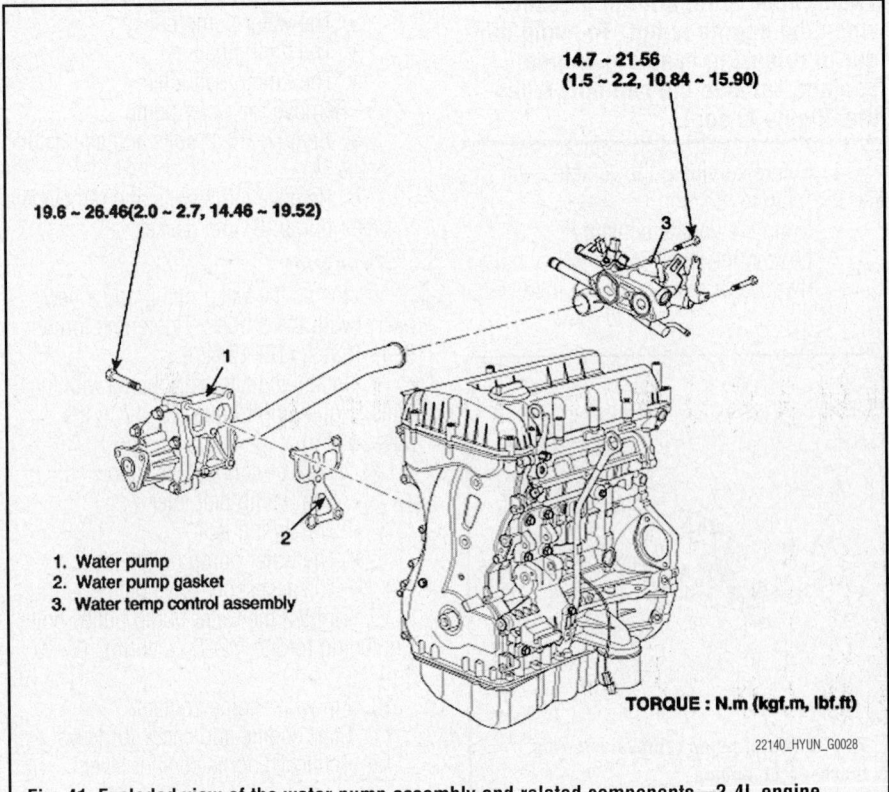

14.7 ~ 21.56
(1.5 ~ 2.2, 10.84 ~ 15.90)

19.6 ~ 26.46(2.0 ~ 2.7, 14.46 ~ 19.52)

1. Water pump
2. Water pump gasket
3. Water temp control assembly

TORQUE : N.m (kgf.m, lbf.ft)

22140_HYUN_G0028

Fig. 41 Exploded view of the water pump assembly and related components—2.4L engine

The system is under high pressure when the engine is hot. To avoid danger of releasing scalding engine coolant, remove the cap only when the engine is cool.

1. Before servicing the vehicle, refer to the Precautions Section.
2. Drain the engine coolant.
3. Remove the accessory drive belt.
4. Remove the exhaust manifold.
5. Remove the water pump:
   a. Remove the 4 bolts and pump pulley.
   b. Remove the water pump and gasket.
6. Remove the water inlet pipe nut.

To install:

7. Install the water pump and a new gasket. Tighten the 5 bolts to 15–20 ft. lbs. (20–27 Nm).

8. Install the 4 bolts and pump pulley.
9. Install the water inlet pipe nut. Tightening torque: 15–20 ft. lbs. (20–27 Nm).
10. Install exhaust manifold.
11. Install accessory drive belt.
12. Fill with engine coolant.
13. Start engine and check for leaks.
14. Recheck engine coolant level.

2.7L Engine

See Figures 42 and 43.

The system is under high pressure when the engine is hot. To avoid danger of releasing scalding engine coolant, remove the cap only when the engine is cool.

1. Before servicing the vehicle, refer to the Precautions Section.

2. Drain the engine coolant.
3. Remove or disconnect the following:
   • The accessory drive belt
   • The timing belt
   • The timing belt idler
   • The water pump (A) gasket (B)

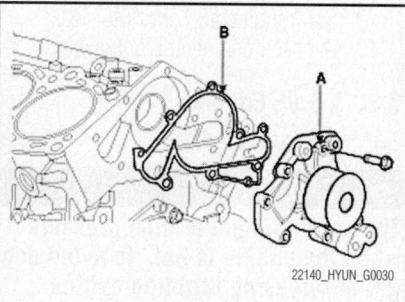

22140_HYUN_G0030

Fig. 42 Remove water pump and gasket— 2.7L engine

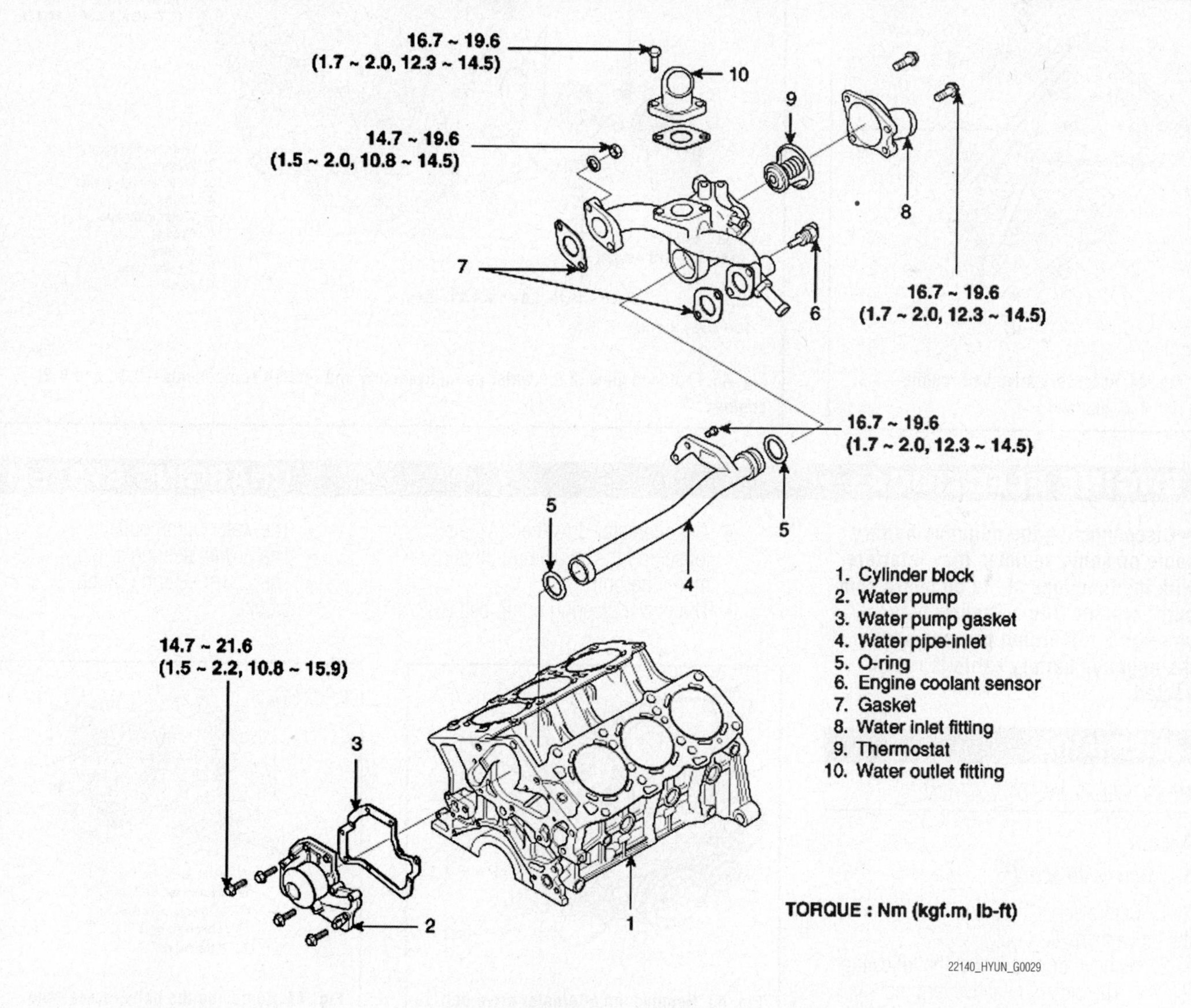

16.7 ~ 19.6
(1.7 ~ 2.0, 12.3 ~ 14.5)

14.7 ~ 19.6
(1.5 ~ 2.0, 10.8 ~ 14.5)

16.7 ~ 19.6
(1.7 ~ 2.0, 12.3 ~ 14.5)

16.7 ~ 19.6
(1.7 ~ 2.0, 12.3 ~ 14.5)

14.7 ~ 21.6
(1.5 ~ 2.2, 10.8 ~ 15.9)

1. Cylinder block
2. Water pump
3. Water pump gasket
4. Water pipe-inlet
5. O-ring
6. Engine coolant sensor
7. Gasket
8. Water inlet fitting
9. Thermostat
10. Water outlet fitting

TORQUE : Nm (kgf.m, lb-ft)

22140_HYUN_G0029

Fig. 43 Exploded view of the water pump assembly and related components—2.7L engine

To install:

4. Install the water pump (A) and a new gasket (B) with the 8 bolts. Tightening torque: 11–16 ft. lbs. (15–22 Nm).

5. Install the timing belt idler.
6. Install the timing belt.
7. Install the accessory drive belt.
8. Fill with engine coolant.
9. Start engine and check for leaks.
10. Recheck engine coolant level.

3.3L & 3.8L Engines

See Figures 44 and 45.

> ✳✳ **CAUTION**
>
> **The system is under high pressure when the engine is hot. To avoid danger of releasing scalding engine coolant, remove the cap only when the engine is cool.**

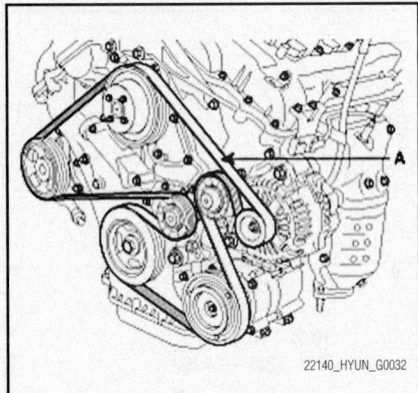

Fig. 44 Accessory drive belt routing—3.3L and 3.8L engines

1. Before servicing the vehicle, refer to the Precautions Section.
2. Drain the engine coolant.
3. Remove or disconnect the following:
   - The accessory drive belt (A)
   - The 4 bolts and pump pulley
   - The water pump and gasket

To install:

➡ **Clean the contact face before assembly.**

4. Install the water pump and a new gasket with 12 bolts. Tightening torque: 16–17 ft. lbs. (22–24 Nm); 84–108 inch lbs. (10–12 Nm).

5. Install the 4 bolts and the pump pulley. Tightening torque: 72–84 inch lbs. (8–10 Nm).

6. Install the accessory drive belt.
7. Fill with engine coolant.
8. Start engine and check for leaks.
9. Recheck engine coolant level.

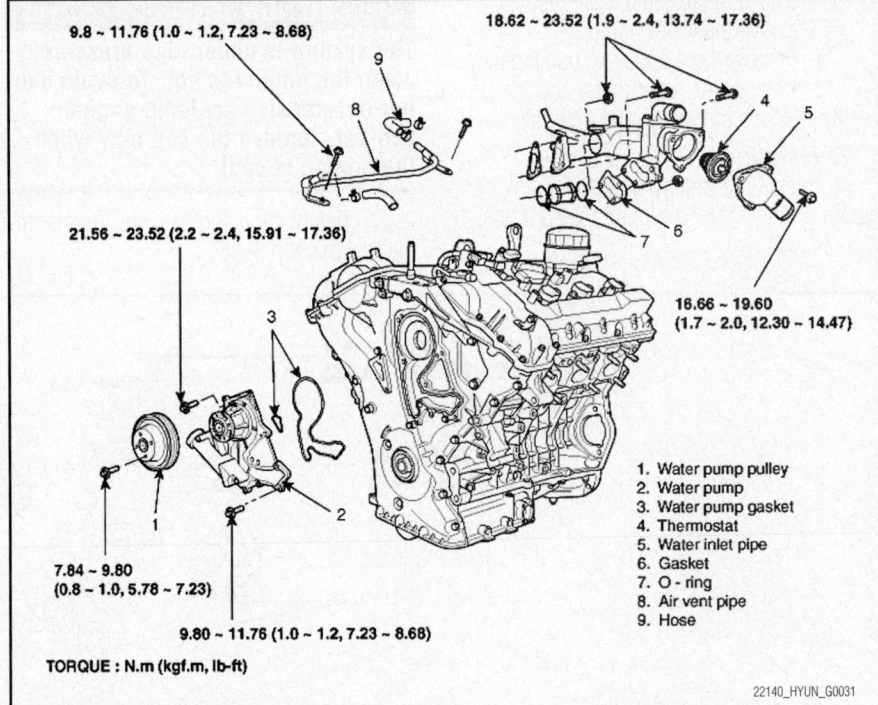

1. Water pump pulley
2. Water pump
3. Water pump gasket
4. Thermostat
5. Water inlet pipe
6. Gasket
7. O - ring
8. Air vent pipe
9. Hose

TORQUE : N.m (kgf.m, lb-ft)

Fig. 45 Exploded view of the water pump assembly and related components—3.3L and 3.8L engines

---

ENGINE ELECTRICAL

➡ **Disconnecting the negative battery cable on some vehicles may interfere with the functions of the on board computer system. The computer may undergo a relearning process once the negative battery cable is reconnected.**

ALTERNATOR

REMOVAL & INSTALLATION

Accent

See Figures 46 and 47.

1. Before servicing the vehicle, refer to the precautions section.
2. Remove or disconnect the following:
   - The battery negative terminal
   - Temporarily loosen the water pump pulley bolts

- The alternator drive belt (A), after loosening the adjusting bolt and mounting bolt
- The power steering pump belt (B)

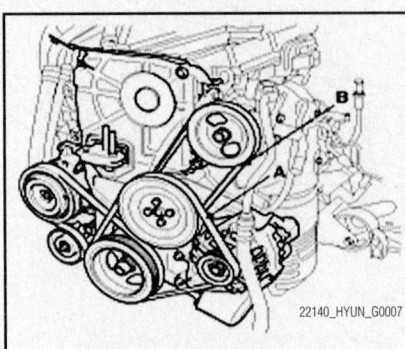

Fig. 46 Remove the alternator drive belt (A) and the power steering pump belt (B)

CHARGING SYSTEM

- The water pump pulley
- The power steering pump
- The power steering pump bracket

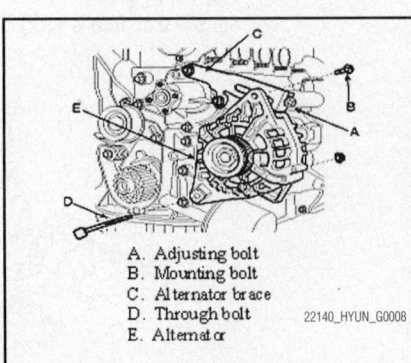

A. Adjusting bolt
B. Mounting bolt
C. Alternator brace
D. Through bolt
E. Alternator

Fig. 47 Removing the bolts, brace, and alternator—1.6L engine

- The alternator connector, and remove the cable from alternator "B" terminal
- The adjusting bolt (A)
- The mounting bolt (B)
- The alternator brace (C)
- Pull out the through bolt (D)

3. Remove the alternator (E).

To install:

4. Install or connect the following:
- The alternator (E)
- The through bolt (D)
- The alternator brace (C)
- The mounting bolt (B)
- The adjusting bolt (A)
- The alternator connector, and install the cable from alternator "B" terminal
- The power steering pump bracket
- The power steering pump
- The water pump pulley
- The power steering pump belt
- The alternator drive belt and tighten the adjusting bolt and mounting bolt

5. Adjust the alternator belt and torque the bolts to the following specifications:
   a. Pivot bolt to 14–18 ft. lbs. (19–25 Nm).
   b. Adjustment bolt to 14–20 ft. lbs. (19–28 Nm).

6. Tighten the water pump pulley bolts to 70–86 inch lbs. (8–10 Nm).

7. Connect the battery negative terminal.

All Models Except Accent

1. Before servicing the vehicle, refer to the precautions section.

2. Remove or disconnect the following:
- Negative battery cable
- Alternator drive belt
- Alternator wiring harness connectors
- Alternator

To install:

3. Install or connect the following:
- Alternator
- Alternator wiring harness connectors
- Alternator drive belt

4. Adjust the alternator belt and torque the bolts to the following specifications:
   a. 2.0L engine: Pivot bolt to 14–18 ft. lbs. (19–25 Nm) and adjustment bolt to 105–132 inch lbs. (12–15 Nm).
   b. 2.4L engine: Pivot bolt to 26–41 ft. lbs. (34–54 Nm) and the adjustment bolt to 14–18 ft. lbs. (19–25 Nm).

c. 3.3L and 3.8L engines: Pivot bolt to 14–18 ft. lbs. (19–25 Nm) and adjustment bolt to 11–16 ft. lbs. (15–22 Nm).

5. Connect the negative battery cable.

VOLTAGE REGULATOR

ADJUSTMENT

See Figures 48 and 49.

1. Before servicing the vehicle, refer to the Precautions Section.

2. Prior to the test, check the following items and correct if necessary.
- Check that the battery installed on the vehicle is fully charged
- Check the alternator drive belt tension
- Turn ignition switch to "OFF"
- Disconnect the battery ground cable
- Connect a digital voltmeter between the "B" terminal of the alternator and ground by connecting the positive (+) lead of the voltmeter to the "B" terminal of the alternator and connecting the negative (-) lead to good ground or the negative (-) battery terminal
- Disconnect the alternator output wire from the alternator "B" terminal
- Connect a DC ammeter (0 to 150 Amps) in series between the "B" terminal and the disconnected output wire
- Connect the negative (-) lead wire of the ammeter to the disconnected output wire
- Attach the engine tachometer and connect the battery ground cable

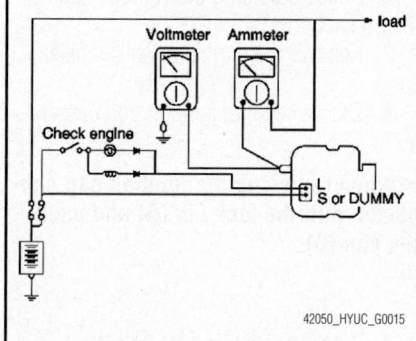

42050_HYUC_G0015

Fig. 48 Diagram of connected Voltmeter and Ammeter

GASOLINE	
Voltage regulator ambient temperature °C (°F)	Regulating voltage (V)
-20 (-4)	14.2 ~ 15.4
20 (68)	14.0 ~ 15.0
60 (140)	13.7 ~ 14.9
80 (176)	13.5 ~ 14.7

42050_HYUC_G0016

Fig. 49 Regulating Voltage table

3. Turn on the ignition switch and check to see that the voltmeter is properly installed and indicates the following value: Voltage reading = Battery voltage of 12.0–14.1 volts.

4. If it reads 0 volts, there is an open circuit in the wiring between the alternator "B" terminal and the battery and the battery negative (-) terminal, or the fusible link is blown.

5. Start the engine keeping all lights and accessories off.

6. Run the engine at a speed of about 2,500 rpm and read the voltmeter when the alternator output current drops to 10 Amps or less.

7. If the voltmeter reading agrees with the value listed in the Regulating Voltage Table below, the voltage regulator is functioning correctly.

8. If the reading is other than the standard value, the voltage regulator or the alternator is faulty.

9. Upon completion of the test, reduce the engine speed to idle, and turn off the ignition switch.

10. Disconnect the battery ground cable.

11. Remove the voltmeter and ammeter and the engine tachometer.

12. Connect the alternator output wire to the alternator "B" terminal.

13. Connect the battery ground cable.

REMOVAL & INSTALLATION

The voltage regulator is an internal component of the alternator. In order to replace the voltage regulator, the entire alternator assembly must be replaced. Refer to Alternator, installation and removal.

ENGINE ELECTRICAL

DISTRIBUTORLESS IGNITION SYSTEM

FIRING ORDER

See Figures 50 through 53.

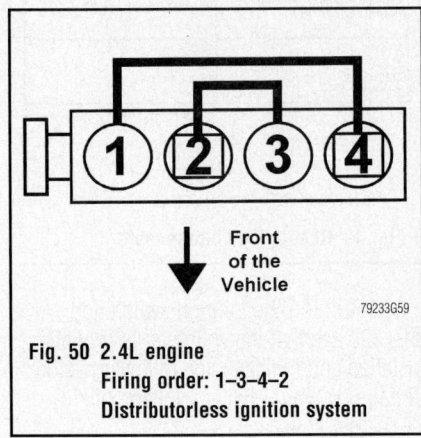

Fig. 50 2.4L engine
Firing order: 1–3–4–2
Distributorless ignition system

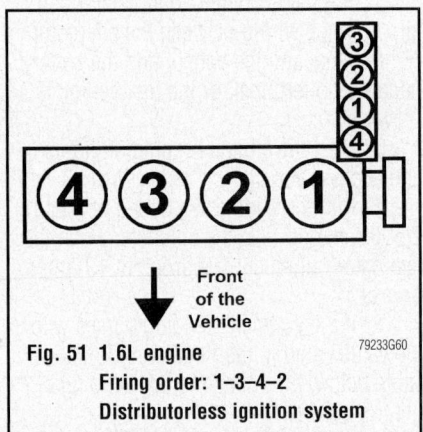

Fig. 51 1.6L engine
Firing order: 1–3–4–2
Distributorless ignition system

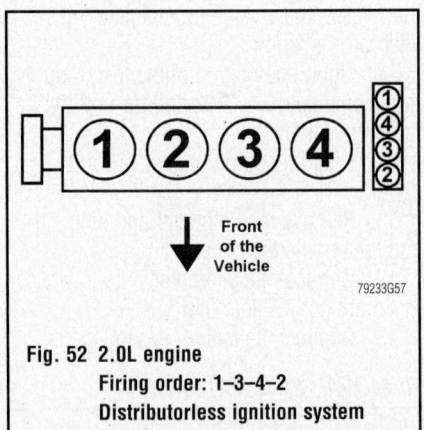

Fig. 52 2.0L engine
Firing order: 1–3–4–2
Distributorless ignition system

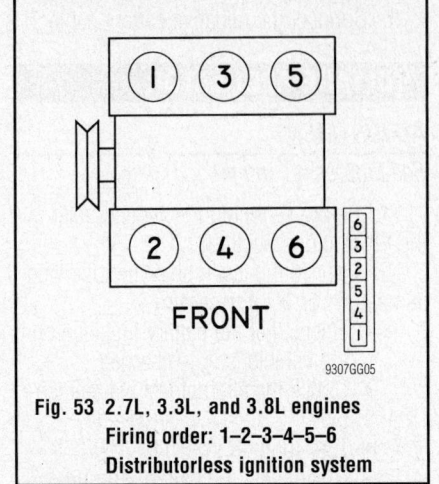

Fig. 53 2.7L, 3.3L, and 3.8L engines
Firing order: 1–2–3–4–5–6
Distributorless ignition system

IGNITION COIL PACK

REMOVAL & INSTALLATION
See Figure 54.

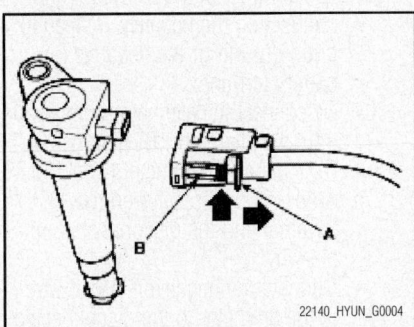

Fig. 54 When removing the ignition coil connector, pull the lock pin (A) and push the clip (B)

1. Before servicing the vehicle, refer to the Precautions Section.
2. Remove the engine cover (as necessary).
3. Disconnect the ignition coil connector.

➡**When removing the ignition coil connector, pull the lock pin (A) and push the clip (B).**

4. Remove the ignition coil.
5. Installation is the reverse of removal.

IGNITION TIMING

ADJUSTMENT

Ignition timing is controlled by the electronic control ignition timing system. The standard reference ignition timing data for the engine operating conditions are pre-programmed in the memory of the Engine Control Module (ECM). The engine operating conditions (speed, load, warm-up condition, etc.) are detected by the various sensors. Based on these sensor signals and the ignition timing data, signals to interrupt the primary current are sent to the ECM. The ignition coil is activated, and timing is controlled.

SPARK PLUGS

REMOVAL & INSTALLATION
See Figure 54.

1. Before servicing the vehicle, refer to the Precautions Section.
2. Remove the engine cover (as necessary).
3. Disconnect the ignition coil connector.

➡**When removing the ignition coil connector, pull the lock pin (A) and push the clip (B).**

4. Remove the ignition coil.
5. Use a spark plug socket and wrench to remove the spark plugs.

❋❋ WARNING

Be careful that no contaminates enter through the spark plug holes.

➡**Check the electrode gap on the spark plugs before installation. Specification: 0.039–0.043 inch (1.0–1.1 mm).**

6. To install, reverse the removal procedure. Tighten the spark plugs to 11 ft. lbs. (15 Nm).

ENGINE ELECTRICAL

STARTING SYSTEM

STARTER

REMOVAL & INSTALLATION

1. Before servicing the vehicle, refer to the Precautions Section.
2. Remove or disconnect the following:
   • Negative battery cable and wait at least 3 minutes
   • Speedometer cable and shift cable from the transaxle
   • Starter motor wiring
   • Starter motor bolts and the starter

To install:

3. Installation is the reverse of removal. Tighten starter motor bolts to 20–25 ft. lbs. (27–34 Nm).

SOLENOID OR RELAY REPLACEMENT

See Figure 55.

1. Remove fuse box cover.
2. Remove the starter relay (A).
3. Install new starter relay.
4. Replace fuse box cover.

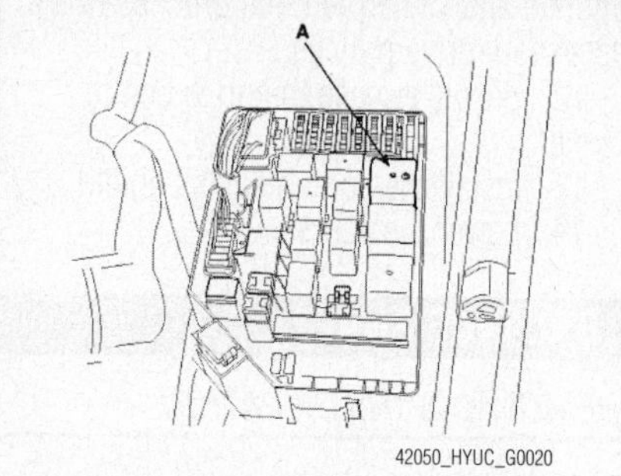

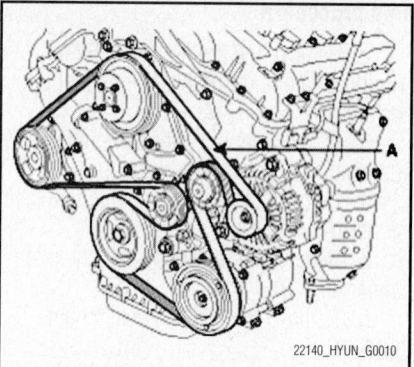

42050_HYUC_G0020

Fig. 55 Starter relay position (A) in the fuse box

ENGINE MECHANICAL

ACCESSORY DRIVE BELTS

ACCESSORY BELT ROUTING

See Figures 56 through 60.

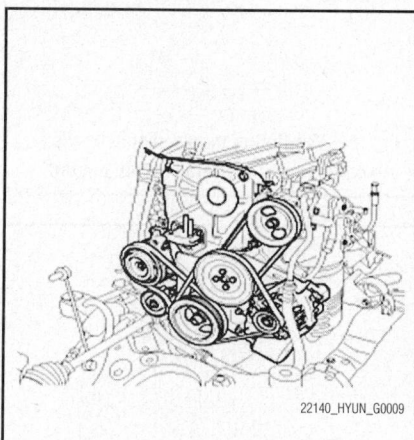

22140_HYUN_G0009

Fig. 56 Accessory drive belt routing—1.6L engine

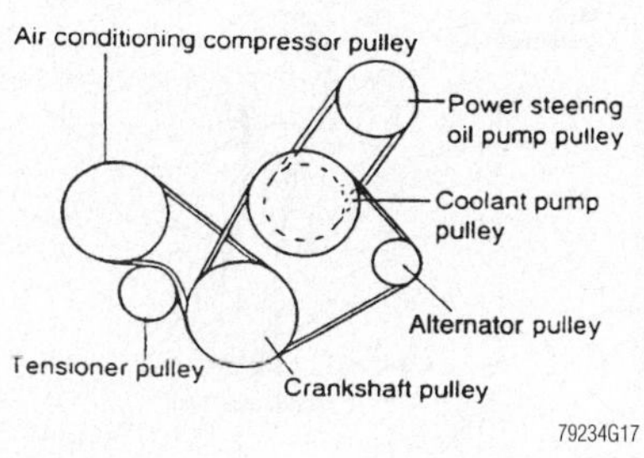

Air conditioning compressor pulley
Power steering oil pump pulley
Coolant pump pulley
Alternator pulley
Tensioner pulley
Crankshaft pulley

79234G17

Fig. 57 Accessory drive belt routing—2.0L engine

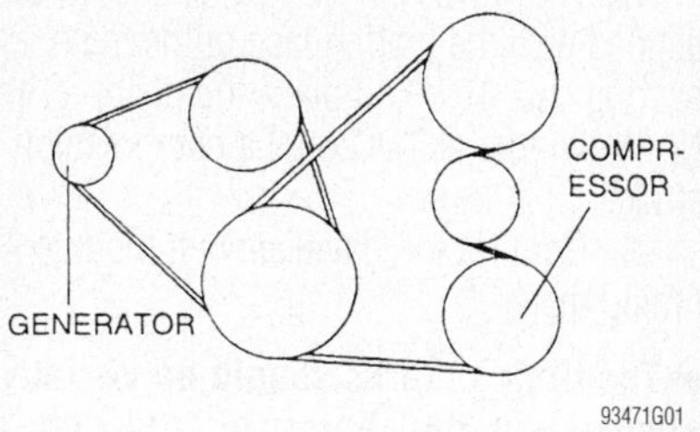

GENERATOR
COMPR-ESSOR

93471G01

Fig. 58 Accessory drive belt routing—2.4L engine

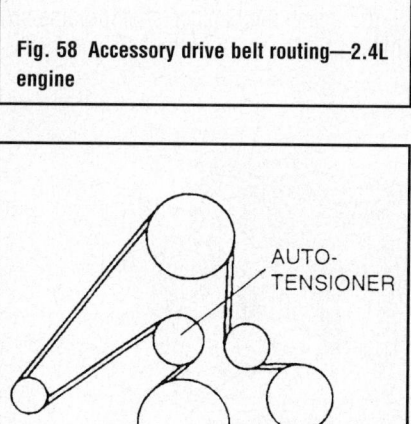

AUTO-TENSIONER

93471G02

Fig. 59 Accessory drive belt routing—2.7L engine

INSPECTION

Inspect the accessory drive belt for signs of glazing or cracking. A glazed belt will be perfectly smooth from slippage, while a good belt will have a slight texture of fabric visible. Cracks will usually start at the inner edge of the belt and run outward. All worn or damaged accessory drive belts should be replaced immediately.

22140_HYUN_G0010

Fig. 60 Accessory drive belt routing—3.3L and 3.8L engines

ADJUSTMENT

1. Loosen the tension mounting bolt.
2. Turn the adjusting bolt to obtain the proper belt tension, then retighten the mounting bolt.
3. Recheck the deflection of the drive belt.

REMOVAL & INSTALLATION

1. Before servicing the vehicle, refer to the Precautions Section.
2. Raise and support the vehicle.
3. Remove the engine splash shield.
4. Rotate the drive belt tensioner clockwise to release the drive belt tension.
5. Remove the drive belt from the alternator.
6. Slowly release the drive belt tensioner.
7. Remove the drive belt from the accessory drive pulleys.

To install:

8. Install the drive belt to the accessory drive pulley.
9. Rotate the drive belt tensioner clockwise.

10. Install the drive belt to the alternator.

11. Ensure the drive belt is properly aligned and seated into the grooves of the accessory drive pulleys.

12. Slowly release the drive belt tensioner.

13. Install the engine splash shield.

14. Lower the vehicle.

BALANCE SHAFT

REMOVAL & INSTALLATION

2.4L Engine

See Figures 61 through 66.

1. Before servicing the vehicle, refer to the Precautions Section.

➡ **Engine removal is not required for this procedure.**

2. Remove the timing chain. Refer to Timing Chain and Sprockets, removal & installation.

3. Install a set pin after compressing the balance shaft chain tensioner.

4. Remove the balance shaft chain tensioner (A).

5. Remove the balance shaft chain tensioner arm (B).

6. Remove the balance shaft chain guide (C).

7. Remove the balance shaft module (A) and balance shaft chain (B).

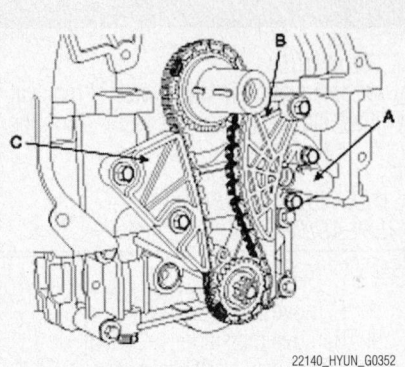

Fig. 62 Remove the balance shaft chain tensioner (A), the balance shaft chain tensioner arm (B), and the balance shaft chain guide (C)—2.4L engine

To install:

8. The key of the crankshaft should be aligned with the mating face of the main bearing cap. This will place the piston of the No. 1 cylinder at TDC on the compression stroke.

9. Confirm the balance shaft module timing mark.

➡ **The timing marks should be visually aligned with the centers of adjacent cast timing notches.**

10. Install the balance shaft module so that the timing mark of the balance shaft

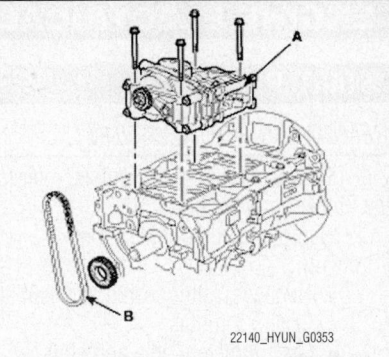

Fig. 63 Remove the balance shaft module (A) and balance shaft chain (B)—2.4L engine

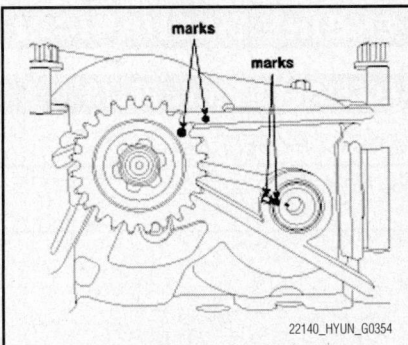

Fig. 64 The timing marks should be visually aligned as illustrated—2.4L engine

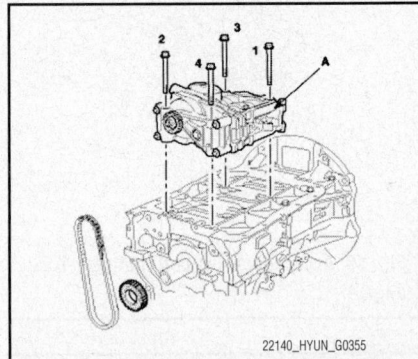

Fig. 65 Tighten the balance shaft module retaining bolts as illustrated—2.4L engine

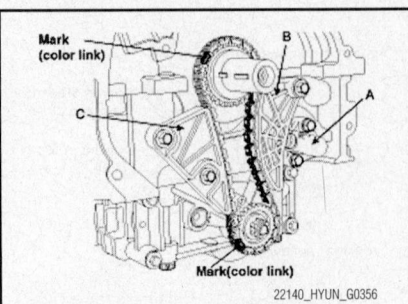

Fig. 66 Install the balance shaft chain guide (C), the balance shaft tensioner arm (B), and the balance shaft tensioner (A)—2.4L engine

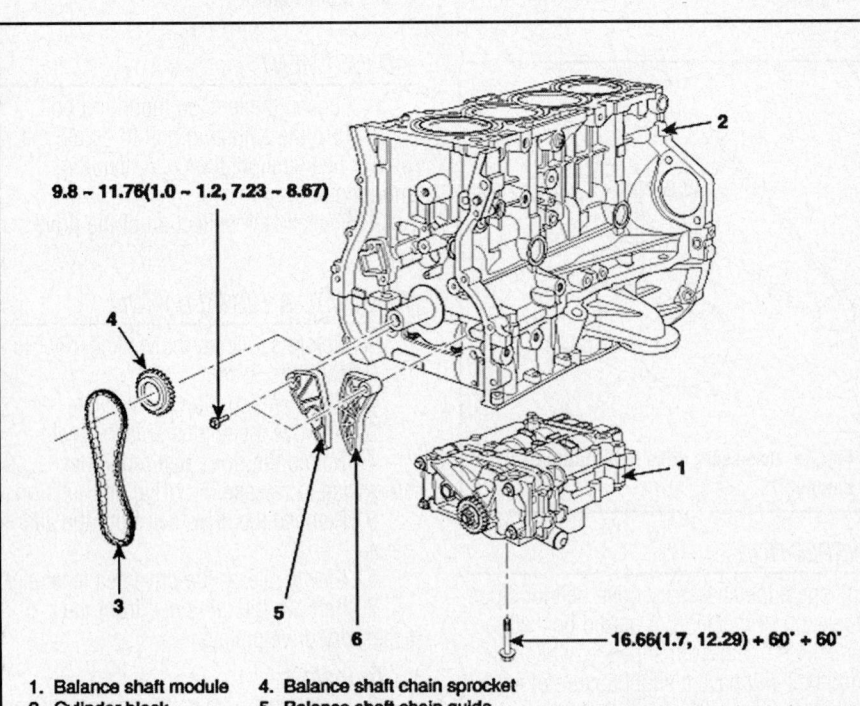

9.8 ~ 11.76(1.0 ~ 1.2, 7.23 ~ 8.67)

16.66(1.7, 12.29) + 60° + 60°

1. Balance shaft module
2. Cylinder block
3. Balance shaft chain
4. Balance shaft chain sprocket
5. Balance shaft chain guide
6. Balance shaft chain tensioner arm

TORQUE : N.m (kgf.m, lbf.ft)

Fig. 61 Exploded view of the balance shaft components—2.4L engine

module sprocket is matched with the timing mark (color link) of the balance shaft chain.

11. Tighten the balance shaft module retaining bolts:
    a. Step 1: 12 ft. lbs. (17 Nm).
    b. Step 2: Add 60°.
    c. Step 3: Add another 60° on the last pass.

12. Install the balance shaft chain guide (C). Tightening torque: 87–104 inch lbs. (10–12 Nm).

13. Install the balance shaft tensioner arm (B). Tightening torque: 87–104 inch lbs. (10–12 Nm).

14. Install the balance shaft tensioner (A) and remove the set pin. Tightening torque: 87–104 inch lbs. (10–12 Nm).

15. Confirm the timing marks.

16. Install the timing chain. Refer to Timing Chain and Sprockets, removal & installation.

CAMSHAFT AND VALVE LIFTERS

INSPECTION

See Figures 67 and 68.

1. Inspect the cam lobes.
    a. Using a micrometer, measure the cam lobe height.
    b. If the cam lobe height is less than specified, replace the camshaft.

2. Inspect the camshaft journal clearance.
    a. Clean the bearing caps and camshaft journals.
    b. Place the camshafts on the cylinder head.
    c. Lay a strip of Plastigage across each of the camshaft journal.
    d. Install the bearing caps and tighten the bolts with specified torque.

➡**Do not turn the camshaft.**

    e. Remove the bearing caps.
    f. Measure the Plastigage at its widest point.
    g. If the oil clearance is greater than specified, replace the camshaft. If necessary, replace the bearing caps and cylinder head as a set.
    h. Completely remove the Plastigage.
    i. Remove the camshafts.

3. Inspect the camshaft end play.
    a. Install the camshafts.
    b. Using a dial indicator, measure the end play while moving the camshaft back and forth.
    c. If the end play is greater than specified, replace the camshaft. If necessary,

replace the bearing caps and cylinder head as a set.
    d. Remove the camshafts.

4. Inspect the Continuous Variable Valve Timing (CVVT) assembly.
    a. Check that the CVVT assembly will not turn.
    b. Apply vinyl tape to all the parts except the one indicated by the arrow in the illustration.
    c. Wrap tape around the tip of the air gun and apply air of approx. 14 psi to the port of the camshaft. Perform this in order to release the lock pin for the maximum delay angle locking.

➡**Wrap a shop rag around the CVVT as the oil may spray out when the air pressure is applied.**

    d. Under the condition of air pressure being applied, turn the CVVT assembly to the advance angle side with your hand.
    • Depending on the air pressure, the CVVT assembly will turn to the advance side
    • If air is leaking from the port and air pressure cannot be maintained, the locking pin will not release

5. Except the position where the lock pin meets at the maximum delay angle, let the

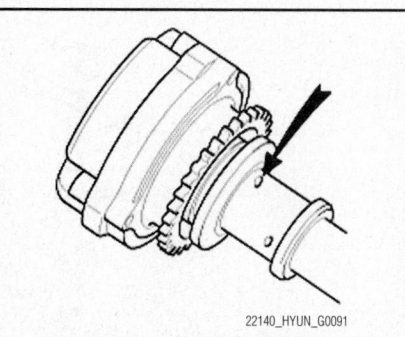

Fig. 67 Apply vinyl tape to the CVVT on all parts except the one indicated by the arrow

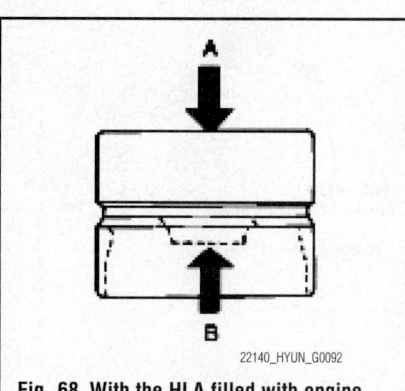

Fig. 68 With the HLA filled with engine oil, hold A and press B by hand

CVVT assembly turn back and forth and check the movable range and that there is no disturbance.
    a. The CVVT should move smoothly in the range of about 20°.
    b. Turn the CVVT assembly with your hand and lock it at the maximum delay angle position.

6. Inspect the Hydraulic Lash Adjuster (HLA).
    a. With the HLA filled with engine oil, hold A and press B by hand.
    b. If B moves, replace the HLA.

REMOVAL & INSTALLATION

1.6L & 2.0L Engines

See Figures 69 through 72.

Engine removal is not required for this procedure. Use a fender cover to avoid damaging painted surfaces. To avoid damage, unplug the wiring connectors carefully while holding the connector portion. Mark all wiring and hoses to avoid misconnection. Inspect the timing belt before removal. Turn the crankshaft pulley so that the No. 1 piston is at Top Dead Center (TDC).

1. Before servicing the vehicle, refer to the Precautions Section.

2. Disconnect the terminals from battery and remove the battery.

3. Remove the engine cover.

4. Remove the undercover.

5. Drain the engine coolant. Remove the radiator cap to speed draining.

6. Remove the intake air hose and air cleaner assembly:
    a. Disconnect the breather hose from intake air hose.
    b. Remove the intake air hose and air cleaner upper cover.
    c. Disconnect the ECM connectors.
    d. Remove the air cleaner element and air cleaner lower cover.

7. Remove the battery tray.

8. Remove the upper radiator hose and lower radiator hose.

9. Remove the heater hoses.

10. Remove the fuel hose.

11. Remove the accelerator cable by loosening the lock-nut, then slip the cable end out of the throttle linkage.

12. Disconnect the Throttle Position Sensor (TPS) connector and the MAP sensor connector.

13. Remove the engine wire harness connectors and wire harness clamps from cylinder head and the intake manifold:
    a. Disconnect the rear oxygen sensor connector.
    b. Disconnect the air conditioner compressor switch connector.

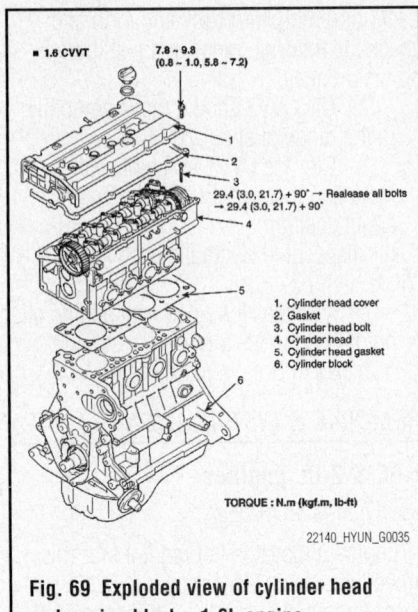

Fig. 69 Exploded view of cylinder head and engine block—1.6L engine

c. Disconnect the knock sensor connector.

d. Disconnect the injector connectors.

e. Remove the wire harness bracket.

f. Disconnect the Idle Speed Actuator (ISA) connector.

g. Disconnect the front oxygen sensor connector.

h. Disconnect the Crankshaft Position Sensor (CKP) connector.

i. Disconnect the Oil Control Valve (OCV) connector.

j. Disconnect the ignition coil connector.

k. Disconnect the ignition coil condenser connector.

l. Disconnect the Camshaft Position Sensor (CMP) connector.

m. Disconnect the ground cable.

n. Remove the wire harness bracket.

14. Remove or disconnect the following:

• The hose of the Purge Control Solenoid Valve (PCSV) side
• The brake booster vacuum hose
• The power steering pump and fix the pump to vehicle with a wire
• The ignition coil
• The exhaust manifold
• The intake manifold
• The timing belt
• The cylinder head cover
• The camshaft sprocket
• The timing chain auto tensioner
• The camshaft bearing caps and camshafts

To install:

Thoroughly clean all parts to be assembled. Rotate the crankshaft, set the No. 1 piston at TDC.

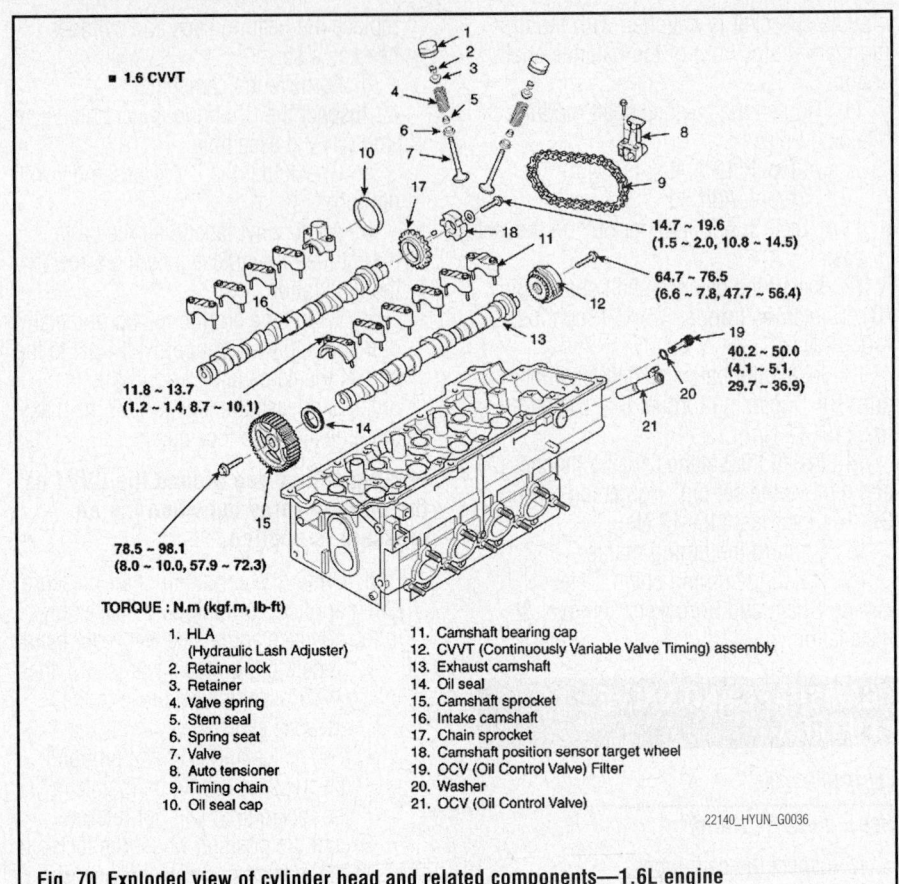

Fig. 70 Exploded view of cylinder head and related components—1.6L engine

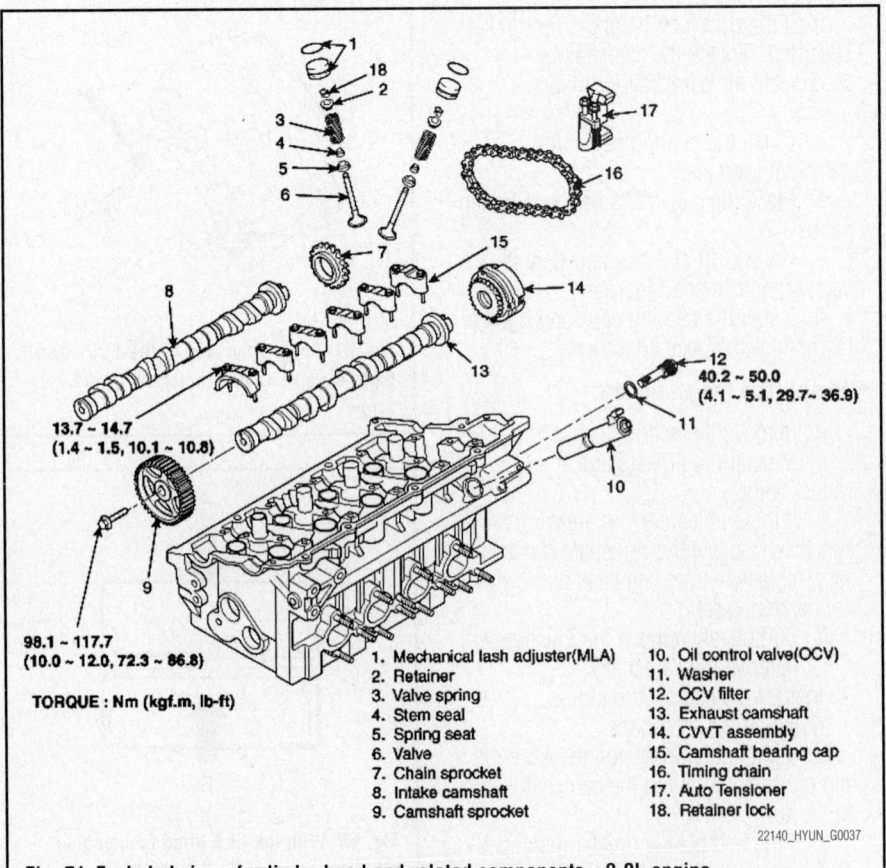

Fig. 71 Exploded view of cylinder head and related components—2.0L engine

15. Install the camshafts:

a. Align the camshaft timing chain with the intake timing chain sprocket and exhaust timing chain sprocket as shown.

b. Install the camshaft and bearing caps and tighten as follows:

- 1.6L engine: 108–120 inch lbs. (12–14 Nm).
- 2.0L engine: 120–132 inch lbs. (14–15 Nm)

c. Install the timing chain auto tensioner and tighten to 72–84 inch lbs. (8–10 Nm).

16. Using the SST (09221 - 21000), install the camshaft bearing oil seal.

17. Install the camshaft sprocket.

18. Install the cylinder head cover.

a. Install the cylinder head cover gasket in the groove of the cylinder head cover.

➡**Before installing the cylinder head cover gasket, thoroughly clean the cylinder head cover and the groove. When installing, make sure the cylinder head cover gasket is seated securely in the corners of the recesses with no gap.**

b. Apply liquid gasket to the head cover gasket at the corners of the recess.

➡**Use liquid gasket, Loctite® No. 5999. Check that the mating surfaces are clean and dry before applying liquid gasket. After assembly, wait at least 30 minutes before filling the engine with oil.**

c. Install the cylinder head cover with bolts:

- Step 1: Pre-tighten all bolts by 36–48 inch lbs. (4–5 Nm)
- Step 2: Tighten by the specified torque 72–84 inch lbs. (8–10 Nm)

19. Install or connect the following:

- The timing belt
- The intake manifold

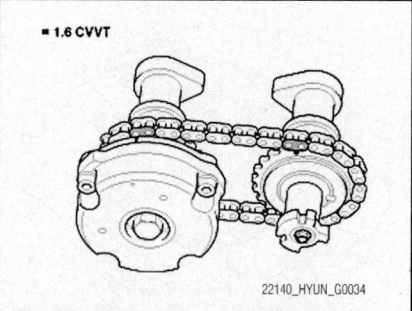

Fig. 72 Align the camshaft timing chain with the intake timing chain sprocket and exhaust timing chain sprocket—1.6L engine

- The exhaust manifold
- The ignition coil
- The power steering pump
- The brake booster hose
- The hose of the PCSV side

20. Install the engine wire harness connectors and wire harness clamps to the cylinder head and the intake manifold:

- The wire harness bracket
- The ground cable
- The CMP(Camshaft position sensor) connector
- The ignition coil condenser connector
- The ignition coil connector
- The OCV connector
- The CKP connector
- The front oxygen sensor connector
- The ISA connector
- The wire harness bracket
- The fuel injector connectors
- The knock sensor connector
- The air conditioner compressor switch connector
- The rear oxygen sensor connector

21. Install or connect the following:

- The TPS connector and the MAP sensor connector
- The accelerator cable
- The fuel hose

- The heater hoses
- The upper radiator hose and lower radiator hose
- The battery tray
- The intake air hose and air cleaner assembly

22. Install the air cleaner element and air cleaner lower cover. Tightening torque: 72–84 inch lbs. (8–10 Nm).

23. Connect the ECM connectors.

24. Install the intake air hose and air cleaner upper cover.

25. Connect the breather hose to intake air hose.

26. Install the undercover.

27. Install the engine cover. Tightening torque: 36–48 inch lbs. (4–6 Nm).

28. Install the battery and connect the battery terminals.

29. Fill with engine coolant.

30. Start the engine and check for leaks.

31. Recheck engine coolant level and oil level.

2.4L Engine

See Figures 73 and 74.

Engine removal is not required for this procedure. Use a fender cover to avoid damaging painted surfaces. To avoid damage, unplug the wiring connectors carefully

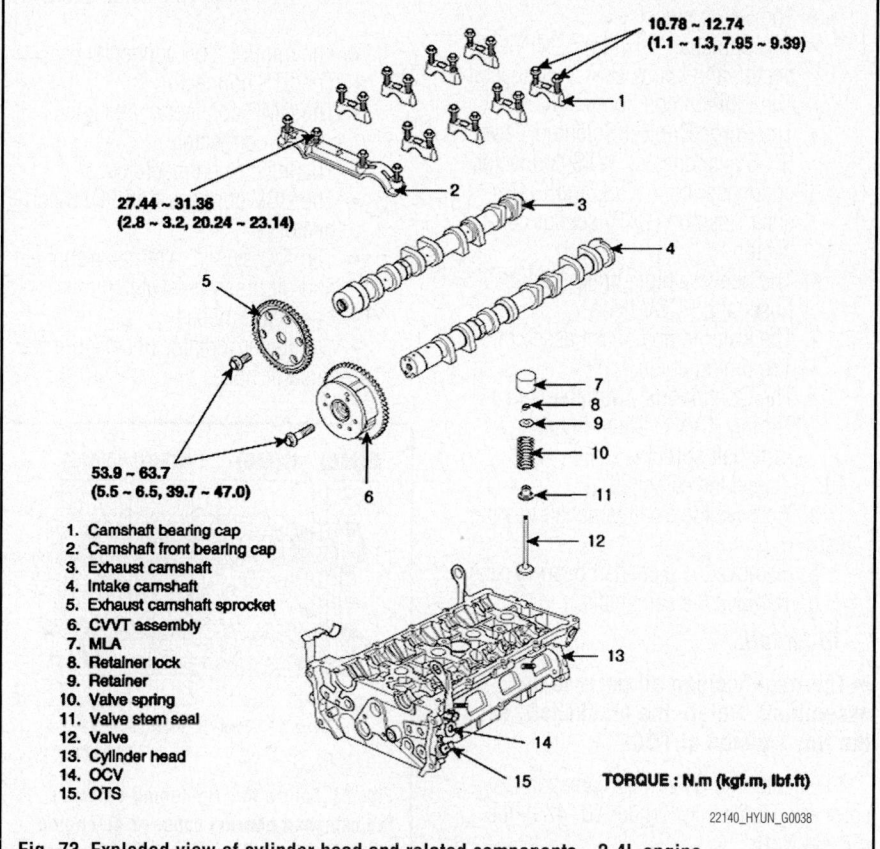

10.78 ~ 12.74
(1.1 ~ 1.3, 7.95 ~ 9.39)

27.44 ~ 31.36
(2.8 ~ 3.2, 20.24 ~ 23.14)

53.9 ~ 63.7
(5.5 ~ 6.5, 39.7 ~ 47.0)

1. Camshaft bearing cap
2. Camshaft front bearing cap
3. Exhaust camshaft
4. Intake camshaft
5. Exhaust camshaft sprocket
6. CVVT assembly
7. MLA
8. Retainer lock
9. Retainer
10. Valve spring
11. Valve stem seal
12. Valve
13. Cylinder head
14. OCV
15. OTS

TORQUE : N.m (kgf.m, lbf.ft)

Fig. 73 Exploded view of cylinder head and related components—2.4L engine

while holding the connector portion. Mark all wiring and hoses to avoid misconnection. Inspect the timing belt before removing. Turn the crankshaft pulley so that the No. 1 piston is at Top Dead Center (TDC).

1. Before servicing the vehicle, refer to the Precautions Section.

2. Disconnect the negative terminal from the battery.

3. Remove the engine cover.

4. Remove the air duct.

5. Remove the intake air hose and air cleaner assembly.

   a. Disconnect the AFS connector.

   b. Disconnect the breather hose from air cleaner hose.

   c. Disconnect the ECM connector.

   d. Remove the intake air hose and air cleaner assembly.

6. Remove front wheels.

7. Remove the undercover.

8. Drain the engine coolant. Remove the radiator cap to speed draining.

9. Remove or disconnect the following:

- The upper and lower radiator hose
- The heater hoses
- The A/C switch, alternator connector, and oil pressure switch
- The Oil Control Valve (OCV) connector and OTS connector
- The injector connectors
- The ETS connector
- The Camshaft Position (CMP) connector, and knock sensor connector
- The ignition coil connectors
- The Purge Control Solenoid Valve (PCSV) connector, WTS connector, condenser connector, and Crankshaft Position (CKP) sensor connector
- The delivery pipe, brake vacuum hose, and PCSV hose
- The water temp control assembly
- The timing chain
- The Continuously Variable Valve Timing (CVVT) assembly and camshaft sprocket

10. Remove the camshaft.

   a. Remove the front camshaft bearing cap.

   b. Remove the camshaft bearing caps.

   c. Remove the camshafts.

To install:

➡**Thoroughly clean all parts to be assembled. Rotate the crankshaft, set the No. 1 piston at TDC.**

11. Install the CVVT and camshaft sprocket. Tightening torque: 40–47 ft. lbs. (54–64 Nm).

➡**Hold the hexagonal head wrench portion of the camshaft with a vise, and install the bolt and CVVT assembly.**

12. Install the camshafts. Apply a light coat of engine oil on camshaft journals.

13. Install the camshaft bearing caps in their proper locations. Follow the illustrated tightening order and the following specifications:

- M6: 8–9 ft. lbs. (11–13 Nm)
- M8: 20–23 ft. lbs. (27–31 Nm)

14. Install the timing chain.

15. Check and adjust the valve clearance.

16. Install the water temp control assembly and tighten as follows:

   a. Bolt: 11–16 ft. lbs. (15–22 Nm)

   b. Nut: 15–20 ft. lbs. (20–27 Nm)

➡**Assemble water temp control assembly and water inlet pipe to water pump assembly before nuts for assembling of water inlet pipe to be tightened. Insert after wetting O-ring or inner surface of thermostat housing. Always use a new O-ring.**

17. Install or connect the following:

- The delivery pipe, brake hose, and PCSV hose
- The PCSV connector, WTS connector, condenser connector, and CKP sensor connector
- The ignition coil connector
- The ETS connector
- The CMP connector, and knock sensor connector
- The injector connectors
- The OCV connector and OTS connector
- The A/C switch, alternator connect, and oil pressure switch
- The heater hoses
- The upper radiator hose and lower radiator hose

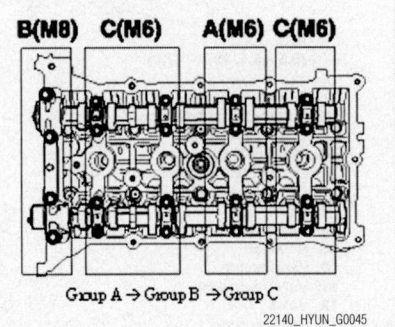

Fig. 74 Follow this tightening order for the camshaft bearing caps—2.4L engine

- The intake air hose and air cleaner assembly
- The engine cover
- The negative terminal to the battery

18. Fill with engine coolant.

19. Start the engine and check for leaks.

20. Recheck the engine coolant level and oil level.

2.7L Engine

See Figures 75 through 78.

Engine removal is not required for this procedure. Use a fender cover to avoid damaging painted surfaces. To avoid damage, unplug the wiring connectors carefully while holding the connector portion. Mark all wiring and hoses to avoid misconnection. Inspect the timing belt before removing. Turn the crankshaft pulley so that the No. 1 piston is at Top Dead Center (TDC).

1. Before servicing the vehicle, refer to the Precautions Section.

2. Remove the side cover (A, B, C) and the engine cover (D).

3. Disconnect the battery terminal and the battery.

4. Remove the radiator drain plug and drain engine coolant. Remove the radiator cap to speed draining.

5. Remove the air cleaner assembly.

   a. Disconnect the Air Flow Sensor (AFS) connector.

   b. Remove the breather hose from intake hose.

   c. Remove the intake hose and air cleaner upper cover.

   d. Remove the air cleaner lower cover.

6. Remove the upper radiator hose and lower radiator hose.

7. Remove the heater hoses.

8. Remove the engine wire harness connectors and wire harness clamps from the cylinder head and the intake manifold:

- Throttle position sensor (TPS) connector

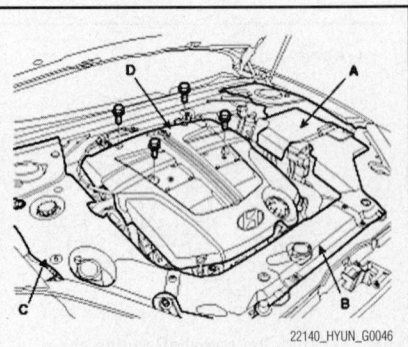

Fig. 75 Remove the side cover and engine cover—2.7L engine

- Idle speed actuator (ISA) connector
- Purge Control Solenoid Valve (PCSV) connector
- Knock sensor connector
- Camshaft Position (CMP) sensor connector
- Engine ground line
- Heated oxygen sensor (Bank 2, Sensor 1) connector
- Engine temperature coolant sensor connector
- Ignition coil connector
- Crankshaft Position (CKP) sensor connector
- Heated oxygen sensor (Bank 1, Sensor 2) connector
- Fuel injector connectors

9. Remove or disconnect the following:
- The fuel inlet from the delivery pipe
- The Purge Control Solenoid Valve (PCSV) hose
- The brake booster vacuum hose
- The accelerator cable by loosening the locknut, then slip the cable end out of the throttle linkage
- The auto-cruise connector and the auto-cruise cable
- The PCV hose
- The power steering pump
- The timing belt
- The cylinder head covers
- The camshaft sprocket

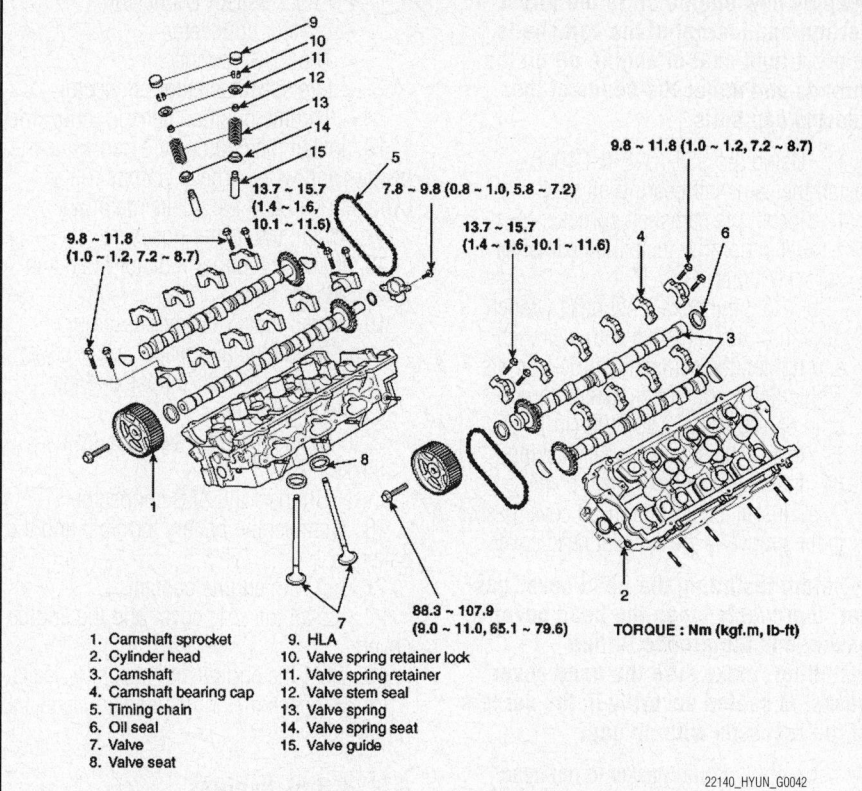

Fig. 77 Exploded view of cylinder head and related components—2.7L engine

1. Camshaft sprocket
2. Cylinder head
3. Camshaft
4. Camshaft bearing cap
5. Timing chain
6. Oil seal
7. Valve
8. Valve seat
9. HLA
10. Valve spring retainer lock
11. Valve spring retainer
12. Valve stem seal
13. Valve spring
14. Valve spring seat
15. Valve guide

22140_HYUN_G0042

- The camshaft bearing caps
- The camshafts

To install:

Thoroughly clean all parts to be assembled. Rotate the crankshaft, set the No. 1 piston at TDC.

10. Install the camshafts.

a. Align the camshaft timing chain with the intake timing chain sprocket and exhaust timing chain sprocket as shown.

b. Install the camshaft.

c. Install the camshaft bearing caps and tighten as follows:

d. M6 (38mm) bolt: 7–9 ft. lbs. (10–13 Nm)

e. M6 (50mm) bolt: 11–12 ft. lbs. (15–17 Nm)

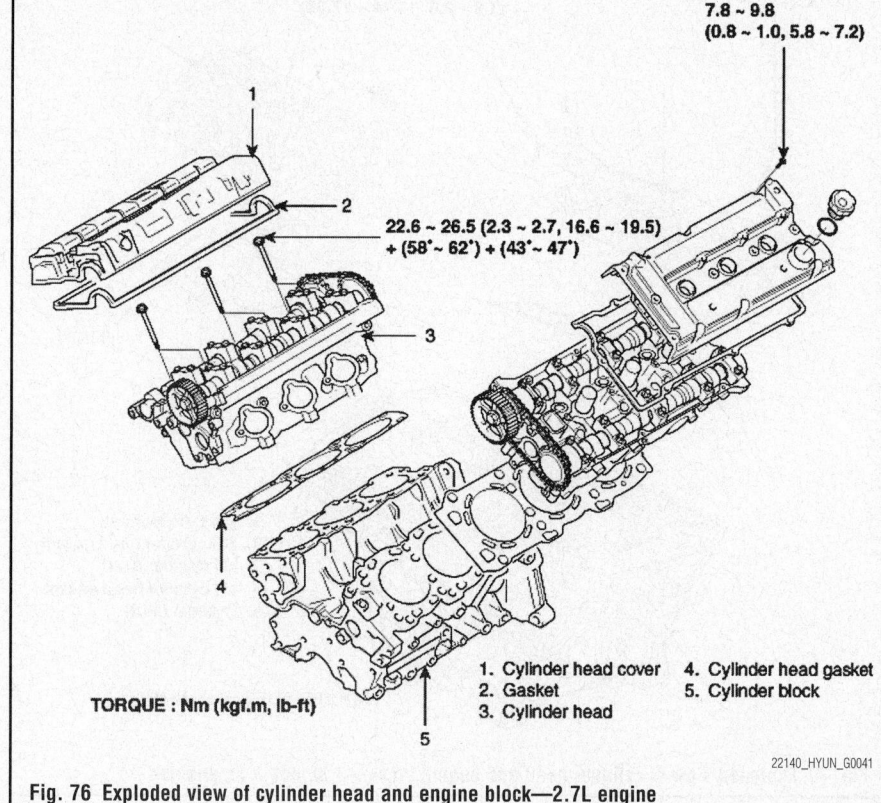

1. Cylinder head cover
2. Gasket
3. Cylinder head
4. Cylinder head gasket
5. Cylinder block

TORQUE : Nm (kgf.m, lb-ft)

22140_HYUN_G0041

Fig. 76 Exploded view of cylinder head and engine block—2.7L engine

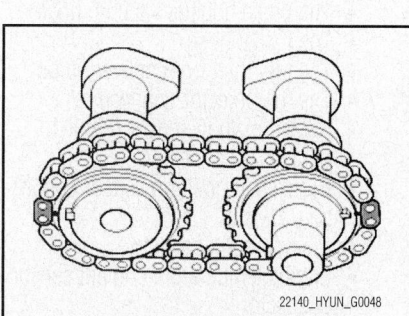

22140_HYUN_G0048

Fig. 78 Align the camshaft timing chain with the intake timing chain sprocket and exhaust timing chain sprocket as shown—2.7L engine

➡**Apply new engine oil to the thrust portion and journal of the camshafts. Apply a light coat of engine oil on the threads and under the heads of the bearing cap bolts.**

11. Using the SST (09214-21000), install the camshaft bearing oil seal.

12. Install the camshaft sprocket.

a. Temporarily install the camshaft sprocket bolts.

b. Hold the hexagonal head wrench portion of the camshaft with a wrench, and tighten the camshaft sprocket bolts. Tightening torque of camshaft sprocket bolt: 65–80 ft. lbs. (88–108 Nm).

13. Install the semi-circular packing.

14. Install the cylinder head cover.

a. Install the cylinder head cover gasket in the groove of the cylinder head cover.

➡**Before installing the head cover gasket, thoroughly clean the head cover gasket and the groove. When installing, make sure the head cover gasket is seated securely in the corners of the recesses with no gap.**

b. Apply liquid gasket to the head cover gasket at the corners of the recess. Use liquid gasket, Loctite® No. 5699. Check that the mating surfaces are clean and dry before applying liquid gasket. After assembly, wait at least 30 minutes before filling the engine with oil.

c. Install the cylinder head covers with the 16 bolts. Uniformly tighten the bolts in several passes. Tightening torque: 6–7 ft. lbs. (8–10 Nm).

15. Install or connect the following:
- The timing belt
- The power steering pump
- The PCV hose
- The auto-cruise connector and the auto- cruise cable
- The accelerator cable by loosening the locknut, then slip the cable end out of the throttle linkage
- The brake booster vacuum hose
- The PCSV hose
- The fuel inlet from delivery pipe
- The fuel injector connectors
- Heated oxygen sensor (Bank 1, Sensor 2) connector
- Crankshaft position sensor connector
- Ignition coil connector
- Engine temperature coolant sensor connector
- Heated oxygen sensor (Bank 2, Sensor 1) connector
- Engine ground line
- Camshaft position sensor connector

- Knock sensor connector
- Injector connector
- The PCSV connector
- Idle speed actuator connector
- Throttle position sensor connector

16. Install the engine wire harness connectors and wire harness clamps to the cylinder head and the intake manifold.

17. Install the heater hoses.

18. Install the upper radiator hose and lower radiator hose.

19. Install the air cleaner assembly.

a. Install the air cleaner lower cover.

b. Install the intake hose and air cleaner upper cover.

c. Install the breather hose from intake hose.

d. Connect the AFS connector.

20. Connect the battery terminal and the battery.

21. Fill with engine coolant.

22. Install the side cover and the engine cover.

23. Start the engine and check for leaks.

24. Recheck the engine coolant level and oil level.

3.3L & 3.8L Engines

See Figures 79 through 82.

Use a fender cover to avoid damaging painted surfaces. To avoid damage, unplug the wiring connectors carefully while holding the connector portion. Mark all wiring and hoses to avoid misconnection. Inspect the timing belt before removing. Turn the crankshaft pulley so that the No. 1 piston is at Top Dead Center (TDC).

1. Before servicing the vehicle, refer to the Precautions Section.

2. Remove or disconnect the following:
- The negative battery connection
- The timing chain
- The water temperature control assembly
- The camshaft bearing cap
- The camshaft assembly

To install:

3. Thoroughly clean all parts to be assembled. Rotate the crankshaft, set the No. 1 piston at TDC.

4. Install the Continuously Variable Valve Timing (CVVT) and camshaft sprocket. Tightening torque: 48–56 ft. lbs. (65–76 Nm).

a. Install camshaft-inlet to dowel pin of CVVT assembly. At this time, do not install to oil hole of camshaft-inlet.

b. Hold the hexagonal head wrench portion of the camshaft with a vise, and install the bolt and CVVT assembly.

c. Do not rotate the CVVT assembly when the camshaft is installed to the dowel pin of the CVVT assembly.

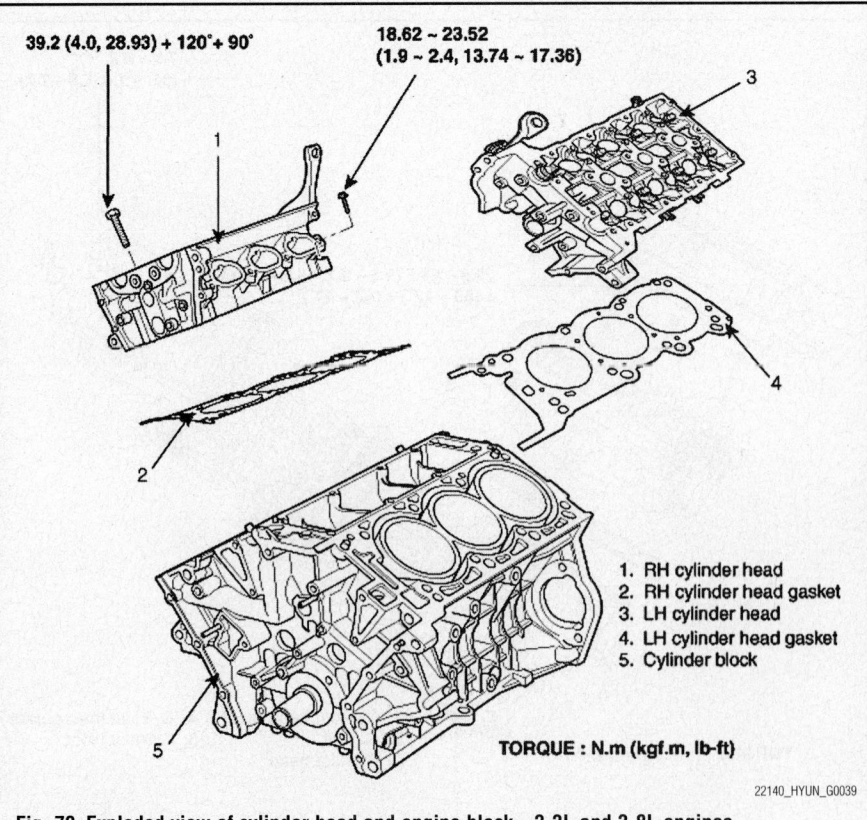

39.2 (4.0, 28.93) + 120°+ 90°

18.62 ~ 23.52 (1.9 ~ 2.4, 13.74 ~ 17.36)

1. RH cylinder head
2. RH cylinder head gasket
3. LH cylinder head
4. LH cylinder head gasket
5. Cylinder block

TORQUE : N.m (kgf.m, lb-ft)

22140_HYUN_G0039

Fig. 79 Exploded view of cylinder head and engine block—3.3L and 3.8L engines

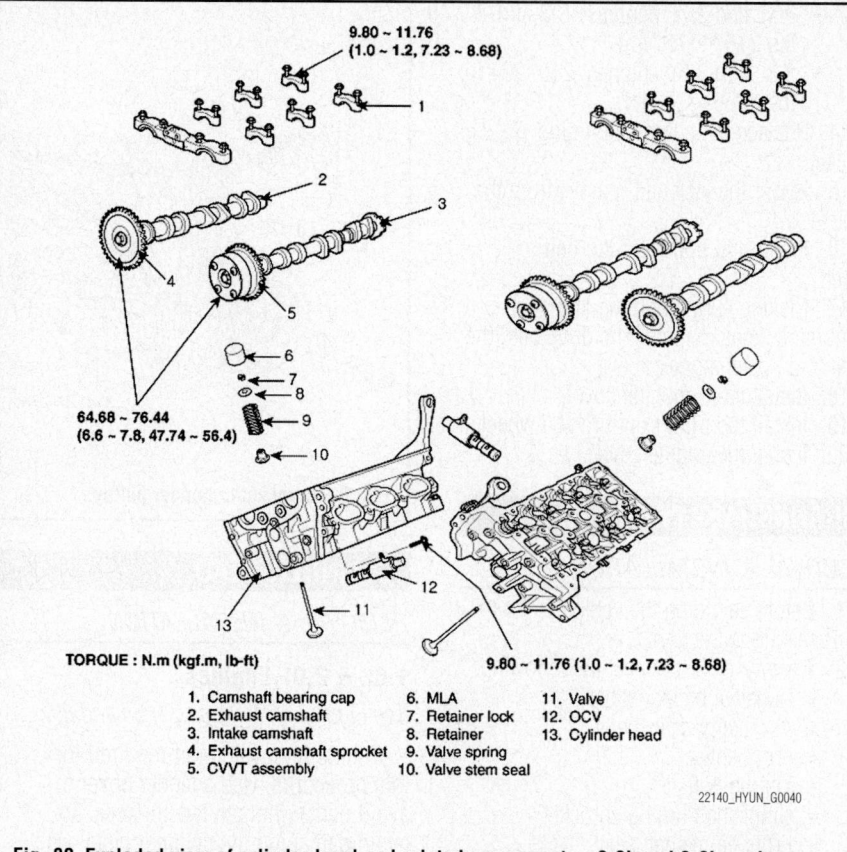

9.80 ~ 11.76
(1.0 ~ 1.2, 7.23 ~ 8.68)

64.68 ~ 76.44
(6.6 ~ 7.8, 47.74 ~ 56.4)

TORQUE : N.m (kgf.m, lb-ft)

9.80 ~ 11.76 (1.0 ~ 1.2, 7.23 ~ 8.68)

1. Camshaft bearing cap
2. Exhaust camshaft
3. Intake camshaft
4. Exhaust camshaft sprocket
5. CVVT assembly
6. MLA
7. Retainer lock
8. Retainer
9. Valve spring
10. Valve stem seal
11. Valve
12. OCV
13. Cylinder head

22140_HYUN_G0040

Fig. 80 Exploded view of cylinder head and related components—3.3L and 3.8L engines

5. Install the camshafts.
   a. Apply a light coat of engine oil on camshaft journals.
   b. Assemble the key groove of camshaft rear side to the same level of head top surface.
   c. Be careful to get the right bank, left bank, intake side, and exhaust side in the correct position before assembling.
6. Install the camshaft bearing caps in the sequence shown:
   a. Step 1—Tightening torque: 48 inch lbs. (6 Nm).
   b. Step 2—Tightening torque: 84–108 inch lbs. (10–12 Nm).

✳✳ WARNING

Be careful to properly position the right bank, left bank, intake side, exhaust side, and front mark on the camshaft bearing caps while assembling.

✳✳ WARNING

Rotate the crankshaft so as not to contact the valves to the pistons by positioning the pistons 0.3937 inch (10mm) below the top of the cylinder block.

7. Install the water temperature control assembly.
8. Install the timing chain.
9. Check and adjust the valve clearance, as necessary.
10. Connect the negative battery cable.
11. Fill with engine coolant.
12. Start the engine and check for leaks.
13. Recheck the engine coolant level and oil level.

CAMSHAFT BEARING REPLACEMENT

Check each bearing for damage. If the bearing surface is excessively damaged, replace the cylinder head assembly or camshaft bearing cap, as necessary.

CRANKSHAFT DAMPER

REMOVAL & INSTALLATION

See Figures 83 through 87.

1. Before servicing the vehicle, refer to the Precautions Section.
2. Remove the engine cover.
3. Remove the front right wheel and tire.
4. Remove the right side cover.
5. Remove the accessory drive belt (A), the idler (B) and the tensioner (C).

➡**In removing the accessory drive belt, fix a tool in the auto tensioner pulley bolt and turn the bolt counter clockwise.**

6. Remove the timing belt upper cover (A).
7. Align the groove of the pulley with the timing mark of the timing belt cover by turning the crankshaft pulley clockwise.
8. Check if the timing mark of the camshaft sprocket is aligned with that of the cylinder head cover with No. 1 cylinder piston at Top Dead Center (TDC).

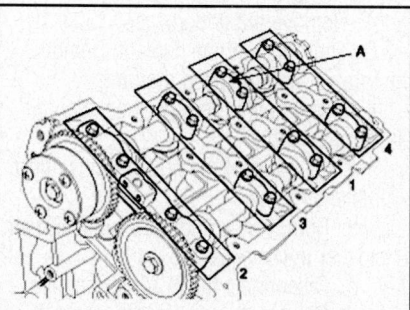

22140_HYUN_G0051

Fig. 81 Install the camshaft bearing caps in the sequence shown—3.3L and 3.8L engines

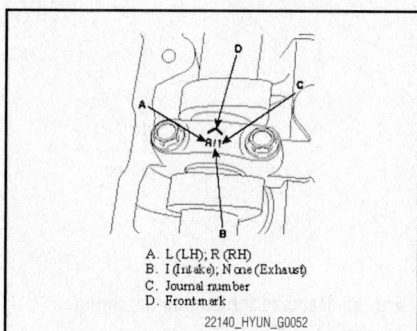

A. L (LH); R (RH)
B. I (Intake); None (Exhaust)
C. Journal number
D. Frontmark

22140_HYUN_G0052

Fig. 82 Be careful to properly position the camshaft bearing caps according to its markings—3.3L and 3.8L engines

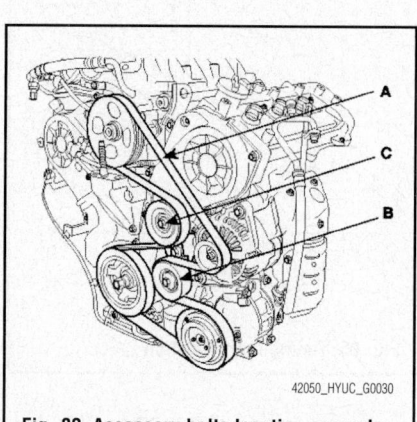

42050_HYUC_G0030

Fig. 83 Accessory belts location example

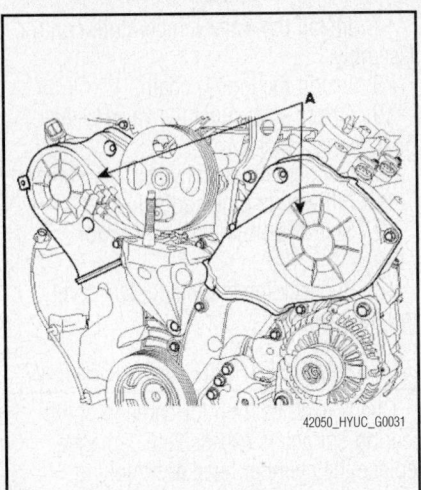

Fig. 84 Timing belt upper cover location example

9. Support the engine oil pan with a jack.

> ☀☀ **WARNING**
>
> **Put a wooden or rubber block between the jack and the engine oil pan.**

10. Remove the engine mounting bracket (A).

11. Remove the crankshaft damper pulley (A).

12. Remove the crankshaft damper.

To install:

13. Install the crankshaft damper. Torque the mounting bolt as follows:
- 1.6L engine: 101–109 ft. lbs. (137–148 Nm)
- 2.0L engine: 116–123 ft. lbs. (157–167 Nm)

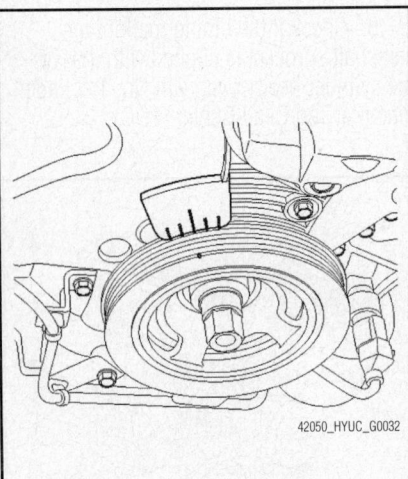

Fig. 85 Timing mark location

- 2.4L and 2.7L engines: 123–130 ft. lbs. (167–176 Nm)
- 3.3L and 3.8L engines: 210–224 ft. lbs. (285–304 Nm)

14. Replace the engine mounting bracket.

15. Lower the jack and remove from the vehicle.

16. Install the upper timing belt cover.

17. Making sure all timing and alignment marks match, install the drive belt, the idler, and the tensioner.

18. Install the right side cover.

19. Install the right front tire and wheel.

20. Install the engine cover.

CRANKSHAFT FRONT SEAL

REMOVAL & INSTALLATION

1. Before servicing the vehicle, refer to the Precautions Section.

2. Remove or disconnect the following:
- Negative battery cable
- Accessory drive belts
- Front cover
- Timing belt
- Crankshaft timing sprocket
- Front crankshaft seal

To install:

3. Install the front crankshaft seal so that it is flush with the oil pump housing.

4. Install or connect the following:
- Crankshaft timing sprocket
- Timing belt
- Front cover
- Accessory drive belts
- Negative battery cable

5. Start the engine and check for leaks.

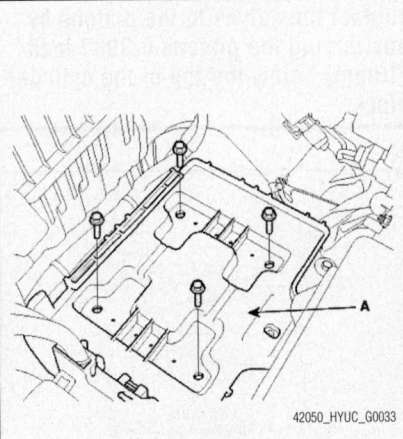

Fig. 86 Remove the engine mounting bracket (A)

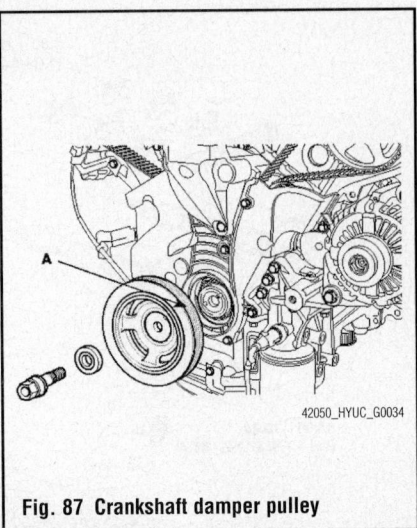

Fig. 87 Crankshaft damper pulley

CYLINDER HEAD

REMOVAL & INSTALLATION

1.6L & 2.0L Engines

See Figures 69 through 71, 88 and 89.

Engine removal is not required for this procedure. Use a fender cover to avoid damaging painted surfaces. To avoid damaging the cylinder head, wait until the engine coolant temperature drops below normal temperature before removing it. When handling a metal gasket, take care not to fold the gasket or damage the contact surface of the gasket. To avoid damage, unplug the wiring connectors carefully while holding the connector portion. Mark all wiring and hoses to avoid misconnection. Inspect the timing belt before removing the cylinder head. Turn the crankshaft pulley so that the No. 1 piston is at Top Dead Center (TDC).

1. Before servicing the vehicle, refer to the Precautions Section.

2. Disconnect the terminals from battery and remove the battery.

3. Remove the engine cover.

4. Remove the undercover.

5. Drain the engine coolant. Remove the radiator cap to speed draining.

6. Remove the intake air hose and air cleaner assembly:

a. Disconnect the breather hose from intake air hose.

b. Remove the intake air hose and air cleaner upper cover.

c. Disconnect the ECM connectors.

d. Remove the air cleaner element and air cleaner lower cover.

7. Remove the battery tray.

8. Remove the upper radiator hose and lower radiator hose.

9. Remove the heater hoses.

10. Remove the fuel hose.

11. Remove the accelerator cable by loosening the lock-nut, then slip the cable end out of the throttle linkage.

12. Disconnect the Throttle Position Sensor (TPS) connector and the MAP sensor connector.

13. Remove the engine wire harness connectors and wire harness clamps from cylinder head and the intake manifold:

a. Disconnect the rear oxygen sensor connector.

b. Disconnect the air conditioner compressor switch connector.

c. Disconnect the knock sensor connector.

d. Disconnect the injector connectors.

e. Remove the wire harness bracket.

f. Disconnect the Idle Speed Actuator (ISA) connector.

g. Disconnect the front oxygen sensor connector.

h. Disconnect the Crankshaft Position Sensor (CKP) connector.

i. Disconnect the Oil Control Valve (OCV) connector.

j. Disconnect the ignition coil connector.

k. Disconnect the ignition coil condenser connector.

l. Disconnect the Camshaft Position Sensor (CMP) connector.

m. Disconnect the ground cable.

n. Remove the wire harness bracket.

14. Remove or disconnect the following:

- The hose of the Purge Control Solenoid Valve (PCSV) side
- The brake booster vacuum hose
- The power steering pump and fix the pump to vehicle with a wire
- The ignition coil
- The exhaust manifold
- The intake manifold
- The timing belt
- The cylinder head cover
- The camshaft sprocket
- The timing chain auto tensioner
- The camshaft bearing caps and camshafts
- The OCV

- The OCV filter
- The engine mounting support bracket fixing bolts

15. Remove the cylinder head bolts, then remove the cylinder head:

a. Using an 8mm hexagon wrench, uniformly loosen and remove the 10 cylinder head bolts, in several passes, in the sequence shown. Head warpage or cracking could result from removing bolts in an incorrect order.

b. Lift the cylinder head from the dowels on the cylinder block and replace the cylinder head on wooden blocks on a bench. Be careful not to damage the contact surfaces of the cylinder head and cylinder block.

To install:

Thoroughly clean all parts to be assembled. Always use a new cylinder head and manifold gasket. Always use a new cylinder head bolt. The cylinder head gasket is a metal gasket. Take care not to bend it. Rotate the crankshaft, set the No. 1 piston at TDC.

16. Install the cylinder head gasket on the cylinder block. Be careful of the installation direction.

17. Place the cylinder head quietly in order not to damage the gasket with the bottom part of the end.

18. Install the cylinder head bolts:

a. Apply a light coat if engine oil on the threads and under the heads of the cylinder head bolts.

b. Using an 8mm and 10mm hexagon wrench, install and tighten the 10 cylinder head bolts and plate washers, in several passes, in the sequence shown.

- 1.6L engine—Step 1: 22 ft. lbs. (29 Nm) plus 90°
- 1.6L engine—Step 2: Release all bolts
- 1.6L engine—Step 3: 22 ft. lbs. (29 Nm) plus 90°
- 2.0L engine—M10 bolts Step 1: 18 ft. lbs. (25 Nm) plus 60°–65°
- 2.0L engine—M10 bolts Step 2: Additional 60°–65°
- 2.0L engine—M12 bolts Step 1: 22 ft. lbs. (29 Nm) plus 60°–65°
- 2.0L engine—M12 bolts Step 2: Additional 60°–65°

19. Install the engine mounting support bracket fixing bolts.

20. Install the OCV filter. Tightening torque: 30–37 ft. lbs. (40–50 Nm).

➡**Always use a new OCV filter gasket and keep the OCV filter clean.**

21. Install the OCV. Tightening torque: 84–108 inch lbs. (10–12 Nm).

➡**Do not reuse the OCV when dropped. Keep the OCV clean. Do not hold the OCV sleeve during servicing. When the OCV is installed on the engine, do not move the engine while holding the OCV yoke.**

22. Install the camshafts:

a. Align the camshaft timing chain with the intake timing chain sprocket and exhaust timing chain sprocket as shown.

b. Install the camshaft and bearing caps.

- 1.6L engine—Tightening torque: 108–120 inch lbs. (12–14 Nm).

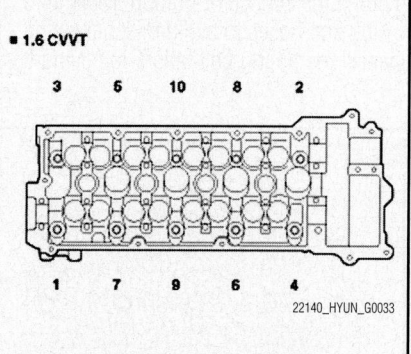

Fig. 88 Cylinder head torque/loosen sequence—1.6L & 2.0L engines

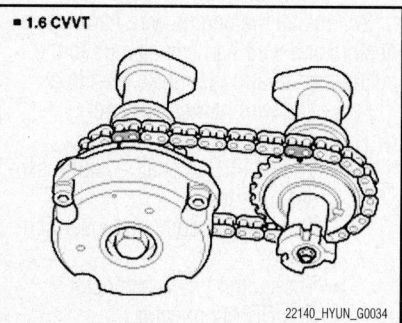

Fig. 89 Align the camshaft timing chain with the intake timing chain sprocket and exhaust timing chain sprocket—1.6L engine

- 2.0L engine—Tightening torque: 120–132 inch lbs. (14–15 Nm)
  c. Install the timing chain auto tensioner. Tightening torque: 72–84 inch lbs. (8–10 Nm).

23. Using the SST (09221 - 21000), install the camshaft bearing oil seal.

24. Install the camshaft sprocket.

25. Install the cylinder head cover.
  a. Install the cylinder head cover gasket in the groove of the cylinder head cover.

➡**Before installing the cylinder head cover gasket, thoroughly clean the cylinder head cover and the groove. When installing, make sure the cylinder head cover gasket is seated securely in the corners of the recesses with no gap.**

  b. Apply liquid gasket to the head cover gasket at the corners of the recess.

➡**Use liquid gasket, Loctite® No. 5999. Check that the mating surfaces are clean and dry before applying liquid gasket. After assembly, wait at least 30 minutes before filling the engine with oil.**

  c. Install the cylinder head cover with bolts.
    - Step 1: Pre-tighten all bolts by 36–48 inch lbs. (4–5 Nm)
    - Step 2: Tighten by the specified torque 72–84 inch lbs. (8–10 Nm)

26. Install or connect the following:
  - The timing belt
  - The intake manifold
  - The exhaust manifold
  - The ignition coil
  - The power steering pump
  - The brake booster hose
  - The hose of the PCSV side

27. Install the engine wire harness connectors and wire harness clamps to the cylinder head and the intake manifold:
  - The wire harness bracket
  - The ground cable
  - The CMP(Camshaft position sensor) connector
  - The ignition coil condenser connector
  - The ignition coil connector
  - The OCV connector
  - The CKP connector
  - The front oxygen sensor connector
  - The ISA connector
  - The wire harness bracket
  - The fuel injector connectors
  - The knock sensor connector
  - The air conditioner compressor switch connector

  - The rear oxygen sensor connector

28. Install or connect the following:
  - The TPS connector and the MAP sensor connector
  - The accelerator cable
  - The fuel hose
  - The heater hoses
  - The upper radiator hose(A) and lower radiator hose
  - The battery tray
  - The intake air hose and air cleaner assembly

29. Install the air cleaner element and air cleaner lower cover. Tightening torque: 72–84 inch lbs. (8–10 Nm).

30. Connect the ECM connectors.

31. Install the intake air hose and air cleaner upper cover.

32. Connect the breather hose to intake air hose.

33. Install the undercover.

34. Install the engine cover. Tightening torque: 36–48 inch lbs. (4–6 Nm).

35. Install the battery and connect the battery terminals.

36. Fill with engine coolant.

37. Start the engine and check for leaks.

38. Recheck engine coolant level and oil level.

2.4L Engine

See Figures 73, 90 through 92.

Engine removal is not required for this procedure. Use a fender cover to avoid damaging painted surfaces. To avoid damaging the cylinder head, wait until the engine coolant temperature drops below normal temperature before removing it. When handling a metal gasket, take care not to fold the gasket or damage the contact surface of the gasket. To avoid damage, unplug the wiring connectors carefully while holding the connector portion. Mark all wiring and hoses to avoid misconnection. Inspect the timing belt before removing the

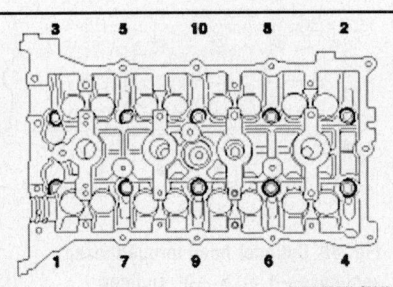

Fig. 90 Uniformly loosen/tighten and remove/install the cylinder head bolts, in several passes, in the sequence shown— 2.4L engine

cylinder head. Turn the crankshaft pulley so that the No. 1 piston is at Top Dead Center (TDC).

1. Before servicing the vehicle, refer to the Precautions Section.

2. Disconnect the negative terminal from the battery.

3. Remove the engine cover(A).

4. Remove the air duct.

5. Remove the intake air hose and air cleaner assembly.
  a. Disconnect the AFS connector.
  b. Disconnect the breather hose from air cleaner hose.
  c. Disconnect the ECM connector.
  d. Remove the intake air hose and air cleaner assembly.

6. Remove front wheels.

7. Remove the undercover.

8. Drain the engine coolant. Remove the radiator cap to speed draining.

9. Remove or disconnect the following:
  - The upper and lower radiator hose
  - The heater hoses
  - The A/C switch, alternator connector, and oil pressure switch
  - The Oil Control Valve (OCV) connector and OTS connector
  - The injector connectors
  - The ETS connector
  - The Camshaft Position (CMP) connector, and knock sensor connector
  - The ignition coil connectors
  - The Purge Control Solenoid Valve (PCSV) connector, WTS connector, condenser connector, and Crankshaft Position (CKP) sensor connector
  - The delivery pipe, brake vacuum hose, and PCSV hose
  - The water temp control assembly
  - The intake manifold
  - The exhaust manifold
  - The timing chain
  - The Continuously Variable Valve Timing (CVVT) assembly and camshaft sprocket

10. Remove the camshaft.
  a. Remove the front camshaft bearing cap.
  b. Remove the camshaft bearing caps.
  c. Remove the camshafts.

11. Remove the OCV and OTS.

12. Remove the cylinder head bolts, then remove the cylinder head:
  a. Using triple square wrench, uniformly loosen and remove the 10 cylinder head bolts, in several passes, in the sequence shown. Remove the 10 cylinder head bolts and plate washers.

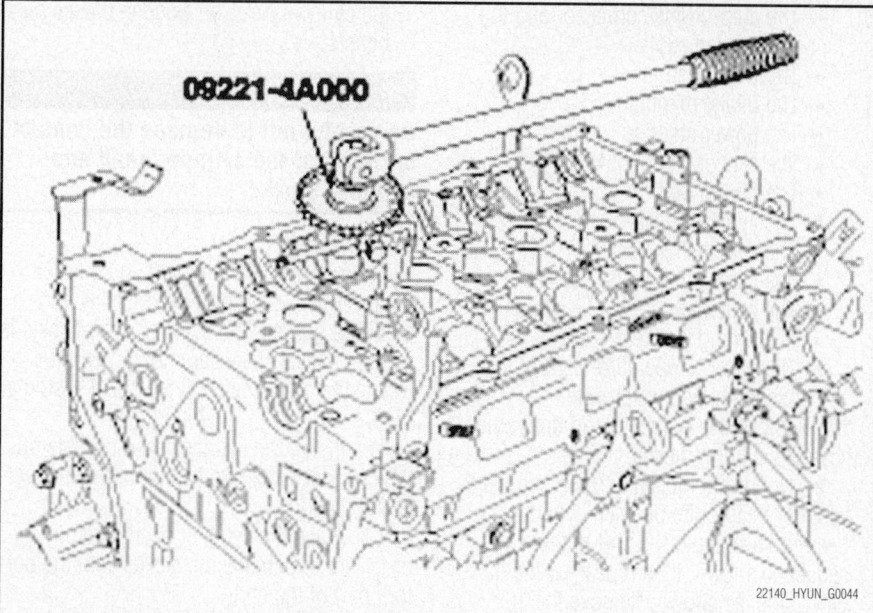

09221-4A000

22140_HYUN_G0044

Fig. 91 Using SST (09221-4A000) to install the cylinder head bolts—2.4L engine

✳✳ WARNING

Head warpage or cracking could result from removing bolts in an incorrect order.

b. Lift the cylinder head from the dowels on the cylinder block and place the cylinder head on wooden blocks on a bench.

✳✳ WARNING

Be careful not to damage the contact surfaces of the cylinder head and cylinder block.

To install:

➥Thoroughly clean all parts to be assembled. Always use a new head and manifold gasket. The cylinder head gasket is a metal gasket. Take care not to bend it. Rotate the

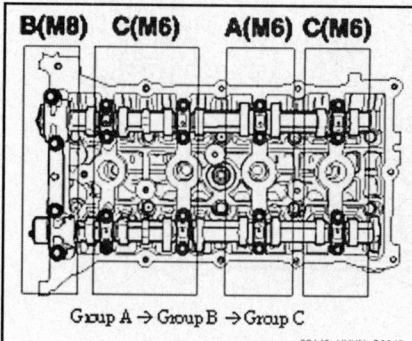

B(M8) C(M6)   A(M6) C(M6)

Group A → Group B → Group C

22140_HYUN_G0045

Fig. 92 Follow this tightening order for the camshaft bearing caps—2.4L engine

crankshaft, set the No. 1 piston at TDC.

13. Install OCV filter. Keep the OCV filter clean.

14. Install the cylinder head gasket on the cylinder block. Be careful of the installation direction.

15. Place the cylinder head carefully in order not to damage the gasket.

16. Using SST (09221-4A000), install the cylinder head bolts.

a. Apply a light coat if engine oil on the threads and under the heads of the cylinder head bolts.

b. Using a wrench, install and tighten the 10 cylinder head bolts and plate washers, in several passes, in the sequence shown. Tightening torque: 25 ft. lbs. (34 Nm) plus 90° and then an additional 90°.

➥**Always use new cylinder head bolts.**

17. Install the OCV. Tightening torque: 84–108 inch lbs. (10–12 Nm).

18. Install the OTS. Tightening torque: 15–17 ft. lbs. (20–24 Nm).

➥**Do not reuse the OCV if it is dropped. Keep the OCV clean. Do not hold onto the OCV sleeve during servicing. When the OCV is installed on the engine, do not move the engine while holding the OCV yoke.**

19. Install the CVVT and camshaft sprocket. Tightening torque: 40–47 ft. lbs. (54–64 Nm).

➥**Hold the hexagonal head wrench portion of the camshaft with a vise, and install the bolt and CVVT assembly.**

20. Install the camshafts. Apply a light coat of engine oil on camshaft journals.

21. Install the camshaft bearing caps in their proper locations. Follow the illustrated tightening order.

• Tightening torque M6: 8–9 ft. lbs. (11–13 Nm)
• Tightening torque M8: 20–23 ft. lbs. (27–31 Nm)

22. Install the timing chain.

23. Check and adjust the valve clearance.

24. Install the exhaust manifold.

25. Install the intake manifold.

26. Install the water temp control assembly, and tighten as follows:

a. Bolt: 11–16 ft. lbs. (15–22 Nm)
b. Nut: 15–20 ft. lbs. (20–27 Nm)

➥**Assemble water temp control assembly and water inlet pipe to water pump assembly before nuts for assembling of water inlet pipe to be tightened. Insert after wetting O-ring or inner surface of thermostat housing. Always use a new O-ring.**

27. Install or connect the following:

• The delivery pipe, brake hose, and PCSV hose
• The PCSV connector, WTS connector, condenser connector, and CKP sensor connector
• The ignition coil connector
• The ETS connector
• The CMP connector, and knock sensor connector
• The injector connectors
• The OCV connector and OTS connector
• The A/C switch, alternator connect, and oil pressure switch
• The heater hoses
• The upper radiator hose and lower radiator hose
• The intake air hose and air cleaner assembly
• The engine cover
• The negative terminal to the battery

28. Fill with engine coolant.

29. Start the engine and check for leaks.

30. Recheck the engine coolant level and oil level.

2.7L Engine

See Figures 76, 77, 21, 93, and 94.

Engine removal is not required for this procedure. Use a fender cover to avoid damaging painted surfaces. To avoid damaging the cylinder head, wait until the engine coolant temperature drops below normal temperature before removing it. When handling a metal gasket, take care not to fold the gasket or damage the contact surface of the gasket. To avoid damage, unplug the wiring connectors carefully while holding the connector portion. Mark all wiring and hoses to avoid misconnection. Inspect the timing belt before removing the cylinder head. Turn the crankshaft pulley so that the No. 1 piston is at Top Dead Center (TDC).

1. Before servicing the vehicle, refer to the Precautions Section.

2. Remove the side cover (A, B, C) and the engine cover (D).

3. Disconnect the battery terminal and the battery.

4. Remove the radiator drain plug and drain engine coolant. Remove the radiator cap to speed draining.

5. Remove the air cleaner assembly.

a. Disconnect the Air Flow Sensor (AFS) connector.

b. Remove the breather hose from intake hose.

c. Remove the intake hose and air cleaner upper cover.

d. Remove the air cleaner lower cover.

6. Remove the upper radiator hose and lower radiator hose.

7. Remove the heater hoses.

8. Remove the engine wire harness connectors and wire harness clamps from the cylinder head and the intake manifold:

- Throttle position sensor (TPS) connector
- Idle speed actuator (ISA) connector
- Purge Control Solenoid Valve (PCSV) connector
- Knock sensor connector
- Camshaft Position (CMP) sensor connector
- Engine ground line
- Heated oxygen sensor (Bank 2, Sensor 1) connector
- Engine temperature coolant sensor connector
- Ignition coil connector
- Crankshaft Position (CKP) sensor connector
- Heated oxygen sensor (Bank 1, Sensor 2) connector
- Fuel injector connectors

9. Remove or disconnect the following:

- The fuel inlet from the delivery pipe
- The Purge Control Solenoid Valve (PCSV) hose
- The brake booster vacuum hose
- The accelerator cable by loosening the locknut, then slip the cable end out of the throttle linkage

- The auto-cruise connector and the auto-cruise cable
- The PCV hose
- The intake manifold
- The power steering pump
- The exhaust manifold
- The timing belt
- The cylinder head covers
- The camshaft sprocket
- The camshaft bearing caps
- The camshafts
- The timing belt rear cover
- The water temperature control assembly and water pipe

10. Remove the cylinder head bolts, then remove the cylinder heads.

a. Uniformly loosen and remove the 8 cylinder head bolts on each cylinder head in several passes and in the sequence shown, then repeat for the other side, as shown. Remove the 16 cylinder head bolts and plate washer.

✳✳ WARNING

Head warpage or cracking could result from removing bolts in an incorrect order.

b. Lift the cylinder head from the dowels on the cylinder block and place

the cylinder head on wooden blocks on a bench.

✳✳ WARNING

Be careful not to damage the contact surfaces of the cylinder head and cylinder block.

To install:

Thoroughly clean all parts to be assembled. Always use a new head gasket and manifold gasket. The cylinder head gasket is a metal gasket. Take care not to bend it. Rotate the crankshaft, set the No. 1 piston at TDC.

11. Install the cylinder head gaskets on the cylinder block. Be careful of the installation direction.

12. Place the cylinder head carefully in order not to damage the gasket with the bottom part of the end.

13. Install cylinder head bolts as follows:

a. Apply a light coat if engine oil on the threads and under the heads of the cylinder head bolts.

b. Install the plate washer to the cylinder head bolt.

c. Install and uniformly tighten the cylinder head bolts on each cylinder head in several passes and in the

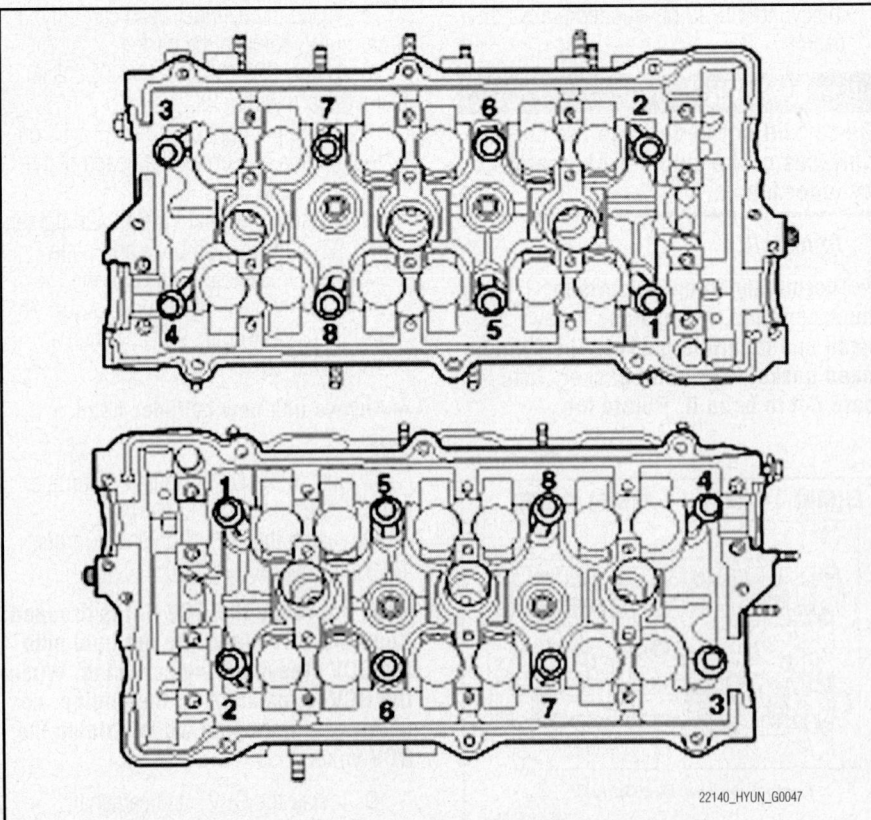

Fig. 93 Uniformly loosen/tighten and remove/install the cylinder head bolts on each cylinder head in several passes and in the sequence shown—2.7L engine

22140_HYUN_G0047

sequence shown, then repeat for the other side, as shown:

➡**If only 1 of the cylinder head bolts does not meet the torque specification, replace the cylinder head bolt.**

- Step 1: Torque bolts to 18 ft. lbs. (25 Nm)
- Step 2: Retighten the cylinder head bolts by 60° in the numerical order shown
- Step 3: Retighten the cylinder head bolts by 45° in the numerical order shown

14. Install the water pipe and water temperature control assembly. Tightening torque: 11–5 ft. lbs. (15–20 Nm).

15. Install the timing belt rear cover. Tightening torque: 7–9 ft. lbs. (10–12 Nm).

16. Install the camshafts.

a. Align the camshaft timing chain with the intake timing chain sprocket and exhaust timing chain sprocket as shown.

b. Install the camshaft.

c. Install the camshaft bearing caps, as follows:

- Tightening torque: M6 (38mm) bolt: 7–9 ft. lbs. (10–13 Nm)
- Tightening torque: M6 (50mm) bolt: 11–12 ft. lbs. (15–17 Nm)

➡**Apply new engine oil to the thrust portion and journal of the camshafts. Apply a light coat of engine oil on the threads and under the heads of the bearing cap bolts.**

17. Using the SST (09214-21000), install the camshaft bearing oil seal.

18. Install the camshaft sprocket.

a. Temporarily install the camshaft sprocket bolts.

b. Hold the hexagonal head wrench portion of the camshaft with a wrench, and tighten the camshaft sprocket bolts. Tightening torque of camshaft sprocket bolt: 65–80 ft. lbs. (88–108 Nm).

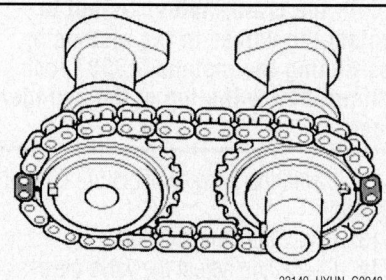

22140_HYUN_G0048

Fig. 94 Align the camshaft timing chain with the intake timing chain sprocket and exhaust timing chain sprocket as shown— 2.7L engine

19. Install the semi-circular packing.

20. Install the cylinder head cover.

a. Install the cylinder head cover gasket in the groove of the cylinder head cover.

➡**Before installing the head cover gasket, thoroughly clean the head cover gasket and the groove. When installing, make sure the head cover gasket is seated securely in the corners of the recesses with no gap.**

b. Apply liquid gasket to the head cover gasket at the corners of the recess. Use liquid gasket, Loctite® No. 5699. Check that the mating surfaces are clean and dry before applying liquid gasket. After assembly, wait at least 30 minutes before filling the engine with oil.

c. Install the cylinder head covers with the 16 bolts. Uniformly tighten the bolts in several passes. Tightening torque: 6–7 ft. lbs. (8–10 Nm).

21. Install or connect the following:

- The timing belt
- The exhaust manifold
- The power steering pump
- The intake manifold
- The PCV hose
- The auto-cruise connector and the auto-cruise cable
- The accelerator cable by loosening the locknut, then slip the cable end out of the throttle linkage
- The brake booster vacuum hose
- The PCSV hose
- The fuel inlet from delivery pipe
- The fuel injector connectors
- Heated oxygen sensor (Bank 1, Sensor 2) connector
- Crankshaft position sensor connector
- Ignition coil connector
- Engine temperature coolant sensor connector
- Heated oxygen sensor (Bank 2, Sensor 1) connector
- Engine ground line
- Camshaft position sensor connector
- Knock sensor connector
- Injector connector
- The PCSV connector
- Idle speed actuator connector
- Throttle position sensor connector

22. Install the engine wire harness connectors and wire harness clamps to the cylinder head and the intake manifold.

23. Install the heater hoses.

24. Install the upper radiator hose and lower radiator hose.

25. Install the air cleaner assembly.

a. Install the air cleaner lower cover.

b. Install the intake hose and air cleaner upper cover.

c. Install the breather hose from intake hose.

d. Connect the AFS connector.

26. Connect the battery terminal and the battery.

27. Fill with engine coolant.

28. Install the side cover and the engine cover.

29. Start the engine and check for leaks.

30. Recheck the engine coolant level and oil level.

3.3L & 3.8L Engines

See Figures 79, 80, 95 through 98.

Use a fender cover to avoid damaging painted surfaces. To avoid damaging the cylinder head, wait until the engine coolant temperature drops below normal temperature before removing it. When handling a metal gasket, take care not to fold the gasket or damage the contact surface of the gasket. To avoid damage, unplug the wiring connectors carefully while holding the connector portion. Mark all wiring and hoses to avoid misconnection. Inspect the timing belt before removing the cylinder head. Turn the crankshaft pulley so that the No. 1 piston is at Top Dead Center (TDC).

1. Before servicing the vehicle, refer to the Precautions Section.

➡**Engine removal is required for this procedure.**

2. Remove or disconnect the following:

- The negative battery connection
- The exhaust manifold
- The intake manifold
- The timing chain
- The water temperature control assembly
- The camshaft bearing cap
- The camshaft assembly

3. The cylinder head bolts, then remove cylinder head.

a. Uniformly loosen and remove the 16 cylinder head bolts, in several passes, in the sequence shown.

b. Remove the 16 cylinder head bolts and plate washers.

✸✸ WARNING

Head warpage or cracking could result from removing bolts in an incorrect order.

c. Lift the cylinder head from the dowels on the cylinder block and place

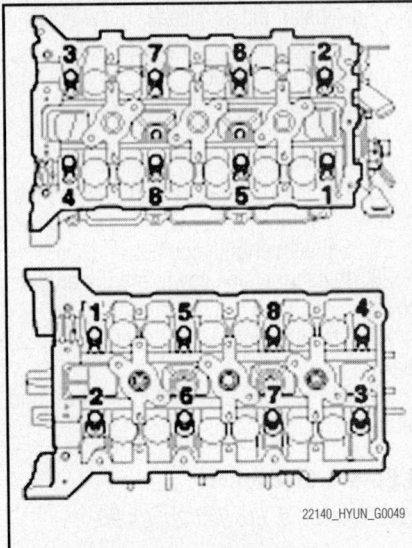

Fig. 95 Uniformly loosen/tighten and remove/install the 16 cylinder head bolts, in several passes, in the sequence shown—3.3L and 3.8L engines

the cylinder head on wooden blocks on a bench.

❊❊ WARNING

Be careful not to damage the contact surfaces of the cylinder head and cylinder block.

To install:

Thoroughly clean all parts to be assembled. Always use a new head and manifold gasket. The cylinder head gasket is a metal gasket. Take care not to bend it. Rotate the crankshaft, set the No. 1 piston at TDC.

4. Ensure the sealant locations on the cylinder head and cylinder block are free of engine oil or any debris.

5. Apply sealant on the cylinder block top face before assembling cylinder head gaskets.

➡**The part must be assembled within 5 minutes after sealant is applied. The bead width should be 0.08–0.12 inch (2–3mm). The sealant location: 0.04–0.06 inch (1.0–1.5mm) from block surface. Recommended sealant: Liquid sealant TB1217H.**

6. Install the cylinder head. Remove any extruded sealant after assembling cylinder heads.

7. Place the cylinder head carefully in order not to damage the gasket with the bottom part of the end.

8. Install cylinder head bolts.
   a. Do not apply engine oil on the threads or under the heads of the cylinder head bolts.

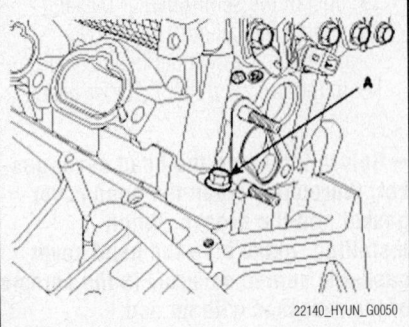

Fig. 96 Tighten bolt (A) to 14–17 ft. lbs. (19–24 Nm)—3.3L and 3.8L engines

   b. Using SST(09221-4A000), install and tighten the cylinder head bolts and plate washers, in several passes, in the sequence shown.
   • Step 1: tighten to 28–30 ft. lbs. (37–41 Nm)
   • Step 2: tighten an additional: 120° plus or minus 2°
   • Step 3: tighten an additional: 90° plus or minus 2°
   c. Tighten bolt (A) to 14–17 ft. lbs. (19–24 Nm).

➡**Always use new cylinder head bolts.**

9. Install the Continuously Variable Valve Timing (CVVT) and camshaft sprocket. Tightening torque: 48–56 ft. lbs. (65–76 Nm).
   a. Install camshaft-inlet to dowel pin of CVVT assembly. At this time, do not install to oil hole of camshaft-inlet.
   b. Hold the hexagonal head wrench portion of the camshaft with a vise, and install the bolt and CVVT assembly.
   c. Do not rotate the CVVT assembly when the camshaft is installed to the dowel pin of the CVVT assembly.
10. Install the camshafts.
   a. Apply a light coat of engine oil on camshaft journals.

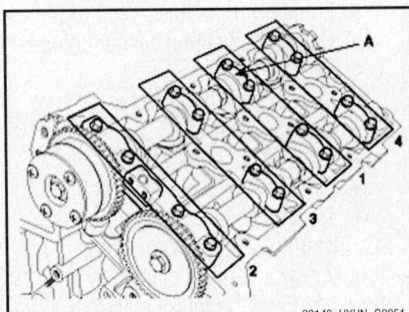

Fig. 97 Install the camshaft bearing caps in the sequence shown—3.3L and 3.8L engines

   b. Assemble the key groove of camshaft rear side to the same level of head top surface.
   c. Be careful to get the right bank, left bank, intake side, and exhaust side in the correct position before assembling.
11. Install the camshaft bearing caps in the sequence shown:
   a. Step 1—Tightening torque: 48 inch lbs. (6 Nm).
   b. Step 2—Tightening torque: 84–108 inch lbs. (10–12 Nm).

❊❊ WARNING

Be careful to properly position the right bank, left bank, intake side, exhaust side, and front mark on the camshaft bearing caps while assembling.

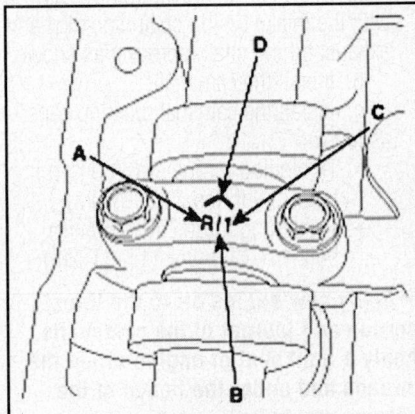

A. L (LH); R (RH)
B. I (Intake); None (Exhaust)
C. Journal number
D. Front mark

Fig. 98 Be careful to properly position the camshaft bearing caps according to its markings—3.3L and 3.8L engines

❊❊ WARNING

Rotate the crankshaft so as not to contact the valves to the pistons by positioning the pistons 0.3937 inch (10mm) below the top of the cylinder block.

12. Install the water temperature control assembly.
13. Install the timing chain.
14. Check and adjust the valve clearance, as necessary.
15. Install the exhaust manifold.
16. Install the intake manifold.
17. Connect the negative battery cable.
18. Fill with engine coolant.

19. Start the engine and check for leaks.
20. Recheck the engine coolant level and oil level.

ENGINE ASSEMBLY

REMOVAL & INSTALLATION

See Figures 99 through 103.

➡**Hyundai recommends that the engine and transaxle be removed as a single unit on all models.**

1. Before servicing the vehicle, refer to the Precautions Section.
2. Drain the cooling system.
3. Drain the transaxle.
4. Drain the engine oil.

5. Relieve fuel system pressure.
6. Remove or disconnect the following:
   - Battery
   - Hood
   - Air intake assembly
   - Accessory drive belts
   - Engine wiring harness connectors
   - Reverse lamp switch connector, if equipped
   - Speedometer cable
   - Alternator harness connectors
   - Oil pressure gauge sender connector
   - Radiator hoses
   - Cooling fan
   - Fuel lines
   - Control cable, if equipped

   - Brake booster vacuum line
   - Intake manifold vacuum lines
   - Heater hoses
   - Accelerator cable
   - Cruise control cable, if equipped
   - Engine ground cable
7. If equipped with a manual transaxle, disconnect or remove the following:
   - Clutch cable
   - Select control valve connector
   - Shift linkage rods
8. If equipped with an automatic transaxle, disconnect or remove the following:
   - Transaxle oil cooler lines
   - Shift cable
   - Transaxle wiring connectors

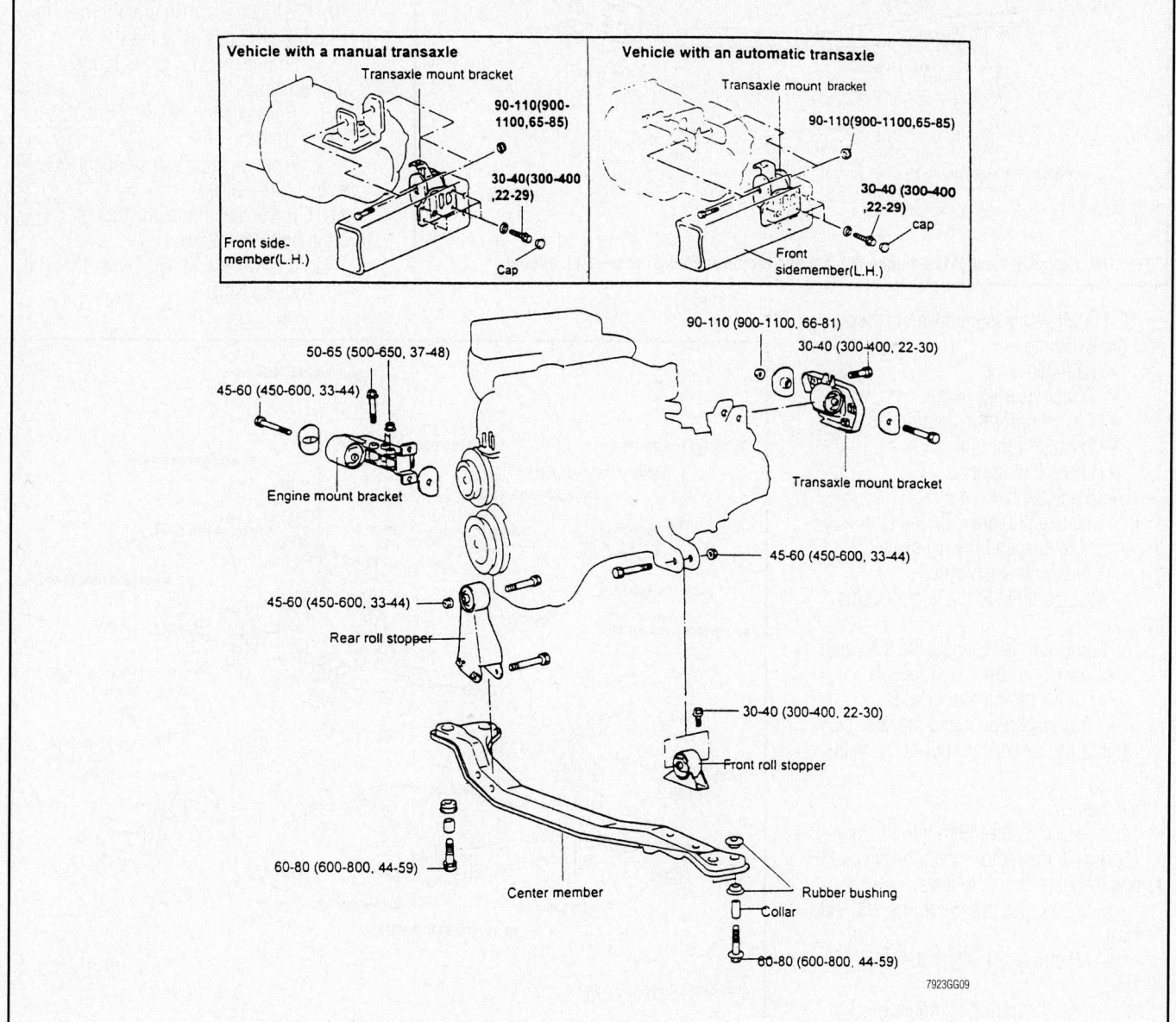

Fig. 99 Exploded view of the engine mounts and torque specifications—1.6L engine

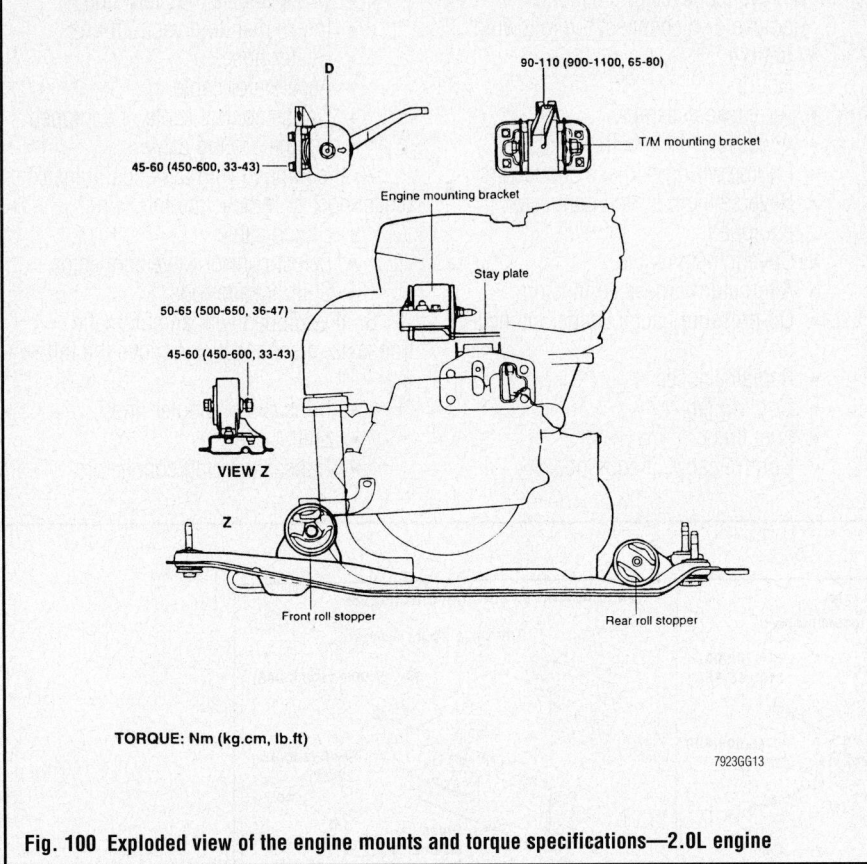

Fig. 100 Exploded view of the engine mounts and torque specifications—2.0L engine

b. Tiburon and Elantra: 33–43 ft. lbs. (45–60 Nm).

c. Accent: 22–30 ft. lbs. (30–40 Nm).

d. Azera: 65–79 ft. lbs. (90–110 Nm).

17. Install or connect the following:
- Front and rear roll stoppers
- Engine mount
- Transaxle mount

18. Remove the engine hoist.

19. For Accent, torque the mount through bolts as follows:

a. Engine mount: 33–43 ft. lbs. (45–60 Nm).

b. Transaxle mount: 65–80 ft. lbs. (90–110 Nm).

c. Front and rear roll stoppers: 33–43 ft. lbs. (45–60 Nm).

20. For Elantra and Tiburon, torque the mount through bolts as follows:

a. Engine mount: 36–47 ft. lbs. (50–65 Nm).

b. Transaxle mount: 65–80 ft. lbs. (90–110 Nm).

c. Front and rear roll stoppers: 33–43 ft. lbs. (45–60 Nm).

21. For Sonata, torque the mount through bolts as follows:

a. 4 cylinder engine mount: 43–58 ft. lbs. (60–80 Nm).

9. For all vehicles, remove or disconnect the following:
- Radiator
- Power steering pump
- A/C compressor, if equipped
- Exhaust front pipe
- Lower ball joints
- Stabilizer bar links

10. Separate the inner CV-joints from the transaxle and suspend the halfshafts out of the work area with safety wire.

11. Attach a hoist to the engine lifting eyes.

12. Remove or disconnect the following:
- Front and rear roll stoppers
- Engine mount and bracket
- Transaxle mount and bracket

13. Lift the powertrain out of the vehicle.

To install:

14. Lower the powertrain into position.

15. Install the motor mount bracket and torque the fasteners as follows:

a. V6 engines: 43–58 ft. lbs. (60–80 Nm).

b. All others: 37–48 ft. lbs. (45–60 Nm).

16. Install the transaxle mount bracket and torque the fasteners as follows:

a. Sonata: 29–36 ft. lbs. (40–50 Nm).

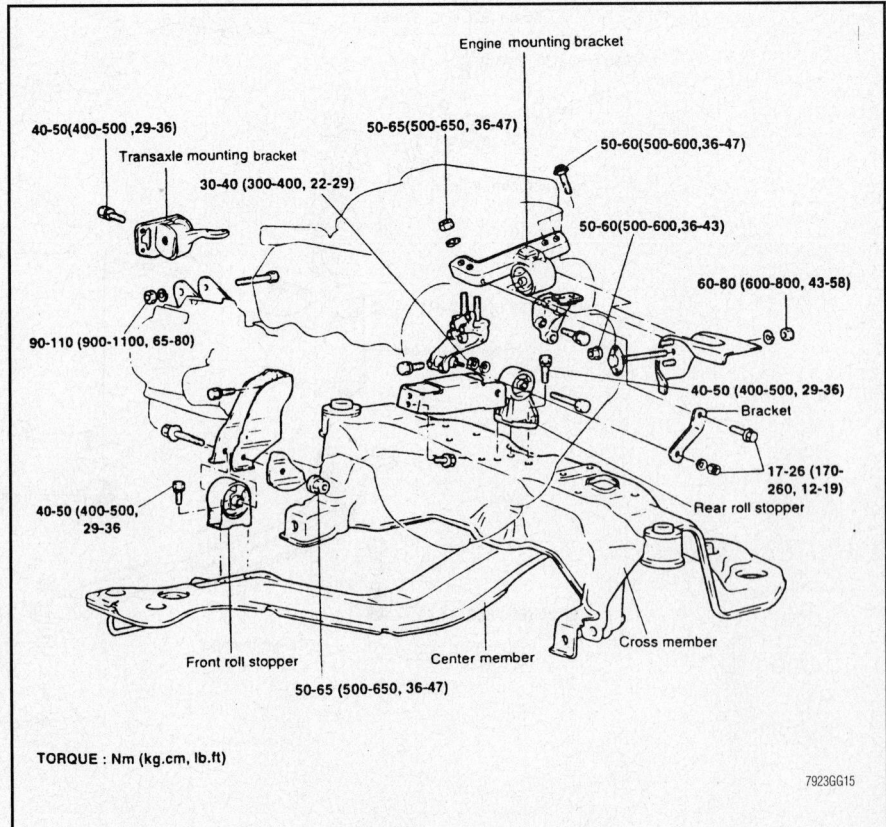

Fig. 101 Exploded view of the engine mounts and torque specifications—2.4L engine

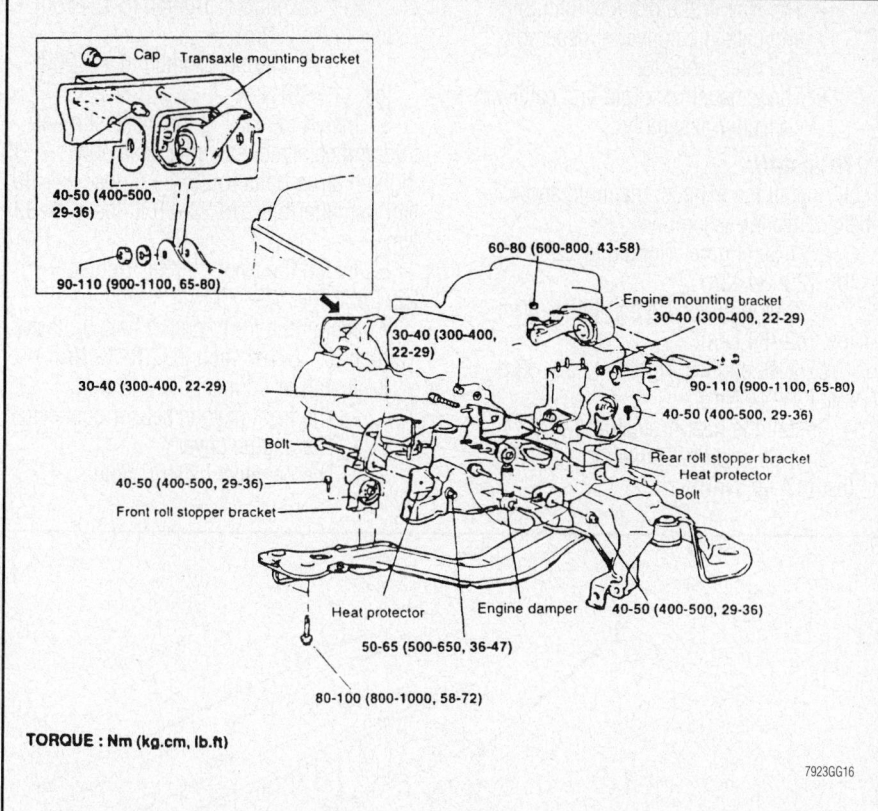

TORQUE : Nm (kg.cm, lb.ft)

7923GG16

Fig. 102 Exploded view of the engine mounts and torque specifications—3.3L engine

• Transaxle wiring connectors

26. For all vehicles, install or connect the following:
  • Engine ground cable
  • Cruise control cable, if equipped
  • Accelerator cable
  • Heater hoses
  • Intake manifold vacuum lines
  • Brake booster vacuum line
  • Fuel lines
  • Cooling fan
  • Radiator hoses
  • Oil pressure gauge sender connector
  • Alternator harness connectors
  • Speedometer cable
  • Reverse lamp switch connector
  • Engine wiring harness connectors
  • Accessory drive belts
  • Air intake assembly
  • Hood
  • Battery

27. Fill the engine with clean oil.
28. Fill the transaxle to the correct level.
29. Fill the cooling system to the proper level.
30. Start the engine and check for leaks.

b. V6 engine mount: 65–80 ft. lbs. (90–110 Nm).
c. Transaxle mount: 65–80 ft. lbs. (90–110 Nm).
d. Front roll stopper: 36–47 ft. lbs. (50–65 Nm).
e. Rear roll stopper: 22–29 ft. lbs. (30–40 Nm).

22. For Azera, torque the mount through bolts as follows:
a. Front roll stopper: 36–47 ft. lbs. (50–65 Nm).
b. Rear roll stopper: 36–47 ft. lbs. (50–65 Nm).

23. Install or connect the following:
  • Axle halfshafts using new circlips
  • Stabilizer bar links
  • Lower ball joints
  • Exhaust front pipe
  • A/C compressor, if equipped
  • Power steering pump
  • Radiator

24. If equipped with a manual transaxle, install or connect the following:
  • Clutch cable
  • Select control valve connector
  • Shift linkage rods

25. If equipped with an automatic transaxle, install or connect the following:
  • Transaxle oil cooler lines
  • Shift cable

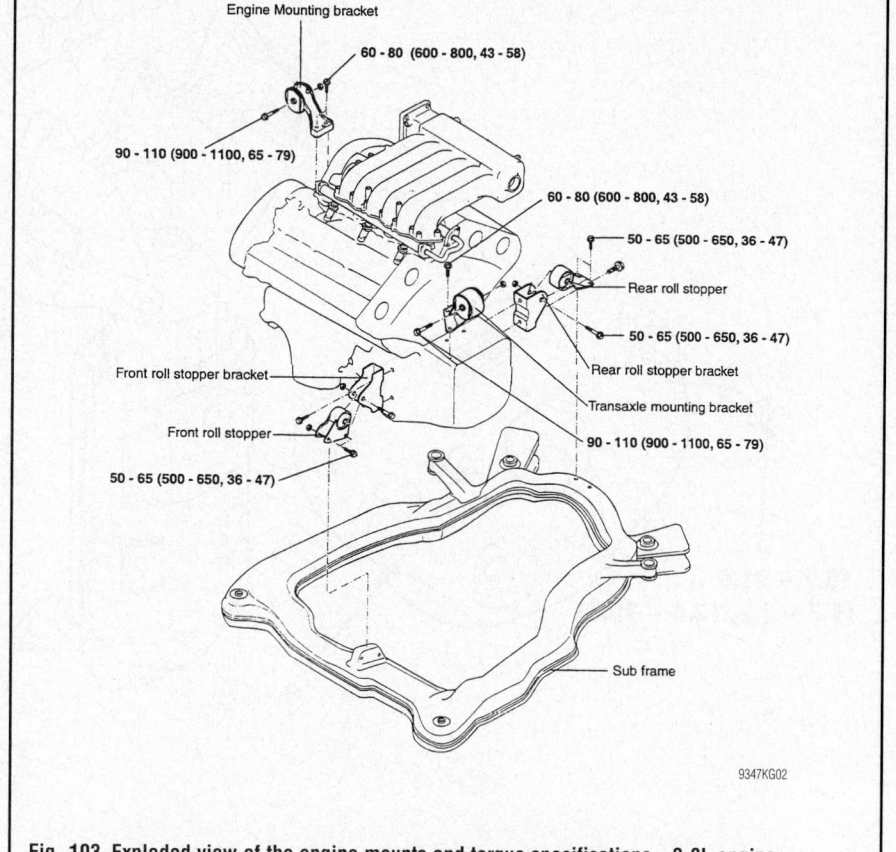

9347KG02

Fig. 103 Exploded view of the engine mounts and torque specifications—3.8L engine

EXHAUST MANIFOLD

REMOVAL & INSTALLATION

1.6L, 2.0L & 2.4L Engines

See Figures 104 through 106.

1. Before servicing the vehicle, refer to the Precautions Section.

2. Remove or disconnect the following:

- The negative battery cable
- The engine cover
- The front oxygen sensor connector
- The front muffler heat protector
- The front muffler
- The stay of the exhaust manifold and catalytic converter assembly
- The heat protector
- The exhaust manifold and catalytic converter assembly

To install:

3. Install the exhaust manifold and catalytic converter assembly:

   a. 1.6L engine: Tighten to 22–25 ft. lbs. (29–34 Nm).

   b. 2.0L engine: Tighten to 31–40 ft lbs. (42–54 Nm).

   c. 2.4L engine: Tighten to 29–33 ft. lbs. (39–44 Nm).

4. Install the heat protector:

   a. 1.6L engine: Tighten to 12–16 ft. lbs. (17–22 Nm).

   b. 2.0L engine: Tighten to 12–16 ft. lbs. (17–22 Nm).

   c. 2.4L engine: Tighten to 14–20 ft. lbs. (19–27 Nm).

5. Install the stay of the exhaust manifold and catalytic converter assembly. Tighten large bolts to 25–29 ft. lbs. (34–39 Nm); smaller bolts to 22–29 ft. lbs. (29–39 Nm).

6. Install the front muffler. Tighten to 22–29 ft. lbs. (29–39 Nm).

7. Install the front muffler heat protector. Tighten to 72–108 inch lbs. (8–12 Nm).

8. Install or connect the following:

- The front oxygen sensor connector
- The engine cover
- The negative battery cable

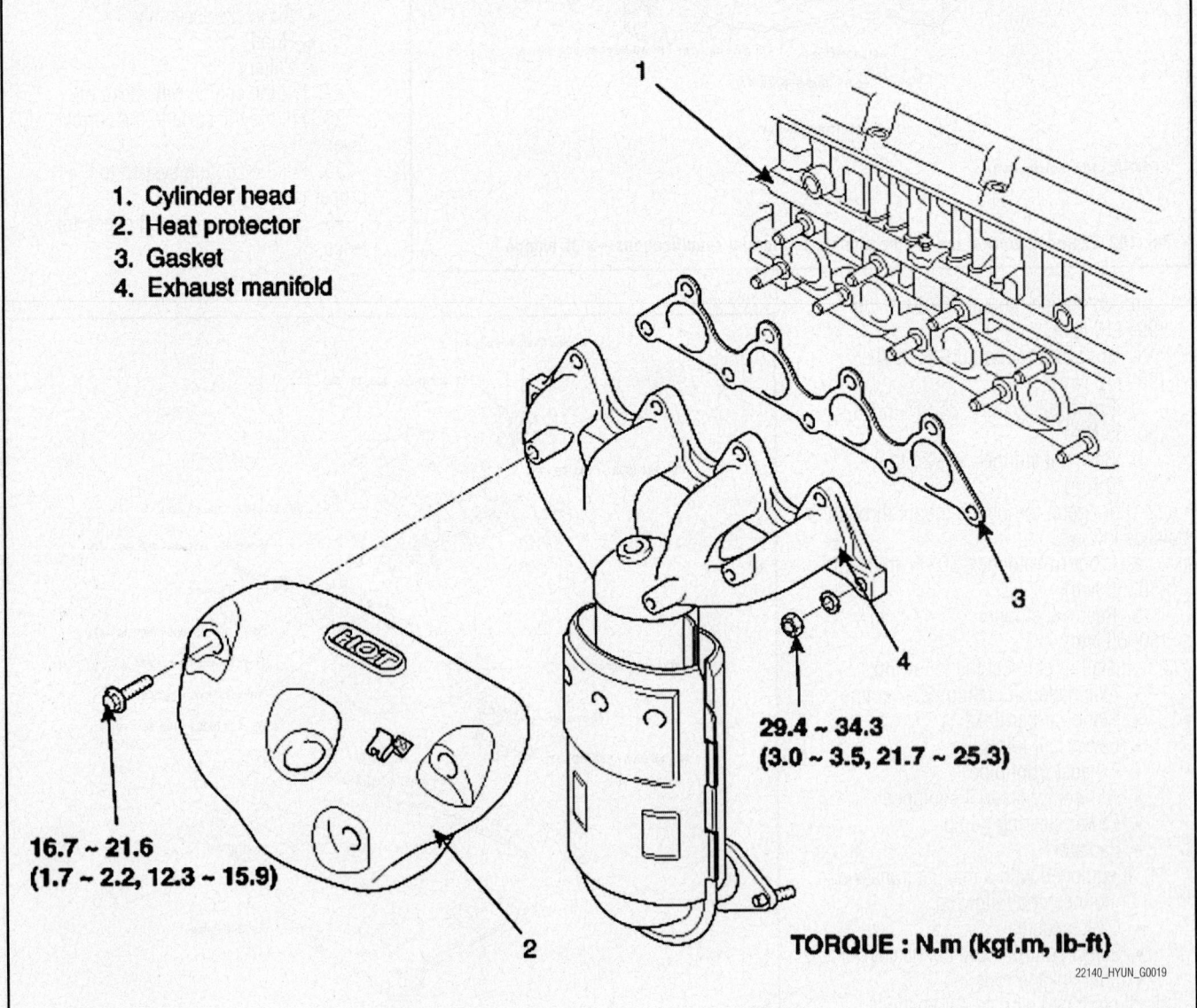

1. Cylinder head
2. Heat protector
3. Gasket
4. Exhaust manifold

16.7 ~ 21.6
(1.7 ~ 2.2, 12.3 ~ 15.9)

29.4 ~ 34.3
(3.0 ~ 3.5, 21.7 ~ 25.3)

TORQUE : N.m (kgf.m, lb-ft)

22140_HYUN_G0019

Fig. 104 Exploded view of the exhaust manifold and related components—1.6L engine

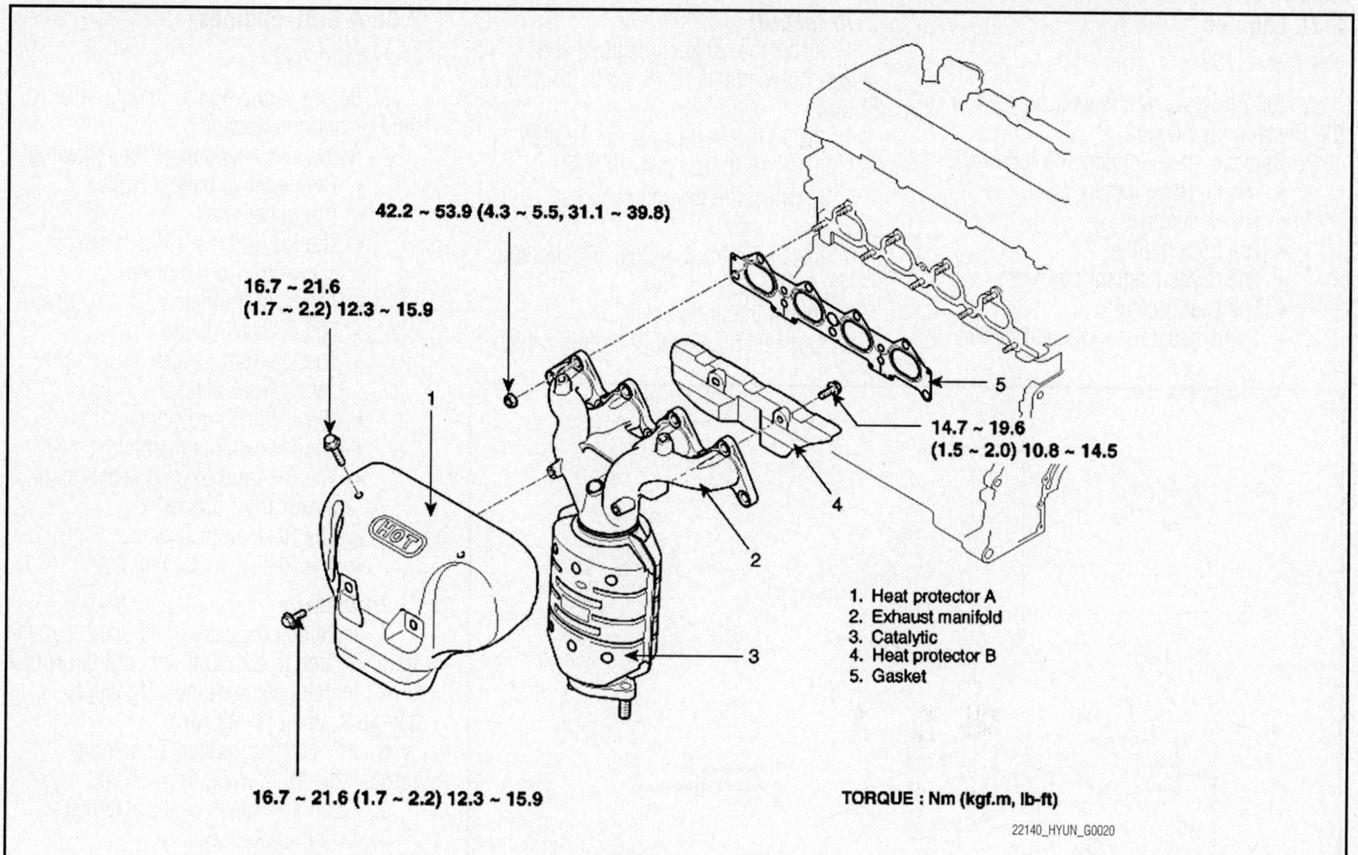

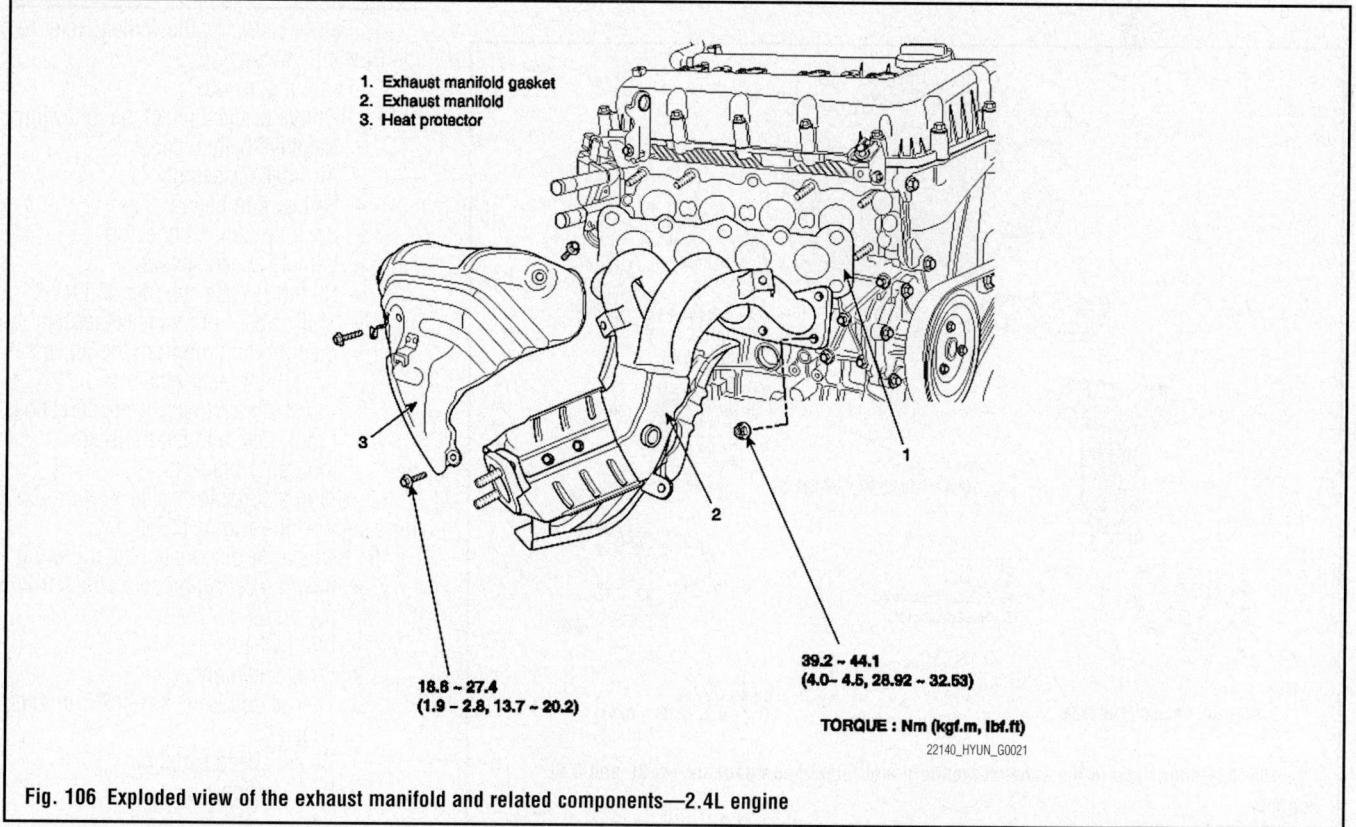

42.2 ~ 53.9 (4.3 ~ 5.5, 31.1 ~ 39.8)

16.7 ~ 21.6
(1.7 ~ 2.2) 12.3 ~ 15.9

14.7 ~ 19.6
(1.5 ~ 2.0) 10.8 ~ 14.5

1. Heat protector A
2. Exhaust manifold
3. Catalytic
4. Heat protector B
5. Gasket

16.7 ~ 21.6 (1.7 ~ 2.2) 12.3 ~ 15.9

TORQUE : Nm (kgf.m, lb-ft)

22140_HYUN_G0020

Fig. 105 Exploded view of the exhaust manifold and related components—2.0L engine

1. Exhaust manifold gasket
2. Exhaust manifold
3. Heat protector

18.6 ~ 27.4
(1.9 ~ 2.8, 13.7 ~ 20.2)

39.2 ~ 44.1
(4.0~ 4.5, 28.92 ~ 32.53)

TORQUE : Nm (kgf.m, lbf.ft)

22140_HYUN_G0021

Fig. 106 Exploded view of the exhaust manifold and related components—2.4L engine

2.7L Engines

See Figure 107.

1. Before servicing the vehicle, refer to the Precautions Section.
2. Remove or disconnect the following:
   - The negative battery cable
   - The undercover
   - The front muffler
   - The oxygen sensor connector
   - The heat protector
   - The exhaust manifold and gasket

To install:

3. Install the exhaust manifold and gasket. Tighten bolts to 18–22 ft. lbs. (25–29 Nm).
4. Install the heat protector. Tighten bolts to 12–16 ft. lbs. (17–22 Nm).
5. Connect the oxygen sensor connector.
6. Install the front muffler. Tighten to 22–29 ft. lbs. (29–39 Nm).
7. Install the undercover.
8. Connect the negative battery cable.

3.3L & 3.8L Engines

See Figure 108.

1. Before servicing the vehicle, refer to the Precautions Section.
2. Remove or disconnect the following:
   - The negative battery cable
   - The undercover
   - The LH, RH rear oxygen sensor connector from bracket
   - The front muffler
   - The oil level gauge
   - The LH front oxygen sensor connector from bracket
   - The LH heat protector
   - The LH exhaust manifold
   - The RH front oxygen sensor connector from bracket
   - The RH heat protector
   - The RH exhaust manifold

To install:

3. Install a new gasket and exhaust manifold. Tighten to 29–33 ft. lbs. (39–44 Nm).
4. Install heat protector. Tighten to 12–16 ft. lbs. (17–22 Nm).
5. Install front muffler. Tightening torque: 29–43 ft. lbs. (39–59 Nm).
6. Connect oxygen sensor connector.
7. Install undercover.
8. Connect the negative battery cable.

FLYWHEEL

REMOVAL & INSTALLATION

1. Before servicing the vehicle, refer to the Precautions Section.
2. Drain the transaxle.
3. Remove or disconnect the following:
   - Negative battery cable
   - Air intake assembly
   - Battery and battery tray
   - Back-up lamp connector
   - Vehicle speed sensor
   - Clutch release cylinder and lever
   - Shift cable from transaxle assembly
   - Steering column from the universal joint in the gear box
   - Clutch housing upper mounting bolts
   - Front, rear, and left transaxle mounting brackets
4. Using a suitable engine support fixture, support the engine assembly.
5. Remove or disconnect the following:
   - Power steering pressure hose from the pump
   - Front wheel
   - Strut assembly
   - Tie rod and sway bar link from the knuckle
   - Wheel speed sensor
   - Brake caliper

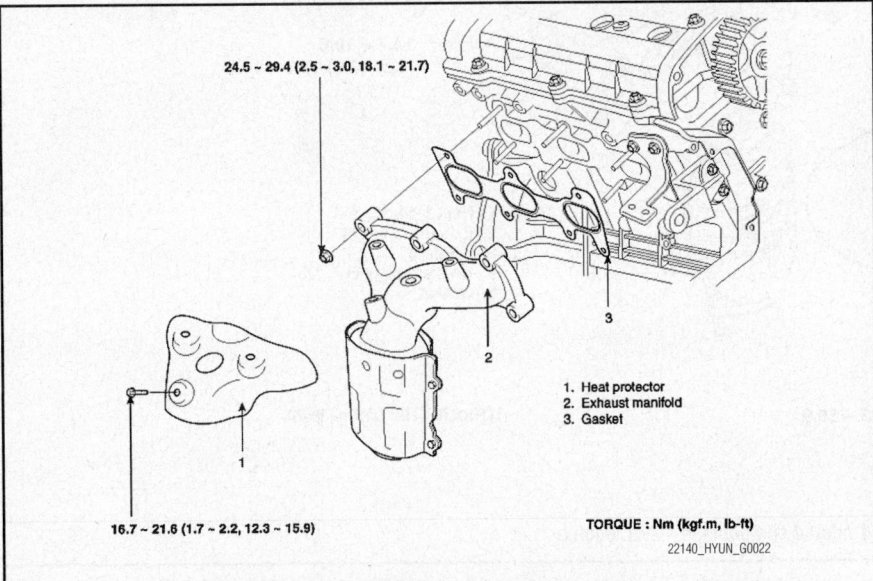

24.5 ~ 29.4 (2.5 ~ 3.0, 18.1 ~ 21.7)

16.7 ~ 21.6 (1.7 ~ 2.2, 12.3 ~ 15.9)

TORQUE : Nm (kgf.m, lb-ft)

1. Heat protector
2. Exhaust manifold
3. Gasket

22140_HYUN_G0022

Fig. 107 Exploded view of the exhaust manifold and related components—2.7L engine

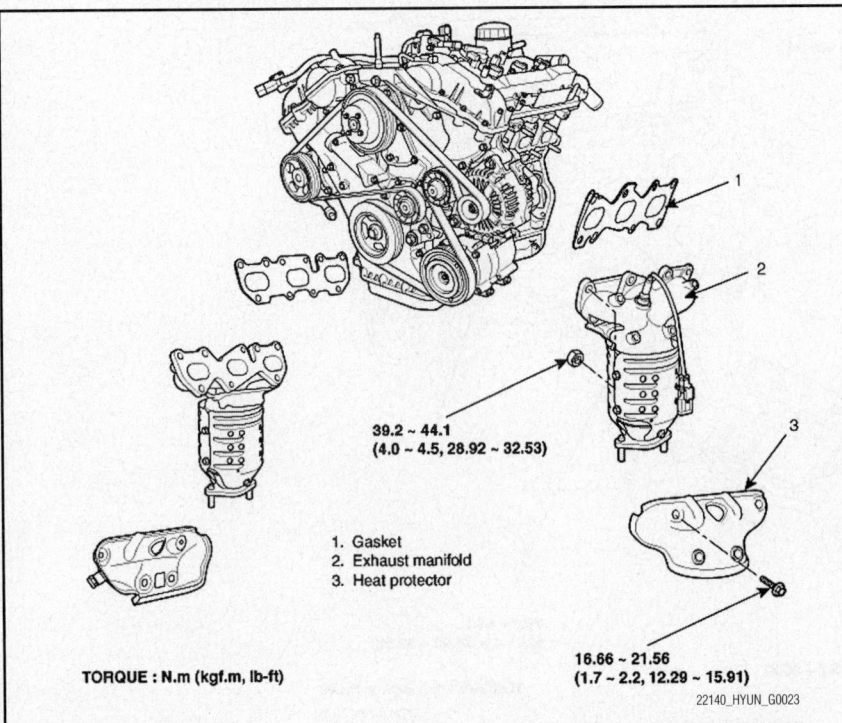

39.2 ~ 44.1
(4.0 ~ 4.5, 28.92 ~ 32.53)

16.66 ~ 21.56
(1.7 ~ 2.2, 12.29 ~ 15.91)

TORQUE : N.m (kgf.m, lb-ft)

1. Gasket
2. Exhaust manifold
3. Heat protector

22140_HYUN_G0023

Fig. 108 Exploded view of the exhaust manifold and related components—3.3L and 3.8L engines

- Engine splash guard
- Front muffler
- Power steering hose on the front cross member

6. Using a suitable jack, support the sub-frame cross member.

7. Remove the cross member mounting bolts, and lower the cross member assembly with the steering gear and stabilizer bar attached.

8. Using a suitable jack, support the transaxle assembly.

9. Remove the front and rear roll stoppers.

10. Remove the engine and transaxle mounting bolts.

11. Slowly lower the transaxle from the vehicle.

12. Insert the special tool (09411-11000) in the clutch disc to prevent the disc from falling.

13. Loosen the bolts which attach the clutch cover to the flywheel in a star pattern. Loosen the bolts in succession, one or two turns at a time, to avoid bending the cover flange.

➡**Do not clean the clutch disc or the release bearing with cleaning solvent.**

14. Remove clutch cover assembly and then the clutch disc assembly.

15. Remove the flywheel.

To install:

16. Installation is the reverse order of removal.

INTAKE MANIFOLD

REMOVAL & INSTALLATION

1.6L, 2.0L & 2.4L Engines
See Figure 109 through 112.

1. Before servicing the vehicle, refer to the Precautions Section.

2. Relieve the fuel system pressure.

3. Drain the cooling system.

4. Remove or disconnect the following:
- The negative battery cable
- The engine cover
- The accelerator cable (A)
- The Throttle Position Sensor (TPS) connector (B) and the MAP sensor connector (F)
- The Idle Speed Actuator (ISA) connector (B)
- The Positive Crankcase Ventilation (PCV) hose (D) and breather hose (E)

5. Disconnect the fuel injector connectors.

6. Remove the heater hose, Purge Control Solenoid Valve (PCSV), and the brake

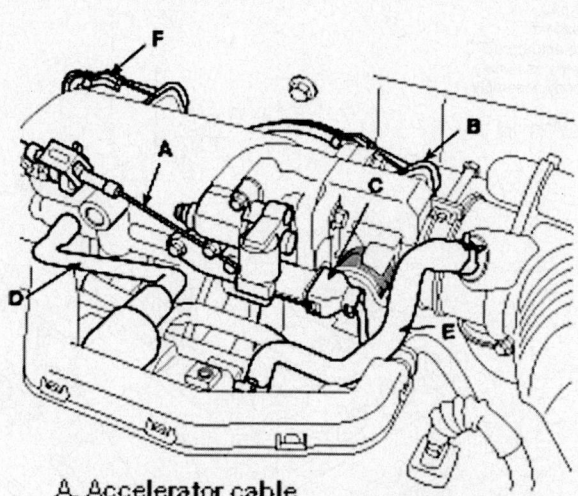

A. Accelerator cable
B. Throttle Position Sensor (TPS) connector
C. Idle Speed Actuator (ISA) connector
D. Positive Crankcase Ventilation (PCV) hose
E. Breather hose
F. MAP sensor connector

22140_HYUN_G0018

Fig. 109 Remove the accelerator cable, the TPS connector, and the MAP sensor connector. Disconnect the ISA connector, the PCV hose, and breather hose—1.6L engine

vacuum hose from the throttle body and intake manifold.

7. Disconnect the PCSV and water temperature sensor connector.

8. Remove the delivery pipe.

9. Remove the intake manifold stay.

10. Remove the intake manifold.

To install:

11. Install or connect the following:
- The intake manifold. Tighten to 11–15 ft. lbs. (15–20 Nm)
- The intake manifold stay. Tighten to 13–18 ft. lbs. (18–25 Nm)

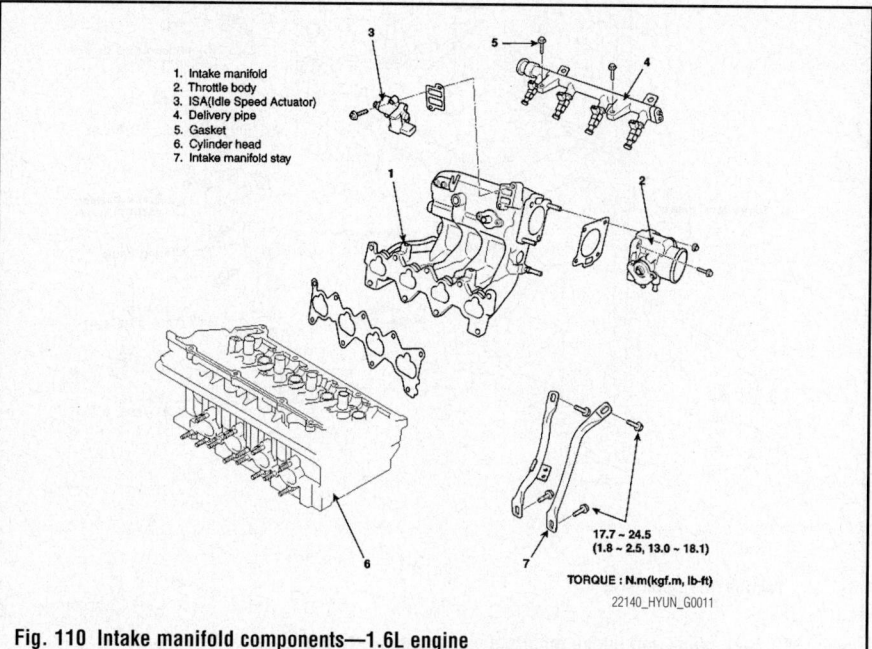

1. Intake manifold
2. Throttle body
3. ISA (Idle Speed Actuator)
4. Delivery pipe
5. Gasket
6. Cylinder head
7. Intake manifold stay

17.7 ~ 24.5
(1.8 ~ 2.5, 13.0 ~ 18.1)

TORQUE : N.m(kgf.m, lb-ft)

22140_HYUN_G0011

Fig. 110 Intake manifold components—1.6L engine

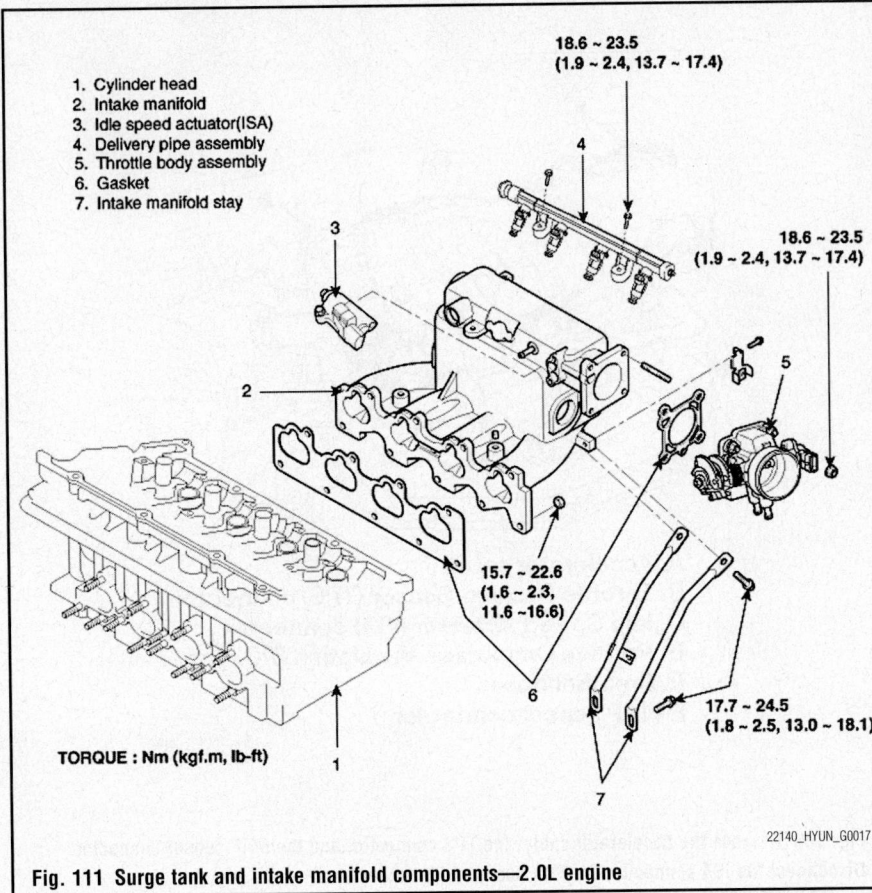

1. Cylinder head
2. Intake manifold
3. Idle speed actuator(ISA)
4. Delivery pipe assembly
5. Throttle body assembly
6. Gasket
7. Intake manifold stay

18.6 ~ 23.5
(1.9 ~ 2.4, 13.7 ~ 17.4)

18.6 ~ 23.5
(1.9 ~ 2.4, 13.7 ~ 17.4)

15.7 ~ 22.6
(1.6 ~ 2.3,
11.6 ~16.6)

17.7 ~ 24.5
(1.8 ~ 2.5, 13.0 ~ 18.1)

TORQUE : Nm (kgf.m, lb-ft)

22140_HYUN_G0017

Fig. 111 Surge tank and intake manifold components—2.0L engine

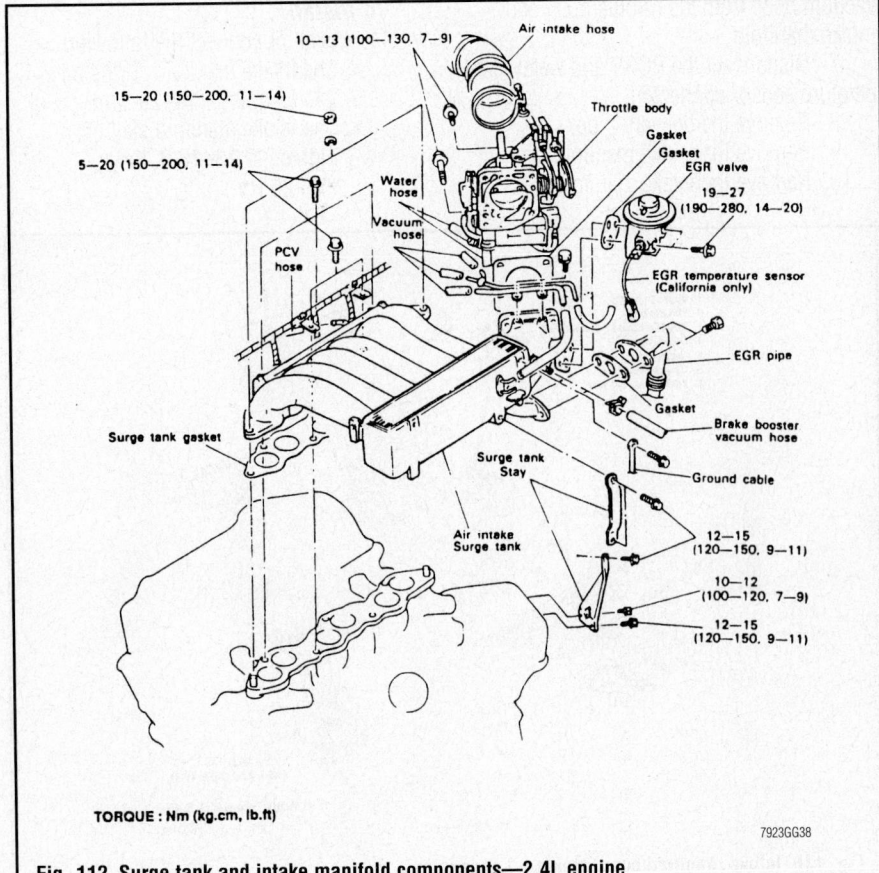

10—13 (100—130, 7—9)

15—20 (150—200, 11—14)

5—20 (150—200, 11—14)

Air intake hose

Throttle body

Gasket
Gasket
EGR valve

19—27
(190—280, 14—20)

Water
hose

Vacuum
hose

PCV
hose

EGR temperature sensor
(California only)

EGR pipe

Gasket

Brake booster
vacuum hose

Surge tank gasket

Surge tank
Stay

Ground cable

Air intake
Surge tank

12—15
(120—150, 9—11)

10—12
(100—120, 7—9)

12—15
(120—150, 9—11)

TORQUE : Nm (kg.cm, lb.ft)

7923GG38

Fig. 112 Surge tank and intake manifold components—2.4L engine

- The delivery pipe. Tighten to 14–20 ft. lbs. (19–28 Nm)
- The PCSV and water temperature sensor connector
- The heater hose, PCSV, and the brake vacuum hose to the throttle body and intake manifold
- The fuel injector connectors
- The Positive Crankcase Ventilation (PCV) hose (D) and breather hose (E)
- The Idle Speed Actuator (ISA) connector (B)
- The Throttle Position Sensor (TPS) connector (B) and the MAP sensor connector (F)
- The accelerator cable (A)
- The engine cover
- The negative battery cable

12. Fill the cooling system.

13. Start the engine and check for leaks.

2.7L Engine

See Figures 113 and 114.

1. Before servicing the vehicle, refer to the Precautions Section.

2. Relieve the fuel system pressure.

3. Drain the cooling system.

4. Remove the negative battery cable.

5. Remove the engine cover.

6. Remove air cleaner hose.

7. Remove surge tank assembly:

   a. Disconnect the accelerator cable.

   b. Disconnect the TPS connector.

   c. Disconnect the ISA connector.

   d. Disconnect the injector connector.

   e. Disconnect the Purge Control Solenoid Valve (PCSV) connector.

   f. Disconnect the PCSV hose.

   g. Disconnect the brake booster vacuum hose.

   h. Disconnect the PCV hose.

   i. Disconnect the IAT sensor connector.

   j. Remove the surge tank stay.

   k. Remove the surge tank assembly.

8. Remove the injector assembly.

9. Remove the intake manifold and gasket.

To install:

10. Install the intake manifold and gasket. Tighten the bolts in the sequence shown to 14–17 ft. lbs. (19–24 Nm).

11. Install the injector assembly.

12. Install the surge tank assembly. Tighten bolts to 14–17 ft. lbs. (19–24 Nm).

   a. Install the surge tank stay. Tighten bolts to 14–17 ft. lbs. (19–24 Nm).

   b. Install the ground cable.

   c. Connect the IAT sensor connector.

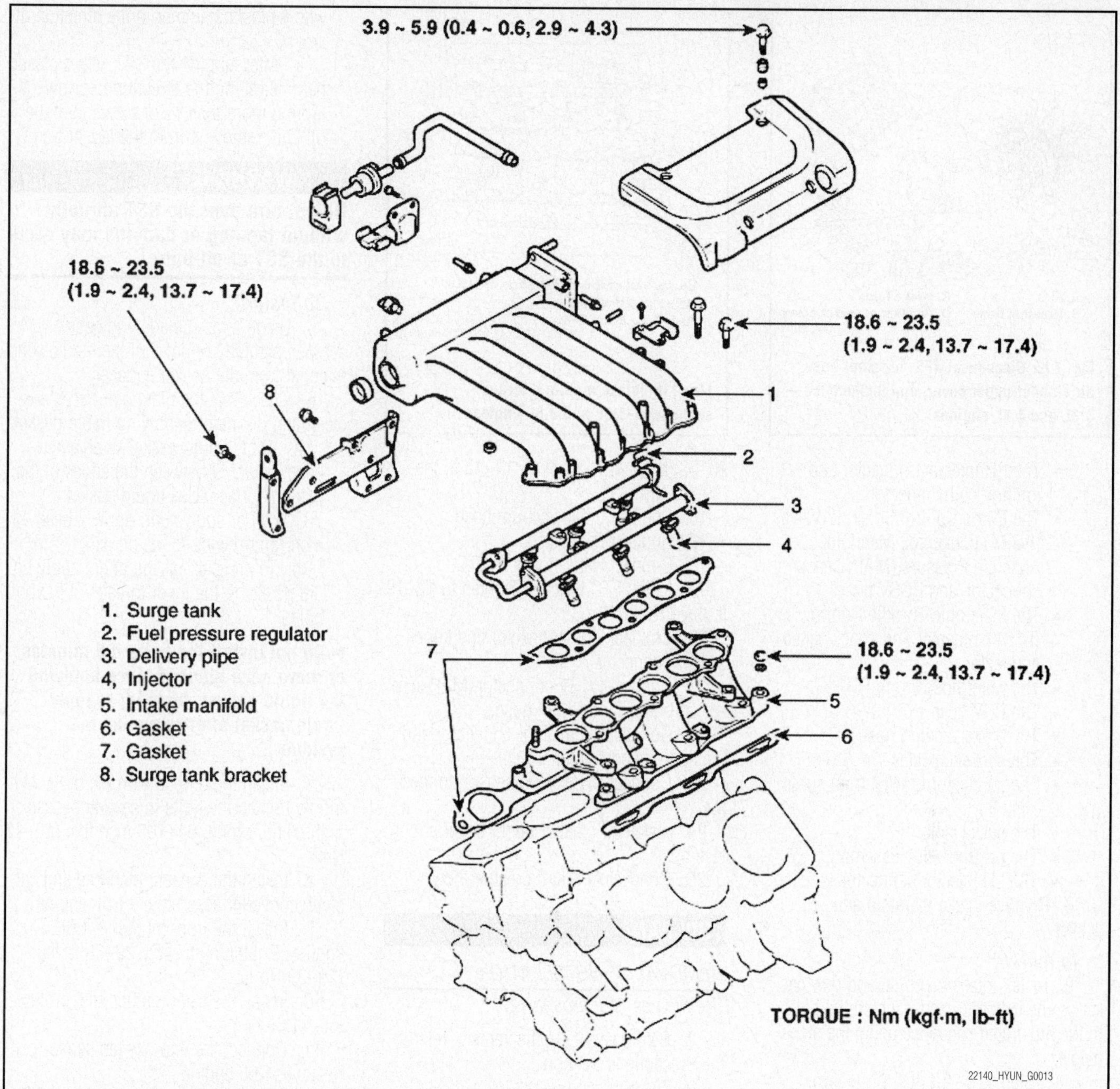

3.9 ~ 5.9 (0.4 ~ 0.6, 2.9 ~ 4.3)

18.6 ~ 23.5
(1.9 ~ 2.4, 13.7 ~ 17.4)

18.6 ~ 23.5
(1.9 ~ 2.4, 13.7 ~ 17.4)

18.6 ~ 23.5
(1.9 ~ 2.4, 13.7 ~ 17.4)

1. Surge tank
2. Fuel pressure regulator
3. Delivery pipe
4. Injector
5. Intake manifold
6. Gasket
7. Gasket
8. Surge tank bracket

TORQUE : Nm (kgf·m, lb-ft)

22140_HYUN_G0013

Fig. 113 Surge tank and intake manifold components—2.7L engine

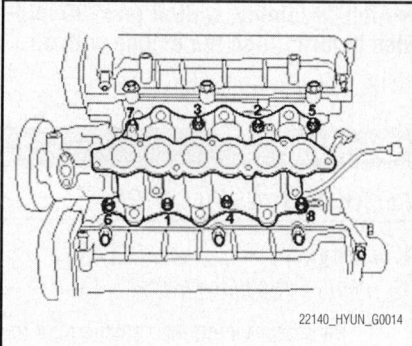

22140_HYUN_G0014

Fig. 114 Intake manifold torque sequence—2.7L engine

d. Connect the PCV hose.
e. Connect the brake booster vacuum hose.
f. Connect the PCSV hose.
g. Connect the PCSV connector.
h. Connect the injector connector.
i. Connector the ISA connector.
j. Connector the TPS connector.
k. Connector the actuator cable.
13. Install the air cleaner hose.
14. Install the engine cover.
15. Install the negative battery cable.
16. Fill the cooling system.
17. Start the engine and check for leaks.

3.3L & 3.8L Engines

See Figures 115 through 117.

1. Before servicing the vehicle, refer to the Precautions Section.
2. Relieve the fuel system pressure.
3. Drain the cooling system.
4. Remove the negative battery cable.
5. Disconnect AFS (A) and breather hose (B).
6. Remove air cleaner upper cover (D) and intake hose (C).
7. Remove or disconnect the following:
• The RH oxygen sensor connector

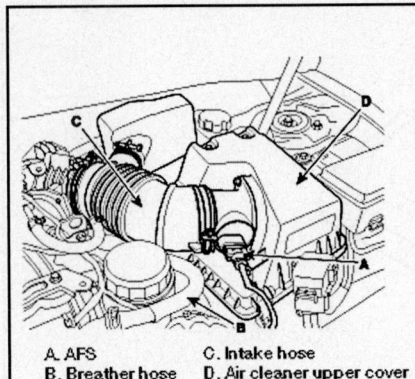

A. AFS
B. Breather hose
C. Intake hose
D. Air cleaner upper cover

22140_HYUN_G0015

Fig. 115 Disconnect AFS, breather hose, air cleaner upper cover, and intake hose—3.3L and 3.8L engines

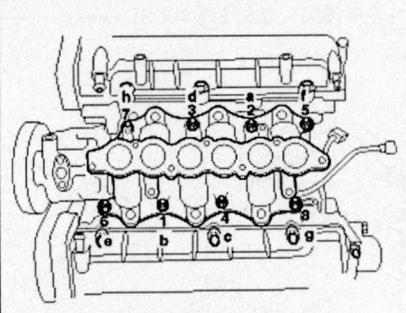

Be careful of the installation order
1st step order: a-h
2nd step order: 1-8

22140_HYUN_G0016

Fig. 116 Intake manifold torque sequence—3.3L and 3.8L engines

- The RH injector connector and ignition coil connector
- The Purge Control Solenoid Valve (PCSV) connector, Manifold Absolute Pressure (MAP) sensor connector, and PCSV hose
- The Electronic Throttle Control (ETC) connector and knock sensor connector
- The water hoses from ETC
- The PCV hose
- The brake vacuum hose
- The surge tank stay
- The connector bracket from surge tank
- The surge tank
- The breather Pipe assembly
- The LH injector connector.

8. Remove intake manifold and gasket

To install:

9. Install intake manifold and new gasket on the cylinder head. Tighten the bolts in the illustrated sequence using the steps below:
   a. Step 1: 3–4 ft. lbs. (4–6 Nm).
   b. Step 2: 14–17 ft. lbs. (19–24 Nm).
   c. Step 3: Repeat 2nd step twice.
10. Install the delivery pipe.
11. Connect the LH injector connector.
12. Connect the breather pipe assembly. Tighten to 84–108 inch lbs. (10–12 Nm).
13. Install the surge tank. Tighten long bolt to 84–108 inch lbs. (10–12 Nm); short bolt and nut to 14–17 ft. lbs. (19–24 Nm).
14. Install the connector bracket on the surge tank. Tighten to 5–8 ft. lbs. (7–11 Nm).
15. Install surge tank stay. Tighten to

20–23 ft. lbs. (27–31 Nm); 14–17 ft. lbs. (19–24 Nm).
16. Connect brake vacuum hose.
17. Connect PCV hose.
18. Connect water hoses to ETC.
19. Install ETC bracket. Tighten to 12–19 ft. lbs. (16–26 Nm).
20. Connect ETC connector and knock sensor connector.
21. Connect PCSV connector, MAP sensor connector and PCSV hose.
22. Connect RH injector connector and ignition coil connector.
23. Connect RH oxygen sensor connector.
24. Install air cleaner upper cover and intake hose.
25. Connect AFS and breather hose.

OIL PAN

REMOVAL & INSTALLATION

See Figures 118 through 122.

1. Before servicing the vehicle, refer to the Precautions Section.
2. Drain the engine oil.
3. Remove or disconnect the following:
   - The rear oxygen sensor connector—1.6L engine
   - The front muffler heat protector—1.6L engine
   - The front muffler—1.6L engine
   - The exhaust manifold and catalytic converter assembly—1.6L engine
   - The front exhaust pipe—2.7L engine
   - The oil pan bolts
4. Using the SST (09215-3C000), remove the oil pan.
   a. Insert the SST between the oil pan and the ladder frame by tapping it

with a plastic hammer in the direction of arrow.
   b. After tapping the SST with a plastic hammer along the direction of arrow around more than ⅔ of the edge of the oil pan, remove it from the ladder frame.

❄❄ WARNING

Do not turn over the SST abruptly without tapping or damage may occur to the SST or oil pan.

To install:

5. Using a razor blade and gasket scraper, carefully remove all the old packing material from the gasket surfaces.
6. Check that the mating surfaces are clean and dry before applying liquid gasket.
   a. Apply liquid gasket as an even bead, centered between the edges of the mating surface. Use liquid gasket: TB1217H or equivalent. Apply a bead ⅛ inch (3mm) wide to the oil pan.
   b. To prevent leakage of oil, apply liquid gasket to the inner threads of the bolt holes.

➡ **Do not install the parts if 5 minutes or more have elapsed since applying the liquid gasket. Instead, reapply liquid gasket after removing the residue.**

7. Install the oil pan with the bolts. Uniformly tighten the bolts in several passes. Tightening torque: 84–108 inch lbs. (10–12 Nm).
8. Install the exhaust manifold and catalytic converter assembly—1.6L engine
9. Install the front muffler—1.6L engine. Tightening torque: 22–29 ft. lbs. (29–39 Nm).
10. Install the front muffler heat protector—1.6L engine.
11. Connect the rear oxygen sensor connector—1.6L engine.
12. Install the front exhaust pipe—2.7L engine.

➡ **After assembly, wait at least 30 minutes before filling the engine with oil.**

13. Fill with engine oil.

OIL PUMP

REMOVAL & INSTALLATION

1.6L Engine

See Figures 123 through 127.

1. Before servicing the vehicle, refer to the Precautions Section.

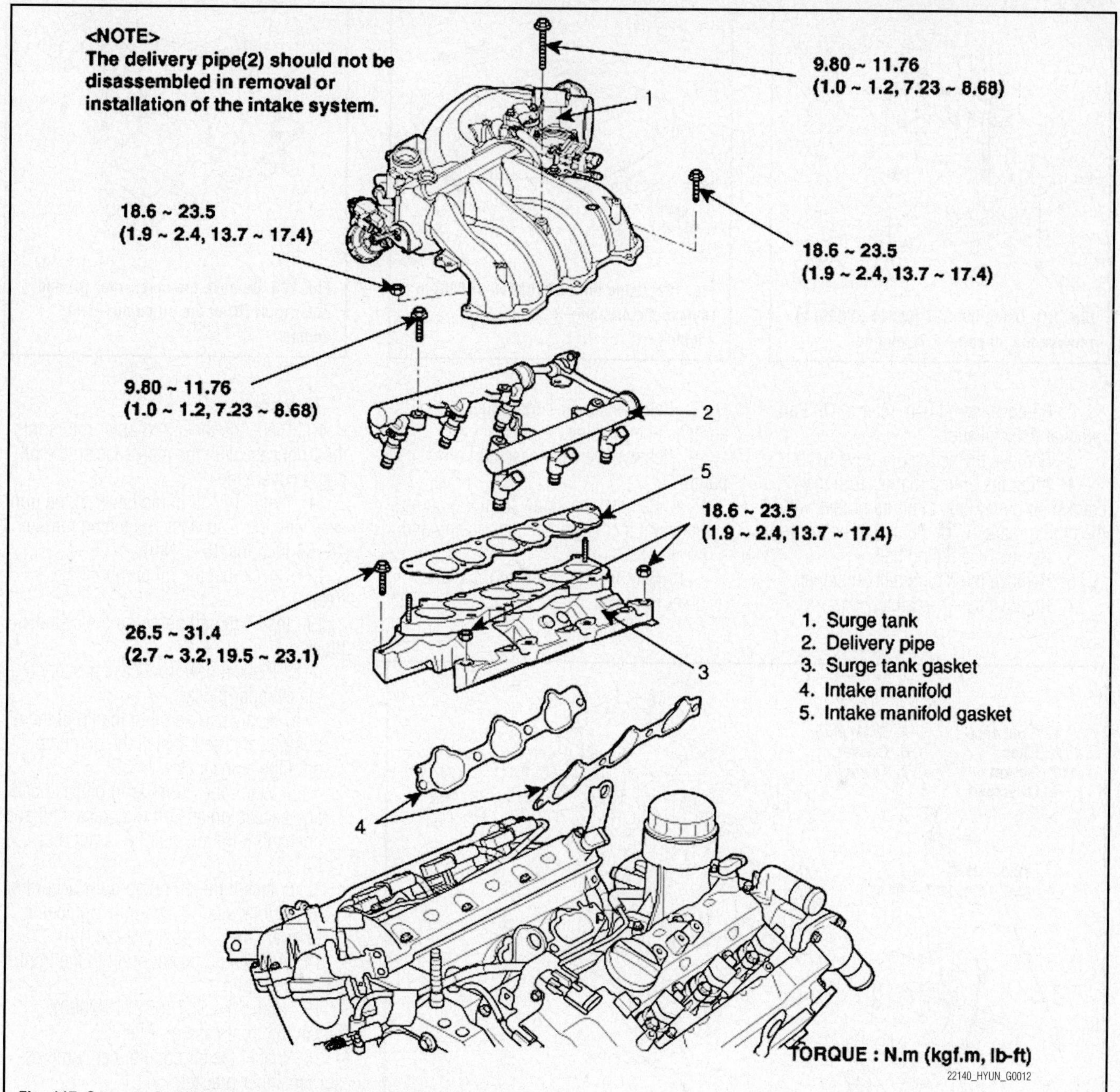

<NOTE>
The delivery pipe(2) should not be disassembled in removal or installation of the intake system.
9.80 ~ 11.76
(1.0 ~ 1.2, 7.23 ~ 8.68)
1
18.6 ~ 23.5
(1.9 ~ 2.4, 13.7 ~ 17.4)
18.6 ~ 23.5
(1.9 ~ 2.4, 13.7 ~ 17.4)
9.80 ~ 11.76
(1.0 ~ 1.2, 7.23 ~ 8.68)
2
5
18.6 ~ 23.5
(1.9 ~ 2.4, 13.7 ~ 17.4)
26.5 ~ 31.4
(2.7 ~ 3.2, 19.5 ~ 23.1)
3
1. Surge tank
2. Delivery pipe
3. Surge tank gasket
4. Intake manifold
5. Intake manifold gasket
4
TORQUE : N.m (kgf.m, lb-ft)
22140_HYUN_G0012
Fig. 117 Surge tank and intake manifold components—3.3L and 3.8L engines

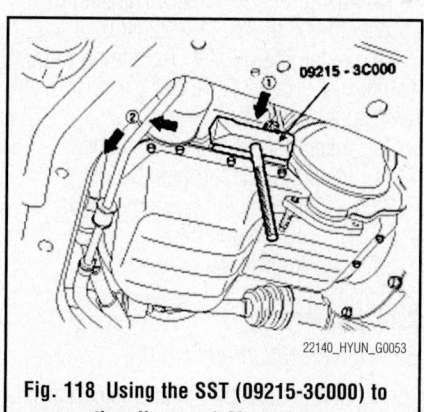

09215 - 3C000
①
②
22140_HYUN_G0053
Fig. 118 Using the SST (09215-3C000) to remove the oil pan—1.6L engine

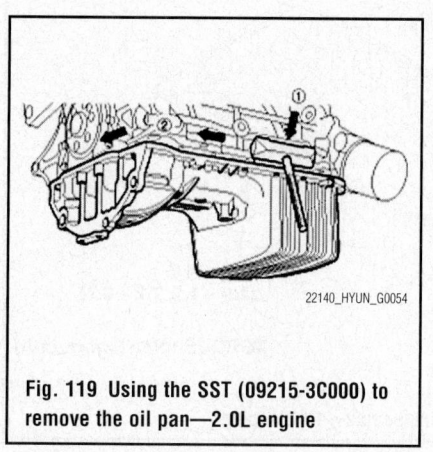

①
②
22140_HYUN_G0054
Fig. 119 Using the SST (09215-3C000) to remove the oil pan—2.0L engine

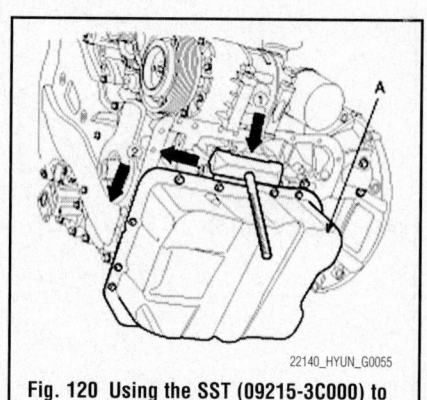

A
22140_HYUN_G0055
Fig. 120 Using the SST (09215-3C000) to remove the oil pan—2.4L engine

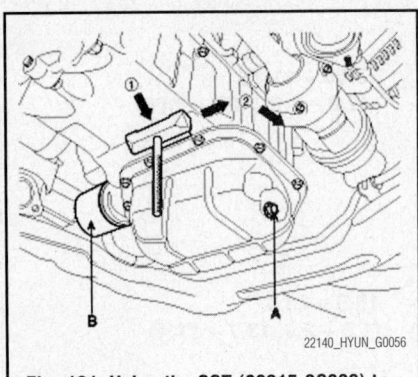

Fig. 121 Using the SST (09215-3C000) to remove the oil pan—2.7L engine

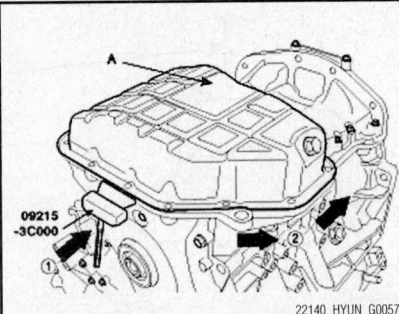

Fig. 122 Using the SST (09215-3C000) to remove the oil pan—3.3L and 3.8L engines

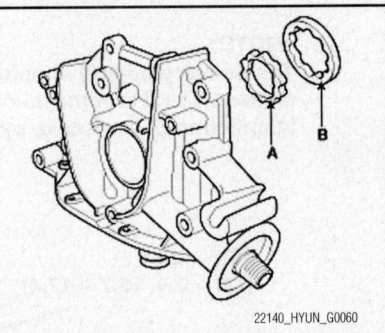

Fig. 123 Remove the inner rotor (A) and outer rotor (B) of the oil pump—1.6L engine

2. Remove the oil pan, refer to Oil Pan, removal & installation.

3. Remove the accessory drive belts.

4. Align the timing marks: Turn the crankshaft pulley, and align its groove with the timing mark "T" of the timing belt cover

5. Remove the timing belt.

6. Remove the timing belt tensioner.

7. Remove the alternator.

8. Remove the air conditioner compressor tensioner bracket.

9. Remove the front case and oil pump.

    a. Remove the screw from the pump housing, then separate the housing and cover.

    b. Remove the inner rotor (A) and outer rotor (B).

To install:

10. Place the inner and outer rotors into the front case with the marks facing the oil pump cover side.

11. Install the oil pump cover to the front case with the 7 screws. Tightening torque: 48–60 inch lbs. (6–7 Nm).

12. Check that the oil pump turns freely.

13. Install the oil pump on the cylinder block.

    a. Place a new front case gasket on the cylinder block.

    b. Apply engine oil to the lip of the oil pump seal. Then, install the oil pump onto the crankshaft.

    c. When the pump is in place, clean any excess grease off the crankshaft and check that the oil seal lip is not distorted.

    d. Install the oil pump bolts according to the following illustration. Tightening torque: 14–17 ft. lbs. (19–24 Nm).

14. Apply a light coat of oil to the front case oil seal lip.

15. Using the SST (09214-32000), install the front case oil seal.

16. Install the air conditioner compressor tensioner bracket.

17. Install the alternator.

18. Install the oil screen. Tightening torque: 11–16 ft. lbs. (15–22 Nm).

19. Install the oil pan. Tightening torque: 7–9 ft. lbs. (10–12 Nm). Refer to Oil Pan, removal & installation.

20. Install the timing belt tensioner.

21. Install the timing belt.

22. Install the accessory drive belts.

23. Fill with engine oil.

24. Check for leaks.

2.0L Engine

See Figures 128 through 130.

1. Before servicing the vehicle, refer to the Precautions Section.

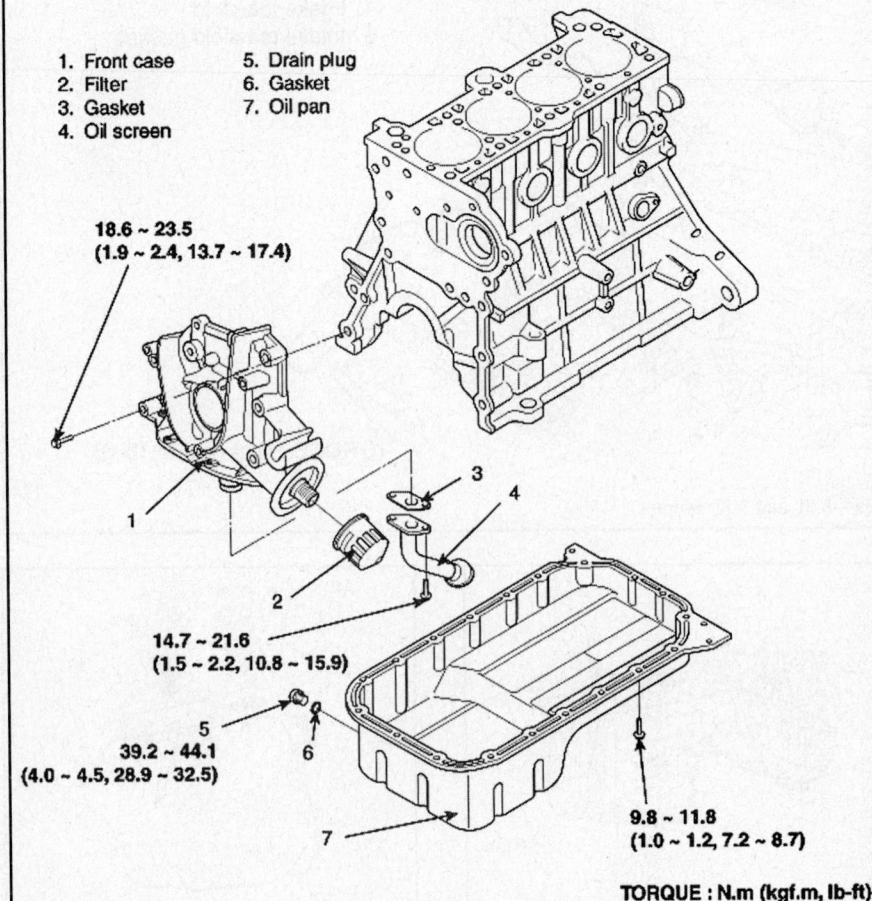

1. Front case
2. Filter
3. Gasket
4. Oil screen
5. Drain plug
6. Gasket
7. Oil pan

18.6 ~ 23.5
(1.9 ~ 2.4, 13.7 ~ 17.4)

14.7 ~ 21.6
(1.5 ~ 2.2, 10.8 ~ 15.9)

39.2 ~ 44.1
(4.0 ~ 4.5, 28.9 ~ 32.5)

9.8 ~ 11.8
(1.0 ~ 1.2, 7.2 ~ 8.7)

TORQUE : N.m (kgf.m, lb-ft)

Fig. 124 Expanded view of lubrication system components—1.6L engine

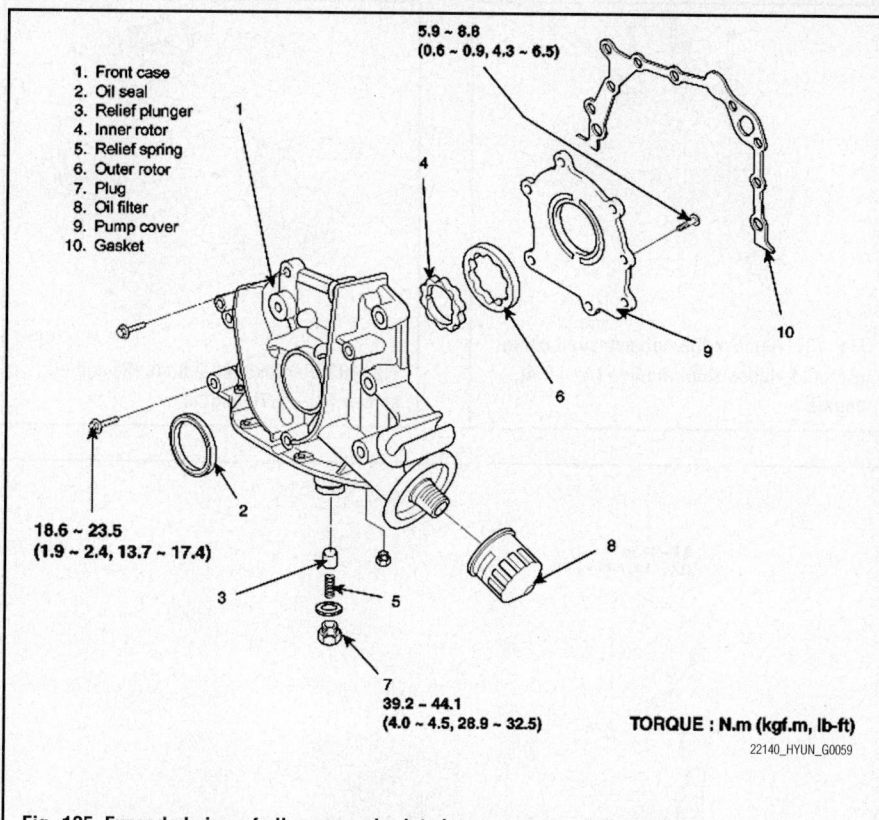

1. Front case
2. Oil seal
3. Relief plunger
4. Inner rotor
5. Relief spring
6. Outer rotor
7. Plug
8. Oil filter
9. Pump cover
10. Gasket

5.9 ~ 8.8
(0.6 ~ 0.9, 4.3 ~ 6.5)

18.6 ~ 23.5
(1.9 ~ 2.4, 13.7 ~ 17.4)

39.2 ~ 44.1
(4.0 ~ 4.5, 28.9 ~ 32.5)

TORQUE : N.m (kgf.m, lb-ft)

22140_HYUN_G0059

Fig. 125 Expanded view of oil pump and related components—1.6L engine

2. Remove the oil pan, refer to Oil Pan, removal & installation.

3. Remove the drive belts.

4. Turn the crankshaft and align the white groove on the crankshaft pulley with the pointer on the lower cover.

5. Remove the timing belt.

6. Remove the front case and oil pump.

a. Remove the screws from the pump housing, then separate the housing and cover.

b. Remove the inner and outer rotors.

To install:

7. Place the inner and outer rotors into the front case with the marks facing the oil pump cover side.

8. Install the oil pump cover to the front case with the 7 screws. Tightening torque: 48–84 inch lbs. (6–9 Nm).

9. Check that the oil pump turns freely.

10. Install the oil pump on the cylinder block.

a. Place a new front case gasket on the cylinder block.

b. Apply engine oil to the lip of the oil pump seal. Then, install the oil pump onto the crankshaft.

c. When the pump is in place, clean any excess grease off the crankshaft and check that the oil seal lip is not distorted.

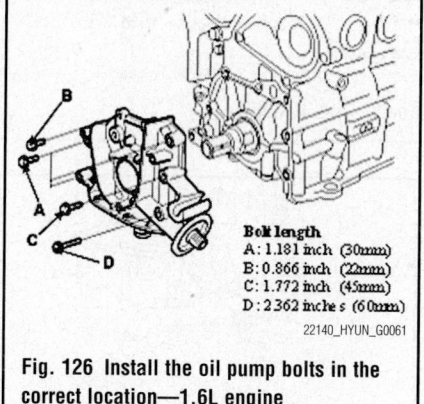

Bolt length
A : 1.181 inch (30mm)
B : 0.866 inch (22mm)
C : 1.772 inch (45mm)
D : 2.362 inches (60mm)

22140_HYUN_G0061

Fig. 126 Install the oil pump bolts in the correct location—1.6L engine

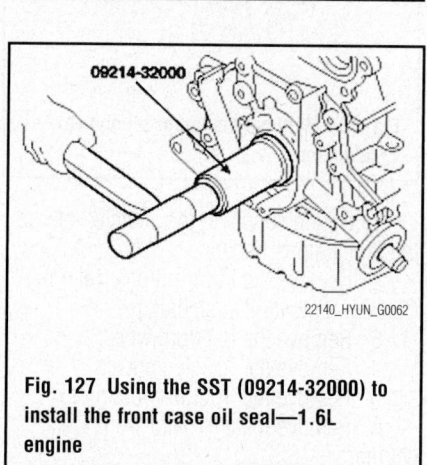

09214-32000

22140_HYUN_G0062

Fig. 127 Using the SST (09214-32000) to install the front case oil seal—1.6L engine

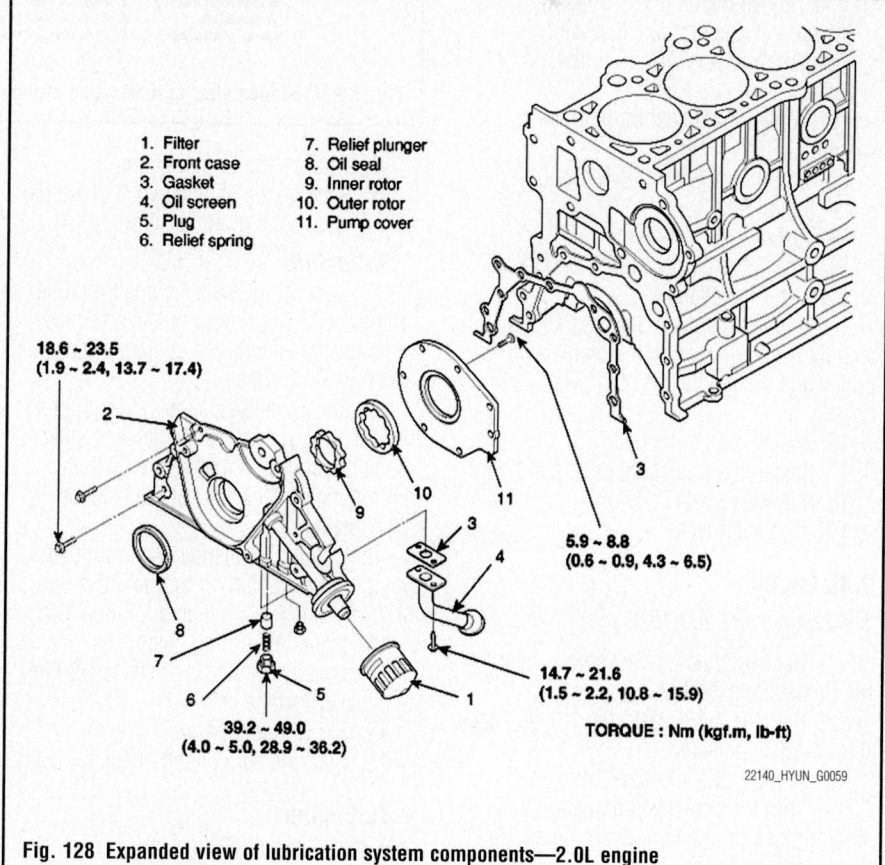

1. Filter
2. Front case
3. Gasket
4. Oil screen
5. Plug
6. Relief spring
7. Relief plunger
8. Oil seal
9. Inner rotor
10. Outer rotor
11. Pump cover

18.6 ~ 23.5
(1.9 ~ 2.4, 13.7 ~ 17.4)

5.9 ~ 8.8
(0.6 ~ 0.9, 4.3 ~ 6.5)

14.7 ~ 21.6
(1.5 ~ 2.2, 10.8 ~ 15.9)

39.2 ~ 49.0
(4.0 ~ 5.0, 28.9 ~ 36.2)

TORQUE : Nm (kgf.m, lb-ft)

22140_HYUN_G0059

Fig. 128 Expanded view of lubrication system components—2.0L engine

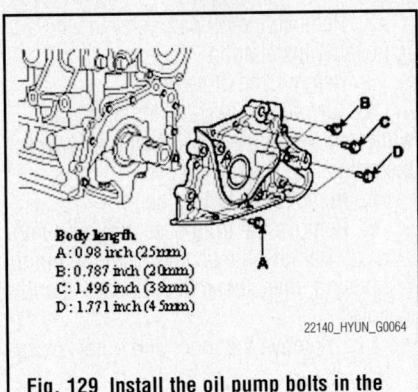

Body length
A : 0.98 inch (25mm)
B : 0.787 inch (20mm)
C : 1.496 inch (38mm)
D : 1.771 inch (45mm)

22140_HYUN_G0064

Fig. 129 Install the oil pump bolts in the correct position—2.0L engine

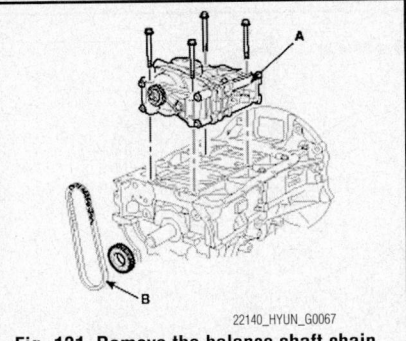

22140_HYUN_G0067

Fig. 131 Remove the balance shaft chain (B) and balance shaft module (A)—2.4L engine

22140_HYUN_G0069

Fig. 133 Remove the 2 bolts (B) and oil screen (A)—2.7L engine

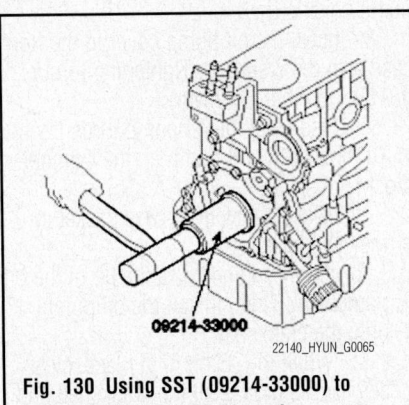

09214-33000

22140_HYUN_G0065

Fig. 130 Using SST (09214-33000) to install the oil seal—2.0L engine

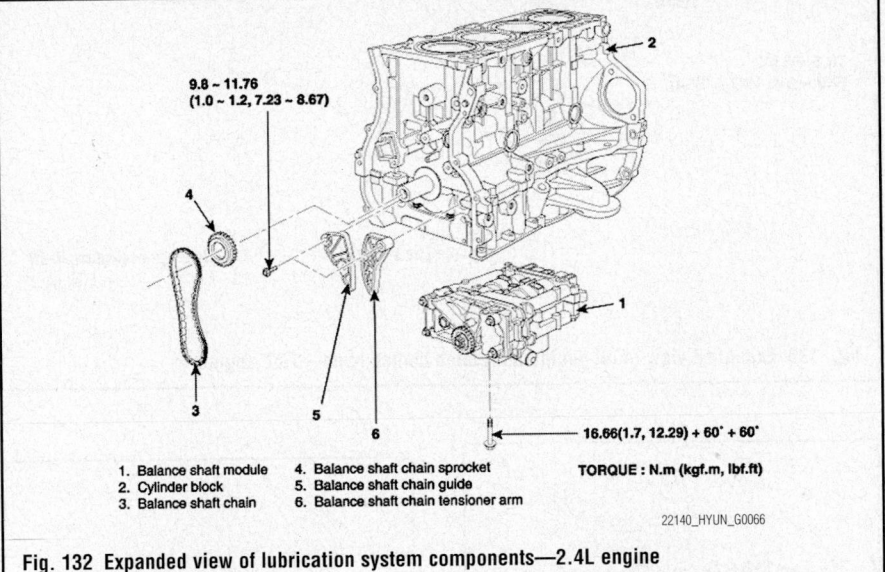

9.8 ~ 11.76
(1.0 ~ 1.2, 7.23 ~ 8.67)

16.66(1.7, 12.29) + 60° + 60°

TORQUE : N.m (kgf.m, lbf.ft)

1. Balance shaft module
2. Cylinder block
3. Balance shaft chain
4. Balance shaft chain sprocket
5. Balance shaft chain guide
6. Balance shaft chain tensioner arm

22140_HYUN_G0066

Fig. 132 Expanded view of lubrication system components—2.4L engine

d. Install the oil pump bolts in the correct position as illustrated. Tightening torque: 15–20 ft. lbs. (20–27 Nm).

11. Apply a light coat of oil to the seal lip.

12. Using the SST (09214-33000), install the oil seal.

13. Install the oil screen.

14. Install the oil pan. Refer to Oil Pan, removal and installation.

15. Ensure that the crankshaft aligns with the white groove on the crankshaft pulley and the pointer on the lower cover.

16. Install the timing belt.

17. Install the accessory drive belts.

18. Fill with engine oil.

19. Check for leaks.

2.4L Engine

See Figures 131 and 132.

1. Before servicing the vehicle, refer to the Precautions Section.

2. Remove the oil pan, refer to Oil Pan, removal & installation.

3. Remove the accessory drive belt.

4. Turn the crankshaft and align the white groove on the crankshaft pulley with the pointer on the lower cover.

5. Remove the timing chain.

6. Remove the balance shaft chain (B) and balance shaft module/oil pump (A).

To install:

7. Confirm the balance shaft module timing mark. The timing marks are to be visually aligned with the centers of adjacent cast timing notches.

8. Install the balance shaft module so that the timing mark of the balance shaft module sprocket is matched with the timing mark (color link) of the balance shaft chain.

9. Install the balance shaft module/oil pump bolts. Tightening torque: 12 ft. lbs. (17 Nm) plus 60° plus an additional 60°.

10. Install the timing chain.

11. Install the oil pan. Refer to Oil Pan, removal & installation.

12. Fill with engine oil.

13. Start the engine and check for leaks.

2.7L Engine

See Figures 133 through 139.

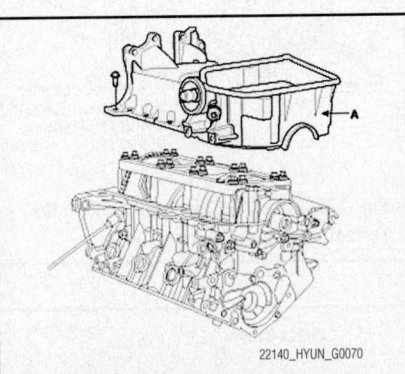

22140_HYUN_G0070

Fig. 134 Remove the upper oil pan (A)—2.7L engine

1. Before servicing the vehicle, refer to the Precautions Section.

2. Remove the lower oil pan, refer to Oil Pan, removal & installation.

3. Remove the RH front wheel.

4. Remove the RH side cover.

5. Remove the front exhaust pipe.

6. Remove the alternator from the engine.

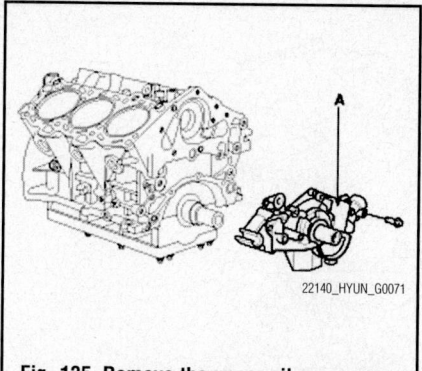

Fig. 135 Remove the upper oil pump case (A)—2.7L engine

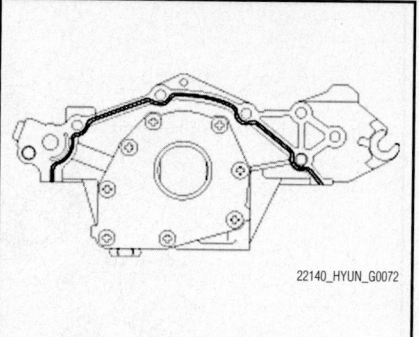

Fig. 137 Apply liquid gasket to the oil pump as shown—2.7L engine

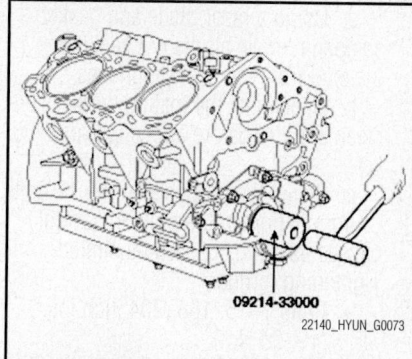

Fig. 138 Using the special tool (09214-33000) to install the oil seal—2.7L engine

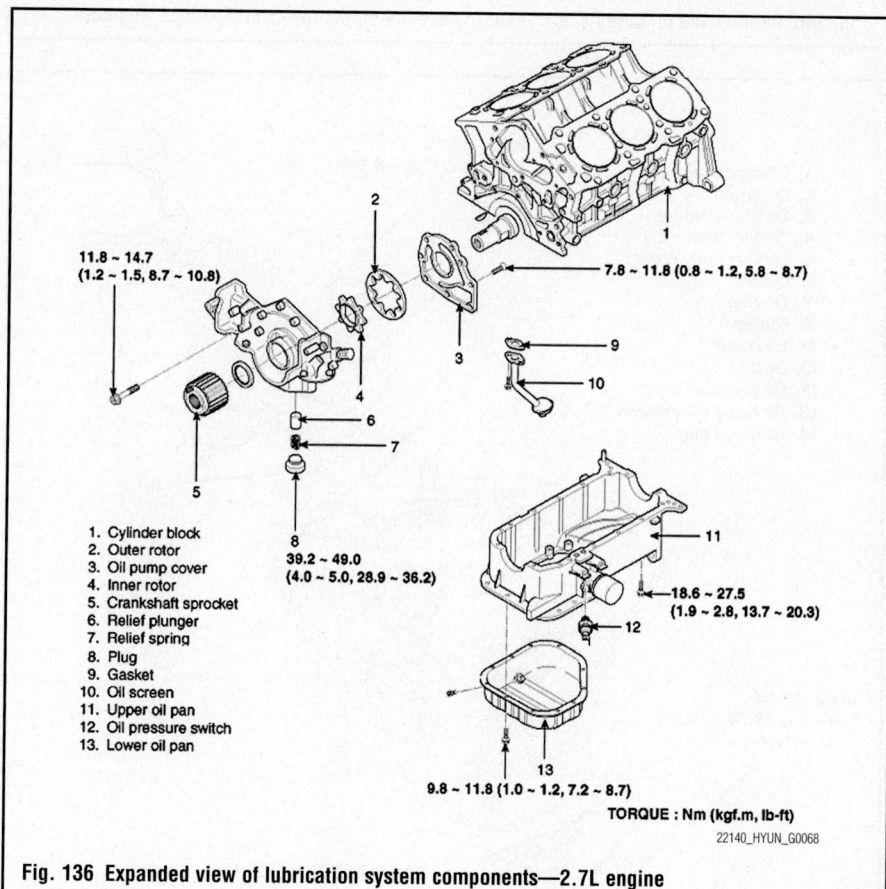

11.8 ~ 14.7
(1.2 ~ 1.5, 8.7 ~ 10.8)

7.8 ~ 11.8 (0.8 ~ 1.2, 5.8 ~ 8.7)

39.2 ~ 49.0
(4.0 ~ 5.0, 28.9 ~ 36.2)

18.6 ~ 27.5
(1.9 ~ 2.8, 13.7 ~ 20.3)

9.8 ~ 11.8 (1.0 ~ 1.2, 7.2 ~ 8.7)

TORQUE : Nm (kgf.m, lb-ft)

1. Cylinder block
2. Outer rotor
3. Oil pump cover
4. Inner rotor
5. Crankshaft sprocket
6. Relief plunger
7. Relief spring
8. Plug
9. Gasket
10. Oil screen
11. Upper oil pan
12. Oil pressure switch
13. Lower oil pan

Fig. 136 Expanded view of lubrication system components—2.7L engine

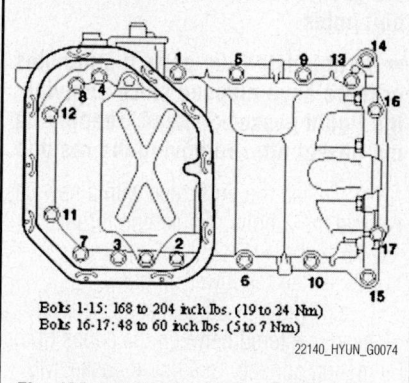

Bolts 1-15: 168 to 204 inch lbs. (19 to 24 Nm)
Bolts 16-17: 48 to 60 inch lbs. (5 to 7 Nm)

Fig. 139 Install the oil pan with the 17 bolts. Uniformly tighten the bolts in several passes—2.7L engine

7. Turn the crankshaft and align the white groove on the crankshaft pulley with the pointer on the lower cover.
8. Remove the timing belt.
9. Remove the oil screen:
   a. Remove the 2 bolts (B).
   b. Remove the oil screen (A) and gasket.
10. Remove the upper oil pan.
11. Remove the oil pump case.
   a. Remove the screws from the pump housing, then separate the housing and cover.
   b. Remove the inner and outer rotors.

To install:
12. Place the inner and outer rotors into the front case with the marks facing the oil pump cover side.
13. Install the oil pump cover to front case with the 8 screws. Tightening torque: 72–108 inch lbs. (8–12 Nm).
14. Check that the oil pump turns freely.
15. Install the oil pump on the cylinder block.
   a. Using a razor blade and gasket scraper, remove all the old liquid gasket from the gasket surfaces and sealing grooves.

   b. Using a non-residue solvent, clean both sealing surfaces.
   c. Apply liquid gasket to the oil pump as shown in the illustration. Use liquid gasket MS 721-40A or equivalent.
   d. To prevent leakage of oil, apply liquid gasket to the inner threads of the bolt holes.
   e. Do not install the parts if 5 minutes or more have elapsed since applying the liquid gasket. Instead, reapply liquid gasket after removing the residue.
   f. Place a new O-ring on the cylinder block.
   g. Engage the spline teeth of the oil pump drive gear with large teeth of the crankshaft, and slide the oil pump on the crankshaft.
   h. Install the oil pump with 5 bolts. Uniformly tighten the bolts in several passes. Tightening torque: 108–132 inch lbs. (12–15 Nm).
16. Apply a light coat of oil to the seal lip.
17. Using the special tool (09214-33000), install the oil seal.
18. Install the upper oil pan.

a. Using a razor blade and gasket scraper, remove all the old packing material from the gasket surfaces.

b. Check that the mating surfaces are clean and dry before applying liquid gasket.

c. Install the oil pan with the 17 bolts. Uniformly tighten the bolts in several passes using the pattern illustrated. Tightening torque:
- Bolts 1–15: 168–204 inch lbs. (19–24 Nm)
- Bolts 16–17: 48–60 inch lbs. (5–7 Nm)

➡To prevent leakage of oil, apply liquid gasket to the inner threads of the bolt holes.

➡Do not install the parts if 5 minutes or more have elapsed since applying the liquid gasket. Instead, reapply liquid gasket after removing the residue.

19. Install the oil screen with a new gasket and the 2 bolts. Tightening torque: 11–16 ft. lbs. (15–22 Nm).

20. Install the lower oil pan.

a. Apply liquid gasket as an even bead, centered between the edges of the mating surface. Use liquid gasket MS 721-40A or equivalent.

b. To prevent leakage of oil, apply liquid gasket to the inner threads of the bolt holes.

c. Do not install the parts if 5 minutes or more have elapsed since applying the liquid gasket. Instead, reapply liquid gasket after removing the residue.

d. Install the lower oil pan 10 bolts. Uniformly tighten the bolts several passes. Tightening torque: 84–108 inch lbs. (10–12 Nm).

21. Install the timing belt.

22. Install the accessory drive belt.

23. Install the alternator.

24. Install the front exhaust pipe.

25. Install the RH front wheel.

26. After assembly, wait at least 30 minutes before filling the engine with oil to allow the gasket material to cure.

27. Fill the engine with oil.

28. Start the engine and check for leaks.

29. Recheck the engine oil level.

3.3L & 3.8L Engines

See Figures 140 and 141.

1. Before servicing the vehicle, refer to the Precautions Section.

2. Remove the lower oil pan, refer to Oil Pan, removal & installation.

3. Remove the oil pump chain cover.

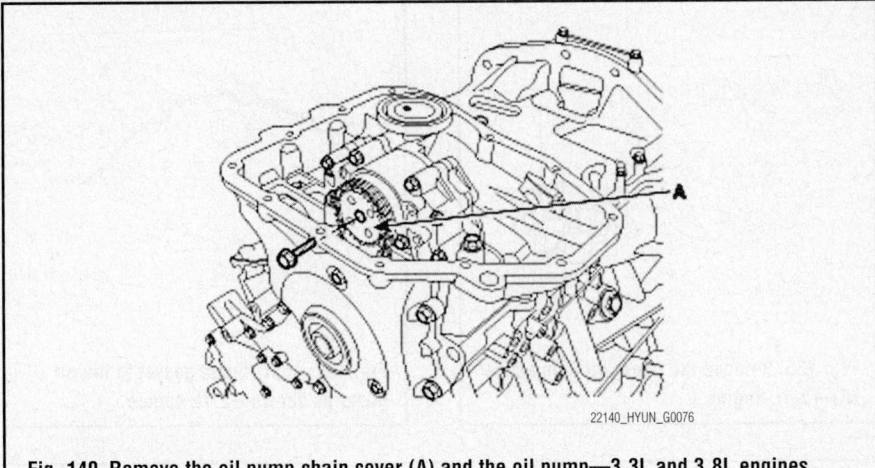

Fig. 140 Remove the oil pump chain cover (A) and the oil pump—3.3L and 3.8L engines

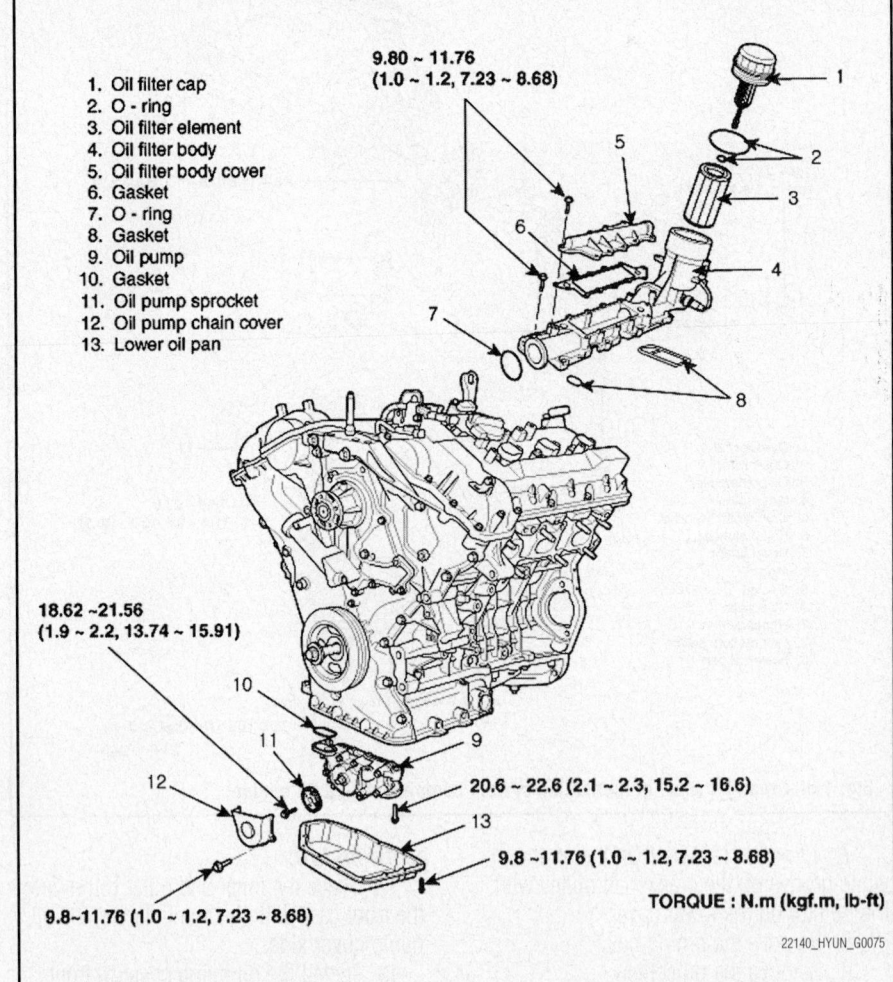

1. Oil filter cap
2. O - ring
3. Oil filter element
4. Oil filter body
5. Oil filter body cover
6. Gasket
7. O - ring
8. Gasket
9. Oil pump
10. Gasket
11. Oil pump sprocket
12. Oil pump chain cover
13. Lower oil pan

9.80 ~ 11.76 (1.0 ~ 1.2, 7.23 ~ 8.68)

18.62 ~21.56 (1.9 ~ 2.2, 13.74 ~ 15.91)

20.6 ~ 22.6 (2.1 ~ 2.3, 15.2 ~ 16.6)

9.8 ~11.76 (1.0 ~ 1.2, 7.23 ~ 8.68)

9.8~11.76 (1.0 ~ 1.2, 7.23 ~ 8.68)

TORQUE : N.m (kgf.m, lb-ft)

Fig. 141 Expanded view of lubrication system components—3.3L and 3.8L engines

4. Remove the oil pump chain sprocket (A).

5. Remove the oil pump.

To install:

6. Install the oil pump using a new O-ring. Tighten the bolts to 15–17 ft. lbs. (21–23 Nm).

➡Always use a new O-ring.

7. Install the oil pump sprocket and oil pump chain on the oil pump. Tightening torque: 14–16 ft. lbs. (19–22 Nm).

8. Install the oil pump chain cover. Tightening torque: 84–108 inch lbs. (10–12 Nm).

9. Install the lower oil pan. Refer to Oil Pan, removal & installation.

10. After assembly, wait at least 30 minutes before filling the engine with oil to allow the gasket material to cure.

11. Fill the engine with oil.

12. Start the engine and check for leaks.

13. Recheck the engine oil level.

INSPECTION

See Figures 142 through 144.

1. Before servicing the vehicle, refer to the Precautions Section.

2. Remove the relief plunger: Remove the plug (A), spring (B), and relief plunger (C).

3. Inspect the relief plunger.
   a. Coat the plunger with engine oil
   b. Check that it falls smoothly into the plunger hole by its own weight.
   c. If it does not, replace the relief plunger.
   d. If necessary, replace the front case.

4. Inspect the relief valve spring.
   a. Inspect for a distorted or broken relief valve spring.
   b. Standard value:
   - Free height: 1.8346 inch (46.6mm)—1.6L engine
   - Load: 13.4 lbs. plus or minus 0.9 lbs./1.5787 inch (6.1 kg plus or minus 0.4kg/40.1mm)—1.6L engine
   - Free height: 1.724 inch (43.8mm)—2.0L engine
   - Load: 8.14 plus or minus 0.88 lbs./1.579 inch (3.7kg plus or minus 0.4kg/40.1mm)—2.0L engine
   - Free height: 1.724 inch (43.8mm)—2.7L engine
   - Load: 10 lbs./1.547 inch (4.6kg/39.1mm)—2.7L engine

5. Inspect the rotor side clearance.
   a. Using a feeler gauge and precision straight edge, measure the clearance between the rotors and precision straight edge.
   b. Side clearance:
   - Inner rotor: 0.0016–0.0033 inch (0.04–0.085mm)—1.6L engine
   - Outer rotor: 0.0016–0.0035 inch (0.04–0.09mm)—1.6L engine
   - Outer gear: 0.0016–0.0035 inch (0.04–0.09mm)—2.0L engine
   - Inner gear: 0.0016–0.0033 inch (0.04–0.085mm)—2.0L engine
   - Side clearance: 0.0016–0.0037 inch (0.04–0.095mm)—2.7L engine
   c. If the side clearance is greater than maximum, replace the rotors as a set. If necessary, replace the front case.

6. Inspect the rotor tip clearance.
   a. Using a feeler gauge, measure the tip clearance between the inner and outer rotor tips.
   b. Tip clearance:
   - 0.0010–0.0027 inch (0.025–0.069mm)—1.6L engine
   - 0.0010–0.0027 inch (0.025–0.069mm)—2.0L engine
   - 0.0010–0.0027 inch (0.025–0.069mm)—2.7L engine
   c. If the tip clearance is greater than specified, replace the rotors as a set.

7. Inspect the rotor body clearance.
   a. Using a feeler gauge, measure the clearance between the outer rotor and body.
   b. Body clearance:
   - 0.0024–0.0035 inch (0.060–0.090mm)—1.6L engine
   - 0.0047–0.0073 inch (0.120–0.185mm)—2.0L engine
   - 0.0039–0.0017 inch (0.100–0.181mm)—2.7L engine
   c. If the body clearance is greater than specified, replace the rotors as a set. If necessary, replace the front case.

8. Install the relief plunger and spring into the front case hole.

9. Install the plug. Tightening torque: 29–36 ft. lbs. (39–49 Nm).

PISTON AND RING

POSITIONING

See Figures 145 through 147.

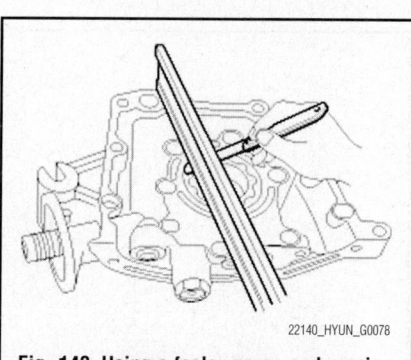

Fig. 143 Using a feeler gauge and precision straight edge to measure the side clearance of the oil pump

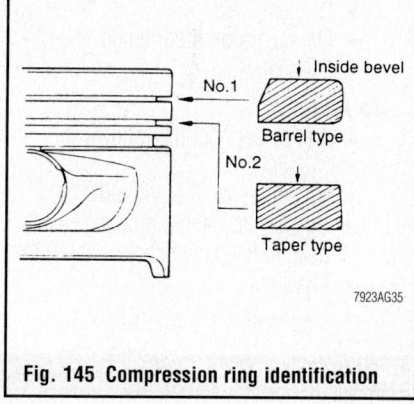

Fig. 145 Compression ring identification

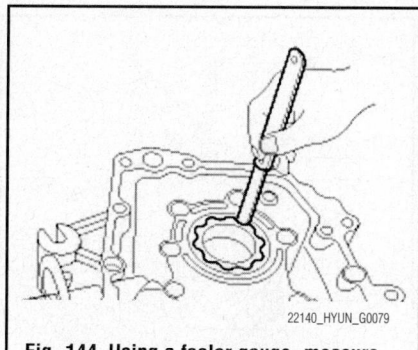

Fig. 142 Remove the relief plunger for inspection—1.6L engine (other engines similar)

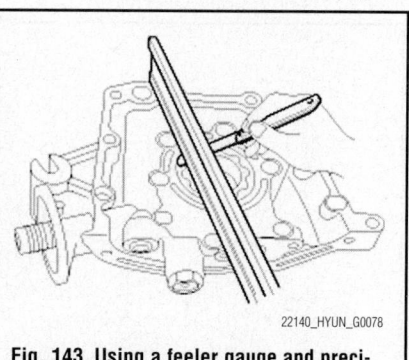

Fig. 144 Using a feeler gauge, measure the tip clearance between the inner and outer rotor tips of the oil pump

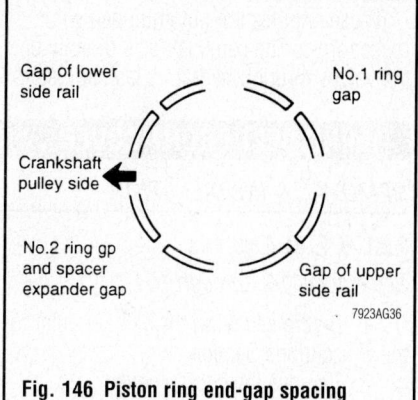

Fig. 146 Piston ring end-gap spacing

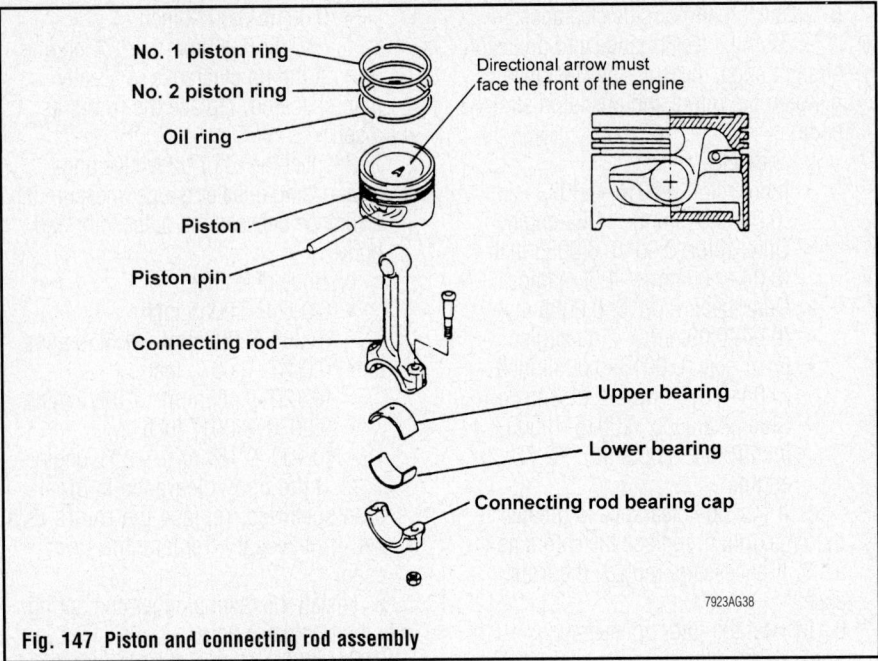

Fig. 147 Piston and connecting rod assembly

REAR MAIN SEAL

REMOVAL & INSTALLATION

1. Before servicing the vehicle, refer to the Precautions Section.
2. Remove or disconnect the following:
   • Transaxle
   • Flywheel
   • Oil seal case
   • Oil separator, if equipped
   • Oil seal

To install:
3. Install or connect the following:
   • Oil seal
   • Oil separator, if equipped
   • Oil seal case and torque the case bolts to 84–108 inch lbs. (8–10 Nm)
   • Flywheel
   • Transaxle

ROCKER ARMS/SHAFTS

REMOVAL & INSTALLATION

These engines are not equipped with rocker arms. The camshaft acts directly on the valves through hydraulic lash adjusters.

TIMING BELT FRONT COVER

REMOVAL & INSTALLATION

1.6L & 2.0L Engines

See Figures 148 through 154.

1. Before servicing the vehicle, refer to the Precautions Section.
2. Remove the engine cover.
3. Remove RH front wheel.

4. Remove 2 bolts (B) and RH side cover (A).
5. Remove the engine mount bracket.
6. Set the jack to the engine oil pan.

➡ **Place wooden block between the jack and engine oil pan.**

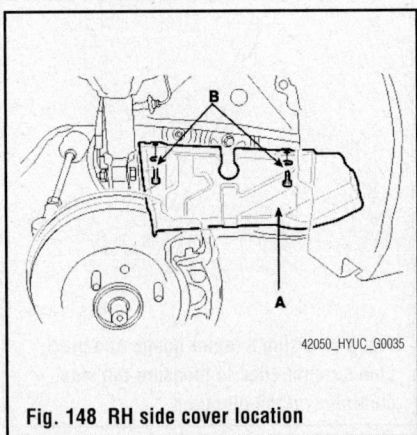

Fig. 148 RH side cover location

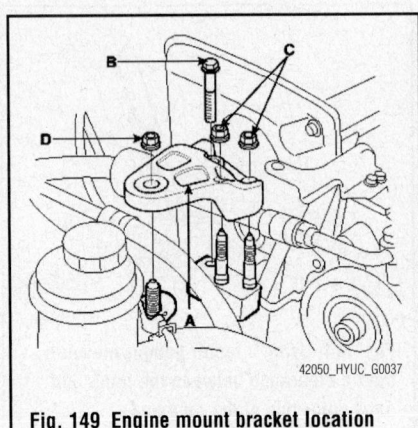

Fig. 149 Engine mount bracket location

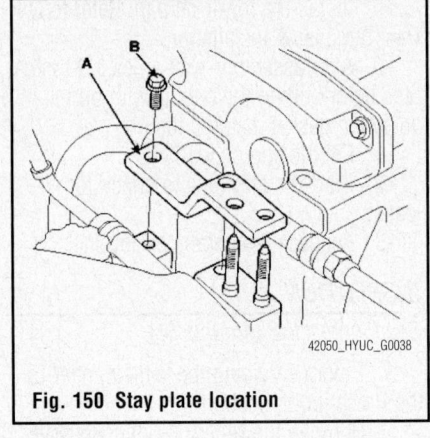

Fig. 150 Stay plate location

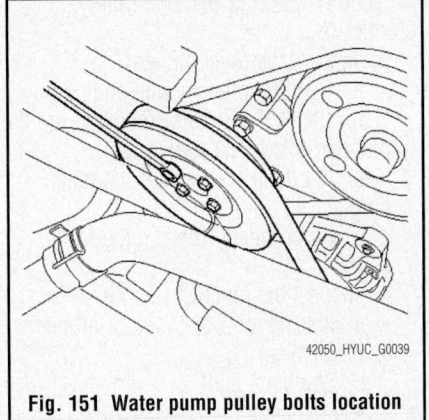

Fig. 151 Water pump pulley bolts location

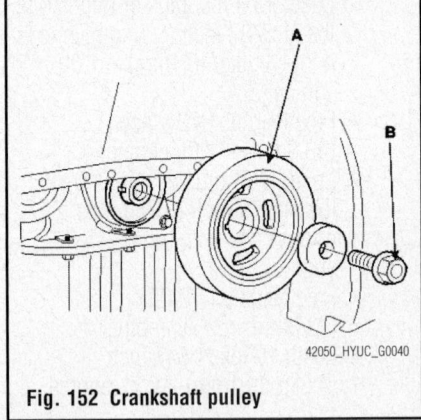

Fig. 152 Crankshaft pulley

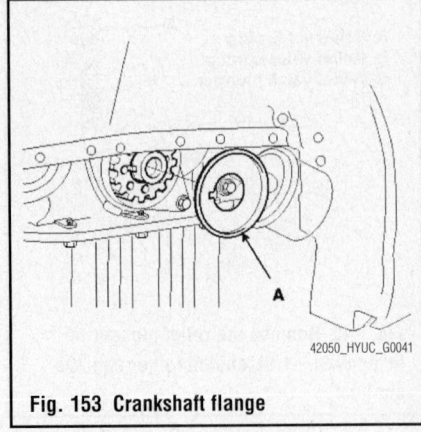

Fig. 153 Crankshaft flange

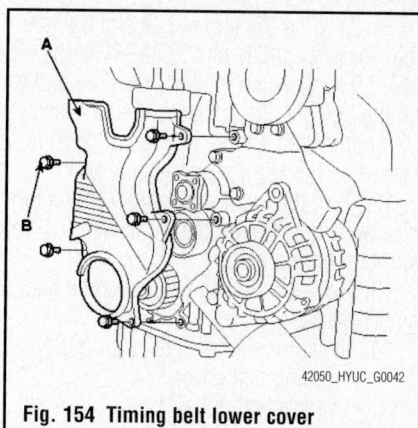

Fig. 154 Timing belt lower cover

7. Remove the bolt (B), 3 nuts (C, D) and engine mount bracket (A).

8. Remove the bolt (B) and stay plate (A).

9. Temporarily loosen the water pump pulley bolts.

10. Remove alternator belt.

11. Remove air compressor belt.

12. Remove power steering belt.

13. Remove 4 bolts and water pump pulley.

14. Remove the 4 bolts and timing belt upper cover.

15. Turn the crankshaft pulley, and align its groove with timing mark "T" of the timing belt cover.

16. Remove the crankshaft pulley bolt (B) and crankshaft pulley (A).

17. Remove the crankshaft flange (A).

18. Remove the 5 bolts (B) and timing belt lower cover (A).

To install:

19. Install timing belt lower cover and secure with the 5 bolts. Tighten and torque to 7 ft. lbs. (10 Nm).

20. Install crankshaft flange.

21. Install crankshaft pulley and secure with the crankshaft pulley bolt. Tighten and torque to 125–133 ft. lbs. (170–180 Nm).

22. Install the timing belt upper cover and secure with the 4 bolts. Tighten and torque to 7 ft. lbs. (10 Nm).

23. Install the water pump pulley and loosely secure with the 4 bolts.

24. Install the power steering, the air compressor and the alternator belts.

25. Tighten all 4 water pump pulley bolts to 7 ft. lbs. (10 Nm).

26. Install the stay plate and bolts. Tighten and torque to 55 ft. lbs. (69 Nm).

27. Install engine mounting brackets, secure with bolts and nuts. Tighten and torque to 80 ft. lbs. (94 Nm).

28. Lower the jack and remove from under the vehicle.

29. Install the RH side cover and secure with 2 bolts.

30. Install the RH front wheel.

31. Install the engine cover.

2.7L Engine

See Figures 155 through 158.

1. Before servicing the vehicle, refer to the Precautions Section.

➡**Engine removal is not required for this procedure.**

2. Remove the engine cover.

3. Remove RH front wheel.

4. Remove the 2 bolts (B) and RH side cover (A).

5. Turn the crankshaft pulley, and align its groove with timing mark "T" of the timing belt cover.

➡**Always turn the crankshaft clockwise.**

6. Remove the auto-tensioner (B) and the drive belt (A).

7. Remove the engine mount bracket.

   a. Set the jack to the engine oil pan.

➡**Place wooden block between the jack and engine oil pan.**

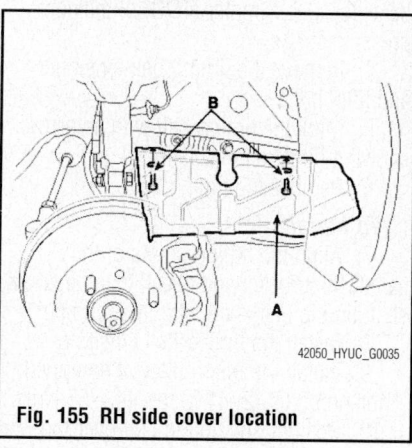

Fig. 155 RH side cover location

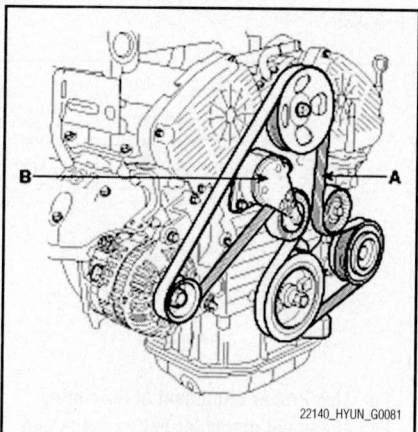

Fig. 156 Location of the auto-tensioner (B) and the drive belt (A)—2.7L engine

Fig. 157 Remove the 7 bolts (B) and timing belt upper cover (A)—2.7L engine

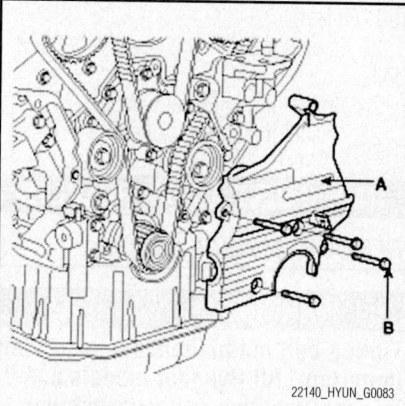

Fig. 158 Remove the 4 bolts (B) and timing belt lower cover (A)—2.7L engine

   b. Remove the 2 bolts, 2 nuts and engine mount bracket.

8. Remove the power steering pump.

9. Remove the 7 bolts (B) and timing belt upper cover (A).

10. Remove the crankshaft damper/pulley bolt and crankshaft damper/pulley.

11. Remove the drive belt idler pulley.

12. Remove the 4 bolts (B) and timing belt lower cover (A).

To install:

13. Install the timing belt lower cover (A) with 4 bolts (B). Tightening torque of timing belt cover bolt: 84–108 inch lbs. (10–12 Nm).

14. Install the drive belt idler pulley. Tightening torque idler pulley bolt: 33–40 ft. lbs. (44–54 Nm).

15. Install the crankshaft damper/pulley. Tightening torque crankshaft pulley bolt: 123–130 ft. lbs. (167–177 Nm).

➡**Make sure that crankshaft sprocket pin fits the small hole in the pulley.**

16. Install the timing belt upper cover with 7 bolts.

17. Install the power steering pump.

18. Install the drive belt tensioner and drive belt.

19. Install the drive belt to the tensioner pulley and install the tensioner as shown. Tightening torque: 18–20 ft. lbs. (25–28 Nm).

20. Install the drive belt following order:
   a. Alternator
   b. Power steering
   c. Crankshaft pulley
   d. A/C pulley, if equipped

21. Rotate the tensioner arm clockwise (about 14°) using a wrench and return it slowly to the original position.

22. Install the engine mount bracket. Tightening torque of engine mount bracket with 2 nuts and 2 bolts: 43–58 ft. lbs. (59–79 Nm).

23. Install the RH side cover with the 2 bolts.

24. Install the RH front wheel.

25. Install the engine cover.

TIMING BELT AND SPROCKETS

REMOVAL & INSTALLATION

✳✳ WARNING

Timing belt maintenance is extremely important. All Hyundai models use interference-type non-freewheeling engines. Should the timing belt break in these engines, the valves in the cylinder head will come in contact with the pistons, causing major engine damage. The recommended replacement interval for timing belts is 60,000 miles.

1.6L Engine

1. Before servicing the vehicle, refer to the Precautions Section.

2. Remove or disconnect the following:
   • Negative battery cable
   • Engine coolant
   • Engine support bracket
   • Accessory drive belts
   • Water pump pulley
   • Crankshaft pulley
   • Timing belt cover

3. Move the timing belt tensioner pulley toward the water pump and secure it.

4. Remove the timing belt.

To install:

5. Install the timing belt and turn the crankshaft sprocket in a reverse direction and align the timing marks.

6. Turn the crankshaft 2 turns in its operating direction and realign the camshaft

sprocket timing mark to Top Dead Center (TDC).

7. Install or connect the following:
   • Timing belt cover. Tighten the bolts to 84 inch lbs. (10 Nm).
   • Crankshaft damper/pulley. Tighten to 111 ft. lbs. (150 Nm).
   • Water pump pulley
   • Accessory drive belts
   • Engine support bracket
   • Negative battery cable

8. Refill the engine with coolant to the correct level.

2.0L Engine

See Figure 159.

1. Before servicing the vehicle, refer to the Precautions Section.

2. Remove or disconnect the following:
   • Negative battery cable
   • Engine coolant
   • Water pump pulley bolts
   • Alternator bolt, loosen only
   • Water pump pulley and drive belts
   • Crankshaft pulley
   • Timing belt cover(s)

3. Rotate the crankshaft clockwise and align the timing marks so No. 1 piston will be at Top Dead Center (TDC) of the compression stroke.

4. Remove the timing belt tensioner and idler pulley.

5. Mark the timing belt with an arrow showing direction of rotation.

6. Remove the timing belt.

To install:

7. Align the timing marks of the camshaft sprocket and check that the crankshaft timing marks are still in alignment.

8. Install the timing belt tensioner.

9. Install the idler pulley, if equipped. Tighten bolt to 32–41 ft. lbs. (43–55 Nm).

10. Position the timing belt over the camshaft sprocket, then over the crankshaft sprocket.

11. Tension the timing belt and tighten the tensioner pulley bolt to 32–41 ft. lbs. (43–55 Nm). When properly tensioned, the timing belt should deflect 0.16–0.24 inch (4–6mm) when a force of 5 lbs. (2.2kg) is placed on the longest span of the belt.

12. Turn the crankshaft sprocket one turn clockwise and realign the crankshaft sprocket timing mark.

13. Recheck the belt tension and adjust as necessary.

14. Install or connect the following:
   • Timing belt cover(s)
   • Crankshaft pulley
   • A/C compressor belt
   • Water pump pulley
   • V belt
   • Negative battery cable
   • Engine with coolant

2.7L Engine

See Figures 160 through 167.

1. Before servicing the vehicle, refer to the Precautions Section.

2. Remove the timing belt front covers (upper and lower). Refer to Timing Belt Front Cover, removal & installation.

3. Remove the engine support bracket (A).

4. Check that timing marks of the camshaft timing pulleys and cylinder head covers are aligned. If not, turn the crankshaft 1 revolution (360°).

5. Remove the timing belt tensioner. Alternately loosen the 2 bolts and remove the tensioner (A).

6. Remove the timing belt.

➡**If the timing belt is to be reused, make an arrow indicating the turning direction to make sure that the belt is reinstalled in the same direction as before.**

7. Remove the tensioner arm assembly (A) and timing belt idler pulley (B).

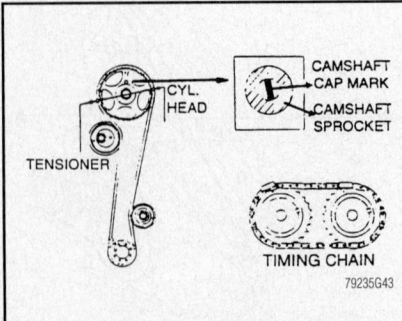

Fig. 159 Proper alignment of the timing belt alignment marks for belt removal and installation—2.0L engine

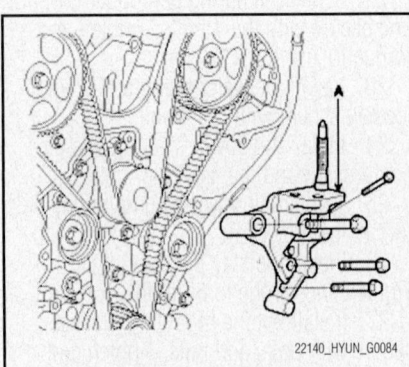

Fig. 160 Remove the engine support bracket (A)—2.7L engine

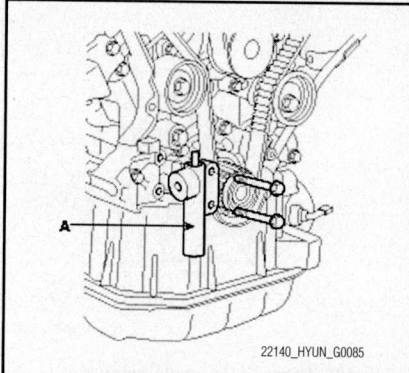

Fig. 161 Alternately loosen the 2 bolts and remove the tensioner (A)—2.7L engine

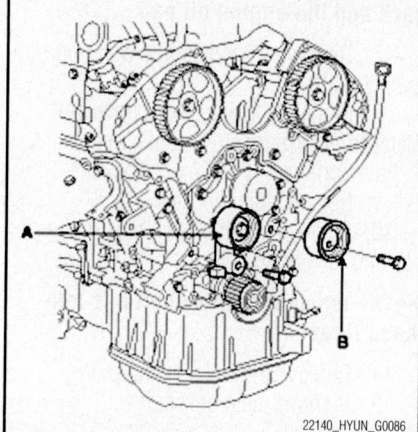

Fig. 162 Remove the tensioner arm assembly (A) and timing belt idler pulley (B)—2.7L engine

8. Remove the crankshaft sprocket.
9. Remove the camshaft sprockets. Hold the hexagonal head wrench portion of the camshaft with a wrench and remove the bolt and camshaft sprocket.

✳✳ WARNING

Be careful not to damage the cylinder head and valve lifter with the wrench.

To install:

10. Install the crankshaft sprocket: Align the pulley set key with the key groove the crankshaft sprocket and slide on the crankshaft sprocket.

11. Install the camshaft sprockets and tighten the bolts to the specified torque.
   a. Temporarily install the camshaft sprocket bolts.
   b. Hold the hexagonal head wrench portion of the camshaft with a wrench, and tighten the camshaft sprocket bolts. Tightening torque: 65–80 ft. lbs. (88–108 Nm).

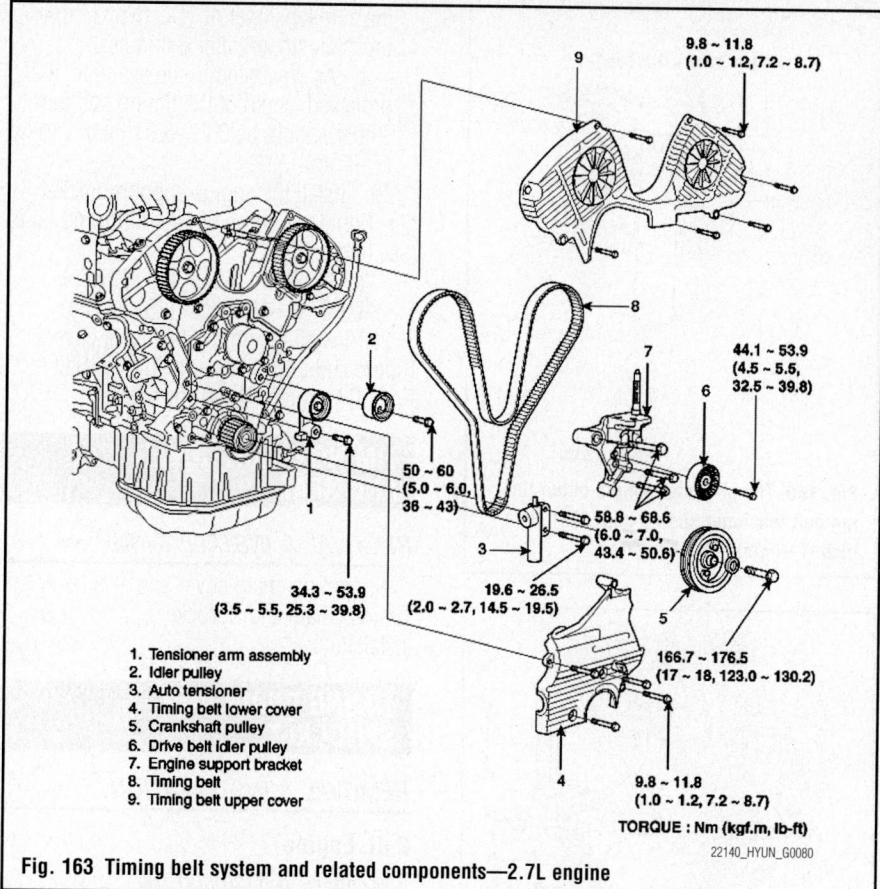

1. Tensioner arm assembly
2. Idler pulley
3. Auto tensioner
4. Timing belt lower cover
5. Crankshaft pulley
6. Drive belt idler pulley
7. Engine support bracket
8. Timing belt
9. Timing belt upper cover

TORQUE: Nm (kgf.m, lb-ft)

Fig. 163 Timing belt system and related components—2.7L engine

12. Install the idler pulley and the tensioner pulley. Tightening torque:
   a. Idler pulley bolt: 36–43 ft. lbs. (49–59 Nm).

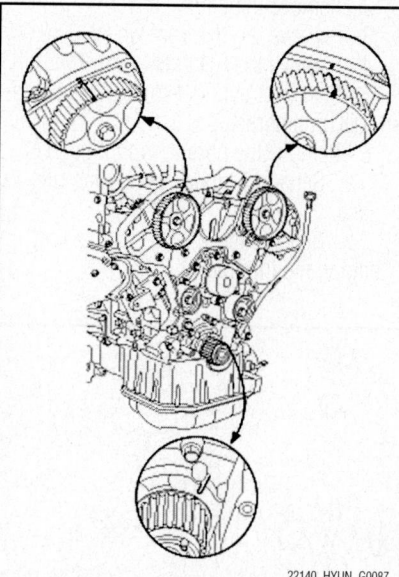

Fig. 164 Align the timing marks of the camshaft sprocket and crankshaft sprocket with the No. 1 piston placed at Top Dead Center (TDC) of its compression stroke—2.7L engine

   b. Tensioner arm fixed bolt: 25–40 ft. lbs. (34–54 Nm).

➡Insert and install the idler pulley to the roll pin that is pressed in the water pump boss.

13. Align the timing marks of the camshaft sprocket and crankshaft sprocket with the No. 1 piston placed at Top Dead Center (TDC) of its compression stroke.

14. Set the timing belt tensioner.
   a. Using a press, slowly press in the push rod.

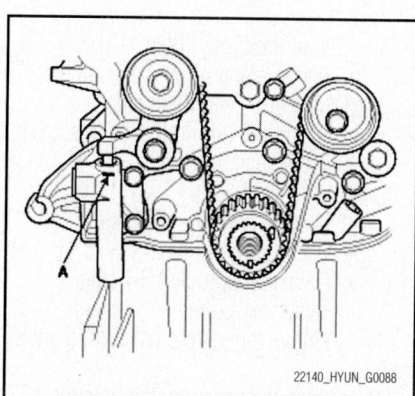

Fig. 165 Remove the set pin (A) from the tensioner—2.7L engine

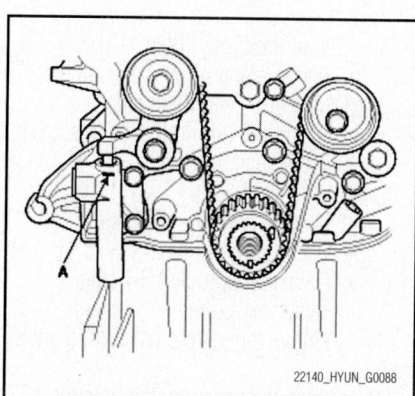

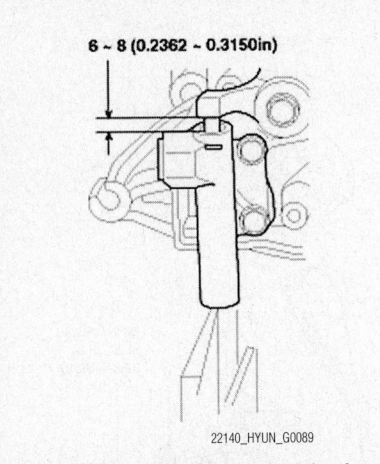

Fig. 166 The projected length of the timing belt tensioner should be 0.27–0.31 inch (7–9mm)—2.7L engine

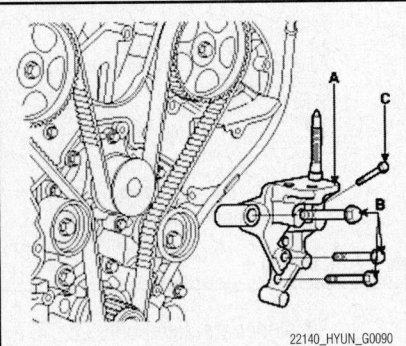

Fig. 167 Install the engine support bracket (A)—2.7L engine

b. Align the holes of the push rod and housing.

c. Pass a set pin through the holes to keep the setting position of the push rod.

d. Release the press.

15. Install the timing belt tensioner.

a. Temporarily install the tensioner with the 2 bolts.

b. Alternately tighten the 2 bolts. Tightening torque: 15–20 ft. lbs. (20–27 Nm).

16. Install the timing belt.

a. Remove any oil or water on the sprockets, and keep them clean.

b. Install the timing belt in this order:

- Crankshaft sprocket
- Idler pulley
- Camshaft sprocket LH side
- Water pump pulley
- Camshaft sprocket RH side
- Tensioner pulley

17. Remove the set pin (A) from the tensioner.

18. Check the timing belt tensioner.

a. Rotate the crankshaft 2 turns clockwise and measure the projected length of

the auto tensioner at TDC (No. 1 compression stroke) after 5 minutes.

b. As illustrated in the example, the projected length of the timing belt tensioner should be 0.27–0.31 inch (7–9mm).

19. Install the engine support bracket (A). Tightening torque of the following bolts (see illustration):

a. B: 43–51 ft. lbs. (59–67 Nm).

b. C: 11–16 ft. lbs. (15–22 Nm).

20. Install the timing belt front covers (upper and lower). Refer to Timing Belt Front Cover, removal & installation.

TIMING CHAIN COVER AND SEAL

REMOVAL & INSTALLATION

For timing chain cover and seal, refer to Timing Chain and Sprockets, removal & installation.

TIMING CHAIN AND SPROCKETS

REMOVAL & INSTALLATION

2.4L Engine

See Figures 168 through 180.

1. Before servicing the vehicle, refer to the Precautions Section.

➡**Engine removal is not required for this procedure.**

2. Remove the engine cover.

3. Remove the RH front wheel.

4. Remove the RH side cover.

5. Set No. 1 cylinder to Top Dead Center (TDC)/compression.

6. Remove the engine mount bracket.

a. Set a jack under the engine oil pan.

b. Remove the 2 bolts, 2 nuts and engine mount bracket (A).

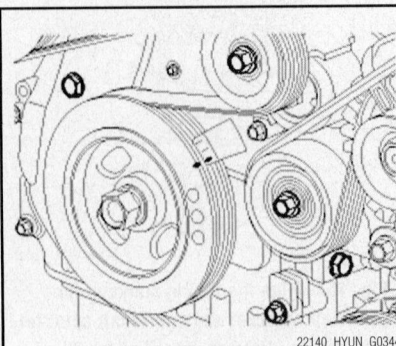

Fig. 168 Set No. 1 cylinder to Top Dead Center (TDC)—2.4L engine

Fig. 169 Remove the engine mount bracket (A)—2.4L engine

➡**Place a wooden block between the jack and the engine oil pan.**

7. Temporarily loosen the water pump pulley bolts.

8. Remove the accessory drive belt. Refer to Accessory Drive Belts, removal & installation.

9. Remove the idler pulley.

10. Remove the drive belt tensioner pulley and tensioner.

➡**The tensioner pulley bolt has left-hand threads.**

11. Remove the water pump pulley.

12. Remove the crankshaft damper/pulley.

13. Remove the engine support bracket.

14. Disconnect the ignition coil connectors.

15. Remove the ignition coils.

16. Remove the PCV hose and breather hose from the cylinder head cover.

17. Remove the cylinder head cover bolts and remove the cylinder head cover (A) and gasket.

18. Remove the A/C compressor lower bolts.

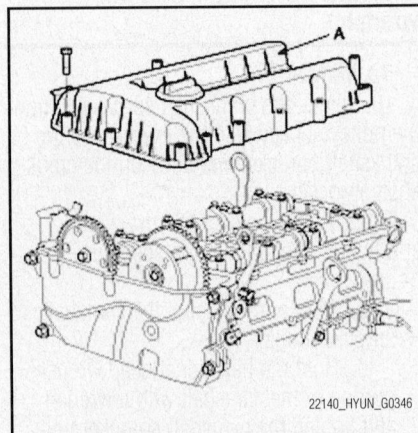

Fig. 170 Remove the cylinder head cover (A) and gasket—2.4L engine

19. Remove the compressor bracket.
20. Drain the engine oil.
21. Remove the oil pan. Refer to Oil Pan, removal & installation.
22. Remove the timing chain cover (A) by prying the portions between the cylinder head and cylinder block with a screwdriver.

➡ **Be careful not to damage the contact surfaces of cylinder block, cylinder head and timing chain cover.**

23. The key of the crankshaft should be aligned with the mating face of the main bearing cap. As a result, the piston of the No. 1 cylinder is placed at TDC on the compression stroke.
24. Install a set pin after compressing the timing chain tensioner.
25. Remove the timing chain tensioner (A).
26. Remove the timing chain tensioner arm (B).
27. Remove the timing chain.

28. Remove the timing chain guide (A).
29. Remove the timing chain oil jet (A).
30. Remove the crankshaft chain sprocket (B).

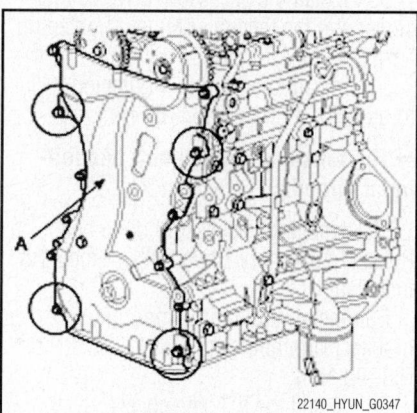

Fig. 171 Remove the timing chain cover (A)—2.4L engine

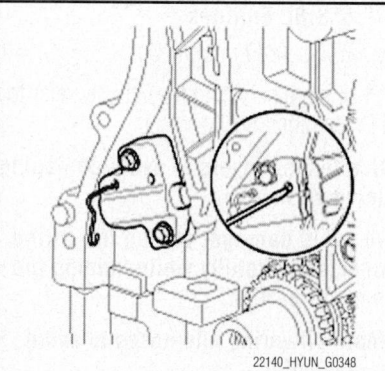

Fig. 172 Install a set pin after compressing the timing chain tensioner—2.4L engine

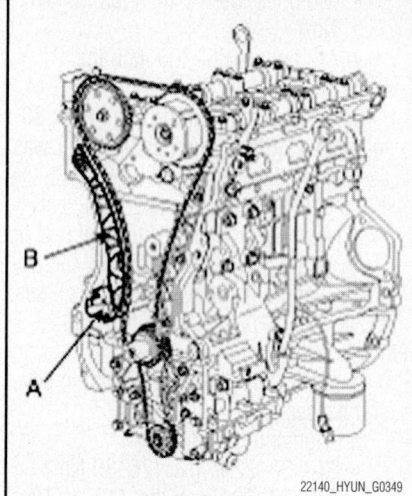

Fig. 173 Remove the timing chain tensioner (A) and the tensioner arm (B)—2.4L engine

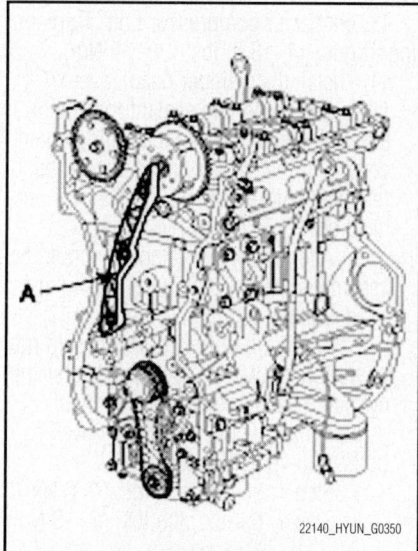

Fig. 174 Remove the timing chain guide (A)—2.4L engine

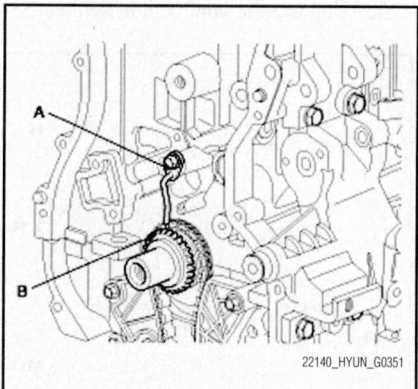

Fig. 175 Remove the timing chain oil jet (A) and the crankshaft chain sprocket (B)—2.4L engine

To install:
31. Install crankshaft chain sprocket.
32. Install timing chain oil jet. Tightening torque: 69–87 inch lbs. (8–10 Nm).
33. Set the crankshaft so that the key of the crankshaft is aligned with the mating surface of main bearing cap. Set the intake and exhaust camshaft assembly so that the TDC mark of the intake sprocket and the exhaust sprocket are aligned with the top surface of the cylinder head. This will place the piston on the No. 1 cylinder at TDC on the compression stroke.
34. Install the timing chain guide (A). Tightening torque: 87–104 inch lbs. (10–12 Nm).
35. Install the timing chain.
   a. To install the timing chain so that there is no slack between the camshaft and crankshaft, use the following procedure.
   b. Place the chain over the crankshaft sprocket (A), then the timing chain guide (B), then the intake camshaft sprocket (C), then the exhaust camshaft sprocket (D).

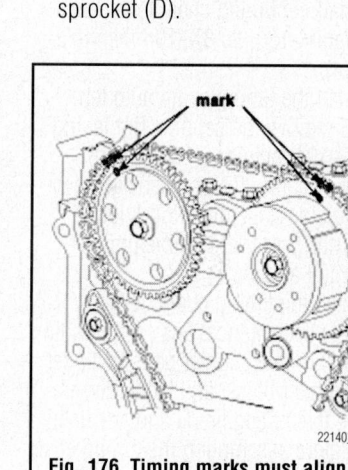

Fig. 176 Timing marks must align with the chain and camshaft sprockets as shown—2.4L engine

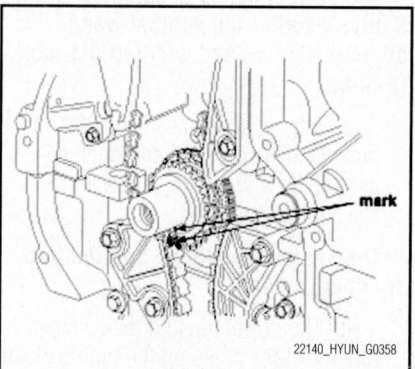

Fig. 177 Timing marks must align with the chain and crankshaft sprocket as shown—2.4L engine

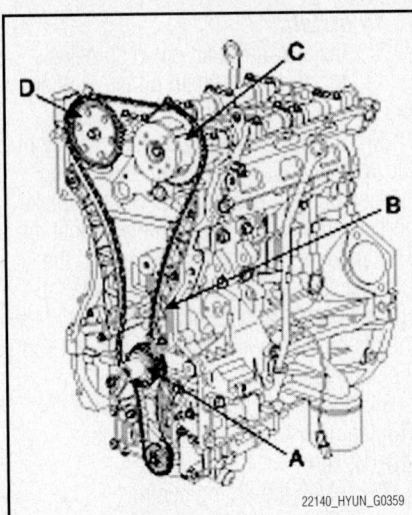

Fig. 178 Install the timing chain according to the order shown—2.4L engine

c. The timing mark of each of the sprockets should be matched with the timing mark (color link) of the timing chain while installing the timing chain.

36. Install the timing chain tensioner arm. Tightening torque: 87–104 inch lbs. (10–12 Nm).

37. Install the timing chain auto tensioner and remove the set pin. Tightening torque: 87–104 inch lbs. (10–12 Nm).

38. After rotating the crankshaft 2 revolutions in a clockwise direction (viewed from the front), confirm that the timing marks are still aligned.

39. Install the timing chain cover.

a. The sealant locations on the chain cover and on the counter parts (cylinder head, cylinder block, and ladder frame) must be free of engine oil and any debris.

b. Before assembling the timing chain cover, the liquid sealant Loctite® 5900 should be applied on the gap between the cylinder head and the cylinder block.

➡**The parts must be assembled within 5 minutes after the sealant was applied. Use a bead width of 0.1 inch (2.5mm).**

c. After applying Loctite® 5900 on the timing chain cover, the part must be assembled within 5 minutes after the sealant was applied.

➡**The sealant should be applied in a continuous bead.**

d. The dowel pins on the cylinder block and the holes on the timing chain cover should be used as a reference in order to assemble the timing chain cover into exact position. Tightening torque:

• M6 bolts: 69–87 inch lbs. (8–10 Nm)
• M8 bolts: 165–200 inch lbs. (19–23 Nm)

e. Wait on starting the engine for 30 minutes after the timing chain cover was assembled.

40. Install timing chain cover oil seal

a. Apply engine oil to a new oil seal lip.

b. Using SST (09214-3K000, 09231-H1100) and a hammer, tap in the oil seal.

41. Install oil pan. Refer to Oil Pan, removal & installation.

a. Uniformly tighten the bolts in several passes.

b. Tightening torque:
• M8: 20–22 ft. lbs. (26–30 Nm)
• M6: 87–104 inch lbs. (10–12 Nm)

c. After assembly, wait at least 30 minutes before filling the engine with oil.

42. Install the air compressor bracket. Tightening torque: 14–17 ft. lbs. (20–24 Nm).

43. Install air compressor bolt. Tightening torque: 14–18 ft. lbs. (20–25 Nm).

44. Install the cylinder head cover.

a. The hardening sealant located on the upper area between the timing chain cover and the cylinder head should be removed before assembling the cylinder head cover.

b. After applying sealant, it should be assembled within 5 minutes. Apply in a bead width of 0.1 inch (2.5mm).

c. The firing of the engine should not be performed within 30 minutes after the cylinder head cover was assembled.

d. Install the cylinder head cover bolts, and tighten as follows:
• Step 1: 35–52 inch lbs. (4–6 Nm)
• Step 2: 69–87 inch lbs. (8–10 Nm)

➡**Do not reuse the cylinder head cover gasket.**

45. Install the ignition coils.
46. Connect the ignition coil connectors.

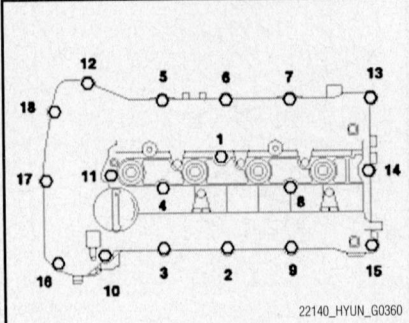

Fig. 179 Cylinder head cover torque sequence shown—2.4L engine

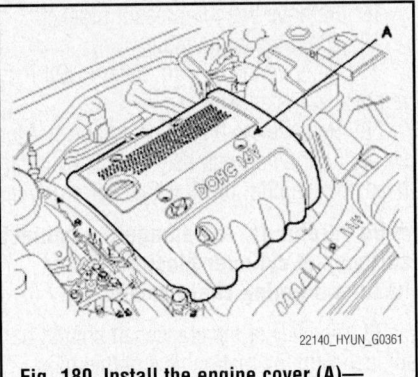

Fig. 180 Install the engine cover (A)—2.4L engine

47. Install the engine support bracket. Tightening torque:

a. M10 bolts: 29–33 ft. lbs. (39–44 Nm).

b. M8 bolts: 15–18 ft. lbs. (20–25 Nm).

48. Using SST (09231-3K000), install the crankshaft damper. Tightening torque: 123–130 ft. lbs. (167–176 Nm).

49. Install the water pump pulley. Tightening torque: 69–87 inch lbs. (8–10 Nm).

50. Install the accessory drive belt tensioner and the tensioner pulley. Tightening torque: 40–47 ft. lbs. (54–64 Nm).

51. Install idler pulley. Tightening torque: 40–47 ft. lbs. (54–64 Nm).

➡**The tensioner pulley bolt has left-hand threads.**

52. Install the accessory drive belt. Refer to Accessory Drive Belts, removal & installation.

53. Install the engine mounting bracket. Tightening torque: 47–61 ft. lbs. (64–83 Nm).

54. Install the RH side cover.
55. Install the RH front wheel.
56. Install the engine cover (A). Tightening torque: 35–52 inch lbs. (4–6 Nm).

3.3L & 3.8L Engines

See Figures 181 through 201.

1. Before servicing the vehicle, refer to the Precautions Section.

➡**Use fender covers to avoid damaging painted surfaces.**

➡**To avoid damage, unplug the wiring connectors carefully while holding the connector portion.**

➡**Mark all wiring and hoses to avoid misconnection.**

➡**Turn the crankshaft pulley so that the No. 1 piston is at Top Dead Center (TDC).**

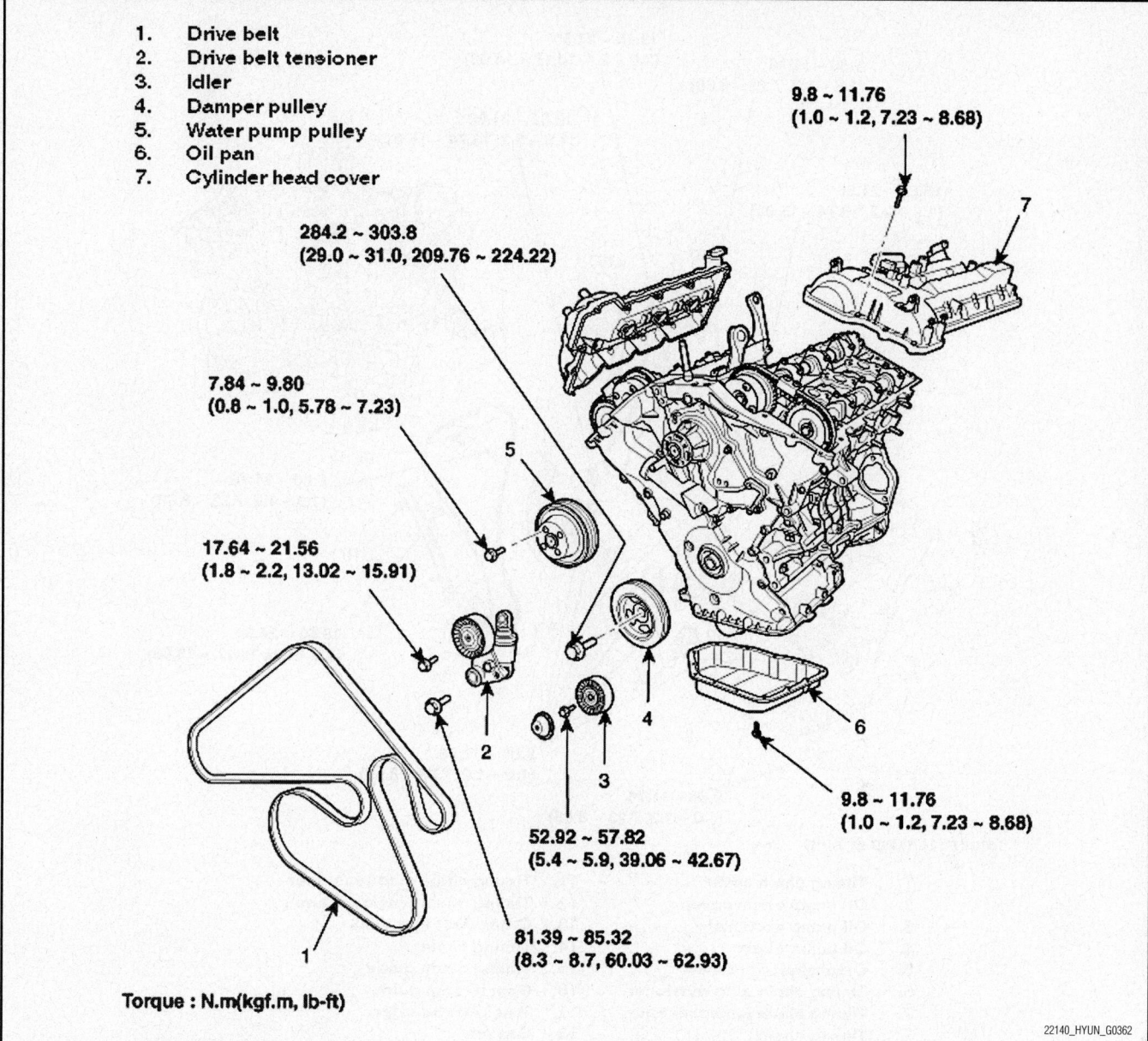

1. Drive belt
2. Drive belt tensioner
3. Idler
4. Damper pulley
5. Water pump pulley
6. Oil pan
7. Cylinder head cover

9.8 ~ 11.76
(1.0 ~ 1.2, 7.23 ~ 8.68)

284.2 ~ 303.8
(29.0 ~ 31.0, 209.76 ~ 224.22)

7.84 ~ 9.80
(0.8 ~ 1.0, 5.78 ~ 7.23)

17.64 ~ 21.56
(1.8 ~ 2.2, 13.02 ~ 15.91)

9.8 ~ 11.76
(1.0 ~ 1.2, 7.23 ~ 8.68)

52.92 ~ 57.82
(5.4 ~ 5.9, 39.06 ~ 42.67)

81.39 ~ 85.32
(8.3 ~ 8.7, 60.03 ~ 62.93)

Torque : N.m(kgf.m, lb-ft)

22140_HYUN_G0362

Fig. 181 Exploded view of cylinder head and accessory drive belt components—3.3L & 3.8L engines

2. Disconnect the negative terminal from the battery.

3. Remove the engine cover.

4. Remove the air duct.

5. Remove the intake air hose and air cleaner assembly.

   a. Disconnect the AFS connector (A).

   b. Disconnect the breather hose (B) from air cleaner hose.

   c. Disconnect the ECM connector.

   d. Remove the intake air hose (C) and air cleaner (D).

6. Remove the RH front wheel.

7. Remove the undercover.

8. Remove the RH side cover.

9. Drain the engine coolant. Remove the radiator cap to speed draining.

10. Remove the upper radiator hose.

11. Drain the engine oil.

12. Remove the lower oil pan. Refer to Oil Pan, removal & installation.

13. Place a jack underneath the upper oil pan.

14. Just loosen the transaxle mounting bolts (A) without removing the transaxle mounting (B).

15. Remove the engine coolant reservoir tank (A).

16. Remove the engine mounting bracket (B).

17. Loosen the A/C pipe bracket mounting bolt.

18. Remove the No. 1 engine mounting (A) through the lower position of the A/C pipe line.

19. Remove the surge tank and engine wiring.

   a. Disconnect the RH oxygen sensor connector.

   b. Disconnect the VIS solenoid valve connector (3.8L only).

   c. Disconnect the power steering oil pressure sensor connector.

   d. Disconnect the RH injector connector and the ignition coil connector.

   e. Disconnect the OCV connector and knock sensor connector.

   f. Disconnect the LH front oxygen sensor connector, alternator connector, and air compressor connector.

   g. Disconnect the LH ignition coil connector, injector connector, condenser connector, and ground

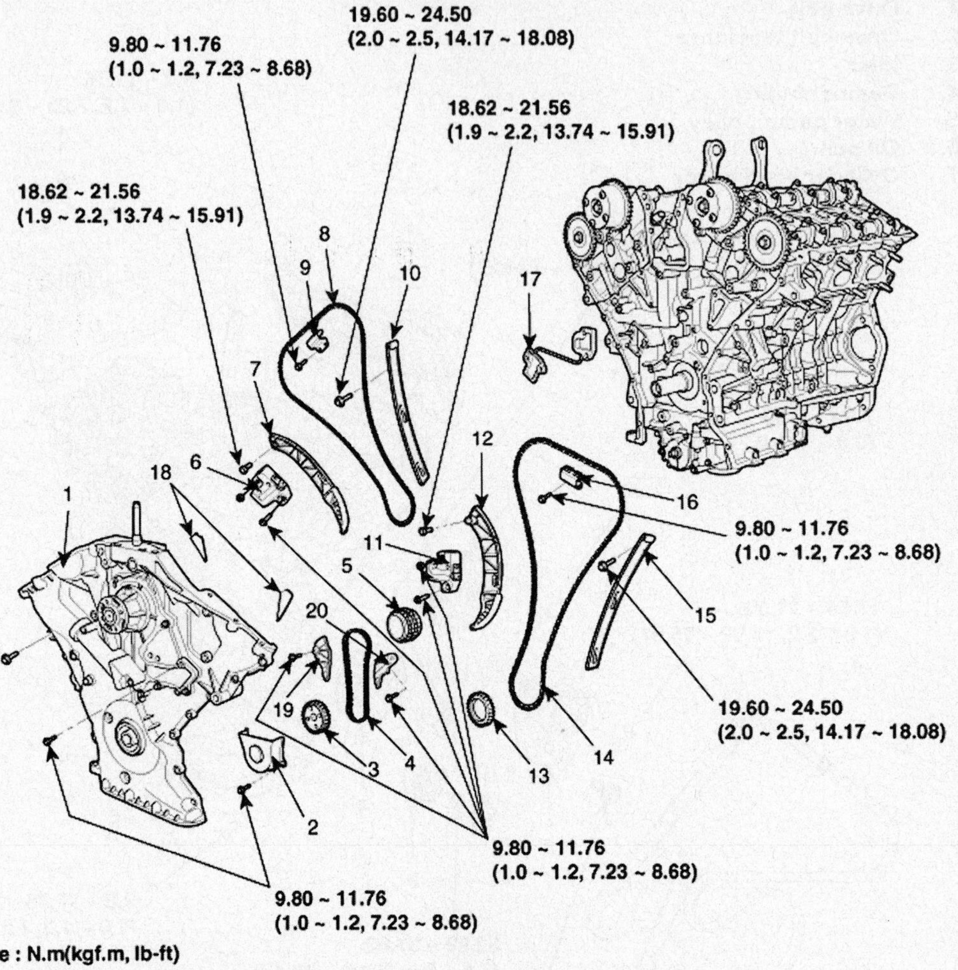

9.80 ~ 11.76
(1.0 ~ 1.2, 7.23 ~ 8.68)

19.60 ~ 24.50
(2.0 ~ 2.5, 14.17 ~ 18.08)

18.62 ~ 21.56
(1.9 ~ 2.2, 13.74 ~ 15.91)

18.62 ~ 21.56
(1.9 ~ 2.2, 13.74 ~ 15.91)

9.80 ~ 11.76
(1.0 ~ 1.2, 7.23 ~ 8.68)

19.60 ~ 24.50
(2.0 ~ 2.5, 14.17 ~ 18.08)

9.80 ~ 11.76
(1.0 ~ 1.2, 7.23 ~ 8.68)

9.80 ~ 11.76
(1.0 ~ 1.2, 7.23 ~ 8.68)

Torque : N.m(kgf.m, lb-ft)

1. Timing chain cover
2. Oil pump chain cover
3. Oil pump sprocket
4. Oil pump chain
5. Crankshaft sprocket
6. Timing chain auto tensioner
7. Timing chain tensioner arm
8. Timing chain
9. Cam to cam guide
10. Timing chain guide
11. Timing chain auto tensioner
12. Timing chain tensioner arm
13. Crankshaft sprocket
14. Timing chain
15. Timing chain guide
16. Cam to cam guide
17. Tensioner adapter
18. Gasket
19. Oil pump chain guide
20. Oil pump tensioner assembly

22140_HYUN_G0363

Fig. 182 Exploded view of timing chain and related components—3.3L & 3.8L engines

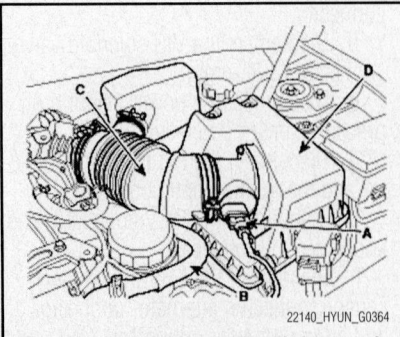

22140_HYUN_G0364

Fig. 183 Remove the intake air hose and air cleaner assembly—3.3L & 3.8L engines

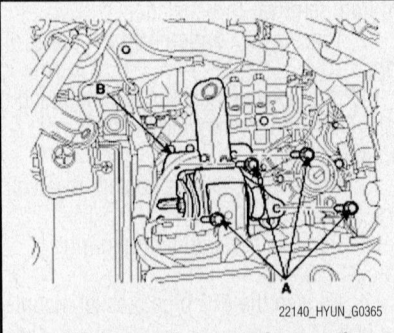

22140_HYUN_G0365

Fig. 184 Loosen the transaxle mounting bolts (A) without removing the transaxle mounting (B)—3.3L & 3.8L engines

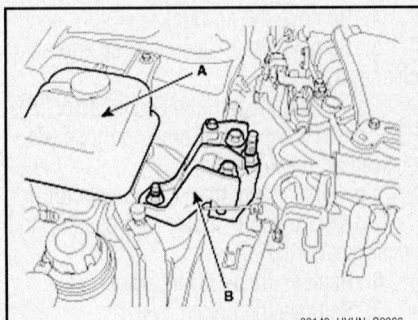

22140_HYUN_G0366

Fig. 185 Remove the engine coolant reservoir tank (A) and the engine mounting bracket (B)—3.3L & 3.8L engines

22140_HYUN_G0367

Fig. 186 Remove the No. 1 engine mounting (A) through the lower position of the A/C pipe line—3.3L & 3.8L engines

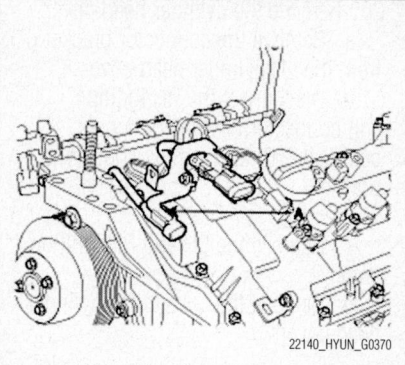

22140_HYUN_G0370

Fig. 189 Remove the connector bracket (A) from the LH cylinder head cover—3.3L and 3.8L engines

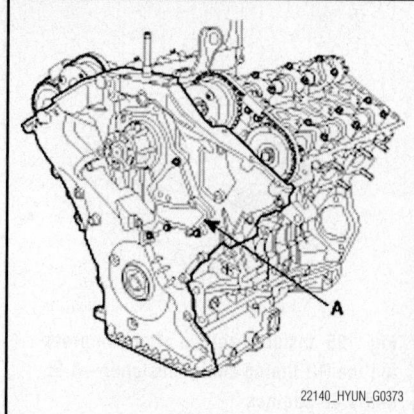

22140_HYUN_G0373

Fig. 192 Remove the timing chain cover (A)—3.3L and 3.8L engines

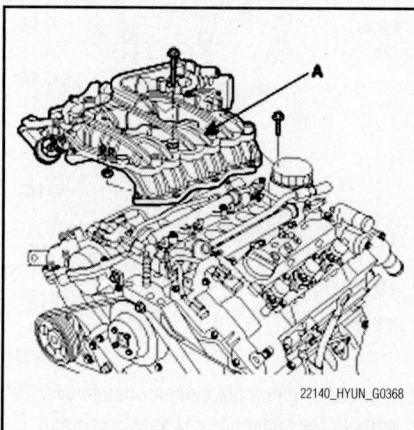

22140_HYUN_G0368

Fig. 187 Remove the surge tank (A)—3.3L engine

22140_HYUN_G0371

Fig. 190 Turn the crankshaft pulley and align its groove with the timing mark "T" of the lower timing chain cover—3.3L and 3.8L engines

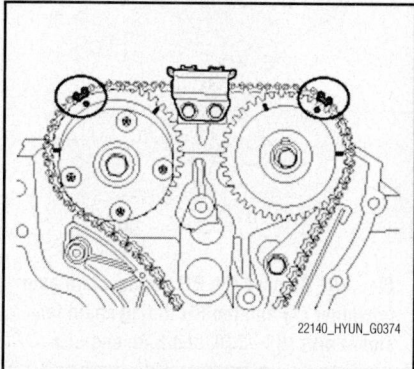

22140_HYUN_G0374

Fig. 193 Timing marks on chain and camshaft sprockets illustrated—3.3L and 3.8L engines

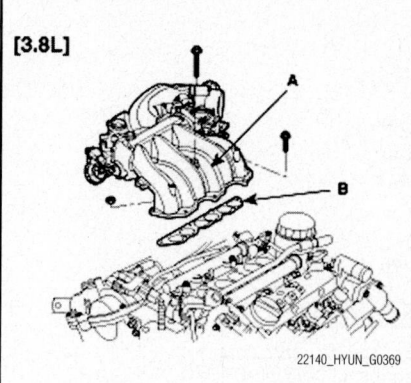

[3.8L]

22140_HYUN_G0369

Fig. 188 Remove the surge tank (A)—3.8L engine

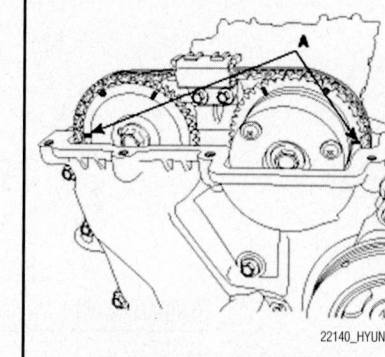

22140_HYUN_G0372

Fig. 191 Check that the mark (A) of the camshaft timing sprockets are in straight line on the cylinder head surface as shown—3.3L and 3.8L engines

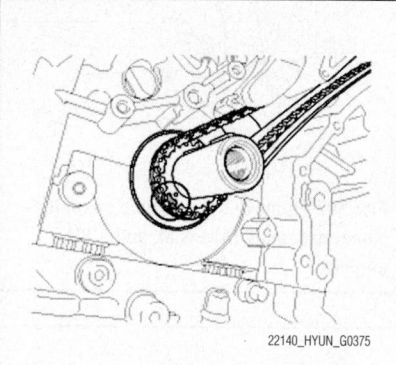

22140_HYUN_G0375

Fig. 194 Timing marks on chain and crankshaft sprocket illustrated—3.3L and 3.8L engines

h. Remove the wiring harness protector.

i. Disconnect the LH CMPS and oil pressure switch connector.

j. Disconnect the ETC connector and the knock sensor connector.

k. Disconnect the PCSV connector, the MAP sensor connector, and the PCSV hose.

l. Remove the Electronic Throttle Control (ETC) bracket.

m. Disconnect the water hoses from the ETC.

n. Disconnect the PCV hose.

o. Disconnect the brake vacuum hose.

p. Remove the surge tank stay.

q. Remove the connector bracket from the surge tank.

r. Remove the surge tank (A).

➡**Cover the inlet of intake manifold with a clean woven stuff or vinyl cover to prevent foreign materials from entering.**

Fig. 195 Install a set pin after compressing the RH timing chain tensioner—3.3L and 3.8L engines

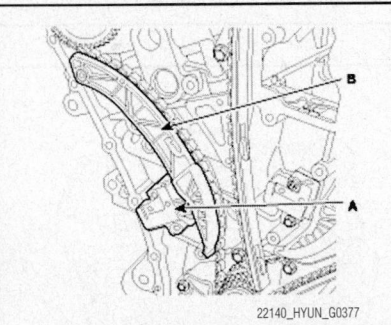

Fig. 196 Remove the RH timing chain auto tensioner (A) and the RH timing chain tensioner arm (B)—3.3L and 3.8L engines

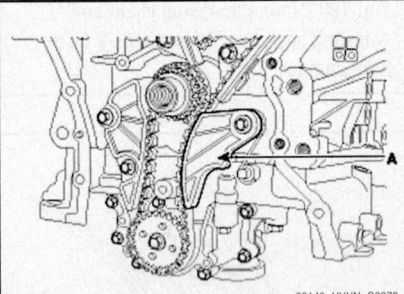

Fig. 197 Remove the oil pump chain tensioner assembly (A)—3.3L and 3.8L engines

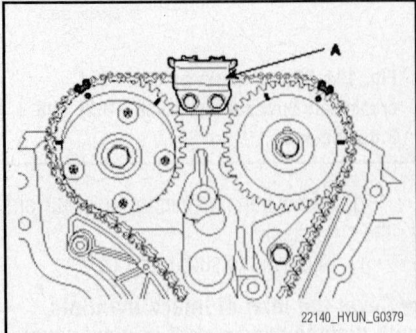

Fig. 198 Remove the LH cam-to-cam guide (A)—3.3L and 3.8L engines

20. Remove the cylinder head cover.

   a. Remove the connector bracket (A) from the LH cylinder head cover.

   b. Disconnect the RH ignition coil connector(A), the condenser connector (B) and remove the wiring bracket (C).

   c. Remove the LH and RH ignition coils.

   d. Remove the LH and RH cylinder head covers.

➡**Cover the upside of engine head with a clean vinyl cover to prevent foreign materials from entering.**

21. Set No. 1 cylinder to TDC/compression.

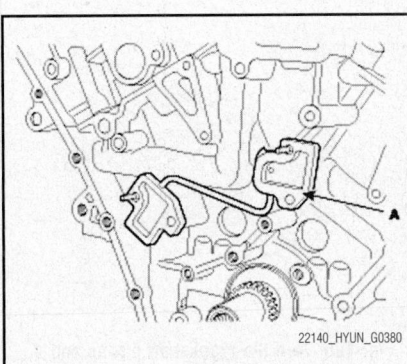

Fig. 199 Remove the tensioner adapter assembly (A)—3.3L and 3.8L engines

   a. Turn the crankshaft pulley and align its groove with the timing mark "T" of the lower timing chain cover.

✳✳ WARNING

Do not rotate engine counterclockwise.

   b. Check that the mark (A) of the camshaft timing sprockets are in straight line on the cylinder head surface as shown in the illustration. If not,

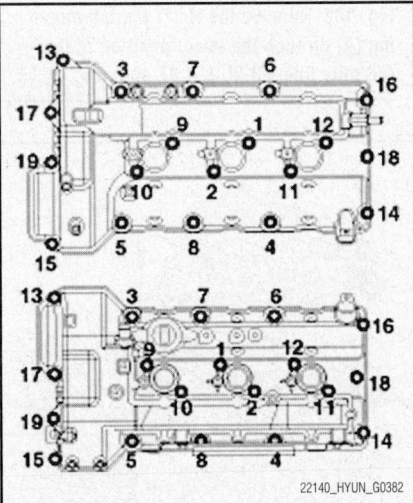

Fig. 201 Tighten the cylinder head cover bolts in the sequence shown—3.3L and 3.8L engines

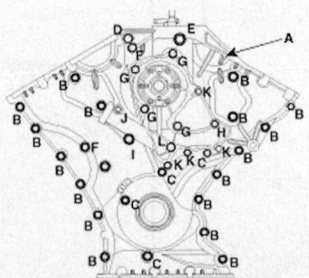

Bolt (No.)	Torque Specification
B (17)	14-16 ft. lbs. (19-22 Nm)
C (4)	87-104 inch lbs. (10-12 Nm)
D (1)	43-51 ft. lbs. (59-69 Nm)
E (1)	43-51 ft. lbs. (59-69 Nm)
F (2)	18-20 ft. lbs. (25-27 Nm)
G (4)	16-17 ft. lbs. (22-24 Nm)
H (1)	87-104 inch lbs. (10-12 Nm)
I (1)	87-104 inch lbs. (10-12 Nm)
J (1)	87-104 inch lbs. (10-12 Nm)
K (4)	87-104 inch lbs. (10-12 Nm)
L (1)	16-20 ft. lbs. (22-27 Nm)

Fig. 200 Tighten the timing chain cover as shown—3.3L and 3.8L engines

turn the crankshaft one revolution (360°). Do not rotate engine counter-clockwise.

22. Remove the accessory drive belt. Refer to Accessory Drive Belts, removal & installation.

23. Remove the crankshaft damper pulley. Refer to Crankshaft Damper, removal & installation.

➡**Use the SST (flywheel stopper, 09231-3C300) to remove the crankshaft pulley bolt, after removing the starter.**

24. Lift up the engine assembly by using the jack.

25. Remove the power steering pump.

26. Remove the air conditioner compressor.

27. Remove the alternator.

28. Remove the drive belt idler.

29. Remove the drive belt auto tensioner.

30. Remove water pump pulley.

31. Remove the timing chain cover (A).

✳ WARNING

Be careful not to damage the contact surfaces of the cylinder block, cylinder head, or timing chain cover.

➡**Before removing the timing chain, mark the RH/LH timing chain with an identification based on the location of the sprocket as the identification mark on the chain for TDC can be erased accidentally.**

32. Install a set pin after compressing the RH timing chain tensioner.

33. Remove the RH cam-to-cam guide.

34. Remove the RH timing chain auto tensioner (A) and the RH timing chain tensioner arm (B).

35. Remove the RH timing chain.

36. Remove the RH timing chain guide.

37. Remove the oil pump chain cover.

38. Remove the oil pump chain tensioner assembly (A).

39. Remove the oil pump chain guide.

40. Remove the oil pump chain sprocket and oil pump chain.

41. Remove the crankshaft sprocket that drives the oil pump.

42. Install a set pin after compressing the LH timing chain tensioner.

43. Remove the LH cam-to-cam guide (A).

44. Remove the LH timing chain auto tensioner and LH timing chain tensioner arm.

45. Remove the LH timing chain.

46. Remove the LH timing chain guide.

47. Remove the crankshaft sprocket.

48. Remove the tensioner adapter assembly (A).

To install:

49. Check the camshaft sprocket and crankshaft sprocket for abnormal wear, cracks, or damage. Replace as necessary.

50. Inspect the tensioner arm and chain guide for abnormal wear, cracks, or damage. Replace as necessary.

51. Check that the tensioner piston moves smoothly when the ratchet pawl is released.

52. Install the jack to the upper oil pan.

53. The key of crankshaft must be aligned with the timing mark of timing chain cover. Then, piston of No. 1 cylinder is placed at the TDC on the compression stroke.

54. Install the tensioner adapter assembly.

55. Install the crankshaft sprocket.

56. Install the LH timing chain guide. Tightening torque: 14–18 ft. lbs. (20–25 Nm).

57. Install the LH timing chain.

　a. To install the timing chain with no slack between the camshaft and crankshaft, use the following procedure.

　b. Place the crankshaft sprocket on first, then the timing chain guide, then the exhaust camshaft sprocket, and finally the intake camshaft sprocket.

　c. The timing mark of each sprockets should be matched with the timing mark (color link) of the timing chain when installing the timing chain.

58. Install the LH timing chain tensioner arm. Tightening torque: 14–16 ft. lbs. (19–22 Nm).

59. Install the chain tensioner. Tightening torque: 87–104 inch lbs. (10–12 Nm).

60. Install the LH cam-to-cam guide. Tightening torque: 87–104 inch lbs. (10–12 Nm).

61. Install the crankshaft sprocket.

62. Install the oil pump chain and the oil pump sprocket. Tightening torque: 14–16 ft. lbs. (19–22 Nm).

63. Install the RH timing chain guide. Tightening torque: 14–18 ft. lbs. (20–25 Nm).

64. Install the RH timing chain.

　a. To install the timing chain with no slack between the camshaft and the crankshaft, use the following procedure.

　b. Place the chain on the crankshaft sprocket first, then the intake camshaft sprocket, then the exhaust camshaft sprocket.

　c. The timing mark of each of the sprockets must be matched with the timing mark (color link) of timing chain when installing the timing chain.

65. Install the RH timing chain tensioner arm. Tightening torque: 14–16 ft. lbs. (19–22 Nm).

66. Install the RH timing chain auto tensioner. Tightening torque: 87–104 inch lbs. (10–12 Nm).

67. Install the RH cam-to-cam guide. Tightening torque: 87–104 inch lbs. (10–12 Nm).

68. Install the oil pump chain guide. Tightening torque: 87–104 inch lbs. (10–12 Nm).

69. Install the oil pump chain tensioner assembly. Tightening torque: 87–104 inch lbs. (10–12 Nm).

70. Pull out the pins of hydraulic tensioner (LH & RH).

71. Install the oil pump chain cover. Tightening torque: 87–104 inch lbs. (10–12 Nm).

72. After rotating the crankshaft 2 revolutions in a clockwise direction (viewed from front), confirm that the timing marks are aligned.

✳ WARNING

Always turn the crankshaft clockwise.

73. Install the timing chain cover.

　a. The sealant locations on the chain cover and on the counter parts (cylinder head, cylinder block, and lower oil pan) must be free of engine oil and any debris.

　b. Before assembling the timing chain cover, the liquid sealant TB 1217H should be applied on the gap between cylinder head and cylinder block.

➡**The part must be assembled within 5 minutes after the sealant is applied. Use a bead width of 0.1 inch (2.5mm).**

　c. After applying liquid sealant TB1217H on the timing chain cover, the part must be assembled within 5 minutes.

➡ **The sealant should be applied in a continuous bead. Bead width: 0.1 inch (2.5mm).**

d. Install a new gasket to the timing chain cover.

e. The dowel pins on the cylinder block and holes on the timing chain cover should be used as a reference in order to aid assembly of the timing chain cover into position.

f. Tighten the timing cover as shown in the illustration.

74. Wait 30 minutes after the timing chain cover was assembled before starting the engine.

75. Install the water pump pulley. Tightening torque: 69–87 inch lbs. (8–10 Nm).

76. Install the drive belt auto tensioner. Tightening torque: Bolt (B): 60–63 ft. lbs. (81–85 Nm). Bolt (C): 13–16 ft. lbs. (18–22 Nm).

77. Install the drive belt idler. Tightening torque: 39–43 ft. lbs. (53–58 Nm).

78. Install the alternator. Tightening torque: 20–25 ft. lbs. (27–33 Nm).

79. Install the air conditioner compressor.

80. Install the power steering pump.

81. Lower the engine assembly using the jack.

82. Using SST (09231-3C100), install the timing chain cover oil seal.

83. Using SST (09231-3C300), install the crankshaft damper pulley. Tightening torque: 210–224 ft. lbs. (284–304 Nm).

84. Install the accessory drive belt. Refer to Accessory Drive Belts, removal & installation.

85. Install the cylinder head cover.

a. The hardening sealant located on the upper area between the timing chain cover and the cylinder head should be removed before assembling the cylinder head cover.

b. After applying sealant (TB1217H), it should be assembled within 5 minutes. Bead width: 0.1 inch (2.5mm).

c. Wait 30 minutes after the cylinder head cover was assembled before starting the engine.

d. Install the cylinder head cover bolts as shown in the illustration. Tightening torque: 87–104 inch lbs. (10–12 Nm).

➡ **Do not reuse the cylinder head cover gasket.**

86. Install the ignition coils.

87. Connect the RH ignition coil connector, the condenser connector, and install the wiring bracket.

88. Install the connector bracket from the LH cylinder head cover.

89. Install the surge tank. Tightening torque: 87–104 inch lbs. (10–12 Nm).

a. Install the connector bracket(A) on the surge tank. Tightening torque: 210–224 ft. lbs. (284–304 Nm).

b. Install the surge tank stay. Tightening torque: 20–23 ft. lbs. (27–31 Nm).

90. Connect the brake vacuum hose.

91. Connect the PCV hose.

92. Connect the water hoses to the ETC.

93. Install the ETC bracket. Tightening torque: 12–19 ft. lbs. (16–26 Nm).

94. Connect the PCSV connector, the MAP sensor connector, and the PCSV hose.

95. Connect the ETC connector and the knock sensor connector.

96. Connect the LH CMPS and oil pressure switch connector.

97. Install the wiring harness protector, and connect the LH ignition connector, injector connector, condenser connector, and ground.

98. Connect the OCV connector and knock sensor connector.

99. Connect the LH front oxygen sensor connector, alternator connector, and air compressor connector.

100. Connect the RH injector connector and the ignition coil connector.

101. Connect the power steering oil pressure sensor connector.

102. Connect the RH oxygen sensor connector and VIS solenoid valve connector.

103. Connect the RH oxygen sensor connector.

104. Install the No. 1 engine mounting through the lower position of A/C pipe line. Tightening torque: 36–47 ft. lbs. (49–64 Nm).

105. Install the A/C pipe bracket mounting bolt.

106. Install the engine coolant reservoir tank.

107. Install the engine mounting bracket. Tightening torque: 47–62 ft. lbs. (64–83 Nm).

108. Install the transaxle mounting bolts. Tightening torque: 36–47 ft. lbs. (49–64 Nm).

109. Remove the jack from the upper oil pan.

110. Install the lower oil pan. Refer to Oil Pan, removal & installation. Uniformly tighten the bolts in several passes. Tightening torque: 87–104 inch lbs. (10–12 Nm).

111. Install the upper radiator hose.

112. Install the side cover.

113. Install the undercover.

114. Install the front wheels.

115. Install the intake air hose and air cleaner assembly.

a. Install the intake air hose and air cleaner.

b. Connect the ECM connector.

c. Connect the breather hose to air cleaner hose.

d. Connect the AFS connector.

e. Install the air duct.

116. Install the engine cover.

117. Connect the negative terminal to the battery.

118. Refill the engine with the proper amount and type of engine oil.

119. Refill the radiator and reservoir tank with engine coolant.

120. Bleed the air from the cooling system.

121. Run the engine and check for leaks.

VALVE COVERS

REMOVAL & INSTALLATION

See Figures 202 through 204.

✳✳ WARNING

Use fender covers to avoid damaging painted surfaces. To avoid damage, unplug the wiring connectors carefully while holding the connector portion.

➡ **Mark all wiring and hoses to avoid misconnection.**

1. Before servicing the vehicle, refer to the Precautions Section.

2. Remove the air duct.

3. Disconnect the negative terminal from the battery.

4. Remove the engine cover.

5. Disconnect the AFS connector.

6. Remove the Positive Crankcase Ventilation (PCV) hose and breather hose.

7. Disconnect the breather hose from air cleaner hose.

8. Remove the intake air hose and air cleaner assembly.

9. Remove the ignition coils. Refer to Ignition Coil Pack, removal & installation.

10. Remove the valve cover (also called cylinder head cover).

11. Remove the valve cover gasket and thoroughly clean the cylinder head surface, where the gasket rests, and the valve cover.

To install:

➡ **Before installing the valve cover gasket, thoroughly clean the gasket and the groove.**

➡ **When installing, make sure the valve cover gasket is seated securely in the corners of the recesses with no gap.**

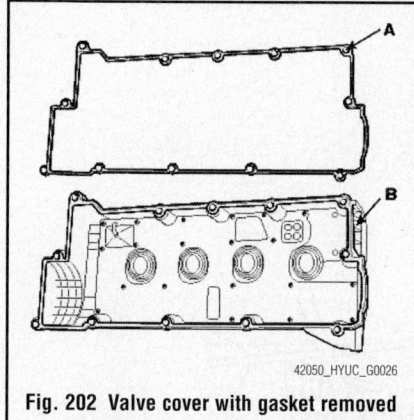

42050_HYUC_G0026

Fig. 202 Valve cover with gasket removed

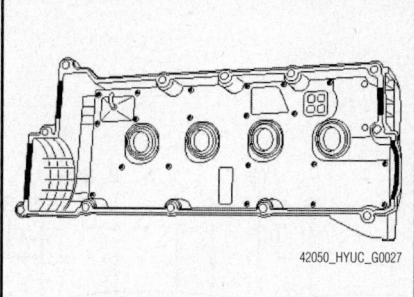

42050_HYUC_G0027

Fig. 203 Valve cover liquid gasket placement

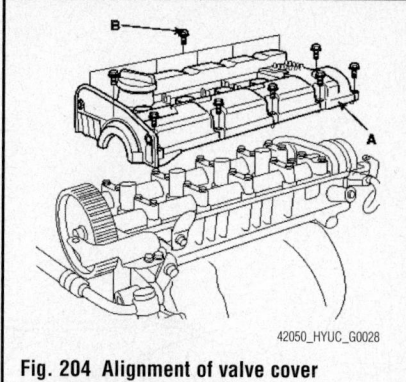

42050_HYUC_G0028

Fig. 204 Alignment of valve cover mounting

12. Install the valve cover gasket (A) in the groove of the valve cover (B).

13. Apply liquid gasket to the valve cover gasket at the corners of the recess.

➠**Use liquid gasket—Loctite® No. 5999.**

14. Check that the mating surfaces are clean and dry before applying liquid gasket.

15. Install the valve cover (A) with the 12 bolts (B).

16. Uniformly tighten the bolts in several passes. Tightening torque: 6–7 ft. lbs. (8–10 Nm).

17. Install the ignition coils. Refer to Ignition Coil Pack, removal & installation.

18. Connect the breather hose from air cleaner hose.

19. Install the PCV hose and breather hose.

20. Connect the AFS connector.

21. Install the intake air hose and air cleaner assembly.

22. Install the engine cover.

23. Connect the negative battery cable to the negative battery terminal.

24. Install the air duct.

VALVE LASH

ADJUSTMENT

All engines use hydraulic valve lash adjusters. Valve lash adjustments are not necessary or possible on these engines.

ENGINE PERFORMANCE & EMISSION CONTROLS

COMPONENT LOCATIONS

See Figures 205 through 211.

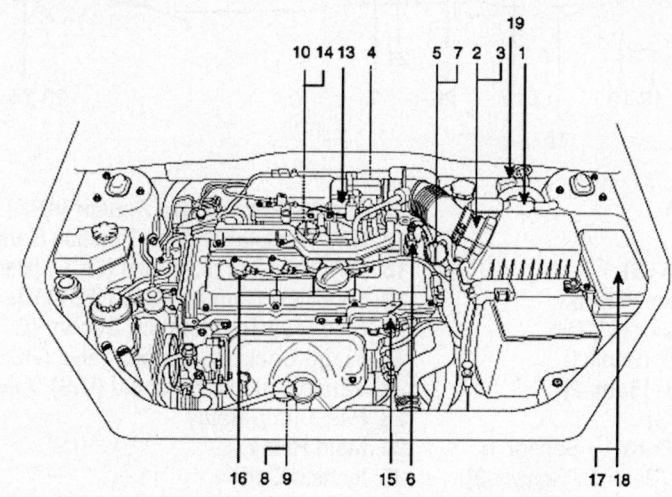

1. ECM (for M/T) / PCM (for A/T)
2. Mass Air Flow Sensor (MAFS)
3. Intake Air Temperature Sensor (IATS)
4. Throttle Position Sensor (TPS)
5. Engine Coolant Temperature Sensor (ECTS)
6. Camshaft Position Sensor (CMPS)
7. Crankshaft Position Sensor (CKPS)
8. Heated Oxygen Sensor (HO2S) [Bank 1/Sensor 1]
9. Heated Oxygen Sensor (HO2S) [Bank 1/Sensor 2]
10. Knock Sensor (KS)
11. Wheel Speed Sensor (WSS)
12. Injector
13. Idle Speed Control Actuator (ISCA)
14. Purge Control Solenoid Valve (PCSV)
15. CVVT Oil Control Valve (OCV)
16. Ignition Coil
17. Main Relay
18. Fuel Pump Relay
19. Multi Purpose Check Connector (20 pin)

22140_HYUN_G0211

Fig. 205 Underhood sensor locations—Accent with 1.6L engine

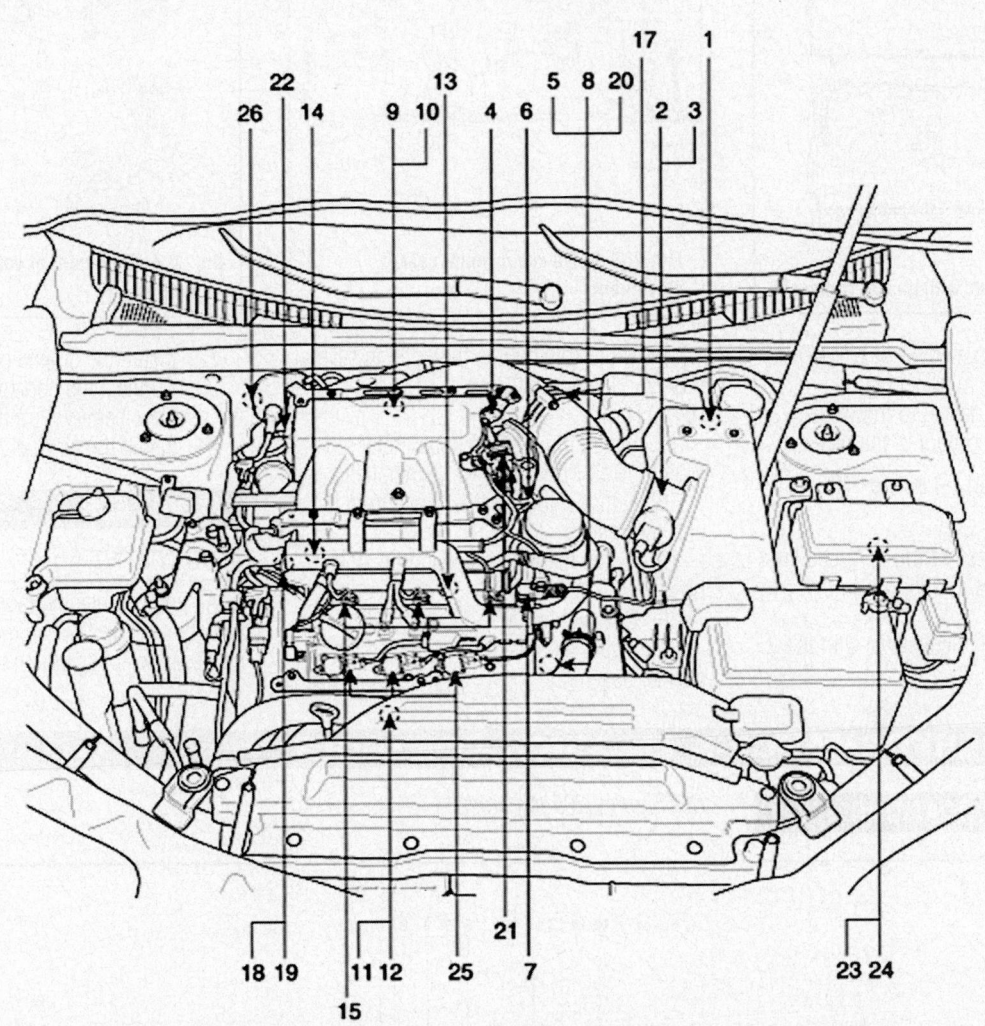

1. PCM (Powertrain Control Module)
2. Mass Air Flow Sensor (MAFS)
3. Intake Air Temperature Sensor (IATS)
4. Manifold Absolute Pressure Sensor (MAPS)
5. Engine Coolant Temperature Sensor (ECTS)
6. Camshaft Position Sensor (CMPS) [Bank 1]
7. Camshaft Position Sensor (CMPS) [Bank 2]
8. Crankshaft Position Sensor (CKPS)
9. Heated Oxygen Sensor (HO2S) [Bank 1 / Sensor 1]
10. Heated Oxygen Sensor (HO2S) [Bank 1 / Sensor 2]
11. Heated Oxygen Sensor (HO2S) [Bank 2 / Sensor 1]
12. Heated Oxygen Sensor (HO2S) [Bank 2 / Sensor 2]
13. Knock Sensor (KS) #1
14. Knock Sensor (KS) #2
15. Injector

16. Accelerator Position Sensor (APS)
17. ETC Module [Throttle Position Sensor (TPS) + ETC Motor]
18. CVVT Oil Control Valve (OCV) [Bank 1]
19. CVVT Oil Control Valve (OCV) [Bank 2]
20. CVVT Oil Temperature Sensor (OTS)
21. Purge Control Solenoid Valve (PCSV)
22. Variable Intake Solenoid (VIS) Valve
23. Fuel Pump Relay
24. Main Relay
25. Ignition Coil
26. Power Steering Pressure Sensor (PSPS)

22140_HYUN_G0212

Fig. 206 Underhood sensor locations—Azera with 3.3L and 3.8L engines

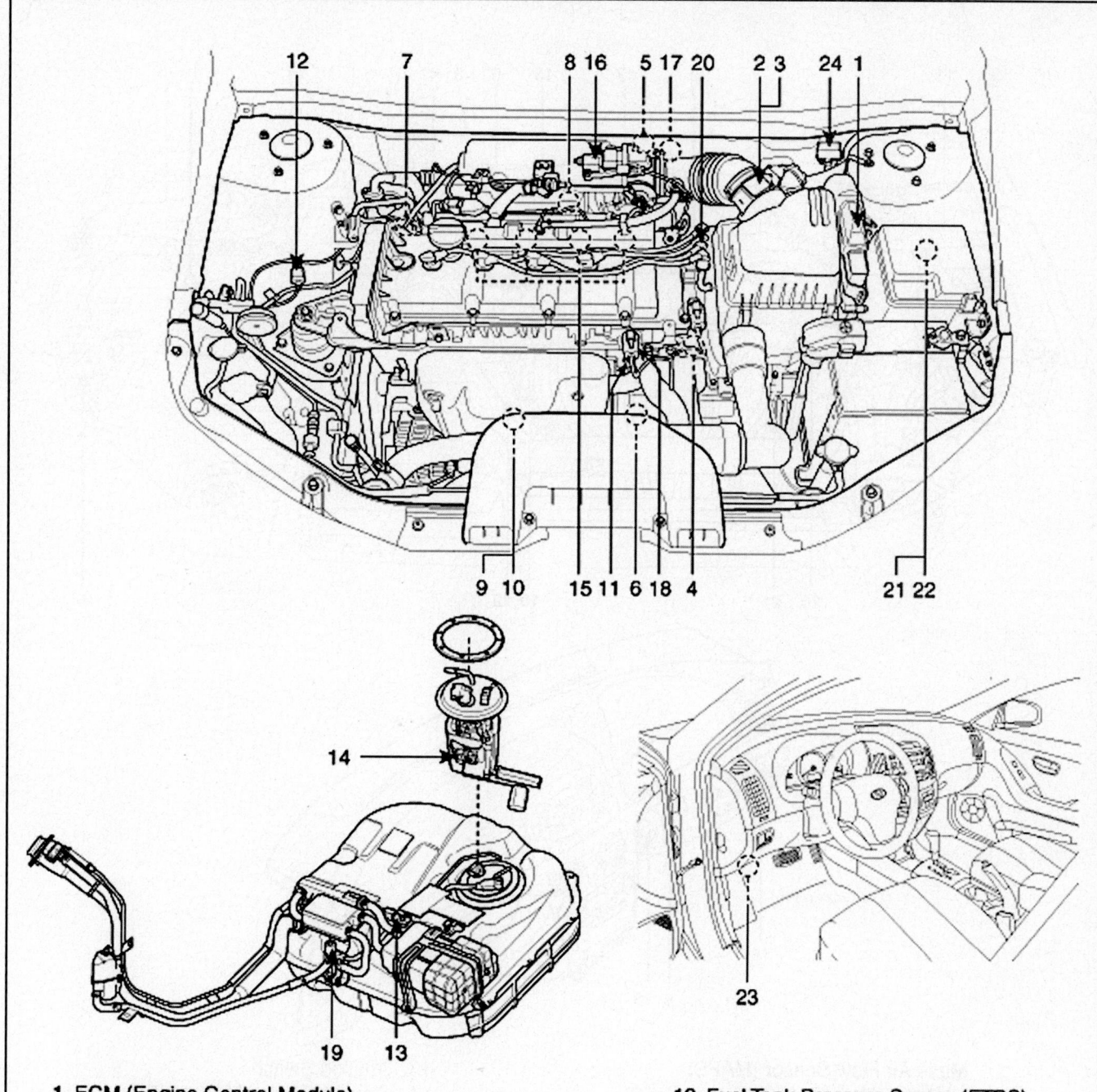

1. ECM (Engine Control Module)
2. Mass Air Flow Sensor (MAFS)
3. Intake Air Temperature Sensor (IATS)
4. Engine Coolant Temperature Sensor (ECTS)
5. Throttle Position Sensor (TPS)
6. Crankshaft Position Sensor (CKPS)
7. Camshaft Position Sensor (CMPS)
8. Knock Sensor (KS)
9. Heated Oxygen Sensor (HO2S) [Bank 1/Sensor 1]
10. Heated Oxygen Sensor (HO2S) [Bank 1/Sensor 2]
11. CVVT Oil Temperature Sensor (OTS)
12. A/C Pressure Transducer (APT)

13. Fuel Tank Pressure Sensor (FTPS)
14. Fuel Level Sensor (FLS)
15. Injector
16. Idle Speed Control Actuator (ISCA)
17. Purge Control Solenoid Valve (PCSV)
18. CVVT Oil Control Valve (OCV)
19. Canister Close Valve (CCV)
20. Ignition Coil
21. Main Relay
22. Fuel Pump Relay
23. Data Link Connector (DLC)
24. Multi-Purpose Connector

22140_HYUN_G0213

Fig. 207 Underhood and fuel tank sensor locations—Elantra with 2.0L engine

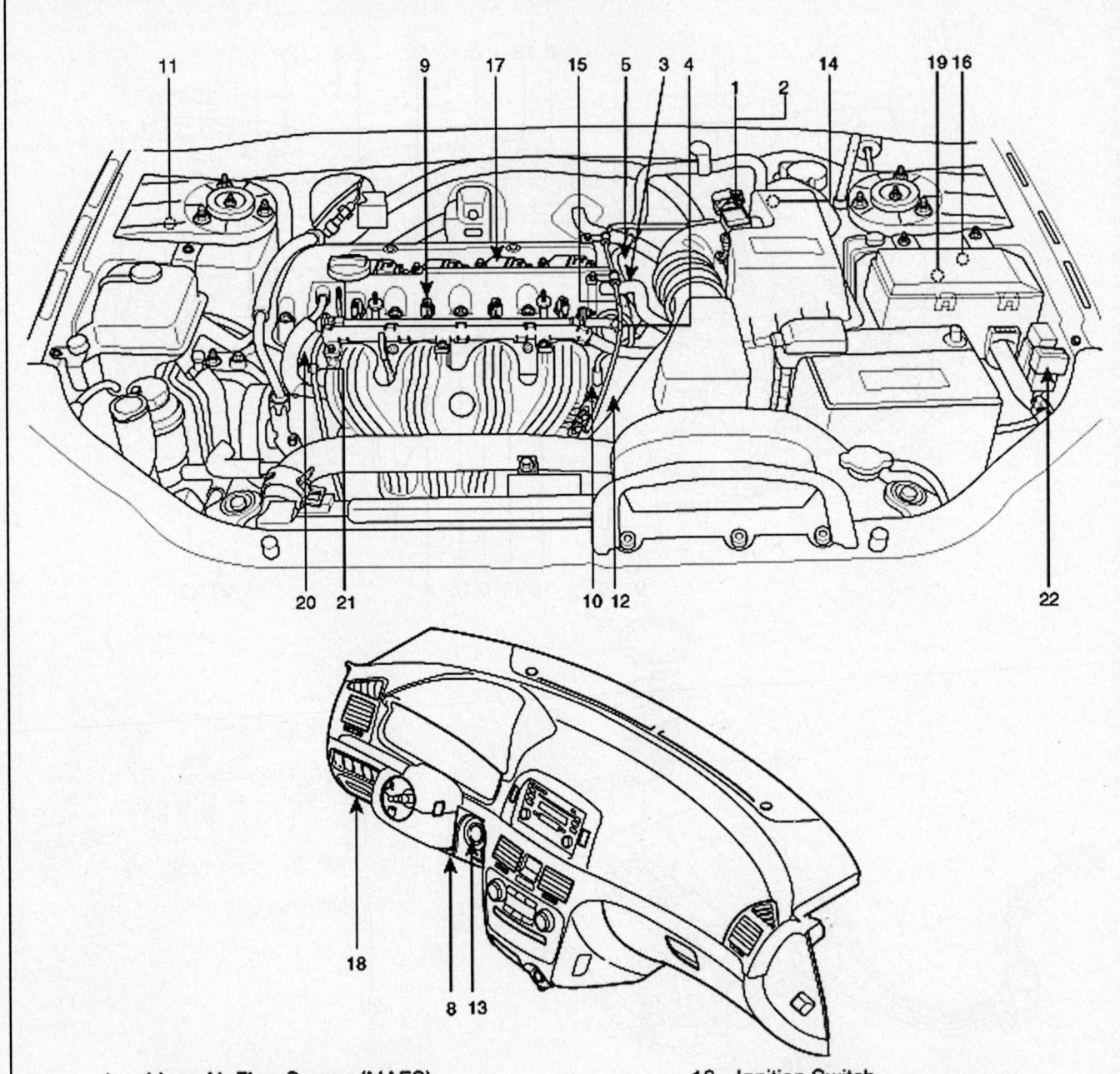

1. Mass Air Flow Sensor (MAFS)
2. Intake Air Temperature Sensor (IATS)
3. Engine Coolant Temperature Sensor (ECTS)
4. Camshaft Position Sensor (CMPS)
5. Crankshaft Position Sensor (CKPS)
6. Heated Oxygen Sensor (Front)
7. Heated Oxygen Sensor (Rear)
8. Accelerator Position Sensor (APS)
9. Injector
10. Electronic Throttle Body
11. Wheel Speed Sensor (WSS)
12. Knock Sensor
13. Ignition Switch
14. Engine Control Moduel (ECM)
15. Purge Control Solenoid Valve (PCSV)
16. Main Relay
17. Ignition Coil
18. Data Link Connector (DLC)
19. Fuel Pump Relay
20. CVVT Oil Control Valve (OCV)
21. CVVT Oil Temperature Sensor (OTS)
22. Multi-purpose Check Connector

22140_HYUN_G0214

Fig. 208 Underhood and instrument panel sensor locations—Sonata with 2.4L engine

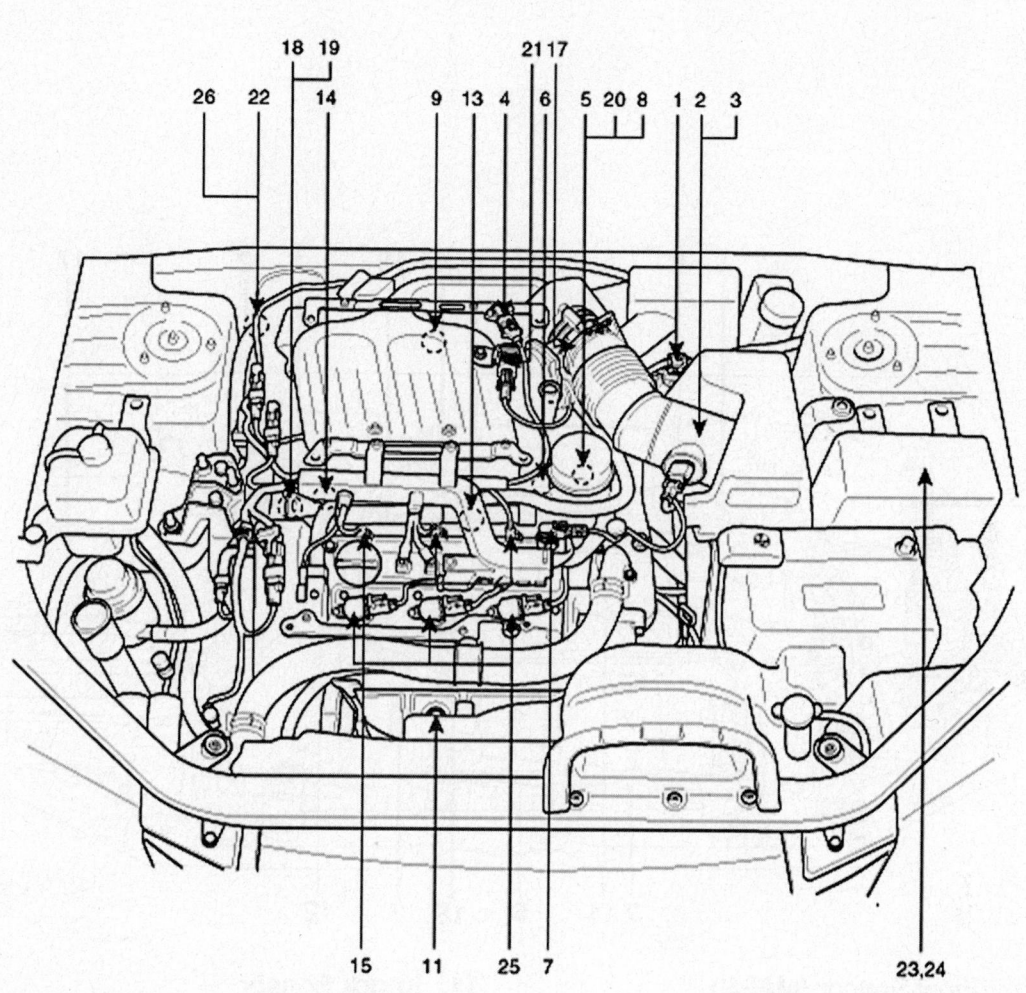

1. PCM (Powertrain Control Module)
2. Mass Air Flow Sensor (MAFS)
3. Intake Air Temperature Sensor (IATS)
4. Manifold Absolute Pressure Sensor (MAPS)
5. Engine Coolant Temperature Sensor (ECTS)
6. Camshaft Position Sensor (CMPS) [Bank 1]
7. Camshaft Position Sensor (CMPS) [Bank 2]
8. Crankshaft Position Sensor (CKPS)
9. Heated Oxygen Sensor (HO2S) [Bank 1 / Sensor 1]
10. Heated Oxygen Sensor (HO2S) [Bank 1 / Sensor 2]
11. Heated Oxygen Sensor (HO2S) [Bank 2 / Sensor 1]
12. Heated Oxygen Sensor (HO2S) [Bank 2 / Sensor 2]
13. Knock Sensor (KS) #1
14. Knock Sensor (KS) #2
15. Injector

16. Accelerator Position Sensor (APS)
17. ETC Module [Throttle Position Sensor (TPS) + ETC Motor]
18. CVVT Oil Control Valve (OCV) [Bank 1]
19. CVVT Oil Control Valve (OCV) [Bank 2]
20. CVVT Oil Temperature Sensor (OTS)
21. Purge Control Solenoid Valve (PCSV)
22. Variable Intake Solenoid (VIS) Valve
23. Fuel Pump Relay
24. Main Relay
25. Ignition Coil
26. Power Steering Pressure Sensor (PSPS)

22140_HYUN_G0215

Fig. 209 Underhood sensor locations—Sonata with 3.3L engine

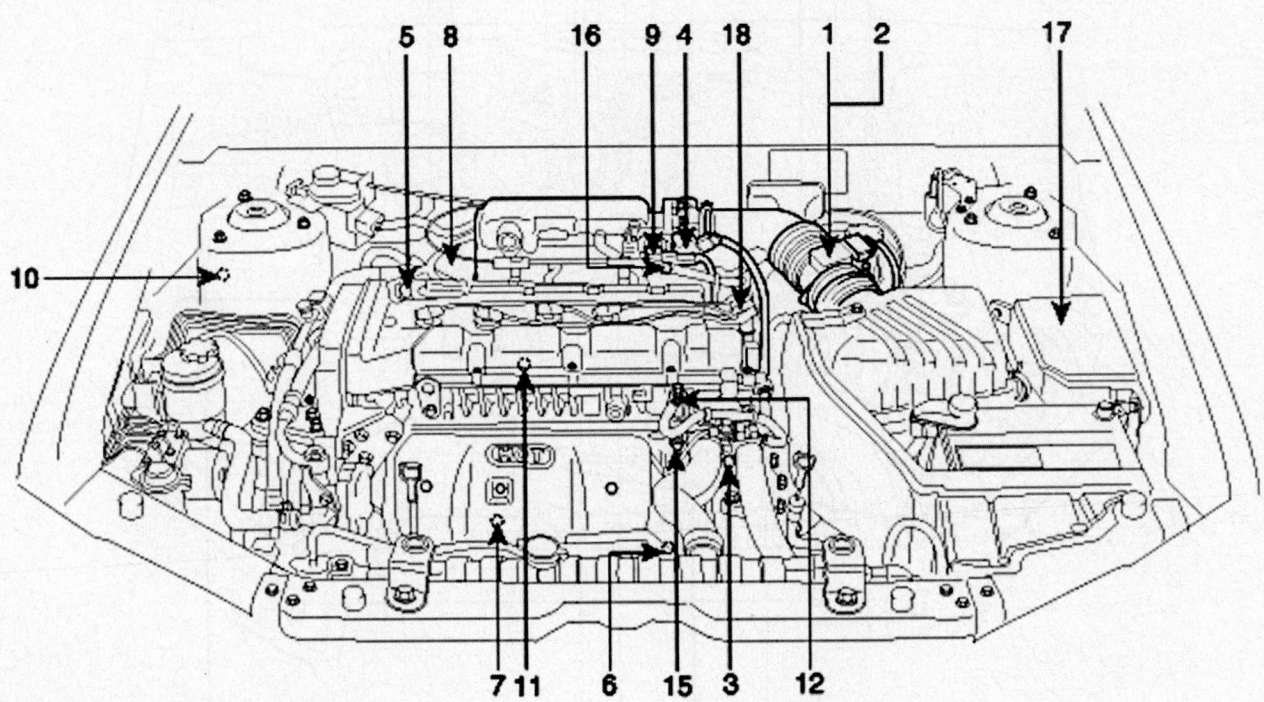

1. Mass Air Flow Sensor (MAFS)
2. Intake Air Temperature Sensor (IATS)
3. Engine Coolant Temperature Sensor (ECTS)
4. Throttle Position Sensor (TPS)
5. Camshaft Position Sensor (CMPS)
6. Crankshaft Position Sensor (CKPS)
7. Heated Oxygen Sensor (HO2S, Sensor 1)
8. Injector
9. Idle Speed Control Actuator (ISCA)
10. Vehicle Speed Sensor (VSS)

11. Knock Sensor
12. CVVT Oil Control Valve (OCV)
13. Ignition Switch
14. Heated Oxygen Sensor (HO2S, Sensor 2)
15. CVVT Oil Temperature Sensor (OTS)
16. Purge Control Solenoid Valve (PCSV)
17. Main Relay
18. Ignition Coil

22140_HYUN_G0216

Fig. 210 Underhood sensor locations—Tiburon with 2.0L engine

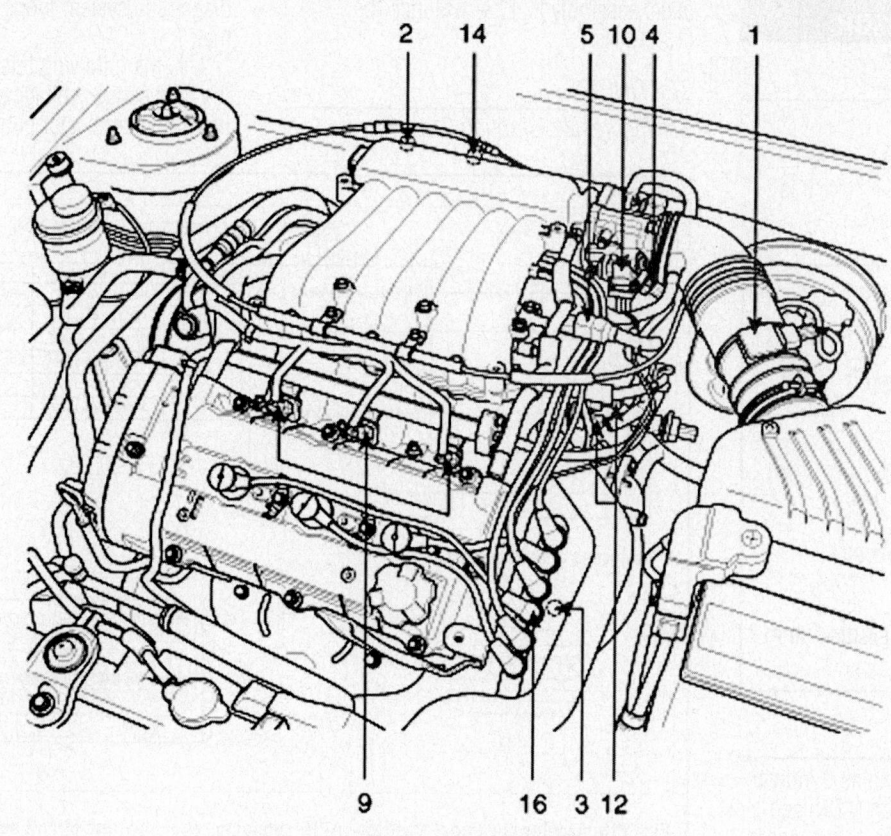

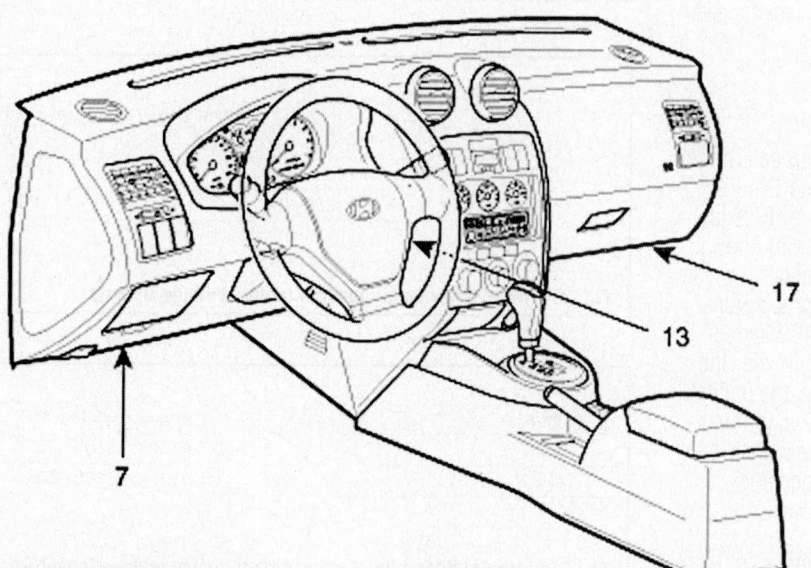

1. Mass Air Flow Sensor (MAFS)
2. Intake Air Temperature Sensor (IATS)
3. Engine Coolant Temperature Sensor (ECTS)
4. Throttle Position Sensor (TPS)
5. Camshaft Position Sensor (CMPS)
6. Crankshaft Position Sensor (CKPS)
7. Diagnostic Link Connector (DLC)
8. Fuel Pump Relay

9. Injector
10. Idle Speed Control Actuator (ISCA)
11. Vehicle Speed Sensor (VSS)
12. Knock Sensor
13. Ignition Switch
14. Purge Control Solenoid Valve (PCSV)
15. Main Relay

16. Ignition Coil
17. ECM

Fig. 211 Underhood and instrument panel sensor locations—Tiburon with 2.7L engine

ACCELERATOR PEDAL POSITION (APP) SENSOR

LOCATION

See Figure 212.

The Accelerator Pedal Position (APP) sensor is located inside the vehicle. It is part of the accelerator pedal assembly.

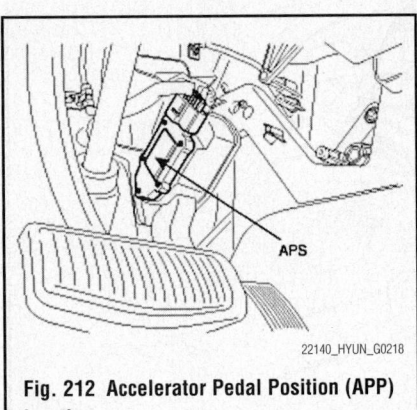

Fig. 212 Accelerator Pedal Position (APP) location

OPERATION

The accelerator pedal contains 2 individual Accelerator Pedal Position (APP) sensors within the assembly. The APP sensors 1 and 2 are potentiometer type sensors each with 3 circuits:

- A 5-volt reference circuit
- A low reference circuit
- A signal circuit

The APP sensors are used to determine the pedal angle. The Powertrain Control Module (PCM) provides each APP sensor with a 5-volt reference circuit and a low reference circuit. The APP sensors provide the PCM with signal voltage proportional to the pedal movement. The APP sensor 1 signal voltage at rest position is near the low reference and increases as the pedal is actuated. The APP sensor 2 signal voltage at rest position is near the 5-volt reference and decreases as the pedal is actuated.

REMOVAL & INSTALLATION

1. Before servicing the vehicle, refer to the Precautions Section.
2. Disconnect the ground cable from the battery.
3. Disconnect the Accelerator Pedal Position (APP) sensor connector.
4. Remove the nut securing accelerator pedal assembly.

To install:

5. Install in the reverse order of removal.

6. Tighten the nut securing accelerator pedal assembly to 113–139 inch lbs. (13–16 Nm).

TESTING

See Figures 213 through 215.

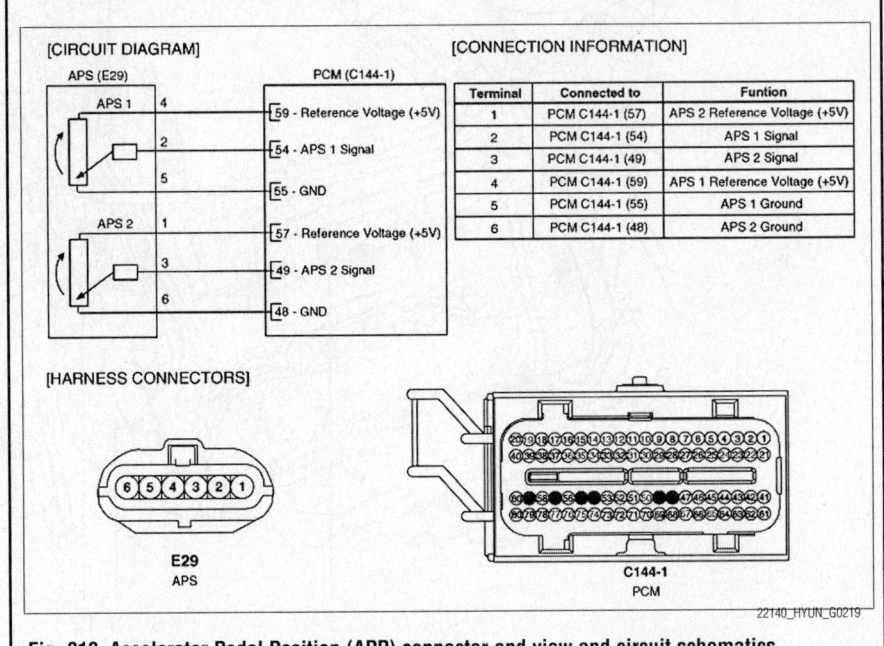

Fig. 213 Accelerator Pedal Position (APP) connector end view and circuit schematics

Pedal Position	Output Voltage (V) [Vref = 5.0V]	
	APS1	APS2
C.T	0.7 ~ 0.8V	0.29 ~ 0.46V
W.O.T	3.85 ~ 4.35V	1.93 ~ 2.18V

Fig. 214 Accelerator pedal position table—Voltage Output

Item	Sensor Resistance
APS1	0.7 ~ 1.3kΩ at 68°F (20°C)
APS2	1.4 ~ 2.6kΩ at 68°F (20°C)

Fig. 215 Accelerator pedal position circuit testing table—Sensor Resistance

The Accelerator Pedal Position (APP) sensor is installed on the accelerator pedal module and detects the rotation angle of the accelerator pedal. The APP is one of the most important sensors in engine control system, so it consists of the 2 sensors which adapt an individual sensor power and ground line. The second sensor monitors the first sensor and its output voltage is half of the first one. If the ratio of sensor 1 and 2 is out of the range (approximately ½), the diagnostic system judges that it is abnormal.

Use the following reference tables to test the APP sensor voltage and resistance values at the connector end points.

CAMSHAFT POSITION (CMP) SENSOR

LOCATION

See Figures 216 through 224.

Refer to the accompanying illustrations for Camshaft Position (CMP) sensor location.

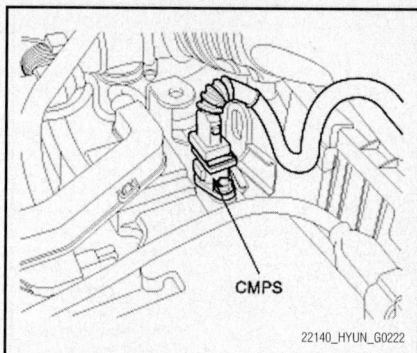

Fig. 216 Camshaft Position (CMP) sensor location—Accent 1.6L engine

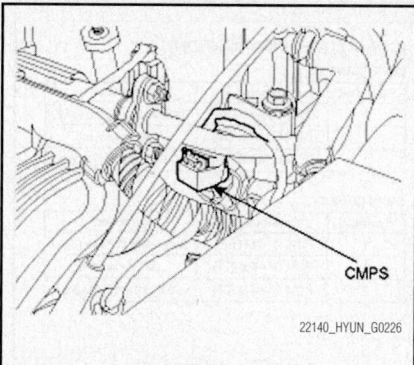

Fig. 220 Camshaft Position (CMP) sensor location—Sonata 2.4L engine

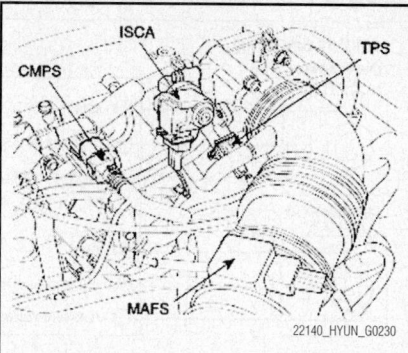

Fig. 224 Camshaft Position (CMP) sensor location—Tiburon 2.7L engine

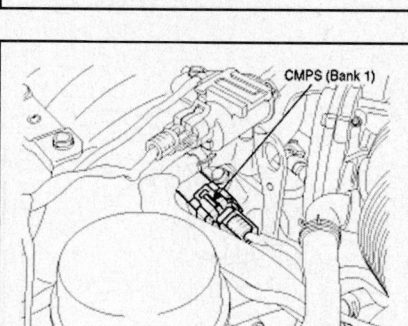

Fig. 217 Camshaft Position (CMP) sensor location (Bank 1)—Azera 3.3L and 3.8L engines

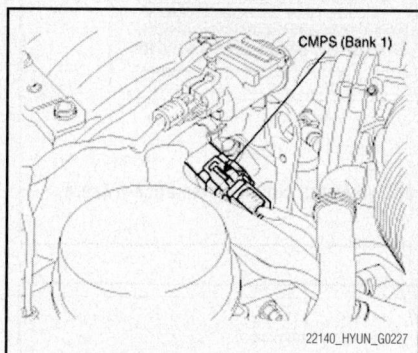

Fig. 221 Camshaft Position (CMP) sensor location (Bank 1)—Sonata 3.3L engine

OPERATION

The Camshaft Position (CMP) sensor is triggered by a notched reluctor wheel built onto the camshaft sprocket. The CMP sensor is connected to the PCM by the following circuits:

- A 5-volt circuit
- A low reference circuit
- A signal circuit

The CMP sensor is a Hall-effect sensor and detects the camshaft position by using a Hall element. It is related with the Crankshaft Position (CKP) sensor and detects the piston position of each cylinder which the CKP sensor can't detect. This sensor has a Hall-effect IC which changes in output voltage when a magnetic field is made on the IC with the current flow.

REMOVAL & INSTALLATION

1. Before servicing the vehicle, refer to the Precautions Section.
2. Disconnect the negative battery cable.
3. Disconnect the connector from the CMP sensor.
4. Remove the bolt that retains the CMP sensor.
5. Remove the CMP sensor.

To install:

6. Installation is the reverse of the removal procedure.
7. Tighten the bolt that retains the CMP sensor to: 61–86 inch lbs. (7–10 Nm).

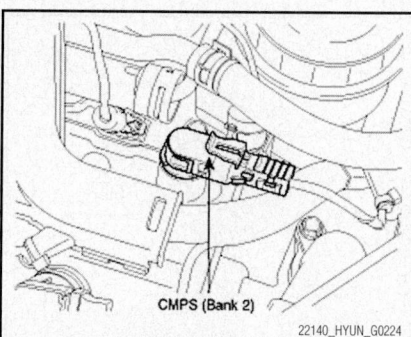

Fig. 218 Camshaft Position (CMP) sensor location (Bank 2)—Azera 3.3L and 3.8L engines

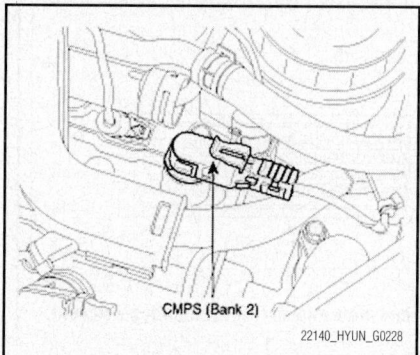

Fig. 222 Camshaft Position (CMP) sensor location (Bank 2)—Sonata 3.3L engine

TESTING

See Figures 225 and 226.

During normal operation the PCM controls all ignition functions. If either the Crankshaft Position (CKP) or Camshaft Position (CMP) sensor signal is lost, the engine will continue to run because the PCM will default to a limp home mode using the remaining sensor input. Diagnostic trouble codes are available to accurately diagnose the ignition system with an OBD2 scan tool.

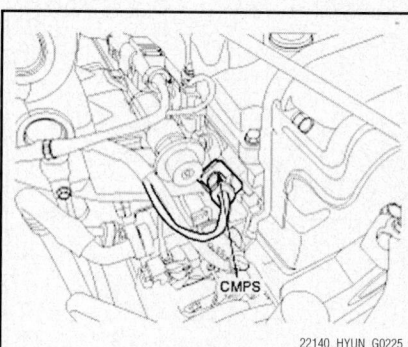

Fig. 219 Camshaft Position (CMP) sensor location—Elantra 2.0L engine

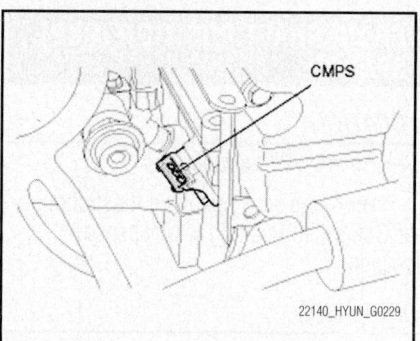

Fig. 223 Camshaft Position (CMP) sensor location—Tiburon 2.0L engine

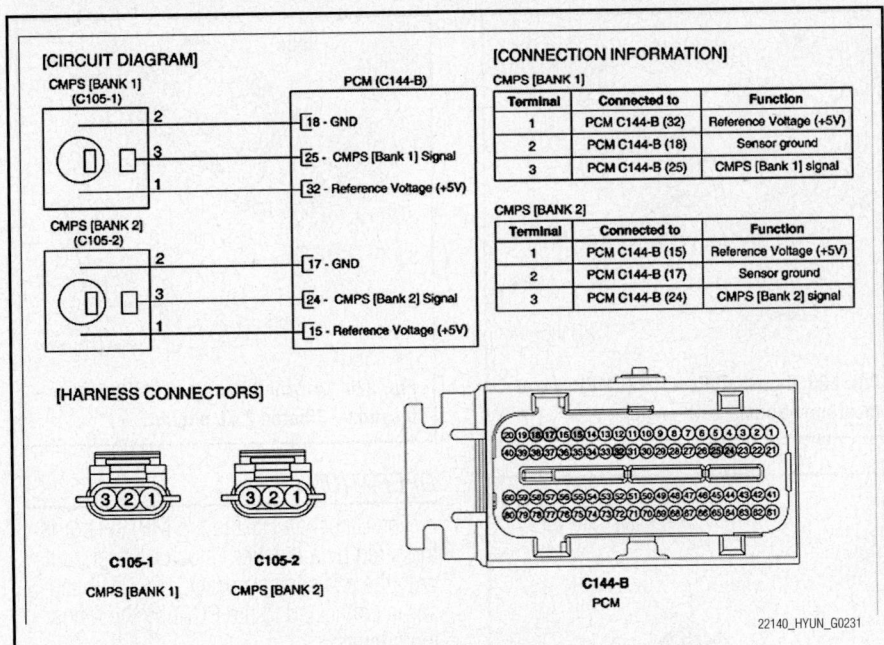

Fig. 225 Camshaft Position (CMP) sensor connector end view and circuit schematics—Azera, Sonata 3.3L engine

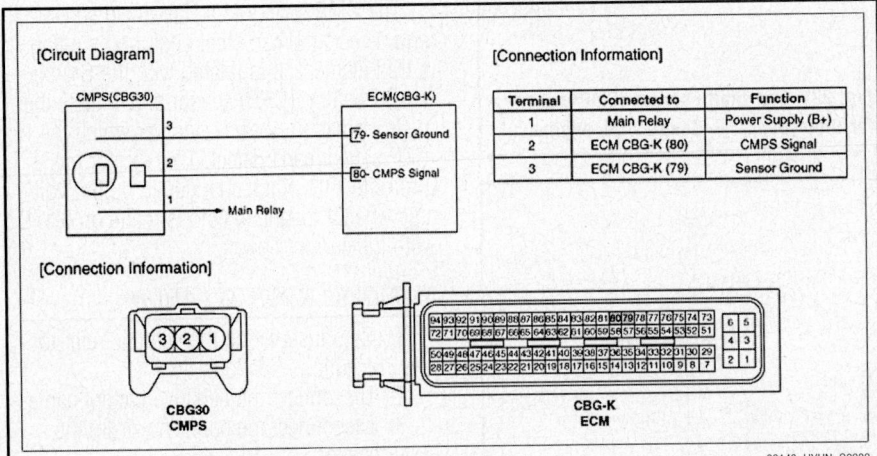

Fig. 226 Camshaft Position (CMP) sensor connector end view and circuit schematics—Accent, Elantra, Sonata 2.4L, and Tiburon

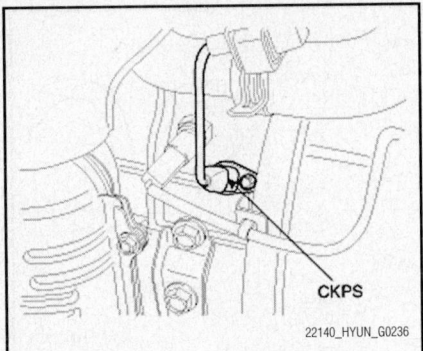

Fig. 227 Crankshaft Position (CKP) sensor location—Accent 1.6L engine

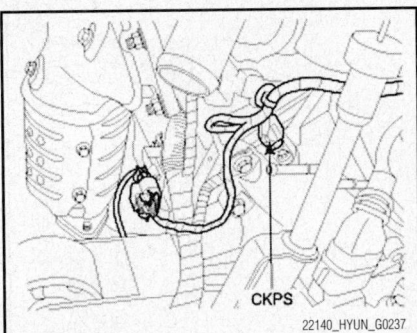

Fig. 228 Crankshaft Position (CKP) sensor location—Azera 3.3L, 3.8L, and Sonata 3.3L engines

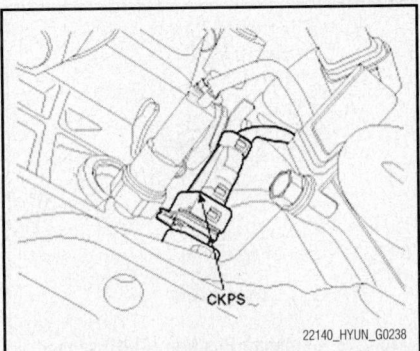

Fig. 229 Crankshaft Position (CKP) sensor location—Elantra 2.0L engine

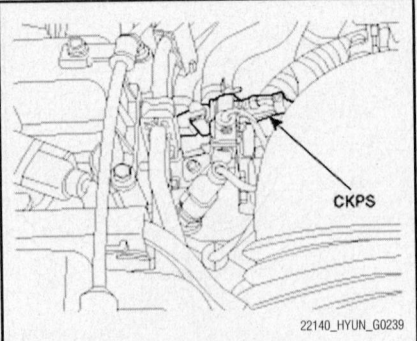

Fig. 230 Crankshaft Position (CKP) sensor location—Sonata 2.4L engine

1. Inspect the CMP sensor for correct installation. Remove the CMP sensor from the engine and inspect the sensor O-ring for damage. If the sensor is loose, incorrectly installed, or damaged, replace the CMP sensor.

2. Engage the CMP sensor harness connector to the CMP sensor.

3. Connect the scan tool to the diagnostic connector.

4. With the ignition **ON**, engine OFF observe the CMP active counter parameter on the scan tool.

5. Pass a flat steel object across the tip of the sensor repeatedly. The CMP active counter parameter should increment with each pass of the steel object.

6. If the parameter does not increment, replace the CMP sensor.

CRANKSHAFT POSITION (CKP) SENSOR

LOCATION

See Figures 227 through 232.

Refer to the accompanying illustrations for Crankshaft Position (CKP) sensor location.

OPERATION

The Crankshaft Position (CKP) sensor is a Hall Effect type sensor that generates voltage using a sensor and a target wheel

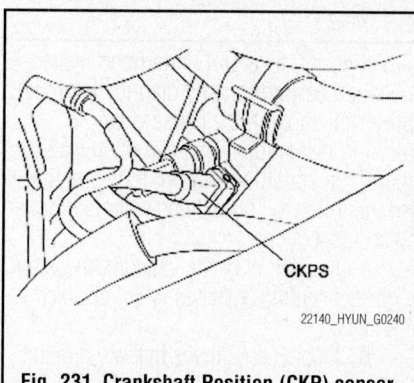

Fig. 231 Crankshaft Position (CKP) sensor location—Tiburon 2.0L engine

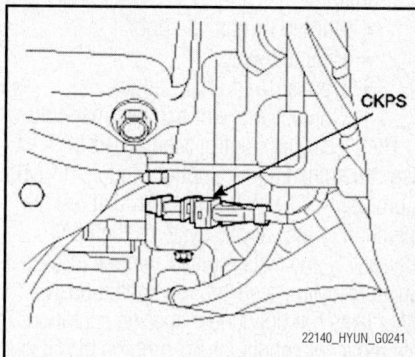

Fig. 232 Crankshaft Position (CKP) sensor location—Tiburon 2.7L engine

mounted on the crankshaft. There are 58 slots in the target wheel where one is longer than the others. When the slot in the wheel aligns with the sensor, the sensor voltage output is low. When the metal tooth in the wheel aligns the sensor, the sensor voltage is high. During one crankshaft rotation there are 58 rectangular signals and one longer signal. The PCM calculates engine RPM by using the sensor's signal and controls the injection duration and ignition timing. Using the signal differences caused by the longer slot, the PCM identifies which cylinder is at TDC.

REMOVAL & INSTALLATION

1. Before servicing the vehicle, refer to the Precautions Section.
2. Disconnect the negative battery cable.
3. Disconnect the connector from the sensor.
4. Remove the bolt that retains the sensor in place.
5. Remove the sensor from its mounting.

  To install:

6. Installation is the reverse of the removal procedure.
7. Tighten the sensor retaining bolt to: 70–104 inch lbs. (8–12 Nm).

TESTING

During normal operation the PCM controls all ignition functions. If either the Crankshaft Position (CKP) or Camshaft Position (CMP) sensor signal is lost, the engine will continue to run because the PCM will default to a limp home mode using the remaining sensor input. Diagnostic trouble codes are available to accurately diagnose the ignition system with an OBD2 scan tool.

1. Inspect the CKP sensor for correct installation. Remove the CKP sensor from the engine and inspect the sensor O-ring for damage. If the sensor is loose, incorrectly installed, or damaged, replace the CKP sensor.
2. Engage the CKP sensor harness connector to the CKP sensor.
3. Connect the scan tool to the diagnostic connector.
4. With the ignition **ON**, engine OFF observe the CKP active counter parameter on the scan tool.
5. Pass a flat steel object across the tip of the sensor repeatedly. The CKP active counter parameter should increment with each pass of the steel object.
6. If the parameter does not increment, replace the CKP sensor.

Component Inspection
See Figure 233.

With a scan tool, check the signal waveform of the Crankshaft Position (CKP) and the Camshaft Position (CMP) sensor. The waveform should appear as the following illustration.

ELECTRIC FAN SWITCH

The engine cooling fan is operated by a relay that is controlled through the Engine Control Module (ECM). The cooling fan relay will activate depending upon the needs of one or more systems including the cooling system, the air conditioning, and the heating systems.

ELECTRONIC CONTROL MODULE (ECM)

For the Accent (automatic transmission), Azera, and Sonata 3.3L vehicles, refer to Powertrain Control Module (PCM). Some manufacturers refer to the Electronic Control Module (ECM) as the Engine Control Module (ECM).

LOCATION
See Figures 234 through 237.

Refer to the accompanying illustrations for Electronic Control Module (ECM) location.

OPERATION

The powertrain has electronic controls to reduce exhaust emissions while maintaining excellent drivability and fuel economy. The Electronic Control Module (ECM) is the control center of this system. The ECM monitors numerous engine and vehicle functions. The ECM constantly looks at the information from various sensors and other inputs, and controls the systems that affect vehicle performance and emissions. The ECM also performs the diagnostic tests on various parts of the system. The ECM can recognize operational problems and alert the driver via the Malfunction Indicator Lamp (MIL). When the ECM detects a malfunction, the ECM stores a Diagnostic Trouble Code (DTC). The problem area isidentified by the particular DTC that is set. The control module supplies a buffered voltage to various sensors and switches. Review the components and wiring diagrams in order to determine which systems are controlled by the ECM. The following are some of the functions that the ECM controls:

- The ignition system
- The Knock Sensor (KS) system
- The Evaporative Emissions (EVAP) system

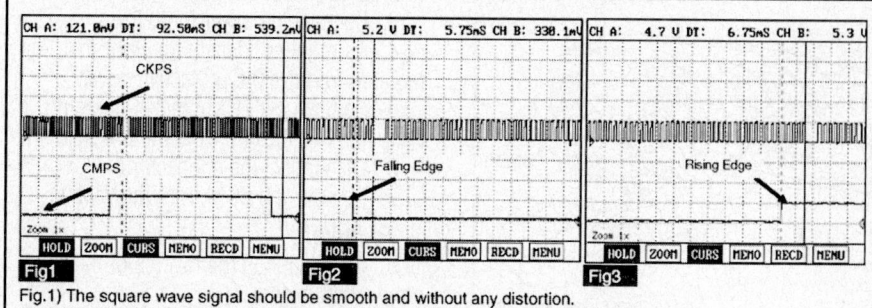

Fig.1) The square wave signal should be smooth and without any distortion.
Fig.2,3)The CMPS falling(rising) edge is coincided with 3~5 tooth of the CKP from one longer signal(missing tooth)

Fig. 233 Waveform graph of the CKP and the CMP sensors

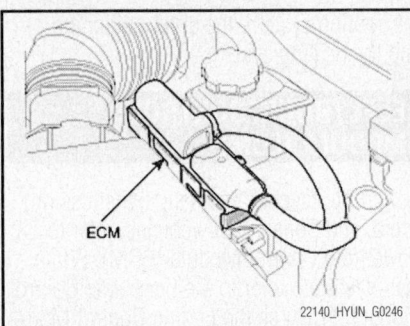

Fig. 234 Electronic Control Module (ECM) location—Accent 1.6L (Manual Transmission)

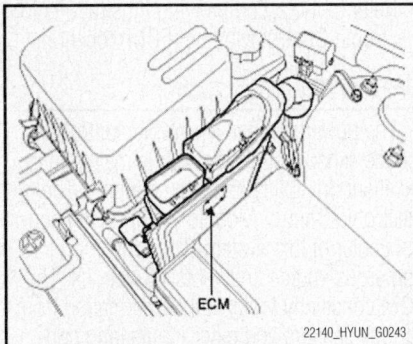

Fig. 235 Electronic Control Module (ECM) location—Elantra 2.0L engine

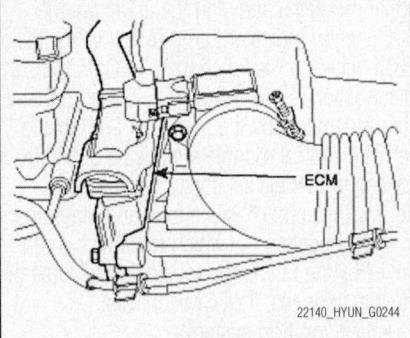

Fig. 236 Electronic Control Module (ECM) location—Sonata 2.4L engine

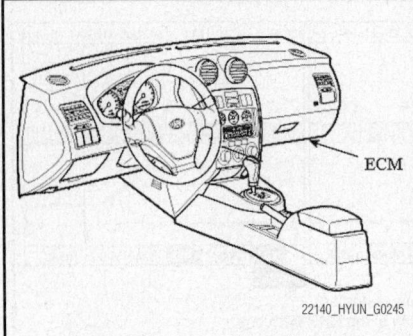

Fig. 237 Electronic Control Module (ECM) location—Tiburon 2.0L and 2.7L engines

- The alternator
- The A/C and fan systems
- The cooling fan control
- The fuel injection system
- The emission control systems
- The on-board diagnostics

REMOVAL & INSTALLATION

Except Tiburon

See Figure 238.

1. Before servicing the vehicle, refer to the Precautions Section.

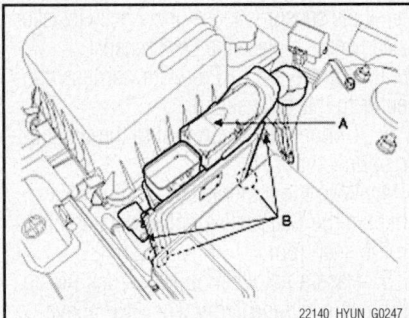

Fig. 238 Disconnect the ECM connector(s) (A) and remove the ECM mounting bolts (B)

2. Turn ignition switch off.
3. Disconnect the negative battery cable from the battery.
4. Disconnect the ECM connector(s) (A).
5. Remove the ECM mounting bolts (B) and remove the ECM from the air cleaner assembly.

To install:

6. Installation is the reverse of the removal.
7. Tighten the ECM mounting bolts to: 86–104 inch lbs. (10–12 Nm).

Tiburon

1. Before servicing the vehicle, refer to the Precautions Section.
2. Disconnect the negative battery cable.
3. Remove the lower inner trim.
4. As required, detach the floor mat. As required, remove the protective cover.
5. Remove the ECM bracket retaining nuts. Remove the clip from the bracket.
6. Disconnect the connectors.
7. Remove the ECM from the vehicle.

To install:

8. Installation is the reverse of the removal procedure.

➡ **When replacing the ECM, be careful to use the right part number, as damage to the injection system could occur.**

TESTING

1. Perform a careful underhood inspection when performing any diagnostic procedure or diagnosing the cause of an emission test failure. This can often lead to repairing a condition without further steps. Use the following guidelines when performing an inspection:

  a. Inspect all of the vacuum hoses for correct routing, pinches, cuts, or disconnects.

  b. Inspect any hoses that are difficult to see.

  c. Inspect all of the wires in the engine compartment for the following conditions:

- Burned or chafed spots
- Pinched wires
- Contact with sharp edges
- Contact with hot exhaust manifold

The Electronic Control Module (ECM), also called the Engine Control Module (ECM), is programmed with test routines that test the operation of the various systems the ECM controls. Some tests monitor internal ECM functions. Many tests are run continuously. Other tests run only under specific conditions, referred to as conditions for running the Diagnostic Trouble Code (DTC). When the vehicle is operating within the conditions for running a particular test, the ECM monitors certain parameters and determines if the values are within an expected range. The parameters and values considered outside the range of normal operation are listed as conditions for setting the DTC. When the conditions for setting the DTC occur, the ECM executes the action taken when the DTC sets. Some DTC's alert the driver via the Malfunction Indicator Lamp (MIL) or a message. Other DTC's do not trigger a driver warning, but are stored in memory. The ECM also saves data and input parameters when most DTC's are set.

The DTC's are categorized by type. The DTC type is determined by the MIL operation and the manner in which the fault data is stored when a particular DTC fails. In some cases, there may be exceptions to this structure. Therefore, when diagnosing the system it is important to read the action taken when the DTC sets and the conditions for clearing the DTC.

Many intermittent open or shorted circuits come and go with harness and connector movement caused by vibration, engine torque, bumps, or rough pavement.

2. Test the wiring harness and connectors by performing the following:

- Move the related ECM connectors and wiring while monitoring the appropriate scan tool data

- With the engine running, move the related connectors and wiring while monitoring engine operation
- If harness or connector movement affects the data displayed, the component and system operation, or the engine operation, inspect and repair the harness or connections as necessary

3. Test the electrical connections and/or wiring by performing the following:

- Inspect for incorrect mating of the connector halves or terminals not fully seated in the connector body
- Inspect for improperly formed or damaged terminals. Test for incorrect terminal tension
- Inspect for poor terminal to wire connections including terminals crimped over insulation. This requires removing the terminal from the connector body
- Inspect for corrosion or water intrusion. Pierced or damaged insulation can allow moisture to enter the wiring. The conductor can corrode inside the insulation with little visible evidence. Look for swollen and stiff sections of wire in the suspect circuits
- Inspect for wires that are broken inside the insulation

4. Test the ECM ground circuit:

a. Measure resistance between the ECM and a good chassis ground. Resistance specification: 1 ohm or less.

b. Use the backside of the ECM harness connector as the ECM side check point.

c. If a problem is found, make the necessary repairs.

5. If a problem is not located through the above procedures, the ECM may be faulty.

6. Hyundai recommends to replace the ECM with a new one and then check the vehicle again. If the vehicle operates normally, then the problem was likely with the ECM.

7. Retest the original ECM, if possible, by installing it into a known-good vehicle and check vehicle operations.

ENGINE COOLANT TEMPERATURE (ECT) SENSOR

LOCATION

See Figure 239.

The Engine Coolant Temperature (ECT) sensor is located in the engine coolant passage of the cylinder head for detecting the engine coolant temperature.

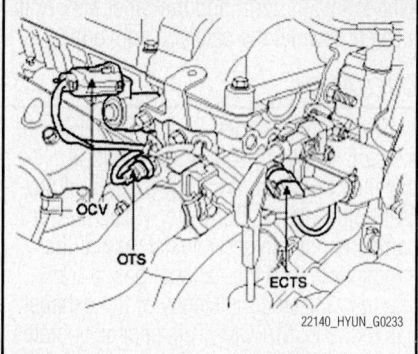

Fig. 239 Engine coolant temperature sensor location

22140_HYUN_G0233

OPERATION

The Engine Coolant Temperature (ECT) sensor uses a thermistor whose resistance changes with the temperature in order to detect the engine coolant temperature. The electrical resistance of the ECT decreases as the temperature increases, and increases as the temperature decreases. The 5-volt reference in the ECM/PCM is supplied to the ECT via a resistor in the ECM/PCM. The resistor in the ECM/PCM and the thermistor in the ECT are connected in series. When the resistance value of the thermistor in the ECT changes according to the engine coolant temperature, the output voltage also changes. During cold engine operation, the ECM/PCM increases the fuel injection duration and controls the ignition timing using the information of engine coolant temperature to avoid engine stalling and improve drivability.

REMOVAL & INSTALLATION

1. Before servicing the vehicle, refer to the Precautions Section.

2. Drain the coolant to a level below the bottom of the sensor.

3. Disconnect the ground cable from the battery and then remove the sensor connector.

4. Remove the coolant temperature sensor.

To install:

5. Reverse the removal procedure.

6. Tighten the sensor to 15–29 ft. lbs. (20–39 Nm).

TESTING

See Figures 240 and 241.

➥ **Clear Diagnostic Trouble Codes (DTC's) if applicable.**

1. Inspect the Engine Coolant Temperature (ECT) sensor electrical connection. Check the terminal for poor connections, loose wires, bent, broken, or corroded pins, and then verify that the connector is securely fastened.

2. Turn ignition switch OFF.

3. Disconnect ECT sensor connector.

4. Remove the ECT sensor.

5. After immersing the thermistor of the sensor into engine coolant, measure the resistance between the ECT terminals 1 and 3.

6. Check that the resistance is within the specification according to the following chart.

7. Replace the ECT sensor if the component is not within specification.

Temperature [°C(°F)]	Resistance (kΩ)
-40(-40)	48.14
-20(-4)	14.13 ~ 16.83
0(32)	5.79
20(68)	2.31 ~ 2.59
40(104)	1.15
60(140)	0.59
80(176)	0.32

22140_HYUN_G0234

Fig. 240 Resistance chart specifications

[Circuit Diagram]

ECTS(CBG11)   ECM(CBG-K)

1 — 15- ECTS Signal
2 → Indicators & Gauges
3 — 14- Sensor Ground

[Connection Information]

Terminal	Connected to	Function
1	ECM CBG-K (15)	ECTS Signal
2	Indicators & Gauges	-
3	ECM CBG-K (14)	Sensor Ground

[Harness Connector]

CBG11
ECTS

CBG-K
ECM

22140_HYUN_G0235

Fig. 241 Engine Coolant Temperature (ECT) sensor connector end view and circuit schematics

FUEL LEVEL SENDING UNIT

LOCATION

See Figure 242.

The fuel level sending unit is located on the fuel pump assembly module.

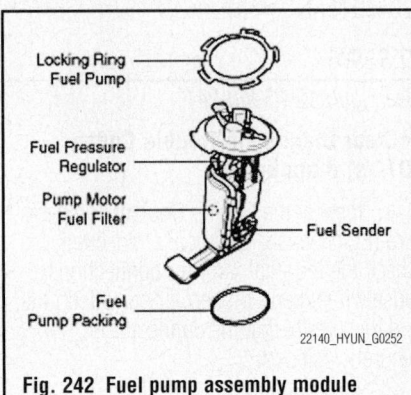

Fig. 242 Fuel pump assembly module

OPERATION

The fuel level sending unit is attached to the side of the fuel pump module. The fuel level sensor utilizes a variable resistor connected to a float that is buoyed up or down by the level of the fuel in the fuel tank.

REMOVAL & INSTALLATION

The fuel level sending unit is integral with the fuel pump module assembly. Refer to Fuel Pump, removal & installation in the Fuel System Section.

TESTING

1. Before servicing the vehicle, refer to the Precautions Section.
2. Check the fuel gauge in the instrument cluster panel for an accurate reading of the fuel level in the tank.
3. If the fuel gauge sticks at one extreme of the meter, no matter the fuel level in the tank, check for a break or short in the wiring from the fuel level sending unit.
4. Check the fuel level sending unit by removing the fuel pump assembly. Refer to Fuel Pump, removal and installation.
5. Using an ohmmeter, measure the resistance between the terminals while moving the float up and down. If the resistance does not change smoothly, replace the fuel level sending unit.

HEATED OXYGEN SENSOR (HO2S)

LOCATION

The Heated Oxygen Sensors (HO2S) are located in the exhaust system. On some vehicles, one sensor is located up at the exhaust manifold(s) and the other sensor is located down at the catalytic converter.

OPERATION

The Heated Oxygen Sensor (HO2S) consists of zirconium and alumina and is installed upstream and downstream of the Manifold Catalyst Converter (MCC). After it compares the oxygen consistency of the atmosphere with the exhaust gas, it transfers the oxygen consistency of the exhaust gas to the ECM/PCM. This sensor operates most efficiently when the temperature of the sensor tip is higher than 698°F (370°C). Therefore, a heater has been integrated with the oxygen sensor that is controlled by the ECM/PCM duty signal. When the exhaust gas temperature is lower than the specified value, the heater warms the sensor tip. The ECM/PCM commands the heater ON or OFF to maintain a specific HO2S operating temperature range. The ECM/PCM monitors the voltage on the HO2S heater low control circuit for heater fault diagnosis. If the ECM/PCM detects that the HO2S heater low control circuit voltage is not within a specified range, a DTC is set.

The HO2S supplies the electronic control assembly with a signal indicating either a rich or lean mixture condition as the engine operates.

REMOVAL & INSTALLATION

1. Before servicing the vehicle, refer to the Precautions Section.
2. Disconnect the electrical connector from the sensor.
3. Remove the oxygen sensor.

To install:

4. Installation is the reverse of the removal procedure.

➡**Apply anti-seize compound to the threaded portion of the sensor, prior to installation. Never apply anti-seize compound to the protector of the sensor.**

TESTING

See Figures 243 through 246.

1. Perform a visual inspection of the sensor as follows:
   a. If the sensor tip has a black/sooty deposit, this may indicate a rich fuel mixture.
   b. If the sensor tip has a white, gritty deposit, this may indicate an internal coolant leak.
   c. If the sensor tip has a brown deposit, this could indicate oil consumption.
2. Check that the heater resistance is according to specification.

Item	Specification
Heater Resistance (Ω)	8.1 ~ 11.1Ω at 69.8°F (21°C)

22140_HYUN_G0255

Fig. 243 Heated oxygen sensor heater resistance specifications

A/F Ratio	Output Voltage (V)
RICH	0.75 ~ 1.00V
LEAN	0 ~ 0.12V

22140_HYUN_G0256

Fig. 244 Heated oxygen sensor rich/lean output voltage specifications

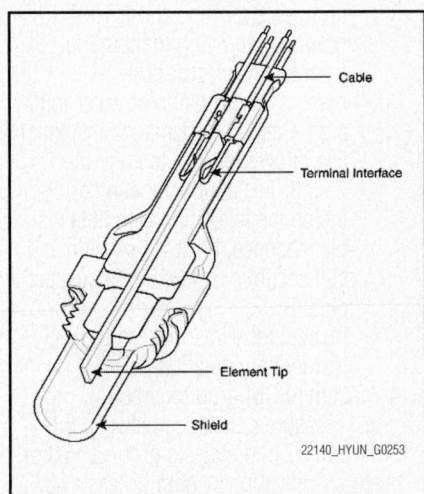

Fig. 245 Heated oxygen sensor cut-away view

3. Check that the output voltage in the rich/lean readings fluctuate according to specifications.
4. Connect an OBD2 scan tool to the Data Link Connector (DLC).
5. With the engine running, the normal waveform for the Heated Oxygen Sensor (HO2S) and the HO2S heater should be similar to those represented in the following waveform illustration.

INTAKE AIR TEMPERATURE (IAT) SENSOR

LOCATION

The Intake Air Temperature (IAT) sensor is mounted in the intake air hose of the air cleaner assembly. The IAT is integrated inside the Mass Air Flow (MAF) sensor.

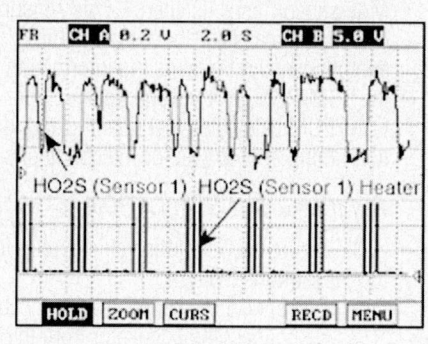 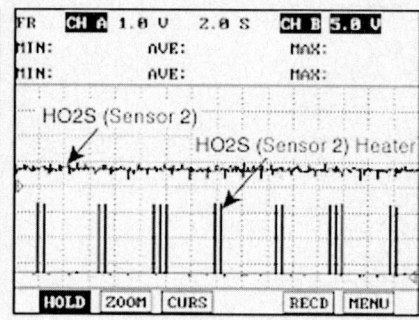 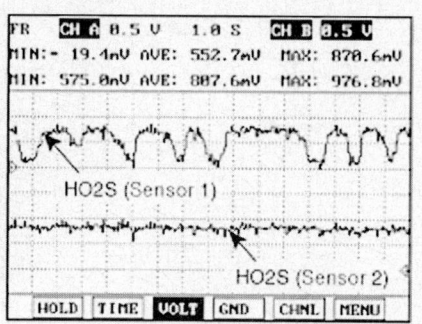

Fig. 246 Waveform of heated oxygen sensors and heater element

OPERATION

See Figure 247.

The Intake Air Temperature (IAT) sensor is a variable resistor that measures the temperature of the air entering the engine intake manifold. The ECM/PCM supplies 5 volts to the IAT signal circuit and a ground for the IAT low reference circuit. When the sensor is cold, the resistance is greater. This results in a greater voltage on the signal circuit that is interpreted by the ECM/PCM as a colder IAT. As the sensor becomes warmer, the resistance decreases. This results in a lesser voltage on the IAT signal circuit that is interpreted by the ECM/PCM as a warmer IAT. If the ECM/PCM detects an IAT sensor signal voltage that is not within a calibrated range of the IAT sensor 1 signal voltage, a DTC is set.

REMOVAL & INSTALLATION

1. Before servicing the vehicle, refer to the Precautions Section.
2. Disconnect the negative battery cable.
3. Disconnect the connector from the sensor.
4. Remove the sensor retaining screws.
5. Remove the sensor from its mounting.

To install:

6. Installation is the reverse of the removal procedure.

TESTING

See Figure 248.

1. Disconnect the Intake Air Temperature (IAT) sensor connector.
2. Measure the resistance between the signal terminal and ground terminal of the IAT connector and compare the resistance to the following Temperature and Resistance Table.
3. If the IAT sensor fails to meet the resistance specifications, replace the IAT sensor.

Temperature		Resistance (kΩ)
°C	°F	
-40	-40	100.87
-20	-4	28.58
0	32	9.40
10	50	5.66
20	68	3.51
40	104	1.47
60	140	0.67
80	176	0.33

Fig. 248 Temperature and Resistance Table

LOCATION

The Knock Sensor (KS) is located in the side of the cylinder block.

OPERATION

See Figure 249.

The Knock Sensor (KS) is used to detect engine vibrations caused by pre-ignition or detonation and provides information to the ECM/PCM. The ECM/PCM will then retard the timing in an attempt to eliminate detonation.

REMOVAL & INSTALLATION

1. Before servicing the vehicle, refer to the Precautions Section.
2. Disconnect the negative battery cable.
3. Disconnect the sensor connector.
4. Remove the sensor from its mounting.

To install:

5. Installation is the reverse of the removal procedure.
6. Tighten the sensor to 12–17 ft. lbs. (16–24 Nm).

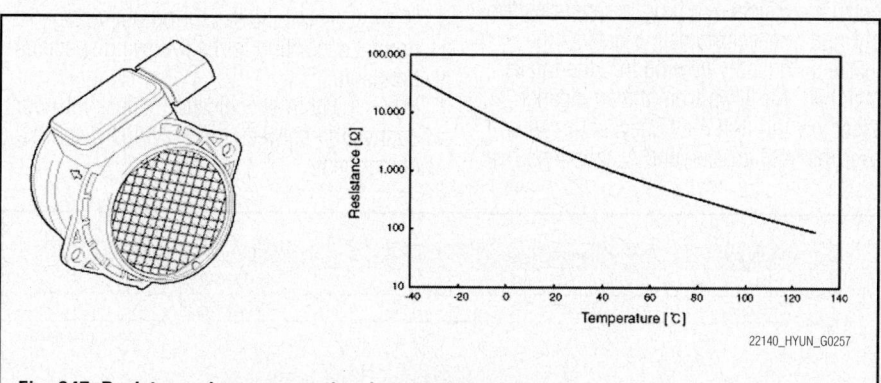

Fig. 247 Resistance decreases as the air temperature increases

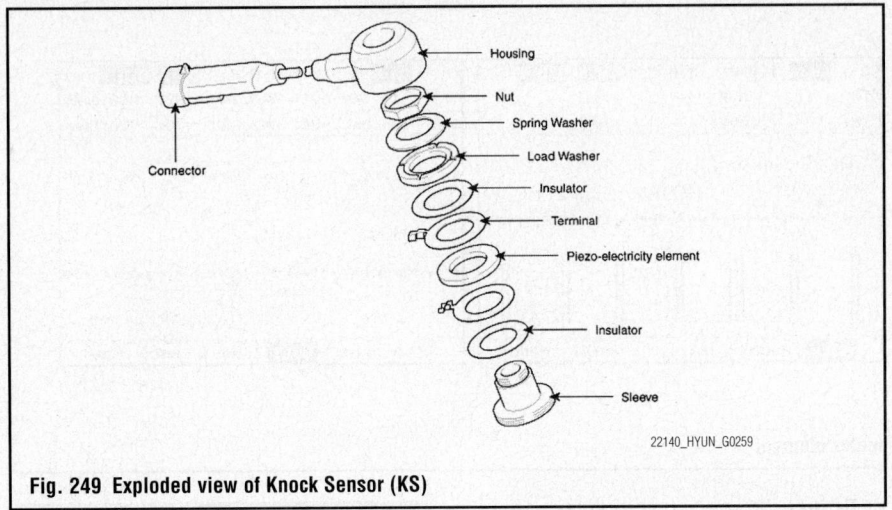

Fig. 249 Exploded view of Knock Sensor (KS)

TESTING

1. Disconnect the Knock Sensor (KS) electrical connector.
2. Connect an ohmmeter.
3. Measure the resistance between terminals 1 and 2. Specification should be 5mohm at 68°F (20°C).
4. If not within specification, replace the KS.

MALFUNCTION INDICATOR LIGHT (MIL)

RESET PROCEDURES

1. Proper operation of the Malfunction Indicator Lamp (MIL):
   - The MIL will illuminate with the ignition switch ON and the engine OFF
   - The MIL will turn OFF when the engine is started
   - The MIL will remain ON if the self-diagnostic system has detected a malfunction
   - The MIL may turn OFF if the malfunction is no longer present
   - If the MIL is illuminated and then the engine stalls, the MIL will remain illuminated as long as the ignition switch is ON
   - If the MIL is not illuminated and the engine stalls, the MIL will not illuminate until the ignition switch is cycled OFF, then ON
2. Resetting the MIL:
   - The control module turns OFF the MIL after 3 consecutive ignition cycles that the diagnostic system runs and does not fail
   - A current Diagnostic Trouble Code (DTC) clears when the diagnostic cycle runs and passes
   - There may still be a history of DTC's stored in the system. These will clear after 40 consecutive warm-up cycles, if no failures are reported by any other related diagnostic system
   - Manual resetting of the MIL and any DTC stored in the system, requires the use of an OBD2 scan tool connected to the Data Link Connector (DLC) for communication with the vehicle. Follow the instructions of the scan tool for both retrieval and resetting of DTC's.

➡️**If the error symptoms causing the MIL to illuminate have been corrected, the MIL will return to normal operation.**

MASS AIR FLOW (MAF) SENSOR

LOCATION

The Mass Air Flow (MAF) sensor is mounted in the intake air hose of the air cleaner assembly. This sensor is combined with the Intake Air Temperature (IAT) sensor.

OPERATION

The Mass Air Flow Sensor (MAF) sensor is a hot-film type sensor and is located in between the air cleaner and the throttle body. It consists of a tube, a sensor assembly, and honeycomb cell. It detects the intake air quantity flowing into the intake manifold. Air flows from the air cleaner assembly through the honeycomb cell and over the hot film element. At this time, heat transfer is generated by convection and the MAF sensor loses its energy. This sensor detects the mass air flow by using the energy loss and transfers the information to the ECM/PCM by a voltage transfer. The ECM/PCM calculates fuel quantity and ignition timing appropriate for the conditions.

REMOVAL & INSTALLATION

1. Before servicing the vehicle, refer to the Precautions Section.
2. Disconnect the negative battery cable.
3. Disconnect the connector from the sensor.
4. Remove the air cleaner and air intake assembly, as required.
5. Remove the sensor from its mounting.

To install:

6. Installation is the reverse of the removal procedure.

TESTING

See Figure 250.

1. Verify the integrity of the air induction system by inspecting for the following conditions:
   - Damaged components
   - Loose or improper installation
   - An air flow restriction
   - Any vacuum leak
   - Water intrusion
2. Verify that any electrical aftermarket devices are properly connected and grounded.
3. With the engine running, observe the scan tool MAF sensor parameter. The reading should be between 1,700–3,200 Hz depending on the Engine Coolant Temperature (ECT).
4. A Wide Open Throttle (WOT) acceleration from a stop should cause the MAF sensor parameter on the scan tool to increase rapidly. This increase should be from 2–6 g/s at idle to greater than 100 g/s at the time of the 1–2 shift.
5. Run the engine and connect the Hi-Scan (Pro)® to the Data Link Connector (DLC).
6. Check if the sensor output voltage is normal according to the following specification table.
7. If within specification, check for poor connection between the ECM/PCM and the component.

Condition	Output Voltage (V)	Intake Air Quantity (kg/h)
Idle	0.6 ~ 1.0	11.66 ~ 19.85
3000 rpm	1.7 ~ 2.0	43.84 ~ 58.79

22140_HYUN_G0260

Fig. 250 Output voltage and intake air quantity table

8. If not within specification, substitute the sensor with a known good component, check for proper operation. If the problem is corrected, replace the sensor.

MANIFOLD ABSOLUTE PRESSURE (MAP) SENSOR

LOCATION

The Manifold Absolute Pressure (MAP) sensor is located on the intake manifold.

OPERATION

The Manifold Absolute Pressure (MAP) sensor is a pressure sensitive variable resistor. It measures the changes in the intake manifold pressure which result from engine load and speed changes, and converts these to a voltage output. This sensor is used to measure the barometric pressure at start up, and under certain conditions, allows the ECM/PCM to automatically adjust for different altitudes. The ECM/PCM supplies 5 volts to the sensor and monitors the voltage on a signal line. The sensor provides a path to ground through its variable resistor. The sensor input affects fuel delivery and ignition timing controls in the ECM/PCM.

REMOVAL & INSTALLATION

1. Before servicing the vehicle, refer to the Precautions Section.
2. Disconnect the negative battery cable.
3. Disconnect the connector from the sensor.
4. Remove the sensor retaining screws.
5. Remove the sensor from its mounting.

To install:

6. Installation is the reverse of the removal procedure.
7. Tighten the MAP sensor installation bolt to: 78–104 inch lbs. (9–12 Nm).

TESTING

See Figures 251 through 253.

1. Measure the voltage between terminal 4 (sensor ground) and terminal 1 (sensor output).
2. Specification should be 4–5 volts, with the ignition switch ON.
3. Specification should be 0.8–2.4 volts at idle.
4. If not within specification, replace the sensor.

OIL PRESSURE SENSOR

LOCATION

The oil pressure sensor is located near the base of the oil filter assembly.

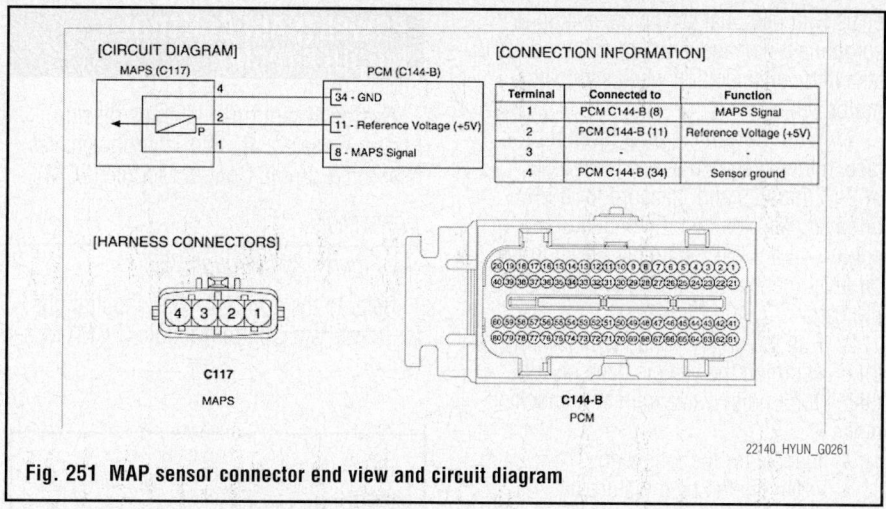
Fig. 251 MAP sensor connector end view and circuit diagram

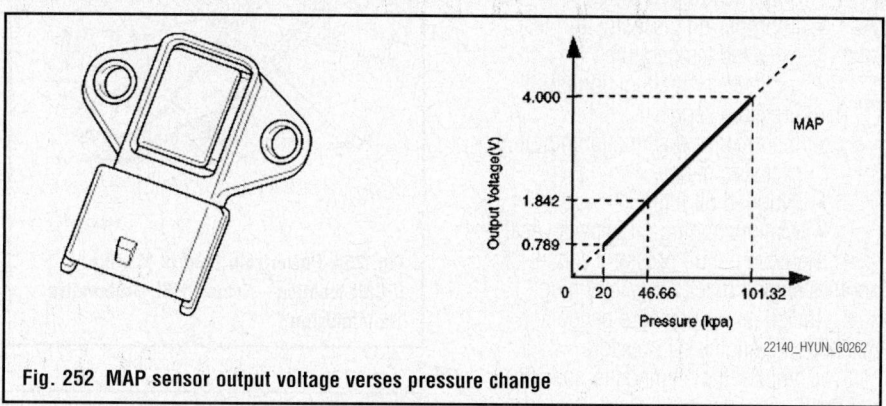
Fig. 252 MAP sensor output voltage verses pressure change

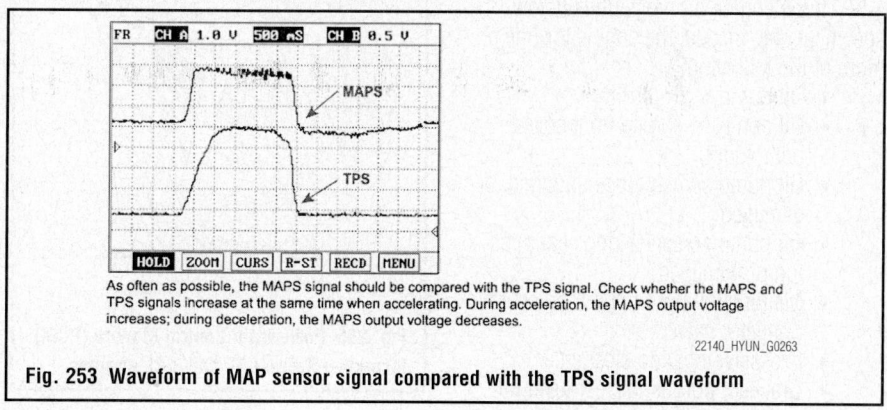
As often as possible, the MAPS signal should be compared with the TPS signal. Check whether the MAPS and TPS signals increase at the same time when accelerating. During acceleration, the MAPS output voltage increases; during deceleration, the MAPS output voltage decreases.

Fig. 253 Waveform of MAP sensor signal compared with the TPS signal waveform

OPERATION

The oil pressure sensor measures the amount of pressure generated from the output of the oil pump. If the oil pressure is too low, mechanical damage and scoring may occur on non-lubricated parts.

REMOVAL & INSTALLATION

✳✳ CAUTION

Do not perform this procedure with the engine hot, wait until the engine is cool or cold.

1. Before servicing the vehicle, refer to the Precautions Section.
2. Turn the ignition **OFF**.
3. Remove the oil pressure sensor electrical connector.
4. Remove the oil pressure sensor.

To install:

5. To install, reverse the removal procedure.
6. Tighten the sensor to 113–130 inch lbs. (13–15 Nm).

TESTING

Before servicing any vehicle, please be sure to read the precautions section, which deals with personal safety, prevention of component damage, and important points to take into consideration when servicing a motor vehicle.

1. With the vehicle on a level surface, allow adequate drain down time of 2–3 minutes and measure for a low oil level. Add the recommended grade engine oil and fill the crankcase until the oil level measures full on the oil level indicator.

2. Run the engine, and verify low, or no oil pressure on the vehicle gage or light. Listen for a noisy valve train or a knocking noise.

3. Inspect for the following:
- Oil diluted by moisture or unburned fuel mixtures
- Improper oil viscosity for the expected temperature
- Incorrect or malfunctioning oil pressure sender
- Incorrect or malfunctioning oil pressure gauge
- Plugged oil filter
- Malfunctioning oil bypass valve

4. Remove the oil pressure sender or another engine block oil gallery plug.

5. Install an oil pressure gauge and measure the engine oil pressure.

6. Compare the readings to specifications.

7. If the engine oil pressure is below specifications, inspect the engine for one or more of the following:
- Oil pump worn or dirty
- Oil pump-to-engine front cover bolts loose
- Oil pump screen loose, plugged, or damaged
- Oil pump screen O-ring seal missing or damaged
- Malfunctioning oil pump pressure regulator valve
- Excessive bearing clearance
- Cracked, porous, or restricted oil galleries
- Oil gallery plugs missing or incorrectly installed
- Broken lash adjusters

8. Higher than recommended oil pressure may be caused by one or more of the following conditions:
- Worn or sticking oil pump pressure relief valve
- Plugged oil filter
- Improper viscosity oil

9. If the oil pressure tests according to specification and the oil pressure sensor gives a faulty reading, replace the oil pressure sensor (switch) as necessary.

POWERTRAIN CONTROL MODULE (PCM)

For the Accent (manual transmission), Elantra, Sonata 2.4L, and Tiburon vehicles, refer to Electronic Control Module (ECM).

LOCATION

See Figures 254 through 256.

Refer to the accompanying illustrations for Powertrain Control Module (PCM) location.

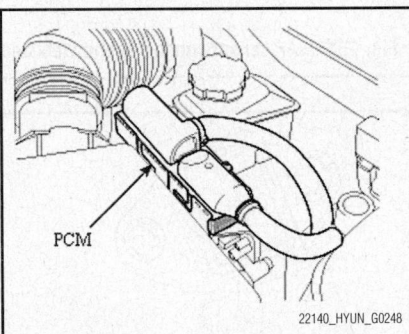

PCM

22140_HYUN_G0248

Fig. 254 Powertrain Control Module (PCM) location—Accent 1.6L (Automatic Transmission)

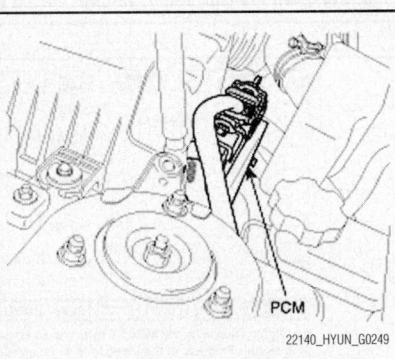

PCM

22140_HYUN_G0249

Fig. 255 Powertrain Control Module (PCM) location—Azera 3.3L and 3.8L engines

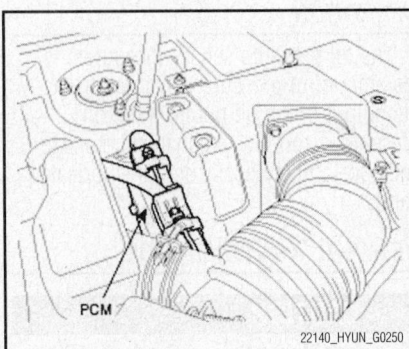

PCM

22140_HYUN_G0250

Fig. 256 Powertrain Control Module (PCM) location—Sonata 3.3L engine

OPERATION

The powertrain has electronic controls to reduce exhaust emissions while maintaining excellent drivability and fuel economy. The Powertrain Control Module (PCM) is the control center of this system. The PCM monitors numerous engine and vehicle functions. The PCM constantly looks at the information from various sensors and other inputs, and controls the systems that affect vehicle performance and emissions. The PCM also performs the diagnostic tests on various parts of the system. The PCM can recognize operational problems and alert the driver via the Malfunction Indicator Lamp (MIL). When the PCM detects a malfunction, the PCM stores a Diagnostic Trouble Code (DTC). The problem area is identified by the particular DTC that is set. The control module supplies a buffered voltage to various sensors and switches. Review the components and wiring diagrams in order to determine which systems are controlled by the PCM. The following are some of the functions that the PCM controls:
- The ignition system
- The Knock Sensor (KS) system
- The Evaporative Emissions (EVAP) system
- The alternator
- The A/C and fan systems
- The cooling fan control
- The fuel injection system
- The emission control systems
- The on-board diagnostics

REMOVAL & INSTALLATION

See Figure 257.

1. Before servicing the vehicle, refer to the Precautions Section.

2. Turn the ignition switch off.

3. Disconnect the negative battery cable from the battery.

4. Disconnect the PCM connector(s) (A).

5. Remove the PCM mounting bolts (B) and remove the PCM from the air cleaner assembly.

To install:

6. Installation is the reverse of the removal.

7. Tighten the PCM mounting bolts to: 86–104 inch lbs. (10–12 Nm).

TESTING

Perform a careful underhood inspection when performing any diagnostic procedure or diagnosing the cause of an emission test failure. This can often lead to repairing a

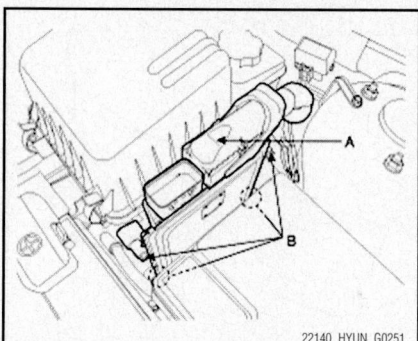

Fig. 257 Disconnect the PCM connector(s) (A) and remove the PCM mounting bolts (B)

condition without further steps. Use the following guidelines when performing an inspection:

1. Inspect all of the vacuum hoses for correct routing, pinches, cuts, or disconnects.

2. Inspect any hoses that are difficult to see.

3. Inspect all of the wires in the engine compartment for the following conditions:
- Burned or chafed spots
- Pinched wires
- Contact with sharp edges
- Contact with hot exhaust manifold

The Powertrain Control Module (PCM) is programmed with test routines that test the operation of the various systems the PCM controls. Some tests monitor internal PCM functions. Many tests are run continuously. Other tests run only under specific conditions, referred to as conditions for running the Diagnostic Trouble Code (DTC). When the vehicle is operating within the conditions for running a particular test, the PCM monitors certain parameters and determines if the values are within an expected range. The parameters and values considered outside the range of normal operation are listed as conditions for setting the DTC. When the conditions for setting the DTC occur, the PCM executes the action taken when the DTC sets. Some DTC's alert the driver via the Malfunction Indicator Lamp (MIL) or a message. Other DTC's do not trigger a driver warning, but are stored in memory. The PCM also saves data and input parameters when most DTC's are set.

The DTC's are categorized by type. The DTC type is determined by the MIL operation and the manner in which the fault data is stored when a particular DTC fails. In some cases, there may be exceptions to this structure. Therefore, when diagnosing the

system it is important to read the action taken when the DTC sets and the conditions for clearing the DTC.

Many intermittent open or shorted circuits come and go with harness and connector movement caused by vibration, engine torque, bumps, or rough pavement.

4. Test the wiring harness and connectors by performing the following:
- Move the related PCM connectors and wiring while monitoring the appropriate scan tool data
- With the engine running, move the related connectors and wiring while monitoring engine operation
- If harness or connector movement affects the data displayed, the component and system operation, or the engine operation, inspect and repair the harness or connections as necessary

5. Test the electrical connections and/or wiring by performing the following:
- Inspect for incorrect mating of the connector halves or terminals not fully seated in the connector body
- Inspect for improperly formed or damaged terminals. Test for incorrect terminal tension
- Inspect for poor terminal to wire connections including terminals crimped over insulation. This requires removing the terminal from the connector body
- Inspect for corrosion or water intrusion. Pierced or damaged insulation can allow moisture to enter the wiring. The conductor can corrode inside the insulation with little visible evidence. Look for swollen and stiff sections of wire in the suspect circuits
- Inspect for wires that are broken inside the insulation

6. Test the PCM ground circuit:
   a. Measure resistance between the PCM and a good chassis ground. Resistance specification: 1 ohm or less.
   b. Use the backside of the PCM harness connector as the PCM side check point.
   c. If a problem is found, make the necessary repairs.

7. If a problem is not located through the above procedures, the PCM may be faulty.

8. Hyundai recommends to replace the PCM with a new one and then check the vehicle again. If the vehicle operates nor-

mally, then the problem was likely with the PCM.

9. Retest the original PCM, if possible, by installing it into a known-good vehicle and check vehicle operations.

THROTTLE POSITION SENSOR (TPS)

LOCATION

See Figure 258.

The Throttle Position Sensor (TPS) is mounted on the throttle body.

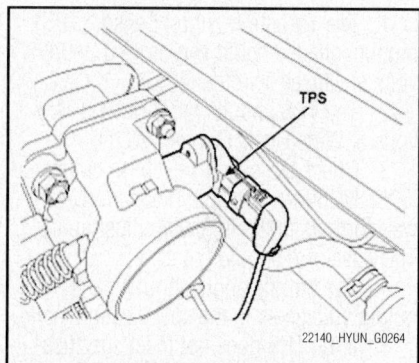

Fig. 258 Throttle Position Sensor (TPS) location

OPERATION

The Throttle Position Sensor (TPS) is mounted on the throttle body and detects the opening angle of the throttle plate. The TPS has a variable resistor (potentiometer) that changes in resistance characteristics according to the throttle angle. During acceleration, the TPS resistance between the reference 5-volt and the signal terminal decreases and the output voltage increases; during deceleration, the TPS resistance increases and the TPS output voltage decreases. The ECM/PCM supplies a reference 5-volt to the TPS and the output voltage increases directly with the opening of the throttle valve. The TPS output voltage will vary from 0.2–0.8 volts at Closed Throttle (CT) to 4.3–4.8 volts at Wide Open Throttle (WOT). The ECM/PCM determines operating conditions such as idle (closed throttle), part load, acceleration/deceleration, and wide-open throttle from the TPS. Also the ECM/PCM uses the Manifold Absolute Pressure (MAP) sensor signal along with the TPS signal to adjust fuel injection duration and ignition timing.

REMOVAL & INSTALLATION

1. Before servicing the vehicle, refer to the Precautions Section.
2. Disconnect the negative battery cable.
3. Disconnect the sensor connector.
4. Remove the sensor retaining screws.
5. Remove the sensor from its mounting.

To install:

6. Installation is the reverse of the removal procedure.

TESTING

See Figures 253 and 259

1. The Throttle Position Sensor (TPS) output voltage should vary from 0.2–0.8 volts at Closed Throttle (CT).
2. The TPS should vary from 4.3–4.8 volts at Wide Open Throttle (WOT).
3. Check the signal waveform compared to the Manifold Absolute Pressure (MAP) waveform as in the following illustration. The waveform should coincide.
4. The throttle angle should match the output voltage as in the specification table.
5. If the TPS does not meet specifications, replace the TPS.

chain sprocket of the intake camshaft. This system controls the intake camshaft to provide the optimal valve timing for every driving condition. The ECM/PCM controls the Variable Camshaft Timing Oil Control Solenoid (VCTOCS)—also known as the Oil Control Valve (OCV)—based on the signal outputs from the Mass Air Flow (MAF) sensor, Throttle Position Sensor (TPS), and Engine Coolant Temperature (ECT) sensor.

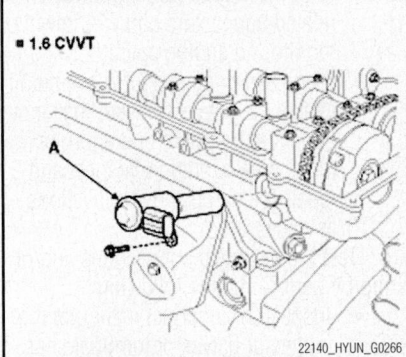

Fig. 260 Variable camshaft timing oil control solenoid location—Accent 1.6L engine

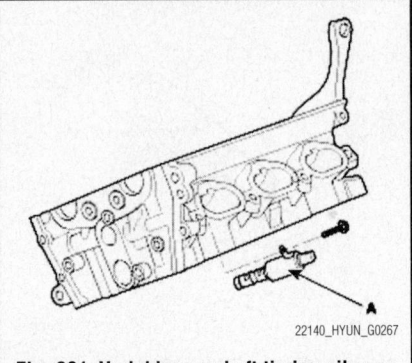

Fig. 261 Variable camshaft timing oil control solenoid location—Azera 3.3L and 3.8L engines

Fig. 262 Variable camshaft timing oil control solenoid location—Elantra 2.0L and Tiburon 2.0L engines

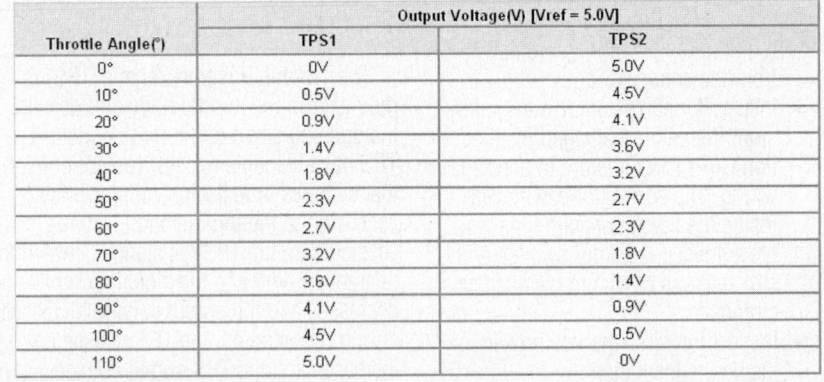

Throttle Angle(°)	Output Voltage(V) [Vref = 5.0V]	
	TPS1	TPS2
0°	0V	5.0V
10°	0.5V	4.5V
20°	0.9V	4.1V
30°	1.4V	3.6V
40°	1.8V	3.2V
50°	2.3V	2.7V
60°	2.7V	2.3V
70°	3.2V	1.8V
80°	3.6V	1.4V
90°	4.1V	0.9V
100°	4.5V	0.5V
110°	5.0V	0V

Fig. 259 Throttle angle and output voltage specification table

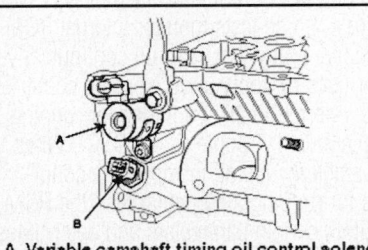

A. Variable camshaft timing oil control solenoid
B. Oil temperature sending unit

Fig. 263 Variable camshaft timing oil control solenoid location—Sonata 2.4L engine

VARIABLE CAMSHAFT TIMING OIL CONTROL SOLENOID

LOCATION

See Figures 260 through 265.

Refer to the accompanying illustrations for variable camshaft timing oil control solenoid location.

OPERATION

See Figure 266.

The Continuously Variable Valve Timing (CVVT) system is installed to the

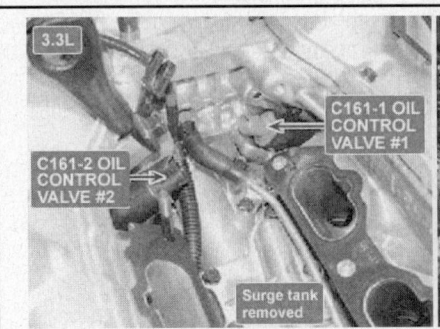

Fig. 264 Variable camshaft timing oil control solenoid location—Sonata 3.3L engine

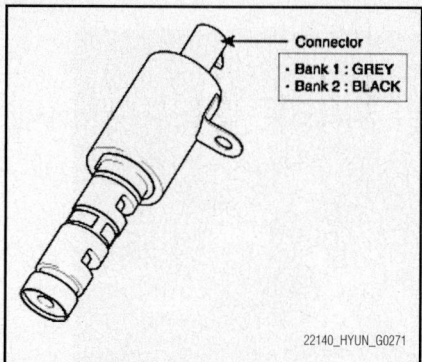

Fig. 265 Variable camshaft timing oil control solenoid component view

The CVVT controller regulates the intake camshaft angle using oil pressure through the VCTOCS. As a result, the relative position between the camshaft and the crankshaft becomes optimal. The engine torque improves, fuel economy improves, and the exhaust emissions decrease within the overall driving conditions.

The CVVT system makes continuous intake valve timing changes based on operating conditions. Intake valve timing is optimized to allow the engine to produce maximum power. Cam angle is advanced to obtain the EGR effect and reduce pumping

loss. The intake valve is closed quickly to reduce the entry of the air/fuel mixture into the intake port and improve the changing effect. Reducing the cam advance at idle stabilizes combustion and reduces engine speed. If a malfunction occurs, the CVVT system control is disabled and the valve timing is fixed at the fully retarded position.

REMOVAL & INSTALLATION

1.6L, 2.0L & 2.4L Engines

1. Before servicing the vehicle, refer to the Precautions Section.
2. Disconnect the ground cable from the battery.
3. Disconnect the connector from Variable Camshaft Timing Oil Control Solenoid (VCTOCS).
4. Remove the bolt retaining the VCTOCS.
5. Remove the VCTOCS.

To install:

➡**Always use a new gasket/O-ring.**

6. Install the VCTOCS.
7. Tighten the mounting bolt to 86–104 inch lbs. (20–39 Nm).
8. Connect the electrical connector to the VCTOCS.

3.3L & 3.8L Engines

See Figure 267.

1. Before servicing the vehicle, refer to the Precautions Section.
2. Disconnect the ground cable from the battery.
3. Disconnect the connector from Variable Camshaft Timing Oil Control Solenoid (VCTOCS) on the right-hand bank and/or left-hand bank.
4. Remove the bolt retaining the VCTOCS.
5. Remove the VCTOCS.

To install:

➡**Always use a new gasket/O-ring.**

6. Install the VCTOCS.
7. Tighten the mounting bolt to 86–104 inch lbs. (20–39 Nm).
8. Connect the electrical connector to the VCTOCS.

TESTING

1. Ensure the vehicle has the proper oil viscosity and that the oil is not overdue maintenance.
2. Observe the engine oil level. The engine oil level should be within the operating range.
3. Allow the engine to reach operating temperature.
4. Connect a suitable OBD2 scan tool to the Data Link Connector (DLC) inside the vehicle.
5. Increase the engine speed to 1,500 RPM.
6. Command each solenoid to 25 percent. The angle desired parameter should match the solenoid actual parameter.

Component Testing

See Figures 268 and 269.

1. Measure the resistance of each variable camshaft timing oil control solenoid valve assembly. Resistance should read according to the resistance specification table shown.
   a. Ignition **OFF**.
   b. Disconnect Variable Camshaft Timing Oil Control Solenoid (VCTOCS) connector.
   c. Measure resistance between terminals 1 and 2 of the VCTOCS connector (component side).
   • Specification: Approx. 6.9–7.9 ohms at 68°F (20°C)
   • If not according to specification, replace the oil control solenoid
2. Check the operation of the VCTOCS.
   a. Start the engine and let it idle.

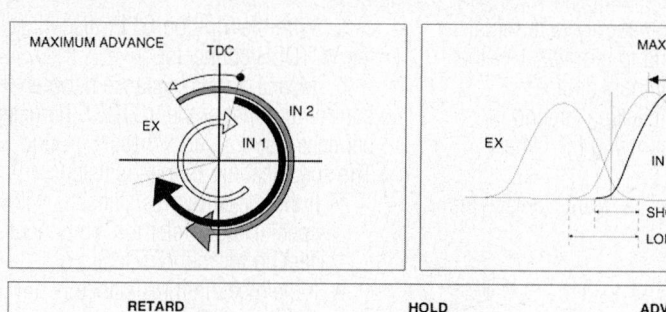

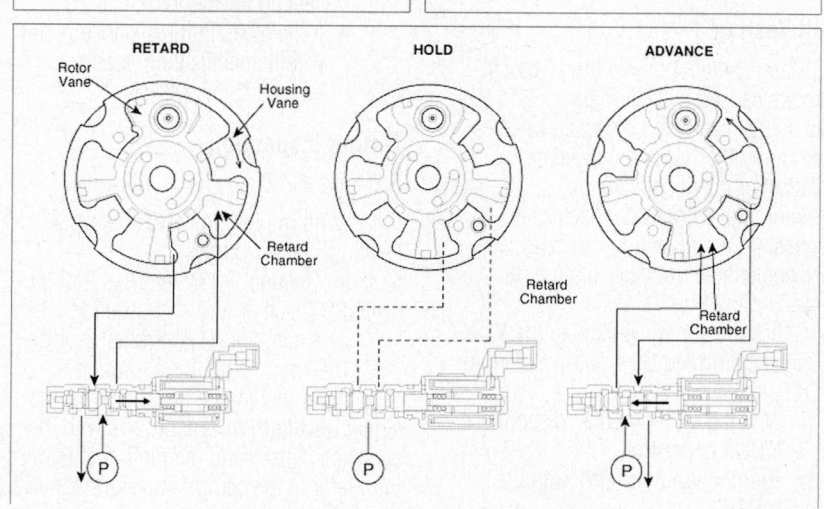

Fig. 266 Continuously Variable Valve Timing (CVVT) system illustrated

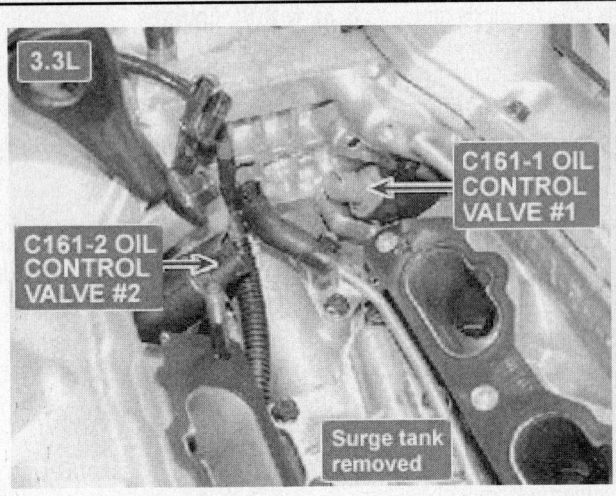

Fig. 267 Variable camshaft timing oil control solenoid location—Sonata 3.3L engine (Azera similar)

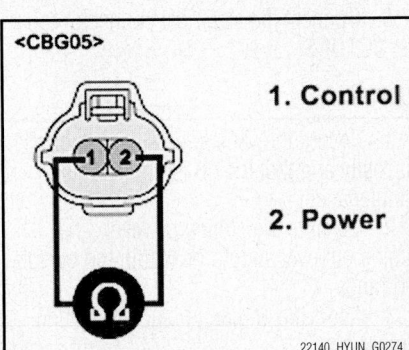

Fig. 268 Use an ohmmeter to check the resistance level of the oil control solenoid

Temperature °C(°F)	Resistance (Ω)
0(32)	6.2 ~ 7.4
20(68)	6.8 ~ 8.0
40(104)	7.4 ~ 8.6
60(140)	8.0 ~ 9.2
80(176)	8.6 ~ 9.8

Fig. 269 Resistance table for the oil control solenoid compared to temperature

b. With VCTOCS connector still disconnected, connect a 12 volt supply to terminal 2 and a ground to terminal 1 of the VCTOCS (component side).
- The engine should suddenly run rough at idle or stall out as the VCTOCS abruptly changes the timing
- If the engine runs rough or stalls when the 12 volt supply is connected, check for connection problems to the VCTOCS or problem with the supply side from the ECM/PCM

- If no change is noted, the VCTOCS may be defective

3. Check the VCTOCS and oil filter assembly.
   a. Turn the ignition **OFF**.
   b. Check VCTOCS filter for sticking or contamination.
   c. Remove the VCTOCS and visually check the spool column of VCTOCS for contamination.
   d. If a problem is found, clean or replace as necessary.
   e. If no problem is found:
   - Apply a 12 volt supply to terminal 2 and a ground to terminal 1 of the VCTOCS (Component side)
   - Verify that a "clicking" sound is heard when applying the battery voltage
   - If no "clicking" is heard, replace the VCTOCS

Circuit Testing

1. Many malfunctions in the electrical system are caused by poor harness and terminals. Faults can also be caused by interference from other electrical systems, mechanical or chemical damage.

2. Thoroughly check connectors for looseness, poor connection, bending, corrosion, contamination, deterioration, or damage.

3. Check the supply voltage to the Variable Camshaft Timing Oil Control Solenoid (VCTOCS).
   a. With the ignition **OFF**, disconnect the VCTOCS connector.
   b. Turn the ignition "ON" with the engine "OFF."
   c. Measure the voltage between the

power terminal of the harness connector and a chassis ground. The specification: battery voltage (12 volts).
   - If the 12 volt supply is not present, check the fuse between the Main Relay and the VCTOCS for an open condition, check for an open in power circuit between the Main Relay and the VCTOCS power circuit
   - Make the repairs as necessary

4. Check for a short to ground in the harness.
   a. With the ignition **OFF**, disconnect the VCTOCS connector.
   b. Measure the resistance between the control terminal of the VCTOCS harness connector and a good chassis ground. The specification: Infinite resistance.
   - If the measured resistance is within specification, refer to Component Testing for the VCTOCS
   - If the measured resistance is not within specification, repair or replace as necessary

Timing Inspection

See Figure 270.

1. With the ignition **OFF**, set up an oscilloscope as follows:
   a. Channel A (+): terminal 2 of the CKPS (back probe), (-): ground.
   b. Channel B (+): terminal 2 of the CMPS (back probe), (-): ground.

2. Start the engine and check for the signal waveform to synchronize with the camshaft sensor and the tooth that is missing. Refer to the sample waveforms in the illustration.

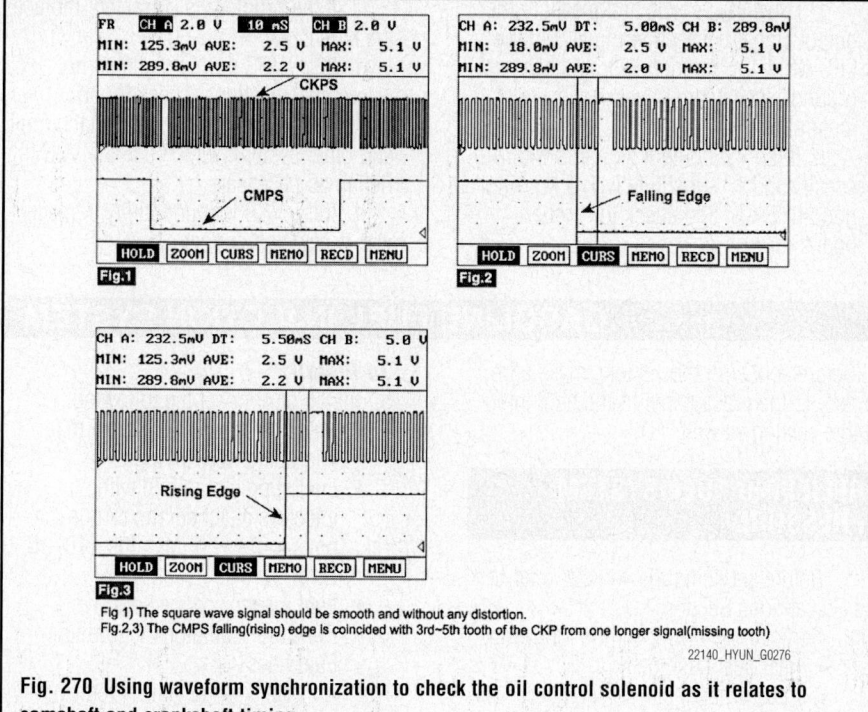

Fig 1) The square wave signal should be smooth and without any distortion.
Fig.2,3) The CMPS falling(rising) edge is coincided with 3rd~5th tooth of the CKP from one longer signal(missing tooth)

22140_HYUN_G0276

Fig. 270 Using waveform synchronization to check the oil control solenoid as it relates to camshaft and crankshaft timing

7. Start the engine.
8. Measure the voltage between terminal 1 of the VSS harness connector and a chassis ground. Specification: Battery voltage (approximately 12 volts).

a. If the measured voltage is within specifications, inspect the ground circuit.

b. If the measured voltage is not within specifications, check for an open/short to ground in the power harness.

9. Repair as necessary.
10. Inspect the ground circuit.

a. Turn the ignition **OFF**.

b. Disconnect VSS connector.

c. Measure the resistance between terminal 2 of the VSS harness connector and the chassis ground. The specification: Approx. 0 ohms.

d. If the measured resistance is within specifications, inspect the signal circuit inspection.

e. If the measured resistance is not within specifications, check for an

VEHICLE SPEED SENSOR (VSS)

LOCATION

The Vehicle Speed Sensor (VSS) is attached to the output shaft of the transaxle.

OPERATION

The Vehicle Speed Sensor (VSS) is a Hall-effect sensor. The sensor converts the transaxle gear revolutions into pulse signals, which are sent to the ECM/PCM.

REMOVAL & INSTALLATION

1. Before servicing the vehicle, refer to the Precautions Section.
2. Raise and support the vehicle safely.
3. Place a drip pan below the Vehicle Speed Sensor (VSS) to catch any spilled fluid when it is removed.
4. Disconnect the VSS connector.
5. Remove the sensor from its mounting.

To install:

6. Installation is the reverse of the removal procedure.
7. Replace any lost transmission fluid.

TESTING

See Figures 271 and 272.

1. Before servicing the vehicle, refer to the Precautions Section.
2. Many malfunctions in the electrical system are caused by poor harness and terminal connections. Faults can also be caused by interference from other electrical systems and mechanical or chemical damage.

3. Thoroughly check all connectors (and connections) for looseness, bending, corrosion, contamination, deterioration, or damage.

4. Make repairs as necessary.
5. Turn the ignition **OFF**.
6. Disconnect the VSS connector.

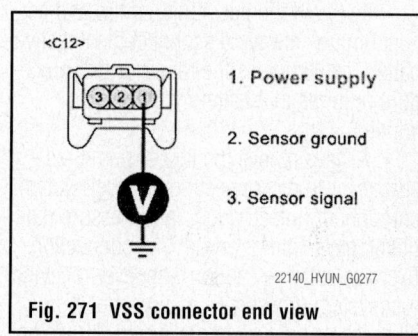

<C12>

1. Power supply
2. Sensor ground
3. Sensor signal

22140_HYUN_G0277

Fig. 271 VSS connector end view

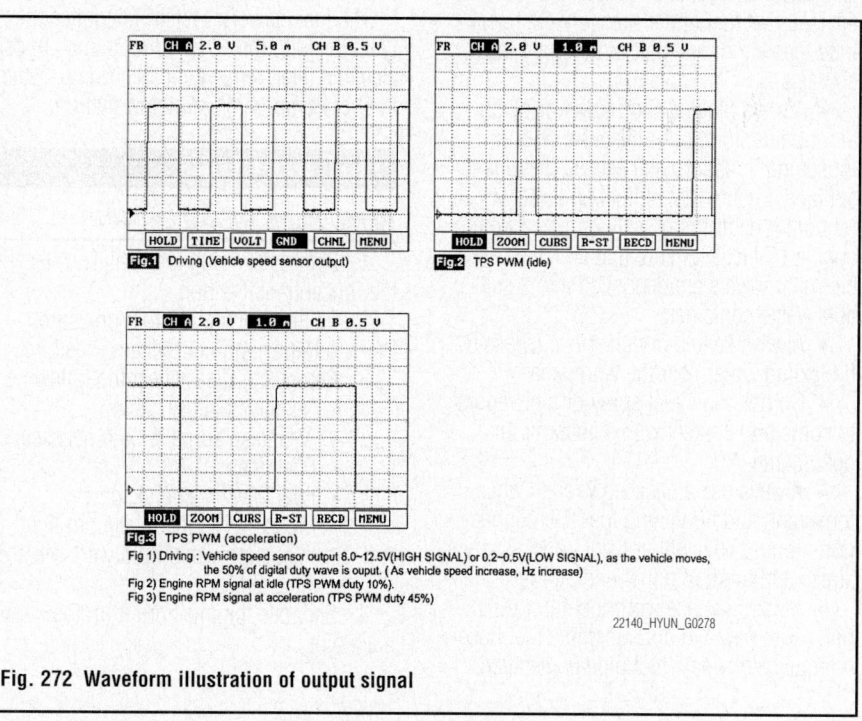

Fig.1 Driving (Vehicle speed sensor output)
Fig.2 TPS PWM (idle)
Fig.3 TPS PWM (acceleration)

Fig 1) Driving : Vehicle speed sensor output 8.0~12.5V(HIGH SIGNAL) or 0.2~0.5V(LOW SIGNAL), as the vehicle moves, the 50% of digital duty wave is ouput. (As vehicle increase, Hz increase)
Fig 2) Engine RPM signal at idle (TPS PWM duty 10%).
Fig 3) Engine RPM signal at acceleration (TPS PWM duty 45%)

22140_HYUN_G0278

Fig. 272 Waveform illustration of output signal

open/short to power in the ground harness. Repair as necessary.

11. Inspect the signal circuit

a. Turn the ignition **OFF**.

b. Connect a scan tool to the Data Link Connector (DLC).

c. Start the engine and select "SCOPEMETER FUNCTION" on the scan tool.

d. Drive the vehicle and measure the output signal between terminal 3 of the EPS CM harness connector and chassis ground. Specification: see waveform illustration.

e. If the VSS output signal is within specifications, substitute with a known-good EPS CM and check for proper operation.

f. If the problem is corrected, replace the EPS CM.

g. If the VSS output signal is not within specifications, check for an open/short to ground in the signal circuit and other systems which use the VSS. Repair as necessary.

h. If the VSS is found faulty, replace with a new component.

FUEL   GASOLINE FUEL INJECTION SYSTEM

FUEL SYSTEM SERVICE PRECAUTIONS

Safety is the most important factor when performing, not only fuel system maintenance, but any type of maintenance. Failure to conduct maintenance and repairs in a safe manner may result in serious personal injury or death. Maintenance and testing of the vehicle's fuel system components can be accomplished safely and effectively by adhering to the following rules and guidelines.

• To avoid the possibility of fire and personal injury, always disconnect the negative battery cable unless the repair or test procedure requires that battery voltage be applied.

• Always relieve the fuel system pressure prior to disconnecting any fuel system component (injector, fuel rail, pressure regulator, etc.), fitting, or fuel line connection. Exercise extreme caution whenever relieving fuel system pressure to avoid exposing skin, face, and eyes to fuel spray. Please be advised that fuel under pressure may penetrate the skin or any part of the body that it contacts.

• Always place a shop towel or cloth around the fitting or connection prior to loosening to absorb any excess fuel due to spillage. Ensure that all fuel spillage (should it occur) is quickly removed from engine surfaces. Ensure that all fuel soaked cloths or towels are deposited into a suitable waste container.

• Always keep a dry chemical (Class B) fire extinguisher near the work area.

• Do not allow fuel spray or fuel vapors to come into contact with a spark or an open flame.

• Always use a back-up wrench when loosening and tightening fuel line connection fittings. This will prevent unnecessary stress and torsion to fuel line piping.

• Always replace worn fuel fitting O-rings with new. Do not substitute fuel hose or equivalent where fuel pipe is installed.

Before servicing the vehicle, make sure to refer to the precautions in the beginning of this section as well.

RELIEVING FUEL SYSTEM PRESSURE

1. Before servicing the vehicle, refer to the Precautions Section.

2. Remove or disconnect the following:
   • Rear seat cushion
   • Access panel
   • Fuel pump module connector

3. Start the engine and allow it to run until it stalls.

4. Turn the ignition switch to the **OFF** position.

5. Disconnect the negative battery cable.

6. Attach the fuel pump harness connector.

FUEL FILTER

REMOVAL & INSTALLATION

The fuel delivery system integrates the fuel filter with the in-tank fuel pump. To service this filter, remove the fuel pump. Refer to Fuel Pump, removal & installation.

FUEL INJECTORS

REMOVAL & INSTALLATION

1. Before servicing the vehicle, refer to the Precautions Section.

2. Relieve the fuel system pressure. Refer to Relieving Fuel System Pressure.

3. Remove or disconnect the following:
   • Negative battery cable
   • Air intake surge tank, if necessary
   • Fuel lines
   • Fuel injector connectors
   • Pressure regulator vacuum line
   • Fuel supply manifold with injectors attached

4. Separate the injectors from the supply manifold.

To install:

5. Install or connect the following:
   • Injectors to the fuel supply manifold using new O-rings
   • Fuel supply manifold with injectors attached and torque the bolts to 84–132 inch lbs. (10–15 Nm)
   • Fuel injector connectors
   • Pressure regulator vacuum line
   • Fuel lines
   • Air intake surge tank, if removed
   • Negative battery cable

6. Start the engine and check for leaks.

FUEL PUMP

REMOVAL & INSTALLATION

See Figures 273 and 274.

1. Before servicing the vehicle, refer to the Precautions Section.

2. Relieve the fuel system pressure. Refer to Relieving Fuel System Pressure.

3. Disconnect the fuel feed line (A) and canister hoses (B).

4. Unscrew the fuel pump mounting bolts (C) and remove the fuel pump assembly.

To install:

5. Install the fuel pump assembly.

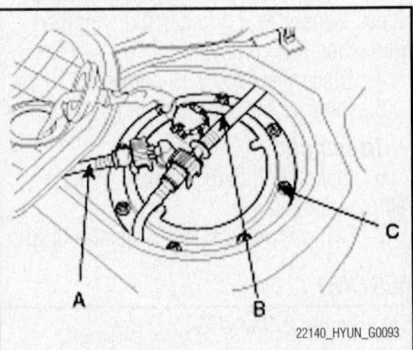

22140_HYUN_G0093

Fig. 273 Disconnect the fuel feed line (A) and canister hoses (B). Unscrew the fuel pump mounting bolts (C)

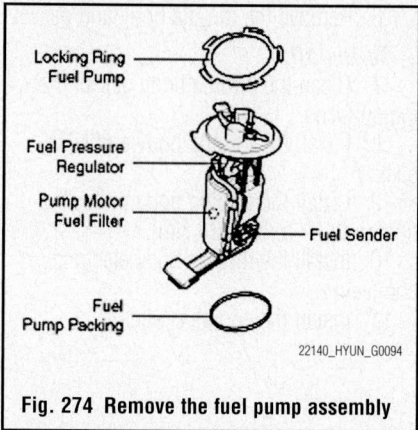

Fig. 274 Remove the fuel pump assembly

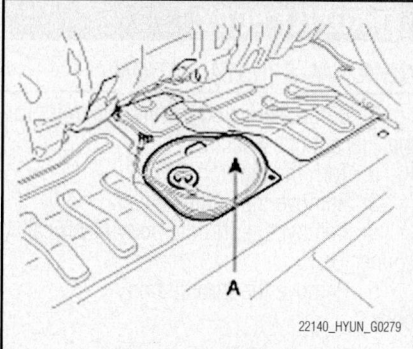

Fig. 275 Remove the service cover (A) under the back seat—except Azera

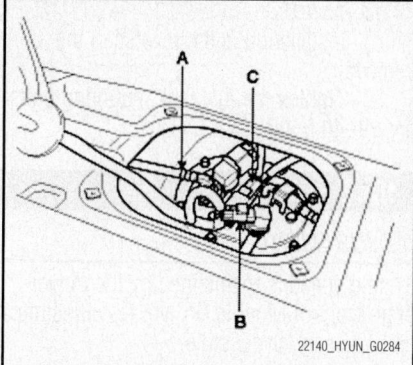

Fig. 279 Disconnect the fuel feed quick-connector (A), fuel tank pressure sensor connector (B), and fuel pump and sub fuel sender connector (C)—Azera

6. Install the fuel pump mounting bolts (C). Tightening the bolts/nuts to: 17–26 inch lbs. (2–3 Nm).

7. Connect the fuel feed line (A) and canister hoses (B).

FUEL PRESSURE REGULATOR

REMOVAL & INSTALLATION

The fuel pressure regulator is built into the fuel pump. Please refer to Fuel Pump, removal & installation.

FUEL TANK

REMOVAL & INSTALLATION

See Figures 275 through 281.

1. Before servicing the vehicle, refer to the Precautions Section.

2. Remove the rear seat cushion—except Azera.

3. Remove the service cover (A) under the back seat—except Azera.

4. Remove the service cover (A) in the trunk—Azera.

5. Disconnect the fuel pump connector (A).

6. Start the engine and wait until the fuel in the fuel line is exhausted.

7. After the engine stalls, turn the ignition switch to the OFF position.

8. Disconnect the negative battery connection.

9. Disconnect the fuel feed line (A) and canister hose (B)—except Azera.

10. Disconnect the fuel feed quick-connector (A), fuel tank pressure sensor connector (B), and fuel pump and sub fuel sender connector (C)—Azera.

11. Raise and safely support the vehicle.

12. Remove the center muffler and main muffler, as needed.

13. Support the fuel tank with a jack.

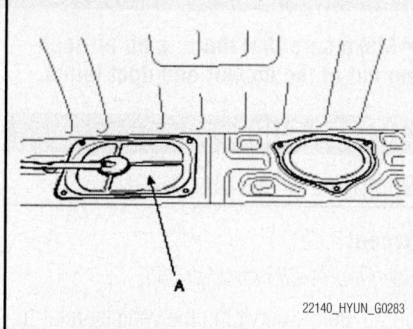

Fig. 276 Remove the service cover (A) in the trunk—Azera

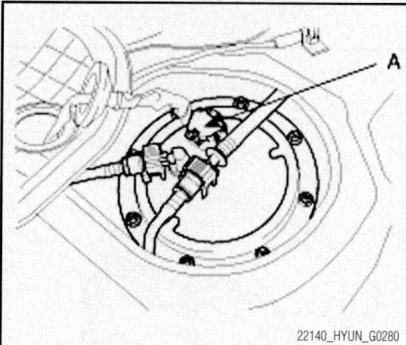

Fig. 277 Disconnect the fuel pump connector (A)

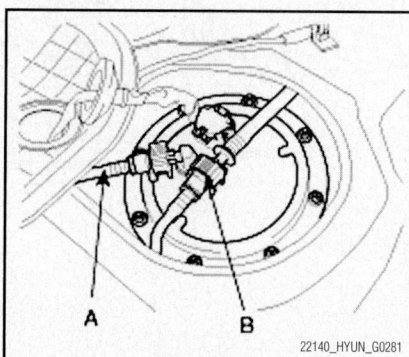

Fig. 278 Disconnect the fuel feed line (A) and canister hose (B)—except Azera

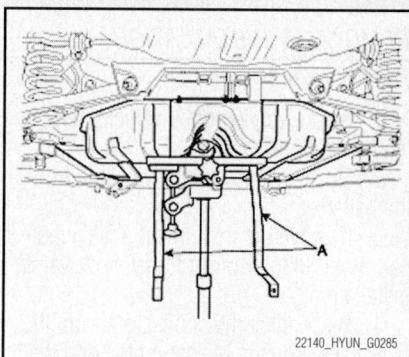

Fig. 280 Support the fuel tank with a jack and remove the fuel tank band (A)

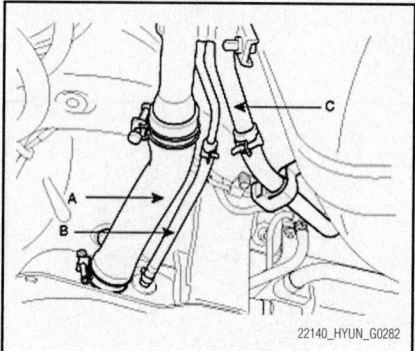

Fig. 281 Disconnect the fuel filler pipe (A), the leveling hose (B), and canister hose (C)

14. Unscrew the brake hose mounting bolts on the left-hand and right-hand sides—Accent only.

15. Remove the fuel tank bands (A).

16. Disconnect the fuel filler pipe (A), the leveling hose (B), and canister hose (C).

17. By moving the jack down slowly, remove the fuel tank from the vehicle.

To install:

18. Installation is the reverse of the removal.

19. Tighten the fuel tank mounting bolts to: 29–40 ft. lbs. (39–54 Nm).

IDLE SPEED

ADJUSTMENT

Idle speed is maintained by the Powertrain Control Module (PCM). No adjustment is necessary or possible.

THROTTLE BODY

REMOVAL & INSTALLATION

1. Before servicing the vehicle, refer to the Precautions Section.
2. Turn the ignition **OFF**.
3. Remove the engine cover.
4. Remove the throttle body electrical connector.
5. Remove the throttle body bolts.

6. Remove the throttle body and gasket.

To install:

7. Clean the throttle body gasket mating surfaces.
8. Install the throttle body and NEW gasket.
9. Install the throttle body bolts and tighten to 21 ft. lbs. (28 Nm).
10. Install the throttle body electrical connector.
11. Install the engine cover.

HEATING & AIR CONDITIONING SYSTEM

BLOWER MOTOR

REMOVAL & INSTALLATION

See Figure 282.

1. Before servicing the vehicle, refer to the Precautions Section.
2. Disconnect the negative cable from the battery.
3. Remove the instrument panel crash pad. Refer to Instrument Panel, removal & installation.
4. Disconnect the connectors from the intake actuator, the blower motor, and the power mosfet.
5. Remove the blower motor (A) after loosening the mounting screws.

To install:

6. Installation is the reverse order of removal.

➡ **Make sure that there is no air leaking out of the blower and duct joints.**

HEATER CORE

REMOVAL & INSTALLATION

Accent

See Figures 283 through 286.

1. Before servicing the vehicle, refer to the Precautions Section.
2. Disconnect the negative battery cable and wait 3 minutes for the SRS memory to drain.

✳✳ **CAUTION**

After disconnecting the negative battery cable, wait for at least 3 minutes for the SRS module to deplete its stored energy.

3. Drain the cooling system into a clean container for reuse.
4. Remove or disconnect the following:

- Heater hoses with the vacuum hose from the heater housing
- Discharge and recover the air conditioning system refrigerant

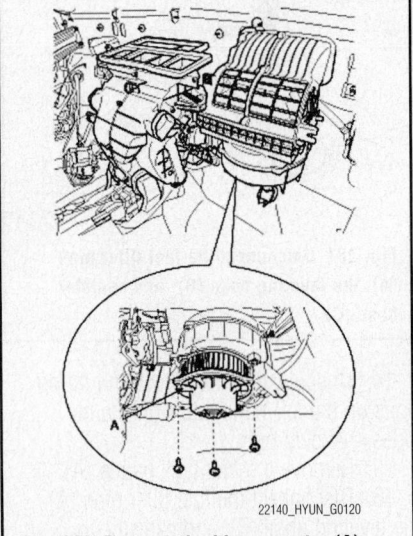

22140_HYUN_G0120

Fig. 282 Remove the blower motor (A) after loosening the mounting screws

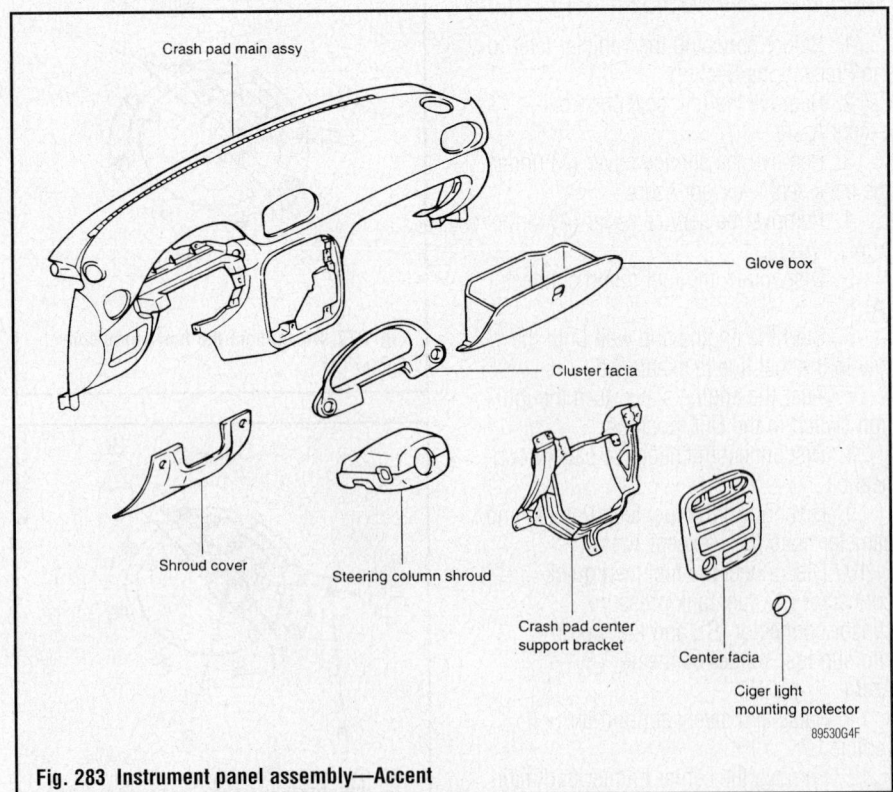

89530G4F

Fig. 283 Instrument panel assembly—Accent

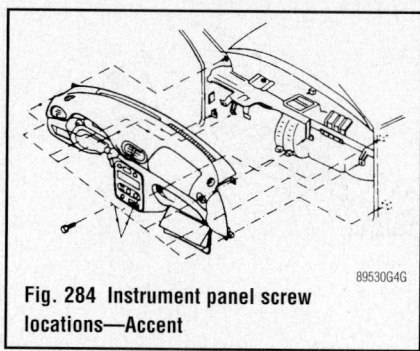

Fig. 284 Instrument panel screw locations—Accent

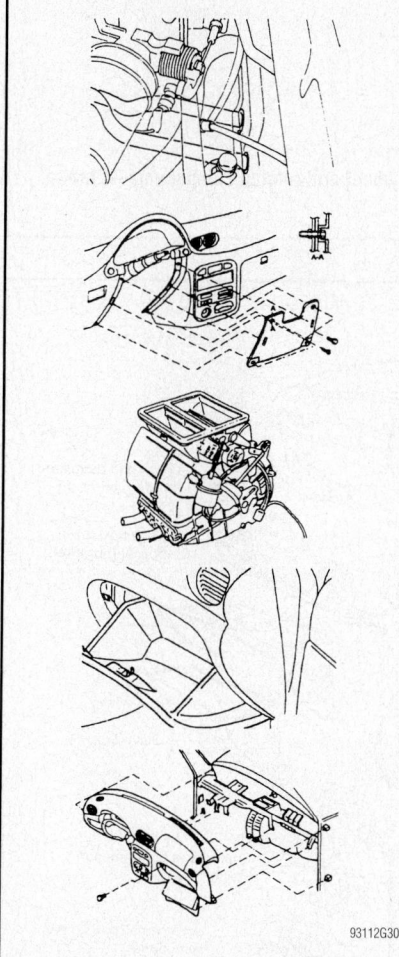

Fig. 285 View of the heater housing assembly and related components—Accent

- Suction and discharge hoses from the evaporator assembly
5. Remove the SRS module and the steering wheel, as follows:
  - Steering wheel-to-SRS module nuts
  - SRS module from the steering wheel and disconnect the electrical connector
  - Steering wheel-to-steering column nut
  - Steering wheel from the steering column

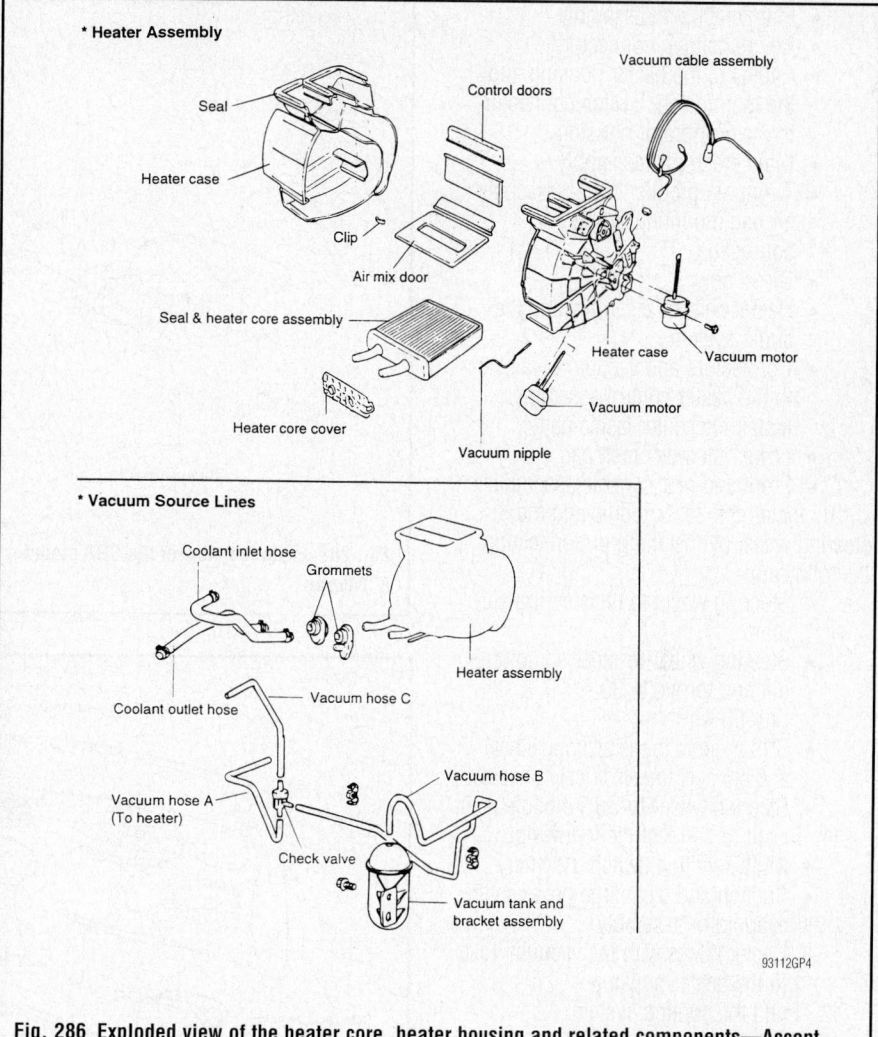

Fig. 286 Exploded view of the heater core, heater housing and related components—Accent

6. Remove or disconnect the following:
- Multi-function switch assembly
- Front and rear console assemblies
- Lower left side crash pad
- Center fascia panel and disconnect the connectors and vacuum connector from the heater control assembly
- Heater control assembly and the audio system
- Glove box
- 4 mounting bolts from the passenger air bag mounting bracket, if equipped
- Main crash pad assembly
- Cables from the heater housing and the thermostatic switch connector from the evaporator housing
- Any remaining connectors
- Main crash pad assembly
- 3 evaporator mounting bolts/nuts
- Evaporator housing
- 3 mounting bolts from the heater housing
- Heater housing

7. Disassemble the heater housing by removing or disconnecting the following:
- Vacuum motor-to-heater housing bolts (2 for each vacuum motor)
- Vacuum motor rod end connection and remove the vacuum motors
- Heater housing cover clips
- Cover and the heater core

To install:
8. Assemble the heater housing by installing or connecting the following:
- Heater core and the cover
- Heater housing cover clips
- Vacuum motor rod end connection and install the vacuum motors
- Vacuum motor-to-heater housing bolts (2 for each vacuum motor)
9. Install or connect the following:
- Heater housing
- 3 mounting bolts to the heater housing
- Evaporator housing
- 3 evaporator mounting bolts (or nuts)

- Main crash pad assembly
- Any remaining connectors
- Cables to the heater housing and the thermostatic switch connector to the evaporator housing
- Main crash pad assembly
- 4 mounting bolts to the passenger air bag mounting bracket, if equipped
- Glove box
- Heater control assembly and the audio system
- Connectors and vacuum connector to the heater control assembly. Install the center fascia panel.
- Lower left side crash pad
- Front and rear console assemblies

10. Install the SRS module and the steering wheel by installing or connecting the following:
- Steering wheel to the steering column
- Steering wheel-to-steering column nut and torque to 30–37 ft. lbs. (40–50 Nm)
- SRS module to the steering wheel and connect the electrical connector
- Steering wheel-to-SRS module nuts

11. Install or connect the following:
- Multi-function switch assembly
- Suction and discharge hoses to the evaporator assembly
- Heater hoses with the vacuum hose to the heater housing

12. Refill the cooling system.
13. Connect the negative battery cable.
14. Evacuate, charge, and leak test the air conditioning system refrigerant.
15. Operate the engine to normal operating temperatures; check the climate control operation and check for leaks.

Elantra

See Figures 287 through 291.

1. Before servicing the vehicle, refer to the Precautions Section.
2. Disconnect the negative battery cable.

✳✳ CAUTION

After disconnecting the negative battery cable, wait for at least 3 minutes for the SRS module to deplete its stored energy.

3. Discharge and recover the air conditioning system refrigerant.
4. Drain the engine coolant into a clean container for reuse.
5. Disconnect the heater hoses from the heater core. Plug the openings.
6. Disconnect the vacuum line from the heater housing vacuum nipple.

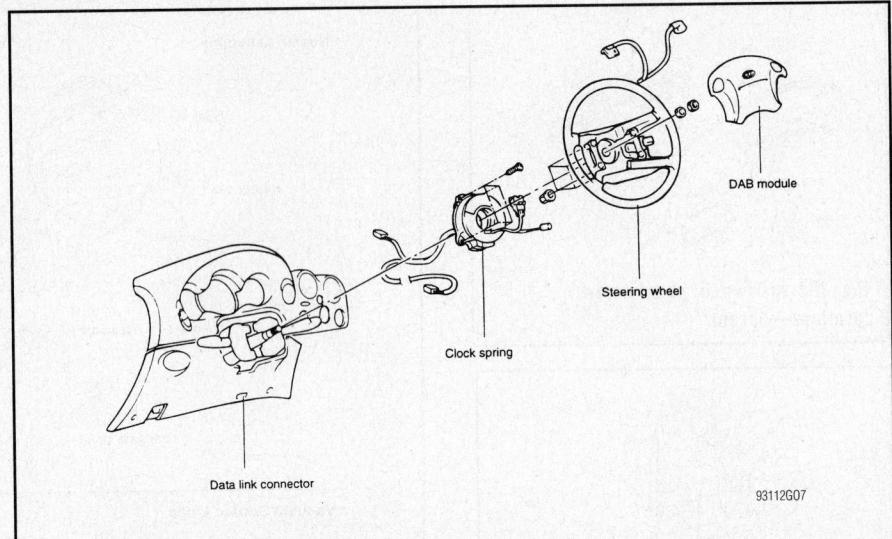

Fig. 287 Exploded view of the SRS module, steering wheel and related components—Elantra & Tiburon

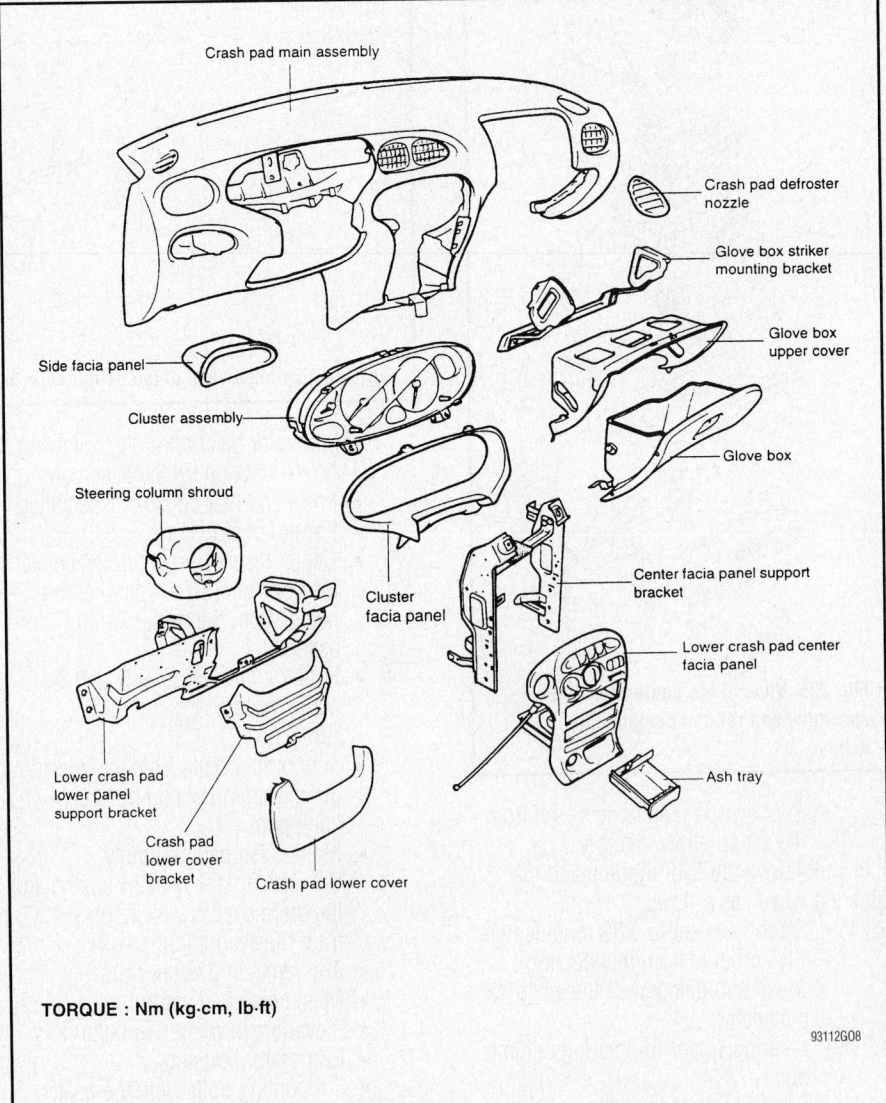

TORQUE : Nm (kg·cm, lb·ft)

93112G08

Fig. 288 Exploded view of the instrument panel and related components—Elantra

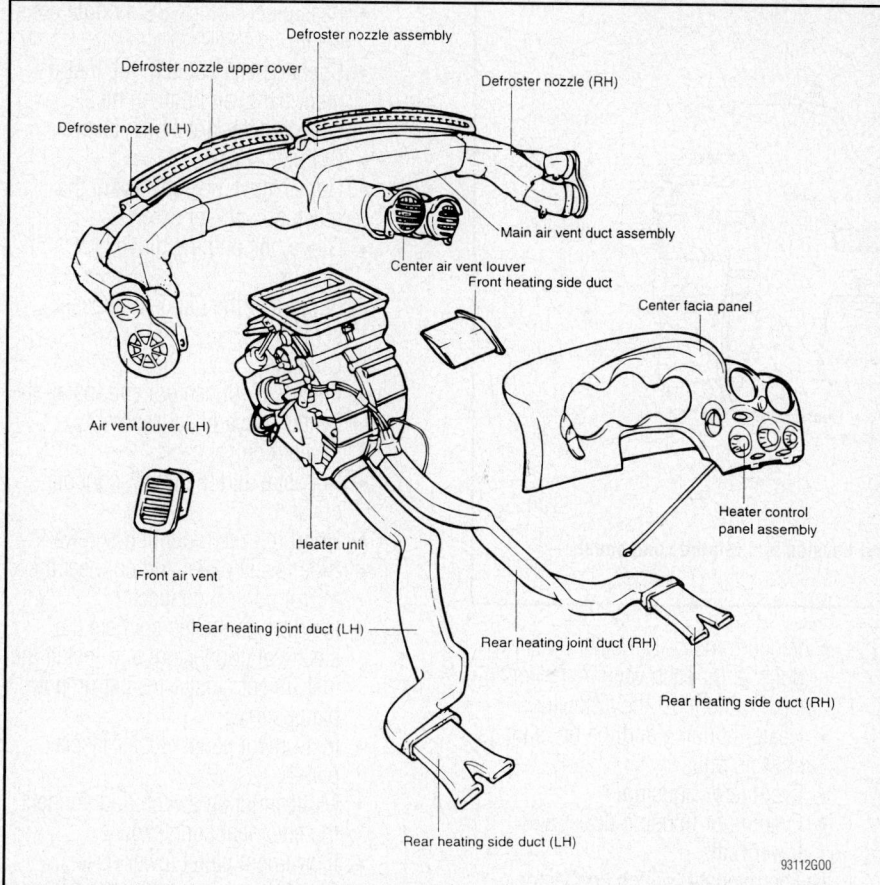

Fig. 289 Exploded view of the heater housing, center fascia, distribution ducts and related components—Elantra Coupe & Tiburon

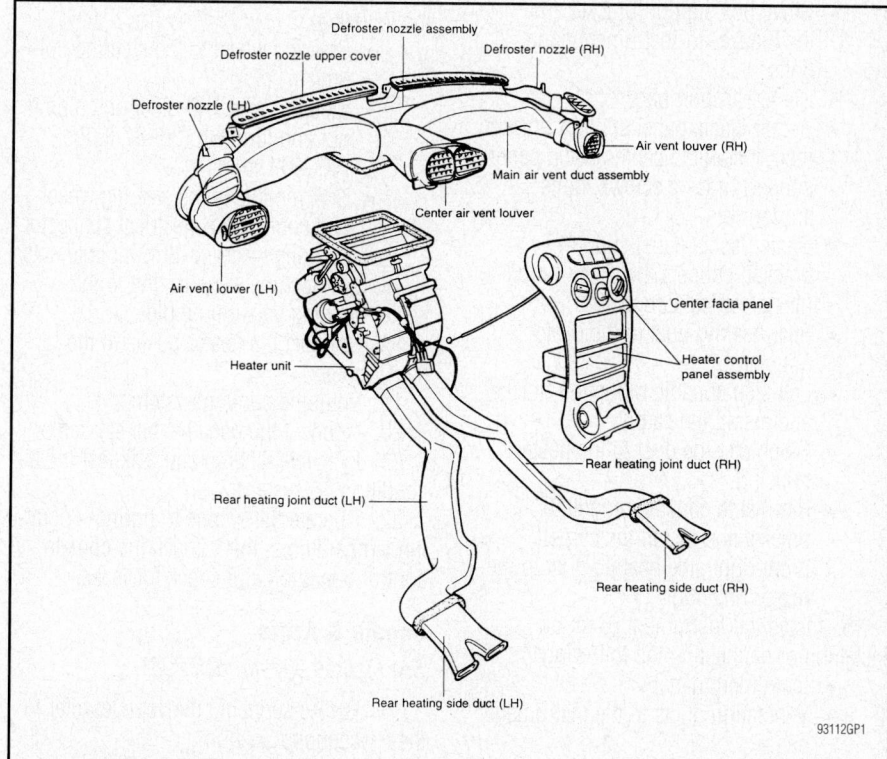

Fig. 290 Exploded view of the heater housing, center fascia, distribution ducts and related components—Elantra Sedan and Wagon & Tiburon

7. Remove the SRS module and the steering wheel by removing or disconnecting the following:
- Steering wheel-to-SRS module nuts
- SRS module from the steering wheel and disconnect the electrical connector
- Steering wheel-to-steering column nut
- Steering wheel from the steering column

8. Remove the instrument panel by removing or disconnecting the following:
- Steering column cover screws and the covers
- Instrument panel lower cover at the driver's side
- Multi-function switch and disconnect the electrical connector at the steering column
- Instrument panel cluster fascia panel
- Instrument cluster-to-instrument panel screws, disconnect the electrical connectors and remove the instrument cluster
- Side fascia panel and disconnect the mirror control connector
- Hood release mounting screws
- Rheostat and the upper console cover
- Heater control cable
- Electrical connectors at the center of the instrument panel and remove the center fascia panel assembly
- Console
- Radio-to-chassis screws and the radio
- Glove box screws and the glove box
- Glove box striker screws and the upper glove box cover
- Defroster nozzle
- Loosen the speedometer drive gear sleeve and disconnect the speedometer cable from the instrument panel
- Passenger's side SRS module connector
- Instrument panel-to-chassis bolts
- Any remaining electrical connectors
- Ventilation ducts from the instrument panel
- Instrument panel

9. Remove or disconnect the following:
- Front right side heating duct from the heater housing
- Pull back the carpet and remove the right side console mounting bracket
- Front left side duct from the heater housing
- Pull back the carpet and remove the left side console mounting bracket

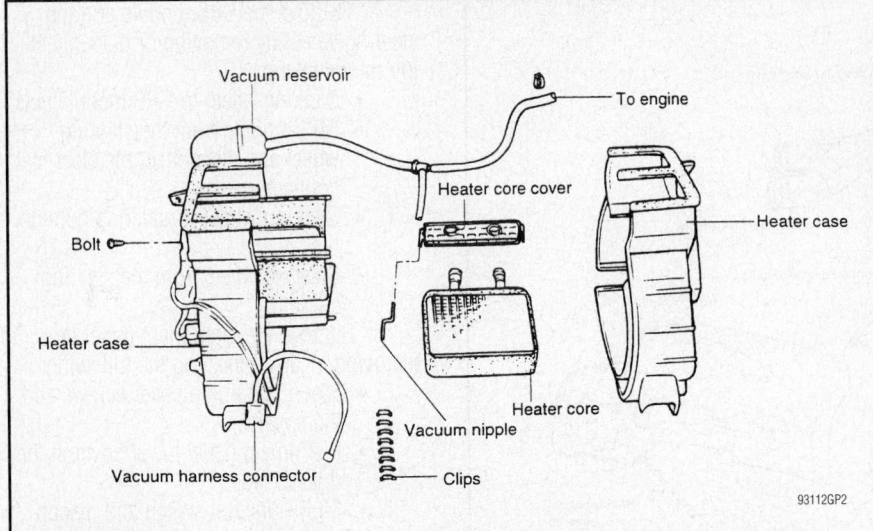

Fig. 291 Exploded view of the heater core and heater housing and related components—Elantra & Tiburon

- Rear heating duct from the heater housing
- Control modules electrical connectors at the center fascia panel support bracket
- Center fascia panel support bracket screws, bolts and/or nuts; then, remove the center fascia panel support bracket
- Center support bars
- Glove box support bracket-to-instrument panel bolts and the bracket

10. Remove the evaporator housing by removing or disconnecting the following:
- Refrigerant lines from the evaporator housing and discard the O-rings
- Thermostatic switch connector
- Evaporator housing upper and lower bolts
- Evaporator housing

11. Remove or disconnect the following:
- Heater housing-to-chassis bolts and the housing
- Vacuum motor-to-heater housing bolts (2 for each vacuum motor)
- Vacuum motor rod end connection and remove the vacuum motors
- Heater housing cover clips
- Cover and the heater core

To install:

12. Assemble the heater housing by installing or connecting the following:
- Heater core and the cover
- Heater housing cover clips
- Vacuum motor rod end connection and install the vacuum motors

- Vacuum motor-to-heater housing bolts (2 for each vacuum motor)

13. Install or connect the following:
- Heater housing and the housing-to-chassis bolts
- Evaporator housing
- Evaporator housing upper and lower bolts
- Thermostatic switch connector
- Connect the refrigerant lines to the evaporator housing

14. Install or connect the following:
- Glove box support bracket and the bracket-to-instrument panel bolts
- Center support bars
- Center fascia panel support bracket; then, install the center fascia panel support bracket screws, bolts and/or nuts
- Center fascia panel support bracket, connect the control modules electrical connectors
- Rear heating duct to the heater housing
- Left side console mounting bracket and install the carpet
- Front left side duct to the heater housing
- Right side console mounting bracket and install the carpet
- Front right side heating duct to the heater housing

15. Install the instrument panel by installing or connecting the following:
- Instrument panel
- Ventilation ducts to the instrument panel
- Instrument panel-to-chassis bolts

- Passenger's side SRS module connector
- Speedometer cable to the instrument panel and tighten the speedometer drive gear sleeve
- Defroster nozzle
- Upper glove box cover and the glove box striker screws
- Glove box and the glove box screws
- Radio and the radio-to-chassis screws
- Console
- Electrical connectors and install the center fascia panel assembly
- Heater control cable
- Rheostat and the upper console cover
- Hood release mounting screws
- Side fascia panel and connect the mirror control connector
- Instrument cluster, connect the electrical connectors and install the instrument cluster-to-instrument panel screws
- Instrument panel cluster fascia panel
- Multi-function switch and connect the electrical connector
- Instrument panel lower cover
- Steering column cover and the cover screws

16. Install the SRS module and the steering wheel by installing or connecting the following:
- Steering wheel to the steering column
- Steering wheel-to-steering column nut and torque to 30–37 ft. lbs. (40–50 Nm)
- SRS module to the steering wheel and connect the electrical connector
- Steering wheel-to-SRS module nuts

17. Connect the vacuum line to the heater housing vacuum nipple.

18. Connect the heater hoses to the heater core.

19. Refill the cooling system.

20. Connect the negative battery cable.

21. Evacuate, charge and leak test the air conditioning system.

22. Operate the engine to normal operating temperatures; then, check the climate control operation and check for leaks.

Sonata & Azera

See Figures 292 through 298.

1. Before servicing the vehicle, refer to the Precautions Section.

2. Disconnect the negative battery cable.

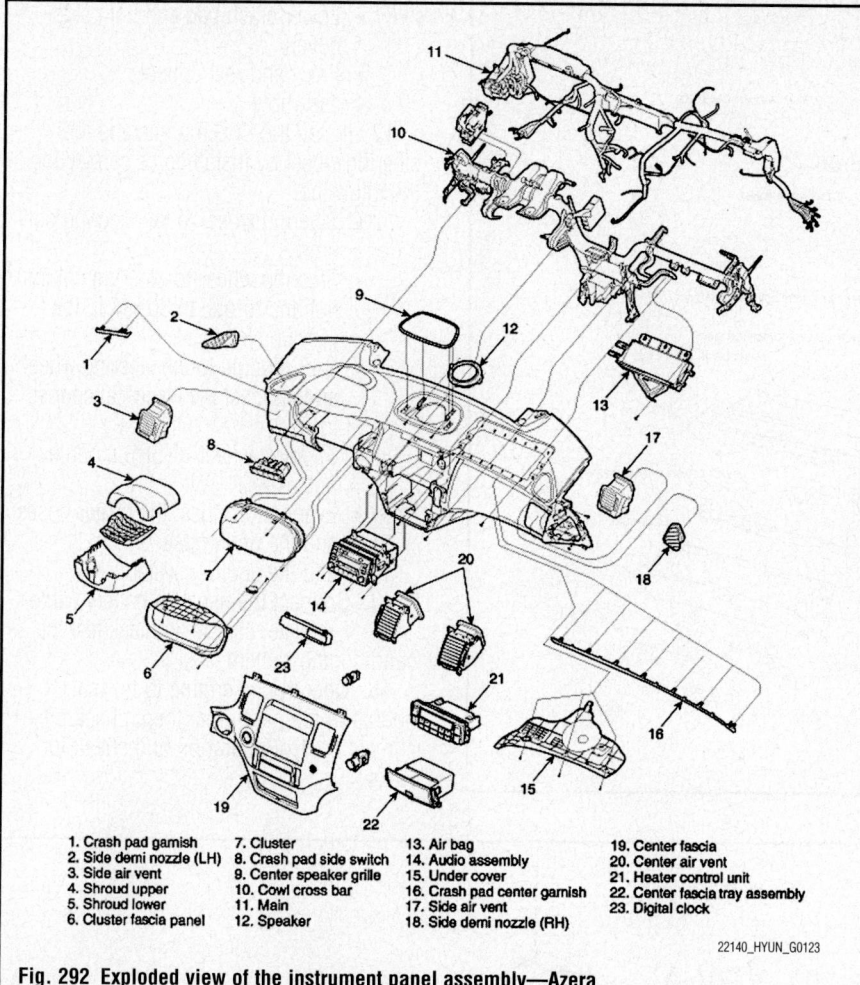

1. Crash pad garnish
2. Side demi nozzle (LH)
3. Side air vent
4. Shroud upper
5. Shroud lower
6. Cluster fascia panel
7. Cluster
8. Crash pad side switch
9. Center speaker grille
10. Cowl cross bar
11. Main
12. Speaker
13. Air bag
14. Audio assembly
15. Under cover
16. Crash pad center garnish
17. Side air vent
18. Side demi nozzle (RH)
19. Center fascia
20. Center air vent
21. Heater control unit
22. Center fascia tray assembly
23. Digital clock

22140_HYUN_G0123

Fig. 292 Exploded view of the instrument panel assembly—Azera

- Steering wheel-to-steering column nut
- Steering wheel from the steering column

9. Remove or disconnect the following:
- Front and rear console assembly and remove both side covers
- Glove box, the center pad cover, the center crash pad and the cassette assembly
- Lower crash pad. Remove the console mounting bracket and the center support bracket
- Rear heater ducts from the heater housing
- Control assembly
- Blower speed control actuator connector and the blend door actuator connector, if equipped with semi-automatic temperature control
- 4 retaining bolts and remove the heater assembly

10. Disassemble the heater housing by removing or disconnecting the following:
- Vacuum motor-to-heater housing bolts (2 for each vacuum motor)
- Vacuum motor rod end connection and remove the vacuum motors

3. Drain the cooling system into a clean container for reuse.

4. Remove the heater hoses from the heater housing.

5. Discharge and recover the air conditioning system refrigerant.

6. Remove the suction and discharge hoses from the evaporator assembly. Cap the hoses to minimize contamination.

7. Remove the evaporator drain hose.

8. Remove the SRS module and the steering wheel by removing or disconnecting the following:
- Steering wheel-to-SRS module nuts
- SRS module from the steering wheel and disconnect the electrical connector

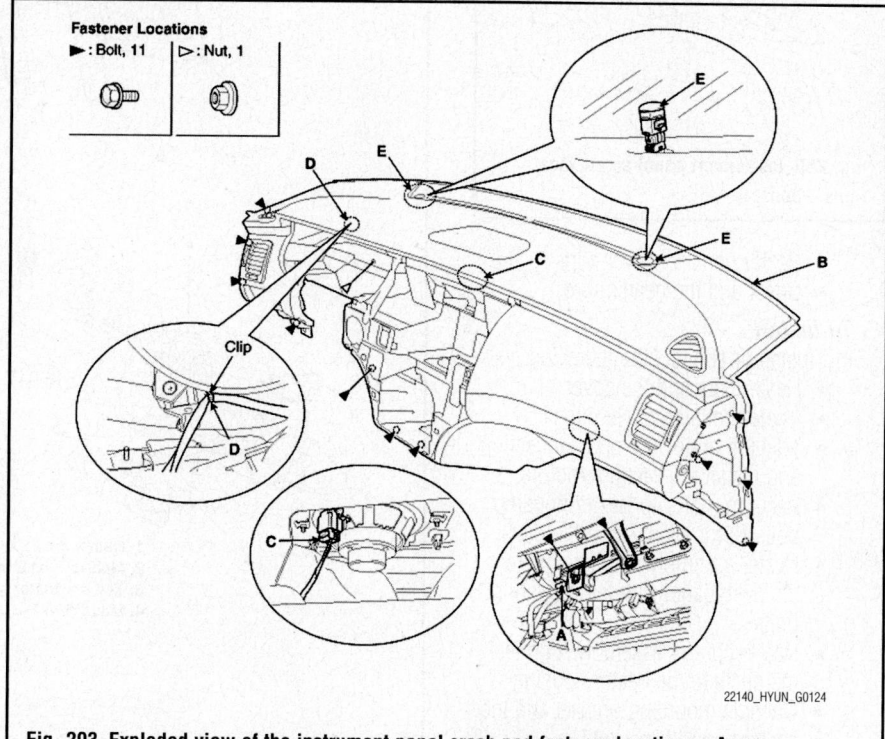

Fastener Locations
▶ : Bolt, 11   ▷ : Nut, 1

22140_HYUN_G0124

Fig. 293 Exploded view of the instrument panel crash pad fastener locations—Azera

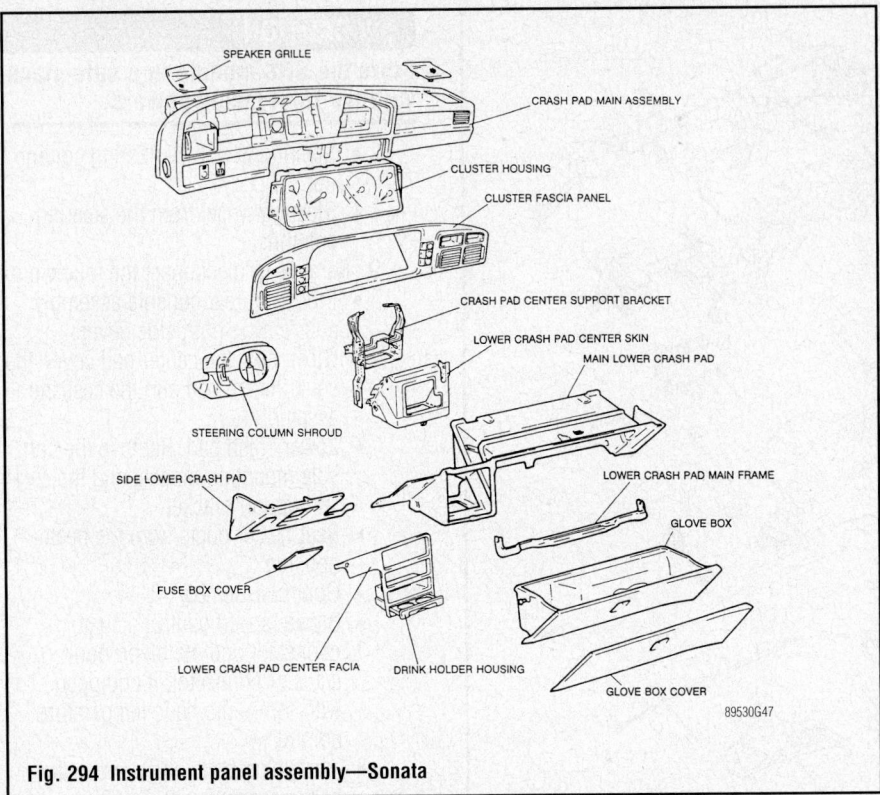

Fig. 294 Instrument panel assembly—Sonata

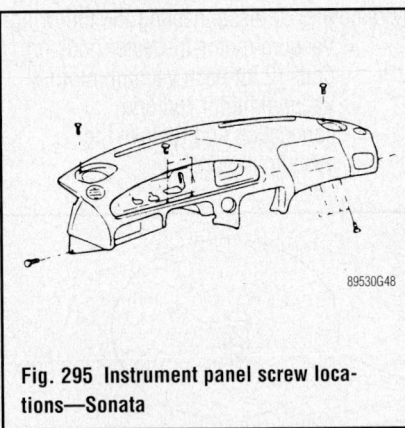

Fig. 295 Instrument panel screw locations—Sonata

- Heater housing cover clips
- Cover and the heater core

To install:
11. Install or connect the following:
- Heater core and the cover
- Heater housing cover clips
- Vacuum motor rod end connection and install the vacuum motors
- Vacuum motor-to-heater housing bolts (2 for each vacuum motor)
- Heater assembly and attach it to the dash panel with the mounting bolts
- Heater control assembly. Connect the ducts to the heater housing
- Console mounting bracket and the center support bracket

- Lower crash pad and both side covers
- Front and rear console assembly

12. Install the SRS module and the steering wheel by installing or connecting the following:
- Steering wheel to the steering column
- Steering wheel-to-steering column nut and torque to 30–37 ft. lbs. (40–50 Nm)
- SRS module to the steering wheel and connect the electrical connector
- Steering wheel-to-SRS module nuts
- Evaporator tubes, the heater hoses and the drain hose

13. Refill the cooling system.
14. Connect the negative battery cable.
15. Evacuate, charge and leak test the air conditioning system.
16. Operate the engine to normal operating temperatures; then, check the climate control operation and check for leaks.

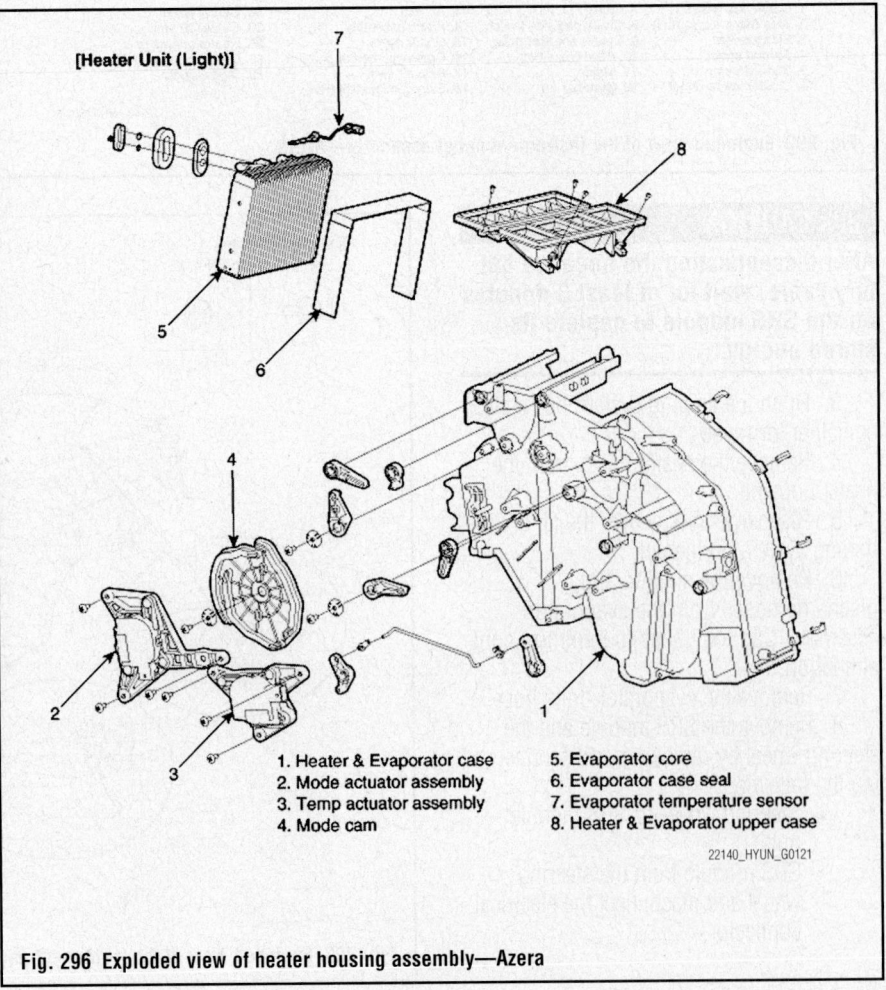

1. Heater & Evaporator case
2. Mode actuator assembly
3. Temp actuator assembly
4. Mode cam
5. Evaporator core
6. Evaporator case seal
7. Evaporator temperature sensor
8. Heater & Evaporator upper case

Fig. 296 Exploded view of heater housing assembly—Azera

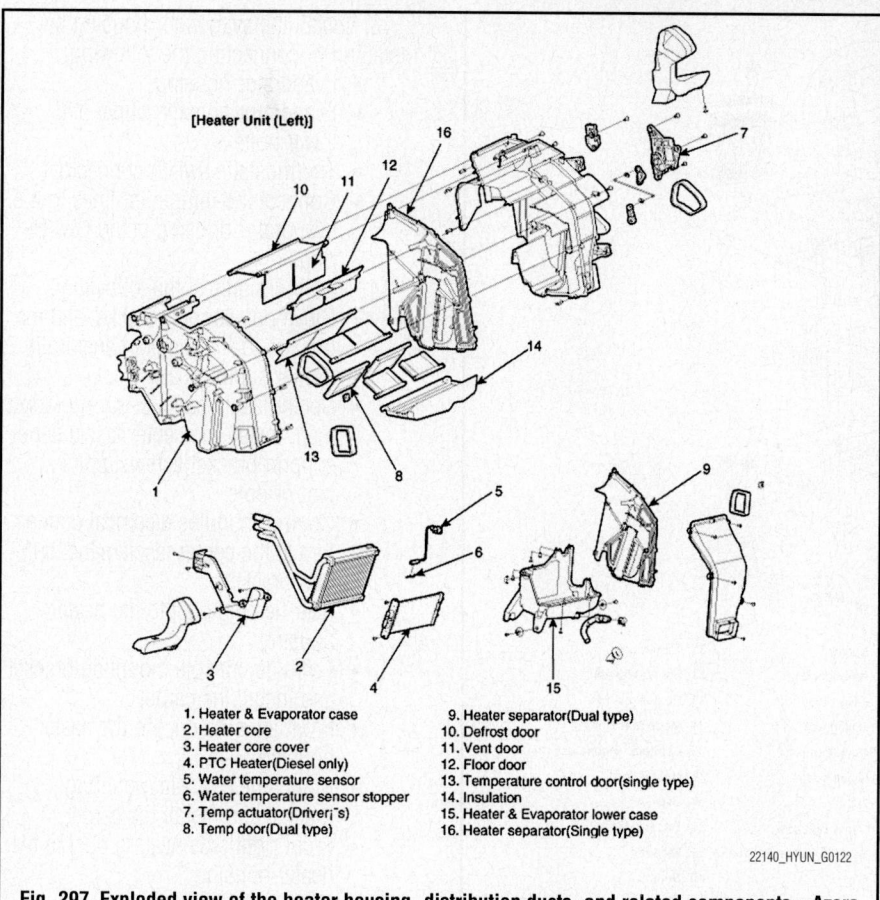

[Heater Unit (Left)]

1. Heater & Evaporator case
2. Heater core
3. Heater core cover
4. PTC Heater(Diesel only)
5. Water temperature sensor
6. Water temperature sensor stopper
7. Temp actuator(Driver¡¯s)
8. Temp door(Dual type)
9. Heater separator(Dual type)
10. Defrost door
11. Vent door
12. Floor door
13. Temperature control door(single type)
14. Insulation
15. Heater & Evaporator lower case
16. Heater separator(Single type)

22140_HYUN_G0122

Fig. 297 Exploded view of the heater housing, distribution ducts, and related components—Azera

Tiburon

See Figures 289, 287, 290, and 299

1. Before servicing the vehicle, refer to the Precautions Section.

2. Disconnect the negative battery cable.

✳✳ CAUTION

After disconnecting the negative battery cable, wait for at least 3 minutes for the SRS module to deplete its stored energy.

3. Discharge and recover the air conditioning system refrigerant.

4. Drain the engine coolant into a clean container for reuse.

5. Disconnect the heater hoses from the heater core. Plug the openings.

6. Disconnect the vacuum line from the heater housing vacuum nipple.

7. Remove the SRS module and the steering wheel by removing or disconnecting the following:

- Steering wheel-to-SRS module nuts
- SRS module from the steering wheel and disconnect the electrical connector
- Steering wheel-to-steering column nut
- Steering wheel from the steering column

8. Remove the instrument panel by removing or disconnecting the following:

- Upper console cover
- Center fascia panel and disconnect the cigar lighter connector
- 3 lower instrument panel screws and the lower instrument panel
- Radio-to-bracket bolts and the radio
- Rheostat switch, the hood release handle and DLC from the lower instrument panel
- 5 cluster fascia panel-to-instrument panel screws; then, disconnect the heater control cable and the cluster electrical connectors and remove the cluster fascia panel
- 4 instrument cluster-to-instrument panel screws and the instrument cluster
- 2 glove box-to-instrument panel bolts and the glove box
- 4 upper glove box cover-to-instrument panel screws, the 2 glove box

striker screws and the upper glove box cover
- Upper instrument panel speaker grille
- 2 upper speaker-to-instrument panel screws
- Instrument panel-to-chassis bolts, disconnect the electrical connectors and remove the instrument panel

9. Remove or disconnect the following:

- Front right side heating duct from the heater housing
- Pull back the carpet and remove the right side console mounting bracket
- Front left side duct from the heater housing
- Pull back the carpet and remove the left side console mounting bracket
- Rear heating duct from the heater housing
- Control modules electrical connectors at the center fascia panel support bracket
- Center fascia panel support bracket screws, bolts and/or nuts; then, remove the center fascia panel support bracket
- Center support bars
- Glove box support bracket-to-instrument panel bolts and the bracket

10. Remove the evaporator housing by removing or disconnecting the following:

- Refrigerant lines (located in the engine compartment), from the evaporator housing and discard the O-rings
- Thermostatic switch connector
- Evaporator housing upper and lower bolts
- Evaporator housing
- Heater housing-to-chassis bolts and the housing

11. Disassemble the heater housing by removing or disconnecting the following:

- Vacuum motor-to-heater housing bolts (2 for each vacuum motor)
- Vacuum motor rod end connection and remove the vacuum motors
- Heater housing cover clips
- Cover and the heater core

To install:

12. Assemble the heater housing by installing or connecting the following:

- Heater core and the cover
- Heater housing cover clips
- Vacuum motor rod end connection and install the vacuum motors
- Vacuum motor-to-heater housing bolts (2 for each vacuum motor)
- Heater housing and the housing-to-chassis bolts

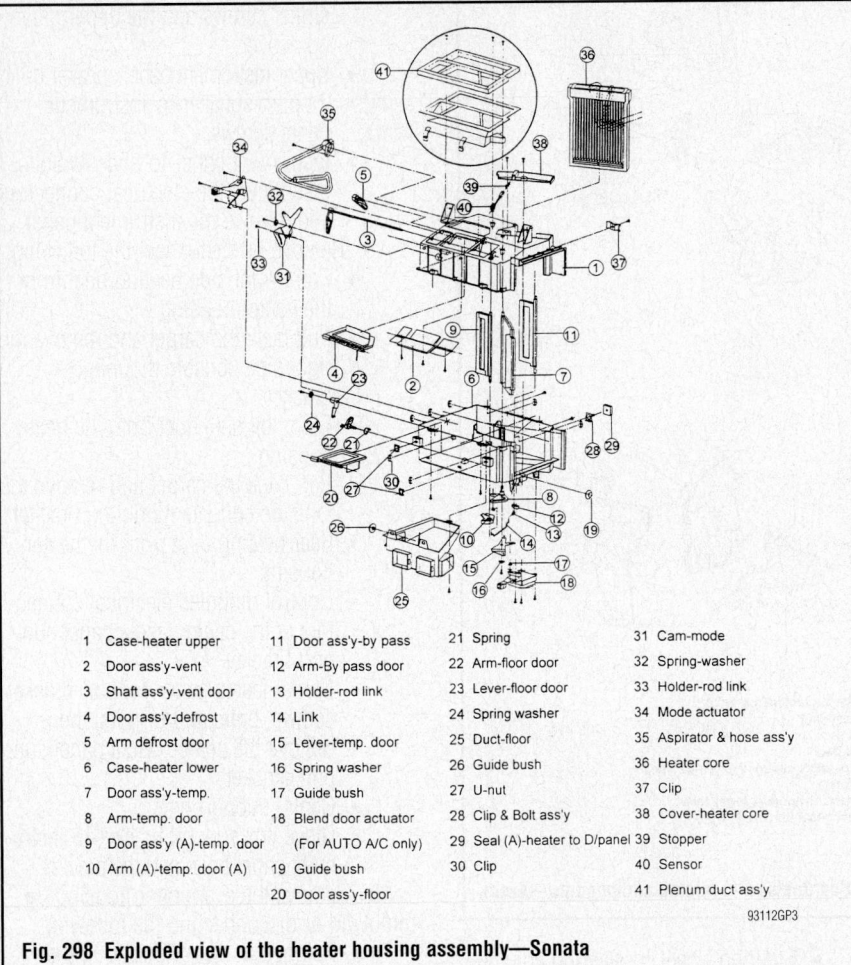

1	Case-heater upper	11	Door ass'y-by pass	21	Spring	31	Cam-mode
2	Door ass'y-vent	12	Arm-By pass door	22	Arm-floor door	32	Spring-washer
3	Shaft ass'y-vent door	13	Holder-rod link	23	Lever-floor door	33	Holder-rod link
4	Door ass'y-defrost	14	Link	24	Spring washer	34	Mode actuator
5	Arm defrost door	15	Lever-temp. door	25	Duct-floor	35	Aspirator & hose ass'y
6	Case-heater lower	16	Spring washer	26	Guide bush	36	Heater core
7	Door ass'y-temp.	17	Guide bush	27	U-nut	37	Clip
8	Arm-temp. door	18	Blend door actuator	28	Clip & Bolt ass'y	38	Cover-heater core
9	Door ass'y (A)-temp. door		(For AUTO A/C only)	29	Seal (A)-heater to D/panel	39	Stopper
10	Arm (A)-temp. door (A)	19	Guide bush	30	Clip	40	Sensor
		20	Door ass'y-floor			41	Plenum duct ass'y

93112GP3

Fig. 298 Exploded view of the heater housing assembly—Sonata

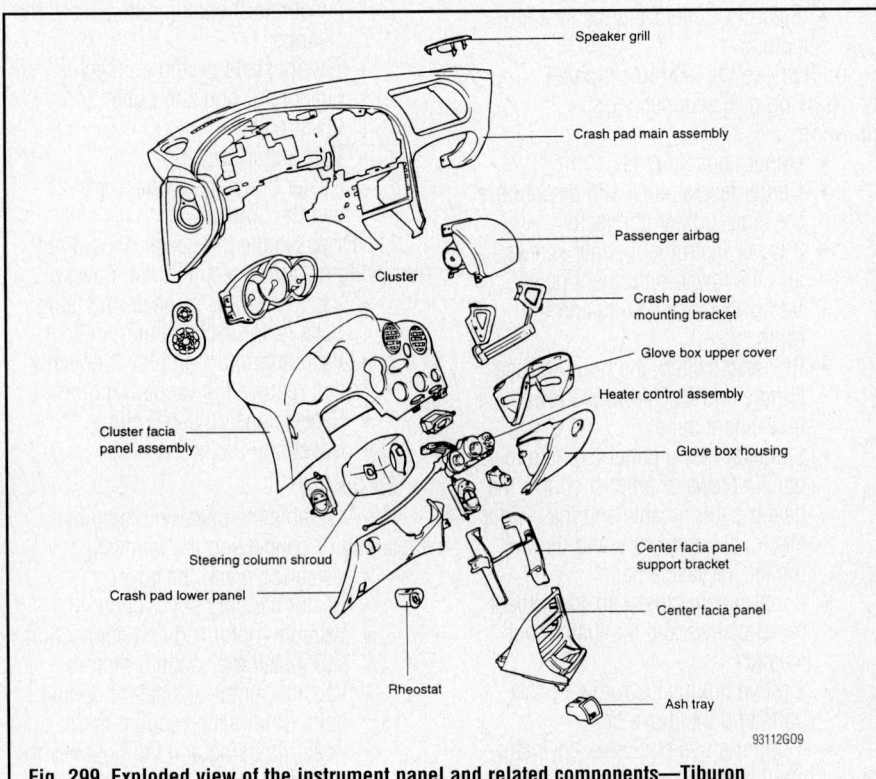

Speaker grill

Crash pad main assembly

Passenger airbag

Cluster

Crash pad lower mounting bracket

Glove box upper cover

Heater control assembly

Cluster facia panel assembly

Glove box housing

Steering column shroud

Center facia panel support bracket

Crash pad lower panel

Center facia panel

Rheostat

Ash tray

93112G09

Fig. 299 Exploded view of the instrument panel and related components—Tiburon

13. Install the evaporator housing by installing or connecting the following:
   - Evaporator housing
   - Evaporator housing upper and lower bolts
   - Thermostatic switch connector
   - Connect the refrigerant lines to the evaporator housing using new O-rings

14. Install or connect the following:
   - Glove box support bracket and the bracket-to-instrument panel bolts
   - Center support bars
   - Center fascia panel support bracket; then, install the center fascia panel support bracket screws, bolts and/or nuts
   - Control modules electrical connectors at the center fascia panel support bracket
   - Rear heating duct to the heater housing
   - Left side console mounting bracket and install the carpet
   - Front left side duct to the heater housing
   - Right side console mounting bracket and install the carpet
   - Front right side heating duct to the heater housing

15. Install the instrument panel by installing or connecting the following:
   - Instrument panel, connect the electrical connectors and install the instrument panel-to-chassis bolts
   - 2 upper speaker-to-instrument panel screws
   - Upper instrument panel speaker grille
   - Upper glove box cover, the 2 glove box striker screws and the 4 upper glove box cover-to-instrument panel screws
   - Glove box and the 2 glove box-to-instrument panel bolts
   - Instrument cluster and the 4 instrument cluster-to-instrument panel screws
   - Cluster fascia panel; then, connect the heater control cable and the cluster electrical connectors and install the 5 cluster fascia panel-to-instrument panel screws
   - Rheostat switch, the hood release handle and DLC to the lower instrument panel
   - Radio and the radio-to-bracket bolts
   - Lower instrument panel and the 3 lower instrument panel screws
   - Cigar lighter connector and install the center fascia panel

- Upper console cover
16. Install the SRS module and the steering wheel by installing or connecting the following:
  - Steering wheel to the steering column
  - Steering wheel-to-steering column nut and torque to 30–37 ft. lbs. (40–50 Nm)

- SRS module to the steering wheel and connect the electrical connector
- Steering wheel-to-SRS module nuts
17. Install or connect the following:
  - Vacuum line to the heater housing vacuum nipple
  - Heater hoses to the heater core

18. Refill the cooling system.
19. Connect the negative battery cable.
20. Evacuate, charge and leak test the air conditioning system.
21. Operate the engine to normal operating temperatures; then, check the climate control operation and check for leaks.

STEERING

POWER RACK & PINION STEERING GEAR

REMOVAL & INSTALLATION

See Figure 300.

1. Before servicing the vehicle, refer to the Precautions Section.
2. Remove or disconnect the following:
  - Negative battery cable
  - Front wheels
  - Outer tie rod ends
  - Steering column flexible coupler

- Power steering fluid hoses, if equipped with power steering
- Sub-frame center beam
- Exhaust front pipe
- Left lower control arm
- Stabilizer bar
- Steering gear

To install:
3. Install or connect the following:
  - Steering gear and torque the bolts to 44–59 ft. lbs. (60–80 Nm)
  - Stabilizer bar
  - Left lower control arm

- Exhaust front pipe
- Sub-frame center beam
- Power steering fluid hoses, if equipped with power steering
- Steering column flexible coupler and torque the bolt to 11–14 ft. lbs. (15–19 Nm)
- Outer tie rod ends and torque the nuts to 17–25 ft. lbs. (23–34 Nm)
- Front wheels
- Negative battery cable
4. Fill the power steering system.
5. Start the engine and check for leaks.

POWER STEERING PUMP

REMOVAL & INSTALLATION

1. Before servicing the vehicle, refer to the Precautions Section.
2. Loosen the bolt fixing the wiring bracket and move the wiring aside.
3. Remove the pressure hose from the oil pump.
4. Disconnect the suction hose from the suction connector and drain the fluid into a container.
5. Loosen the tension adjusting bolt on the power steering "V" belt.
6. Remove the "V" belt from the power steering oil pump pulley.
7. Loosen the power steering oil pump mounting bolt and the tension adjusting bolt.
8. Remove the steering oil pump assembly.
9. Installation is the reverse of the removal procedure.

BLEEDING

☀ CAUTION

The fluid level should be checked with the engine OFF to prevent injury from moving components. Use only PSF-3 power steering fluid, or equivalent. Do not overfill.

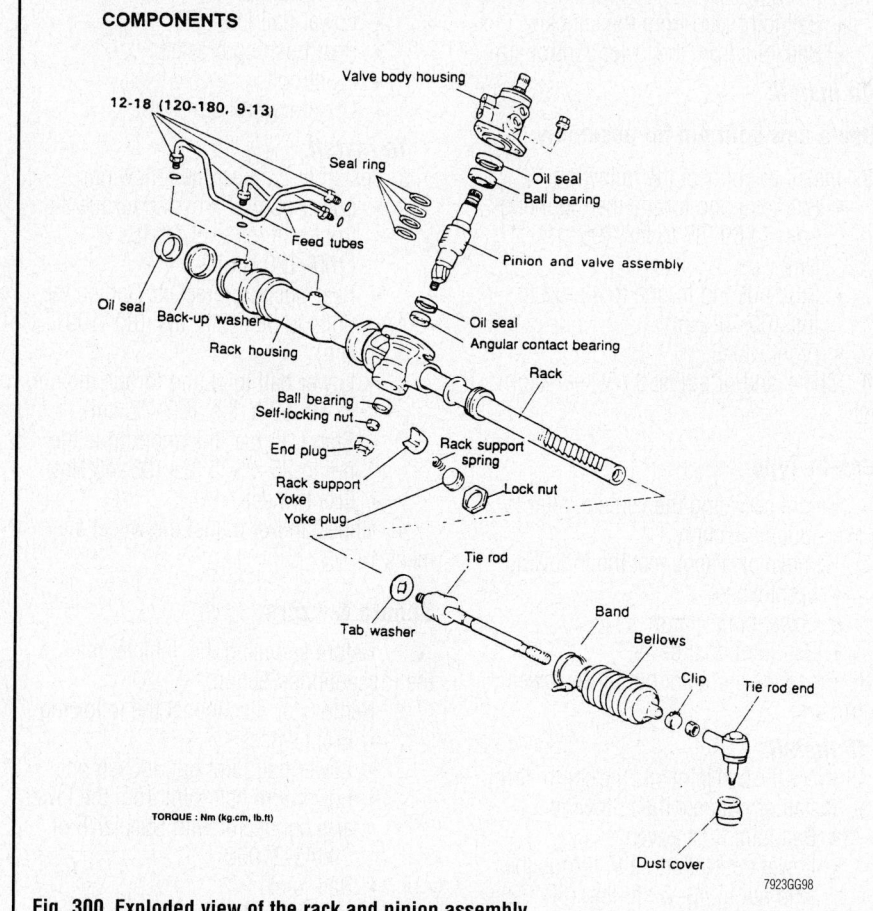

COMPONENTS

12-18 (120-180, 9-13)

Valve body housing
Seal ring
Oil seal
Ball bearing
Feed tubes
Pinion and valve assembly
Oil seal
Back-up washer
Rack housing
Oil seal
Angular contact bearing
Rack
Ball bearing
Self-locking nut
End plug
Rack support spring
Rack support Yoke
Lock nut
Yoke plug
Tie rod
Tab washer
Band
Bellows
Clip
Tie rod end
Dust cover

TORQUE : Nm (kg.cm, lb.ft)

7923GG98

Fig. 300 Exploded view of the rack and pinion assembly

1. Before servicing the vehicle, refer to the Precautions Section.

2. Wipe the reservoir fill cap clean before removal.

3. Fill the pump fluid reservoir to the proper level. The fluid level should be within the "FILL RANGE" listed on the exterior of the reservoir when the fluid is at normal ambient temperature, approximately 70–80°F (21–27°C).

4. Let the fluid settle in the system for at least 2 minutes.

5. Start the engine and let it run for a few seconds. Then turn the engine **OFF**.

6. Add fluid if necessary. Repeat the above procedure until the fluid level remains constant after running the engine.

7. Raise the front wheels off the ground.

8. Start the engine. Slowly turn the steering wheel right and left, lightly contacting the wheel stops.

9. Turn the engine off. Check the fluid level and add power steering fluid if necessary.

10. Start the engine. Lower the vehicle and turn the steering wheel slowly from lock to lock.

11. Stop the engine. Check the fluid level and refill as required.

12. If the fluid is extremely foamy, allow the vehicle to stabilize a few minutes, then repeat the above procedure.

SUSPENSION

CONTROL LINKS

REMOVAL & INSTALLATION

See Figure 301.

1. Before servicing the vehicle, refer to the Precautions Section.

2. Raise and safely support the vehicle.

3. Remove the front wheel and tire from the front hub.

☀ WARNING

Be careful not to damage the hub bolts when removing the front wheel and tire.

4. Remove the stabilizer bar control link (A) by removing the upper and lower mounting nuts (B).

5. Remove the stabilizer bar control link.

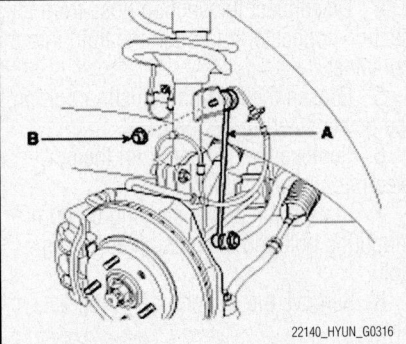

22140_HYUN_G0316

Fig. 301 Stabilizer control link and mounting nuts shown

To install:

6. Install the stabilizer bar control link.

7. Tighten the stabilizer bar control link upper and lower mounting nuts:

   a. 25–32 ft. lbs. (35–45 Nm)—Accent and Tiburon.

   b. 72–87 ft. lbs. (100–120 Nm)—Azera, Elantra, and Sonata.

8. Install the front wheels, and lower the vehicle. Tighten the wheel nuts to: 65–80 ft. lbs. (90–110 Nm).

LOWER BALL JOINT

REMOVAL & INSTALLATION

Bolt-On Type

1. Before servicing the vehicle, refer to the Precautions Section.

2. Remove or disconnect the following:
   • Front wheel
   • Ball joint stud from the knuckle
   • Ball joint from the lower control arm

To install:

➡ **Use a new split pin for assembly.**

3. Install or connect the following:
   • Ball joint and torque the mounting bolts to 69–87 ft. lbs. (95–120 Nm)
   • Stud nut and torque to 43–52 ft. lbs. (60–72 Nm)
   • Front wheel

4. Check and/or adjust the wheel alignment.

Press-In Type

1. Before servicing the vehicle, refer to the Precautions Section.

2. Remove or disconnect the following:
   • Front wheel
   • Lower control arm
   • Ball joint dust cover

3. Press the ball joint out of the lower control arm.

To install:

4. Press the ball joint into the control arm.

5. Install or connect the following:
   • Ball joint dust cover
   • Lower control arm and torque the stud nut to 43–52 ft. lbs. (60–72 Nm)
   • Front wheel

FRONT SUSPENSION

6. Check and/or adjust the wheel alignment.

LOWER CONTROL ARM

REMOVAL & INSTALLATION

Except Sonata & Azera

1. Before servicing the vehicle, refer to the Precautions Section.

2. Remove or disconnect the following:
   • Front wheel
   • Stabilizer bar link
   • Lower ball joint
   • Rear bushing bracket
   • Front bolt
   • Lower control arm

To install:

3. Install or connect the following:
   • Lower control arm and torque the front bolt to 72–87 ft. lbs. (100–120 Nm)
   • Rear bushing bracket. Tighten the bolts to 58–72 ft. lbs. (80–100 Nm)
   • Lower ball joint and torque the nut to 43–52 ft. lbs. (60–72 Nm)
   • Stabilizer bar link and torque the nut to 25–33 ft. lbs. (35–45 Nm)
   • Front wheel

4. Check and/or adjust the wheel alignment.

Sonata & Azera

1. Before servicing the vehicle, refer to the Precautions Section.

2. Remove or disconnect the following:
   • Front wheel
   • Lower ball joint nut, loosen only
   • Lower arm ball joint from the lower arm connector with Special Tool 09445-21000
   • Ball joint
   • Fork from the lower arm connector
   • Stabilizer bar link

- Control arm inner bushing bolts
- Lower control arm

To install:

3. Install or connect the following:
   - Lower control arm and torque the front bushing bolts to 74–88 ft. lbs. (100–120 Nm) and the rear bushing bolt to 88–103 ft. lbs. (120–140 Nm)
   - Stabilizer bar link and torque the nut to 26–33 ft. lbs. (35–45 Nm)
   - Damper fork lower bolt and torque the nut to 74–88 ft. lbs. (100–120 Nm)
   - Lower ball joint and torque the nut to 55–66 ft. lbs. (75–90 Nm)
   - Front wheel

4. Check and/or adjust the wheel alignment.

MACPHERSON STRUT

REMOVAL & INSTALLATION

Accent, Elantra & Tiburon

See Figure 302.

1. Before servicing the vehicle, refer to the Precautions Section.

2. Remove the strut from the vehicle and install a spring compressor.

3. Compress the coil spring so that the end of the spring comes away from the spring seat.

4. Remove or disconnect the following:
   - Upper strut mount
   - Upper spring seat
   - Compressed spring from the strut
   - Spring from the spring compressor

To install:

5. Compress the spring and install it on the strut.

6. Install or connect the following:
   - Upper spring seat and the upper strut mount and torque the nut to 29–36 ft. lbs. (40–50 Nm)
   - Strut to the vehicle

7. Check and/or adjust the wheel alignment.

Sonata

See Figure 303.

1. Before servicing the vehicle, refer to the Precautions Section.

2. Remove or disconnect the following:
   - Front wheel
   - Brake hose bracket from the mounting fork
   - Mounting fork and lower arm connecting bolt
   - Strut upper mounting nuts
   - Strut assembly

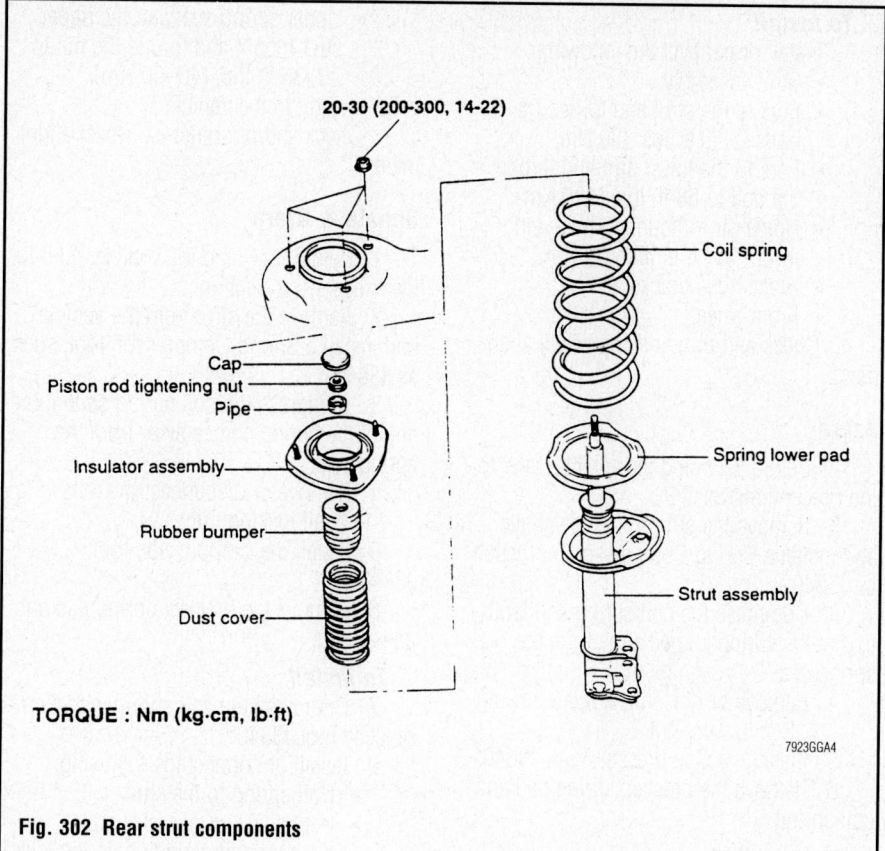

TORQUE : Nm (kg·cm, lb·ft)

7923GGA4

Fig. 302 Rear strut components

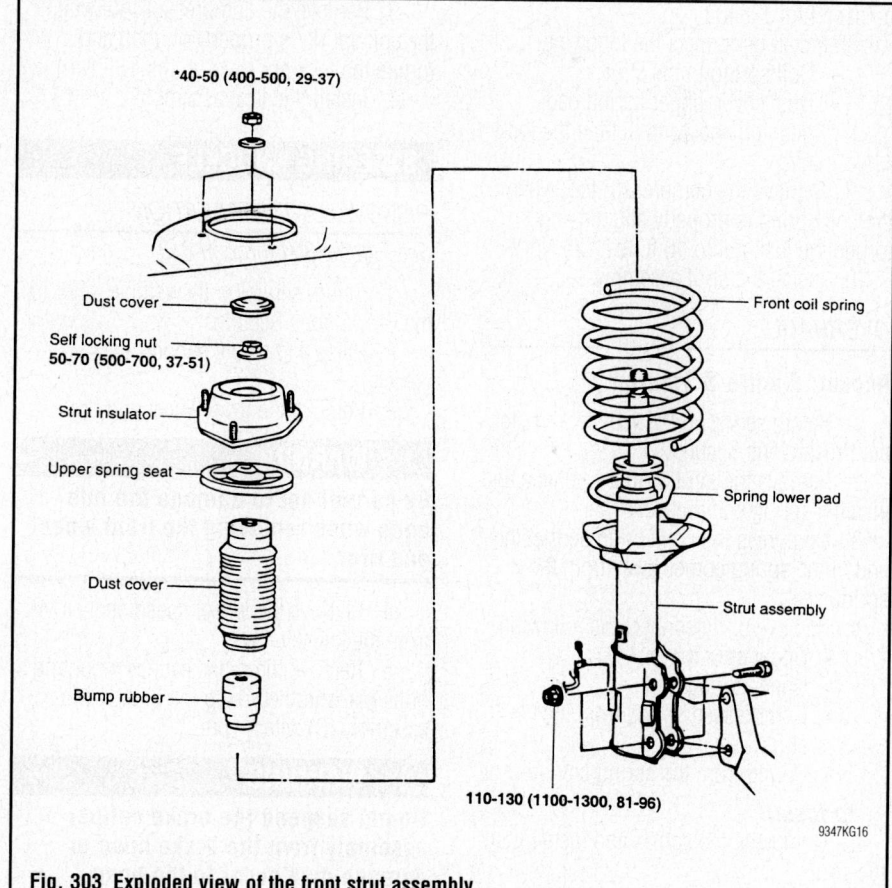

9347KG16

Fig. 303 Exploded view of the front strut assembly

To install:

3. Install or connect the following:
- Strut assembly
- Fork to the strut and torque the bolts to 59 ft. lbs. (80 Nm)
- Fork to the lower arm and torque the bolt to 88 ft. lbs. (120 Nm)
- Upper strut mounting nuts and torque to 36 ft. lbs. (50 Nm)
- Brake hose bracket
- Front wheel

4. Check and/or adjust the wheel alignment.

Azera

1. Before servicing the vehicle, refer to the Precautions Section.

2. Remove the strut from the vehicle and install a Spring Compressor Tool, such as J38402.

3. Compress the coil spring so that the end of the spring comes away from the spring seat.

4. Remove or disconnect the following:
- Self locking nut

5. Install the Compressor Tool, J38402.

6. Remove the bracket, spring pad and coil spring.

To install:

7. Compress the coil spring with Compressor Tool J38402.

8. Install or connect the following:
- Coil spring to the strut
- Dust cover, upper spring pad, bushing and hand tighten the lock nut

9. Remove the compressor tool when the coil spring is properly aligned and torque the lock nut to 18 ft. lbs. (25 Nm).

10. Install the strut assembly.

OVERHAUL

Accent, Elantra & Tiburon

1. Before servicing the vehicle, refer to the Precautions Section.

2. Remove the strut from the vehicle and install a spring compressor.

3. Compress the coil spring so that the end of the spring comes away from the spring seat.

4. Remove or disconnect the following:
- Upper strut mount
- Upper spring seat
- Compressed spring from the strut
- Spring from the spring compressor

To install:

5. Compress the spring and install it on the strut.

6. Install or connect the following:

- Upper spring seat and the upper strut mount and torque the nut to 29–36 ft. lbs. (40–50 Nm)
- Strut to the vehicle

7. Check and/or adjust the wheel alignment.

Sonata & Azera

1. Before servicing the vehicle, refer to the Precautions Section.

2. Remove the strut from the vehicle and install a Spring Compressor Tool, such as J38402.

3. Compress the coil spring so that the end of the spring comes away from spring seat.

4. Remove or disconnect the following:
- Self locking nut

5. Install the Compressor Tool, J38402.

6. Remove the bracket, spring pad and coil spring.

To install:

7. Compress the coil spring with Compressor Tool J38402.

8. Install or connect the following:
- Coil spring to the strut
- Dust cover, upper spring pad, bushing and hand tighten the lock nut

9. Remove the compressor tool when the coil spring is properly aligned and torque the lock nut to 18 ft. lbs. (25 Nm).

10. Install the strut assembly.

STEERING KNUCKLE

REMOVAL & INSTALLATION

See Figures 304 through 318.

1. Before servicing the vehicle, refer to the Precautions Section.

2. Raise and safely support the vehicle.

3. Remove the front wheel and tire.

✱✱ WARNING

Be careful not to damage the hub bolts when removing the front wheel and tire.

4. Remove the wheel speed sensor (A) from the knuckle.

5. Remove the brake caliper mounting bolts (A), and then hang the brake caliper assembly (B) with a wire.

✱✱ WARNING

Do not suspend the brake caliper assembly from the brake hose or damage may occur to the hose.

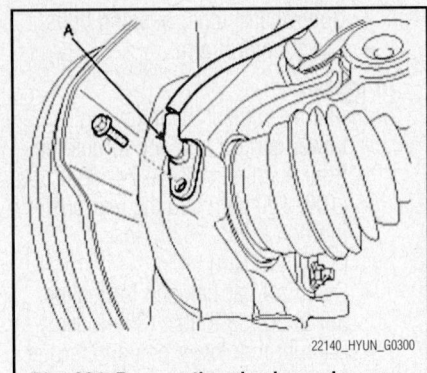

Fig. 304 Remove the wheel speed sensor (A) from the knuckle

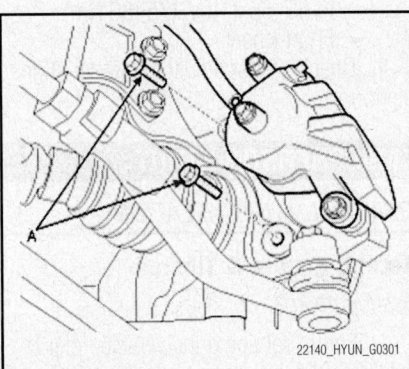

Fig. 305 Remove the brake caliper mounting bolts (A)

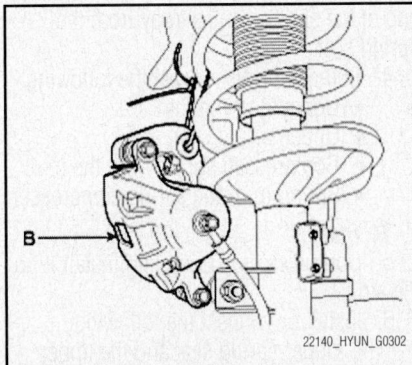

Fig. 306 Hang the brake caliper assembly (B) with a wire

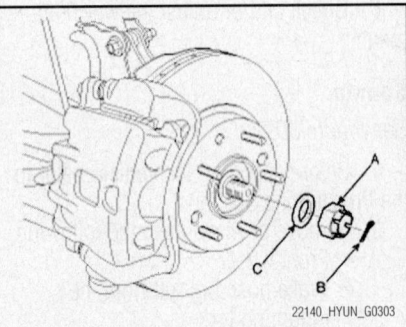

Fig. 307 Remove the split pin (B), the castle nut (A), and the washer (C) from the front hub

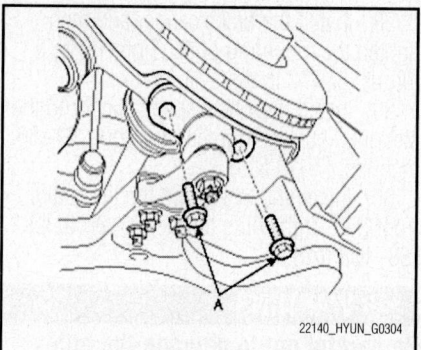

Fig. 308 Remove the ball joint assembly mounting bolt (A) from the steering knuckle

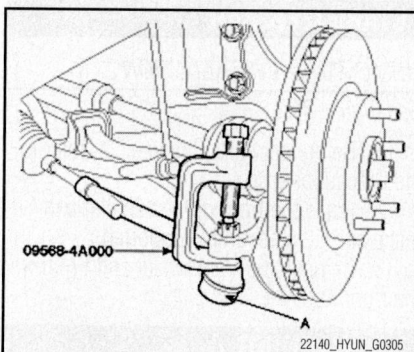

Fig. 309 Disconnect the ball joint (A) from the steering knuckle using the special tool (09568-4A000)

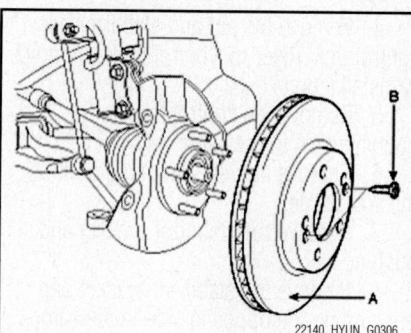

Fig. 310 Remove the brake disc (A) from the front hub assembly after removing the screws (B)

6. Remove the split pin (B), then remove castle nut (A) and washer (C) from the front hub.

7. Remove the ball joint assembly mounting bolt (A) from the steering knuckle.

8. Remove the tie rod end ball joint from the knuckle.

   a. Remove the split pin and castle nut.

   b. Disconnect the ball joint (A) from the steering knuckle using the special tool (09568-4A000).

➡**Apply a few drops of oil to the special tool at the boot contact end.**

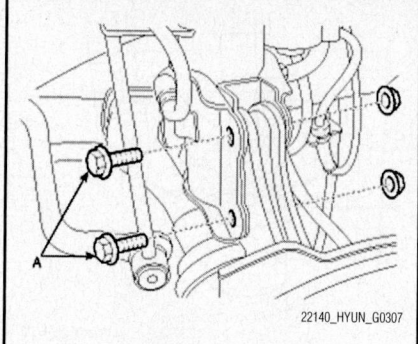

Fig. 311 Remove the strut assembly mounting bolts (A) and nuts

9. Remove the brake disc (A) from the front hub assembly after removing the screws (B).

10. Remove the strut assembly mounting bolts (A) and nuts.

11. Remove the hub and knuckle assembly.

✳✳ WARNING

Be careful not to damage the boot and rotor teeth.

➡**Disassemble the hub and steering knuckle assembly.**

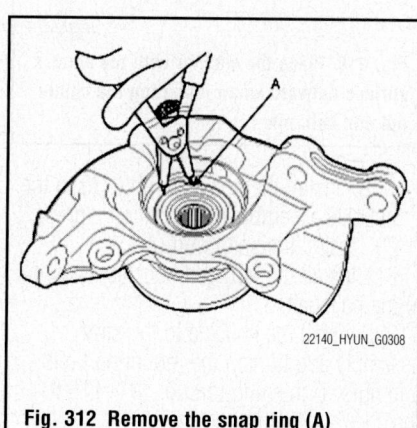

Fig. 312 Remove the snap ring (A)

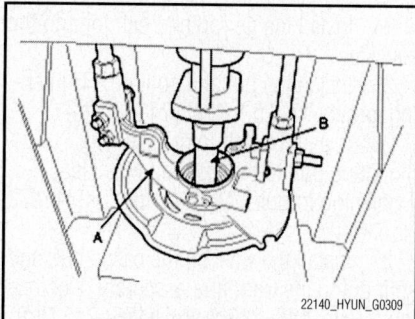

Fig. 313 Install the front knuckle assembly (A) on a press with an adapter (B) upon the hub assembly shaft

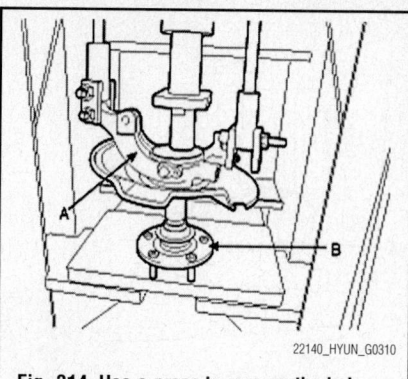

Fig. 314 Use a press to remove the hub assembly (B) from the knuckle assembly (A)

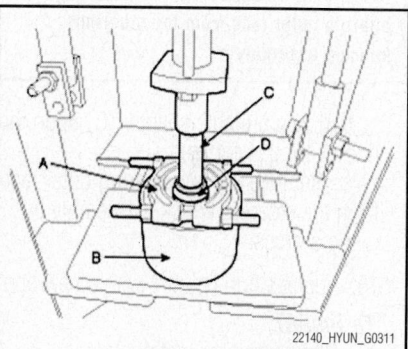

Fig. 315 Install tool (A) for removing the hub bearing inner race on the hub assembly. Use adapters (B) and (C) for the hub assembly shaft. Using a press, remove the hub bearing inner race (D) from the hub assembly

12. Remove the snap ring (A).

13. Remove the hub assembly from the knuckle assembly.

   a. Install the front knuckle assembly (A) on a press.

   b. Lay a suitable adapter (B) upon the hub assembly shaft.

   c. Remove the hub assembly (B) from the knuckle assembly (A) using a press.

14. Remove the hub bearing inner race from the hub assembly.

   a. Install a suitable tool (A) for removing the hub bearing inner race on the hub assembly.

   b. Lay the hub assembly and tool (A) upon a suitable adapter (B).

   c. Lay a suitable adapter (C) upon the hub assembly shaft.

   d. Remove the hub bearing inner race (D) from the hub assembly using a press.

15. Remove the hub bearing outer race from the knuckle assembly.

   a. Lay the hub assembly (A) upon a suitable adapter (B).

22140_HYUN_G0312

Fig. 316 Lay the hub assembly (A) on adapters (B) and (C) upon the hub bearing outer race. Using a press, remove the hub bearing outer race from the steering knuckle assembly

b. Lay a suitable adapter (C) upon the hub bearing outer race.

c. Remove the hub bearing outer race from the steering knuckle assembly by using a press.

16. Replace hub bearing with a new one.

To install:

17. Preparation for installation:

a. Check the hub for cracks and the splines for wear.

b. Check the brake disc for scoring and damage.

c. Check the steering knuckle for cracks.

d. Check the bearing for cracks or damage.

18. Install the hub bearing to the knuckle assembly.

a. Lay the knuckle assembly on a press.

b. Lay a new hub bearing upon the steering knuckle assembly.

c. Lay a suitable adapter upon the hub bearing.

d. Install the hub bearing to the steering knuckle assembly by using a press.

✳✳ WARNING

Do not press against the inner race of the hub bearing or damage may occur to the bearing assembly.

19. Install the hub assembly to the knuckle assembly.

a. Lay the hub assembly (A) upon a suitable adapter (B).

b. Lay the knuckle assembly (C) upon the hub assembly (A).

c. Lay a suitable adapter (D) upon the hub bearing.

22140_HYUN_G0313

Fig. 317 Lay the hub assembly (A) upon adapter (B). Lay the knuckle assembly (C) upon the hub assembly (A). Place an adapter (D) upon the hub bearing. Using a press, install the hub assembly (A) to the knuckle assembly (C)

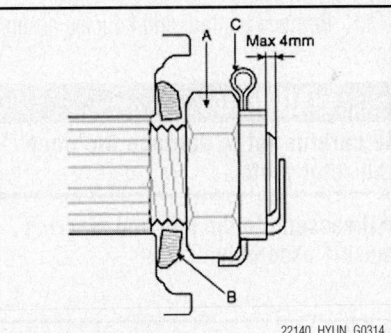

22140_HYUN_G0314

Fig. 318 Place the washer with the convex surface outward when installing the castle nut and split pin

d. Install the hub assembly (A) to the knuckle assembly (C) by using a press.

20. Install the snap ring (A).

21. Install the hub and knuckle assembly to the halfshaft.

22. Install the knuckle to the strut assembly and tighten the mounting bolts and nuts. Tightening torque: 101–116 ft. lbs. (137–157 Nm).

23. Install the brake disc to the front hub assembly, and tighten the screws.

24. Install the tie rod end ball joint to the knuckle.

25. Install the nut and split pin. Tightening torque: 17–25 ft. lbs. (24–33 Nm).

26. Install the ball joint assembly mounting bolt to the steering knuckle. Tightening torque: 72–87 ft. lbs. (98–118 Nm).

27. Install the washer, the castle nut, and split pin to the front hub assembly. Tightening torque: 145–203 ft. lbs. (196–275 Nm).

➡**The washer should be assembled with the convex surface outward when installing the castle nut and split pin.**

28. Install the brake caliper and then tighten the mounting bolts. Tightening torque: 58–72 ft. lbs. (79–98 Nm).

29. Install the wheel speed sensor to the steering knuckle. Tightening torque: 61–86 inch lbs. (7–10 Nm).

30. Install the wheel and the tire to the front hub. Tightening torque: 65–80 ft. lbs. (88–108 Nm).

✳✳ WARNING

Be careful not to damage the hub bolts when installing the front wheel and tire.

STABILIZER BAR

REMOVAL & INSTALLATION

See Figure 319.

1. Before servicing the vehicle, refer to the Precautions Section.

2. Raise the front of the vehicle, and make sure it is securely supported.

3. Remove the front wheel and tire from the front hub.

✳✳ WARNING

Be careful not to damage the hub bolts when removing the front wheel and tire.

4. Remove the nut and stabilizer bar control link. Refer to Control Links, removal & installation.

5. Remove the control link on the opposite side in the same way.

6. Remove the rear mounting bolts of the sub-frame.

7. Remove the stabilizer bracket and bushing.

8. Remove the stabilizer bracket and bushing on the opposite side in the same way.

9. Remove the stabilizer bar.

✳✳ WARNING

Be careful not to damage the power steering pressure tube.

To install:

10. Installation is the reverse of the removal.

11. Tighten the stabilizer bracket bolts to: 32–39 ft. lbs. (45–55 Nm).

12. Tighten the sub-frame mounting bolts to: 68–86 ft. lbs. (95–120 Nm).

13. Tighten the stabilizer bar control link upper and lower mounting nuts:

a. Accent and Tiburon: 25–32 ft. lbs. (35–45 Nm).

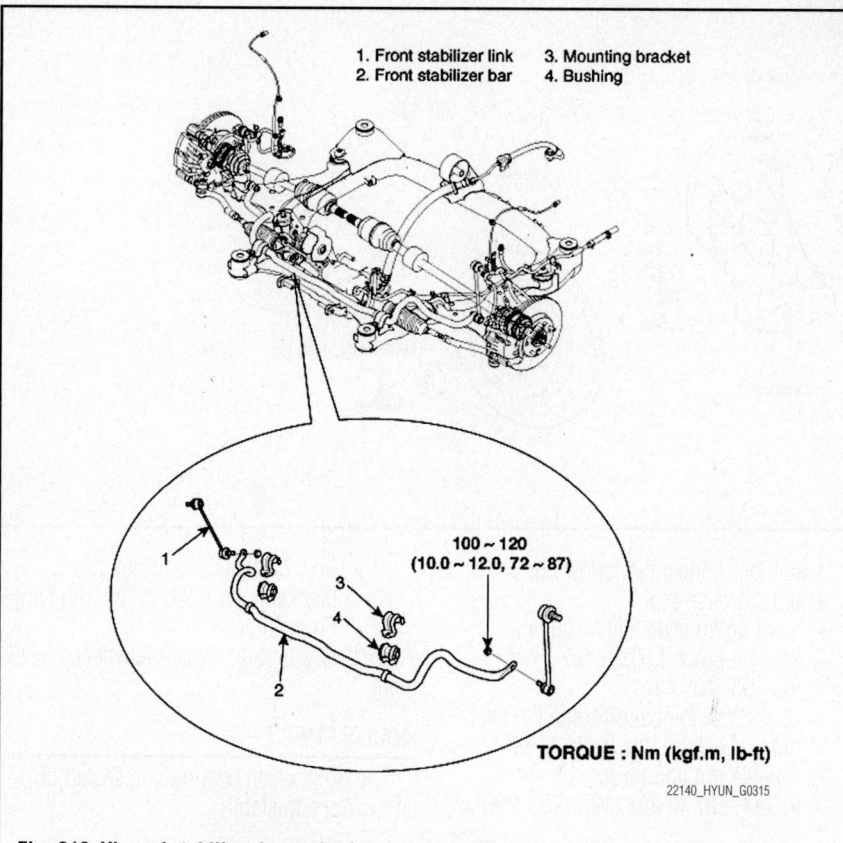

1. Front stabilizer link 3. Mounting bracket
2. Front stabilizer bar 4. Bushing

100 ~ 120
(10.0 ~ 12.0, 72 ~ 87)

TORQUE : Nm (kgf.m, lb-ft)

22140_HYUN_G0315

Fig. 319 View of stabilizer bar and related components

b. Azera, Elantra and Sonata: 72–87 ft. lbs. (100–120 Nm).

14. Tighten the wheel nuts to: 65–80 ft. lbs. (90–110 Nm).

UPPER BALL JOINT

REMOVAL & INSTALLATION

The upper ball joints are replaced with the upper control arms as an assembly.

UPPER CONTROL ARM

REMOVAL & INSTALLATION

Sonata & Azera

See Figure 320.

1. Before servicing the vehicle, refer to the Precautions Section.

2. Support the lower control arm assembly with a floor jack.

3. Remove or disconnect the following:
- Front wheel
- Ball joint nut, loosen only
- Upper arm ball joint from the steering knuckle with Special Tool 09568-34000
- Wheel house panel nuts
- Upper arm assembly
- Upper arm shaft

To install:

4. Install or connect the following:
- Upper control arm shaft
- Upper control arm assembly and torque the bolts to 73 ft. lbs. (100 Nm)
- Wheel house panel nuts and torque the nuts to 48 ft. lbs. (65 Nm)
- Upper arm ball joint to the steering knuckle and torque the bolts to 33 ft. lbs. (45 Nm)
- Front wheel

WHEEL HUB AND BEARING

REMOVAL & INSTALLATION

See Figure 321.

1. Before servicing the vehicle, refer to the Precautions Section.

2. Remove or disconnect the following:
- Front wheel
- Brake caliper
- Lower ball joint
- Spindle nut
- Knuckle pinch bolts
- Steering knuckle

3. Press the hub out of the wheel bearing.

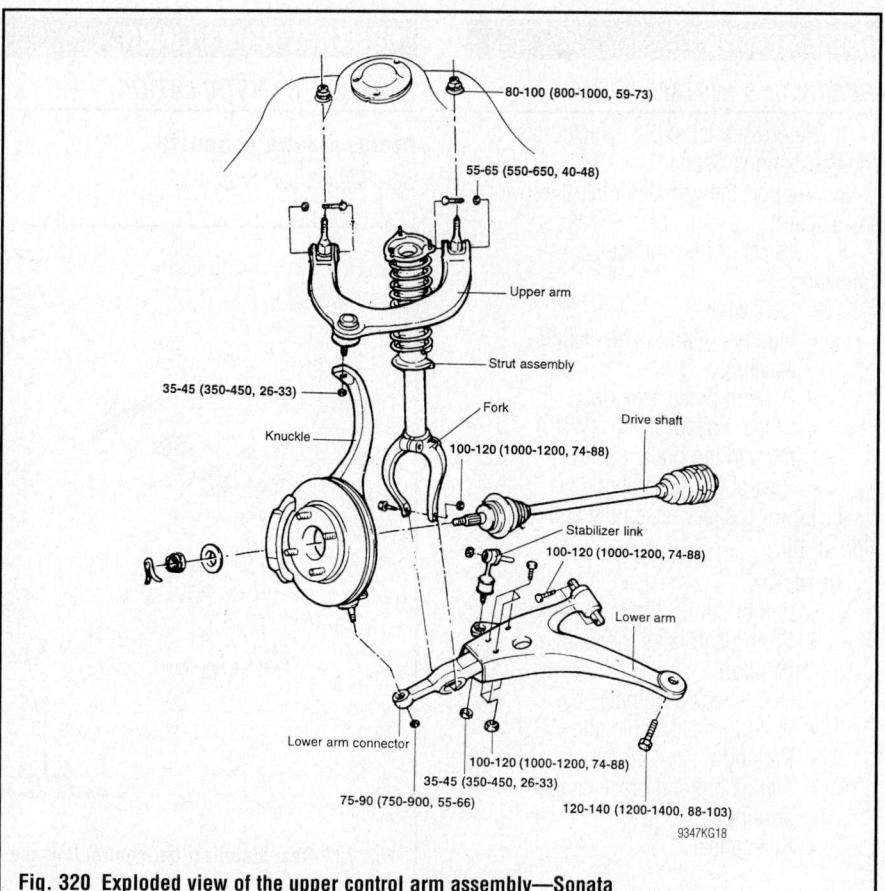

80-100 (800-1000, 59-73)

55-65 (550-650, 40-48)

Upper arm

Strut assembly

35-45 (350-450, 26-33)

Fork

Drive shaft

Knuckle

100-120 (1000-1200, 74-88)

Stabilizer link

100-120 (1000-1200, 74-88)

Lower arm

Lower arm connector

100-120 (1000-1200, 74-88)

35-45 (350-450, 26-33)

75-90 (750-900, 55-66)

120-140 (1200-1400, 88-103)

9347KG18

Fig. 320 Exploded view of the upper control arm assembly—Sonata

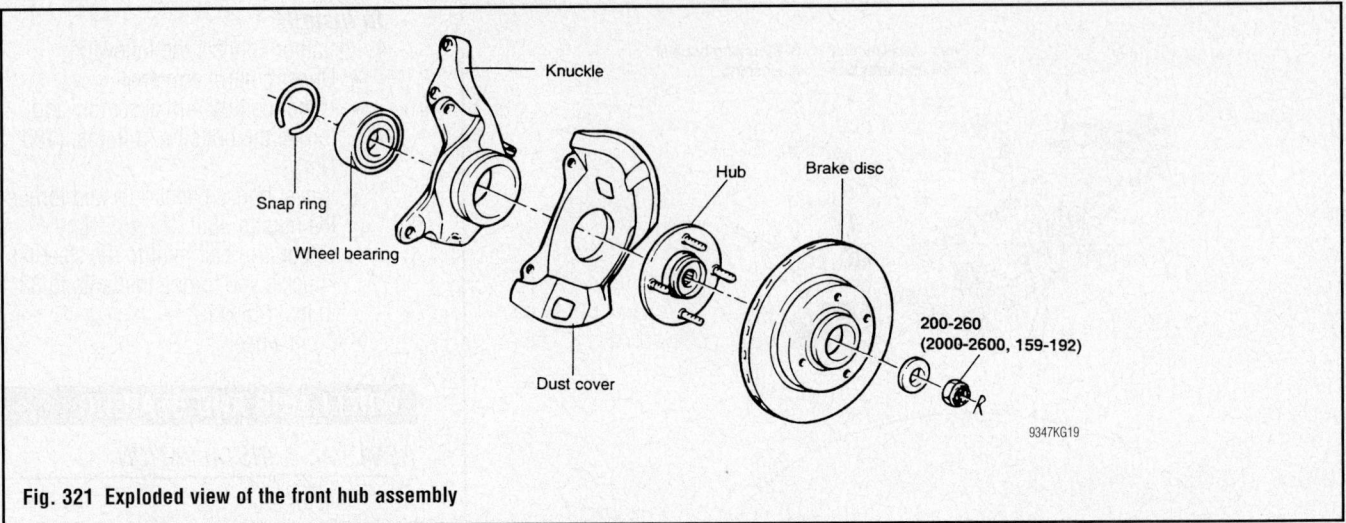

Fig. 321 Exploded view of the front hub assembly

4. Press the wheel bearings out of the steering knuckle.

5. If necessary, press the inner race off the hub.

To install:

6. Press the wheel bearings into the steering knuckle.

7. Install the outer grease seal and press the hub into the wheel bearings.

8. Install or connect the following:
- Inner grease seal
- Steering knuckle and torque the knuckle pinch bolts to 65–76 ft. lbs. (95–105 Nm)
- Lower ball joint and torque the stud nut to 43–52 ft. lbs. (60–72 Nm)
- Spindle nut and torque the nut to 144–187 ft. lbs. (195–253 Nm)

- Brake caliper and torque the bracket bolts to 50 ft. lbs. (68 Nm)
- Front wheel

9. Check and/or adjust the wheel alignment.

ADJUSTMENT

The front wheel bearing is a sealed unit and is not adjustable.

SUSPENSION

REAR SUSPENSION

COIL SPRING

REMOVAL & INSTALLATION

1. Before servicing the vehicle, refer to the Precautions Section.

2. Support the vehicle under the lower control arm.

3. Remove or disconnect the following:
- Rear wheel
- Flange nut and brake caliper assembly
- Parking brake assembly
- Wheel Speed Sensor (WSS) and the parking brake cable
- Rear shock assembly

4. Lower the jack assembly and remove the spring.

To install:

5. Install or connect the following:
- Spring and raise the jack into position
- Rear shock assembly
- Parking brake cable and WSS
- Parking brake assembly
- Flange nut and brake caliper assembly
- Rear wheel

CONTROL ARMS/LINKS

REMOVAL & INSTALLATION

Azera, Elantra & Sonata
See Figures 322 and 323.

1. Before servicing the vehicle, refer to the Precautions Section.

2. Remove the rear wheel and tire.

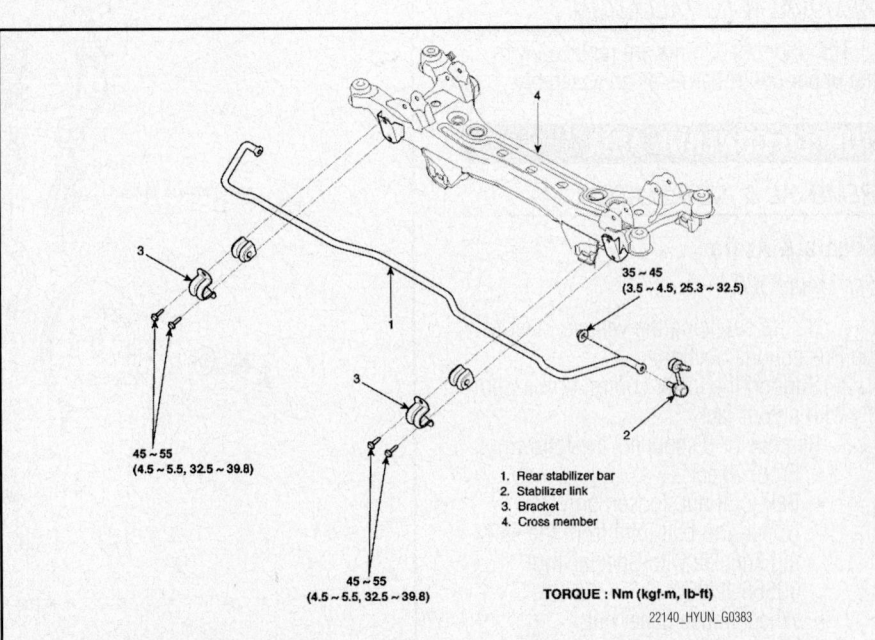

Fig. 322 Rear stabilizer bar control link and related rear suspension components—Azera and Sonata

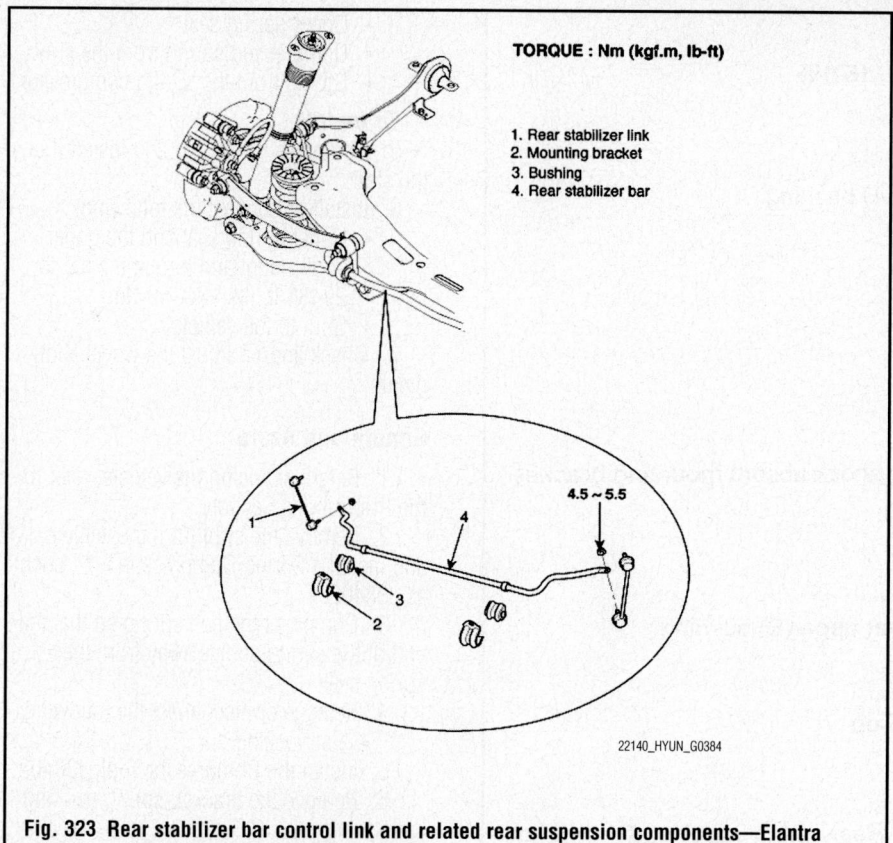

TORQUE : Nm (kgf.m, lb-ft)

1. Rear stabilizer link
2. Mounting bracket
3. Bushing
4. Rear stabilizer bar

4.5 ~ 5.5

22140_HYUN_G0384

Fig. 323 Rear stabilizer bar control link and related rear suspension components—Elantra

✲✲ WARNING

Be careful not to damage the hub bolts when removing the wheel and tire.

3. Remove the mounting nuts from the stabilizer bar control link.

4. Remove the stabilizer link.

To install:

5. Connect the rear stabilizer bar control link.

6. Install the stabilizer bar control link nuts and tighten to specification:

    a. Azera and Sonata: 25–33 ft. lbs. (35–45 Nm).

    b. Elantra: 33–40 ft. lbs. (45–55 Nm).

7. Install the rear wheel and tire and tighten the wheel nuts to 65–80 ft. lbs. (90–110 Nm).

Tiburon

See Figure 324.

1. Before servicing the vehicle, refer to the Precautions Section.

2. Remove the mounting nuts from the rear stabilizer bar control link.

3. Remove the stabilizer bar control link from the rear strut assembly and the stabilizer bar.

To install:

4. Install the mounting nuts on the stabilizer bar control link. Tighten to: 25–33 ft. lbs. (35–45 Nm).

5. Install the rear wheel and tire. Tighten the wheel nuts to: 65–80 ft. lbs. (90–110 Nm).

MACPHERSON STRUTS

REMOVAL & INSTALLATION

Accent, Elantra & Tiburon

1. Before servicing the vehicle, refer to the Precautions Section.

2. Remove or disconnect the following:

- Rear seatback assembly and wheel house cover
- Rear wheel
- Upper mounting nuts
- Brake hose and wheel speed sensor connectors
- Carrier mounting nuts
- Strut assembly

To install:

3. Install or connect the following:

- Strut assembly and torque the carrier mounting nuts to 66 ft. lbs. (90 Nm)
- Brake hose and wheel speed sensor connectors
- Upper mounting nuts and torque the nuts to 22 ft. lbs. (30 Nm) on the Accent and to 37 ft. lbs. (50 Nm) for the Elantra
- Wheel house cover and wheel
- Rear seatback

Sonata & Azera

See Figure 325.

1. Before servicing the vehicle, refer to the Precautions Section.

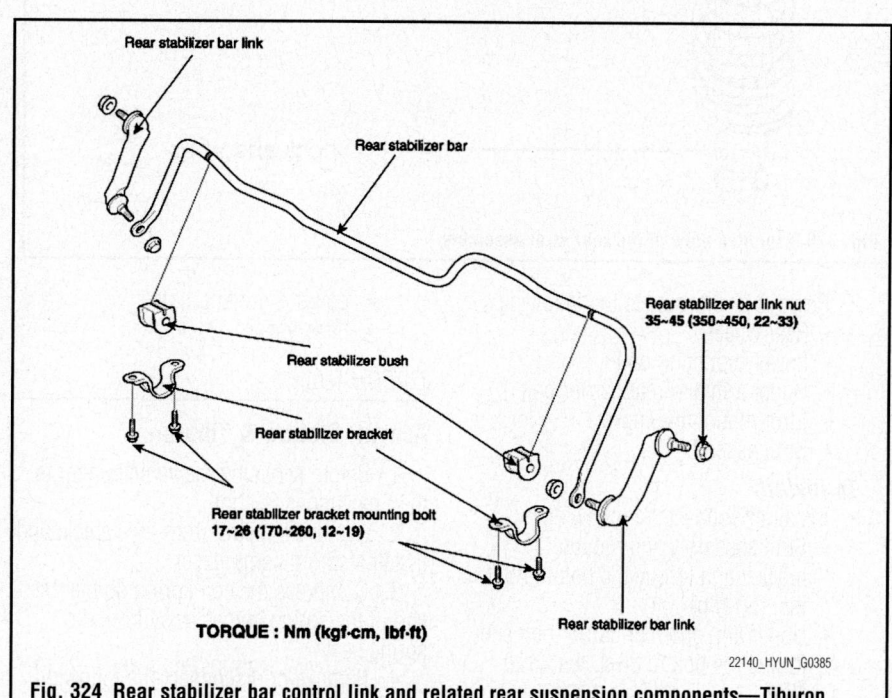

Rear stabilizer bar link

Rear stabilizer bar

Rear stabilizer bar link nut
35~45 (350~450, 22~33)

Rear stabilizer bush

Rear stabilizer bracket

Rear stabilizer bracket mounting bolt
17~26 (170~260, 12~19)

Rear stabilizer bar link

TORQUE : Nm (kgf-cm, lbf-ft)

22140_HYUN_G0385

Fig. 324 Rear stabilizer bar control link and related rear suspension components—Tiburon

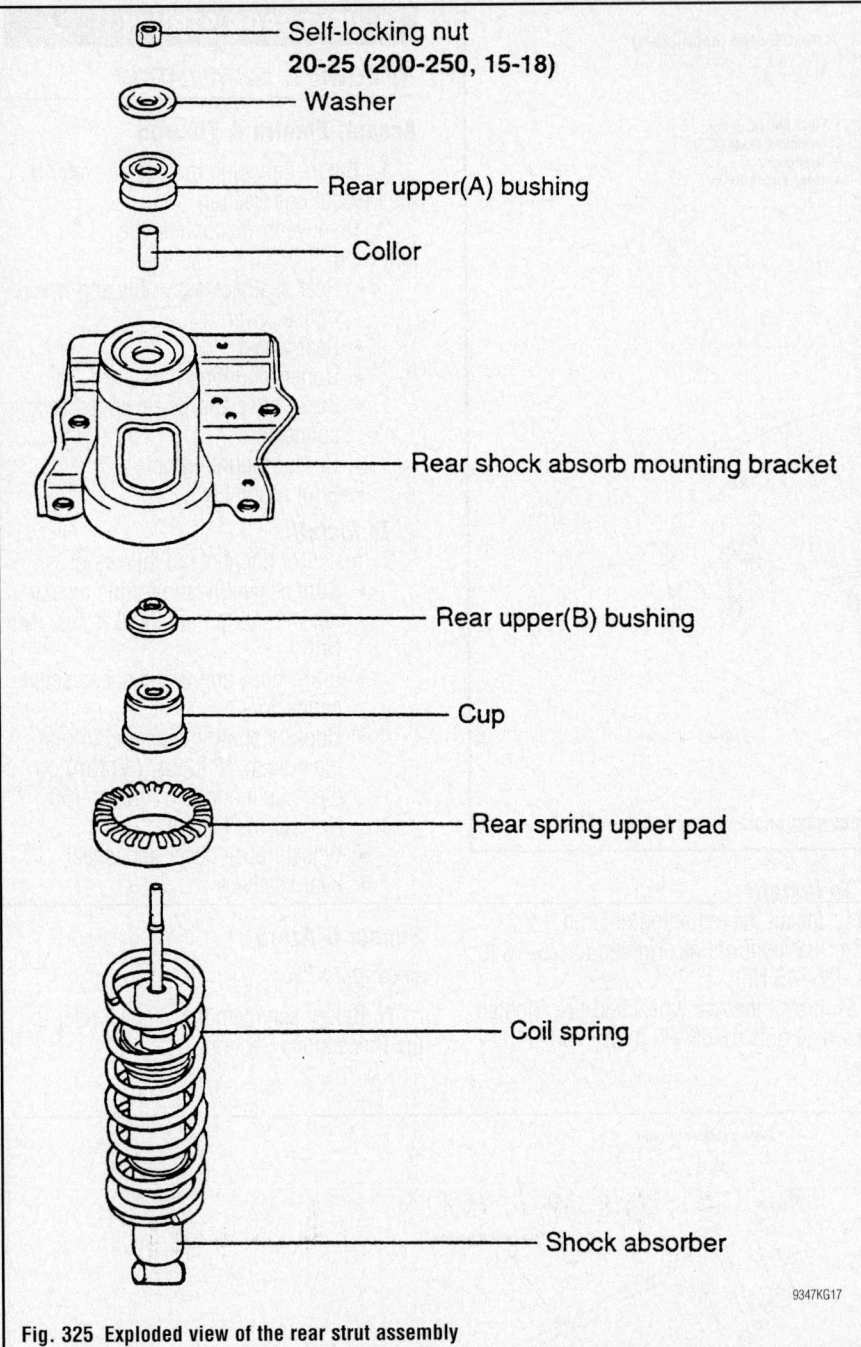

Self-locking nut
20-25 (200-250, 15-18)

Washer

Rear upper(A) bushing

Collor

Rear shock absorb mounting bracket

Rear upper(B) bushing

Cup

Rear spring upper pad

Coil spring

Shock absorber

9347KG17

Fig. 325 Exploded view of the rear strut assembly

2. Remove or disconnect the following:
- Rear wheel
- Lower mounting bolt
- Upper arm and rear carrier bolt
- Strut mounting bracket
- Strut assembly

To install:

3. Install or connect the following:
- Strut assembly and mounting bracket and torque the bolt to 36 ft. lbs. (50 Nm)
- Upper arm and rear carrier bolt and torque the bolt to 88 ft. lbs. (120 Nm)

- Lower mounting bolt
- Rear wheel

OVERHAUL

Accent, Elantra & Tiburon

1. Before servicing the vehicle, refer to the Precautions Section.

2. Remove the strut from the vehicle and install a spring compressor.

3. Compress the coil spring so that the end of the spring comes away from the spring seat.

4. Remove or disconnect the following:
- Upper strut mount

- Upper spring seat
- Compressed spring from the strut
- Spring from the spring compressor

To install:

5. Compress the spring and install it on the strut.

6. Install or connect the following:
- Upper spring seat and the upper strut mount and torque the nut to 29–36 ft. lbs. (40–50 Nm)
- Strut to the vehicle

7. Check and/or adjust the wheel alignment.

Sonata and Azera

1. Before servicing the vehicle, refer to the Precautions Section.

2. Remove the strut from the vehicle and install a Spring Compressor Tool, such as J38402.

3. Compress the coil spring so that the end of the spring comes away from the spring seat.

4. Remove or disconnect the following:
- Self locking nut

5. Install the Compressor Tool, J38402.

6. Remove the bracket, spring pad and coil spring.

To install:

7. Compress the coil spring with Compressor Tool J38402.

8. Install or connect the following:
- Coil spring to the strut
- Dust cover, upper spring pad, bushing and hand tighten the lock nut

9. Remove the compressor tool when the coil spring is properly aligned and torque the lock nut to 18 ft. lbs. (25 Nm).

10. Install the strut assembly.

SHOCK ABSORBER

REMOVAL & INSTALLATION

1. Before servicing the vehicle, refer to the Precautions Section.

2. Support the vehicle under the lower control arm.

3. Remove or disconnect the following:
- Rear wheel
- Upper shock absorber upper nut
- Lower shock absorber nut
- Shock absorber

To install:

4. Install or connect the following:
- Shock absorber. Tighten the upper nut to 15–22 ft. lbs. (20–30 Nm)
- Tighten the lower nut to 88–103 ft. lbs. (120–140 Nm)
- Rear wheel.

TESTING

1. Check the rubber parts for damage or deterioration.

2. Check for correct height and proper return of shock absorber to original height.

3. Check the shock absorber for abnormal resistance or unusual sounds.

4. Check for oil leakage around seals.

5. Replace if necessary.

WHEEL HUB AND BEARING

REMOVAL & INSTALLATION

With Rear Drum Brakes

See Figure 326.

1. Before servicing the vehicle, refer to the Precautions Section.

2. Remove or disconnect the following:
   - Rear wheel
   - Speed sensor, if equipped
   - Grease cap
   - Flange nut
   - Outer bearing
   - Brake drum
   - Inner grease seal
   - Inner bearing

3. Drive the bearing races out of the drum hub.

To install:

4. Install the inner and outer bearing races.

5. Apply grease to the bearings and to the cavity in the hub.

6. Install or connect the following:
   - Inner bearing
   - Inner grease seal
   - Brake drum
   - Outer bearing
   - Flange nut and torque the nut to 159–192 ft. lbs. (200–260 Nm)
   - Grease cap
   - Wheel speed sensor, if equipped
   - Rear wheel

With Rear Disc Brakes

See Figure 327.

1. Before servicing the vehicle, refer to the Precautions Section.

2. Release the parking brake.

3. Remove or disconnect the following:
   - Rear wheel
   - Wheel speed sensor, if equipped

- Brake caliper and rotor
- Rear axle hub bolts
- Tone wheel with Tool 09445-21000
- Carrier assembly
- Nut after unstaking it

4. Press out the rear axle hub.

5. Remove the bearing inner race with Tool 09445-21000.

6. Remove the bushings from the carrier with Tools 09453-33000B and 09545-21100.

To install:

7. Press in the bushings to the carrier with Tools 09453-33000B and 09545-21100.

8. Press in the bearing to the hub with Tool 09221-21000.

9. Tighten the flange nut to meet the concave portion of the spindle.

10. Press in the tone wheel with Tool 09221-21000. Torque the nut to 191 ft. lbs. (260 Nm).

11. Install the hub and bearing assembly to the backing plate and torque the bolts to 88 ft. lbs. (120 Nm).

12. Install the brake caliper and rotor.

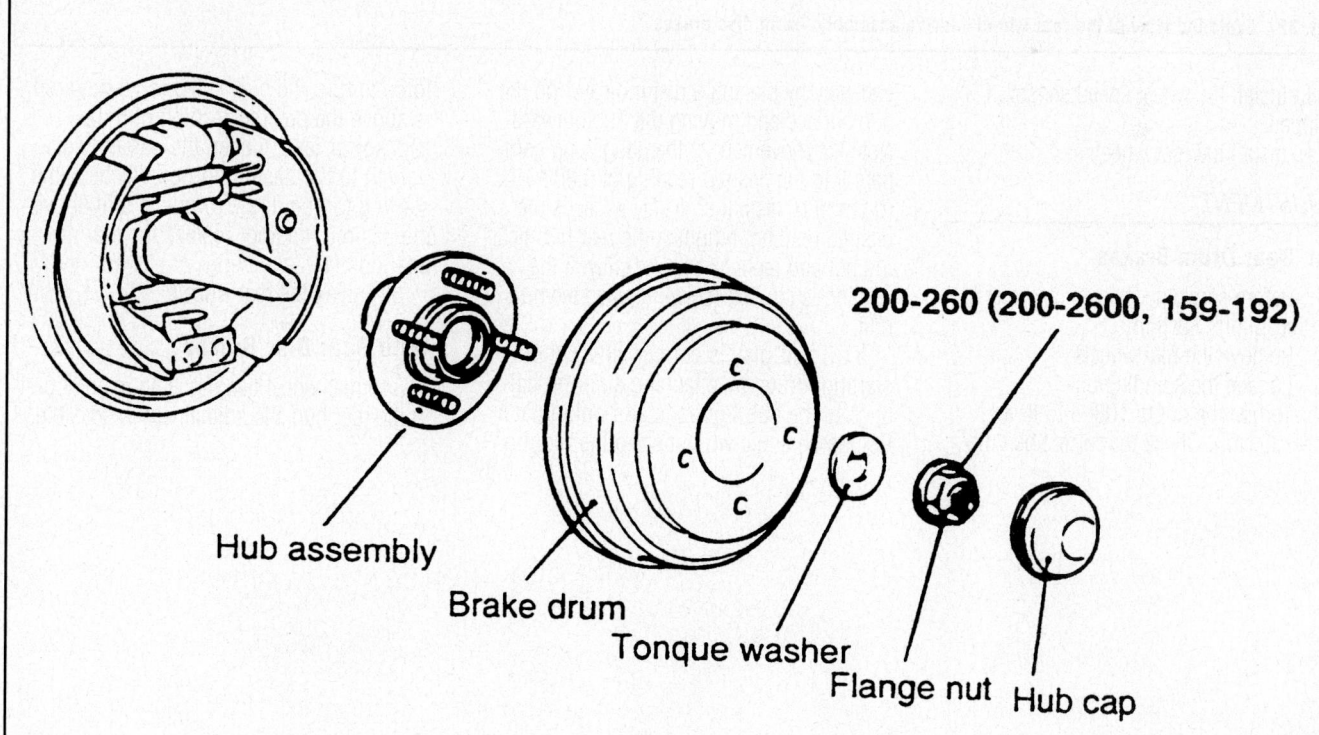

200-260 (200-2600, 159-192)

Hub assembly

Brake drum

Tonque washer

Flange nut Hub cap

9347KG20

Fig. 326 Exploded view of the rear hub assembly—with drum brakes

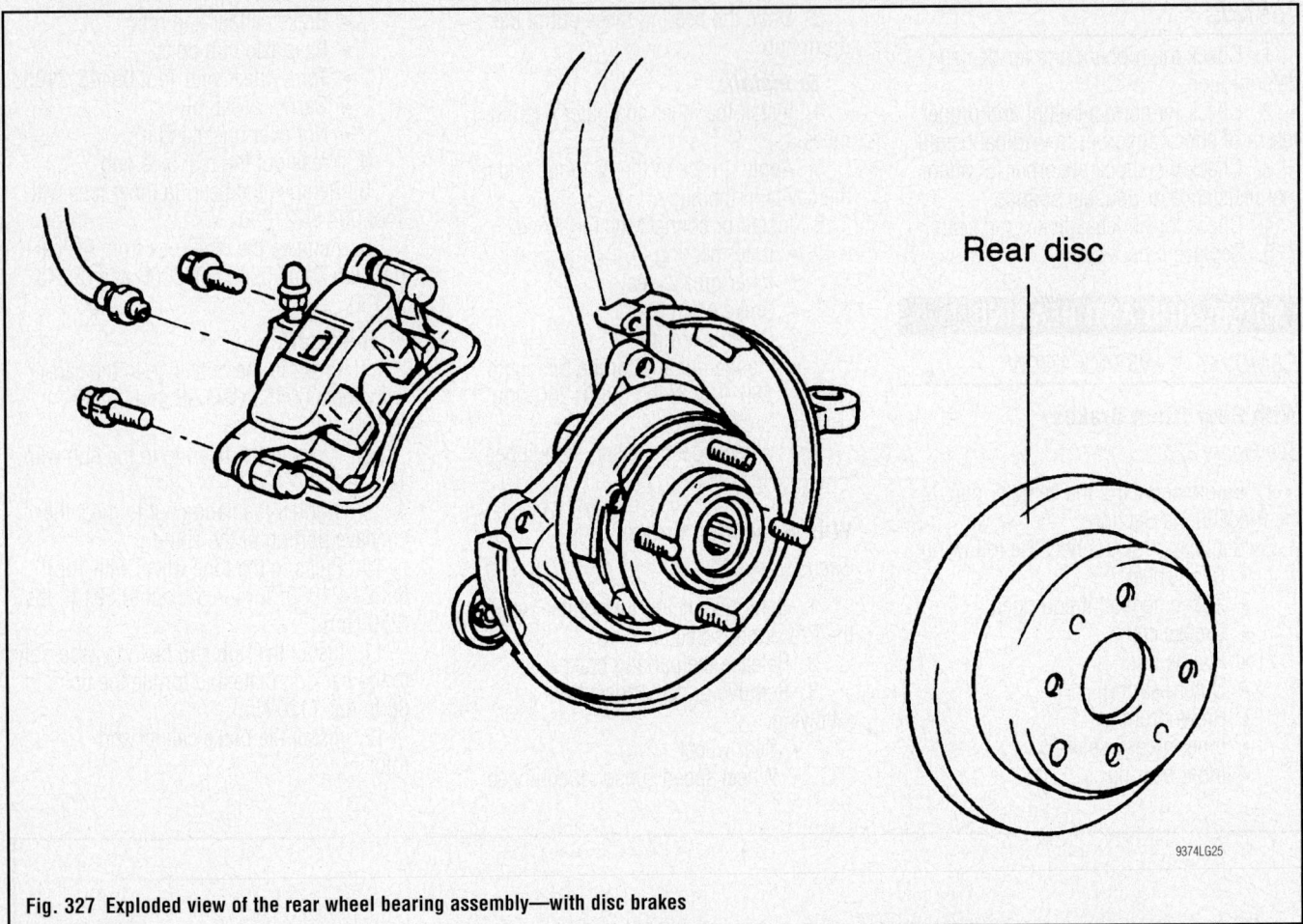

Rear disc

Fig. 327 Exploded view of the rear wheel bearing assembly—with disc brakes

13. Install the wheel speed sensor, if equipped.

14. Install the rear wheel.

ADJUSTMENT

With Rear Drum Brakes

1. Before servicing the vehicle, refer to the Precautions Section.

2. Remove the rear wheels.

3. Loosen the spindle nut.

4. Torque the nut to 108–145 ft. lbs. (150–200 Nm). Check for correct bearing end-play by placing a dial indicator on the hub surface and moving the hub outward. Note the movement of the gauge and compare it to the desired reading of 0.008 in. (0.2mm) or less. If end-play exceeds the desired reading, retighten the rear hub bearing nut and recheck the end-play. If the reading is still excessive, replace the hub unit.

5. If end-play is correct, check the starting torque by attaching a spring balance to the hub lug bolts and pulling at a 90 degree angle while noting the required force to turn the hub. If the force required is above the desired reading of 5 lbs. (2.3 kg) or less, loosen the nut and again tighten to the desired torque. Recheck the starting torque. If the torque is still above the desired reading, replace the rear bearings.

6. Install the rear wheels.

With Rear Disc Brakes

The rear wheel bearing is an integral part of the rear hub. No adjustment is possible.

HYUNDAI

Entourage

2

SPECIFICATIONS AND MAINTENANCE CHARTS

ENGINE AND VEHICLE IDENTIFICATION

Engine							Model Year	
Code ①	Liters (cc)	Cu. In.	Cyl.	Fuel Sys.	Engine Type	Eng. Mfg.	Code ②	Year
F	3.8 (3778)	230.55	V6	MPFI	DOHC	KIA	7	2007
							8	2008

MPFI: Multi-Point Fuel Injection

DOHC: Double Overhead Camshafts

① 8th digit of VIN

② 10th digit of VIN

22140_ENTO_C0001

GENERAL ENGINE SPECIFICATIONS

All measurements are given in inches.

Years	Model	Engine Displacement Liters	Engine Series VIN	Net Horsepower @ rpm	Net Torque @ rpm (ft. lbs.)	Bore x Stroke (in.)	Com- pression Ratio	Oil Pressure @ rpm
2007	Entourage	3.8	3	250@6000	253@3500	3.78x3.43	10.4:1	18.8@1000
2008	Entourage	3.8	3	250@6000	255@4500	3.78x3.43	10.4:1	18.8@1000

22140_ENTO_C0002

GASOLINE ENGINE TUNE-UP SPECIFICATIONS

Year	Engine Displacement Liters	Engine VIN	Spark Plug Gap (in.)	Ignition Timing (deg.) MT	AT	Fuel Pump (psi)	Idle Speed (rpm) MT	AT	Valve Clearance In.	Ex.
2007	3.8	3	0.039-0.043	①	①	54.3-55.8	②	②	HYD	HYD
2008	3.8	3	0.039-0.043	①	①	54.3-55.8	②	②	HYD	HYD

NOTE: The Vehicle Emission Control Information label reflects specification changes made during production.

Follow the figures on the label if they differ from those in this chart.

HYD: Hydraulic

① Ignition timing is preset and cannot be adjusted

② Idle speed is maintained by the Electronic Control Module (ECM)

22140_ENTO_C0003

CAPACITIES

Year	Model	Engine Displacement Liters	Engine VIN	Engine Oil with Filter (qts.)	Transmission (pts.) Manual	Transmission (pts.) Auto. ①	Fuel Tank (gal.)	Cooling System (qts.)
2007	Entourage	3.8	3	5.5	—	23.0	21.1	9.1
2008	Entourage	3.8	3	5.5	—	23.0	21.1	9.1

NOTE: All capacities are approximate. Add fluid gradually and check to be sure a proper fluid level is obtained.

① Drain and refill

22140_ENTO_C0004

FLUID SPECIFICATIONS

Year	Model	Engine Displacement Liters	Engine ID/VIN	Engine Oil	Auto. Trans.	Manual Trans.	Power Steering Fluid	Brake Master Cylinder	Cooling System
2007	Entourage	3.8	3	5W-20	①	—	PSF-3	②	③
2008	Entourage	3.8	3	5W-20	①	—	PSF-3	②	③

DOT: Department Of Transportation

① DIAMOND ATF SP-III, SK ATF SP-III

② DOT 3, DOT 4, or equivalent

③ Ethylene glycol base for aluminum radiator

22140_ENTO_C0005

VALVE SPECIFICATIONS

Year	Engine Displacement Liters	Engine VIN	Seat Angle (deg.)	Face Angle (deg.)	Spring Test Pressure (lbs. @ in.)	Spring Installed Height (in.)	Stem-to-Guide Clearance (in.) Intake	Stem-to-Guide Clearance (in.) Exhaust	Stem Diameter (in.) Intake	Stem Diameter (in.) Exhaust
2007	3.8	3	44.75-45.20	45.25-45.75	90.4-96.2 @0.953	1.7267	0.0008-0.0019	0.0012-0.0021	0.2151-0.2157	0.2149-0.2153
2008	3.8	3	44.75-45.20	45.25-45.75	90.4-96.2 @0.953	1.7267	0.0008-0.0019	0.0012-0.0021	0.2151-0.2157	0.2149-0.2153

22140_ENTO_C0006

CAMSHAFT AND BEARING SPECIFICATIONS CHART

All measurements are given in inches.

Year	Engine Displ. Liters	Engine ID/VIN	Journal Dia.	Brg. Oil Clearance	Shaft End-play	Runout	Journal Bore	Lobe Height Intake	Lobe Height Exhaust
2007	3.8	3	①	②	0.0008-0.0071	NA	NA	1.8425	1.8031
2008	3.8	3	①	②	0.0008-0.0071	NA	NA	1.8425	1.8031

NA: Not Available

① Intake No. 1 is 1.1009-1.1016 inch

　Intake No. 2, 3, 4 are 0.9430-0.9437 inch

　Exhaust No.1 is 1.1009-1.1016 inch

　Exhaust No. 2, 3, 4 are 0.9430-0.9437 inch

② Intake No. 1 is 0.0008-0.0022 inch

　Intake No. 2, 3, 4 are 0.0012-0.0026 inch

　Exhaust No.1 is 0.0008-0.0022 inch

　Exhaust No. 2, 3, 4 are 0.0012-0.0026 inch

22140_ENTO_C0007

CRANKSHAFT AND CONNECTING ROD SPECIFICATIONS

All measurements are given in inches.

Year	Engine Displacement Liters	Engine VIN	Crankshaft Main Brg. Journal Dia.	Crankshaft Main Brg. Oil Clearance	Crankshaft Shaft End-play	Crankshaft Thrust on No.	Connecting Rod Journal Diameter	Connecting Rod Oil Clearance	Connecting Rod Side Clearance
2007	3.8	3	2.7142-2.7149	0.0008-0.0016	0.0039-0.0110	3	2.2834-2.2842	0.0015-0.0022	0.0039-0.0098
2008	3.8	3	2.7142-2.7149	0.0008-0.0016	0.0039-0.0110	3	2.2834-2.2842	0.0015-0.0022	0.0039-0.0098

22140_ENTO_C0008

PISTON AND RING SPECIFICATIONS

All measurements are given in inches.

Year	Engine Displ. Liters	Engine VIN	Piston Clearance	Ring Gap Top Compression	Ring Gap Bottom Compression	Ring Gap Oil Control	Ring Side Clearance Top Compression	Ring Side Clearance Bottom Compression	Ring Side Clearance Oil Control
2007	3.8	3	0.0012-0.0020	0.0067-0.0126	0.0126-0.0185	0.0078-0.0275	0.0012-0.0027	0.0012-0.0027	0.0024-0.0059
2008	3.8	3	0.0012-0.0020	0.0067-0.0126	0.0126-0.0185	0.0078-0.0275	0.0012-0.0027	0.0012-0.0027	0.0024-0.0059

22140_ENTO_C0009

TORQUE SPECIFICATIONS
All readings in ft. lbs.

Year	Engine Displacement Liters	Engine VIN	Cylinder Head Bolts	Main Bearing Bolts	Rod Bearing Bolts	Crankshaft Damper Bolts	Flywheel Bolts	Manifold Intake	Manifold Exhaust	Spark Plugs	Oil Pan Drain Plug
2007	3.8	3	①	②	③	210-224	53-56	14-17	29-33	15-22	25-33
2008	3.8	3	①	②	③	210-224	53-56	14-17	29-33	15-22	25-33

① Step 1: 29 ft. lbs.

  Step 2: Plus 120 degrees

  Step 3: Plus 90 degrees

② M11 bolts (inner) Step 1: 36 ft. lbs.

  M11 bolts (inner) Step 2: Plus 90 degrees

  M8 bolts (outer) Step 1: 15 ft. lbs.

  M8 bolts (outer) Step 2: Plus 120 degrees

  M8 bolts (side): 22-23 ft. lbs.

③ Step 1: 15 ft. lbs.

  Step 2: Plus 90 degrees

22140_ENTO_C0010

WHEEL ALIGNMENT

Year	Model		Caster Range (+/-Deg.)	Caster Preferred Setting (Deg.)	Camber Range (+/-Deg.)	Camber Preferred Setting (Deg.)	Toe-in (Deg.)
2007	Entourage	Front	—	4 05' +/- 30'	—	0 +/- 30'	0 +/- 0.08
		Rear	—	—	—	-20 +/- 30'	0.14 +/- 0.08
2008	Entourage	Front	—	4 05' +/- 30'	—	0 +/- 30'	0 +/- 0.08
		Rear	—	—	—	-20 +/- 30'	0.14 +/- 0.08

22140_ENTO_C0011

TIRE, WHEEL AND BALL JOINT SPECIFICATIONS

Year	Model	OEM Tires Standard	OEM Tires Optional	Tire Pressures (psi) Front	Tire Pressures (psi) Rear	Wheel Size	Ball Joint Inspection	Lug Nut Torque (ft. lbs.)
2007	Entourage	P225/70R16	P235/60R17	35	35	6.5J x 16 6.5J x 17	①	65-80
2008	Entourage	P225/70R16	P235/60R17	35	35	6.5J x 16 6.5J x 17	①	65-80

① Replace if any measurable movement is found.

22140_ENTO_C0012

BRAKE SPECIFICATIONS
All measurements in inches unless noted

Year	Model		Brake Disc Original Thickness	Brake Disc Minimum Thickness	Brake Disc Maximum Runout	Brake Drum Diameter Original Inside Diameter	Brake Drum Diameter Max. Wear Limit	Brake Drum Diameter Maximum Machine Diameter	Minimum Lining Thickness	Brake Caliper Bracket Bolts (ft. lbs.)	Brake Caliper Mounting Bolts (ft. lbs.)
2007	Entourage	F	1.100	1.020	0.001	—	—	—	0.079	58-72	16-23
		R	0.470	0.410	0.002	—	—	—	0.079	36-43	16-23
2008	Entourage	F	1.100	1.020	0.001	—	—	—	0.079	58-72	16-23
		R	0.470	0.410	0.002	—	—	—	0.079	36-43	16-23

22140_ENTO_C0013

SCHEDULED MAINTENANCE INTERVALS
HYUNDAI—Entourage

TO BE SERVICED	TYPE OF SERVICE	VEHICLE MILEAGE INTERVAL (x1000)												
		7.5	15	22.5	30	37.5	45	52.5	60	67.5	75	82.5	90	97.5
Accessory drive belts	S/I	✓	✓	✓	✓	✓	✓	✓	✓	✓	✓	✓	✓	✓
Air cleaner element (engine)	R				✓				✓				✓	
Air conditioner system	S/I	Inspect the system operation annually												
Automatic transaxle fluid	S/I		✓		✓		✓		✓		✓		✓	
Automatic transaxle fluid & filter	R	105,000 miles (under normal usage)												
Brake lines, hoses, and connections	S/I		✓		✓		✓		✓		✓		✓	
Brake pads, calipers, & rotors	S/I		✓		✓		✓		✓		✓		✓	
Cabin air filter	R	Every 12 months or 10,000 miles												
Chassis and body fasteners	S/I				✓				✓				✓	
Cooling system hoses and coolant level	S/I		✓		✓		✓		✓		✓		✓	
Crankcase ventilation hose, vapor hose, and fuel filter cap	S/I				✓				✓					
Driveshafts and CV-boots	S/I		✓		✓		✓		✓		✓		✓	
Electronic throttle control	S/I		✓		✓		✓		✓		✓		✓	
Engine coolant	R								✓					
Engine oil and filter	R	✓	✓	✓	✓	✓	✓	✓	✓	✓	✓	✓	✓	✓
Exhaust pipe connections, muffler, and suspension bolts	S/I				✓				✓				✓	
Front and rear brakes	S/I				✓				✓				✓	
Fuel filter	R							✓						
Fuel lines, fuel hoses, and connections	S/I				✓				✓				✓	
Locks and hinges	S/I	✓	✓	✓	✓	✓	✓	✓	✓	✓	✓	✓	✓	✓
Spark plugs (standard)	R				✓				✓				✓	
Spark plugs (Iridium coated) 100,000 mile replacement	R													
Spark plugs (Platinum coated)	R								✓					
Steering gear rack, linkage, and boots	S/I				✓				✓				✓	
Upper and lower arm ball joints	S/I		✓		✓		✓		✓		✓		✓	
Wheel bearings	S/I				✓				✓				✓	

R: Replace   S/I: Service or Inspect

FREQUENT OPERATION MAINTENANCE (SEVERE SERVICE)

If a vehicle is operated under any of the following conditions it is considered severe service:

- Extremely dusty areas.

- 50% or more of the vehicle operation is in 90°F (32°C) or higher temperatures, or constant operation in temperatures below 32°F (0°C).

- Prolonged idling (vehicle operation in stop and go traffic).

- Frequent short running periods (engine does not warm to normal operating temperatures).

- Police, taxi, delivery usage or trailer towing usage.

Oil & oil filter: change every 3,000 miles.

Brake pads, calipers & rotors: service or inspect every 7,500 miles.

Driveshaft boots: service or inspect every 7,500 miles

Steering gear rack, linkage & boots: service or inspect every 7,500 miles.

Air cleaner filter: service or inspect every 15,000 miles.

Automatic transaxle fluid & filter: replace every 30,000 miles.

Rear brake drums & linings: service or inspect every 15,000 miles.

Spark plugs: service or inspect every 24,000 miles.

PRECAUTIONS

Before servicing any vehicle, please be sure to read all of the following precautions, which deal with personal safety, prevention of component damage, and important points to take into consideration when servicing a motor vehicle:

• Never open, service or drain the radiator or cooling system when the engine is hot; serious burns can occur from the steam and hot coolant.

• Observe all applicable safety precautions when working around fuel. Whenever servicing the fuel system, always work in a well-ventilated area. Do not allow fuel spray or vapors to come in contact with a spark, open flame, or excessive heat (a hot drop light, for example). Keep a dry chemical fire extinguisher near the work area. Always keep fuel in a container specifically designed for fuel storage; also, always properly seal fuel containers to avoid the possibility of fire or explosion. Refer to the additional fuel system precautions later in this section.

• Fuel injection systems often remain pressurized, even after the engine has been turned **OFF**. The fuel system pressure must be relieved before disconnecting any fuel lines. Failure to do so may result in fire and/or personal injury.

• Brake fluid often contains polyglycol ethers and polyglycols. Avoid contact with the eyes and wash your hands thoroughly after handling brake fluid. If you do get brake fluid in your eyes, flush your eyes with clean, running water for 15 minutes. If eye irritation persists, or if you have taken brake fluid internally, IMMEDIATELY seek medical assistance.

• The EPA warns that prolonged contact with used engine oil may cause a number of skin disorders, including cancer. You should make every effort to minimize your exposure to used engine oil. Protective gloves should be worn when changing oil. Wash your hands and any other exposed skin areas as soon as possible after exposure to used engine oil. Soap and water, or waterless hand cleaner should be used.

• All new vehicles are now equipped with an air bag system, often referred to as a Supplemental Restraint System (SRS) or Supplemental Inflatable Restraint (SIR) system. The system must be disabled before performing service on or around system components, steering column, instrument panel components, wiring and sensors. Failure to follow safety and disabling procedures could result in accidental air bag deployment, possible personal injury and unnecessary system repairs.

• Always wear safety goggles when working with, or around, the air bag system. When carrying a non-deployed air bag, be sure the bag and trim cover are pointed away from your body. When placing a non-deployed air bag on a work surface, always face the bag and trim cover upward, away from the surface. This will reduce the motion of the module if it is accidentally deployed. Refer to the additional air bag system precautions later in this section.

• Clean, high quality brake fluid from a sealed container is essential to the safe and proper operation of the brake system. You should always buy the correct type of brake fluid for your vehicle. If the brake fluid becomes contaminated, completely flush the system with new fluid. Never reuse any brake fluid. Any brake fluid that is removed from the system should be discarded. Also, do not allow any brake fluid to come in contact with a painted surface; it will damage the paint.

• Never operate the engine without the proper amount and type of engine oil; doing so WILL result in severe engine damage.

• Timing belt maintenance is extremely important. Many models utilize an interference-type, non-freewheeling engine. If the timing belt breaks, the valves in the cylinder head may strike the pistons, causing potentially serious (also time-consuming and expensive) engine damage. Refer to the maintenance interval charts for the recommended replacement interval for the timing belt, and to the timing belt section for belt replacement and inspection.

• Disconnecting the negative battery cable on some vehicles may interfere with the functions of the on-board computer system(s) and may require the computer to undergo a relearning process once the negative battery cable is reconnected.

• When servicing drum brakes, only disassemble and assemble one side at a time, leaving the remaining side intact for reference.

• Only an MVAC-trained, EPA-certified automotive technician should service the air conditioning system or its components.

BRAKES

GENERAL INFORMATION

See Figure 1.

The Anti-Lock Brake System (ABS) controls the hydraulic brake pressure of all four wheels during sudden braking and braking on hazardous road surfaces, preventing the wheels from locking. The ABS provides the following benefits:

• Enables steering around obstacles with a greater degree of certainty during panic braking

• Enables stopping during panic braking while allowing stability and control, even on curves

• If a malfunction occurs in the ABS, the system will operate as a normal brake (fail safe mode). A diagnostic function and a fail-safe system have been included for serviceability.

PRECAUTIONS

• Certain components within the ABS system are not intended to be serviced or repaired individually.

• Do not use rubber hoses or other parts not specifically specified for and ABS system. When using repair kits, replace all parts included in the kit. Partial or incorrect repair may lead to functional problems and require the replacement of components.

• Lubricate rubber parts with clean, fresh brake fluid to ease assembly. Do not

ANTI-LOCK BRAKE SYSTEM (ABS)

use shop air to clean parts; damage to rubber components may result.

• Use only DOT 3 brake fluid from an unopened container.

• If any hydraulic component or line is removed or replaced, it may be necessary to bleed the entire system.

• A clean repair area is essential. Always clean the reservoir and cap thoroughly before removing the cap. The slightest amount of dirt in the fluid may plug an orifice and impair the system function. Perform repairs after components have been thoroughly cleaned; use only denatured alcohol to clean components. Do not allow ABS components to come into contact with any substance

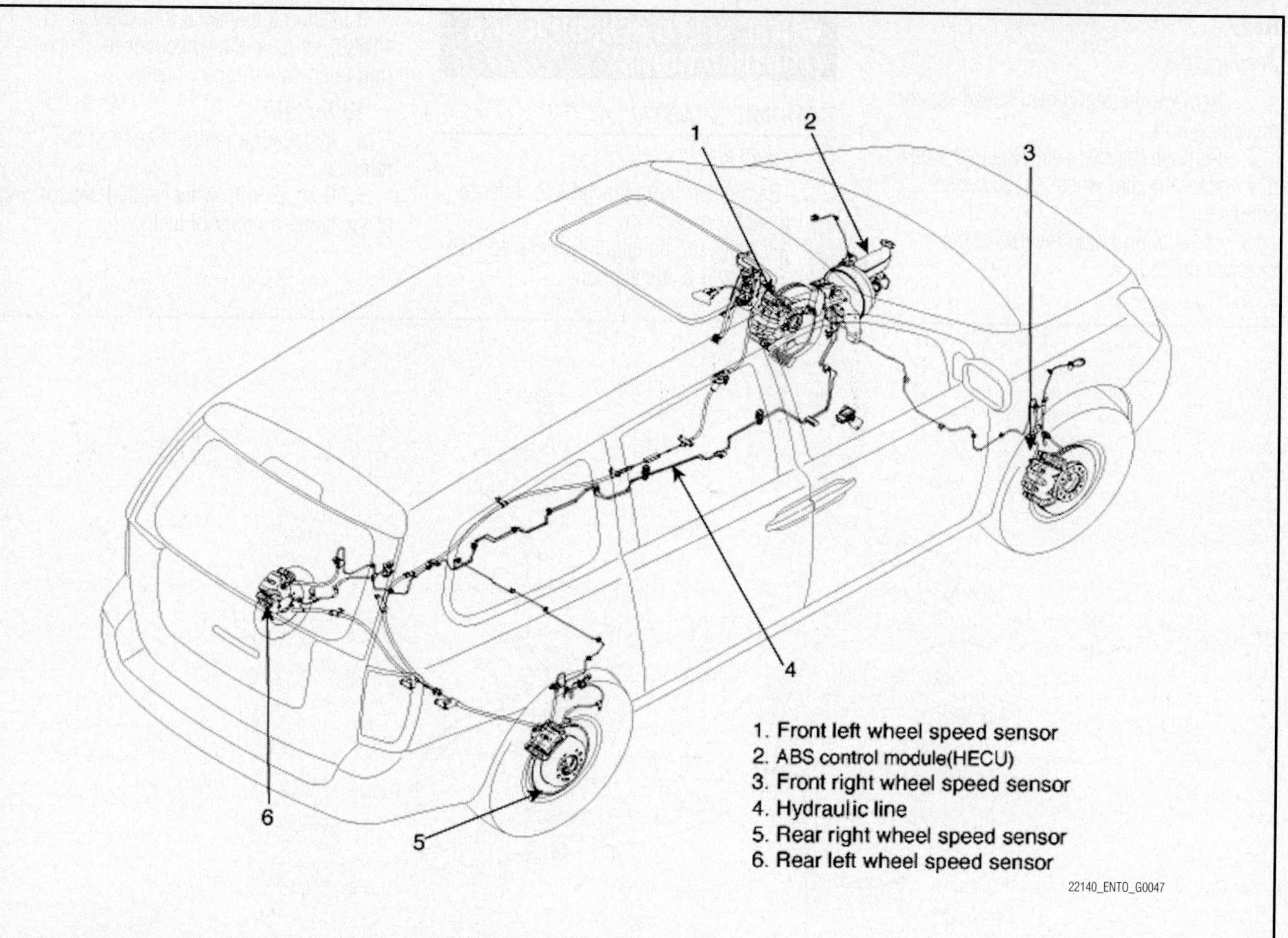

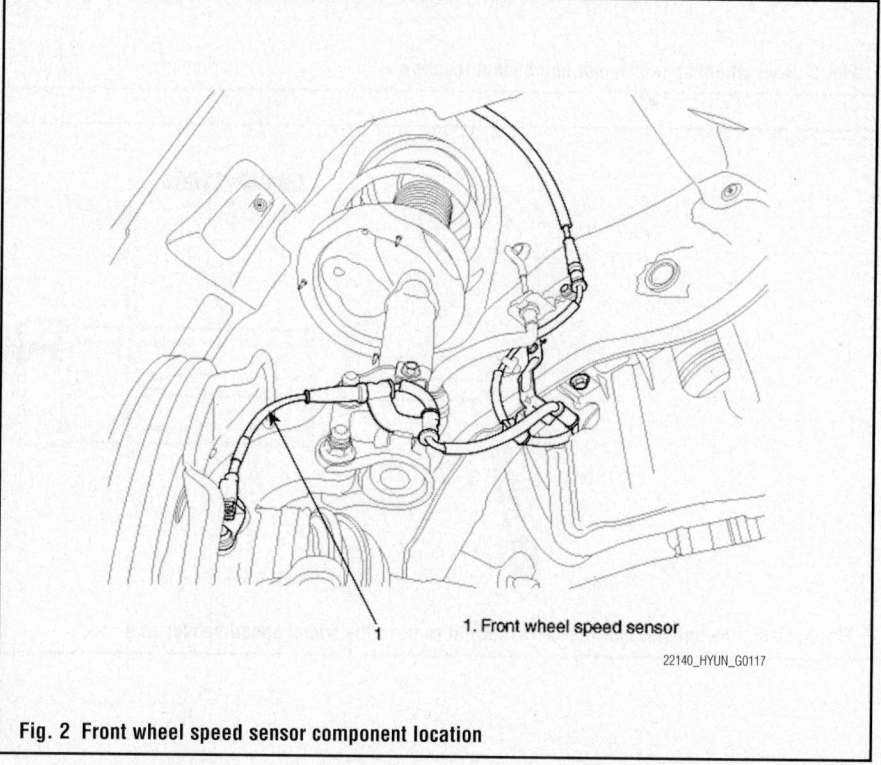

1. Front left wheel speed sensor
2. ABS control module(HECU)
3. Front right wheel speed sensor
4. Hydraulic line
5. Rear right wheel speed sensor
6. Rear left wheel speed sensor

22140_ENTO_G0047

Fig. 1 Components and component locations of the ABS system

containing mineral oil; this includes used shop rags.

• The Anti-Lock control unit is a microprocessor similar to other computer units in the vehicle. Ensure that the ignition switch is **OFF** before removing or installing controller harnesses. Avoid static electricity discharge at or near the controller.

• If any arc welding is to be done on the vehicle, the control unit should be unplugged before welding operations begin.

WHEEL SPEED SENSORS

REMOVAL & INSTALLATION

Front

See Figure 2.

1. Remove the front wheel speed sensor mounting bolt.
2. Remove the front wheel guard.
3. Disconnect the wheel speed sensor connector.
4. Remove the front wheel speed sensor.
5. Installation is the reverse of the removal procedure.

1. Front wheel speed sensor

22140_HYUN_G0117

Fig. 2 Front wheel speed sensor component location

Rear

See Figure 3.

1. Remove the rear wheel speed sensor mounting bolt.

2. Remove the rear seat side pad then disconnect the rear wheel speed sensor connector.

3. Installation is the reverse of the removal procedure.

WHEEL SPEED SENSOR RINGS (TOOTHED RINGS)

REMOVAL & INSTALLATION

See Figure 4.

1. Before servicing the vehicle, refer to the Precautions Section.

2. Remove the halfshaft(s). Refer to Halfshafts, removal & installation.

3. Using the special tool (09432-11000), remove the wheel speed sensor ring (also called tone wheel).

To install:

4. Installation is the reverse of the removal.

5. Replace with wheel speed sensor ring of the same number of teeth.

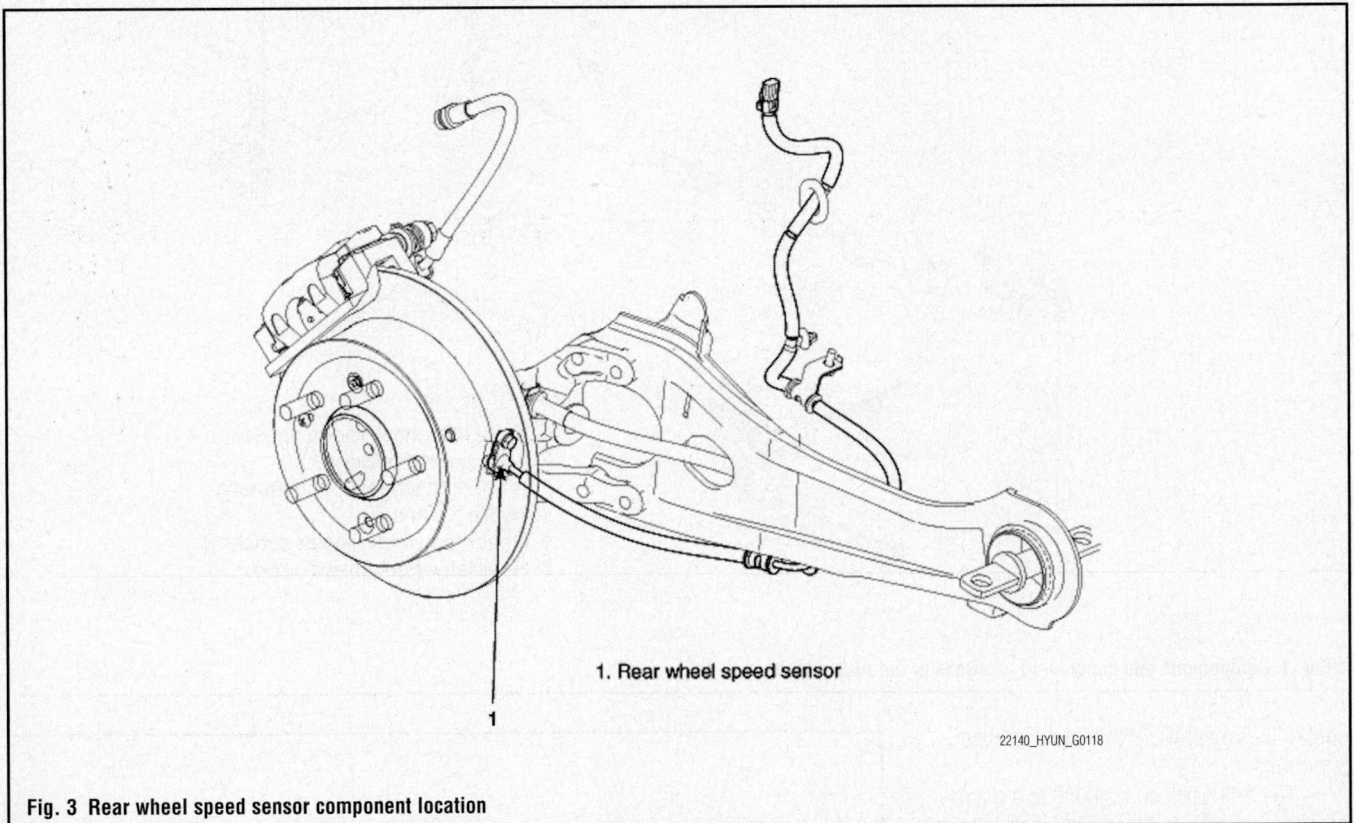

1. Rear wheel speed sensor

22140_HYUN_G0118

Fig. 3 Rear wheel speed sensor component location

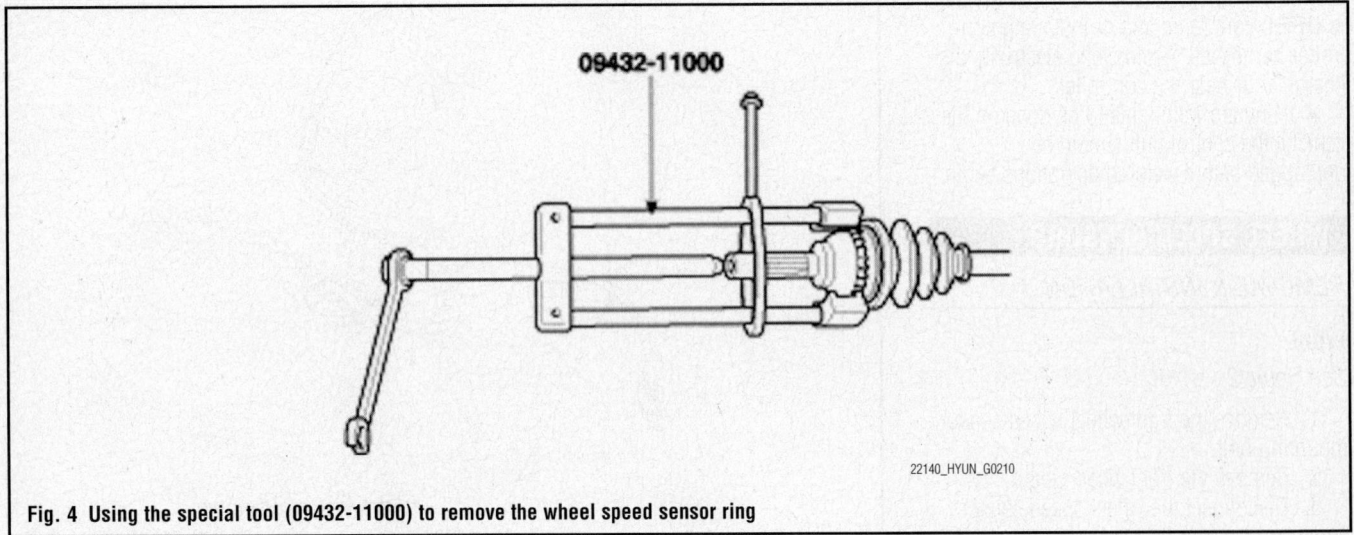

09432-11000

22140_HYUN_G0210

Fig. 4 Using the special tool (09432-11000) to remove the wheel speed sensor ring

BRAKES — BLEEDING THE BRAKE SYSTEM

BLEEDING PROCEDURE

BLEEDING PROCEDURE

These vehicles come standard with a 4-wheel Anti-Lock Braking System (ABS). Please refer to the section, Bleeding the ABS System.

BLEEDING THE ABS SYSTEM

This procedure should be followed to ensure adequate bleeding of air and the filling of the ABS unit, the brake lines, and the master cylinder with brake fluid.

1. Before servicing the vehicle, refer to the Precautions Section.
2. Remove the reservoir cap and fill the brake reservoir with brake fluid.

✳✳ WARNING

If there is any brake fluid on any painted surface, wash it off immediately.

➡ **When pressure bleeding, do not depress the brake pedal. Recommended brake fluid: DOT3 or DOT4.**

3. Connect a clear plastic tube to the wheel cylinder bleeder plug and insert the other end of the tube into a clear plastic bottle that is half filled with clean brake fluid.
4. Connect the Hi-Scan Pro® to the data link connector located underneath the dash panel.
5. Select and operate according to the instructions on the Hi-Scan Pro® screen.

✳✳ CAUTION

You must obey the maximum operating time of the ABS motor with the Hi-Scan Pro® to prevent the motor pump from burning.

6. Select Hyundai vehicle diagnosis.
7. Select vehicle name.
8. Select Anti-Lock Brake system.
9. Select air bleeding mode.
10. Press "YES" to operate motor pump and solenoid valve.

✳✳ WARNING

Wait 60 seconds before operating the air bleeding or damage to the motor may occur.

11. Wait 60 seconds before operating the air bleeding.
12. Pump the brake pedal several times, and then loosen the bleeder screw until fluid starts to run out without bubbles. Then, close the bleeder screw.
13. Repeat until there are no more bubbles in the fluid for each wheel.

BRAKES — FRONT DISC BRAKES

✳✳ CAUTION

Dust and dirt accumulating on brake parts during normal use may contain asbestos fibers from production or aftermarket brake linings. Breathing excessive concentrations of asbestos fibers can cause serious bodily harm. Exercise care when servicing brake parts. Do not sand or grind brake lining unless equipment used is designed to contain the dust residue. Do not clean brake parts with compressed air or by dry brushing. Cleaning should be done by dampening the brake components with a fine mist of water, then wiping the brake components clean with a dampened cloth. Dispose of cloth and all residue containing asbestos fibers in an impermeable container with the appropriate label. Follow practices prescribed by the Occupational Safety and Health Administration (OSHA) and the Environmental Protection Agency (EPA) for the handling, processing, and disposing of dust or debris that may contain asbestos fibers.

BRAKE CALIPER

REMOVAL & INSTALLATION

See Figure 5.

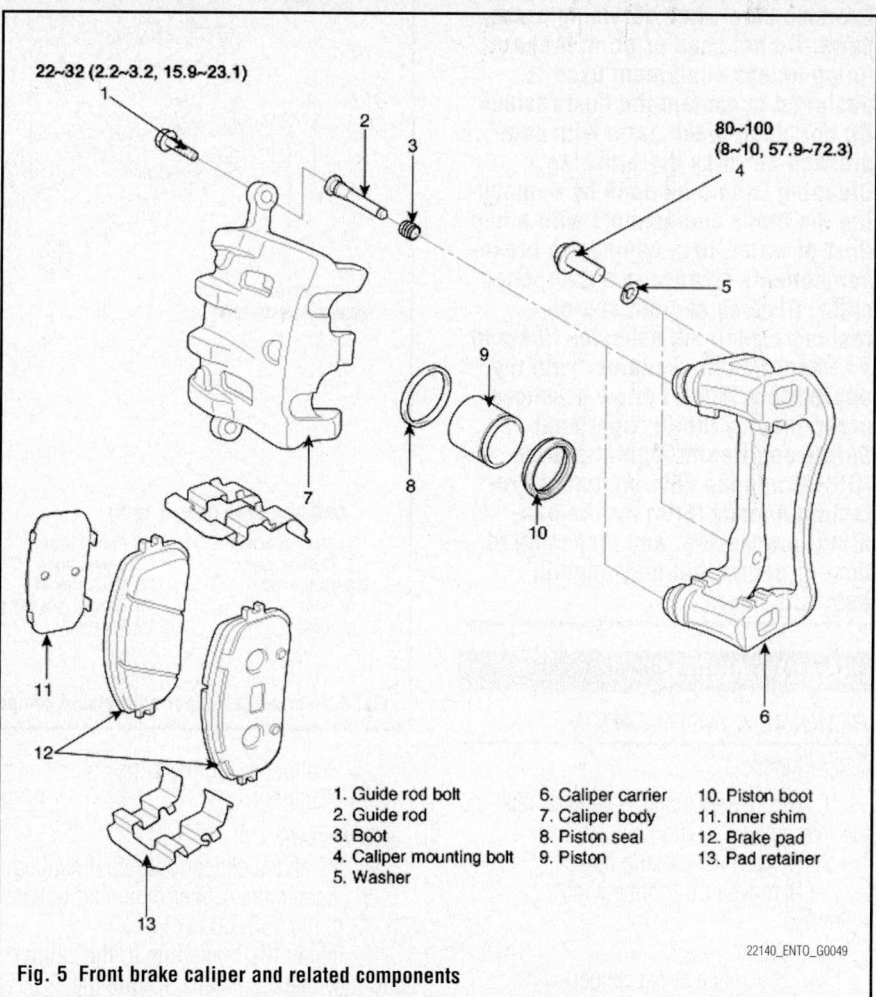

22~32 (2.2~3.2, 15.9~23.1)

80~100 (8~10, 57.9~72.3)

1. Guide rod bolt	6. Caliper carrier
2. Guide rod	7. Caliper body
3. Boot	8. Piston seal
4. Caliper mounting bolt	9. Piston
5. Washer	10. Piston boot
	11. Inner shim
	12. Brake pad
	13. Pad retainer

22140_ENTO_G0049

Fig. 5 Front brake caliper and related components

1. Before servicing the vehicle, refer to the Precautions Section.
2. Remove or disconnect the following:
   - Wheel
   - Brake hose from the caliper
   - Caliper mounting bolts
   - Caliper

To install:
3. Install or connect the following:
   - Caliper
   - Caliper mounting bolts and tighten to 58–73 ft. lbs. (80–100 Nm)
   - Brake hose to the caliper and tighten the fitting to 18–22 ft. lbs. (25–30 Nm)
   - Front wheel

4. Bleed the brake system.
5. Before attempting to move the vehicle, pump the brake pedal to seat the pads against the rotors. Make sure the vehicle has a firm brake pedal. Check the level of the brake fluid and add fluid if necessary.

DISC BRAKE PADS

REMOVAL & INSTALLATION
See Figure 5.

1. Before servicing the vehicle, refer to the Precautions Section.
2. Remove or disconnect the following:
   - Wheel

- Caliper mounting bolts
- Caliper and support to one side with a wire.

✳✳ WARNING
Do not let the caliper hang by the hose.

- Pads and shims

To install:
3. Install the pads, clips, and shims.
4. Bottom the caliper piston using tool 09581-11000 or a C-clamp.
5. Install the caliper mounting bolts and tighten to 16–24 ft. lbs. (22–32 Nm).
6. Install the wheel.

BRAKES ▮ REAR DISC BRAKES

✳✳ CAUTION

Dust and dirt accumulating on brake parts during normal use may contain asbestos fibers from production or aftermarket brake linings. Breathing excessive concentrations of asbestos fibers can cause serious bodily harm. Exercise care when servicing brake parts. Do not sand or grind brake lining unless equipment used is designed to contain the dust residue. Do not clean brake parts with compressed air or by dry brushing. Cleaning should be done by dampening the brake components with a fine mist of water, then wiping the brake components clean with a dampened cloth. Dispose of cloth and all residue containing asbestos fibers in an impermeable container with the appropriate label. Follow practices prescribed by the Occupational Safety and Health Administration (OSHA) and the Environmental Protection Agency (EPA) for the handling, processing, and disposing of dust or debris that may contain asbestos fibers.

BRAKE CALIPER

REMOVAL & INSTALLATION
See Figure 6.

1. Before servicing the vehicle, refer to the Precautions Section.
2. Release the parking brake.
3. Remove or disconnect the following:
   - Wheel
   - Brake line at the caliper

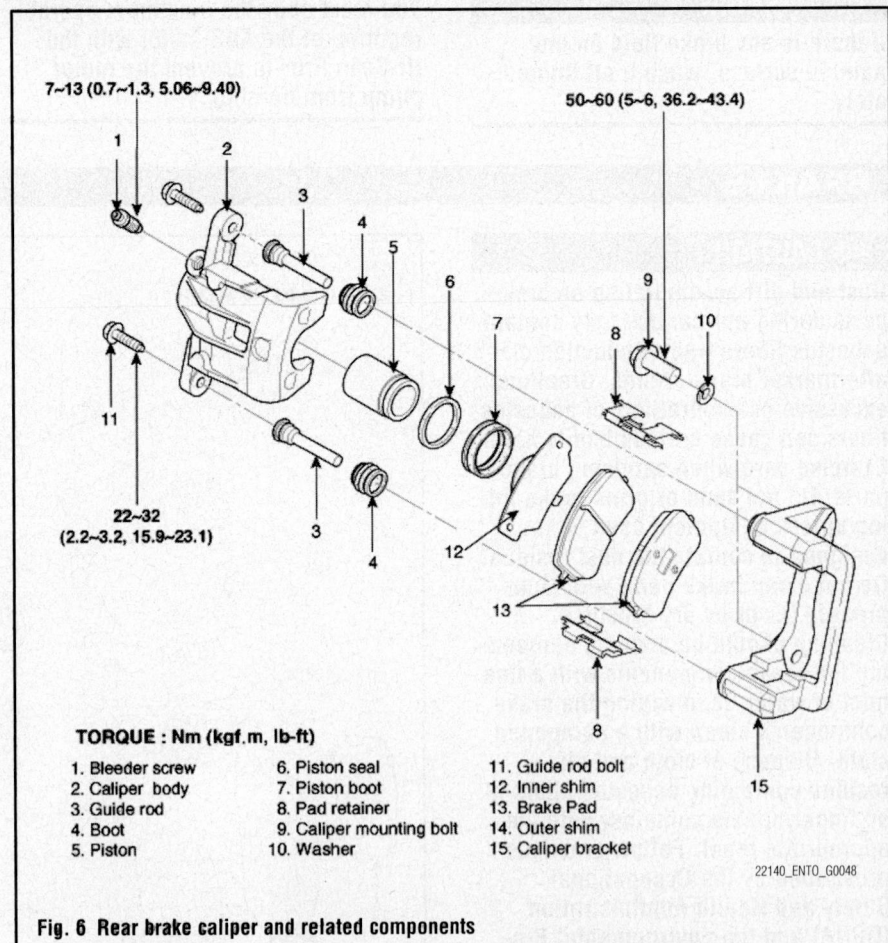

7~13 (0.7~1.3, 5.06~9.40)
50~60 (5~6, 36.2~43.4)
22~32 (2.2~3.2, 15.9~23.1)

TORQUE : Nm (kgf.m, lb-ft)

1. Bleeder screw
2. Caliper body
3. Guide rod
4. Boot
5. Piston
6. Piston seal
7. Piston boot
8. Pad retainer
9. Caliper mounting bolt
10. Washer
11. Guide rod bolt
12. Inner shim
13. Brake Pad
14. Outer shim
15. Caliper bracket

22140_ENTO_G0048

Fig. 6 Rear brake caliper and related components

- Caliper mounting bolts
- Caliper

To install:
4. Install the caliper onto its mounting.
5. Tighten the caliper mounting bolts: 36–43 ft. lbs. (50–60 Nm).
6. Install the brake line to the caliper with new metal gaskets. Torque the

brake line union bolt to 18–22 ft. lbs. (24–30 Nm).
7. Bleed the system.
8. Install the wheel.
9. Pump the brake pedal until the brake pads are seated and a firm pedal is achieved before attempting to move the vehicle.

※ CAUTION

Do not move the vehicle until a firm pedal is obtained.

10. Road test the vehicle to check for proper brake operation.

DISC BRAKE PADS

REMOVAL & INSTALLATION

See Figure 6.

1. Before servicing the vehicle, refer to the Precautions Section.

PARKING BRAKE CABLES

ADJUSTMENT

See Figure 7.

The parking brake cable adjustment must be performed AFTER adjusting the parking brake shoes. Refer to Parking Brake Shoes, adjustment.

➡**After adjusting the parking brake, make sure there is clearance between the adjusting nut and pin and that the brake is not dragging.**

1. Fully apply and release the parking brake 3 times.

2. Adjust the adjusting nut (A) so the parking brake pedal stroke is 3.46–3.86 inches (88–98mm) when 44 lbs. (200N) of force is applied.

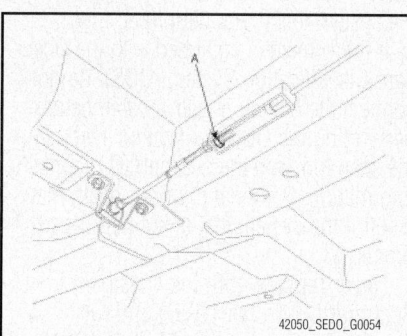

42050_SEDO_G0054

Fig. 7 Adjust the adjusting nut (A) so the parking brake pedal stroke is according to specification

PARKING BRAKE SHOES

REMOVAL & INSTALLATION

See Figures 8 through 12.

1. Before servicing the vehicle, refer to the Precautions Section.

2. Raise and safely support the vehicle.

2. Remove or disconnect the following:
   - Rear wheels
   - Lower caliper mounting bolt and rotate the caliper upward
   - Pads from the caliper support
   - Pad retainers, if necessary

To install:

3. Install or connect the following:
   - Pad retainers, if removed
   - Pads onto the pad retainers

4. Compress the caliper piston using a C-clamp.

3. Remove the rear wheel and tire assembly.

4. Remove the hub cap and rotor.

5. Remove the hub nut (A) and washer (B), then remove the rear hub (C).

6. Remove the shoe hold-down pin and spring (A) by pressing then rotating the spring.

7. Remove the adjuster assembly (B) and the lower return spring (C).

8. Remove the strut (B) and the upper return spring (A).

42050_SEDO_G0048

Fig. 8 Remove the hub nut (A) and washer (B), then remove the rear hub (C)

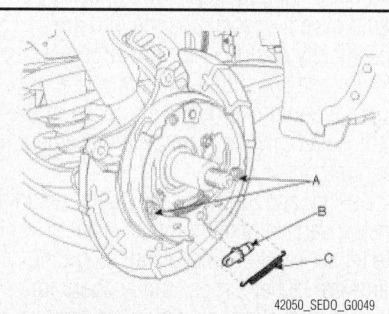

42050_SEDO_G0049

Fig. 9 Remove the shoe hold-down pin and spring (A) by pressing then rotating the spring. Then, remove the adjuster assembly (B) and lower return spring (C)

5. Rotate the caliper downward and install the mounting bolt. Tighten to 16–24 ft. lbs. (22–32 Nm).

6. Install the wheel.

7. Pump the brake pedal until the brake pads are seated and a firm pedal is achieved before attempting to move the vehicle.

※ CAUTION

Do not move the vehicle until a firm pedal is obtained.

8. Road test the vehicle to check for proper brake operation.

9. Remove the retaining ring (B) of the parking brake wire (A) from the back of the backing plate.

10. Disconnect the parking brake wire from the brake shoe.

To install:

11. Install the backing plate.

12. Connect the parking brake wire to the brake shoe.

13. Install the shoe hold-down pin and spring to hold the brake shoe.

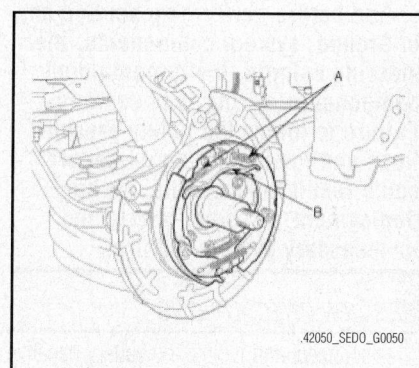

.42050_SEDO_G0050

Fig. 10 Remove the strut (B) and the upper return spring (A)

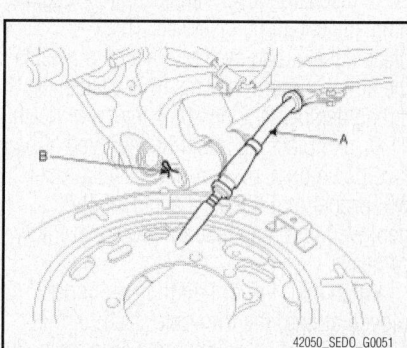

42050_SEDO_G0051

Fig. 11 Remove the retaining ring (B) of the parking brake wire (A) from the back of the backing plate

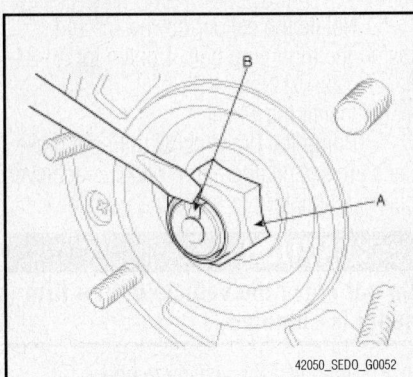

Fig. 12 After installing the hub nut, lock the nut to the spindle as shown (B)

14. Install the adjuster assembly and the lower return spring.

15. Install the upper return spring and strut.

16. Lubricate the hub as needed, then install the rear hub.

17. Install the hub nut (A), then lock the nut to the spindle as shown in the accompanying illustration.

18. Install the hub cap and the rotor.

19. Install the rear wheel and tire assembly.

20. Tighten the parking brake adjusting nut as needed.

ADJUSTMENT

See Figure 13.

1. Before servicing the vehicle, refer to the Precautions Section.

2. Raise and safely support the vehicle.

3. Remove the rear wheel and tire.

4. After removing the plug from the disc, rotate the toothed wheel with a screwdriver until the disc does not move, then back it up 5 notches.

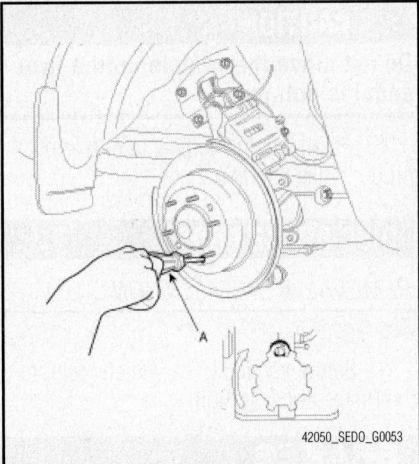

Fig. 13 After removing the plug from the disc, rotate the toothed wheel with a screwdriver (A) until the disc does not move, then back it up 5 notches

CHASSIS ELECTRICAL   AIR BAG (SUPPLEMENTAL RESTRAINT SYSTEM)

GENERAL INFORMATION

✳✳ CAUTION

This vehicle is equipped with an air bag system. The system must be disarmed before performing service on, or around, system components, the steering column, instrument panel components, wiring, and sensors. Failure to follow the safety precautions and the disarming procedure could result in accidental air bag deployment, possible injury, and unnecessary system repairs.

SERVICE PRECAUTIONS

Disconnect and isolate the battery negative cable before beginning any airbag system component diagnosis, testing, removal, or installation procedures. Allow system capacitor to discharge for 3 minutes before beginning any component service. This will disable the airbag system. Failure to disable the airbag system may result in accidental airbag deployment, personal injury, or death.

Do not place an intact undeployed airbag face down on a solid surface. The airbag will propel into the air if accidentally deployed and may result in personal injury or death.

When carrying or handling an undeployed airbag, the trim side (face) of the airbag should be pointing towards the body to minimize possibility of injury if accidental deployment occurs. Failure to do this may result in personal injury or death.

Replace airbag system components with OEM replacement parts. Substitute parts may appear interchangeable, but internal differences may result in inferior occupant protection. Failure to do so may result in occupant personal injury or death.

Wear safety glasses, rubber gloves, and long sleeved clothing when cleaning powder residue from vehicle after an airbag deployment. Powder residue emitted from a deployed airbag can cause skin irritation. Flush affected area with cool water if irritation is experienced. If nasal or throat irritation is experienced, exit the vehicle for fresh air until the irritation ceases. If irritation continues, see a physician.

Do not use a replacement airbag that is not in the original packaging. This may result in improper deployment, personal injury, or death.

The factory installed fasteners, screws and bolts used to fasten airbag components have a special coating and are specifically designed for the airbag system. Do not use substitute fasteners. Use only original equipment fasteners listed in the parts catalog when fastener replacement is required.

During, and following, any child restraint anchor service, due to impact event or vehicle repair, carefully inspect all mounting hardware, tether straps, and anchors for proper installation, operation, or damage. If a child restraint anchor is found damaged in any way, the anchor must be replaced. Failure to do this may result in personal injury or death.

Deployed and non-deployed airbags may or may not have live pyrotechnic material within the airbag inflator.

Do not dispose of driver/passenger/curtain airbags or seat belt tensioners unless you are sure of complete deployment. Refer to the Hazardous Substance Control System for proper disposal.

Dispose of deployed airbags and tensioners consistent with state, provincial, local, and federal regulations.

After any airbag component testing or service, do not connect the battery negative cable. Personal injury or death may result if the system test is not performed first.

If the vehicle is equipped with the Occupant Classification System (OCS), do not connect the battery negative cable before performing the OCS Verification Test using the scan tool and the appropriate diagnostic information. Personal injury or death may result if the system test is not performed properly.

Never replace both the Occupant Restraint Controller (ORC) and the Occupant Classification Module (OCM) at the same time. If both require replacement, replace one, then perform the Airbag System test before replacing the other.

Both the ORC and the OCM store Occupant Classification System (OCS) calibration data, which they transfer to one another when one of them is replaced. If both are replaced at the same time, an irreversible fault will be set in both modules and the OCS may malfunction and cause personal injury or death.

If equipped with OCS, the Seat Weight Sensor is a sensitive, calibrated unit and must be handled carefully. Do not drop or handle roughly. If dropped or damaged, replace with another sensor. Failure to do so may result in occupant injury or death.

If equipped with OCS, the front passenger seat must be handled carefully as well. When removing the seat, be careful when it setting on the floor not to drop it. If dropped, the sensor may be inoperative, could result in occupant injury, or possibly death.

If equipped with OCS, when the passenger front seat is on the floor, no one should sit in the front passenger seat. This uneven force may damage the sensing ability of the seat weight sensors. If sat on and damaged, the sensor may be inoperative, could result in occupant injury, or possibly death.

Several precautions must be observed when handling the inflator module to avoid accidental deployment and possible personal injury.

• Never carry the inflator module by the wires or connector on the underside of the module

• When carrying a live inflator module, hold it securely with both hands, and ensure that the bag and trim cover are pointed away

• Place the inflator module on a bench or other surface with the bag and trim cover facing up

• With the inflator module on the bench, never place anything on or close to the module which may be thrown in the event of an accidental deployment

Before servicing the vehicle, make sure to refer to the precautions in the beginning of this section as well.

DISARMING THE SYSTEM

1. Before servicing the vehicle, refer to the Precautions Section.
2. Record the radio anti-theft code data. Remove the ignition key from the vehicle.
3. Disconnect the negative battery cable.
4. Wait at least 3 minutes for the system capacitor to discharge before performing any service.

ARMING THE SYSTEM

1. Before servicing the vehicle, refer to the Precautions Section.

2. Reconnect the negative battery cable.
3. To confirm proper system operation, turn the ignition switch to the ON position. The SRS indicator light will be lit for at least 6 seconds and then go off.

CLOCKSPRING CENTERING

See Figures 14 through 16.

1. Disconnect and isolate the battery negative cable. Allow the system capacitor to discharge for 3 minutes before beginning any component service. This will disable the airbag system.

✳✳ CAUTION

Failure to disable the airbag system may result in accidental airbag deployment, personal injury, or death.

2. Remove the ignition key from the vehicle.

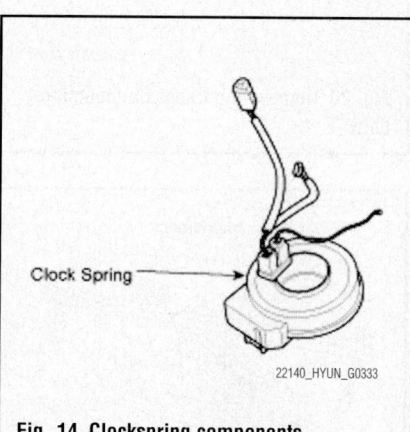

Fig. 14 Clockspring components

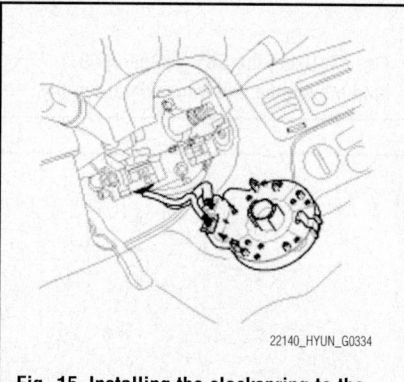

Fig. 15 Installing the clockspring to the steering column

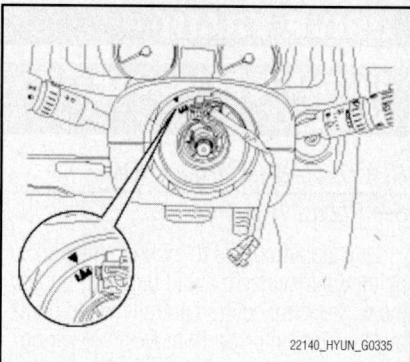

Fig. 16 Aligning the marks to center the clockspring

3. Connect the clockspring harness connector and horn harness connector to the clockspring.
4. Set the clockspring on neutral position and after turning the front wheels to the straight-ahead position, install the clockspring.
   a. Check connectors and protective tube for damage, and terminals for deformities.
   b. If even one abnormal point is discovered, replace the clockspring with a new one.
5. Connect the clockspring harness connector and the steering switch harness connector to the clockspring.
6. Set the center position by getting the marks between the clockspring and the cover into line.
7. Turn the clockspring clockwise to the stop and then 2.4 revolutions counterclockwise. See the illustration of marks in line.
8. Install the steering wheel column cover and the steering wheel.
9. Connect the Driver Airbag (DAB) module connector and the horn connector, then install the DAB module on the steering wheel.
10. Secure the DAB with the new mounting bolts and tighten to 70–96 inch lbs. (8–11 NM).
11. Connect the battery negative cable.
12. After installing the airbag, confirm proper system operation:
    a. Turn the ignition switch ON; the SRS indicator light should be turned on for about 6 seconds and then go off.
    b. Make sure the horn functions properly.

DRIVE TRAIN

AUTOMATIC TRANSAXLE ASSEMBLY

REMOVAL & INSTALLATION

See Figures 17 through 25.

Use fender covers to avoid damaging painted surfaces. To avoid damage, unplug the wiring connectors carefully while holding the connector portion. Mark all wiring and hoses for reassembly.

1. Before servicing the vehicle, refer to the Precautions Section.
2. Record the radio anti-theft code data.
3. Disconnect the negative terminal from the battery, then disconnect the positive terminal from the battery.
4. Remove the battery.
5. Disconnect the AFS connector.
6. Remove the air cleaner upper cover (B) by loosening the clips (A).
7. Remove the air cleaner assembly.
8. Disconnect the air cleaner hose by loosening the clamp.
9. Remove the battery tray by removing the 4 mounting bolts.
10. Disconnect the transaxle wire harness connectors:
    a. Remove the inhibiter switch connector.
    b. Remove the solenoid valve connector.

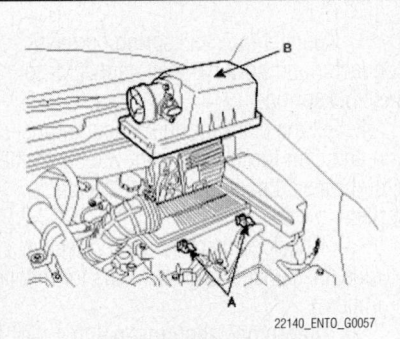

Fig. 17 Remove the air cleaner upper cover (B) by loosening the clips (A)

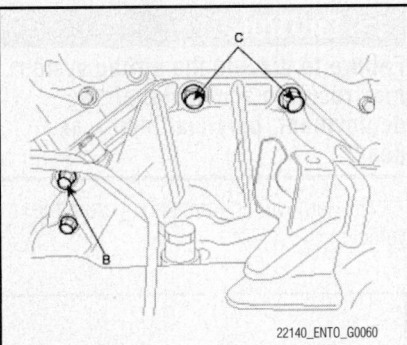

Fig. 20 Remove the transaxle mounting bolts (B, C)

Fig. 23 Remove the roll stopper mounting bolt (A)

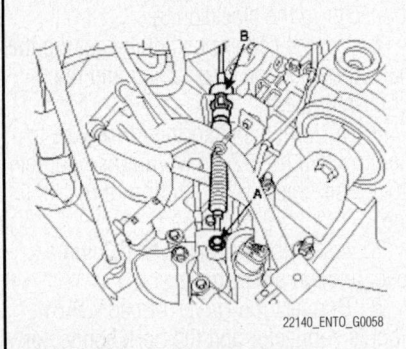

Fig. 18 Remove the shift cable by removing the bolt (A) and clip (B)

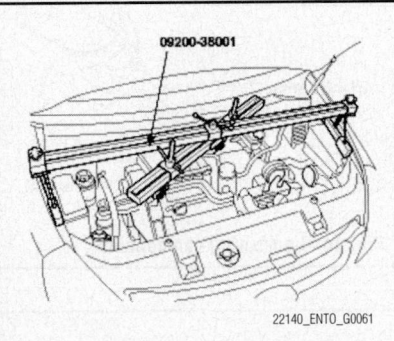

Fig. 21 Using the SST (09200-38001), hold the engine and transaxle assembly

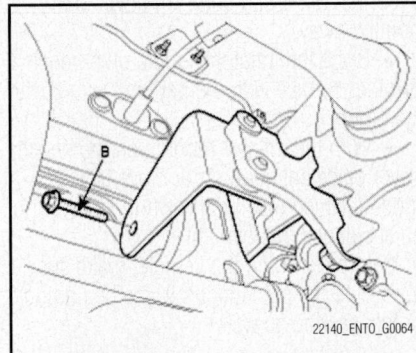

Fig. 24 Remove the roll stopper mounting bolt (B)

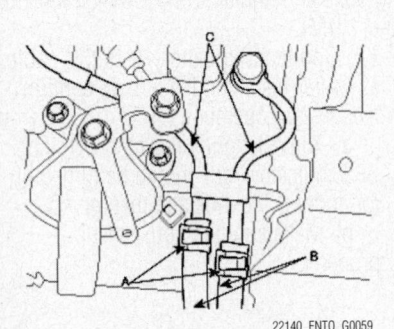

Fig. 19 Disconnect the transaxle oil cooler hoses (A) from the tubes (C) by loosening the clamps (B)

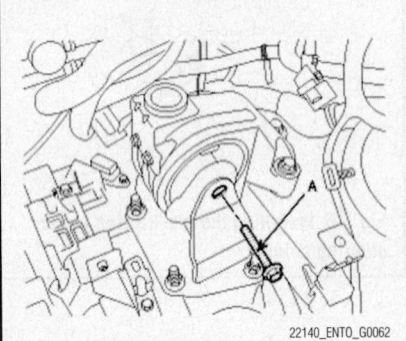

Fig. 22 Remove the transaxle insulator mounting bolt (A)

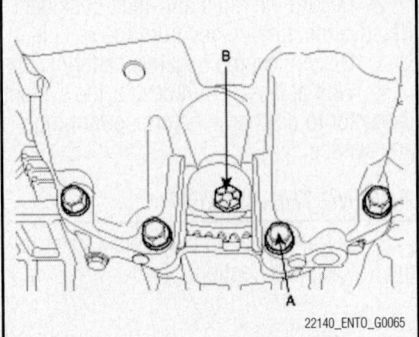

Fig. 25 Remove the transaxle under mounting bolts (A) and the drive plate bolts (B)

c. Remove the input speed sensor connector.

d. Remove the output speed sensor connector.

e. Remove the vehicle speed sensor connector.

f. Remove the CKP sensor connector.

11. Remove the ground circuit line from the transaxle.

12. Remove the shift cable by removing the bolt (A) and clip (B).

13. Disconnect the transaxle oil cooler hoses (A) from the tubes (C) by loosening the clamps (B).

14. Remove the transaxle mounting bolts (B, C).

15. Using the SST(09200-38001), hold the engine and transaxle assembly safely.

16. Remove the transaxle insulator mounting bolt (A).

17. Remove the front wheels.

18. Remove the power steering column joint bolt.

19. Raise and safely support the vehicle.

20. Remove the undercover.

21. Drain transaxle oil.

22. Drain the power steering oil through the return tube.

23. Disconnect the power steering pressure tube from the power steering oil pump.

24. Disconnect the lower arm, the tie rod end ball joint, and the stabilizer bar control link from the front knuckle.

25. Remove the roll stopper mounting bolts (A) and (B).

26. Remove the mounting bolts from the sub frame by supporting the sub frame with a jack.

27. Remove halfshafts from the transaxle. Refer to Halfshafts, removal & installation.

28. Install a jack for supporting the transaxle assembly.

29. Remove the transaxle under mounting bolts (A) and the drive plate bolts (B).

30. Lifting the vehicle up and lowering the jack slowly, remove the transaxle assembly.

To install:

31. Lower the vehicle or lift up the jack to install the transaxle assembly.

32. Tighten the transaxle under mounting bolts to: 47–62 ft. lbs. (65–85 Nm).

33. Install the starter motor, if removed. Tighten the bolts to: 47–62 ft. lbs. (65–85 Nm).

34. Tighten the transaxle under mounting bolts to: 33–38 ft. lbs. (46–53 Nm).

35. After removing the jack, install the halfshafts. Refer to Halfshafts, removal & installation.

36. Install the sub frame.

37. Tighten the roll stopper mounting bolts. Tighten to 65–80 ft. lbs. (90–110 Nm).

38. Connect the return tube with a clamp.

39. Connect the lower arm, the tie rod end ball joint, and the stabilizer bar control link to the front knuckle.

40. Connect the power steering pressure tube to the power steering oil pump.

41. Install the undercover.

42. Install the steering column joint bolt.

43. Install the front wheels and tires.

44. Tighten the transaxle insulator mounting bolt to: 65–80 ft. lbs. (90–110 Nm).

45. Tighten the transaxle mounting bolts to: 47–62 ft. lbs. (65–85 Nm)

46. Remove the SST (09200-38001) holding the engine and transaxle assembly.

47. Connect the transaxle oil cooler hoses to the tubes by fastening the clamps.

48. Install the shift cable by tightening the bolt and clip. Tighten the bolt to: 84–120 inch lbs. (10–14 Nm).

49. Connect the transaxle wire harness connectors:

a. Install the inhibiter switch connector.

b. Install the solenoid valve connector.

c. Install the input speed sensor connector.

d. Install the output speed sensor connector.

e. Install the vehicle speed sensor connector.

f. Install the CKP sensor connector.

50. Install the ground circuit line to the transaxle.

51. Install the battery tray by tightening the 4 mounting bolts.

52. Connect the air cleaner hose by fastening the clamp.

53. Install the air cleaner assembly.

54. Install the air cleaner upper cover by fastening the clips.

55. Connect the AFS connector.

56. Install the battery.

57. Refill the power steering fluid and bleed the power steering system.

58. Fill the transaxle with the proper grade and type of transaxle fluid.

59. Run the engine and recheck the transaxle fluid. Road test the vehicle.

60. Reprogram the radio anti-theft codes.

61. When replacing the automatic transaxle, reset the automatic transaxle values by using the High-Scan Pro tool:

a. Connect the Hi-Scan Pro connector to the Data Link Connector (DLC) under the crash pad and the power cable to the cigarette jack under the center fascia.

b. Turn the ignition switch ON and power on the Hi-Scan Pro.

c. Select the vehicle's name.

d. Select AUTOMATIC TRANSAXLE.

e. Select RESETTING AUTO T/A VALUES and perform the procedure.

f. Perform the procedure by pressing F1 (REST).

HALFSHAFTS

REMOVAL & INSTALLATION

See Figure 26.

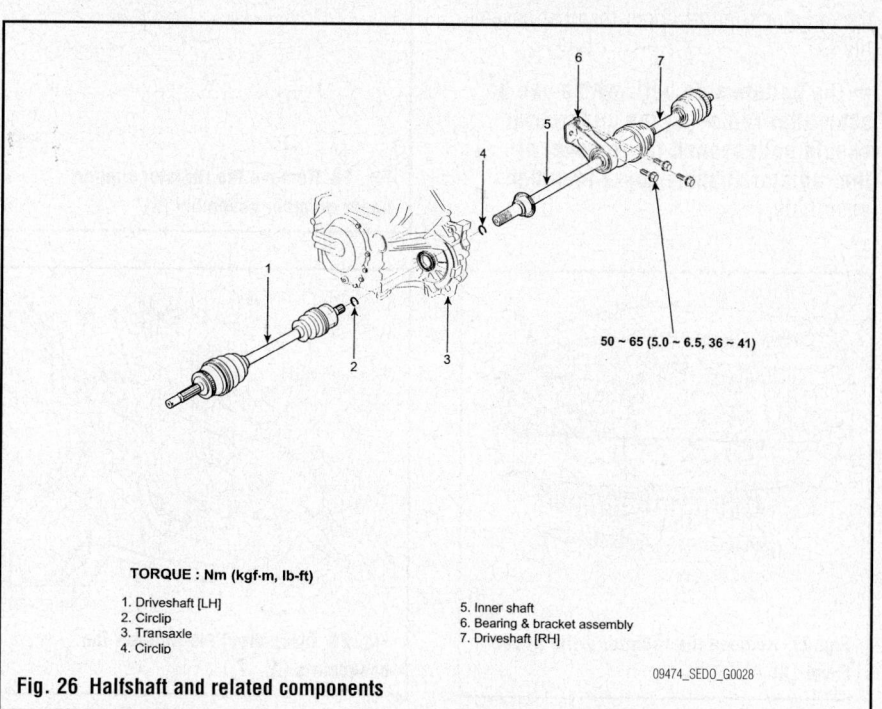

TORQUE : Nm (kgf·m, lb-ft)

1. Driveshaft [LH]
2. Circlip
3. Transaxle
4. Circlip
5. Inner shaft
6. Bearing & bracket assembly
7. Driveshaft [RH]

50 ~ 65 (5.0 ~ 6.5, 36 ~ 41)

Fig. 26 Halfshaft and related components

09474_SEDO_G0028

1. Before servicing the vehicle, refer to the Precautions Section.
2. Record the radio anti-theft code data.
3. Disconnect the negative battery cable.
4. Raise and safely support the vehicle.
5. Remove the front tires.
6. Drain the transaxle fluid.
7. Unstake the halfshaft lock nut. Remove the lock nut.
8. Remove the split pin and castle nut from the tie rod end ball joint. Disconnect the tie rod end from the knuckle, using tool SST (09568-4A000).
9. Remove the split pin and lower arm ball and nut. Using a plastic hammer, disconnect the halfshaft from the front hub assembly.

10. To remove the right halfshaft, remove the right halfshaft heat protector. Remove the inner shaft bearing bracket assembly mounting bolts.
11. Insert a pry bar between the transaxle case and halfshaft joint; separate the halfshaft from the transaxle. Pull the halfshaft from the transaxle case.

To install:

→ **Use new circlips for assembly.**

12. Installation is the reverse of the removal procedure.
13. Torque the inner shaft bearing bracket assembly bolts to 36–47 ft. lbs. (50–65 Nm).

14. Torque the split pin and lower arm bolt and nut to 65–87 ft. lbs. (90–110 Nm).
15. Torque the tie rod end ball joint split pin and castle nut to 43–58 ft. lbs. (60–80 Nm).
16. Torque the halfshaft lock nut to 177–199 ft. lbs. (245–275 Nm).
17. Torque the tire and wheel assembly to 65–80 ft. lbs. (90–110 Nm).
18. Fill the transaxle with the proper grade and type of transaxle fluid.
19. Run the engine and recheck the transaxle fluid. Road test the vehicle.
20. Reprogram the radio anti-theft codes.

ENGINE COOLING

ENGINE FAN

REMOVAL & INSTALLATION

This procedure involves the removal of the radiator assembly. Refer to Radiator, removal & installation.

RADIATOR

REMOVAL & INSTALLATION

See Figures 27 through 33.

1. Before servicing the vehicle, refer to the Precautions Section.
2. Drain the engine coolant.
3. Remove the radiator grille upper cover (A).
4. For convenience sake, remove the radiator support upper member assembly (A).

→ **The bottom side bolt, which can be seen after removing the undercover, should be loosened for removal of the radiator support upper member assembly.**

5. Disconnect the radiator upper and lower hoses.
6. Disconnect the transaxle oil cooler hoses.
7. Disconnect the radiator fan connectors (A).
8. Disconnect the pressure lines from the radiator assembly.
9. Separate the condenser (A) from the radiator assembly (B) by removing the bolts (C).

10. Remove the radiator bracket.
11. Remove the radiator assembly.
12. Remove the radiator cooling fans (A).

To install:

13. Install the radiator fans to the radiator. Tighten to 48–72 inch lbs. (5–8 Nm).
14. Install the radiator assembly into the vehicle.
15. Install the radiator bracket.

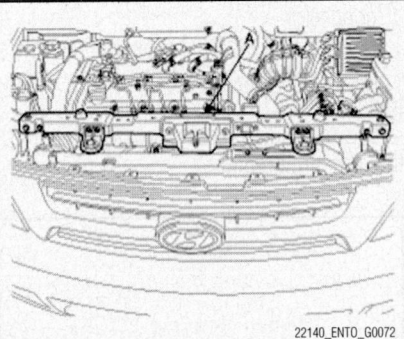

Fig. 28 Remove the radiator support upper member assembly (A)

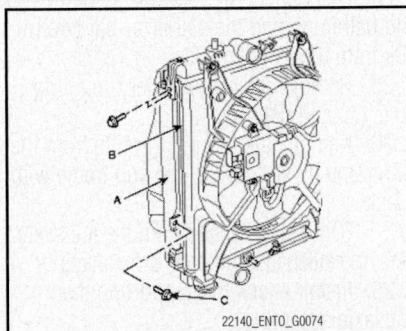

Fig. 30 Separate the condenser (A) from the radiator assembly (B) by removing the bolts (C)

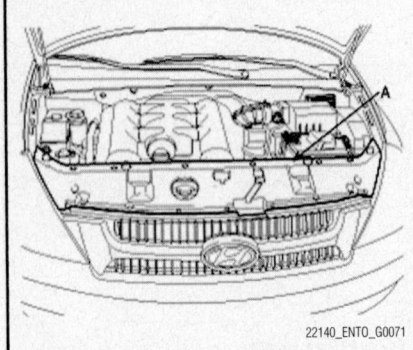

Fig. 27 Remove the radiator grille upper cover (A)

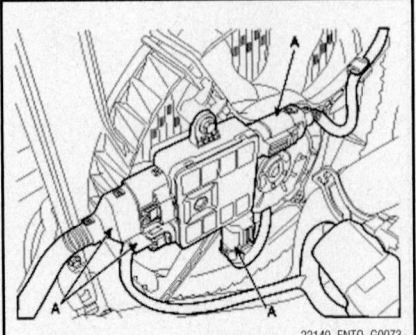

Fig. 29 Disconnect the radiator fan connectors (A)

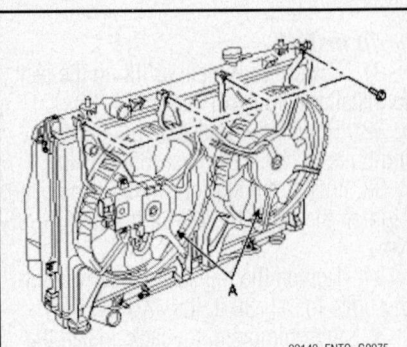

Fig. 31 Remove the radiator cooling fans (A)

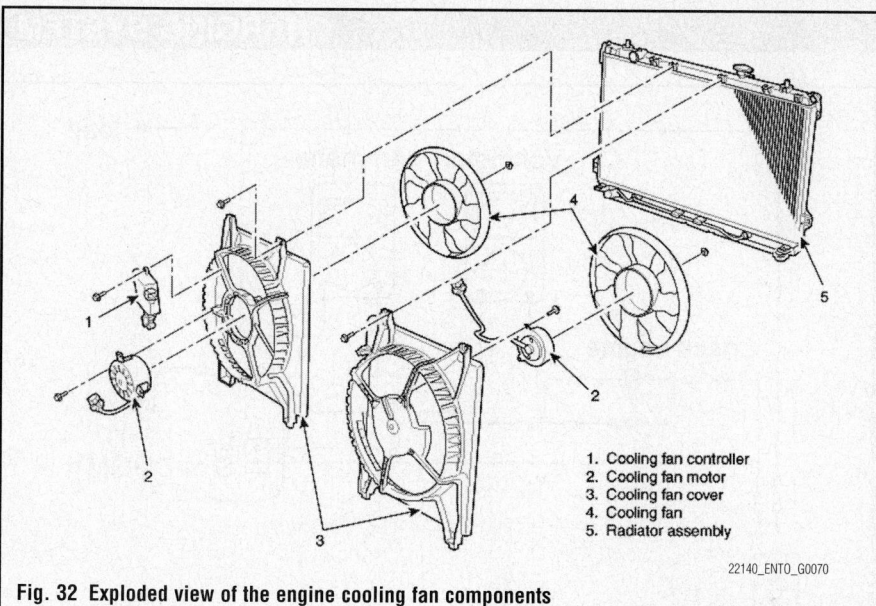

1. Cooling fan controller
2. Cooling fan motor
3. Cooling fan cover
4. Cooling fan
5. Radiator assembly

22140_ENTO_G0070

Fig. 32 Exploded view of the engine cooling fan components

16. Fix the condenser to the radiator assembly and tighten as follows:
   a. Bolts (D): 48–72 inch lbs. (5–8 Nm).
   b. Bolts (C): 60–84 inch lbs. (7–10 Nm).
17. Connect the radiator fan connectors.
18. Connect the pressure lines to the radiator assembly.
19. Connect transaxle oil cooler hoses.
20. Connect radiator upper and lower hoses.
21. Install the radiator support upper member assembly.
22. Install the radiator grille upper cover.
23. Fill with engine coolant.
24. Start engine and check for leaks.
25. Recheck engine coolant level and add coolant as necessary.

THERMOSTAT

REMOVAL & INSTALLATION

See Figure 34.

1. Before servicing the vehicle, refer to the Precautions Section.

➡**Removal of the thermostat would have an adverse effect, causing a lowering of cooling efficiency. Do not remove the thermostat, even if the engine tends to overheat.**

2. Drain engine coolant just below the level of the thermostat.
3. Remove the water inlet (A) and thermostat (B).

To install:

4. Place the thermostat in the thermostat housing.

a. Install the thermostat with the jiggle valve upward.
   b. Install a new thermostat (B).
5. Install the water inlet (A) and tighten to 12–15 ft. lbs. (17–20 Nm).
6. Fill with engine coolant.
7. Start the engine and check for leaks.

WATER PUMP

REMOVAL & INSTALLATION

See Figure 35.

1. Before servicing the vehicle, refer to the Precautions Section.
2. Record the radio anti-theft code data.
3. Drain the cooling system.
4. Disconnect the negative battery cable.
5. Remove the accessory drive belt.
6. Remove the water pump pulley.
7. Remove the water pump retaining bolts.
8. Remove the water pump from the vehicle.

To install:

9. Installation is the reverse of the removal procedure.
10. Reprogram the radio anti-theft code data.
11. Install the water pump (A) and a new gasket (B) with 12 bolts.
12. Be sure to use a new water pump gasket. Torque the retaining bolts to 16–17 ft. lbs. (22–23 Nm).
13. Torque the water pump pulley bolts to 6–7 ft. lbs. (8–10 Nm).
14. Be sure to refill the cooling system with the proper grade and type coolant.
15. Start the engine and check for leaks.

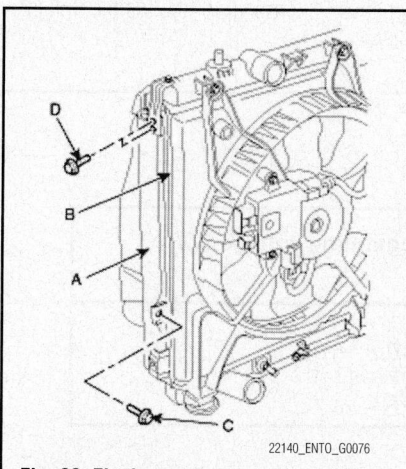

22140_ENTO_G0076

Fig. 33 Fix the condenser to the radiator assembly

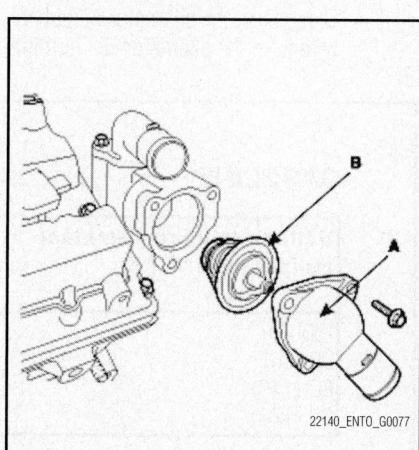

22140_ENTO_G0077

Fig. 34 Remove the water inlet (A) and thermostat (B)

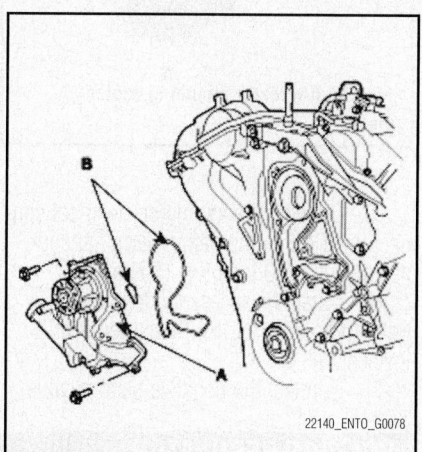

22140_ENTO_G0078

Fig. 35 Install the water pump (A) and a new gasket (B) with 12 bolts

➡Disconnecting the negative battery cable on some vehicles may interfere with the functions of the on board computer system. The computer may undergo a relearning process once the negative battery cable is reconnected.

ALTERNATOR

REMOVAL & INSTALLATION

See Figure 36.

1. Before servicing the vehicle, refer to the Precautions Section.
2. Disconnect the negative battery cable.
3. Disconnect the electrical connectors from the alternator.
4. Remove the accessory drive belt. Refer to Accessory Drive Belts, removal & installation.
5. Remove the alternator mounting bolts.
6. Remove the alternator.

To install:

7. Install the alternator.
8. Tighten the pivot bolt to 14–18 ft. lbs. (19–25 Nm) and the adjustment bolt to 11–16 ft. lbs. (15–22 Nm).

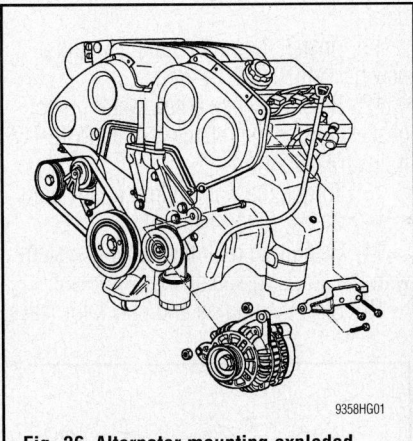

Fig. 36 Alternator mounting exploded view

9. Connect the alternator electrical connectors. Tighten the battery terminal connector nut to 60 inch lbs. (7 Nm).
10. Install the accessory drive belt. Refer to Accessory Drive Belts, removal & installation.
11. Connect the negative battery cable.

VOLTAGE REGULATOR

ADJUSTMENT

See Figures 37 and 38.

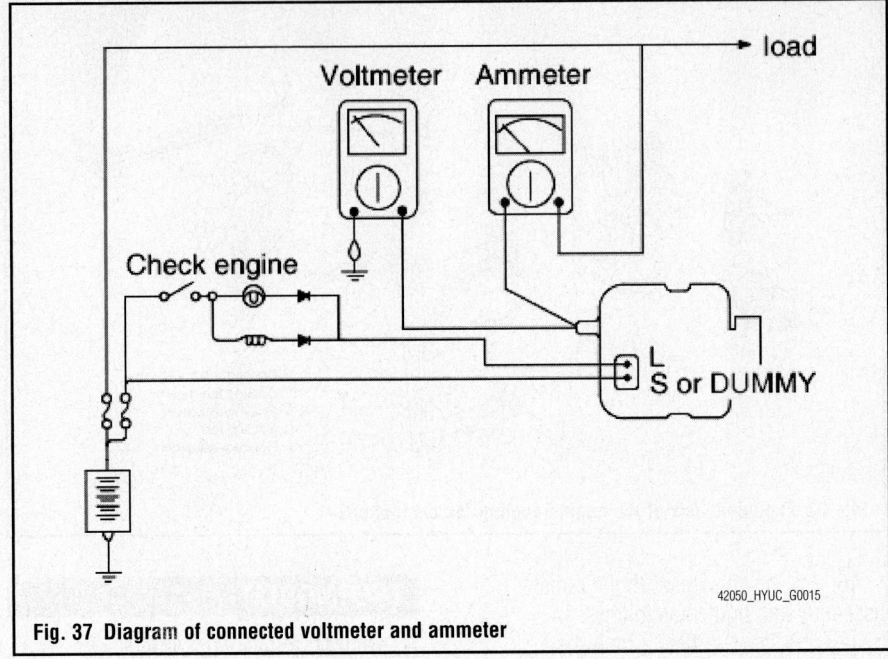

Fig. 37 Diagram of connected voltmeter and ammeter

1. Before servicing the vehicle, refer to the Precautions Section.
2. Prior to the test, check the following items and correct if necessary.
   - Check that the battery installed on the vehicle is fully charged
   - Check the alternator drive belt tension
   - Turn ignition switch to **OFF**
   - Disconnect the battery ground cable
   - Connect a digital voltmeter between the "B" terminal of the alternator and ground by connecting the positive (+) lead of the voltmeter to the "B" terminal of the alternator and connecting the negative (-) lead to good ground or the negative (-) battery terminal
   - Disconnect the alternator output wire from the alternator "B" terminal
   - Connect a DC ammeter (0 to 150 Amps) in series between the "B" terminal and the disconnected output wire
   - Connect the negative (-) lead wire of the ammeter to the disconnected output wire
   - Attach the engine tachometer and connect the battery ground cable

3. Turn on the ignition switch and check to see that the voltmeter is properly installed and indicates the following value: Voltage reading = Battery voltage of 12.0–14.1 volts.
4. If it reads 0 volts, there is an open circuit in the wiring between the alternator "B" terminal and the battery and the battery negative (-) terminal, or the fusible link is blown.

GASOLINE

Voltage regulator ambient temperature °C (°F)	Regulating voltage (V)
-20 (-4)	14.2 ~ 15.4
20 (68)	14.0 ~ 15.0
60 (140)	13.7 ~ 14.9
80 (176)	13.5 ~ 14.7

Fig. 38 Regulating voltage table

5. Start the engine keeping all lights and accessories off.

6. Run the engine at a speed of about 2,500 rpm and read the voltmeter when the alternator output current drops to 10 Amps or less.

7. If the voltmeter reading agrees with the value listed in the Regulating Voltage Table below, the voltage regulator is functioning correctly.

8. If the reading is other than the standard value, the voltage regulator or the alternator is faulty.

9. Upon completion of the test, reduce the engine speed to idle, and turn off the ignition switch.

10. Disconnect the battery ground cable.

11. Remove the voltmeter and ammeter and the engine tachometer.

12. Connect the alternator output wire to the alternator "B" terminal.

13. Connect the battery ground cable.

REMOVAL & INSTALLATION

The voltage regulator is an internal component of the alternator. In order to replace the voltage regulator, the entire alternator assembly must be replaced. Refer to Alternator, installation and removal.

ENGINE ELECTRICAL

IGNITION SYSTEM

FIRING ORDER

See Figure 39.

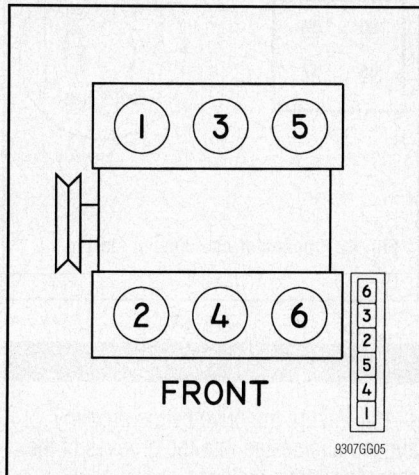

Fig. 39 Firing order: 1–2–3–4–5–6 Distributorless ignition system—3.8L engine

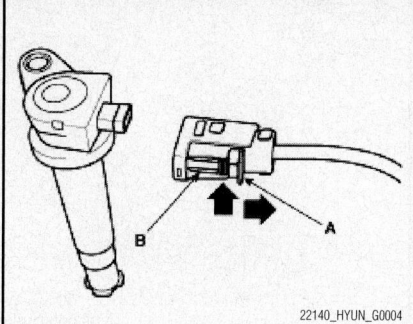

22140_HYUN_G0004

Fig. 40 When removing the ignition coil connector, pull the lock pin (A) and push the clip (B)

➡**When removing the ignition coil connector, pull the lock pin (A) and push the clip (B).**

4. Remove the bolt from the ignition coil.
5. Remove the ignition coil.
6. Installation is the reverse of removal.

condition, etc.) are detected by the various sensors. Based on these sensor signals and the ignition timing data, signals to interrupt the primary current are sent to the ECM. The ignition coil is activated, and timing is controlled.

SPARK PLUGS

REMOVAL & INSTALLATION
See Figure 40.

1. Before servicing the vehicle, refer to the Precautions Section.
2. Remove the engine cover (as necessary).
3. Disconnect the ignition coil connector.

➡**When removing the ignition coil connector, pull the lock pin (A) and push the clip (B).**

4. Remove the ignition coil.
5. Use a spark plug socket and wrench to remove the spark plugs.

✳✳ WARNING
Be careful that no contaminates enter through the spark plug holes.

➡**Check the electrode gap on the spark plugs before installation. Specification: 0.039–0.043 inch (1.0–1.1mm).**

6. To install, reverse the removal procedure. Tighten the spark plugs to 11 ft. lbs. (15 Nm).

IGNITION COIL PACK

REMOVAL & INSTALLATION
See Figure 40.

1. Before servicing the vehicle, refer to the Precautions Section.
2. Remove the engine cover (as necessary).
3. Disconnect the ignition coil connector.

IGNITION TIMING

ADJUSTMENT

Ignition timing is controlled by the electronic control ignition timing system. The standard reference ignition timing data for the engine operating conditions are pre-programmed in the memory of the Engine Control Module (ECM). The engine operating conditions (speed, load, warm-up

ENGINE ELECTRICAL

STARTER

REMOVAL & INSTALLATION

1. Before servicing the vehicle, refer to the Precautions Section.
2. Record the radio anti-theft code data.
3. Remove or disconnect the following:
   - Negative battery cable
   - Starter motor electrical connectors
   - Starter motor

To install:

4. Install or connect the following:
   - Starter motor. Tighten the bolts to 20–24 ft. lbs. (27–33 Nm).
   - Starter motor electrical connectors. Tighten the battery erminal nut to 106–141 inch lbs. (12–16 Nm).
   - Negative battery cable
   - Reprogram the radio anti-theft codes

SOLENOID OR RELAY REPLACEMENT

See Figures 41 and 42.

1. Before servicing the vehicle, refer to the Precautions Section.

STARTING SYSTEM

2. Remove the fuse box cover.
3. Remove the starter relay (A).
4. Using an ohmmeter, check that there is continuity between each terminal.
   a. No continuity between terminals 30–87.
   b. Continuity between terminals 85–86.

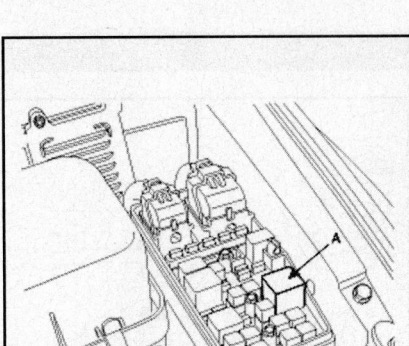

Fig. 41 Remove the starter relay (A)

5. Apply 12 volts to terminal 85 and a ground to terminal 86.
   a. Check for continuity between terminals 30 and 87.
   b. If there is no continuity, replace the starter relay.
6. Install the starter relay.
7. Install the fuse box cover.

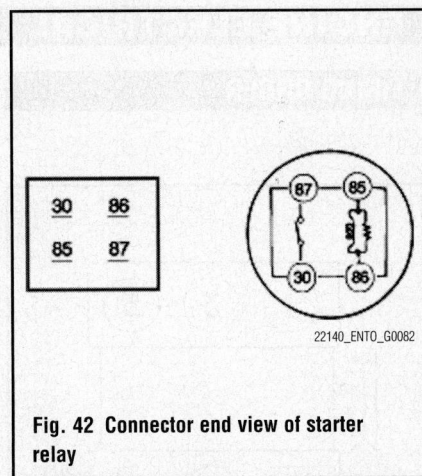

Fig. 42 Connector end view of starter relay

ENGINE MECHANICAL

ACCESSORY DRIVE BELTS

ACCESSORY BELT ROUTING

See Figure 43.

Refer to the accompanying illustration for accessory belt routing.

INSPECTION

Inspect the accessory drive belt for signs of glazing or cracking. A glazed belt will be perfectly smooth from slippage, while a good belt will have a slight texture of fabric visible. Cracks will usually start at the inner edge of the belt and run outward. All worn

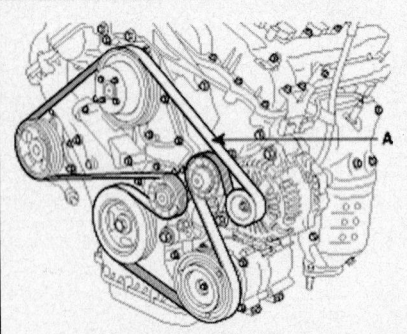

Fig. 43 Accessory drive belt routing—3.8L engine

or damaged accessory drive belts should be replaced immediately.

ADJUSTMENT

1. Loosen the tension mounting bolt.
2. Turn the adjusting bolt to obtain the proper belt tension, then retighten the mounting bolt.
3. Recheck the deflection of the drive belt.

REMOVAL & INSTALLATION

See Figure 43.

1. Before servicing the vehicle, refer to the Precautions Section.
2. Raise and support the vehicle.
3. Remove the engine splash shield.
4. Rotate the drive belt tensioner clockwise to release the drive belt tension.
5. Remove the drive belt from the alternator.
6. Slowly release the drive belt tensioner.
7. Remove the drive belt from the accessory drive pulleys.

To install:

8. Install the drive belt to the accessory drive pulley.
9. Rotate the drive belt tensioner clockwise.
10. Install the drive belt to the alternator.

11. Ensure the drive belt is properly aligned and seated into the grooves of the accessory drive pulleys.
12. Slowly release the drive belt tensioner.
13. Install the engine splash shield.
14. Lower the vehicle.

CAMSHAFT AND VALVE LIFTERS

INSPECTION

See Figures 44 and 45.

1. Inspect the cam lobes.
   a. Using a micrometer, measure the cam lobe height.
   b. If the cam lobe height is less than specified, replace the camshaft.
2. Inspect the camshaft journal clearance.
   a. Clean the bearing caps and camshaft journals.
   b. Place the camshafts on the cylinder head.
   c. Lay a strip of Plastigage® across each of the camshaft journal.
   d. Install the bearing caps and tighten the bolts with specified torque.

➡**Do not turn the camshaft.**

   e. Remove the bearing caps.

f. Measure the Plastigage® at its widest point.

g. If the oil clearance is greater than specified, replace the camshaft. If necessary, replace the bearing caps and cylinder head as a set.

h. Completely remove the Plastigage®.

i. Remove the camshafts.

3. Inspect the camshaft end play.

a. Install the camshafts.

b. Using a dial indicator, measure the end play while moving the camshaft back and forth.

c. If the end play is greater than specified, replace the camshaft. If necessary, replace the bearing caps and cylinder head as a set.

d. Remove the camshafts.

4. Inspect the Continuous Variable Valve Timing (CVVT) assembly.

a. Check that the CVVT assembly will not turn.

b. Apply vinyl tape to all the parts except the one indicated by the arrow in the illustration.

c. Wrap tape around the tip of the air gun and apply air of approx. 14 psi to the port of the camshaft. Perform this in order to release the lock pin for the maximum delay angle locking.

➡ **Wrap a shop rag around the CVVT as the oil may spray out when the air pressure is applied.**

d. Under the condition of air pressure being applied, turn the CVVT assembly to the advance angle side with your hand.

• Depending on the air pressure, the CVVT assembly will turn to the advance side

• If air is leaking from the port and air pressure cannot be maintained, the locking pin will not release

5. Except the position where the lock pin

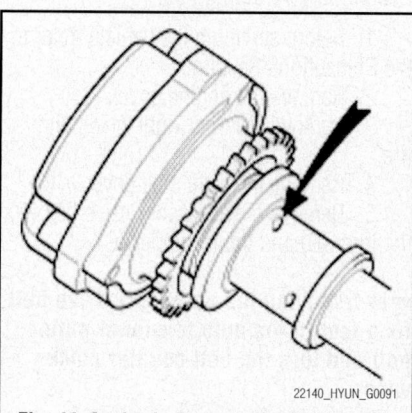

Fig. 44 Apply vinyl tape to the CVVT on all parts except the one indicated by the arrow

22140_HYUN_G0091

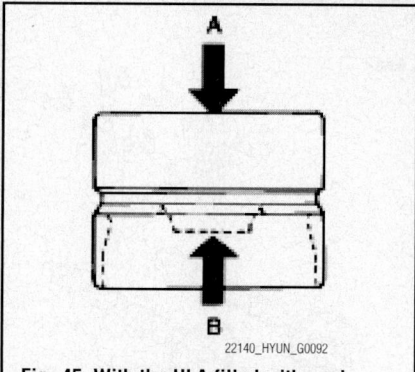

22140_HYUN_G0092

Fig. 45 With the HLA filled with engine oil, hold A and press B by hand

meets at the maximum delay angle, let the CVVT assembly turn back and forth and check the movable range and that there is no disturbance.

a. The CVVT should move smoothly in the range of about 20°.

b. Turn the CVVT assembly with your hand and lock it at the maximum delay angle position.

6. Inspect the Hydraulic Lash Adjuster (HLA).

a. With the HLA filled with engine oil, hold A and press B by hand.

b. If B moves, replace the HLA.

REMOVAL & INSTALLATION

See Figures 46 through 49.

Use fender covers to avoid damaging painted surfaces. To avoid damage, unplug the wiring connectors carefully while holding the connector portion. Mark all wiring and hoses to avoid misconnection. Inspect the timing belt before removing. Turn the crankshaft pulley so that the No. 1 piston is at Top Dead Center (TDC).

1. Before servicing the vehicle, refer to the Precautions Section.

2. Remove or disconnect the following:

• The negative battery connection
• The timing chain
• The water temperature control assembly
• The camshaft bearing cap
• The camshaft assembly

To install:

Thoroughly clean all parts to be assembled. Rotate the crankshaft, set the No. 1 piston at TDC.

3. Install the Continuously Variable Valve Timing (CVVT) and camshaft sprocket. Tighten to 48–56 ft. lbs. (65–76 Nm).

a. Install camshaft-inlet to dowel pin of CVVT assembly. At this time, do not install to oil hole of camshaft-inlet.

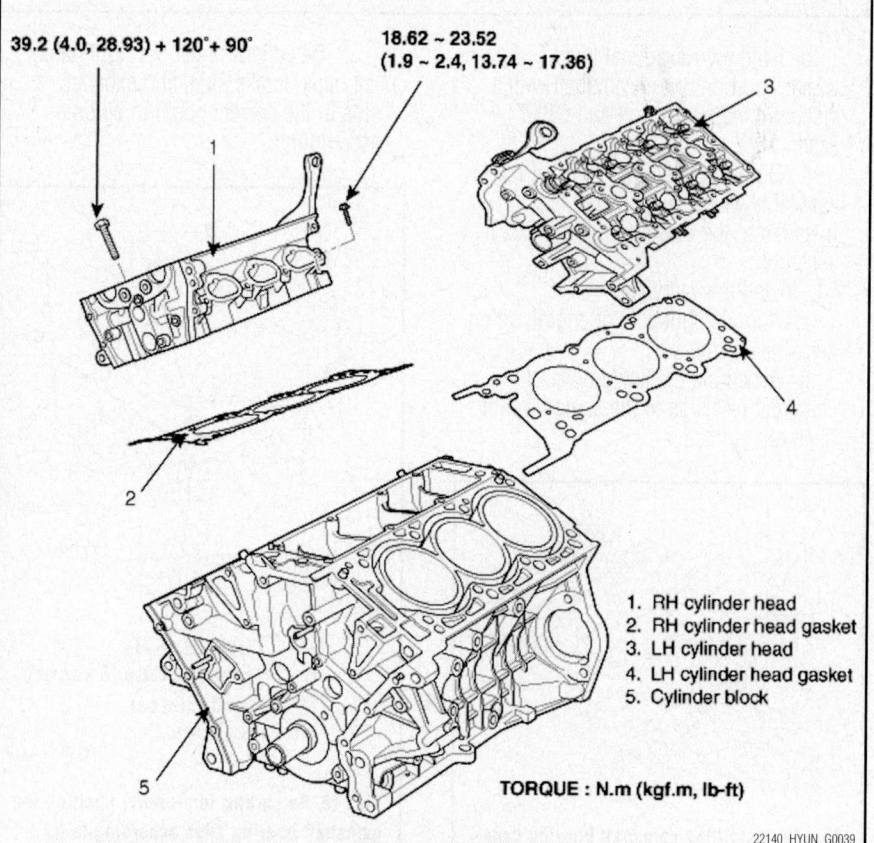

39.2 (4.0, 28.93) + 120° + 90°

18.62 ~ 23.52 (1.9 ~ 2.4, 13.74 ~ 17.36)

1. RH cylinder head
2. RH cylinder head gasket
3. LH cylinder head
4. LH cylinder head gasket
5. Cylinder block

TORQUE : N.m (kgf.m, lb-ft)

22140_HYUN_G0039

Fig. 46 Exploded view of cylinder head and engine block

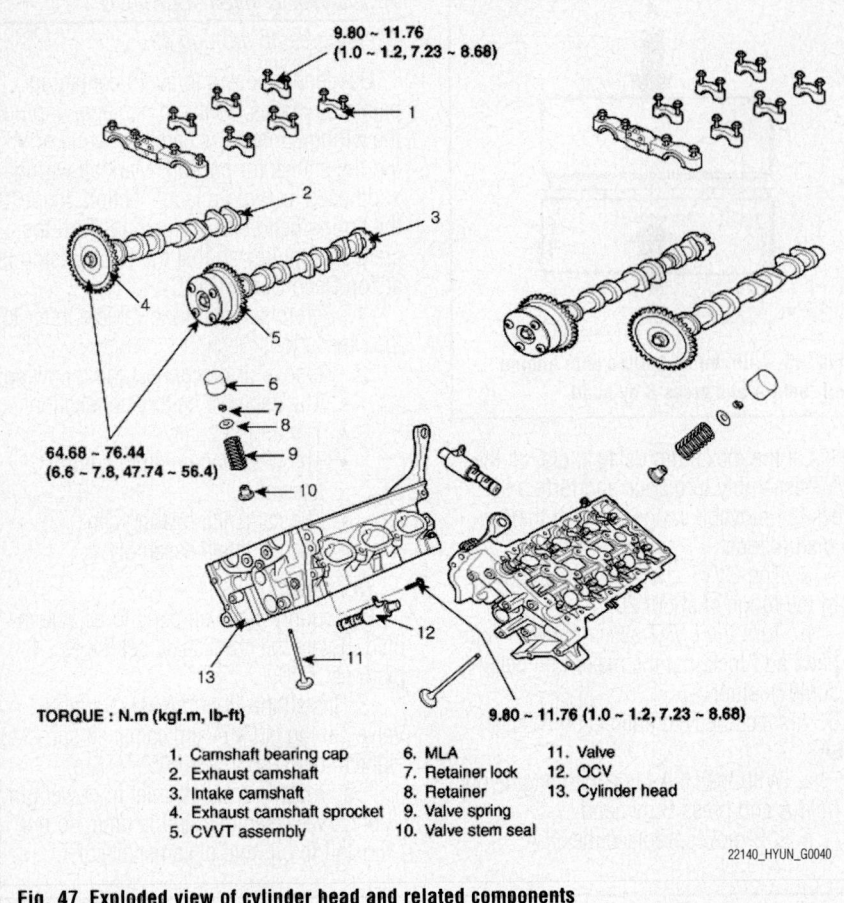

9.80 ~ 11.76
(1.0 ~ 1.2, 7.23 ~ 8.68)

64.68 ~ 76.44
(6.6 ~ 7.8, 47.74 ~ 56.4)

9.80 ~ 11.76 (1.0 ~ 1.2, 7.23 ~ 8.68)

TORQUE : N.m (kgf.m, lb-ft)

1. Camshaft bearing cap
2. Exhaust camshaft
3. Intake camshaft
4. Exhaust camshaft sprocket
5. CVVT assembly

6. MLA
7. Retainer lock
8. Retainer
9. Valve spring
10. Valve stem seal

11. Valve
12. OCV
13. Cylinder head

22140_HYUN_G0040

Fig. 47 Exploded view of cylinder head and related components

b. Hold the hexagonal head wrench portion of the camshaft with a vise, and install the bolt and CVVT assembly.

c. Do not rotate the CVVT assembly when the camshaft is installed to the dowel pin of the CVVT assembly.

4. Install the camshafts.

a. Apply a light coat of engine oil on camshaft journals.

b. Assemble the key groove of camshaft rear side to the same level of head top surface.

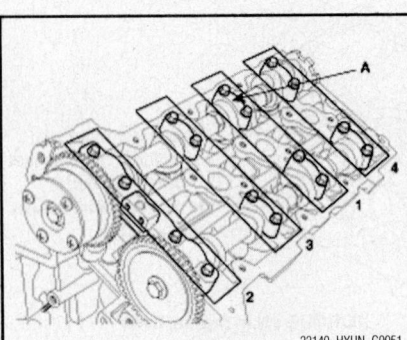

22140_HYUN_G0051

Fig. 48 Install the camshaft bearing caps in the sequence shown

c. Be careful to get the right bank, left bank, intake side, and exhaust side in the correct position before assembling.

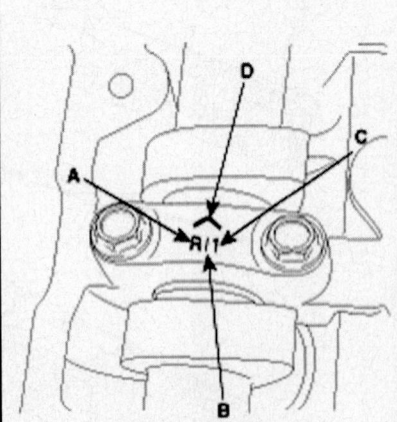

A. L (LH); R (RH)
B. I (Intake); None (Exhaust)
C. Journal number
D. Front mark

22140_HYUN_G0052

Fig. 49 Be careful to properly position the camshaft bearing caps according to its markings

5. Install the camshaft bearing caps in the sequence shown, and tighten to the following specifications:

a. Step 1: 48 inch lbs. (6 Nm)
b. Step 2: 84–108 inch lbs. (10–12 Nm)

✳✳ WARNING

Be careful to properly position the right bank, left bank, intake side, exhaust side, and front mark on the camshaft bearing caps while assembling.

✳✳ WARNING

Rotate the crankshaft so as not to contact the valves to the pistons by positioning the pistons 0.3937 inch (10mm) below the top of the cylinder block.

6. Install the water temperature control assembly.

7. Install the timing chain.

8. Check and adjust the valve clearance, as necessary.

9. Connect the negative battery cable.

10. Fill with engine coolant.

11. Start the engine and check for leaks.

12. Recheck the engine coolant level and oil level.

CAMSHAFT BEARING REPLACEMENT

Check each bearing for damage. If the bearing surface is excessively damaged, replace the cylinder head assembly or camshaft bearing cap, as necessary.

CRANKSHAFT DAMPER

REMOVAL & INSTALLATION

See Figures 50 through 55.

1. Before servicing the vehicle, refer to the Precautions Section.

2. Remove the engine cover.

3. Remove the front right wheel and tire.

4. Remove the right side cover.

5. Remove the accessory drive belt (A), the idler (B) and the tensioner (C).

➡In removing the accessory drive belt, fix a tool in the auto tensioner pulley bolt and turn the bolt counter clockwise.

6. Remove the timing belt upper over (A).

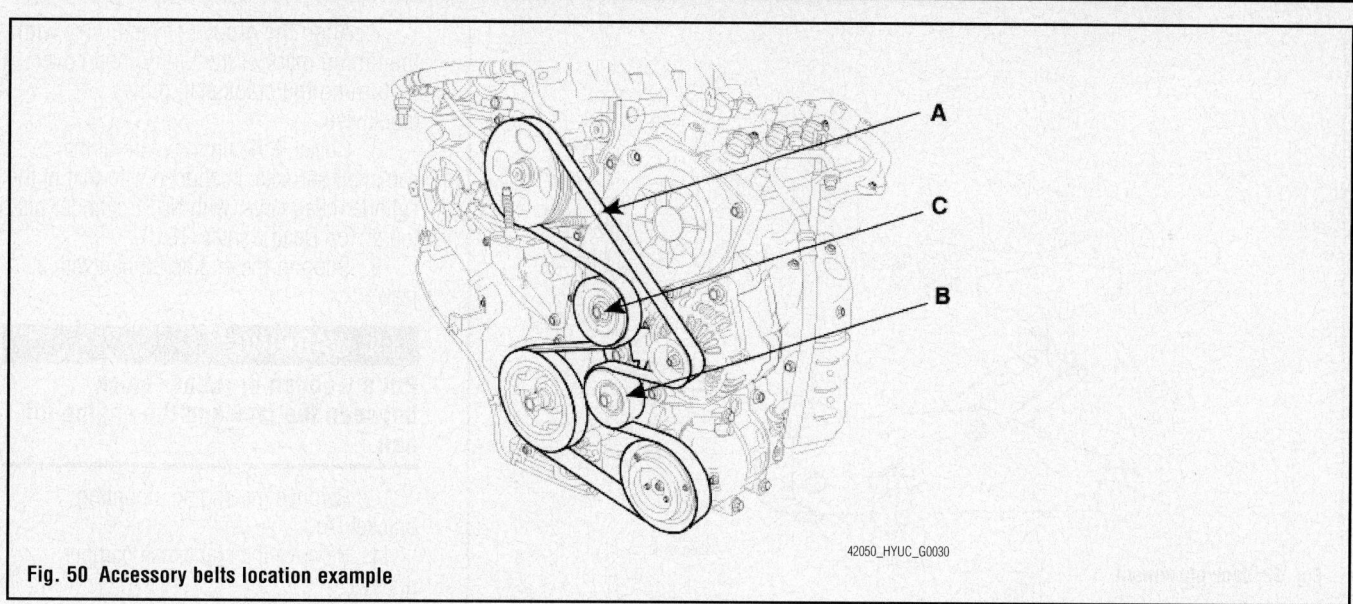

Fig. 50 Accessory belts location example

42050_HYUC_G0030

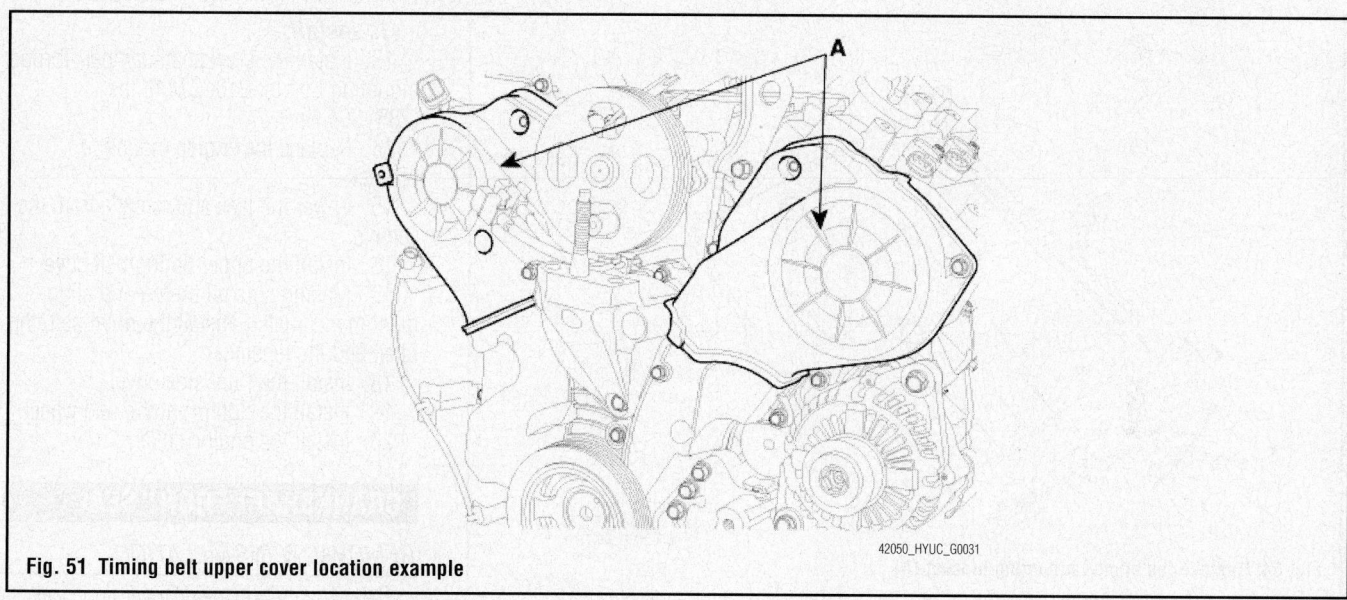

Fig. 51 Timing belt upper cover location example

42050_HYUC_G0031

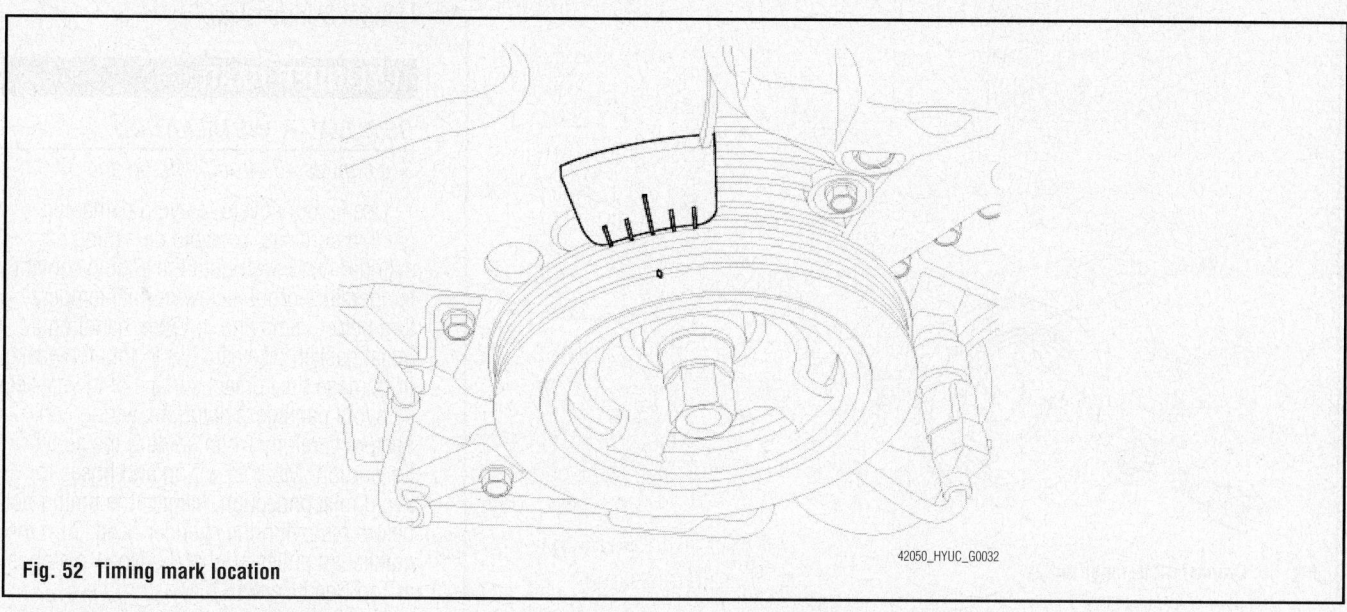

Fig. 52 Timing mark location

42050_HYUC_G0032

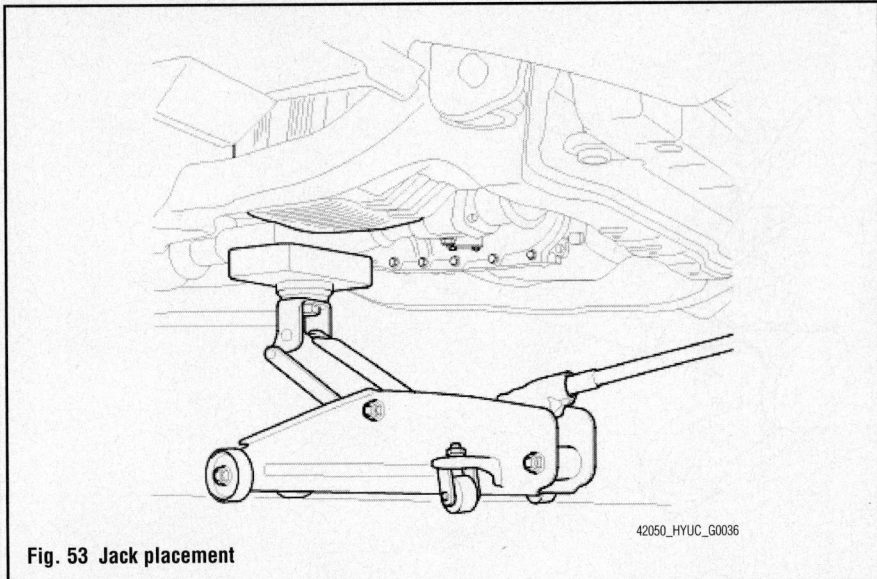

Fig. 53 Jack placement

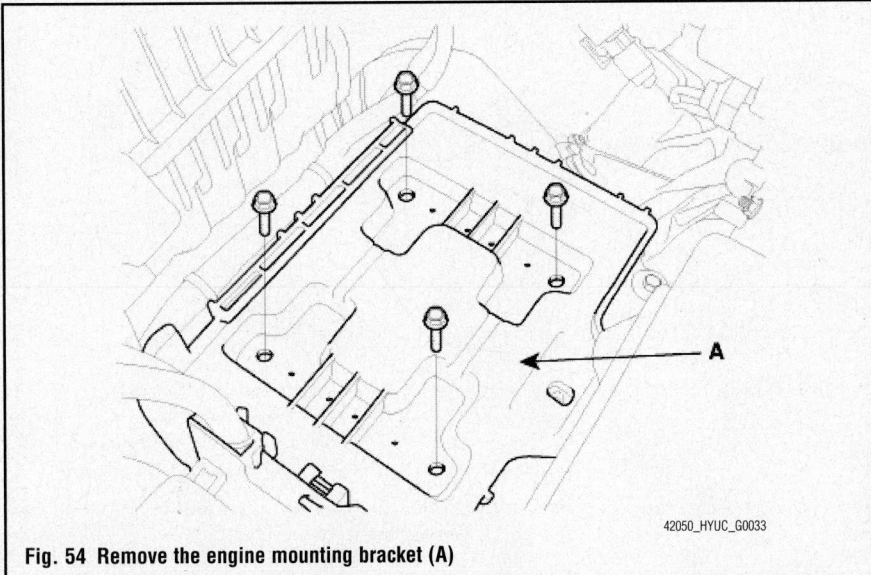

Fig. 54 Remove the engine mounting bracket (A)

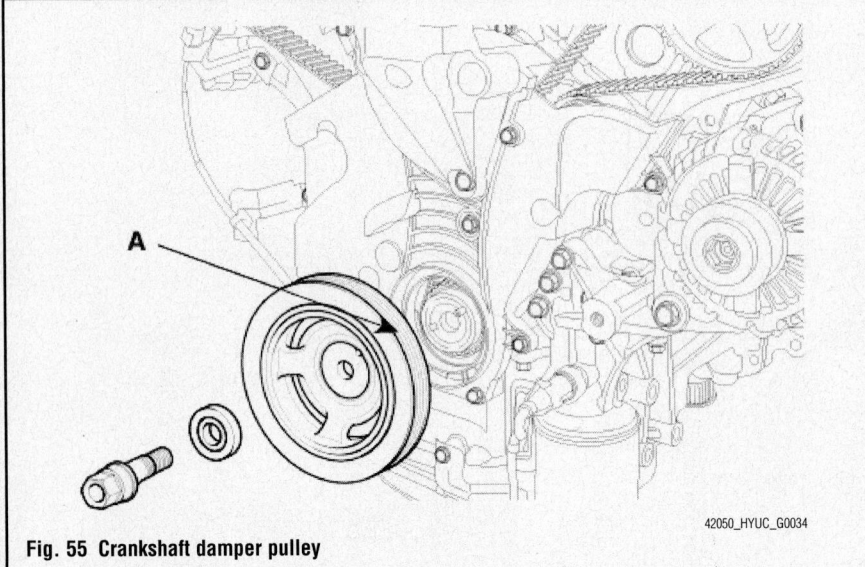

Fig. 55 Crankshaft damper pulley

7. Align the groove of the pulley with the timing mark of the timing belt cover by turning the crankshaft pulley clockwise.

8. Check if the timing mark of the camshaft sprocket is aligned with that of the cylinder head cover with No. 1 cylinder piston at Top Dead Center (TDC).

9. Support the engine oil pan with a jack.

✳✳ WARNING

Put a wooden or rubber block between the jack and the engine oil pan.

10. Remove the engine mounting bracket (A).

11. Remove the crankshaft damper pulley (A).

12. Remove the crankshaft damper.

To install:

13. Install the crankshaft damper. Torque mounting bolt to: 210–224 ft. lbs. (285–304 Nm).

14. Replace the engine mounting bracket.

15. Lower the jack and remove from the vehicle.

16. Install the upper timing belt cover.

17. Making sure all timing and alignment marks match, install the drive belt, the idler, and the tensioner.

18. Install the right side cover.

19. Install the right front tire and wheel.

20. Install the engine cover.

CRANKSHAFT FRONT SEAL

REMOVAL & INSTALLATION

Refer to Timing Chain Cover and Seal, removal & installation.

CYLINDER HEAD

REMOVAL & INSTALLATION

See Figures 46 through 49, 56 and 57.

Use fender covers to avoid damaging painted surfaces. To avoid damaging the cylinder head, wait until the engine coolant temperature drops below normal temperature before removing it. When handling a metal gasket, take care not to fold the gasket or damage the contact surface of the gasket. To avoid damage, unplug the wiring connectors carefully while holding the connector portion. Mark all wiring and hoses to avoid misconnection. Inspect the timing belt before removing the cylinder head. Turn the crankshaft pulley so that the No. 1 piston is at Top Dead Center (TDC).

1. Before servicing the vehicle, refer to the Precautions Section.

➡ **Engine removal is required for this procedure.**

2. Remove or disconnect the following:
- The negative battery connection
- The exhaust manifold
- The intake manifold
- The timing chain
- The water temperature control assembly
- The camshaft bearing cap
- The camshaft assembly

3. The cylinder head bolts, then remove cylinder head.

a. Uniformly loosen and remove the 16 cylinder head bolts, in several passes, in the sequence shown.

b. Remove the 16 cylinder head bolts and plate washers.

✳✳ WARNING

Head warpage or cracking could result from removing bolts in an incorrect order.

c. Lift the cylinder head from the dowels on the cylinder block and place the cylinder head on wooden blocks on a bench.

✳✳ WARNING

Be careful not to damage the contact surfaces of the cylinder head and cylinder block.

To install:

Thoroughly clean all parts to be assembled. Always use a new head and manifold gasket. The cylinder head gasket is a metal gasket. Take care not to bend it. Rotate the crankshaft, set the No. 1 piston at TDC.

4. Ensure the sealant locations on the cylinder head and cylinder block are free of engine oil or any debris.

5. Apply sealant on the cylinder block top face before assembling cylinder head gaskets.

➡ **The part must be assembled within 5 minutes after sealant is applied. The bead width should be 0.08–0.12 inch (2–3mm). The sealant location: 0.04–0.06 inch (1.0–1.5mm) from block surface. Recommended sealant: Liquid sealant TB1217H.**

6. Install the cylinder head. Remove any extruded sealant after assembling cylinder heads.

7. Place the cylinder head carefully in order not to damage the gasket with the bottom part of the end.

8. Install cylinder head bolts and tighten as follows:

a. Do not apply engine oil on the threads or under the heads of the cylinder head bolts.

b. Using SST(09221-4A000), install and tighten the cylinder head bolts and plate washers, in several passes, in the sequence shown.

c. Step 1: Tighten to 28–30 ft. lbs. (37–41 Nm)

d. Step 2: Tighten an additional: 120° plus or minus 2°

e. Step 3: Tighten an additional: 90° plus or minus 2°

f. Tighten bolt (A) to: 14–17 ft. lbs. (19–24 Nm).

➡ **Always use new cylinder head bolts.**

9. Install the Continuously Variable Valve Timing (CVVT) and camshaft sprocket. Tighten to 48–56 ft. lbs. (65–76 Nm).

a. Install camshaft-inlet to dowel pin of CVVT assembly. At this time, do not install to oil hole of camshaft-inlet.

b. Hold the hexagonal head wrench portion of the camshaft with a vise, and install the bolt and CVVT assembly.

c. Do not rotate the CVVT assembly when the camshaft is installed to the dowel pin of the CVVT assembly.

10. Install the camshafts.

a. Apply a light coat of engine oil on camshaft journals.

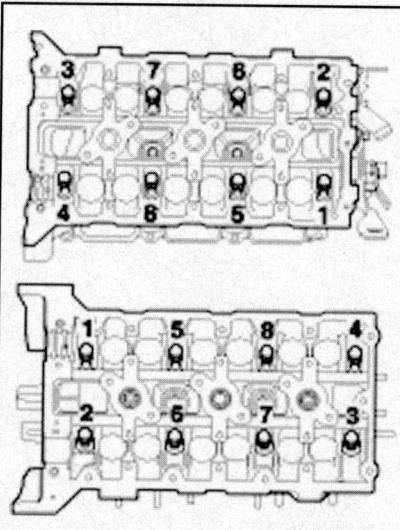

Fig. 56 Uniformly loosen/tighten and remove/install the 16 cylinder head bolts, in several passes, in the sequence shown

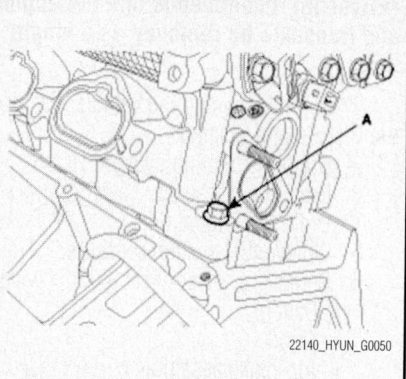

22140_HYUN_G0050

Fig. 57 Tighten bolt (A) to 14–17 ft. lbs. (19–24 Nm)

b. Assemble the key groove of camshaft rear side to the same level of head top surface.

c. Be careful to get the right bank, left bank, intake side, and exhaust side in the correct position before assembling.

11. Install the camshaft bearing caps in the sequence shown, to the following specifications:

a. Step 1: 48 inch lbs. (6 Nm)

b. Step 2: 84–108 inch lbs. (10–12 Nm)

✳✳ WARNING

Be careful to properly position the right bank, left bank, intake side, exhaust side, and front mark on the camshaft bearing caps while assembling.

✳✳ WARNING

Rotate the crankshaft so as not to contact the valves to the pistons by positioning the pistons 0.3937 inch (10mm) below the top of the cylinder block.

12. Install the water temperature control assembly.

13. Install the timing chain.

14. Check and adjust the valve clearance, as necessary.

15. Install the exhaust manifold.

16. Install the intake manifold.

17. Connect the negative battery cable.

18. Fill with engine coolant.

19. Start the engine and check for leaks.

20. Recheck the engine coolant level and oil level.

ENGINE ASSEMBLY

REMOVAL & INSTALLATION

See Figure 58.

➡**Hyundai recommends that the engine and transaxle be removed as a single unit.**

1. Before servicing the vehicle, refer to the Precautions Section.
2. Drain the cooling system.
3. Drain the transaxle.
4. Drain the engine oil.
5. Relieve fuel system pressure.
6. Remove or disconnect the following:
   - Battery
   - Hood
   - Air intake assembly
   - Accessory drive belts
   - Engine wiring harness connectors
   - Reverse lamp switch connector
   - Speedometer cable
   - Alternator harness connectors
   - Oil pressure gauge sender connector
   - Radiator hoses
   - Cooling fan
   - Fuel lines
   - Control cable, if equipped
   - Brake booster vacuum line
   - Intake manifold vacuum lines
   - Heater hoses
   - Accelerator cable
   - Cruise control cable, if equipped
   - Engine ground cable
   - Transaxle oil cooler lines
   - Shift cable
   - Transaxle wiring connectors
   - Radiator
   - Power steering pump
   - A/C compressor
   - Exhaust front pipe
   - Lower ball joints
   - Stabilizer bar control links
7. Separate the inner CV-joints from the transaxle and suspend the halfshafts out of the work area with safety wire.
8. Attach a hoist to the engine lifting eyes.
9. Remove or disconnect the following:
   - Front and rear roll stoppers
   - Engine mount and bracket
   - Transaxle mount and bracket
10. Lift the powertrain out of the vehicle.

To install:
11. Lower the powertrain into position.
12. Install the motor mount bracket and torque the fasteners to: 43–58 ft. lbs. (60–80 Nm).
13. Install the transaxle mount bracket and torque the fasteners to: 65–79 ft. lbs. (90–110 Nm).
14. Install or connect the following:
   - Front and rear roll stoppers
   - Engine mount
   - Transaxle mount
15. Remove the engine hoist.
16. Torque the mount through bolts as follows:
   a. Engine mount: 65–80 ft. lbs. (90–110 Nm).
   b. Transaxle mount: 65–80 ft. lbs. (90–110 Nm).

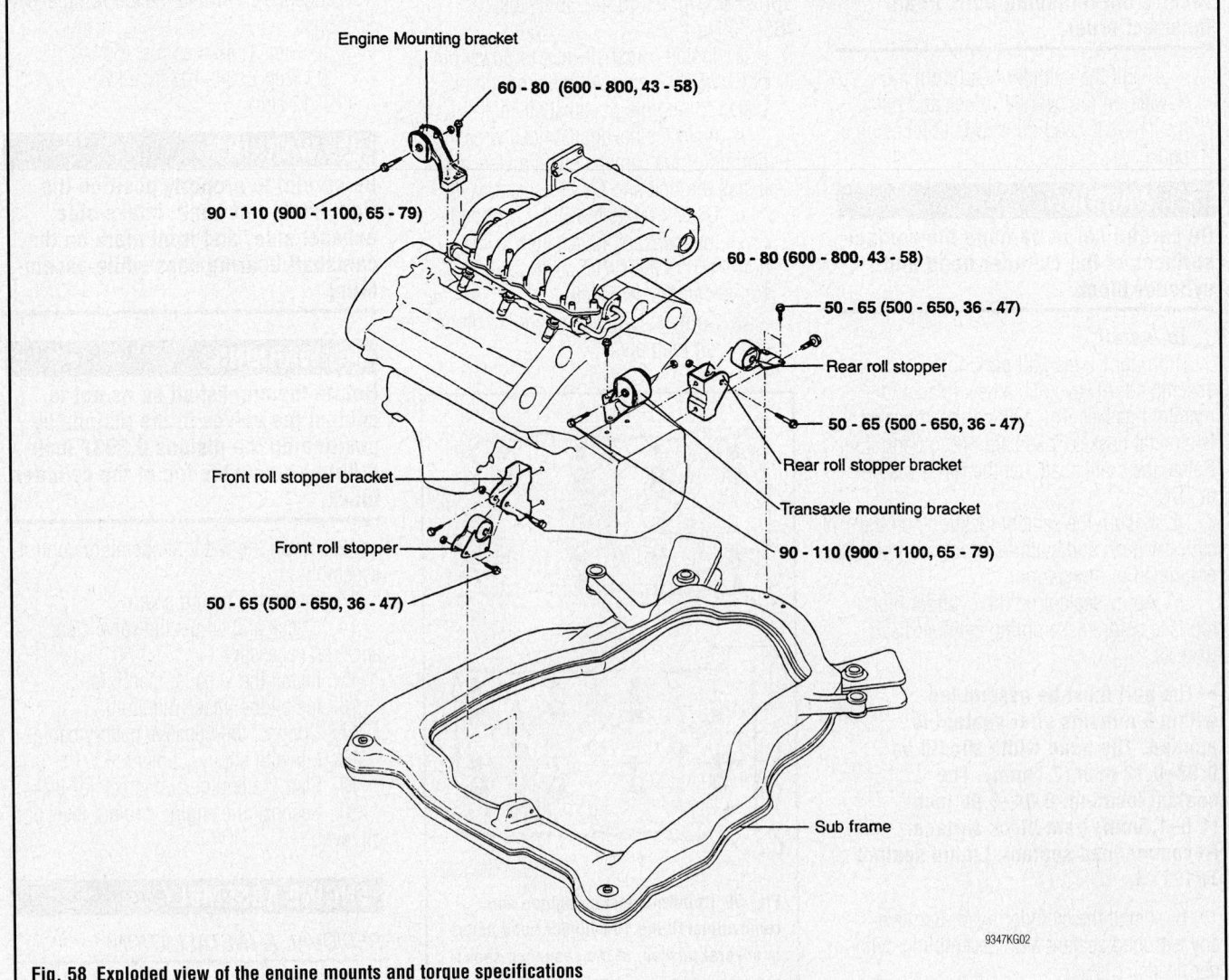

Fig. 58 Exploded view of the engine mounts and torque specifications

9347KG02

c. Front roll stopper: 36–47 ft. lbs.
(50–65 Nm).

d. Rear roll stopper: 36–47 ft. lbs.
(50–65 Nm).

17. Install or connect the following:
- Axle halfshafts using new circlips
- Stabilizer bar control links
- Lower ball joints
- Exhaust front pipe
- A/C compressor
- Power steering pump
- Radiator
- Transaxle oil cooler lines
- Shift cable
- Transaxle wiring connectors
- Engine ground cable
- Cruise control cable
- Accelerator cable
- Heater hoses
- Intake manifold vacuum lines
- Brake booster vacuum line
- Fuel lines
- Cooling fan
- Radiator hoses
- Oil pressure gauge sender connector
- Alternator harness connectors
- Speedometer cable
- Reverse lamp switch connector
- Engine wiring harness connectors
- Accessory drive belts
- Air intake assembly
- Hood
- Battery

18. Fill the engine with clean oil.
19. Fill the transaxle to the correct level.
20. Fill the cooling system to the proper level.
21. Start the engine and check for leaks.

EXHAUST MANIFOLD

REMOVAL & INSTALLATION

See Figure 59.

1. Before servicing the vehicle, refer to the Precautions Section.
2. Remove or disconnect the following:
- The negative battery cable
- The under cover
- The LH, RH rear oxygen sensor connector from bracket
- The front muffler
- The oil level gauge
- The LH front oxygen sensor connector from bracket
- The LH heat protector
- The LH exhaust manifold
- The RH front oxygen sensor connector from bracket
- The RH heat protector
- The RH exhaust manifold

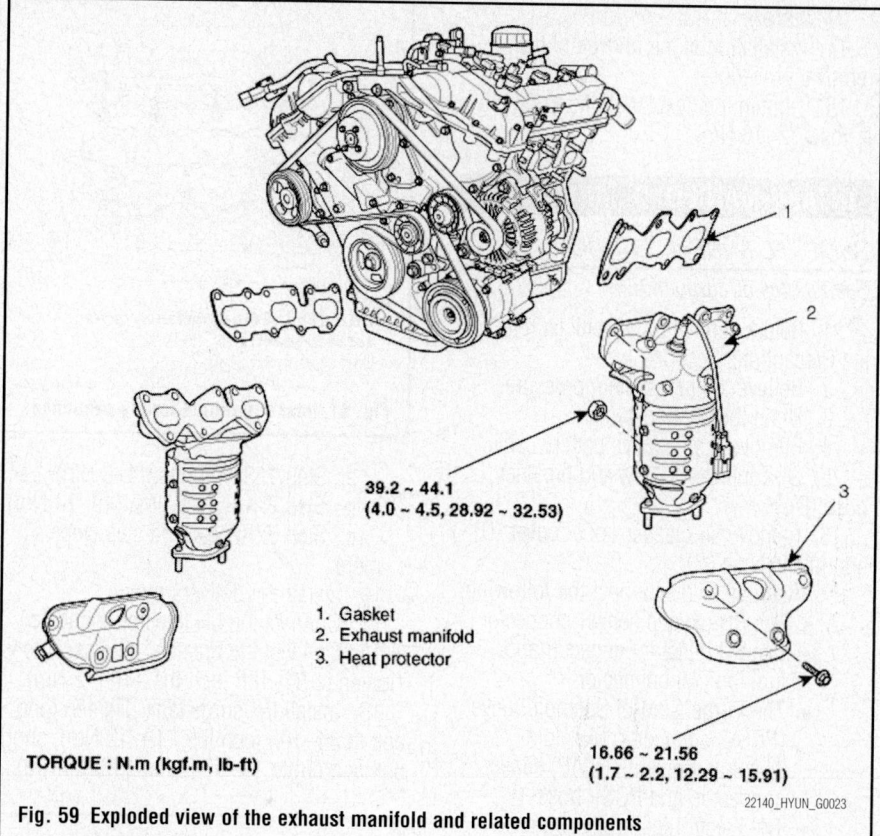

39.2 ~ 44.1
(4.0 ~ 4.5, 28.92 ~ 32.53)

16.66 ~ 21.56
(1.7 ~ 2.2, 12.29 ~ 15.91)

1. Gasket
2. Exhaust manifold
3. Heat protector

TORQUE : N.m (kgf.m, lb-ft)

22140_HYUN_G0023

Fig. 59 Exploded view of the exhaust manifold and related components

To install:

3. Install a new gasket and exhaust manifold. Tighten to 29–33 ft. lbs. (39–44 Nm).
4. Install heat protector. Tighten to 12–16 ft. lbs. (17–22 Nm).
5. Install front muffler. Tightening to 29–43 ft. lbs. (39–59 Nm).
6. Connect oxygen sensor connector.
7. Install under cover.
8. Connect the negative battery cable.

FLEXPLATE

REMOVAL & INSTALLATION

1. Before servicing the vehicle, refer to the Precautions Section.
2. Drain the transaxle.
3. Remove or disconnect the following:
- Negative battery cable
- Air intake assembly
- Battery and battery tray
- Back-up lamp connector
- Vehicle speed sensor
- Clutch release cylinder and lever
- Shift cable from transaxle assembly
- Steering column from the universal joint in the gear box
- Clutch housing upper mounting bolts
- Front, rear, and left transaxle mounting brackets

4. Using a suitable engine support fixture, support the engine assembly.
5. Remove or disconnect the following:
- Power steering pressure hose from the pump
- Front wheel
- Strut assembly
- Tie rod and sway bar link from the knuckle
- Wheel speed sensor
- Brake caliper
- Engine splash guard
- Front muffler
- Power steering hose on the front cross member

6. Using a suitable jack, support the sub-frame cross member.
7. Remove the cross member mounting bolts, and lower the cross member assembly with the steering gear and stabilizer bar attached.
8. Using a suitable jack, support the transaxle assembly.
9. Remove the front and rear roll stoppers.
10. Remove the engine and transaxle mounting bolts.
11. Slowly lower the transaxle from the vehicle.
12. Loosen the bolts that attach the flexplate in a star pattern. Loosen the bolts in succession, 1 or 2 turns at a time.
13. Remove the flexplate.

To install:

14. Installation is the reverse of the removal procedure.

15. Tighten the flexplate bolts to 53–56 ft. lbs. (72–76 Nm).

INTAKE MANIFOLD

REMOVAL & INSTALLATION

See Figures 60 through 62.

1. Before servicing the vehicle, refer to the Precautions Section.
2. Relieve the fuel system pressure.
3. Drain the cooling system.
4. Remove the negative battery cable.
5. Disconnect AFS (A) and breather hose (B).
6. Remove air cleaner upper cover (D) and intake hose (C).
7. Remove or disconnect the following:
   - The RH oxygen sensor connector
   - The RH injector connector and ignition coil connector
   - The Purge Control Solenoid Valve (PCSV) connector, Manifold Absolute Pressure (MAP) sensor connector, and PCSV hose
   - The Electronic Throttle Control (ETC) connector and knock sensor connector
   - The water hoses from ETC
   - The PCV hose
   - The brake vacuum hose
   - The surge tank stay
   - The connector bracket from surge tank
   - The surge tank
   - The breather Pipe assembly
   - The LH injector connector.
8. Remove intake manifold and gasket

To install:

9. Install intake manifold and new gasket on the cylinder head. Tighten the bolts in the illustrated sequence using the steps below:

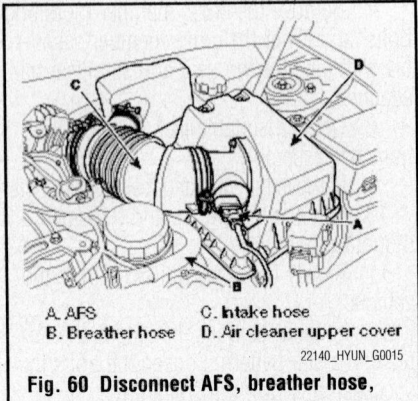

A. AFS
B. Breather hose
C. Intake hose
D. Air cleaner upper cover

22140_HYUN_G0015

Fig. 60 Disconnect AFS, breather hose, air cleaner upper cover, and intake hose

Be careful of the installation order
1st step order: a-h
2nd step order: 1-8

22140_HYUN_G0016

Fig. 61 Intake manifold torque sequence

a. Step 1: 3–4 ft. lbs. (4–6 Nm).
b. Step 2: 14–17 ft. lbs. (19–24 Nm).
c. Step 3: Repeat 2nd step twice or more.

10. Install the delivery pipe.
11. Connect the LH injector connector.
12. Connect the breather pipe assembly. Tighten to 84–108 inch lbs. (10–12 Nm).
13. Install the surge tank. Tighten long bolt to 84–108 inch lbs. (10–12 Nm); short bolt and nut to 14–17 ft. lbs. (19–24 Nm).

14. Install the connector bracket on the surge tank. Tighten to 5–8 ft. lbs. (7–11 Nm).
15. Install surge tank stay. Tighten to 20–23 ft. lbs. (27–31 Nm); 14–17 ft. lbs. (19–24 Nm).
16. Connect brake vacuum hose.
17. Connect PCV hose.
18. Connect water hoses to ETC.
19. Install ETC bracket. Tighten to 12–19 ft. lbs. (16–26 Nm).
20. Connect ETC connector and knock sensor connector.
21. Connect PCSV connector, MAP sensor connector and PCSV hose.
22. Connect RH injector connector and ignition coil connector.
23. Connect RH oxygen sensor connector.
24. Install air cleaner upper cover and intake hose.
25. Connect AFS and breather hose.

OIL PAN

REMOVAL & INSTALLATION

See Figure 63.

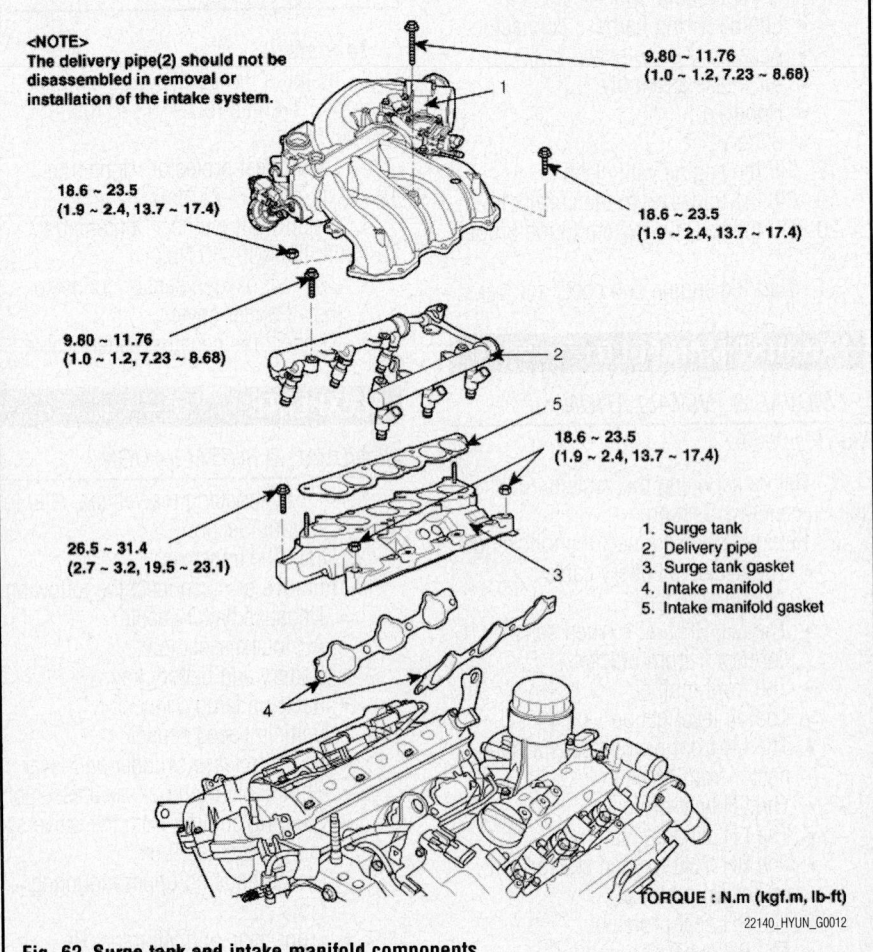

<NOTE>
The delivery pipe(2) should not be disassembled in removal or installation of the intake system.

9.80 ~ 11.76
(1.0 ~ 1.2, 7.23 ~ 8.68)

18.6 ~ 23.5
(1.9 ~ 2.4, 13.7 ~ 17.4)

18.6 ~ 23.5
(1.9 ~ 2.4, 13.7 ~ 17.4)

9.80 ~ 11.76
(1.0 ~ 1.2, 7.23 ~ 8.68)

18.6 ~ 23.5
(1.9 ~ 2.4, 13.7 ~ 17.4)

26.5 ~ 31.4
(2.7 ~ 3.2, 19.5 ~ 23.1)

1. Surge tank
2. Delivery pipe
3. Surge tank gasket
4. Intake manifold
5. Intake manifold gasket

TORQUE : N.m (kgf.m, lb-ft)

22140_HYUN_G0012

Fig. 62 Surge tank and intake manifold components

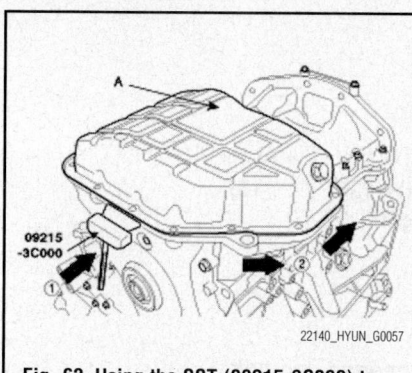

Fig. 63 Using the SST (09215-3C000) to remove the oil pan

1. Before servicing the vehicle, refer to the Precautions Section.
2. Drain the engine oil.
3. Remove the oil pan bolts.
4. Using the SST (09215-3C000), remove the oil pan.

   a. Insert the SST between the oil pan and the ladder frame by tapping it with a plastic hammer in the direction of arrow.

   b. After tapping the SST with a plastic hammer along the direction of arrow around more than ⅔ of the edge of the oil pan, remove it from the ladder frame.

❊❊ WARNING

Do not turn over the SST abruptly without tapping. Damage may occur to the SST or the oil pan.

To install:

5. Using a razor blade and gasket scraper, carefully remove all the old packing material from the gasket surfaces.
6. Check that the mating surfaces are clean and dry before applying liquid gasket.

   a. Apply liquid gasket as an even bead, centered between the edges of the mating surface. Use liquid gasket: TB1217H or equivalent. Apply a bead ⅛ inch (3mm) wide to the oil pan.

   b. To prevent leakage of oil, apply liquid gasket to the inner threads of the bolt holes.

➡ **Do not install the parts if 5 minutes or more have elapsed since applying the liquid gasket. Instead, reapply liquid gasket after removing the residue.**

7. Install the oil pan with the bolts. Uniformly tighten the bolts in several passes to a final torque of 84–108 inch lbs. (10–12 Nm).

➡ **After assembly, wait at least 30 minutes before filling the engine with oil.**

8. Fill the engine with the proper type and amount of engine oil.

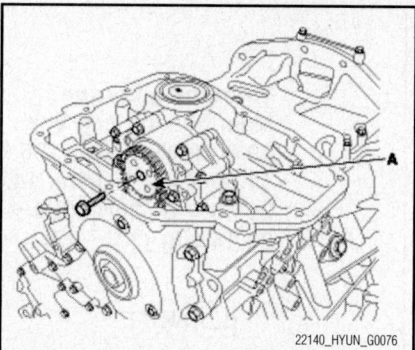

Fig. 64 Remove the oil pump chain cover (A) and the oil pump

OIL PUMP

REMOVAL & INSTALLATION

See Figures 64 and 65.

1. Before servicing the vehicle, refer to the Precautions Section.
2. Remove the lower oil pan, refer to Oil Pan, removal & installation.
3. Remove the oil pump chain cover.
4. Remove the oil pump chain sprocket (A).
5. Remove the oil pump.

To install:

6. Install the oil pump using a new O-ring. Tighten the bolts to 15–17 ft. lbs. (21–23 Nm).

➡ **Always use a new O-ring.**

7. Install the oil pump sprocket and oil pump chain on the oil pump. Tighten to 14–16 ft. lbs. (19–22 Nm).
8. Install the oil pump chain cover. Tighten to 84–108 inch lbs. (10–12 Nm).
9. Install the lower oil pan. Refer to Oil Pan, removal & installation.
10. After assembly, wait at least 30 minutes before filling the engine with oil to allow the gasket material to cure.
11. Fill the engine with the proper type and amount of engine oil.
12. Start the engine and check for leaks.
13. Recheck the engine oil level.

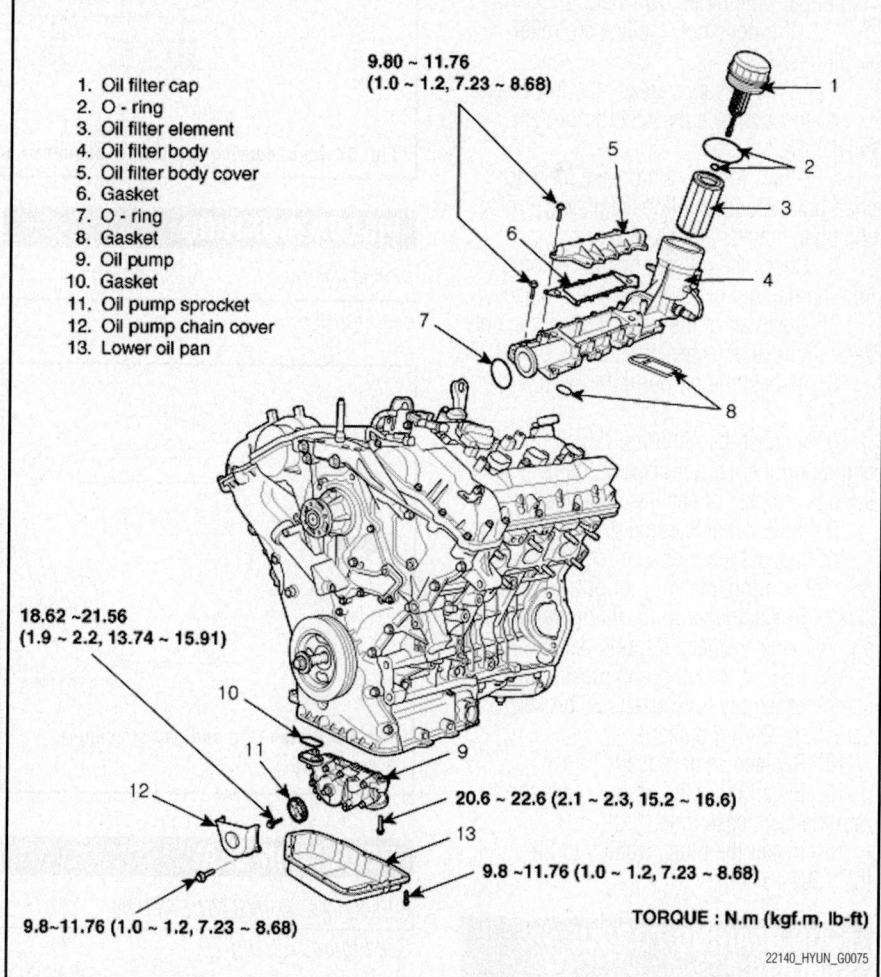

1. Oil filter cap
2. O - ring
3. Oil filter element
4. Oil filter body
5. Oil filter body cover
6. Gasket
7. O - ring
8. Gasket
9. Oil pump
10. Gasket
11. Oil pump sprocket
12. Oil pump chain cover
13. Lower oil pan

9.80 ~ 11.76 (1.0 ~ 1.2, 7.23 ~ 8.68)

18.62 ~21.56 (1.9 ~ 2.2, 13.74 ~ 15.91)

20.6 ~ 22.6 (2.1 ~ 2.3, 15.2 ~ 16.6)

9.8 ~11.76 (1.0 ~ 1.2, 7.23 ~ 8.68)

9.8~11.76 (1.0 ~ 1.2, 7.23 ~ 8.68)

TORQUE : N.m (kgf.m, lb-ft)

Fig. 65 Expanded view of lubrication system components

INSPECTION

See Figure 66.

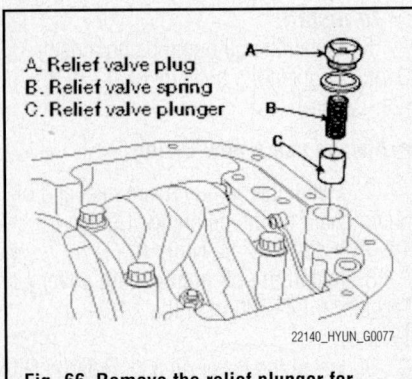

A. Relief valve plug
B. Relief valve spring
C. Relief valve plunger

22140_HYUN_G0077

Fig. 66 Remove the relief plunger for inspection

1. Before servicing the vehicle, refer to the Precautions Section.
2. Remove the relief plunger: remove the plug (A), spring (B), and relief plunger (C).
3. Inspect the relief plunger.
   a. Coat the plunger with engine oil
   b. Check that it falls smoothly into the plunger hole by its own weight.
   c. If it does not, replace the relief plunger.
   d. If necessary, replace the front case.
4. Inspect for a distorted or broken relief valve spring.
5. Check the oil pump case for worn shaft hole, clogged oil passage, worn rotor chamber, cracks, and other faults.
6. Check the oil seal lips for deformation, hardening, or wear. Replace if defective.
7. Clean all of the parts in cleaning solvent. Remove any varnish, sludge, or dirt.
8. Inspect the oil pump for wear and scoring.
9. Inspect the oil pump housing and engine front cover for scoring, damaged threads, cracks, or casting imperfections.
10. Inspect the oil pump gears for damage.
11. Inspect the pressure regulator valve area for scoring, sticking, or burrs.
12. Check the valve for fitting condition and damage. Replace if found defective.
13. Inspect the oil pump pickup tube and screen assembly for looseness, broken screen, or O-ring damage.
14. Replace as necessary.
15. Install the relief plunger and spring into the front case hole.
16. Install the plug. Tighten to 29–36 ft. lbs. (39–49 Nm).

MAIN BEARING TORQUE SEQUENCE

See Figure 67.

Tightening torque
Main bearing cap bolt
49.00Nm(5.0 kgf.m, 36.16lb-ft) + 90° (1 ~ 8)
19.60 Nm(2.0 kgf.m, 14.46lb-ft)+ 120° (9 ~ 16)
29.40 ~ 31.36Nm(3.0 ~ 3.2 kgf.m, 21.70 ~ 23.14lb-ft) (17 ~ 22)

⊍NOTE

- Always use new main bearing cap bolt.
- If any of the bearing cap bolts in broken or deformed, replace it.

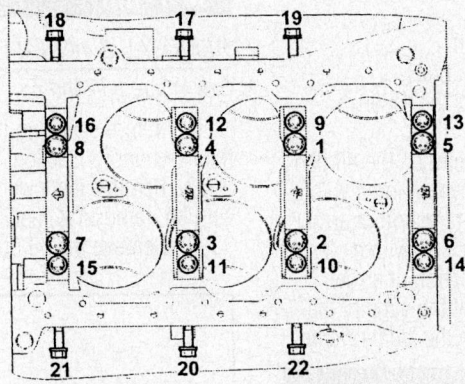

09474_SEDO_G0018

Fig. 67 Main bearing cap torque sequence and specification

PISTON AND RING

POSITIONING

See Figure 68.

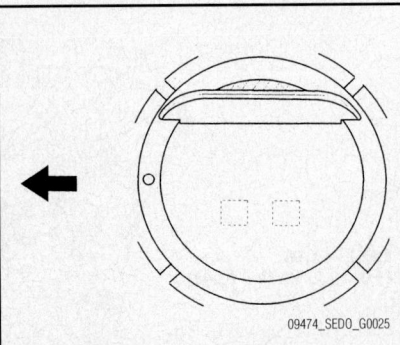

09474_SEDO_G0025

Fig. 68 Piston ring end gap spacing—3.8L Engine

REAR MAIN SEAL

REMOVAL & INSTALLATION

See Figures 69 and 70.

1. Before servicing the vehicle, refer to the Precautions Section.
2. Record the radio anti-theft code data.
3. Remove the engine from the vehicle.
4. Remove the flexplate from the engine.
5. Remove the rear main oil seal from its mounting.

To install:

6. Install the oil seal to the oil seal housing using tool 09231-3C200 and tool 09231-H1100.
7. Continue the installation in the reverse order of the removal procedure.

ROCKER ARMS/SHAFTS

REMOVAL & INSTALLATION

These engines are not equipped with rocker arms. The camshaft acts directly on the valves through hydraulic lash adjusters.

TIMING CHAIN COVER AND SEAL

REMOVAL & INSTALLATION

For timing chain cover and seal, refer to Timing Chain and Sprockets, removal & installation.

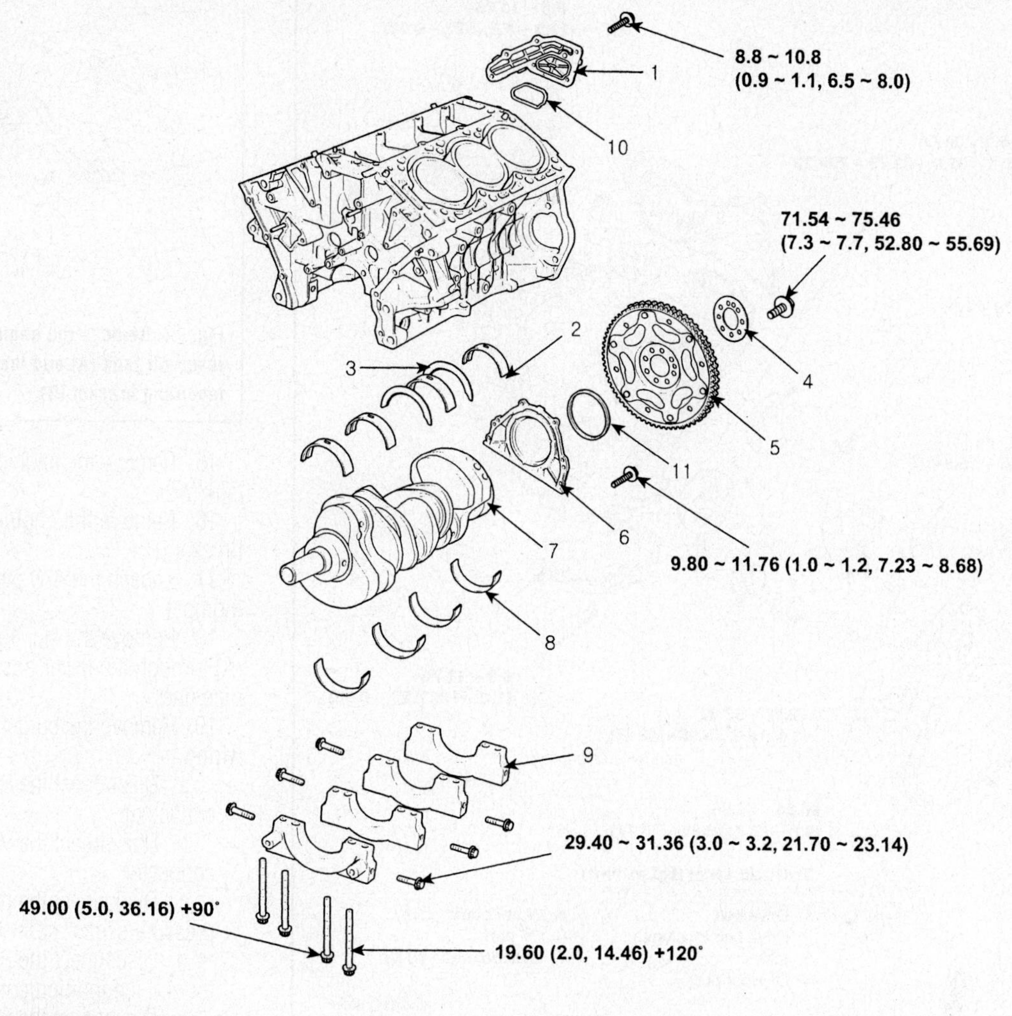

**8.8 ~ 10.8
(0.9 ~ 1.1, 6.5 ~ 8.0)**

**71.54 ~ 75.46
(7.3 ~ 7.7, 52.80 ~ 55.69)**

9.80 ~ 11.76 (1.0 ~ 1.2, 7.23 ~ 8.68)

29.40 ~ 31.36 (3.0 ~ 3.2, 21.70 ~ 23.14)

49.00 (5.0, 36.16) +90°

19.60 (2.0, 14.46) +120°

TORQUE : N.m (kgf.m, lb-ft)

1. Oil drain cover
2. Crankshaft upper bearing
3. Thrust bearing
4. Plate adapter
5. Drive plate
6. Rear oil seal case
7. Crankshaft
8. Crankshaft lower bearing
9. Main bearing cap
10. Oil drain cover gasket
11. Rear oil seal

09474_SEDO_G0016

Fig. 69 Engine block and related components

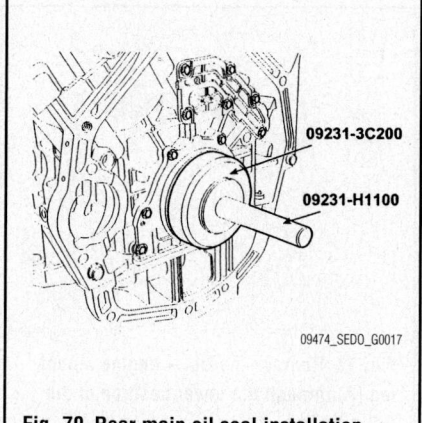

09231-3C200

09231-H1100

09474_SEDO_G0017

Fig. 70 Rear main oil seal installation

TIMING CHAIN AND SPROCKETS

REMOVAL & INSTALLATION

See Figures 71 through 90.

1. Before servicing the vehicle, refer to the Precautions Section.

➡ **Use fender covers to avoid damaging painted surfaces.**

➡ **To avoid damage, unplug the wiring connectors carefully while holding the connector portion.**

➡ **Mark all wiring and hoses to avoid misconnection.**

➡ **Turn the crankshaft pulley so that the No. 1 piston is at Top Dead Center (TDC).**

2. Disconnect the negative terminal from the battery.
3. Remove the engine cover.
4. Remove the air duct.
5. Remove the intake air hose and air cleaner assembly.
   a. Disconnect the AFS connector (A).
   b. Disconnect the breather hose (B) from air cleaner hose.
   c. Disconnect the ECM connector.
   d. Remove the intake air hose (C) and air cleaner (D).

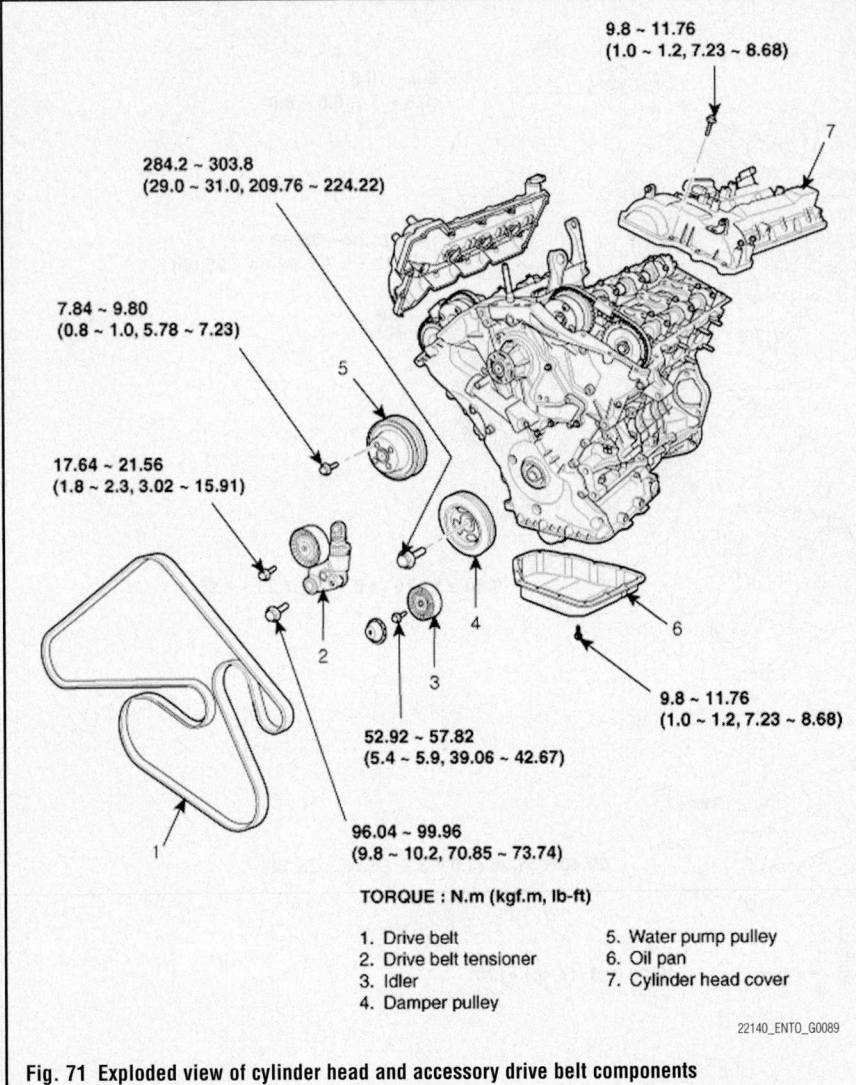

9.8 ~ 11.76
(1.0 ~ 1.2, 7.23 ~ 8.68)

284.2 ~ 303.8
(29.0 ~ 31.0, 209.76 ~ 224.22)

7.84 ~ 9.80
(0.8 ~ 1.0, 5.78 ~ 7.23)

17.64 ~ 21.56
(1.8 ~ 2.3, 3.02 ~ 15.91)

9.8 ~ 11.76
(1.0 ~ 1.2, 7.23 ~ 8.68)

52.92 ~ 57.82
(5.4 ~ 5.9, 39.06 ~ 42.67)

96.04 ~ 99.96
(9.8 ~ 10.2, 70.85 ~ 73.74)

TORQUE : N.m (kgf.m, lb-ft)

1. Drive belt
2. Drive belt tensioner
3. Idler
4. Damper pulley
5. Water pump pulley
6. Oil pan
7. Cylinder head cover

22140_ENTO_G0089

Fig. 71 Exploded view of cylinder head and accessory drive belt components

6. Remove the RH front wheel.
7. Remove the undercover.
8. Remove the RH side cover.
9. Drain the engine coolant. Remove the radiator cap to speed draining.
10. Remove the upper radiator hose.
11. Drain the engine oil.

12. Remove the lower oil pan. Refer to Oil Pan, removal & installation.
13. Place a jack underneath the upper oil pan.
14. Just loosen the transaxle mounting bolts (A) without removing the transaxle mounting (B).

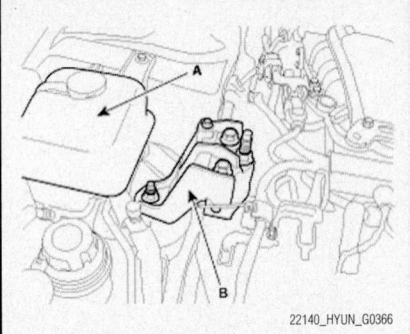

22140_HYUN_G0366

Fig. 74 Remove the engine coolant reservoir tank (A) and the engine mounting bracket (B)

15. Remove the engine coolant reservoir tank (A).
16. Remove the engine mounting bracket (B).
17. Loosen the A/C pipe bracket mounting bolt.
18. Remove the No. 1 engine mounting (A) through the lower position of the A/C pipe line.
19. Remove the surge tank and engine wiring.
   a. Disconnect the RH oxygen sensor connector.
   b. Disconnect the VIS solenoid valve connector.
   c. Disconnect the power steering oil pressure sensor connector.
   d. Disconnect the RH injector connector and the ignition coil connector.
   e. Disconnect the OCV connector and knock sensor connector.
   f. Disconnect the LH front oxygen sensor connector, alternator connector, and air compressor connector.
   g. Disconnect the LH ignition coil connector, injector connector, condenser connector, and ground
   h. Remove the wiring harness protector.

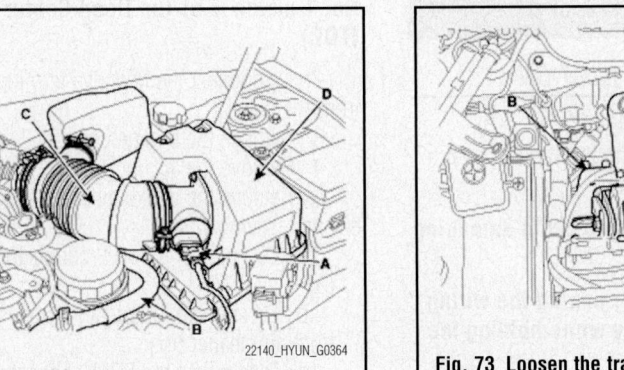

22140_HYUN_G0364

Fig. 72 Remove the intake air hose and air cleaner assembly

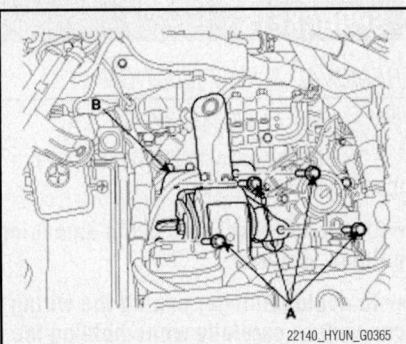

22140_HYUN_G0365

Fig. 73 Loosen the transaxle mounting bolts (A) without removing the transaxle mounting (B)

22140_HYUN_G0367

Fig. 75 Remove the No. 1 engine mounting (A) through the lower position of the A/C pipe line

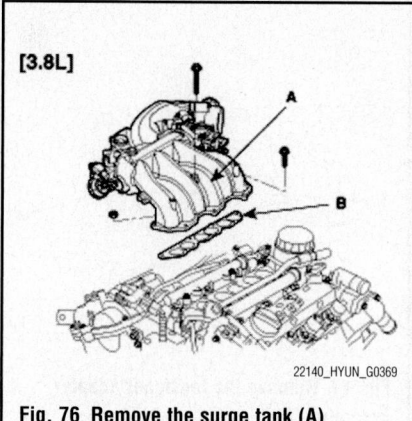

Fig. 76 Remove the surge tank (A)

Fig. 78 Turn the crankshaft pulley and
align its groove with the timing mark "T"
of the lower timing chain cover

➡Use the SST (flywheel stopper,
09231-3C300) to remove the crankshaft
pulley bolt, after removing the starter.

24. Lift up the engine assembly by using
the jack.
25. Remove the power steering pump.
26. Remove the air conditioner compressor.
27. Remove the alternator.
28. Remove the drive belt idler.
29. Remove the drive belt auto tensioner.
30. Remove water pump pulley.
31. Remove the timing chain cover (A).

i. Disconnect the LH CMPS and oil
pressure switch connector.
j. Disconnect the ETC connector and
the knock sensor connector.
k. Disconnect the PCSV connector,
the MAP sensor connector, and the
PCSV hose.
l. Remove the Electronic Throttle
Control (ETC) bracket.
m. Disconnect the water hoses from
the ETC.
n. Disconnect the PCV hose.
o. Disconnect the brake vacuum hose.
p. Remove the surge tank stay.
q. Remove the connector bracket from
the surge tank.
r. Remove the surge tank (A).

**Cover the inlet of intake manifold with a
clean woven stuff or vinyl cover to prevent foreign materials from entering.**

20. Remove the cylinder head cover.
a. Remove the connector bracket (A)
from the LH cylinder head cover.
b. Disconnect the RH ignition coil
connector(A), the condenser connector
(B) and remove the wiring bracket (C).
c. Remove the LH and RH ignition
coils.

d. Remove the LH and RH cylinder
head covers.

➡**Cover the upside of engine head with
a clean vinyl cover to prevent foreign
materials from entering.**

21. Set No. 1 cylinder to TDC/
compression.
a. Turn the crankshaft pulley and
align its groove with the timing mark "T"
of the lower timing chain cover.

❊❊ WARNING

**Do not rotate engine counterclock-
wise.**

b. Check that the mark (A) of the
camshaft timing sprockets are in
straight line on the cylinder head surface as shown in the illustration. If not,
turn the crankshaft one revolution
(360°). Do not rotate engine counterclockwise.
22. Remove the accessory drive belt.
Refer to Accessory Drive Belts, removal &
installation.
23. Remove the crankshaft damper pulley. Refer to Crankshaft Damper, removal &
installation.

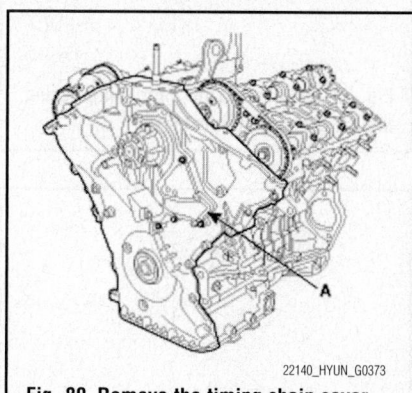

Fig. 80 Remove the timing chain cover
(A)

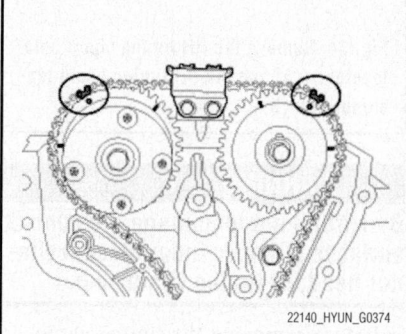

Fig. 81 Timing marks on chain and
camshaft sprockets illustrated

Fig. 77 Remove the connector bracket (A)
from the LH cylinder head cover

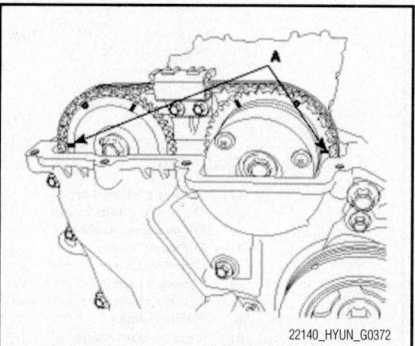

Fig. 79 Check that the mark (A) of the
camshaft timing sprockets are in straight
line on the cylinder head surface as shown

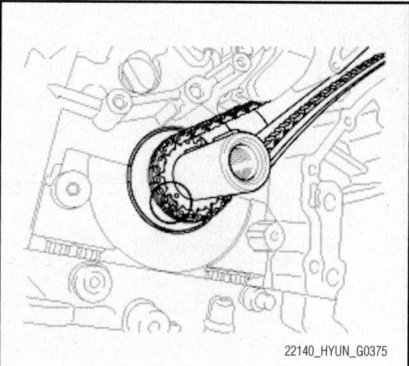

Fig. 82 Timing marks on chain and
crankshaft sprocket illustrated

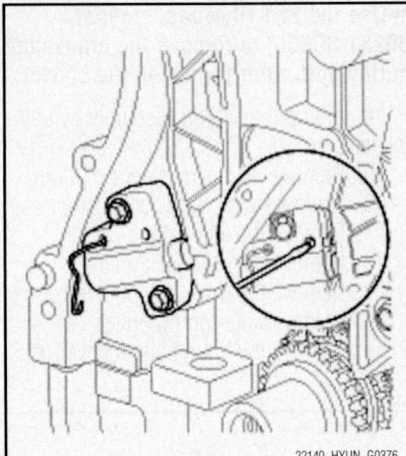

Fig. 83 Install a set pin after compressing the RH timing chain tensioner

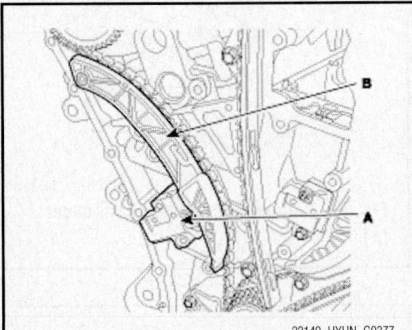

Fig. 84 Remove the RH timing chain auto tensioner (A) and the RH timing chain tensioner arm (B)

✳✳ WARNING

Be careful not to damage the contact surfaces of the cylinder block, cylinder head, or timing chain cover.

→Before removing the timing chain, mark the RH/LH timing chain with an identification based on the location of the sprocket as the identification mark

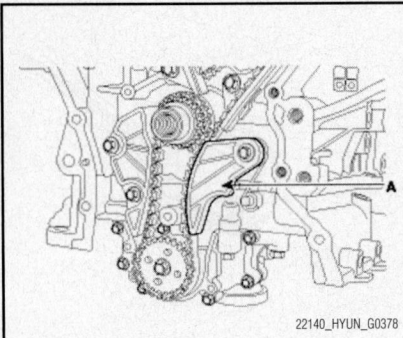

Fig. 85 Remove the oil pump chain tensioner assembly (A)

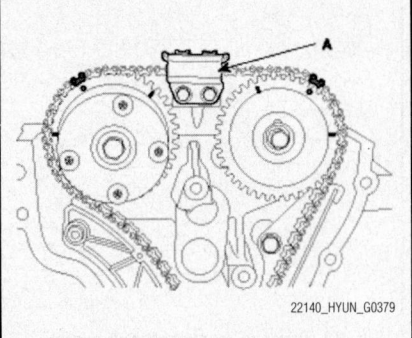

Fig. 86 Remove the LH cam-to-cam guide (A)

Fig. 87 Remove the tensioner adapter assembly (A)

on the chain for TDC can be erased accidentally.

32. Install a set pin after compressing the RH timing chain tensioner.
33. Remove the RH cam-to-cam guide.
34. Remove the RH timing chain auto tensioner (A) and the RH timing chain tensioner arm (B).
35. Remove the RH timing chain.
36. Remove the RH timing chain guide.
37. Remove the oil pump chain cover.

38. Remove the oil pump chain tensioner assembly (A).
39. Remove the oil pump chain guide.
40. Remove the oil pump chain sprocket and oil pump chain.
41. Remove the crankshaft sprocket that drives the oil pump.
42. Install a set pin after compressing the LH timing chain tensioner.
43. Remove the LH cam-to-cam guide (A).

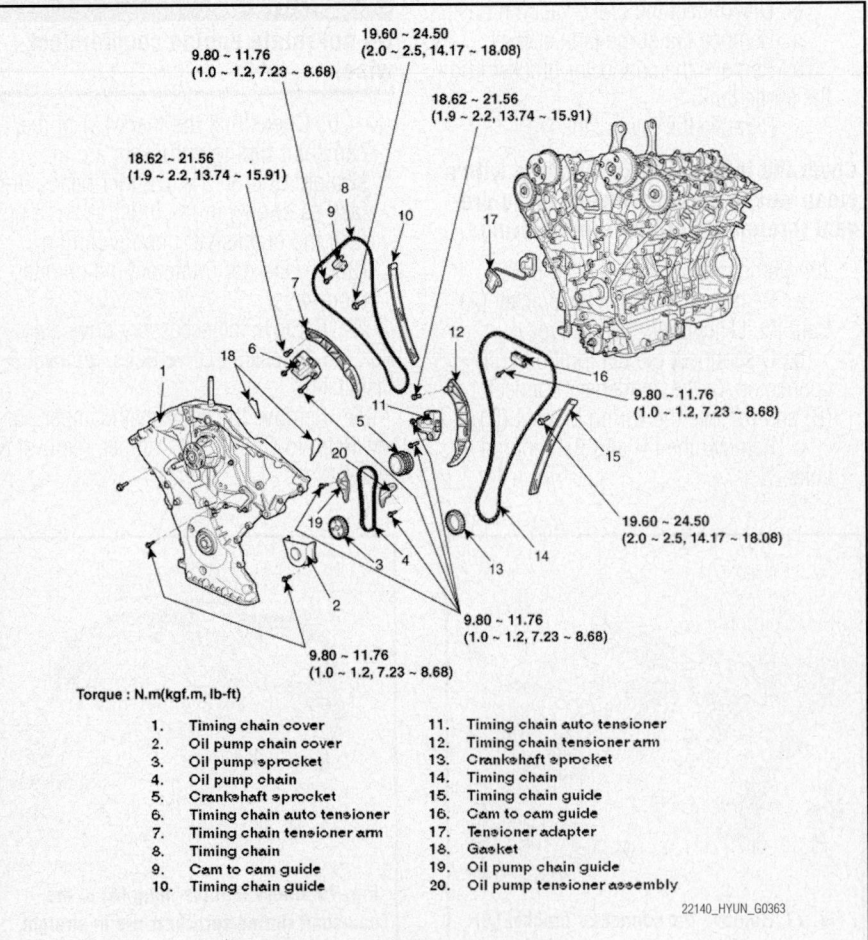

Torque : N.m(kgf.m, lb-ft)

1. Timing chain cover	11. Timing chain auto tensioner
2. Oil pump chain cover	12. Timing chain tensioner arm
3. Oil pump sprocket	13. Crankshaft sprocket
4. Oil pump sprocket	14. Timing chain
5. Crankshaft sprocket	15. Timing chain guide
6. Timing chain auto tensioner	16. Cam to cam guide
7. Timing chain tensioner arm	17. Tensioner adapter
8. Timing chain	18. Gasket
9. Cam to cam guide	19. Oil pump chain guide
10. Timing chain guide	20. Oil pump tensioner assembly

Fig. 88 Exploded view of timing chain and related components

44. Remove the LH timing chain auto tensioner and LH timing chain tensioner arm.

45. Remove the LH timing chain.

46. Remove the LH timing chain guide.

47. Remove the crankshaft sprocket.

48. Remove the tensioner adapter assembly (A).

To install:

49. Check the camshaft sprocket and crankshaft sprocket for abnormal wear, cracks, or damage. Replace as necessary.

50. Inspect the tensioner arm and chain guide for abnormal wear, cracks, or damage. Replace as necessary.

51. Check that the tensioner piston moves smoothly when the ratchet pawl is released.

52. Install the jack to the upper oil pan.

53. The key of crankshaft must be aligned with the timing mark of timing chain cover. Then, piston of No. 1 cylinder is placed at the TDC on the compression stroke.

54. Install the tensioner adapter assembly.

55. Install the crankshaft sprocket.

56. Install the LH timing chain guide. Tighten to 14–18 ft. lbs. (20–25 Nm).

57. Install the LH timing chain.

a. To install the timing chain with no slack between the camshaft and crankshaft, use the following procedure.

b. Place the crankshaft sprocket on first, then the timing chain guide, then the exhaust camshaft sprocket, and finally the intake camshaft sprocket.

c. The timing mark of each sprockets should be matched with the timing mark (color link) of the timing chain when installing the timing chain.

58. Install the LH timing chain tensioner arm. Tighten to 14–16 ft. lbs. (19–22 Nm).

59. Install the chain tensioner. Tighten to 87–104 inch lbs. (10–12 Nm).

60. Install the LH cam-to-cam guide. Tighten to 87–104 inch lbs. (10–12 Nm).

61. Install the crankshaft sprocket.

62. Install the oil pump chain and the oil pump sprocket. Tighten to 14–16 ft. lbs. (19–22 Nm).

63. Install the RH timing chain guide. Tighten to 14–18 ft. lbs. (20–25 Nm).

64. Install the RH timing chain.

a. To install the timing chain with no slack between the camshaft and the crankshaft, use the following procedure.

b. Place the chain on the crankshaft sprocket first, then the intake camshaft sprocket, then the exhaust camshaft sprocket.

c. The timing mark of each of the sprockets must be matched with the timing mark (color link) of timing chain when installing the timing chain.

65. Install the RH timing chain tensioner arm. Tighten to 14–16 ft. lbs. (19–22 Nm).

66. Install the RH timing chain auto tensioner. Tighten to 87–104 inch lbs. (10–12 Nm).

67. Install the RH cam-to-cam guide. Tighten to 87–104 inch lbs. (10–12 Nm).

68. Install the oil pump chain guide. Tighten to 87–104 inch lbs. (10–12 Nm).

69. Install the oil pump chain tensioner assembly. Tighten to 87–104 inch lbs. (10–12 Nm).

70. Pull out the pins of hydraulic tensioner (LH & RH).

71. Install the oil pump chain cover. Tighten to 87–104 inch lbs. (10–12 Nm).

72. After rotating the crankshaft 2 revolutions in a clockwise direction (viewed from front), confirm that the timing marks are aligned.

✷✷ WARNING

Always turn the crankshaft clockwise.

73. Install the timing chain cover.

a. The sealant locations on the chain cover and on the counter parts (cylinder head, cylinder block, and lower oil pan) must be free of engine oil and any debris.

b. Before assembling the timing chain cover, the liquid sealant TB 1217H should be applied on the gap between cylinder head and cylinder block.

➡**The part must be assembled within 5 minutes after the sealant is applied. Use a bead width of 0.1 inch (2.5mm).**

c. After applying liquid sealant TB1217H on the timing chain cover, the part must be assembled within 5 minutes.

➡**The sealant should be applied in a continuous bead. Bead width: 0.1 inch (2.5mm).**

d. Install a new gasket to the timing chain cover.

e. The dowel pins on the cylinder block and holes on the timing chain cover should be used as a reference in order to aid assembly of the timing chain cover into position.

f. Tighten the timing cover as shown in the illustration.

74. Wait 30 minutes after the timing chain cover was assembled before starting the engine.

75. Install the water pump pulley. Tighten to 69–87 inch lbs. (8–10 Nm).

76. Install the drive belt auto tensioner. Tighten to Bolt (B): 60–63 ft. lbs. (81–85 Nm).

Bolt (C): 13–16 ft. lbs. (18–22 Nm).

77. Install the drive belt idler. Tighten to 39–43 ft. lbs. (53–58 Nm).

78. Install the alternator. Tighten to 20–25 ft. lbs. (27–33 Nm).

79. Install the air conditioner compressor.

80. Install the power steering pump.

81. Lower the engine assembly using the jack.

82. Using SST (09231-3C100), install the timing chain cover oil seal.

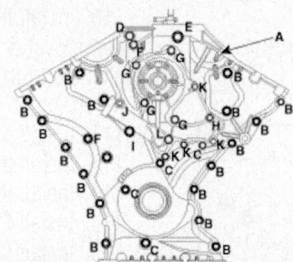

Bolt (No.)	Torque Specification
B (17)	14-16 ft. lbs. (19-22 Nm)
C (4)	87-104 inch lbs. (10-12 Nm)
D (1)	43-51 ft. lbs. (59-69 Nm)
E (1)	43-51 ft. lbs. (59-69 Nm)
F (2)	18-20 ft. lbs. (25-27 Nm)
G (4)	16-17 ft. lbs. (22-24 Nm)
H (1)	87-104 inch lbs. (10-12 Nm)
I (1)	87-104 inch lbs. (10-12 Nm)
J (1)	87-104 inch lbs. (10-12 Nm)
K (4)	87-104 inch lbs. (10-12 Nm)
L (1)	16-20 ft. lbs. (22-27 Nm)

22140_HYUN_G0381

Fig. 89 Tighten the timing chain cover as shown

83. Using SST (09231-3C300), install the crankshaft damper pulley. Tighten to 210–224 ft. lbs. (284–304 Nm).

84. Install the accessory drive belt. Refer to Accessory Drive Belts, removal & installation.

85. Install the cylinder head cover.

a. The hardening sealant located on the upper area between the timing chain cover and the cylinder head should be removed before assembling the cylinder head cover.

b. After applying sealant (TB1217H), it should be assembled within 5 minutes. Bead width: 0.1 inch (2.5mm).

c. Wait 30 minutes after the cylinder head cover was assembled before starting the engine.

d. Install the cylinder head cover bolts as shown in the illustration. Tighten to 87–104 inch lbs. (10–12 Nm).

➡Do not reuse the cylinder head cover gasket.

86. Install the ignition coils.

87. Connect the RH ignition coil connector, the condenser connector, and install the wiring bracket.

88. Install the connector bracket from the LH cylinder head cover.

89. Install the surge tank. Tighten to 87–104 inch lbs. (10–12 Nm).

a. Install the connector bracket(A) on the surge tank. Tighten to 210–224 ft. lbs. (284–304 Nm).

b. Install the surge tank stay. Tighten to 20–23 ft. lbs. (27–31 Nm).

90. Connect the brake vacuum hose.

91. Connect the PCV hose.

92. Connect the water hoses to the ETC.

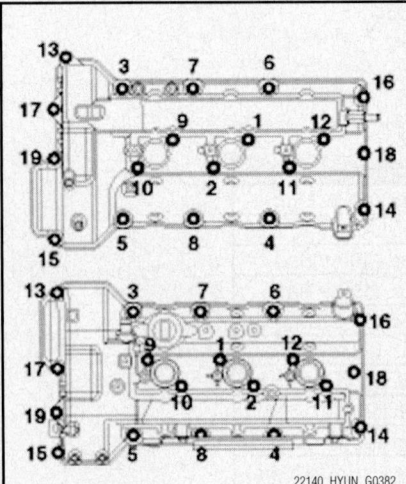

Fig. 90 Tighten the cylinder head cover bolts in the sequence shown

22140_HYUN_G0382

93. Install the ETC bracket. Tighten to 12–19 ft. lbs. (16–26 Nm).

94. Connect the PCSV connector, the MAP sensor connector, and the PCSV hose.

95. Connect the ETC connector and the knock sensor connector.

96. Connect the LH CMPS and oil pressure switch connector.

97. Install the wiring harness protector, and connect the LH ignition connector, injector connector, condenser connector, and ground.

98. Connect the OCV connector and knock sensor connector.

99. Connect the LH front oxygen sensor connector, alternator connector, and air compressor connector.

100. Connect the RH injector connector and the ignition coil connector.

101. Connect the power steering oil pressure sensor connector.

102. Connect the RH oxygen sensor connector and VIS solenoid valve connector.

103. Connect the RH oxygen sensor connector.

104. Install the No. 1 engine mounting through the lower position of A/C pipe line. Tighten to 36–47 ft. lbs. (49–64 Nm).

105. Install the A/C pipe bracket mounting bolt.

106. Install the engine coolant reservoir tank.

107. Install the engine mounting bracket. Tighten to 47–62 ft. lbs. (64–83 Nm).

108. Install the transaxle mounting bolts. Tighten to 36–47 ft. lbs. (49–64 Nm).

109. Remove the jack from the upper oil pan.

110. Install the lower oil pan. Refer to Oil Pan, removal & installation. Uniformly tighten the bolts in several passes. Tighten to 87–104 inch lbs. (10–12 Nm).

111. Install the upper radiator hose.

112. Install the side cover.

113. Install the undercover.

114. Install the front wheels.

115. Install the intake air hose and air cleaner assembly.

a. Install the intake air hose and air cleaner.

b. Connect the ECM connector.

c. Connect the breather hose to air cleaner hose.

d. Connect the AFS connector.

e. Install the air duct.

116. Install the engine cover.

117. Connect the negative terminal to the battery.

118. Refill the engine with the proper amount and type of engine oil.

119. Refill the radiator and reservoir tank with engine coolant.

120. Bleed the air from the cooling system.

121. Run the engine and check for leaks.

VALVE COVERS

REMOVAL & INSTALLATION

See Figures 72 through 77 and 90.

1. Before servicing the vehicle, refer to the Precautions Section.

✳✳ WARNING

Use fender covers to avoid damaging painted surfaces.

✳✳ WARNING

To avoid damage, unplug the wiring connectors carefully while holding the connector portion.

➡Mark all wiring and hoses to avoid misconnection.

2. Disconnect the negative terminal from the battery.

3. Remove the engine cover.

4. Remove the air duct.

5. Remove the intake air hose and air cleaner assembly.

a. Disconnect the AFS connector (A).

b. Disconnect the breather hose (B) from air cleaner hose.

c. Disconnect the ECM connector.

d. Remove the intake air hose (C) and air cleaner (D).

6. Remove the RH front wheel.

7. Remove the undercover.

8. Remove the RH side cover.

9. Drain the engine coolant. Remove the radiator cap to speed draining.

10. Remove the upper radiator hose.

11. Drain the engine oil.

12. Remove the lower oil pan. Refer to Oil Pan, removal & installation.

13. Place a jack underneath the upper oil pan.

14. Just loosen the transaxle mounting bolts (A) without removing the transaxle mounting (B).

15. Remove the engine coolant reservoir tank (A).

16. Remove the engine mounting bracket (B).

17. Loosen the A/C pipe bracket mounting bolt.

18. Remove the No. 1 engine mounting (A) through the lower position of the A/C pipe line.

19. Remove the surge tank and engine wiring.

a. Disconnect the RH oxygen sensor connector.

b. Disconnect the VIS solenoid valve connector.

c. Disconnect the power steering oil pressure sensor connector.

d. Disconnect the RH injector connector and the ignition coil connector.

e. Disconnect the OCV connector and knock sensor connector.

f. Disconnect the LH front oxygen sensor connector, alternator connector, and air compressor connector.

g. Disconnect the LH ignition coil connector, injector connector, condenser connector, and ground

h. Remove the wiring harness protector.

i. Disconnect the LH CMPS and oil pressure switch connector.

j. Disconnect the ETC connector and the knock sensor connector.

k. Disconnect the PCSV connector, the MAP sensor connector, and the PCSV hose.

l. Remove the Electronic Throttle Control (ETC) bracket.

m. Disconnect the water hoses from the ETC.

n. Disconnect the PCV hose.

o. Disconnect the brake vacuum hose.

p. Remove the surge tank stay.

q. Remove the connector bracket from the surge tank.

r. Remove the surge tank (A).

➡**Cover the inlet of intake manifold with a clean woven stuff or vinyl cover to prevent foreign materials from entering.**

20. Remove the valve cover(s), also called cylinder head cover(s).

a. Remove the connector bracket (A) from the LH cylinder head cover.

b. Disconnect the RH ignition coil connector(A), the condenser connector (B) and remove the wiring bracket (C).

c. Remove the LH and RH ignition coils.

d. Remove the LH and RH cylinder head covers.

➡**Cover the upside of engine head with a clean vinyl cover to prevent foreign materials from entering.**

To install:

21. Install the valve cover, also called cylinder head cover.

a. The sealant located on the upper area between the timing chain cover and the cylinder head should be removed before assembling the cylinder head cover.

b. After applying sealant (TB1217H), it should be assembled within 5 minutes. Bead width: 0.1 inch (2.5mm).

c. Wait 30 minutes after the cylinder head cover was assembled before starting the engine.

d. Install the cylinder head cover bolts as shown in the illustration. Tighten to 87–104 inch lbs. (10–12 Nm).

➡**Do not reuse the cylinder head cover gasket.**

22. Install the ignition coils.

23. Connect the RH ignition coil connector, the condenser connector, and install the wiring bracket.

24. Install the connector bracket from the LH cylinder head cover.

25. Install the surge tank. Tighten to 87–104 inch lbs. (10–12 Nm).

a. Install the connector bracket(A) on the surge tank. Tighten to 210–224 ft. lbs. (284–304 Nm).

b. Install the surge tank stay. Tighten to 20–23 ft. lbs. (27–31 Nm).

26. Connect the brake vacuum hose.

27. Connect the PCV hose.

28. Connect the water hoses to the ETC.

29. Install the ETC bracket. Tighten to 12–19 ft. lbs. (16–26 Nm).

30. Connect the PCSV connector, the MAP sensor connector, and the PCSV hose.

31. Connect the ETC connector and the knock sensor connector.

32. Connect the LH CMPS and oil pressure switch connector.

33. Install the wiring harness protector, and connect the LH ignition connector, injector connector, condenser connector, and ground.

34. Connect the OCV connector and knock sensor connector.

35. Connect the LH front oxygen sensor connector, alternator connector, and air compressor connector.

36. Connect the RH injector connector and the ignition coil connector.

37. Connect the power steering oil pressure sensor connector.

38. Connect the RH oxygen sensor connector and VIS solenoid valve connector.

39. Connect the RH oxygen sensor connector.

40. Install the No. 1 engine mounting through the lower position of A/C pipe line. Tighten to 36–47 ft. lbs. (49–64 Nm).

41. Install the A/C pipe bracket mounting bolt.

42. Install the engine coolant reservoir tank.

43. Install the engine mounting bracket. Tighten to 47–62 ft. lbs. (64–83 Nm).

44. Install the transaxle mounting bolts. Tighten to 36–47 ft. lbs. (49–64 Nm).

45. Remove the jack from the upper oil pan.

46. Install the lower oil pan. Refer to Oil Pan, removal & installation. Uniformly tighten the bolts in several passes. Tighten to 87–104 inch lbs. (10–12 Nm).

47. Install the upper radiator hose.

48. Install the side cover.

49. Install the undercover.

50. Install the front wheel(s).

51. Install the intake air hose and air cleaner assembly.

a. Install the intake air hose and air cleaner.

b. Connect the ECM connector.

c. Connect the breather hose to air cleaner hose.

d. Connect the AFS connector.

e. Install the air duct.

52. Install the engine cover.

53. Connect the negative terminal to the battery.

54. Refill the engine with the proper amount and type of engine oil.

55. Refill the radiator and reservoir tank with engine coolant.

56. Bleed the air from the cooling system.

57. Run the engine and check for leaks.

VALVE LASH

ADJUSTMENT

All engines use hydraulic valve lash adjusters. Valve lash adjustments are not necessary or possible on these engines.

ENGINE PERFORMANCE & EMISSION CONTROL

COMPONENT LOCATIONS

See Figure 91.

ACCELERATOR PEDAL POSITION (APP) SENSOR

LOCATION
See Figure 92.

The Accelerator Pedal Position (APP) sensor is located inside the vehicle. It is part of the accelerator pedal assembly.

OPERATION

The accelerator pedal contains 2 individual Accelerator Pedal Position (APP) sensors within the assembly. The APP sensors 1 and 2 are potentiometer type sensors each with 3 circuits:

- A 5-volt reference circuit
- A low reference circuit
- A signal circuit

The APP sensors are used to determine the pedal angle. The Powertrain Control Module (PCM) provides each APP sensor with a 5-volt reference circuit and a low reference circuit. The APP sensors provide the PCM with signal voltage proportional to the pedal movement. The APP sensor 1

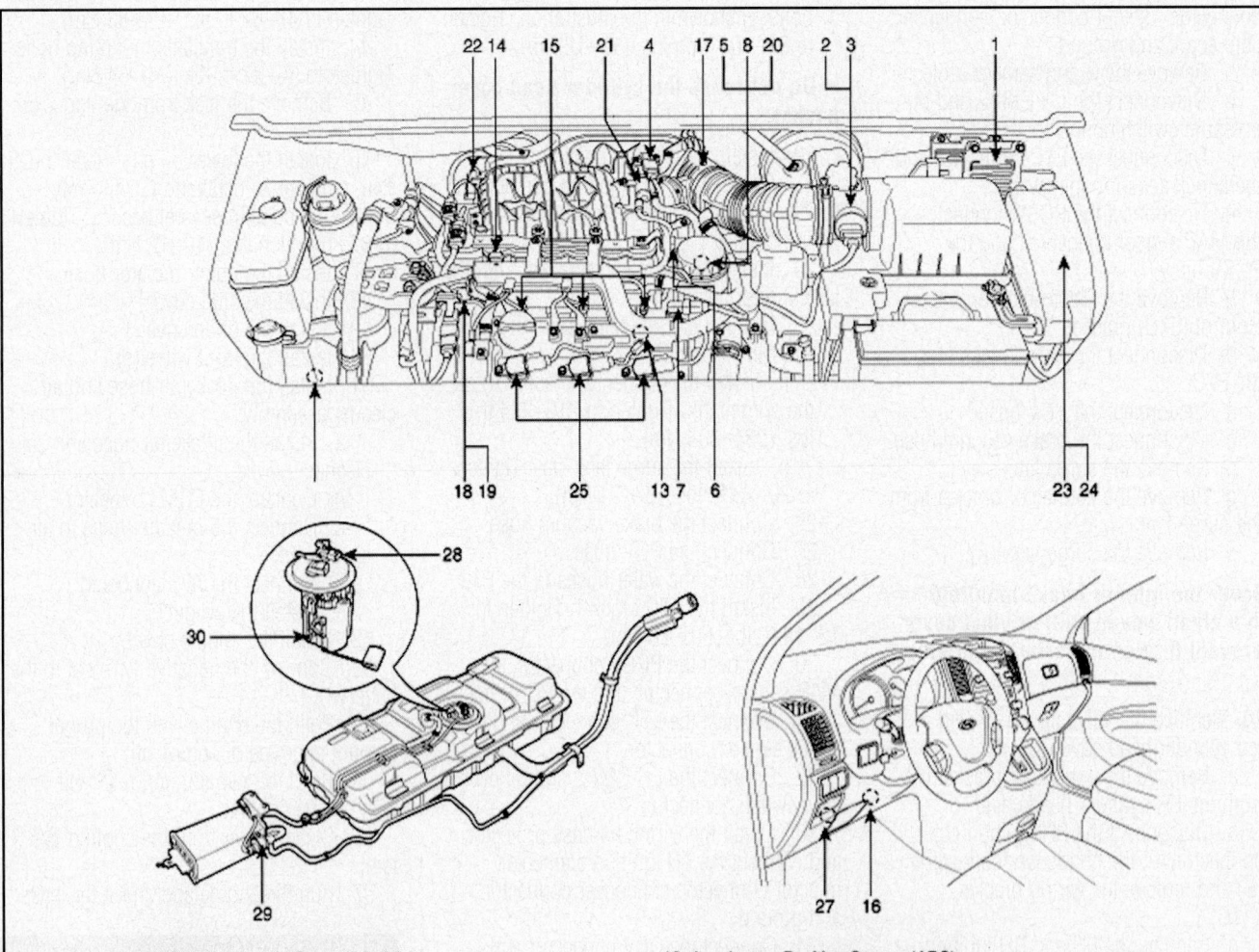

1. PCM (Powertrain Control Module)
2. Mass Air Flow Sensor (MAFS)
3. Intake Air Temperature Sensor (IATS)
4. Manifold Absolute Pressure Sensor (MAPS)
5. Engine Coolant Temperature Sensor (ECTS)
6. Camshaft Position Sensor (CMPS) [Bank 1]
7. Camshaft Position Sensor (CMPS) [Bank 2]
8. Crankshaft Position Sensor (CKPS)
9. Heated Oxygen Sensor (HO2S) [Bank 1 / Sensor 1]
10. Heated Oxygen Sensor (HO2S) [Bank 1 / Sensor 2]
11. Heated Oxygen Sensor (HO2S) [Bank 2 / Sensor 1]
12. Heated Oxygen Sensor (HO2S) [Bank 2 / Sensor 2]
13. Knock Sensor (KS) #1
14. Knock Sensor (KS) #2
15. Injector

16. Accelerator Position Sensor (APS)
17. ETC Module [Throttle Position Sensor (TPS) + ETC Motor]
18. CVVT Oil Control Valve (OCV) [Bank 1]
19. CVVT Oil Control Valve (OCV) [Bank 2]
20. CVVT Oil Temperature Sensor (OTS)
21. Purge Control Solenoid Valve (PCSV)
22. Variable Intake Solenoid (VIS) Valve
23. Fuel Pump Relay
24. Main Relay
25. Ignition Coil
26. A/C Pressure Transducer (APT)
27. Data Link Connector (DLC)
28. Fuel Tank Pressure Sensor (FTPS)
29. Canister Close Valve (CCV)
30. Fuel Level Sensor (FLS)

22140_ENTO_G0083

Fig. 91 Component and component sensor locations

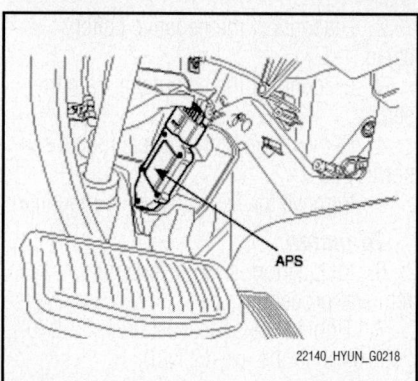

Fig. 92 Accelerator Pedal Position (APP) location

signal voltage at rest position is near the low reference and increases as the pedal is actuated. The APP sensor 2 signal voltage at rest position is near the 5-volt reference and decreases as the pedal is actuated.

REMOVAL & INSTALLATION

1. Before servicing the vehicle, refer to the Precautions Section.
2. Disconnect the ground cable from the battery.
3. Disconnect the Accelerator Pedal Position (APP) sensor connector.
4. Remove the nut securing accelerator pedal assembly.

To install:

5. Install in the reverse order of removal.
6. Tighten the nut securing accelerator pedal assembly to 113–139 inch lbs. (13–16 Nm).

TESTING

See Figures 93 through 95.

The Accelerator Pedal Position (APP) sensor is installed on the accelerator pedal module and detects the rotation angle of the accelerator pedal. The APP is one of the most important sensors in engine control system, so it consists of the 2 sensors which

adapt an individual sensor power and ground line. The second sensor monitors the first sensor and its output voltage is half of the first one. If the ratio of sensor 1 and 2 is out of the range (approximately ½), the diagnostic system judges that it is abnormal.

Use the following reference tables to test the APP sensor voltage and resistance values at the connector end points.

CAMSHAFT POSITION (CMP) SENSOR

LOCATION

See Figures 96 and 97.

Pedal Position	Output Voltage (V) [Vref = 5.0V]	
	APS1	APS2
C.T	0.7 ~ 0.8V	0.29 ~ 0.46V
W.O.T	3.85 ~ 4.35V	1.93 ~ 2.18V

Fig. 94 Accelerator pedal position table—Voltage Output

Item	Sensor Resistance
APS1	0.7 ~ 1.3kΩ at 68°F (20°C)
APS2	1.4 ~ 2.6kΩ at 68°F (20°C)

Fig. 95 Accelerator pedal position circuit testing table—Sensor Resistance

Refer to the accompanying illustration for Camshaft Position (CMP) sensor locations.

OPERATION

The Camshaft Position (CMP) sensor is triggered by a notched reluctor wheel built onto the camshaft sprocket. The CMP sensor is connected to the PCM by the following circuits:

- A 5-volt circuit
- A low reference circuit
- A signal circuit

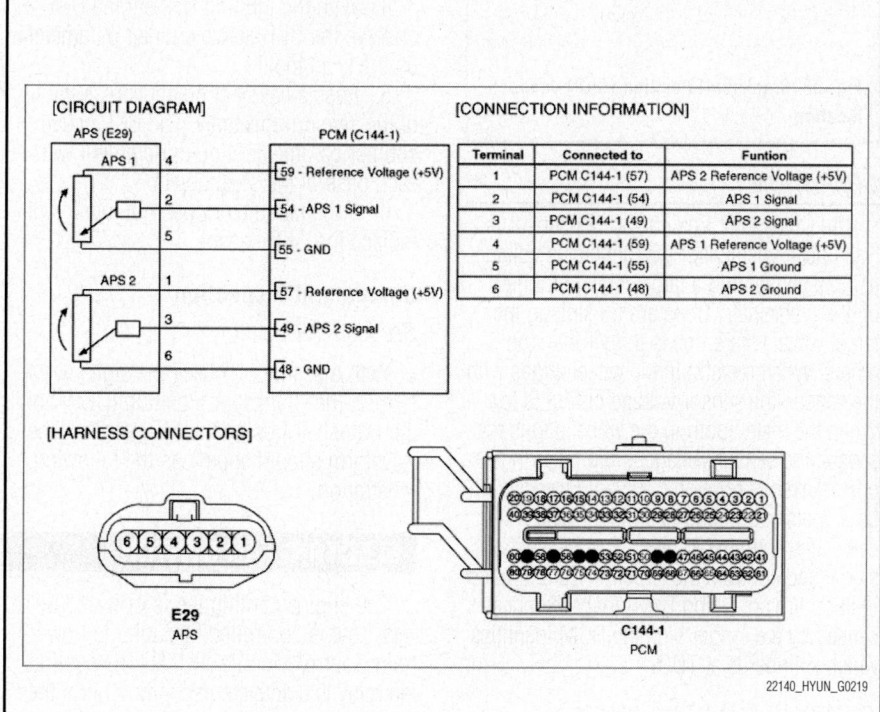

[CIRCUIT DIAGRAM]
APS (E29)   PCM (C144-1)

APS 1   4 — [59 - Reference Voltage (+5V)]
        2 — [54 - APS 1 Signal]
        5 — [55 - GND]

APS 2   1 — [57 - Reference Voltage (+5V)]
        3 — [49 - APS 2 Signal]
        6 — [48 - GND]

[CONNECTION INFORMATION]

Terminal	Connected to	Funtion
1	PCM C144-1 (57)	APS 2 Reference Voltage (+5V)
2	PCM C144-1 (54)	APS 1 Signal
3	PCM C144-1 (49)	APS 2 Signal
4	PCM C144-1 (59)	APS 1 Reference Voltage (+5V)
5	PCM C144-1 (55)	APS 1 Ground
6	PCM C144-1 (48)	APS 2 Ground

[HARNESS CONNECTORS]

E29
APS

C144-1
PCM

Fig. 93 Accelerator Pedal Position (APP) connector end view and circuit schematics

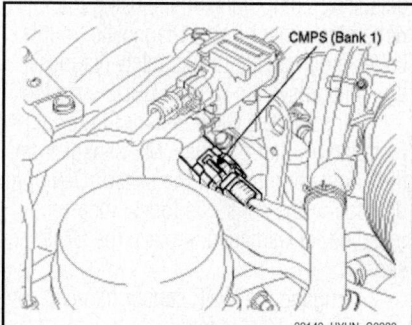

Fig. 96 Camshaft Position (CMP) sensor location—Bank 1

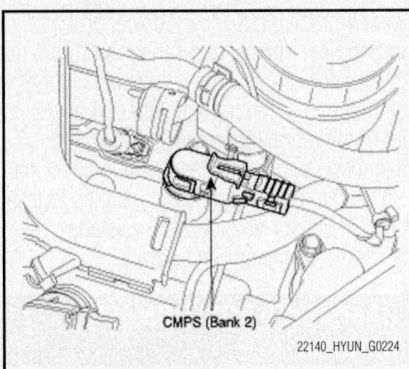

Fig. 97 Camshaft Position (CMP) sensor location—Bank 2

The CMP sensor is a Hall-effect sensor and detects the camshaft position by using a Hall element. It is related with the Crankshaft Position (CKP) sensor and detects the piston position of each cylinder which the CKP sensor can't detect. This sensor has a Hall-effect IC which changes in output voltage when a magnetic field is made on the IC with the current flow.

REMOVAL & INSTALLATION

1. Before servicing the vehicle, refer to the Precautions Section.
2. Disconnect the negative battery cable.
3. Disconnect the connector from the CMP sensor.
4. Remove the bolt that retains the CMP sensor.
5. Remove the CMP sensor.

To install:
6. Installation is the reverse of the removal procedure.
7. Tighten the bolt that retains the CMP sensor to: 61–86 inch lbs. (7–10 Nm).

TESTING

During normal operation the PCM controls all ignition functions. If either the Crankshaft Position (CKP) or Camshaft Position (CMP) sensor signal is lost, the engine will continue to run because the PCM will default to a limp home mode using the remaining sensor input. Diagnostic trouble codes are available to accurately diagnose the ignition system with an OBD2 scan tool.

1. Inspect the CMP sensor for correct installation. Remove the CMP sensor from the engine and inspect the sensor O-ring for damage. If the sensor is loose, incorrectly installed, or damaged, replace the CMP sensor.
2. Engage the CMP sensor harness connector to the CMP sensor.
3. Connect the scan tool to the diagnostic connector.

4. With the ignition ON, engine OFF observe the CMP active counter parameter on the scan tool.
5. Pass a flat steel object across the tip of the sensor repeatedly. The CMP active counter parameter should increment with each pass of the steel object.
6. If the parameter does not increment, replace the CMP sensor.

COOLANT TEMPERATURE SENSOR

Refer to Engine Coolant Temperature (ECT) Sensor.

CRANKSHAFT POSITION (CKP) SENSOR

LOCATION

See Figure 98.

Refer to the accompanying illustration for Crankshaft Position (CKP) sensor location.

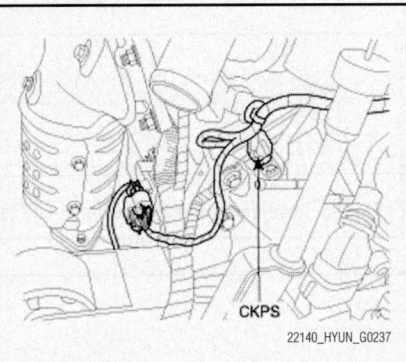

Fig. 98 Crankshaft Position (CKP) sensor location

OPERATION

The Crankshaft Position (CKP) sensor is a Hall Effect type sensor that generates voltage using a sensor and a target wheel mounted on the crankshaft. There are 58 slots in the target wheel where one is longer than the others. When the slot in the wheel aligns with the sensor, the sensor voltage output is low. When the metal tooth in the wheel aligns the sensor, the sensor voltage is high. During one crankshaft rotation there are 58 rectangular signals and one longer signal. The PCM calculates engine RPM by using the sensor's signal and controls the injection duration and ignition timing. Using the signal differences caused by the longer slot, the PCM identifies which cylinder is at TDC.

REMOVAL & INSTALLATION

1. Before servicing the vehicle, refer to the Precautions Section.

2. Disconnect the negative battery cable.
3. Disconnect the connector from the sensor.
4. Remove the bolt that retains the sensor in place.
5. Remove the sensor from its mounting.

To install:
6. Installation is the reverse of the removal procedure.
7. Tighten the sensor retaining bolt to: 70–104 inch lbs. (8–12 Nm).

TESTING

During normal operation the PCM controls all ignition functions. If either the Crankshaft Position (CKP) or Camshaft Position (CMP) sensor signal is lost, the engine will continue to run because the PCM will default to a limp home mode using the remaining sensor input. Diagnostic trouble codes are available to accurately diagnose the ignition system with an OBD2 scan tool.

1. Inspect the CKP sensor for correct installation. Remove the CKP sensor from the engine and inspect the sensor O-ring for damage. If the sensor is loose, incorrectly installed, or damaged, replace the CKP sensor.
2. Engage the CKP sensor harness connector to the CKP sensor.
3. Connect the scan tool to the diagnostic connector.
4. With the ignition ON, engine OFF observe the CKP active counter parameter on the scan tool.
5. Pass a flat steel object across the tip of the sensor repeatedly. The CKP active counter parameter should increment with each pass of the steel object.
6. If the parameter does not increment, replace the CKP sensor.

Component Inspection
See Figure 99.

With a scan tool, check the signal waveform of the Crankshaft Position (CKP) and the Camshaft Position (CMP) sensor. The waveform should appear as the following illustration.

ELECTRIC FAN SWITCH

The engine cooling fan is operated by a relay that is controlled through the Powertrain Control Module (PCM). The cooling fan relay will activate depending upon the needs of one or more systems including the cooling system, the air conditioning, and the heating systems.

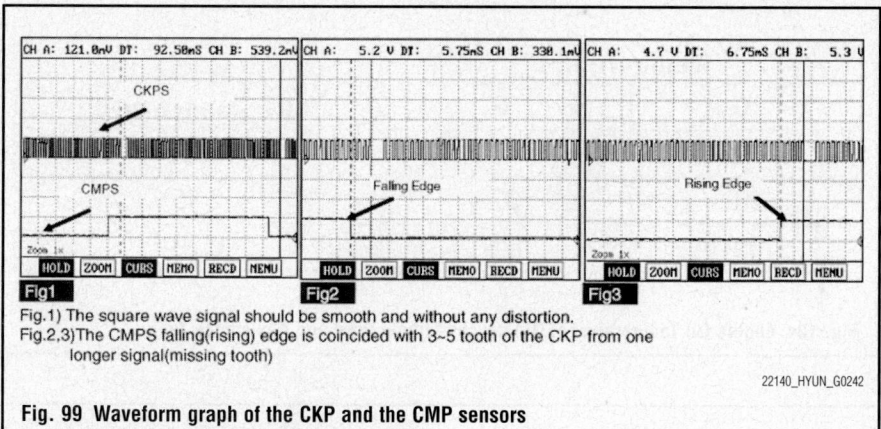

Fig.1) The square wave signal should be smooth and without any distortion.
Fig.2,3)The CMPS falling(rising) edge is coincided with 3~5 tooth of the CKP from one longer signal(missing tooth)

22140_HYUN_G0242

Fig. 99 Waveform graph of the CKP and the CMP sensors

Temperature [°C(°F)]	Resistance (kΩ)
–40(–40)	48.14
–20(–4)	14.13 ~ 16.83
0(32)	5.79
20(68)	2.31 ~ 2.59
40(104)	1.15
60(140)	0.59
80(176)	0.32

22140_HYUN_G0234

Fig. 101 Resistance chart specifications

ELECTRONIC CONTROL MODULE (ECM)

Refer to Powertrain Control Module (PCM).

ENGINE COOLANT TEMPERATURE (ECT) SENSOR

LOCATION

See Figure 100.

The Engine Coolant Temperature (ECT) sensor is located in the engine coolant passage of the cylinder head for detecting the engine coolant temperature. Refer to the accompanying illustration for sensor location.

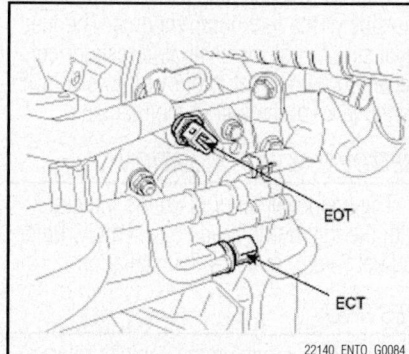

22140_ENTO_G0084

Fig. 100 Engine Coolant Temperature (ECT) sensor and Engine Oil Temperature (EOT) sensor locations

OPERATION

The Engine Coolant Temperature (ECT) sensor uses a thermistor whose resistance changes with the temperature in order to detect the engine coolant temperature. The electrical resistance of the ECT decreases as the temperature increases, and increases as the temperature decreases. The 5-volt refer-
ence in the PCM is supplied to the ECT via a resistor in the PCM. The resistor in the PCM and the thermistor in the ECT are connected in series. When the resistance value of the thermistor in the ECT changes according to the engine coolant temperature, the output voltage also changes. During cold engine operation, the PCM increases the fuel injection duration and controls the ignition timing using the information of engine coolant temperature to avoid engine stalling and improve drivability.

REMOVAL & INSTALLATION

1. Before servicing the vehicle, refer to the Precautions Section.
2. Drain the coolant to a level below the bottom of the sensor.
3. Disconnect the ground cable from the battery and then remove the sensor connector.
4. Remove the coolant temperature sensor.

To install:
5. Reverse the removal procedure.

6. Tighten the sensor to 15–29 ft. lbs. (20–39 Nm).

TESTING

See Figures 101 and 102.

➡**Clear Diagnostic Trouble Codes (DTC's) if applicable.**

1. Inspect the Engine Coolant Temperature (ECT) sensor electrical connection. Check the terminal for poor connections, loose wires, bent, broken, or corroded pins, and then verify that the connector is securely fastened.
2. Turn ignition switch OFF.
3. Disconnect ECT sensor connector.
4. Remove the ECT sensor.
5. After immersing the thermistor of the sensor into engine coolant, measure the resistance between the ECT terminals 1 and 3.
6. Check that the resistance is within the specification according to the following chart.
7. Replace the ECT sensor if the component is not within specification.

ENGINE OIL TEMPERATURE (EOT) SENSOR

LOCATION

See Figure 100.

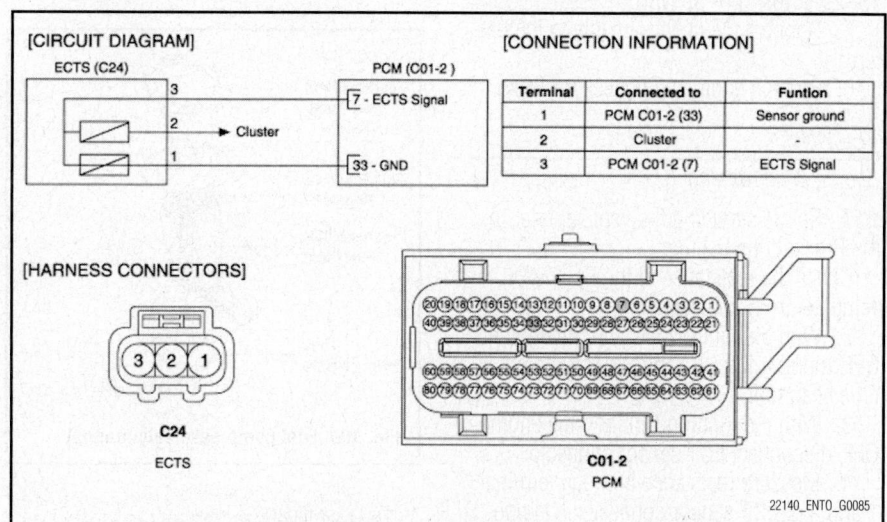

22140_ENTO_G0085

Fig. 102 Engine Coolant Temperature (ECT) sensor connector end view and circuit schematics

Refer to the illustration for Engine Oil Temperature (EOT) sensor location.

OPERATION

See Figure 103.

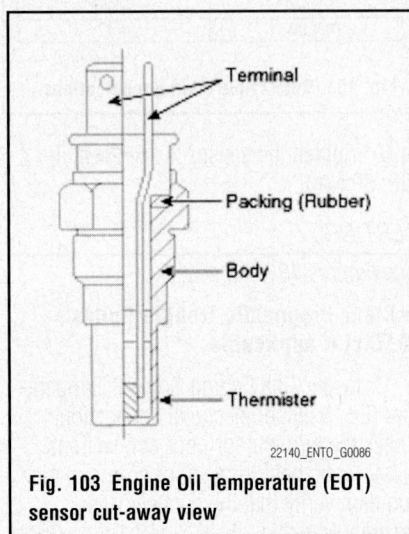

Fig. 103 Engine Oil Temperature (EOT) sensor cut-away view

The Engine Oil Temperature (EOT) sensor is a negative coefficient thermistor used by the PCM to measure engine oil temperature for the purpose of adjusting Continuously Variable Valve Timing (CVVT) calculations.

REMOVAL & INSTALLATION

1. Before servicing the vehicle, refer to the Precautions Section.
2. Disconnect the ground cable from the battery and then remove the sensor connector.
3. Remove Engine Oil Temperature (EOT) sensor.

To install:

4. Install the EOT sensor and tighten to 15–29 ft. lbs. (20–39 Nm).
5. Connect the negative cable to the battery.
6. Run the engine and check for leaks.

TESTING

See Figures 104 and 105.

1. Before servicing the vehicle, refer to the Precautions Section.
Check the resistance of the Engine Oil Temperature (EOT) sensor.
2. With the ignition ON and the engine OFF, monitor the Oil Temperature parameter on a scan tool.
3. With the ignition OFF and the engine OFF, disconnect EOT sensor connector.
4. Measure resistance between terminal 1 and 2 of EOT sensor connector (Component Side).

Temperature		Resistance(kΩ)
°C	°F	
-20	-4	16.52
20	32	2.45
80	176	0.29

22140_ENTO_G0087

Fig. 104 Engine Oil Temperature (EOT) sensor Temperature and Resistance Table

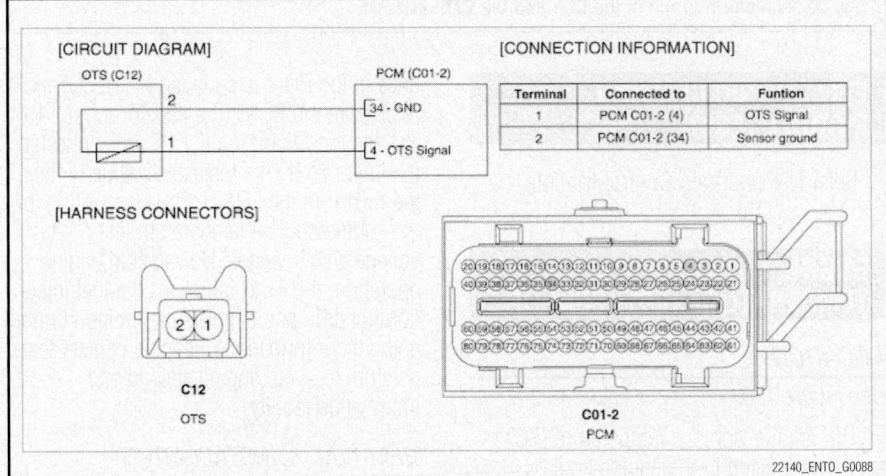

Fig. 105 Engine Oil Temperature (EOT) sensor connector end view and schematic diagram

5. The specifications should be as listed in the following table.
6. If the EOT sensor does not meet the specifications, replace the EOT sensor.

FUEL LEVEL SENDING UNIT

LOCATION

See Figure 106.

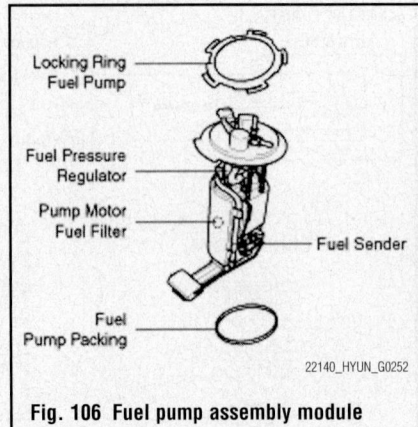

22140_HYUN_G0252

Fig. 106 Fuel pump assembly module

The fuel level sending unit is located on the fuel pump assembly module.

OPERATION

The fuel level sending unit is attached to the side of the fuel pump module. The fuel level sensor utilizes a variable resistor connected to a float that is buoyed up or down by the level of the fuel in the fuel tank.

REMOVAL & INSTALLATION

The fuel level sending unit is integral with the fuel pump module assembly. Refer to Fuel Pump, removal & installation.

TESTING

1. Before servicing the vehicle, refer to the Precautions Section.
2. Check the fuel gauge in the instrument cluster panel for an accurate reading of the fuel level in the tank.
3. If the fuel gauge sticks at one extreme of the meter, no matter the fuel level in the tank, check for a break or short in the wiring from the fuel level sending unit.
4. Check the fuel level sending unit by removing the fuel pump assembly. Refer to Fuel Pump, removal and installation.
5. Using an ohmmeter, measure the resistance between the terminals while moving the float up and down. If the resistance

does not change smoothly, replace the fuel level sending unit.

HEATED OXYGEN SENSOR (HO2S)

LOCATION

See Figures 107 through 110.

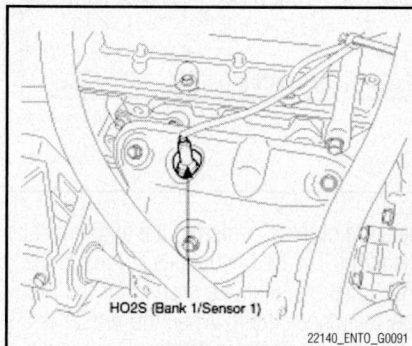

HO2S (Bank 1/Sensor 1)

22140_ENTO_G0091

Fig. 107 Heated Oxygen (HO2S) sensor location—Bank 1, Sensor 1

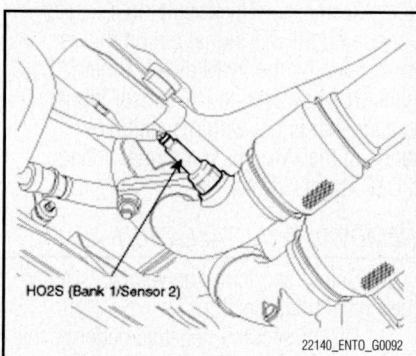

HO2S (Bank 1/Sensor 2)

22140_ENTO_G0092

Fig. 108 Heated Oxygen (HO2S) sensor location—Bank 1, Sensor 2

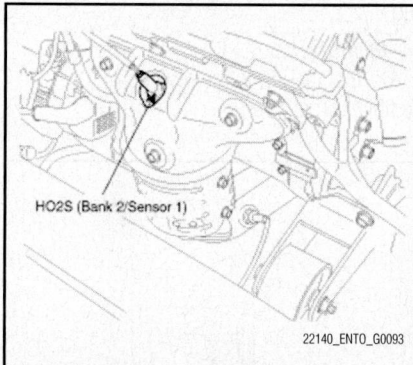

HO2S (Bank 2/Sensor 1)

22140_ENTO_G0093

Fig. 109 Heated Oxygen (HO2S) sensor location—Bank 2, Sensor 1

The Heated Oxygen Sensors (HO2S) sensors are located in the exhaust system. One sensor is located up at the exhaust manifold(s) and the other sensor is located down at the catalytic converter.

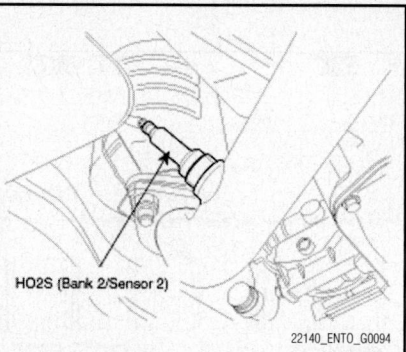

HO2S (Bank 2/Sensor 2)

22140_ENTO_G0094

Fig. 110 Heated Oxygen (HO2S) sensor location—Bank 2, Sensor 2

OPERATION

The Heated Oxygen Sensor (HO2S) consists of zirconium and alumina and is installed upstream and downstream of the Manifold Catalyst Converter (MCC). After it compares the oxygen consistency of the atmosphere with the exhaust gas, it transfers the oxygen consistency of the exhaust gas to the PCM. This sensor operates most efficiently when the temperature of the sensor tip is higher than 698°F (370°C). Therefore, a heater has been integrated with the oxygen sensor that is controlled by the PCM duty signal. When the exhaust gas temperature is lower than the specified value, the heater warms the sensor tip. The PCM commands the heater ON or OFF to maintain a specific HO2S operating temperature range. The PCM monitors the voltage on the HO2S heater low control circuit for heater fault diagnosis. If the PCM detects that the HO2S heater low control circuit voltage is not within a specified range, a DTC is set.

The HO2S supplies the electronic control assembly with a signal indicating either a rich or lean mixture condition as the engine operates.

REMOVAL & INSTALLATION

1. Before servicing the vehicle, refer to the Precautions Section.
2. Disconnect the electrical connector from the sensor.
3. Remove the oxygen sensor.

To install:

4. Installation is the reverse of the removal procedure.

➡**Apply anti-seize compound to the threaded portion of the sensor, prior to installation. Never apply anti-seize compound to the protector of the sensor.**

TESTING

See Figures 111 through 115.

1. Perform a visual inspection of the sensor as follows:

   a. If the sensor tip has a black/sooty deposit, this may indicate a rich fuel mixture.

   b. If the sensor tip has a white, gritty deposit, this may indicate an internal coolant leak.

   c. If the sensor tip has a brown deposit, this could indicate oil consumption.

2. Check that the heater resistance is according to specification.
3. Check that the output voltage in the rich/lean readings fluctuate according to specifications.
4. Connect an OBD2 scan tool to the Data Link Connector (DLC).
5. With the engine running, the normal waveform for the Heated Oxygen Sensor (HO2S) and the HO2S heater should be similar to those represented in the following waveform illustration.

Item	Specification
Heater Resistance (Ω)	8.1 ~ 11.1Ω at 69.8°F (21°C)

22140_HYUN_G0255

Fig. 111 Heated oxygen sensor heater resistance specifications

A/F Ratio	Output Voltage (V)
RICH	0.75 ~ 1.00V
LEAN	0 ~ 0.12V

22140_HYUN_G0256

Fig. 112 Heated oxygen sensor rich/lean output voltage specifications

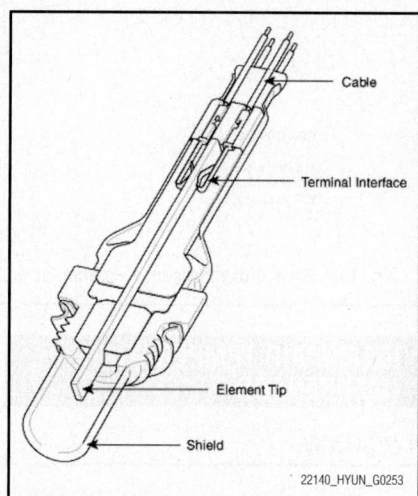

Cable

Terminal Interface

Element Tip

Shield

22140_HYUN_G0253

Fig. 113 Heated oxygen sensor cutaway view

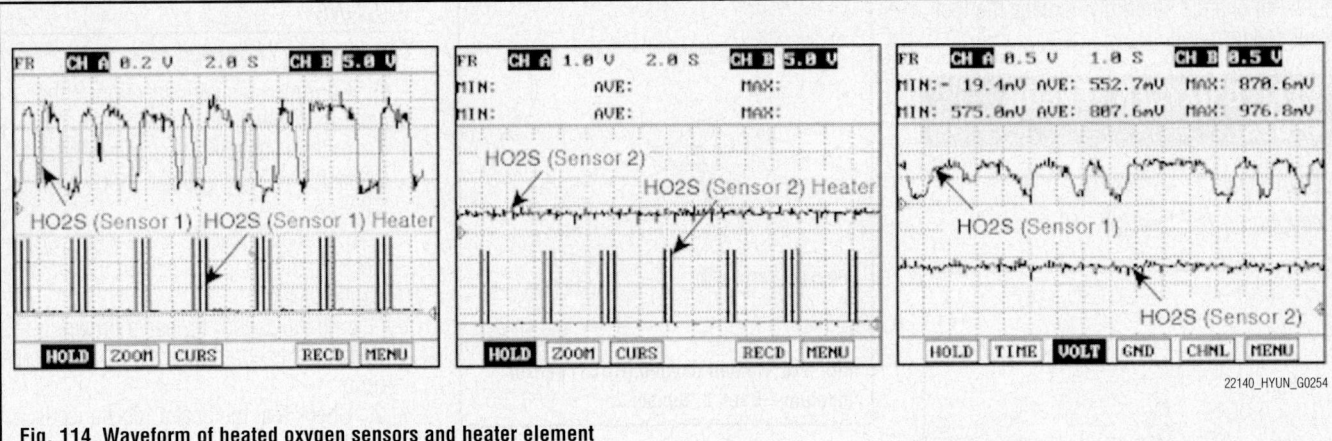

Fig. 114 Waveform of heated oxygen sensors and heater element

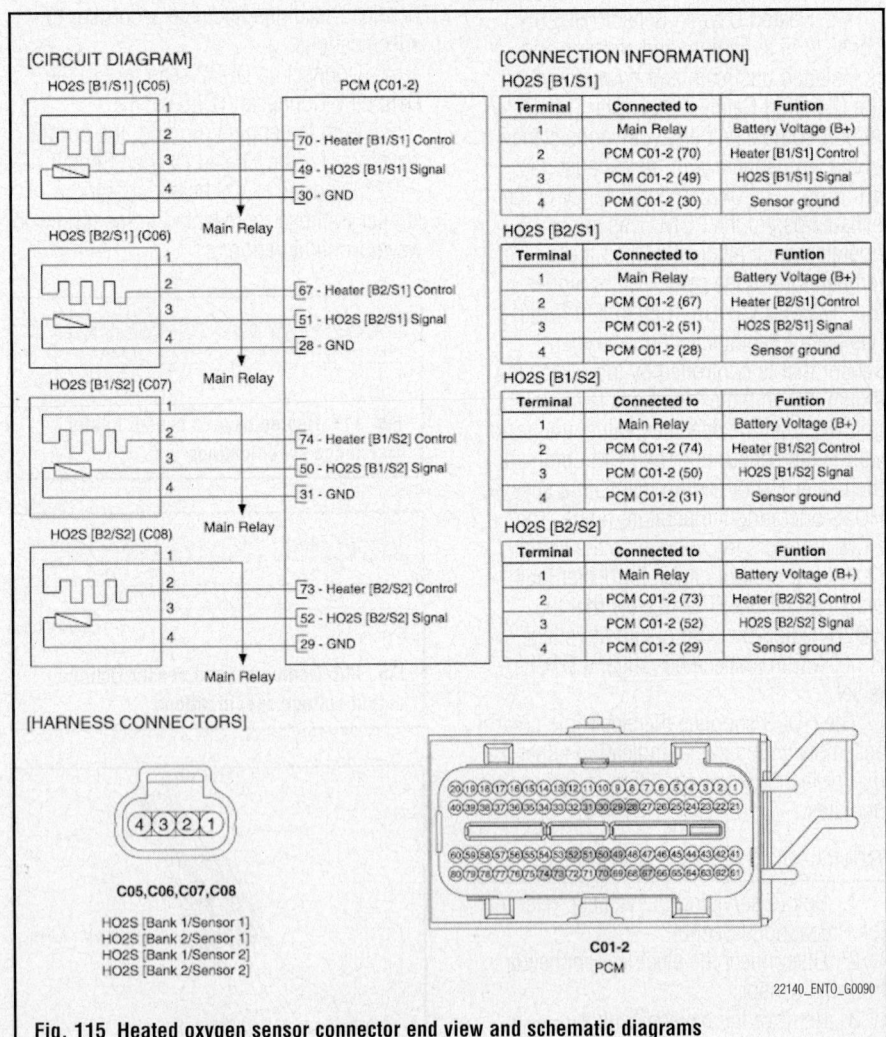

[CIRCUIT DIAGRAM]

HO2S [B1/S1] (C05) .... PCM (C01-2)

70 - Heater [B1/S1] Control
49 - HO2S [B1/S1] Signal
30 - GND

HO2S [B2/S1] (C06) .... Main Relay

67 - Heater [B2/S1] Control
51 - HO2S [B2/S1] Signal
28 - GND

HO2S [B1/S2] (C07) .... Main Relay

74 - Heater [B1/S2] Control
50 - HO2S [B1/S2] Signal
31 - GND

HO2S [B2/S2] (C08) .... Main Relay

73 - Heater [B2/S2] Control
52 - HO2S [B2/S2] Signal
29 - GND

Main Relay

[CONNECTION INFORMATION]

HO2S [B1/S1]

Terminal	Connected to	Funtion
1	Main Relay	Battery Voltage (B+)
2	PCM C01-2 (70)	Heater [B1/S1] Control
3	PCM C01-2 (49)	HO2S [B1/S1] Signal
4	PCM C01-2 (30)	Sensor ground

HO2S [B2/S1]

Terminal	Connected to	Funtion
1	Main Relay	Battery Voltage (B+)
2	PCM C01-2 (67)	Heater [B2/S1] Control
3	PCM C01-2 (51)	HO2S [B2/S1] Signal
4	PCM C01-2 (28)	Sensor ground

HO2S [B1/S2]

Terminal	Connected to	Funtion
1	Main Relay	Battery Voltage (B+)
2	PCM C01-2 (74)	Heater [B1/S2] Control
3	PCM C01-2 (50)	HO2S [B1/S2] Signal
4	PCM C01-2 (31)	Sensor ground

HO2S [B2/S2]

Terminal	Connected to	Funtion
1	Main Relay	Battery Voltage (B+)
2	PCM C01-2 (73)	Heater [B2/S2] Control
3	PCM C01-2 (52)	HO2S [B2/S2] Signal
4	PCM C01-2 (29)	Sensor ground

[HARNESS CONNECTORS]

C05,C06,C07,C08

HO2S [Bank 1/Sensor 1]
HO2S [Bank 2/Sensor 1]
HO2S [Bank 1/Sensor 2]
HO2S [Bank 2/Sensor 2]

C01-2
PCM

Fig. 115 Heated oxygen sensor connector end view and schematic diagrams

INTAKE AIR TEMPERATURE (IAT) SENSOR

LOCATION

See Figure 116.

The Intake Air Temperature (IAT) sensor is mounted in the intake air hose of the air cleaner assembly. The IAT is integrated inside the Mass Air Flow (MAF) sensor.

OPERATION

See Figure 117.

The Intake Air Temperature (IAT) sensor is a variable resistor that measures the temperature of the air entering the engine intake manifold. The PCM supplies 5 volts to the IAT signal circuit and a ground for the IAT low reference circuit. When the sensor is cold, the resistance is greater. This results in a greater voltage on the signal circuit that is interpreted by the PCM as a colder IAT. As the sensor becomes warmer, the resistance decreases. This results in a lesser voltage on the IAT signal circuit that is interpreted by the PCM as a warmer IAT. If the PCM detects an IAT sensor signal voltage that is not within a calibrated range of the IAT sensor 1 signal voltage, a DTC is set.

REMOVAL & INSTALLATION

1. Before servicing the vehicle, refer to the Precautions Section.
2. Disconnect the negative battery cable.
3. Disconnect the connector from the sensor.
4. Remove the sensor retaining screws.
5. Remove the sensor from its mounting.

To install:

6. Installation is the reverse of the removal procedure.

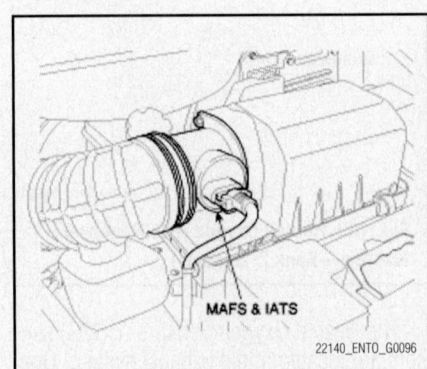

Fig. 116 Intake Air Temperature (IAT) sensor and Mass Air Flow (MAF) sensor location

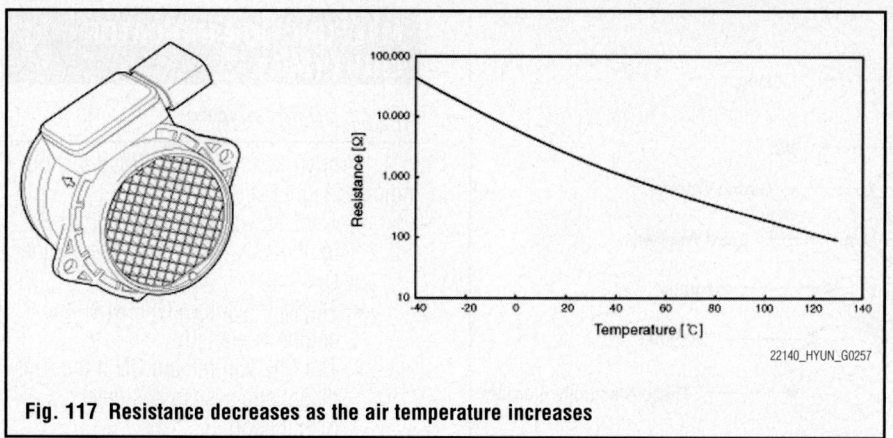

Fig. 117 Resistance decreases as the air temperature increases

TESTING

See Figures 118 and 119.

1. Disconnect the Intake Air Temperature (IAT) sensor connector.

2. Measure the resistance between the signal terminal and ground terminal of the IAT connector and compare the resistance to the following Temperature and Resistance Table.

Temperature		Resistance (kΩ)
°C	°F	
-40	-40	100.87
-20	-4	28.58
0	32	9.40
10	50	5.66
20	68	3.51
40	104	1.47
60	140	0.67
80	176	0.33

Fig. 118 Temperature and Resistance Table

3. If the IAT sensor fails to meet the resistance specifications, replace the IAT sensor.

KNOCK SENSOR (KS)

LOCATION

See Figure 120.

The knock sensor is located in the side of the cylinder block.

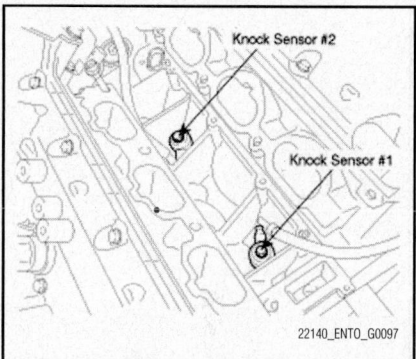

Fig. 120 Location of Knock Sensor (KS) No. 1 and No. 2

OPERATION

See Figure 121.

The Knock Sensor (KS) is used to detect engine vibrations caused by pre-ignition or detonation and provides information to the PCM. The PCM will then retard the timing in an attempt to eliminate detonation.

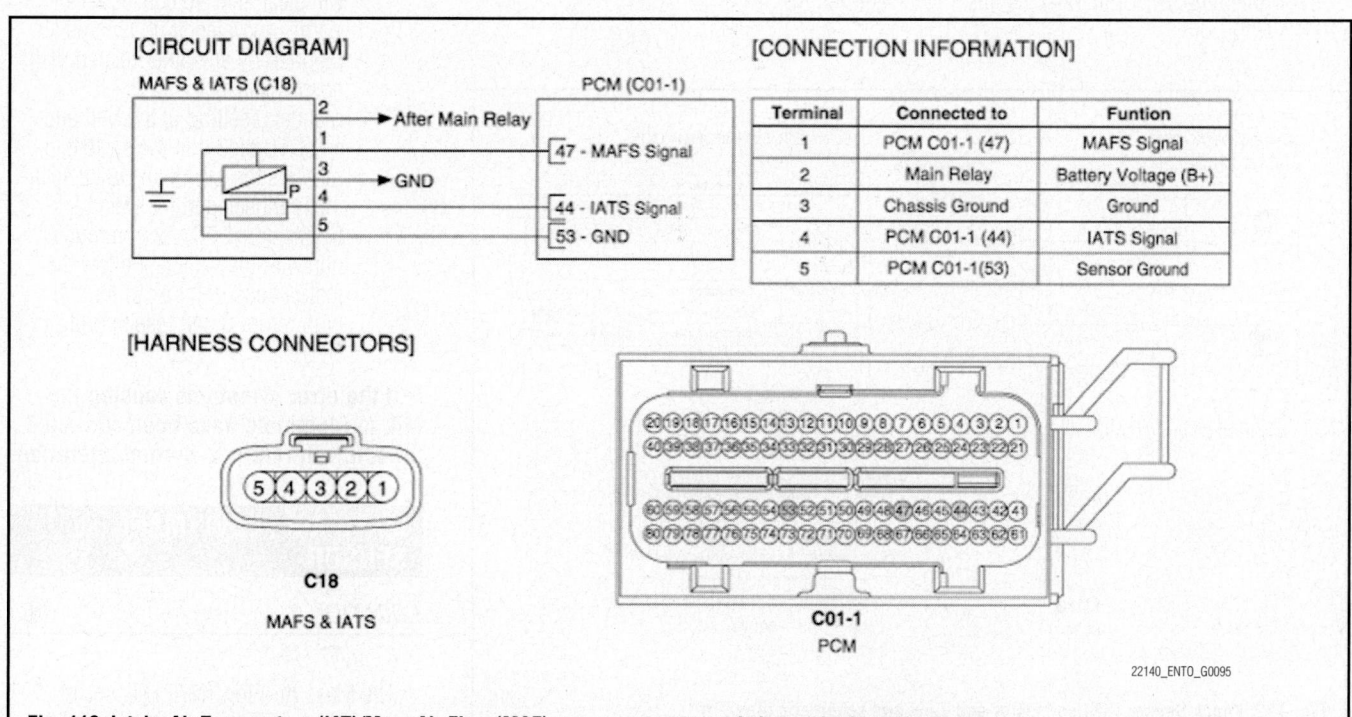

Fig. 119 Intake Air Temperature (IAT)/Mass Air Flow (MAF) sensor connector end view and schematic diagram

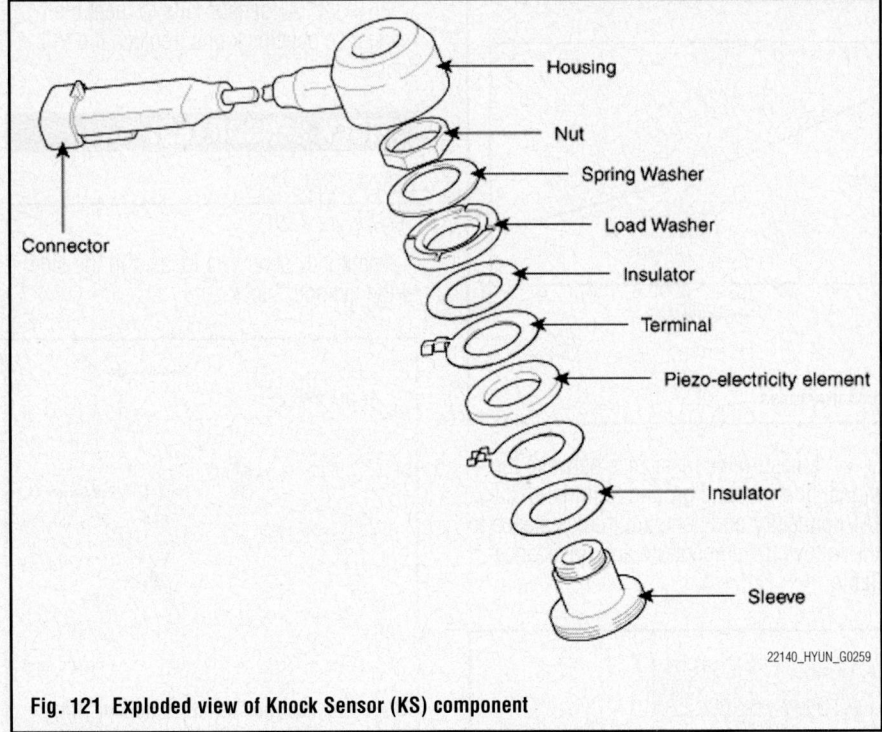

Fig. 121 Exploded view of Knock Sensor (KS) component

REMOVAL & INSTALLATION

1. Before servicing the vehicle, refer to the Precautions Section.
2. Disconnect the negative battery cable.
3. Disconnect the sensor connector.
4. Remove the sensor from its mounting.

To install:

5. Installation is the reverse of the removal procedure.
6. Tighten the sensor to 12–17 ft. lbs. (16–24 Nm).

TESTING

See Figure 122.

1. Disconnect the Knock Sensor (KS) electrical connector.
2. Connect an ohmmeter.
3. Measure the resistance between terminals 1 and 2. Specification should be 5mohm at 68°F (20°C).
4. If not within specification, replace the effected KS.

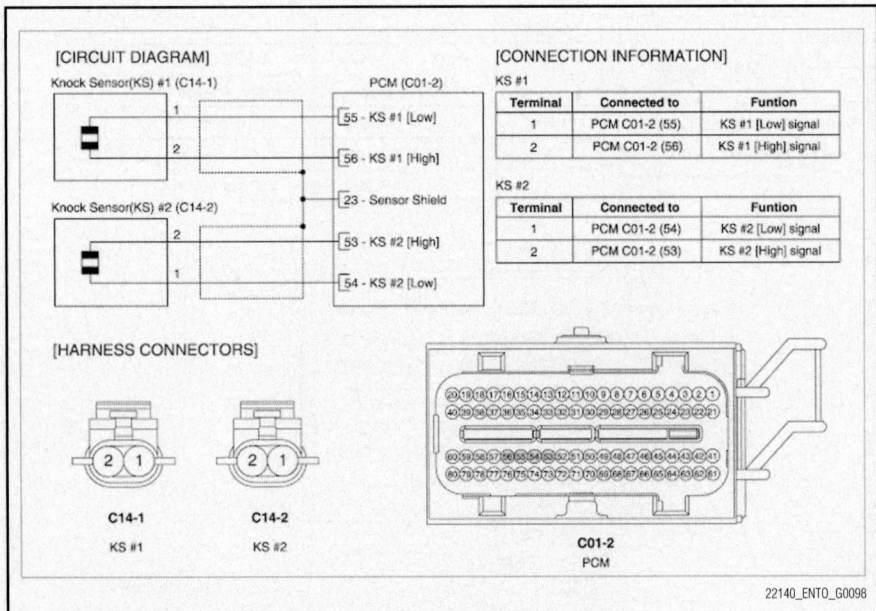

Fig. 122 Knock Sensor (KS) connector end view and schematic diagram

MALFUNCTION INDICATOR LIGHT (MIL)

RESET PROCEDURES

1. Proper operation of the Malfunction Indicator Light (MIL):
   - The MIL will illuminate with the ignition switch ON and the engine OFF
   - The MIL will turn OFF when the engine is started
   - The MIL will remain ON if the self-diagnostic system has detected a malfunction
   - The MIL may turn OFF if the malfunction is no longer present
   - If the MIL is illuminated and then the engine stalls, the MIL will remain illuminated as long as the ignition switch is ON
   - If the MIL is not illuminated and the engine stalls, the MIL will not illuminate until the ignition switch is cycled OFF, then ON
2. Resetting the MIL:
   - The control module turns OFF the MIL after 3 consecutive ignition cycles that the diagnostic system runs and does not fail
   - The control module turns OFF the MIL after a current Diagnostic Trouble Code (DTC) clears when the diagnostic cycle runs and passes
   - There may still be a history of DTC's stored in the system. These will clear after 40 consecutive warm-up cycles, if no failures are reported by any other related diagnostic system
   - Manual resetting of the MIL and any DTC stored in the system, requires the use of an OBD2 scan tool connected to the Data Link Connector (DLC) for communication with the vehicle. Follow the instructions of the scan tool for both retrieval and resetting of DTC's.

➡️**If the error symptoms causing the MIL to illuminate have been corrected, the MIL will return to normal operation.**

MASS AIR FLOW (MAF) SENSOR

LOCATION

See Figure 116.

The Mass Air Flow (MAF) sensor is mounted in the intake air hose of the air cleaner assembly. This sensor is combined with the Intake Air Temperature (IAT) sensor.

OPERATION

The Mass Air Flow Sensor (MAF) sensor is a hot-film type sensor and is located in between the air cleaner and the throttle body. It consists of a tube, a sensor assembly, and honeycomb cell. It detects the intake air quantity flowing into the intake manifold. Air flows from the air cleaner assembly through the honeycomb cell and over the hot film element. At this time, heat transfer is generated by convection and the MAF sensor loses its energy. This sensor detects the mass air flow by using the energy loss and transfers the information to the PCM by a voltage transfer. The PCM calculates fuel quantity and ignition timing appropriate for the conditions.

REMOVAL & INSTALLATION

1. Before servicing the vehicle, refer to the Precautions Section.
2. Disconnect the negative battery cable.
3. Disconnect the connector from the sensor.
4. Remove the air cleaner and air intake assembly, as required.
5. Remove the sensor from its mounting.

To install:
6. Installation is the reverse of the removal procedure.

TESTING

See Figures 119, 123 and 124.

1. Verify the integrity of the air induction system by inspecting for the following conditions:
   - Damaged components
   - Loose or improper installation
   - An air flow restriction
   - Any vacuum leak
   - Water intrusion
2. Verify that any electrical aftermarket devices are properly connected and grounded.
3. With the engine running, observe the scan tool MAF sensor parameter. The reading should be between 1,700–3,200 Hz depending on the Engine Coolant Temperature (ECT).
4. A Wide Open Throttle (WOT) acceleration from a stop should cause the MAF sensor parameter on the scan tool to increase rapidly. This increase should be from 2–6 g/s at idle to greater than 100 g/s at the time of the 1–2 shift.
5. Run the engine and connect the Hi-Scan (Pro)® to the Data Link Connector (DLC).

Condition	Output Voltage (V)	Intake Air Quantity (kg/h)
Idle	0.6 ~ 1.0	11.66 ~ 19.85
3000 rpm	1.7 ~ 2.0	43.84 ~ 58.79

22140_HYUN_G0260

Fig. 123 Output Voltage and Intake Air Quantity Table

Air Flow (kg/h)	Output Frequency (Hz)
12.6	2,617
18.0	2,958
23.4	3,241
32.4	3,653
43.2	4,024
57.6	4,399
72.0	4,704
108.0	5,329
144.0	5,897
198.0	6,553
270.0	7,240
360.0	7,957
486.0	8,738
666.0	9,644
900.0	10,590

22140_ENTO_G0099

Fig. 124 Air Flow and Output Frequency Table

6. Check if the sensor output voltage is normal according to the following specification table.
7. If within specification, check for poor connection between the PCM and the component.
8. If not within specification, substitute the sensor with a known good component, check for proper operation. If the problem is corrected, replace the sensor.

MANIFOLD ABSOLUTE PRESSURE (MAP) SENSOR

LOCATION

See Figure 125.

The MAP sensor is located on the intake manifold.

OPERATION

This sensor is a pressure sensitive variable resistor. It measures the changes in the intake manifold pressure which result from engine load and speed changes, and converts these to a voltage output. This sensor is used to measure the barometric pressure at start up, and under certain conditions, allows the PCM to automatically adjust for different altitudes. The

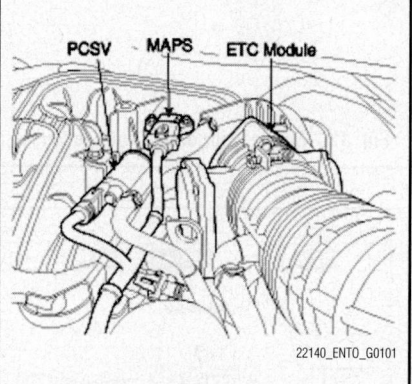

22140_ENTO_G0101

Fig. 125 Location of the MAP sensor, Purge Control Solenoid Valve (PCSV), and the Electronic Throttle Control (ETC) module

PCM supplies 5 volts to the sensor and monitors the voltage on a signal line. The sensor provides a path to ground through its variable resistor. The sensor input affects fuel delivery and ignition timing controls in the PCM.

REMOVAL & INSTALLATION

1. Before servicing the vehicle, refer to the Precautions Section.
2. Disconnect the negative battery cable.
3. Disconnect the connector from the sensor.
4. Remove the sensor retaining screws.
5. Remove the sensor from its mounting.

To install:
6. Installation is the reverse of the removal procedure.
7. Tighten the MAP sensor installation bolt to: 78–104 inch lbs. (9–12 Nm).

TESTING

See Figures 126 through 128.

1. Measure the voltage between terminal 4 (sensor ground) and terminal 1 (sensor output).
2. Specification should be 4–5 volts, with the ignition switch ON.
3. Specification should be 0.8–2.4 volts at idle.
4. If not within specification, replace the sensor.

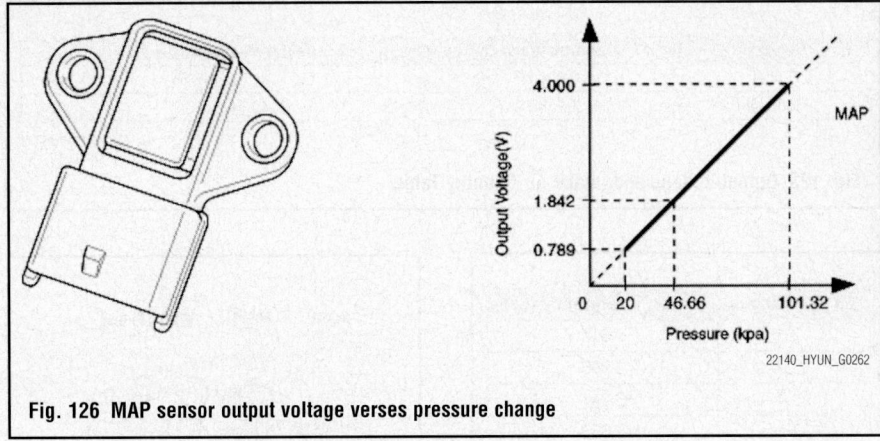

Fig. 126 MAP sensor output voltage verses pressure change

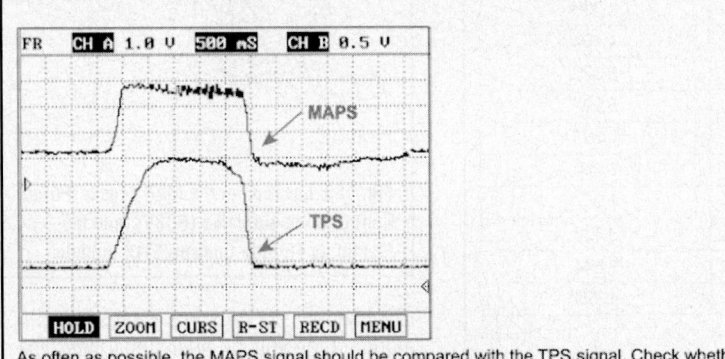

As often as possible, the MAPS signal should be compared with the TPS signal. Check whether the MAPS and TPS signals increase at the same time when accelerating. During acceleration, the MAPS output voltage increases; during deceleration, the MAPS output voltage decreases.

22140_HYUN_G0263

Fig. 127 Waveform of MAP sensor signal compared with the TPS signal waveform

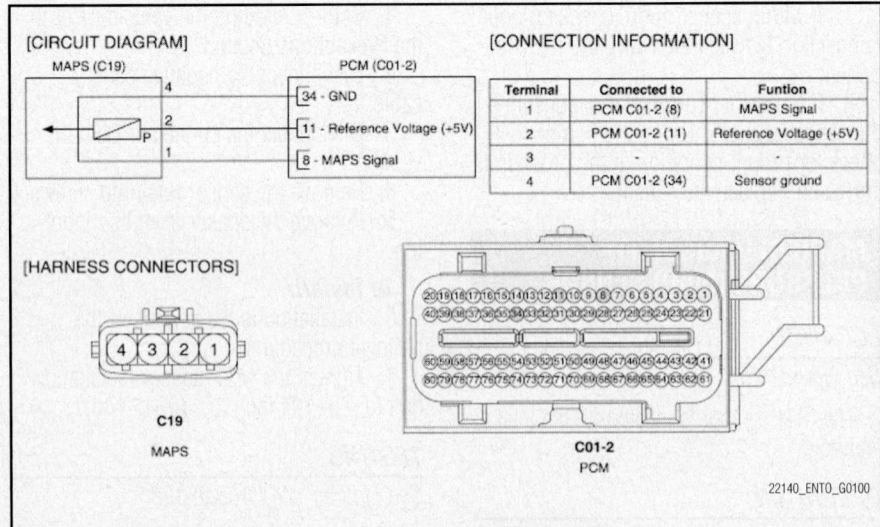

Fig. 128 MAP sensor connector end view and schematic diagram

OIL PRESSURE SENSOR

LOCATION

The oil pressure sensor is located near the base of the oil filter assembly.

OPERATION

The oil pressure sensor measures the amount of pressure generated from the output of the oil pump. If the oil pressure is too low, mechanical damage and scoring may occur on non-lubricated parts.

REMOVAL & INSTALLATION

✱✱ CAUTION

Do not perform this procedure with the engine hot, wait until the engine is cool or cold.

1. Before servicing the vehicle, refer to the Precautions Section.
2. Turn the ignition OFF.
3. Remove the oil pressure sensor electrical connector.
4. Remove the oil pressure sensor.

To install:

5. To install, reverse the removal procedure.
6. Tighten the sensor to 113–130 inch lbs. (13–15 Nm).

TESTING

Before servicing any vehicle, please be sure to read the precautions section, which deals with personal safety, prevention of component damage, and important points to take into consideration when servicing a motor vehicle.

1. With the vehicle on a level surface, allow adequate drain down time of 2–3 minutes and measure for a low oil level. Add the recommended grade engine oil and fill the crankcase until the oil level measures full on the oil level indicator.
2. Run the engine, and verify low, or no oil pressure on the vehicle gage or light. Listen for a noisy valve train or a knocking noise.
3. Inspect for the following:
   - Oil diluted by moisture or unburned fuel mixtures
   - Improper oil viscosity for the expected temperature
   - Incorrect or malfunctioning oil pressure sender
   - Incorrect or malfunctioning oil pressure gauge
   - Plugged oil filter
   - Malfunctioning oil bypass valve
4. Remove the oil pressure sender or another engine block oil gallery plug.
5. Install an oil pressure gauge and measure the engine oil pressure.
6. Compare the readings to specifications.
7. If the engine oil pressure is below specifications, inspect the engine for one or more of the following:
   - Oil pump worn or dirty
   - Oil pump-to-engine front cover bolts loose
   - Oil pump screen loose, plugged, or damaged

- Oil pump screen O-ring seal missing or damaged
- Malfunctioning oil pump pressure regulator valve
- Excessive bearing clearance
- Cracked, porous, or restricted oil galleries
- Oil gallery plugs missing or incorrectly installed
- Broken lash adjusters

8. Higher than recommended oil pressure may be caused by one or more of the following conditions:

- Worn or sticking oil pump pressure relief valve
- Plugged oil filter
- Improper viscosity oil

9. If the oil pressure tests according to specification and the oil pressure sensor gives a faulty reading, replace the oil pressure sensor (switch) as necessary.

POWERTRAIN CONTROL MODULE (PCM)

LOCATION

See Figure 129.

Refer to the accompanying illustration for Powertrain Control Module (PCM) location.

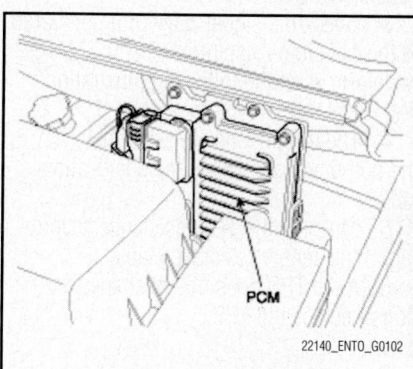

22140_ENTO_G0102

Fig. 129 Powertrain Control Module (PCM) location

OPERATION

The powertrain has electronic controls to reduce exhaust emissions while maintaining excellent drivability and fuel economy. The Powertrain Control Module (PCM) is the control center of this system. The PCM monitors numerous engine and vehicle functions. The PCM constantly looks at the information from various sensors and other inputs, and controls the systems that affect vehicle performance and emissions. The PCM also performs the diagnostic tests on various parts of the system. The PCM can recognize operational problems and alert

the driver via the Malfunction Indicator Lamp (MIL). When the PCM detects a malfunction, the PCM stores a Diagnostic Trouble Code (DTC). The problem area is identified by the particular DTC that is set. The control module supplies a buffered voltage to various sensors and switches. Review the components and wiring diagrams in order to determine which systems are controlled by the PCM. The following are some of the functions that the PCM controls:

- The ignition system
- The Knock Sensor (KS) system
- The Evaporative Emissions (EVAP) system
- The alternator
- The A/C and fan systems
- The cooling fan control
- The fuel injection system
- The emission control systems
- The on-board diagnostics

REMOVAL & INSTALLATION

1. Before servicing the vehicle, refer to the Precautions Section.
2. Turn the ignition switch **OFF**.
3. Disconnect the negative battery cable from the battery.
4. Disconnect the PCM connectors.
5. Remove the PCM mounting bolts.
6. Remove the PCM from the engine compartment.

To install:

7. Installation is the reverse of the removal procedure.
8. Tighten the PCM mounting bolts to: 86–104 inch lbs. (10–12 Nm).

TESTING

1. Perform a careful underhood inspection when performing any diagnostic procedure or diagnosing the cause of an emission test failure. This can often lead to repairing a condition without further steps. Use the following guidelines when performing an inspection:

a. Inspect all of the vacuum hoses for correct routing, pinches, cuts, or disconnects.

b. Inspect any hoses that are difficult to see.

c. Inspect all of the wires in the engine compartment for the following conditions:

- Burned or chafed spots
- Pinched wires
- Contact with sharp edges
- Contact with hot exhaust manifold

The Powertrain Control Module (PCM) is programmed with test routines that test the

operation of the various systems the PCM controls. Some tests monitor internal PCM functions. Many tests are run continuously. Other tests run only under specific conditions, referred to as conditions for running the Diagnostic Trouble Code (DTC). When the vehicle is operating within the conditions for running a particular test, the PCM monitors certain parameters and determines if the values are within an expected range. The parameters and values considered outside the range of normal operation are listed as conditions for setting the DTC. When the conditions for setting the DTC occur, the PCM executes the action taken when the DTC sets. Some DTC's alert the driver via the Malfunction Indicator Lamp (MIL) or a message. Other DTC's do not trigger a driver warning, but are stored in memory. The PCM also saves data and input parameters when most DTC's are set.

The DTC's are categorized by type. The DTC type is determined by the MIL operation and the manner in which the fault data is stored when a particular DTC fails. In some cases, there may be exceptions to this structure. Therefore, when diagnosing the system it is important to read the action taken when the DTC sets and the conditions for clearing the DTC.

Many intermittent open or shorted circuits come and go with harness and connector movement caused by vibration, engine torque, bumps, or rough pavement.

2. Test the wiring harness and connectors by performing the following:

- Move the related PCM connectors and wiring while monitoring the appropriate scan tool data
- With the engine running, move the related connectors and wiring while monitoring engine operation
- If harness or connector movement affects the data displayed, the component and system operation, or the engine operation, inspect and repair the harness or connections as necessary

3. Test the electrical connections and/or wiring by performing the following:

- Inspect for incorrect mating of the connector halves or terminals not fully seated in the connector body
- Inspect for improperly formed or damaged terminals. Test for incorrect terminal tension
- Inspect for poor terminal to wire connections including terminals crimped over insulation. This requires removing the terminal from the connector body

- Inspect for corrosion or water intrusion. Pierced or damaged insulation can allow moisture to enter the wiring. The conductor can corrode inside the insulation with little visible evidence. Look for swollen and stiff sections of wire in the suspect circuits
- Inspect for wires that are broken inside the insulation

4. Test the PCM ground circuit:

a. Measure resistance between the PCM and a good chassis ground. Resistance specification: 1 ohm or less.

b. Use the backside of the PCM harness connector as the PCM side check point.

c. If a problem is found, make the necessary repairs.

5. If a problem is not located through the above procedures, the PCM may be faulty.

6. Hyundai recommends to replace the PCM with a new one and then check the vehicle again. If the vehicle operates normally, then the problem was likely with the PCM.

7. Retest the original PCM, if possible, by installing it into a known-good vehicle and check vehicle operations.

THROTTLE POSITION SENSOR (TPS)

LOCATION

See Figure 125.

The Throttle Position Sensor (TPS) is integrated in the Electronic Throttle Control (ETC) module.

OPERATION

The Electronic Throttle Control (ETC) system is an electronically controlled throttle device which controls the throttle valve. It consists of the ETC motor, throttle body, and Throttle Position Sensor (TPS). A mechanical throttle control system receives a driver's intention via a wire cable between the accelerator and the throttle valve, while the ETC system uses the signal from the Accelerator Pedal Position (APP) sensor installed on the accelerator pedal. After the PCM receives the APP sensor signal and calculates the throttle opening angle, it activates the throttle valve by using the ETC motor. Additionally, it can handle the cruise control function without any special devices.

REMOVAL & INSTALLATION

See Figure 130.

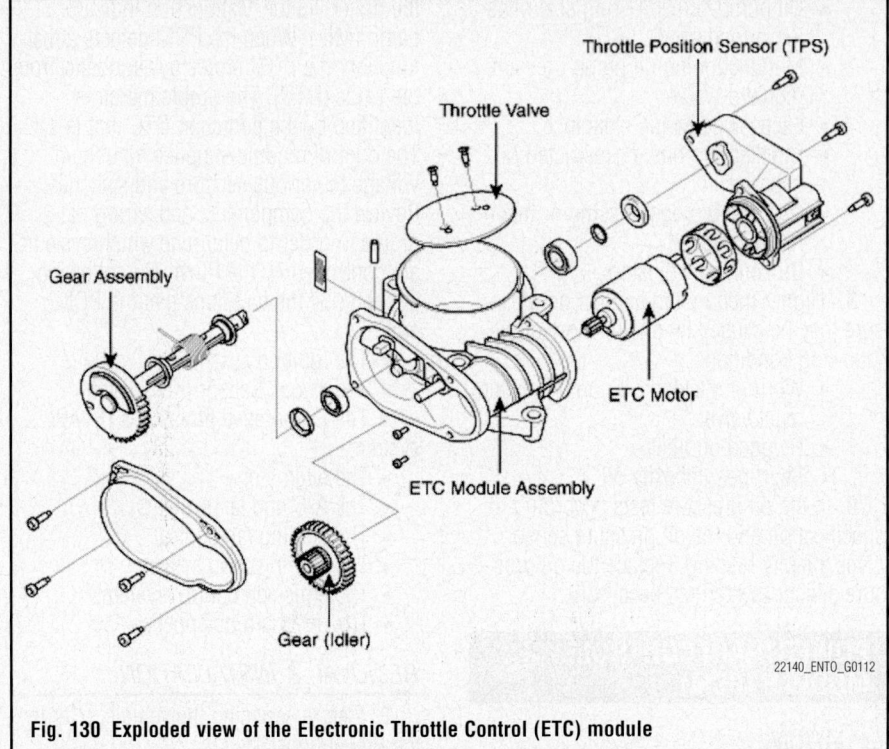

Fig. 130 Exploded view of the Electronic Throttle Control (ETC) module

1. Before servicing the vehicle, refer to the Precautions Section.
2. Disconnect the negative battery cable.
3. Disconnect the sensor connector.
4. Remove the sensor retaining screws.
5. Remove the sensor from its mounting.

To install:

6. Installation is the reverse of the removal procedure.

TESTING

See Figures 127 and 131 through 134.

1. The Throttle Position Sensor (TPS) output voltage within the Electronic Throttle Control (ETC) module should vary from 0.2–0.8 volts at Closed Throttle (CT).

2. The TPS should vary from 4.3–4.8 volts at Wide Open Throttle (WOT).

3. Check the signal waveform compared to the Manifold Absolute Pressure (MAP) waveform as in the following illustration. The waveform should coincide.

4. The throttle angle should match the output voltage as in the specification table.

5. Measure the TPS resistance according to the sensor resistance table.

6. If the TPS does not meet specifications, replace the TPS.

Throttle Angle(°)	Output Voltage(V) [Vref = 5.0V]	
	TPS1	TPS2
0°	0V	5.0V
10°	0.5V	4.5V
20°	0.9V	4.1V
30°	1.4V	3.6V
40°	1.8V	3.2V
50°	2.3V	2.7V
60°	2.7V	2.3V
70°	3.2V	1.8V
80°	3.6V	1.4V
90°	4.1V	0.9V
100°	4.5V	0.5V
110°	5.0V	0V

Fig. 131 Throttle angle and output voltage specification table

Item	Sensor Resistance
TPS1	4.0 ~ 6.0kΩ at 68°F (20°C)
TPS2	2.72 ~ 4.08kΩ at 68°F (20°C)

22140_ENTO_G0103

Fig. 132 TPS resistance table

VARIABLE CAMSHAFT TIMING OIL CONTROL SOLENOID

LOCATION

See Figures 135 and 136.

Refer to the accompanying illustrations for variable camshaft timing oil control solenoid location.

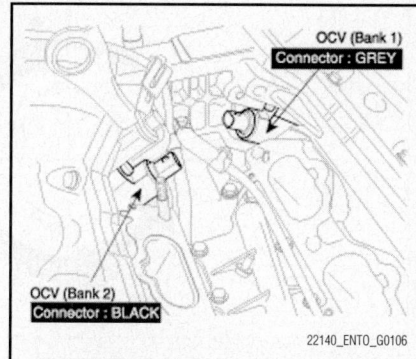

Fig. 135 Variable camshaft timing oil control solenoid location

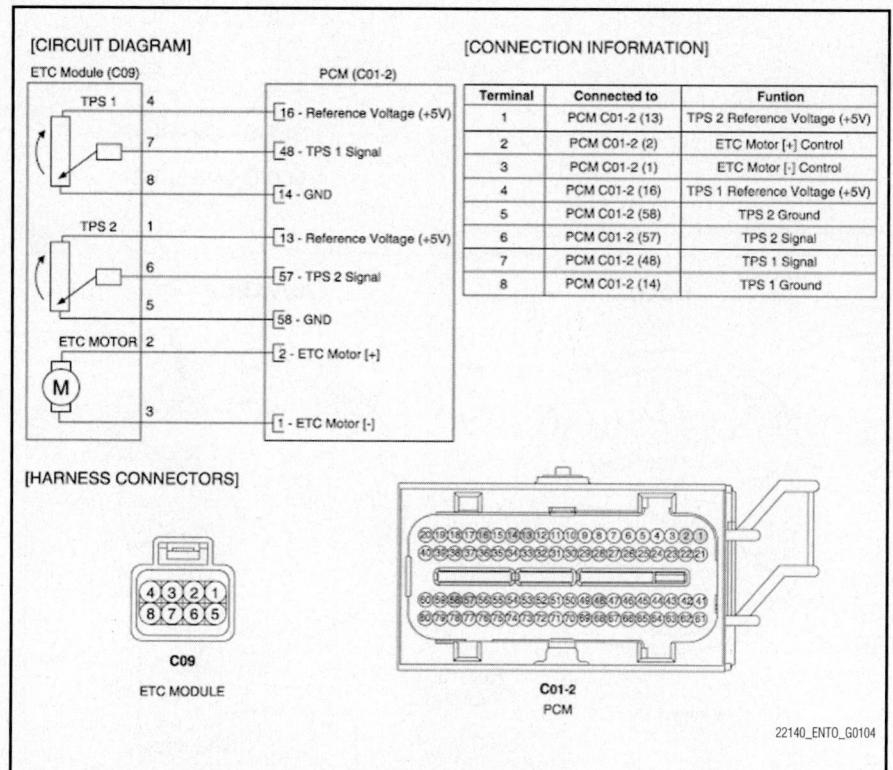

[CONNECTION INFORMATION]

Terminal	Connected to	Funtion
1	PCM C01-2 (13)	TPS 2 Reference Voltage (+5V)
2	PCM C01-2 (2)	ETC Motor [+] Control
3	PCM C01-2 (1)	ETC Motor [-] Control
4	PCM C01-2 (16)	TPS 1 Reference Voltage (+5V)
5	PCM C01-2 (58)	TPS 2 Ground
6	PCM C01-2 (57)	TPS 2 Signal
7	PCM C01-2 (48)	TPS 1 Signal
8	PCM C01-2 (14)	TPS 1 Ground

22140_ENTO_G0104

Fig. 133 TPS connector end view and schematic diagram

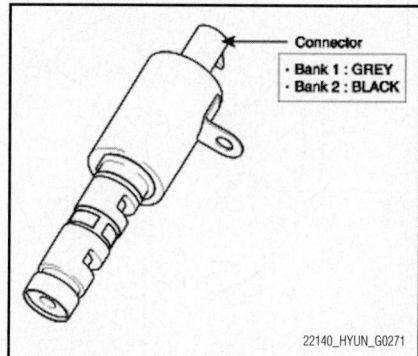

Fig. 136 Variable camshaft timing oil control solenoid component view

OPERATION

See Figure 137.

The Continuously Variable Valve Timing (CVVT) system is installed to the chain sprocket of the intake camshaft. This system controls the intake camshaft to provide the optimal valve timing for every driving condition. The PCM controls the Variable Camshaft Timing Oil Control Solenoid (VCTOCS)—also known as the Oil Control Valve (OCV)—based on the signal outputs from the Mass Air Flow (MAF) sensor, Throttle Position Sensor (TPS), and Engine Coolant Temperature (ECT) sensor. The CVVT controller regulates the intake camshaft angle using oil pressure through the VCTOCS. As a result, the relative position between the camshaft and the crankshaft becomes optimal. The engine torque improves, fuel economy improves, and the exhaust emissions decrease within overall driving conditions.

The CVVT system makes continuous intake valve timing changes based on operating conditions. Intake valve timing is

Mode	Description	Symptom	Possible Cause
MODE 1	FORCED ENGINE SHUTDOWN	Engine stop	• ETC system can't proceed reliable algorithm procedure •• Fatal PCM internal programming error •• Faulty intake system or throttle body
MODE 2	FORCED IDLE & POWER MANAGEMENT	Forced idle state controlled by fuel quantity regulation and ignition timing adjustment	• ETC system can't control engine power via throttle device • Disabled throttle control or broken throttle position information
MODE 3	FORCED IDLE	Forced idle state and no response for accelerator activation	• No information about the accelerator position •• Malfunctioning APS 1 and 2, faulty A/D converter or internal controller
MODE 4	LIMIT PERFORMANCE & POWER MANAGEMENT	Engine power is determined by accelerator position and idle power requirement (Limited vehicle running)	• ETC system can't securely control engine power
MODE 5	LIMIT PERFORMANCE	1. Engine power varies with accelerator position, but driver 2. MIL ON (Normal vehicle running)	• Not reliable accelerator position signal or bad maximum power generation •• Faulty APS, ignition voltage or internal controller
MODE 6	NORMAL	Normal	

22140_ENTO_G0105

Fig. 134 TPS and ETC module mode of operation and troubleshooting chart

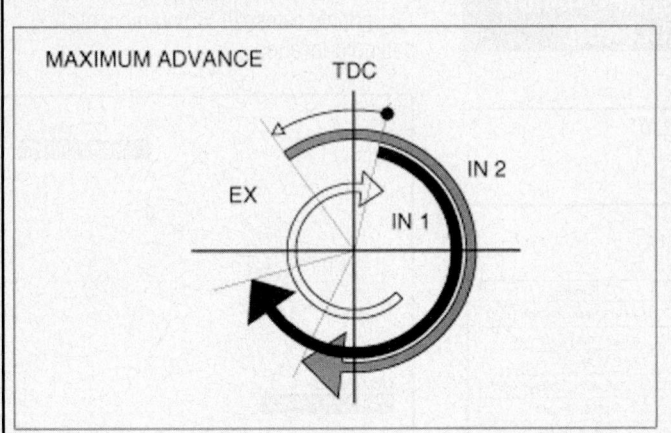

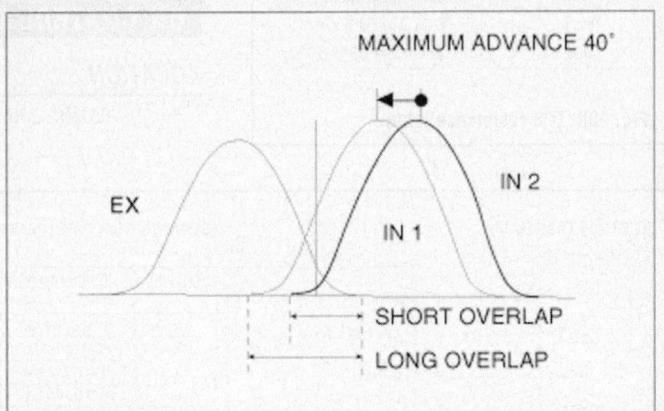

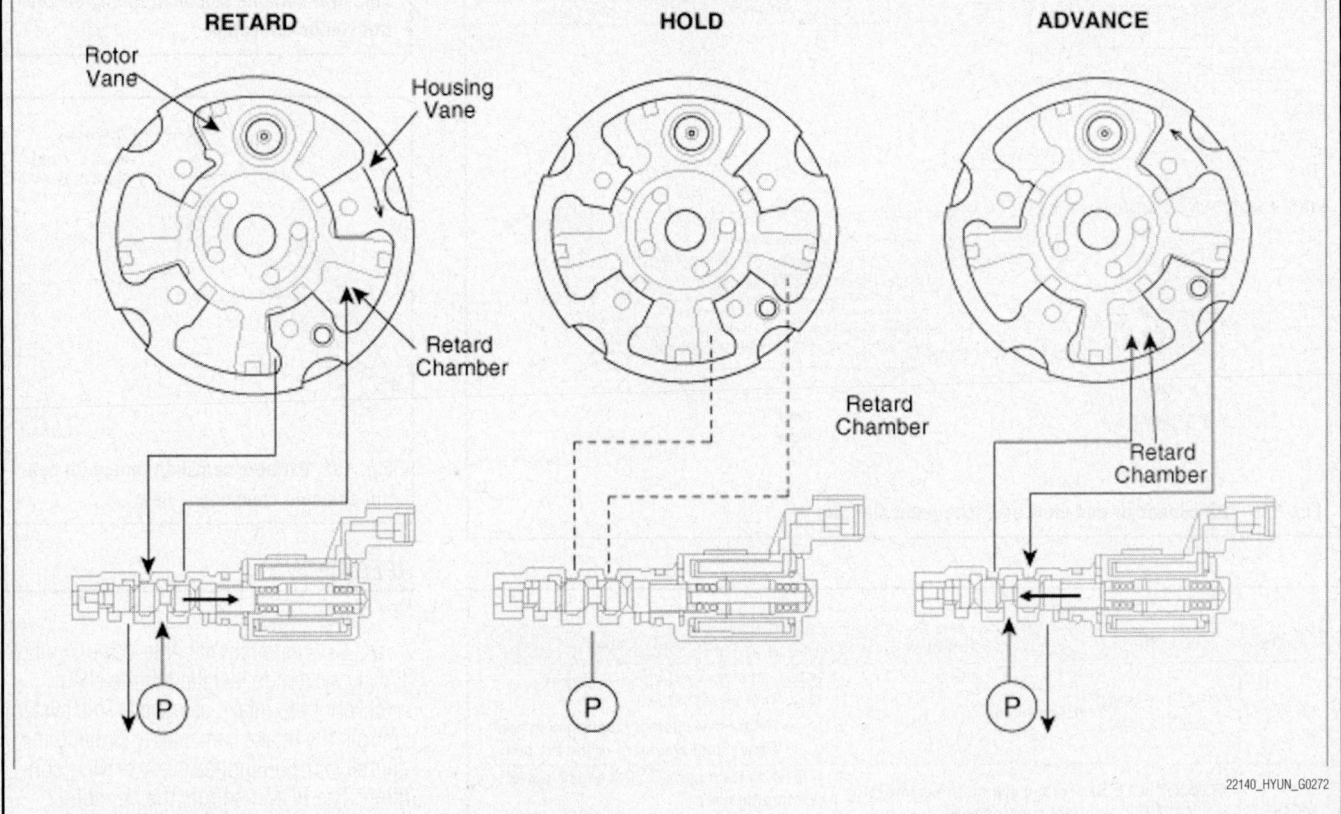

Fig. 137 Continuously Variable Valve Timing (CVVT) system illustrated

optimized to allow the engine to produce maximum power. Cam angle is advanced to obtain the EGR effect and reduce pumping loss. The intake valve is closed quickly to reduce the entry of the air/fuel mixture into the intake port and improve the changing effect. Reducing the cam advance at idle stabilizes combustion and reduces engine speed. If a malfunction occurs, the CVVT system control is disabled and the valve timing is fixed at the fully retarded position.

REMOVAL & INSTALLATION

See Figure 138.

1. Before servicing the vehicle, refer to the Precautions Section.

Bank	Component side	Harness side
Bank 1 (RH)	Grey	Grey
Bank 2 (LH)	Black	Black

22140_ENTO_G0107

Fig. 138 Bank and color match table for the variable camshaft timing oil control solenoids

2. Disconnect the ground cable from the battery.

3. Disconnect the connector from Variable Camshaft Timing Oil Control Solenoid (VCTOCS) on the right-hand bank and/or left-hand bank.

4. Remove the bolt retaining the VCTOCS.

5. Remove the VCTOCS.

To install:

➡**Always use a new gasket/O-ring.**

6. Install the VCTOCS.

❄❄ WARNING

If the VCTOCS, also called Oil Control Valves (OCV), are installed incorrectly, the engine may be damaged. When installing them, ensure the VCTOCS/OCV and harness connector colors match (components and harness side) according to the following table.

7. Tighten the mounting bolt to 86–104 inch lbs. (20–39 Nm).

8. Connect the electrical connector to the VCTOCS.

TESTING

1. Ensure the vehicle has the proper oil viscosity and that the oil is not overdue maintenance.

2. Observe the engine oil level. The engine oil level should be within the operating range.

3. Allow the engine to reach operating temperature.

4. Connect a suitable OBD2 scan tool to the Data Link Connector (DLC) inside the vehicle.

5. Increase the engine speed to 1,500 RPM.

6. Command each solenoid to 25 percent. The angle desired parameter should match the solenoid actual parameter.

Component Testing

See Figures 139 and 140.

1. Measure the resistance of each variable camshaft timing oil control solenoid valve assembly. Resistance should read according to the resistance specification table shown.

a. Ignition **OFF**.

b. Disconnect Variable Camshaft Timing Oil Control Solenoid (VCTOCS) connector.

c. Measure resistance between terminals 1 and 2 of the VCTOCS connector (component side).

- Specification: Approx. 6.7–7.7 ohms at 68°F (20°C)
- If not according to specification, replace the oil control solenoid

2. Check the operation of the VCTOCS.

a. Start the engine and let it idle.

b. With VCTOCS connector still disconnected, connect a 12 volt supply to terminal 2 and a ground to terminal 1 of the VCTOCS (component side).

- The engine should suddenly run rough at idle or stall out as the VCTOCS abruptly changes the timing

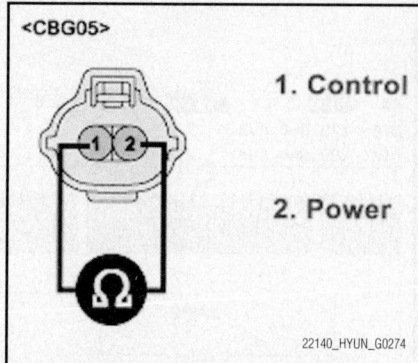

<CBG05>

1. Control

2. Power

22140_HYUN_G0274

Fig. 139 Use an ohmmeter to check the resistance level of the oil control solenoid

Temperature °C(°F)	Resistance (Ω)
0(32)	6.2 ~ 7.4
20(68)	6.8 ~ 8.0
40(104)	7.4 ~ 8.6
60(140)	8.0 ~ 9.2
80(176)	8.6 ~ 9.8

22140_HYUN_G0275

Fig. 140 Resistance table for the oil control solenoid compared to temperature

- If the engine runs rough or stalls when the 12 volt supply is connected, check for connection problems to the VCTOCS or problem with the supply side from the PCM
- If no change is noted, the VCTOCS may be defective

3. Check the VCTOCS and oil filter assembly.

a. Turn the ignition **OFF**.

b. Check VCTOCS filter for sticking or contamination.

c. Remove the VCTOCS and visually check the spool column of VCTOCS for contamination.

d. If a problem is found, clean or replace as necessary.

e. If no problem is found:

- Apply a 12 volt supply to terminal 2 and a ground to terminal 1 of the VCTOCS (Component side)
- Verify that a "clicking" sound is heard when applying the battery voltage
- If no "clicking" is heard, replace the VCTOCS

Circuit Testing

See Figure 141.

1. Many malfunctions in the electrical system are caused by poor harness and terminals. Faults can also be caused by interference from other electrical systems, mechanical or chemical damage.

2. Thoroughly check connectors for looseness, poor connection, bending, corrosion, contamination, deterioration, or damage.

3. Check the supply voltage to the Variable Camshaft Timing Oil Control Solenoid (VCTOCS).

a. With the ignition "OFF," disconnect the VCTOCS connector.

b. Turn the ignition "ON" with the engine "OFF."

c. Measure the voltage between the power terminal of the harness connector

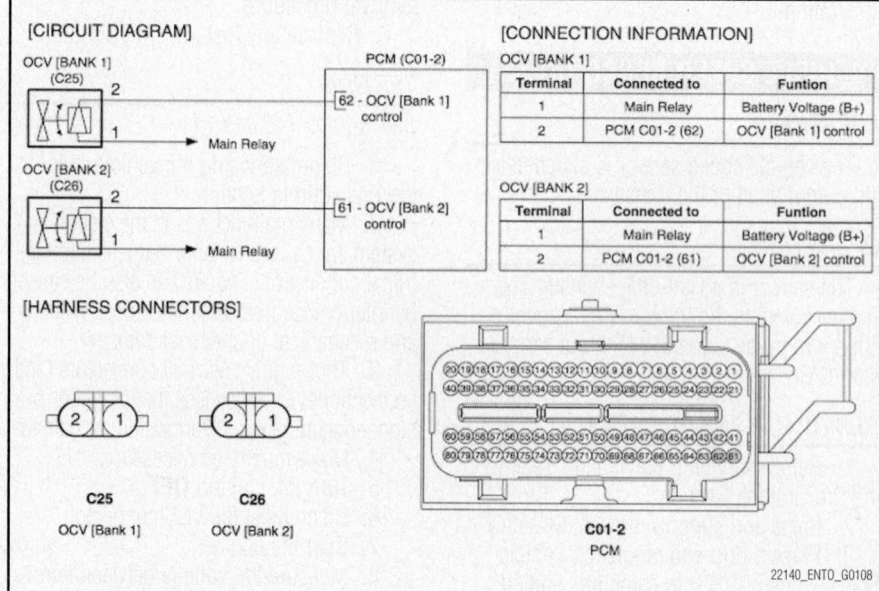

[CIRCUIT DIAGRAM]

OCV [BANK 1] (C25)

OCV [BANK 2] (C26)

PCM (C01-2)

62 - OCV [Bank 1] control

61 - OCV [Bank 2] control

Main Relay

[CONNECTION INFORMATION]

OCV [BANK 1]

Terminal	Connected to	Funtion
1	Main Relay	Battery Voltage (B+)
2	PCM C01-2 (62)	OCV [Bank 1] control

OCV [BANK 2]

Terminal	Connected to	Funtion
1	Main Relay	Battery Voltage (B+)
2	PCM C01-2 (61)	OCV [Bank 2] control

[HARNESS CONNECTORS]

C25
OCV [Bank 1]

C26
OCV [Bank 2]

C01-2
PCM

22140_ENTO_G0108

Fig. 141 Variable camshaft timing oil control solenoid connector end view and schematic diagram

and a chassis ground. The specification: battery voltage (12 volts).

- If the 12 volt supply is not present, check the fuse between the Main Relay and the VCTOCS for an open condition, check for an open in power circuit between the Main Relay and the VCTOCS power circuit
- Make the repairs as necessary

4. Check for a short to ground in the harness.

   a. With the ignition "OFF," disconnect the VCTOCS connector.

   b. Measure the resistance between the control terminal of the VCTOCS harness connector and a good chassis ground. The specification: Infinite resistance.

- If the measured resistance is within specification, refer to Component Testing for the VCTOCS
- If the measured resistance is not within specification, repair or replace as necessary

Timing Inspection

See Figure 142.

1. With the ignition "OFF," set up an oscilloscope as follows:

   a. Channel A (+): terminal 2 of the CKPS (back probe), (-): ground.

   b. Channel B (+): terminal 2 of the CMPS (back probe), (-): ground.

2. Start the engine and check for the signal waveform to synchronize with the camshaft sensor and the tooth that is missing. Refer to the sample waveforms in the illustration.

VEHICLE SPEED SENSOR (VSS)

LOCATION

The vehicle speed sensor is attached to the output shaft of the transaxle.

OPERATION

The sensor is a Hall-effect sensor. The sensor converts the transaxle gear revolutions into pulse signals, which are sent to the PCM.

REMOVAL & INSTALLATION

1. Before servicing the vehicle, refer to the Precautions Section.
2. Raise and support the vehicle safely.
3. Place a drip pan below the Vehicle Speed Sensor (VSS) to catch any spilled fluid when it is removed.
4. Disconnect the VSS connector.

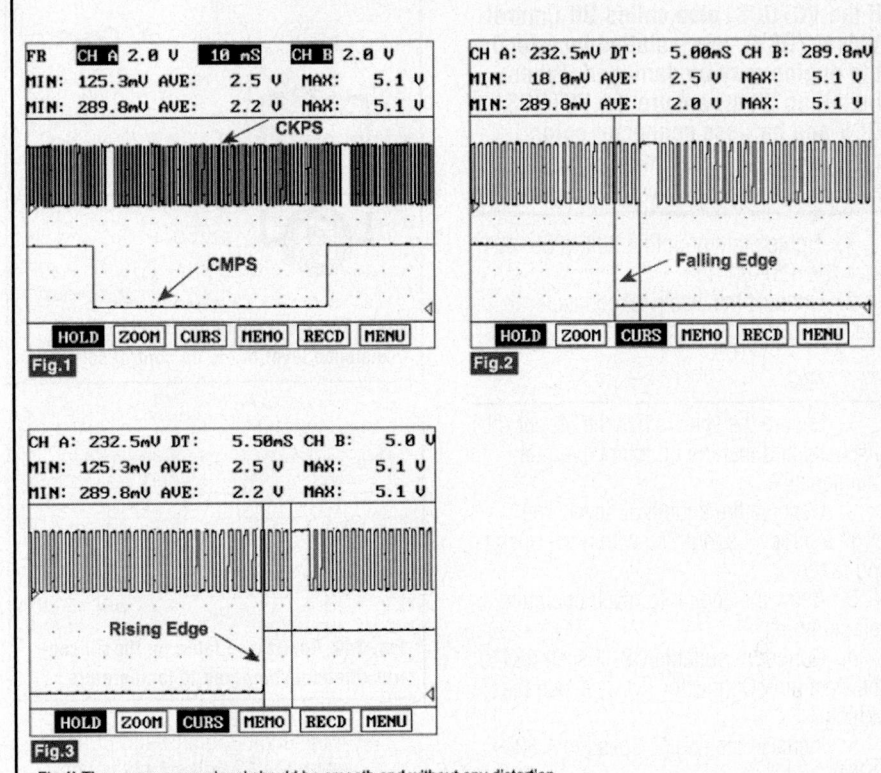

Fig 1) The square wave signal should be smooth and without any distortion.
Fig.2,3) The CMPS falling(rising) edge is coincided with 3rd~5th tooth of the CKP from one longer signal(missing tooth)

22140_HYUN_G0276

Fig. 142 Illustration using the waveform synchronization to check the oil control solenoid as it relates to camshaft and crankshaft timing

5. Remove the sensor from its mounting.

To install:

6. Installation is the reverse of the removal procedure.
7. Replace any lost transmission fluid.

TESTING

See Figures 143 and 144.

1. Before servicing the vehicle, refer to the Precautions Section.
2. Many malfunctions in the electrical system are caused by poor harness and terminal connections. Faults can also be caused by interference from other electrical systems and mechanical or chemical damage.
3. Thoroughly check all connectors (and connections) for looseness, bending, corrosion, contamination, deterioration, or damage.
4. Make repairs as necessary.
5. Turn the ignition OFF.
6. Disconnect the VSS connector.
7. Start the engine.
8. Measure the voltage between terminal 1 of the VSS harness connector and a chassis ground. Specification: Battery voltage (approximately 12 volts).

   a. If the measured voltage is within specifications, inspect the ground circuit.

   b. If the measured voltage is not within specifications, check for an open/short to ground in the power harness.

9. Repair as necessary.
10. Inspect the ground circuit.

   a. Turn the ignition OFF.

   b. Disconnect VSS connector.

   c. Measure the resistance between terminal 2 of the VSS harness connector and the chassis ground. The specification: Approx. 0 ohms.

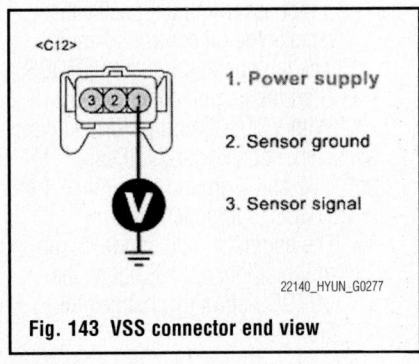

1. Power supply
2. Sensor ground
3. Sensor signal

22140_HYUN_G0277

Fig. 143 VSS connector end view

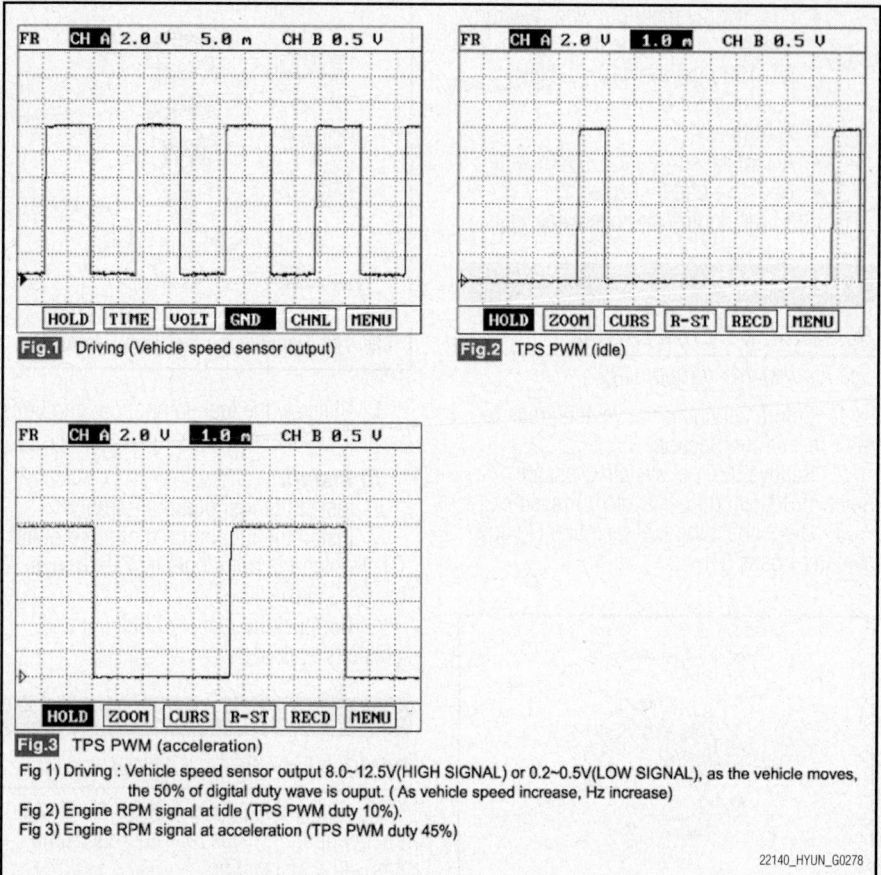

Fig.1 Driving (Vehicle speed sensor output)

Fig.2 TPS PWM (idle)

Fig.3 TPS PWM (acceleration)

Fig 1) Driving : Vehicle speed sensor output 8.0~12.5V(HIGH SIGNAL) or 0.2~0.5V(LOW SIGNAL), as the vehicle moves, the 50% of digital duty wave is ouput. (As vehicle speed increase, Hz increase)
Fig 2) Engine RPM signal at idle (TPS PWM duty 10%).
Fig 3) Engine RPM signal at acceleration (TPS PWM duty 45%)

22140_HYUN_G0278

Fig. 144 Waveform illustration of output signal

d. If the measured resistance is within specifications, inspect the signal circuit inspection.

e. If the measured resistance is not within specifications, check for an open/short to power in the ground harness. Repair as necessary.

11. Inspect the signal circuit

a. Turn the ignition **OFF**.

b. Connect a scan tool to the Data Link Connector (DLC).

c. Start the engine and select "SCOPEMETER FUNCTION" on the scan tool.

d. Drive the vehicle and measure the output signal between terminal 3 of the EPS CM harness connector and chassis ground. Specification: see waveform illustration.

e. If the VSS output signal is within specifications, substitute with a known-good EPS CM and check for proper operation.

f. If the problem is corrected, replace the EPS CM.

g. If the VSS output signal is not within specifications, check for an open/short to ground in the signal circuit and other systems which use the VSS. Repair as necessary.

h. If the VSS is found faulty, replace with a new component.

FUEL SYSTEMS

FUEL

GASOLINE FUEL INJECTION SYSTEM

FUEL SYSTEM SERVICE PRECAUTIONS

Safety is the most important factor when performing, not only fuel system maintenance, but any type of maintenance. Failure to conduct maintenance and repairs in a safe manner may result in serious personal injury or death. Maintenance and testing of the vehicle's fuel system components can be accomplished safely and effectively by adhering to the following rules and guidelines.

• To avoid the possibility of fire and personal injury, always disconnect the negative battery cable unless the repair or test procedure requires that battery voltage be applied.

• Always relieve the fuel system pressure prior to disconnecting any fuel system component (injector, fuel rail, pressure regulator, etc.), fitting, or fuel line connection. Exercise extreme caution whenever relieving fuel system pressure to avoid exposing

skin, face, and eyes to fuel spray. Please be advised that fuel under pressure may penetrate the skin or any part of the body that it contacts.

• Always place a shop towel or cloth around the fitting or connection prior to loosening to absorb any excess fuel due to spillage. Ensure that all fuel spillage (should it occur) is quickly removed from engine surfaces. Ensure that all fuel soaked cloths or towels are deposited into a suitable waste container.

• Always keep a dry chemical (Class B) fire extinguisher near the work area.

• Do not allow fuel spray or fuel vapors to come into contact with a spark or an open flame.

• Always use a back-up wrench when loosening and tightening fuel line connection fittings. This will prevent unnecessary stress and torsion to fuel line piping.

• Always replace worn fuel fitting O-rings with new. Do not substitute fuel

hose or equivalent where fuel pipe is installed.

Before servicing the vehicle, make sure to refer to the precautions in the beginning of this section as well.

RELIEVING FUEL SYSTEM PRESSURE

See Figures 145 and 146.

1. Before servicing the vehicle, refer to the Precautions Section.

2. Remove the second seats.

3. Remove the access panel (A).

4. Disconnect the fuel pump module connector (A).

5. Start the engine and allow it to run until it stalls.

6. Turn the ignition switch to the **OFF** position.

7. Disconnect the negative battery cable.

8. Attach the fuel pump module connector.

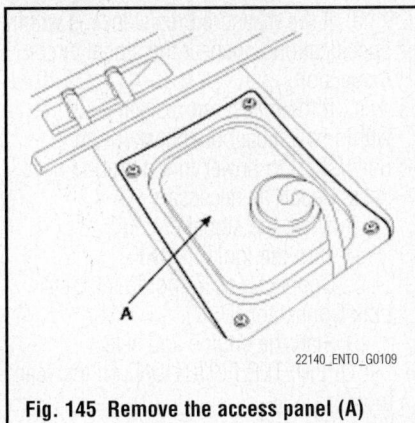

Fig. 145 Remove the access panel (A)

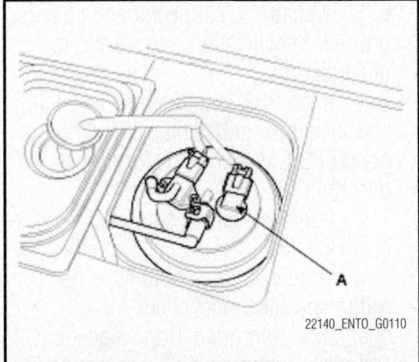

Fig. 146 Disconnect the fuel pump module connector (A)

FUEL FILTER

REMOVAL & INSTALLATION

The fuel delivery system integrates the fuel filter with the in-tank fuel pump. To service this filter, remove the fuel pump. Refer to Fuel Pump, removal & installation.

FUEL INJECTORS

REMOVAL & INSTALLATION

1. Before servicing the vehicle, refer to the Precautions Section.
2. Relieve the fuel system pressure. Refer to Relieving Fuel System Pressure.
3. Remove or disconnect the following:
   - Negative battery cable
   - Air intake surge tank, if necessary
   - Fuel lines
   - Fuel injector connectors
   - Fuel supply manifold with injectors attached
4. Separate the injectors from the supply manifold.

To install:
5. Install or connect the following:
   - Injectors to the fuel supply manifold using new O-rings

- Fuel supply manifold with injectors attached and torque the bolts to 84–132 inch lbs. (10–15 Nm)
- Fuel injector connectors
- Fuel lines
- Air intake surge tank, if removed
- Negative battery cable
6. Start the engine and check for leaks.

FUEL PUMP

REMOVAL & INSTALLATION

See Figures 147 through 149.

1. Before servicing the vehicle, refer to the Precautions Section.
2. Relieve the fuel system pressure. Refer to Relieving Fuel System Pressure.
3. Disconnect the fuel feed line (A) and canister hoses (B).

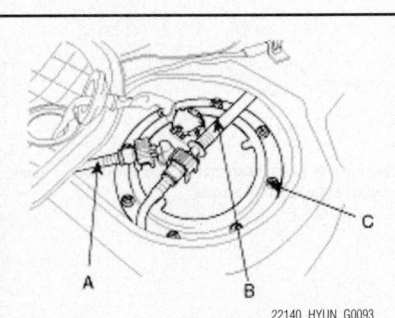

Fig. 147 Disconnect the fuel feed line (A) and canister hoses (B). Unscrew the fuel pump mounting bolts (C)

4. Unscrew the fuel pump mounting bolts (C) and remove the fuel pump assembly.

To install:
5. Install the fuel pump assembly.
6. Install the fuel pump mounting bolts (C). Tighten the bolts/nuts to 17–26 inch lbs. (2–3 Nm).
7. Connect the fuel feed line (A) and canister hoses (B).

FUEL PRESSURE REGULATOR

REMOVAL & INSTALLATION

The fuel pressure regulator is built into the fuel pump. Please refer to Fuel Pump, removal & installation.

FUEL TANK

REMOVAL & INSTALLATION

See Figures 150 through 153.

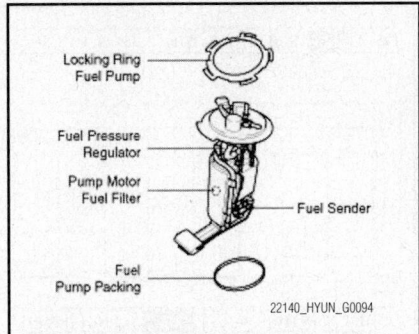

Fig. 148 Remove the fuel pump assembly

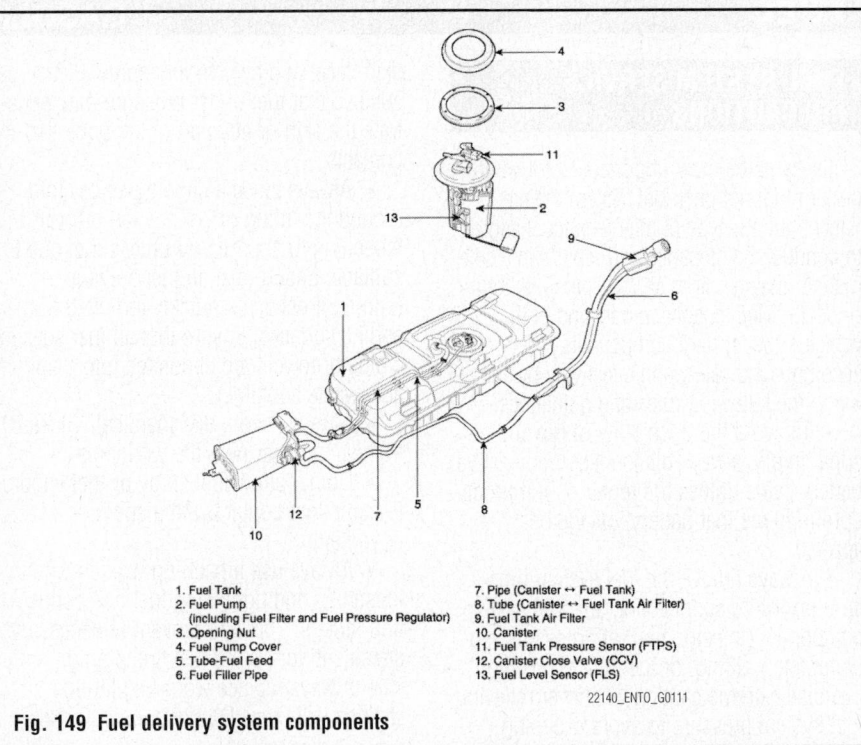

1. Fuel Tank
2. Fuel Pump (including Fuel Filter and Fuel Pressure Regulator)
3. Opening Nut
4. Fuel Pump Cover
5. Tube-Fuel Feed
6. Fuel Filler Pipe
7. Pipe (Canister ↔ Fuel Tank)
8. Tube (Canister ↔ Fuel Tank Air Filter)
9. Fuel Tank Air Filter
10. Canister
11. Fuel Tank Pressure Sensor (FTPS)
12. Canister Close Valve (CCV)
13. Fuel Level Sensor (FLS)

Fig. 149 Fuel delivery system components

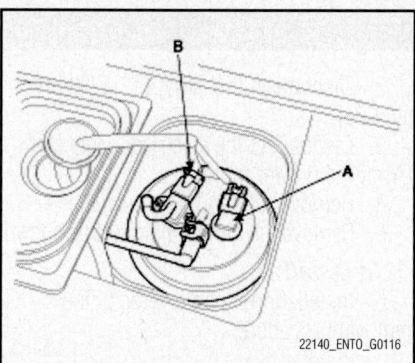

Fig. 150 Disconnect the fuel pump connector (A) and the fuel tank pressure sensor connector (B)

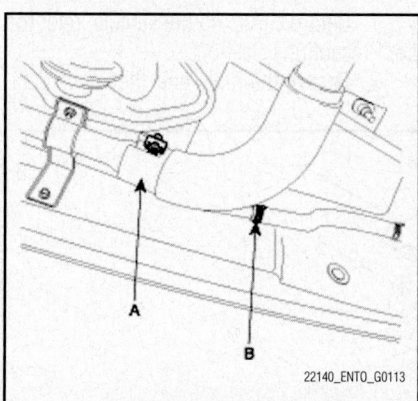

Fig. 151 Disconnect the fuel filler hose (A) and the leveling hose (B)

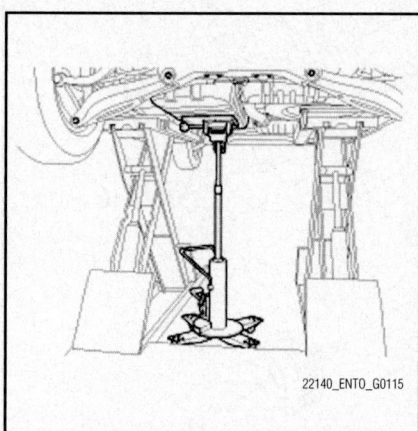

Fig. 152 Support the fuel tank with a jack

1. Before servicing the vehicle, refer to the Precautions Section.
2. Relieve the fuel system pressure. Refer to Relieving Fuel System Pressure.
3. Disconnect the fuel pump connector (A) and the fuel tank pressure sensor connector (B).
4. Raise and safely support the vehicle.

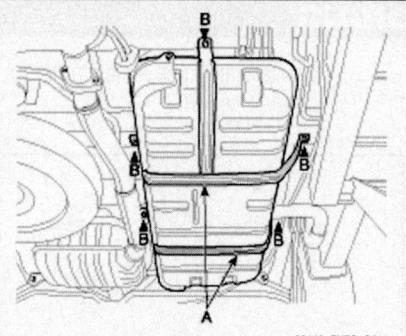

Fig. 153 Remove the fuel tank band (A), mounting bolts (B), and remove the fuel tank from the vehicle

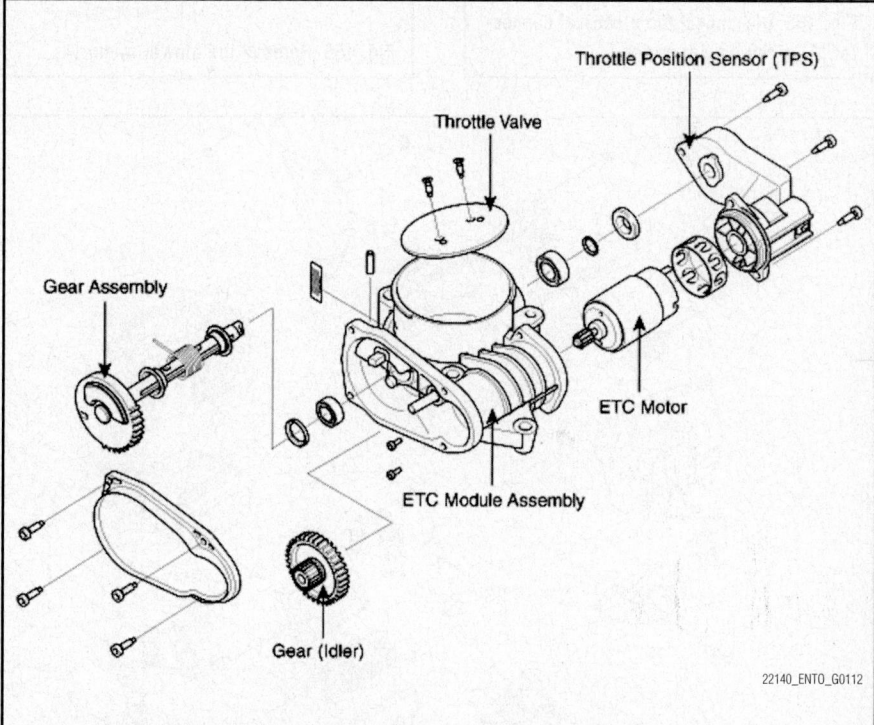

Fig. 154 Exploded view of the Electronic Throttle Control (ETC) module

5. Disconnect the fuel filler hose (A) and the leveling hose (B).
6. Disconnect the fuel feed quick connector near the canister.
7. Support the fuel tank with a suitable jack.
8. Remove the fuel tank band (A) by unscrewing the mounting bolts (B).
9. Remove the fuel tank from the vehicle.

To install:
10. Installation is the reverse of the removal procedure.
11. Tighten the fuel tank band bolts (B) to: 29–40 ft. lbs. (39–54 Nm).

IDLE SPEED

Idle speed is maintained by the Powertrain Control Module (PCM). No adjustment is necessary or possible.

THROTTLE BODY

REMOVAL & INSTALLATION

See Figure 154.

The throttle body may also be referred to as the Electronic Throttle Control (ETC) module.
1. Before servicing the vehicle, refer to the Precautions Section.
2. Turn the ignition OFF.
3. Remove the engine cover.
4. Remove the throttle body electrical connector.
5. Remove the throttle body bolts.
6. Remove the throttle body and gasket.

To install:
7. Clean the throttle body gasket mating surfaces.
8. Install the throttle body and NEW gasket.
9. Install the throttle body bolts and tighten to 21 ft. lbs. (28 Nm).
10. Install the throttle body electrical connector.
11. Install the engine cover.

HEATING & AIR CONDITIONING

BLOWER MOTOR

REMOVAL & INSTALLATION

See Figures 155 and 156.

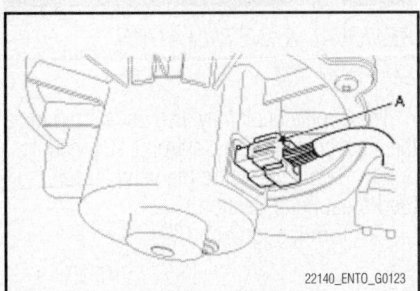

Fig. 155 Disconnect the electrical connector (A) of the blower motor

1. Before servicing the vehicle, refer to the Precautions Section.

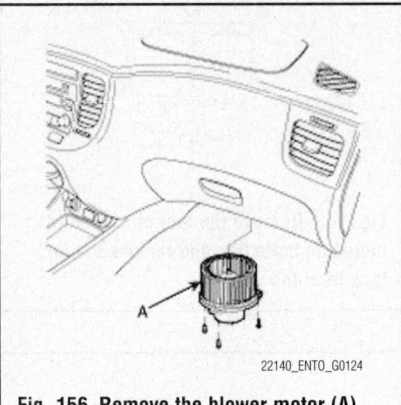

Fig. 156 Remove the blower motor (A)

2. Disconnect the negative (-) battery terminal.

3. Disconnect the electrical connector (A) of the blower motor.

4. Remove the mounting screws.

5. Remove the blower motor (A) as shown.

To install:

6. Installation is the reverse of the removal procedure.

HEATER CORE

REMOVAL & INSTALLATION

See Figures 157 and 158.

1. Before servicing the vehicle, refer to the Precautions Section.

2. Record the radio anti-theft code data.

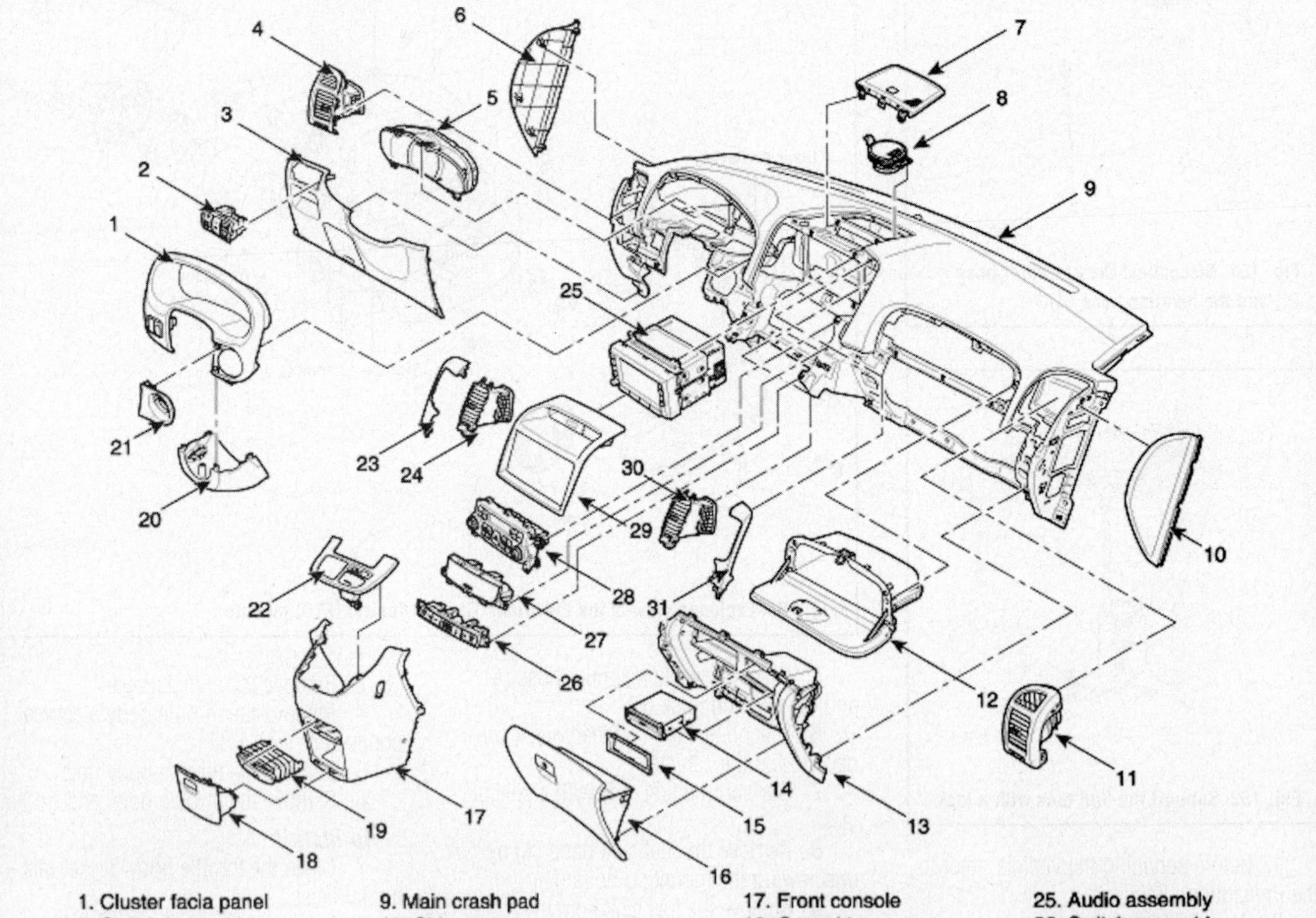

1. Cluster facia panel	9. Main crash pad
2. Switch assembly	10. Side cover
3. Lower panel	11. Side air vent
4. Side air vent	12. Multi box
5. Cluster	13. Lower crash pad panel
6. Side cover	14. DVD
7. Center speaker cover	15. DVD cover
8. Center speaker	16. Glove box

17. Front console	25. Audio assembly
18. Consol tray	26. Switch assembly
19. Cup holders	27. Center tray
20. Shroud	28. Heater control unit
21. Key box cover	29. Center facia panel
22. Console upper cover	30. Center air vent
23. Center garnish	31. Center garnish
24. Center air vent	

Fig. 157 Instrument panel exploded view

3. Disconnect the negative battery cable.

4. Drain the cooling system.

5. Discharge the air conditioning system.

6. Disconnect and plug the heater hoses.

7. Remove the front seat.

8. Tilt the steering column down.

9. Remove the screws and detach the clips from the cluster fascia panel.

10. Disconnect the connector and remove the cluster fascia panel.

11. Remove the screws and detach the clips from the center fascia panel.

12. Disconnect the connector and remove the center fascia panel.

13. Remove the radio retaining screws.

14. Disconnect the connectors and remove the radio.

15. Disconnect the damper from the glovebox lid.

16. Remove the glovebox lid from the lift.

17. Disconnect the retaining pins and remove the glovebox assembly.

18. Remove the dash pad side cover, center cover, and under cover.

19. Remove the front A-pillar trim.

20. Remove the photo sensor.

21. Remove the speaker connector.

22. Disconnect the passenger's air bag connector.

23. Loosen the bolt and nut. Remove the dash pad.

24. Disconnect the electrical connectors from the cross bar assembly.

25. Loosen the bolts and nuts that retain the assembly in place.

26. Remove the cross bar assembly.

27. Disconnect the connectors from the temperature control actuator, the mode control actuator, and the evaporator temperature sensor.

28. Remove the heater/blower unit from the vehicle, after removing the 3 retaining screws.

29. Separate the blower unit from the heater unit, after removing the 2 retaining screws.

30. Remove the heater core cover.

31. Remove the heater core from its mounting.

To install:

32. Install the heater core in the heater unit.

33. Attach the heater unit to the blower housing.

34. Position the assembly in the vehicle. Attach the retaining screws.

35. Continue the installation in the reverse order of the removal procedure.

36. Make sure that you reprogram the anti-theft code for the radio.

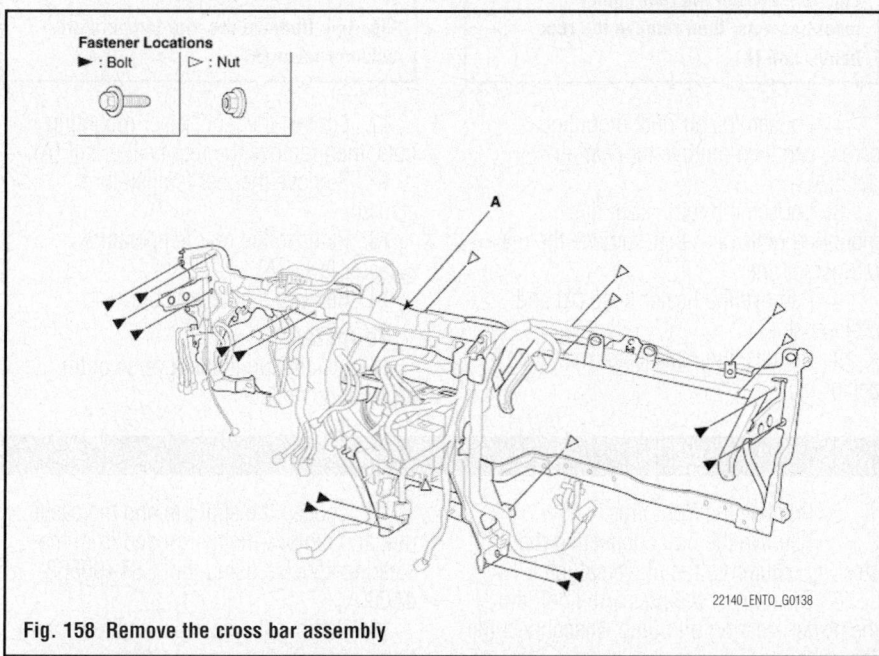

Fig. 158 Remove the cross bar assembly

AUXILIARY HEATING & AIR CONDITIONING SYSTEM

BLOWER MOTOR

REMOVAL & INSTALLATION
See Figure 159.

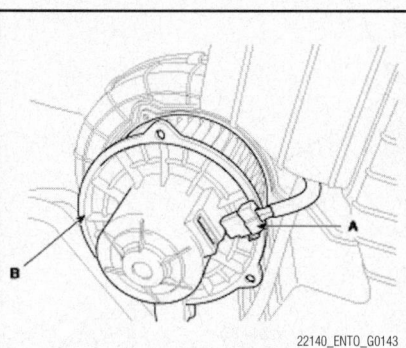

Fig. 159 Detach the rear blower motor connector (A), then remove the screws and the blower motor (B)

1. Before servicing the vehicle, refer to the Precautions Section.

2. Disconnect the negative (-) battery terminal.

3. Detach the connector (A) from the blower motor.

4. Remove the mounting screws, then remove the blower motor (B).

5. Installation is the reverse of the removal procedure.

HEATER CORE

REMOVAL & INSTALLATION
See Figures 160 through 164.

1. Before servicing the vehicle, refer to the Precautions Section.

2. Disconnect the negative (-) battery terminal.

3. Recover the refrigerant with a recover/recycling/charging station.

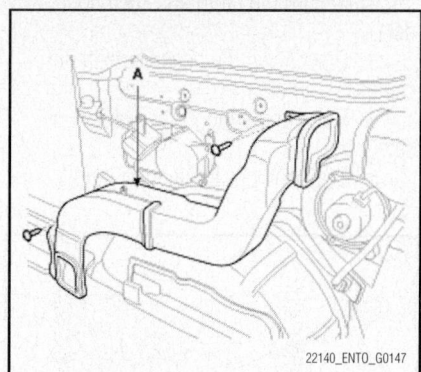

Fig. 160 Loosen the air duct mounting screw, and then remove the rear air duct (A)

4. When the engine is cool, drain the engine coolant from the radiator.

5. Remove the luggage side trim.

6. Remove the rear speaker.

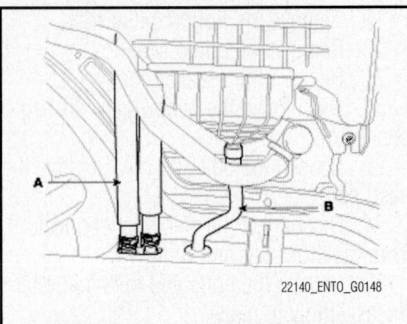

Fig. 161 Remove the heater hose (A) and drain hose (B)

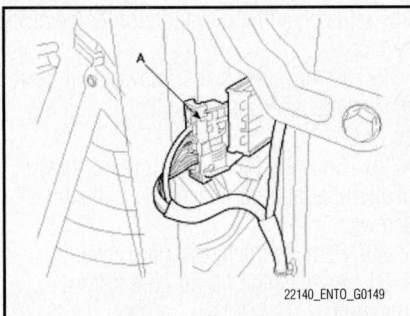

Fig. 162 Remove the rear heater main connector (A)

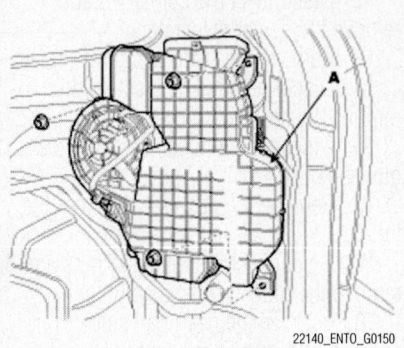

Fig. 163 Loosen the rear heater mounting nuts, then remove the rear heater unit (A)

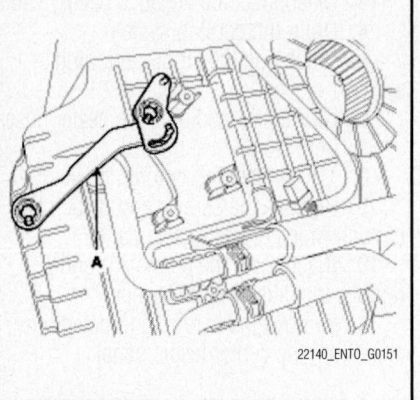

Fig. 164 Remove the rear temperature actuator lever (A)

7. Loosen the air duct mounting screw, and then remove the rear air duct (A).

8. Loosen the refrigerant line mounting bolts, and then remove the rear refrigerant line.

9. Remove the heater hose (A) and drain hose (B).

10. Remove the rear heater main connector (A).

11. Loosen the rear heater mounting nuts, then remove the rear heater unit (A).

12. Remove the rear temperature actuator.

13. Remove the rear temperature actuator lever (A).

14. Remove the heater core.

To install:

15. Installation is the reverse of the removal procedure.

STEERING

POWER RACK & PINION STEERING GEAR

REMOVAL & INSTALLATION

See Figures 165 through 167.

1. Before servicing the vehicle, refer to the Precautions Section.

2. Record the radio anti-theft code data.

3. Disconnect the negative battery cable.

4. Raise and support the vehicle safely.

5. Drain the power steering fluid.

6. Remove the front tires.

7. Remove the bolt connecting the steering column to the universal joint.

8. Disconnect the pressure line from the power steering oil pump. Disconnect the return hose.

9. Loosen the split pin and the castle nut, and remove the tie rod end from the steering knuckle using tool SST (09568-4A000).

10. Remove the split pin and lower arm bolts and nut.

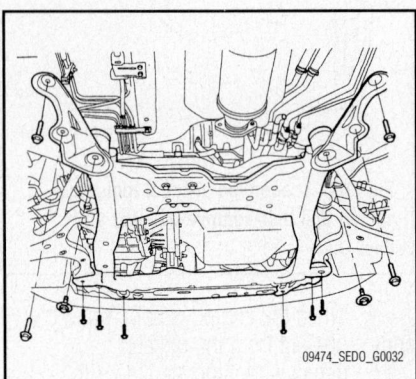

Fig. 165 Subframe attaching bolts

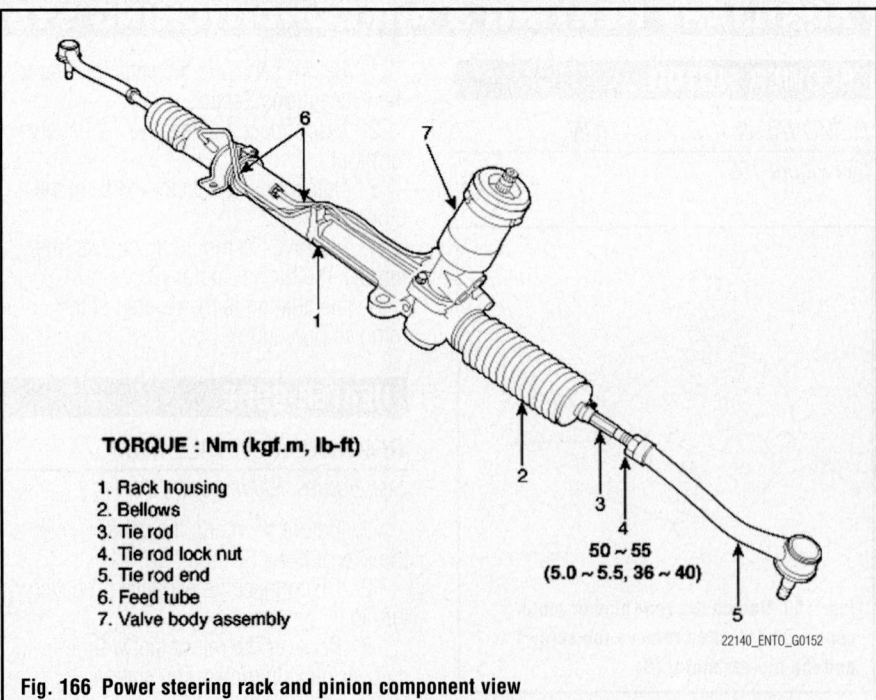

TORQUE : Nm (kgf.m, lb-ft)

1. Rack housing
2. Bellows
3. Tie rod
4. Tie rod lock nut
5. Tie rod end
6. Feed tube
7. Valve body assembly

50 ~ 55
(5.0 ~ 5.5, 36 ~ 40)

Fig. 166 Power steering rack and pinion component view

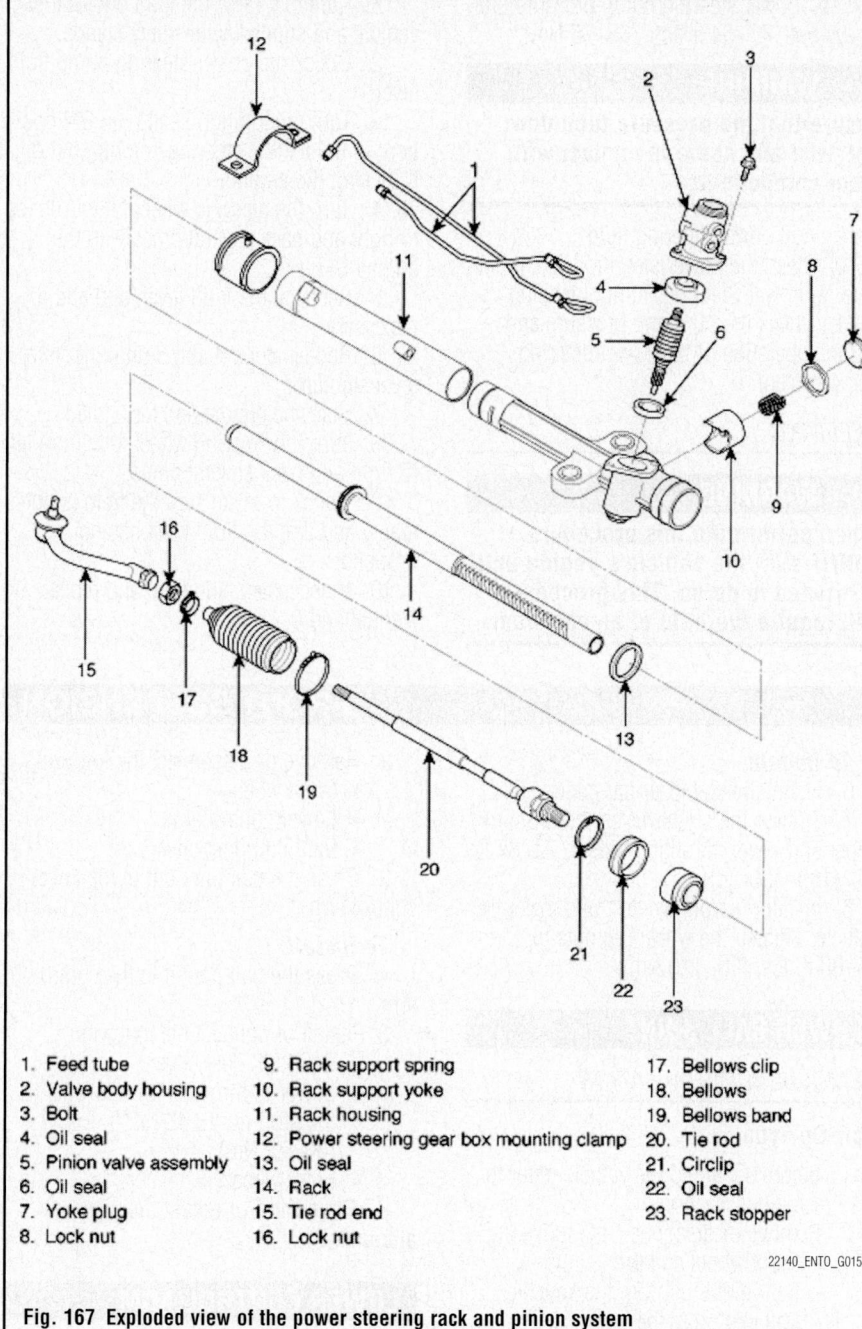

Fig. 167 Exploded view of the power steering rack and pinion system

1. Feed tube
2. Valve body housing
3. Bolt
4. Oil seal
5. Pinion valve assembly
6. Oil seal
7. Yoke plug
8. Lock nut
9. Rack support spring
10. Rack support yoke
11. Rack housing
12. Power steering gear box mounting clamp
13. Oil seal
14. Rack
15. Tie rod end
16. Lock nut
17. Bellows clip
18. Bellows
19. Bellows band
20. Tie rod
21. Circlip
22. Oil seal
23. Rack stopper

22140_ENTO_G0153

a. Remove the fuel pump fuse.

b. Start the engine and allow it to stall.

c. While operating the starting motor intermittently (not more than 15 seconds at a time), turn the steering wheel all the way to the left and then all the way to the right. Repeat the procedure 5–6 times. Do not hold the steering wheel in the full turn position for more than 10 seconds.

d. Check the fluid level to be sure it does not fall below the lower position on the filter.

→**Be sure to follow the procedure outlined for bleeding the system of air. If bleeding is done at engine idle, air will be broken up and absorbed into the fluid.**

23. Check the wheel alignment. Adjust as needed.

POWER STEERING PUMP

REMOVAL & INSTALLATION

See Figures 168 and 169.

1. Before servicing the vehicle, refer to the Precautions Section.

2. Disconnect the pressure tube (A) from the oil pump by loosening the eye bolt.

3. Disconnect the suction hose (B) from the suction pipe.

4. Remove the accessory drive belt. Refer to Accessory Drive Belts, removal & installation.

5. Loosen the power steering pump mounting bolt and nut.

6. Remove the power steering pump assembly (A) from the pump bracket.

→**Be careful not to spill fluid from the power steering oil pump.**

To install:

7. Installation is the reverse of the removal procedure.

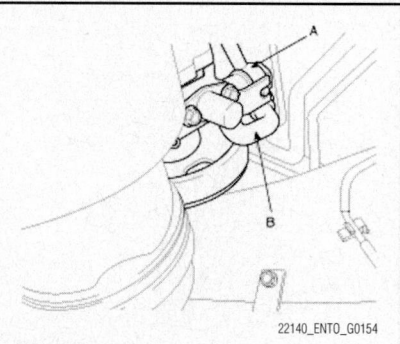

Fig. 168 Disconnect the pressure line (A) and suction hose (B) from the power steering pump

22140_ENTO_G0154

11. Disconnect the stabilizer control link from the strut assembly. Repeat on the other side of the vehicle.

12. Remove the front and rear roll stopper bolts and nuts.

13. Remove the subframe retaining bolts and nuts.

14. Remove the rear roll stopper from the subframe.

15. Disconnect the pressure line and the return line from the steering gear valve body housing.

16. Remove the steering gear assembly from the subframe, after removing the retaining bolts.

To install:

17. Installation is the reverse of the removal procedure.

18. Torque the steering gear to subframe mounting bolts to 65–80 ft. lbs. (90–110 Nm).

19. Torque the split pin and lower arm bolt and nut to 65–87 ft. lbs. (90–110 Nm).

20. Torque the bolt connecting the steering column to the universal joint to 108–156 inch lbs. (13–18 Nm).

21. Fill the power steering pump with the proper grade and type of fluid.

22. To bleed the system of air:

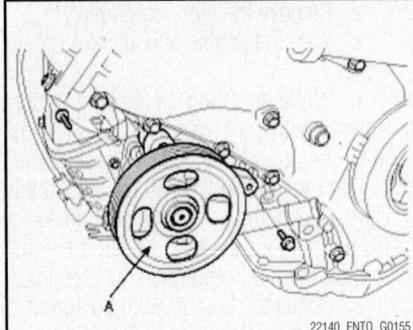

Fig. 169 After loosening the mounting bolt and nut, remove the power steering pump (A) from the bracket

8. Installation is the reverse of the removal procedure. Tighten the retainers as follows:

   a. Power steering pump bracket mounting bolt and nut: 25–36 ft. lbs (35–50 Nm).

   b. Power steering pump pressure line eye bolt: 47–54 ft. lbs. (65–75 Nm).

✳✳ WARNING

Ensure that the pressure tube does not twist and come in contact with other components.

9. Add power steering fluid
10. Bleed the power steering system. Refer to Power Steering Pump, bleeding.
11. Check the oil pump pressure and test the operation of the power steering pump system.

BLEEDING

✳✳ WARNING

When performing this procedure, doNOT start the vehicle's engine until instructed to do so. This procedure will require the help of an assistant.

1. Carefully raise the front end of the vehicle and support with safety stands.
2. Check the power steering pump fluid level.
3. Turn the ignition key to the **ON** position, to unlock the steering column, but do NOT start the engine.
4. Turn the steering wheel fully from left to right and back several times with the engine OFF.
5. Recheck the fluid level, and add as necessary.
6. Repeat steps 4 and 5 until the fluid level stabilizes.
7. Start the engine and let it idle.
8. Turn the steering wheel fully from left to right and back several times.
9. Check to make sure the fluid is not foamy and that the fluid level has not dropped.
10. If necessary, add fluid and repeat steps 8 and 9.

SUSPENSION

CONTROL LINKS

REMOVAL & INSTALLATION

See Figure 170.

1. Before servicing the vehicle, refer to the Precautions Section.
2. Raise and safely support the vehicle.
3. Remove the front wheel and tire from the front hub.

✳✳ WARNING

Be careful not to damage the hub bolts when removing the front wheel and tire.

4. Remove the stabilizer bar control link (A) by removing the upper and lower mounting nuts (B).
5. Remove the stabilizer bar control link.

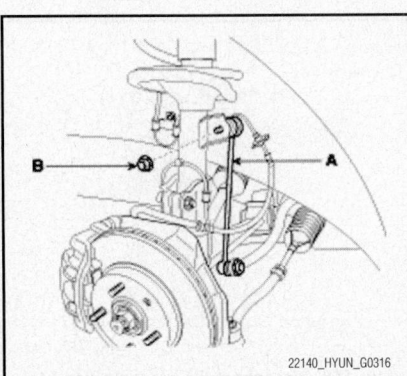

Fig. 170 Stabilizer control link and mounting nuts shown

To install:

6. Install the stabilizer bar control link.
7. Tighten the stabilizer bar control link upper and lower mounting nuts to 72–87 ft. lbs. (100–120 Nm).
8. Install the front wheels, and lower the vehicle. Tighten the wheel lug nuts to 65–80 ft. lbs. (90–110 Nm).

LOWER BALL JOINT

REMOVAL & INSTALLATION

Bolt-On Type

1. Before servicing the vehicle, refer to the Precautions Section.
2. Remove or disconnect the following:
   - Front wheel and tire
   - Ball joint stud from the knuckle
   - Ball joint from the lower control arm

To install:

➡**Use a new split pin for assembly.**

3. Install or connect the following:
   - Ball joint and torque the mounting bolts to 69–87 ft. lbs. (95–120 Nm)
   - Stud nut and torque to 43–52 ft. lbs. (60–72 Nm)
   - Front wheel
4. Check and/or adjust the wheel alignment.

Press-In Type

1. Before servicing the vehicle, refer to the Precautions Section.

FRONT SUSPENSION

2. Remove or disconnect the following:
   - Front wheel
   - Lower control arm
   - Ball joint dust cover
3. Press the ball joint out of the lower control arm.

To install:

4. Press the ball joint into the control arm.
5. Install or connect the following:
   - Ball joint dust cover
   - Lower control arm and torque the stud nut to 43–52 ft. lbs. (60–72 Nm)
   - Front wheel
6. Check and/or adjust the wheel alignment.

LOWER CONTROL ARM

REMOVAL & INSTALLATION

See Figures 171 and 172.

1. Before servicing the vehicle, refer to the Precautions Section.
2. Raise and safely support the vehicle.
3. Remove the front wheel and tire from the front hub.

✳✳ WARNING

Be careful not to damage the hub bolts when removing the wheel and tire.

4. Remove the lower arm (A) mounting bolt (B) from the knuckle.

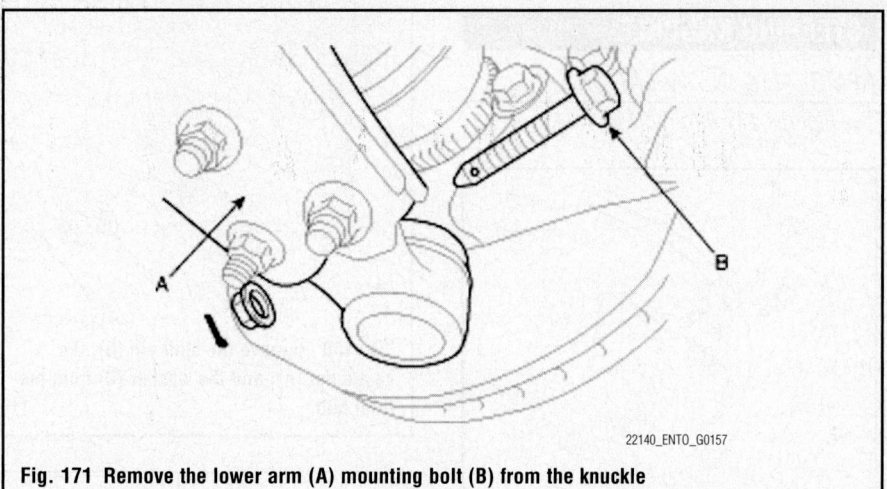

Fig. 171 Remove the lower arm (A) mounting bolt (B) from the knuckle

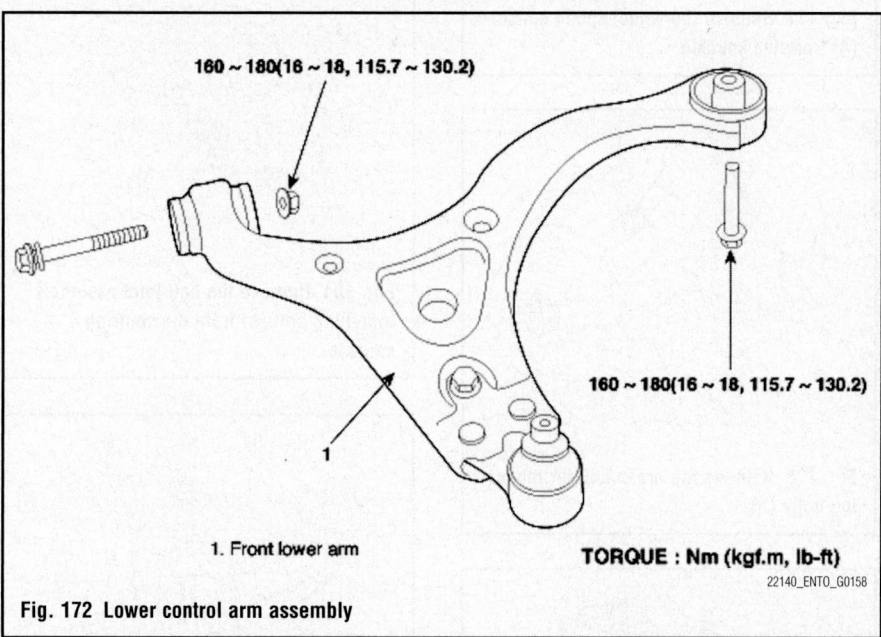

160 ~ 180(16 ~ 18, 115.7 ~ 130.2)

160 ~ 180(16 ~ 18, 115.7 ~ 130.2)

1. Front lower arm

TORQUE : Nm (kgf.m, lb-ft)

Fig. 172 Lower control arm assembly

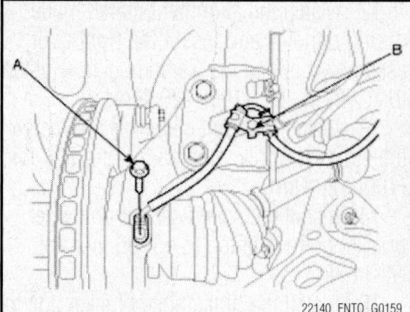

Fig. 173 Remove the speed sensor (A) and wire bracket bolts (B) from the front steering knuckle

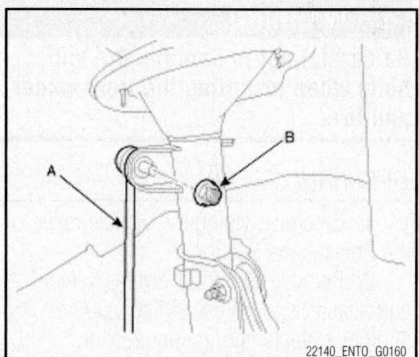

Fig. 174 Remove the front stabilizer control link (A) and nut (B) from the strut

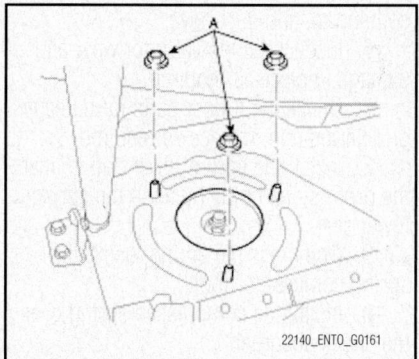

Fig. 175 Remove the upper strut mounting nuts (A)

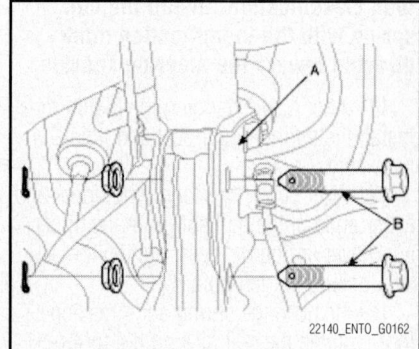

Fig. 176 Remove the strut assembly (A) and bolts (B) from its mounting

5. Remove the lower arm mounting bolts.

6. Remove the lower arm from the vehicle.

To install:

7. Installation is the reverse of the removal procedure.

8. Torque the lower arm mounting bolts to 116–130 ft. lbs. (160–180 Nm).

9. Torque the front lower arm ball joint mounting bolt to 65–87 ft. lbs. (90–120 Nm).

10. Check the front end alignment and adjust as necessary.

MACPHERSON STRUT

REMOVAL & INSTALLATION

See Figures 173 through 176.

1. Before servicing the vehicle, refer to the Precautions Section.

2. Raise and safely support the vehicle.

3. Remove the front wheels and tires.

4. Remove the brake hose bracket bolts from the strut assembly.

5. Remove the speed sensor (A) and wire bracket bolts (B) from the front steering knuckle.

6. Make an alignment marking on the camber adjusting bolt and strut for installation alignment approximation later.

7. Remove the front stabilizer control link (A) and nut (B) from the strut.

8. Remove the upper strut mounting nuts (A).

9. Remove the front strut mounting bolts from the knuckle.

10. Remove the strut assembly (A) and bolts (B) from its mounting in the steering knuckle.

To install:

11. Install the strut upper mounting nuts. Tighten to 33–43 ft. lbs. (45–60 Nm).

12. Match the alignment marks made during removal and install the front strut assembly bolts to the front knuckle. Tighten to 72–87 ft. lbs. (100–120 Nm).

13. Install the front stabilizer link nut to the strut assembly. Tighten to 72–87 ft. lbs. (100–120 Nm).

14. Install the speed sensor and wire bracket bolts. Tighten to 60–96 inch lbs. (7–11 Nm).

15. Install the brake hose bracket bolt to the axle assembly.

16. Install the wheel and the tire to the front hub. Tighten to 65–80 ft. lbs. (90–110 Nm).

✳✳ WARNING

Be careful not to damage the hub bolts when installing the front wheel and tire.

OVERHAUL

1. Before servicing the vehicle, refer to the Precautions Section.

2. Remove the strut from the vehicle and attach Service Tool 09546-2600 or another suitable spring compressor.

3. Compress the coil spring.

4. Remove the self-locking nut from the strut.

5. Remove the insulator, spring seat, coil spring, and dust cover.

6. Inspect the insulator for wear and damage, replace as required.

7. Check the rubber parts for damage or deterioration, replace as required.

8. Install the spring lower pad so that the protrusions fit the holes in the spring lower seat.

9. Compress the spring, using the spring compressor tool.

10. Install the compressed spring over the shock absorber.

➡There are 2 color identification marks on the coil spring, one indicates model option and the other indicates load classification. Install the coil spring with the identification mark directed toward the steering knuckle.

11. After fully extending the piston rod, install the spring upper seat and insulator assembly.

12. After correctly seating the upper and lower ends of the coil spring in the upper and lower spring grooves, tighten the new self locking nut temporarily.

13. Remove the spring compression tool. Tighten the self locking nut to 43–51 ft. lbs. (60–70 Nm).

14. Install the strut to the vehicle.

STEERING KNUCKLE

REMOVAL & INSTALLATION

See Figures 177 through 191.

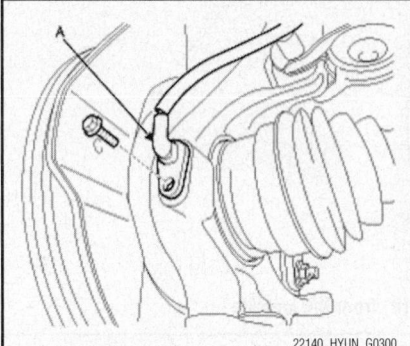

Fig. 177 Remove the wheel speed sensor (A) from the knuckle

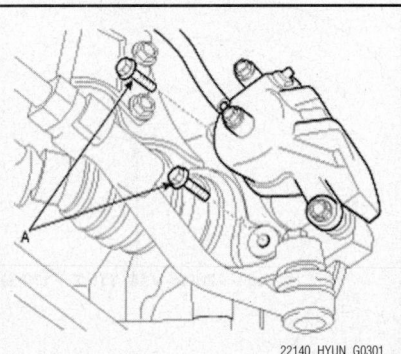

Fig. 178 Remove the brake caliper mounting bolts (A)

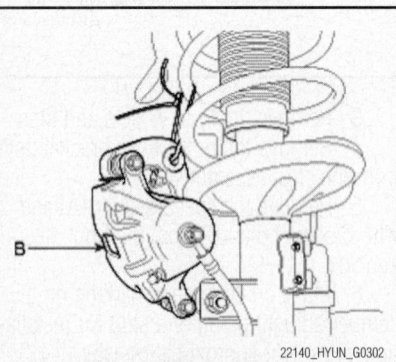

Fig. 179 Hang the brake caliper assembly (B) with a wire

1. Before servicing the vehicle, refer to the Precautions Section.

2. Raise and safely support the vehicle.

3. Remove the front wheel and tire.

✳✳ WARNING

Be careful not to damage the hub bolts when removing the front wheel and tire.

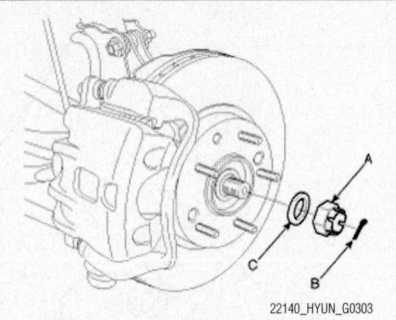

Fig. 180 Remove the split pin (B), the castle nut (A), and the washer (C) from the front hub

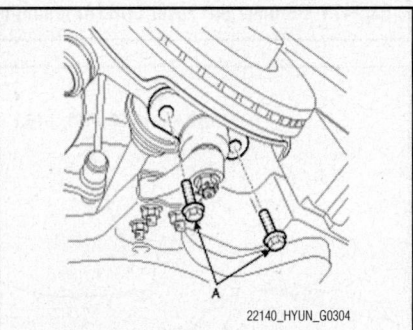

Fig. 181 Remove the ball joint assembly mounting bolt (A) from the steering knuckle

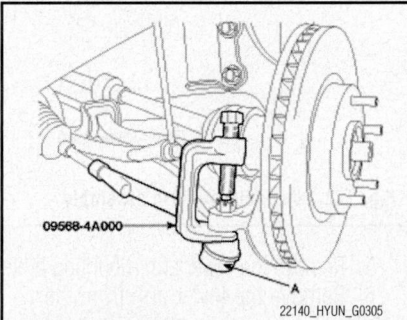

Fig. 182 Disconnect the ball joint (A) from the steering knuckle using the special tool (09568-4A000)

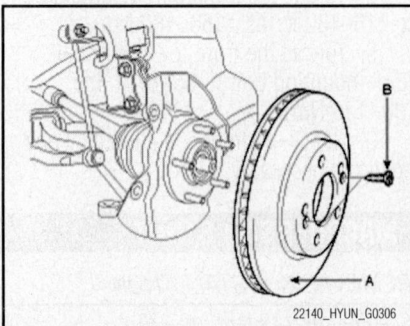

Fig. 183 Remove the brake disc (A) from the front hub assembly after removing the screws (B)

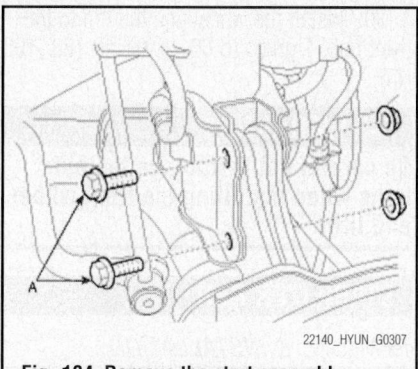

Fig. 184 Remove the strut assembly mounting bolts (A) and nuts

4. Remove the wheel speed sensor (A) from the knuckle.

5. Remove the brake caliper mounting bolts (A), and then hang the brake caliper assembly (B) with a wire.

✳✳ WARNING

Do not suspend the brake caliper assembly from the brake hose or damage may occur to the hose.

6. Remove the split pin (B), then remove castle nut (A) and washer (C) from the front hub.

7. Remove the ball joint assembly mounting bolt (A) from the steering knuckle.

8. Remove the tie rod end ball joint from the knuckle.

a. Remove the split pin and castle nut.

b. Disconnect the ball joint (A) from the steering knuckle using the special tool (09568-4A000).

➡Apply a few drops of oil to the special tool at the boot contact end.

9. Remove the brake disc (A) from the front hub assembly after removing the screws (B).

10. Remove the strut assembly mounting bolts (A) and nuts.

11. Remove the hub and knuckle assembly.

✳✳ WARNING

Be careful not to damage the boot and rotor teeth.

➡**Disassemble the hub and steering knuckle assembly.**

12. Remove the snap ring (A).

13. Remove the hub assembly from the knuckle assembly.

a. Install the front knuckle assembly (A) on a press.

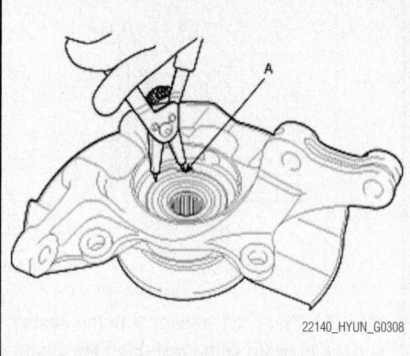

Fig. 185 Remove the snap ring (A)

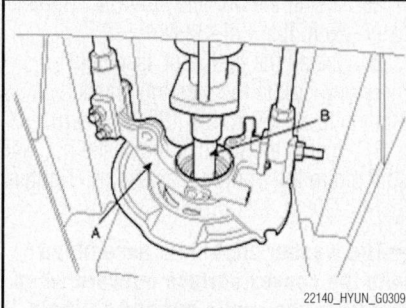

Fig. 186 Install the front knuckle assembly (A) on a press with an adapter (B) upon the hub assembly shaft

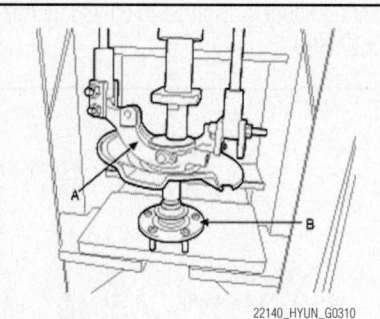

Fig. 187 Use a press to remove the hub assembly (B) from the knuckle assembly (A)

b. Lay a suitable adapter (B) upon the hub assembly shaft.

c. Remove the hub assembly (B) from the knuckle assembly (A) using a press.

14. Remove the hub bearing inner race from the hub assembly.

a. Install a suitable tool (A) for removing the hub bearing inner race on the hub assembly.

b. Lay the hub assembly and tool (A) upon a suitable adapter (B).

c. Lay a suitable adapter (C) upon the hub assembly shaft.

d. Remove the hub bearing inner race (D) from the hub assembly using a press.

15. Remove the hub bearing outer race from the knuckle assembly.

a. Lay the hub assembly (A) upon a suitable adapter (B).

b. Lay a suitable adapter (C) upon the hub bearing outer race.

c. Remove the hub bearing outer race from the steering knuckle assembly by using a press.

16. Replace hub bearing with a new one.

To install:

17. Preparation for installation:

a. Check the hub for cracks and the splines for wear.

b. Check the brake disc for scoring and damage.

c. Check the steering knuckle for cracks.

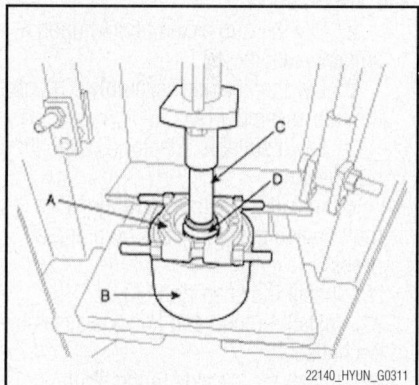

Fig. 188 Install tool (A) for removing the hub bearing inner race on the hub assembly. Use adapters (B) and (C) for the hub assembly shaft. Using a press, remove the hub bearing inner race (D) from the hub assembly

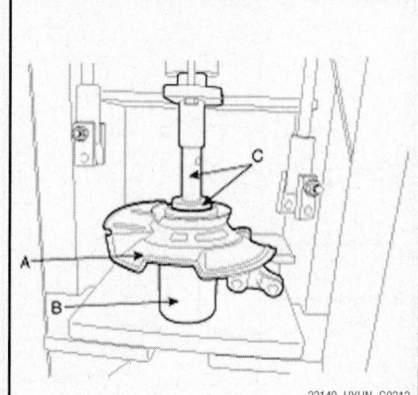

Fig. 189 Lay the hub assembly (A) on adapters (B) and (C) upon the hub bearing outer race. Using a press, remove the hub bearing outer race from the steering knuckle assembly

d. Check the bearing for cracks or damage.

18. Install the hub bearing to the knuckle assembly.

a. Lay the knuckle assembly on a press.

b. Lay a new hub bearing upon the steering knuckle assembly.

c. Lay a suitable adapter upon the hub bearing.

d. Install the hub bearing to the steering knuckle assembly by using a press.

✳✳ WARNING
Do not press against the inner race of the hub bearing or damage may occur to the bearing assembly.

19. Install the hub assembly to the knuckle assembly.

a. Lay the hub assembly (A) upon a suitable adapter (B).

b. Lay the knuckle assembly (C) upon the hub assembly (A).

c. Lay a suitable adapter (D) upon the hub bearing.

d. Install the hub assembly (A) to the knuckle assembly (C) by using a press.

20. Install the snap ring (A).

21. Install the hub and knuckle assembly to the halfshaft.

22. Install the knuckle to the strut assembly and tighten the mounting bolts and nuts. Tighten to 101–116 ft. lbs. (137–157 Nm).

23. Install the brake disc to the front hub assembly, and tighten the screws.

24. Install the tie rod end ball joint to the knuckle.

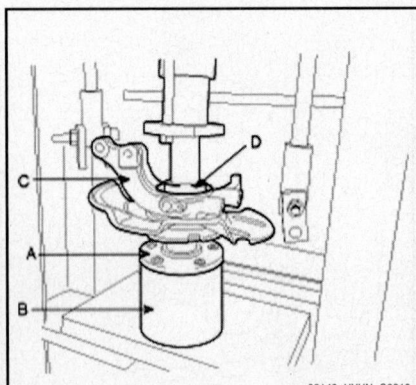

Fig. 190 Lay the hub assembly (A) upon adapter (B). Lay the knuckle assembly (C) upon the hub assembly (A). Place an adapter (D) upon the hub bearing. Using a press, install the hub assembly (A) to the knuckle assembly (C)

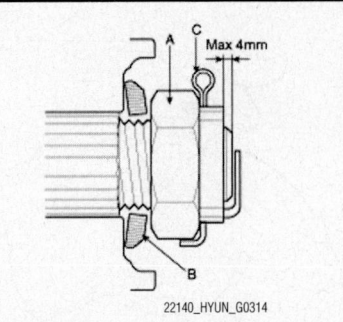

Fig. 191 Place the washer with the convex surface outward when installing the castle nut and split pin

25. Install the nut and split pin. Tighten to 17–25 ft. lbs. (24–33 Nm).

26. Install the ball joint assembly mounting bolt to the steering knuckle. Tighten to 72–87 ft. lbs. (98–118 Nm).

27. Install the washer, the castle nut, and split pin to the front hub assembly. Tighten to 145–203 ft. lbs. (196–275 Nm).

➡ **The washer should be assembled with the convex surface outward when installing the castle nut and split pin.**

28. Install the brake caliper and then tighten the mounting bolts. Tighten to 58–72 ft. lbs. (79–98 Nm).

29. Install the wheel speed sensor to the steering knuckle. Tighten to 61–86 inch lbs. (7–10 Nm).

30. Install the wheel and the tire to the front hub. Tighten to 65–80 ft. lbs. (88–108 Nm).

✳✳ WARNING
Be careful not to damage the hub bolts when installing the front wheel and tire.

STABILIZER BAR

REMOVAL & INSTALLATION
See Figure 192.

1. Before servicing the vehicle, refer to the Precautions Section.

2. Raise the front of the vehicle, and make sure it is securely supported.

3. Remove the front wheel and tire from the front hub.

✳✳ WARNING
Be careful not to damage the hub bolts when removing the front wheel and tire.

4. Remove the nut and stabilizer bar control link. Refer to Control Links, removal & installation.

5. Remove the control link on the opposite side in the same way.

6. Remove the rear mounting bolts of the sub-frame.

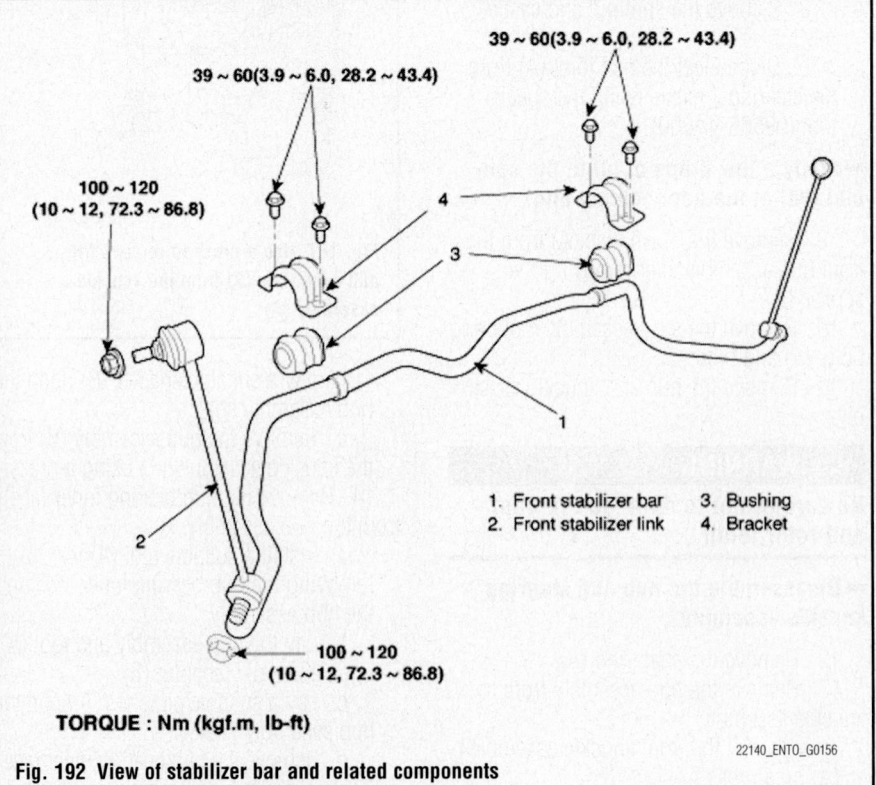

1. Front stabilizer bar
2. Front stabilizer link
3. Bushing
4. Bracket

TORQUE : Nm (kgf.m, lb-ft)

Fig. 192 View of stabilizer bar and related components

7. Remove the stabilizer bracket and bushing.

8. Remove the stabilizer bracket and bushing on the opposite side in the same way.

9. Remove the stabilizer bar.

❊❊ WARNING

Be careful not to damage the power steering pressure tube.

To install:

10. Installation is the reverse of the removal.

11. Tighten the stabilizer bracket bolts to: 32–39 ft. lbs. (45–55 Nm).

12. Tighten the sub-frame mounting bolts to: 68–86 ft. lbs. (95–120 Nm).

13. Tighten the stabilizer bar control link upper and lower mounting nuts to: 72–87 ft. lbs. (100–120 Nm).

14. Tighten the wheel nuts to: 65–80 ft. lbs. (90–110 Nm).

WHEEL HUB AND BEARING (SEALED UNIT)

REMOVAL & INSTALLATION

See Figure 193.

1. Before servicing the vehicle, refer to the Precautions Section.

2. Remove or disconnect the following:
   - Front wheel
   - Brake caliper
   - Lower ball joint

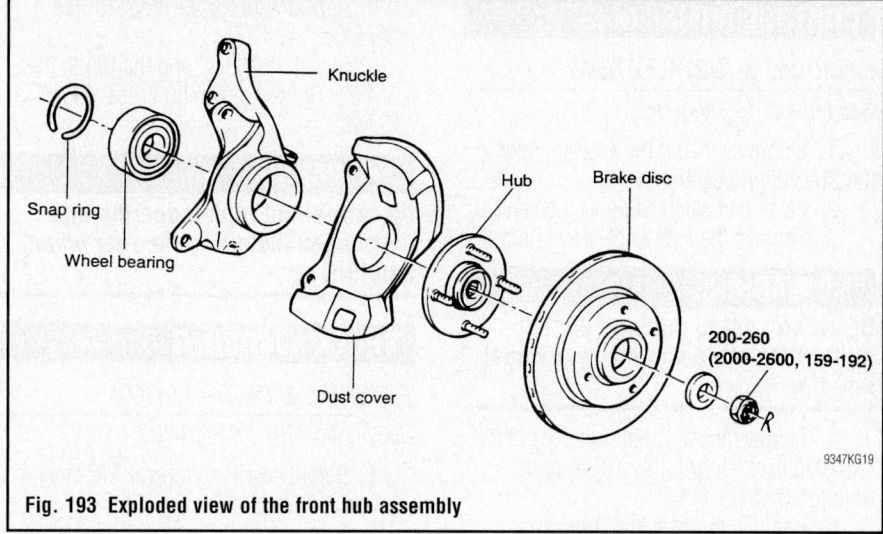

Fig. 193 Exploded view of the front hub assembly

- Spindle nut
- Knuckle pinch bolts
- Steering knuckle

3. Press the hub out of the wheel bearing.

4. Press the wheel bearings out of the steering knuckle.

5. If necessary, press the inner race off the hub.

To install:

6. Press the wheel bearings into the steering knuckle.

7. Install the outer grease seal and press the hub into the wheel bearings.

8. Install or connect the following:
   - Inner grease seal

- Steering knuckle and torque the knuckle pinch bolts to 65–76 ft. lbs. (95–105 Nm).
- Lower ball joint and torque the stud nut to 43–52 ft. lbs. (60–72 Nm).
- Spindle nut and torque the nut to 144–187 ft. lbs. (195–253 Nm).
- Brake caliper and torque the bracket bolts to 50 ft. lbs. (68 Nm).
- Front wheel

9. Check and/or adjust the wheel alignment.

ADJUSTMENT

The front wheel bearing is a sealed unit and is not adjustable.

SUSPENSION   REAR SUSPENSION

COIL SPRING

REMOVAL & INSTALLATION

See Figure 194.

1. Before servicing the vehicle, refer to the Precautions Section.

2. Raise and safely support the vehicle.

3. Remove the rear wheels and tires.

4. While supporting the lower arm with a jack, remove the lower arm bolt from the rear knuckle.

5. Loosen the lower arm bolt from the crossmember.

6. Remove the spring, lower seat, and the upper pad.

To install:

7. Installation is the reverse of the removal procedure.

8. Torque the lower arm mounting bolt to knuckle to 87–116 ft. lbs. (120–160 Nm).

9. Torque the lower arm to crossmember bolt to 145–195 ft. lbs. (200–270 Nm).

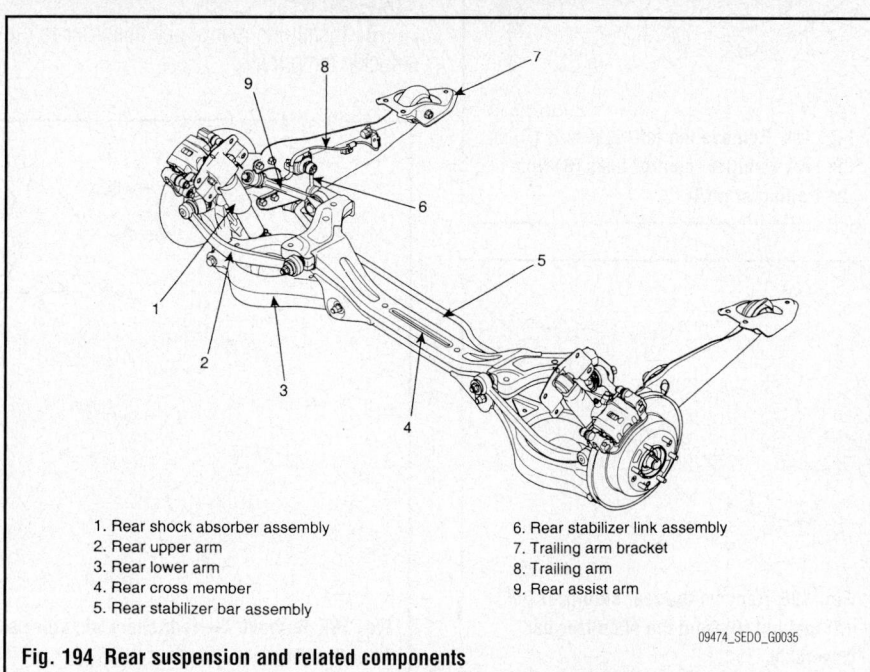

1. Rear shock absorber assembly
2. Rear upper arm
3. Rear lower arm
4. Rear cross member
5. Rear stabilizer bar assembly
6. Rear stabilizer link assembly
7. Trailing arm bracket
8. Trailing arm
9. Rear assist arm

Fig. 194 Rear suspension and related components

CONTROL ARMS/LINKS

REMOVAL & INSTALLATION

See Figures 195 and 196.

1. Before servicing the vehicle, refer to the Precautions Section.
2. Raise and safely support the vehicle.
3. Remove the rear wheels and tires.

✳✳ WARNING

Be careful not to damage the hub bolts when removing the rear wheel and tire.

4. Remove the left/right nuts (C) of the rear stabilizer control links (B) from the trailing arm (A).
5. Remove the rear stabilizer link (A) and nut (B) from the stabilizer bar assembly.
6. Remove the stabilizer control links from the vehicle.

To install:

7. Install the rear stabilizer control link (A) and nut (B) to the stabilizer bar assembly. Tighten to 36–47 ft. lbs. (50–65 Nm).
8. Install the stabilizer control link nut to the trailing arm. Tighten to 36–47 ft. lbs. (50–65 Nm).

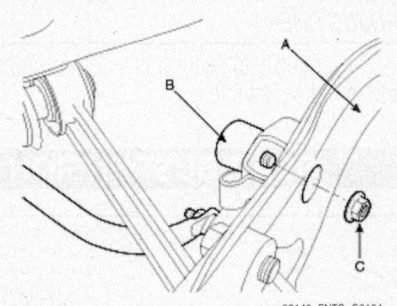

Fig. 195 Remove the left/right nuts (C) of the rear stabilizer control links (B) from the trailing arm (A)

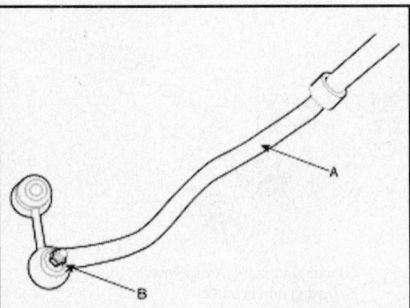

Fig. 196 Remove the rear stabilizer link (A) and nut (B) from the stabilizer bar assembly

9. Repeat appropriate steps for the other side.
10. Install the wheel and the tire to the rear hub. Tighten to 65–80 ft. lbs. (90–110 Nm).

✳✳ WARNING

Be careful not to damage the hub bolts when installing the rear wheel and tire.

SHOCK ABSORBER

REMOVAL & INSTALLATION

See Figures 197 through 200.

1. Before servicing the vehicle, refer to the Precautions Section.
2. Raise and safely support the vehicle.
3. Remove the rear wheels and tires.

✳✳ WARNING

Be careful not to damage the hub bolts when removing the rear wheel and tire.

4. Remove the rear shock absorber assembly mounting bolts (A) and nut (B) from the body.
5. Remove the rear shock absorber assembly nut (B) from the rear knuckle, then remove the shock absorber assembly (A).
6. Remove the rear shock absorber bracket bolt (A).

To install:

7. Install the connecting bolt (A) between the rear shock absorber and the bracket. Tighten to 116–130 ft. lbs. (160–180 Nm).
8. Install the rear shock absorber to the knuckle temporarily.

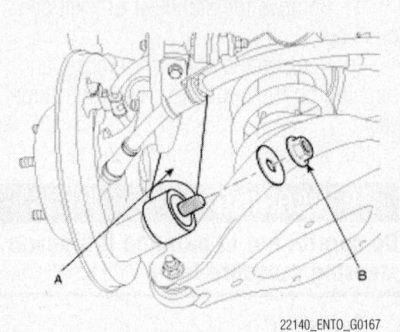

Fig. 198 Remove the rear shock absorber assembly nut (B) from the rear knuckle, then remove the shock absorber assembly (A)

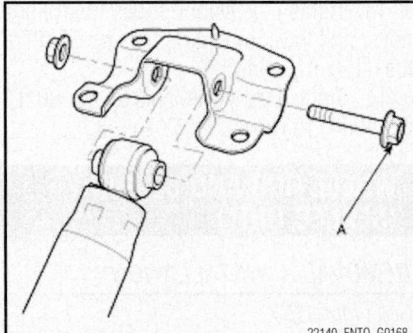

Fig. 199 Remove the rear shock absorber bracket bolt (A)

9. Install the rear shock absorber bracket mounting bolts and nut. Tightening bolt to: 60–80 ft. lbs. (80–110 Nm); tighten nut to: 65–87 ft. lbs. (90–120 Nm).
10. Install the rear shock absorber nut to the knuckle. Tighten to 116–130 ft. lbs. (160–180 Nm).

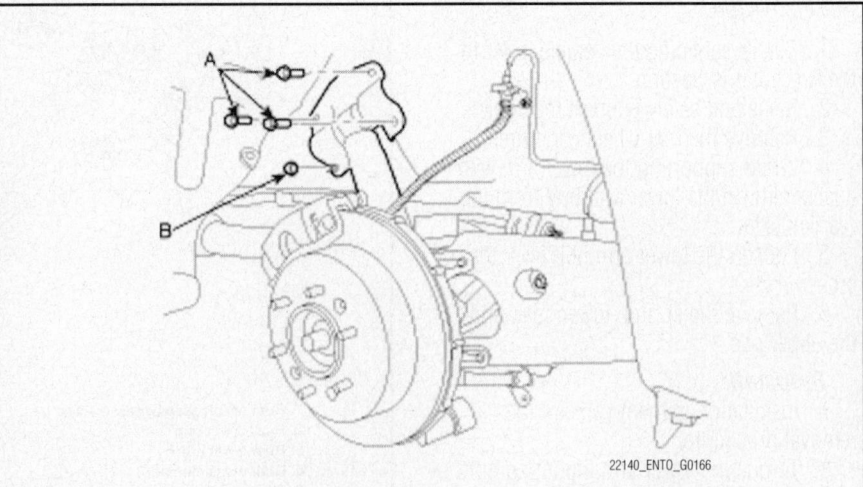

Fig. 197 Remove the rear shock absorber assembly mounting bolts (A) and nut (B) from the body

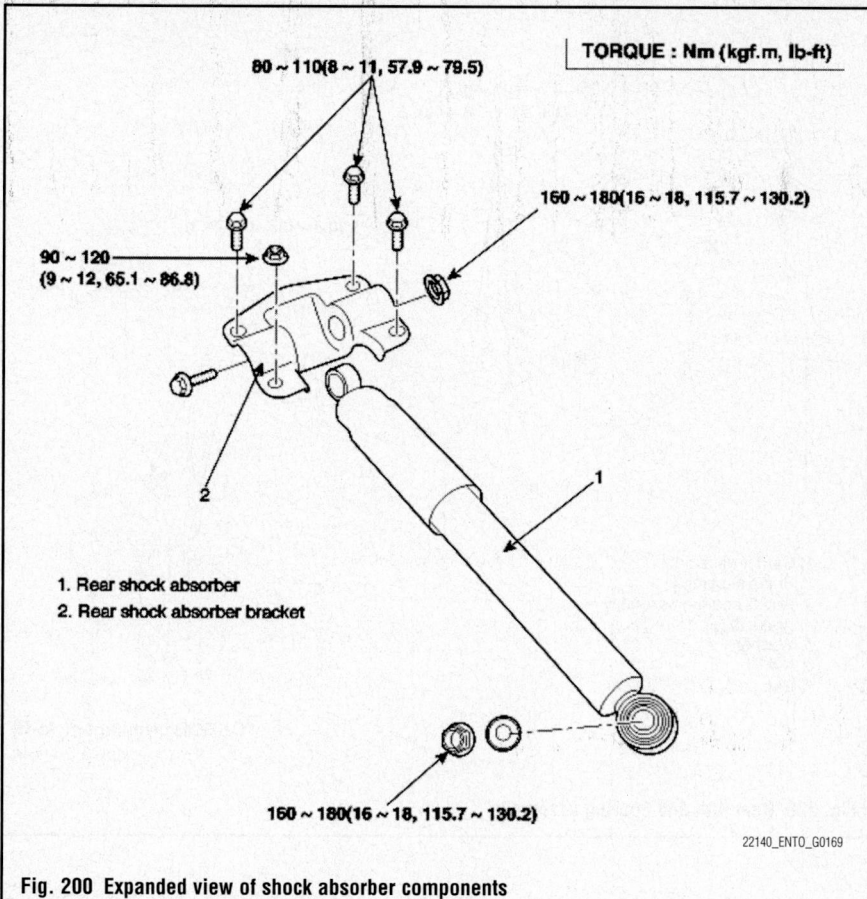

TORQUE : Nm (kgf.m, lb-ft)

80 ~ 110(8 ~ 11, 57.9 ~ 79.5)

160 ~ 180(16 ~ 18, 115.7 ~ 130.2)

90 ~ 120
(9 ~ 12, 65.1 ~ 86.8)

1. Rear shock absorber
2. Rear shock absorber bracket

160 ~ 180(16 ~ 18, 115.7 ~ 130.2)

22140_ENTO_G0169

Fig. 200 Expanded view of shock absorber components

11. Install the wheel and the tire to the rear hub. Tighten to 65–80 ft. lbs. (90–110 Nm).

❊❊ WARNING

Be careful not to damage the hub bolts when installing the rear wheel and tire.

TESTING

1. Check the rubber parts for damage or deterioration.
2. Check for correct height and proper return of shock absorber to original height.
3. Check the shock absorber for abnormal resistance or unusual sounds.
4. Check for oil leakage around the seals.
5. Replace if necessary.

WHEEL HUB AND BEARING (SEALED UNIT)

REMOVAL & INSTALLATION

See Figures 201 through 206.

1. Before servicing the vehicle, refer to the Precautions Section.
2. Release the parking brake.

3. Raise and safely support the vehicle.
4. Remove the rear wheels and tires.
5. Support the lower part of the lower arm (A) using a jack and remove the bolt and nut (B).
6. Remove the coil spring and upper pad.
7. Remove the wheel speed sensor.
8. Remove the rear brake caliper assembly from the carrier assembly and suspend it with a wire.
9. Remove the rear brake disc by loosening the screws.

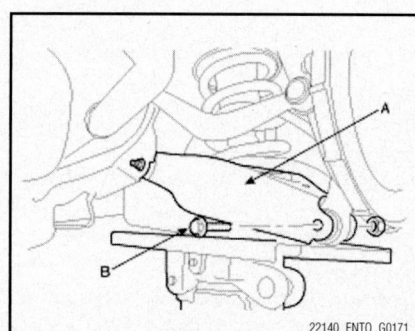

22140_ENTO_G0171

Fig. 201 Support the lower part of the lower arm (A) using a jack and remove the bolt and nut (B)

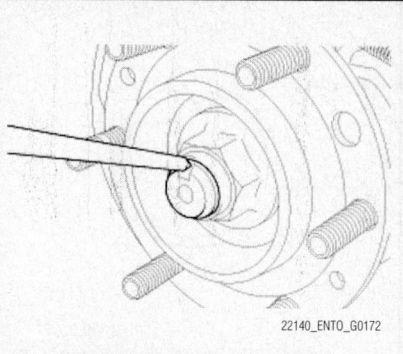

22140_ENTO_G0172

Fig. 202 Unstake the lock nut using a chisel and hammer

10. Unstake the lock nut using a chisel and hammer.
11. Remove the lock nut and washer.
12. Remove the hub and bearing assembly (A).

To install:

13. Installation is the reverse of the removal procedure.

➡**The rear hub lock nuts should be replaced with new ones.**

14. Torque the hub lock nut to 145–188 ft. lbs. (200–260 Nm).

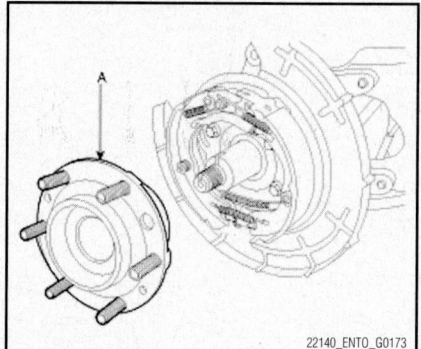

22140_ENTO_G0173

Fig. 203 Remove the hub and bearing assembly (A)

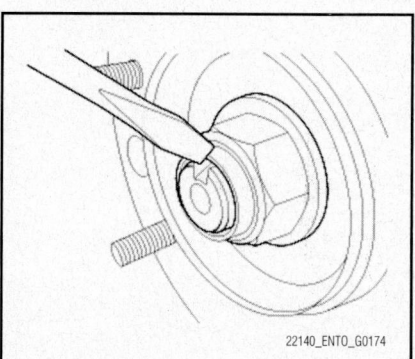

22140_ENTO_G0174

Fig. 204 Stake the lock nut using a chisel and hammer

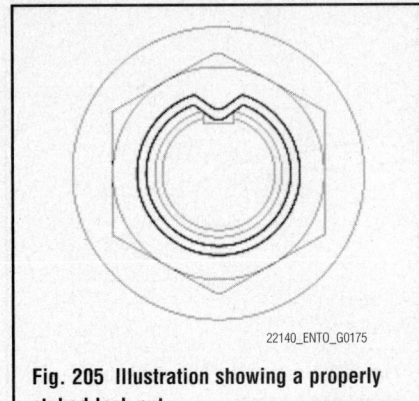

Fig. 205 Illustration showing a properly staked lock nut

15. After installation of the lock nut, stake the lock nut using a chisel and hammer as shown in the illustration.

16. Torque the caliper retaining bolts to 36–43 ft. lbs. (50–60 Nm).

17. Torque the lower arm retaining nut and bolt to 116–130 ft. lbs. (160–180 Nm).

18. Install the rear wheels and tires. Tighten to 65–80 ft. lbs. (90–110 Nm).

ADJUSTMENT

The rear wheel bearing is an integral part of the rear hub. No adjustment is possible.

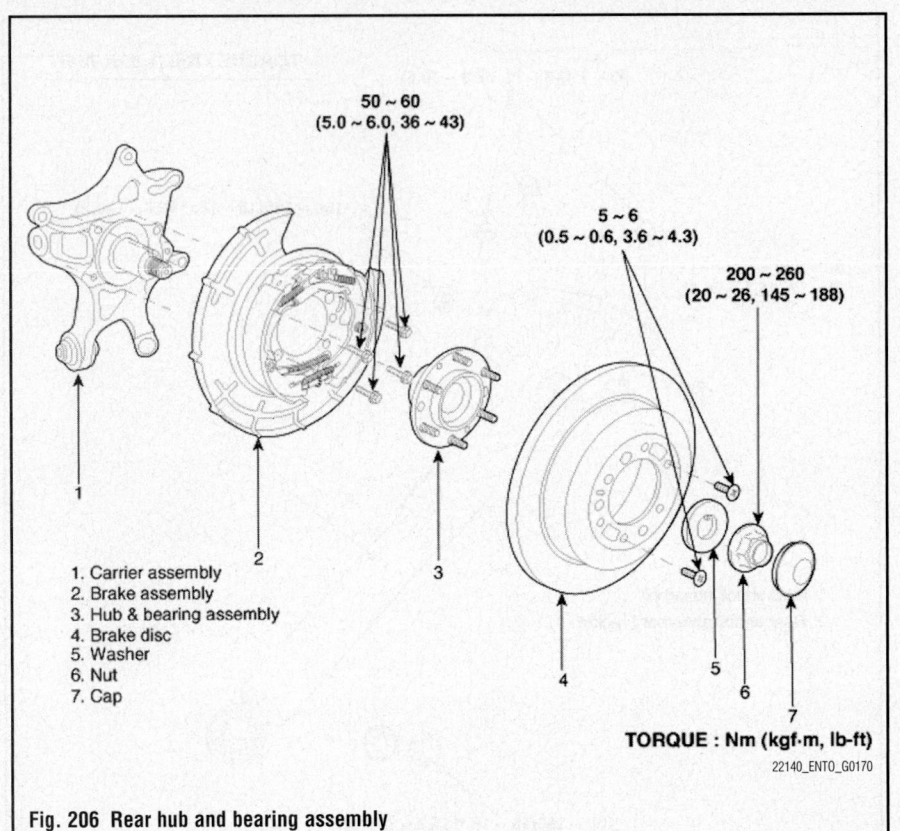

1. Carrier assembly
2. Brake assembly
3. Hub & bearing assembly
4. Brake disc
5. Washer
6. Nut
7. Cap

TORQUE : Nm (kgf·m, lb-ft)

Fig. 206 Rear hub and bearing assembly

HYUNDAI

3

Santa Fe • Veracruz

SPECIFICATIONS AND MAINTENANCE CHARTS

ENGINE AND VEHICLE IDENTIFICATION

Engine							Model Year	
Code ①	Liters (cc)	Cu. In.	Cyl.	Fuel Sys.	Engine Type	Eng. Mfg.	Code ②	Year
B	2.4 (2351)	143.50	I4	MPFI	DOHC	Hyundai	6	2006
D	2.7 (2656)	162.00	V6	MPFI	DOHC	Hyundai	7	2007
E	3.3 (3342)	203.86	V6	MPFI	DOHC	Hyundai	8	2008
E	3.5 (3497)	213.30	V6	MPFI	DOHC	Hyundai		
C	3.8 (3778)	230.55	V6	MPFI	DOHC	Hyundai		

MPFI: Multi-Point Fuel Injection

DOHC: Double Overhead Camshafts

① 8th digit of VIN

② 10th digit of VIN

22140_SANT_C0001

GENERAL ENGINE SPECIFICATIONS

All measurements are given in inches.

Year	Model	Engine Displacement Liters	Engine Series VIN	Net Horsepower @ rpm	Net Torque @ rpm (ft. lbs.)	Bore x Stroke (in.)	Com-pression Ratio	Oil Pressure @ rpm
2006	Santa Fe	2.4	B	149@5500	156@3000	3.410 x 3.940	10.0:1	11.6@Idle
	Santa Fe	2.7	D	173@6000	182@4000	3.413 x 2.953	10.0:1	7.3@Idle
	Santa Fe	3.5	E	200@5500	219@3500	3.661 x 3.779	10.0:1	11.4@700
2007	Santa Fe	2.7	D	185@6000	183@4000	3.413 x 2.953	10.4:1	18.8@1000
	Santa Fe	3.3	E	242@6000	226@4500	3.622 x 3.299	10.4:1	18.8@1000
	Veracruz	3.8	C	260@6000	257@4500	3.780 x 3.425	10.4:1	18.8@1000
2008	Santa Fe	2.7	D	185@6000	183@4000	3.413 x 2.953	10.4:1	18.8@1000
	Santa Fe	3.3	E	242@6000	226@4500	3.622 x 3.299	10.4:1	18.8@1000
	Veracruz	3.8	C	260@6000	257@4500	3.780 x 3.425	10.4:1	18.8@1000

22140_SANT_C0002

GASOLINE ENGINE TUNE-UP SPECIFICATIONS

Year	Engine Displacement Liters	Engine VIN	Spark Plug Gap (in.)	Ignition Timing (deg.) MT	Ignition Timing (deg.) AT	Fuel Pump (psi)	Idle Speed (rpm) MT	Idle Speed (rpm) AT	Valve Clearance In.	Valve Clearance Ex.
2006	2.4	B	0.039-0.043	①	①	38.0	②	②	HYD	HYD
	2.7	D	0.039-0.043	①	①	38.0	②	②	HYD	HYD
	3.5	E	0.039-0.043	①	①	45.5-50.0	②	②	HYD	HYD
2007	2.7	D	0.039-0.043	①	①	54.3-55.8	②	②	HYD	HYD
	3.3	E	0.039-0.043	①	①	54.3-55.8	②	②	HYD	HYD
	3.8	C	0.039-0.043	①	①	54.3-55.8	②	②	HYD	HYD
2008	2.7	D	0.039-0.043	①	①	54.3-55.8	②	②	HYD	HYD
	3.3	E	0.039-0.043	①	①	54.3-55.8	②	②	HYD	HYD
	3.8	C	0.039-0.043	①	①	54.3-55.8	②	②	HYD	HYD

NOTE: The Vehicle Emission Control Information label reflects specification changes made during production.

Follow the figures on the label if they differ from those in this chart.

HYD: Hydraulic

① Ignition timing is preset and cannot be adjusted

② Idle speed is maintained by the Electronic Control Module (ECM)

22140_SANT_C0003

CAPACITIES

Year	Model	Engine Displacement Liters	Engine VIN	Engine Oil with Filter (qts.)	Transmission (pts.) Manual	Transmission (pts.) Auto. ①	Fuel Tank (gal.)	Cooling System (qts.)
2006	Santa Fe	2.4	B	4.5	4.6	16.6	17.1	7.7
	Santa Fe	2.7	D	4.8	4.6	16.6	17.1	9.1
	Santa Fe	3.5	E	4.6	4.6	18.0	19.1	8.7
2007	Santa Fe	2.7	D	4.8	4.6	16.4	19.8	8.2
	Santa Fe	3.3	E	6.8	—	23.0	21.1	9.1
	Veracruz	3.8	C	6.8	—	23.0	20.6	9.1
2008	Santa Fe	2.7	D	4.8	4.6	16.4	19.8	8.2
	Santa Fe	3.3	E	6.8	—	23.0	21.1	9.1
	Veracruz	3.8	C	6.8	—	23.0	20.6	9.1

NOTE: All capacities are approximate. Add fluid gradually and check to be sure a proper fluid level is obtained.

① Drain and refill

22140_SANT_C0004

FLUID SPECIFICATIONS

Year	Model	Engine Displacement Liters	Engine ID/VIN	Engine Oil	Manual Trans.	Auto. Trans.	Drive Axle	Power Steering Fluid	Brake Master Cylinder	Cooling System
2006	Santa Fe	2.4	B	5W-20	①	②	③	PSF-3	④	⑤
	Santa Fe	2.7	D	5W-20	①	②	③	PSF-3	④	⑤
	Santa Fe	3.5	E	5W-20	①	②	③	PSF-3	④	⑤
2007	Santa Fe	2.7	D	5W-20	⑥	②	⑦	PSF-3	④	⑤
	Santa Fe	3.3	E	5W-20	—	②	⑦	PSF-3	④	⑤
	Veracruz	3.8	C	5W-20	—	⑧	⑦	PSF-3	④	⑤
2008	Santa Fe	2.7	D	5W-20	⑥	②	⑦	PSF-3	④	⑤
	Santa Fe	3.3	E	5W-20	—	②	⑦	PSF-3	④	⑤
	Veracruz	3.8	C	5W-20	—	⑧	⑦	PSF-3	④	⑤

DOT: Department Of Transportation

① HYUNDAI GENUINE PART MTF 75W/90 (API GL-4)

② DIAMOND ATF SP-3, SK ATF SP-3

③ Hypoid gear oil (API GL-5, SAE 80W/90, SHELL SPIRAX AX or equiv.)

④ DOT 3, DOT 4, or equivalent

⑤ Ethylene glycol base for aluminum radiator

⑥ HYUNDAI GENUINE PART MTF 75W/85 (API GL-4)

⑦ Hypoid gear oil (API GL-5, SAE 75W/90)

⑧ HYUNDAI GENUINE PART TFF ATF T-IV JWS-3309, Mobil ATF 3309

22140_SANT_C0005

VALVE SPECIFICATIONS

Year	Engine Displacement Liters	Engine VIN	Seat Angle (deg.)	Face Angle (deg.)	Spring Test Pressure (lbs. @ in.)	Spring Installed Height (in.)	Stem-to-Guide Clearance (in.)		Stem Diameter (in.)	
							Intake	Exhaust	Intake	Exhaust
2006	2.4	B	44.0-44.5	45.0-45.5	55.8 @1.57	1.804	0.0008-0.0019	0.0020-0.0030	0.2585-0.2591	0.2571-0.2579
	2.7	D	NA	45.0-45.5	48.4 @1.378	1.673	0.0008-0.0020	0.0014-0.0026	0.2350-0.2354	0.2340-0.2350
	3.5	E	43.5-44.0	45.0-45.5	53.0 @1.492	1.827	0.0009-0.0020	0.0020-0.0033	0.2580-0.2590	0.2570-0.2580
2007	2.7	D	NA	45.0-45.5	40.6-44.8 @1.379	NA	0.0008-0.0020	0.0014-0.0026	0.2348-0.2354	0.2343-0.2348
	3.3	E	44.75-45.20	45.25-45.75	90.4-96.2 @0.953	NA	0.0008-0.0019	0.0012-0.0021	0.2151-0.2157	0.2149-0.2153
	3.8	C	44.75-45.20	45.25-45.75	90.4-96.2 @0.953	NA	0.0008-0.0019	0.0012-0.0021	0.2151-0.2157	0.2149-0.2153
2008	2.7	D	NA	45.0-45.5	40.6-44.8 @1.379	NA	0.0008-0.0020	0.0014-0.0026	0.2348-0.2354	0.2343-0.2348
	3.3	E	44.75-45.20	45.25-45.75	90.4-96.2 @0.953	NA	0.0008-0.0019	0.0012-0.0021	0.2151-0.2157	0.2149-0.2153
	3.8	C	44.75-45.20	45.25-45.75	90.4-96.2 @0.953	NA	0.0008-0.0019	0.0012-0.0021	0.2151-0.2157	0.2149-0.2153

NA: Not Available

22140_SANT_C0006

CAMSHAFT AND BEARING SPECIFICATIONS CHART
All measurements are given in inches.

Year	Engine Displ. Liters	Engine ID/VIN	Journal Dia.	Brg. Oil Clearance	Shaft End-play	Runout	Journal Bore	Lobe Height Intake	Lobe Height Exhaust
2006	2.4	B	1.0200	0.0020-0.0030	0.0040-0.0060	NA	NA	1.3974-1.3776	1.3904-1.3707
	2.7	D	1.0222-1.0228	0.0007-0.0024	0.0039-0.0059	NA	NA	1.7303-1.7382	1.7303-1.7382
	3.5	E	1.0220-1.0224	0.0007-0.0024	0.0039-0.0059	NA	NA	1.3818-1.3897	1.3705-1.3783
2007	2.7	D	1.1009-1.1016	0.0012-0.0022	0.0039-0.0079	NA	NA	1.7520	1.7520
	3.3	E	①	②	0.0008-0.0071	NA	NA	1.8242	1.8045
	3.8	C	③	④	0.0008-0.0071	NA	NA	1.8425	1.8031
2008	2.7	D	1.1009-1.1016	0.0012-0.0022	0.0039-0.0079	NA	NA	1.7520	1.7520
	3.3	E	①	②	0.0008-0.0071	NA	NA	1.8242	1.8045
	3.8	C	③	④	0.0008-0.0071	NA	NA	1.8425	1.8031

NA: Not Available

① For LH and RH camshafts:
   Intake No. 1 is 1.1009-1.1016 inch
   Intake No. 2, 3, 4 are 0.9430-0.9437 inch
   Exhaust No.1 is 1.1009-1.1016 inch
   Exhaust No. 2, 3, 4 are 0.9430-0.9437 inch

② For LH and RH camshafts:
   Intake No. 1 is 0.0008-0.0022 inch
   Intake No. 2, 3, 4 are 0.0012-0.0026 inch
   Exhaust No. 1 is 0.0008-0.0022 inch
   Exhaust No. 2, 3, 4 are 0.0012-0.0026 inch

③ For LH and RH camshafts:
   Intake No. 1 is 1.1009-1.1015 inch
   Intake No. 2, 3, 4 are 0.9430-0.9437 inch
   Exhaust No.1 is 1.1009-1.1015 inch
   Exhaust No. 2, 3, 4 are 0.9430-0.9437 inch

④ For LH and RH camshafts:
   Intake No. 1 is 0.0011-0.0022 inch
   Intake No. 2, 3, 4 are 0.0012-0.0026 inch
   Exhaust No.1 is 0.0011-0.0022 inch
   Exhaust No. 2, 3, 4 are 0.0012-0.0026 inch

22140_SANT_C0007

CRANKSHAFT AND CONNECTING ROD SPECIFICATIONS
All measurements are given in inches.

Year	Engine Displacement Liters	Engine VIN	Crankshaft				Connecting Rod		
			Main Brg. Journal Dia.	Main Brg. Oil Clearance	Shaft End-play	Thrust on No.	Journal Diameter	Oil Clearance	Side Clearance
2006	2.4	B	2.2434-2.2441	①	0.0020-0.0098	3	1.8900-1.8903	0.0006-0.0019	0.0040-0.0098
	2.7	D	2.4402-2.4409	0.0002-0.0009	0.0028-0.0098	3	1.8891-1.8898	0.0007-0.0014	0.0039-0.0098
	3.5	E	2.5190-2.5197	0.0009-0.0016	0.0020-0.0098	3	2.1650-2.1653	0.0010-0.0017	0.0039-0.0098
2007	2.7	D	2.4402-2.4409	0.0002-0.0009	0.0028-0.0098	3	1.8891-1.8898	0.0007-0.0014	0.0039-0.0098
	3.3	E	2.7142-2.7149	0.0008-0.0016	0.0039-0.0110	3	2.1635-2.1642	0.0015-0.0022	0.0039-0.0098
	3.8	C	2.7142-2.7149	0.0008-0.0016	0.0039-0.0110	3	2.1635-2.1642	0.0015-0.0022	0.0039-0.0098
2008	2.7	D	2.4402-2.4409	0.0002-0.0009	0.0028-0.0098	3	1.8891-1.8898	0.0007-0.0014	0.0039-0.0098
	3.3	E	2.7142-2.7149	0.0008-0.0016	0.0039-0.0110	3	2.1635-2.1642	0.0015-0.0022	0.0039-0.0098
	3.8	C	2.7142-2.7149	0.0008-0.0016	0.0039-0.0110	3	2.1635-2.1642	0.0015-0.0022	0.0039-0.0098

① No. 1, 2, 4, 5 are 0.0007-0.0014 inch

No. 3 is 0.0009-0.0017 inch

22140_SANT_C0008

PISTON AND RING SPECIFICATIONS
All measurements are given in inches.

Year	Engine Displ. Liters	Engine VIN	Piston Clearance	Ring Gap			Ring Side Clearance		
				Top Compression	Bottom Compression	Oil Control	Top Compression	Bottom Compression	Oil Control
2006	2.4	B	0.0008-0.0016	0.0098-0.0138	0.0157-0.0216	0.0039-0.0157	0.0012-0.0027	0.0008-0.0024	0.0024-0.0059
	2.7	D	0.0004-0.0012	0.0079-0.0138	0.0146-0.0205	0.0079-0.0276	0.0016-0.0031	0.0012-0.0028	NA
	3.5	E	0.0012-0.0020	0.0078-0.0118	0.0157-0.0216	0.0079-0.0276	0.0016-0.0031	0.0008-0.0024	NA
2007	2.7	D	0.0008-0.0020	0.0059-0.0118	0.0118-0.0177	0.0078-0.0275	0.0016-0.0031	0.0012-0.0027	0.0024-0.0059
	3.3	E	0.0008-0.0016	0.0067-0.0126	0.0126-0.0185	0.0078-0.0275	0.0016-0.0031	0.0012-0.0027	0.0024-0.0059
	3.8	C	0.0012-0.0020	0.0067-0.0126	0.0126-0.0185	0.0078-0.0275	0.0012-0.0027	0.0012-0.0027	0.0024-0.0059
2008	2.7	D	0.0008-0.0020	0.0059-0.0118	0.0118-0.0177	0.0078-0.0275	0.0016-0.0031	0.0012-0.0027	0.0024-0.0059
	3.3	E	0.0008-0.0016	0.0067-0.0126	0.0126-0.0185	0.0078-0.0275	0.0016-0.0031	0.0012-0.0027	0.0024-0.0059
	3.8	C	0.0012-0.0020	0.0067-0.0126	0.0126-0.0185	0.0078-0.0275	0.0012-0.0027	0.0012-0.0027	0.0024-0.0059

NA: Not Available

22140_SANT_C0009

TORQUE SPECIFICATIONS
All readings in ft. lbs.

Year	Engine Displacement Liters	Engine VIN	Cylinder Head Bolts	Main Bearing Bolts	Rod Bearing Bolts	Crankshaft Damper Bolts	Flywheel Bolts	Manifold Intake	Manifold Exhaust	Spark Plugs	Oil Pan Drain Plug
2006	2.4	B	①	②	③	58-72	94-101	④	⑤	15-22	25-33
	2.7	D	⑥	⑦	⑧	65-80	53-56	14-15	18-22	15-22	25-33
	3.5	E	75-82	52-59	⑨	130-138	NA	14-16	29-33	15-22	25-33
2007	2.7	D	⑩	⑪	②	123-130	53-56	14-17	22-25	15-22	25-33
	3.3	E	⑫	⑬	②	210-224	53-56	14-17	29-33	15-22	25-33
	3.8	C	⑫	⑬	②	210-224	53-56	14-17	29-33	15-22	25-33
2008	2.7	D	⑩	⑪	②	123-130	53-56	14-17	22-25	15-22	25-33
	3.3	E	⑫	⑬	②	210-224	53-56	14-17	29-33	15-22	25-33
	3.8	C	⑫	⑬	②	210-224	53-56	14-17	29-33	15-22	25-33

NOTE: Dip main bearing bolts and crankshaft damper bolt in clean engine oil prior to tightening.

NA: Not Available

① If using used parts:
   Step 1: 14 ft. lbs. (20 Nm)
   Step 2: Plus an additional 90 degrees
   Step 3: Plus an additional 90 degrees
   If using new parts:
   Step 1: 46 ft. lbs. (64 Nm)
   Step 2: Release the bolts
   Step 3: 14 ft. lbs. (20 Nm)
   Step 4: Plus an additional 90 degrees
   Step 5: Plus an additional 90 degrees

② 15 ft. lbs., plus 90 degrees

③ 14 ft. lbs., plus 90 degrees

④ Bolt (M8): 11-14 ft. lbs.
   Nut: 22-30 ft. lbs.

⑤ Bolt (M8): 18-22 ft. lbs.
   Bolt (M10): 25-40 ft. lbs.

⑥ Step 1: 18 ft. lbs.
   Step 2: Plus an additional 58-62 degrees
   Step 3: Plus an additional 43-47 degrees

⑦ Bolt (M10): 20-24 ft. lbs., plus 90-94 degrees
   Bolt (M8): 10-14 ft. lbs., plus 90-94 degrees

⑧ 12-15 ft. lbs., plus 90-94 degrees

⑨ 26 ft. lbs., plus 90 degrees

⑩ Step 1: 18 ft. lbs.
   Step 3: Plus an additional 60 degrees
   Step 3: Plus an additional 45 degrees

⑪ Bolt (M10): 20-24 ft. lbs., plus 90 degrees, plus 5 degrees
   Bolt (M8): 9-14 ft. lbs., plus 90 degrees, plus 5 degrees

⑫ Step 1: 29 ft. lbs.
   Step 2: Plus 120 degrees
   Step 3: Plus 90 degrees

⑬ M11 bolts (inner) Step 1: 36 ft. lbs.
   M11 bolts (inner) Step 2: Plus 90 degrees
   M8 bolts (outer) Step 1: 15 ft. lbs.
   M8 bolts (outer) Step 2: Plus 120 degrees
   M8 bolts (side): 22-23 ft. lbs.

22140_SANT_C0010

WHEEL ALIGNMENT

Year	Model		Caster Range (+/-Deg.)	Caster Preferred Setting (Deg.)	Camber Range (+/-Deg.)	Camber Preferred Setting (Deg.)	Toe-in (Deg.)
2006	Santa Fe	Front	①	②	③	0.00	④
		Rear	—	—	⑤	-0.00	⑥
2007	Santa Fe	Front	①	⑦	①	⑧	⑨
		Rear	—	—	①	-1.00	⑩
	Veracruz	Front	①	⑪	①	⑫	⑥
		Rear	—	—	①	-1.00	⑬
2008	Santa Fe	Front	①	⑦	①	⑧	⑨
		Rear	—	—	①	-1.00	⑩
	Veracruz	Front	①	⑪	①	⑫	⑥
		Rear	—	—	①	-1.00	⑬

① +/- 30'

② 2 degrees 30'

③ +/- 30' (Maximum difference between LH and RH: 30')

④ -0.08 +/- 0.08 inch

⑤ +/- 30' (Maximum difference between LH and RH: 45')

⑥ 0 +/- 0.08 inch

⑦ 4 degrees 25'

⑧ -30'

⑨ 0 +/- 2 inches

⑩ 2 +/- 2 inches

⑪ 4 degrees 20' and 4 degrees 41'

⑫ -0 degrees 30'

⑬ 0.08 +/- 0.08 inch

22140_SANT_C0011

TIRE, WHEEL AND BALL JOINT SPECIFICATIONS

Year	Model	OEM Tires Standard	OEM Tires Optional	Tire Pressures (psi) Front	Tire Pressures (psi) Rear	Wheel Size	Ball Joint Inspection	Lug Nut Torque (ft. lbs.)
2006	Santa Fe	P225/70R16	—	30	30	6.5J x 16	①	67-81
2007	Santa Fe	P235/70R16 P235/60R18	—	30	30	7.0J x 16 7.0J x 18	①	65-80
	Veracruz	P245/65R17 P245/60R18	—	30	30	7.0J x 17 7.0J x 18	①	65-80
2008	Santa Fe	P235/70R16 P235/60R18	—	30	30	7.0J x 16 7.0J x 18	①	65-80
	Veracruz	P245/65R17 P245/60R18	—	30	30	7.0J x 17 7.0J x 18	①	65-80

NA: Not Available

① Replace the ball joint if rotating torque exceeds specification: 13-31 inch lbs.

22140_SANT_C0012

BRAKE SPECIFICATIONS

All measurements in inches unless noted

Year	Model		Brake Disc			Brake Drum Diameter			Minimum Lining Thickness	Brake Caliper	
			Original Thickness	Minimum Thickness	Maximum Runout	Original Inside Diameter	Max. Wear Limit	Maximum Machine Diameter		Bracket Bolts (ft. lbs.)	Mounting Bolts (ft. lbs.)
2006	Santa Fe	F	1.024	0.961	0.002	—	—	—	0.079	59-74	16-24
		R	0.390	0.330	0.002	—	—	—	0.079	37-44	16-24
2007	Santa Fe	F	1.100	1.040	0.001	—	—	—	0.120-0.160	58-72	16-23
		R	0.430	0.370	0.001	—	—	—	0.120	36-47	16-23
	Veracruz	F	1.100	1.040	0.001	—	—	—	0.120-0.160	54-62	19-28
		R	0.470	0.410	0.001	—	—	—	0.120	47-54	16-23
2008	Santa Fe	F	1.100	1.040	0.001	—	—	—	0.120-0.160	58-72	16-23
		R	0.430	0.370	0.001	—	—	—	0.120	36-47	16-23
	Veracruz	F	1.100	1.040	0.001	—	—	—	0.120-0.160	54-62	19-28
		R	0.470	0.410	0.001	—	—	—	0.120	47-54	16-23

F: Front

R: Rear

22140_SANT_C0013

SCHEDULED MAINTENANCE INTERVALS
HYUNDAI—SANTA FE & VERACRUZ

TO BE SERVICED	TYPE OF SERVICE	VEHICLE MILEAGE INTERVAL (x1000)												
		7.5	15	22.5	30	37.5	45	52.5	60	67.5	75	82.5	90	97.5
Accessory drive belts	S/I	✓	✓	✓	✓	✓	✓	✓	✓	✓	✓	✓	✓	✓
Air cleaner element (engine)	R				✓				✓				✓	
Air conditioner system	S/I	Inspect the system operation annually												
Automatic transaxle fluid	S/I		✓		✓		✓		✓		✓		✓	
Automatic transaxle fluid &	R	105,000 miles (under normal usage)												
Brake lines, hoses, and connections	S/I		✓		✓		✓		✓		✓		✓	
Brake pads, calipers, & rotors	S/I		✓		✓		✓		✓		✓		✓	
Cabin air filter	R	Every 12 months or 10,000 miles												
Chassis and body fasteners	S/I				✓				✓				✓	
Cooling system hoses and coolant level	S/I		✓		✓		✓		✓		✓		✓	
Crankcase ventilation hose, vapor hose, and fuel filter cap	S/I				✓				✓					
Driveshafts and CV-boots	S/I		✓		✓		✓		✓		✓		✓	
Electronic throttle control	S/I		✓		✓		✓		✓		✓		✓	
Engine coolant	R								✓					
Engine oil and filter	R	✓	✓	✓	✓	✓	✓	✓	✓	✓	✓	✓	✓	✓
Exhaust pipe connections, muffler, and suspension bolts	S/I				✓				✓				✓	
Front and rear brakes	S/I				✓				✓				✓	
Fuel filter	R							✓						
Fuel lines, fuel hoses, and connections	S/I				✓				✓				✓	
Locks and hinges	S/I	✓	✓	✓	✓	✓	✓	✓	✓	✓	✓	✓	✓	✓
Spark plugs (standard)	R				✓				✓				✓	
Spark plugs (Iridium coated) 100,000 mile replacement	R													
Spark plugs (Platinum coated)	R								✓					
Steering gear rack, linkage, and boots	S/I				✓				✓				✓	
Upper and lower arm ball joints	S/I		✓		✓		✓		✓		✓		✓	
Wheel bearings	S/I				✓				✓				✓	

R: Replace   S/I: Service or Inspect

FREQUENT OPERATION MAINTENANCE (SEVERE SERVICE)

If a vehicle is operated under any of the following conditions it is considered severe service:

- Extremely dusty areas.

- 50% or more of the vehicle operation is in 90°F (32°C) or higher temperatures, or constant operation in temperatures below 32°F (0°C).

- Prolonged idling (vehicle operation in stop and go traffic).

- Frequent short running periods (engine does not warm to normal operating temperatures).

- Police, taxi, delivery usage or trailer towing usage.

Oil & oil filter: change every 3,000 miles.

Brake pads, calipers & rotors: service or inspect every 7,500 miles.

Driveshaft boots: service or inspect every 7,500 miles

Steering gear rack, linkage & boots: service or inspect every 7,500 miles.

Air cleaner filter: service or inspect every 15,000 miles.

Automatic transaxle fluid & filter: replace every 30,000 miles.

Rear brake drums & linings: service or inspect every 15,000 miles.

Spark plugs: service or inspect every 24,000 miles.

GENERAL INFORMATION

Before servicing any vehicle, please be sure to read all of the following precautions, which deal with personal safety, prevention of component damage, and important points to take into consideration when servicing a motor vehicle:

• Never open, service or drain the radiator or cooling system when the engine is hot; serious burns can occur from the steam and hot coolant.

• Observe all applicable safety precautions when working around fuel. Whenever servicing the fuel system, always work in a well-ventilated area. Do not allow fuel spray or vapors to come in contact with a spark, open flame, or excessive heat (a hot drop light, for example). Keep a dry chemical fire extinguisher near the work area. Always keep fuel in a container specifically designed for fuel storage; also, always properly seal fuel containers to avoid the possibility of fire or explosion. Refer to the additional fuel system precautions later in this section.

• Fuel injection systems often remain pressurized, even after the engine has been turned **OFF**. The fuel system pressure must be relieved before disconnecting any fuel lines. Failure to do so may result in fire and/or personal injury.

• Brake fluid often contains polyglycol ethers and polyglycols. Avoid contact with the eyes and wash your hands thoroughly after handling brake fluid. If you do get brake fluid in your eyes, flush your eyes with clean, running water for 15 minutes. If eye irritation persists, or if you have taken

brake fluid internally, IMMEDIATELY seek medical assistance.

• The EPA warns that prolonged contact with used engine oil may cause a number of skin disorders, including cancer. You should make every effort to minimize your exposure to used engine oil. Protective gloves should be worn when changing oil. Wash your hands and any other exposed skin areas as soon as possible after exposure to used engine oil. Soap and water, or waterless hand cleaner should be used.

• All new vehicles are now equipped with an air bag system, often referred to as a Supplemental Restraint System (SRS) or Supplemental Inflatable Restraint (SIR) system. The system must be disabled before performing service on or around system components, steering column, instrument panel components, wiring and sensors. Failure to follow safety and disabling procedures could result in accidental air bag deployment, possible personal injury and unnecessary system repairs.

• Always wear safety goggles when working with, or around, the air bag system. When carrying a non-deployed air bag, be sure the bag and trim cover are pointed away from your body. When placing a non-deployed air bag on a work surface, always face the bag and trim cover upward, away from the surface. This will reduce the motion of the module if it is accidentally deployed. Refer to the additional air bag system precautions later in this section.

• Clean, high quality brake fluid from a sealed container is essential to the safe and

proper operation of the brake system. You should always buy the correct type of brake fluid for your vehicle. If the brake fluid becomes contaminated, completely flush the system with new fluid. Never reuse any brake fluid. Any brake fluid that is removed from the system should be discarded. Also, do not allow any brake fluid to come in contact with a painted surface; it will damage the paint.

• Never operate the engine without the proper amount and type of engine oil; doing so WILL result in severe engine damage.

• Timing belt maintenance is extremely important. Many models utilize an interference-type, non-freewheeling engine. If the timing belt breaks, the valves in the cylinder head may strike the pistons, causing potentially serious (also time-consuming and expensive) engine damage. Refer to the maintenance interval charts for the recommended replacement interval for the timing belt, and to the timing belt section for belt replacement and inspection.

• Disconnecting the negative battery cable on some vehicles may interfere with the functions of the on-board computer system(s) and may require the computer to undergo a relearning process once the negative battery cable is reconnected.

• When servicing drum brakes, only disassemble and assemble one side at a time, leaving the remaining side intact for reference.

• Only an MVAC-trained, EPA-certified automotive technician should service the air conditioning system or its components.

BRAKES

GENERAL INFORMATION

The Anti-Lock Brake System (ABS) controls the hydraulic brake pressure of all four wheels during sudden braking and braking on hazardous road surfaces, preventing the wheels from locking. The ABS provides the following benefits:

• Enables steering around obstacles with a greater degree of certainty during panic braking

• Enables stopping during panic braking while allowing stability and control, even on curves

• If a malfunction occurs in the ABS, the system will operate as a normal brake (fail safe mode). A diagnostic function and a fail-safe system have been included for serviceability.

WHEEL SPEED SENSORS

REMOVAL & INSTALLATION

Front
See Figure 1.

1. Remove the front wheel speed sensor mounting bolt.
2. Remove the front wheel guard after removing the mud ground.
3. Remove the front wheel speed sensor after disconnecting the wheel speed sensor connector.
4. Installation is the reverse of the removal procedure.

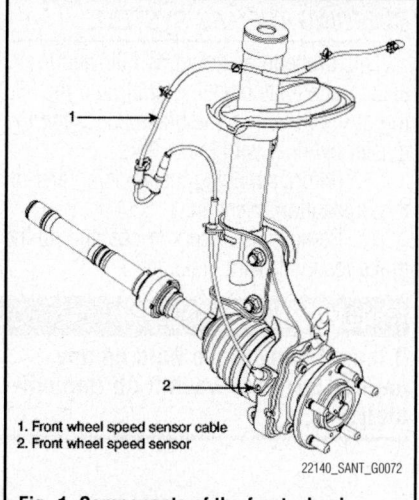

1. Front wheel speed sensor cable
2. Front wheel speed sensor

22140_SANT_G0072

Fig. 1 Components of the front wheel speed sensor

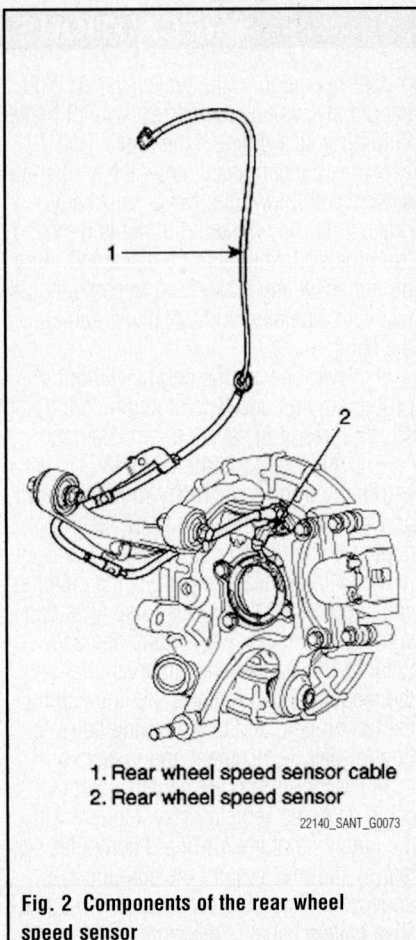

1. Rear wheel speed sensor cable
2. Rear wheel speed sensor

22140_SANT_G0073

Fig. 2 Components of the rear wheel speed sensor

Rear

See Figure 2.

1. Remove the rear wheel speed sensor mounting bolt.

2. Remove the rear seat side pad then disconnect the rear wheel speed sensor connector.

3. Installation is the reverse of the removal procedure.

WHEEL SPEED SENSOR RINGS (TOOTHED RINGS)

REMOVAL & INSTALLATION

See Figure 3.

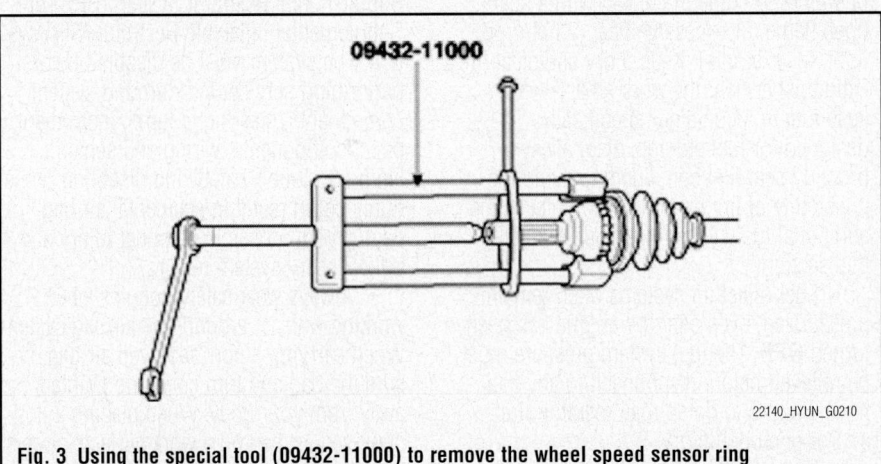

09432-11000

22140_HYUN_G0210

Fig. 3 Using the special tool (09432-11000) to remove the wheel speed sensor ring

1. Before servicing the vehicle, refer to the Precautions Section.

2. Remove the halfshaft(s). Refer to Halfshafts, removal & installation.

3. Using the special tool (09432-11000), remove the wheel speed sensor ring (also called tone wheel).

To install:

4. Installation is the reverse of the removal.

5. Replace with wheel speed sensor ring of the same number of teeth.

BRAKES

BLEEDING THE BRAKE SYSTEM

BLEEDING PROCEDURE

These vehicles come standard with a 4-wheel Anti-Lock Braking System (ABS). Please refer to the section, Bleeding the ABS System.

BLEEDING THE ABS SYSTEM

This procedure should be followed to ensure adequate bleeding of air and the filling of the ABS unit, the brake lines, and the master cylinder with brake fluid.

1. Before servicing the vehicle, refer to the Precautions Section.

2. Remove the reservoir cap and fill the brake reservoir with brake fluid.

✳ WARNING

If there is any brake fluid on any painted surface, wash it off immediately.

➡ **When pressure bleeding, do not depress the brake pedal. Recommended brake fluid: DOT3 or DOT4.**

3. Connect a clear plastic tube to the wheel cylinder bleeder plug and insert the other end of the tube into a clear plastic bottle that is half filled with clean brake fluid.

4. Connect the Hi-Scan Pro® to the data link connector located underneath the dash panel.

5. Select and operate according to the instructions on the Hi-Scan Pro® screen.

✳✳ CAUTION

You must obey the maximum operating time of the ABS motor with the Hi-Scan Pro® to prevent the motor pump from burning.

6. Select Hyundai vehicle diagnosis.

7. Select vehicle name.

8. Select Anti-Lock Brake system.

9. Select air bleeding mode.

10. Press "YES" to operate motor pump and solenoid valve.

✳✳ WARNING

Wait 60 seconds before operating the air bleeding or damage to the motor may occur.

11. Wait 60 seconds before operating the air bleeding.

12. Pump the brake pedal several times, and then loosen the bleeder screw until fluid starts to run out without bubbles. Then, close the bleeder screw.

13. Repeat until there are no more bubbles in the fluid for each wheel.

BRAKES

FRONT DISC BRAKES

✳✳ CAUTION

Dust and dirt accumulating on brake parts during normal use may contain asbestos fibers from production or aftermarket brake linings. Breathing excessive concentrations of asbestos fibers can cause serious bodily harm. Exercise care when servicing brake parts. Do not sand or grind brake lining unless equipment used is designed to contain the dust residue. Do not clean brake parts with compressed air or by dry brushing. Cleaning should be done by dampening the brake components with a fine mist of water, then wiping the brake components clean with a dampened cloth. Dispose of cloth and all residue containing asbestos fibers in an impermeable container with the appropriate label. Follow practices prescribed by the Occupational Safety and Health Administration (OSHA) and the Environmental Protection Agency (EPA) for the handling, processing, and disposing of dust or debris that may contain asbestos fibers.

BRAKE CALIPER

REMOVAL & INSTALLATION

1. Before servicing the vehicle, refer to the Precautions Section.
2. Remove or disconnect the following:
   • Wheel
   • Brake hose from the caliper
   • Caliper mounting bolts
   • Caliper

To install:
3. Install or connect the following:
   • Caliper
   • Caliper mounting bolts and tighten to 58–73 ft. lbs. (80–100 Nm)
   • Brake hose to the caliper and tighten the fitting to 18–22 ft. lbs. (25–30 Nm)
   • Wheel
4. Bleed the brake system.
5. Before attempting to move the vehicle, pump the brake pedal to seat the pads against the rotors. Make sure the vehicle has a firm brake pedal. Check the level of the brake fluid and add fluid if necessary.

DISC BRAKE PADS

REMOVAL & INSTALLATION

1. Before servicing the vehicle, refer to the Precautions Section.
2. Remove or disconnect the following:
   • Wheel
   • Caliper mounting bolts
   • Caliper and support it to one side with a wire
   • Pads and shims

✳✳ WARNING

To prevent damage to the caliper assembly or brake hose, use a short piece of wire to hang the caliper from the undercarriage.

To install:
3. Install the pads, clips, and shims.
4. Bottom the caliper piston using tool 09581-11000 or a C-clamp.
5. Install the caliper mounting bolts and tighten to 58–73 ft. lbs. (80–100 Nm).
6. Install the wheel.

BRAKES

REAR DISC BRAKES

✳✳ CAUTION

Dust and dirt accumulating on brake parts during normal use may contain asbestos fibers from production or aftermarket brake linings. Breathing excessive concentrations of asbestos fibers can cause serious bodily harm. Exercise care when servicing brake parts. Do not sand or grind brake lining unless equipment used is designed to contain the dust residue. Do not clean brake parts with compressed air or by dry brushing. Cleaning should be done by dampening the brake components with a fine mist of water, then wiping the brake components clean with a dampened cloth. Dispose of cloth and all residue containing asbestos fibers in an impermeable container with the appropriate label. Follow practices prescribed by the Occupational Safety and Health Administration (OSHA) and the Environmental Protection Agency (EPA) for the handling, processing, and disposing of dust or debris that may contain asbestos fibers.

BRAKE CALIPER

REMOVAL & INSTALLATION

See Figure 4.

1. Before servicing the vehicle, refer to the Precautions Section.
2. Release the parking brake.

3. Remove or disconnect the following:
   • Wheel
   • Brake line at the caliper
   • Caliper mounting bolts
   • Caliper

To install:
4. Install the caliper onto its mounting.

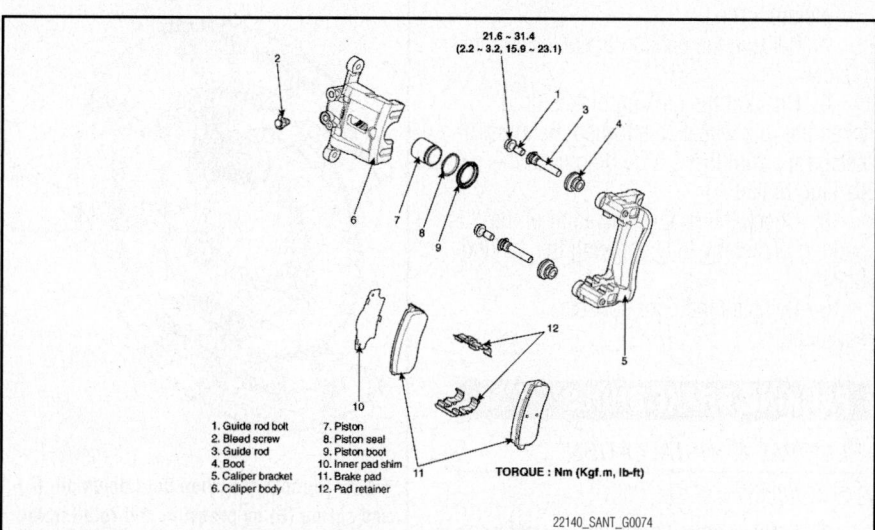

1. Guide rod bolt
2. Bleed screw
3. Guide rod
4. Boot
5. Caliper bracket
6. Caliper body
7. Piston
8. Piston seal
9. Piston boot
10. Inner pad shim
11. Brake pad
12. Pad retainer

TORQUE : Nm (Kgf.m, lb-ft)

21.6 ~ 31.4
(2.2 ~ 3.2, 15.9 ~ 23.1)

22140_SANT_G0074

Fig. 4 Rear brake components

5. Tighten the caliper mounting bolts: 36–43 ft. lbs. (50–60 Nm).

6. Install the brake line to the caliper with new metal gaskets. Torque the brake line union bolt to 18–22 ft. lbs. (24–30 Nm).

7. Bleed the system.

8. Install the wheel.

9. Pump the brake pedal until the brake pads are seated and a firm pedal is achieved before attempting to move the vehicle.

✷✷ CAUTION

Do not move the vehicle until a firm pedal is obtained.

10. Road test the vehicle to check for proper brake operation.

DISC BRAKE PADS

REMOVAL & INSTALLATION

See Figure 4.

1. Before servicing the vehicle, refer to the Precautions Section.

2. Remove or disconnect the following:
- Rear wheels
- Lower caliper mounting bolt and rotate the caliper upward
- Pads from the caliper support
- Pad retainers, if necessary

To install:

3. Install or connect the following:
- Pad retainers, if removed
- Pads onto the pad retainers

4. Compress the caliper piston using a C-clamp.

5. Rotate the caliper downward and install the mounting bolt. Tighten to: 16–24 ft. lbs. (22–32 Nm).

6. Install the wheel.

7. Pump the brake pedal until the brake pads are seated and a firm pedal is achieved before attempting to move the vehicle.

✷✷ CAUTION

Do not move the vehicle until a firm pedal is obtained.

8. Road test the vehicle to check for proper brake operation.

BRAKES

PARKING BRAKE CABLES

ADJUSTMENT

1. Before servicing the vehicle, refer to the Precautions Section.

2. After servicing the rear brake assembly, loosen the parking brake adjusting nut, start the engine, and depress the brake pedal several times in order to set the self-adjusting brake system before adjusting the parking brake.

3. Block the front wheels, then raise the rear of the vehicle and make sure it is securely supported.

4. Set the parking brake 1 click toward engagement.

5. Remove the floor console, if equipped.

6. Tighten the adjusting nut until the parking brakes drag slightly when the rear wheels are turned.

7. Release the parking brake lever completely.

8. Check if the parking brakes drag when the rear wheels are turned. Readjust if necessary until there is no drag from the parking brakes.

9. Check the proper operation of the parking brakes by fully applying the parking brakes.

10. Reinstall the floor console, if equipped.

PARKING BRAKE SHOES

REMOVAL & INSTALLATION

See Figures 5 and 6.

1. Before servicing the vehicle, refer to the Precautions Section.

2. Raise and safely support the vehicle.

3. Remove the rear wheel and tire assembly.

4. Remove the brake caliper. Refer to Rear Disc Brakes, Brake Caliper, removal & installation.

5. Remove the brake rotor. Refer to Rear Disc Brakes, Rotor, removal & installation.

6. Remove the clip retainer and the parking brake cable.

7. Remove the shoe hold down pin (A) and spring (B) by pressing and rotating the spring.

8. Remove the adjuster assembly (B) and the lower return spring (A).

9. Remove the upper return spring (C) and the brake shoes (D).

10. Remove the operating lever assembly (E).

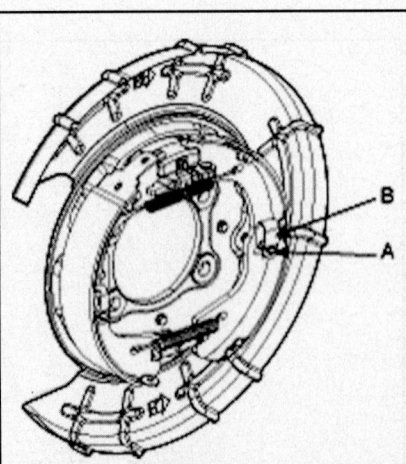

Fig. 5 Remove the shoe hold down pin (A) and spring (B) by pressing and rotating the spring

22140_SANT_G0075

PARKING BRAKE

To install:

11. Install the operating lever assembly (E).

12. Install the upper return spring (C) and the brake shoes (D).

13. Install the adjuster assembly (B) and the lower return spring (A).

14. Install the shoe hold down pin and spring by pressing and rotating the spring.

15. Install the parking brake cable and then install the clip.

16. Install the rear brake disc rotor. Refer to Rear Disc Brakes, Rotor, removal & installation.

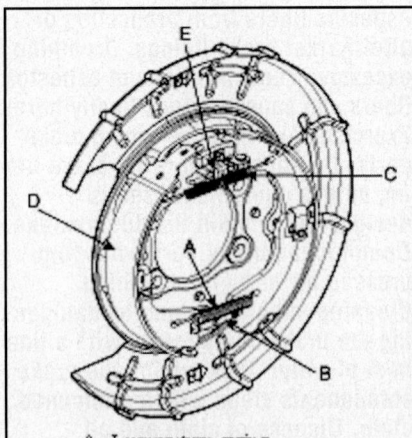

A. Lower return spring
B. Adjuster assembly
C. Upper return spring
D. Brake shoes
E. Operating lever assembly

22140_SANT_G0076

Fig. 6 Remove the adjuster assembly, the lower return spring, the upper return spring, the brake shoes, and the operating lever assembly

17. Adjust the rear brake shoe clearance:
   a. Remove the plug from the disc.
   b. Rotate the toothed wheel of adjuster with a screw driver until the disc is not moving.
   c. Back off 5 notches in the opposite direction.
18. Install the brake caliper. Refer to Rear Disc Brakes, Brake Caliper, removal & installation.
19. Install the tire and wheel.
20. If the parking brake shoe or the brake disc are replaced with a new one, perform the brake shoe brake-in procedure:
   a. While operating the parking brake engaged with a 15 lb. (69 N) effort, drive the vehicle 0.31 miles (500 meters) at the speed of about 37 mph (60 kph).
   b. Repeat the above procedure at least 2 times.

21. The parking brake must be able to operate at 220 lbs. (981 N) with no difficulty.
22. The parking brake should hold on a 30 percent grade.
23. Ensure that all parts move smoothly.
24. The parking brake indicator lamp must turn ON when the parking brake is applied and turn OFF when the parking brake is released.

ADJUSTMENT

See Figure 7.

1. Before servicing the vehicle, refer to the Precautions Section.
2. Raise and safely support the vehicle.
3. Remove the rear wheel and tire assembly.
4. Adjust the rear brake shoe clearance:
   a. Remove the plug from the disc.
   b. Rotate the toothed wheel of

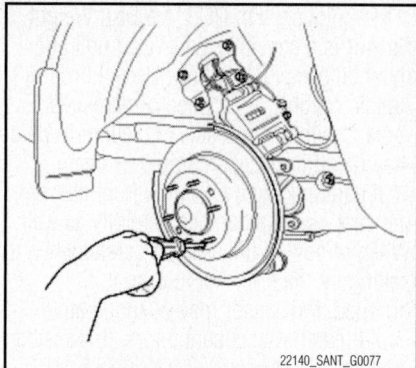

22140_SANT_G0077

Fig. 7 After removing the plug from the disc, rotate the toothed wheel with a screwdriver until the disc does not move, then back it off 5 notches

adjuster with a screw driver until the disc is not moving.
   c. Back off 5 notches in the opposite direction.

CHASSIS ELECTRICAL

AIR BAG (SUPPLEMENTAL RESTRAINT SYSTEM)

GENERAL INFORMATION

✳✳ CAUTION

This vehicle is equipped with an air bag system. The system must be disarmed before performing service on, or around, system components, the steering column, instrument panel components, wiring, and sensors. Failure to follow the safety precautions and the disarming procedure could result in accidental air bag deployment, possible injury, and unnecessary system repairs.

SERVICE PRECAUTIONS

Disconnect and isolate the battery negative cable before beginning any airbag system component diagnosis, testing, removal, or installation procedures. Allow system capacitor to discharge for 3 minutes before beginning any component service. This will disable the airbag system. Failure to disable the airbag system may result in accidental airbag deployment, personal injury, or death.

Do not place an intact undeployed airbag face down on a solid surface. The airbag will propel into the air if accidentally deployed and may result in personal injury or death.

When carrying or handling an undeployed airbag, the trim side (face) of the airbag should be pointing towards the body to minimize possibility of injury if accidental deployment occurs. Failure to do this may result in personal injury or death.

Replace airbag system components with OEM replacement parts. Substitute parts may appear interchangeable, but internal differences may result in inferior occupant protection. Failure to do so may result in occupant personal injury or death.

Wear safety glasses, rubber gloves, and long sleeved clothing when cleaning powder residue from vehicle after an airbag deployment. Powder residue emitted from a deployed airbag can cause skin irritation. Flush affected area with cool water if irritation is experienced. If nasal or throat irritation is experienced, exit the vehicle for fresh air until the irritation ceases. If irritation continues, see a physician.

Do not use a replacement airbag that is not in the original packaging. This may result in improper deployment, personal injury, or death.

The factory installed fasteners, screws and bolts used to fasten airbag components have a special coating and are specifically designed for the airbag system. Do not use substitute fasteners. Use only original equipment fasteners listed in the parts catalog when fastener replacement is required.

During, and following, any child restraint anchor service, due to impact event or vehicle repair, carefully inspect all mounting hardware, tether straps, and anchors for proper installation, operation, or damage. If a child restraint anchor is found damaged in any way, the anchor must be replaced. Failure to do this may result in personal injury or death.

Deployed and non-deployed airbags may or may not have live pyrotechnic material within the airbag inflator.

Do not dispose of driver/passenger/curtain airbags or seat belt tensioners unless you are sure of complete deployment. Refer to the Hazardous Substance Control System for proper disposal.

Dispose of deployed airbags and tensioners consistent with state, provincial, local, and federal regulations.

After any airbag component testing or service, do not connect the battery negative cable. Personal injury or death may result if the system test is not performed first.

If the vehicle is equipped with the Occupant Classification System (OCS), do not connect the battery negative cable before performing the OCS Verification Test using the scan tool and the appropriate diagnostic information. Personal injury or death may result if the system test is not performed properly.

Never replace both the Occupant Restraint Controller (ORC) and the Occupant Classification Module (OCM) at the same time. If both require replacement, replace one, then perform the Airbag System test before replacing the other.

Both the ORC and the OCM store Occupant Classification System (OCS) calibration data, which they transfer to one another when one of them is replaced. If both are replaced at the same time, an irreversible fault will be set in both modules and the OCS may malfunction and cause personal injury or death.

If equipped with OCS, the Seat Weight Sensor is a sensitive, calibrated unit and must be handled carefully. Do not drop or handle roughly. If dropped or damaged, replace with another sensor. Failure to do so may result in occupant injury or death.

If equipped with OCS, the front passenger seat must be handled carefully as well. When removing the seat, be careful when it setting on the floor not to drop it. If dropped, the sensor may be inoperative, could result in occupant injury, or possibly death.

If equipped with OCS, when the passenger front seat is on the floor, no one should sit in the front passenger seat. This uneven force may damage the sensing ability of the seat weight sensors. If sat on and damaged, the sensor may be inoperative, could result in occupant injury, or possibly death.

Several precautions must be observed when handling the inflator module to avoid accidental deployment and possible personal injury.

• Never carry the inflator module by the wires or connector on the underside of the module

• When carrying a live inflator module, hold it securely with both hands, and ensure that the bag and trim cover are pointed away

• Place the inflator module on a bench or other surface with the bag and trim cover facing up

• With the inflator module on the bench, never place anything on or close to the module which may be thrown in the event of an accidental deployment

Before servicing the vehicle, make sure to refer to the precautions in the beginning of this section as well.

DISARMING THE SYSTEM

1. Before servicing the vehicle, refer to the Precautions Section.
2. Record the radio anti-theft code data. Remove the ignition key from the vehicle.
3. Disconnect the negative battery cable.
4. Wait at least 3 minutes for the system capacitor to discharge before performing any service.

ARMING THE SYSTEM

1. Before servicing the vehicle, refer to the Precautions Section.
2. Reconnect the negative battery cable.
3. To confirm proper system operation, turn the ignition switch to the ON position. The SRS indicator light will be lit for at least 6 seconds and then go off.

CLOCKSPRING CENTERING (IF APPLICABLE)

See Figures 8 through 10.

1. Disconnect and isolate the battery negative cable. Allow the system capacitor to discharge for 3 minutes before beginning any component service. This will disable the airbag system.

❊❊ CAUTION

Failure to disable the airbag system may result in accidental airbag deployment, personal injury, or death.

2. Remove the ignition key from the vehicle.
3. Connect the clockspring harness connector and horn harness connector to the clockspring.
4. Set the clockspring on neutral position and after turning the front wheels to the straight-ahead position, install the clockspring.
   a. Check connectors and protective tube for damage, and terminals for deformities.
   b. If even one abnormal point is discovered, replace the clockspring with a new one.
5. Connect the clockspring harness connector and the steering switch harness connector to the clockspring.
6. Set the center position by getting the marks between the clockspring and the cover into line.

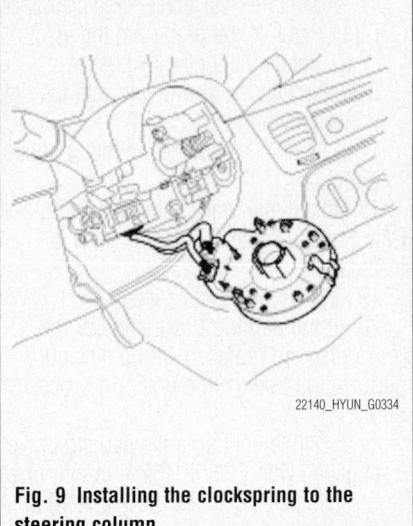
22140_HYUN_G0334

Fig. 9 Installing the clockspring to the steering column

7. Turn the clockspring clockwise to the stop and then 2.4 revolutions counterclockwise. See the illustration of marks in line.
8. Install the steering wheel column cover and the steering wheel.
9. Connect the Driver Airbag (DAB) module connector and the horn connector, then install the DAB module on the steering wheel.
10. Secure the DAB with the new mounting bolts. Tightening torque of DAB mounting bolt: 70–96 inch lbs. (8–11 NM).
11. Connect the battery negative cable.
12. After installing the airbag, confirm proper system operation:
    a. Turn the ignition switch ON; the SRS indicator light should be turned on for about 6 seconds and then go off.
    b. Make sure the horn functions properly.

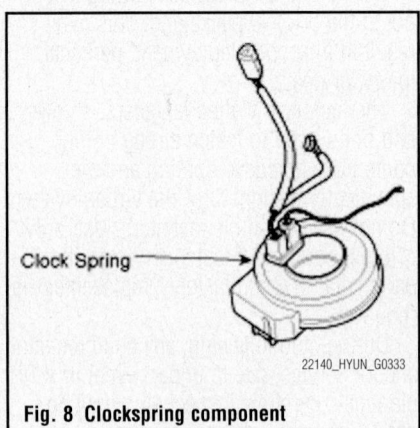
Clock Spring
22140_HYUN_G0333

Fig. 8 Clockspring component

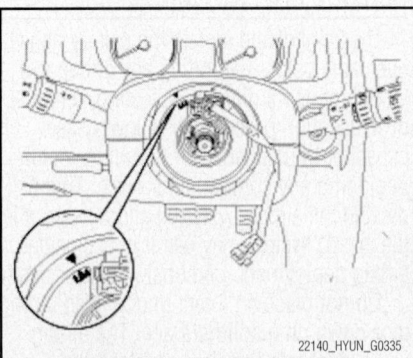
22140_HYUN_G0335

Fig. 10 Aligning the marks to center the clockspring

DRIVETRAIN

AUTOMATIC TRANSAXLE ASSEMBLY

REMOVAL & INSTALLATION

2.4L Engine

See Figure 11.

Use fender covers to avoid damaging painted surfaces. To avoid damage, unplug the wiring connectors carefully while holding the connector portion. Mark all wiring and hoses to avoid misconnection.

1. Before servicing the vehicle, refer to the Precautions Section.
2. Remove the engine cover.
3. Remove the battery after removing the battery terminal.
4. Remove the air duct assembly.
5. Remove the air cleaner assembly by disconnecting the Air Flow Sensor (AFS) connector, the clamp, and the ECM connector.
6. Remove the battery tray.
7. Remove the ground cable from the transaxle.
8. Disconnect the inhibiter switch connector, solenoid valve connector, and the input shaft speed sensor connector.
9. Disconnect the output shaft speed sensor connector.
10. Remove the control cable assembly.
11. Remove the oil cooler hoses.
12. Install the special tools (09200-38001), the engine support fixture and the adapter on the engine assembly.
13. Remove the transaxle upper mounting bolts and the starter motor mounting bolts.
14. Remove the 4 bolts and take off the transaxle support bracket.
15. Remove the steering joint assembly bolt.

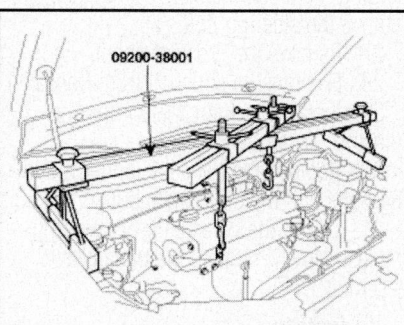

Fig. 11 Install the special tools (09200-38001), the engine support fixture and the adapter on the engine assembly—2.4L engine

16. Remove the front wheels and tires.
17. Remove the side mud cover.
18. Remove the under shield cover.
19. Drain the transaxle fluid by removing the oil drain plug.
20. Remove the lower arm ball joint mounting nut, the stabilizer link mounting nut, and the tie rod end mounting nut from the front knuckles.
21. Remove the roll stopper mounting bolts.
22. Remove the muffler hanger rubber.
23. Supporting the sub frame with a jack and the Special tool (09624-38000), remove the mounting bolts.
24. Disconnect the halfshafts from the transaxle.
25. Remove the drive plate mounting bolts.
26. Supporting the transaxle with a jack, remove the transaxle lower mounting bolts.
27. Lowering the jack slowly, remove the transaxle.

❊❊ WARNING

When removing the transaxle assembly, be careful not to damage any surrounding parts or body components.

To install:

28. Install the transaxle lower mounting bolts after fitting the transaxle assembly into the engine assembly. Tighten to: 31–40 ft. lbs. (43–55 Nm).
29. Install the drive plate mounting bolts. Tighten to: 33–38 ft. lbs. (46–53 Nm).
30. Connect the halfshafts to the transaxle.
31. Supporting the sub frame with a jack and the Special tool(09624-38000), install the mounting bolts. Tighten the bolts to: 101–118 ft. lbs. (140–160 Nm).
32. Install the muffler hanger rubber.
33. Install the roll stopper mounting bolts. Tighten to: 36–47 ft. lbs. (50–65 Nm).
34. Install the lower arm ball joint mounting nut, the stabilizer link mounting nut, and the tie rod end mounting nut to the front knuckles.
35. Install the under shield cover.
36. Install the side mud cover.
37. Install the front wheels and tires.
38. Install the steering joint assembly bolt.
39. Install the transaxle support bracket bolts. Tighten to: 43–58 ft. lbs. (60–80 Nm).

40. Install the transaxle upper mounting bolts. Tighten to: 43–58 ft. lbs. (60–80 Nm).
41. Install the starter motor mounting bolts. Tighten to: 28–43 ft. lbs. (39–60 Nm).
42. Remove the special tool (09200-38001).
43. Connect the transaxle oil cooler hoses to the tubes by fastening the clamps.
44. Install the control cable assembly.
45. Install the output speed sensor connector.
46. Connect the inhibiter switch connector, solenoid valve connector and the input shaft speed sensor connector.
47. Install the ground cable to transaxle.
48. Install the battery tray.
49. Install the air cleaner assembly by connecting the AFS connector, the clamp, and the ECM connector.
50. Install the air duct assembly.
51. Install the battery and the battery terminal.
52. Install the engine cover.
53. After completing the installation perform the following procedure:
   a. Adjust the shift cable.
   b. Refill the transaxle fluid.
   c. Clean the battery posts and cable terminals with sandpaper and grease them to prevent corrosion before installing.
   d. When replacing the automatic transaxle, reset the automatic transaxle's values by using the High-Scan Pro, as follows:
   • Connect the Hi-Scan Pro connector to the data link connector under the crash pad and power cable to the cigar jack under the center fascia
   • Turn the ignition switch on and power on the Hi-Scan Pro
   • Select the vehicle name
   • Select AUTOMATIC TRANSAXLE
   • Select RESETTING AUTO T/A VALUES and perform the procedure
   • Perform the procedure by pressing F1 (REST)
54. Install the transaxle control cable and adjust as follows.
   a. Move the shift lever and the transaxle range switch to the "N" Position, and install the control cable.
   b. When connecting the control cable to the transaxle mounting bracket, install the clip until it contacts the control cable.
   c. Remove any free-play in the control cable by adjusting the nut and then check to see that the select lever moves smoothly.

d. Check to see that the control cable has been adjusted correctly.

55. Fill the transaxle fluid to the proper level.

56. Start the vehicle, check for leaks and repair if necessary.

3.5L Engine

1. Before servicing the vehicle, refer to the Precautions Section.
2. Drain the transaxle.
3. Remove the battery and tray.
4. Remove the air cleaner duct.
5. Remove the air cleaner and air flow hose.
6. Disconnect the back-up light connector, Crankshaft Position (CKP) sensor and oil pressure switch connectors.
7. Disconnect the speedometer cable.
8. Remove the cotter pin from the shift cable at the transaxle.
9. Remove the clip from the shift cable on the transaxle side.
10. Remove the clip from the select cable at the transaxle side.
11. Separate the steering column shaft joint.
12. Separate the power steering pump hose.
13. Separate the hose after removing the clip from the power steering return hose.
14. Remove the upper transaxle bolt.
15. Remove the starter motor.
16. Install an engine support fixture.
17. Remove the front wheels and calipers.
18. Disconnect the tie rod end, wheel speed sensor, and knuckle mounting bolt.
19. Remove the transaxle mounting bracket and insulator.
20. Remove the front roll stopper insulator bolt, upper and lower stopper bolts and the front roll stopper.
21. Remove the rear roll stopper insulator bolt, and stopper bolt and the roll stopper.
22. Remove the drive shaft.
23. Remove the subframe bolts and subframe.
24. Install a transaxle jack.
25. Remove the transaxle lower bolts.
26. Remove the transaxle-to-engine bolts.
27. Remove the transaxle.

To install:

28. Installation is the reverse of removal keeping in mind the following steps and torques:

a. Tighten the transaxle case bolts to: 15–20 ft. lbs. (20–27 Nm).

b. Tighten the transaxle mounting bracket bolts to: 43–58 ft. lbs. (60–80 Nm).

c. Tighten the transaxle mounting bracket bolts to: 29–40 ft. lbs. (40–55 Nm).

d. Tighten the transaxle mounting insulator bolt to: 65–80 ft. lbs. (90–110 Nm).

e. Tighten the front roll stopper-to-transaxle bolts to: 43–58 ft. lbs. (60–80 Nm).

f. Tighten the front roll stopper insulator bolt and nut to: 36–47 ft. lbs. (50–65 Nm).

g. Tighten the front roll stopper-to-subframe bolts to: 43–58 ft. lbs. (60–80 Nm).

h. Tighten the rear roll stopper-to-subframe bolts to: 43–58 ft. lbs. (60–80 Nm).

i. Tighten the rear roll stopper-to-transaxle bolt and nut to: 36–47 ft. lbs. (50–65 Nm).

j. Tighten the rear roll stopper insulator bolt and nut to: 36–47 ft. lbs. (50–65 Nm).

k. Move the shift lever and the transaxle range switch to the **N** position and install the cable.

l. When attaching the control cable to the bracket, make sure the clip installs so it contacts the cable.

m. Adjust the nut to remove any free-play in the cable and make sure the lever moves freely.

29. Fill the transaxle to the correct level.
30. Start the engine and check for leaks.
31. Check the wheel alignment and adjust as necessary.

2.7L, 3.3L & 3.8L Engines

See Figures 12 and 13.

Use fender covers to avoid damaging painted surfaces. To avoid damage, unplug the wiring connectors carefully while holding the connector portion. Mark all wiring and hoses to avoid misconnection.

1. Before servicing the vehicle, refer to the Precautions Section.
2. Disconnect the negative terminal from the battery.
3. Remove the engine cover.
4. Remove the intake air hose and the air cleaner assembly.

a. Disconnect the AFS connector.

b. Disconnect the breather hose from air cleaner hose.

c. Disconnect the PCM connectors.

d. Remove the intake air hose and air cleaner.

5. Disconnect the positive terminal from the battery and remove the battery.
6. Remove the transaxle oil cooler hoses.
7. Remove engine wiring.

a. Disconnect the RH rear oxygen sensor connector.

b. Disconnect the LH rear oxygen sensor connector and the CPS connector.

8. Disconnect the transaxle wire harness connector and remove transaxle control cable.

a. Remove the wiring brackets.

b. After removing a transaxle bracket, remove the inhibiter switch connector and shift cable.

c. Remove the solenoid valve connector.

d. Remove the input speed sensor, output speed sensor and vehicle speed sensor connector.

e. Disconnect the ground wire.

9. Disconnect the power steering pressure sensor connector.
10. Remove the power steering hose mounting bolts.
11. Remove the front wheels.
12. Disconnect the EPS connector around the left hand side front wheel.
13. Remove the transaxle mounting bolts.
14. Using the SST (09200-38001), hold the engine and transaxle assembly safely.
15. Remove the transaxle insulator mounting bolt.
16. Raise and safely support the vehicle.
17. Remove the undercover.
18. Drain the transaxle oil.
19. Disconnect the power steering pump hose.
20. Disconnect the lower arm assembly from the knuckle.
21. Disconnect the tie rod end ball joint from the knuckle after removing the split pin.
22. Disconnect the stabilizer bar link.
23. Remove the front roll stopper mounting bolt.
24. Remove the front exhaust pipe.
25. Remove the rear roll stopper mounting bolt.
26. Using the SST (09624-38000) and holding the cross member with a jack, remove the steering bolt.
27. Remove the cross member.
28. Remove the halfshaft from transaxle.
29. Install a jack for supporting the transaxle assembly.
30. Remove the transaxle under mounting bolts and the drive plate bolts.
31. Lifting the vehicle up or lowering the jack slowly, remove the transaxle assembly.

To install:

32. Lowering the vehicle or lifting up a jack, install the transaxle assembly.
33. Tighten the transaxle under mounting bolts. Tighten to:

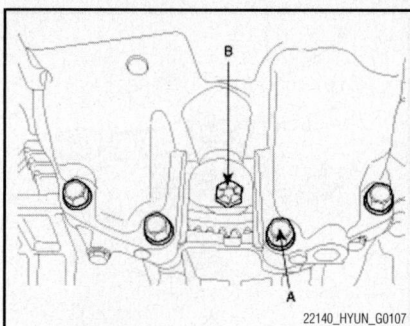

Fig. 12 Tighten the transaxle under mounting bolts (A, B)—2.7L, 3.3L, and 3.8L engines

   a. Bolts (A): 25–30 ft. lbs. (34–41.2 Nm).

   b. Bolt (B): 33–38 ft. lbs. (45–52 Nm).

34. Remove the jack and insert the half-shafts.

35. Supporting the cross member with the SST (09624-38000), tighten the steering column bolt and the cross member mounting bolts.

36. Tighten the rear roll stopper mounting bolt to: 36–47 ft. lbs. (49–64 Nm).

37. Install the front exhaust pipe.

38. Tighten the front roll stopper mounting bolt to: 36–47 ft. lbs. (49–64 Nm).

39. Install the steering bar tie rod, the stabilizer bar link, and the lower arm assembly.

40. Clamp the power steering pump hose.

41. Install the undercover.

42. After lowering the vehicle, tighten the transaxle insulator mounting bolt(A). Tighten to: 47–62 ft. lbs. (63–83 Nm).

43. Tighten the transaxle mounting bolts (B, C).

   a. Bolts (B): 24–36 ft. lbs. (32–49 Nm).

   b. Bolts (C): 47–62 ft. lbs. (64–83 Nm).

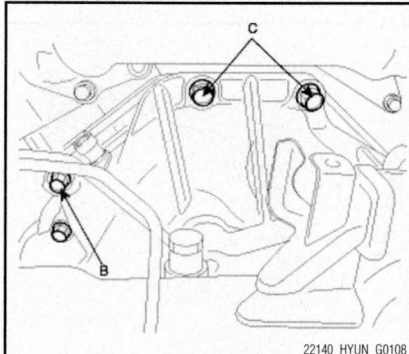

Fig. 13 Tighten the transaxle under mounting bolts (B, C)—2.7L, 3.3L, and 3.8L engines

44. Remove the SST (09200-38001) holding the engine and transaxle assembly.

45. Connect the EPS connector and install the front wheels and tires.

46. Install the power steering hose mounting bolts.

47. Connect the power steering pressure sensor connector.

48. Connect the transaxle wire harness connector and the control cable.

   a. Install the wiring brackets.

   b. Connect the inhibitor switch connector and the shift cable and install the transaxle bracket.

   c. Connect the solenoid valve connector.

   d. Connect the input/output speed sensor connectors and vehicle speed sensor connector.

   e. Connect the ground wire.

49. Connect the engine wiring.

   a. Connect the RH rear oxygen sensor connector.

   b. Connect the LH rear oxygen sensor connector and the CPS connector.

50. Clamp the transaxle oil cooler hoses(A).

51. After disconnecting the positive terminal from the battery, remove the battery.

52. Install the intake air hose and the air cleaner assembly.

   a. Connect the AFS connector.

   b. Clamp the breather hose from the air cleaner hose.

   c. Connect the PCM connectors.

   d. Install the intake air hose and the air cleaner assembly.

53. Install the engine cover.

54. Connect the negative terminal on the battery.

55. Fill the transaxle fluid to the proper level.

56. Start the vehicle, check for leaks and repair if necessary.

MANUAL TRANSAXLE ASSEMBLY

REMOVAL & INSTALLATION

2.4L Engine

See Figures 14 through 16.

Use fender covers to avoid damaging painted surfaces. To avoid damage, unplug the wiring connectors carefully while holding the connector portion. Mark all wiring and hoses to avoid misconnection.

1. Before servicing the vehicle, refer to the Precautions Section.

2. Remove the engine cover.

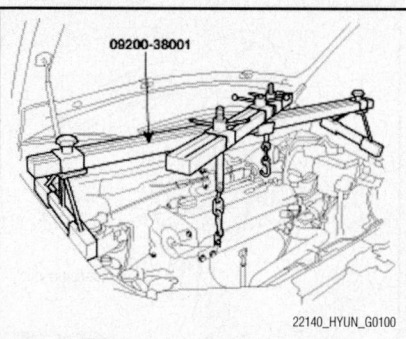

Fig. 14 Using the special tool (09200-38001) to support the engine assembly—2.4L engine

3. Remove the battery after removing the battery terminal connections.

4. Remove the air duct assembly.

5. Remove the air cleaner assembly by disconnecting the clamp and the ECM connector.

6. Remove the ground cable from the transaxle.

7. Disconnect the vehicle speed sensor and the back lamp switch integrated connector.

8. Remove the control cable assembly by removing the snap pins and clips.

9. Remove the control cable bracket.

10. Using the special tool (09200-38001), support the engine assembly safely.

11. Remove the transaxle upper mounting bolts and the starter motor mounting bolts.

12. After removing the bolts, take the transaxle insulator mounting bracket off.

13. Remove the steering joint assembly bolt.

14. Raise and safely support the vehicle.

15. Remove the front wheels and tires.

16. Remove the lower arm ball joint mounting nut, the stabilizer link mounting nut, and the tie rod end mounting nut from the front knuckles.

17. Remove the under shield cover.

18. Remove the roll stopper mounting bolts.

19. Disconnect the muffler hanger rubber.

20. Supporting the sub frame (A) with a jack and the Special tool (09624-38000), remove the mounting bolts.

21. Disconnect the halfshafts from the transaxle.

22. Remove the clutch release cylinder assembly.

23. Supporting the transaxle with a jack, remove the transaxle lower mounting bolts.

24. Lowering the jack slowly, remove the transaxle.

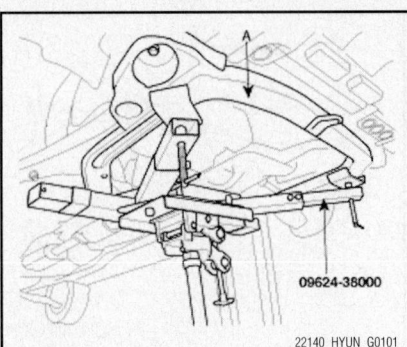

Fig. 15 Support the sub frame (A) with a jack and the Special tool (09624-38000) and remove the mounting bolts—2.4L engine

To install:

25. Fit the transaxle assembly to the engine assembly.

26. Install the transaxle lower mounting bolts. Tighten to:

    a. Bolts (A): 22–30 ft. lbs. (30–42 Nm).

    b. Bolts (B): 31–40 ft. lbs. (43–55 Nm).

27. Install the clutch release cylinder assembly.

28. Connect the halfshafts to the transaxle.

29. Supporting the sub frame with a jack and the Special tool(09624-38000), install the mounting bolts. Tighten to: 101–118 ft. lbs. (140–160 Nm).

30. Install the muffler hanger rubber.

31. Install the roll stopper bracket bolts. Tighten to: 36–47 ft. lbs. (50–65 Nm).

32. Install the under shield cover.

33. Install the lower arm ball joint mounting nut, the stabilizer link mounting nut, and the tie rod end mounting nut to the front knuckles.

34. Install the front wheels and tires.

35. Install the steering joint assembly bolt. Tighten to: 13–18 ft. lbs. (18–25 Nm).

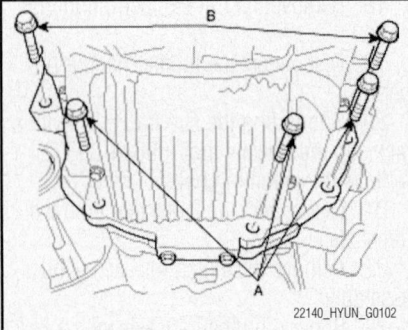

Fig. 16 Install the transaxle lower mounting bolts—2.4L engine

36. Install the transaxle insulator mounting bracket bolts. Tighten to: 43–58 ft. lbs. (60–80 Nm).

37. Remove the special tool (09200-38001).

38. Install the transaxle upper mounting bolts. Tighten to: 43–60 ft. lbs. (60–80 Nm).

39. Install the starter motor mounting bolts. Tighten to: 28–43 ft. lbs. (39–60 Nm).

40. Install the control cable bracket. Tighten to: 11–16 ft. lbs. (15–22 Nm).

41. Connect the vehicle speed sensor and the back lamp switch integrated connector.

42. Install the control cable assembly by installing the clips and pins.

43. Install the ground cable to the transaxle.

44. Install the air cleaner assembly and the ECM connector.

45. Install the air duct assembly.

46. Install the battery and connect the battery terminals.

47. Install the engine cover.

48. Fill the transaxle to the correct level.

49. Start the engine and check for leaks.

50. Check the wheel alignment and adjust as necessary.

2.7L Engine

See Figures 17 through 19.

Use fender covers to avoid damaging painted surfaces. To avoid damage, unplug the wiring connectors carefully while holding the connector portion. Mark all wiring and hoses to avoid misconnection.

1. Before servicing the vehicle, refer to the Precautions Section.

2. Remove the inter-cooler assembly and the engine cover.

3. Remove the battery.

4. Disconnect the AFS connector.

5. Remove the air cleaner hose and loosen the clamp bolt.

6. Remove the air cleaner upper cover by removing the clips.

7. Remove the air cleaner assembly by removing the 2 mounting bolts.

8. Remove the battery tray by removing the 4 mounting bolts.

9. Remove the ground wire from the transaxle case.

10. Disconnect the vehicle speed sensor and the backup lamp switch integrated connector.

11. Remove the control cable assemblies (A) by removing the snap pins (B) and clips (C).

12. Remove the Concentric Slave Cylinder (CSC) tube (A), which is being clamped, by loosening the nut (B).

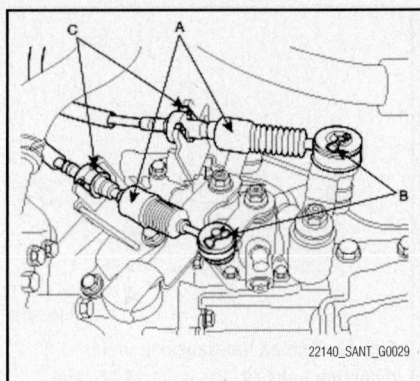

Fig. 17 Remove the control cable assemblies (A) by removing the snap pins (B) and clips (C)—2.7L engine

13. Remove the four mounting bolts of upper part of the transaxle.

14. Support the engine and transaxle by using the special tool (09200-38001).

15. Remove the transaxle insulator bracket by removing the bolts.

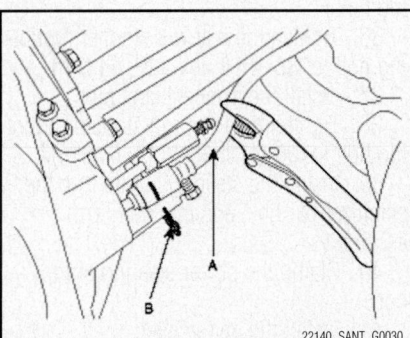

Fig. 18 Remove Concentric Slave Cylinder (CSC) tube (A), which is being clamped, by loosening the nut (B)—2.7L engine

16. Remove the front wheels and tires.

17. Lift up the vehicle.

18. Remove the steering column joint bolt.

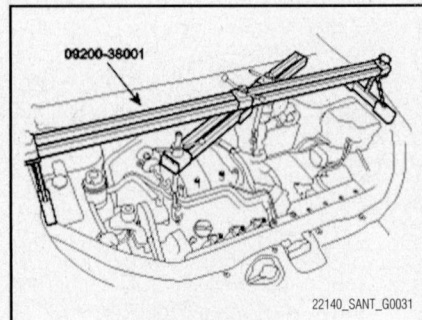

Fig. 19 Using the special tool (09200-38001) to support the engine and transaxle—2.7L engine

19. Remove the undercover.

20. Drain power steering oil through the return tube.

21. Disconnect the power steering pressure hose from the power steering oil pump.

22. Drain the transaxle fluid through the drain plug.

23. Disconnect the lower arm, the tie rod end ball joint, the stabilizer bar link from the front knuckle.

24. Remove the roll stopper mounting bolts.

25. Remove the mounting bolts from the sub frame by supporting the sub frame with a jack.

26. Remove the halfshafts from the transaxle.

27. Disconnect the starter motor connector and remove the starter motor.

28. With 4WD, remove the transfer case assembly. Refer to Transfer Case, removal & installation.

29. Remove the mounting bolts of the lower part of the transaxle, and the left side cover.

30. Remove the transaxle assembly while supporting it with a jack.

To install:

31. Lowering the vehicle or lifting up a jack, install the transaxle assembly.

32. Tighten the transaxle under mounting bolts to: 47–62 ft. lbs. (65–85 Nm).

33. Install the starter motor and connect the starter motor connector.

34. With 4WD, install the transfer case assembly. Refer to Transfer Case, removal & installation.

35. Install the halfshafts to the transaxle.

36. Install the transaxle insulator and mounting bracket by tightening the bolts to: 44–59 ft. lbs. (60–80 Nm).

37. Install the sub frame and tighten the bolts to: 44–59 ft. lbs. (60–80 Nm).

38. Connect the power steering return hose.

39. Connect the lower arm, the rod end ball joint, and the stabilizer bar control link to the front knuckle.

40. Install the steering column joint bolt.

41. Install the CSC tube by tightening the nut.

42. Refill transaxle oil through the inlet hole and tighten the filler plug to: 22–25 ft. lbs. (30–35 Nm).

43. Install the undercover.

44. Install the front wheels and tires.

45. Install the transaxle mounting bolts and remove the SST (09200-38001) holding the engine and transaxle assembly. Tighten the mounting bolts to: 47–62 ft. lbs. (65–85 Nm).

46. Connect the power steering pressure hose to the power steering oil pump.

47. Install the control cable assemblies by installing the clips and snap pins.

➡ **Do not reuse the control cable clips. Replace with new ones.**

48. Install the vehicle speed sensor and the backup lamp integrated connector.

49. Install the ground wire from the transaxle case.

50. Install the battery tray and the 4 mounting bolts.

51. Install the air cleaner assembly by tightening the 2 mounting bolts.

52. Install the air cleaner upper cover.

53. Install the air cleaner hose and tighten the clamp bolt.

54. Connect the AFS connector.

55. Install the battery.

56. Refill the power steering fluid.

57. Install the engine cover and the intercooler assembly.

➡ **Bleed any air from the power steering system.**

58. Start the engine and check for leaks.

59. Check the wheel alignment and adjust as necessary.

3.5L Engine

1. Before servicing the vehicle, refer to the Precautions Section.

2. Drain the transaxle.

3. Remove the battery and tray.

4. Remove the air cleaner duct.

5. Remove the air cleaner and air flow hose.

6. Disconnect the back-up light connector, Crankshaft Position (CKP) sensor and oil pressure switch connectors.

7. Disconnect the speedometer cable.

8. Remove the clutch release cylinder bolt.

9. Remove the cotter pin from the shift cable at the transaxle.

10. Remove the clip from the shift cable on the transaxle side.

11. Remove the clip from the select cable at the transaxle side.

12. Separate the steering column shaft joint.

13. Separate the power steering pump hose.

14. Separate the hose after removing the clip from the power steering return hose.

15. Remove the upper transaxle bolt.

16. Remove the starter motor.

17. Install an engine support fixture.

18. Remove the front wheels and calipers.

19. Disconnect the tie rod end, wheel speed sensor, and knuckle mounting bolt.

20. Remove the transaxle mounting bracket and insulator.

21. Remove the front roll stopper insulator bolt, upper and lower stopper bolts and the front roll stopper.

22. Remove the rear roll stopper insulator bolt, and stopper bolt and the roll stopper.

23. Remove the drive shaft.

24. Remove the subframe bolts and subframe.

25. Install a transaxle jack.

26. Remove the transaxle lower bolts.

27. Remove the transaxle-to-engine bolts.

28. Remove the manual transaxle.

To install:

29. Installation is the reverse of removal keeping in mind the following steps and torques:

   a. Transaxle case bolts to: 15–20 ft. lbs. (20–27 Nm).

   b. Tighten the transaxle mounting bracket bolts to: 43–58 ft. lbs. (60–80 Nm).

   c. Transaxle mounting insulator bolt to: 65–80 ft. lbs. (90–110 Nm).

   d. Tighten the front roll stopper-to-transaxle bolts to: 43–58 ft. lbs. (60–80 Nm).

   e. Tighten the front roll stopper insulator bolt and nut to: 36–47 ft. lbs. (50–65 Nm).

   f. Tighten the front roll stopper-to-subframe bolts to: 43–58 ft. lbs. (60–80 Nm).

   g. Tighten the rear roll stopper-to-subframe bolts to: 43–58 ft. lbs. (60–80 Nm).

   h. Tighten the rear roll stopper-to-transaxle bolt and nut to: 36–47 ft. lbs. (50–65 Nm).

   i. Tighten the rear roll stopper insulator bolt and nut to: 36–47 ft. lbs. (50–65 Nm).

   j. Tighten the clutch release cylinder retainers to: 11–16 ft. lbs. (15–22 Nm).

30. Fill the transaxle to the correct level.

31. Start the engine and check for leaks.

32. Check the wheel alignment and adjust as necessary.

CLUTCH DRIVEN DISC & PRESSURE PLATE

REMOVAL & INSTALLATION

See Figures 20 through 25.

1. Before servicing the vehicle, refer to the Precautions Section.

2. Remove the transaxle assembly. Refer to Manual Transaxle Assembly, removal & installation.

Fig. 20 Special tool placement diagram

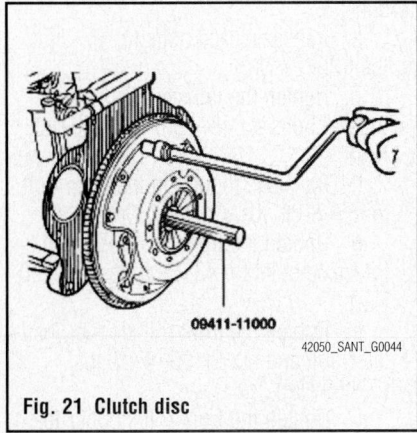

Fig. 21 Clutch disc

3. Insert the special tool (09411-11000) in the clutch disc to prevent the disc from falling.

4. Loosen the bolts which attach the clutch cover to the flywheel in a star pattern. Loosen the bolts in succession, 1–2 turns at a time, to avoid bending the cover flange.

→Do not clean the clutch disc or the release bearing with cleaning solvent.

5. Remove the release fork shaft and bushing.

6. Remove clutch cover assembly and then the driven disc and pressure plate.

To install:

✱✱ WARNING

When installing the clutch, apply grease to each part, but be careful not to apply excessive grease. It can cause clutch slippage and shudder.

7. Install the driven disc and pressure plate in to the clutch cover assembly.

8. Install the clutch disc assembly to the flywheel using the special tool (09411-11000).

Fig. 22 Clutch grease application

9. Install the clutch cover assembly to the flywheel and temporarily tighten the bolts 1 or 2 steps at a time in a star pattern.

10. Tightening torque of the clutch cover bolt: 11–16 ft. lbs. (15–22 Nm).

11. Align the bearing (A) to the release fork (B) and then install it to the sleeve of the housing.

12. Apply multipurpose grease (CAS-MOLY® L9508) to the bearing sleeve, contact point of the release fork (B), and the bushing inner surface (C).

13. Install the release lever to the release fork.

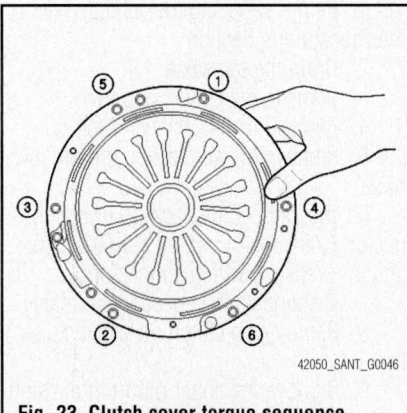

Fig. 23 Clutch cover torque sequence

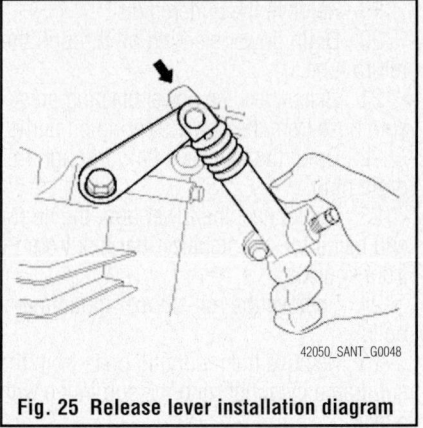

Fig. 25 Release lever installation diagram

14. Install the transaxle assembly to the engine. Refer to Manual Transaxle Assembly, removal & installation.

15. Test the clutch for proper operation.

ADJUSTMENTS

The clutch system is hydraulic and requires no adjustment.

CLUTCH MASTER CYLINDER

REMOVAL & INSTALLATION

See Figures 26 and 27.

1. Before servicing the vehicle, refer to the Precautions Section.

✱✱ WARNING

Do not spill brake fluid on the vehicle; it may damage the paint; if brake fluid does contact the paint, wash it off immediately with water.

2. Remove the brake fluid from the clutch master cylinder reservoir with a syringe.

3. Clamp the clutch master cylinder hose (A). If there is no enough room for clamping, you can also clamp the hose (B) from the brake master cylinder side.

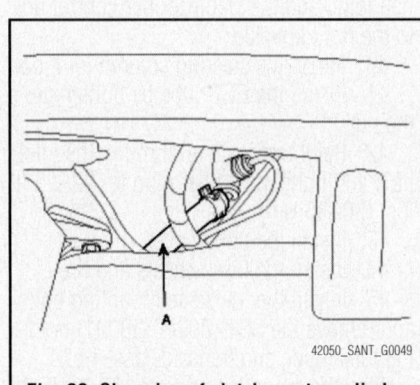

Fig. 26 Clamping of clutch master cylinder hose diagram

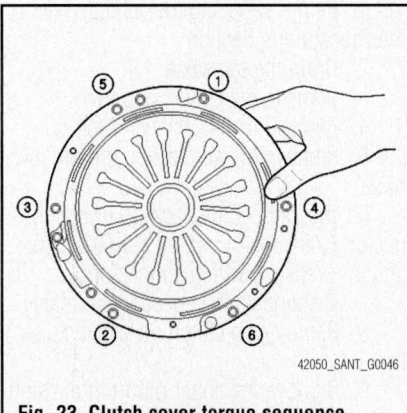

Fig. 24 Release fork grease application

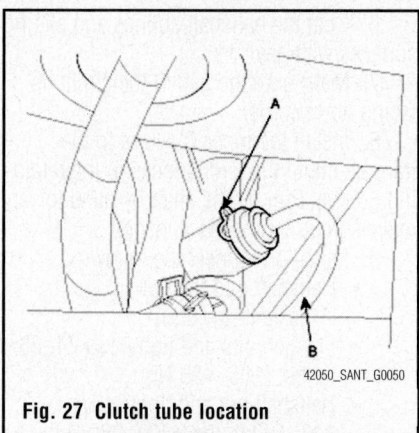

Fig. 27 Clutch tube location

4. Disconnect the hose from the cylinder by releasing the clutch master cylinder clamp.

5. Remove the clip (A) and disconnect the clutch tube (B).

6. Remove the pin and washer which connect the clutch pedal with the clutch master cylinder.

7. After loosening the clutch master cylinder assembly mounting bolts under the driver's seat, remove the clutch master cylinder. It can be helpful to do this step after removing the clutch pedal mounting bracket.

To install:

8. Installation is the reverse of removal.

CLUTCH SLAVE CYLINDER

REMOVAL & INSTALLATION

See Figures 28 and 29.

1. Before servicing the vehicle, refer to the Precautions Section.

2. Disconnect the clutch tube.

3. Remove the 2 clutch slave cylinder mounting bolts (A).

4. Remove the clutch slave cylinder (also called the clutch release cylinder).

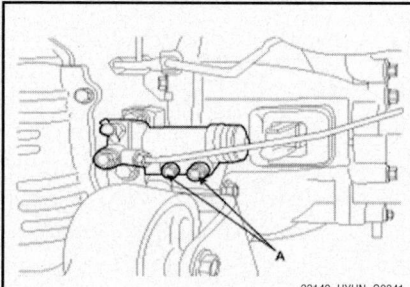

Fig. 28 Disconnect the clutch tube and remove the 2 clutch release slave cylinder mounting bolts (A)

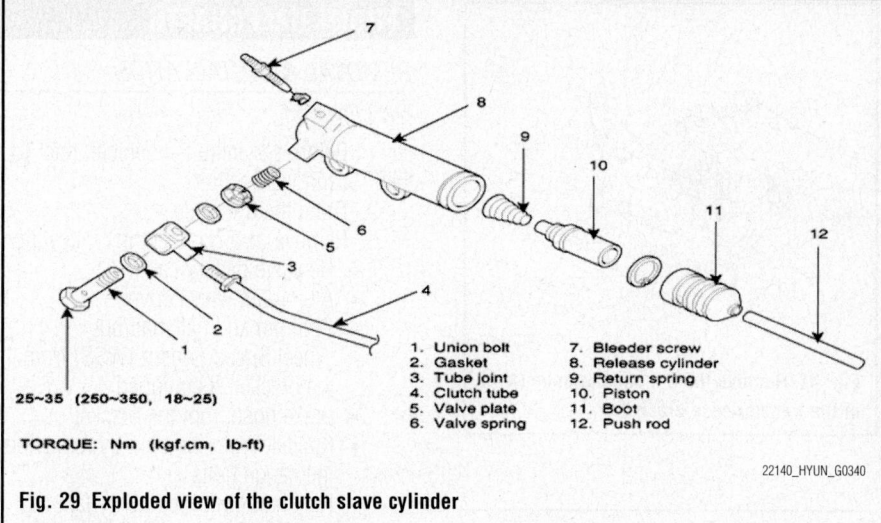

25~35 (250~350, 18~25)

TORQUE: Nm (kgf.cm, lb-ft)

1. Union bolt
2. Gasket
3. Tube joint
4. Clutch tube
5. Valve plate
6. Valve spring
7. Bleeder screw
8. Release cylinder
9. Return spring
10. Piston
11. Boot
12. Push rod

Fig. 29 Exploded view of the clutch slave cylinder

To install:

5. Check the clutch release cylinder for fluid leakage.

6. Check the clutch release cylinder boots for damage.

7. Coat the clutch clevis push rod with specified grease: CASMOLY® L9508.

8. Install the release cylinder (A) to the transaxle. Tighten the bolts to: 11–16 ft. lbs. (15–22 Nm).

9. Install the clutch tube.

CLUTCH HYDRAULIC SYSTEM BLEEDING

Bleeding air from the hydraulic clutch system is necessary whenever any part of the system has been disconnected or the fluid level (in the reservoir) has been allowed to fall so low that air has been drawn into the master cylinder.

❊❊ WARNING

NEVER use fluid that has been bled from a clutch system to fill the master cylinder reservoir, as it may be aerated, contain excessive moisture and/or be contaminated in some other way.

1. Before servicing the vehicle, refer to the Precautions Section.

2. Fill the clutch master cylinder reservoir with new hydraulic clutch fluid.

3. Attach a hose to the bleeder on the clutch actuator and submerge the other end of the hose in a container of hydraulic clutch fluid.

4. Have an assistant slowly depress and hold the clutch pedal.

5. Loosen the bleeder to purge air.

6. Tighten the bleeder.

7. Repeat the above 3 steps until all air is completely purged from the system.

8. Refill the clutch master cylinder reservoir.

TRANSFER CASE ASSEMBLY

REMOVAL & INSTALLATION

See Figures 30 through 32.

1. Before servicing the vehicle, refer to the Precautions Section.

2. Disconnect the negative (-) terminal from the battery.

3. Raise and safely support the vehicle.

4. Remove the front muffler.

5. Disconnect the driveshaft (A) (also called the propeller shaft) by removing the 3 bolts.

6. Disconnect the right halfshaft from the transfer case.

7. Drain the fluid by removing the oil drain plug.

8. After completely draining the fluid, install the oil drain plug. Torque: 29–43 ft. lbs. (40–60 Nm).

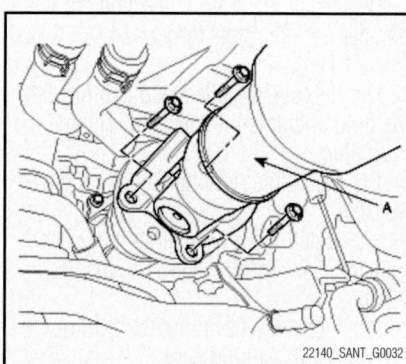

Fig. 30 Disconnect the driveshaft (A) by removing the 3 bolts

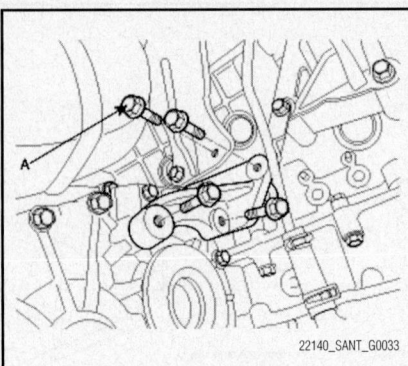

Fig. 31 Remove the 4 mounting bolts (A) of the transfer case bracket

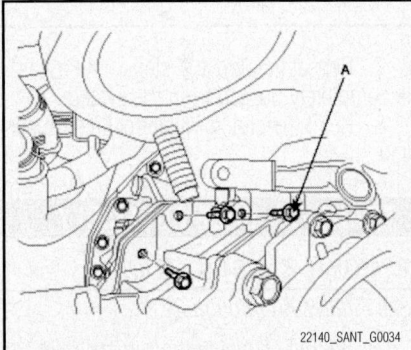

Fig. 32 Remove the transfer case assembly by removing the 6 mounting bolts (A)

9. Remove the 4 mounting bolts (A) of the transfer case bracket.

10. Remove the transfer case assembly by removing the 6 mounting bolts (A).

To install:

➡**For installation, use a new O-ring (47354-39300).**

11. Temporarily install the transfer case assembly to the transaxle and then tighten the 6 mounting bolts (A). Torque: 46–49 ft. lbs. (62–67 Nm).

12. Install the 4 mounting bolts of the transfer case bracket. Torque: 34–37 ft. lbs. (47–51 Nm).

13. Remove the filler plug and fill with the fluid and amount specified. Tighten the filler plug.

14. Install the right halfshaft to the transfer case.

15. Install the driveshaft to the transfer case. Torque: 36–51 ft. lbs. (50–70 Nm).

16. Install the front muffler. Torque: 29–43 ft. lbs. (40–60 Nm).

17. Connect the negative (-) terminal to the battery.

FRONT HALFSHAFTS

REMOVAL & INSTALLATION

See Figure 33.

1. Before servicing the vehicle, refer to the Precautions Section.

2. Drain the transaxle.

3. Remove or disconnect the following:
- Negative battery cable
- Aluminum wheel cover
- Split pin and halfshaft nut
- Wheel Speed Sensor (WSS) from the bracket, if equipped
- Brake hose from the bracket
- Knuckle from the strut by removing the flange bolts
- Halfshaft from the hub by tapping the end with a plastic mallet
- Halfshaft from the transaxle using a pry bar.

4. Insert a plug in the transaxle opening.

To install:

➡**Use new circlips, split pins, and self-locking nuts for assembly.**

5. Remove the plug from the transaxle opening.

6. Coat the halfshaft splines and sliding surfaces with gear oil.

7. Make sure the gap of the circlip is facing downwards.

8. Install the inner CV-joint to the transaxle until the circlip locks in the retaining groove. Pull on the shaft by hand to make sure it is properly engaged.

9. Install or connect the following:
- Halfshaft into the hub
- Knuckle to the strut
- Flange bolts and tighten to 74–88 ft. lbs. (100–120 Nm)
- Halfshaft nut and tighten to 146–190 ft. lbs. (200–260 Nm)
- New split pin
- Brake hose to the bracket
- WSS to the bracket, if equipped
- Aluminum wheel cover
- Negative battery cable

10. Fill the transaxle to the correct level and check for leaks.

FRONT PINION SEAL

REMOVAL & INSTALLATION

See Figures 34 through 40.

1. Before servicing the vehicle, refer to the Precautions Section.

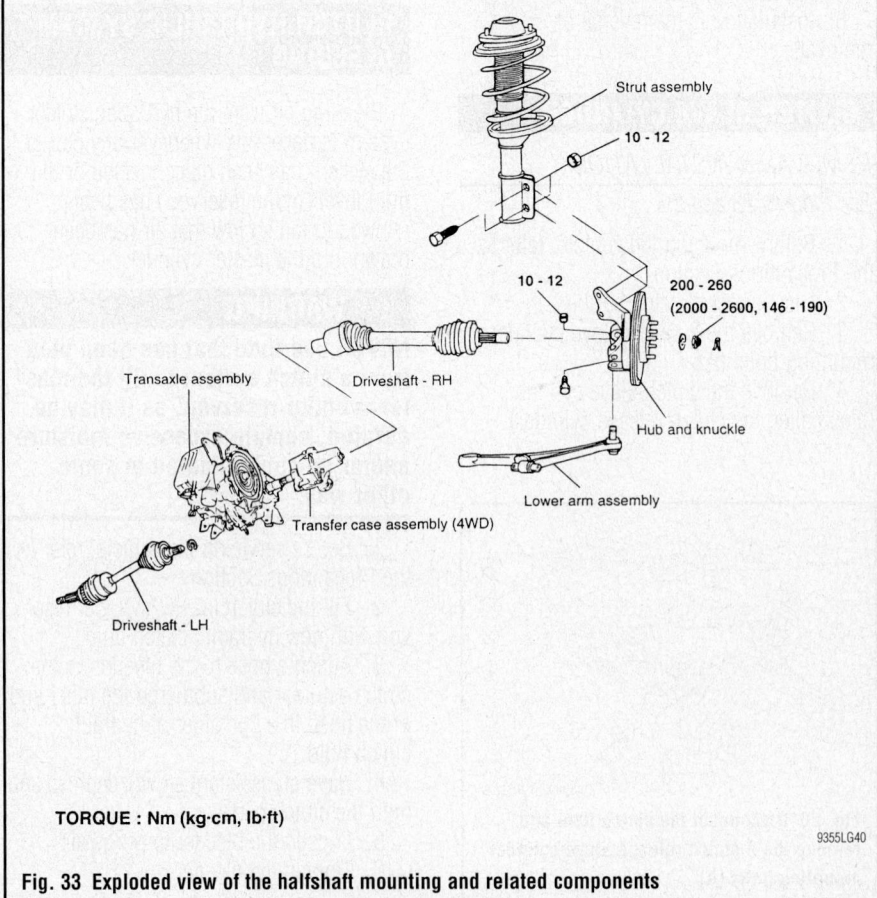

TORQUE : Nm (kg·cm, lb·ft)

Fig. 33 Exploded view of the halfshaft mounting and related components

2. Remove the propeller shaft.

3. Drain the transfer oil through drain plug hole.

4. Remove the 7 pinion assembly mounting bolts (B) and then remove the pinion assembly (A).

5. Remove the O-ring (A) from the pinion shaft and case assembly.

6. Remove the spacer (A).

7. Remove the HEX lock nut (A).

8. Remove the rear flange assembly (A).

9. Using the SST(09455-32200), remove the pinion oil seal (A) from the pinion case.

To install:

10. Install the oil seal and the rear flange assembly.

11. Tighten down the hex lock nut until there is an axial clearance **B** of 0.0099–0.0296 inch (0.25mm–0.75mm) between pinion and pinion case.

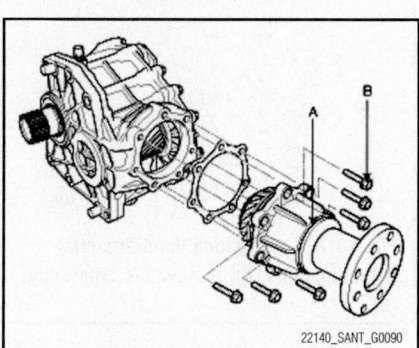

Fig. 34 Remove the 7 pinion assembly mounting bolts (B) and then remove the pinion assembly (A)

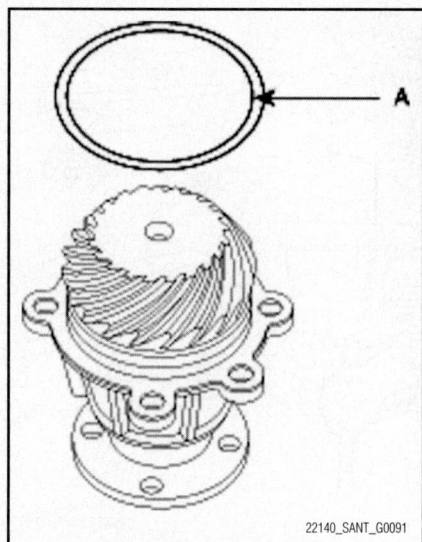

Fig. 35 Remove the O-ring (A) from the pinion shaft and case assembly

Fig. 36 Remove the spacer (A)

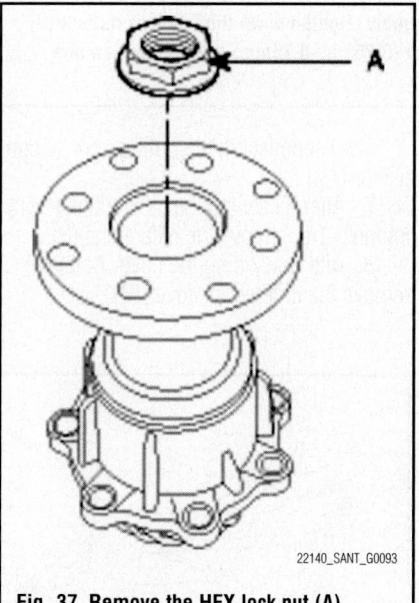

Fig. 37 Remove the HEX lock nut (A)

a. Measurement of **B**:
• Apply an additional 200N tension load (F1) to the pinion head according to arrangement Y. Rotate the pinion 10 times and measure dimension C
• Apply a 200N load (F2) to the rear flange. Rotate the pinion 10 times and measure dimension D
• B = C minus D
b. Measure dimension Ev.
• Tighten HEX lock nut
• Measure the drag torque of the pinion assembly. Target is 16–19 inch lbs. (1.8–2.1 Nm)

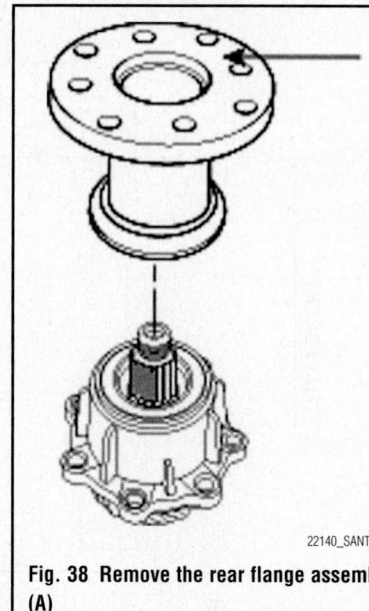

Fig. 38 Remove the rear flange assembly (A)

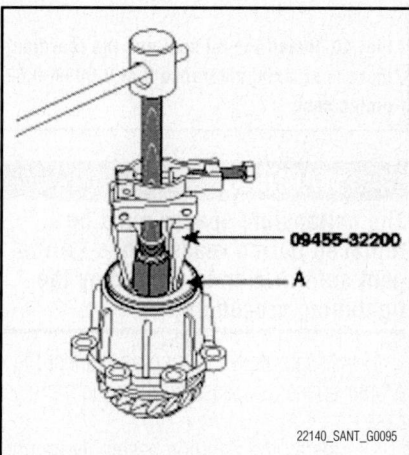

Fig. 39 Using the SST(09455-32200), remove the pinion oil seal (A) from the pinion case

➡To achieve the target preload torque, it is permissible to gradually increase the tightening torque. If the permissible preload torque of 19 inch lbs. (2.1 Nm) is exceeded, it is not allowed to loosen the Hex lock nut. Disassemble the whole pinion assembly. Replace the collapsible spacer and repeat the mounting procedure.

c. Measure dimension En.
• Check preload. A preload torque of 16–19 inch lbs. (1.8–2.1 Nm) corresponds to a bearing preload of 0.0047–0.0063 inch (0.12–0.16mm)
• Ev minus En minus B = 0.0047–0.0063 inch (0.12–0.16mm)

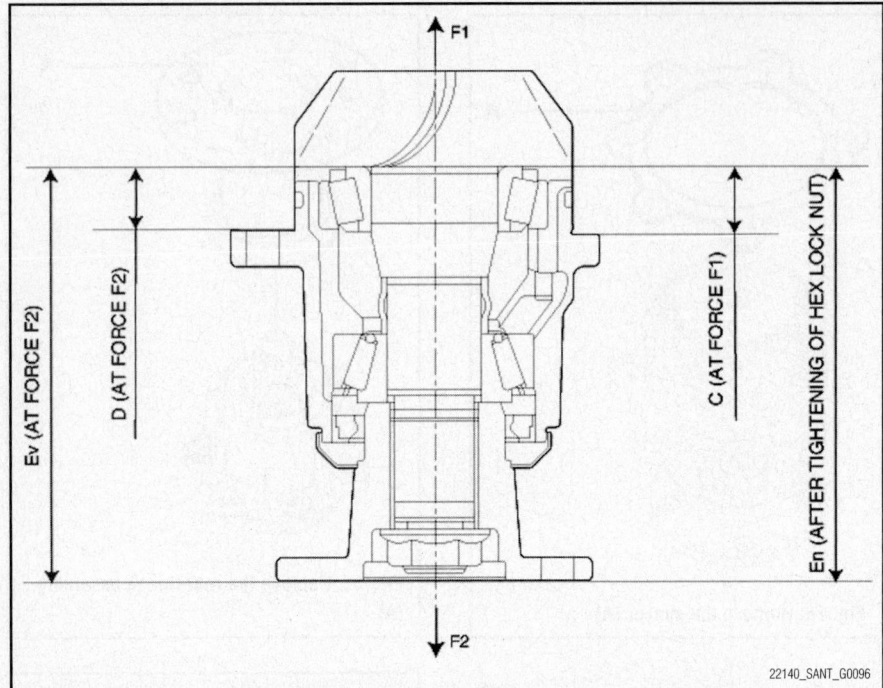

Fig. 40 Install the oil seal and the rear flange assembly. Tighten down the hex lock nut until there is an axial clearance B of 0.0099–0.0296 inch (0.25mm–0.75mm) between pinion and pinion case

22140_SANT_G0096

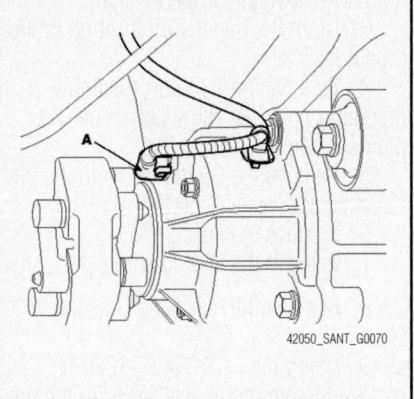

Fig. 42 Disconnect the coupling control connector (A)

42050_SANT_G0070

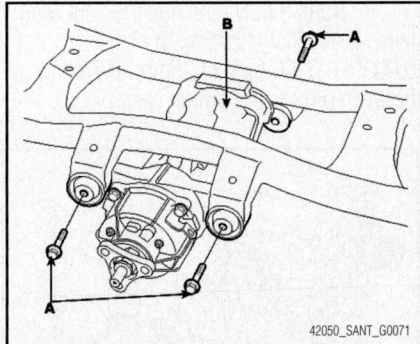

Fig. 43 After loosening the differential mounting bolts (A), remove the differential (B)

42050_SANT_G0071

❋ WARNING

The collapsible spacer must be replaced during reassembly. Permanent deformation is caused by the tightening procedure.

12. Assemble the O-ring and select the proper spacer for reassembling the transfer case.

13. Install the 7 pinion assembly mounting bolts. Tighten the bolts to: 15–22 ft. lbs. (37–40 Nm).

14. Install the pinion assembly.

15. Install the propeller shaft to the transfer case. Tighten to: 36–51 ft. lbs. (50–70 Nm).

16. Fill the transfer case with the proper amount and type of oil.

17. Check for leaks.

REAR AXLE HOUSING

REMOVAL & INSTALLATION

See Figures 41 through 44.

1. Before servicing the vehicle, refer to the Precautions Section.

2. Drain the differential gear oil.

3. Remove the rear drive shaft.

4. Remove the propeller shaft.

5. Support the differential assembly (A) with the jack (B).

6. Disconnect the coupling control connector (A).

7. After loosening the differential mounting bolts (A), remove the differential (B).

8. After loosening the cover bolts (A), remove the differential cover (B).

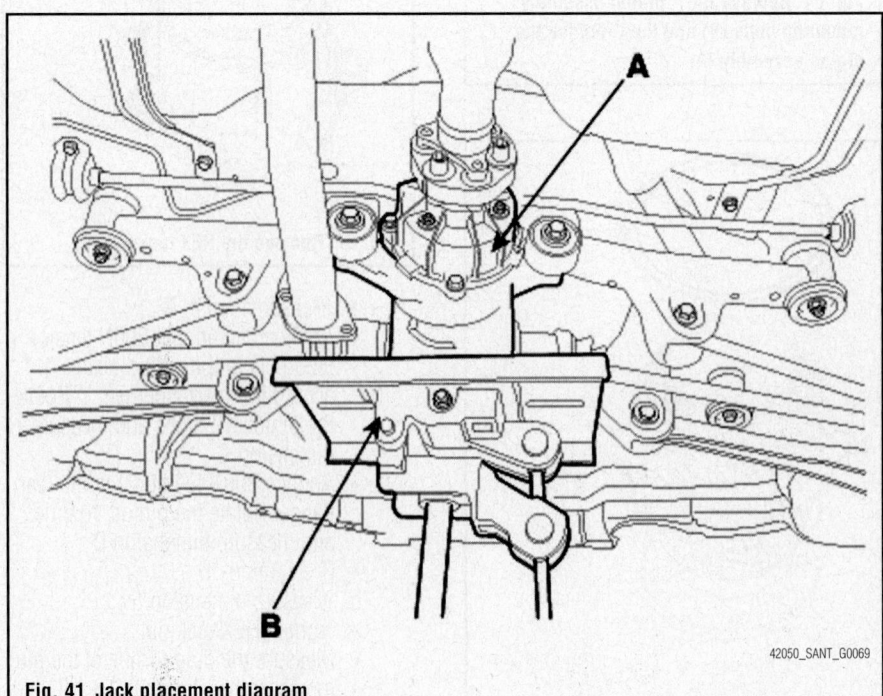

Fig. 41 Jack placement diagram

42050_SANT_G0069

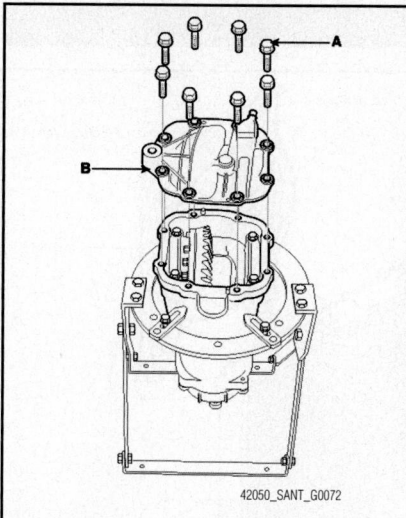

Fig. 44 After loosening the cover bolts (A), remove the differential cover (B)

To install:

9. Apply liquid gasket to the differential cover (B).

10. Install the mounting bolts (A). Tighten to: 29–36 ft. lbs. (39–49 Nm).

11. Install the differential and tighten the mounting bolts to: 51–65 ft. lbs. (69–88 Nm).

12. Using the transaxle jack, install the differential assembly.

13. Connect the coupling control connector.

14. Install the propeller shaft.

15. Install the rear drive shaft.

16. Fill the gear oil with the proper amount and type of gear oil.

REAR AXLE SHAFT, BEARING & SEAL

REMOVAL & INSTALLATION

See Figures 45 and 46.

1. Before servicing the vehicle, refer to the Precautions Section.

2. Remove the brake disc rotor from the rear axle carrier assembly. Refer to Rear Disc Brakes, Rotor, removal & installation.

3. Remove the hub assembly mounting bolts (B) from the rear axle carrier (A).

4. Remove the hub assembly (B) and the parking brake assembly (C) from the rear axle carrier (A).

➡**Do not disassemble the hub assembly.**

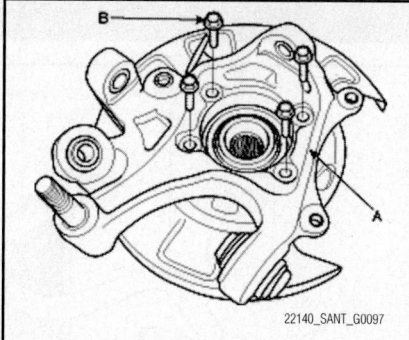

Fig. 45 Remove the hub assembly mounting bolts (B) from the rear axle carrier (A)

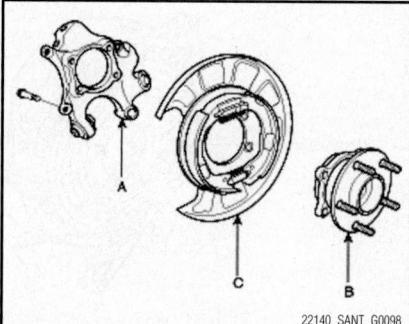

Fig. 46 Remove the hub assembly (B) and the parking brake assembly (C) from the rear axle carrier (A)

5. Check the hub for cracks and the splines for wear.

6. Check the brake disc for scoring and damage.

7. Check the rear axle carrier for cracks

8. Check the bearing for cracks or damage.

9. Replace the hub assembly if problems are found.

To install:

10. Install the parking brake assembly and the hub assembly to the rear axle carrier.

11. Install the hub assembly to the rear axle carrier and then tighten the mounting bolt to: 58–65 ft. lbs. (79–88 Nm).

12. Install the brake disc to the rear axle carrier assembly. Tighten screw to: 43–52 inch lbs. (5–6 Nm).

REAR HALFSHAFTS

REMOVAL & INSTALLATION

1. Before servicing the vehicle, refer to the Precautions Section.

2. Remove or disconnect the following:
- Rear wheel
- Split pin and nut
- Spare tire and support the hanger of the main muffler to avoid interfering with the carrier during right hand shaft removal

3. Matchmark the propeller rubber coupling and differential flange, then remove the bolts and nuts.

4. Support the differential carrier with a jack and remove the differential carrier mounting nuts and bolts.
- Shaft from the carrier by inserting a prybar between the carrier and the shaft
- Differential carrier to the rear after lowering the jack
- Shaft from the axle hub using a plastic mallet
- Shaft from the vehicle

To install:

5. Install or connect the following:
- Shaft into the axle hub
- Differential carrier to the rear using a jack
- Shaft to the carrier
- Differential carrier mounting nuts and bolts. Tighten the carrier nuts and bolts to 51–58 ft. lbs. (70–80 Nm) and the carrier rear bracket bolt to 58–73 ft. lbs. (80–100 Nm)
- Propeller shaft
- Spare tire and support for the main muffler hanger
- New shaft nut and tighten to 146–253 ft. lbs. (200–260 Nm)
- New split pin
- Rear wheel

REAR PINION SEAL

REMOVAL & INSTALLATION

1. Before servicing the vehicle, refer to the Precautions Section.

2. Remove the differential case assembly.

3. Remove the side bearing inner races.

4. Remove the drive gear.

5. Remove the lock pin.

6. Remove the self-locking nut.

7. Remove the drive pinion.

8. Remove the drive pinion bearing inner race.

9. Remove the pinion oil seal.

10. Installation is the reverse of the removal procedure.

ENGINE COOLING

ENGINE FAN

REMOVAL & INSTALLATION

1. Before servicing the vehicle, refer to the Precautions Section.

2. Disconnect the ground cable from the battery cable.

3. Disconnect the connectors from the fan motor and the harness from the shroud.

4. For vehicles with automatic transaxles, remove the oil cooler hose from the shroud.

5. Remove the four bolts holding the shroud.

6. Remove the shroud with the fan motor.

7. Remove the fan mounting clip and detach the fan from the fan motor.

8. Remove the 3 screws and detach the fan motor.

To install:

9. Reattach the fan motor to the fan shroud and insert the 3 screws to secure.

10. Reattach the fan to the fan motor and install the fan mounting clip to secure.

11. Install the fan shroud with attached fan and fan motor to the radiator.

12. Install and tighten the four bolts holding the fan shroud in place to the radiator.

13. Install the oil cooler hose back on the shroud if previously removed.

14. Reconnect the connectors from the fan motor and the harness from the shroud.

15. Reconnect the battery ground or negative cable.

RADIATOR

REMOVAL & INSTALLATION

See Figures 47 and 48.

1. Before servicing the vehicle, refer to the Precautions Section.

2. Remove the air duct.

3. Disconnect the negative battery cable.

4. Remove the air cleaner assembly.

5. Drain the cooling system. Remove the radiator cap to speed draining.

6. Remove the inlet hose from the radiator.

7. Remove the transaxle oil cooler lines from the retainer clip at the bottom of the cooling fan shroud.

8. Remove the fan shroud clip from the condenser tubes.

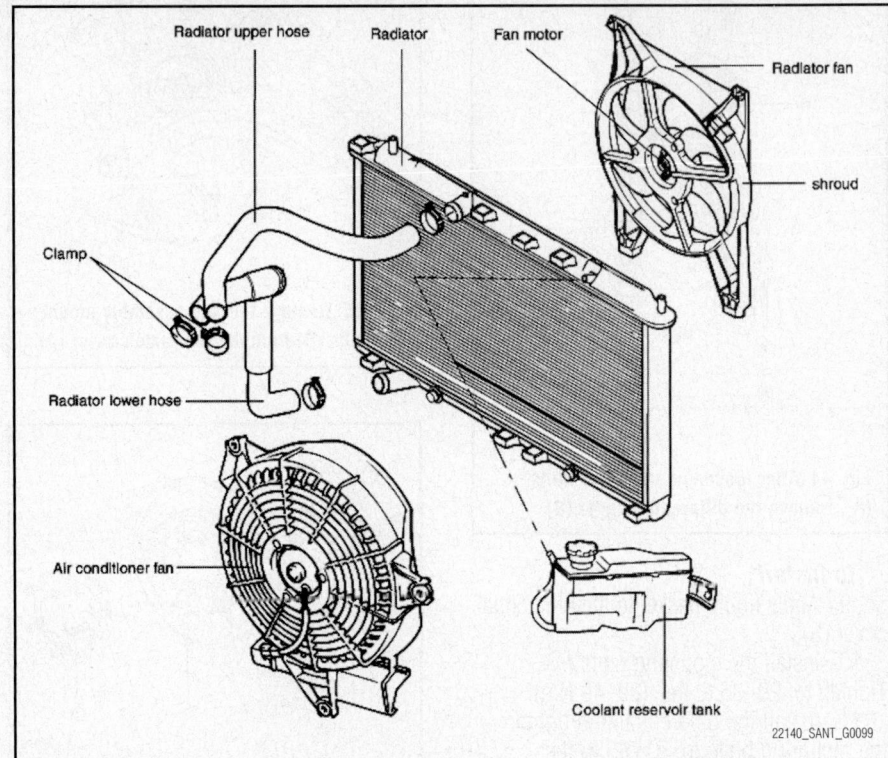

Fig. 47 Engine cooling system component location—2006 2.4L, 2.7L and 3.5L engines

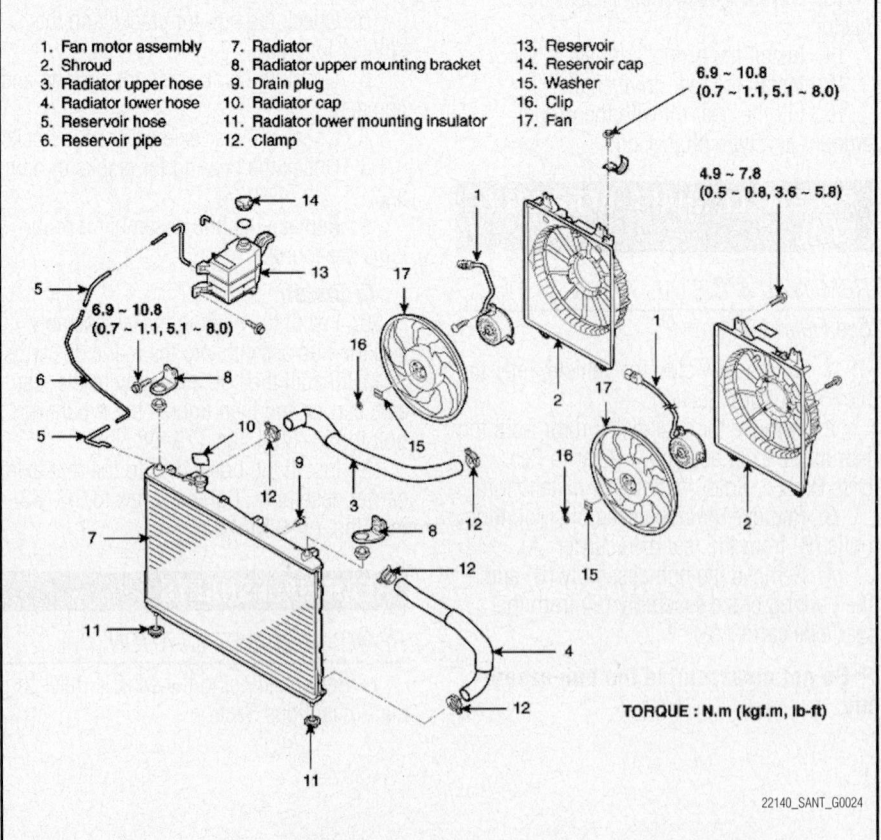

1. Fan motor assembly
2. Shroud
3. Radiator upper hose
4. Radiator lower hose
5. Reservoir hose
6. Reservoir pipe
7. Radiator
8. Radiator upper mounting bracket
9. Drain plug
10. Radiator cap
11. Radiator lower mounting insulator
12. Clamp
13. Reservoir
14. Reservoir cap
15. Washer
16. Clip
17. Fan

6.9 ~ 10.8
(0.7 ~ 1.1, 5.1 ~ 8.0)

4.9 ~ 7.8
(0.5 ~ 0.8, 3.6 ~ 5.8)

6.9 ~ 10.8
(0.7 ~ 1.1, 5.1 ~ 8.0)

TORQUE : N.m (kgf.m, lb-ft)

Fig. 48 Engine cooling system component location—2007–08 2.7L, 3.3L and 3.8L engines

9. Remove the bolt that connects the fan shroud to the condenser hold down bracket.

10. Remove the air deflectors from the top of the radiator.

11. Remove the cooling fan shroud bolts.

12. Remove the coolant reservoir hose from the radiator overflow neck.

13. Remove the radiator upper support brackets and bolts that connect to the fan shroud.

14. Disconnect the engine cooling fan motors electrical connectors.

15. Remove the cooling fan motors electrical harness from the fan shroud clips.

16. Remove the cooling fan shroud.

17. Remove the outlet hose from the radiator.

18. Disconnect the transaxle oil cooler pipes from the radiator.

19. Tilt the top of the radiator rearward.

20. Remove the condenser hold down bracket from the radiator.

21. Lift the condenser from the mounting tabs on the radiator and position the condenser aside.

22. Remove the radiator.

To install:

23. Install the radiator to the lower mounts.

➡**Verify that the condenser is fully seated in the radiator mounting tabs.**

24. Install the condenser to the mounting tabs on the radiator.

25. Install the condenser hold down bracket to the radiator and condenser.

26. Install the outlet hose to the radiator.

27. Connect the transaxle oil cooler pipes to the radiator.

➡**Ensure the lower edge of the fan shroud engages the clip at the bottom of the radiator.**

28. Install the cooling fan shroud.

29. Install the cooling fan motors electrical harness to the fan shroud clips.

30. Connect the engine cooling fan motors electrical connectors.

31. Install the fan shroud clip to the condenser tubes.

32. Install the cooling fan shroud bolts and tighten to 89 inch lbs. (10 Nm).

33. Install the radiator upper support brackets and bolts that connect to the fan shroud.

34. Install the air deflectors to the top of the radiator.

35. Install the bolt that connects the fan shroud to the condenser hold down bracket.

36. Install the inlet hose to the radiator.

37. Install the air cleaner assembly.

38. Install the coolant reservoir hose to the radiator overflow neck.

39. Install the transaxle oil cooler lines to the retainer clip at the bottom of the cooling fan shroud.

40. Fill the cooling system.

41. Connect the negative battery cable.

THERMOSTAT

REMOVAL & INSTALLATION

✳✳ CAUTION

Make sure the engine and radiator are cool to the touch prior to beginning this procedure in order to prevent scalding or burns.

1. Before servicing the vehicle, refer to the Precautions Section.

2. Drain the coolant down to thermostat level or below.

3. Remove the coolant outlet fitting and gasket.

4. Remove the thermostat.

To install:

5. Check that the flange of the thermostat is correctly seated in the socket of the thermostat housing.

6. Install the inlet fitting. Tighten the bolt to: 84–132 inch lbs. (10–15 Nm).

7. Refill the coolant with the proper amount and type of fluid.

8. Run the engine through warm-up and check for leaks.

WATER PUMP

REMOVAL & INSTALLATION

2.4L Engine

See Figure 49.

✳✳ CAUTION

The system is under high pressure when the engine is hot. To avoid danger of releasing scalding engine coolant, remove the cap only when the engine is cool.

1. Before servicing the vehicle, refer to the Precautions Section.

2. Drain the engine coolant.

3. Remove the accessory drive belt.

4. Remove the exhaust manifold.

5. Remove the water pump:

a. Remove the 4 bolts and pump pulley.

b. Remove the water pump and gasket.

6. Remove the water inlet pipe nut.

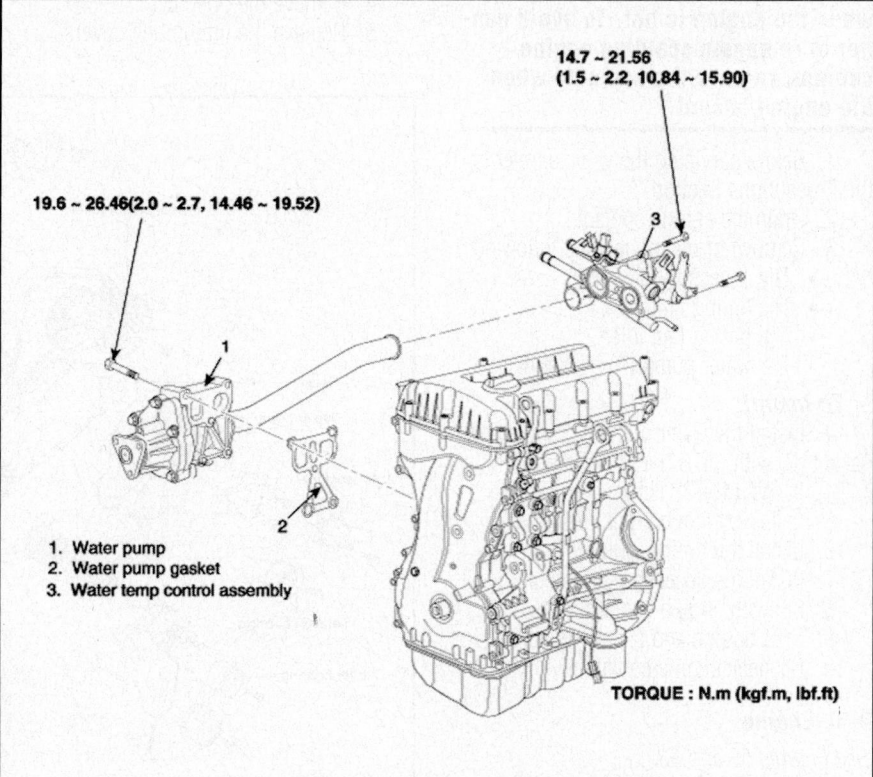

14.7 ~ 21.56
(1.5 ~ 2.2, 10.84 ~ 15.90)

19.6 ~ 26.46(2.0 ~ 2.7, 14.46 ~ 19.52)

1. Water pump
2. Water pump gasket
3. Water temp control assembly

TORQUE : N.m (kgf.m, lbf.ft)

Fig. 49 Exploded view of the water pump assembly and related components—2.4L engine

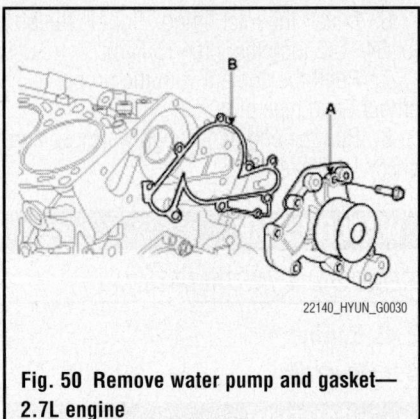

Fig. 50 Remove water pump and gasket—2.7L engine

To install:

7. Install the water pump and a new gasket. Tighten the 5 bolts to 15–20 ft. lbs. (20–27 Nm).

8. Install the 4 bolts and pump pulley.

9. Install the water inlet pipe nut. Tighten to 15–20 ft. lbs. (20–27 Nm).

10. Install exhaust manifold.

11. Install accessory drive belt.

12. Fill with engine coolant.

13. Start engine and check for leaks.

14. Recheck engine coolant level.

2.7L Engine

See Figures 50 and 51.

❊❊ CAUTION

The system is under high pressure when the engine is hot. To avoid danger of releasing scalding engine coolant, remove the cap only when the engine is cool.

1. Before servicing the vehicle, refer to the Precautions Section.

2. Drain the engine coolant.

3. Remove or disconnect the following:
   - The accessory drive belt
   - The timing belt
   - The timing belt idler
   - The water pump (A) gasket (B)

To install:

4. Install the water pump (A) and a new gasket (B) with the 8 bolts. Tighten to 11–16 ft. lbs. (15–22 Nm).

5. Install the timing belt idler.

6. Install the timing belt.

7. Install the accessory drive belt.

8. Fill with engine coolant.

9. Start engine and check for leaks.

10. Recheck engine coolant level.

3.5L Engine

See Figures 52 and 53.

1. Before servicing the vehicle, refer to the Precautions Section.

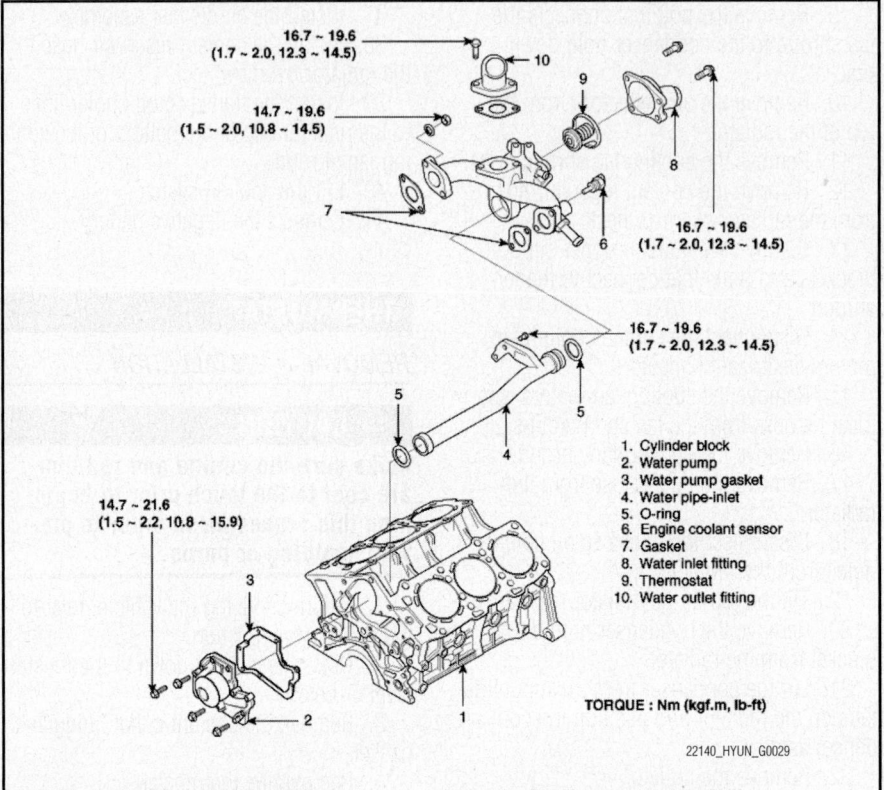

16.7 ~ 19.6
(1.7 ~ 2.0, 12.3 ~ 14.5)

14.7 ~ 19.6
(1.5 ~ 2.0, 10.8 ~ 14.5)

16.7 ~ 19.6
(1.7 ~ 2.0, 12.3 ~ 14.5)

16.7 ~ 19.6
(1.7 ~ 2.0, 12.3 ~ 14.5)

14.7 ~ 21.6
(1.5 ~ 2.2, 10.8 ~ 15.9)

1. Cylinder block
2. Water pump
3. Water pump gasket
4. Water pipe-inlet
5. O-ring
6. Engine coolant sensor
7. Gasket
8. Water inlet fitting
9. Thermostat
10. Water outlet fitting

TORQUE : Nm (kgf.m, lb-ft)

Fig. 51 Exploded view of the water pump assembly and related components—2.7L engine

2. Drain the cooling system.

3. Disconnect the negative battery cable.

4. Remove the drive belt.

5. Remove the timing belt covers.

6. Remove the timing belt tensioner.

7. Remove the idler pulley.

8. Remove the water pump bolts.

9. Remove the water pump and gasket.

10. Clean the gasket mating surfaces.

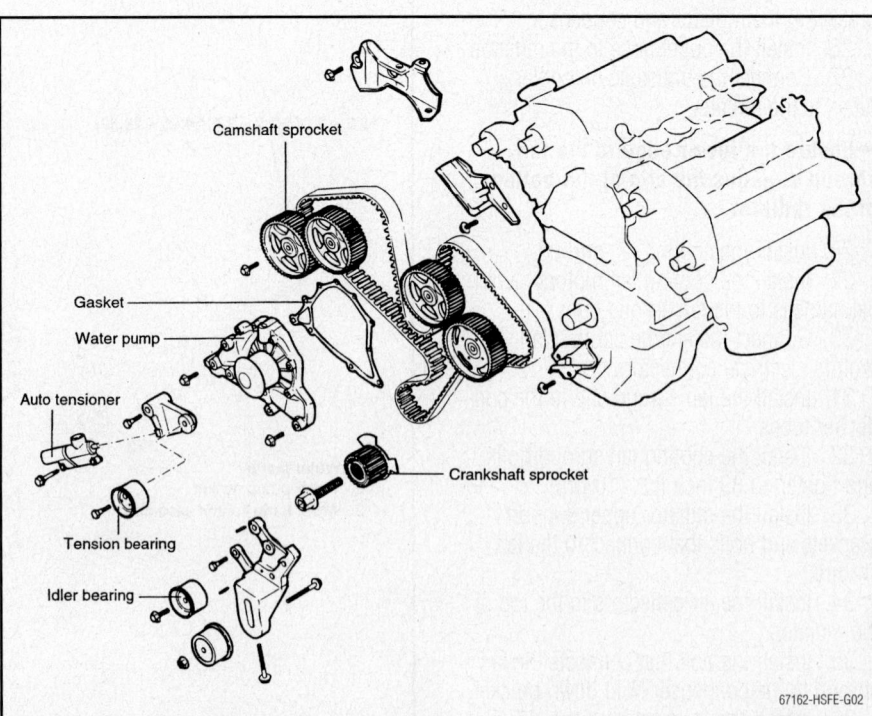

Camshaft sprocket

Gasket

Water pump

Auto tensioner

Tension bearing

Idler bearing

Crankshaft sprocket

Fig. 52 Exploded view of the water pump mounting and related components—3.5L engine

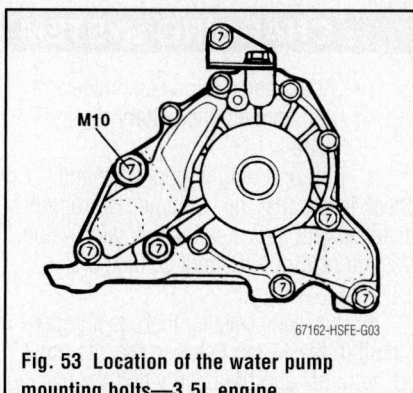

Fig. 53 Location of the water pump mounting bolts—3.5L engine

To install:

11. Install the water pump and gasket.
12. Install the bolts. Tighten the M8 bolts to 11–16 ft. lbs. (15–22 Nm) and the M10 bolt to 24–36 ft. lbs. (33–50 Nm).
13. Install the idler pulley.
14. Install the timing belt tensioner.
15. Install the timing belt covers.
16. Install the drive belt.
17. Connect the negative battery cable.
18. Fill the cooling system.
19. Start the engine and check for leaks.

3.3L & 3.8L Engines

See Figures 54 and 55.

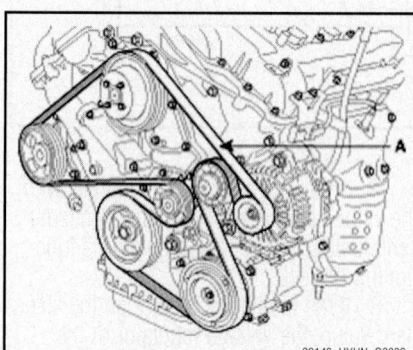

Fig. 54 Accessory drive belt routing—3.3L and 3.8L engines

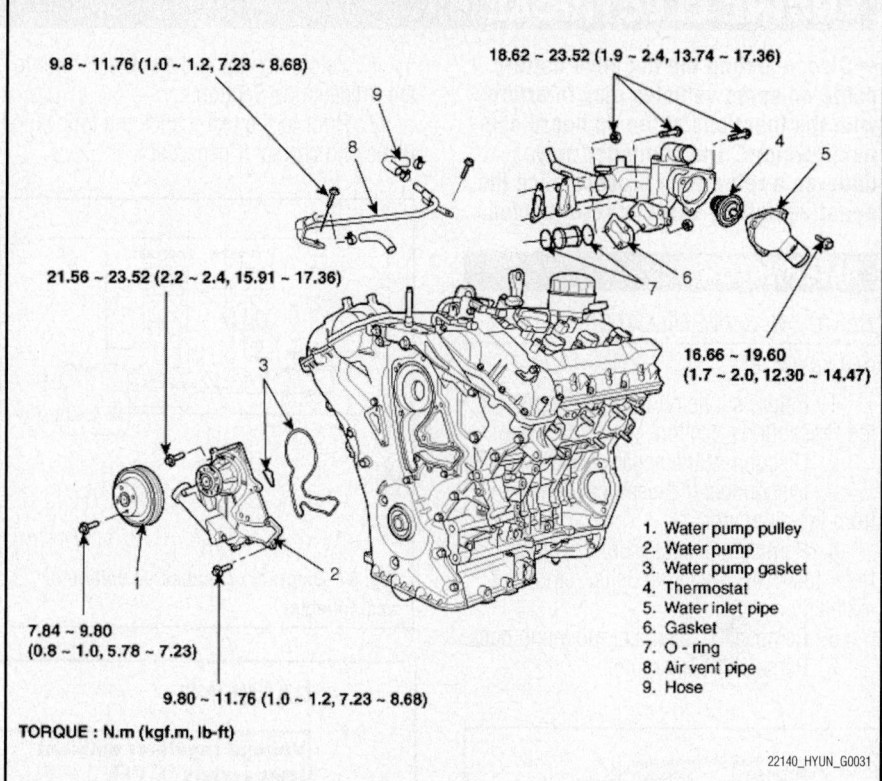

9.8 ~ 11.76 (1.0 ~ 1.2, 7.23 ~ 8.68)

18.62 ~ 23.52 (1.9 ~ 2.4, 13.74 ~ 17.36)

21.56 ~ 23.52 (2.2 ~ 2.4, 15.91 ~ 17.36)

16.66 ~ 19.60 (1.7 ~ 2.0, 12.30 ~ 14.47)

7.84 ~ 9.80 (0.8 ~ 1.0, 5.78 ~ 7.23)

9.80 ~ 11.76 (1.0 ~ 1.2, 7.23 ~ 8.68)

TORQUE : N.m (kgf.m, lb-ft)

1. Water pump pulley
2. Water pump
3. Water pump gasket
4. Thermostat
5. Water inlet pipe
6. Gasket
7. O - ring
8. Air vent pipe
9. Hose

Fig. 55 Exploded view of the water pump assembly and related components—3.3L and 3.8L engines

❊❊ CAUTION

The system is under high pressure when the engine is hot. To avoid danger of releasing scalding engine coolant, remove the cap only when the engine is cool.

1. Before servicing the vehicle, refer to the Precautions Section.
2. Drain the engine coolant.
3. Remove or disconnect the following:
   - The accessory drive belt (A)
   - The 4 bolts and pump pulley
   - The water pump and gasket

To install:

→Clean the contact face before assembly.

4. Install the water pump and a new gasket with 12 bolts. Tighten to 16–17 ft. lbs. (22–24 Nm); 84–108 inch lbs. (10–12 Nm).
5. Install the 4 bolts and the pump pulley. Tighten to 72–84 inch lbs. (8–10 Nm).
6. Install the accessory drive belt.
7. Fill with engine coolant.
8. Start engine and check for leaks.
9. Recheck engine coolant level.

➡ **Disconnecting the negative battery cable on some vehicles may interfere with the functions of the on board computer system. The computer may undergo a relearning process once the negative battery cable is reconnected.**

ALTERNATOR

REMOVAL & INSTALLATION

See Figure 56.

1. Before servicing the vehicle, refer to the Precautions Section.
2. Disconnect the negative battery cable.
3. Disconnect the electrical connectors from the alternator.
4. Remove the accessory drive belt. Refer to Accessory Drive Belts, removal & installation.
5. Remove the alternator mounting bolts.
6. Remove the alternator.

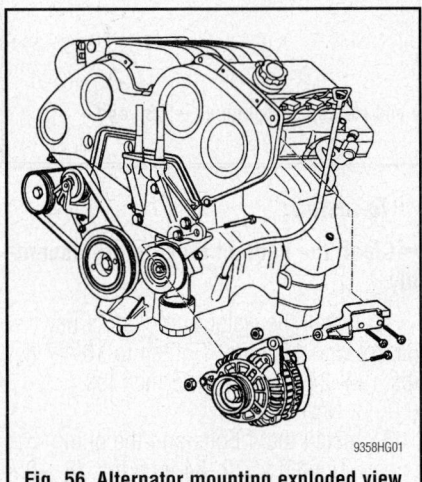

Fig. 56 Alternator mounting exploded view

To install:
7. Install the alternator.
8. Tighten the pivot bolt to 14–18 ft. lbs. (19–25 Nm) and the adjustment bolt to 11–16 ft. lbs. (15–22 Nm).
9. Connect the alternator electrical connectors. Tighten the battery terminal connector nut to 60 inch lbs. (7 Nm).
10. Install the accessory drive belt. Refer to Accessory Drive Belts, removal & installation.
11. Connect the negative battery cable.

VOLTAGE REGULATOR

ADJUSTMENT

See Figures 57 and 58.

1. Before servicing the vehicle, refer to the Precautions Section.
2. Prior to the test, check the following items and correct if necessary.

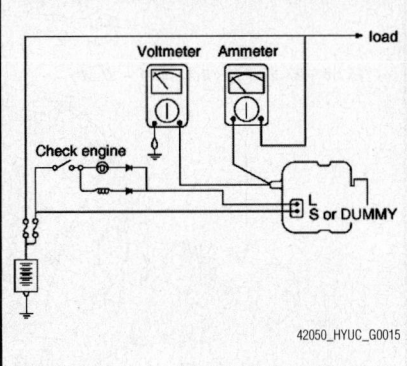

Fig. 57 Diagram of connected voltmeter and ammeter

GASOLINE	
Voltage regulator ambient temperature °C (°F)	**Regulating voltage (V)**
-20 (-4)	14.2 ~ 15.4
20 (68)	14.0 ~ 15.0
60 (140)	13.7 ~ 14.9
80 (176)	13.5 ~ 14.7

42050_HYUC_G0016

Fig. 58 Regulating voltage table

- Check that the battery installed on the vehicle is fully charged
- Check the alternator drive belt tension
- Turn ignition switch to "OFF"
- Disconnect the battery ground cable
- Connect a digital voltmeter between the "B" terminal of the alternator and ground by connecting the positive (+) lead of the voltmeter to the "B" terminal of the alternator and connecting the negative (-) lead to good ground or the negative (-) battery terminal
- Disconnect the alternator output wire from the alternator "B" terminal
- Connect a DC ammeter (0 to 150 Amps) in series between the "B" terminal and the disconnected output wire
- Connect the negative (-) lead wire of the ammeter to the disconnected output wire

- Attach the engine tachometer and connect the battery ground cable
3. Turn on the ignition switch and check to see that the voltmeter is properly installed and indicates the following value: Voltage reading = Battery voltage of 12.0–14.1 volts.
4. If it reads 0 volts, there is an open circuit in the wiring between the alternator "B" terminal and the battery and the battery negative (-) terminal, or the fusible link is blown.
5. Start the engine keeping all lights and accessories off.
6. Run the engine at a speed of about 2,500 rpm and read the voltmeter when the alternator output current drops to 10 Amps or less.

7. If the voltmeter reading agrees with the value listed in the Regulating Voltage Table below, the voltage regulator is functioning correctly.
8. If the reading is other than the standard value, the voltage regulator or the alternator is faulty.
9. Upon completion of the test, reduce the engine speed to idle, and turn off the ignition switch.
10. Disconnect the battery ground cable.
11. Remove the voltmeter and ammeter and the engine tachometer.
12. Connect the alternator output wire to the alternator "B" terminal.
13. Connect the battery ground cable.

REMOVAL & INSTALLATION

The voltage regulator is an internal component of the alternator. In order to replace the voltage regulator, the entire alternator assembly must be replaced. Refer to Alternator, installation and removal.

ENGINE ELECTRICAL **IGNITION SYSTEM**

FIRING ORDER

See Figures 59 and 60.

3.5L engine—Firing order:
1–2–3–4–5–6.

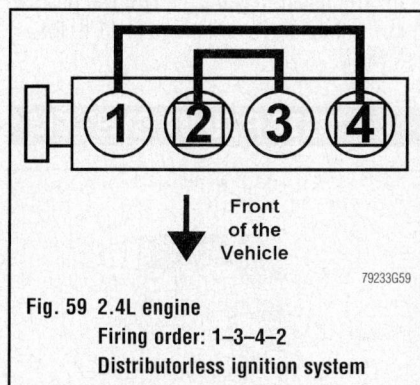

Fig. 59 2.4L engine
Firing order: 1–3–4–2
Distributorless ignition system

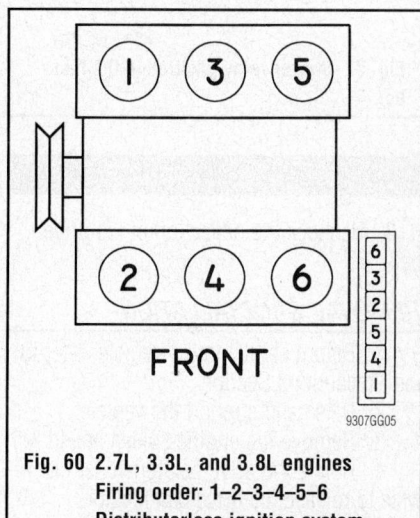

Fig. 60 2.7L, 3.3L, and 3.8L engines
Firing order: 1–2–3–4–5–6
Distributorless ignition system

IGNITION COIL PACK

REMOVAL & INSTALLATION

2006 2.4L, 2.7L & 3.5L Engines

1. Before servicing the vehicle, refer to the Precautions Section.
2. Remove the engine cover.
3. Disconnect the spark plug wires and ignition coil connectors.
4. Remove the ignition coil pack.
5. Installation is the reverse of the removal procedure.

2007–08 2.7L, 3.3L & 3.8L Engines
See Figure 61.

1. Before servicing the vehicle, refer to the Precautions Section.

2. Remove the engine cover (as necessary).
3. Disconnect the ignition coil connector.

→**When removing the ignition coil connector, pull the lock pin (A) and push the clip (B).**

4. Remove the bolt from the ignition coil.
5. Remove the ignition coil.
6. Installation is the reverse of the removal procedure.

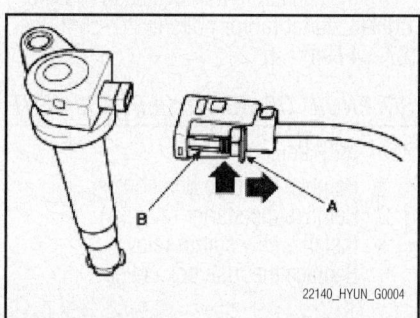

Fig. 61 When removing the ignition coil connector, pull the lock pin (A) and push the clip (B)

IGNITION TIMING

Ignition timing is controlled by the electronic control ignition timing system. The standard reference ignition timing data for the engine operating conditions are pre-programmed in the memory of the Engine Control Module (ECM). The engine operating conditions (speed, load, warm-up condition, etc.) are detected by the various sensors. Based on these sensor signals and the ignition timing data, signals to interrupt the primary current are sent to the ECM. The ignition coil is activated, and timing is controlled.

SPARK PLUGS

REMOVAL & INSTALLATION

2.4L Engine

1. Before servicing the vehicle, refer to the Precautions Section.
2. Remove the spark plug cables.

→**When removing the spark plug cable, pull on the spark plug cable boot (not the cable), as it may be damaged.**

3. Using a spark plug socket, remove the spark plugs.

✳✳ **CAUTION**

Be careful that no contaminants enter through the spark plug holes.

4. Check the electrode gap of the new spark plugs. Specification: 0.039–0.043 inch (1.0–1.1mm.

To install:
5. Installation is the reverse of the removal procedure.
6. Tighten the spark plugs to 11 ft. lbs. (15 Nm).

2006 2.7L & 3.5L Engines

1. Before servicing the vehicle, refer to the Precautions Section.
2. Remove the engine cover.
3. Disconnect the Variable Intake System (VIS) actuator connectors and the fuel injector connectors.
4. Remove the accelerator cable.
5. Remove surge tank sub assembly.
6. Remove the spark plug cable.
7. Remove the spark plug.
8. Check the electrode gap of the new spark plugs. Specification: 0.039–0.043 inch (1.0–1.1mm).

To install:
9. Installation is the reverse of the removal procedure.
10. Tighten the spark plugs to 11 ft. lbs. (15 Nm).

2007–08 2.7L Engines
See Figure 61.

1. Before servicing the vehicle, refer to the Precautions Section.
2. Remove the engine cover (as necessary).
3. Disconnect the ignition coil connector.

→**When removing the ignition coil connector, pull the lock pin (A) and push the clip (B).**

4. Remove the ignition coil.
5. Use a spark plug socket and wrench to remove the spark plugs.

✳✳ **WARNING**

Be careful that no contaminates enter through the spark plug holes.

→**Check the electrode gap on the spark plugs before installation. Specification: 0.039–0.043 inch (1.0–1.1 mm).**

6. To install, reverse the removal procedure. Tighten the spark plugs to 11 ft. lbs. (15 Nm).

3.3L & 3.8L Engines

See Figure 61.

1. Before servicing the vehicle, refer to the Precautions Section.
2. Remove the engine cover.

3. Disconnect the ignition coil connector.

➡**When removing the ignition coil connector, pull the lock pin (A) and push the clip (B).**

4. Remove the ignition coil.
5. Use a spark plug socket and wrench to remove the spark plugs.

To install:

3. Installation is the reverse of removal. Tighten starter motor bolts to 20–25 ft. lbs. (27–34 Nm).

✳✳ **WARNING**

Be careful that no contaminates enter through the spark plug holes.

➡**Check the electrode gap on the spark plugs before installation. Specification: 0.039–0.043 inch (1.0–1.1 mm).**

6. To install, reverse the removal procedure. Tighten the spark plugs to 11 ft. lbs. (15 Nm).

ENGINE ELECTRICAL

STARTER

REMOVAL & INSTALLATION

1. Before servicing the vehicle, refer to the Precautions Section.
2. Remove or disconnect the following:
   - Negative battery cable and wait at least 3 minutes
   - Speedometer cable and shift cable from the transaxle
   - Starter motor wiring
   - Starter motor bolts and the starter

SOLENOID OR RELAY REPLACEMENT

See Figure 62.

1. Remove the fuse box cover.
2. Remove the starter relay (A).
3. Install a new starter relay.
4. Replace the fuse box cover.

STARTING SYSTEM

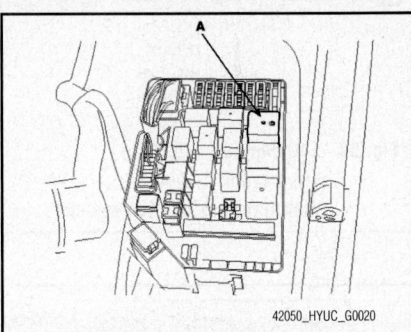

42050_HYUC_G0020

Fig. 62 Starter relay position in the fuse box.

ENGINE MECHANICAL

➡**Disconnecting the negative battery cable may interfere with the functions of the on board computer systems and may require the computer to undergo a relearning process, once the negative battery cable is reconnected.**

ACCESSORY DRIVE BELTS

ACCESSORY BELT ROUTING

See Figures 63 through 66.

Refer to the accompanying illustrations for accessory drive belt routing.

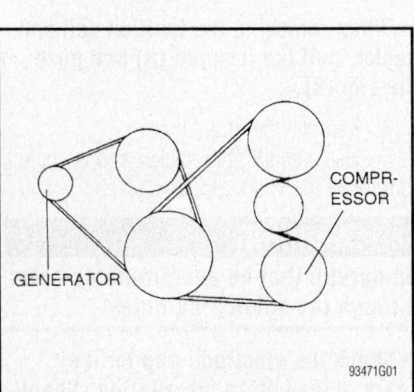

93471G01

Fig. 63 Accessory drive belt routing—2.4L engine

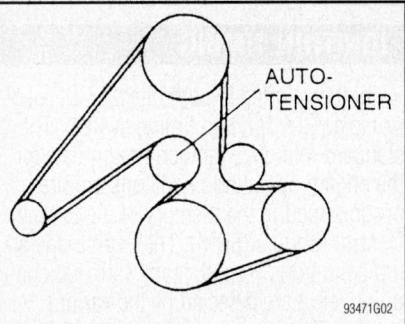

93471G02

Fig. 64 Accessory drive belt routing—2.7L engine

INSPECTION

Inspect the accessory drive belt for signs of glazing or cracking. A glazed belt will be perfectly smooth from slippage, while a good belt will have a slight texture of fabric visible. Cracks will usually start at the inner edge of the belt and run outward. All worn or damaged accessory drive belts should be replaced immediately.

ADJUSTMENT

1. Loosen the tension mounting bolt.
2. Turn the adjusting bolt to obtain the proper belt tension, then retighten the mounting bolt.

3. Recheck the deflection of the drive belt.

REMOVAL & INSTALLATION

1. Before servicing the vehicle, refer to the Precautions Section.
2. Raise and support the vehicle.
3. Remove the engine splash shield.
4. Rotate the drive belt tensioner clockwise to release the drive belt tension.
5. Remove the drive belt from the alternator.
6. Slowly release the drive belt tensioner.
7. Remove the drive belt from the accessory drive pulleys.

To install:

8. Install the drive belt to the accessory drive pulley.
9. Rotate the drive belt tensioner clockwise.
10. Install the drive belt to the alternator.
11. Ensure the drive belt is properly aligned and seated into the grooves of the accessory drive pulleys.
12. Slowly release the drive belt tensioner.
13. Install the engine splash shield.
14. Lower the vehicle.

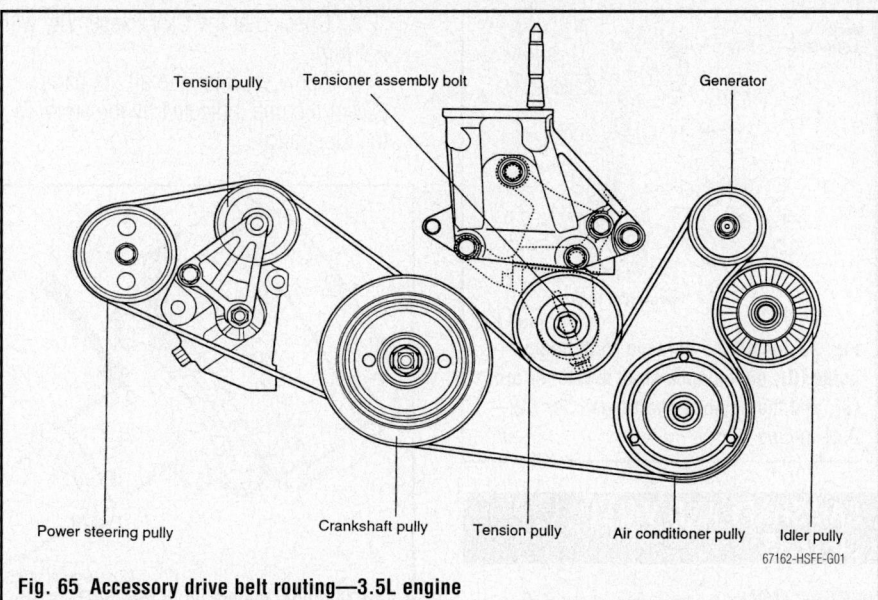

Fig. 65 Accessory drive belt routing—3.5L engine

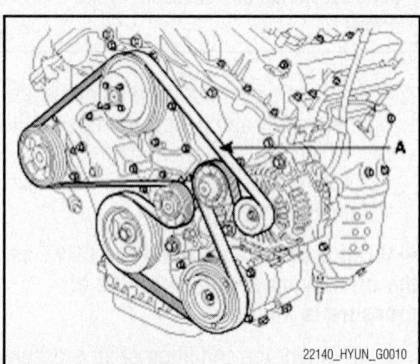

Fig. 66 Accessory drive belt routing—3.3L and 3.8L engines

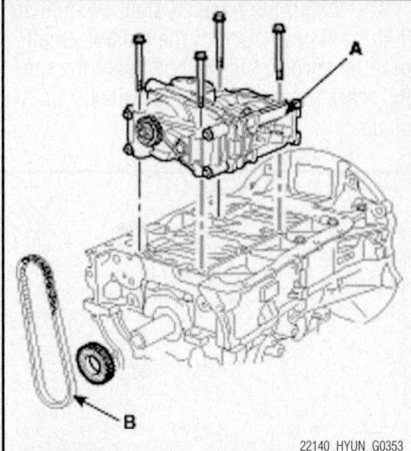

Fig. 68 Remove the balance shaft module (A) and balance shaft chain (B)—2.4L engine

BALANCE SHAFT

REMOVAL & INSTALLATION

2.4L Engine

See Figures 67 through 72.

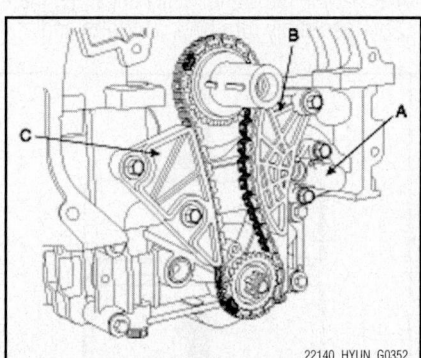

Fig. 67 Remove the balance shaft chain tensioner (A), the balance shaft chain tensioner arm (B), and the balance shaft chain guide (C)—2.4L engine

1. Before servicing the vehicle, refer to the Precautions Section.

➡**Engine removal is not required for this procedure.**

2. Remove the timing chain. Refer to Timing Chain and Sprockets, removal & installation.

3. Install a set pin after compressing the balance shaft chain tensioner.

4. Remove the balance shaft chain tensioner (A).

5. Remove the balance shaft chain tensioner arm (B).

6. Remove the balance shaft chain guide (C).

7. Remove the balance shaft module (A) and balance shaft chain (B).

To install:

8. The key of the crankshaft should be aligned with the mating face of the main bearing cap. This will place the piston of the No. 1 cylinder at TDC on the compression stroke.

9. Confirm the balance shaft module timing mark.

➡**The timing marks should be visually aligned with the centers of adjacent cast timing notches.**

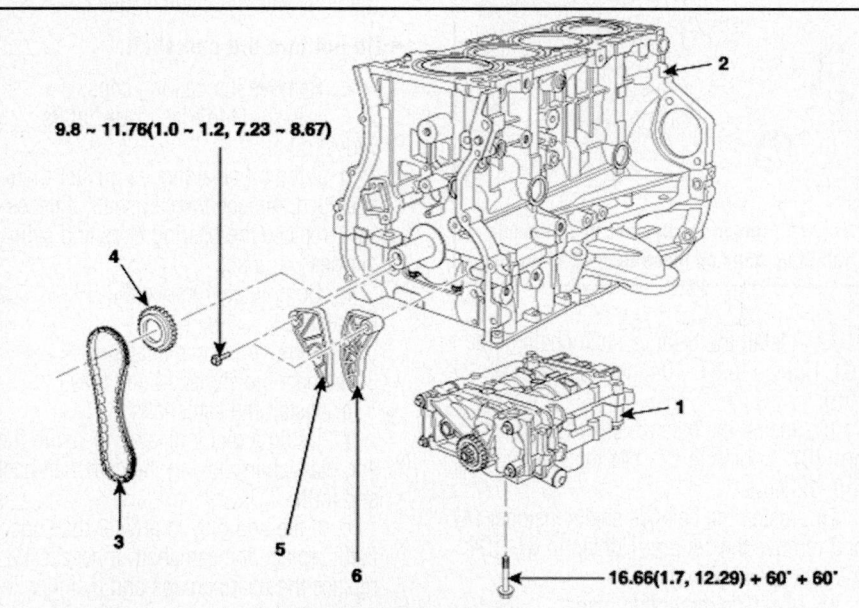

1. Balance shaft module
2. Cylinder block
3. Balance shaft chain
4. Balance shaft chain sprocket
5. Balance shaft chain guide
6. Balance shaft chain tensioner arm

TORQUE : N.m (kgf.m, lbf.ft)

Fig. 69 Exploded view of the balance shaft components—2.4L engine

10. Install the balance shaft module so that the timing mark of the balance shaft module sprocket is matched with the timing mark (color link) of the balance shaft chain.

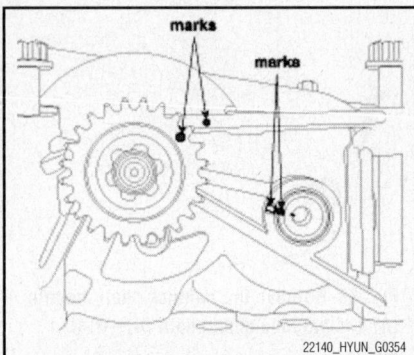

Fig. 70 The timing marks should be visually aligned as illustrated—2.4L engine

11. Tighten the balance shaft module retaining bolts:
   a. Step 1: 12 ft. lbs. (17 Nm).
   b. Step 2: Add 60°.
   c. Step 3: Add another 60° on the last pass.

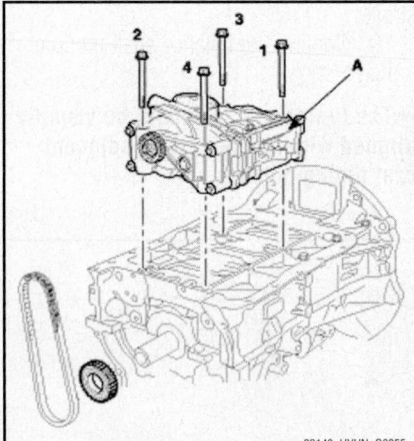

Fig. 71 Tighten the balance shaft module retaining bolts as illustrated—2.4L engine

12. Install the balance shaft chain guide (C). Tighten to 87–104 inch lbs. (10–12 Nm).

13. Install the balance shaft tensioner arm (B). Tighten to 87–104 inch lbs. (10–12 Nm).

14. Install the balance shaft tensioner (A) and remove the set pin. Tighten to 87–104 inch lbs. (10–12 Nm).

15. Confirm the timing marks.

16. Install the timing chain. Refer to Timing Chain and Sprockets, removal & installation.

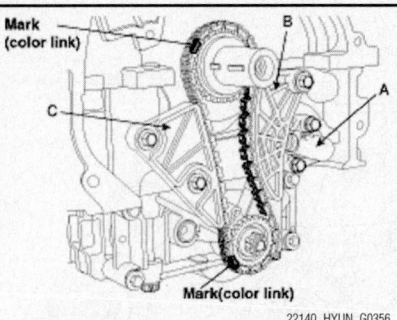

Fig. 72 Install the balance shaft chain guide (C), the balance shaft tensioner arm (B), and the balance shaft tensioner (A)—2.4L engine

CAMSHAFT AND VALVE LIFTERS

INSPECTION

See Figures 73 and 74.

1. Inspect the cam lobes.
   a. Using a micrometer, measure the cam lobe height.
   b. If the cam lobe height is less than specified, replace the camshaft.
2. Inspect the camshaft journal clearance.
   a. Clean the bearing caps and camshaft journals.
   b. Place the camshafts on the cylinder head.
   c. Lay a strip of Plastigage® across each of the camshaft journal.
   d. Install the bearing caps and tighten the bolts with specified torque.

➡**Do not turn the camshaft.**

   e. Remove the bearing caps.
   f. Measure the Plastigage® at its widest point.
   g. If the oil clearance is greater than specified, replace the camshaft. If necessary, replace the bearing caps and cylinder head as a set.
   h. Completely remove the Plastigage®.
   i. Remove the camshafts.
3. Inspect the camshaft end play.
   a. Install the camshafts.
   b. Using a dial indicator, measure the end play while moving the camshaft back and forth.
   c. If the end play is greater than specified, replace the camshaft. If necessary, replace the bearing caps and cylinder head as a set.
   d. Remove the camshafts.
4. Inspect the Continuous Variable Valve Timing (CVVT) assembly.

   a. Check that the CVVT assembly will not turn.
   b. Apply vinyl tape to all the parts except the one indicated by the arrow in the illustration.

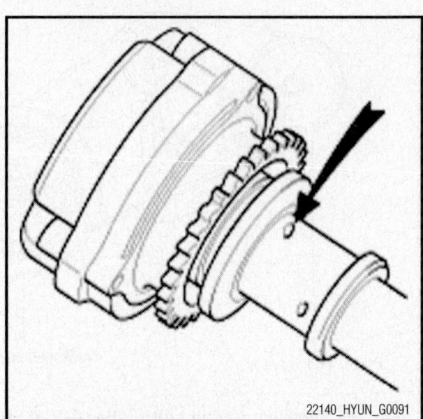

Fig. 73 Apply vinyl tape to the CVVT on all parts except the one indicated by the arrow

   c. Wrap tape around the tip of the air gun and apply air of approx. 14 psi to the port of the camshaft. Perform this in order to release the lock pin for the maximum delay angle locking.

➡**Wrap a shop rag around the CVVT as the oil may spray out when the air pressure is applied.**

   d. Under the condition of air pressure being applied, turn the CVVT assembly to the advance angle side with your hand:
   • Depending on the air pressure, the CVVT assembly will turn to the advance side
   • If air is leaking from the port and air pressure cannot be maintained, the locking pin will not release
5. Except the position where the lock pin meets at the maximum delay angle, let the CVVT assembly turn back and forth and

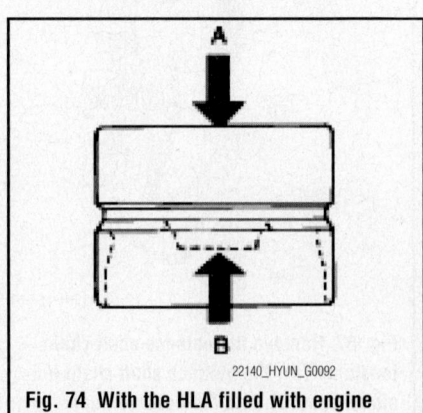

Fig. 74 With the HLA filled with engine oil, hold A and press B by hand

check the movable range and that there is no disturbance.

    a. The CVVT should move smoothly in the range of about 20°.

    b. Turn the CVVT assembly with your hand and lock it at the maximum delay angle position.

6. Inspect the Hydraulic Lash Adjuster (HLA).

    a. With the HLA filled with engine oil, hold A and press B by hand.

    b. If B moves, replace the HLA.

REMOVAL & INSTALLATION

2.4L Engine

See Figures 75 and 76.

Engine removal is not required for this procedure. Use fender covers to avoid damaging painted surfaces. To avoid damage, unplug the wiring connectors carefully while holding the connector portion. Mark all wiring and hoses to avoid misconnection. Inspect the timing belt before removing. Turn the crankshaft pulley so that the No. 1 piston is at Top Dead Center (TDC).

1. Before servicing the vehicle, refer to the Precautions Section.

2. Disconnect the negative terminal from the battery.

3. Remove the engine cover.

4. Remove the air duct.

5. Remove the intake air hose and air cleaner assembly.

    a. Disconnect the AFS connector.

    b. Disconnect the breather hose from air cleaner hose.

    c. Disconnect the ECM connector.

    d. Remove the intake air hose and air cleaner assembly.

6. Remove front wheels.

7. Remove the undercover.

8. Drain the engine coolant. Remove the radiator cap to speed draining.

9. Remove or disconnect the following:
- The upper and lower radiator hose
- The heater hoses
- The A/C switch, alternator connector, and oil pressure switch
- The Oil Control Valve (OCV) connector and OTS connector
- The injector connectors
- The ETS connector
- The Camshaft Position (CMP) connector, and knock sensor connector
- The ignition coil connectors
- The Purge Control Solenoid Valve (PCSV) connector, WTS connector, condenser connector, and Crankshaft Position (CKP) sensor connector
- The delivery pipe, brake vacuum hose, and PCSV hose

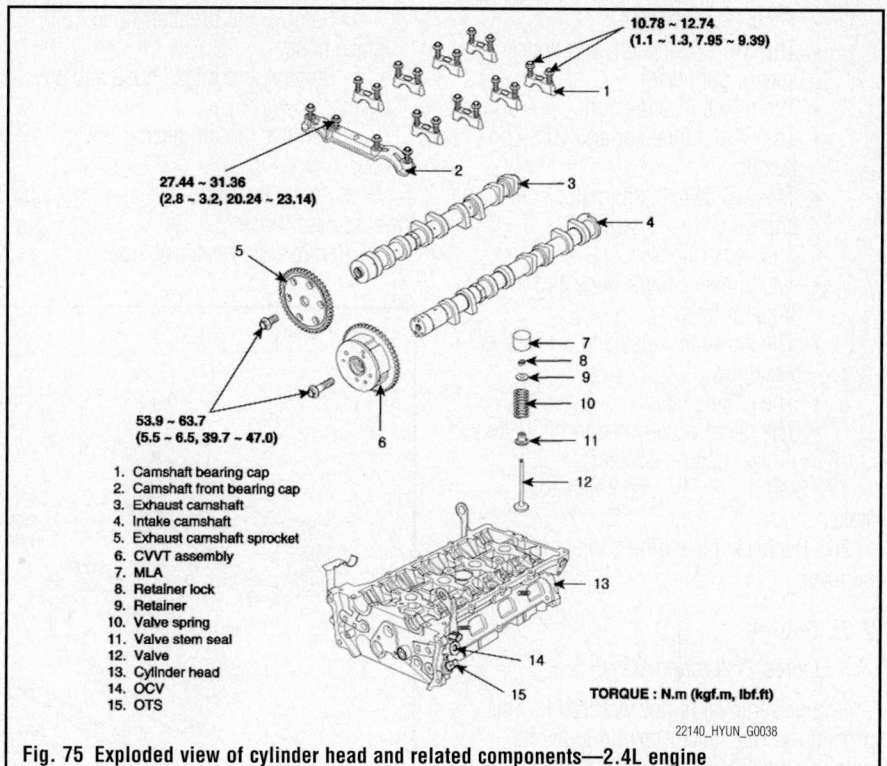

Fig. 75 Exploded view of cylinder head and related components—2.4L engine

1. Camshaft bearing cap
2. Camshaft front bearing cap
3. Exhaust camshaft
4. Intake camshaft
5. Exhaust camshaft sprocket
6. CVVT assembly
7. MLA
8. Retainer lock
9. Retainer
10. Valve spring
11. Valve stem seal
12. Valve
13. Cylinder head
14. OCV
15. OTS

TORQUE : N.m (kgf.m, lbf.ft)

22140_HYUN_G0038

- The water temp control assembly
- The timing chain
- The Continuously Variable Valve Timing (CVVT) assembly and camshaft sprocket

10. Remove the camshaft.

    a. Remove the front camshaft bearing cap.

    b. Remove the camshaft bearing caps.

    c. Remove the camshafts.

To install:

➡**Thoroughly clean all parts to be assembled. Rotate the crankshaft, set the No. 1 piston at TDC.**

11. Install the CVVT and camshaft sprocket. Tighten to 40–47 ft. lbs. (54–64 Nm).

➡**Hold the hexagonal head wrench portion of the camshaft with a vise, and install the bolt and CVVT assembly.**

12. Install the camshafts. Apply a light coat of engine oil on camshaft journals.

13. Install the camshaft bearing caps in their proper locations. Follow the illustrated tightening order.
- Tightening torque M6: 8–9 ft. lbs. (11–13 Nm)
- Tightening torque M8: 20–23 ft. lbs. (27–31 Nm)

14. Install the timing chain.

15. Check and adjust the valve clearance.

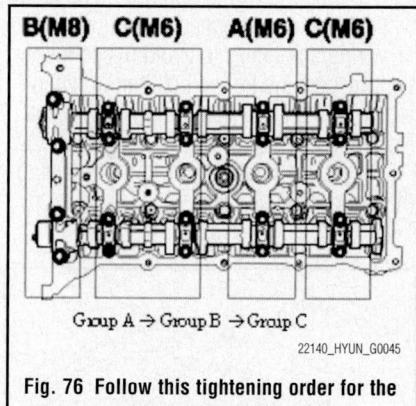

B(M8) C(M6) A(M6) C(M6)

Group A → Group B → Group C

22140_HYUN_G0045

Fig. 76 Follow this tightening order for the camshaft bearing caps—2.4L engine

16. Install the water temp control assembly and tighten as follows:
- Bolt: 11–16 ft. lbs. (15–22 Nm)
- Nut: 15–20 ft. lbs. (20–27 Nm)

➡**Assemble water temp control assembly and water inlet pipe to water pump assembly before nuts for assembling of water inlet pipe to be tightened. Insert after wetting O-ring or inner surface of thermostat housing. Always use a new O-ring.**

17. Install or connect the following:
- The delivery pipe, brake hose, and PCSV hose
- The PCSV connector, WTS connector, condenser connector, and CKP sensor connector
- The ignition coil connector

- The ETS connector
- The CMP connector, and knock sensor connector
- The injector connectors
- The OCV connector and OTS connector
- The A/C switch, alternator connect, and oil pressure switch
- The heater hoses
- The upper radiator hose and lower radiator hose
- The intake air hose and air cleaner assembly
- The engine cover
- The negative terminal to the battery

18. Fill with engine coolant.

19. Start the engine and check for leaks.

20. Recheck the engine coolant level and oil level.

2.7L Engine

See Figures 77 through 80.

Engine removal is not required for this procedure. Use fender covers to avoid damaging painted surfaces. To avoid damage, unplug the wiring connectors carefully while holding the connector portion. Mark all wiring and hoses to avoid misconnection. Inspect the timing belt before removing. Turn the crankshaft pulley so that the No. 1 piston is at Top Dead Center (TDC).

1. Before servicing the vehicle, refer to the Precautions Section.

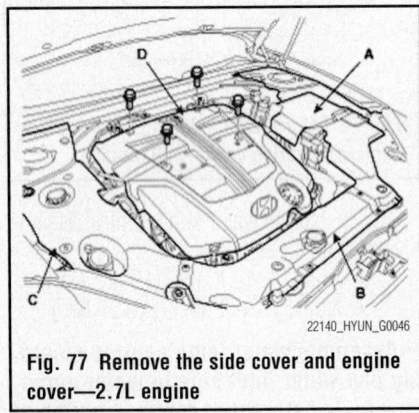

Fig. 77 Remove the side cover and engine cover—2.7L engine

2. Remove the side cover (A, B, C) and the engine cover (D).

3. Disconnect the battery terminal and the battery.

4. Remove the radiator drain plug and drain engine coolant. Remove the radiator cap to speed draining.

5. Remove the air cleaner assembly.

   a. Disconnect the Air Flow Sensor (AFS) connector.

b. Remove the breather hose from intake hose.

   c. Remove the intake hose and air cleaner upper cover.

   d. Remove the air cleaner lower cover.

6. Remove the upper radiator hose and lower radiator hose.

7. Remove the heater hoses.

8. Remove the engine wire harness connectors and wire harness clamps from the cylinder head and the intake manifold:

- Throttle position sensor (TPS) connector
- Idle speed actuator (ISA) connector
- Purge Control Solenoid Valve (PCSV) connector
- Knock sensor connector

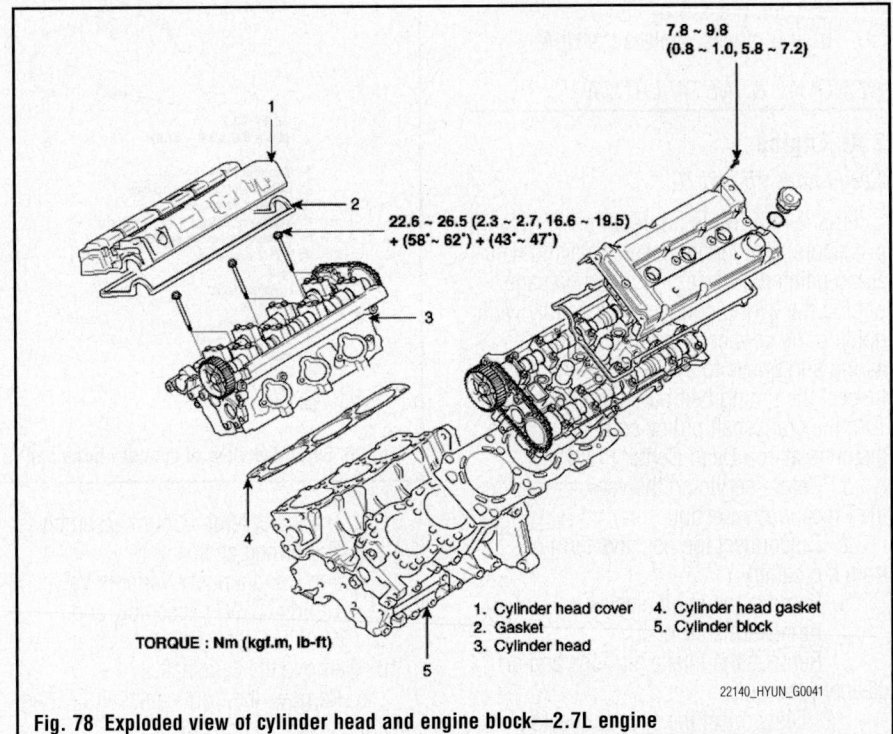

7.8 ~ 9.8 (0.8 ~ 1.0, 5.8 ~ 7.2)

22.6 ~ 26.5 (2.3 ~ 2.7, 16.6 ~ 19.5) + (58° ~ 62°) + (43° ~ 47°)

TORQUE : Nm (kgf.m, lb-ft)

1. Cylinder head cover
2. Gasket
3. Cylinder head
4. Cylinder head gasket
5. Cylinder block

22140_HYUN_G0041

Fig. 78 Exploded view of cylinder head and engine block—2.7L engine

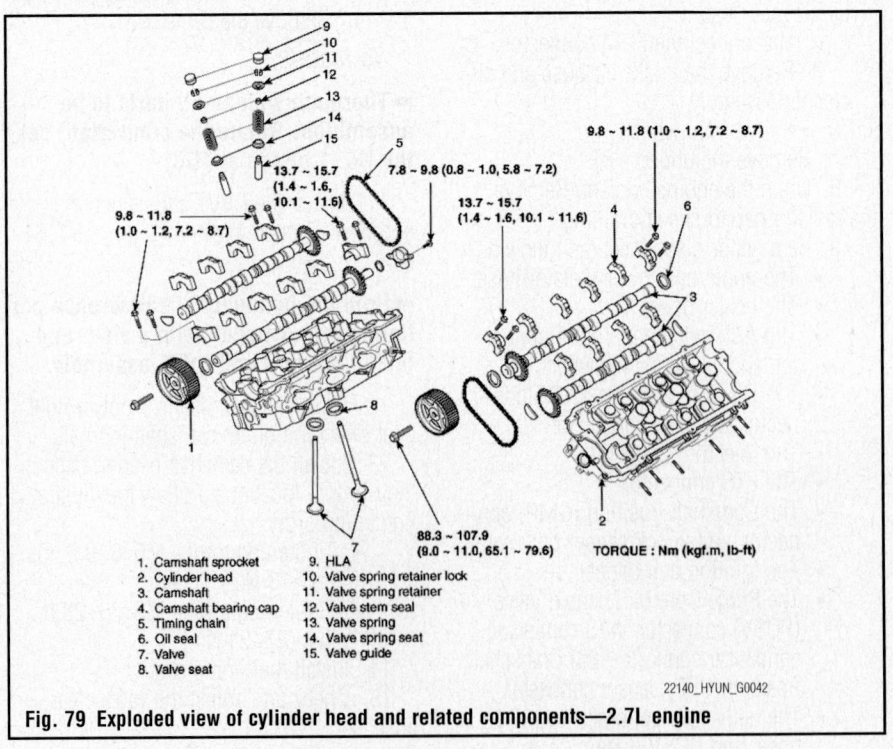

9.8 ~ 11.8 (1.0 ~ 1.2, 7.2 ~ 8.7)

13.7 ~ 15.7 (1.4 ~ 1.6, 10.1 ~ 11.6)

7.8 ~ 9.8 (0.8 ~ 1.0, 5.8 ~ 7.2)

13.7 ~ 15.7 (1.4 ~ 1.6, 10.1 ~ 11.6)

9.8 ~ 11.8 (1.0 ~ 1.2, 7.2 ~ 8.7)

88.3 ~ 107.9 (9.0 ~ 11.0, 65.1 ~ 79.6)

TORQUE : Nm (kgf.m, lb-ft)

1. Camshaft sprocket
2. Cylinder head
3. Camshaft
4. Camshaft bearing cap
5. Timing chain
6. Oil seal
7. Valve
8. Valve seat
9. HLA
10. Valve spring retainer lock
11. Valve spring retainer
12. Valve stem seal
13. Valve spring
14. Valve spring seat
15. Valve guide

22140_HYUN_G0042

Fig. 79 Exploded view of cylinder head and related components—2.7L engine

- Camshaft Position (CMP) sensor connector
- Engine ground line
- Heated oxygen sensor (Bank 2, Sensor 1) connector
- Engine temperature coolant sensor connector
- Ignition coil connector
- Crankshaft Position (CKP) sensor connector
- Heated oxygen sensor (Bank 1, Sensor 2) connector
- Fuel injector connectors

9. Remove or disconnect the following:
- The fuel inlet from the delivery pipe
- The Purge Control Solenoid Valve (PCSV) hose
- The brake booster vacuum hose
- The accelerator cable by loosening the locknut, then slip the cable end out of the throttle linkage
- The auto-cruise connector and the auto-cruise cable
- The PCV hose
- The power steering pump
- The timing belt
- The cylinder head covers

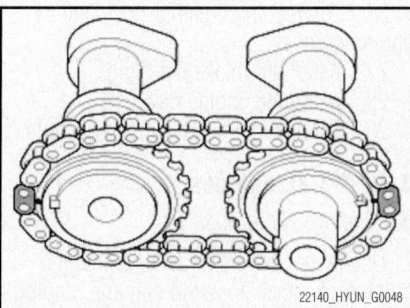

Fig. 80 Align the camshaft timing chain with the intake timing chain sprocket and exhaust timing chain sprocket as shown— 2.7L engine

- The camshaft sprocket
- The camshaft bearing caps
- The camshafts

To install:

Thoroughly clean all parts to be assembled. Rotate the crankshaft, set the No. 1 piston at TDC.

10. Install the camshafts.
   a. Align the camshaft timing chain with the intake timing chain sprocket and exhaust timing chain sprocket as shown.
   b. Install the camshaft.
   c. Install the camshaft bearing caps, and tighten as follows:

- M6 (38mm) bolt: 7–9 ft. lbs. (10–13 Nm)
- M6 (50mm) bolt: 11–12 ft. lbs. (15–17 Nm)

➡**Apply new engine oil to the thrust portion and journal of the camshafts. Apply a light coat of engine oil on the threads and under the heads of the bearing cap bolts.**

11. Using the SST (09214-21000), install the camshaft bearing oil seal.
12. Install the camshaft sprocket.
   a. Temporarily install the camshaft sprocket bolts.
   b. Hold the hexagonal head wrench portion of the camshaft with a wrench, and tighten the camshaft sprocket bolts. Tightening torque of camshaft sprocket bolt: 65–80 ft. lbs. (88–108 Nm).
13. Install the semi-circular packing.
14. Install the cylinder head cover.
   a. Install the cylinder head cover gasket in the groove of the cylinder head cover.

➡**Before installing the head cover gasket, thoroughly clean the head cover gasket and the groove. When installing, make sure the head cover gasket is seated securely in the corners of the recesses with no gap.**

   b. Apply liquid gasket to the head cover gasket at the corners of the recess. Use liquid gasket, Loctite® No. 5699. Check that the mating surfaces are clean and dry before applying liquid gasket. After assembly, wait at least 30 minutes before filling the engine with oil.
   c. Install the cylinder head covers with the 16 bolts. Uniformly tighten the bolts in several passes. Tighten to 6–7 ft. lbs. (8–10 Nm).
15. Install or connect the following:
- The timing belt
- The power steering pump
- The PCV hose
- The auto-cruise connector and the auto-cruise cable
- The accelerator cable by loosening the locknut, then slip the cable end out of the throttle linkage
- The brake booster vacuum hose
- The PCSV hose
- The fuel inlet from delivery pipe
- The fuel injector connectors
- Heated oxygen sensor (Bank 1, Sensor 2) connector
- Crankshaft position sensor connector
- Ignition coil connector
- Engine temperature coolant sensor connector

- Heated oxygen sensor (Bank 2, Sensor 1) connector
- Engine ground line
- Camshaft position sensor connector
- Knock sensor connector
- Injector connector
- The PCSV connector
- Idle speed actuator connector
- Throttle position sensor connector

16. Install the engine wire harness connectors and wire harness clamps to the cylinder head and the intake manifold.
17. Install the heater hoses.
18. Install the upper radiator hose and lower radiator hose.
19. Install the air cleaner assembly.
   a. Install the air cleaner lower cover.
   b. Install the intake hose and air cleaner upper cover.
   c. Install the breather hose from intake hose.
   d. Connect the AFS connector.
20. Connect the battery terminal and the battery.
21. Fill with engine coolant.
22. Install the side cover and the engine cover.
23. Start the engine and check for leaks.
24. Recheck the engine coolant level and oil level.

3.5L Engine

See Figures 81 through 87.

1. Before servicing the vehicle, refer to the Precautions Section.
2. Disconnect the negative battery cable.
3. Remove the engine cover.
4. Remove the intake manifold.
5. Disconnect the breather hose and engine harness.
6. Remove the timing belt.
7. Disconnect the spark plug wires.
8. Remove the rocker arm cover.
9. Remove the camshaft sprockets.

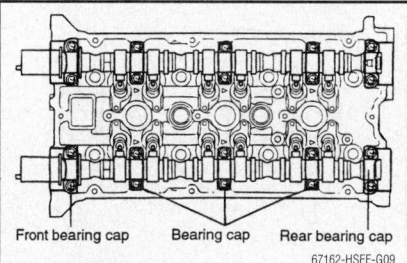

Front bearing cap   Bearing cap   Rear bearing cap

Fig. 81 Remove the camshaft bearing caps—3.5L engine

10. Remove the camshaft bearing caps. Make sure to note the cap location prior to removal.

11. Remove the camshafts.

12. If necessary, remove the rocker arms and lifters.

To install:

13. Rotate the crankshaft until the number one cylinder is at Top Dead Center (TDC).

14. Check the position of the rocker arm to make sure it is proper installed on the lash adjuster and valve.

15. Install the camshaft dowel pin as illustrated.

16. When installing the camshafts be careful to place them in their correct positions. They are marked as follows:

- Left bank intake is marked with an (I)
- Left bank exhaust is marked with an (E)
- Right bank intake is marked with an (J)
- Right bank exhaust is marked with an (H)

17. Make sure the camshaft caps are installed in their original locations. Bearing

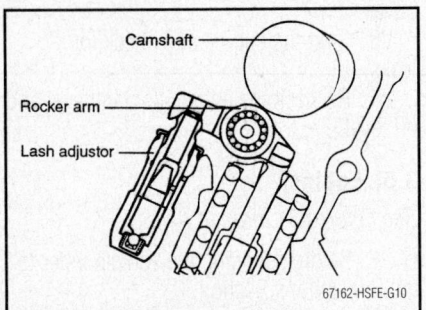

Fig. 82 Camshaft and related component positioning—3.5L engine

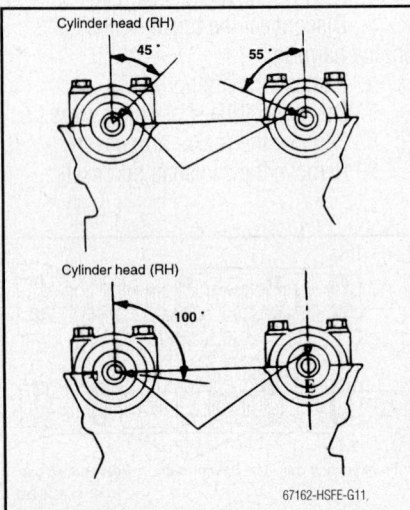

Fig. 83 Install the camshaft dowel pin as shown—3.5L engine

caps 3, 4, and 5 have the front mark on them. All caps are marked with an **I** for Intake or an **E** for Exhaust.

18. Install the caps, check once more they are installed in the original locations and tighten the caps using 2 –3 passes. Tighten the outer cap bolts (identified by an * mark in the illustration) to 14 ft. lbs. (19 Nm) and the inner cap bolts to 84 inch lbs. (10 Nm).

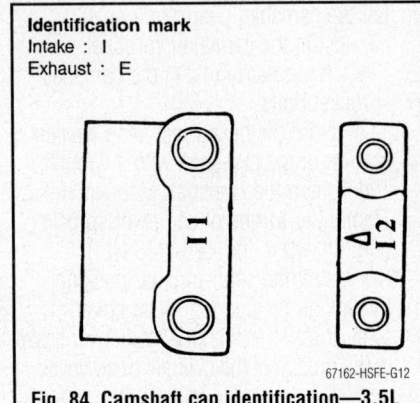

Identification mark
Intake : I
Exhaust : E

Fig. 84 Camshaft cap identification—3.5L engine

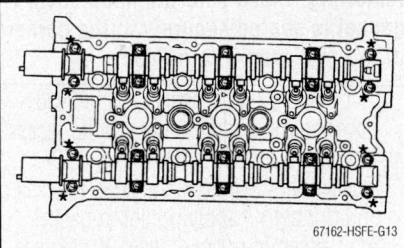

Fig. 85 Camshaft cap bolt location (the * mark on the illustration identifies the outer cap bolts)—3.5L engine

19. Install a new rocker cover gasket.

20. Clean the camshaft cap sealing surface using a plastic scraper to avoid damage.

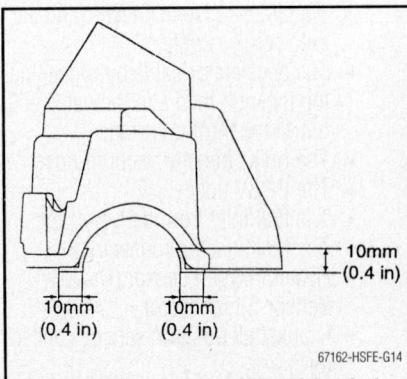

Fig. 86 Apply a 0.4 inch (10mm) bead of sealant to the locations shown—3.5L engine

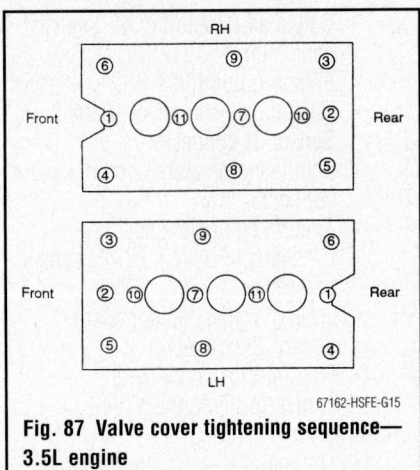

Fig. 87 Valve cover tightening sequence—3.5L engine

21. Apply a 0.4 inch (10mm) bead of LT 5900 sealant to the locations illustrated.

22. Be careful that the gasket is positioned properly when installing the rocker arm cover and that it stays in its positioned when the cover bolts are tightened.

23. Install the cover, making sure to use the washers when installing the bolts and tighten in the sequence illustrated to 72 inch lbs. (8 Nm).

24. Connect the spark plug wires.

25. Install the timing belt.

26. Connect the breather hose and engine harness.

27. Install the intake manifold.

28. Install the engine cover.

29. Connect the negative battery cable.

3.3L & 3.8L Engines

See Figures 88 through 91.

Use fender covers to avoid damaging painted surfaces. To avoid damage, unplug the wiring connectors carefully while holding the connector portion. Mark all wiring and hoses to avoid misconnection. Inspect the timing belt before removing. Turn the crankshaft pulley so that the No. 1 piston is at Top Dead Center (TDC).

1. Before servicing the vehicle, refer to the Precautions Section.

2. Remove or disconnect the following:

- The negative battery connection
- The timing chain
- The water temperature control assembly
- The camshaft bearing cap
- The camshaft assembly

To install:

Thoroughly clean all parts to be assembled. Rotate the crankshaft, set the No. 1 piston at TDC.

3. Install the Continuously Variable Valve Timing (CVVT) and camshaft

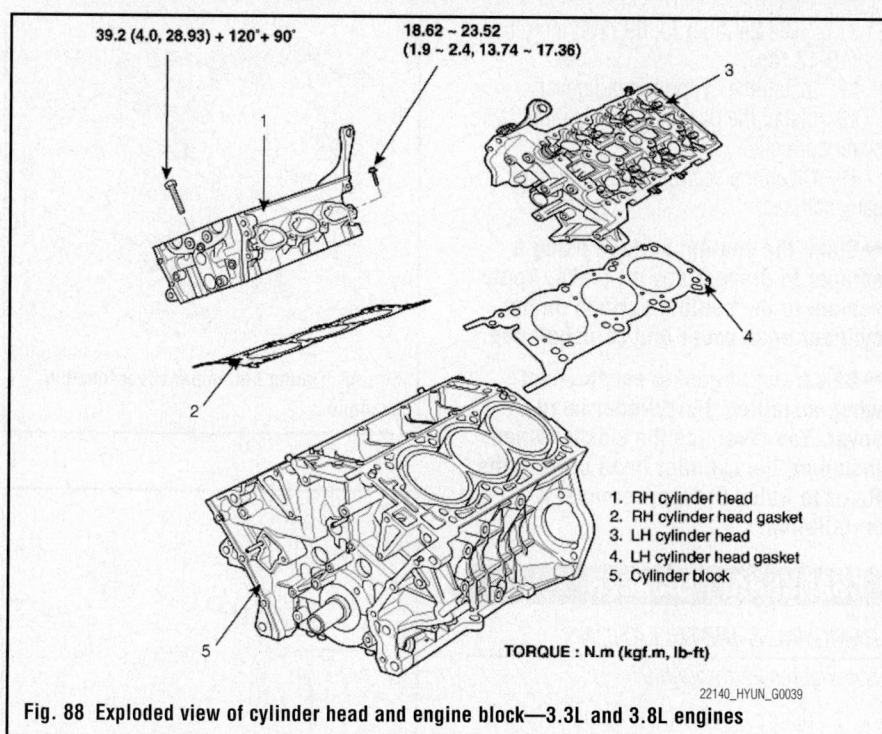

Fig. 88 Exploded view of cylinder head and engine block—3.3L and 3.8L engines

1. RH cylinder head
2. RH cylinder head gasket
3. LH cylinder head
4. LH cylinder head gasket
5. Cylinder block

TORQUE : N.m (kgf.m, lb-ft)

22140_HYUN_G0039

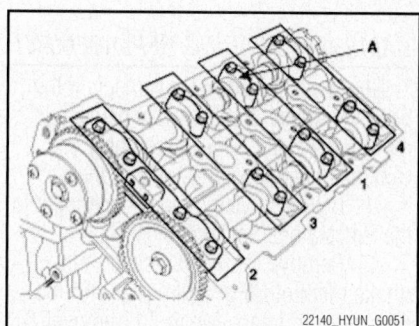

Fig. 89 Exploded view of cylinder head and related components—3.3L and 3.8L engines

1. Camshaft bearing cap
2. Exhaust camshaft
3. Intake camshaft
4. Exhaust camshaft sprocket
5. CVVT assembly
6. MLA
7. Retainer lock
8. Retainer
9. Valve spring
10. Valve stem seal
11. Valve
12. OCV
13. Cylinder head

TORQUE : N.m (kgf.m, lb-ft)

22140_HYUN_G0040

sprocket. Tighten to 48–56 ft. lbs. (65–76 Nm).

a. Install camshaft-inlet to dowel pin of CVVT assembly. At this time, do not install to oil hole of camshaft-inlet.

b. Hold the hexagonal head wrench portion of the camshaft with a vise, and install the bolt and CVVT assembly.

c. Do not rotate the CVVT assembly when the camshaft is installed to the dowel pin of the CVVT assembly.

4. Install the camshafts.

a. Apply a light coat of engine oil on camshaft journals.

b. Assemble the key groove of camshaft rear side to the same level of head top surface.

c. Be careful to get the right bank, left bank, intake side, and exhaust side in the correct position before assembling.

5. Install the camshaft bearing caps in the sequence shown:

a. Step 1—Tighten to 48 inch lbs. (6 Nm).

b. Step 2—Tighten to 84–108 inch lbs. (10–12 Nm).

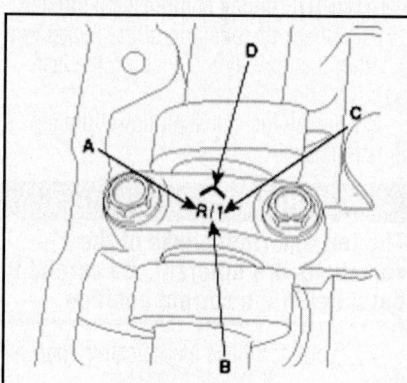

Fig. 90 Install the camshaft bearing caps in the sequence shown—3.3L and 3.8L engines

22140_HYUN_G0051

✳✳ WARNING

Be careful to properly position the right bank, left bank, intake side, exhaust side, and front mark on the camshaft bearing caps while assembling.

A. L (LH); R (RH)
B. I (Intake); None (Exhaust)
C. Journal number
D. Front mark

22140_HYUN_G0052

Fig. 91 Be careful to properly position the camshaft bearing caps according to its markings—3.3L and 3.8L engines

✳✳ WARNING

Rotate the crankshaft so as not to contact the valves to the pistons by positioning the pistons 0.3937 inch (10mm) below the top of the cylinder block.

6. Install the water temperature control assembly.

7. Install the timing chain.

8. Check and adjust the valve clearance, as necessary.

9. Connect the negative battery cable.

10. Fill with engine coolant.

11. Start the engine and check for leaks.

12. Recheck the engine coolant level and oil level.

CAMSHAFT BEARING REPLACEMENT

Check each bearing for damage. If the bearing surface is excessively damaged, replace the cylinder head assembly or camshaft bearing cap, as necessary.

1. Before servicing the vehicle, refer to the Precautions Section.

2. Remove the engine cover and intake manifold. For additional information, refer to Intake Manifold, removal & installation.

3. Disconnect the breather hose and the engine harness.

4. Remove the timing belt.

5. Remove the spark plug cables.

6. Loosen the cylinder head cover bolts and then remove the cylinder head cover (also called the valve cover).

7. Remove the camshaft sprockets.

8. Remove the camshaft bearing caps.

9. Remove the camshafts.

10. Rotate the crankshaft so No. 1 cylinder is in TDC on the compression stroke.

11. Check the position of the rocker arm whether it is exactly installed on the lash adjuster and valve or not.

12. Install the camshaft dowel pin as illustration.

❋❋ WARNING

The left and right banks of the camshafts are different. Be careful to put them in the correct position.

13. Look for these identification markings:

    a. Left bank Intake (IN): I. Exhaust (Ex): E.

    b. Right bank Intake (IN): J. Exhaust (Ex): H.

14. Confirm the identification mark and the number. Bearing caps for No. 3, No. 4, and No. 5 should have a front mark. Arrange the front mark upon the cylinder head while installing the bearing caps.

15. Use the identification mark—Intake **I** and Exhaust **E**.

16. Tighten each bearing cap in 2–3 steps:

    a. The 16 outer (*) to: 14–15 ft. lbs. (19–21 Nm).

    b. The 24 inner to: 89–106 inch lbs. (10-12 Nm).

17. Install the cylinder head cover.

18. Install the gasket to the cylinder head cover correctly.

19. Clean the sealing surface on the camshaft cap.

➥**Clean the sealing surface using a scraper to prevent any oil leaks. Apply sealant to the sealing surface on the cylinder head cover and camshaft cap.**

➥**Be careful of gasket escapement when installing the cylinder head cover. You must use the washer when installing the cylinder head cover bolts. Refer to Valve Covers, removal & installation.**

CRANKSHAFT DAMPER

REMOVAL & INSTALLATION

See Figures 92 through 97.

1. Before servicing the vehicle, refer to the Precautions Section.

2. Remove the engine cover.

3. Remove the front right wheel and tire.

4. Remove the right side cover.

5. Remove the accessory drive belt (A), the idler (B), and the tensioner (C).

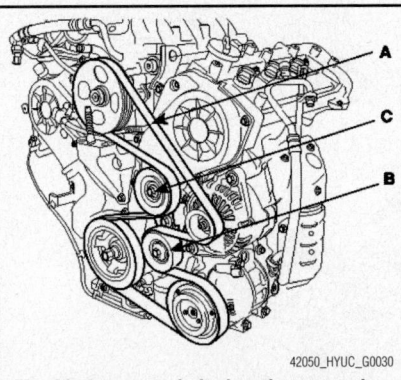

Fig. 92 Accessory belts location example

➥**In removing the accessory drive belt, fix a tool in the auto tensioner pulley bolt and turn the bolt counter clockwise.**

6. Remove the timing belt upper cover (A).

7. Align the groove of the pulley with the timing mark of the timing belt cover by turning the crankshaft pulley clockwise.

8. Check if the timing mark of the camshaft sprocket is aligned with that of the cylinder head cover with No. 1 cylinder piston at Top Dead Center (TDC).

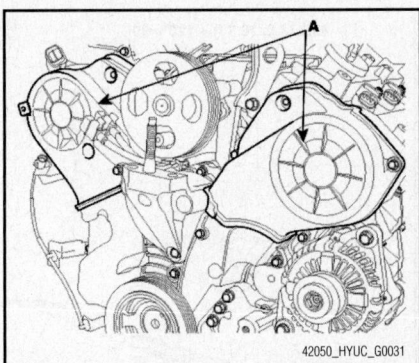

Fig. 93 Timing belt upper cover location example

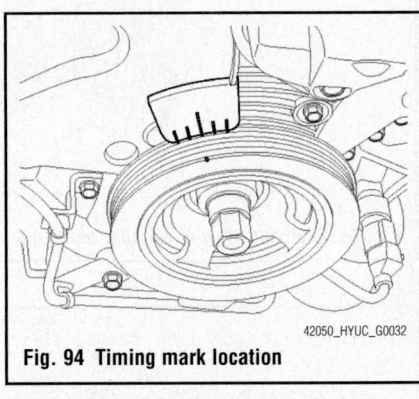

Fig. 94 Timing mark location

9. Support the engine oil pan with a jack.

❋❋ WARNING

Put a wooden or rubber block between the jack and the engine oil pan.

10. Remove the engine mounting bracket (A).

11. Remove the crankshaft damper pulley (A).

To install:

12. Install the crankshaft damper. Torque mounting bolt to:

- 123–130 ft. lbs. (167–176 Nm)— 2.4L and 2.7L engines

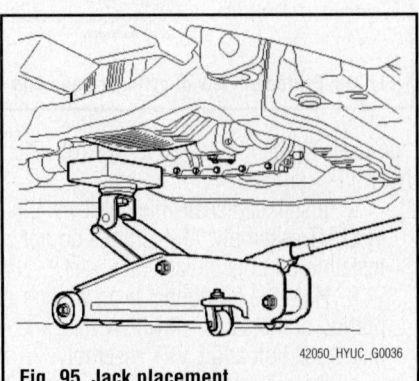

Fig. 95 Jack placement

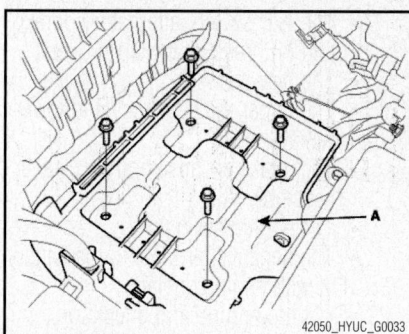

Fig. 96 Remove the engine mounting bracket (A)

- 130–138 ft. lbs. (176–187 Nm)—3.5L engine
- 210–224 ft. lbs. (285–304 Nm)—3.3L and 3.8L engines
13. Replace the engine mounting bracket.
14. Lower the jack and remove from the vehicle.
15. Install the upper timing belt cover.
16. Making sure all timing and alignment marks match, install the drive belt, the idler, and the tensioner.
17. Install the right side cover.
18. Install the right front tire and wheel.
19. Install the engine cover.

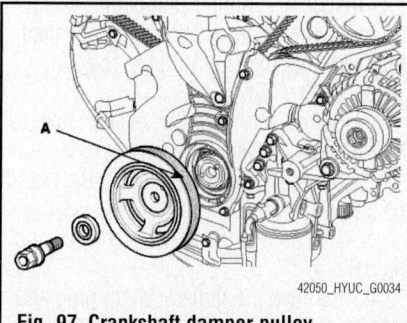

Fig. 97 Crankshaft damper pulley

CRANKSHAFT FRONT SEAL

REMOVAL & INSTALLATION

1. Before servicing the vehicle, refer to the Precautions Section.
2. Remove or disconnect the following:
   - Negative battery cable
   - Accessory drive belts
   - Front cover
   - Timing belt
   - Crankshaft timing sprocket
   - Front crankshaft seal

To install:
3. Install the front crankshaft seal so that it is flush with the oil pump housing.
4. Install or connect the following:
   - Crankshaft timing sprocket
   - Timing belt

- Front cover
- Accessory drive belts
- Negative battery cable
5. Start the engine and check for leaks.

CYLINDER HEAD

REMOVAL & INSTALLATION

2.4L Engine

See Figures 75, 76, 98 and 99.

Engine removal is not required for this procedure. Use fender covers to avoid damaging painted surfaces. To avoid damaging the cylinder head, wait until the engine coolant temperature drops below normal temperature before removing it. When handling a metal gasket, take care not to fold the gasket or damage the contact surface of the gasket. To avoid damage, unplug the wiring connectors carefully while holding the connector portion. Mark all wiring and hoses to avoid misconnection. Inspect the timing belt before removing the cylinder head. Turn the crankshaft pulley so that the No. 1 piston is at Top Dead Center (TDC).

1. Before servicing the vehicle, refer to the Precautions Section.
2. Disconnect the negative terminal from the battery.
3. Remove the engine cover(A).
4. Remove the air duct.
5. Remove the intake air hose and air cleaner assembly.
   a. Disconnect the AFS connector.
   b. Disconnect the breather hose from air cleaner hose.
   c. Disconnect the ECM connector.
   d. Remove the intake air hose and air cleaner assembly.
6. Remove front wheels.
7. Remove the undercover.
8. Drain the engine coolant. Remove the radiator cap to speed draining.
9. Remove or disconnect the following:
   - The upper and lower radiator hose
   - The heater hoses
   - The A/C switch, alternator connector, and oil pressure switch
   - The Oil Control Valve (OCV) connector and OTS connector
   - The injector connectors
   - The ETS connector
   - The Camshaft Position (CMP) connector, and knock sensor connector
   - The ignition coil connectors
   - The Purge Control Solenoid Valve (PCSV) connector, WTS connector, condenser connector, and Crankshaft Position (CKP) sensor connector

- The delivery pipe, brake vacuum hose, and PCSV hose
- The water temp control assembly
- The intake manifold
- The exhaust manifold
- The timing chain
- The Continuously Variable Valve Timing (CVVT) assembly and camshaft sprocket
10. Remove the camshaft.
    a. Remove the front camshaft bearing cap.
    b. Remove the camshaft bearing caps.
    c. Remove the camshafts.
11. Remove the OCV and OTS.
12. Remove the cylinder head bolts, then remove the cylinder head:
    a. Using triple square wrench, uniformly loosen and remove the 10 cylinder head bolts, in several passes, in the sequence shown. Remove the 10 cylinder head bolts and plate washers.

✳✳ WARNING

Head warpage or cracking could result from removing bolts in an incorrect order.

    b. Lift the cylinder head from the dowels on the cylinder block and place the cylinder head on wooden blocks on a bench.

✳✳ WARNING

Be careful not to damage the contact surfaces of the cylinder head and cylinder block.

To install:

➡**Thoroughly clean all parts to be assembled. Always use a new head and manifold gasket. The cylinder head gasket is a metal gasket. Take care not to bend it. Rotate the crankshaft, set the No. 1 piston at TDC.**

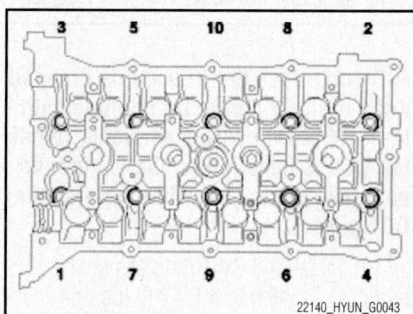

Fig. 98 Uniformly loosen/tighten and remove/install the cylinder head bolts, in several passes, in the sequence shown—2.4L engine

13. Install OCV filter. Keep the OCV filter clean.

14. Install the cylinder head gasket on the cylinder block. Be careful of the installation direction.

15. Place the cylinder head carefully in order not to damage the gasket.

16. Using SST (09221-4A000), install the cylinder head bolts.

a. Apply a light coat if engine oil on the threads and under the heads of the cylinder head bolts.

b. Using a wrench, install and tighten the 10 cylinder head bolts and plate washers, in several passes, in the sequence shown. Tighten to 25 ft. lbs. (34 Nm) plus 90° and then an additional 90°.

➡**Always use new cylinder head bolts.**

17. Install the OCV. Tighten to 84–108 inch lbs. (10–12 Nm).

18. Install the OTS. Tighten to 15–17 ft. lbs. (20–24 Nm).

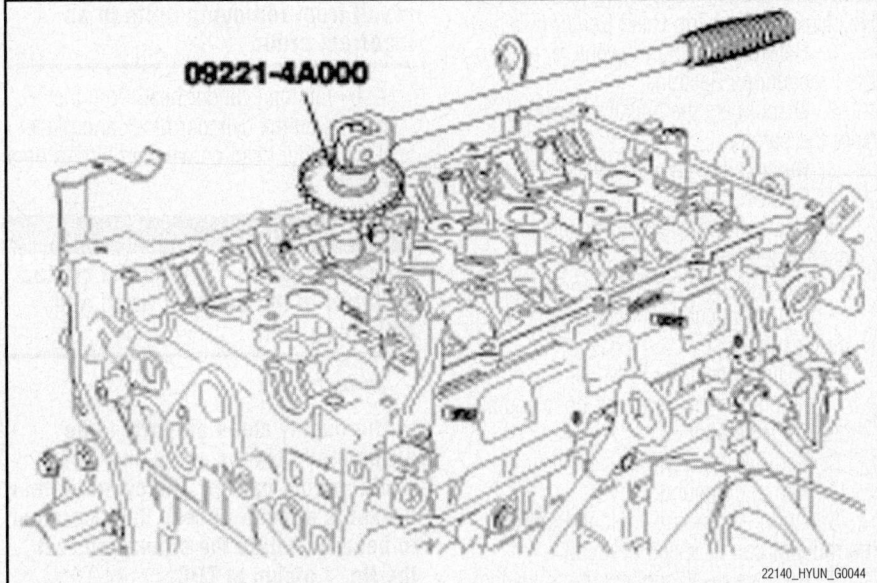

09221-4A000

22140_HYUN_G0044

Fig. 99 Using SST (09221-4A000) to install the cylinder head bolts—2.4L engine

➡**Do not reuse the OCV if it is dropped. Keep the OCV clean. Do not hold onto the OCV sleeve during servicing. When the OCV is installed on the engine, do not move the engine while holding the OCV yoke.**

19. Install the CVVT and camshaft sprocket. Tighten to 40–47 ft. lbs. (54–64 Nm).

➡**Hold the hexagonal head wrench portion of the camshaft with a vise, and install the bolt and CVVT assembly.**

20. Install the camshafts. Apply a light coat of engine oil on camshaft journals.

21. Install the camshaft bearing caps in their proper locations. Follow the illustrated tightening order and tighten to:
- M6: 8–9 ft. lbs. (11–13 Nm)
- M8: 20–23 ft. lbs. (27–31 Nm)

22. Install the timing chain.

23. Check and adjust the valve clearance.

24. Install the exhaust manifold.

25. Install the intake manifold.

26. Install the water temp control assembly, and tighten as follows:
- Bolt: 11–16 ft. lbs. (15–22 Nm)
- Nut: 15–20 ft. lbs. (20–27 Nm)

➡**Assemble water temp control assembly and water inlet pipe to water pump assembly before nuts for assembling of water inlet pipe to be tightened. Insert after wetting O-ring or inner surface of thermostat housing. Always use a new O-ring.**

27. Install or connect the following:
- The delivery pipe, brake hose, and PCSV hose
- The PCSV connector, WTS connector, condenser connector, and CKP sensor connector
- The ignition coil connector
- The ETS connector
- The CMP connector, and knock sensor connector
- The injector connectors
- The OCV connector and OTS connector

- The A/C switch, alternator connect, and oil pressure switch
- The heater hoses
- The upper radiator hose and lower radiator hose
- The intake air hose and air cleaner assembly
- The engine cover
- The negative terminal to the battery

28. Fill with engine coolant.

29. Start the engine and check for leaks.

30. Recheck the engine coolant level and oil level.

2.7L Engine

See Figures 76 through 80 and 100.

Engine removal is not required for this procedure. Use fender covers to avoid damaging painted surfaces. To avoid damaging the cylinder head, wait until the engine coolant temperature drops below normal temperature before removing it. When handling a metal gasket, take care not to fold the gasket or damage the contact surface of the gasket. To avoid damage, unplug the wiring connectors carefully while holding the connector portion. Mark all wiring and hoses to avoid misconnection. Inspect the timing belt before removing the cylinder head. Turn the crankshaft pulley so that the No. 1 piston is at Top Dead Center (TDC).

1. Before servicing the vehicle, refer to the Precautions Section.

2. Remove the side cover (A, B, C) and the engine cover (D).

3. Disconnect the battery terminal and the battery.

4. Remove the radiator drain plug and drain engine coolant. Remove the radiator cap to speed draining.

5. Remove the air cleaner assembly.

a. Disconnect the Air Flow Sensor (AFS) connector.

b. Remove the breather hose from intake hose.

c. Remove the intake hose and air cleaner upper cover.

d. Remove the air cleaner lower cover.

6. Remove the upper radiator hose and lower radiator hose.

7. Remove the heater hoses.

8. Remove the following engine wire harness connectors and wire harness clamps from the cylinder head and the intake manifold:
- Throttle position sensor (TPS) connector
- Idle speed actuator (ISA) connector

- Purge Control Solenoid Valve (PCSV) connector
- Knock sensor connector
- Camshaft Position (CMP) sensor connector
- Engine ground line
- Heated oxygen sensor (Bank 2, Sensor 1) connector
- Engine temperature coolant sensor connector
- Ignition coil connector
- Crankshaft Position (CKP) sensor connector
- Heated oxygen sensor (Bank 1, Sensor 2) connector
- Fuel injector connectors

9. Remove or disconnect the following:
- The fuel inlet from the delivery pipe
- The Purge Control Solenoid Valve (PCSV) hose
- The brake booster vacuum hose
- The accelerator cable by loosening the locknut, then slip the cable end out of the throttle linkage
- The auto-cruise connector and the auto-cruise cable
- The PCV hose
- The intake manifold
- The power steering pump
- The exhaust manifold
- The timing belt
- The cylinder head covers
- The camshaft sprocket
- The camshaft bearing caps
- The camshafts
- The timing belt rear cover
- The water temperature control assembly and water pipe

10. Remove the cylinder head bolts, then remove the cylinder heads, as follows:

a. Uniformly loosen and remove the 8 cylinder head bolts on each cylinder head in several passes and in the sequence shown, then repeat for the other side, as shown. Remove the 16 cylinder head bolts and plate washer.

❊❊ WARNING

Head warpage or cracking could result from removing bolts in an incorrect order.

b. Lift the cylinder head from the dowels on the cylinder block and place the cylinder head on wooden blocks on a bench.

❊❊ WARNING

Be careful not to damage the contact surfaces of the cylinder head and cylinder block.

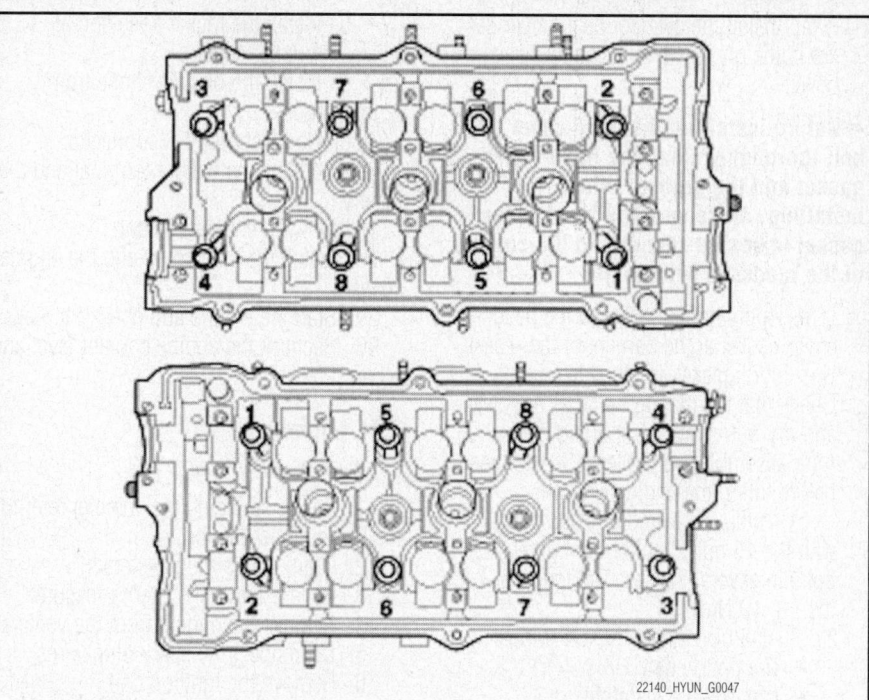

Fig. 100 Uniformly loosen/tighten and remove/install the cylinder head bolts on each cylinder head in several passes and in the sequence shown—2.7L engine

22140_HYUN_G0047

To install:

Thoroughly clean all parts to be assembled. Always use a new head gasket and manifold gasket. The cylinder head gasket is a metal gasket. Take care not to bend it. Rotate the crankshaft, set the No. 1 piston at TDC.

11. Install the cylinder head gaskets on the cylinder block. Be careful of the installation direction.

12. Place the cylinder head carefully in order not to damage the gasket with the bottom part of the end.

13. Install cylinder head bolts:

a. Apply a light coat if engine oil on the threads and under the heads of the cylinder head bolts.

b. Install the plate washer to the cylinder head bolt.

c. Install and uniformly tighten the cylinder head bolts on each cylinder head in several passes and in the sequence shown, then repeat for the other side, as shown:

➡**If only 1 of the cylinder head bolts does not meet the torque specification, replace the cylinder head bolt.**

- Step 1: Torque bolts to 18 ft. lbs. (25 Nm)
- Step 2: Retighten the cylinder head bolts by 60° in the numerical order shown
- Step 3: Retighten the cylinder head bolts by 45° in the numerical order shown

14. Install the water pipe and water temperature control assembly. Tighten to 11–5 ft. lbs. (15–20 Nm).

15. Install the timing belt rear cover. Tighten to 7–9 ft. lbs. (10–12 Nm).

16. Install the camshafts, as follows:

a. Align the camshaft timing chain with the intake timing chain sprocket and exhaust timing chain sprocket as shown.

b. Install the camshaft.

c. Install the camshaft bearing caps and tighten as follows:

- M6 (38mm) bolt: 7–9 ft. lbs. (10–13 Nm)
- M6 (50mm) bolt: 11–12 ft. lbs. (15–17 Nm)

➡**Apply new engine oil to the thrust portion and journal of the camshafts. Apply a light coat of engine oil on the threads and under the heads of the bearing cap bolts.**

17. Using the SST (09214-21000), install the camshaft bearing oil seal.

18. Install the camshaft sprocket, as follows:

a. Temporarily install the camshaft sprocket bolts.

b. Hold the hexagonal head wrench portion of the camshaft with a wrench, and tighten the camshaft sprocket bolts. Tightening torque of camshaft sprocket bolt: 65–80 ft. lbs. (88–108 Nm).

19. Install the semi-circular packing.

20. Install the cylinder head cover.

a. Install the cylinder head cover gasket in the groove of the cylinder head cover.

➡ **Before installing the head cover gasket, thoroughly clean the head cover gasket and the groove. When installing, make sure the head cover gasket is seated securely in the corners of the recesses with no gap.**

b. Apply liquid gasket to the head cover gasket at the corners of the recess. Use liquid gasket, Loctite® No. 5699. Check that the mating surfaces are clean and dry before applying liquid gasket. After assembly, wait at least 30 minutes before filling the engine with oil.

c. Install the cylinder head covers with the 16 bolts. Uniformly tighten the bolts in several passes. Tighten to 6–7 ft. lbs. (8–10 Nm).

21. Install or connect the following:
- The timing belt
- The exhaust manifold
- The power steering pump
- The intake manifold
- The PCV hose
- The auto-cruise connector and the auto- cruise cable
- The accelerator cable by loosening the locknut, then slip the cable end out of the throttle linkage
- The brake booster vacuum hose
- The PCSV hose
- The fuel inlet from delivery pipe
- The fuel injector connectors
- Heated oxygen sensor (Bank 1, Sensor 2) connector
- Crankshaft position sensor connector
- Ignition coil connector
- Engine temperature coolant sensor connector
- Heated oxygen sensor (Bank 2, Sensor 1) connector
- Engine ground line
- Camshaft position sensor connector
- Knock sensor connector
- Injector connector
- The PCSV connector
- Idle speed actuator connector
- Throttle position sensor connector

22. Install the engine wire harness connectors and wire harness clamps to the cylinder head and the intake manifold.

23. Install the heater hoses.

24. Install the upper radiator hose and lower radiator hose.

25. Install the air cleaner assembly, as follows:

a. Install the air cleaner lower cover.

b. Install the intake hose and air cleaner upper cover.

c. Install the breather hose from intake hose.

d. Connect the AFS connector.

26. Connect the battery terminal and the battery.

27. Fill with engine coolant.

28. Install the side cover and the engine cover.

29. Start the engine and check for leaks.

30. Recheck the engine coolant level and oil level.

3.5L Engine

See Figures 101 through 103.

1. Before servicing the vehicle, refer to the Precautions Section.

2. Drain the cooling system.

3. Relieve the fuel system pressure.

4. Remove the engine from the vehicle.

5. Disconnect the spark plug wires.

6. Remove the ignition coil.

7. Remove the timing belt covers.

8. Remove the timing belt and camshaft sprockets.

9. Remove the heat shield and exhaust manifold.

10. Remove the water pump pulley and valve cover.

11. Remove the intake and exhaust camshafts.

12. Loosen the cylinder head bolts in the sequence shown using a 12mm socket in 2–3 steps.

13. Remove the cylinder head and gasket.

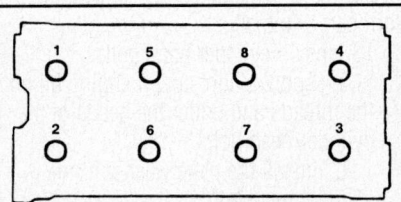

Fig. 101 Cylinder head bolt loosening sequence—3.5L engine

To install:

14. Clean the gasket mating surfaces

15. Install the cylinder head gasket so the surface with the identification mark faces towards the head.

16. Install the cylinder head.

17. Tighten the cylinder head bolts to 75–82 ft. lbs. (105–115 Nm).

18. Install the intake and exhaust camshafts.

19. Install the water pump pulley and valve cover.

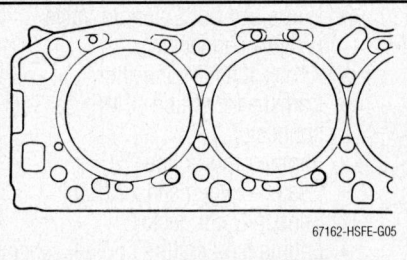

Fig. 102 Location of the cylinder head gasket identification mark—3.5L engine

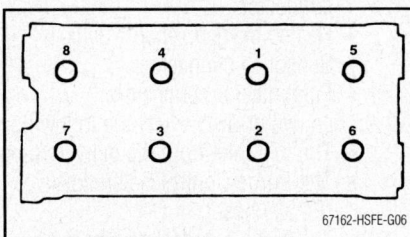

Fig. 103 Cylinder head bolt tightening sequence—3.5L engine

20. Install the heat shield and exhaust manifold.

21. Install the timing belt and camshaft sprockets.

22. Install the timing belt covers.

23. Install the ignition coil.

24. Connect the spark plug wires.

25. Install the engine in the vehicle.

26. Fill the cooling system.

27. Change the oil and filter.

28. Start the engine and check for leaks.

3.3L & 3.8L Engines

See Figures 88 through 90, 104 and 105.

Use fender covers to avoid damaging painted surfaces. To avoid damaging the cylinder head, wait until the engine coolant temperature drops below normal temperature before removing it. When handling a metal gasket, take care not to fold the gasket or damage the contact surface of the gasket. To avoid damage, unplug the wiring connectors carefully while holding the connector portion. Mark all wiring and hoses to avoid misconnection. Inspect the timing belt before removing the cylinder head. Turn the crankshaft pulley so that the No. 1 piston is at Top Dead Center (TDC).

1. Before servicing the vehicle, refer to the Precautions Section.

➡ **Engine removal is required for this procedure.**

2. Remove or disconnect the following:
- The negative battery connection
- The exhaust manifold
- The intake manifold

- The timing chain
- The water temperature control assembly
- The camshaft bearing cap
- The camshaft assembly

3. The cylinder head bolts, then remove cylinder head.

a. Uniformly loosen and remove the 16 cylinder head bolts, in several passes, in the sequence shown.

b. Remove the 16 cylinder head bolts and plate washers.

> ✳✳ **WARNING**
>
> **Head warpage or cracking could result from removing bolts in an incorrect order.**

c. Lift the cylinder head from the dowels on the cylinder block and place the cylinder head on wooden blocks on a bench.

> ✳✳ **WARNING**
>
> **Be careful not to damage the contact surfaces of the cylinder head and cylinder block.**

To install:

Thoroughly clean all parts to be assembled. Always use a new head and manifold gasket. The cylinder head gasket is a metal gasket. Take care not to bend it. Rotate the crankshaft, set the No. 1 piston at TDC.

4. Ensure the sealant locations on the cylinder head and cylinder block are free of engine oil or any debris.

5. Apply sealant on the cylinder block top face before assembling cylinder head gaskets.

➡**The part must be assembled within 5 minutes after sealant is applied. The bead width should be 0.08–0.12 inch (2–3mm). The sealant location: 0.04–0.06 inch (1.0–1.5mm) from block surface. Recommended sealant: Liquid sealant TB1217H.**

6. Install the cylinder head. Remove any extruded sealant after assembling cylinder heads.

7. Place the cylinder head carefully in order not to damage the gasket with the bottom part of the end.

8. Install cylinder head bolts.

a. Do not apply engine oil on the threads or under the heads of the cylinder head bolts.

b. Using SST(09221-4A000), install and tighten the cylinder head bolts and plate washers, in several passes, in the sequence shown.

- Step 1—Tighten to 28–30 ft. lbs. (37–41 Nm)
- Step 2—Tighten an additional: 120° plus or minus 2°
- Step 3—Tighten an additional: 90° plus or minus 2°

c. Tighten bolt (A) to: 14–17 ft. lbs. (19–24 Nm).

➡**Always use new cylinder head bolts.**

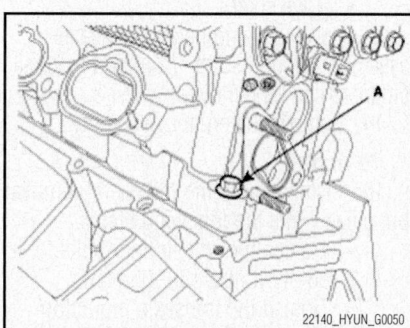

Fig. 104 Tighten bolt (A) to 14–17 ft. lbs. (19–24 Nm)—3.3L and 3.8L engines

9. Install the Continuously Variable Valve Timing (CVVT) and camshaft sprocket. Tighten to 48–56 ft. lbs. (65–76 Nm).

a. Install camshaft-inlet to dowel pin of CVVT assembly. At this time, do not install to oil hole of camshaft-inlet.

b. Hold the hexagonal head wrench portion of the camshaft with a vise, and install the bolt and CVVT assembly.

c. Do not rotate the CVVT assembly when the camshaft is installed to the dowel pin of the CVVT assembly.

10. Install the camshafts.

a. Apply a light coat of engine oil on camshaft journals.

b. Assemble the key groove of camshaft rear side to the same level of head top surface.

c. Be careful to get the right bank, left bank, intake side, and exhaust side in the correct position before assembling.

11. Install the camshaft bearing caps, in the sequence shown, to:

a. Step 1: Tighten to 48 inch lbs. (6 Nm).

b. Step 2: Tighten to 84–108 inch lbs. (10–12 Nm).

> ✳✳ **WARNING**
>
> **Be careful to properly position the right bank, left bank, intake side, exhaust side, and front mark on the camshaft bearing caps while assembling.**

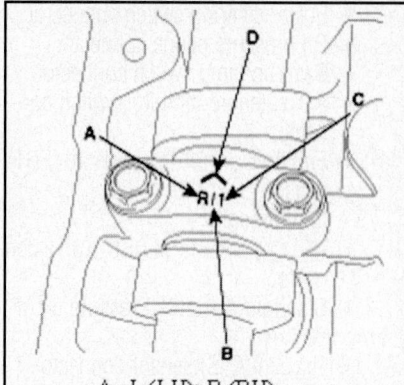

A. L (LH); R (RH)
B. I (Intake); None (Exhaust)
C. Journal number
D. Front mark

Fig. 105 Be careful to properly position the camshaft bearing caps according to its markings—3.3L and 3.8L engines

> ✳✳ **WARNING**
>
> **Rotate the crankshaft so as not to contact the valves to the pistons by positioning the pistons 0.3937 inch (10mm) below the top of the cylinder block.**

12. Install the water temperature control assembly.

13. Install the timing chain.

14. Check and adjust the valve clearance, as necessary.

15. Install the exhaust manifold.

16. Install the intake manifold.

17. Connect the negative battery cable.

18. Fill with engine coolant.

19. Start the engine and check for leaks.

20. Recheck the engine coolant level and oil level.

ENGINE ASSEMBLY

REMOVAL & INSTALLATION

2.4L & 2.7L Engines

See Figure 106.

1. Before servicing the vehicle, refer to the Precautions Section.

2. Remove the battery and air cleaner assembly.

3. Drain the cooling system.

4. Drain the engine oil.

5. Drain the transaxle fluid.

6. Relieve the fuel system pressure.

7. Disconnect the following electrical connections:

- Starter
- Alternator
- Throttle Position Sensor (TPS)

- Power steering switch connector
- Oil pressure gauge connector
- Back-up lamp switch connector
- A/T solenoid inhibitor switch connector
- Engine Coolant Temperature (ECT) sensor
- Ignition coil
- Idle Speed Control (ISC) valve connector
- Manifold Absolute Pressure (MAP) sensor
- Oxygen (O$_2$S) sensor connector

8. If equipped with an automatic transaxle, disconnect the oil cooler lines.

9. Remove or disconnect the following:
- Radiator hoses from the engine
- Radiator
- Engine ground
- Brake vacuum hose
- Heater hoses at the engine
- Throttle cable at the engine
- Cruise control cable at the engine, if equipped
- Main fuel line at the supply/return pipe
- Speedometer cable at the transaxle
- Clutch or control cable from the transaxle
- Power steering hoses from the pump
- Steering dust cover in the engine compartment
- Gear box universal joint bolt
- Front wheel
- Brake caliper and support with wire
- Strut lower bolt
- Front muffler bolts
- Transaxle control rod and extension rod, if equipped with a manual transaxle

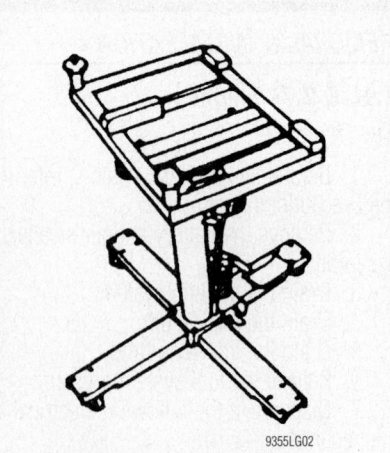

Fig. 106 Attach the special tool to the transaxle jack and support the transaxle

10. Support the transaxle with a jack using the special attachment shown in the accompanying illustration.

11. Make sure all cable, harness connectors and hoses are disconnected from the engine and transaxle.

12. Remove or disconnect the following:
- Engine and transaxle mounting brackets
- Sub frame bolts
- Drive shaft

13. Lower the engine and transaxle assembly enough so the front and rear roll stoppers can be removed.

14. Remove the engine assembly.

To install:

15. Installation is the reverse of removal but please note the following steps:
- Tighten the roll stopper bolts to 36–47 ft. lbs. (50–65 Nm)
- Tighten the transaxle mounting bracket bolts to 65–80 ft. lbs. (90–110 Nm)
- Tighten the engine mount bracket bolts to 43–58 ft. lbs. (60–80 Nm)

16. Fill the engine crankcase to the correct level.

17. Fill the transaxle to the correct level.

18. Fill the cooling system.

19. Fill the power steering system.

20. Start the engine and check for leaks.

21. Check the wheel alignment and adjust as necessary.

3.5L Engine

See Figure 106.

1. Before servicing the vehicle, refer to the Precautions Section.

2. Remove the battery and engine cover assembly.

3. Remove the battery stay.

4. Remove the air cleaner.

5. Drain the cooling system.

6. Drain the engine oil.

7. Drain the transaxle fluid.

8. Relieve the fuel system pressure.

9. Disconnect the following electrical connections:
- Alternator
- Starter
- Power steering switch connector
- Oil pressure gauge connector
- A/C switch
- Fuel injector connectors
- Back-up lamp switch connector
- A/T solenoid inhibitor switch connector
- Ignition coils
- Power TR selector connector
- Idle Speed Control (ISC) valve connector

- AFS and ATS connectors
- Oxygen (O$_2$S) sensor connector

10. Remove any remaining electrical connections that would interfere with engine removal.

11. If equipped with an automatic transaxle, disconnect the oil cooler lines.

12. Disconnect the radiator hoses from the engine.

13. Disconnect the engine and transaxle grounds.

14. Disconnect the brake booster vacuum hose.

15. Disconnect the heater hoses from the engine.

16. Disconnect the fuel delivery and return lines.

17. Disconnect the speedometer cable from the transaxle.

18. Disconnect the control cable from the transaxle.

19. Disconnect the power steering hose from the engine mount bracket.

20. Disconnect the steering dust cover in the engine compartment.

21. Disconnect the gear box universal joint bolt. Mark the locations prior to removal to aid in installation.

22. Remove the front wheel.

23. Remove the brake caliper and support with wire.

24. Remove the strut lower bolt and disconnect the strut from the knuckle.

25. Remove the wheel speed sensor from the knuckle.

26. Remove the front muffler bolts.

27. Support the transaxle with a jack using the special attachment shown in the accompanying illustration.

28. Make sure all cable, harness connectors and hoses are disconnected from the engine and transaxle.

29. Remove the engine and transaxle mounting brackets.

30. Remove the sub frame bolts.

31. Lower the transaxle side down, then lift the engine and transaxle assembly from the vehicle.

To install:

32. Installation is the reverse of removal but please note the following steps:

a. Tighten the front roll stopper-to-transaxle bolts to 43–58 ft. lbs. (60–80 Nm).

b. Tighten the front roll stopper insulator bolt and nut to 36–47 ft. lbs. (50–65 Nm).

c. Tighten the front roll stopper-to-sub-frame bolts to 43–58 ft. lbs. (60–80 Nm).

d. Tighten the rear roll stopper-to-sub-frame bolts to 43–58 ft. lbs. (60–80 Nm).

e. Tighten the rear roll stopper-to-transaxle bolt and nut to 36–47 ft. lbs. (50–65 Nm).

f. Tighten the rear roll stopper insulator bolt and nut to 36–47 ft. lbs. (50–65 Nm).

g. Tighten the transaxle mounting sub-bracket bolts to 43–58 ft. lbs. (60–80 Nm).

h. Tighten the transaxle mounting bracket bolts to 43–58 ft. lbs. (60–80 Nm).

i. Tighten the transaxle mounting insulator bolt to 65–80 ft. lbs. (90–110 Nm).

j. Tighten the engine mount bracket bolts to 43–58 ft. lbs. (60–80 Nm).

k. Tighten the engine mount insulator bolt to 65–80 ft. lbs. (90–110 Nm).

33. Fill the engine crankcase to the correct level.

34. Fill the transaxle to the correct level.

35. Fill the cooling system.

36. Fill the power steering system.

37. Start the engine and check for leaks.

38. Check the wheel alignment and adjust as necessary.

3.3L & 3.8L Engines

See Figure 107.

➡**Hyundai recommends that the engine and transaxle be removed as a single unit.**

1. Before servicing the vehicle, refer to the Precautions Section.

2. Drain the cooling system.

3. Drain the transaxle.

4. Drain the engine oil.

5. Relieve fuel system pressure.

6. Remove or disconnect the following:
- Battery
- Hood
- Air intake assembly
- Accessory drive belts
- Engine wiring harness connectors
- Reverse lamp switch connector
- Speedometer cable
- Alternator harness connectors
- Oil pressure gauge sender connector
- Radiator hoses
- Cooling fan
- Fuel lines
- Control cable, if equipped
- Brake booster vacuum line
- Intake manifold vacuum lines
- Heater hoses
- Accelerator cable
- Cruise control cable, if equipped
- Engine ground cable
- Transaxle oil cooler lines
- Shift cable

- Transaxle wiring connectors
- Radiator
- Power steering pump
- A/C compressor
- Exhaust front pipe
- Lower ball joints
- Stabilizer bar control links

7. Separate the inner CV-joints from the transaxle and suspend the halfshafts out of the work area with safety wire.

8. Attach a hoist to the engine lifting eyes.

9. Remove or disconnect the following:
- Front and rear roll stoppers
- Engine mount and bracket
- Transaxle mount and bracket

10. Lift the powertrain out of the vehicle.

To install:

11. Lower the powertrain into position.

12. Install the motor mount bracket and torque the fasteners to: 43–58 ft. lbs. (60–80 Nm).

13. Install the transaxle mount bracket and torque the fasteners to: 65–79 ft. lbs. (90–110 Nm).

14. Install or connect the following:
- Front and rear roll stoppers
- Engine mount
- Transaxle mount

15. Remove the engine hoist.

16. Torque the mount through bolts as follows:

a. Engine mount: 65–80 ft. lbs. (90–110 Nm).

b. Transaxle mount: 65–80 ft. lbs. (90–110 Nm).

c. Front roll stopper: 36–47 ft. lbs. (50–65 Nm).

d. Rear roll stopper: 36–47 ft. lbs. (50–65 Nm).

17. Install or connect the following:
- Axle halfshafts using new circlips
- Stabilizer bar control links
- Lower ball joints
- Exhaust front pipe
- A/C compressor
- Power steering pump
- Radiator
- Transaxle oil cooler lines
- Shift cable
- Transaxle wiring connectors

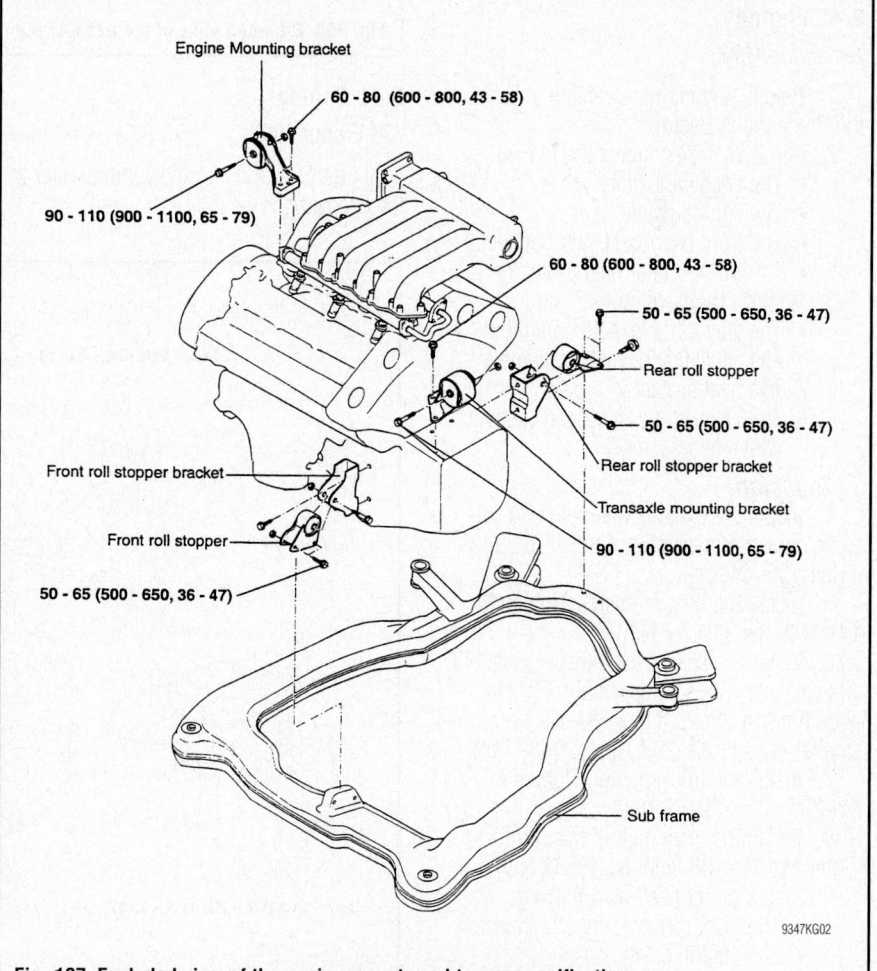

Fig. 107 Exploded view of the engine mounts and torque specifications

- Engine ground cable
- Cruise control cable
- Accelerator cable
- Heater hoses
- Intake manifold vacuum lines
- Brake booster vacuum line
- Fuel lines
- Cooling fan
- Radiator hoses
- Oil pressure gauge sender connector
- Alternator harness connectors
- Speedometer cable
- Reverse lamp switch connector
- Engine wiring harness connectors
- Accessory drive belts
- Air intake assembly
- Hood
- Battery

18. Fill the engine with clean oil.
19. Fill the transaxle to the correct level.
20. Fill the cooling system to the proper level.
21. Start the engine and check for leaks.

EXHAUST MANIFOLD

REMOVAL & INSTALLATION

2.4L Engine

See Figure 108.

1. Before servicing the vehicle, refer to the Precautions Section.
2. Remove or disconnect the following:
   - The negative battery cable
   - The engine cover
   - The front oxygen sensor connector
   - The front muffler heat protector
3. Remove the front muffler
   - The stay of the exhaust manifold and catalytic converter assembly
   - The heat protector
   - The exhaust manifold and catalytic converter assembly

To install:

4. Install the exhaust manifold and catalytic converter assembly. Tighten to 29–33 ft. lbs. (39–44 Nm).
5. Install the heat protector. Tighten to 14–20 ft. lbs. (19–27 Nm).
6. Install the stay of the exhaust manifold and catalytic converter assembly. Tighten large bolts to 25–29 ft. lbs. (34–39 Nm); smaller bolts to 22–29 ft. lbs. (29–39 Nm).
7. Install the front muffler. Tighten to 22–29 ft. lbs. (29–39 Nm).
8. Install the front muffler heat protector. Tighten to 72–108 inch lbs. (8–12 Nm).
9. Install or connect the following:
   - The front oxygen sensor connector
   - The engine cover
   - The negative battery cable

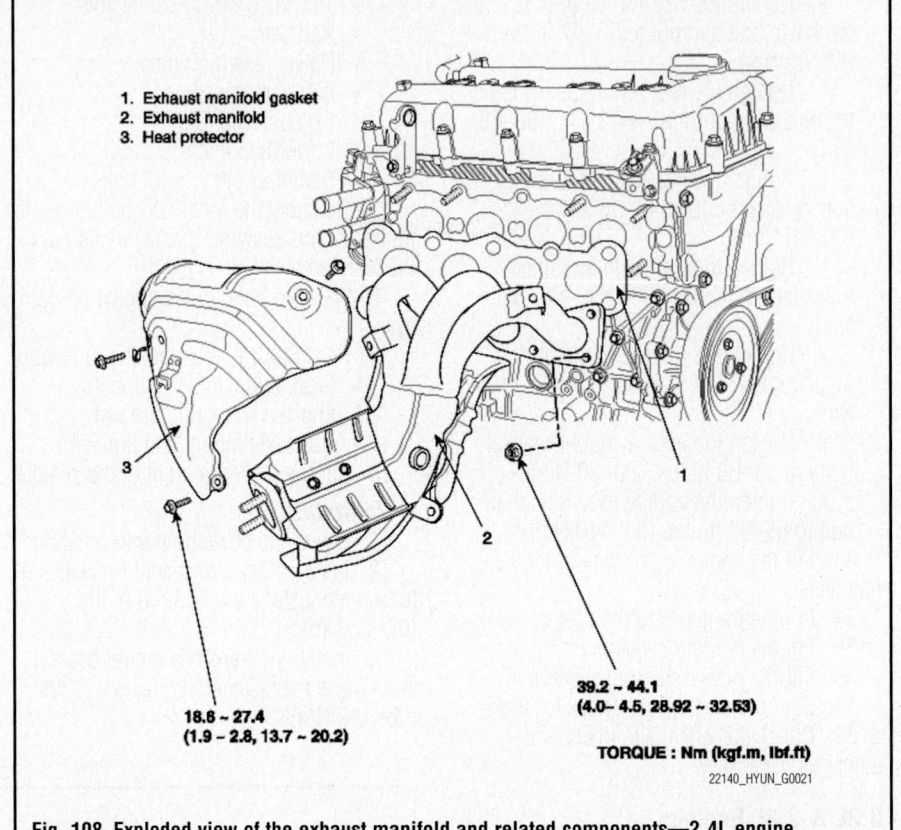

1. Exhaust manifold gasket
2. Exhaust manifold
3. Heat protector

18.6 ~ 27.4
(1.9 ~ 2.8, 13.7 ~ 20.2)

39.2 ~ 44.1
(4.0~ 4.5, 28.92 ~ 32.53)

TORQUE : Nm (kgf.m, lbf.ft)

22140_HYUN_G0021

Fig. 108 Exploded view of the exhaust manifold and related components—2.4L engine

2.7L Engine

See Figure 109.

1. Before servicing the vehicle, refer to the Precautions Section.

2. Remove or disconnect the following:
   - The negative battery cable
   - The under cover
   - The front muffler
   - The oxygen sensor connector

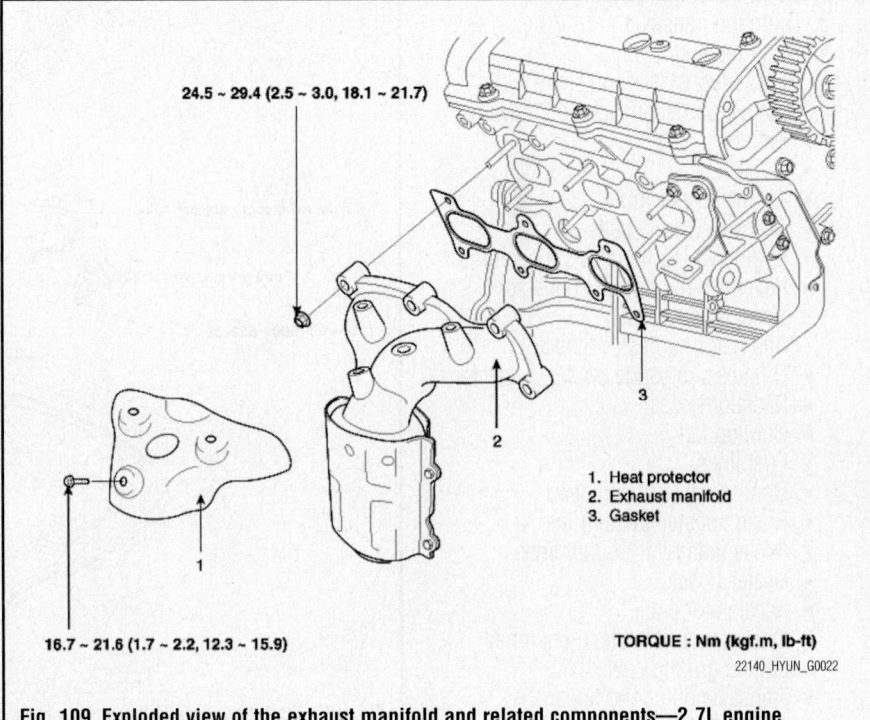

24.5 ~ 29.4 (2.5 ~ 3.0, 18.1 ~ 21.7)

16.7 ~ 21.6 (1.7 ~ 2.2, 12.3 ~ 15.9)

1. Heat protector
2. Exhaust manifold
3. Gasket

TORQUE : Nm (kgf.m, lb-ft)

22140_HYUN_G0022

Fig. 109 Exploded view of the exhaust manifold and related components—2.7L engine

- The heat protector
- The exhaust manifold and gasket

To install:

3. Install the exhaust manifold and gasket. Tighten bolts to 18–22 ft. lbs. (25–29 Nm).

4. Install the heat protector. Tighten bolts to 12–16 ft. lbs. (17–22 Nm).

5. Connect the oxygen sensor connector.

6. Install the front muffler. Tighten to 22–29 ft. lbs. (29–39 Nm).

7. Install the undercover.

8. Connect the negative battery cable.

3.5L Engine

1. Before servicing the vehicle, refer to the Precautions Section.

2. If necessary, disconnect the Oxygen ($O_2$2) sensor connector and remove the sensor.

3. Remove the front muffler.

4. Remove the heat shield.

5. Remove the exhaust manifold retainers.

6. Remove the exhaust manifold and gasket.

To install:

7. Install a new gasket and the manifold. Tighten the manifold bolts to 29–33 ft. lbs. (40–45 Nm).

8. Install the heat shield.

9. If removed, install the O_2 sensor, tighten to 29–36 ft. lbs. (40–50 Nm) and attach the electrical connector.

10. Install the front muffler.

3.3L & 3.8L Engines

See Figure 110.

1. Before servicing the vehicle, refer to the Precautions Section.

2. Remove or disconnect the following:
- The negative battery cable
- The under cover
- The LH, RH rear oxygen sensor connector from bracket
- The front muffler
- The oil level gauge
- The LH front oxygen sensor connector from bracket
- The LH heat protector
- The LH exhaust manifold
- The RH front oxygen sensor connector from bracket
- The RH heat protector
- The RH exhaust manifold

To install:

3. Install a new gasket and exhaust manifold. Tighten to 29–33 ft. lbs. (39–44 Nm).

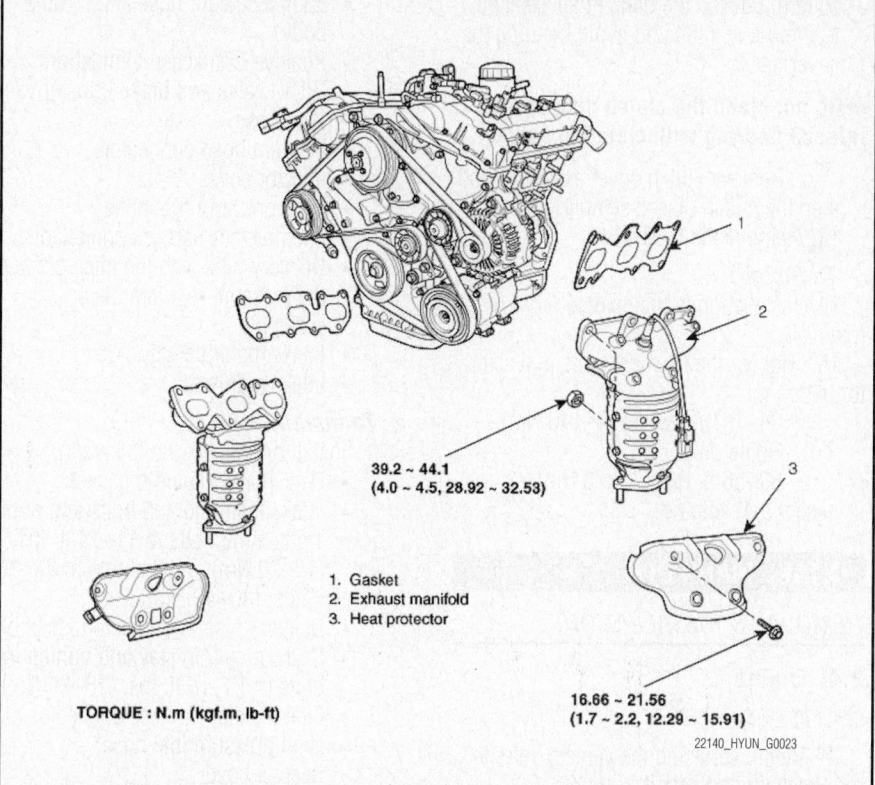

39.2 ~ 44.1
(4.0 ~ 4.5, 28.92 ~ 32.53)

1. Gasket
2. Exhaust manifold
3. Heat protector

TORQUE : N.m (kgf.m, lb-ft)

16.66 ~ 21.56
(1.7 ~ 2.2, 12.29 ~ 15.91)

22140_HYUN_G0023

Fig. 110 Exploded view of the exhaust manifold and related components—3.3L and 3.8L engines

4. Install heat protector. Tighten to 12–16 ft. lbs. (17–22 Nm).

5. Install front muffler. Tighten to 29–43 ft. lbs. (39–59 Nm).

6. Connect oxygen sensor connector.

7. Install under cover.

8. Connect the negative battery cable.

FLYWHEEL

REMOVAL & INSTALLATION

1. Before servicing the vehicle, refer to the Precautions Section.

2. Drain the transaxle.

3. Remove or disconnect the following:
- Negative battery cable
- Air intake assembly
- Battery and battery tray
- Back-up lamp connector
- Vehicle speed sensor
- Clutch release cylinder and lever
- Shift cable from transaxle assembly
- Steering column from the universal joint in the gear box
- Clutch housing upper mounting bolts
- Front, rear and left transaxle mounting brackets

4. Using a suitable engine support fixture, support the engine assembly.

5. Remove or disconnect the following:

- Power steering pressure hose from the pump
- Front wheel
- Strut assembly
- Tie rod and sway bar link from the knuckle
- Wheel speed sensor
- Brake caliper
- Engine splash guard
- Front muffler
- Power steering hose on the front cross member

6. Using a suitable jack, support the sub-frame cross member.

7. Remove the cross member mounting bolts, and lower the cross member assembly with the steering gear and stabilizer bar attached.

8. Using a suitable jack, support the transaxle assembly.

9. Remove the front and rear roll stoppers.

10. Remove the engine and transaxle mounting bolts.

11. Slowly lower the transaxle from the vehicle.

12. If equipped with a clutch:

a. Insert the special tool (09411-11000) in the clutch disc to prevent the disc from falling.

b. Loosen the bolts which attach the clutch cover to the flywheel in a star pattern. Loosen the bolts in succession, 1–2 turns at a time, to avoid bending the cover flange.

➡**Do not clean the clutch disc or the release bearing with cleaning solvent.**

c. Remove clutch cover assembly and then the clutch disc assembly.

13. Remove the flywheel.

To install:

14. Installation is the reverse order of removal.

15. Tighten the flywheel bolts in a star formation to:

a. 94–101 ft. lbs. (130–140 Nm)—2.4L engine only.

b. 53–56 ft. lbs. (72–76 Nm)—except 2.4L engine.

INTAKE MANIFOLD

REMOVAL & INSTALLATION

2.4L Engine

See Figure 111.

1. Before servicing the vehicle, refer to the Precautions Section.

2. Remove or disconnect the following:
- Negative battery cable
- Air breather hose from the throttle body
- Throttle cable

- Engine coolant hose and throttle body
- Positive Crankcase Ventilation (PCV) valve and brake booster vacuum hose
- Vacuum hose connector
- Injector cover
- High pressure fuel hose
- Fuel injector harness connector
- Delivery pipe with the injectors and the pressure regulator as an assembly
- Intake manifold stay
- Intake manifold

To install:

3. Install or connect the following:
- New intake manifold gasket
- Intake manifold and bolts and nuts. Tighten the bolts to 11–14 ft. lbs. (15–20 Nm) and the nuts to 22–30 ft. lbs. (30–42 Nm).
- Delivery pipe and injector assembly
- Intake manifold stay and tighten the bolts to 13–18 ft. lbs. (18–25 Nm)
- Fuel injector harness connector
- High pressure fuel hose
- Injector cover
- Vacuum hose connector
- PCV valve and brake booster vacuum hose
- Throttle body and engine coolant hose
- Throttle cable

- Air breather hose from the throttle body
- Negative battery cable

4. Start the engine and check for proper operation.

2.7L Engine

See Figures 112 and 113.

1. Before servicing the vehicle, refer to the Precautions Section.

2. Remove or disconnect the following:
- Negative battery cable
- Air breather hose from the throttle body
- Throttle and cruise control cables
- Engine coolant hose and throttle body
- Positive Crankcase Ventilation (PCV) valve and brake booster vacuum hose
- Vacuum hose connector
- Surge tank stay
- High pressure fuel hose
- Surge tank and gasket
- Fuel injector harness connector
- Delivery pipe with the injectors and the pressure regulator as an assembly
- Engine Coolant Temperature (ECT) sensor electrical connector
- Intake manifold

To install:

3. Install or connect the following:
- New intake manifold gasket
- Intake manifold and tighten the bolts to 14–15 ft. lbs. (19–21 Nm)
- ECT electrical connector
- Delivery pipe with the injectors and the pressure regulator as an assembly
- Fuel injector harness connector
- Surge tank and gasket
- High pressure fuel hose
- Surge tank stay
- Vacuum hose connector
- PCV valve and brake booster vacuum hose
- Engine coolant hose and throttle body
- Throttle and cruise control cables
- Air breather hose from the throttle body
- Negative battery cable

4. Start the engine and check for proper operation.

3.5L Engine

See Figures 114 and 115.

1. Before servicing the vehicle, refer to the Precautions Section.

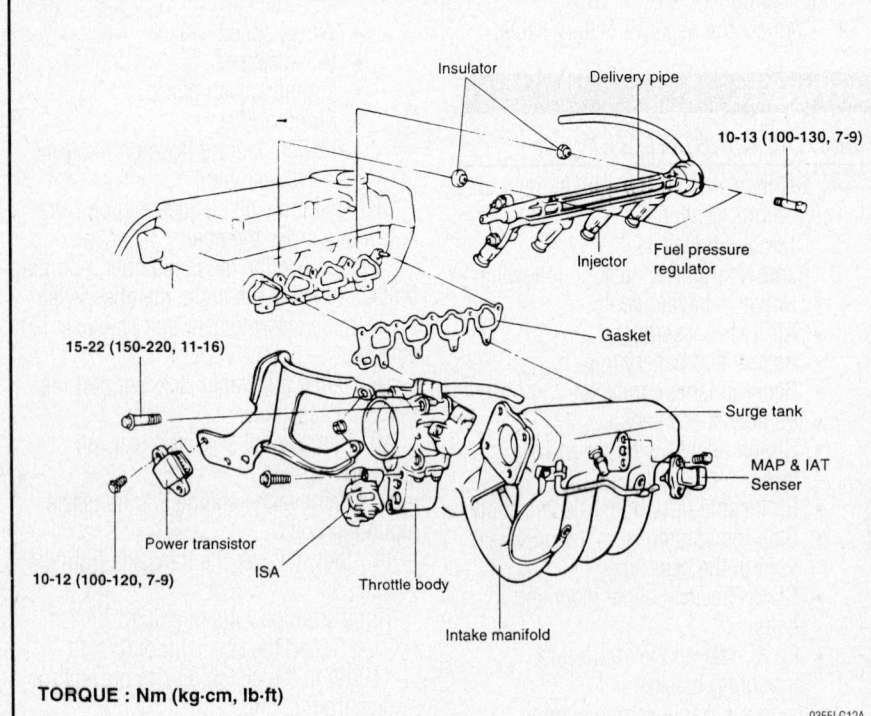

Insulator
Delivery pipe
10-13 (100-130, 7-9)
Fuel pressure regulator
Injector
15-22 (150-220, 11-16)
Gasket
Surge tank
MAP & IAT Senser
Power transistor
ISA
10-12 (100-120, 7-9)
Throttle body
Intake manifold

TORQUE : Nm (kg·cm, lb·ft)

9355LG12A

Fig. 111 Exploded view of the intake manifold—2.4L engine

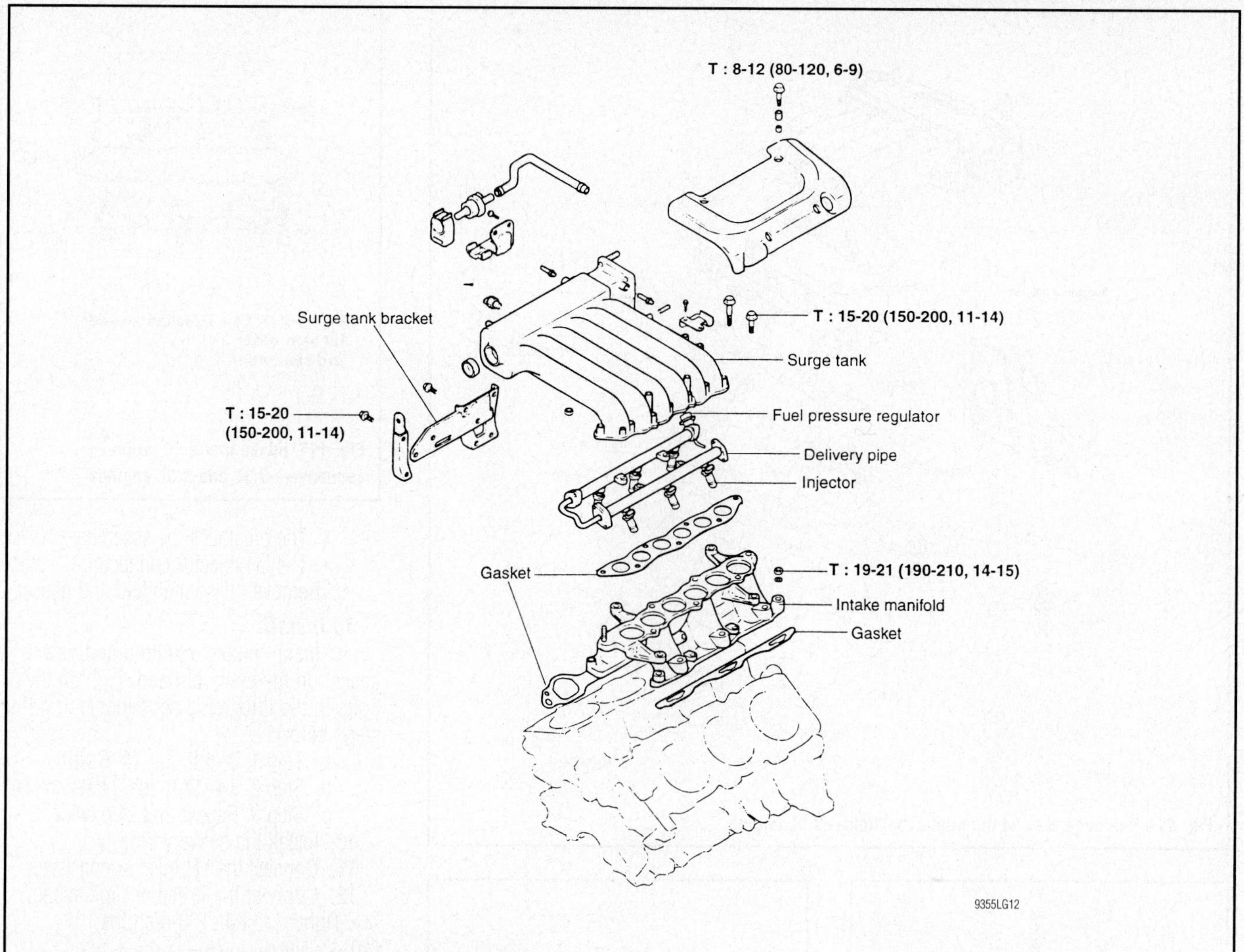

T : 8-12 (80-120, 6-9)

Surge tank bracket

T : 15-20 (150-200, 11-14)

Surge tank

T : 15-20
(150-200, 11-14)

Fuel pressure regulator

Delivery pipe

Injector

Gasket

T : 19-21 (190-210, 14-15)

Intake manifold

Gasket

9355LG12

Fig. 112 Exploded view of the intake manifold—2.7L engine

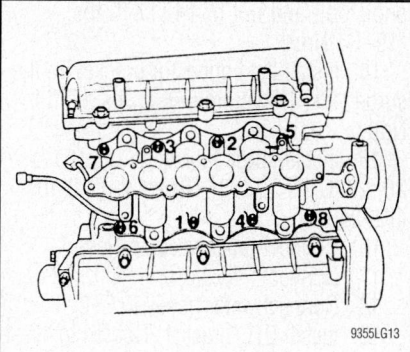

9355LG13

Fig. 113 Intake manifold torque sequence—2.7L engine

2. Disconnect the negative battery cable.
3. Remove the air breather hose from the throttle body.
4. Remove the Positive Crankcase Ventilation (PCV) valve and brake booster vacuum hoses.
5. Disconnect the vacuum hose connections.

6. Remove the surge tank stay.
7. Remove the surge tank and gasket.
8. Disconnect the fuel injector harness connector.
9. Remove the delivery pipe with the injectors and the pressure regulator as an assembly.
10. Disconnect the Engine Coolant Temperature (ECT) sensor electrical connector.
11. Remove the intake manifold bolts, manifold and gasket.

To install:
12. Install a new intake manifold gasket.
13. Install the intake manifold and tighten the bolts in the sequence illustrated to: 14–16 ft. lbs. (20–23 Nm).
14. Attach the CTS electrical connector.
15. Install the delivery pipe with the injectors and the pressure regulator as an assembly.
16. Attach the fuel injector harness connector.
17. Install the surge tank and gasket.
18. Install the surge tank stay and

tighten the retainers to: 11–14 ft. lbs. (15–20 Nm).
19. Attach the vacuum hose connectors.
20. Connect the PCV valve and brake booster vacuum hoses.
21. Connect the engine coolant hose to the throttle body.
22. Connect the air breather hose to the throttle body.
23. Connect the negative battery cable.
24. Start the engine and check for proper operation.

3.3L & 3.8L Engines
See Figures 116 through 118.

1. Before servicing the vehicle, refer to the Precautions Section.
2. Relieve the fuel system pressure.
3. Drain the cooling system.
4. Remove the negative battery cable.
5. Disconnect AFS (A) and breather hose (B).
6. Remove air cleaner upper cover (D) and intake hose (C).

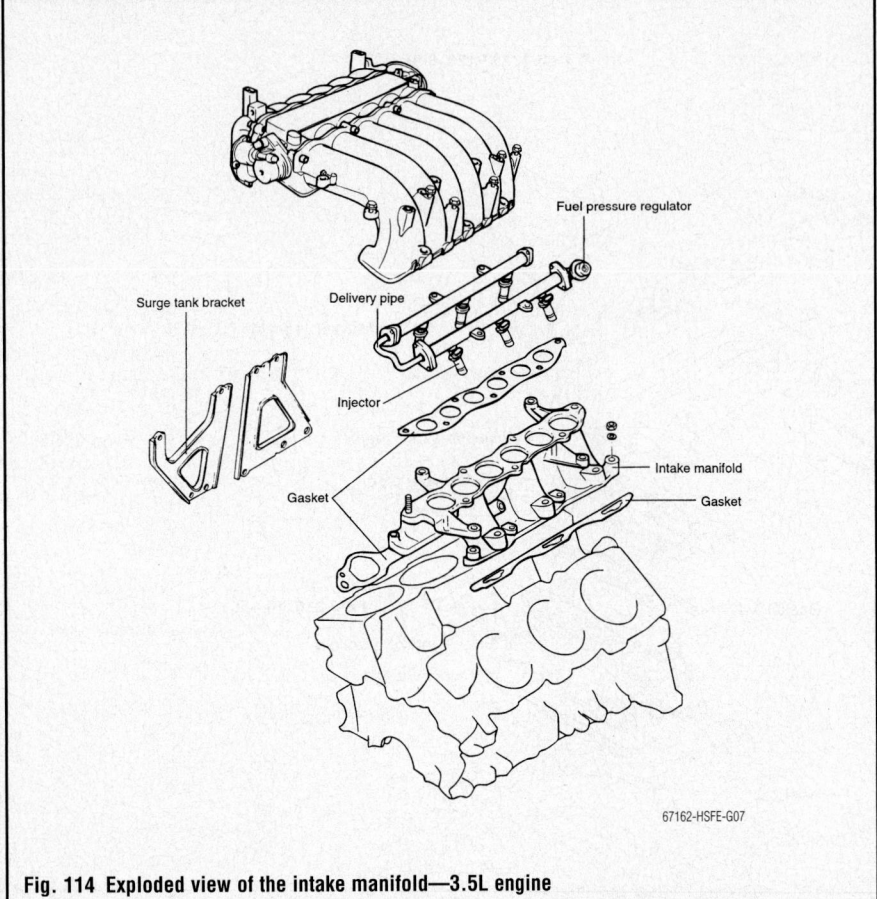

Fig. 114 Exploded view of the intake manifold—3.5L engine

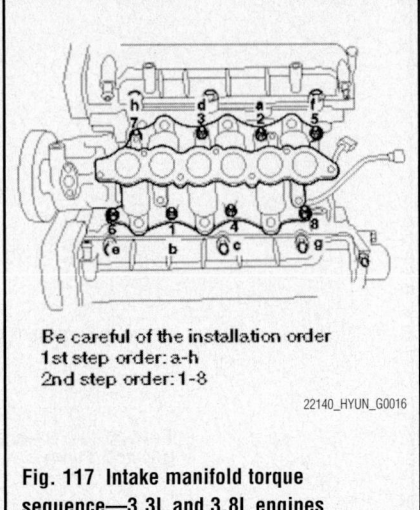

Be careful of the installation order
1st step order: a-h
2nd step order: 1-8

Fig. 117 Intake manifold torque
sequence—3.3L and 3.8L engines

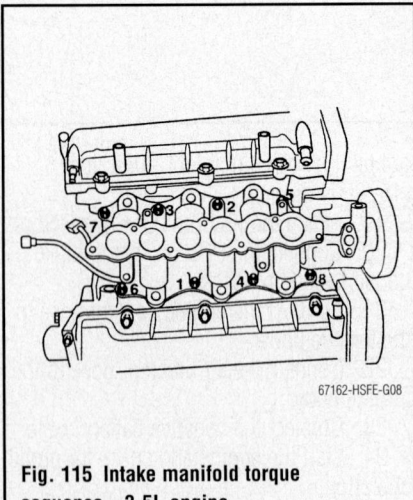

Fig. 115 Intake manifold torque
sequence—3.5L engine

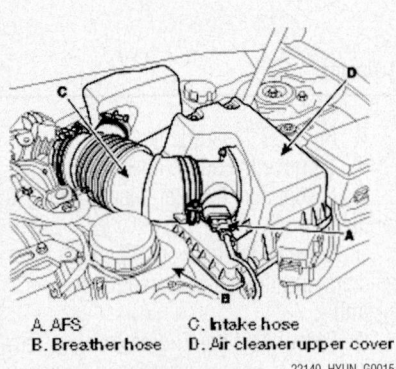

A. AFS
B. Breather hose
C. Intake hose
D. Air cleaner upper cover

Fig. 116 Disconnect AFS, breather hose,
air cleaner upper cover, and intake hose—
3.3L and 3.8L engines

7. Remove or disconnect the following:
- The RH oxygen sensor connector
- The RH injector connector and ignition coil connector
- The Purge Control Solenoid Valve (PCSV) connector, Manifold Absolute Pressure (MAP) sensor connector, and PCSV hose
- The Electronic Throttle Control (ETC) connector and knock sensor connector
- The water hoses from ETC
- The PCV hose
- The brake vacuum hose
- The surge tank stay
- The connector bracket from surge tank
- The surge tank
- The breather Pipe assembly
- The LH injector connector.

8. Remove intake manifold and gasket

To install:

9. Install intake manifold and new gasket on the cylinder head. Tighten the bolts in the illustrated sequence using the steps below:
   a. Step 1: 3–4 ft. lbs. (4–6 Nm).
   b. Step 2: 14–17 ft. lbs. (19–24 Nm).
   c. Step 3: Repeat 2nd step twice.

10. Install the delivery pipe.

11. Connect the LH injector connector.

12. Connect the breather pipe assembly. Tighten to 84–108 inch lbs. (10–12 Nm).

13. Install the surge tank. Tighten long bolt to 84–108 inch lbs. (10–12 Nm); short bolt and nut to 14–17 ft. lbs. (19–24 Nm).

14. Install the connector bracket on the surge tank. Tighten to 5–8 ft. lbs. (7–11 Nm).

15. Install surge tank stay. Tighten to 20–23 ft. lbs. (27–31 Nm); 14–17 ft. lbs. (19–24 Nm).

16. Connect brake vacuum hose.

17. Connect PCV hose.

18. Connect water hoses to ETC.

19. Install ETC bracket. Tighten to 12–19 ft. lbs. (16–26 Nm).

20. Connect ETC connector and knock sensor connector.

21. Connect PCSV connector, MAP sensor connector and PCSV hose.

22. Connect RH injector connector and ignition coil connector.

23. Connect RH oxygen sensor connector.

24. Install air cleaner upper cover and intake hose.

25. Connect AFS and breather hose.

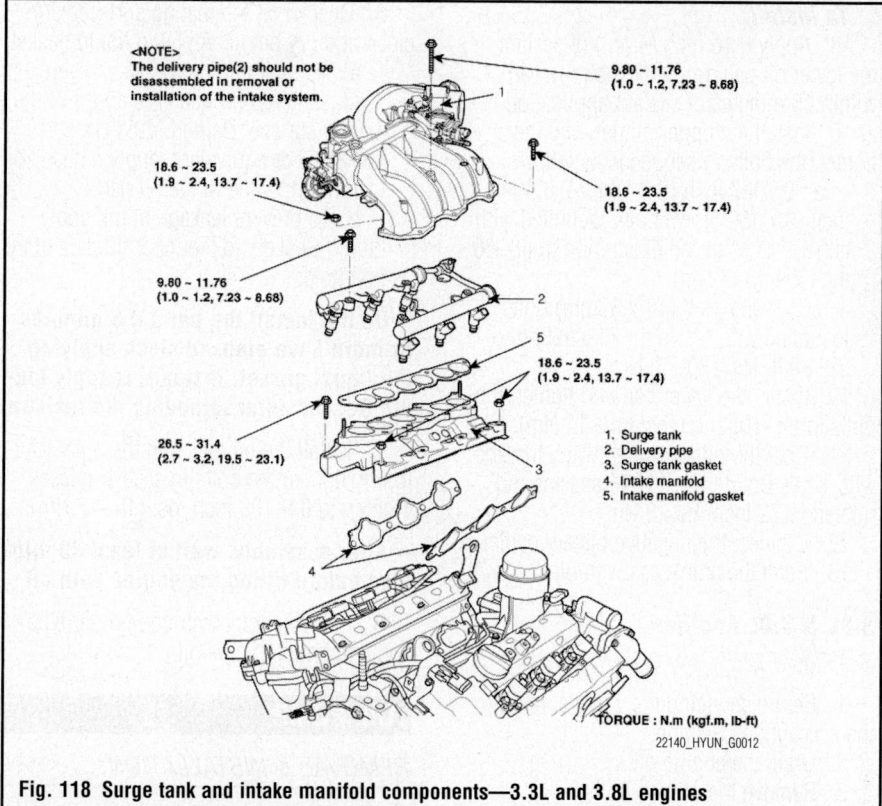

<NOTE>
The delivery pipe(2) should not be disassembled in removal or installation of the intake system.

9.80 ~ 11.76
(1.0 ~ 1.2, 7.23 ~ 8.68)

18.6 ~ 23.5
(1.9 ~ 2.4, 13.7 ~ 17.4)

18.6 ~ 23.5
(1.9 ~ 2.4, 13.7 ~ 17.4)

9.80 ~ 11.76
(1.0 ~ 1.2, 7.23 ~ 8.68)

18.6 ~ 23.5
(1.9 ~ 2.4, 13.7 ~ 17.4)

26.5 ~ 31.4
(2.7 ~ 3.2, 19.5 ~ 23.1)

1. Surge tank
2. Delivery pipe
3. Surge tank gasket
4. Intake manifold
5. Intake manifold gasket

TORQUE : N.m (kgf.m, lb-ft)

22140_HYUN_G0012

Fig. 118 Surge tank and intake manifold components—3.3L and 3.8L engines

OIL PAN

REMOVAL & INSTALLATION

2.4L Engine

See Figure 119.

1. Before servicing the vehicle, refer to the Precautions Section.
2. Drain the engine oil.
3. Disconnect the negative battery cable.
4. Remove the oil pan bolts, note the bolts length and location.
5. Tap the oil pan with a rubber mallet and remove the upper and lower pan components.

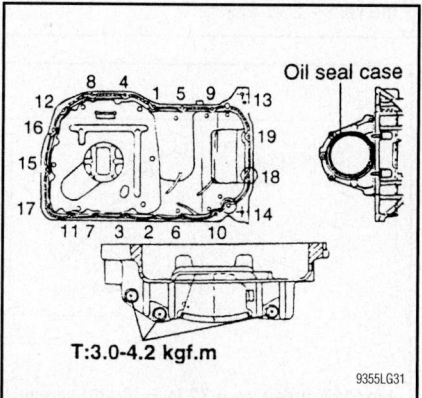

T:3.0-4.2 kgf.m

9355LG31

Fig. 119 Oil pan torque sequence and sealant application points—2.4L engine

6. Clean the gasket mating surfaces.

To install:

7. Apply 0.16 inch (4mm) of sealant to the oil pan groove as illustrated. Install the oil pan within 15 minutes of sealant application.
8. Install the upper and lower oil pans and the bolts making sure the proper length is installed in its original position. Tighten the bolts in sequence to 84–108 inch lbs. (10–12 Nm).
9. Connect the negative battery cable.
10. Refill the crankcase with oil.

2.7L Engine

See Figures 120 and 121.

1. Before servicing the vehicle, refer to the Precautions Section.
2. Drain the engine oil.
3. Remove or disconnect the following:
   • Negative battery cable
   • Lower oil pan bolts and the pan
   • Upper oil pan bolts and the pan
4. Clean the gasket mating surfaces.

To install:

5. Apply 0.16 inch (4mm) of sealant to the lower oil pan groove. Install the pan within 15 minutes of sealant installation.
6. Install the upper oil pan and the tighten the bolts in sequence as follows:
   a. 0.937 x 1.4961 inch (10 x 38mm) bolt to 22–30 ft. lbs. (30–42 Nm).

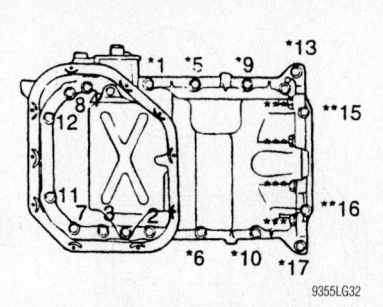

9355LG32

Fig. 120 Upper oil pan torque sequence (tighten bolts indicated with * to 14–20 inch (19–28 Nm), bolts indicated with a ** to 48–60 inch lbs. (5–7 Nm) and bolts indicated with a * to 22–30 ft. lbs. (30–42 Nm)—2.7L engine**

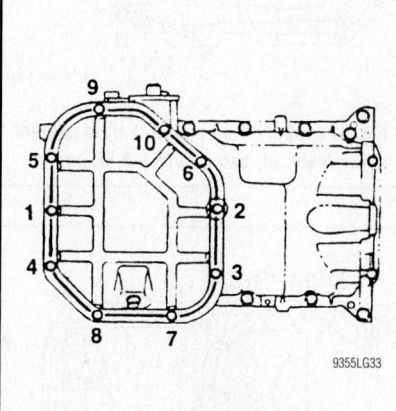

9355LG33

Fig. 121 Lower oil pan bolt torque sequence—2.7L engine

   b. 0.3150 x 0.866a inch (8 x 22mm) bolt 14–20 inch (19–28 Nm).
   c. 6.7519 inch (171.5mm) bolt to 48–60 inch lbs. (5–7 Nm).
   d. 6.7520 inch (152.5mm) bolt to 48–60 inch lbs. (5–7 Nm).
7. Install the lower pan and tighten the bolts to 84–108 inch lbs. (10–12 Nm).
8. Connect the negative battery cable.
9. Refill the crankcase with oil.

3.5L Engine

See Figures 122 through 124.

1. Before servicing the vehicle, refer to the Precautions Section.
2. Drain the engine oil.
3. Remove the oil pressure switch.
4. Disconnect the negative battery cable.
5. Remove the lower oil pan bolts and the pan.
6. Remove the upper oil pan bolts and the pan.
7. Clean the gasket mating surfaces

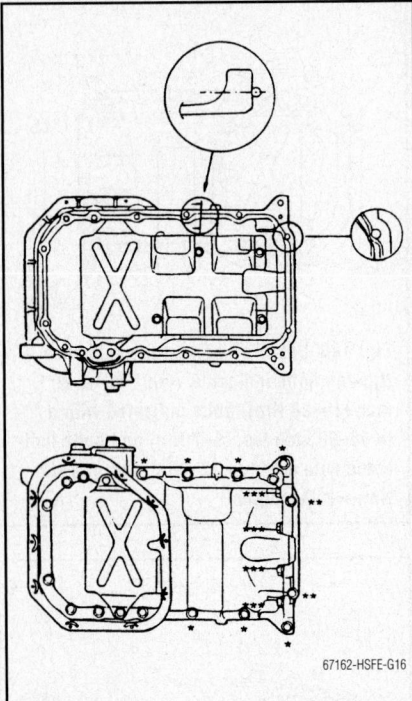

Fig. 122 Apply 0.16 inch (4mm) of sealant to the lower oil pan groove—3.5L engine

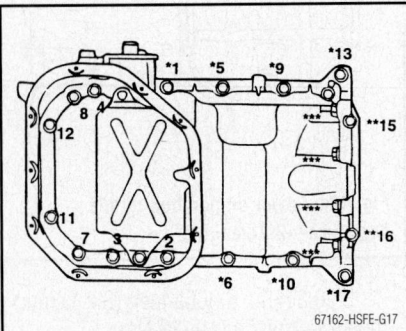

Fig. 123 Upper oil pan torque sequence (tighten bolts indicated with * or ** in the illustration to 48–60 inch lbs. (5–7 Nm), and bolts indicated with a * to 22–30 ft. lbs. (30–42 Nm)—3.5L engine**

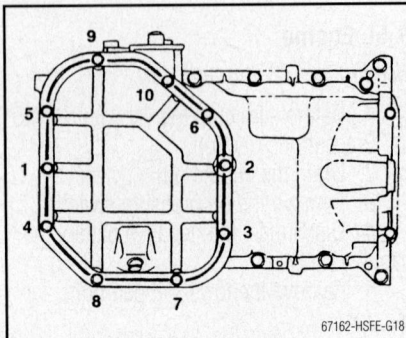

Fig. 124 Lower oil pan bolt torque sequence—3.5L engine

To install:

8. Apply 0.16 inch (4mm) of sealant to the lower oil pan groove. Install the pan within 15 minutes of sealant application.

9. Install the upper oil pan and the tighten the bolts in sequence as follows:

a. 0.7087 inch (6 x 18mm), 6.004 inch (6 x 152.5mm) bolts identified with either * or ** in the illustration to 48–60 inch lbs. (5–7 Nm).

b. 1.4961 inch (10 x 38mm) bolts identified with *** in the illustration to 22–30 ft. lbs. (30–42 Nm).

10. Install the lower pan and tighten the bolts to 84–108 inch lbs. (10–12 Nm).

11. Coat the oil pressure switch threads with Three Bond® No. 1104E sealant and tighten to 72 inch lbs. (8 Nm).

12. Connect the negative battery cable.

13. Refill the crankcase with oil.

3.3L & 3.8L Engines

See Figure 125.

1. Before servicing the vehicle, refer to the Precautions Section.

2. Drain the engine oil.

3. Remove the oil pan bolts.

4. Using the SST (09215-3C000), remove the oil pan.

a. Insert the SST between the oil pan and the ladder frame by tapping it with a plastic hammer in the direction of arrow.

b. After tapping the SST with a plastic hammer along the direction of arrow around more than ⅔ of the edge of the oil pan, remove it from the ladder frame.

✳✳ WARNING

Do not turn over the SST abruptly without tapping. Damage may occur to the SST or the oil pan.

To install:

5. Using a razor blade and gasket scraper, carefully remove all the old packing material from the gasket surfaces.

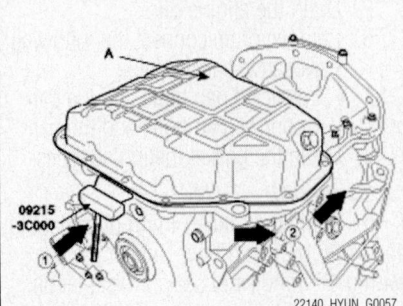

Fig. 125 Using the SST (09215-3C000) to remove the oil pan`

6. Check that the mating surfaces are clean and dry before applying liquid gasket.

a. Apply liquid gasket as an even bead, centered between the edges of the mating surface. Using Liquid Gasket® TB1217H or equivalent, apply a bead ⅛ inch (3mm) wide to the oil pan.

b. To prevent leakage of oil, apply Liquid Gasket® to the inner threads of the bolt holes.

➡ **Do not install the parts if 5 minutes or more have elapsed since applying the liquid gasket. Instead, reapply Liquid Gasket® after removing the residue.**

7. Install the oil pan with the bolts. Uniformly tighten the bolts in several passes. Tighten to 84–108 inch lbs. (10–12 Nm).

➡ **After assembly, wait at least 30 minutes before filling the engine with oil.**

8. Fill the engine with the proper type and amount of engine oil.

OIL PUMP

REMOVAL & INSTALLATION

2.4L Engine

See Figures 126 through 132.

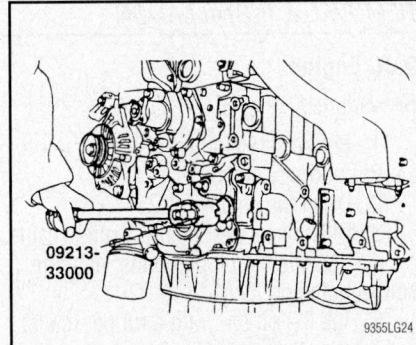

Fig. 126 Use Tool 09213-33000, remove the plug cap from the oil pump portion of the case—2.4L engine

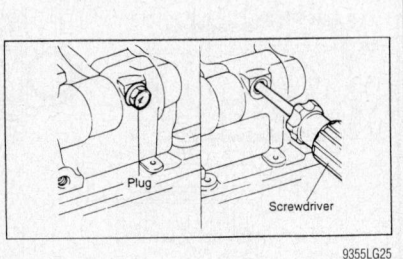

Fig. 127 Insert an 0.32 inch (8mm) screwdriver at least 2.4 inch (60mm) into the plug hole—2.4L engine

1. Before servicing the vehicle, refer to the Precautions Section.
2. Drain the engine oil.
3. Remove or disconnect the following:
   - Negative battery cable
   - Timing belt
   - Oil pan
   - Oil screen and gasket
   - Oil pressure switch
   - Oil filter bracket and gasket
4. Using Tool 09213-33000, remove the plug cap from the oil pump portion of the case.
   - Plug from the left side of the block and insert a 0.32 inch (8mm) screwdriver into the plug hole. The screwdriver must be inserted at least 2½ inches (60mm).
   - Pump driven gear the left counter balance shaft bolt
   - Front case bolts (noting the bolt length and location), the case and gasket.
   - Two counter balance shafts from the block
   - Oil pump cover from the case
   - Oil pump gears from the case
   - Screwdriver from the plug hole

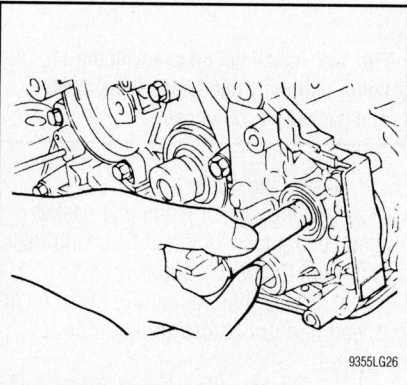

Fig. 128 Remove the pump driven gear the left counter balance shaft bolt—2.4L engine

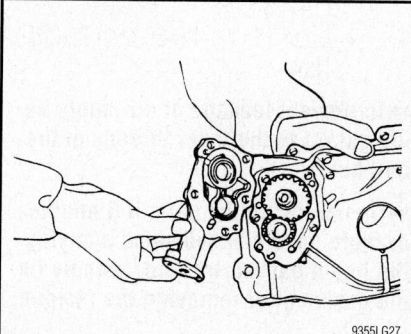

Fig. 129 Remove the oil pump cover from the case—2.4L engine

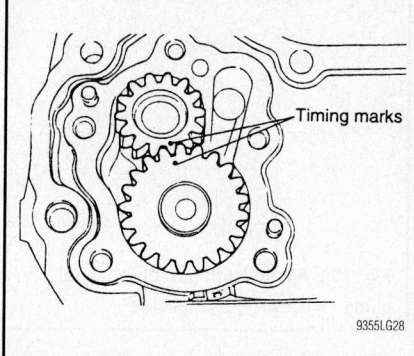

Fig. 130 Apply engine oil to both gears and align the gear timing marks—2.4L engine

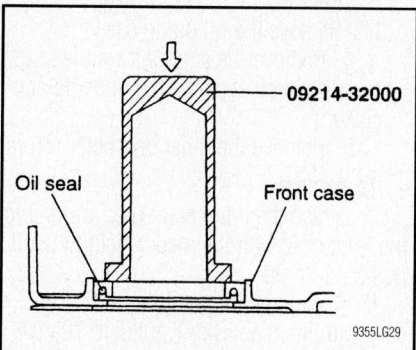

Fig. 131 Using crankshaft front oil seal install Tool 09214-32000, install the oil seal into the case—2.4L engine

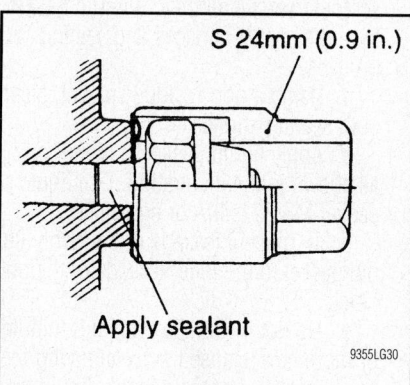

Fig. 132 Front case bolt length and location—2.4L engine

To install:

5. Install the oil pump gears.
6. Inspect the tip clearance of the gears using a feeler gauge. The specifications are as follows:
   a. Standard value drive gear: 0.0063–0.0083 inch (0.16–0.21mm).
   b. Standard value driven gear: 0.0071–0.0083 inch (0.18–0.21mm).
   c. Limit drive gear: 0.0098 inch (0.25mm).

d. Limit driven gear: 0.0098 inch (0.25mm).
7. Inspect the side clearance of the gears using a feeler gauge. The specifications are as follows:
   a. Standard value drive gear: 0.0031–0.0055 inch (0.08–0.14mm).
   b. Standard value driven gear: 0.0024–0.0047 inch (0.06–0.12mm).
   c. Limit drive gear: 0.0098 inch (0.25mm).
   d. Limit driven gear: 0.0098 inch (0.25mm).
8. Apply engine oil to both gears and align the gear timing marks.
9. Install the oil pump case.
10. Using crankshaft front oil seal install Tool 09214-32000, install the oil seal into the case.
11. Place special tool 09214-32100 on the front of the crankshaft and apply a coat of oil to the outside of the tool to aid in case installation.
12. Install or connect the following:
    - New front case gasket and temporarily tighten the flange bolts
    - Front case and tighten the bolts to 14–20 ft. lbs. (20–27 Nm), making sure the correct length bolt is installed in the correct location
13. Insert a 0.32 inch (8mm) screwdriver into the plug hole. The screwdriver must be inserted at least 2½ inches (60mm). Verify that the shaft is in place and install the bolt.
14. Install or connect the following:
    - New O-ring on the groove on the front case
    - Plug case and tighten to 14–20 ft. lbs. (20–27 Nm)
    - Oil screen and gasket
    - Oil pan
    - Oil pressure switch using a 24mm deep socket. Apply Three Bond® 1104 sealant to the threads before installation and tighten to 72–108 inch lbs. (8–12 Nm)
    - Timing belt
    - Negative battery cable
15. Fill the crankcase to the correct level.
16. Start the engine and check for leaks.

2.7L Engine

See Figures 133 through 139.

1. Before servicing the vehicle, refer to the Precautions Section.
2. Remove the lower oil pan, refer to Oil Pan, removal & installation.
3. Remove the RH front wheel.
4. Remove the RH side cover.
5. Remove the front exhaust pipe.

6. Remove the alternator from the engine.

7. Turn the crankshaft and align the white groove on the crankshaft pulley with the pointer on the lower cover.

8. Remove the timing belt.

9. Remove the oil screen:

Fig. 133 Remove the 2 bolts (B) and oil screen (A)—2.7L engine

Fig. 134 Remove the upper oil pan (A)—2.7L engine

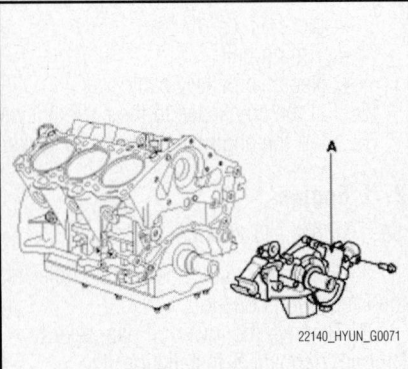

Fig. 135 Remove the upper oil pump case (A)—2.7L engine

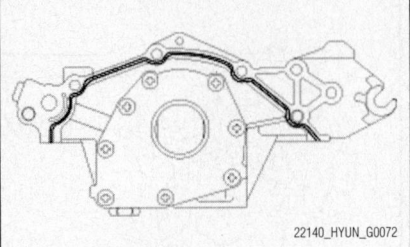

Fig. 136 Apply liquid gasket to the oil pump as shown—2.7L engine

a. Remove the 2 bolts (B).

b. Remove the oil screen (A) and gasket.

10. Remove the upper oil pan.

11. Remove the oil pump case.

a. Remove the screws from the pump housing, then separate the housing and cover.

b. Remove the inner and outer rotors.

To install:

12. Place the inner and outer rotors into the front case with the marks facing the oil pump cover side.

13. Install the oil pump cover to front case with the 8 screws. Tighten to 72–108 inch lbs. (8–12 Nm).

14. Check that the oil pump turns freely.

15. Install the oil pump on the cylinder block.

a. Using a razor blade and gasket scraper, remove all the old liquid gasket from the gasket surfaces and sealing grooves.

b. Using a non-residue solvent, clean both sealing surfaces.

c. Apply liquid gasket to the oil pump as shown in the illustration. Use liquid gasket MS 721-40A or equivalent.

d. To prevent leakage of oil, apply liquid gasket to the inner threads of the bolt holes.

e. Do not install the parts if 5 minutes or more have elapsed since applying the liquid gasket. Instead, reapply liquid gasket after removing the residue.

f. Place a new O-ring on the cylinder block.

g. Engage the spline teeth of the oil pump drive gear with large teeth of the crankshaft, and slide the oil pump on the crankshaft.

h. Install the oil pump with 5 bolts. Uniformly tighten the bolts in several passes. Tighten to 108–132 inch lbs. (12–15 Nm).

16. Apply a light coat of oil to the seal lip.

17. Using the special tool (09214-33000), install the oil seal.

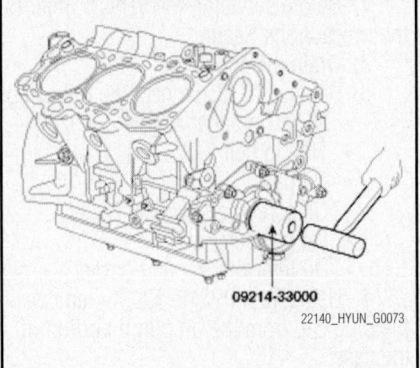

Fig. 137 Using the special tool (09214-33000) to install the oil seal—2.7L engine

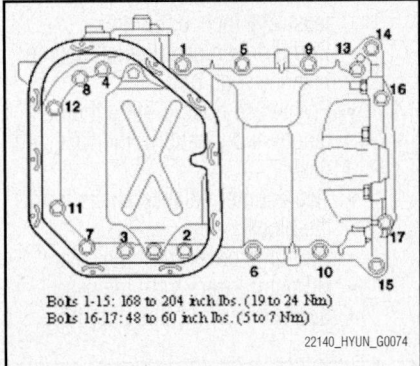

Bolts 1-15: 168 to 204 inch lbs. (19 to 24 Nm)
Bolts 16-17: 48 to 60 inch lbs. (5 to 7 Nm)

Fig. 138 Install the oil pan with the 17 bolts. Uniformly tighten the bolts in several passes—2.7L engine

18. Install the upper oil pan.

a. Using a razor blade and gasket scraper, remove all the old packing material from the gasket surfaces.

b. Check that the mating surfaces are clean and dry before applying liquid gasket.

c. Install the oil pan with the 17 bolts. Uniformly tighten the bolts in several passes using the pattern illustrated. Tightening torque:

- Bolts 1–15: 168–204 inch lbs. (19–24 Nm)
- Bolts 16–17: 48–60 inch lbs. (5–7 Nm)

➡**To prevent leakage of oil, apply liquid gasket to the inner threads of the bolt holes.**

➡**Do not install the parts if 5 minutes or more have elapsed since applying the liquid gasket. Instead, reapply liquid gasket after removing the residue.**

19. Install the oil screen with a new gasket and the 2 bolts. Tighten to 11–16 ft. lbs. (15–22 Nm).

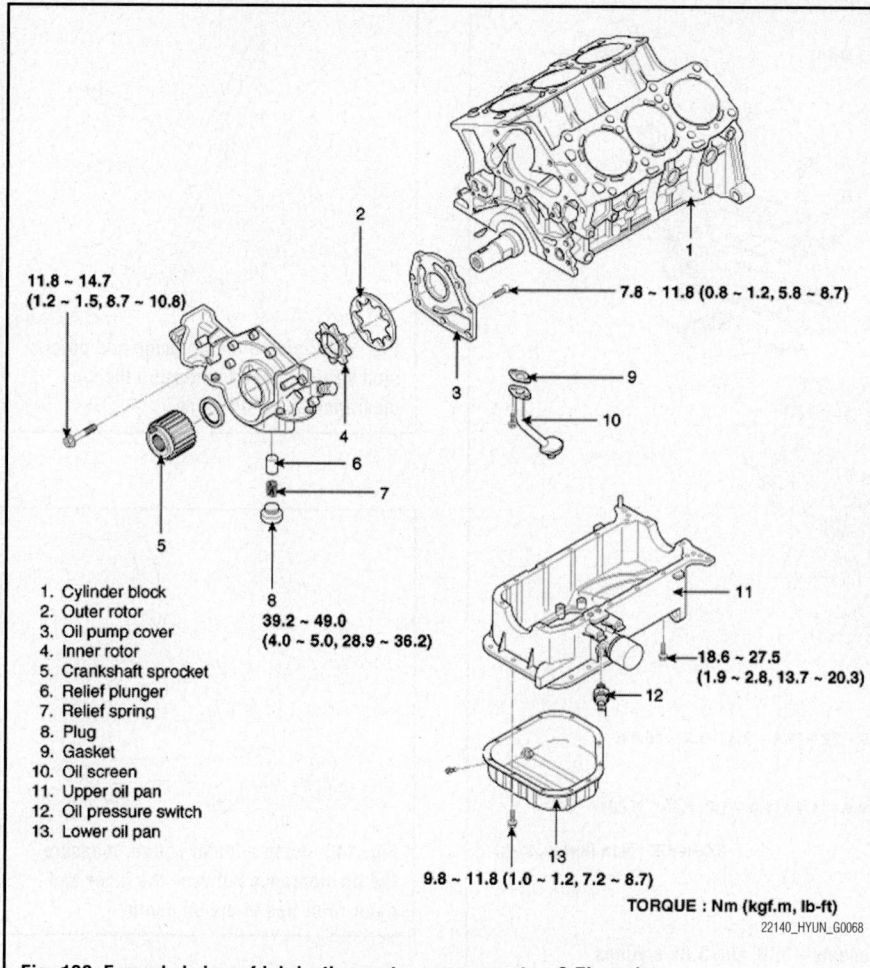

11.8 ~ 14.7
(1.2 ~ 1.5, 8.7 ~ 10.8)

7.8 ~ 11.8 (0.8 ~ 1.2, 5.8 ~ 8.7)

1. Cylinder block
2. Outer rotor
3. Oil pump cover
4. Inner rotor
5. Crankshaft sprocket
6. Relief plunger
7. Relief spring
8. Plug
9. Gasket
10. Oil screen
11. Upper oil pan
12. Oil pressure switch
13. Lower oil pan

39.2 ~ 49.0
(4.0 ~ 5.0, 28.9 ~ 36.2)

18.6 ~ 27.5
(1.9 ~ 2.8, 13.7 ~ 20.3)

9.8 ~ 11.8 (1.0 ~ 1.2, 7.2 ~ 8.7)

TORQUE : Nm (kgf.m, lb-ft)

22140_HYUN_G0068

Fig. 139 Expanded view of lubrication system components—2.7L engine

20. Install the lower oil pan.

a. Apply liquid gasket as an even bead, centered between the edges of the mating surface. Use liquid gasket MS 721-40A or equivalent.

b. To prevent leakage of oil, apply liquid gasket to the inner threads of the bolt holes.

c. Do not install the parts if 5 minutes or more have elapsed since applying the liquid gasket. Instead, reapply liquid gasket after removing the residue.

d. Install the lower oil pan 10 bolts. Uniformly tighten the bolts several passes. Tighten to 84–108 inch lbs. (10–12 Nm).

21. Install the timing belt.

22. Install the accessory drive belt.

23. Install the alternator.

24. Install the front exhaust pipe.

25. Install the RH front wheel.

26. After assembly, wait at least 30 minutes before filling the engine with oil to allow the gasket material to cure.

27. Fill the engine with oil.

28. Start the engine and check for leaks.

29. Recheck the engine oil level.

3.5L Engine

See Figure 140.

1. Before servicing the vehicle, refer to the Precautions Section.

2. Drain the engine oil.

3. Disconnect the negative battery cable.

4. Remove the oil pressure switch.

5. Remove the oil filter and pans.

6. Remove the oil screen and gasket.

7. Remove the oil filter bracket and gasket.

8. Remove the oil relief valve plug from the pump case.

9. Remove the oil pump case.

To install:

10. Install the oil pump gears.

11. Inspect the side clearance of the gears using a feeler gauge. The specifications are as follows:

a. Standard body clearance: 0.0039–0.0071 inch (0.100–0.181mm).

b. Standard side clearance: 0.0016–0.0037 inch (0.040–0.095mm).

c. Oil tip clearance: 0.0024–0.0071 inch (0.06–0.18mm).

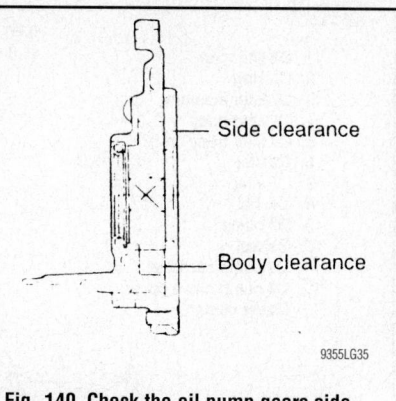

Side clearance

Body clearance

9355LG35

Fig. 140 Check the oil pump gears side and body clearance–3.5L

12. Install the oil pump case with a new gasket. Tighten the bolt to 11–14 ft. lbs. (15–20 Nm) and the screw to 72–108 inch lbs. (8–12 Nm).

13. Install a new oil seal into the pump as tightly as possible.

14. Using crankshaft front oil seal install Tool 09214-33000, install the oil seal into the case.

15. Install the relief plunger and spring and tighten the valve plug to 29–36 ft. lbs. (40–50 Nm).

16. Install the oil screen and a new gasket.

17. Install the oil pans and filter.

18. Connect the negative battery cable.

19. Fill the crankcase to the correct level.

20. Start the engine and check for leaks.

3.3L & 3.8L Engines

See Figures 141 and 142.

1. Before servicing the vehicle, refer to the Precautions Section.

2. Remove the lower oil pan, refer to Oil Pan, removal & installation.

3. Remove the oil pump chain cover.

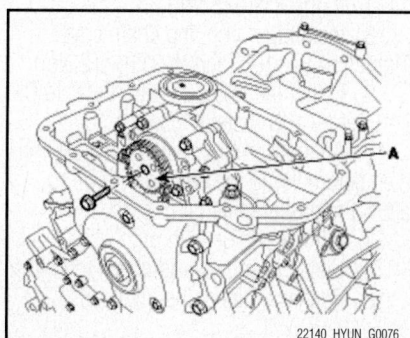

A

22140_HYUN_G0076

Fig. 141 Remove the oil pump chain cover (A) and the oil pump—3.3L and 3.8L engines

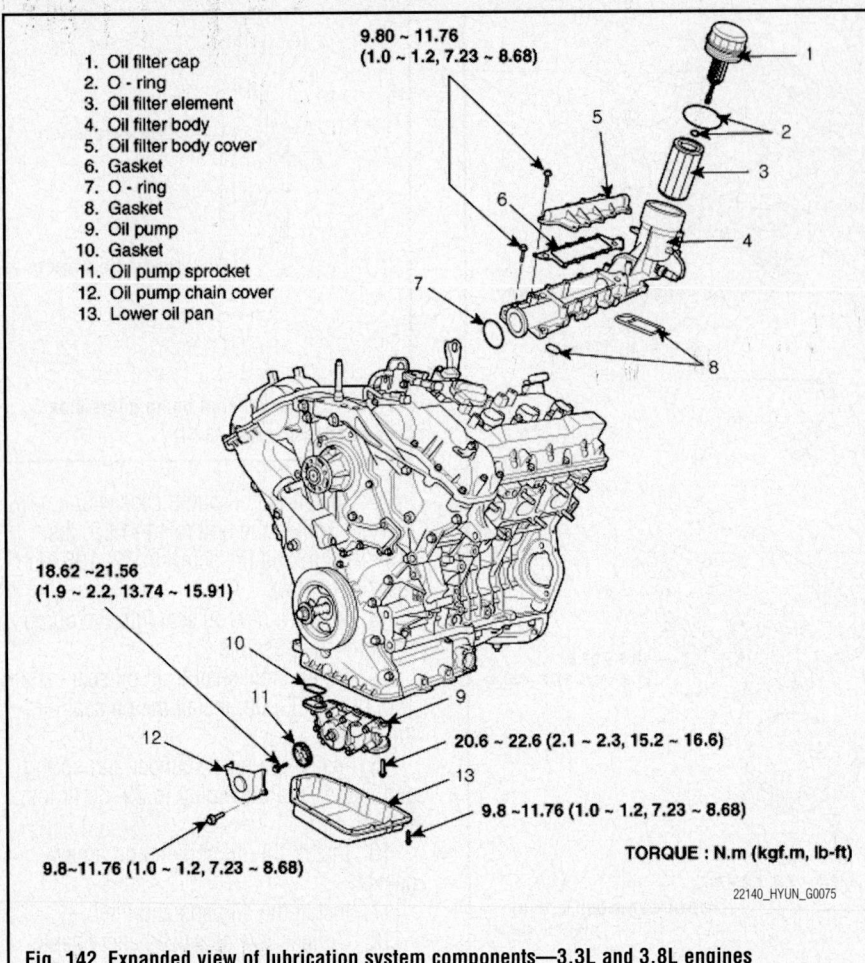

1. Oil filter cap
2. O - ring
3. Oil filter element
4. Oil filter body
5. Oil filter body cover
6. Gasket
7. O - ring
8. Gasket
9. Oil pump
10. Gasket
11. Oil pump sprocket
12. Oil pump chain cover
13. Lower oil pan

9.80 ~ 11.76
(1.0 ~ 1.2, 7.23 ~ 8.68)

18.62 ~ 21.56
(1.9 ~ 2.2, 13.74 ~ 15.91)

20.6 ~ 22.6 (2.1 ~ 2.3, 15.2 ~ 16.6)

9.8 ~11.76 (1.0 ~ 1.2, 7.23 ~ 8.68)

9.8~11.76 (1.0 ~ 1.2, 7.23 ~ 8.68)

TORQUE : N.m (kgf.m, lb-ft)

22140_HYUN_G0075

Fig. 142 Expanded view of lubrication system components—3.3L and 3.8L engines

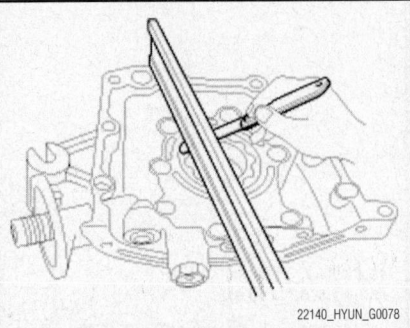

22140_HYUN_G0078

Fig. 144 Using a feeler gauge and precision straight edge to measure the side clearance of the oil pump

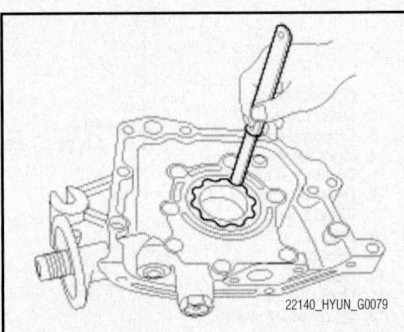

22140_HYUN_G0079

Fig. 145 Using a feeler gauge, measure the tip clearance between the inner and outer rotor tips of the oil pump

4. Remove the oil pump chain sprocket (A).

5. Remove the oil pump.

To install:

6. Install the oil pump using a new O-ring. Tighten the bolts to 15–17 ft. lbs. (21–23 Nm).

➡**Always use a new O-ring.**

7. Install the oil pump sprocket and oil pump chain on the oil pump. Tighten to 14–16 ft. lbs. (19–22 Nm).

8. Install the oil pump chain cover. Tighten to 84–108 inch lbs. (10–12 Nm).

9. Install the lower oil pan. Refer to Oil Pan, removal & installation.

10. After assembly, wait at least 30 minutes before filling the engine with oil to allow the gasket material to cure.

11. Fill the engine with oil.

12. Start the engine and check for leaks.

13. Recheck the engine oil level.

INSPECTION

See Figures 143 through 145.

1. Before servicing the vehicle, refer to the Precautions Section.

2. Remove the relief plunger: Remove the plug (A), spring (B), and relief plunger (C).

3. Inspect the relief plunger.
   a. Coat the plunger with engine oil
   b. Check that it falls smoothly into the plunger hole by its own weight.
   c. If it does not, replace the relief plunger.

d. If necessary, replace the front case.

4. Inspect the relief valve spring.

5. Inspect for a distorted or broken relief valve spring.

6. Inspect the rotor side clearance.
   a. Using a feeler gauge and precision straight edge, measure the clearance between the rotors and precision straight edge.

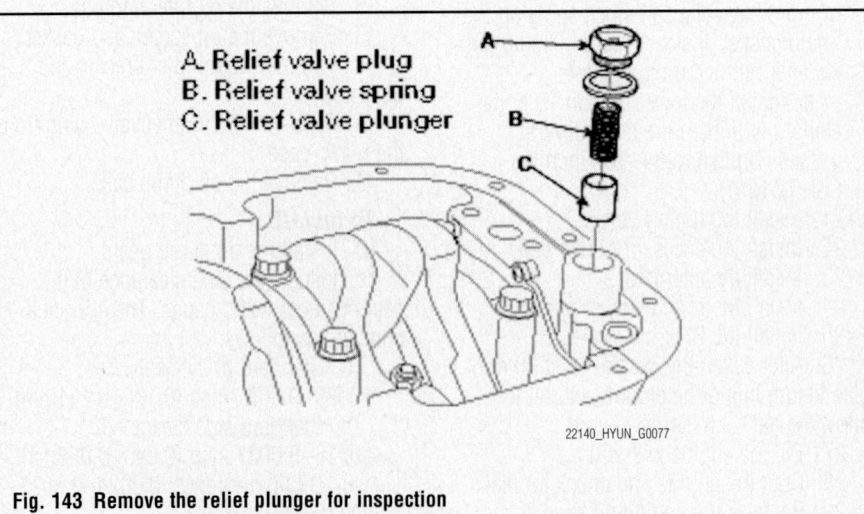

A. Relief valve plug
B. Relief valve spring
C. Relief valve plunger

22140_HYUN_G0077

Fig. 143 Remove the relief plunger for inspection

b. If the side clearance is greater than maximum, replace the rotors as a set. If necessary, replace the front case.

7. Inspect the rotor tip clearance.

a. Using a feeler gauge, measure the tip clearance between the inner and outer rotor tips.

b. If the tip clearance is greater than specified, replace the rotors as a set.

8. Inspect the rotor body clearance, as follows:

a. Using a feeler gauge, measure the clearance between the outer rotor and body.

b. If the body clearance is greater than specified, replace the rotors as a set. If necessary, replace the front case.

9. Install the relief plunger and spring into the front case hole.

10. Install the plug. Tighten to 29–36 ft. lbs. (39–49 Nm).

MAIN BEARING TORQUE SEQUENCE

See Figures 146 and 147.

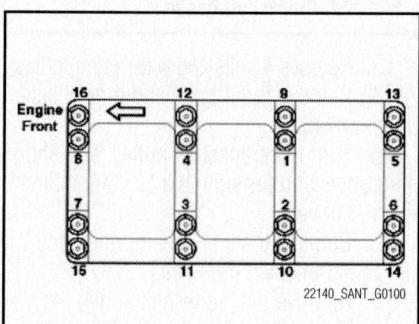

Fig. 146 Main bearing torque sequence—2.7L engine

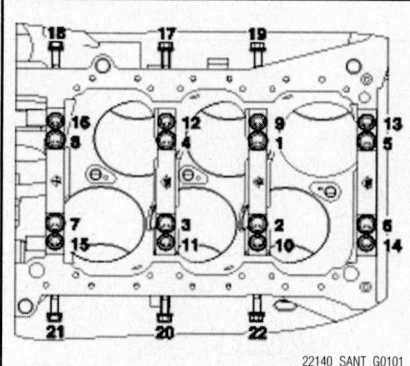

Fig. 147 Main bearing torque sequence—3.3L and 3.8L engines

PISTON AND RING

POSITIONING

See Figure 148.

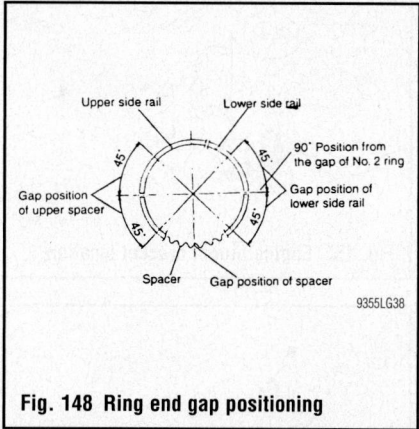

Fig. 148 Ring end gap positioning

REAR MAIN SEAL

REMOVAL & INSTALLATION

2.4L Engine

See Figure 149.

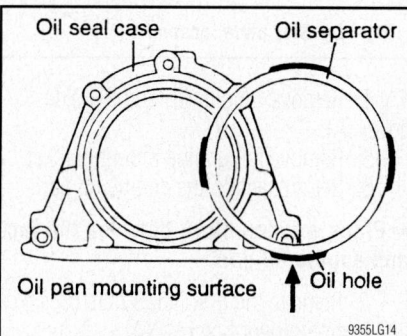

Fig. 149 Position the oil seal case so that the oil hole in the separator may be directed downwards—2.4L engine

1. Before servicing the vehicle, refer to the Precautions Section.

2. Remove or disconnect the following:
- Transaxle
- Clutch pressure plate and disc, if equipped
- Flywheel
- Rear main oil seal case bolts and the case
- Rear main oil seal

To install:

3. Install the rear main oil seal. Drive the seal square into the seal case.

4. Install the oil seal case so that the oil hole in the separator may be directed downwards. Tighten the bolts to 84–108 inch lbs. (10–12 Nm).

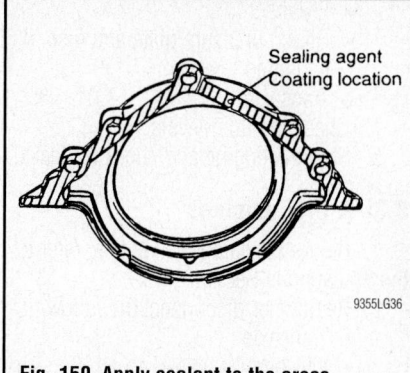

Fig. 150 Apply sealant to the areas shown—2.7L and 3.5L engines

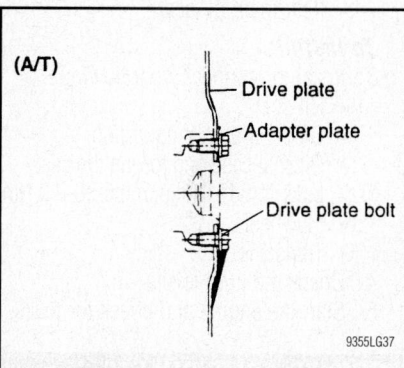

Fig. 151 Install the drive plate and adapter plate—2.7L and 3.5L engines

5. Install the flywheel.

6. Install the clutch pressure plate and disc, if equipped.

7. Install the transaxle.

8. Check the fluid levels.

9. Start the engine and check for leaks.

2.7L & 3.5L Engines

See Figures 150 and 151.

1. Before servicing the vehicle, refer to the Precautions Section.

2. Remove or disconnect the following:
- Transaxle
- Clutch pressure plate and disc, if equipped
- Flywheel
- Drive plate and adapter plate
- Oil seal case bolts and the case
- Rear main oil seal

To install:

3. Install or connect the following:
- Oil seal. Drive the seal square into the seal case
- Oil seal case so that the oil hole in the separator may be directed downwards. Tighten the bolts to 84–108 inch lbs. (10–12 Nm).
- Drive plate and adapter plate. Tighten the bolt to 53–56 ft. lbs. (73–77 Nm).

- Flywheel
- Clutch pressure plate and disc, if equipped
- Transaxle

4. Check the fluid levels.
5. Start the engine and check for leaks.

3.3L & 3.8L Engines

1. Before servicing the vehicle, refer to the Precautions Section.
2. Remove or disconnect the following:
   - Transaxle
   - Flywheel
   - Oil seal case
   - Oil separator, if equipped
   - Rear main oil seal

To install:

3. Install or connect the following:
   - Oil seal
   - Oil separator, if equipped
   - Oil seal case and torque the case bolts to 84–108 inch lbs. (8–10 Nm)
   - Flywheel
   - Transaxle
4. Check the fluid levels.
5. Start the engine and check for leaks.

ROCKER ARMS/SHAFTS

REMOVAL & INSTALLATION

These engines are not equipped with rocker arms. The camshaft acts directly on the valves through hydraulic/mechanical lash adjusters.

TIMING BELT FRONT COVER

REMOVAL & INSTALLATION

2.4L Engine

See Figures 152 through 158.

1. Before servicing the vehicle, refer to the Precautions Section.
2. Remove the engine cover.
3. Remove RH front wheel.

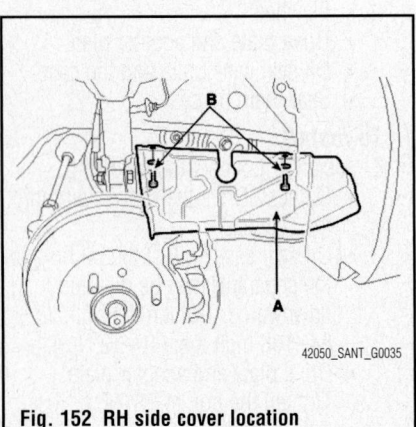

Fig. 152 RH side cover location

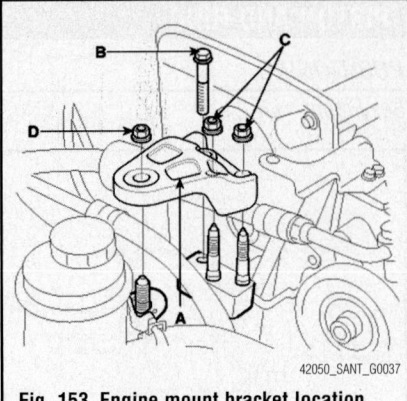

Fig. 153 Engine mount bracket location

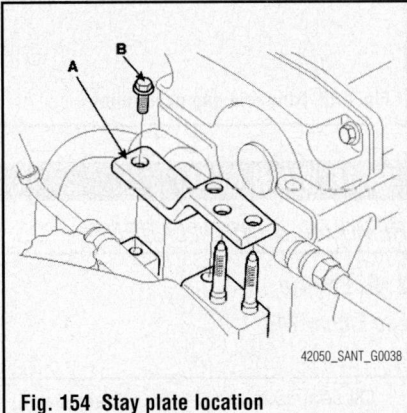

Fig. 154 Stay plate location

4. Remove 2 bolts (B) and RH side cover (A).
5. Remove the engine mount bracket.
6. Set the jack to the engine oil pan.

➡**Place wooden block between the jack and engine oil pan.**

7. Remove the bolt (B), 3 nuts (C, D) and engine mount bracket (A).
8. Remove the bolt (B) and stay plate (A).
9. Temporarily loosen the water pump pulley bolts.
10. Remove alternator belt.
11. Remove air compressor belt.
12. Remove power steering belt.

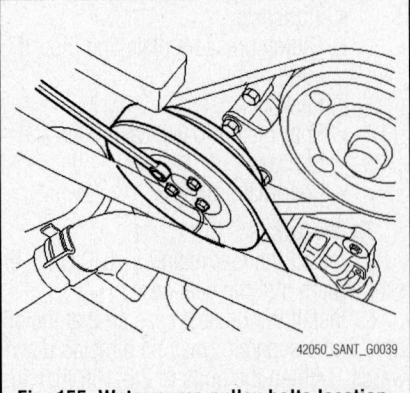

Fig. 155 Water pump pulley bolts location

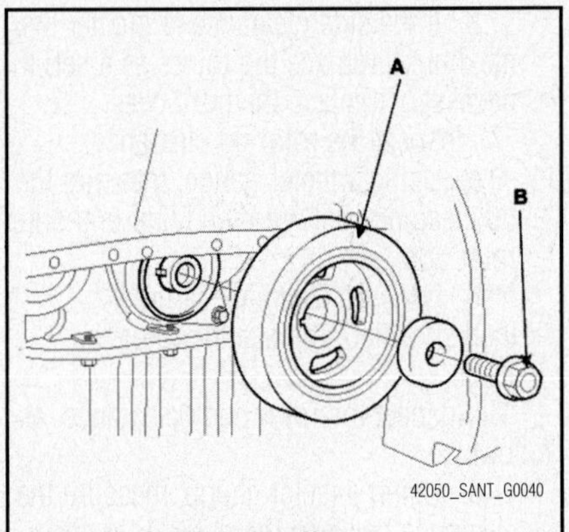

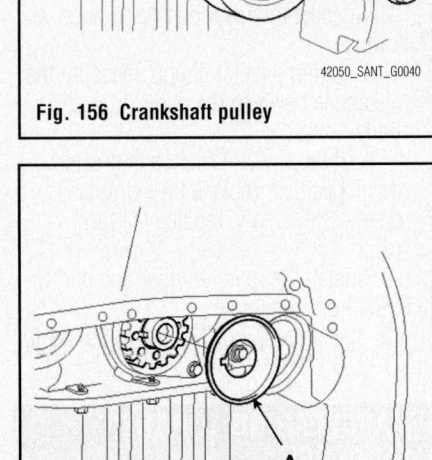

Fig. 156 Crankshaft pulley

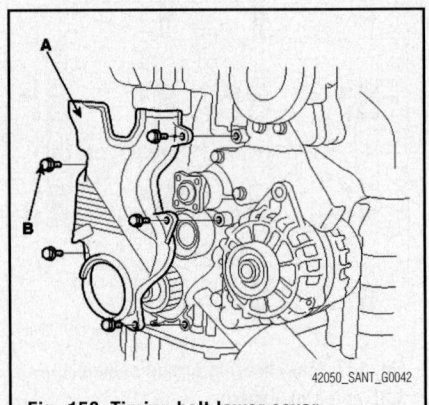

Fig. 157 Crankshaft flange

13. Remove 4 bolts and water pump pulley.
14. Remove the 4 bolts and timing belt upper cover.
15. Turn the crankshaft pulley, and align its groove with timing mark "T" of the timing belt cover.
16. Remove the crankshaft pulley bolt (B) and crankshaft pulley (A).
17. Remove the crankshaft flange (A).
18. Remove the 5 bolts (B) and timing belt lower cover (A).

To install:

19. Install timing belt lower cover and secure with the 5 bolts. Tighten and torque to 84 inch lbs. (10 Nm).

Fig. 158 Timing belt lower cover

20. Install crankshaft flange.

21. Install crankshaft pulley and secure with the crankshaft pulley bolt. Tighten and torque to 125–133 ft. lbs. (170–180 Nm).

22. Install the timing belt upper cover and secure with the 4 bolts. Tighten and torque to 84 inch lbs. (10 Nm).

23. Install the water pump pulley and loosely secure with the 4 bolts.

24. Install the power steering, the air compressor and the alternator belts.

25. Tighten all 4 water pump pulley bolts to 84 inch lbs. (10 Nm).

26. Install the stay plate and bolts. Tighten and torque to 55 ft. lbs. (69 Nm).

27. Install engine mounting brackets, secure with bolts and nuts. Tighten and torque to 80 ft. lbs. (94 Nm).

28. Lower the jack and remove from under the vehicle.

29. Install the RH side cover and secure with 2 bolts.

30. Install the RH front wheel.

31. Install the engine cover.

2.7L Engine

See Figures 159 through 163.

1. Before servicing the vehicle, refer to the Precautions Section.

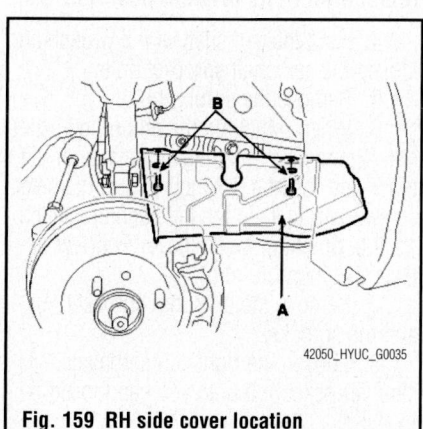

Fig. 159 RH side cover location

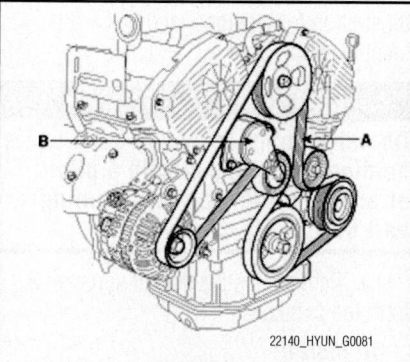

Fig. 160 Location of the auto-tensioner (B) and the drive belt (A)—2.7L engine

➡**Engine removal is not required for this procedure.**

2. Remove the engine cover.

3. Remove RH front wheel.

4. Remove the 2 bolts (B) and RH side cover (A).

5. Turn the crankshaft pulley, and align its groove with timing mark "T" of the timing belt cover.

➡**Always turn the crankshaft clockwise.**

6. Remove the auto-tensioner (B) and the drive belt (A).

7. Remove the engine mount bracket.

a. Set the jack to the engine oil pan.

➡**Place wooden block between the jack and engine oil pan.**

b. Remove the 2 bolts, 2 nuts and engine mount bracket.

8. Remove the power steering pump.

9. Remove the 7 bolts (B) and timing belt upper cover (A).

10. Remove the crankshaft damper/pulley bolt and crankshaft damper/pulley.

11. Remove the drive belt idler pulley.

12. Remove the 4 bolts (B) and timing belt lower cover (A).

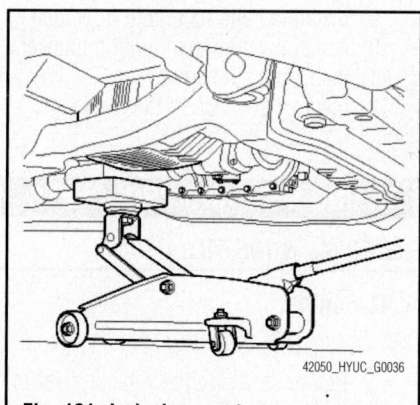

Fig. 161 Jack placement

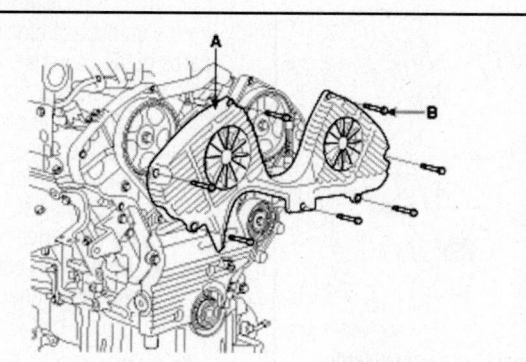

Fig. 162 Remove the 7 bolts (B) and timing belt upper cover (A)—2.7L engine

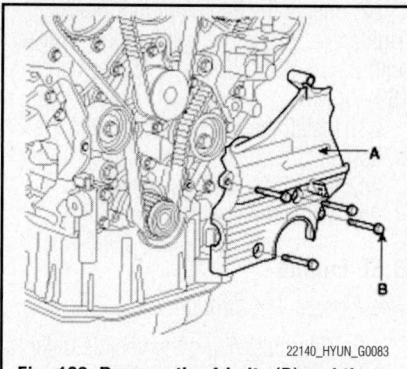

Fig. 163 Remove the 4 bolts (B) and timing belt lower cover (A)—2.7L engine

To install:

13. Install the timing belt lower cover (A) with 4 bolts (B). Tightening torque of timing belt cover bolt: 84–108 inch lbs. (10–12 Nm).

14. Install the drive belt idler pulley. Tightening torque idler pulley bolt: 33–40 ft. lbs. (44–54 Nm).

15. Install the crankshaft damper/pulley. Tightening torque crankshaft pulley bolt: 123–130 ft. lbs. (167–177 Nm).

➡**Make sure that crankshaft sprocket pin fits the small hole in the pulley.**

16. Install the timing belt upper cover with 7 bolts.

17. Install the power steering pump.

18. Install the drive belt tensioner and drive belt.

19. Install the drive belt to the tensioner pulley and install the tensioner as shown. Tighten to 18–20 ft. lbs. (25–28 Nm).

20. Install the drive belt following order:

a. Alternator

b. Power steering

c. Crankshaft pulley

d. A/C pulley, if equipped

21. Rotate the tensioner arm clockwise (about 14°) using a wrench and return it slowly to the original position.

22. Install the engine mount bracket. Tightening torque of engine mount bracket with 2 nuts and 2 bolts: 43–58 ft. lbs. (59–79 Nm).

23. Install the RH side cover with the 2 bolts.

24. Install the RH front wheel.

25. Install the engine cover.

3.5L Engine

See Figures 164 through 166.

1. Before servicing the vehicle, refer to the Precautions Section.

2. Remove the wheel of passenger side.

3. Remove the side cover.

4. Remove the engine cover.

5. Disconnect the power steering hose retainer on the engine mounting bracket.

6. Remove the engine mounting bracket.

➡ **Before removing the engine mounting bracket, support the engine oil pan with a jack, or equivalent, in order to support the engine when the bracket is removed.**

7. Disconnect the connectors from the timing belt upper cover and remove the upper cover.

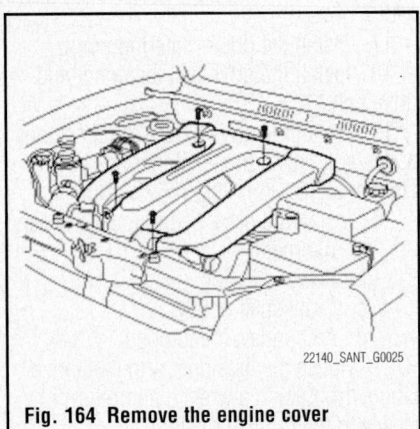

Fig. 164 Remove the engine cover

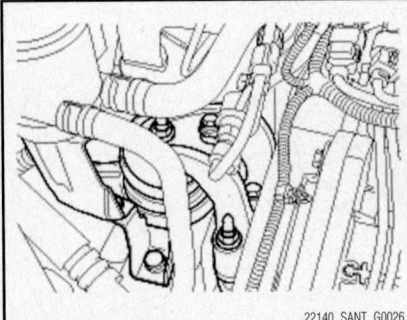

Fig. 165 Disconnect the power steering hose retainer on the engine mounting bracket

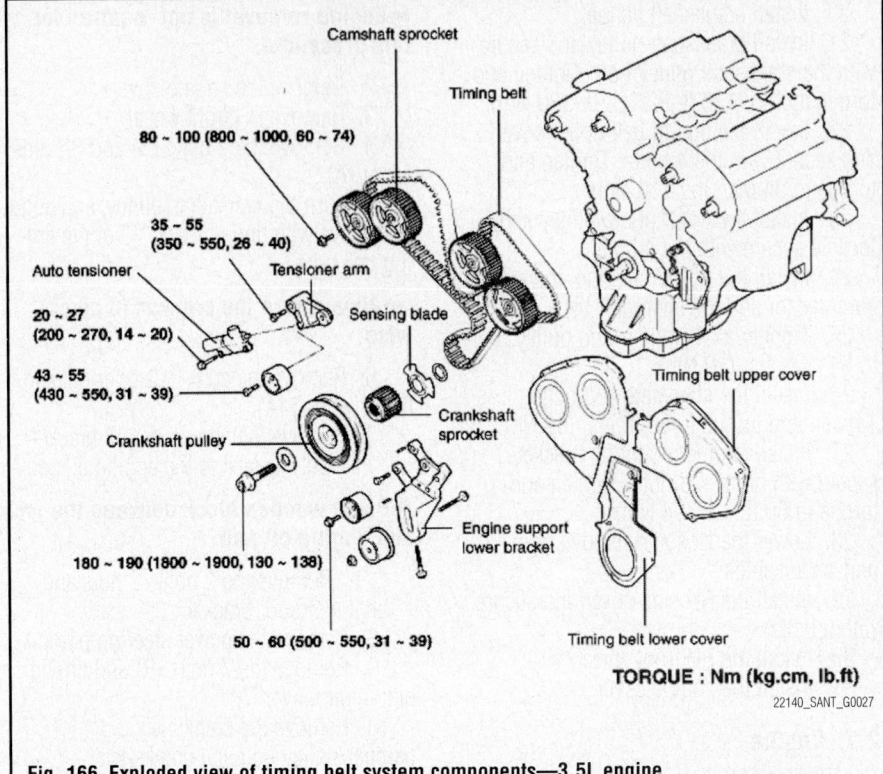

Fig. 166 Exploded view of timing belt system components—3.5L engine

Camshaft sprocket

Timing belt

80 ~ 100 (800 ~ 1000, 60 ~ 74)

35 ~ 55 (350 ~ 550, 26 ~ 40)

Auto tensioner

Tensioner arm

20 ~ 27 (200 ~ 270, 14 ~ 20)

Sensing blade

43 ~ 55 (430 ~ 550, 31 ~ 39)

Crankshaft sprocket

Crankshaft pulley

180 ~ 190 (1800 ~ 1900, 130 ~ 138)

Engine support lower bracket

Timing belt upper cover

50 ~ 60 (500 ~ 550, 31 ~ 39)

Timing belt lower cover

TORQUE : Nm (kg.cm, lb.ft)

To install:

8. Install the timing belt cover.

9. Install the alternator and drive belt.

10. Install the engine mounting bracket.

11. Install the side cover.

12. Install the wheel on the passenger side.

TIMING BELT AND SPROCKETS

REMOVAL & INSTALLATION

2.4L Engine

See Figures 167 through 185.

1. Before servicing the vehicle, refer to the Precautions Section.

2. Align the timing marks to set the No. 1 piston to Top Dead Center (TDC) by rotating the crankshaft clockwise. The timing marks of the camshaft sprocket and the cylinder head cover should be aligned and the dowel pin of the camshaft sprocket should be at the upper side.

3. Remove or disconnect the following:

- Crankshaft pulley, water pump pulley and drive belt
- Timing belt cover
- Auto tensioner
- Timing belt

➡ **Mark the timing belt if it is being reused; mark an arrow on the belt noting the direction of rotation or the front**

of the engine to make sure the belt is reinstalled in its original position.

4. Hold the camshaft with a wrench and loosen the camshaft sprocket bolts.

5. Remove the sprockets.

6. When removing the oil pump socket nut, first remove the plug at the side of the block and insert a 0.3 inch (8mm) diameter screwdriver to keep the left counterbalance shaft in position. Insert the screwdriver at least 2½ inch (60 mm).

7. Remove the oil pump sprocket nut and the sprocket.

8. Loosen the right counterbalance shaft sprocket bolt until you can loosen it by hand.

9. Remove tensioner **B** and timing belt **B**. Refer to the accompanying illustration for tensioner and belt identification.

❊❊ CAUTION

Do not attempt to loosen bolts while holding the sprocket with a pliers or any tool after removing timing belt B.

10. Remove the crankshaft sprocket **B** from the crankshaft.

To install:

11. Install the crankshaft sprocket **B** to the crankshaft.

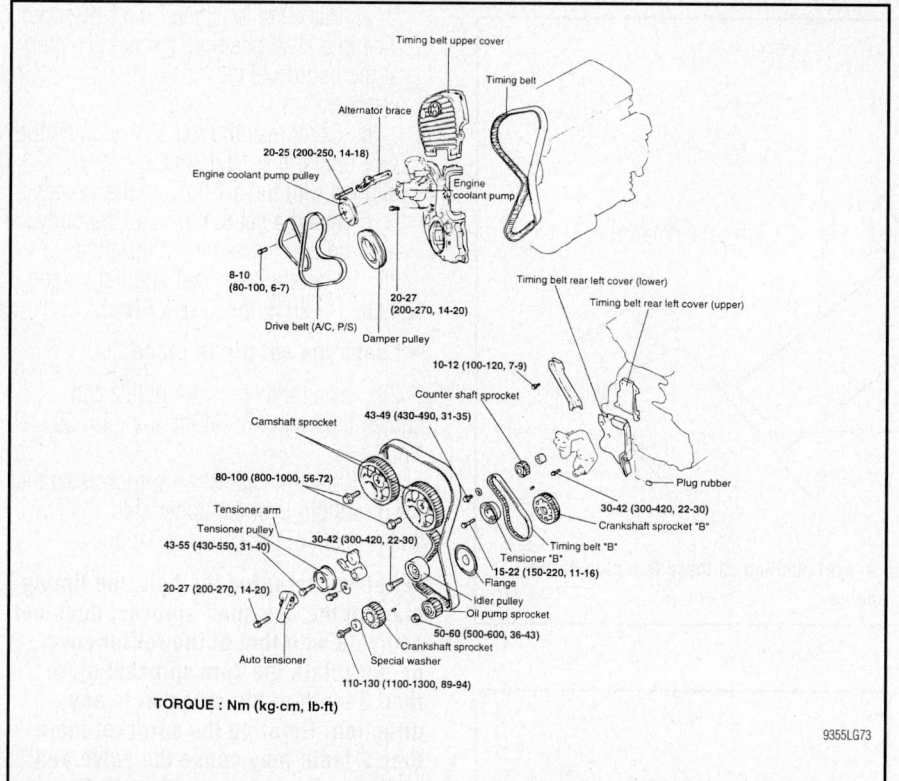

Fig. 167 Exploded view of the timing belt assembly and related components—2.4L engine

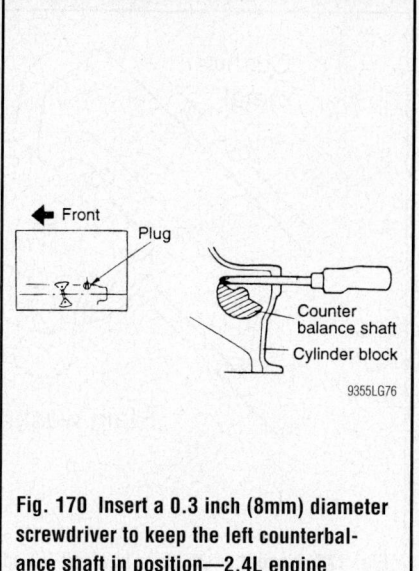

Fig. 170 Insert a 0.3 inch (8mm) diameter screwdriver to keep the left counterbalance shaft in position—2.4L engine

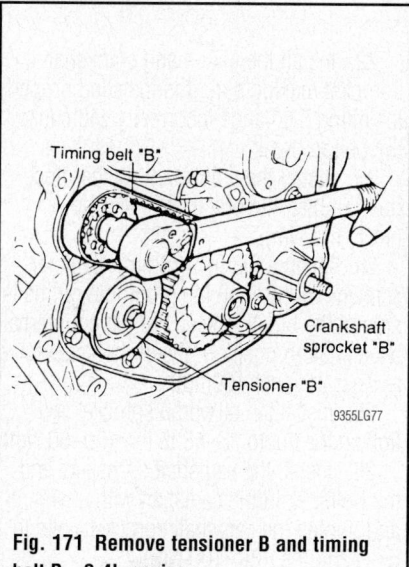

Fig. 171 Remove tensioner B and timing belt B—2.4L engine

✻✻ WARNING

Pay attention to the direction of the flange. If it is installed in the wrong direction, the belt will break.

12. Apply engine oil lightly to the outer surface of the spacer and install the spacer to the right counterbalance shaft. Be sure to install the spacer correctly.

13. Install the counterbalance shaft sprocket onto the right counterbalance shaft and then tighten the flange bolt by hand until it is tight.

14. Align the timing mark on each sprocket with its corresponding timing mark on the front case.

15. Install the timing belt **B** and make sure there is no slack.

16. Install the tensioner **B** so that the center of the pulley is located on the left side of the mounting bolt and the pulley flange faces the front of the engine.

17. Align the timing mark on the right counterbalance shaft sprocket with the timing mark on the front case.

18. Lift the tensioner **B** to tighten tensioner **B** so that its tension side is pulled tight.

19. Tighten the bolt on tensioner **B**.

➡ As the bolt is being tightened, make sure the shaft does not turn. If the shaft turns, the belt will be over tightened.

20. Make sure the timing marks are aligned.

21. Check the belt tension by depressing the center of the belt span with an index finger. The deflection should be 0.20–0.28 inch (5–7mm).

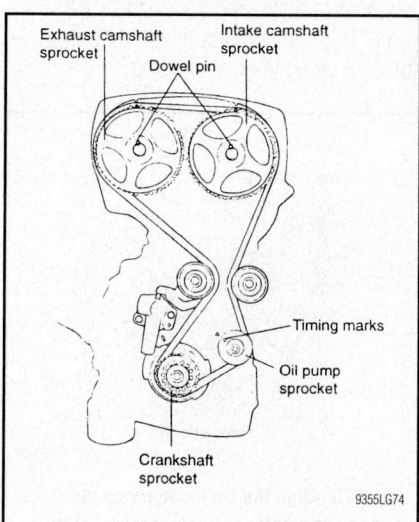

Fig. 168 Correct sprocket alignment when the belt is installed—2.4L engine

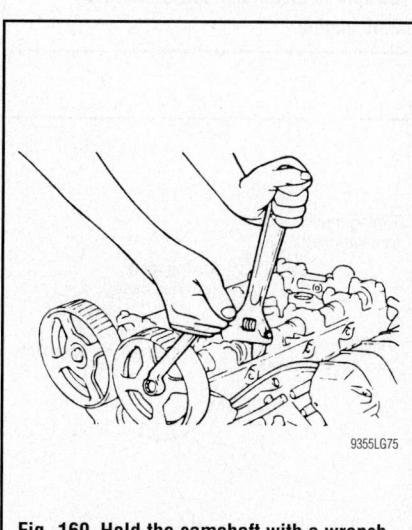

Fig. 169 Hold the camshaft with a wrench and loosen the camshaft sprocket bolts—2.4L engine

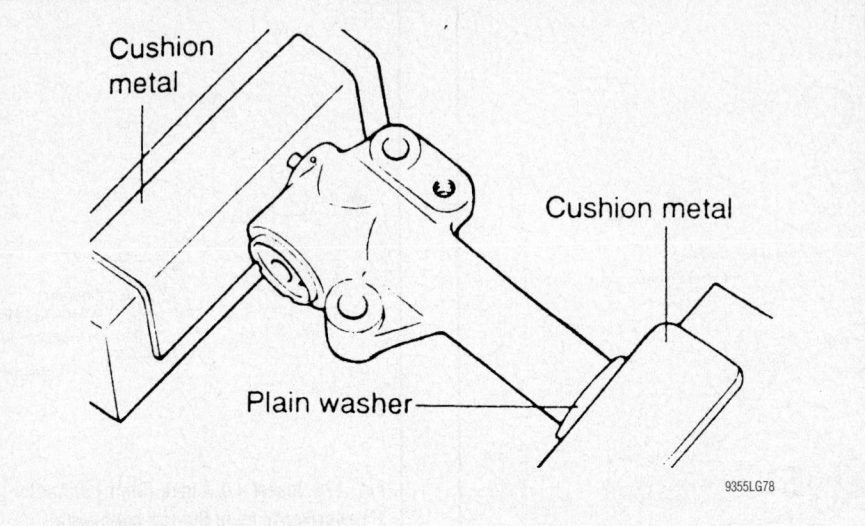

Fig. 172 Place the tensioner in a soft-jawed vise in a level position. If there is a plug at the bottom of the tensioner, use a plain washer—2.4L engine

22. Install the flange and crankshaft sprocket making sure it is installed properly. Installing the flange incorrectly will cause the belt to break.

23. Install the crankshaft washer and bolt. Tighten the bolt to 80–94 ft. lbs. (110–130 Nm).

24. Insert a 0.3 inch (8mm) diameter screwdriver through the plug hole on the left side of the block to keep the left counterbalance shaft in position. Insert the screwdriver at least 2½ inch (60mm).

25. Install the oil pump sprocket and tighten the nut to 36–43 ft. lbs. (50–60 Nm).

26. Install the camshaft sprockets and the bolts: Hold the camshaft with a wrench and tighten the camshaft sprocket bolts to 56–72 ft. lbs. (80–100 Nm).

27. Reset the auto tensioner as follows:

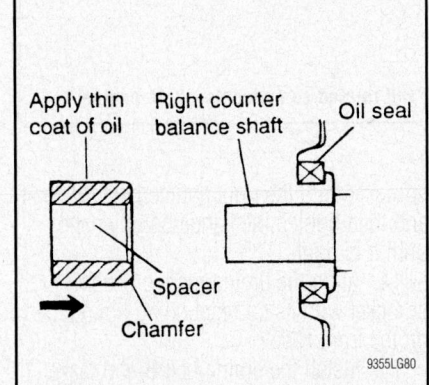

Fig. 174 Lightly apply engine oil to the outer surface of the spacer and install the spacer to the right counterbalance shaft. Be sure to install the spacer correctly—2.4L engine

a. Place the tensioner in a soft-jawed vice in a level position. If there is a plug at the bottom of the tensioner, use a plain washer.

b. Compress the rod slowly using the vice until the set hole in the rod is aligned with the set hole on the cylinder.

c. Insert a set pin through the body and rod and leave the pin installed.

28. Install the tensioner and tighten the bolts to 14–20 ft. lbs. (20–27 Nm).

➡**Leave the set pin in place.**

29. Install the tensioner pulley and tighten the bolt to 31–40 ft. lbs. (43–55 Nm).

30. Rotate the camshaft sprockets so that the dowel pin is at the upper side. Set the timing mark of the sprocket correctly.

➡**Before installing the belt, the timing mark of the camshaft sprocket does not coincide with that of the rocker cover, do not rotate the cam sprocket more than 2 teeth of the sprocket in any direction. Rotating the sprocket more than 2 teeth may cause the valve and piston to contact each other. If it is necessary to rotate the sprocket more that 2 teeth, rotate the crankshaft sprocket counterclockwise first, based on the timing mark. After the camshaft sprocket is properly timed, return the crankshaft to TDC.**

31. Align the crankshaft sprocket timing marks.

32. Align the pump sprocket timing marks.

33. Install the timing belt counterclockwise around the tensioner pulley and crankshaft sprocket. Hold the belt onto the tensioner pulley using your hand.

34. Pull the belt around the oil pump sprocket using your other hand.

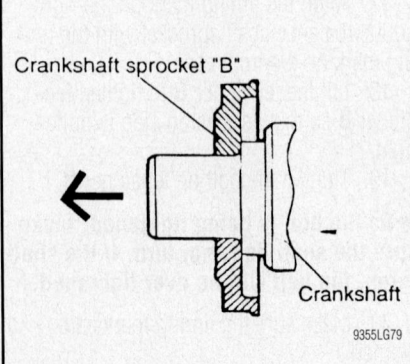

Fig. 173 Pay attention to the direction of the flange, if it is installed in the wrong direction, the belt will break—2.4L engine

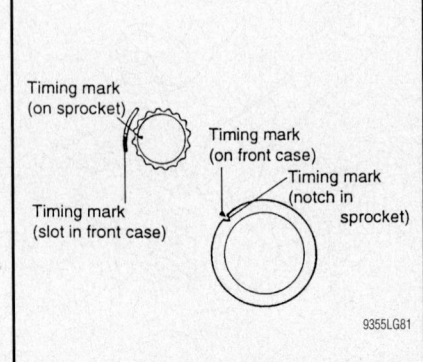

Fig. 175 Align the timing mark on each sprocket with its corresponding timing mark on the front case—2.4L engine

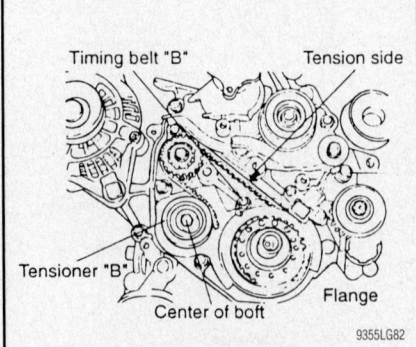

Fig. 176 Align the timing mark on the right counterbalance shaft sprocket with the timing mark on the front case—2.4L engine

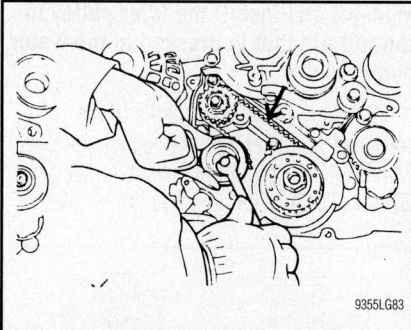

Fig. 177 Lift the tensioner B to tighten tensioner B so that its tension side is pulled tight—2.4L engine

35. Install the belt around the right-hand idler pulley, then the intake camshaft sprocket.

36. Turn the exhaust camshaft sprocket 1 tooth clockwise to align its timing mark with the cylinder top surface, then pull the belt around the exhaust camshaft sprocket.

37. Raise the tensioner pulley gently so that the belt does not sag and temporarily tighten the pulley center bolt.

38. Recheck that all timing marks are correct.

39. Remove the set pin from the auto tensioner.

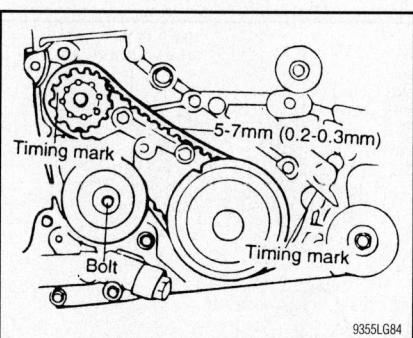

Fig. 178 Make sure the timing belt B marks are aligned and the belt tensioner is correct—2.4L engine

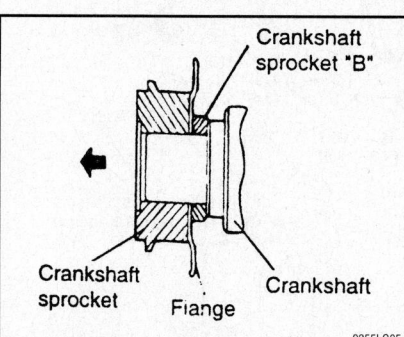

Fig. 179 Install the flange and crankshaft sprocket making sure it is installed properly—2.4L engine

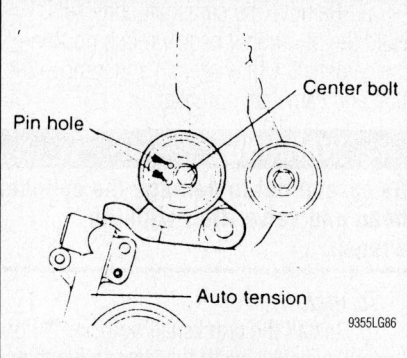

Fig. 180 Install the tensioner pulley—2.4L engine

40. Rotate the crankshaft 2 turns clockwise and let it sit for around 15 minutes. After 15 minutes, measure the auto tensioner protrusion **A** (the distance between the tensioner arm and tensioner) as shown in the accompanying illustration. The specification should be 0.24–0.35 inch (6–9mm).

41. Install the timing covers. Tighten the bolts as shown in the accompanying illustration to specification as shown in the accompanying illustration.

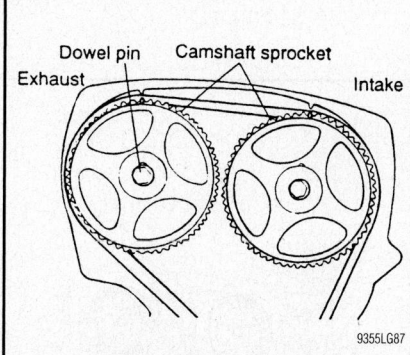

Fig. 181 Rotate the camshaft sprockets so that the dowel pin is at the upper side, set the timing mark of the sprocket correctly—2.4L engine

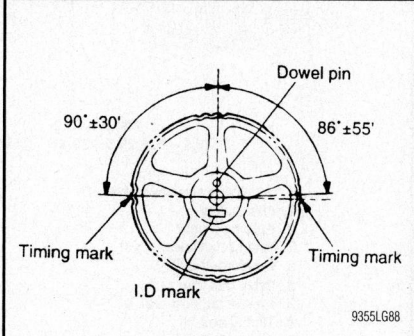

Fig. 182 Camshaft sprocket alignment—2.4L engine

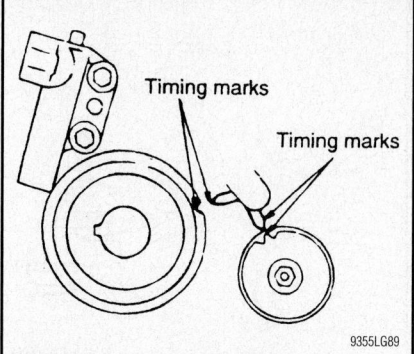

Fig. 183 Align the oil pump sprocket timing marks—2.4L engine

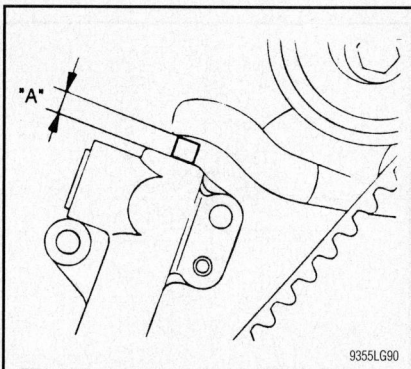

Fig. 184 Measure the auto tensioner protrusion A—2.4L engine

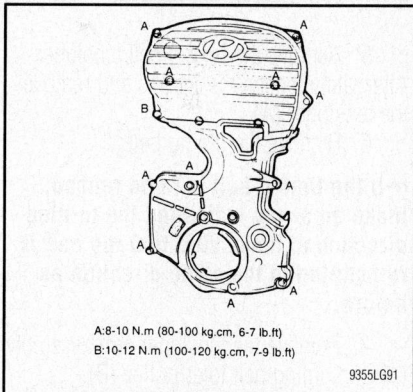

A:8-10 N.m (80-100 kg.cm, 6-7 lb.ft)
B:10-12 N.m (100-120 kg.cm, 7-9 lb.ft)

Fig. 185 Timing cover bolt location and torque specifications A—2.4L engine

2.7L Engine

See Figures 186 through 193.

1. Before servicing the vehicle, refer to the Precautions Section.

2. Remove the timing belt front covers (upper and lower). Refer to Timing Belt Front Cover, removal & installation.

3. Remove the engine support bracket (A).

4. Check that timing marks of the camshaft timing pulleys and cylinder head covers are aligned. If not, turn the crankshaft 1 revolution (360°).

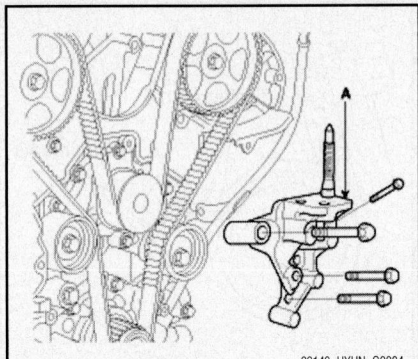

Fig. 186 Remove the engine support bracket (A)—2.7L engine

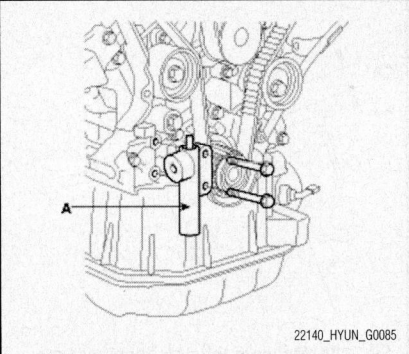

Fig. 187 Alternately loosen the 2 bolts and remove the tensioner (A)—2.7L engine

5. Remove the timing belt tensioner. Alternately loosen the 2 bolts and remove the tensioner (A).

6. Remove the timing belt.

➡If the timing belt is to be reused, make an arrow indicating the turning direction to make sure that the belt is reinstalled in the same direction as before.

7. Remove the tensioner arm assembly (A) and timing belt idler pulley (B).

8. Remove the crankshaft sprocket.

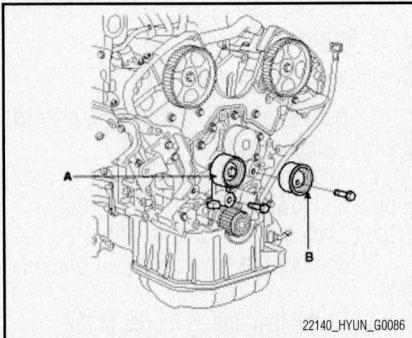

Fig. 188 Remove the tensioner arm assembly (A) and timing belt idler pulley (B)—2.7L engine

9. Remove the camshaft sprockets. Hold the hexagonal head wrench portion of the camshaft with a wrench and remove the bolt and camshaft sprocket.

❊❊ WARNING

Be careful not to damage the cylinder head and valve lifter with the wrench.

To install:

10. Install the crankshaft sprocket: Align the pulley set key with the key groove the crankshaft sprocket and slide on the crankshaft sprocket.

11. Install the camshaft sprockets and tighten the bolts to the specified torque.

a. Temporarily install the camshaft sprocket bolts.

b. Hold the hexagonal head wrench portion of the camshaft with a wrench, and tighten the camshaft sprocket bolts. Tighten to 65–80 ft. lbs. (88–108 Nm).

12. Install the idler pulley and the tensioner pulley. Tightening torque:

a. Idler pulley bolt: 36–43 ft. lbs. (49–59 Nm).

b. Tensioner arm fixed bolt: 25–40 ft. lbs. (34–54 Nm).

➡Insert and install the idler pulley to the roll pin that is pressed in the water pump boss.

13. Align the timing marks of the camshaft sprocket and crankshaft sprocket with the No. 1 piston placed at Top Dead Center (TDC) of its compression stroke.

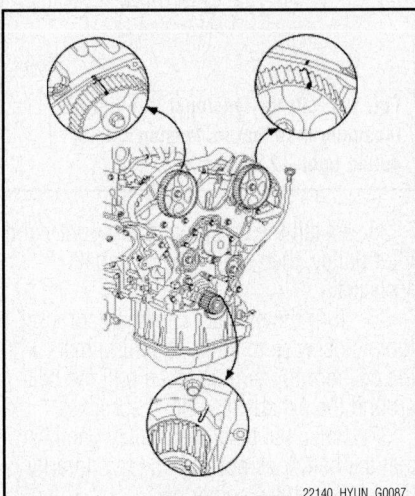

Fig. 190 Align the timing marks of the camshaft sprocket and crankshaft sprocket with the No. 1 piston placed at Top Dead Center (TDC) of its compression stroke—2.7L engine

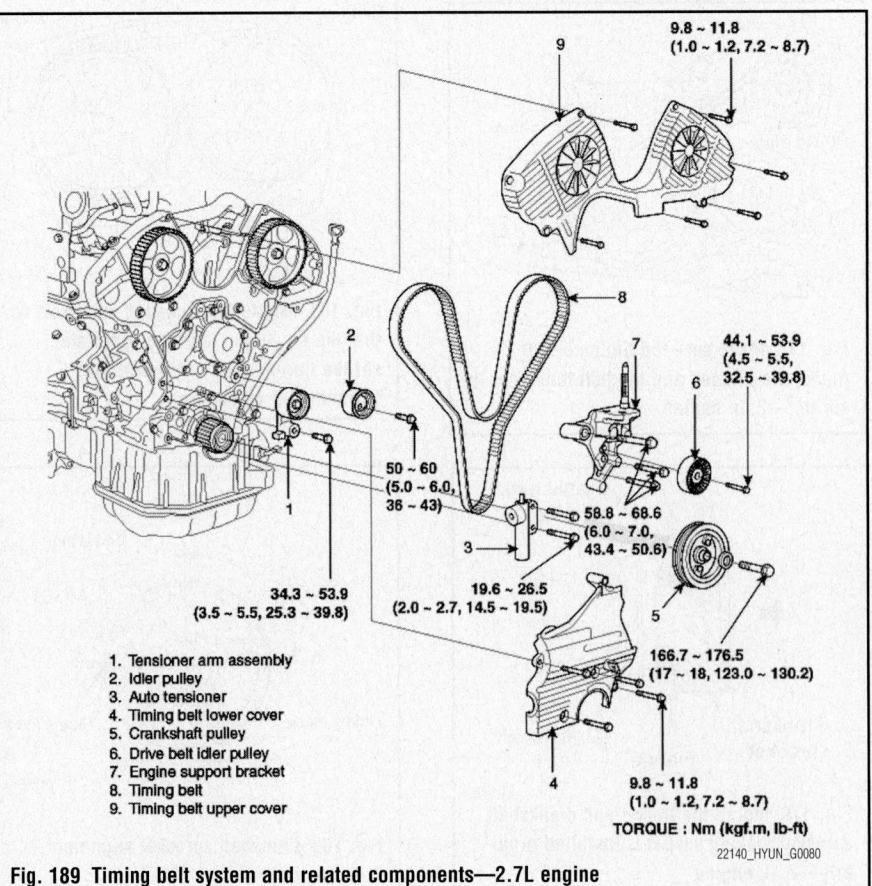

1. Tensioner arm assembly
2. Idler pulley
3. Auto tensioner
4. Timing belt lower cover
5. Crankshaft pulley
6. Drive belt idler pulley
7. Engine support bracket
8. Timing belt
9. Timing belt upper cover

9.8 ~ 11.8 (1.0 ~ 1.2, 7.2 ~ 8.7)

44.1 ~ 53.9 (4.5 ~ 5.5, 32.5 ~ 39.8)

50 ~ 60 (5.0 ~ 6.0, 36 ~ 43)

58.8 ~ 68.6 (6.0 ~ 7.0, 43.4 ~ 50.6)

34.3 ~ 53.9 (3.5 ~ 5.5, 25.3 ~ 39.8)

19.6 ~ 26.5 (2.0 ~ 2.7, 14.5 ~ 19.5)

166.7 ~ 176.5 (17 ~ 18, 123.0 ~ 130.2)

9.8 ~ 11.8 (1.0 ~ 1.2, 7.2 ~ 8.7)

TORQUE : Nm (kgf.m, lb-ft)

Fig. 189 Timing belt system and related components—2.7L engine

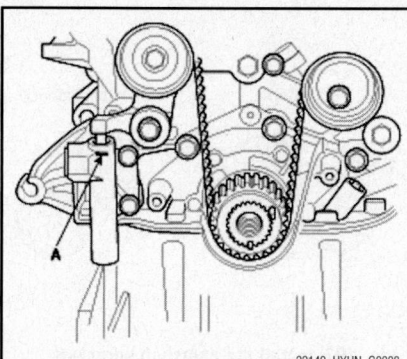

Fig. 191 Remove the set pin (A) from the tensioner—2.7L engine

14. Set the timing belt tensioner.
    a. Using a press, slowly press in the push rod.
    b. Align the holes of the push rod and housing.
    c. Pass a set pin through the holes to keep the setting position of the push rod.
    d. Release the press.
15. Install the timing belt tensioner.
    a. Temporarily install the tensioner with the 2 bolts.
    b. Alternately tighten the 2 bolts. Tighten to 15–20 ft. lbs. (20–27 Nm).

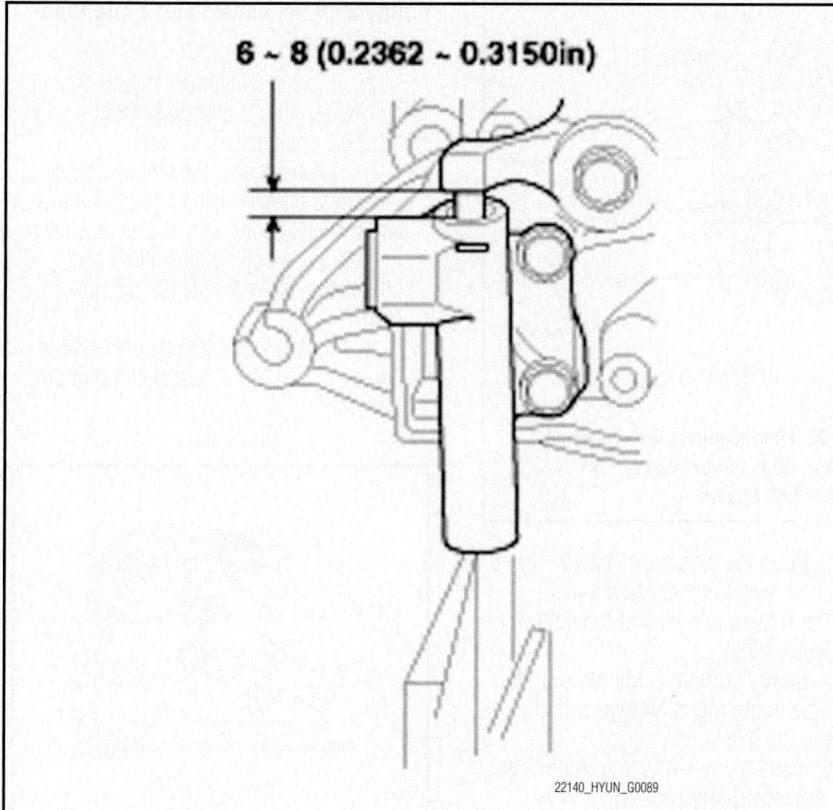

Fig. 192 The projected length of the timing belt tensioner should be 0.27–0.31 inch (7–9mm)—2.7L engine

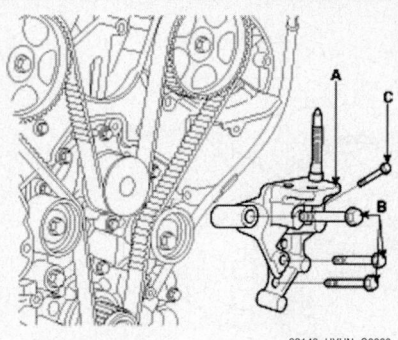

Fig. 193 Install the engine support bracket (A)—2.7L engine

16. Install the timing belt.
    a. Remove any oil or water on the sprockets, and keep them clean.
    b. Install the timing belt in this order:
    • Crankshaft sprocket
    • Idler pulley
    • Camshaft sprocket LH side
    • Water pump pulley
    • Camshaft sprocket RH side
    • Tensioner pulley
17. Remove the set pin (A) from the tensioner.
18. Check the timing belt tensioner.

    a. Rotate the crankshaft 2 turns clockwise and measure the projected length of the auto tensioner at TDC (No. 1 compression stroke) after 5 minutes.
    b. As illustrated in the example, the projected length of the timing belt tensioner should be 0.27–0.31 inch (7–9mm).
19. Install the engine support bracket (A). Tightening torque of the following bolts (see illustration):
    a. B: 43–51 ft. lbs. (59–67 Nm).
    b. C: 11–16 ft. lbs. (15–22 Nm).
20. Install the timing belt front covers (upper and lower). Refer to Timing Belt Front Cover, removal & installation.

3.5L Engine

See Figures 194 through 200.

1. Before servicing the vehicle, refer to the Precautions Section.
2. Remove the passenger side wheel.
3. Remove the side cover.
4. Remove the engine cover.
5. Disconnect the power steering hose from the engine mount bracket.
6. Support the engine with a jack and a wooden block and remove the engine mount bracket.
7. Disconnect the connectors from the upper timing belt cover.
8. Remove the upper timing belt cover.
9. Remove the drive belt.
10. Remove the 2 alternator mounting nuts and 2 bolts attaching the engine support bracket.
11. Loosen the 5 engine support bracket bolts.
12. Remove the alternator.
13. Remove the engine support bracket by moving the engine up and down slightly and remove the bracket upwards.
14. Remove the auto tensioner.
15. Rotate the crankshaft clockwise and align the timing mark to set the No. 1 cylinder to Top Dead Center (TDC). Make sure the timing marks of the camshaft sprocket and cylinder head cover should align with each other.

➡ **If reusing the belt, mark the direction of rotation on the belt to ensure proper belt installation.**

16. Unbolt the tensioner and remove the belt.
17. Hold the flange of the camshaft with a wrench, unfasten the sprocket bolts and remove the sprockets.

To install:
18. Install the idler pulley to the engine support bracket.

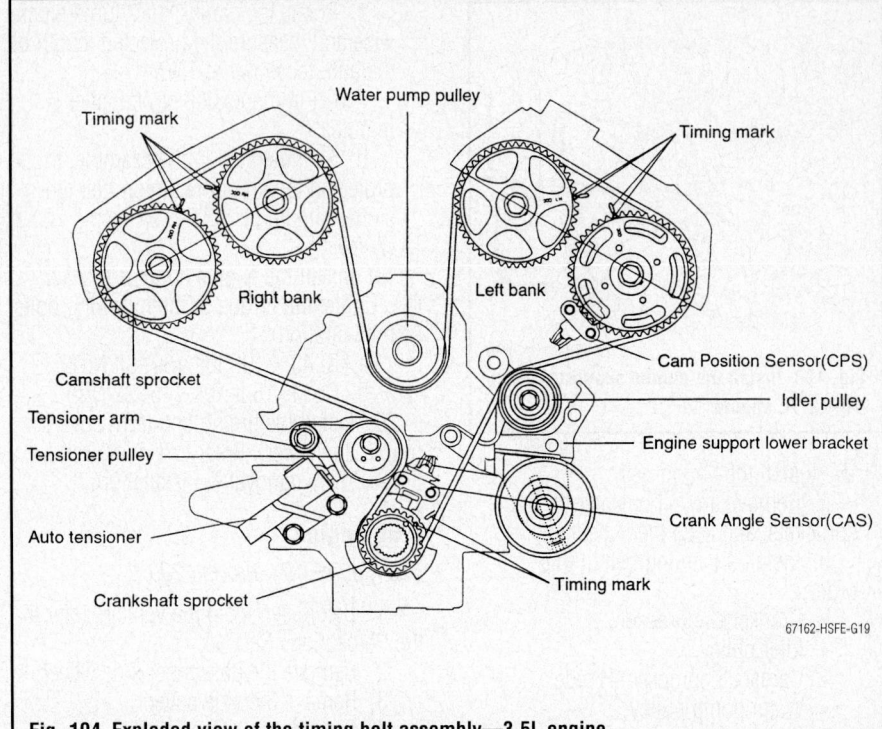

Fig. 194 Exploded view of the timing belt assembly—3.5L engine

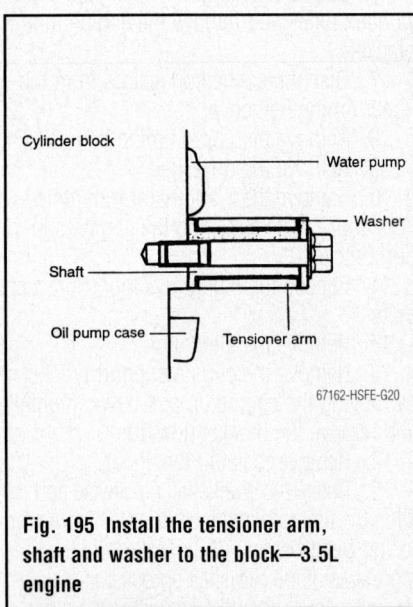

Fig. 195 Install the tensioner arm, shaft and washer to the block—3.5L engine

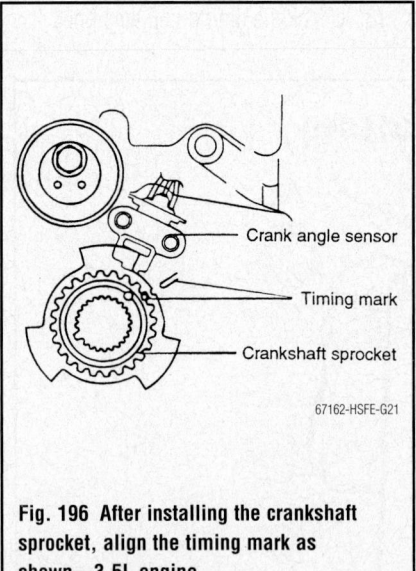

Fig. 196 After installing the crankshaft sprocket, align the timing mark as shown—3.5L engine

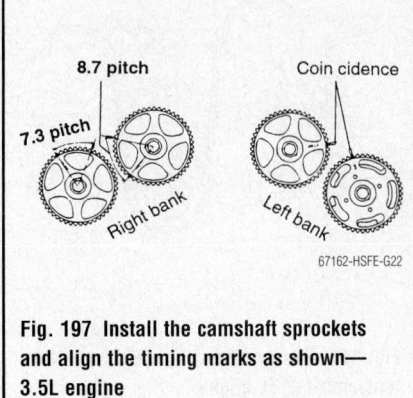

Fig. 197 Install the camshaft sprockets and align the timing marks as shown—3.5L engine

- Crankshaft sprocket
- Idler pulley
- Left hand exhaust camshaft sprocket
- Right hand intake camshaft sprocket
- Right hand exhaust camshaft sprocket
- Tensioner pulley

23. Make sure the engine is at TDC or remove the belt and reset to TDC, then, reinstall the belt as outlined above.

24. Adjust the belt tension as follows with the tensioner pin still installed:

➡ **Be careful not to turn the tensioner pulley with the center bolt while tightening the bolt.**

a. Turn the crankshaft ¼ turn counterclockwise. Turn it clockwise to fit it into the TDC position.

b. Release the center bolt and apply tension to the belt using a torque wrench and the tensioner pulley socket as shown in the accompanying illustration and tighten the center bolt to 45 inch lbs. (5 Nm).

c. Tighten the auto tensioner bolt to 31–39 ft. lbs. (43–55 Nm) and pull out the pin.

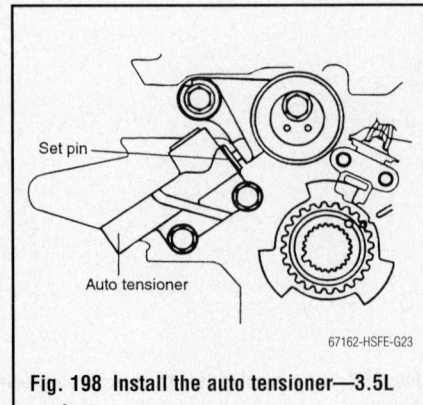

Fig. 198 Install the auto tensioner—3.5L engine

19. Install the tensioner arm, shaft and washer to the block. Tighten to 25–40 ft. lbs. (35–55 Nm).

20. Install the crankshaft sprocket and align the timing mark as illustrated. Tighten the bolts to 60–74 ft. lbs. (80–100 Nm). Be careful not to bend the crankshaft sensing blade.

21. Install the auto tensioner to the oil pump case. If the auto tensioner is in its fully extended position, perform the following:

a. Place the tensioner in a soft jawed vise in a level position. Use a plain washer if there is a plug at the bottom of the tensioner.

b. Slowly compress the rod until the set hole in the rod is aligned with the set hole in the tensioner.

c. Insert a 3.8—4.5mm pin through the tensioner body and rod.

22. Align the timing marks of each sprocket and install the timing belt in this order:

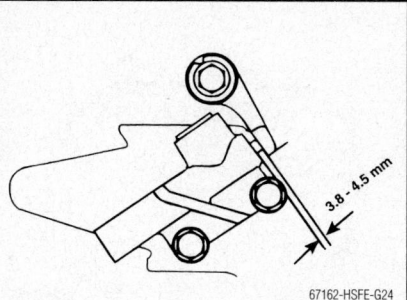

Fig. 199 Insert a 3.8—4.5mm pin through the tensioner body and rod when resetting the tensioner—3.5L engine

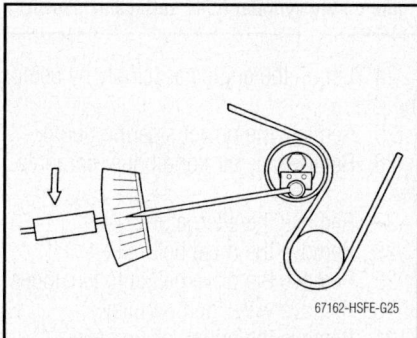

Fig. 200 Release the center bolt and apply tension to the belt using a torque wrench and the tensioner pulley socket as shown—3.5L engine

25. Adjust the belt tension as follows with the tensioner pin removed:

➡ **Be careful not to turn the tensioner pulley with the center bolt as you tighten the bolt.**

a. Turn the crankshaft 2 rotations clockwise and measure the projected load of the auto tensioner in the TDC position after 5 minutes. The projected length should be 3.8–4.5mm.

26. Check that all timing marks are aligned.
27. Install the engine support bracket.
28. Install the timing belt cover.
29. Install the alternator and drive belt.
30. Install the engine mount bracket.
31. Install the side cover.
32. Install the engine cover.
33. Install the wheel.

TIMING CHAIN COVER AND SEAL

REMOVAL & INSTALLATION

3.3L & 3.8L Engines

See Figures 201 through 211.

1. Before servicing the vehicle, refer to the Precautions Section.

➡ **Use fender covers to avoid damaging painted surfaces.**

➡ **To avoid damage, unplug the wiring connectors carefully while holding the connector portion.**

➡ **Mark all wiring and hoses to avoid misconnection.**

➡ **Turn the crankshaft pulley so that the No. 1 piston is at Top Dead Center (TDC).**

2. Disconnect the negative terminal from the battery.
3. Remove the engine cover.
4. Remove the air duct.
5. Remove the intake air hose and air cleaner assembly.
   a. Disconnect the AFS connector (A).
   b. Disconnect the breather hose (B) from air cleaner hose.
   c. Disconnect the ECM connector.
   d. Remove the intake air hose (C) and air cleaner (D).
6. Remove the RH front wheel.
7. Remove the undercover.

8. Remove the RH side cover.
9. Drain the engine coolant. Remove the radiator cap to speed draining.
10. Remove the upper radiator hose.
11. Drain the engine oil.
12. Remove the lower oil pan. Refer to Oil Pan, removal & installation.
13. Place a jack underneath the upper oil pan.
14. Loosen the transaxle mounting bolts (A) without removing the transaxle mounting (B).
15. Remove the engine coolant reservoir tank (A).
16. Remove the engine mounting bracket (B).
17. Loosen the A/C pipe bracket mounting bolt.
18. Remove the No. 1 engine mounting (A) through the lower position of the A/C pipe line.
19. Remove the surge tank and engine wiring.
   a. Disconnect the RH oxygen sensor connector.
   b. Disconnect the VIS solenoid valve connector.
   c. Disconnect the power steering oil pressure sensor connector.

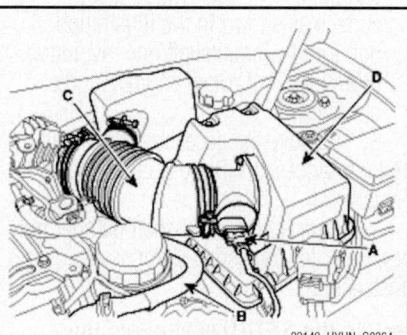

Fig. 201 Remove the intake air hose (C) and air cleaner (D) assembly

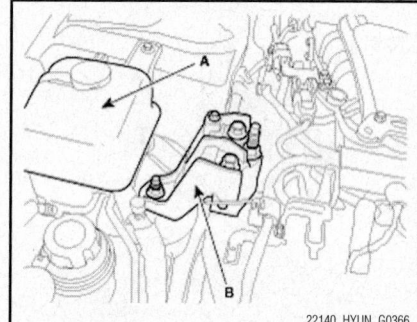

Fig. 203 Remove the engine coolant reservoir tank (A) and the engine mounting bracket (B)

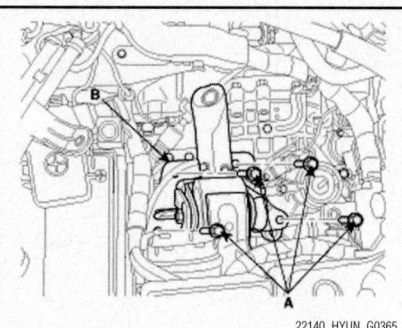

Fig. 202 Loosen the transaxle mounting bolts (A) without removing the transaxle mounting (B)

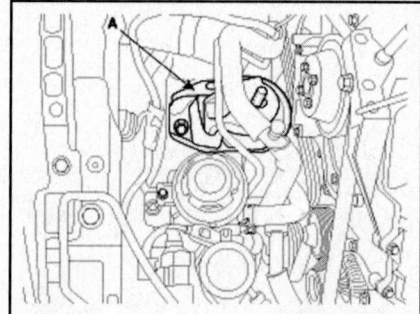

Fig. 204 Remove the No. 1 engine mounting (A) through the lower position of the A/C pipe line

d. Disconnect the RH injector connector and the ignition coil connector.

e. Disconnect the OCV connector and knock sensor connector.

f. Disconnect the LH front oxygen sensor connector, alternator connector, and air compressor connector.

g. Disconnect the LH ignition coil connector, injector connector, condenser connector, and ground

h. Remove the wiring harness protector.

i. Disconnect the LH CMPS and oil pressure switch connector.

j. Disconnect the ETC connector and the knock sensor connector.

k. Disconnect the PCSV connector, the MAP sensor connector, and the PCSV hose.

l. Remove the Electronic Throttle Control (ETC) bracket.

m. Disconnect the water hoses from the ETC.

n. Disconnect the PCV hose.

o. Disconnect the brake vacuum hose.

p. Remove the surge tank stay.

q. Remove the connector bracket from the surge tank.

r. Remove the surge tank (A).

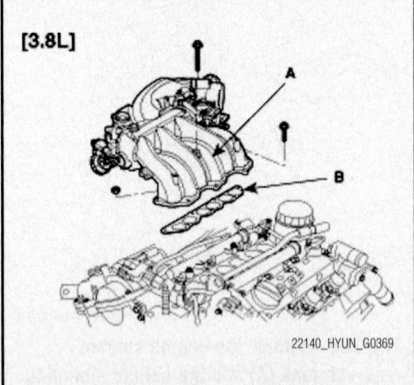

Fig. 205 Remove the surge tank (A)

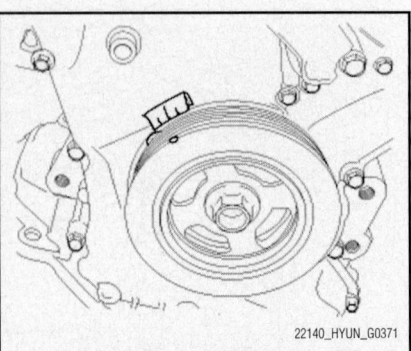

Fig. 206 Remove the connector bracket (A) from the LH cylinder head cover

➡Cover the inlet of intake manifold with a clean woven stuff or vinyl cover to prevent foreign materials from entering.

20. Remove the cylinder head cover.

a. Remove the connector bracket (A) from the LH cylinder head cover.

b. Disconnect the RH ignition coil connector(A), the condenser connector (B) and remove the wiring bracket (C).

c. Remove the LH and RH ignition coils.

d. Remove the LH and RH cylinder head covers.

➡Cover the upside of engine head with a clean vinyl cover to prevent foreign materials from entering.

21. Set No. 1 cylinder to TDC/compression.

a. Turn the crankshaft pulley and align its groove with the timing mark "T" of the lower timing chain cover.

✳✳ WARNING

Do not rotate engine counterclockwise.

b. Check that the mark (A) of the camshaft timing sprockets are in straight line on the cylinder head surface as shown in the illustration. If not, turn the crankshaft one revolution (360°). Do not rotate engine counterclockwise.

22. Remove the accessory drive belt. Refer to Accessory Drive Belts, removal & installation.

23. Remove the crankshaft damper pulley. Refer to Crankshaft Damper, removal & installation.

➡Use the SST (flywheel stopper, 09231-3C300) to remove the crankshaft pulley bolt, after removing the starter.

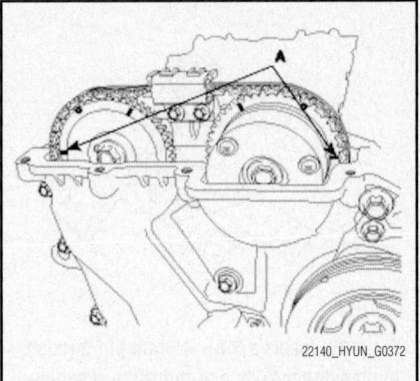

Fig. 208 Check that the mark (A) of the camshaft timing sprockets are in straight line on the cylinder head surface as shown

24. Lift up the engine assembly by using the jack.

25. Remove the power steering pump.

26. Remove the air conditioner compressor.

27. Remove the alternator.

28. Remove the drive belt idler.

29. Remove the drive belt auto tensioner.

30. Remove water pump pulley.

31. Remove the timing chain cover (A).

✳✳ WARNING

Be careful not to damage the contact surfaces of the cylinder block, cylinder head, or timing chain cover.

To install:

32. Install the timing chain cover.

a. The sealant locations on the chain cover and on the counter parts (cylinder head, cylinder block, and lower oil pan) must be free of engine oil and any debris.

b. Before assembling the timing chain cover, the liquid sealant TB 1217H should be applied on the gap between cylinder head and cylinder block.

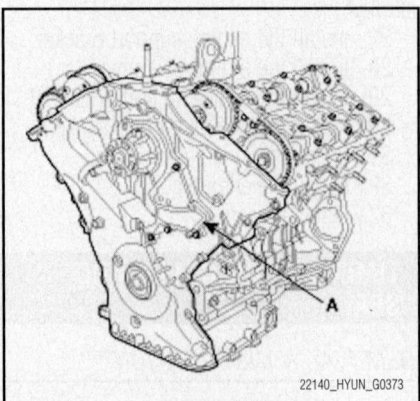

Fig. 209 Remove the timing chain cover (A)

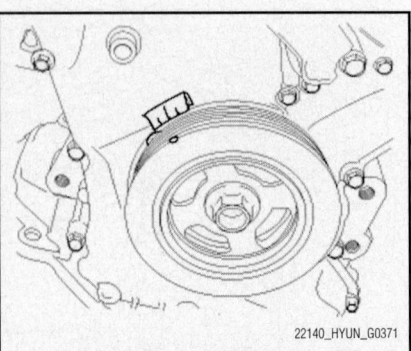

Fig. 207 Turn the crankshaft pulley and align its groove with the timing mark "T" of the lower timing chain cover

➡**The part must be assembled within 5 minutes after the sealant is applied. Use a bead width of 0.1 inch (2.5mm).**

c. After applying liquid sealant TB1217H on the timing chain cover, the part must be assembled within 5 minutes.

➡**The sealant should be applied in a continuous bead. Bead width: 0.1 inch (2.5mm).**

d. Install a new gasket to the timing chain cover.

e. The dowel pins on the cylinder block and holes on the timing chain cover should be used as a reference in order to aid assembly of the timing chain cover into position.

f. Tighten the timing cover as shown in the illustration.

33. Wait 30 minutes after the timing chain cover was assembled before starting the engine.

34. Install the water pump pulley. Tighten to 69–87 inch lbs. (8–10 Nm).

35. Install the drive belt auto tensioner. Tighten to Bolt (B): 60–63 ft. lbs. (81–85 Nm).Bolt (C): 13–16 ft. lbs. (18–22 Nm).

36. Install the drive belt idler. Tighten to 39–43 ft. lbs. (53–58 Nm).

37. Install the alternator. Tighten to 20–25 ft. lbs. (27–33 Nm).

38. Install the air conditioner compressor.

39. Install the power steering pump.

40. Lower the engine assembly using the jack.

41. Using SST (09231-3C100), install the timing chain cover oil seal.

42. Using SST (09231-3C300), install the crankshaft damper pulley. Tighten to 210–224 ft. lbs. (284–304 Nm).

43. Install the accessory drive belt. Refer to Accessory Drive Belts, removal & installation.

44. Install the cylinder head cover.

a. The hardening sealant located on the upper area between the timing chain cover and the cylinder head should be removed before assembling the cylinder head cover.

b. After applying sealant (TB1217H), it should be assembled within 5 minutes. Bead width: 0.1 inch (2.5mm).

c. Wait 30 minutes after the cylinder head cover was assembled before starting the engine.

d. Install the cylinder head cover bolts as shown in the illustration. Tighten to 87–104 inch lbs. (10–12 Nm).

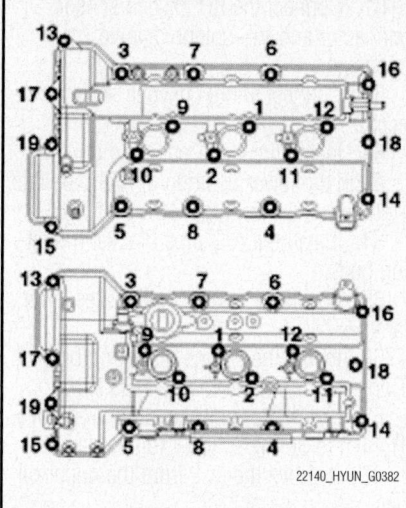

Fig. 211 Tighten the cylinder head cover bolts in the sequence shown

➡**Do not reuse the cylinder head cover gasket.**

45. Install the ignition coils.
46. Connect the RH ignition coil connector, the condenser connector, and install the wiring bracket.
47. Install the connector bracket from the LH cylinder head cover.
48. Install the surge tank. Tighten to 87–104 inch lbs. (10–12 Nm).

a. Install the connector bracket(A) on the surge tank. Tighten to 210–224 ft. lbs. (284–304 Nm).

b. Install the surge tank stay. Tighten to 20–23 ft. lbs. (27–31 Nm).

49. Connect the brake vacuum hose.
50. Connect the PCV hose.
51. Connect the water hoses to the ETC.
52. Install the ETC bracket. Tighten to 12–19 ft. lbs. (16–26 Nm).
53. Connect the PCSV connector, the MAP sensor connector, and the PCSV hose.
54. Connect the ETC connector and the knock sensor connector.
55. Connect the LH CMPS and oil pressure switch connector.
56. Install the wiring harness protector, and connect the LH ignition connector, injector connector, condenser connector, and ground.
57. Connect the OCV connector and knock sensor connector.
58. Connect the LH front oxygen sensor connector, alternator connector, and air compressor connector.
59. Connect the RH injector connector and the ignition coil connector.
60. Connect the power steering oil pressure sensor connector.

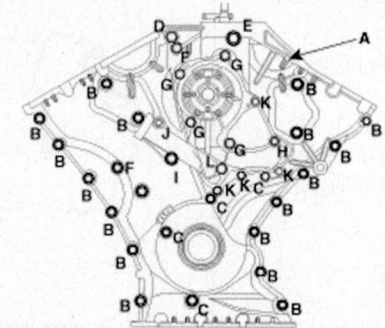

Bolt (No.)	Torque Specification
B (17)	14-16 ft. lbs. (19-22 Nm)
C (4)	87-104 inch lbs. (10-12 Nm)
D (1)	43-51 ft. lbs. (59-69 Nm)
E (1)	43-51 ft. lbs. (59-69 Nm)
F (2)	18-20 ft. lbs. (25-27 Nm)
G (4)	16-17 ft. lbs. (22-24 Nm)
H (1)	87-104 inch lbs. (10-12 Nm)
I (1)	87-104 inch lbs. (10-12 Nm)
J (1)	87-104 inch lbs. (10-12 Nm)
K (4)	87-104 inch lbs. (10-12 Nm)
L (1)	16-20 ft. lbs. (22-27 Nm)

Fig. 210 Tighten the timing chain cover as shown

61. Connect the RH oxygen sensor connector and VIS solenoid valve connector.

62. Connect the RH oxygen sensor connector.

63. Install the No. 1 engine mounting through the lower position of A/C pipe line. Tighten to 36–47 ft. lbs. (49–64 Nm).

64. Install the A/C pipe bracket mounting bolt.

65. Install the engine coolant reservoir tank.

66. Install the engine mounting bracket. Tighten to 47–62 ft. lbs. (64–83 Nm).

67. Install the transaxle mounting bolts. Tighten to 36–47 ft. lbs. (49–64 Nm).

68. Remove the jack from the upper oil pan.

69. Install the lower oil pan. Refer to Oil Pan, removal & installation. Uniformly tighten the bolts in several passes. Tighten to 87–104 inch lbs. (10–12 Nm).

70. Install the upper radiator hose.

71. Install the side cover.

72. Install the undercover.

73. Install the front wheels.

74. Install the intake air hose and air cleaner assembly.

   a. Install the intake air hose and air cleaner.

   b. Connect the ECM connector.

   c. Connect the breather hose to air cleaner hose.

   d. Connect the AFS connector.

   e. Install the air duct.

75. Install the engine cover.

76. Connect the negative terminal to the battery.

77. Refill the engine with the proper amount and type of engine oil.

78. Refill the radiator and reservoir tank with engine coolant.

79. Bleed the air from the cooling system.

80. Run the engine and check for leaks.

TIMING CHAIN AND SPROCKETS

REMOVAL & INSTALLATION

3.3L & 3.8L Engines
See Figures 212 through 219.

1. Before servicing the vehicle, refer to the Precautions Section.

➡**Use fender covers to avoid damaging painted surfaces.**

➡**To avoid damage, unplug the wiring connectors carefully while holding the connector portion.**

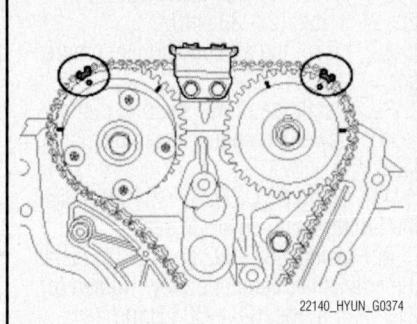

Fig. 212 Timing marks on chain and camshaft sprockets illustrated

➡**Mark all wiring and hoses to avoid misconnection.**

➡**Turn the crankshaft pulley so that the No. 1 piston is at Top Dead Center (TDC).**

2. Disconnect the negative terminal from the battery.

3. Remove the timing chain cover. Refer to Timing Chain Cover and Seal, removal & installation.

➡**Before removing the timing chain, mark the RH/LH timing chain with an identification based on the location of the sprocket as the identification mark on the chain for the TDC can be erased accidentally.**

4. Install a set pin after compressing the RH timing chain tensioner.

5. Remove the RH cam-to-cam guide.

6. Remove the RH timing chain auto tensioner (A) and the RH timing chain tensioner arm (B).

7. Remove the RH timing chain.

8. Remove the RH timing chain guide.

9. Remove the oil pump chain cover.

10. Remove the oil pump chain tensioner assembly (A).

11. Remove the oil pump chain guide.

12. Remove the oil pump chain sprocket and oil pump chain.

13. Remove the crankshaft sprocket that drives the oil pump.

14. Install a set pin after compressing the LH timing chain tensioner.

15. Remove the LH cam-to-cam guide (A).

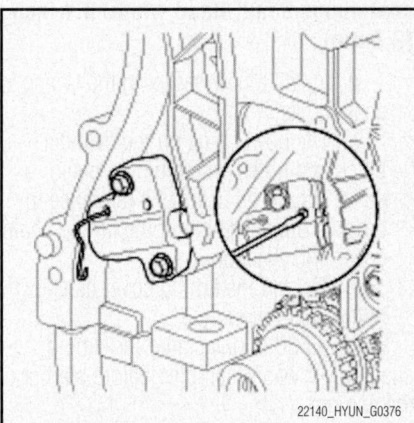

Fig. 214 Install a set pin after compressing the RH timing chain tensioner

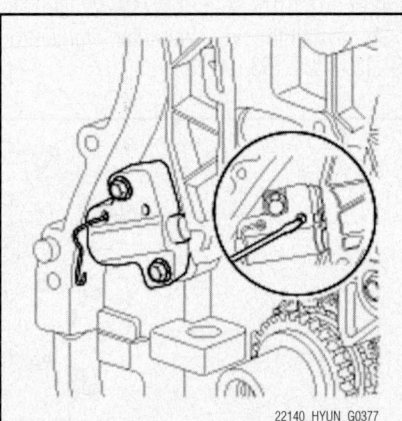

Fig. 215 Remove the RH timing chain auto tensioner (A) and the RH timing chain tensioner arm (B)

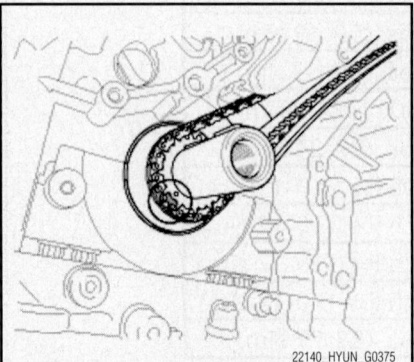

Fig. 213 Timing marks on chain and crankshaft sprocket illustrated

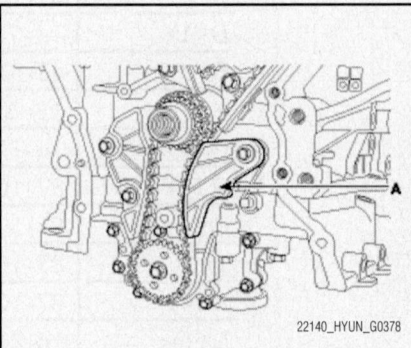

Fig. 216 Remove the oil pump chain tensioner assembly (A)

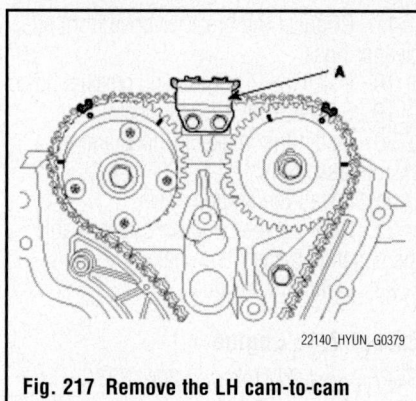

Fig. 217 Remove the LH cam-to-cam guide (A)

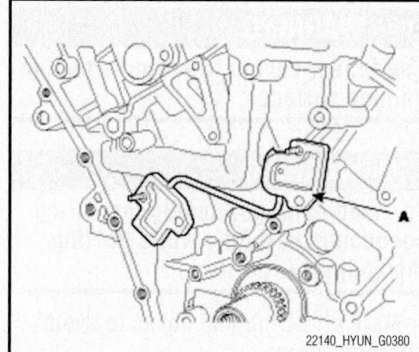

Fig. 218 Remove the tensioner adapter assembly (A)

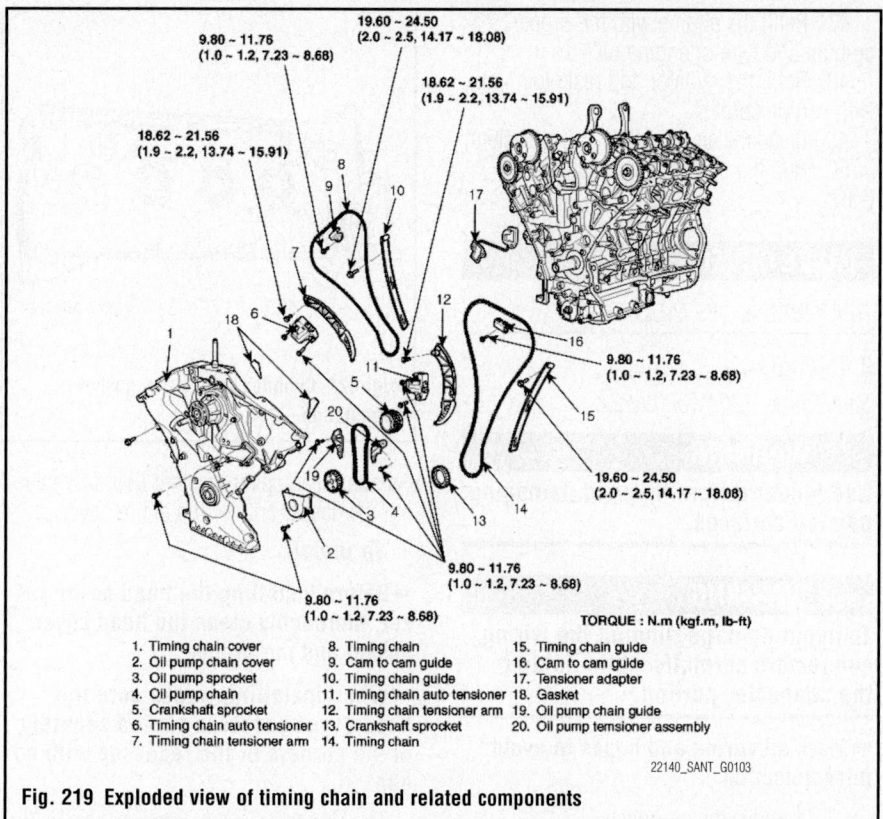

TORQUE : N.m (kgf.m, lb-ft)

1. Timing chain cover	8. Timing chain	15. Timing chain guide
2. Oil pump chain cover	9. Cam to cam guide	16. Cam to cam guide
3. Oil pump sprocket	10. Timing chain guide	17. Tensioner adapter
4. Oil pump chain	11. Timing chain auto tensioner	18. Gasket
5. Crankshaft sprocket	12. Timing chain tensioner arm	19. Oil pump chain guide
6. Timing chain auto tensioner	13. Crankshaft sprocket	20. Oil pump temsioner assembly
7. Timing chain tensioner arm	14. Timing chain	

Fig. 219 Exploded view of timing chain and related components

16. Remove the LH timing chain auto tensioner and LH timing chain tensioner arm.

17. Remove the LH timing chain.

18. Remove the LH timing chain guide.

19. Remove the crankshaft sprocket.

20. Remove the tensioner adapter assembly (A).

To install:

21. Check the camshaft sprocket and crankshaft sprocket for abnormal wear, cracks, or damage. Replace as necessary.

22. Inspect the tensioner arm and chain guide for abnormal wear, cracks, or damage. Replace as necessary.

23. Check that the tensioner piston moves smoothly when the ratchet pawl is released.

24. Install the jack to the upper oil pan.

25. The key of crankshaft must be aligned with the timing mark of timing chain cover. Then, piston of No. 1 cylinder is placed at the TDC on the compression stroke.

26. Install the tensioner adapter assembly.

27. Install the crankshaft sprocket.

28. Install the LH timing chain guide. Tighten to 14–18 ft. lbs. (20–25 Nm).

29. Install the LH timing chain.

a. To install the timing chain with no slack between the camshaft and crankshaft, use the following procedure.

b. Place the crankshaft sprocket on first, then the timing chain guide, then the exhaust camshaft sprocket, and finally the intake camshaft sprocket.

c. The timing mark of each sprockets should be matched with the timing mark (color link) of the timing chain when installing the timing chain.

30. Install the LH timing chain tensioner arm. Tighten to 14–16 ft. lbs. (19–22 Nm).

31. Install the chain tensioner. Tighten to 87–104 inch lbs. (10–12 Nm).

32. Install the LH cam-to-cam guide. Tighten to 87–104 inch lbs. (10–12 Nm).

33. Install the crankshaft sprocket.

34. Install the oil pump chain and the oil pump sprocket. Tighten to 14–16 ft. lbs. (19–22 Nm).

35. Install the RH timing chain guide. Tighten to 14–18 ft. lbs. (20–25 Nm).

36. Install the RH timing chain.

a. To install the timing chain with no slack between the camshaft and the crankshaft, use the following procedure.

b. Place the chain on the crankshaft sprocket first, then the intake camshaft sprocket, then the exhaust camshaft sprocket.

c. The timing mark of each of the sprockets must be matched with the timing mark (color link) of timing chain when installing the timing chain.

37. Install the RH timing chain tensioner arm. Tighten to 14–16 ft. lbs. (19–22 Nm).

38. Install the RH timing chain auto tensioner. Tighten to 87–104 inch lbs. (10–12 Nm).

39. Install the RH cam-to-cam guide. Tighten to 87–104 inch lbs. (10–12 Nm).

40. Install the oil pump chain guide. Tighten to 87–104 inch lbs. (10–12 Nm).

41. Install the oil pump chain tensioner assembly. Tighten to 87–104 inch lbs. (10–12 Nm).

42. Pull out the pins of hydraulic tensioner (LH & RH).

43. Install the oil pump chain cover. Tighten to 87–104 inch lbs. (10–12 Nm).

44. After rotating the crankshaft 2 revolutions in a clockwise direction (viewed from front), confirm that the timing marks are aligned.

✳✳ WARNING
Always turn the crankshaft clockwise.

45. Install the timing chain cover. Refer to Timing Chain Cover and Seal, removal & installation.

46. Connect the negative terminal to the battery.

47. Refill the engine with the proper amount and type of engine oil.
48. Refill the radiator and reservoir tank with engine coolant.
49. Bleed the air from the cooling system.
50. Run the engine and check for leaks.

VALVE COVERS

REMOVAL & INSTALLATION

2.4L Engine

See Figures 220 through 222.

✳✳ WARNING

Use fender covers to avoid damaging painted surfaces.

✳✳ WARNING

To avoid damage, unplug the wiring connectors carefully while holding the connector portion.

➡Mark all wiring and hoses to avoid misconnection.

1. Remove the air duct.
2. Disconnect the negative terminal from the battery.
3. Remove the engine cover.
4. Disconnect the AFS connector .
5. Remove the Positive Crankcase Ventilation (PCV) hose and breather hose.
6. Disconnect the breather hose from air cleaner hose.
7. Remove the intake air hose and air cleaner assembly.
8. Remove the spark plug wires.
9. Remove the cylinder head cover (also called valve cover).
10. Remove the cylinder head cover gasket and thoroughly clean the cylinder head surface (where the gasket rests) and the

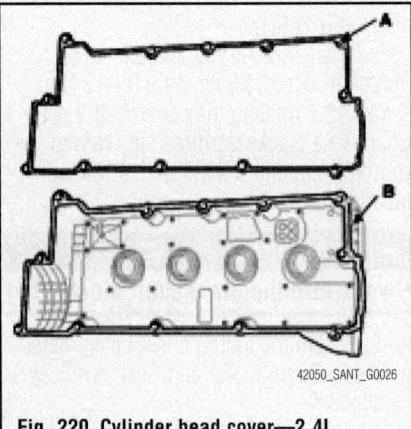

Fig. 220 Cylinder head cover—2.4L engine

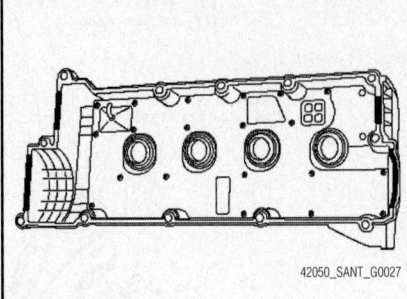

Fig. 221 Cylinder head cover gasket installation—2.4L engine

cylinder head cover whether you are replacing the gasket only or the entire cover.

To install:

➡Before installing the head cover gasket, thoroughly clean the head cover gasket and the groove.

➡When installing, make sure the head cover gasket is seated securely in the corners of the recesses with no gap.

11. Install the cylinder head cover gasket (A) in the groove of the cylinder head cover (B).
12. Apply liquid gasket to the head cover gasket at the corners of the recess.

➡Use liquid gasket—Loctite® No. 5999.

13. Check that the mating surfaces are clean and dry before applying liquid gasket
14. Install the cylinder head cover (A) with the 12 bolts (B).
15. Uniformly tighten the bolts in several passes. Tighten to 72–84 inch lbs. (8–10 Nm).
16. Install the spark plug wires.

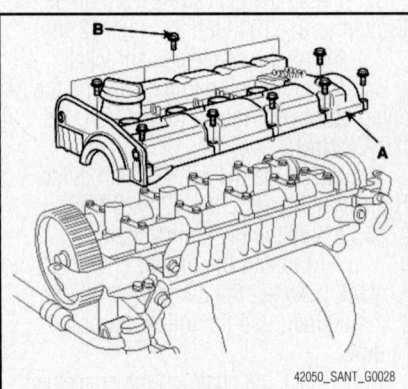

Fig. 222 Alignment of cylinder head cover mounting—2.4L engine

17. Connect the breather hose from air cleaner hose.
18. Install the PCV hose and breather hose.
19. Connect the AFS connector.
20. Install the intake air hose and air cleaner assembly.
21. Install the engine cover.
22. Connect the negative battery cable to the negative battery terminal.
23. Install the air duct.

Except 2.4L Engine

See Figures 201 through 206, 223 and 224.

1. Before servicing the vehicle, refer to the Precautions Section.

✳✳ WARNING

Use fender covers to avoid damaging painted surfaces.

✳✳ WARNING

To avoid damage, unplug the wiring connectors carefully while holding the connector portion.

➡Mark all wiring and hoses to avoid misconnection.

2. Disconnect the negative terminal from the battery.
3. Remove the engine cover.
4. Remove the air duct.
5. Remove the intake air hose and air cleaner assembly.
   a. Disconnect the AFS connector (A).
   b. Disconnect the breather hose (B) from air cleaner hose.
   c. Disconnect the ECM connector.
   d. Remove the intake air hose (C) and air cleaner (D).
6. Remove the RH front wheel.
7. Remove the undercover.
8. Remove the RH side cover.
9. Drain the engine coolant. Remove the radiator cap to speed draining.
10. Remove the upper radiator hose.
11. Drain the engine oil.
12. Remove the lower oil pan. Refer to Oil Pan, removal & installation.
13. Place a jack underneath the upper oil pan.
14. Just loosen the transaxle mounting bolts (A) without removing the transaxle mounting (B).
15. Remove the engine coolant reservoir tank (A).
16. Remove the engine mounting bracket (B).
17. Loosen the A/C pipe bracket mounting bolt.

18. Remove the No. 1 engine mounting (A) through the lower position of the A/C pipe line.

19. Remove the surge tank and engine wiring.

    a. Disconnect the RH oxygen sensor connector.

    b. Disconnect the VIS solenoid valve connector.

    c. Disconnect the power steering oil pressure sensor connector.

    d. Disconnect the RH injector connector and the ignition coil connector.

    e. Disconnect the OCV connector and knock sensor connector.

    f. Disconnect the LH front oxygen sensor connector, alternator connector, and air compressor connector.

    g. Disconnect the LH ignition coil connector, injector connector, condenser connector, and ground

    h. Remove the wiring harness protector.

    i. Disconnect the LH CMPS and oil pressure switch connector.

    j. Disconnect the ETC connector and the knock sensor connector.

    k. Disconnect the PCSV connector, the MAP sensor connector, and the PCSV hose.

    l. Remove the Electronic Throttle Control (ETC) bracket.

    m. Disconnect the water hoses from the ETC.

    n. Disconnect the PCV hose.

    o. Disconnect the brake vacuum hose.

    p. Remove the surge tank stay.

    q. Remove the connector bracket from the surge tank.

    r. Remove the surge tank (A).

➡**Cover the inlet of intake manifold with a clean woven stuff or vinyl cover to prevent foreign materials from entering.**

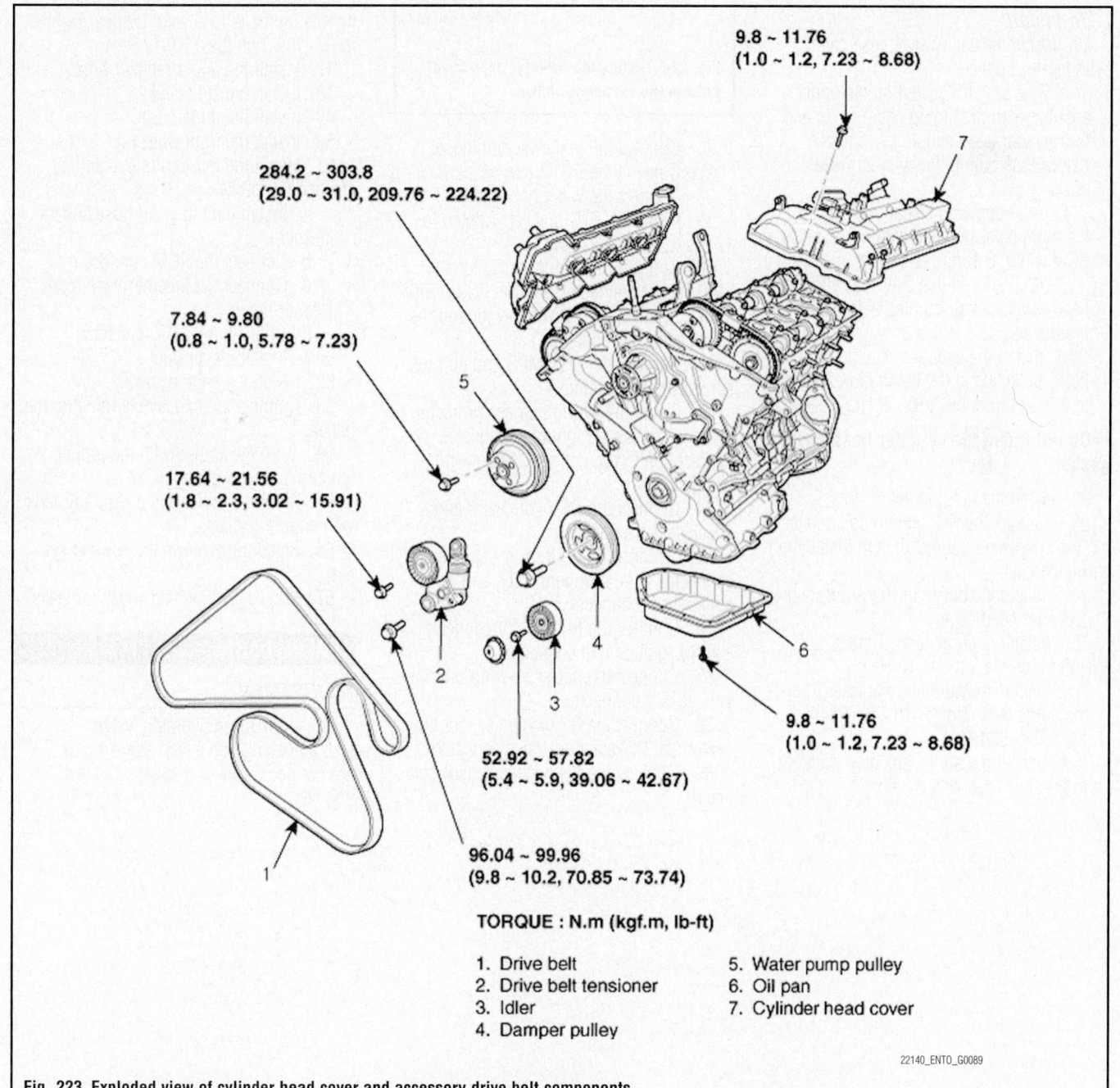

9.8 ~ 11.76
(1.0 ~ 1.2, 7.23 ~ 8.68)

284.2 ~ 303.8
(29.0 ~ 31.0, 209.76 ~ 224.22)

7.84 ~ 9.80
(0.8 ~ 1.0, 5.78 ~ 7.23)

17.64 ~ 21.56
(1.8 ~ 2.3, 3.02 ~ 15.91)

9.8 ~ 11.76
(1.0 ~ 1.2, 7.23 ~ 8.68)

52.92 ~ 57.82
(5.4 ~ 5.9, 39.06 ~ 42.67)

96.04 ~ 99.96
(9.8 ~ 10.2, 70.85 ~ 73.74)

TORQUE : N.m (kgf.m, lb-ft)

1. Drive belt
2. Drive belt tensioner
3. Idler
4. Damper pulley
5. Water pump pulley
6. Oil pan
7. Cylinder head cover

22140_ENTO_G0089

Fig. 223 Exploded view of cylinder head cover and accessory drive belt components

20. Remove the valve cover(s), also called cylinder head cover(s).

   a. Remove the connector bracket (A) from the LH cylinder head cover.

   b. Disconnect the RH ignition coil connector(A), the condenser connector (B) and remove the wiring bracket (C).

   c. Remove the LH and RH ignition coils.

   d. Remove the LH and RH cylinder head covers.

→**Cover the upside of engine head with a clean vinyl cover to prevent foreign materials from entering.**

To install:

21. Install the valve cover, also called cylinder head cover.

   a. The sealant located on the upper area between the timing chain cover and the cylinder head should be removed before assembling the cylinder head cover.

   b. After applying sealant (TB1217H), it should be assembled within 5 minutes. Bead width: 0.1 inch (2.5mm).

   c. Wait 30 minutes after the cylinder head cover was assembled before starting the engine.

   d. Install the cylinder head cover bolts as shown in the illustration. Tighten to 87–104 inch lbs. (10–12 Nm).

→**Do not reuse the cylinder head cover gasket.**

22. Install the ignition coils.

23. Connect the RH ignition coil connector, the condenser connector, and install the wiring bracket.

24. Install the connector bracket from the LH cylinder head cover.

25. Install the surge tank. Tighten to 87–104 inch lbs. (10–12 Nm).

   a. Install the connector bracket(A) on the surge tank. Tighten to 210–224 ft. lbs. (284–304 Nm).

   b. Install the surge tank stay. Tighten to 20–23 ft. lbs. (27–31 Nm).

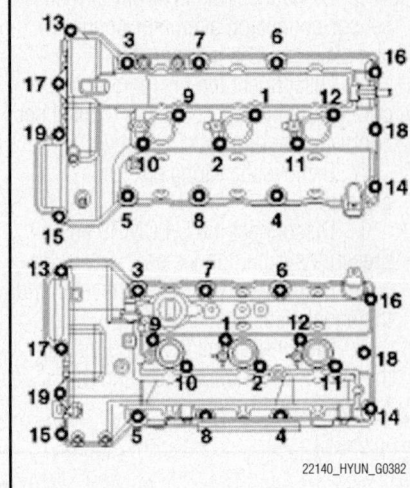

Fig. 224 Tighten the cylinder head cover bolts in the sequence shown

26. Connect the brake vacuum hose.

27. Connect the PCV hose.

28. Connect the water hoses to the ETC.

29. Install the ETC bracket. Tighten to 12–19 ft. lbs. (16–26 Nm).

30. Connect the PCSV connector, the MAP sensor connector, and the PCSV hose.

31. Connect the ETC connector and the knock sensor connector.

32. Connect the LH CMPS and oil pressure switch connector.

33. Install the wiring harness protector, and connect the LH ignition connector, injector connector, condenser connector, and ground.

34. Connect the OCV connector and knock sensor connector.

35. Connect the LH front oxygen sensor connector, alternator connector, and air compressor connector.

36. Connect the RH injector connector and the ignition coil connector.

37. Connect the power steering oil pressure sensor connector.

38. Connect the RH oxygen sensor connector and VIS solenoid valve connector.

39. Connect the RH oxygen sensor connector.

40. Install the No. 1 engine mounting through the lower position of A/C pipe line. Tighten to 36–47 ft. lbs. (49–64 Nm).

41. Install the A/C pipe bracket mounting bolt.

42. Install the engine coolant reservoir tank.

43. Install the engine mounting bracket. Tighten to 47–62 ft. lbs. (64–83 Nm).

44. Install the transaxle mounting bolts. Tighten to 36–47 ft. lbs. (49–64 Nm).

45. Remove the jack from the upper oil pan.

46. Install the lower oil pan. Refer to Oil Pan, removal & installation. Uniformly tighten the bolts in several passes. Tighten to 87–104 inch lbs. (10–12 Nm).

47. Install the upper radiator hose.

48. Install the side cover.

49. Install the undercover.

50. Install the front wheel(s).

51. Install the intake air hose and air cleaner assembly.

   a. Install the intake air hose and air cleaner.

   b. Connect the ECM connector.

   c. Connect the breather hose to air cleaner hose.

   d. Connect the AFS connector.

   e. Install the air duct.

52. Install the engine cover.

53. Connect the negative terminal to the battery.

54. Refill the engine with the proper amount and type of engine oil.

55. Refill the radiator and reservoir tank with engine coolant.

56. Bleed the air from the cooling system.

57. Run the engine and check for leaks.

VALVE LASH

ADJUSTMENT

All engines use hydraulic valve lash adjusters. Valve lash adjustments are not necessary or possible on these engines.

ENGINE PERFORMANCE & EMISSION CONTROL

COMPONENT LOCATIONS

See Figures 225 through 230.

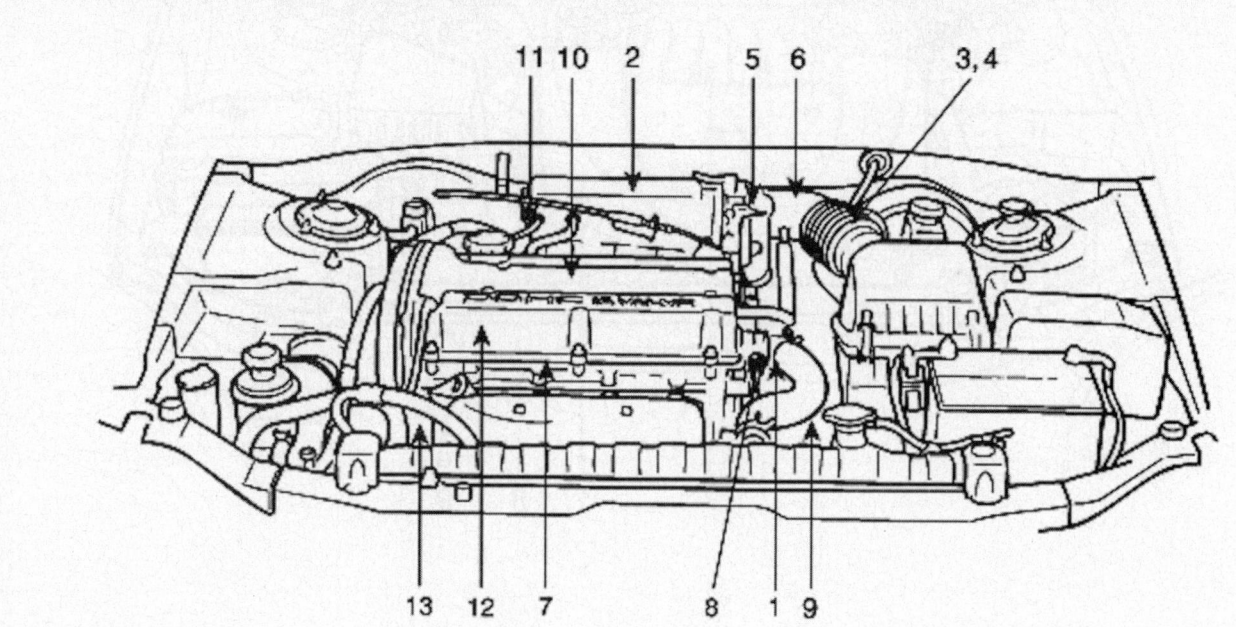

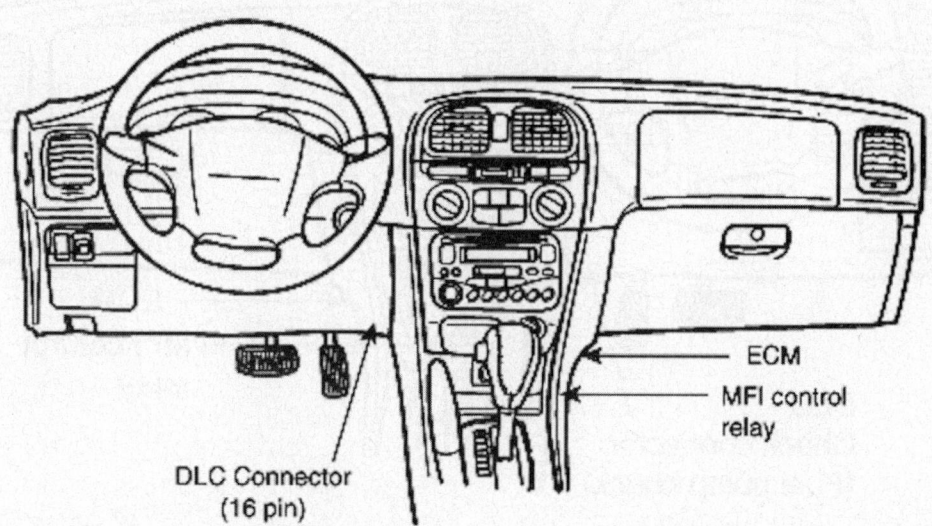

1. Engine Coolant Temperature Sensor (ECTS)
2. Manifold Absolute Pressure (MAP) Sensor
3. Mass Air Flow (MAF) sensor
4. Intake Air Temperature Sensor (ATS)
   - Bulit in MAF Sensor
5. Throttle Position Sensor (TPS)
6. Idle Speed Control Actuator (ISA)
7. Heated Oxygen Sensor (HO2S)
8. Camshaft Position (TDC) Sensor
9. Crankshaft Position (CKP) Sensor
10. Injector
11. Purge Control Solenoid Valve (PCSV)
12. Knock Sensor
13. Power Steering Oil Pressure Switch

22140_SANT_G0104

Fig. 225 Underhood and related sensor locations—2006 Santa Fe with 2.4L engine

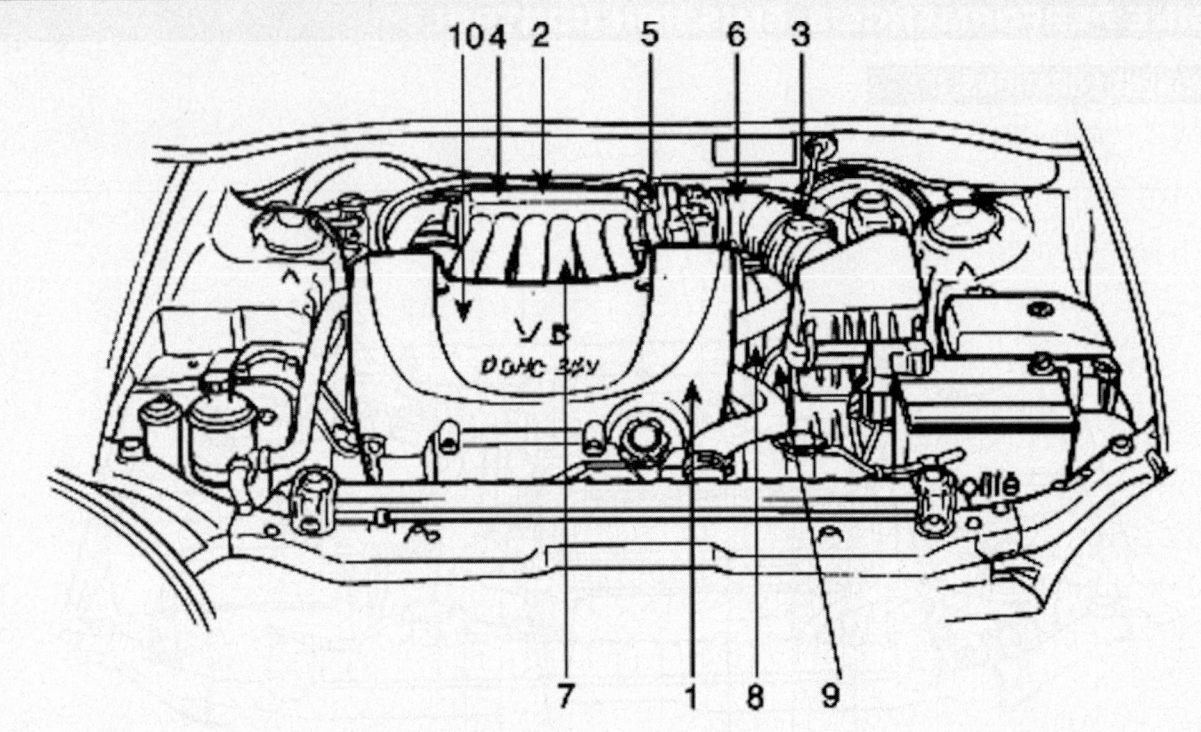

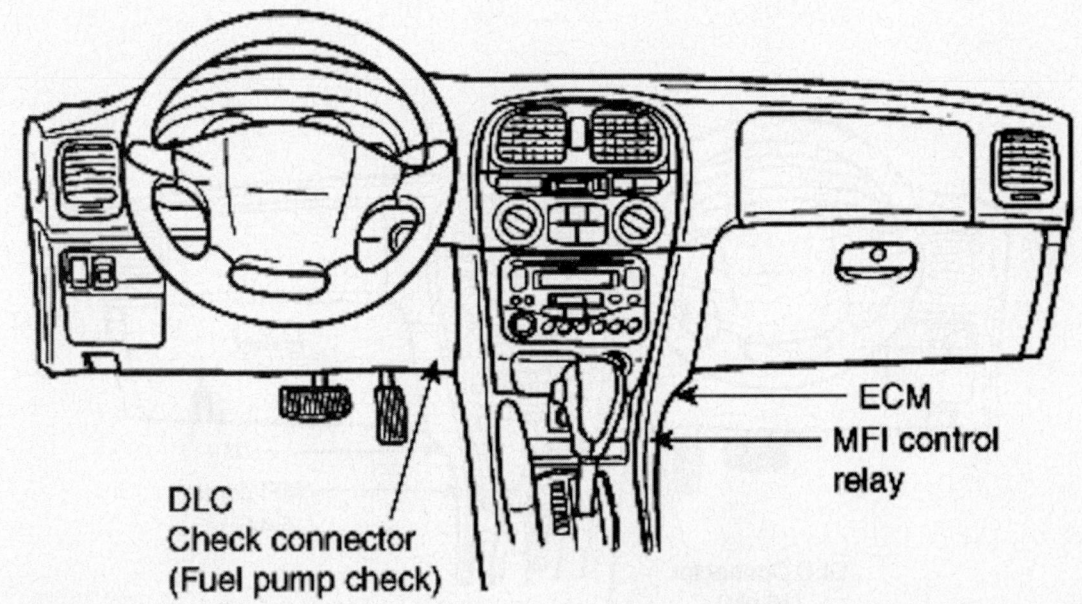

DLC
Check connector
(Fuel pump check)

ECM

MFI control
relay

1. Engine Coolant Temperature Sensor (ECTS)
2. Purge Control Solenoid Valve (PCSV)
3. Mass Air Flow (MAF) sensor
4. Intake Air Temperature Sensor (ATS)
5. Throttle Position Sensor (TPS)
6. Idle Speed Control Actuator (ISA)
7. Heated Oxygen Sensor (HO2S)
8. Camshaft Position (TDC) Sensor
9. Crankshaft Position (CKP) Sensor
10. Injector

22140_SANT_G0105

Fig. 226 Underhood and related sensor locations—2006 Santa Fe with 2.7L engine

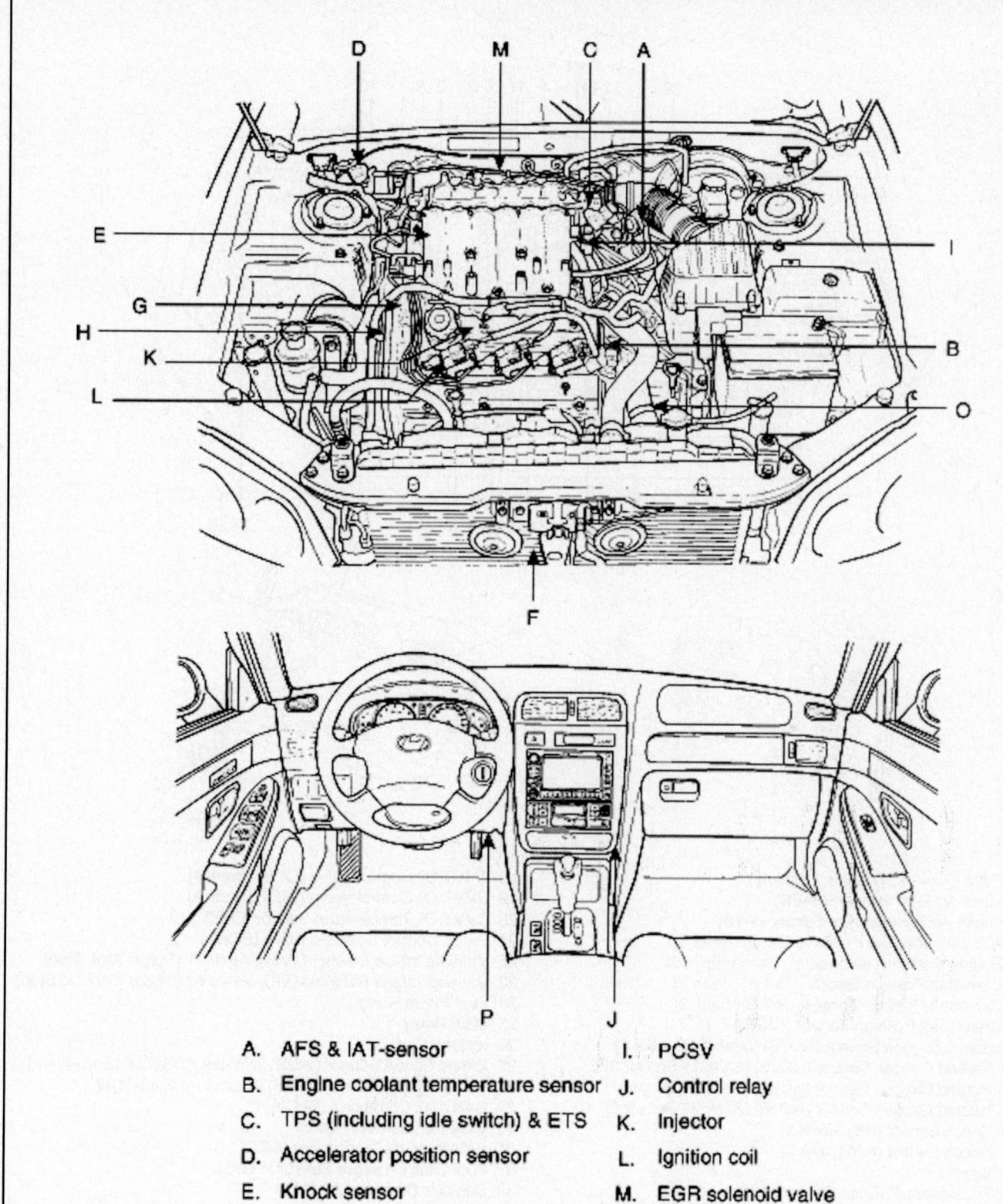

A. AFS & IAT-sensor
B. Engine coolant temperature sensor
C. TPS (including idle switch) & ETS
D. Accelerator position sensor
E. Knock sensor
F. O2 sensor
G. Crankshaft position sensor
.H. Camshaft position sensor
I. PCSV
J. Control relay
K. Injector
L. Ignition coil
M. EGR solenoid valve
N. Ignition power transistor
O. Inhibitor switch
P. DLC connector

22140_SANT_G0106

Fig. 227 Underhood and related sensor locations—2006 Santa Fe with 3.5L engine

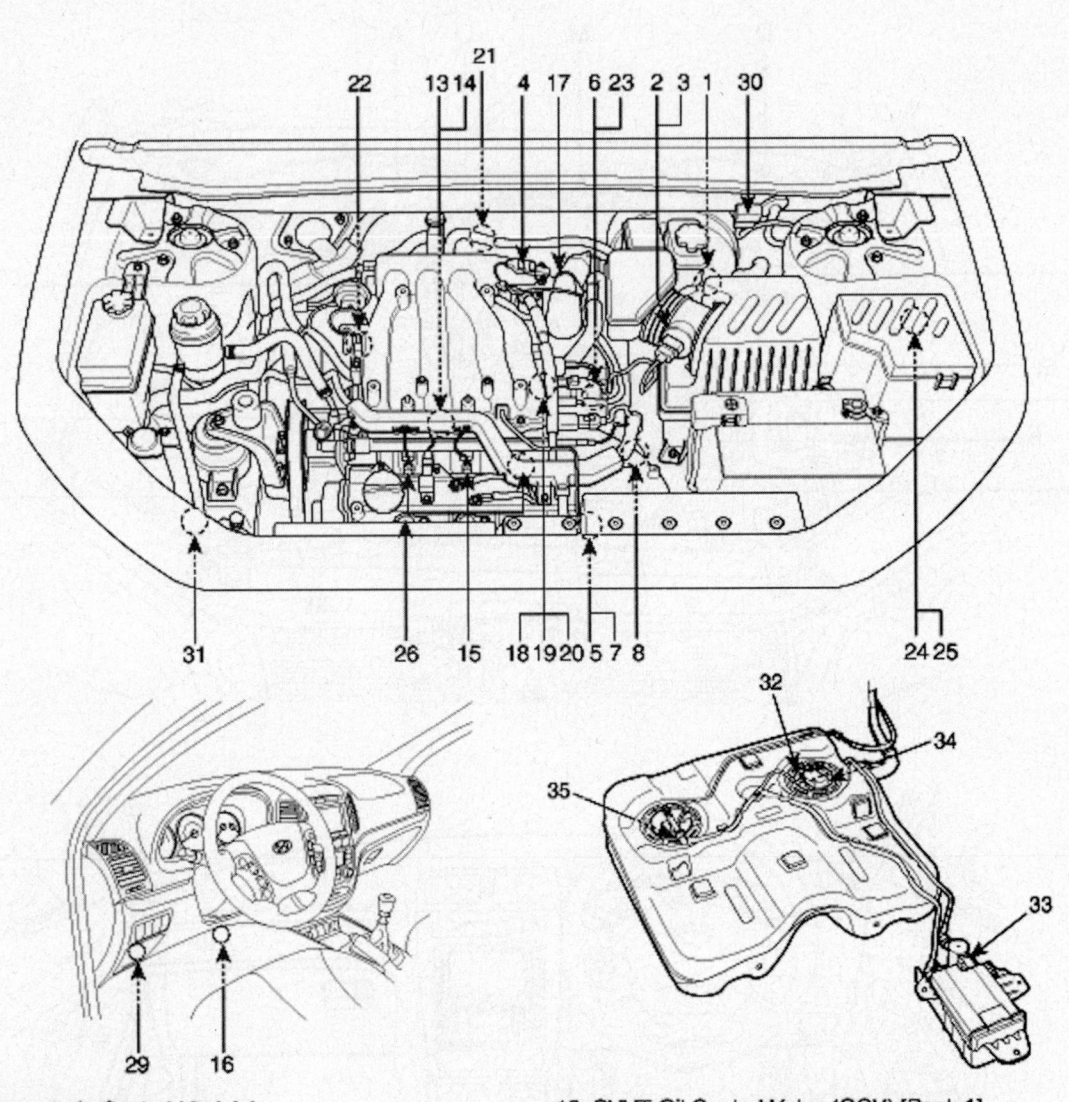

1. PCM (Powertrain Control Module)
2. Mass Air Flow Sensor (MAFS)
3. Intake Air Temperature Sensor (IATS)
4. Manifold Absolute Pressure Sensor (MAPS)
5. Engine Coolant Temperature Sensor (ECTS)
6. Camshaft Position Sensor (CMPS) [Bank 1]
7. Camshaft Position Sensor (CMPS) [Bank 2]
8. Crankshaft Position Sensor (CKPS)
9. Heated Oxygen Sensor (HO2S) [Bank 1 / Sensor 1]
10. Heated Oxygen Sensor (HO2S) [Bank 1 / Sensor 2]
11. Heated Oxygen Sensor (HO2S) [Bank 2 / Sensor 1]
12. Heated Oxygen Sensor (HO2S) [Bank 2 / Sensor 2]
13. Knock Sensor (KS) [Bank 1]
14. Knock Sensor (KS) [Bank 2]
15. Injector
16. Accelerator Position Sensor (APS)
17. ETC Module [Throttle Position Sensor (TPS) + ETC Motor]

18. CVVT Oil Control Valve (OCV) [Bank 1]
19. CVVT Oil Control Valve (OCV) [Bank 2]
20. CVVT Oil Temperature Sensor (OTS)
21. Purge Control Solenoid Valve (PCSV)
22. Variable Intake Solenoid (VIS) Valve #1 (Surge Tank Side)
23. Variable Intake Solenoid (VIS) Valve #2 (Intake Manifold Side)
24. Fuel Pump Relay
25. Main Relay
26. Ignition Coil
27. Wheel Speed Sensor (WSS) [Without ABS/ESP (Euro-III/IV)]
28. Vehicle Speed Sensor (VSS) [Except for Euro-III/IV]
29. Data Link Connector (DLC)
30. Multi-Purpose Connector
31. A/C Pressure Transducer (APT)
32. Fuel Tank Pressure Sensor (FTPS)
33. Canister Close Valve (CCV)
34. Fuel Level Sensor (FLS) 1
35. Fuel Level Sensor (FLS) 2

22140_SANT_G0107

Fig. 228 Underhood and related sensor locations—2007–08 Santa Fe with 2.7L engine

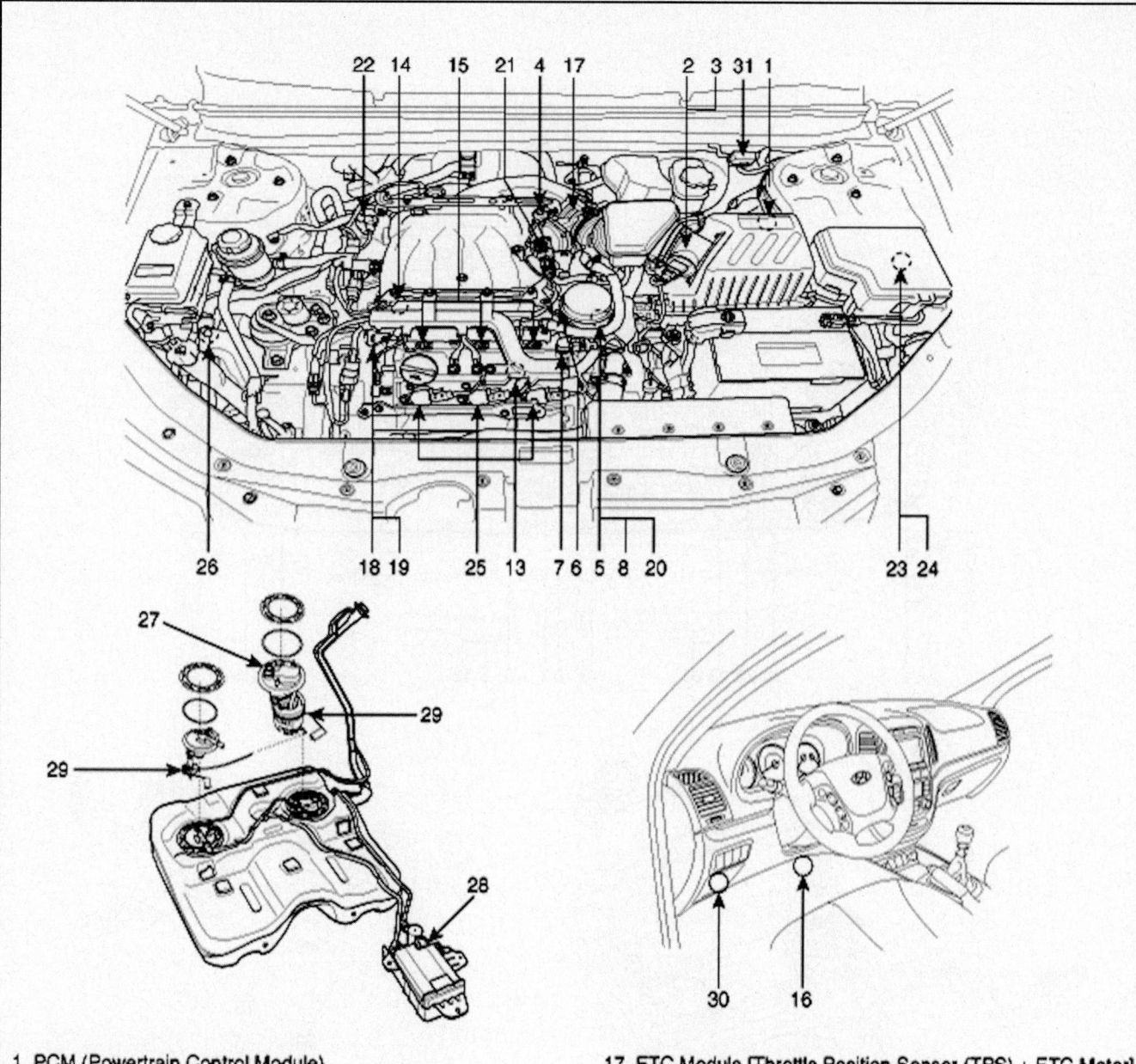

1. PCM (Powertrain Control Module)
2. Mass Air Flow Sensor (MAFS)
3. Intake Air Temperature Sensor (IATS)
4. Manifold Absolute Pressure Sensor (MAPS)
5. Engine Coolant Temperature Sensor (ECTS)
6. Camshaft Position Sensor (CMPS) [Bank 1]
7. Camshaft Position Sensor (CMPS) [Bank 2]
8. Crankshaft Position Sensor (CKPS)
9. Heated Oxygen Sensor (HO2S) [Bank 1 / Sensor 1]
10. Heated Oxygen Sensor (HO2S) [Bank 1 / Sensor 2]
11. Heated Oxygen Sensor (HO2S) [Bank 2 / Sensor 1]
12. Heated Oxygen Sensor (HO2S) [Bank 2 / Sensor 2]
13. Knock Sensor (KS) #1
14. Knock Sensor (KS) #2
15. Injector
16. Accelerator Position Sensor (APS)

17. ETC Module [Throttle Position Sensor (TPS) + ETC Motor]
18. CVVT Oil Control Valve (OCV) [Bank 1]
19. CVVT Oil Control Valve (OCV) [Bank 2]
20. CVVT Oil Temperature Sensor (OTS)
21. Purge Control Solenoid Valve (PCSV)
22. Variable Intake Solenoid (VIS) Valve
23. Fuel Pump Relay
24. Main Relay
25. Ignition Coil
26. A/C Pressure Transducer (APT)
27. Fuel Tank Pressure Sensor (FTPS)
28. Canister Close Valve (CCV)
29. Fuel Level Sensor (FLS)
30. Wheel Speed Sensor (WSS)
31. Data Link Connector (DLC)
32. Multi-Purpose Check Connector

22140_SANT_G0108

Fig. 229 Underhood and related sensor locations—2007–08 Santa Fe with 3.3L engine

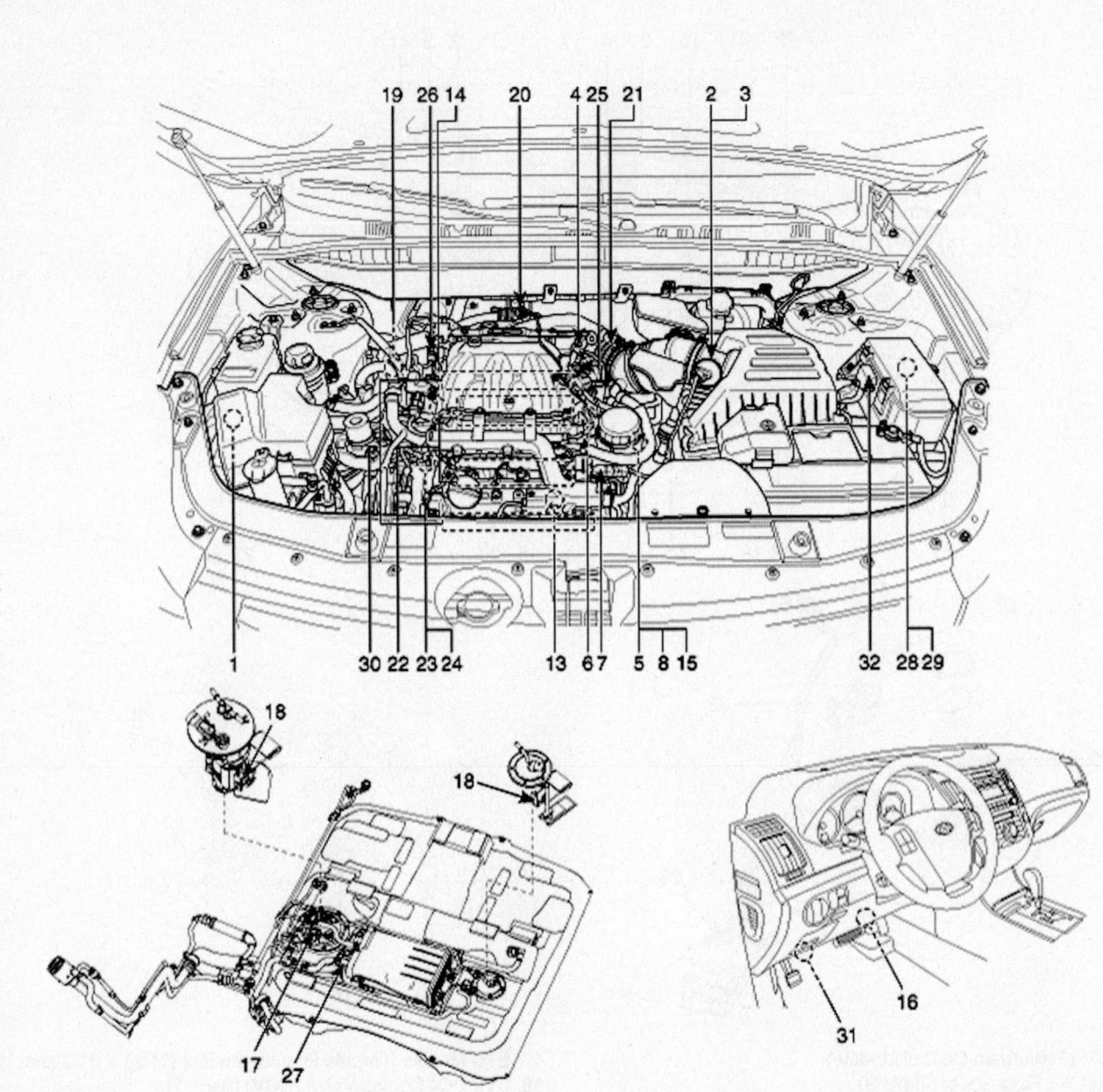

1. PCM (Powertrain Control Module)
2. Mass Air Flow Sensor (MAFS)
3. Intake Air Temperature Sensor (IATS)
4. Manifold Absolute Pressure Sensor (MAPS)
5. Engine Coolant Temperature Sensor (ECTS)
6. Camshaft Position Sensor (CMPS) [Bank 1]
7. Camshaft Position Sensor (CMPS) [Bank 2]
8. Crankshaft Position Sensor (CKPS)
9. Heated Oxygen Sensor (HO2S)
   [Bank 1 / Sensor 1]
10. Heated Oxygen Sensor (HO2S)
    [Bank 1 / Sensor 2]
11. Heated Oxygen Sensor (HO2S)
    [Bank 2 / Sensor 1]
12. Heated Oxygen Sensor (HO2S)
    [Bank 2 / Sensor 2]
13. Knock Sensor (KS) #1
14. Knock Sensor (KS) #2
15. CVVT Oil Temperature Sensor (OTS)
16. Accelerator Position Sensor (APS)
17. Fuel Tank Pressure Sensor (FTPS)
18. Fuel Level Sensor (FLS)
19. A/C Pressure Transducer (APT)
20. Power Steering Pressure Sensor (PSPS)
21. ETC Module [Throttle Position Sensor
    (TPS) + ETC Motor]
22. Injector
23. CVVT Oil Control Valve (OCV) [Bank 1]
24. CVVT Oil Control Valve (OCV) [Bank 2]
25. Purge Control Solenoid Valve (PCSV)
26. Variable Intake Solenoid (VIS) Valve
27. Canister Close Valve (CCV)
28. Main Relay
29. Fuel Pump Relay
30. Ignition Coil
31. Data Link Connector (DLC)
32. Multi-Purpose Check Connector

22140_SANT_G0109

Fig. 230 Underhood and related sensor locations—2007–08 Veracruz with 3.8L engine

ACCELERATOR PEDAL POSITION (APP) SENSOR

LOCATION

See Figure 231.

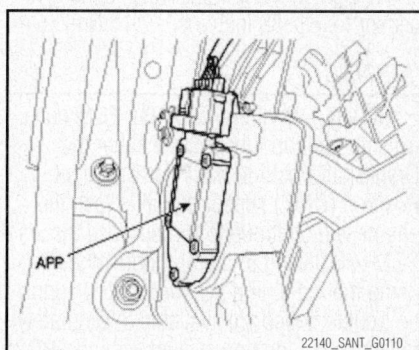

Fig. 231 Accelerator Pedal Position (APP) location

The Accelerator Pedal Position (APP) sensor is located inside the vehicle. It is part of the accelerator pedal assembly.

OPERATION

The accelerator pedal contains 2 individual Accelerator Pedal Position (APP) sensors within the assembly. The APP sensors 1 and 2 are potentiometer type sensors each with 3 circuits:

- A 5-volt reference circuit
- A low reference circuit
- A signal circuit

The APP sensors are used to determine the pedal angle. The Powertrain Control Module (PCM) provides each APP sensor with a 5-volt reference circuit and a low reference circuit. The APP sensors provide the PCM with signal voltage proportional to the pedal movement. The APP sensor 1 signal voltage at rest position is near the low reference and increases as the pedal is actuated. The APP sensor 2 signal voltage at rest position is near the 5-volt reference and decreases as the pedal is actuated.

REMOVAL & INSTALLATION

1. Before servicing the vehicle, refer to the Precautions Section.
2. Disconnect the ground cable from the battery.
3. Disconnect the Accelerator Pedal Position (APP) sensor connector.
4. Remove the nut securing accelerator pedal assembly.

To install:

5. Install in the reverse order of removal.
6. Tighten the nut securing accelerator

pedal assembly to 113–139 inch lbs. (13–16 Nm).

TESTING

See Figures 232 through 234.

The Accelerator Pedal Position (APP) sensor is installed on the accelerator pedal module and detects the rotation angle of the accelerator pedal. The APP is one of the most important sensors in engine control system, so it consists of the 2 sensors which adapt an individual sensor power and ground line. The second sensor monitors the first sensor and its output voltage is half of the first one. If the ratio of sensor 1 and 2 is out of the range (approximately ½), the diagnostic system judges that it is abnormal.

Use the following reference tables to test the APP sensor voltage and resistance values at the connector end points.

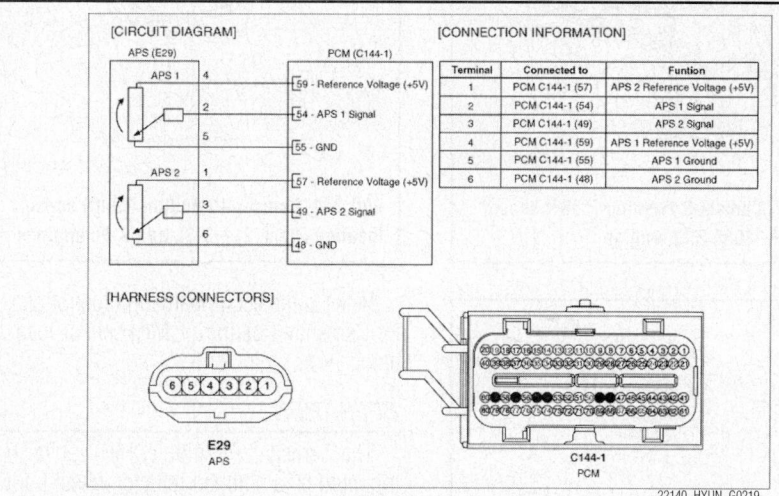

Fig. 232 Accelerator Pedal Position (APP) connector end view and circuit schematics

Pedal Position	Output Voltage (V) [Vref = 5.0V]	
	APS1	APS2
C.T	0.7 ~ 0.8V	0.29 ~ 0.46V
W.O.T	3.85 ~ 4.35V	1.93 ~ 2.18V

Fig. 233 Accelerator pedal position table—Voltage Output

Item	Sensor Resistance
APS1	0.7 ~ 1.3kΩ at 68°F (20°C)
APS2	1.4 ~ 2.6kΩ at 68°F (20°C)

Fig. 234 Accelerator pedal position circuit testing table—Sensor Resistance

CAMSHAFT POSITION (CMP) SENSOR

LOCATION

See Figures 235 through 241.

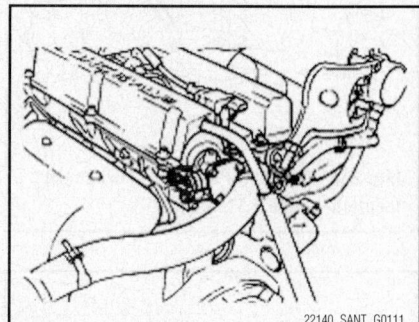

Fig. 235 Camshaft Position (CMP) sensor location—2006 2.4L engine

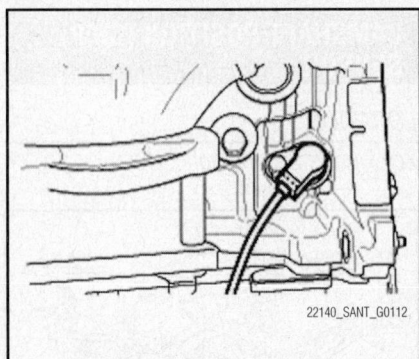

Fig. 236 Camshaft Position (CMP) sensor location—2006 2.7L engine

Fig. 237 Camshaft Position (CMP) sensor location—2006 3.5L engine

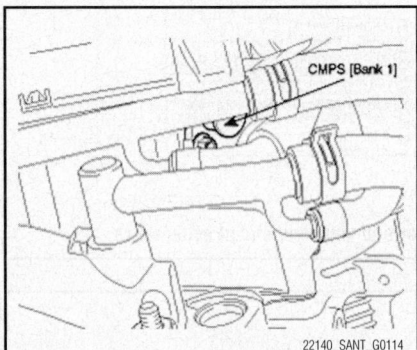

Fig. 238 Camshaft Position (CMP) sensor location (Bank 1)—2007-08 2.7L engine

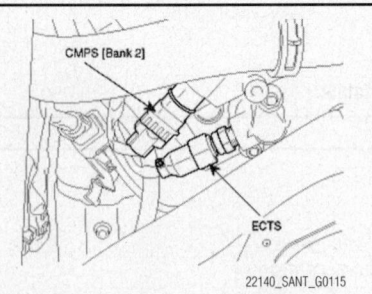

Fig. 239 Camshaft Position (CMP) sensor location (Bank 2) and Engine Coolant Temperature (ECT) Sensor—2007-08 2.7L engine

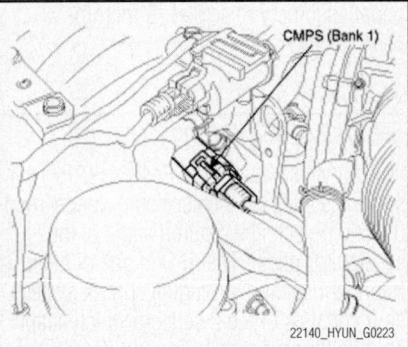

Fig. 240 Camshaft Position (CMP) sensor location (Bank 1)—3.3L and 3.8L engines

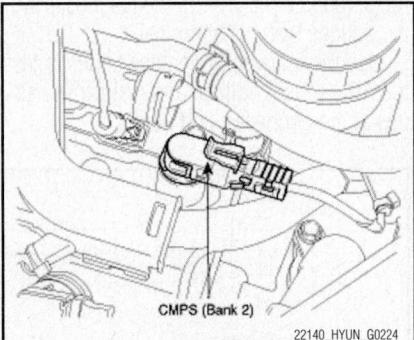

Fig. 241 Camshaft Position (CMP) sensor location (Bank 2)—3.3L and 3.8L engines

Refer to the accompanying illustrations for Camshaft Position (CMP) sensor location.

OPERATION

The Camshaft Position (CMP) sensor is triggered by a notched reluctor wheel built onto the camshaft sprocket. The CMP sensor is connected to the PCM by the following circuits:

- A 5-volt circuit
- A low reference circuit
- A signal circuit

The CMP sensor is a Hall-effect sensor and detects the camshaft position by using a Hall element. It is related with the Crankshaft Position (CKP) sensor and detects the piston position of each cylinder which the CKP sensor can't detect. This sensor has a Hall-effect IC which changes in output voltage when a magnetic field is made on the IC with the current flow.

REMOVAL & INSTALLATION

1. Before servicing the vehicle, refer to the Precautions Section.
2. Disconnect the negative battery cable.
3. Disconnect the connector from the CMP sensor.

4. Remove the bolt that retains the CMP sensor.
5. Remove the CMP sensor.

To install:

6. Installation is the reverse of the removal procedure.
7. Tighten the bolt that retains the CMP sensor to: 61-86 inch lbs. (7-10 Nm).

TESTING

During normal operation the PCM controls all ignition functions. If either the Crankshaft Position (CKP) or Camshaft Position (CMP) sensor signal is lost, the engine will continue to run because the PCM will default to a limp home mode using the remaining sensor input. Diagnostic trouble codes are available to accurately diagnose the ignition system with an OBD2 scan tool.

1. Inspect the CMP sensor for correct installation. Remove the CMP sensor from the engine and inspect the sensor O-ring for damage. If the sensor is loose, incorrectly installed, or damaged, replace the CMP sensor.
2. Engage the CMP sensor harness connector to the CMP sensor.
3. Connect the scan tool to the diagnostic connector.
4. With the ignition **ON**, engine OFF observe the CMP active counter parameter on the scan tool.
5. Pass a flat steel object across the tip of the sensor repeatedly. The CMP active counter parameter should increment with each pass of the steel object.
6. If the parameter does not increment, replace the CMP sensor.

COOLANT TEMPERATURE SENSOR

Refer to Engine Coolant Temperature (ECT) Sensor

CRANKSHAFT POSITION (CKP) SENSOR

LOCATION

See Figures 242 through 246.

OPERATION

The crankshaft position sensor is a Hall Effect type sensor that generates voltage using a sensor and a target wheel mounted on the crankshaft. There are 58 slots in the target wheel where one is longer than the others. When the slot in the wheel aligns with the sensor, the sensor voltage output is

Fig. 242 Crankshaft Position (CKP) sensor location—2006 2.4L engine

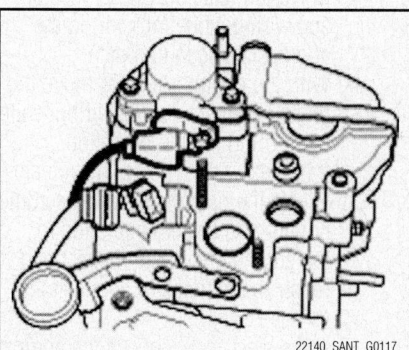

Fig. 243 Crankshaft Position (CKP) sensor location—2006 2.7L engine

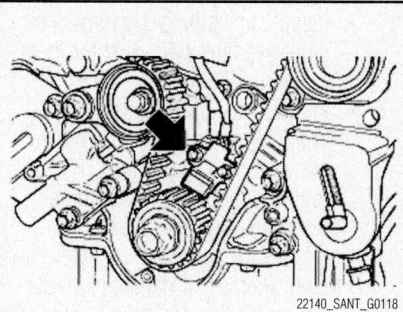

Fig. 244 Crankshaft Position (CKP) sensor location—2006 3.5L engine

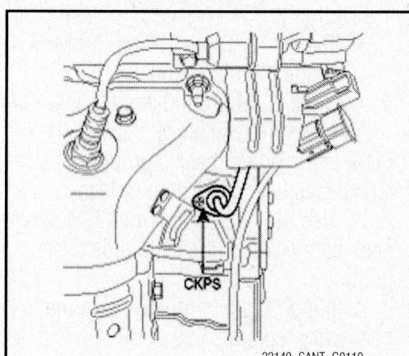

Fig. 245 Crankshaft Position (CKP) sensor location—2007–08 2.7L engine

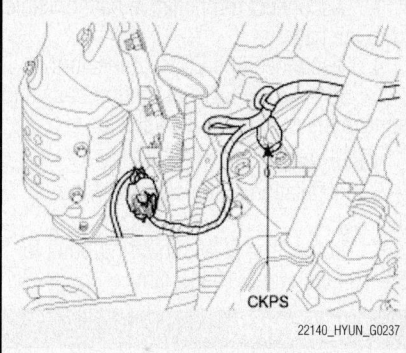

Fig. 246 Crankshaft Position (CKP) sensor location—3.3L and 3.8L engines

low. When the metal tooth in the wheel aligns the sensor, the sensor voltage is high. During one crankshaft rotation there are 58 rectangular signals and one longer signal. The PCM calculates engine RPM by using the sensor's signal and controls the injection duration and ignition timing. Using the signal differences caused by the longer slot, the PCM identifies which cylinder is at TDC.

REMOVAL & INSTALLATION

1. Before servicing the vehicle, refer to the Precautions Section.
2. Disconnect the negative battery cable.
3. Disconnect the connector from the sensor.
4. Remove the bolt that retains the sensor in place.
5. Remove the sensor from its mounting.

To install:

6. Installation is the reverse of the removal procedure.

7. Tighten the sensor retaining bolt to: 70–104 inch lbs. (8–12 Nm).

TESTING

During normal operation the PCM controls all ignition functions. If either the Crankshaft Position (CKP) or Camshaft Position (CMP) sensor signal is lost, the engine will continue to run because the PCM will default to a limp home mode using the remaining sensor input. Diagnostic trouble codes are available to accurately diagnose the ignition system with an OBD2 scan tool.

1. Inspect the CKP sensor for correct installation. Remove the CKP sensor from the engine and inspect the sensor O-ring for damage. If the sensor is loose, incorrectly installed, or damaged, replace the CKP sensor.
2. Engage the CKP sensor harness connector to the CKP sensor.
3. Connect the scan tool to the diagnostic connector.
4. With the ignition **ON**, engine OFF observe the CKP active counter parameter on the scan tool.
5. Pass a flat steel object across the tip of the sensor repeatedly. The CKP active counter parameter should increment with each pass of the steel object.
6. If the parameter does not increment, replace the CKP sensor.

Component Inspection

See Figure 247.

With a scan tool, check the signal waveform of the Crankshaft Position (CKP) and the Camshaft Position (CMP) sensor. The waveform should appear as the following illustration.

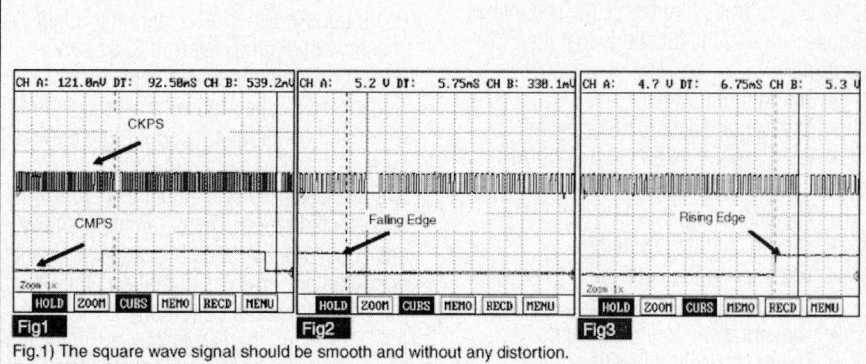

Fig.1) The square wave signal should be smooth and without any distortion.
Fig.2,3)The CMPS falling(rising) edge is coincided with 3~5 tooth of the CKP from one
   longer signal(missing tooth)

Fig. 247 Waveform graph of the CKP and the CMP sensors

ELECTRIC FAN SWITCH

The engine cooling fan is operated by a relay that is controlled through the Engine Control Module (ECM). The cooling fan relay will activate depending upon the needs of one or more systems including the cooling system, the air conditioning, and the heating systems.

ELECTRONIC CONTROL MODULE (ECM)

LOCATION

The Electronic Control Module (ECM) is located in the engine compartment near the air filter assembly or on the passenger's side near the front of the vehicle. In some models, the ECM is located near the passenger front seat under the glove box area.

OPERATION

This vehicle has electronic controls to reduce exhaust emissions while maintaining excellent drivability and fuel economy. The Electronic Control Module (ECM) is the control center of this system. The ECM monitors numerous engine and vehicle functions. The ECM constantly looks at the information from various sensors and other inputs, and controls the systems that affect vehicle performance and emissions. The ECM also performs the diagnostic tests on various parts of the system. The ECM can recognize operational problems and alert the driver via the Malfunction Indicator Lamp (MIL). When the ECM detects a malfunction, the ECM stores a Diagnostic Trouble Code (DTC). The problem area is identified by the particular DTC that is set. The control module supplies a buffered voltage to various sensors and switches. Review the components and wiring diagrams in order to determine which systems are controlled by the ECM. The following are some of the functions that the ECM controls:

- The ignition system
- The Knock Sensor (KS) system
- The Evaporative Emissions (EVAP) system
- The alternator
- The A/C and fan systems
- The cooling fan control
- The fuel injection system
- The emission control systems
- The on-board diagnostics

REMOVAL & INSTALLATION

1. Before servicing the vehicle, refer to the Precautions Section.
2. Turn ignition switch off.

3. Disconnect the negative battery cable from the battery.
4. Disconnect the ECM connectors.
5. Remove the ECM mounting bolts and remove the ECM from the vehicle.

To install:

6. Installation is the reverse of the removal.
7. Tighten the ECM mounting bolts to: 86–104 inch lbs. (10–12 Nm).

✳✳ WARNING

When replacing the ECM, be careful to use the right part number, as damage to the injection system could occur.

TESTING

1. Perform a careful underhood inspection when performing any diagnostic procedure or diagnosing the cause of an emission test failure. This can often lead to repairing a condition without further steps. Use the following guidelines when performing an inspection:

a. Inspect all of the vacuum hoses for correct routing, pinches, cuts, or disconnects.

b. Inspect any hoses that are difficult to see.

c. Inspect all of the wires in the engine compartment for the following conditions:

- Burned or chafed spots
- Pinched wires
- Contact with sharp edges
- Contact with hot exhaust manifold

The Electronic Control Module (ECM), also called the Engine Control Module (ECM), is programmed with test routines that test the operation of the various systems the ECM controls. Some tests monitor internal ECM functions. Many tests are run continuously. Other tests run only under specific conditions, referred to as conditions for running the Diagnostic Trouble Code (DTC). When the vehicle is operating within the conditions for running a particular test, the ECM monitors certain parameters and determines if the values are within an expected range. The parameters and values considered outside the range of normal operation are listed as conditions for setting the DTC. When the conditions for setting the DTC occur, the ECM executes the action taken when the DTC sets. Some DTC's alert the driver via the Malfunction Indicator Lamp (MIL) or a message. Other DTC's do not trigger a driver warning, but are stored in memory. The ECM also saves data and input parameters when most DTC's are set.

The DTC's are categorized by type. The DTC type is determined by the MIL operation and the manner in which the fault data is stored when a particular DTC fails. In some cases, there may be exceptions to this structure. Therefore, when diagnosing the system it is important to read the action taken when the DTC sets and the conditions for clearing the DTC.

Many intermittent open or shorted circuits come and go with harness and connector movement caused by vibration, engine torque, bumps, or rough pavement.

2. Test the wiring harness and connectors by performing the following:
- Move the related ECM connectors and wiring while monitoring the appropriate scan tool data
- With the engine running, move the related connectors and wiring while monitoring engine operation
- If harness or connector movement affects the data displayed, the component and system operation, or the engine operation, inspect and repair the harness or connections as necessary

3. Test the electrical connections and/or wiring by performing the following:
- Inspect for incorrect mating of the connector halves or terminals not fully seated in the connector body
- Inspect for improperly formed or damaged terminals. Test for incorrect terminal tension
- Inspect for poor terminal to wire connections including terminals crimped over insulation. This requires removing the terminal from the connector body
- Inspect for corrosion or water intrusion. Pierced or damaged insulation can allow moisture to enter the wiring. The conductor can corrode inside the insulation with little visible evidence. Look for swollen and stiff sections of wire in the suspect circuits
- Inspect for wires that are broken inside the insulation

4. Test the ECM ground circuit:
a. Measure resistance between the ECM and a good chassis ground. Resistance specification: 1 ohm or less.
b. Use the backside of the ECM harness connector as the ECM side check point.
c. If a problem is found, make the necessary repairs.

5. If a problem is not located through the above procedures, the ECM may be faulty.

6. Hyundai recommends to replace the ECM with a new one and then check the vehicle again. If the vehicle operates normally, then the problem was likely with the ECM.

7. Retest the original ECM, if possible, by installing it into a known-good vehicle and check vehicle operations.

ENGINE COOLANT TEMPERATURE (ECT) SENSOR

LOCATION

The Engine Coolant Temperature (ECT) sensor is located in the engine coolant passage near the thermostat.

OPERATION

The Engine Coolant Temperature (ECT) sensor uses a thermistor whose resistance changes with the temperature in order to detect the engine coolant temperature. The electrical resistance of the ECT decreases as the temperature increases, and increases as the temperature decreases. The 5-volt reference in the PCM is supplied to the ECT via a resistor in the PCM. The resistor in the PCM and the thermistor in the ECT are connected in series. When the resistance value of the thermistor in the ECT changes according to the engine coolant temperature, the output voltage also changes. During cold engine operation, the PCM increases the fuel injection duration and controls the ignition timing using the information of engine coolant temperature to avoid engine stalling and improve drivability.

REMOVAL & INSTALLATION

1. Before servicing the vehicle, refer to the Precautions Section.
2. Drain the coolant to a level below the bottom of the sensor.
3. Disconnect the ground cable from the battery and then remove the sensor connector.
4. Remove the Engine Coolant Temperature (ECT) sensor.

To install:
5. Reverse the removal procedure.
6. Tighten the ECT sensor to 15–29 ft. lbs. (20–39 Nm).

TESTING

See Figures 248 and 249.

➡**Clear Diagnostic Trouble Codes (DTC's) if applicable.**

1. Inspect the Engine Coolant Temperature (ECT) sensor electrical connection. Check the terminal for poor connections,

Temperature [°C(°F)]	Resistance (kΩ)
-40(-40)	48.14
-20(-4)	14.13 ~ 16.83
0(32)	5.79
20(68)	2.31 ~ 2.59
40(104)	1.15
60(140)	0.59
80(176)	0.32

22140_HYUN_G0234

Fig. 248 Resistance chart specifications

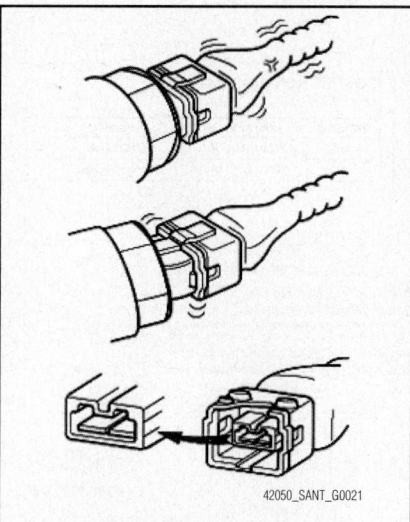

42050_SANT_G0021

Fig. 249 Inspecting the Engine Coolant Temperature (ECT) sensor connector

loose wires, bent, broken, or corroded pins, and then verify that the connector is securely fastened.
2. Turn ignition switch OFF.
3. Disconnect ECT sensor connector.
4. Remove the ECT sensor.
5. After immersing the thermistor of the sensor into engine coolant, measure the resistance between the ECT terminals 1 and 3.
6. Check that the resistance is within the specification according to the following chart.
7. Replace the ECT sensor if the component is not within specification.

ENGINE OIL TEMPERATURE (EOT) SENSOR

LOCATION

See Figure 250.

Refer to the accompanying illustration for Engine Oil Temperature (EOT) sensor location.

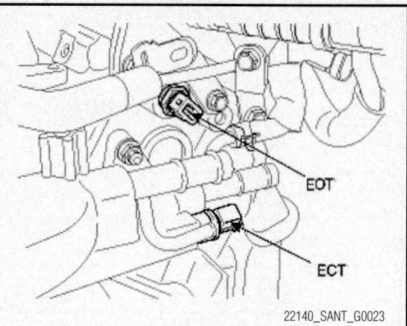

22140_SANT_G0023

Fig. 250 Engine Coolant Temperature (ECT) sensor and Engine Oil Temperature (EOT) sensor locations

OPERATION

See Figure 251.

The Engine Oil Temperature (EOT) sensor is a negative coefficient thermistor used by the PCM to measure engine oil temperature for the purpose of adjusting Continuously Variable Valve Timing (CVVT) calculations.

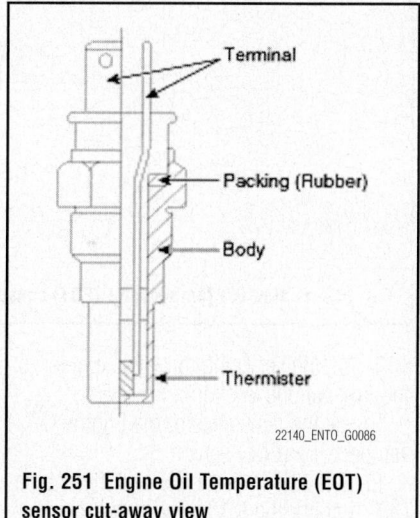

22140_ENTO_G0086

Fig. 251 Engine Oil Temperature (EOT) sensor cut-away view

REMOVAL & INSTALLATION

1. Before servicing the vehicle, refer to the Precautions Section.
2. Disconnect the ground cable from the battery and then remove the sensor connector.
3. Remove Engine Oil Temperature (EOT) sensor.

To install:
4. Install the EOT sensor and tighten to 15–29 ft. lbs. (20–39 Nm).
5. Connect the negative cable to the battery.
6. Run the engine and check for leaks.

TESTING

See Figures 252 and 253.

Temperature		Resistance(kΩ)
°C	°F	
-20	-4	16.52
20	32	2.45
80	176	0.29

22140_ENTO_G0087

Fig. 252 Engine Oil Temperature (EOT) sensor Temperature and Resistance Table

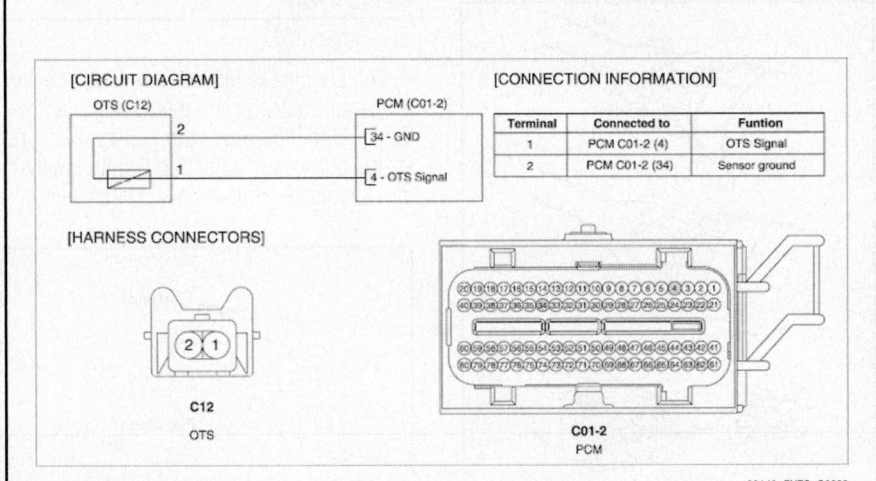

22140_ENTO_G0088

Fig. 253 Engine Oil Temperature (EOT) sensor connector end view and schematic diagram

1. Before servicing the vehicle, refer to the Precautions Section.

Check the resistance of the Engine Oil Temperature (EOT) sensor.

2. With the ignition **ON** and the engine OFF, monitor the Oil Temperature parameter on a scan tool.

3. With the ignition **OFF** and the engine OFF, disconnect EOT sensor connector.

4. Measure resistance between terminal 1 and 2 of EOT sensor connector (Component Side).

5. The specifications should be as listed in the following table.

6. If the EOT sensor does not meet the specifications, replace the EOT sensor.

FUEL LEVEL SENDING UNIT

LOCATION

See Figure 254.

The fuel level sending unit is located on the fuel pump assembly module.

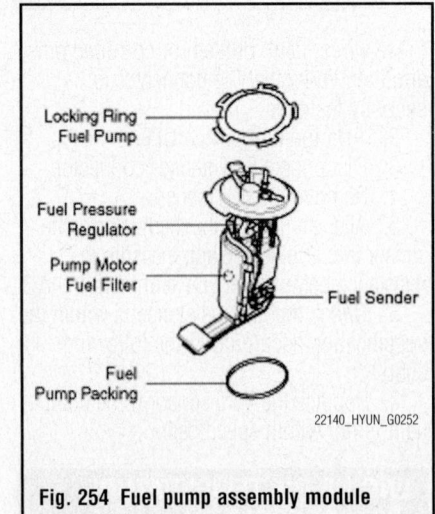

22140_HYUN_G0252

Fig. 254 Fuel pump assembly module

OPERATION

The fuel level sending unit is attached to the side of the fuel pump module. The fuel level sensor utilizes a variable resistor connected to a float that is buoyed up or down by the level of the fuel in the fuel tank.

REMOVAL & INSTALLATION

The fuel level sending unit is integral with the fuel pump module assembly. Refer to Fuel Pump, removal & installation.

TESTING

1. Before servicing the vehicle, refer to the Precautions Section.

2. Check the fuel gauge in the instrument cluster panel for an accurate reading of the fuel level in the tank.

3. If the fuel gauge sticks at one extreme of the meter, no matter the fuel level in the tank, check for a break or short in the wiring from the fuel level sending unit.

4. Check the fuel level sending unit by removing the fuel pump assembly. Refer to Fuel Pump, removal and installation.

5. Using an ohmmeter, measure the resistance between the terminals while moving the float up and down. If the resistance does not change smoothly, replace the fuel level sending unit.

HEATED OXYGEN (HO2S) SENSOR

LOCATION

The Heated Oxygen Sensors (HO2S) are located in the exhaust system. On some vehicles, one sensor is located up at the exhaust manifold(s) and the other sensor is located down at the catalytic converter.

OPERATION

The Heated Oxygen Sensor (HO2S) consists of zirconium and alumina and is installed upstream and downstream of the Manifold Catalyst Converter (MCC). After it compares the oxygen consistency of the atmosphere with the exhaust gas, it transfers the oxygen consistency of the exhaust gas to the PCM. This sensor operates most efficiently when the temperature of the sensor tip is higher than 698°F (370°C). Therefore, a heater has been integrated with the oxygen sensor that is controlled by the PCM duty signal. When the exhaust gas temperature is lower than the specified value, the heater warms the sensor tip. The PCM commands the heater ON or OFF to maintain a specific HO2S operating temperature range. The PCM monitors the voltage on the HO2S heater low control circuit for heater fault diagnosis. If the PCM detects that the HO2S heater low control circuit voltage is not within a specified range, a DTC is set.

The HO2S supplies the electronic control assembly with a signal indicating either a rich or lean mixture condition as the engine operates.

REMOVAL & INSTALLATION

1. Before servicing the vehicle, refer to the Precautions Section.

2. Disconnect the electrical connector from the sensor.

3. Remove the Heated Oxygen Sensor (HO2S).

To install:

4. Installation is the reverse of the removal procedure.

➡**Apply anti-seize compound to the threaded portion of the sensor, prior to installation. Never apply anti-seize compound to the protector of the sensor.**

TESTING

See Figures 255 through 258.

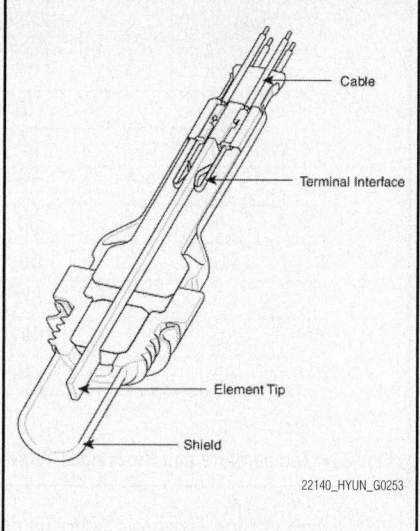

Fig. 257 Heated oxygen sensor cut-away view

1. Perform a visual inspection of the sensor as follows:

   a. If the sensor tip has a black/sooty deposit, this may indicate a rich fuel mixture.

   b. If the sensor tip has a white, gritty deposit, this may indicate an internal coolant leak.

   c. If the sensor tip has a brown deposit, this could indicate oil consumption.

2. Check that the heater resistance is according to specification.

3. Check that the output voltage in the rich/lean readings fluctuate according to specifications.

4. Connect an OBD2 scan tool to the Data Link Connector (DLC).

5. With the engine running, the normal waveform for the Heated Oxygen Sensor (HO2S) and the HO2S heater should be similar to those represented in the following waveform illustration.

INTAKE AIR TEMPERATURE (IAT) SENSOR

LOCATION

See Figure 259.

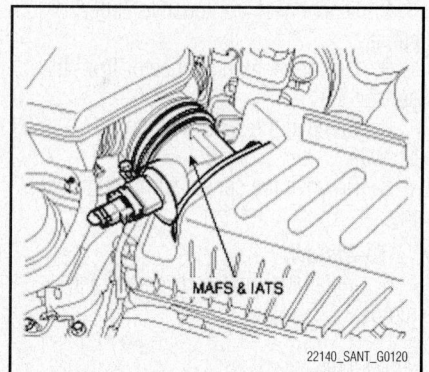

Fig. 259 Intake Air Temperature (IAT) sensor and Mass Air Flow (MAF) sensor location

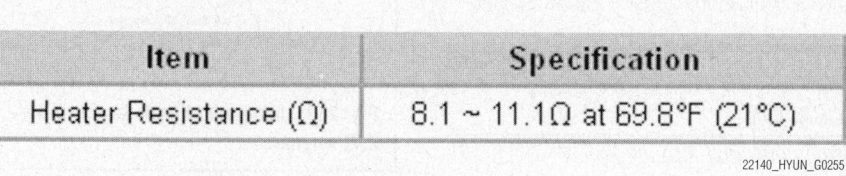

Item	Specification
Heater Resistance (Ω)	8.1 ~ 11.1Ω at 69.8°F (21°C)

Fig. 255 Heated oxygen sensor heater resistance specifications

A/F Ratio	Output Voltage (V)
RICH	0.75 ~ 1.00V
LEAN	0 ~ 0.12V

Fig. 256 Heated oxygen sensor rich/lean output voltage specifications

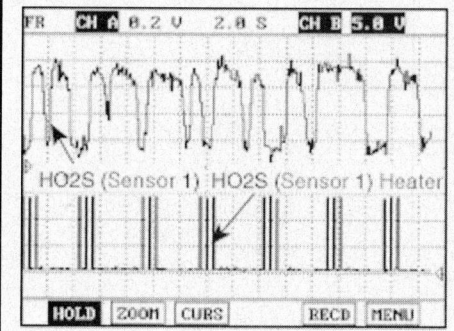

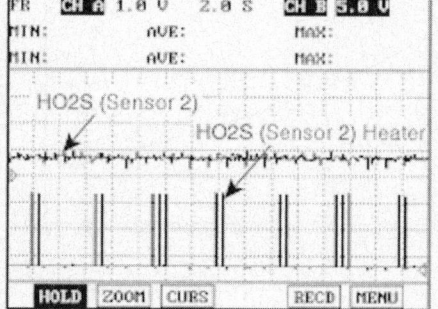

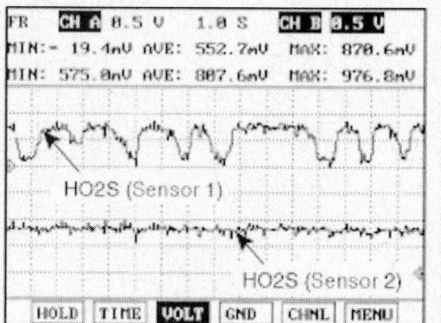

Fig. 258 Waveform of heated oxygen sensors and heater element

The Intake Air Temperature (IAT) sensor is mounted in the intake air hose of the air cleaner assembly. The IAT is integrated inside the Mass Air Flow (MAF) sensor.

OPERATION

See Figure 260.

The Intake Air Temperature (IAT) sensor is a variable resistor that measures the temperature of the air entering the engine intake manifold. The PCM supplies 5 volts to the IAT signal circuit and a ground for the IAT low reference circuit. When the sensor is cold, the resistance is greater. This results in a greater voltage on the signal circuit that is interpreted by the PCM as a colder IAT. As the sensor becomes warmer, the resistance decreases. This results in a lesser voltage on the IAT signal circuit that is interpreted by the PCM as a warmer IAT. If the PCM detects an IAT sensor signal voltage that is not within a calibrated range of the IAT sensor 1 signal voltage, a DTC is set.

REMOVAL & INSTALLATION

1. Before servicing the vehicle, refer to the Precautions Section.
2. Disconnect the negative battery cable.
3. Disconnect the connector from the sensor.
4. Remove the sensor retaining screws.
5. Remove the sensor from its mounting.

To install:

6. Installation is the reverse of the removal procedure.

TESTING

See Figure 261.

Temperature		Resistance (kΩ)
°C	°F	
-40	-40	100.87
-20	-4	28.58
0	32	9.40
10	50	5.66
20	68	3.51
40	104	1.47
60	140	0.67
80	176	0.33

22140_HYUN_G0258

Fig. 261 Temperature and Resistance Table

1. Disconnect the Intake Air Temperature (IAT) sensor connector.
2. Measure the resistance between the signal terminal and ground terminal of the IAT connector and compare the resistance to the following Temperature and Resistance Table.
3. If the IAT sensor fails to meet the resistance specifications, replace the IAT sensor.

KNOCK SENSOR (KS)

LOCATION

See Figures 262 through 265.

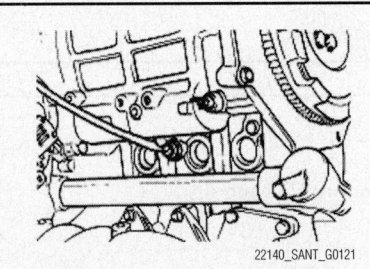

22140_SANT_G0121

Fig. 262 Location of Knock Sensor (KS)— 2.4L engine

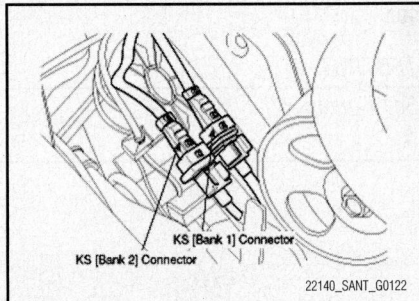

KS [Bank 1] Connector
KS [Bank 2] Connector

22140_SANT_G0122

Fig. 263 Location of Knock Sensor (KS)— 2.7L engine

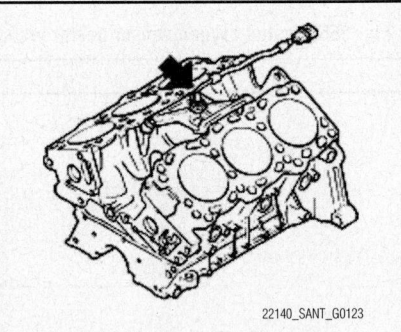

22140_SANT_G0123

Fig. 264 Location of Knock Sensor (KS)— 3.5L engine

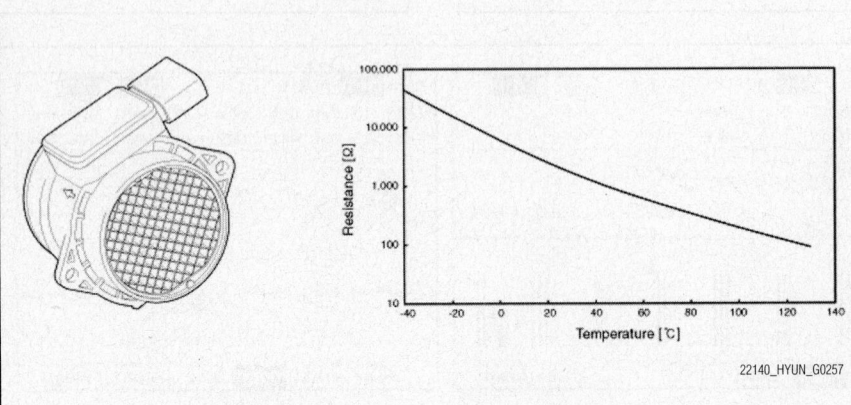

22140_HYUN_G0257

Fig. 260 Resistance decreases as the air temperature increases

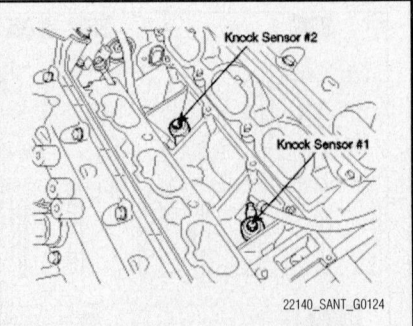

Knock Sensor #2
Knock Sensor #1

22140_SANT_G0124

Fig. 265 Location of Knock Sensor (KS) No. 1 and No. 2—3.3L and 3.8L engines

The knock sensor is located in the side of the cylinder block.

OPERATION

See Figure 266.

The Knock Sensor (KS) is used to detect engine vibrations caused by pre-ignition or detonation and provides information to the PCM. The PCM will then retard the timing in an attempt to eliminate detonation.

REMOVAL & INSTALLATION

1. Before servicing the vehicle, refer to the Precautions Section.
2. Disconnect the negative battery cable.
3. Disconnect the sensor connector.
4. Remove the sensor from its mounting.

To install:

5. Installation is the reverse of the removal procedure.
6. Tighten the sensor to 12–17 ft. lbs. (16–24 Nm).

TESTING

1. Disconnect the Knock Sensor (KS) electrical connector.
2. Connect an ohmmeter.
3. Measure the resistance between terminals 1 and 2. Specification should be 5mohm at 68°F (20°C).
4. If not within specification, replace the effected KS.

MALFUNCTION INDICATOR LIGHT (MIL)

RESET PROCEDURES

1. Proper operation of the Malfunction Indicator Light (MIL):
   - The MIL will illuminate with the ignition switch ON and the engine OFF
   - The MIL will turn OFF when the engine is started
   - The MIL will remain ON if the self-diagnostic system has detected a malfunction
   - The MIL may turn OFF if the malfunction is no longer present
   - If the MIL is illuminated and then the engine stalls, the MIL will remain illuminated as long as the ignition switch is ON
   - If the MIL is not illuminated and the engine stalls, the MIL will not illuminate until the ignition switch is cycled OFF, then ON
2. Resetting the MIL:
   - The control module turns OFF the MIL after 3 consecutive ignition cycles that the diagnostic system runs and does not fail
   - The control module turns OFF the MIL after a current Diagnostic Trouble Code (DTC) clears when the diagnostic cycle runs and passes
   - There may still be a history of DTC's

stored in the system. These will clear after 40 consecutive warm-up cycles, if no failures are reported by any other related diagnostic system
- Manual resetting of the MIL and any DTC stored in the system, requires the use of an OBD2 scan tool connected to the Data Link Connector (DLC) for communication with the vehicle. Follow the instructions of the scan tool for both retrieval and resetting of DTC's.

➡️**If the error symptoms causing the MIL to illuminate have been corrected, the MIL will return to normal operation.**

MASS AIR FLOW (MAF) SENSOR

LOCATION

See Figure 259.

The Mass Air Flow (MAF) sensor is mounted in the intake air hose of the air cleaner assembly. This sensor is combined with the Intake Air Temperature (IAT) sensor.

OPERATION

The Mass Air Flow Sensor (MAF) sensor is a hot-film type sensor and is located in between the air cleaner and the throttle body. It consists of a tube, a sensor assembly, and honeycomb cell. It detects the intake air quantity flowing into the intake manifold. Air flows from the air cleaner assembly through the honeycomb cell and over the hot film element. At this time, heat transfer is generated by convection and the MAF sensor loses its energy. This sensor detects the mass air flow by using the energy loss and transfers the information to the PCM by a voltage transfer. The PCM calculates fuel quantity and ignition timing appropriate for the conditions.

REMOVAL & INSTALLATION

1. Before servicing the vehicle, refer to the Precautions Section.
2. Disconnect the negative battery cable.
3. Disconnect the connector from the sensor.
4. Remove the air cleaner and air intake assembly, as required.
5. Remove the sensor from its mounting.

To install:

6. Installation is the reverse of the removal procedure.

TESTING

See Figures 267 and 268.

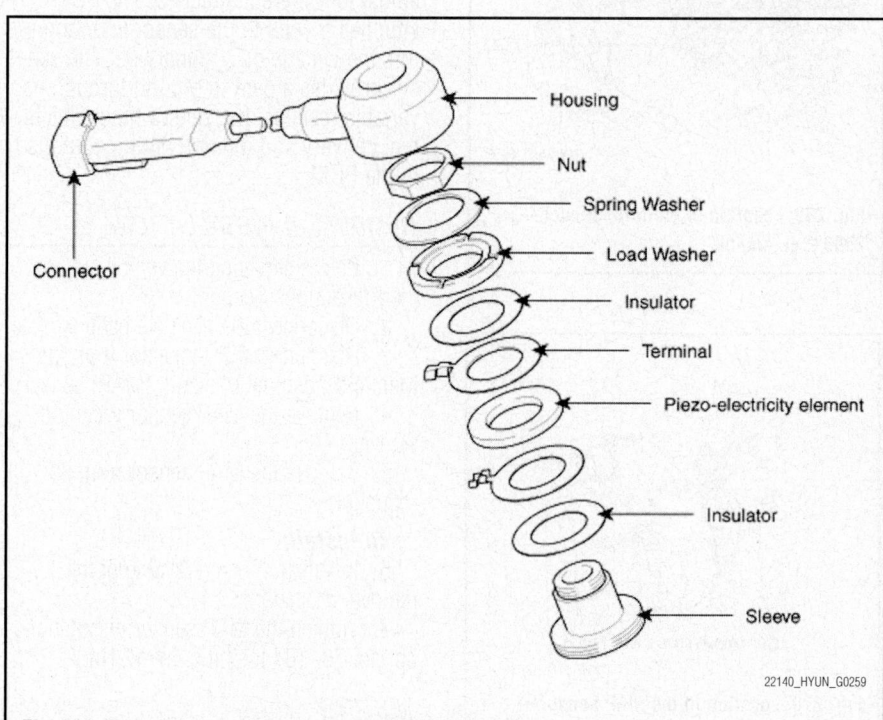

Housing
Nut
Spring Washer
Load Washer
Insulator
Terminal
Piezo-electricity element
Insulator
Sleeve

Connector

22140_HYUN_G0259

Fig. 266 Exploded view of Knock Sensor (KS) component

Condition	Output Voltage (V)	Intake Air Quantity (kg/h)
Idle	0.6 ~ 1.0	11.66 ~ 19.85
3000 rpm	1.7 ~ 2.0	43.84 ~ 58.79

22140_HYUN_G0260

Fig. 267 Output Voltage and Intake Air Quantity Table

1. Verify the integrity of the air induction system by inspecting for the following conditions:
   - Damaged components
   - Loose or improper installation
   - An air flow restriction
   - Any vacuum leak
   - Water intrusion

2. Verify that any electrical aftermarket devices are properly connected and grounded.

3. With the engine running, observe the scan tool MAF sensor parameter. The reading should be between 1,700–3,200 Hz depending on the Engine Coolant Temperature (ECT).

4. A Wide Open Throttle (WOT) acceleration from a stop should cause the MAF sensor parameter on the scan tool to increase rapidly. This increase should be from 2–6 g/s at idle to greater than 100 g/s at the time of the 1–2 shift.

5. Run the engine and connect the Hi-Scan (Pro)® to the Data Link Connector (DLC).

Air Flow (kg/h)	Output Frequency (Hz)
12.6	2,617
18.0	2,958
23.4	3,241
32.4	3,653
43.2	4,024
57.6	4,399
72.0	4,704
108.0	5,329
144.0	5,897
198.0	6,553
270.0	7,240
360.0	7,957
486.0	8,738
666.0	9,644
900.0	10,590

22140_ENTO_G0099

Fig. 268 Air Flow and Output Frequency Table

6. Check if the sensor output voltage is normal according to the following specification table.

7. If within specification, check for poor connection between the PCM and the component.

8. If not within specification, substitute the sensor with a known good component, check for proper operation. If the problem is corrected, replace the sensor.

MANIFOLD ABSOLUTE PRESSURE (MAP) SENSOR

LOCATION

See Figures 269 through 271.

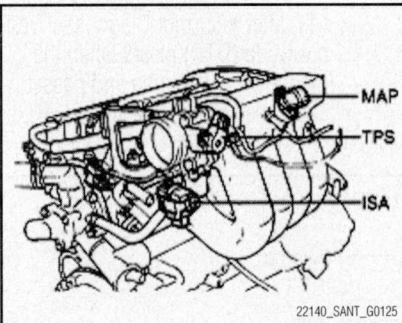

22140_SANT_G0125

Fig. 269 Location of the MAP sensor— 2006 2.4L engine

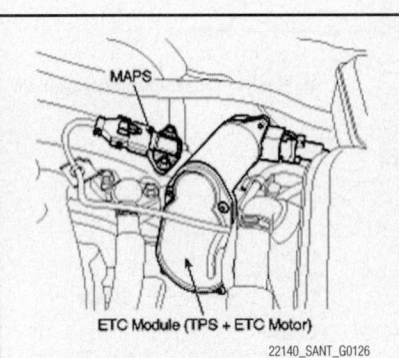

22140_SANT_G0126

Fig. 270 Location of the MAP sensor— 2.7L engine

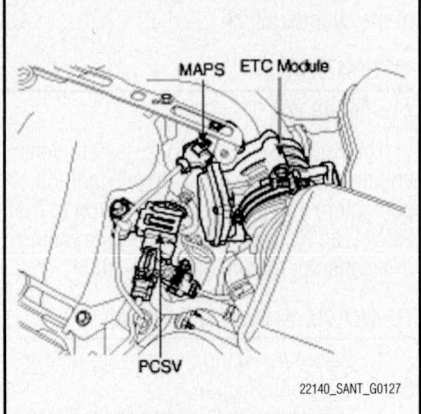

22140_SANT_G0127

Fig. 271 Location of the MAP sensor, Purge Control Solenoid Valve (PCSV), and the Electronic Throttle Control (ETC) module—3.3L and 3.8L engines

The Manifold Absolute Pressure (MAP) sensor is located on the intake manifold.

OPERATION

The Manifold Absolute Pressure (MAP) sensor is a pressure sensitive variable resistor. It measures the changes in the intake manifold pressure which result from engine load and speed changes, and converts these to a voltage output. This sensor is used to measure the barometric pressure at start up, and under certain conditions, allows the PCM to automatically adjust for different altitudes. The PCM supplies 5 volts to the sensor and monitors the voltage on a signal line. The sensor provides a path to ground through its variable resistor. The sensor input affects fuel delivery and ignition timing controls in the PCM.

REMOVAL & INSTALLATION

1. Before servicing the vehicle, refer to the Precautions Section.

2. Disconnect the negative battery cable.

3. Disconnect the connector from the Manifold Absolute Pressure (MAP) sensor.

4. Remove the MAP sensor retaining screws.

5. Remove the MAP sensor from its mounting.

To install:

6. Installation is the reverse of the removal procedure.

7. Tighten the MAP sensor installation bolt to: 78–104 inch lbs. (9–12 Nm).

TESTING

See Figures 272 and 273.

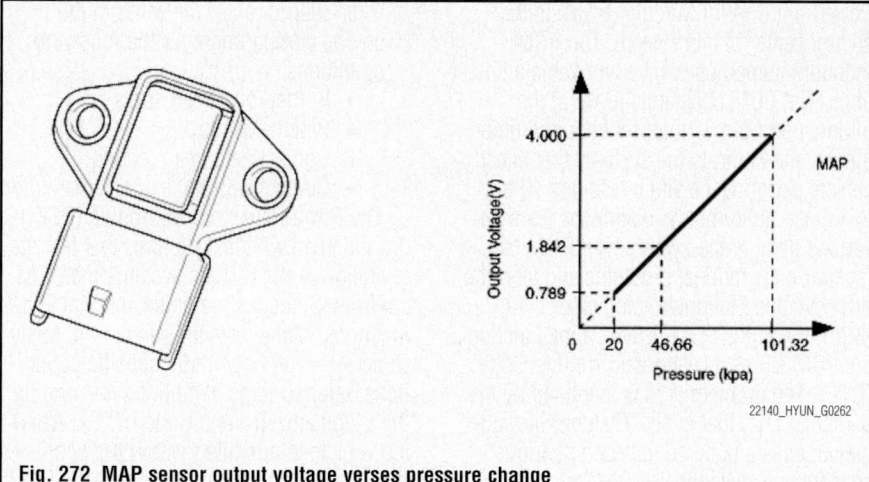

Fig. 272 MAP sensor output voltage verses pressure change

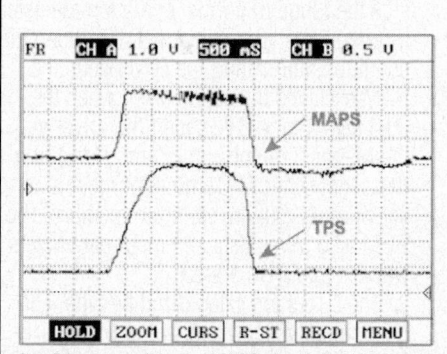

As often as possible, the MAPS signal should be compared with the TPS signal. Check whether the MAPS and TPS signals increase at the same time when accelerating. During acceleration, the MAPS output voltage increases; during deceleration, the MAPS output voltage decreases.

22140_HYUN_G0263

Fig. 273 Waveform of MAP sensor signal compared with the TPS signal waveform

1. Measure the voltage between terminal 4 (sensor ground) and terminal 1 (sensor output).

2. Specification should be 4–5 volts, with the ignition switch ON.

3. Specification should be 0.8–2.4 volts at idle.

4. If not within specification, replace the sensor.

OIL PRESSURE SENSOR

LOCATION

The oil pressure sensor is located near the base of the oil filter assembly.

OPERATION

The oil pressure sensor measures the amount of pressure generated from the output of the oil pump. If the oil pressure is too low, mechanical damage and scoring may occur.

REMOVAL & INSTALLATION

✸✸ CAUTION

Do not perform this procedure with the engine hot, wait until the engine is cool or cold.

1. Before servicing the vehicle, refer to the Precautions Section.

2. Turn the ignition **OFF**.

3. Remove the oil pressure sensor electrical connector.

4. Remove the oil pressure sensor.

To install:

5. To install, reverse the removal procedure.

6. Tighten the sensor to 113–130 inch lbs. (13–15 Nm).

TESTING

Before servicing any vehicle, please be sure to read the precautions section, which deals with personal safety, prevention of component damage, and important points to take into consideration when servicing a motor vehicle.

1. With the vehicle on a level surface, allow adequate drain down time of 2–3 minutes and measure for a low oil level. Add the recommended grade engine oil and fill the crankcase until the oil level measures full on the oil level indicator.

2. Run the engine, and verify low, or no oil pressure on the vehicle gage or light. Listen for a noisy valve train or a knocking noise.

3. Inspect for the following:
- Oil diluted by moisture or unburned fuel mixtures
- Improper oil viscosity for the expected temperature
- Incorrect or malfunctioning oil pressure sender
- Incorrect or malfunctioning oil pressure gauge
- Plugged oil filter
- Malfunctioning oil bypass valve

4. Remove the oil pressure sender or another engine block oil gallery plug.

5. Install an oil pressure gauge and measure the engine oil pressure.

6. Compare the readings to specifications.

7. If the engine oil pressure is below specifications, inspect the engine for one or more of the following:
- Oil pump worn or dirty
- Oil pump-to-engine front cover bolts loose
- Oil pump screen loose, plugged, or damaged
- Oil pump screen O-ring seal missing or damaged
- Malfunctioning oil pump pressure regulator valve
- Excessive bearing clearance
- Cracked, porous, or restricted oil galleries
- Oil gallery plugs missing or incorrectly installed
- Broken lash adjusters

8. Higher than recommended oil pressure may be caused by one or more of the following conditions:
- Worn or sticking oil pump pressure relief valve
- Plugged oil filter
- Improper viscosity oil

9. If the oil pressure tests according to specification and the oil pressure sensor gives a faulty reading, replace the oil pressure sensor (switch) as necessary.

POWERTRAIN CONTROL MODULE (PCM)

LOCATION

See Figures 274 through 276.

Refer to the accompanying illustrations for Powertrain Control Module (PCM) locations.

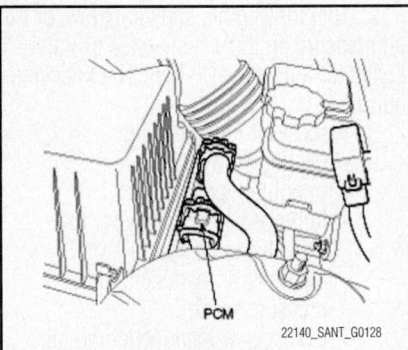

Fig. 274 Powertrain Control Module (PCM) location—2007–08 2.7L engine

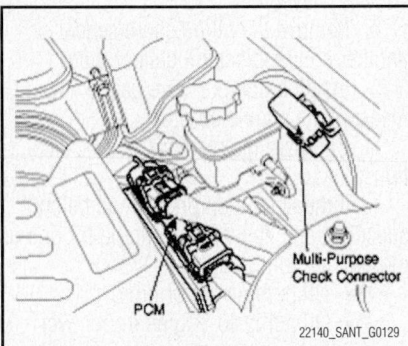

Fig. 275 Powertrain Control Module (PCM) location—2007–08 3.3L engine

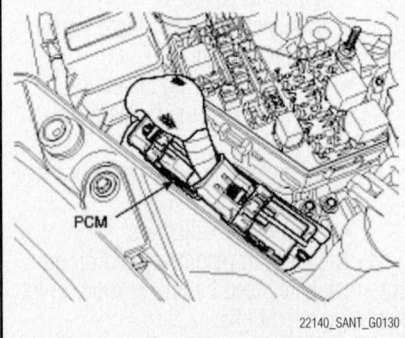

Fig. 276 Powertrain Control Module (PCM) location—2007–08 3.8L engine

OPERATION

The powertrain has electronic controls to reduce exhaust emissions while maintaining excellent drivability and fuel economy. The Powertrain Control Module (PCM) is the control center of this system. The PCM monitors numerous engine and vehicle functions. The PCM constantly looks at the information from various sensors and other inputs, and controls the systems that affect vehicle performance and emissions. The PCM also performs the diagnostic tests on various parts of the system. The PCM can recognize operational problems and alert the driver via the Malfunction Indicator Lamp (MIL). When the PCM detects a malfunction, the PCM stores a Diagnostic Trouble Code (DTC). The problem area is identified by the particular DTC that is set. The control module supplies a buffered voltage to various sensors and switches. Review the components and wiring diagrams in order to determine which systems are controlled by the PCM. The following are some of the functions that the PCM controls:

- The ignition system
- The Knock Sensor (KS) system
- The Evaporative Emissions (EVAP) system
- The alternator
- The A/C and fan systems
- The cooling fan control
- The fuel injection system
- The emission control systems
- The on-board diagnostics

REMOVAL & INSTALLATION

1. Before servicing the vehicle, refer to the Precautions Section.
2. Turn the ignition switch **OFF**.
3. Disconnect the negative battery cable from the battery.
4. Disconnect the PCM connectors.
5. Remove the PCM mounting bolts.
6. Remove the PCM from the engine compartment.

To install:

7. Installation is the reverse of the removal procedure.
8. Tighten the PCM mounting bolts to: 86–104 inch lbs. (10–12 Nm).

TESTING

1. Perform a careful underhood inspection when performing any diagnostic procedure or diagnosing the cause of an emission test failure. This can often lead to repairing a condition without further steps. Use the following guidelines when performing an inspection:

   a. Inspect all of the vacuum hoses for correct routing, pinches, cuts, or disconnects.

   b. Inspect any hoses that are difficult to see.

   c. Inspect all of the wires in the engine compartment for the following conditions:
   - Burned or chafed spots
   - Pinched wires
   - Contact with sharp edges
   - Contact with hot exhaust manifold

The Powertrain Control Module (PCM) is programmed with test routines that test the operation of the various systems the PCM controls. Some tests monitor internal PCM functions. Many tests are run continuously. Other tests run only under specific conditions, referred to as conditions for running the Diagnostic Trouble Code (DTC). When the vehicle is operating within the conditions for running a particular test, the PCM monitors certain parameters and determines if the values are within an expected range. The parameters and values considered outside the range of normal operation are listed as conditions for setting the DTC. When the conditions for setting the DTC occur, the PCM executes the action taken when the DTC sets. Some DTC's alert the driver via the Malfunction Indicator Lamp (MIL) or a message. Other DTC's do not trigger a driver warning, but are stored in memory. The PCM also saves data and input parameters when most DTC's are set.

The DTC's are categorized by type. The DTC type is determined by the MIL operation and the manner in which the fault data is stored when a particular DTC fails. In some cases, there may be exceptions to this structure. Therefore, when diagnosing the system it is important to read the action taken when the DTC sets and the conditions for clearing the DTC.

Many intermittent open or shorted circuits come and go with harness and connector movement caused by vibration, engine torque, bumps, or rough pavement.

2. Test the wiring harness and connectors by performing the following:
   - Move the related PCM connectors and wiring while monitoring the appropriate scan tool data
   - With the engine running, move the related connectors and wiring while monitoring engine operation
   - If harness or connector movement affects the data displayed, the component and system operation, or the engine operation, inspect and repair the harness or connections as necessary

3. Test the electrical connections and/or wiring by performing the following:
   - Inspect for incorrect mating of the connector halves or terminals not fully seated in the connector body

- Inspect for improperly formed or damaged terminals. Test for incorrect terminal tension
- Inspect for poor terminal to wire connections including terminals crimped over insulation. This requires removing the terminal from the connector body
- Inspect for corrosion or water intrusion. Pierced or damaged insulation can allow moisture to enter the wiring. The conductor can corrode inside the insulation with little visible evidence. Look for swollen and stiff sections of wire in the suspect circuits
- Inspect for wires that are broken inside the insulation

4. Test the PCM ground circuit:

a. Measure resistance between the PCM and a good chassis ground. Resistance specification: 1 ohm or less.

b. Use the backside of the PCM harness connector as the PCM side check point.

c. If a problem is found, make the necessary repairs.

5. If a problem is not located through the above procedures, the PCM may be faulty.

6. Hyundai recommends to replace the PCM with a new one and then check the vehicle again. If the vehicle operates normally, then the problem was likely with the PCM.

7. Retest the original PCM, if possible, by installing it into a known-good vehicle and check vehicle operations.

THROTTLE POSITION SENSOR (TPS)

LOCATION

See Figures 277 through 281.

The Throttle Position Sensor (TPS) is mounted on the throttle body.

Fig. 277 Throttle Position Sensor (TPS) location—2006 2.4L engine

Fig. 278 Throttle Position Sensor (TPS) location—2006 2.7L engine

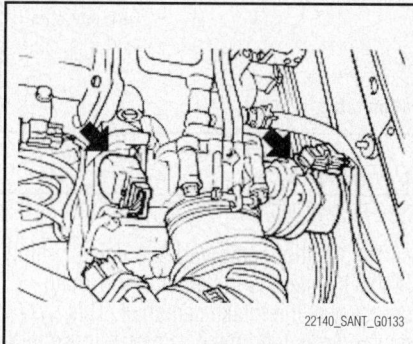

Fig. 279 Throttle Position Sensor (TPS) location—2006 3.5L engine

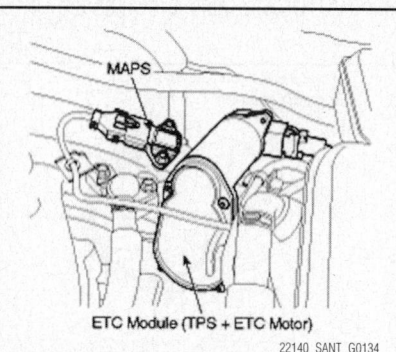

Fig. 280 Throttle Position Sensor (TPS) location—2007–08 2.7L engine

OPERATION

The Throttle Position Sensor (TPS) is mounted on the throttle body and detects the opening angle of the throttle plate. The TPS has a variable resistor (potentiometer) that changes in resistance characteristics according to the throttle angle. During acceleration, the TPS resistance between the reference 5-volt and the signal terminal decreases and the output voltage increases; during deceleration, the TPS resistance increases and the TPS output voltage decreases. The ECM/PCM supplies a reference 5-volt to the TPS and the

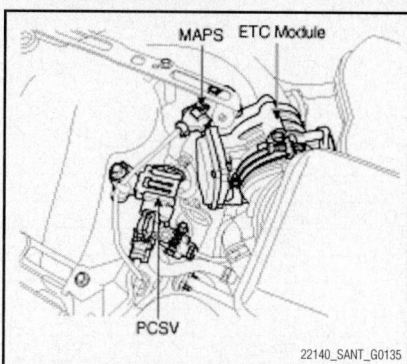

Fig. 281 Throttle Position Sensor (TPS) location—3.3L and 3.8L engines 2007–08

output voltage increases directly with the opening of the throttle valve. The TPS output voltage will vary from 0.2–0.8 volts at Closed Throttle (CT) to 4.3–4.8 volts at Wide Open Throttle (WOT). The ECM/PCM determines operating conditions such as idle (closed throttle), part load, acceleration/deceleration, and wide-open throttle from the TPS. Also the ECM/PCM uses the Manifold Absolute Pressure (MAP) sensor signal along with the TPS signal to adjust fuel injection duration and ignition timing.

REMOVAL & INSTALLATION

1. Before servicing the vehicle, refer to the Precautions Section.

2. Disconnect the negative battery cable.

3. Disconnect the Throttle Position Sensor (TPS) connector.

4. Remove the TPS retaining screws.

5. Remove the TPS from its mounting.

To install:

6. Installation is the reverse of the removal procedure.

TESTING

See Figures 273 and 282.

1. The Throttle Position Sensor (TPS) output voltage should vary from 0.2–0.8 volts at Closed Throttle (CT).

2. The TPS should vary from 4.3–4.8 volts at Wide Open Throttle (WOT).

3. Check the signal waveform compared to the Manifold Absolute Pressure (MAP) waveform as in the following illustration. The waveform should coincide.

4. The throttle angle should match the output voltage as in the specification table.

5. If the TPS does not meet specifications, replace the TPS.

Throttle Angle(°)	Output Voltage(V) [Vref = 5.0V]	
	TPS1	TPS2
0°	0V	5.0V
10°	0.5V	4.5V
20°	0.9V	4.1V
30°	1.4V	3.6V
40°	1.8V	3.2V
50°	2.3V	2.7V
60°	2.7V	2.3V
70°	3.2V	1.8V
80°	3.6V	1.4V
90°	4.1V	0.9V
100°	4.5V	0.5V
110°	5.0V	0V

22140_HYUN_G0265

Fig. 282 Throttle angle and output voltage specification table

VARIABLE CAMSHAFT TIMING OIL CONTROL SOLENOID

LOCATION

See Figures 283 and 284.

Refer to the accompanying illustrations for variable camshaft timing oil control solenoid location.

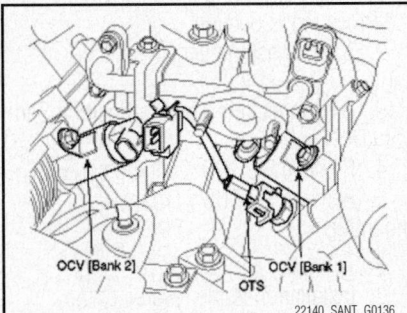

22140_SANT_G0136

Fig. 283 Variable camshaft timing oil control solenoid location (Bank 1, Bank 2)—2007–08 2.7L engine

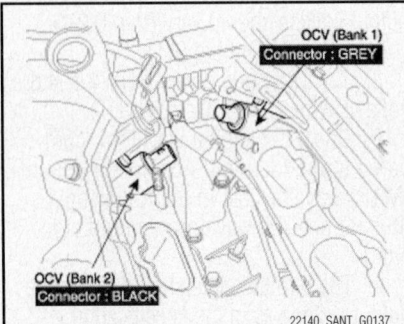

OCV (Bank 1)
Connector : GREY

OCV (Bank 2)
Connector : BLACK

22140_SANT_G0137

Fig. 284 Variable camshaft timing oil control solenoid location (Bank 1, Bank 2)—2007–08 3.3L and 3.8L engines

OPERATION

See Figure 285.

The Continuously Variable Valve Timing (CVVT) system is installed to the chain sprocket of the intake camshaft. This system controls the intake camshaft to provide the optimal valve timing for every driving condition. The ECM/PCM controls the Variable Camshaft Timing Oil Control Solenoid (VCTOCS)—also known as the Oil

Control Valve (OCV)—based on the signal outputs from the Mass Air Flow (MAF) sensor, Throttle Position Sensor (TPS), and Engine Coolant Temperature (ECT) sensor. The CVVT controller regulates the intake camshaft angle using oil pressure through the VCTOCS. As a result, the relative position between the camshaft and the crankshaft becomes optimal. The engine torque improves, fuel economy improves, and the exhaust emissions decrease within the overall driving conditions.

The CVVT system makes continuous intake valve timing changes based on operating conditions. Intake valve timing is optimized to allow the engine to produce maximum power. Cam angle is advanced to obtain the EGR effect and reduce pumping loss. The intake valve is closed quickly to reduce the entry of the air/fuel mixture into the intake port and improve the changing effect. Reducing the cam advance at idle stabilizes combustion and reduces engine speed. If a malfunction occurs, the CVVT system control is disabled and the valve timing is fixed at the fully retarded position.

REMOVAL & INSTALLATION

1. Before servicing the vehicle, refer to the Precautions Section.
2. Disconnect the ground cable from the battery.

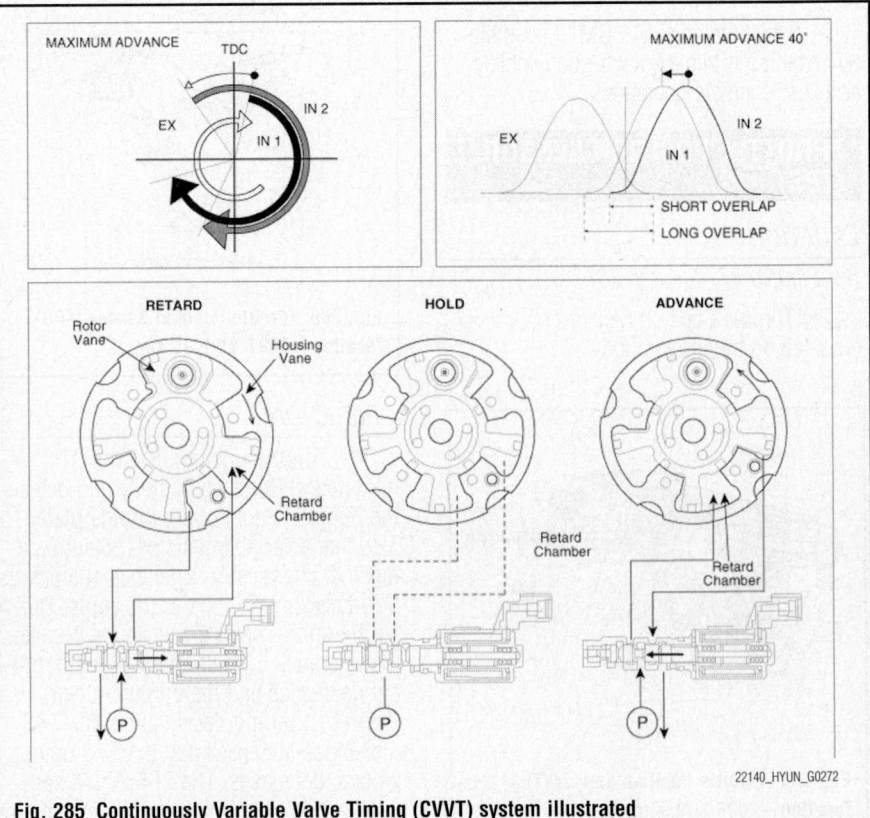

22140_HYUN_G0272

Fig. 285 Continuously Variable Valve Timing (CVVT) system illustrated

3. Disconnect the connector from Variable Camshaft Timing Oil Control Solenoid (VCTOCS) on the right-hand bank and/or left-hand bank.

4. Remove the bolt retaining the VCTOCS.

5. Remove the VCTOCS.

To install:

➥**Always use a new gasket/O-ring.**

6. Install the VCTOCS.

7. Tighten the mounting bolt to 86–104 inch lbs. (20–39 Nm).

8. Connect the electrical connector to the VCTOCS.

TESTING

1. Ensure the vehicle has the proper oil viscosity and that the oil is not overdue maintenance.

2. Observe the engine oil level. The engine oil level should be within the operating range.

3. Allow the engine to reach operating temperature.

4. Connect a suitable OBD2 scan tool to the Data Link Connector (DLC) inside the vehicle.

5. Increase the engine speed to 1,500 RPM.

6. Command each solenoid to 25 percent. The angle desired parameter should match the solenoid actual parameter.

Component Testing

See Figures 286 and 287.

1. Measure the resistance of each variable camshaft timing oil control solenoid valve assembly. Resistance should read according to the resistance specification table shown.

　a. Ignition "OFF."

　b. Disconnect Variable Camshaft Timing Oil Control Solenoid (VCTOCS) connector.

　c. Measure resistance between terminals 1 and 2 of the VCTOCS connector (component side).

　　• Specification: Approx. 6.9–7.9 ohms at 68°F (20°C)

　　• If not according to specification, replace the oil control solenoid

2. Check the operation of the VCTOCS.

　a. Start the engine and let it idle.

　b. With VCTOCS connector still disconnected, connect a 12 volt supply to terminal 2 and a ground to terminal 1 of the VCTOCS (component side).

<CBG05>

1. Control

2. Power

22140_HYUN_G0274

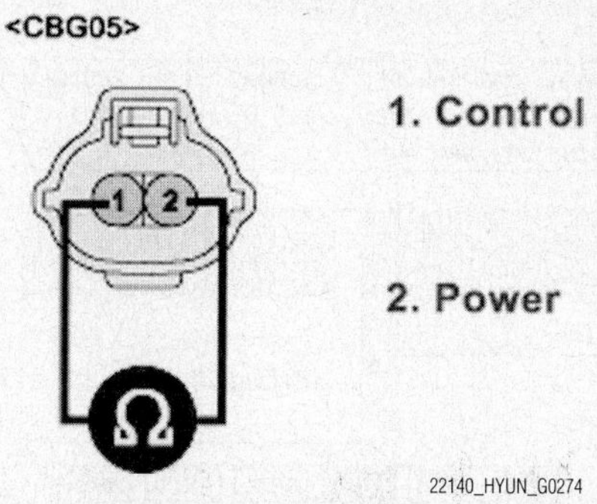

Fig. 286 Use an ohmmeter to check the resistance level of the oil control solenoid

• The engine should suddenly run rough at idle or stall out as the VCTOCS abruptly changes the timing

• If the engine runs rough or stalls when the 12 volt supply is connected, check for connection problems to the VCTOCS or problem with the supply side from the ECM/PCM

• If no change is noted, the VCTOCS may be defective

3. Check the VCTOCS and oil filter assembly.

　a. Turn the ignition "OFF."

　b. Check VCTOCS filter for sticking or contamination.

　c. Remove the VCTOCS and visually check the spool column of VCTOCS for contamination.

　d. If a problem is found, clean or replace as necessary.

　e. If no problem is found:

　　• Apply a 12 volt supply to terminal 2 and a ground to terminal 1 of the VCTOCS (Component side)

　　• Verify that a "clicking" sound is heard when applying the battery voltage

　　• If no "clicking" is heard, replace the VCTOCS

Temperature °C(°F)	Resistance (Ω)
0(32)	6.2 ~ 7.4
20(68)	6.8 ~ 8.0
40(104)	7.4 ~ 8.6
60(140)	8.0 ~ 9.2
80(176)	8.6 ~ 9.8

22140_HYUN_G0275

Fig. 287 Resistance table for the oil control solenoid compared to temperature

Circuit Testing

1. Many malfunctions in the electrical system are caused by poor harness and terminals. Faults can also be caused by interference from other electrical systems, mechanical or chemical damage.

2. Thoroughly check connectors for looseness, poor connection, bending, corrosion, contamination, deterioration, or damage.

3. Check the supply voltage to the Variable Camshaft Timing Oil Control Solenoid (VCTOCS).

　a. With the ignition "OFF," disconnect the VCTOCS connector.

　b. Turn the ignition "ON" with the engine "OFF."

　c. Measure the voltage between the power terminal of the harness connector and a chassis ground. The specification: battery voltage (12 volts).

　　• If the 12 volt supply is not present, check the fuse between the Main Relay and the VCTOCS for an open condition, check for an open in power circuit between the Main Relay and the VCTOCS power circuit

　　• Make the repairs as necessary

4. Check for a short to ground in the harness.

　a. With the ignition "OFF," disconnect the VCTOCS connector.

　b. Measure the resistance between the control terminal of the VCTOCS harness connector and a good chassis ground. The specification: Infinite resistance.

　　• If the measured resistance is within specification, refer to Component Testing for the VCTOCS

　　• If the measured resistance is not within specification, repair or replace as necessary

Timing Inspection

See Figure 288.

1. With the ignition "OFF," set up an oscilloscope as follows:

　a. Channel A (+): terminal 2 of the CKPS (back probe), (-): ground.

　b. Channel B (+): terminal 2 of the CMPS (back probe), (-): ground.

2. Start the engine and check for the signal waveform to synchronize with the camshaft sensor and the tooth that is missing. Refer to the sample waveforms in the illustration.

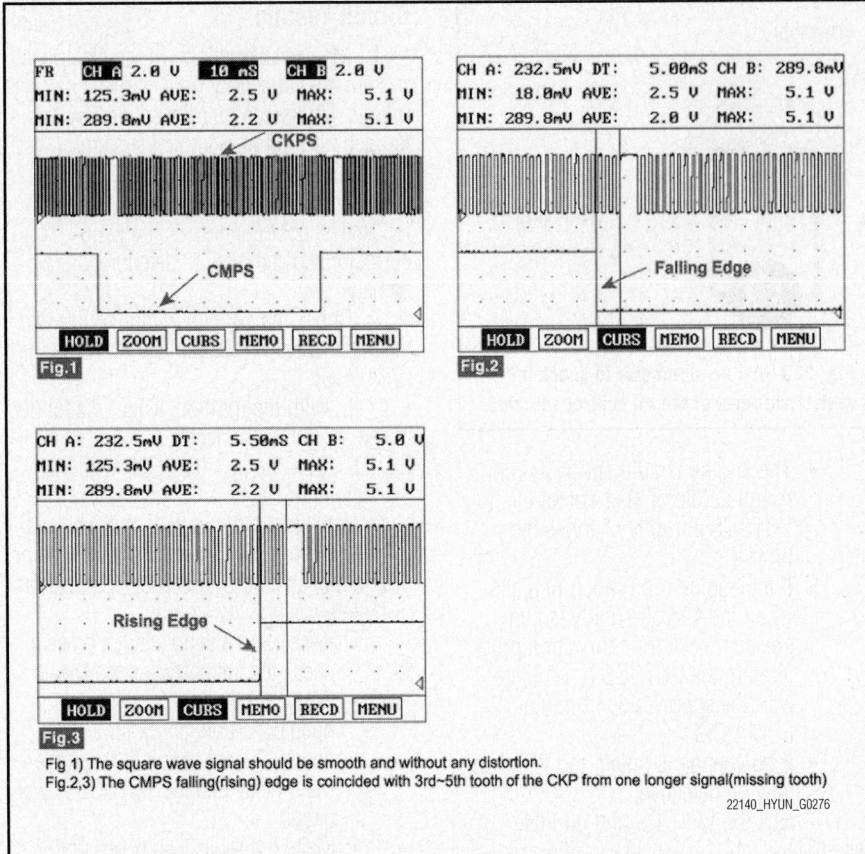

Fig 1) The square wave signal should be smooth and without any distortion.
Fig.2,3) The CMPS falling(rising) edge is coincided with 3rd~5th tooth of the CKP from one longer signal(missing tooth)

22140_HYUN_G0276

Fig. 288 Using waveform synchronization to check the oil control solenoid as it relates to camshaft and crankshaft timing

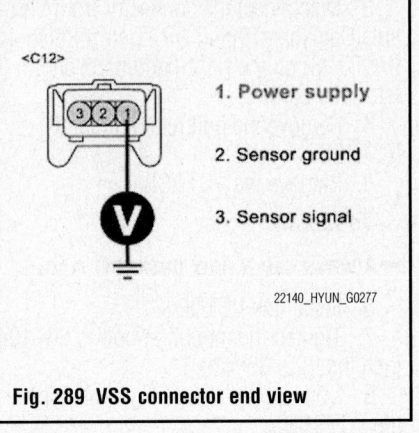

Fig. 289 VSS connector end view

TESTING

See Figures 289 and 290.

1. Before servicing the vehicle, refer to the Precautions Section.

2. Many malfunctions in the electrical system are caused by poor harness and terminal connections. Faults can also be caused by interference from other electrical systems and mechanical or chemical damage.

3. Thoroughly check all connectors (and connections) for looseness, bending, corrosion, contamination, deterioration, or damage.

4. Make repairs as necessary.

5. Turn the ignition **OFF**.

6. Disconnect the VSS connector.

VEHICLE SPEED SENSOR (VSS)

LOCATION

The Vehicle Speed Sensor (VSS) is attached to the output shaft of the transaxle.

OPERATION

The sensor is a Hall-effect sensor. The sensor converts the transaxle gear revolutions into pulse signals, which are sent to the PCM.

REMOVAL & INSTALLATION

1. Before servicing the vehicle, refer to the Precautions Section.

2. Raise and support the vehicle safely.

3. Place a drip pan below the Vehicle Speed Sensor (VSS) to catch any spilled fluid when it is removed.

4. Disconnect the VSS connector.

5. Remove the sensor from its mounting.

To install:

6. Installation is the reverse of the removal procedure.

7. Replace any lost transaxle fluid.

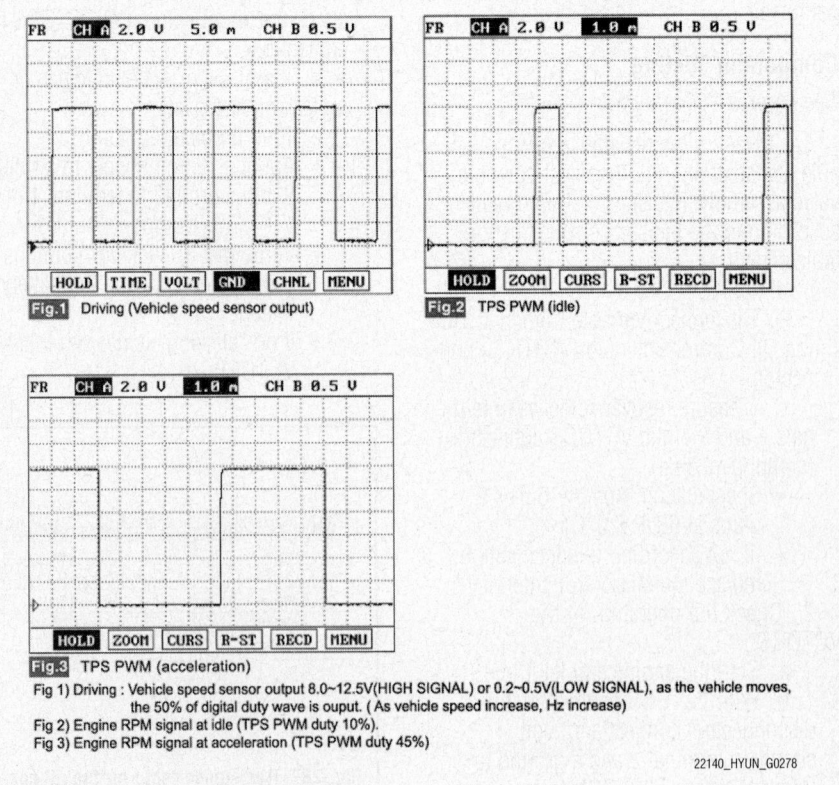

Fig 1) Driving : Vehicle speed sensor output 8.0~12.5V(HIGH SIGNAL) or 0.2~0.5V(LOW SIGNAL), as the vehicle moves, the 50% of digital duty wave is ouput. (As vehicle speed increase, Hz increase)
Fig 2) Engine RPM signal at idle (TPS PWM duty 10%).
Fig 3) Engine RPM signal at acceleration (TPS PWM duty 45%)

22140_HYUN_G0278

Fig. 290 Waveform illustration of output signal

7. Start the engine.

8. Measure the voltage between terminal 1 of the VSS harness connector and a chassis ground. Specification: Battery voltage (approximately 12 volts).

    a. If the measured voltage is within specifications, inspect the ground circuit.

    b. If the measured voltage is not within specifications, check for an open/short to ground in the power harness.

9. Repair as necessary.

10. Inspect the ground circuit.

    a. Turn the ignition **OFF**.

    b. Disconnect VSS connector.

    c. Measure the resistance between terminal 2 of the VSS harness connector and the chassis ground. The specification: Approx. 0 ohms.

    d. If the measured resistance is within specifications, inspect the signal circuit inspection.

    e. If the measured resistance is not within specifications, check for an open/short to power in the ground harness. Repair as necessary.

11. Inspect the signal circuit

    a. Turn the ignition **OFF**.

    b. Connect a scan tool to the Data Link Connector (DLC).

    c. Start the engine and select "SCOPEMETER FUNCTION" on the scan tool.

    d. Drive the vehicle and measure the output signal between terminal 3 of the EPS CM harness connector and chassis ground. Specification: see waveform illustration.

    e. If the VSS output signal is within specifications, substitute with a known-good EPS CM and check for proper operation.

    f. If the problem is corrected, replace the EPS CM.

    g. If the VSS output signal is not within specifications, check for an open/short to ground in the signal circuit and other systems which use the VSS. Repair as necessary.

    h. If the VSS is found faulty, replace with a new component.

FUEL

GASOLINE FUEL INJECTION SYSTEM

FUEL SYSTEM SERVICE PRECAUTIONS

Safety is the most important factor when performing, not only fuel system maintenance, but any type of maintenance. Failure to conduct maintenance and repairs in a safe manner may result in serious personal injury or death. Maintenance and testing of the vehicle's fuel system components can be accomplished safely and effectively by adhering to the following rules and guidelines.

• To avoid the possibility of fire and personal injury, always disconnect the negative battery cable unless the repair or test procedure requires that battery voltage be applied.

• Always relieve the fuel system pressure prior to disconnecting any fuel system component (injector, fuel rail, pressure regulator, etc.), fitting, or fuel line connection. Exercise extreme caution whenever relieving fuel system pressure to avoid exposing skin, face, and eyes to fuel spray. Please be advised that fuel under pressure may penetrate the skin or any part of the body that it contacts.

• Always place a shop towel or cloth around the fitting or connection prior to loosening to absorb any excess fuel due to spillage. Ensure that all fuel spillage (should it occur) is quickly removed from engine surfaces. Ensure that all fuel soaked cloths or towels are deposited into a suitable waste container.

• Always keep a dry chemical (Class B) fire extinguisher near the work area.

• Do not allow fuel spray or fuel vapors to come into contact with a spark or an open flame.

• Always use a back-up wrench when loosening and tightening fuel line connection fittings. This will prevent unnecessary stress and torsion to fuel line piping.

• Always replace worn fuel fitting O-rings with new. Do not substitute fuel hose or equivalent where fuel pipe is installed.

Before servicing the vehicle, make sure to refer to the precautions in the beginning of this section as well.

RELIEVING FUEL SYSTEM PRESSURE

1. Before servicing the vehicle, refer to the Precautions Section.

2. Remove the fuel filler cap.

3. Remove the fuel pump fuse.

4. Start and run the engine until it stalls.

5. Disconnect the negative battery cable.

6. Replace the fuse.

FUEL FILTER

REMOVAL & INSTALLATION

The fuel delivery system integrates the fuel filter with the in-tank fuel pump. To service this filter, remove the fuel pump. Refer to Fuel Pump, removal & installation.

FUEL INJECTORS

REMOVAL & INSTALLATION

See Figure 291.

1. Before servicing the vehicle, refer to the Precautions Section.

2. Relieve the fuel system pressure. Refer to Relieving Fuel System Pressure.

3. Remove or disconnect the following:
   • Negative battery cable
   • Air breather hose from the throttle body
   • Throttle cable
   • Engine coolant hose and throttle body
   • Positive Crankcase Ventilation (PCV) valve and brake booster vacuum hose
   • Vacuum hose connector
   • Injector cover
   • High pressure fuel hose
   • Fuel injector harness connector
   • Delivery pipe with the injectors and the pressure regulator as an assembly
   • Injector from the delivery pipe

4. Remove and discard the injector O-ring and grommet.

To install:

5. Install a new grommet and O-ring.

6. Apply a coating of spindle oil or gasoline to the injector O-ring.

7. Install the injector into the delivery pipe while turning the injector left and right making sure the injector turns smoothly. If the injector does not turn smoothly check

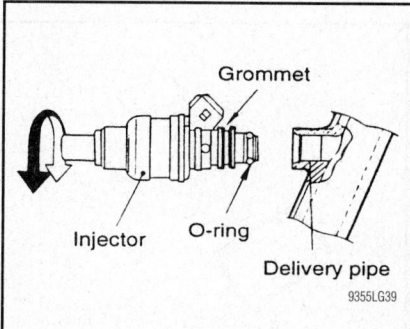

Fig. 291 Install the injector using a twisting motion

for a jammed O-ring, remove the injector and reinsert it again.

8. Install or connect the following:
   - Delivery pipe (fuel rail) and injector assembly. Torque the bolts to 84–132 inch lbs. (10–15 Nm)
   - Intake manifold stay and tighten the bolts to 13–18 ft. lbs. (18–25 Nm)
   - Fuel injector harness connector
   - High pressure fuel hose
   - Injector cover
   - Vacuum hose connector
   - PCV valve and brake booster vacuum hose
   - Throttle body and engine coolant hose
   - Throttle cable
   - Air breather hose from the throttle body
   - Negative battery cable

9. Start the engine and check for proper operation.

FUEL PUMP

REMOVAL & INSTALLATION

See Figures 292 through 294.

1. Before servicing the vehicle, refer to the Precautions Section.
2. Relieve the fuel system pressure. Refer to Relieving Fuel System Pressure.
3. Remove the second seat.
4. Open the carpet over the fuel pump.
5. Remove the service cover (A) from on top of the fuel pump.
6. Disconnect the fuel feed line (A) and canister hoses (B).
7. Unscrew the fuel pump mounting bolts (C) and remove the fuel pump assembly.

To install:

8. Install the fuel pump assembly.
9. Install the fuel pump mounting bolts

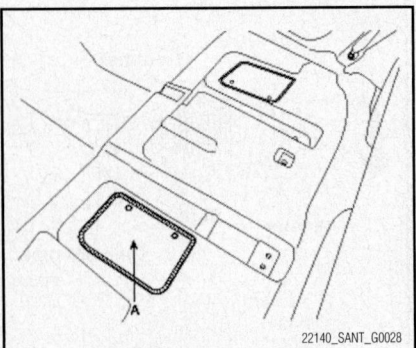

Fig. 292 Remove the service cover (A) from on top of the fuel pump

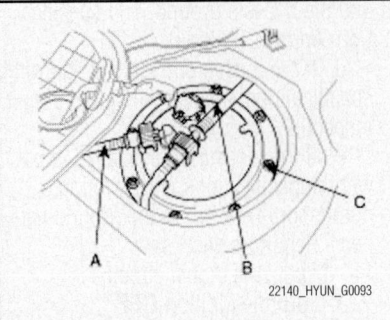

Fig. 293 Disconnect the fuel feed line (A) and canister hoses (B). Unscrew the fuel pump mounting bolts (C)

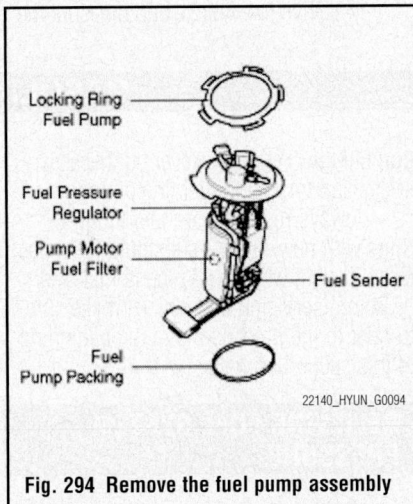

Fig. 294 Remove the fuel pump assembly

(C). Tighten the bolts/nuts to 17–26 inch lbs. (2–3 Nm).

10. Connect the fuel feed line (A) and canister hoses (B).

FUEL PRESSURE REGULATOR

REMOVAL & INSTALLATION

The fuel pressure regulator is built into the fuel pump. Please refer to Fuel Pump, removal & installation.

FUEL TANK

REMOVAL & INSTALLATION

See Figures 295 through 302.

1. Before servicing the vehicle, refer to the Precautions Section.
2. Remove the 2nd seat.
3. Open the carpet above the fuel pump.
4. Remove the service cover of the fuel pump.
5. Disconnect the fuel pump connector (A).
6. Start the engine and wait until the fuel in the fuel line is exhausted.

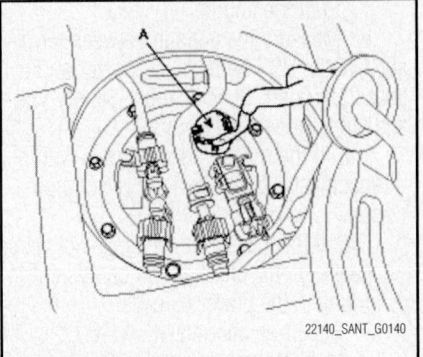

Fig. 295 Disconnect the fuel pump connector (A)

7. After the engine stops, Turn the ignition switch **OFF**.
8. Disconnect the fuel tank pressure sensor connector (A).
9. Open the service cover of the sub fuel sender.
10. Disconnect the sub fuel sender connector (A) and the canister close valve connector (B).
11. Raise and safely support the vehicle.
12. Remove the muffler assembly.
13. Remove the propeller shaft (4WD only).
14. Support the fuel tank with a jack.

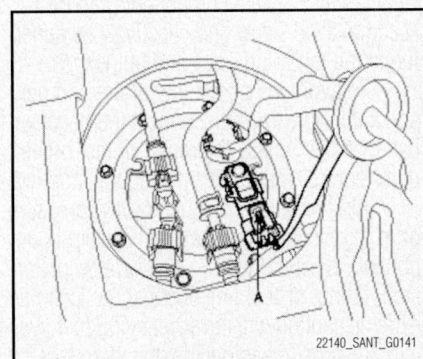

Fig. 296 Disconnect the fuel tank pressure sensor connector (A)

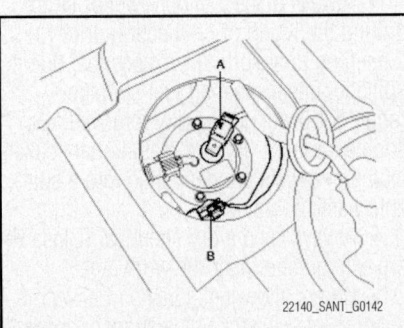

Fig. 297 Disconnect the sub fuel sender connector (A) and the canister close valve connector (B)

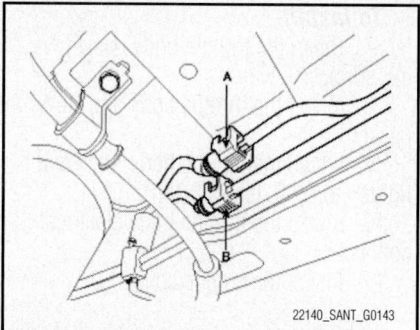

Fig. 298 Disconnect the fuel feed tube quick-connector (A) and the vacuum tube quick-connector (B)

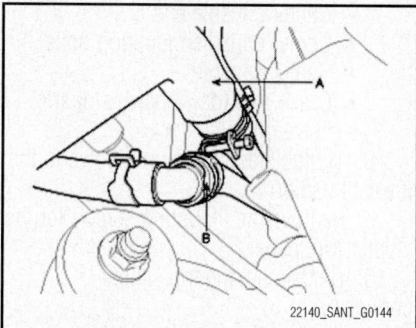

Fig. 299 Disconnect the fuel filler hose (A) and the ventilation hose quick-connector (B)

15. Disconnect the fuel feed tube quick-connector (A) and the vacuum tube quick-connector (B).

16. Remove the bracket near the fuel tank air filter.

17. Disconnect the fuel filler hose (A) and the ventilation hose quick-connector (B).

18. Disconnect the leveling tube quick-connector.

19. Remove the fuel tank cover.

20. Remove the fuel tank band mounting nuts (A) and remove the fuel tank from the vehicle.

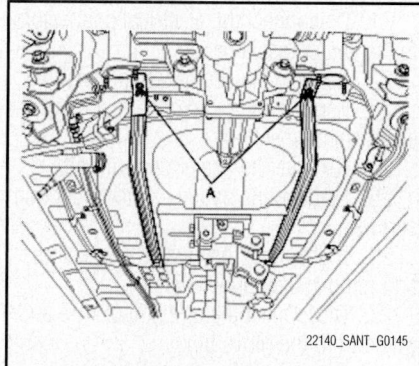

Fig. 300 Remove the fuel tank band mounting nuts (A) and remove the fuel tank from the vehicle

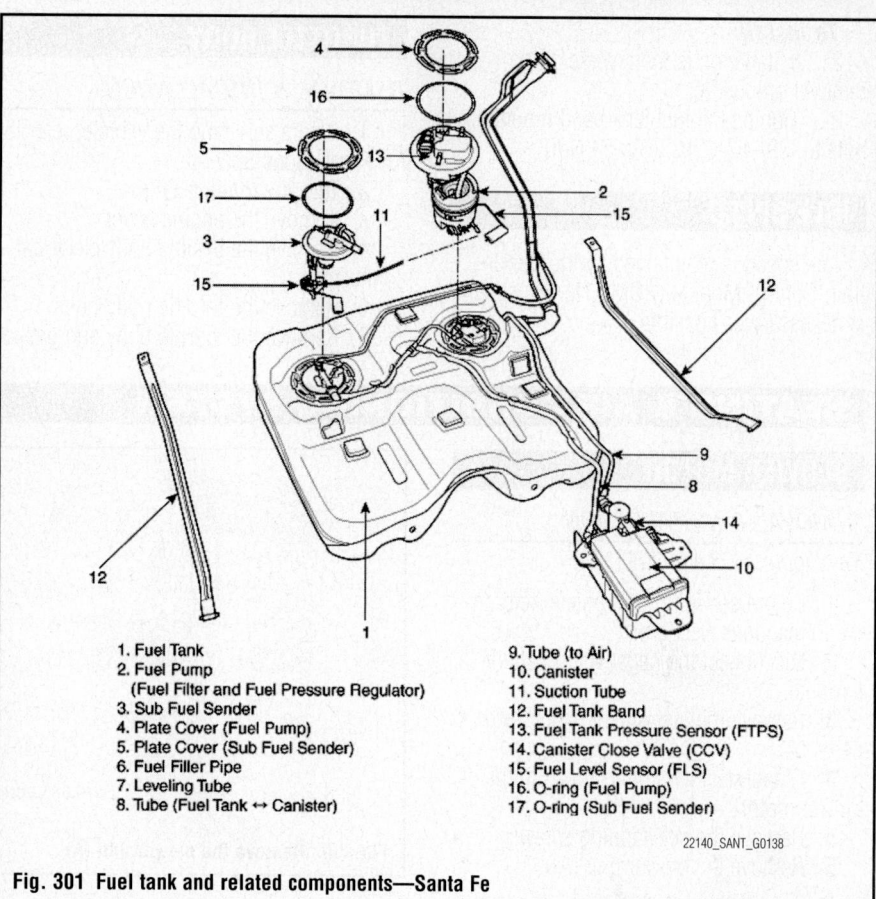

1. Fuel Tank
2. Fuel Pump
   (Fuel Filter and Fuel Pressure Regulator)
3. Sub Fuel Sender
4. Plate Cover (Fuel Pump)
5. Plate Cover (Sub Fuel Sender)
6. Fuel Filler Pipe
7. Leveling Tube
8. Tube (Fuel Tank ↔ Canister)

9. Tube (to Air)
10. Canister
11. Suction Tube
12. Fuel Tank Band
13. Fuel Tank Pressure Sensor (FTPS)
14. Canister Close Valve (CCV)
15. Fuel Level Sensor (FLS)
16. O-ring (Fuel Pump)
17. O-ring (Sub Fuel Sender)

Fig. 301 Fuel tank and related components—Santa Fe

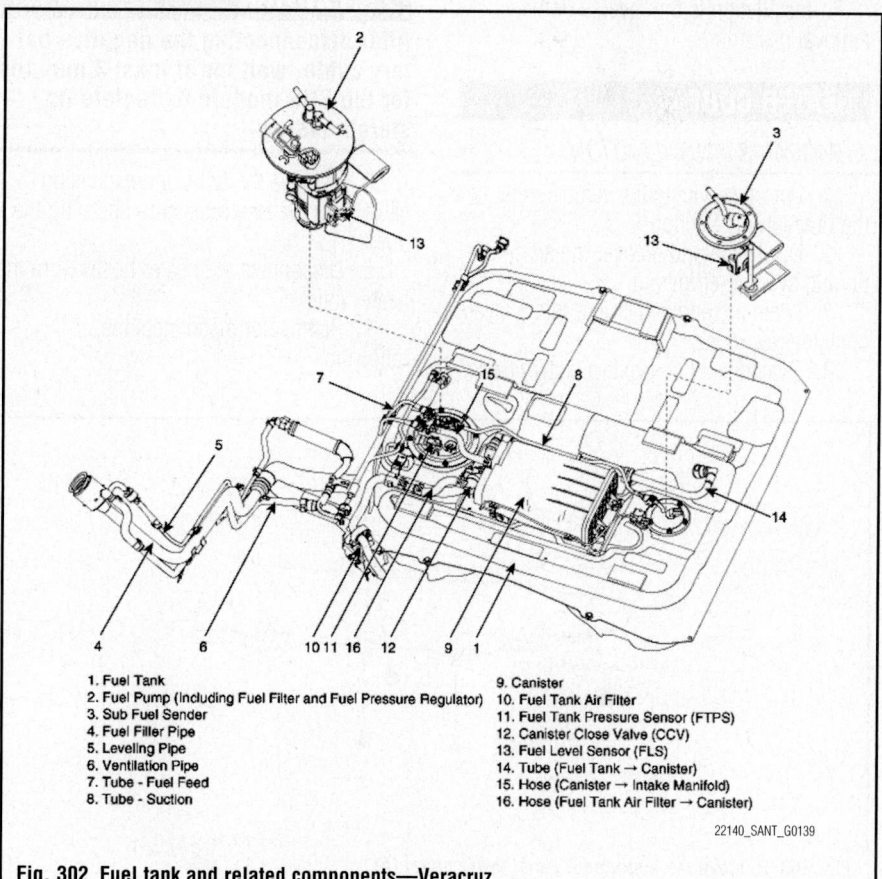

1. Fuel Tank
2. Fuel Pump (Including Fuel Filter and Fuel Pressure Regulator)
3. Sub Fuel Sender
4. Fuel Filler Pipe
5. Leveling Pipe
6. Ventilation Pipe
7. Tube - Fuel Feed
8. Tube - Suction

9. Canister
10. Fuel Tank Air Filter
11. Fuel Tank Pressure Sensor (FTPS)
12. Canister Close Valve (CCV)
13. Fuel Level Sensor (FLS)
14. Tube (Fuel Tank → Canister)
15. Hose (Canister → Intake Manifold)
16. Hose (Fuel Tank Air Filter → Canister)

Fig. 302 Fuel tank and related components—Veracruz

To install:

21. Installation is the reverse of the removal procedure.

22. Tighten the fuel tank band mounting nuts to: 29–40 ft. lbs. (39–54 Nm).

IDLE SPEED

Idle speed is maintained by the Powertrain Control Module (PCM). No adjustment is necessary or possible.

THROTTLE BODY

REMOVAL & INSTALLATION

1. Before servicing the vehicle, refer to the Precautions Section.
2. Turn the ignition **OFF**.
3. Remove the engine cover.
4. Remove the throttle body electrical connector.
5. Remove the throttle body bolts.
6. Remove the throttle body and gasket.

To install:

7. Clean the throttle body gasket mating surfaces.
8. Install the throttle body and NEW gasket.
9. Install the throttle body bolts and tighten to 21 ft. lbs. (28 Nm).
10. Install the throttle body electrical connector.
11. Install the engine cover.

HEATING & AIR CONDITIONING SYSTEM

BLOWER MOTOR

REMOVAL & INSTALLATION

See Figures 303 and 304.

1. Before servicing the vehicle, refer to the Precautions Section.
2. Disconnect the negative (-) battery terminal.
3. Remove the instrument panel lower panel (A).
4. Disconnect the connector from the blower motor.
5. Remove the self-tapping screws.
6. Remove the blower unit (A).

To install:

7. Installation is the reverse of the removal procedure.

HEATER CORE

REMOVAL & INSTALLATION

1. Before servicing the vehicle, refer to the Precautions Section.
2. Discharge and recover the air conditioning system refrigerant.
3. Drain the engine coolant into a clean container for reuse.
4. Disconnect the negative battery cable.

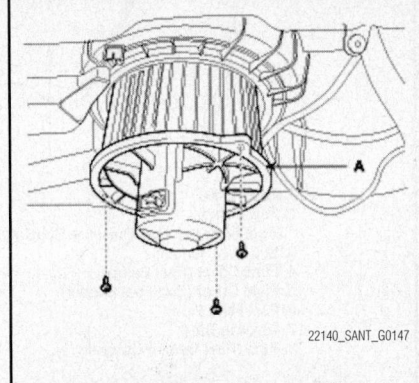

22140_SANT_G0147

Fig. 304 Remove the blower unit (A)

✳✳ CAUTION

After disconnecting the negative battery cable, wait for at least 3 minutes for the SRS module to deplete its stored energy.

5. Remove the bolts and expansion valve from the evaporate core and plug the lines.
6. Disconnect the heater hoses from the heater unit.
7. Remove or disconnect the following:

- Front seats
- Center console end cover
- Center console mounting bolts
- Center console
- Crash pad (dashboard) side trim
- Front pillar trim

8. Tilt the steering column down to the lowest position.
9. Remove the mounting screws for the cluster trim panel.
10. Disconnect the trip sensor connector.
11. Remove the gauge cluster fascia panel.
12. Disconnect the hood release cable from the hood release handle.
13. Remove the screws, bolts, and clips from the lower crash pad panels.
14. Disconnect the self-diagnosis connector from the lower crash pad panel.
15. Remove the lower crash pad panel.
16. Remove the front console side trim.
17. Remove the front console upper cover.
18. Remove the front console mounting screws.
19. Remove the front console.
20. Remove the mounting screws and clips for the center fascia panel.
21. Disconnect the electrical connectors to the center fascia panel.
22. Remove the center fascia panel.
23. Remove the radio and disconnect the electrical connectors.
24. Remove the heater control unit.
25. Remove the gauge cluster mounting screws.
26. Disconnect the cluster connectors and remove the gauge cluster.
27. Disconnect the air damper wire and guide from the glove box.
28. Disengage the hinge pins and remove the glove box.
29. Remove the driver airbag module by removing the 2 mounting bolts.

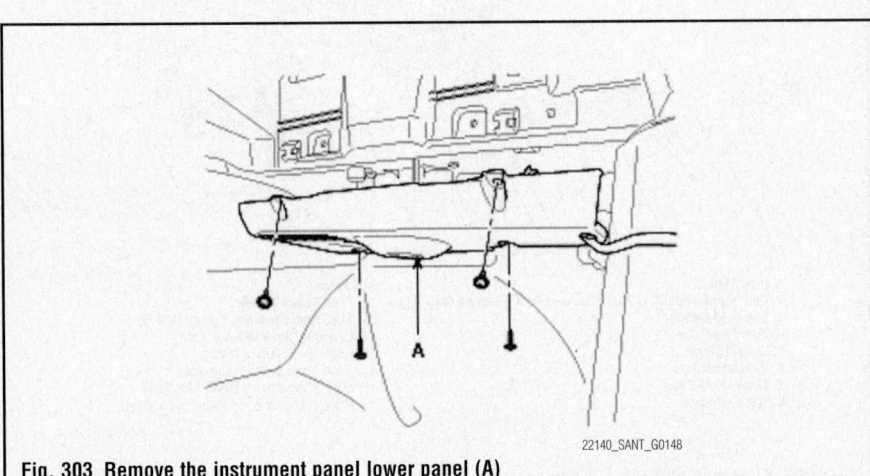

22140_SANT_G0148

Fig. 303 Remove the instrument panel lower panel (A)

30. Remove the steering wheel center lock nut.

31. Align the marks on the steering shaft and wheel.

32. Install Special Tool 09561-11002 to remove the steering wheel.

33. Remove the upper and lower steering column shrouds.

34. Disconnect the 2 lower tightening bolts and remove the lower crash pad.

35. Disconnect the passenger airbag connector.

36. Remove the main crash pad.

37. Remove the cross member.

38. Remove the heater and evaporator unit.

39. Remove the side bracket and lower cover for the unit to access the heater core.

To install:

40. Installation is the reverse of the removal procedure.

41. Observe the following:
- Ensure the crash pad fits on the guide pins correctly and no wiring harnesses are pinched.
- Any damaged trim clips must be replaced.

AUXILIARY HEATING & AIR CONDITIONING SYSTEM

BLOWER MOTOR

REMOVAL & INSTALLATION

See Figure 305.

1. Before servicing the vehicle, refer to the Precautions Section.

2. Disconnect the negative (-) battery terminal.

3. Remove the luggage side trim.

4. Disconnect the connector (A) of the blower motor.

5. Remove the blower motor (B) after loosening the mounting screws.

To install:

6. Installation is the reverse of the removal procedure.

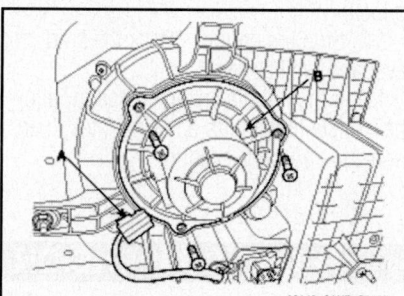

Fig. 305 Disconnect the connector (A) of the blower motor and remove the blower motor (B) after loosening the mounting screws

HEATER CORE

REMOVAL & INSTALLATION

See Figures 306 through 309.

1. Before servicing the vehicle, refer to the Precautions Section.

2. Disconnect the negative (-) battery terminal.

3. Recover the refrigerant with a recover/recycling/charging station.

4. When the engine is cool, drain the engine coolant from the radiator.

5. Remove luggage side trim.

6. Loosen the refrigerant line mounting nuts and remove the rear refrigerant line.

7. Remove the air duct mounting screws and remove the air duct.

8. Loosen the refrigerant line mounting bolts, and then remove the rear refrigerant line bracket (A).

9. Remove the heater core cover (A) and disconnect the heater hose.

10. Remove the rear heater (A) unit after loosening the mounting bolts and nuts.

11. Remove the heater core (B).

To install:

12. Installation is the reverse of the removal procedure.

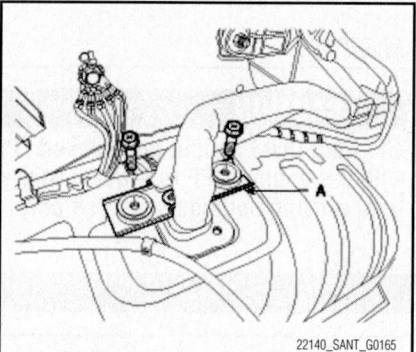

Fig. 306 Remove the rear refrigerant line bracket (A)

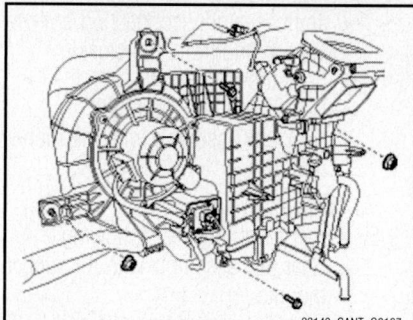

Fig. 308 Remove the rear heater (A) unit after loosening the mounting bolts and nuts

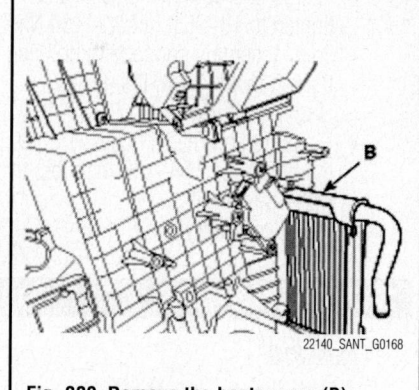

Fig. 309 Remove the heater core (B)

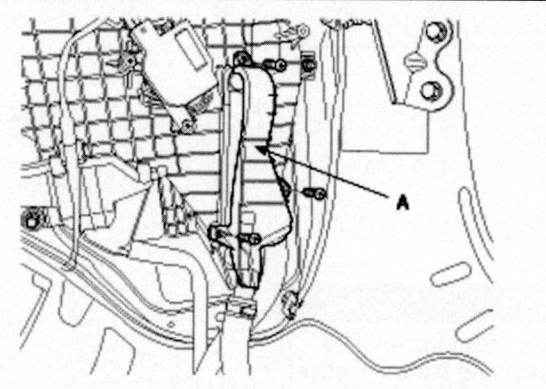

Fig. 307 Remove the heater core cover (A) and disconnect the heater hose

STEERING

POWER RACK & PINION STEERING GEAR

REMOVAL & INSTALLATION

1. Before servicing the vehicle, refer to the Precautions Section.
2. Center the steering wheel and lock it in position.
3. Drain the power steering system.
4. Remove or disconnect the following:
   - Negative battery cable
   - Pressure and return hoses
   - Joint assembly connecting bolt
   - Tie rod end from the knuckle
   - Feed tube
   - Gear box mounting bolts
   - Gear box assembly with the rubber mounts

To install:

5. Install or connect the following:
   - Gear box assembly with the rubber mounts
   - Gear box mounting bolts and tighten to 66–81 ft. lbs. (90–110 Nm)
   - Feed tube and tighten to 84–132 inch lbs. (10–16 Nm)
   - Tie rod end from the knuckle and tighten to 18–25 ft. lbs. (24–34 Nm)
   - Joint assembly connecting bolt and tighten to 11–14 ft. lbs. (15–20 Nm)
   - Pressure and return hoses. Tighten the fittings to 108–156 inch lbs. (12–18 Nm).
   - Negative battery cable

6. Fill the power steering system to the correct level with the appropriate fluid type.
7. Check the wheel alignment and adjust as necessary.

POWER STEERING PUMP

REMOVAL & INSTALLATION

1. Before servicing the vehicle, refer to the Precautions Section.
2. Loosen the bolt fixing the wiring bracket and move the wiring aside.
3. Remove the pressure hose from the oil pump.
4. Disconnect the suction hose from the suction connector and drain the fluid into a container.
5. Loosen the tension adjusting bolt on the power steering belt.
6. Remove the belt from the power steering oil pump pulley.
7. Loosen the power steering oil pump mounting bolt and the tension adjusting bolt.
8. Remove the steering oil pump assembly.
9. Installation is the reverse of the removal procedure.

BLEEDING

❈❈ CAUTION

The fluid level should be checked with the engine OFF to prevent injury from moving components. Use only

PSF-3 power steering fluid, or equivalent. Do not overfill.

1. Before servicing the vehicle, refer to the Precautions Section.
2. Wipe the reservoir fill cap clean before removal.
3. Fill the pump fluid reservoir to the proper level. The fluid level should be within the "FILL RANGE" listed on the exterior of the reservoir when the fluid is at normal ambient temperature, approximately 70–80°F (21–27°C).
4. Let the fluid settle in the system for at least 2 minutes.
5. Start the engine and let it run for a few seconds. Then turn the engine **OFF**.
6. Add fluid if necessary. Repeat the above procedure until the fluid level remains constant after running the engine.
7. Raise the front wheels off the ground.
8. Start the engine. Slowly turn the steering wheel right and left, lightly contacting the wheel stops.
9. Turn the engine off. Check the fluid level and add power steering fluid if necessary.
10. Start the engine. Lower the vehicle and turn the steering wheel slowly from lock to lock.
11. Stop the engine. Check the fluid level and refill as required.
12. If the fluid is extremely foamy, allow the vehicle to stabilize a few minutes, then repeat the above procedure.

SUSPENSION

See Figure 310.

CONTROL LINKS

REMOVAL & INSTALLATION

See Figures 311 and 312.

1. Before servicing the vehicle, refer to the Precautions Section.
2. Raise and safely support the vehicle.
3. Remove the front wheel and tire from the front hub.

❈❈ WARNING

Be careful not to damage the hub bolts when removing the front wheel and tire.

FRONT SUSPENSION

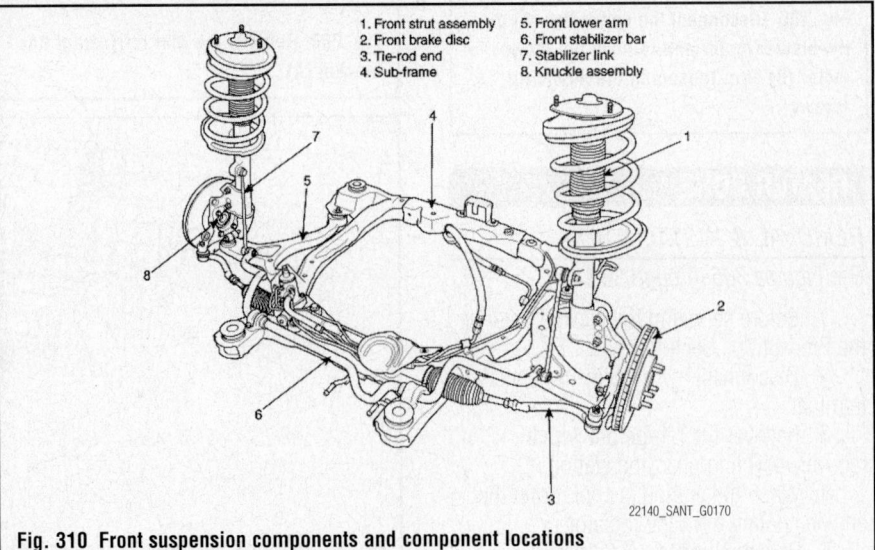

1. Front strut assembly
2. Front brake disc
3. Tie-rod end
4. Sub-frame
5. Front lower arm
6. Front stabilizer bar
7. Stabilizer link
8. Knuckle assembly

22140_SANT_G0170

Fig. 310 Front suspension components and component locations

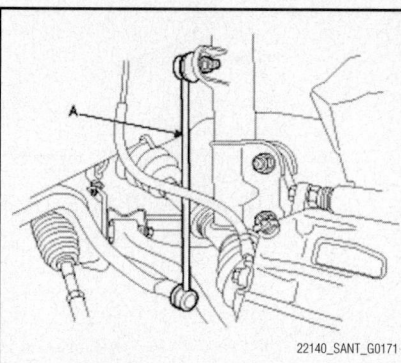

Fig. 311 Stabilizer control link and mounting nuts shown

4. Remove the stabilizer bar control link (A) by removing the upper and lower mounting nuts.

5. Remove the stabilizer bar control link.

To install:

6. Install the stabilizer bar control link.

7. Tighten the stabilizer bar control link upper and lower mounting nuts to 72–87 ft. lbs. (100–120 Nm).

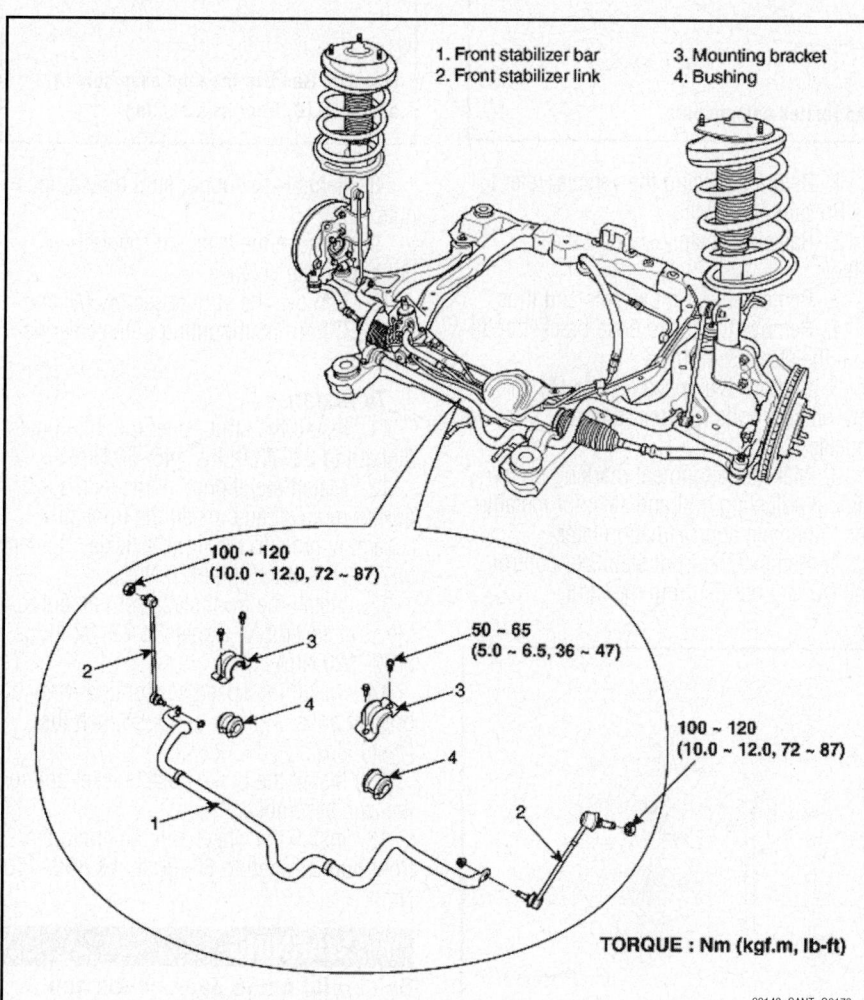

1. Front stabilizer bar
2. Front stabilizer link
3. Mounting bracket
4. Bushing

100 ~ 120
(10.0 ~ 12.0, 72 ~ 87)

50 ~ 65
(5.0 ~ 6.5, 36 ~ 47)

100 ~ 120
(10.0 ~ 12.0, 72 ~ 87)

TORQUE : Nm (kgf.m, lb-ft)

Fig. 312 View of stabilizer bar and related components

8. Install the front wheels, and lower the vehicle. Tighten the wheel lug nuts to 65–80 ft. lbs. (90–110 Nm).

LOWER BALL JOINT

REMOVAL & INSTALLATION

See Figures 313 and 314.

1. Before servicing the vehicle, refer to the Precautions Section.

2. Raise and safely support the vehicle.

3. Remove the front wheel and tire from the front hub.

❊❊ WARNING

Be careful not to damage the hub bolts when removing the front wheel and tire.

4. Remove the tie rod end ball joint from the knuckle.

   a. Remove the split pin (A).

   b. Remove the castle nut (B).

   c. Disconnect the ball joint (C) from

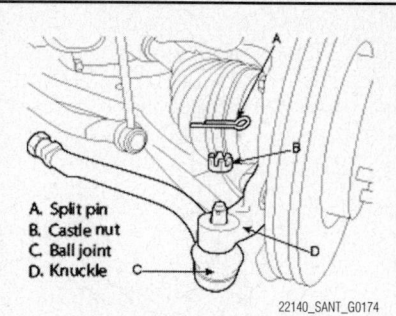

A. Split pin
B. Castle nut
C. Ball joint
D. Knuckle

Fig. 313 Remove the split pin and castle nut. Disconnect the ball joint from the knuckle

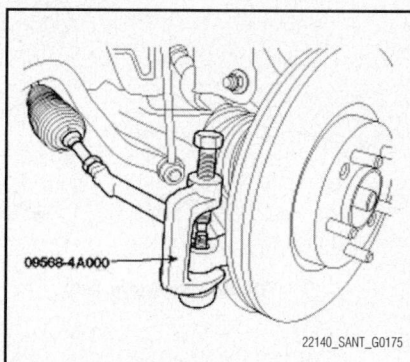

Fig. 314 Using the special tool (09568-4A000) to remove the ball joint

knuckle (D) using the special tool (09568-4A000).

➡**Apply a few drops of oil to the special tool (boot contact part).**

To install:

5. Install the tie rod end ball joint (C) to the knuckle (D).

6. Install the castle nut (B) and the split pin (A). Tighten the nut to: 17–25 ft. lbs. (24–33 Nm).

7. Install the front wheel and tire to the front hub.

LOWER CONTROL ARM

REMOVAL & INSTALLATION

See Figure 315.

1. Before servicing the vehicle, refer to the Precautions Section.

2. Remove or disconnect the following:

   • Front wheel

   • Ball joint-to-knuckle bolt

   • Sub frame bolts and the frame

   • Lower arm bolts and the lower control arm

To install:

3. Installation is the reverse of removal. Tighten the fasteners as follows:

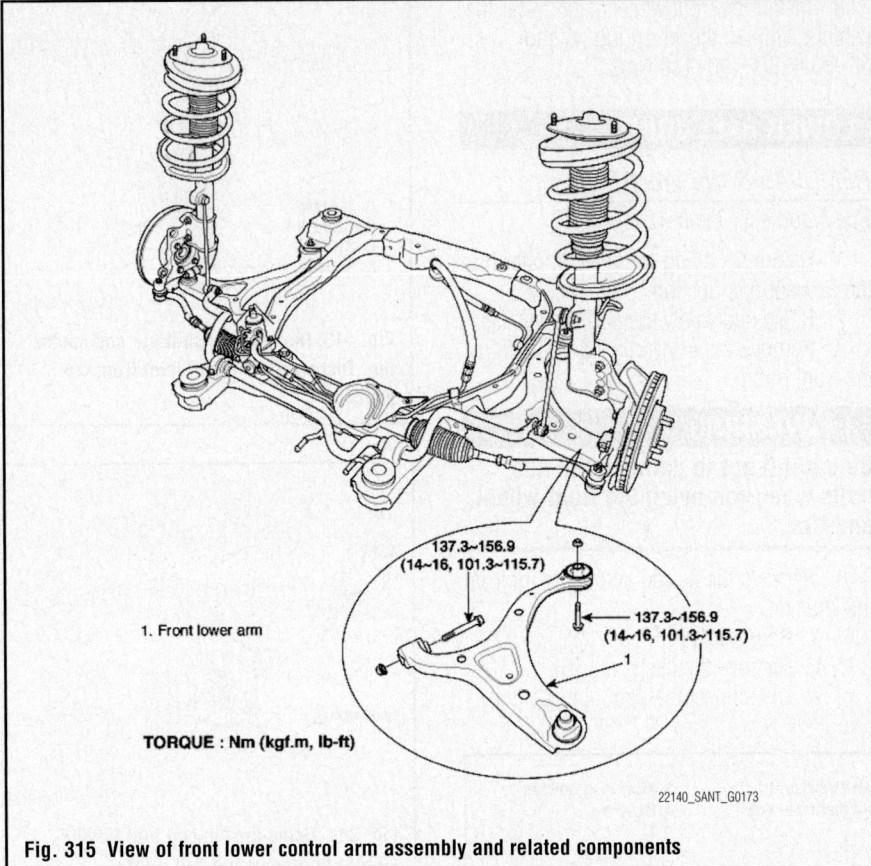

1. Front lower arm

137.3~156.9
(14~16, 101.3~115.7)

137.3~156.9
(14~16, 101.3~115.7)

1

TORQUE : Nm (kgf.m, lb-ft)

22140_SANT_G0173

Fig. 315 View of front lower control arm assembly and related components

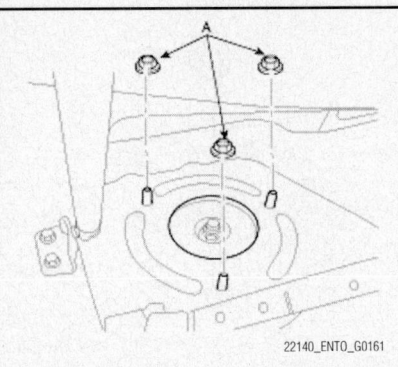

22140_ENTO_G0161

Fig. 318 Remove the upper strut mounting nuts (A)

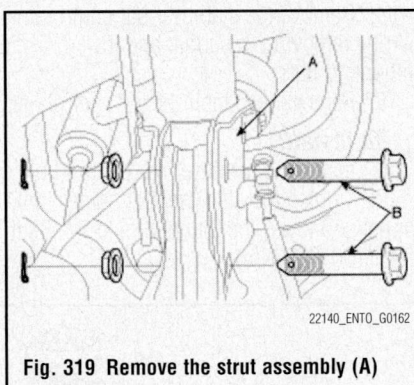

22140_ENTO_G0162

Fig. 319 Remove the strut assembly (A) and bolts (B) from its mounting

a. Tighten the lower control arm bolt (A) to 74–88 ft. lbs. (100–120 Nm).

b. Tighten bolt (B) to 74–88 ft. lbs. (100–120 Nm).

c. Tighten the sub frame bolts to 118–148 ft. lbs. (160–200 Nm).

d. Tighten the ball joint-to-knuckle bolt to 74–88 ft. lbs. (100–120 Nm).

4. Check the wheel alignment and adjust as necessary.

MACPHERSON STRUT

REMOVAL & INSTALLATION

See Figures 316 through 319.

1. Before servicing the vehicle, refer to the Precautions Section.

2. Raise and safely support the vehicle.

3. Remove the front wheels and tires.

4. Remove the brake hose bracket bolts from the strut assembly.

5. Remove the speed sensor (A) and wire bracket bolts (B) from the front steering knuckle.

6. Make an alignment marking on the camber adjusting bolt and strut for installation alignment approximation later.

7. Remove the front stabilizer control link (A) and nut (B) from the strut.

8. Remove the upper strut mounting nuts (A).

9. Remove the front strut mounting bolts from the knuckle.

10. Remove the strut assembly (A) and bolts (B) from its mounting in the steering knuckle.

To install:

11. Install the strut upper mounting nuts. Tighten to 33–43 ft. lbs. (45–60 Nm).

12. Match the alignment marks made during removal and install the front strut assembly bolts to the front knuckle. Tighten to 72–87 ft. lbs. (100–120 Nm).

13. Install the front stabilizer link nut to the strut assembly. Tighten to 72–87 ft. lbs. (100–120 Nm).

14. Install the speed sensor and wire bracket bolts. Tighten to 60–96 inch lbs. (7–11 Nm).

15. Install the brake hose bracket bolt to the axle assembly.

16. Install the wheel and the tire to the front hub. Tighten to 65–80 ft. lbs. (90–110 Nm).

❊❊ WARNING

Be careful not to damage the hub bolts when installing the front wheel and tire.

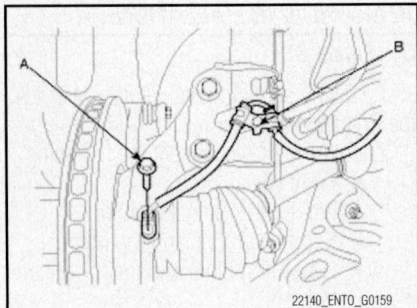

22140_ENTO_G0159

Fig. 316 Remove the speed sensor (A) and wire bracket bolts (B) from the front steering knuckle

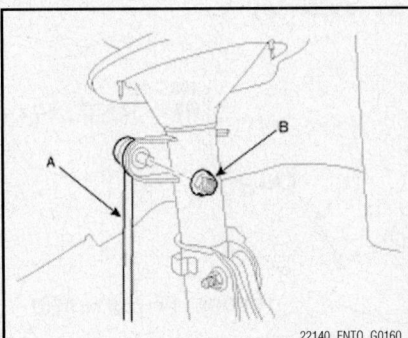

22140_ENTO_G0160

Fig. 317 Remove the front stabilizer control link (A) and nut (B) from the strut

17. Check the wheel alignment and adjust as necessary.

OVERHAUL

1. Before servicing the vehicle, refer to the Precautions Section.

2. Remove the strut from the vehicle and attach Service Tool 09546-2600 or another suitable spring compressor.

3. Compress the coil spring.

4. Remove the self-locking nut from the strut.

5. Remove the insulator, spring seat, coil spring, and dust cover.

6. Inspect the insulator for wear and damage, replace as required.

7. Check the rubber parts for damage or deterioration, replace as required.

8. Install the spring lower pad so that the protrusions fit the holes in the spring lower seat.

9. Compress the spring, using the spring compressor tool.

10. Install the compressed spring over the shock absorber.

➡ **There are 2 color identification marks on the coil spring, one indicates model option and the other indicates load classification. Install the coil spring with the identification mark directed toward the steering knuckle.**

11. After fully extending the piston rod, install the spring upper seat and insulator assembly.

12. After correctly seating the upper and lower ends of the coil spring in the upper and lower spring grooves, tighten the NEW self-locking nut temporarily.

13. Remove the spring compression tool. Tighten the self-locking nut to 43–51 ft. lbs. (60–70 Nm).

14. Install the strut to the vehicle.

15. Check the wheel alignment and adjust as necessary.

STEERING KNUCKLE

REMOVAL & INSTALLATION

See Figure 320.

1. Before servicing the vehicle, refer to the Precautions Section.

2. Raise and safely support the vehicle.

3. Remove the front wheels and tires.

4. Remove the brake rotor. Refer to Front Disc Brakes, Rotor, removal & installation.

5. Remove the lower ball joint. Refer to Lower Ball Joint, removal & installation.

6. Remove the wheel speed sensor (B), the strut lower mounting bolt (C), and the

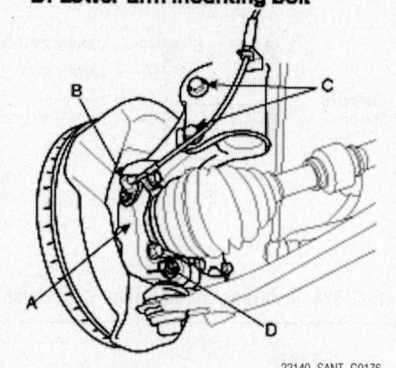

A. Steering knuckle
B. Wheel speed sensor
C. Strut lower mounting bolt
D. Lower arm mounting bolt

22140_SANT_G0176

Fig. 320 Remove the wheel speed sensor, strut lower mounting bolt, and lower arm mounting bolt from the steering knuckle

lower arm mounting bolt (D) from the steering knuckle (A).

7. Remove the hub and knuckle assembly.

❊❊ WARNING

Be careful not to damage the boot and rotor teeth.

To install:

8. Install the hub and knuckle assembly.

9. Install the wheel speed sensor (B), the strut lower mounting bolt (C), and the lower arm mounting bolt (D) to the knuckle (A) and tighten:

a. The wheel speed sensor (B) to: 61–96 inch lbs. (7–11 Nm).

b. Bolts (C) to: 112–127 ft. lbs. (152–172 Nm).

c. Bolt (D) to: 72–87 ft. lbs. (98–118 Nm).

10. Install the lower ball joint. Refer to Lower Ball Joint, removal & installation.

11. Install the brake rotor. Refer to Front Disc Brakes, Rotor, removal & installation.

12. Install the front wheels and tires.

13. Check alignment and adjust as necessary.

STABILIZER BAR

REMOVAL & INSTALLATION

See Figure 321.

1. Before servicing the vehicle, refer to the Precautions Section.

2. Raise the front of the vehicle, and make sure it is securely supported.

3. Remove the front wheel and tire from the front hub.

❊❊ WARNING

Be careful not to damage the hub bolts when removing the front wheel and tire.

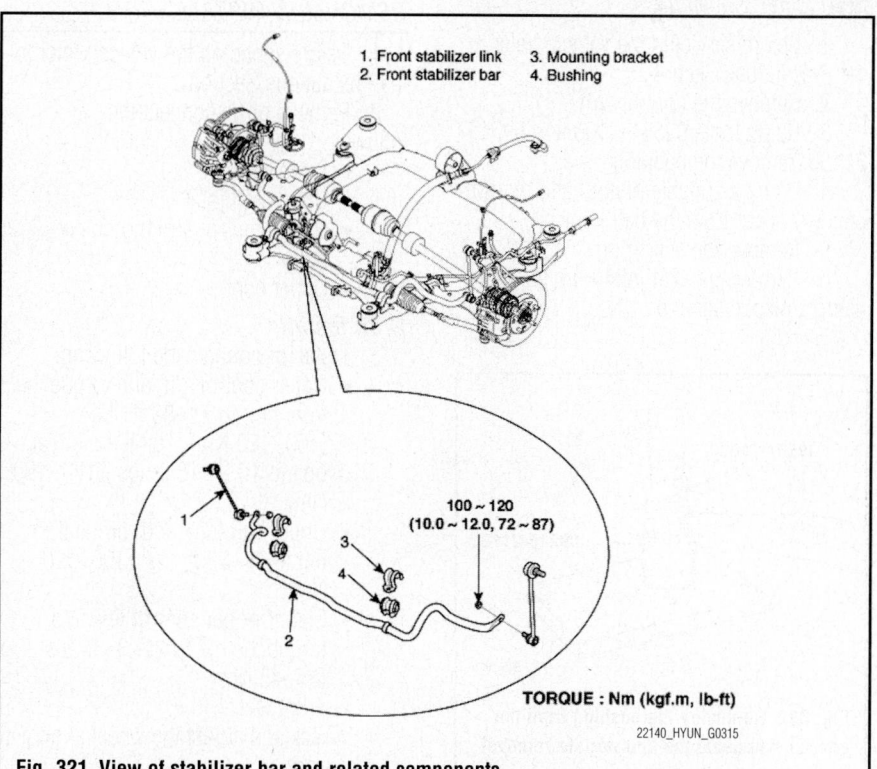

1. Front stabilizer link
2. Front stabilizer bar
3. Mounting bracket
4. Bushing

100 ~ 120
(10.0 ~ 12.0, 72 ~ 87)

TORQUE : Nm (kgf.m, lb-ft)

22140_HYUN_G0315

Fig. 321 View of stabilizer bar and related components

4. Remove the nut and stabilizer bar control link. Refer to Control Links, removal & installation.

5. Remove the control link on the opposite side in the same way.

6. Remove the rear mounting bolts of the sub-frame.

7. Remove the stabilizer bracket and bushing.

8. Remove the stabilizer bracket and bushing on the opposite side in the same way.

9. Remove the stabilizer bar.

❊ WARNING

Be careful not to damage the power steering pressure tube.

To install:

10. Installation is the reverse of the removal.

11. Tighten the stabilizer bracket bolts to: 32–39 ft. lbs. (45–55 Nm).

12. Tighten the sub-frame mounting bolts to: 68–86 ft. lbs. (95–120 Nm).

13. Tighten the stabilizer bar control link upper and lower mounting nuts to 72–87 ft. lbs. (100–120 Nm).

14. Tighten the wheel nuts to: 65–80 ft. lbs. (90–110 Nm).

UPPER BALL JOINT

REMOVAL & INSTALLATION

See Figures 322 and 323.

1. Before servicing the vehicle, refer to the Precautions Section.

2. Remove the control arm.

3. Using tools 09551-3100 and 09216-21100, remove the bushing.

4. Using a suitable prytool and remove the dust cover from the ball joint.

5. Remove the snap-ring.

6. Remove the ball joint from the arm using a plastic hammer.

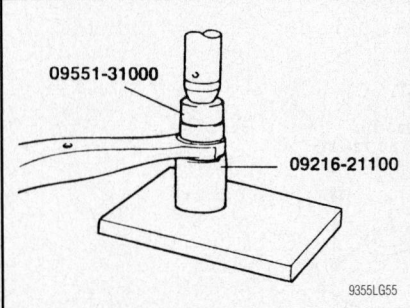

Fig. 322 Removing the bushing from the control arm using the appropriate removal and installation tools

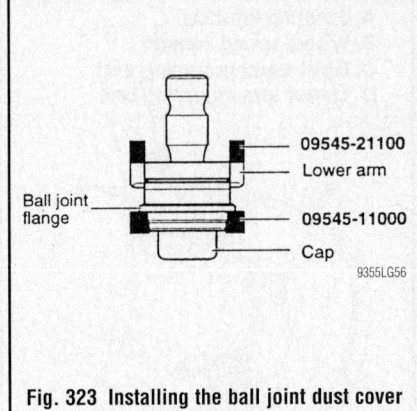

Fig. 323 Installing the ball joint dust cover

To install:

7. Install the ball joint.

8. Install the bushings into the arm using the appropriate tools. Make sure that the ball joint flange is supported while pressing down on the bushing until the flange touches the arm surface.

9. Install the ball joint snap-ring. Be careful to keep the snap-ring expansion as small as possible during installation.

10. Apply multi-purpose grease to the dust cover lip and inside of the cover.

11. Using tool 09545-11000, install the dust cover until it is completely seated on the snap ring.

12. Install the control arm.

UPPER CONTROL ARM

REMOVAL & INSTALLATION

1. Before servicing the vehicle, refer to the Precautions Section.

2. Remove or disconnect the following:
- Front wheel
- Upper ball joint
- Upper control arm mounting bolts
- Upper control arm

To install:

3. Install or connect the following:
- Upper control arm and torque the front bolt to 74–89 ft. lbs. (100–120 Nm). Tighten the rear bolt to 103–118 ft. lbs. (140–160 Nm)
- Upper ball joint and torque the nut to 74–89 ft. lbs. (100–120 Nm)
- Stabilizer bar control link and torque the nut to 25–33 ft. lbs. (35–45 Nm)
- Front wheel

4. Check and adjust the wheel alignment as necessary.

WHEEL HUB AND BEARING (SEALED UNIT)

REMOVAL & INSTALLATION

See Figures 324 through 331.

1. Before servicing the vehicle, refer to the Precautions Section.

2. Remove or disconnect the following:
- Front wheel
- Wheel Speed Sensor (WSS) from the knuckle
- Brake caliper and suspend it to one side using wire
- Split pin and nut from the axle
- Strut from the knuckle
- Tie rod end from the knuckle
- Lower ball joint bolt
- Axle shaft from the knuckle using a plastic hammer
- Brake disc
- Knuckle assembly
- Snap-ring from the hub
- Hub from the knuckle by installing tools 09517-3A00 and 09517-2900, then tighten the nut of the tool to separate the tool from the knuckle
- Wheel bearing inner race from the hub using tools 09455-2100 and 09545-34100

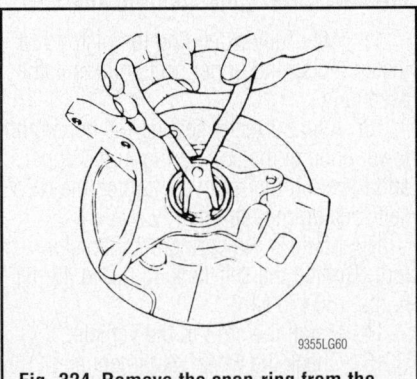

Fig. 324 Remove the snap-ring from the hub

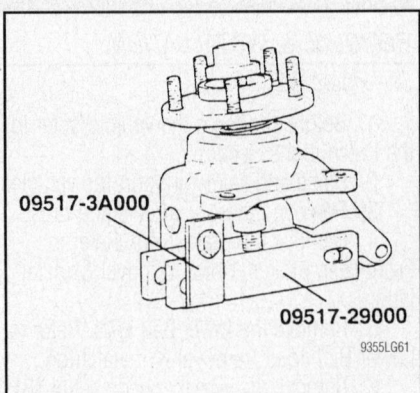

Fig. 325 Remove the hub from the knuckle

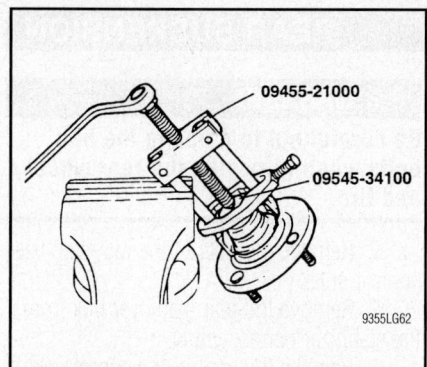

Fig. 326 Remove the wheel bearing inner race from the hub

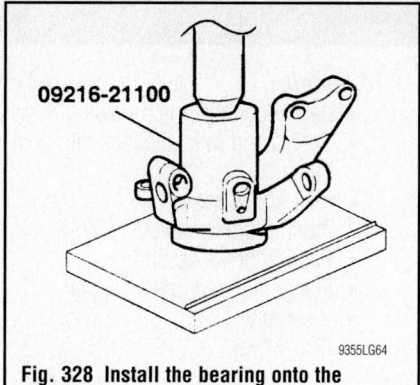

Fig. 328 Install the bearing onto the knuckle

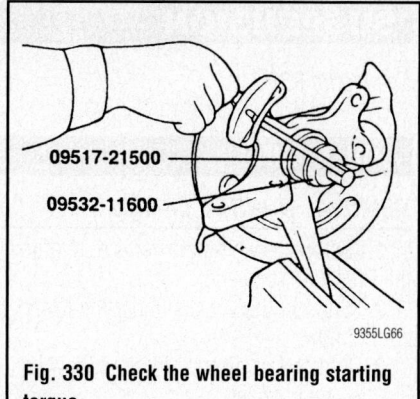

Fig. 330 Check the wheel bearing starting torque

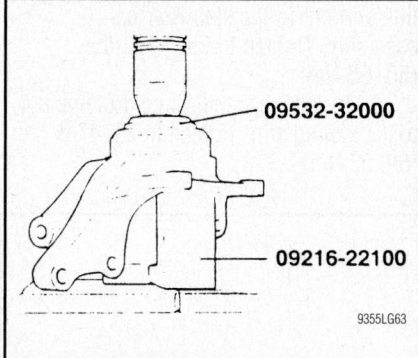

Fig. 327 Remove the wheel bearing outer race from the knuckle

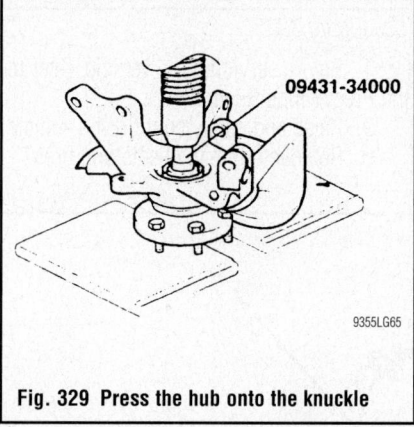

Fig. 329 Press the hub onto the knuckle

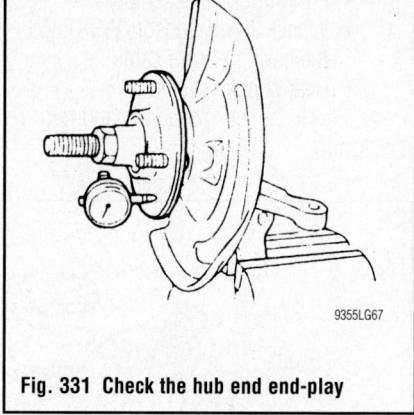

Fig. 331 Check the hub end end-play

- Wheel bearing outer race from the knuckle using tools 09532-3200 and 09216-22100

3. Check all components for wear or damage. Replace as necessary.

To install:

4. Apply a thin coat of multi-purpose grease to the surface on the knuckle and bearing.

5. Install or connect the following:
- Bearing onto the knuckle using tool 09216-21100
- Snap-ring into the groove of the knuckle

- Backing plate onto the knuckle

✳✳ WARNING

Do not press against the outer race of the bearing as this can cause bearing damage. Always use a new bearing kit.

- Hub onto the knuckle by pressing it into position using tool 09431-3400

6. Rotate the bearing several times to seat the bearing.

7. Measure the wheel bearing torque using a torque wrench. The standard measurement is 17 inch lbs. (2 Nm).

8. Measure the end-play of the hub using a dial gauge. The specification is 0.0025–0.0035 inch (0.064–0.088mm).

9. Install the remaining components in the reverse order of removal.

10. Check the wheel alignment and adjust as necessary.

ADJUSTMENT

The wheel bearings are sealed units and are not adjustable.

SUSPENSION

See Figure 332.

COIL SPRING

REMOVAL & INSTALLATION

1. Before servicing the vehicle, refer to the Precautions Section.
2. Support the vehicle under the lower control arm.
3. Remove or disconnect the following:
   - Rear wheel
   - Flange nut and brake caliper assembly
   - Parking brake assembly
   - Wheel Speed Sensor (WSS) and the parking brake cable
   - Rear shock assembly
4. Lower the jack assembly and remove the spring.

To install:
5. Install or connect the following:
   - Spring and raise the jack into position
   - Rear shock assembly
   - Parking brake cable and WSS
   - Parking brake assembly
   - Flange nut and brake caliper assembly
   - Rear wheel

CONTROL ARMS/LINKS

REMOVAL & INSTALLATION

See Figure 333.

1. Before servicing the vehicle, refer to the Precautions Section.
2. Raise and safely support the vehicle.
3. Remove the rear wheels and tires.

✴✴ WARNING

Be careful not to damage the hub bolts when removing the rear wheel and tire.

4. Remove the nuts of the rear stabilizer control links.
5. Remove the rear stabilizer link from the stabilizer bar assembly.
6. Remove the control links from the vehicle.

To install:
7. Install the rear stabilizer control link and nut to the stabilizer bar assembly. Tighten to 36–47 ft. lbs. (50–65 Nm).
8. Install the stabilizer control link nut to the trailing arm. Tighten to 36–47 ft. lbs. (50–65 Nm).

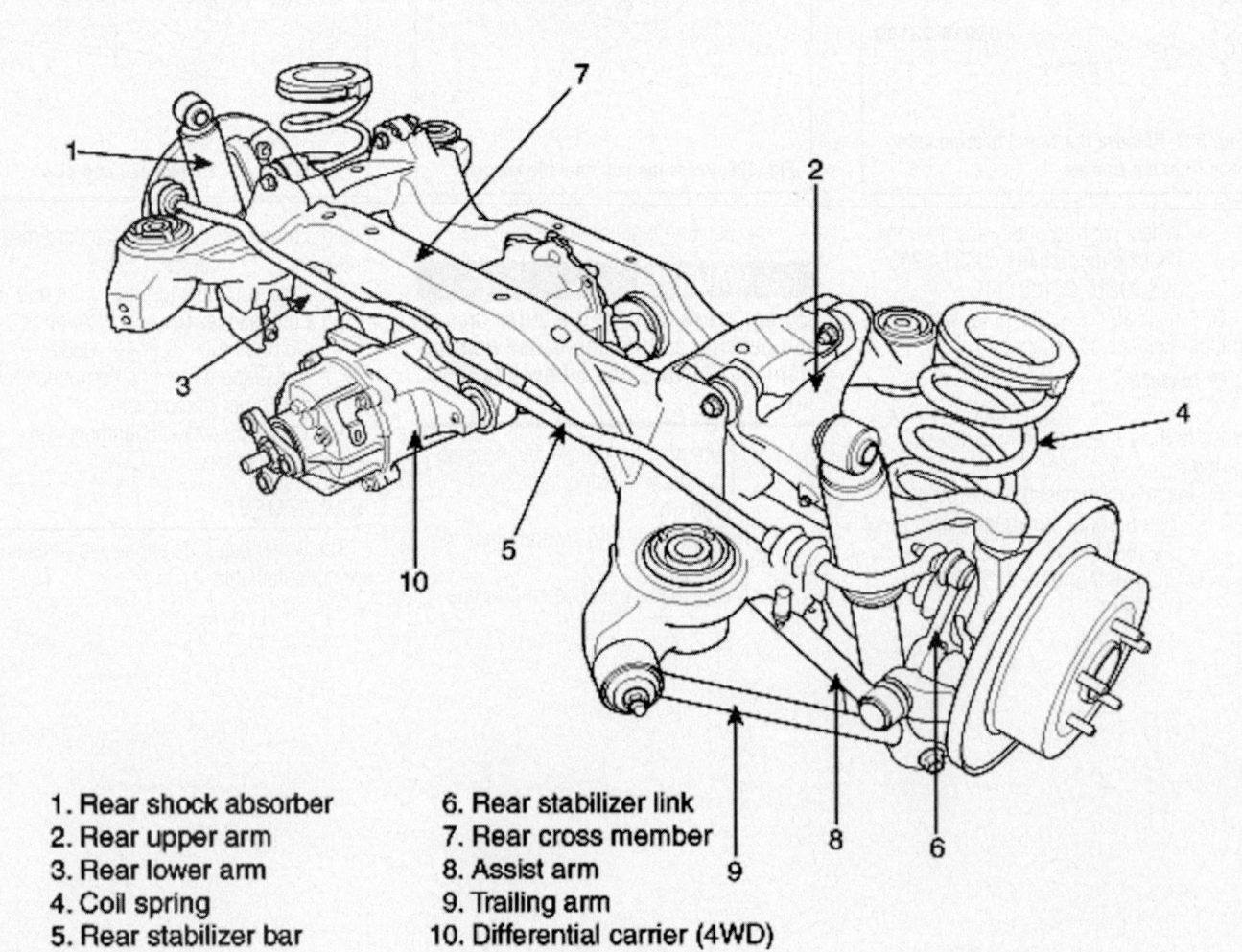

1. Rear shock absorber
2. Rear upper arm
3. Rear lower arm
4. Coil spring
5. Rear stabilizer bar
6. Rear stabilizer link
7. Rear cross member
8. Assist arm
9. Trailing arm
10. Differential carrier (4WD)

22140_SANT_G0177

Fig. 332 Rear suspension assembly and related components

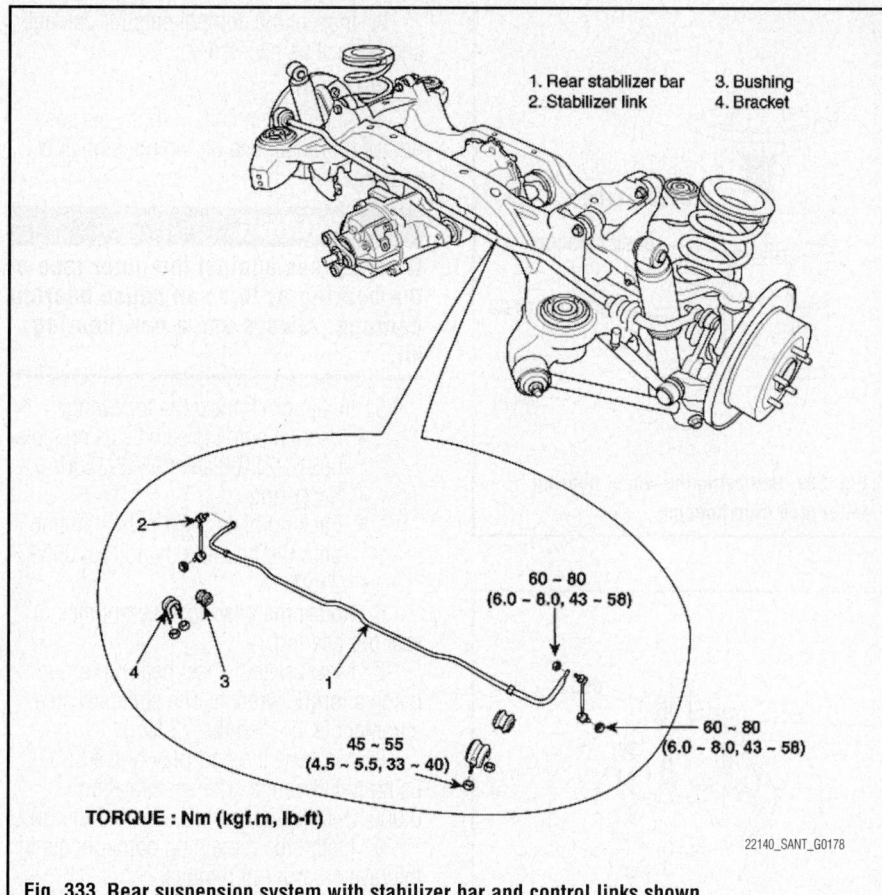

1. Rear stabilizer bar   3. Bushing
2. Stabilizer link   4. Bracket

60 – 80
(6.0 – 8.0, 43 – 58)

60 ~ 80
(6.0 ~ 8.0, 43 ~ 58)

45 ~ 55
(4.5 ~ 5.5, 33 ~ 40)

TORQUE : Nm (kgf.m, lb-ft)

22140_SANT_G0178

Fig. 333 Rear suspension system with stabilizer bar and control links shown

9. Repeat appropriate steps for the other side.

10. Install the wheel and the tire to the rear hub. Tighten to 65–80 ft. lbs. (90–110 Nm).

❋❋ WARNING

Be careful not to damage the hub bolts when installing the rear wheel and tire.

SHOCK ABSORBER

REMOVAL & INSTALLATION

1. Before servicing the vehicle, refer to the Precautions Section.

2. Support the vehicle under the lower control arm.

3. Remove or disconnect the following:
   • Rear wheel
   • Upper shock absorber upper nut
   • Lower shock absorber nut
   • Shock absorber

To install:

4. Install the shock absorber. Tighten the upper nut to 15–22 ft. lbs. (20–30 Nm). Tighten the lower nut to 88–103 ft. lbs. (120–140 Nm).

5. Install the rear wheel.

TESTING

1. Check the rubber parts for damage or deterioration.

2. Check for correct height and proper return of shock absorber to original height.

3. Check the shock absorber for abnormal resistance or unusual sounds.

4. Check for oil leakage around seals.

5. Replace as necessary.

WHEEL HUB AND BEARING (SEALED UNIT)

REMOVAL & INSTALLATION

See Figures 334 through 338.

1. Before servicing the vehicle, refer to the Precautions Section.

2. Remove or disconnect the following:
   • Rear wheel
   • Flange nut and washer
   • Drum or rotor
   • Brake line
   • Parking brake assembly
   • Parking brake cable
   • Spindle bolts and the spindle
   • Rear hub from the housing using tool 09517-43001
   • Wheel bearing snap-ring
   • Wheel bearing inner race from the housing using tools 09500-2100, 09527-33000, and 09216-22100

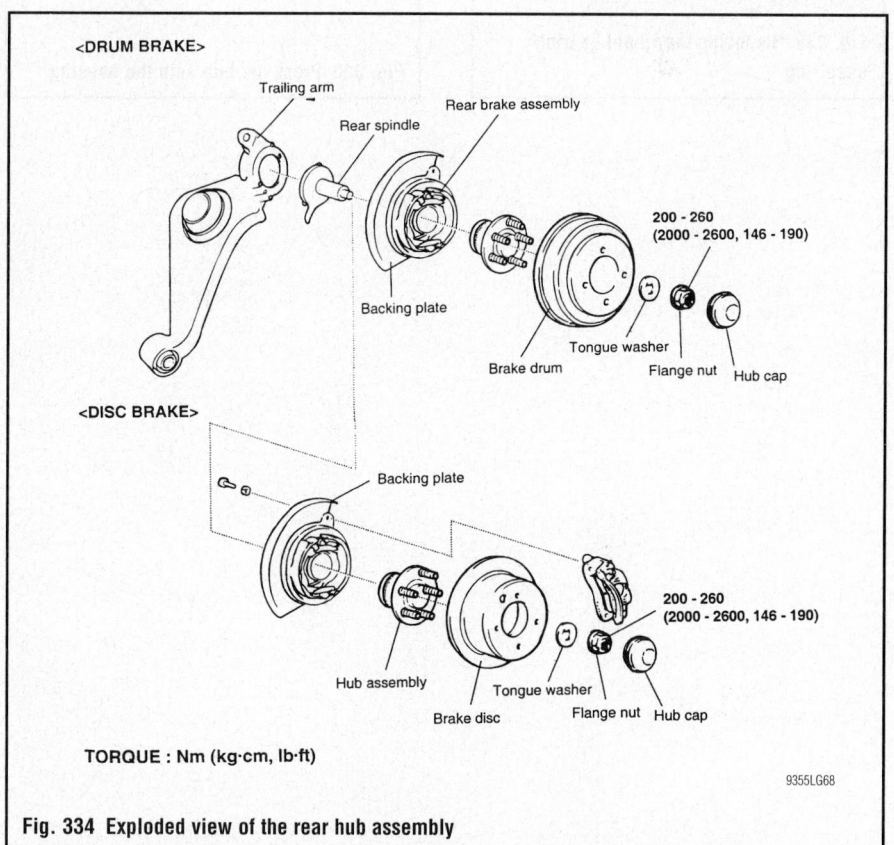

<DRUM BRAKE>

Trailing arm

Rear spindle

Rear brake assembly

Backing plate

200 - 260
(2000 - 2600, 146 - 190)

Tongue washer

Brake drum   Flange nut   Hub cap

<DISC BRAKE>

Backing plate

200 - 260
(2000 - 2600, 146 - 190)

Hub assembly

Brake disc   Tongue washer

Flange nut   Hub cap

TORQUE : Nm (kg·cm, lb-ft)

9355LG68

Fig. 334 Exploded view of the rear hub assembly

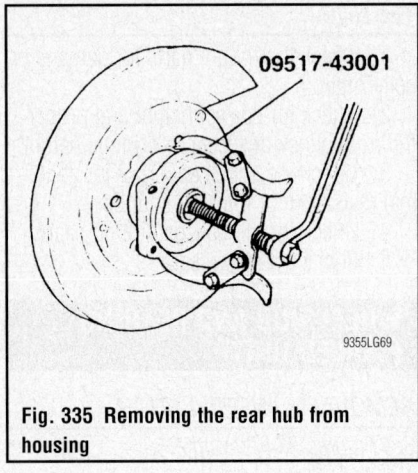

Fig. 335 Removing the rear hub from housing

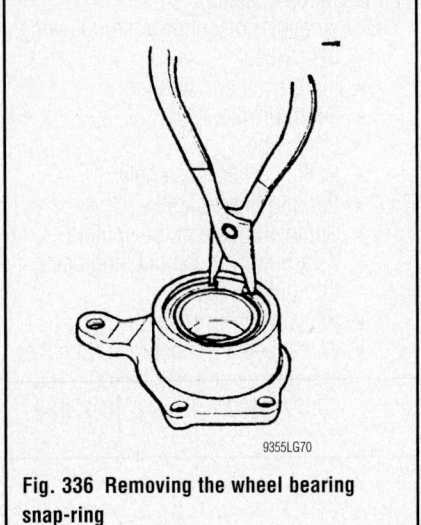

Fig. 336 Removing the wheel bearing snap-ring

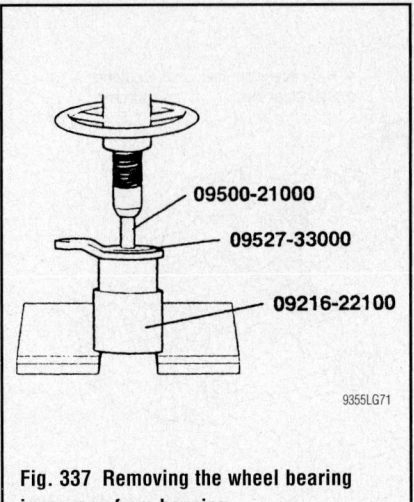

Fig. 337 Removing the wheel bearing inner race from housing

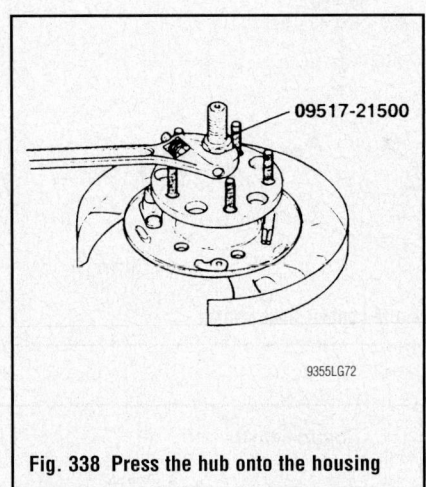

Fig. 338 Press the hub onto the housing

3. Inspect the components for damage and replace as necessary.

To install:

4. Apply a thin coat of multi-purpose grease to the surface on the housing and bearing.

> ### ✳✳ WARNING
>
> **Do not press against the outer race of the bearing as this can cause bearing damage. Always use a new bearing kit.**

5. Install or connect the following:
   - Bearing onto the spindle using tools 09216-21100 and 09532-3200
   - Snap-ring
   - Backing plate, then press the hub onto the housing using tool 09517-21500

6. Rotate the bearing several times to seat the bearing.

7. Measure the wheel bearing torque using a torque wrench. The standard measurement is 17 inch lbs. (2 Nm).

8. Measure the end-play of the hub using a dial gauge. The specification is 0.00012–0.00032 inch (0.003–0.008mm).

9. Install the remaining components in the reverse order of removal.

10. Check the wheel alignment and adjust as necessary.

ADJUSTMENT

The wheel bearings are sealed units and are not adjustable.

SPECIFICATIONS AND MAINTENANCE CHARTS

ENGINE AND VEHICLE IDENTIFICATION

		Engine						Model Year	
Code ①	Liters (cc)	Cu. In.	Cyl.	Fuel Sys.	Engine Type	Eng. Mfg.	Code ②		Year
B	2.0 (1975)	120.52	I4	MPFI	DOHC	Hyundai	6		2006
D	2.7 (2656)	164.30	V6	MPFI	DOHC	Hyundai	7		2007
							8		2008

MPFI: Multi-Point Fuel Injection

DOHC: Double Overhead Camshafts

① 8th digit of VIN

② 10th digit of VIN

22140_TUCS_C0001

GENERAL ENGINE SPECIFICATIONS

All measurements are given in inches.

Year	Model	Engine Displacement Liters	Engine Series VIN	Net Horsepower @ rpm	Net Torque @ rpm (ft. lbs.)	Bore x Stroke (in.)	Com-pression Ratio	Oil Pressure @ rpm
2006	Tucson	2.0	B	140@6000	136@4500	3.23 x 3.68	10.1:1	36@1500
	Tucson	2.7	D	173@6000	178@4000	3.41 x 2.95	10.0:1	NA
2007	Tucson	2.0	B	140@6000	136@4500	3.23 x 3.68	10.1:1	36@1500
	Tucson	2.7	D	173@6000	178@4000	3.41 x 2.95	10.0:1	NA
2008	Tucson	2.0	B	140@6000	136@4500	3.23 x 3.68	10.1:1	36@1500
	Tucson	2.7	D	173@6000	178@4000	3.41 x 2.95	10.0:1	NA

NA: Not Available

22140_TUCS_C0002

GASOLINE ENGINE TUNE-UP SPECIFICATIONS

Year	Engine Displacement Liters	Engine VIN	Spark Plug Gap (in.)	Ignition Timing (deg.) MT	Ignition Timing (deg.) AT	Fuel Pump (psi)	Idle Speed (rpm) MT	Idle Speed (rpm) AT	Valve Clearance (in.) In.	Valve Clearance (in.) Ex.
2006	2.0	B	0.039-0.043	①	①	49.8	②	②	0.0079	0.0110
	2.7	D	0.039-0.043	①	①	49.8	②	②	HYD	HYD
2007	2.0	B	0.039-0.043	①	①	49.8	②	②	0.0079	0.0110
	2.7	D	0.039-0.043	①	①	49.8	②	②	HYD	HYD
2008	2.0	B	0.039-0.043	①	①	49.8	②	②	0.0079	0.0110
	2.7	D	0.039-0.043	①	①	49.8	②	②	HYD	HYD

NOTE: The Vehicle Emission Control Information label reflects specification changes made during production.

Follow the figures on the label if they differ from those in this chart.

HYD: Hydraulic

① Ignition timing is preset and cannot be adjusted

② Idle speed is maintained by the Electronic Control Module (ECM)

22140_TUCS_C0003

CAPACITIES

Year	Model	Engine Displacement Liters	Engine VIN	Engine Oil with Filter (qts.)	Transmission (pts.)		Fuel Tank (gal.)	Cooling System (qts.)
					Manual	Auto. ①		
2006	Tucson	2.0	B	4.2	4.5	16.4	15.3	6.4
	Tucson	2.7	D	4.8	—	16.4	17.7	8.2
2007	Tucson	2.0	B	4.2	4.5	16.4	15.3	6.4
	Tucson	2.7	D	4.8	—	16.4	17.7	8.2
2008	Tucson	2.0	B	4.2	4.5	16.4	15.3	6.4
	Tucson	2.7	D	4.8	—	16.4	17.7	8.2

NOTE: All capacities are approximate. Add fluid gradually and check to be sure a proper fluid level is obtained.

① Drain and refill

22140_TUCS_C0004

FLUID SPECIFICATIONS

Year	Model	Engine Displacement Liters	Engine ID/VIN	Engine Oil	Manual Trans.	Auto. Trans.	Drive Axle	Power Steering Fluid	Brake Master Cylinder	Cooling System
2006	Tucson	2.0	B	5W-20	①	②	③	PSF-3	④	⑤
	Tucson	2.7	D	5W-20	—	②	③	PSF-3	④	⑤
2007	Tucson	2.0	B	5W-20	①	②	③	PSF-3	④	⑤
	Tucson	2.7	D	5W-20	—	②	③	PSF-3	④	⑤
2008	Tucson	2.0	B	5W-20	①	②	③	PSF-3	④	⑤
	Tucson	2.7	D	5W-20	—	②	③	PSF-3	④	⑤

DOT: Department Of Transportation

① HYUNDAI GENUINE PART MTF 75W/90 (API GL-4)
② DIAMOND ATF SP-3, SK ATF SP-3
③ Hypoid gear oil (API GL-5, SAE 80W/90, SHELL SPIRAX AX or equivalent)
④ DOT 3, DOT 4, or equivalent
⑤ Ethylene glycol base for aluminum radiator

22140_TUCS_C0005

VALVE SPECIFICATIONS

Year	Engine Displacement Liters	Engine VIN	Seat Angle (deg.)	Face Angle (deg.)	Spring Test Pressure (lbs. @ in.)	Spring Installed Height (in.)	Stem-to-Guide Clearance (in.)		Stem Diameter (in.)	
							Intake	Exhaust	Intake	Exhaust
2006	2.0	B	45	45	41.5 @1.535	1.535	0.0008-0.0019	0.0014-0.0026	0.2348-0.2354	0.2343-0.2348
	2.7	D	NA	45.0-45.5	48.4 @1.378	NA	0.0008-0.0020	0.0012-0.0026	0.2350-0.2354	0.2340-0.2350
2007	2.0	B	45	45	41.5 @1.535	1.535	0.0008-0.0019	0.0014-0.0026	0.2348-0.2354	0.2343-0.2348
	2.7	D	NA	45.0-45.5	48.4 @1.378	NA	0.0008-0.0020	0.0012-0.0026	0.2350-0.2354	0.2340-0.2350
2008	2.0	B	45	45	41.5 @1.535	1.535	0.0008-0.0019	0.0014-0.0026	0.2348-0.2354	0.2343-0.2348
	2.7	D	NA	45.0-45.5	48.4 @1.378	NA	0.0008-0.0020	0.0012-0.0026	0.2350-0.2354	0.2340-0.2350

NA: Not Available

22140_TUCS_C0006

CAMSHAFT AND BEARING SPECIFICATIONS CHART

All measurements are given in inches.

Year	Engine Displ. Liters	Engine ID/VIN	Journal Dia.	Brg. Oil Clearance	Shaft End-play	Runout	Journal Bore	Lobe Height Intake	Lobe Height Exhaust
2006	2.0	B	1.1023	0.0008-0.0024	0.0040-0.0080	NA	NA	1.7527-1.7566	1.7487-1.7527
	2.7	D	1.0222-1.0228	0.0007-0.0024	0.0039-0.0059	NA	NA	1.7303-1.7382	1.7303-1.7382
2007	2.0	B	1.1023	0.0008-0.0024	0.0040-0.0080	NA	NA	1.7527-1.7566	1.7487-1.7527
	2.7	D	1.0222-1.0228	0.0007-0.0024	0.0039-0.0059	NA	NA	1.7303-1.7382	1.7303-1.7382
2008	2.0	B	1.1023	0.0008-0.0024	0.0040-0.0080	NA	NA	1.7527-1.7566	1.7487-1.7527
	2.7	D	1.0222-1.0228	0.0007-0.0024	0.0039-0.0059	NA	NA	1.7303-1.7382	1.7303-1.7382

NA: Not Available

22140_TUCS_C0007

CRANKSHAFT AND CONNECTING ROD SPECIFICATIONS

All measurements are given in inches.

Year	Engine Displacement Liters	Engine VIN	Crankshaft Main Brg. Journal Dia.	Crankshaft Main Brg. Oil Clearance	Crankshaft Shaft End-play	Crankshaft Thrust on No.	Connecting Rod Journal Diameter	Connecting Rod Oil Clearance	Connecting Rod Side Clearance
2006	2.0	B	2.2440	0.0011-0.0018	0.0023-0.0100	3	1.7700	0.0009-0.0016	0.0039-0.0100
	2.7	D	2.4402-2.4409	0.0002-0.0009	0.0028-0.0098	3	1.8891-1.8898	0.0007-0.0014	0.0039-0.0098
2007	2.0	B	2.2440	0.0011-0.0018	0.0023-0.0100	3	1.7700	0.0009-0.0016	0.0039-0.0100
	2.7	D	2.4402-2.4409	0.0002-0.0009	0.0028-0.0098	3	1.8891-1.8898	0.0007-0.0014	0.0039-0.0098
2008	2.0	B	2.2440	0.0011-0.0018	0.0023-0.0100	3	1.7700	0.0009-0.0016	0.0039-0.0100
	2.7	D	2.4402-2.4409	0.0002-0.0009	0.0028-0.0098	3	1.8891-1.8898	0.0007-0.0014	0.0039-0.0098

22140_TUCS_C0008

PISTON AND RING SPECIFICATIONS

All measurements are given in inches.

Year	Engine Displ. Liters	Engine VIN	Piston Clearance	Ring Gap			Ring Side Clearance		
				Top Compression	Bottom Compression	Oil Control	Top Compression	Bottom Compression	Oil Control
2006	2.0	B	0.0008-0.0016	0.0090-0.0149	0.0130-0.0189	0.0078-0.0236	0.0015-0.0031	0.0012-0.0027	NA
	2.7	D	0.0004-0.0012	0.0079-0.0138	0.0146-0.0205	0.0079-0.0276	0.0016-0.0031	0.0012-0.0028	NA
2007	2.0	B	0.0008-0.0016	0.0090-0.0149	0.0130-0.0189	0.0078-0.0236	0.0015-0.0031	0.0012-0.0027	NA
	2.7	D	0.0004-0.0012	0.0079-0.0138	0.0146-0.0205	0.0079-0.0276	0.0016-0.0031	0.0012-0.0028	NA
2008	2.0	B	0.0008-0.0016	0.0090-0.0149	0.0130-0.0189	0.0078-0.0236	0.0015-0.0031	0.0012-0.0027	NA
	2.7	D	0.0004-0.0012	0.0079-0.0138	0.0146-0.0205	0.0079-0.0276	0.0016-0.0031	0.0012-0.0028	NA

NA: Not Available

22140_TUCS_C0009

TORQUE SPECIFICATIONS

All readings in ft. lbs.

Year	Engine Displacement Liters	Engine VIN	Cylinder Head Bolts	Main Bearing Bolts	Rod Bearing Bolts	Crankshaft Damper Bolts	Flywheel Bolts	Manifold		Spark Plugs	Oil Pan Drain Plug
								Intake	Exhaust		
2006	2.0	B	①	②	37-39	125-133	88-95	12-17	32-40	15-22	30-33
	2.7	D	③	④	⑤	130-138	53-56	14-15	22-26	15-22	25-33
2007	2.0	B	①	②	37-39	125-133	88-95	12-17	32-40	15-22	30-33
	2.7	D	③	④	⑤	130-138	53-56	14-15	22-26	15-22	25-33
2008	2.0	B	①	②	37-39	125-133	88-95	12-17	32-40	15-22	30-33
	2.7	D	③	④	⑤	130-138	53-56	14-15	22-26	15-22	25-33

NOTE: Dip main bearing bolts and crankshaft damper bolt in clean engine oil prior to tightening.

① Step 1: M10 bolts to 18 ft. lbs.; M12 bolts to 22 ft. lbs.
   Step 2: Plus 60-65 degrees
   Step 3: Plus 60-65 degrees

② Step 1: 20-24 ft. lbs.
   Step 2: Plus 60-65 degrees

③ Step 1: 18 ft. lbs.
   Step 2: Plus 58-62 degrees
   Step 3: Plus 43-47 degrees

④ Step 1: M10 bolts to 20-24 ft. lbs.; M8 bolts to 10-14 ft. lbs.
   Step 2: Plus 90-94 degrees

⑤ Step 1: 12-15 ft. lbs.
   Step 2: Plus 90-94 degrees

22140_TUCS_C0010

WHEEL ALIGNMENT

Year	Model		Caster Range (+/-Deg.)	Caster Preferred Setting (Deg.)	Camber Range (+/-Deg.)	Camber Preferred Setting (Deg.)	Toe-in (Deg.)
2006	Tucson	Front	①	②	①	0.00	0 +/- 0.08
		Rear	—	—	①	③	④
2007	Tucson	Front	①	②	①	0.00	0 +/- 0.08
		Rear	—	—	①	③	④
2008	Tucson	Front	①	②	①	0.00	0 +/- 0.08
		Rear	—	—	①	③	④

① +/- 30'

② 3 degrees 32'

③ -0 degrees 55'

④ 4.6 +3, -1

22140_TUCS_C0011

TIRE, WHEEL AND BALL JOINT SPECIFICATIONS

Year	Model	OEM Tires Standard	OEM Tires Optional	Tire Pressures (psi) Front	Tire Pressures (psi) Rear	Wheel Size	Ball Joint Inspection	Lug Nut Torque (ft. lbs.)
2006	Tucson	P215/65R16	P235/60R16	30	30	6.5J x 16	①	66-81
2007	Tucson	P215/65R16	P235/60R16	30	30	6.5J x 16	①	66-81
2008	Tucson	P215/65R16	P235/60R16	30	30	6.5J x 16	①	66-81

OEM: Original Equipment Manufacturer

PSI: Pounds Per Square Inch

① Replace the ball joint if too loose or if rotating torque exceeds specification: 4-21 inch lbs.

22140_TUCS_C0012

BRAKE SPECIFICATIONS

All measurements in inches unless noted

Year	Model		Brake Disc Original Thickness	Brake Disc Minimum Thickness	Brake Disc Maximum Runout	Brake Drum Diameter Original Inside Diameter	Max. Wear Limit	Brake Drum Diameter Maximum Machine Diameter	Minimum Lining Thickness	Brake Caliper Bracket Bolts (ft. lbs.)	Brake Caliper Mounting Bolts (ft. lbs.)
2006	Tucson	F	1.014	0.961	0.002	—	—	—	0.079	59-74	16-24
		R	0.394	0.315	0.001	—	—	—	0.079	59-74	16-24
2007	Tucson	F	1.014	0.961	0.002	—	—	—	0.079	59-74	16-24
		R	0.394	0.315	0.001	—	—	—	0.079	59-74	16-24
2008	Tucson	F	1.014	0.961	0.002	—	—	—	0.079	59-74	16-24
		R	0.394	0.315	0.001	—	—	—	0.079	59-74	16-24

F: Front

R: Rear

22140_TUCS_C0013

SCHEDULED MAINTENANCE INTERVALS
HYUNDAI—TUCSON

TO BE SERVICED	TYPE OF SERVICE	VEHICLE MILEAGE INTERVAL (x1000)												
		7.5	15	22.5	30	37.5	45	52.5	60	67.5	75	82.5	90	97.5
Accessory drive belts	S/I	✓	✓	✓	✓	✓	✓	✓	✓	✓	✓	✓	✓	✓
Air cleaner element (engine)	R				✓				✓				✓	
Air conditioner system	S/I	Inspect the system operation annually												
Automatic transaxle fluid	S/I		✓		✓		✓		✓		✓		✓	
Automatic transaxle fluid &	R	105,000 miles (under normal usage)												
Brake lines, hoses, and connections	S/I		✓		✓		✓		✓		✓		✓	
Brake pads, calipers, & rotors	S/I		✓		✓		✓		✓		✓		✓	
Cabin air filter	R	Every 12 months or 10,000 miles												
Chassis and body fasteners	S/I				✓				✓				✓	
Cooling system hoses and coolant level	S/I		✓		✓		✓		✓		✓		✓	
Crankcase ventilation hose, vapor hose, and fuel filter cap	S/I				✓				✓					
Driveshafts and CV-boots	S/I		✓		✓		✓		✓		✓		✓	
Electronic throttle control	S/I		✓		✓		✓		✓		✓		✓	
Engine coolant	R								✓					
Engine oil and filter	R	✓	✓	✓	✓	✓	✓	✓	✓	✓	✓	✓	✓	✓
Exhaust pipe connections, muffler, and suspension bolts	S/I				✓				✓				✓	
Front and rear brakes	S/I				✓				✓				✓	
Fuel filter	R							✓						
Fuel lines, fuel hoses, and connections	S/I				✓				✓				✓	
Locks and hinges	S/I	✓	✓	✓	✓	✓	✓	✓	✓	✓	✓	✓	✓	✓
Spark plugs (standard)	R				✓				✓				✓	
Spark plugs (Iridium coated) 100,000 mile replacement	R													
Spark plugs (Platinum coated)	R								✓					
Steering gear rack, linkage, and boots	S/I				✓				✓				✓	
Timing Belt	R								✓					
Upper and lower arm ball joints	S/I		✓		✓		✓		✓		✓		✓	
Wheel bearings	S/I				✓				✓				✓	

R: Replace   S/I: Service or Inspect

FREQUENT OPERATION MAINTENANCE (SEVERE SERVICE)

If a vehicle is operated under any of the following conditions it is considered severe service:

- Extremely dusty areas.

- 50% or more of the vehicle operation is in 90°F (32°C) or higher temperatures, or constant operation in temperatures below 32°F (0°C).

- Prolonged idling (vehicle operation in stop and go traffic).

- Frequent short running periods (engine does not warm to normal operating temperatures).

- Police, taxi, delivery usage or trailer towing usage.

Oil & oil filter: change every 3,000 miles.

Brake pads, calipers & rotors: service or inspect every 7,500 miles.

Driveshaft boots: service or inspect every 7,500 miles

Steering gear rack, linkage & boots: service or inspect every 7,500 miles.

Air cleaner filter: service or inspect every 15,000 miles.

Automatic transaxle fluid & filter: replace every 30,000 miles.

Rear brake drums & linings: service or inspect every 15,000 miles.

Spark plugs: service or inspect every 24,000 miles.

PRECAUTIONS

Before servicing any vehicle, please be sure to read all of the following precautions, which deal with personal safety, prevention of component damage, and important points to take into consideration when servicing a motor vehicle:

• Never open, service or drain the radiator or cooling system when the engine is hot; serious burns can occur from the steam and hot coolant.

• Observe all applicable safety precautions when working around fuel. Whenever servicing the fuel system, always work in a well-ventilated area. Do not allow fuel spray or vapors to come in contact with a spark, open flame, or excessive heat (a hot drop light, for example). Keep a dry chemical fire extinguisher near the work area. Always keep fuel in a container specifically designed for fuel storage; also, always properly seal fuel containers to avoid the possibility of fire or explosion. Refer to the additional fuel system precautions later in this section.

• Fuel injection systems often remain pressurized, even after the engine has been turned **OFF**. The fuel system pressure must be relieved before disconnecting any fuel lines. Failure to do so may result in fire and/or personal injury.

• Brake fluid often contains polyglycol ethers and polyglycols. Avoid contact with the eyes and wash your hands thoroughly after handling brake fluid. If you do get brake fluid in your eyes, flush your eyes with clean, running water for 15 minutes. If eye irritation persists, or if you have taken brake fluid internally, IMMEDIATELY seek medical assistance.

• The EPA warns that prolonged contact with used engine oil may cause a number of skin disorders, including cancer. You should make every effort to minimize your exposure to used engine oil. Protective gloves should be worn when changing oil. Wash your hands and any other exposed skin areas as soon as possible after exposure to used engine oil. Soap and water, or waterless hand cleaner should be used.

• All new vehicles are now equipped with an air bag system, often referred to as a Supplemental Restraint System (SRS) or Supplemental Inflatable Restraint (SIR) system. The system must be disabled before performing service on or around system components, steering column, instrument panel components, wiring and sensors. Failure to follow safety and disabling procedures could result in accidental air bag deployment, possible personal injury and unnecessary system repairs.

• Always wear safety goggles when working with, or around, the air bag system. When carrying a non-deployed air bag, be sure the bag and trim cover are pointed away from your body. When placing a non-deployed air bag on a work surface, always face the bag and trim cover upward, away from the surface. This will reduce the motion of the module if it is accidentally deployed. Refer to the additional air bag system precautions later in this section.

• Clean, high quality brake fluid from a sealed container is essential to the safe and proper operation of the brake system. You should always buy the correct type of brake fluid for your vehicle. If the brake fluid becomes contaminated, completely flush the system with new fluid. Never reuse any brake fluid. Any brake fluid that is removed from the system should be discarded. Also, do not allow any brake fluid to come in contact with a painted surface; it will damage the paint.

• Never operate the engine without the proper amount and type of engine oil; doing so WILL result in severe engine damage.

• Timing belt maintenance is extremely important. Many models utilize an interference-type, non-freewheeling engine. If the timing belt breaks, the valves in the cylinder head may strike the pistons, causing potentially serious (also time-consuming and expensive) engine damage. Refer to the maintenance interval charts for the recommended replacement interval for the timing belt, and to the timing belt section for belt replacement and inspection.

• Disconnecting the negative battery cable on some vehicles may interfere with the functions of the on-board computer system(s) and may require the computer to undergo a relearning process once the negative battery cable is reconnected.

• When servicing drum brakes, only disassemble and assemble one side at a time, leaving the remaining side intact for reference.

• Only an MVAC-trained, EPA-certified automotive technician should service the air conditioning system or its components.

BRAKES

GENERAL INFORMATION

See Figure 1.

The Anti-Lock Brake System (ABS) controls the hydraulic brake pressure of all four wheels during sudden braking and braking on hazardous road surfaces, preventing the wheels from locking. The ABS provides the following benefits:

• Enables steering around obstacles with a greater degree of certainty during panic braking

• Enables stopping during panic braking while allowing stability and control, even on curves

• If a malfunction occurs in the ABS, the system will operate as a normal brake (fail safe mode). A diagnostic function and a fail-safe system have been included for serviceability.

ANTI-LOCK BRAKE SYSTEM (ABS)

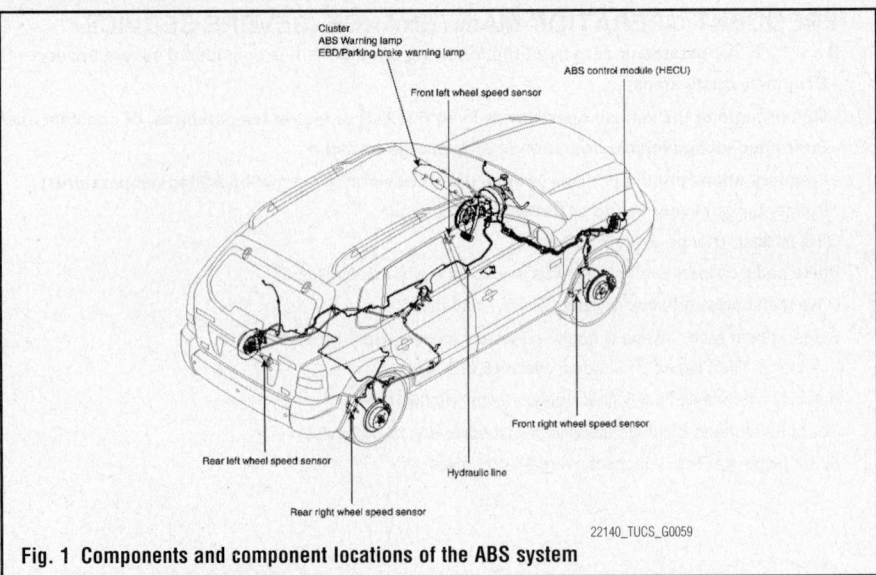

Fig. 1 Components and component locations of the ABS system

22140_TUCS_G0059

WHEEL SPEED SENSORS

REMOVAL & INSTALLATION

Front
See Figure 2.

1. Remove the front wheel speed sensor mounting bolt.
2. Remove the front wheel guard after removing the mud ground.

3. Remove the front wheel speed sensor after disconnecting the wheel speed sensor connector.
4. Installation is the reverse of the removal procedure.

Rear
See Figure 2.

1. Remove the rear wheel speed sensor mounting bolt.

2. Remove the rear seat side pad then disconnect the rear wheel speed sensor connector.
3. Installation is the reverse of the removal procedure.

WHEEL SPEED SENSOR RINGS (TOOTHED RINGS)

REMOVAL & INSTALLATION
See Figure 3.

1. Before servicing the vehicle, refer to the Precautions Section.
2. Remove the halfshaft(s). Refer to Halfshafts, removal & installation.
3. Using the special tool (09432-11000), remove the wheel speed sensor ring (also called the tone wheel).

To install:
4. Install new wheel speed sensor ring(s).

➡**Replace with a wheel speed sensor ring of the same number of teeth as the original equipment.**

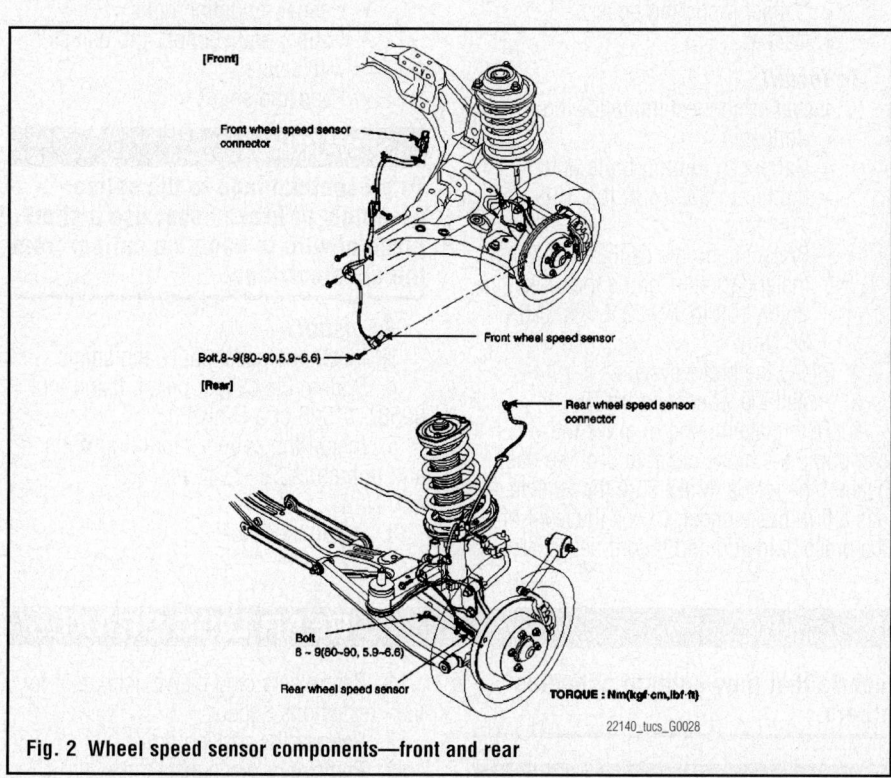

Fig. 2 Wheel speed sensor components—front and rear

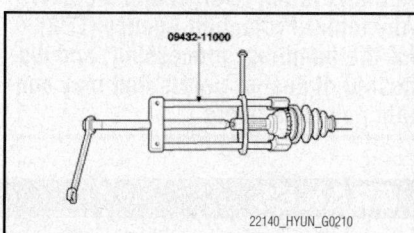

Fig. 3 Using the special tool (09432-11000) to remove the wheel speed sensor ring

BRAKES

BLEEDING PROCEDURE

This vehicle comes standard with a 4-wheel Anti-Lock Braking System (ABS). Please refer to the section, Bleeding the ABS System.

BLEEDING THE ABS SYSTEM

This procedure should be followed to ensure adequate bleeding of air and the filling of the ABS unit, the brake lines, and the master cylinder with brake fluid.
1. Before servicing the vehicle, refer to the Precautions Section.
2. Remove the reservoir cap and fill the brake reservoir with brake fluid.

✳✳ WARNING

If there is any brake fluid on any painted surface, wash it off immediately.

➡**When pressure bleeding, do not depress the brake pedal. Recommended brake fluid: DOT3 or DOT4.**

3. Connect a clear plastic tube to the wheel cylinder bleeder plug and insert the other end of the tube into a clear plastic bottle that is half filled with clean brake fluid.
4. Connect the Hi-Scan Pro® to the data link connector located underneath the dash panel.
5. Select and operate according to the instructions on the Hi-Scan Pro® screen.

✳✳ CAUTION

You must obey the maximum operating time of the ABS motor with the Hi-Scan Pro® to prevent the motor pump from burning.

BLEEDING THE BRAKE SYSTEM

6. Select Hyundai vehicle diagnosis.
7. Select vehicle name.
8. Select Anti-Lock Brake system.
9. Select air bleeding mode.
10. Press "YES" to operate motor pump and solenoid valve.

✳✳ WARNING

Wait 60 seconds before operating the air bleeding or damage to the motor may occur.

11. Wait 60 seconds before operating the air bleeding.
12. Pump the brake pedal several times, and then loosen the bleeder screw until fluid starts to run out without bubbles. Then, close the bleeder screw.
13. Repeat until there are no more bubbles in the fluid for each wheel.

BRAKES

FRONT DISC BRAKES

❊❊ CAUTION

Dust and dirt accumulating on brake parts during normal use may contain asbestos fibers from production or aftermarket brake linings. Breathing excessive concentrations of asbestos fibers can cause serious bodily harm. Exercise care when servicing brake parts. Do not sand or grind brake lining unless equipment used is designed to contain the dust residue. Do not clean brake parts with compressed air or by dry brushing. Cleaning should be done by dampening the brake components with a fine mist of water, then wiping the brake components clean with a dampened cloth. Dispose of cloth and all residue containing asbestos fibers in an impermeable container with the appropriate label. Follow practices prescribed by the Occupational Safety and Health Administration (OSHA) and the Environmental Protection Agency (EPA) for the handling, processing, and disposing of dust or debris that may contain asbestos fibers.

BRAKE CALIPER

REMOVAL & INSTALLATION

1. Before servicing the vehicle, refer to the Precautions Section.
2. Remove or disconnect the following:
   - Wheel
   - Brake hose from the caliper
   - Caliper mounting bolts
   - Caliper

To install:

3. Install or connect the following:
   - Caliper
   - Caliper mounting bolts and tighten to 59–74 ft. lbs. (80–100 Nm)
   - Brake line to the caliper with 2 new metal gaskets. Torque the brake line union bolt to 18–22 ft. lbs. (25–30 Nm)
4. Bleed the brake system.
5. Install the wheel and tire.
6. Before attempting to move the vehicle, pump the brake pedal to seat the pads against the rotors. Make sure the vehicle has a firm brake pedal. Check the level of the brake fluid and add fluid if necessary.

DISC BRAKE PADS

REMOVAL & INSTALLATION

1. Before servicing the vehicle, refer to the Precautions Section.
2. Remove or disconnect the following:
   - Wheel
   - Caliper mounting bolts
   - Caliper and support it to one side with a wire
   - Pads and shims

❊❊ WARNING

To prevent damage to the caliper assembly or brake hose, use a short piece of wire to hang the caliper from the undercarriage.

To install:

3. Install the pads, clips, and shims.
4. Bottom the caliper piston using tool 09581-11000 or a C-clamp.
5. Install the caliper mounting bolts and tighten to 58–73 ft. lbs. (80–100 Nm).
6. Install the wheel.

BRAKES

REAR DISC BRAKES

❊❊ CAUTION

Dust and dirt accumulating on brake parts during normal use may contain asbestos fibers from production or aftermarket brake linings. Breathing excessive concentrations of asbestos fibers can cause serious bodily harm. Exercise care when servicing brake parts. Do not sand or grind brake lining unless equipment used is designed to contain the dust residue. Do not clean brake parts with compressed air or by dry brushing. Cleaning should be done by dampening the brake components with a fine mist of water, then wiping the brake components clean with a dampened cloth. Dispose of cloth and all residue containing asbestos fibers in an impermeable container with the appropriate label. Follow practices prescribed by the Occupational Safety and Health Administration (OSHA) and the Environmental Protection Agency (EPA) for the handling, processing, and disposing of dust or

debris that may contain asbestos fibers.

BRAKE CALIPER

REMOVAL & INSTALLATION

See Figure 4.

1. Before servicing the vehicle, refer to the Precautions Section.
2. Release the parking brake.
3. Remove or disconnect the following:
   - Wheel
   - Brake line at the caliper

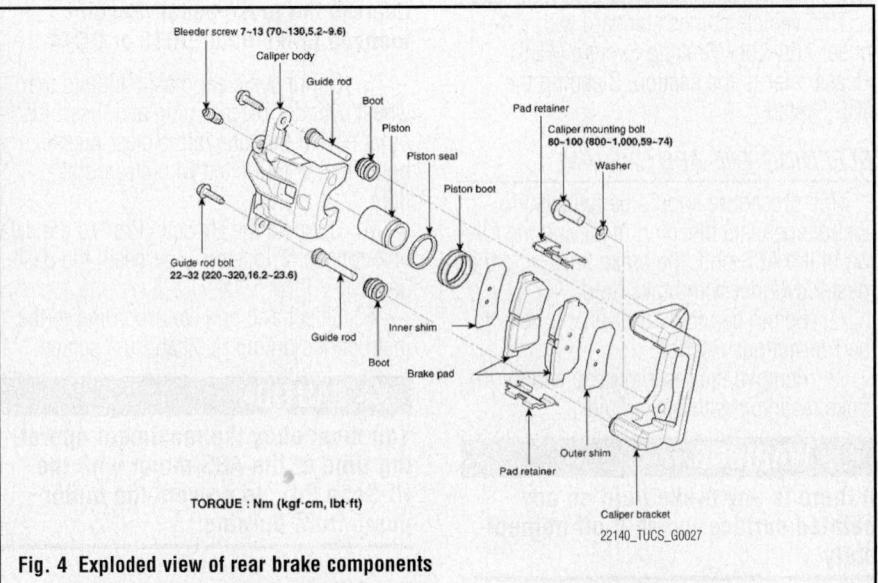

Bleeder screw 7~13 (70~130,5.2~9.6)
Caliper body
Guide rod
Boot
Piston
Pad retainer
Caliper mounting bolt 80~100 (800~1,000,59~74)
Washer
Piston seal
Piston boot
Guide rod bolt 22~32 (220~320,16.2~23.6)
Guide rod
Inner shim
Boot
Brake pad
Outer shim
Pad retainer
Caliper bracket

TORQUE : Nm (kgf·cm, lbf·ft)

22140_TUCS_G0027

Fig. 4 Exploded view of rear brake components

- Brake pads
- Caliper mounting bolts
- Caliper

To install:

4. Install or connect the following:
- Caliper onto its mounting
- Mounting bolts. Torque the bolts to 59–74 ft. lbs. (80–100 Nm)
- Brake pads
- Brake line to the caliper with 2 new metal gaskets. Torque the brake line union bolt to 18–22 ft. lbs. (24–30 Nm)

5. Bleed the system.
6. Install the wheel.
7. Pump the brake pedal until the brake pads are seated and a firm pedal is achieved before attempting to move the vehicle.

✳✳ CAUTION

Do not move the vehicle until a firm pedal is obtained.

8. Road test the vehicle to check for proper brake operation.

DISC BRAKE PADS

REMOVAL & INSTALLATION
See Figure 4.

1. Before servicing the vehicle, refer to the Precautions Section.
2. Remove or disconnect the following:
- Rear wheels
- Lower caliper mounting bolt and rotate the caliper upward
- Pads from the caliper support
- Pad retainers, if necessary

To install:

3. Install or connect the following:
- Pad retainers, if removed
- Pads onto the pad retainers

4. Compress the caliper piston using a C-clamp.
5. Rotate the caliper downward and install the mounting bolt. Tighten to: 16–24 ft. lbs. (22–32 Nm).
6. Install the wheel.
7. Pump the brake pedal until the brake pads are seated and a firm pedal is achieved before attempting to move the vehicle.

✳✳ CAUTION

Do not move the vehicle until a firm pedal is obtained.

8. Road test the vehicle to check for proper brake operation.

BRAKES
PARKING BRAKE

PARKING BRAKE CABLES

ADJUSTMENT

1. Before servicing the vehicle, refer to the Precautions Section.
2. After servicing the rear brake assembly, loosen the parking brake adjusting nut, start the engine, and depress the brake pedal several times in order to set the self-adjusting brake system before adjusting the parking brake.
3. Block the front wheels, then raise the rear of the vehicle and make sure it is securely supported.
4. Set the parking brake 1 click toward engagement.
5. Remove the floor console.
6. Tighten the adjusting nut until the parking brakes drag slightly when the rear wheels are turned.
7. Release the parking brake lever completely.
8. Check if the parking brakes drag when the rear wheels are turned. Readjust if necessary until there is no drag from the parking brakes.
9. Check the proper operation of the parking brakes by fully applying the parking brakes.
10. Reinstall the floor console.

PARKING BRAKE SHOES

REMOVAL & INSTALLATION
See Figures 5 and 6.

1. Before servicing the vehicle, refer to the Precautions Section.

2. Raise and safely support the vehicle.
3. Remove the rear wheel and tire assembly.
4. Remove the brake caliper. Refer to Rear Disc Brakes, Brake Caliper, removal & installation.
5. Remove the brake rotor. Refer to Rear Disc Brakes, Rotor, removal & installation.
6. Remove the clip retainer and the parking brake cable.
7. Remove the shoe hold down pin (A) and spring (B) by pressing and rotating the spring.
8. Remove the adjuster assembly (B) and the lower return spring (A).
9. Remove the upper return spring (C) and the brake shoes (D).
10. Remove the operating lever assembly (E).

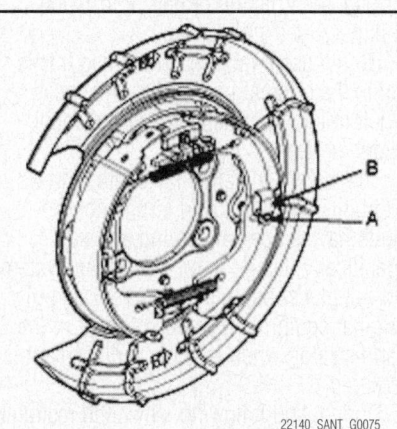

22140_SANT_G0075

Fig. 5 Remove the shoe hold down pin (A) and spring (B) by pressing and rotating the spring

To install:

11. Install the operating lever assembly (E).
12. Install the upper return spring (C) and the brake shoes (D).
13. Install the adjuster assembly (B) and the lower return spring (A).
14. Install the shoe hold down pin and spring by pressing and rotating the spring.
15. Install the parking brake cable and then install the clip.
16. Install the rear brake disc rotor. Refer to Rear Disc Brakes, Rotor, removal & installation.
17. Adjust the rear brake shoe clearance:

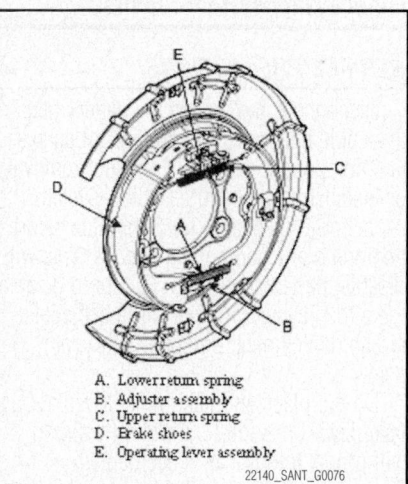

A. Lower return spring
B. Adjuster assembly
C. Upper return spring
D. Brake shoes
E. Operating lever assembly

22140_SANT_G0076

Fig. 6 Remove the adjuster assembly, the lower return spring, the upper return spring, the brake shoes, and the operating lever assembly

a. Remove the plug from the disc.

b. Rotate the toothed wheel of adjuster with a screw driver until the disc is not moving.

c. Back off 5 notches in the opposite direction.

18. Install the brake caliper. Refer to Rear Disc Brakes, Brake Caliper, removal & installation.

19. Install the tire and wheel.

20. If the parking brake shoe or the brake disc are replaced with a new one, perform the brake shoe brake-in procedure:

a. While operating the parking brake engaged with a 15 lb. (69 N) effort, drive the vehicle 0.31 miles (500 meters) at the speed of about 37 mph (60 kph).

b. Repeat the above procedure at least 2 times.

21. The parking brake must be able to operate at 220 lbs. (981 N) with no difficulty.

22. The parking brake should hold on a 30 percent grade.

23. Ensure that all parts move smoothly.

24. The parking brake indicator lamp must turn ON when the parking brake is applied and turn OFF when the parking brake is released.

ADJUSTMENT

See Figure 7.

1. Before servicing the vehicle, refer to the Precautions Section.

2. Raise and safely support the vehicle.

3. Remove the rear wheel and tire assembly.

4. Adjust the rear brake shoe clearance:

a. Remove the plug from the disc.

b. Rotate the toothed wheel of adjuster with a screw driver until the disc is not moving.

c. Back off 5 notches in the opposite direction.

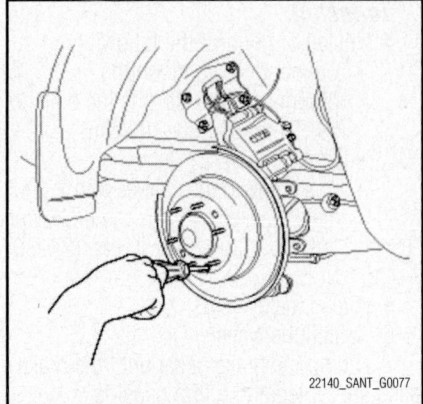

22140_SANT_G0077

Fig. 7 After removing the plug from the disc, rotate the toothed wheel with a screwdriver until the disc does not move, then back it off 5 notches

---

CHASSIS ELECTRICAL

AIR BAG (SUPPLEMENTAL RESTRAINT SYSTEM)

GENERAL INFORMATION

✳✳ CAUTION

This vehicle is equipped with an air bag system. The system must be disarmed before performing service on, or around, system components, the steering column, instrument panel components, wiring, and sensors. Failure to follow the safety precautions and the disarming procedure could result in accidental air bag deployment, possible injury, and unnecessary system repairs.

SERVICE PRECAUTIONS

Disconnect and isolate the battery negative cable before beginning any airbag system component diagnosis, testing, removal, or installation procedures. Allow system capacitor to discharge for 3 minutes before beginning any component service. This will disable the airbag system. Failure to disable the airbag system may result in accidental airbag deployment, personal injury, or death.

Do not place an intact undeployed airbag face down on a solid surface. The airbag will propel into the air if accidentally deployed and may result in personal injury or death.

When carrying or handling an undeployed airbag, the trim side (face) of the airbag should be pointing towards the body to minimize possibility of injury if accidental deployment occurs. Failure to do this may result in personal injury or death.

Replace airbag system components with OEM replacement parts. Substitute parts may appear interchangeable, but internal differences may result in inferior occupant protection. Failure to do so may result in occupant personal injury or death.

Wear safety glasses, rubber gloves, and long sleeved clothing when cleaning powder residue from vehicle after an airbag deployment. Powder residue emitted from a deployed airbag can cause skin irritation. Flush affected area with cool water if irritation is experienced. If nasal or throat irritation is experienced, exit the vehicle for fresh air until the irritation ceases. If irritation continues, see a physician.

Do not use a replacement airbag that is not in the original packaging. This may result in improper deployment, personal injury, or death.

The factory installed fasteners, screws and bolts used to fasten airbag components have a special coating and are specifically designed for the airbag system. Do not use substitute fasteners. Use only original equipment fasteners listed in the parts catalog when fastener replacement is required.

During, and following, any child restraint anchor service, due to impact event or vehicle repair, carefully inspect all mounting hardware, tether straps, and anchors for proper installation, operation, or damage. If a child restraint anchor is found damaged in any way, the anchor must be replaced. Failure to do this may result in personal injury or death.

Deployed and non-deployed airbags may or may not have live pyrotechnic material within the airbag inflator.

Do not dispose of driver/passenger/curtain airbags or seat belt tensioners unless you are sure of complete deployment. Refer to the Hazardous Substance Control System for proper disposal.

Dispose of deployed airbags and tensioners consistent with state, provincial, local, and federal regulations.

After any airbag component testing or service, do not connect the battery negative cable. Personal injury or death may result if the system test is not performed first.

If the vehicle is equipped with the Occupant Classification System (OCS), do not connect the battery negative cable before performing the OCS Verification Test using the scan tool and the appropriate diagnostic information. Personal injury or death may result if the system test is not performed properly.

Never replace both the Occupant Restraint Controller (ORC) and the Occupant Classification Module (OCM) at the same time. If both require replacement, replace one, then perform the Airbag System test before replacing the other.

Both the ORC and the OCM store Occupant Classification System (OCS) calibration data, which they transfer to one another when one of them is replaced. If both are replaced at the same time, an irreversible fault will be set in both modules and the OCS may malfunction and cause personal injury or death.

If equipped with OCS, the Seat Weight Sensor is a sensitive, calibrated unit and must be handled carefully. Do not drop or handle roughly. If dropped or damaged, replace with another sensor. Failure to do so may result in occupant injury or death.

If equipped with OCS, the front passenger seat must be handled carefully as well. When removing the seat, be careful when it setting on the floor not to drop it. If dropped, the sensor may be inoperative, could result in occupant injury, or possibly death.

If equipped with OCS, when the passenger front seat is on the floor, no one should sit in the front passenger seat. This uneven force may damage the sensing ability of the seat weight sensors. If sat on and damaged, the sensor may be inoperative, could result in occupant injury, or possibly death.

Several precautions must be observed when handling the inflator module to avoid accidental deployment and possible personal injury.

• Never carry the inflator module by the wires or connector on the underside of the module

• When carrying a live inflator module, hold it securely with both hands, and ensure that the bag and trim cover are pointed away

• Place the inflator module on a bench or other surface with the bag and trim cover facing up

• With the inflator module on the bench, never place anything on or close to the module which may be thrown in the event of an accidental deployment

Before servicing the vehicle, make sure to refer to the precautions in the beginning of this section as well.

DISARMING THE SYSTEM

1. Before servicing the vehicle, refer to the Precautions Section.
2. Record the radio anti-theft code data. Remove the ignition key from the vehicle.
3. Disconnect the negative battery cable.
4. Wait at least 3 minutes for the system capacitor to discharge before performing any service.

ARMING THE SYSTEM

1. Before servicing the vehicle, refer to the Precautions Section.
2. Reconnect the negative battery cable.
3. To confirm proper system operation, turn the ignition switch to the ON position. The SRS indicator light will be lit for at least 6 seconds and then go off.

CLOCKSPRING CENTERING

See Figures 8 through 10.

1. Disconnect and isolate the battery negative cable. Allow the system capacitor to discharge for 3 minutes before beginning any component service. This will disable the airbag system.

❊❊ CAUTION

Failure to disable the airbag system may result in accidental airbag deployment, personal injury, or death.

2. Remove the ignition key from the vehicle.
3. Connect the clockspring harness connector and horn harness connector to the clockspring.
4. Set the clockspring on neutral position and after turning the front wheels to the straight-ahead position, install the clockspring.

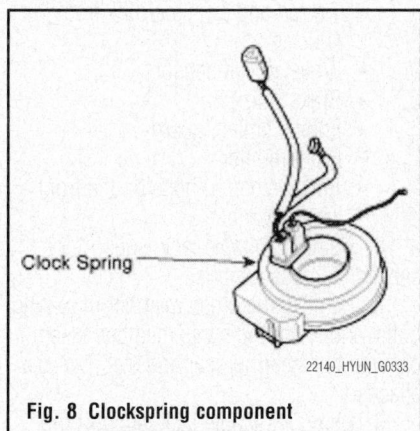

Fig. 8 Clockspring component

a. Check connectors and protective tube for damage, and terminals for deformities.
b. If even one abnormal point is discovered, replace the clockspring with a new one.
5. Connect the clockspring harness connector and the steering switch harness connector to the clockspring.

6. Set the center position by getting the marks between the clockspring and the cover into line.
7. Turn the clockspring clockwise to the stop and then 2.4 revolutions counter-clockwise. See the illustration of marks in line.

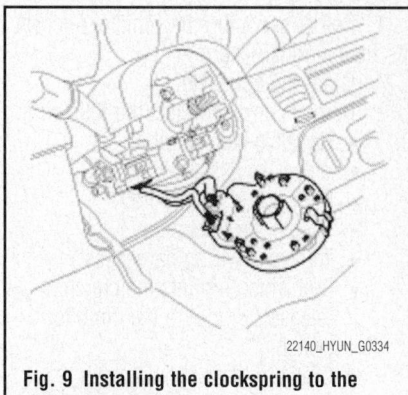

Fig. 9 Installing the clockspring to the steering column

8. Install the steering wheel column cover and the steering wheel.
9. Connect the Driver Airbag (DAB) module connector and the horn connector, then install the DAB module on the steering wheel.
10. Secure the DAB with the new mounting bolts. Tightening torque of DAB mounting bolt: 70–96 inch lbs. (8–11 NM).
11. Connect the battery negative cable.

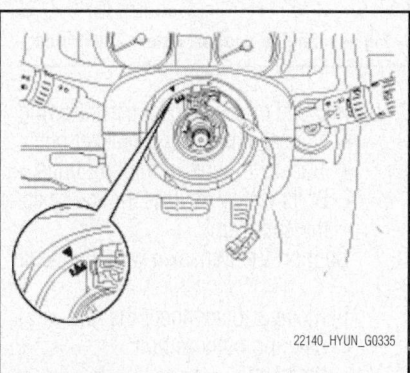

Fig. 10 Aligning the marks to center the clockspring

12. After installing the airbag, confirm proper system operation:
a. Turn the ignition switch **ON**; the SRS indicator light should be turned on for about 6 seconds and then go off.
b. Make sure the horn functions properly.

DRIVETRAIN

AUTOMATIC TRANSAXLE ASSEMBLY

REMOVAL & INSTALLATION

See Figure 11.

1. Before servicing the vehicle, refer to the Precautions Section.
2. Drain the transaxle.
3. Remove or disconnect the following:
   • Negative battery cable
   • Air intake assembly
   • Battery and battery tray
   • Intercooler inlet pipe
   • Transaxle wiring harnesses
   • Bolt which mounts the clutch release cylinder to the inhibiter switch
   • Clutch release cylinder clip
   • Oil cooler hose clamps
4. Use Special Tool 09200-38001 to support the engine.

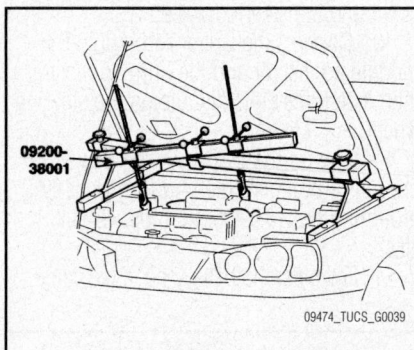

09474_TUCS_G0039

Fig. 11 Engine support fixture illustrated

5. Remove or disconnect the following:
   • Transaxle mounting bracket bolts
   • Transaxle upper mounting bolts
   • Bolts which mounts the transaxle to the sub-frame
6. Support the transaxle with a suitable jack.
7. Remove or disconnect the following:
   • Steering column bolt
   • Halfshafts
   • Bolt which mounts the transaxle to the rear sub-frame
   • Drive shaft, if equipped with 4WD
   • Lower transaxle mounting bolts
   • Transaxle assembly

To install:

8. Before servicing the vehicle, refer to the Precautions Section.
9. After installation, reset the adaptive learning value of the TCM, if necessary.

MANUAL TRANSAXLE ASSEMBLY

REMOVAL & INSTALLATION

2.0L Engine

1. Before servicing the vehicle, refer to the Precautions Section.
2. Drain the transaxle.
3. Remove or disconnect the following:

   • Negative battery cable
   • Air intake assembly
   • Battery and battery tray
   • Back-up lamp connector
   • Vehicle speed sensor
   • Clutch release cylinder and lever
   • Shift cable from transaxle assembly
   • Steering column from the universal joint in the gear box
   • Clutch housing upper mounting bolts
   • Front, rear, and left transaxle mounting brackets
4. Using a suitable engine support fixture, support the engine assembly.
5. Remove or disconnect the following:

   • Power steering pressure hose from the pump
   • Front wheel
   • Strut assembly
   • Tie rod and sway bar link from the knuckle
   • Wheel speed sensor
   • Brake caliper
   • Engine splash guard
   • Front muffler
   • Power steering hose on the front cross member
6. Using a suitable jack, support the sub-frame cross member.
7. Remove the cross member mounting bolts, and lower the cross member assembly with the steering gear and stabilizer bar attached.
8. Using a suitable jack, support the transaxle assembly.
9. Remove the front and rear roll stoppers.
10. Remove the engine and transaxle mounting bolts.
11. Slowly lower the transaxle from the vehicle.
12. Installation is the reverse order of removal.

CLUTCH DRIVEN DISC & PRESSURE PLATE

REMOVAL & INSTALLATION

See Figures 12 through 17.

1. Before servicing the vehicle, refer to the Precautions Section.
2. Remove the transaxle assembly. Refer to Manual Transaxle Assembly, removal & installation.
3. Insert the special tool (09411-11000) in the clutch disc to prevent the disc from falling.

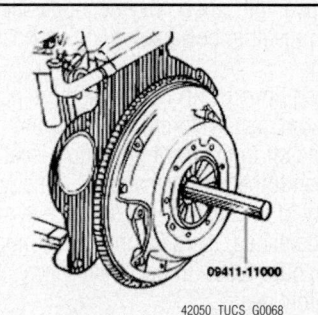

42050_TUCS_G0068

Fig. 12 Insert the special tool (09411-11000) in the clutch disc to prevent the disc from falling

4. Loosen the bolts which attach the clutch cover to the flywheel in a star pattern. Loosen the bolts in succession, 1–2 turns at a time, to avoid bending the cover flange.

➡ **Do not clean the clutch disc or the release bearing with cleaning solvent.**

5. Remove the release fork shaft and bushing.
6. Remove clutch cover assembly and then driven disc and pressure plate.

To install:

➡ **When installing the clutch, apply grease to each part, but be careful not to apply excessive grease. It can cause clutch slippage and shudder.**

7. Install the driven disc and pressure plate in to the clutch cover assembly.
8. Install the clutch disc assembly to the flywheel using the special tool (09411-11000).
9. Install the clutch cover assembly to the flywheel and temporarily tighten the bolts 1–2 steps at a time in a star pattern.
10. Tighten the clutch cover bolts to: 11–16 ft. lbs. (15–22 Nm).

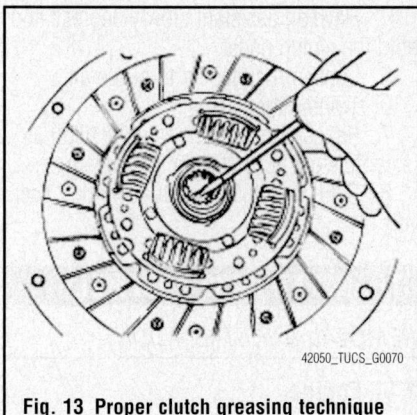

Fig. 13 Proper clutch greasing technique

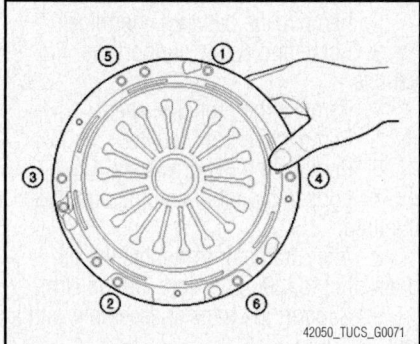

Fig. 14 Clutch cover installation torque pattern

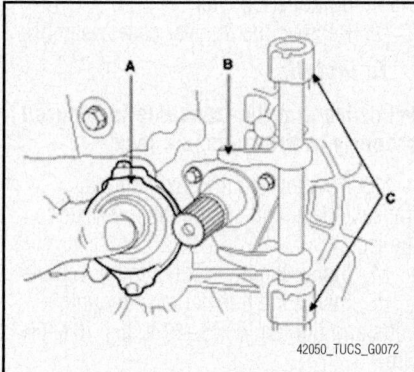

Fig. 15 Grease application illustrated

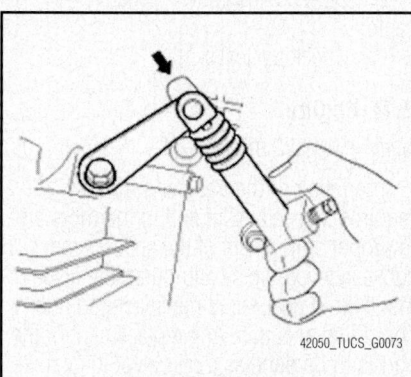

Fig. 16 Release lever installation

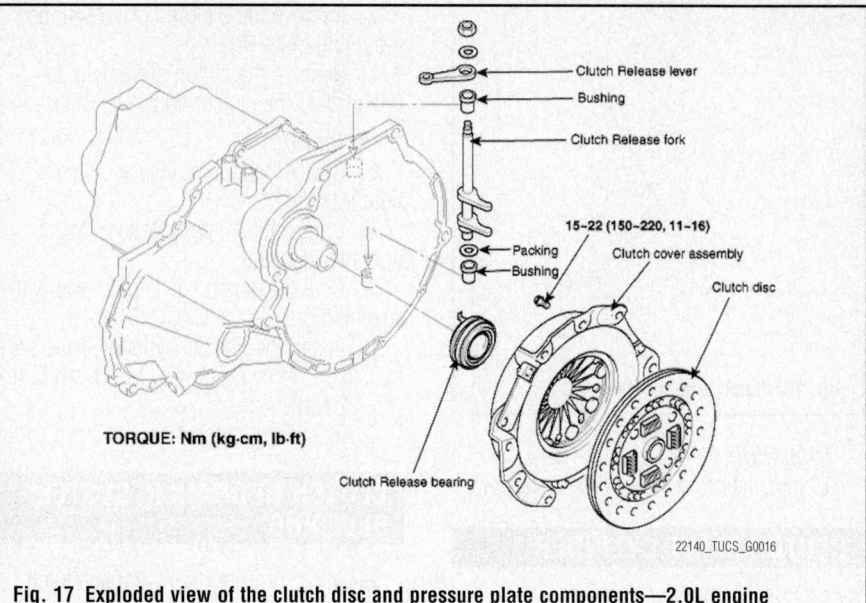

TORQUE: Nm (kg-cm, lb-ft)

Fig. 17 Exploded view of the clutch disc and pressure plate components—2.0L engine

11. Align the bearing (A) to the release fork (B) and then install it to the sleeve of the housing.

12. Apply multipurpose grease (CAS-MOLY® L9508) to the bearing sleeve, contact point of the release fork (B), and the bushing inner surface (C).

13. Install the release lever to the release fork.

14. Install the transaxle assembly to the engine. Refer to Manual Transaxle Assembly, removal & installation.

15. Test the clutch for proper operation.

➡ **If the transaxle assembly is installed to the engine without installing the release lever to the release fork, the release bearing can be separated, as the release fork rotates freely.**

ADJUSTMENTS

The clutch system is hydraulic and requires no adjustment.

CLUTCH MASTER CYLINDER

REMOVAL & INSTALLATION

See Figures 18 and 19.

❊❋ WARNING

Do not spill brake fluid on the vehicle; it may damage the paint; if brake fluid does contact the paint, wash it off immediately with water.

1. Before servicing the vehicle, refer to the Precautions Section.

2. Remove the brake fluid from the clutch master cylinder reservoir with a syringe.

3. Clamp the clutch master cylinder hose (A). If there is not enough room for clamping, you can also clamp the hose (B) from the brake master cylinder side.

4. Disconnect the hose from the cylinder by releasing the clutch master cylinder clamp.

5. Remove the clip (A) and disconnect the clutch tube (B).

6. Remove the pin and washer which connect the clutch pedal with the clutch master cylinder.

7. Remove the clutch master cylinder assembly mounting bolts under the driver's seat.

8. Remove the clutch master cylinder.

➡ **It can be helpful to do this step after removing the clutch pedal mounting bracket.**

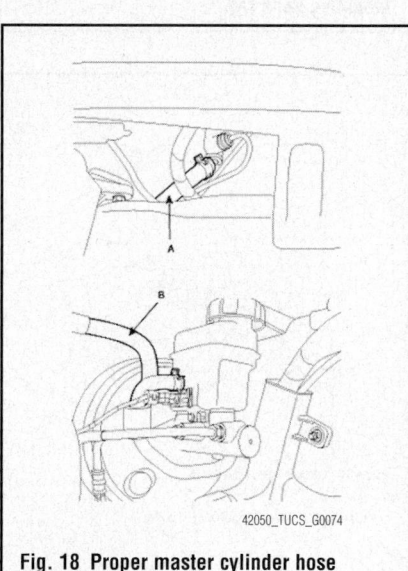

Fig. 18 Proper master cylinder hose clamping placement

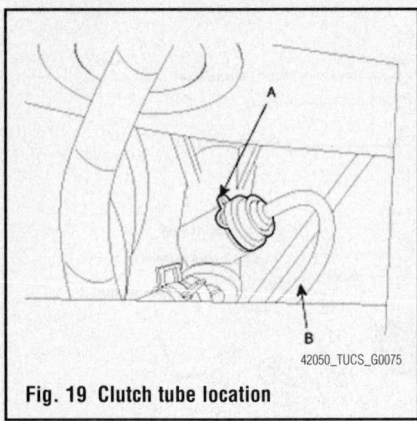

Fig. 19 Clutch tube location

To install:

9. Installation is the reverse of removal.

CLUTCH SLAVE CYLINDER

REMOVAL & INSTALLATION

See Figures 20 and 21.

1. Before servicing the vehicle, refer to the Precautions Section.
2. Disconnect the clutch tube.

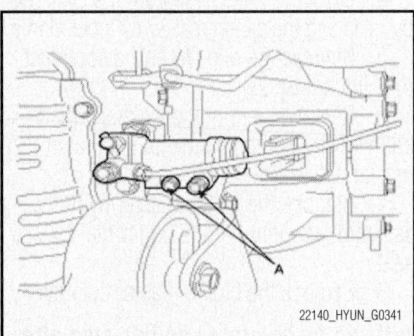

Fig. 20 Disconnect the clutch tube and remove the 2 clutch release slave cylinder mounting bolts (A)

3. Remove the 2 clutch slave cylinder mounting bolts (A).
4. Remove the clutch slave cylinder (also called the clutch release cylinder).

To install:

5. Check the clutch release cylinder for fluid leakage.
6. Check the clutch release cylinder boots for damage.
7. Coat the clutch clevis push rod with specified grease: CASMOLY® L9508.
8. Install the release cylinder (A) to the transaxle. Tighten the bolts to: 11–16 ft. lbs. (15–22 Nm).
9. Install the clutch tube.

CLUTCH HYDRAULIC SYSTEM BLEEDING

Bleeding air from the hydraulic clutch system is necessary whenever any part of the system has been disconnected or the fluid level (in the reservoir) has been allowed to fall so low that air has been drawn into the master cylinder.

✻ WARNING

NEVER use fluid that has been bled from a clutch system to fill the master cylinder reservoir, as it may be aerated, contain excessive moisture and/or be contaminated in some other way.

1. Before servicing the vehicle, refer to the Precautions Section.
2. Fill the clutch master cylinder reservoir with new hydraulic clutch fluid.
3. Attach a hose to the bleeder on the clutch actuator and submerge the other end of the hose in a container of hydraulic clutch fluid.

4. Have an assistant slowly depress and hold the clutch pedal.
5. Loosen the bleeder to purge air.
6. Tighten the bleeder.
7. Repeat the above 3 steps until all air is completely purged from the system.
8. Refill the clutch master cylinder reservoir.

TRANSFER CASE ASSEMBLY

REMOVAL & INSTALLATION

2.0L Engine

1. Before servicing the vehicle, refer to the Precautions Section.
2. Remove the battery (-) terminal.
3. Raise and safely support the vehicle.
4. Remove the propeller shaft.
5. Remove the front muffler.
6. Remove the RH driveshaft.
7. Loosen the oil drain plug and drain the fluid.
8. After draining, re-tighten the oil drain plug to: 29–43 ft. lbs. (39–58 Nm).
9. Support the transfer assembly with a jack.
10. Remove the transfer case assembly by loosening the mounting bolts.
11. Remove the 2 transfer bracket mounting bolts together.
12. Remove the transfer case assembly.

To install:

➡**Ensure that the transaxle is secured properly to the transaxle jack.**

13. Position the transaxle onto a transaxle jack and secure the transaxle to the jack.
14. Install the transaxle into the vehicle.
15. Install the transfer case mounting bolts and tighten to: 45–49 ft. lbs. (61–66 Nm).
16. Fill the transfer case assembly with the proper type and amount of lubricant.
17. Connect the battery negative (-) cable.
18. Check for leaks.

2.7L Engine

See Figures 22 through 24.

The repair of the transfer assembly requires a special skill and furthermore an improper adjustment of the spacers may cause a severe noise and durability issue. The hypoid gear set is manufactured and controlled as a pair. Any replacement of the part is to be done as a pair, hypoid gear shaft assembly 47308-39200 and pinion shaft 47311-39000.

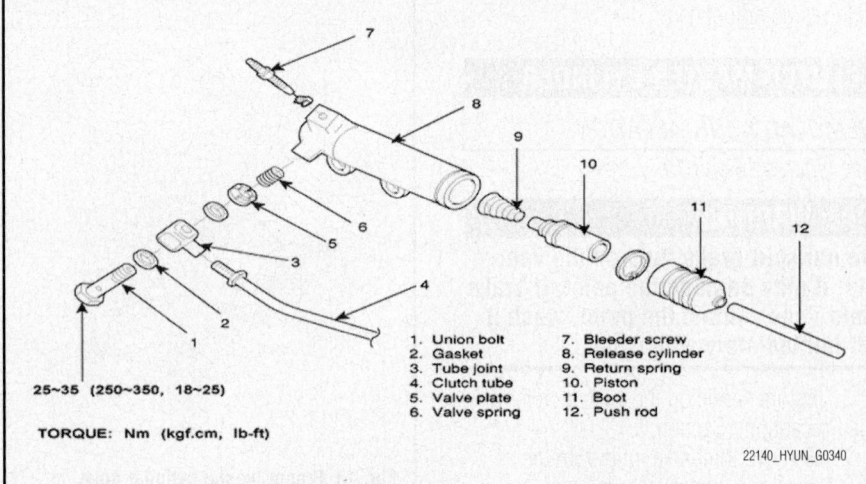

1. Union bolt	7. Bleeder screw
2. Gasket	8. Release cylinder
3. Tube joint	9. Return spring
4. Clutch tube	10. Piston
5. Valve plate	11. Boot
6. Valve spring	12. Push rod

25~35 (250~350, 18~25)

TORQUE: Nm (kgf.cm, lb-ft)

Fig. 21 Exploded view of the clutch slave cylinder

1. Remove the battery negative (-) cable.

2. Remove the air intake hose and air cleaner cover.

3. Disconnect the oxygen sensor connector left bank and right bank.

4. Remove the wheel and tire (RH).

5. Remove the engine side cover (RH).

6. Raise and safely support the vehicle.

7. Remove the lower arm ball joint mounting bolts.

8. Remove the lower arm ball joint from the front axle steering knuckle.

9. Remove the steering bar tie rod ball joint from the knuckle.

10. Remove the drive shaft from the transfer case assembly.

11. Remove the front exhaust pipe.

12. Drain the transfer oil through drain plug hole.

13. Remove the pinion case mounting bolts (A).

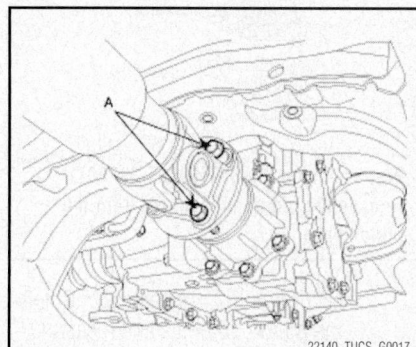

Fig. 22 Remove the pinion case mounting bolts (A)

14. Remove the alternator drive belt using the tensioner and then alternator assembly by loosening the mounting bolts.

15. Remove the alternator wire terminal mounting nut and connector.

16. Remove the heat protector (A) using a hexagonal socket.

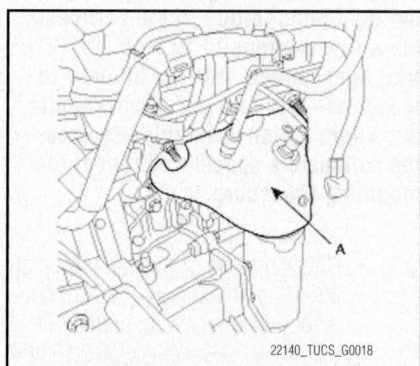

Fig. 23 Remove the heat protector (A) using a hexagonal socket

17. Remove the 7 exhaust manifold mounting nuts.

18. Remove the transfer mounting bracket (A).

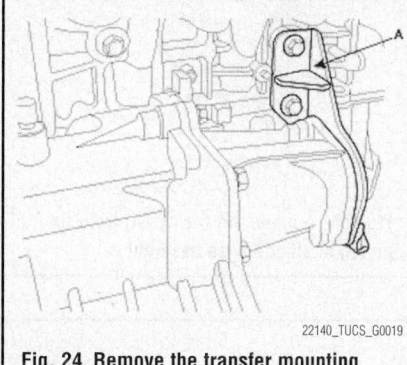

Fig. 24 Remove the transfer mounting bracket (A)

19. Remove the 4 transfer mounting bolts.

20. Using a flat head screw driver, carefully remove the transfer case assembly from the transaxle.

To install:

21. Install the transfer case assembly to the transaxle.

➡**To aid in installation of the transfer case, install it by moving the pinion in left and right directions, and slightly rotating the inner drive shaft of transfer assembly.**

22. Install the 4 transfer mounting bolts. Tighten to: 45–49 ft. lbs. (61–66 Nm).

23. Install the transfer mounting bracket. Tighten to:

a. 2 bolts to: 34–37 ft. lbs. (46–50 Nm).

b. 3 bolts to: 17–20 ft. lbs. (24–28 Nm).

24. Tighten the 7 exhaust manifold mounting nuts to: 22–25 ft. lbs. (29–34 Nm).

25. Tighten the 3 heat protector mounting nuts to: 104–130 inch lbs. (12–15 Nm).

26. Install the alternator and wire connectors. Tighten the mounting bolts to: 15–22 ft. lbs. (20–29 Nm).

27. Tighten the 6 pinion case mounting bolts to: 15–22 ft. lbs. (36–39 Nm).

28. Refill the transfer oil through the filler plug.

29. Install the front exhaust pipe.

30. Install the drive shaft (RH) to the transfer assembly.

31. Install the lower arm ball joint mounting and steering bar tie rod ball joint.

32. Lower the vehicle.

33. Install the engine side cover (RH) and the wheel and tire (RH).

34. Connect the oxygen sensor connectors.

35. Install the air intake hose and air cleaner cover and then connect the oxygen sensor connector.

36. Connect the battery negative (-) cable.

37. Check for leaks.

FRONT HALFSHAFT

REMOVAL & INSTALLATION

See Figure 25.

1. Before servicing the vehicle, refer to the Precautions Section.

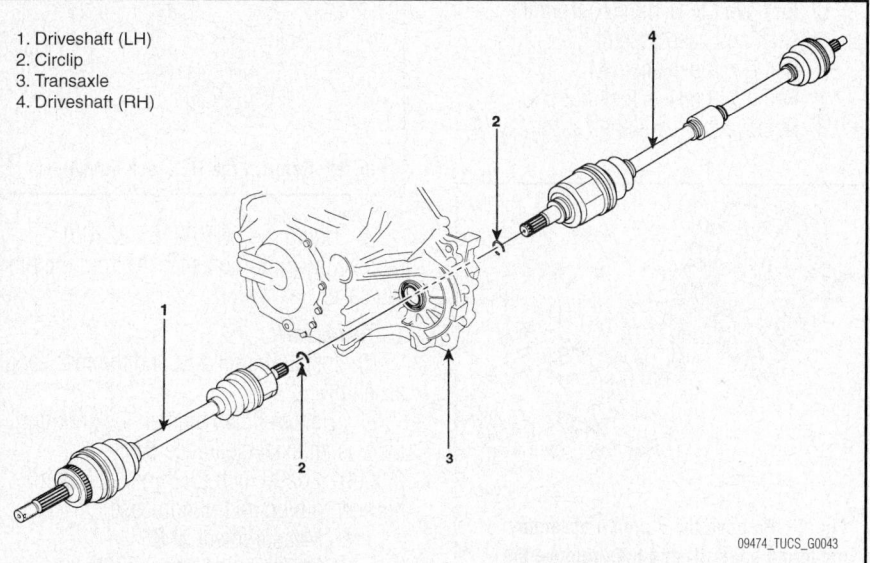

1. Driveshaft (LH)
2. Circlip
3. Transaxle
4. Driveshaft (RH)

Fig. 25 Exploded view of the halfshaft mounting and related components—Front

2. Remove or disconnect the following:
- Front wheel
- Spindle nut
- Wheel speed sensor, if equipped
- Lower ball joint
- Lower arm ball joint mounting nuts

3. Press the stub shaft out of the hub.

4. Using a suitable pry bar, pry the inner joint out of the transaxle.

To install:

➡**Use new circlips, split pins, and self-locking nuts for assembly.**

5. Install the inner joint so that the circlip is felt to seat in the retaining groove.

6. Guide the stub shaft into the hub.

7. Install or connect the following:
- Lower ball joint. Tighten the mounting bolts to 74–89 ft. lbs. (100–120 Nm)
- Wheel speed sensor, if equipped
- Spindle nut and torque the nut to 148–207 ft. lbs. (200–280 Nm)
- Front wheel

8. Check and/or adjust the wheel alignment.

FRONT PINION SEAL

REMOVAL & INSTALLATION

See Figures 26 through 32.

1. Before servicing the vehicle, refer to the Precautions Section.

2. Remove the propeller shaft.

3. Drain the transfer oil through drain plug hole.

4. Remove the 7 pinion assembly mounting bolts (B) and then remove the pinion assembly (A).

5. Remove the O-ring (A) from the pinion shaft and case assembly.

6. Remove the spacer (A).

7. Remove the HEX lock nut (A).

8. Remove the rear flange assembly (A).

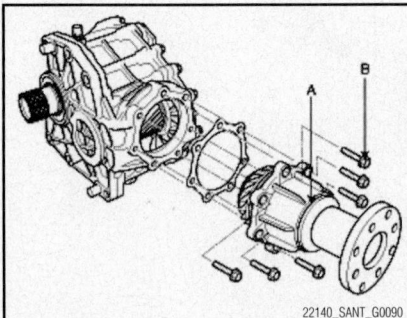

Fig. 26 Remove the 7 pinion assembly mounting bolts (B) and then remove the pinion assembly (A)

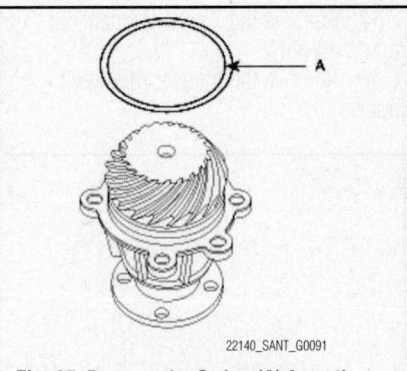

Fig. 27 Remove the O-ring (A) from the pinion shaft and case assembly

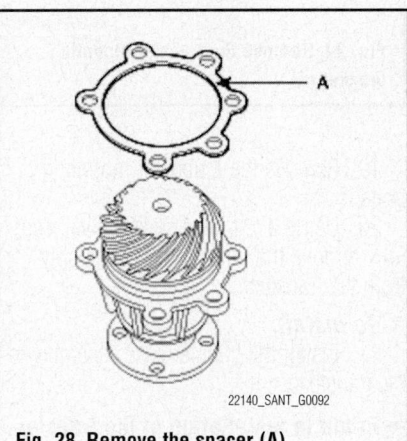

Fig. 28 Remove the spacer (A)

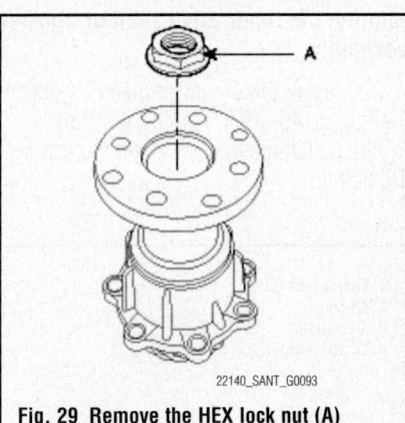

Fig. 29 Remove the HEX lock nut (A)

9. Using the SST(09455-32200), remove the pinion oil seal (A) from the pinion case.

To install:

10. Install the oil seal and the rear flange assembly.

11. Tighten down the hex lock nut until there is an axial clearance **B** of 0.0099–0.0296 inch (0.25mm–0.75mm) between pinion and pinion case.

a. Measurement of **B**:
- Apply an additional 200N tension load (F1) to the pinion head

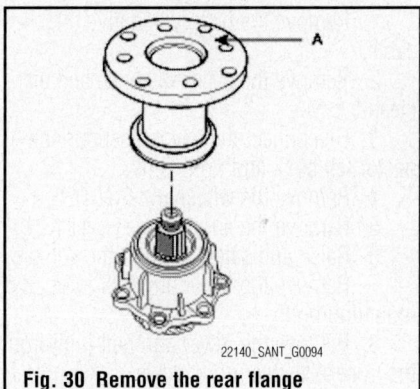

Fig. 30 Remove the rear flange assembly (A)

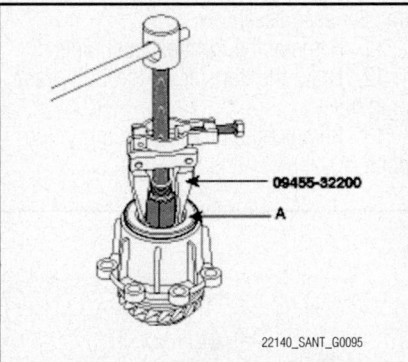

Fig. 31 Using the SST(09455-32200), remove the pinion oil seal (A) from the pinion case

according to arrangement Y. Rotate the pinion 10 times and measure dimension C
- Apply a 200N load (F2) to the rear flange. Rotate the pinion 10 times and measure dimension D
- B = C minus D

b. Measure dimension Ev.
- Tighten HEX lock nut
- Measure the drag torque of the pinion assembly. Target is 16–19 inch lbs. (1.8–2.1 Nm)

➡**To achieve the target preload torque, it is permissible to gradually increase the tightening torque. If the permissible preload torque of 19 inch lbs. (2.1 Nm) is exceeded, it is not allowed to loosen the Hex lock nut. Disassemble the whole pinion assembly. Replace the collapsible spacer and repeat the mounting procedure.**

c. Measure dimension En.
- Check preload. A preload torque of 16–19 inch lbs. (1.8–2.1 Nm) corresponds to a bearing preload of 0.0047–0.0063 inch (0.12–0.16mm)
- Ev minus En minus B = 0.0047–0.0063 inch (0.12–0.16mm)

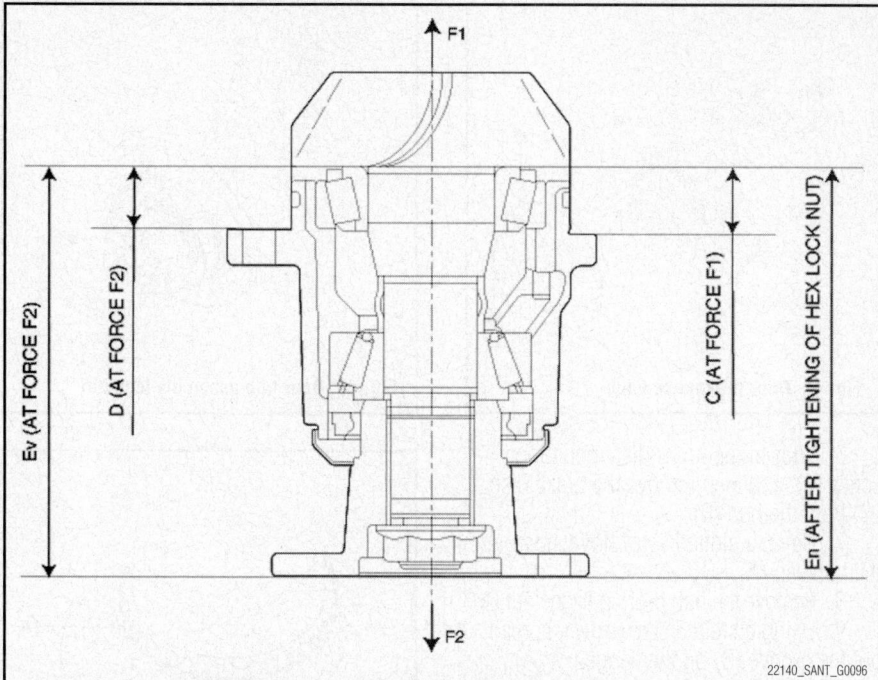

Fig. 32 Install the oil seal and the rear flange assembly. Tighten down the hex lock nut until there is an axial clearance B of 0.0099–0.0296 inch (0.25mm–0.75mm) between pinion and pinion case

✳✳ WARNING

The collapsible spacer must be replaced during reassembly. Permanent deformation is caused by the tightening procedure.

12. Assemble the O-ring and select the proper spacer for reassembling the transfer case.

13. Install the 7 pinion assembly mounting bolts. Tighten the bolts to: 15–22 ft. lbs. (37–40 Nm).

14. Install the pinion assembly.

15. Install the propeller shaft to the transfer case. Tighten to: 36–51 ft. lbs. (50–70 Nm).

16. Fill the transfer case with the proper amount and type of oil.

17. Check for leaks.

REAR AXLE HOUSING

REMOVAL & INSTALLATION

See Figures 33 through 36.

1. Before servicing the vehicle, refer to the Precautions Section.

2. Drain the differential gear oil.

3. Remove the rear drive shaft.

4. Remove the propeller shaft.

5. Support the differential assembly (A) with the jack (B).

6. Disconnect the coupling control connector (A).

7. After loosening the differential mounting bolts (A), remove the differential (B).

8. After loosening the cover bolts (A), remove the differential cover (B).

To install:

9. Apply liquid gasket to the differential cover (B).

10. Install the mounting bolts (A). Tighten to: 29–36 ft. lbs. (39–49 Nm).

11. Install the differential and tighten the mounting bolts to: 51–65 ft. lbs. (69–88 Nm).

12. Using the transaxle jack, install the differential assembly.

13. Connect the coupling control connector.

14. Install the propeller shaft.

15. Install the rear drive shaft.

16. Fill the gear oil with the proper amount and type of gear oil.

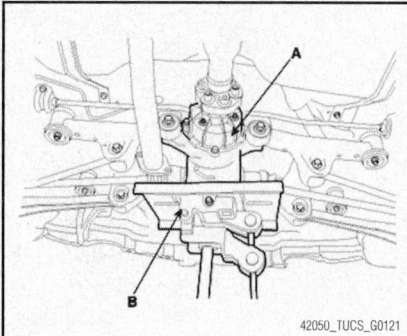

Fig. 33 Jack placement

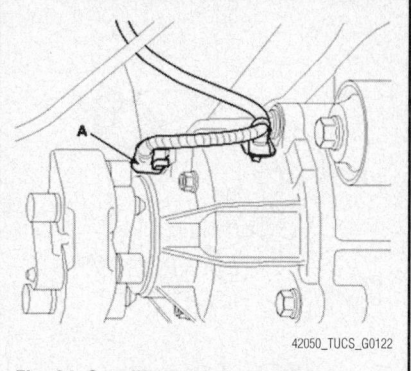

Fig. 34 Coupling control connector

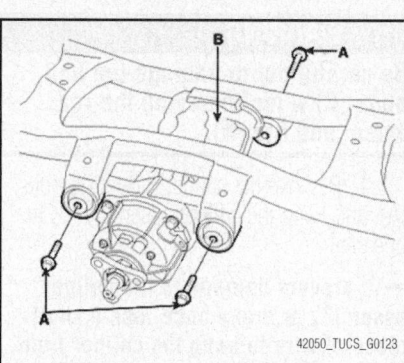

Fig. 35 Differential mounting bolt locations

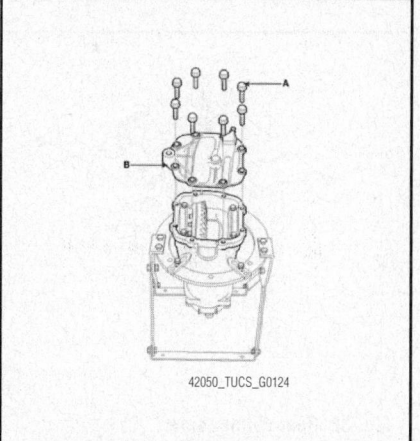

Fig. 36 Exploded view of removed rear differential cover

REAR AXLE SHAFT, BEARING & SEAL

REMOVAL & INSTALLATION

See Figures 37 through 46.

1. Before servicing the vehicle, refer to the Precautions Section.

2. Raise and safely support the vehicle.

3. Remove the rear wheel and tire (A) from rear hub (B).

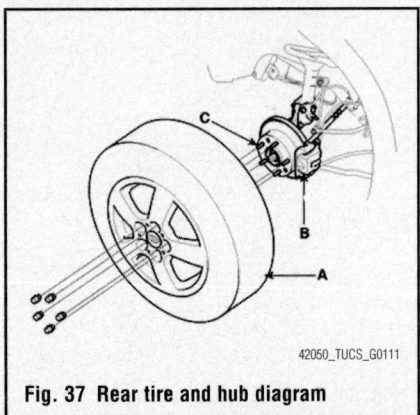

Fig. 37 Rear tire and hub diagram

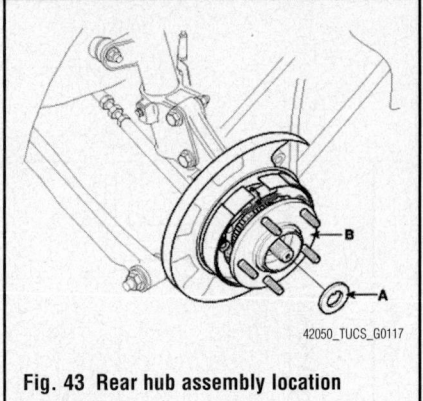

Fig. 43 Rear hub assembly location

✳✳ WARNING

Be careful not to damage the hub bolts (C) when removing the rear wheel and tire (A).

4. Remove the caliper mounting bolts (A), and hang the caliper assembly (B) to one side.

➡**To prevent damage to the caliper assembly or brake hose, use a short piece of wire to hang the caliper from the undercarriage.**

5. Remove the wheel speed sensor (B) from the axle carrier (A).

6. Loosen the brake disc mounting screw (A), and then remove the brake disc (C) from the hub (B).

7. Using a slotted screwdriver, remove the hub cap (A).

8. Remove the hub bearing flange nut (A).

9. Using a slotted screwdriver, spread out the groove (B) on the flange nut (A).

10. Loosen the hub bearing flange nut (A).

11. Remove the rear hub washer (A) and rear hub assembly (B).

✳✳ CAUTION

Be careful not to disassemble the rear hub assembly.

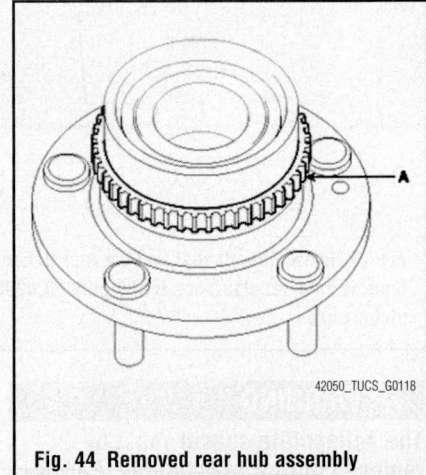

Fig. 44 Removed rear hub assembly

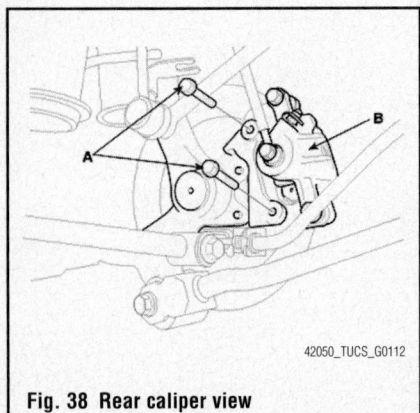

Fig. 38 Rear caliper view

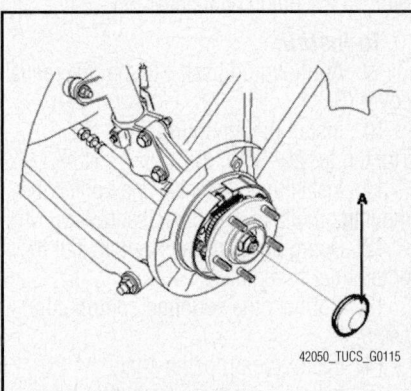

Fig. 41 Removing the hub cap

✳✳ WARNING

Care must be taken not to scratch or damage the teeth of the rotor. The rotor must never be dropped. If the teeth of the rotor are chipped or deformed in any way, wheel rotation speed will not be detected accurately and the system will react abnormally.

12. Loosen the rear dust cover mounting bolts (A) and then remove the rear parking brake assembly (B).

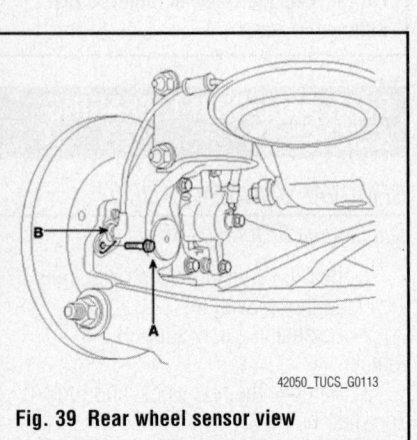

Fig. 39 Rear wheel sensor view

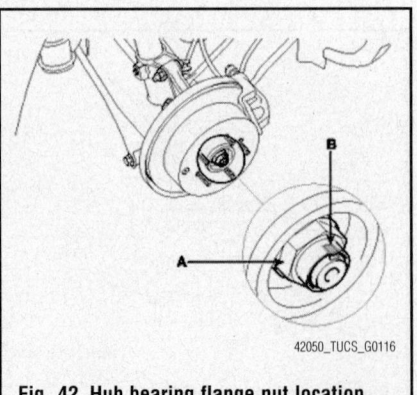

Fig. 42 Hub bearing flange nut location

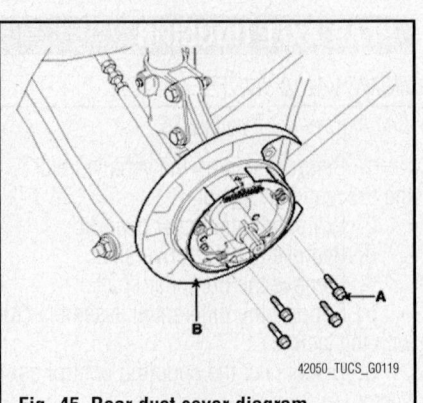

Fig. 45 Rear dust cover diagram

13. Remove the rear axle carrier (A).

14. Remove the trailing arm mounting bolt (B).

15. Remove the suspension arm mounting nut (C).

16. Remove the strut mounting nuts (D).

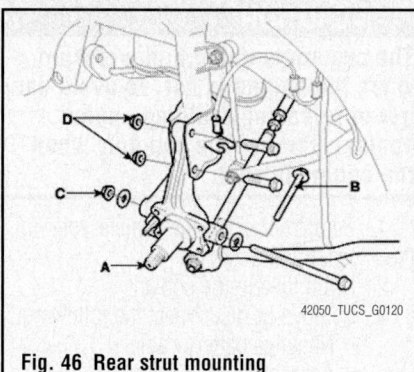

Fig. 46 Rear strut mounting

42050_TUCS_G0120

To install:

17. Before installation, check the following:

    a. The hub for cracks and the splines for wear.

    b. The brake disc for scoring and damage.

    c. The rear axle carrier for cracks

    d. The bearing for cracks or damage.

    e. Replace the hub as an assembly if problems are found.

18. Installation is the reverse of the removal procedure.

REAR HALFSHAFT

REMOVAL & INSTALLATION

1. Before servicing the vehicle, refer to the Precautions Section.

2. Remove or disconnect the following:
- Rear wheel
- Wheel speed sensor, if equipped
- Spindle nut
- Trailing arm mounting bolt
- Lower arm mounting nuts

3. Press the stub shaft out of the hub.

4. Using a suitable pry bar, pry the inner joint out of the differential.

To install:

➡**Use new circlips, split pins and self-locking nuts for assembly.**

5. Install the inner joint to the differential so that the circlip is felt to seat in the retaining groove and the halfshaft cannot be removed by hand.

6. Install or connect the following:
- Lower arm mounting nuts. Tighten to 104–118 ft. lbs. (140–160 Nm)
- Trailing arm mounting bolt. Tighten to 74–89 ft. lbs. (100–120 Nm)
- Spindle nut. Tighten to 148–207 ft. lbs. (200–280 Nm)
- Wheel speed sensor, if equipped
- Rear wheel

REAR PINION SEAL

REMOVAL & INSTALLATION

1. Before servicing the vehicle, refer to the Precautions Section.

2. Remove the differential case assembly.

3. Remove the side bearing inner races.

4. Remove the drive gear.

5. Remove the lock pin.

6. Remove the self-locking nut.

7. Remove the drive pinion.

8. Remove the drive pinion bearing inner race.

9. Remove the pinion oil seal.

10. Installation is the reverse of the removal procedure.

ENGINE COOLING

ENGINE FAN

REMOVAL & INSTALLATION

1. Before servicing the vehicle, refer to the Precautions Section.

2. Disconnect the ground cable from the battery cable.

3. Disconnect the connectors from the fan motor and the harness from the shroud.

4. For vehicles with automatic transaxles, remove the oil cooler hose from the shroud.

5. Remove the 4 bolts holding the shroud.

6. Remove the shroud with the fan motor.

7. Remove the fan mounting clip and detach the fan from the fan motor.

8. Remove the 3 screws and detach the fan motor.

To install:

9. Reattach the fan motor to the fan shroud and insert the 3 screws to secure.

10. Reattach the fan to the fan motor and install the fan mounting clip to secure.

11. Install the fan shroud with attached fan and fan motor to the radiator.

12. Install and tighten the 4 bolts holding the fan shroud in place with the radiator.

13. Install the oil cooler hose back on the shroud if previously removed.

14. Reconnect the connectors from the fan motor and the harness from the shroud.

15. Reconnect the battery ground or negative cable.

RADIATOR

REMOVAL & INSTALLATION

See Figure 47.

1. Before servicing the vehicle, refer to the Precautions Section.

2. Remove the air duct.

3. Disconnect the negative battery cable.

4. Remove the air cleaner assembly.

5. Drain the cooling system. Remove the radiator cap to speed draining.

6. Remove the inlet hose from the radiator.

7. Remove the transaxle oil cooler lines from the retainer clip at the bottom of the cooling fan shroud.

8. Remove the fan shroud clip from the condenser tubes.

9. Remove the bolt that connects the fan shroud to the condenser hold down bracket.

10. Remove the air deflectors from the top of the radiator.

11. Remove the cooling fan shroud bolts.

12. Remove the coolant reservoir hose from the radiator overflow neck.

13. Remove the radiator upper support brackets and bolts that connect to the fan shroud.

14. Disconnect the engine cooling fan motors electrical connectors.

15. Remove the cooling fan motors electrical harness from the fan shroud clips.

16. Remove the cooling fan shroud.

17. Remove the outlet hose from the radiator.

18. Disconnect the transaxle oil cooler pipes from the radiator.

19. Tilt the top of the radiator rearward.

20. Remove the condenser hold down bracket from the radiator.

21. Lift the condenser from the mounting tabs on the radiator and position the condenser aside.

22. Remove the radiator.

To install:

23. Install the radiator to the lower mounts.

➡**Verify that the condenser is fully seated in the radiator mounting tabs.**

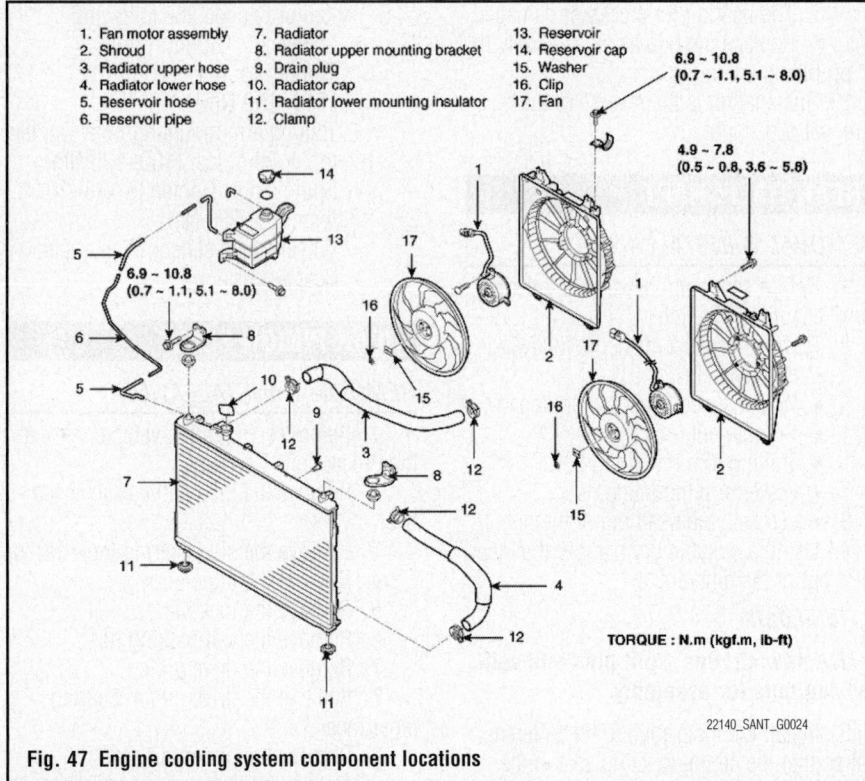

Fig. 47 Engine cooling system component locations

1. Fan motor assembly
2. Shroud
3. Radiator upper hose
4. Radiator lower hose
5. Reservoir hose
6. Reservoir pipe
7. Radiator
8. Radiator upper mounting bracket
9. Drain plug
10. Radiator cap
11. Radiator lower mounting insulator
12. Clamp
13. Reservoir
14. Reservoir cap
15. Washer
16. Clip
17. Fan

6.9 ~ 10.8
(0.7 ~ 1.1, 5.1 ~ 8.0)

4.9 ~ 7.8
(0.5 ~ 0.8, 3.6 ~ 5.8)

6.9 ~ 10.8
(0.7 ~ 1.1, 5.1 ~ 8.0)

TORQUE : N.m (kgf.m, lb-ft)

22140_SANT_G0024

24. Install the condenser to the mounting tabs on the radiator.

25. Install the condenser hold down bracket to the radiator and condenser.

26. Install the outlet hose to the radiator.

27. Connect the transaxle oil cooler pipes to the radiator.

➡**Ensure the lower edge of the fan shroud engages the clip at the bottom of the radiator.**

28. Install the cooling fan shroud.

29. Install the cooling fan motors electrical harness to the fan shroud clips.

30. Connect the engine cooling fan motors electrical connectors.

31. Install the fan shroud clip to the condenser tubes.

32. Install the cooling fan shroud bolts and tighten to 89 inch lbs. (10 Nm).

33. Install the radiator upper support brackets and bolts that connect to the fan shroud.

34. Install the air deflectors to the top of the radiator.

35. Install the bolt that connects the fan shroud to the condenser hold down bracket.

36. Install the inlet hose to the radiator.

37. Install the air cleaner assembly.

38. Install the coolant reservoir hose to the radiator overflow neck.

39. Install the transaxle oil cooler lines to the retainer clip at the bottom of the cooling fan shroud.

40. Fill the cooling system.

41. Connect the negative battery cable.

42. Run the engine through warm-up and check for leaks.

THERMOSTAT

REMOVAL & INSTALLATION

✳✳ CAUTION

Make sure the engine and radiator are cool to the touch prior to beginning this procedure in order to prevent scalding or burns.

1. Before servicing the vehicle, refer to the Precautions Section.

2. Drain the coolant down to thermostat level or below.

3. Remove the coolant outlet fitting and gasket.

4. Remove the thermostat.

To install:

5. Check that the flange of the thermostat is correctly seated in the socket of the thermostat housing.

6. Install the inlet fitting. Tighten the bolt to: 84–132 inch lbs. (10–15 Nm).

7. Refill the coolant with the proper amount and type of fluid.

8. Run the engine through warm-up and check for leaks.

WATER PUMP

REMOVAL & INSTALLATION

2.0L Engine

See Figure 48.

✳✳ CAUTION

The system is under high pressure when the engine is hot. To avoid danger of releasing scalding engine coolant, remove the cap only when the engine is cool.

1. Before servicing the vehicle, refer to the Precautions Section.

2. Drain the engine coolant.

3. Remove or disconnect the following:
- Negative battery cable
- Accessory drive belts
- Timing belt
- Timing belt idler

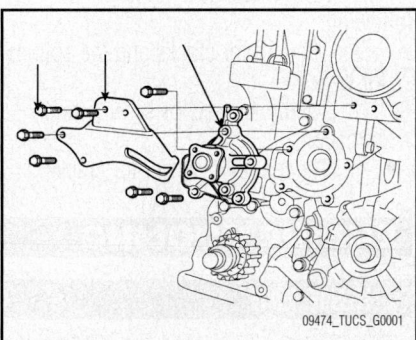

Fig. 48 Exploded view of water pump and alternator bracket—2.0L engine

09474_TUCS_G0001

- Water pump pulley
- Alternator bracket
- Water pump and gasket

➡**The water pump bolts are different lengths. Note the bolt location for assembly.**

To install:

4. Install the water pump with a new gasket. Tighten the bolts 15–20 ft. lbs. (20–27 Nm).

5. Install or connect the following:
- Alternator bracket
- Water pump pulley
- Timing belt idler
- Timing belt
- Accessory drive belts
- Negative battery cable

6. Fill the cooling system to the proper level.

7. Start the engine and check for leaks.

2.7L Engine

See Figures 49 and 50.

> ✳✳ **CAUTION**
>
> **The system is under high pressure when the engine is hot. To avoid danger of releasing scalding engine coolant, remove the cap only when the engine is cool.**

1. Before servicing the vehicle, refer to the Precautions Section.
2. Drain the engine coolant.
3. Remove or disconnect the following:
   - The accessory drive belt
   - The timing belt
   - The timing belt idler
   - The water pump (A) gasket (B)

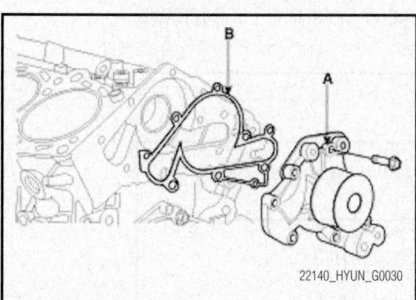

Fig. 49 Remove water pump and gasket—2.7L engine

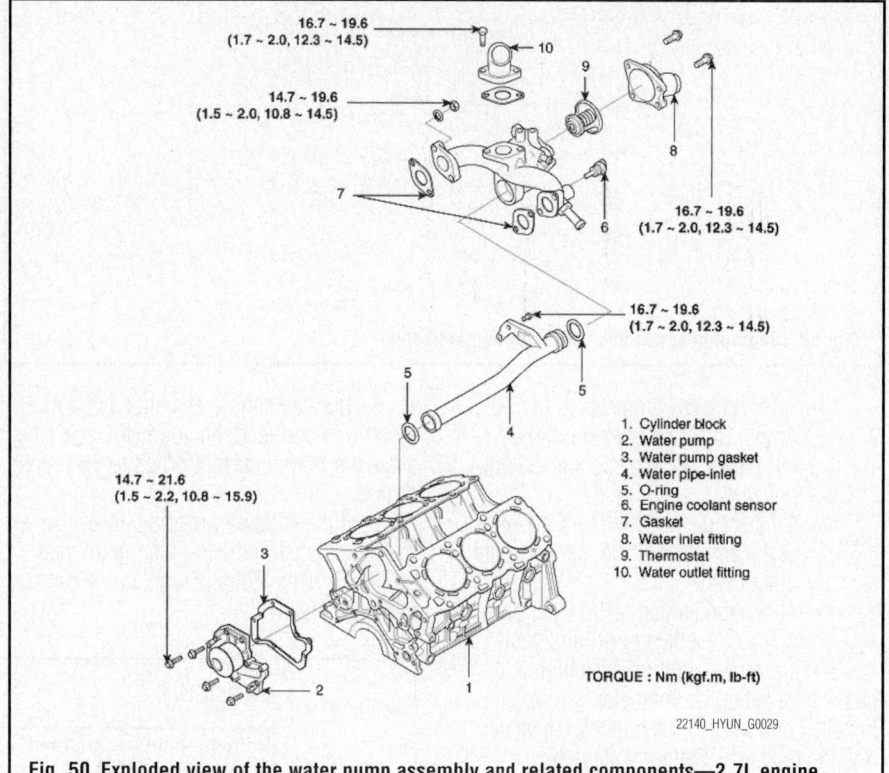

TORQUE : Nm (kgf.m, lb-ft)

1. Cylinder block
2. Water pump
3. Water pump gasket
4. Water pipe-inlet
5. O-ring
6. Engine coolant sensor
7. Gasket
8. Water inlet fitting
9. Thermostat
10. Water outlet fitting

Fig. 50 Exploded view of the water pump assembly and related components—2.7L engine

To install:

4. Install the water pump (A) and a new gasket (B) with the 8 bolts. Tightening torque: 11–16 ft. lbs. (15–22 Nm).
5. Install the timing belt idler.
6. Install the timing belt.
7. Install the accessory drive belt.
8. Fill with engine coolant.
9. Start engine and check for leaks.
10. Recheck engine coolant level.

ENGINE ELECTRICAL

➡Disconnecting the negative battery cable on some vehicles may interfere with the functions of the on board computer system. The computer may undergo a relearning process once the negative battery cable is reconnected.

ALTERNATOR

REMOVAL & INSTALLATION

See Figure 51.

1. Before servicing the vehicle, refer to the Precautions Section.
2. Disconnect the negative battery cable.
3. Disconnect the electrical connectors from the alternator.
4. Remove the accessory drive belt. Refer to Accessory Drive Belts, removal & installation.
5. Remove the alternator mounting bolts.
6. Remove the alternator.

To install:

7. Install the alternator.
8. Tighten the pivot bolt to 14–18 ft.

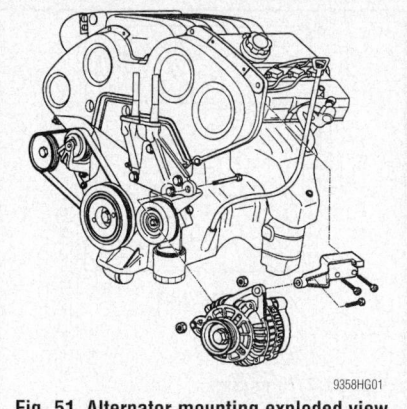

Fig. 51 Alternator mounting exploded view

lbs. (19–25 Nm) and the adjustment bolt to 11–16 ft. lbs. (15–22 Nm).

9. Connect the alternator electrical connectors. Tighten the battery terminal connector nut to 60 inch lbs. (7 Nm).
10. Install the accessory drive belt. Refer to Accessory Drive Belts, removal & installation.
11. Connect the negative battery cable.

CHARGING SYSTEM

VOLTAGE REGULATOR

ADJUSTMENT

See Figures 52 and 53.

1. Before servicing the vehicle, refer to the Precautions Section.
2. Prior to the test, check the following items and correct if necessary.
   - Check that the battery installed on the vehicle is fully charged
   - Check the alternator drive belt tension
   - Turn ignition switch to **OFF**
   - Disconnect the battery ground cable
   - Connect a digital voltmeter between the "B" terminal of the alternator and ground by connecting the positive (+) lead of the voltmeter to the "B" terminal of the alternator and connecting the negative (-) lead to good ground or the negative (-) battery terminal
   - Disconnect the alternator output wire from the alternator "B" terminal

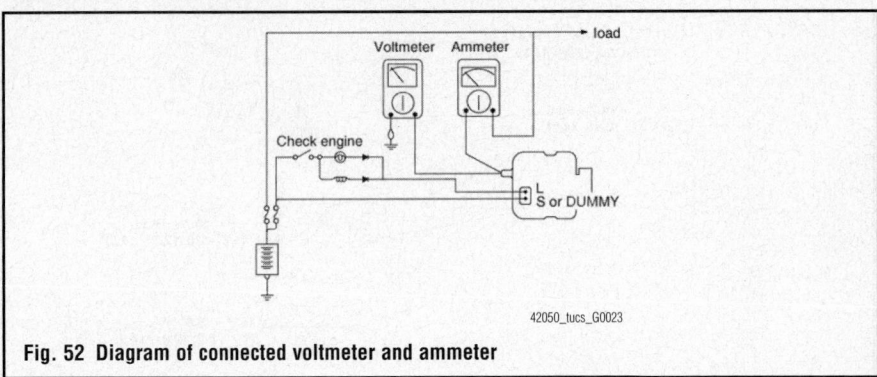

Fig. 52 Diagram of connected voltmeter and ammeter

- Connect a DC ammeter (0 to 150 Amps) in series between the "B" terminal and the disconnected output wire
- Connect the negative (-) lead wire of the ammeter to the disconnected output wire
- Attach the engine tachometer and connect the battery ground cable

3. Turn on the ignition switch and check to see that the voltmeter is properly installed and indicates the following value: Voltage reading = Battery voltage of 12.0–14.1 volts.

4. If it reads 0 volts, there is an open circuit in the wiring between the alternator "B" terminal and the battery and the battery negative (-) terminal, or the fusible link is blown.

5. Start the engine keeping all lights and accessories off.

6. Run the engine at a speed of about 2,500 rpm and read the voltmeter when the alternator output current drops to 10 Amps or less.

7. If the voltmeter reading agrees with the value listed in the Regulating Voltage Table below, the voltage regulator is functioning correctly.

8. If the reading is other than the standard value, the voltage regulator or the alternator is faulty.

9. Upon completion of the test, reduce the engine speed to idle, and turn off the ignition switch.

10. Disconnect the battery ground cable.

11. Remove the voltmeter and ammeter and the engine tachometer.

12. Connect the alternator output wire to the alternator "B" terminal.

13. Connect the battery ground cable.

REMOVAL & INSTALLATION

The voltage regulator is an internal component of the alternator. In order to replace the voltage regulator, the entire alternator assembly must be replaced. Refer to Alternator, installation and removal.

GASOLINE	
Voltage regulator ambient temperature °C (°F)	Regulating voltage (V)
-20 (-4)	14.2 ~ 15.4
20 (68)	14.0 ~ 15.0
60 (140)	13.7 ~ 14.9
80 (176)	13.5 ~ 14.7

Fig. 53 Regulating voltage table

ENGINE ELECTRICAL

ADJUSTMENT

These engines are equipped with a Distributorless Ignition System (DIS). No adjustment is necessary.

FIRING ORDERS

See Figures 54 and 55.

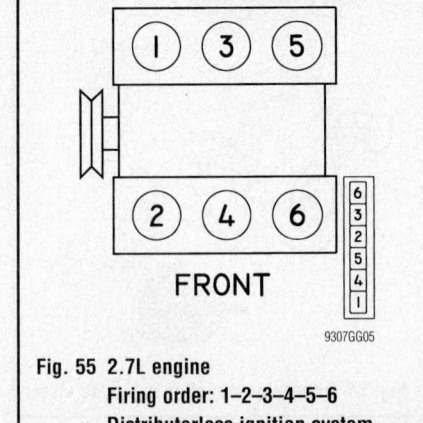

**Fig. 55 2.7L engine
Firing order: 1–2–3–4–5–6
Distributorless ignition system**

IGNITION COIL PACK

REMOVAL & INSTALLATION

See Figures 56 and 57.

1. Before servicing the vehicle, refer to the Precautions Section.

IGNITION SYSTEM

2. Remove the engine cover, if necessary.

3. Disconnect the spark plug cables and ignition coil connectors.

4. Remove the ignition coil pack (A).

5. Installation is the reverse of removal.

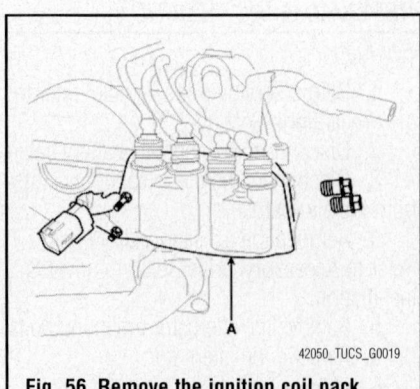

Fig. 56 Remove the ignition coil pack (A)—2.0L engine

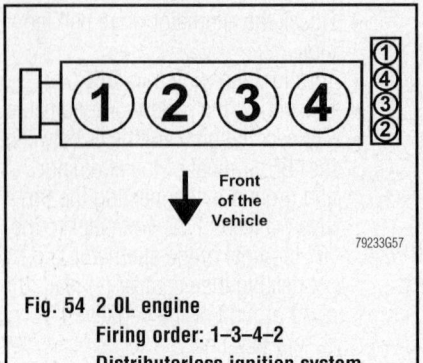

**Fig. 54 2.0L engine
Firing order: 1–3–4–2
Distributorless ignition system**

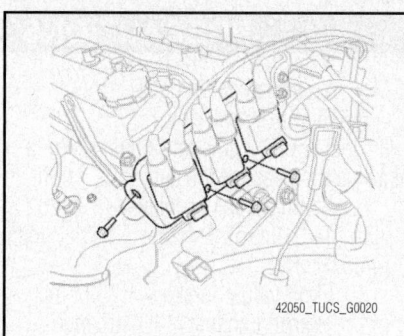

**Fig. 57 Remove the ignition coil pack—
2.7L engine**

IGNITION TIMING

ADJUSTMENT

Ignition timing is controlled by the electronic control ignition timing system. The standard reference ignition timing data for the engine operating conditions are pre-programmed in the memory of the Engine Control Module (ECM). The engine operating conditions (speed, load, warm-up condition, etc.) are detected by the various sensors. Based on these sensor signals and the ignition timing data, signals to interrupt the primary current are sent to the ECM. The ignition coil is activated, and timing is controlled.

SPARK PLUGS

REMOVAL & INSTALLATION

2.0L Engine

1. Before servicing the vehicle, refer to the Precautions Section.
2. Remove the spark plug cables.

✳✳ WARNING

When removing the spark plug cable, pull on the spark plug cable boot (not the cable), as it may be damaged.

3. Using a spark plug socket, remove the spark plugs.

✳✳ WARNING

Be careful that no contaminants enter through the spark plug holes.

4. Check the electrode gap of the new spark plugs. Specification: 0.039–0.043 inch (1.0–1.1mm).

To install:

5. Installation is the reverse of the removal procedure.
6. Tighten the spark plugs to 15–22 ft. lbs. (20–30 Nm).

2.7L Engine

1. Before servicing the vehicle, refer to the Precautions Section.
2. Remove the engine cover.
3. Disconnect the Variable Intake System (VIS) actuator connectors and the fuel injector connectors.
4. Remove the accelerator cable.
5. Remove surge tank sub assembly.
6. Remove the spark plug cable.
7. Remove the spark plug.
8. Check the electrode gap of the new spark plugs. Specification: 0.039–0.043 inch (1.0–1.1mm).

To install:

9. Installation is the reverse of the removal procedure.
10. Tighten the spark plugs to 15–22 ft. lbs. (20–30 Nm).

ENGINE ELECTRICAL

STARTER

REMOVAL & INSTALLATION

See Figure 58.

1. Before servicing the vehicle, refer to the Precautions Section.
2. Remove the negative battery cable and wait at least 3 minutes

3. Disconnect the starter cable (A) from the B terminal (B) on the solenoid (C).
4. Disconnect the connector (D) from the S terminal (E).
5. Remove the 2 bolts holding the starter.
6. Remove the starter motor assembly.

STARTING SYSTEM

To install:

7. Installation is the reverse of the removal procedure.
8. Tighten the starter motor bolts to 20–25 ft. lbs. (27–34 Nm).

SOLENOID OR RELAY REPLACEMENT

See Figure 59.

1. Before servicing the vehicle, refer to the Precautions Section.
2. Remove the fuse box cover.
3. Remove the starter relay (A).
4. Install a new starter relay.
5. Replace the fuse box cover.

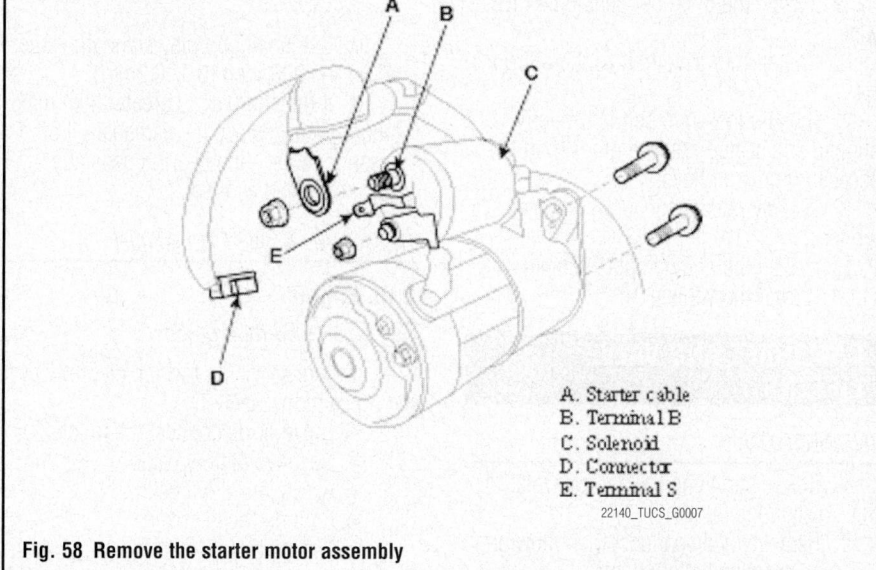

A. Starter cable
B. Terminal B
C. Solenoid
D. Connector
E. Terminal S

Fig. 58 Remove the starter motor assembly

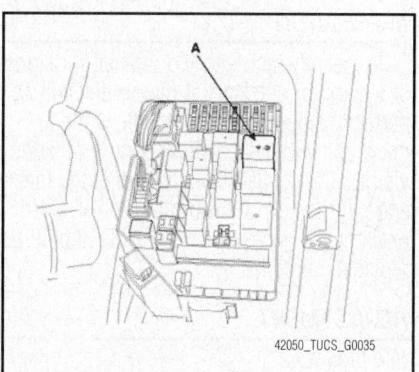

Fig. 59 Starter relay position in the fuse box.

ENGINE MECHANICAL

→Disconnecting the negative battery cable may interfere with the functions of the on board computer systems and may require the computer to undergo a relearning process, once the negative battery cable is reconnected.

ACCESSORY DRIVE BELTS

ACCESSORY BELT ROUTING

See Figures 60 and 61.

Refer to the accompanying illustrations for belt routing.

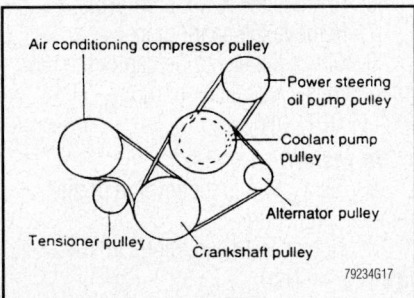

Fig. 60 Accessory drive belt routing—2.0L engines

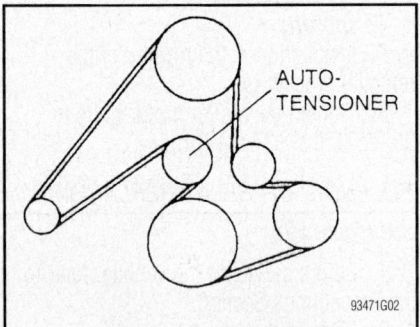

Fig. 61 Accessory drive belt routing—2.7L engine

INSPECTION

Inspect the accessory drive belt for signs of glazing or cracking. A glazed belt will be perfectly smooth from slippage, while a good belt will have a slight texture of fabric visible. Cracks will usually start at the inner edge of the belt and run outward. All worn or damaged accessory drive belts should be replaced immediately.

ADJUSTMENT

See Figure 62.

1. Before servicing the vehicle, refer to the Precautions Section.

2. Loosen the tension mounting bolt (B).
3. Turn the adjusting bolt to obtain the proper belt tension, then retighten the mounting bolt.
4. Recheck the deflection of the drive belt.

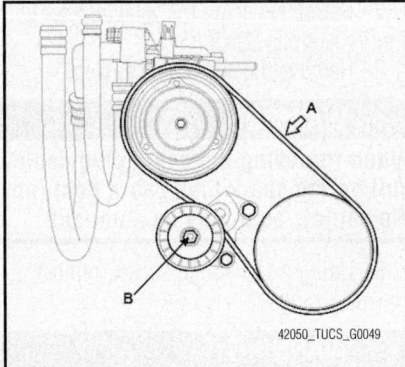

42050_TUCS_G0049

Fig. 62 Drive belt adjustment diagram

REMOVAL & INSTALLATION

1. Before servicing the vehicle, refer to the Precautions Section.
2. Raise and support the vehicle.
3. Remove the engine splash shield.
4. Rotate the drive belt tensioner clockwise to release the drive belt tension.
5. Remove the drive belt from the alternator.
6. Slowly release the drive belt tensioner.
7. Remove the drive belt from the accessory drive pulleys.

To install:
8. Install the drive belt to the accessory drive pulley.
9. Rotate the drive belt tensioner clockwise.
10. Install the drive belt to the alternator.
11. Ensure the drive belt is properly aligned and seated into the grooves of the accessory drive pulleys.
12. Slowly release the drive belt tensioner.
13. Install the engine splash shield.
14. Lower the vehicle.

CAMSHAFT AND VALVE LIFTERS

INSPECTION

1. Before servicing the vehicle, refer to the Precautions Section.
2. Remove or disconnect the following:
   - Negative battery cable

PRECAUTIONS

- Accessory drive belts
- Cylinder head cover
3. Inspect the camshaft lobes. Using a micrometer, measure the cam lobe height.
   a. Standard value: Intake = 1.7566 inch (44.618mm), Exhaust = 1.7527 inch (44.518mm).
   b. Limit value: Intake = 1.7527 inch (44.518mm), Exhaust = 1.7487 inch (44.418mm).
4. If the cam lobe height is less than the minimum limit value, replace the camshaft.
5. Inspect the camshaft journal clearance.
   a. Clean the bearing caps and camshaft journals.
   b. Place the camshafts on the cylinder head.
   c. Lay a strip of Plastigage® across each of the camshaft journals.
   d. Install the bearing caps.

→Do not turn the camshaft.

   e. Remove the bearing caps.
   f. Measure the Plastigage® at its widest point. The bearing oil clearance should be within specification.
   - Standard value: 0.0008–0.0024 inch (0.020–0.061mm)
   - Limit value: 0.0039 inch (0.1mm)
   g. If the oil clearance is greater than the maximum limit, replace the camshaft. If necessary, replace the bearing caps and cylinder head as a set.
6. Completely remove the Plastigage®.
7. Remove the camshafts.
8. Inspect camshaft end play.
   a. Install the camshafts.
   b. Using a dial indicator, measure the end play while moving the camshaft back and forth.
   c. Camshaft end play standard value: 0.004–0.008 inch (0.1–0.2mm).
   d. If the end play is greater than maximum limit, replace the camshaft. If necessary, replace the bearing caps and cylinder head as a set.

REMOVAL & INSTALLATION

2.0L Engine

See Figures 63 through 65.

1. Before servicing the vehicle, refer to the Precautions Section.
2. Remove or disconnect the following:
   - Negative battery cable
   - Accessory drive belts
   - Cylinder head cover
   - Timing belt

- Camshaft sprocket
- Timing chain auto tensioner
- Camshaft bearing caps and camshaft timing chain
- Intake and exhaust camshafts
- Mechanical lash adjusters

➡**Keep all valvetrain components in order for assembly.**

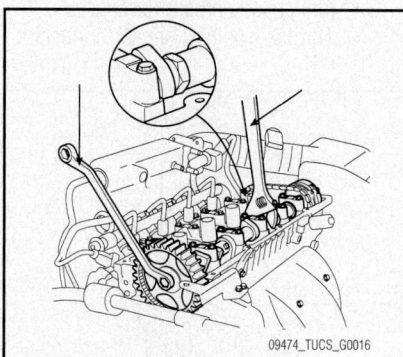

Fig. 63 Removing the camshaft sprocket—2.0L engine

To install:

3. Install or connect the following:
- Mechanical lash adjusters in their original positions
- Intake and exhaust camshafts with the timing chain aligned as shown

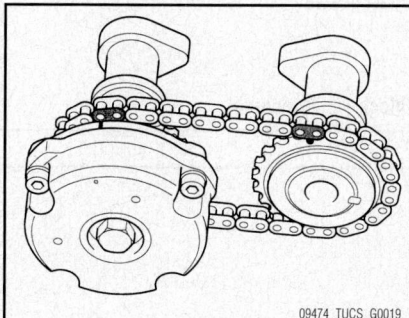

Fig. 64 Camshaft timing chain alignment marks—2.0L engine

- Camshaft bearing caps. Tighten to 10 ft. lbs. (14 Nm).
- Timing chain auto tensioner. Tighten to 72–84 inch lbs. (8–10 Nm).

4. Using Special Tool 09221-21000 seal installer, install the camshaft bearing oil seal.

5. Install or connect the following:
- Camshaft sprocket and torque the bolt to 74–89 ft. lbs. (100–120 Nm)
- Timing belt

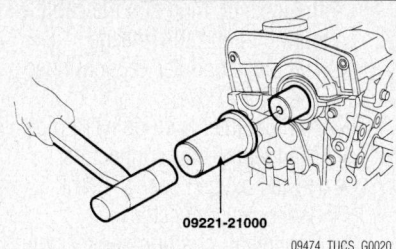

Fig. 65 Camshaft bearing oil seal install tool—2.0L engine

- Cylinder head cover
- Accessory drive belts
- Negative battery cable

6. Refill the cooling system to the correct level.

7. Refill the engine oil to the correct level.

8. Start the engine and check for leaks.

2.7L Engine

See Figures 66 through 69.

Engine removal is not required for this procedure. Use Fender cover to avoid damaging painted surfaces. To avoid damage, unplug the wiring connectors carefully while holding the connector portion. Mark all wiring and hoses to avoid misconnection. Inspect the timing belt before removing. Turn the crankshaft pulley so that the No. 1 piston is at Top Dead Center (TDC).

1. Before servicing the vehicle, refer to the Precautions Section.

2. Remove the side cover (A, B, C) and the engine cover (D).

3. Disconnect the battery terminal and the battery.

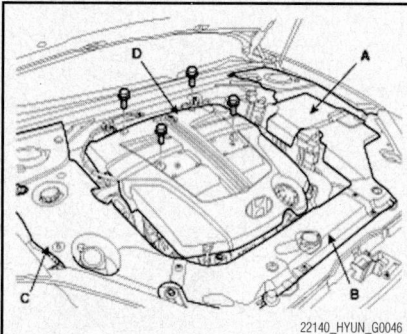

Fig. 66 Remove the side cover and engine cover—2.7L engine

4. Remove the radiator drain plug and drain engine coolant. Remove the radiator cap to speed draining.

5. Remove the air cleaner assembly.
a. Disconnect the Air Flow Sensor (AFS) connector.

b. Remove the breather hose from intake hose.
c. Remove the intake hose and air cleaner upper cover.
d. Remove the air cleaner lower cover.
6. Remove the upper radiator hose and lower radiator hose.
7. Remove the heater hoses.
8. Remove the engine wire harness connectors and wire harness clamps from the cylinder head and the intake manifold:
- Throttle position sensor (TPS) connector
- Idle speed actuator (ISA) connector
- Purge Control Solenoid Valve (PCSV) connector
- Knock sensor connector
- Camshaft Position (CMP) sensor connector
- Engine ground line
- Heated oxygen sensor (Bank 2, Sensor 1) connector
- Engine temperature coolant sensor connector
- Ignition coil connector
- Crankshaft Position (CKP) sensor connector
- Heated oxygen sensor (Bank 1, Sensor 2) connector
- Fuel injector connectors
9. Remove or disconnect the following:
- The fuel inlet from the delivery pipe
- The Purge Control Solenoid Valve (PCSV) hose
- The brake booster vacuum hose
- The accelerator cable by loosening the locknut, then slip the cable end out of the throttle linkage
- The auto-cruise connector and the auto-cruise cable
- The PCV hose
- The power steering pump
- The timing belt
- The cylinder head covers
- The camshaft sprocket
- The camshaft bearing caps
- The camshafts

To install:

Thoroughly clean all parts to be assembled. Rotate the crankshaft, set the No. 1 piston at TDC.

10. Install the camshafts.
a. Align the camshaft timing chain with the intake timing chain sprocket and exhaust timing chain sprocket as shown.
b. Install the camshaft.
c. Install the camshaft bearing caps.
- Tightening torque: M6 (38mm) bolt: 7–9 ft. lbs. (10–13 Nm)
- Tightening torque: M6 (50mm) bolt: 11–12 ft. lbs. (15–17 Nm)

➡Apply new engine oil to the thrust portion and journal of the camshafts. Apply a light coat of engine oil on the threads and under the heads of the bearing cap bolts.

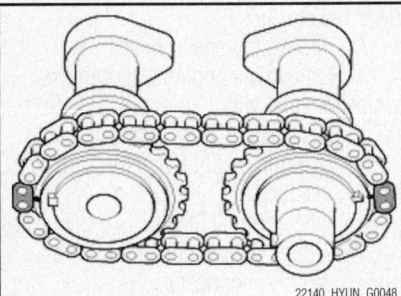

Fig. 67 Align the camshaft timing chain with the intake timing chain sprocket and exhaust timing chain sprocket as shown—2.7L engine

11. Using the SST (09214-21000), install the camshaft bearing oil seal.

12. Install the camshaft sprocket.

a. Temporarily install the camshaft sprocket bolts.

b. Hold the hexagonal head wrench portion of the camshaft with a wrench, and tighten the camshaft sprocket bolts. Tightening torque of camshaft sprocket bolt: 65–80 ft. lbs. (88–108 Nm).

13. Install the semi-circular packing.

14. Install the cylinder head cover.

a. Install the cylinder head cover gasket in the groove of the cylinder head cover.

➡Before installing the head cover gasket, thoroughly clean the head cover gasket and the groove. When installing, make sure the head cover gasket is seated securely in the corners of the recesses with no gap.

b. Apply liquid gasket to the head cover gasket at the corners of the recess. Use liquid gasket, Loctite® No. 5699. Check that the mating surfaces are clean and dry before applying liquid gasket. After assembly, wait at least 30 minutes before filling the engine with oil.

c. Install the cylinder head covers with the 16 bolts. Uniformly tighten the bolts in several passes. Tightening torque: 6–7 ft. lbs. (8–10 Nm).

15. Install or connect the following:
- The timing belt
- The power steering pump
- The PCV hose
- The auto-cruise connector and the auto- cruise cable
- The accelerator cable by loosening

the locknut, then slip the cable end out of the throttle linkage
- The brake booster vacuum hose
- The PCSV hose
- The fuel inlet from delivery pipe
- The fuel injector connectors
- Heated oxygen sensor (Bank 1, Sensor 2) connector
- Crankshaft position sensor connector
- Ignition coil connector

- Engine temperature coolant sensor connector
- Heated oxygen sensor (Bank 2, Sensor 1) connector
- Engine ground line
- Camshaft position sensor connector
- Knock sensor connector
- Injector connector
- The PCSV connector
- Idle speed actuator connector
- Throttle position sensor connector

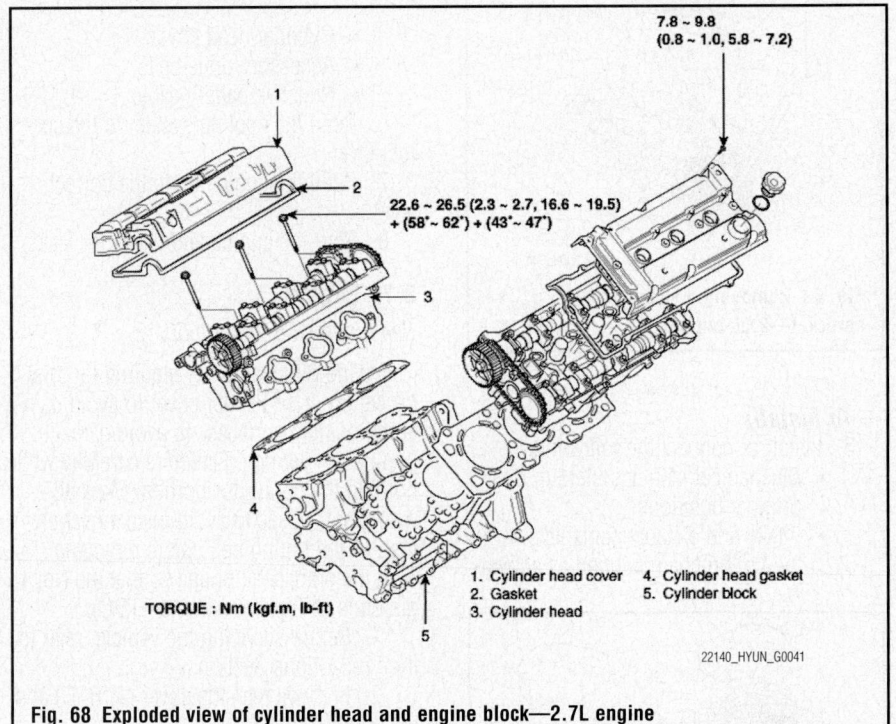

7.8 ~ 9.8
(0.8 ~ 1.0, 5.8 ~ 7.2)

22.6 ~ 26.5 (2.3 ~ 2.7, 16.6 ~ 19.5)
+ (58°~ 62°) + (43°~ 47°)

1. Cylinder head cover
2. Gasket
3. Cylinder head
4. Cylinder head gasket
5. Cylinder block

TORQUE : Nm (kgf.m, lb-ft)

Fig. 68 Exploded view of cylinder head and engine block—2.7L engine

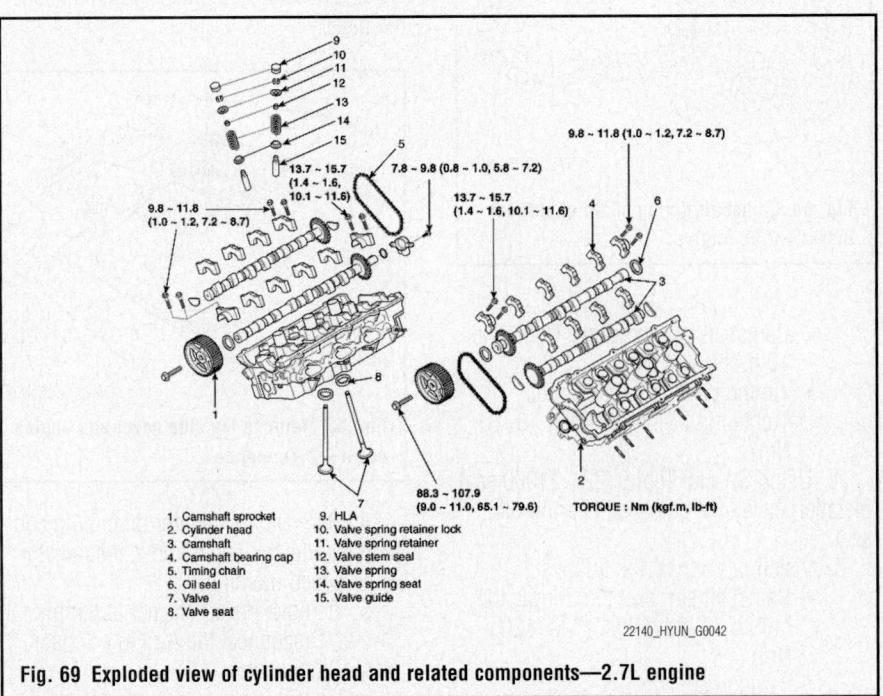

9.8 ~ 11.8 (1.0 ~ 1.2, 7.2 ~ 8.7)

13.7 ~ 15.7
(1.4 ~ 1.6,
10.1 ~ 11.6)

7.8 ~ 9.8 (0.8 ~ 1.0, 5.8 ~ 7.2)

13.7 ~ 15.7
(1.4 ~ 1.6, 10.1 ~ 11.6)

9.8 ~ 11.8
(1.0 ~ 1.2, 7.2 ~ 8.7)

88.3 ~ 107.9
(9.0 ~ 11.0, 65.1 ~ 79.6)

TORQUE : Nm (kgf.m, lb-ft)

1. Camshaft sprocket
2. Cylinder head
3. Camshaft
4. Camshaft bearing cap
5. Timing chain
6. Oil seal
7. Valve
8. Valve seat
9. HLA
10. Valve spring retainer lock
11. Valve spring retainer
12. Valve stem seal
13. Valve spring
14. Valve spring seat
15. Valve guide

Fig. 69 Exploded view of cylinder head and related components—2.7L engine

16. Install the engine wire harness connectors and wire harness clamps to the cylinder head and the intake manifold.
17. Install the heater hoses.
18. Install the upper radiator hose and lower radiator hose.
19. Install the air cleaner assembly.
   a. Install the air cleaner lower cover.
   b. Install the intake hose and air cleaner upper cover.
   c. Install the breather hose from intake hose.
   d. Connect the AFS connector.
20. Connect the battery terminal and the battery.
21. Fill with engine coolant.
22. Install the side cover and the engine cover.
23. Start the engine and check for leaks.
24. Recheck the engine coolant level and oil level.

CAMSHAFT BEARING REPLACEMENT

Check each bearing for damage. If the bearing surface is excessively damaged, replace the cylinder head assembly or camshaft bearing cap, as necessary.

1. Before servicing the vehicle, refer to the Precautions Section.
2. Remove the engine cover and intake manifold. For additional information, refer to Intake Manifold, removal & installation.
3. Disconnect the breather hose and the engine harness.
4. Remove the timing belt.
5. Remove the spark plug cables.
6. Loosen the cylinder head cover bolts and then remove the cylinder head cover (also called the valve cover).
7. Remove the camshaft sprockets.
8. Remove the camshaft bearing caps.
9. Remove the camshafts.
10. Rotate the crankshaft so No. 1 cylinder is at Top Dead Center (TDC) on the compression stroke).
11. Check the position of the rocker arm whether it is exactly installed on the lash adjuster and valve or not.
12. Install the camshaft dowel pin.

⁂ WARNING

The left and right banks of the camshafts are different. Be careful to put them in the correct position.

13. Look for identification markings for the left and right banks of the camshafts:
   a. Left bank Intake (IN): I. Exhaust (Ex): E.
   b. Right bank Intake (IN): J. Exhaust (Ex): H.
14. Confirm the identification mark and the number. Bearing caps for No. 3, No. 4, and No. 5 should have a front mark. Arrange the front mark upon the cylinder head while installing the bearing caps.
15. Use the identification mark—Intake **I** and Exhaust **E**.
16. Tighten each bearing cap in 2–3 steps:
   a. The 16 outer to: 14–15 ft. lbs. (19–21 Nm).
   b. The 24 inner to: 89–106 inch lbs. (10–12 Nm).
17. Install the cylinder head cover.
18. Install the gasket to the cylinder head cover correctly.
19. Clean the sealing surface on the camshaft cap.

➡Clean the sealing surface using a scraper to prevent any oil leaks. Apply sealant to the sealing surface on the cylinder head cover and camshaft cap.

➡Be careful of gasket escapement when installing the cylinder head cover. You must use the washer when installing the cylinder head cover bolts. Refer to Valve Covers, removal & installation.

CRANKSHAFT DAMPER

REMOVAL & INSTALLATION

See Figures 70 through 74.

1. Before servicing the vehicle, refer to the Precautions Section.
2. Remove the engine cover.
3. Remove the front right wheel and tire.
4. Remove the right side cover.
5. Remove the drive belt (A), the idler (B), and the tensioner (C).

➡In removing the drive belt, fix a tool in the auto tensioner pulley bolt and turn the bolt counter clockwise.

6. Remove the timing belt upper cover (A).

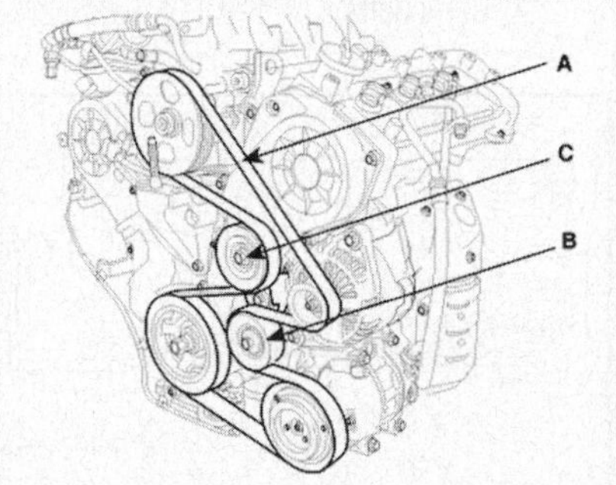

Fig. 70 Drive belt, idler, and tensioner locations

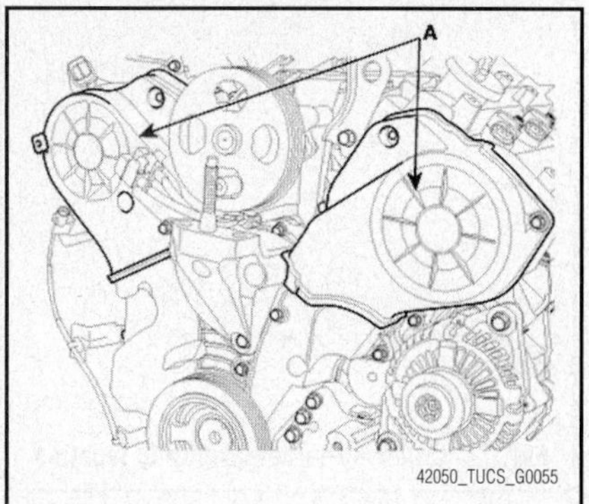

Fig. 71 Timing belt upper cover location

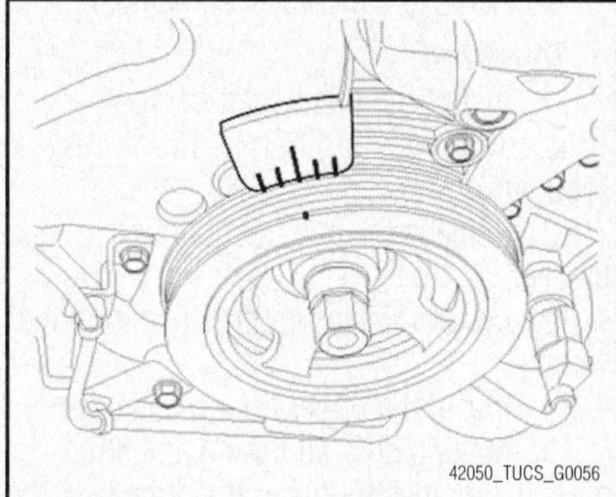

Fig. 72 Timing mark location

Align the groove of the pulley with the timing mark of the timing belt cover by turning the crankshaft pulley clockwise. Check if the timing mark of the camshaft sprocket is aligned with that of the cylinder head cover at the moment. No. 1 cylinder piston at Top Dead Center (TDC).

7. Support the engine oil pan with a jack.

⁂ CAUTION

Put a wooden or rubber block between the jack and the engine oil pan.

8. Remove the engine mounting bracket (A).

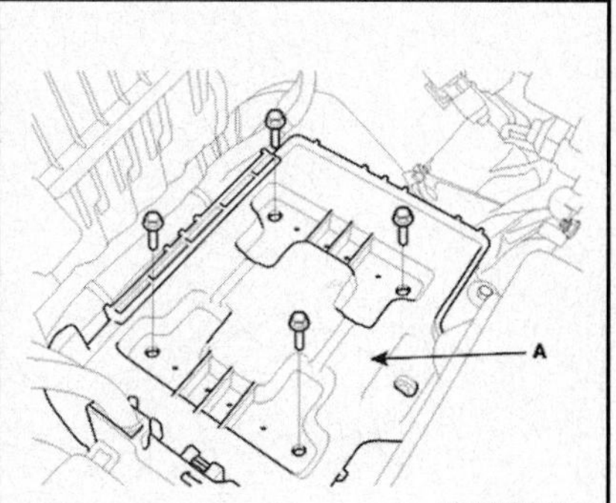

Fig. 73 Engine mounting bracket location

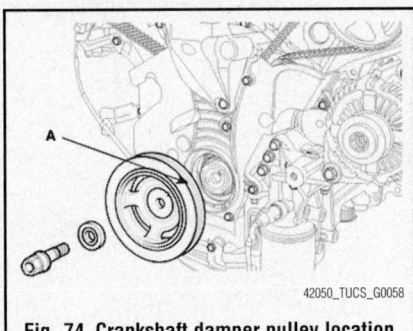

Fig. 74 Crankshaft damper pulley location

9. Remove the crankshaft damper pulley (A).

10. Remove the crankshaft damper.

To install:

11. Install the crankshaft damper. Torque the mounting bolt to 133 ft. lbs. (180 Nm).

12. Replace the engine mounting bracket.

13. Lower the jack and remove from the vehicle.

14. Install the upper timing belt cover.

15. Making sure all timing and alignment marks match, install the drive belt, the idler, and the tensioner.

16. Install the right side cover.

17. Install the right front tire and wheel.

18. Install the engine cover.

CRANKSHAFT FRONT SEAL

REMOVAL & INSTALLATION

See Figure 75.

1. Before servicing the vehicle, refer to the Precautions Section.

2. Remove or disconnect the following:
   - Negative battery cable
   - Accessory drive belts
   - Front covers
   - Timing belt
   - Crankshaft timing sprocket
   - Front crankshaft seal

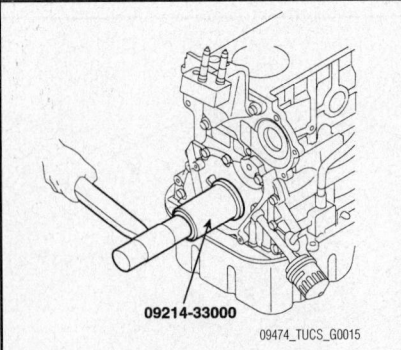

Fig. 75 Installing the front crankshaft seal—2.0L engine, 2.7L engine similar

To install:

3. Install the front crankshaft seal using Special Tool 09214–33000 seal installer

4. Install or connect the following:
   - Crankshaft timing sprocket
   - Timing belt
   - Front covers
   - Accessory drive belts
   - Negative battery cable

5. Start the engine and check for leaks.

CYLINDER HEAD

REMOVAL & INSTALLATION

2.0L Engine

See Figures 76 and 77.

1. Before servicing the vehicle, refer to the Precautions Section.

2. Drain the cooling system.

3. Relieve the fuel system pressure. Refer to Relieving Fuel System Pressure.

4. Remove or disconnect the following:
   - Negative battery cable
   - Engine cover
   - Air intake assembly
   - Upper and lower radiator hose
   - Heater hoses
   - Engine wiring harnesses connectors and clamps from the cylinder head and intake manifold
   - Fuel supply hose
   - Purge Control Solenoid Valve (PCSV) hose
   - Brake booster vacuum hose
   - Throttle cable
   - Power steering pump and mounting bracket
   - Spark plug cables
   - PCV hose
   - Cylinder head cover
   - Timing belt
   - Exhaust manifold
   - Intake manifold
   - Camshaft sprocket
   - Timing chain auto tensioner
   - Camshaft bearing caps
   - Camshafts

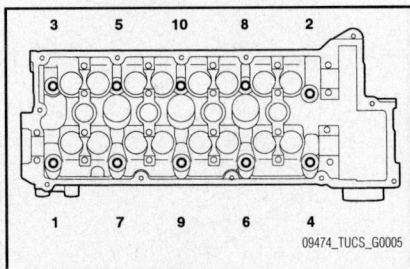

Fig. 76 Cylinder head removal sequence—2.0L engine

- Oil Control Valve (OCV)
- OCV filter
- Coolant hose from coolant pipe

5. Remove the cylinder head mounting bolts in sequence as shown.

6. Remove the cylinder head.

To install:

7. Install the cylinder head using a new gasket.

8. Apply a light coat of engine oil to the mounting bolts and tighten in sequence as shown:

   a. Step 1: M10 bolts to 18 ft. lbs. (25 Nm). M12 bolts to 22 ft. lbs. (30 Nm).

   b. Step 2: Plus 60–65 degrees.

   c. Step 3: Plus an additional 60–65 degrees.

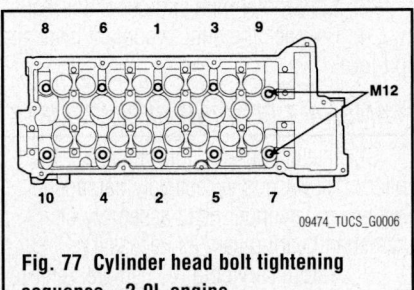

Fig. 77 Cylinder head bolt tightening sequence—2.0L engine

9. Install or connect the following:
   - OCV filter and tighten to 30–38 ft. lbs. (41–51 Nm)
   - OCV and tighten to 84–108 inch lbs. (10–12 Nm)
   - Camshafts and bearing caps
   - Timing chain auto tensioner
   - Timing belt

10. Install the cylinder head cover and new gasket.

   a. Apply liquid gasket to the head cover gasket at the corners of the recesses.

   b. Tighten the cylinder head cover mounting bolts uniformly in several passes to 72–89 inch lbs. (8–10 Nm).

11. Install or connect the following:
   - Intake manifold
   - Exhaust manifold
   - PCV
   - Spark plug wires
   - Power steering pump mounting bracket. Tighten to 26–37 ft. lbs. (35–50 Nm)
   - Power steering pump
   - Throttle cable
   - Brake booster hose
   - PCSV hose
   - Fuel supply hose
   - Engine wiring harnesses to the cylinder head and intake manifold

- Heater hoses
- Upper and lower radiator hoses
- Air intake assembly
- Engine cover
- Negative battery cable

12. Fill the engine with coolant to the correct level.

13. Start the engine and check for leaks.

2.7L Engine

See Figures 78 through 80.

1. Before servicing the vehicle, refer to the Precautions Section.

2. Drain the cooling system.

3. Relieve the fuel system pressure. Refer to Relieving Fuel System Pressure.

4. Remove or disconnect the following:
- Negative battery cable
- Engine cover
- Air intake assembly
- Upper and lower radiator hoses
- Heater hoses
- Engine wiring harnesses from the cylinder head and intake manifold
- Fuel supply hoses
- Purge Control Solenoid Valve (PCSV) hose
- Brake booster vacuum hose
- Throttle cable
- PCV hose
- Intake manifold
- Power steering pump
- Exhaust manifold
- Timing belt
- Spark plug cable
- Cylinder head covers
- Camshaft bearing caps
- Camshafts
- Timing belt rear cover
- Water temperature control assembly and water pipe

5. Loosen the cylinder head bolts in several passes in the sequence shown.

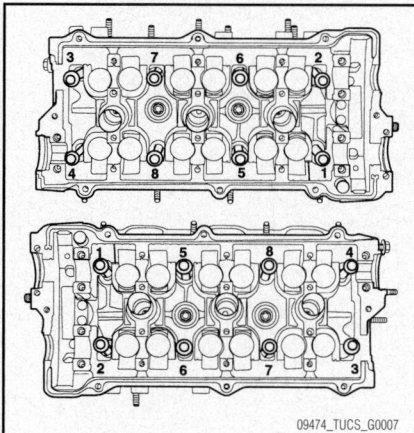

Fig. 78 Cylinder head bolt removal sequence—2.7L engine

09474_TUCS_G0007

6. Remove the cylinder bolts and plate washer.

7. Remove the cylinder heads.

To install:

8. Install the cylinder heads with new gaskets.

9. Apply a light coat of engine oil the cylinder head bolts and tighten in sequence as follows:
   a. Step 1: Tighten to 18 ft. lbs. (25 Nm).
   b. Step 2: Plus 60 degrees.
   c. Step 3: Plus 45 degrees.

10. Install or connect the following:
- Water pipe and water temperature control assembly. Tighten to 11–14 ft. lbs. (15–20 Nm)
- Timing belt rear cover. Tighten to 84–108 inch lbs. (10–12 Nm)

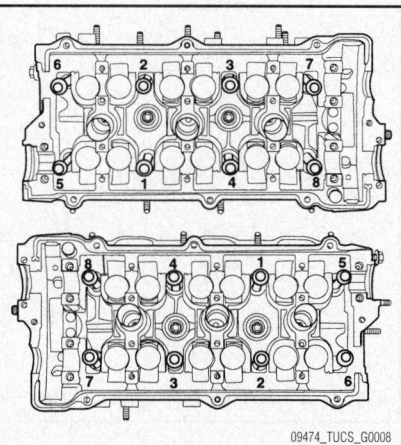

Fig. 79 Cylinder head bolt tightening sequence—2.7L engine

09474_TUCS_G0008

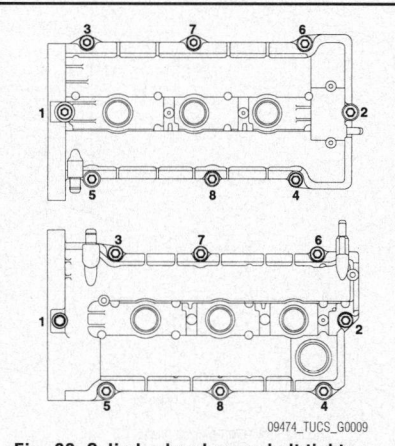

Fig. 80 Cylinder head cover bolt tightening sequence—2.7L engine

09474_TUCS_G0009

- Camshafts and camshaft bearing caps
- Cylinder head cover gaskets. Apply liquid gasket material to the gasket at the corners of the recess
- Cylinder head covers. Tighten bolts in sequence to 72–89 inch lbs. (8–10 Nm).

11. Install or connect the following:
- Spark plug cable
- Timing belt
- Exhaust manifold
- Power steering pump
- Intake manifold
- PCV hose
- Throttle cable
- Brake booster vacuum hose
- PCSV hose
- Fuel supply hose
- Engine wiring harnesses to the cylinder head and intake manifold
- Heater hoses
- Upper and lower radiator hoses
- Air intake assembly
- Engine cover
- Negative battery cable

12. Refill the engine with coolant to the correct level.

13. Start the engine and check for leaks.

ENGINE ASSEMBLY

REMOVAL & INSTALLATION

➡**Hyundai recommends that the engine and transaxle be removed as a single unit on all models.**

1. Before servicing the vehicle, refer to the Precautions Section.

2. Drain the cooling system.

3. Drain the transaxle.

4. Drain the engine oil.

5. Relieve the fuel system pressure.

6. Remove or disconnect the following:
- Negative battery cable
- Air intake assembly
- Engine cover
- Upper and lower radiator hoses
- Heater hoses

7. Remove the following engine wiring harnesses:
- Oil control valve (OCV) connector
- Oil temperature sensor connector
- Engine coolant temperature (ECT) sensor
- Ignition coil connector
- Throttle position sensor (TPS) connector
- Idle Speed Actuator (ISA) connector
- Camshaft position sensor (CMP) connector
- Fuel injector connectors

- Ground cable from intake manifold
- Compressor switch
- Front heated oxygen sensors
- Crankshaft position (CKP) connector
- Oil pressure switch connector
- Purge control solenoid valve (PCSV) connector

8. Remove or disconnect the following:
- Fuel supply hose
- PCSV hose
- Brake booster vacuum hose
- Throttle cable
- Power steering pump
- Battery and battery bracket
- Transaxle wiring harnesses
- Transaxle cooler hose
- Engine splash guard
- Front exhaust pipe
- ABS wheel speed sensor from front knuckles
- Front strut lower mounting bolts
- Front calipers
- Steering U-joint mounting bolt

9. Using a suitable jack, support the engine and transaxle assembly.

10. Remove or disconnect the following:
- Engine mounting bracket
- Transaxle mounting bracket
- Sub-frame mounting bolts

11. Jack up the vehicle and remove the engine/transaxle assembly.

To install:

12. Move the powertrain assembly into position and lower the vehicle into place.

13. Install or connect the following:
- Sub-frame mounting bolts and nuts. Tighten the nuts to 118–133 ft. lbs. (160–180 Nm) and the bolts to 52–66 ft. lbs. (70–90 Nm)
- Transaxle mounting bracket. Tighten to 118–133 ft. lbs. (160–180 Nm)
- Engine mounting bracket. Tighten to 118–133 ft. lbs. (160–180 Nm)
- Steering U-joint mounting bolt
- Front calipers
- Front strut lower mounting bolts
- ABS wheel speed sensors
- Front exhaust pipe
- Engine splash guard
- Transaxle cooler hose
- Transaxle wiring harness
- Battery and battery bracket
- Power steering pump
- Throttle cable
- Brake booster vacuum hose
- PCSV hose
- Fuel supply hose
- Engine wiring harnesses
- Heater hoses
- Upper and lower radiator hoses
- Air intake assembly

- Engine cover
- Negative battery cable

14. Fill the engine with clean oil.
15. Fill the transaxle to the correct level.
16. Fill the cooling system to the proper level.
17. Start the engine and check for leaks.

EXHAUST MANIFOLD

REMOVAL & INSTALLATION

See Figures 81 and 82.

1. Before servicing the vehicle, refer to the Precautions Section.

2. Remove or disconnect the following:
- Negative battery cable
- Engine cover
- Front Oxygen Sensor (O$_2$S) connector

- Front muffler
- Manifold heat shield
- Exhaust manifold

To install:

3. Installation is the reverse of the removal procedure.

4. Observe the following torques:
   a. 2.0L Engine
   - Exhaust manifold: 32–40 ft. lbs. (43–55 Nm)
   - Manifold heat shield: 12–16 ft. lbs. (17–22 Nm)
   b. 2.7L Engine
   - Exhaust manifold: 22–26 ft. lbs. (30–35 Nm)
   - Manifold heat shield: 12–16 ft. lbs. (17–22 Nm)
   - Front exhaust pipe: 22–30 ft. lbs. (30–40 Nm)

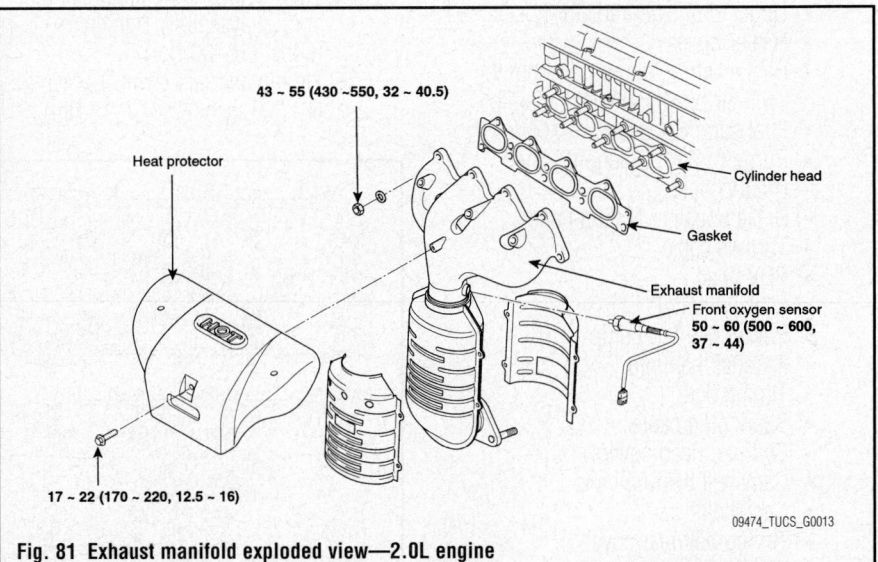

Fig. 81 Exhaust manifold exploded view—2.0L engine

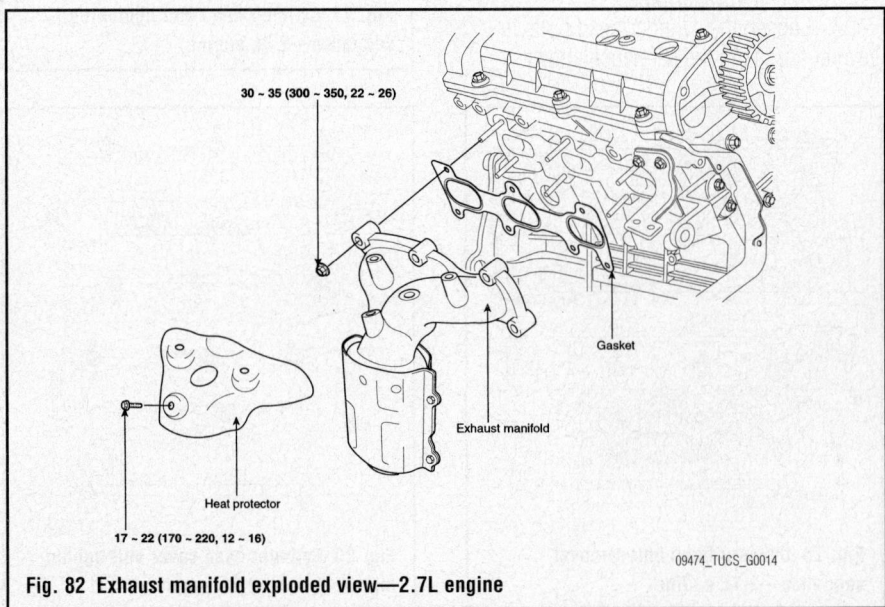

Fig. 82 Exhaust manifold exploded view—2.7L engine

FLYWHEEL

REMOVAL & INSTALLATION

1. Before servicing the vehicle, refer to the Precautions Section.
2. Drain the transaxle.
3. Remove or disconnect the following:
- Negative battery cable
- Air intake assembly
- Battery and battery tray
- Back-up lamp connector
- Vehicle speed sensor
- Clutch release cylinder and lever
- Shift cable from transaxle assembly
- Steering column from the universal joint in the gear box
- Clutch housing upper mounting bolts
- Front, rear and left transaxle mounting brackets

4. Using a suitable engine support fixture, support the engine assembly.
5. Remove or disconnect the following:
- Power steering pressure hose from the pump
- Front wheel
- Strut assembly
- Tie rod and sway bar link from the knuckle
- Wheel speed sensor
- Brake caliper
- Engine splash guard
- Front muffler
- Power steering hose on the front cross member

6. Using a suitable jack, support the sub-frame cross member.
7. Remove the cross member mounting bolts, and lower the cross member assembly with the steering gear and stabilizer bar attached.
8. Using a suitable jack, support the transaxle assembly.
9. Remove the front and rear roll stoppers.
10. Remove the engine and transaxle mounting bolts.
11. Slowly lower the transaxle from the vehicle.
12. If equipped with a clutch:

    a. Insert the special tool (09411-11000) in the clutch disc to prevent the disc from falling.

    b. Loosen the bolts which attach the clutch cover to the flywheel in a star pattern. Loosen the bolts in succession, 1–2 turns at a time, to avoid bending the cover flange.

➡**Do not clean the clutch disc or the release bearing with cleaning solvent.**

c. Remove clutch cover assembly and then the clutch disc assembly.
13. Remove the flywheel.

To install:

14. Installation is the reverse order of removal.
15. Tighten the flywheel bolts in a star formation to: 53–56 ft. lbs. (72–76 Nm).

INTAKE MANIFOLD

REMOVAL & INSTALLATION

2.0L Engine

See Figure 83.

1. Before servicing the vehicle, refer to the Precautions Section.
2. Relieve the fuel system pressure. Refer to Relieving Fuel System Pressure.
3. Drain the cooling system.
4. Remove or disconnect the following:
- Negative battery cable
- Engine cover
- Air intake assembly
- TPS and ISA connectors
- PCV and breather hoses
- Throttle cable
- Fuel rail
- Heater hoses
- PCSV hose
- Brake booster vacuum hose
- Intake manifold braces
- Intake manifold and gasket

To install:

5. Installation is the reverse of the removal procedure.
6. Tighten the intake manifold nuts to 12–17 ft. lbs. (16–23 Nm).

2.7L Engine

See Figures 84 and 85.

1. Before servicing the vehicle, refer to the Precautions Section.
2. Relieve the fuel system pressure. Refer to Relieving Fuel System Pressure.
3. Drain the cooling system.
4. Remove or disconnect the following:
- Negative battery cable
- Throttle cable
- All electrical connectors to the surge tank assembly
- PCSV hose
- Brake booster vacuum hose
- PCV hose
- Surge tank brace
- Surge tank assembly
- Fuel rail
- Intake manifold and gasket

To install:

5. Install the intake manifold and gasket.
6. Tighten the bolts in sequence to 14–15 ft. lbs. (19–21 Nm).
7. Install the surge tank assembly and tighten to 11–15 ft. lbs. (15–20 Nm).
8. The remainder of the installation is the reverse of removal.

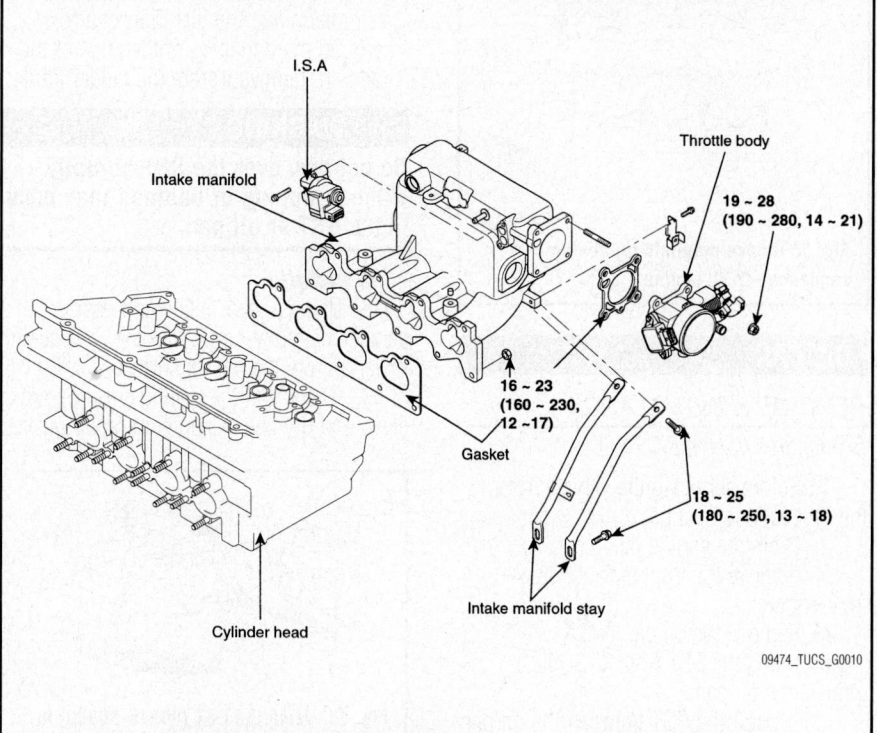

Fig. 83 Exploded view of intake manifold—2.0L engine

09474_TUCS_G0010

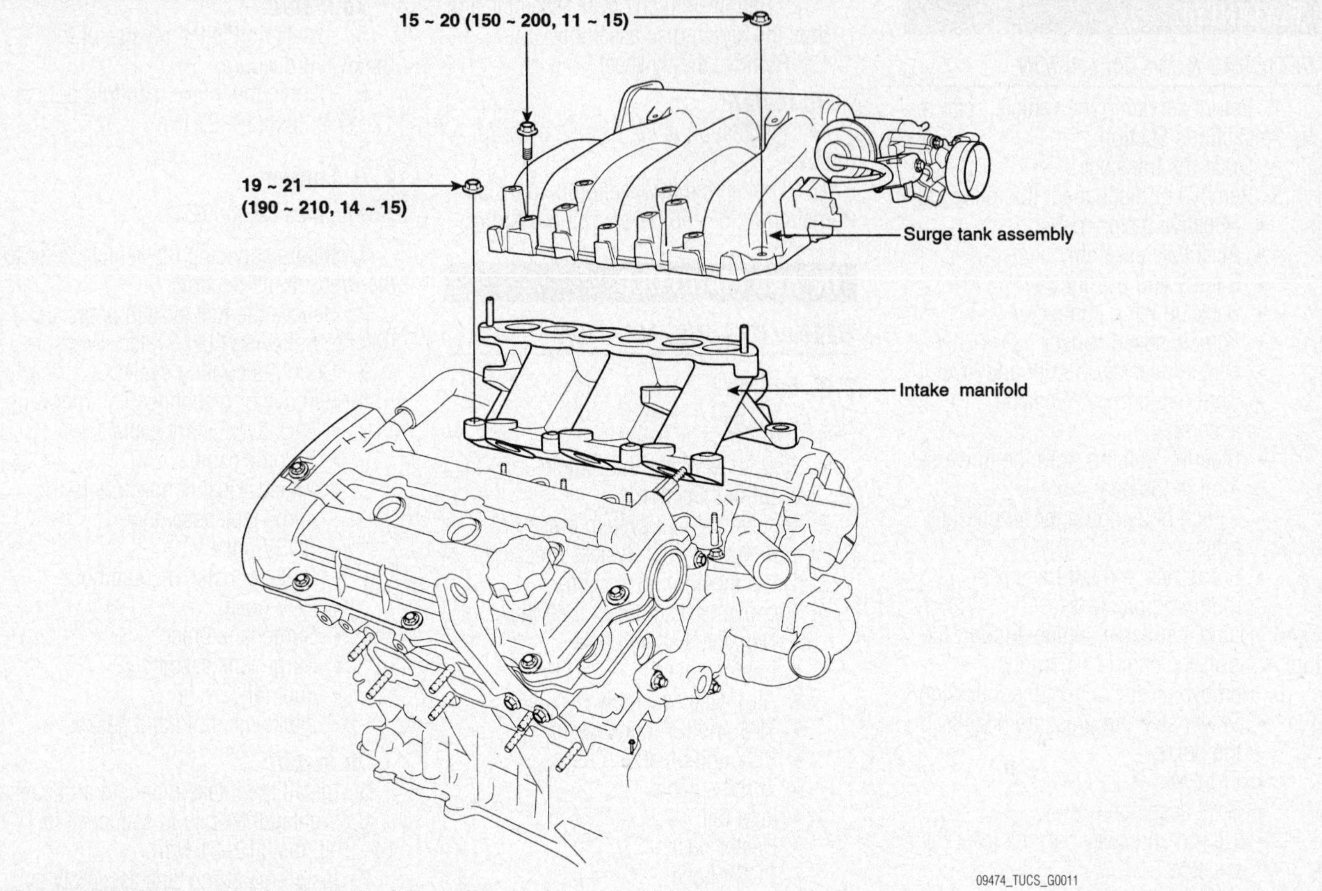

15 ~ 20 (150 ~ 200, 11 ~ 15)

19 ~ 21
(190 ~ 210, 14 ~ 15)

Surge tank assembly

Intake manifold

09474_TUCS_G0011

Fig. 84 Exploded view of intake manifold—2.7L engine

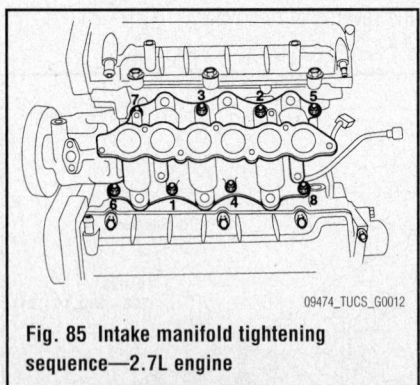

09474_TUCS_G0012

Fig. 85 Intake manifold tightening sequence—2.7L engine

OIL PAN

REMOVAL & INSTALLATION

See Figures 86 and 87.

1. Before servicing the vehicle, refer to the Precautions Section.
2. Drain the engine oil.
3. Remove the front exhaust pipe—2.7L engine.
4. Remove the oil pan bolts.
5. Using the SST (09215-3C000), remove the oil pan:
   a. Insert the SST between the oil pan and the ladder frame by tapping it with a

plastic hammer in the direction of arrow.
   b. After tapping the SST with a plastic hammer along the direction of arrow around more than ⅔ of the edge of the oil pan, remove it from the ladder frame.

❉❉ WARNING

Do not turn over the SST abruptly without tapping or damage may occur to the SST or oil pan.

To install:

6. Using a razor blade and gasket scraper, carefully remove all the old packing material from the gasket surfaces.
7. Check that the mating surfaces are clean and dry before applying liquid gasket:

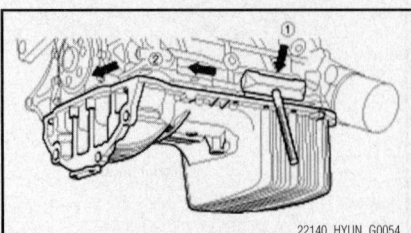

22140_HYUN_G0054

Fig. 86 Using the SST (09215-3C000) to remove the oil pan—2.0L engine

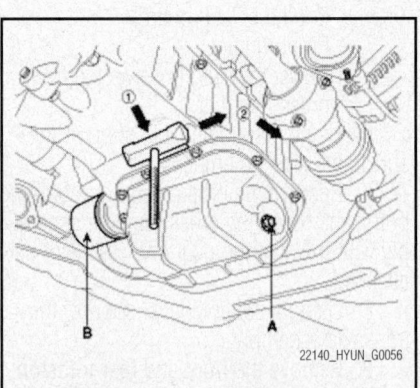

22140_HYUN_G0056

Fig. 87 Using the SST (09215-3C000) to remove the oil pan—2.7L engine

   a. Apply liquid gasket as an even bead, centered between the edges of the mating surface. Use liquid gasket: TB1217H or equivalent. Apply a bead ⅛ inch (3mm) wide to the oil pan.
   b. To prevent leakage of oil, apply liquid gasket to the inner threads of the bolt holes.

➡**Do not install the parts if 5 minutes or more have elapsed since applying the liquid gasket. Instead, reapply liquid gasket after removing the residue.**

8. Install the oil pan with the bolts. Uniformly tighten the bolts in several passes. Tightening torque: 84–108 inch lbs. (10–12 Nm).

9. Install the front exhaust pipe—2.7L engine.

➡**After assembly, wait at least 30 minutes before filling the engine with oil.**

10. Fill with engine oil.
11. Run engine and check for leaks.

OIL PUMP

REMOVAL & INSTALLATION

2.0L Engine

See Figures 88 through 90.

1. Before servicing the vehicle, refer to the Precautions Section.

2. Remove the oil pan. Refer to Oil Pan, removal & installation.

3. Remove the drive belts.

4. Turn the crankshaft and align the white groove on the crankshaft pulley with the pointer on the lower cover.

5. Remove the timing belt.

6. Remove the front case and oil pump.
   a. Remove the screws from the pump housing, then separate the housing and cover.
   b. Remove the inner and outer rotors.

To install:

7. Place the inner and outer rotors into the front case with the marks facing the oil pump cover side.

8. Install the oil pump cover to the front case with the 7 screws. Tightening torque: 48–84 inch lbs. (6–9 Nm).

9. Check that the oil pump turns freely.

10. Install the oil pump on the cylinder block.
    a. Place a new front case gasket on the cylinder block.
    b. Apply engine oil to the lip of the oil

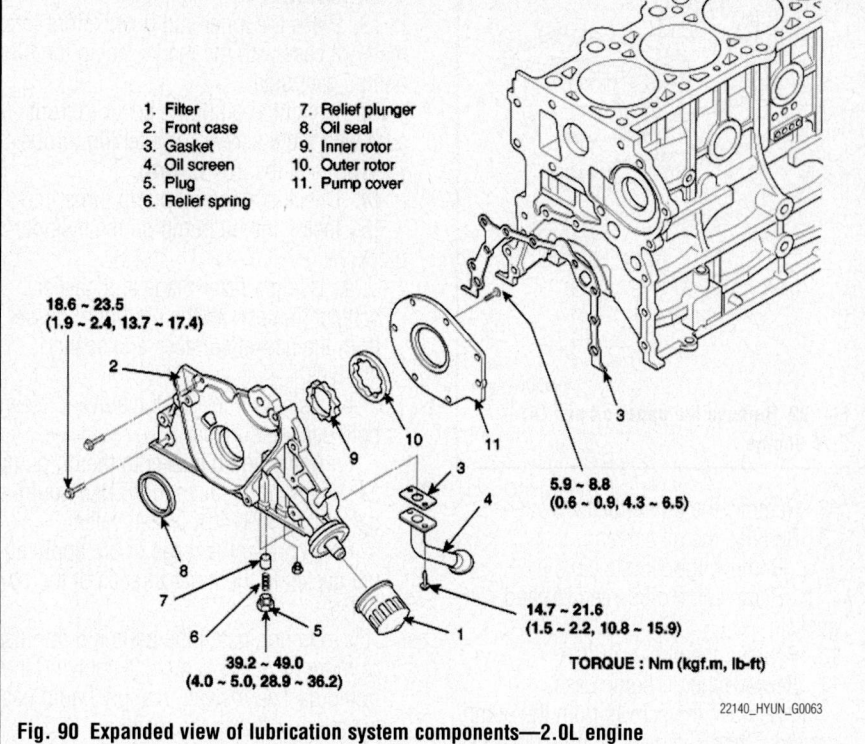

1. Filter
2. Front case
3. Gasket
4. Oil screen
5. Plug
6. Relief spring
7. Relief plunger
8. Oil seal
9. Inner rotor
10. Outer rotor
11. Pump cover

18.6 ~ 23.5
(1.9 ~ 2.4, 13.7 ~ 17.4)

5.9 ~ 8.8
(0.6 ~ 0.9, 4.3 ~ 6.5)

14.7 ~ 21.6
(1.5 ~ 2.2, 10.8 ~ 15.9)

39.2 ~ 49.0
(4.0 ~ 5.0, 28.9 ~ 36.2)

TORQUE : Nm (kgf.m, lb-ft)

22140_HYUN_G0063

Fig. 90 Expanded view of lubrication system components—2.0L engine

pump seal. Then, install the oil pump onto the crankshaft.

c. When the pump is in place, clean any excess grease off the crankshaft and check that the oil seal lip is not distorted.

d. Install the oil pump bolts in the correct position as illustrated. Tightening torque: 15–20 ft. lbs. (20–27 Nm).

11. Apply a light coat of oil to the seal lip.

12. Using the SST (09214-33000), install the oil seal.

13. Install the oil screen.

14. Install the oil pan. Refer to Oil Pan, removal and installation.

15. Ensure that the crankshaft aligns with the white groove on the crankshaft pulley and the pointer on the lower cover.

16. Install the timing belt.

17. Install the accessory drive belts.
18. Fill with engine oil.
19. Check for leaks.

2.7L Engine

See Figures 91 through 96.

1. Before servicing the vehicle, refer to the Precautions Section.

2. Remove the lower oil pan, refer to Oil Pan, removal & installation.

3. Remove the RH front wheel.

4. Remove the RH side cover.

5. Remove the front exhaust pipe.

6. Remove the alternator from the engine.

7. Turn the crankshaft and align the white groove on the crankshaft pulley with the pointer on the lower cover.

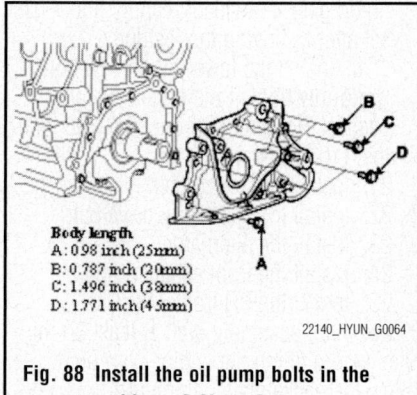

Body length
A : 0.98 inch (25mm)
B : 0.787 inch (20mm)
C : 1.496 inch (38mm)
D : 1.771 inch (45mm)

22140_HYUN_G0064

Fig. 88 Install the oil pump bolts in the correct position—2.0L engine

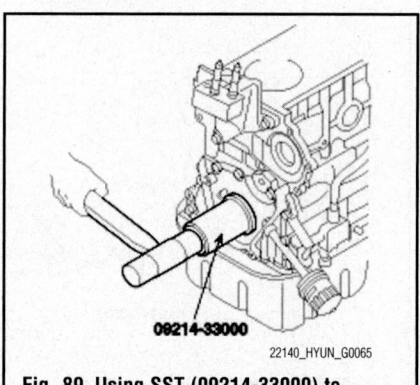

09214-33000

22140_HYUN_G0065

Fig. 89 Using SST (09214-33000) to install the oil seal—2.0L engine

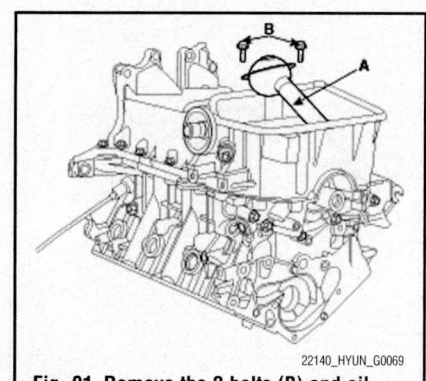

22140_HYUN_G0069

Fig. 91 Remove the 2 bolts (B) and oil screen (A)—2.7L engine

Fig. 92 Remove the upper oil pan (A)—2.7L engine

8. Remove the timing belt.
9. Remove the oil screen:
   a. Remove the 2 bolts (B).
   b. Remove the oil screen (A) and gasket.
10. Remove the upper oil pan.
11. Remove the oil pump case.
    a. Remove the screws from the pump housing, then separate the housing and cover.
    b. Remove the inner and outer rotors.

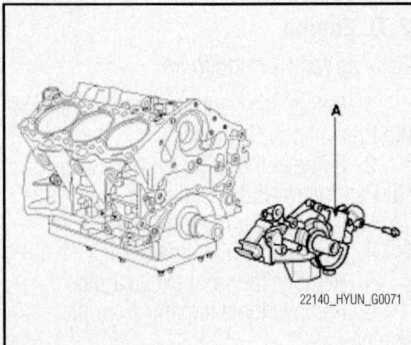

Fig. 93 Remove the upper oil pump case (A)—2.7L engine

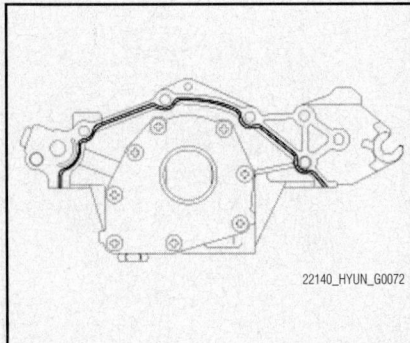

Fig. 94 Apply liquid gasket to the oil pump as shown—2.7L engine

To install:

12. Place the inner and outer rotors into the front case with the marks facing the oil pump cover side.
13. Install the oil pump cover to front case with the 8 screws. Tightening torque: 72–108 inch lbs. (8–12 Nm).
14. Check that the oil pump turns freely.
15. Install the oil pump on the cylinder block.
    a. Using a razor blade and gasket scraper, remove all the old liquid gasket from the gasket surfaces and sealing grooves.
    b. Using a non-residue solvent, clean both sealing surfaces.
    c. Apply liquid gasket to the oil pump as shown in the illustration. Use liquid gasket MS 721-40A or equivalent.
    d. To prevent leakage of oil, apply liquid gasket to the inner threads of the bolt holes.
    e. Do not install the parts if 5 minutes or more have elapsed since applying the liquid gasket. Instead, reapply liquid gasket after removing the residue.
    f. Place a new O-ring on the cylinder block.
    g. Engage the spline teeth of the oil pump drive gear with large teeth of the crankshaft, and slide the oil pump on the crankshaft.
    h. Install the oil pump with 5 bolts. Uniformly tighten the bolts in several passes. Tightening torque: 108–132 inch lbs. (12–15 Nm).
16. Apply a light coat of oil to the seal lip.
17. Using the special tool (09214-33000), install the oil seal.
18. Install the upper oil pan.
    a. Using a razor blade and gasket scraper, remove all the old packing material from the gasket surfaces.
    b. Check that the mating surfaces are clean and dry before applying liquid gasket.

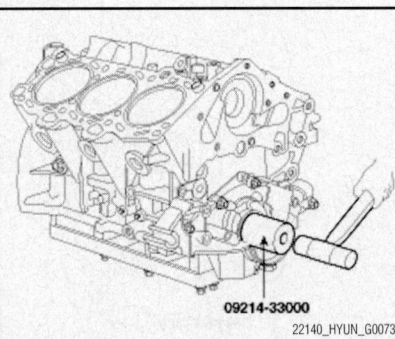

Fig. 95 Using the special tool (09214-33000) to install the oil seal—2.7L engine

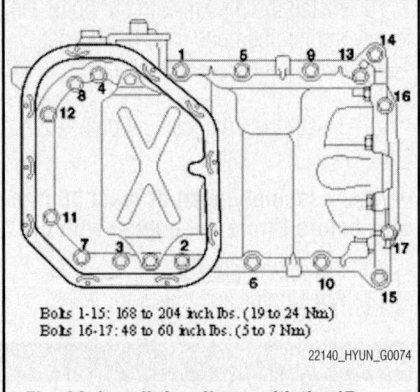

Bolts 1-15: 168 to 204 inch lbs. (19 to 24 Nm)
Bolts 16-17: 48 to 60 inch lbs. (5 to 7 Nm)

Fig. 96 Install the oil pan with the 17 bolts. Uniformly tighten the bolts in several passes—2.7L engine

   c. Install the oil pan with the 17 bolts. Uniformly tighten the bolts in several passes using the pattern illustrated. Tightening torque:
   - Bolts 1–15: 168–204 inch lbs. (19–24 Nm)
   - Bolts 16–17: 48–60 inch lbs. (5–7 Nm)

➡**To prevent leakage of oil, apply liquid gasket to the inner threads of the bolt holes.**

➡**Do not install the parts if 5 minutes or more have elapsed since applying the liquid gasket. Instead, reapply liquid gasket after removing the residue.**

19. Install the oil screen with a new gasket and the 2 bolts. Tightening torque: 11–16 ft. lbs. (15–22 Nm).
20. Install the lower oil pan.
    a. Apply liquid gasket as an even bead, centered between the edges of the mating surface. Use liquid gasket MS 721-40A or equivalent.
    b. To prevent leakage of oil, apply liquid gasket to the inner threads of the bolt holes.
    c. Do not install the parts if 5 minutes or more have elapsed since applying the liquid gasket. Instead, reapply liquid gasket after removing the residue.
    d. Install the lower oil pan 10 bolts. Uniformly tighten the bolts several passes. Tightening torque: 84–108 inch lbs. (10–12 Nm).
21. Install the timing belt.
22. Install the accessory drive belt.
23. Install the alternator.
24. Install the front exhaust pipe.
25. Install the RH front wheel.
26. After assembly, wait at least 30 minutes before filling the engine with oil to allow the gasket material to cure.
27. Fill the engine with oil.

28. Start the engine and check for leaks.
29. Recheck the engine oil level.

INSPECTION

See Figures 97 through 99.

1. Before servicing the vehicle, refer to the Precautions Section.
2. Remove the relief plunger: Remove the plug (A), spring (B), and relief plunger (C).
3. Inspect the relief plunger.
   a. Coat the plunger with engine oil
   b. Check that it falls smoothly into the plunger hole by its own weight.
   c. If it does not, replace the relief plunger.
   d. If necessary, replace the front case.
4. Inspect the relief valve spring.
5. Inspect for a distorted or broken relief valve spring.
6. Inspect the rotor side clearance.
   a. Using a feeler gauge and precision straight edge, measure the clearance between the rotors and precision straight edge.
   b. If the side clearance is greater than maximum, replace the rotors as a set. If necessary, replace the front case.
7. Inspect the rotor tip clearance.
   a. Using a feeler gauge, measure the

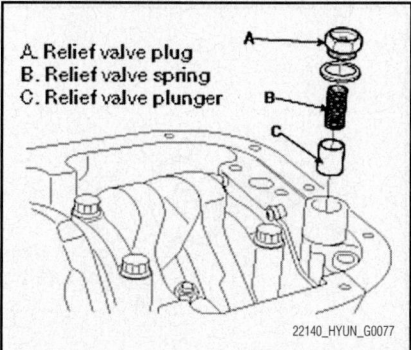

A. Relief valve plug
B. Relief valve spring
C. Relief valve plunger

22140_HYUN_G0077

Fig. 97 Remove the relief plunger for inspection

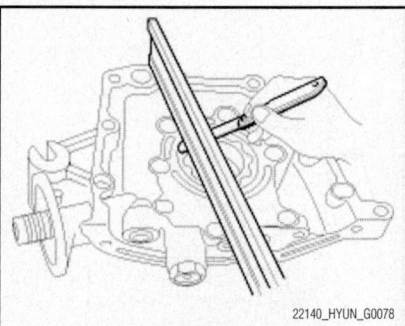

22140_HYUN_G0078

Fig. 98 Using a feeler gauge and precision straight edge to measure the side clearance of the oil pump

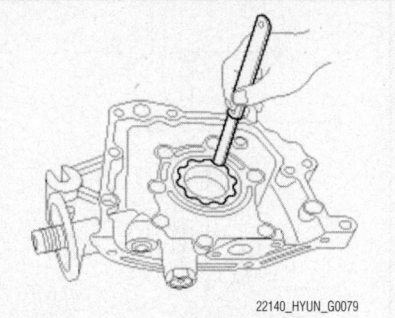

22140_HYUN_G0079

Fig. 99 Using a feeler gauge, measure the tip clearance between the inner and outer rotor tips of the oil pump

tip clearance between the inner and outer rotor tips.
   b. If the tip clearance is greater than specified, replace the rotors as a set.
8. Inspect the rotor body clearance.
   a. Using a feeler gauge, measure the clearance between the outer rotor and body.
   b. If the body clearance is greater than specified, replace the rotors as a set. If necessary, replace the front case.
9. Install the relief plunger and spring into the front case hole.
10. Install the plug. Tightening torque: 29–36 ft. lbs. (39–49 Nm).

PISTON AND RING

POSITIONING

See Figures 100 and 101.

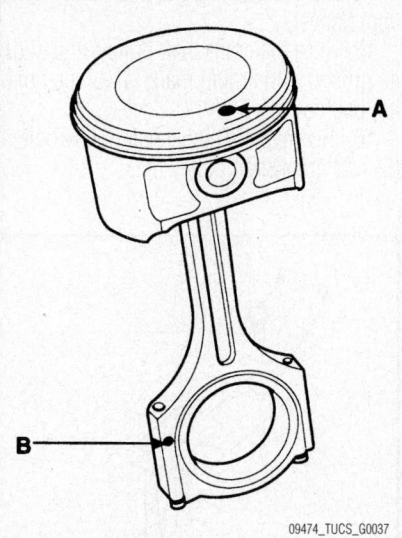

09474_TUCS_G0037

Fig. 100 The piston front mark and connecting rod front mark must face the timing belt side of the engine.

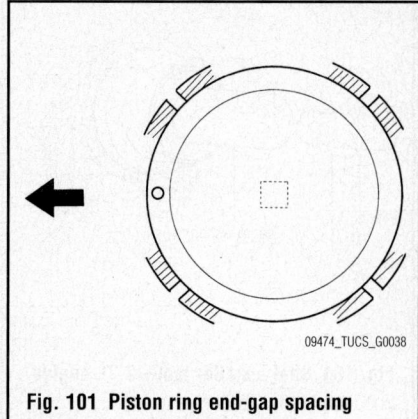

09474_TUCS_G0038

Fig. 101 Piston ring end-gap spacing

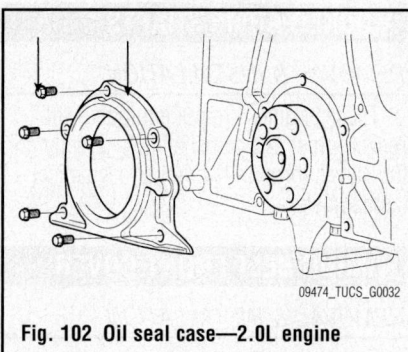

09474_TUCS_G0032

Fig. 102 Oil seal case—2.0L engine

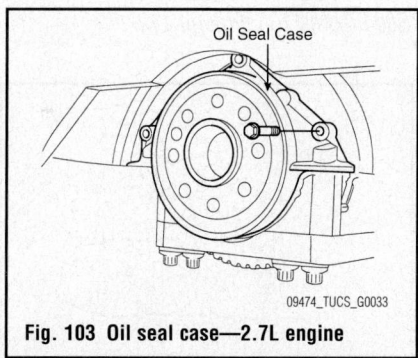

Oil Seal Case

09474_TUCS_G0033

Fig. 103 Oil seal case—2.7L engine

REAR MAIN SEAL

REMOVAL & INSTALLATION

See Figures 102 through 104.

1. Before servicing the vehicle, refer to the Precautions Section.
2. Remove or disconnect the following:
   - Transaxle
   - Flywheel
   - Oil seal case
   - Oil seal

To install:

3. Install or connect the following:
   - Oil seal, using Seal Installer tool
   - Oil seal case and torque the case bolts to 84–108 inch lbs. (10–12 Nm)
   - Flywheel
   - Transaxle

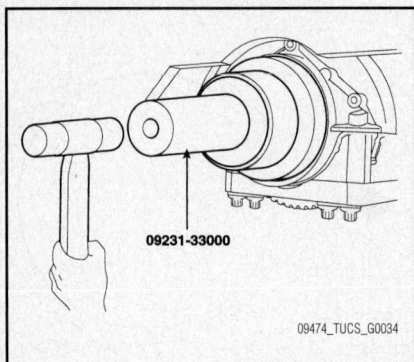

Fig. 104 Seal installer tool—2.7L engine, 2.0L engine is similar

ROCKER ARMS/SHAFTS

REMOVAL & INSTALLATION

These engines are not equipped with rocker arms. The camshaft acts directly on the valves through hydraulic/mechanical lash adjusters.

TIMING BELT FRONT COVER

REMOVAL & INSTALLATION

2.0L Engine

See Figures 105 through 112.

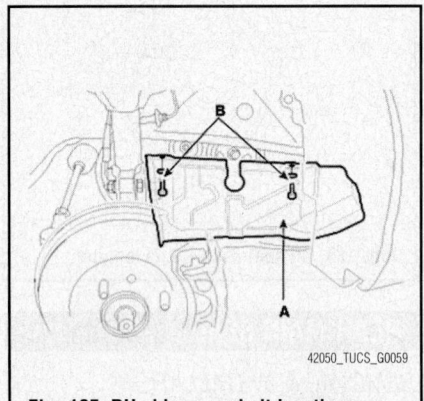

Fig. 105 RH side cover bolt locations

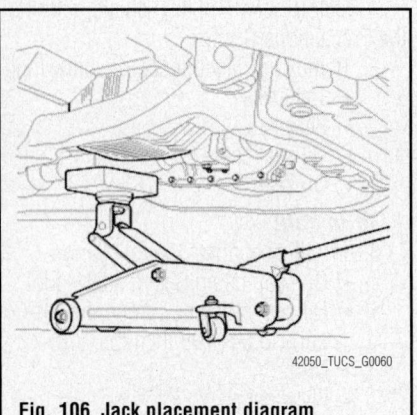

Fig. 106 Jack placement diagram

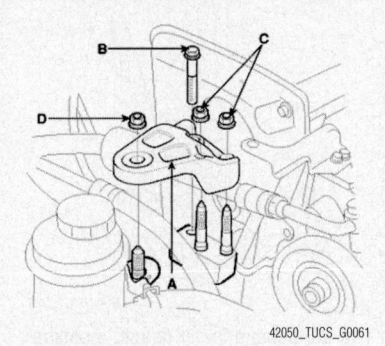

Fig. 107 Engine mounting bracket bolt locations

1. Before servicing the vehicle, refer to the Precautions Section.
2. Remove the engine cover.
3. Remove the Right Hand (RH) front wheel.
4. Remove 2 bolts (B) and RH side cover (A).
5. Remove the engine mount bracket.
6. Set a jack under the engine oil pan.

➡**Place a wooden block between the jack and engine oil pan.**

7. Remove the bolt (B), 3 nuts (C, D) and engine mount bracket (A).
8. Remove the bolt (B) and stay plate (A).
9. Temporarily loosen the water pump pulley bolts.
10. Remove alternator belt.
11. Remove air compressor belt.
12. Remove power steering belt.
13. Remove 4 bolts and water pump pulley.
14. Remove the 4 bolts and timing belt upper cover.
15. Turn the crankshaft pulley, and align its groove with timing mark "T" of the timing belt cover.
16. Remove the crankshaft pulley bolt (B) and crankshaft pulley (A).

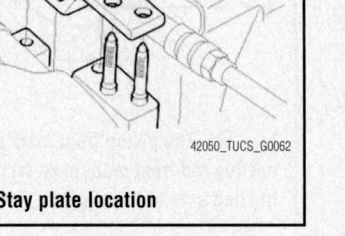

Fig. 108 Stay plate location

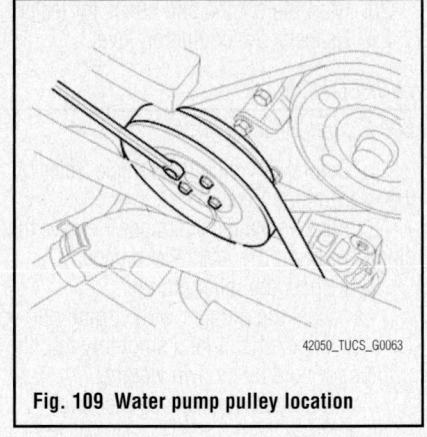

Fig. 109 Water pump pulley location

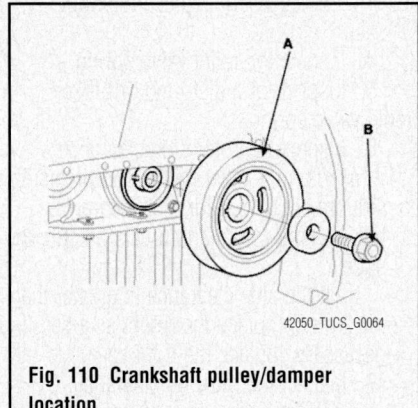

Fig. 110 Crankshaft pulley/damper location

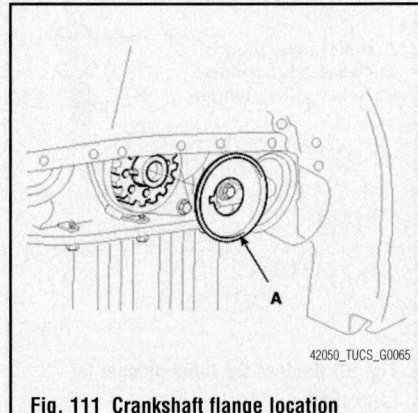

Fig. 111 Crankshaft flange location

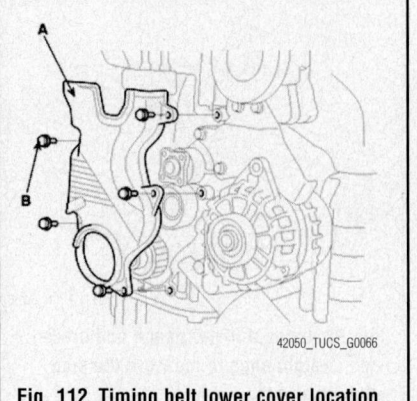

Fig. 112 Timing belt lower cover location

17. Remove the crankshaft flange (A).
18. Remove the 5 bolts (B) and timing belt lower cover (A).

To install:

19. Install timing belt lower cover and secure with the 5 bolts. Tighten and torque to 84 inch lbs. (10 Nm).
20. Install crankshaft flange.
21. Install crankshaft pulley and secure with the crankshaft pulley bolt. Tighten and torque to 125–133 ft. lbs. (170–180 Nm).
22. Install the timing belt upper cover and secure with the 4 bolts. Tighten and torque to 84 inch lbs. (10 Nm).
23. Install the water pump pulley and loosely secure with the 4 bolts.
24. Install the power steering, the air compressor and the alternator belts.
25. Tighten all 4 water pump pulley bolts to 84 inch lbs. (10 Nm).
26. Install the stay plate and bolts. Tighten and torque to 55 ft. lbs. (69 Nm).
27. Install engine mounting brackets, secure with bolts and nuts. Tighten and torque to 80 ft. lbs. (94 Nm).
28. Lower the jack and remove from under the vehicle.
29. Install the RH side cover and secure with 2 bolts.
30. Install the RH front wheel.
31. Install the engine cover.

2.7L Engine

See Figures 113 through 117.

1. Before servicing the vehicle, refer to the Precautions Section.

➡**Engine removal is not required for this procedure.**

2. Remove the engine cover.
3. Remove the Right Hand (RH) front wheel.
4. Remove the 2 bolts (B) and RH side cover (A).
5. Turn the crankshaft pulley, and align

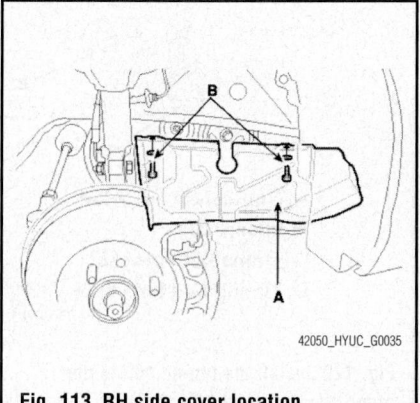

Fig. 113 RH side cover location

its groove with timing mark "T" of the timing belt cover.

➡**Always turn the crankshaft clockwise.**

6. Remove the auto-tensioner (B) and the drive belt (A).
7. Remove the engine mount bracket:
   a. Set the jack to the engine oil pan.

➡**Place wooden block between the jack and engine oil pan.**

   b. Remove the 2 bolts, 2 nuts and engine mount bracket.

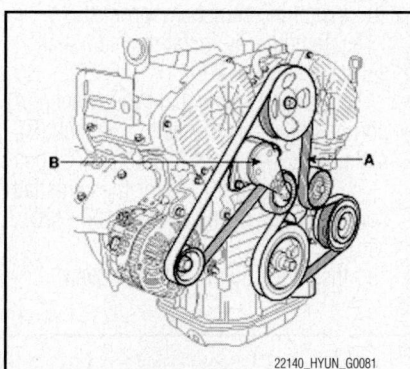

Fig. 114 Location of the auto-tensioner (B) and the drive belt (A)—2.7L engine

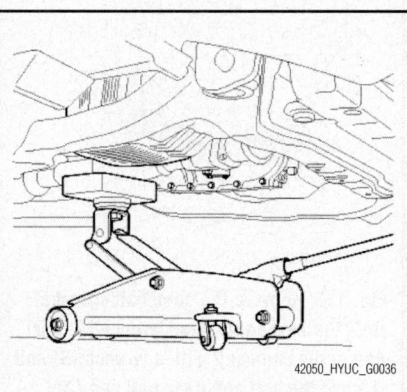

Fig. 115 Jack placement

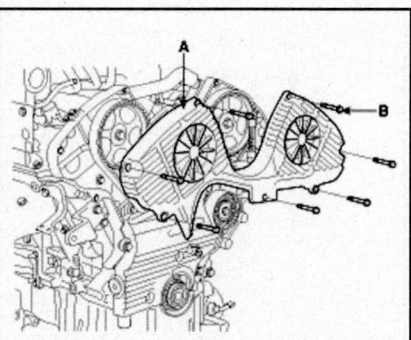

Fig. 116 Remove the 7 bolts (B) and timing belt upper cover (A)—2.7L engine

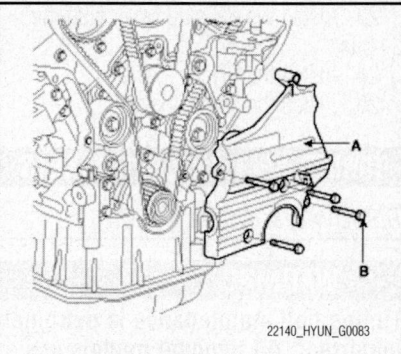

Fig. 117 Remove the 4 bolts (B) and timing belt lower cover (A)—2.7L engine

8. Remove the power steering pump.
9. Remove the 7 bolts (B) and timing belt upper cover (A).
10. Remove the crankshaft damper/pulley bolt and crankshaft damper/pulley.
11. Remove the drive belt idler pulley.
12. Remove the 4 bolts (B) and timing belt lower cover (A).

To install:

13. Install the timing belt lower cover (A) with 4 bolts (B). Tightening torque of timing belt cover bolt: 84–108 inch lbs. (10–12 Nm).
14. Install the drive belt idler pulley. Tightening torque of idler pulley bolt: 33–40 ft. lbs. (44–54 Nm).
15. Install the crankshaft damper/pulley. Tightening torque for crankshaft pulley bolt: 123–130 ft. lbs. (167–177 Nm).

➡**Make sure that the crankshaft sprocket pin fits the small hole in the pulley.**

16. Install the timing belt upper cover with 7 bolts.
17. Install the power steering pump.
18. Install the drive belt tensioner and drive belt.
19. Install the drive belt to the tensioner pulley and install the tensioner as shown. Tightening torque: 18–20 ft. lbs. (25–28 Nm).
20. Install the drive belt in the following order:
    a. Alternator
    b. Power steering
    c. Crankshaft pulley
    d. A/C pulley
21. Using a wrench, rotate the tensioner arm clockwise (about 14°) and return it slowly to the original position.
22. Install the engine mount bracket. Tightening torque of engine mount bracket with 2 nuts and 2 bolts: 43–58 ft. lbs. (59–79 Nm).

23. Install the RH side cover with the 2 bolts.
24. Install the RH front wheel.
25. Install the engine cover.

TIMING BELT AND SPROCKETS

REMOVAL & INSTALLATION

✳✳ WARNING

Timing belt maintenance is extremely important. All Hyundai models use interference-type non-freewheeling engines. Should the timing belt break in these engines, the valves in the cylinder head will come in contact with the pistons, causing major engine damage. The recommended replacement interval for a timing belt is 60,000 miles.

2.0L Engine

See Figures 118 through 123.

1. Before servicing the vehicle, refer to the Precautions Section.
2. Remove the timing belt front cover. Refer to Timing Belt Front Cover, removal & installation.
3. Remove the timing belt tensioner and timing belt.

➡ **If the timing belt is to be reused, make an arrow indicating the turning direction to make sure that the belt is reinstalled in the same direction as before.**

4. Remove the bolt (B) and timing belt idler (A).
5. Remove the crankshaft sprocket.
6. Remove the cylinder head cover.
   a. Remove the spark plug cables.
   b. Remove the accelerator cable from the cylinder head cover.
   c. Remove the Positive Crankcase Ventilation (PCV) hose and breather hose.

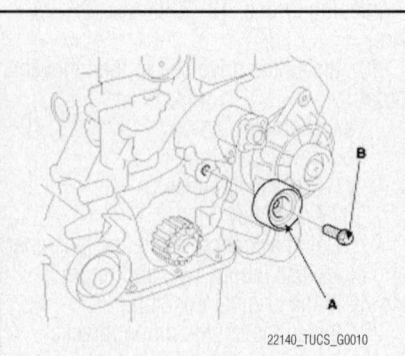

Fig. 118 Remove the bolt (B) and timing belt idler (A)—2.0L engine

d. Remove the 12 bolts and cylinder head cover.
7. Remove the camshaft sprocket.
   a. Hold the hexagonal head wrench (A) portion of the camshaft with a wrench (B).
   b. Remove the bolt and camshaft sprocket (C).

✳✳ WARNING

Be careful not to damage the cylinder head and valve lifter with the wrench.

To install:

8. Install the camshaft sprocket.
   a. Temporarily install the camshaft sprocket bolt.
   b. Hold the hexagonal head wrench (A) portion of the camshaft with a wrench (B), and tighten the camshaft sprocket (C) bolt.
   c. Tightening torque for the camshaft sprocket bolt: 74–89 ft. lbs. (100–120 Nm).
9. Install the cylinder head cover.

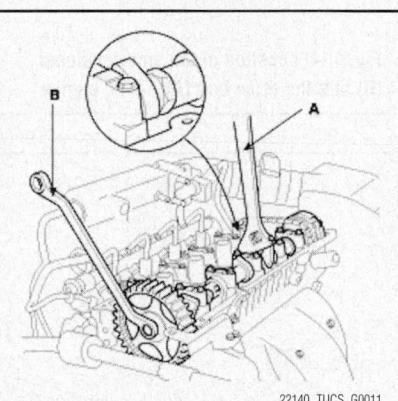

Fig. 119 Remove the camshaft sprocket. Hold the hexagonal head wrench (A) portion of the camshaft with a wrench (B) and remove the bolt and camshaft sprocket (C)—2.0L engine

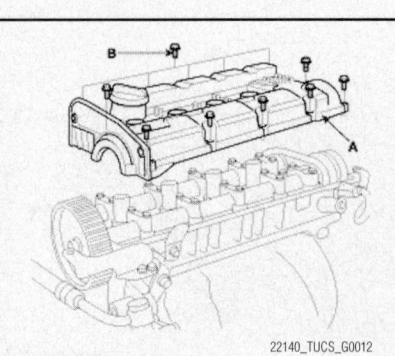

Fig. 120 Install cylinder head cover (A) and 12 bolts (B)—2.0L engine

a. Install cylinder head cover (A) and 12 bolts (B).
b. Install the PCV hose and breather hose.

Fig. 121 Align the timing marks of the camshaft sprocket (A) and crankshaft sprocket (B) with the No. 1 piston placed at TDC and its compression stroke—2.0L engine

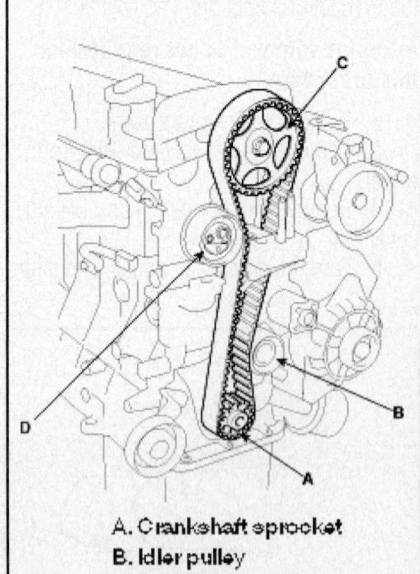

A. Crankshaft sprocket
B. Idler pulley
C. Camshaft sprocket
D. Timing belt tensioner

Fig. 122 Install the timing belt in the order illustrated—2.0L engine

c. Install the accelerator cable to the cylinder head cover.

d. Install the spark plug cables.

10. Install the crankshaft sprocket.

11. Align the timing marks of the camshaft sprocket (A) and crankshaft sprocket (B) with the No. 1 piston placed at Top Dead Center (TDC) and its compression stroke.

12. Install the idler pulley and tighten the bolt to: 32–40 ft. lbs. (43–55 Nm).

13. Install the timing belt tensioner loosely enough for the adjuster to rotate. Make sure that the stopper of the base is leaning against the lower sealing cap on the cylinder head.

14. Install the timing belt so there is no slack in the order illustrated.

➡**The tensioner can be installed after the timing belt.**

15. Check the alignment of the timing marks on each sprocket.

16. Remove the pin fixing the tensioner arm.

17. Using a hex wrench, turn the adjuster counterclockwise to make the indicator on the arm (A) align at the center of the base notch.

❊❊ **WARNING**

Do not rotate the adjuster clockwise. It will result in damage to the auto tensioner.

18. Tightening the tensioner bolt so the indicator does not move. Tightening torque: 17–21 ft. lbs. (23–28 Nm).

19. Turn the crankshaft 2 revolutions in the operating direction (clockwise) and check that the indicator is in the center of the base.

20. If the indicator is not located at the

center of the base, slacken the bolt and repeat the above procedure.

21. Install the timing belt front cover. Refer to Timing Belt Front Cover, removal & installation.

2.7L Engine

See Figures 124 through 131.

1. Before servicing the vehicle, refer to the Precautions Section.

2. Remove the timing belt front covers (upper and lower). Refer to Timing Belt Front Cover, removal & installation.

3. Remove the engine support bracket (A).

4. Check that timing marks of the camshaft timing pulleys and cylinder head covers are aligned. If not, turn the crankshaft 1 revolution (360°).

5. Remove the timing belt tensioner. Alternately loosen the 2 bolts and remove the tensioner (A).

6. Remove the timing belt.

➡**If the timing belt is to be reused, make an arrow indicating the turning**

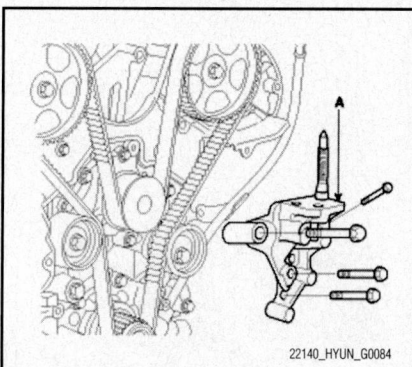

Fig. 124 Remove the engine support bracket (A)—2.7L engine

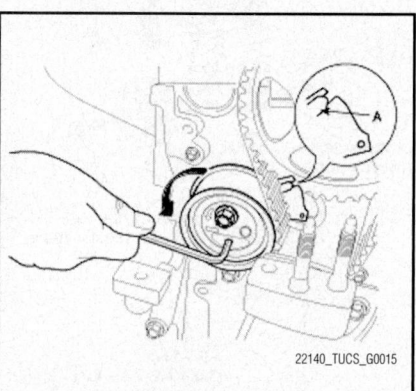

Fig. 123 Using a hex wrench, turn the adjuster counterclockwise to align the indicator on the arm (A) to the center of the base notch—2.0L engine

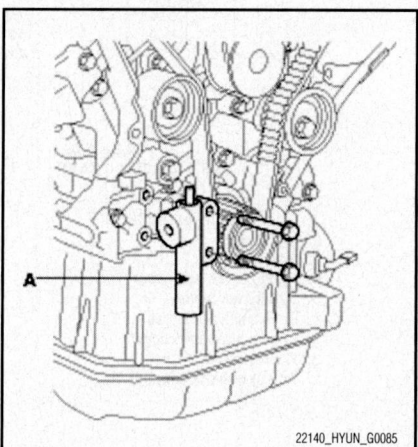

Fig. 125 Alternately loosen the 2 bolts and remove the tensioner (A)—2.7L engine

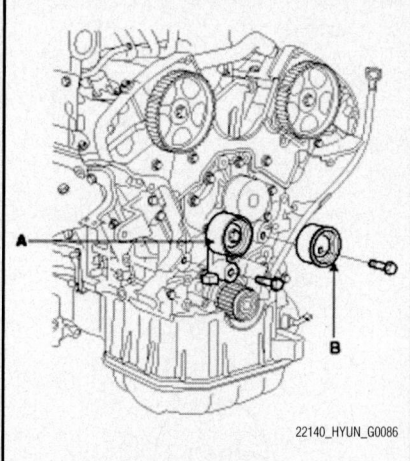

Fig. 126 Remove the tensioner arm assembly (A) and timing belt idler pulley (B)—2.7L engine

direction to make sure that the belt is reinstalled in the same direction as before.

7. Remove the tensioner arm assembly (A) and timing belt idler pulley (B).

8. Remove the crankshaft sprocket.

9. Remove the camshaft sprockets. Hold the hexagonal head wrench portion of the camshaft with a wrench and remove the bolt and camshaft sprocket.

❊❊ **WARNING**

Be careful not to damage the cylinder head and valve lifter with the wrench.

To install:

10. Install the crankshaft sprocket: Align the pulley set key with the key groove the crankshaft sprocket and slide on the crankshaft sprocket.

11. Install the camshaft sprockets and tighten the bolts to the specified torque.

a. Temporarily install the camshaft sprocket bolts.

b. Hold the hexagonal head wrench portion of the camshaft with a wrench, and tighten the camshaft sprocket bolts. Tightening torque: 65–80 ft. lbs. (88–108 Nm).

12. Install the idler pulley and the tensioner pulley. Tightening torque:

a. Idler pulley bolt: 36–43 ft. lbs. (49–59 Nm).

b. Tensioner arm fixed bolt: 25–40 ft. lbs. (34–54 Nm).

➡**Insert and install the idler pulley to the roll pin that is pressed in the water pump boss.**

13. Align the timing marks of the camshaft sprocket and crankshaft sprocket

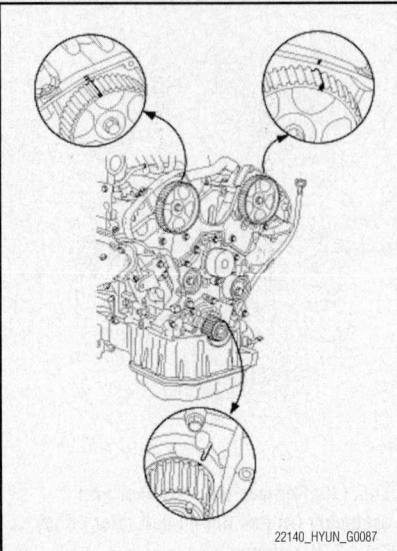

Fig. 127 Align the timing marks of the camshaft sprocket and crankshaft sprocket with the No. 1 piston placed at Top Dead Center (TDC) of its compression stroke—2.7L engine

with the No. 1 piston placed at Top Dead Center (TDC) of its compression stroke.

14. Set the timing belt tensioner.

a. Using a press, slowly press in the push rod.

b. Align the holes of the push rod and housing.

c. Pass a set pin through the holes to keep the setting position of the push rod.

d. Release the press.

15. Install the timing belt tensioner.

a. Temporarily install the tensioner with the 2 bolts.

b. Alternately tighten the 2 bolts. Tightening torque: 15–20 ft. lbs. (20–27 Nm).

16. Install the timing belt.

a. Remove any oil or water on the sprockets, and keep them clean.

b. Install the timing belt in this order:

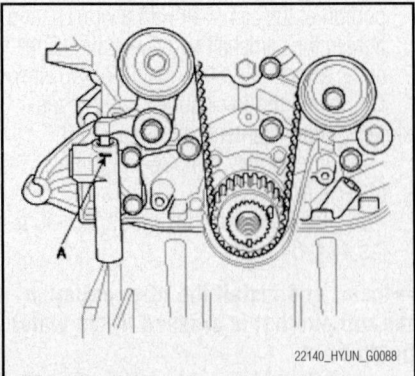

Fig. 128 Remove the set pin (A) from the tensioner—2.7L engine

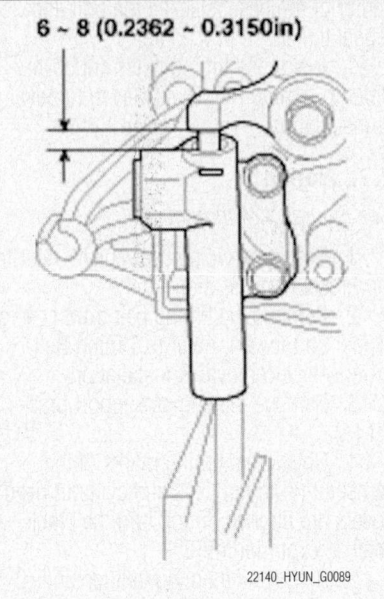

6 ~ 8 (0.2362 ~ 0.3150in)

Fig. 129 The projected length of the timing belt tensioner should be 0.27–0.31 inch (7–9mm)—2.7L engine

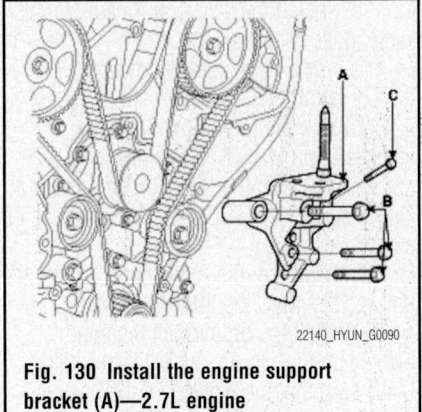

Fig. 130 Install the engine support bracket (A)—2.7L engine

- Crankshaft sprocket
- Idler pulley
- Camshaft sprocket LH side
- Water pump pulley
- Camshaft sprocket RH side
- Tensioner pulley

17. Remove the set pin (A) from the tensioner.

18. Check the timing belt tensioner.

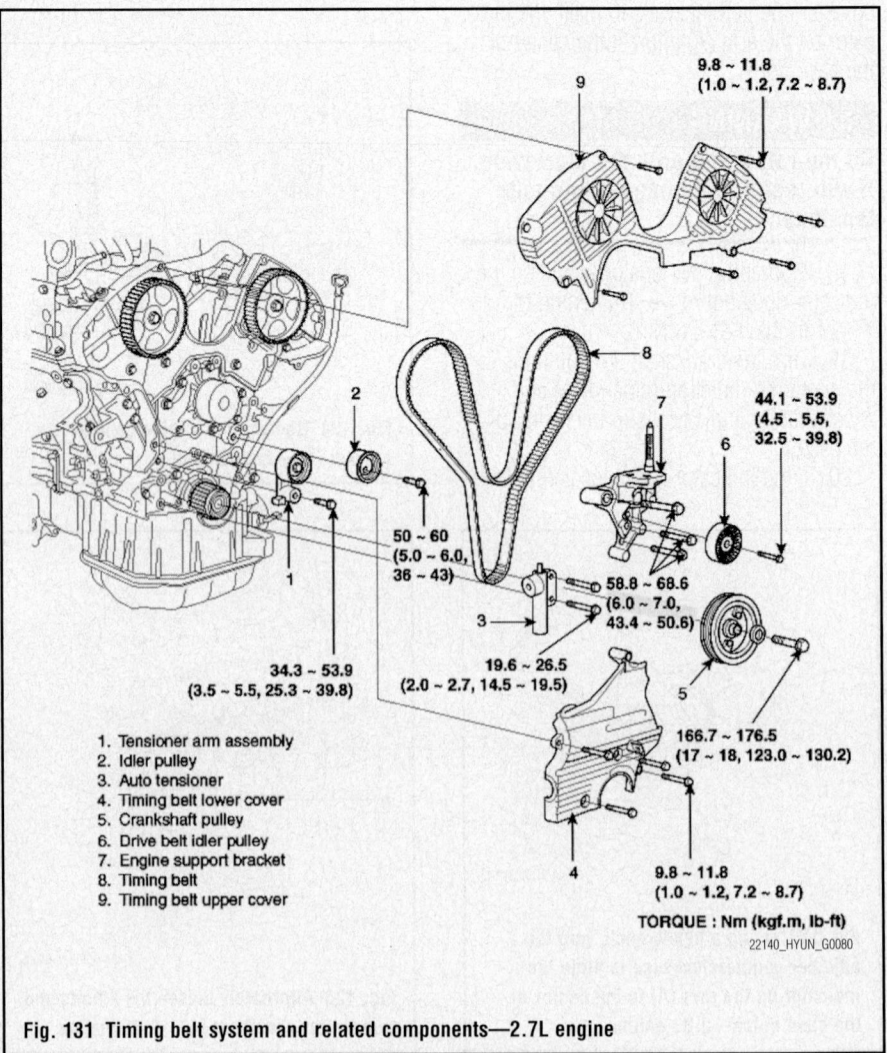

9.8 ~ 11.8
(1.0 ~ 1.2, 7.2 ~ 8.7)

44.1 ~ 53.9
(4.5 ~ 5.5, 32.5 ~ 39.8)

50 ~ 60
(5.0 ~ 6.0, 36 ~ 43)

58.8 ~ 68.6
(6.0 ~ 7.0, 43.4 ~ 50.6)

34.3 ~ 53.9
(3.5 ~ 5.5, 25.3 ~ 39.8)

19.6 ~ 26.5
(2.0 ~ 2.7, 14.5 ~ 19.5)

166.7 ~ 176.5
(17 ~ 18, 123.0 ~ 130.2)

9.8 ~ 11.8
(1.0 ~ 1.2, 7.2 ~ 8.7)

1. Tensioner arm assembly
2. Idler pulley
3. Auto tensioner
4. Timing belt lower cover
5. Crankshaft pulley
6. Drive belt idler pulley
7. Engine support bracket
8. Timing belt
9. Timing belt upper cover

TORQUE : Nm (kgf.m, lb-ft)

Fig. 131 Timing belt system and related components—2.7L engine

a. Rotate the crankshaft 2 turns clockwise and measure the projected length of the auto tensioner at TDC (No. 1 compression stroke) after 5 minutes.

b. As illustrated in the example, the projected length of the timing belt tensioner should be 0.27–0.31 inch (7–9mm).

19. Install the engine support bracket (A). Tightening torque of the following bolts (see illustration):

a. B: 43–51 ft. lbs. (59–67 Nm).

b. C: 11–16 ft. lbs. (15–22 Nm).

20. Install the timing belt front covers (upper and lower). Refer to Timing Belt Front Cover, removal & installation.

VALVE COVERS

REMOVAL & INSTALLATION

See Figures 132 through 134.

➡**The valve covers are also referred to as cylinder head covers.**

✲✲ WARNING

Use fender covers to avoid damaging painted surfaces.

✲✲ WARNING

To avoid damage, unplug the wiring connectors carefully while holding the connector portion.

➡**Mark all wiring and hoses to avoid misconnection.**

1. Remove the air duct.

2. Disconnect the negative terminal from the battery.

3. Remove the engine cover.

4. Disconnect the AFS connector .

5. Remove the Positive Crankcase Ventilation (PCV) hose and breather hose.

6. Disconnect the breather hose from air cleaner hose.

7. Remove the intake air hose and air cleaner assembly.

8. Remove the spark plug wires.

9. Remove the cylinder head cover.

10. Remove the cylinder head cover gasket and thoroughly clean the cylinder head surface (where the gasket rests) and the cylinder head cover whether you are replacing the gasket only or the entire cover.

To install:

➡**Before installing the head cover gasket, thoroughly clean the head cover gasket and the groove.**

➡**When installing, make sure the head cover gasket is seated securely in the corners of the recesses with no gap.**

11. Install the cylinder head cover gasket (A) in the groove of the cylinder head cover (B).

12. Apply liquid gasket to the head cover gasket at the corners of the recess.

➡**Use liquid gasket—Loctite® No. 5999.**

13. Check that the mating surfaces are clean and dry before applying liquid gasket.

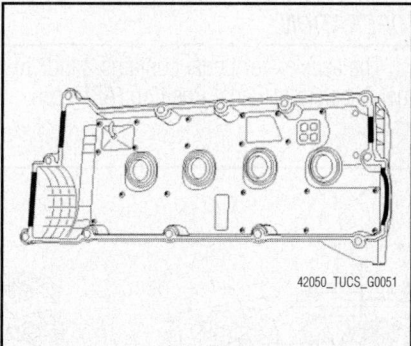

Fig. 133 Cylinder head cover gasket installation

14. Install the cylinder head cover (A) with the 12 bolts (B).

15. Uniformly tighten the bolts in several passes. Tightening torque: 72–84 inch lbs. (8–10 Nm).

16. Install the spark plug wires.

17. Connect the breather hose from air cleaner hose.

18. Install the PCV hose and breather hose.

19. Connect the AFS connector.

20. Install the intake air hose and air cleaner assembly.

21. Install the engine cover.

22. Connect the negative battery cable to the negative battery terminal.

23. Install the air duct.

VALVE LASH

ADJUSTMENT

2.0L Engine

See Figures 135 through 138.

1. Before servicing the vehicle, refer to the Precautions Section.

2. Remove or disconnect the following:
- Negative battery cable
- Engine cover
- Upper timing belt cover
- Cylinder head cover

3. Set the No. 1 cylinder to Top Dead Center (TDC) as follows:

a. Turn the crankshaft pulley and align its groove with the timing mark "T" of the lower timing belt cover.

b. Ensure the hole of the camshaft timing pulley is aligned with the timing mark of the bearing cap.

4. Inspect the clearance of the valves as shown.

5. Turn the crankshaft one revolution and align the timing mark "T" of the lower timing belt cover to set the No. 4 cylinder and TDC.

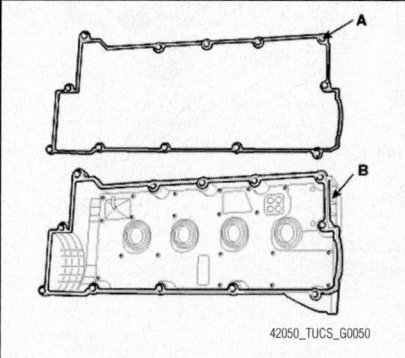

Fig. 132 Cylinder head (valve) cover and gasket diagram

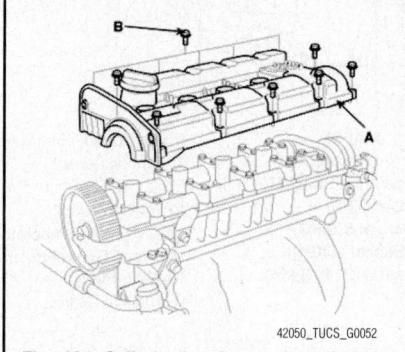

Fig. 134 Cylinder head cover installation diagram

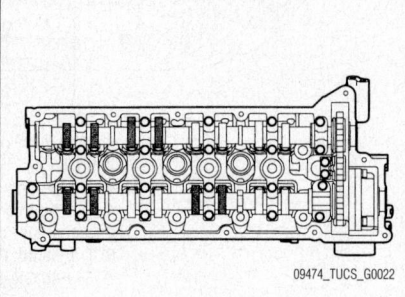

Fig. 135 Valves to inspect with no. 1 cylinder at TDC

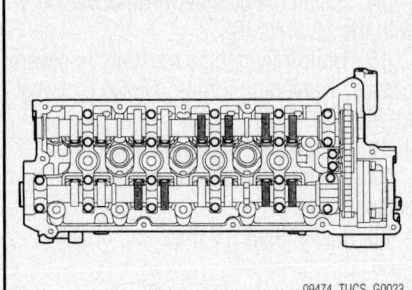

Fig. 136 Valves to inspect with no. 4 at TDC

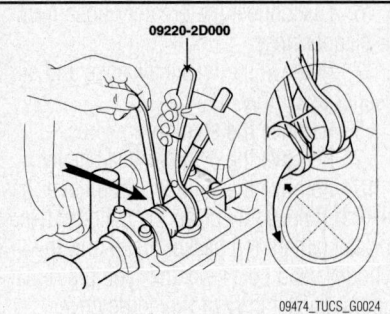

Fig. 137 Press down on the valve lifter with special tool and place a stopper between the camshaft and valve lifter.

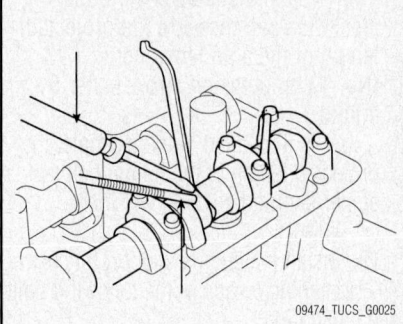

Fig. 138 Remove adjusting shim with a screwdriver and magnet

6. Inspect the clearance of the valves as shown.

7. If adjustment is necessary:

a. Turn the crankshaft so that the cam lobe on the adjusting valve is upward.

b. Using Special Tool 09220-2D000, press down on the valve lifter and place a stopper between the camshaft and valve lifter and remove the special tool.

c. Remove the adjusting shim with a small screwdriver and magnet.

d. Measure the thickness of the removed shim using a micrometer.

e. Calculate the thickness of a new shim so that the valve clearance comes within the specified valve.

f. Place a new adjusting shim on the valve lifter.

g. Using Special Tool 09220-2D000, press down on the valve lifter and remove the stopper.

h. Recheck the valve clearance.

2.7L Engine

The 2.7L engine utilizes hydraulic valve lash adjusters. Valve lash adjustments are not necessary or possible on this engine.

ENGINE PERFORMANCE & EMISSION CONTROL

See Figures 139 and 140.

ACCELERATOR PEDAL POSITION (APP) SENSOR

LOCATION

See Figure 141.

The Accelerator Pedal Position (APP) sensor is located inside the vehicle. It is part of the accelerator pedal assembly.

OPERATION

The accelerator pedal contains 2 individual Accelerator Pedal Position (APP) sen-

COMPONENT LOCATIONS

sors within the assembly. The APP sensors 1 and 2 are potentiometer type sensors each with 3 circuits:

- A 5-volt reference circuit
- A low reference circuit
- A signal circuit

The APP sensors are used to determine the pedal angle. The Electronic Control

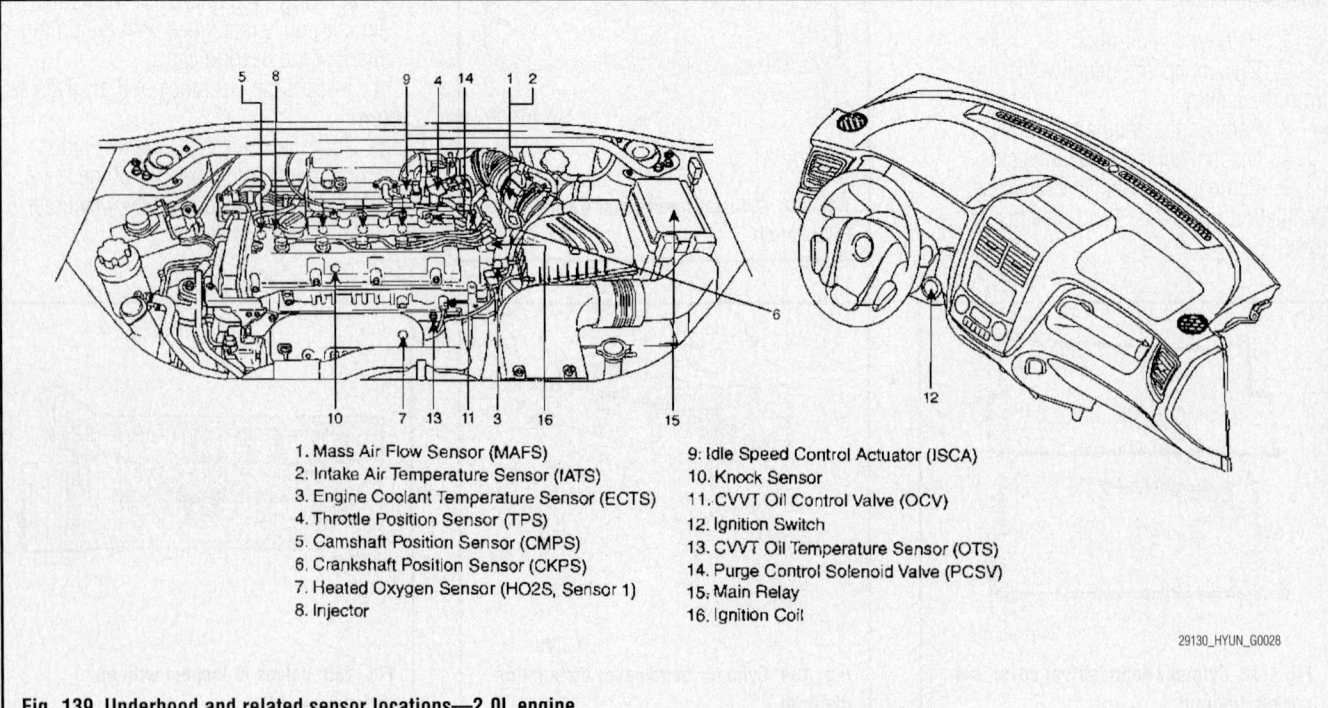

1. Mass Air Flow Sensor (MAFS)
2. Intake Air Temperature Sensor (IATS)
3. Engine Coolant Temperature Sensor (ECTS)
4. Throttle Position Sensor (TPS)
5. Camshaft Position Sensor (CMPS)
6. Crankshaft Position Sensor (CKPS)
7. Heated Oxygen Sensor (HO2S, Sensor 1)
8. Injector
9. Idle Speed Control Actuator (ISCA)
10. Knock Sensor
11. CVVT Oil Control Valve (OCV)
12. Ignition Switch
13. CVVT Oil Temperature Sensor (OTS)
14. Purge Control Solenoid Valve (PCSV)
15. Main Relay
16. Ignition Coil

Fig. 139 Underhood and related sensor locations—2.0L engine

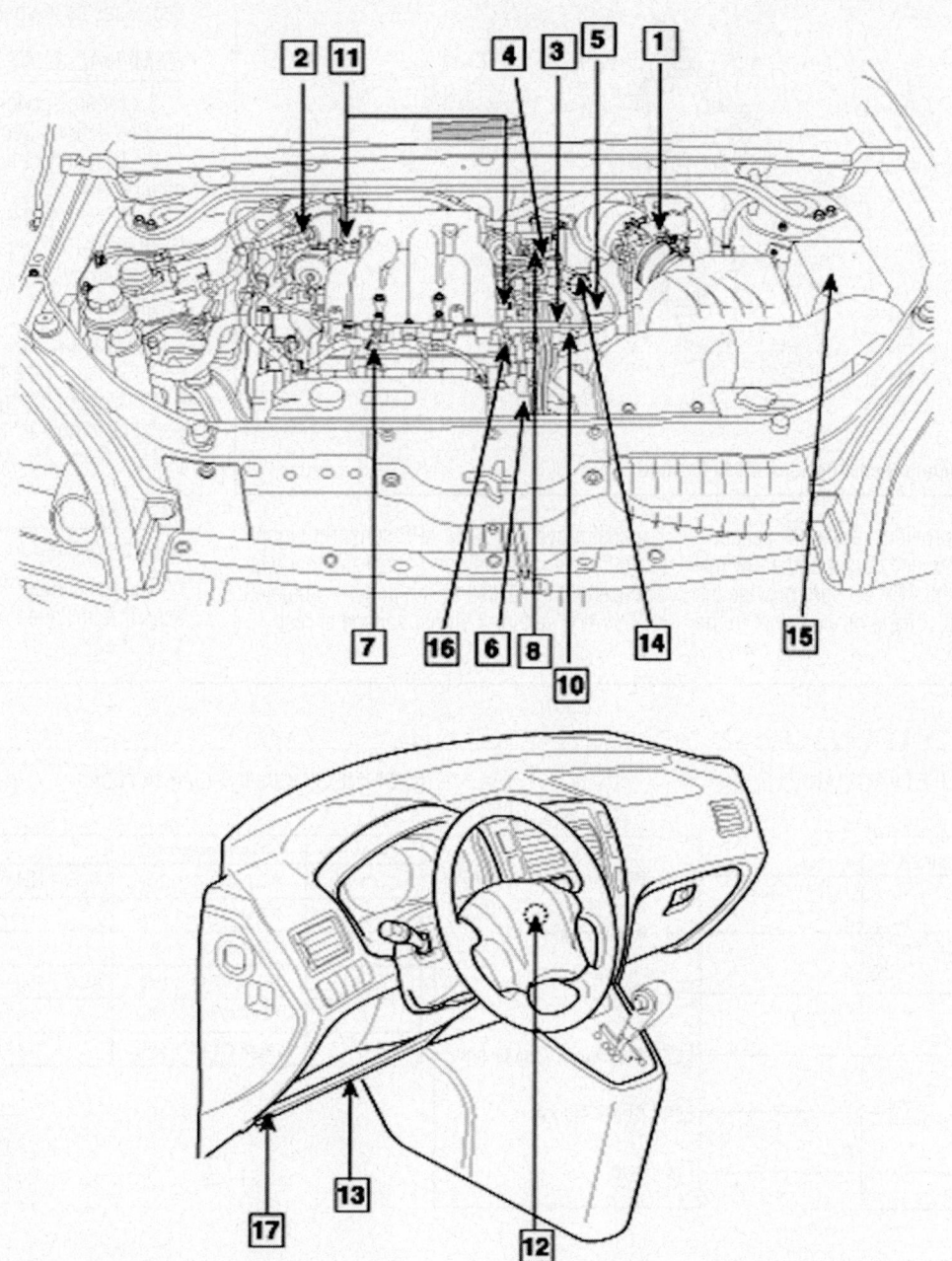

Fig. 140 Underhood and related sensor locations—2.7L engine

1. Mass Air Flow Sensor (MAFS) [With CVVT]
2. Intake Air Temperature Sensor (IATS)
3. Engine Coolant Temperature Sensor (ECTS)
4. Throttle Position Sensor (TPS)
5. Camshaft Position Sensor (CMPS)
6. Crankshaft Position Sensor (CKPS)
7. Injector
8. Idle Speed Control Actuator (ISCA)
9. Vehicle Speed Sensor (VSS)
10. Knock Sensor
11. VIS Control solenoid valve
12. Ignition Switch
13. ECM
14. Purge Control Solenoid Valve (PCSV)
15. Main Relay
16. Ignition Coil
17. DLC (Diagnostic Link Connector)
18. Heated Oxygen Sensor (Bank1, Sensor1)
19. Heated Oxygen seneor (Bank1, Sensor2)
20. Heated Oxygen seneor (Bank2, Sensor1)
21. Heated Oxygen seneor (Bank2, Sensor2)

22140_TUCS_G0073

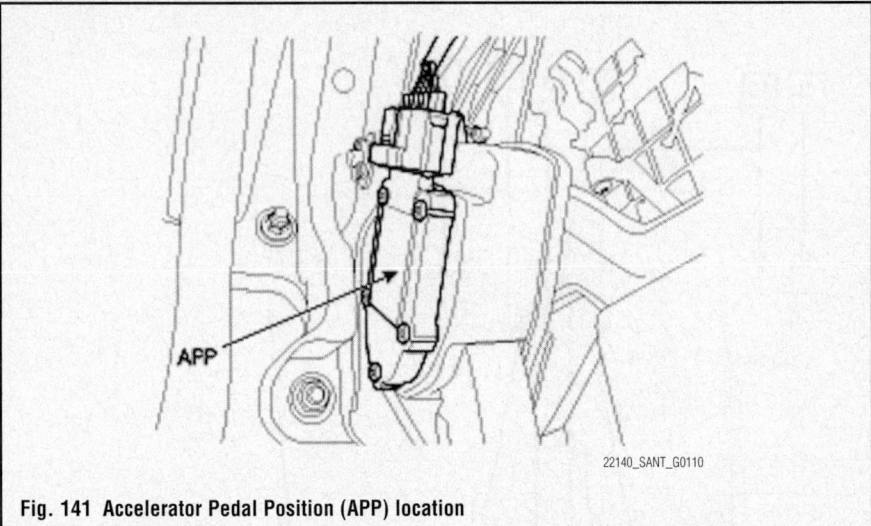

Fig. 141 Accelerator Pedal Position (APP) location

22140_SANT_G0110

Module (ECM) provides each APP sensor with a 5-volt reference circuit and a low reference circuit. The APP sensors provide the ECM with signal voltage proportional to the pedal movement. The APP sensor 1 signal voltage at rest position is near the low reference and increases as the pedal is actuated. The APP sensor 2 signal voltage at rest position is near the 5-volt reference and decreases as the pedal is actuated.

REMOVAL & INSTALLATION

1. Before servicing the vehicle, refer to the Precautions Section.
2. Disconnect the ground cable from the battery.
3. Disconnect the Accelerator Pedal Position (APP) sensor connector.
4. Remove the nut securing accelerator pedal assembly.

To install:

5. Install in the reverse order of removal.
6. Tighten the nut securing accelerator pedal assembly to 113–139 inch lbs. (13–16 Nm).

TESTING

See Figures 142 through 144.

The Accelerator Pedal Position (APP) sensor is installed on the accelerator pedal

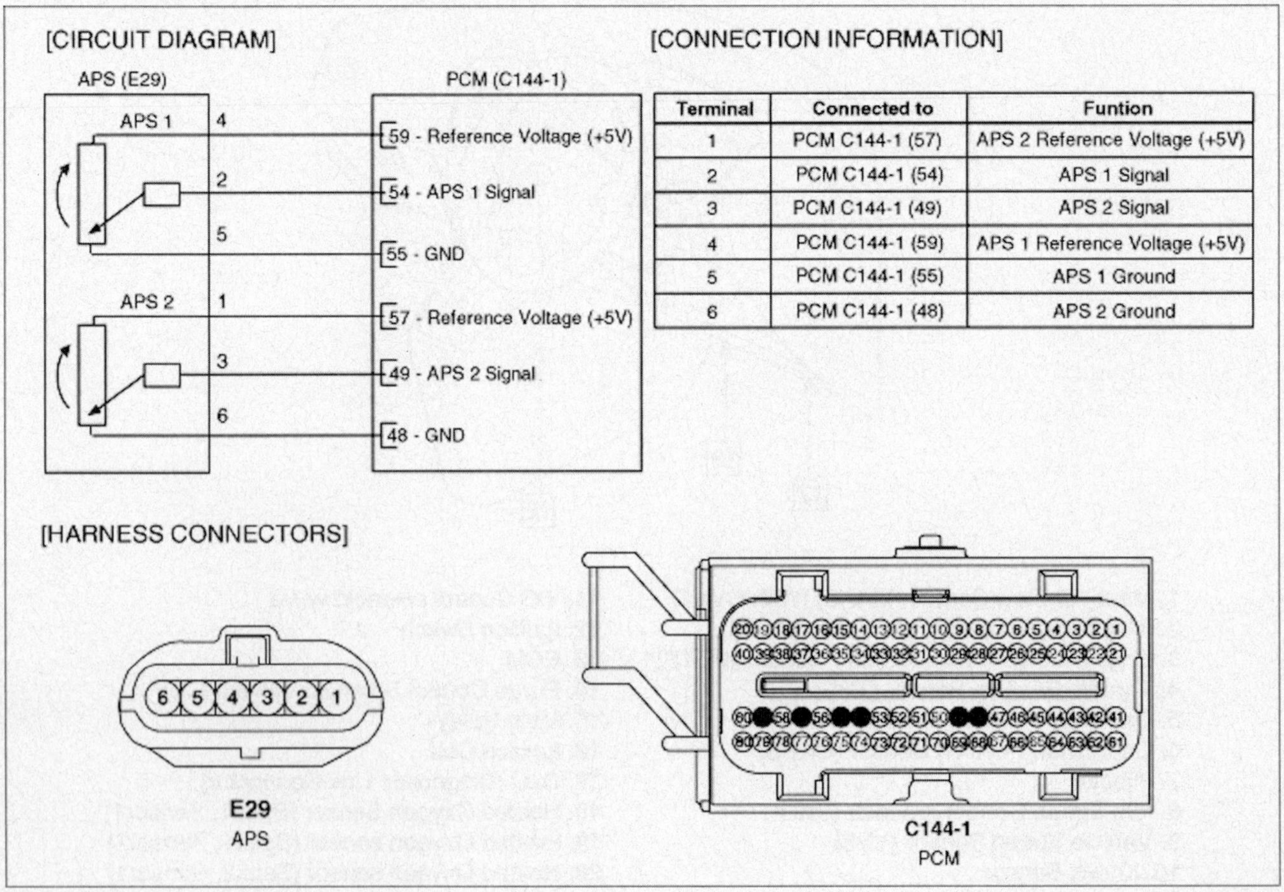

[CIRCUIT DIAGRAM]

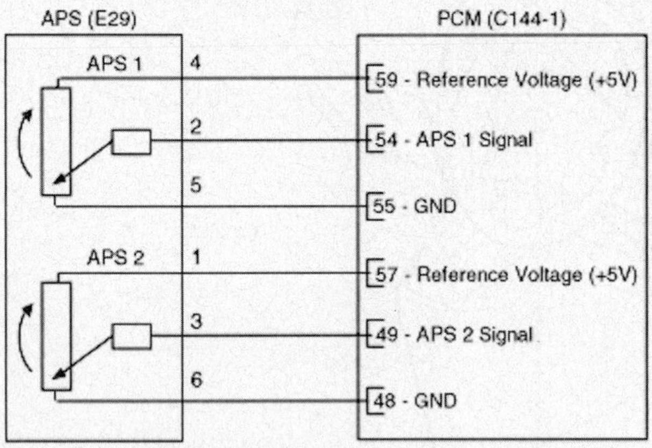

APS (E29)

APS 1
4 — 59 - Reference Voltage (+5V)
2 — 54 - APS 1 Signal
5 — 55 - GND

APS 2
1 — 57 - Reference Voltage (+5V)
3 — 49 - APS 2 Signal
6 — 48 - GND

PCM (C144-1)

[CONNECTION INFORMATION]

Terminal	Connected to	Funtion
1	PCM C144-1 (57)	APS 2 Reference Voltage (+5V)
2	PCM C144-1 (54)	APS 1 Signal
3	PCM C144-1 (49)	APS 2 Signal
4	PCM C144-1 (59)	APS 1 Reference Voltage (+5V)
5	PCM C144-1 (55)	APS 1 Ground
6	PCM C144-1 (48)	APS 2 Ground

[HARNESS CONNECTORS]

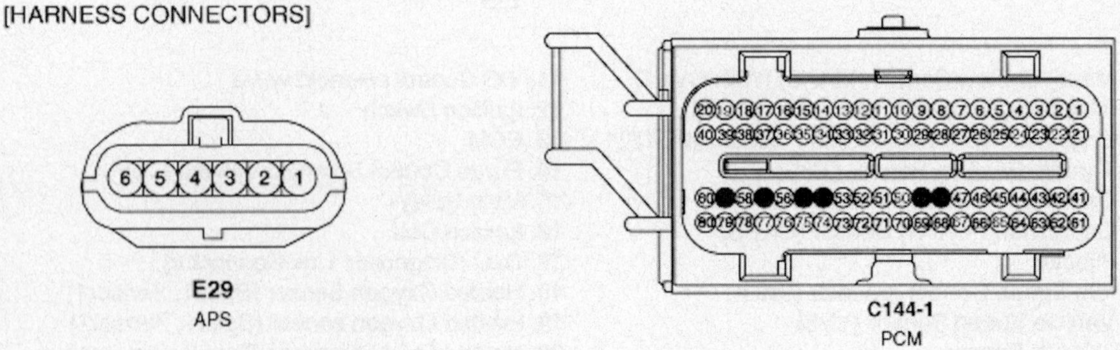

E29
APS

C144-1
PCM

22140_HYUN_G0219

Fig. 142 Accelerator Pedal Position (APP) connector end view and circuit schematics

| Pedal Position | Output Voltage (V) [Vref = 5.0V] | |
	APS1	APS2
C.T	0.7 ~ 0.8V	0.29 ~ 0.46V
W.O.T	3.85 ~ 4.35V	1.93 ~ 2.18V

22140_HYUN_G0220

Fig. 143 Accelerator pedal position table—Voltage Output

Item	Sensor Resistance
APS1	0.7 ~ 1.3kΩ at 68°F (20°C)
APS2	1.4 ~ 2.6kΩ at 68°F (20°C)

22140_HYUN_G0221

Fig. 144 Accelerator pedal position circuit testing table—Sensor Resistance

module and detects the rotation angle of the accelerator pedal. The APP is one of the most important sensors in engine control system, so it consists of the 2 sensors which adapt an individual sensor power and ground line. The second sensor monitors the first sensor and its output voltage is half of the first one. If the ratio of sensor 1 and 2 is out of the range (approximately ½), the diagnostic system judges that it is abnormal.

Use the following reference tables to test the APP sensor voltage and resistance values at the connector end points.

CAMSHAFT POSITION (CMP) SENSOR

LOCATION

On the 2.0L engine, the Camshaft Position (CMP) sensor is located near the top of the engine, on the left side of the engine. On the 2.7L engine, the CMP sensor is located near the top of the engine, on the right side of the engine near the ignition coils.

OPERATION

The Camshaft Position (CMP) sensor is triggered by a notched reluctor wheel built onto the camshaft sprocket. The CMP sensor is connected to the ECM by the following circuits:
- A 5-volt circuit
- A low reference circuit
- A signal circuit

The CMP sensor is a Hall-effect sensor and detects the camshaft position by using a Hall element. It is related with the Crankshaft Position (CKP) sensor and detects the piston position of each cylinder which the CKP sensor can't detect. This sensor has a Hall-effect IC which changes in output voltage when a magnetic field is made on the IC with the current flow.

REMOVAL & INSTALLATION

1. Before servicing the vehicle, refer to the Precautions Section.
2. Disconnect the negative battery cable.
3. Disconnect the connector from the CMP sensor.
4. Remove the bolt that retains the CMP sensor.
5. Remove the CMP sensor.

To install:
6. Installation is the reverse of the removal procedure.
7. Tighten the bolt that retains the CMP sensor to: 61–86 inch lbs. (7–10 Nm).

TESTING

2.0L Engine

During normal operation the ECM controls all ignition functions. If either the Crankshaft Position (CKP) or Camshaft Position (CMP) sensor signal is lost, the engine will continue to run because the ECM will default to a limp home mode using the remaining sensor input. Diagnostic trouble codes are available to accurately diagnose the ignition system with an OBD2 scan tool.

1. Inspect the CMP sensor for correct installation. Remove the CMP sensor from the engine and inspect the sensor O-ring for damage. If the sensor is loose, incorrectly installed, or damaged, replace the CMP sensor.
2. Engage the CMP sensor harness connector to the CMP sensor.
3. Set up an oscilloscope as follows:
   a. Channel A (+) to terminal 2 of the CKPS, (-) to ground.
   b. Channel B (+) to terminal 2 of the CMPS, (-) to ground.
4. Start the engine and check the signal waveform. Check whether the waveform synchronizes with crankshaft sensor or not and a tooth is missing.

5. The square wave signal should be smooth and without distortion.

6. The CMPS falling (rising) edge should coincide with the 3–5 tooth of the CKP from one longer signal (missing tooth).

7. If the waveform signal is normal, check for a poor connection between the ECM and the components.

8. If the waveform signal is not normal, remove the sensor and calculate the air gap between the sensor and the flywheel/torque converter.

9. The air gap should be 0.07 inch. Measure the distance of the housing to the teeth on the flywheel/torque converter (measurement "A") and from the mounting surface on the sensor to sensor tip (measurement "B"), then subtract "B" from "A".

10. Check the sensor for contamination, deterioration, or damage.

11. Substitute the sensor with a known good component, check for proper operation. If the problem is corrected, replace the sensor.

2.7L Engine

During normal operation the ECM controls all ignition functions. If either the Crankshaft Position (CKP) or Camshaft Position (CMP) sensor signal is lost, the engine will continue to run because the ECM will default to a limp home mode using the remaining sensor input. Diagnostic trouble codes are available to accurately diagnose the ignition system with an OBD2 scan tool.

1. Inspect the CMP sensor for correct installation. Remove the CMP sensor from the engine and inspect the sensor O-ring for damage. If the sensor is loose, incorrectly installed, or damaged, replace the CMP sensor.

2. Engage the CMP sensor harness connector to the CMP sensor.

3. Connect the Hi-Scan tool to the data link connector.

4. Start the engine and let it idle.

5. Monitor "CKP T/WHEELS-LO CMP" and "CKP T/WHEELS-HI CMP" parameters on the scan tool data list.

6. The specification should be "CKP T/WHEELS-LO CMP": 38 plus/minus 4 tooth, and "CKP T/WHEELS-HI CMP": 98 plus/minus 4 tooth.

7. If the specification is normal, check for a poor connection between the ECM and the components.

8. If the specification is not normal, remove the sensor and calculate the air gap.

9. Check the sensor for contamination, deterioration, or damage.

10. Substitute the sensor with a known good component, check for proper operation. If the problem is corrected, replace the sensor.

CRANKSHAFT POSITION (CKP) SENSOR

LOCATION

See Figures 145 and 146.

Refer to the accompanying illustrations for Crankshaft Position (CKP) sensor location.

OPERATION

The Crankshaft Position (CKP) sensor is a Hall Effect type sensor that generates voltage using a sensor and a target wheel mounted on the crankshaft. There are 58 slots in the target wheel where one is longer than the others. When the slot in the wheel aligns with the sensor, the sensor voltage output is low. When the metal tooth in the wheel aligns the sensor, the sensor voltage is high. During one crankshaft rotation there are 58 rectangular signals and one longer signal. The ECM calculates engine RPM by using the sensor's signal and controls the injection duration and ignition timing. Using the signal differences caused by the longer slot, the ECM identifies which cylinder is at TDC.

REMOVAL & INSTALLATION

1. Before servicing the vehicle, refer to the Precautions Section.

2. Disconnect the negative battery cable.

3. Disconnect the connector from the sensor.

4. Remove the bolt that retains the sensor in place.

5. Remove the sensor from its mounting.

To install:

6. Installation is the reverse of the removal procedure.

7. Tighten the sensor retaining bolt to: 70–104 inch lbs. (8–12 Nm).

8. Clearance between the sensor and the sensor wheel should be 0.020–0.059 inch.

TESTING

Component Inspection

See Figure 147.

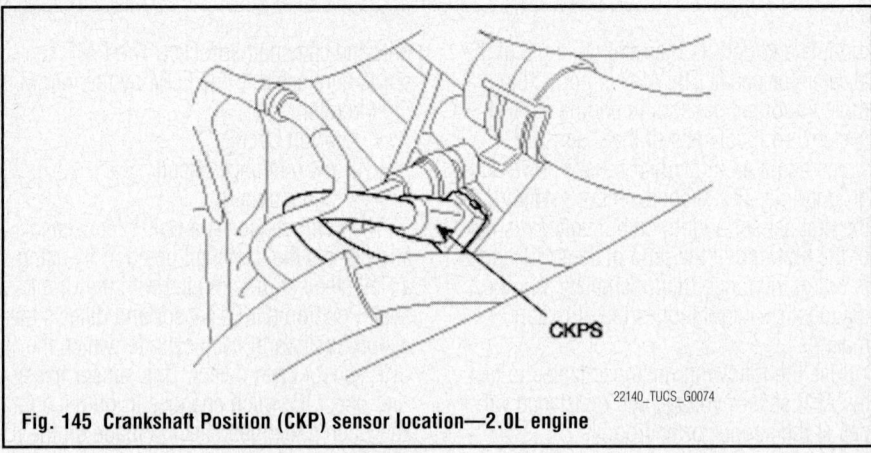

Fig. 145 Crankshaft Position (CKP) sensor location—2.0L engine

22140_TUCS_G0074

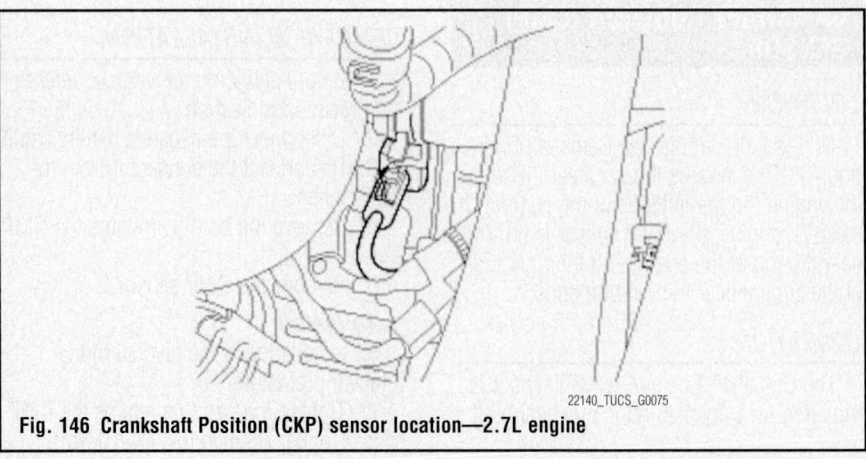

Fig. 146 Crankshaft Position (CKP) sensor location—2.7L engine

22140_TUCS_G0075

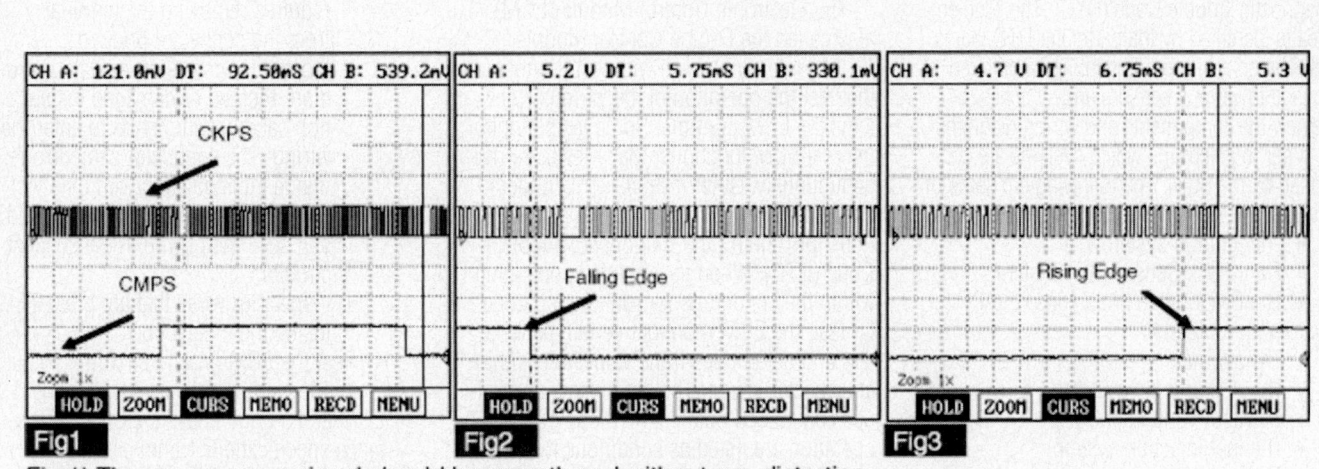

Fig.1) The square wave signal should be smooth and without any distortion.

Fig.2,3) The CMPS falling(rising) edge is coincided with 3~5 tooth of the CKP from one longer signal(missing tooth)

22140_HYUN_G0242

Fig. 147 Waveform graph of the CKP and the CMP sensors

With a scan tool, check the signal waveform of the Crankshaft Position (CKP) and the Camshaft Position (CMP) sensor. The waveform should appear as the following illustration.

2.0L Engine

1. Be sure that the CKP and ECM connectors are connected.

2. Set up an oscilloscope as follows:

a. Channel A (+) to terminal 2 of the CKP, (-) to ground.

b. Channel B (+) to terminal 2 of the CMP, (-) to ground.

3. Start the engine and check the signal waveform. Check whether the waveform synchronizes with the camshaft sensor or not and a tooth is missing.

4. The square wave signal should be smooth and without distortion.

5. The CMP falling (rising) edge should coincide with 3–5 tooth of the CKP from one longer signal (missing tooth).

6. If the waveform signal is normal, check for a poor connection between the ECM and the components.

7. If the waveform signal is not normal, remove the sensor and calculate the air gap between the sensor and the flywheel/torque converter.

8. The air gap should be 0.012–0.067 inch. Measure the distance of the housing to the teeth on the flywheel/torque converter (measurement "A") and from the mounting surface on the sensor to the sensor tip (measurement "B"), then subtract "B" from "A".

9. Check the sensor for contamination, deterioration, or damage.

10. Substitute the sensor with a known good component, check for proper operation. If the problem is corrected, replace the sensor.

2.7L Engine

1. Connect the Hi-Scan tool to the data link connector.

2. Start the engine and let it idle.

3. Monitor "CKP T/WHEELS-LO CMP" and "CKP T/WHEELS-HI CMP" parameters on the scan tool data list.

4. Specification should be "CKP T/WHEELS-LO CMP": 38 plus/minus 4 tooth, and "CKP T/WHEELS-HI CMP": 98 plus/minus 4 tooth.

5. If the reading is normal, check for a poor connection between the ECM and the components.

6. If the reading is not normal, remove the sensor and calculate the air gap between the sensor and the flywheel/torque converter.

7. The air gap should be 0.012–0.067 inch. Measure from the distance of the housing to the teeth on the flywheel/torque converter (measurement "A") and from the mounting surface on the sensor to the sensor tip (measurement "B"), then subtract "B" from "A".

8. Check the sensor for contamination, deterioration, or damage.

9. Substitute the sensor with a known good component and check for proper oper-

ation. If the problem is corrected, replace the sensor.

ELECTRIC FAN SWITCH

The engine cooling fan is operated by a relay that is controlled through the Electronic Control Module (ECM). The cooling fan relay will activate depending upon the needs of one or more systems including the cooling system, the air conditioning, and the heating systems.

ELECTRONIC CONTROL MODULE (ECM)

LOCATION

The Electronic Control Module (ECM) is located in the passenger compartment on the driver's side near the steering column.

OPERATION

This vehicle has electronic controls to reduce exhaust emissions while maintaining excellent drivability and fuel economy. The Electronic Control Module (ECM) is the control center of this system. The ECM monitors numerous engine and vehicle functions. The ECM constantly looks at the information from various sensors and other inputs, and controls the systems that affect vehicle performance and emissions. The ECM also performs the diagnostic tests on various parts of the system. The ECM can recognize operational problems and alert the driver via the Malfunction Indicator Lamp (MIL). When the

ECM detects a malfunction, the ECM stores a Diagnostic Trouble Code (DTC). The problem area is identified by the particular DTC that is set. The control module supplies a buffered voltage to various sensors and switches. Review the components and wiring diagrams in order to determine which systems are controlled by the ECM. The following are some of the functions that the ECM controls:

- The ignition system
- The Knock Sensor (KS) system
- The Evaporative Emissions (EVAP) system
- The alternator
- The A/C and fan systems
- The cooling fan control
- The fuel injection system
- The emission control systems
- The on-board diagnostics

REMOVAL & INSTALLATION

1. Before servicing the vehicle, refer to the Precautions Section.
2. Turn ignition switch **OFF**.
3. Disconnect the negative battery cable from the battery.
4. Disconnect the ECM connectors.
5. Remove the ECM mounting bolts and remove the ECM from the vehicle.

To install:

6. Installation is the reverse of the removal.
7. Tighten the ECM mounting bolts to: 86–104 inch lbs. (10–12 Nm).

❊❊ WARNING

When replacing the ECM, be careful to use the right part number, as damage to the injection system could occur.

TESTING

1. Perform a careful underhood inspection when performing any diagnostic procedure or diagnosing the cause of an emission test failure. This can often lead to repairing a condition without further steps. Use the following guidelines when performing an inspection:

a. Inspect all of the vacuum hoses for correct routing, pinches, cuts, or disconnects.

b. Inspect any hoses that are difficult to see.

c. Inspect all of the wires in the engine compartment for the following conditions:

- Burned or chafed spots
- Pinched wires
- Contact with sharp edges

- Contact with hot exhaust manifold

The Electronic Control Module (ECM), also called the Engine Control Module (ECM), is programmed with test routines that test the operation of the various systems the ECM controls. Some tests monitor internal ECM functions. Many tests are run continuously. Other tests run only under specific conditions, referred to as conditions for running the Diagnostic Trouble Code (DTC). When the vehicle is operating within the conditions for running a particular test, the ECM monitors certain parameters and determines if the values are within an expected range. The parameters and values considered outside the range of normal operation are listed as conditions for setting the DTC. When the conditions for setting the DTC occur, the ECM executes the action taken when the DTC sets. Some DTC's alert the driver via the Malfunction Indicator Lamp (MIL) or a message. Other DTC's do not trigger a driver warning, but are stored in memory. The ECM also saves data and input parameters when most DTC's are set.

The DTC's are categorized by type. The DTC type is determined by the MIL operation and the manner in which the fault data is stored when a particular DTC fails. In some cases, there may be exceptions to this structure. Therefore, when diagnosing the system it is important to read the action taken when the DTC sets and the conditions for clearing the DTC.

Many intermittent open or shorted circuits come and go with harness and connector movement caused by vibration, engine torque, bumps, or rough pavement.

2. Test the wiring harness and connectors by performing the following:

- Move the related ECM connectors and wiring while monitoring the appropriate scan tool data
- With the engine running, move the related connectors and wiring while monitoring engine operation
- If harness or connector movement affects the data displayed, the component and system operation, or the engine operation, inspect and repair the harness or connections as necessary

3. Test the electrical connections and/or wiring by performing the following:

- Inspect for incorrect mating of the connector halves or terminals not fully seated in the connector body
- Inspect for improperly formed or damaged terminals. Test for incorrect terminal tension
- Inspect for poor terminal to wire connections including terminals

crimped over insulation. This requires removing the terminal from the connector body

- Inspect for corrosion or water intrusion. Pierced or damaged insulation can allow moisture to enter the wiring. The conductor can corrode inside the insulation with little visible evidence. Look for swollen and stiff sections of wire in the suspect circuits
- Inspect for wires that are broken inside the insulation

4. Test the ECM ground circuit:

a. Measure resistance between the ECM and a good chassis ground. Resistance specification: 1 ohm or less.

b. Use the backside of the ECM harness connector as the ECM side check point.

c. If a problem is found, make the necessary repairs.

5. If a problem is not located through the above procedures, the ECM may be faulty.

6. Hyundai recommends to replace the ECM with a new one and then check the vehicle again. If the vehicle operates normally, then the problem was likely with the ECM.

7. Retest the original ECM, if possible, by installing it into a known-good vehicle and check vehicle operations.

ENGINE COOLANT TEMPERATURE (ECT) SENSOR

LOCATION

See Figures 148 and 149.

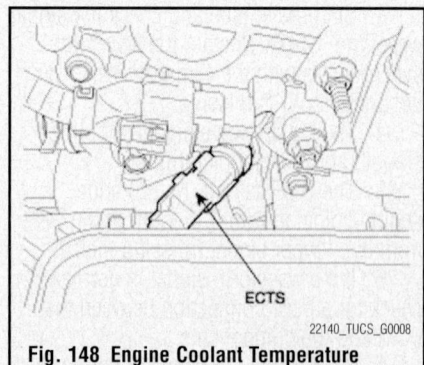

ECTS

22140_TUCS_G0008

Fig. 148 Engine Coolant Temperature (ECT) sensor location—2.0L engine

The Engine Coolant Temperature (ECT) sensor is located in the engine coolant passage near the thermostat.

OPERATION

The Engine Coolant Temperature (ECT) sensor uses a thermistor whose resistance

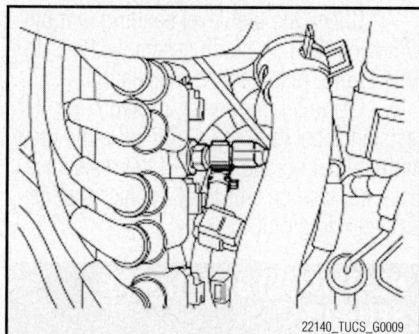

Fig. 149 Engine Coolant Temperature (ECT) sensor location—2.7L engine

Temperature [°C(°F)]	Resistance (kΩ)
-40(-40)	48.14
-20(-4)	14.13 ~ 16.83
0(32)	5.79
20(68)	2.31 ~ 2.59
40(104)	1.15
60(140)	0.59
80(176)	0.32

Fig. 150 Resistance chart specifications

changes with the temperature in order to detect the engine coolant temperature. The electrical resistance of the ECT decreases as the temperature increases, and increases as the temperature decreases. The 5-volt reference in the ECM is supplied to the ECT via a resistor in the ECM. The resistor in the ECM and the thermistor in the ECT are connected in series. When the resistance value of the thermistor in the ECT changes according to the engine coolant temperature, the output voltage also changes. During cold engine operation, the ECM increases the fuel injection duration and controls the ignition timing using the information of engine coolant temperature to avoid engine stalling and improve drivability.

REMOVAL & INSTALLATION

1. Before servicing the vehicle, refer to the Precautions Section.
2. Drain the coolant to a level below the bottom of the sensor.
3. Disconnect the ground cable from the battery and then remove the sensor connector.
4. Remove the Engine Coolant Temperature (ECT) sensor.

To install:

5. Reverse the removal procedure.
6. Tighten the sensor to 15–29 ft. lbs. (20–39 Nm).

TESTING

See Figures 150 and 151.

➡**Clear Diagnostic Trouble Codes (DTC's) if applicable.**

1. Inspect the Engine Coolant Temperature (ECT) sensor electrical connection. Check the terminal for poor connections, loose wires, bent, broken, or corroded pins, and then verify that the connector is securely fastened.
2. Turn ignition switch **OFF**.
3. Disconnect ECT sensor connector.
4. Remove the ECT sensor.

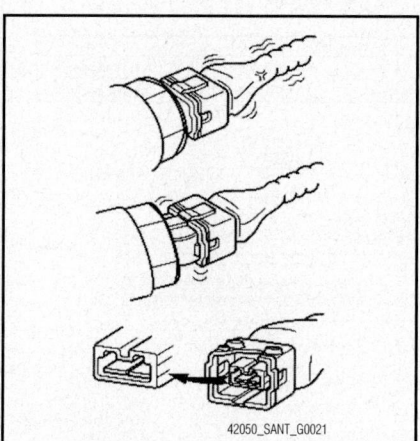

Fig. 151 Inspecting the Engine Coolant Temperature (ECT) sensor connector

5. After immersing the thermistor of the sensor into engine coolant, measure the resistance between the ECT terminals 1 and 3.
6. Check that the resistance is within the specification according to the following chart.
7. Replace the ECT sensor if the component is not within specification.

ENGINE OIL TEMPERATURE (EOT) SENSOR

LOCATION

See Figure 152.

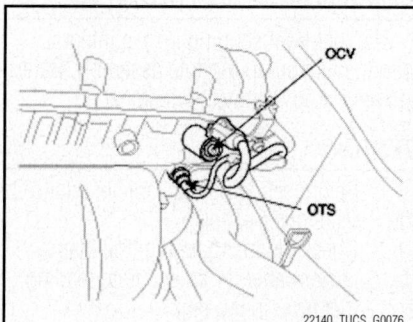

Fig. 152 Oil Control Valve (OCV) and Engine Oil Temperature Sensor (OTS)/(EOT) locations—2.0L engine

Refer to the accompanying illustration for sensor location.

OPERATION

See Figure 153.

The Engine Oil Temperature (EOT) sensor is a negative coefficient thermistor used by the ECM to measure engine oil temperature for the purpose of adjusting Continuously Variable Valve Timing (CVVT) calculations.

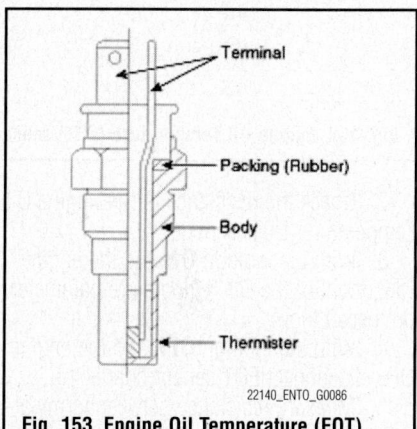

Fig. 153 Engine Oil Temperature (EOT) sensor cut-away view

REMOVAL & INSTALLATION

1. Before servicing the vehicle, refer to the Precautions Section.
2. Disconnect the ground cable from the battery and then remove the sensor connector.
3. Remove Engine Oil Temperature (EOT) sensor.

To install:

4. Install the EOT sensor and tighten to 15–29 ft. lbs. (20–39 Nm).
5. Connect the negative cable to the battery.
6. Run the engine and check for leaks.

TESTING

See Figures 154 and 155.

1. Before servicing the vehicle, refer to the Precautions Section.

Temperature		Resistance(kΩ)
°C	°F	
-20	-4	16.52
20	32	2.45
80	176	0.29

22140_ENTO_G008

Fig. 154 Engine Oil Temperature (EOT) sensor Temperature and Resistance Table

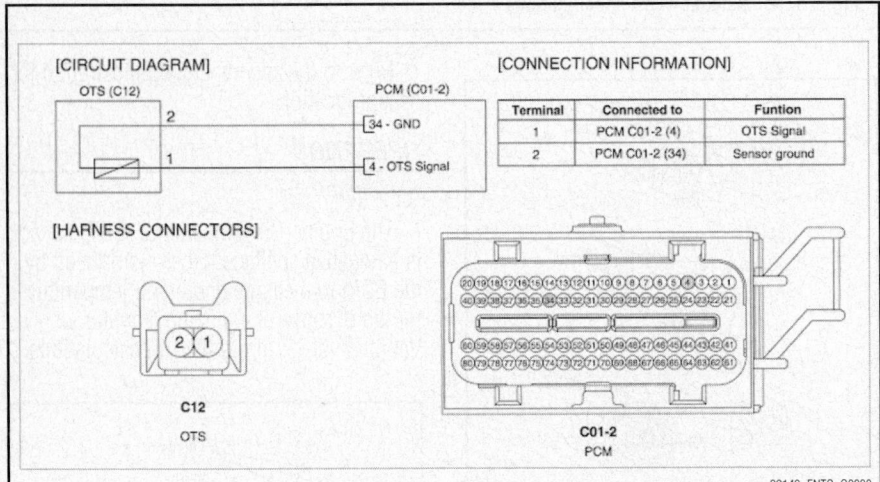

22140_ENTO_G0088

Fig. 155 Engine Oil Temperature (EOT) sensor connector end view and schematic diagram

2. Check the resistance of the Engine Oil Temperature (EOT) sensor.

3. With the ignition **ON** and the engine OFF, monitor the Oil Temperature parameter on a scan tool.

4. With the ignition **OFF** and the engine OFF, disconnect EOT sensor connector.

5. Measure resistance between terminal 1 and 2 of EOT sensor connector (Component Side).

6. The specifications should be as listed in the following table.

7. If the EOT sensor does not meet the specifications, replace the EOT sensor.

FUEL LEVEL SENDING UNIT

LOCATION

See Figure 156.

The fuel level sending unit is located on the fuel pump assembly module.

OPERATION

The fuel level sending unit is attached to the side of the fuel pump module. The fuel level sensor utilizes a variable resistor connected to a float that is buoyed up or down by the level of the fuel in the fuel tank.

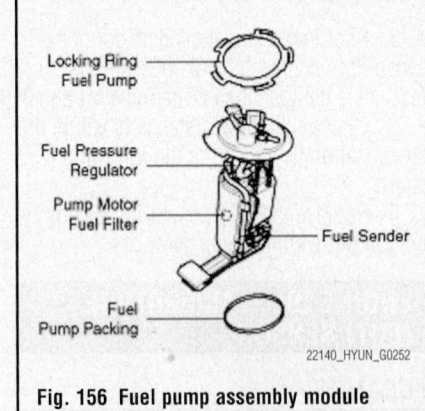

22140_HYUN_G0252

Fig. 156 Fuel pump assembly module

REMOVAL & INSTALLATION

The fuel level sending unit is integral with the fuel pump module assembly. Refer to Fuel Pump, removal & installation.

TESTING

1. Before servicing the vehicle, refer to the Precautions Section.

2. Check the fuel gauge in the instrument cluster panel for an accurate reading of the fuel level in the tank.

3. If the fuel gauge sticks at one extreme of the meter, no matter the fuel level in the tank, check for a break or short in the wiring from the fuel level sending unit.

4. Check the fuel level sending unit by removing the fuel pump assembly. Refer to Fuel Pump, removal and installation.

5. Using an ohmmeter, measure the resistance between the terminals while moving the float up and down. If the resistance does not change smoothly, replace the fuel level sending unit.

HEATED OXYGEN (HO2S) SENSOR

LOCATION

The Heated Oxygen Sensors (HO2S) are located in the exhaust system. On some vehicles, one sensor is located up at the exhaust manifold(s) and the other sensor is located down at the catalytic converter.

OPERATION

The Heated Oxygen Sensor (HO2S) consists of zirconium and alumina and is installed upstream and downstream of the Manifold Catalyst Converter (MCC). After it compares the oxygen consistency of the atmosphere with the exhaust gas, it transfers the oxygen consistency of the exhaust gas to the ECM. This sensor operates most efficiently when the temperature of the sensor tip is higher than 698°F (370°C). Therefore, a heater has been integrated with the oxygen sensor that is controlled by the ECM duty signal. When the exhaust gas temperature is lower than the specified value, the heater warms the sensor tip. The ECM commands the heater ON or OFF to maintain a specific HO2S operating temperature range. The ECM monitors the voltage on the HO2S heater low control circuit for heater fault diagnosis. If the ECM detects that the HO2S heater low control circuit voltage is not within a specified range, a DTC is set.

The HO2S supplies the electronic control assembly with a signal indicating either a rich or lean mixture condition as the engine operates.

REMOVAL & INSTALLATION

1. Before servicing the vehicle, refer to the Precautions Section.

2. Disconnect the electrical connector from the sensor.

3. Remove the oxygen sensor.

To install:

4. Installation is the reverse of the removal procedure.

➡**Apply anti-seize compound to the threaded portion of the sensor, prior to installation. Never apply anti-seize compound to the protector of the sensor.**

TESTING

See Figures 157 through 160.

1. Perform a visual inspection of the sensor as follows:

a. If the sensor tip has a black/sooty deposit, this may indicate a rich fuel mixture.

b. If the sensor tip has a white, gritty deposit, this may indicate an internal coolant leak.

c. If the sensor tip has a brown deposit, this could indicate oil consumption.

2. Check that the heater resistance is according to specification.

3. Check that the output voltage in the rich/lean readings fluctuate according to specifications.

4. Connect an OBD2 scan tool to the Data Link Connector (DLC).

5. With the engine running, the normal waveform for the Heated Oxygen Sensor (HO2S) and the HO2S heater should be similar to those represented in the following waveform illustration.

INTAKE AIR TEMPERATURE (IAT) SENSOR

LOCATION

See Figures 161 and 162.

The Intake Air Temperature (IAT) sensor is mounted in the intake air hose of the air cleaner assembly.

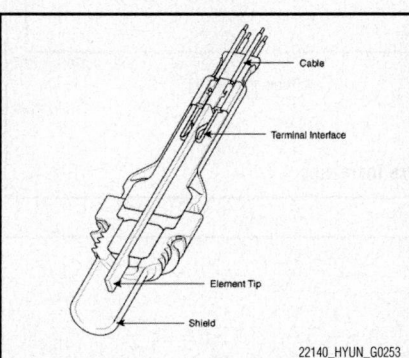

A/F Ratio	Output Voltage (V)
RICH	0.75 ~ 1.00V
LEAN	0 ~ 0.12V

22140_HYUN_G0256

Fig. 158 Heated oxygen sensor rich/lean output voltage specifications

22140_HYUN_G0253

Fig. 159 Heated oxygen sensor cut-away view

OPERATION

See Figure 163.

The Intake Air Temperature (IAT) sensor is a variable resistor that measures the temperature of the air entering the engine intake manifold. The ECM supplies 5 volts to the IAT signal circuit and a ground for the IAT low reference circuit. When the sensor is cold, the resistance is greater. This results in a greater voltage on the signal circuit that is interpreted by the ECM as a colder IAT. As

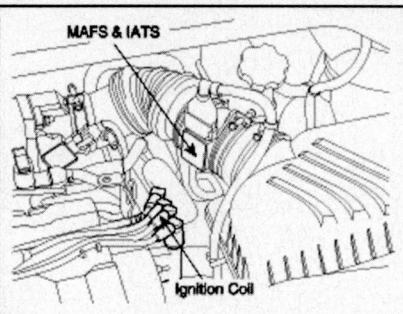

22140_TUCS_G0077

Fig. 161 Intake Air Temperature (IAT) sensor and Mass Air Flow (MAF) sensor locations—2.0L engine

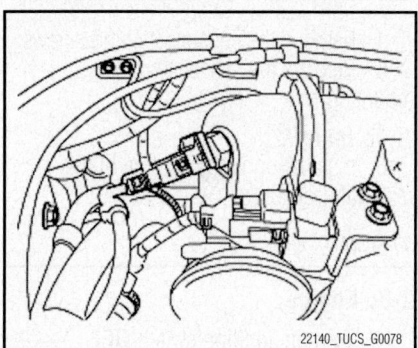

22140_TUCS_G0078

Fig. 162 Intake Air Temperature (IAT) sensor location—2.7L engine

the sensor becomes warmer, the resistance decreases. This results in a lesser voltage on the IAT signal circuit that is interpreted by the ECM as a warmer IAT. If the ECM detects an IAT sensor signal voltage that is not within a calibrated range of the IAT sensor 1 signal voltage, a DTC is set.

REMOVAL & INSTALLATION

1. Before servicing the vehicle, refer to the Precautions Section.

2. Disconnect the negative battery cable.

3. Disconnect the connector from the sensor.

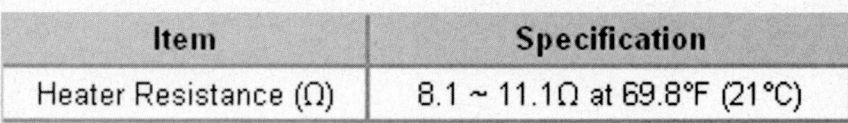

Item	Specification
Heater Resistance (Ω)	8.1 ~ 11.1Ω at 69.8°F (21°C)

22140_HYUN_G0255

Fig. 157 Heated oxygen sensor heater resistance specifications

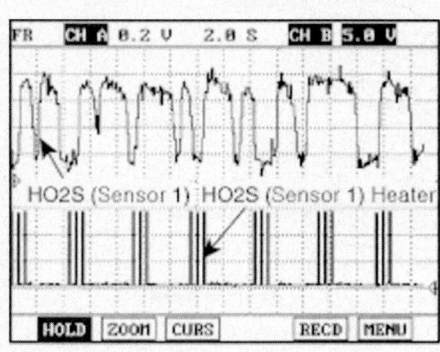

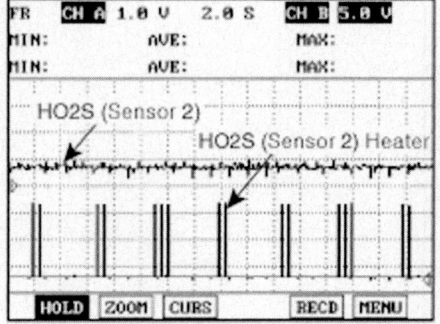

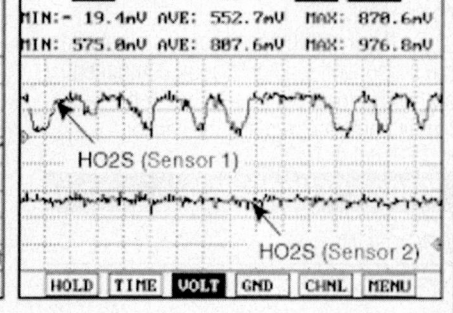

22140_HYUN_G0254

Fig. 160 Waveform of heated oxygen sensors and heater element

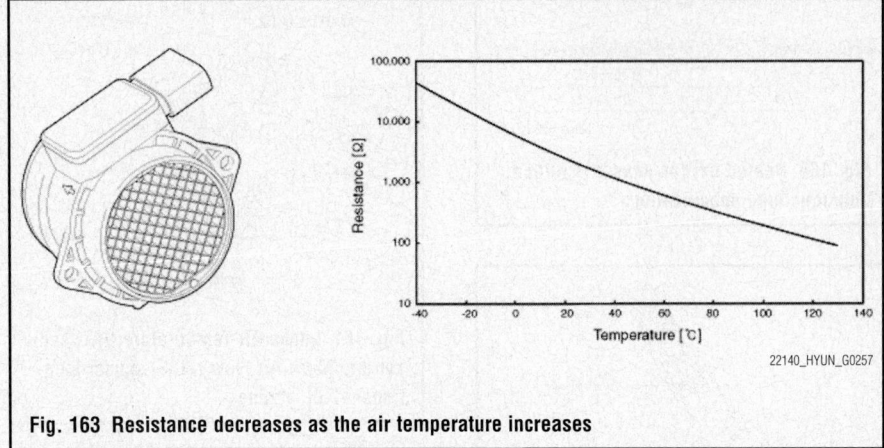

Fig. 163 Resistance decreases as the air temperature increases

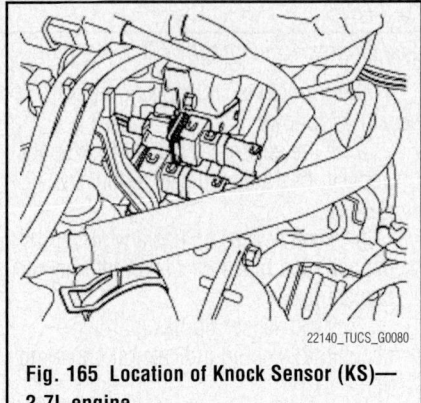

Fig. 165 Location of Knock Sensor (KS)—2.7L engine

4. Remove the sensor retaining screws.
5. Remove the sensor from its mounting.

To install:

6. Installation is the reverse of the removal procedure.

TESTING

2.0L Engine

1. Turn the ignition switch **OFF**.
2. Disconnect the sensor connector.
3. Measure the resistance between terminals 1 and 5 (component side).
4. Specification should be 2.35–3.54 kohms at 68°F.
5. If within specification, check for a poor connection between the ECM and the component.
6. If not within specification, substitute the sensor with a known good component and check for proper operation. If the problem is corrected, replace the sensor.

2.7L Engine

1. Remove the sensor connector.
2. Measure the voltage between the sensor terminal 1 and 2.
3. Specification should be 2.35–2.54 volts at 68°F.
4. If within specification, check for a poor connection between the ECM and the component.
5. If not within specification, substitute the sensor with a known good component and check for proper operation. If the problem is corrected, replace the sensor.

KNOCK SENSOR (KS)

LOCATION

See Figures 164 and 165.

The Knock Sensor (KS) is located in the side of the cylinder block.

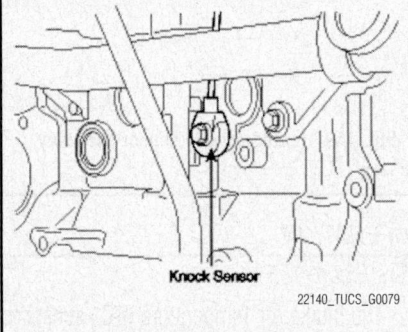

Fig. 164 Location of Knock Sensor (KS)—2.0L engine

OPERATION

See Figure 166.

The Knock Sensor (KS) is used to detect engine vibrations caused by pre-ignition or detonation and provides information to the ECM. The ECM will then retard the timing in an attempt to eliminate detonation.

REMOVAL & INSTALLATION

1. Before servicing the vehicle, refer to the Precautions Section.
2. Disconnect the negative battery cable.
3. Disconnect the sensor connector.
4. Remove the sensor from its mounting.

To install:

5. Installation is the reverse of the removal procedure.
6. Tighten the sensor to 12–17 ft. lbs. (16–24 Nm).

TESTING

1. Disconnect the Knock Sensor (KS) electrical connector.
2. Connect an ohmmeter.
3. Measure the resistance between terminals 1 and 2. Specification should be 5mohm at 68°F (20°C).
4. If not within specification, replace the effected KS.

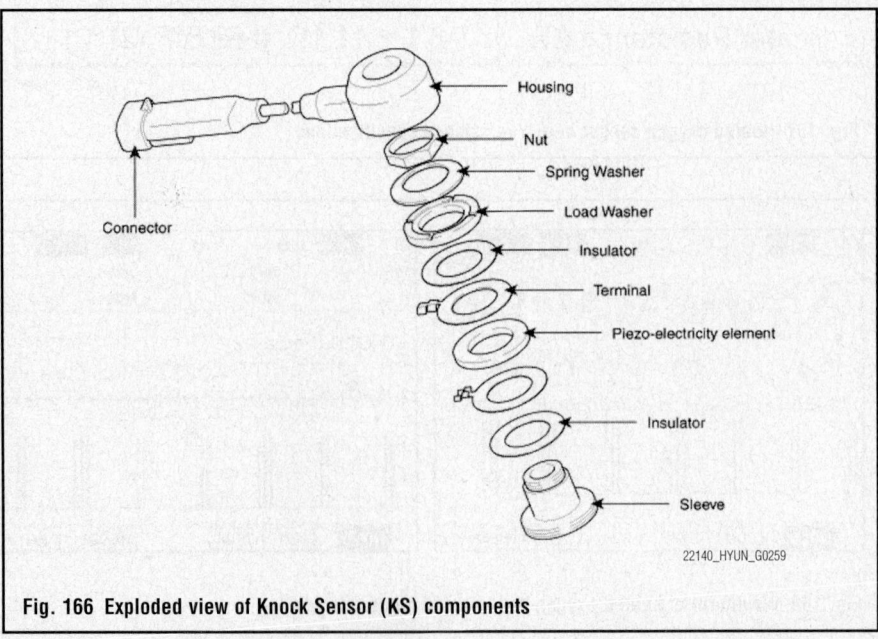

Fig. 166 Exploded view of Knock Sensor (KS) components

MALFUNCTION INDICATOR LIGHT (MIL)

RESET PROCEDURES

1. Proper operation of the Malfunction Indicator Light (MIL):
- The MIL will illuminate with the ignition switch **ON** and the engine OFF
- The MIL will turn OFF when the engine is started
- The MIL will remain ON if the self-diagnostic system has detected a malfunction
- The MIL may turn OFF if the malfunction is no longer present
- If the MIL is illuminated and then the engine stalls, the MIL will remain illuminated as long as the ignition switch is ON
- If the MIL is not illuminated and the engine stalls, the MIL will not illuminate until the ignition switch is cycled OFF, then ON

2. Resetting the MIL:
- The control module turns OFF the MIL after 3 consecutive ignition cycles that the diagnostic system runs and does not fail
- The control module turns OFF the MIL after a current Diagnostic Trouble Code (DTC) clears when the diagnostic cycle runs and passes
- There may still be a history of DTC's stored in the system. These will clear after 40 consecutive warm-up cycles, if no failures are reported by any other related diagnostic system
- Manual resetting of the MIL and any DTC stored in the system, requires the use of an OBD2 scan tool connected to the Data Link Connector (DLC) for communication with the vehicle. Follow the instructions of the scan tool for both retrieval and resetting of DTC's.

➡**If the error symptoms causing the MIL to illuminate have been corrected, the MIL will return to normal operation.**

MASS AIR FLOW (MAF) SENSOR

LOCATION

See Figures 161 and 167.

The Mass Air Flow (MAF) sensor is mounted in the intake air hose of the air cleaner assembly.

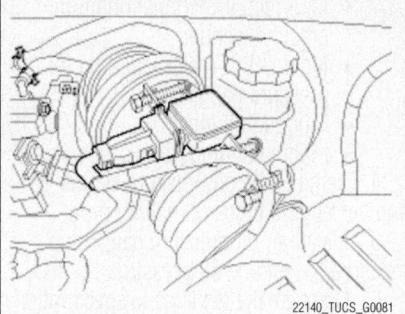

Fig. 167 Mass Air Flow (MAF) sensor location—2.7L engine

OPERATION

The Mass Air Flow Sensor (MAF) sensor is a hot-film type sensor and is located in between the air cleaner and the throttle body. It consists of a tube, a sensor assembly, and honeycomb cell. It detects the intake air quantity flowing into the intake manifold. Air flows from the air cleaner assembly through the honeycomb cell and over the hot film element. At this time, heat transfer is generated by convection and the MAF sensor loses its energy. This sensor detects the mass air flow by using the energy loss and transfers the information to the ECM by a voltage transfer. The ECM calculates fuel quantity and ignition timing appropriate for the conditions.

REMOVAL & INSTALLATION

1. Before servicing the vehicle, refer to the Precautions Section.
2. Disconnect the negative battery cable.
3. Disconnect the connector from the sensor.
4. Remove the air cleaner and air intake assembly, as required.
5. Remove the sensor from its mounting.

To install:

6. Installation is the reverse of the removal procedure.

TESTING

See Figures 168 and 169.

1. Verify the integrity of the air induction

system by inspecting for the following conditions:
- Damaged components
- Loose or improper installation
- An air flow restriction
- Any vacuum leak
- Water intrusion

2. Verify that any electrical aftermarket devices are properly connected and grounded.
3. With the engine running, observe the scan tool MAF sensor parameter. The reading should be between 1,700–3,200 Hz depending on the Engine Coolant Temperature (ECT).
4. A Wide Open Throttle (WOT) acceleration from a stop should cause the MAF sensor parameter on the scan tool to increase rapidly. This increase should be from 2–6 g/s at idle to greater than 100 g/s at the time of the 1–2 shift.
5. Run the engine and connect the Hi-Scan (Pro)® to the Data Link Connector (DLC).
6. Check if the sensor output voltage is normal according to the following specification table.
7. If within specification, check for poor

Air Flow (kg/h)	Output Frequency (Hz)
12.6	2,617
18.0	2,958
23.4	3,241
32.4	3,653
43.2	4,024
57.6	4,399
72.0	4,704
108.0	5,329
144.0	5,897
198.0	6,553
270.0	7,240
360.0	7,957
486.0	8,738
666.0	9,644
900.0	10,590

22140_ENTO_G0099

Fig. 169 Air Flow and Output Frequency Table

Condition	Output Voltage (V)	Intake Air Quantity (kg/h)
Idle	0.6 ~ 1.0	11.66 ~ 19.85
3000 rpm	1.7 ~ 2.0	43.84 ~ 58.79

22140_HYUN_G0260

Fig. 168 Output Voltage and Intake Air Quantity Table

connection between the ECM and the component.

8. If not within specification, substitute the sensor with a known good component, check for proper operation. If the problem is corrected, replace the sensor.

OIL PRESSURE SENSOR

LOCATION

The oil pressure sensor is located near the base of the oil filter assembly.

OPERATION

The oil pressure sensor measures the amount of pressure generated from the output of the oil pump. If the oil pressure is too low, mechanical damage and scoring may occur.

REMOVAL & INSTALLATION

❊❊ CAUTION

Do not perform this procedure with the engine hot, wait until the engine is cool or cold.

1. Before servicing the vehicle, refer to the Precautions Section.
2. Turn the ignition **OFF**.
3. Remove the oil pressure sensor electrical connector.
4. Remove the oil pressure sensor.

To install:
5. To install, reverse the removal procedure.
6. Tighten the sensor to 16 ft. lbs. (22 Nm).

TESTING

Before servicing any vehicle, please be sure to read the precautions section, which deals with personal safety, prevention of component damage, and important points to take into consideration when servicing a motor vehicle.

1. With the vehicle on a level surface, allow adequate drain down time of 2–3 minutes and measure for a low oil level. Add the recommended grade engine oil and fill the crankcase until the oil level measures full on the oil level indicator.
2. Run the engine, and verify low, or no oil pressure on the vehicle gage or light. Listen for a noisy valve train or a knocking noise.
3. Inspect for the following:
   - Oil diluted by moisture or unburned fuel mixtures
   - Improper oil viscosity for the expected temperature

- Incorrect or malfunctioning oil pressure sender
- Incorrect or malfunctioning oil pressure gauge
- Plugged oil filter
- Malfunctioning oil bypass valve

4. Remove the oil pressure sender or another engine block oil gallery plug.
5. Install an oil pressure gauge and measure the engine oil pressure.
6. Compare the readings to specifications.
7. If the engine oil pressure is below specifications, inspect the engine for one or more of the following:
   - Oil pump worn or dirty
   - Oil pump-to-engine front cover bolts loose
   - Oil pump screen loose, plugged, or damaged
   - Oil pump screen O-ring seal missing or damaged
   - Malfunctioning oil pump pressure regulator valve
   - Excessive bearing clearance
   - Cracked, porous, or restricted oil galleries
   - Oil gallery plugs missing or incorrectly installed
   - Broken lash adjusters

8. Higher than recommended oil pressure may be caused by one or more of the following conditions:
   - Worn or sticking oil pump pressure relief valve
   - Plugged oil filter
   - Improper viscosity oil

9. If the oil pressure tests according to specification and the oil pressure sensor gives a faulty reading, replace the oil pressure sensor (switch) as necessary.

THROTTLE POSITION SENSOR (TPS)

LOCATION
See Figures 170 and 171.

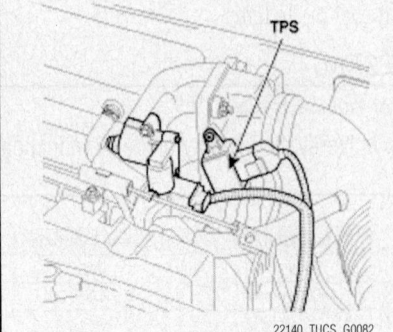

Fig. 170 Throttle Position Sensor (TPS) location—2.0L engine

22140_TUCS_G0082

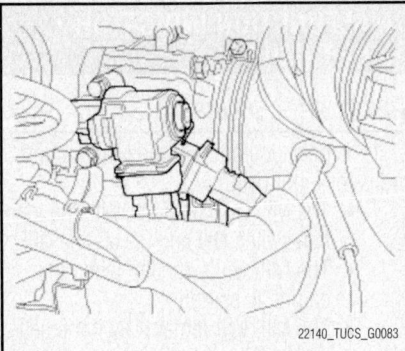

22140_TUCS_G0083

Fig. 171 Throttle Position Sensor (TPS) location—2.7L engine

The Throttle Position Sensor (TPS) is mounted on the throttle body.

OPERATION

The Throttle Position Sensor (TPS) is mounted on the throttle body and detects the opening angle of the throttle plate. The TPS has a variable resistor (potentiometer) that changes in resistance characteristics according to the throttle angle. During acceleration, the TPS resistance between the reference 5-volt and the signal terminal decreases and the output voltage increases; during deceleration, the TPS resistance increases and the TPS output voltage decreases. The ECM supplies a reference 5-volt to the TPS and the output voltage increases directly with the opening of the throttle valve. The TPS output voltage will vary from 0.2–0.8 volts at Closed Throttle (CT) to 4.3–4.8 volts at Wide Open Throttle (WOT). The ECM determines operating conditions such as idle (closed throttle), part load, acceleration/deceleration, and wide-open throttle from the TPS. Also the ECM uses the Manifold Absolute Pressure (MAP) sensor signal along with the TPS signal to adjust fuel injection duration and ignition timing.

REMOVAL & INSTALLATION

1. Before servicing the vehicle, refer to the Precautions Section.
2. Disconnect the negative battery cable.
3. Disconnect the sensor connector.
4. Remove the sensor retaining screws.
5. Remove the sensor from its mounting.

To install:
6. Installation is the reverse of the removal procedure.

TESTING
See Figures 172 and 173.

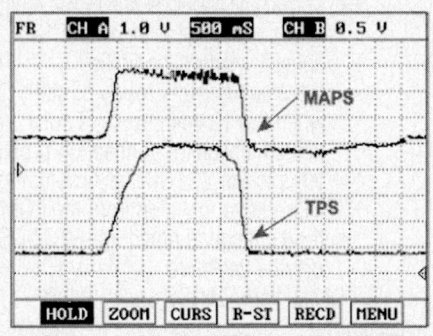

As often as possible, the MAPS signal should be compared with the TPS signal. Check whether the MAPS and TPS signals increase at the same time when accelerating. During acceleration, the MAPS output voltage increases; during deceleration, the MAPS output voltage decreases.

22140_HYUN_G0263

Fig. 172 Waveform of MAP sensor signal compared with the TPS signal waveform

Throttle Angle(°)	Output Voltage(V) [Vref = 5.0V]	
	TPS1	TPS2
0°	0V	5.0V
10°	0.5V	4.5V
20°	0.9V	4.1V
30°	1.4V	3.6V
40°	1.8V	3.2V
50°	2.3V	2.7V
60°	2.7V	2.3V
70°	3.2V	1.8V
80°	3.6V	1.4V
90°	4.1V	0.9V
100°	4.5V	0.5V
110°	5.0V	0V

22140_HYUN_G0265

Fig. 173 Throttle angle and output voltage specification table

1. The Throttle Position Sensor (TPS) output voltage should vary from 0.2–0.8 volts at Closed Throttle (CT).

2. The TPS should vary from 4.3–4.8 volts at Wide Open Throttle (WOT).

3. Check the signal waveform compared to the Manifold Absolute Pressure (MAP) waveform as in the following illustration. The waveform should coincide.

4. The throttle angle should match the output voltage as in the specification table.

5. If the TPS does not meet specifications, replace the TPS.

VARIABLE CAMSHAFT TIMING OIL CONTROL SOLENOID

LOCATION

See Figure 152.

Refer to the accompanying illustration for location.

OPERATION

See Figure 174.

The Continuously Variable Valve Timing (CVVT) system is installed to the chain sprocket of the intake camshaft. This system controls the intake camshaft to provide the optimal valve timing for every driving condition. The ECM controls the Variable

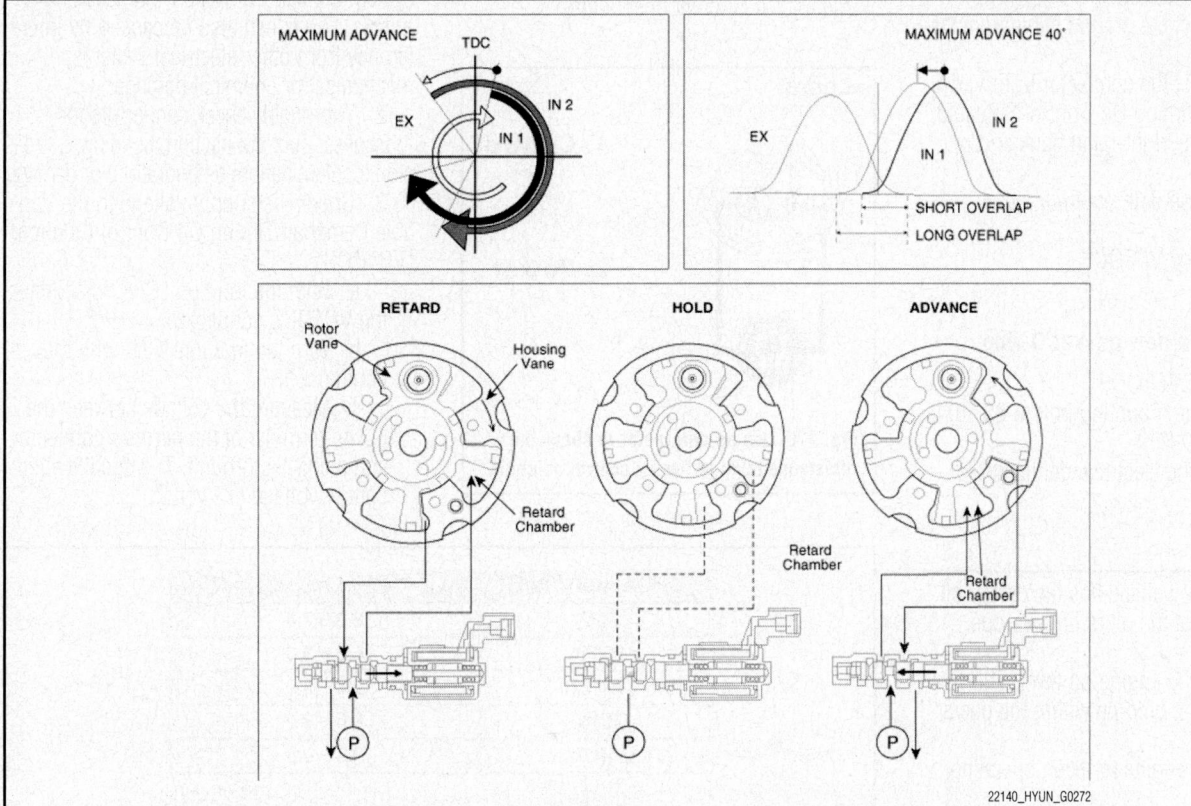

22140_HYUN_G0272

Fig. 174 Continuously Variable Valve Timing (CVVT) system illustrated

Camshaft Timing Oil Control Solenoid (VCTOCS)—also known as the Oil Control Valve (OCV)—based on the signal outputs from the Mass Air Flow (MAF) sensor, Throttle Position Sensor (TPS), and Engine Coolant Temperature (ECT) sensor. The CVVT controller regulates the intake camshaft angle using oil pressure through the VCTOCS. As a result, the relative position between the camshaft and the crankshaft becomes optimal. The engine torque improves, fuel economy improves, and the exhaust emissions decrease within the overall driving conditions.

The CVVT system makes continuous intake valve timing changes based on operating conditions. Intake valve timing is optimized to allow the engine to produce maximum power. Cam angle is advanced to obtain the EGR effect and reduce pumping loss. The intake valve is closed quickly to reduce the entry of the air/fuel mixture into the intake port and improve the changing effect. Reducing the cam advance at idle stabilizes combustion and reduces engine speed. If a malfunction occurs, the CVVT system control is disabled and the valve timing is fixed at the fully retarded position.

REMOVAL & INSTALLATION

1. Before servicing the vehicle, refer to the Precautions Section.
2. Disconnect the ground cable from the battery.
3. Disconnect the connector from Variable Camshaft Timing Oil Control Solenoid (VCTOCS) on the right-hand bank and/or left-hand bank.
4. Remove the bolt retaining the VCTOCS.
5. Remove the VCTOCS.

To install:

➡**Always use a new gasket/O-ring.**

6. Install the VCTOCS.
7. Tighten the mounting bolt to 86–104 inch lbs. (20–39 Nm).
8. Connect the electrical connector to the VCTOCS.

TESTING

1. Ensure the vehicle has the proper oil viscosity and that the oil is not overdue maintenance.
2. Observe the engine oil level. The engine oil level should be within the operating range.
3. Allow the engine to reach operating temperature.
4. Connect a suitable OBD2 scan tool to the Data Link Connector (DLC) inside the vehicle.
5. Increase the engine speed to 1,500 RPM.
6. Command each solenoid to 25 percent. The angle desired parameter should match the solenoid actual parameter.

Component Testing

See Figures 175 and 176.

1. Measure the resistance of each variable camshaft timing oil control solenoid valve assembly. Resistance should read according to the resistance specification table shown.
   a. Ignition "OFF."
   b. Disconnect Variable Camshaft Timing Oil Control Solenoid (VCTOCS) connector.
   c. Measure resistance between terminals 1 and 2 of the VCTOCS connector (component side).
   • Specification: Approx. 6.9–7.9 ohms at 68°F (20°C)
   • If not according to specification, replace the oil control solenoid
2. Check the operation of the VCTOCS.
   a. Start the engine and let it idle.
   b. With VCTOCS connector still disconnected, connect a 12 volt supply to terminal 2 and a ground to terminal 1 of the VCTOCS (component side).

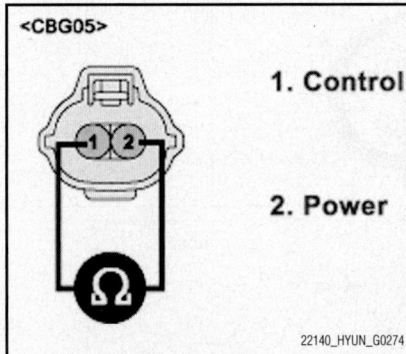

<CBG05>

1. Control

2. Power

22140_HYUN_G0274

Fig. 175 Use an ohmmeter to check the resistance level of the oil control solenoid

• The engine should suddenly run rough at idle or stall out as the VCTOCS abruptly changes the timing
• If the engine runs rough or stalls when the 12 volt supply is connected, check for connection problems to the VCTOCS or problem with the supply side from the ECM
• If no change is noted, the VCTOCS may be defective

3. Check the VCTOCS and oil filter assembly.
   a. Turn the ignition **OFF.**
   b. Check VCTOCS filter for sticking or contamination.
   c. Remove the VCTOCS and visually check the spool column of VCTOCS for contamination.
   d. If a problem is found, clean or replace as necessary.
   e. If no problem is found:
   • Apply a 12 volt supply to terminal 2 and a ground to terminal 1 of the VCTOCS (Component side)
   • Verify that a "clicking" sound is heard when applying the battery voltage
   • If no "clicking" is heard, replace the VCTOCS

Circuit Testing

1. Many malfunctions in the electrical system are caused by poor harness and terminals. Faults can also be caused by interference from other electrical systems, mechanical or chemical damage.
2. Thoroughly check connectors for looseness, poor connection, bending, corrosion, contamination, deterioration, or damage.
3. Check the supply voltage to the Variable Camshaft Timing Oil Control Solenoid (VCTOCS).
   a. With the ignition "OFF," disconnect the VCTOCS connector.
   b. Turn the ignition "ON" with the engine "OFF."
   c. Measure the voltage between the power terminal of the harness connector and a chassis ground. The specification: battery voltage (12 volts).

Temperature °C(°F)	Resistance (Ω)
0(32)	6.2 ~ 7.4
20(68)	6.8 ~ 8.0
40(104)	7.4 ~ 8.6
60(140)	8.0 ~ 9.2
80(176)	8.6 ~ 9.8

22140_HYUN_G0275

Fig. 176 Resistance table for the oil control solenoid compared to temperature

- If the 12 volt supply is not present, check the fuse between the Main Relay and the VCTOCS for an open condition, check for an open in power circuit between the Main Relay and the VCTOCS power circuit
- Make the repairs as necessary

4. Check for a short to ground in the harness.

a. With the ignition "OFF," disconnect the VCTOCS connector.

b. Measure the resistance between the control terminal of the VCTOCS harness connector and a good chassis ground. The specification: Infinite resistance.

- If the measured resistance is within specification, refer to Component Testing for the VCTOCS
- If the measured resistance is not within specification, repair or replace as necessary

Timing Inspection

See Figure 177.

1. With the ignition "OFF," set up an oscilloscope as follows:

a. Channel A (+): terminal 2 of the CKPS (back probe), (-): ground.

b. Channel B (+): terminal 2 of the CMPS (back probe), (-): ground.

2. Start the engine and check for the signal waveform to synchronize with the camshaft sensor and the tooth that is missing. Refer to the sample waveforms in the illustration.

VEHICLE SPEED SENSOR (VSS)

LOCATION

The Vehicle Speed Sensor (VSS) is attached to the output shaft of the transaxle.

OPERATION

The Vehicle Speed Sensor (VSS) is a Hall-effect sensor. The sensor converts the transaxle gear revolutions into pulse signals, which are sent to the ECM.

REMOVAL & INSTALLATION

1. Before servicing the vehicle, refer to the Precautions Section.

2. Raise and support the vehicle safely.

3. Place a drip pan below the Vehicle Speed Sensor (VSS) to catch any spilled fluid when it is removed.

4. Disconnect the VSS connector.

5. Remove the sensor from its mounting.

To install:

6. Installation is the reverse of the removal procedure.

7. Replace any lost transaxle fluid.

TESTING

See Figures 178 and 179.

1. Before servicing the vehicle, refer to the Precautions Section.

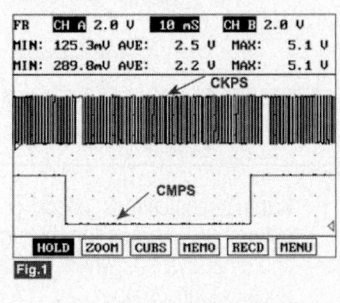

<C12>

1. Power supply

2. Sensor ground

3. Sensor signal

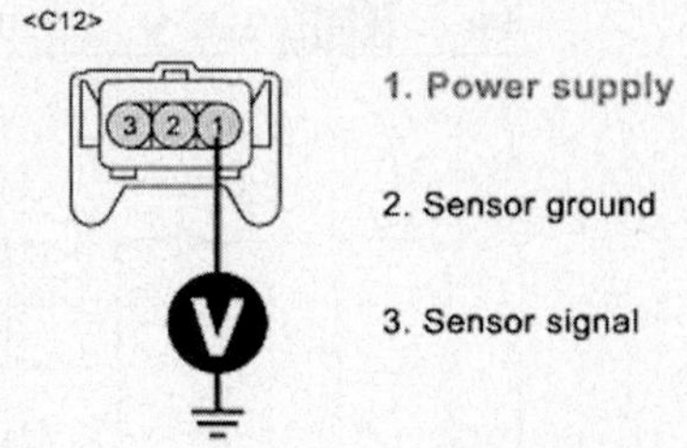

22140_HYUN_G0277

Fig. 178 VSS connector end view

2. Many malfunctions in the electrical system are caused by poor harness and terminal connections. Faults can also be caused by interference from other electrical systems and mechanical or chemical damage.

3. Thoroughly check all connectors (and connections) for looseness, bending, corrosion, contamination, deterioration, or damage.

4. Make repairs as necessary.

5. Turn the ignition **OFF**.

6. Disconnect the VSS connector.

7. Start the engine.

8. Measure the voltage between terminal 1 of the VSS harness connector and a chassis ground. Specification: Battery voltage (approximately 12 volts).

a. If the measured voltage is within specifications, inspect the ground circuit.

b. If the measured voltage is not within specifications, check for an open/short to ground in the power harness.

9. Repair as necessary.

10. Inspect the ground circuit.

a. Turn the ignition **OFF.**

b. Disconnect VSS connector.

c. Measure the resistance between terminal 2 of the VSS harness connector and the chassis ground. The specification: Approx. 0 ohms.

d. If the measured resistance is within specifications, inspect the signal circuit inspection.

e. If the measured resistance is not within specifications, check for an open/short to power in the ground harness. Repair as necessary.

11. Inspect the signal circuit

a. Turn the ignition **OFF.**

b. Connect a scan tool to the Data Link Connector (DLC).

c. Start the engine and select "SCOPEMETER FUNCTION" on the scan tool.

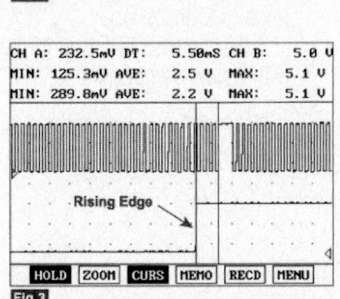

Fig 1) The square wave signal should be smooth and without any distortion.
Fig.2,3) The CMPS falling(rising) edge is coincided with 3rd~5th tooth of the CKP from one longer signal(missing tooth)

22140_HYUN_G0276

Fig. 177 Using waveform synchronization to check the oil control solenoid as it relates to camshaft and crankshaft timing

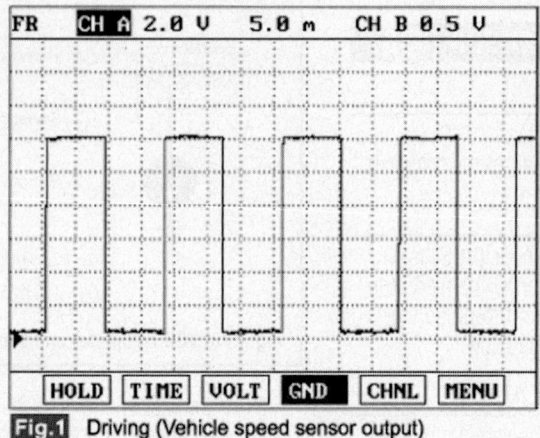

Fig.1 Driving (Vehicle speed sensor output)

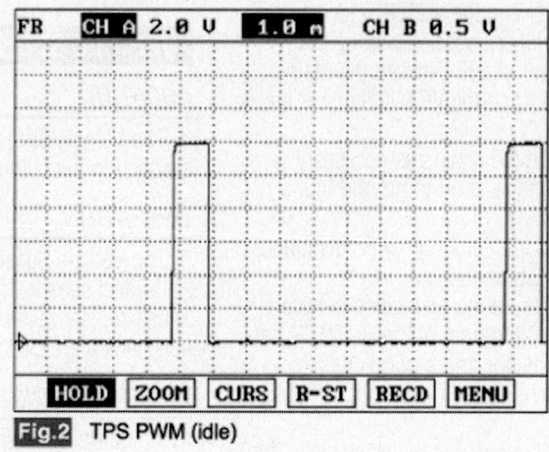

Fig.2 TPS PWM (idle)

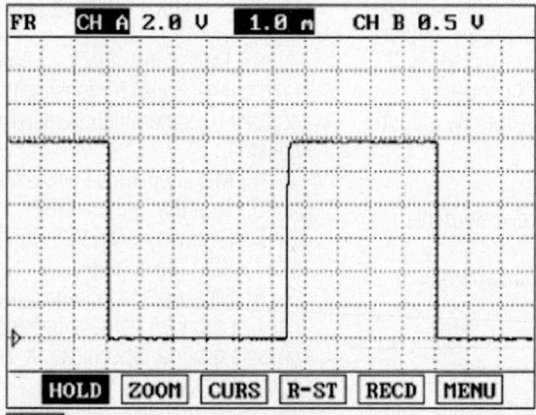

Fig.3 TPS PWM (acceleration)

Fig 1) Driving : Vehicle speed sensor output 8.0~12.5V(HIGH SIGNAL) or 0.2~0.5V(LOW SIGNAL), as the vehicle moves,
the 50% of digital duty wave is ouput. (As vehicle speed increase, Hz increase)
Fig 2) Engine RPM signal at idle (TPS PWM duty 10%).
Fig 3) Engine RPM signal at acceleration (TPS PWM duty 45%)

22140_HYUN_G0278

Fig. 179 Waveform illustration of output signal

d. Drive the vehicle and measure the output signal between terminal 3 of the EPS CM harness connector and chassis ground. Specification: see waveform illustration.

e. If the VSS output signal is within specifications, substitute with a known-good EPS CM and check for proper operation.

f. If the problem is corrected, replace the EPS CM.

g. If the VSS output signal is not within specifications, check for an open/short to ground in the signal circuit and other systems which use the VSS. Repair as necessary.

h. If the VSS is found faulty, replace with a new component.

FUEL GASOLINE FUEL INJECTION SYSTEM

FUEL SYSTEM SERVICE PRECAUTIONS

Safety is the most important factor when performing, not only fuel system maintenance, but any type of maintenance. Failure to conduct maintenance and repairs in a safe manner may result in serious personal injury or death. Maintenance and testing of the vehicle's fuel system components can be accomplished safely and effectively by adhering to the following rules and guidelines.

• To avoid the possibility of fire and personal injury, always disconnect the negative battery cable unless the repair or test procedure requires that battery voltage be applied.

• Always relieve the fuel system pressure prior to disconnecting any fuel system component (injector, fuel rail, pressure regulator, etc.), fitting, or fuel line connection. Exercise extreme caution whenever relieving fuel system pressure to avoid exposing skin, face, and eyes to fuel spray. Please be advised that fuel under pressure may penetrate the skin or any part of the body that it contacts.

• Always place a shop towel or cloth around the fitting or connection prior to loosening to absorb any excess fuel due to spillage. Ensure that all fuel spillage (should it occur) is quickly removed from engine surfaces. Ensure that all fuel soaked cloths or towels are deposited into a suitable waste container.

• Always keep a dry chemical (Class B) fire extinguisher near the work area.

• Do not allow fuel spray or fuel vapors to come into contact with a spark or an open flame.

• Always use a back-up wrench when loosening and tightening fuel line connection fittings. This will prevent unnecessary stress and torsion to fuel line piping.

• Always replace worn fuel fitting O-rings with new. Do not substitute fuel hose or equivalent where fuel pipe is installed.

Before servicing the vehicle, make sure to refer to the precautions in the beginning of this section as well.

RELIEVING FUEL SYSTEM PRESSURE

1. Before servicing the vehicle, refer to the Precautions Section.
2. Remove or disconnect the following:

• Rear seat cushion
• Access panel
• Fuel pump module connector
3. Start the engine and allow it to run until it stalls.
4. Turn the ignition switch to the **OFF-** position.
5. Disconnect the negative battery cable.
6. Attach the fuel pump harness connector.

FUEL FILTER

REMOVAL & INSTALLATION

The fuel delivery system integrates the fuel filter with the in-tank fuel pump. To service this filter, remove the fuel pump. Refer to Fuel Pump, removal & installation.

FUEL PUMP

REMOVAL & INSTALLATION

See Figures 180 and 181.

1. Before servicing the vehicle, refer to the Precautions Section.
2. Relieve the fuel system pressure. Refer to Relieving Fuel System Pressure.
3. Remove the second seat.
4. Open the carpet over the fuel pump.
5. Remove the service cover from on top of the fuel pump.
6. Disconnect the fuel feed line (A) and canister hoses (B).
7. Unscrew the fuel pump mounting bolts (C) and remove the fuel pump assembly.

To install:
8. Install the fuel pump assembly.
9. Install the fuel pump mounting bolts (C). Tighten the bolts/nuts to 17–26 inch lbs. (2–3 Nm).

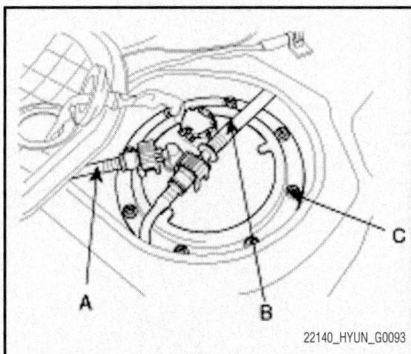

Fig. 180 Disconnect the fuel feed line (A) and canister hoses (B). Unscrew the fuel pump mounting bolts (C)

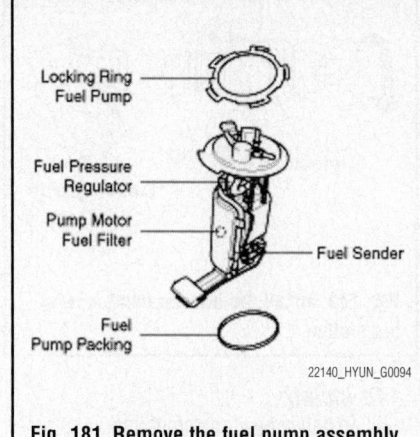

22140_HYUN_G0094

Fig. 181 Remove the fuel pump assembly

10. Connect the fuel feed line (A) and canister hoses (B).
11. Start the vehicle and check for leaks.
12. Install the second seat.

FUEL PRESSURE REGULATOR

REMOVAL & INSTALLATION

The fuel pressure regulator is built into the fuel pump. Please refer to Fuel Pump, removal & installation.

FUEL RAIL & INJECTORS

REMOVAL & INSTALLATION

See Figure 182.

1. Before servicing the vehicle, refer to the Precautions Section.
2. Relieve the fuel system pressure. Refer to Relieving Fuel System Pressure.
3. Remove or disconnect the following:

• Negative battery cable
• Air breather hose from the throttle body
• Throttle cable
• Engine coolant hose and throttle body
• Positive Crankcase Ventilation (PCV) valve and brake booster vacuum hose
• Vacuum hose connector
• Injector cover
• High pressure fuel hose
• Fuel injector harness connector
• Delivery pipe with the injectors and the pressure regulator as an assembly
• Injector from the delivery pipe
4. Remove and discard the injector O-ring and grommet.

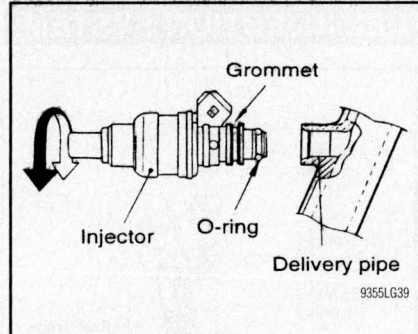

Fig. 182 Install the injector using a twisting motion

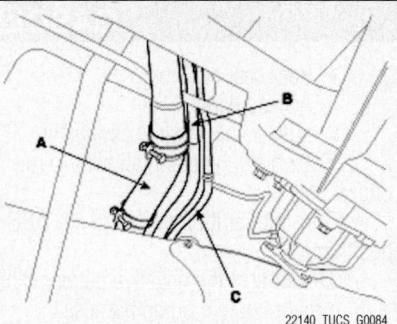

Fig. 183 Disconnect the fuel filler hose (A), fuel leveling hose (B), and ventilation hose (C)

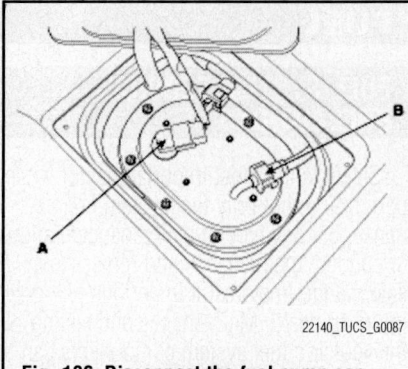

Fig. 186 Disconnect the fuel pump connector (A) and fuel feed hose (B)

To install:

5. Install a new grommet and O-ring.

6. Apply a coating of spindle oil or gasoline to the injector O-ring.

7. Install the injector into the delivery pipe while turning the injector left and right making sure the injector turns smoothly. If the injector does not turn smoothly, check for a jammed O-ring, remove the injector and reinsert it again.

8. Install or connect the following:
   • Delivery pipe (fuel rail) and injector assembly. Torque the bolts to 84–132 inch lbs. (10–15 Nm)
   • Intake manifold stay and tighten the bolts to 13–18 ft. lbs. (18–25 Nm)
   • Fuel injector harness connector
   • High pressure fuel hose
   • Injector cover
   • Vacuum hose connector
   • PCV valve and brake booster vacuum hose
   • Throttle body and engine coolant hose
   • Throttle cable
   • Air breather hose from the throttle body
   • Negative battery cable

9. Start the engine and check for leaks and for proper operation.

FUEL TANK

REMOVAL & INSTALLATION

See Figures 183 through 188.

1. Before servicing the vehicle, refer to the Precautions Section.

2. Relieve the fuel system pressure. Refer to Relieving Fuel System Pressure.

3. Raise and safely support the vehicle.

4. Remove the front and main muffler assembly.

5. Remove the propeller-shaft (4WD only).

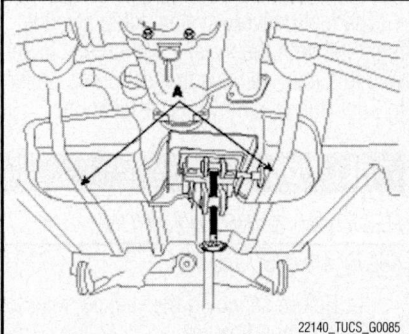

Fig. 184 Support the fuel tank with a jack and then remove the 2 fuel tank bands (A)

6. Disconnect the fuel filler hose (A), fuel leveling hose (B), and ventilation hose (C).

7. Support the fuel tank with a jack and then remove the 2 fuel tank bands (A).

8. Remove the parking brake mounting bolts.

9. Slowly lower the fuel tank just enough to provide work space to disconnect the fuel pump connector (A) and fuel feed hose (B).

10. Disconnect the sub fuel sender connector (A).

11. Remove the fuel tank.

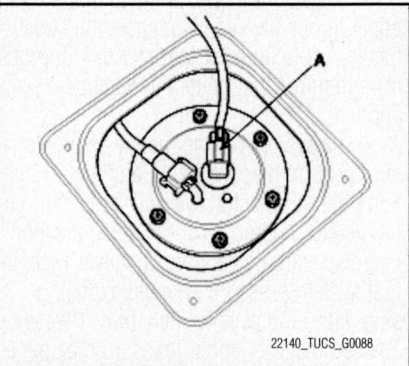

Fig. 187 Disconnect the sub fuel sender connector (A)

To install:

12. Installation is the reverse of the removal procedure.

13. Tighten the fuel tank band mounting nuts to: 29–40 ft. lbs. (39–54 Nm).

IDLE SPEED

ADJUSTMENT

Idle speed is maintained by the Electronic Control Module (ECM). No adjustment is necessary or possible.

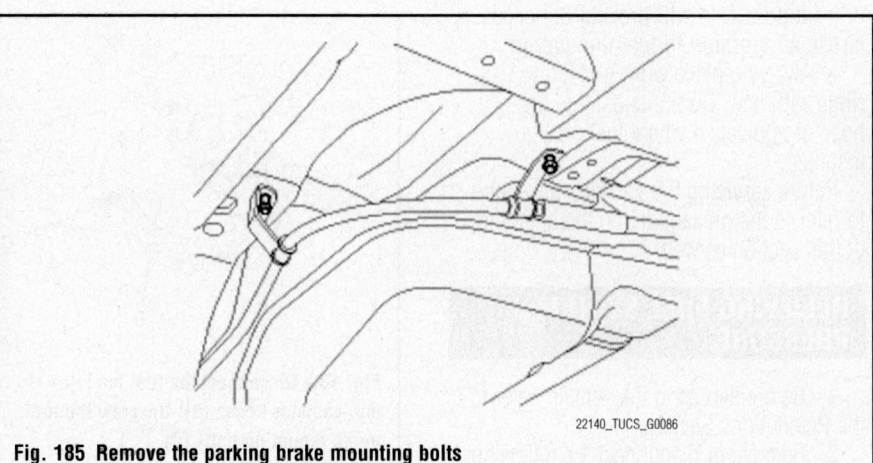

Fig. 185 Remove the parking brake mounting bolts

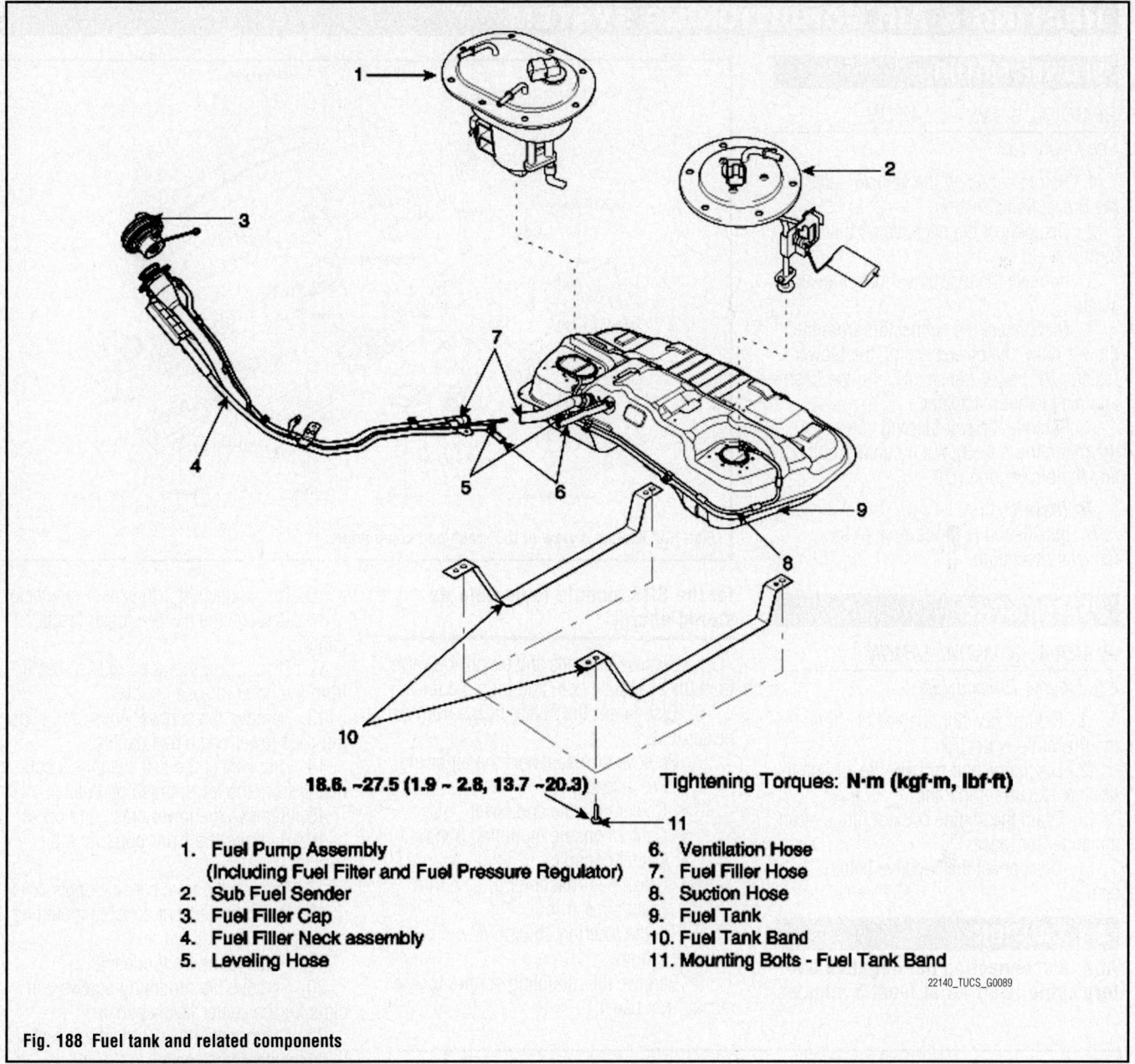

18.6. ~27.5 (1.9 ~ 2.8, 13.7 ~20.3)   Tightening Torques: **N·m (kgf·m, lbf·ft)**

1. Fuel Pump Assembly
   (Including Fuel Filter and Fuel Pressure Regulator)
2. Sub Fuel Sender
3. Fuel Filler Cap
4. Fuel Filler Neck assembly
5. Leveling Hose

6. Ventilation Hose
7. Fuel Filler Hose
8. Suction Hose
9. Fuel Tank
10. Fuel Tank Band
11. Mounting Bolts - Fuel Tank Band

22140_TUCS_G0089

Fig. 188 Fuel tank and related components

THROTTLE BODY

REMOVAL & INSTALLATION

1. Before servicing the vehicle, refer to the Precautions Section.
2. Turn the ignition **OFF**.
3. Remove the engine cover.
4. Remove the throttle body electrical connector.
5. Remove the throttle body bolts.
6. Remove the throttle body and gasket.

To install:

7. Clean the throttle body gasket mating surfaces.
8. Install the throttle body and NEW gasket.
9. Install the throttle body bolts and tighten to 21 ft. lbs. (28 Nm).
10. Install the throttle body electrical connector.
11. Install the engine cover.

HEATING & AIR CONDITIONING SYSTEM

BLOWER MOTOR

REMOVAL & INSTALLATION

See Figure 189.

1. Before servicing the vehicle, refer to the Precautions Section.

2. Disconnect the negative (-) battery terminal.

3. Remove the instrument panel lower section.

4. Disconnect the connectors from the blower relay, the blower motor, the blower resistor (or power transistor), and the fresh and recirculation actuator.

5. Remove the self-tapping screws (A), the mounting nut (B), the mounting bolt (C), and the blower unit (D).

To install:

6. Installation is the reverse of the removal procedure.

HEATER CORE

REMOVAL & INSTALLATION

See Figures 190 and 191.

1. Before servicing the vehicle, refer to the Precautions Section.

2. Discharge and recover the air conditioning system refrigerant.

3. Drain the engine coolant into a clean container for reuse.

4. Disconnect the negative battery cable.

❈❈ CAUTION

After disconnecting the negative battery cable, wait for at least 3 minutes

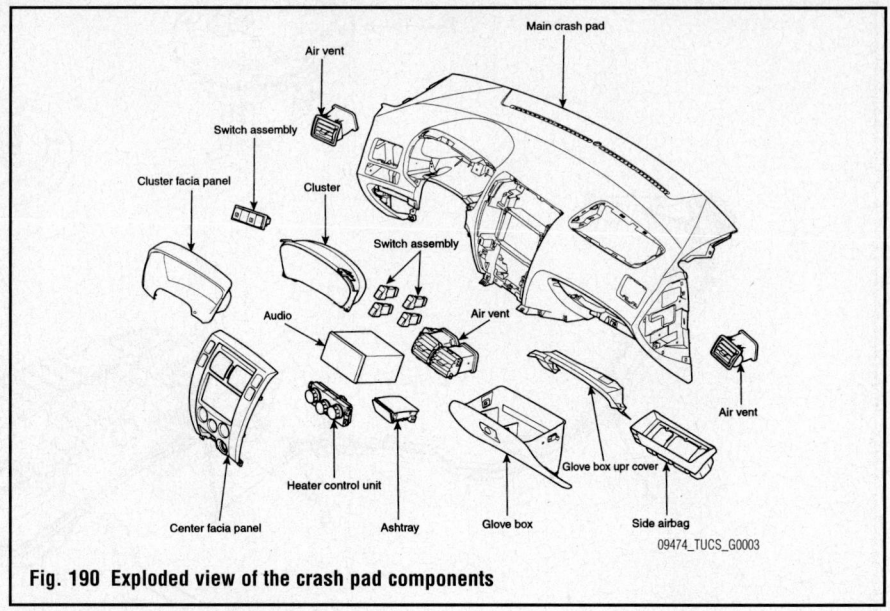

Fig. 190 Exploded view of the crash pad components

09474_TUCS_G0003

for the SRS module to deplete its stored energy.

5. Remove the bolts and expansion valve from the evaporate core and plug the lines.

6. Disconnect the heater hoses from the heater unit.

7. Remove or disconnect the following:
• Front seats
• Center console end cover
• Center console mounting bolts
• Center console
• Crash pad (dashboard) side trim
• Front pillar trim

8. Tilt the steering column down to the lowest position.

9. Remove the mounting screws for the cluster trim panel.

10. Disconnect the trip sensor connector.

11. Remove the gauge cluster fascia panel.

12. Disconnect the hood release cable from the hood release handle.

13. Remove the screws, bolts, and clips from the lower crash pad panels.

14. Disconnect the self-diagnosis connector from the lower crash pad panel.

15. Remove the lower crash pad panel.

16. Remove the front console side trim.

17. Remove the front console upper cover.

18. Remove the front console mounting screws.

19. Remove the front console.

20. Remove the mounting screws and clips for the center fascia panel.

21. Disconnect the electrical connectors to the center fascia panel.

22. Remove the center fascia panel.

23. Remove the radio and disconnect the electrical connectors.

24. Remove the heater control unit.

25. Remove the gauge cluster mounting screws.

26. Disconnect the cluster connectors and remove the gauge cluster.

27. Disconnect the air damper wire and guide from the glove box.

28. Disengage the hinge pins and remove the glove box.

29. Remove the driver airbag module by removing the 2 mounting bolts.

30. Remove the steering wheel center lock nut.

31. Align the marks on the steering shaft and wheel.

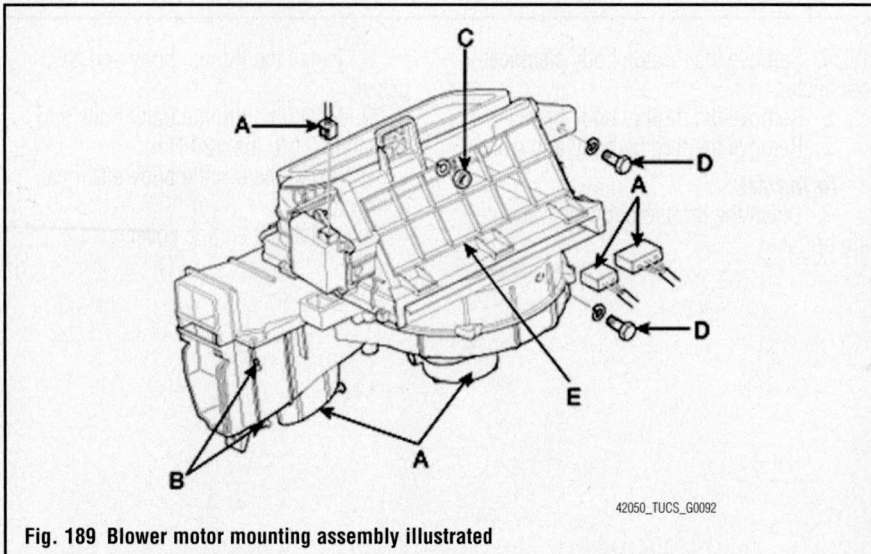

Fig. 189 Blower motor mounting assembly illustrated

42050_TUCS_G0092

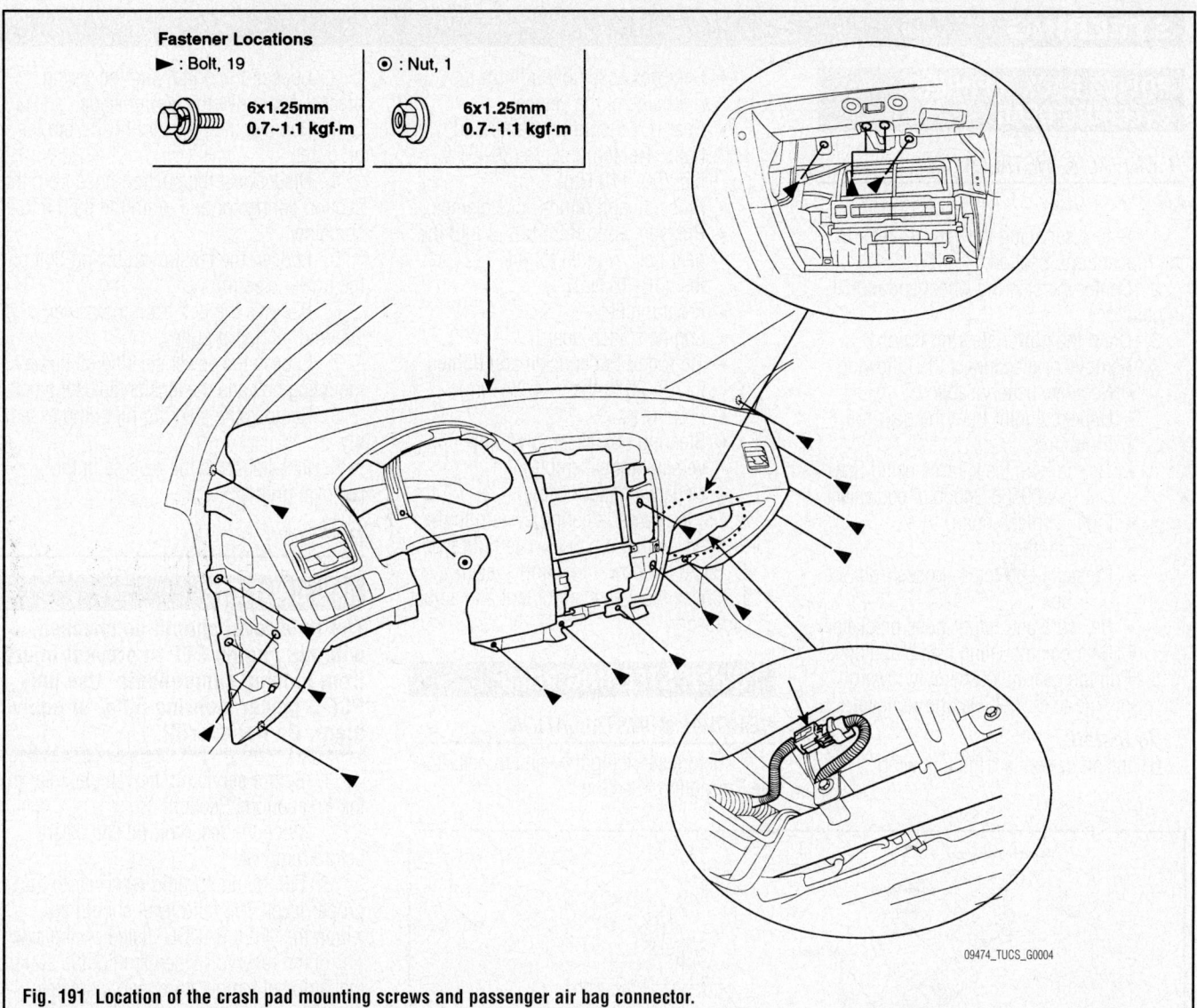

Fastener Locations

► : Bolt, 19

6x1.25mm
0.7~1.1 kgf·m

⊙ : Nut, 1

6x1.25mm
0.7~1.1 kgf·m

09474_TUCS_G0004

Fig. 191 Location of the crash pad mounting screws and passenger air bag connector.

32. Install Special Tool 09561-11002 to remove the steering wheel.

> ❊❊ **WARNING**
>
> **Do not hammer on the steering wheel to remove. Damage to the steering column may occur.**

33. Remove the upper and lower steering column shrouds.

34. Disconnect the 2 lower tightening bolts and remove the lower crash pad.

35. Disconnect the passenger airbag connector.

36. Remove the main crash pad.

37. Remove the cross member.

38. Remove the heater and evaporator unit.

39. Remove the side bracket and lower cover for the unit to access the heater core.

To install:

40. Installation is the reverse of the removal procedure.

41. Observe the following:

• Ensure the crash pad fits on the guide pins correctly and no wiring harnesses are pinched

• Any damaged trim clips must be replaced

42. Fill the engine coolant with the proper amount and type of fluid. Check for leaks.

43. Connect the negative battery cable.

44. Evacuate and recharge the A/C system. Check for leaks.

STEERING

POWER RACK & PINION STEERING GEAR

REMOVAL & INSTALLATION

See Figure 192.

1. Before servicing the vehicle, refer to the Precautions Section.
2. Center the steering wheel and lock it in position.
3. Drain the power steering fluid.
4. Remove or disconnect the following:
   - Negative battery cable.
   - Universal joint from the gear box
   - Front tires
   - Tie rod from the knuckle using Special Tool 09568-34000 or equivalent
   - Engine splash guard
   - Front muffler
   - Pressure and return hoses from the gear box
   - Pressure and return hose brackets
   - Gear box mounting clamp and bolts
5. Pull the gear box assembly toward the right side of the vehicle and remove.

To install:

6. Install or connect the following:
   - Gear box assembly in from the right side of the vehicle
   - Gear box mounting clamp and bolts. Tighten bolts to: 66–81 ft. lbs. (90–110 Nm)
   - Pressure and return hose clamps
   - Pressure and return hoses into the gear box. Tighten to: 84–132 inch lbs. (10–16 Nm)
   - Front muffler
   - Engine splash guard
   - Tie rod to the knuckle and tighten to: 18–25 ft. lbs. (24–34 Nm)
   - Front tires
   - Steering box assembly to the universal joint assembly
   - Negative battery cable
7. Fill the power steering system to the correct level with the appropriate fluid type.
8. Bleed the power steering system.
9. Check the wheel alignment and adjust as necessary.

POWER STEERING PUMP

REMOVAL & INSTALLATION

1. Before servicing the vehicle, refer to the Precautions Section.

2. Loosen the bolt fixing the wiring bracket and move the wiring aside.
3. Remove the pressure hose from the oil pump.
4. Disconnect the suction hose from the suction connector and drain the fluid into a container.
5. Loosen the tension adjusting bolt on the power steering belt.
6. Remove the belt from the power steering oil pump pulley.
7. Loosen the power steering oil pump mounting bolt and the tension adjusting bolt.
8. Remove the steering oil pump assembly.
9. Installation is the reverse of the removal procedure.

BLEEDING

✳✳ CAUTION

The fluid level should be checked with the engine OFF to prevent injury from moving components. Use only PSF-3 power steering fluid, or equivalent. Do not overfill.

1. Before servicing the vehicle, refer to the Precautions Section.
2. Wipe the reservoir fill cap clean before removal.
3. Fill the pump fluid reservoir to the proper level. The fluid level should be within the "FILL RANGE" listed on the exterior of the reservoir when the fluid is at normal ambient temperature, approximately 70–80°F (21–27°C).
4. Let the fluid settle in the system for at least 2 minutes.
5. Start the engine and let it run for a few seconds. Then turn the engine **OFF**.
6. Add fluid if necessary. Repeat the above procedure until the fluid level remains constant after running the engine.
7. Raise the front wheels off the ground.
8. Start the engine. Slowly turn the steering wheel right and left, lightly contacting the wheel stops.
9. Turn the engine off. Check the fluid level and add power steering fluid if necessary.
10. Start the engine. Lower the vehicle and turn the steering wheel slowly from lock to lock.
11. Stop the engine. Check the fluid level and refill as required.
12. If the fluid is extremely foamy, allow the vehicle to stabilize a few minutes, then repeat the above procedure.

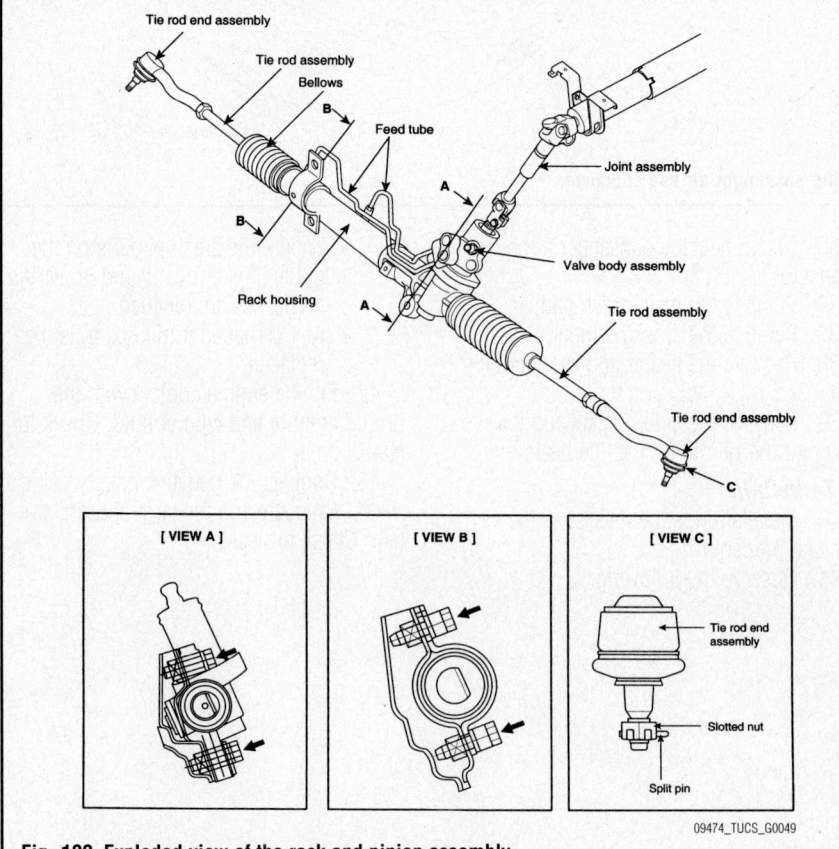

09474_TUCS_G0049

Fig. 192 Exploded view of the rack and pinion assembly

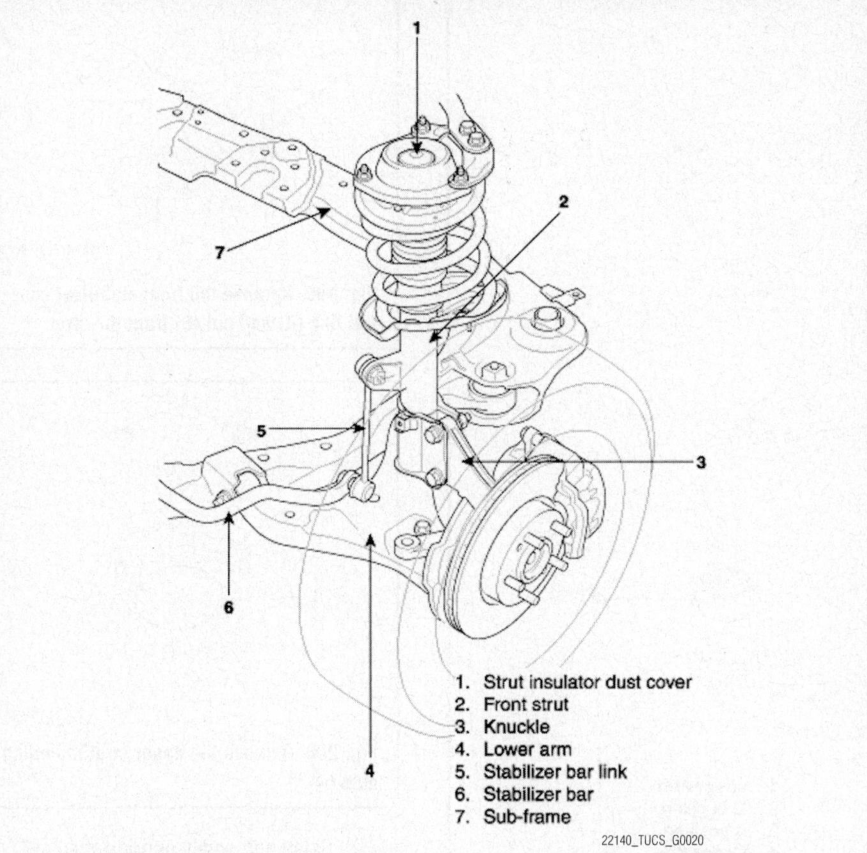

SUSPENSION

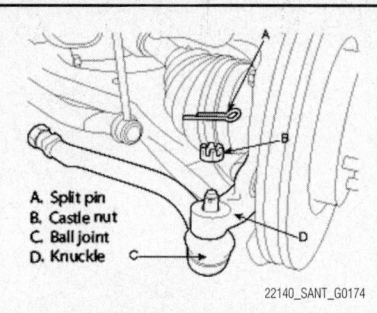
Fig. 195 Remove the split pin and castle nut. Disconnect the ball joint from the knuckle

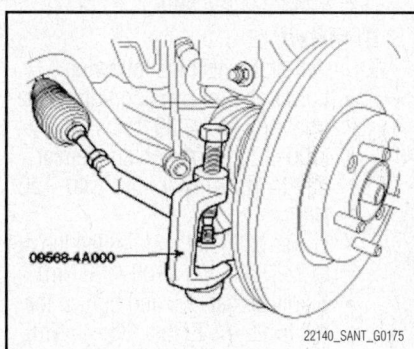

1. Strut insulator dust cover
2. Front strut
3. Knuckle
4. Lower arm
5. Stabilizer bar link
6. Stabilizer bar
7. Sub-frame

Fig. 193 View of front suspension components

See Figure 193.

CONTROL LINKS

REMOVAL & INSTALLATION

See Figure 194.

1. Before servicing the vehicle, refer to the Precautions Section.
2. Raise and safely support the vehicle.
3. Remove the front wheel and tire from the front hub.

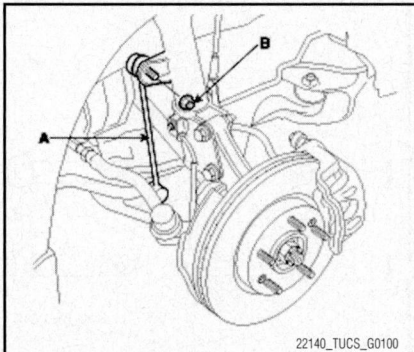
Fig. 194 Stabilizer control link and mounting nuts shown

✳✳ WARNING

Be careful not to damage the hub bolts when removing the front wheel and tire.

4. Remove the stabilizer bar control link (A) by removing the upper (B) and lower mounting nuts.
5. Remove the stabilizer bar control link.

To install:
6. Install the stabilizer bar control link.
7. Tighten the stabilizer bar control link upper and lower mounting nuts to 74–89 ft. lbs. (100–120 Nm).
8. Install the front wheels. Tighten the wheel lug nuts to 65–80 ft. lbs. (90–110 Nm).

LOWER BALL JOINT

REMOVAL & INSTALLATION

See Figures 195 and 196.

1. Before servicing the vehicle, refer to the Precautions Section.
2. Raise and safely support the vehicle.
3. Remove the front wheel and tire from the front hub.

FRONT SUSPENSION

A. Split pin
B. Castle nut
C. Ball joint
D. Knuckle

✳✳ WARNING

Be careful not to damage the hub bolts when removing the front wheel and tire.

4. Remove the tie rod end ball joint from the knuckle.
   a. Remove the split pin (A).
   b. Remove the castle nut (B).
   c. Disconnect the ball joint (C) from knuckle (D) using the special tool (09568-4A000).

➡**Apply a few drops of oil to the special tool (boot contact part).**

To install:
5. Install the tie rod end ball joint (C) to the knuckle (D).
6. Install the castle nut (B) and the split pin (A). Tighten the nut to: 17–25 ft. lbs. (24–33 Nm).
7. Install the front wheel and tire to the front hub.

LOWER CONTROL ARM

REMOVAL & INSTALLATION

See Figure 197.

Fig. 196 Using the special tool (09568-4A000) to remove the ball joint

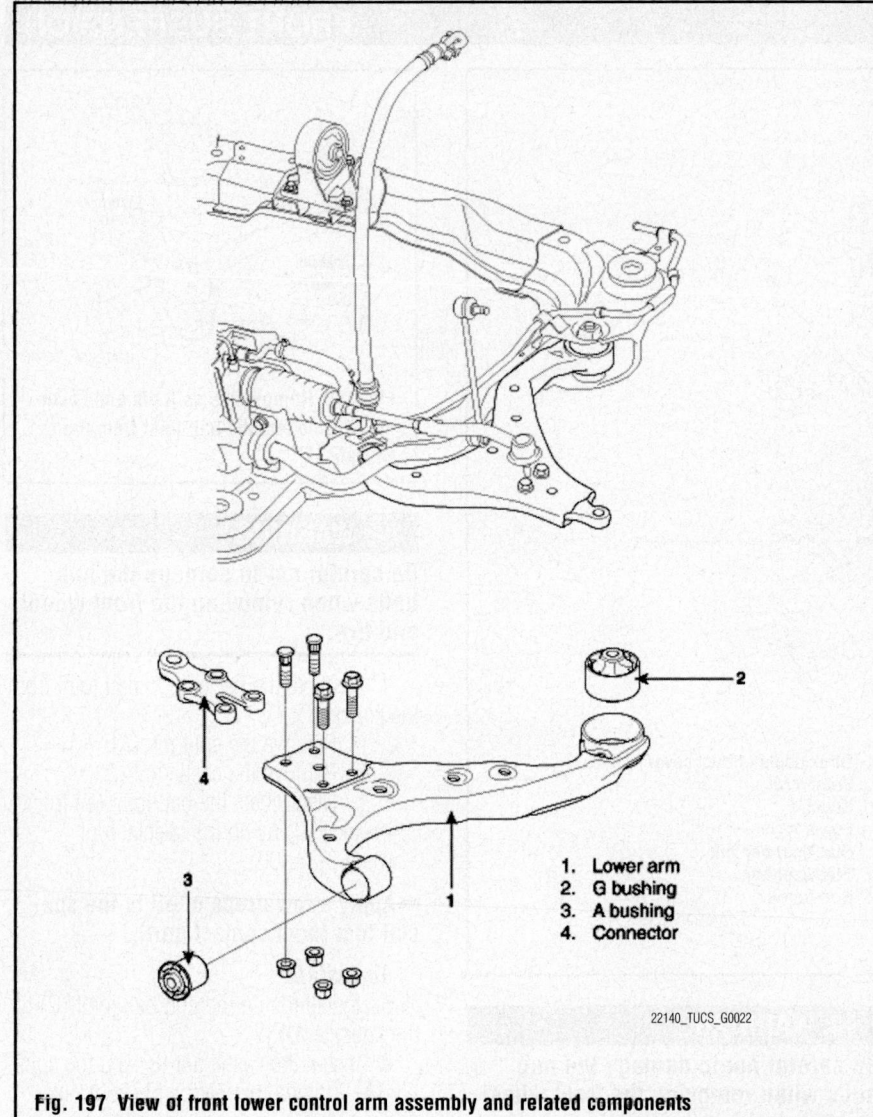

Fig. 197 View of front lower control arm assembly and related components

1. Lower arm
2. G bushing
3. A bushing
4. Connector

22140_TUCS_G0022

1. Before servicing the vehicle, refer to the Precautions Section.

2. Remove or disconnect the following:
- Front wheel
- Lower ball joint
- Lower control arm mounting bolts
- Lower control arm

To install:

3. Install or connect the following:
- Lower control arm and tighten the front bolt to 74–89 ft. lbs. (100–120 Nm). Tighten the rear bolt to 103–118 ft. lbs. (140–160 Nm)
- Lower ball joint and tighten the nut to 74–89 ft. lbs. (100–120 Nm)
- Stabilizer bar link and tighten the nut to 25–33 ft. lbs. (35–45 Nm)
- Front wheel

4. Check and/or adjust the wheel alignment.

STRUT

REMOVAL & INSTALLATION

See Figures 198 through 201.

1. Before servicing the vehicle, refer to the Precautions Section.

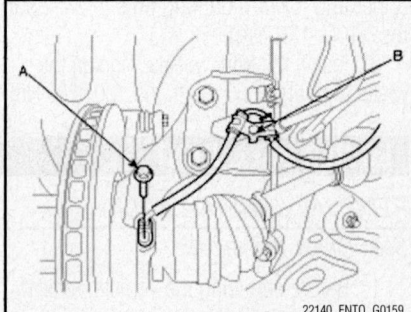

Fig. 198 Remove the speed sensor (A) and wire bracket bolts (B) from the front steering knuckle

22140_ENTO_G0159

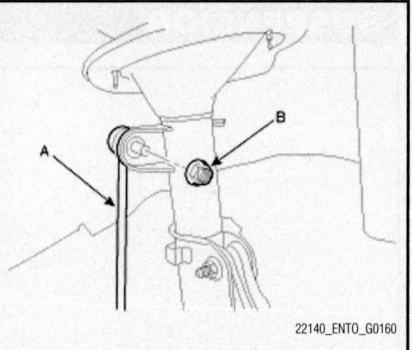

22140_ENTO_G0160

Fig. 199 Remove the front stabilizer control link (A) and nut (B) from the strut

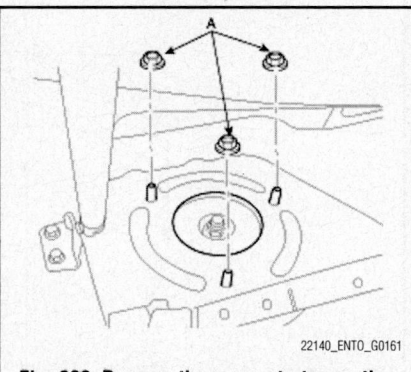

22140_ENTO_G0161

Fig. 200 Remove the upper strut mounting nuts (A)

2. Raise and safely support the vehicle.

3. Remove the front wheels and tires.

4. Remove the brake hose bracket bolts from the strut assembly.

5. Remove the speed sensor (A) and wire bracket bolts (B) from the front steering knuckle.

6. Make an alignment marking on the camber adjusting bolt and strut for installation alignment approximation later.

7. Remove the front stabilizer control link (A) and nut (B) from the strut.

8. Remove the upper strut mounting nuts (A).

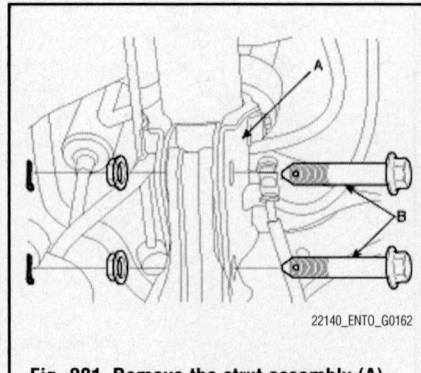

22140_ENTO_G0162

Fig. 201 Remove the strut assembly (A) and bolts (B) from its mounting

9. Remove the front strut mounting bolts from the knuckle.

10. Remove the strut assembly (A) and bolts (B) from its mounting in the steering knuckle.

To install:

11. Install the strut upper mounting nuts and tighten to: 33–44 ft. lbs. (45–60 Nm).

12. Match the alignment marks made during removal and install the front strut assembly bolts to the front knuckle. Tighten the lower mounting bolts to: 103–118 ft. lbs. (140–160 Nm).

13. Install the stabilizer bar control link mounting nut to the strut assembly and tighten to: 74–89 ft. lbs. (100–120 Nm).

14. Install the speed sensor and wire bracket bolts. Tightening torque: 60–96 inch lbs. (7–11 Nm).

15. Install the brake hose bracket bolt to the axle assembly.

16. Install the wheel and the tire to the front hub. Tightening torque: 65–80 ft. lbs. (90–110 Nm).

✳✳ WARNING

Be careful not to damage the hub bolts when installing the front wheel and tire.

17. Check the wheel alignment and adjust as necessary.

OVERHAUL

See Figure 202.

1. Before servicing the vehicle, refer to the Precautions Section.

2. Remove the strut from the vehicle and attach Service Tool 09546-2600 or another suitable spring compressor.

3. Compress the coil spring.

4. Remove the self-locking nut from the strut.

5. Remove the insulator, spring seat, coil spring, and dust cover.

6. Inspect the insulator for wear and damage, replace as required.

7. Check the rubber parts for damage or deterioration, replace as required.

8. Install the spring lower pad so that the protrusions fit the holes in the spring lower seat.

9. Compress the spring, using the spring compressor tool.

10. Install the compressed spring over the shock absorber.

➥**There are 2 color identification marks on the coil spring, one indicates model option and the other indicates load classification. Install the coil**

spring with the identification mark directed toward the steering knuckle.

11. After fully extending the piston rod, install the spring upper seat and insulator assembly.

12. After correctly seating the upper and lower ends of the coil spring in the upper and lower spring grooves, tighten the NEW self-locking nut temporarily.

13. Remove the spring compression tool. Tighten the self-locking nut to 43–51 ft. lbs. (60–70 Nm).

14. Install the strut to the vehicle.

15. Check the wheel alignment and adjust as necessary.

STEERING KNUCKLE

REMOVAL & INSTALLATION
See Figure 203.

1. Before servicing the vehicle, refer to the Precautions Section.

2. Raise and safely support the vehicle.

3. Remove the front wheels and tires.

4. Remove the brake rotor. Refer to

Front Disc Brakes, Rotor, removal & installation.

5. Remove the lower ball joint. Refer to Lower Ball Joint, removal & installation.

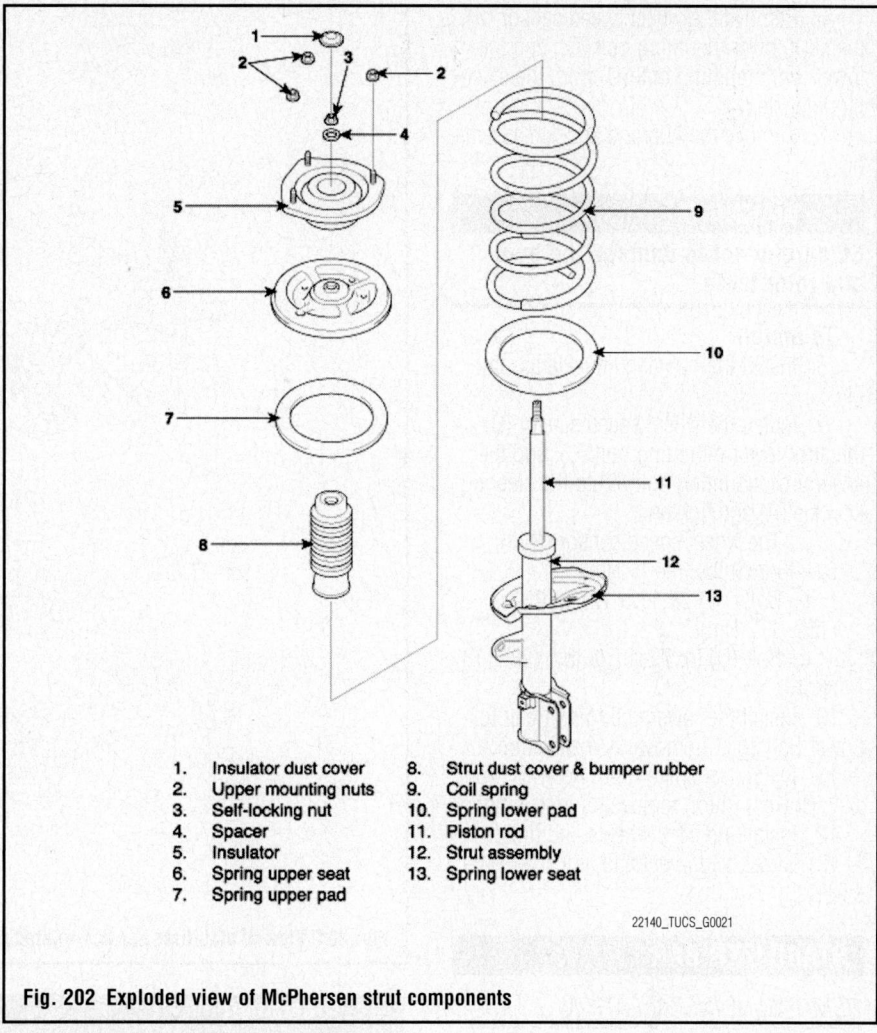

1.	Insulator dust cover	8.	Strut dust cover & bumper rubber
2.	Upper mounting nuts	9.	Coil spring
3.	Self-locking nut	10.	Spring lower pad
4.	Spacer	11.	Piston rod
5.	Insulator	12.	Strut assembly
6.	Spring upper seat	13.	Spring lower seat
7.	Spring upper pad		

22140_TUCS_G0021

Fig. 202 Exploded view of McPhersen strut components

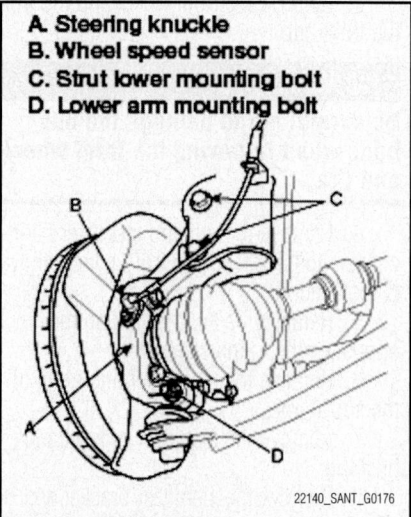

A. Steering knuckle
B. Wheel speed sensor
C. Strut lower mounting bolt
D. Lower arm mounting bolt

22140_SANT_G0176

Fig. 203 Remove the wheel speed sensor, strut lower mounting bolt, and lower arm mounting bolt from the steering knuckle

6. Remove the wheel speed sensor (B), the strut lower mounting bolt (C), and the lower arm mounting bolt (D) from the steering knuckle (A).

7. Remove the hub and knuckle assembly.

✳✳ WARNING

Be careful not to damage the boot and rotor teeth.

To install:

8. Install the hub and knuckle assembly.

9. Install the wheel speed sensor (B), the strut lower mounting bolt (C), and the lower arm mounting bolt (D) to the steering knuckle (A) and tighten:

   a. The wheel speed sensor (B) to: 61–96 inch lbs. (7–11 Nm).

   b. Bolts (C) to: 112–127 ft. lbs. (152–172 Nm).

   c. Bolt (D) to: 72–87 ft. lbs. (98–118 Nm).

10. Install the lower ball joint. Refer to Lower Ball Joint, removal & installation.

11. Install the brake rotor. Refer to Front Disc Brakes, Rotor, removal & installation.

12. Install the front wheels and tires.

13. Check alignment and adjust as necessary.

STABILIZER BAR

REMOVAL & INSTALLATION

See Figure 204.

1. Before servicing the vehicle, refer to the Precautions Section.

2. Raise the front of the vehicle, and make sure it is securely supported.

3. Remove the front wheel and tire from the front hub.

✳✳ WARNING

Be careful not to damage the hub bolts when removing the front wheel and tire.

4. Remove the nut and stabilizer bar control link. Refer to Control Links, removal & installation.

5. Remove the control link on the opposite side in the same way.

6. Remove the rear mounting bolts of the sub-frame.

7. Remove the stabilizer bracket and bushing.

8. Remove the stabilizer bracket and bushing on the opposite side in the same way.

9. Remove the stabilizer bar.

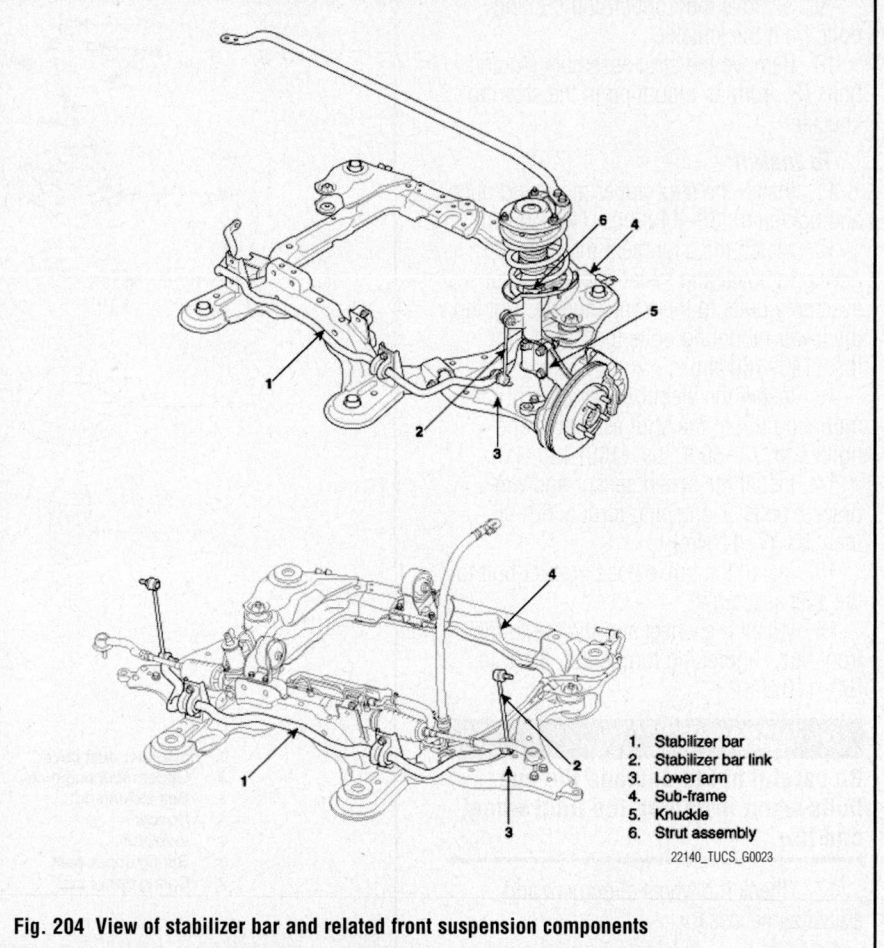

1. Stabilizer bar
2. Stabilizer bar link
3. Lower arm
4. Sub-frame
5. Knuckle
6. Strut assembly

22140_TUCS_G0023

Fig. 204 View of stabilizer bar and related front suspension components

✳✳ WARNING

Be careful not to damage the power steering pressure tube.

To install:

10. Installation is the reverse of the removal procedure.

11. Tighten the stabilizer bracket bolts to: 37–48 ft. lbs. (50–65 Nm).

12. Tighten the sub-frame mounting bolts to: 68–86 ft. lbs. (95–120 Nm).

13. Tighten the stabilizer bar control link upper and lower mounting nuts to 72–87 ft. lbs. (100–120 Nm).

14. Tighten the wheel nuts to: 65–80 ft. lbs. (90–110 Nm).

WHEEL HUB AND BEARING (SEALED UNIT)

REMOVAL & INSTALLATION

See Figures 205 through 212.

1. Before servicing the vehicle, refer to the Precautions Section.

2. Remove or disconnect the following:

   • Front wheel

   • Wheel Speed Sensor (WSS) from the knuckle

   • Brake caliper and suspend it to one side using wire

   • Split pin and nut from the axle

   • Strut from the knuckle

   • Tie rod end from the knuckle

   • Lower ball joint bolt

   • Axle shaft from the knuckle using a plastic hammer

   • Brake disc

   • Knuckle assembly

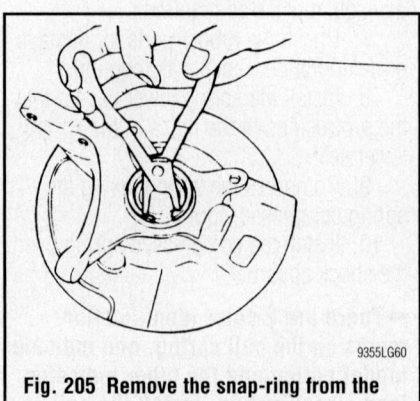

9355LG60

Fig. 205 Remove the snap-ring from the hub

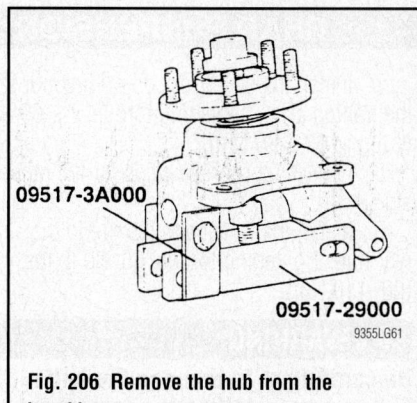

Fig. 206 Remove the hub from the knuckle

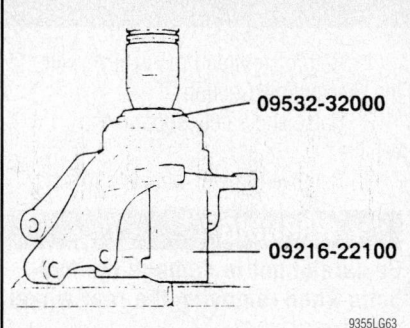

Fig. 208 Remove the wheel bearing outer race from the knuckle

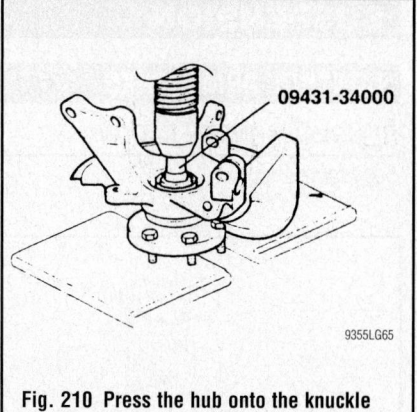

Fig. 210 Press the hub onto the knuckle

- Snap-ring from the hub
- Hub from the knuckle by installing tools 09517-3A00 and 09517-2900, then tighten the nut of the tool to separate the tool from the knuckle
- Wheel bearing inner race from the hub using tools 09455-2100 and 09545-34100
- Wheel bearing outer race from the knuckle using tools 09532-3200 and 09216-22100

3. Check all components for wear or damage. Replace as necessary.

To install:

4. Apply a thin coat of multi-purpose grease to the surface on the knuckle and bearing.

5. Install or connect the following:
- Bearing onto the knuckle using tool 09216-21100
- Snap-ring into the groove of the knuckle

- Backing plate onto the knuckle
- Hub onto the knuckle by pressing it into position using tool 09431-3400

✳✳ WARNING

Do not press against the outer race of the bearing as this can cause bearing damage. Always use a new bearing kit.

6. Rotate the bearing several times to seat the bearing.

7. Measure the wheel bearing torque using a torque wrench. The standard measurement is 17 inch lbs. (2 Nm).

8. Measure the end-play of the hub using a dial gauge. The specification is 0.0025–0.0035 inch (0.064–0.088mm).

9. Install the remaining components in the reverse order of removal.

10. Check the wheel alignment and adjust as necessary.

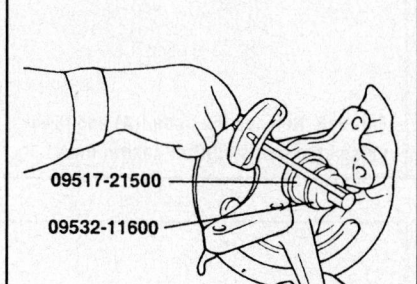

Fig. 211 Check the wheel bearing starting torque

ADJUSTMENT

The wheel bearings are sealed units and are not adjustable.

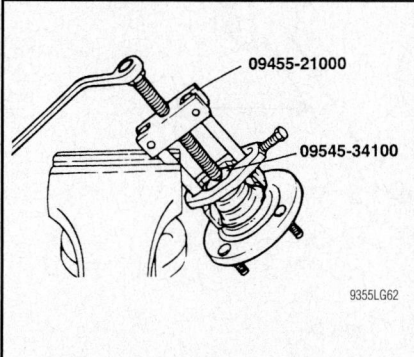

Fig. 207 Remove the wheel bearing inner race from the hub

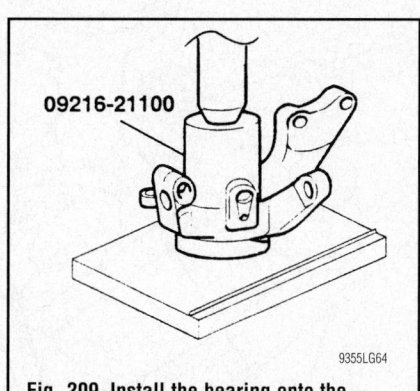

Fig. 209 Install the bearing onto the knuckle

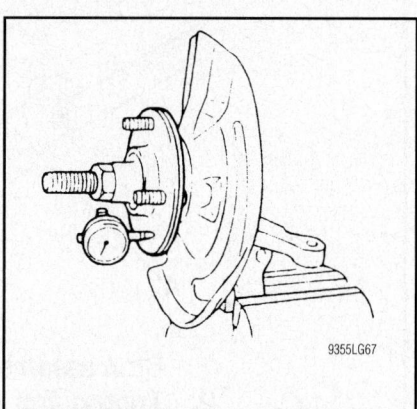

Fig. 212 Check the hub end end-play

CONTROL ARMS/LINKS

REMOVAL & INSTALLATION

See Figure 213.

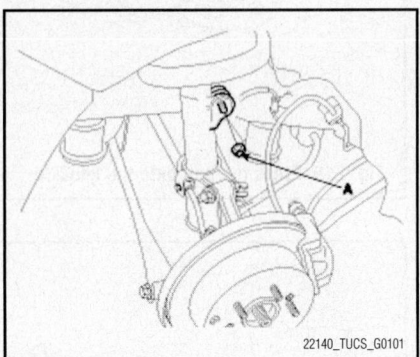

**Fig. 213 Remove the upper (A) and lower
nuts of the rear stabilizer control links**

1. Before servicing the vehicle, refer to the Precautions Section.
2. Raise and safely support the vehicle.
3. Remove the rear wheels and tires.

❊❊ WARNING

Be careful not to damage the hub bolts when removing the rear wheel and tire.

4. Remove the upper (A) and lower nuts of the rear stabilizer control links.
5. Remove the rear stabilizer control link from the stabilizer bar assembly.

To install:

6. Install the rear stabilizer control link and nut to the stabilizer bar assembly. Tightening torque: 74–89 ft. lbs. (100–120 Nm).

7. Install the stabilizer control link nut to the trailing arm. Tightening torque: 74–89 ft. lbs. (100–120 Nm).
8. Repeat appropriate steps for the other side.
9. Install the wheel and the tire to the rear hub. Tightening torque: 65–80 ft. lbs. (90–110 Nm).

❊❊ WARNING

Be careful not to damage the hub bolts when installing the rear wheel and tire.

STRUT

REMOVAL & INSTALLATION

See Figures 214 and 215.

1. Before servicing the vehicle, refer to the Precautions Section.

[2WD]

1. Strut assembly
2. Trailing arm
3. Suspension arm
4. Cross member
5. Carrier
6. Disc brake assembly

22140_TUCS_G0024

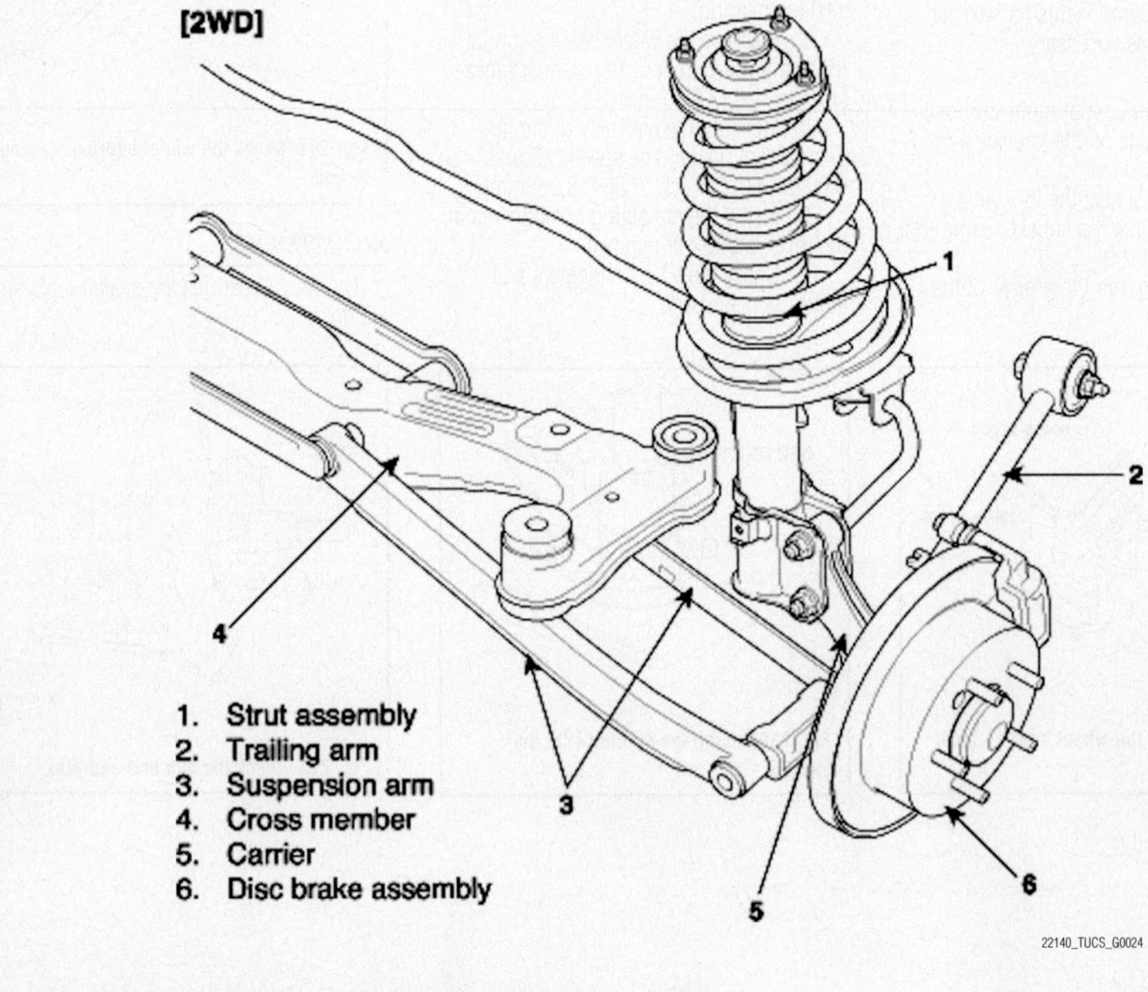

Fig. 214 View of strut assembly and related rear suspension components—2WD

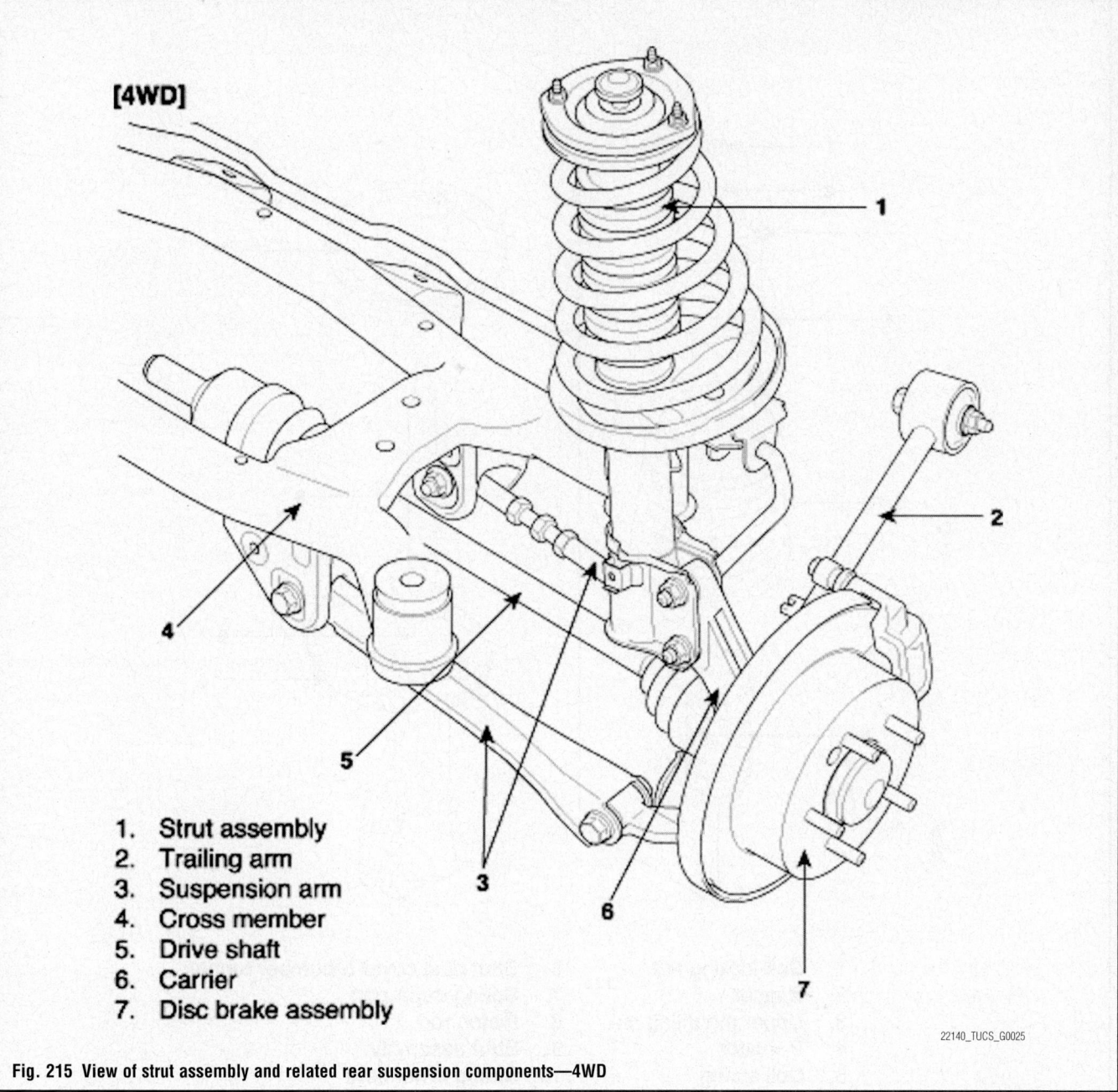

[4WD]

1. Strut assembly
2. Trailing arm
3. Suspension arm
4. Cross member
5. Drive shaft
6. Carrier
7. Disc brake assembly

22140_TUCS_G0025

Fig. 215 View of strut assembly and related rear suspension components—4WD

2. Remove or disconnect the following:
- Rear wheel
- Speed sensor
- Stabilizer bar control link mounting nut
- Upper strut mounting nuts
- Lower strut mounting bolts
- Strut assembly

To install:
3. Install or connect the following:
- Strut assembly and tighten the lower mounting bolts to 103–118 ft. lbs. (140–160 Nm)
- Upper strut mounting bolts. Tighten to 22–30 ft. lbs. (30–40 Nm)

- Stabilizer bar link mounting nut. Tighten to 74–89 ft. lbs. (100–120 Nm)
- Speed sensor. Tighten the mounting bolt to 60–96 inch lbs. (7–11 Nm).
- Rear wheel. Tighten the lug nuts to 66–81 ft. lbs. (90–110 Nm).
4. Check the alignment and adjust as necessary.

OVERHAUL
See Figure 216.

1. Before servicing the vehicle, refer to the Precautions Section.

2. Remove the strut from the vehicle and attach Service Tool 09546-2600 or another suitable spring compressor.
3. Compress the coil spring.
4. Remove the self-locking nut from the strut.
5. Remove the insulator, spring seat, coil spring, and dust cover.
6. Inspect the insulator for wear and damage, replace as required.
7. Check the rubber parts for damage or deterioration, replace as required.

To reassemble:
8. Install the spring lower pad so that the protrusions fit the holes in the spring lower seat.

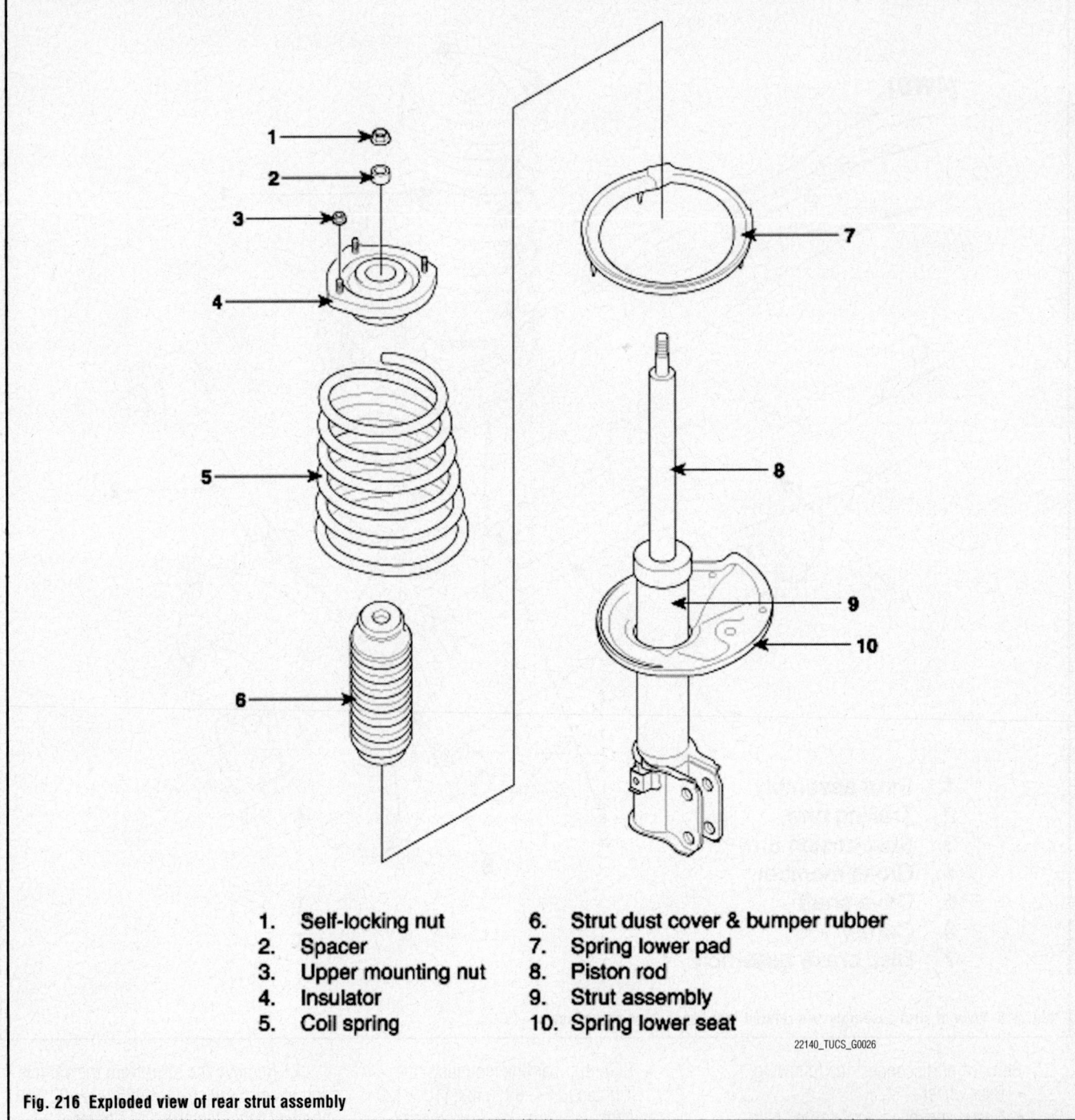

1. Self-locking nut
2. Spacer
3. Upper mounting nut
4. Insulator
5. Coil spring
6. Strut dust cover & bumper rubber
7. Spring lower pad
8. Piston rod
9. Strut assembly
10. Spring lower seat

22140_TUCS_G0026

Fig. 216 Exploded view of rear strut assembly

9. Compress the spring, using the spring compressor tool.

10. Install the compressed spring over the shock absorber.

➡**There are 2 color identification marks on the coil spring, one indicates model option and the other indicates load classification. Install the coil spring with the identification mark directed toward the steering knuckle.**

11. After fully extending the piston rod, install the spring upper seat and insulator assembly.

12. After correctly seating the upper and lower ends of the coil spring in the upper and lower spring grooves, tighten the NEW self-locking nut temporarily.

13. Remove the spring compression tool. Tighten the self-locking nut to 43–51 ft. lbs. (60–70 Nm).

14. Install the strut to the vehicle.

15. Check the wheel alignment and adjust as necessary.

WHEEL HUB AND BEARING (SEALED UNIT)

REMOVAL & INSTALLATION

See Figures 217 through 221.

1. Before servicing the vehicle, refer to the Precautions Section.

2. Remove or disconnect the following:
   - Rear wheel
   - Flange nut and washer
   - Drum or rotor

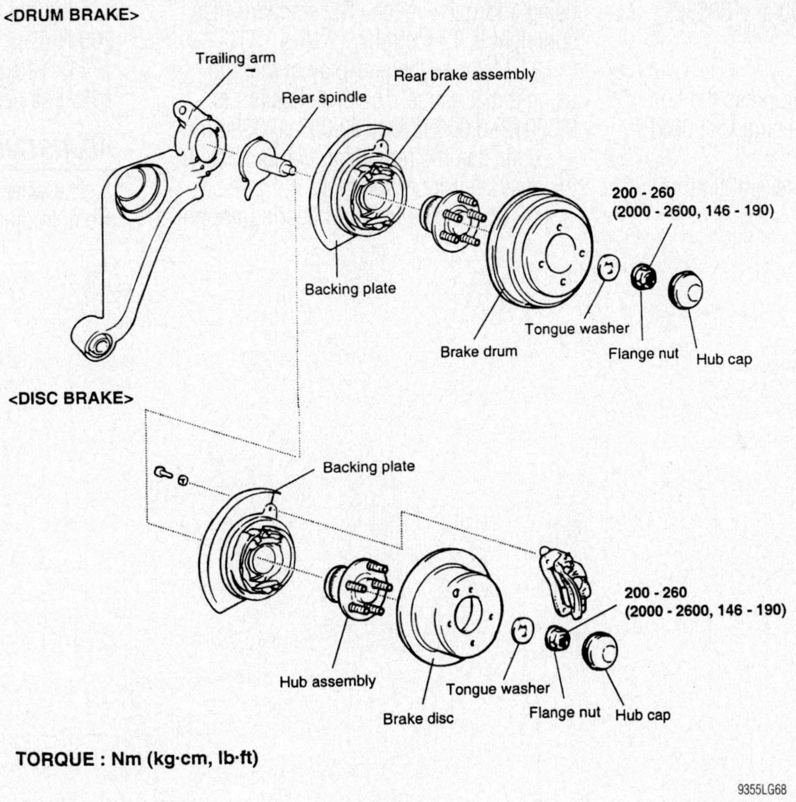

<DRUM BRAKE>

Trailing arm
Rear spindle
Rear brake assembly
200 - 260
(2000 - 2600, 146 - 190)
Backing plate
Tongue washer
Brake drum
Flange nut
Hub cap

<DISC BRAKE>

Backing plate
200 - 260
(2000 - 2600, 146 - 190)
Hub assembly
Tongue washer
Brake disc
Flange nut
Hub cap

TORQUE : Nm (kg·cm, lb·ft)

9355LG68

Fig. 217 Exploded view of the rear hub assembly

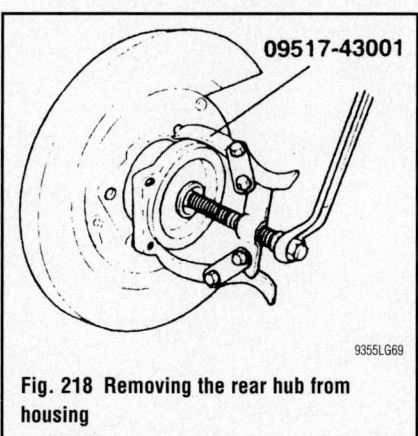

Fig. 218 Removing the rear hub from housing

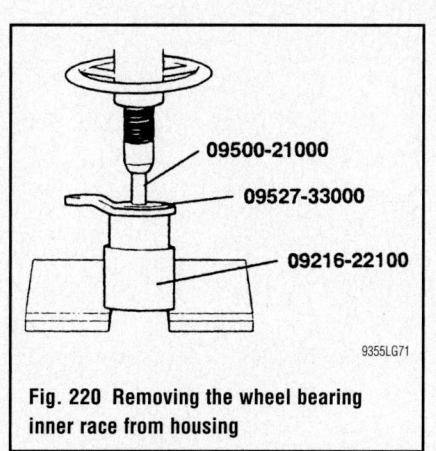

Fig. 220 Removing the wheel bearing inner race from housing

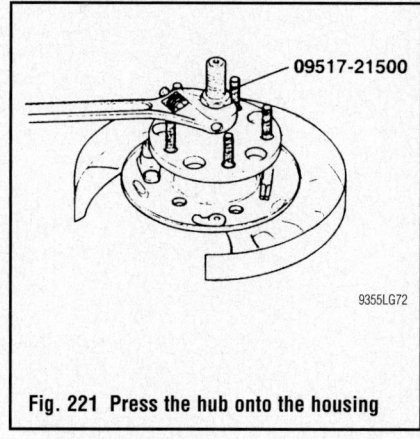

Fig. 221 Press the hub onto the housing

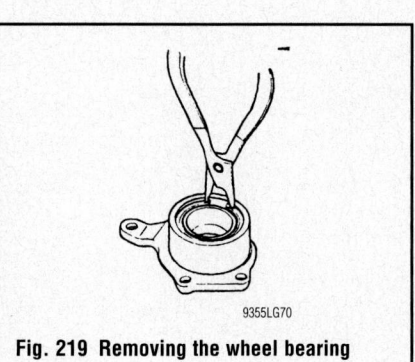

Fig. 219 Removing the wheel bearing snap-ring

- Brake line
- Parking brake assembly
- Parking brake cable
- Spindle bolts and the spindle
- Rear hub from the housing using tool 09517-43001
- Wheel bearing snap-ring
- Wheel bearing inner race from the housing using tools 09500-2100, 09527-33000, and 09216-22100

3. Inspect the components for damage and replace as necessary.

To install:

4. Apply a thin coat of multi-purpose grease to the surface on the housing and bearing.

✳✳ WARNING

Do not press against the outer race of the bearing as this can cause bearing damage. Always use a new bearing kit.

5. Install or connect the following:

- Bearing onto the spindle using tools 09216-21100 and 09532-3200
- Snap-ring
- Backing plate, then press the hub onto the housing using tool 09517-21500

6. Rotate the bearing several times to seat the bearing.

7. Measure the wheel bearing torque using a torque wrench. The standard measurement is 17 inch lbs. (2 Nm).

8. Measure the end-play of the hub using a dial gauge. The specification is 0.00012–0.00032 inch (0.003–0.008mm).

9. Install the remaining components in the reverse order of removal.

10. Tighten the hub and bearing assembly to the backing plate. Torque the hub bearing nut to 148–192 ft. lbs. (200–260 Nm).

11. Check the wheel alignment and adjust as necessary.

ADJUSTMENT

The wheel bearings are sealed units and are not adjustable.

HYUNDAI

Diagnostic Trouble Codes

DIAGNOSTIC TROUBLE CODES

OBD II VEHICLE APPLICATIONS

HYUNDAI

Accent
2006–2008
• 1.6L I4 MPFI (DOHC) .... VIN C
Azera
2006
• 3.8L V6 MPFI (DOHC) .... VIN F
2007–2008
• 3.3L V6 MPFI (DOHC) .... VIN D
• 3.8L V6 MPFI (DOHC) .... VIN F
Elantra
2006–2008
• 2.0L I4 MPFI (DOHC) .... VIN D
Entourage
2007–2008
• 3.8L V6 MPFI (DOHC) .... VIN 3
Santa Fe
2006
• 2.4L I4 MPFI (DOHC) .... VIN B
• 2.7L V6 MPFI (DOHC) .... VIN D
• 3.5L V6 MPFI (DOHC) .... VIN E
2007–2008
• 2.7L V6 MPFI (DOHC) .... VIN D
• 3.3L V6 MPFI (DOHC) .... VIN E

Sonata
2006–2008
• 2.4L I4 MPFI (DOHC) .... VIN C
• 3.3L V6 MPFI (DOHC) .... VIN F
Tiburon
2006–2008
• 2.0L I4 MPFI (DOHC) .... VIN D
• 2.7L V6 MPFI (DOHC) .... VIN F
Tucson
2006–2008
• 2.0L I4 MPFI (DOHC) .... VIN B
• 2.7L V6 MPFI (DOHC) .... VIN D
Veracruz
2007–2008
• 3.8L V6 MPFI (DOHC) .... VIN C

HYUNDAI REFERENCE INFORMATION

OBD II TROUBLE CODE LIST

To use this information, first read and record All codes in memory along with Freeze Frame data. *If a PCM Reset function is done prior to recording this data,* All *codes and freeze frame data are lost!*

Look up the appropriate trouble code in the list on the following pages. The left hand column includes the code number, the number of trips to set the code (e.g., **1T or 2T**), the year, model description, and type of OBD II Monitor that failed (e.g., **CCM or O2S**). This data can be used to determine how to drive a vehicle after a repair in order to validate the repair has been completed.

The **(N/MIL)** designator in the left hand column indicates the trouble code does not turn on the Malfunction Indicator Lamp or MIL. The **(STS Lamp)** indicator in the left column indicates a code that turns on the Service Transmission Soon lamp. This code may or may not turn "on" the MIL.

Gas Engine OBD II Trouble Code List (P0xxx Codes)

DTC	Trouble Code Title, Conditions & Possible Causes
DTC: P0011 **2T CCM, MIL: Yes** **Years:** 2006, 2007, 2008 **Models:** Accent, Azera, Elantra, Entourage, Santa Fe, Sonata, Tiburon, Tucson, Veracruz **Engines:** 1.6L VIN C, 2.0L VIN B, 2.0L VIN D, 2.4L VIN C, 2.7L VIN D, 3.3L VIN D, 3.3L VIN E, 3.3L VIN F, 3.8L VIN C, 3.8L VIN F, 3.8L VIN 3 **Transmissions:** All	**Camshaft Position Timing Over-Advanced or System Performance (Bank 1)** The Engine Control Module (ECM) has enabled the CMP actuator. The system voltage is more than 11 volts. DTC P0011 is set when the control module has enabled the CMP actuator and the difference between the desired CMP actuator angle and the actual CMP actuator angle is greater than specification. **Possible Causes:** • CMP actuator control circuit is open or shorted to ground • CMP actuator control circuit is shorted to voltage • CMP actuator control circuit has high resistance • Engine oil pressure is low, improper oil viscosity, engine oil level is low • Excessive timing chain play • CMP actuator control or solenoid has failed • Recent engine mechanical repairs with an incorrectly installed camshaft, camshaft actuator, or timing chain • Poor connection(s) at ECM • ECM has failed
DTC: P0012 **2T CCM, MIL: Yes** **Years:** 2006, 2007, 2008 **Models:** Accent, Azera, Entourage, Santa Fe, Sonata, Veracruz **Engines:** 1.6L VIN C, 2.7L VIN D, 3.3L VIN D, 3.3L VIN E, 3.3L VIN F, 3.8L VIN C, 3.8L VIN F, 3.8L VIN 3 **Transmissions:** All	**Camshaft Position Timing Over-Retarded (Bank 1)** The Engine Control Module (ECM) has enabled the CMP actuator. The system voltage is more than 11 volts. DTC P0012 is set when the control module has enabled the CMP actuator and there is no change or a steady error is present. **Possible Causes:** • Engine oil pressure is low, improper oil viscosity, engine oil level is low • Excessive timing chain play • CMP actuator control or solenoid is stuck or has failed • Recent engine mechanical repairs with an incorrectly installed camshaft, camshaft actuator, or timing chain • Poor connection(s) at ECM • ECM has failed
DTC: P0016 **2T CCM, MIL: Yes** **Years:** 2006, 2007, 2008 **Models:** Accent, Azera, Elantra, Entourage, Santa Fe, Sonata, Tiburon, Tucson, Veracruz **Engines:** 1.6L VIN C, 2.0L VIN B, 2.0L VIN D, 2.4L VIN B, 2.4L VIN C, 2.7L VIN D, 3.3L VIN D, 3.3L VIN E, 3.3L VIN F, 3.5L VIN E, 3.8L VIN C, 3.8L VIN F, 3.8L VIN 3 **Transmissions:** All	**Crankshaft Position (CKP) - Intake Camshaft Position (CMP) Correlation (Bank 1)** The engine is at full operating temperature at idle and no active faults are present. DTC P0016 is set when the ECM detects one of the following conditions: (1) The ECM detects a deviation in the relationship between a camshaft and the crankshaft; (2) A camshaft is more than 12 degrees advanced in relationship to the crankshaft; (3) A camshaft is more than 12 degrees retarded in relationship to the crankshaft. The condition exists continuously for 1 minute. **Possible Causes:** • Engine oil pressure is low • Improper crankshaft balancer torque • Failed CKP solenoid • Incorrect valve timing • Incorrectly installed camshaft, crankshaft, or timing chain
DTC: P0018 **2T CCM, MIL: Yes** **Years:** 2006, 2007, 2008 **Models:** Azera, Entourage, Santa Fe, Sonata, Veracruz **Engines:** 2.7L VIN D, 3.3L VIN D, 3.3L VIN E, 3.3L VIN F, 3.8L VIN C, 3.8L VIN F, 3.8L VIN 3 **Transmissions:** All	**Crankshaft Position (CKP) - Intake Camshaft Position (CMP) Correlation Bank 2** The engine is at full operating temperature at idle and no active faults are present. DTC P0018 is set when the ECM detects that the timing is misaligned. **Possible Causes:** • Failed CKP solenoid • Incorrect valve timing • CKP is loosened or not installed correctly • Incorrectly installed camshaft, crankshaft, or timing chain
DTC: P0021 **2T CCM, MIL: Yes** **Years:** 2006, 2007, 2008 **Models:** Azera, Entourage, Santa Fe, Sonata, Veracruz **Engines:** 2.7L VIN D, 3.3L VIN D, 3.3L VIN E, 3.3L VIN F, 3.8L VIN 3, 3.8L VIN C, 3.8L VIN F **Transmissions:** All	**Camshaft Position (CMP) Timing Over-Advanced or System Performance (Bank 2)** The engine is at full operating temperature at idle and no active faults are present. DTC P0021 is set when the ECM detects that the phaser is moving at an unexpected rate. **Possible Causes:** • CMP actuator control circuit is open or shorted to ground • CMP actuator control circuit is shorted to voltage • CMP actuator control circuit has high resistance • CMP actuator control or solenoid has failed • Engine oil in poor condition has a major impact on the camshaft actuator • Engine oil pressure is low or oil level is low • Recent engine mechanical repairs with an incorrectly installed camshaft, camshaft actuator, or timing chain • Poor connection(s) at ECM • ECM has failed

DTC	Trouble Code Title, Conditions & Possible Causes
DTC: P0022 **2T CCM, MIL: Yes** **Years:** 2006, 2007, 2008 **Models:** Azera, Entourage, Santa Fe, Sonata, Veracruz **Engines:** 2.7L VIN D, 3.3L VIN D, 3.3L VIN E, 3.3L VIN F, 3.8L VIN 3, 3.8L VIN C, 3.8L VIN F **Transmissions:** All	**Camshaft Position (CMP) Timing Over-Retarded (Bank 2)** The engine is at full operating temperature at idle and no active faults are present. DTC P0022 is set when the ECM detects that the phaser is stuck or has a steady state error. **Possible Causes:** • CMP actuator control circuit is open or shorted to ground • CMP actuator control circuit is shorted to voltage • CMP actuator control circuit has high resistance • CMP actuator control or solenoid is stuck or has failed • Engine oil in poor condition has a major impact on the camshaft actuator • Engine oil pressure is low or oil level is low • Recent engine mechanical repairs with an incorrectly installed camshaft, camshaft actuator, or timing chain • Poor connection(s) at ECM • ECM has failed
DTC: P0026 **2T CCM, MIL: Yes** **Years:** 2006, 2007, 2008 **Models:** Azera, Entourage, Santa Fe, Sonata, Veracruz **Engines:** 2.7L VIN D, 3.3L VIN D, 3.3L VIN E, 3.3L VIN F, 3.8L VIN 3, 3.8L VIN C, 3.8L VIN F **Transmissions:** All	**Intake Valve Control Solenoid Circuit Range/Performance (Bank 1)** The engine is at full operating temperature at idle and no active faults are present. DTC P0026 is set when the ECM detects that the oil control valve is stuck. Valve cleaning is not in progress. Off sets available. **Possible Causes:** • Oil pressure loss has occurred • Oil Control Valve (OCV) has seized • ECM has failed
DTC: P0028 **2T CCM, MIL: Yes** **Years:** 2006, 2007, 2008 **Models:** Azera, Entourage, Santa Fe, Sonata, Veracruz **Engines:** 2.7L VIN D, 3.3L VIN D, 3.3L VIN E, 3.3L VIN F, 3.8L VIN 3, 3.8L VIN C, 3.8L VIN F **Transmissions:** All	**Intake Valve Control Solenoid Circuit Range/Performance (Bank 2)** The engine is at full operating temperature at idle and no active faults are present. DTC P0028 is set when the ECM detects that the oil control valve is stuck. Valve cleaning is not in progress. Off sets available. **Possible Causes:** • Oil pressure loss has occurred • Oil Control Valve (OCV) has seized • ECM has failed
DTC: P0030 **2T CCM, MIL: Yes** **Years:** 2006, 2007, 2008 **Models:** Accent, Azera, Elantra, Entourage, Santa Fe, Sonata, Tiburon, Tucson, Veracruz **Engines:** 1.6L VIN C, 2.0L VIN B, 2.0L VIN D, 2.4L VIN C, 2.7L VIN D, 2.7L VIN F, 3.3L VIN D, 3.3L VIN E, 3.3L VIN F, 3.8L VIN 3, 3.8L VIN C, 3.8L VIN F **Transmissions:** All	**O2 Sensor Heater Control Circuit (Bank 1/Sensor 1)** The engine is running. DTC P0030 is set when the ECM detects that the heater current is below a certain threshold for a predetermined time. **Possible Causes:** • HO2S heater low side control circuit open or shorted to ground • HO2S heater low side control circuit short to voltage • HO2S heater power circuit is open (check the PRE O2 fuse) • HO2S heater is damaged or it has failed • Harness or connector is not secure or is corroded • Contaminated, deteriorated, or aged sensor • ECM has failed
DTC: P0031 **2T CCM, MIL: Yes** **Years:** 2006, 2007, 2008 **Models:** Accent, Azera, Elantra, Entourage, Santa Fe, Sonata, Tiburon, Tucson, Veracruz **Engines:** 1.6L VIN C, 2.0L VIN B, 2.0L VIN D, 2.4L VIN B, 2.4L VIN C, 2.7L VIN D, 2.7L VIN F, 3.3L VIN D, 3.3L VIN E, 3.3L VIN F, 3.5L VIN E, 3.8L VIN 3, 3.8L VIN C, 3.8L VIN F **Transmissions:** All	**HO2S Heater Control Circuit Low Voltage (Bank 1/Sensor 1)** The engine is running. The Ignition 1 Signal parameter is more than 10 volts. DTC P0031 is set when the ECM detects that the difference between the predicted voltage and the actual voltage is below a calibrated limit. **Possible Causes:** • HO2S heater low side control circuit open or shorted to ground • HO2S heater power circuit is open (check the PRE O2 fuse) • HO2S heater is damaged or it has failed • Harness or connector not secure or corroded • ECM has failed

DTC	Trouble Code Title, Conditions & Possible Causes
DTC: P0032 **2T CCM, MIL: Yes** **Years:** 2006, 2007, 2008 **Models:** Accent, Azera, Elantra, Entourage, Santa Fe, Sonata, Tiburon, Tucson, Veracruz **Engines:** 1.6L VIN C, 2.0L VIN B, 2.0L VIN D, 2.4L VIN B, 2.4L VIN C, 2.7L VIN D, 2.7L VIN F, 3.3L VIN D, 3.3L VIN E, 3.3L VIN F, 3.5L VIN E, 3.8L VIN 3, 3.8L VIN C, 3.8L VIN F **Transmissions:** All	**HO2S Heater Control Circuit High Voltage (Bank 1/Sensor 1)** The engine is running. The Ignition 1 Signal parameter is more than 10 volts. DTC P0032 is set when the ECM detects that the difference between the predicted voltage and the actual voltage exceeds a calibrated limit. **Possible Causes:** • HO2S heater low side control circuit short to voltage • HO2S heater is damaged or it has failed • Harness or connector not secure or corroded • ECM has failed
DTC: P0036 **2T CCM, MIL: Yes** **Years:** 2006, 2007, 2008 **Models:** Accent, Azera, Elantra, Entourage, Santa Fe, Sonata, Tiburon, Tucson, Veracruz **Engines:** 1.6L VIN C, 2.0L VIN B, 2.0L VIN D, 2.4L VIN C, 2.7L VIN D, 2.7L VIN F, 3.3L VIN D, 3.3L VIN E, 3.3L VIN F, 3.8L VIN 3, 3.8L VIN C, 3.8L VIN F **Transmissions:** All	**HO2S Heater Control Circuit (Bank 1/Sensor 2)** The engine is running. DTC P0036 is set when the ECM detects that the difference between the expected and the actual HO2S 2 heater low control circuit voltage is below a calibrated value for more than a predetermined time. **Possible Causes:** • HO2S heater low side control circuit open or shorted to ground • HO2S heater low side control circuit short to voltage • HO2S heater power circuit is open (check the PRE O2 fuse) • HO2S heater is damaged or it has failed • Heater resistance is out of range • Contaminated, deteriorated, or aged sensor • Harness or connector not secure or corroded • ECM has failed
DTC: P0037 **2T CCM, MIL: Yes** **Years:** 2006, 2007, 2008 **Models:** Accent, Azera, Elantra, Entourage, Santa Fe, Sonata, Tiburon, Tucson, Veracruz **Engines:** 1.6L VIN C, 2.0L VIN B, 2.0L VIN D, 2.4L VIN B, 2.4L VIN C, 2.7L VIN D, 2.7L VIN F, 3.3L VIN D, 3.3L VIN E, 3.3L VIN F, 3.5L VIN E, 3.8L VIN 3, 3.8L VIN C, 3.8L VIN F **Transmissions:** All	**HO2S Heater Control Circuit Low Voltage (Bank 1/Sensor 2)** The engine is running. DTC P0037 is set when the ECM detects that the difference between the expected and the actual HO2S 2 heater low control circuit voltage is below a calibrated value for more than a predetermined time. A short to ground or open in heater circuit is detected. **Possible Causes:** • HO2S heater low side control circuit open or shorted to ground • HO2S heater power circuit is open (check the PRE O2 fuse) • HO2S heater is damaged or it has failed • Heater resistance is out of range • Contaminated, deteriorated, or aged sensor • Harness or connector not secure or corroded • ECM has failed
DTC: P0038 **2T CCM, MIL: Yes** **Years:** 2006, 2007, 2008 **Models:** Accent, Azera, Elantra, Entourage, Santa Fe, Sonata, Tiburon, Tucson, Veracruz **Engines:** 1.6L VIN C, 2.0L VIN B, 2.0L VIN D, 2.4L VIN B, 2.4L VIN C, 2.7L VIN D, 2.7L VIN F, 3.3L VIN D, 3.3L VIN E, 3.3L VIN F, 3.5L VIN E, 3.8L VIN 3, 3.8L VIN C, 3.8L VIN F **Transmissions:** All	**HO2S Heater Control Circuit High Voltage (Bank 1/Sensor 2)** The engine is running. DTC P0038 is set when the ECM detects that the difference between the expected and the actual HO2S 2 heater low control circuit voltage exceeds a calibrated value for more than a predetermined time. A short to battery in heater circuit is detected. **Possible Causes:** • HO2S heater low side control circuit short to voltage • HO2S heater is damaged or it has failed • Harness or connector not secure or corroded • ECM has failed
DTC: P0050 **2T CCM, MIL: Yes** **Years:** 2006, 2007, 2008 **Models:** Azera, Entourage, Santa Fe, Sonata, Tiburon, Tucson, Veracruz **Engines:** 2.7L VIN D, 2.7L VIN F, 3.3L VIN D, 3.3L VIN E, 3.3L VIN F, 3.8L VIN 3, 3.8L VIN C, 3.8L VIN F **Transmissions:** All	**HO2S Heater Control Circuit (Bank 1/Sensor 2)** The engine is running. DTC P0050 is set when the control module detects that the affected HO2S heater low control circuit is not within a specified range. **Possible Causes:** • HO2S heater low side control circuit open or shorted to ground • HO2S heater low side control circuit short to voltage • HO2S heater power circuit is open (check the PRE O2 fuse) • HO2S heater is damaged or it has failed • Harness or connector not secure or corroded • ECM has failed

DTC	Trouble Code Title, Conditions & Possible Causes
DTC: P0051 **2T CCM, MIL: Yes** **Years:** 2006, 2007, 2008 **Models:** Azera, Entourage, Santa Fe, Sonata, Tiburon, Tucson, Veracruz **Engines:** 2.7L VIN D, 2.7L VIN F, 3.3L VIN D, 3.3L VIN E, 3.3L VIN F, 3.5L VIN E, 3.8L VIN 3, 3.8L VIN C, 3.8L VIN F **Transmissions:** All	**HO2S Heater Circuit Low (Bank 2/Sensor 1)** The engine is running and there are no disabling faults. The battery voltage is between 11 to 16 volts. DTC P0051 is set when the control module detects that the affected HO2S heater low control circuit is not within a specified range. **Possible Causes:** • HO2S heater low side control circuit open or shorted to ground • HO2S heater power circuit is open (check the PRE O2 fuse) • HO2S heater is damaged or it has failed • Harness or connector not secure or corroded • ECM has failed
DTC: P0052 **2T CCM, MIL: Yes** **Years:** 2006, 2007, 2008 **Models:** Azera, Entourage, Santa Fe, Sonata, Tiburon, Tucson, Veracruz **Engines:** 2.7L VIN D, 2.7L VIN F, 3.3L VIN D, 3.3L VIN E, 3.3L VIN F, 3.5L VIN E, 3.8L VIN 3, 3.8L VIN C, 3.8L VIN F **Transmissions:** All	**HO2S Heater Circuit High (Bank 2/Sensor 1)** The engine is running and there are no disabling faults. The battery voltage is between 11 to 16 volts. DTC P0052 is set when the control module detects that the affected HO2S heater low control circuit is not within a specified range. **Possible Causes:** • HO2S heater low side control circuit open or shorted to battery voltage • HO2S heater is damaged or it has failed • Harness or connector not secure or corroded • ECM has failed
DTC: P0053 **2T CCM, MIL: Yes** **Years:** 2006 **Models:** Santa Fe **Engines:** 2.4L VIN B, 3.5L VIN E **Transmissions:** All	**HO2S Heater Resistance (Bank 1/Sensor 1)** The engine is running and there are no disabling faults. The battery voltage is between 11 to 16 volts. DTC P0053 is set when the control module detects that the affected HO2S heater low control circuit is not within a specified range. **Possible Causes:** • HO2S heater low side control circuit open or shorted to ground • HO2S heater power circuit is open • An incorrect or out of range resistance value at engine start-up • HO2S heater is damaged or it has failed • ECM has failed
DTC: P0054 **2T CCM, MIL: Yes** **Years:** 2006 **Models:** Santa Fe **Engines:** 2.4L VIN B, 3.5L VIN E **Transmissions:** All	**HO2S Heater Resistance (Bank 1/Sensor 2)** The engine is running and there are no disabling faults. The battery voltage is between 11 to 16 volts. DTC P0054 is set when the ECM detects that the affected HO2S heater low control circuit is not within a specified range at engine start-up. **Possible Causes:** • HO2S heater low side control circuit open or shorted to ground • HO2S heater power circuit is open • An incorrect or out of range resistance value at engine start-up • HO2S heater is damaged or it has failed • ECM has failed
DTC: P0056 **2T CCM, MIL: Yes** **Years:** 2006, 2007, 2008 **Models:** Azera, Entourage, Santa Fe, Sonata, Tiburon, Tucson, Veracruz **Engines:** 2.7L VIN D, 2.7L VIN F, 3.3L VIN D, 3.3L VIN E, 3.3L VIN F, 3.8L VIN 3, 3.8L VIN C, 3.8L VIN F **Transmissions:** All	**HO2S Heater Control Circuit (Bank 2/Sensor 2)** The engine is running and there are no disabling faults. The battery voltage is between 11 to 16 volts. DTC P0056 is set when the ECM detects that the affected HO2S heater low control circuit is not within a specified range at engine start-up. **Possible Causes:** • HO2S heater low side control circuit open or shorted to ground • HO2S heater low side control circuit short to voltage • HO2S heater power circuit is open (check the PRE O2 fuse) • HO2S heater is damaged or it has failed • Harness or connector not secure or corroded • ECM has failed
DTC: P0057 **2T CCM, MIL: Yes** **Years:** 2006, 2007, 2008 **Models:** Azera, Entourage, Santa Fe, Sonata, Tiburon, Tucson, Veracruz **Engines:** 2.7L VIN D, 2.7L VIN F, 3.3L VIN D, 3.3L VIN E, 3.3L VIN F, 3.5L VIN E, 3.8L VIN 3, 3.8L VIN C, 3.8L VIN F **Transmissions:** All	**HO2S Heater Control Circuit Low (Bank 2/Sensor 2)** The engine is running and there are no disabling faults. The battery voltage is between 11 to 16 volts. DTC P0057 is set when the ECM detects that the affected HO2S heater low control circuit is not within a specified range. **Possible Causes:** • HO2S heater low side control circuit open or shorted to ground • HO2S heater power circuit is open (check the PRE O2 fuse) • HO2S heater is damaged or it has failed • Harness or connector not secure or corroded • ECM has failed

DTC	Trouble Code Title, Conditions & Possible Causes
DTC: P0058 **2T CCM, MIL: Yes** **Years:** 2006, 2007, 2008 **Models:** Azera, Entourage, Santa Fe, Sonata, Tiburon, Tucson, Veracruz **Engines:** 2.7L VIN D, 2.7L VIN F, 3.3L VIN D, 3.3L VIN E, 3.3L VIN F, 3.5L VIN E, 3.8L VIN 3, 3.8L VIN C, 3.8L VIN F **Transmissions:** All	**HO2S Heater Control Circuit High (Bank 2/Sensor 2)** The engine is running and there are no disabling faults. The battery voltage is between 11 to 16 volts. DTC P0058 is set when the ECM detects that the affected HO2S heater low control circuit is not within a specified range. **Possible Causes:** • HO2S heater low side control circuit short to battery voltage • HO2S heater is damaged or it has failed • Harness or connector not secure or corroded • ECM has failed
DTC: P0059 **2T CCM, MIL: Yes** **Years:** 2006 **Models:** Santa Fe **Engines:** 3.5L VIN E **Transmissions:** All	**HO2S Heater Resistance (Bank 2/Sensor 1)** The engine is started. The ignition voltage is less than 16 volts. DTC P0059 is set when the ECM detects that the affected HO2S heater control circuit is not within a specified range. **Possible Causes:** • HO2S heater voltage supply circuit has a short to ground or an open/high resistance • HO2S heater voltage supply circuit fuse is open • HO2S heater low control circuit has a short to ground • HO2S heater low control circuit has a short to voltage or an open/high resistance • ECM has failed
DTC: P0060 **2T CCM, MIL: Yes** **Years:** 2006 **Models:** Santa Fe **Engines:** 3.5L VIN E **Transmissions:** All	**HO2S Heater Resistance (Bank 2/Sensor 2)** The engine is started. The ignition voltage is less than 16 volts. DTC P0060 is set when the ECM detects that the affected HO2S heater control circuit is not within a specified range. **Possible Causes:** • HO2S heater voltage supply circuit has a short to ground or an open/high resistance • HO2S heater voltage supply circuit fuse is open • HO2S heater low control circuit has a short to ground • HO2S heater low control circuit has a short to voltage or an open/high resistance • ECM has failed
DTC: P0068 **1T CCM, MIL: Yes** **Years:** 2006, 2007, 2008 **Models:** Accent **Engines:** 1.6L VIN C **Transmissions:** All	**MAFS/MAPS-TPS Correlation** The engine is running. DTC P0068 is set if the sensor input value of the TPS is lower or higher than the threshold value, which is dependent on MAF (MAP). The condition exist for greater than 300 seconds. The rationality check is run to correlate the actual and secondary load. **Possible Causes:** • Vacuum leaks • Dirty throttle body • Throttle body damage • Faulty MAP sensor • Faulty TPS • Faulty throttle body assembly
DTC: P0075 **1T CCM, MIL: Yes** **Years:** 2006, 2007, 2008 **Models:** Accent **Engines:** 1.6L VIN C **Transmissions:** All	**Intake Valve Control Solenoid Circuit (Bank 1)** The engine is running. DTC P0075 is set when the ECM detects an open in the Oil Control Valve (OCV) control circuit. **Possible Causes:** • Circuit continuity check is open • An open or short to ground in the power circuit • An open in the control circuit • Faulty OCV • Faulty ECM
DTC: P0076 **1T CCM, MIL: Yes** **Years:** 2006, 2007, 2008 **Models:** Accent, Azera, Elantra, Entourage, Santa Fe, Sonata, Tiburon, Tucson, Veracruz **Engines:** 1.6L VIN C, 2.0L VIN B, 2.0L VIN D, 2.4L VIN C, 2.7L VIN D, 3.3L VIN D, 3.3L VIN E, 3.3L VIN F, 3.8L VIN 3, 3.8L VIN C, 3.8L VIN F **Transmissions:** All	**Intake Valve Control Solenoid Circuit Low (Bank 1)** The engine is running and no disabling faults are current. The ignition voltage is between 11 and 16 volts. An Oil Control Valve (OCV) failure is detected. A short to ground or open circuit of the OCV is detected. **Possible Causes:** • An open or short to ground in the control circuit • An open in the power circuit • Poor connection is causing a fault • Faulty OCV • Faulty ECM

DTC	Trouble Code Title, Conditions & Possible Causes

DTC: P0077
1T CCM, MIL: Yes
Years: 2006, 2007, 2008
Models: Accent, Azera, Elantra, Entourage, Santa Fe, Sonata, Tiburon, Tucson, Veracruz
Engines: 1.6L VIN C, 2.0L VIN B, 2.0L VIN D, 2.4L VIN C, 2.7L VIN D, 3.3L VIN D, 3.3L VIN E, 3.3L VIN F, 3.8L VIN 3, 3.8L VIN C, 3.8L VIN F
Transmissions: All
	Intake Valve Control Solenoid Circuit High (Bank 1)
	The engine is running and no disabling faults are current. The ignition voltage is between 11 and 16 volts. An Oil Control Valve (OCV) output failure is detected. A short to battery voltage is detected.
	Possible Causes:
	  • A short to battery voltage in the control circuit
	  • Poor connection is causing a fault
	  • Faulty OCV
	  • Faulty ECM

DTC: P0082
1T CCM, MIL: Yes
Years: 2006, 2007, 2008
Models: Azera, Entourage, Santa Fe, Sonata, Veracruz
Engines: 2.7L VIN D, 3.3L VIN D, 3.3L VIN E, 3.3L VIN F, 3.8L VIN 3, 3.8L VIN C, 3.8L VIN F
Transmissions: All
	Intake Valve Control Solenoid Circuit Low (Bank 2)
	The engine is running and no disabling faults are current. The ignition voltage is between 11 and 16 volts. An Oil Control Valve (OCV) output failure is detected. A short to ground or open circuit of the OCV is detected.
	Possible Causes:
	  • An open in the power circuit
	  • Open or short to ground in the control circuit
	  • Poor connection is causing a fault
	  • Faulty OCV
	  • Faulty ECM

DTC: P0083
1T CCM, MIL: Yes
Years: 2006, 2007, 2008
Models: Azera, Entourage, Santa Fe, Sonata, Veracruz
Engines: 2.7L VIN D, 3.3L VIN D, 3.3L VIN E, 3.3L VIN F, 3.8L VIN 3, 3.8L VIN C, 3.8L VIN F
Transmissions: All
	Intake Valve Control Solenoid Circuit High (Bank 2)
	The engine is running and no disabling faults are current. The ignition voltage is between 11 and 16 volts. An Oil Control Valve (OCV) output failure is detected. A short to battery voltage of the OCV circuit is detected.
	Possible Causes:
	  • A short to battery voltage in the control circuit
	  • Poor connection is causing a fault
	  • Faulty OCV
	  • Faulty ECM

DTC: P0100
2T CCM, MIL: Yes
Years: 2006
Models: Accent
Engines: 1.6L VIN C
Transmissions: All
	Mass or Volume Airflow Sensor Circuit Malfunction
	The engine is started with an engine runtime of over 5 seconds. The PCM detects an unexpected voltage condition on the MAF sensor circuit.
	Possible Causes:
	  • MAF sensor signal circuit is shorted to ground
	  • MAF sensor signal circuit is shorted to VREF or system power
	  • MAF sensor ground circuit is open between sensor and ground
	  • MAF sensor is damaged or has failed
	  • ECM has failed

DTC: P0101
2T CCM, MIL: Yes
Years: 2006, 2007, 2008
Models: Azera, Elantra, Entourage, Santa Fe, Sonata, Tiburon, Tucson, Veracruz
Engines: 2.0L VIN B, 2.0L VIN D, 2.4L VIN B, 2.4L VIN C, 2.7L VIN D, 2.7L VIN F, 3.3L VIN D, 3.3L VIN E, 3.3L VIN F, 3.5L VIN E, 3.8L VIN 3, 3.8L VIN C, 3.8L VIN F
Transmissions: All
	Mass or Volume Air Flow Circuit Range / Performance
	The engine is started and running at idle speed. The PCM detects the MAF sensor signal is less than 0.5 volt. Or with the engine speed more than 3,000 RPM, the MAF sensor signal is more than 4.5 volts.
	Possible Causes:
	  • Base engine vacuum leak, PCV valve leaking or stuck open
	  • An air flow restriction
	  • Water intrusion (in cold climates, inspect for any snow or ice buildup)
	  • Verify that any electrical aftermarket devices are properly connected and grounded
	  • MAF sensor element (wire) is contaminated or dirty
	  • MAF sensor signal circuit is shorted to ground
	  • MAF sensor signal circuit is shorted to VREF or system power
	  • MAF sensor ground circuit is open between sensor and ground
	  • MAF sensor has failed
	  • ECM has failed

DTC	Trouble Code Title, Conditions & Possible Causes
DTC: P0102 **2T CCM, MIL: Yes** **Years:** 2006, 2007, 2008 **Models:** Azera, Elantra, Entourage, Santa Fe, Sonata, Tiburon, Tucson, Veracruz **Engines:** 2.0L VIN B, 2.0L VIN D, 2.4L VIN B, 2.4L VIN C, 2.7L VIN D, 2.7L VIN F, 3.3L VIN D, 3.3L VIN E, 3.3L VIN F, 3.5L VIN E, 3.8L VIN 3, 3.8L VIN C, 3.8L VIN F **Transmissions:** All	**Mass or Volume Airflow Sensor Circuit Low Input** The engine is started and running for over 5 seconds. The PCM detected the MAF sensor signal was less than 0.5 volt during the test. **Possible Causes:** • Base engine vacuum leak, PCV valve leaking or stuck open • An air flow restriction • Water intrusion (in cold climates, inspect for any snow or ice buildup) • Verify that any electrical aftermarket devices are properly connected and grounded • MAF sensor signal circuit is open or shorted to ground • MAF sensor power (VREF) circuit is open or shorted to ground • MAF sensor element (wire) is contaminated or dirty • MAF sensor has failed • ECM has failed
DTC: P0103 **2T CCM, MIL: Yes** **Years:** 2006, 2007, 2008 **Models:** Azera, Elantra, Entourage, Santa Fe, Sonata, Tiburon, Tucson, Veracruz **Engines:** 2.0L VIN B, 2.0L VIN D, 2.4L VIN B, 2.4L VIN C, 2.7L VIN D, 2.7L VIN F, 3.3L VIN D, 3.3L VIN E, 3.3L VIN F, 3.5L VIN E, 3.8L VIN 3, 3.8L VIN C, 3.8L VIN F **Transmissions:** All	**Mass or Volume Air Flow Circuit High Input** The engine is running for over 5 seconds and the PCM detected the MAF sensor input was out of range on the high end. **Possible Causes:** • Base engine vacuum leak, PCV valve leaking or stuck open • An air flow restriction • Water intrusion (in cold climates, inspect for any snow or ice buildup) • Verify that any electrical aftermarket devices are properly connected and grounded • MAF sensor signal circuit is open between the sensor and PCM • MAF sensor signal circuit is shorted to VREF or system power • MAF sensor ground circuit is open between sensor and ground • MAF sensor element (wire) is contaminated or dirty • MAF sensor or has failed • ECM has failed
DTC: P0105 **2T CCM, MIL: Yes** **Years:** 2006, 2007, 2008 **Models:** Azera, Entourage, Santa Fe, Sonata, Veracruz **Engines:** 2.7L VIN D, 3.3L VIN D, 3.3L VIN E, 3.3L VIN F, 3.8L VIN 3, 3.8L VIN C, 3.8L VIN F **Transmissions:** All	**Manifold Absolute Pressure/Barometric Pressure Circuit** The engine is started and running over 5 seconds. The PCM detected an unexpected voltage condition (i.e., more than 4.50 volts or less than 1.95 volts) on the MAP sensor circuit for greater than 4 seconds. **Possible Causes:** • Base engine vacuum leak, PCV valve leaking or stuck open • An air flow restriction • Water intrusion (in cold climates, inspect for any snow or ice buildup) • Verify that any electrical aftermarket devices are properly connected and grounded • MAP sensor signal circuit is open or shorted to ground • MAP sensor ground circuit open between sensor and ground • MAP sensor power (VREF) circuit is open • MAF sensor element (wire) is contaminated or dirty • MAF sensor has failed • ECM has failed
DTC: P0106 **2T CCM, MIL: Yes** **Years:** 2006, 2007, 2008 **Models:** Accent, Azera, Azera, Entourage, Santa Fe, Sonata, Veracruz **Engines:** 1.6L VIN C, 2.4L VIN B, 2.7L VIN D, 3.3L VIN D, 3.3L VIN E, 3.3L VIN F, 3.5L VIN E, 3.8L VIN 3, 3.8L VIN C, 3.8L VIN F **Transmissions:** All	**Manifold Absolute Pressure/Barometric Pressure Circuit Range/Performance** The engine is fully warmed to operating temperature and is running between 600 and 3,000 RPM. The MAP sensor output voltage is out of the threshold value during a rationality check. **Possible Causes:** • Loose or improper installation of air induction system • An air flow restriction • Any vacuum leak • Improperly routed vacuum hoses • In cold climates, inspect for snow or ice buildup • MAP sensor signal circuit high resistance or shorted to system power • MAP sensor VREF circuit high resistance or shorted to system power • MAP sensor is damaged or has failed • MAP sensor Low reference circuit open • ECM has failed

DTC	Trouble Code Title, Conditions & Possible Causes
DTC: P0107 **2T CCM, MIL: Yes** **Years:** 2006, 2007, 2008 **Models:** Accent, Azera, Azera, Entourage, Santa Fe, Sonata, Veracruz **Engines:** 1.6L VIN C, 2.4L VIN B, 2.7L VIN D, 3.3L VIN D, 3.3L VIN E, 3.3L VIN F, 3.5L VIN E, 3.8L VIN 3, 3.8L VIN C, 3.8L VIN F **Transmissions:** All	**Manifold Absolute Pressure/Barometric Pressure Circuit Low Input** The ignition is ON or the engine is running. This DTC runs continuously within the enabling conditions. DTC P0107 is set when the ECM detects that the MAP sensor voltage is less than the threshold value. **Possible Causes:** • MAP sensor VREF circuit high resistance or shorted to ground • MAP sensor signal circuit high resistance shorted to ground or open condition • MAP sensor is damaged or has failed • Intake manifold vacuum leaks • Damage to the MAP sensor housing • Damage to the MAP sensor seal • Loose or improperly installed MAP sensor • Restriction in the vacuum source of the MAP sensor • ECM has failed
DTC: P0108 **2T CCM, MIL: Yes** **Years:** 2006, 2007, 2008 **Models:** Accent, Azera, Azera, Entourage, Santa Fe, Sonata, Veracruz **Engines:** 1.6L VIN C, 2.4L VIN B, 2.7L VIN D, 3.3L VIN D, 3.3L VIN E, 3.3L VIN F, 3.5L VIN E, 3.8L VIN 3, 3.8L VIN C, 3.8L VIN F **Transmissions:** All	**Manifold Absolute Pressure/Barometric Pressure Circuit High Input** The ignition is ON or the engine is running. This DTC runs continuously within the enabling conditions. DTC P0108 is set when the ECM detects that the MAP sensor voltage is greater than 4.5 volts for more than 2.5 seconds. **Possible Cause:** • MAP sensor VREF circuit high resistance or shorted to ground • MAP sensor signal circuit high resistance shorted to ground or open condition • MAP sensor is damaged or has failed • Intake manifold vacuum leaks • Damage to the MAP sensor housing • Damage to the MAP sensor seal • Water intrusion (in cold climates, inspect for any snow or ice buildup) • Loose or improperly installed MAP sensor • Restriction in the vacuum source of the MAP sensor • ECM has failed
DTC: P0109 **2T CCM, MIL: Yes** **Years:** 2006, 2007, 2008 **Models:** Azera, Entourage, Santa Fe, Sonata, Veracruz **Engines:** 2.7L VIN D, 3.3L VIN D, 3.3L VIN F, 3.8L VIN 3, 3.8L VIN C, 3.8L VIN F **Transmissions:** All	**Manifold Absolute Pressure/Barometric Pressure Circuit Intermittent** The engine is running. This DTC runs continuously within the enabling conditions. DTC P0109 is set when the ECM detects that the MAP sensor voltage is lower than the threshold value for 2 minutes. **Possible Cause:** • MAP sensor VREF circuit high resistance or shorted to ground • MAP sensor signal circuit high resistance shorted to ground or open condition • MAP sensor is damaged or has failed • Intake manifold vacuum leaks • Damage to the MAP sensor housing • Damage to the MAP sensor seal • Water intrusion (in cold climates, inspect for any snow or ice buildup) • Loose or improperly installed MAP sensor • Restriction in the vacuum source of the MAP sensor • ECM has failed
DTC: P0110 **2T CCM, MIL: Yes** **Years:** 2006, 2007, 2008 **Models:** Azera, Entourage, Santa Fe, Sonata, Veracruz **Engines:** 2.7L VIN D, 3.3L VIN D, 3.3L VIN E, 3.3L VIN F, 3.8L VIN 3, 3.8L VIN C, 3.8L VIN F **Transmissions:** All	**Intake Air Temperature Sensor 1 Circuit** The ignition is ON or the engine is running. The PCM detected an unexpected voltage condition on the IAT sensor signal circuit during the test. **Possible Cause:** • IAT sensor signal circuit is open or shorted to ground • IAT sensor ground circuit is open between sensor and the PCM • IAT sensor signal circuit is shorted to VREF or system power • IAT sensor is damaged or has failed • PCM has failed

DTC	Trouble Code Title, Conditions & Possible Causes
DTC: P0111 **2T CCM, MIL: Yes** **Years:** 2006, 2007, 2008 **Models:** Accent, Azera, Elantra, Entourage, Santa Fe, Sonata, Tiburon, Tucson, Veracruz **Engines:** 1.6L VIN C, 2.0L VIN B, 2.0L VIN D, 2.4L VIN B, 2.4L VIN C, 2.7L VIN D, 2.7L VIN F, 3.3L VIN D, 3.3L VIN E, 3.3L VIN F, 3.5L VIN E, 3.8L VIN 3, 3.8L VIN C, 3.8L VIN F **Transmissions:** All	**Intake Air Temperature Sensor 1 Circuit Range/Performance** If the sensor is out of specification, a code is set. The output voltage is monitored. If the PCM detects that the Intake Air Temperature (IAT), correlated to coolant temperature, does not change, the PCM determines that a fault exists and a DTC is stored. **Possible Cause:** • IAT sensor signal circuit is shorted to sensor or chassis ground • IAT sensor is damaged or has failed • ECM has failed
DTC: P0112 **2T CCM, MIL: Yes** **Years:** 2006, 2007, 2008 **Models:** Accent, Azera, Elantra, Entourage, Santa Fe, Sonata, Tiburon, Tucson, Veracruz **Engines:** 1.6L VIN C, 2.0L VIN B, 2.0L VIN D, 2.4L VIN B, 2.4L VIN C, 2.7L VIN D, 2.7L VIN F, 3.3L VIN D, 3.3L VIN E, 3.3L VIN F, 3.5L VIN E, 3.8L VIN 3, 3.8L VIN C, 3.8L VIN F **Transmissions:** All	**Intake Air Temperature Sensor 1 Circuit Low Input** The engine is running. DTC P0112 is set when the ECM detects that the IAT output signal is less than 0.1 volt for more than 10 seconds. **Possible Causes:** • IAT sensor signal circuit is shorted to sensor or chassis ground • IAT sensor is damaged or has failed • ECM has failed
DTC: P0113 **2T CCM, MIL: Yes** **Years:** 2006, 2007, 2008 **Models:** Accent, Azera, Elantra, Entourage, Santa Fe, Sonata, Tiburon, Tucson, Veracruz **Engines:** 1.6L VIN C, 2.0L VIN B, 2.0L VIN D, 2.4L VIN B, 2.4L VIN C, 2.7L VIN D, 2.7L VIN F, 3.3L VIN D, 3.3L VIN E, 3.3L VIN F, 3.5L VIN E, 3.8L VIN 3, 3.8L VIN C, 3.8L VIN F **Transmissions:** All	**Intake Air Temperature Sensor 1 Circuit High Input** The engine is running. DTC P0113 is set when the ECM detects that the IAT output signal is greater than 4.9 volts for more than 10 seconds. **Possible Causes:** • IAT sensor signal circuit is open between the sensor and ECM • IAT sensor signal circuit is shorted to VREF or system power • IAT sensor is damaged or has failed • ECM has failed
DTC: P0115 **1T CCM, MIL: No** **Years:** 2006, 2007, 2008 **Models:** Accent, Azera, Entourage, Santa Fe, Sonata, Veracruz **Engines:** 1.6L VIN C, 2.7L VIN D, 3.3L VIN D, 3.3L VIN E, 3.3L VIN F, 3.8L VIN 3, 3.8L VIN C, 3.8L VIN F **Transmissions:** All	**Engine Coolant Temperature (ECT) Sensor Circuit** The ignition is ON or the engine is running and the PCM detected an unexpected voltage condition on the ECT sensor signal circuit during the test. **Possible Causes:** • ECT sensor signal circuit is open or shorted ground • ECT sensor ground circuit is open between sensor and PCM • ECT sensor signal circuit is shorted to VREF or system power • ECT sensor is damaged or has failed • PCM has failed
DTC: P0116 **2T CCM, MIL: Yes** **Years:** 2006, 2007, 2008 **Models:** Accent, Azera, Elantra, Entourage, Santa Fe, Sonata, Tiburon, Tucson, Veracruz **Engines:** 1.6L VIN C, 2.0L VIN B, 2.0L VIN D, 2.4L VIN B, 2.4L VIN C, 2.7L VIN D, 2.7L VIN F, 3.3L VIN D, 3.3L VIN E, 3.3L VIN F, 3.5L VIN E, 3.8L VIN 3, 3.8L VIN C, 3.8L VIN F **Transmissions:** All	**Engine Coolant Temperature (ECT) Sensor Performance** The engine is running for over 20 minutes. The PCM detected the ECT sensor signal is more than 68°F from the model curve stored in memory. This could be an intermittent fault. Check for a problem related to the Cooling system. **Possible Causes:** • ECT sensor circuit is open or shorted ground (intermittent fault) • ECT sensor ground circuit is open (an intermittent fault) • ECT sensor has drifted out of calibration or has failed • PCM has failed

DTC	Trouble Code Title, Conditions & Possible Causes
DTC: P0117 **2T CCM, MIL: Yes** **Years:** 2006, 2007, 2008 **Models:** Accent, Azera, Elantra, Entourage, Santa Fe, Sonata, Tiburon, Tucson, Veracruz **Engines:** 1.6L VIN C, 2.0L VIN B, 2.0L VIN D, 2.4L VIN B, 2.4L VIN C, 2.7L VIN D, 2.7L VIN F, 3.3L VIN D, 3.3L VIN E, 3.3L VIN F, 3.5L VIN E, 3.8L VIN 3, 3.8L VIN C, 3.8L VIN F **Transmissions:** All	**Engine Coolant Temperature (ECT) Sensor Circuit Low Input** The ignition is ON or the engine is running. DTC P0117 is set when the ECM detects that the ECT sensor output signal was less than 0.1 volt for more than 40 seconds. **Possible Causes:** • ECT sensor connector is damaged or shorted • ECT sensor signal circuit shorted to ground • ECT sensor is damaged or has failed • ECM has failed
DTC: P0118 **2T CCM, MIL: Yes** **Years:** 2006, 2007, 2008 **Models:** Accent, Azera, Elantra, Entourage, Santa Fe, Sonata, Tiburon, Tucson, Veracruz **Engines:** 1.6L VIN C, 2.0L VIN B, 2.0L VIN D, 2.4L VIN B, 2.4L VIN C, 2.7L VIN D, 2.7L VIN F, 3.3L VIN D, 3.3L VIN E, 3.3L VIN F, 3.5L VIN E, 3.8L VIN 3, 3.8L VIN C, 3.8L VIN F **Transmissions:** All	**Engine Coolant Temperature (ECT) Sensor Circuit High Input** The engine is running. DTC P0118 is set when the ECM detects that the ECT output signal is greater than 4.9 volts for more than 40 seconds. **Possible Causes:** • ECT sensor connector is damaged, loose, or open • ECT sensor signal circuit is open or high resistance • ECT sensor signal circuit is shorted to VREF • ECT sensor is damaged or has failed • ECM has failed
DTC: P0119 **2T CCM, MIL: Yes** **Years:** 2006, 2007, 2008 **Models:** Santa Fe, Sonata, Tiburon, Tucson **Engines:** 2.4L VIN C, 2.7L VIN D, 2.7L VIN F **Transmissions:** All	**Engine Coolant Temperature (ECT) Sensor Circuit Intermittent** The ignition is ON or the engine is running. DTC P0119 is set when the ECM detects that the ECT sensor signal is intermittent or has abruptly changed for greater than 4 seconds. **Possible Causes:** • ECT sensor connector is damaged, loose, or open • ECT sensor signal circuit is open or high resistance • ECT sensor signal circuit is shorted to VREF • ECT sensor is damaged or has failed • ECM has failed
DTC: P0120 **1T CCM, MIL: Yes** **Years:** 2007, 2008 **Models:** Santa Fe **Engines:** 2.7L VIN D, 3.3L VIN E **Transmissions:** All	**Throttle Position (TP) Sensor Switch Circuit** DTC P0120 is set when the TP sensor voltage is less than 0.27 volt or more than 4.67 volts for more than 0.5 second. **Possible Causes:** • TP Sensor 1 circuit is open, shorted to ground, or has a high resistance condition • Throttle body assembly is damaged or it has failed • ECM has failed
DTC: P0121 **2T CCM, MIL: Yes** **Years:** 2006, 2007, 2008 **Models:** Accent, Elantra, Santa Fe, Sonata, Tiburon, Tucson **Engines:** 1.6L VIN C, 2.0L VIN B, 2.0L VIN D, 2.4L VIN B, 2.4L VIN C, 2.7L VIN D, 2.7L VIN F, 3.5L VIN E **Transmissions:** All	**Throttle Position (TP) Sensor Range/Performance** The engine is running. The PCM detected the TP sensor input was incorrect when it was compared to the MAF sensor signal. **Possible Causes:** • TP sensor connector is damaged, open or shorted • TP Sensor low reference circuit is open or has high resistance • TP Sensor 1 signal is open, shorted to ground or to 5v VREF • TP Sensor 1 VREF circuit is open or shorted to ground • Throttle body assembly is damaged or it has failed • ECM has failed
DTC: P0122 **1T CCM, MIL: Yes** **Years:** 2006, 2007, 2008 **Models:** Accent, Azera, Elantra, Entourage, Santa Fe, Sonata, Tiburon, Tucson, Veracruz **Engines:** 1.6L VIN C, 2.0L VIN B, 2.0L VIN D, 2.4L VIN B, 2.4L VIN C, 2.7L VIN D, 2.7L VIN F, 3.3L VIN D, 3.3L VIN E, 3.3L VIN F, 3.5L VIN E, 3.8L VIN 3, 3.8L VIN C, 3.8L VIN F **Transmissions:** All	**Throttle Position (TP) Sensor/Switch Circuit Low Input** The engine is running for over 5 seconds and the PCM detected the TP sensor signal was 0.2 volts or lower during the test. **Possible Causes:** • TP Sensor low reference circuit is open or has high resistance • TP Sensor 1 signal is open, shorted to ground or low reference • TP Sensor 1 VREF circuit is open or shorted to ground • Throttle body assembly is damaged or it has failed • ECM has failed

DTC	Trouble Code Title, Conditions & Possible Causes
DTC: P0123 **1T CCM, MIL: Yes** **Years:** 2006, 2007, 2008 **Models:** Accent, Azera, Elantra, Entourage, Santa Fe, Sonata, Tiburon, Tucson, Veracruz **Engines:** 1.6L VIN C, 2.0L VIN B, 2.0L VIN D, 2.4L VIN B, 2.4L VIN C, 2.7L VIN D, 2.7L VIN F, 3.3L VIN D, 3.3L VIN E, 3.3L VIN F, 3.5L VIN E, 3.8L VIN 3, 3.8L VIN C, 3.8L VIN F **Transmissions:** All	**Throttle Position (TP) Sensor 1 Circuit High Input** The engine is running for over 5 seconds and the PCM detected the TP sensor signal was 4.96 volts or higher during the test. **Possible Causes:** • TP Sensor 1 VREF shorted to voltage • Throttle body assembly is damaged or it has failed • TP Sensor low reference circuit is open or has high resistance • TP Sensor 1 signal is shorted to voltage • ECM has failed
DTC: P0124 **2T CCM, MIL: Yes** **Years:** 2006, 2007, 2008 **Models:** Accent **Engines:** 1.6L VIN C **Transmissions:** All	**Throttle/Pedal Position Sensor/Switch "A" Circuit Intermittent** The engine is running at a speed greater than 600 RPM for more than 20 seconds. The engine coolant temperature is greater than 167°F (75°C). The PCM detected the difference between the modeled relative load and the measured relative load as over the threshold value. **Possible Causes:** • Poor connection at Throttle Position Sensor (TPS) • TPS has failed
DTC: P0125 **2T CCM, MIL: Yes** **Years:** 2006, 2007, 2008 **Models:** Azera, Elantra, Entourage, Santa Fe, Sonata, Tiburon, Tucson, Veracruz **Engines:** 2.0L VIN B, 2.0L VIN D, 2.4L VIN B, 2.4L VIN C, 2.7L VIN D, 2.7L VIN F, 3.3L VIN D, 3.3L VIN E, 3.3L VIN F, 3.5L VIN E, 3.8L VIN 3, 3.8L VIN C, 3.8L VIN F **Transmissions:** All	**Engine Coolant Temperature (ECT) Insufficient For Closed Loop Fuel Control** The engine is running. No disabling faults are set that are related to the MAFS/MAPS, catalyst, fuel system, or engine oil temperature sensor. The engine coolant temperature does not reach the coolant temperature threshold within a predetermined period of time. **Possible Causes:** • Check the operation of the thermostat (it may be stuck open) • Coolant level is too low or the coolant mixture is incorrect • ECT sensor signal circuit has a high resistance condition • ECT sensor is damaged or it has failed
DTC: P0128 **2T CCM, MIL: Yes** **Years:** 2006, 2007, 2008 **Models:** Accent, Azera, Elantra, Entourage, Santa Fe, Sonata, Tiburon, Tucson, Veracruz **Engines:** 1.6L VIN C, 2.0L VIN B, 2.0L VIN D, 2.4L VIN B, 2.4L VIN C, 2.7L VIN D, 2.7L VIN F, 3.3L VIN D, 3.3L VIN E, 3.3L VIN F, 3.5L VIN E, 3.8L VIN 3, 3.8L VIN C, 3.8L VIN F **Transmissions:** All	**Engine Coolant Temperature (ECT) Below Thermostat Regulating Temperature** DTC P0128 is set when the PCM detects that the ECT sensor input is less than 40°F at startup, engine started, and the PCM detected the ECT sensor did not reach 167°F after a normal warm up period had expired during the CCM Rationality test. **Possible Causes:** • Check the operation of the thermostat (it may be stuck open) • Coolant level is too low, or the coolant mixture is incorrect • ECT sensor signal circuit short to ground or high resistance • ECT sensor is damaged or it has failed • ECM has failed
DTC: P0130 **2T CCM, MIL: Yes** **Years:** 2006, 2007, 2008 **Models:** Accent, Santa Fe, Sonata, Tiburon, Tucson **Engines:** 1.6L VIN C, 2.0L VIN B, 2.0L VIN D, 2.4L VIN C, 2.7L VIN D, 2.7L VIN F **Transmissions:** All	**HO2S-11 Circuit Malfunction (Bank 1/Sensor 1)** The engine is running. DTC P0130 is set when the ECM detects that the Loop Status parameter is open. DTC P0130 sets within 50 seconds when the above condition is met. **Possible Causes:** • Heated Oxygen Sensor (HO2S) 1 high signal circuit short to voltage • HO2S 1 high signal circuit open/high resistance or a short to ground • HO2S 1 has failed • ECM has failed

DTC	Trouble Code Title, Conditions & Possible Causes
DTC: P0131 **2T CCM, MIL: Yes** **Years:** 2006, 2007, 2008 **Models:** Accent, Azera, Elantra, Entourage, Santa Fe, Sonata, Tiburon, Tucson, Veracruz **Engines:** 1.6L VIN C, 2.0L VIN B, 2.0L VIN D, 2.4L VIN B, 2.4L VIN C, 2.7L VIN D, 2.7L VIN F, 3.3L VIN D, 3.3L VIN E, 3.3L VIN F, 3.5L VIN E, 3.8L VIN 3, 3.8L VIN C, 3.8L VIN F **Transmissions:** All	**HO2S-11 Circuit Low Voltage/Input (Bank 1/Sensor 1)** The engine is running for more than 60 seconds and the battery voltage is greater than 10 volts. DTC P0131 is set when the ECM detects that the HO2S-11 is less than 0.04 volts for more than 12:5 seconds. **Possible Causes:** • Low fuel pressure, fuel filter restricted, or fuel injectors plugged • HO2S signal circuit is shorted to ground (an intermittent fault) • HO2S is damaged or contaminated • PCM has failed
DTC: P0132 **2T CCM, MIL: Yes** **Years:** 2006, 2007, 2008 **Models:** Accent, Azera, Elantra, Entourage, Santa Fe, Sonata, Tiburon, Tucson, Veracruz **Engines:** 1.6L VIN C, 2.0L VIN B, 2.0L VIN D, 2.4L VIN B, 2.4L VIN C, 2.7L VIN D, 2.7L VIN F, 3.3L VIN D, 3.3L VIN E, 3.3L VIN F, 3.5L VIN E, 3.8L VIN 3, 3.8L VIN C, 3.8L VIN F **Transmissions:** All	**HO2S-11 Circuit High Voltage/Input (Bank 1/Sensor 1)** The engine is running in closed loop. DTC P0132 is set when the ECM detects that the HO2S-11 parameter is more than 1.3 volts for a predetermined time. **Possible Causes:** • Fuel pressure regulator leaking or fuel injectors leaking • HO2S signal circuit is shorted to the heater power circuit • HO2S may be contaminated or it has failed • HO2S heater is damaged or has failed • PCM has failed
DTC: P0133 **2T CCM, MIL: Yes** **Years:** 2006, 2007, 2008 **Models:** Accent, Azera, Elantra, Entourage, Santa Fe, Sonata, Tiburon, Tucson, Veracruz **Engines:** 1.6L VIN C, 2.0L VIN B, 2.0L VIN D, 2.4L VIN B, 2.4L VIN C, 2.7L VIN D, 2.7L VIN F, 3.3L VIN D, 3.3L VIN E, 3.3L VIN F, 3.5L VIN E, 3.8L VIN 3, 3.8L VIN C, 3.8L VIN F **Transmissions:** All	**HO2S-11 Slow Response (Bank 1/Sensor 1)** The engine is running greater than 60 seconds and has reached an operating temperature of at least 158°F (70°C). No disabling faults are present. The calculated response rate rich-to-lean or from lean-to-rich is too slow (out of the threshold in the PCM). **Possible Causes:** • HO2S signal circuit is open or shorted to ground • HO2S element is contaminated or it has failed • HO2S heater is damaged or has failed • Intake air leaks, exhaust manifold leaks or PCV system leaks • MAF sensor out of calibration (it may be dirty or contaminated) • Exhaust system is leaking or severely restricted • ECM has failed
DTC: P0134 **2T CCM, MIL: Yes** **Years:** 2006, 2007, 2008 **Models:** Accent, Azera, Elantra, Entourage, Santa Fe, Sonata, Tiburon, Tucson, Veracruz **Engines:** 1.6L VIN C, 2.0L VIN B, 2.0L VIN D, 2.4L VIN B, 2.4L VIN C, 2.7L VIN D, 2.7L VIN F, 3.3L VIN D, 3.3L VIN E, 3.3L VIN F, 3.5L VIN E, 3.8L VIN 3, 3.8L VIN C, 3.8L VIN F **Transmissions:** All	**HO2S-11 Circuit Insufficient Activity (Bank 1/Sensor 1)** The engine is running in closed loop with a battery voltage greater than 10 volts. No disabling faults are present. The PCM detected the HO2S-11 signal remained fixed from 0.415 to 0.515 volts for more than 1 minute during the CCM test. **Possible Causes:** • Exhaust system is leaking or severely restricted • HO2S signal circuit is open or shorted to ground • HO2S element is contaminated or it has failed • HO2S heater is damaged or has failed • PCM has failed
DTC: P0135 **2T O2S HTR, MIL: Yes** **Years:** 2006, 2007, 2008 **Models:** Accent **Engines:** 1.6L VIN C **Transmissions:** All	**HO2S-11 Heater Circuit Malfunction (Bank 1/Sensor 1)** The engine is started and running in closed loop and the PCM detected the HO2S-11 heater current was less than 0.2 amps, or that is was more than 3.5 amps during the test period. **Possible Causes:** • HO2S heater voltage supply circuit open or short to ground • HO2S heater low control circuit open or short to voltage • HO2S heater element is damaged or has failed • ECM has failed

DTC	Trouble Code Title, Conditions & Possible Causes
DTC: P0136 **2T CCM, MIL: Yes** **Years:** 2006, 2007, 2008 **Models:** Accent, Elantra, Santa Fe, Sonata, Tiburon, Tucson **Engines:** 1.6L VIN C, 2.0L VIN B, 2.0L VIN D, 2.4L VIN C, 2.7L VIN D, 2.7L VIN F **Transmissions:** All	**HO2S-12 Circuit Malfunction (Bank 1/Sensor 2)** The engine is started and running in closed loop and the PCM detected an unexpected high voltage on the HO2S circuit; or the HO2S signal was fixed at mid-range (350–550 mV) or not switching properly. **Possible Causes:** • Fuel system lean • HO2S signal circuit is open between the sensor and the PCM • HO2S signal circuit is shorted to sensor or chassis ground • HO2S signal circuit is shorted to VREF or system power (B+) • HO2S is damaged, contaminated or it has failed • PCM has failed
DTC: P0137 **2T CCM, MIL: Yes** **Years:** 2006, 2007, 2008 **Models:** Accent, Azera, Elantra, Entourage, Santa Fe, Sonata, Tiburon, Tucson, Veracruz **Engines:** 1.6L VIN C, 2.0L VIN B, 2.0L VIN D, 2.4L VIN B, 2.4L VIN C, 2.7L VIN D, 2.7L VIN F, 3.3L VIN D, 3.3L VIN E, 3.3L VIN F, 3.5L VIN E, 3.8L VIN 3, 3.8L VIN C, 3.8L VIN F **Transmissions:** All	**HO2S-12 Circuit Low Input (Bank 1/Sensor 2)** The engine is started and running in closed loop and the PCM detected the HO2S-12 signal of less than 0.16 volt during the CCM test. **Possible Causes:** • Fuel system lean • Low fuel pressure, fuel filter restricted or fuel injectors plugged • HO2S signal circuit is shorted to ground (an intermittent fault) • HO2S may be contaminated or it has failed • HO2S heater is damaged or has failed • PCM has failed
DTC: P0138 **2T CCM, MIL: Yes** **Years:** 2006, 2007, 2008 **Models:** Accent, Azera, Elantra, Entourage, Santa Fe, Sonata, Tiburon, Tucson, Veracruz **Engines:** 1.6L VIN C, 2.0L VIN B, 2.0L VIN D, 2.4L VIN B, 2.4L VIN C, 2.7L VIN D, 2.7L VIN F, 3.3L VIN D, 3.3L VIN E, 3.3L VIN F, 3.5L VIN E, 3.8L VIN 3, 3.8L VIN C, 3.8L VIN F **Transmissions:** All	**HO2S-12 Circuit High Input (Bank 1/Sensor 2)** The engine is running in closed loop and the PCM detected the HO2S-12 signal of more than 1.3 volts during the CCM test. **Possible Causes:** • Exhaust system is leaking or severely restricted • HO2S high signal circuit short to voltage • HO2S low signal circuit is open high resistance or short to voltage • HO2S is damaged or contaminated • ECM has failed
DTC: P0139 **2T CCM, MIL: Yes** **Years:** 2006, 2007, 2008 **Models:** Accent, Azera, Elantra, Entourage, Santa Fe, Sonata, Tiburon, Tucson, Veracruz **Engines:** 1.6L VIN C, 2.0L VIN B, 2.0L VIN D, 2.4L VIN B, 2.4L VIN C, 2.7L VIN D, 2.7L VIN F, 3.3L VIN D, 3.3L VIN E, 3.3L VIN F, 3.5L VIN E, 3.8L VIN 3, 3.8L VIN C, 3.8L VIN F **Transmissions:** All	**HO2S-12 Slow Response (Bank 1/Sensor 2)** The engine is running in closed loop, ECT sensor signal more than 158°F, and the ECM detected the number of HO2S-12 rich-to-lean or lean-to-rich switches was less than a calibrated amount. **Possible Causes:** • HO2S signal circuit is open or shorted to ground • HO2S element is contaminated or it has failed • HO2S heater is damaged or has failed • Intake air leaks, exhaust manifold leaks or PCV system leaks • MAF sensor out of calibration (it may be dirty or contaminated) • ECM has failed
DTC: P0140 **2T CCM, MIL: Yes** **Years:** 2006, 2007, 2008 **Models:** Accent, Azera, Elantra, Entourage, Santa Fe, Sonata, Tiburon, Tucson, Veracruz **Engines:** 1.6L VIN C, 2.0L VIN B, 2.0L VIN D, 2.4L VIN B, 2.4L VIN C, 2.7L VIN D, 2.7L VIN F, 3.3L VIN D, 3.3L VIN E, 3.3L VIN F, 3.5L VIN E, 3.8L VIN 3, 3.8L VIN C, 3.8L VIN F **Transmissions:** All	**HO2S-12 Circuit Insufficient Activity (Bank 1/Sensor 2)** The engine is running in closed loop and the PCM detected the HO2S-12 signal remained fixed from 0.415 to 0.515 volts for more than 1 minute during the CCM test. **Possible Causes:** • HO2S signal circuit is open or shorted to ground • Exhaust system is leaking or severely restricted • HO2S element is contaminated, damaged or has failed • PCM has failed

DTC	Trouble Code Title, Conditions & Possible Causes
DTC: P0141 **2T O2S HTR, MIL: Yes** **Years:** 2006, 2007, 2008 **Models:** Accent **Engines:** 1.6L VIN C **Transmissions:** All	**HO2S-12 Heater Circuit Performance (Bank 1/Sensor 2)** DTC P0141 is set when the ECM detects that the HO2S 2 heater current is more than 2.5 amps or less than 0.25 amp. DTC P0141 sets within 30 seconds during the heater current test when the above condition is met. **Possible Causes:** • HO2S heater voltage supply circuit open short to ground • HO2S heater low control circuit open or short to voltage • HO2S heater element is damaged or has failed • ECM has failed
DTC: P0150 **2T CCM, MIL: Yes** **Years:** 2006, 2007, 2008 **Models:** Santa Fe, Tiburon, Tucson **Engines:** 2.7L VIN D, 2.7L VIN F **Transmissions:** All	**HO2S-21 Sensor Circuit Malfunction (Bank 2/Sensor 1)** The engine is running in closed loop and the PCM detected failure condition: the HO2S signal was too high; or the HO2S signal was fixed without variance; or the HO2S signal was too long in switching between rich-to-lean. **Possible Causes:** • HO2S signal circuit is open between the sensor and the PCM • HO2S signal circuit is shorted to sensor or chassis ground • HO2S signal circuit is shorted to VREF or system power (B+) • HO2S is damaged, contaminated or it has failed • PCM has failed
DTC: P0151 **2T CCM, MIL: Yes** **Years:** 2006, 2007, 2008 **Models:** Azera, Entourage, Santa Fe, Sonata, Tiburon, Tucson, Veracruz **Engines:** 2.7L VIN D, 2.7L VIN F, 3.3L VIN D, 3.3L VIN E, 3.3L VIN F, 3.5L VIN E, 3.8L VIN 3, 3.8L VIN C, 3.8L VIN F **Transmissions:** All	**HO2S-21 Circuit Low Input (Bank 2/Sensor 1)** The engine is running and the Loop Status parameter is Closed. The Ignition 1 Signal parameter is between 10–18 volts. DTC P0151 is set when the control module detects that the HO2S-21 voltage parameter is less than 100 mV. DTC P0151 sets within 3 seconds when the above condition is met. **Possible Causes:** • Fuel system lean • HO2S high signal circuit open high resistance or short to ground • HO2S low signal circuit open or high resistance • HO2S is damaged or contaminated • ECM has failed
DTC: P0152 **2T CCM, MIL: Yes** **Years:** 2006, 2007, 2008 **Models:** Azera, Entourage, Santa Fe, Sonata, Tiburon, Tucson, Veracruz **Engines:** 2.7L VIN D, 2.7L VIN F, 3.3L VIN D, 3.3L VIN E, 3.3L VIN F, 3.5L VIN E, 3.8L VIN 3, 3.8L VIN C, 3.8L VIN F **Transmissions:** All	**HO2S-21 Circuit High Input (Bank 2/Sensor 1)** The engine is running and the Loop Status parameter is Closed. The Ignition 1 Signal parameter is between 10–18 volts. DTC P0152 is set when the control module detects that the HO2S Bank 2 Sensor 1 voltage parameter is more than 1.2 volts. DTC P0151 sets within 3 seconds when the above condition is met. **Possible Causes:** • Air leaks in the exhaust system, intake manifold, vacuum lines • Fuel system rich condition • HO2S high signal circuit high resistance open or shorted to ground • HO2S high signal circuit shorted to voltage • HO2S low signal circuit high resistance or open • ECM has failed
DTC: P0153 **2T CCM, MIL: Yes** **Years:** 2006, 2007, 2008 **Models:** Azera, Entourage, Santa Fe, Sonata, Tiburon, Tucson, Veracruz **Engines:** 2.7L VIN D, 2.7L VIN F, 3.3L VIN D, 3.3L VIN E, 3.3L VIN F, 3.5L VIN E, 3.8L VIN 3, 3.8L VIN C, 3.8L VIN F **Transmissions:** All	**HO2S-21 Sensor Circuit Slow Response (Bank 2/Sensor 1)** The engine is running and the Engine Coolant Temperature (ECT) Sensor parameter is more than 158°F (70°C). The Ignition 1 Signal parameter is between 10–18 volts. The Loop Status parameter is Closed. DTC P0153 is set when the control module detects that the HO2S 1 rich-to-lean or lean-to-rich average response time is over 1 second. **Possible Causes:** • Exhaust system is leaking or severely restricted • HO2S high signal circuit is open or high resistance • HO2S high signal circuit is shorted to voltage • HO2S low signal circuit is open or high resistance • HO2S is damaged or contaminated • ECM has failed
DTC: P0154 **2T CCM, MIL: Yes** **Years:** 2006, 2007, 2008 **Models:** Azera, Entourage, Santa Fe, Sonata, Tiburon, Tucson, Veracruz **Engines:** 2.7L VIN D, 2.7L VIN F, 3.3L VIN D, 3.3L VIN E, 3.3L VIN F, 3.5L VIN E, 3.8L VIN 3, 3.8L VIN C, 3.8L VIN F **Transmissions:** All	**HO2S-21 Circuit Insufficient Activity (Bank 2/Sensor 1)** The Ignition 1 Signal parameter is between 10–18 volts. DTC P0154 is set when the control module detects that the HO2S-21 parameter is between 400–550 mV for more than 1 minute during the CCM test. **Possible Causes:** • Exhaust system is leaking or severely restricted • HO2S high signal circuit is open or high resistance • HO2S high signal circuit is shorted to voltage • HO2S low signal circuit is open or high resistance • HO2S low signal circuit is shorted to voltage • HO2S is damaged or contaminated • ECM has failed

DTC	Trouble Code Title, Conditions & Possible Causes
DTC: P0155 **2T O2S HTR, MIL: Yes** **Years:** 2006 **Models:** Sonata **Engines:** 2.4L VIN C **Transmissions:** All	**HO2S-21 Heater Performance (Bank 2/Sensor 1)** The engine is running. The ECT Sensor parameter is more than 140°F (60°C). The Ignition 1 Signal parameter is between 10–18 volts. The Engine Speed parameter is between 500–3,000 RPM. DTC P0155 is set when the control module detects that the HO2S-21 heater current parameter is more than 3.125 amps or less than 0.25 amp. **Possible Causes:** • HO2S heater voltage supply circuit short to ground or open • HO2S heater low control circuit open or short to voltage • HO2S heater element is damaged or has failed • ECM has failed
DTC: P0156 **2T O2S HTR, MIL: Yes** **Years:** 2007, 2008 **Models:** Tiburon, Tucson **Engines:** 2.7L VIN D, 2.7L VIN F **Transmissions:** All	**HO2S-22 Sensor Circuit Malfunction (Bank 2/Sensor 2)** The engine is running. The ECT Sensor parameter is more than 140°F (60°C). The Ignition 1 Signal parameter is between 10–18 volts. The Engine Speed parameter is between 500–3,000 RPM. DTC P0156 is set when the control module detects that the HO2S-22 heater signal is too high, is fixed without variance, or takes too long to switch between rich and lean parameters. **Possible Causes:** • HO2S signal circuit is open between the sensor and the PCM • HO2S signal circuit is shorted to sensor or chassis ground • HO2S signal circuit is shorted to VREF or system power (B+) • HO2S is damaged, contaminated or it has failed • PCM has failed
DTC: P0157 **2T O2S HTR, MIL: Yes** **Years:** 2007, 2008 **Models:** Azera, Entourage, Santa Fe, Sonata, Tiburon, Tucson, Veracruz **Engines:** 2.7L VIN D, 2.7L VIN F, 3.3L VIN D, 3.3L VIN E, 3.3L VIN F, 3.5L VIN E, 3.8L VIN 3, 3.8L VIN C, 3.8L VIN F **Transmissions:** All	**HO2S-22 Circuit Low Input (Bank 2/Sensor 2)** The engine is running The Loop Status parameter is Closed. The Ignition 1 Signal parameter is between 10–18 volts. The control module detects that the sensor voltage parameter is less than 80 mV. DTC P0157 sets within 100 seconds during the lean test when the above condition is met. **Possible Causes:** • Fuel system lean • HO2S high signal circuit high resistance short to ground or open condition • HO2S low signal circuit is open or high resistance • HO2S is damaged or contaminated • ECM has failed
DTC: P0158 **2T O2S HTR, MIL: Yes** **Years:** 2007, 2008 **Models:** Azera, Entourage, Santa Fe, Sonata, Tiburon, Tucson, Veracruz **Engines:** 2.7L VIN D, 2.7L VIN F, 3.3L VIN D, 3.3L VIN E, 3.3L VIN F, 3.5L VIN E, 3.8L VIN 3, 3.8L VIN C, 3.8L VIN F **Transmissions:** All	**HO2S-22 Circuit High Input (Bank 2/Sensor 2)** The engine is running. The Loop Status parameter is Closed. The Ignition 1 Signal parameter is between 10–18 volts. The control module detects that the sensor voltage parameter is more than 1.2 volts. **Possible Causes:** • Exhaust system is leaking or severely restricted • HO2S high signal circuit short to voltage • HO2S low signal circuit is open high resistance or shorted to voltage • HO2S is damaged or contaminated • ECM has failed
DTC: P0159 **2T O2S HTR, MIL: Yes** **Years:** 2007, 2008 **Models:** Azera, Entourage, Santa Fe, Sonata, Tiburon, Tucson, Veracruz **Engines:** 2.7L VIN D, 2.7L VIN F, 3.3L VIN D, 3.3L VIN E, 3.3L VIN F, 3.5L VIN E, 3.8L VIN 3, 3.8L VIN C, 3.8L VIN F **Transmissions:** All	**HO2S-22 Sensor Circuit Slow Response (Bank 2/Sensor 2)** The engine is running. The Loop Status parameter is Closed. The control module detects that the response time to switch from rich-to-lean or from lean-to-rich was over 1 second. **Possible Causes:** • HO2S signal circuit is open or shorted to ground • HO2S element is contaminated or it has failed • HO2S heater is damaged or has failed • Intake air leaks, exhaust manifold leaks or PCV system leaks • MAF sensor out of calibration (it may be dirty or contaminated) • PCM has failed

DTC	Trouble Code Title, Conditions & Possible Causes
DTC: P0160 **2T O2S HTR, MIL: Yes** **Years:** 2007, 2008 **Models:** Azera, Entourage, Santa Fe, Sonata, Tiburon, Tucson, Veracruz **Engines:** 2.7L VIN D, 2.7L VIN F, 3.3L VIN D, 3.3L VIN E, 3.3L VIN F, 3.5L VIN E, 3.8L VIN 3, 3.8L VIN C, 3.8L VIN F **Transmissions:** All	**HO2S-22 Insufficient Activity (Bank 2/Sensor 2)** The engine is running. The Loop Status parameter is closed. DTC P0160 runs once per drive cycle when the above conditions are met. The control module detects that the sensor parameter has remained fixed at 0.515 volts or less from more than 1 minute during the CCM test. **Possible Causes:** • HO2S signal circuit is open or shorted to ground • HO2S element is contaminated or it has failed • HO2S heater is damaged or has failed • PCM has failed
DTC: P0161 **2T O2S HTR, MIL: Yes** **Years:** 2006 **Models:** Sonata **Engines:** 3.3L VIN F **Transmissions:** All	**HO2S-22 Sensor Heater Circuit Malfunction (Bank 2/Sensor 2)** The engine is running in closed loop and the PCM detected the HO2S-22 heater current was less than 0.2 amps, or that is was more than 3.5 amps during the test period. **Possible Causes:** • HO2S heater high voltage supply circuit open or shorted to ground • HO2S heater low control circuit open or high resistance • HO2S heater low control circuit shorted to voltage • HO2S heater element is damaged • ECM has failed
DTC: P0170 **2T FUEL, MIL: Yes** **Years:** 2006, 2007, 2008 **Models:** Elantra, Santa Fe, Sonata, Tiburon, Tucson **Engines:** 2.0L VIN B, 2.0L VIN D, 2.4L VIN C, 2.7L VIN D, 2.7L VIN F **Transmissions:** All	**HO2S Fuel Trim/Sensor System Lambda Controller at the Limit (Bank 1)** The is running in closed loop for 3 to 5 minutes and the PCM detected the Fuel Trim system control was too rich or too lean under these conditions in the Fuel System Monitor test. **Possible Causes:** • Air leaks after the MAF sensor or PCV system • Air leaks in intake manifold, exhaust pipes, or exhaust manifold • Base engine mechanical fault affecting one or more cylinders • Exhaust leaks located in front of the HO2S location • Fuel control sensor is out of calibration (i.e., ECT, IAT, or MAP) • Fuel delivery system supplying too little fuel during cruise or idle periods (e.g., faulty fuel pump or dirty, restricted fuel filter) • Fuel injector (one or more) dirty or pressure regulator has failed • Vacuum hose is disconnected, broken, leaking, or loose • HO2S is contaminated, deteriorated, or it has failed • Vehicle driven low on fuel or until it ran out of fuel
DTC: P0171 **2T FUEL, MIL: Yes** **Years:** 2006, 2007, 2008 **Models:** Accent, Azera, Elantra, Entourage, Santa Fe, Sonata, Tiburon, Tucson, Veracruz **Engines:** 1.6L VIN C, 2.0L VIN B, 2.0L VIN D, 2.4L VIN B, 2.7L VIN D, 2.7L VIN F, 3.3L VIN D, 3.3L VIN E, 3.3L VIN F, 3.5L VIN E, 3.8L VIN 3, 3.8L VIN C, 3.8L VIN F **Transmissions:** All	**Fuel Trim Malfunction System Too Lean (Bank 1)** The engine is running in closed loop status. DTC P0171 is set when the PCM detects the Fuel system is too lean (i.e., it was beyond a calibrated value stored in the PCM memory). **Possible Causes:** • Air leaks in intake manifold, exhaust pipes, or exhaust manifold • Fuel control sensor is out of calibration (ECT, IAT, or MAF) • Low fuel pressure (fuel filter clogged, pressure regulator failure) • One or more injectors restricted or pressure regulator has failed • HO2S element is contaminated, deteriorated, or has failed • Vacuum hose is disconnected, broken, leaking, or loose
DTC: P0172 **2T FUEL, MIL: Yes** **Years:** 2006, 2007, 2008 **Models:** Accent, Azera, Elantra, Entourage, Santa Fe, Sonata, Tiburon, Tucson, Veracruz **Engines:** 1.6L VIN C, 2.0L VIN B, 2.0L VIN D, 2.4L VIN B, 2.7L VIN D, 2.7L VIN F, 3.3L VIN D, 3.3L VIN E, 3.3L VIN F, 3.5L VIN E, 3.8L VIN 3, 3.8L VIN C, 3.8L VIN F **Transmissions:** All	**Fuel Trim Malfunction System Too Rich (Bank 1)** The engine is running in closed loop status and the PCM detected the Fuel system was too rich (i.e., it was beyond a calibrated value stored in the PCM memory). **Possible Causes:** • Base engine mechanical fault affecting one or more cylinders • Excess fuel vapors in crankcase (the oil needs to be changed) • EVAP system component has failed or canister fuel saturated • Fuel control sensor is out of calibration (i.e., ECT, IAT, or MAF) • Fuel delivery system supplying too much fuel during cruise or idle periods (e.g. faulty fuel pump or faulty pressure regulator) • Fuel injector(s) is leaking or stuck partially open (one or more) • HO2S is contaminated, deteriorated or it has failed

DTC	Trouble Code Title, Conditions & Possible Causes
DTC: P0173 **2T FUEL, MIL: Yes** **Years:** 2006, 2007, 2008 **Models:** Santa Fe, Tiburon, Tucson **Engines:** 2.7L VIN D, 2.7L VIN F **Transmissions:** All	**HO2S Sensor System Lambda Controller at the Limit (Bank 2)** The engine is running in close loop status and the PCM detected the Fuel system was too rich or too lean during 2 or more consecutive trips. **Possible Causes:** • Base engine mechanical fault affecting one or more cylinders • EVAP system component has failed or canister is fuel saturated • Exhaust leaks located in front of the HO2S location • Fuel control sensor is out of calibration (i.e., ECT, IAT or MAF) • Fuel delivery system supplying too much fuel during cruise or idle periods (e.g., faulty fuel pump, or faulty pressure regulator) • Fuel injector(s) is leaking or stuck partially open (one or more) • HO2S is contaminated, deteriorated or it has failed
DTC: P0174 **2T FUEL, MIL: Yes** **Years:** 2006, 2007, 2008 **Models:** Azera, Entourage, Santa Fe, Sonata, Tiburon, Tucson, Veracruz **Engines:** 2.7L VIN D, 2.7L VIN F, 3.3L VIN D, 3.3L VIN E, 3.3L VIN F, 3.5L VIN E, 3.8L VIN 3, 3.8L VIN C, 3.8L VIN F **Transmissions:** All	**Fuel Trim Malfunction System Too Lean (Bank 2)** The engine is running in closed loop status and the PCM detected the Fuel system was too lean (i.e., it was below a calibrated value stored in the PCM memory). **Possible Causes:** • Air leaks in intake manifold, exhaust pipes, or exhaust manifold • Fuel control sensor is out of calibration (ECT, IAT, or MAF) • Low fuel pressure (fuel filter clogged, pressure regulator failure) • One or more injectors restricted or pressure regulator has failed • HO2S element is contaminated, deteriorated, or has failed • Vacuum hose is disconnected, broken, leaking, or loose
DTC: P0175 **2T FUEL, MIL: Yes** **Years:** 2006, 2007, 2008 **Models:** Azera, Entourage, Santa Fe, Sonata, Tiburon, Tucson, Veracruz **Engines:** 2.7L VIN D, 2.7L VIN F, 3.3L VIN D, 3.3L VIN E, 3.3L VIN F, 3.5L VIN E, 3.8L VIN 3, 3.8L VIN C, 3.8L VIN F **Transmissions:** All	**Fuel Trim Malfunction System Too Rich (Bank 2)** The engine is running in closed loop status and the PCM detected the Fuel system was too rich (i.e., it was beyond a calibrated value stored in the PCM memory). **Possible Causes:** • Base engine mechanical fault affecting one or more cylinders • Excess fuel vapors in crankcase (the oil needs to be changed) • EVAP system component has failed or canister fuel saturated • Fuel control sensor is out of calibration (i.e., ECT, IAT, or MAF) • Fuel delivery system supplying too much fuel during cruise or idle periods (e.g. faulty fuel pump or faulty pressure regulator) • Fuel injector(s) is leaking or stuck partially open (one or more) • HO2S is contaminated, deteriorated, or it has failed
DTC: P0181 **2T FUEL, MIL: Yes** **Years:** 2006 **Models:** Santa Fe **Engines:** 2.4L VIN B, 3.5L VIN E **Transmissions:** All	**Fuel Temp Sensor A Circuit Range/Performance** If the voltage difference between the fuel tank temperature and the Engine Coolant Temperature (ECT) at starting is greater than the threshold value (above 59°F), a fault exists. Rationality check. ECT sensor temperature is 14 to 122°F. **Possible Causes:** • Poor connection • Faulty fuel temperature sensor • Faulty PCM
DTC: P0182 **2T FUEL, MIL: Yes** **Years:** 2006 **Models:** Santa Fe **Engines:** 2.4L VIN B, 3.5L VIN E **Transmissions:** All	**Fuel Temperature Sensor A Circuit Low Input** If the fuel temperature sensor signal is less than 0.1 volt after starting, P0182 is set. Output voltage is monitored. **Possible Causes:** • Poor connection • Short to ground in fuel temperature sensor circuit • Faulty fuel temperature sensor • Faulty PCM
DTC: P0183 **2T FUEL, MIL: Yes** **Years:** 2006 **Models:** Santa Fe **Engines:** 2.4L VIN B, 3.5L VIN E **Transmissions:** All	**Fuel Temp Sensor A Circuit High Input** If the sensor signal is above 4.6 volts after starting, P0183 is set. Output voltage is monitored. **Possible Causes:** • Poor connection • Open in fuel temperature sensor circuit • Faulty fuel temperature sensor • Faulty PCM

DTC	Trouble Code Title, Conditions & Possible Causes
DTC: P0196 **1T FUEL, MIL: Yes** **Years:** 2006, 2007, 2008 **Models:** Azera, Elantra, Entourage, Santa Fe, Sonata, Tiburon, Tucson, Veracruz **Engines:** 2.0L VIN B, 2.4L VIN C, 2.7L VIN D, 3.3L VIN D, 3.3L VIN E, 3.3L VIN F, 3.8L VIN 3, 3.8L VIN C, 3.8L VIN F **Transmissions:** All	**Engine Oil Temperature Sensor Range/Performance** Stuck oil temperature sensor signal or unusual low or high signal. Condition 1 (signal high or low), engine coolant temperature more than 158°F and oil temperature less than 68°F. Condition 2 (signal high or low), engine coolant temperature less than 158°F and oil temperature above 212°F. Condition 3 (stuck signal) engine coolant temperature less than 104°F. **Possible Causes:** • Contact resistance in connectors • Faulty oil temperature sensor
DTC: P0197 **1T FUEL, MIL: Yes** **Years:** 2006, 2007, 2008 **Models:** Azera, Elantra, Entourage, Santa Fe, Sonata, Tiburon, Tucson, Veracruz **Engines:** 2.0L VIN B, 2.4L VIN C, 2.7L VIN D, 3.3L VIN D, 3.3L VIN E, 3.3L VIN F, 3.8L VIN 3, 3.8L VIN C, 3.8L VIN F **Transmissions:** All	**Engine Oil Temperature Sensor Low Input** Signal voltage lower than the possible range of a properly operating OTS. Voltage range check. Engine coolant temperature less than 212°F. Oil temperature above 309°F. **Possible Causes:** • Short circuit to ground • Contact resistance in connectors • Faulty oil temperature sensor
DTC: P0198 **1T FUEL, MIL: Yes** **Years:** 2006, 2007, 2008 **Models:** Azera, Elantra, Entourage, Santa Fe, Sonata, Tiburon, Tucson, Veracruz **Engines:** 2.0L VIN B, 2.4L VIN C, 2.7L VIN D, 3.3L VIN D, 3.3L VIN E, 3.3L VIN F, 3.8L VIN 3, 3.8L VIN C, 3.8L VIN F **Transmissions:** All	**Engine Oil Temperature Sensor High Input** Signal voltage higher than the possible range of a properly operating OTS. Voltage range check. Five minutes after engine start if engine coolant temperature less than 14°F. Oil temperature -33°F. **Possible Causes:** • Open circuit to battery • Contact resistance in connectors • Faulty oil temperature sensor

DTC: P0201
2T FUEL, MIL: Yes
Years: 2006, 2007, 2008
Models: Accent, Santa Fe
Engines: 1.6L VIN C, 2.4L VIN B, 3.5L VIN E
Transmissions: All
	Fuel Injector 1 Control Circuit Malfunction
	The engine speed is more than 80 RPM. DTC P0201 is set when the control module detects an incorrect voltage on the high voltage supply circuit or the high voltage control circuit.
	Note: Drive the vehicle at off-idle speeds and have an assistant monitor the misfire current counters. Observe if more than one cylinder is misfiring. This may not be apparent until after a repair is completed. If an injector fuse is open for one cylinder bank, the Scan Tool may only display 2 or 3 cylinders misfiring.
	Possible Causes:
	    • Fuel injector power circuit (B+) is open (check the power fuse)
	    • Check the fuel injector harness connector for proper connection, proper terminal contact, and proper terminal retention force
	    • Fuel injector control circuit is open between injector and ECM
	    • Fuel injector control circuit is grounded between injector and ECM
	    • Fuel injector is damaged or has failed
	    • ECM is damaged

DTC: P0202
2T FUEL, MIL: Yes
Years: 2006, 2007, 2008
Models: Accent, Santa Fe
Engines: 1.6L VIN C, 2.4L VIN B, 3.5L VIN E
Transmissions: All
	Fuel Injector 2 Control Circuit Malfunction
	The engine speed is more than 80 RPM. DTC P0202 is set when the control module detects an incorrect voltage on the high voltage supply circuit or the high voltage control circuit.
	Note: Drive the vehicle at off-idle speeds and have an assistant monitor the misfire current counters. Observe if more than one cylinder is misfiring. This may not be apparent until after a repair is completed. If an injector fuse is open for one cylinder bank, the Scan Tool may only display 2 or 3 cylinders misfiring.
	Possible Causes:
	    • Fuel injector power circuit (B+) is open (check the power fuse)
	    • Check the fuel injector harness connector for proper connection, proper terminal contact, and proper terminal retention force
	    • Fuel injector control circuit is open between injector and ECM
	    • Fuel injector control circuit is grounded between injector and ECM
	    • Fuel injector is damaged or has failed
	    • ECM is damaged

DTC	Trouble Code Title, Conditions & Possible Causes
DTC: P0203 **2T FUEL, MIL: Yes** **Years:** 2006, 2007, 2008 **Models:** Accent, Santa Fe **Engines:** 1.6L VIN C, 2.4L VIN B, 3.5L VIN E **Transmissions:** All	**Fuel Injector 3 Control Circuit Malfunction** The engine speed is more than 80 RPM. DTC P0203 is set when the control module detects an incorrect voltage on the high voltage supply circuit or the high voltage control circuit. **Note: Drive the vehicle at off-idle speeds and have an assistant monitor the misfire current counters. Observe if more than one cylinder is misfiring. This may not be apparent until after a repair is completed. If an injector fuse is open for one cylinder bank, the Scan Tool may only display 2 or 3 cylinders misfiring.** **Possible Causes:** • Fuel injector power circuit (B+) is open (check the power fuse) • Check the fuel injector harness connector for proper connection, proper terminal contact, and proper terminal retention force • Fuel injector control circuit is open between injector and ECM • Fuel injector control circuit is grounded between injector and ECM • Fuel injector is damaged or has failed • ECM is damaged
DTC: P0204 **2T FUEL, MIL: Yes** **Years:** 2006, 2007, 2008 **Models:** Accent, Santa Fe **Engines:** 1.6L VIN C, 2.4L VIN B, 3.5L VIN E **Transmissions:** All	**Fuel Injector 4 Control Circuit Malfunction** The engine speed is more than 80 RPM. DTC P0204 is set when the control module detects an incorrect voltage on the high voltage supply circuit or the high voltage control circuit. **Note: Drive the vehicle at off-idle speeds and have an assistant monitor the misfire current counters. Observe if more than one cylinder is misfiring. This may not be apparent until after a repair is completed. If an injector fuse is open for one cylinder bank, the Scan Tool may only display 2 or 3 cylinders misfiring.** **Possible Causes:** • Fuel injector power circuit (B+) is open (check the power fuse) • Check the fuel injector harness connector for proper connection, proper terminal contact, and proper terminal retention force • Fuel injector control circuit is open between injector and ECM • Fuel injector control circuit is grounded between injector and ECM • Fuel injector is damaged or has failed • ECM is damaged
DTC: P0205 **2T FUEL, MIL: Yes** **Years:** 2006 **Models:** Santa Fe **Engines:** 3.5L VIN E **Transmissions:** All	**Fuel Injector 5 Control Circuit Malfunction** The engine speed is more than 80 RPM. DTC P0205 is set when the control module detects an incorrect voltage on the high voltage supply circuit or the high voltage control circuit. **Note: Drive the vehicle at off-idle speeds and have an assistant monitor the misfire current counters. Observe if more than one cylinder is misfiring. This may not be apparent until after a repair is completed. If an injector fuse is open for one cylinder bank, the Scan Tool may only display 2 or 3 cylinders misfiring.** **Possible Causes:** • Fuel injector power circuit (B+) is open (check the power fuse) • Check the fuel injector harness connector for proper connection, proper terminal contact, and proper terminal retention force • Fuel injector control circuit is open between injector and ECM • Fuel injector control circuit is grounded between injector and ECM • Fuel injector is damaged or has failed • ECM is damaged
DTC: P0206 **2T FUEL, MIL: Yes** **Years:** 2006 **Models:** Santa Fe **Engines:** 3.5L VIN E **Transmissions:** All	**Fuel Injector 6 Control Circuit Malfunction** The engine speed is more than 80 RPM. DTC P0206 is set when the control module detects an incorrect voltage on the high voltage supply circuit or the high voltage control circuit. **Note: Drive the vehicle at off-idle speeds and have an assistant monitor the misfire current counters. Observe if more than one cylinder is misfiring. This may not be apparent until after a repair is completed. If an injector fuse is open for one cylinder bank, the Scan Tool may only display 2 or 3 cylinders misfiring.** **Possible Causes:** • Fuel injector power circuit (B+) is open (check the power fuse) • Check the fuel injector harness connector for proper connection, proper terminal contact, and proper terminal retention force • Fuel injector control circuit is open between injector and ECM • Fuel injector control circuit is grounded between injector and ECM • Fuel injector is damaged or has failed • ECM is damaged

DTC	Trouble Code Title, Conditions & Possible Causes
DTC: P0217 **2T FUEL, MIL: No** **Years:** 2006, 2007, 2008 **Models:** Azera, Entourage, Santa Fe, Sonata, Veracruz **Engines:** 2.7L VIN D, 3.3L VIN D, 3.3L VIN E, 3.3L VIN F, 3.8L VIN 3, 3.8L VIN C, 3.8L VIN F **Transmissions:** A/T	**Engine Coolant Over Temperature Condition** This diagnostic introduces a delay and also looks for excessive engine loads. Once the delay period passes and an excessive load was not experienced, the diagnostic checks whether the coolant temperature has exceeded a maximum threshold in order to make a pass/fail determination. No disabling faults are present. The coolant sensor is within range. The coolant temperature is equal or greater than 122°F. The Intake Air Temperature (IAT) is equal to or greater than 95°F. **Possible Causes:** • Poor connection • Lack of engine coolant • Water pump problems • Faulty Engine Coolant Temperature (ECT) sensor • Faulty PCM
DTC: P0220 **1T FUEL, MIL: Yes** **Years:** 2006 **Models:** Santa Fe **Engines:** 3.5L VIN E **Transmissions:** All	**Throttle/Pedal Position Sensor/Switch B Circuit** The ignition is ON or the engine runtime is over 5 seconds. The PCM detected the Electronic Throttle System (ETS) throttle position sensor input was out of range (i.e., it was too high or too low) during the CCM test. **Possible Causes:** • TP Sensor B circuit is open, shorted to ground, or has a high resistance condition • Throttle body assembly is damaged or it has failed • ECM has failed
DTC: P0221 **2T FUEL, MIL: Yes** **Years:** 2007, 2008 **Models:** Sonata **Engines:** 2.4L VIN C **Transmissions:** All	**Throttle/Pedal Position Sensor/Switch B Circuit Range/Performance** The system voltage is more than 7 volts. DTC P0221 is set when the TP sensor 1 disagrees more than 6.3 percent from TP sensor 2. The TP sensor 2 disagrees more than 9 percent from the throttle position calculated from Mass Air Flow (MAF) signal. **Possible Causes:** • TP Sensor 2 circuit is open, shorted to ground, or has a high resistance condition • Throttle body assembly is damaged, or it has failed • ECM is damaged
DTC: P0222 **2T FUEL, MIL: Yes** **Years:** 2007, 2008 **Models:** Azera, Entourage, Santa Fe, Sonata, Veracruz **Engines:** 2.4L VIN C, 2.7L VIN D, 3.3L VIN D, 3.3L VIN E, 3.3L VIN F, 3.5L VIN E, 3.8L VIN 3, 3.8L VIN C, 3.8L VIN F **Transmissions:** All	**Throttle/Pedal Position Sensor/Switch B Circuit Low Input** The system voltage is more than 7 volts. DTC P0222 is set when the TP sensor 2 voltage is less than 0.16 volt. **Possible Causes:** • APP signal 2 circuit is shorted to ground • TP Sensor 2 signal circuit is shorted to ground • TP Sensor 5-volt REF circuit is open or shorted to ground • ECM is damaged
DTC: P0223 **1T FUEL, MIL: Yes** **Years:** 2007, 2008 **Models:** Azera, Entourage, Santa Fe, Sonata, Veracruz **Engines:** 2.4L VIN C, 2.7L VIN D, 3.3L VIN D, 3.3L VIN E, 3.3L VIN F, 3.5L VIN E, 3.8L VIN 3, 3.8L VIN C, 3.8L VIN F **Transmissions:** All	**Throttle/Pedal Position Sensor/Switch B Circuit High Input** The system voltage is more than 7 volts. DTC P0223 is set when the TP sensor 2 voltage is more than 4.88 volts. **Possible Causes:** • APP Sensor 2 signal circuit is shorted to the 5-volt REF circuit • TP Sensor 2 signal circuit is shorted to the 5-volt REF circuit • TP Sensor 5-volt REF circuit is open or shorted to ground • ECM is damaged
DTC: P0224 **1T FUEL, MIL: Yes** **Years:** 2006 **Models:** Santa Fe **Engines:** 3.5L VIN E **Transmissions:** All	**Throttle Position Sensor B Linearity** The system voltage is more than 7 volts. DTC P0224 is set when the TP sensor 2 voltage is too low as compared to the threshold. **Possible Causes:** • APP Sensor 2 signal circuit is shorted to the 5-volt REF circuit • TP Sensor 2 signal circuit is shorted to the 5-volt REF circuit • TP Sensor 5-volt REF circuit is open or shorted to ground • ECM is damaged

DTC	Trouble Code Title, Conditions & Possible Causes
DTC: P0230 **2T CCM, MIL:** Yes **Years:** 2007, 2008 **Models:** Accent, Azera, Elantra, Entourage, Santa Fe, Sonata, Tiburon, Tucson, Veracruz **Engines:** 1.6L VIN C, 2.0L VIN B, 2.0L VIN D, 2.4L VIN C, 2.7L VIN D, 2.7L VIN F, 3.3L VIN D, 3.3L VIN E, 3.3L VIN F, 3.8L VIN 3, 3.8L VIN C, 3.8L VIN F **Transmissions:** All	**Fuel Pump Circuit Malfunction** The ignition is ON and the PCM detected an unexpected voltage condition on the fuel pump circuit through the fuel pump monitoring input. **Possible Causes:** • Fuel pump control circuit is open or shorted to ground • Fuel pump relay power circuit from ignition switch is open • Fuel pump relay is damaged or has failed • PCM has failed
DTC: P0231 **2T CCM, MIL:** Yes **Years:** 2006, 2007, 2008 **Models:** Accent **Engines:** 1.6L VIN C **Transmissions:** All	**Electric Fuel Pump Relay Open or Short Circuit** DTC P0231 is set when the ECM detects that the circuit continuity check is high. **Possible Causes:** • Poor connection • Short to power in control circuit • Fuel pump relay has failed • Faulty PCM
DTC: P0232 **2T CCM, MIL:** Yes **Years:** 2006, 2007, 2008 **Models:** Accent **Engines:** 1.6L VIN C **Transmissions:** All	**Electric Fuel Pump Relay Short Circuit** DTC P0232 is set when the ECM detects that the circuit continuity check is low. **Possible Causes:** • Poor connection • Short to ground in control circuit • Fuel pump relay has failed • Faulty PCM
DTC: P0261 **2T CCM, MIL:** Yes **Years:** 2006, 2007, 2008 **Models:** Accent, Azera, Elantra, Entourage, Santa Fe, Sonata, Tiburon, Tucson, Veracruz **Engines:** 1.6L VIN C, 2.0L VIN B, 2.0L VIN D, 2.4L VIN C, 2.7L VIN D, 2.7L VIN F, 3.3L VIN D, 3.3L VIN E, 3.3L VIN F, 3.8L VIN 3, 3.8L VIN C, 3.8L VIN F **Transmissions:** All	**Fuel Injector 1 Control Circuit Low Voltage** The engine speed is more than 80 RPM. DTC P0261 is set when the control module detects an incorrect voltage on the high voltage supply circuit or the high voltage control circuit. **Possible Causes:** • Fuel injector 1 connector is damaged or shorted • Fuel injector 1 power circuit is open (check the INJ/Coil fuse) • Fuel injector 1 control circuit is shorted to ground • Fuel injector 1 is damaged or it has failed • ECM is damaged
DTC: P0262 **2T CCM, MIL:** Yes **Years:** 2006, 2007, 2008 **Models:** Accent, Azera, Elantra, Entourage, Santa Fe, Sonata, Tiburon, Tucson, Veracruz **Engines:** 1.6L VIN C, 2.0L VIN B, 2.0L VIN D, 2.4L VIN C, 2.7L VIN D, 2.7L VIN F, 3.3L VIN D, 3.3L VIN E, 3.3L VIN F, 3.8L VIN 3, 3.8L VIN C, 3.8L VIN F **Transmissions:** All	**Fuel Injector 1 Control Circuit High Voltage** The engine speed is more than 80 RPM. DTC P0262 is set when the control module detects an incorrect voltage on the high voltage supply circuit or the high voltage control circuit. **Possible Causes:** • Fuel injector 1 connector is damaged or shorted • Fuel injector 1 control circuit is shorted to system power (B+) • Fuel injector 1 is damaged or it has failed • ECM is damaged

DTC	Trouble Code Title, Conditions & Possible Causes
DTC: P0264 **2T CCM, MIL: Yes** **Years:** 2006, 2007, 2008 **Models:** Accent, Azera, Elantra, Entourage, Santa Fe, Sonata, Tiburon, Tucson, Veracruz **Engines:** 1.6L VIN C, 2.0L VIN B, 2.0L VIN D, 2.4L VIN C, 2.7L VIN D, 2.7L VIN F, 3.3L VIN D, 3.3L VIN E, 3.3L VIN F, 3.8L VIN 3, 3.8L VIN C, 3.8L VIN F **Transmissions:** All	**Fuel Injector 2 Control Circuit Low Voltage** The engine speed is more than 80 RPM. DTC P0264 is set when the control module detects an incorrect voltage on the high voltage supply circuit or the high voltage control circuit. **Possible Causes:** • Fuel injector 2 connector is damaged or shorted • Fuel injector 2 power circuit is open (check the INJ/Coil fuse) • Fuel injector 2 control circuit is shorted to ground • Fuel injector 2 is damaged or it has failed • ECM is damaged
DTC: P0265 **2T CCM, MIL: Yes** **Years:** 2006, 2007, 2008 **Models:** Accent, Azera, Elantra, Entourage, Santa Fe, Sonata, Tiburon, Tucson, Veracruz **Engines:** 1.6L VIN C, 2.0L VIN B, 2.0L VIN D, 2.4L VIN C, 2.7L VIN D, 2.7L VIN F, 3.3L VIN D, 3.3L VIN E, 3.3L VIN F, 3.8L VIN 3, 3.8L VIN C, 3.8L VIN F **Transmissions:** All	**Fuel Injector 2 Control Circuit High Voltage** The engine speed is more than 80 RPM. DTC P0265 is set when the control module detects an incorrect voltage on the high voltage supply circuit or the high voltage control circuit. **Possible Causes:** • Fuel injector 2 connector is damaged or shorted • Fuel injector 2 control circuit is shorted to system power (B+) • Fuel injector 2 is damaged or it has failed • ECM is damaged
DTC: P0267 **2T CCM, MIL: Yes** **Years:** 2006, 2007, 2008 **Models:** Accent, Azera, Elantra, Entourage, Santa Fe, Sonata, Tiburon, Tucson, Veracruz **Engines:** 1.6L VIN C, 2.0L VIN B, 2.0L VIN D, 2.4L VIN C, 2.7L VIN D, 2.7L VIN F, 3.3L VIN D, 3.3L VIN E, 3.3L VIN F, 3.8L VIN 3, 3.8L VIN C, 3.8L VIN F **Transmissions:** All	**Fuel Injector 3 Control Circuit Low Voltage** The engine speed is more than 80 RPM. DTC P0267 is set when the control module detects an incorrect voltage on the high voltage supply circuit or the high voltage control circuit. **Possible Causes:** • Fuel injector 3 connector is damaged or shorted • Fuel injector 3 power circuit is open (check the INJ/Coil fuse) • Fuel injector 3 control circuit is shorted to ground • Fuel injector 3 is damaged or it has failed • ECM is damaged
DTC: P0268 **2T CCM, MIL: Yes** **Years:** 2006, 2007, 2008 **Models:** Accent, Azera, Elantra, Entourage, Santa Fe, Sonata, Tiburon, Tucson, Veracruz **Engines:** 1.6L VIN C, 2.0L VIN B, 2.0L VIN D, 2.4L VIN C, 2.7L VIN D, 2.7L VIN F, 3.3L VIN D, 3.3L VIN E, 3.3L VIN F, 3.8L VIN 3, 3.8L VIN C, 3.8L VIN F **Transmissions:** All	**Fuel Injector 3 Control Circuit High Voltage** The engine speed is more than 80 RPM. DTC P0268 is set when the control module detects an incorrect voltage on the high voltage supply circuit or the high voltage control circuit. **Possible Causes:** • Fuel injector 3 connector is damaged or shorted • Fuel injector 3 control circuit is shorted to system power (B+) • Fuel injector 3 is damaged or it has failed • ECM is damaged
DTC: P0270 **2T CCM, MIL: Yes** **Years:** 2006, 2007, 2008 **Models:** Accent, Azera, Elantra, Entourage, Santa Fe, Sonata, Tiburon, Tucson, Veracruz **Engines:** 1.6L VIN C, 2.0L VIN B, 2.0L VIN D, 2.4L VIN C, 2.7L VIN D, 2.7L VIN F, 3.3L VIN D, 3.3L VIN E, 3.3L VIN F, 3.8L VIN 3, 3.8L VIN C, 3.8L VIN F **Transmissions:** All	**Fuel Injector 4 Control Circuit Low Voltage** The engine speed is more than 80 RPM. DTC P0270 is set when the control module detects an incorrect voltage on the high voltage supply circuit or the high voltage control circuit. **Possible Causes:** • Fuel injector 4 connector is damaged or shorted • Fuel injector 4 power circuit is open (check the INJ/Coil fuse) • Fuel injector 4 control circuit is shorted to ground • Fuel injector 4 is damaged or it has failed • ECM is damaged

DTC	Trouble Code Title, Conditions & Possible Causes
DTC: P0271 **2T CCM, MIL: Yes** **Years:** 2006, 2007, 2008 **Models:** Accent, Azera, Elantra, Entourage, Santa Fe, Sonata, Tiburon, Tucson, Veracruz **Engines:** 1.6L VIN C, 2.0L VIN B, 2.0L VIN D, 2.4L VIN C, 2.7L VIN D, 2.7L VIN F, 3.3L VIN D, 3.3L VIN E, 3.3L VIN F, 3.8L VIN 3, 3.8L VIN C, 3.8L VIN F **Transmissions:** All	**Fuel Injector 4 Control Circuit High Voltage** The engine speed is more than 80 RPM. DTC P0271 is set when the control module detects an incorrect voltage on the high voltage supply circuit or the high voltage control circuit. **Possible Causes:** • Fuel injector 4 connector is damaged or shorted • Fuel injector 4 control circuit is shorted to system power (B+) • Fuel injector 4 is damaged or it has failed • ECM is damaged
DTC: P0273 **2T CCM, MIL: Yes** **Years:** 2006, 2007, 2008 **Models:** Azera, Entourage, Santa Fe, Sonata, Tiburon, Tucson, Veracruz **Engines:** 2.7L VIN D, 2.7L VIN F, 3.3L VIN D, 3.3L VIN E, 3.3L VIN F, 3.8L VIN 3, 3.8L VIN C, 3.8L VIN F **Transmissions:** All	**Fuel Injector 5 Control Circuit Low Voltage** The engine speed is more than 80 RPM. DTC P0273 is set when the ECM detects a grounded fuel injector circuit. **Possible Causes:** • Fuel injector 5 connector is damaged or shorted • Fuel injector 5 power circuit is open (check the INJ/Coil fuse) • Fuel injector 5 control circuit is shorted to ground • Fuel injector 5 is damaged or it has failed • ECM is damaged
DTC: P0274 **2T CCM, MIL: Yes** **Years:** 2006, 2007, 2008 **Models:** Azera, Entourage, Santa Fe, Sonata, Tiburon, Tucson, Veracruz **Engines:** 2.7L VIN D, 2.7L VIN F, 3.3L VIN D, 3.3L VIN E, 3.3L VIN F, 3.8L VIN 3, 3.8L VIN C, 3.8L VIN F **Transmissions:** All	**Fuel Injector 5 Control Circuit High Voltage** The engine speed is more than 80 RPM. DTC P0274 is set when the ECM detects a short to voltage on the fuel injector circuits. **Possible Causes:** • Fuel injector 5 connector is damaged or shorted • Fuel injector 5 control circuit is shorted to system power (B+) • Fuel injector 5 is damaged or it has failed • ECM is damaged
DTC: P0276 **2T CCM, MIL: Yes** **Years:** 2006, 2007, 2008 **Models:** Azera, Entourage, Santa Fe, Sonata, Tiburon, Tucson, Veracruz **Engines:** 2.7L VIN D, 2.7L VIN F, 3.3L VIN D, 3.3L VIN E, 3.3L VIN F, 3.8L VIN 3, 3.8L VIN C, 3.8L VIN F **Transmissions:** All	**Fuel Injector 6 Control Circuit Low Voltage** The engine speed is more than 80 RPM. DTC P0276 is set when the ECM detects a grounded fuel injector circuit. **Possible Causes:** • Fuel injector 6 connector is damaged or shorted • Fuel injector 6 power circuit is open (check the INJ/Coil fuse) • Fuel injector 6 control circuit is shorted to ground • Fuel injector 6 is damaged or it has failed • ECM is damaged
DTC: P0277 **2T CCM, MIL: Yes** **Years:** 2006, 2007, 2008 **Models:** Azera, Entourage, Santa Fe, Sonata, Tiburon, Tucson, Veracruz **Engines:** 2.7L VIN D, 2.7L VIN F, 3.3L VIN D, 3.3L VIN E, 3.3L VIN F, 3.8L VIN 3, 3.8L VIN C, 3.8L VIN F **Transmissions:** All	**Fuel Injector 6 Control Circuit High Voltage** The engine speed is more than 80 RPM. DTC P0277 is set when the ECM detects a short to voltage on the fuel injector circuits. **Possible Causes:** • Fuel injector 6 connector is damaged or shorted • Fuel injector 6 control circuit is shorted to system power (B+) • Fuel injector 6 is damaged or it has failed • ECM is damaged

DTC	Trouble Code Title, Conditions & Possible Causes
DTC: P0300 **2T MISFIRE, MIL: Yes** **Years:** 2006, 2007, 2008 **Models:** Accent, Azera, Elantra, Entourage, Santa Fe, Sonata, Tiburon, Tucson, Veracruz **Engines:** 1.6L VIN C, 2.0L VIN B, 2.0L VIN D, 2.4L VIN B, 2.4L VIN C, 2.7L VIN D, 2.7L VIN F, 3.3L VIN D, 3.3L VIN E, 3.3L VIN F, 3.5L VIN E, 3.8L VIN 3, 3.8L VIN C, 3.8L VIN F **Transmissions:** All	**Random/Multiple Cylinder Misfire Detected** The engine speed is between 400–3,500 RPM. DTC P0300 is set when the ECM is detecting a crankshaft rotation speed variation indicating a misfire sufficient to cause emission or catalytic converter damaging levels to exceed mandated standards. **Note: When the MIL is flashing, the injector may be disabled for the misfiring cylinder to protect the catalytic converter.** **Possible Causes:** • A misfire DTC could be caused by an excessive vibration from sources other than the engine. Inspect for the following possible sources: A tire or wheel that is out of round or out of balance, variable thickness brake rotors, an unbalanced drive shaft, certain rough road conditions, a damaged accessory drive component or belt, a damaged reluctor wheel • Base engine mechanical fault • Check for vacuum leaks in the throttle body, in the vacuum hoses, and in the PCV valve/hoses • Check ECM power grounds (verify they are clean and secure) • Fuel metering problem • Fuel injector wire harness electrical connectors connected to wrong fuel injectors • Fuel pressure that is too low or too high • Contaminated fuel (alcohol/other contaminants) • Restricted exhaust system • Ignition system fault (coil, plug, or wire) • Fuel injector failure
DTC: P0301 **2T MISFIRE, MIL: Yes** **Years:** 2006, 2007, 2008 **Models:** Accent, Azera, Elantra, Entourage, Santa Fe, Sonata, Tiburon, Tucson, Veracruz **Engines:** 1.6L VIN C, 2.0L VIN B, 2.0L VIN D, 2.4L VIN B, 2.4L VIN C, 2.7L VIN D, 2.7L VIN F, 3.3L VIN D, 3.3L VIN E, 3.3L VIN F, 3.5L VIN E, 3.8L VIN 3, 3.8L VIN C, 3.8L VIN F **Transmissions:** All	**Cylinder 1 Misfire Detected** The engine speed is between 400–3,500 RPM. DTC P0301 is set when the ECM is detecting a crankshaft rotation speed variation indicating a misfire sufficient to cause emission or catalytic converter damaging levels to exceed mandated standards. **Note: When the MIL is flashing, the injector may be disabled for the misfiring cylinder to protect the catalytic converter.** **Possible Causes:** • A misfire DTC could be caused by an excessive vibration from sources other than the engine. Inspect for the following possible sources: A tire or wheel that is out of round or out of balance, variable thickness brake rotors, an unbalanced drive shaft, certain rough road conditions, a damaged accessory drive component or belt, a damaged reluctor wheel • Vacuum leaks in the throttle body, intake manifold, vacuum hoses, or PCV valve/hoses • Check ECM power grounds (verify they are clean and secure) • Fuel metering problem • Fuel injector wire harness electrical connectors connected to wrong fuel injectors • Fuel pressure that is too low or too high • Contaminated fuel (alcohol/other contaminants) • Restricted exhaust system • Fuel injector failure • Air leak in the intake manifold, or in the EGR or PCV system • Base engine mechanical fault that affects only Cylinder 1 • Fuel component fault that affects only Cylinder 1 (i.e., dirty/failed fuel injector) • Ignition system fault (coil or plug) that affects only Cylinder 1
DTC: P0302 **2T MISFIRE, MIL: Yes** **Years:** 2006, 2007, 2008 **Models:** Accent, Azera, Elantra, Entourage, Santa Fe, Sonata, Tiburon, Tucson, Veracruz **Engines:** 1.6L VIN C, 2.0L VIN B, 2.0L VIN D, 2.4L VIN B, 2.4L VIN C, 2.7L VIN D, 2.7L VIN F, 3.3L VIN D, 3.3L VIN E, 3.3L VIN F, 3.5L VIN E, 3.8L VIN 3, 3.8L VIN C, 3.8L VIN F **Transmissions:** All	**Cylinder 2 Misfire Detected** The engine speed is between 400–3,500 RPM. DTC P0302 is set when the ECM is detecting a crankshaft rotation speed variation indicating a misfire sufficient to cause emission or catalytic converter damaging levels to exceed mandated standards. **Note: When the MIL is flashing, the injector may be disabled for the misfiring cylinder to protect the catalytic converter.** **Possible Causes:** • A misfire DTC could be caused by an excessive vibration from sources other than the engine. Inspect for the following possible sources: A tire or wheel that is out of round or out of balance, variable thickness brake rotors, an unbalanced drive shaft, certain rough road conditions, a damaged accessory drive component or belt, a damaged reluctor wheel • Vacuum leaks in the throttle body, intake manifold, vacuum hoses, or PCV valve/hoses • Check ECM power grounds (verify they are clean and secure) • Fuel metering problem • Fuel injector wire harness electrical connectors connected to wrong fuel injectors • Fuel pressure that is too low or too high • Contaminated fuel (alcohol/other contaminants) • Restricted exhaust system • Fuel injector failure • Air leak in the intake manifold, or in the EGR or PCV system • Base engine mechanical fault that affects only Cylinder 2 • Fuel component fault that affects only Cylinder 2 (i.e., dirty/failed fuel injector) • Ignition system fault (coil or plug) that affects only Cylinder 2

DTC	Trouble Code Title, Conditions & Possible Causes
DTC: P0303 **2T MISFIRE, MIL: Yes** **Years:** 2006, 2007, 2008 **Models:** Accent, Azera, Elantra, Entourage, Santa Fe, Sonata, Tiburon, Tucson, Veracruz **Engines:** 1.6L VIN C, 2.0L VIN B, 2.0L VIN D, 2.4L VIN B, 2.4L VIN C, 2.7L VIN D, 2.7L VIN F, 3.3L VIN D, 3.3L VIN E, 3.3L VIN F, 3.5L VIN E, 3.8L VIN 3, 3.8L VIN C, 3.8L VIN F **Transmissions:** All	**Cylinder 3 Misfire Detected** The engine speed is between 400–3,500 RPM. DTC P0303 is set when the ECM is detecting a crankshaft rotation speed variation indicating a misfire sufficient to cause emission or catalytic converter damaging levels to exceed mandated standards. **Note: When the MIL is flashing, the injector may be disabled for the misfiring cylinder to protect the catalytic converter.** **Possible Causes:** • A misfire DTC could be caused by an excessive vibration from sources other than the engine. Inspect for the following possible sources: A tire or wheel that is out of round or out of balance, variable thickness brake rotors, an unbalanced drive shaft, certain rough road conditions, a damaged accessory drive component or belt, a damaged reluctor wheel • Vacuum leaks in the throttle body, intake manifold, vacuum hoses, or PCV valve/hoses • Check ECM power grounds (verify they are clean and secure) • Fuel metering problem • Fuel injector wire harness electrical connectors connected to wrong fuel injectors • Fuel pressure that is too low or too high • Contaminated fuel (alcohol/other contaminants) • Restricted exhaust system • Fuel injector failure • Air leak in the intake manifold, or in the EGR or PCV system • Base engine mechanical fault that affects only Cylinder 3 • Fuel component fault that affects only Cylinder 3 (i.e., dirty/failed fuel injector) • Ignition system fault (coil or plug) that affects only Cylinder 3
DTC: P0304 **2T MISFIRE, MIL: Yes** **Years:** 2006, 2007, 2008 **Models:** Accent, Azera, Elantra, Entourage, Santa Fe, Sonata, Tiburon, Tucson, Veracruz **Engines:** 1.6L VIN C, 2.0L VIN B, 2.0L VIN D, 2.4L VIN B, 2.4L VIN C, 2.7L VIN D, 2.7L VIN F, 3.3L VIN D, 3.3L VIN E, 3.3L VIN F, 3.5L VIN E, 3.8L VIN 3, 3.8L VIN C, 3.8L VIN F **Transmissions:** All	**Cylinder 4 Misfire Detected** The engine speed is between 400–3,500 RPM. DTC P0304 is set when the ECM is detecting a crankshaft rotation speed variation indicating a misfire sufficient to cause emission or catalytic converter damaging levels to exceed mandated standards. **Note: When the MIL is flashing, the injector may be disabled for the misfiring cylinder to protect the catalytic converter.** **Possible Causes:** • A misfire DTC could be caused by an excessive vibration from sources other than the engine. Inspect for the following possible sources: A tire or wheel that is out of round or out of balance, variable thickness brake rotors, an unbalanced drive shaft, certain rough road conditions, a damaged accessory drive component or belt, a damaged reluctor wheel • Vacuum leaks in the throttle body, intake manifold, vacuum hoses, or PCV valve/hoses • Check ECM power grounds (verify they are clean and secure) • Fuel metering problem • Fuel injector wire harness electrical connectors connected to wrong fuel injectors • Fuel pressure that is too low or too high • Contaminated fuel (alcohol/other contaminants) • Restricted exhaust system • Fuel injector failure • Air leak in the intake manifold, or in the EGR or PCV system • Base engine mechanical fault that affects only Cylinder 4 • Fuel component fault that affects only Cylinder 4 (i.e., dirty/failed fuel injector) • Ignition system fault (coil or plug) that affects only Cylinder 4
DTC: P0305 **2T MISFIRE, MIL: Yes** **Years:** 2006, 2007, 2008 **Models:** Azera, Azera, Entourage, Santa Fe, Sonata, Tiburon, Tucson, Veracruz **Engines:** 2.7L VIN D, 2.7L VIN F, 3.3L VIN D, 3.3L VIN E, 3.3L VIN F, 3.5L VIN E, 3.8L VIN 3, 3.8L VIN C, 3.8L VIN F **Transmissions:** All	**Cylinder 5 Misfire Detected** The engine speed is between 400–3,500 RPM. DTC P0305 is set when the ECM is detecting a crankshaft rotation speed variation indicating a misfire sufficient to cause emission or catalytic converter damaging levels to exceed mandated standards. **Note: When the MIL is flashing, the injector may be disabled for the misfiring cylinder to protect the catalytic converter.** **Possible Causes:** • A misfire DTC could be caused by an excessive vibration from sources other than the engine. Inspect for the following possible sources: A tire or wheel that is out of round or out of balance, variable thickness brake rotors, an unbalanced drive shaft, certain rough road conditions, a damaged accessory drive component or belt, a damaged reluctor wheel • Vacuum leaks in the throttle body, intake manifold, vacuum hoses, or PCV valve/hoses • Check ECM power grounds (verify they are clean and secure) • Fuel metering problem • Fuel injector wire harness electrical connectors connected to wrong fuel injectors • Fuel pressure that is too low or too high • Contaminated fuel (alcohol/other contaminants) • Restricted exhaust system • Fuel injector failure • Air leak in the intake manifold, or in the EGR or PCV system • Base engine mechanical fault that affects only Cylinder 5 • Fuel component fault that affects only Cylinder 5 (i.e., dirty/failed fuel injector) • Ignition system fault (coil or plug) that affects only Cylinder 5

DTC	Trouble Code Title, Conditions & Possible Causes
DTC: P0306 **2T MISFIRE, MIL: Yes** **Years:** 2006, 2007, 2008 **Models:** Azera, Azera, Entourage, Santa Fe, Sonata, Tiburon, Tucson, Veracruz **Engines:** 2.7L VIN D, 2.7L VIN F, 3.3L VIN D, 3.3L VIN E, 3.3L VIN F, 3.5L VIN E, 3.8L VIN 3, 3.8L VIN C, 3.8L VIN F **Transmissions:** All	**Cylinder 6 Misfire Detected** The engine speed is between 400–3,500 RPM. DTC P0306 is set when the ECM is detecting a crankshaft rotation speed variation indicating a misfire sufficient to cause emission or catalytic converter damaging levels to exceed mandated standards. **Note: When the MIL is flashing, the injector may be disabled for the misfiring cylinder to protect the catalytic converter.** **Possible Causes:** • A misfire DTC could be caused by an excessive vibration from sources other than the engine. Inspect for the following possible sources: A tire or wheel that is out of round or out of balance, variable thickness brake rotors, an unbalanced drive shaft, certain rough road conditions, a damaged accessory drive component or belt, a damaged reluctor wheel • Vacuum leaks in the throttle body, intake manifold, vacuum hoses, or PCV valve/hoses • Check ECM power grounds (verify they are clean and secure) • Fuel metering problem • Fuel injector wire harness electrical connectors connected to wrong fuel injectors • Fuel pressure that is too low or too high • Contaminated fuel (alcohol/other contaminants) • Restricted exhaust system • Fuel injector failure • Air leak in the intake manifold, or in the EGR or PCV system • Base engine mechanical fault that affects only Cylinder 6 • Fuel component fault that affects only Cylinder 6 (i.e., dirty/failed fuel injector) • Ignition system fault (coil or plug) that affects only Cylinder 6
DTC: P0315 **1T CCM, MIL: Yes** **Years:** 2006, 2007, 2008 **Models:** Azera, Elantra, Entourage, Santa Fe, Sonata, Tiburon, Tucson, Veracruz **Engines:** 2.0L VIN B, 2.4L VIN C, 2.7L VIN D, 2.7L VIN F, 3.3L VIN D, 3.3L VIN E, 3.3L VIN F, 3.8L VIN 3, 3.8L VIN C, 3.8L VIN F **Transmissions:** All	**Crankshaft Position (CKP) System Variation Not Learned/Segment Time Acquisition Incorrect** The engine is running. DTC P0315 is set when the CKP system variation values are not stored in the ECM memory. A misfire induces a decrease in the engine speed and causes a variation in the segment period. **Possible causes:** • Crankshaft Position System Variation not Learned • Interference in the signal circuit of the CKP sensor • Debris between the CKP sensor and the reluctor wheel • A damaged or misaligned reluctor wheel • Mechanical damage (worn crankshaft main bearings, excessive crankshaft runout, a damaged crankshaft) • Contact resistance in connectors • Faulty Control Module
DTC: P0320 **2T CCM, MIL: Yes** **Years:** 2006 **Models:** Santa Fe **Engines:** 2.4L VIN B, 3.5L VIN E **Transmissions:** All	**Ignition/Distributor Engine Speed Input Circuit Malfunction** The engine is started and running 2 seconds. The PCM detected a problem in the Ignition Failure Sensor or its circuit. **Possible Causes:** • Ignition failure sensor circuit is open or shorted to ground • Ignition failure sensor is damaged or has failed • PCM has failed
DTC: P0325 **2T CCM, MIL: Yes** **Years:** 2006, 2007, 2008 **Models:** Azera, Elantra, Entourage, Santa Fe, Sonata, Tiburon, Tucson, Veracruz **Engines:** 2.0L VIN B, 2.0L VIN D, 2.4L VIN B, 2.4L VIN C, 2.7L VIN D, 2.7L VIN F, 3.3L VIN D, 3.3L VIN E, 3.3L VIN F, 3.5L VIN E, 3.8L VIN 3, 3.8L VIN C, 3.8L VIN F **Transmissions:** All	**Knock Sensor 1 Circuit (Bank 1)** The engine speed is between 1,000–5,000 RPM. The Throttle Position (TP) indicated angle is more than 10 percent. The engine load is more than 40 percent. The Engine Coolant Temperature (ECT) is more than 140°F (60°C). DTC P0325 is set when the control module detects a malfunction in the KS diagnostic circuitry that will not allow proper diagnosis of the KS system. **Possible Causes:** • KS connector is damaged or shorted • KS signal circuit is shorted to ground • KS has physical damage (may have been dropped) • KS improper installation (too loose or over-tightened may cause DTC to set) • KS mounting surface contains burs, casting flash, or other foreign material • ECM has failed
DTC: P0326 **2T CCM, MIL: Yes** **Years:** 2006, 2007, 2008 **Models:** Accent, Azera, Entourage, Santa Fe, Sonata, Veracruz **Engines:** 1.6L VIN C, 2.0L VIN B, 2.0L VIN D, 2.4L VIN B, 2.4L VIN C, 2.7L VIN D, 2.7L VIN F, 3.3L VIN D, 3.3L VIN E, 3.3L VIN F, 3.5L VIN E, 3.8L VIN 3, 3.8L VIN C, 3.8L VIN F **Transmissions:** All	**Knock Sensor 1 Circuit Range/Performance (Bank 1)** DTC P0326 runs continuously when the engine is running. DTC P0326 is set when the KS signal indicates an excessive engine knock is present. The control module commanded spark retard at a given engine load and speed is more than the calibrated value. **Possible Causes:** • Knock Sensor signal circuits are open or shorted • Knock Sensor circuit is shorted to system power • Knock Sensor is damaged or it has failed • ECM has failed

DTC	Trouble Code Title, Conditions & Possible Causes
DTC: P0327 **2T CCM, MIL: Yes** **Years:** 2006, 2007, 2008 **Models:** Accent **Engines:** 1.6L VIN C **Transmissions:** All	**Knock Sensor (KS) 1 Circuit Low Input** DTC P0327 runs continuously when the Engine Coolant Temperature (ECT) is greater than -104°F. The engine run time is greater than 1 second. DTC P0327 is set when the KS signal circuits are shorted to ground or power. **Possible Causes:** • KS 1 connector is damaged or shorted • KS 1 signal circuit is shorted to ground • KS 1 has physical damage (may have been dropped) • KS 1 improper installation (too loose or over-tightened may cause DTC to set) • KS 1 mounting surface contains burs, casting flash, or other foreign material • ECM has failed
DTC: P0328 **2T CCM, MIL: Yes** **Years:** 2006, 2007, 2008 **Models:** Accent **Engines:** 1.6L VIN C **Transmissions:** All	**Knock Sensor (KS) 1 Circuit High Input** DTC P0328 runs continuously when the Engine Coolant Temperature (ECT) is greater than -104°F. The engine run time is greater than 1 second. DTC P0328 is set when the KS signal circuits are shorted to ground or power. **Possible Causes:** • Electromagnetic Interference (EMI): KS circuits routed too close to high load circuits or components • KS 1 connector is damaged or shorted • KS 1 signal circuit is shorted to ground • KS 1 has physical damage (may have been dropped) • KS 1 improper installation (too loose or over-tightened may cause DTC to set) • KS 1 mounting surface contains burs, casting flash, or other foreign material • ECM has failed
DTC: P0330 **2T CCM, MIL: Yes** **Years:** 2006, 2007, 2008 **Models:** Azera, Entourage, Santa Fe, Sonata, Tiburon, Tucson, Veracruz **Engines:** 2.7L VIN D, 2.7L VIN F, 3.3L VIN D, 3.3L VIN E, 3.3L VIN F, 3.8L VIN 3, 3.8L VIN C, 3.8L VIN F **Transmissions:** All	**Knock Sensor (KS) 2 Circuit (Bank 2)** DTC P0330 is set when the KS signal circuits are shorted to ground or power. When the DTC is set, the spark timing will be retarded to reduce spark knock and may cause reduced engine power. **Possible Causes:** • Knock Sensor 2 connector is damaged or open • Knock Sensor 2 circuit is shorted to ground • Knock Sensor 2 is damaged or it has failed • ECM has failed
DTC: P0331 **2T CCM, MIL: Yes** **Years:** 2006, 2007, 2008 **Models:** Azera, Entourage, Santa Fe, Sonata, Tucson, Veracruz **Engines:** 2.7L VIN D, 3.3L VIN D, 3.3L VIN E, 3.3L VIN F, 3.8L VIN 3, 3.8L VIN C, 3.8L VIN F **Transmissions:** All	**Knock Sensor (KS) 2 Circuit Range/Performance (Bank 2)** DTC P0331 is set when the KS signal indicates an excessive engine knock is present. The control module commanded spark retard at a given engine load and speed is more than the calibrated value. **Possible Causes:** • Knock Sensor 2 connector is damaged or open • Knock Sensor 2 circuit is shorted to ground • Knock Sensor 2 is damaged or it has failed • ECM has failed
DTC: P0335 **1T CCM, MIL: Yes** **Years:** 2006, 2007, 2008 **Models:** Accent, Azera, Elantra, Entourage, Santa Fe, Sonata, Tiburon, Tucson, Veracruz **Engines:** 1.6L VIN C, 2.0L VIN B, 2.0L VIN D, 2.4L VIN C, 2.7L VIN D, 2.7L VIN F, 3.3L VIN D, 3.3L VIN E, 3.3L VIN F, 3.8L VIN 3, 3.8L VIN C, 3.8L VIN F **Transmissions:** All	**Crankshaft Position (CKP) Sensor A Circuit Malfunction** The engine is cranking or operating. The ECM has detected more than 12 camshaft revolutions. DTC P0335 is set when the ECM does not detect a signal from the CKP sensor. OR The ECM detects a CKP signal without reference pulse for more than 3 revolutions. **Possible Causes:** • CKP sensor connector is damaged, open or shorted • CKP sensor positive (+) circuit or (−) circuit is open or shorted to ground • CKP sensor is physically damaged or it is improperly installed • CKP sensor has failed • Electromagnetic Interference (EMI) in the CKP sensor circuits • Excessive air gap between the CKP sensor and reluctor ring • Foreign material lodged between CKP sensor and reluctor ring • ECM has failed

DTC	Trouble Code Title, Conditions & Possible Causes
DTC: P0336 **1T CCM, MIL: Yes** **Years:** 2006, 2007, 2008 **Models:** Accent, Azera, Entourage, Santa Fe, Sonata, Veracruz **Engines:** 1.6L VIN C, 2.4L VIN C, 2.7L VIN D, 3.3L VIN D, 3.3L VIN E, 3.3L VIN F, 3.8L VIN 3, 3.8L VIN C, 3.8L VIN F **Transmissions:** All	**Crankshaft Position (CKP) Sensor A Circuit Range/Performance** The engine is cranking or operating. The ECM has detected more than 12 camshaft revolutions. DTC P0336 is set when the ECM re-syncs the engine position 6 or more times during an ignition cycle. OR The ECM detects 14 or more interruptions in the engine speed signal during an ignition cycle. **Possible Causes:** • CKP sensor wires routed close to other wiring or components • CKP sensor wires routed close to after-market add-on devices • CKP sensor wires routed close to solenoids, relays, and motors • CKP sensor is physically damaged or it is improperly installed • CKP sensor has failed • Electromagnetic Interference (EMI) in CKP sensor circuit • Excessive air gap between the CKP sensor and reluctor ring • Foreign material lodged between CKP sensor and reluctor ring • ECM has failed
DTC: P0337 **1T CCM, MIL: Yes** **Years:** 2006, 2007, 2008 **Models:** Accent, Santa Fe **Engines:** 1.6L VIN C, 2.4L VIN B, 3.5L VIN E **Transmissions:** All	**Crankshaft Position (CKP) Sensor A Circuit Low Input** The output voltage of the CKP sensor remains low for more than 2 seconds when the change of the CMP sensor output voltage is zero, then the PCM determines a fault and stores a code. Change in output voltage is monitored. **Possible Causes:** • CKP sensor wires routed close to other wiring or components • CKP sensor resistance out of specification • CKP sensor is damaged or it has failed • Excessive air gap between the CKP sensor and reluctor ring • Foreign material lodged between CKP sensor and reluctor ring • ECM has failed
DTC: P0338 **1T CCM, MIL: Yes** **Years:** 2006, 2007, 2008 **Models:** Accent, Santa Fe **Engines:** 1.6L VIN C, 2.4L VIN B, 3.5L VIN E **Transmissions:** All	**Crankshaft Position (CKP) Sensor A Circuit High Input** The engine is cranking or operating. If the output voltage of the CKP sensor remains high for more than 2 seconds, when the change of the CMP sensor output voltage is zero, the PCM determines a fault and stores a code. Change in output voltage is monitored. **Possible Causes:** • CKP sensor wires routed close to other wiring or components • CKP sensor resistance out of specification • CKP sensor is damaged or it has failed • Excessive air gap between the CKP sensor and reluctor ring • Foreign material lodged between CKP sensor and reluctor ring • ECM has failed
DTC: P0339 **1T CCM, MIL: Yes** **Years:** 2006, 2007, 2008 **Models:** Accent **Engines:** 1.6L VIN C **Transmissions:** All	**Crankshaft Position Sensor A Circuit** The engine is cranking or operating. The ECM has detected more than 12 camshaft revolutions. The DTC P0339 is set when the ECM detects a difference of more than 8 teeth between reference gap position pulses for 4 consecutive crankshaft revolutions in which the same number of pulses are detected. **Possible Causes:** • CKP sensor wires routed close to other wiring or components • CKP sensor resistance out of specification • CKP sensor is damaged or it has failed • Excessive air gap between the CKP sensor and reluctor ring • Foreign material lodged between CKP sensor and reluctor ring • ECM has failed
DTC: P0340 **2T CCM, MIL: Yes** **Years:** 2006, 2007, 2008 **Models:** Accent, Azera, Elantra, Entourage, Santa Fe, Sonata, Tiburon, Tucson, Veracruz **Engines:** 1.6L VIN C, 2.0L VIN B, 2.0L VIN D, 2.4L VIN C, 2.7L VIN D, 2.7L VIN F, 3.3L VIN D, 3.3L VIN E, 3.3L VIN F, 3.8L VIN 3, 3.8L VIN C, 3.8L VIN F **Transmissions:** All	**Camshaft Position (CMP) Sensor Circuit Malfunction (Bank 1 or Single Sensor)** The engine is cranking. OR The engine is running for more than 3 seconds. This diagnostic runs continuously when either condition is met. DTC P0340 is set when the PCM detects an invalid or irregular CMP signal, or it did not detect any CMP signals. **Possible Causes:** • CMP sensor circuit shorted to ground • CMP sensor circuit open or high resistance • VREF short to ground or open • Excessive air gap between reluctor wheel and sensor magnet • CMP sensor is damaged (cracked), or it has failed • ECM has failed

DTC	Trouble Code Title, Conditions & Possible Causes
DTC: P0341 **2T CCM, MIL: Yes** **Years:** 2006, 2007, 2008 **Models:** Accent, Azera, Entourage, Santa Fe, Sonata, Veracruz **Engines:** 1.6L VIN C, 2.4L VIN C, 2.7L VIN D, 3.3L VIN D, 3.3L VIN E, 3.3L VIN F, 3.8L VIN 3, 3.8L VIN C, 3.8L VIN F **Transmissions:** All	**Camshaft Position (CMP) Sensor A Circuit Range/Performance (Bank 1 or Single Sensor)** The battery voltage is between 10–16 volts. The crankshaft sensor tests normal. DTC P0341 is set when the ECM detects the incorrect number of CMP sensor pulses in 10 revolutions of the crankshaft. **Possible Causes:** • CMP sensor wires routed close to other wiring or components • Camshaft reluctor wheel damage, incorrect sensor installation • CMP sensor is contacting the reluctor wheel • CMP sensor is damaged (cracked), or it has failed • Electromagnetic interference in CMP sensor circuit (due to the sensor wires routed to close to ignition cables or motors) • Excessive air gap between reluctor wheel and sensor magnet • Foreign material lodged between sensor and the reluctor wheel • ECM has failed
DTC: P0342 **2T CCM, MIL: Yes** **Years:** 2006, 2007, 2008 **Models:** Accent, Santa Fe **Engines:** 1.6L VIN C, 2.4L VIN B, 3.5L VIN E **Transmissions:** All	**Camshaft Position Sensor A Circuit Low Input** The engine is running at a speed over 600 RPM. DTC P0342 is set when the ECM detects the incorrect number of CMP sensor pulses in 10 revolutions of the crankshaft. **Possible Causes:** • CMP sensor connector is damaged or open • CMP sensor signal circuit is shorted to ground • CMP sensor signal circuit is open or it has high resistance • CMP sensor reluctor wheel is damaged or loose • ECM has failed
DTC: P0343 **2T CCM, MIL: Yes** **Years:** 2006, 2007, 2008 **Models:** Accent, Santa Fe **Engines:** 1.6L VIN C, 2.4L VIN B, 3.5L VIN E **Transmissions:** All	**Camshaft Position (CMP) Sensor Circuit High Input** The engine is running at a speed over 600 RPM. DTC P0343 is set when the ECM detects the incorrect number of CMP sensor pulses in 10 revolutions of the crankshaft. **Possible Causes:** • CMP sensor connector is damaged or shorted • CMP sensor signal circuit is open • CMP sensor signal circuit is shorted to system power • CMP sensor is damaged or it has failed • CMP sensor reluctor wheel is damaged or loose • ECM has failed
DTC: P0346 **2T CCM, MIL: Yes** **Years:** 2006, 2007, 2008 **Models:** Azera, Entourage, Santa Fe, Sonata, Veracruz **Engines:** 2.7L VIN D, 3.3L VIN D, 3.3L VIN E, 3.3L VIN F, 3.8L VIN 3, 3.8L VIN F, 3.8 VIN C **Transmissions:** All	**Camshaft Position (CMP) Sensor A Circuit Range/Performance (Bank 2)** DTC P0346 is set when the ECM detects a signal from the CMP sensor, but the number of pulses are at least 5 less than, or more than, what is expected for 1 crankshaft revolution. OR The CMP sensor does NOT correlate to the crankshaft position. **Possible Causes:** • CMP sensor wires routed close to other wiring or components • Camshaft reluctor wheel damage, incorrect sensor installation • CMP sensor is contacting the reluctor wheel • CMP sensor is damaged (cracked), or it has failed • Electromagnetic interference in CMP sensor circuit (due to the sensor wires routed to close to ignition cables or motors) • Excessive air gap between reluctor wheel and sensor magnet • Foreign material lodged between sensor and the reluctor wheel • ECM has failed
DTC: P0350 **2T CCM, MIL: Yes** **Years:** 2006, 2007, 2008 **Models:** Santa Fe, Sonata, Tiburon, Tucson **Engines:** 2.4L VIN B, 2.7L VIN D, 2.7L VIN F, 3.5L VIN E **Transmissions:** All	**Ignition Coil Primary/Secondary Circuit Malfunction** The engine is running or cranking. DTC P0350 is set when the ECM detects a fault in the IC timing control circuit. **Possible Causes:** • High resistance in the ground circuits for the ignition coil • Ignition control circuit for an open • Poor connection at the ignition coil or ECM • Faulty ignition coil • ECM has failed
DTC: P0351 **2T CCM, MIL: Yes** **Years:** 2006, 2007, 2008 **Models:** Azera, Entourage, Santa Fe, Sonata, Tiburon, Tucson, Veracruz **Engines:** 2.7L VIN D, 2.7L VIN F, 3.3L VIN D, 3.3L VIN E, 3.3L VIN F, 3.8L VIN 3, 3.8L VIN C, 3.8L VIN F **Transmissions:** All	**Ignition Coil A Primary/Secondary Circuit Malfunction** The engine is running. DTC P0351 is set when the ECM detects an open on the circuit or the ignition coil/module for less than 1 second. OR The ECM detects a short to ground on the circuit or the ignition coil/module for less than 1 second. OR The ECM detects a short to voltage on the circuit or the ignition coil/module for less than 1 second. **Possible Causes:** • High resistance in the ground circuits for the ignition coil • Ignition control circuit for an open • Poor connection at the ignition coil or ECM • Faulty ignition coil • ECM has failed

DTC	Trouble Code Title, Conditions & Possible Causes
DTC: P0352 **2T CCM, MIL: Yes** **Years:** 2006, 2007, 2008 **Models:** Azera, Entourage, Santa Fe, Sonata, Tiburon, Tucson, Veracruz **Engines:** 2.7L VIN D, 2.7L VIN F, 3.3L VIN D, 3.3L VIN E, 3.3L VIN F, 3.8L VIN 3, 3.8L VIN C, 3.8L VIN F **Transmissions:** All	**Ignition Coil B Primary/Secondary Circuit Malfunction** The engine is running. DTC P0352 is set when the ECM detects an open on the circuit or the ignition coil/module for less than 1 second. OR The ECM detects a short to ground on the circuit or the ignition coil/module for less than 1 second. OR The ECM detects a short to voltage on the circuit or the ignition coil/module for less than 1 second. **Possible Causes:** • High resistance in the ground circuits for the ignition coil • Ignition control circuit for an open • Poor connection at the ignition coil or ECM • Faulty ignition coil • ECM has failed
DTC: P0353 **2T CCM, MIL: Yes** **Years:** 2006, 2007, 2008 **Models:** Azera, Entourage, Santa Fe, Sonata, Tiburon, Tucson, Veracruz **Engines:** 2.7L VIN D, 2.7L VIN F, 3.3L VIN D, 3.3L VIN E, 3.3L VIN F, 3.8L VIN 3, 3.8L VIN C, 3.8L VIN F **Transmissions:** All	**Ignition Coil C Primary/Secondary Circuit Malfunction** The engine is running. DTC P0353 is set when the ECM detects an open on the circuit or the ignition coil/module for less than 1 second. OR The ECM detects a short to ground on the circuit or the ignition coil/module for less than 1 second. OR The ECM detects a short to voltage on the circuit or the ignition coil/module for less than 1 second. **Possible Causes:** • High resistance in the ground circuits for the ignition coil • Ignition control circuit for an open • Poor connection at the ignition coil or ECM • Faulty ignition coil • ECM has failed
DTC: P0354 **2T CCM, MIL: Yes** **Years:** 2006, 2007, 2008 **Models:** Azera, Entourage, Santa Fe, Sonata, Tiburon, Tucson, Veracruz **Engines:** 2.7L VIN D, 2.7L VIN F, 3.3L VIN D, 3.3L VIN E, 3.3L VIN F, 3.8L VIN 3, 3.8L VIN C, 3.8L VIN F **Transmissions:** All	**Ignition Coil D Primary/Secondary Circuit Malfunction** The engine is running. DTC P0354 is set when the ECM detects one of the following failures on the IC circuit for up to 6 seconds: An open, a short to ground, or a short to voltage. **Possible Causes:** • High resistance in the ground circuits for the ignition coil • Ignition control circuit for an open • Poor connection at the ignition coil or ECM • Faulty ignition coil • ECM has failed
DTC: P0355 **2T CCM, MIL: Yes** **Years:** 2006, 2007, 2008 **Models:** Azera, Entourage, Santa Fe, Sonata, Tiburon, Tucson, Veracruz **Engines:** 2.7L VIN D, 2.7L VIN F, 3.3L VIN D, 3.3L VIN E, 3.3L VIN F, 3.8L VIN 3, 3.8L VIN C, 3.8L VIN F **Transmissions:** All	**Ignition Coil E Primary/Secondary Circuit Malfunction** The engine is running. DTC P0355 is set when the ECM detects one of the following failures on the IC circuit for up to 6 seconds: An open, a short to ground, or a short to voltage. **Possible Causes:** • High resistance in the ground circuits for the ignition coil • Ignition control circuit for an open • Poor connection at the ignition coil or ECM • Faulty ignition coil • ECM has failed
DTC: P0356 **2T CCM, MIL: Yes** **Years:** 2006, 2007, 2008 **Models:** Azera, Entourage, Santa Fe, Sonata, Tiburon, Tucson, Veracruz **Engines:** 2.7L VIN D, 2.7L VIN F, 3.3L VIN D, 3.3L VIN E, 3.3L VIN F, 3.8L VIN 3, 3.8L VIN C, 3.8L VIN F **Transmissions:** All	**Ignition Coil F Primary/Secondary Circuit Malfunction** The engine is running. DTC P0356 is set when the ECM detects one of the following failures on the IC circuit for up to 6 seconds: An open, a short to ground, or a short to voltage. **Possible Causes:** • High resistance in the ground circuits for the ignition coil • Ignition control circuit for an open • Poor connection at the ignition coil or ECM • Faulty ignition coil • ECM has failed

DTC	Trouble Code Title, Conditions & Possible Causes
DTC: P0401 **1T CCM, MIL: Yes** **Years:** 2006 **Models:** Santa Fe **Engines:** 2.4L VIN B, 3.5L VIN E **Transmissions:** All	**Exhaust Gas Recirculation (EGR) Flow Insufficient Detected** The engine run time may need to be more than 3 minutes. The Ignition 1 Signal parameter is between 11–18 volts. The Engine Coolant Temperature (ECT) sensor parameter is between 167–302°F (75–150°C). The Engine Speed parameter is between 1,000–1,500 RPM. As the EGR flow test is running, you will see the desired EGR Position parameter and the EGR Position Sensor parameter on the scan tool momentarily change from 0 to a calibrated value above 0. DTC P0401 is set when the MAP changes monitored by the ECM during the EGR flow tests indicate an insufficient amount of EGR flow. **Possible Causes:** • MAP sensor stuck, skewed, or faulty vacuum supply • Restrictions in EGR passages or valve caused by carbon deposits or casting flash • Incorrect EGR valve for engine application • Vacuum or exhaust leaks between the EGR valve and the intake manifold • Leaking, restricted, or modified exhaust system • Engine mechanical failure • Faulty EGR valve
DTC: P0420 **1T CAT, MIL: Yes** **Years:** 2006, 2007, 2008 **Models:** Accent, Azera, Elantra, Entourage, Santa Fe, Sonata, Tiburon, Tucson, Veracruz **Engines:** 1.6L VIN C, 2.0L VIN B, 2.0L VIN D, 2.4L VIN B, 2.4L VIN C, 2.7L VIN D, 2.7L VIN F, 3.3L VIN D, 3.3L VIN E, 3.3L VIN F, 3.5L VIN E, 3.8L VIN 3, 3.8L VIN C, 3.8L VIN F **Transmissions:** All	**Catalyst System Efficiency Below Threshold (Bank 1)** The vehicle is in Closed Loop. The calculated catalyst temperature is between 852–1,562°F (400–850°C), and stable. The rear HO2S has exceeded the dew point for more than 60 seconds. DTC P0420 is set when the ECM has determined the catalyst efficiency has degraded below a calibrated threshold. **Note: A new converter with less than 100 miles on it may set DTC P0420 due to out-gassing of the internal matting. Operating the vehicle at highway speeds for approximately 1 hour may correct the condition.** **Possible Causes:** • Air leaks at the exhaust manifold or in the exhaust pipes • Base engine problems (i.e., high engine oil or coolant usage) • Catalytic converter is damaged, contaminated, or has failed • Continuous engine misfire conditions, or weak/low coil output • Front HO2S or rear HO2S is contaminated with fuel or moisture • Rear HO2S is loose in the mounting hole (check it for a leak)
DTC: P0430 **2T CAT, MIL: Yes** **Years:** 2006, 2007, 2008 **Models:** Azera, Entourage, Santa Fe, Sonata, Tiburon, Tucson, Veracruz **Engines:** 2.7L VIN D, 2.7L VIN F, 3.3L VIN D, 3.3L VIN E, 3.3L VIN F, 3.5L VIN E, 3.8L VIN 3, 3.8L VIN C, 3.8L VIN F **Transmissions:** All	**Catalyst System Efficiency Below Threshold (Bank 2)** The engine has been running for more than 10 minutes. The vehicle has been driven at more than 1,000 RPM for more than 1 minute. The vehicle is in Closed Loop. The vehicle has Fuel Trim Learn enabled. The Engine Coolant Temperature (ECT) is between 156–257°F (70–125°C). The catalytic converter calculated temperature is greater than or equal to 842°F (450°C). DTC P0420 is set when the ECM has determined the catalyst efficiency has degraded below a calibrated threshold. **Possible Causes:** • Air leaks at the exhaust manifold or in the exhaust pipes • Base engine problems (i.e., high engine oil or coolant usage) • Catalytic converter is damaged, contaminated, or has failed • Continuous engine misfire conditions, or weak or low coil output • Front HO2S or rear HO2S is contaminated with fuel or moisture • Rear HO2S is loose in the mounting hole (check it for a leak)
DTC: P0441 **1T EVAP, MIL: Yes** **Years:** 2006, 2007, 2008 **Models:** Azera, Elantra, Entourage, Santa Fe, Sonata, Tiburon, Tucson, Veracruz **Engines:** 2.0L VIN B, 2.0L VIN D, 2.4L VIN B, 2.4L VIN C, 2.7L VIN D, 2.7L VIN F, 3.3L VIN D, 3.3L VIN E, 3.3L VIN F, 3.5L VIN E, 3.8L VIN 3, 3.8L VIN C, 3.8L VIN F **Transmissions:** All	**Evaporative Emission (EVAP) System Incorrect Purge Flow** The engine is running and the vehicle is driven at a speed of 35–40 mph for at least 5–10 minutes under light engine load conditions. Then with the Purge solenoid commanded ON and then OFF, the PCM detected the Purge solenoid valve remained OPEN during the EVAP Monitor flow test. **Possible Causes:** • Fuel filler cap is loose, cross-threaded, damaged, or wrong part • Fuel tank, fuel filler neck, or fuel sending unit O-ring is leaking • Fuel tank vapor line(s) is clogged, damaged, or disconnected • Fuel tank pressure sensor low reference circuit is open • Fuel tank pressure sensor is damaged or it has failed • EVAP charcoal canister is clogged or loaded with fuel or water • EVAP purge or EVAP vent valve is damaged or it has failed • ECM has failed

DTC	Trouble Code Title, Conditions & Possible Causes
DTC: P0442 **1T EVAP, MIL: Yes** **Years:** 2006, 2007, 2008 **Models:** Accent, Azera, Elantra, Entourage, Santa Fe, Sonata, Tiburon, Tucson, Veracruz **Engines:** 1.6L VIN C, 2.0L VIN B, 2.0L VIN D, 2.4L VIN B, 2.4L VIN C, 2.7L VIN D, 2.7L VIN F, 3.3L VIN D, 3.3L VIN E, 3.3L VIN F, 3.5L VIN E, 3.8L VIN 3, 3.8L VIN C, 3.8L VIN F **Transmissions:** All	**Evaporative Emission (EVAP) System Small Leak Detected** The engine run time minimum is 10 minutes. The vehicle has traveled more than 5 miles (8 km) this trip. The ignition is OFF. A refueling event is not detected. DTC P0442 runs once per drive cycle when the above conditions are met. One test occurs at ignition OFF after a drive cycle and may require up to 45 minutes to complete. DTC P0442 is set when the control module detects a pressure change that is less than a calibrated amount. **Possible Causes:** • Fuel filler cap is loose, cross-threaded, damaged, or wrong part • Fuel tank, fuel filler neck, or fuel sending unit O-ring is leaking • Fuel tank vapor line(s) is clogged, damaged, or disconnected • Fuel tank pressure sensor low reference circuit is open • Fuel tank pressure sensor is damaged or it has failed • EVAP charcoal canister is clogged or loaded with fuel or water • EVAP purge or EVAP vent valve is damaged or it has failed • ECM has failed
DTC: P0444 **2T EVAP, MIL: Yes** **Years:** 2006, 2007, 2008 **Models:** Accent, Azera, Elantra, Entourage, Santa Fe, Sonata, Tiburon, Tucson, Veracruz **Engines:** 1.6L VIN C, 2.0L VIN B, 2.0L VIN D, 2.4L VIN B, 2.4L VIN C, 2.7L VIN D, 2.7L VIN F, 3.3L VIN D, 3.3L VIN E, 3.3L VIN F, 3.5L VIN E, 3.8L VIN 3, 3.8L VIN C, 3.8L VIN F **Transmissions:** All	**Evaporative Emission (EVAP) System Purge Control Valve Circuit Open** Engine started and running at idle speed. The PCM detected an unexpected low voltage condition on the EVAP Purge solenoid circuit as the solenoid was commanded ON and OFF in the test. **Possible Causes:** • Purge solenoid control circuit open between solenoid and PCM • Purge solenoid power circuit is open (check the power source) • Purge control solenoid is damaged or has failed • PCM has failed
DTC: P0445 **2T EVAP, MIL: Yes** **Years:** 2006, 2007, 2008 **Models:** Azera, Elantra, Entourage, Santa Fe, Sonata, Tiburon, Tucson, Veracruz **Engines:** 2.0L VIN B, 2.0L VIN D, 2.4L VIN B, 2.4L VIN C, 2.7L VIN D, 2.7L VIN F, 3.3L VIN D, 3.3L VIN E, 3.3L VIN F, 3.5L VIN E, 3.8L VIN 3, 3.8L VIN C, 3.8L VIN F **Transmissions:** All	**Evaporative Emission (EVAP) System Purge Control Valve Circuit Shorted** Engine started **and** running at idle speed. **The PCM detected an unexpected** high voltage condition on the EVAP Purge solenoid circuit as the solenoid was commanded **ON** and **OFF** in the test. **Possible Causes:** • Purge solenoid control circuit is shorted to system power • Purge control solenoid is damaged or has failed (short circuit) • PCM has failed
DTC: P0446 **2T EVAP, MIL: Yes** **Years:** 2006, 2007, 2008 **Models:** Accent **Engines:** 1.6L VIN C **Transmissions:** All	**Evaporative Emission (EVAP) System Vent Control Circuit** Engine started and running at cruise speed for 2–3 minutes. The PCM detected the EVAP Vent Control solenoid was in a closed position continuously during the CCM test. **Possible Causes:** • Vent solenoid control circuit is shorted to ground • Vent solenoid power circuit is open (check the power source) • Vent control solenoid is damaged or has failed • PCM has failed
DTC: P0447 **2T EVAP, MIL: Yes** **Years:** 2006, 2007, 2008 **Models:** Azera, Elantra, Entourage, Santa Fe, Sonata, Tiburon, Tucson, Veracruz **Engines:** 2.0L VIN B, 2.0L VIN D, 2.4L VIN B, 2.4L VIN C, 2.7L VIN D, 2.7L VIN F, 3.3L VIN D, 3.3L VIN E, 3.3L VIN F, 3.5L VIN E, 3.8L VIN 3, 3.8L VIN C, 3.8L VIN F **Transmissions:** All	**Evaporative Emission (EVAP) System Vent Control Circuit Open** Engine started **and** running at cruise speed for 2–3 minutes. **T**he PCM detected the EVAP Vent Control solenoid control circuit was in a continuous low state during the CCM test. **Possible Causes:** • Purge solenoid control circuit is shorted to ground • Purge solenoid power circuit is open (check the power source) • Purge solenoid is damaged or has failed • PCM has failed

DTC	Trouble Code Title, Conditions & Possible Causes
DTC: P0448 **2T EVAP, MIL: Yes** **Years:** 2006, 2007, 2008 **Models:** Azera, Elantra, Entourage, Santa Fe, Sonata, Tiburon, Tucson, Veracruz **Engines:** 2.0L VIN B, 2.0L VIN D, 2.4L VIN B, 2.4L VIN C, 2.7L VIN D, 2.7L VIN F, 3.3L VIN D, 3.3L VIN E, 3.3L VIN F, 3.5L VIN E, 3.8L VIN 3, 3.8L VIN C, 3.8L VIN F **Transmissions:** All	**Evaporative Emission (EVAP) System Vent Control Circuit Shorted** Engine started and running at cruise speed for 2–3 minutes. The PCM detected the EVAP Vent Control solenoid control circuit was in a continuous high state during the CCM test. **Possible Causes:** • Purge solenoid control circuit is shorted to system power • Purge solenoid is damaged or has failed • PCM has failed
DTC: P0449 **2T CCM, MIL: Yes** **Years:** 2006, 2007, 2008 **Models:** Accent, Elantra, Santa Fe, Sonata, Tiburon, Tucson **Engines:** 1.6L VIN C, 2.0L VIN B, 2.0L VIN D, 2.4L VIN C, 2.7L VIN D, 2.7L VIN F **Transmissions:** All	**Evaporative Emission (EVAP) Control System Vent Valve/Solenoid Circuit** The PCM measures pressure in the fuel tank, by means of a sensor, during all engine operating conditions, except for start and stop. If the pressure is lower than the threshold, a DTC is set. DTC P0449 is set when the control module detects that the commanded state of the driver and the actual state of the control circuit do not match. **Possible Causes:** • Ignition 1 voltage Purge Supply open or shorted to ground • Purge solenoid control circuit is open or shorted to ground • Purge solenoid control circuit is shorted to voltage • Purge solenoid is damaged or it has failed • ECM has failed
DTC: P0450 **2T EVAP, MIL: Yes** **Years:** 2006, 2007, 2008 **Models:** Accent **Engines:** 1.6L VIN C **Transmissions:** All	**Evaporative Emission (EVAP) System Pressure Sensor/Switch** Engine started and running, the vehicle is not moving. With the EVAP Vapor sensor commanded ON, the PCM detected an unexpected voltage condition on the EVAP Pressure sensor circuit. **Possible Causes:** • Pressure sensor signal circuit is open or shorted to ground • Pressure sensor signal circuit is shorted to VREF or power • Pressure sensor power (VREF) circuit is open • Pressure sensor is damaged or has failed • PCM has failed
DTC: P0451 **2T CCM, MIL: Yes** **Years:** 2006, 2007, 2008 **Models:** Accent, Azera, Elantra, Entourage, Santa Fe, Sonata, Tiburon, Tucson, Veracruz **Engines:** 1.6L VIN C, 2.0L VIN B, 2.0L VIN D, 2.4L VIN B, 2.4L VIN C, 2.7L VIN D, 2.7L VIN F, 3.3L VIN D, 3.3L VIN E, 3.3L VIN F, 3.5L VIN E, 3.8L VIN 3, 3.8L VIN C, 3.8L VIN F **Transmissions:** All	**Evaporative Emission (EVAP) System Pressure Sensor Range/Performance** Engine started and running with the vehicle not moving. With the EVAP Vapor sensor commanded ON, the PCM detected the EVAP pressure sensor signal was not plausible during the test. **Note: This condition (code) can be due to a fuel sloshing condition.** **Possible Causes:** • Pressure sensor vacuum hoses loose or damaged • Pressure sensor is damaged or out-of-calibration • VSV for the EVAP pressure sensor is damaged or has failed • PCM has failed
DTC: P0452 **2T CCM, MIL: Yes** **Years:** 2006, 2007, 2008 **Models:** Accent, Azera, Elantra, Entourage, Santa Fe, Sonata, Tiburon, Tucson, Veracruz **Engines:** 1.6L VIN C, 2.0L VIN B, 2.0L VIN D, 2.4L VIN B, 2.4L VIN C, 2.7L VIN D, 2.7L VIN F, 3.3L VIN D, 3.3L VIN E, 3.3L VIN F, 3.5L VIN E, 3.8L VIN 3, 3.8L VIN C, 3.8L VIN F **Transmissions:** All	**Evaporative Emission (EVAP) System Pressure Sensor/Switch Low** Engine started and running over 5 seconds. The PCM detected an unexpected low voltage condition on the EVAP Pressure sensor circuit during the CCM test. **Possible Causes:** • Pressure sensor signal circuit is shorted to ground • Pressure sensor power (VREF) circuit is open • Pressure sensor is damaged or has failed • PCM has failed

DTC	Trouble Code Title, Conditions & Possible Causes
DTC: P0453 **2T CCM, MIL: Yes** **Years:** 2006, 2007, 2008 **Models:** Accent, Azera, Elantra, Entourage, Santa Fe, Sonata, Tiburon, Tucson, Veracruz **Engines:** 1.6L VIN C, 2.0L VIN B, 2.0L VIN D, 2.4L VIN B, 2.4L VIN C, 2.7L VIN D, 2.7L VIN F, 3.3L VIN D, 3.3L VIN E, 3.3L VIN F, 3.5L VIN E, 3.8L VIN 3, 3.8L VIN C, 3.8L VIN F **Transmissions:** All	**Evaporative Emission (EVAP) System Pressure Sensor/Switch High** The engine is running over 5 seconds. The PCM detected an unexpected high voltage condition on the EVAP Pressure sensor circuit during the CCM test. **Possible Causes:** • Pressure sensor signal circuit is shorted to power • Pressure sensor ground circuit open between sensor and PCM • Pressure sensor is damaged or has failed • PCM has failed
DTC: P0454 **2T CCM, MIL: Yes** **Years:** 2006, 2007, 2008 **Models:** Azera, Elantra, Entourage, Santa Fe, Sonata, Tiburon, Tucson, Veracruz **Engines:** 2.0L VIN B, 2.0L VIN D, 2.4L VIN B, 2.4L VIN C, 2.7L VIN D, 2.7L VIN F, 3.3L VIN D, 3.3L VIN E, 3.3L VIN F, 3.5L VIN E, 3.8L VIN 3, 3.8L VIN C, 3.8L VIN F **Transmissions:** All	**Evaporative Emission (EVAP) System Pressure Sensor/Switch Intermittent** The PCM measures pressure stability in the fuel tank, by means of a sensor, for a predetermined duration. If a fluctuation is larger than the threshold, DTC P0454 is set. **Possible Causes:** • Fuel Tank Pressure (FTP) sensor signal performance • Restriction in EVAP canister or vent lines • FTP sensor is damaged or it has failed • ECM has Failed
DTC: P0455 **2T CCM, MIL: Yes** **Years:** 2006, 2007, 2008 **Models:** Accent, Azera, Elantra, Entourage, Santa Fe, Sonata, Tiburon, Tucson, Veracruz **Engines:** 1.6L VIN C, 2.0L VIN B, 2.0L VIN D, 2.4L VIN B, 2.4L VIN C, 2.7L VIN D, 2.7L VIN F, 3.3L VIN D, 3.3L VIN E, 3.3L VIN F, 3.5L VIN E, 3.8L VIN 3, 3.8L VIN C, 3.8L VIN F **Transmissions:** All	**Evaporative Emission (EVAP) System Large Leak Detected** The engine is started, the ECT sensor is more than 185°F, the IAT sensor is between 14–122°F, and the fuel level is from 25–75 percent. The engine is running at cruise speed. The PCM detected a large change in the fuel tank pressure (due to a large leak) during the EVAP Monitor leak test. **Possible Causes:** • Canister Vent (CV) solenoid may be stuck in open position • EVAP canister tube, EVAP canister purge outlet tube, or EVAP return tube disconnected or cracked, or canister is damaged • EVAP canister purge valve stuck closed, or canister damaged • Fuel filler cap missing, loose (not tightened) or the wrong part • Fuel vapor hoses/tubes blocked or restricted, or fuel vapor control valve tube or fuel vapor vent valve assembly blocked • Fuel Tank Pressure (FTP) sensor has failed (mechanical fault) • Fuel tank control valve is contaminated, damaged, or has failed
DTC: P0456 **2T CCM, MIL: Yes** **Years:** 2006, 2007, 2008 **Models:** Accent, Azera, Elantra, Entourage, Santa Fe, Sonata, Tiburon, Tucson, Veracruz **Engines:** 1.6L VIN C, 2.0L VIN B, 2.0L VIN D, 2.4L VIN B, 2.4L VIN C, 2.7L VIN D, 2.7L VIN F, 3.3L VIN D, 3.3L VIN E, 3.3L VIN F, 3.5L VIN E, 3.8L VIN 3, 3.8L VIN C, 3.8L VIN F **Transmissions:** All	**Evaporative Emission (EVAP) System Leak detected (very small leak)** The engine is started, the ECT sensor is more than 185°F, the IAT sensor is between 14–122°F, and the fuel level is from 25–75 percent. The engine is running at cruise speed. The PCM detected a small change in the fuel tank pressure (due to a small leak) during the EVAP Monitor leak test. **Possible Causes:** • Leakage in EVAP system line • Faulty Charcoal Canister Vent (CCV), Purge Control Solenoid Valve (PCSV), or Fuel Tank Pressure Sensor (FTPS)
DTC: P0457 **2T CCM, MIL: Yes** **Years:** 2006, 2007, 2008 **Models:** Accent **Engines:** 1.6L VIN C **Transmissions:** All	**Evaporative Emission (EVAP) System Leak Detected (Fuel Tank Cap Loose/Off)** The engine is started, the ECT sensor is more than 185°F, the IAT sensor is between 14–122°F, and the fuel level is from 25–75 percent. The PCM detected a large change in fuel tank pressure during the EVAP System Monitor leak test. **Possible Causes:** • Canister Vent (CV) solenoid may be stuck in open position • EVAP canister tube, EVAP canister purge outlet tube, or EVAP return tube disconnected or cracked, or canister is damaged • Fuel filler cap missing, loose (not tightened), or the wrong part • Fuel Tank Pressure (FTP) sensor has failed (mechanical fault) • Fuel tank control valve is contaminated, damaged, or has failed

DTC	Trouble Code Title, Conditions & Possible Causes
DTC: P0458 **2T CCM, MIL: Yes** **Years:** 2006, 2007, 2008 **Models:** Accent **Engines:** 1.6L VIN C **Transmissions:** All	**Evaporative Emission (EVAP) System Purge Control Valve Circuit Low** The engine RPM is greater than 80. The system voltage is between 10–18 volts. DTC P0458 is set when the control module detects that the commanded state of the driver and the actual state of the control circuit do not match. Circuit continuity check is low. **Possible Causes:** • Ignition 1 voltage circuit is open or shorted to ground (Check Fuse) • EVAP canister purge valve • ECM has failed
DTC: P0459 **2T CCM, MIL: Yes** **Years:** 2006, 2007, 2008 **Models:** Accent **Engines:** 1.6L VIN C **Transmissions:** All	**Evaporative Emission (EVAP) System Purge Control Valve Circuit High** The engine RPM is greater than 80. The system voltage is between 10–18 volts. DTC P0459 is set when the control module detects that the commanded state of the driver and the actual state of the control circuit do not match. Circuit continuity check is high. **Possible Causes:** • Purge Solenoid Control Circuit short to voltage • EVAP canister purge Solenoid valve • ECM has failed
DTC: P0460 **1T CCM, MIL: No** **Years:** 2006 **Models:** Santa Fe **Engines:** 2.4L VIN B, 3.5L VIN E **Transmissions:** All	**Fuel Level Sensor Circuit** The engine is running. DTC P0461 is set when the ECM does not detect a change in fuel level of at least 1.6 percent over a distance of 120 mi (193 km). **Possible Causes:** • Fuel level sensor signal circuit is shorted to system power • Fuel level sensor ground circuit is open • Fuel level sender is damaged, binding, or not aligned properly • ECM has failed
DTC: P0461 **1T CCM, MIL: No** **Years:** 2006, 2007, 2008 **Models:** Accent, Azera, Entourage, Santa Fe, Sonata, Tiburon, Tucson, Veracruz **Engines:** 1.6L VIN C, 2.4L VIN B, 2.4L VIN C, 2.7L VIN D, 2.7L VIN F, 3.3L VIN D, 3.3L VIN E, 3.3L VIN F, 3.5L VIN E, 3.8L VIN 3, 3.8L VIN C, 3.8L VIN F **Transmissions:** All	**Fuel Level Sensor A Circuit Range/Performance** The engine is running. DTC P0461 is set when the ECM does not detect a change in fuel level of at least 1.6 percent over a distance of 120 mi (193 km). **Possible Causes:** • Fuel level sensor signal circuit is shorted to system power • Fuel level sensor ground circuit is open • Fuel level sender is damaged, binding, or not aligned properly • ECM has failed
DTC: P0462 **1T CCM, MIL: No** **Years:** 2006, 2007, 2008 **Models:** Accent, Azera, Entourage, Santa Fe, Sonata, Tiburon, Tucson, Veracruz **Engines:** 1.6L VIN C, 2.4L VIN B, 2.4L VIN C, 2.7L VIN D, 2.7L VIN F, 3.3L VIN D, 3.3L VIN E, 3.3L VIN F, 3.5L VIN E, 3.8L VIN 3, 3.8L VIN C, 3.8L VIN F **Transmissions:** All	**Fuel Level Sensor A Circuit Low Input** The ignition is ON, with the engine running. The system voltage is between 9–16 volts. DTC P0462 is set when the sender output is less than 0.39 volt for greater than 30 seconds. **Possible Causes:** • Fuel level sensor signal circuit is shorted to ground • Fuel level sensor ground circuit is open • Fuel level sender is damaged, binding, or not aligned properly • ECM has failed
DTC: P0463 **1T CCM, MIL: No** **Years:** 2006, 2007, 2008 **Models:** Accent, Azera, Entourage, Santa Fe, Sonata, Tiburon, Tucson, Veracruz **Engines:** 1.6L VIN C, 2.4L VIN B, 2.4L VIN C, 2.7L VIN D, 2.7L VIN F, 3.3L VIN D, 3.3L VIN E, 3.3L VIN F, 3.5L VIN E, 3.8L VIN 3, 3.8L VIN C, 3.8L VIN F **Transmissions:** All	**Fuel Level Sensor A Circuit High Input** The ignition is ON, with the engine running. The system voltage is between 9–16 volts. DTC P0463 is set when the sensor output is greater than 2.9 volts for greater than 30 seconds. **Possible Causes:** • Fuel level sensor low reference circuit shorted to voltage or high resistance • Fuel level sensor signal circuit shorted to voltage or high resistance • Fuel level sender is damaged, binding, or not aligned properly • ECM has failed

DTC	Trouble Code Title, Conditions & Possible Causes
DTC: P0464 **1T CCM, MIL: No** **Years:** 2006, 2007, 2008 **Models:** Azera, Entourage, Santa Fe, Sonata, Tiburon, Tucson, Veracruz **Engines:** 2.4L VIN C, 2.7L VIN D, 2.7L VIN F, 3.3L VIN D, 3.3L VIN E, 3.3L VIN F, 3.8L VIN 3, 3.8L VIN C, 3.8L VIN F **Transmissions:** All	**Fuel Level Sensor A Circuit Intermittent** The ignition is OFF. DTC P0464 runs and fails 2 out of 3 test cycles. DTC P0464 is set when the fuel level change is greater than 10 percent for less than 30 seconds. **Possible Causes:** • Fuel level sensor signal circuit is shorted to system power • Fuel level sensor ground circuit is open or tests for high resistance • Fuel level sender is damaged, binding, or not aligned properly • ECM has failed
DTC: P0480 **2T CCM, MIL: Yes** **Years:** 2006, 2007, 2008 **Models:** Azera, Entourage, Santa Fe, Sonata, Veracruz **Engines:** 2.7L VIN D, 3.3L VIN D, 3.3L VIN E, 3.3L VIN F, 3.8L VIN 3, 3.8L VIN C, 3.8L VIN F **Transmissions:** All	**Fan 1 Control Circuit Malfunction** The engine is running. No disabling faults are present. Enable time delay is equal to or greater than 0.5 seconds. A short to ground, short to battery, or open circuit of fan relay output is detected. Fault information is provided by an output driver chip. **Possible Causes:** • Poor connection at fan module • Open in power circuit to cooling fan • Open or short in control circuit to PCM • Faulty fan relay • Faulty cooling fan module • Faulty PCM
DTC: P0481 **2T CCM, MIL: Yes** **Years:** 2007, 2008 **Models:** Santa Fe, Veracruz **Engines:** 2.7L VIN D, 3.3L VIN E, 3.8L VIN C **Transmissions:** All	**Fan 2 Control Circuit Malfunction** The engine is running. No disabling faults are present. Enable time delay is equal to or greater than 0.5 seconds. A short to ground, short to battery, or open circuit of fan relay output is detected. Fault information is provided by an output driver chip. **Possible Causes:** • Poor connection at fan module • Open in power circuit to cooling fan • Open or short in control circuit to PCM • Faulty fan relay • Faulty cooling fan module • Faulty PCM
DTC: P0489 **2T CCM, MIL: Yes** **Years:** 2006 **Models:** Santa Fe **Engines:** 2.4L VIN B, 3.5L VIN E **Transmissions:** All	**Exhaust Gas Recirculation (EGR) Control Circuit Low Input** The engine is running. No disabling faults are present. The PCM detects that the surge voltage and output level voltage monitored are below a predetermined threshold. **Possible Causes:** • Poor connection at EGR solenoid • Open or short to ground in EGR solenoid valve circuit • Faulty EGR solenoid valve • Faulty PCM
DTC: P0490 **2T CCM, MIL: Yes** **Years:** 2006 **Models:** Santa Fe **Engines:** 2.4L VIN B, 3.5L VIN E **Transmissions:** All	**Exhaust Gas Recirculation (EGR) Control Circuit High Input** The engine is running. No disabling faults are present. The PCM detects that the surge voltage and output level voltage monitored are above a predetermined threshold. **Possible Causes:** • Poor connection at EGR solenoid • Short to battery in EGR solenoid valve circuit • Faulty EGR solenoid valve • Faulty PCM
DTC: P0496 **2T CCM, MIL: Yes** **Years:** 2006, 2007, 2008 **Models:** Accent **Engines:** 1.6L VIN C **Transmissions:** All	**Evaporative Emission (EVAP) System High Purge Flow** The engine is running for at least 600 seconds. The idle speed controller is activated. The engine coolant temperature at start is above 12°F. The PCM detects fuel tank pressure behavior related to a stuck canister purge valve. Mixture adaptation activated. Tank ventilation must be active for 10 seconds. **Possible Causes:** • Leakage at the fuel evaporative system • Purge Control Solenoid Valve (PCSV) has failed • Faulty PCM

DTC	Trouble Code Title, Conditions & Possible Causes
DTC: P0497 **2T CCM, MIL: Yes** **Years:** 2006, 2007, 2008 **Models:** Accent **Engines:** 1.6L VIN C **Transmissions:** All	**Evaporative Emission (EVAP) System Low Purge Flow** The engine is running for at least 600 seconds. The idle speed controller is activated. The engine coolant temperature at start is above 12°F. The PCM detects fuel tank pressure behavior related to a stuck canister purge valve. Mixture adaptation activated. Tank ventilation must be active for 10 seconds. **Possible Causes:** • Clog in the fuel evaporative system • Purge Control Solenoid Valve (PCSV) has failed • Faulty PCM
DTC: P0498 **2T CCM, MIL: Yes** **Years:** 2006, 2007, 2008 **Models:** Accent **Engines:** 1.6L VIN C **Transmissions:** All	**Evaporative Emission (EVAP) System Vent Valve Control Circuit Low** Circuit continuity check, low. **Possible Causes:** • Poor connection • Short to ground in control circuit • Canister Close Valve (CCV) has failed • Faulty PCM
DTC: P0499 **2T CCM, MIL: Yes** **Years:** 2006, 2007, 2008 **Models:** Accent **Engines:** 1.6L VIN C **Transmissions:** All	**Evaporative Emission (EVAP) System Vent Valve Control Circuit High** Circuit continuity check, high. **Possible Causes:** • Poor connection • Short to power in control circuit • Canister Close Valve (CCV) has failed • Faulty PCM
DTC: P0500 **2T CCM, MIL: Yes** **Years:** 2006, 2007, 2008 **Models:** Santa Fe, Tiburon, Tucson **Engines:** 2.0L VIN B, 2.0L VIN D, 2.7L VIN D, 2.7L VIN F, 3.5L VIN E **Transmissions:** All	**Vehicle Speed Sensor (VSS) Open Or Short** The engine is started and the vehicle is driven with an engine speed over 3,000 RPM at an engine load over 70 percent. The Closed Throttle switch is indicating OFF and the PCM did not detect any VSS signals for 4 seconds. **Possible Causes:** • VSS signal circuit is open or shorted to ground • VSS power or ground circuit is open • Excessive VSS to rotor gap • Incorrect VSS rotor alignment • VSS rotor damage • Incorrect VSS sensor • Faulty VSS sensor • PCM has failed
DTC: P0501 **2T CCM, MIL: Yes** **Years:** 2006, 2007, 2008 **Models:** Accent, Azera, Elantra, Entourage, Santa Fe, Sonata, Tiburon, Tucson, Veracruz **Engines:** 1.6L VIN C, 2.0L VIN B, 2.0L VIN D, 2.7L VIN D, 2.7L VIN F, 3.3L VIN D, 3.3L VIN E, 3.3L VIN F, 3.8L VIN 3, 3.8L VIN C, 3.8L VIN F **Transmissions:** All	**Vehicle Speed Sensor A Range/Performance** The engine is running and the vehicle is driven with an engine speed over 2,000 RPM. The PCM did not detect any VSS signals. **Possible Causes:** • VSS signal circuit from the sensor to the I/P Cluster to the PCM is open, shorted to ground, or to system power • VSS (Magnetic) signal (+) or (−) circuit is open or shorted • VSS (Magnetic) is damaged or has failed • Excessive VSS to rotor gap • Incorrect VSS rotor alignment • VSS rotor damage • Incorrect VSS sensor • Faulty VSS sensor • PCM has failed
DTC: P0502 **2T CCM, MIL: Yes** **Years:** 2006 **Models:** Santa Fe **Engines:** 2.4L VIN B **Transmissions:** All	**Vehicle Speed Sensor (VSS) Circuit Low** DTC P0502 is set when the Vehicle Speed Sensor (VSS) signal is less than threshold value. Signal check. The threshold value is determined when the vehicle holds a constant speed for more than 20 seconds. **Possible Causes:** • Open in signal circuit • Open in battery and ground circuit • Short to ground in signal circuit • Faulty VSS • Faulty PCM

DTC	Trouble Code Title, Conditions & Possible Causes
DTC: P0503 **2T CCM, MIL: Yes** **Years:** 2006 **Models:** Santa Fe **Engines:** 2.4L VIN B **Transmissions:** All	**Vehicle Speed Sensor Circuit Malfunction High** TP sensor DTC's P0120 or P0121 are not set. The TP sensor angle is 15 percent or more. The engine speed is 1,000–5,000 RPM. DTC P0503 is set when the transmission output speed is 100 RPM or less for 3 seconds. When the DTC is set, the ECM disables Cruise Control. **Possible Causes:** • Open in signal circuit • Open in battery and ground circuit • Short to ground in signal circuit • Short to battery in signal circuit • Faulty VSS • VSS rotor damage • Incorrect VSS sensor • Faulty VSS sensor • Faulty PCM
DTC: P0504 **2T CCM, MIL: Yes** **Years:** 2006, 2007, 2008 **Models:** Azera, Entourage, Santa Fe, Sonata, Veracruz **Engines:** 2.4L VIN C, 2.7L VIN D, 3.3L VIN D, 3.3L VIN E, 3.3L VIN F, 3.8L VIN 3, 3.8L VIN C, 3.8L VIN F **Transmissions:** All	**Brake Switch A/B Correlation** During driving, the 2 brake signals are compared. Case 1: Engine works. Vehicle speed sensor is abnormal. Case 2: Engine works. Vehicle speed sensor is normal. Vehicle speed is over 12 mph (20 kph), for at least 1 second. **Possible Causes:** • Poor connection • Brake Pedal Position Sensor out of adjustment • Stop lamp voltage supply circuit to the PCM has an open, short to ground, or a short to battery positive voltage • Faulty PCM
DTC: P0505 **2T CCM, MIL: Yes** **Years:** 2006, 2007, 2008 **Models:** Accent, Elantra, Sonata **Engines:** 1.6L VIN C, 2.0L VIN D, 2.4L VIN C **Transmissions:** All	**Idle Air Control System** The engine is started and running at hot idle speed in closed loop for 30 seconds. The PCM detected the real engine speed is lower or higher than the threshold value of the desired engine speed during catalyst heating. **Possible Causes:** • Poor connection • Leak or clog in intake air system • Carbon pile • ISCA is faulty • ECM has failed
DTC: P0506 **2T CCM, MIL: Yes** **Years:** 2006, 2007, 2008 **Models:** Accent, Azera, Elantra, Entourage, Santa Fe, Sonata, Tiburon, Tucson, Veracruz **Engines:** 1.6L VIN C, 2.0L VIN B, 2.0L VIN D, 2.4L VIN B, 2.7L VIN D, 2.7L VIN F, 3.3L VIN D, 3.3L VIN E, 3.3L VIN F, 3.5L VIN E, 3.8L VIN 3, 3.8L VIN C, 3.8L VIN F **Transmissions:** All	**Idle Air Control System RPM Lower Than Expected** The engine is started and running at idle speed under these conditions: Intake Air Temperature (IAT) sensor is less than 114°F during the last ignition cycle, Long Term fuel trim from -8 percent to +8 percent, the Engine Coolant Temperature (ECT) sensor is more than 176°F, the Intake Air Temperature (IAT) sensor is more than 14°F, the system voltage is over 10.0 volts, and the PCM detected the Actual idle speed was over 200 RPM lower than the Target idle speed for 10 seconds during the CCM Rationality test. **Possible Causes:** • ISC motor "open" circuit is open or shorted to ground • ISC motor "close" circuit is open or shorted to ground • ISC motor is damaged or has failed • Base engine problem (i.e., compression or misfire condition) • Vacuum leak(s) • Restricted exhaust system • A faulty Positive Crankcase Ventilation (PCV) valve • PCM is damaged

DTC	Trouble Code Title, Conditions & Possible Causes
DTC: P0507 **2T CCM, MIL: Yes** **Years:** 2006, 2007, 2008 **Models:** Accent, Azera, Elantra, Entourage, Santa Fe, Sonata, Tiburon, Tucson, Veracruz **Engines:** 1.6L VIN C, 2.0L VIN B, 2.0L VIN D, 2.4L VIN B, 2.4L VIN C, 2.7L VIN D, 2.7L VIN F, 3.3L VIN D, 3.3L VIN E, 3.3L VIN F, 3.5L VIN E, 3.8L VIN 3, 3.8L VIN C, 3.8L VIN F **Transmissions:** All	**Idle Air Control System RPM Higher Than Expected** The engine is running at idle speed in closed loop, under these conditions: Condition 1: The Intake Air Temperature (IAT) sensor input is less than 114°F during the last drive cycle, the Long Term fuel trim from -8 percent to +8 percent, the Engine Coolant Temperature (ECT) sensor input is more than 176°F, the Intake Air Temperature (IAT) sensor input is more than 14°F, and the system voltage is over 10.0 volts. The PCM detected the Actual idle speed was more than 200 RPM higher than the Target idle speed for 10 seconds. Condition 2: Power steering pressure switch signal indicating OFF, engine load less than 40 percent, the IAT sensor input more than 14°F, and the PCM detected the Actual idle speed was more than 120 RPM higher than the Target idle speed for 10 seconds. **Possible Causes:** • Base engine problem (i.e., compression or misfire condition) • Vacuum leak(s) • Throttle valves binding open or binding closed • Throttle body assembly is damaged or it has failed • A faulty Positive Crankcase Ventilation (PCV) valve • ISC motor control circuit(s) open or shorted to ground • ISC motor is damaged or has failed • PCM is damaged
DTC: P050B **2T CCM, MIL: Yes** **Years:** 2006, 2007, 2008 **Models:** Azera, Entourage, Santa Fe, Sonata, Veracruz **Engines:** 2.7L VIN D, 3.3L VIN D, 3.3L VIN E, 3.3L VIN F, 3.8L VIN 3, 3.8L VIN C, 3.8L VIN F **Transmissions:** All	**Cold Start Ignition Timing Performance/Spark Timing Error** The engine is running. The vehicle is not rapidly accelerating or decelerating. The battery voltage is between 11.0–16.0 volts. No DTC's are set related to the CKP sensor, Ignition coil, or Misfire. The PCM checks the spark timing under detecting conditions. If the actual spark timing differs from the commanded spark timing, the PCM sets DTC P050B. **Possible Causes:** • Faulty Ignition Coil • Faulty PCM
DTC: P0510 **1T CCM, MIL: No** **Years:** 2006 **Models:** Santa Fe **Engines:** 3.5L VIN E **Transmissions:** All	**Closed Throttle Position (CTP) Switch Malfunction** The engine is started and the vehicle is driven at over 30 mph then back to a stop at least 15 times. The TP sensor signal is over 2.0 volts at least once, and the PCM detected the CTP switch remained OFF for over 2 seconds. **Possible Causes:** • Closed throttle position switch signal circuit is open or grounded • Closed throttle position switch signal circuit is shorted to power • Closed throttle position switch or TP sensor damaged or failed • PCM has failed
DTC: P0532 **1T CCM, MIL: No** **Years:** 2006, 2007, 2008 **Models:** Accent, Azera, Entourage, Santa Fe, Sonata, Veracruz **Engines:** 1.6L VIN C, 2.4L VIN C, 2.7L VIN D, 3.3L VIN D, 3.3L VIN E, 3.3L VIN F, 3.8L VIN 3, 3.8L VIN C, 3.8L VIN F **Transmissions:** All	**A/C Refrigerant Pressure Sensor A Circuit Low Input** Engine is running. Battery voltage is between 11–16 volts. DTC P0532 is set when the ECM detects that the A/C pressure sensor signal is shorted to ground or open. The sensor output is 0.05 volt. **Possible Causes:** • Poor connection • Open in power circuit • Open or short to ground in signal circuit • Faulty A/C pressure sensor • Faulty PCM
DTC: P0533 **1T CCM, MIL: No** **Years:** 2006, 2007, 2008 **Models:** Accent, Azera, Entourage, Santa Fe, Sonata, Veracruz **Engines:** 1.6L VIN C, 2.4L VIN C, 2.7L VIN D, 3.3L VIN D, 3.3L VIN E, 3.3L VIN F, 3.8L VIN 3, 3.8L VIN C, 3.8L VIN F **Transmissions:** All	**A/C Refrigerant Pressure Sensor A Circuit High Input** Engine is running. Battery voltage is between 11–16 volts. DTC P0533 is set when the ECM detects that the sensor signal has a short to high voltage. Sensor output is 4.65 volts. **Possible Causes:** • Poor connection • Open in signal circuit open • Open in ground circuit • Faulty A/C pressure sensor • Short to voltage in the 5-volt reference circuit • Faulty PCM

DTC	Trouble Code Title, Conditions & Possible Causes
DTC: P0551 **1T CCM, MIL: Yes** **Years:** 2006, 2007, 2008 **Models:** Santa Fe, Sonata, Tiburon, Tucson **Engines:** 2.4L VIN B, 2.4L VIN C, 2.7L VIN D, 2.7L VIN F, 3.3L VIN F, 3.5L VIN E **Transmissions:** All	**Power Steering Pressure Sensor/Switch Circuit Range/Performance** The engine is running. If a power steering switch signal is ON when the engine speed is more than 2,500 RPM, load value is greater than 55 percent and engine coolant temperature is above 50°F, DTC P0551 will set. The signal of the power steering pressure switch is monitored. **Possible Causes:** • Poor connection • Faulty power steering switch • Open or short in power steering switch • Faulty PCM
DTC: P0552 **1T CCM, MIL: Yes** **Years:** 2006, 2007, 2008 **Models:** Azera, Sonata, Veracruz **Engines:** 2.4L VIN C, 3.3L VIN F, 3.8L VIN C, 3.8L VIN F **Transmissions:** All	**Power Steering Pressure Sensor/Switch Circuit Low Input** The engine is running. The PCM detects the sensor signal has a short to low voltage. The sensor output 0.25 volt. **Possible Causes:** • Poor connection • Open in power circuit • Open or short to ground in signal circuit • Faulty P/S pressure sensor • Faulty PCM
DTC: P0553 **1T CCM, MIL: Yes** **Years:** 2006, 2007, 2008 **Models:** Azera, Sonata, Veracruz **Engines:** 2.4L VIN C, 3.3L VIN F, 3.8L VIN C, 3.8L VIN F **Transmissions:** All	**Power Steering Pressure Sensor/Switch Circuit High Input** The engine is running. The PCM detects the sensor signal has a short to low voltage. The sensor output is 4.65 volts. **Possible Causes:** • Poor connection • Short in signal circuit • Open in ground circuit • Faulty P/S pressure sensor • Faulty PCM
DTC: P0560 **1T CCM, MIL: No** **Years:** 2006, 2007, 2008 **Models:** Accent, Elantra, Santa Fe, Sonata, Tiburon, Tucson **Engines:** 1.6L VIN C, 2.0L VIN B, 2.0L VIN D, 2.4L VIN B, 2.4L VIN C, 2.7L VIN D, 2.7L VIN F, 3.5L VIN E **Transmissions:** All	**System Voltage Malfunction** DTC P0560 is set when the ECM detects a system voltage out of range for 10 seconds. When the DTC is set, the ECM will command the charge indicator and or warning message to be illuminated on the Instrument Panel Cluster (IPC) and the Driver Information Center (DIC), if equipped. **Possible Causes:** • Check for high resistance at battery connections • Battery terminals are corroded, dirty, or loose • Check the drive belt for excessive wear and the proper tension • Test the operation of the Alternator (it may be undercharging) • ECM has failed
DTC: P0561 **1T CCM, MIL: No** **Years:** 2006, 2007, 2008 **Models:** Accent **Engines:** 1.6L VIN C **Transmissions:** All	**System Voltage Unstable** The vehicle speed is below 25 mph (40 km/h). The system voltage is below 11 volts or above 16 volts. Engine speed is above 1,500 RPM. DTC P0560 is set when the ECM detects a system voltage out of range for 2 seconds. **Possible Causes:** • Check for high resistance at battery connections • Check the drive belt for excessive wear and the proper tension • Test the operation of the Alternator (it may be undercharging) • ECM has failed
DTC: P0562 **1T CCM, MIL: No** **Years:** 2006, 2007, 2008 **Models:** Accent, Azera, Elantra, Entourage, Santa Fe, Sonata, Tiburon, Tucson, Veracruz **Engines:** 1.6L VIN C, 2.0L VIN B, 2.0L VIN D, 2.4L VIN C, 2.7L VIN D, 2.7L VIN F, 3.3L VIN D, 3.3L VIN E, 3.3L VIN F, 3.5L VIN E, 3.8L VIN 3, 3.8L VIN C, 3.8L VIN F **Transmissions:** All	**System Voltage Low** The vehicle speed is above 5 mph (8 km/h). The system voltage is between 9.5–18 volts. DTC P0562 is set when the ECM detects an unexpected low voltage condition on the ignition circuit. **Possible Causes:** • Check for high resistance at battery connections • Check the drive belt for excessive wear and the proper tension • Test the operation of the Alternator (it may be undercharging) • ECM has failed

DTC	Trouble Code Title, Conditions & Possible Causes
DTC: P0563 **1T CCM, MIL: No** **Years:** 2006, 2007, 2008 **Models:** Accent, Azera, Elantra, Entourage, Santa Fe, Sonata, Tiburon, Tucson, Veracruz **Engines:** 1.6L VIN C, 2.0L VIN B, 2.0L VIN D, 2.4L VIN C, 2.7L VIN D, 2.7L VIN F, 3.3L VIN D, 3.3L VIN E, 3.3L VIN F, 3.5L VIN E, 3.8L VIN 3, 3.8L VIN C, 3.8L VIN F **Transmissions:** All	**System Voltage High** The vehicle speed is above 5 mph (8 km/h). DTC P0563 is set when the ECM detects an unexpected high voltage condition on the ignition circuit. **Possible Causes:** • Condition of the battery • Test the operation of the Alternator (it may be overcharging) • ECM has failed
DTC: P0564 **1T ECM, MIL: No** **Years:** 2006, 2007, 2008 **Models:** Azera, Azera, Entourage, Santa Fe, Sonata, Veracruz **Engines:** 2.4L VIN C, 2.7L VIN D, 3.3L VIN D, 3.3L VIN E, 3.3L VIN F, 3.5L VIN E, 3.8L VIN 3, 3.8L VIN C, 3.8L VIN F **Transmissions:** All	**Cruise Control Multi-Function Input A Circuit** The ignition is ON. The cruise control on/off switch is ON. DTC P0564 is set when the ECM detects an invalid voltage signal on the cruise control set/coast and resume/accelerate switch signal circuit for greater than 1.5 seconds. The ECM runs this diagnostic every 0.05 seconds. When the DTC is set, the Cruise Control System is disabled. **Possible Causes:** • Internally shorted inflatable restraint steering wheel module coil • Cruise control set/coast and resume/accelerate switch signal circuit high resistance, short to voltage, or short to ground • Poor connections at the cruise control switch • Poor connections at the harness connector of the ECM • Faulty cruise control switch • ECM has failed
DTC: P0565 **1T ECM, MIL: No** **Years:** 2006, 2007, 2008 **Models:** Azera, Azera, Entourage, Santa Fe, Sonata, Veracruz **Engines:** 2.7L VIN D, 3.3L VIN D, 3.3L VIN E, 3.3L VIN F, 3.8L VIN 3, 3.8L VIN C, 3.8L VIN F **Transmissions:** All	**Cruise Control ON Signal** The engine is running. The ignition voltage is at or greater than 9 volts. The cruise control system type is learned. DTC P0565 is set when the PCM detects that the main signal is switching too frequently or stuck for too long. When the DTC is set, the Cruise Control System is disabled. **Possible Causes:** • Poor connections at the cruise control switch • Poor connections at the harness connector of the ECM • Faulty cruise switch • Faulty PCM
DTC: P0566 **1T ECM, MIL: No** **Years:** 2006, 2007, 2008 **Models:** Azera, Azera, Entourage, Santa Fe, Sonata, Veracruz **Engines:** 2.7L VIN D, 3.3L VIN D, 3.3L VIN E, 3.3L VIN F, 3.5L VIN E, 3.8L VIN 3, 3.8L VIN C, 3.8L VIN F **Transmissions:** All	**Cruise Control OFF Signal** The engine is running. The ignition voltage is at or greater than 9 volts. The cruise control system type is learned. DTC P0566 is set when the PCM detects that the cancel switch signal is switching too frequently or stuck for too long. When the DTC is set, the Cruise Control System is disabled. **Possible Causes:** • Poor connections at the cruise control switch • Poor connections at the harness connector of the ECM • Faulty cruise switch • Faulty PCM
DTC: P0567 **1T ECM, MIL: No** **Years:** 2006, 2007, 2008 **Models:** Azera, Azera, Entourage, Santa Fe, Sonata, Veracruz **Engines:** 2.7L VIN D, 3.3L VIN D, 3.3L VIN E, 3.3L VIN F, 3.8L VIN 3, 3.8L VIN C, 3.8L VIN F **Transmissions:** All	**Cruise Control RESUME Signal** The engine is running. The ignition voltage is at or greater than 9 volts. The cruise control system type is learned. DTC P0567 is set when the PCM detects that the Resume switch signal is switching too frequently or stuck for too long. When the DTC is set, the Cruise Control System is disabled. **Possible Causes:** • Poor connections at the cruise control switch • Poor connections at the harness connector of the ECM • Faulty cruise switch • Faulty PCM
DTC: P0568 **1T ECM, MIL: No** **Years:** 2006, 2007, 2008 **Models:** Azera, Azera, Entourage, Santa Fe, Sonata, Veracruz **Engines:** 2.7L VIN D, 3.3L VIN D, 3.3L VIN E, 3.3L VIN F, 3.8L VIN 3, 3.8L VIN C, 3.8L VIN F **Transmissions:** All	**Cruise Control SET Signal** The engine is running. The ignition voltage is at or greater than 9 volts. The cruise control system type is learned. DTC P0568 is set when the PCM detects that the SET switch signal is switching too frequently or stuck for too long. When the DTC is set, the Cruise Control System is disabled. **Possible Causes:** • Poor connections at the cruise control switch • Poor connections at the harness connector of the ECM • Faulty cruise switch • Faulty PCM

DTC	Trouble Code Title, Conditions & Possible Causes
DTC: P0571 **2T ECM, MIL: Yes** **Years:** 2006, 2007, 2008 **Models:** Azera, Azera, Entourage, Santa Fe, Sonata, Veracruz **Engines:** 2.7L VIN D, 3.3L VIN D, 3.3L VIN E, 3.3L VIN F, 3.5L VIN E, 3.8L VIN 3, 3.8L VIN C, 3.8L VIN F **Transmissions:** All	**Brake Switch A Circuit** The engine is running normally. The Vehicle Speed Sensor (VSS) signal is normal. The vehicle speed is greater than 12 mph (20 kph). The brake lamp is OFF and not changing the brake lamp signal for more than 3 seconds. The PCM detects the brake lamp input signal when the vehicle stops. **Possible Causes:** • Poor connection • Open or short to ground in signal circuit • Faulty PCM
DTC: P0600 **1T ECM, MIL: Yes** **Years:** 2007, 2008 **Models:** Sonata, Tucson **Engines:** 2.0L VIN B, 2.4L VIN C **Transmissions:** All	**Serial Communication Link Malfunction** The PCM determines there is a CAN communication error and sets DTC P0600 if communication with other control devices (e.g. ABS) via CAN is impossible or the PCM detects that the communication time via CAN exceeds the threshold value. **Possible Causes:** • Open or short in CAN line • Contact resistance in connectors • Faulty PCM
DTC: P0601 **1T ECM, MIL: Yes** **Years:** 2006, 2007, 2008 **Models:** Azera, Entourage, Santa Fe, Sonata, Veracruz **Engines:** 2.7L VIN D, 3.3L VIN D, 3.3L VIN E, 3.3L VIN F, 3.8L VIN 3, 3.8L VIN C, 3.8L VIN F **Transmissions:** All	**Internal Control Module Memory Check Sum Error** Key on or engine running for 1 second, and the PCM detected an internal checksum data error during the initial Self-Test. **Possible Causes:** • Ground circuits open high resistance or short • Voltage supply circuits open high resistance or short • ECM is not programmed or has failed
DTC: P0602 **1T ECM, MIL: Yes** **Years:** 2006, 2007, 2008 **Models:** Azera, Entourage, Santa Fe, Sonata, Veracruz **Engines:** 2.7L VIN D, 3.3L VIN D, 3.3L VIN E, 3.3L VIN F, 3.8L VIN 3, 3.8L VIN C, 3.8L VIN F **Transmissions:** All	**Control Module Programming Error** The ignition switch is in Run or Crank. DTC P0602 runs once per ignition cycle. P0602 is set when the ECM detects an internal failure or incomplete programming for more than 10 seconds. **Possible Causes:** • ECM is not programmed • Ground circuits open high resistance or short • Voltage supply circuits open high resistance or short • ECM has failed
DTC: P0603 **1T ECM, MIL: Yes** **Years:** 2006, 2007, 2008 **Models:** Veracruz **Engines:** 3.8L VIN C **Transmissions:** All	**ECM(EEPROM)-KAM Error** The ignition switch is ON. DTC P0603 is set when input is not available or the ROM I.D. has been changed. **Possible Causes:** • Ground circuits open high resistance or short • Voltage supply circuits open high resistance or short • TCM has failed
DTC: P0604 **1T ECM, MIL: Yes** **Years:** 2006, 2007, 2008 **Models:** Azera, Entourage, Santa Fe, Sonata, Veracruz **Engines:** 2.7L VIN D, 3.3L VIN D, 3.3L VIN E, 3.3L VIN F, 3.8L VIN 3, 3.8L VIN C, 3.8L VIN F **Transmissions:** All	**Internal Control Module Random Access Memory (RAM) Error** The ignition switch is ON. DTC P0604 is set when the ECM detects an internal failure or incomplete programming for more than 10 seconds. **Possible Causes:** • Ground circuits open high resistance or short • Voltage supply circuits open high resistance or short • PCM has failed
DTC: P0605 **1T ECM, MIL: Yes** **Years:** 2006, 2007, 2008 **Models:** Accent, Elantra, Santa Fe, Sonata, Tiburon, Tucson **Engines:** 1.6L VIN C, 2.0L VIN B, 2.0L VIN D, 2.4L VIN C, 2.7L VIN D, 2.7L VIN F **Transmissions:** All	**Internal Control Module Read Only Memory (ROM) Error** The ignition switch is in the ON position. DTC P0605 is set when the PCM detects an internal failure or incomplete programming for more than 10 seconds. **Possible Causes:** • Ground circuits open high resistance or short • Voltage supply circuits open high resistance or short • ECM incomplete programming or has failed

DTC	Trouble Code Title, Conditions & Possible Causes
DTC: P0606 **1T ECM, MIL: Yes** **Years:** 2006, 2007, 2008 **Models:** Azera, Elantra, Entourage, Santa Fe, Sonata, Veracruz **Engines:** 2.0L VIN D, 2.7L VIN D, 3.3L VIN D, 3.3L VIN E, 3.3L VIN F, 3.8L VIN 3, 3.8L VIN C, 3.8L VIN F **Transmissions:** All	**PCM/PCM Processor (PCM SELF TEST Failed)** **Control Module Internal Performance** The ignition switch is in the ON position. DTC P0606 is set when the ECM detects an internal failure or incomplete programming for more than 10 seconds. **Possible Causes:** • Ground circuits open high resistance or short • Voltage supply circuits open high resistance or short • ECM incomplete programming or has failed
DTC: P061B **1T ECM, MIL: Yes** **Years:** 2006, 2007, 2008 **Models:** Azera, Entourage, Santa Fe, Sonata, Veracruz **Engines:** 2.7L VIN D, 3.3L VIN D, 3.3L VIN E, 3.3L VIN F, 3.8L VIN 3, 3.8L VIN C, 3.8L VIN F **Transmissions:** All	**Internal Control Module Torque Calculation Performance** Desired torque error has been detected. **Possible Causes:** • Ground circuits open high resistance or short • Voltage supply circuits open high resistance or short • ECM incomplete programming or has failed
DTC: P0624 **1T ECM, MIL: No** **Years:** 2006, 2007, 2008 **Models:** Accent **Engines:** 1.6L VIN C **Transmissions:** All	**Fuel Cap Lamp Control Circuit** The ignition switch is ON. The PCM detects that the circuit continuity check is: high, low, or open. **Possible Causes:** • Poor connection • Open or short • Instrument cluster • Faulty PCM
DTC: P0625 **1T ECM, MIL: No** **Years:** 2006, 2007, 2008 **Models:** Elantra, Sonata **Engines:** 2.0L VIN D, 2.4L VIN C **Transmissions:** All	**Generator Field/F Terminal Circuit Low Input** No generator, Crankshaft Position (CKP) sensors, or Camshaft Position (CMP) sensor DTC's are set. The engine is less than 3,000 RPM. The generator has not been commanded OFF by the ECM or scan tool. DTC P0625 is set when the ECM detects a Pulse Width Modulation (PWM) signal less than 5 percent for at least 15 seconds. **Possible Causes:** • Generator Field Duty Cycle Signal short to ground or open/high resistance • GENF Terminal Signal parameter not within the specified range • Poor connections at the harness connector of the ECM • Faulty charging system • ECM has failed
DTC: P0626 **1T ECM, MIL: No** **Years:** 2006, 2007, 2008 **Models:** Elantra, Sonata **Engines:** 2.0L VIN D, 2.4L VIN C **Transmissions:** All	**Generator Field/F Terminal Circuit High Input** The engine is not running. The ignition is in the ON position. DTC P0626 is set when the ECM detects a Pulse Width Modulated (PWM) signal greater than 65 percent for at least 15 seconds. When the DTC is set, the ECM will command the charge indicator and or warning message to be illuminated on the Instrument Panel Cluster (IPC) and the Driver Information Center (DIC), if equipped. **Possible Causes:** • Generator Field Duty Cycle Signal short to voltage • GENF Terminal Signal parameter not within the specified range • Poor connections at the harness connector of the ECM • ECM has failed
DTC: P0630 **1T CCM, MIL: Yes** **Years:** 2006, 2007, 2008 **Models:** Accent, Azera, Elantra, Entourage, Santa Fe, Sonata, Tiburon, Tucson, Veracruz **Engines:** 1.6L VIN C, 2.0L VIN B, 2.0L VIN D, 2.4L VIN B, 2.4L VIN C, 2.7L VIN D, 2.7L VIN F, 3.3L VIN D, 3.3L VIN E, 3.3L VIN F, 3.5L VIN E, 3.8L VIN 3, 3.8L VIN C, 3.8L VIN F **Transmissions:** All	**VIN not Programmed or Incompatible ECM/PCM** The ignition is ON. The PCM internal check is run and does not find the VIN in the boot area. **Possible Causes:** • PCM is new and has not yet been programmed • PCM is from another vehicle or has failed

DTC	Trouble Code Title, Conditions & Possible Causes
DTC: P0638 **1T CCM, MIL: Yes** **Years:** 2006, 2007, 2008 **Models:** Azera, Entourage, Santa Fe, Sonata, Veracruz **Engines:** 2.4L VIN C, 2.7L VIN D, 3.3L VIN D, 3.3L VIN E, 3.3L VIN F, 3.8L VIN 3, 3.8L VIN C, 3.8L VIN F **Transmissions:** All	**Throttle Actuator Control (TAC) Range/Performance (Bank 1)** The ignition is ON. The ignition voltage is more than 5 volts. DTC P0638 is set when the Electronic Throttle System (ETS) control malfunctions. **Possible Causes:** • TAC motor control circuit is open, shorted to ground, or shorted to system power (B+) • Throttle valves not in their rest position • Throttle valves binding open or binding closed • Throttle valves moving open or closed without spring pressure • Throttle body assembly is damaged or it has failed • ECM is damaged
DTC: P0641 **1T CCM, MIL: Yes** **Years:** 2006, 2007, 2008 **Models:** Azera, Entourage, Santa Fe, Sonata, Veracruz **Engines:** 2.7L VIN D, 3.3L VIN D, 3.3L VIN E, 3.3L VIN F, 3.8L VIN 3, 3.8L VIN C, 3.8L VIN F **Transmissions:** All	**Sensor Reference Voltage A Circuit/Open** The ignition switch is ON. The PCM runs a reference voltage check detecting a short in the power supply line. DTC P0641 is set when the PCM detects a voltage out of tolerance condition. **Possible Causes:** • A short to voltage on signal circuit of certain components may cause this DTC to set • Short in sensor power supply line • ECM has failed
DTC: P0642 **1T CCM, MIL: Yes** **Years:** 2006, 2007, 2008 **Models:** Accent, Sonata **Engines:** 1.6L VIN C, 2.4L VIN C **Transmissions:** All	**Sensor Reference Voltage A Circuit Low** The ignition switch is in the ON position. DTC P0642 is set when the ECM detects a low voltage condition on the 5-volt reference for more than 3 seconds. **Possible Causes:** • Air Conditioning (A/C) Refrigerant Pressure Sensor has failed • Fuel Tank Pressure (FTP) Sensor has failed • APP sensor 2 5-volt reference 1 circuit short to ground • A/C pressure sensor 5-volt reference 1 circuit short to ground • FTP sensor 5-volt reference 1 circuit short to ground • MAP sensor 5-volt reference 1 circuit or short to ground • Engine oil pressure (EOP) switch 5-volt reference 1 circuit short to ground • MAP sensor has failed • ECM has failed
DTC: P0643 **1T CCM, MIL: Yes** **Years:** 2006, 2007, 2008 **Models:** Accent, Sonata **Engines:** 1.6L VIN C, 2.4L VIN C **Transmissions:** All	**Sensor Reference Voltage A Circuit Low** The ignition is in the ON position. DTC P0643 is set when the ECM detects a high voltage condition on the 5-volt reference for more than 3 seconds. **Possible Causes:** • Air Conditioning (A/C) Refrigerant Pressure Sensor has failed • Fuel Tank Pressure (FTP) Sensor has failed • APP sensor 2 5-volt reference 1 circuit for a short to voltage • A/C pressure sensor 5-volt reference 1 circuit for a short to voltage • FTP sensor 5-volt reference 1 circuit for a short to voltage • MAP sensor 5-volt reference 1 circuit for a short to voltage • Engine oil pressure (EOP) switch 5-volt reference 1 circuit for a short to voltage • MAP sensor has failed • ECM has failed
DTC: P0645 **1T CCM, MIL: Yes** **Years:** 2006, 2007, 2008 **Models:** Accent **Engines:** 1.6L VIN C **Transmissions:** All	**A/C Clutch Relay Control Circuit** The ECM A/C Compressor Clutch Relay Control driver transitions from ON to OFF or from OFF to ON. DTC P0645 is set when the ECM detects an open on the control circuit of the A/C compressor clutch relay when commanded OFF with the engine in crank or run status. **Possible Causes:** • A/C compressor clutch supply voltage circuit for short to ground • Open circuit (check A/C Fuse) • A/C compressor clutch relay short to ground, short to voltage, or an open condition • Faulty ground(s) on ECM • Faulty A/C clutch relay • ECM is damaged

DTC	Trouble Code Title, Conditions & Possible Causes
DTC: P0646 **1T CCM, MIL: No** **Years:** 2006, 2007, 2008 **Models:** Accent, Azera, Entourage, Santa Fe, Sonata, Veracruz **Engines:** 1.6L VIN C, 2.4L VIN C, 2.7L VIN D, 3.3L VIN D, 3.3L VIN E, 3.3L VIN F, 3.8L VIN 3, 3.8L VIN C, 3.8L VIN F **Transmissions:** All	**A/C Clutch Relay Control Circuit Low** The ECM driver transitions from ON to OFF or from OFF to ON. DTC P0646 is set when the ECM detects a short to ground on the control circuit of the A/C compressor clutch relay. **Possible Causes:** • A/C compressor clutch supply voltage circuit for short to ground • Open circuit (check A/C Fuse) • A/C compressor clutch relay short to ground, short to voltage, or an open condition • Faulty ground(s) on ECM • Faulty A/C clutch relay • ECM is damaged
DTC: P0647 **1T CCM, MIL: No** **Years:** 2006, 2007, 2008 **Models:** Accent, Azera, Entourage, Santa Fe, Sonata, Veracruz **Engines:** 1.6L VIN C, 2.4L VIN C, 2.7L VIN D, 3.3L VIN D, 3.3L VIN E, 3.3L VIN F, 3.8L VIN 3, 3.8L VIN C, 3.8L VIN F **Transmissions:** All	**A/C Clutch Relay Control Circuit High** The ECM driver transitions from ON to OFF or from OFF to ON. DTC P0647 is set when the ECM detects a short to voltage on the control circuit of the A/C compressor clutch relay. **Possible Causes:** • A/C compressor clutch relay short to voltage • Faulty A/C clutch relay • ECM is damaged
DTC: P0650 **2T CCM, MIL: Yes** **Years:** 2006, 2007, 2008 **Models:** Accent, Azera, Elantra, Entourage, Santa Fe, Sonata, Tiburon, Tucson, Veracruz **Engines:** 1.6L VIN C, 2.0L VIN B, 2.0L VIN D, 2.4L VIN C, 2.7L VIN D, 2.7L VIN F, 3.3L VIN D, 3.3L VIN E, 3.3L VIN F, 3.8L VIN 3, 3.8L VIN C, 3.8L VIN F **Transmissions:** All	**Malfunction Indicator Lamp (MIL) Control Circuit** DTC P0650 runs continuously when the ignition is ON and the ignition voltage is between 9–18 volts. DTC P0650 is set when the control module detects that the commanded state of the MIL driver and the actual state of the control circuit do not match for more than 1 second. **Possible Causes:** • MIL control circuit is open or shorted to ground • MIL control circuit is shorted to system power • MIL control power circuit is open in the Instrument Panel Cluster • MIL (lamp) is damaged or has failed • Instrument Panel Cluster or the ECM has failed
DTC: P0651 **1T CCM, MIL: Yes** **Years:** 2006, 2007, 2008 **Models:** Azera, Entourage, Santa Fe, Sonata, Veracruz **Engines:** 2.7L VIN D, 3.3L VIN D, 3.3L VIN E, 3.3L VIN F, 3.8L VIN 3, 3.8L VIN C, 3.8L VIN F **Transmissions:** All	**Sensor Reference Voltage B Circuit/Open** The ignition is in the ON position. DTC P0651 is set when the ECM detects a voltage out of tolerance condition on the 5-volt reference bus for more than 3 seconds. **Possible Causes:** • Short in sensor power supply line • ECM has failed
DTC: P0652 **1T CCM, MIL: Yes** **Years:** 2007, 2008 **Models:** Sonata **Engines:** 2.4L VIN C **Transmissions:** All	**Sensor Reference Voltage B Circuit Low** The ignition is in the ON position. DTC P0652 is set when the ECM detects a low voltage condition on the 5-volt reference bus for at least 0.04 seconds. **Possible Causes:** • Open or short to ground in power circuit • Poor connection or damaged harness • ECM has failed
DTC: P0653 **1T CCM, MIL: Yes** **Years:** 2007, 2008 **Models:** Sonata **Engines:** 2.4L VIN C **Transmissions:** All	**Sensor Reference Voltage B Circuit High** The ignition is in the ON position. DTC P0653 is set when the ECM detects a high voltage condition on the 5-volt reference bus for at least 0.04 seconds. **Possible Causes:** • Open or short to ground in power circuit • Poor connection or damage harness • ECM has failed

DTC	Trouble Code Title, Conditions & Possible Causes
DTC: P0660 **2T CCM, MIL: Yes** **Years:** 2006, 2007, 2008 **Models:** Azera, Entourage, Santa Fe, Sonata, Veracruz **Engines:** 2.7L VIN D, 3.3L VIN D, 3.3L VIN E, 3.3L VIN F, 3.8L VIN 3, 3.8L VIN C, 3.8L VIN F **Transmissions:** All	**Intake Manifold Tuning Valve Control Circuit/Open (Bank 1)** The engine is working normally. The battery voltage is between 11–16 volts. The PCM checks the Variable Intake System (VIS) every 10 seconds. DTC P0660 is set when the PCM detects an open or short in the VIS circuit. **Possible Causes:** • Poor connection • Open or short in the VIS circuit • Faulty VIS • Faulty PCM
DTC: P0661 **1T CCM, MIL: No** **Years:** 2007, 2008 **Models:** Tucson **Engines:** 2.7L VIN D **Transmissions:** All	**Intake Manifold Tuning Valve No. 1 (IV) Control Circuit Low** DTC P0661 is set if the ECM detects that the valve control circuit is shorted to ground. Driver stage check. **Possible Causes:** • Open in power supply harness • Short to ground in control harness • Contact resistance in connectors • Faulty intake manifold tuning valve
DTC: P0662 **1T CCM, MIL: No** **Years:** 2007, 2008 **Models:** Tucson **Engines:** 2.7L VIN D **Transmissions:** All	**Intake Manifold Tuning Valve No. 1 (IV) Control Circuit High** DTC P0662 is set if the ECM detects that the valve control circuit is open or shorted to battery voltage. Driver stage check. **Possible Causes:** • Open or short to battery in control harness • Contact resistance in connectors • Faulty intake manifold tuning valve
DTC: P0664 **1T CCM, MIL: No** **Years:** 2007, 2008 **Models:** Tucson **Engines:** 2.7L VIN D **Transmissions:** All	**Intake Manifold Tuning Valve No. 2 (MV) Control Circuit Low** DTC P0664 is set if the ECM detects that the valve control circuit is shorted to ground. Driver stage check. **Possible Causes:** • Open in power supply harness • Short to ground in control harness • Contact resistance in connectors • Faulty intake manifold tuning valve
DTC: P0665 **1T CCM, MIL: No** **Years:** 2007, 2008 **Models:** Tucson **Engines:** 2.7L VIN D **Transmissions:** All	**Intake Manifold Tuning Valve No. 2 (MV) Control Circuit High** DTC P0665 is set if the ECM detects that the valve control circuit is open or shorted to battery voltage. Driver stage check. **Possible Causes:** • Open or short to battery in control harness • Contact resistance in connectors • Faulty intake manifold tuning valve
DTC: P0685 **1T CCM, MIL: No** **Years:** 2006, 2007, 2008 **Models:** Azera, Entourage, Santa Fe, Sonata, Veracruz **Engines:** 2.7L VIN D, 3.3L VIN D, 3.3L VIN E, 3.3L VIN F, 3.8L VIN 3, 3.8L VIN C, 3.8L VIN F **Transmissions:** All	**ECM/PCM Power Relay Control Circuit (Open)** The engine is running and the battery voltage is between 11–16 volts. DTC P0685 is set when the PCM detects a short to ground, short to battery, or an open circuit on the Main Relay output. **Possible Causes:** • Poor connection • Open or short in the control circuit • PCM is damaged
DTC: P0698 **1T CCM, MIL: Yes** **Years:** 2007, 2008 **Models:** Sonata **Engines:** 2.4L VIN C **Transmissions:** All	**Sensor Reference Voltage C Circuit Low** The ignition switch is in the ON position. DTC P0698 is set when the PCM detects the APS1 voltage is 0.7 volt, for at least 0.1 second. **Possible Causes:** • Open or short to ground in power circuit • Poor connection or damaged harness • Faulty ECM
DTC: P0699 **1T CCM, MIL: Yes** **Years:** 2007, 2008 **Models:** Sonata **Engines:** 2.4L VIN C **Transmissions:** All	**Sensor Reference Voltage C Circuit High** The ignition switch is in the ON position. DTC P0699 is set when the PCM detects the APS1 voltage is 5.5 volts, for at least 0.01 second. **Possible Causes:** • Open or short to ground in power circuit • Poor connection or damaged harness • Faulty ECM

DTC	Trouble Code Title, Conditions & Possible Causes
DTC: P0700 **1T CCM, MIL: Yes** **Years:** 2006, 2007, 2008 **Models:** Accent, Elantra, Santa Fe, Sonata, Tiburon, Tucson **Engines:** 1.6L VIN C, 2.0L VIN B, 2.0L VIN D, 2.4L VIN C, 2.7L VIN D, 2.7L VIN F **Transmissions:** All	**Transmission Control Module (TCM) Requested MIL Illumination** The ignition is ON or the engine is running. DTC P0700 is set when the TCM is requesting MIL illumination. **Possible Causes:** • MIL control circuit is shorted to ground • Check the TCM for any trouble codes in memory that are responsible for the request to turn on the MIL • TCM has failed
DTC: P0703 **2T CCM, MIL: No** **Years:** 2006, 2007, 2008 **Models:** Santa Fe, Sonata, Tiburon, Tucson **Engines:** 2.0L VIN B, 2.0L VIN D, 2.4L VIN C, 2.7L VIN D, 2.7L VIN F, 3.5L VIN E **Transmissions:** All	**Stop Lamp Switch Circuit Malfunction** The engine is started and running at cruise speed. The PCM detected the Brake Switch signal did not cycle from HIGH to LOW as the brake pedal was pressed and released during the CCM test. **Possible Causes:** • Brake switch signal circuit is open or shorted to ground • Brake switch power circuit is open (check power from the relay) • Brake switch is damaged or has failed • TCM has failed
DTC: P0706 **2T CCM, MIL: No** **Years:** 2007, 2008 **Models:** Veracruz **Engines:** 3.8L VIN C **Transmissions:** All	**Transaxle Range Sensor Range/Performance** The battery voltage is between 6–15.5 volts. DTC P0706 is set when the system detects there is a stuck signal from the shift lever switch. When DTC P0706 is set, the self learning control is inhibited and the transaxle is locked in 3rd gear. **Possible Causes:** • TCU has failed
DTC: P0707 **2T CCM, MIL: No** **Years:** 2006, 2007, 2008 **Models:** Azera, Elantra, Entourage, Santa Fe, Sonata, Tiburon, Tucson, Veracruz **Engines:** 2.0L VIN B, 2.0L VIN D, 2.4L VIN C, 2.7L VIN D, 2.7L VIN F, 3.3L VIN D, 3.3L VIN E, 3.3L VIN F, 3.5L VIN E, 3.8L VIN 3, 3.8L VIN C, 3.8L VIN F **Transmissions:** All	**Transaxle Range Switch Circuit Low Input** The engine is started and running at cruise speed with VSS signals present. The PCM detected an unexpected low voltage condition on the Transaxle Range (TR) switch circuit during the CCM test. **Possible Causes:** • TR sensor signal circuit is open or shorted to ground between the switch and PCM • TR sensor is damaged or has failed • TCM has failed
DTC: P0708 **2T CCM, MIL: No** **Years:** 2006, 2007, 2008 **Models:** Azera, Elantra, Entourage, Santa Fe, Sonata, Tiburon, Tucson, Veracruz **Engines:** 2.0L VIN B, 2.0L VIN D, 2.4L VIN C, 2.7L VIN D, 2.7L VIN F, 3.3L VIN D, 3.3L VIN E, 3.3L VIN F, 3.5L VIN E, 3.8L VIN 3, 3.8L VIN C, 3.8L VIN F **Transmissions:** All	**Transaxle Range Switch Circuit High Input** The engine is started and running at cruise speed with VSS signals present. The PCM detected an unexpected high voltage condition on the Transaxle Range (TR) switch circuit during the CCM test. **Possible Causes:** • TR sensor signal circuit is shorted to voltage • TR sensor is damaged or has failed • TCM has failed
DTC: P0711 **2T CCM, MIL: No** **Years:** 2006, 2007, 2008 **Models:** Azera, Elantra, Entourage, Santa Fe, Sonata, Tiburon, Tucson, Veracruz **Engines:** 2.0L VIN B, 2.0L VIN D, 2.4L VIN C, 2.7L VIN D, 2.7L VIN F, 3.3L VIN D, 3.3L VIN E, 3.3L VIN F, 3.5L VIN E, 3.8L VIN 3, 3.8L VIN C, 3.8L VIN F **Transmissions:** All	**Transaxle Fluid Temperature (TFT) Sensor Rationality** The engine is running. The Intake air temperature is equal to or greater than -13°F -25°C. No other fault codes are present. This DTC code is set when the ATF temperature output voltage is lower than a value generated by thermistor resistance, in a normal operating range, for approximately 1 second or longer. The TCM regards the ATF temperature as fixed at a value of 176°F (80°C). **Possible Causes:** • Sensor signal circuit short to ground • Faulty sensor • Faulty PCM

DTC	Trouble Code Title, Conditions & Possible Causes
DTC: P0712 **2T CCM, MIL: No** **Years:** 2006, 2007, 2008 **Models:** Azera, Elantra, Entourage, Santa Fe, Sonata, Tiburon, Tucson, Veracruz **Engines:** 2.0L VIN B, 2.0L VIN D, 2.4L VIN C, 2.7L VIN D, 2.7L VIN F, 3.3L VIN D, 3.3L VIN E, 3.3L VIN F, 3.5L VIN E, 3.8L VIN 3, 3.8L VIN C, 3.8L VIN F **Transmissions:** All	**Transaxle Fluid Temperature (TFT) Sensor Circuit Low Input** The engine is started and running. The gear selector is in any position except for Neutral. The PCM detected the TFT sensor indicated less than 0.50 volt (Scan Tool reads 315°F) during the CCM test. **Possible Causes:** • TFT sensor signal circuit is shorted to ground • TFT sensor ground circuit is open between sensor and PCM • TFT sensor is damaged or has failed • PCM has failed
DTC: P0713 **2T CCM, MIL: No** **Years:** 2006, 2007, 2008 **Models:** Azera, Elantra, Entourage, Santa Fe, Sonata, Tiburon, Tucson, Veracruz **Engines:** 2.0L VIN B, 2.0L VIN D, 2.4L VIN C, 2.7L VIN D, 2.7L VIN F, 3.3L VIN D, 3.3L VIN E, 3.3L VIN F, 3.5L VIN E, 3.8L VIN 3, 3.8L VIN C, 3.8L VIN F **Transmissions:** All	**Transaxle Fluid Temperature (TFT) Sensor Circuit High Input** The engine is started and running. The gear selector is in any position except for Neutral. The PCM detected the TFT sensor indicated more than 4.90 volts (Scan Tool reads -40°F) during the CCM test. **Possible Causes:** • TFT sensor signal circuit is open or shorted to voltage • TFT sensor is damaged or has failed • PCM has failed
DTC: P0715 **1T CCM, MIL: Yes** **Years:** 2006, 2007, 2008 **Models:** Santa Fe, Tiburon, Tucson **Engines:** 2.0L VIN B, 2.0L VIN D, 2.7L VIN D, 2.7L VIN F, 3.5L VIN E **Transmissions:** A/T	**Input Speed Sensor (ISS) Circuit** The engine is started. The vehicle is driven at cruise speed for 3–5 minutes. The TCM detected too large a change in the gear/speed ratio from the Input Speed sensor signal during the CCM Rationality test. **Possible Causes:** • ISS signal circuit is open or shorted to ground • ISS is damaged or has failed • ISS rotor damage or excessive air gap • TCM has failed
DTC: P0717 **2T CCM, MIL: Yes** **Years:** 2006, 2007, 2008 **Models:** Azera, Elantra, Entourage, Santa Fe, Sonata, Veracruz **Engines:** 2.0L VIN D, 2.4L VIN C, 2.7L VIN D, 3.3L VIN D, 3.3L VIN E, 3.3L VIN F, 3.8L VIN 3, 3.8L VIN C, 3.8L VIN F **Transmissions:** A/T	**Input Speed Sensor (ISS) Circuit Open or Short (GND)** The engine is started. The vehicle is driven at a speed of over 30 mph. The PCM detected an unexpected low voltage condition during the CCM test. **Possible Causes:** • ISS signal circuit is open or shorted to ground • ISS is damaged or has failed • ISS rotor damage or excessive air gap • TCM has failed
DTC: P0720 **1T CCM, MIL: Yes** **Years:** 2006, 2007, 2008 **Models:** Santa Fe, Sonata, Tiburon, Tucson **Engines:** 2.0L VIN B, 2.0L VIN D, 2.7L VIN D, 2.7L VIN F, 3.5L VIN E **Transmissions:** A/T	**A/T Output Speed Sensor (OSS) Circuit Open or Short (GND)** Engine started, vehicle driven at a speed of over 30 mph, and the PCM detected the OSS signal was less than 50 percent of the VSS signal for 1 second at a speed of 6 mph, and with the Stop lamp switch indicating ON during the CCM test. **Possible Causes:** • OSS high and low signal circuit is open or shorted to ground • OSS terminals are corroded or damaged • OSS unit is damaged or it has failed • TCM has failed
DTC: P0721 **1T CCM, MIL: Yes** **Years:** 2007, 2008 **Models:** Sonata **Engines:** 2.4L VIN C **Transmissions:** A/T	**A/T Output Speed Sensor (OSS) Fail** The engine is started and running. The selected range is not PARK or NEUTRAL. The vehicle speed is greater than 10 mph (16 km/h). DTC P0721 is set when the TCM detects no output shaft speed when there is vehicle speed. **Possible Causes:** • OSS high and low signal circuit is open, shorted to ground or to power • OSS terminals are corroded or damaged • OSS unit is damaged or it has failed • TCM has failed

DTC	Trouble Code Title, Conditions & Possible Causes
DTC: P0722 **1T CCM, MIL: Yes** **Years:** 2006, 2007, 2008 **Models:** Azera, Elantra, Entourage, Santa Fe, Sonata, Veracruz **Engines:** 2.0L VIN D, 2.4L VIN C, 2.7L VIN D, 3.3L VIN D, 3.3L VIN E, 3.3L VIN F, 3.8L VIN 3, 3.8L VIN C, 3.8L VIN F **Transmissions:** A/T	**Output Speed Sensor (OSS) Circuit Open or Short (GND)** The engine is started and running. The selected range is not PARK or NEUTRAL. The vehicle speed is greater than 10 mph (16 km/h). DTC P0722 is set when the TCM detects no output shaft speed when there is vehicle speed. **Possible Causes:** • OSS high and low signal circuit is open or shorted to ground • OSS terminals are corroded or damaged • OSS unit is damaged or it has failed • TCM has failed
DTC: P0729 **2T CCM, MIL: Yes** **Years:** 2006, 2007, 2008 **Models:** Veracruz **Engines:** 3.8L VIN C **Transmissions:** A/T	**6th Gear Incorrect Ratio** The engine is running with a throttle position greater than 10 percent. The output speed of the transaxle is greater than 500 RPM and the shift lever switch is in "D" range. The shift gear is currently 6th gear. DTC P0729 is set when the value of input shaft speed is not equal to the value of the output shaft, when multiplied by the 6th gear ratio, while the transaxle is engaged in 6th gear. This malfunction is mainly caused by mechanical trouble, such as a control valve sticking or a solenoid valve malfunctioning, rather than an electrical issue. **Possible Causes:** • C2 clutch • B1 brake • Valve body (C2, B1 pressure system) • TCU has failed • Line pressure solenoid valve (SLT) • Shift control solenoid valve (SLC1, SLC3)
DTC: P0730 **2T CCM, MIL: Yes** **Years:** 2006, 2007, 2008 **Models:** Veracruz **Engines:** 3.8L VIN C **Transmissions:** A/T	**Incorrect Gear Ratio** The engine is running with a throttle position greater than 10 percent. The shift lever switch is in "D" range. The shift gear is currently 1st gear. DTC P0730 is set when the value of the input shaft speed is not equal to the value of the output shaft, when multiplied by the 1st gear ratio, while the transaxle is engaged in 1st gear. This malfunction is mainly caused by mechanical trouble, such as a control valve sticking or a solenoid valve malfunctioning, rather than an electrical issue. **Possible Causes:** • C1 clutch • B2 brake • Valve body(C1, B2 pressure system) • TCU has failed • Line pressure solenoid valve (SLT) • Shift control solenoid valve (SLC2, SLC3, SLB1) • 3-way solenoid valve (S1, S2)
DTC: P0731 **2T CCM, MIL: Yes** **Years:** 2006, 2007, 2008 **Models:** Azera, Elantra, Entourage, Santa Fe, Sonata, Tiburon, Tucson, Veracruz **Engines:** 2.0L VIN B, 2.0L VIN D, 2.4L VIN C, 2.7L VIN D, 2.7L VIN F, 3.3L VIN D, 3.3L VIN E, 3.3L VIN F, 3.5L VIN E, 3.8L VIN 3, 3.8L VIN C, 3.8L VIN F **Transmissions:** A/T	**Gear 1 Incorrect Ratio** Engine started, vehicle driven at a speed of over 5 mph, then after a gear shift, the PCM detected the OSS signal times the gear ratio of the new gear ratio did not match the ISS signal during the CCM test. **Possible Causes:** • Solenoid or related pressure switch is damaged or has failed • Problems related to the Input Speed or Output Speed sensor • Low/Reverse clutch is damaged, leaking, or has failed • Problems related to the transmission valve body • Transmission fluid level incorrect or level is too low
DTC: P0732 **2T CCM, MIL: Yes** **Years:** 2006, 2007, 2008 **Models:** Azera, Elantra, Entourage, Santa Fe, Sonata, Tiburon, Tucson, Veracruz **Engines:** 2.0L VIN B, 2.0L VIN D, 2.4L VIN C, 2.7L VIN D, 2.7L VIN F, 3.3L VIN D, 3.3L VIN E, 3.3L VIN F, 3.5L VIN E, 3.8L VIN 3, 3.8L VIN C, 3.8L VIN F **Transmissions:** A/T	**Gear 2 Incorrect Ratio** The engine is started. The vehicle is driven at a speed of over 5 mph. After a gear shift, the PCM detected the OSS signal times the gear ratio of the new gear ratio did not match the ISS signal during the CCM test. **Possible Causes:** • Solenoid or related pressure switch is damaged or has failed • Problems related to the Input Speed or Output Speed sensor • Low/Reverse clutch is damaged, leaking or has failed • Problems related to the transmission valve body • Transmission fluid level incorrect or level is too low

DTC	Trouble Code Title, Conditions & Possible Causes
DTC: P0733 **2T CCM, MIL: Yes** **Years:** 2006, 2007, 2008 **Models:** Azera, Elantra, Entourage, Santa Fe, Sonata, Tiburon, Tucson, Veracruz **Engines:** 2.0L VIN B, 2.0L VIN D, 2.4L VIN C, 2.7L VIN D, 2.7L VIN F, 3.3L VIN D, 3.3L VIN E, 3.3L VIN F, 3.5L VIN E, 3.8L VIN 3, 3.8L VIN C, 3.8L VIN F **Transmissions:** A/T	**Gear 3 Incorrect Ratio** The engine is started. The vehicle is driven at a speed of over 5 mph. After a gear shift, the PCM detected the OSS signal times the gear ratio of the new gear ratio did not match the ISS signal during the CCM test. **Possible Causes:** • Solenoid or related pressure switch is damaged or has failed • Problems related to the Input Speed or Output Speed sensor • Low/Reverse clutch is damaged, leaking or has failed • Problems related to the transmission valve body • Transmission fluid level incorrect or level is too low
DTC: P0734 **2T CCM, MIL: Yes** **Years:** 2006, 2007, 2008 **Models:** Azera, Elantra, Entourage, Santa Fe, Sonata, Tiburon, Tucson, Veracruz **Engines:** 2.0L VIN B, 2.0L VIN D, 2.4L VIN C, 2.7L VIN D, 2.7L VIN F, 3.3L VIN D, 3.3L VIN E, 3.3L VIN F, 3.5L VIN E, 3.8L VIN 3, 3.8L VIN C, 3.8L VIN F **Transmissions:** A/T	**Gear 4 Incorrect Ratio** The engine is started. The vehicle is driven at a speed of over 5 mph. After a gear shift, the PCM detected the OSS signal times the gear ratio of the new gear ratio did not match the ISS signal during the CCM test. **Possible Causes:** • Solenoid or related pressure switch is damaged or has failed • Problems related to the Input Speed or Output Speed sensor • Low/Reverse clutch is damaged, leaking or has failed • Problems related to the transmission valve body • Transmission fluid level incorrect or level is too low
DTC: P0735 **2T CCM, MIL: Yes** **Years:** 2006, 2007, 2008 **Models:** Azera, Entourage, Santa Fe, Sonata, Veracruz **Engines:** 3.3L VIN D, 3.3L VIN E, 3.3L VIN F, 3.5L VIN E, 3.8L VIN 3, 3.8L VIN C, 3.8L VIN F **Transmissions:** A/T	**Gear 5 Incorrect Ratio** The engine is started. The vehicle is driven at a speed of over 5 mph. After a gear shift, the PCM detected the OSS signal times the gear ratio of the new gear ratio did not match the ISS signal during the CCM test. **Possible Causes:** • Solenoid or related pressure switch is damaged or has failed • Problems related to the Input Speed or Output Speed sensor • Low/Reverse clutch is damaged, leaking or has failed • Problems related to the transmission valve body • Transmission fluid level incorrect or level is too low
DTC: P0736 **2T CCM, MIL: Yes** **Years:** 2006, 2007, 2008 **Models:** Azera, Santa Fe, Sonata, Tiburon, Tucson **Engines:** 2.0L VIN B, 2.0L VIN D, 2.4L VIN C, 2.7L VIN D, 2.7L VIN F, 3.3L VIN D, 3.3L VIN F, 3.5L VIN E, 3.8L VIN F **Transmissions:** A/T	**Reverse Incorrect Ratio** The engine is started and the vehicle is driven at a speed of over 3 mph with Reverse Gear commanded ON. DTC P0736 is set when the PCM detects an incorrect Reverse Gear ratio during the CCM Rationality test. **Possible Causes:** • Reverse Gear is damaged or has failed • Low/Reverse clutch is damaged, leaking, or has failed • Problems related to the Input Speed or Output Speed sensor • Problems related to the transmission valve body • Transmission fluid level incorrect or level is too low
DTC: P0741 **2T CCM, MIL: Yes** **Years:** 2006, 2007, 2008 **Models:** Azera, Elantra, Entourage, Santa Fe, Sonata, Tiburon, Tucson, Veracruz **Engines:** 2.0L VIN B, 2.0L VIN D, 2.4L VIN C, 2.7L VIN D, 2.7L VIN F, 3.3L VIN D, 3.3L VIN E, 3.3L VIN F, 3.5L VIN E, 3.8L VIN 3, 3.8L VIN C, 3.8L VIN F **Transmissions:** A/T	**Torque Converter Clutch (TCC) System Stuck Off** The PCM/TCM increases the duty ratio to engage the Damper Clutch by monitoring slip RPM (difference value between engine speed and turbine speed). To decrease the slip of the Damper Clutch, the PCM/TCM increases the duty ratio by applying more hydraulic pressure. When slip RPM does not drop under some value with a 100 percent duty ratio, the PCM/TCM determines that the Torque Converter Clutch is stuck OFF and sets this code. **Possible Causes:** • Faulty TCC or oil pressure system • Faulty TCC solenoid valve • Faulty body control valve • Faulty PCM/TCM

DTC	Trouble Code Title, Conditions & Possible Causes
DTC: P0742 **2T CCM, MIL: Yes** **Years:** 2006, 2007, 2008 **Models:** Azera, Elantra, Entourage, Santa Fe, Sonata, Tiburon, Tucson, Veracruz **Engines:** 2.0L VIN B, 2.0L VIN D, 2.7L VIN D, 2.7L VIN F, 3.3L VIN D, 3.3L VIN E, 3.3L VIN F, 3.5L VIN E, 3.8L VIN 3, 3.8L VIN C, 3.8L VIN F **Transmissions:** A/T	**Torque Converter Clutch (TCC) System Stuck On** The TCM increases the duty ratio to engage the Damper Clutch by monitoring the slip RPM (difference value between engine speed and turbine speed). If a very small amount of slip RPM is maintained, though the TCM applies 0 percent duty ratio value, then the TCM determines that the Torque Converter Clutch is stuck ON and sets this code. **Possible Causes:** • Faulty TCC or oil pressure system • Faulty TCC solenoid valve • Faulty body control valve • Faulty PCM/TCM
DTC: P0743 **2T CCM, MIL: Yes** **Years:** 2006, 2007, 2008 **Models:** Azera, Elantra, Entourage, Santa Fe, Sonata, Tiburon, Tucson **Engines:** 2.0L VIN B, 2.0L VIN D, 2.4L VIN C, 2.7L VIN D, 2.7L VIN F, 3.3L VIN D, 3.3L VIN E, 3.3L VIN F, 3.5L VIN E, 3.8L VIN 3, 3.8L VIN F **Transmissions:** A/T	**Torque Converter Clutch (TCC) Control Solenoid Valve Open or Short (GND)** Engine started, vehicle driven to a speed over 30 mph, and the PCM detected an unexpected low voltage condition on the Damper Clutch Control circuit during the CCM test. **Possible Causes:** • Damper Clutch solenoid control circuit is open or shorted to ground • Damper Clutch solenoid valve is damaged or has failed • TCM has failed
DTC: P0746 **2T CCM, MIL: No** **Years:** 2006, 2007, 2008 **Models:** Azera, Entourage, Santa Fe, Sonata **Engines:** 3.3L VIN D, 3.3L VIN E, 3.3L VIN F, 3.8L VIN 3, 3.8L VIN F **Transmissions:** A/T	**VFS Solenoid Valve Malfunction** The TCM checks the Variable Forced Solenoid (VFS) Control Signal by monitoring the feedback signal from the solenoid valve drive circuit. If an unexpected signal is monitored (for example, high voltage is detected when low voltage is expected, or low voltage is detected when high voltage is expected), the TCM judges that the Low and Reverse control solenoid circuit is malfunctioning and sets this code. **Possible Causes:** • Open or short in circuit • Faulty VFS valve • Faulty PCM/TCM
DTC: P0748 **2T CCM, MIL: No** **Years:** 2006, 2007, 2008 **Models:** Azera, Elantra, Entourage, Santa Fe, Sonata **Engines:** 2.0L VIN D, 2.4L VIN C, 2.7L VIN D, 3.3L VIN D, 3.3L VIN E, 3.3L VIN F, 3.8L VIN 3, 3.8L VIN F **Transmissions:** A/T	**VFS Solenoid Valve Circuit Open or Short (GND)** Engine started, vehicle driven to over 3 mph, and the PCM detected an unexpected low voltage condition on the PCS (solenoid) circuit. **Possible Causes:** • PCS (solenoid) control circuit is shorted to ground • PCS (solenoid) is damaged or has failed • TCM is damaged or has failed
DTC: P0750 **2T CCM, MIL: Yes** **Years:** 2006, 2007, 2008 **Models:** Azera, Elantra, Entourage, Santa Fe, Sonata, Tiburon, Tucson **Engines:** 2.0L VIN B, 2.0L VIN D, 2.4L VIN C, 2.7L VIN D, 2.7L VIN F, 3.3L VIN D, 3.3L VIN E, 3.3L VIN F, 3.5L VIN E, 3.8L VIN 3, 3.8L VIN F **Transmissions:** A/T	**Low and Reverse Solenoid Valve Circuit Open or Short (GND)** The engine is running. The battery voltage is between 11–16 volts. The transaxle is in gear. The TCM checks the Low and Reverse Control Signal by monitoring the feedback signal from the solenoid valve drive circuit. If an unexpected signal is monitored (for example, high voltage is detected when low voltage is expected, or low voltage is detected when high voltage is expected), the TCM judges that the Low and Reverse control solenoid circuit is malfunctioning and sets this code. **Possible Causes:** • Open or short to ground in circuit • Faulty LR SOLENOID VALVE • Faulty PCM/TCM
DTC: P0755 **2T CCM, MIL: Yes** **Years:** 2006, 2007, 2008 **Models:** Azera, Elantra, Entourage, Santa Fe, Sonata, Tiburon, Tucson **Engines:** 2.0L VIN B, 2.0L VIN D, 2.4L VIN C, 2.7L VIN D, 2.7L VIN F, 3.3L VIN D, 3.3L VIN E, 3.3L VIN F, 3.5L VIN E, 3.8L VIN 3, 3.8L VIN F **Transmissions:** A/T	**Underdrive Solenoid Valve Circuit Open or Short (GND)** The TCM checks the Under Drive Clutch Control Signal by monitoring the feedback signal from the solenoid valve drive circuit. If an unexpected signal is monitored (for example, high voltage is detected when low voltage is expected, or low voltage is detected when high voltage is expected), the TCM judges that Under Drive Clutch control solenoid circuit is malfunctioning and sets this code. **Possible Causes:** • Open or short to ground in circuit • Faulty UD SOLENOID VALVE • Faulty PCM/TCM

DTC	Trouble Code Title, Conditions & Possible Causes
DTC: P0760 **2T CCM, MIL: Yes** **Years:** 2006, 2007, 2008 **Models:** Azera, Elantra, Entourage, Santa Fe, Sonata, Tiburon, Tucson **Engines:** 2.0L VIN B, 2.0L VIN D, 2.4L VIN C, 2.7L VIN D, 2.7L VIN F, 3.3L VIN D, 3.3L VIN E, 3.3L VIN F, 3.5L VIN E, 3.8L VIN 3, 3.8L VIN F **Transmissions:** A/T	**Second Solenoid Valve Circuit Open or Short (GND)** The TCM checks the Under Drive Clutch Control Signal by monitoring the feedback signal from the solenoid valve drive circuit .If an unexpected signal is monitored, (for example, high voltage is detected when low voltage is expected or low voltage is detected when high voltage is expected) the TCM judges that 2nd Brake drive control solenoid circuit is malfunctioning and sets this code. **Possible Causes:** • Open or short to ground in circuit • Faulty 2nd SOLENOID VALVE • Faulty PCM/TCM
DTC: P0765 **2T CCM, MIL: Yes** **Years:** 2006, 2007, 2008 **Models:** Azera, Elantra, Entourage, Santa Fe, Sonata, Tiburon, Tucson **Engines:** 2.0L VIN B, 2.0L VIN D, 2.4L VIN C, 2.7L VIN D, 2.7L VIN F, 3.3L VIN D, 3.3L VIN E, 3.3L VIN F, 3.5L VIN E, 3.8L VIN 3, 3.8L VIN F **Transmissions:** A/T	**Overdrive Solenoid Valve Circuit Open or Short (GND)** The TCM checks the Under Drive Clutch Control Signal by monitoring the feedback signal from the solenoid valve drive circuit. If an unexpected signal is monitored (for example, high voltage is detected when low voltage is expected or low voltage is detected when high voltage is expected), the TCM judges that the OVER DRIVE CLUTCH drive control solenoid circuit is malfunctioning and sets this code. **Possible Causes:** • Open or short to ground in circuit • Faulty OD SOLENOID VALVE • Faulty PCM/TCM
DTC: P0770 **1T CCM, MIL: Yes** **Years:** 2006, 2007, 2008 **Models:** Azera, Entourage, Santa Fe, Sonata **Engines:** 3.3L VIN D, 3.3L VIN E, 3.3L VIN F, 3.5L VIN E, 3.8L VIN 3, 3.8L VIN F **Transmissions:** All	**Reduction Solenoid Valve Circuit Open or Short (GND)** The TCM checks the Reduction Control Signal by monitoring the feedback signal from the solenoid valve drive circuit. If an unexpected signal is monitored (for example, high voltage is detected when low voltage is expected, or low voltage is detected when high voltage is expected), the TCM judges that the Reduction control solenoid circuit is malfunctioning and sets this code. **Possible Causes:** • Open or short to ground in circuit • Faulty reduction solenoid valve • Faulty PCM/TCM
DTC: P0780 **1T CCM, MIL: Yes** **Years:** 2007, 2008 **Models:** Veracruz **Engines:** 3.8L VIN C **Transmissions:** A/T	**Shift Error** The engine is running with the transaxle fluid temperature greater than 149°F (65°C) and the shift lever switch in the "D" range. The transaxle output speed is greater than 300 RPM. DTC P0780 is set when the TCU detects abnormal shifting. **Possible Causes:** • S1, S2 solenoid-valve • TCU has failed • Line pressure solenoid valve (SLT) • Shift solenoid valve (SLC1, SLC2, SLC3, SLB1)
DTC: P0880 **1T CCM, MIL: No** **Years:** 2006, 2007, 2008 **Models:** Elantra **Engines:** 2.0L VIN D **Transmissions:** A/T	**TCM Power Signal Error** The input voltage to the TCM is between 9–22 volts after 0.5 seconds have passed from IGN ON. DTC P0880 is set when the threshold voltage is higher or lower than specification (input voltage greater than 24.5 volts or less than 7 volts). **Possible Causes:** • Open or Short in harness • Faulty TCM
DTC: P0882 **1T CCM, MIL: No** **Years:** 2007, 2008 **Models:** Veracruz **Engines:** 3.8L VIN C **Transmissions:** A/T	**Battery Voltage Low** The ignition is in the ON position and the battery voltage is greater than 9 volts. DTC P0882 is set when the TCU detects that the input is a lower voltage than the specified value. **Possible Causes:** • Battery is not fully charged or past its serviceable life • Alternator is not producing a sufficient charge • TCU has failed • Poor connection in the wiring harness
DTC: P0883 **1T CCM, MIL: No** **Years:** 2007, 2008 **Models:** Veracruz **Engines:** 3.8L VIN C **Transmissions:** A/T	**Battery Voltage High** The ignition is in the ON position and the battery voltage is less than 18 volts. DTC P0883 is set when the TCU detects that the input is a higher voltage than the specified value. **Possible Causes:** • Battery is past its serviceable life • Alternator is over-charging • TCU has failed • Faulty connection in the wiring harness

DTC	Trouble Code Title, Conditions & Possible Causes
DTC: P0885 **1T CCM, MIL: No** **Years:** 2007, 2008 **Models:** Azera, Entourage, Santa Fe, Sonata, Tiburon, Tucson **Engines:** 2.0L VIN B, 2.0L VIN D, 2.4L VIN C, 2.7L VIN D, 2.7L VIN D, 2.7L VIN F, 3.3L VIN D, 3.3L VIN E, 3.3L VIN F, 3.8L VIN 3, 3.8L VIN F **Transmissions:** A/T	**A/T Control Relay Malfunction** The TCM checks the A/T control relay signal by monitoring the control signal. If, after the ignition key is turned ON, an unexpected voltage value, which is lower than battery voltage is detected, the TCM sets this code. **Possible Causes:** • Open or short in circuit • Faulty A/T control relay • Faulty PCM/TCM
DTC: P0890 **1T CCM, MIL: No** **Years:** 2006, 2007, 2008 **Models:** Azera, Entourage, Santa Fe, Sonata **Engines:** 2.7L VIN D, 3.3L VIN D, 3.3L VIN E, 3.3L VIN F, 3.8L VIN 3, 3.8L VIN F **Transmissions:** A/T	**TCM Power Relay Sense Circuit Low** The TCM checks the A/T control relay signal by monitoring the control signal. If, after the ignition key is turned ON, an unexpected voltage value, which is lower than battery voltage, is detected, the TCM sets this code. **Possible Causes:** • Open or short in circuit • Faulty A/T control relay • Faulty PCM/TCM
DTC: P0891 **1T CCM, MIL: No** **Years:** 2006, 2007, 2008 **Models:** Azera, Entourage, Santa Fe, Sonata **Engines:** 2.7L VIN D, 3.3L VIN D, 3.3L VIN E, 3.3L VIN F, 3.8L VIN 3, 3.8L VIN F **Transmissions:** A/T	**TCM Power Relay Sense Circuit High** The TCM checks the A/T control relay signal by monitoring the control signal. If, after the ignition key is turned ON, an unexpected voltage value, which is higher than battery voltage is detected, the TCM sets this code. **Possible Causes:** • Open or short in circuit • Faulty A/T control relay • Faulty PCM/TCM
DTC: P0942 **1T CCM, MIL: No** **Years:** 2007, 2008 **Models:** Veracruz **Engines:** 3.8L VIN C **Transmissions:** A/T	**Hydraulic Pressure Unit** The Transmission Control Unit (TCU) monitors the status of the SHIFT SOLENOID VALVE C (SLC1) in order to maintain optimum gear shift. The TCU sets this code when the Solenoid valves are not working normally. **Possible Causes:** • C1 Clutch • Valve body (C1 pressure system) • TCU has failed • Shift solenoid valve (SLC1) • S1, S2 solenoid valve
DTC: P0961 **1T CCM, MIL: No** **Years:** 2007, 2008 **Models:** Veracruz **Engines:** 3.8L VIN C **Transmissions:** A/T	**Line Pressure Control (PC) Solenoid System Range/Performance** The Transmission Control Unit (TCU) monitors the status of the PCSV-A (SLT) in order to maintain optimum gear shift. The TCU sets this code when the Solenoid valves are not working normally. **Possible Causes:** • Line pressure solenoid valve (SLT) • Wiring harness (SLT) • TCU has failed
DTC: P0962 **1T CCM, MIL: No** **Years:** 2007, 2008 **Models:** Veracruz **Engines:** 3.8L VIN C **Transmissions:** A/T	**Line Pressure Control (PC) Solenoid Control Circuit Low (Short to GND or Open)** The engine is running with a speed greater than 400 RPM. The battery voltage is between 10.2–15.5 volts. TCU communication is normal. The TCU monitors the status of the PCSV-A (SLT) in order to maintain optimum gear shift. The TCU sets this code when the Solenoid valves are not working normally. **Possible Causes:** • Line pressure solenoid valve (SLT) • Wiring harness (SLT) • PC Solenoid Control Circuit open or short to ground • TCU

DTC	Trouble Code Title, Conditions & Possible Causes
DTC: P0963 **1T CCM, MIL: No** **Years:** 2007, 2008 **Models:** Veracruz **Engines:** 3.8L VIN C **Transmissions:** A/T	**Line Pressure Control (PC) Solenoid Control Circuit High (Short to +B)** The engine is running with an engine speed of greater than 400 RPM. The battery voltage is between 10.2–15.5 volts. The TCU communication is normal. The TCU monitors the status of the PCSV-A (SLT) in order to maintain the optimum gear shift. The TCU sets this code when the Solenoid valves are not working normally. **Possible Causes:** • Line pressure solenoid valve (SLT) • PC Solenoid Control Circuit short to battery voltage • Wiring harness (SLT) • TCU
DTC: P0973 **1T CCM, MIL: No** **Years:** 2007, 2008 **Models:** Veracruz **Engines:** 3.8L VIN C **Transmissions:** A/T	**Shift Solenoid (SS) "A" Circuit Low (S1) (Short to GND)** The engine is running at a speed greater than 400 RPM. The battery voltage is between 10.2–15.5 volts and the TCU communication is normal. The shift solenoid output signal is ON. The TCU monitors the status of the SCSV-A (S1) in order to maintain optimum gear shift. The TCU sets this code when the Solenoid valves are not working normally. **Possible Causes:** • Solenoid valve (S1) • Wiring harness (S1) • Shift Solenoid (SS) 1 Control circuit short to ground • TCU
DTC: P0974 **1T CCM, MIL: No** **Years:** 2007, 2008 **Models:** Veracruz **Engines:** 3.8L VIN C **Transmissions:** A/T	**Shift Solenoid (SS) "A" Circuit High (S1) (Short to +B or Open)** The engine is running at a speed greater than 400 RPM. The battery voltage is between 10.2–15.5 volts and the TCU communication is normal. The shift solenoid output signal is ON. The TCU monitors the status of the SCSV-A (S1) in order to maintain optimum gear shift. The TCU sets this code when the Solenoid valves are not working normally. **Possible Causes:** • Solenoid valve (S1) • Wiring harness (S1) • Shift Solenoid (SS) 1 Control circuit open or short to battery voltage • TCU
DTC: P0976 **1T CCM, MIL: No** **Years:** 2007, 2008 **Models:** Veracruz **Engines:** 3.8L VIN C **Transmissions:** A/T	**Shift Solenoid (SS) "B" Circuit Low (S2) (Short to GND)** The engine is running at a speed greater than 400 RPM. The battery voltage is between 10.2–15.5 volts and the TCU communication is normal. The shift solenoid output signal is ON. The TCU monitors the status of the SCSV-B (S2) in order to maintain optimum gear shift. The TCU sets this code when the Solenoid valves are not working normally. **Possible Causes:** • Solenoid valve (S2) • Wiring harness (S2) • Shift Solenoid (SS) 2 Control circuit short to ground • TCU
DTC: P0977 **1T CCM, MIL: No** **Years:** 2007, 2008 **Models:** Veracruz **Engines:** 3.8L VIN C **Transmissions:** A/T	**Shift Solenoid (SS) "B" Circuit High (S2) (Short to +B or Open)** The engine is running at a speed greater than 400 RPM. The battery voltage is between 10.2–15.5 volts and the TCU communication is normal. The shift solenoid output signal is ON. The TCU monitors the status of the SCSV-B (S2) in order to maintain optimum gear shift. The TCU sets this code when the Solenoid valves are not working normally. **Possible Causes:** • Solenoid valve (S2) • Wiring harness (S2) • Shift Solenoid (SS) 2 Control circuit open or short to battery voltage • TCU
DTC: P0978 **1T CCM, MIL: No** **Years:** 2007, 2008 **Models:** Veracruz **Engines:** 3.8L VIN C **Transmissions:** A/T	**Shift Solenoid (SS) "C" (SLC1) Range/Performance** The engine is running at a speed greater than 400 RPM. The battery voltage is between 10.2–15.5 volts and the TCU communication is normal. The shift solenoid output signal is ON. The TCU monitors the status of the SCSV-C (SLC1) in order to maintain optimum gear shift. The TCU sets this code when the Solenoid valves are not working normally. **Possible Causes:** • Solenoid valve (SLC1) • Wiring harness (SLC1) • Shift Solenoid (SS) Control circuit short to voltage, open/high resistance • TCU

DTC	Trouble Code Title, Conditions & Possible Causes
DTC: P0979 **1T CCM, MIL: No** **Years:** 2007, 2008 **Models:** Veracruz **Engines:** 3.8L VIN C **Transmissions:** A/T	**Shift Solenoid (SS) "C" (SLC1) Circuit Low (Short to GND or Open)** The engine is running at a speed greater than 400 RPM. The battery voltage is between 10.2–15.5 volts and the TCU communication is normal. The TCU monitors the status of the SCSV-C (SLC1) in order to maintain optimum gear shift. The TCU sets this code when the Solenoid valves are not working normally. **Possible Causes:** • Solenoid valve (SLC1) • Wiring harness (SLC1) • Shift Solenoid (SS) Control circuit open or short to ground • TCU
DTC: P0980 **1T CCM, MIL: No** **Years:** 2007, 2008 **Models:** Veracruz **Engines:** 3.8L VIN C **Transmissions:** A/T	**Shift Solenoid (SS) "C" (SLC1) Circuit High (Short to +B)** The engine is running at a speed greater than 400 RPM. The battery voltage is between 10.2–15.5 volts and the TCU communication is normal. The TCU monitors the status of the SCSV-C (SLC1) in order to maintain optimum gear shift. The TCU sets this code when the Solenoid valves are not working normally. **Possible Causes:** • Solenoid valve (SLC1) • Wiring harness (SLC1) • Shift Solenoid (SS) Control circuit open or short to battery voltage • TCU
DTC: P0981 **1T CCM, MIL: No** **Years:** 2007, 2008 **Models:** Veracruz **Engines:** 3.8L VIN C **Transmissions:** A/T	**Shift Solenoid (SS) "D" (SLC2) Range/Performance** The engine is running at a speed greater than 400 RPM. The battery voltage is between 10.2–15.5 volts and the TCU communication is normal. The TCU monitors the status of the SCSV-D (SLC2) in order to maintain optimum gear shift. The TCU sets this code when the Solenoid valves are not working normally. **Possible Causes:** • Solenoid valve (SLC2) • Wiring harness (SLC2) • Shift Solenoid (SS) Control circuit open or short to battery voltage or ground • TCU
DTC: P0982 **1T CCM, MIL: No** **Years:** 2007, 2008 **Models:** Veracruz **Engines:** 3.8L VIN C **Transmissions:** A/T	**Shift Solenoid (SS) "D" (SLC2) Circuit Low (Short to GND or Open)** The engine is running at a speed greater than 400 RPM. The battery voltage is between 10.2–15.5 volts and the TCU communication is normal. The TCU monitors the status of the SCSV-D (SLC2) in order to maintain optimum gear shift. The TCU sets this code when the Solenoid valves are not working normally. **Possible Causes:** • Solenoid valve (SLC2) • Wiring harness (SLC2) • Shift Solenoid (SS) Control circuit open or short to ground • TCU
DTC: P0983 **1T CCM, MIL: No** **Years:** 2007, 2008 **Models:** Veracruz **Engines:** 3.8L VIN C **Transmissions:** A/T	**Shift Solenoid (SS) "D" (SLC2) Circuit High (Short to +B)** The engine is running at a speed greater than 400 RPM. The battery voltage is between 10.2–15.5 volts and the TCU communication is normal. The TCU monitors the status of the SCSV-D (SLC2) in order to maintain optimum gear shift. The TCU sets this code when the Solenoid valves are not working normally. **Possible Causes:** • Solenoid valve (SLC2) • Wiring harness (SLC2) • Shift Solenoid (SS) Control circuit short to battery voltage • TCU
DTC: P0984 **1T CCM, MIL: No** **Years:** 2007, 2008 **Models:** Veracruz **Engines:** 3.8L VIN C **Transmissions:** A/T	**Shift Solenoid (SS) "E" (SLC3) Range/Performance** The engine is running at a speed greater than 400 RPM. The battery voltage is between 10.2–15.5 volts and the TCU communication is normal. The TCU monitors the status of the SCSV-E (SLC3) in order to maintain optimum gear shift. The TCU sets this code when the Solenoid valves are not working normally. **Possible Causes:** • Solenoid valve (SLC3) • Wiring harness (SLC3) • Shift Solenoid (SS) Control circuit open or short to ground or battery voltage • TCU

DTC	Trouble Code Title, Conditions & Possible Causes
DTC: P0985 **1T CCM, MIL: No** **Years:** 2007, 2008 **Models:** Veracruz **Engines:** 3.8L VIN C **Transmissions:** A/T	**Shift Solenoid (SS) "E" (SLC3) Circuit Low (Short to GND or Open)** The engine is running at a speed greater than 400 RPM. The battery voltage is between 10.2–15.5 volts and the TCU communication is normal. The TCU monitors the status of the SCSV-E (SLC3) in order to maintain optimum gear shift. The TCU sets this code when the Solenoid valves are not working normally. **Possible Causes:** • Solenoid valve (SLC3) • Wiring harness (SLC3) • Shift Solenoid (SS) Control circuit open or short to ground • TCU
DTC: P0986 **1T CCM, MIL: No** **Years:** 2007, 2008 **Models:** Veracruz **Engines:** 3.8L VIN C **Transmissions:** A/T	**Shift Solenoid (SS) "E" (SLC3) Circuit High (Short to +B)** The engine is running at a speed greater than 400 RPM. The battery voltage is between 10.2–15.5 volts and the TCU communication is normal. The TCU monitors the status of the SCSV-E (SLC3) in order to maintain optimum gear shift. The TCU sets this code when the Solenoid valves are not working normally. **Possible Causes:** • Solenoid valve (SLC3) • Wiring harness (SLC3) • Shift Solenoid (SS) Control circuit short to battery voltage • TCU
DTC: P0997 **1T CCM, MIL: No** **Years:** 2007, 2008 **Models:** Veracruz **Engines:** 3.8L VIN C **Transmissions:** A/T	**Shift Solenoid (SS) "F" (SLB1) Range/Performance** The engine is running at a speed greater than 400 RPM. The battery voltage is between 10.2–15.5 volts and the TCU communication is normal. The TCU monitors the status of the SCSV-F (SLB1) in order to maintain optimum gear shift. The TCU sets this code when the Solenoid valves are not working normally. **Possible Causes:** • Solenoid valve (SLB1) • Wiring harness (SLB1) • Shift Solenoid (SS) Control circuit open or short to ground or battery voltage • TCU
DTC: P0998 **1T CCM, MIL: No** **Years:** 2007, 2008 **Models:** Veracruz **Engines:** 3.8L VIN C **Transmissions:** A/T	**Shift Solenoid (SS)"F" (SLB1) Circuit Low (Short to GND or Open)** The engine is running at a speed greater than 400 RPM. The battery voltage is between 10.2–15.5 volts and the TCU communication is normal. The TCU monitors the status of the SCSV-F (SLB1) in order to maintain optimum gear shift. The TCU sets this code when the Solenoid valves are not working normally. **Possible Causes:** • Solenoid valve (SLB1) • Wiring harness (SLB1) • Shift Solenoid (SS) Control circuit open or short to ground • TCU
DTC: P0999 **1T CCM, MIL: No** **Years:** 2007, 2008 **Models:** Veracruz **Engines:** 3.8L VIN C **Transmissions:** A/T	**Shift Solenoid (SS) "F" (SLB1) Circuit High (Short to +B)** The engine is running at a speed greater than 400 RPM. The battery voltage is between 10.2–15.5 volts and the TCU communication is normal. The TCU monitors the status of the SCSV-F (SLB1) in order to maintain optimum gear shift. The TCU sets this code when the Solenoid valves are not working normally. **Possible Causes:** • Solenoid valve (SLB1) • Wiring harness (SLB1) • Shift Solenoid (SS) Control circuit short to battery voltage • TCU

OBD II Trouble Code List (P1xxx Codes)

DTC	Trouble Code Title, Conditions & Possible Causes
DTC: P1106 **2T CCM, MIL: No** **Years:** 2006, 2007, 2008 **Models:** Azera, Entourage, Santa Fe, Sonata, Veracruz **Engines:** 2.7L VIN D, 3.3L VIN D, 3.3L VIN E, 3.3L VIN F, 3.8L VIN 3, 3.8L VIN C, 3.8L VIN F **Transmissions:** All	**MAP Sensor Circuit Intermittent High** The engine is running for more than 10 seconds. No active faults are present. DTC P1106 is set when the PCM detects that the output signals of the MAP sensor are intermittently above 4.5 volts. **Possible Causes:** • Poor connection at sensor • Short to battery in signal circuit • Open in ground circuit • Faulty MAP sensor • Faulty PCM

DTC	Trouble Code Title, Conditions & Possible Causes
DTC: P1107 **2T CCM, MIL: Yes** **Years:** 2006, 2007, 2008 **Models:** Azera, Entourage, Santa Fe, Sonata, Veracruz **Engines:** 2.7L VIN D, 3.3L VIN D, 3.3L VIN E, 3.3L VIN F, 3.8L VIN 3, 3.8L VIN C, 3.8L VIN F **Transmissions:** All	**MAP Sensor Circuit Intermittent Low** The engine is running. The ignition voltage is at or above 11 volts. No Throttle Position Sensor (TPS) faults are active. DTC P01107 is set when the PCM detects that the MAP sensor output signal is intermittently below 0.25 volt. **Possible Causes:** • Poor connection • Open or short to ground in the power circuit • Open or short to ground in the signal circuit • Faulty MAPS • Faulty PCM
DTC: P1111 **2T CCM, MIL: No** **Years:** 2006, 2007, 2008 **Models:** Azera, Entourage, Santa Fe, Sonata, Veracruz **Engines:** 2.7L VIN D, 3.3L VIN D, 3.3L VIN E, 3.3L VIN F, 3.8L VIN 3, 3.8L VIN C, 3.8L VIN F **Transmissions:** All	**Intake Air Temperature (IAT) Sensor Circuit Intermittent Low Input** The engine is running and is in a warmed condition. There are no VSS, ECT, or MAF faults active. DTC P1111 is set when the PCM detects that the IAT sensor output signal is intermittently over 4.9 volts. **Possible Causes:** • Poor connection • Open or short in signal circuit • Open in ground circuit • Faulty IAT sensor • Faulty PCM
DTC: P1112 **2T CCM, MIL: No** **Years:** 2006, 2007, 2008 **Models:** Azera, Entourage, Santa Fe, Sonata, Veracruz **Engines:** 2.7L VIN D, 3.3L VIN D, 3.3L VIN E, 3.3L VIN F, 3.8L VIN 3, 3.8L VIN C, 3.8L VIN F **Transmissions:** All	**Intake Air Temperature (IAT) Sensor Circuit Intermittent High Input** The engine is running. DTC P1112 is set when the PCM detects the IAT sensor output signal is intermittently below 0.1 volt. **Possible Causes:** • Poor connection • Short to ground in the signal circuit • Open in ground circuit • Faulty IAT sensor • Faulty PCM
DTC: P1114 **2T CCM, MIL: Yes** **Years:** 2006, 2007, 2008 **Models:** Azera, Entourage, Santa Fe, Sonata, Veracruz **Engines:** 2.7L VIN D, 3.3L VIN D, 3.3L VIN E, 3.3L VIN F, 3.8L VIN 3, 3.8L VIN C, 3.8L VIN F **Transmissions:** All	**Engine Coolant Temperature (ECT) Sensor Circuit Intermittent Low** DTC P1114 is set when the PCM detects the ECT sensor output signal is intermittently below 0.1 volt. **Possible Causes:** • Poor connection • Short to ground in signal circuit • Open in ground circuit • Faulty ECT sensor • Faulty PCM
DTC: P1115 **2T CCM, MIL: Yes** **Years:** 2006, 2007, 2008 **Models:** Azera, Entourage, Santa Fe, Sonata, Veracruz **Engines:** 2.7L VIN D, 3.3L VIN D, 3.3L VIN E, 3.3L VIN F, 3.8L VIN 3, 3.8L VIN C, 3.8L VIN F **Transmissions:** All	**Engine Coolant Temperature (ECT) Sensor Circuit Intermittent High** DTC P1115 is set when the PCM detects the ECT sensor output signal is intermittently above 4.9 volts. **Possible Causes:** • Poor connection • Open or short to battery in signal circuit • Open in ground circuit • Faulty ECT sensor • Faulty PCM
DTC: P1192 **2T CCM, MIL: Yes** **Years:** 2006 **Models:** Santa Fe **Engines:** 3.5L VIN E **Transmissions:** All	**ETS LimpHome (Target Following Malfunction)** The ignition switch is in the ON position. The ETS motor relay is ON. The battery voltage is greater than 11 volts. TPS 1 is normal. DTC 1192 is set when the PCM detects a fault in the ETS motor circuit. **Possible Causes:** • Poor connection • Short in ETS motor circuit • Faulty ETS motor • Faulty PCM

DTC	Trouble Code Title, Conditions & Possible Causes
DTC: P1193 **2T CCM, MIL: Yes** **Years:** 2006 **Models:** Santa Fe **Engines:** 3.5L VIN E **Transmissions:** All	**ETS LimpHome (Low RPM)** The ignition switch is in the ON position. The ETS motor relay is ON. The battery voltage is greater than 11 volts. The TPS 1 is normal. DTC P1193 is set when the PCM detects an abnormally low RPM under detecting conditions. **Possible Causes:** • Poor connector • Intake/exhaust system blockage • Check throttle plate for carbon deposits • Faulty ETS system • Faulty TPS • Faulty ETS motor • Faulty PCM
DTC: P1194 **2T CCM, MIL: Yes** **Years:** 2006 **Models:** Santa Fe **Engines:** 3.5L VIN E **Transmissions:** All	**ETS LimpHome (TPS2 Malfunction)** The ignition switch is in the ON position. The ETS motor relay is ON. The battery voltage is greater than 11 volts. The TPS 1 is normal. The engine coolant temperature is above 158°F. DTC P1194 is set when the PCM detects the TPS 2 voltage of less than or equal to 0.7 volt. **Possible Causes:** • Poor connector • Short in ETS motor circuit • Faulty ETS motor • Faulty PCM
DTC: P1195 **2T CCM, MIL: Yes** **Years:** 2006 **Models:** Santa Fe **Engines:** 3.5L VIN E **Transmissions:** All	**ETS LimpHome (Target Following Delay)** The ignition switch is in the ON position. The ETS motor relay is ON. The battery voltage is greater than 11 volts. The TPS 1 is normal. The engine coolant temperature is above 158°F. DTC P1195 is set when the PCM detects the TPS 1 voltage of less than or equal to 0.2 volt. **Possible Causes:** • Poor connector • Short in ETS motor circuit • Faulty ETS motor • Faulty PCM
DTC: P1196 **2T CCM, MIL: Yes** **Years:** 2006 **Models:** Santa Fe **Engines:** 3.5L VIN E **Transmissions:** All	**ETS LimpHome (Closed Throttle Stuck)** The ignition switch is in the ON position. The ETS motor relay is ON. The battery voltage is greater than 11 volts. The engine coolant temperature is above 158°F. DTC P1196 is set when the PCM detects no response to changes in throttle position. **Possible Causes:** • Poor connector • Short in ETS motor circuit • Faulty ETS motor • Faulty PCM
DTC: P1295 **1T CCM, MIL: Yes** **Years:** 2006, 2007, 2008 **Models:** Azera, Entourage, Santa Fe, Sonata, Veracruz **Engines:** 2.7L VIN D, 3.3L VIN D, 3.3L VIN E, 3.3L VIN F, 3.8L VIN 3, 3.8L VIN C, 3.8L VIN F **Transmissions:** All	**Throttle Actuator Control System Power Management** The ignition switch is in the ON position. DTC P1295 is set when the PCM detects a problem in the power management mode. **Possible Causes:** • TPS malfunction • TPS malfunction plus MAF sensor malfunction • MAP malfunction plus TPS malfunction • Faulty PCM
DTC: P1330 **2T CCM, MIL: Yes** **Years:** 2006 **Models:** Santa Fe **Engines:** 3.5L VIN E **Transmissions:** All	**Spark Timing Adjust** The engine is running. DTC P1330 is set when the PCM detects an unexpected (invalid) signal on the Spark Timing Adjust circuit during the CCM test. **Possible Causes:** • Spark timing adjust circuit is open or shorted to ground • Spark timing adjust circuit is shorted to VREF or system power • PCM has failed

DTC	Trouble Code Title, Conditions & Possible Causes
DTC: P1505 **1T CCM, MIL: Yes** **Years:** 2006, 2007, 2008 **Models:** Accent, Elantra, Santa Fe, Tiburon, Tucson **Engines:** 1.6L VIN C, 2.0L VIN B, 2.0L VIN D, 2.7L VIN D, 2.7L VIN F **Transmissions:** All	**Idle Charge Actuator Signal Low (Coil No. 1)** DTC P1505 is set when the PCM detects the Idle Charge Actuator (ICA) circuit is open or shorted to ground. Driver stage check. **Possible Causes:** • Short to ground or open in harness • Contact resistance in connectors • Faulty ICA valve
DTC: P1506 **1T CCM, MIL: Yes** **Years:** 2006, 2007, 2008 **Models:** Accent, Elantra, Santa Fe, Tiburon, Tucson **Engines:** 1.6L VIN C, 2.0L VIN B, 2.0L VIN D, 2.7L VIN D, 2.7L VIN F **Transmissions:** All	**Idle Charge Actuator Signal High (Coil No. 1)** DTC P1506 is set when the PCM detects the idle charge actuator control circuit is shorted to battery voltage. Driver stage check. **Possible Causes:** • Short to battery in harness • Contact resistance in connectors • Faulty ICA valve
DTC: P1507 **1T CCM, MIL: Yes** **Years:** 2006, 2007, 2008 **Models:** Accent, Elantra, Santa Fe, Tiburon, Tucson **Engines:** 1.6L VIN C, 2.0L VIN B, 2.0L VIN D, 2.7L VIN D, 2.7L VIN F **Transmissions:** All	**Idle Charge Actuator Signal Low (Coil No. 2)** DTC P1507 is set when the PCM detects the idle charge actuator control circuit is open or shorted to ground. Driver stage check. **Possible Causes:** • Open or short to ground in harness • Contact resistance in connectors • Faulty ICA valve
DTC: P1508 **1T CCM, MIL: Yes** **Years:** 2006, 2007, 2008 **Models:** Accent, Elantra, Santa Fe, Tiburon, Tucson **Engines:** 1.6L VIN C, 2.0L VIN B, 2.0L VIN D, 2.7L VIN D, 2.7L VIN F **Transmissions:** All	**Idle Charge Actuator Signal High (Coil No. 2)** DTC P1508 is set when the PCM detects idle charge actuator control circuit is shorted to battery voltage. Driver stage check. **Possible Causes:** • Short to battery in harness • Contact resistance in connectors • Faulty ICA valve
DTC: P1523 **1T CCM, MIL: Yes** **Years:** 2006, 2007, 2008 **Models:** Azera, Entourage, Santa Fe, Sonata, Veracruz **Engines:** 2.7L VIN D, 3.3L VIN D, 3.3L VIN E, 3.3L VIN F, 3.8L VIN 3, 3.8L VIN C, 3.8L VIN F **Transmissions:** All	**Throttle Actuator Control (TAC) System Throttle Valve Stuck** The throttle actuation from the previous mode is not OFF. The throttle actuator mode is OFF. The ETC power control mode is normal. TPS1 and 2 are normal. The sensor supply voltage is normal. DTC P1523 is set when the PCM detects the throttle fails to return to an unpowered default position when the power to the ETC motor is turned OFF. A fault is set for failure to return to default position within a specified time. **Possible Causes:** • Carbon in throttle • Broken throttle return spring • Throttle sticky or icy (in cold weather climate) • Faulty PCM
DTC: P1550 **1T CCM, MIL: Yes** **Years:** 2006, 2007, 2008 **Models:** Accent **Engines:** 1.6L VIN C **Transmissions:** All	**Knock Sensor Evaluation IC** Circuit continuity check, pulse test. DTC P1550 is set when the PCM detects the knock sensor signal is outside the acceptable parameters. **Possible Causes:** • Poor connection • Open or short in control circuit • Faulty knock sensor • Faulty PCM
DTC: P1560 **1T CCM, MIL: Yes** **Years:** 2006, 2007, 2008 **Models:** Accent **Engines:** 1.6L VIN C **Transmissions:** All	**Knock Control Serial Port Interface (SPI) Check** The SPI communication check runs continuously. The knock sensor signal is sent to the CPU through the SPI. DTC P1560 is set when the PCM detects a malfunction between the SPI and the CPU. **Possible Causes:** • Poor connection • Faulty PCM

DTC	Trouble Code Title, Conditions & Possible Causes
DTC: P1603 **1T CCM, MIL: No** **Years:** 2006, 2007, 2008 **Models:** Santa Fe, Tiburon, Tucson **Engines:** 2.0L VIN B, 2.0L VIN D, 2.7L VIN D, 2.7L VIN F **Transmissions:** All	**Can Communication Bus Off** The ignition switch is in the ON position. The TCM reads data on the CAN-BUS line and checks whether the data is equal to the data that the TCM sent earlier. If the data is not the same, the TCM decides that either the CAN-BUS line or TCM are malfunctioning and sets this code. **Possible Causes:** • Open or short in CAN communication harness • Faulty TCM
DTC: P1604 **1T CCM, MIL: No** **Years:** 2006, 2007, 2008 **Models:** Santa Fe, Tiburon, Tucson **Engines:** 2.0L VIN B, 2.0L VIN D, 2.7L VIN D, 2.7L VIN F **Transmissions:** All	**NO ID from ECU** When the TCM cannot read the data from the ECM through the CAN-BUS line, the TCM sets this code. CAN-BUS circuit malfunctioning or a faulty ECM can be a possible cause of this DTC. **Possible Causes:** • Open or short in CAN communication harness • Faulty ECM • Faulty TCM
DTC: P161B **1T CCM, MIL: Yes** **Years:** 2006, 2007, 2008 **Models:** Azera, Entourage, Santa Fe, Sonata, Tucson, Veracruz **Engines:** 2.7L VIN D, 3.3L VIN D, 3.3L VIN E, 3.3L VIN F, 3.8L VIN 3, 3.8L VIN C, 3.8L VIN F **Transmissions:** All	**ECM/PCM Internal Error Torque Calculation** The engine is running with an engine speed greater than 600 RPM. No faults are present. DTC P161B is set if the delivered torque is grossly different from the desired torque. **Possible Causes:** • Intake air leakage • Faulty ETS System • Clogged exhaust system • Faulty PCM
DTC: P1610 **1T CCM, MIL: No** **Years:** 2006, 2007, 2008 **Models:** Accent, Azera, Entourage, Santa Fe, Tucson, Veracruz **Engines:** 1.6L VIN C, 2.0L VIN B, 2.7L VIN D, 3.3L VIN D, 3.3L VIN E, 3.8L VIN 3, 3.8L VIN C, 3.8L VIN F **Transmissions:** All	**Non-Immobilizer EMS connected to an Immobilizer** The ignition is ON. The ECM sets DTC P1610 if Non-Immobilizer EMS is installed on the vehicle equipped with an Immobilizer. **Possible Causes:** • Invalid PCM installed • Non-immobilizer PCM connected
DTC: P1674 **2T CCM, MIL: Yes** **Years:** 2006, 2007, 2008 **Models:** Accent, Azera, Entourage, Santa Fe, Tucson, Veracruz **Engines:** 1.6L VIN C, 2.0L VIN B, 2.7L VIN D, 3.3L VIN D, 3.3L VIN E, 3.8L VIN 3, 3.8L VIN C, 3.8L VIN F **Transmissions:** All	**Transponder Status Error** The ignition is ON and on the registering TP procedure. The ECM sets DTC P1674 if a transponder key is inserted for registration procedure that can't be registered (TP not in the password mode or whose transport data has been changed). **Possible Causes:** • Invalid transponder
DTC: P1675 **2T CCM, MIL: Yes** **Years:** 2006, 2007, 2008 **Models:** Accent, Azera, Entourage, Santa Fe, Tucson, Veracruz **Engines:** 1.6L VIN C, 2.0L VIN B, 2.7L VIN D, 3.3L VIN D, 3.3L VIN E, 3.8L VIN 3, 3.8L VIN C, 3.8L VIN F **Transmissions:** All	**Transponder Programming Error** The ECM sets DTC P1675 if characteristic data of the transponder doesn't coincide with that of the ECM because of a transponder programming error. **Possible Causes:** • Invalid transponder
DTC: P1676 **2T CCM, MIL: Yes** **Years:** 2006, 2007, 2008 **Models:** Accent, Azera, Entourage, Santa Fe, Tucson, Veracruz **Engines:** 1.6L VIN C, 2.0L VIN B, 2.7L VIN D, 3.3L VIN D, 3.3L VIN E, 3.8L VIN 3, 3.8L VIN C, 3.8L VIN F **Transmissions:** All	**SMARTRA Message Error** The ignition switch is in the ON position. DTC P1676 is set when the PCM detects there is any fault in the message from SMARTRA to the ECU. **Possible Causes:** • Faulty SMARTRA

DTC	Trouble Code Title, Conditions & Possible Causes
DTC: P169A **2T CCM, MIL: Yes** **Years:** 2006, 2007, 2008 **Models:** Accent, Azera, Entourage, Santa Fe, Tucson, Veracruz **Engines:** 1.6L VIN C, 2.0L VIN B, 2.7L VIN D, 3.3L VIN D, 3.3L VIN E, 3.8L VIN 3, 3.8L VIN C, 3.8L VIN F **Transmissions:** All	**SMARTRA Authentication Fail** The ignition switch is in the ON position. The PCM sets DTC P169A if authentication between the PCM and SMARTRA has failed. **Possible Causes:** • Virgin or neutral SMARTRA mismatched with a learned EMS • Locking of SMARTRA
DTC: P1690 **1T CCM, MIL: No** **Years:** 2006, 2007, 2008 **Models:** Accent, Azera, Entourage, Santa Fe, Tiburon, Tucson, Veracruz **Engines:** 1.6L VIN C, 2.0L VIN B, 2.0L VIN D, 2.7L VIN D, 2.7L VIN F, 3.3L VIN D, 3.3L VIN E, 3.8L VIN 3, 3.8L VIN C, 3.8L VIN F **Transmissions:** All	**SMARTRA No Response** The ignition switch is in the ON position. DTC P1690 is set when there is no answer from SMARTRA or an invalid message from SMARTRA to the ECM. **Possible Causes:** • Open or short in antenna or SMARTRA circuit • Antenna • SMARTRA • Faulty transponder • Faulty ECM
DTC: P1691 **1T CCM, MIL: No** **Years:** 2006, 2007, 2008 **Models:** Accent, Azera, Entourage, Santa Fe, Tiburon, Tucson, Veracruz **Engines:** 1.6L VIN C, 2.0L VIN B, 2.0L VIN D, 2.7L VIN D, 2.7L VIN F, 3.3L VIN D, 3.3L VIN E, 3.8L VIN 3, 3.8L VIN C, 3.8L VIN F **Transmissions:** All	**Immobilizer Antenna Coil Error** The ignition switch is in the ON position. The ECM sets DTC P1691 if there is any fault in the immobilizer antenna coil. **Possible Causes:** • Open or short in antenna or SMARTRA circuit • Antenna • SMARTRA • Faulty transponder • Faulty ECM
DTC: P1692 **2T CCM, MIL: Yes** **Years:** 2006, 2007, 2008 **Models:** Accent, Azera, Entourage, Santa Fe, Tucson, Veracruz **Engines:** 1.6L VIN C, 2.0L VIN B, 2.7L VIN D, 3.3L VIN D, 3.3L VIN E, 3.8L VIN 3, 3.8L VIN C, 3.8L VIN F **Transmissions:** All	**Immobilizer Indicator Lamp Error** The ignition switch is in the ON position. The ECM sets DTC P1692 if there is a short circuit in the immobilizer lamp circuit. **Possible Causes:** • Short Circuit in immobilizer lamp circuit • Open or short to ground in the control harness • Faulty PCM/ECM
DTC: P1693 **1T CCM, MIL: No** **Years:** 2006, 2007, 2008 **Models:** Accent, Azera, Entourage, Santa Fe, Tiburon, Tucson, Veracruz **Engines:** 1.6L VIN C, 2.0L VIN B, 2.0L VIN D, 2.7L VIN D, 2.7L VIN F, 3.3L VIN D, 3.3L VIN E, 3.8L VIN 3, 3.8L VIN C, 3.8L VIN F **Transmissions:** All	**Immobilizer Transponder Error** The ignition switch is in the ON position. The ECM sets DTC P1693 if there is an abnormal response from the transponder. **Possible Causes:** • Invalid transponder • Corrupted data from the transponder • More than 1 TP in the magnetic field • No TP (key without TP) in the magnetic field
DTC: P1694 **1T CCM, MIL: No** **Years:** 2006, 2007, 2008 **Models:** Accent, Azera, Entourage, Santa Fe, Tucson, Veracruz **Engines:** 1.6L VIN C, 2.0L VIN B, 2.0L VIN D, 2.7L VIN D, 2.7L VIN F, 3.3L VIN D, 3.3L VIN E, 3.8L VIN 3, 3.8L VIN C, 3.8L VIN F **Transmissions:** All	**EMS message error** The ignition switch is in the ON position. The ECM sets DTC P1694 if a request from the EMS is invalid. The ignition is ON. DTC P1689 is set when the Electronic Brake Control Module (EBCM) receives an invalid delivered torque signal for 2 seconds. When the DTC is set, the EBCM disables the Traction Control Switch (TCS)/Vehicle Stability Enhancement System (VSES) for the duration of the ignition cycle. The Traction Control Off is turned ON. The ABS remains functional. The Stability Off indicator is turned ON. **Possible Causes:** • Open or short in antenna or SMARTRA circuit • Antenna • SMARTRA • Faulty transponder • Protocol layer violation (invalid request or check sum error) • Faulty ECM

DTC	Trouble Code Title, Conditions & Possible Causes
DTC: P1695 **1T CCM, MIL: No** **Years:** 2006, 2007, 2008 **Models:** Accent, Azera, Entourage, Santa Fe, Tiburon, Tucson, Veracruz **Engines:** 1.6L VIN C, 2.0L VIN B, 2.0L VIN D, 2.7L VIN D, 2.7L VIN F, 3.3L VIN D, 3.3L VIN E, 3.8L VIN 3, 3.8L VIN C, 3.8L VIN F **Transmissions:** All	**EMS memory error** The ignition switch is in the ON position. Inconsistent data from EEPROM. Invalid write operation from EEPROM. Not plausible immobilizer indicator store in ECM. No valid data from SMARTRA after 3 attempts from the ECM. Invalid tester message or unexpected request from tester. **Possible Causes:** • Open or short in antenna or SMARTRA circuit • Antenna • SMARTRA • Faulty transponder • Protocol layer violation (invalid request or check sum error) • EMS has failed • Faulty ECM
DTC: P1696 **1T CCM, MIL: No** **Years:** 2006, 2007, 2008 **Models:** Accent, Azera, Entourage, Santa Fe, Tiburon, Tucson, Veracruz **Engines:** 1.6L VIN C, 2.0L VIN B, 2.0L VIN D, 2.7L VIN D, 2.7L VIN F, 3.3L VIN D, 3.3L VIN E, 3.8L VIN 3, 3.8L VIN C, 3.8L VIN F **Transmissions:** All	**Transponder Authentication Fail** The ignition switch is in the ON position. The ECM sets DTC P1696 if an invalid key is inserted into the key hole for Authentication **Possible Causes:** • Virgin TP mismatched with a learned ECM • Learned TP mismatched with a learned ECM • Invalid transponder
DTC: P1697 **1T CCM, MIL: No** **Years:** 2006, 2007, 2008 **Models:** Azera, Entourage, Santa Fe, Veracruz **Engines:** 2.7L VIN D, 3.3L VIN E, 3.8L VIN 3, 3.8L VIN C, 3.8L VIN F **Transmissions:** All	**Immobilizer Tool Message Error** The ignition switch is in the ON position. The ECM sets DTC P1697 if a Request from a Tester is Invalid. **Possible Causes:** • Protocol layer violation (invalid request or check sum error) • Poor connection between scanner and diagnostic connector • Scanner program not up-to-date
DTC: P1699 **1T CCM, MIL: No** **Years:** 2006, 2007, 2008 **Models:** Accent, Azera, Entourage, Santa Fe, Tucson, Veracruz **Engines:** 1.6L VIN C, 2.0L VIN B, 2.7L VIN D, 3.3L VIN D, 3.3L VIN E, 3.8L VIN 3, 3.8L VIN C, 3.8L VIN F **Transmissions:** All	**Immobilizer Twice Overtrial** The ignition switch is in the ON position. This is a special function for the engine start by the vehicle manufacturer. The engine can be started for moving from the production line to an area where the key teaching is done. The ECM sets DTC P1699 if the number of times Twice IGN is tried exceeds the maximum limit (greater than or equal to 32 times). **Possible Causes:** • Over time trial of Twice IGN
DTC: P1716 **2T CCM, MIL: Yes** **Years:** 2006, 2007, 2008 **Models:** Tucson **Engines:** 2.0L VIN B, 2.7L VIN D **Transmissions:** A/T	**Steering Wheel Angle Sensor signal (CAN error)** This code is related to the communication line between the ESP and the TCCU and is set when the CAN bus off or ECU internal error occurs. If a failure is detected, the vehicle is controlled with 2WD by intercepting the EMC current. **Possible Causes:** • Steering Angle sensor, ESP MODULE, TCCU connector loose or poor terminal to wire connection • Open/short in CAN line • Steering Angle sensor faulty • Faulty ESP MODULE • Faulty TCCU
DTC: P1717 **2T CCM, MIL: Yes** **Years:** 2006, 2007, 2008 **Models:** Santa Fe, Tucson **Engines:** 2.0L VIN B, 2.4L VIN B, 2.7L VIN D, 3.5L VIN E **Transmissions:** A/T	**Steering Wheel Angle Sensor 1 (Input Signal)** DTC P1717 is set when the voltage of the signal is above 4.5 volts or a loss of signal has occurred for more than 1 second. If the ESP system applied this code, it is set when the CAN bus off or ECU internal error occurred. If a failure is detected, the vehicle is controlled with 2WD by intercepting the EMC current. **Possible Causes:** • Steering angle sensor, TCCU connector loose or poor terminal to wire connection • Steering angle sensor circuit open/short • Faulty Steering angle sensor • Faulty TCCU

DTC	Trouble Code Title, Conditions & Possible Causes
DTC: P1718 **2T CCM, MIL: Yes** **Years:** 2006, 2007, 2008 **Models:** Santa Fe, Tucson **Engines:** 2.0L VIN B, 2.4L VIN B, 2.7L VIN D, 3.5L VIN E **Transmissions:** A/T	**Steering Wheel Angle Sensor 2 (Input Signal)** DTC P1718 is set when the voltage of the signal is above 4.5 volts or a loss of signal has occurred for more than 1 second. If the ESP system applied this code, it is set when the CAN bus off or ECU internal error occurred. If a failure is detected, the vehicle is controlled with 2WD by intercepting the EMC current. **Possible Causes:** • Steering angle sensor, TCCU connector loose or poor terminal to wire connection • Steering angle sensor circuit open/short • Faulty Steering angle sensor • Faulty TCCU
DTC: P1719 **2T CCM, MIL: Yes** **Years:** 2006, 2007, 2008 **Models:** Santa Fe, Tucson **Engines:** 2.0L VIN B, 2.4L VIN B, 2.7L VIN D, 3.5L VIN E **Transmissions:** A/T	**Steering Wheel Angle Sensor C (Input Signal)** DTC P1719 is set when the voltage of the signal is above 4.5 volts or a loss of signal has occurred for more than 1 second. If the ESP system applied this code, it is set when the CAN bus off or ECU internal error occurred. If a failure is detected, the vehicle is controlled with 2WD by intercepting the EMC current. **Possible Causes:** • Steering angle sensor, TCCU connector loose or poor terminal to wire connection • Steering angle sensor circuit open/short • Faulty Steering angle sensor • Faulty TCCU
DTC: P1726 **1T CCM, MIL: Yes** **Years:** 2006, 2007, 2008 **Models:** Santa Fe, Tucson **Engines:** 2.0L VIN B, 2.4L VIN B, 2.7L VIN D, 3.5L VIN E **Transmissions:** A/T	**Throttle Position Sensor (TPS) Loss of Signal** This code related to the communication line between the ECU and the TCCU. DTC P1726 is set when the CAN signal from the ECU cannot be received for more than 1 second or an ECU internal error occurred. If a failure is detected, the TCCU prohibits the ITM control. **Possible Causes:** • ECM, TCCU connector loose or poor terminal to wire connection • CAN HIGH/LOW circuit open/short • APS (TPS) faulty • Faulty TCCU • Faulty ECM
DTC: P1728 **2T CCM, MIL: Yes** **Years:** 2006, 2007, 2008 **Models:** Santa Fe, Tucson **Engines:** 2.0L VIN B, 2.4L VIN B, 2.7L VIN D, 3.3L VIN E, 3.5L VIN E **Transmissions:** A/T	**EMC Open/Short To Battery** This DTC code is related to the EMC and is set if the control harness has an open or short to a battery source. If failure is detected, the TCCU prohibits the ITM control and the vehicle is controlled with 2WD by intercepting the EMC current. **Possible Causes:** • Power supply malfunction • EMC, TCCU connector loose or poor terminal to wire connection • EMC circuit open or short to battery • Faulty EMC motor • Faulty TCCU
DTC: P1729 **2T CCM, MIL: Yes** **Years:** 2006, 2007, 2008 **Models:** Santa Fe, Tucson **Engines:** 2.0L VIN B, 2.4L VIN B, 2.7L VIN D, 3.5L VIN E **Transmissions:** A/T	**EMC Short To Ground** This DTC is related to the EMC and is set if the control harness has an open or a short to ground. If a failure is detected, the TCCU prohibits the ITM control and the vehicle is controlled with 2WD by intercepting current EMC current. **Possible Causes:** • EMC, TCCU connector loose or poor terminal to wire connection • EMC circuit open or short to ground • Faulty EMC motor • Faulty TCCU
DTC: P1738 **2T CCM, MIL: Yes** **Years:** 2006, 2007, 2008 **Models:** Santa Fe, Tucson **Engines:** 2.0L VIN B, 2.4L VIN B, 2.7L VIN D, 3.5L VIN E **Transmissions:** A/T	**4WD ECU Invalid Part Number** This DTC is set when an inappropriate 4WD-ECU (TCCU) is detected. If a failure is detected, the TCCU prohibits the ITM control. **Possible Causes:** • Part Number invalid for 4WD ECU • Faulty TCCU
DTC: P1745 **2T CCM, MIL: Yes** **Years:** 2007, 2008 **Models:** Santa Fe **Engines:** 2.7L VIN D, 3.3L VIN E **Transmissions:** A/T	**Invalid Part Number** This DTC is set when an inappropriate 4WD-ECU (TCCU) is detected. If a failure is detected, the TCCU prohibits the ITM control. **Possible Causes:** • Part Number invalid for 4WD ECU • Faulty TCCU

DTC	Trouble Code Title, Conditions & Possible Causes
DTC: P1750 **2T CCM, MIL: Yes** **Years:** 2007, 2008 **Models:** Santa Fe, Veracruz **Engines:** 2.7L VIN D, 3.3L VIN E, 3.8L VIN C **Transmissions:** A/T	**Front Left Wheel Speed Sensor** No faults are current in the system. The battery voltage is between 10.2–15.5 volts. DTC P1750 is set when the TCM cannot read the data from the ABS ECU through the CAN-BUS line. CAN-BUS circuit malfunctioning or ABS ECU can be a possible cause of this DTC. **Possible Causes:** • Short or open in CAN communication circuit • Faulty ECU
DTC: P1750A **1T CCM, MIL: Yes** **Years:** 2006, 2007, 2008 **Models:** Santa Fe, Tucson **Engines:** 2.0L VIN B, 2.4L VIN B, 2.7L VIN D, 3.5L VIN E **Transmissions:** A/T	**Front Left Speed Sensor (With ABS)** This code is related to the Wheel Speed Sensor and is set when the difference between the front and the rear wheel speed is greater than 19 mph (30 kph) for more than 30 seconds or a loss of signal has occurred. If the ESP (or FTCS) system applied this code, the signal going through the CAN line is not received by the TCCU for more than 1 second. If a failure is detected, the TCCU prohibits the ITM control. **Possible Causes:** • ABS MODULE, TCCU connector loose or poor terminal to wire connection • Wheel sensor communication line circuit open/short • Faulty Wheel sensor • Faulty ABS MODULE • Faulty TCCU
DTC: P1750B **2T CCM, MIL: Yes** **Years:** 2006, 2007, 2008 **Models:** Santa Fe, Tucson **Engines:** 2.0L VIN B, 2.4L VIN B, 2.7L VIN D, 3.5L VIN E **Transmissions:** A/T	**Front Left Speed Sensor (Without ABS)** This code is related to the Wheel Speed Sensor and is set when the difference between the front and the rear wheel speed is above 19 mph (30 kph) for more than 30 seconds or loss of signal has occurred. If the ESP (or FTCS) system applied this code, the signal going through the CAN line is not received by the TCCU for more than 1 second. If a failure is detected, the TCCU prohibits the ITM control. **Possible Causes:** • ABS MODULE, TCCU connector loose or poor terminal to wire connection • Wheel sensor communication line circuit open/short • Faulty Wheel sensor • Faulty ABS MODULE • Faulty TCCU
DTC: P1751 **2T CCM, MIL: Yes** **Years:** 2007, 2008 **Models:** Santa Fe, Veracruz **Engines:** 2.7L VIN D, 3.3L VIN E, 3.8L VIN C **Transmissions:** A/T	**Front Right Wheel Speed Sensor** No faults are current in the system. The battery voltage is between 10.2–15.5 volts. When the TCM cannot read the data from the ABS ECU through the CAN-BUS line, the TCM sets this code. CAN-BUS circuit malfunctioning or ABS ECU can be a possible cause of this DTC. **Possible Causes:** • Short or open in the CAN communication circuit • Faulty ECU
DTC: P1751A **2T CCM, MIL: Yes** **Years:** 2006, 2007, 2008 **Models:** Santa Fe, Tucson **Engines:** 2.0L VIN B, 2.4L VIN B, 2.7L VIN D, 3.5L VIN E **Transmissions:** A/T	**Front Right Speed Sensor (With ABS)** This code is related to the Wheel Speed Sensor and is set when the difference between the front and the rear wheel speed is above 19 mph (30 kph) for more than 30 seconds or loss of signal has occurred. If the ESP (or FTCS) system applied this code, the signal going through the CAN line is not received by the TCCU for more than 1 second. If a failure is detected, the TCCU prohibits the ITM control. **Possible Causes:** • ABS MODULE, TCCU connector loose or poor terminal to wire connection • Wheel sensor communication line circuit open/short • Faulty Wheel sensor • Faulty ABS MODULE • Faulty TCCU
DTC: P1751B **2T CCM, MIL: Yes** **Years:** 2006, 2007, 2008 **Models:** Santa Fe, Tucson **Engines:** 2.0L VIN B, 2.4L VIN B, 2.7L VIN D, 3.5L VIN E **Transmissions:** A/T	**Front Right Speed Sensor (Without ABS)** This code is related to the Wheel Speed Sensor and is set when the difference between the front and the rear wheel speed is greater than 19 mph (30 kph) for more than 30 seconds or a loss of signal has occurred. If a failure is detected, the TCCU prohibits the ITM control. **Possible Causes:** • ABS MODULE, TCCU connector loose or poor terminal to wire connection • Wheel sensor communication line circuit open/short to battery • Wheel sensor air gap is incorrect • Faulty wheel speed sensor • Faulty TCCU

DTC	Trouble Code Title, Conditions & Possible Causes
DTC: P1752 **2T CCM, MIL: Yes** **Years:** 2007, 2008 **Models:** Santa Fe, Veracruz **Engines:** 2.7L VIN D, 3.3L VIN E, 3.8L VIN C **Transmissions:** A/T	**Rear Left Wheel Speed Sensor** When the TCM cannot read the data from the ABS ECU through the CAN-BUS line, the TCM sets this code. CAN-BUS circuit malfunctioning or ABS ECU can be a possible cause of this DTC. **Possible Causes:** • Short or open in the CAN communication circuit • Faulty ECU
DTC: P1752A **2T CCM, MIL: Yes** **Years:** 2006, 2007, 2008 **Models:** Santa Fe, Tucson **Engines:** 2.0L VIN B, 2.4L VIN B, 2.7L VIN D, 3.5L VIN E **Transmissions:** A/T	**Rear Left speed Sensor (With ABS)** This code is related to the Wheel Speed Sensor and is set when the difference between the front and the rear wheel speed is above 19 mph (30 kph) for more than 30 seconds or loss of signal has occurred. If the ESP (or FTCS) system applied this code, the signal going through the CAN line is not received by the TCCU for more than 1 second. If a failure is detected, the TCCU prohibits the ITM control. **Possible Causes:** • ABS MODULE, TCCU connector loose or poor terminal to wire connection • Wheel sensor communication line circuit open/short • Faulty Wheel sensor • Faulty ABS MODULE • Faulty TCCU
DTC: P1752B **2T CCM, MIL: Yes** **Years:** 2006, 2007, 2008 **Models:** Santa Fe, Tucson **Engines:** 2.0L VIN B, 2.4L VIN B, 2.7L VIN D, 3.5L VIN E **Transmissions:** A/T	**Rear Left Speed Sensor (Without ABS)** This code is related to the Wheel Speed Sensor and is set when the difference between the front and the rear wheel speed is greater than 19 mph (30 kph) for more than 30 seconds or a loss of signal has occurred. If a failure is detected, the TCCU prohibits the ITM control. **Possible Causes:** • ABS MODULE, TCCU connector loose or poor terminal to wire connection • Wheel sensor communication line circuit open/short to battery • Wheel sensor air gap is incorrect • Faulty wheel speed sensor • Faulty TCCU
DTC: P1753 **2T CCM, MIL: Yes** **Years:** 2007, 2008 **Models:** Santa Fe, Veracruz **Engines:** 2.7L VIN D, 3.3L VIN E, 3.8L VIN C **Transmissions:** A/T	**Rear Right Wheel Speed Sensor** When the TCM cannot read the data from the ABS ECU through the CAN-BUS line, the TCM sets this code. CAN-BUS circuit malfunctioning or ABS ECU can be a possible cause of this DTC. **Possible Causes:** • Short or open in the CAN communication circuit • Faulty ECU
DTC: P1753A **2T CCM, MIL: Yes** **Years:** 2006, 2007, 2008 **Models:** Santa Fe, Tucson **Engines:** 2.0L VIN B, 2.4L VIN B, 2.7L VIN D, 3.5L VIN E **Transmissions:** A/T	**Rear Right Speed Sensor (With ABS)** This code is related to the Wheel Speed Sensor and is set when the difference between the front and the rear wheel speed is above 19 mph (30 kph) for more than 30 seconds or loss of signal has occurred. If the ESP (or FTCS) system applied this code, the signal going through the CAN line is not received by the TCCU for more than 1 second. If a failure is detected, the TCCU prohibits the ITM control. **Possible Causes:** • ABS MODULE, TCCU connector loose or poor terminal to wire connection • Wheel sensor communication line circuit open/short • Faulty Wheel sensor • Faulty ABS MODULE • Faulty TCCU
DTC: P1753B **2T CCM, MIL: Yes** **Years:** 2006, 2007, 2008 **Models:** Santa Fe, Tucson **Engines:** 2.0L VIN B, 2.4L VIN B, 2.7L VIN D, 3.5L VIN E **Transmissions:** A/T	**Rear Right Speed Sensor (Without ABS)** This code is related to the Wheel Speed Sensor and is set when the difference between the front and the rear wheel speed is greater than 19 mph (30 kph) for more than 30 seconds or a loss of signal has occurred. If a failure is detected, the TCCU prohibits the ITM control. **Possible Causes:** • ABS MODULE, TCCU connector loose or poor terminal to wire connection • Wheel sensor communication line circuit open/short to battery • Wheel sensor air gap is incorrect • Faulty wheel speed sensor • Faulty TCCU

DTC	Trouble Code Title, Conditions & Possible Causes
DTC: P1760 **2T CCM, MIL: Yes** **Years:** 2007, 2008 **Models:** Veracruz **Engines:** 3.8L VIN C **Transmissions:** A/T	**Neutral Condition AT R Range SLC3 Stick** The engine is running with a speed greater than 400 RPM. The shift lever switch is in the "R" range. The gear position is in reverse. No fault is current. The battery voltage is between 10.2–15.5 volts. The TCU monitors the status of the SHIFT SOLENOID VALVE C (SLC3) in order to maintain an optimum gear shift. DTC P1760 is set when the Solenoid valves are not working normally. **Possible Causes:** • C3 Clutch • B2 Brake • Valve-body (C3, B2 pressure system) • TCU has failed • Solenoid valve (SLC3) • S1, S2 Solenoid valve
DTC: P1764 **2T CCM, MIL: Yes** **Years:** 2006, 2007, 2008 **Models:** Santa Fe, Tiburon, Tucson **Engines:** 2.0L VIN B, 2.0L VIN D, 2.4L VIN B, 2.7L VIN D, 2.7L VIN F, 3.5L VIN E **Transmissions:** A/T	**ECU-ITM CAN Error** This code is related to the communication line between the ECU and the TCCU. DTC 1764 is set when the CAN signal from the ECU is not received for more than 1 second or an ECU internal error occurred. If a failure is detected, the TCCU prohibits the ITM control. **Possible Causes:** • ECM, TCCU connector loose or poor terminal to wire connection • CAN HIGH/LOW circuit open/short • Faulty ECM • Faulty TCCU
DTC: P1765 **2T CCM, MIL: Yes** **Years:** 2006, 2007, 2008 **Models:** Santa Fe, Tucson **Engines:** 2.0L VIN B, 2.4L VIN B, 2.7L VIN D, 3.5L VIN E **Transmissions:** A/T	**TCS-ITM CAN Error** This code is related to the communication line between the TCS, ESP (or FTCS) and the TCCU. DTC 1765 is set when the CAN signal from the ECU is not received for more than 1 second or an ECU internal error occurred. If a failure is detected, the TCCU prohibits the ITM control. **Possible Causes:** • ECM, TCCU connector loose or poor terminal to wire connection • CAN HIGH/LOW circuit open/short • Faulty ECM • Faulty TCCU
DTC: P1766 **2T CCM, MIL: Yes** **Years:** 2006, 2007, 2008 **Models:** Santa Fe, Tucson **Engines:** 2.0L VIN B, 2.4L VIN B, 2.7L VIN D, 3.5L VIN E **Transmissions:** A/T	**CAN Communication BUS OFF** This code is related to the communication line between the ECM and the TCCU. DTC 1766 is set when the CAN signal from the ECU is not received for more than 1 second or an ECU internal error occurred. If a failure is detected, the TCCU prohibits the ITM control. **Possible Causes:** • CAN Bus • ABS MODULE, TCCU connector loose or poor terminal to wire connection • CAN communication line circuit open/short • Faulty ECU • Faulty ABS MODULE • Faulty TCCU
DTC: P1767 **2T CCM, MIL: Yes** **Years:** 2006, 2007, 2008 **Models:** Santa Fe, Tucson **Engines:** 2.0L VIN B, 2.4L VIN B, 2.7L VIN D, 3.5L VIN E **Transmissions:** A/T	**ABS Active Signal (CAN Error)** This code is related to the communication line between the ECM and the TCCU. DTC 1766 is set when the CAN signal from the ECU is not received for more than 1 second or an ECU internal error occurred. If a failure is detected, the TCCU prohibits the ITM control. **Possible Causes:** • CAN Bus • ABS MODULE, TCCU connector loose or poor terminal to wire connection • CAN communication line circuit open/short • Faulty ECU • Faulty ABS MODULE • Faulty TCCU
DTC: P1769 **2T CCM, MIL: No** **Years:** 2007, 2008 **Models:** Santa Fe **Engines:** 2.7L VIN D, 3.3L VIN E **Transmissions:** A/T	**Mini Spare Detected** DTC 1769 is set when the ITM-ECU detects that the tire size is 10 percent bigger than the normal tire by calculating each wheel speed. **Possible Causes:** • Tire size has been changed • Faulty TCCU
DTC: P1770 **1T CCM, MIL: No** **Years:** 2007, 2008 **Models:** Santa Fe **Engines:** 2.7L VIN D, 3.3L VIN E **Transmissions:** A/T	**Plate And Oil Temperature Thresholds Exceeded (LOCK MODE)** DTC P1770 is set when the ITM-ECU decides that the clutch (EMC) and oil are over-heated (above specified value) for more than 1 second. **Possible Causes:** • Faulty SWITCH (4WD-LOCK) • Faulty EMC • Faulty TCCU

DTC	Trouble Code Title, Conditions & Possible Causes
DTC: P1771 **1T CCM, MIL: No** **Years:** 2007, 2008 **Models:** Santa Fe **Engines:** 2.7L VIN D, 3.3L VIN E **Transmissions:** A/T	**Plate And Oil Temperature Thresholds Exceeded (SHUT DOWN)** DTC P1771 is set when the ITM-ECU decides that the clutch (EMC) and oil are over-heated (above specified value) for more than 1 second. **Possible Causes:** • Faulty EMC • Faulty TCCU
DTC: P1780 **2T CCM, MIL: Yes** **Years:** 2007, 2008 **Models:** Santa Fe **Engines:** 2.7L VIN D, 3.3L VIN E **Transmissions:** A/T	**Torque Control Signal Failure** DTC P1780 is set when the ITM-ECU detects invalid engine information like engine displacement or torque. If a failure is detected, the TCCU prohibits the ITM control and cuts the current to the control coil. **Possible Causes:** • Invalid engine size received

OBD II Trouble Code List (P2xxx Codes)

DTC	Trouble Code Title, Conditions & Possible Causes
DTC: P2015 **1T CCM, MIL: No** **Years:** 2006 **Models:** Santa Fe **Engines:** 3.5L VIN E **Transmissions:** All	**V. Intake Motor Rotation SNS** DTC P2015 is set if the VIS system could not approach a target position. The engine speed is equal to or greater than 3,750 RPM. **Possible Causes:** • Poor connection • Open or short to battery in harness • Faulty VIS motor or VIS motor rotation sensor • Faulty PCM
DTC: P2065 **2T CCM, MIL: Yes** **Years:** 2006, 2007, 2008 **Models:** Azera, Santa Fe, Veracruz **Engines:** 2.7L VIN D, 3.3L VIN D, 3.3L VIN E, 3.8L VIN C, 3.8L VIN F **Transmissions:** All	**Fuel Level Sensor B Circuit** The engine is running. The ignition voltage is greater than 11 volts. DTC P2065 is set if the difference between fuel level A and B is less than 2 percent. **Possible Causes:** • Poor connection • Short in signal circuit • Faulty Fuel Level sender A/B • Faulty PCM
DTC: P2066 **2T CCM, MIL: Yes** **Years:** 2006, 2007, 2008 **Models:** Azera, Santa Fe, Veracruz **Engines:** 2.7L VIN D, 3.3L VIN D, 3.3L VIN E, 3.8L VIN C, 3.8L VIN F **Transmissions:** All	**Fuel Level Sensor B Performance** The engine is running. No fuel level fault is current. DTC P2066 is set if the fuel level difference between the current and the previous is lower than 10 percent while the odometer difference between the present and the previous is higher than 106 miles (170km). **Possible Causes:** • Poor connection • Faulty fuel level sender (stuck)
DTC: P2067 **2T CCM, MIL: Yes** **Years:** 2006, 2007, 2008 **Models:** Azera, Santa Fe, Veracruz **Engines:** 2.7L VIN D, 3.3L VIN D, 3.3L VIN E, 3.8L VIN C, 3.8L VIN F **Transmissions:** All	**Fuel Level Sensor B Circuit Low** The engine is running and the ignition voltage is greater than 11 volts. DTC P2067 is set if the fuel level is too low (fuel level sender signal is below 0.9 percent). **Possible Causes:** • Poor connection • Open or short to ground in signal circuit • Faulty Fuel Level sender A • Faulty PCM
DTC: P2068 **2T CCM, MIL: Yes** **Years:** 2006, 2007, 2008 **Models:** Azera, Santa Fe, Veracruz **Engines:** 2.7L VIN D, 3.3L VIN D, 3.3L VIN E, 3.8L VIN C, 3.8L VIN F **Transmissions:** All	**Fuel Level Sensor B Circuit High** The engine is running and the ignitions voltage is greater than 11 volts. DTC P2068 is set if the fuel level is too high (fuel level sensor signal greater than 43 percent). **Possible Causes:** • Poor connection • Short to battery in signal circuit • Faulty Fuel Level sender A • Faulty PCM

DTC	Trouble Code Title, Conditions & Possible Causes
DTC: P2096 **2T CCM, MIL: Yes** **Years:** 2006, 2007, 2008 **Models:** Accent, Azera, Elantra, Entourage, Santa Fe, Sonata, Veracruz **Engines:** 1.6L VIN C, 2.0L VIN D, 2.4L VIN C, 2.7L VIN D, 3.3L VIN D, 3.8L VIN 3, 3.8L VIN C, 3.8L VIN F **Transmissions:** All	**Post Catalyst Fuel Trim System B1 Too Lean** The engine is running in closed loop. There are no disabling faults present. The system voltage is greater than or equal to 11 volts. The Intake Air Temperature (IAT) is greater than 14°F (-10°C). DTC P2096 is set when the fuel trim rear sensor control limits are exceeded. **Possible Causes:** • HO2S is contaminated or deteriorated due to poor fuel usage • Damaged wiring between the HO2S and the ECM • Terminal corrosion or water intrusion in the HO2S harness connectors • Incorrect terminal tension • HO2S is not securely installed • Exhaust leaks • HO2S 1 circuits shorted together between the HO2S connector and the ECM • Shorted terminals or poor connections at the HO2S • Shorted terminals or poor connections at the ECM • HO2S has failed • ECM has failed
DTC: P2097 **2T CCM, MIL: Yes** **Years:** 2006, 2007, 2008 **Models:** Accent, Azera, Elantra, Entourage, Santa Fe, Sonata, Veracruz **Engines:** 1.6L VIN C, 2.0L VIN D, 2.4L VIN C, 2.7L VIN D, 3.3L VIN D, 3.8L VIN 3, 3.8L VIN C, 3.8L VIN F **Transmissions:** All	**Post Catalyst Fuel Trim System B1 Too Rich** The engine is running in closed loop. There are no disabling faults present. The system voltage is greater than or equal to 11 volts. The Intake Air Temperature (IAT) is greater than 14°F (-10°C). DTC P2097 is set when the fuel trim rear sensor control limits are exceeded. **Possible Causes:** • HO2S is contaminated or deteriorated due to poor fuel usage • Damaged wiring between the HO2S and the ECM • Terminal corrosion or water intrusion in the HO2S harness connectors • Incorrect terminal tension • HO2S is not securely installed • Exhaust leaks • HO2S 1 circuits shorted together between the HO2S connector and the ECM • Shorted terminals or poor connections at the HO2S • Shorted terminals or poor connections at the ECM • HO2S has failed • ECM has failed
DTC: P2098 **2T CCM, MIL: Yes** **Years:** 2006, 2007, 2008 **Models:** Azera, Entourage, Santa Fe, Veracruz **Engines:** 2.7L VIN D, 3.3L VIN D, 3.8L VIN 3, 3.8L VIN C, 3.8L VIN F **Transmissions:** All	**Post Catalyst Fuel Trim System B2 Too Lean** The engine is running in closed loop. There are no disabling faults present. The system voltage is greater than or equal to 11 volts. The Intake Air Temperature (IAT) is greater than 14°F (-10°C). DTC P2098 is set when the fuel trim rear sensor control limits are exceeded. **Possible Causes:** • HO2S is contaminated or deteriorated due to poor fuel usage • Damaged wiring between the HO2S and the ECM • Terminal corrosion or water intrusion in the HO2S harness connectors • Incorrect terminal tension • HO2S is not securely installed • Exhaust leaks • HO2S 1 circuits shorted together between the HO2S connector and the ECM • Shorted terminals or poor connections at the HO2S • Shorted terminals or poor connections at the ECM • HO2S has failed • ECM has failed
DTC: P2099 **2T CCM, MIL: Yes** **Years:** 2006, 2007, 2008 **Models:** Azera, Entourage, Santa Fe, Veracruz **Engines:** 2.7L VIN D, 3.3L VIN D, 3.8L VIN 3, 3.8L VIN C, 3.8L VIN F **Transmissions:** All	**Post Catalyst Fuel Trim System B2 Too Rich** The engine is running in closed loop. There are no disabling faults present. The system voltage is greater than or equal to 11 volts. The Intake Air Temperature (IAT) is greater than 14°F (-10°C). DTC P2098 is set when the fuel trim rear sensor control limits are exceeded. **Possible Causes:** • HO2S is contaminated or deteriorated due to poor fuel usage • Damaged wiring between the HO2S and the ECM • Terminal corrosion or water intrusion in the HO2S harness connectors • Incorrect terminal tension • HO2S is not securely installed • Exhaust leaks • HO2S 1 circuits shorted together between the HO2S connector and the ECM • Shorted terminals or poor connections at the HO2S • Shorted terminals or poor connections at the ECM • HO2S has failed • ECM has failed

DTC	Trouble Code Title, Conditions & Possible Causes
DTC: P2100 **1T CCM, MIL: Yes** **Years:** 2006 **Models:** Santa Fe **Engines:** 3.5L VIN E **Transmissions:** All	**ETS Relay-Malfunction** The ignition switch is in the ON position. The motor relay is ON. DTC P2100 is set when the indicated throttle position does not match the predicted throttle position for a predetermined time. **Possible Causes:** • Poor connection • Open in ETS relay circuit • Faulty ETS relay/fuse • Faulty PCM
DTC: P2101 **1T CCM, MIL: Yes** **Years:** 2007, 2008 **Models:** Sonata **Engines:** 2.4L VIN C **Transmissions:** All	**Throttle Actuator Control (TAC) Motor Circuit Range/Performance** The ignition switch is in the ON position. DTC P2101 is set when the indicated throttle position does not match the predicted throttle position for a predetermined time. **Possible Causes:** • TAC motor control circuit is open, shorted to ground or to voltage • Throttle body assembly is damaged or it has failed • ECM has failed
DTC: P2102 **1T CCM, MIL: Yes** **Years:** 2006 **Models:** Santa Fe **Engines:** 3.5L VIN E **Transmissions:** All	**Throttle Actuator Control (TAC) Motor Circuit Low** The ignition switch is in the ON position. DTC P2102 is set when the motor circuit voltage is lower than expected. **Possible Causes:** • Poor connection • Short to ground in the ETS motor circuit • Faulty ETS motor • Faulty PCM
DTC: P2103 **1T CCM, MIL: Yes** **Years:** 2006 **Models:** Santa Fe **Engines:** 3.5L VIN E **Transmissions:** All	**Throttle Actuator Control (TAC) Motor Circuit High** The ignition switch is in the ON position. DTC P2103 is set when the motor circuit voltage is higher than expected. **Possible Causes:** • Poor connection • Short to battery voltage in the ETS motor circuit • Faulty ETS motor • Faulty PCM
DTC: P2104 **1T CCM, MIL: Yes** **Years:** 2006, 2007, 2008 **Models:** Azera, Entourage, Santa Fe, Sonata, Veracruz **Engines:** 2.4L VIN C, 2.7L VIN D, 3.3L VIN D, 3.3L VIN E, 3.3L VIN F, 3.8L VIN 3, 3.8L VIN C, 3.8L VIN F **Transmissions:** All	**Throttle Actuator Control (TAC) System Forced Idle** The ignition switch is in the ON position. DTC P2104 is set when the motor circuit was forced into an idle mode. **Possible Causes:** • Faulty Accelerator Pedal Position Sensor (APS) • Faulty APS plus brake • Faulty APS plus vehicle speed sensor • Faulty APS plus brake plus vehicle speed sensor • Faulty PCM
DTC: P2105 **1T CCM, MIL: Yes** **Years:** 2006, 2007, 2008 **Models:** Azera, Entourage, Santa Fe, Sonata, Veracruz **Engines:** 2.4L VIN C, 2.7L VIN D, 3.3L VIN D, 3.3L VIN E, 3.3L VIN F, 3.8L VIN 3, 3.8L VIN C, 3.8L VIN F **Transmissions:** All	**Throttle Actuator Control (TAC) System Forced Engine Shutdown** The ignition switch is in the ON position. DTC P2105 is set when the PCM detects an internal failure or incomplete programming and moves into shutdown mode. **Possible Causes:** • Faulty APS plus MAPS plus ETS • PCM fuse open • Poor connection at the PCM • Faulty PCM
DTC: P2106 **1T CCM, MIL: Yes** **Years:** 2006, 2007, 2008 **Models:** Azera, Entourage, Santa Fe, Sonata, Veracruz **Engines:** 2.4L VIN C, 2.7L VIN D, 3.3L VIN D, 3.3L VIN E, 3.3L VIN F, 3.5L VIN E, 3.8L VIN 3, 3.8L VIN C, 3.8L VIN F **Transmissions:** All	**Throttle Actuator Control (TAC) System Forced Limited Power** The ignition switch is in the ON position. DTC P2106 is set when the PCM detects that the system is in Limit Performance Mode. **Possible Causes:** • Faulty APS • Faulty APS plus Brake • Faulty APS plus Vehicle speed sensor • Faulty APS plus Vehicle speed sensor plus Brake • Faulty ETC Items (TPS or ETC motor) • Faulty PCM

DTC	Trouble Code Title, Conditions & Possible Causes
DTC: P2107 **1T CCM, MIL: Yes** **Years:** 2006 **Models:** Santa Fe **Engines:** 3.5L VIN E **Transmissions:** All	**ETS-ECM Malfunction (EEPROM R/W)** The DTC will set if the PCM cannot read or write on the EEPROM. Check reading and writing. Ignition switch ON. Reading or writing error has occurred. **Possible Causes:** • Poor connection • PCM has failed
DTC: P2108 **1T CCM, MIL: Yes** **Years:** 2006 **Models:** Santa Fe **Engines:** 3.5L VIN E **Transmissions:** All	**ETS-ECM Malfunction** The ignition switch is in the ON position. DTC P2108 is set when the PCM detects an unexpected error in the ETS system. Case 1: PCM/ETS error in the communication line between the PCM to the ETS. Case 2: ETS/PCM error in the communication line between the ETS to the PCM. **Possible Causes:** • Faulty PCM
DTC: P2110 **1T CCM, MIL: Yes** **Years:** 2007, 2008 **Models:** Sonata **Engines:** 2.4L VIN C **Transmissions:** All	**Throttle Actuator Control (TAC) System Forced Limited RPM** DTC P2110 is set when the ECM detects an ETC system malfunction has occurred and the engine is in Limp-Home mode. The ECM limits engine speed to 1,500 RPM after setting the DTC, as an Emergency Operation (Limp-Home) mode. **Possible Causes:** • Throttle body is damaged or it has failed
DTC: P2111 **1T CCM, MIL: Yes** **Years:** 2006 **Models:** Santa Fe **Engines:** 3.5L VIN E **Transmissions:** All	**Throttle Actuator Control System Stuck Open** The ignition switch is in the ON position. The motor relay is ON. DTC P2111 is set when the throttle actuator control system valve is stuck open. **Possible Causes:** • Poor connector • Faulty throttle valve • Faulty ETS motor • Faulty PCM
DTC: P2112 **1T CCM, MIL: Yes** **Years:** 2006 **Models:** Santa Fe **Engines:** 3.5L VIN E **Transmissions:** All	**Throttle Actuator Control System Stuck Closed** The ignition switch is in the OFF position. DTC P2112 is set when the throttle actuator control system valve is stuck closed. **Possible Causes:** • Poor connector • Faulty throttle valve • Faulty ETS motor • Faulty PCM
DTC: P2118 **1T CCM, MIL: Yes** **Years:** 2006, 2007, 2008 **Models:** Santa Fe, Sonata **Engines:** 2.4L VIN C, 3.5L VIN E **Transmissions:** All	**Throttle Actuator Control Motor Current Range/Performance** The battery voltage is greater than 10 volts. No relevant faults are current. DTC P2118 is set when the ECM detects that the Pulse Width Modulation (PWM) signal of the TAC motor exceeds the threshold value. **Possible Causes:** • Open in control circuit • Poor connection or damaged harness • Faulty ETC motor
DTC: P2119 **1T CCM, MIL: No** **Years:** 2006, 2007, 2008 **Models:** Santa Fe, Sonata **Engines:** 2.4L VIN C, 3.5L VIN E **Transmissions:** All	**Throttle Actuator Control Throttle Body Range/Performance** DTC P2119 is set when the ECM detects that the TPS adaptation procedure is abnormal. **Possible Causes:** • Poor connection or damaged harness • Faulty ETC motor
DTC: P2122 **1T CCM, MIL: Yes** **Years:** 2006, 2007, 2008 **Models:** Azera, Entourage, Santa Fe, Sonata, Veracruz **Engines:** 2.4L VIN C, 2.7L VIN D, 3.3L VIN D, 3.3L VIN E, 3.3L VIN F, 3.5L VIN E, 3.8L VIN 3, 3.8L VIN C, 3.8L VIN F **Transmissions:** All	**Throttle/Pedal Position Sensor/Switch D Circuit Low Input** DTC P2122 is set when the PCM detects that the Accelerator Position Sensor (APS 1) output signals are below the predetermined threshold (less than 0.125 volt). **Possible Causes:** • Poor connection • Open or short to ground in the Power circuit • Open or short to ground in the Signal Circuit • Faulty APS • Faulty PCM

DTC	Trouble Code Title, Conditions & Possible Causes
DTC: P2123 **12T CCM, MIL: Yes** **Years:** 2006, 2007, 2008 **Models:** Azera, Entourage, Santa Fe, Sonata, Veracruz **Engines:** 2.4L VIN C, 2.7L VIN D, 3.3L VIN D, 3.3L VIN E, 3.3L VIN F, 3.5L VIN E, 3.8L VIN 3, 3.8L VIN C, 3.8L VIN F **Transmissions:** All	**Throttle/Pedal Position Sensor/Switch D Circuit High Input** DTC P2123 is set when the PCM detects that the output signals of the APP 1 are above the threshold (greater than 4.5 volts). **Possible Causes:** • Poor connection • Short to battery in signal circuit • Open in ground circuit • Faulty APS • Faulty PCM
DTC: P2125 **1T CCM, MIL: Yes** **Years:** 2006 **Models:** Santa Fe **Engines:** 3.5L VIN E **Transmissions:** All	**Acceleration Position Sensor 2 Circuit** The ETS/PCM communication is normal. The idle switch is ON. DTC P2125 is set when the PCM detects that the acceleration position sensor voltage readings are above or below the predetermined threshold. **Possible Causes:** • Poor connector • Faulty APS1 • Open or short in APS1 circuit • Faulty PCM
DTC: P2127 **1T CCM, MIL: Yes** **Years:** 2006, 2007, 2008 **Models:** Azera, Entourage, Santa Fe, Sonata, Veracruz **Engines:** 2.4L VIN C, 2.7L VIN D, 3.3L VIN D, 3.3L VIN E, 3.3L VIN F, 3.5L VIN E, 3.8L VIN 3, 3.8L VIN C, 3.8L VIN F **Transmissions:** All	**Throttle/Pedal Position Sensor/Switch E Circuit Low** The ignition switch is in the ON position. This code detects a continuous short to ground or open in either the circuit or the sensor. **Possible Causes:** • Poor connection • Open or short to ground in the Power circuit • Open or short to ground in the signal circuit • Faulty APS • Faulty PCM
DTC: P2128 **2T CCM, MIL: Yes** **Years:** 2006, 2007, 2008 **Models:** Azera, Entourage, Santa Fe, Sonata, Veracruz **Engines:** 2.4L VIN C, 2.7L VIN D, 3.3L VIN D, 3.3L VIN E, 3.3L VIN F, 3.5L VIN E, 3.8L VIN 3, 3.8L VIN C, 3.8L VIN F **Transmissions:** All	**Throttle/Pedal Position Sensor/Switch E Circuit High** The ignition is in the ON position. DTC P2128 is set when the PCM detects throttle position output signals about the threshold value (greater than 3 volts). **Possible Causes:** • Poor connection • Short to battery voltage in the Power circuit • Open in ground circuit • Faulty APS • Faulty PCM
DTC: P2135 **1T CCM, MIL: Yes** **Years:** 2006, 2007, 2008 **Models:** Azera, Entourage, Santa Fe, Sonata, Veracruz **Engines:** 2.7L VIN D, 3.3L VIN D, 3.3L VIN E, 3.3L VIN F, 3.5L VIN E, 3.8L VIN 3, 3.8L VIN C, 3.8L VIN F **Transmissions:** All	**Throttle/Pedal Position Sensor/Switch A/B Voltage Correlation** The ignition switch is in the ON position. DTC P2135 is set when the PCM detects that the output signals from TPS 1 and 2 are more than 4.5 percent different from each other for a specified number of times. **Possible Causes:** • Poor connection • Open or short in the TPS circuit • Faulty TPS • Faulty PCM
DTC: P2138 **2T CCM, MIL: Yes** **Years:** 2006, 2007, 2008 **Models:** Azera, Entourage, Santa Fe, Sonata, Veracruz **Engines:** 2.4L VIN C, 2.7L VIN D, 3.3L VIN D, 3.3L VIN E, 3.3L VIN F, 3.5L VIN E, 3.8L VIN 3, 3.8L VIN C, 3.8L VIN F **Transmissions:** All	**Throttle/Pedal Position Sensor/Switch D/E Voltage Correlation** The ignition switch is in the ON position. DTC P2138 is set when the PCM detects that the output signals from APS 1 and 2 are more than 4.5 percent different from each other for a specified number of times. **Possible Causes:** • Poor connection • Open or short in the APS circuit • Faulty APS • Faulty PCM

DTC	Trouble Code Title, Conditions & Possible Causes
DTC: P2159 **1T CCM, MIL: Yes** **Years:** 2007, 2008 **Models:** Sonata **Engines:** 2.4L VIN C **Transmissions:** All	**Vehicle Speed Sensor B Range/Performance** The engine is running at a speed greater than 2,100 RPM. The coolant temperature is greater than 140°F (60°C). The ECM evaluates the engine speed and mass air flow if there is no vehicle speed signal. This evaluation of both values will detect an open circuit or a short circuit error on the wheel speed sensor. DTC P2159 is set if there is no vehicle speed signal from the wheel speed sensor while both engine speed and mass air flow are higher than a predetermined threshold. **Possible Causes:** • Open or short in harness • Poor connection or damaged harness • VSS is faulty
DTC: P2173 **1T CCM, MIL: Yes** **Years:** 2006, 2007, 2008 **Models:** Azera, Entourage, Santa Fe, Sonata, Veracruz **Engines:** 2.7L VIN D, 3.3L VIN D, 3.3L VIN E, 3.3L VIN F, 3.8L VIN 3, 3.8L VIN C, 3.8L VIN F **Transmissions:** All	**Throttle Actuator Control System High Airflow Detected** The engine is running and no faults are current. DTC P2173 is set when comparing the real intake air flow to the intake air flow calculated by the ETS and the air flow is more than the threshold value for more than 19 seconds. **Possible Causes:** • Air leakage between the TP and MAF sensors • Faulty throttle body or intake manifold • Faulty MAF sensor
DTC: P2176 **1T CCM, MIL: Yes** **Years:** 2006 **Models:** Santa Fe **Engines:** 3.5L VIN E **Transmissions:** All	**Throttle Actuator Control System Stuck Closed** DTC P2176 is set when the PCM detects throttle actuator system is stuck closed. TPS output as the throttle valve is closed is less than 0.025 volt. **Possible Causes:** • Poor connector • Faulty throttle valve • Faulty ETS motor • Faulty PCM
DTC: P2187 **2T CCM, MIL: Yes** **Years:** 2006, 2007, 2008 **Models:** Azera, Entourage, Santa Fe, Sonata, Veracruz **Engines:** 2.4L VIN C, 2.7L VIN D, 3.3L VIN D, 3.3L VIN E, 3.3L VIN F, 3.8L VIN 3, 3.8L VIN C, 3.8L VIN F **Transmissions:** All	**Fuel Trim System Too Lean at Idle (Bank 1)** The engine has been running under idle state over 5 minutes. The engine coolant temperature is between 140°F (60°C)–239°F (115°C). No disabling faults are present (DTC's related to the HO2S, purge valve, or catalyst). DTC P2187 is set when the PCM detects the average fuel trim exceeds the limit for a predetermined time. **Possible Causes:** • Air leakage • Improper fuel pressure • PCV valve stuck • Clogging of injector • Leak in exhaust system • Faulty MAP, TPS, ECTS • Faulty front HO2S • Contaminated fuel • Faulty PCM
DTC: P2188 **2T CCM, MIL: Yes** **Years:** 2006, 2007, 2008 **Models:** Azera, Entourage, Santa Fe, Sonata, Veracruz **Engines:** 2.4L VIN C, 2.7L VIN D, 3.3L VIN D, 3.3L VIN E, 3.3L VIN F, 3.8L VIN 3, 3.8L VIN C, 3.8L VIN F **Transmissions:** All	**Fuel Trim System Too Rich at Idle (Bank 1)** The engine has been running under idle state over 5 minutes. The engine coolant temperature is between 140°F (60°C)–239°F (115°C). No disabling faults are present (DTC's related to the HO2S, purge valve, or catalyst). DTC P2188 is set when the PCM detects the average fuel trim exceeds the limit for a predetermined time. **Possible Causes:** • Faulty ignition system • EVAP PCSV malfunction • Faulty fuel injectors (leakage) • Leak in exhaust system • Faulty MAP, TPS, ECTS • Contaminated fuel • Improper fuel pressure • Faulty front HO2S • Faulty PCM

DTC	Trouble Code Title, Conditions & Possible Causes
DTC: P2189 **2T CCM, MIL: Yes** **Years:** 2006, 2007, 2008 **Models:** Azera, Entourage, Santa Fe, Sonata, Veracruz **Engines:** 2.7L VIN D, 3.3L VIN D, 3.3L VIN E, 3.3L VIN F, 3.8L VIN 3, 3.8L VIN C, 3.8L VIN F **Transmissions:** All	**Fuel Trim System Too Lean at Idle (Bank 2)** The engine has been running under idle state over 5 minutes. The engine coolant temperature is between 140°F (60°C)–239°F (115°C). No disabling faults are present (DTC's related to the HO2S, purge valve, or catalyst). DTC P2189 is set when the PCM detects the average fuel trim exceeds the limit for a predetermined time. **Possible Causes:** • Contaminated fuel • Air leakage • Improper fuel pressure • Clogging of injector • Heated Oxygen Sensor (HO2S)
DTC: P2190 **2T CCM, MIL: Yes** **Years:** 2006, 2007, 2008 **Models:** Azera, Entourage, Santa Fe, Sonata, Veracruz **Engines:** 2.7L VIN D, 3.3L VIN D, 3.3L VIN E, 3.3L VIN F, 3.8L VIN 3, 3.8L VIN C, 3.8L VIN F **Transmissions:** All	**Fuel Trim System Too Rich at Idle (Bank 2)** The engine has been running under idle state over 5 minutes. The engine coolant temperature is between 140°F (60°C)–239°F (115°C). No disabling faults are present (DTC's related to the HO2S, purge valve, or catalyst). DTC P2190 is set when the PCM detects the average fuel trim exceeds the limit for a predetermined time. **Possible Causes:** • Blocking of intake system • Fuel leakage in injector • Contaminated fuel • Improper fuel pressure • Heated Oxygen Sensor (HO2S)
DTC: P2191 **1T CCM, MIL: Yes** **Years:** 2007, 2008 **Models:** Sonata **Engines:** 2.4L VIN C **Transmissions:** All	**System Too Lean at Higher Load (Multiple) (Bank 1)** No relevant faults are present. Lambda adaptation is active. The engine coolant temperature is greater than 163°F (73°C). DTC P2191 is set when the lambda controller reaches the maximum or minimum threshold, then feedback control is no longer possible and emissions will be increased. The ECM sets DTC P2191 if no proportional post catalyst fuel trim adaptation occurs for a defined time after the lambda controller has reached its maximum threshold at partial load. **Possible Causes:** • Air leakage in intake, exhaust, or EVAP system • Faulty PCV system • Faulty sensor signals • Fuel system restriction (dirty fuel filter)
DTC: P2192 **1T CCM, MIL: Yes** **Years:** 2007, 2008 **Models:** Sonata **Engines:** 2.4L VIN C **Transmissions:** All	**System Too Rich at Higher Load (Bank 1)** No relevant faults are present. Lambda adaptation is active. The engine coolant temperature is greater than 163°F (73°C). DTC P2192 is set when the lambda controller reaches the maximum or minimum threshold, then feedback control is no longer possible and emissions will be increased. The ECM sets DTC P21912 if no proportional post catalyst fuel trim adaptation occurs for a defined time after the lambda controller has reached its maximum threshold at partial load. **Possible Causes:** • Air restriction in intake or exhaust system • Front HO2S or MAFS contamination • Faulty sensor signals • EVAP system • Fuel system
DTC: P2195 **2T CCM, MIL: Yes** **Years:** 2006, 2007, 2008 **Models:** Azera, Elantra, Entourage, Santa Fe, Sonata, Veracruz **Engines:** 2.0L VIN D, 2.7L VIN D, 3.3L VIN D, 3.3L VIN E, 3.3L VIN F, 3.8L VIN 3, 3.8L VIN C, 3.8L VIN F **Transmissions:** All	**HO2S-11 Signal Biased/Stuck Lean (Bank 1/Sensor 1)** The engine is running at least 60 seconds. The battery voltage is at or above 10 volts. DTC P2195 is set when the PCM detects the HO2S output signal is lean during power enrichment conditions. **Possible Causes:** • Damaged wiring between the HO2S and the ECM • Terminal corrosion or water intrusion in the HO2S harness connectors • Clogging of fuel filter in fuel pump • Faulty HO2S • PCM has failed
DTC: P2196 **2T CCM, MIL: Yes** **Years:** 2006, 2007, 2008 **Models:** Azera, Elantra, Entourage, Santa Fe, Sonata, Veracruz **Engines:** 2.0L VIN D, 2.7L VIN D, 3.3L VIN D, 3.3L VIN E, 3.3L VIN F, 3.8L VIN 3, 3.8L VIN C, 3.8L VIN F **Transmissions:** All	**HO2S-11 Signal Biased/Stuck Rich (Bank 1/Sensor 1)** The engine is running at least 60 seconds. The battery voltage is at or above 10 volts. DTC P2196 is set when the PCM detects the HO2S signal is rich during fuel cut-off conditions. **Possible Causes:** • Damaged wiring between the HO2S and the PCM • Terminal corrosion or water intrusion in the HO2S harness connectors • Exhaust leaks • Faulty HO2S • PCM has failed

DTC	Trouble Code Title, Conditions & Possible Causes
DTC: P2197 **2T CCM, MIL: Yes** **Years:** 2006, 2007, 2008 **Models:** Azera, Entourage, Santa Fe, Sonata, Veracruz **Engines:** 2.7L VIN D, 3.3L VIN D, 3.3L VIN E, 3.3L VIN F, 3.8L VIN 3, 3.8L VIN C, 3.8L VIN F **Transmissions:** All	**HO2S-21 Signal Biased/Stuck Lean (Bank 2/Sensor 1)** The engine is running at least 60 seconds. The battery voltage is at or above 10 volts. DTC P2197 is set when the PCM detects the output signals from the HO2S are lean during power enrichment conditions. **Possible Causes:** • Poor Connection • Faulty HO2S • Clogging of fuel filter in fuel pump • Faulty PCM
DTC: P2198 **2T CCM, MIL: Yes** **Years:** 2006, 2007, 2008 **Models:** Azera, Entourage, Santa Fe, Sonata, Veracruz **Engines:** 2.7L VIN D, 3.3L VIN D, 3.3L VIN E, 3.3L VIN F, 3.8L VIN 3, 3.8L VIN C, 3.8L VIN F **Transmissions:** All	**HO2S-21 Signal Biased/Stuck Rich (Bank 2/Sensor 1)** The engine is running at least 60 seconds. The battery voltage is at or above 10 volts. DTC P2198 is set when the PCM detects the HO2S signal is rich during fuel cut-off conditions. **Possible Causes:** • Poor Connection • Faulty HO2S • Faulty PCM
DTC: P2226 **2T CCM, MIL: Yes** **Years:** 2006, 2007, 2008 **Models:** Accent **Engines:** 1.6L VIN C **Transmissions:** All	**Barometric Pressure (BARO) Sensor Circuit** A barometric pressure sensor is installed within the ECM to permit altitude corrections to be applied to fuel injection quantity calculations. If the change of ambient pressure value is over 50 hPa during 20 seconds or more, PCU sets DTC P2226. **Possible Causes:** • Clog at the sensing hole • Faulty ECM
DTC: P2227 **2T CCM, MIL: Yes** **Years:** 2006, 2007, 2008 **Models:** Accent **Engines:** 1.6L VIN C **Transmissions:** All	**Barometric Pressure (BARO) Sensor Circuit Range/Performance** A barometric pressure sensor is installed within the ECM to permit altitude corrections to be applied to fuel injection quantity calculations. If the output of the ambient pressure sensor is shown to be an abnormal value, the PCU sets DTC P2227. **Possible Causes:** • Clog at the sensing hole • Faulty ECM
DTC: P2228 **2T CCM, MIL: Yes** **Years:** 2006, 2007, 2008 **Models:** Accent **Engines:** 1.6L VIN C **Transmissions:** All	**Barometric Pressure (BARO) Sensor Circuit Low Input** A barometric pressure sensor is installed within the ECM to permit altitude corrections to be applied to fuel injection quantity calculations. If the output of the ambient pressure sensor is below 0.2 volts, the PCU sets DTC P2228. **Possible Causes:** • 5-volt reference circuit open short to ground or a high resistance • BARO signal circuit short to ground • Poor connection at the BARO sensor • Poor connection at the ECM • Faulty BARO sensor • ECM has failed
DTC: P2229 **2T CCM, MIL: Yes** **Years:** 2006, 2007, 2008 **Models:** Accent **Engines:** 1.6L VIN C **Transmissions:** All	**Barometric Pressure (BARO) Sensor Circuit High Input** A barometric pressure sensor is installed within the ECM to permit altitude corrections to be applied to fuel injection quantity calculations. If the output of the ambient pressure sensor is over 4.8 volts, the PCU sets DTC P2229. **Possible Causes:** • BARO signal circuit open or short to voltage • BARO low reference circuit open or high resistance • 5-volt reference circuit short to voltage • Poor connection at the BARO sensor • Poor connection at the ECM • Faulty BARO sensor • ECM has failed
DTC: P2231 **2T CCM, MIL: Yes** **Years:** 2006, 2007, 2008 **Models:** Elantra **Engines:** 2.0L VIN D **Transmissions:** All	**HO2S-11 Sensor Signal Circuit Shorted to Heater Circuit (Bank 1/Sensor 1)** The engine is running in a stable driving condition. No lambda controller stop mode. The exhaust temperature is greater than 752°F (400°C). No relevant faults are present. The battery voltage is between 11–16 volts. DTC P2231 is set if the ECM detects that the front HO2S heater control circuit is shorted to the output circuit. **Possible Causes:** • Short in heater control circuit • Poor connection or damaged harness

DTC	Trouble Code Title, Conditions & Possible Causes
DTC: P2232 **2T CCM, MIL: Yes** **Years:** 2006, 2007, 2008 **Models:** Accent **Engines:** 1.6L VIN C **Transmissions:** All	**HO2S-12 Signal Circuit Shorted to Heater Circuit (Bank 1/Sensor 2)** HO2S-12 is in the rear side of Catalytic Converter to check the proper operation of the catalyst. Oxygen density after the catalytic converter has to be within a specific range (around 0.5 volt when there is no acceleration or deceleration). If the oxygen density changes in accordance with HO2S-12, it is an indication of catalytic converter poor performance. If the counter that records the rapid signal voltage changes is greater than 5, DTC P2232 is set. **Possible Causes:** • Poor connection • Short to power in the signal circuit • Faulty HO2S • ECM has failed
DTC: P2237 **2T CCM, MIL: Yes** **Years:** 2006, 2007, 2008 **Models:** Elantra **Engines:** 2.0L VIN D **Transmissions:** All	**HO2S-11 Pumping Current Control Circuit/Open (Bank 1/Sensor 1)** No relevant faults are present. The ECM monitors the front HO2S tip temperature and signal at the same time to check for an open circuit in the Pumping Current Circuit. The PCM sets DTC P2237 if the front O2 tip temperature is higher than the predetermined threshold and the front HO2S signal is unchanged. **Possible Causes:** • Poor connection or damaged harness • Open in HO2S control circuit • Faulty HO2S • ECM has failed
DTC: P2243 **2T CCM, MIL: Yes** **Years:** 2006, 2007, 2008 **Models:** Elantra **Engines:** 2.0L VIN D **Transmissions:** All	**HO2S-11 Reference Voltage Circuit/Open (Bank 1/Sensor 1)** The battery voltage is between 11–16 volts. The ECM monitors the front O2 sensor control circuit and sets DTC P2243 if the front O2 sensor signal is unchanged. **Possible Causes:** • Poor connection or damaged harness • Open in HO2S control circuit • Faulty HO2S • ECM has failed
DTC: P2251 **2T CCM, MIL: Yes** **Years:** 2006, 2007, 2008 **Models:** Elantra **Engines:** 2.0L VIN D **Transmissions:** All	**HO2S-11 Sensor Reference Ground Circuit/Open (Bank 1/Sensor 1)** The ECM monitors the front O2 sensor signal level during the fuel cut-off operation. This signal monitoring will determine if there is an open circuit in the Reference Ground Circuit. The PCM sets DTC P2251 if the front O2 signal during fuel cut-off is higher than a predetermined threshold. **Possible Causes:** • Poor connection or damaged harness • Open in HO2S control circuit • Faulty HO2S • ECM has failed
DTC: P2252 **2T CCM, MIL: Yes** **Years:** 2006 **Models:** Santa Fe **Engines:** 2.4L VIN B, 3.5L VIN E **Transmissions:** All	**HO2S Reference Ground Circuit Low** The engine is running for more than 2 seconds. DTC P2252 is set when an electrical check finds the ground circuit low (voltage is less than 0.4 volt). **Possible Causes:** • Poor connection or damaged harness • Open or short to ground in reference ground circuit • Faulty HO2S sensor
DTC: P2253 **2T CCM, MIL: Yes** **Years:** 2006 **Models:** Santa Fe **Engines:** 2.4L VIN B, 3.5L VIN E **Transmissions:** All	**HO2S Reference Ground Circuit High** The engine is running for more than 2 seconds. DTC P22523 is set when an electrical check finds the ground circuit high (voltage is greater than 3.7 volts). **Possible Causes:** • Poor connection or damaged harness • Short to battery in reference ground circuit • Faulty HO2S sensor
DTC: P2270 **2T CCM, MIL: Yes** **Years:** 2006, 2007, 2008 **Models:** Azera, Entourage, Santa Fe, Sonata, Veracruz **Engines:** 2.0L VIN D, 2.7L VIN D, 3.3L VIN D, 3.3L VIN E, 3.3L VIN F, 3.8L VIN 3, 3.8L VIN C, 3.8L VIN F **Transmissions:** All	**HO2S-12 Signal Stuck Lean (Bank 1/Sensor 2)** The engine is running for at least 60 seconds. The battery voltage is at or above 10 volts. DTC P2270 is set when the PCM detects the HO2S signal is lean during power enrichment conditions. **Possible Causes:** • Poor connection or damaged harness • Faulty HO2S • Clogging of fuel filter in fuel pump • PCM has failed

DTC	Trouble Code Title, Conditions & Possible Causes
DTC: P2271 **2T CCM, MIL: Yes** **Years:** 2006, 2007, 2008 **Models:** Azera, Elantra, Entourage, Santa Fe, Sonata, Veracruz **Engines:** 2.0L VIN D, 2.7L VIN D, 3.3L VIN D, 3.3L VIN E, 3.3L VIN F, 3.8L VIN 3, 3.8L VIN C, 3.8L VIN F **Transmissions:** All	**HO2S-12 Signal Stuck Rich (Bank 1/Sensor 2)** The engine is running for at least 60 seconds. The battery voltage is at or above 10 volts. DTC P2271 is set when the PCM detects the HO2S signal is rich during fuel cut-off conditions. **Possible Causes:** • Poor connection or damaged harness • Faulty HO2S • PCM has failed
DTC: P2272 **2T CCM, MIL: Yes** **Years:** 2006, 2007, 2008 **Models:** Azera, Entourage, Santa Fe, Sonata, Veracruz **Engines:** 2.7L VIN D, 3.3L VIN D, 3.3L VIN E, 3.3L VIN F, 3.8L VIN 3, 3.8L VIN C, 3.8L VIN F **Transmissions:** All	**HO2S-22 Signal Stuck Lean (Bank 2/Sensor 2)** The engine is running for at least 60 seconds. The battery voltage is at or above 10 volts. DTC P2272 is set when the PCM detects the HO2S signal is lean during power enrichment conditions. **Possible Causes:** • Poor connection or damaged harness • Faulty HO2S • Clogging of fuel filter in fuel pump • PCM has failed
DTC: P2273 **2T CCM, MIL: Yes** **Years:** 2006, 2007, 2008 **Models:** Azera, Entourage, Santa Fe, Sonata, Veracruz **Engines:** 2.7L VIN D, 3.3L VIN D, 3.3L VIN E, 3.3L VIN F, 3.8L VIN 3, 3.8L VIN C, 3.8L VIN F **Transmissions:** All	**HO2S-22 Signal Stuck Rich (Bank 2/Sensor 2)** The engine is running for at least 60 seconds. The battery voltage is at or above 10 volts. DTC P2273 is set when the PCM detects the HO2S signal is rich during fuel cut-off conditions. **Possible Causes:** • Poor connection or damaged harness • Faulty HO2S • PCM has failed
DTC: P2414 **2T CCM, MIL: Yes** **Years:** 2006, 2007, 2008 **Models:** Elantra **Engines:** 2.0L VIN D **Transmissions:** All	**HO2S-11 Sensor Exhaust Sample Error (Bank 1/Sensor 1)** The battery voltage is between 11–16 volts. The sensor tip temperature is greater than 1,202°F (650°C). No relevant fault is present. The ECM monitors the front HO2S signal level during partial load or full load operation to check if the front HO2S is mechanically attached to the exhaust gas pipe while the vehicle is running. The ECM sets DTC P2414 if the front HO2S signal is within the predetermined threshold. **Possible Causes:** • Poor connection or damaged harness • Faulty HO2S • PCM has failed
DTC: P2422 **2T CCM, MIL: Yes** **Years:** 2006, 2007, 2008 **Models:** Azera, Entourage, Santa Fe, Sonata, Veracruz **Engines:** 2.4L VIN B, 2.7L VIN D, 3.3L VIN D, 3.3L VIN E, 3.3L VIN F, 3.5L VIN E, 3.8L VIN 3, 3.8L VIN C, 3.8L VIN F **Transmissions:** All	**Evaporative Emission (EVAP) System Canister Clogging** The battery voltage is between 10–16 volts. The barometric pressure is greater than 72 kPa. The engine coolant temperature at startup is 40–95°F (5–35°C). The intake air temperature at startup is 40–95°F (5–35°C). The fuel level is 15–85 percent. While checking the output signals from the fuel tank pressure sensor at purging, if the fuel tank vacuum is higher than a prescribed threshold, the PCM sets DTC P2422. **Possible Causes:** • Faulty canister close valve • Clogging of canister air filter
DTC: P2507 **1T CCM, MIL: Yes** **Years:** 2006, 2007, 2008 **Models:** Azera, Entourage, Santa Fe, Sonata, Veracruz **Engines:** 2.7L VIN D, 3.3L VIN D, 3.3L VIN E, 3.3L VIN F, 3.8L VIN 3, 3.8L VIN C, 3.8L VIN F **Transmissions:** All	**PCM/PCM Input Signal Low** When the ignition switch is turned ON, battery voltage is applied from the battery to the PCM through the main relay. When the ignition switch is turned OFF, the PCM is supplied with power through the battery power input line to control the basic operation of vehicle. If the battery power input line has a problem, the PCM sets DTC P2507. **Possible Causes:** • Poor connection or damaged harness • Open or short to ground in line • PCM has failed

DTC	Trouble Code Title, Conditions & Possible Causes
DTC: P2610 **2T CCM, MIL: Yes** **Years:** 2006, 2007, 2008 **Models:** Azera, Entourage, Santa Fe, Sonata, Veracruz **Engines:** 2.7L VIN D, 3.3L VIN D, 3.3L VIN E, 3.3L VIN F, 3.8L VIN 3, 3.8L VIN C, 3.8L VIN F **Transmissions:** All	**ECM/PCM-Engine Off Timer Performance** The engine is running for greater than 10 seconds with a battery voltage above 8 volts. No memory failure has occurred. The PCM continues to calculate data of several sensors despite turning the ignition OFF. When the ignition turns ON, the PCM uses the calculated data for ease of vehicle start. If abnormal engine off timer performance is detected for a calibrated time, the PCM sets DTC P2610. **Possible Causes:** • Poor battery condition • Connecting condition • PCM has failed
DTC: P2626 **2T CCM, MIL: Yes** **Years:** 2006, 2007, 2008 **Models:** Elantra **Engines:** 2.0L VIN D **Transmissions:** All	**HO2S-11 Sensor Pumping Current Trim Circuit/Open (Bank 1/Sensor 1)** No relevant faults are present. The ECM monitors the front O2 sensor control circuit and sets DTC P2243 if the front O2 sensor signal is stuck. **Possible Causes:** • Poor connection or damaged harness • Open in HO2S control circuit • Faulty HO2S
DTC: P2762 **2T CCM, MIL: Yes** **Years:** 2007, 2008 **Models:** Veracruz **Engines:** 3.8L VIN C **Transmissions:** All	**Torque Converter Clutch (TCC) Pressure Control Solenoid System Range/Performance** The engine is running at a speed greater than 400 RPM. The battery voltage is 10.2–15.5 volts. The TCU communication is normal. The PCM/TCM increases the duty ratio to engage the Damper Clutch by monitoring slip RPM (difference value between engine speed and turbine speed). To decrease the slip of the Damper Clutch, the PCM/TCM increases the duty ratio by applying more hydraulic pressure. When slip RPM does not drop under some value with a 100 percent duty ratio, the PCM/TCM determines that the Torque Converter Clutch is stuck OFF and sets DTC P2762. **Possible Causes:** • TCC PC Solenoid open, short to volts, short to ground • The Control Solenoid connectors and pins have the following conditions: damage, bent pins, debris, broken retaining tab, and/or contamination • Faulty solenoid valve (SLU) • Faulty TCU
DTC: P2763 **2T CCM, MIL: Yes** **Years:** 2007, 2008 **Models:** Veracruz **Engines:** 3.8L VIN C **Transmissions:** All	**Torque Converter Clutch (TCC) Pressure Control Solenoid Control Circuit Low (Short to GND or Open)** The engine is running at a speed greater than 400 RPM. The battery voltage is 10.2–15.5 volts. The TCU communication is normal. The PCM/TCM increases the duty ratio to engage the Damper Clutch by monitoring slip RPM (difference value between engine speed and turbine speed). To decrease the slip of the Damper Clutch, the PCM/TCM increases the duty ratio by applying more hydraulic pressure. When slip RPM does not drop under some value with a 100 percent duty ratio, the PCM/TCM determines that the Torque Converter Clutch is stuck OFF and sets DTC P2763. **Possible Causes:** • TCC PC Solenoid open or short to ground • The Control Solenoid connectors and pins have the following conditions: damage, bent pins, debris, broken retaining tab, and/or contamination • Faulty solenoid valve (SLU) • Faulty TCU
DTC: P2764 **2T CCM, MIL: Yes** **Years:** 2007, 2008 **Models:** Veracruz **Engines:** 3.8L VIN C **Transmissions:** All	**Torque Converter Clutch (TCC) Pressure Control Solenoid Control Circuit High (Short to +B)** The engine is running at a speed greater than 400 RPM. The battery voltage is 10.2–15.5 volts. The TCU communication is normal. The PCM/TCM increases the duty ratio to engage the Damper Clutch by monitoring slip RPM (difference value between engine speed and turbine speed). To decrease the slip of the Damper Clutch, the PCM/TCM increases the duty ratio by applying more hydraulic pressure. When slip RPM does not drop under some value with a 100 percent duty ratio, the PCM/TCM determines that the Torque Converter Clutch is stuck OFF and sets DTC P2764. **Possible Causes:** • TCC PC Solenoid short to voltage • The Control Solenoid connectors and pins have the following conditions: damage, bent pins, debris, broken retaining tab, and/or contamination • Faulty solenoid valve (SLB1) • Faulty TCU
DTC: P2A00 **2T CCM, MIL: Yes** **Years:** 2006, 2007, 2008 **Models:** Azera, Entourage, Santa Fe, Sonata, Veracruz **Engines:** 2.7L VIN D, 3.3L VIN D, 3.3L VIN E, 3.3L VIN F, 3.8L VIN 3, 3.8L VIN C, 3.8L VIN F **Transmissions:** All	**HO2S-11 Not Ready (Bank 1/Sensor 1)** The engine is running at idle for more than 20 seconds. DTC P2A00 is set when the PCM detects that the HO2S is not ready. **Possible Causes:** • Poor connection • HO2S has failed • PCM has failed

DTC	Trouble Code Title, Conditions & Possible Causes
DTC: P2A01 **2T CCM, MIL: Yes** **Years:** 2006, 2007, 2008 **Models:** Azera, Entourage, Santa Fe, Sonata, Veracruz **Engines:** 2.7L VIN D, 3.3L VIN D, 3.3L VIN E, 3.3L VIN F, 3.8L VIN 3, 3.8L VIN C, 3.8L VIN F **Transmissions:** All	**HO2S-12 Not Ready (Bank 1/Sensor 2)** The engine is running and no other disabling faults are present. While checking output signals from the HO2S during deceleration fuel cut-off, DTC P2A01 is set if the HO2S response time is too long. **Possible Causes:** • Poor connection • HO2S has failed • PCM has failed
DTC: P2A03 **2T CCM, MIL: Yes** **Years:** 2006, 2007, 2008 **Models:** Azera, Entourage, Santa Fe, Sonata, Veracruz **Engines:** 2.7L VIN D, 3.3L VIN D, 3.3L VIN E, 3.3L VIN F, 3.8L VIN 3, 3.8L VIN C, 3.8L VIN F **Transmissions:** All	**HO2S-12 Not Ready (Bank 1/Sensor 2)** The engine is running for more than 20 seconds. DTC P2A03 is set when the PCM detects that the HO2S is not ready. **Possible Causes:** • Poor connection • HO2S has failed • PCM has failed
DTC: P2A04 **2T CCM, MIL: Yes** **Years:** 2006, 2007, 2008 **Models:** Azera, Entourage, Santa Fe, Sonata, Veracruz **Engines:** 2.7L VIN D, 3.3L VIN D, 3.3L VIN E, 3.3L VIN F, 3.8L VIN 3, 3.8L VIN C, 3.8L VIN F **Transmissions:** All	**HO2S-22 Not Ready (Bank 2/Sensor 2)** The engine is running and in deceleration fuel cut-off state. No other disabling faults are present. DTC P2A04 is set when the PCM detects the HO2S response time is too long during deceleration fuel cut-off. **Possible Causes:** • Poor connection • HO2S has failed • PCM has failed

OBD II Trouble Code List (U1xxx Codes)

DTC	Trouble Code Title, Conditions & Possible Causes
DTC: U0001 **2T ECM, MIL: Yes** **Years:** 2006, 2007, 2008 **Models:** Accent, Azera, Elantra, Entourage, Santa Fe, Sonata, Tiburon, Veracruz **Engines:** 1.6L VIN C, 2.0L VIN D, 2.4L VIN C, 2.7L VIN D, 3.3L VIN D, 3.3L VIN E, 3.3L VIN F, 3.5L VIN E, 3.8L VIN 3, 3.8L VIN C, 3.8L VIN F **Transmissions:** All	**High Speed CAN Communication Bus** The engine is running for more than 2 seconds with the ignition voltage at or above 11 volts. DTC U0001 is set when the PCM detects a failure in communication between the PCM and other modules in the vehicle which are on the CAN serial bus. The condition exists for more than 1.5 seconds. **Possible Causes:** • CAN BUS • CAN Communication module component
DTC: U0100 **1T ECM, MIL: Yes** **Years:** 2006, 2007, 2008 **Models:** Elantra, Santa Fe, Sonata, Veracruz **Engines:** 2.0L VIN D, 2.4L VIN C, 2.7L VIN D, 3.3L VIN E, 3.8L VIN C **Transmissions:** All	**Lost Communication With ECM/PCM "A"** Voltage supplied to the modules is in the normal operating voltage range. The vehicle power mode requires serial data communication to occur. DTC U0100 is set when a supervised periodic message that includes the transmitter module availability has not been received. When the DTC is set, the module uses a default value for the missing parameter. **Note: If more than one module is not communicating, use Data Communication Schematics to determine which module is closest to the Data Link Connector (DLC). Start diagnostics with that module.** **Possible Causes:** • Short or open in CAN communication circuit • Poor module connections • ECM input/output voltage and ground circuits • Faulty ECU • Faulty TCU

DTC	Trouble Code Title, Conditions & Possible Causes
DTC: U0101 **1T ECM, MIL: Yes** **Years:** 2006, 2007, 2008 **Models:** Accent, Elantra, Santa Fe, Tiburon, Tucson **Engines:** 1.6L VIN C, 2.0L VIN D, 2.7L VIN D, 2.7L VIN F, 3.3L VIN E **Transmissions:** All	**Lost Communication With Transmission Control Module (TCM)** Voltage supplied to the modules is in the normal operating voltage range. The vehicle power mode requires serial data communication to occur. DTC U0101 is set when a supervised periodic message that includes the transmitter module availability has not been received. When the DTC is set, the module uses a default value for the missing parameter. **Note: If more than one module is not communicating, use Data Communication Schematics to determine which module is closest to the Data Link Connector (DLC). Start diagnostics with that module.** **Possible Causes:** • Poor connection or damaged harness • Faulty module in communication link • Faulty ECM
DTC: U0121 **1T ECM, MIL: Yes** **Years:** 2006, 2007, 2008 **Models:** Santa Fe, Veracruz **Engines:** 2.7L VIN D, 3.3L VIN E, 3.8L VIN C **Transmissions:** All	**Lost Communication With Anti-Lock Brake System (ABS) Control Module** Voltage supplied to the modules is in the normal operating voltage range. The vehicle power mode requires serial data communication to occur. TCU communication is normal with the ABS ECU. No faults are current in the system. DTC U0121 is set when a supervised periodic message that includes the transmitter module availability has not been received. When the DTC is set, the module uses a default value for the missing parameter. **Possible Causes:** • Faulty module in communication link • Short or open in CAN communication circuit • Faulty ABS ECU • Faulty TCU
DTC: U0122 **1T ECM, MIL: Yes** **Years:** 2006, 2007, 2008 **Models:** Santa Fe, Veracruz **Engines:** 2.7L VIN D, 3.3L VIN E, 3.8L VIN C **Transmissions:** All	**Lost Communication With Vehicle Dynamics Control Module** Voltage supplied to the modules is in the normal operating voltage range. The vehicle power mode requires serial data communication to occur. TCU communication is normal with the ABS ECU. No faults are current in the system. DTC U0122 is set when a supervised periodic message that includes the transmitter module availability has not been received. When the DTC is set, the module uses a default value for the missing parameters. **Note: If more than one module is not communicating, use Data Communication Schematics to determine which module is closest to the Data Link Connector (DLC). Start diagnostics with that module.** **Possible Causes:** • Faulty module in communication link • Short or open in CAN communication circuit • Faulty ESP ECU • Faulty TCU
DTC: U0126 **1T ECM, MIL: Yes** **Years:** 2007, 2008 **Models:** Santa Fe **Engines:** 2.7L VIN D, 3.3L VIN E **Transmissions:** All	**Lost Communication With Steering Angle Sensor Module** This code is related to the communication line between the ECU and the TCCU. DTC U0126 is set when the CAN signal from the ECU is not received for more than 1 second or an ECU internal error occur. When the DTC is set, the TCCU prohibits the ITM control. **Possible Causes:** • CAN HIGH/LOW circuit open/short • Faulty SAS (ESP only) • Faulty TCCU

SPECIFICATIONS AND MAINTENANCE CHARTS

ENGINE AND VEHICLE IDENTIFICATION

			Engine					Model Year	
Code ①	Liters (cc)	Cu. In.	Cyl.	Fuel Sys.	Engine Type	Eng. Mfg.		Code ②	Year
VQ35DE	3.5 (3498)	213.5	6	SFI	DOHC	Nissan		6	2006
VK45DE	4.5 (4494)	274.2	8	SFI	DOHC	Nissan		7	2007
								8	2008

SFI: Sequential Fuel Injection

DOHC: Double Overhead Camshaft

① Stamped on the upper rear of the engine block, just behind one of the cylinder heads

② 10th digit of the Vehicle Identification Number (VIN)

22140_FX35_C0001

GENERAL ENGINE SPECIFICATIONS

All measurements are given in inches.

Year	Model	Engine Displacement Liters	Engine Series ID/VIN	Net Horsepower @ rpm	Net Torque @ rpm (ft. lbs.)	Bore x Stroke (in.)	Compression Ratio	Oil Pressure @ rpm
2006	FX35	3.5	VQ35DE	275@6200	268@4800	3.76 x 3.21	10.3:1	43@2000
	FX45	4.5	VK45DE	320@6000	335@4000	3.66 x 3.26	10.5:1	43@2000
2007	FX35	3.5	VQ35DE	275@6200	268@4800	3.76 x 3.21	10.3:1	43@2000
	FX45	4.5	VK45DE	320@6000	335@4000	3.66 x 3.26	10.5:1	43@2000
2008	FX35	3.5	VQ35DE	275@6200	268@4800	3.76 x 3.21	10.3:1	43@2000
	FX45	4.5	VK45DE	320@6000	335@4000	3.66 x 3.26	10.5:1	43@2000

22140_FX35_C0002

GASOLINE ENGINE TUNE-UP SPECIFICATIONS

Year	Engine Displacement Liters	Engine Series ID/VIN	Spark Plug Gap (in.)	Ignition Timing (deg.) ①		Fuel Pump (psi)	Idle Speed (rpm)		Valve Clearance (in.) ②	
				MT	AT		MT	AT	Intake	Exhaust
2006	3.5	VQ35DE	0.043	—	10-20B	51	—	600-700	0.010-0.013	0.011-0.015
	4.5	VK45DE	0.043	—	7-17B	51	—	600-700	0.010-0.013	0.011-0.015
2007	3.5	VQ35DE	0.043	—	10-20B	51	—	600-700	0.010-0.013	0.011-0.015
	4.5	VK45DE	0.043	—	7-17B	51	—	600-700	0.010-0.013	0.011-0.015
2008	3.5	VQ35DE	0.043	—	10-20B	51	—	600-700	0.010-0.013	0.011-0.015
	4.5	VK45DE	0.043	—	7-17B	51	—	600-700	0.010-0.013	0.011-0.015

NOTE: The Vehicle Emission Control Information label reflects specification changes made during production.

Follow the figures on the label if they differ from those in this chart.

B: Before top dead center

① With terminals TC and CG of DLC3 connected

② Engine cold - approximately 68°F (20°C)

22140_FX35_C0003

CAPACITIES

Year	Model	Engine Displacement Liters	Engine Series ID/VIN	Engine Oil with Filter (qts.)	Transmission (pts.) Manual	Transmission (pts.) Auto. ①	Transfer Case (pts.)	Drive Axle Front (pts.)	Drive Axle Rear (pts.)	Fuel Tank (gal.)	Cooling System (qts.)
2006	FX35	3.5	VQ35DE	5.0	—	21.8	2.6	1.4	3.0	23.8	9.1
	FX45	4.5	VK45DE	6.1	—	21.8	2.6	1.4	3.0	23.8	10.6
2007	FX35	3.5	VQ35DE	5.0	—	21.8	2.6	1.4	3.0	23.8	9.1
	FX45	4.5	VK45DE	6.1	—	21.8	2.6	1.4	3.0	23.8	10.6
2008	FX35	3.5	VQ35DE	5.0	—	21.8	2.6	1.4	3.0	23.8	9.1
	FX45	4.5	VK45DE	6.1	—	21.8	2.6	1.4	3.0	23.8	10.6

NOTE: All capacities are approximate. Add fluid gradually and check to be sure a proper fluid level is obtained.

① Drain and refill

22140_FX35_C0004

FLUID SPECIFICATIONS

Year	Model	Engine Displacement Liters	Engine Series ID/VIN	Engine Oil	Manual Trans.	Auto. Trans.	Drive Axle	Transfer Case	Power Steering Fluid	Brake Master Cylinder	Cooling System
2006	FX35	3.5	VQ35DE	5W-30	—	①	②	③	④	⑤	⑥
	FX45	4.5	VK45DE	5W-30	—	①	②	③	④	⑤	⑥
2007	FX35	3.5	VQ35DE	5W-30	—	①	②	③	④	⑤	⑥
	FX45	4.5	VK45DE	5W-30	—	①	②	③	④	⑤	⑥
2008	FX35	3.5	VQ35DE	5W-30	—	①	②	③	④	⑤	⑥
	FX45	4.5	VK45DE	5W-30	—	①	②	③	④	⑤	⑥

DOT: Department Of Transportation

① Genuine NISSAN Matic J ATF

② Genuine NISSAN Differential Oil Hypoid Super GL-5 80W-90 or API GL-5 Viscosity SAE 80W-90

③ Genuine NISSAN Matic D ATF (Continental U.S. and Alaska) or Canada NISSAN Automatic Transmission Fluid or equivalent (if available)

④ Genuine NISSAN PSF or equivalent

⑤ Genuine NISSAN Super Heavy Duty Brake Fluid or equivalent DOT 3

⑥ Genuine NISSAN Long Life Antifreeze/Coolant or equivalent

22140_FX35_C0005

VALVE SPECIFICATIONS

Year	Engine Displacement Liters	Engine Series ID/VIN	Seat Angle (deg.)	Face Angle (deg.)	Spring Test Pressure (lbs. @ in.)	Spring Installed Height (in.)	Stem-to-Guide Clearance (in.)		Stem Diameter (in.)	
							Intake	Exhaust	Intake	Exhaust
2006	3.5	VQ35DE	45.15-45.45	44.23-45.08	84-95 @1.071	1.457	0.0008-0.0021	0.0012-0.0025	0.2348-0.2354	0.2344-0.2350
	4.5	VK45DE	45.15-45.45	44.23-45.08	65-74 @0.961	1.331	0.0008-0.0018	0.0012-0.0022	0.2351-0.2354	0.2347-0.2350
2007	3.5	VQ35DE	45.15-45.45	44.23-45.08	84-95 @1.071	1.457	0.0008-0.0021	0.0012-0.0025	0.2348-0.2354	0.2344-0.2350
	4.5	VK45DE	45.15-45.45	44.23-45.08	65-74 @0.961	1.331	0.0008-0.0018	0.0012-0.0022	0.2351-0.2354	0.2347-0.2350
2008	3.5	VQ35DE	45.15-45.45	44.23-45.08	84-95 @1.071	1.457	0.0008-0.0021	0.0012-0.0025	0.2348-0.2354	0.2344-0.2350
	4.5	VK45DE	45.15-45.45	44.23-45.08	65-74 @0.961	1.331	0.0008-0.0018	0.0012-0.0022	0.2351-0.2354	0.2347-0.2350

22140_FX35_C0006

CAMSHAFT AND BEARING SPECIFICATIONS CHART
All measurements are given in inches.

Year	Engine Displ. Liters	Engine Series ID/VIN	Journal Dia.	Brg. Oil Clearance	Shaft End-play	Runout	Journal Bore	Lobe Height	
								Intake	Exhaust
2006	3.5	VQ35DE	①	②	0.0045-0.0074	0.0020 max.	NA	1.7633-1.7738	1.7633-1.7738
	4.5	VK45DE	③	④	0.0045-0.0074	0.0010 max.	NA	1.7633-1.7738	1.7293-1.7368
2007	3.5	VQ35DE	①	②	0.0045-0.0074	0.0020 max.	NA	1.7633-1.7738	1.7633-1.7738
	4.5	VK45DE	③	④	0.0045-0.0074	0.0010 max.	NA	1.7633-1.7738	1.7293-1.7368
2008	3.5	VQ35DE	①	②	0.0045-0.0074	0.0020 max.	NA	1.7633-1.7738	1.7633-1.7738
	4.5	VK45DE	③	④	0.0045-0.0074	0.0010 max.	NA	1.7633-1.7738	1.7293-1.7368

NA: Not Available

① No. 1: 1.0211-1.0218 in.
  No. 2, 3, 4: 0.9230-0.9238 in.

② No. 1: 0.0018-0.0034 in.
  No. 2, 3, 4: 0.0014-0.0030 in.

③ No. 1: 1.0212-1.0218 in.
  No. 2, 3, 4, 5: 1.0218-1.0224 in.

④ No. 1: 0.0018-0.0033 in.
  No. 2, 3, 4, 5: 0.0012-0.0027 in.

22140_FX35_C0007

CRANKSHAFT AND CONNECTING ROD SPECIFICATIONS

All measurements are given in inches.

Year	Engine Displacement Liters	Engine Series ID/VIN	Crankshaft				Connecting Rod		
			Main Brg. Journal Dia.	Main Brg. Oil Clearance	Shaft End-play	Thrust on No.	Journal Diameter	Oil Clearance	Side Clearance
2006	3.5	VQ35DE	①	0.0014-0.0018	0.0039-0.0098	3	④	0.0013-0.0023	0.0079-0.0138
	4.5	VK45DE	②	③	0.0039-0.0098	3	④	0.0008-0.0018	0.0079-0.0138
2007	3.5	VQ35DE	①	0.0014-0.0018	0.0039-0.0098	3	④	0.0013-0.0023	0.0079-0.0138
	4.5	VK45DE	②	③	0.0039-0.0098	3	④	0.0008-0.0018	0.0079-0.0138
2008	3.5	VQ35DE	①	0.0014-0.0018	0.0039-0.0098	3	④	0.0013-0.0023	0.0079-0.0138
	4.5	VK45DE	②	③	0.0039-0.0098	3	④	0.0008-0.0018	0.0079-0.0138

① Depends on the grade of the crankshaft. The nominal range is: 2.3603-2.3612 in.
② Depends on the grade of the crankshaft. The nominal range is: 2.5173-2.5183 in.
③ Journals 1 and 5: 0.00004-0.0004 in.
　 Journals 2, 3 & 4: 0.0003-0.0007 in.

④ Grade 0: 2.0460-2.0462 in.
　 Grade 1: 2.0457-2.0460 in.
　 Grade 2: 2.0455-2.0457 in.

22140_FX35_C0008

PISTON AND RING SPECIFICATIONS

All measurements are given in inches.

Year	Engine Displ. Liters	Engine Series ID/VIN	Piston Clearance	Ring Gap			Ring Side Clearance		
				Top Compression	Bottom Compression	Oil Control	Top Compression	Bottom Compression	Oil Control
2006	3.5	VQ35DE	0.0004-0.0012	0.0091-0.0130	0.0130-0.0189	0.0079-0.0197	0.0018-0.0031	0.0012-0.0028	0.0026-0.0053
	4.5	VK45DE	0.0004-0.0012	0.0087-0.0126	0.0087-0.0126	0.0079-0.0197	0.0018-0.0031	0.0012-0.0028	0.0026-0.0053
2007	3.5	VQ35DE	0.0004-0.0012	0.0091-0.0130	0.0130-0.0189	0.0079-0.0197	0.0018-0.0031	0.0012-0.0028	0.0026-0.0053
	4.5	VK45DE	0.0004-0.0012	0.0087-0.0126	0.0087-0.0126	0.0079-0.0197	0.0018-0.0031	0.0012-0.0028	0.0026-0.0053
2008	3.5	VQ35DE	0.0004-0.0012	0.0091-0.0130	0.0130-0.0189	0.0079-0.0197	0.0018-0.0031	0.0012-0.0028	0.0026-0.0053
	4.5	VK45DE	0.0004-0.0012	0.0087-0.0126	0.0087-0.0126	0.0079-0.0197	0.0018-0.0031	0.0012-0.0028	0.0026-0.0053

22140_FX35_C0009

TORQUE SPECIFICATIONS
All readings in ft. lbs.

Year	Engine Displacement Liters	Engine Series ID/VIN	Cylinder Head Bolts	Main Bearing Bolts	Rod Bearing Bolts	Crankshaft Damper Bolts	Flywheel Bolts	Manifold Intake	Manifold Exhaust	Spark Plugs	Oil Pan Drain Plug
2006	3.5	VQ35DE	①	②	③	④	65	⑤	22	18	25
	4.5	VK45DE	⑥	⑦	⑧	⑨	65	21	21	18	25
2007	3.5	VQ35DE	①	②	③	④	65	⑤	22	18	25
	4.5	VK45DE	⑥	⑦	⑧	⑨	65	21	21	18	25
2008	3.5	VQ35DE	①	②	③	④	65	⑤	22	18	25
	4.5	VK45DE	⑥	⑦	⑧	⑨	65	21	21	18	25

NOTE: Dip main bearing bolts, crankshaft damper bolt, and flywheel bolts in clean engine oil prior to tightening.

① Step 1: Tighten in sequence to 72 ft. lbs.
Step 2: Completely loosen all in reverse sequence
Step 3: Tighten in sequence to 29 ft. lbs.
Step 4: Tighten 90 degrees
Step 5: Tighten another 90 degrees

② Step 1: Tighten in sequence to 10 ft. lbs.
Step 2: Tighten in sequence to 26 ft. lbs.
Step 3: Tighten 90 degrees

③ Step 1: Tighten to 14 ft. lbs.
Step 2: Tighten 90 degrees

④ Step 1: Tighten to 69 ft. lbs.
Step 2: Tighten 90 degrees

⑤ Step 1: Tighten to 60 inch lbs.
Step 2: Tighten to 21 ft. lbs.

⑥ Step 1: Tighten in sequence to 72 ft. lbs.
Step 2: Completely loosen all in reverse sequence
Step 3: Tighten in sequence to 32 ft. lbs.
Step 4: Tighten 60 degrees
Step 5: Tighten another 60 degrees

⑦ Step 1: Tighten M12 bolts in sequence 1-10 to 29 ft. lbs.
Step 2: Tighten M9 bolts in sequence 11-20 to 22 ft. lbs.
Step 3: Tighten M12 bolts in sequence 1-10 another 40 degrees
Step 4: Tighten M9 bolts in sequence 11-20 another 30 degrees
Step 5: Tighten M10 bolts in sequence 21-30 to 36 ft. lbs.

⑧ Step 1: Tighten to 11 ft. lbs.
Step 2: Tighten 60 degrees

⑨ Step 1: Tighten to 33 ft. lbs.
Step 2: Tighten 90 degrees

22140_FX35_C0010

WHEEL ALIGNMENT

Year	Model		Caster Range (+/-Deg.)	Caster Preferred Setting (Deg.)	Camber Range (+/-Deg.)	Camber Preferred Setting (Deg.)	Toe-in (Deg.)
2006	FX35	Front	0.75	+3.78	0.75	-0.73	0.060+/-0.040
		Rear	—	—	0.50	-0.80	0.170+/-0.080
	FX45	Front	0.75	+3.78	0.75	-0.73	0.060+/-0.040
		Rear	—	—	0.50	-0.80	0.170+/-0.080
2007	FX35	Front	0.75	+3.78	0.75	-0.73	0.060+/-0.040
		Rear	—	—	0.50	-0.80	0.170+/-0.080
	FX45	Front	0.75	+3.78	0.75	-0.73	0.060+/-0.040
		Rear	—	—	0.50	-0.80	0.170+/-0.080
2008	FX35	Front	0.75	+3.78	0.75	-0.73	0.060+/-0.040
		Rear	—	—	0.50	-0.80	0.170+/-0.080
	FX45	Front	0.75	+3.78	0.75	-0.73	0.060+/-0.040
		Rear	—	—	0.50	-0.80	0.170+/-0.080

NOTE: Measurements are given for unladen vehicle: fuel, engine coolant, and fluid levels are full. Spare tire, jack, hand tools, and mats are in designated positions.

22140_FX35_C0011

TIRE, WHEEL AND BALL JOINT SPECIFICATIONS

| Year | Model | OEM Tires | | Tire Pressures (psi) | | Wheel | Ball Joint | Lug Nut |
		Standard	Optional	Front	Rear	Size	Inspection	Torque (ft. lbs.)
2006	FX35	P265/60R18	P265/50R20	32	32	8.0JJ x 18 8.0JJ x 20	①	80
	FX45	P265/50R20	—	32	32	8.0JJ x 20	①	80
2007	FX35	P265/60R18	P265/50R20	32	32	8.0JJ x 18 8.0JJ x 20	①	80
	FX45	P265/50R20	—	32	32	8.0JJ x 20	①	80
2008	FX35	P265/60R18	P265/50R20	32	32	8.0JJ x 18 8.0JJ x 20	①	80
	FX45	P265/50R20	—	32	32	8.0JJ x 20	①	80

OEM: Original Equipment Manufacturer

PSI: Pounds Per Square Inch

① Replace if any measurable axial end play is found.

22140_FX35_C0012

BRAKE SPECIFICATIONS
All measurements in inches unless noted

| Year | Model | | Brake Disc | | | Brake Drum Diameter | | | Minimum Lining Thickness | Brake Caliper | |
			Original Thickness	Minimum Thickness	Maximum Runout	Original Inside Diameter	Max. Wear Limit	Maximum Machine Diameter		Bracket Bolts (ft. lbs.)	Mounting Bolts (ft. lbs.)
2006	FX35	F	1.339	1.260	0.0016	—	—	—	0.079	34	122
		R	0.630	0.551	0.0020	—	—	—	0.079	32	62
	FX45	F	1.339	1.260	0.0016	—	—	—	0.079	34	122
		R	0.630	0.551	0.0020	—	—	—	0.079	32	62
2007	FX35	F	1.339	1.260	0.0016	—	—	—	0.079	34	122
		R	0.630	0.551	0.0020	—	—	—	0.079	32	62
	FX45	F	1.339	1.260	0.0016	—	—	—	0.079	34	122
		R	0.630	0.551	0.0020	—	—	—	0.079	32	62
2008	FX35	F	1.339	1.260	0.0016	—	—	—	0.079	34	122
		R	0.630	0.551	0.0020	—	—	—	0.079	32	62
	FX45	F	1.339	1.260	0.0016	—	—	—	0.079	34	122
		R	0.630	0.551	0.0020	—	—	—	0.079	32	62

F: Front

R: Rear

22140_FX35_C0013

SCHEDULED MAINTENANCE INTERVALS
INFINITI—FX35 & FX45

TO BE SERVICED	TYPE OF SERVIC	VEHICLE MILEAGE INTERVAL (x1000)													
		7.5	15	22.5	30	37.5	45	52.5	60	67.5	75	82.5	90	97.5	105	120
Accessory drive belts ①	S/I								✓							✓
Air cleaner element (engine)	R				✓				✓				✓			✓
Air conditioner system	S/I	Inspect system operation annually														
Automatic transaxle fluid	S/I		✓		✓		✓		✓		✓		✓		✓	✓
Brake lines, hoses, cables, and connections	S/I		✓		✓		✓		✓		✓		✓		✓	✓
Brake pads, calipers, & rotors	S/I		✓		✓		✓		✓		✓		✓		✓	✓
Differential gear oil	S/I		✓		✓		✓		✓		✓		✓		✓	✓
Driveshafts and CV-boots	S/I		✓		✓		✓		✓		✓		✓		✓	✓
Engine coolant	R								✓				✓			✓
Engine oil and filter	R	✓	✓	✓	✓	✓	✓	✓	✓	✓	✓	✓	✓	✓	✓	✓
Exhaust pipe connections, muffler, and suspension bolts	S/I				✓				✓				✓			✓
EVAP vapor lines	S/I				✓				✓				✓			✓
Fuel lines and connections	S/I				✓				✓				✓			✓
In-cabin microfilter	S/I		✓		✓		✓		✓		✓		✓		✓	✓
Spark plugs (Platinum-tipped)	R	105,000 miles (under normal usage)														
Steering system	S/I				✓				✓				✓			✓
Suspension system	S/I				✓				✓				✓			✓
Transfer case fluid	S/I		✓		✓		✓		✓		✓		✓		✓	✓
Valve clearance	S/I	Whenever valve noise increases														

R: Replace   S/I: Service or Inspect

① Replace if worn or damaged or if the auto-tensioner has reached its limit (V8)

FREQUENT OPERATION MAINTENANCE (SEVERE SERVICE)

If a vehicle is operated under any of the following conditions it is considered severe service:

- Extremely dusty areas.

- 50% or more of the vehicle operation is in 90°F (32°C) or higher temperatures, or constant operation in temperatures below 32°F (0°C).

- Prolonged idling (vehicle operation in stop and go traffic).

- Frequent short running periods (engine does not warm to normal operating temperatures).

- Police, taxi, delivery usage, or trailer towing usage.

Automatic transaxle fluid (and filter), transfer case fluid, and differential gear oil: check every 15,000 miles, replace every 30,000 miles

Brake pads, calipers & rotors: service or inspect every 7,500 miles

Driveshafts and CV-boots inspect every 7,500 miles

Exhaust system inspect every 7,500 miles

Oil and oil filter: change every 3,750 miles

Steering system and suspension components inspect for looseness and damage every 7,500 miles

22140_FX35_C0014

PRECAUTIONS

Before servicing any vehicle, please be sure to read all of the following precautions, which deal with personal safety, prevention of component damage, and important points to take into consideration when servicing a motor vehicle:

• Never open, service or drain the radiator or cooling system when the engine is hot; serious burns can occur from the steam and hot coolant.

• Observe all applicable safety precautions when working around fuel. Whenever servicing the fuel system, always work in a well-ventilated area. Do not allow fuel spray or vapors to come in contact with a spark, open flame, or excessive heat (a hot drop light, for example). Keep a dry chemical fire extinguisher near the work area. Always keep fuel in a container specifically designed for fuel storage; also, always properly seal fuel containers to avoid the possibility of fire or explosion. Refer to the additional fuel system precautions later in this section.

• Fuel injection systems often remain pressurized, even after the engine has been turned **OFF**. The fuel system pressure must be relieved before disconnecting any fuel lines. Failure to do so may result in fire and/or personal injury.

• Brake fluid often contains polyglycol ethers and polyglycols. Avoid contact with the eyes and wash your hands thoroughly after handling brake fluid. If you do get brake fluid in your eyes, flush your eyes with clean, running water for 15 minutes. If eye irritation persists, or if you have taken brake fluid internally, IMMEDIATELY seek medical assistance.

• The EPA warns that prolonged contact with used engine oil may cause a number of skin disorders, including cancer. You should make every effort to minimize your exposure to used engine oil. Protective gloves should be worn when changing oil. Wash your hands and any other exposed skin areas as soon as possible after exposure to used engine oil. Soap and water, or waterless hand cleaner should be used.

• All new vehicles are now equipped with an air bag system, often referred to as a Supplemental Restraint System (SRS) or Supplemental Inflatable Restraint (SIR) system. The system must be disabled before performing service on or around system components, steering column, instrument panel components, wiring and sensors. Failure to follow safety and disabling procedures could result in accidental air bag deployment, possible personal injury and unnecessary system repairs.

• Always wear safety goggles when working with, or around, the air bag system. When carrying a non-deployed air bag, be sure the bag and trim cover are pointed away from your body. When placing a non-deployed air bag on a work surface, always face the bag and trim cover upward, away from the surface. This will reduce the motion of the module if it is accidentally deployed. Refer to the additional air bag system precautions later in this section.

• Clean, high quality brake fluid from a sealed container is essential to the safe and proper operation of the brake system. You should always buy the correct type of brake fluid for your vehicle. If the brake fluid becomes contaminated, completely flush the system with new fluid. Never reuse any brake fluid. Any brake fluid that is removed from the system should be discarded. Also, do not allow any brake fluid to come in contact with a painted surface; it will damage the paint.

• Never operate the engine without the proper amount and type of engine oil; doing so WILL result in severe engine damage.

• Timing belt maintenance is extremely important. Many models utilize an interference-type, non-freewheeling engine. If the timing belt breaks, the valves in the cylinder head may strike the pistons, causing potentially serious (also time-consuming and expensive) engine damage. Refer to the maintenance interval charts for the recommended replacement interval for the timing belt, and to the timing belt section for belt replacement and inspection.

• Disconnecting the negative battery cable on some vehicles may interfere with the functions of the on-board computer system(s) and may require the computer to undergo a relearning process once the negative battery cable is reconnected.

• When servicing drum brakes, only disassemble and assemble one side at a time, leaving the remaining side intact for reference.

• Only an MVAC-trained, EPA-certified automotive technician should service the air conditioning system or its components.

BRAKES

GENERAL INFORMATION

PRECAUTIONS

• Certain components within the ABS system are not intended to be serviced or repaired individually.

• Do not use rubber hoses or other parts not specifically specified for and ABS system. When using repair kits, replace all parts included in the kit. Partial or incorrect repair may lead to functional problems and require the replacement of components.

• Lubricate rubber parts with clean, fresh brake fluid to ease assembly. Do not use shop air to clean parts; damage to rubber components may result.

• Use only DOT 3 brake fluid from an unopened container.

• If any hydraulic component or line is removed or replaced, it may be necessary to bleed the entire system.

• A clean repair area is essential. Always clean the reservoir and cap thoroughly before removing the cap. The slightest amount of dirt in the fluid may plug an orifice and impair the system function. Perform repairs after components have been thoroughly cleaned; use only denatured alcohol

ANTI-LOCK BRAKE SYSTEM (ABS)

to clean components. Do not allow ABS components to come into contact with any substance containing mineral oil; this includes used shop rags.

• The Anti-Lock control unit is a microprocessor similar to other computer units in the vehicle. Ensure that the ignition switch is **OFF** before removing or installing controller harnesses. Avoid static electricity discharge at or near the controller.

• If any arc welding is to be done on the vehicle, the control unit should be unplugged before welding operations begin.

BRAKES

BLEEDING PROCEDURE

BLEEDING PROCEDURE

When any part of the hydraulic system has been disconnected for repair or replacement, air may get into the lines and cause spongy pedal action (because air can be compressed and brake fluid cannot). To correct this condition, it is necessary to bleed the hydraulic system so to be sure all air is purged.

When bleeding the brake system, bleed one brake cylinder at a time, beginning at the cylinder with the longest hydraulic line (farthest from the master cylinder) first. ALWAYS keep the master cylinder reservoir filled with brake fluid during the bleeding operation. Never use brake fluid that has been drained from the hydraulic system, no matter how clean it is.

The primary and secondary hydraulic brake systems are separate and are bled independently. During the bleeding operation, do not allow the reservoir to run dry. Keep the master cylinder reservoir filled with brake fluid.

1. Clean all dirt from around the master cylinder fill cap, remove the cap and fill the master cylinder with brake fluid until the level is within ¼ inch (6mm) of the top edge of the reservoir.

2. Clean the bleeder screws at all 4 wheels. The bleeder screws are located on the top of the brake calipers.

3. Attach a length of rubber hose over the bleeder screw and place the other end of the hose in a glass jar, submerged in brake fluid.

4. Open the bleeder screw ½–¾ turn. Have an assistant slowly depress the brake pedal.

※ CAUTION

Brake fluid contains polyglycol ethers and polyglycols. Avoid contact with the eyes and wash your hands thoroughly after handling brake fluid. If you do get brake fluid in your eyes, flush your eyes with clean, running water for 15 minutes. If eye irritation persists, or if you have taken brake fluid internally, IMMEDIATELY seek medical assistance.

5. Close the bleeder screw and tell your assistant to allow the brake pedal to return slowly. Continue this process to purge all air from the system.

6. When bubbles cease to appear at the end of the bleeder hose, close the bleeder screw and remove the hose. Tighten the bleeder screw to the proper torque.

7. Check the master cylinder fluid level and add fluid accordingly. Do this after bleeding each wheel.

8. Repeat the bleeding operation at the remaining 3 wheels, ending with the one closet to the master cylinder.

9. Fill the master cylinder reservoir to the proper level.

MASTER CYLINDER BLEEDING

1. Before servicing the vehicle, refer to the Precautions Section.

2. If removed from the vehicle, clamp the master cylinder in a vise with soft-jaw caps.

3. Attach the special tools for bleeding the master cylinder in the following fashion:

   a. Thread the bleeder tube adapters into the primary and secondary outlet ports of the master cylinder and tighten the adapters.

   b. Thread a bleeder tube into each adapter and tighten the tube nuts.

   c. Flex each bleeder tube and place the open ends into the neck of the master cylinder reservoir. Position the open ends of the tubes into the reservoir so their outlets are below the surface of the brake fluid in the reservoir when filled.

➡**Make sure the ends of the bleeder tubes stay below the surface of the brake fluid in the reservoir at all times during the bleeding procedure.**

4. Fill the brake fluid reservoir with fresh brake fluid (DOT 3).

5. Using an appropriately sized wooden dowel as a pushrod, slowly press the pistons inward discharging brake fluid through the bleeder tubes, then release the pressure, allowing the pistons to return to the released position. Repeat this several times until all air bubbles are expelled from the master cylinder bore and bleeder tubes.

6. Remove the bleeder tubes and adapters from the master cylinder and plug the master cylinder outlet ports.

7. Install the fill cap on the reservoir.

8. Remove the master cylinder from the vise.

9. Install the master cylinder on the vehicle.

BRAKE LINE BLEEDING

Refer to Bleeding the Brake System, Bleeding Procedure.

BRAKES

※ CAUTION

Dust and dirt accumulating on brake parts during normal use may contain asbestos fibers from production or aftermarket brake linings. Breathing excessive concentrations of asbestos fibers can cause serious bodily harm. Exercise care when servicing brake parts. Do not sand or grind brake lining unless equipment used is designed to contain the dust residue. Do not clean brake parts with compressed air or by dry brushing. Cleaning should be done by dampen-ing the brake components with a fine mist of water, then wiping the brake components clean with a dampened cloth. Dispose of cloth and all residue containing asbestos fibers in an impermeable container with the appropriate label. Follow practices prescribed by the Occupational Safety and Health Administration (OSHA) and the Environmental Protection Agency (EPA) for the handling, processing, and disposing of dust or debris that may contain asbestos fibers.

BRAKE CALIPER

REMOVAL & INSTALLATION

See Figure 1.

1. Before servicing the vehicle, refer to the Precautions Section.

2. Raise and safely support the vehicle.

3. Remove the front wheels from the vehicle.

4. Drain the brake fluid.

5. Remove the union bolts and torque member bolts, and remove the brake caliper assembly from the vehicle.

6. If necessary, remove the rotor.

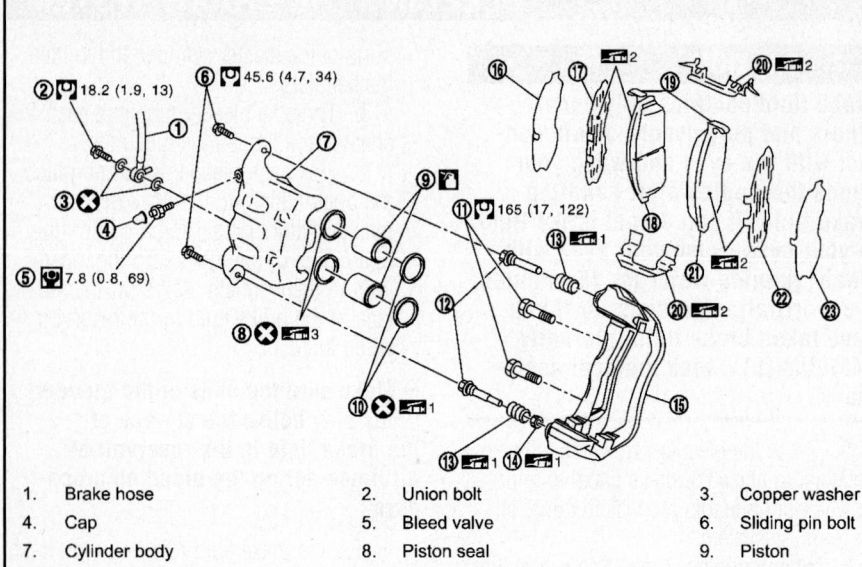

1. Brake hose
2. Union bolt
3. Copper washer
4. Cap
5. Bleed valve
6. Sliding pin bolt
7. Cylinder body
8. Piston seal
9. Piston
10. Piston boot
11. Torque member mounting bolt
12. Sliding pin
13. Sliding pin boot
14. Bushing
15. Torque member
16. Inner shim cover
17. Inner shim
18. Inner pad
19. Pad wear sensor
20. Pad retainer
21. Outer pad

22140_FX35_G0001

Fig. 1 Exploded view of the front disc brake mounting and components

DISC BRAKE PADS

REMOVAL & INSTALLATION

1. Before servicing the vehicle, refer to the Precautions Section.
2. Raise and safely support the vehicle.
3. Remove the front wheels.
4. Remove the lower sliding pin bolt.
5. Suspend the cylinder body with strong cord or wire, then remove the pad and shim from the torque member.

To install:

6. Position the inner shim and shim cover onto the inner pad, and outer shim onto the outer pad.
7. Push the caliper piston in so that the brake pad is firmly installed.
8. Position the cylinder body on the torque member.

➡ **The use of a disc brake piston tool can make it easier to push in the piston.**

✻✻ WARNING

By pushing in the piston, brake fluid returns to the master cylinder reservoir tank. Watch the level of the surface of reservoir tank. Brake fluid is corrosive to painted surfaces.

9. Reattach the pad retainer to the torque member. When attaching the pad retainer, attach it firmly so that it does not float up higher than the torque member.
10. Install the lower sliding pin bolt and tighten it to 34 ft. lbs. (46 Nm).
11. Check the brake assembly for drag.
12. Install the wheels.
13. Before attempting to move the vehicle, pump the brake pedal to seat the pads against the rotors. Make sure the vehicle has a firm brake pedal. Check the level of the brake fluid and add fluid if necessary.

To install:

➡ **Use Only new DOT-3 brake fluid. Do not reuse drained brake fluid.**

7. If removed, install the rotor.
8. Install the caliper assembly onto the vehicle. Tighten the sliding pin bolts to 34 ft. lbs. (46 Nm) and the torque member bolts to 122 ft. lbs. (165 Nm).

➡ **When attaching the caliper assembly to the vehicle, wipe any oil off the knuckle spindle, washers, and caliper assembly attachment surfaces.**

9. Reattach the brake hose to the brake caliper assembly, and tighten the union bolt to 13 ft. lbs. (18 Nm).

➡ **Do not reuse the old copper washers for the union bolt. Attach the brake hose to the caliper assembly together only using the specified union bolt and washers.**

10. Refill the brake system with new brake fluid and bleed the system.
11. Install the wheels.
12. Before attempting to move the vehicle, pump the brake pedal to seat the pads against the rotors. Make sure the vehicle has a firm brake pedal. Check the level of the brake fluid and add fluid if necessary.

BRAKES

✻✻ CAUTION

Dust and dirt accumulating on brake parts during normal use may contain asbestos fibers from production or aftermarket brake linings. Breathing excessive concentrations of asbestos fibers can cause serious bodily harm. Exercise care when servicing brake parts. Do not sand or grind brake lining unless equipment used is designed to contain the dust residue. Do not clean brake parts with compressed air or by dry brushing. Cleaning should be done by dampening the brake components with a fine mist of water, then wiping the brake components clean with a dampened cloth. Dispose of cloth and all residue containing asbestos fibers in an impermeable container with the appropriate label. Follow practices prescribed by the Occupational Safety and Health Administration (OSHA) and the Environmental Protection Agency (EPA) for the handling, processing, and disposing of dust or debris that may contain asbestos fibers.

REAR DISC BRAKES

BRAKE CALIPER

REMOVAL & INSTALLATION

See Figure 2.

1. Before servicing the vehicle, refer to the Precautions Section.
2. Raise and safely support the vehicle.
3. Remove the wheels.

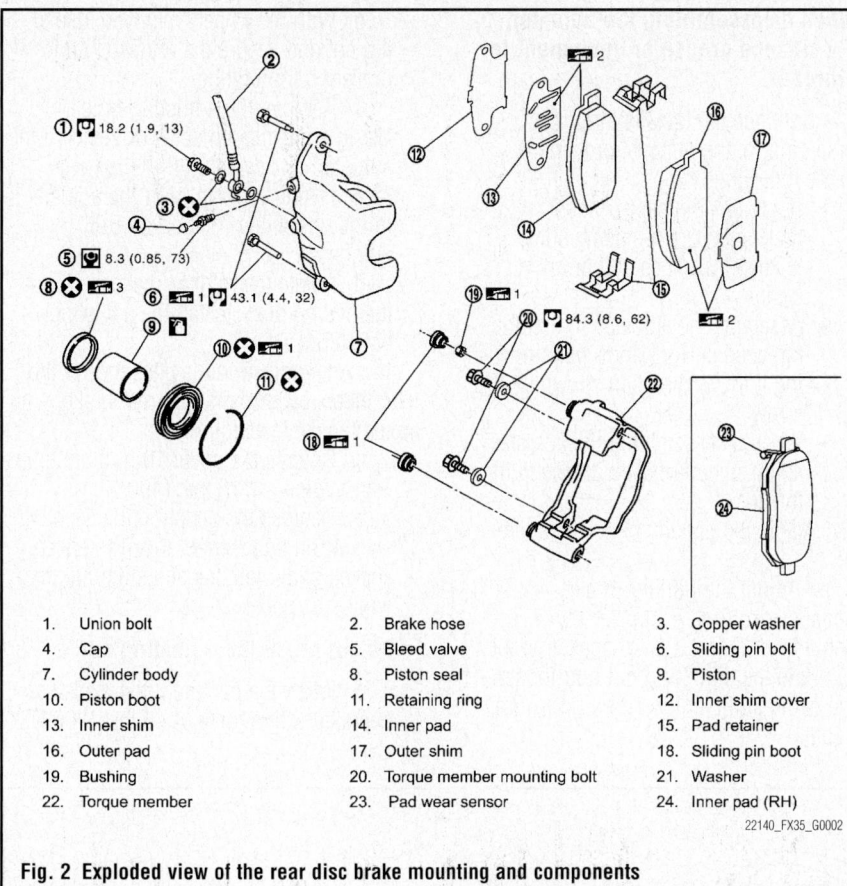

1. Union bolt	2. Brake hose	3. Copper washer
4. Cap	5. Bleed valve	6. Sliding pin bolt
7. Cylinder body	8. Piston seal	9. Piston
10. Piston boot	11. Retaining ring	12. Inner shim cover
13. Inner shim	14. Inner pad	15. Pad retainer
16. Outer pad	17. Outer shim	18. Sliding pin boot
19. Bushing	20. Torque member mounting bolt	21. Washer
22. Torque member	23. Pad wear sensor	24. Inner pad (RH)

22140_FX35_G0002

Fig. 2 Exploded view of the rear disc brake mounting and components

4. Drain the brake fluid.

5. Remove the union bolts and torque member bolts, and remove the brake caliper assembly from the vehicle.

6. If necessary, remove the disc rotor.

To install:

➡**Only use new DOT-3 brake fluid. Do not reuse drained brake fluid.**

7. If removed, install the disc rotor.

8. Install the caliper assembly onto the vehicle. Tighten the sliding pin bolts to 32 ft. lbs. (43 Nm) and the torque member bolts to 62 ft. lbs. (84 Nm).

➡**When attaching the caliper assembly to the vehicle, wipe any oil off the knuckle spindle, washers, and caliper assembly attachment surfaces.**

9. Reattach the brake hose to the brake caliper assembly, and tighten the union bolt to 13 ft. lbs. (18 Nm).

➡**Do not reuse the old copper washers for the union bolt. Attach the brake hose to the caliper assembly together only using the specified union bolt and washers.**

10. Refill the brake system with new brake fluid and bleed the system.

11. Install the wheels.

12. Before attempting to move the vehicle, pump the brake pedal to seat the pads against the rotors. Make sure the vehicle has a firm brake pedal. Check the level of the brake fluid and add fluid if necessary.

DISC BRAKE PADS

REMOVAL & INSTALLATION

1. Before servicing the vehicle, refer to the Precautions Section.

2. Raise and safely support the vehicle.

3. Remove the wheels.

4. Remove the upper sliding pin bolt.

5. Suspend the cylinder body with strong cord or wire, then remove the pad and shim from the torque member.

To install:

6. Apply silicon-based grease to the backside of the pad and to both sides of the shim, then attach the inner shim and shim cover to the inner pad. Attach the outer shim and outer shim cover to the outer pad.

7. Install the pad retainer and mount the pad onto the torque member.

8. Position the cylinder body on the torque member.

➡**Using a disc brake piston tool can make it easier to push in the piston.**

✳✳ WARNING

By pushing in the piston, brake fluid returns to the master cylinder reservoir tank. Watch the level of the surface of the reservoir tank. Brake fluid is corrosive to painted surfaces.

9. Install the top sliding pin bolt and tighten it to 32 ft. lbs. (43 Nm).

10. Check the brake assembly for drag.

11. Install the wheels

BRAKES

PARKING BRAKE SHOES

REMOVAL & INSTALLATION

See Figures 3 through 5.

1. Before servicing the vehicle, refer to the Precautions Section.

2. Raise and safely support the vehicle.

3. Release the parking brake.

4. Remove the rear wheels.

5. Remove the rotor. If the rotor cannot be removed:

a. Secure the rotor in place with lug nuts and remove the adjuster hole plug.

b. Using a flat-bladed screwdriver, rotate the adjuster in direction **B** to retract and loosen the brake shoes.

6. Remove the anti-rattle pins, retainers, anti-rattle springs, and return springs.

7. Remove the parking brake shoes, adjuster assembly, adjuster spring, and toggle lever.

To install:

8. Check the thickness of the lining.

PARKING BRAKE

Standard thickness **A**: 3.2mm (0.126 inch). Repair limit thickness **A**: 1.5mm (0.059 inch).

9. Check the drum inner diameter. Standard inner diameter: 7.48 inch (190mm). Maximum inner diameter: 7.52 inch (191mm).

10. Check the following:

• Shoes for excessive wear, damage, and peeling

• Shoe sliding surfaces for excessive wear and damage

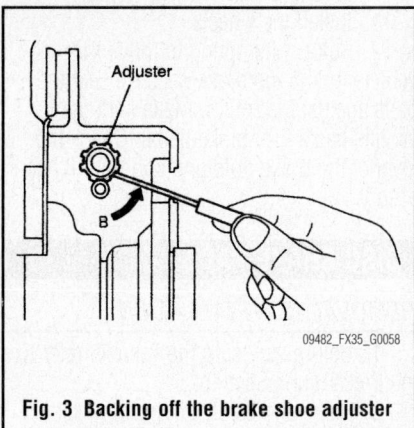

Fig. 3 Backing off the brake shoe adjuster

- Anti-rattle pins for excessive wear and corrosion
- Return springs for sagging
- Check that the adjuster moves smoothly
- Visually check the inside of the drum for excessive wear, cracks, and damage. Check the inside diameter of the drum

11. Replace any suspect parts.

➡**When disassembling the adjuster, apply silicone grease or equivalent to the threads.**

12. Continue the installation in the reverse order of the removal procedure. Note the following:
- Apply brake grease to the brake mechanism, being sure to keep the shoes and drum clean
- Assemble the adjuster so that the threaded part expands when rotating it in the direction shown by the arrow
- When disassembling the adjuster, apply silicone-based grease to the threads

13. Adjust the parking brake shoe tension:

a. Adjust the parking brake pedal stroke to 4–5 clicks fully depressed. Insert a deep socket wrench to rotate the adjusting nut and loosen the cable sufficiently. Then, return the pedal.

b. With the wheels removed, use a lug nut and secure the rotor to hub to prevent it from tilting.

c. Remove the adjusting hole plug. Using a flat-bladed screwdriver, turn the adjuster clockwise until the rotor is locked. After locking, turn the adjuster in the opposite direction by 5–6 notches.

d. Rotate the rotor to make sure that there is no drag. Install the adjusting hole plug.

14. After adjusting the clearance of the rear shoes, with no drag on rear brake, adjust the cable as follows:

a. Operate the pedal 10 or more times with a force of 110 lbs. (490 N).

b. Depress the pedal until a deep socket can be inserted. Insert the deep socket, and rotate the adjusting nut to adjust the pedal stroke.

➡**Do not reuse the adjusting nut.**

c. When the parking brake pedal is operated with a force of 45 lbs. (200 N),

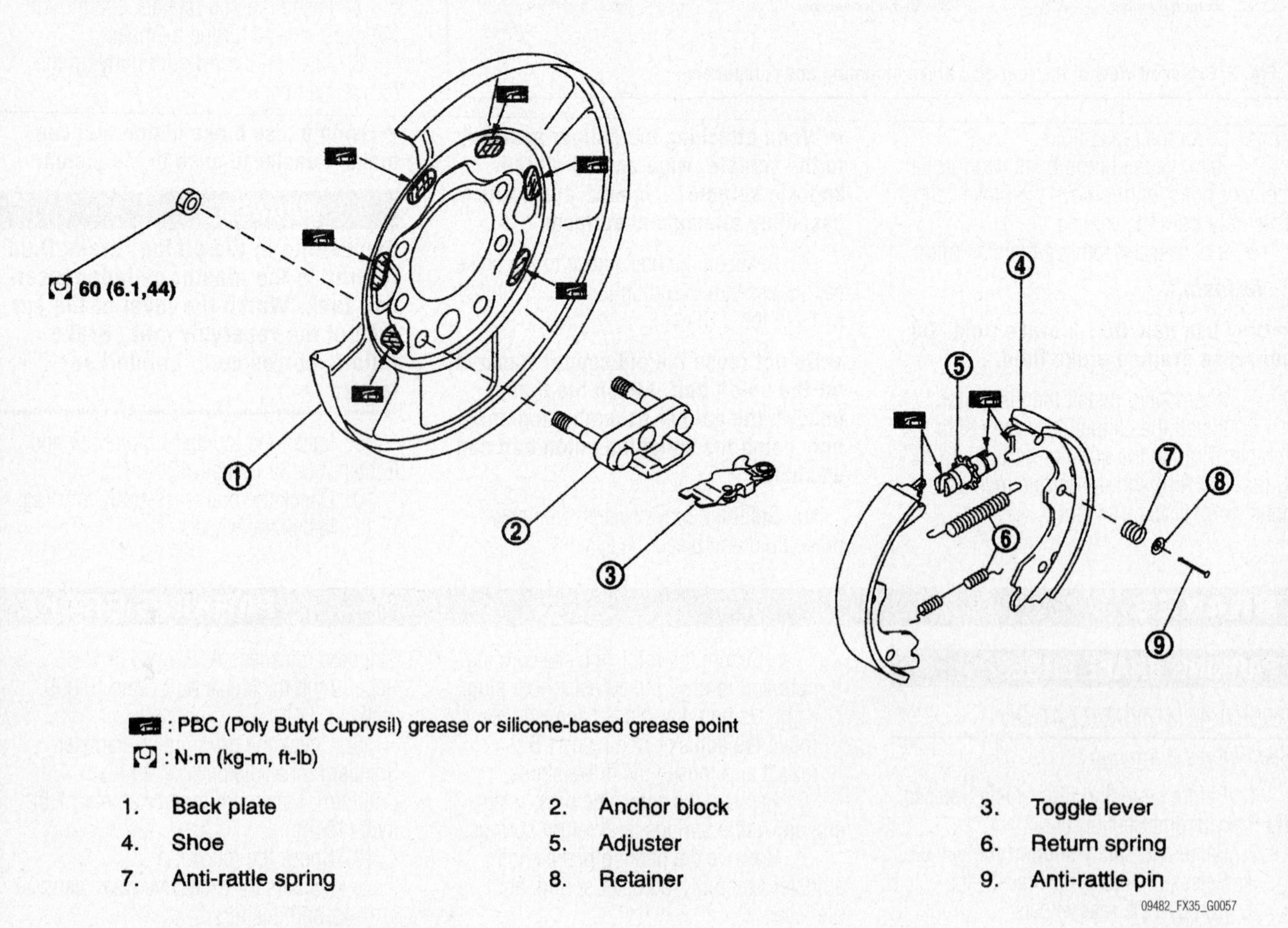

60 (6.1,44)

: PBC (Poly Butyl Cuprysil) grease or silicone-based grease point

: N·m (kg-m, ft-lb)

1. Back plate	2. Anchor block	3. Toggle lever
4. Shoe	5. Adjuster	6. Return spring
7. Anti-rattle spring	8. Retainer	9. Anti-rattle pin

Fig. 4 Parking brake shoes and related components

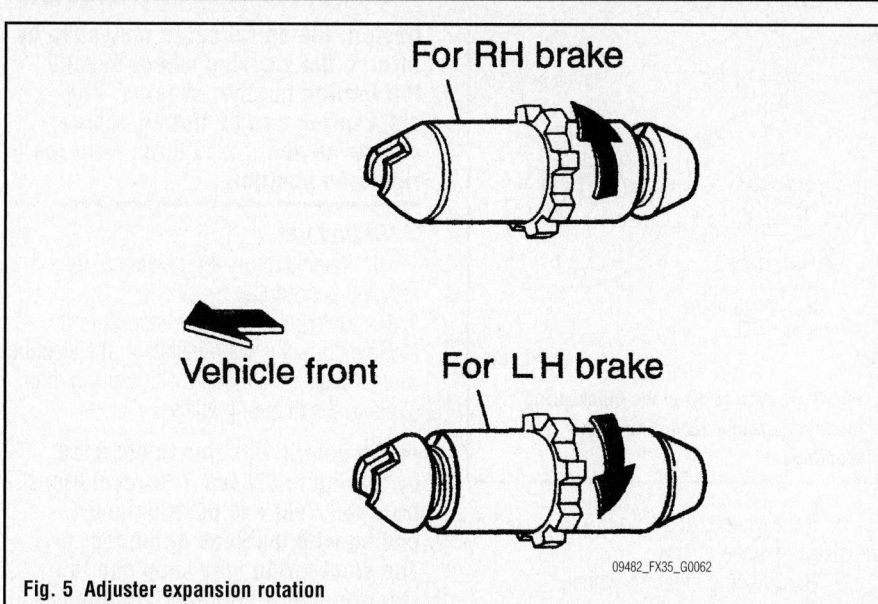

Fig. 5 Adjuster expansion rotation

make sure the stroke is 4–5 notches. (Check it by listening and counting the ratchet clicks).

d. With the parking brake pedal completely returned, make sure there is no drag on the rear brake.

15. Perform the parking brake break-in operation as follows: Safely, drive forward at approximately 25 mph (40 km/h) with the parking brake set with a force of approximately 45 lbs. (200 N) for about 30 seconds.

16. After the break-in operation, check the pedal stroke of parking brake. Readjust if necessary.

➡**To prevent the brake lining from getting too hot, allow a cool off period of approximately 5 minutes after every break-in operation.**

CHASSIS ELECTRICAL

AIR BAG (SUPPLEMENTAL RESTRAINT SYSTEM)

GENERAL INFORMATION

✳✳ CAUTION

These vehicles are equipped with an air bag system. The system must be disarmed before performing service on, or around, system components, the steering column, instrument panel components, wiring and sensors. Failure to follow the safety precautions and the disarming procedure could result in accidental air bag deployment, possible injury and unnecessary system repairs.

SERVICE PRECAUTIONS

Disconnect and isolate the battery negative cable before beginning any airbag system component diagnosis, testing, removal, or installation procedures. Allow system capacitor to discharge for two minutes before beginning any component service. This will disable the airbag system. Failure to disable the airbag system may result in accidental airbag deployment, personal injury, or death.

Do not place an intact undeployed airbag face down on a solid surface. The airbag will propel into the air if accidentally deployed and may result in personal injury or death.

When carrying or handling an undeployed airbag, the trim side (face) of the airbag should be pointing towards the body to minimize possibility of injury if accidental deployment occurs. Failure to do this may result in personal injury or death.

Replace airbag system components with OEM replacement parts. Substitute parts may appear interchangeable, but internal differences may result in inferior occupant protection. Failure to do so may result in occupant personal injury or death.

Wear safety glasses, rubber gloves, and long sleeved clothing when cleaning powder residue from vehicle after an airbag deployment. Powder residue emitted from a deployed airbag can cause skin irritation. Flush affected area with cool water if irritation is experienced. If nasal or throat irritation is experienced, exit the vehicle for fresh air until the irritation ceases. If irritation continues, see a physician.

Do not use a replacement airbag that is not in the original packaging. This may result in improper deployment, personal injury, or death.

The factory installed fasteners, screws and bolts used to fasten airbag components have a special coating and are specifically designed for the airbag system. Do not use substitute fasteners. Use only original equipment fasteners listed in the parts catalog when fastener replacement is required.

During, and following, any child restraint anchor service, due to impact event or vehicle repair, carefully inspect all mounting hardware, tether straps, and anchors for proper installation, operation, or damage. If a child restraint anchor is found damaged in any way, the anchor must be replaced. Failure to do this may result in personal injury or death.

Deployed and non-deployed airbags may or may not have live pyrotechnic material within the airbag inflator.

Do not dispose of driver/passenger/curtain airbags or seat belt tensioners unless you are sure of complete deployment. Refer to the Hazardous Substance Control System for proper disposal.

Dispose of deployed airbags and tensioners consistent with state, provincial, local, and federal regulations.

After any airbag component testing or service, do not connect the battery negative cable. Personal injury or death may result if the system test is not performed first.

If the vehicle is equipped with the Occupant Classification System (OCS), do not connect the battery negative cable before performing the OCS Verification Test using the scan tool and the appropriate diagnostic information. Personal injury or death may result if the system test is not performed properly.

Never replace both the Occupant Restraint Controller (ORC) and the Occupant Classification Module (OCM) at the same time. If both require replacement, replace one, then perform the Airbag System test before replacing the other.

Both the ORC and the OCM store Occupant Classification System (OCS) calibration data, which they transfer to one another when one of them is replaced. If both are replaced at the same time, an irreversible fault will be set in both modules and the OCS may malfunction and cause personal injury or death.

If equipped with OCS, the Seat Weight Sensor is a sensitive, calibrated unit and must be handled carefully. Do not drop or handle roughly. If dropped or damaged, replace with another sensor. Failure to do so may result in occupant injury or death.

If equipped with OCS, the front passenger seat must be handled carefully as well. When removing the seat, be careful when setting on floor not to drop. If dropped, the sensor may be inoperative, could result in occupant injury, or possibly death.

If equipped with OCS, when the passenger front seat is on the floor, no one should sit in the front passenger seat. This uneven force may damage the sensing ability of the seat weight sensors. If sat on and damaged, the sensor may be inoperative, could result in occupant injury, or possibly death.

DISARMING THE SYSTEM

1. Before servicing the vehicle, refer to the Precautions Section.
2. Turn the ignition switch to **OFF**.
3. Disconnect the negative battery cable and isolate it from accidental reconnection. Insulate the cable end with high-quality electrical tape or a similar non-conductive wrapping.
4. Wait at least 3 minutes for the system capacitor to discharge before performing any service. The air bag system is designed to retain enough voltage to deploy the air bag for a short period of time after the battery has been disconnected.

ARMING THE SYSTEM

1. Before servicing the vehicle, refer to the Precautions Section.
2. Reconnect the negative battery cable.
3. To confirm proper system operation, turn the ignition switch to the **ON** position. The SRS indicator light should light for at least 7 seconds and then go off.

CLOCKSPRING CENTERING

See Figures 6 and 7.

✳✳ CAUTION

Before servicing, or working around, the SRS system, turn the ignition switch OFF, disconnect both battery cables and wait at least 3 minutes. When servicing, or working around, the SRS system, do not work directly in front of the air bag module.

1. Before servicing the vehicle, refer to the Precautions Section.
2. Position the front wheels in the straight ahead position.
3. Disconnect the negative and positive battery cables.

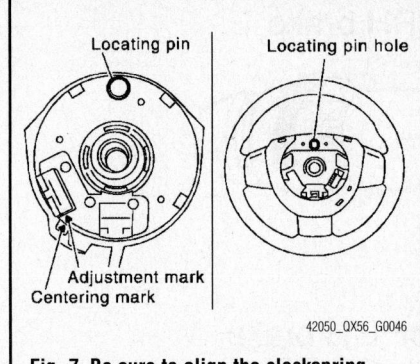

Fig. 7 Be sure to align the clockspring module correctly according to the markings

4. Remove the air bag module affixed to the steering wheel.
5. Remove the steering wheel.
6. Remove the upper and lower steering column covers.
7. Remove the retaining screws of the clockspring module (known by this manufacturer as the spiral cable). Remove the clockspring module.

➡**Do not disassemble the clockspring module. Do not apply lubricant to the clockspring module.**

8. Remove the clockspring module connectors.

✳✳ WARNING

With the steering linkage disconnected, the spiral cable may snap by turning the steering wheel beyond the limited number of turns. The clockspring can be turned counterclockwise about 2 ½ turns from the right end position.

To install:

9. Installation is the reverse of the removal procedure.
10. Be sure to align the clockspring module correctly when installing the steering wheel. Make sure that the clockspring module is in the neutral position.

➡**The neutral position is detected by turning to the left 2 ½ revolutions from the right end position and ending with the knob at the top. The clockspring may snap due to steering operation if installed incorrectly. Also, with the steering linkage disconnected the cable may snap by turning the steering wheel beyond the limited number of turns (2 ½ from the neutral position to both the left and right).**

11. Use the CONSULT-III tool and perform self-diagnosis to ensure no malfunction is detected.
12. Tighten the steering wheel retaining nut to 25 ft. lbs. (34 Nm).
13. When reinstalling the air bag module, be sure to use new bolts. Tighten the bolts to 8 ft. lbs.

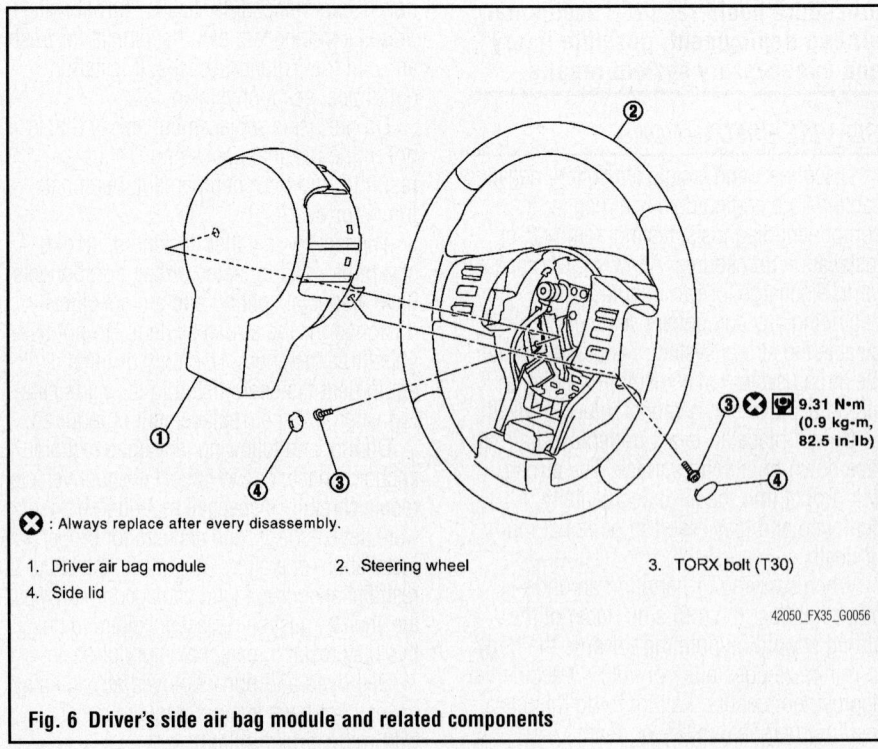

✖ : Always replace after every disassembly.

1. Driver air bag module
2. Steering wheel
3. TORX bolt (T30)
4. Side lid

Fig. 6 Driver's side air bag module and related components

DRIVETRAIN

AUTOMATIC TRANSMISSION ASSEMBLY

REMOVAL & INSTALLATION

See Figures 8 through 13.

1. Before servicing the vehicle, refer to the Precautions Section.

※※ CAUTION

When removing the automatic transmission assembly from the engine, **first remove the Crankshaft Position (CKP) sensor from the A/T assembly. Be careful not to damage the sensor edge.**

2. Disconnect the negative battery terminal.
3. Remove the engine cover.
4. Remove the A/T fluid level gauge.
5. Raise and safely support the vehicle.
6. Remove the engine undercover.
7. Remove the exhaust front tube and center muffler.

8. Remove the driveshaft.
9. Detach the transmission shifter control rod by removing the snap pin retaining the control rod bracket to the control device assembly lever.
10. Remove the CKP sensor from A/T assembly.
11. When handling the CKP sensor, heed the following:
    - Do not subject it to impact by dropping or hitting it
    - Do not disassemble it

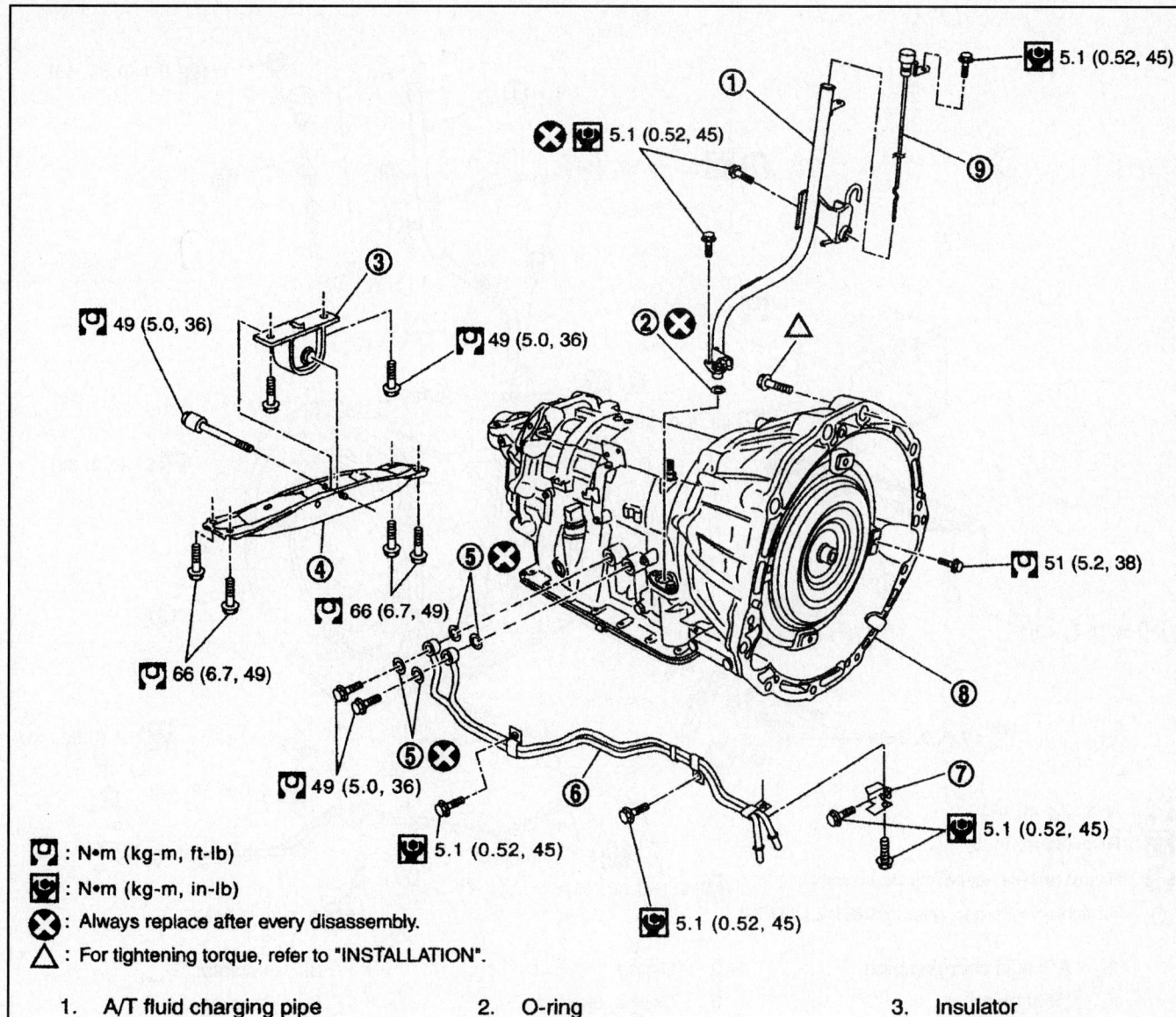

- : N•m (kg-m, ft-lb)
- : N•m (kg-m, in-lb)
- ✖ : Always replace after every disassembly.
- ▲ : For tightening torque, refer to "INSTALLATION".

1. A/T fluid charging pipe
2. O-ring
3. Insulator
4. Rear member
5. Copper washer
6. Fluid cooler tube
7. Bracket
8. Transmission assembly
9. A/T fluid level gauge

09482_FX35_G0005

Fig. 8 Exploded view of transmission mounting—2WD Models

- Do not allow metal filings or debris get on the sensor's front edge magnetic area
- Do not place in an area affected by magnetism

12. Remove the starter motor.

13. Disconnect the A/T fluid cooler tube from the transmission assembly.

14. Remove the dust cover from the converter housing part.

15. Turn the crankshaft, and remove the 4 bolts from the driveplate and torque converter.

➡**When turning the crankshaft, turn it clockwise as viewed from the front of the engine.**

16. For AWD models with the 3.5L engine, remove the dynamic damper.

17. Support the transmission assembly with a transmission jack.

➡**When setting the transmission jack, be careful not to allow it to collide against the drain plug.**

18. Remove the engine rear member.

19. Tilt the transmission slightly to keep the clearance between the body and the transmission and then disconnect the air breather hose from the charging pipe.

20. Disconnect the A/T assembly connector.

21. Remove the A/T fluid charging pipe from the A/T assembly.

22. Plug up the fluid charge pipe hole and other openings.

23. Remove the bolts fixing the transmission assembly to the engine.

24. Secure the torque converter to prevent it from dropping.

25. Secure the transmission assembly to the jack.

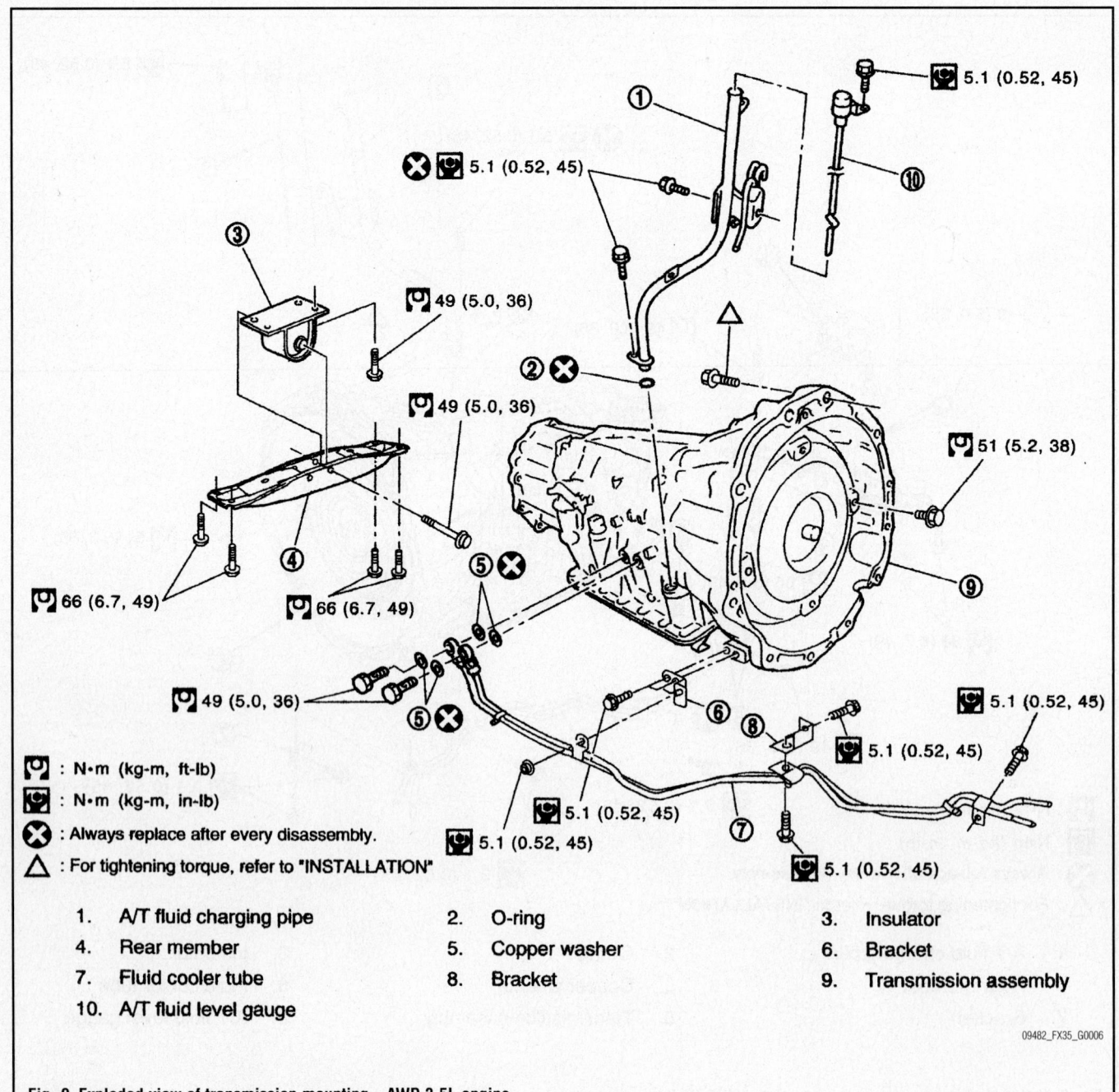

: N·m (kg-m, ft-lb)

: N·m (kg-m, in-lb)

✕ : Always replace after every disassembly.

△ : For tightening torque, refer to "INSTALLATION"

1.	A/T fluid charging pipe	2.	O-ring	3.	Insulator
4.	Rear member	5.	Copper washer	6.	Bracket
7.	Fluid cooler tube	8.	Bracket	9.	Transmission assembly
10.	A/T fluid level gauge				

09482_FX35_G0006

Fig. 9 Exploded view of transmission mounting—AWD 3.5L engine

26. For 2WD models, remove the transmission assembly from the vehicle with the transmission jack.

27. For AWD models, remove the transmission assembly with the transfer case from the vehicle with the transmission jack. Remove the transfer mounting bolts and separate the transfer from the transmission.

To install:

28. After fitting the torque converter to the transmission, be sure to check dimension **A** to ensure it is within the reference value limit:

a. VQ35DE models: 0.98 inch (25mm) or more.

b. VK45DE models: 0.87 inch (22mm) or more.

✳✳ WARNING

When setting the transmission jack, be careful not to allow it to collide against the drain plug.

29. For AWD models, install the transfer case to the transmission assembly.

30. Install the transmission assembly into the vehicle with the transmission jack.

31. Install the bolts fixing the transmission assembly to the engine. Use the accompanying illustrations for proper bolt placement and installation torques.

32. Unplug the fluid holes.

33. Install the A/T fluid charging pipe onto the A/T assembly.

34. Connect the A/T assembly connector.

35. Install the air breather hose.

36. Install the engine rear member.

37. Carefully, remove the transmission jack.

✳✳ CAUTION

When turning the crankshaft, turn it clockwise as viewed from the front of the engine.

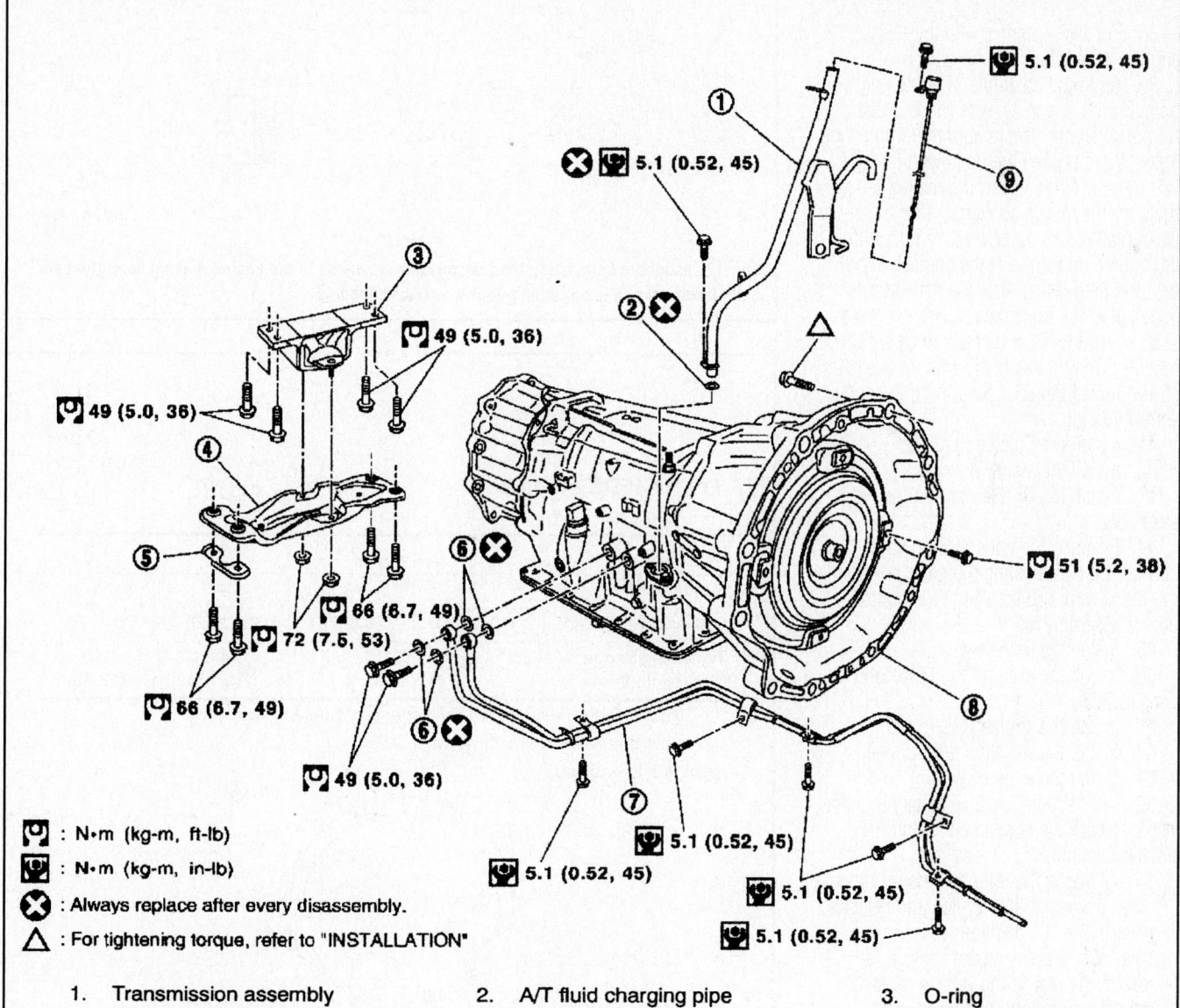

🔧 : N•m (kg-m, ft-lb)

🔧 : N•m (kg-m, in-lb)

❌ : Always replace after every disassembly.

△ : For tightening torque, refer to "INSTALLATION"

1.	Transmission assembly	2.	A/T fluid charging pipe	3.	O-ring
4.	Fluid cooler tube	5.	Copper washer	6.	A/T fluid level gauge
7.	Engine rear member	8.	Insulator		

09482_FX35_G0007

Fig. 10 Exploded view of transmission mounting—AWD 4.5L engine

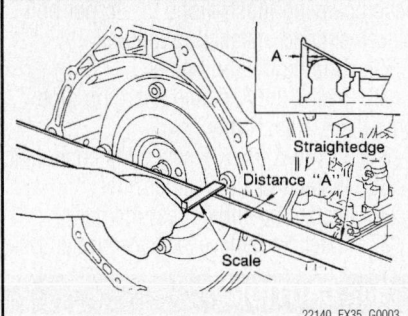

Fig. 11 After fitting the torque converter to the transmission, be sure to check dimension A to ensure it is within the reference limit

38. For AWD models with the 3.5L engine, install the dynamic damper.

39. Align the positions of tightening bolts for the driveplate with those of the torque converter, and temporarily tighten the bolts. Then, tighten the bolts with the specified torque. When tightening the tightening bolts for the torque converter, after fixing the crankshaft pulley bolts, be sure to confirm the tightening torque of the crankshaft pulley mounting bolts. After the converter is installed to the driveplate, rotate the crankshaft several turns and check to be sure that transmission rotates freely without binding.

40. Install the dust cover onto the converter housing.

41. Install the fluid cooler tube.

42. Install the starter motor.

43. Install the CKP sensor on the A/T assembly.

44. Reattach the transmission shifter control rod by installing the snap pin retaining the control rod bracket to the control device assembly lever.

45. Install the driveshaft.

46. Install the exhaust front tube and center muffler.

47. Install the engine undercover.

48. Install the A/T fluid level gauge.

49. Install the engine cover.

50. Connect the negative battery terminal.

51. Check the adjustment of the A/T position, as follows:

    a. Loosen the nut of the control rod.

    b. Place the PNP switch and selector lever in the "P" position.

    c. While pressing the lower lever toward the rear of the vehicle (in "P" position direction), tighten the nut.

➡**Do not push the bracket.**

52. Check the A/T position, as follows:

    a. Place the selector lever in "P" position, and turn ignition switch **ON** (engine **OFF**).

For VQ35DE models

Bolt No.	1	2	3	4
Number of bolts	1	5	2	1
Bolt length " ℓ "mm (in)	55 (2.17)	65 (2.56)	35 (1.38)	40 (1.57)
Tightening torque N·m (kg-m, ft-lb)	75 (7.7, 55)		47 (4.8, 35)	34 (3.5, 25)

◉ Transmission to engine

⊗ Engine to Transmission

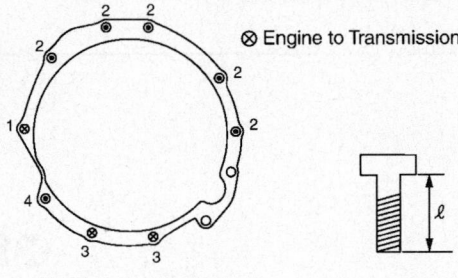

View from vehicle front

Fig. 12 Install the bolts fixing the transmission assembly to the engine making sure of proper bolt placement and installation torques—VQ35DE models

For VK45DE models

Bolt No.	1	2*1	3	4*2
Number of bolts	4	1	4	1
Bolt length mm (in)	70 (2.76)	70 (2.76)	65 (2.56)	70 (2.76)
Tightening torque N·m (kg-m, ft-lb)	113 (12, 83)		74.0 (7.5, 55)	113 (12,83)

*1 : No.2 bolt also secures A/T fluid charging pipe and washer.

*2 : No.4 bolt also secures bracket.

(A) : A/T to engine

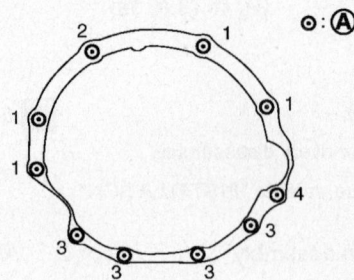

Fig. 13 Install the bolts fixing the transmission assembly to the engine making sure of proper bolt placement and installation torques—VK45DE models

b. Make sure the selector lever can be shifted to positions other than "P" when the brake pedal is depressed. Also, make sure the selector lever can be shifted from "P" position only when the brake pedal is depressed.

c. Move the selector lever and check for excessive effort, sticking, noise, or rattle.

d. Confirm the selector lever stops at each position with the feel of engagement when it is moved through all of the positions. Check whether or not the actual position the selector lever matches the position shown by the shift position indicator and the transmission body.

e. Confirm the back-up lamps illuminate only when the lever is placed in the "R" position. Confirm the back-up lamps do not illuminate when the selector lever is pushed against the "R" position while in the "P" or "N" positions.

f. Confirm the engine can only be started with the selector lever is in the "P" and "N" positions.

g. Make sure the transmission is locked completely in the "P" position.

h. When the selector lever is set to manual shift gate, make sure the manual mode is displayed on the combination meter. Shift the selector lever to "+" and "−" sides, and make sure set shift position changes.

TRANSFER CASE ASSEMBLY

REMOVAL & INSTALLATION

AWD Models

See Figure 14.

1. Before servicing the vehicle, refer to the Precautions Section.
2. Raise and safely support the vehicle.
3. Remove fixing bolts and nuts of the tunnel stay and the member stay and remove from the vehicle.
4. Remove the exhaust front tube.
5. Remove the front and rear driveshafts.
6. Disconnect the transfer assembly wiring harness connector and separate the wiring harness from the transfer assembly.
7. Remove the air breather hose.
8. Support the transfer assembly with a jack.
9. Remove the engine rear mounting.
10. Remove the transfer mounting bolts and separate the transfer from the transmission.

✳✳ CAUTION

Secure the transfer assembly to a jack or similar support.

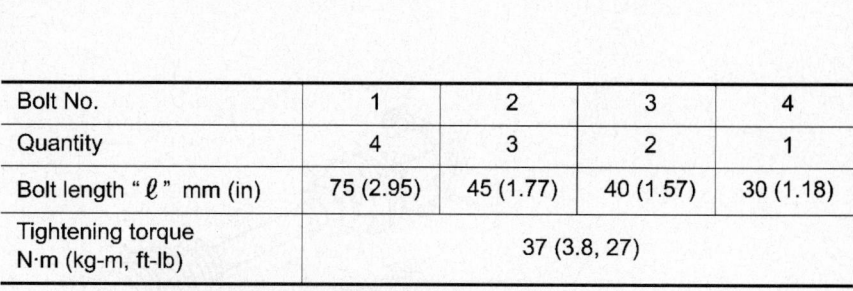

Bolt No.	1	2	3	4
Quantity	4	3	2	1
Bolt length "ℓ" mm (in)	75 (2.95)	45 (1.77)	40 (1.57)	30 (1.18)
Tightening torque N·m (kg-m, ft-lb)	37 (3.8, 27)			

⊙ : Transfer to Transmission
⊗ : Transmission to transfer

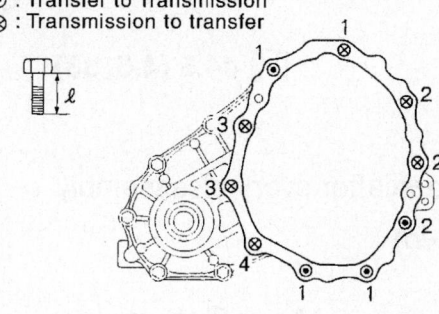

Fig. 14 Transfer case-to-transmission mounting bolt length and tightening specifications illustrated

To install:

11. Install the transfer mounting bolts and mount the transfer onto the transmission. Tighten the bolts in the proper order and torque specification as illustrated.
12. Install the engine rear mounting.
13. Install the air breather hose.
14. Connect the transfer assembly wiring harness connector and attach the wiring harness to the transfer assembly.
15. Install the front and rear driveshafts.
16. Install the exhaust front tube.
17. Install the fixing bolts and nuts of the tunnel stay and the member stay.
18. Check the fluid level and for fluid leakage.

FRONT HALFSHAFT

REMOVAL & INSTALLATION

AWD Models

See Figures 15 through 17.

1. Before servicing the vehicle, refer to the Precautions Section.
2. Raise and safely support the vehicle.
3. Remove the wheels.
4. Remove the undercover.
5. Remove the cotter pin. Then remove the locknut from the driveshaft.
6. Remove the wheel sensor wiring harness from the strut assembly.

✳✳ CAUTION

Do not pull on the wheel sensor wiring harness.

7. Remove the brake hose lock plate. Then remove the brake hose from the strut assembly.
8. Remove the mounting bolts and nuts between the strut assembly and the steering knuckle.
9. Separate the halfshaft from the steering knuckle.

✳✳ CAUTION

When removing the halfshaft, do not apply an excessive angle to the halfshaft joint. Also, be careful not to excessively extend the slide joint.

10. For the left-hand halfshaft, remove the fixing bolts of the front final drive side assembly halfshaft, then remove the halfshaft from the vehicle.
11. For the right-hand halfshaft, pry off the halfshaft from the front final drive assembly.
12. Inspect the halfshaft, as follows:
    a. Move the joint up/down, left/right, and in the axial direction. Check for any rough movement or significant looseness.
    b. Check the boot for cracks or other damage, and also for grease leakage.

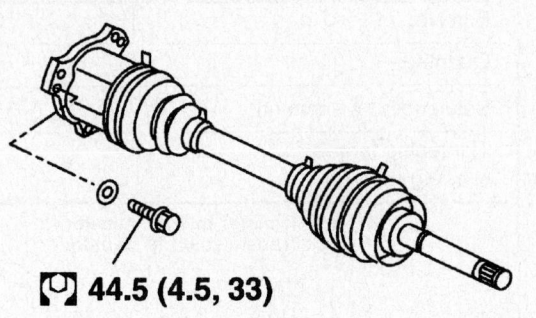
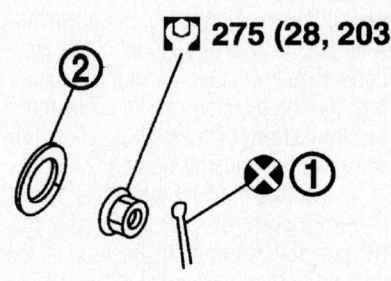

275 (28, 203)

44.5 (4.5, 33)

⊗ : Always replace after every disassembly

🔧 : N·m(kg-m,ft-lb)

1. Cotter pin   2. Washer

67162-FX35-G199

Fig. 15 View of left-hand front halfshaft

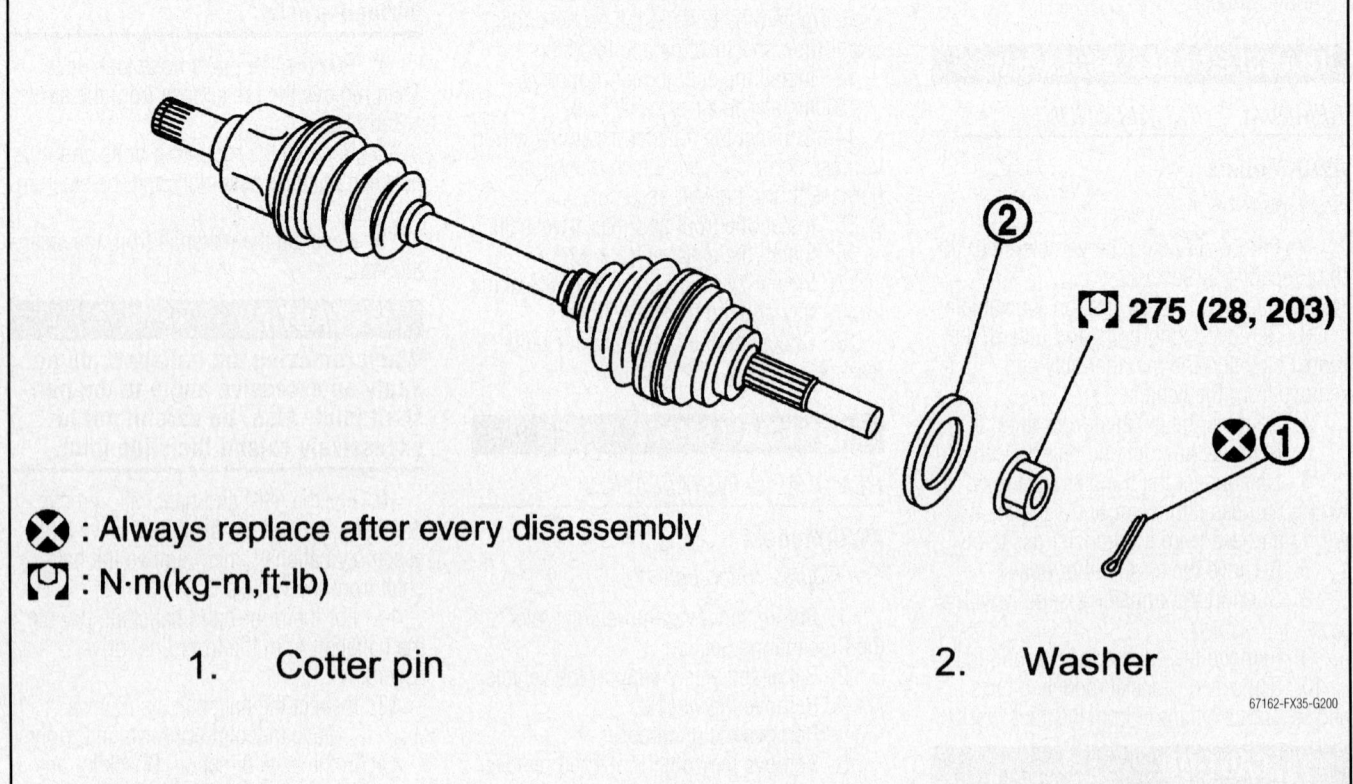

275 (28, 203)

⊗ : Always replace after every disassembly

🔧 : N·m(kg-m,ft-lb)

1. Cotter pin   2. Washer

67162-FX35-G200

Fig. 16 View of right-hand front halfshaft

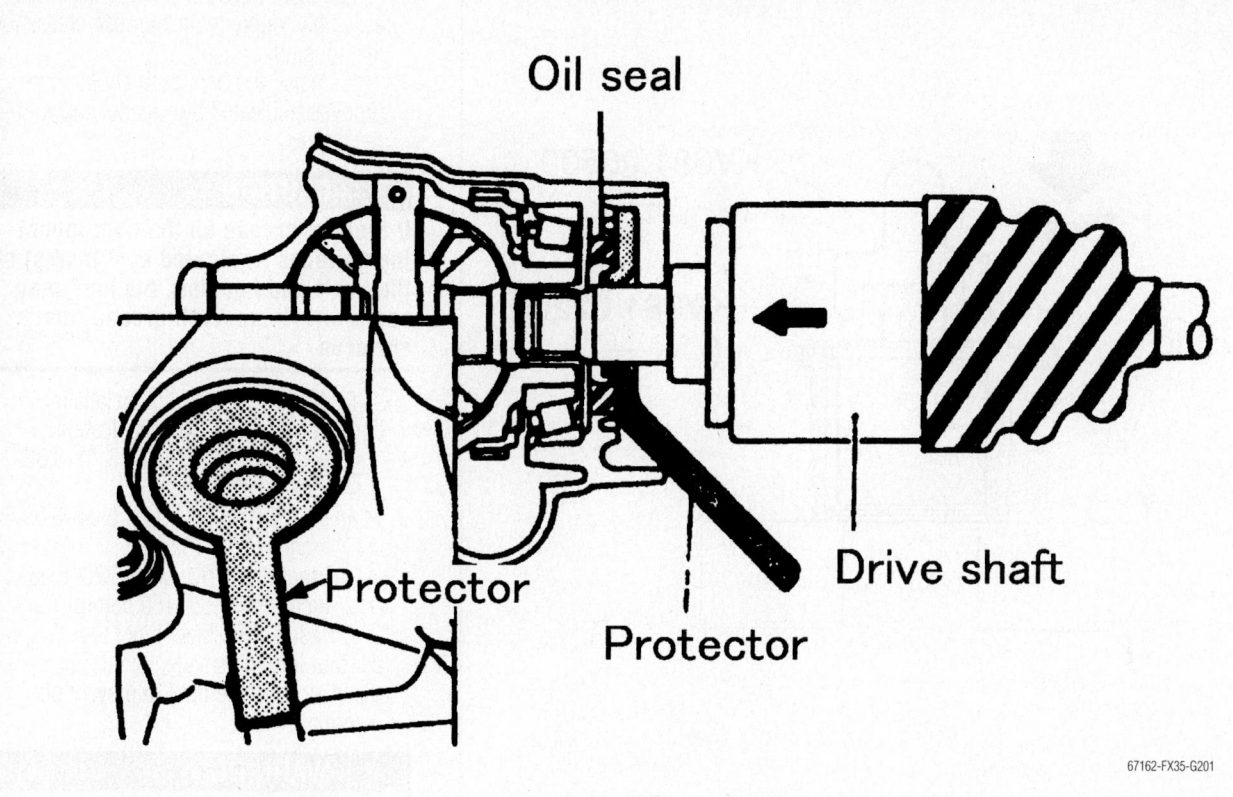

Fig. 17 Place the protector SST: KV38107900 onto front final drive assembly to prevent damage to the oil seal while inserting drive shaft

c. If damage is found, disassemble the halfshaft and replace the defective part(s) with new one(s).

To install:

➡**Refer to component parts location and do not reuse non-reusable parts.**

13. Installation is the reverse of removal. During installation, tighten the fasteners to the following specifications:
   a. Halfshaft-to-hub locknut: 203 ft. lbs. (275 Nm).
   b. Left-hand side halfshaft-to-drive unit bolts: 33 ft. lbs. (45 Nm).

14. Right-hand side, in order to prevent damage to the front final drive assembly side oil seal, first fit protector SST: KV38107900 onto the oil seal before inserting the halfshaft. Slide the halfshaft into the slide joint and tap it with a hammer to install it securely.

➡**Be sure to check that the circular clip is securely fastened after inserting the halfshaft.**

15. Check the condition of the wheel sensor wiring harness. Repair if necessary.

CV-JOINTS OVERHAUL

See Figures 18 through 21.

1. Before servicing the vehicle, refer to the Precautions Section.

2. Raise and safely support the vehicle.

3. For the front final drive assembly side, perform the following:
   a. Press the halfshaft in a vice.

➡**When retaining the shaft in a vice, always use copper or aluminum plates between the vise and shaft.**

   b. Remove the boot bands.
   c. If the plug needs to be removed, move the boot toward the wheel side, and drive it out with a plastic hammer.
   d. Put matching marks on the spider assembly and shaft.

➡**Use paint for the matching mark, but don't damage the spider assembly and halfshaft.**

   e. Remove the snapring and remove the spider assembly from the shaft.
   f. Remove the boot from the shaft.
   g. Remove the old grease on the slide joint assembly.

4. For the wheel side, perform the following:
   a. Place the halfshaft in a vice.

➡**When retaining the halfshaft in a vice, always use copper or aluminum plates between a vise and halfshaft.**

   b. Remove the boot bands.
   c. Remove the boot from the joint sub-assembly.
   d. Screw a halfshaft puller or equivalent 1.18 inch (30mm) or more into the threaded part of the joint sub-assembly. Pull the joint sub-assembly out of the halfshaft.

➡**If the joint sub-assembly cannot be removed after 5 or more unsuccessful attempts, replace the shaft and joint subassembly as a set. Use a sliding hammer on the halfshaft and remove the halfshaft.**

   e. Remove the boot from the half-shaft.
   f. Remove the circular clip from the halfshaft.
   g. While rotating the ball cage, remove the old grease from the joint sub-assembly.

5. Replace the halfshaft if there is any run-out, cracking, or other damage.

6. Make sure there is no rough rotation or unusual axial looseness of the joint sub-assembly.

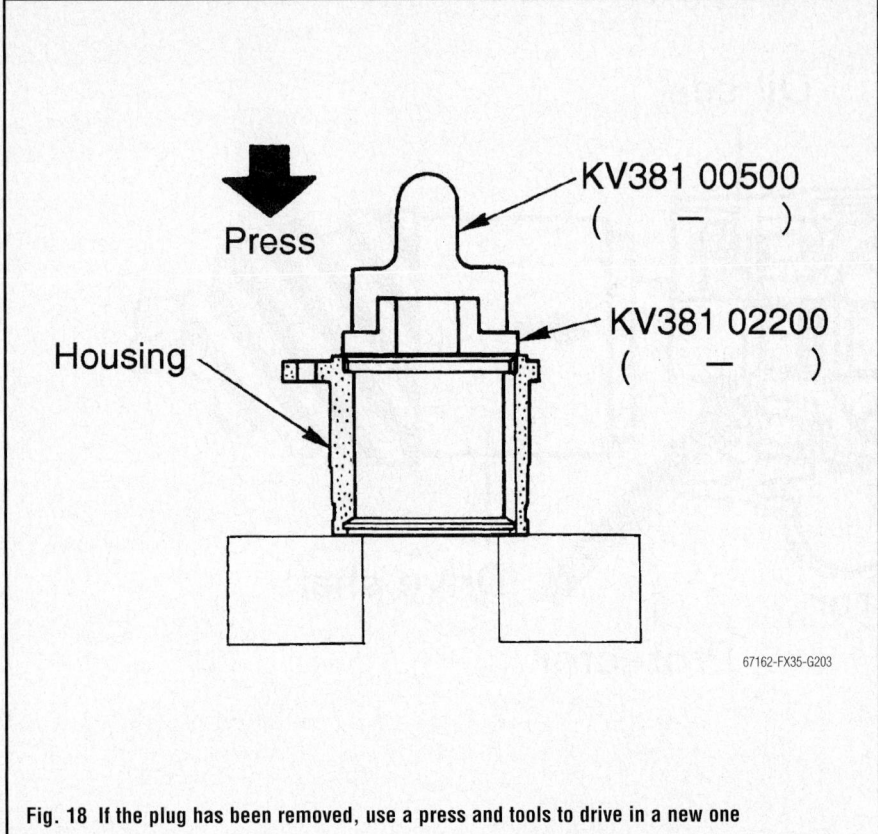

Fig. 18 If the plug has been removed, use a press and tools to drive in a new one

KV381 00500
(—)

KV381 02200
(—)

Press

Housing

67162-FX35-G203

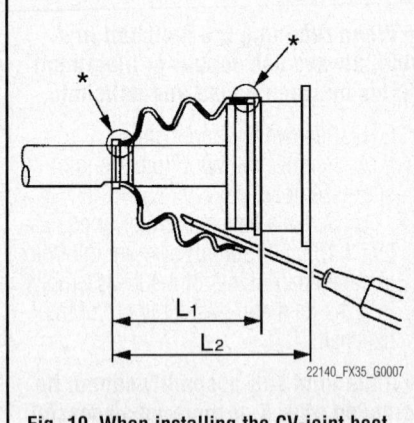

Fig. 19 When installing the CV-joint boot, make sure the installation length meets specification

22140_FX35_G0007

7. Make sure there is no foreign material inside the joint sub-assembly.

8. Check joint sub-assembly for compression scar, cracks, or fractures.

➡**If there are any irregular conditions of joint sub-assembly components, replace the entire joint sub-assembly.**

9. Inspect the housing and spider assembly. If the roller or roller surface of the spider assembly have scratches or other wear, replace the housing and the spider assembly.

➡**The housing and spider assembly are components which are used as a set.**

To assemble:
10. To assemble the front final drive assembly side, perform the following:

a. If the plug has been removed, use a drift to press in a new one.

➡**Discard the old plug, and replace it with a new one.**

b. Wind the serrated part of the shaft with tape. Install the boot band and boot on the shaft. Be careful not to damage the boot.

➡**Discard the old boot band and boot, and replace them with new ones.**

c. Remove the protective tape wound around the serrated part of the shaft.

d. Align the alignment marks, which were made when the spider assembly was removed. Install the spider assembly with the serration chamfer facing the shaft.

e. Secure the spider assembly with a snapring.

➡**Discard the old snapring, and replace it with a new one.**

f. Apply multi-purpose grease to the spider assembly and sliding surface.

g. Install the housing onto the spider assembly. Apply multi-purpose grease to the housing.

h. Install the boot securely into the grooves (indicated by * marks shown in the figure).

✳✳ CAUTION

If there is grease on the boot mounting surfaces (indicated by * marks) of the shaft and housing, the boot may come off. Remove all grease from surfaces.

i. Make sure the boot installation length is according to specification.
- VK45DE models (L1): 3.74–3.82 inches (95–97mm)
- VQ35DE models (L2): 5.94–6.02 inches (151–153mm)
- Front final drive side: 6.20–6.28 inches (157.55–159.55mm)

j. Insert a flat-bladed pry tool or similar tool into the smaller side of the boot. Bleed air from the boot to prevent boot deformation.

✳✳ WARNING

The boot may break if the boot installation length is less than the standard value. Take care not to touch the tip of the pry tool to the inside surface of the boot.

k. Install the new larger and smaller boot bands securely with a suitable tool.

➡**Discard the old boot bands, and replace them with new ones.**

l. After installing the housing and shaft, rotate the boot to check whether or not the actual position is correct. If the boot position is not correct, secure the boot with new boot bands again.

11. Assemble the wheel side of the half-shaft as follows:

a. Insert grease (Nissan genuine grease or equivalent) into the joint sub-assembly serration hole until grease begins to ooze from the ball groove and serration hole. After inserting the grease, use a shop cloth to wipe off the old grease that has oozed out.

b. Wind the serrated part of the shaft with tape for protection of the boot, then install the boot band and boot on the shaft. Be careful not to damage the boot.

➡**Discard the old boot band and boot, and replace them with new ones.**

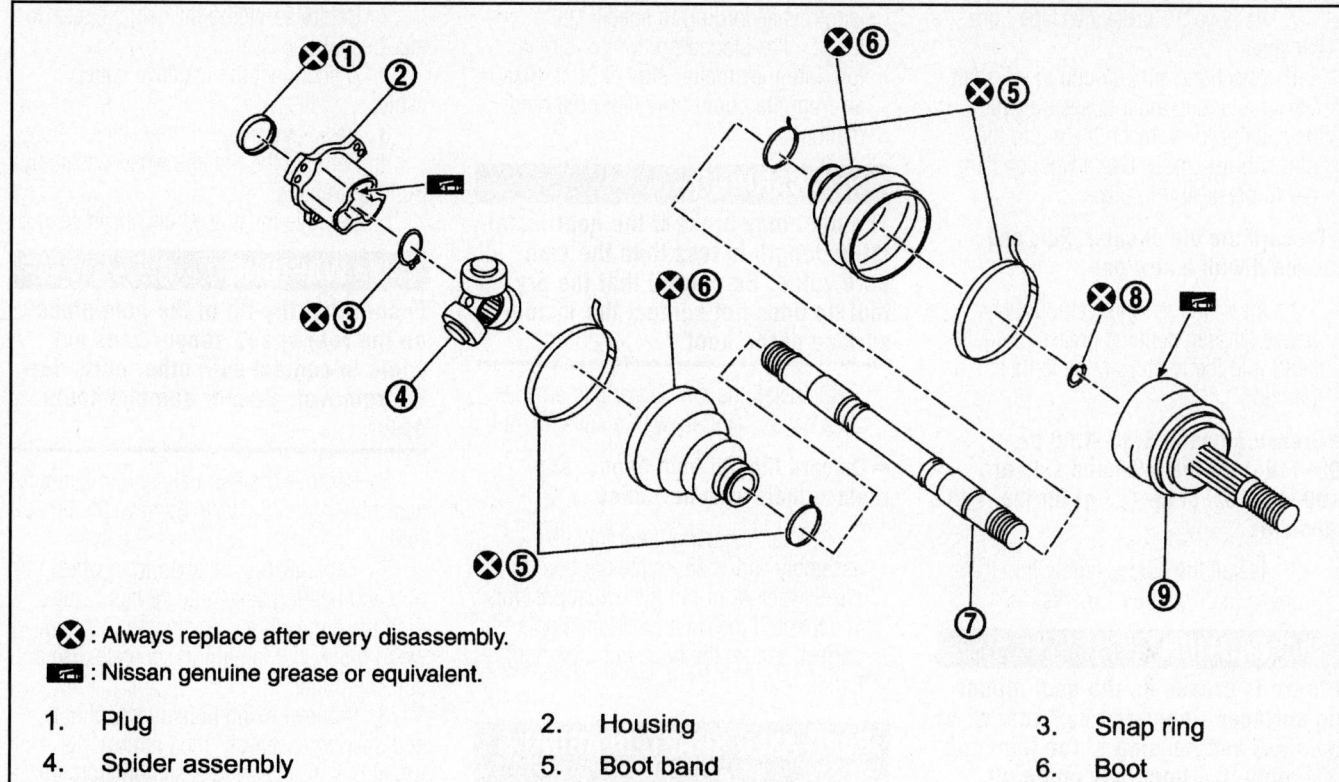

⊗ : Always replace after every disassembly.

▭ : Nissan genuine grease or equivalent.

1.	Plug	2.	Housing	3.	Snap ring
4.	Spider assembly	5.	Boot band	6.	Boot
7.	Shaft	8.	Circular clip	9.	Joint sub-assembly

67162-FX35-G202

Fig. 20 Exploded view of the left-hand halfshaft and CV-joint

⊗ : Always replace after every disassembly.

▭ : Nissan genuine grease or equivalent.

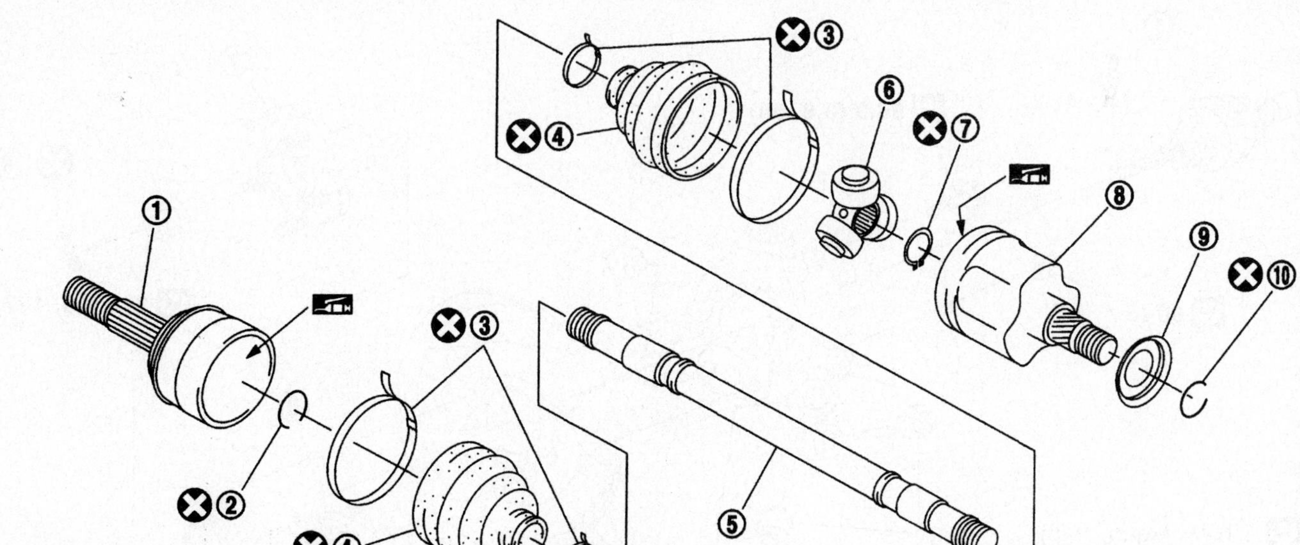

1.	Joint sub-assembly	2.	Circular clip	3.	Boot band
4.	Boot	5.	Shaft	6.	Spider assembly
7.	Snap ring	8.	Housing	9.	Dust shield
10.	Circular clip				

67162-FX35-G206

Fig. 21 Exploded view of the right-hand halfshaft and CV-joint

c. Remove the protective tape from the shaft.

d. Attach the circular clip to the shaft. The circular clip must fit securely into the shaft groove. Attach the nut to the joint sub-assembly. Use a wooden hammer to press-fit it in place.

➡ **Discard the old circular clip, and replace it with a new one.**

e. Insert the specified amount of grease (Nissan genuine grease or equivalent) into the boot from the large end of the boot.

➡ **Grease amount: 3.35–4.06 oz. (95–115g) for the left-hand side or 3.99–4.34 oz. (113–123 g) for the right-hand side.**

f. Install the boot securely into the grooves (indicated by * marks).

✳ CAUTION

If there is grease on the boot mounting surfaces (indicated by * marks) of the shaft and housing of the joint sub assembly, the boot may come off. Remove all grease from the surfaces.

g. Make sure the boot installation length is according to specification. Insert a flat-bladed pry tool or similar tool into the smaller side of boot. Bleed air from the boot to prevent boot deformation.

✳ WARNING

The boot may brake if the boot installation length is less than the standard value. Be careful that the pry tool tip does not contact the inside surface of the boot.

h. Install the new larger and smaller boot bands securely with a suitable tool.

➡ **Discard the old boot bands, and replace them with new ones.**

i. After installing the joint sub-assembly and shaft, rotate the boot to check whether or not the actual position is correct. If the boot position is not correct, secure the boot with new boot bands again.

REAR AXLE SHAFT, BEARING & SEAL

REMOVAL & INSTALLATION

See Figure 22.

1. Before servicing the vehicle, refer to the Precautions Section.
2. Disconnect the negative battery cable.
3. Raise and support the vehicle safely.
4. Remove the tire and wheel assembly from the vehicle.
5. Remove the rear wheel speed sensor.

✳ WARNING

Ensure that the tip of the pole piece on the rear speed sensor does not come in contact with other parts during removal. Sensor damage could occur.

6. Remove the rear caliper and support assembly out of the way. Remove the brake rotor.
7. Separate the halfshaft from wheel hub and bearing assembly by lightly tapping the end with a suitable hammer and wood block. If it is hard to separate, use a suitable puller.
8. Remove fixing bolts of wheel hub and bearing assembly, then remove the wheel hub and bearing assembly from the axle.
9. Remove the parking brake cable and parking brake shoe from the back plate.

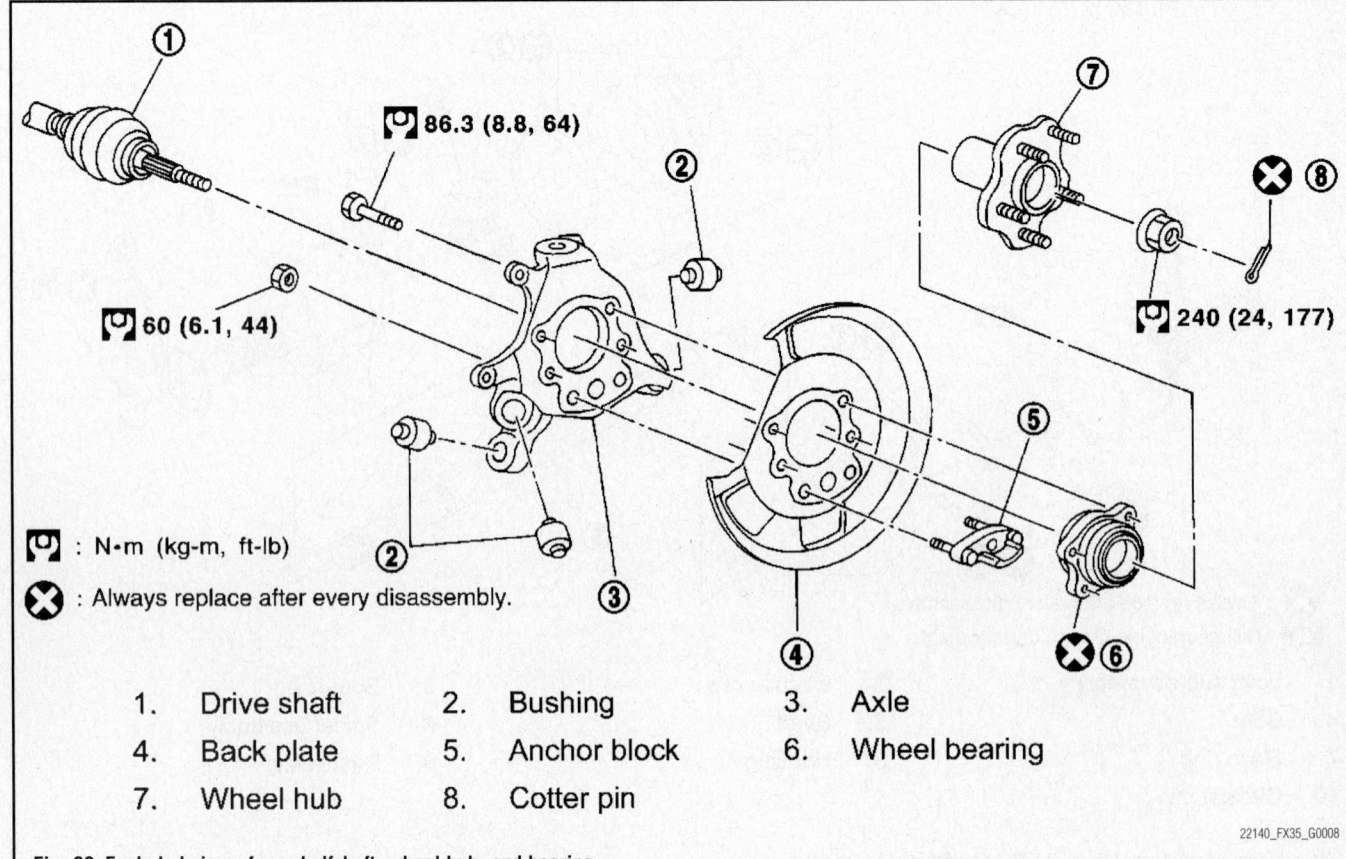

86.3 (8.8, 64)
60 (6.1, 44)
240 (24, 177)

: N•m (kg-m, ft-lb)

: Always replace after every disassembly.

1.	Drive shaft	2.	Bushing	3.	Axle
4.	Back plate	5.	Anchor block	6.	Wheel bearing
7.	Wheel hub	8.	Cotter pin		

Fig. 22 Exploded view of rear halfshaft, wheel hub, and bearing

22140_FX35_G0008

10. Remove the fixing nuts of the anchor block, then remove the anchor block and back plate from the axle.

11. Loosen the fixing bolts and nuts of the front lower link, radius rod, and rear lower link in the side of the suspension member.

12. Set a jack under the rear lower link. Then remove the fixing bolt in the front lower link side of the shock absorber.

13. Remove the bolt and nut in the axle side of the rear lower link. Then remove the coil spring.

14. Remove the fixing bolts and nuts in the axle side of the front lower link and radius rod.

15. Remove the suspension arm and cotter pin at the axle, then loosen the mounting nut.

16. Use a ball joint remover to remove the suspension arm from the axle. Be careful not to damage the ball joint boot.

➡**Temporarily tighten the mounting nut to prevent damage to the threads and to prevent the ball joint remover from coming off.**

17. Remove the axle/halfshaft from the vehicle.

18. Remove wheel bearing fixing bolts and anchor block fixing nuts, and remove the wheel hub and bearing assembly, back plate, and anchor block from the axle/halfshaft.

19. Using a drift and a puller, press the wheel hub out to remove from wheel bearing.

20. Using a drift and a puller, press the wheel bearing outer side inner race out to remove from wheel hub.

21. Using a suitable drift, remove each bushing from the axle.

To install:

22. Press fit a wheel hub into wheel bearing with a drift.

➡**Press fit a drift while holding it against the wheel bearing inner side inner race. The wheel bearing cannot be reused.**

23. Install the back plate and wheel hub and bearing assembly.

24. Install the anchor block onto axle.

25. Install the rear axle shaft. Install the companion flange to the rear axle shaft, then install a new self-locking nut.

26. While holding the rear axle shaft in position, tighten a new self-locking nut to 177 ft. lbs. (240 Nm).

27. Install the rear brake disc, caliper assembly and parking brake.

28. Install the tire and wheel assembly and lower the vehicle. Check the parking brake stroke and adjust as required.

29. Before moving the vehicle, pump the brakes until a firm pedal is achieved.

REAR HALFSHAFT

REMOVAL & INSTALLATION

See Figure 23.

1. Before servicing the vehicle, refer to the Precautions Section.

2. Raise and safely support the vehicle.

3. Remove the wheels.

4. Remove the cotter pin. Then, remove the locknut from the outer end of the half-shaft.

5. Remove the fixing nuts and bolts between the side flange and halfshaft.

6. Separate the halfshaft from the wheel hub and bearing assembly by lightly tapping the end with a suitable hammer and wood block. If it is hard to separate, use a suitable puller.

7. Remove the halfshaft from the axle and from the vehicle.

✳✳ CAUTION

When removing the halfshaft, do not apply an excessive angle to the halfshaft joint. Also be careful not to excessively extend the slide joint.

To install:

8. Inspect the halfshaft, as follows:

a. Move the joint up/down, left / right, and in the axial direction. Check for any rough movement or significant freeplay.

b. Check the boot for cracks or other damage, and also for grease leak-age.

c. If damage is found, disassemble the halfshaft and replace the defective part(s) with new one(s).

➡**Refer to component parts location and do not reuse non-reusable parts.**

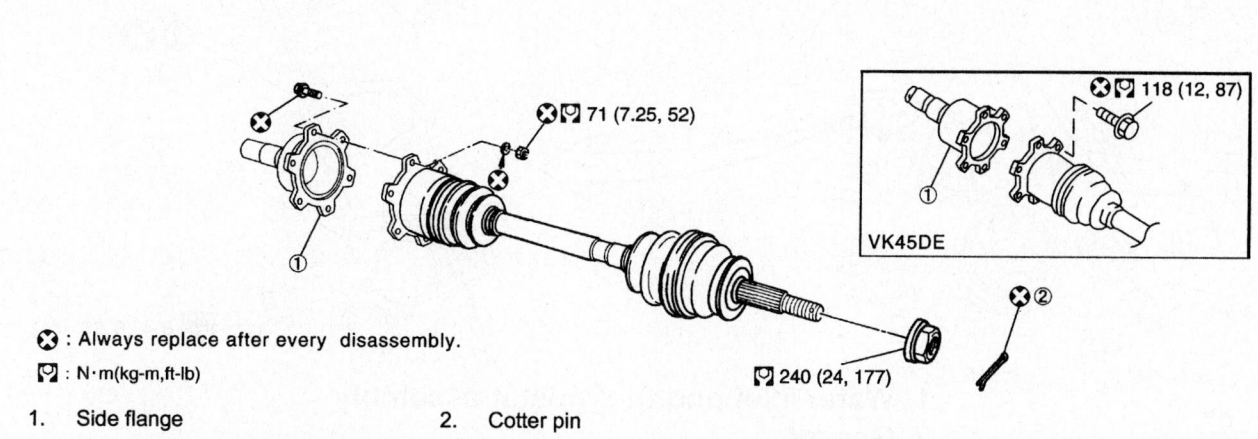

✖ : Always replace after every disassembly.

🔧 : N·m(kg-m,ft-lb)

1. Side flange   2. Cotter pin

71 (7.25, 52)

118 (12, 87)

VK45DE

240 (24, 177)

67162-FX35-G207

Fig. 23 View of the rear halfshaft with torque specifications

9. Installation is the reverse of removal. During installation, tighten the fasteners to the following specifications:

    a. Halfshaft-to-hub locknut: 177 ft. lbs. (240 Nm).

    b. Halfshaft-to-side flange bolts: 52 ft. lbs. (71 Nm).

REAR PINION SEAL

REMOVAL & INSTALLATION

1. Before servicing the vehicle, refer to the Precautions Section.

2. Raise and safely support the vehicle.

3. Remove the rear driveshaft.

4. Put a matching mark on the end of the drive pinion corresponding to the matching mark on the final drive companion flange.

➡**Use paint for the matching mark; never damage the drive pinion. The matching mark on the final drive companion flange indicates the maximum vertical run out position.**

5. Using the drive pinion flange wrench KV40104000 or equivalent, remove the drive pinion locknut.

6. Remove the companion flange using the puller.

7. Remove the front oil seal using the side bearing outer race puller ST33290001 (J34286) or equivalent.

To install:

8. Apply multi-purpose grease to sealing lips of oil seal. Press the front oil seal into the carrier with tool ST30720000 (J25405) or equivalent.

➡**When installing the side oil seal, be careful not to install it crooked.**

Do not reuse the old side oil seal. Always replace the oil seal with a new one.

9. Align the matching mark of the drive pinion with the matching mark of the companion flange, then install the companion flange.

10. Apply oil on the threaded part of the drive pinion and the seating surface of the drive pinion locknut.

11. Install the drive pinion nut with tool KV40104000 or equivalent. Tighten to 109–238 ft. lbs. (147–323 Nm).

➡**The drive pinion locknut is not reusable. Never reuse the drive pinion nut.**

12. Install the rear driveshaft.

ENGINE COOLING

THERMOSTAT

REMOVAL & INSTALLATION

3.5L Engine

See Figure 24.

✳✳ CAUTION

Never remove the radiator cap when the engine is hot. Serious burns could occur from high-pressure engine coolant escaping from the radiator.

1. Before servicing the vehicle, refer to the Precautions Section.

2. Be sure the engine is cold.

3. Disconnect the negative battery cable.

4. Remove the front engine cover.

5. Drain the engine coolant using the radiator drain plug and the water drain plug at the front of the cylinder block. Properly dispose of used coolant.

6. Remove the air duct (inlet).

7. Remove the water inlet and thermostat housing retaining bolts.

8. Remove the thermostat assembly from the engine.

➡**Do not disassemble the water inlet and thermostat assembly. Replace them as a unit, if required.**

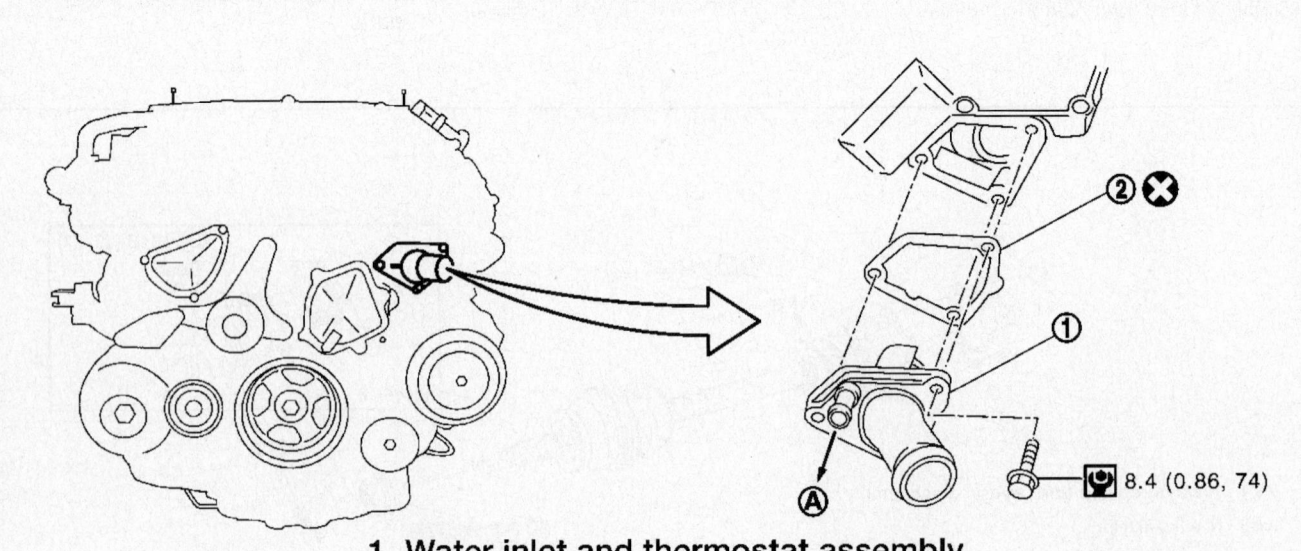

1. Water inlet and thermostat assembly
2. Gasket

22140_FX35_G0009

Fig. 24 Thermostat assembly and related components—3.5L engine

To install:

9. Installation is the reverse of the removal procedure.

10. Tighten the water inlet and thermostat housing retaining bolts to 74 inch lbs. (8 Nm).

11. Be sure to refill the cooling system using the proper grade and type engine coolant.

12. Start the engine and check for leaks.

13. Start the engine and allow it to reach operation temperature. Recheck the coolant level, fill as required.

4.5L Engine

See Figures 25 and 26.

➡**Never remove the radiator cap when the engine is hot. Serious burns could occur from high-pressure engine coolant escaping from the radiator.**

1. Before servicing the vehicle, refer to the Precautions Section.

2. Be sure the engine is cold.

3. Disconnect the negative battery cable.

4. Drain the engine coolant using the radiator drain plug and the water drain plug at the front of the cylinder block. Properly dispose of used coolant.

5. Remove the front engine cover.

6. Remove the air duct (inlet).

7. Disconnect the water suction hose from the water inlet.

8. Remove the thermostat housing, water outlet pipe, water connector, water control valve, water outlet, and heater pipe.

➡**Do not disassemble the water control valve.**

To install:

9. Installation is the reverse of the removal procedure.

10. Install the thermostat and water control valve with the whole circumference of each flange part fit securely inside the rubber ring.

11. Install the thermostat with the jiggle valve facing upwards. The position deviation may be within plus or minus 10° as illustrated.

12. Install the water control valve with the UP mark facing upward and the frame center part facing upward. The position deviation may be within plus or minus 10°.

13. Be sure to refill the cooling using the proper grade and type engine coolant.

14. Start the engine and check for leaks.

15. Start the engine and allow it to reach full operating temperature.

16. Once the vehicle has cooled, recheck the coolant level.

WATER PUMP

REMOVAL & INSTALLATION

3.5L Engine

See Figures 27 through 30.

The water pump cannot be disassembled and should be replaced as a unit when found defective.

1. Before servicing the vehicle, refer to the Precautions Section.

➡**During service, be sure to prevent engine coolant from contacting the accessory drive belt.**

2. Remove the front engine undercover.

3. Remove the drive belts.

4. Drain the engine coolant from the radiator.

❊❊ CAUTION

Make sure the engine is cold before draining the coolant.

5. Remove the air duct (inlet), power duct, and air cleaner case assembly.

6. Remove the cylinder block drain plug (front side) of engine.

7. Remove the chain tensioner cover and water pump cover. Use a seal cutter to separate the two mating surfaces.

❊❊ WARNING

Be careful not to damage the mating surfaces.

8. Remove the timing chain tensioner (primary), as follows:

   a. Pull the lever down and release the plunger stopper tab.

➡**The plunger stopper tab can be pushed up to release the coaxial structure with a lever.**

   b. Insert a stopper pin into the tensioner body hole to hold the lever and keep the plunger stopper tab released.

➡**An Allen wrench can be used for a stopper pin.**

   c. Insert the plunger into the tensioner body by pressing the timing chain slack guide.

   d. Keep the slack guide depressed and hold the plunger in by pushing the stopper pin deeper through the lever and into the tensioner body hole.

   e. Turn the crankshaft pulley approximately 20° clockwise, so that the timing chain on the timing chain tensioner (primary) side is loose.

   f. Remove the mounting bolts and timing chain tensioner (primary).

❊❊ WARNING

Be careful not to drop mounting bolts inside the chain case.

9. Remove the 3 water pump fixing bolts. Secure a gap between the water pump gear and timing chain, by turning the crankshaft pulley counterclockwise until timing chain slack on the water pump sprocket is at its maximum.

10. Insert M8 bolts into the water pump upper and lower mounting bolt holes until they contact the timing chain case. Then, alternately tighten each bolt ½ turn, and pull out the water pump.

➡**The M8 bolts should be 0.0492 x 1.97 inch (1.25mm x 50mm).**

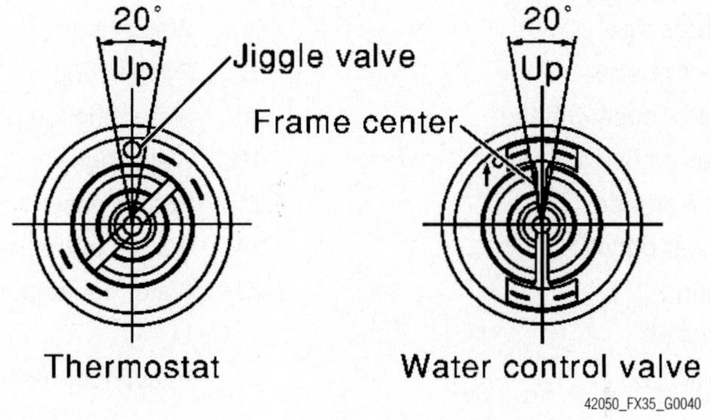

Fig. 25 Thermostat and water control valve installation alignment—4.5L engine

42050_FX35_G0040

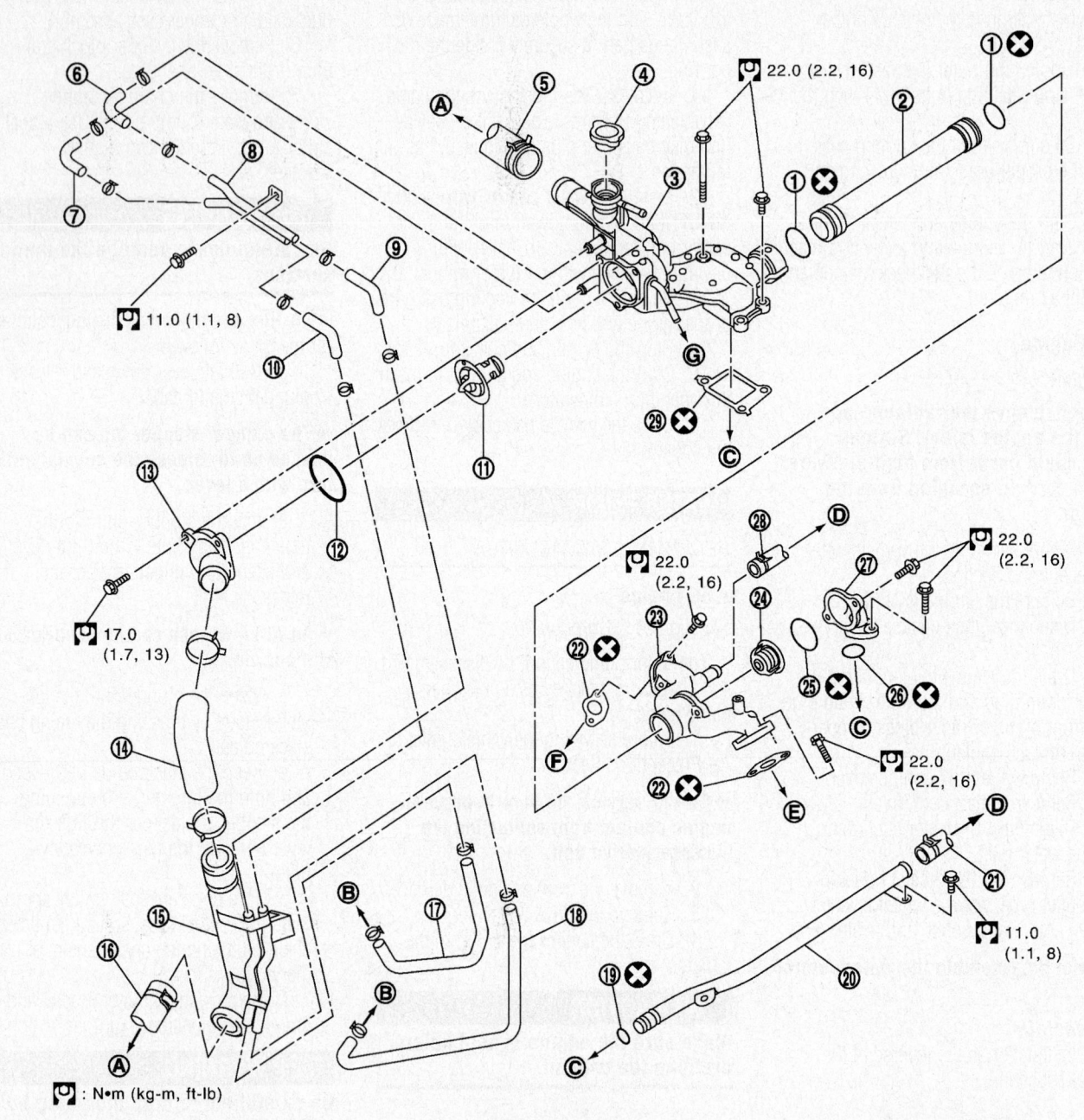

: N•m (kg-m, ft-lb)

1.	O-ring	2.	Water outlet pipe	3.	Thermostat housing
4.	Radiator cap	5.	Radiator hose (upper)	6.	Water hose
7.	Water hose	8.	Water pipe	9.	Water hose
10.	Water hose	11.	Thermostat	12.	Rubber ring
13.	Water inlet	14.	Water suction hose	15.	Water suction pipe
16.	Radiator hose (lower)	17.	Water hose	18.	Water hose
19.	O-ring	20.	Heater pipe	21.	Heater hose
22.	Gasket	23.	Water outlet	24.	Water control valve
25.	Rubber ring	26.	O-ring	27.	Water connector
28.	Heater hose	29.	Gasket		
A.	To radiator	B.	To oil cooler	C.	To cylinder block

22140_FX35_G0010

Fig. 26 Thermostat assembly and related components—4.5L engine

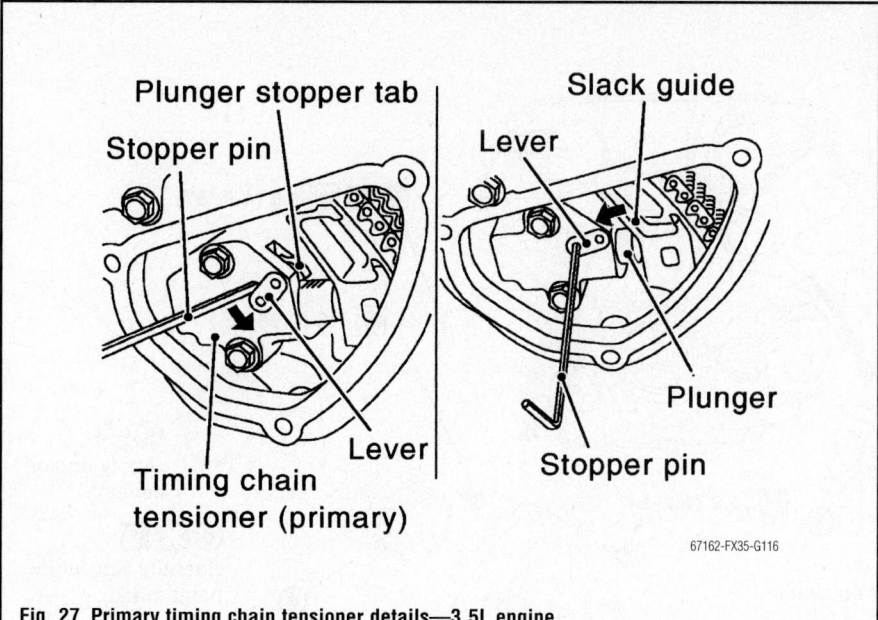

Fig. 27 Primary timing chain tensioner details—3.5L engine

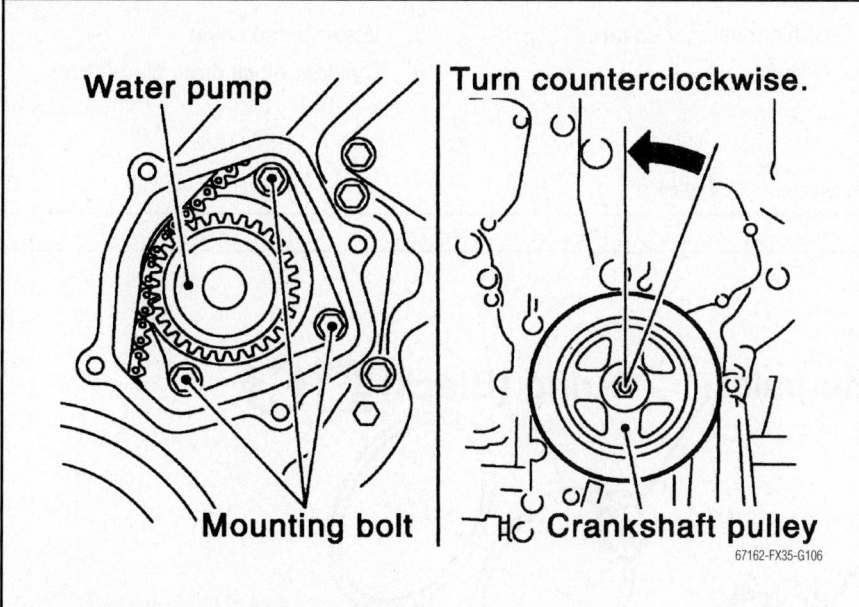

Fig. 28 Water pump mounting bolt locations—3.5L engine

✳✳ WARNING

Pull the water pump straight out while preventing the vane from contacting the socket in the installation area. Remove the water pump without allowing the sprocket to contact the timing chain.

11. Remove the M8 bolts and O-rings from the water pump.
12. Do not disassemble the water pump.
13. Check for badly rusted or corroded water pump body assembly.
14. Check for rough operation due to excessive end play.

15. If any defects are found, replace the water pump.

To install:
16. Install new O-rings on the water pump.
17. Apply engine oil and engine coolant to the water pump O-rings as illustrated.
18. Position the O-ring with the white paint mark toward the engine front side.
19. Install the water pump.

✳✳ CAUTION

Do not allow the cylinder block to damage the O-rings during installation.

20. Make sure the timing chain and water pump sprocket are engaged properly.
21. Install the water pump by alternately and evenly tightening the mounting bolts.
22. Install the timing chain tensioner (primary), as follows:
   a. Remove all dust and foreign material completely from the backside of the chain tensioner and from the installation area of the rear timing chain case.
   b. Turn the crankshaft pulley clockwise so that the timing chain on the timing chain tensioner (primary) side is loose.
   c. Apply engine oil to the oil hole and tensioner when installing the timing chain tensioner.
   d. Install the timing chain tensioner (primary).
   e. Remove the stopper pin.
23. Install the chain tensioner cover and water pump cover.

➡**Before installing, remove all traces of liquid gasket from the mating surface of the water pump cover and chain tensioner cover using a scraper. Also, remove traces of liquid gasket from the mating surface of the front timing chain case.**

24. Apply a continuous bead of liquid gasket 0.091–0.130 inch (2–3mm) thick to the mating surface of the chain tensioner cover and water pump cover.

➡**Use RTV Silicone Sealant or equivalent.**

25. Install the cylinder block drain plug (front side).
26. Apply thread sealant to the thread of cylinder block drain plug.

➡**Use Genuine Thread Sealant or equivalent.**

27. Install the air duct (inlet), power duct and air cleaner case assembly.
28. Fill the engine with coolant.
29. Install the drive belts.
30. Install the front engine undercover.
31. Check for engine coolant leaks using a radiator cap tester.
32. Start the engine and let it idle for 3 minutes, then raise the engine RPM up to 3,000 RPM under no load to purge air from the high-pressure chamber of the chain tensioner. The engine may produce a rattling noise. This indicates that air remains in the chamber, but it is not a matter of concern.
33. With the engine idling, visually make sure that there are no leaks of engine coolant.

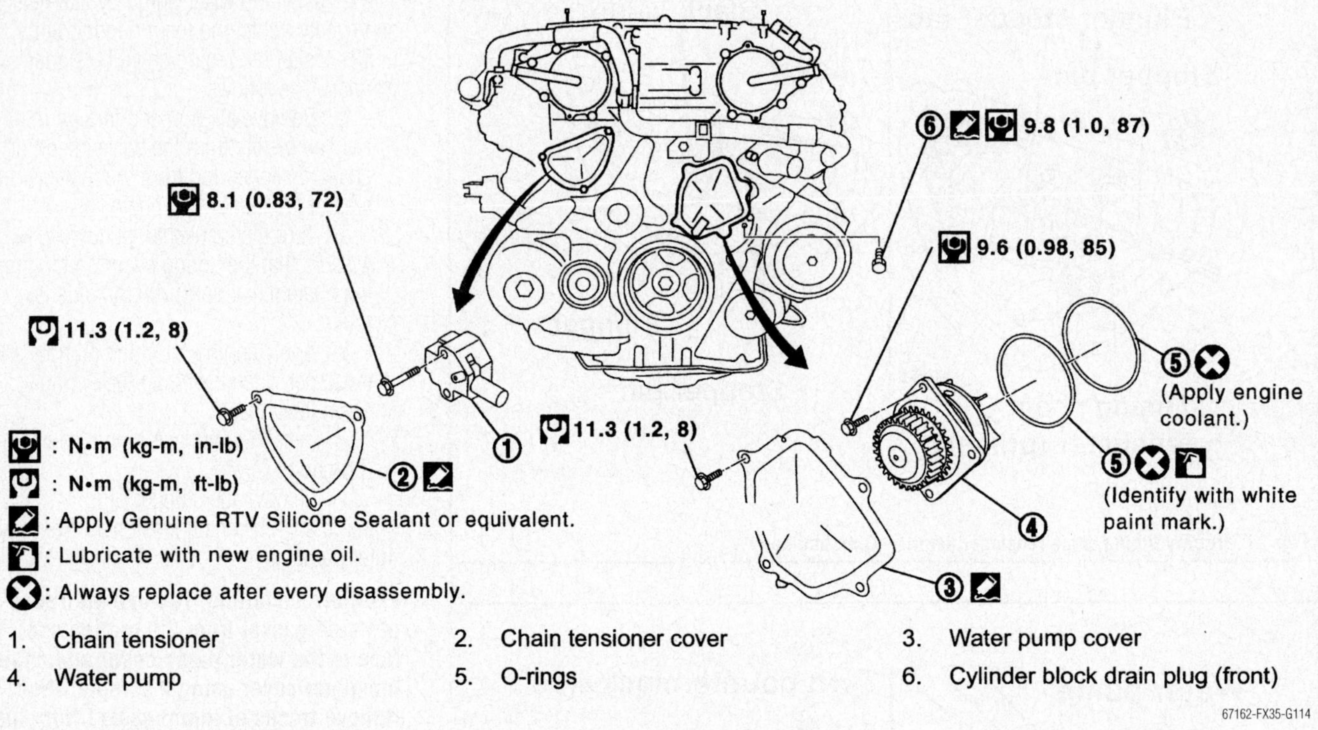

⑥ ✏ 🔧 9.8 (1.0, 87)

🔧 8.1 (0.83, 72)

🔧 9.6 (0.98, 85)

💿 11.3 (1.2, 8)

⑤ ❌ (Apply engine coolant.)

💿 11.3 (1.2, 8)

⑤ ❌ 🛢 (Identify with white paint mark.)

🔧 : N•m (kg-m, in-lb)

💿 : N•m (kg-m, ft-lb)

✏ : Apply Genuine RTV Silicone Sealant or equivalent.

🛢 : Lubricate with new engine oil.

❌ : Always replace after every disassembly.

①

② ✏

③ ✏

④

1. Chain tensioner
2. Chain tensioner cover
3. Water pump cover
4. Water pump
5. O-rings
6. Cylinder block drain plug (front)

67162-FX35-G114

Fig. 29 Exploded view of water pump and related components—3.5L engine

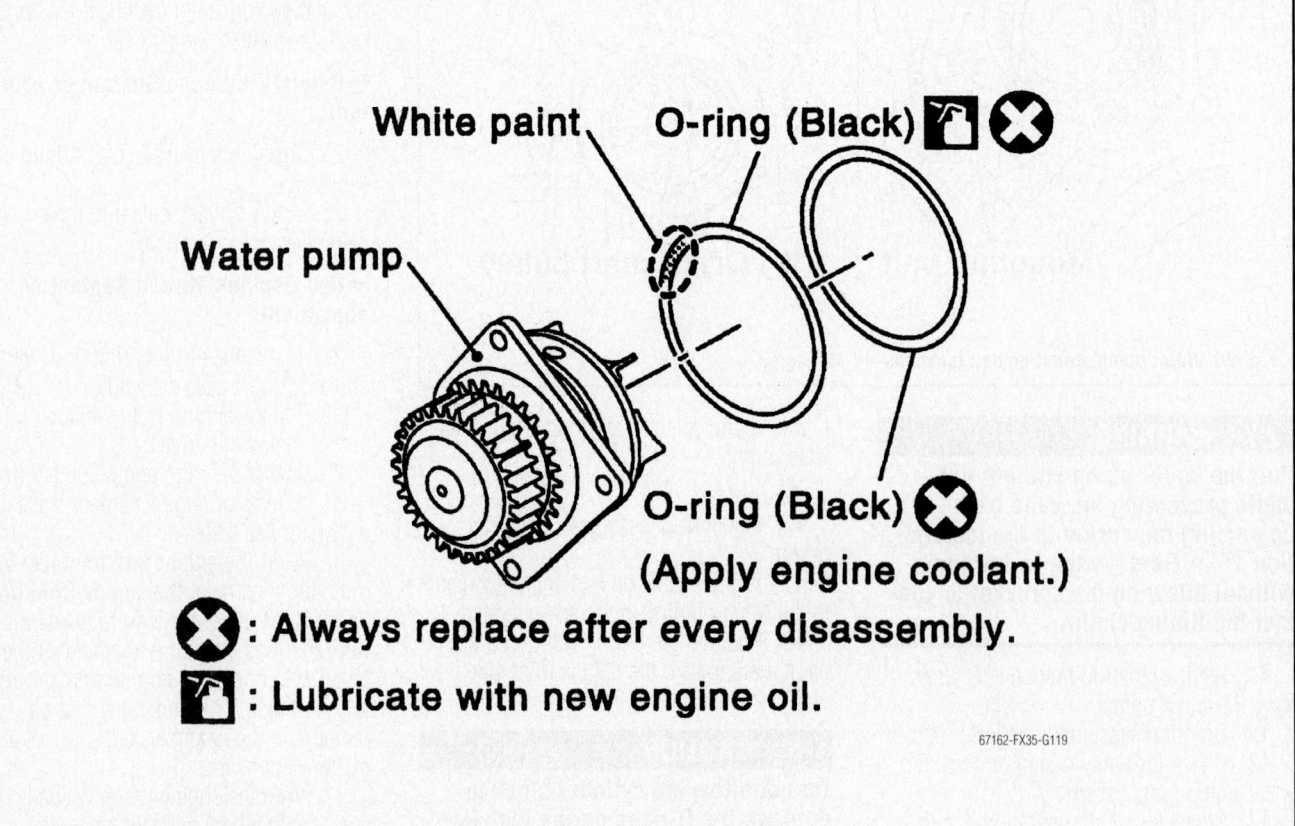

White paint   O-ring (Black) 🛢 ❌

Water pump

O-ring (Black) ❌
(Apply engine coolant.)

❌ : Always replace after every disassembly.

🛢 : Lubricate with new engine oil.

67162-FX35-G119

Fig. 30 Water pump O-ring positioning for installation—3.5L engine

4.5L Engine

See Figure 31.

1. Before servicing the vehicle, refer to the Precautions Section.

❋❋ CAUTION

When removing water pump, be careful not to get engine coolant on the drive belt.

➡**The water pump cannot be disassembled and should be replaced as a unit if it is found faulty.**

2. Drain the engine coolant from the drain plugs on the radiator and both sides of the engine block.

❋❋ WARNING

Make sure the engine is cold before draining the coolant.

3. Remove the engine front undercover.
4. Remove the air duct (inlet).
5. Remove the accessory drive belt. Refer to Accessory Drive Belts, removal & installation.
6. Remove the fan coupling with the cooling fan, and then the water pump pulley.
7. Remove the water pump.

➡**Engine coolant will leak from the cylinder block, so have a receptacle ready under the vehicle.**

❋❋ WARNING

Handle the water pump vane so that it does not contact any other parts.

8. Inspect the water pump after removal for the following:
   a. Significant dirt or rusting on water pump body and vane.
   b. Free play in the vane shaft.

c. Ensure that the water pump turns smoothly when rotated by hand.

9. If found faulty, replace the water pump.

To install:

10. Install the water pump. Tighten the bolts to 21 ft. lbs. (28 Nm).

11. Install the water pump pulley, then the fan coupling with the cooling fan.

12. Install the accessory drive belt. Refer to Accessory Drive Belts, removal & installation.

13. Install the air duct (inlet).

14. Install the engine front undercover.

15. Tighten the drain plugs on the radiator and both sides of the engine block.

16. Fill the cooling system with coolant.

17. Check for leaks of engine coolant using a radiator cap tester.

18. Start and warm up the engine. Visually make sure that there are no leaks of engine coolant.

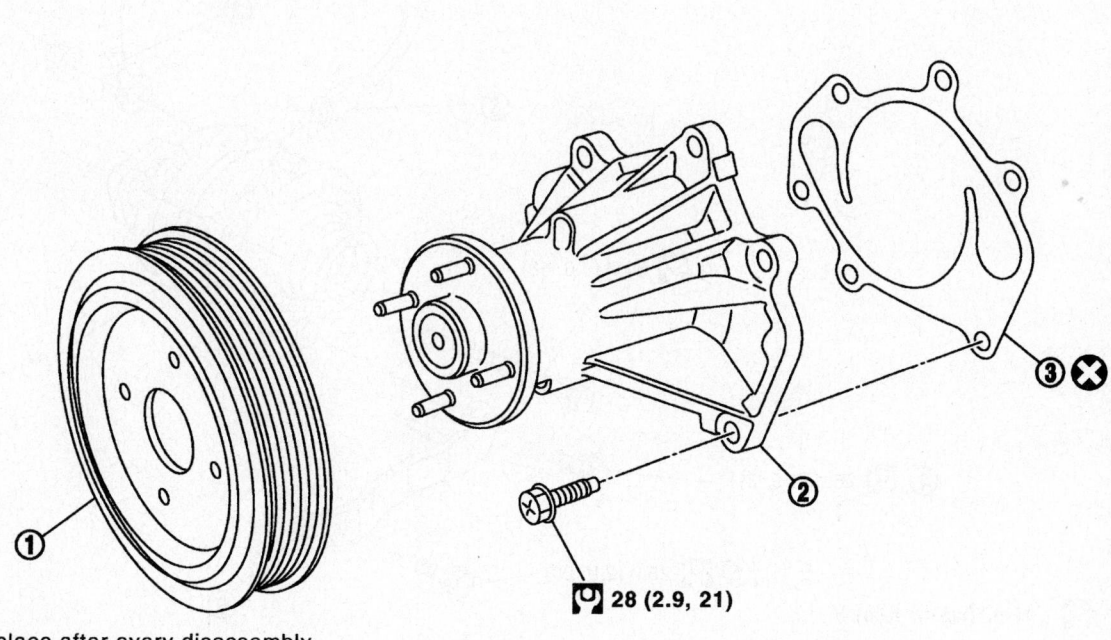

❌ : Always replace after every disassembly.

🔧 : N•m (kg-m, ft-lb)

28 (2.9, 21)

1. Water pump pulley
2. Water pump
3. Gasket

67162-FX35-G121

Fig. 31 Exploded view of the water pump mounting—4.5L engine

ALTERNATOR

REMOVAL & INSTALLATION

3.5L Engine

See Figures 32 and 33.

1. Before servicing the vehicle, refer to the Precautions Section.
2. Disconnect the negative battery cable.
3. Remove engine front undercover.
4. Remove the alternator and power steering oil pump belt.
5. Disconnect the alternator connector. Remove the **B** terminal nut.

6. Remove the harness clip and water hose bracket from the alternator.
7. For 2WD models:
   a. Remove the oil pressure switch harness clip from the alternator stay.
   b. Disconnect the oil pressure switch connector.
8. Remove the alternator stay mounting bolts and alternator stay.
9. Remove the alternator mounting bolt.
10. Remove the alternator assembly, by lowering it out of the bottom of the engine compartment.

To install:

11. Reposition the alternator in place on the engine and tighten the mounting bolts.

Tighten the long alternator bolt 48 ft. lbs. (65 Nm) and the short alternator bracket bolts to 21 ft. lbs. (28 Nm).

12. Tighten the **B** terminal nut carefully to 84 inch lbs. (10 Nm).
13. For 2WD models:
   a. Connect the oil pressure switch connector.
   b. Install the oil pressure switch harness clip to the alternator stay.
14. Reconnect the alternator wiring.
15. Install the accessory drive belt. Refer to Accessory Drive Belts, removal & installation.
16. Install the front engine undercover.
17. Reconnect the negative battery cable.

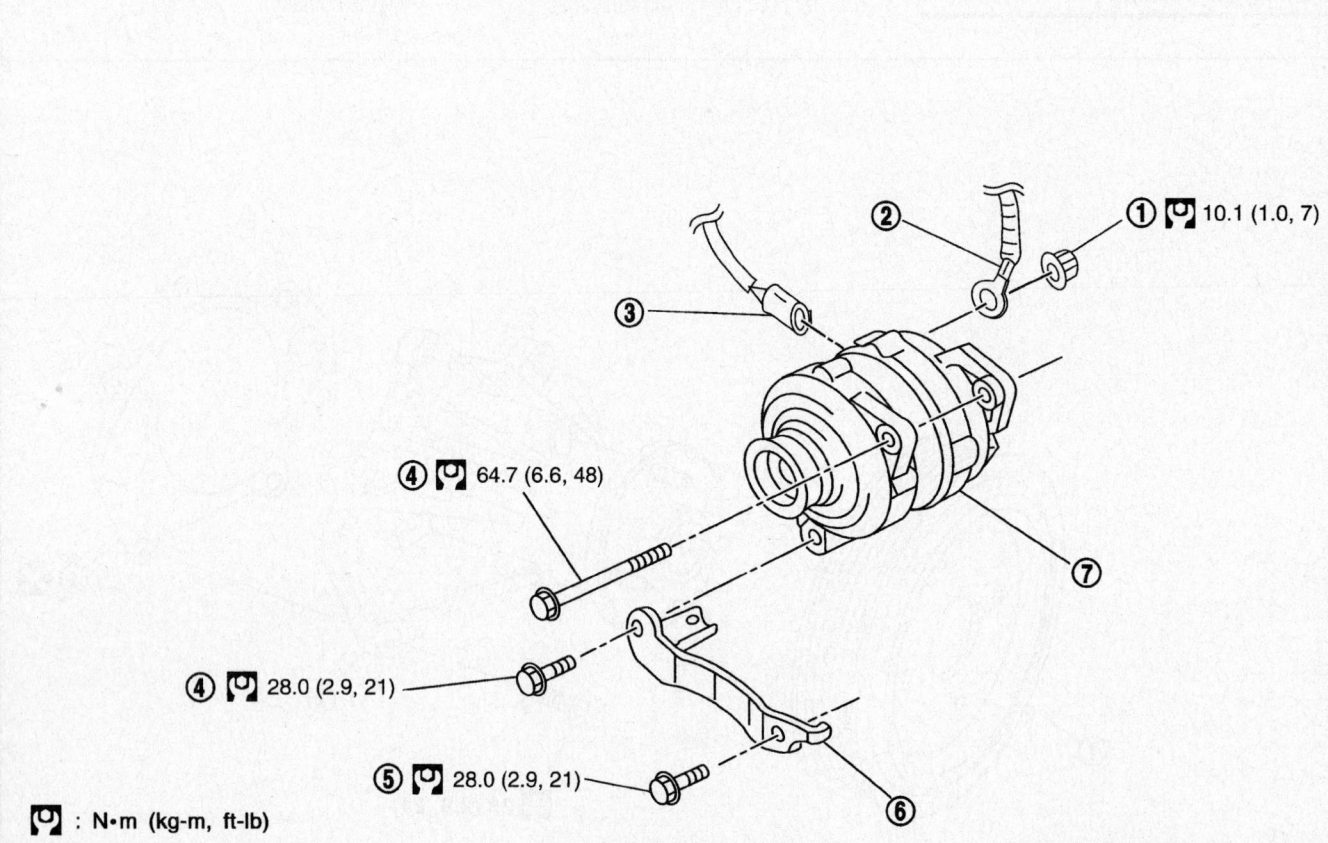

① 🔧 10.1 (1.0, 7)

④ 🔧 64.7 (6.6, 48)

④ 🔧 28.0 (2.9, 21)

⑤ 🔧 28.0 (2.9, 21)

🔧 : N•m (kg-m, ft-lb)

1.	B terminal nut	2.	Alternator B terminal harness	3.	Alternator connector
4.	Alternator mounting bolt	5.	Alternator stay mounting bolt	6.	Alternator stay
7.	Alternator				

67162-FX35-G100

Fig. 32 Exploded view of alternator mounting—3.5L engine

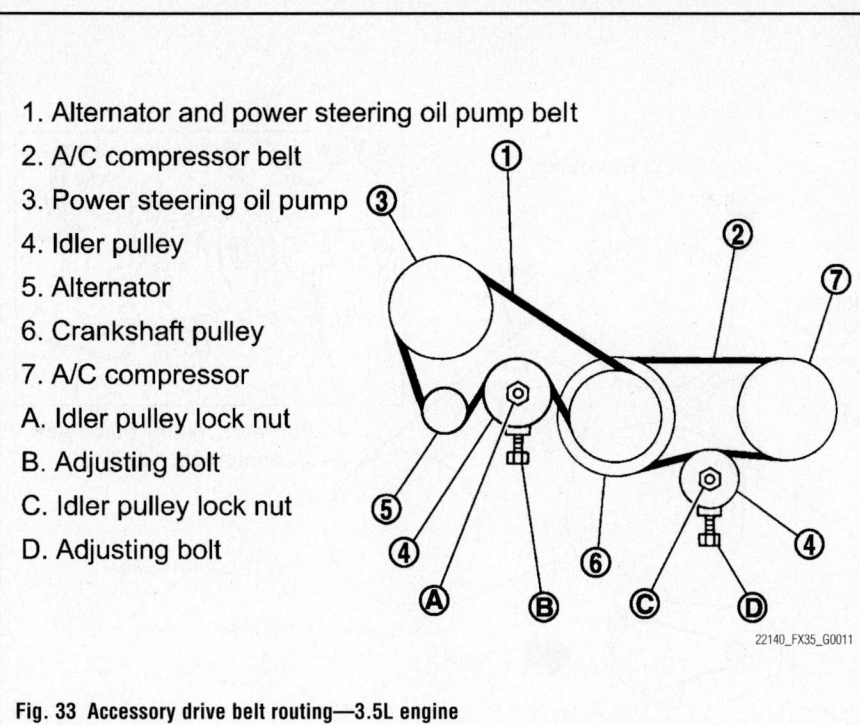

1. Alternator and power steering oil pump belt
2. A/C compressor belt
3. Power steering oil pump
4. Idler pulley
5. Alternator
6. Crankshaft pulley
7. A/C compressor
A. Idler pulley lock nut
B. Adjusting bolt
C. Idler pulley lock nut
D. Adjusting bolt

22140_FX35_G0011

Fig. 33 Accessory drive belt routing—3.5L engine

4.5L Engine

See Figures 34 and 35.

1. Before servicing the vehicle, refer to the Precautions Section.

2. Disconnect the negative battery cable.

3. Remove the front engine undercover.

4. Remove the lower cooling fan shroud.

5. Remove the alternator, water pump, A/C belt. Refer to Accessory Drive Belts, removal & installation.

6. Remove the alternator mounting bolts.

7. Disconnect the alternator wiring connector.

8. Remove the alternator assembly, by lowering it out of the bottom of the engine compartment.

To install:

9. Reposition the alternator in place on the engine and tighten the mounting bolts. Tighten the long alternator bolt 37 ft. lbs. (50 Nm), the alternator-to-bracket bolt to 21 ft. lbs. (28 Nm), and the bracket-to-engine bolts to 17 ft. lbs. (24 Nm).

10. Tighten the **B** terminal nut carefully to 84 inch lbs. (10 Nm).

11. Reconnect the alternator wiring.

12. Install the alternator, water pump, A/C drive belt. Refer to Accessory Drive Belts, removal & installation.

13. Install the lower cooling fan shroud.

14. Install the front engine undercover.

15. Reconnect the negative battery cable.

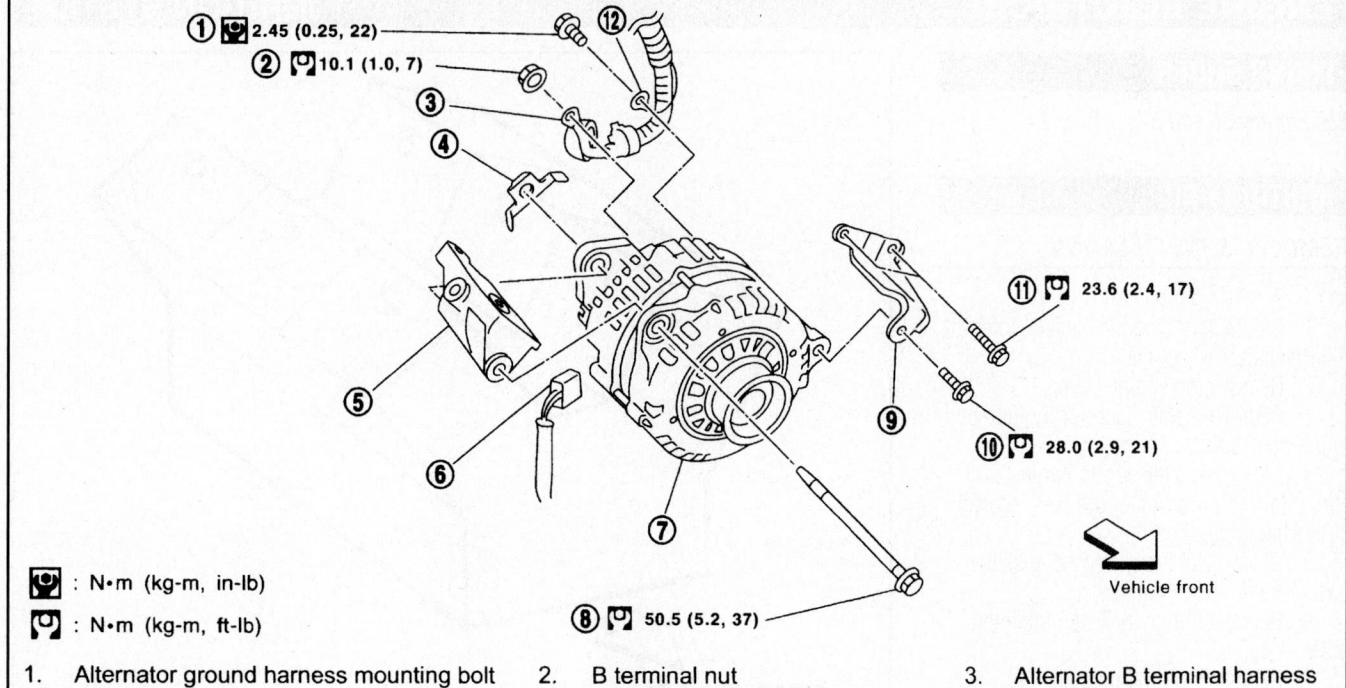

① ⬚ 2.45 (0.25, 22)
② ⬚ 10.1 (1.0, 7)
⑪ ⬚ 23.6 (2.4, 17)
⑩ ⬚ 28.0 (2.9, 21)
⑧ ⬚ 50.5 (5.2, 37)

Vehicle front

⬚ : N•m (kg-m, in-lb)
⬚ : N•m (kg-m, ft-lb)

1.	Alternator ground harness mounting bolt	2.	B terminal nut	3.	Alternator B terminal harness
4.	Alternator Nut	5.	Alternator bracket	6.	Alternator connector
7.	Alternator	8.	Alternator mounting bolt	9.	Alternator stay
10.	Alternator mounting bolt	11.	Alternator stay mounting bolt	12.	Alternator ground harness

Fig. 34 Exploded view of alternator mounting—4.5L engine

67162-FX35-G102

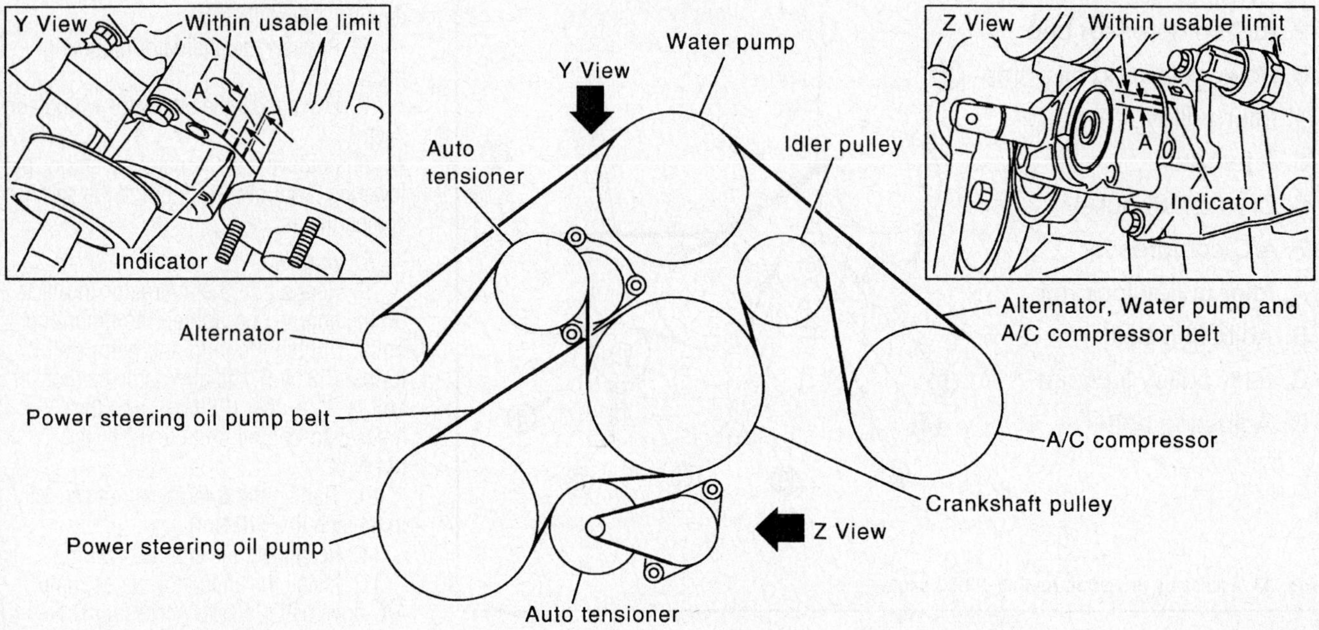

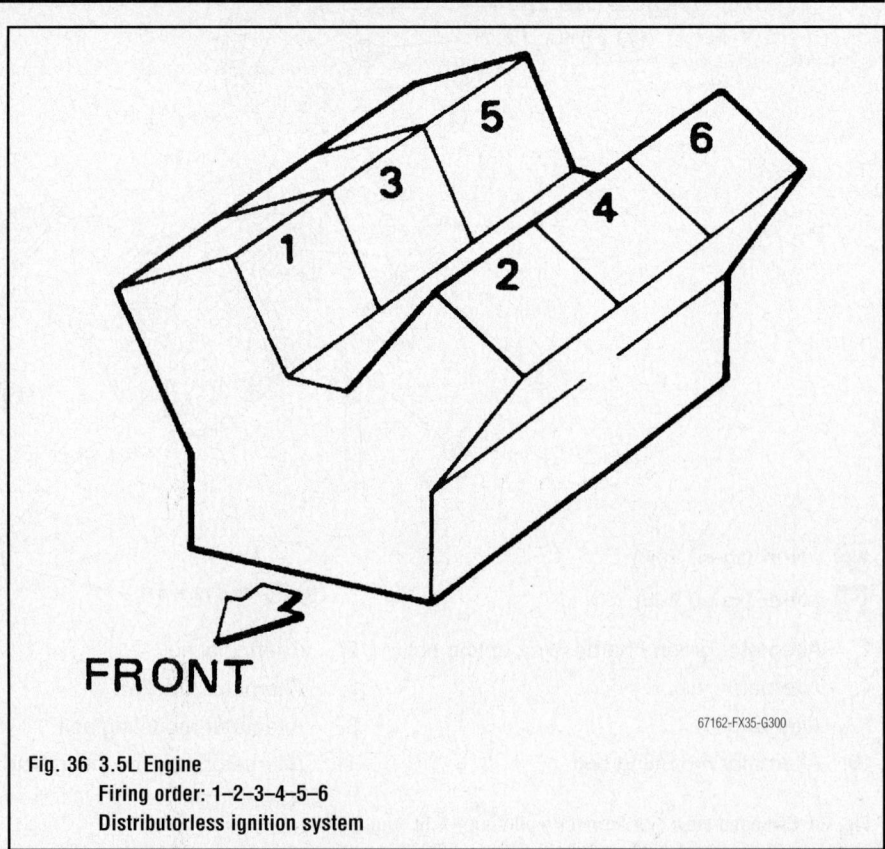

Fig. 35 Accessory drive belt routing—4.5L engine

67162-FX35-G103

ENGINE ELECTRICAL   IGNITION SYSTEM

FIRING ORDER

See Figures 36 and 37.

IGNITION COIL

REMOVAL & INSTALLATION

See Figures 38 and 39.

1. Before servicing the vehicle, refer to the Precautions Section.
2. Remove the engine cover.
3. Remove the air duct (for ignition coil of left bank side).
4. Move aside the wiring harness, wiring harness bracket, and hoses located above the ignition coil.
5. Disconnect the wiring harness connector from the ignition coil.
6. Remove the ignition coil retaining bolt.
7. Remove the ignition coil.

✺ WARNING

Do not subject the ignition coils to excessive shock or vibration.

67162-FX35-G300

Fig. 36 3.5L Engine
Firing order: 1–2–3–4–5–6
Distributorless ignition system

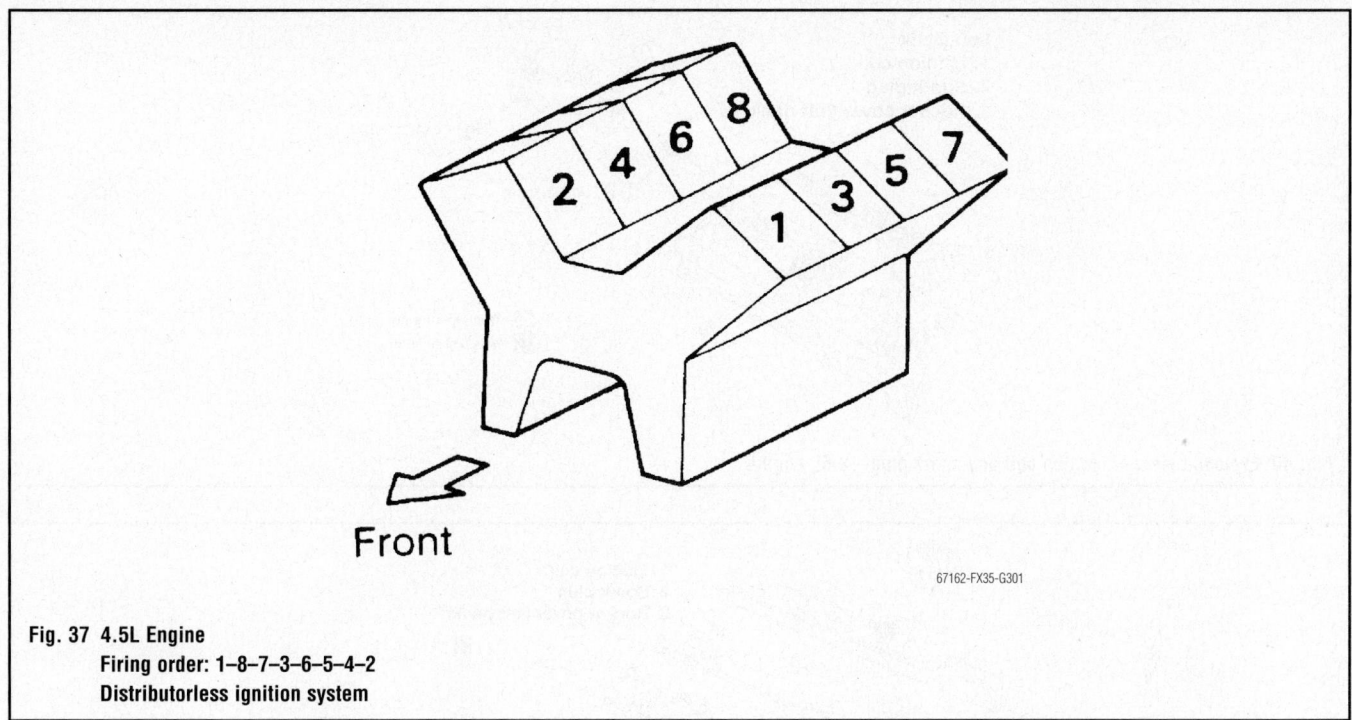

Fig. 37 4.5L Engine
Firing order: 1–8–7–3–6–5–4–2
Distributorless ignition system

67162-FX35-G301

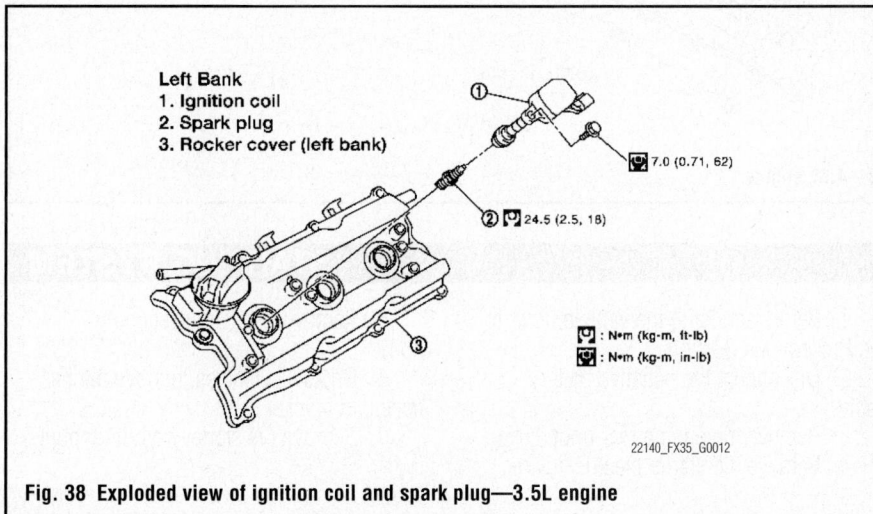

Left Bank
1. Ignition coil
2. Spark plug
3. Rocker cover (left bank)

7.0 (0.71, 62)

24.5 (2.5, 18)

: N•m (kg-m, ft-lb)
: N•m (kg-m, in-lb)

22140_FX35_G0012

Fig. 38 Exploded view of ignition coil and spark plug—3.5L engine

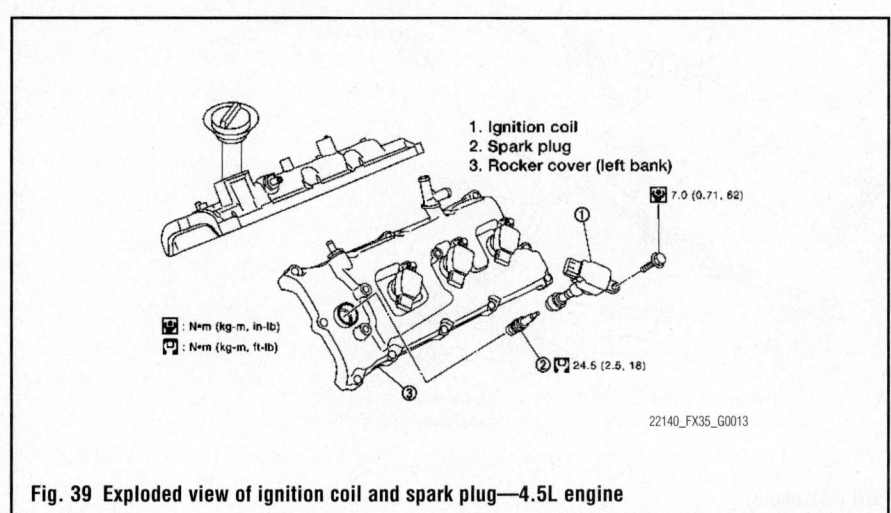

1. Ignition coil
2. Spark plug
3. Rocker cover (left bank)

7.0 (0.71, 62)

: N•m (kg-m, in-lb)
: N•m (kg-m, ft-lb)

24.5 (2.5, 18)

22140_FX35_G0013

Fig. 39 Exploded view of ignition coil and spark plug—4.5L engine

To install:

8. Install the ignition coil on the engine. Tighten the retaining bolt to 62 inch lbs. (7 Nm).

9. Reconnect the wiring harness to the coil.

10. Reposition the wiring harness, bracket and hoses.

11. Install the air duct and the engine cover.

IGNITION TIMING

The ignition timing is controlled by the Electronic Control Module (ECM). No adjustment is necessary or possible.

SPARK PLUGS

REMOVAL & INSTALLATION

See Figures 40 and 41.

1. Disconnect the negative battery cable.
2. Remove the engine cover.
3. Remove the ignition coil retaining bolt.
4. Remove the ignition coil.
5. Remove the spark plug using a spark plug socket and wrench.

To install:

6. Be sure the spark plug gap is to specification (0.043 inch).

7. Carefully install the spark plug and torque to specification: 18 ft. lbs. (25 Nm).

8. Install the ignition coil, torque the retaining bolt to 62 inch lbs. (7 Nm).

9. Install the engine cover.

10. Connect the negative battery cable.

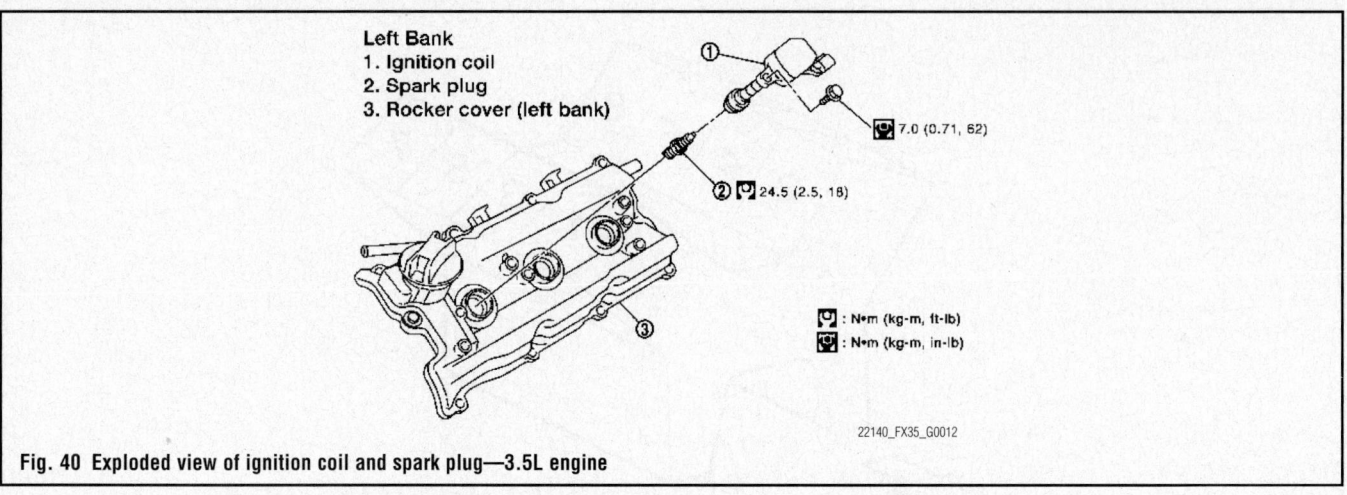

Fig. 40 Exploded view of ignition coil and spark plug—3.5L engine

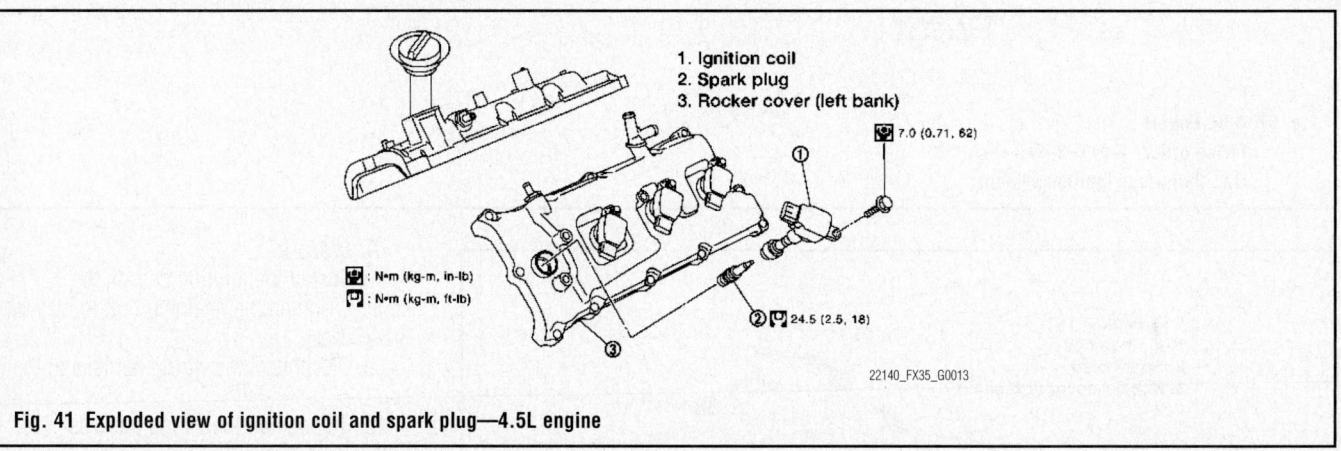

Fig. 41 Exploded view of ignition coil and spark plug—4.5L engine

ENGINE ELECTRICAL   STARTING SYSTEM

STARTER

REMOVAL & INSTALLATION

3.5L Engine

See Figures 42 and 43.

1. Before servicing the vehicle, refer to the Precautions Section.
2. Disconnect the negative battery cable.
3. Remove the engine rear undercover.
4. Remove the starter electrical wires.
5. Remove the starter retaining bolts.
6. On 2WD vehicles, remove the harness clip bracket.
7. Remove the starter from its mounting.

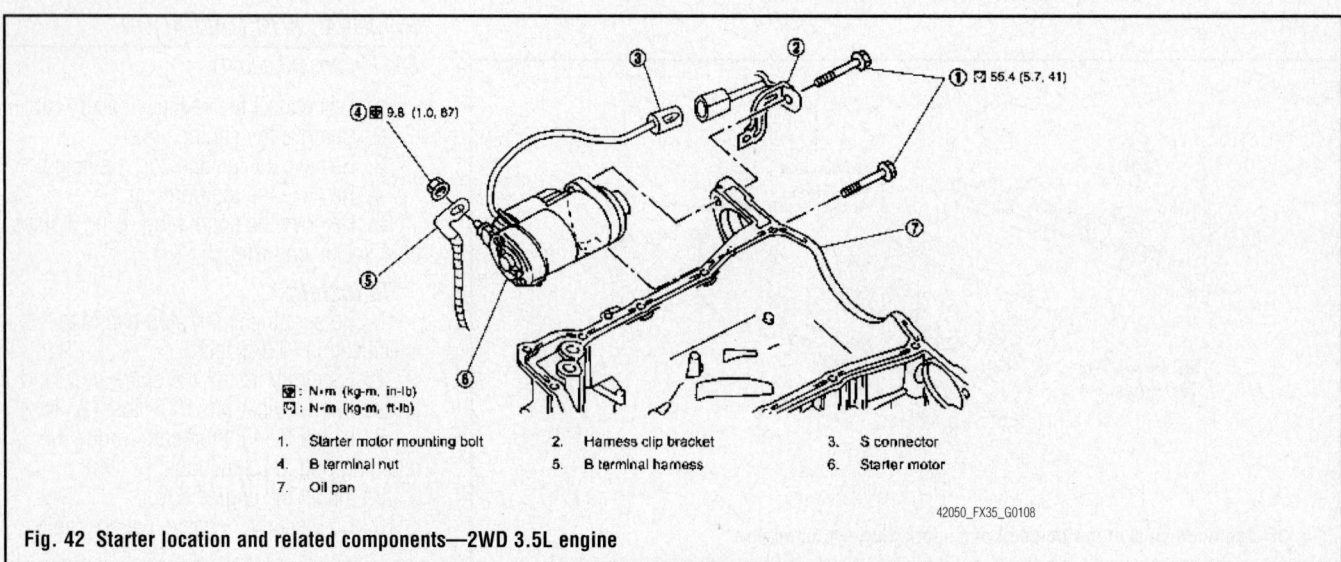

1. Starter motor mounting bolt
2. Harness clip bracket
3. S connector
4. B terminal nut
5. B terminal harness
6. Starter motor
7. Oil pan

Fig. 42 Starter location and related components—2WD 3.5L engine

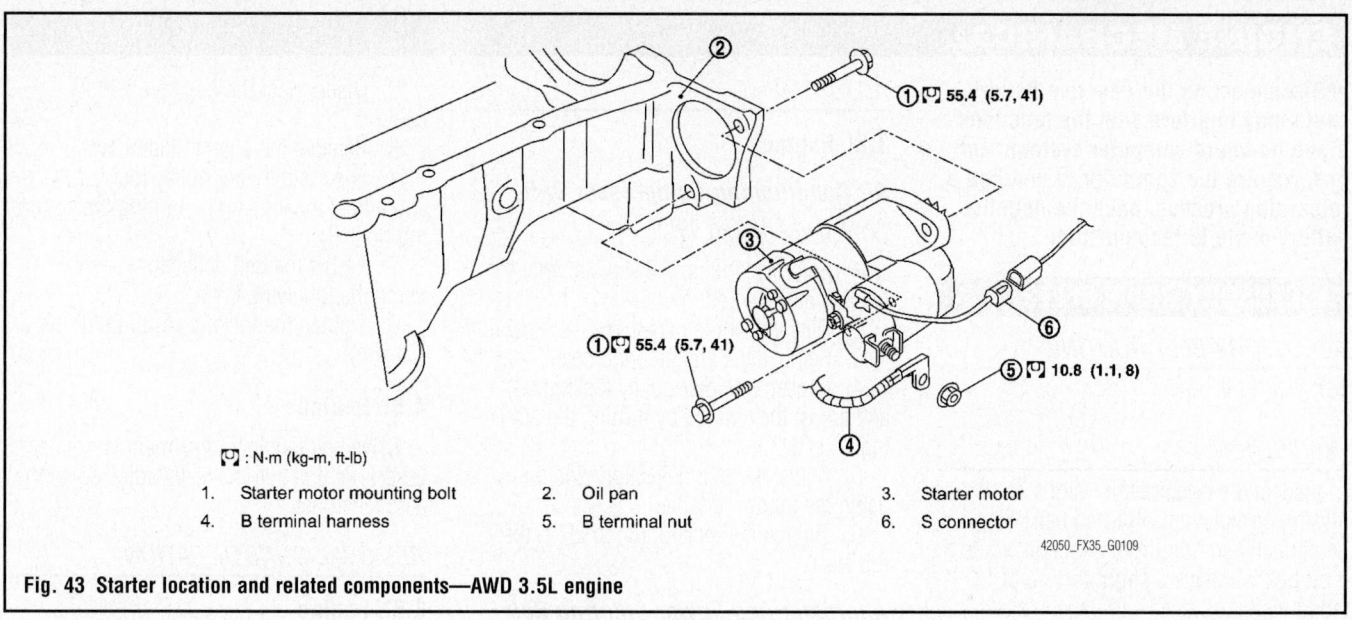

: N·m (kg-m, ft-lb)

1. Starter motor mounting bolt
2. Oil pan
3. Starter motor
4. B terminal harness
5. B terminal nut
6. S connector

42050_FX35_G0109

Fig. 43 Starter location and related components—AWD 3.5L engine

To install:

8. Installation is the reverse of the removal procedure.

9. Tighten the retaining bolts to 41 ft. lbs. (55 Nm).

10. Tighten the terminal nut to 87 inch lbs. (10 Nm).

4.5L Engine

See Figures 44 and 45.

1. Before servicing the vehicle, refer to the Precautions Section.

2. Disconnect the negative battery cable.

3. Remove the front and rear engine undercover.

4. Remove the starter electrical wires.

5. Remove the starter retaining bolts.

6. Loosen the automatic transmission fluid cooler tube clip bolts.

7. Remove the starter from its mounting.

To install:

8. Installation is the reverse of the removal procedure.

9. Tighten the retaining bolts to 34 ft. lbs. (47 Nm).

10. Tighten the terminal nut to 87 inch lbs. (10 Nm).

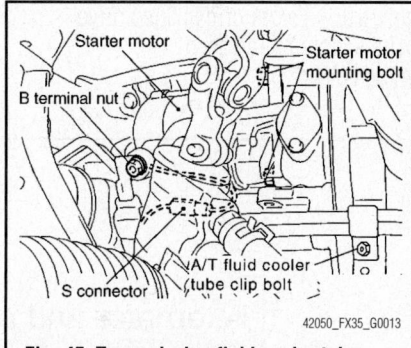

42050_FX35_G0013

Fig. 45 Transmission fluid cooler tube bolt location—4.5L engine

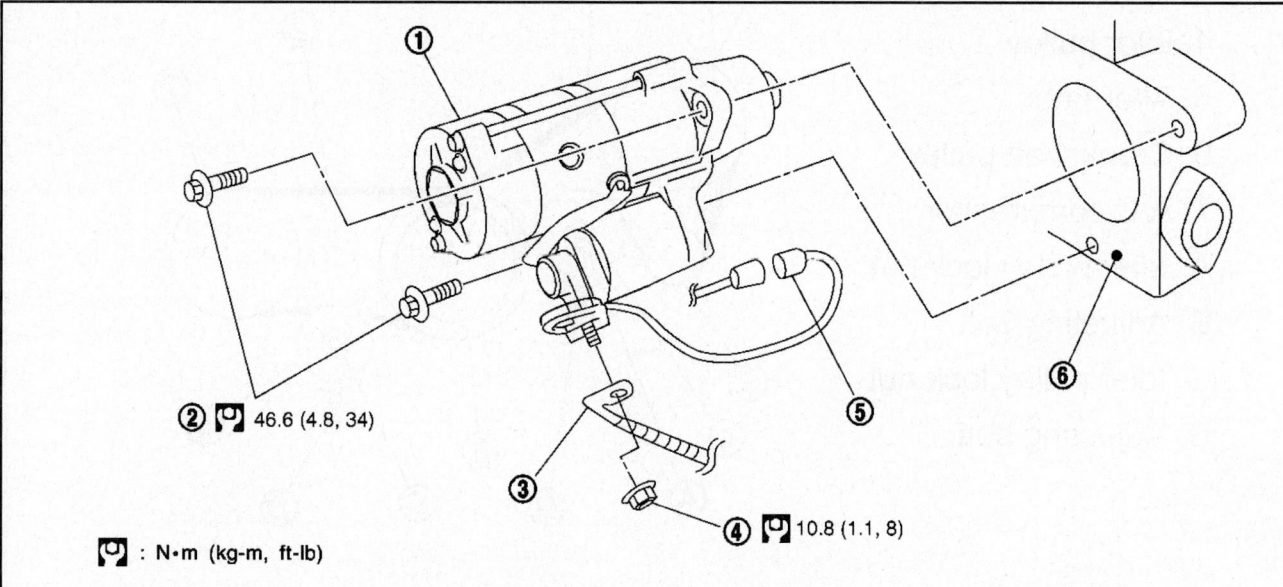

: N•m (kg-m, ft-lb)

1. Starter motor
2. Starter motor mounting bolt
3. B terminal harness
4. B terminal nut
5. S connector
6. Cylinder block

42050_FX35_G0012

Fig. 44 Starter location and related components—4.5L engine

ENGINE MECHANICAL

➡**Disconnecting the negative battery cable may interfere with the functions of the on board computer systems and may require the computer to undergo a relearning process, once the negative battery cable is reconnected.**

ACCESSORY DRIVE BELTS

ACCESSORY BELT ROUTING

See Figures 46 and 47.

INSPECTION

Inspect the drive belt for signs of glazing or cracking. A glazed belt will be perfectly smooth from slippage, while a g od belt will have a slight texture of fabric visible. Cracks will usually start at the inner edge of the belt and run outward. All worn or damaged drive belts should be replaced immediately.

ADJUSTMENT

3.5L Engine

Air Conditioning Compressor Belt

See Figures 48 and 49.

1. Before servicing the vehicle, refer to the Precautions Section.
2. Disconnect the negative battery cable.
3. Remove the engine undercover.
4. Loosen the idler pulley locknut (C) and adjust the tension by turning the adjusting bolt (D).
5. Adjust the belt deflection/tension using the following table.
6. Tighten the locknut (C) to 26 ft. lbs. (35 Nm).

Alternator And Power Steering Belt

See Figures 50 and 49.

1. Before servicing the vehicle, refer to the Precautions Section.

2. Disconnect the negative battery cable.
3. Remove the engine undercover.
4. Loosen the idler pulley locknut (A) and adjust the tension by turning the adjusting bolt (B).
5. Adjust the belt deflection/tension using the following table.
6. Tighten the locknut (A) to 26 ft. lbs. (35 Nm).

4.5L Engine

Drive belt tension adjustment is not necessary, as it is automatically adjusted by the auto tensioner.

REMOVAL & INSTALLATION

3.5L Engine

1. Before servicing the vehicle, refer to the Precautions Section.
2. Disconnect the negative battery cable.
3. Remove the engine undercover.

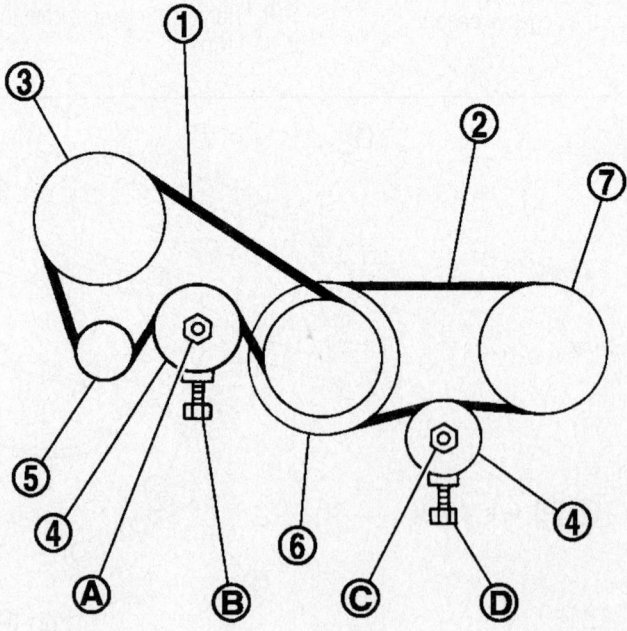

1. Alternator and power steering oil pump belt
2. A/C compressor belt
3. Power steering oil pump
4. Idler pulley
5. Alternator
6. Crankshaft pulley
7. A/C compressor
A. Idler pulley lock nut
B. Adjusting bolt
C. Idler pulley lock nut
D. Adjusting bolt

22140_FX35_G0011

Fig. 46 Accessory drive belt routing—3.5L engine

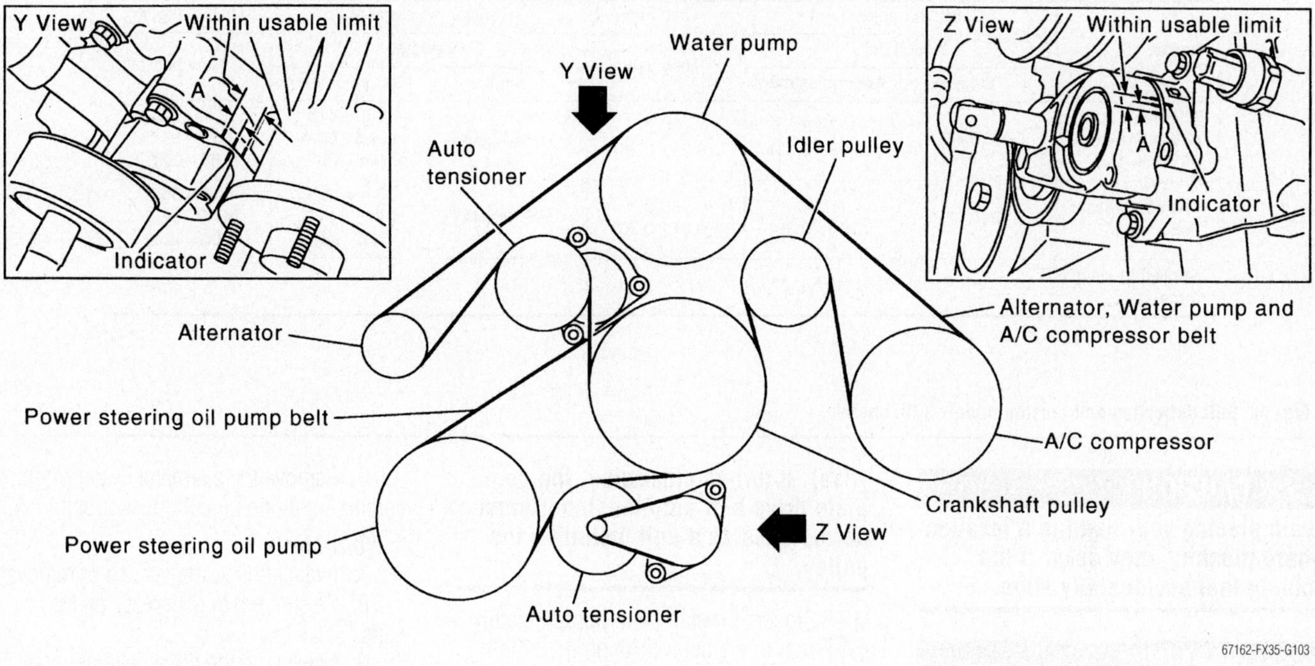

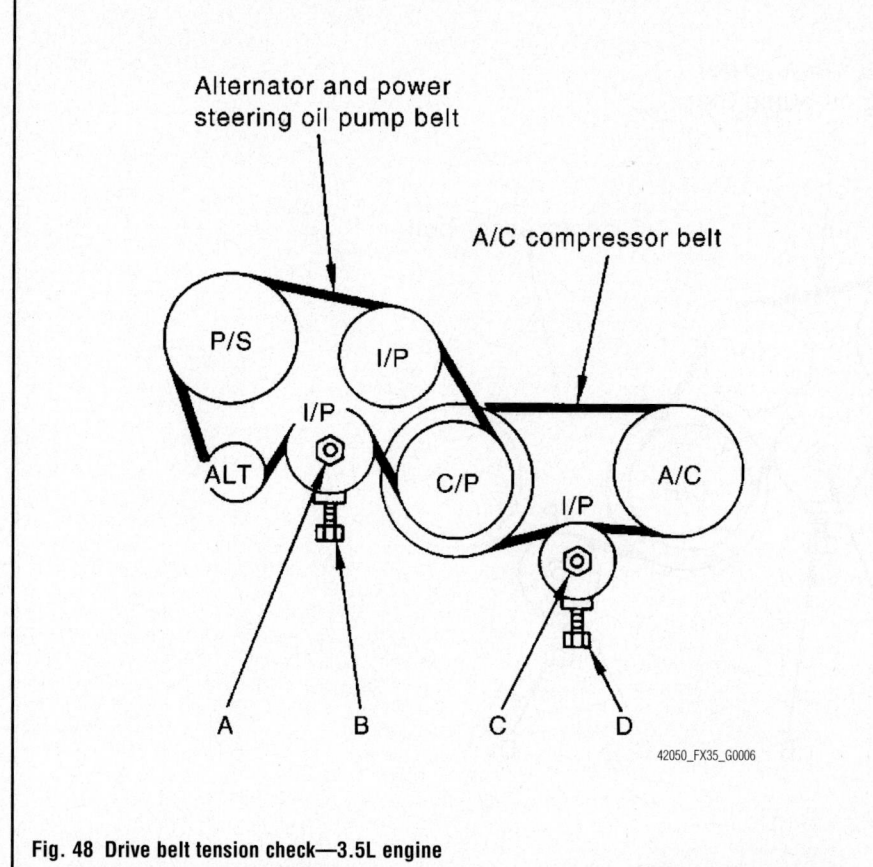

Fig. 47 Accessory drive belt routing—4.5L engine

Fig. 48 Drive belt tension check—3.5L engine

4. Remove the alternator and power steering belt.

5. Remove the air conditioning compressor belt.

To install:

6. Installation is the reverse of removal.

7. Be sure not to get grease or oil on the belts.

8. Make sure the drive belts are correctly engaged with the pulley groove

9. Torque and adjust the drive belts to specification. Refer to Accessory Drive Belts, Adjustment.

4.5L Engine

Except Power Steering Belt

See Figure 52.

1. Before servicing the vehicle, refer to the Precautions Section.

2. Disconnect the negative battery cable.

3. Remove the air duct.

4. Holding the hexagonal part in the pulley center of the auto tensioner with a wrench, move the wrench handle in the direction of the arrow shown in the illustration.

Belt Deflection and Tension

Items	Deflection adjustment		Unit: mm (in)	Tension adjustment		Unit: N (kg, lb)
	Used belt		New belt	Used belt		New belt
	Limit	After adjustment		Limit	After adjustment	
Alternator and power steering oil pump belt	12 (0.47)	7 - 8 (0.28 - 0.31)	6 - 7 (0.24 - 0.28)	294 (30, 66)	730 - 818 (74.5 - 83.4, 164 - 184)	838 - 926 (85.5 - 94.5, 188 - 208)
A/C compressor belt	12 (0.47)	9 - 10 (0.35 - 0.39)	8 - 9 (0.31 - 0.35)	196 (20, 44)	348 - 436 (35.5 - 44.5, 78 - 98)	470 - 559 (47.9 - 57.0, 106 - 126)
Applied pushing force	98 N (10 kg, 22 lb)			—		

22140_FX35_G0014

Fig. 49 Belt deflection and tension table—3.5L engine

wise). If turned clockwise, the complete drive belt auto tensioner must be replaced as a unit including the pulley.

5. Insert a metallic bar approximately 0.24 inch in diameter through the holding boss to lock the auto tensioner pulley arm.

➥**Leave the auto tensioner pulley arm locked until the belt is reinstalled.**

6. Remove the alternator, water pump, and air conditioning compressor belt.

To install:

7. Installation is the reverse of removal.

8. Be sure not to get grease on the belts.

9. Make sure the drive belts are correctly engaged with the pulley groove

10. Check that belt tension is within specification.

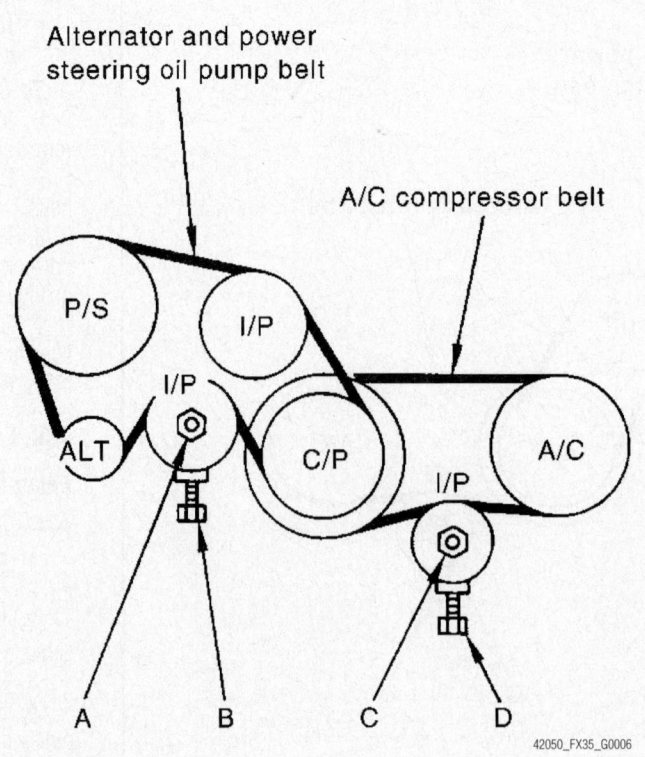

Alternator and power steering oil pump belt

A/C compressor belt

42050_FX35_G0006

Fig. 50 Drive belt tension check—3.5L engine

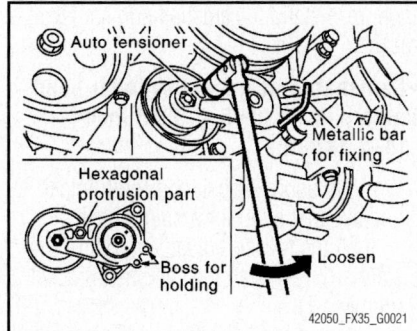

Fig. 51 Alternator, water pump, and air conditioning compressor belt removal—4.5L engine

Power Steering Belt

See Figure 53.

1. Before servicing the vehicle, refer to the Precautions Section.
2. Disconnect the negative battery cable.
3. Remove the air duct.
4. Remove the alternator, water pump, and air conditioning compressor belt.
5. Holding the hexagonal protrusion part of the auto tensioner pulley with a box wrench and move pulley in the direction of the arrow shown in the illustration.

✸ CAUTION

Avoid placing your hand in a location where pinching may occur if the holding tool accidentally slips.

6. Insert a metallic bar approximately 0.24 inch in diameter through the holding boss to lock the auto tensioner pulley arm.

➡ **Leave the auto tensioner pulley arm locked until the belt is reinstalled.**

7. Remove the power steering belt.

To install:

8. Installation is the reverse of removal.
9. Be sure not to get grease on the belts.

Fig. 52 Power steering belt removal—4.5L engine

10. Make sure the drive belts are correctly engaged with the pulley groove
11. Check that belt tension is within specification.

CAMSHAFT AND VALVE LIFTERS

REMOVAL & INSTALLATION

3.5L Engine

See Figures 53 through 61.

1. Before servicing the vehicle, refer to the Precautions Section.
2. Remove the front timing chain case, camshaft sprocket, timing chain, and rear timing chain case.
3. Remove the camshaft position sensor (PHASE) (right and left banks) from the cylinder head back side.

✸✸ WARNING

Handle the camshaft position sensor carefully to avoid dropping and shocks. Do not disassemble and do not allow metal powder to adhere to magnetic part at the sensor tip. Do not place sensors in a location where they are exposed to magnetism.

4. Remove the intake valve timing control solenoid valves.
5. Discard the intake valve timing control solenoid valve gaskets.
6. Remove the camshaft brackets. Equally loosen the camshaft bracket bolts in several steps in reverse order of camshaft bearing cap mounting bolt tightening sequence shown in the figure.

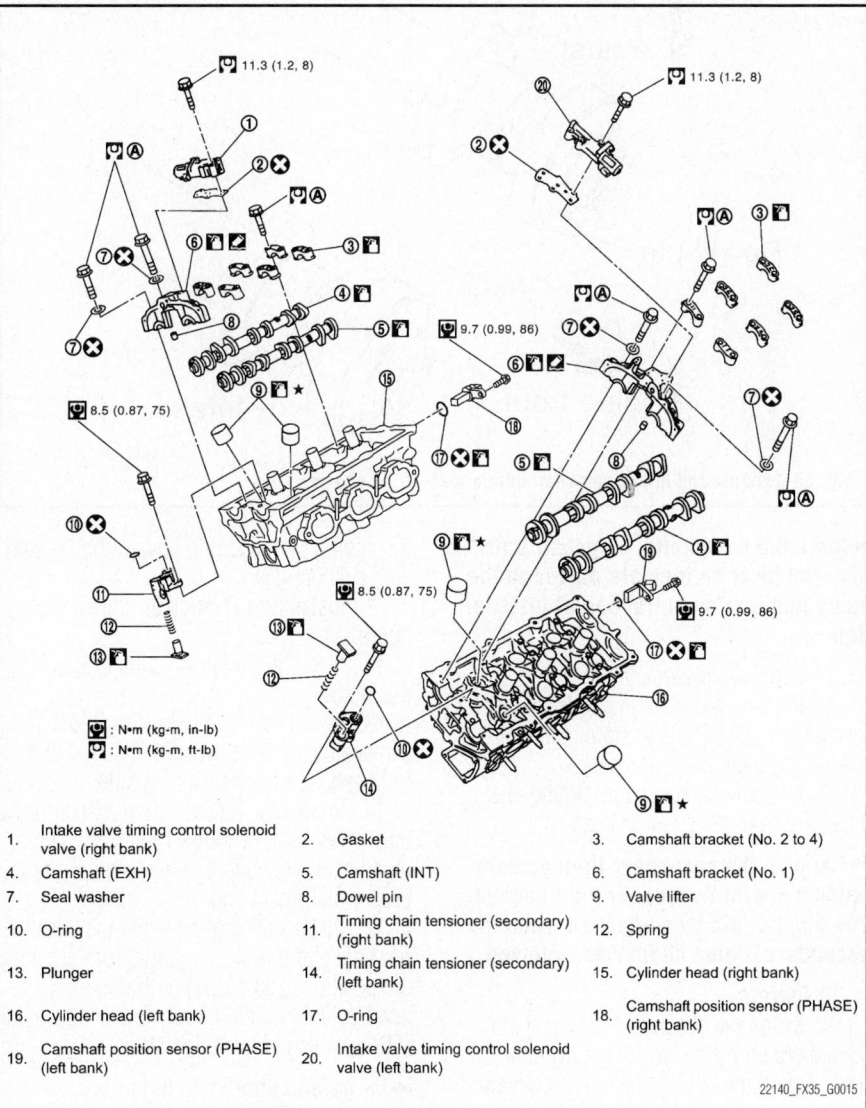

1.	Intake valve timing control solenoid valve (right bank)	2.	Gasket	3.	Camshaft bracket (No. 2 to 4)
4.	Camshaft (EXH)	5.	Camshaft (INT)	6.	Camshaft bracket (No. 1)
7.	Seal washer	8.	Dowel pin	9.	Valve lifter
10.	O-ring	11.	Timing chain tensioner (secondary) (right bank)	12.	Spring
13.	Plunger	14.	Timing chain tensioner (secondary) (left bank)	15.	Cylinder head (right bank)
16.	Cylinder head (left bank)	17.	O-ring	18.	Camshaft position sensor (PHASE) (right bank)
19.	Camshaft position sensor (PHASE) (left bank)	20.	Intake valve timing control solenoid valve (left bank)		

Fig. 53 Exploded view of camshaft and related components—3.5L engine

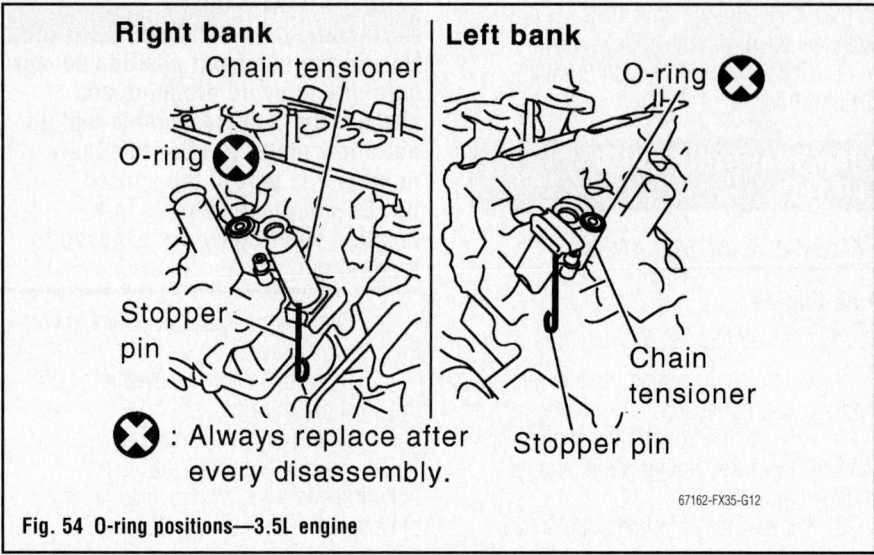

Fig. 54 O-ring positions—3.5L engine

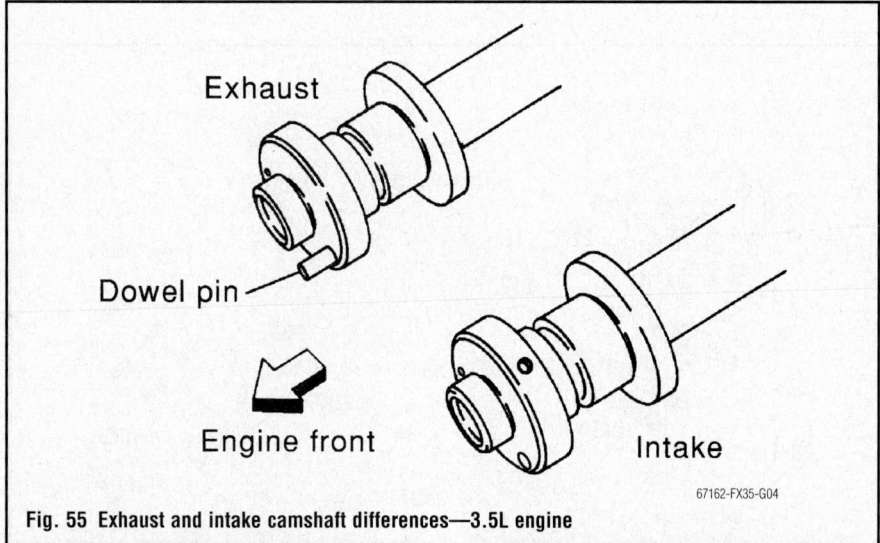

Fig. 55 Exhaust and intake camshaft differences—3.5L engine

➡**Mark the camshafts, camshaft brackets, and bolts so they are placed in the same position and direction at installation.**

7. Remove the camshaft(s).

8. Remove the valve lifters. Identify installation positions and store them without mixing them up.

9. Remove the secondary timing chain tensioner from the cylinder head.

➡**Remove the secondary timing chain tensioner with its stopper pin attached. The stopper pin was attached when the secondary timing chain was removed.**

To install:

10. Install the secondary timing chain tensioners on both sides of the cylinder head. Install the timing chain tensioner with its stopper pin attached and with the timing chain tensioner sliding part facing downward on the right-side cylinder head while the sliding part is facing upward on the left-side cylinder head.

11. Install new O-rings as shown in the figure.

12. Install valve lifters in the original positions.

13. Install the camshafts. Install the camshaft with the dowel pin attached to its front end face on the exhaust side.

14. Follow the identification marks made during removal, or follow the identification marks that are present on new camshafts for proper placement and direction.

15. Install camshafts so that the dowel pin hole and dowel pin on the front end face are positioned as shown in the figure. Number 1 cylinder should be in Top Dead Center (TDC) on its compression stroke.

➡**Large and small pin holes are located on the front end face of the intake camshaft at intervals of 180°.**

Face the small diameter side pin hole upward (in cylinder head upper face direction). Though the camshaft does not stop at the portion as shown in the figure, for the placement of the cam nose, it is generally accepted that the camshaft is placed in the same direction of the figure.

16. Install the camshaft brackets. Remove foreign material completely from the camshaft bracket backside and from the cylinder head installation face. Install the camshaft bracket in the original position and direction as shown in figure.

17. Install camshaft brackets (numbers 2–4) aligning the stamp marks as shown in the figure.

➡**There are no identification marks indicating left and right for camshaft bracket (number 1).**

18. Apply liquid gasket to mating surface of camshaft bracket (number 1) as shown on both right and left banks. Use Genuine RTV Silicone Sealant or equivalent.

19. Tighten the camshaft bracket bolts in the following steps, in numerical order as shown.

    a. Tighten numbers 7–10 to 12 inch lbs. (2 Nm).

    b. Tighten numbers 1–6 to 12 inch lbs. (2 Nm).

    c. Tighten numbers 1–10 to 48 inch lbs. (6 Nm).

    d. Tighten numbers 1–10 in numerical order to 96 inch lbs. (10 Nm).

➡**After tightening the mounting bolts of camshaft brackets (number 1), be sure to wipe off excessive liquid gasket from the mating surface of the rocker cover and the mating surface of the rear timing chain case.**

20. Measure difference in levels between the front end faces of the number 1 camshaft bracket and the cylinder head. If the measurement is outside the specified range, re-install the camshaft and camshaft brackets.

➡**Camshaft bracket and cylinder head standard is: -0.0055–0.0055 inch (-0.14–0.14mm).**

    a. Measure 2 positions (both intake and exhaust side) for a single bank.

    b. If the measured value is out of the standard, re-install the camshaft bracket (number 1).

21. Inspect and adjust the valve clearance.

22. The remainder of installation is the reverse of removal.

• Follow your identification marks made during removal, or follow the identification marks that are present on new camshafts for proper placement and direction.

Bank	INT/EXH	Dowel pin	Paint marks		Identification mark
			M1	M2	
RH	EXH	Yes	No	Orange	RE
	INT	No	Pink	No	RE
LH	INT	No	Pink	No	LH
	EXH	Yes	No	Orange	LH

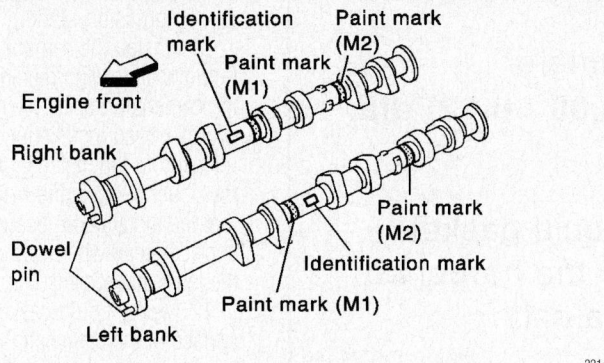

Fig. 56 View of camshaft identification and paint markings—3.5L engine

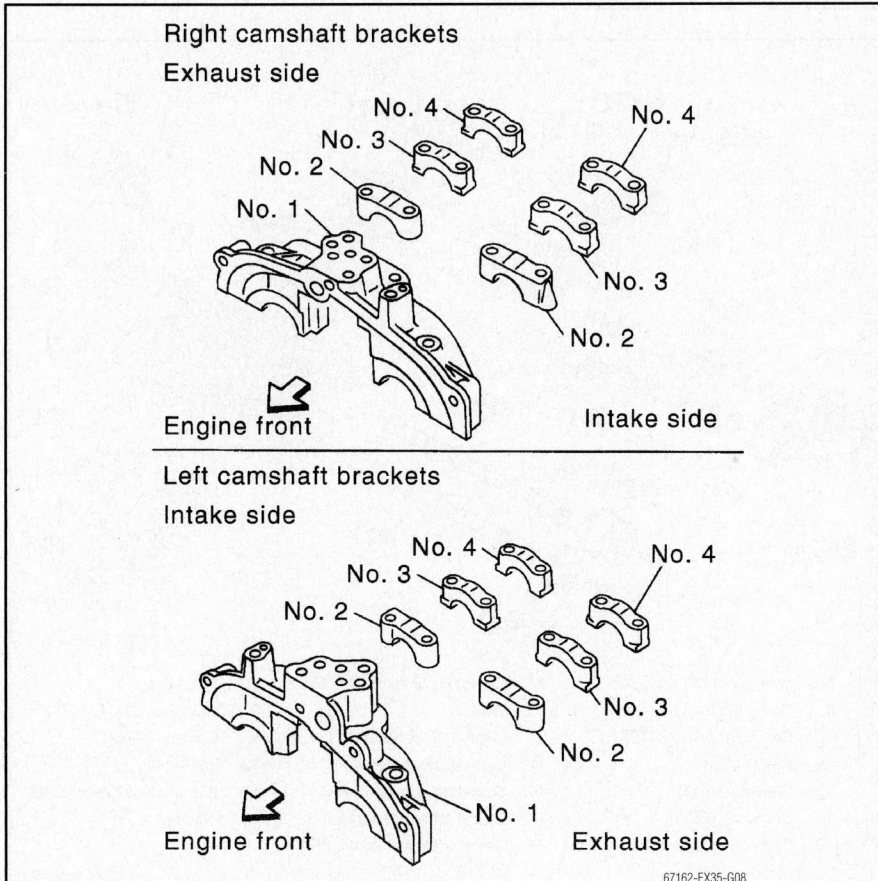

Fig. 58 Camshaft bracket positions—3.5L engine

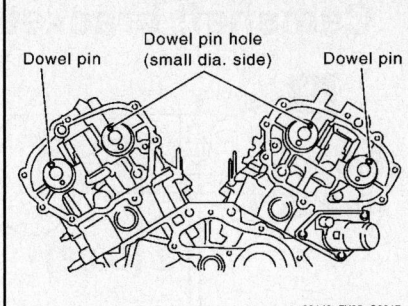

Fig. 57 Install camshafts so that the dowel pin hole and dowel pin on the front end face are positioned as shown—3.5L engine

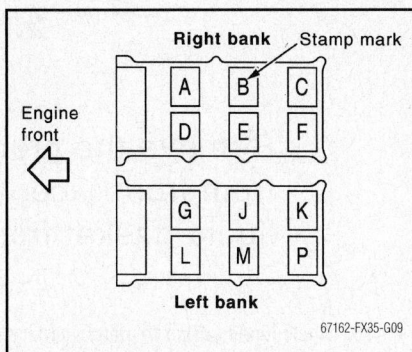

Fig. 59 Camshaft bracket identification mark positions—3.5L engine

4.5L Engine

See Figures 63 through 70.

1. Before servicing the vehicle, refer to the Precautions Section.
2. Remove the engine assembly from the vehicle. Refer to Engine Assembly, removal & installation.
3. Remove the timing chain. Refer to Timing Chain and Sprockets, removal & installation.
4. With the hexagonal part of the camshaft locked with a wrench, loosen the bolts securing the camshaft sprocket, then remove the camshaft sprocket.

❈❈ WARNING

After removing the timing chain, do not turn the crankshaft or camshaft separately, otherwise the valves may strike the piston head.

5. Remove the intake and exhaust camshaft brackets. Mark the camshafts, camshaft brackets, and bolts so that they can be installed in their original positions and direction. Equally loosen the camshaft brackets and bolts in several steps in the reverse order of the tightening sequence.

Camshaft bracket (No. 1)

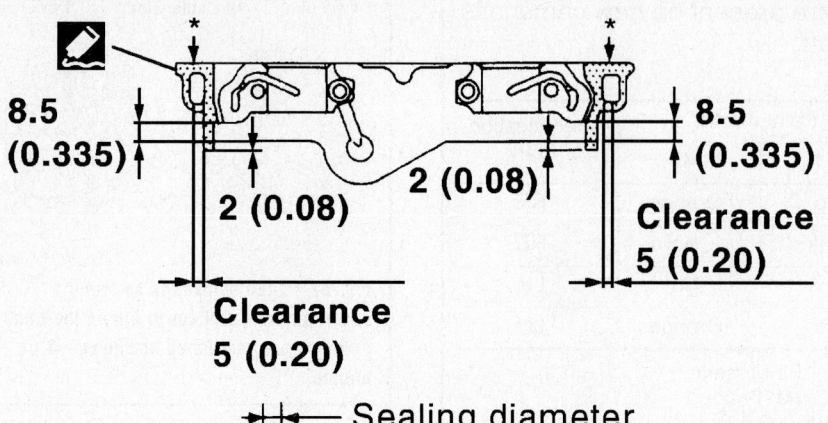

8.5 (0.335)

8.5 (0.335)

2 (0.08)

2 (0.08)

Clearance 5 (0.20)

Clearance 5 (0.20)

Sealing diameter
2.0 - 3.0 (0.08 - 0.12) dia.

* : Remove the protruding liquid gasket from front face. (Remove the hardened liquid gasket from surface only.)

22140_FX35_G0018

Fig. 60 Apply liquid gasket to mating surface of camshaft bracket (number 1) as shown—3.5L engine

8. Identify installation positions and store all parts without mixing them up.

To install:

9. Install the valve lifters and adjusting shims. Ensure they are installed in their original positions.

10. Install the camshafts. Follow identification marks made during removal, or follow the identification marks that are present on the new camshafts for proper placement and direction. Install the camshafts so that the dowel pins on the front end faces are positioned as shown (number 1 cylinder at TDC should be on its compression stroke).

11. Install the camshaft brackets. Remove all foreign material completely from the camshaft bracket backsides and from the cylinder head installation faces. Install by referring to the installation location mark on the upper surface and front mark. Install so that the installation location mark can be correctly read when viewed from the side of the left exhaust bank.

12. Apply liquid gasket to the mating surface of the camshaft bracket (number 1) as shown.

➡**Use Genuine RTV Silicone Sealant or equivalent.**

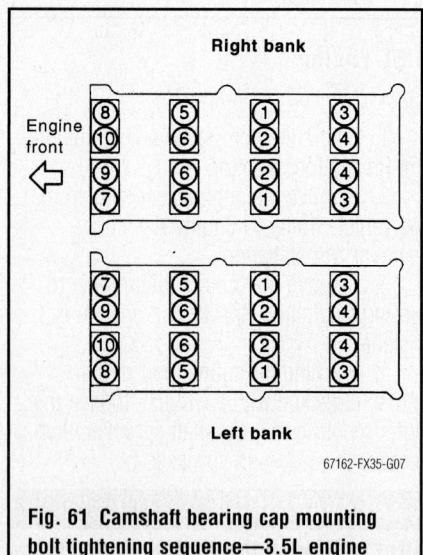

67162-FX35-G07

Fig. 61 Camshaft bearing cap mounting bolt tightening sequence—3.5L engine

Lightly tap with a plastic hammer to remove camshaft brackets numbers 1 and 6.

➡**The bottom surface of each bracket may adhere to the cylinder head because of the liquid gasket material that was applied.**

6. Remove the camshaft.

7. Remove the adjusting shims and valve lifters.

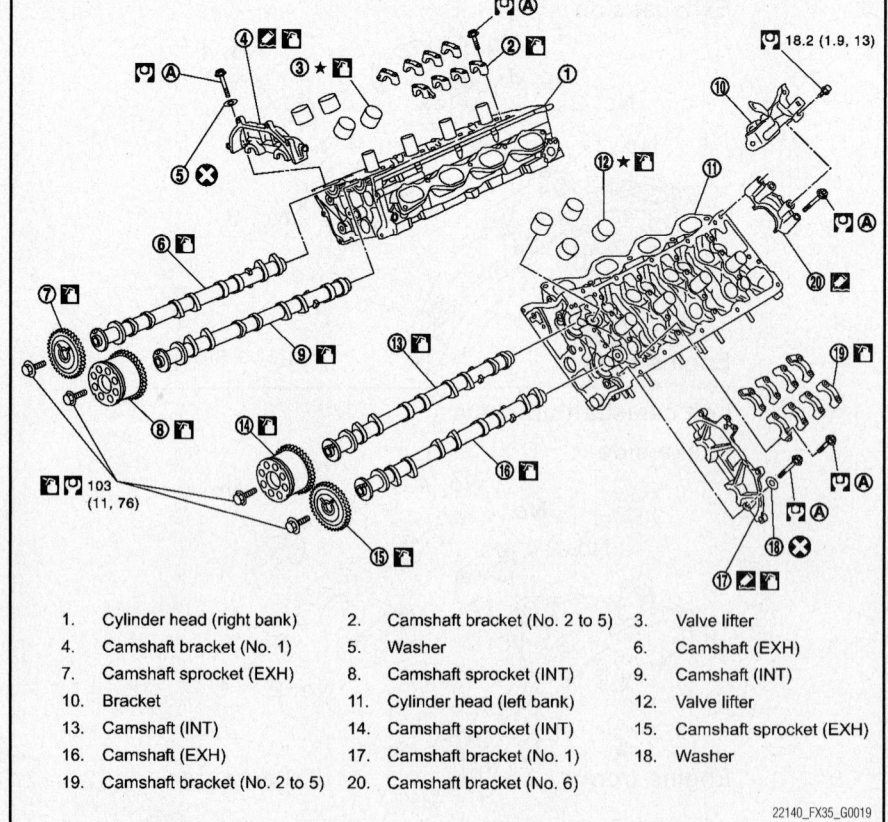

1.	Cylinder head (right bank)	2.	Camshaft bracket (No. 2 to 5)	3.	Valve lifter
4.	Camshaft bracket (No. 1)	5.	Washer	6.	Camshaft (EXH)
7.	Camshaft sprocket (EXH)	8.	Camshaft sprocket (INT)	9.	Camshaft (INT)
10.	Bracket	11.	Cylinder head (left bank)	12.	Valve lifter
13.	Camshaft (INT)	14.	Camshaft sprocket (INT)	15.	Camshaft sprocket (EXH)
16.	Camshaft (EXH)	17.	Camshaft bracket (No. 1)	18.	Washer
19.	Camshaft bracket (No. 2 to 5)	20.	Camshaft bracket (No. 6)		

22140_FX35_G0019

Fig. 62 Exploded view of the camshaft and related components—4.5L engine

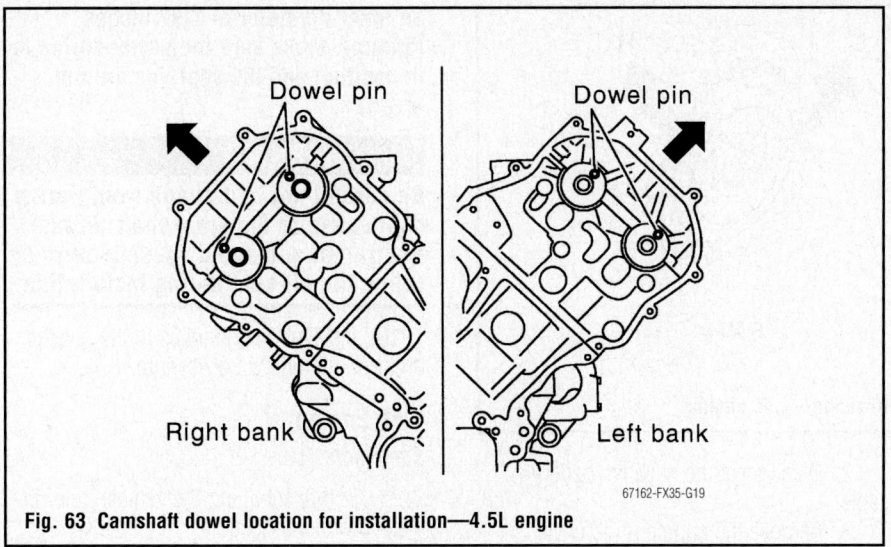

Fig. 63 Camshaft dowel location for installation—4.5L engine

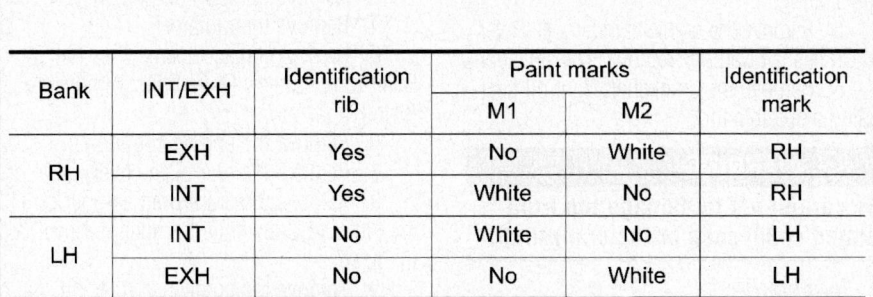

Bank	INT/EXH	Identification rib	Paint marks		Identification mark
			M1	M2	
RH	EXH	Yes	No	White	RH
	INT	Yes	White	No	RH
LH	INT	No	White	No	LH
	EXH	No	No	White	LH

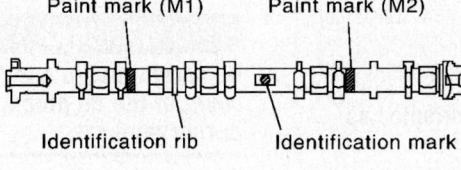

Fig. 64 Location and type of identification mark for camshafts—4.5L engine

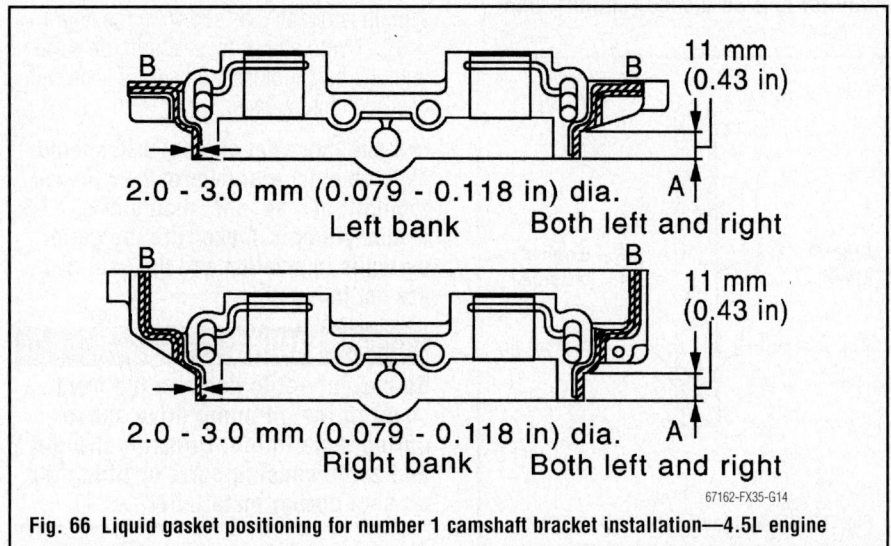

Fig. 66 Liquid gasket positioning for number 1 camshaft bracket installation—4.5L engine

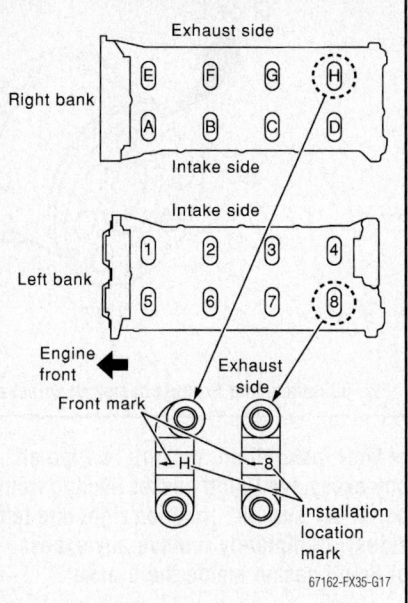

Fig. 65 Camshaft bearing cap identification—4.5L engine

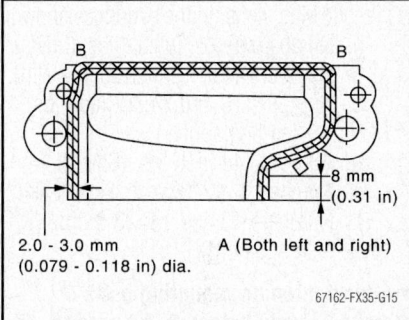

Fig. 67 Liquid gasket positioning for number 6 camshaft bracket installation—4.5L engine.

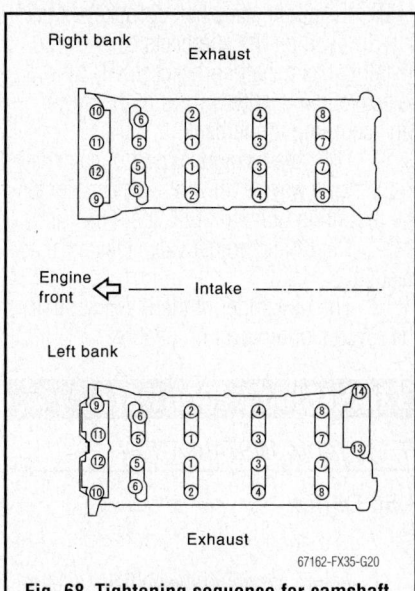

Fig. 68 Tightening sequence for camshaft bearing caps—4.5L engine

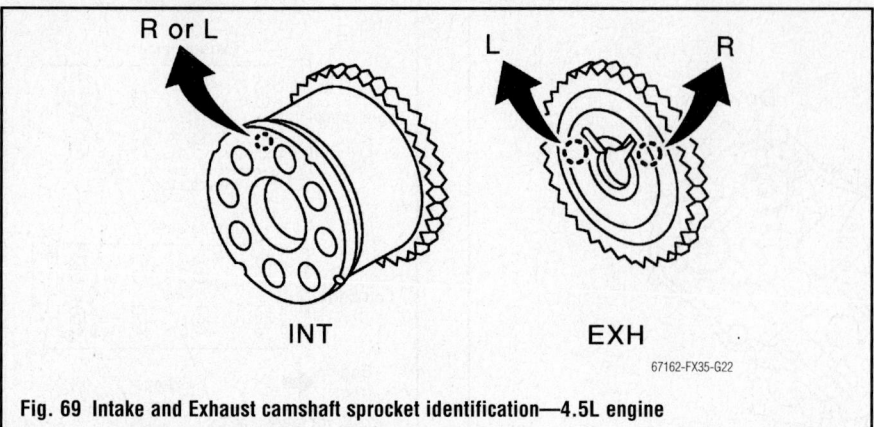

Fig. 69 Intake and Exhaust camshaft sprocket identification—4.5L engine

➡**After installation, be sure to wipe off any excessive liquid gasket leaking from parts "A" and "B" (both on right and left sides). Completely remove any excess of liquid gasket inside the bracket.**

13. Apply liquid gasket to the mating surface of camshaft bracket (number 6) on the left bank intake as shown.

14. Tighten the camshaft bracket (cap) bolts in the following steps, in the sequence shown.

    a. Numbers 9–12: 12 inch lbs. (2 Nm).

    b. Numbers 1–8: 12 inch lbs. (2 Nm).

    c. Numbers 13 and 14: 12 inch lbs. (2 Nm)—left bank only.

    d. All bolts: 48 inch lbs. (6 Nm).

    e. Numbers 1–12: 96 inch lbs. (10 Nm).

    f. Numbers 13 and 14: 23 ft. lbs. (31 Nm)—left bank only.

➡**After tightening mounting bolts of camshaft brackets (caps), be sure to wipe off excessive liquid gasket from the mating surface of the rocker cover and the mating surface of the front cover.**

15. Install the camshaft sprockets. Make sure to position the sprockets as shown. Install the camshaft sprocket (EXH) by selectively using the groove of the dowel pin according to the bank.

16. Lock the hexagonal part of camshaft in the same way as for removal, and tighten the mounting bolts to 76 ft. lbs. (103 Nm).

17. Check and adjust valve clearance as needed.

18. The remainder of the installation is the reverse of removal procedure.

CRANKSHAFT FRONT SEAL

REMOVAL & INSTALLATION

3.5L Engine

See Figure 70.

1. Before servicing the vehicle, refer to the Precautions Section.

2. Disconnect the negative battery cable.

3. Remove the engine undercover.

4. Remove the accessory drive belts. Refer to Accessory Drive Belts, removal & installation.

5. Remove the crankshaft damper. Refer to Crankshaft Damper, removal & installation.

6. Remove the crankshaft front oil seal using a suitable tool.

> **✲✲ WARNING**
>
> **Be careful not to damage the front timing chain case or the crankshaft.**

To install:

7. Apply new engine oil to the oil seal lip and the dust seal lip of new crankshaft front oil seal.

8. Install the crankshaft front oil seal.

➡**The oil seal must be oriented as shown in the figure.**

9. Using a suitable drift, press-fit until the height of front oil seal is level with the mounting surface.

➡**A suitable drift should have an outer diameter of 2.36 inches (60mm), and**

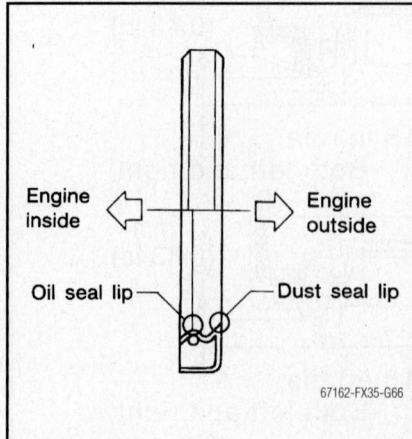

Fig. 70 Oil seal installation orientation

an inner diameter of 1.97 inches (50mm). Make sure the garter spring is in position and the seal lips are not inverted.

> **✲✲ WARNING**
>
> **Be careful not to damage front timing chain case or the crankshaft. Press-fit straight and avoid causing burrs or tilting the oil seal during installation.**

10. Installation continues in the reverse order of the removal procedure.

4.5L Engine

See Figure 70.

1. Before servicing the vehicle, refer to the Precautions Section.

2. Disconnect the negative battery cable.

3. Remove the front engine undercover.

4. Remove the radiator.

5. Remove the accessory drive belt. Refer to Accessory Drive Belts, removal & installation.

6. Remove the engine cooling fan.

7. Remove the rear plate cover.

8. Remove the crankshaft damper pulley. Refer to Crankshaft Damper, removal & installation.

9. Remove the crankshaft front oil seal using a suitable tool.

> **✲✲ WARNING**
>
> **Be careful not to damage the front cover or the oil pump drive spacer during removal.**

To install:

10. Apply new engine oil to the oil seal lip and dust seal lip of the new crankshaft front oil seal.

11. Install the front oil seal so that each seal lip is oriented as shown in the figure.

12. Using a front oil seal drift, press-fit until the height of front oil seal is level with the mounting surface.

➡**A suitable front oil seal drift should have an outer diameter of 2.20 inches (56mm), and an inner diameter of 1.93 inches (49mm). Make sure the garter spring is in position and the seal lips are not inverted.**

> **✲✲ WARNING**
>
> **Be careful not to damage the front cover or the oil pump drive spacer during installation. Press-fit straight and avoid causing burrs or tilting the oil seal during installation.**

13. Installation continues in the reverse order of the removal procedure.

CYLINDER HEAD

REMOVAL & INSTALLATION

3.5L Engine

See Figures 71 through 75.

1. Before servicing the vehicle, refer to the Precautions Section.

2. Remove the camshafts. Refer to Camshaft and Valve Lifters, removal & installation.

3. Temporarily support the front suspension member to support the engine.

➡**Temporary support means that the engine is adequately stable although the weight supported by the hoist may be released. The front suspension member is removed and the cylinder head is hung by the hoist with an engine slinger installed.**

4. Release the hoist from hanging, then remove the engine slinger.

5. Remove the fuel tube and fuel injector assembly.

6. Remove the intake manifold.

7. Remove the exhaust manifold.

8. Remove the water inlet and thermostat housing.

9. Remove the water outlet and water piping.

10. Loosen the cylinder head bolts in the reverse of the tightening order.

11. Remove the cylinder head.

12. Remove the cylinder head gaskets.

To install:

13. Inspect the cylinder head bolt diameters. The cylinder head bolts are tightened by the plastic zone tightening method. Whenever the size difference between d1 and d2 exceeds the limit, replace the bolt with a new one. The specification for d1 minus d2 for the cylinder head bolts is 0.0043 inch (0.11mm).

➡**If the reduction of the outer diameter appears in a position other than at d2, use it as the d2 point value.**

14. Check the cylinder head for distortion. Using a scraper, wipe off oil, scale, gasket, sealant, and carbon deposits from the surface of the cylinder head. At each of several locations on the bottom surface of the cylinder head, measure distortion in 6 directions. If cylinder head distortion exceeds the recommended limit of 0.004 inch (0.1mm), replace the cylinder head.

✳✳ WARNING

Do not allow gasket fragments to enter engine oil or engine coolant passages.

15. Install a new cylinder head gasket.

16. Turn the crankshaft until number 1 piston is set at Top Dead Center (TDC) on the compression stroke. The crankshaft key

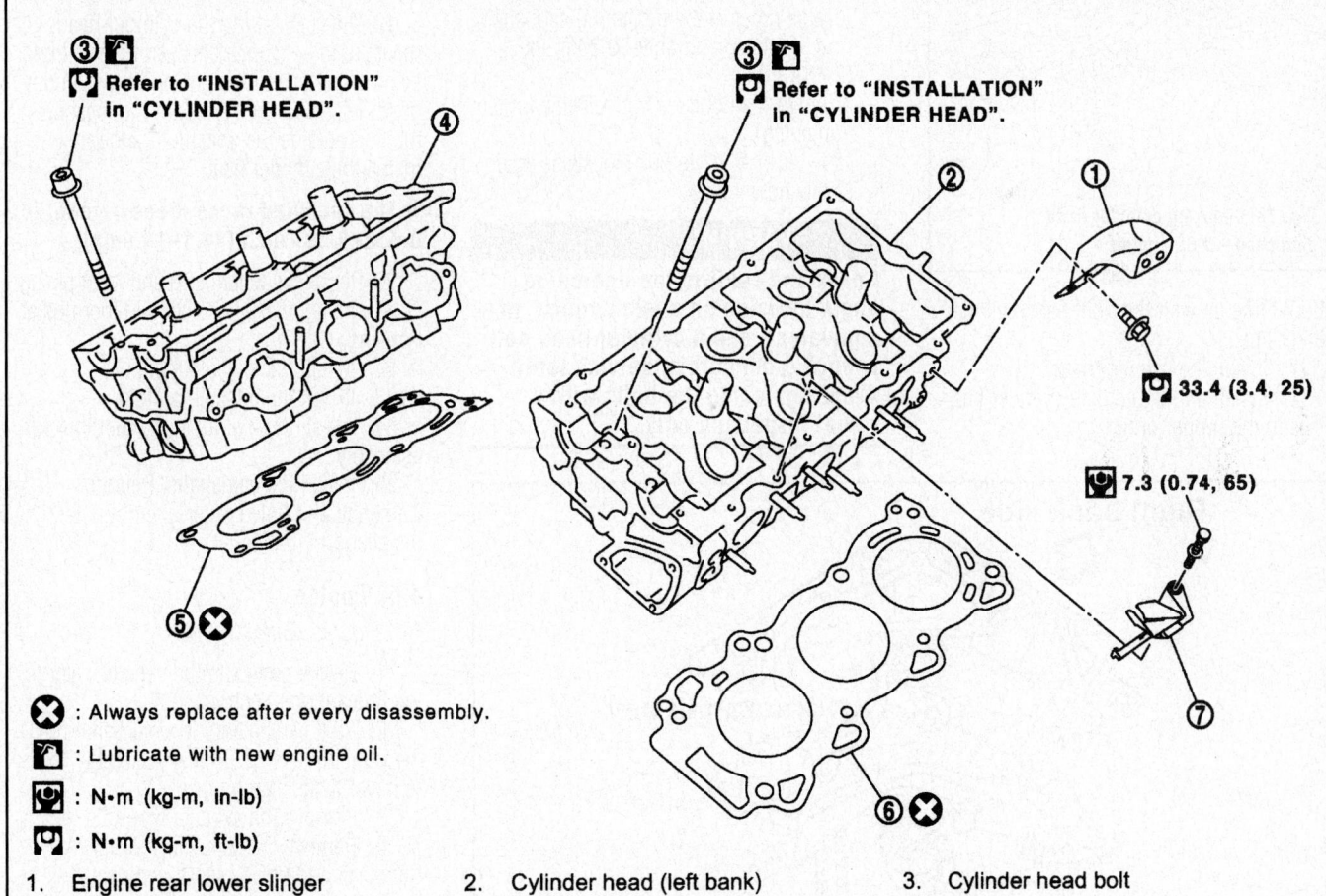

❌ : Always replace after every disassembly.

🛢 : Lubricate with new engine oil.

📷 : N•m (kg-m, in-lb)

📷 : N•m (kg-m, ft-lb)

1. Engine rear lower slinger
2. Cylinder head (left bank)
3. Cylinder head bolt
4. Cylinder head (right bank)
5. Cylinder head gasket (right bank)
6. Cylinder head gasket (left bank)
7. Oil level gauge guide

67162-FX35-G128

Fig. 71 Exploded view of cylinder head and related components—3.5L engine

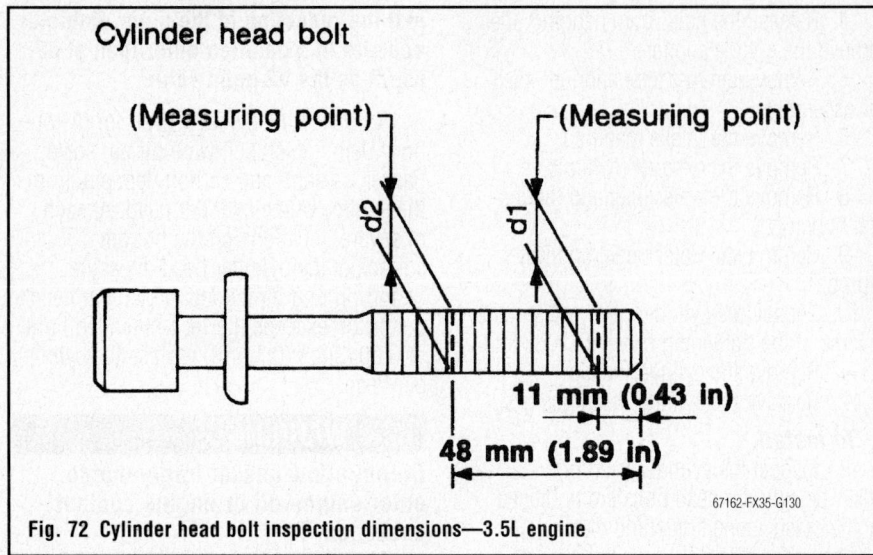

Fig. 72 Cylinder head bolt inspection dimensions—3.5L engine

11 mm (0.43 in)
48 mm (1.89 in)

67162-FX35-G130

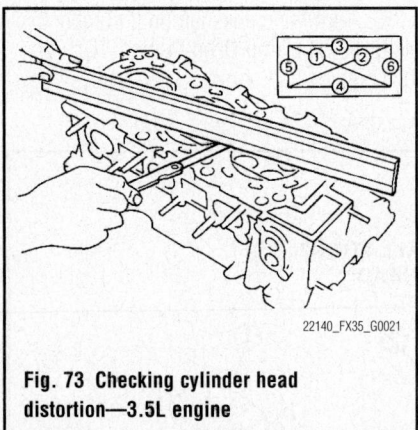

Fig. 73 Checking cylinder head distortion—3.5L engine

should line up with the right bank cylinder center line.

17. Install the cylinder head.

18. Install and tighten the cylinder head bolts in the proper order:

a. Apply new engine oil to the threads and seat surfaces of the cylinder head bolts.

b. Tighten all bolts to 72 ft. lbs. (98 Nm).

c. Completely loosen all bolts in the reverse order of the tightening sequence.

d. Retighten all bolts to 29 ft. lbs. (39 Nm).

e. Turn all bolts 90° clockwise (angle tightening).

f. Turn all bolts 90° clockwise again (angle tightening).

※ WARNING

Check and confirm the tightening angle by using an angle wrench, or equivalent, and a cylinder head bolt wrench (commercial service tool). Avoid tightening the bolts with a visual inspection only.

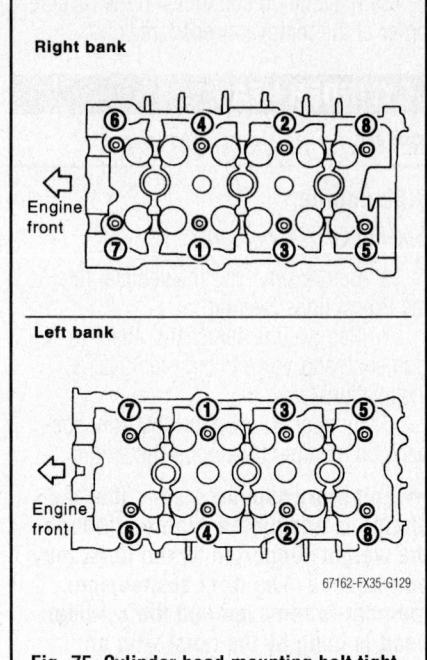

Fig. 75 Cylinder head mounting bolt tightening sequence—3.5L engine

67162-FX35-G129

19. After installing the cylinder head, measure the distance between the front end faces of the cylinder block and the cylinder head (left and right banks). If the measurement is outside the specified range, re-install the cylinder head.

➡ **The specified range measurement is 0.555–0.587 inch (14.1–14.9mm).**

20. Install the water outlet and water piping.

21. Install the water inlet and thermostat housing.

22. Install the exhaust manifold.

23. Install the intake manifold.

24. Install the fuel tube and fuel injector assembly.

25. Install the camshafts. Refer to Camshaft and Valve Lifters, removal & installation.

4.5L Engine

See Figures 76 through 80.

1. Before servicing the vehicle, refer to the Precautions Section.

2. Remove the engine assembly from the vehicle. Refer to Engine Assembly, removal & installation.

3. Remove the exhaust manifold.

4. Remove the camshafts. Refer to Camshaft and Valve Lifters, removal & installation.

5. Loosen the cylinder head bolts in the reverse of the tightening order.

6. Remove the cylinder head.

7. Remove the cylinder head gaskets.

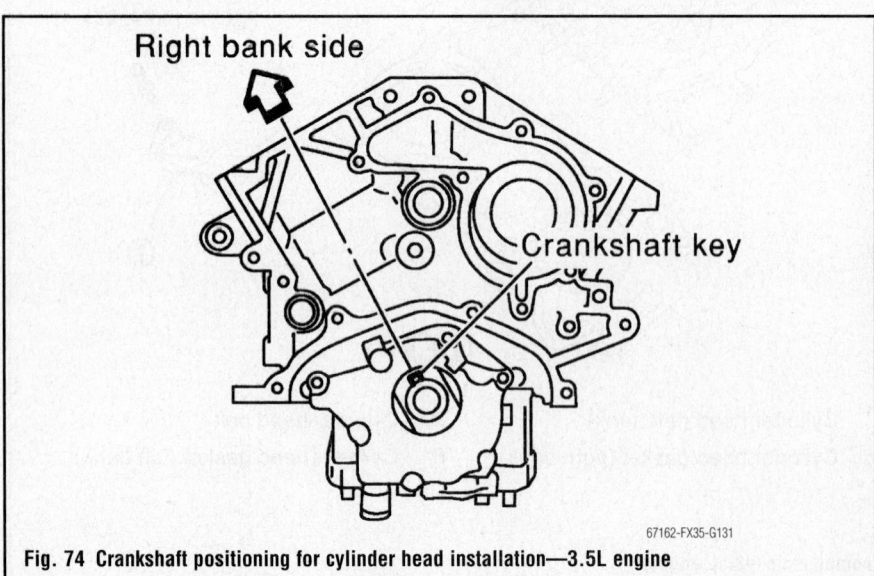

Fig. 74 Crankshaft positioning for cylinder head installation—3.5L engine

67162-FX35-G131

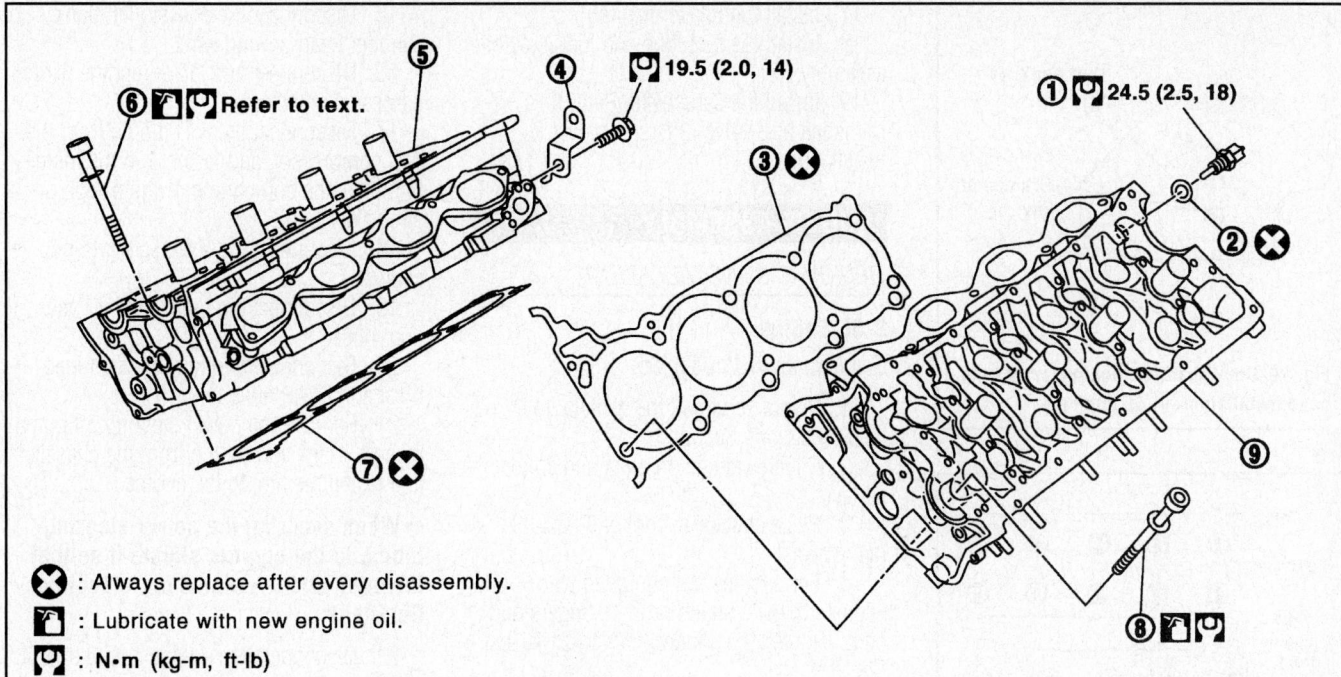

⊗ : Always replace after every disassembly.

🛢 : Lubricate with new engine oil.

🔧 : N•m (kg-m, ft-lb)

I. Engine coolant temperature sensor	2. Washer	3. Cylinder head gasket (left bank)
4. Harness bracket	5. Cylinder head (right bank)	6. Cylinder head bolt
7. Cylinder head gasket (right bank)	8. Cylinder head bolt	9. Cylinder head (left bank)

67162-FX35-G132

Fig. 76 Exploded view of the cylinder head mounting and related components—4.5L engine

To install:

8. Inspect the cylinder head bolt diameters. The cylinder head bolts are tightened by the plastic zone tightening method. Whenever the size difference between d1 and d2 exceeds the limit, replace the bolt with a new one. The specification for d1 minus d2 for the cylinder head bolts is 0.0071 inch (0.18mm).

➡If the reduction of the outer diameter appears in a position other than at d2, use it as the d2 point value.

9. Check the cylinder head for distortion. Using a scraper, wipe off oil, scale, gasket, sealant and carbon deposits from the surface of the cylinder head. At each of several locations on the bottom surface of the cylinder head, measure distortion in 6 directions. If cylinder head distortion exceeds the recommended limit of 0.004 inch (0.1mm), replace the cylinder head.

✳✳ WARNING

Do not allow gasket fragments to enter engine oil or engine coolant passages.

10. Install a new cylinder head gasket.
11. Turn the crankshaft until number 1 piston is set at Top Dead Center (TDC) on the compression stroke. The crankshaft key should line up with the left bank cylinder center line.

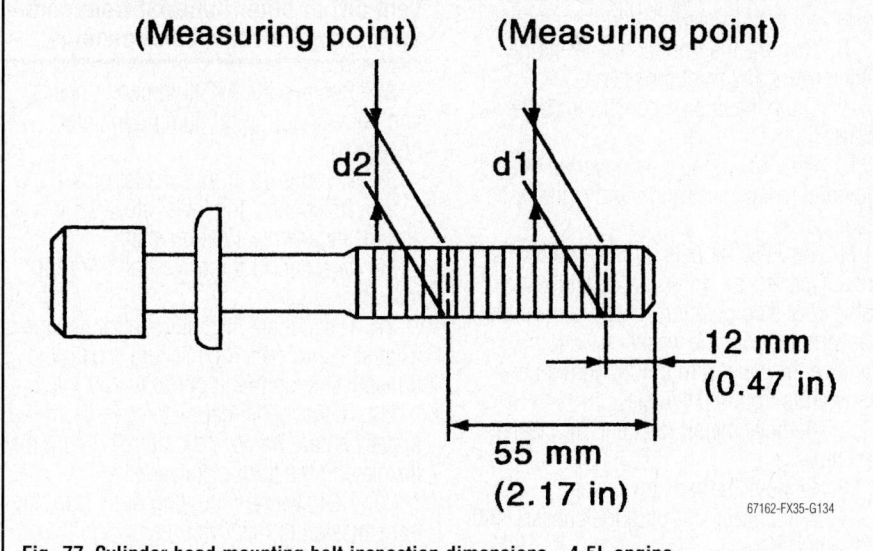

67162-FX35-G134

Fig. 77 Cylinder head mounting bolt inspection dimensions—4.5L engine

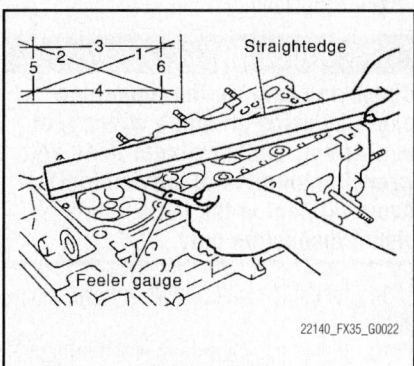

22140_FX35_G0022

Fig. 78 Checking cylinder head distortion—4.5L engine

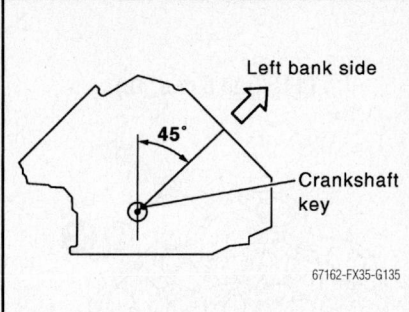

Fig. 79 Crankshaft positioning for cylinder head installation—4.5L engine

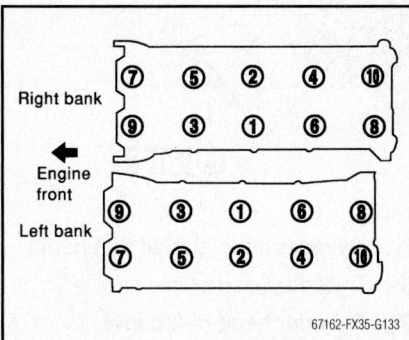

Fig. 80 Cylinder head mounting bolt tightening sequence—4.5L engine

12. Install the cylinder head.

13. Install and tighten the cylinder head bolts in the proper order as follows:

   a. Apply new engine oil to the threads and seating surface of the cylinder head bolts.

   b. Tighten all bolts to 72 ft. lbs. (98 Nm).

   c. Completely loosen all bolts in the reverse order of the tightening sequence.

   d. Retighten all bolts to 32 ft. lbs. (44 Nm)

   e. Turn all bolts 60° clockwise (angle tightening).

   f. Turn all bolts 60° clockwise again (angle tightening).

> **⁕⁕ WARNING**
>
> **Check and confirm the tightening angle by using an angle wrench, or equivalent, and a cylinder head bolt wrench (commercial service tool). Avoid tightening the bolts with a visual inspection only.**

14. Install the water outlet and water piping.

15. Install the water inlet and thermostat housing.

16. Install the exhaust manifold.

17. Install the intake manifold.

18. Install the fuel tube and fuel injector assembly.

19. Install the camshafts. Refer to Camshaft and Valve Lifters, removal & installation.

ENGINE ASSEMBLY

REMOVAL & INSTALLATION

3.5L Engine

See Figures 81 through 86.

1. Before servicing the vehicle, refer to the Precautions Section.

2. Situate vehicle on a flat and solid surface.

3. Place chocks at front and back of rear wheels.

4. For engines not equipped with engine slingers, attach proper slingers and bolts. Tighten the slinger bolts to 21 ft. lbs. (28 Nm).

➡**Always use the support point specified for lifting. Use either a 2-point lift type or a separate type lift. If a board-on type is used for unavoidable reasons, support the rear axle jacking point with a transmission jack or similar tool before starting work. The rear axle jacking point needs to be supported since the center of gravity will shift rearward once the engine assembly is removed.**

The engine and transmission assembly are removed from the vehicle with the suspension member from the underside of the vehicle. After removal, separate the engine from the transmission.

5. Release the fuel system pressure.

6. Disconnect both battery terminals.

7. Remove the engine cover, battery cover and both front wheel/tire assemblies.

8. Remove the front and rear engine undercovers and front cross bar.

9. Drain the engine coolant from the radiator.

10. On AWD models, remove the clips of the hood ledge cover and remove the hood ledge cover.

11. On AWD models, remove both wiper arms. Operate the wiper motor, and stop it at the auto stop position. Then, remove the washer tube from the washer tube joint. Remove the wiper arm mounting nuts and wiper arms from the vehicle.

12. Remove the air duct and air cleaner assembly.

13. Remove the radiator hoses.

14. Disconnect and plug the chassis-end of the heater hose.

15. Disconnect the chassis/left bank cylinder head ground wire.

16. Disconnect and tag all engine wiring harness connectors.

17. Disconnect the A/C tubing from the A/C compressor, and fasten it to the inside of the engine compartment with a rope or strong cord.

18. Disconnect the 2 chassis ground cables.

19. Disconnect the brake booster vacuum hose.

20. Disconnect and plug the fuel feed hose and EVAP hose.

21. Remove the power steering oil pump reservoir tank and tubing from the chassis, and fasten them onto the engine.

➡**When securing the power steering tubing to the engine, situate it so that the open end is pointed up to avoid a fluid leak.**

22. Disconnect the engine compartment wiring harness connectors from the passenger compartment, as follows:

   a. Remove the passenger-side kick plate, dashboard side trim, and glove box.

   b. Disconnect the engine compartment wiring harness connectors from the ECM.

   c. Disengage the intermediate fixing point. Pull the engine compartment wiring harnesses through to the engine compartment and fasten them to the engine.

> **⁕⁕ WARNING**
>
> **Be careful not to damage the wiring harness when pulling it out of the passenger compartment. Also, cover the wiring harness connectors to prevent dirt or other material from contaminating the connector openings.**

23. Remove the A/T fluid cooler hoses and power steering oil pump oil cooler hoses.

24. Remove the front exhaust pipe.

25. Disconnect the lower steering joint, and disengage the steering shaft.

26. Disconnect the driveshaft from the transmission.

27. Disconnect the shift control linkage from selector lever, then secure it onto the transmission so that it doesn't hang free.

28. Remove the rear plate cover from the upper oil pan. Remove the bolts holding the flexplate to the torque converter.

29. Remove the mounting bolts from the transmission to the lower rear side of the upper oil pan.

30. Remove the front stabilizer.

31. Detach the left-hand and right-hand side tie-rod ends from the steering knuckles.

32. Remove the lower ends of the left-hand and right-hand struts from the lower arms.

33. Disconnect the left-hand and right-hand lower arms from the suspension member.

34. On AWD models, disconnect both front halfshafts from the knuckles.

35. Use a manual lift table caddy or equivalently rigid tool such as a transmis-sion jack and securely support the bottom of the suspension member and transmission.

➡**Put a piece of wood or something similar as the supporting surface and secure in a completely stable condition.**

36. Remove the rear member mounting bolt.

37. Remove the suspension member mounting bolt and nut.

38. Carefully lower the jack to remove engine, transmission, and suspension member assembly. While performing this, observe the following:

- Confirm there is no interference with vehicle
- Make sure all connection points have been disconnected
- Keep in mind the center of vehicle gravity changes
- If necessary, use jackstands to support the vehicle at the rear jacking point(s) to prevent it from falling off the lift

39. Install engine slingers/supports into the front of the right bank cylinder head and the rear of the left bank cylinder head.

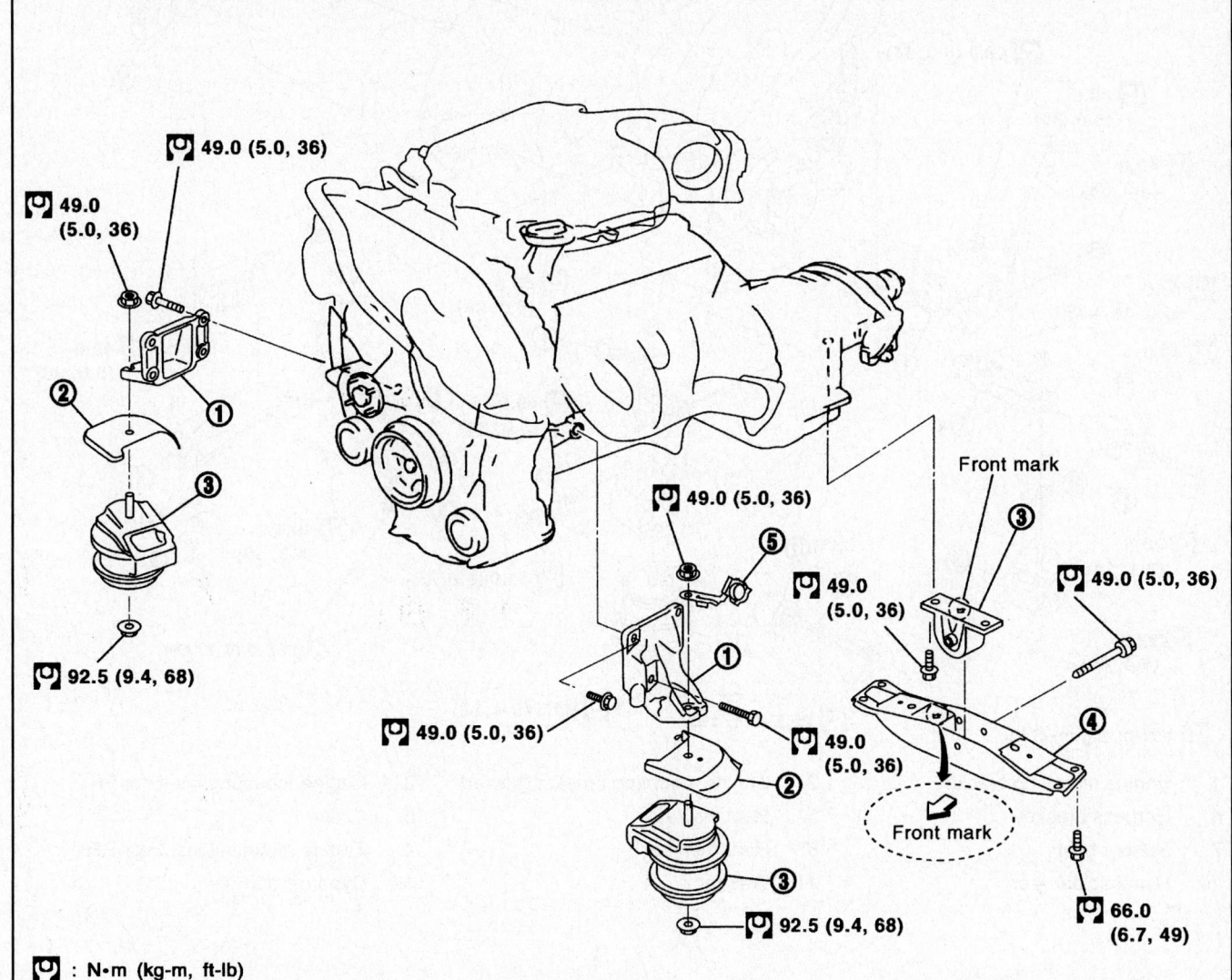

: N•m (kg-m, ft-lb)

1. Engine mounting bracket	2. Heat insulator	3. Engine mounting insulator
4. Rear member	5. Harness bracket	

67162-FX35-G105

Fig. 81 Exploded view of engine mounting—2WD 3.5L engine

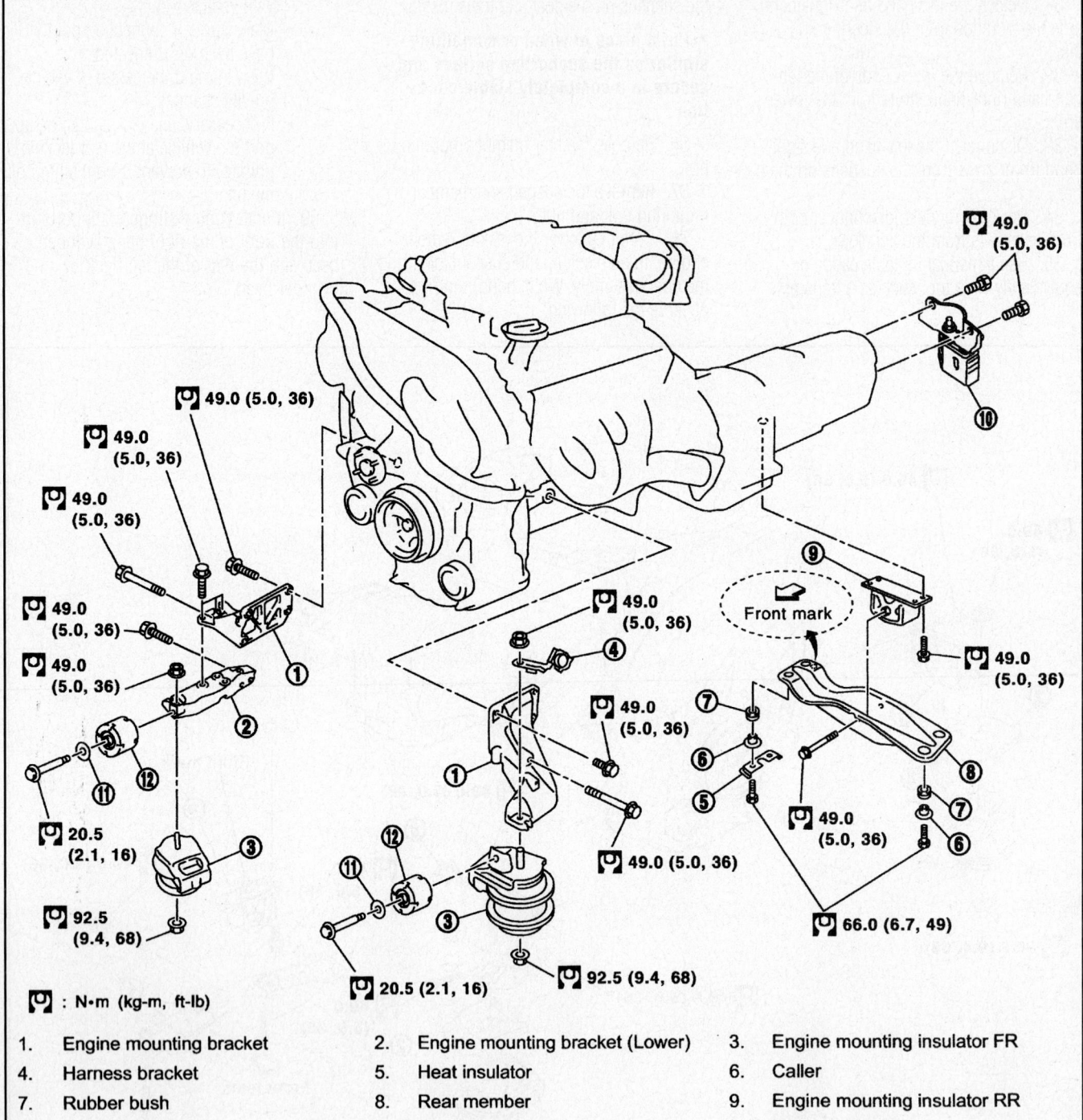

49.0 (5.0, 36)

49.0 (5.0, 36)

49.0 (5.0, 36)

49.0 (5.0, 36)

49.0 (5.0, 36)

49.0 (5.0, 36)

20.5 (2.1, 16)

92.5 (9.4, 68)

49.0 (5.0, 36)

49.0 (5.0, 36)

49.0 (5.0, 36)

20.5 (2.1, 16)

92.5 (9.4, 68)

Front mark

49.0 (5.0, 36)

49.0 (5.0, 36)

66.0 (6.7, 49)

: N•m (kg-m, ft-lb)

1. Engine mounting bracket
2. Engine mounting bracket (Lower)
3. Engine mounting insulator FR
4. Harness bracket
5. Heat insulator
6. Caller
7. Rubber bush
8. Rear member
9. Engine mounting insulator RR
10. Dynamic damper
11. Washer
12. Dynamic damper

67162-FX35-G109

Fig. 82 Exploded view of engine mounting—AWD 3.5L engine

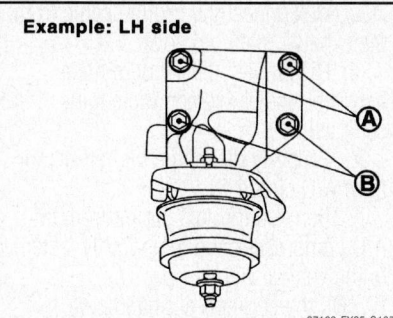

Fig. 83 When installing engine mounting bracket on cylinder block, tighten 2 upper bolts (A), then tighten 2 lower bolts (B)—2WD 3.5L engine

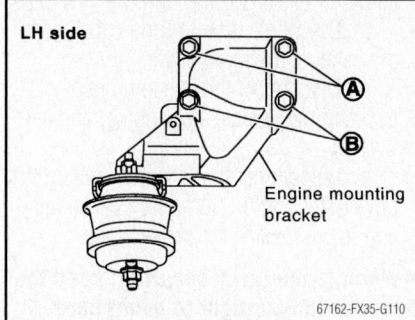

Fig. 84 Left-hand engine mounting bracket bolt identification—AWD 3.5L engine

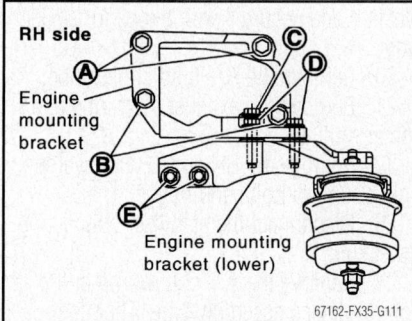

Fig. 85 Right-hand engine mounting bracket bolt identification—AWD 3.5L engine

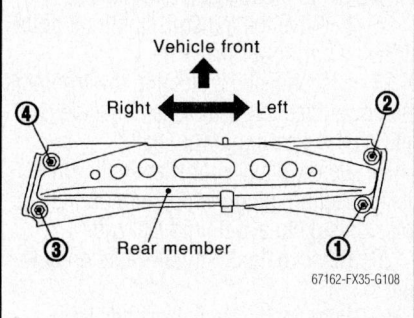

Fig. 86 Rear member mounting bolt tightening sequence—2WD 3.5L engine

40. Remove the power steering oil pump from the engine.

41. Remove the engine mounting insulator bottom nut.

42. Lifting with a hoist, separate the engine and transmission assembly from the suspension member.

✳✳ WARNING

Before and during lifting, check whether any wiring harnesses are still connected. Avoid damaging the engine mounting insulator or getting oil/grease contact on it.

43. On AWD models, remove both front halfshafts.

44. Remove the alternator.

45. Remove the starter motor.

46. On AWD models, detach the front driveshaft from the front final drive assembly.

47. Separate the engine from the transmission assembly.

48. Remove the engine mounting insulator and bracket.

49. On AWD models, remove the front final drive assembly from the oil pan (upper).

To install:

50. On AWD models, install the front final drive assembly onto the upper oil pan.

➡**When installing the engine mounting bracket on the cylinder block, tighten the 2 upper bolts (shown as A) first. Then tighten the 2 lower bolts (shown as B).**

51. Install the engine mounting insulator and bracket.

52. Join the engine and transmission assemblies.

53. On AWD models, reattach the front driveshaft to the front final drive assembly.

54. Install the starter motor.

55. Install the alternator.

56. On AWD models, install both front halfshafts.

57. Join the engine and transmission assembly to the suspension member.

58. Install the engine mounting insulator bottom nut.

59. Install the power steering oil pump.

60. Remove the engine slingers/supports.

61. Carefully raise the jack to install the engine, transmission, and suspension member assembly.

62. Install the suspension member mounting bolt and nut.

63. Install the rear member mounting bolts in the order shown.

64. On AWD models, reconnect both front halfshafts to the knuckles.

65. Connect the left-hand and right-hand lower arms to the suspension member.

66. Reinstall the lower ends of the left-hand and right-hand struts from the lower arms.

67. Reattach the left-hand and right-hand side tie-rod ends to the steering knuckles.

68. Install the front stabilizer.

69. Install the mounting bolts from the transmission to the lower rear side of the upper oil pan.

70. Install the bolts holding the flexplate to the torque converter.

71. Install the rear plate cover onto the upper oil pan.

72. Reconnect the shift control linkage to the selector lever.

73. Install the driveshaft to the transmission.

74. Connect the lower steering joint, and engage the steering shaft.

75. Install the front exhaust pipe.

76. Reattach the A/T fluid cooler hoses and power steering oil pump oil cooler hoses.

77. Route the engine compartment wiring harnesses through to the passenger compartment.

78. Connect the engine compartment wiring harness connectors to the ECM.

79. Install the passenger-side kick plate, dashboard side trim, and glove box.

80. Install the power steering oil pump reservoir tank and tubing.

81. Reconnect the fuel feed and EVAP hoses.

82. Reconnect the brake booster vacuum hose.

83. Reattach the 2 chassis ground cables.

84. Reconnect the A/C tubing to the A/C compressor.

85. Reconnect all engine wiring harness connectors.

86. Reconnect the chassis/left bank cylinder head ground wire.

87. Reconnect the heater hose.

88. Install the radiator hoses.

89. Install the air duct and air cleaner assembly.

90. Install the top right-hand cowl cover.

91. Fill the engine cooling system.

92. Install the front crossbar and the front and rear lower engine covers.

93. Install the engine cover, the battery cover, and both front wheel assemblies.

94. Connect both battery cables.

95. Before starting the engine, check oil/fluid levels including engine coolant and engine oil. If less than required quantity, fill to the specified level.

96. Turn the ignition switch **ON** with the engine **OFF**). With fuel pressure applied to the fuel piping, check for fuel leakage at the connection points.

97. Start the engine. With engine speed increased, check again for fuel leakage at connection points.

98. Run the engine to check for unusual noise and vibration.

99. Warm the engine thoroughly to make sure there is no leakage of fuel, exhaust gases, or any oil/fluids including engine oil and engine coolant.

100. Bleed air from lines and hoses of applicable lines, such as in the cooling system.

101. After cooling down the engine, again check oil/fluid levels including engine oil and engine coolant. Refill to the specified level, if necessary.

4.5L Engine

See Figures 89 through 92.

1. Before servicing the vehicle, refer to the Precautions Section.

2. Situate the vehicle on a flat and solid surface.

3. Place chocks at front and back of rear wheels.

4. For engines not equipped with engine slingers, attach proper slingers and

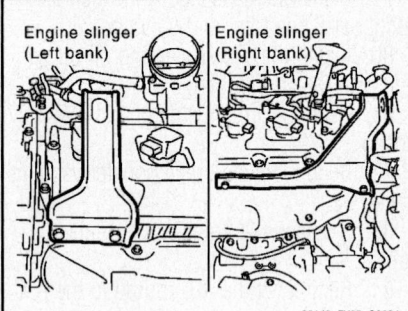

Fig. 87 Install engine slingers into the front of each cylinder head—4.5L engine

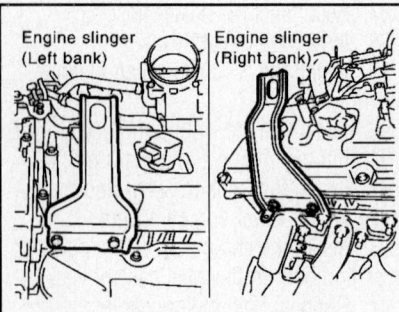

Fig. 88 Change the engine slinger installed to the cylinder head (right bank)—4.5L engine

bolts. Tighten the slinger bolts to 21 ft. lbs. (28 Nm).

➡**Always use the support point specified for lifting. Use either a 2-point lift type or a separate type lift. If a board-on type is used for unavoidable reasons, support the rear axle jacking point with a transmission jack or similar tool before starting work. The rear axle jacking point needs to be supported since the center of gravity will shift rearward once the engine assembly is removed.**

The engine, transmission assembly, and front final drive are removed from the vehicle with the front suspension member from the underside of the vehicle. After removal, separate the engine from the transmission.

5. Relieve the fuel pressure.

✳✳ CAUTION

The fuel injection system remains under pressure after the engine has been OFF. Properly relieve fuel pressure before disconnecting any fuel lines. Failure to do so may result in fire or personal injury.

6. Drain the engine coolant from the radiator after the engine is cold. Take care not to spill engine coolant on the accessory drive belt.

7. Disconnect both battery terminals.

8. Remove the Crankshaft Position (CKP) sensor.

✳✳ WARNING

Handle carefully to avoid dropping and shocks. Do not disassemble. Do not allow metal powder to adhere to the magnetic part at sensor tip. Do not place sensors in a location where they are exposed to magnetism.

9. Remove the hood assembly
10. Remove the engine cover.
11. Remove the front and rear engine undercover.
12. Remove the air duct (inlet), air duct, and air cleaner case assembly.
13. Remove the accessory drive belts.
14. Remove the radiator and radiator hoses (upper and lower).
15. Remove the front wheels and tires.
16. On the engine left hand side:
    a. Disconnect the engine room harness from the engine side and set it aside for easier work.
    b. Disconnect the heater hoses and install plugs to avoid leakage of engine coolant.

c. Disconnect the ground cable from the exhaust manifold cover.

d. Disconnect the vacuum hose, between the vehicle and the engine, and set it aside.

e. Properly discharge the refrigerant from A/C circuit.

f. Remove the A/C piping from the A/C compressor and temporarily fasten it to the vehicle with a rope.

17. On the engine right hand side:
    a. Disconnect the fuel feed hose and EVAP hose. Fit plugs onto disconnected hose to prevent fuel leakage.
    b. Disconnect the engine room harness from the engine side and set it aside for easier work.
    c. Disconnect the ground cable from the exhaust manifold cover.
    d. Disconnect the vacuum hose, between the vehicle and engine, and set it aside.
    e. Disconnect the reservoir tank of the power steering oil pump from the engine and move it aside for easier work.

➡**When temporarily securing, keep the reservoir tank upright to avoid fluid leakage.**

18. Remove the front cross bar.
19. Disconnect the power steering oil pump from the engine. Move it from its location and secure it with a rope for easier work.
20. Remove the A/T fluid cooler tube.
21. Remove the exhaust front tube and center muffler.
22. Remove the RH and LH transverse link mounting bolts and nuts.
23. Disconnect the stabilizer connecting lower rod.
24. Remove the A/T control rod at the control device assembly side. Then temporarily secure it on the transmission, so that it does not sag.
25. Remove the rear plate cover from the oil pan. Then remove the bolts fixing the drive plate to the torque converter.
26. Remove the transmission joint bolts at the oil pan lower rear side.
27. Disconnect the steering lower joint at the power steering gear assembly side and release the steering lower shaft.
28. Remove rear propeller shaft. After disconnection, plug the opening on transmission side to avoid fluid leakage.
29. Remove the front halfshafts on both sides.
30. Remove the front propeller shaft.
31. Remove the 3-way catalyst from both banks.
32. Install engine slingers into the

front of each cylinder head. Tighten the slinger bolts to 25 ft. lbs. (33 Nm).

33. Lift using the engine slingers and a hoist and secure the engine into position.

34. Use a manual lift table caddy, or an equivalently rigid tool such as transmission jack, to securely support the bottom of the suspension member and transmission.

➡️**Put a piece of wood, or something similar, to use as the supporting surface and secure a completely stable condition for the assembly.**

35. Remove the engine rear member mounting bolts.

36. Remove the front suspension member mounting nuts.

37. Carefully lower the jack, or raise the lift to remove the engine, transmission, front final drive, and front suspension member assembly. When performing this work, observe the following:

a. Confirm there is no interference with the vehicle.

b. Make sure that all the connection points have been disconnected.

c. Keep in mind that the center of vehicle gravity changes. If necessary, use jack(s) to support the vehicle at rear jacking point(s) to prevent it from falling off the lift.

38. In order to separate the components from the engine, change the engine slinger installed to the cylinder head (right bank). Tighten the slinger bolts to 25 ft. lbs. (33 Nm).

39. Remove the engine mounting insulators underside nut (RH and LH).

40. Lift with hoist and separate engine and transmission assembly from the front suspension member.

⁑ **WARNING**

Ensure there are no harnesses left connected. Avoid damage to and oil/grease smearing or spills onto the engine mounting insulator.

41. Remove the alternator.
42. Remove the starter motor.
43. Separate the engine from the transmission assembly.
44. Remove the front final drive from the engine.

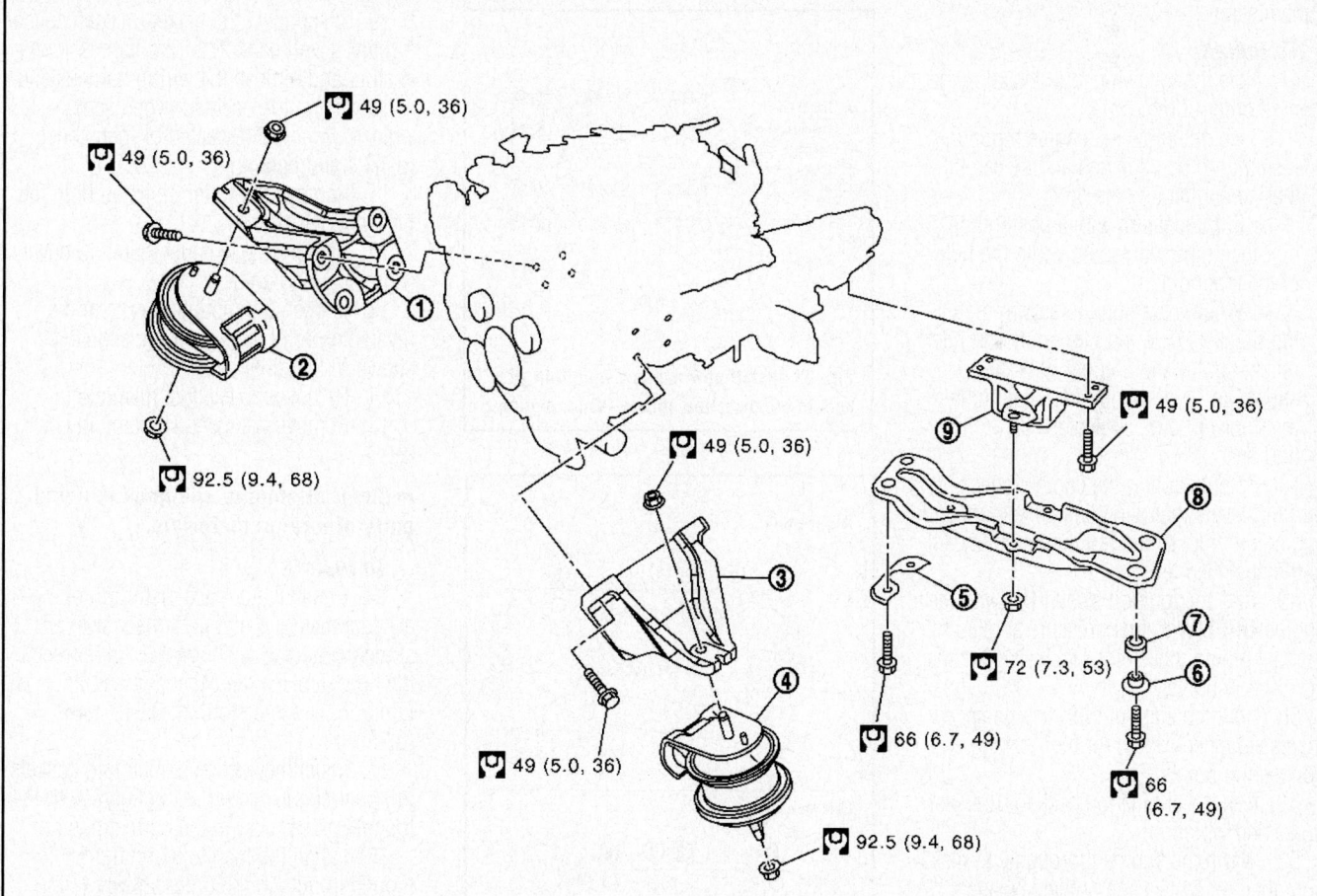

1. Engine mounting bracket (RH)
2. Engine mounting insulator (RH)
3. Engine mounting bracket (LH)
4. Engine mounting insulator (LH)
5. Plate
6. Collar
7. Grommet
8. Engine rear member
9. Engine mounting insulator (rear)

22140_FX35_G0023

Fig. 89 Exploded view of engine mounting—4.5L engine

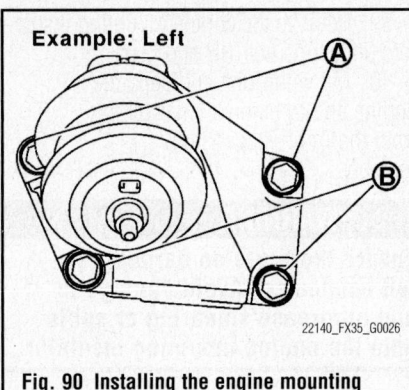

Fig. 90 Installing the engine mounting brackets—4.5L engine

45. Remove the engine mounting insulators (RH and LH) and brackets (RH and LH) from the engine.

46. Remove the engine rear member and engine mounting insulator (rear) from the transmission.

To install:

47. Note the following, and install in the reverse order of removal.

a. Do not allow the engine mounting insulator to be damaged and be careful that no engine oil gets on it.

b. For a location with a positioning pin, insert the pin securely into the hole of the mating part.

c. When installing the engine mounting brackets (RH and LH) on the cylinder block, tighten the 2 upper bolts first, shown as "A" in the figure. Then tighten the 2 lower bolts, shown as "B" in the figure.

48. Before starting the engine, check oil/fluid levels including engine coolant and engine oil. If less than required quantity, fill to the specified level.

49. Turn the ignition switch **ON** with the engine **OFF**). With fuel pressure applied to the fuel piping, check for fuel leakage at the connection points.

50. Start the engine. With engine speed increased, check again for fuel leakage at connection points.

51. Run the engine to check for unusual noise and vibration.

52. Warm the engine thoroughly to make sure there is no leakage of fuel, exhaust gases, or any oil/fluids including engine oil and engine coolant.

53. Bleed air from lines and hoses of applicable lines, such as in the cooling system.

54. After cooling down the engine, again check oil/fluid levels including engine oil and engine coolant. Refill to the specified level, if necessary.

EXHAUST MANIFOLD

REMOVAL & INSTALLATION

3.5L Engine

See Figures 91 through 93.

1. Before servicing the vehicle, refer to the Precautions Section.

✳✳ CAUTION

Perform the work when the exhaust and cooling system have completely cooled down.

2. Remove the engine cover.

3. Remove the air cleaner case and air duct.

4. Remove the front and rear engine undercover and front cross bar.

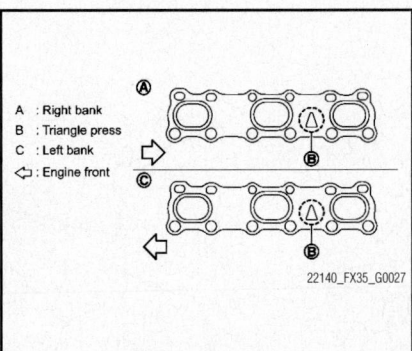

Fig. 91 Install new exhaust manifold gaskets in the direction shown—3.5L engine

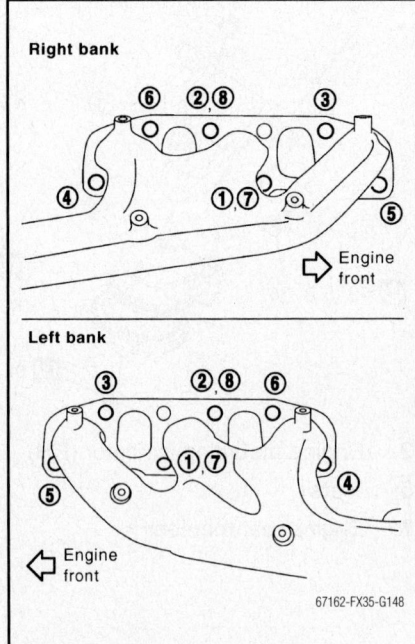

Fig. 92 Exhaust manifold mounting bolt tightening sequence—3.5L engine

5. Disconnect the heated oxygen sensor 2 wiring harness connectors (bank 1 and bank 2).

6. Using a heated oxygen sensor wrench, remove the heated oxygen sensors (bank 1 and bank 2).

✳✳ WARNING

Be careful not to damage heated oxygen sensor. Discard any heated oxygen sensor which has been dropped from a height of more than 20 inches (51cm) onto a hard surface such as a concrete floor; replace with a new sensor.

7. Remove the exhaust mounting bracket between the right/left catalytic converter and transmission.

8. Remove the 3-way catalyst (right and left bank).

9. Disconnect the heated oxygen sensor 1 (bank 1 and bank 2) wiring harness connectors and remove the wiring harness clip.

10. Using the heated oxygen sensor wrench, remove the heated oxygen sensor 1 (bank 1 and bank 2).

11. Remove the water pipes on both the right and left side.

12. Remove the exhaust manifold cover (right and left bank).

13. Loosen the mounting nuts in the reverse order of the tightening sequence shown in the illustration.

14. Remove the exhaust manifold.

15. Remove the exhaust manifold gaskets.

➡ **Cover all engine openings to avoid entry of foreign materials.**

To install:

16. Check the surface distortion of the exhaust manifold mating surface with a straightedge and feeler gauge. If it exceeds the limit, replace the exhaust manifold. Limit of surface distortion: 0.012 inch (0.3mm).

17. Install new exhaust manifold gaskets in the direction shown in the figure with the triangle press mark in the correct position.

18. Install the manifold and tighten the mounting nuts in the order shown. If the stud bolts were removed, install them and tighten them to 11 ft. lbs. (15 Nm).

19. Tighten nuts number 1 and 2 in two steps.

20. Tighten all exhaust manifold nuts-to-engine nuts to 22 ft. lbs. (31 Nm).

21. Install the exhaust manifold cover (right and left bank).

22. Install the water pipes on both the right and left side.

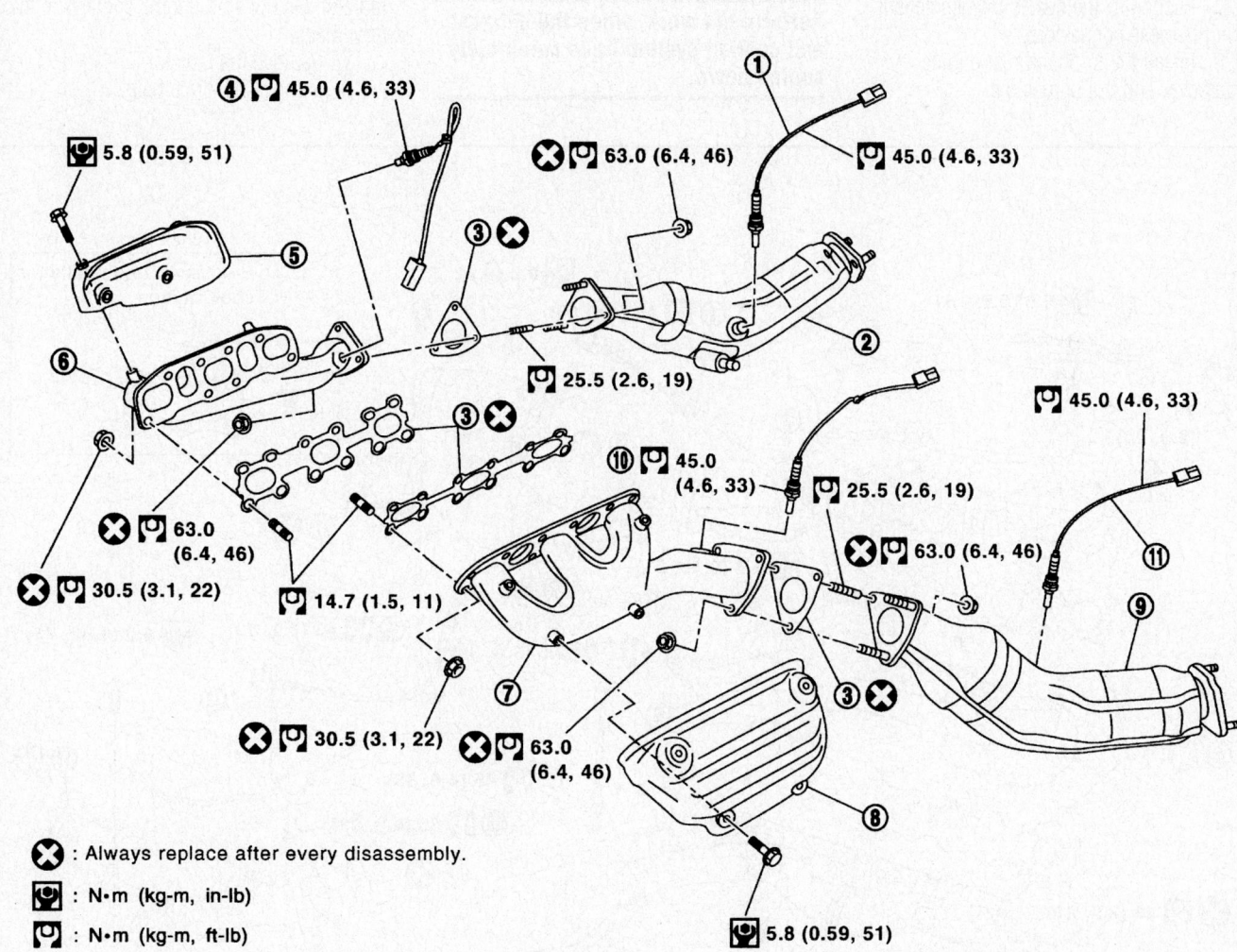

④ ⬚ 45.0 (4.6, 33)

⬚ 5.8 (0.59, 51)

❌ ⬚ 63.0 (6.4, 46)

⬚ 45.0 (4.6, 33)

⑤

③ ❌

⑥

⬚ 25.5 (2.6, 19)

⑩ ⬚ 45.0 (4.6, 33)

⬚ 25.5 (2.6, 19)

⬚ 45.0 (4.6, 33)

③ ❌

❌ ⬚ 63.0 (6.4, 46)

❌ ⬚ 30.5 (3.1, 22)

⬚ 14.7 (1.5, 11)

❌ ⬚ 63.0 (6.4, 46)

⑪

⑦

⑨

❌ ⬚ 30.5 (3.1, 22)

③ ❌

⑧

❌ ⬚ 63.0 (6.4, 46)

⬚ 5.8 (0.59, 51)

❌ : Always replace after every disassembly.

⬚ : N•m (kg-m, in-lb)

⬚ : N•m (kg-m, ft-lb)

1. Heated oxygen sensor 2 (bank 1)	2. Three way catalyst (right bank)	3. Gasket
4. heated oxygen sensor 1 (bank 1)	5. Exhaust manifold cover (right bank)	6. Exhaust manifold (right bank)
7. Exhaust manifold (left bank)	8. Exhaust manifold cover (left bank)	9. Three way catalyst (left bank)
10. heated oxygen sensor 1 (bank 2)	11. Heated oxygen sensor 2 (bank 2)	

67162-FX35-G146

Fig. 93 Exploded view of the exhaust manifold mounting and related components—3.5L engine

23. Install the heated oxygen sensor 1 (bank 1 and bank 2).

24. Install the 3-way catalyst (right and left bank).

25. Install the exhaust mounting bracket between the right/left catalytic converter and transmission.

26. Install the heated oxygen sensors 2 (bank 1 and bank 2).

27. Reconnect the heated oxygen sensor wiring harness connectors.

28. Install the front and rear engine undercover and front cross bar.

29. Install the air cleaner case and air duct.

30. Install the engine cover.

4.5L Engine

See Figures 94 and 95.

1. Before servicing the vehicle, refer to the Precautions Section.

> ※ **CAUTION**
>
> **Perform the work, when the exhaust and cooling system have completely cooled down.**

2. Remove the engine cover.

3. Remove the front and rear engine undercovers.

4. Remove the air duct (inlet), air cleaner case, and mass air flow sensor assembly, air duct and resonator assembly.

5. Remove the front cross bar.

6. Drain the engine coolant from the radiator. Do not spill engine coolant on the drive belts.

7. Remove the radiator.

8. Remove the drive belts.

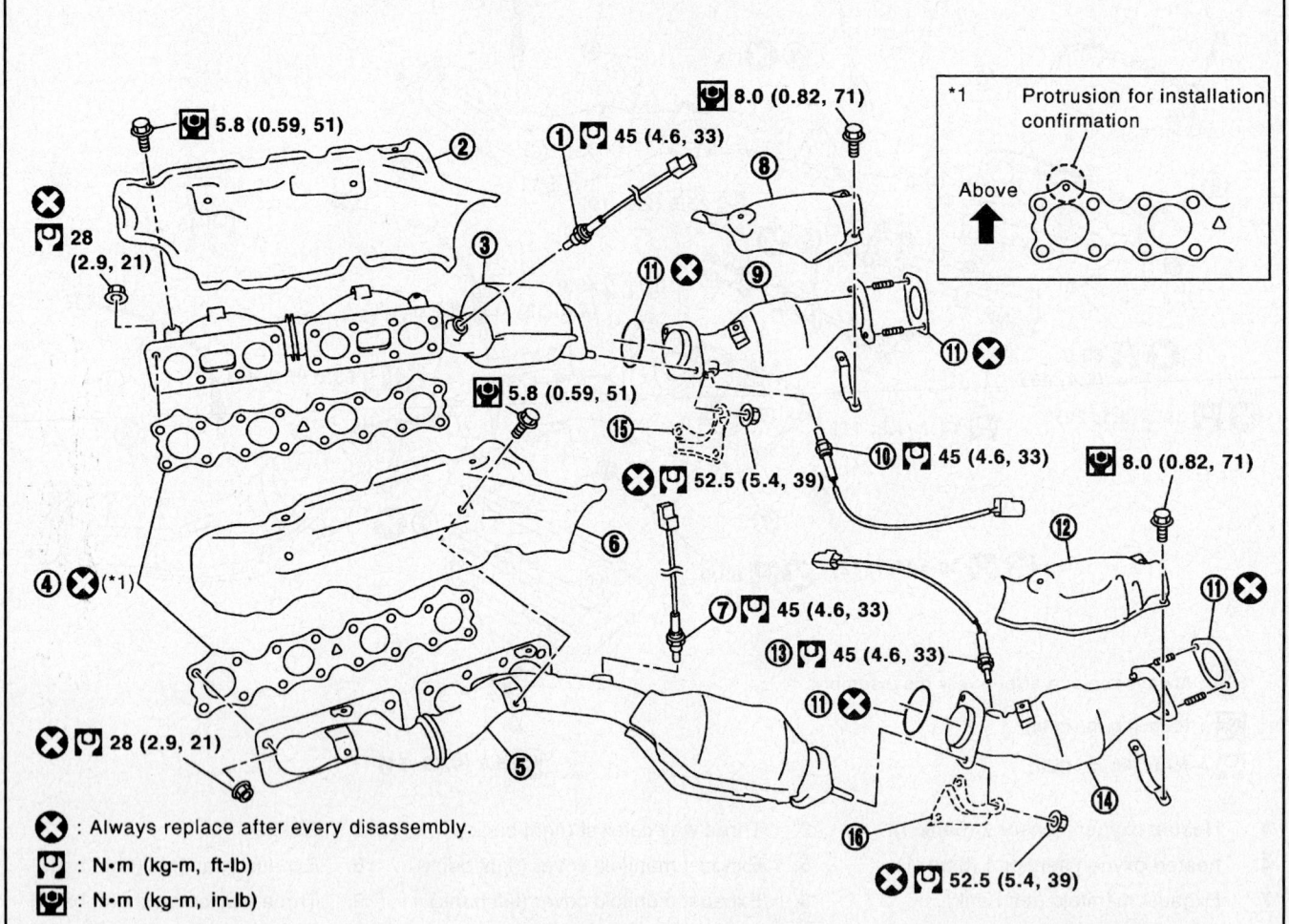

❌ : Always replace after every disassembly.

🔧 : N·m (kg-m, ft-lb)

🔧 : N·m (kg-m, in-lb)

1. Heated oxygen sensor 1 (bank 2)
2. Exhaust manifold cover (right bank)
3. Exhaust manifold (right bank)
4. Gasket
5. Exhaust manifold (left bank)
6. Exhaust manifold cover (left bank)
7. Heated oxygen sensor 1 (bank 1)
8. Three way catalyst cover (right bank)
9. Three way catalyst (right bank)
10. Heated oxygen sensor 2 (bank 2)
11. Gasket
12. Three way catalyst cover (left bank)
13. Heated oxygen sensor 2 (bank 1)
14. Three way catalyst (left bank)
15. Mounting bracket
16. Mounting bracket

67162-FX35-G150

Fig. 94 Exploded view of the exhaust manifold mounting and related components—4.5L engine

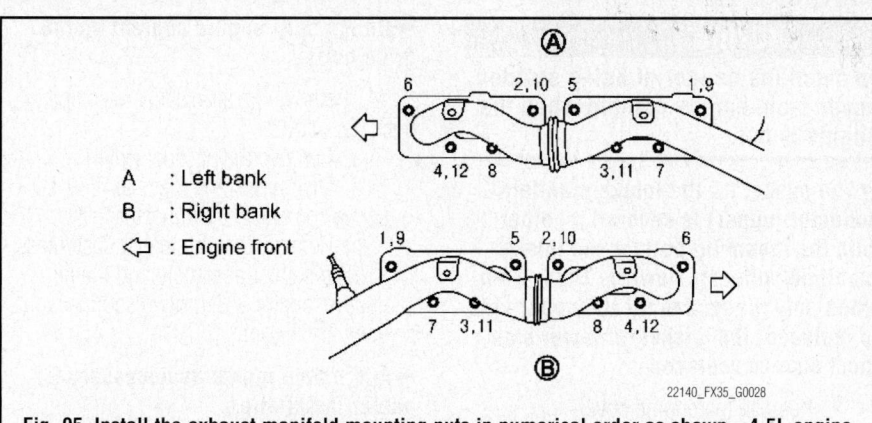

A : Left bank
B : Right bank
◁ : Engine front

22140_FX35_G0028

Fig. 95 Install the exhaust manifold mounting nuts in numerical order as shown—4.5L engine

9. Remove the heated oxygen sensors, as follows:

a. Disconnect the wiring harness connector of each heated oxygen sensor.

b. Remove the heated oxygen sensor 1 and 2 on both banks with a heated oxygen sensor wrench.

✳✳ WARNING

Be careful not to damage the heated oxygen sensor. Discard any heated oxygen sensor which has been dropped from a height of more than 20 inches (51cm) onto a hard surface such as a concrete floor; replace with a new sensor.

10. Remove the exhaust mounting bracket between the 3-way catalysts (right and left bank) and the transmission.

11. Evacuate the A/C system. Disconnect the A/C piping from the A/C compressor, then remove the A/C compressor.

12. Remove the alternator and bracket.

13. Remove the exhaust front tube.

14. Remove the steering lower joint at the power steering gear assembly side, and release the steering lower shaft.

15. Remove the 3-way catalysts (right and left bank).

16. Remove the exhaust manifold covers (right and left bank).

17. Loosen the nuts in the reverse order as shown in the figure.

18. Remove the exhaust manifold.

➥**Disregard the numerical order for numbers 9–12 in removal.**

19. Remove the exhaust manifold gaskets.

➥**Cover the engine openings to avoid entry of foreign materials.**

To install:

20. Check the surface distortion of each exhaust manifold flange mating surface with a straightedge and feeler gauge. If it exceeds the limit, replace exhaust manifold. Limit of surface distortion: 0.012 inch (0.3mm).

21. Install new exhaust manifold gaskets. Install each exhaust manifold gasket with its directional protrusion set upward. Refer to the illustration.

22. Install the exhaust manifold mounting nuts in numerical order as shown in the figure. Tighten nuts numbers 1–4 in two steps.

23. Install the exhaust manifold covers (right and left bank).

24. Install the 3-way catalysts (right and left bank).

25. Install the steering lower joint at the power steering gear assembly side.

26. Install the exhaust front tube.

27. Install the alternator and bracket.

28. Install the A/C compressor, and reconnect the A/C piping to the A/C compressor.

29. Install the exhaust mounting bracket between the 3-way catalysts (right and left bank) and the transmission.

✳✳ WARNING

Be careful not to damage heated oxygen sensor. Discard any heated oxygen sensor which has been dropped from a height of more than 20 inches (51cm) onto a hard surface such as a concrete floor; replace with a new sensor.

30. Install the heated oxygen sensors and tighten to 37 ft. lbs. (50 Nm).

31. Install the accessory drive belts.

32. Install the radiator.

33. Install the front cross bar.

34. Install the air duct (inlet), air cleaner case and mass air flow sensor assembly, air duct and resonator assembly.

35. Install the front and rear engine undercovers.

36. Install the engine cover.

37. Fill the engine cooling system with coolant.

38. Recharge the A/C system.

INTAKE MANIFOLD

REMOVAL & INSTALLATION

3.5L Engine

Upper Intake Manifold

See Figures 96 through 100.

1. Before servicing the vehicle, refer to the Precautions Section.

The upper intake manifold is constructed of 2 halves: the upper intake manifold collector and the lower intake manifold collector. Together these two components create the upper intake manifold. The 3.5L engine also uses a lower intake manifold plenum.

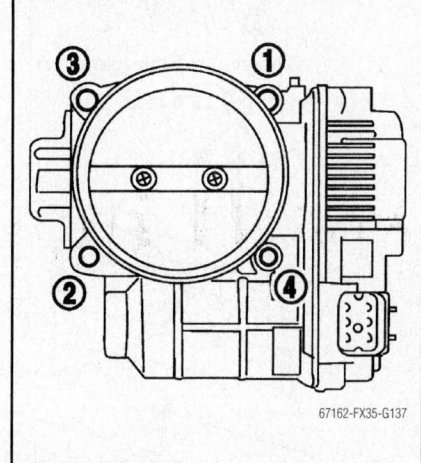

67162-FX35-G137

Fig. 96 Throttle body mounting bolt tightening sequence—3.5L engine

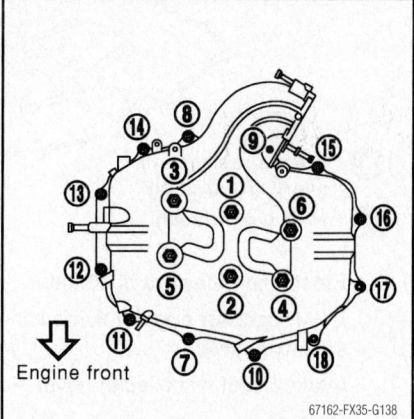

Engine front

67162-FX35-G138

Fig. 97 Tightening sequence for the upper half of the upper intake manifold—3.5L engine

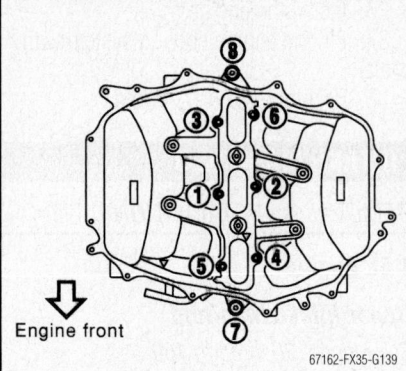

Engine front

67162-FX35-G139

Fig. 98 Tightening sequence for the lower half of the upper intake manifold—3.5L engine

☀☀ CAUTION

To avoid the danger of being scalded, never drain engine coolant when the engine is hot.

➡**The gasket for the intake manifold collector (upper) is secured together with the mounting bolt for the intake manifold collector (lower). Thus, even when only the gasket for the upper side is replaced, the gasket for lower side must also be replaced.**

2. Remove the engine cover.
3. Disconnect and plug the water hoses from the intake manifold collector (upper).

➡**Do not spill engine coolant on the drive belts.**

4. Remove the air cleaner case and air duct, as follows:
   a. Remove the air duct (inlet).
   b. Disconnect the mass air flow sensor wiring harness connector.
   c. Remove the air cleaner case/mass air flow sensor assembly and the air duct/resonator assembly disconnecting them at the joints.

➡**Add match marks as necessary for easier installation.**

   d. Remove the mass air flow sensor from air cleaner case.

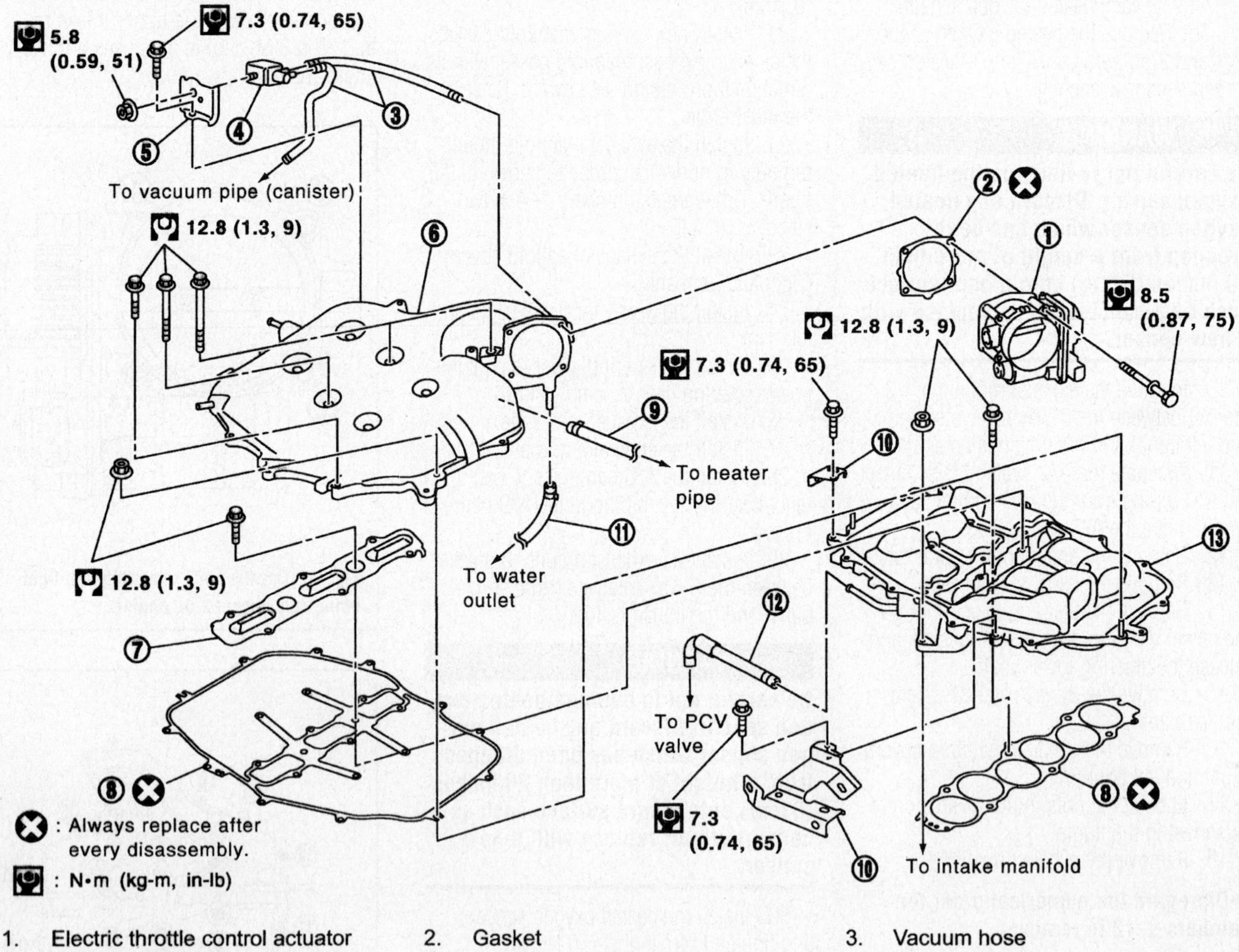

5.8 (0.59, 51)

7.3 (0.74, 65)

To vacuum pipe (canister)

12.8 (1.3, 9)

12.8 (1.3, 9)

7.3 (0.74, 65)

8.5 (0.87, 75)

9

To heater pipe

10

11

To water outlet

12.8 (1.3, 9)

13

7

12

To PCV valve

7.3 (0.74, 65)

8 ✖

10

To intake manifold

8 ✖

✖ : Always replace after every disassembly.

🔧 : N·m (kg-m, in-lb)

1.	Electric throttle control actuator	2.	Gasket	3.	Vacuum hose
4.	EVAP canister purge volume control solenoid valve	5.	Bracket	6.	Intake manifold collector (upper)
7.	Intake manifold collector cover	8.	Gasket	9.	Water hose
10.	Bracket	11.	Water hose	12.	PCV hose
13.	Intake manifold collector (lower)				

67162-FX35-G136

Fig. 99 Exploded view of the upper and lower halves of the upper intake manifold—3.5L engine

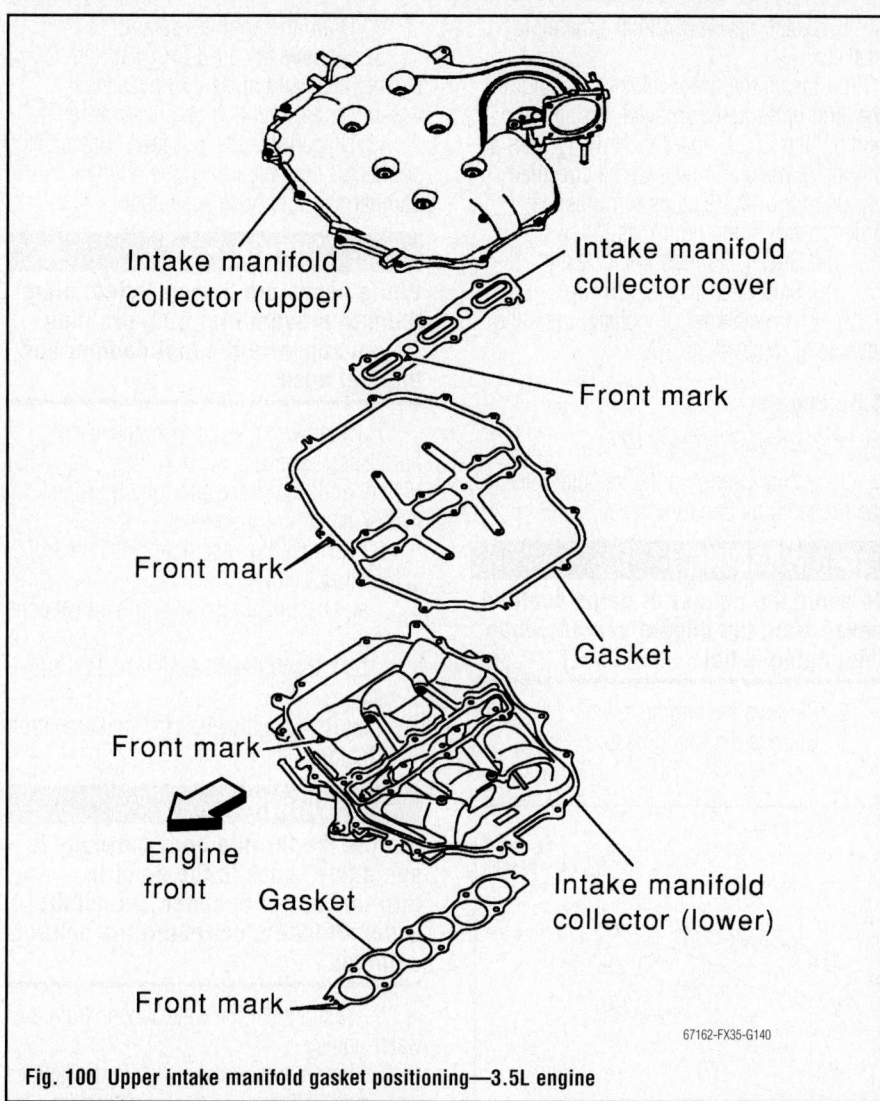

Intake manifold collector (upper)

Intake manifold collector cover

Front mark

Front mark

Gasket

Front mark

Engine front

Gasket

Intake manifold collector (lower)

Front mark

67162-FX35-G140

Fig. 100 Upper intake manifold gasket positioning—3.5L engine

✲✲ WARNING

Handle the mass air flow sensor with care. Do not expose it to harsh vibration or shock. Do not disassemble it or touch its sensor.

e. Remove the resonator in the fender, lifting the left fender protector.

5. Remove the electric throttle control actuator, as follows:

a. Disconnect the wiring harness connector.

b. Loosen the bolts in the reverse order as shown in the figure.

✲✲ WARNING

Handle the throttle body carefully to avoid any shock to the electric throttle control actuator. Do not disassemble.

6. Remove the fuel sub-tube mounting bolt to disconnect it from the rear of the intake manifold collector (lower).

7. Disconnect the vacuum hose and water hose from the intake manifold collector (upper).

8. Disconnect the EVAP canister purge volume control solenoid valve bracket mounting bolt from the intake manifold collector (upper).

9. Loosen the bolts in the reverse order of the illustration to remove the intake manifold collector (upper).

10. Remove the PCV hose between the intake manifold collector and the right-hand rocker cover.

11. Loosen the bolts in the reverse order of the illustration, and remove the intake manifold collector cover, gasket, intake manifold collector (lower) and gasket.

✲✲ CAUTION

Cover all engine openings to avoid entry of foreign materials.

To install:

12. Check the surface distortion of both the intake manifold collector (upper and lower) mating surfaces with a straightedge and feeler gauge. If it exceeds 0.004 inch (0.1mm), replace the intake manifold collector (upper and/or lower).

13. Use the following illustration referring to the gasket front marks, as a guide to installing the parts.

14. Install the intake manifold collector (lower). Tighten the mounting bolts in numerical order as shown in the figure.

➡**Tighten mounting bolts to secure gasket (lower), intake manifold collector (lower), gasket (upper), and intake manifold collector cover.**

15. Reconnect the PCV hose between the intake manifold collector and the right-hand rocker cover.

16. Install the lower manifold. If the stud bolts were removed, install them and tighten them to the specified torque of 52 inch lbs. (6 Nm).

17. The shank length from under the bolt head varies with bolt location. Install the bolts while referring to the numbers shown below and in the illustration of the tightening sequence for the upper half of the upper intake manifold.

➡**The bolt length does not include pilot portion. Make sure to tighten each bolt in the numerical order as shown in the figure.**

- M6 bolt (length 25mm): Positions 7, 8, 10, 11, 13, 14, 15, 16, and 18
- M6 bolt (length 45mm): Positions 2, 4, 5
- M6 bolt (length 60mm): Positions 1, 3, 6, and 9
- M6 nut: Positions 12 and 17

18. Reattach the EVAP canister purge volume control solenoid valve bracket to the intake manifold collector (upper).

19. Reattach the vacuum hose and water hose to the intake manifold collector (upper).

20. Reattach the fuel sub-tube to the rear of the intake manifold collector (lower).

21. Install the electric throttle control actuator.

22. Install the mass air flow sensor in the air cleaner case.

23. Install the air cleaner case and air duct.

24. Reconnect the hoses to the intake manifold collector (upper).

25. Install the engine cover.

Lower Intake Manifold

See Figures 101 and 102.

1. Before servicing the vehicle, refer to the Precautions Section.

2. Relieve the fuel pressure.

3. Remove the intake manifold collector (upper) and (lower). Refer to Upper Intake Manifold procedure.

4. Remove the fuel tube and fuel injector assembly.

5. Loosen the mounting bolts and nuts in the reverse order of the illustration to remove the lower intake manifold.

6. Remove the intake manifold gaskets.

✳✳ CAUTION

Cover all engine openings to avoid entry of foreign materials.

To install:

7. Check the surface distortion of the intake manifold mating surface with a straightedge and feeler gauge. If it exceeds the limit of 0.04 inch (0.1mm), replace the lower intake manifold.

8. Install new lower intake manifold gaskets.

9. Install the lower intake manifold. If the stud bolts were removed, install them and tighten to 8 ft. lbs. (11 Nm). Tighten all mounting bolts and nuts to the specified torque in 2 or more steps in numerical order shown in the figure, as follows:

    a. Step 1: 60 inch lbs. (7 Nm).

    b. Step 2: 21 ft. lbs. (29 Nm).

10. The remainder of installation is the reverse of removal.

4.5L Engine

See Figures 103 through 105.

1. Before servicing the vehicle, refer to the Precautions Section.

✳✳ WARNING

To avoid the danger of being scalded, never drain the engine coolant when the engine is hot.

2. Remove the engine cover.

3. Release the fuel pressure.

4. Drain the engine coolant.

5. Remove the air duct (inlet), air cleaner case and mass air flow sensor assembly, air duct and resonator assembly.

6. Disconnect the fuel feed hose quick connector on the engine side, the fuel damper and fuel hose assembly.

✳✳ CAUTION

While hoses are disconnected, plug them to prevent fuel from draining. Do not separate the fuel damper and the fuel hose.

7. Remove or disconnect the wiring harnesses, brackets, vacuum hose, vacuum gallery and PCV hose and tube from the intake manifold (upper).

8. Remove the electric throttle control actuator as follows:

    a. Disconnect the wiring harness connector.

    b. Loosen the mounting bolts diagonally.

    c. Remove the electric throttle control actuator.

✳✳ WARNING

Handle the throttle body carefully to avoid any shock to the electric throttle control actuator. Do not disassemble the electric throttle control actuator.

9. Disconnect the water hoses from the water gallery.

10. Remove the intake manifold adaptor and water gallery.

11. Loosen the bolts in the reverse order of the tightening sequence, as shown in the figure, to remove the intake manifold (upper).

12. Remove the vacuum tank from the intake manifold (lower).

13. Remove the fuel injector and fuel tube assembly. Refer to the Fuel Injectors, removal & installation.

14. Loosen the bolts in the reverse order of the tightening sequence, as shown in the figure, to remove the intake manifold (lower).

15. Remove the intake manifold gaskets.

✳✳ CAUTION

Cover all engine openings to avoid entry of foreign materials.

To install:

16. Check the surface distortion of both the intake manifold (upper and lower) mating surfaces with a straightedge and feeler gauge. If it exceeds the limit of 0.004 inch

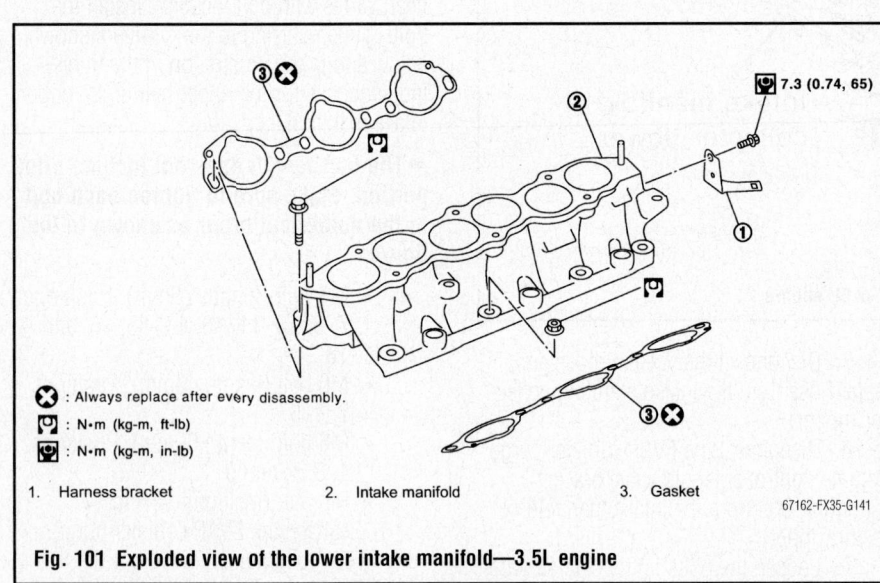

⊗ : Always replace after every disassembly.

⬚ : N•m (kg-m, ft-lb)

⬚ : N•m (kg-m, in-lb)

1. Harness bracket   2. Intake manifold   3. Gasket

67162-FX35-G141

Fig. 101 Exploded view of the lower intake manifold—3.5L engine

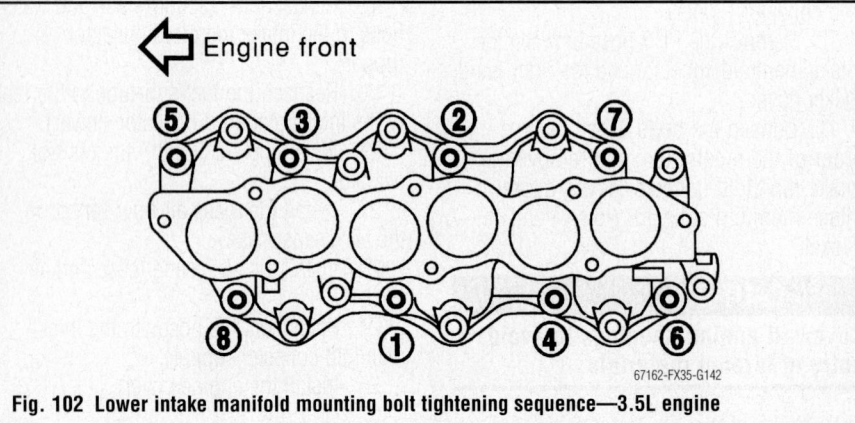

← Engine front

67162-FX35-G142

Fig. 102 Lower intake manifold mounting bolt tightening sequence—3.5L engine

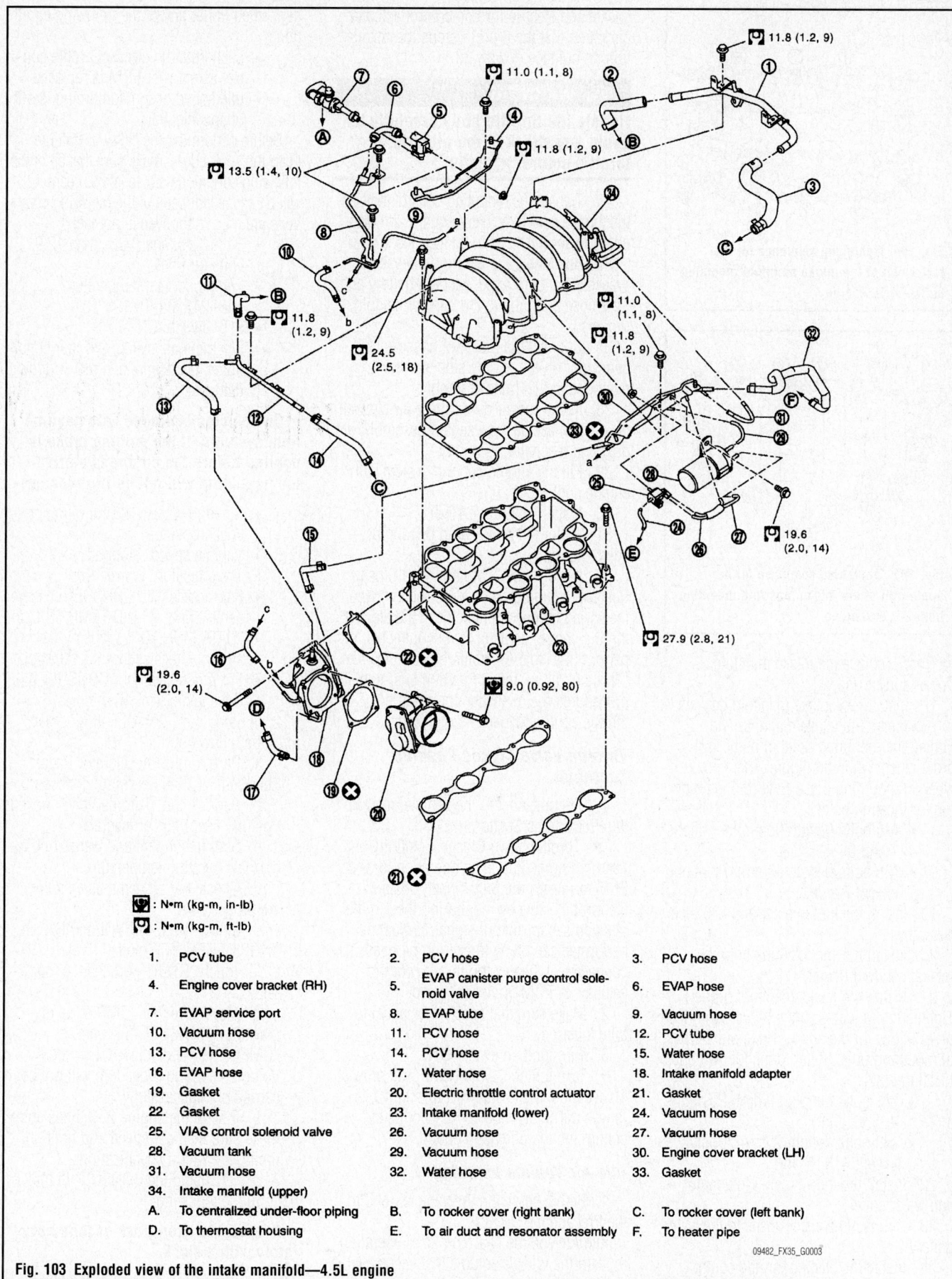

: N•m (kg-m, in-lb)

: N•m (kg-m, ft-lb)

1.	PCV tube	2.	PCV hose	3.	PCV hose
4.	Engine cover bracket (RH)	5.	EVAP canister purge control solenoid valve	6.	EVAP hose
7.	EVAP service port	8.	EVAP tube	9.	Vacuum hose
10.	Vacuum hose	11.	PCV hose	12.	PCV tube
13.	PCV hose	14.	PCV hose	15.	Water hose
16.	EVAP hose	17.	Water hose	18.	Intake manifold adapter
19.	Gasket	20.	Electric throttle control actuator	21.	Gasket
22.	Gasket	23.	Intake manifold (lower)	24.	Vacuum hose
25.	VIAS control solenoid valve	26.	Vacuum hose	27.	Vacuum hose
28.	Vacuum tank	29.	Vacuum hose	30.	Engine cover bracket (LH)
31.	Vacuum hose	32.	Water hose	33.	Gasket
34.	Intake manifold (upper)				
A.	To centralized under-floor piping	B.	To rocker cover (right bank)	C.	To rocker cover (left bank)
D.	To thermostat housing	E.	To air duct and resonator assembly	F.	To heater pipe

09482_FX35_G0003

Fig. 103 Exploded view of the intake manifold—4.5L engine

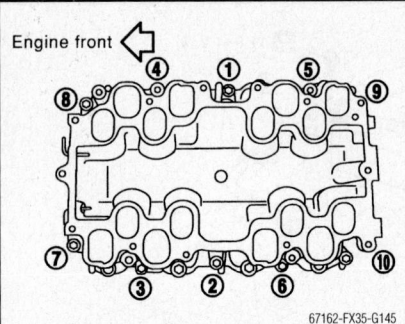

Fig. 104 Tightening sequence for the lower half of the intake manifold mounting bolts—4.5L engine

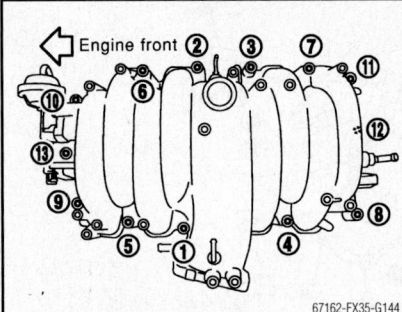

Fig. 105 Tightening sequence for the upper half of the intake manifold mounting bolts—4.5L engine

(0.1mm), replace the intake manifolds (lower and/or upper).

17. Install new intake manifold gaskets.

18. Install the intake manifold and tighten the mounting bolts in numerical order as shown in the figure. There are 2 types of mounting bolts. Refer to the following for locating bolts:

- M8 bolts (length 90mm): Positions 7 and 8
- M8 bolts (length 35mm): Positions except 7 and 8

19. Install the fuel injector and fuel tube assembly.

20. Install the vacuum tank onto the intake manifold (lower).

21. Install the intake manifold (upper). Tighten the mounting bolts in the numerical order shown in the figure. There are 2 types of mounting bolts. Refer to the following for locating bolts:

- M8 bolts (length 80mm): Positions 4, 5, 6, and 7
- M8 bolts (length 25mm): Positions except 4, 5, 6, and 7

22. Install the intake manifold adaptor and water gallery.

23. Connect the water hoses to the water gallery.

24. Install the electric throttle control actuator. Install the intake manifold adapter gasket and electric throttle control actuator gasket so that the 3 protrusions for installation do not face downward.

✳✳ WARNING

Handle the throttle body carefully to avoid any shock to the electric throttle control actuator.

25. Tighten the mounting bolts of the electric throttle control actuator equally and diagonally in several steps to 80 inch lbs. (9 Nm).

26. Reconnect the wiring harnesses, brackets, vacuum hose, vacuum gallery and PCV hose and tube to the intake manifold (upper).

27. Connect the fuel feed hose quick connector on the engine side, the fuel damper and fuel hose assembly.

28. Install the air duct (inlet), air cleaner case and mass air flow sensor assembly, air duct and resonator assembly.

29. Fill the engine cooling system with engine coolant.

30. Install the engine cover.

31. Perform the following drivability adjustments:

32. Perform the "Throttle Valve Closed Position Learning" procedure (below) when the wiring harness connector of the electric throttle control actuator is disconnected, or perform the "Idle Air Volume Learning" and "Throttle Valve Closed Position Learning" procedures (below) when the electric throttle control actuator is replaced.

Throttle Valve Closed Position Learning

1. Before servicing the vehicle, refer to the Precautions Section.

The Throttle Valve Closed Position Learning procedure is an operation for the ECM to relearn the fully closed position of the throttle valve by monitoring the throttle position sensor output signal. It must be performed each time the wiring harness connector of the electric throttle control actuator or ECM is disconnected.

2. Make sure that accelerator pedal is fully released.

3. Turn ignition switch **ON**.

4. Turn ignition switch **OFF** wait at least 10 seconds. Make sure that throttle valve moves during the above 10 seconds by confirming the operating sound.

Idle Air Volume Learning

1. Before servicing the vehicle, refer to the Precautions Section.

Idle Air Volume Learning is an operation to learn the idle air volume that keeps each engine within the specific range. It must be performed under any of the following conditions:

- Each time the electric throttle control actuator or ECM is replaced
- Idle speed or ignition timing is out of specification

Before performing the "Idle Air Volume Learning" procedure, make sure that all of the following conditions are satisfied. Learning will be cancelled if any of the following conditions are missed for even a moment.

- Battery voltage: More than 12.9 volts (at idle)
- Engine coolant temperature: 158–212°F (70–100°C)
- PNP switch: ON
- Electric load switch: OFF (air conditioner, headlamp, and rear window defogger)

➡ **On vehicles equipped with daytime light systems, if the parking brake is applied before the engine is started, the headlamp will not be illuminated.**

- Steering wheel: Neutral (straight-ahead position)
- Vehicle speed: Stopped
- Transmission: Warmed-up
- For models with CONSULT-III, drive vehicle until "FLUID TEMP SE 1" in "DATA MONITOR" mode of "A/T" system indicates less than 0.9 volts.
- For models without CONSULT-III, drive vehicle for 10 minutes.

2. If using the CONSULT-III tool, perform the following:

a. Perform the "Accelerator Pedal Released Position Learning" procedure.

b. Perform the "Throttle Valve Closed Position Learning" procedure.

c. Start the engine and warm it up to normal operating temperature.

d. Check that all items listed above are properly set.

e. Select "IDLE AIR VOL LEARN" in "WORK SUPPORT" mode.

f. Touch "START" and wait 20 seconds.

g. Make sure that "CMPLT" is displayed on CONSULT-III screen. If "CMPLT" is not displayed, the Idle Air Volume Learning procedure will not be carried out successfully.

h. Rev up the engine 2–3 times and make sure that idle speed and ignition timing are within specifications.

3. If NOT using the CONSULT-III tool, perform the following:

➡ **It is best to keep track of time accurately with a clock.**

➡ **It is impossible to switch the diag-**

nostic mode when an accelerator pedal position sensor circuit has a malfunction.

    a. Perform the "Accelerator Pedal Released Position Learning" procedure.
    b. Perform the "Throttle Valve Closed Position Learning" procedure.
    c. Start the engine and warm it up to normal operating temperature.
    d. Check that all items listed above are properly set.
    e. Turn the ignition switch OFF and wait at least 10 seconds.
    f. Confirm that the accelerator pedal is fully released, turn the ignition switch ON and wait 3 seconds.

➡**Repeat the following 2 steps quickly 5 times within 5 seconds.**

    g. Fully depress the accelerator pedal.
    h. Fully release the accelerator pedal.
    i. Wait 7 seconds, fully depress the accelerator pedal and keep it for approx. 20 seconds until the MIL stops blinking and remains ON.
    j. Fully release the accelerator pedal within 3 seconds after the MIL turned ON.
    k. Start the engine and let it idle.
    l. Wait 20 seconds.
    m. Rev up the engine 2–3 times and make sure that idle speed and ignition timing are within specifications.
    n. If the idle speed and ignition timing are not within specification, the Idle Air Volume Learning procedure will not be successful.

Accelerator Pedal Released Position Learning

1. Before servicing the vehicle, refer to the Precautions Section.

The "Accelerator Pedal Released Position Learning" procedure is an operation for the ECM to relearn the fully released position of the accelerator pedal by monitoring the accelerator pedal position sensor output signal. It must be performed each time the wiring harness connector of the accelerator pedal position sensor or ECM is disconnected.

2. Make sure that the accelerator pedal is fully released.
3. Turn the ignition switch **ON** and wait at least 2 seconds.
4. Turn the ignition switch **OFF** wait at least 10 seconds.
5. Turn the ignition switch **ON** and wait at least 2 seconds.
6. Turn the ignition switch **OFF** wait at least 10 seconds.

OIL PAN

REMOVAL & INSTALLATION

3.5L Engine

See Figures 106 through 114.

1. Before servicing the vehicle, refer to the Precautions Section.

✳✳ CAUTION

To avoid the danger of being scalded, never drain the engine oil or engine coolant when the engine is hot.

➡**To remove only the lower oil pan, drain the engine oil and skip to step 26.**

2. Remove the front wheels and tires.
3. Remove the hood assembly.
4. Remove the front and rear engine undercover.
5. Remove the front cross bar.
6. Drain the engine oil.
7. Drain the engine coolant.
8. Remove the engine cover.
9. Remove the air hose from the air duct to the mass air flow and the electric throttle control actuator side.
10. Remove the alternator, power steering pump, and A/C compressor belt.
11. On AWD models, remove the front left and right halfshafts, and side shaft.
12. Remove the engine rear lower slinger, and install the engine rear slinger tool SST: 10006 31U00 (or equivalent) to hold the engine assembly in position. Tighten the engine rear slinger tool mounting bolts to 21 ft. lbs. (28 Nm).
13. Remove the front suspension member.
14. On AWD models, remove the engine mounting bracket, lower engine mounting bracket and insulator.
15. On AWD models, remove the front driveshaft.
16. On AWD models, remove the oil filter and oil filter bracket.
17. Remove the alternator stay.
18. Remove the starter motor.
19. Remove the alternator and power steering pump and A/C compressor idler pulley and bracket assembly.
20. Disconnect the A/T fluid cooler hoses, and remove the oil cooler water pipe mounting bolt.
21. Disconnect the A/T fluid cooler tube.
22. On AWD models, remove the front final drive assembly.
23. Remove the crankshaft position sensor.

✳✳ WARNING

Handle the crankshaft position sensor carefully to avoid dropping it or exposing it to abrupt shocks. Do not disassemble it, do not allow metal powder to adhere to the magnetic part at the sensor tip, and do not place the sensor in a location where it may be exposed to magnetism.

24. Remove the oil filter, as necessary.
25. Remove the oil cooler, as necessary.
26. Remove the oil pan (lower), as follows:
    a. Loosen the mounting bolts in the reverse order of the tightening sequence.
    b. Insert a seal cutter SST: KV10111100 (J37228) or equivalent, between the upper oil pan and lower oil pan.
    c. Slide the seal cutter by tapping on the side of the tool with a hammer.
    d. Remove the lower oil pan.

✳✳ WARNING

Be careful not to damage the mating surface. Do not use a flat-bladed screwdriver as this could damage the mating surfaces.

27. Remove the oil strainer.
28. Remove the transmission joint bolts which pass through the upper oil pan.
29. On 2WD models, remove the rear cover plate.
30. Loosen the upper oil pan bolts in the reverse order of the tightening sequence.
31. Insert a seal cutter SST: KV10111100 (J37228) between the upper oil pan and cylinder block. Slide the seal cutter by tapping on the side of the tool with a hammer.
32. Remove the upper oil pan.

✳✳ WARNING

Be careful not to damage the mating surface. Do not use a flat-bladed screwdriver as this could damage the mating surfaces.

33. Remove the O-rings from the bottom of the cylinder block and oil pump.
34. Remove the oil pan gaskets.
35. For AWD models, remove the axle pipe from the upper oil pan using a suitable drift, if necessary.
36. Clean the oil strainer, if necessary.

To install:
37. On AWD models, install the axle pipe to the oil pan, if removed:
    a. Lubricate the O-ring groove of the axle pipe, O-ring, and O-ring joint of the oil pan with new engine oil.

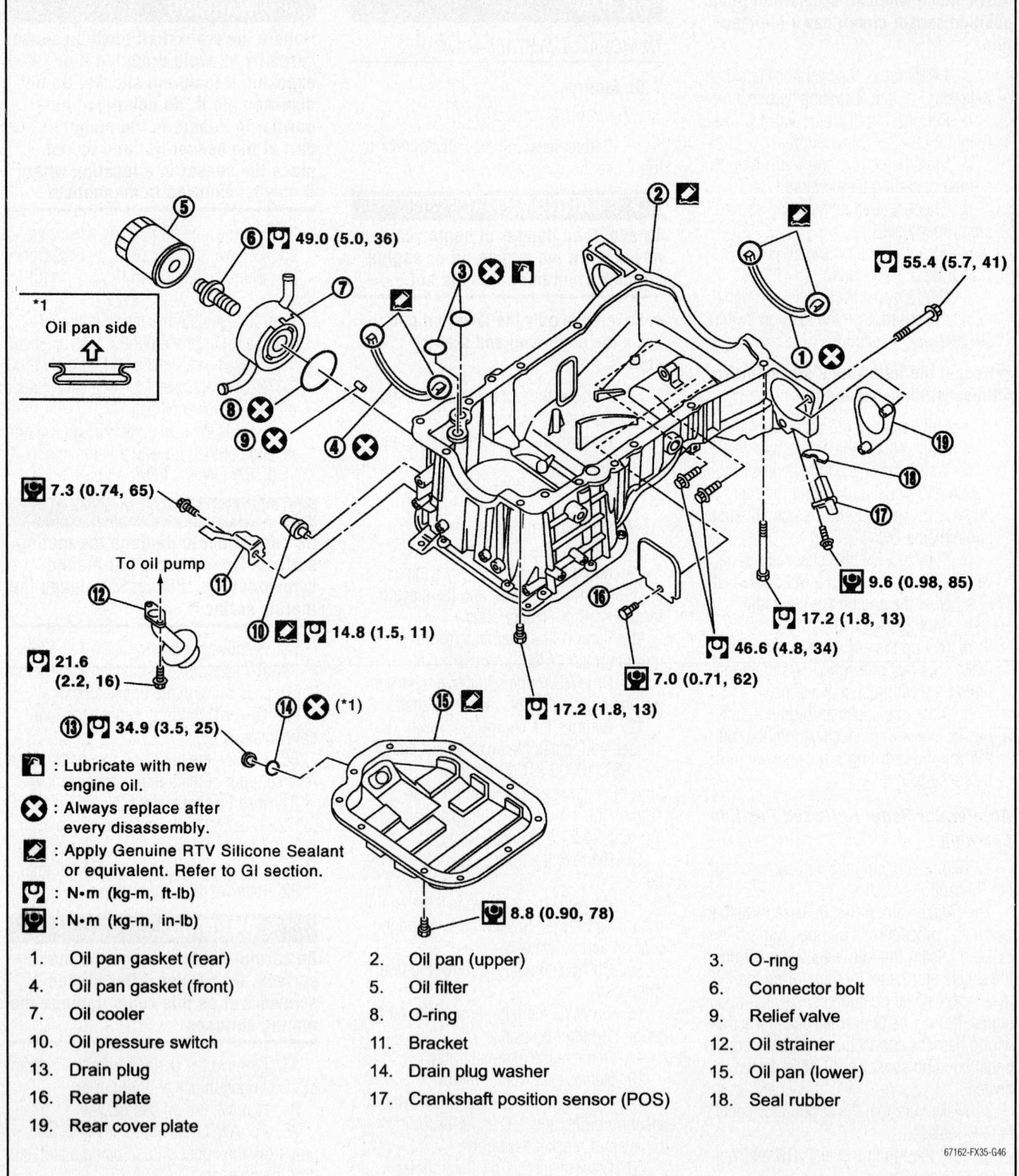

*1
Oil pan side

6 49.0 (5.0, 36)

55.4 (5.7, 41)

7.3 (0.74, 65)

To oil pump

21.6
(2.2, 16)

9.6 (0.98, 85)

17.2 (1.8, 13)

46.6 (4.8, 34)

14.8 (1.5, 11)

7.0 (0.71, 62)

17.2 (1.8, 13)

13 34.9 (3.5, 25)

(*1)

: Lubricate with new
engine oil.

: Always replace after
every disassembly.

: Apply Genuine RTV Silicone Sealant
or equivalent. Refer to GI section.

: N•m (kg-m, ft-lb)

: N•m (kg-m, in-lb)

8.8 (0.90, 78)

1. Oil pan gasket (rear)	2. Oil pan (upper)	3. O-ring
4. Oil pan gasket (front)	5. Oil filter	6. Connector bolt
7. Oil cooler	8. O-ring	9. Relief valve
10. Oil pressure switch	11. Bracket	12. Oil strainer
13. Drain plug	14. Drain plug washer	15. Oil pan (lower)
16. Rear plate	17. Crankshaft position sensor (POS)	18. Seal rubber
19. Rear cover plate		

67162-FX35-G46

Fig. 106 Exploded view of oil pan and related components—2WD 3.5L engine

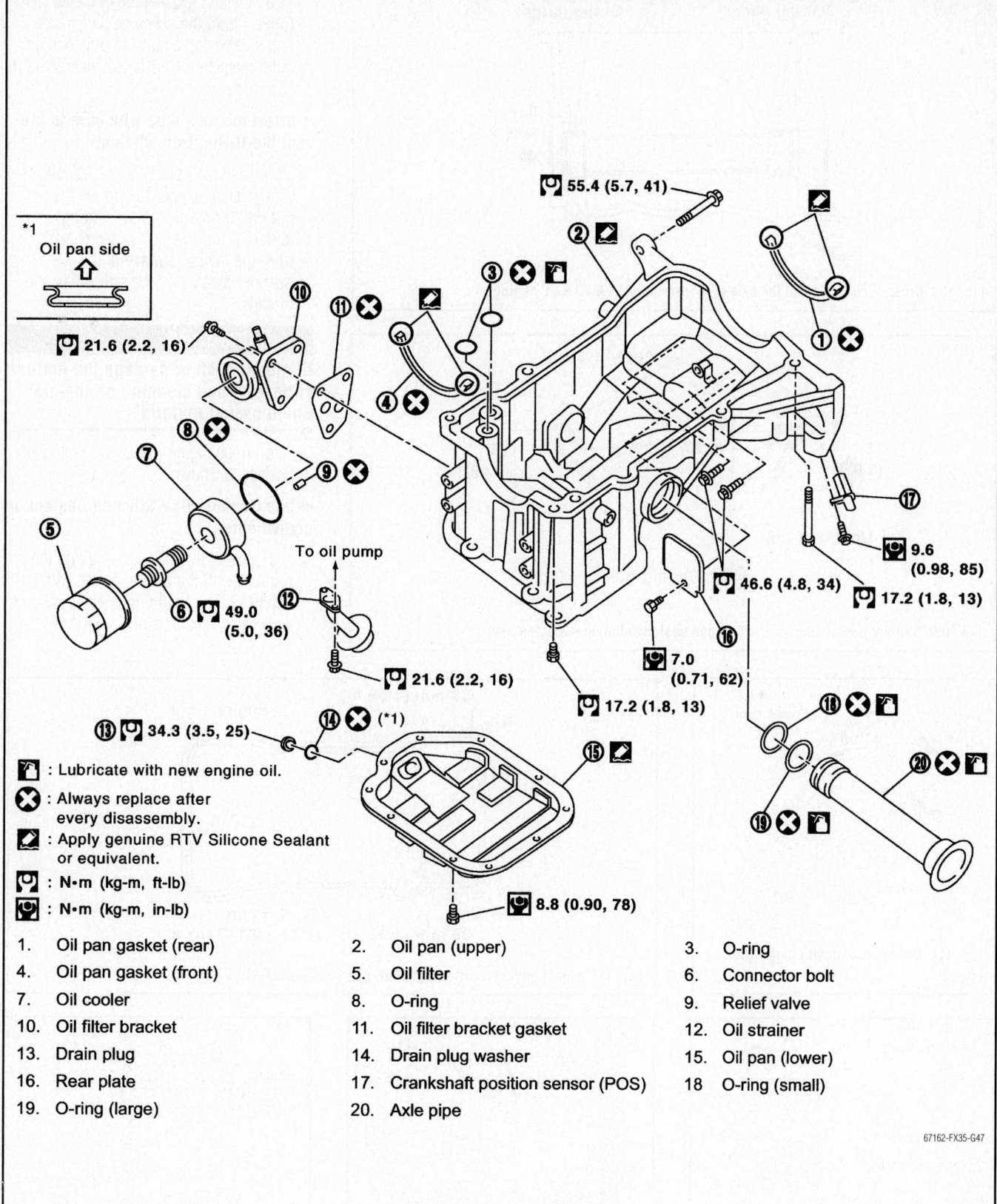

Fig. 107 Exploded view of oil pan and related components—AWD 3.5L engine

*1 Oil pan side

55.4 (5.7, 41)

21.6 (2.2, 16)

To oil pump

49.0 (5.0, 36)

21.6 (2.2, 16)

34.3 (3.5, 25)

(*1)

9.6 (0.98, 85)

46.6 (4.8, 34)

17.2 (1.8, 13)

7.0 (0.71, 62)

17.2 (1.8, 13)

8.8 (0.90, 78)

: Lubricate with new engine oil.

: Always replace after every disassembly.

: Apply genuine RTV Silicone Sealant or equivalent.

: N•m (kg-m, ft-lb)

: N•m (kg-m, in-lb)

1. Oil pan gasket (rear)
2. Oil pan (upper)
3. O-ring
4. Oil pan gasket (front)
5. Oil filter
6. Connector bolt
7. Oil cooler
8. O-ring
9. Relief valve
10. Oil filter bracket
11. Oil filter bracket gasket
12. Oil strainer
13. Drain plug
14. Drain plug washer
15. Oil pan (lower)
16. Rear plate
17. Crankshaft position sensor (POS)
18. O-ring (small)
19. O-ring (large)
20. Axle pipe

67162-FX35-G47

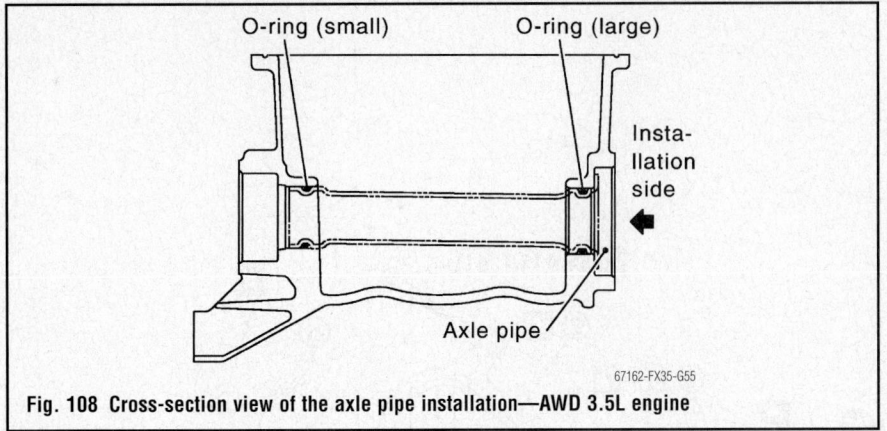

Fig. 108 Cross-section view of the axle pipe installation—AWD 3.5L engine

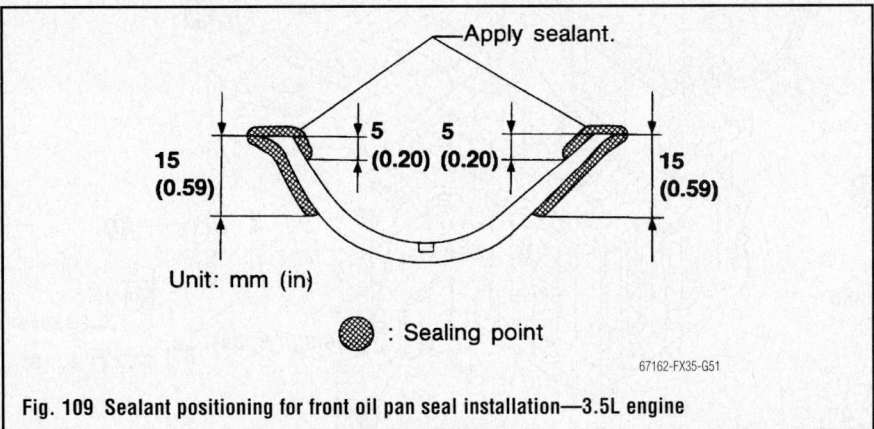

Fig. 109 Sealant positioning for front oil pan seal installation—3.5L engine

b. Install the axle pipe to the oil pan (upper) from the axle pipe flange side (left side) using a suitable drift with an outer diameter of 1.7 to 2.2 inches (43 to 57mm).

➡ **Insert the axle pipe with care to prevent the O-ring from sliding.**

38. Install the upper oil pan, as follows:
   a. Use a scraper to remove the old liquid gasket from all mating surfaces. Remove old liquid gasket from the mating surface of the cylinder block, and the bolt holes and threads.

✳✳ WARNING

Do not scratch or damage the mating surfaces when cleaning off the old liquid gasket material.

b. Apply liquid gasket to the oil pan gaskets as shown.

➡ **Use Genuine RTV Silicone Sealant or equivalent.**

c. Install the new gasket. Align the protrusion of the oil pan gasket with the notches of the front timing chain case and rear oil seal retainer.

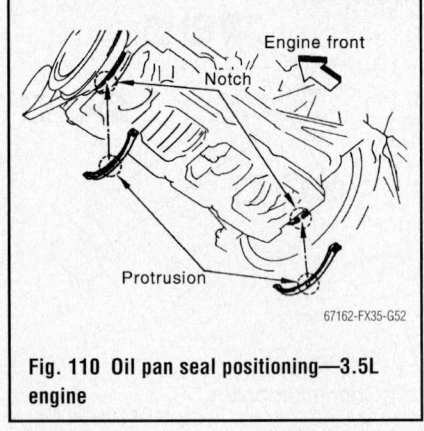

Fig. 110 Oil pan seal positioning—3.5L engine

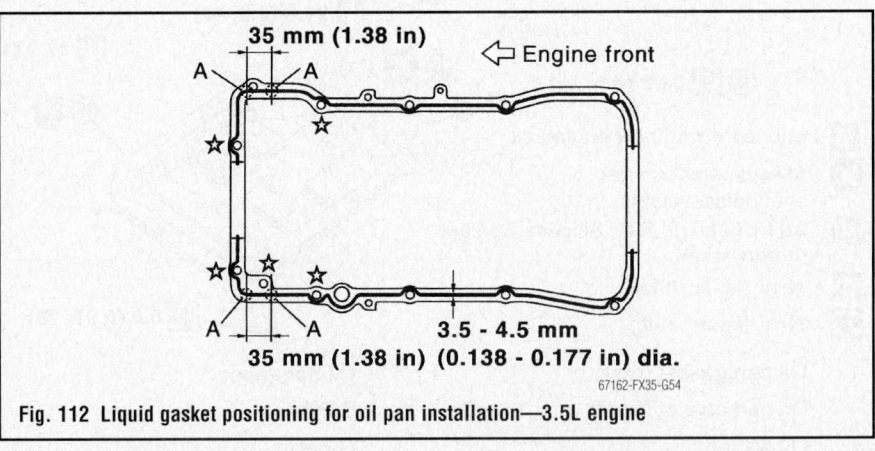

Fig. 112 Liquid gasket positioning for oil pan installation—3.5L engine

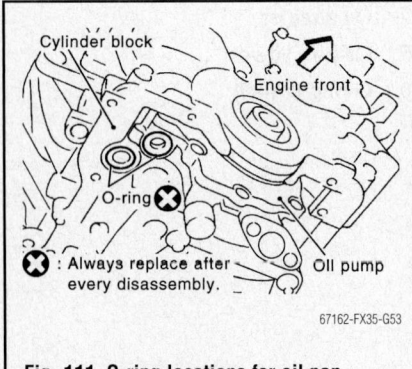

Fig. 111 O-ring locations for oil pan service—3.5L engine

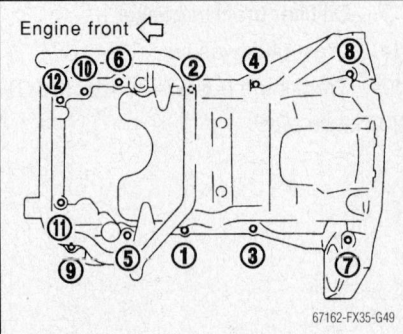

Fig. 113 Upper oil pan mounting bolt tightening sequence—3.5L engine

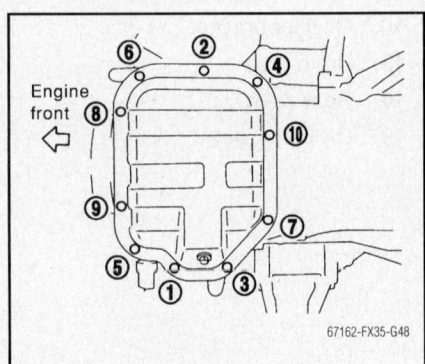

Fig. 114 Lower oil pan mounting bolt tightening sequence—3.5L engine

d. Install the oil pan gasket with the smaller arc to the front timing chain case side.

e. Install new O-rings on the cylinder block and oil pump.

f. Apply a continuous bead of liquid gasket to the cylinder block mating surface of the upper oil pan to a limited portion as shown.

➡**Use Genuine RTV Silicone Sealant or equivalent.**

- For bolt holes with star marks in illustration (5 locations), apply liquid gasket outside the holes.
- Apply a bead of 0.18–0.22 inch (4.5–5.5mm) in diameter to designated area **A**.
- Installation should be done within 5 minutes after coating.

g. Install the upper oil pan. Tighten the mounting bolts in the order shown. There are 2 types of mounting bolts. Refer to the following for locating the bolt positions:

- M8 x 100mm (3.97 inches): positions 5, 7, 8, and 11
- M8 x 25mm (0.98 inches): positions except 5, 7, 8, and 11

h. Tighten the transmission joint bolts.

39. Install the oil strainer onto the oil pump.
40. Install the lower oil pan, as follows:

a. Use a scraper to remove all old liquid gasket material from the mating surfaces.

b. Apply new liquid gasket.

➡**Use Genuine RTV Silicone Sealant or equivalent. Installation should be done within 5 minutes after coating.**

c. Tighten the mounting bolts in numerical order as shown.

41. Install the oil pan drain plug.
42. The remainder of installation is the reverse of removal.

➡**Wait at least 30 minutes after the oil pan is installed before filling the engine with new oil.**

43. Start the engine and check that there is no leakage of engine oil.
44. Stop the engine and wait 10 minutes.
45. Check the engine oil level again.

4.5L Engine

See Figures 115 through 118.

1. Before servicing the vehicle, refer to the Precautions Section.

✳✳ CAUTION

To avoid the danger of being scalded, do not drain engine oil or coolant when the engine is hot.

2. Remove the front wheels and tires.
3. Remove the hood assembly.
4. Remove the engine cover.
5. Remove the front and rear engine undercovers.
6. Drain the engine oil and engine coolant.
7. Remove the accessory drive belts.
8. Remove the auto tensioner for the power steering oil pump belt.
9. Remove the power steering oil pump with the piping connected, and temporarily secure it aside with ropes or equivalent.
10. Remove the A/C compressor with the piping connected, and temporarily secure it aside with ropes or equivalent.
11. Remove the A/C compressor fitting bolts, and install the A/C compressor temporarily on the vehicle side with ropes or equivalent.
12. Remove the wiring harness of the lower side of oil pan.
13. Remove the crankshaft position sensor from the transmission.

✳✳ WARNING

Handle the crankshaft position sensor carefully to avoid dropping it and exposing it to abrupt shocks. Do not disassemble it. Do not allow metal powder to adhere to magnetic part at the sensor tip and do not place sensor in a location where it may be exposed to magnetism.

14. Install an engine slinger to hold the engine assembly in a secure position. Tighten the slinger mounting bolts to 25 ft. lbs. (33.5 Nm).
15. Remove the front suspension member.
16. Remove the front final drive assembly.
17. Remove the oil filter.
18. Disconnect the oil cooler water hoses and remove the oil cooler water pipe and oil cooler.
19. Remove the oil pan as follows:

a. Remove the rear plate cover.

b. Remove the transmission joint bolts which pass through the oil pan.

c. Loosen the oil pan bolts in the reverse order of the tightening sequence.

➡**Disregard the tightening sequence on numbers 11 and 17 during removal.**

d. Insert a seal cutter SST: KV10111100 (J37228), or equivalent, between the oil pan and the cylinder block. Slide the seal cutter by tapping on the side of seal cutter with a hammer.

e. Remove the oil pan.

✳✳ WARNING

Be careful not to damage the mating surfaces. Do not use a flat-bladed screwdriver as this could cause damage to the mating surfaces.

f. Remove the O-rings from the bottom of the oil pump and front cover.

20. As necessary, pull the axle pipe from the oil pan. Hold the pipes and pull them out to the left side.
21. Remove the oil strainer.
22. Clean the oil strainer, if necessary.

To install:

23. Install the oil strainer.
24. Install the axle pipe to the oil pan, if removed:

a. Lubricate the O-ring groove of the axle pipe, O-ring, and O-ring joint of the oil pan with new engine oil.

➡**The right and left O-ring diameters differ from each other. The O-ring with an identification paint mark must be installed on the left front halfshaft side.**

b. Install the axle pipe to the oil pan on the left side.

➡**Insert the axle pipe with care to prevent the O-ring from sliding.**

25. Install the oil pan as follows:

a. Install new O-rings onto the oil pump and the side of the front cover.

b. Apply a continuous bead of RTV liquid gasket to the cylinder block mating surface of the oil pan as shown.

➡**Use Genuine RTV Silicone Sealant or equivalent. Installation should be done within 5 minutes after coating, otherwise the liquid gasket may not seal properly.**

c. Install the oil pan and tighten the mounting bolts in order as shown. There are 3 types of mounting bolts. Refer to the following for locating bolts.

- M6 x 30mm: positions 18 and 19
- M8 x 100mm: positions 5 and 9
- M8 x 45mm: positions except 5, 9, 18 and 19

➡**Tighten bolts number 1 and number 2 in two steps. Bolt number 1 is shown as numbers 1 and 11 and bolt number 2 is shown as 2 and 17 in the illustration. Numbers 11 and 17 are the second steps for numbers 1 and 2.**

d. Tighten the transmission joint bolts to 55 ft. lbs. (74 Nm).

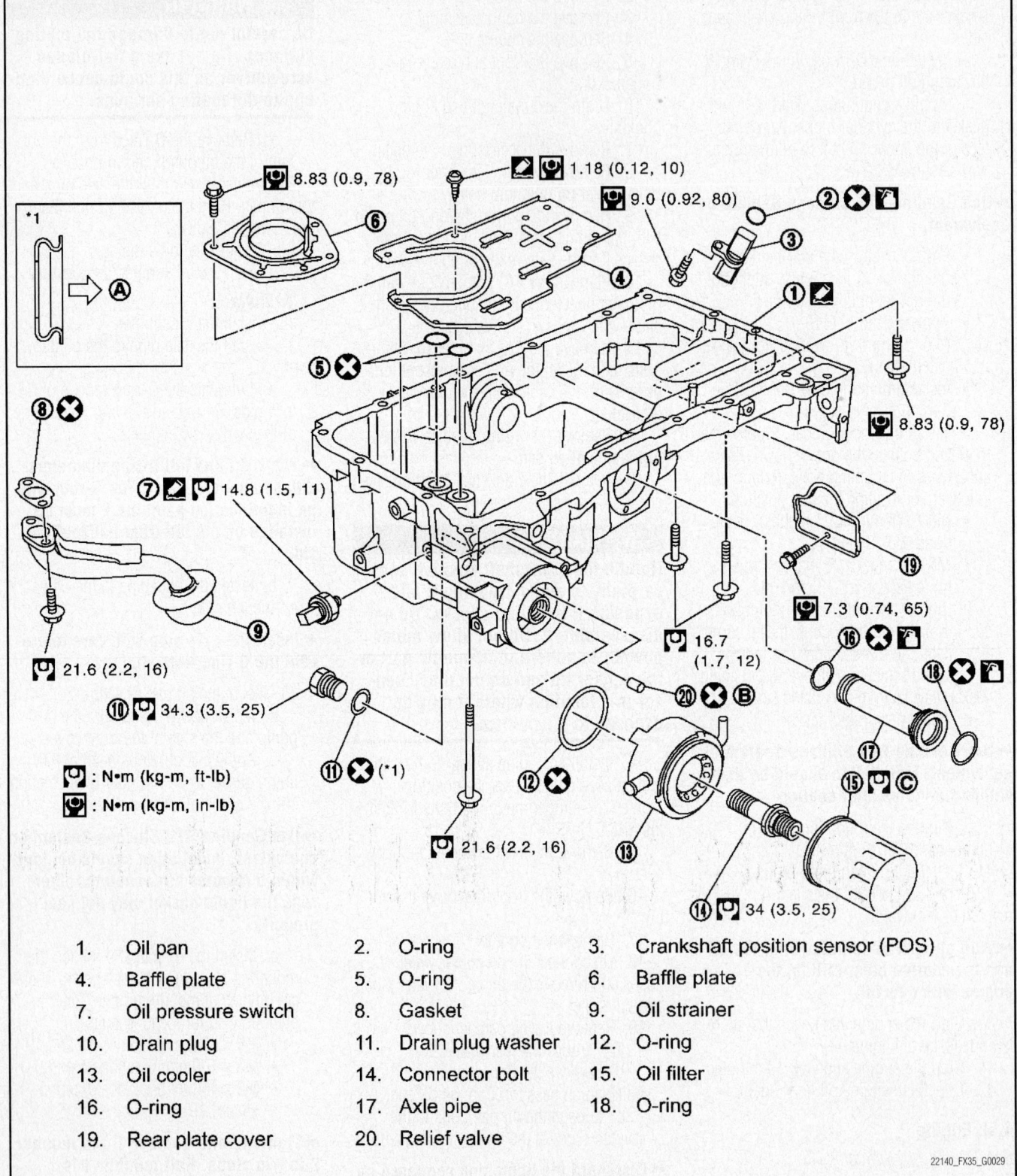

1. Oil pan
2. O-ring
3. Crankshaft position sensor (POS)
4. Baffle plate
5. O-ring
6. Baffle plate
7. Oil pressure switch
8. Gasket
9. Oil strainer
10. Drain plug
11. Drain plug washer
12. O-ring
13. Oil cooler
14. Connector bolt
15. Oil filter
16. O-ring
17. Axle pipe
18. O-ring
19. Rear plate cover
20. Relief valve

22140_FX35_G0029

Fig. 115 Exploded view of the oil pan and related components—4.5L engine

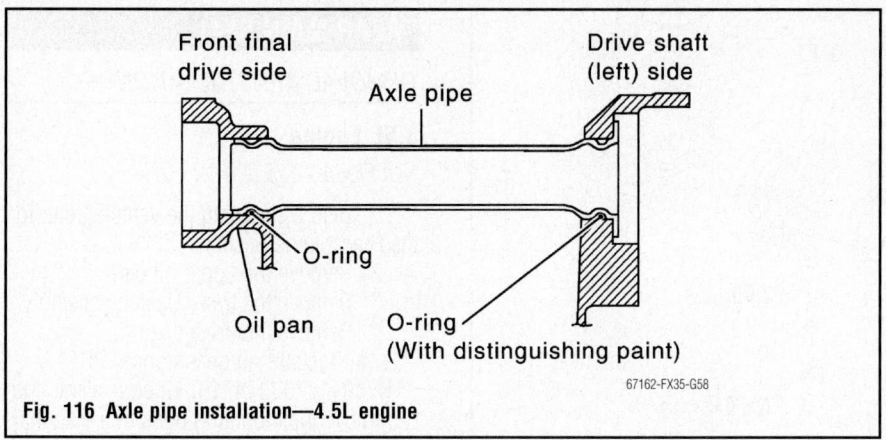

Fig. 116 Axle pipe installation—4.5L engine

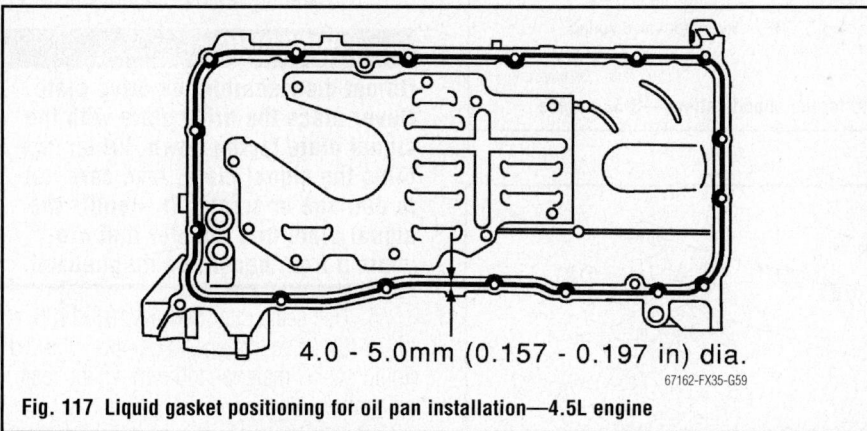

4.0 - 5.0mm (0.157 - 0.197 in) dia.

67162-FX35-G59

Fig. 117 Liquid gasket positioning for oil pan installation—4.5L engine

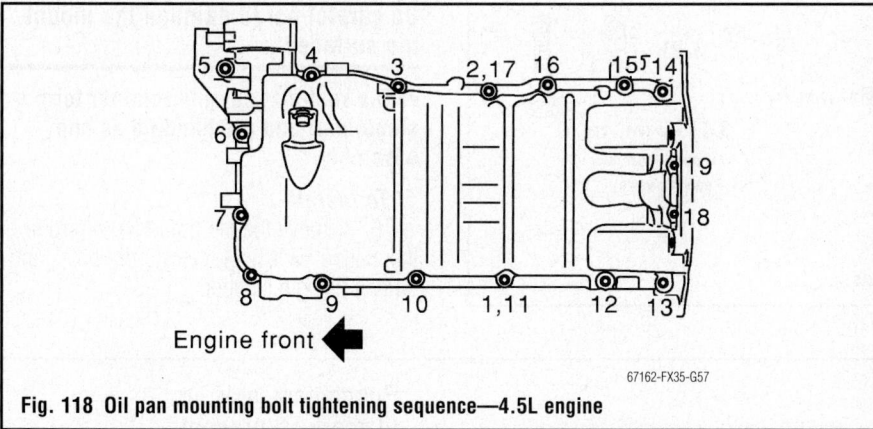

Engine front ◀

67162-FX35-G57

Fig. 118 Oil pan mounting bolt tightening sequence—4.5L engine

e. Install the rear plate cover and tighten the mounting bolt to 65 inch lbs. (7 Nm).

26. Install the oil pan drain plug with a new drain plug washer. Tighten it to 25 ft. lbs. (34 Nm).

27. The remainder of installation is the reverse of the removal procedure.

➡**Wait at least 30 minutes after the oil pan is installed to fill the engine with new oil.**

28. Add the proper amount and type of engine oil.

29. Start the engine and check for leakage of engine oil.

30. Stop the engine and wait for 15 minutes.

31. Check the engine oil level again.

OIL PUMP

REMOVAL & INSTALLATION

3.5L Engine

See Figure 119.

1. Before servicing the vehicle, refer to the Precautions Section.

2. Remove the oil pan (lower and upper) and the oil strainer.

3. Remove the front timing chain case and the timing chain (primary).

4. Remove the oil pump assembly.

To install:

5. Before installation, apply new engine oil to the parts as illustrated in the figure.

6. For pump installation, align the crankshaft flat faces with the oil pump inner rotor flat faces.

7. Installation is the reverse of the removal procedure.

8. After warming up the engine, check for engine oil leakage.

9. Check the engine oil level and add engine oil, as needed.

4.5L Engine

See Figures 120 and 121.

1. Before servicing the vehicle, refer to the Precautions Section.

2. Remove the front cover.

3. Remove the oil pump drive spacer.

4. Set bolts in the 2 bolt holes (M6 x 1.0mm) on the front surface. Using a suitable puller, pull the oil pump drive spacer off the crankshaft.

5. Remove the oil pump.

To install:

6. Install the oil pump.

7. Install the oil pump drive spacer as follows:

a. Insert the oil pump drive spacer so that the crankshaft key and flat surfaces of the oil pump inner rotor mesh properly.

➡**If the positional relationship does not allow the insertion, rotate the oil pump inner rotor to facilitate installation.**

b. After confirming that the position of each part is in correct position for the spacer, force fit the spacer by lightly tapping it with a plastic hammer until it makes contact and goes no further.

8. Installation is the reverse of the removal procedure.

9. After warming up the engine, check for engine oil leakage.

10. Check the engine oil level, and add more oil, if necessary.

PISTON AND RING

POSITIONING

See Figures 122 through 124.

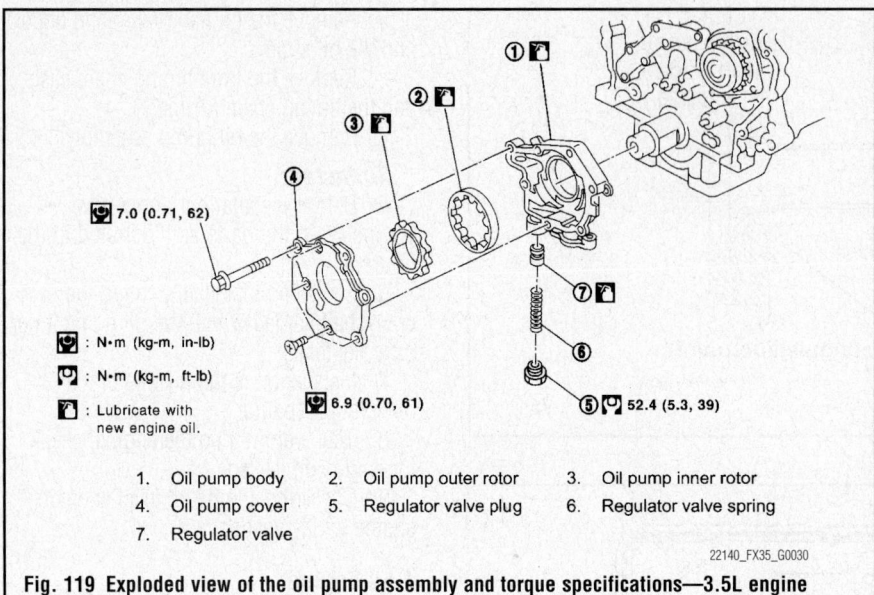

N•m (kg-m, in-lb)

N•m (kg-m, ft-lb)

Lubricate with new engine oil.

7.0 (0.71, 62)

6.9 (0.70, 61)

52.4 (5.3, 39)

1. Oil pump body	2. Oil pump outer rotor	3. Oil pump inner rotor
4. Oil pump cover	5. Regulator valve plug	6. Regulator valve spring
7. Regulator valve		

Fig. 119 Exploded view of the oil pump assembly and torque specifications—3.5L engine

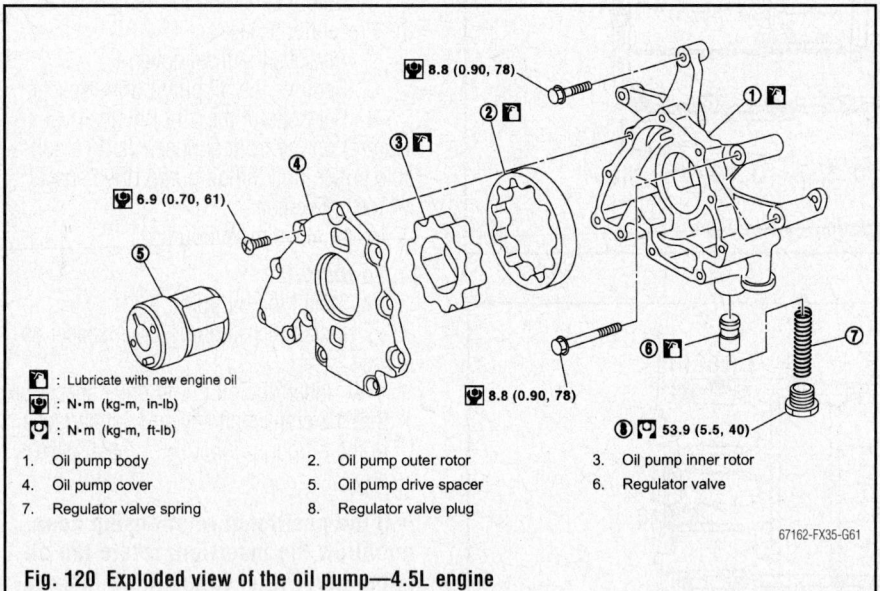

8.8 (0.90, 78)

6.9 (0.70, 61)

8.8 (0.90, 78)

53.9 (5.5, 40)

Lubricate with new engine oil

N•m (kg-m, in-lb)

N•m (kg-m, ft-lb)

1. Oil pump body	2. Oil pump outer rotor	3. Oil pump inner rotor
4. Oil pump cover	5. Oil pump drive spacer	6. Regulator valve
7. Regulator valve spring	8. Regulator valve plug	

Fig. 120 Exploded view of the oil pump—4.5L engine

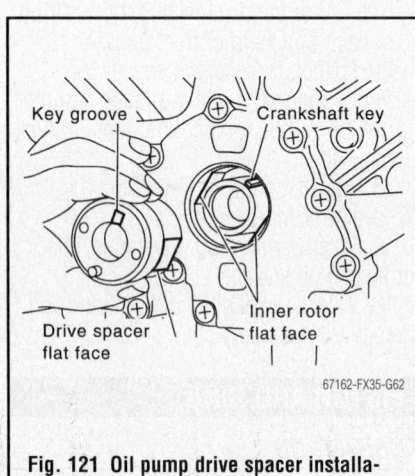

Key groove

Crankshaft key

Inner rotor flat face

Drive spacer flat face

Fig. 121 Oil pump drive spacer installation orientation—4.5L engine

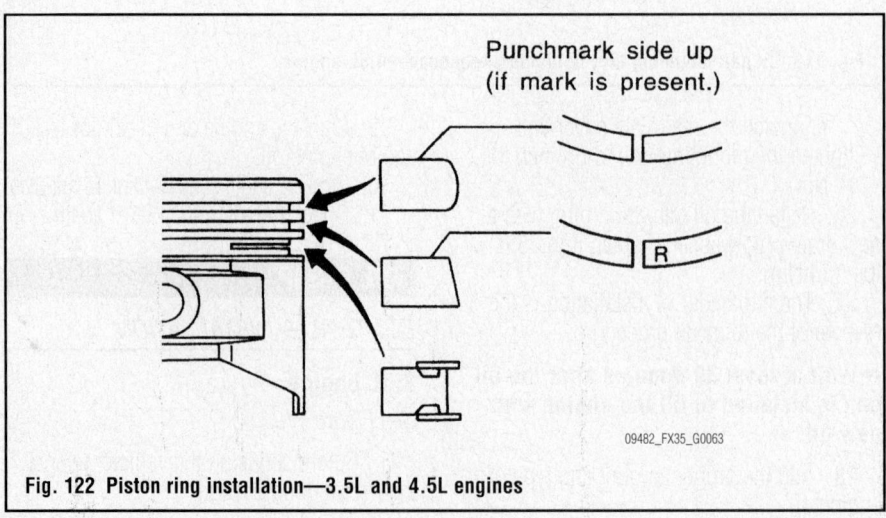

Punchmark side up (if mark is present.)

Fig. 122 Piston ring installation—3.5L and 4.5L engines

REAR MAIN SEAL

REMOVAL & INSTALLATION

3.5L Engine

See Figures 125 and 126.

1. Before servicing the vehicle, refer to the Precautions Section.
2. Remove the upper oil pan.
3. Remove the transmission assembly.
4. Remove the drive plate.
   a. Install ring gear stopper SST: KV1011770 (J44716), or equivalent, and remove the mounting bolts in a diagonal order.
   b. Carefully remove the drive plate.

❊❊ WARNING

Do not disassemble the drive plate. Never place the drive plate with the signal plate facing down. When handling the signal plate, take care not to damage or scratch it. Handle the signal plate in a manner that prevents it from becoming magnetized.

5. Use seal cutter SST: KV10111100 (J37228), or equivalent, to cut away the old liquid gasket material and remove the rear oil seal retainer.

❊❊ WARNING

Be careful not to damage the mounting surfaces.

➡The rear oil seal and retainer form a single part and are handled as one assembly.

To install:

6. Remove the old liquid gasket from the mating surface of the cylinder block and oil pan using a scraper.

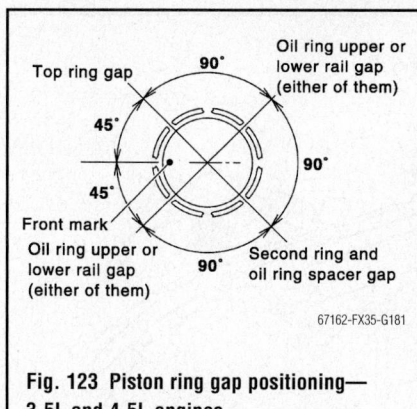

Fig. 123 Piston ring gap positioning—3.5L and 4.5L engines

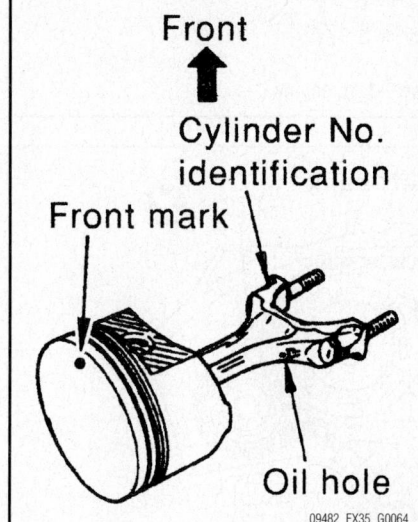

Fig. 124 Piston and connecting rod installation—3.5L and 4.5L engines

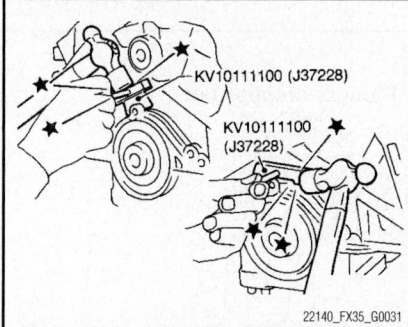

Fig. 125 Use seal cutter SST: KV10111100 (J37228) to cut away the old gasket and remove the rear oil seal retainer—3.5L engine

7. Apply new engine oil to the oil and dust seal lips.

8. Apply liquid gasket to the rear oil seal retainer as illustrated.

➡**Use Genuine RTV Silicone Sealant or equivalent. Installation should be done**

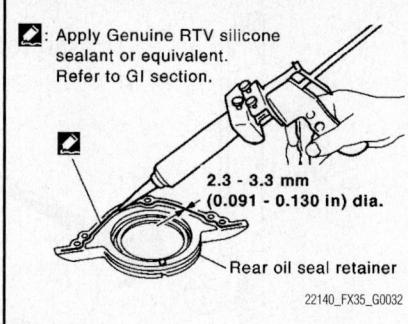

Fig. 126 Apply liquid gasket to the rear oil seal retainer as illustrated—3.5L engine

within 5 minutes after coating, otherwise the liquid gasket may not seal properly.

9. Install the rear oil seal retainer onto the cylinder block.

➡**Make sure the garter spring is in position and the seal lips are not inverted.**

10. The remainder of installation is the reverse of the removal procedure.

4.5L Engine

See Figure 127.

1. Before servicing the vehicle, refer to the Precautions Section.

2. Remove the transmission and transfer assembly.

3. Remove the drive plate.

   a. Install ring gear stopper SST: J-45476, or equivalent.

   b. Holding the ring gear with the ring gear stopper, remove the mounting bolts in a diagonal order.

   c. Carefully remove the drive plate.

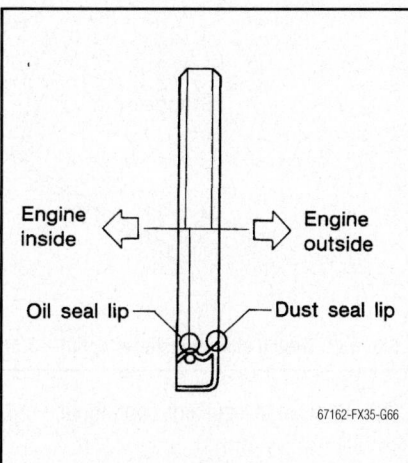

Fig. 127 Rear main seal installation orientation—4.5L engine

※※ **WARNING**

Do not disassemble the drive plate. Never place the drive plate with the signal plate facing down. When handling the signal plate, take care not to damage or scratch it. Handle the signal plate in a manner that prevents it from becoming magnetized.

4. Remove the engine rear plate.

5. Remove the rear oil seal using a suitable tool.

※※ **CAUTION**

Be careful not to damage the crankshaft and oil seal retainer surface.

To install:

6. Apply engine oil to both the oil seal lip and dust seal lip.

7. Install rear oil seal so that each seal lip is oriented as shown in the figure.

8. Using a suitable drift, press fit the oil seal until the height of the oil seal is level with the mounting surface. Rear oil seal drift—outer diameter: 4.02 inches (102mm), inner diameter: 3.39 inches (86mm).

➡**Make sure the garter spring is in position and the seal lips are not inverted. Press fit the seal straight and avoid causing burrs or tilting.**

9. The remainder of installation is the reverse of the removal procedure.

TIMING CHAIN, SPROCKETS, FRONT COVER AND SEAL

REMOVAL & INSTALLATION

3.5L Engine

With Oil Pan Removal

See Figures 128 through 155.

1. Before servicing the vehicle, refer to the Precautions Section.

This section describes procedures for removing/installing the front timing chain case and timing chain related parts, and the rear timing chain case, when the upper oil pan needs to be removed/installed for engine overhaul, etc.

When the upper oil pan needs to be removed or installed, or when the rear timing chain case is removed or installed, remove the oil pans (upper and lower) first. Then, remove the front timing chain case, timing chain related parts, and the rear timing chain case, and install in the reverse order of removal.

2. Place the vehicle on a lift.

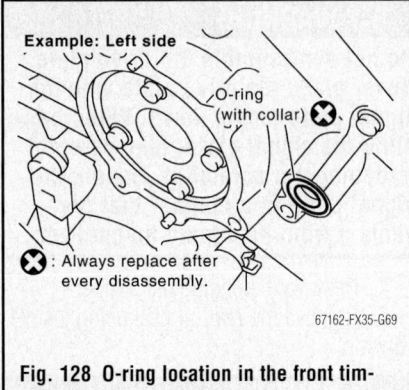

Fig. 128 O-ring location in the front timing chain case—3.5L engine

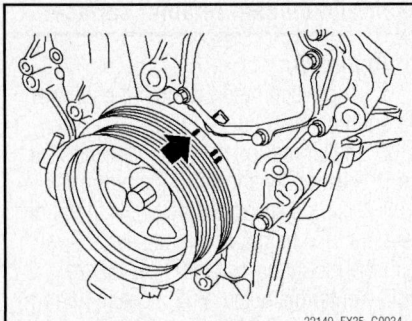

Fig. 129 Rotate the crankshaft pulley clockwise to align the timing with the timing indicator—3.5L engine

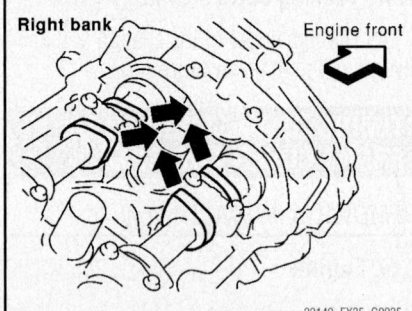

Fig. 130 Make sure the intake and exhaust cam lobes on the number 1 cylinder are located so that they point inward and upward compared to the cylinder head—3.5L engine

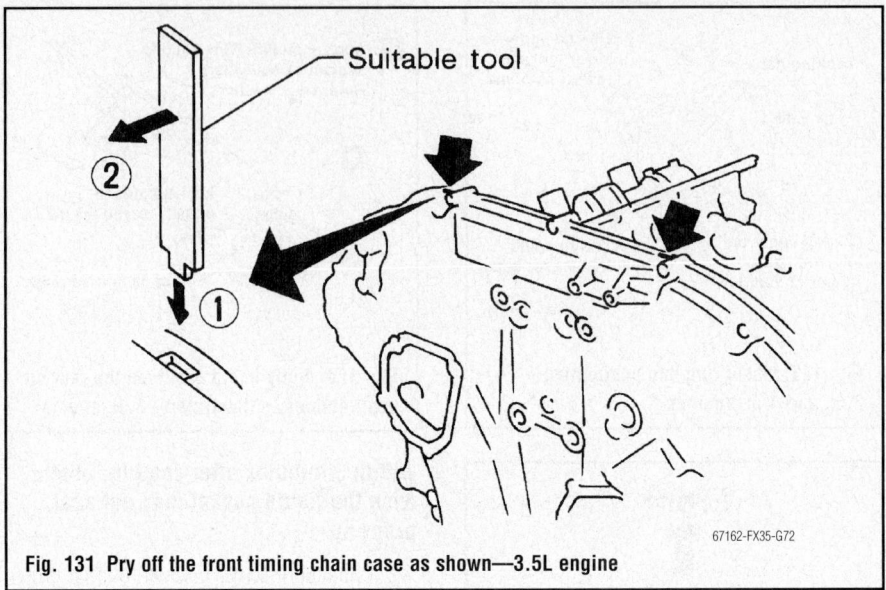

Fig. 131 Pry off the front timing chain case as shown—3.5L engine

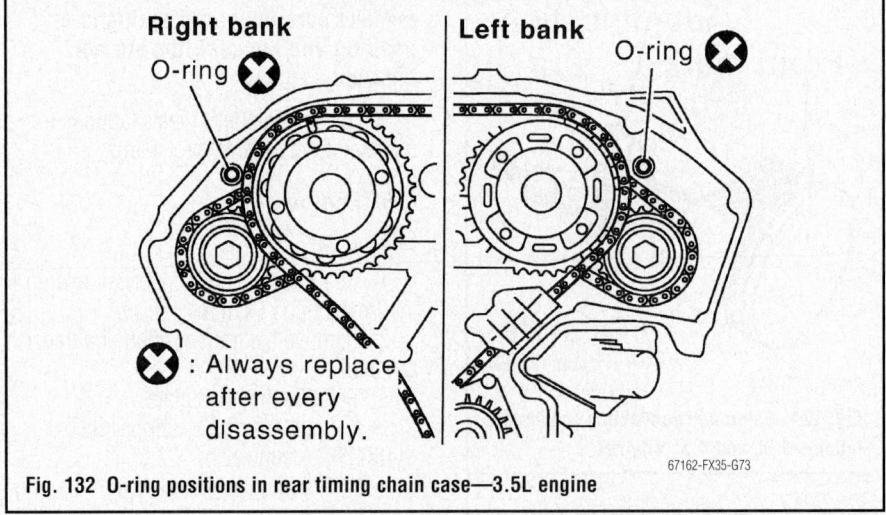

Fig. 132 O-ring positions in rear timing chain case—3.5L engine

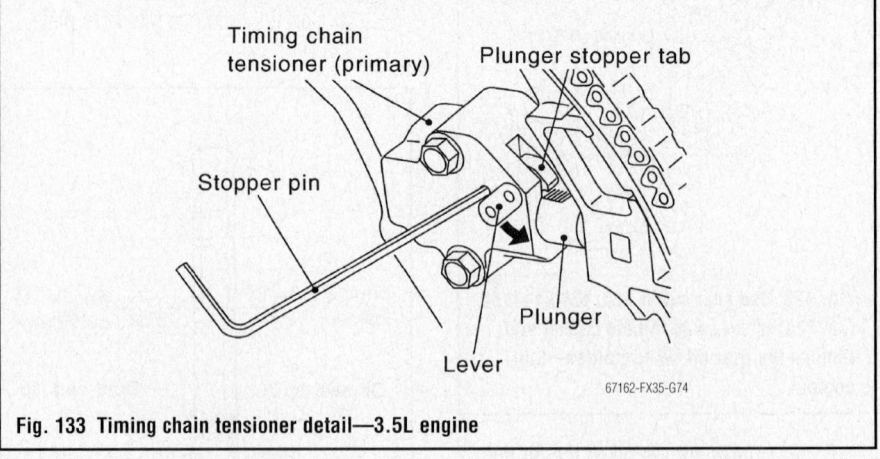

Fig. 133 Timing chain tensioner detail—3.5L engine

3. Remove the front tires.

4. Disconnect the negative battery terminal.

5. Remove the engine cover.

6. Remove the air cleaner case assembly.

7. Remove the front and rear engine undercovers.

8. Drain the engine coolant from the radiator.

9. Drain the engine oil from the oil pan.

10. Remove the engine wiring harnesses.

11. Remove the upper and lower intake manifold collectors.

12. Remove the radiator cooling fan assembly.

13. Remove the A/C compressor from the bracket with its piping connected, and temporarily secure it aside.

14. Remove the power steering oil pump from the bracket with its piping connected, and temporarily secure it aside.

15. Remove the power steering oil pump bracket.

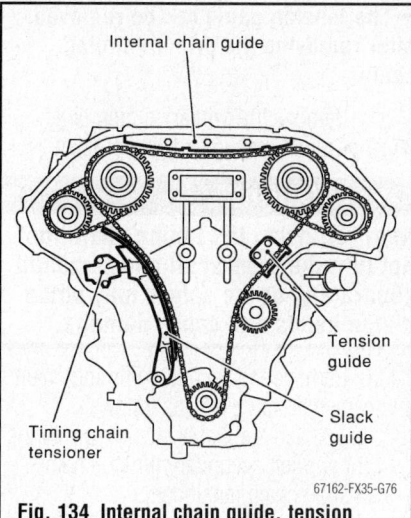

Fig. 134 Internal chain guide, tension guide and slack guide positions—3.5L engine

16. Remove the alternator.

17. Remove the water bypass hose, water hose clamp, and idler pulley bracket from the front timing chain case.

18. Remove the upper and lower oil pan.

19. Remove the right and left intake valve timing control covers by loosening the bolts in the reverse order of the tightening sequence.

20. Use seal cutter SST: KV10111100 (J37228), or equivalent, to cut the liquid gasket for removal.

➡ **The shaft is internally joined to the intake camshaft sprocket center hole. During removal, keep the shaft horizontal until it is completely disconnected.**

21. Remove the collared O-ring from the front timing chain case (left and right side).

22. Remove the right and left rocker covers.

23. Position the engine at compression Top Dead Center (TDC) of number 1 cylinder as follows:

a. Rotate the crankshaft pulley clockwise to align the timing mark (grooved line without color) with the timing indicator.

b. Make sure the intake and exhaust cam lobes on the number 1 cylinder (engine front side of right bank) are located so that they point inward and upward compared to the cylinder head, see illustration.

c. If the cam lobes are not positioned pointing inward and upward as illustrated, rotate the crankshaft 1 full revolution (360°) until they are.

24. Remove the crankshaft pulley, as follows:

a. For 2WD models, remove the rear cover plate.

b. For AWD models, remove the starter motor.

c. Set the ring gear stopper to hold the crankshaft in position.

d. Loosen the crankshaft damper pulley bolt until the bolt seating surface is approximately 0.39 inch (10mm) from its original position.

➡ **Do not completely remove the crankshaft damper pulley bolt, since it will be used as a supporting point for a suitable puller.**

e. Place a suitable puller tab on the holes of the crankshaft damper pulley and pull the crankshaft damper pulley until it releases.

✳✳ WARNING

Do not position the suitable puller tab on the outer edges of the crankshaft pulley since this can damage the internal damper.

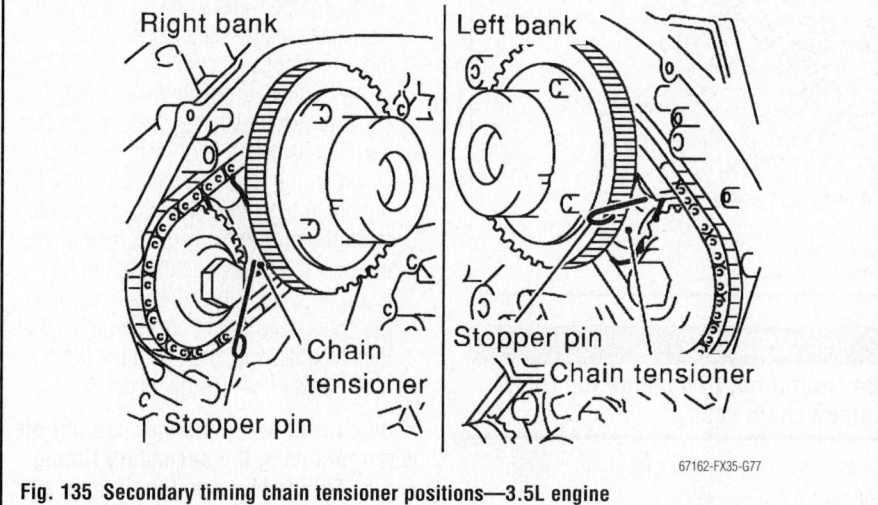

Fig. 135 Secondary timing chain tensioner positions—3.5L engine

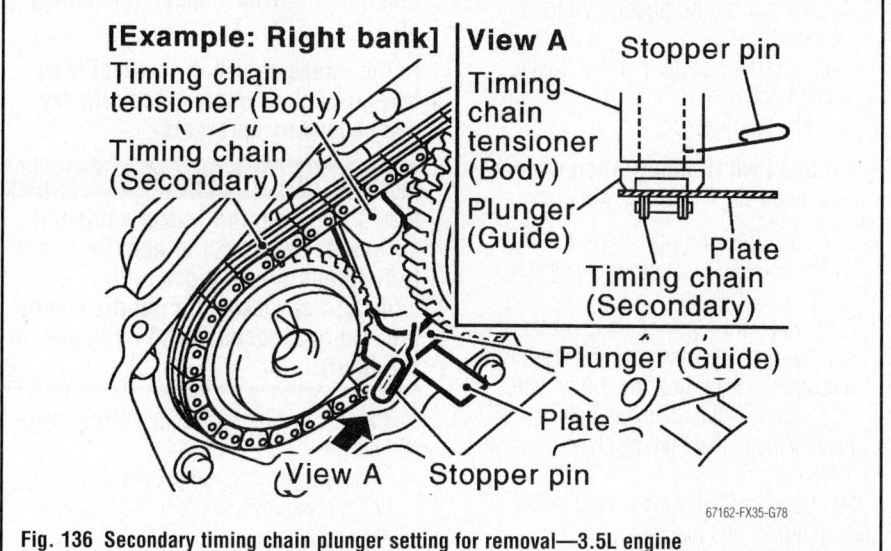

Fig. 136 Secondary timing chain plunger setting for removal—3.5L engine

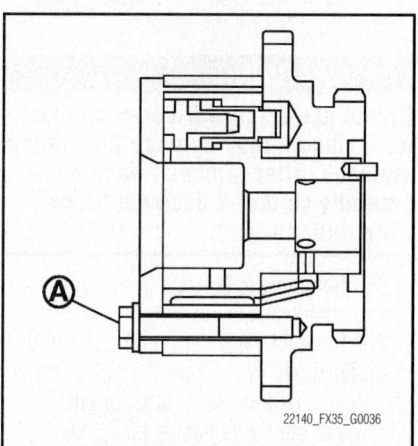

Fig. 137 Cross-section of intake camshaft sprocket—Do NOT loosen bolt A—3.5L engine

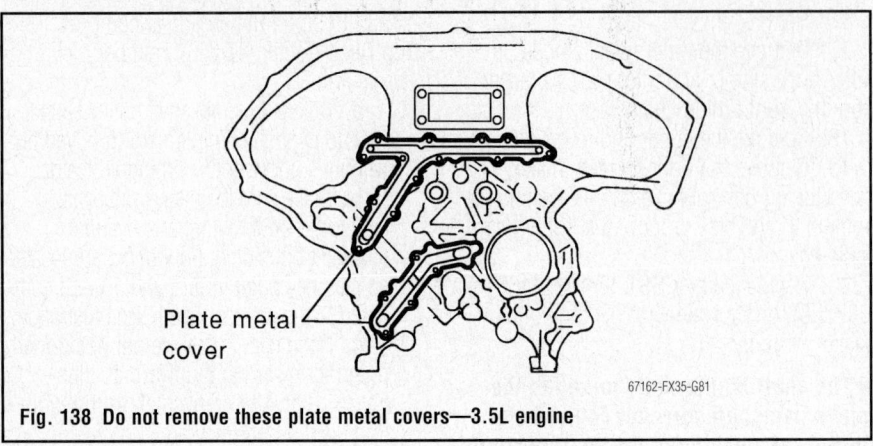

Fig. 138 Do not remove these plate metal covers—3.5L engine

Plate metal cover

67162-FX35-G81

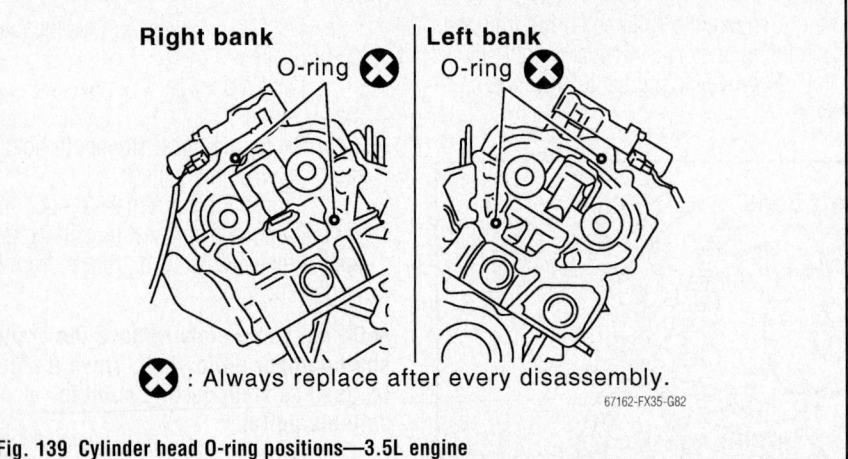

Right bank
O-ring ✗

Left bank
O-ring ✗

✗ : Always replace after every disassembly.

67162-FX35-G82

Fig. 139 Cylinder head O-ring positions—3.5L engine

25. Remove the front timing chain case, as follows:

   a. Loosen the mounting bolts in the reverse order of the tightening sequence.

   b. Insert a suitable tool into the notch at the top of the front timing chain case.

   c. Pry off the case by moving the pry tool as shown. Use a seal cutter to cut the liquid gasket for removal.

✳✳ WARNING
Do not use a screwdriver or similar tool, since it may damage the mating surfaces. After removal, handle it carefully so that it does not tilt or warp under a load.

26. Remove the O-rings from the rear timing chain case.

27. Remove the water pump cover and chain tensioner cover from the front timing chain case. Use the seal cutter, or equivalent, to cut the liquid gasket for removal.

28. Remove the front oil seal from the front timing chain case using a suitable tool.

✳✳ WARNING
Be careful not to damage the front timing chain case.

29. Remove the primary timing chain tensioner, as follows:

   a. Pull the lever down and release the plunger stopper tab. The plunger stopper tab can be pushed up to release.

   b. Insert a stopper pin into the tensioner body hole to hold the lever and keep the tab released.

➡ A 0.098 inch (2.5mm) Allen wrench can be used for a stopper pin.

   c. Insert the plunger into the tensioner body by pressing on the slack guide.

   d. Keep the slack guide depressed and hold it by pushing the stopper pin through the lever hole and body hole.

   e. Remove the mounting bolts and remove the primary timing chain tensioner.

30. Remove the internal chain guide, tension guide, and slack guide.

➡ The tension guide can be removed after removing the primary timing chain.

31. Remove the primary timing chain, tension guide, and crankshaft sprocket.

✳✳ WARNING
After removing the timing chain, do not turn the crankshaft and camshaft separately, or the valves may strike piston heads and cause damage.

32. Remove the secondary timing chain and camshaft sprockets, as follows:

   a. Attach a suitable stopper pin to the right and left secondary timing chain camshaft chain tensioners.

   b. Remove the intake and exhaust camshaft sprocket bolts. Apply paint to the timing chain and camshaft sprockets for alignment during installation. Secure the hexagonal portion of the camshaft using an open-end wrench to hold the camshaft steady while loosening the mounting bolts.

   c. Remove the secondary timing chain together with camshaft sprockets. Turn the camshaft slightly to create slack in the timing chain on the secondary timing chain tensioner side. Insert a 0.020 inch (0.5mm) thick metal or resin plate between the timing chain and timing chain tensioner plunger guide. Remove the secondary timing chain together with the camshaft sprockets with the timing chain loose from the guide groove.

➡ Be careful of the plunger coming-off when removing the secondary timing chain. This is because the plunger of the secondary timing chain tensioner moves during operation, which can result in the fixed stopper pin coming off.

➡ The intake camshaft sprocket is a two-for-one assembly of the primary and secondary sprockets.

✳✳ WARNING
When handling the intake camshaft sprocket, be careful to handle it carefully to avoid any shock to the camshaft sprocket. Do not disassemble. Do not loosen bolt A as shown in the figure.

33. Remove the rear timing chain case, as follows:

   a. Loosen and remove the mounting bolts in the reverse order of the tightening sequence.

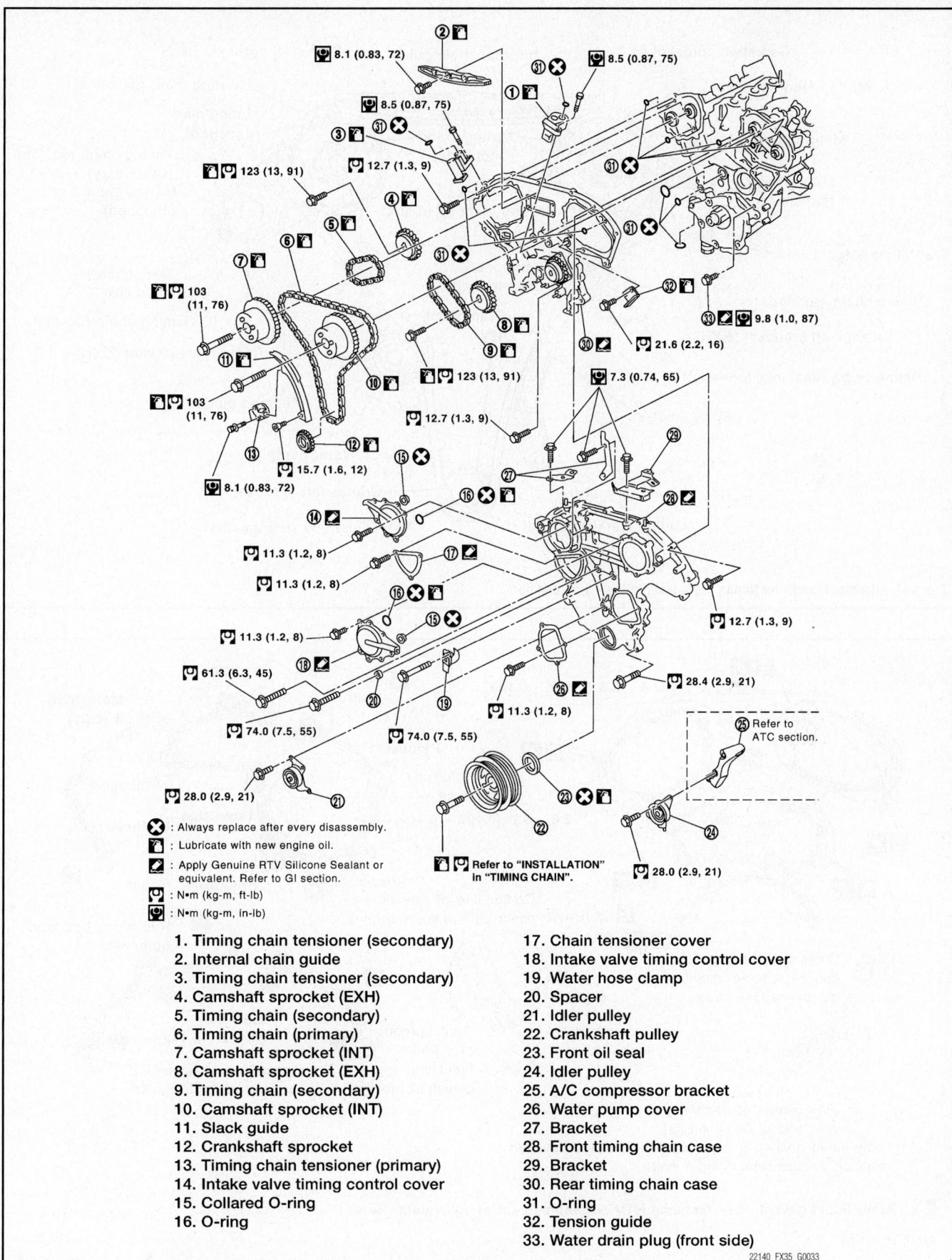

8.1 (0.83, 72)

8.5 (0.87, 75)

8.5 (0.87, 75)

12.7 (1.3, 9)

123 (13, 91)

103 (11, 76)

103 (11, 76)

8.1 (0.83, 72)

15.7 (1.6, 12)

123 (13, 91)

12.7 (1.3, 9)

21.6 (2.2, 16)

9.8 (1.0, 87)

7.3 (0.74, 65)

11.3 (1.2, 8)

11.3 (1.2, 8)

11.3 (1.2, 8)

12.7 (1.3, 9)

61.3 (6.3, 45)

74.0 (7.5, 55)

74.0 (7.5, 55)

11.3 (1.2, 8)

28.4 (2.9, 21)

28.0 (2.9, 21)

28.0 (2.9, 21)

Refer to ATC section.

Refer to "INSTALLATION" in "TIMING CHAIN".

✗ : Always replace after every disassembly.

: Lubricate with new engine oil.

: Apply Genuine RTV Silicone Sealant or equivalent. Refer to GI section.

: N•m (kg-m, ft-lb)

: N•m (kg-m, in-lb)

1. Timing chain tensioner (secondary)	17. Chain tensioner cover
2. Internal chain guide	18. Intake valve timing control cover
3. Timing chain tensioner (secondary)	19. Water hose clamp
4. Camshaft sprocket (EXH)	20. Spacer
5. Timing chain (secondary)	21. Idler pulley
6. Timing chain (primary)	22. Crankshaft pulley
7. Camshaft sprocket (INT)	23. Front oil seal
8. Camshaft sprocket (EXH)	24. Idler pulley
9. Timing chain (secondary)	25. A/C compressor bracket
10. Camshaft sprocket (INT)	26. Water pump cover
11. Slack guide	27. Bracket
12. Crankshaft sprocket	28. Front timing chain case
13. Timing chain tensioner (primary)	29. Bracket
14. Intake valve timing control cover	30. Rear timing chain case
15. Collared O-ring	31. O-ring
16. O-ring	32. Tension guide
	33. Water drain plug (front side)

Fig. 140 Exploded view of timing chain and related components—3.5L engine

22140_FX35_G0033

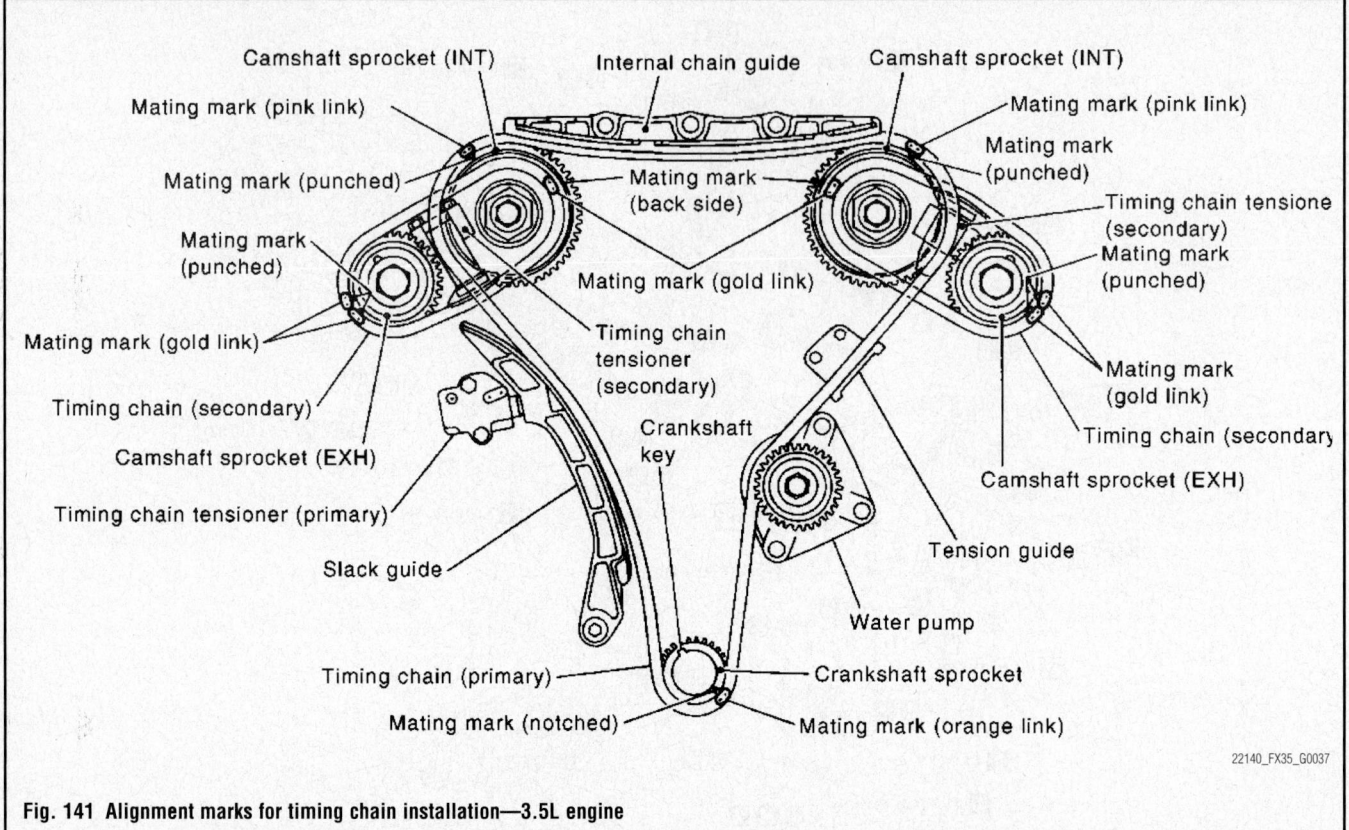

Fig. 141 Alignment marks for timing chain installation—3.5L engine

Cross both ends as shown and be sure to minimize the overlapped area.

Protrusions at beginning and end of liquid gasket

* : Apply liquid gasket to the chamfered surface between camshaft bracket and cylinder head.

Do not protrude in this area.

2.6 - 3.6 (0.102 - 0.142)

More than 8 (0.31)

: Run along bolt hole outer side

Protrusions at beginning and end of liquid gasket

2.6 - 3.6 (0.102 - 0.142)

E Camshaft axis area
Center line of rear timing chain case sealant groove
5 (0.20)
Center line of liquid gasket
2 (0.08)
Joint portion of cylinder head and camshaft bracket

: Apply liquid gasket. (Use Genuine RTV silicone sealant or equivalent. Refer to GI section.)

Unit: mm (in)

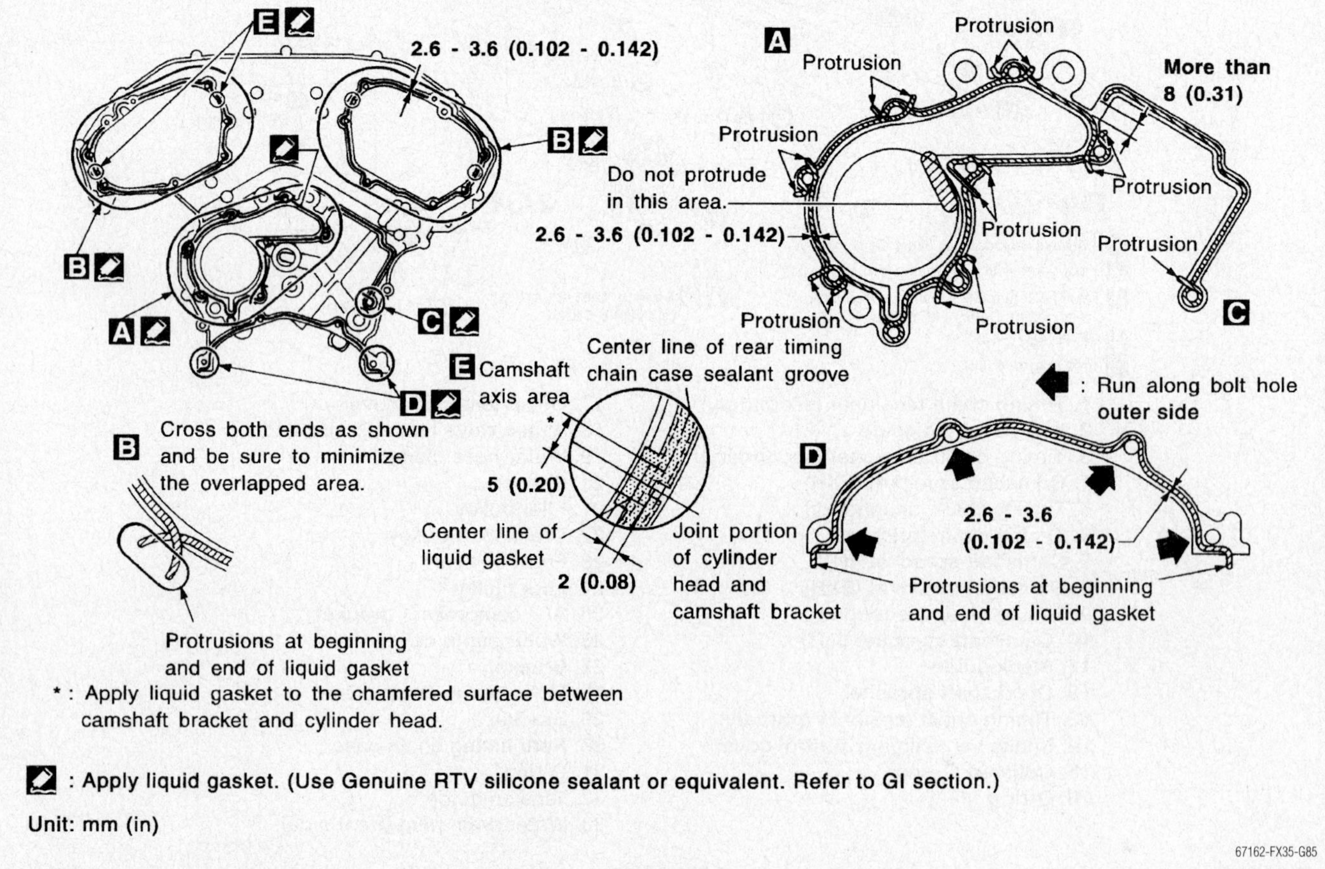

Fig. 142 Liquid gasket positions for timing chain case installation—3.5L engine

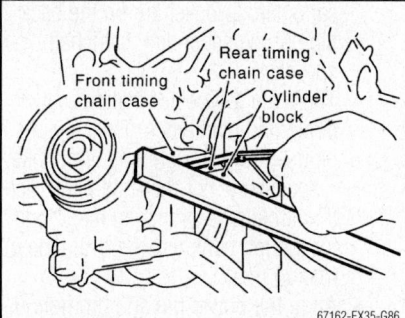

Fig. 143 Tightening sequence for the rear timing chain case mounting bolts—3.5L engine

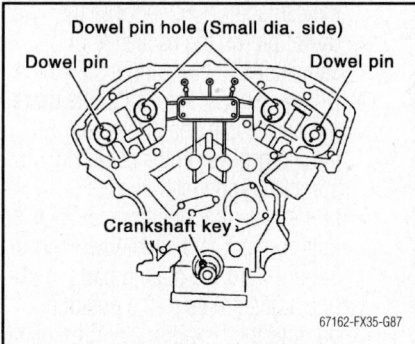

Fig. 144 Check the surface height difference between the rear timing chain case and the cylinder block—3.5L engine

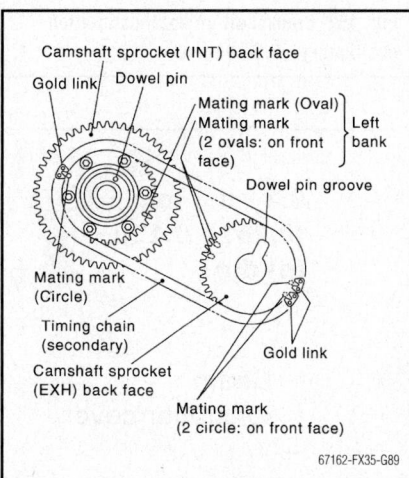

Fig. 145 Camshaft positioning for timing chain installation—3.5L engine

   b. Cut the sealant using a seal cutter and remove the rear timing chain case.

❊❊ WARNING

Do not remove plate metal cover of engine oil passage. After removing the chain case, do not apply any load on the case which could cause warping.

34. Remove the O-rings from the cylinder head.

35. Remove the O-rings from the cylinder block.

36. If necessary, remove the secondary timing chain tensioners from the cylinder head, as follows:

   a. Remove the number 1 camshaft brackets.

   b. Remove the secondary timing chain tensioners with the stopper pins attached.

37. Use a scraper to remove all traces of liquid gasket from the front and rear timing chain cases and opposite mating surfaces.

❊❊ WARNING

Be careful not to allow gasket fragments to enter the oil pan.

38. Remove the old liquid gasket from the bolt holes and threads.

39. Use a scraper to remove all traces of liquid gasket from the water pump cover, chain tensioner cover and intake valve timing control covers.

40. Inspect the timing chain for cracks and any excessive wear at link plates and roller links of the timing chain. Replace the timing chain, as necessary.

To install:

➡**The accompanying illustration shows the relationship between the mating mark on each timing chain and that on the corresponding sprocket, with the components installed.**

41. If removed, install the secondary timing chain tensioners on the cylinder head, as follows:

   a. Install the chain tensioners with stopper pins attached and new O-rings.

   b. Install the number 1 camshaft brackets. Refer to Camshaft and Valve Lifters, removal & installation.

42. Install new O-rings onto the cylinder block.

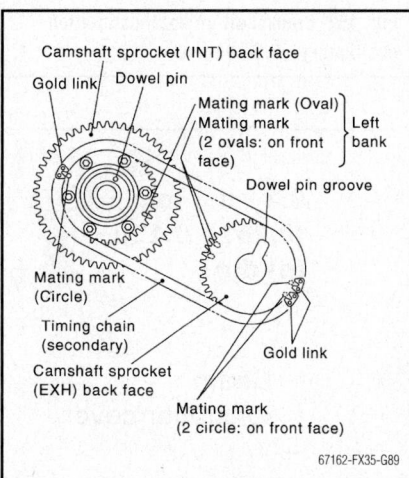

Fig. 146 Secondary timing chain installation alignment—3.5L engine

43. Install new O-rings on the cylinder head.

44. Apply liquid gasket to the rear timing chain case backside, as shown.

➡**Use Genuine RTV Silicone Sealant or equivalent.**

45. For **A** in the illustration, completely wipe out the liquid gasket extended on a portion touching at engine coolant.

46. Apply liquid gasket on the installation position of the water pump and cylinder head completely.

47. Align the rear timing chain case and water pump assembly with right and left dowel pins on cylinder block and install the case.

➡**Make sure the O-rings stay in place during installation on the cylinder block and cylinder head.**

48. Tighten the mounting bolts in the sequence shown. After all bolts are temporarily tightened, retighten them to 108 inch lbs. (13 Nm) in the tightening sequence. There are 2 bolt lengths used for the timing chain case:

   a. 20mm length: Bolt positions 1, 2, 3, 6, 7, 8, 9, and 10.

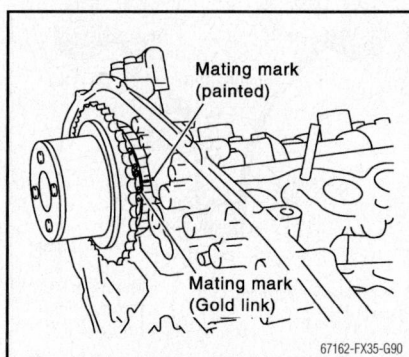

Fig. 147 Intake camshaft sprocket and secondary timing chain alignment with mating mark painted—3.5L engine

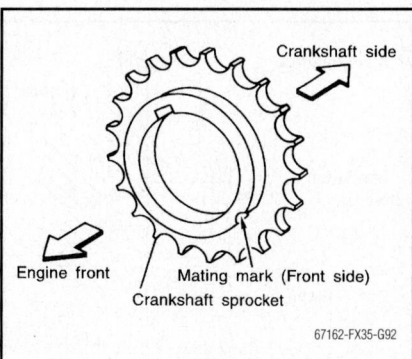

Fig. 148 Crankshaft sprocket installation position—3.5L engine

b. 16mm length: All except bolt positions 1, 2, 3, 6, 7, 8, 9, and 10.

➡ **If RTV Silicone Sealant protrudes beyond the sealing surfaces, wipe it off immediately.**

49. After installing the rear timing chain case, check the surface height difference between the rear timing chain case to the cylinder block at the oil pan mounting surface. If the difference is not within −0.009–0.006 inch (-0.24–0.14mm), repeat the installation procedure.

50. Position the crankshaft so the number 1 piston is set at TDC on the compression stroke. Make sure that the dowel pin hole, dowel pin, and crankshaft key are located as shown.

➡ **Though the camshaft does not stop at the position as shown in the figure, for the placement of the cam nose, it is generally accepted the camshaft is**

placed in the same direction of the figure and as follows:

- Camshaft dowel pin hole (intake side): At the cylinder head upper face side in each bank
- Camshaft dowel pin (exhaust side): At the cylinder head upper face side in each bank
- Crankshaft key: At the cylinder head side of the right bank

✳✳ WARNING
The hole on the small diameter side must be used for the intake side dowel pin hole.

51. Install the secondary timing chains and camshaft sprockets, as follows:

✳✳ WARNING
The matching marks between the timing chain and sprockets slip

easily. Confirm all matching mark positions repeatedly during the installation process.

a. Push the plunger of the secondary chain tensioner and keep it pressed in with a stopper pin.

b. Install the secondary timing chains and camshaft sprockets. Align the mating marks on the secondary timing chain (gold link) with the ones on the intake and exhaust camshaft sprockets (stamped), and install them.

52. Ensure that the sprockets are properly positioned by heeding the following items:

- The mating marks for the intake camshaft sprocket are on the back side of the secondary camshaft sprocket.
- There are 2 types of mating marks, circle and oval types. They should be used for the right and left banks, respectively. For the right bank, use the circle type of mating mark, and for the left bank use the oval type of mating mark.
- Align the dowel pin and pin hole on the camshaft with the groove and dowel pin on the sprocket, and install them.
- On the intake side, align the pin hole on the small diameter side of the camshaft front end with the dowel pin on the back side of camshaft sprocket, and install them.
- On the exhaust side, align the dowel pin on the camshaft front end with the pin groove on the camshaft sprocket, and install them.
- In the case that positions of each mating mark and each dowel pin are not fit to the mating parts, make fine adjustments to the position holding the hexagonal portion of the camshaft with a wrench or equivalent.

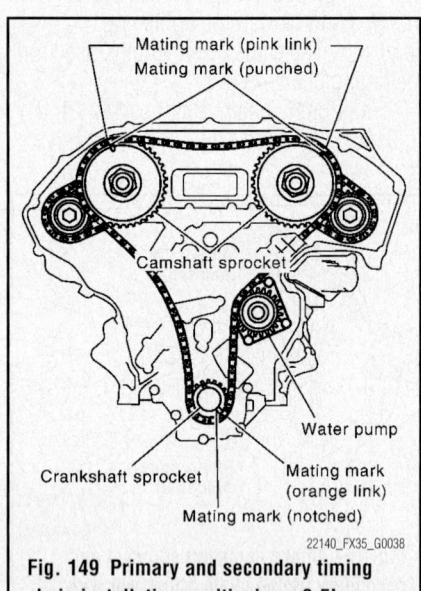

Fig. 149 Primary and secondary timing chain installation positioning—3.5L engine

22140_FX35_G0038

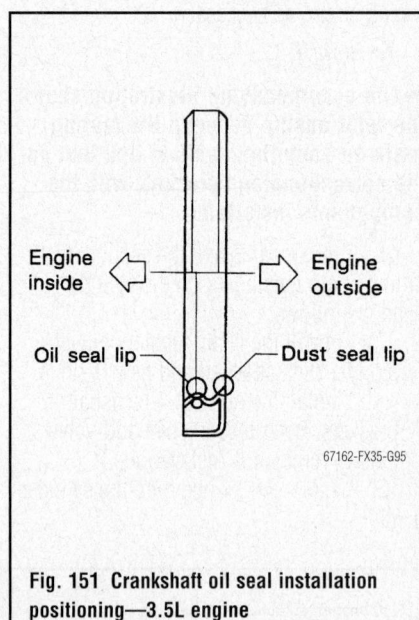

67162-FX35-G95

Fig. 151 Crankshaft oil seal installation positioning—3.5L engine

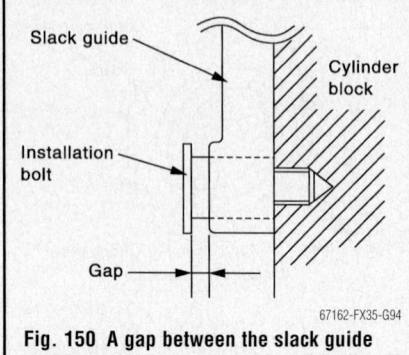

67162-FX35-G94

Fig. 150 A gap between the slack guide bolt head and the slack guide is normal—3.5L engine

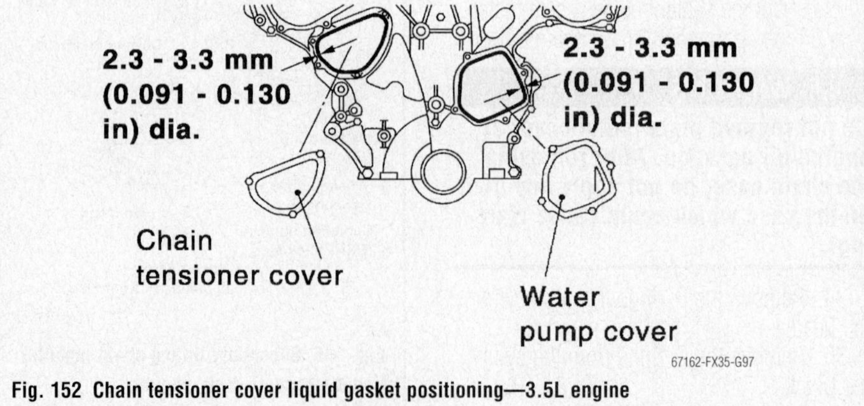

2.3 - 3.3 mm (0.091 - 0.130 in) dia.

2.3 - 3.3 mm (0.091 - 0.130 in) dia.

Chain tensioner cover

Water pump cover

67162-FX35-G97

Fig. 152 Chain tensioner cover liquid gasket positioning—3.5L engine

- Mounting bolts for the camshaft sprockets must be tightened in the next step. Tightening them by hand is enough to prevent the dislocation of dowel pins.
- It may be difficult to visually check the dislocation of the mating marks during and after installation. To make the matching easier, make a mating mark on the top of the sprocket teeth and its extended line in advance with paint.

53. After confirming the mating marks are aligned, tighten the camshaft sprocket mounting bolts, while securing the camshaft using a wrench on a hexagonal portion of the camshaft.

54. Pull the stopper pins out from the secondary timing chain tensioners.

55. Install the primary timing chain, as follows:

➡**During alignment, be careful to prevent dislocation of the mating mark alignments of the secondary timing chains.**

   a. Install the crankshaft sprocket, making sure the mating marks on the crankshaft sprocket face the front of the engine.

   b. Install the primary timing chain, so that the mating mark (punched) on the camshaft sprocket is aligned with the pink link on the timing chain, while the mating mark (notched) on the crankshaft

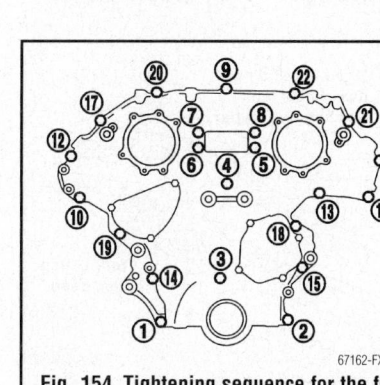

Front timing chain case

2.6 - 3.6 mm
(0.102 -
0.142 in) dia.

Ⓐ Ⓐ Ⓑ

Bolt hole

Protrusion

Liquid gasket protrusion away from bolt hole

✎ : Apply Genuine RTV silicone sealant or equivalent. Refer to GI section.

22140_FX35_G0039

Fig. 153 Apply liquid gasket to front timing chain case back side as shown—3.5L engine

sprocket is aligned with the orange one on the timing chain, as shown.

➡**If it is difficult to align mating marks of the primary timing chain with each sprocket, gradually turn the camshaft using a wrench on the hexagonal portion of the camshaft to align it with the mating marks.**

56. Install the internal chain guide and primary timing chain tensioner.

57. Install the slack guide.

➡**Do not over tighten the slack guide mounting bolts. It is normal for a gap to exist under the bolt seats when the mounting bolts are tightened to specification.**

58. Remove all dirt and foreign materials completely from the back and mounting surfaces of the chain tensioner.

59. Install chain tensioner for slack guide. When installing the chain tensioner, push in the sleeve and keep it depressed with a stopper pin.

60. After chain tensioner installation, pull out the stopper pin by pressing in the slack guide.

61. Reconfirm that the mating marks on sprockets and timing chains have not slipped out of alignment.

62. Install new O-rings on the rear timing chain case.

63. Install the front oil seal on the front timing chain case, as follows:

   a. Apply new engine oil to the oil seal edges.

   b. Make sure the garter spring is in position and the seal lip is not inverted and so that each seal lip is oriented as shown in the figure.

   c. Using a suitable drift, press-fit the oil seal until it is flush with the front timing chain case end face.

64. Install the water pump cover and chain tensioner cover to the front timing

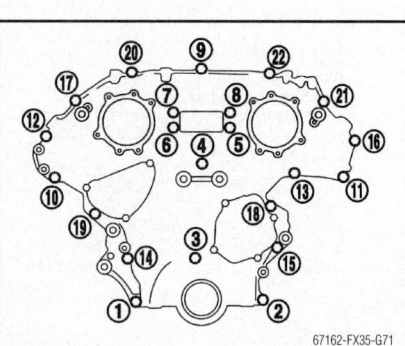

67162-FX35-G71

Fig. 154 Tightening sequence for the front timing chain case mounting bolts—3.5L engine

chain case. Apply liquid gasket to the front timing chain case front side as shown.

➡**Use Genuine RTV Silicone Sealant or equivalent.**

65. Install the front timing chain case as follows:

   a. Apply liquid gasket to front timing chain case back side as shown.

➡**Use Genuine RTV Silicone Sealant or equivalent.**

   b. Install the dowel pin on the rear timing chain case into the dowel pin hole on the front timing chain case.

   c. Tighten the bolts to the specified torque in the order shown. Refer to the following for locating the bolts.
- 8mm bolts (positions 1 and 2): 21 ft. lbs. (28 Nm)
- 6mm bolts (except positions 1 and 2): 108 inch lbs. (13 Nm)

   d. After tightening, retighten them again to the specified torque in the numerical order shown.

66. After installing the front timing chain case, check the surface height difference between the front timing chain case to rear timing chain case on the oil pan mounting surface. The allowable height difference is – 0.006–0.006 inch (– 0.14–0.14mm). If not within specification, repeat the installation procedure.

67. Install the right and left intake valve timing control covers, as follows:

   a. Install the seal rings in the shaft grooves.

   b. Apply a continuous bead of liquid gasket to the intake valve timing control covers.

➡**Use Genuine RTV Silicone Sealant or equivalent.**

   c. Install the collared O-ring in the front cover engine oil hole (left and right sides).

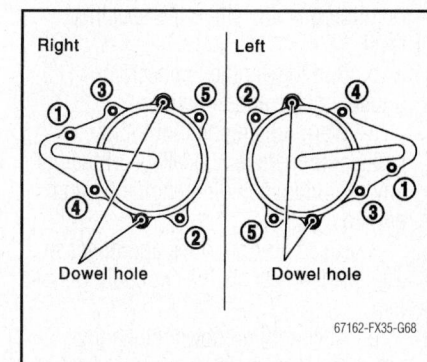

Right   **Left**

Dowel hole   Dowel hole

67162-FX35-G68

Fig. 155 Tightening sequence for the timing control covers—3.5L engine

d. Being careful not to move the seal ring from the installation groove, align the dowel pins on the chain case with the holes to install the intake valve timing control covers. Tighten the bolts in the order shown.

68. Install the crankshaft damper pulley, as follows:

a. Fix the crankshaft in position using ring gear stopper SST: KV10117700 (J44716) or equivalent.

b. Install the crankshaft pulley, taking care not to damage the front oil seal.

✷✷ WARNING

When press-fitting the crankshaft pulley with a plastic hammer, tap on its center portion, not on the circumference.

c. Tighten the crankshaft bolt to 33 ft. lbs. (44 Nm).

d. Put a paint mark on the crankshaft pulley aligned with the angle mark on the crankshaft pulley bolt. Then, further tighten the bolt 90°.

69. Rotate the crankshaft pulley in the normal direction (clockwise when viewed from front) to confirm that it turns smoothly.

70. The remainder of installation is the reverse of the removal procedure.

➡**If the hydraulic pressure inside the chain tensioner drops after removal/installation, the slack in the guide may generate a pounding noise during and just after engine start. However, this does not indicate a problem. The noise will stop after hydraulic pressure rises.**

71. Perform the following once installation is complete:

a. Before starting the engine, check the oil/fluid levels including the engine coolant and engine oil. If less than required quantity, fill to the specified level.

b. Run the engine to check for unusual noise and vibration.

c. Warm up engine thoroughly to make sure there is no leakage of fuel, or any oil/fluids including engine oil and engine coolant.

d. Bleed air from lines and hoses of applicable lines, such as in cooling system.

e. After cooling down the engine, again check oil/fluid levels including engine oil and engine coolant. Refill to the specified level, if necessary.

Without Oil Pan Removal

See Figures 130 through 144, 146, 147 through 157, 158 and 159.

1. Before servicing the vehicle, refer to the Precautions Section.

This section describes the removal/installation procedure of the front timing chain case and timing chain related parts without removing the upper oil pan.

2. Position the vehicle onto a lift or support it in a safe manner so that work can be performed on the underside of the vehicle.

3. Disconnect the negative battery terminal.

4. Remove the engine cover.

5. Remove the air cleaner case assembly.

6. Remove the front and rear engine undercover.

7. Drain the engine coolant from the radiator.

8. Drain the engine oil from oil pan.

9. Remove and label the engine wiring harnesses.

10. Remove the upper and lower intake manifold collectors.

11. Remove the power steering oil pump from the bracket with the piping connected, then temporarily secure it aside.

12. Remove the power steering oil pump bracket.

13. Remove the alternator.

14. Remove the water bypass hose, water hose clamp, idler pulley bracket, and accessory drive belt tensioner from the front timing chain case.

15. Remove the right and left intake valve timing control covers, by loosening the bolts in the reverse order as shown.

16. Use seal cutter SST: KV10111100 (J37228), or equivalent, to cut the liquid gasket for removal.

➡**The shaft is internally joined to the intake camshaft sprocket center hole. During removal, keep the shaft**

horizontal until it is completely disconnected.

17. Remove the collared O-ring from the front timing chain case left and right sides.

18. Remove the right and left rocker covers.

➡**When the secondary timing chain is not removed/installed, the following step and associated sub-steps are not required.**

19. Position the engine with cylinder number 1 on Top Dead Center (TDC) of its compression stroke, as follows:

a. Rotate the crankshaft pulley clockwise to align the timing mark (grooved line without color) with the timing indicator.

b. Make sure the intake and exhaust cam lobes on the number 1 cylinder (engine front side of right bank) are located so that they point inward and upward compared to the cylinder head, see illustration.

c. If the cam lobes are not positioned pointing inward and upward as illustrated, rotate the crankshaft 1 full revolution (360°) until they are.

➡**When only the primary timing chain is removed, the rocker cover does not need to be removed. To confirm that the number 1 cylinder is at its compression TDC, remove the front timing chain case first. Then, check the mating marks on the camshaft sprockets.**

20. Remove the crankshaft pulley, as follows:

a. Remove the rear cover plate (2WD models) or the starter motor (AWD models) and install the ring gear stopper or equivalent.

b. Loosen the crankshaft pulley bolt until the bolt seating surface is 0.39 inch (10mm) from its original position.

➡**Do not completely remove the crankshaft damper pulley bolt, since it will**

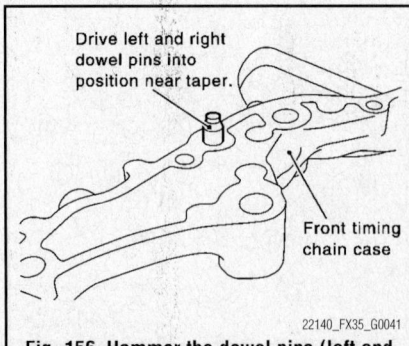

Fig. 156 Hammer the dowel pins (left and right) into the front timing chain case up to a point close to the taper—3.5L engine

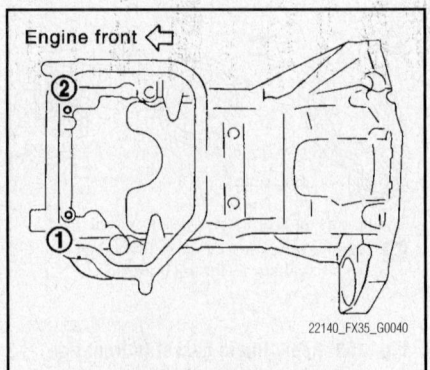

Fig. 157 Tightening sequence for the upper oil pan—3.5L engine

be used as a supporting point for a suitable puller.

    c. Place a suitable puller tab on the holes of the crankshaft damper pulley and pull the crankshaft damper pulley until it releases.

✹✹ WARNING

Do not position the suitable puller tab on the outer edges of the crankshaft pulley since this can damage the internal damper.

21. Remove the lower oil pan.
22. Loosen the 2 mounting bolts in the front of the upper oil pan in the reverse order shown.
23. Remove the front timing chain case, as follows:
    a. Loosen the mounting bolts in the reverse order of the tightening sequence.
    b. Insert a suitable tool into the notch at the top of the front timing chain case, as shown.

✹✹ WARNING

Do not use a screwdriver or similar tool, since it may damage the mating surfaces. After removal, handle it carefully so that it does not tilt or warp under a load.

    c. Pry off the case.
24. Remove the O-rings from the rear timing chain case.
25. Remove the water pump cover and the chain tensioner cover from the front timing chain case.
26. Use a seal cutter to cut the liquid gasket.
27. Remove front oil seal from front timing chain case using a suitable pry tool.

✹✹ WARNING

Be careful not to damage the front timing chain case.

28. Remove the primary timing chain tensioner, as follows:
    a. Pull the lever down and release the plunger stopper tab. The plunger stopper tab can be pushed up to release.
    b. Insert a stopper pin into the tensioner body hole to hold the lever and keep the tab released.

➡ **A 0.098 inch (2.5mm) Allen wrench can be used for a stopper pin.**

    c. Insert the plunger into the tensioner body by pressing on the slack guide.
    d. Keep the slack guide depressed

and hold it by pushing the stopper pin through the lever hole and body hole.
    e. Remove the mounting bolts and remove the primary timing chain tensioner.
29. Remove the internal chain guide, tension guide, and slack guide.

➡ **The tension guide can be removed after removing the primary timing chain.**

30. Remove the primary timing chain, tension guide, and crankshaft sprocket.

✹✹ WARNING

After removing the timing chain, do not turn the crankshaft and camshaft separately, or the valves may strike piston heads and cause damage.

31. Remove the secondary timing chain and camshaft sprockets, as follows:
    a. Attach a suitable stopper pin to the right and left secondary timing chain camshaft chain tensioners.
    b. Remove the intake and exhaust camshaft sprocket bolts. Apply paint to the timing chain and camshaft sprockets for alignment during installation. Secure the hexagonal portion of the camshaft using an open-end wrench to hold the camshaft steady while loosening the mounting bolts.
    c. Remove the secondary timing chain together with camshaft sprockets. Turn the camshaft slightly to create slack in the timing chain on the secondary timing chain tensioner side. Insert a 0.020 inch (0.5mm) thick metal or resin plate between the timing chain and timing chain tensioner plunger guide. Remove the secondary timing chain together with the camshaft sprockets with the timing chain loose from the guide groove.

➡ **Be careful of the plunger coming-off when removing the secondary timing chain. This is because the plunger of the secondary timing chain tensioner moves during operation, which can result in the fixed stopper pin coming off.**

➡ **The intake camshaft sprocket is a two-for-one assembly of the primary and secondary sprockets.**

✹✹ WARNING

When handling the intake camshaft sprocket, be careful to handle it carefully to avoid any shock to the camshaft sprocket. Do not disassemble. Do not loosen bolt A as shown in the figure.

32. Remove the rear timing chain case, as follows:
    a. Loosen and remove the mounting bolts in the reverse order of the tightening sequence.
    b. Cut the sealant using a seal cutter and remove the rear timing chain case.

✹✹ WARNING

Do not remove plate metal cover of engine oil passage. After removing the chain case, do not apply any load on the case which could cause warping.

33. Remove the O-rings from the cylinder head.
34. Remove the O-rings from the cylinder block.
35. If necessary, remove the secondary timing chain tensioners from the cylinder head, as follows:
    a. Remove the number 1 camshaft brackets.
    b. Remove the secondary timing chain tensioners with the stopper pins attached.
36. Use a scraper to remove all traces of liquid gasket from the front and rear timing chain cases and opposite mating surfaces.

✹✹ WARNING

Be careful not to allow gasket fragments to enter the oil pan.

37. Remove the old liquid gasket from the bolt holes and threads.
38. Use a scraper to remove all traces of liquid gasket from the water pump cover, chain tensioner cover and intake valve timing control covers.
39. Inspect the timing chain for cracks and any excessive wear at link plates and roller links of the timing chain. Replace the timing chain, as necessary.

To install:

➡ **Throughout the installation procedure, whenever liquid gasket is to be used make sure to use Genuine RTV Silicone Sealant or equivalent.**

➡ **The accompanying illustration shows the relationship between the mating mark on each timing chain and that on the corresponding sprocket, with the components installed.**

40. If removed, install the secondary timing chain tensioners on the cylinder head, as follows:
    a. Install the chain tensioners with stopper pins attached and new O-rings.

b. Install the number 1 camshaft brackets. Refer to Camshaft and Valve Lifters, removal & installation.

41. Install new O-rings onto the cylinder block.

42. Install new O-rings on the cylinder head.

43. Apply liquid gasket to the rear timing chain case backside, as shown.

➡ **Use Genuine RTV Silicone Sealant or equivalent.**

44. For **A** in the illustration, completely wipe out the liquid gasket extended on a portion touching at engine coolant.

45. Apply liquid gasket on the installation position of the water pump and cylinder head completely.

46. Align the rear timing chain case and water pump assembly with right and left dowel pins on cylinder block and install the case.

➡ **Make sure the O-rings stay in place during installation on the cylinder block and cylinder head.**

47. Tighten the mounting bolts in the sequence shown. After all bolts are temporarily tightened, retighten them to 108 inch lbs. (13 Nm) in the tightening sequence. There are 2 bolt lengths used for the timing chain case:

a. 20mm length: Bolt positions 1, 2, 3, 6, 7, 8, 9, and 10.

b. 16mm length: All except bolt positions 1, 2, 3, 6, 7, 8, 9, and 10.

➡ **If RTV Silicone Sealant protrudes beyond the sealing surfaces, wipe it off immediately.**

48. After installing the rear timing chain case, check the surface height difference between the rear timing chain case to the cylinder block at the oil pan mounting surface. If the difference is not within − 0.009–0.006 inch (− 0.24–0.14mm), repeat the installation procedure.

49. Position the crankshaft so the number 1 piston is set at TDC on the compression stroke. Make sure that the dowel pin hole, dowel pin, and crankshaft key are located as shown.

➡ **Though the camshaft does not stop at the position as shown in the figure, for the placement of the cam nose, it is generally accepted the camshaft is placed in the same direction of the figure and as follows:**

- Camshaft dowel pin hole (intake side): At the cylinder head upper face side in each bank
- Camshaft dowel pin (exhaust side):

At the cylinder head upper face side in each bank

- Crankshaft key: At the cylinder head side of the right bank

> ❄ **WARNING**
>
> **The hole on the small diameter side must be used for the intake side dowel pin hole.**

50. Install the secondary timing chains and camshaft sprockets, as follows:

> ❄ **WARNING**
>
> **The matching marks between the timing chain and sprockets slip easily. Confirm all matching mark positions repeatedly during the installation process.**

a. Push the plunger of the secondary chain tensioner and keep it pressed in with a stopper pin.

b. Install the secondary timing chains and camshaft sprockets. Align the mating marks on the secondary timing chain (gold link) with the ones on the intake and exhaust camshaft sprockets (stamped), and install them.

51. Ensure that the sprockets are properly positioned by heeding the following items:

- The mating marks for the intake camshaft sprocket are on the back side of the secondary camshaft sprocket.
- There are 2 types of mating marks, circle and oval types. They should be used for the right and left banks, respectively. For the right bank, use the circle type of mating mark, and for the left bank use the oval type of mating mark.
- Align the dowel pin and pin hole on the camshaft with the groove and dowel pin on the sprocket, and install them.
- On the intake side, align the pin hole on the small diameter side of the camshaft front end with the dowel pin on the back side of camshaft sprocket, and install them.
- On the exhaust side, align the dowel pin on the camshaft front end with the pin groove on the camshaft sprocket, and install them.
- In the case that positions of each mating mark and each dowel pin are not fit to the mating parts, make fine adjustments to the position holding the hexagonal portion of the camshaft with a wrench or equivalent.

- Mounting bolts for the camshaft sprockets must be tightened in the next step. Tightening them by hand is enough to prevent the dislocation of dowel pins.
- It may be difficult to visually check the dislocation of the mating marks during and after installation. To make the matching easier, make a mating mark on the top of the sprocket teeth and its extended line in advance with paint.

52. After confirming the mating marks are aligned, tighten the camshaft sprocket mounting bolts, while securing the camshaft using a wrench on a hexagonal portion of the camshaft.

53. Pull the stopper pins out from the secondary timing chain tensioners.

54. Install the primary timing chain, as follows:

➡ **During alignment, be careful to prevent dislocation of the mating mark alignments of the secondary timing chains.**

a. Install the crankshaft sprocket, making sure the mating marks on the crankshaft sprocket face the front of the engine.

b. Install the primary timing chain, so that the mating mark (punched) on the camshaft sprocket is aligned with the pink link on the timing chain, while the mating mark (notched) on the crankshaft sprocket is aligned with the orange one on the timing chain, as shown.

➡ **If it is difficult to align mating marks of the primary timing chain with each sprocket, gradually turn the camshaft using a wrench on the hexagonal portion of the camshaft to align it with the mating marks.**

55. Install the internal chain guide and primary timing chain tensioner.

56. Install the slack guide.

➡ **Do not over tighten the slack guide mounting bolts. It is normal for a gap to exist under the bolt seats when the mounting bolts are tightened to specification.**

57. Remove all dirt and foreign materials completely from the back and mounting surfaces of the chain tensioner.

58. Install chain tensioner for slack guide. When installing the chain tensioner, push in the sleeve and keep it depressed with a stopper pin.

59. After chain tensioner installation, pull out the stopper pin by pressing in the slack guide.

60. Reconfirm that the mating marks on sprockets and timing chains have not slipped out of alignment.

61. Install new O-rings on the rear timing chain case.

62. Hammer the dowel pins (right and left) into the front timing chain case up to a point close to the taper in order to shorten the protrusion length.

63. Install the front oil seal on the front timing chain case, as follows:

    a. Apply new engine oil to the oil seal edges.

    b. Make sure the garter spring is in position and the seal lip is not inverted and so that each seal lip is oriented as shown in the figure.

    c. Using a suitable drift, press-fit the oil seal until it is flush with the front timing chain case end face.

64. Install the water pump cover and chain tensioner cover to the front timing chain case. Apply liquid gasket to the front timing chain case front side as shown.

➡**Use Genuine RTV Silicone Sealant or equivalent.**

65. Install the front timing chain case, as follows:

    a. Apply liquid gasket to the front timing chain case back side as shown.

    b. Apply liquid gasket to the oil pan gasket as shown.

    c. Install the new oil pan gasket, and heed the following:

- Align the notch of the front timing chain case with the protrusion of the oil pan gasket
- Apply liquid gasket to the top surface of the upper oil pan as shown

➡**Be careful that oil pan gasket is in place.**

    d. Assemble the front timing chain case by fitting the lower end of the front timing chain case tightly onto the top face of the upper oil pan. From the fitting point, make the entire front timing chain case contact the rear timing chain case completely. Then, while pressing the front timing chain case from its front and top as shown in figure, install the bolts and temporarily tighten them by hand. Hammer the dowel pin until the outer end becomes flush with the surface.

    e. After hand tightening the mounting bolts, retighten them to specified torque in numerical order shown. Refer to the following for locating the bolts.

- 8mm bolts (positions 1 and 2): 21 ft. lbs. (28 Nm)

- 6mm bolts (except positions 1 and 2): 108 inch lbs. (13 Nm)

    f. After tightening, retighten them again to the specified torque in the numerical order shown.

66. Install 2 mounting bolts in the front of the upper oil pan in numerical order shown to 13 ft. lbs. (17 Nm).

67. Install the lower oil pan.

68. Install the right and left intake valve timing control covers, as follows:

    a. Install seal rings in the shaft grooves.

    b. Apply a continuous bead of liquid gasket to the intake valve timing control covers.

➡**Use Genuine RTV Silicone Sealant or equivalent.**

    c. Install the collared O-ring in the front timing chain case oil hole (left and right sides).

    d. Being careful not to move the seal ring from the installation groove, align the dowel pins on the chain case with the holes in the intake valve timing control covers.

    e. Tighten the bolts in the numerical order shown to 96 inch lbs. (11 Nm).

69. Install the crankshaft damper pulley, as follows:

    a. Fix the crankshaft in position using ring gear stopper SST: KV10117700 (J44716) or equivalent.

    b. Install the crankshaft pulley, taking care not to damage the front oil seal. When press-fitting crankshaft pulley with a plastic hammer, tap on its center portion (not the circumference).

    c. Tighten the crankshaft pulley bolt to 33 ft. lbs. (44 Nm).

    d. Put a paint mark on the crankshaft pulley aligned with the angle mark on the crankshaft pulley bolt. Then, further tighten the bolt by 90°.

70. Rotate the crankshaft pulley in the normal direction (clockwise when viewed from the front of the engine) to confirm it turns smoothly.

71. The remainder of installation is the reverse of removal.

➡**If hydraulic pressure inside the chain tensioner drops after removal/installation, slack in the guide may generate a pounding noise during and just after engine start. However, this is normal and the noise will stop after hydraulic pressure rises.**

72. Perform the following once installation is complete:

    a. Before starting the engine, check the oil/fluid levels including the engine coolant and engine oil. If less than required quantity, fill to the specified level.

    b. Run the engine to check for unusual noise and vibration.

    c. Warm up engine thoroughly to make sure there is no leakage of fuel, or any oil/fluids including engine oil and engine coolant.

    d. Bleed air from lines and hoses of applicable lines, such as in cooling system.

    e. After cooling down the engine, again check oil/fluid levels including engine oil and engine coolant. Refill to the specified level, if necessary.

4.5L Engine

See Figures 160 through 179.

1. Before servicing the vehicle, refer to the Precautions Section.

2. Remove the engine assembly from the vehicle.

3. Remove the front engine undercover.

4. Remove the accessory drive belts.

5. Remove the drive belt auto tensioner and idler pulley.

6. Remove the thermostat housing and hoses.

7. Remove the rocker cover, as follows:

    a. Remove the engine cover.

    b. Remove the inlet air duct, air cleaner case and mass air flow sensor assembly, air duct and resonator assembly.

    c. Move the wiring harness on the upper rocker cover and its peripheral aside.

    d. Remove the wiring harness brackets from the No. 6 camshaft bracket.

    e. Remove the electric throttle control actuator.

    f. Remove the ignition coils.

    g. Remove the PCV hose from the PCV valve.

    h. Move the wiring harness on the upper rocker cover.

    i. Remove the ignition coil.

    j. Remove the PCV hose from the PCV valve.

    k. Remove the grommets from the right and left cowl top panel. For the right side grommet, remove the battery, battery tray, then the grommet.

    l. Loosen the rocker cover bolts in reverse order as shown.

❈❈ **WARNING**

Do not hold onto the oil filler neck on the right bank or it may be damaged.

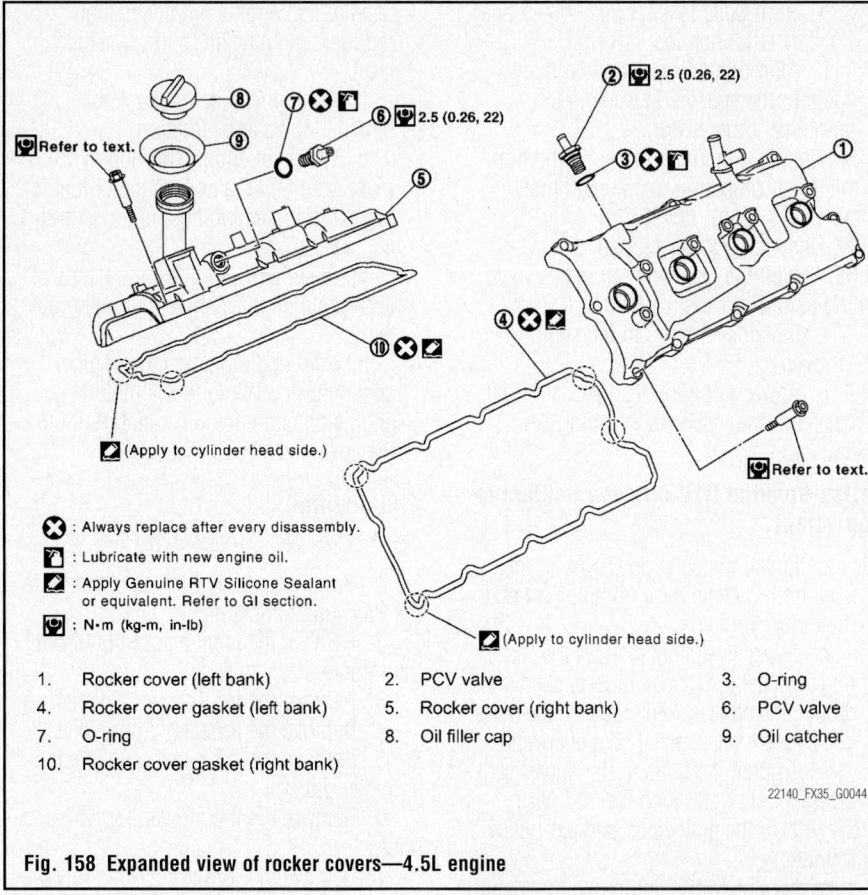

Fig. 158 Expanded view of rocker covers—4.5L engine

❌ : Always replace after every disassembly.

🛢 : Lubricate with new engine oil.

🔧 : Apply Genuine RTV Silicone Sealant or equivalent. Refer to GI section.

⚙ : N•m (kg-m, in-lb)

1. Rocker cover (left bank)
2. PCV valve
3. O-ring
4. Rocker cover gasket (left bank)
5. Rocker cover (right bank)
6. PCV valve
7. O-ring
8. Oil filler cap
9. Oil catcher
10. Rocker cover gasket (right bank)

22140_FX35_G0044

8. Loosen number 10 bolt of the right bank and number 10 and 12 bolts at the left bank from the cowl top panel hole.

   a. Use a scraper to remove all traces of the liquid gasket from the cylinder head and camshaft bracket.

9. If necessary, remove the intake valve timing control position sensor (right and left banks) and the camshaft position sensor (PHASE) from the intake valve timing control cover and front cover.

✳ WARNING

Handle the sensors carefully to avoid dropping and/or shocking them. Also, do not disassemble them. Do not allow metal powder to adhere to the magnetic part at the sensor tip. Do not place sensors in a location where they are exposed to magnetism.

10. If necessary, remove the intake valve timing control solenoid valve from the intake valve timing control cover.

11. Remove the intake valve timing control cover, as follows:

   a. Loosen and remove the mounting bolts in the reverse order shown.

   b. Use a seal cutter to cut the liquid gasket for removal.

✳ WARNING

Exercise care not to damage the mating surfaces. Pull out the cover keeping it level, since an inner part of the cover is engaged with the center of the intake camshaft sprocket.

12. Remove the O-rings from the front cover.

13. Position the crankshaft with number 1 cylinder at Top Dead Center (TDC) of its compression stroke as follows:

   a. Rotate the crankshaft pulley clockwise to align the TDC identification notch (without paint mark) with the timing indicator on the front cover.

   b. Make sure that both intake and exhaust camshaft lobes of the number 1 cylinder (engine front side of left bank) are located as shown in the figure.

   c. If the camshaft lobes are not positioned appropriately, turn the crankshaft pulley one revolution (360°) and align it as shown.

14. Remove the crankshaft pulley as follows:

   a. Remove the rear plate cover and install the ring gear stopper or equivalent.

   b. Loosen the crankshaft damper pulley bolt, and then pull the crankshaft pulley with both hands to remove it.

✳ WARNING

Do not completely remove the crankshaft pulley bolt. Keep the loosened crankshaft pulley bolt in place to

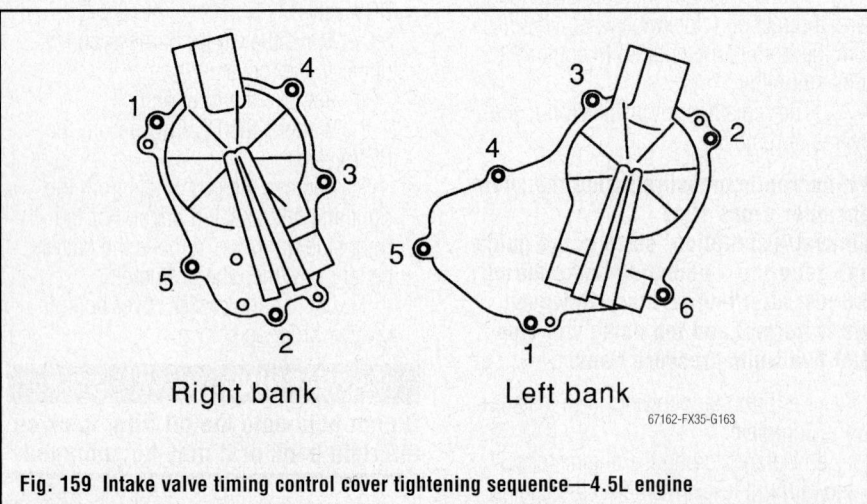

Right bank   Left bank

67162-FX35-G163

Fig. 159 Intake valve timing control cover tightening sequence—4.5L engine

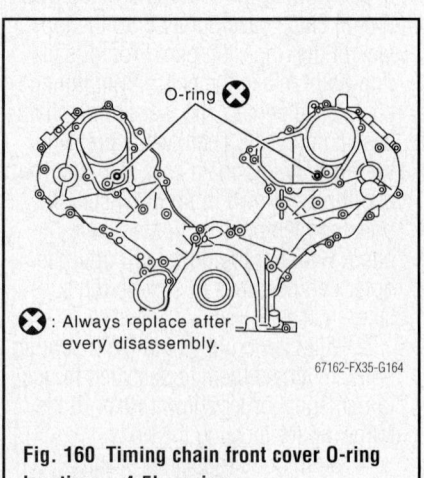

O-ring ❌

❌ : Always replace after every disassembly.

67162-FX35-G164

Fig. 160 Timing chain front cover O-ring locations—4.5L engine

protect the removed crankshaft pulley from dropping. Do not remove the balance weight (inner hexagonal bolt) at the front of the crankshaft pulley.

15. Remove the oil pan and oil strainer.
16. Remove the front cover, as follows:
   a. Loosen the mounting bolts in the reverse order shown.
   b. Use a seal cutter to cut the liquid gasket.

✳✳ WARNING

Exercise care not to damage mating surfaces. After removal, handle the front cover carefully so it does not warp under load.

17. Remove the front oil seal from the front cover using a suitable thin-bladed pry tool.

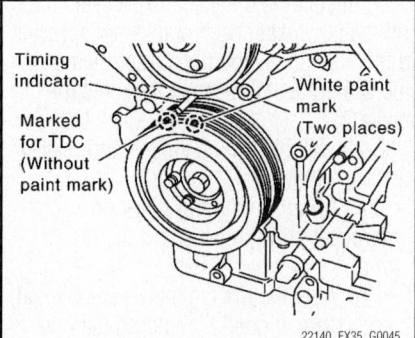

Timing indicator
Marked for TDC (Without paint mark)
White paint mark (Two places)

22140_FX35_G0045

Fig. 161 Rotate the crankshaft pulley clockwise to align the TDC identification notch (without paint mark) with the timing indicator on the front cover—4.5L engine

✳✳ WARNING

Be careful not to damage the front cover.

18. Remove the O-rings from both cylinder heads and cylinder block.
19. Remove the chain tensioner cover from the front cover. Use a seal cutter to cut the liquid gasket for removal.
20. Remove the oil pump drive spacer, by setting bolts in the 2 bolt holes (M6 pitch) on the front surface. Using a suitable puller, pull the oil pump drive spacer off the crankshaft.

➡ **The dimension between the centers of the 2 bolt holes is 1.30 inch (33mm).**

21. Remove the oil pump.
22. Remove the chain tensioner from the left bank, as follows:

➡ **To remove timing chain and related parts, start with those on the left bank. The procedure for removing parts on the right bank is omitted because it is the same as that for the left bank.**

   a. While lightly pressing the tensioner plunger, depress the tensioner tab in (or turn the lever in the direction of the arrow in the accompanying illustration) to unlock the mechanism that stops the tensioner plunger.
   b. Push in the tensioner plunger to align the hole on the lever and that on the pump main body. If you push in the tensioner too far, the holes will not align. Therefore, push in the plunger to the degree at which the start of stopper groove and tab engages.

   c. Insert a stopper pin (hard wire approximately 0.020 inch/0.5mm thick or similar tool) to hold the plunger in position. With the plunger held, remove the chain tensioner.
23. Remove the chain tension guide and timing chain slack guide.
24. Remove the timing chain and crankshaft sprocket.

✳✳ WARNING

After removing the timing chain, do not turn the crankshaft and camshaft separately, or the valves may strike the piston head and cause damage.

25. With the hexagonal part of the camshaft held with a wrench, loosen the bolts securing the camshaft sprocket.
26. Perform the same procedure on the right side.
27. Use a scraper to remove all traces of old liquid gasket from the front cover and opposite mating surfaces. Remove the oil liquid gasket from the bolt holes and threads.
28. Use scraper to remove all trace of liquid gasket from the chain tensioner cover and the intake valve timing control covers.
29. Check the timing chain for cracks and any excessive wear at the roller links and link plates. Replace the timing chain as necessary.

To install:

➡ **The accompanying illustration shows the relationship between the mating mark on each timing chain and that on the corresponding sprocket, with the components installed. Parts with an identification mark (R or L) should be installed on the corresponding bank according to the mark. Parts with an identification mark include:**

- Camshaft sprocket (INT)
- Dowel pin groove of camshaft sprocket (EXH) (camshaft sprocket is the same part on both banks)
- Chain tension guide
- Chain slack guide

➡ **To install the timing chain and related parts, start with those on right bank. The procedure for installing parts on left bank is omitted because it is the same as that for the right bank.**

30. Make sure that the crankshaft key and dowel pin of each camshaft are located as shown in the figure with the crankshaft at number 1 cylinder at compression TDC. The positioning is as follows:
- Camshaft dowel pin—At the cylinder head upper face side in each bank

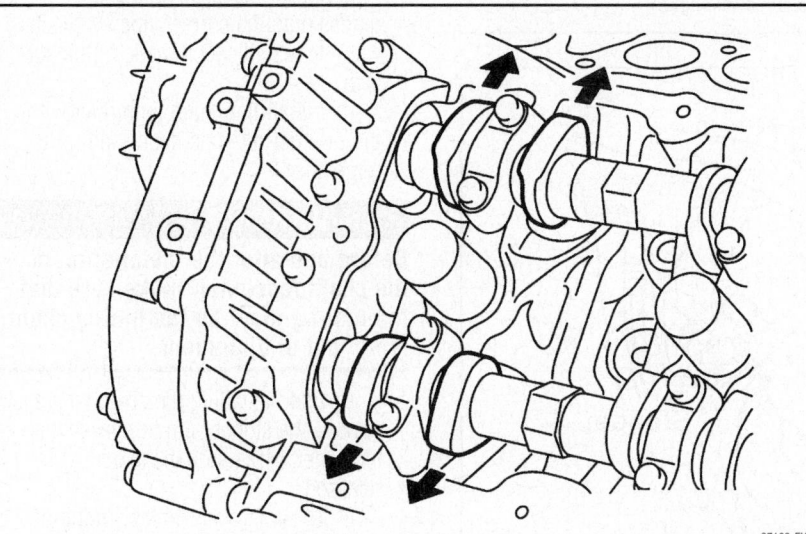

67162-FX35-G166

Fig. 162 With the engine positioned at TDC on number 1 cylinder, the camshaft lobes should be positioned as shown—4.5L engine

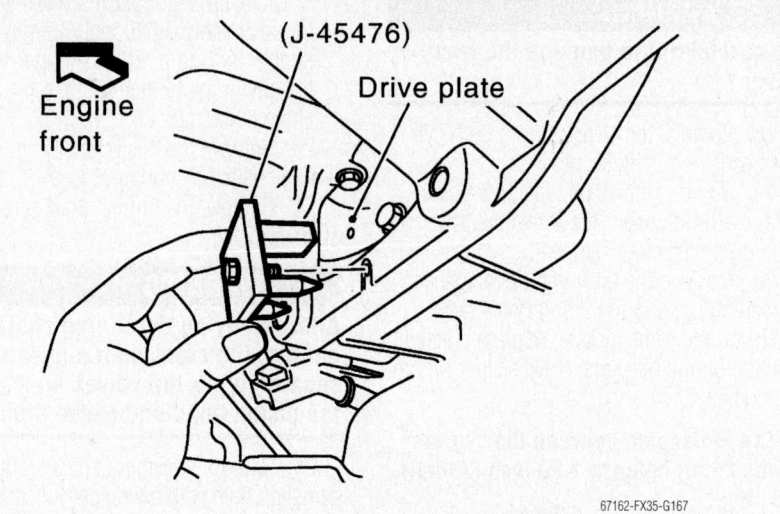

Fig. 163 Use a ring gear holder to fix the crankshaft to loosen the crankshaft pulley bolt—4.5L engine

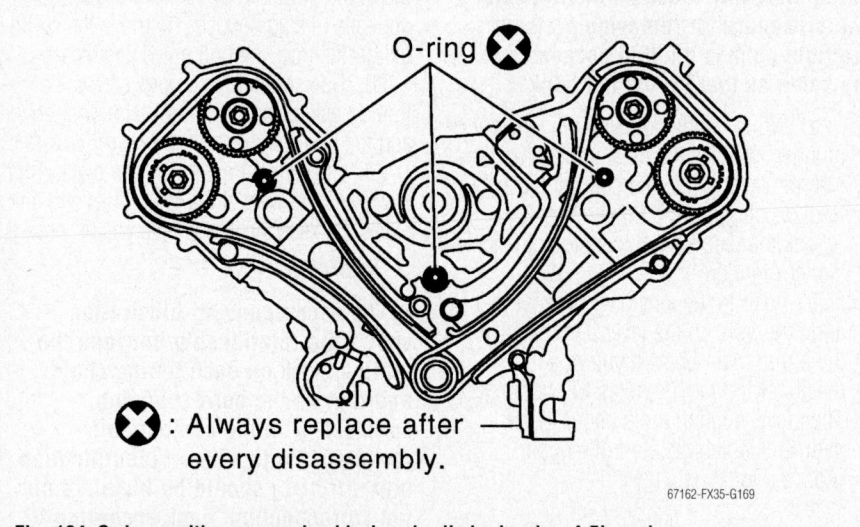

Fig. 164 O-ring positions on engine block and cylinder heads—4.5L engine

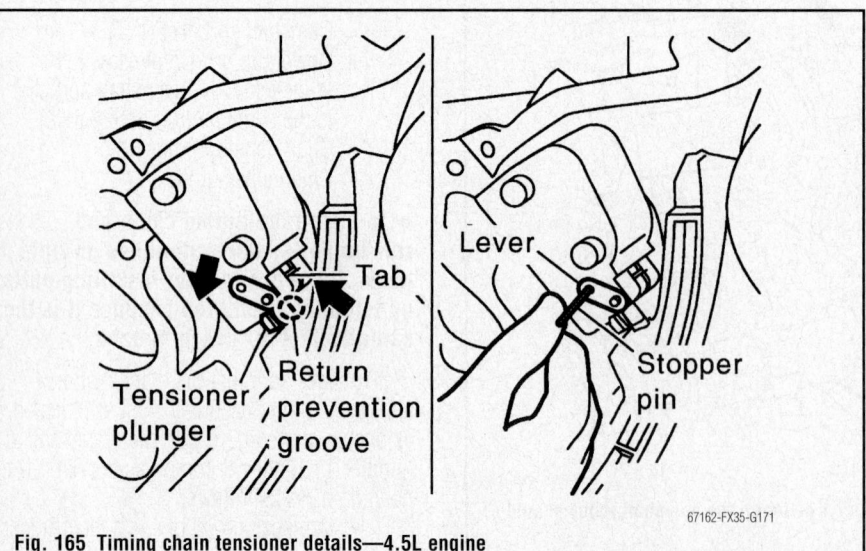

Fig. 165 Timing chain tensioner details—4.5L engine

- Crankshaft key—At the cylinder head side of left bank

➡Though the camshaft may not stop at the position as shown in the figure, for the placement of camshaft lobe nose, it is generally accepted that the camshaft is placed in the same direction of the figure.

31. Install the camshaft sprockets, while heeding the following items:

  a. Install the sprockets onto the correct side by checking with the identification mark on the surface.

  b. Install the camshaft sprocket (EXH) by selectively using the groove of the dowel pin according to the bank. This is a common part used for both banks.

  c. Lock the hexagonal part of the camshaft in the same procedure as removal, and tighten the mounting bolts.

32. Install the crankshaft sprockets for both banks. Install each crankshaft sprocket so that its flange side (the larger diameter side without teeth) faces in the direction shown.

33. Install the timing chains and related parts, as follows:

  a. Align the mating mark on each sprocket and timing chain for installation.

  b. After the mating marks are aligned, keep them aligned by holding them in position by hand.

➡Before installing the chain tensioner, it is possible to change the position of the mating mark on the timing chain for that on each sprocket for alignment.

  c. Install the slack guides and tension guides onto the correct side by checking with the identification mark on the surface.

  d. Install the chain tensioner with the plunger fixed as described in its removal.

✳✳ WARNING

Before and after the installation of the chain tensioner, make sure that the mating mark on the timing chain is not out of alignment.

  e. After installing the chain tensioner, remove the stopper pin to release the tensioner. Make sure the tensioner is released.

  f. To avoid chain-link skipping of timing chain teeth, do not move the crankshaft or camshafts until the front cover is installed.

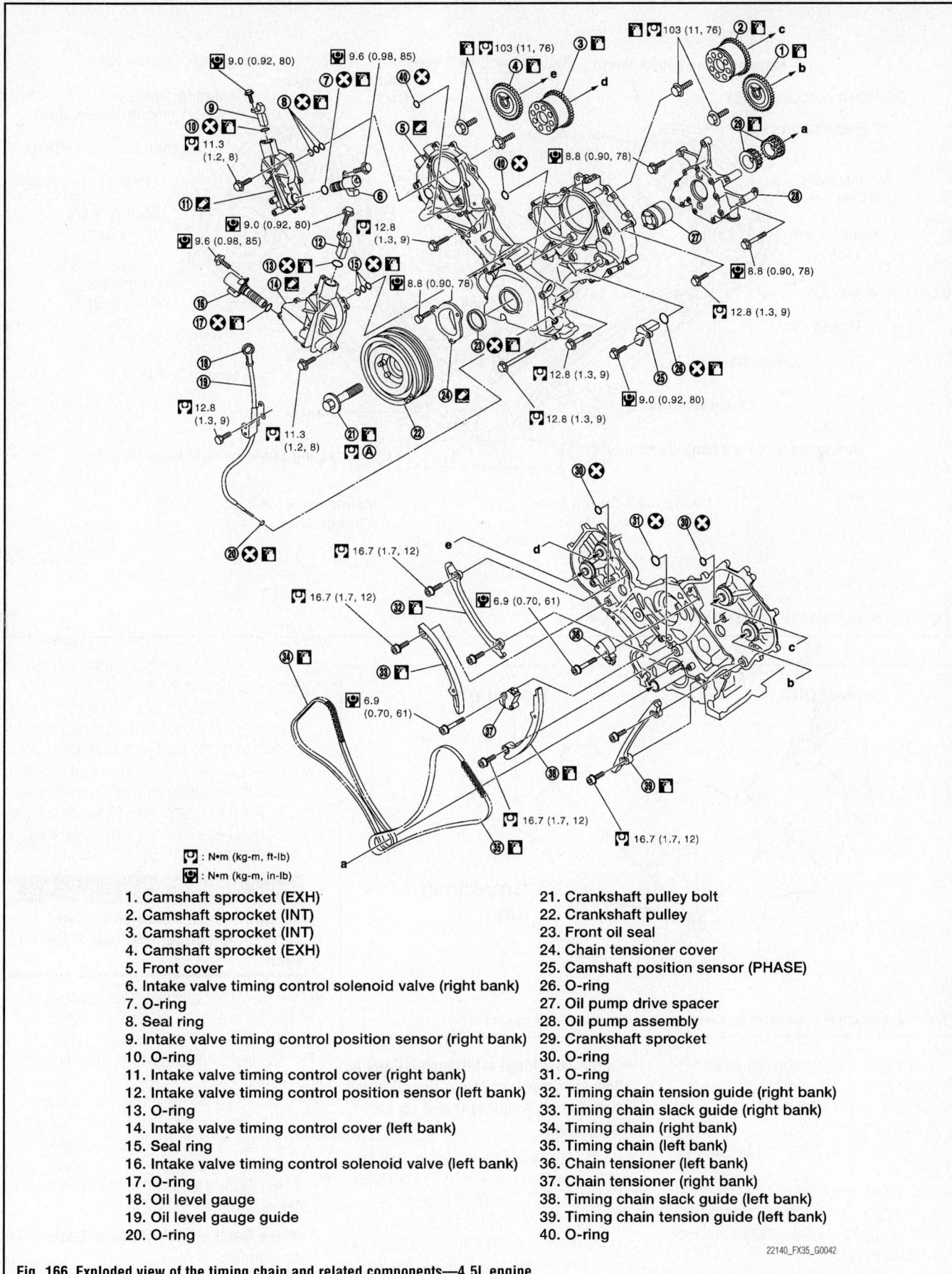

: N•m (kg-m, ft-lb)
: N•m (kg-m, in-lb)

1. Camshaft sprocket (EXH)
2. Camshaft sprocket (INT)
3. Camshaft sprocket (INT)
4. Camshaft sprocket (EXH)
5. Front cover
6. Intake valve timing control solenoid valve (right bank)
7. O-ring
8. Seal ring
9. Intake valve timing control position sensor (right bank)
10. O-ring
11. Intake valve timing control cover (right bank)
12. Intake valve timing control position sensor (left bank)
13. O-ring
14. Intake valve timing control cover (left bank)
15. Seal ring
16. Intake valve timing control solenoid valve (left bank)
17. O-ring
18. Oil level gauge
19. Oil level gauge guide
20. O-ring

21. Crankshaft pulley bolt
22. Crankshaft pulley
23. Front oil seal
24. Chain tensioner cover
25. Camshaft position sensor (PHASE)
26. O-ring
27. Oil pump drive spacer
28. Oil pump assembly
29. Crankshaft sprocket
30. O-ring
31. O-ring
32. Timing chain tension guide (right bank)
33. Timing chain slack guide (right bank)
34. Timing chain (right bank)
35. Timing chain (left bank)
36. Chain tensioner (left bank)
37. Chain tensioner (right bank)
38. Timing chain slack guide (left bank)
39. Timing chain tension guide (left bank)
40. O-ring

22140_FX35_G0042

Fig. 166 Exploded view of the timing chain and related components—4.5L engine

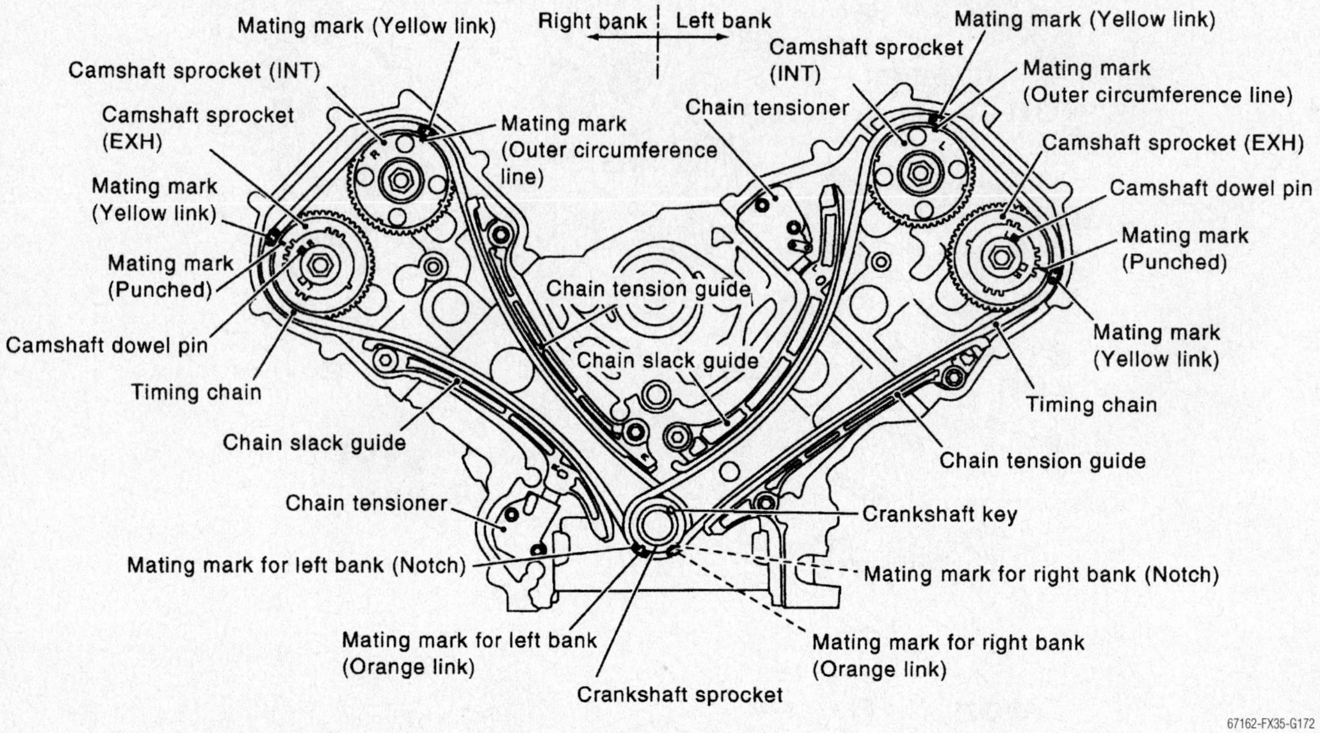

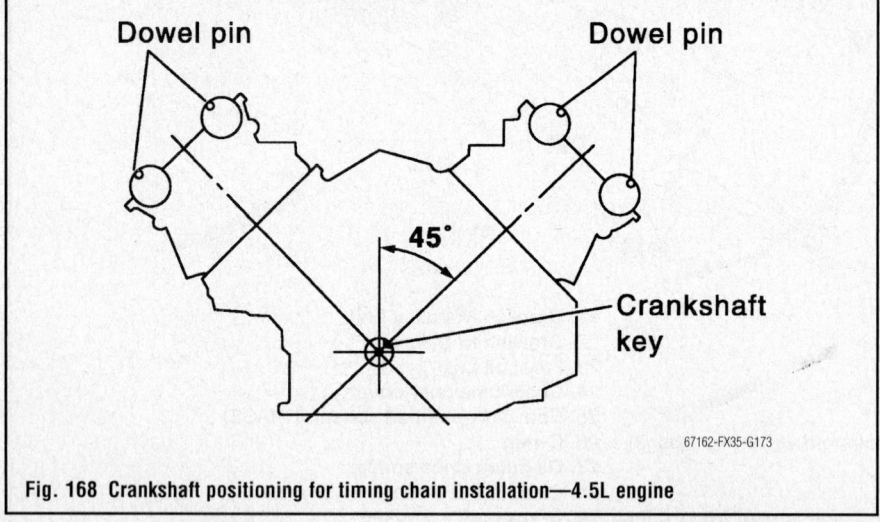

Fig. 167 Timing chain positioning for installation—4.5L engine

Fig. 168 Crankshaft positioning for timing chain installation—4.5L engine

34. Perform the same procedure on the left bank, installing the timing chain and related parts on the left side.

35. Install the oil pump.

36. Install the oil pump drive spacer, as follows:

a. Insert the oil pump drive spacer according to the directions of the crankshaft key and the 2 flat surfaces of the oil pump inner rotor.

➡**If the positional relationship does not allow for insertion, rotate the oil pump inner rotor to properly line it up for installation.**

b. After confirming that the position of each part is in correct condition to allow for the spacer, force fit the spacer by lightly tapping it with a plastic hammer until it makes contact and goes no further.

37. Install the front oil seal in the front cover, as follows:

a. Apply new engine oil to both the oil seal lip and the dust seal lip.

b. Position the seal so that each seal is oriented as shown, then, using a suitable drift, press the front oil seal into the front cover until the face of the oil seal is flush with the mounting surface. The drift used to seat the oil seat should have an outside diameter of 2.20 inch (56mm) and an inner diameter of 1.93 inch (49mm).

✳✱ WARNING

Be careful not to scratch or make burrs on the circumference of the oil seal.

c. Make sure the garter spring is in position and the seal lips are not inverted.

38. Install the chain tensioner cover on the front cover.

39. Install the front cover, as follows:

a. Install new O-rings onto the right and left cylinder heads and cylinder block.

40. Apply a continuous bead of liquid gasket to the front cover as shown.

➡**Use Genuine RTV Silicone Sealant or equivalent.**

a. Make sure again that the mating marks on the timing chain and that on each sprocket are aligned. Then, install the front cover.

✳✳ WARNING

Be careful to avoid interference with the front end of the oil pump drive spacer. Such interference may damage the front oil seal.

b. Tighten the mounting bolts in numerical order shown. There are 4 types of mounting bolts. The bolts are as follows:
- Position A: M6 x 20mm
- Position B: M6 x 45mm
- Position C: M6 x 80mm
- Position D: M6 x 25mm

c. After all bolts are tightened, retighten them in numerical order shown.

✳✳ CAUTION

Be sure to wipe off any excessive liquid gasket leaking onto the surface mating with the oil pan.

41. Install the intake valve timing control cover, as follows:
a. At the back of the intake valve timing control cover, install new seal rings (3 for each bank) to the area to be inserted into the camshaft sprocket (INT).

➡**Do not spread the seal ring excessively to avoid breaks and deformation.**

b. Install the new O-rings on the front cover.

c. Apply a continuous bead of liquid gasket to the intake valve timing control covers as shown.

➡**Use Genuine RTV Silicone Sealant or equivalent.**

d. Tighten the mounting bolts in the numerical order shown.

42. Install the intake valve timing control

position sensor, intake valve timing control solenoid valve, and camshaft position sensor (PHASE) on the intake valve timing control cover and front cover, if removed. Be sure to tighten the bolts with flanges completely seated.

43. Install the oil pan and oil strainer.

44. Install the crankshaft pulley, as follows:
a. Hold the crankshaft with ring gear stopper SST: J-45476 or equivalent.

b. Install the crankshaft pulley, taking care not to damage the front oil seal. Install it according to the dowel pin of the oil pump drive spacer. Lightly tapping its center with plastic hammer, insert the pulley.

✳✳ WARNING

Do not tap the pulley on the side surface where the belt is installed (outer circumference) or damage to the inner damper may occur.

c. Apply engine oil onto the threaded parts of crankshaft pulley bolt and seating area.

d. Tighten the crankshaft pulley bolt to 69 ft. lbs. (93 Nm).

e. Mark the crankshaft pulley to align it with the angle mark on the crankshaft pulley bolt.

f. Further tighten the pulley bolt 90° (angle tightening). Check the tightening angle by referencing it to the notches. The angle between 2 notches is 90°.

45. Rotate the crankshaft pulley in the normal direction of rotation (clockwise when viewed from the front of the engine) to confirm it turns smoothly.

46. The remainder of installation is the reverse of removal.

➡**If hydraulic pressure inside the chain tensioner drops after removal/installation, slack in the guide**

may generate a pounding noise during and just after engine start. This is normal and the noise should stop after the hydraulic pressure rises.

47. For rocker arm cover installation, perform the following:
a. Apply liquid gasket to the joint part of the cylinder head and camshaft bracket as shown.

➡**The accompanying illustration shows an example of the left bank side.**

b. Refer to "a" in the illustration to apply liquid gasket to the joint part of the camshaft bracket (both numbers 1 and 6) and the cylinder head.

c. Refer to "b" in the illustration to apply liquid gasket in 90° to "a."

➡**Use Genuine RTV Silicone Sealant or equivalent.**

d. Install the rocker cover. Check that the rocker cover gasket does not fall from the installation groove of the rocker cover.

e. Tighten the bolts in 2 separate steps in the numerical order shown, as follows:
- 1st Step: 18 inch lbs. (2 Nm)
- 2nd step: 73 inch lbs. (8.3 Nm)

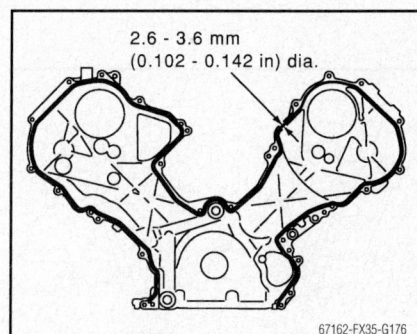

Fig. 171 Liquid gasket positioning for timing chain cover installation—4.5L engine

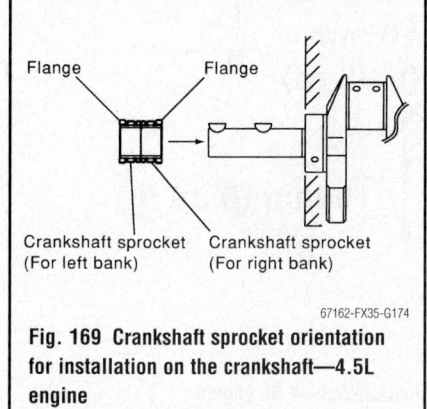

Fig. 169 Crankshaft sprocket orientation for installation on the crankshaft—4.5L engine

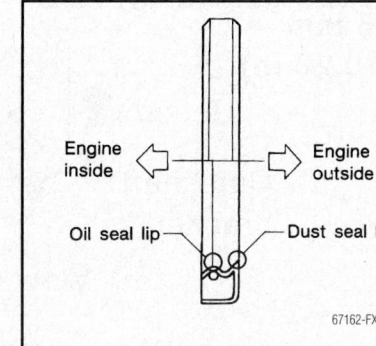

Fig. 170 Install the front timing cover front seal as shown—4.5L engine

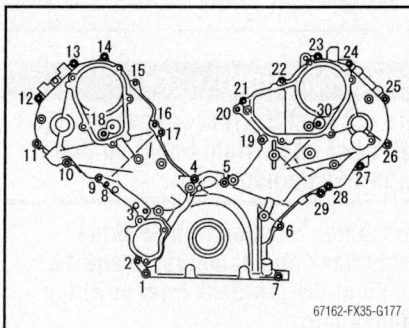

Fig. 172 Front timing chain cover mounting bolt tightening sequence—4.5L engine

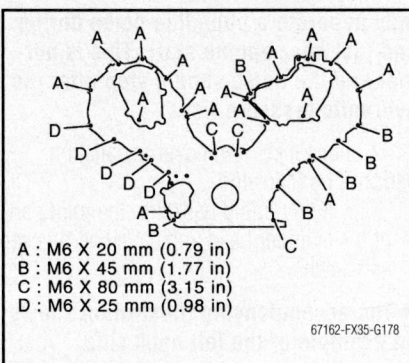

A : M6 X 20 mm (0.79 in)
B : M6 X 45 mm (1.77 in)
C : M6 X 80 mm (3.15 in)
D : M6 X 25 mm (0.98 in)

67162-FX35-G178

Fig. 173 Front timing chain cover mounting bolt identification—4.5L engine

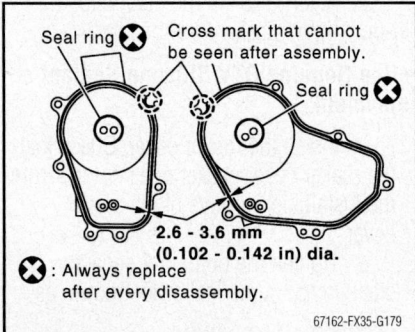

2.6 - 3.6 mm
(0.102 - 0.142 in) dia.

✕ : Always replace after every disassembly.

67162-FX35-G179

Fig. 174 Intake valve timing control cover liquid gasket positioning for installation—4.5L engine

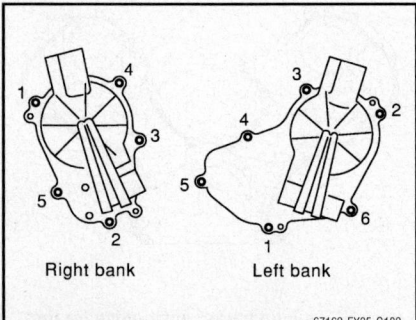

Right bank   Left bank

67162-FX35-G180

Fig. 175 Intake valve timing control cover mounting bolt tightening sequence—4.5L engine.

✳✳ CAUTION

Do not hold the rocker cover by the oil filler neck (right bank) so that it won't be damaged.

➡ Tighten number 10 bolt of the right bank and numbers 10 and 12 bolts of the left bank from cowl top panel hole.

48. Perform the following once installation is complete:
    a. Before starting the engine, check

the oil/fluid levels including the engine coolant and engine oil. If less than required quantity, fill to the specified level.

  b. Run the engine to check for unusual noise and vibration.

  c. Warm up engine thoroughly to make sure there is no leakage of fuel, or any oil/fluids including engine oil and engine coolant.

  d. Bleed air from lines and hoses of applicable lines, such as in cooling system.

  e. After cooling down the engine, again check oil/fluid levels including

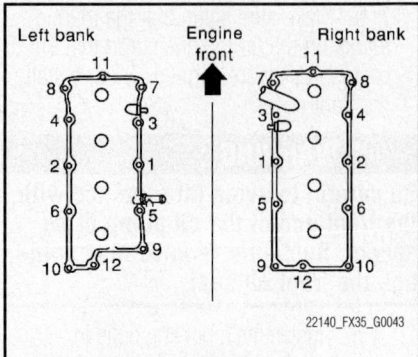

22140_FX35_G0043

Fig. 177 Tightening sequence for rocker cover bolts—4.5L engine

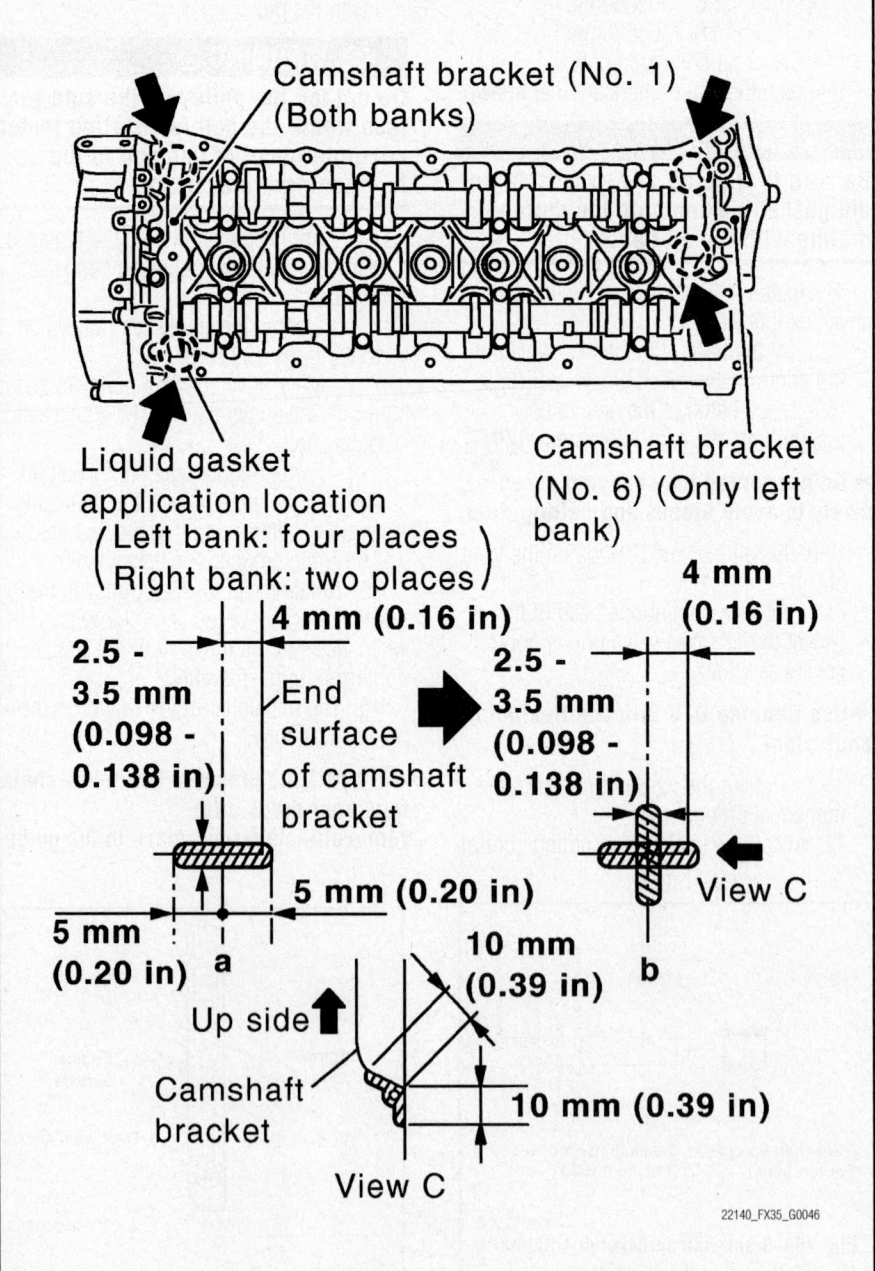

22140_FX35_G0046

Fig. 176 Liquid gasket application for rocker cover installation—4.5L engine

engine oil and engine coolant. Refill to the specified level, if necessary.

VALVE LASH

ADJUSTMENT

3.5L Engine

See Figures 178 through 182.

1. Before servicing the vehicle, refer to the Precautions Section.

Perform inspection after removal, installation, or replacement of camshaft or valve-related parts, or if there is unusual engine condition regarding valve clearance.

2. Remove the right and left rocker covers.

3. Set the number 1 cylinder at Top Dead Center (TDC) of its compression stroke, as follows:

   a. Rotate the crankshaft pulley clockwise until the timing mark (grooved line without color) is aligned with the timing indicator.

   b. Make sure the number 1 cylinder intake and exhaust cam noses are facing inward and upward from the cylinder head, as shown. If they are not positioned as shown, rotate the crankshaft pulley 360° clockwise until they are appropriately positioned.

4. Using a feeler gauge, measure the valve lash clearance as illustrated.

5. Measure the valve clearance for the following cylinders: Cylinder 1 Intake, Cylinder 2 Exhaust, Cylinder 3 Exhaust, Cylinder 6 Intake:

   a. Valve clearance cold (68°F/20°C):
   • Intake: 0.010–0.013 inch (0.26–0.34mm)
   • Exhaust: 0.011–0.015 inch (0.29–0.37mm)
   b. Valve clearance hot (176°F/80°C):

67162-FX35-G23

Fig. 178 Position the crankshaft at TDC for number 1 cylinder—3.5L engine

• Intake: 0.012–0.016 inch (0.304–0.416mm)
• Exhaust: 0.012–0.017 inch (0.308–0.432mm)

6. Rotate the crankshaft by 240° clockwise (when viewed from front) to align the number 3 cylinder at TDC of its compression stroke.

➡ **The crankshaft damper pulley mounting bolt flange has a stamped line every 60°. This can be used as a guide to the rotation angle.**

7. Using a feeler gauge, measure the valve clearance for the following cylinders: Cylinder 2 Intake, Cylinder 3 Intake, Cylinder 4 Exhaust, Cylinder 5 Exhaust:

   a. Valve clearance standard cold (68°F/20°C):
   • Intake: 0.010–0.013 inch (0.26–0.34mm)
   • Exhaust: 0.011–0.015 inch (0.29–0.37mm)
   b. Valve clearance hot (176°F/80°C):
   • Intake: 0.012–0.016 inch (0.304–0.416mm)
   • Exhaust: 0.012–0.017 inch (0.308–0.432mm)

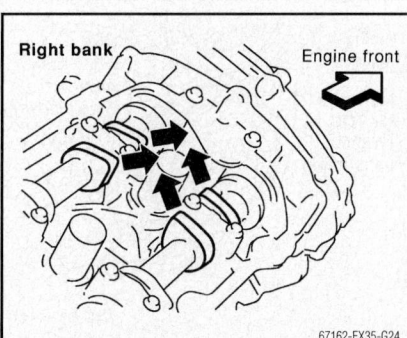

67162-FX35-G24

Fig. 179 When at TDC with number 1 cylinder, the camshaft lobes should point as shown—3.5L engine

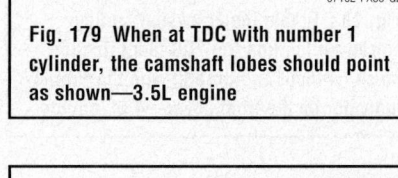

67162-FX35-G25

Fig. 180 Measure the valve lash with the camshaft lobe positioned as shown—3.5L engine

8. Rotate the crankshaft by 240° clockwise (when viewed from front) to align the number 5 cylinder at TDC of its compression stroke.

9. Using a feeler gauge, measure the valve clearance for the following cylinders: Cylinder 1 Exhaust, Cylinder 4 Intake, Cylinder 5 Intake, Cylinder 6 Exhaust:

   a. Valve clearance standard cold (68°F/20°C):
   • Intake: 0.010–0.013 inch (0.26–0.34mm)
   • Exhaust: 0.011–0.015 inch (0.29–0.37mm)
   b. Valve clearance hot (176°F/80°C):
   • Intake: 0.012–0.016 inch (0.304–0.416mm)
   • Exhaust: 0.012–0.017 inch (0.308–0.432mm)

➡ **If the inspection was carried out with a cold engine, make sure the values with a fully warmed up engine are still within specifications.**

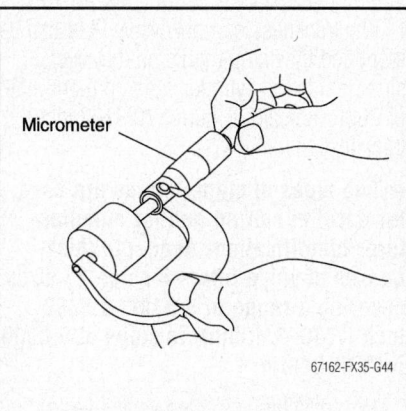

67162-FX35-G44

Fig. 181 Measure the valve lifter height as shown—3.5L engine

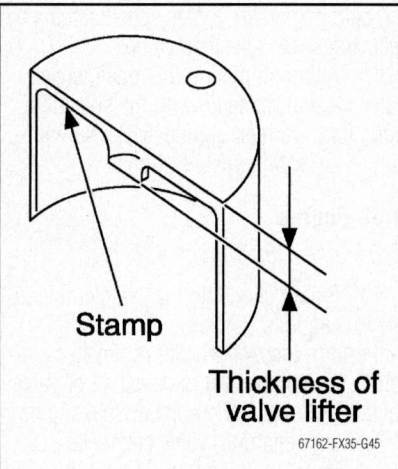

67162-FX35-G45

Fig. 182 Valve lifter identification stamp location—3.5L engine

10. For all valve lifters that are found to be outside the specified range, perform the following steps.

   a. Perform adjustment depending on selected head thickness of valve lifter.

   b. The specified valve lifter thickness is the dimension at normal temperatures. Ignore dimensional differences caused by temperature. Use the specifications for hot engine condition to adjust.

11. Remove the camshaft.

12. Remove the valve lifters at the locations that are outside the standard.

13. Measure the center thickness of the removed valve lifters with a micrometer.

14. Use the following equation to calculate valve lifter thickness for the replacement lifters.

➡**Valve lifter thickness calculation: thickness of replacement valve lifter = t1 + (C1 - C2). t1 = Thickness of removed valve lifter, C1 = measured valve clearance, C2 = standard valve clearance.**

The thickness of a new valve lifter can be identified by stamp marks on the reverse side (inside the cylinder). Stamp mark 788U or 788R indicates 7.88mm (0.3102 inch) in thickness.

➡**Two types of stamp marks are used for parallel setting and for manufacturer identification. Available thicknesses of valve lifters include 27 sizes covering a range of 0.3102–0.3307 inch (7.88–8.40mm) in steps of 0.0008 inch (0.02mm).**

15. Install the selected valve lifter(s).

16. Install the camshaft.

17. Manually turn crankshaft pulley a few turns.

18. Make sure the valve clearances for the cold engine are within specifications by referring to the specified values.

19. After completing the repair, check valve clearances again with the specifications for a warmed engine. Make sure the values are within specifications.

4.5L Engine

See Figures 183 through 191.

1. Before servicing the vehicle, refer to the Precautions Section.

   Perform inspection after removal, installation, or replacement of camshaft or valve-related parts, or if there is unusual engine conditions regarding valve clearance.

2. Warm up the engine. Then turn it OFF.

3. Remove the rocker covers (right and left bank).

4. Set the number 1 cylinder at Top Dead Center (TDC) of its compression stroke, as follows:

   a. Rotate the crankshaft pulley clockwise to align the TDC identification notch (without the paint mark) with the timing indicator on the front cover.

   b. Make sure that both intake and exhaust cam noses of the number 1 cylinder (engine front side of the left engine bank) are located as shown. If not, turn the crankshaft 1 revolution (360°) and align as shown.

5. Using a feeler gauge, measure for specified valve clearance according to the following illustrations:

➡**If inspection was carried out with a cold engine, make sure values with a fully warmed up engine are still within specification.**

6. For any valves that measure out of the standard allowable range, perform the following adjustment steps.

7. Thoroughly wipe off all engine oil around the adjusting shim.

8. Rotate the crankshaft to position the cam nose upward for the camshaft for the valve that must be adjusted.

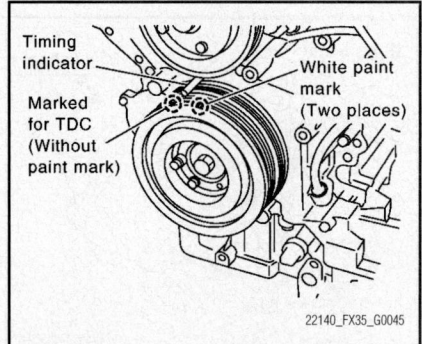

Fig. 183 Rotate the crankshaft pulley clockwise to align the TDC identification notch (without paint mark) with the timing indicator on the front cover—4.5L engine

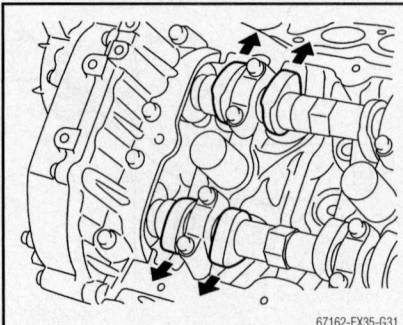

Fig. 184 When at TDC number 1 cylinder, the camshaft lobes should point as shown—4.5L engine

9. Using a small screwdriver or pick, turn the round hole of the adjusting shim in the direction of the arrow (toward the center of the cylinder head).

10. For all valves except the exhaust side of No. 7 and No. 8 cylinders, install the lifter stopper SST: 10115120 (J38972-2) or equivalent, as follows:

   a. Place the camshaft pliers around camshaft as shown in the figure.

   b. Rotate the camshaft pliers so that the valve lifter is pushed down.

✸✸ WARNING

Be careful not to damage the cam surface, valve lifter, or cylinder head with the camshaft pliers.

   c. Place the lifter stopper between the camshaft and the edge of the valve lifter to retain the valve lifter.

✸✸ WARNING

The lifter stopper must be placed as close to the camshaft bracket as possible. Be careful not to damage the cam surface, valve lifter, or cylinder head with the lifter stopper.

   d. Remove the camshaft pliers.

✸✸ WARNING

The camshaft pliers should be removed by rotating it slowly because the lifter stopper hits and damages the journal portion when rotating the camshaft pliers quickly.

11. For the exhaust side of number 7 and 8 cylinders, perform the following:

➡**Exhaust side of number 7 and 8 cylinders do not have space for installing with camshaft pliers SST: KV10115110 (J38972-1). Install lifter stopper SST: KV10115120 (J38972-2)**

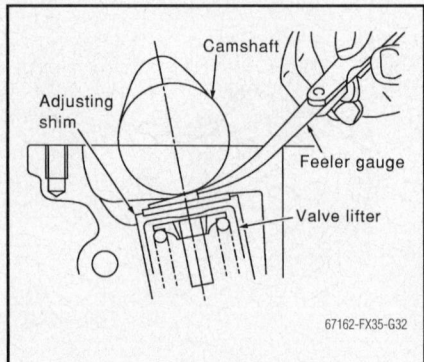

Fig. 185 Checking valve clearance—4.5L engine

Valve clearance: Unit: mm (in)

	Cold	Hot * (reference data)
Intake	0.26 - 0.34 (0.010 - 0.013)	0.304 - 0.416 (0.012 - 0.016)
Exhaust	0.29 - 0.37 (0.011 - 0.015)	0.308 - 0.432 (0.012 - 0.017)

*:Approximately 80°C (176°F)

22140_FX35_G0047

Fig. 186 Valve clearance specifications—4.5L engine

- By referring to the figure, measure the valve clearances at locations marked "×" as shown in the table below (locations indicated with black arrow in figure).
  NOTE:
  Firing order 1-8-7-3-6-5-4-2
- No. 1 cylinder at compression TDC

Measuring position (right bank)		No. 2 CYL.	No. 4 CYL.	No. 6 CYL.	No. 8 CYL.
No. 1 cylinder at compression TDC	EXH				×
	INT	×	×		
Measuring position (left bank)		No. 1 CYL.	No. 3 CYL.	No. 5 CYL.	No. 7 CYL.
No. 1 cylinder at compression TDC	INT	×		×	
	EXH	×			×

↑ (filled) : Measurable at No. 1 cylinder compression TDC

↑ (open) : Measurable at No. 3 cylinder compression TDC

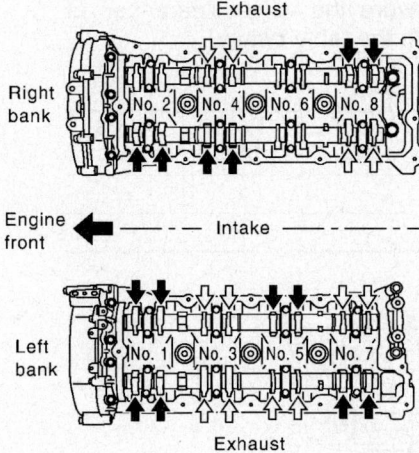

22140_FX35_G0048

Fig. 187 Valve lash inspection positions condition 1—4.5L engine

or equivalent according to the following instructions:

a. Rotate the crankshaft to press the cam nose onto the adjusting part of the valve lifter.

b. Place the lifter stopper between the camshaft and the edge of the valve lifter to retain the valve lifter.

✴✴ WARNING

The lifter stopper must be placed as close to the camshaft bracket as possible. Be careful not to damage the cam surface, valve lifter, or cylinder head with the lifter stopper.

c. Rotate the crankshaft slowly 180° clockwise.

✴✴ WARNING

Rotate the crankshaft slowly because the lifter stopper hits and damages the journal portion by rotating the crankshaft quickly.

12. Blow air into the round hole to separate the adjusting shim from the valve lifter.

✴✴ CAUTION

When blowing, use goggles for eye protection.

13. Remove the adjusting shim with a magnetic tool.

14. Use the following to calculate the adjusting shim thickness for replacement:

a. Use a micrometer to determine the thickness of the removed shim measured at the center.

b. Calculate the thickness of the new adjusting shim so that the valve clearance falls within the specified acceptable range. Valve lifter thickness calculation:

$t = t1 + (C1 − C2)$

t = Valve lifter thickness to be replaced

$t1$ = Removed valve lifter thickness

$C1$ = Measured valve clearance

$C2$ = Standard valve clearance (Intake: 0.012 inch/0.30mm, Exhaust: 0.013 inch/0.33mm at 68°F/20°C)

➡**Shims are available in 64 sizes from 0.0913–0.1161 inch (2.32–2.95mm) in steps of 0.0004 inch (0.01mm). And the thickness of new adjusting shims can be identified by stamp marks on the underside (inside the cylinder).**

- By referring to the figure, measure the valve clearances at locations marked "×" as shown in the table below (locations indicated with white arrow in figure).
- No. 3 cylinder at compression TDC

Measuring position (right bank)		No. 2 CYL.	No. 4 CYL.	No. 6 CYL.	No. 8 CYL.
No. 3 cylinder at compression TDC	EXH		×		
	INT				×
Measuring position (left bank)		No. 1 CYL.	No. 3 CYL.	No. 5 CYL.	No. 7 CYL.
No. 3 cylinder at compression TDC	INT		×		×
	EXH		×	×	

↑ : Measurable at No. 1 cylinder compression TDC

⇧ : Measurable at No. 3 cylinder compression TDC

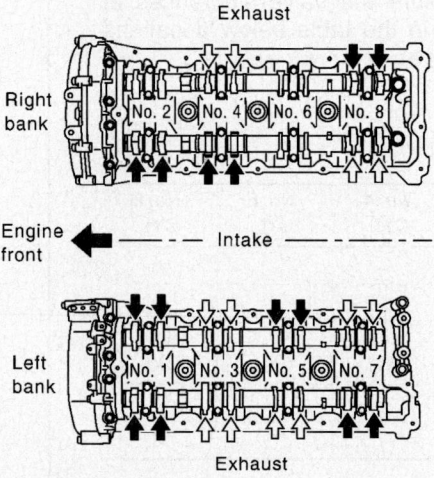

Fig. 188 Valve lash inspection positions condition 2—4.5L engine

- By referring to the figure, measure the valve clearances at locations marked "×" as shown in the table below.
- No. 6 cylinder at compression TDC

Measuring position (right bank)		No. 2 CYL.	No. 4 CYL.	No. 6 CYL.	No. 8 CYL.
No. 6 cylinder at compression TDC	EXH	×		×	
	INT			×	

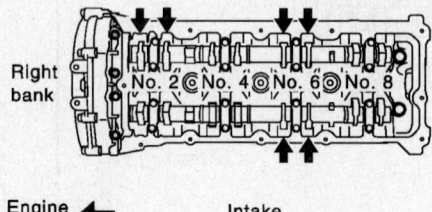

Fig. 189 Valve lash inspection positions condition 3—4.5L engine

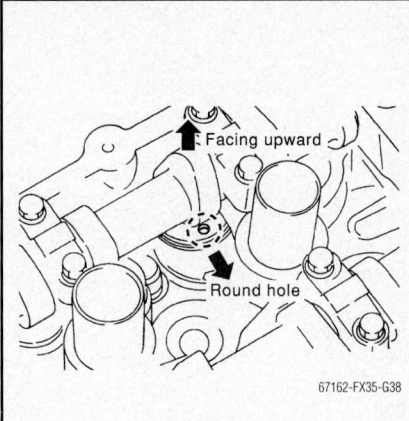

Fig. 190 Position the camshaft lobe as shown for shim replacement—4.5L engine

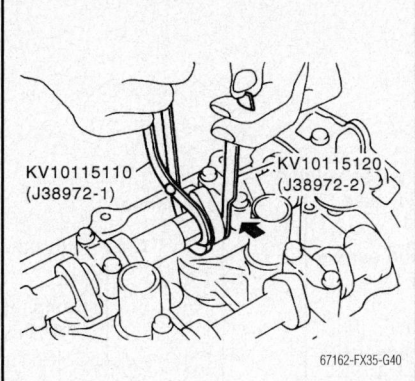

Fig. 191 Use the tools shown to depress the valve spring for shim replacement—4.5L engine

15. Install the new adjusting shim using a suitable tool, with the surface on which the thickness is stamped facing down.

16. For all valve lifters, except exhaust side of number 7 and 8 cylinders, remove the lifter stopper as follows:
   a. Perform the same procedure as described for removal using the camshaft pliers.
   b. Remove the lifter stopper.
   c. Remove the camshaft pliers.

17. For exhaust side of number 7 and 8 cylinder valve lifters, rotate the crankshaft slowly 180° clockwise, then remove the lifter stopper.

18. Manually rotate the crankshaft pulley a few turns.

19. Make sure that the valve clearance is within specifications for cold and hot settings.

20. Valve clearance specifications:
- Cold Intake (68°F/20°C): 0.010–0.013 inch (0.26–0.34mm)
- Cold Exhaust: 0.011–0.015 inch (0.29–0.37mm)
- Hot Intake (176°F/80°C): 0.012–0.016 inch (0.304–0.416mm)
- Hot Exhaust: 0.012–0.016 inch (0.308–0.432mm)

ENGINE PERFORMANCE & EMISSION CONTROL

See Figures 192 and 193.

CAMSHAFT POSITION (CMP) SENSOR

LOCATION

See Figures 194 and 195.

REMOVAL & INSTALLATION

1. Before servicing the vehicle, refer to the Precautions Section.
2. Disconnect the negative battery cable.
3. Disconnect the connector from the CMP sensor.
4. Remove the bolt that retains the CMP sensor.
5. Remove the CMP sensor.

To install:
6. Installation is the reverse of the removal procedure.
7. Tighten the bolt that retains the CMP sensor.
8. Connect the sensor connector.

CRANKSHAFT POSITION (CKP) SENSOR

LOCATION

See Figures 196 and 197.

REMOVAL & INSTALLATION

1. Before servicing the vehicle, refer to the Precautions Section.
2. Disconnect the negative battery cable.
3. Disconnect the connector from the sensor.
4. Remove the bolt that retains the sensor in place.
5. Remove the sensor from its mounting.

To install:
6. Installation is the reverse of the removal procedure.
7. Tighten the sensor retaining bolt.

ELECTRONIC CONTROL MODULE (ECM)

LOCATION

See Figures 198 and 199.

REMOVAL & INSTALLATION

1. Before servicing the vehicle, refer to the Precautions Section.
2. Turn the ignition switch **OFF**.
3. Disconnect the negative battery cable from the battery.
4. Disconnect the ECM connectors.
5. Remove the ECM mounting bolts and remove the ECM from the vehicle.

To install:
6. Installation is the reverse of the removal.
7. Tighten the ECM mounting bolts.

⁂ WARNING

When replacing the ECM, be careful to use the right part number, as damage to the injection system could occur.

ENGINE COOLANT TEMPERATURE (ECT) SENSOR

LOCATION

See Figures 200 and 201.

REMOVAL & INSTALLATION

1. Before servicing the vehicle, refer to the Precautions Section.
2. Drain the coolant to a level below the bottom of the sensor.
3. Disconnect the ground cable from the battery and then remove the sensor connector.
4. Remove the coolant temperature sensor.

To install:
5. Coat the threads of the sensor with a suitable sealant and thread into the housing.
6. Tighten the sensor.
7. Refill the cooling system to the proper level.
8. Attach the electrical connector to the sensor securely.
9. Connect the negative battery cable.

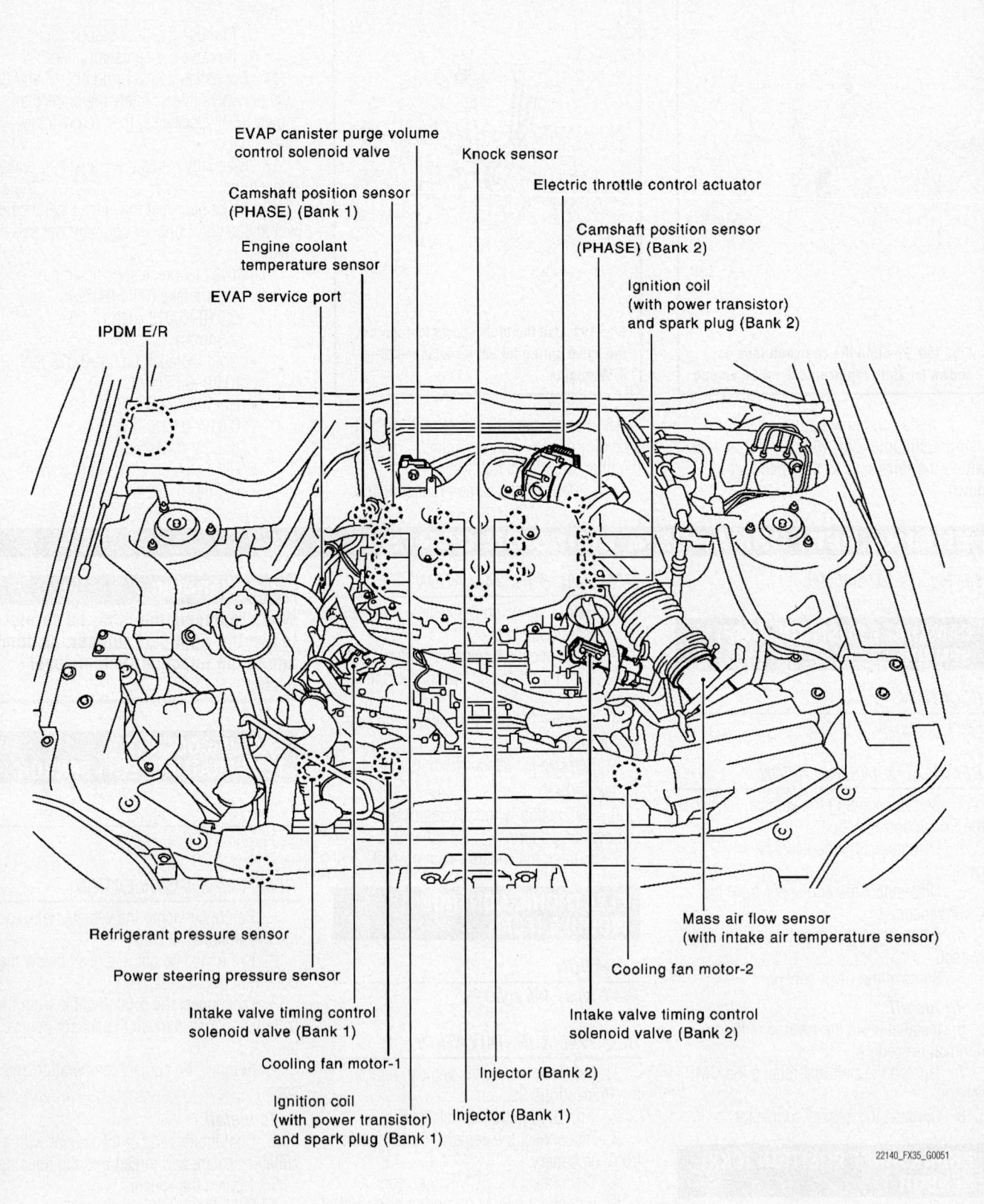

EVAP canister purge volume
control solenoid valve

Knock sensor

Camshaft position sensor
(PHASE) (Bank 1)

Electric throttle control actuator

Engine coolant
temperature sensor

Camshaft position sensor
(PHASE) (Bank 2)

EVAP service port

Ignition coil
(with power transistor)
and spark plug (Bank 2)

IPDM E/R

Refrigerant pressure sensor

Power steering pressure sensor

Intake valve timing control
solenoid valve (Bank 1)

Cooling fan motor-1

Ignition coil
(with power transistor)
and spark plug (Bank 1)

Injector (Bank 1)

Injector (Bank 2)

Intake valve timing control
solenoid valve (Bank 2)

Cooling fan motor-2

Mass air flow sensor
(with intake air temperature sensor)

22140_FX35_G0051

Fig. 192 Underhood and related sensor locations—3.5L engine

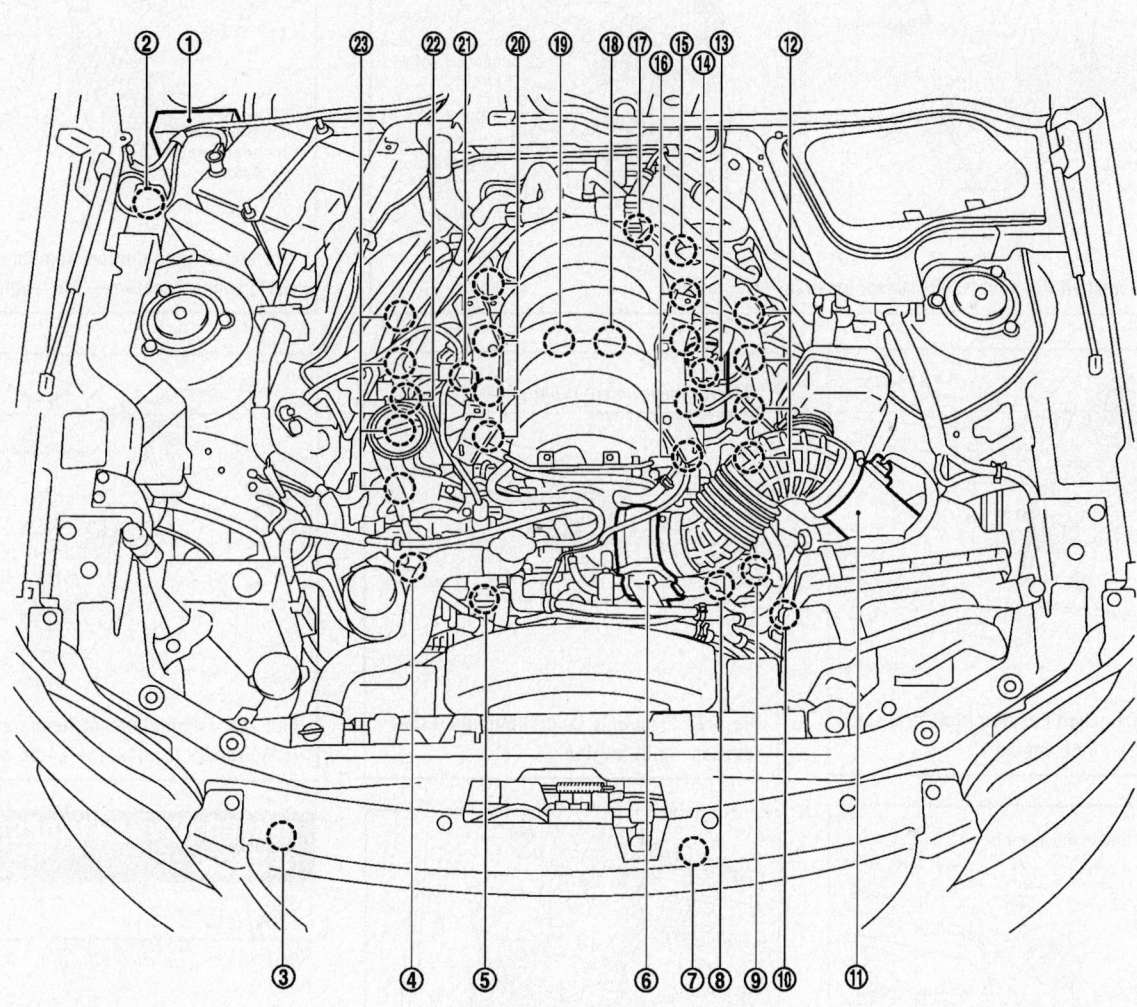

PBIB3224E

1.	IPDM E/R	2.	ICC brake hold relay (ICC models only)	3.	Refrigerant pressure sensor
4.	Intake valve timing control position sensor (Bank 2)	5.	Intake valve timing control solenoid valve (Bank 2)	6.	Electric throttle control actuator
7.	Cooling fan motor	8.	Intake valve timing control solenoid valve (Bank 1)	9.	Intake valve timing control position sensor (Bank 1)
10.	Camshaft position sensor (PHASE)	11.	Mass air flow sensor (with intake air temperature sensor)	12.	Ignition coil (with power transistor) and spark plug (Bank 1)
13.	Vacuum tank	14.	VIAS control solenoid valve	15.	Engine coolant temperature sensor
16.	Fuel injector (Bank 1)	17.	Power valve actuator	18.	Knock sensor (Bank 1)
19.	Knock sensor (Bank 2)	20.	Fuel injector (Bank 2)	21.	EVAP canister purge volume control solenoid valve
22.	EVAP service port	23.	Ignition coil (with power transistor) and spark plug (Bank 2)		

22140_FX35_G0052

Fig. 193 Underhood and related sensor locations—4.5L engine

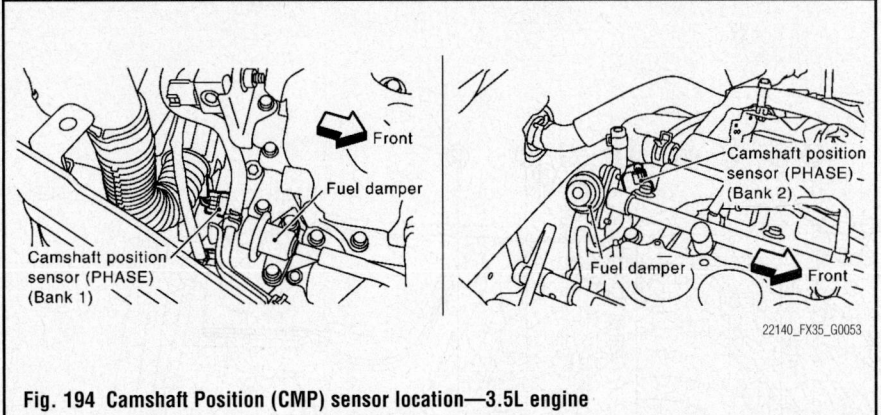

Fig. 194 Camshaft Position (CMP) sensor location—3.5L engine

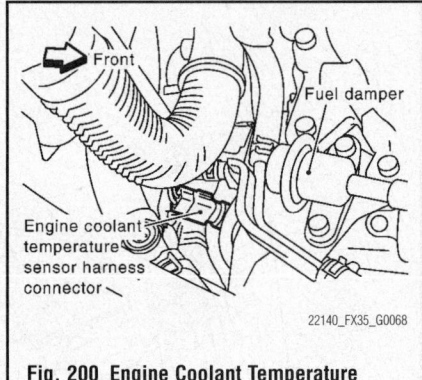

Fig. 200 Engine Coolant Temperature (ECT) sensor location—3.5L engine

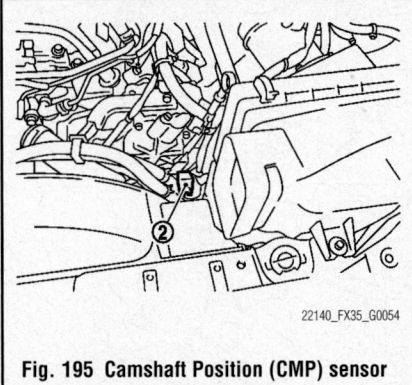

Fig. 195 Camshaft Position (CMP) sensor location (2)—4.5L engine

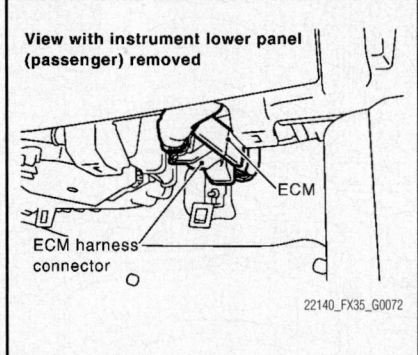

Fig. 198 Electronic Control Module (ECM) location—3.5L engine

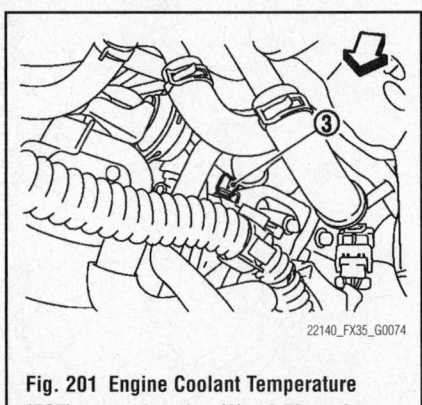

Fig. 201 Engine Coolant Temperature (ECT) sensor location (3)—4.5L engine

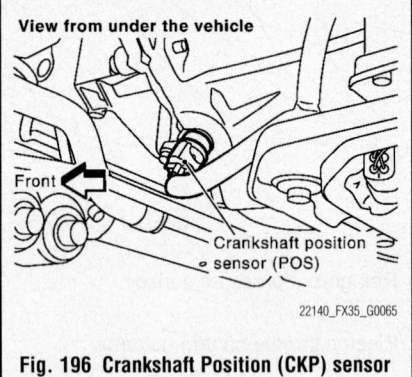

Fig. 196 Crankshaft Position (CKP) sensor location—3.5L engine

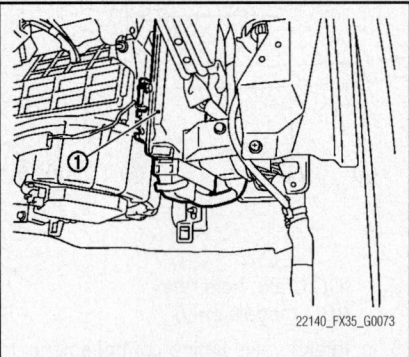

Fig. 199 Electronic Control Module (ECM) location (1)—4.5L engine

HEATED OXYGEN (HO2S) SENSOR

LOCATION

See Figures 202 through 204.

REMOVAL & INSTALLATION

✳✳ CAUTION

The temperature of the exhaust system is extremely high after the engine has been run. To prevent personal injury, allow the exhaust system to

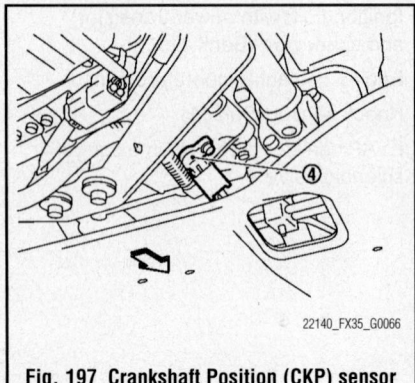

Fig. 197 Crankshaft Position (CKP) sensor location (4)—4.5L engine

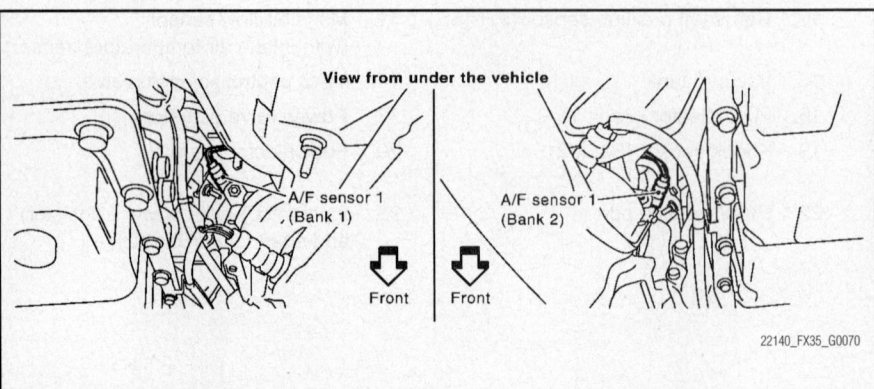

Fig. 202 Location of Heated Oxygen (HO2S) Sensor (bank 1 and bank 2)—3.5L engine

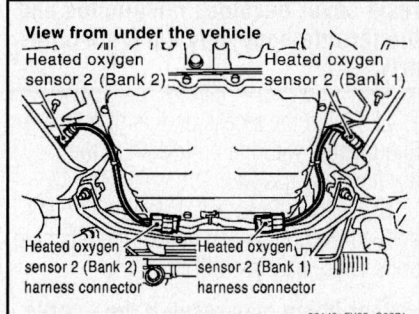

Fig. 203 Location of Heated Oxygen (HO2S) Sensor (bank 1, sensor 2 and bank 2, sensor 2)—3.5L engine

REMOVAL & INSTALLATION

1. Before servicing the vehicle, refer to the Precautions Section.

2. Disconnect the negative battery cable.

3. Disconnect the connector from the sensor.

4. Remove the sensor retaining screws.

5. Remove the sensor from its mounting.

To install:

6. Installation is the reverse of the removal procedure.

7. Handle the sensor assembly carefully, protecting it from impact, extremes of temperature and/or exposure to shop chemicals.

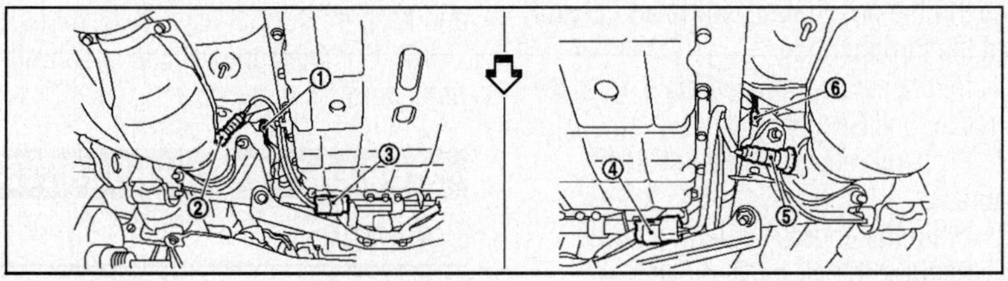

22140_FX35_G0075

⇦ : Vehicle front

1. Air fuel ratio (A/F) sensor 1 (Bank 1)
2. Heated oxygen sensor 2 (Bank 1)
3. Heated oxygen sensor 2 (Bank 1) harness connector
4. Heated oxygen sensor 2 (Bank 2) harness connector
5. Heated oxygen sensor 2 (Bank 2)
6. Air fuel ratio (A/F) sensor 1 (Bank 2)

Fig. 204 Location of Heated Oxygen (HO2S) Sensors—4.5L engine

cool before removing the sensor from the exhaust system.

1. Before servicing the vehicle, refer to the Precautions Section.

2. Disconnect the negative battery cable.

3. Raise and safely support the vehicle, as needed.

4. Detach the electrical connector from the oxygen sensor.

5. Using an oxygen sensor socket, remove the heated oxygen sensor.

To install:

6. If installing the old oxygen sensor, coat the threads with anti-seize compound. New sensors are already coated. Take care not to contaminate the oxygen sensor probe with the anti-seize compound.

7. Install the oxygen sensor. Using the correct tool, tighten the sensor.

8. Attach the wiring to the sensor.

9. Connect the negative battery cable.

INTAKE AIR TEMPERATURE (IAT) SENSOR

LOCATION

See Figures 205 and 206.

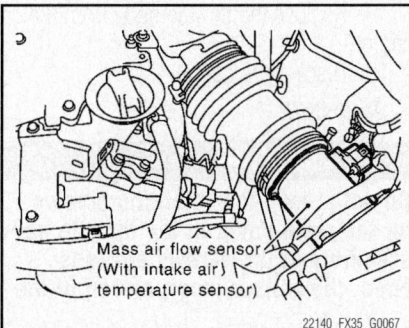

Fig. 205 Location of Mass Air Flow (MAF) and Intake Air Temperature (IAT) sensor—3.5L engine

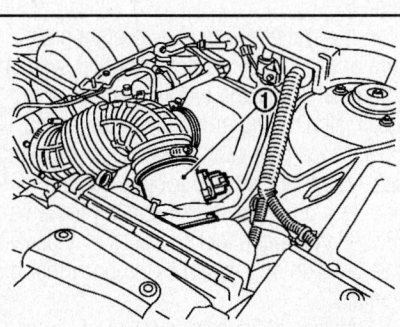

Fig. 206 Location of Mass Air Flow (MAF) and Intake Air Temperature (IAT) sensor (1)—4.5L engine

KNOCK SENSOR (KS)

LOCATION

See Figures 207 and 208.

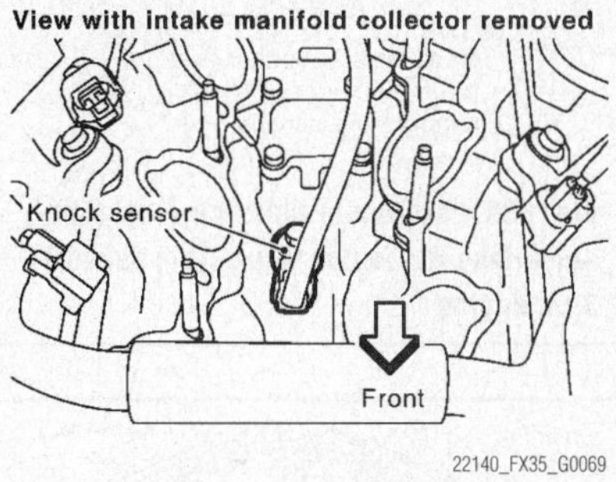

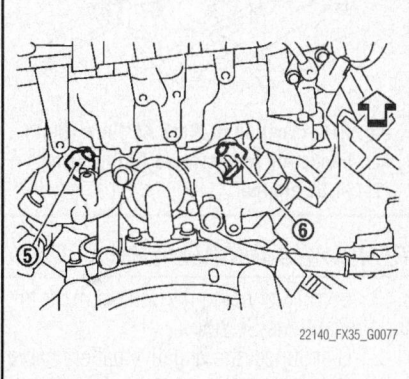

Fig. 207 Location of Knock Sensor (KS)—3.5L engine

Fig. 208 Location of Knock Sensor (KS)—bank 1 (5), bank 2 (6)—4.5L engine

REMOVAL & INSTALLATION

1. Before servicing the vehicle, refer to the Precautions Section.

2. Disconnect the negative battery cable.

3. Disconnect the sensor connector.

4. Remove the sensor from its mounting.

To install:

5. Installation is the reverse of the removal procedure.

6. Tighten the sensor to 16 ft. lbs. (21 Nm).

MASS AIR FLOW (MAF) SENSOR

LOCATION

See Figures 209 and 210.

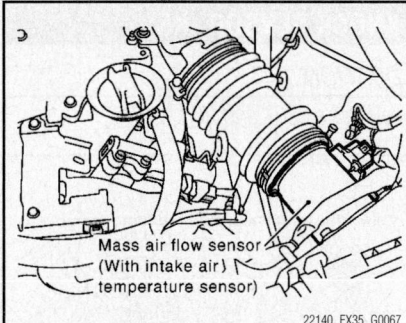

Fig. 209 Location of Mass Air Flow (MAF) and Intake Air Temperature (IAT) sensor—3.5L engine

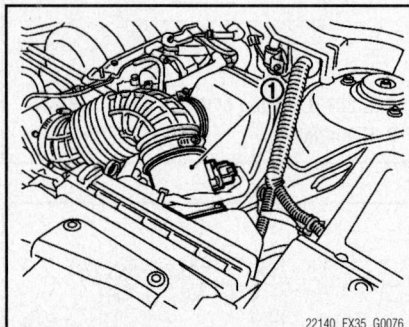

Fig. 210 Location of Mass Air Flow (MAF) and Intake Air Temperature (IAT) sensor (1)—4.5L engine

REMOVAL & INSTALLATION

1. Before servicing the vehicle, refer to the Precautions Section.
2. Disconnect the negative battery cable.
3. Disconnect the connector from the sensor.
4. Remove the air cleaner and air intake assembly, as required.
5. Remove the sensor from its mounting.

To install:

6. Installation is the reverse of the removal procedure.
7. Handle the sensor assembly carefully, protecting it from impact, extremes of temperature, and exposure to shop chemicals.

THROTTLE POSITION SENSOR (TPS)

LOCATION

The Throttle Position Sensor (TPS) is mounted on the throttle body and is incorporated into the throttle body assembly.

REMOVAL & INSTALLATION

The throttle position sensor is an integral part of the throttle body.

1. Before servicing the vehicle, refer to the Precautions Section.
2. Properly relieve the fuel system pressure.
3. Drain the engine coolant.
4. Remove the air intake hose.
5. Remove the battery.
6. Disconnect the throttle position sensor connector.
7. Disconnect the water hose connection.
8. Remove the throttle body retaining bolts.
9. Remove the throttle body from the engine.
10. Discard the gasket.

To install:

✳✳ WARNING

Do not loosen the retaining screws for the resin cover of the throttle body assembly. If the screws are loosened, the sensor incorporated in the resin cover becomes misaligned and the throttle body may not work properly.

11. Align the recess on the intake manifold plenum with the projection of the throttle body gasket.
12. Install the gasket. Install the throttle body to the engine and tighten the retaining bolts to 80 inch lbs. (9 Nm).

➡**Poor idling may result if the throttle body gasket is not installed properly.**

13. Continue the installation in the reverse order of the removal procedure.
14. Connect the negative battery cable.
15. Turn the ignition **ON** and then **OFF**, and keep it off for at least 10 seconds.
16. Complete the vehicle initialization procedure.

VEHICLE SPEED SENSOR (VSS)

LOCATION

The vehicle speed sensor is installed on the transmission.

REMOVAL & INSTALLATION

1. Before servicing the vehicle, refer to the Precautions Section.
2. Raise and support the vehicle safely.
3. Place a drip pan below the Vehicle Speed Sensor (VSS) to catch any spilled transmission fluid when it is removed.
4. Disconnect the VSS connector.
5. Remove the sensor from its mounting.

To install:

6. Install the sensor and tighten the attaching bolt.
7. Replace any lost transmission fluid.
8. Connect the sensor electrical connector.

FUEL

GASOLINE FUEL INJECTION SYSTEM

FUEL SYSTEM SERVICE PRECAUTIONS

Safety is the most important factor when performing not only fuel system maintenance but any type of maintenance. Failure to conduct maintenance and repairs in a safe manner may result in serious personal injury or death. Maintenance and testing of the vehicle's fuel system components can be accomplished safely and effectively by adhering to the following rules and guidelines.

• To avoid the possibility of fire and personal injury, always disconnect the negative battery cable unless the repair or test procedure requires that battery voltage be applied.

• Always relieve the fuel system pressure prior to disconnecting any fuel system component (injector, fuel rail, pressure regulator, etc.), fitting or fuel line connection. Exercise extreme caution whenever relieving fuel system pressure to avoid exposing skin, face and eyes to fuel spray. Please be advised that fuel under pressure may penetrate the skin or any part of the body that it contacts.

• Always place a shop towel or cloth around the fitting or connection prior to loosening to absorb any excess fuel due to spillage. Ensure that all fuel spillage (should it occur) is quickly removed from engine surfaces. Ensure that all fuel soaked cloths or towels are deposited into a suitable waste container.

• Always keep a dry chemical (Class B) fire extinguisher near the work area.

• Do not allow fuel spray or fuel vapors to come into contact with a spark or open flame.

• Always use a back-up wrench when loosening and tightening fuel line connection fittings. This will prevent unnecessary stress and torsion to fuel line piping.

• Always replace worn fuel fitting O-rings with new Do not substitute fuel hose or equivalent where fuel pipe is installed.

Before servicing the vehicle, make sure to also refer to the precautions in the beginning of this section as well.

RELIEVING FUEL SYSTEM PRESSURE

WITH THE CONSULT-III TOOL

1. Before servicing the vehicle, refer to the Precautions Section.
2. Turn the ignition switch **ON**.
3. Perform the "FUEL PRESSURE RELEASE" in "WORK SUPPORT" mode with the CONSULT-III.
4. Start the engine.
5. After engine stalls, crank it over 2–3 times to release all fuel pressure.
6. Turn the ignition switch **OFF**.

WITHOUT THE CONSULT-III TOOL

See Figure 211.

1. Before servicing the vehicle, refer to the Precautions Section.
2. Remove the fuel pump fuse located in IPDM E/R.
3. Start the engine.
4. After the engine stalls, crank it over 2–3 times to release all fuel pressure.
5. Turn the ignition switch **OFF**.
6. Reinstall the fuel pump fuse after servicing the fuel system.

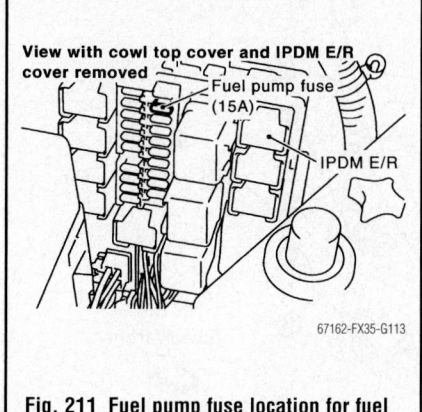

Fig. 211 Fuel pump fuse location for fuel pressure release

FUEL FILTER

REMOVAL & INSTALLATION

The fuel delivery system integrates the fuel filter with the in-tank fuel pump. To service this filter, remove the fuel pump. Refer to Fuel Pump, removal & installation.

FUEL INJECTORS

REMOVAL & INSTALLATION

3.5L Engine

See Figures 212 through 218.

1. Before servicing the vehicle, refer to the Precautions Section.
2. Remove the engine cover.
3. Relieve the fuel pressure.
4. Remove the fuel feed hose (with damper) from the fuel sub-tube.

➥ **There is no fuel return route.**

✳ CAUTION

While the hoses are disconnected, plug them to prevent fuel from draining. Also, do not separate the damper and hose.

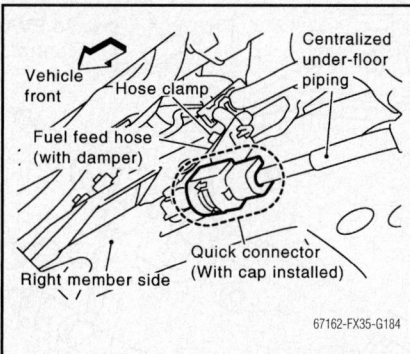

Fig. 212 Quick connect coupling detail for under-floor piping—3.5L engine

5. When separating the fuel feed hose (with damper) and the centralized under-floor piping connection, disconnect the quick connector, as follows:
   a. Remove the quick connector cap from the quick connector connection on the right member side.
   b. Disconnect the fuel feed hose (with damper) from the bracket hose clamp.

➥ **Disconnect the quick connector by using quick connector release tool SST: J-45488, or equivalent.**

   c. With the sleeve side of the quick connector release facing the quick connector, install the quick connector release onto the centralized under-floor piping.

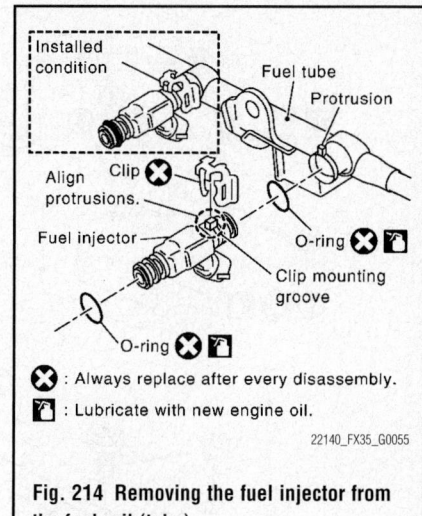

Fig. 214 Removing the fuel injector from the fuel rail (tube)

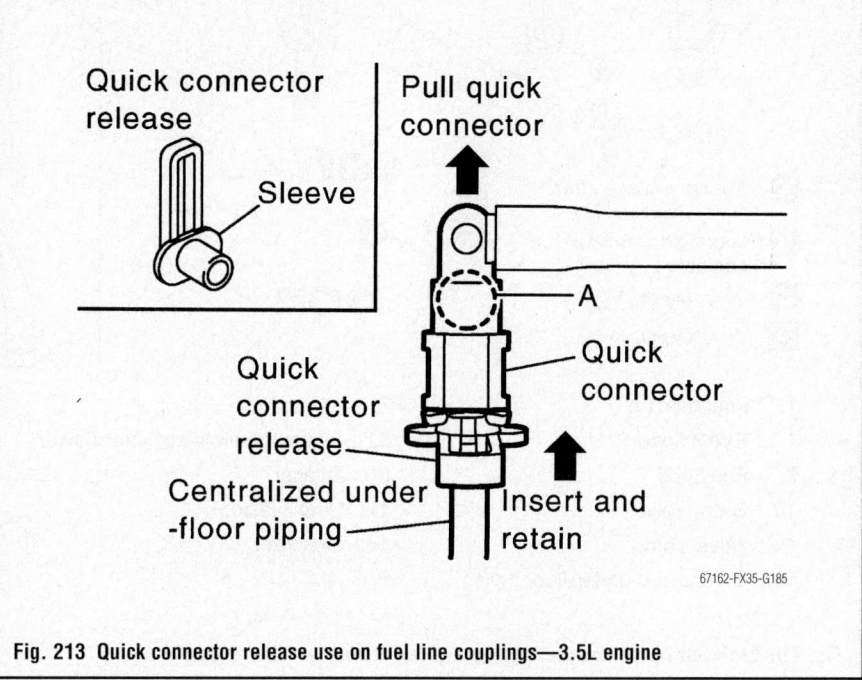

Fig. 213 Quick connector release use on fuel line couplings—3.5L engine

d. Insert the quick connector release into the quick connector until the sleeve contacts and goes no further. Hold the quick connector release at that position.

> **⁕⁕ CAUTION**
>
> **Inserting the quick connector release hard will not disconnect the quick connector. Hold the quick connector release where it contacts and goes no further.**

e. Draw and pull out the quick connector straight from the centralized under-floor piping.

When disconnecting the fuel line, heed the following:

- Pull the quick connector holding **A** position as shown in the figure. Do not pull it with lateral force applied. The O-ring inside the quick connector may be damaged
- Prepare a container and cloth beforehand as fuel will leak out
- Avoid fire and sparks
- Keep parts away from all heat sources. Especially, be careful when welding is performed
- Do not expose the parts to battery electrolyte or other acids
- Do not bend or twist the connection between the quick connector and the fuel feed hose (with damper) during installation/removal
- To keep the connecting portion clean and to avoid damage from foreign materials, cover them completely with plastic bags or a similar material

6. Remove the upper and lower intake manifold collectors. Refer to Intake Manifold, Upper Intake Manifold, removal & installation.

7. Disconnect the wiring harness connector from the fuel injector.

8. Loosen the mounting bolts in the reverse order shown, and remove the fuel tube and fuel injector assembly.

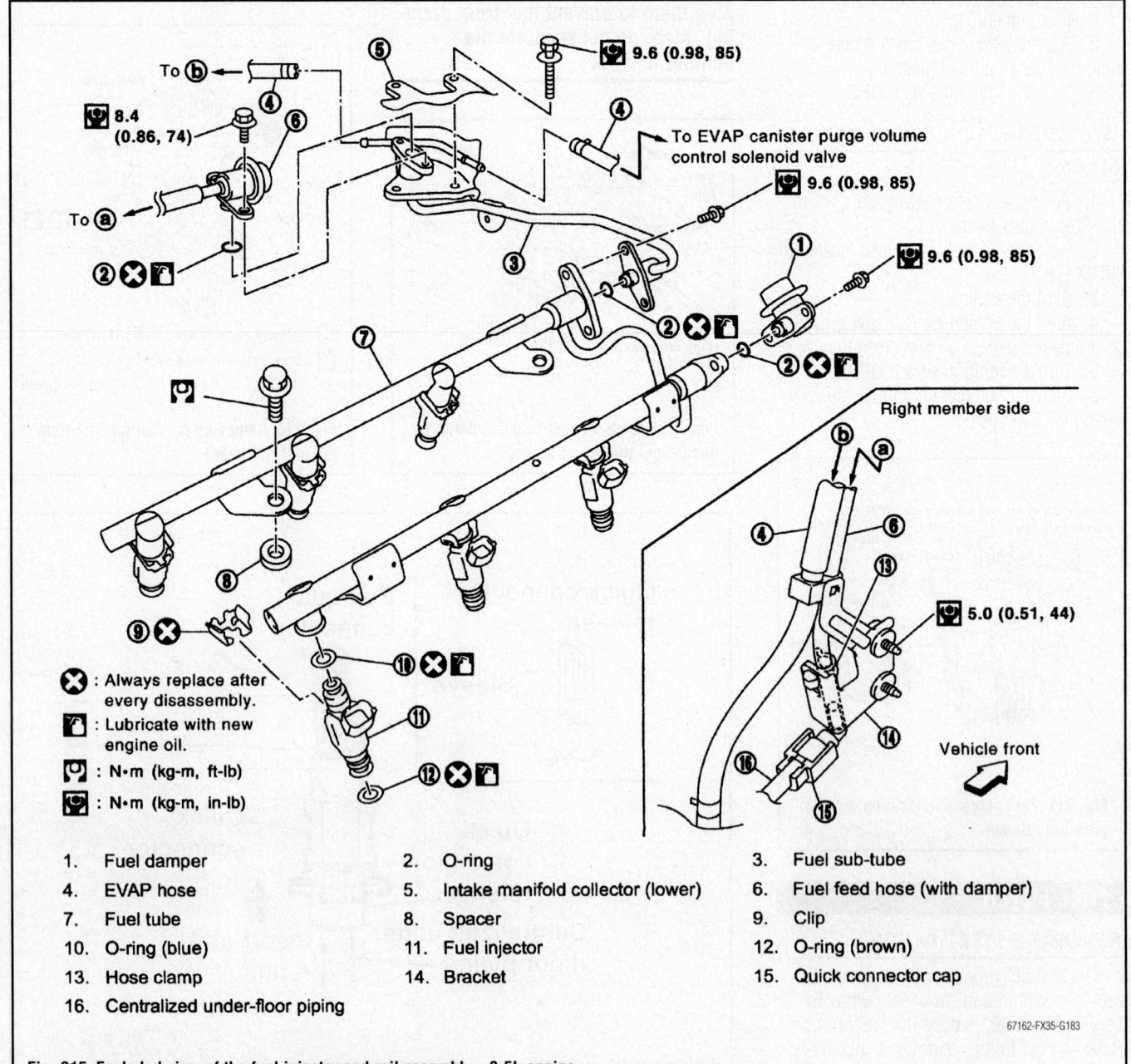

1. Fuel damper
2. O-ring
3. Fuel sub-tube
4. EVAP hose
5. Intake manifold collector (lower)
6. Fuel feed hose (with damper)
7. Fuel tube
8. Spacer
9. Clip
10. O-ring (blue)
11. Fuel injector
12. O-ring (brown)
13. Hose clamp
14. Bracket
15. Quick connector cap
16. Centralized under-floor piping

Fig. 215 Exploded view of the fuel injector and rail assembly—3.5L engine

67162-FX35-G183

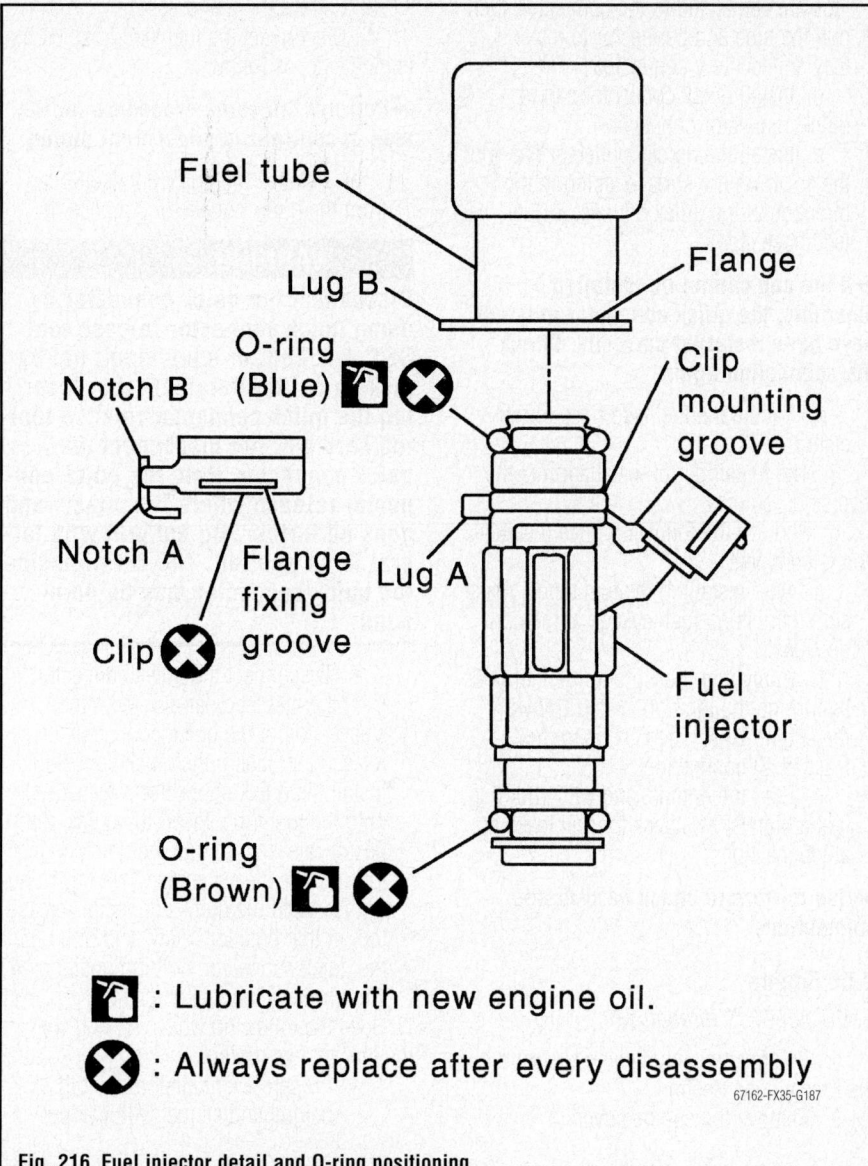

Fig. 216 Fuel injector detail and O-ring positioning

Fuel tube

Flange

Lug B

O-ring (Blue)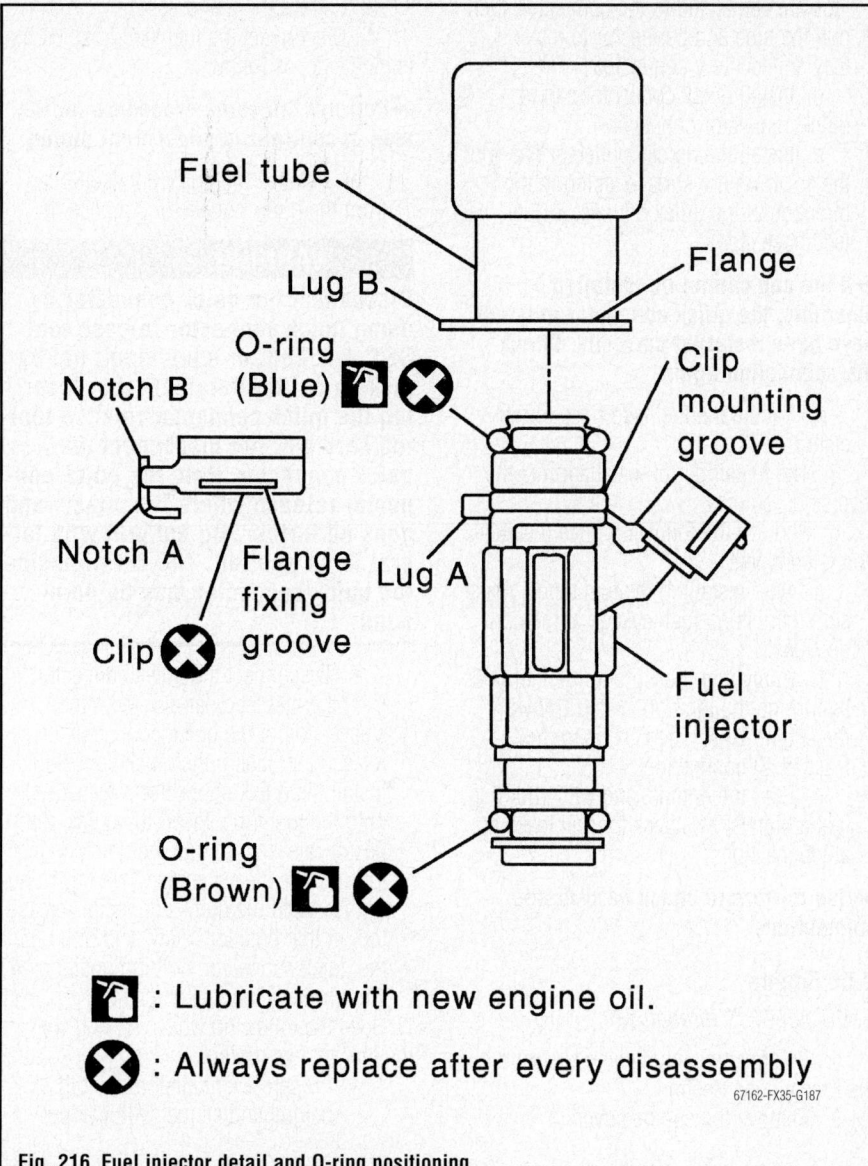

Notch B

Clip mounting groove

Notch A

Flange fixing groove

Lug A

Clip

Fuel injector

O-ring (Brown)

🛢 : Lubricate with new engine oil.

✖ : Always replace after every disassembly

67162-FX35-G187

※※ CAUTION

Do not tilt the assembly or the remaining fuel may leak.

9. Remove the fuel injectors from the fuel rail (tube) with the following procedure:

a. Open and remove the clip.

b. Remove the fuel injector from the fuel tube by pulling straight.

During injector removal, heed the following items:

- Be careful with the remaining fuel that leaks from the fuel tube
- Be careful not to damage the injector nozzles during removal
- Do not bump or drop the fuel injectors
- Do not disassemble the fuel injectors

10. Remove the fuel sub-tube and fuel damper.

To install:

When handling all O-rings in this procedure, heed the following:

- Handle the O-ring with bare hands. Never wear gloves
- Lubricate the O-ring with new engine oil
- Do not clean the O-ring with solvent
- Make sure that the O-ring and its mating part are free of foreign material
- When installing the O-ring, be careful not to scratch it with a tool or fingernails
- Be careful not to twist or stretch the O-ring
- Insert the O-ring straight into the fuel tube

11. Install the fuel damper and fuel sub-tube.

12. Insert the fuel damper and fuel sub-tube straight into the fuel tube.

13. Tighten the mounting bolts.

14. After tightening the mounting bolts, make sure that there is no gap between the flange and fuel tube.

15. Install O-rings onto the fuel injector—the upper and lower O-rings are

---

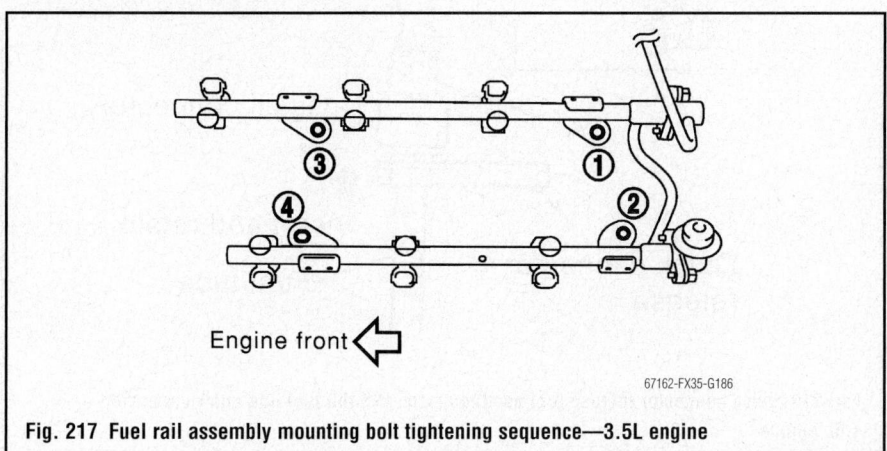

Engine front ⇐

67162-FX35-G186

Fig. 217 Fuel rail assembly mounting bolt tightening sequence—3.5L engine

---

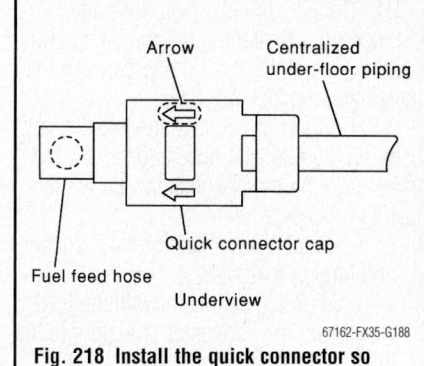

Arrow

Centralized under-floor piping

Quick connector cap

Fuel feed hose

Underview

67162-FX35-G188

Fig. 218 Install the quick connector so that the arrows face the fuel feed tube—3.5L engine

different. The O-rings are identified as follows:

- Fuel tube side O-ring: Blue
- Nozzle side O-ring: Brown

16. Install each fuel injector onto the fuel tube, as follows:

a. Insert the clip into the clip mounting groove on the fuel injector.

b. Insert the clip so that lug "A" of the fuel injector matches notch "A" of the clip.

✱✱ CAUTION

Do not reuse old clips. Replace them with new ones. Be careful to keep the clip from interfering with the O-ring. If interference occurs, replace the O-ring with a new one.

c. Insert the fuel injector into the fuel tube, matching it to the axial center, with the clip attached. Insert the fuel injector so that lug "B" of fuel tube matches notch "B" of the clip. Make sure that the fuel tube flange is securely fixed in the groove on the clip.

d. Make sure that installation is complete by checking that the fuel injector does not rotate or come off.

17. Install the fuel tube and fuel injector assembly onto the intake manifold, and tighten the mounting bolts in 2 steps as shown.

a. 1st Step: 84 inch lbs. (10 Nm).
b. 2nd Step: 17 ft. lbs. (24 Nm).

✱✱ WARNING

Be careful not to let the tips of the injector nozzles come into contact with other parts.

18. Connect the injector sub-wiring harness.

19. Install the upper and lower intake manifold collectors. Refer to Intake Manifold, Upper Intake Manifold, removal & installation.

20. Install the fuel sub-tube on the rear end of the lower intake manifold collector.

21. Connect the fuel feed hose and damper. After tightening the mounting bolts, make sure that there is no gap between the flange and the fuel sub-tube.

22. Connect the quick connector between the fuel feed hose and centralized under-floor piping connection with the following procedure:

a. Check the connection for damage and foreign materials.

b. Align the connector with the tube, then insert the connector straight into the tube until a click is heard.

c. After connecting the quick connector, visually confirm that the 2 retainer

tabs are connected to the connector, then pull the tube and connector to make sure they are securely connected.

d. Install quick connector cap to quick connector connection.

e. Install the quick connector cap with the arrow on the surface facing in the direction of the quick connector (fuel feed hose side).

➡**If the cap cannot be installed smoothly, the quick connector may not have been installed correctly. Check the connection again.**

f. Secure the fuel feed hose to the clamp.

23. The remainder of installation is the reverse of removal.

24. Perform the following once installation is complete.

a. After installing the fuel tubes, make sure there is no fuel leakage at the connections.

b. Apply fuel pressure to the fuel lines by turning the ignition switch **ON** with the engine **OFF**. Then check for fuel leaks at all connections.

c. Start the engine, and while holding it at a high RPM, check for fuel leaks at all connections.

➡**Use mirrors to check hard-to-see connections.**

4.5L Engine

See Figures 219 through 224.

1. Before servicing the vehicle, refer to the Precautions Section.

2. Remove the engine cover.

3. Relieve the fuel pressure.

4. Disconnect the fuel feed hose on the engine side, as follows:

➡**Perform the same procedure for the side of centralized under-floor piping.**

a. Remove the quick connector cap from the quick connector connection.

✱✱ WARNING

Disconnect the quick connector by using quick connector release tool SST: J-45488, or equivalent; not by picking out the retainer tabs. Inserting the quick connector release tool too hard will not disconnect the quick connector. Hold the quick connector release where it contacts and goes no further. Do not pull with lateral force applied. The O-ring inside the quick connector may be damaged.

b. Disconnect the quick connector from the fuel feed damper. With the sleeve side of the quick connector release tool facing the quick connector, install the quick connector release tool onto the fuel tube. Insert the quick connector release tool into the quick connector until the sleeve contacts and goes no further. Hold the quick connector release tool in that position. Draw and pull out the quick connector, holding position "A" straight from the fuel damper.

Heed the following when working with the fuel system tubing:

- Prepare a container and cloth beforehand as fuel will leak out

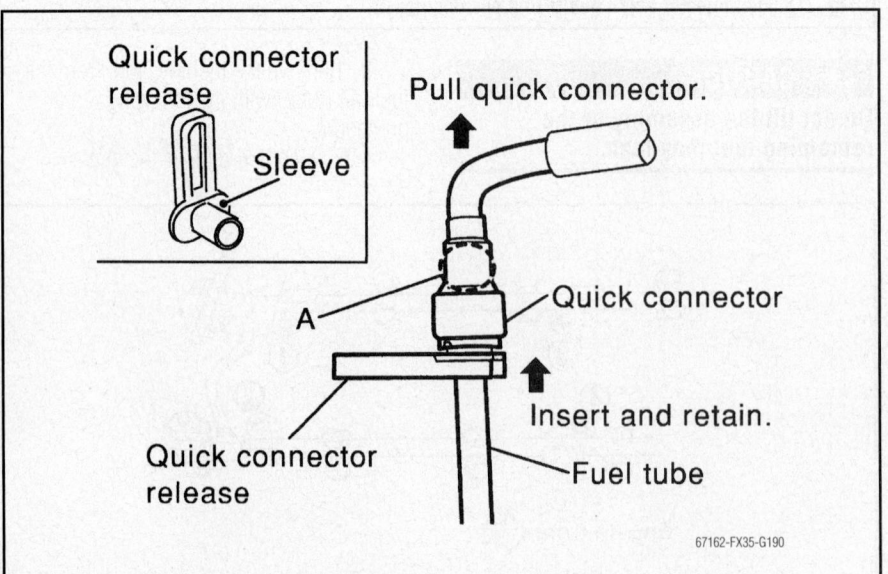

Fig. 219 Quick connector release tool used to disconnect the fuel line quick connector—4.5L engine

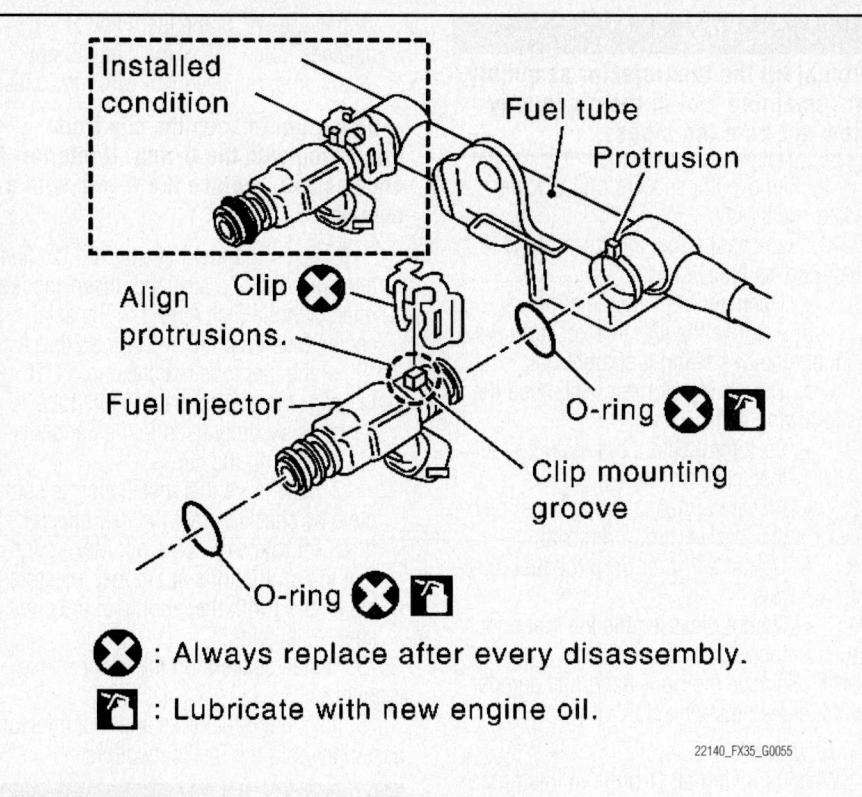

Fig. 220 Removing the fuel injector from the fuel rail (tube)

Installed condition

Fuel tube

Protrusion

Align protrusions.

Clip ⊗

Fuel injector

O-ring ⊗ 🛢

Clip mounting groove

O-ring ⊗ 🛢

⊗ : Always replace after every disassembly.

🛢 : Lubricate with new engine oil.

22140_FX35_G0055

- Avoid fire and sparks
- Keep parts away from a heat source. Especially, be careful when welding is performed.
- Do not expose parts to battery electrolyte or other acids
- Do not bend or twist the connection between the quick connector and the fuel feed hose during installation/removal
- To keep the connecting portion clean and to avoid damage and foreign materials, cover them completely with plastic bags or a similar material.

5. Disconnect the fuel damper and fuel hose assembly from the left-hand and right-hand fuel tubes.

⁑ CAUTION

While the hoses are disconnected, plug them to prevent fuel from leaking and do not separate the fuel damper and fuel hose.

6. Remove the upper intake manifold. Refer to Intake Manifold, removal & installation.

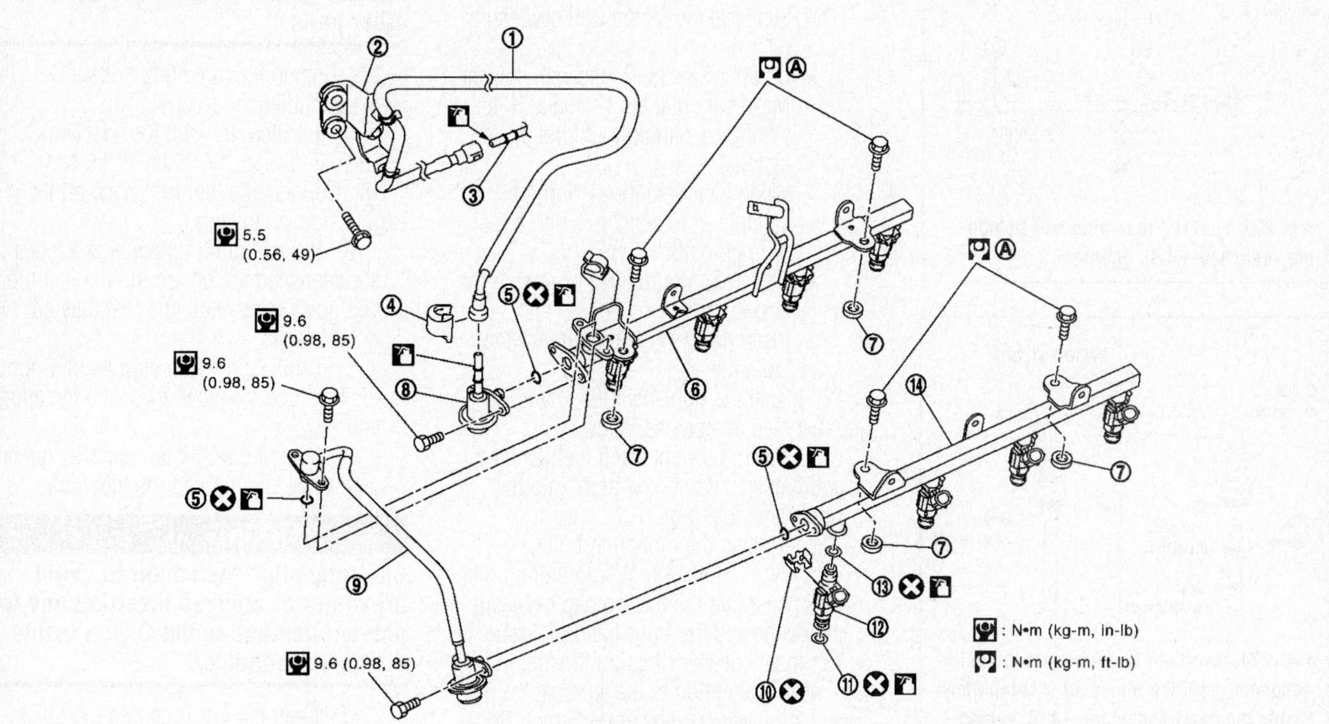

Fig. 221 Exploded view of fuel rail and injector mounting and related components—4.5L engine

5.5 (0.56, 49)

9.6 (0.98, 85)

9.6 (0.98, 85)

⑤⊗🛢

⑤⊗🛢

9.6 (0.98, 85)

⑤⊗🛢

⑬⊗🛢

⑫

⑩⊗   ⑪⊗🛢

🔧 : N•m (kg-m, in-lb)
🔧 : N•m (kg-m, ft-lb)

1. Fuel feed hose	2. Fuel feed hose bracket	3. Centralized under-floor piping
4. Quick connector cap	5. O-ring	6. Fuel tube (RH)
7. Spacer	8. Fuel feed damper	9. Fuel damper and fuel hose assembly
10. Clip	11. O-ring (Green)	12. Fuel injector
13. O-ring (Black)	14. Fuel tube (LH)	

22140_FX35_G0056

Fig. 221 Exploded view of fuel rail and injector mounting and related components—4.5L engine

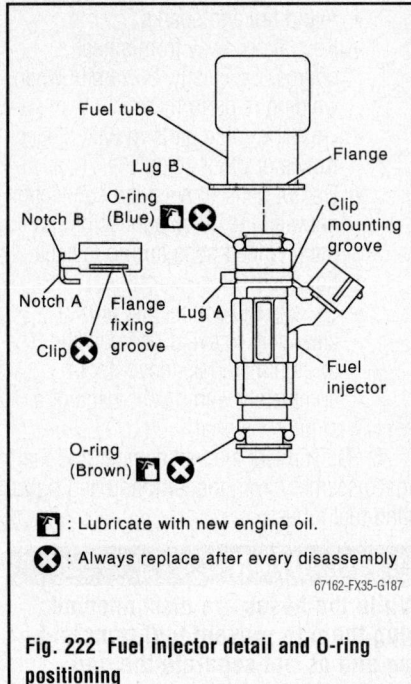

Fig. 222 Fuel injector detail and O-ring positioning

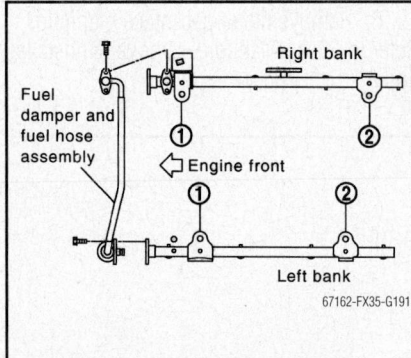

Fig. 223 Fuel rail mounting bolt tightening sequence—4.5L engine

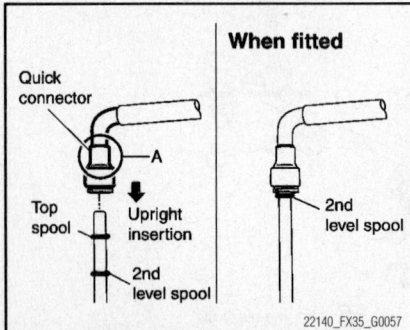

Fig. 224 Insert the fuel tube into the quick connector until the top spool is completely inside the quick connector—4.5L engine

7. Disconnect the wiring harness connector from the fuel injector.

8. Loosen the mounting bolts in the reverse order of the tightening sequence and remove the fuel tube and fuel injector assembly.

⁂ CAUTION

Do not tilt the fuel injector assembly, or remaining fuel in the pipes may flow out from the pipes.

9. Remove the spacers on the lower intake manifold.

10. Remove the fuel injector from the fuel tube, as follows:

　a. Open and remove the clip.

　b. Remove the fuel injector from the fuel tube by pulling it straight out.

　c. During injector removal, heed the following items:

- Be careful with the remaining fuel that leaks from the fuel tube
- Be careful not to damage the injector nozzles during removal
- Do not bump or drop the fuel injectors
- Do not disassemble the fuel injectors

11. Remove the right-hand fuel damper and fuel feed damper.

To install:

When handling all O-rings in this procedure, heed the following:

- Handle the O-ring with bare hands. Never wear gloves
- Lubricate the O-ring with new engine oil
- Do not clean the O-ring with solvent
- Make sure that the O-ring and its mating part are free of foreign material
- When installing the O-ring, be careful not to scratch it with a tool or fingernails
- Be careful not to twist or stretch the O-ring
- Insert the O-ring straight into the fuel tube

12. Install the right-hand fuel damper and fuel feed damper, as follows:

　a. Insert the right-hand fuel damper and fuel feed damper straight into the right-hand fuel tube.

　b. Tighten the mounting bolts.

　c. After tightening the mounting bolts, make sure that there is no gap between the flange and the right-hand fuel tube.

13. Install new O-rings on the fuel injectors, paying attention to the fact that the upper and lower O-rings are different. Be careful not to confuse them. The fuel tube side O-ring is black, whereas the nozzle side O-ring is green.

14. Install the fuel injector onto the fuel tube, as follows:

　a. Insert the clip into clip mounting groove on the fuel injector. Insert the clip

so that lug "A" of the fuel injector matches notch "A" of the clip. Do not reuse clips; replace them with new ones.

➡**Be careful to keep the clip from interfering with the O-ring. If interference occurs, replace the O-ring with a new one.**

　b. Insert the fuel injector into the fuel tube with the clip attached. Insert the fuel injector while matching it to the axial center. Insert the fuel injector so that lug "B" of the fuel tube matches notch "B" of the clip. Make sure that the fuel tube flange is securely fixed in the flange fixing groove on the clip.

　c. Make sure that installation is complete by checking that the fuel injector does not rotate or come off. Make sure that the protrusions of the fuel injectors are aligned with the cutouts of the clips after installation.

15. Install spacers on the lower intake manifold.

16. Install the fuel tube and fuel injector assembly onto the intake manifold.

⁂ WARNING

Be careful not to let the tip of the injector nozzle come in contact with other parts.

17. Tighten the mounting bolts in 2 steps, in numerical order shown:

　a. 1st Step: 84 inch lbs. (10 Nm).

　b. 2nd Step: 17 ft. lbs. (23.5 Nm).

18. Connect the fuel feed hose on the engine side, as follows:

　a. Make sure no foreign substances are deposited in and around the fuel tube and quick connector, and that they are not damaged.

　b. Thinly apply new engine oil around the fuel tube from the tip end to the spool end.

　c. Align the center to insert the quick connector straight into the fuel tube.

⁂ WARNING

Carefully align the center to avoid off-center or crooked insertion and to prevent damage to the O-ring inside the quick connector.

　d. Insert the fuel tube into the quick connector until the top spool is completely inside the quick connector, and the 2nd level spool is exposed right below the quick connector. Hold position "A" when inserting the fuel tube into the quick connector. Insert until you hear a "click" sound and actually feel the engagement. To avoid misidentification of engagement with a similar sound, be sure to perform the next step.

e. Pull the quick connector by hand holding position "A". Make sure it is completely engaged (connected) so that it does not disconnect from fuel tube.

f. Install the quick connector cap on quick connector connection on the engine side only.

g. Install the fuel feed hose to the hose clamps.

19. Perform the preceding step and associated sub steps for the side of the centralized under-floor piping as well.

20. The remainder of installation is the reverse of removal.

21. Perform the following once installation is complete.

a. After installing the fuel tubes, make sure there is no fuel leakage at the connections.

b. Apply fuel pressure to the fuel lines by turning the ignition switch **ON** with the engine **OFF**. Then check for fuel leaks at all connections.

c. Start the engine, and while holding it at a high RPM, check for fuel leaks at all connections.

➡**Use mirrors to check hard-to-see connections.**

FUEL PUMP

REMOVAL & INSTALLATION

See Figures 225 through 227.

1. Before servicing the vehicle, refer to the Precautions Section.

2. Check the fuel level on the fuel gauge. If the fuel gauge indicates full or almost full, drain the fuel from the fuel tank until the gauge indicates a level near ¾ of a tank.

✳✳ CAUTION

Fuel will be spilled when removing the main and sub fuel level sensor units if the level of the fuel in the tank is higher than about ¾ of a tank.

3. In the case that the fuel pump does not operate, perform the following procedure:

a. Insert a hose of less than 1 inch (25mm) in diameter into the fuel filler tube through the fuel filler opening to draw fuel from the fuel filler tube.

b. Disconnect the fuel filler hose from the fuel filler tube.

c. Insert the fuel tube into the fuel tank through the fuel filler hose to draw the fuel from the fuel tank.

4. Release the fuel pressure from the fuel lines.

5. Open the fuel filler lid.

6. Open the filler cap and release the pressure inside the fuel tank.

7. Remove the rear seat cushion, as follows:

a. Pull the lock at the front bottom of the seat cushion forward (1 for each side).

b. Pull the seat cushion upward to release the retaining wire from the plastic hook.

c. Pull the seat cushion forward to remove.

8. Lift up the floor carpet, then remove the inspection hole cover for the main and sub fuel level sensor units by turning the retaining clips clockwise by 90°.

9. Disconnect the wiring harness connector and fuel feed tube.

10. Disconnect the fuel line quick connector, as follows:

a. Hold the sides of the connector, push in the tabs and pull out the tube.

➡**If the quick connector sticks to the tube of the main fuel level sensor unit, push and pull the quick connector several times until they start to move. Then, disconnect them by pulling.**

When dealing with the fuel line quick connector, heed the following:

- The quick connector can be disconnected when the tabs are completely depressed. Do not twist it more than necessary
- Do not use any tools to disconnect the quick connector
- Keep the resin tube away from heat. Be especially careful when welding near the resin tube
- Prevent acidic liquid such as battery electrolyte from getting on the resin tube
- Do not bend or twist the resin tube during connection and disconnection
- Do not remove the remaining retainer on the hard tube (or the equivalent) except when the resin tube or retainer is replaced
- When the resin tube or hard tube (or the equivalent) is replaced, also replace the retainer with a new one
- To keep the connecting portion clean and to avoid damage and foreign materials, cover them completely with plastic bags or a similar material.

✳✳ WARNING

Make sure to not bend the float arm during removal, and avoid impacts, such as falling, when handling components.

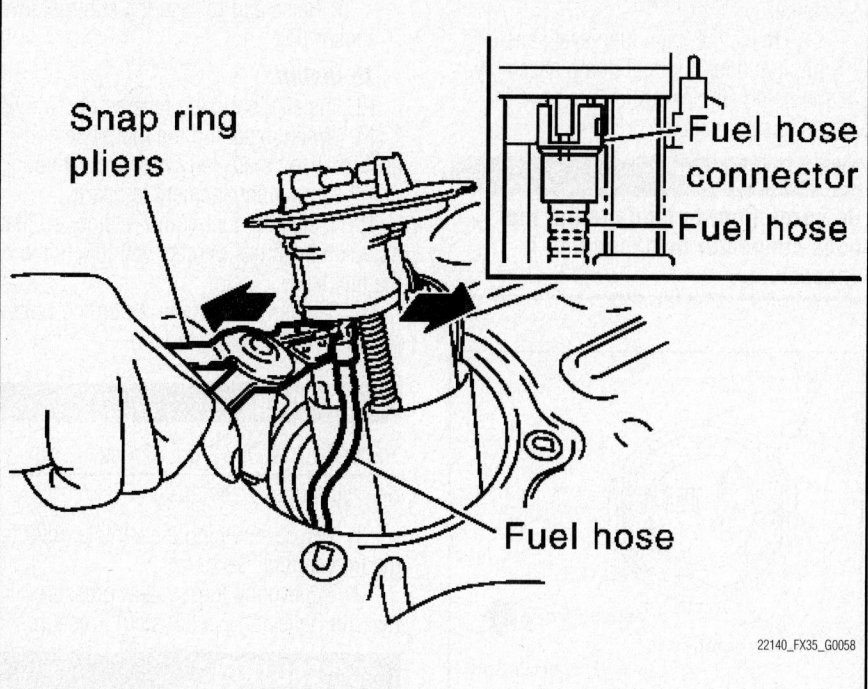

22140_FX35_G0058

Fig. 225 Raise the fuel pump assembly and using snapring pliers, remove the fuel hose connector

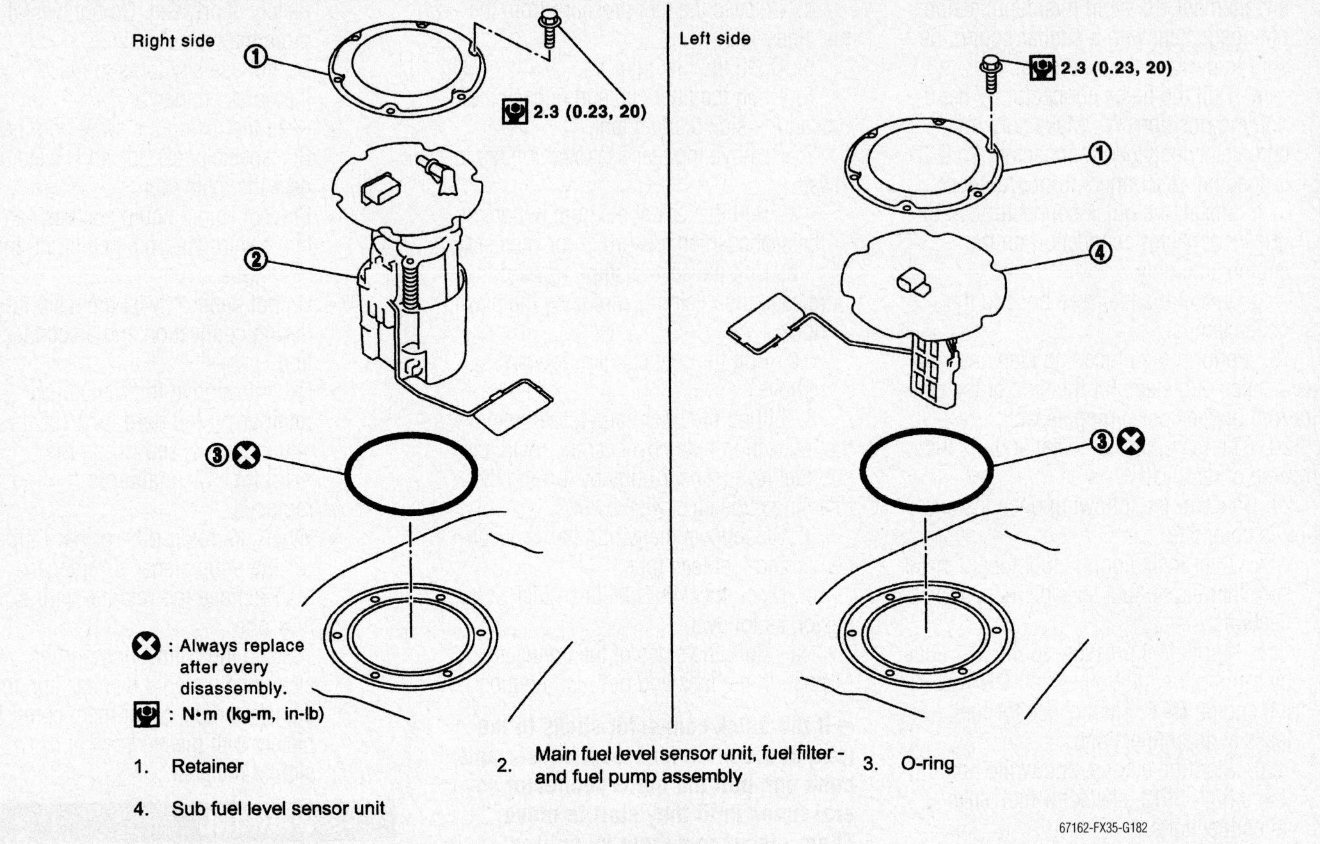

Right side ① Left side

⬛ 2.3 (0.23, 20)

⬛ 2.3 (0.23, 20)

① ④

②

③✕ ③✕

✕ : Always replace after every disassembly.

⬛ : N•m (kg-m, in-lb)

1. Retainer
2. Main fuel level sensor unit, fuel filter - and fuel pump assembly
3. O-ring
4. Sub fuel level sensor unit

67162-FX35-G182

Fig. 226 Exploded view of fuel pump and filter assembly and related components

11. Remove the main fuel level sensor unit, fuel filter, and fuel pump assembly, and sub fuel level sensor unit, as follows:

a. Remove the main fuel sensor unit retainer.

b. Raise the main fuel level sensor unit, fuel filter and fuel pump assembly, and using snapring pliers, remove the fuel hose connector.

※※ WARNING

Be careful not to damage the fuel hose connector by expanding it excessively.

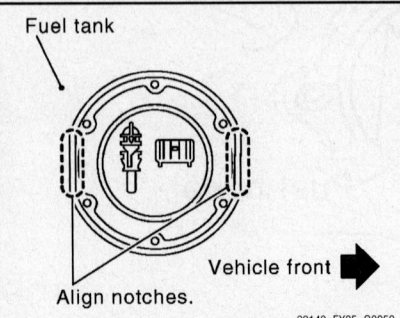

Fuel tank

Vehicle front ➡

Align notches.

22140_FX35_G0059

Fig. 227 Install the fuel pump retainer so that its notch becomes parallel with the notch on the fuel tank

12. Removal of sub fuel level sensor unit:

a. Remove the sub fuel level sensor unit retainer.

b. Raise and release the sub fuel level sensor unit.

To install:

13. Installation is the reverse of removal.

14. When installing the fuel hose connectors insert them fully until a click sound of full stopper engagement is heard.

15. Install the fuel pump retainer so that its notch becomes parallel with the notch on the fuel tank.

16. Tighten the retainer mounting bolts evenly to 20 inch lbs. (2 Nm).

FUEL TANK

REMOVAL & INSTALLATION

See Figures 228 and 229.

1. Before servicing the vehicle, refer to the Precautions Section.

2. Relieve the fuel system pressure. Refer to Relieving Fuel System Pressure.

※※ CAUTION

The fuel injection system remains under pressure even after the engine

has been turned OFF. Properly relieve fuel pressure before disconnecting any fuel lines. Failure to do so may result in fire or personal injury. Do not allow fuel spray or fuel vapors to come in contact with a spark or an open flame. Keep a dry chemical fire extinguisher nearby. Never store fuel in an open container due to risk of fire or explosion.

3. Drain the fuel from the tank.
4. Remove the negative battery cable.
5. Remove the rear seat cushion assembly.

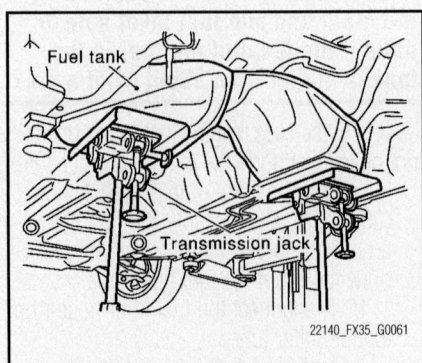

Fuel tank

Transmission jack

22140_FX35_G0061

Fig. 228 Support the lower part of fuel tank with a transmission jack

6. Remove the fuel pump module. Refer to Fuel Pump, removal & installation.

7. Remove the tunnel stay.

8. Remove the exhaust front tube, center muffler, and main muffler.

9. Remove the insulator.

10. Remove the propeller shaft.

11. Remove the parking rear brake cables.

12. Compress the coil springs and remove them.

13. Remove the rear suspension assembly.

14. Remove the fuel tank protector.

15. Disconnect the fuel filler hose, vent hose, and EVAP hoses at the fuel tank side.

16. Support the lower part of fuel tank with a transmission jack.

➡ **Support the fuel tank in a position that fuel tank mounting bands do not engage.**

17. Remove the fuel tank mounting bands.

18. Help support the fuel tank and lower the transmission jack carefully to remove the fuel tank.

➡ **Make sure that all the connection points have been disconnected. Con-** firm there is no interference with the vehicle while lowering the fuel tank.

19. Remove fuel filler tube protector and fuel filler tube, if necessary.

To install:

20. Installation is the reverse of the removal procedure.

21. Tighten the retaining bolts and fasteners to specification as illustrated.

22. Perform the following once installation is complete.

a. After installing the fuel tubes, make sure there is no fuel leakage at the connections.

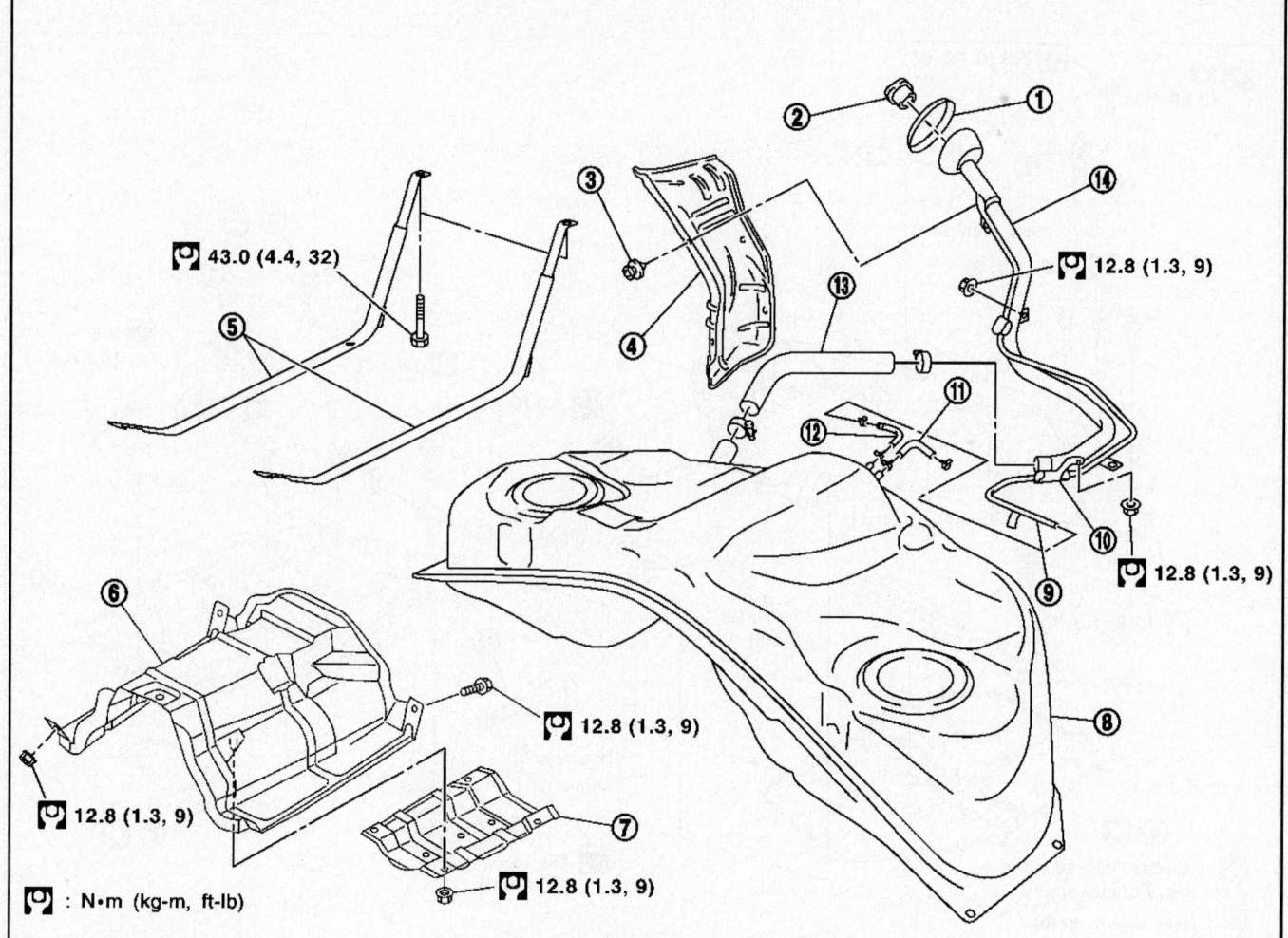

43.0 (4.4, 32)

12.8 (1.3, 9)

12.8 (1.3, 9)

12.8 (1.3, 9)

12.8 (1.3, 9)

12.8 (1.3, 9)

: N•m (kg-m, ft-lb)

1. Grommet	2. Fuel filler cap	3. Clip
4. Fuel filler tube protector	5. Fuel tank mounting band	6. Fuel tank protector
7. Insulator	8. Fuel tank	9. Vent tube
10. Vent hose	11. EVAP hose	12. Vent tube
13. Fuel filler hose	14. Fuel filler tube	

22140_FX35_G0060

Fig. 229 Exploded view of fuel tank and related components

b. Apply fuel pressure to the fuel lines by turning the ignition switch **ON** with the engine **OFF**. Then check for fuel leaks at all connections.

c. Start the engine, and while holding it at a high RPM, check for fuel leaks at all connections.

➡**Use mirrors to check hard-to-see connections.**

IDLE SPEED

ADJUSTMENT

Idle speed is maintained by the Elec-tronic Control Module (ECM). No adjust-ment is necessary or possible.

THROTTLE BODY

REMOVAL & INSTALLATION

3.5L Engine

See Figures 230 and 231.

1. Before servicing the vehicle, refer to the Precautions Section.
2. Disconnect the negative battery cable.
3. Remove the engine cover.

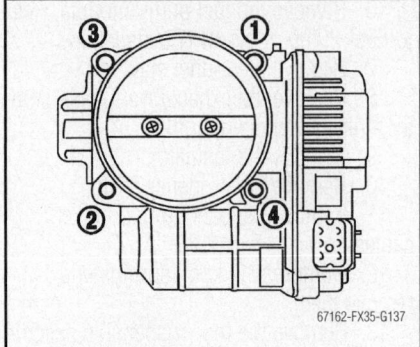

67162-FX35-G137

Fig. 230 Throttle body mounting bolt tightening sequence—3.5L engine

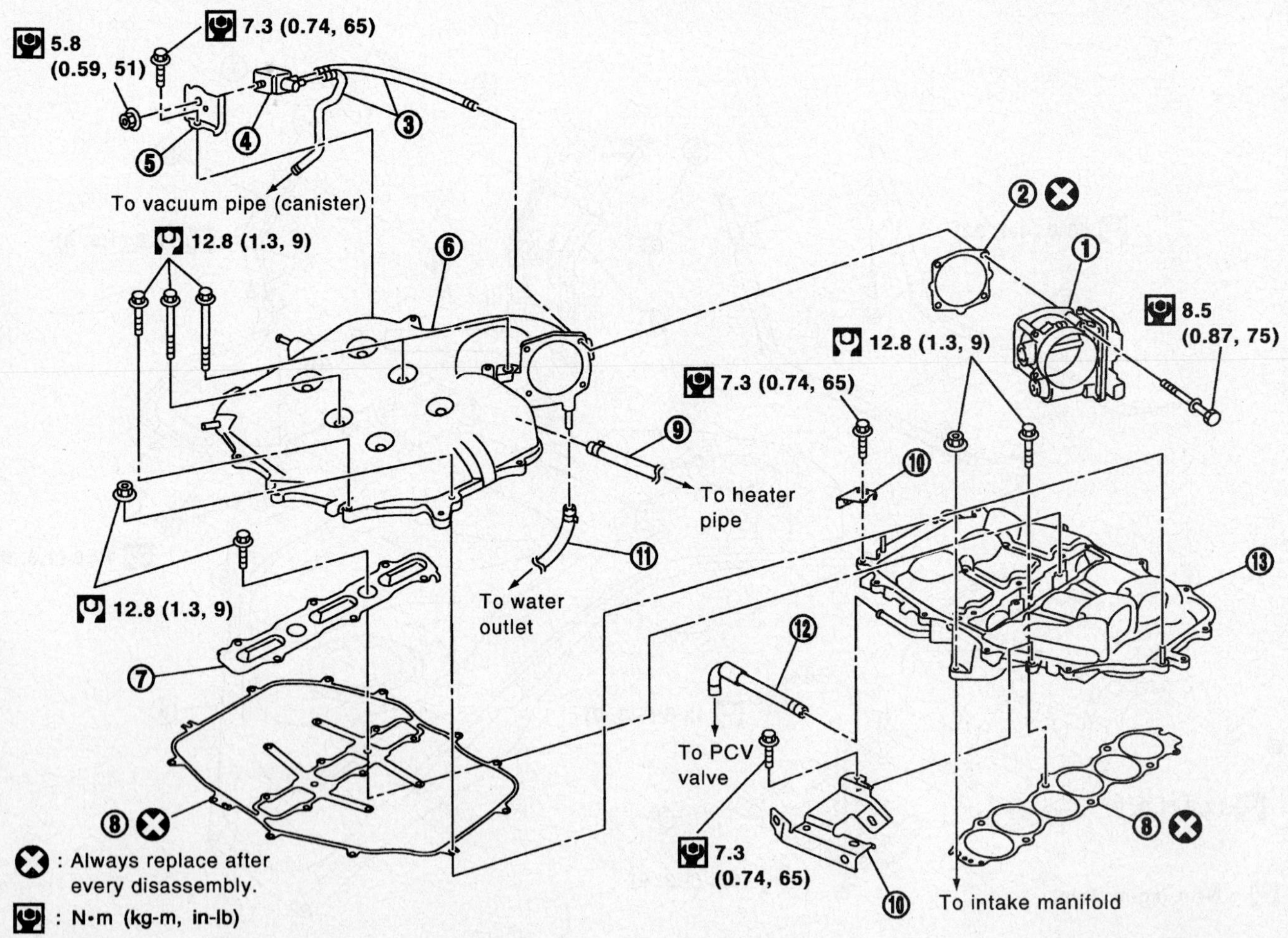

⊗ : Always replace after every disassembly.

🔧 : N•m (kg-m, in-lb)

1. Electric throttle control actuator
2. Gasket
3. Vacuum hose
4. EVAP canister purge volume control solenoid valve
5. Bracket
6. Intake manifold collector (upper)
7. Intake manifold collector cover
8. Gasket
9. Water hose
10. Bracket
11. Water hose
12. PCV hose
13. Intake manifold collector (lower)

67162-FX35-G136

Fig. 231 Exploded view of the upper and lower halves of the upper intake manifold—3.5L engine

4. Disconnect and plug the water hoses from the intake manifold collector (upper).

➡**Do not spill engine coolant on the drive belts.**

5. Remove the air cleaner case and air duct, as follows:

    a. Remove the air duct (inlet).

    b. Disconnect the mass air flow sensor wiring harness connector.

    c. Remove the air cleaner case/mass air flow sensor assembly and the air duct/resonator assembly disconnecting them at the joints.

➡**Add match marks as necessary for easier installation.**

    d. Remove the mass air flow sensor from air cleaner case.

❄❄ WARNING

Handle the mass air flow sensor with care. Do not expose it to harsh vibration or shock. Do not disassemble it or touch its sensor.

    e. Remove the resonator in the fender, lifting the left fender protector.

6. Remove the electric throttle body control actuator, as follows:

    a. Disconnect the wiring harness connector.

    b. Loosen the bolts in the reverse order as shown in the figure.

❄❄ WARNING

Handle the throttle body carefully to avoid any shock to the electric throttle control actuator. Do not disassemble.

To install:

7. Install the electric throttle control actuator. Tighten the bolts to 75 inch lbs. (9 Nm).

8. Install the mass air flow sensor in the air cleaner case.

9. Install the air cleaner case and air duct.

10. Reconnect the hoses to the intake manifold collector (upper).

11. Install the engine cover.

12. Connect the negative battery cable.

13. Perform the following drivability adjustments:

14. Perform the "Throttle Valve Closed Position Learning" procedure (below) when the wiring harness connector of the electric throttle control actuator is disconnected, or perform the "Idle Air Volume Learning" and "Throttle Valve Closed Position Learning" procedures (below) when the electric throttle control actuator is replaced.

Throttle Valve Closed Position Learning

1. Before servicing the vehicle, refer to the Precautions Section.

The Throttle Valve Closed Position Learning procedure is an operation for the ECM to relearn the fully closed position of the throttle valve by monitoring the throttle position sensor output signal. It must be performed each time the wiring harness connector of the electric throttle control actuator or ECM is disconnected.

2. Make sure that accelerator pedal is fully released.

3. Turn ignition switch **ON**.

4. Turn ignition switch **OFF** wait at least 10 seconds. Make sure that throttle valve moves during the above 10 seconds by confirming the operating sound.

Idle Air Volume Learning

1. Before servicing the vehicle, refer to the Precautions Section.

Idle Air Volume Learning is an operation to learn the idle air volume that keeps each engine within the specific range. It must be performed under any of the following conditions:

- Each time the electric throttle control actuator or ECM is replaced
- Idle speed or ignition timing is out of specification

Before performing the "Idle Air Volume Learning" procedure, make sure that all of the following conditions are satisfied. Learning will be cancelled if any of the following conditions are missed for even a moment.

- Battery voltage: More than 12.9 volts (at idle)
- Engine coolant temperature: 158–212°F (70–100°C)
- PNP switch: ON
- Electric load switch: OFF (air conditioner, headlamp, and rear window defogger)

➡**On vehicles equipped with daytime light systems, if the parking brake is applied before the engine is started, the headlamp will not be illuminated.**

- Steering wheel: Neutral (straight-ahead position)
- Vehicle speed: Stopped
- Transmission: Warmed-up
- For models with CONSULT-III, drive vehicle until "FLUID TEMP SE 1" in "DATA MONITOR" mode of "A/T" system indicates less than 0.9 volts.
- For models without CONSULT-III, drive vehicle for 10 minutes.

2. If using the CONSULT-III tool, perform the following:

    a. Perform the "Accelerator Pedal Released Position Learning" procedure.

    b. Perform the "Throttle Valve Closed Position Learning" procedure.

    c. Start the engine and warm it up to normal operating temperature.

    d. Check that all items listed above are properly set.

    e. Select "IDLE AIR VOL LEARN" in "WORK SUPPORT" mode.

    f. Touch "START" and wait 20 seconds.

    g. Make sure that "CMPLT" is displayed on CONSULT-III screen. If "CMPLT" is not displayed, the Idle Air Volume Learning procedure will not be carried out successfully.

    h. Rev up the engine 2–3 times and make sure that idle speed and ignition timing are within specifications.

3. If NOT using the CONSULT-III tool, perform the following:

➡**It is best to keep track of time accurately with a clock.**

➡**It is impossible to switch the diagnostic mode when an accelerator pedal position sensor circuit has a malfunction.**

    a. Perform the "Accelerator Pedal Released Position Learning" procedure.

    b. Perform the "Throttle Valve Closed Position Learning" procedure.

    c. Start the engine and warm it up to normal operating temperature.

    d. Check that all items listed above are properly set.

    e. Turn the ignition switch OFF and wait at least 10 seconds.

    f. Confirm that the accelerator pedal is fully released, turn the ignition switch ON and wait 3 seconds.

➡**Repeat the following 2 steps quickly 5 times within 5 seconds.**

    g. Fully depress the accelerator pedal.

    h. Fully release the accelerator pedal.

    i. Wait 7 seconds, fully depress the accelerator pedal and keep it for approx. 20 seconds until the MIL stops blinking and remains ON.

    j. Fully release the accelerator pedal within 3 seconds after the MIL turned ON.

    k. Start the engine and let it idle.

    l. Wait 20 seconds.

    m. Rev up the engine 2–3 times and make sure that idle speed and ignition timing are within specifications.

n. If the idle speed and ignition timing are not within specification, the Idle Air Volume Learning procedure will not be successful.

Accelerator Pedal Released Position Learning

1. Before servicing the vehicle, refer to the Precautions Section.

The "Accelerator Pedal Released Position Learning" procedure is an operation for the ECM to relearn the fully released position of the accelerator pedal by monitoring the accelerator pedal position sensor output signal. It must be performed each time the wiring harness connector of the accelerator pedal position sensor or ECM is disconnected.

2. Make sure that the accelerator pedal is fully released.

3. Turn the ignition switch **ON** and wait at least 2 seconds.

4. Turn the ignition switch **OFF** wait at least 10 seconds.

5. Turn the ignition switch **ON** and wait at least 2 seconds.

6. Turn the ignition switch **OFF** wait at least 10 seconds.

4.5L Engine
See Figure 232.

1. Before servicing the vehicle, refer to the Precautions Section.

2. Remove the engine cover.

3. Release the fuel pressure.

4. Remove the air duct (inlet), air cleaner case and mass air flow sensor assembly, air duct and resonator assembly.

5. Disconnect the fuel feed hose quick connector on the engine side, the fuel damper and fuel hose assembly.

❊❊ CAUTION

While hoses are disconnected, plug them to prevent fuel from draining. Do not separate the fuel damper and the fuel hose.

6. Remove the electric throttle body control actuator as follows:

a. Disconnect the wiring harness connector.

b. Loosen the mounting bolts diagonally.

c. Remove the electric throttle control actuator.

❊❊ WARNING

Handle the throttle body carefully to avoid any shock to the electric

throttle control actuator. Do not disassemble the electric throttle body control actuator.

To install:

7. Install the electric throttle body control actuator. Install the intake manifold adapter gasket and electric throttle control actuator gasket so that the 3 protrusions for installation do not face downward.

❊❊ WARNING

Handle the throttle body carefully to avoid any shock to the electric throttle control actuator.

8. Tighten the mounting bolts of the electric throttle control actuator equally and diagonally in several steps to 80 inch lbs. (9 Nm).

9. Connect the fuel feed hose quick connector on the engine side, the fuel damper and fuel hose assembly.

10. Install the air duct (inlet), air cleaner case and mass air flow sensor assembly, air duct and resonator assembly.

11. Install the engine cover.

12. Perform the following drivability adjustments:

13. Perform the "Throttle Valve Closed Position Learning" procedure (below) when the wiring harness connector of the electric throttle control actuator is disconnected, or perform the "Idle Air Volume Learning" and "Throttle Valve Closed Position Learning" procedures (below) when the electric throttle control actuator is replaced.

Throttle Valve Closed Position Learning

1. Before servicing the vehicle, refer to the Precautions Section.

The Throttle Valve Closed Position Learning procedure is an operation for the ECM to relearn the fully closed position of the throttle valve by monitoring the throttle position sensor output signal. It must be performed each time the wiring harness connector of the electric throttle control actuator or ECM is disconnected.

2. Make sure that accelerator pedal is fully released.

3. Turn ignition switch **ON**.

4. Turn ignition switch **OFF** wait at least 10 seconds. Make sure that throttle valve moves during the above 10 seconds by confirming the operating sound.

Idle Air Volume Learning

1. Before servicing the vehicle, refer to the Precautions Section.

Idle Air Volume Learning is an operation to learn the idle air volume that keeps each engine within the specific range. It must be performed under any of the following conditions:

- Each time the electric throttle control actuator or ECM is replaced
- Idle speed or ignition timing is out of specification
- Before performing the "Idle Air Volume Learning" procedure, make sure that all of the following conditions are satisfied. Learning will be cancelled if any of the following conditions are missed for even a moment.
- Battery voltage: More than 12.9 volts (at idle)
- Engine coolant temperature: 158–212°F (70–100°C)
- PNP switch: ON
- Electric load switch: OFF (air conditioner, headlamp, and rear window defogger)

➡ **On vehicles equipped with daytime light systems, if the parking brake is applied before the engine is started, the headlamp will not be illuminated.**

- Steering wheel: Neutral (straight-ahead position)
- Vehicle speed: Stopped
- Transmission: Warmed-up
- For models with CONSULT-III, drive vehicle until "FLUID TEMP SE 1" in "DATA MONITOR" mode of "A/T" system indicates less than 0.9 volts.
- For models without CONSULT-III, drive vehicle for 10 minutes.

2. If using the CONSULT-III tool, perform the following:

a. Perform the "Accelerator Pedal Released Position Learning" procedure.

b. Perform the "Throttle Valve Closed Position Learning" procedure.

c. Start the engine and warm it up to normal operating temperature.

d. Check that all items listed above are properly set.

e. Select "IDLE AIR VOL LEARN" in "WORK SUPPORT" mode.

f. Touch "START" and wait 20 seconds.

g. Make sure that "CMPLT" is displayed on CONSULT-III screen. If "CMPLT" is not displayed, the Idle Air Volume Learning procedure will not be carried out successfully.

h. Rev up the engine 2–3 times and make sure that idle speed and ignition timing are within specifications.

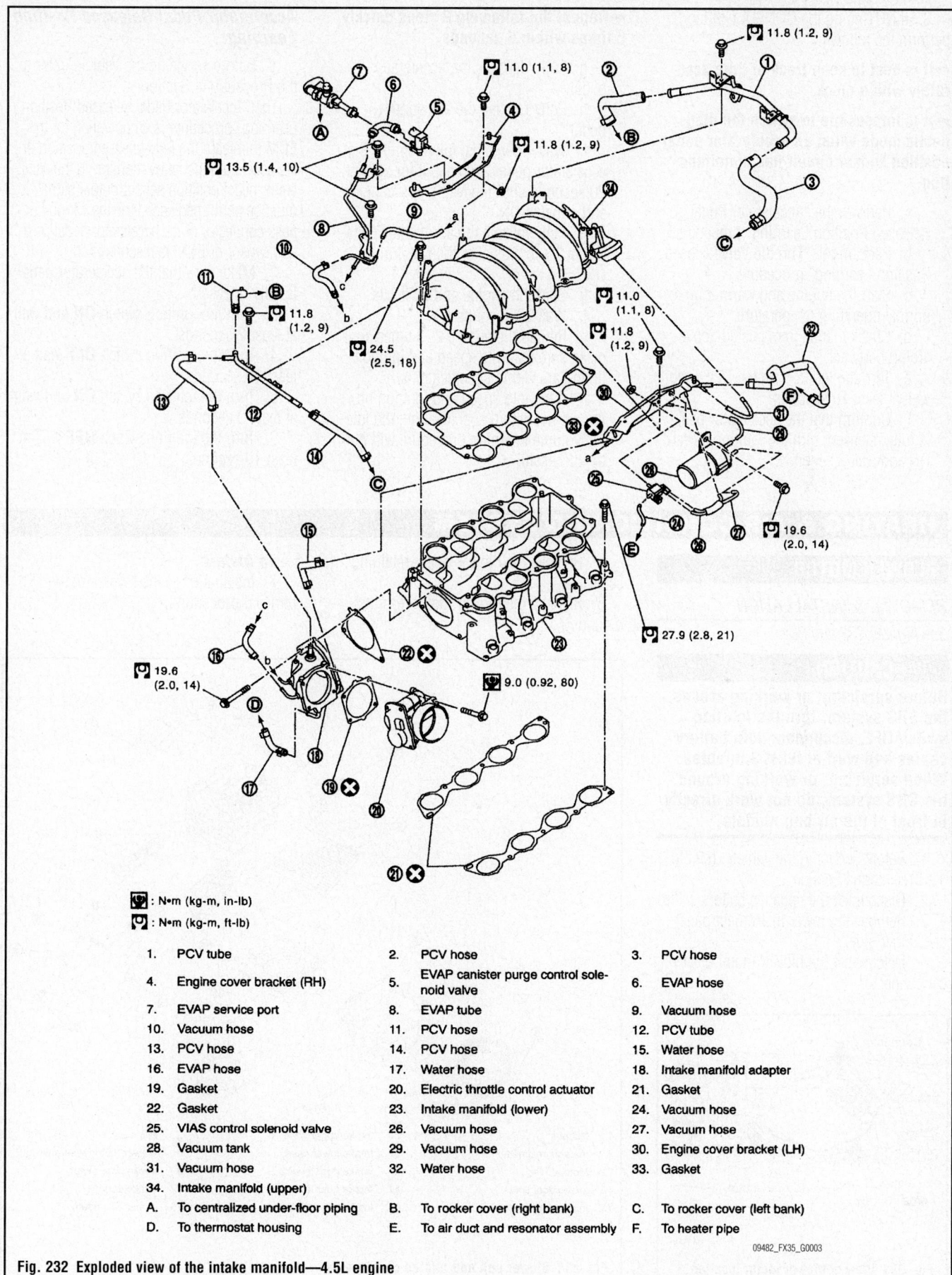

: N•m (kg-m, in-lb)

: N•m (kg-m, ft-lb)

1.	PCV tube	2.	PCV hose	3.	PCV hose
4.	Engine cover bracket (RH)	5.	EVAP canister purge control solenoid valve	6.	EVAP hose
7.	EVAP service port	8.	EVAP tube	9.	Vacuum hose
10.	Vacuum hose	11.	PCV hose	12.	PCV tube
13.	PCV hose	14.	PCV hose	15.	Water hose
16.	EVAP hose	17.	Water hose	18.	Intake manifold adapter
19.	Gasket	20.	Electric throttle control actuator	21.	Gasket
22.	Gasket	23.	Intake manifold (lower)	24.	Vacuum hose
25.	VIAS control solenoid valve	26.	Vacuum hose	27.	Vacuum hose
28.	Vacuum tank	29.	Vacuum hose	30.	Engine cover bracket (LH)
31.	Vacuum hose	32.	Water hose	33.	Gasket
34.	Intake manifold (upper)				
A.	To centralized under-floor piping	B.	To rocker cover (right bank)	C.	To rocker cover (left bank)
D.	To thermostat housing	E.	To air duct and resonator assembly	F.	To heater pipe

09482_FX35_G0003

Fig. 232 Exploded view of the intake manifold—4.5L engine

3. If NOT using the CONSULT-III tool, perform the following:

➡️ **It is best to keep track of time accurately with a clock.**

➡️ **It is impossible to switch the diagnostic mode when an accelerator pedal position sensor circuit has a malfunction.**

a. Perform the "Accelerator Pedal Released Position Learning" procedure.

b. Perform the "Throttle Valve Closed Position Learning" procedure.

c. Start the engine and warm it up to normal operating temperature.

d. Check that all items listed above are properly set.

e. Turn the ignition switch OFF and wait at least 10 seconds.

f. Confirm that the accelerator pedal is fully released, turn the ignition switch ON and wait 3 seconds.

➡️ **Repeat the following 2 steps quickly 5 times within 5 seconds.**

g. Fully depress the accelerator pedal.

h. Fully release the accelerator pedal.

i. Wait 7 seconds, fully depress the accelerator pedal and keep it for approx. 20 seconds until the MIL stops blinking and remains ON.

j. Fully release the accelerator pedal within 3 seconds after the MIL turned ON.

k. Start the engine and let it idle.

l. Wait 20 seconds.

m. Rev up the engine 2–3 times and make sure that idle speed and ignition timing are within specifications.

n. If the idle speed and ignition timing are not within specification, the Idle Air Volume Learning procedure will not be successful.

Accelerator Pedal Released Position Learning

1. Before servicing the vehicle, refer to the Precautions Section.

The "Accelerator Pedal Released Position Learning" procedure is an operation for the ECM to relearn the fully released position of the accelerator pedal by monitoring the accelerator pedal position sensor output signal. It must be performed each time the wiring harness connector of the accelerator pedal position sensor or ECM is disconnected.

2. Make sure that the accelerator pedal is fully released.

3. Turn the ignition switch **ON** and wait at least 2 seconds.

4. Turn the ignition switch **OFF** wait at least 10 seconds.

5. Turn the ignition switch **ON** and wait at least 2 seconds.

6. Turn the ignition switch **OFF** wait at least 10 seconds.

HEATING & AIR CONDITIONING SYSTEM

BLOWER MOTOR

REMOVAL & INSTALLATION

See Figures 233 and 234.

✳✳ CAUTION

Before servicing, or working around, the SRS system, turn the ignition switch OFF, disconnect both battery cables and wait at least 3 minutes. When servicing, or working around, the SRS system, do not work directly in front of the air bag module.

1. Before servicing the vehicle, refer to the Precautions Section.

2. Disconnect the negative battery cable.

3. Remove the lower instrument panel, passenger side.

4. Disconnect the blower motor electrical connector.

5. Remove the blower motor retaining screws.

6. Remove the blower motor from its mounting.

To install:

7. Installation is the reverse of the removal procedure.

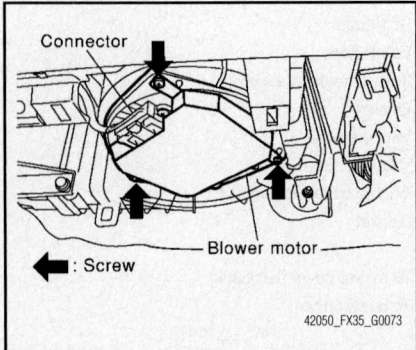

42050_FX35_G0073

Fig. 233 View of blower motor location

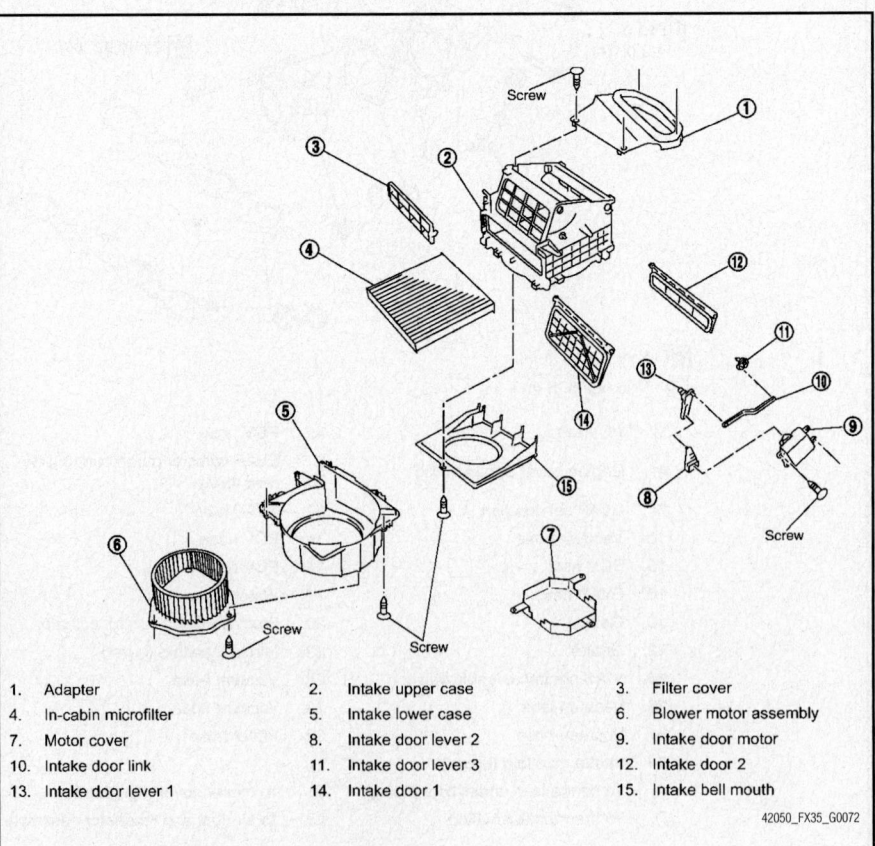

1.	Adapter	2.	Intake upper case	3.	Filter cover
4.	In-cabin microfilter	5.	Intake lower case	6.	Blower motor assembly
7.	Motor cover	8.	Intake door lever 2	9.	Intake door motor
10.	Intake door link	11.	Intake door lever 3	12.	Intake door 2
13.	Intake door lever 1	14.	Intake door 1	15.	Intake bell mouth

42050_FX35_G0072

Fig. 234 Blower unit and related components

HEATER CORE

REMOVAL & INSTALLATION

See Figures 235 through 243.

1. Before servicing the vehicle, refer to the Precautions Section.

2. Use approved refrigerant collecting equipment to discharge refrigerant.

✳✳ CAUTION

Make sure the engine is cold before draining the coolant.

3. Drain the coolant from the cooling system.

4. Remove the cowl top cover.

5. Remove the 2 high-pressure pipe mounting clips.

6. Remove the low-pressure flexible hose bracket mounting bolts.

7. Disconnect the evaporator-side one touch joint, as follows:

   a. Set disconnect SST: 9253089908 (high-pressure side) and SST: 9253089916 (low-pressure side) on the A/C piping.

   b. Slide the disconnect tool toward the front of the vehicle until it clicks.

   c. Slide the A/C pipe toward the front of the vehicle front and disconnect it.

➡Seal the connection opening of the pipe with a cap or vinyl tape to avoid exposure to atmosphere.

8. On 3.5L engines, remove the electronic control throttle assembly.

9. Disconnect the 2 heater hoses from the heater core.

10. Remove the instrument panel assembly, as follows:

   a. Remove the front kicking plate on both sides of the vehicle.

   b. Remove dash side finisher plastic nuts, then remove the dash side finisher.

   c. Pull to the inside of the vehicle, disengage the metal clips and remove the front pillar garnish.

   d. To remove the A/T Select Lever Knob, pull down the knob cover. Remove the lock-pin of the select lever knob. Then, lift up the select lever knob and remove it.

   e. Insert a remover into the side between the gaps of the instrument clock finisher and pull back to the side.

   f. Disconnect the clips and wiring harness connector, then remove the instrument clock finisher.

   g. Insert a remover into the side between the gaps of the A/T console fin-

isher and remove it by lifting the A/T console finisher.

   h. Disconnect the wiring harness connector.

   i. Remove the console finisher screws.

   j. Remove the console finishers.

   k. To remove the center console, remove the mounting screws, then remove the console sub-wiring harness.

✳✳ WARNING

When removing console, be careful not to pull the wiring harness.

   l. Remove the instrument lower cover by pulling down on the front instrument lower cover and disconnecting clips. Pull it horizontally, and remove it from the lower cover pawls.

   m. Remove the instrument passenger lower panel screws, disconnect the wiring harness connector, and remove the lower panel.

   n. Remove the instrument driver lower panel bolt and screws, detach the data link connector, pull to disengage the clip and pawl by removing panel in a horizontal direction. Then, disconnect the in-vehicle sensor and all electrical parts. Remove the grommet and remove the hood lock cable.

   o. Remove the steering column front lower cover screw, disengage the tab, and then remove the steering column front lower cover. Move the steering column telescopic to the rear most position, and move the steering column tilt to the top position.

   p. Remove the steering column lower cover screws, then disengage the tab and remove steering column lower cover.

   q. Remove the steering column upper cover.

   r. Remove the wiper and washer switch

   s. Remove the lighting and turn signal switch.

   t. Pull the steering lock escutcheon back and remove it.

   u. Remove the Combination Meter Assembly by removing the bolts and disconnecting the connector bracket. Remove the bolts and disconnect the wiring harness connector.

✳✳ WARNING

To prevent it from being damaged by interference with the combination meter assembly, protect the combination meter assembly with cloths.

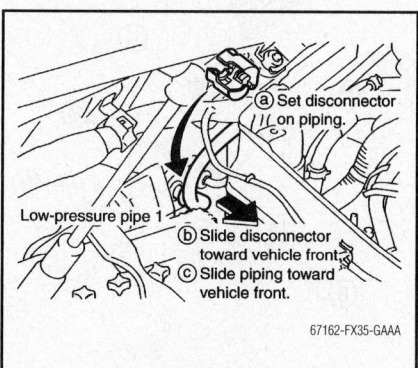

Fig. 235 Slide the disconnect tool toward the front of the vehicle until it clicks

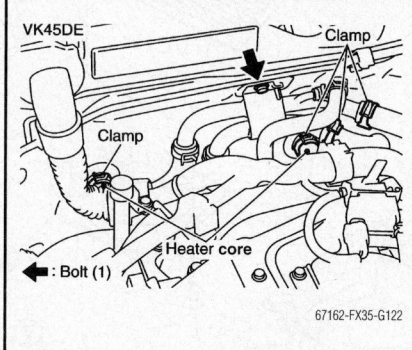

Fig. 237 Heater hose clamp locations for heater core service—4.5L engine

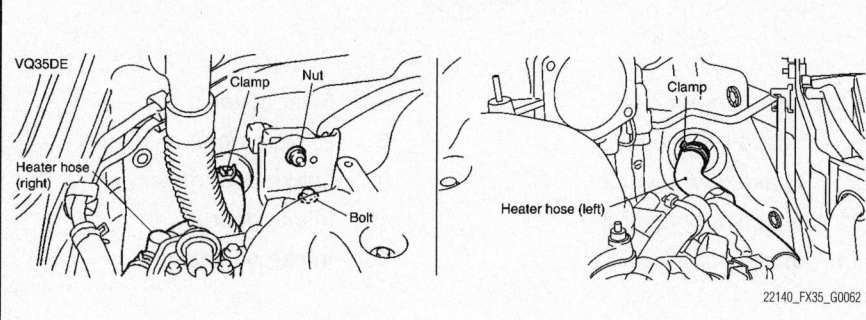

Fig. 236 Heater hose clamp locations for heater core service—3.5L engine

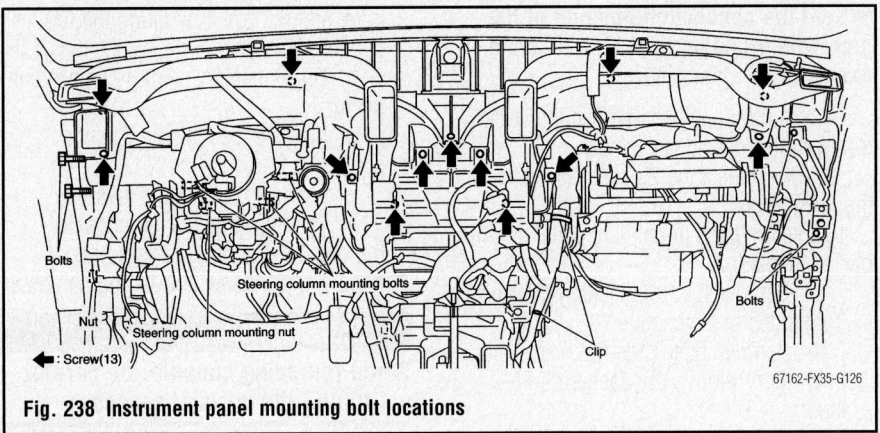

Fig. 238 Instrument panel mounting bolt locations

67162-FX35-G126

v. Remove the instrument panel side panel screws, then pull the panels to the side, disconnect the clip and pawls, and remove the instrument side panels. Perform for both right-hand and left-hand panels.

w. To remove the cluster lid, insert a pry tool into the gap between the instrument panel and pad, pull back towards you, and disconnect the metal clips. Then, disconnect the wiring harness connectors, and remove the cluster lid.

➡️**Cover surroundings with cloth to avoid making scratches or causing damage.**

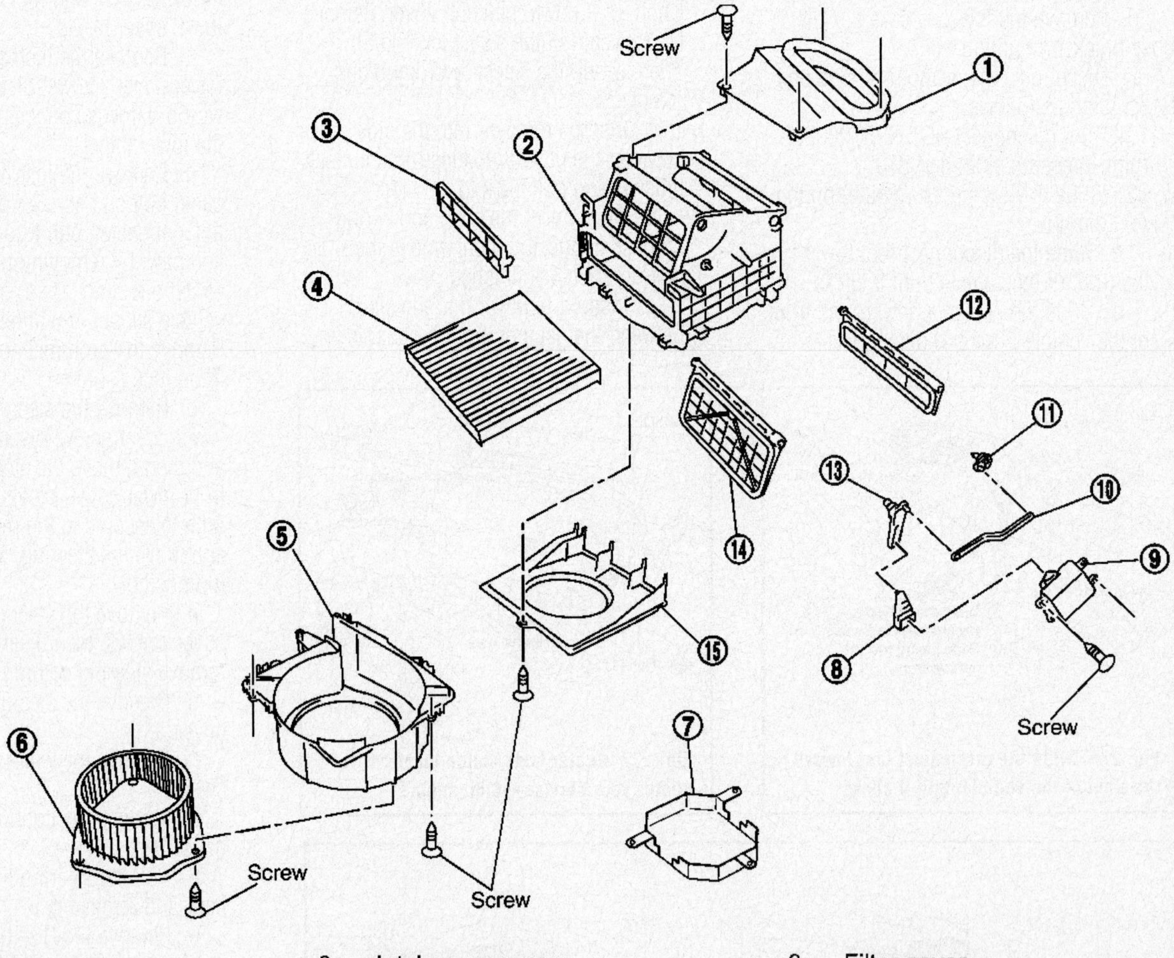

1.	Adapter	2.	Intake upper case	3.	Filter cover
4.	In-cabin microfilter	5.	Intake lower case	6.	Blower motor assembly
7.	Motor cover	8.	Intake door lever 2	9.	Intake door motor
10.	Intake door link	11.	Intake door lever 3	12.	Intake door 2
13.	Intake door lever 1	14.	Intake door 1	15.	Intake bell mouth

42050_FX35_G0072

Fig. 239 Blower unit and related components

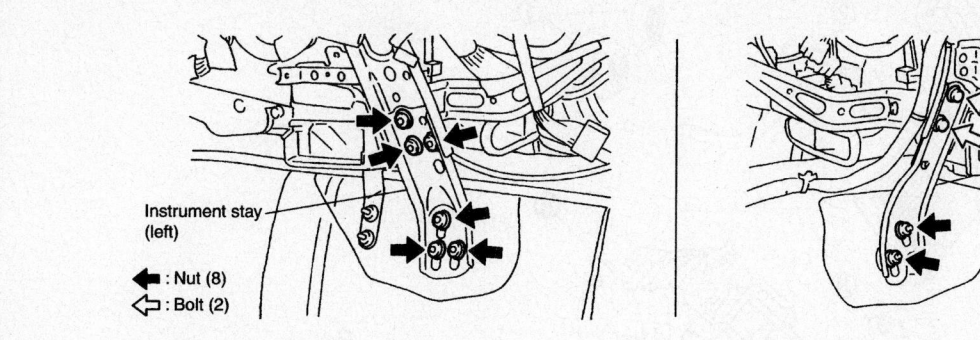

Instrument stay (left)

◀ : Nut (8)
◁ : Bolt (2)

Instrument stay (right)

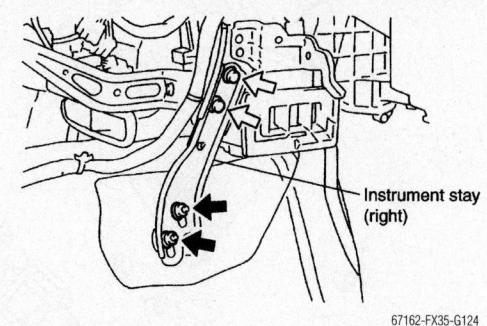

67162-FX35-G124

Fig. 240 Instrument panel stay mounting bolt locations

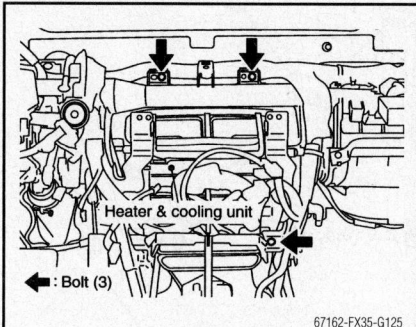

Heater & cooling unit

◀ : Bolt (3)

67162-FX35-G125

Fig. 241 Heater core assembly mounting bolt locations

x. Remove the display unit and audio unit by removing the screws, disconnecting the wiring harness connector, and removing the display unit and audio unit.

❋❋ CAUTION

The Unit is heavy, so be careful not to pinch your fingers when working.

y. Insert a thin pry tool into the gaps between the front defroster grille and instrument panel and pad, lift the front defroster grille upward, and remove the front defroster grille. Perform this task for both right-hand and left-hand grilles.

z. Remove the combination meter bracket bolts and remove the bracket from the vehicle.

aa. Once the mounting bolts of the wiring harness clip and steering column assembly are removed, pull the steering column assembly backward, and free the combination meter bracket from the instrument panel and pad.

bb. Remove the side ventilations by inserting a thin pry tool into the gaps between the instrument panel and pad, pull back to disconnect the metal retaining clips. Then, disconnect the door mirror switch wiring harness connectors, and remove the side ventilations.

cc. Remove the instrument panel and pad by removing the bolts and screws. Then, remove the front passenger air bag module, disconnect the wiring harness connectors, and remove the instrument panel and pad from the passenger door opening.

11. Remove the blower unit, as follows:
a. Remove the ECM with the bracket attached.
b. Disconnect the intake door motor connector and blower fan motor connector.
c. Remove the wiring harness clip from the blower unit.
d. Remove the mounting bolt and screws from the blower unit.

❋❋ CAUTION

Move the blower unit rightward, and remove the locating pin and joint. Then, remove the blower unit downward.

e. Remove the blower unit.
12. Remove the instrument stays (driver-side and passenger-side).
13. Remove the mounting bolts from the heater and cooling unit.
14. Disconnect the drain hose.
15. Remove the ventilator ducts, defroster nozzle and ducts.
16. Remove the steering member mounting bolts, nut and wiring harness clips.
17. Remove the steering member.

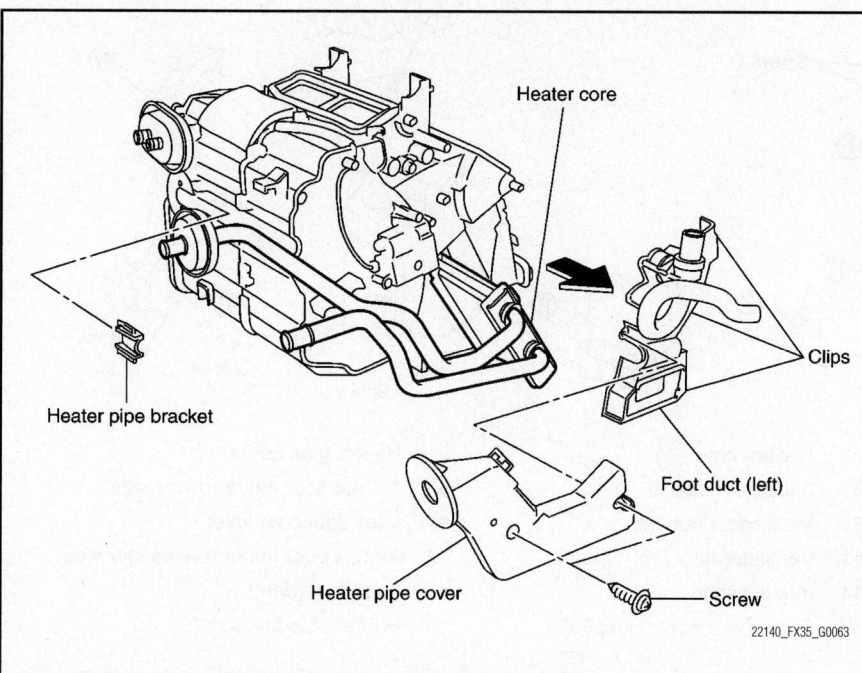

Heater core

Heater pipe bracket

Clips

Foot duct (left)

Heater pipe cover

Screw

22140_FX35_G0063

Fig. 242 Slide the heater core toward the driver's side to remove

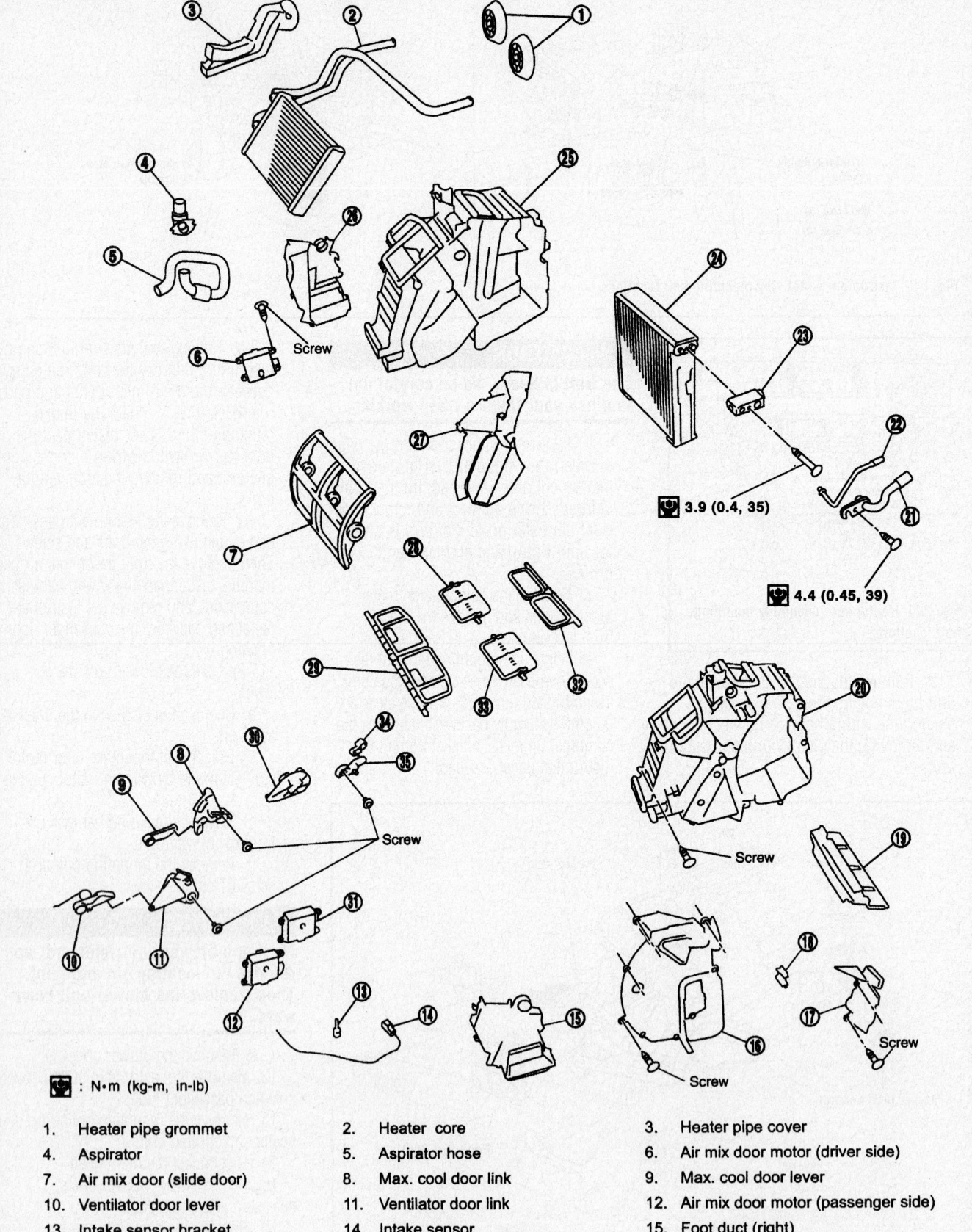

Screw

3.9 (0.4, 35)

4.4 (0.45, 39)

Screw

Screw

Screw

Screw

: N·m (kg-m, in-lb)

1. Heater pipe grommet	2. Heater core	3. Heater pipe cover
4. Aspirator	5. Aspirator hose	6. Air mix door motor (driver side)
7. Air mix door (slide door)	8. Max. cool door link	9. Max. cool door lever
10. Ventilator door lever	11. Ventilator door link	12. Air mix door motor (passenger side)
13. Intake sensor bracket	14. Intake sensor	15. Foot duct (right)
16. Evaporator cover	17. Evaporator cover adaptor	18. Heater pipe bracket

67162-FX35-G127

Fig. 243 Exploded view of the Heater & Cooling Unit, which contains the heater core

18. Remove the mounting screws and then remove the heater pipe cover.

19. Remove the heater pipe bracket.

20. Slide the heater core (shown in the figure) toward the driver's side.

21. Remove the heater core.

To install:

22. Install the heater core into the heating/cooling unit.

23. Install the heater and cooling unit. Tighten the mounting bolts to 60 inch lbs. (7 Nm).

24. Install the steering member.

25. Install the steering member mounting bolts, nut and wiring harness clips. Tighten the steering member mounting bolts to 108 inch lbs. (12 Nm).

26. Install the ventilator ducts, defroster nozzle and ducts.

27. Reconnect the drain hose.

28. Install the mounting bolts for the heater and cooling unit.

29. Install the instrument stays (driver-side and passenger-side).

✳✳ CAUTION

Make sure the locating pin and joint are securely inserted.

30. Install the blower unit, as follows:

a. Install the blower unit.

b. Install the mounting bolt and screws for the blower unit.

c. Install the wiring harness clip for the blower unit.

d. Reconnect the intake door motor connector and blower fan motor connector.

e. Install the ECM with its bracket attached.

31. Install the instrument panel assembly, as follows:

a. Install the front passenger air bag module and instrument panel and pad.

b. Install the side ventilations.

c. Install the combination meter bracket and reinstall the steering column assembly.

d. Install the front defroster grilles.

e. Install the display unit and audio unit.

f. Install the cluster lid.

g. Install the instrument side panels..

h. Install the combination meter assembly.

i. Install the steering lock escutcheon.

j. Install the lighting and turn signal switch.

k. Install the wiper and washer switch

l. Install the steering column upper cover.

m. Install the steering column lower cover.

n. Install the steering column front lower cover.

o. Install the instrument driver lower panel.

p. Reattach the data link connector.

q. Reconnect the in-vehicle sensor and all electrical parts.

r. Install the grommet and hood lock cable.

s. Install the instrument passenger lower panel.

t. Install the instrument lower cover.

u. Connect the center console sub-wiring harness.

v. Install the console finishers.

w. Install the console finisher screws.

x. Reconnect the wiring harness connector.

y. Install the A/T console finisher.

z. Install instrument clock finisher.

aa. Install the A/T select lever knob.

bb. Install the front pillar garnish.

cc. Install the dash side finisher and plastic nuts.

dd. Install the front kicking plate on both sides of the vehicle.

32. Reconnect the two heater hoses to the heater core.

33. On 3.5L engines, install the electronic control throttle assembly.

➡**Replace the O-rings for A/C piping with new ones, then apply compressor oil to them when installing them.**

34. Reconnect the evaporator-side one touch joint. The connection point for the female-side piping is thin, so when inserting the male-side piping, take care not to deform the female-side piping. Slowly insert it in the axial direction. Insert the one-touch joint connection point securely until it clicks. After the piping has been connected, pull on the male-side piping by hand to make sure the piping does not come off.

35. Install the low-pressure flexible hose bracket mounting bolts.

36. Install the 2 high-pressure pipe mounting clips.

37. Install the cowl top cover.

38. Fill the cooling system with the proper amount and type of fluid.

39. Recharge the vehicle A/C system and check for leaks.

STEERING

POWER STEERING GEAR

REMOVAL & INSTALLATION

See Figures 244 through 246.

1. Before servicing the vehicle, refer to the Precautions Section.

✳✳ CAUTION

The spiral cable may snap due to steering operation if the steering column is separated from the steering gear assembly. Therefore fix the steering wheel with a string to avoid turning it too far.

2. Set the wheels in the straight-ahead position.

3. Raise and safely support the vehicle.

4. Remove the tires from vehicle.

5. Remove the undercover.

6. Confirm the slit of the lower joint fits with the projection on the rear cover cap

marking the position on the steering gear assembly nearly fits with the projection on the rear cover cap.

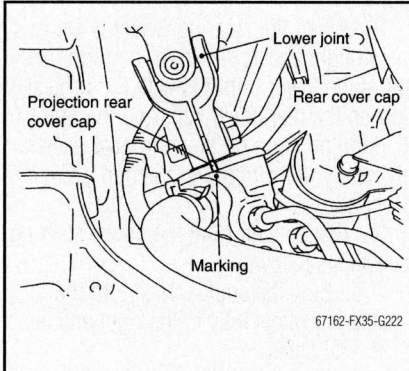

67162-FX35-G222

Fig. 244 Steering gear projection position

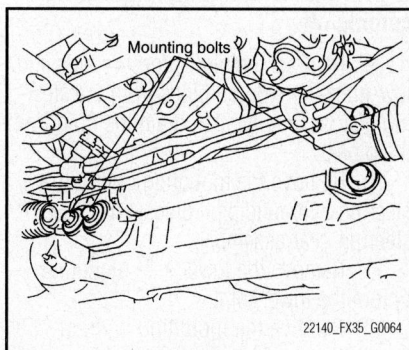

22140_FX35_G0064

Fig. 245 Remove the mounting bolts of the steering gear assembly

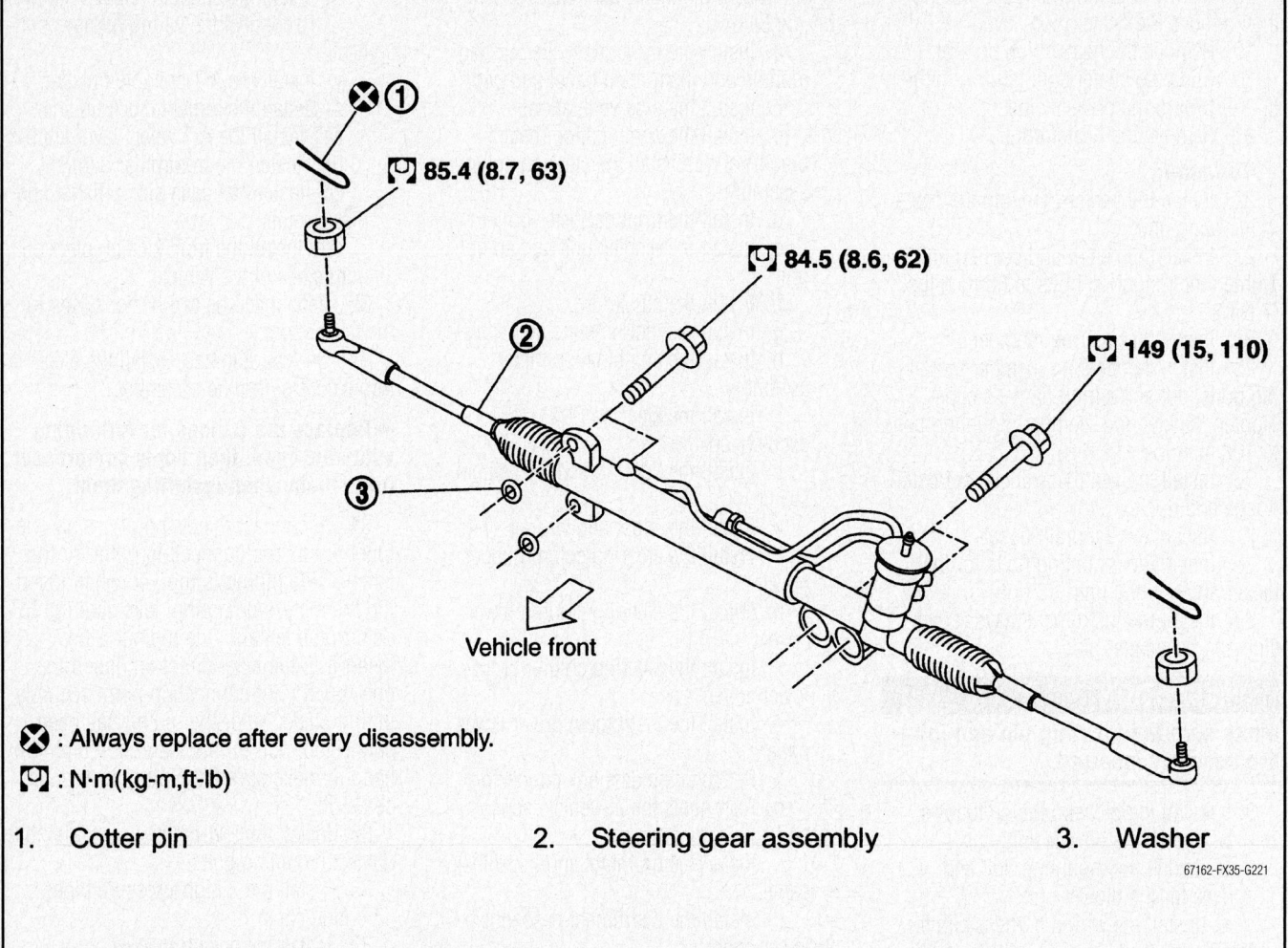

85.4 (8.7, 63)

84.5 (8.6, 62)

149 (15, 110)

Vehicle front

⊗ : Always replace after every disassembly.

🔧 : N·m(kg-m,ft-lb)

1. Cotter pin   2. Steering gear assembly   3. Washer

67162-FX35-G221

Fig. 246 Exploded view of the power steering gear mounting

7. Remove the cotter pin at steering outer socket, then loosen the mounting nut.

8. Use a ball joint remover to remove the steering outer socket from the steering knuckle. Be careful not to damage the ball joint boot.

➡**Temporarily tighten the mounting nut to prevent damage to the threads and to prevent the ball joint remover from coming off.**

9. Remove the high-pressure side and low-pressure side oil pipes from the steering gear assembly, then drain the fluid from the pipes.

10. Remove the mounting bolt of the steering hydraulic pipe bracket from the steering gear assembly.

11. Remove the lower side mounting bolt of the lower joint.

12. Remove the mounting bolts of the steering gear assembly, then remove the steering gear assembly from the vehicle.

To install:

13. Installation is the reverse of the removal procedure.

➡**Refer to component parts location and do not reuse non-reusable parts.**

14. After installation, check wheel alignment.

15. After adjusting wheel alignment, adjust the neutral position of the steering angle sensor.

16. When the steering wheel is set in the straight ahead direction, confirm the slit of the lower joint fits with the projection on the rear cover cap, and that the marking position on steering gear assembly nearly fits with the projection on rear cover cap.

17. Bleed all air from the power steering system, as follows:

a. Stop the engine, then turn the steering wheel fully to the right and left several times.

➡**Do not allow the steering fluid reservoir tank level to drop below the low-**

level line. Check the tank frequently and add fluid as needed.

b. Run the engine at idle speed. Turn the steering wheel fully to the right and then fully to the left, and keep hold for about 3 seconds. Then check whether any fluid leaks have occurred.

c. Repeat sub-step b several times at about 3 second intervals.

❋❋ WARNING

Do not hold the steering wheel in the locked position for more than 10 seconds. There is the possibility that the oil pump may be damaged.

d. Check for air bubbles and cloudiness in the fluid.

e. If air bubbles and/or cloudiness don't fade, stop the engine, and stop air bleeding until the air bubbles and cloudiness fade.

f. Perform until all bubbles and cloudiness are gone.

g. Stop the engine and check the fluid level.

Incomplete air bleeding causes the following. When this happens, bleed the system again:

- Generation of air bubbles in the reservoir tank
- Generation of clicking noise in the oil pump
- Excessive buzzing in the oil pump

➡ **When the vehicle is stationary or while the steering wheel is being turned slowly, some noise may be heard from the oil pump or gear. This noise is normal.**

18. Check that the steering wheel turns smoothly when it is turned several times fully to the end of the left and right.

POWER STEERING PUMP

REMOVAL & INSTALLATION

3.5L Engine

See Figure 247.

1. Before servicing the vehicle, refer to the Precautions Section.
2. Disconnect the negative battery cable.
3. Remove the undercover.

4. Remove the drive belt. Refer to Accessory Drive Belts, removal & installation.
5. Drain the power steering fluid from the reservoir tank into a suitable container. Properly discard the used fluid.
6. Remove the high pressure and the low pressure lines from the power steering fluid pump.
7. Remove the pump mounting bolts.
8. Remove the pump from the vehicle.

To install:

9. Installation is the reverse of the removal procedure.
10. Bleed the power steering system.
11. Adjust the belt tension.

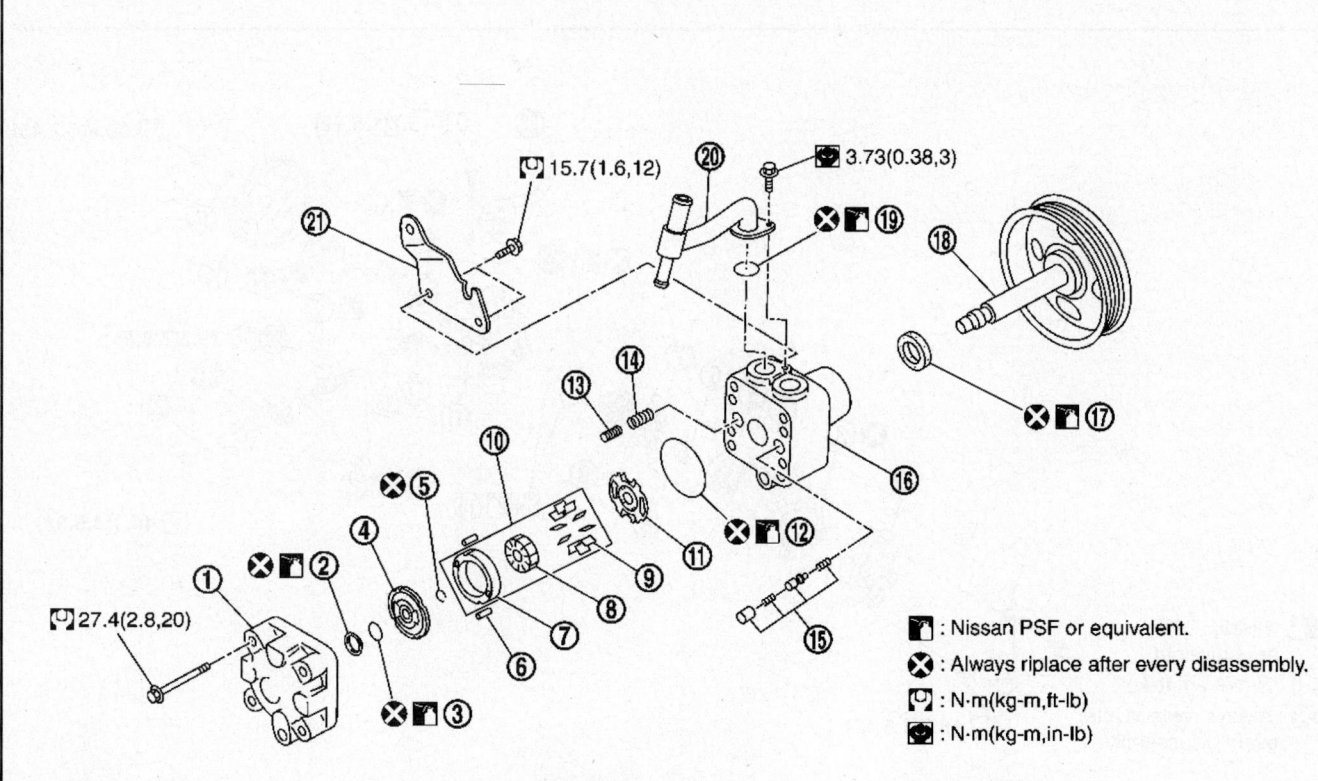

15.7(1.6,12)

3.73(0.38,3)

27.4(2.8,20)

■ : Nissan PSF or equivalent.
✖ : Always riplace after every disassembly.
⬜ : N·m(kg-m,ft-lb)
⬛ : N·m(kg-m,in-lb)

1.	Rear cover	2.	Teflon ring	3.	O-ring
4.	Rear side plate	5.	Rotor snap ring	6.	Dowel pin
7.	Cam ring	8.	Rotor	9.	Vane
10.	Cartridge	11.	Front side plate	12.	O-ring
13.	Flow control valve A	14.	Spring	15.	Flow control valve B assembly
16.	Body assembly	17.	Oil seal	18.	Pulley
19.	O-ring	20.	Suction pipe	21.	Bracket

42050_FX35_G0062

Fig. 247 Exploded view of power steering pump—3.5L engine

4.5L Engine

See Figure 248.

1. Before servicing the vehicle, refer to the Precautions Section.
2. Disconnect the negative battery cable.
3. Remove the undercover.
4. Remove the power steering belt from the auto tensioner.
5. Drain the power steering fluid from the reservoir tank into a suitable container. Properly discard the used fluid.
6. Remove the high pressure and the low pressure lines from the power steering fluid pump.
7. Remove the pump mounting bolts.
8. Remove the pump from the vehicle.

To install:

9. Installation is the reverse of the removal procedure.
10. Bleed the power steering system.

➡**The drive belt tension is automatic and requires no adjustment.**

BLEEDING

1. Before servicing the vehicle, refer to the Precautions Section.
2. Stop the engine.
3. Turn the steering wheel fully to the right and left several times.

➡**Do not allow the fluid level in the reservoir tank to go below the MIN level line. Check and add fluid as needed.**

4. Run the engine at idle speed. Turn the steering wheel fully to the right and then fully to the left. Hold for about 3 seconds. Check for fluid leakage.
5. Repeat the above step several times at 3 second intervals.

✳✳ WARNING

Do not hold the steering wheel in the locked position for more than 10 seconds. Damage to the pump may occur.

6. Check for air bubbles or cloudy fluid. If found, repeat the bleeding procedure.
7. Stop the engine and check the fluid level. Correct as required.

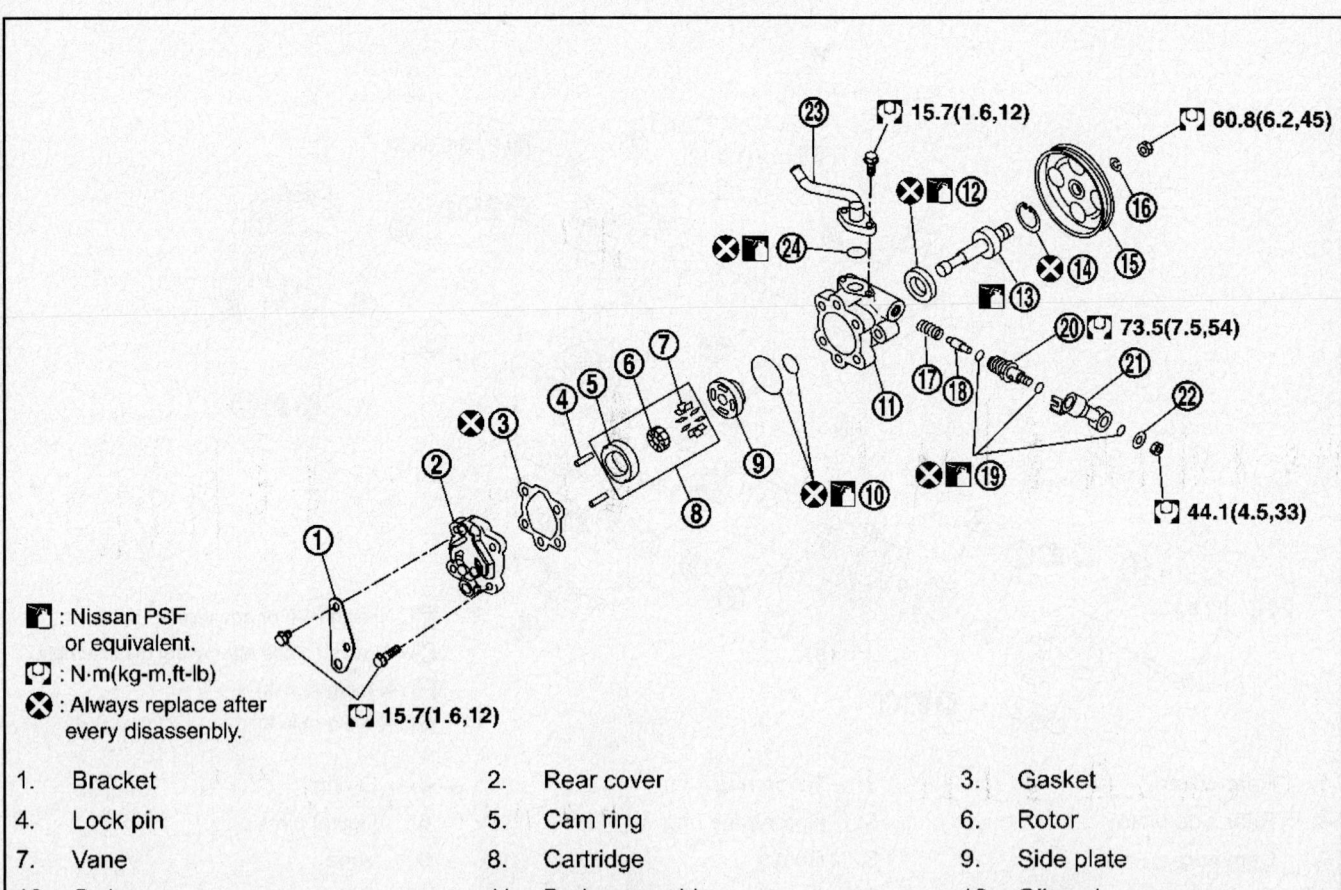

🔹 : Nissan PSF or equivalent.
🔧 : N·m(kg-m,ft-lb)
❌ : Always replace after every disassembly.

🔧 15.7(1.6,12)

1.	Bracket	2.	Rear cover	3.	Gasket
4.	Lock pin	5.	Cam ring	6.	Rotor
7.	Vane	8.	Cartridge	9.	Side plate
10.	O-ring	11.	Body assembly	12.	Oil seal
13.	Drive shaft assembly	14.	Snap ring	15.	Pulley
16.	Spring washer	17.	Spring	18.	Flow control valve
19.	O-ring	20.	Connector bolt	21.	Joint
22.	Washer	23.	Suction pipe	24.	O-ring

42050_FX35_G0063

Fig. 248 Power steering pump—exploded view 4.5L engine

See Figure 249.

LOWER BALL JOINT

REMOVAL & INSTALLATION

The front lower control arm ball joints are not separately replaceable from the control arms themselves. If the joints are found to be defective, the entire assembly must be replaced.

LOWER CONTROL ARM

REMOVAL & INSTALLATION

1. Before servicing the vehicle, refer to the Precautions Section.
2. Raise and safely support the vehicle.
3. Remove the wheels from vehicle.
4. Remove the undercover.
5. Remove the front cross bar.
6. Remove the cotter pin at the transverse link, then loosen the mounting nut.
7. Use a ball joint remover to remove the transverse link from the steering knuckle. Be careful not to damage the ball joint boot.

➡**Temporarily tighten the mounting nut to prevent damage to the threads and to prevent the ball joint remover from coming off.**

8. Remove the mounting bolts which are at the back of the transverse link (mounting part with body), then separate the transverse link.
9. Remove the mounting bolts which are at the front of the transverse link (mounting part with the front suspension member), then separate the transverse link.
10. Remove the transverse link from the vehicle.
11. Check transverse link and bushing for deformation, cracks, or damage. If any non-standard condition is found, replace it.
12. Check the boot of the ball joint for cracks, or other damage, and also for grease leakage. If any non-standard condition is found, replace it.
13. Manually move ball stud to confirm it moves smoothly with no binding.

➡**Before measurement, move ball joint at least ten times by hand to check for smooth movement.**

14. Hook a spring scale onto the ball stud tip. Confirm that the spring scale measurement value is within specifications when the ball stud begins to move. If it is outside the specified range, replace the transverse link assembly.

➡**Swing torque specification: Less than 5–43 inch lbs. (1–5 Nm), measure value of spring scale: less than 5–43 inch lbs. (1–5 Nm).**

15. Attach the mounting nut onto the ball stud. Check that the rotating torque is within specifications with a preload gauge. If it is outside the specified range, replace transverse link assembly.

➡**Rotating torque specification: Less than 5–43 inch lbs. (1–5 Nm).**

16. Move the tip of ball joint in axial direction to check for looseness. If it is outside the specified range, replace transverse link assembly.

➡**Axial end play specification: 0.004 inch (0.1mm).**

To install:

17. Install the transverse link on the vehicle.
18. Install and tighten the mounting bolts at the front of the transverse link (mounting part with the front suspension member) to 89 ft. lbs. (120 Nm), then install and tighten the mounting bolts which are at the back of the transverse link (mounting part with body) to 118 ft. lbs. (160 Nm).
19. Reattach the transverse link to the steering knuckle, and tighten the ball joint nut to 105 ft. lbs. (142 Nm). Be careful not to damage the ball joint boot.
20. Install a new cotter pin.
21. Install the front cross bar. Tighten the inner 2 bolts on each end to 33 ft. lbs. (45 Nm) and the outer 2 bolts on each end to 41 ft. lbs. (55 Nm).
22. Install the undercover.
23. Install the tire.
24. Check the wheel alignment.
25. After adjusting wheel alignment, adjust the neutral position of the steering angle sensor.

MACPHERSON STRUT

REMOVAL & INSTALLATION

See Figure 250.

1. Before servicing the vehicle, refer to the Precautions Section.
2. Raise and safely support the vehicle.
3. Make an alignment marking on the camber adjusting bolt and strut for approximate installation alignment later.
4. Remove the wheels from the vehicle.
5. Remove the brake hose lock plate. Then, remove the brake hose from the strut assembly.
6. Remove the wheel sensor wiring harness from the strut assembly.

✳✳ WARNING

Do not pull on the wheel sensor wiring harness.

7. Remove the stabilizer connecting rod upper nut, separate the stabilizer connecting rod and strut assembly.
8. Remove the attaching bolts and nuts between the strut assembly and the steering knuckle.
9. Remove the mounting nuts on the mounting insulator bracket, then remove the strut upper plate, strut spacer and the strut from the vehicle.

To install:

➡**Attach strut upper plate as shown in the accompanying illustration.**

10. Install the strut upper plate, strut spacer, and the strut onto the vehicle.
11. Install and tighten the mounting nuts on the mounting insulator bracket to 35 ft. lbs. (47 Nm).
12. Install and tighten the attaching bolts and nuts between the strut assembly and the steering knuckle to 134 ft. lbs. (182 Nm).

➡**Use the matchmarks made earlier to approximate the front end alignment during installation of the strut assembly.**

13. Reattach the stabilizer connecting rod to the strut and tighten the upper nut to 75 ft. lbs. (102 Nm).
14. Install the wheel sensor wiring harness onto the strut assembly.
15. Install the brake hose and the hose lock plate.
16. Install the wheel and tire to the vehicle.
17. After installation, check wheel alignment.
18. After adjusting the wheel alignment, adjust the neutral position of steering angle sensor.
19. Double-check to ensure that the wheel sensor wiring harness is properly routed.

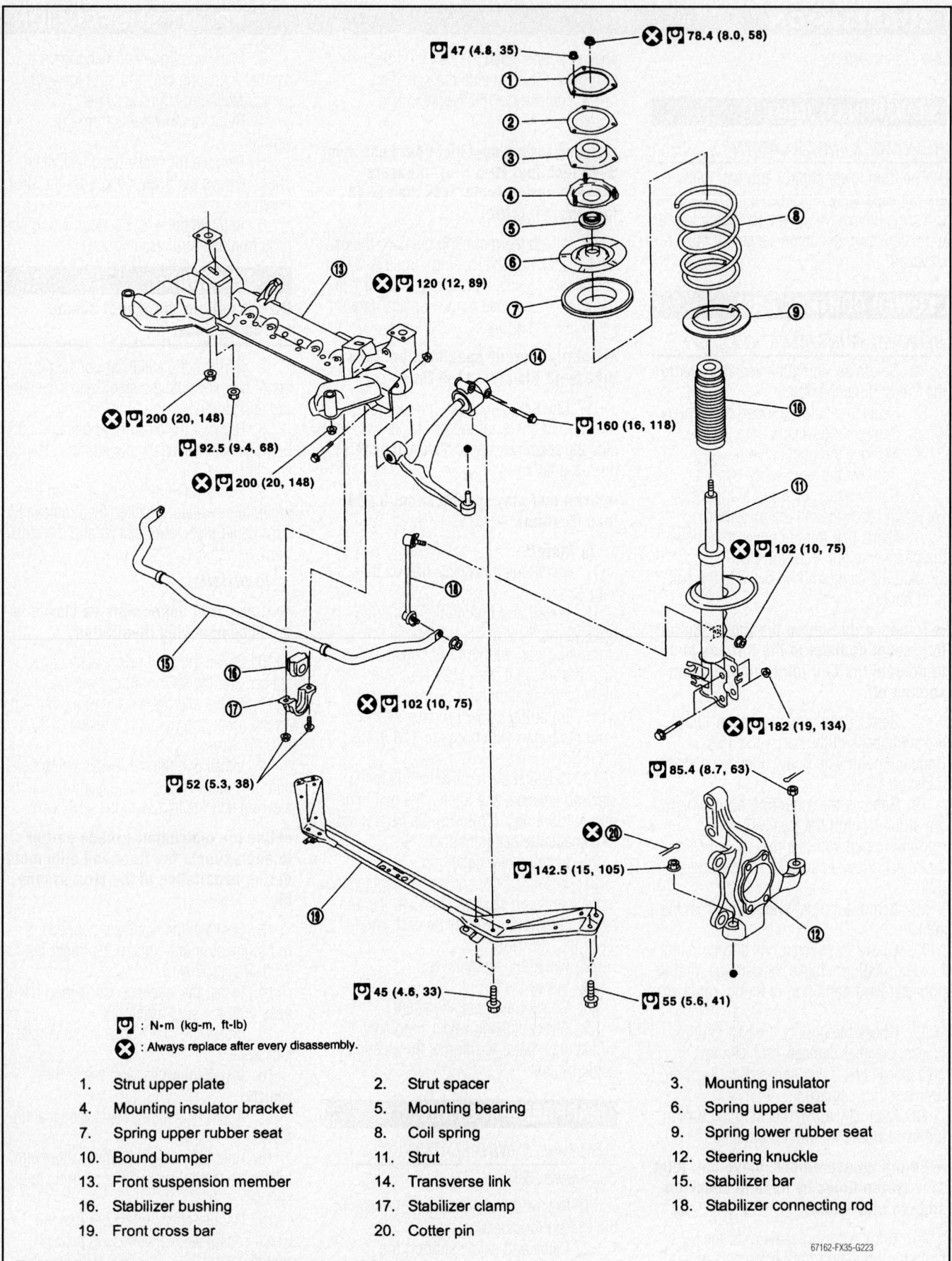

47 (4.8, 35)
78.4 (8.0, 58)
120 (12, 89)
160 (16, 118)
200 (20, 148)
92.5 (9.4, 68)
200 (20, 148)
102 (10, 75)
102 (10, 75)
182 (19, 134)
85.4 (8.7, 63)
52 (5.3, 38)
142.5 (15, 105)
45 (4.6, 33)
55 (5.6, 41)

: N•m (kg-m, ft-lb)

: Always replace after every disassembly.

1. Strut upper plate	2. Strut spacer	3. Mounting insulator
4. Mounting insulator bracket	5. Mounting bearing	6. Spring upper seat
7. Spring upper rubber seat	8. Coil spring	9. Spring lower rubber seat
10. Bound bumper	11. Strut	12. Steering knuckle
13. Front suspension member	14. Transverse link	15. Stabilizer bar
16. Stabilizer bushing	17. Stabilizer clamp	18. Stabilizer connecting rod
19. Front cross bar	20. Cotter pin	

67162-FX35-G223

Fig. 249 Exploded view of the front suspension components

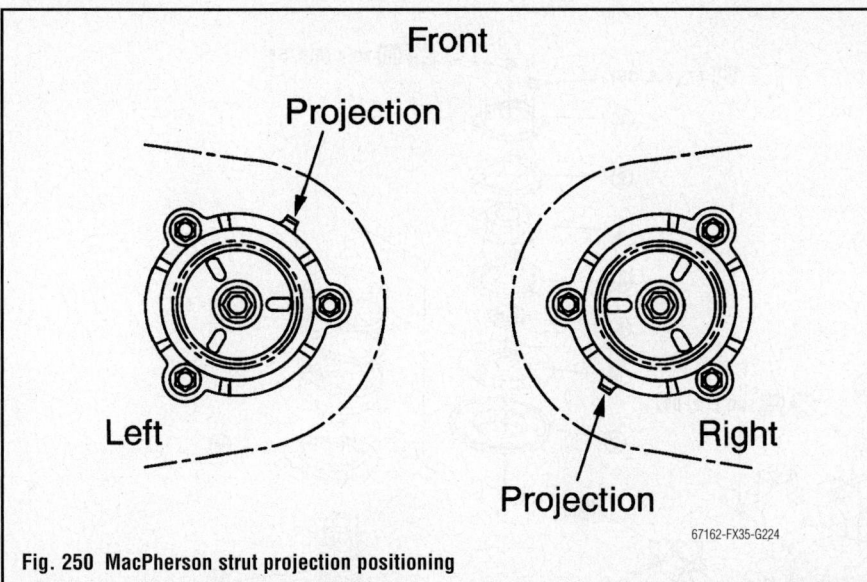

Front

Projection

Left

Right

Projection

67162-FX35-G224

Fig. 250 MacPherson strut projection positioning

STABILIZER BAR

REMOVAL & INSTALLATION

See Figure 251.

1. Before servicing the vehicle, refer to the Precautions Section.
2. Raise and safely support the vehicle.
3. Remove the wheels.
4. Remove the fixing bolts and remove the stabilizer connecting rod mount bracket from the suspension arm.
5. Remove the lower side fixing nut on the stabilizer connecting rod and remove the stabilizer connecting rod from the stabilizer bar.
6. Remove the fixing nuts on the stabilizer clamps and remove the stabilizer from the vehicle.
7. Check the stabilizer bar, stabilizer bushings, stabilizer clamps, stabilizer connecting rod, and stabilizer connecting rod mounting bracket for any deformation, cracks, or damage. Replace if necessary.

To install:
8. Refer to the exploded view for tightening torques. Installation is the reverse of removal.

➡**Do not reuse non-reusable parts during assembly.**

9. The stabilizer bar uses pillow ball type connecting rod, position the ball joint with the case on pillow ball head parallel to the stabilizer bar.
10. When the bushing and clamp are installed to the stabilizer bar, position the bushing and clamp inside of the side slip prevention clamp.

STEERING KNUCKLE

REMOVAL & INSTALLATION

See Figures 252 and 253.

1. Before servicing the vehicle, refer to the Precautions Section.
2. Raise and safely support the vehicle.
3. Remove the appropriate wheel.
4. Remove the brake caliper. Support it in a place where it will not interfere with work.

➡**Avoid depressing brake pedal while brake caliper is removed.**

5. Remove the disc rotor.
6. Remove the wheel sensor from the wheel hub and bearing assembly.

➡**Do not pull on wheel sensor wiring harness.**

7. Remove the cotter pin from the steering outer socket, then loosen the mounting nut.
8. Use a ball joint remover to separate the steering outer socket from the steering knuckle. Be careful not to damage the ball joint boot.

➡**Temporarily tighten the mounting nut to prevent damage to the threads and to prevent the ball joint remover from coming off.**

9. Remove the cotter pin at the transverse link, then loosen the mounting nut.
10. Use a ball joint remover to separate the transverse link from the steering knuckle. Be careful not to damage ball joint boot.

✳✳ CAUTION

Temporarily tighten the mounting nut to prevent damage to the threads and to prevent the ball joint remover from coming off.

11. On AWD models, perform the following:
   a. Remove the cotter pin, then remove the lock nut from the halfshaft.
   b. Remove the steering knuckle from the halfshaft.

✳✳ WARNING

When removing the steering knuckle, do not apply an excessive angle to the halfshaft joint. Also be careful not to excessively extend the slide joint. Do not hang over the halfshaft without proper support.

12. Remove the mounting bolts and nuts between the strut assembly and the steering knuckle.
13. Remove the steering knuckle from the vehicle.
14. Remove the mounting bolts between the steering knuckle and the wheel hub/bearing assembly.
15. Remove the splash guard and wheel hub/bearing assembly from the steering knuckle.
16. Check for deformities, cracks and damage on all parts and replace if necessary.
17. Inspect the ball joint for boot breakage, axial looseness, and torque of transverse link and steering outer socket ball joint. Maximum of allowable axial end play is 0.002 inch (0.05mm) or less.

To install:
18. Install the splash guard and wheel hub/bearing assembly onto the steering knuckle.
19. Install the mounting bolts between the steering knuckle and the wheel hub/bearing assembly. Tighten the bolts to 77 ft. lbs. (104 Nm).
20. Install the steering knuckle on the vehicle.
21. Install the mounting bolts and nuts to the strut assembly and the steering knuckle. Tighten the bolts to 134 ft. lbs. (182 Nm).
22. On AWD models, perform the following:
   a. Install the steering knuckle onto the halfshaft.
   b. Install and tighten the lock nut on the halfshaft to 203 ft. lbs. (275 Nm). Install a new cotter pin.

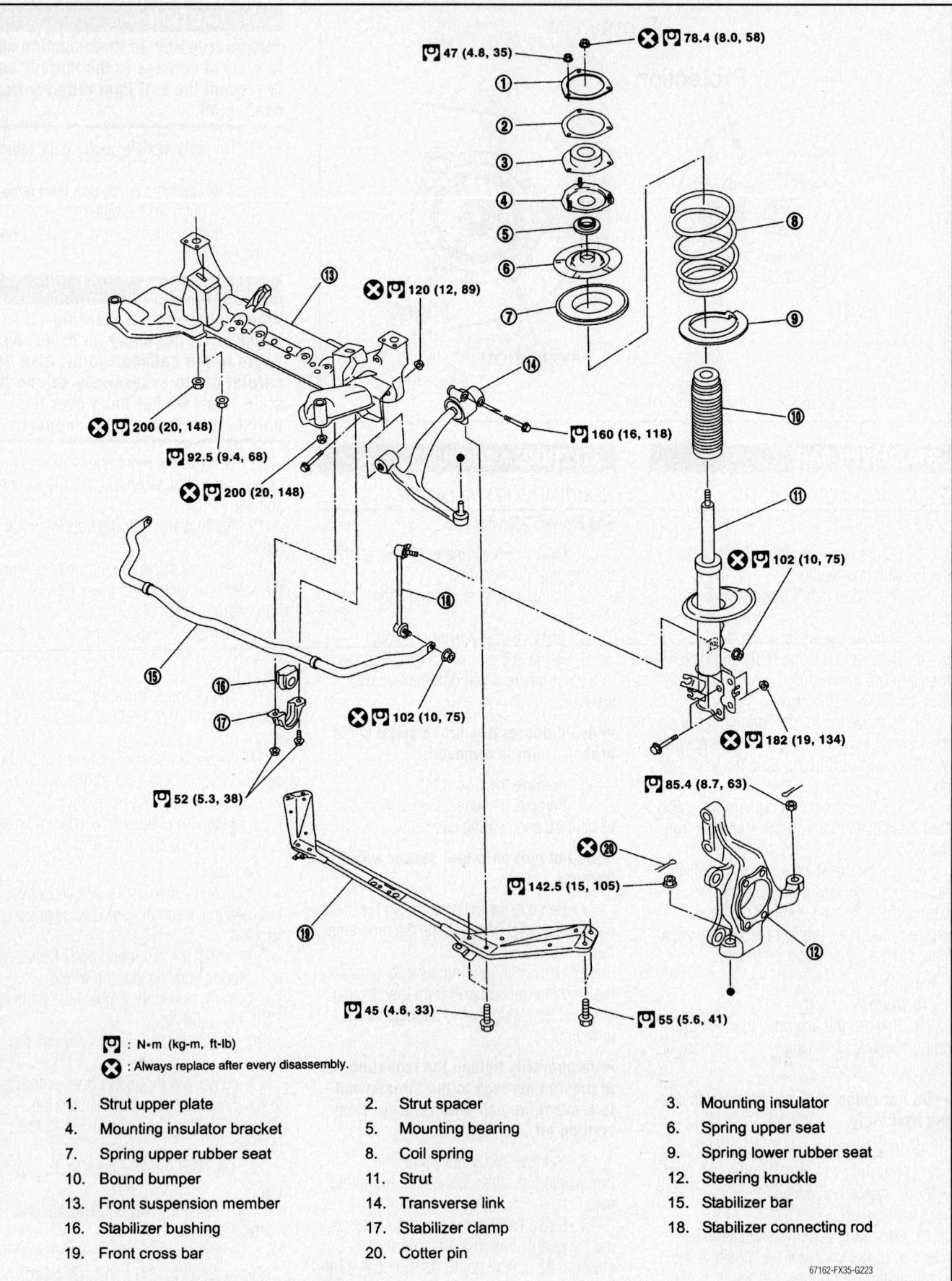

Fig. 251 Exploded view of the front suspension components

[:] : N•m (kg-m, ft-lb)

[X] : Always replace after every disassembly.

1. Strut upper plate
2. Strut spacer
3. Mounting insulator
4. Mounting insulator bracket
5. Mounting bearing
6. Spring upper seat
7. Spring upper rubber seat
8. Coil spring
9. Spring lower rubber seat
10. Bound bumper
11. Strut
12. Steering knuckle
13. Front suspension member
14. Transverse link
15. Stabilizer bar
16. Stabilizer bushing
17. Stabilizer clamp
18. Stabilizer connecting rod
19. Front cross bar
20. Cotter pin

67162-FX35-G223

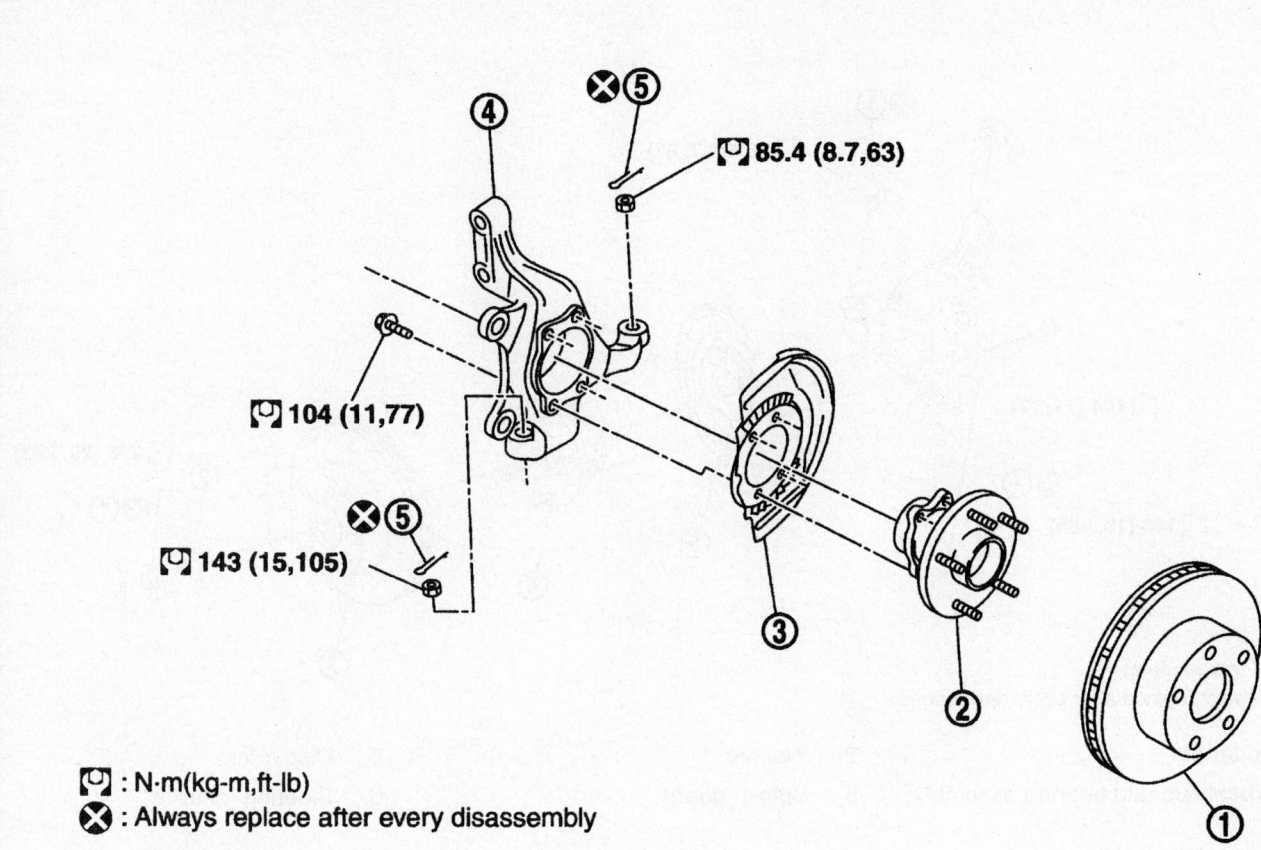

$\boxed{\Box}$ 85.4 (8.7,63)

$\boxed{\Box}$ 104 (11,77)

$\boxed{\Box}$ 143 (15,105)

$\boxed{\Box}$: N·m(kg-m,ft-lb)

$\otimes$: Always replace after every disassembly

1. Disc rotor
2. Wheel hub and bearing assembly
3. Splash guard
4. Steering knuckle
5. Cotter pin

67162-FX35-G228

Fig. 252 Exploded view of front wheel hub mounting—2WD

23. Reattach the transverse link to the steering knuckle. Tighten the ball joint nut to 105 ft. lbs. (143 Nm).
24. Install a new cotter pin.
25. Reconnect the steering outer socket to the steering knuckle. Tighten the ball joint nut to 63 ft. lbs. (86 Nm).
26. Install a new cotter pin.

27. Install the wheel sensor onto the wheel hub and bearing assembly.
28. Install the disc rotor.
29. Install the brake caliper.
30. Install the wheel.
31. Check wheel alignment.
32. After adjusting wheel alignment, adjust the neutral position of the steering angle sensor.

33. Check the installation condition of the wheel sensor wiring harness.

WHEEL BEARINGS

REMOVAL & INSTALLATION

The wheel hub and bearing are an assembly. If a bearing fails specification, replace the wheel hub and bearing as an assembly.

SUSPENSION

See Figure 254.

COIL SPRING

REMOVAL & INSTALLATION

1. Before servicing the vehicle, refer to the Precautions Section.
2. Raise and safely support the vehicle.

3. Remove the rear tire.
4. Position a jack under the rear lower link for support.
5. Loosen the fixing bolt and nut of the rear lower link in the side of the suspension member, and then remove the fixing bolt and nut in the side of the axle.
6. Slowly lower the jack, then remove the upper seat, coil spring and rubber sheet from the rear lower link.

REAR SUSPENSION

7. Remove the fixing bolt and nut in the side of the rear suspension member to remove the rear lower link.
8. Check the rear lower link, bushing, and coil spring for deformation, cracks, and damage. Replace the rear lower link and coil spring, if necessary.

To install:

9. Position the rear lower link on the vehicle.

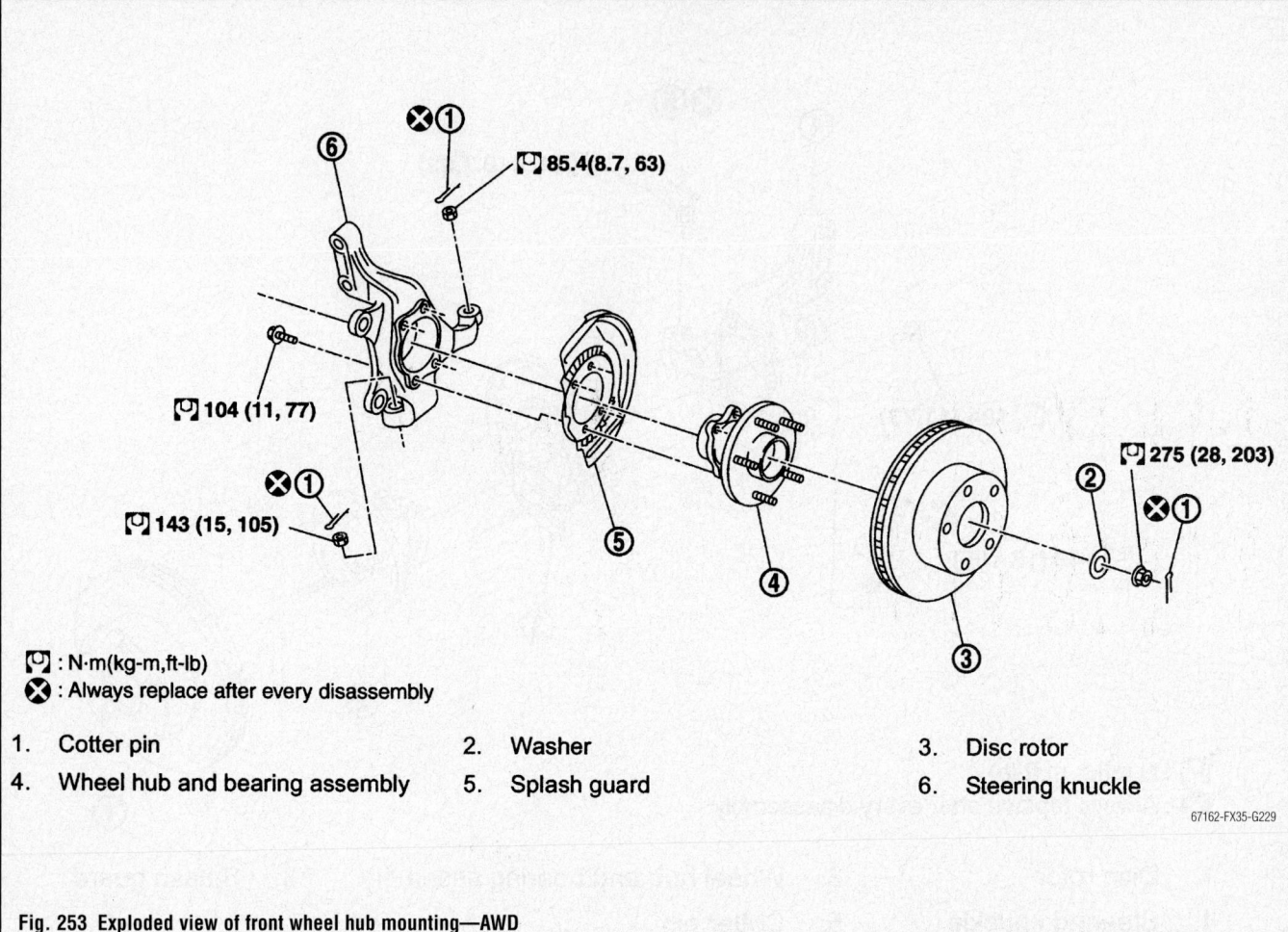

⊡ 85.4(8.7, 63)

⊡ 104 (11, 77)

⊡ 143 (15, 105)

⊡ 275 (28, 203)

⊡ : N·m(kg-m,ft-lb)

✕ : Always replace after every disassembly

1. Cotter pin
2. Washer
3. Disc rotor
4. Wheel hub and bearing assembly
5. Splash guard
6. Steering knuckle

67162-FX35-G229

Fig. 253 Exploded view of front wheel hub mounting—AWD

10. Install and tighten the fixing bolt and nut in the side of the rear suspension member to 48 ft. lbs. (65 Nm).

11. Position the upper seat, coil spring and rubber sheet in place, and then slowly raise the jack under the rear lower link.

➡**Match up the rubber seat indentions and rear lower link grooves. Also, make sure the spring is not upside down. The top and bottom are indicated by paint color.**

12. Install and tighten the fixing bolt and nut in the side of the axle to 77 ft. lbs. (105 Nm).

13. Slowly lower the jack from under the rear lower link.

14. Install the rear tire.

✳✳ CAUTION

Perform the final tightening of the rear suspension member and axle installation position (rubber bushing) under unladen conditions with the tires on level ground.

15. Check the wheel alignment.

16. After adjusting wheel alignment, adjust the neutral position of the steering angle sensor.

LOWER CONTROL ARM

REMOVAL & INSTALLATION

Front Lower Link

1. Before servicing the vehicle, refer to the Precautions Section.

2. Raise and safely support the vehicle.

3. Remove the rear tire.

4. Position a jack under the rear lower link for support.

5. Remove the front lower link protector.

6. Remove the shock absorber assembly from the vehicle.

7. Remove the mounting nut and bolt between the front lower link and the axle.

8. Remove the mounting nut and bolt between the front lower link and the rear suspension member.

9. Remove the front lower link from the vehicle.

10. Check the front lower link and bushing for any deformation, cracks, or damage. Replace it if necessary.

To install:

11. Position the front lower link on the vehicle.

12. Install and tighten the mounting nut and bolt between the front lower link and the rear suspension member to 74 ft. lbs. (101 Nm).

13. Install and tighten the mounting nut and bolt between the front lower link and the axle to 77 ft. lbs. (105 Nm).

14. Install the shock absorber assembly.

15. Install the front lower link protector.

16. Slowly lower the jack from under the rear lower link.

17. Install the rear tire.

✳✳ CAUTION

Perform final tightening of the rear suspension member and axle installation position (rubber bushing) under unladen conditions with the tires on level ground.

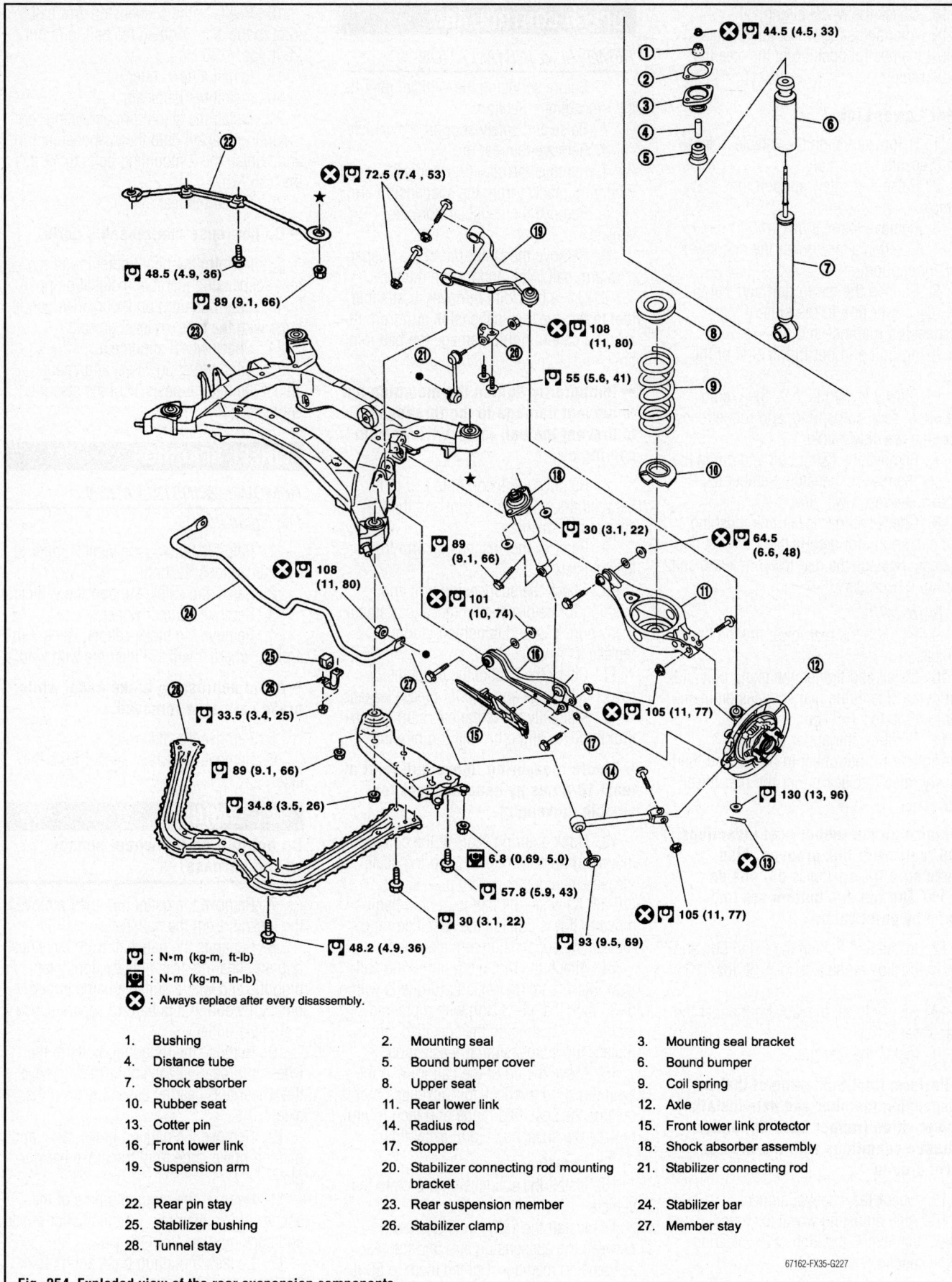

⊗ 44.5 (4.5, 33)

72.5 (7.4 , 53)

48.5 (4.9, 36)

89 (9.1, 66)

⊗ 108 (11, 80)

55 (5.6, 41)

30 (3.1, 22)

⊗ 64.5 (6.6, 48)

89 (9.1, 66)

⊗ 108 (11, 80)

⊗ 101 (10, 74)

⊗ 105 (11, 77)

33.5 (3.4, 25)

89 (9.1, 66)

34.8 (3.5, 26)

6.8 (0.69, 5.0)

57.8 (5.9, 43)

30 (3.1, 22)

48.2 (4.9, 36)

93 (9.5, 69)

⊗ 105 (11, 77)

130 (13, 96)

⊗ 13

[🔧] : N•m (kg-m, ft-lb)

[🔧] : N•m (kg-m, in-lb)

⊗ : Always replace after every disassembly.

1.	Bushing	2.	Mounting seal	3.	Mounting seal bracket
4.	Distance tube	5.	Bound bumper cover	6.	Bound bumper
7.	Shock absorber	8.	Upper seat	9.	Coil spring
10.	Rubber seat	11.	Rear lower link	12.	Axle
13.	Cotter pin	14.	Radius rod	15.	Front lower link protector
16.	Front lower link	17.	Stopper	18.	Shock absorber assembly
19.	Suspension arm	20.	Stabilizer connecting rod mounting bracket	21.	Stabilizer connecting rod
22.	Rear pin stay	23.	Rear suspension member	24.	Stabilizer bar
25.	Stabilizer bushing	26.	Stabilizer clamp	27.	Member stay
28.	Tunnel stay				

67162-FX35-G227

Fig. 254 Exploded view of the rear suspension components

18. Check the wheel alignment.

19. After adjusting wheel alignment, adjust the neutral position of the steering angle sensor.

Rear Lower Link

1. Before servicing the vehicle, refer to the Precautions Section.

2. Raise and safely support the vehicle.

3. Remove the rear tire.

4. Position a jack under the rear lower link for support.

5. Loosen the fixing bolt and nut of the rear lower link in the side of the suspension member and then remove the fixing bolt and nut in the side of the axle.

6. Slowly lower the jack, then remove the upper seat, coil spring, and rubber sheet from the rear lower link.

7. Remove the fixing bolt and nut in the side of the rear suspension member to remove the rear lower link.

8. Check the rear lower link, bushing, and coil spring for deformation, cracks, and damage. Replace the rear lower link and coil spring, if necessary.

To install:

9. Position the rear lower link on the vehicle.

10. Install and tighten the fixing bolt and nut in the side of the rear suspension member to 48 ft. lbs. (65 Nm).

11. Position the upper seat, coil spring and rubber sheet in place, and then slowly raise the jack under the rear lower link.

➡**Match up the rubber seat indentions and rear lower link grooves. Also, make sure the spring is not upside down. The top and bottom are indicated by paint color.**

12. Install and tighten the fixing bolt and nut in the side of the axle to 77 ft. lbs. (105 Nm).

13. Slowly lower the jack from under the rear lower link.

14. Install the rear tire.

➡**Perform final tightening of the rear suspension member and axle installation position (rubber bushing) under unladen conditions with the tires on level ground.**

15. Check the wheel alignment.

16. After adjusting wheel alignment, adjust the neutral position of the steering angle sensor.

UPPER CONTROL ARM

REMOVAL & INSTALLATION

1. Before servicing the vehicle, refer to the Precautions Section.

2. Raise and safely support the vehicle.

3. Remove the rear tire.

4. Remove the stabilizer connecting rod mounting bracket from the suspension arm.

5. Remove the halfshaft from the vehicle.

6. Remove the cotter pin of the suspension arm ball joint, and loosen the nut.

7. Use a ball joint remover or suitable tool to remove the suspension arm from the axle. Be careful not to damage the ball joint boot.

➡**Temporarily tighten the mounting nut to prevent damage to the threads and to prevent the ball joint remover from coming off.**

8. Remove the fixing nuts and bolts between the suspension arm and the rear suspension member.

9. Remove the suspension arm from the vehicle.

10. Check the suspension arm and bushing for deformation, cracks, or damage. If any non-standard condition is found, replace it.

11. Check the boot of the ball joint for cracks or damage and also for grease leakage.

12. Manually move the ball stud to confirm it moves smoothly with no binding.

➡**Before measuring, move ball joint at least 10 times by hand to check for smooth movement.**

13. Hook a spring scale at the cotter pin mounting hole. Confirm the spring scale measurement value is within 2–15 lbs. (10–66 N) when the ball joint stud begins moving. If it is outside the specified range, replace the suspension arm assembly.

14. Attach the mounting nut to the ball stud. Make sure the rotating torque is within 5–30 inch lbs. (1–3 Nm) with a preload gauge . If it is outside the specified range, replace the suspension arm assembly.

15. Move the tip of the ball joint in the axial direction to check for looseness. If it is outside the specified range of 0 inch (0mm), replace the suspension arm assembly.

To install:

16. Install the suspension arm onto the vehicle.

17. Install the fixing nuts and bolts between the suspension arm and the rear suspension member. Tighten them to 53 ft. lbs. (73 Nm).

18. Reattach the suspension arm ball joint to the axle. Tighten the ball joint nut to 96 ft. lbs. (130 Nm).

19. Install a new cotter pin.

20. Install the halfshaft.

21. Install the stabilizer connecting rod mounting bracket onto the suspension arm, and tighten the 2 mounting bolts to 41 ft. lbs. (55 Nm).

22. Install the rear tire.

➡**Do not reuse non-reusable parts.**

23. Perform the final tightening of the rear suspension member installation position (rubber bushing) under unladen conditions with the tires on level ground.

24. Check wheel alignment.

25. After adjusting wheel alignment, adjust the neutral position of the steering angle sensor.

WHEEL BEARINGS

REMOVAL & INSTALLATION

See Figure 255.

1. Before servicing the vehicle, refer to the Precautions Section.

2. Raise and safely support the vehicle.

3. Remove the rear wheel.

4. Remove the brake caliper. Hang it in a place where it will not interfere with work.

➡**Avoid depressing brake pedal while brake caliper is removed.**

5. Remove the disc rotor.

6. Remove the wheel sensor from the axle.

❈❈ WARNING

Do not pull on the wheel sensor wiring harness.

7. Remove the cotter pin, then remove the locknut from the halfshaft.

8. Separate the halfshaft from the wheel hub and bearing assembly by lightly tapping the end with a suitable hammer and block of wood. If it is hard to separate, use a suitable puller.

9. Remove the mounting bolts of the wheel hub/bearing assembly, then remove the wheel hub/bearing assembly from the axle.

10. Remove the parking brake cable and parking brake shoe from the brake backing plate.

11. Remove the mounting nuts of the anchor block, then remove the anchor block and backing plate from the axle.

12. Loosen mounting bolts and nuts of the front lower link, radius rod, and rear

lower link on the side of the suspension member.

13. Set a jack under the rear lower link, then remove the mounting bolt in the front lower link side of the shock absorber.

14. Remove the bolt and nut in the axle side of the rear lower link, then remove the coil spring.

15. Remove the mounting bolts and nuts in the axle side of the front lower link and radius rod.

16. Remove the suspension arm and cotter pin at the axle, then loosen the mounting nut.

17. Use a ball joint remover (or equivalent) to remove the suspension arm from the axle. Be careful not to damage the ball joint boot.

➡**Temporarily tighten the mounting nut to prevent damage to the threads and to prevent the ball joint remover from coming off.**

18. Remove the axle from the vehicle.

19. Inspect the Ball Joint for boot breakage, axial looseness, and torque of suspension arm ball joint. Maximum of allowable axial end play is 0.00 inch (0.0mm).

To install:

20. Install the axle in the vehicle.

21. Reattach the upper control arm to the axle.

22. Tighten the upper control arm mounting nut to 96 ft. lbs. (130 Nm), and install a new cotter pin at the axle.

23. Install the mounting bolts and nuts

in the axle side of the front lower link and radius rod.

24. Install the coil spring, then install the bolt and nut in the axle side of the rear lower link.

25. Install the mounting bolt in the front lower link side of the shock absorber.

26. Tighten the mounting bolts and nuts of the front lower link, radius rod, and rear lower link on the side of the suspension member.

27. Install the anchor block and backing plate onto the axle.

28. Install and tighten the mounting nuts of the anchor block to 44 ft. lbs. (60 Nm).

29. Install the parking brake cable and parking brake shoe from the brake backing plate.

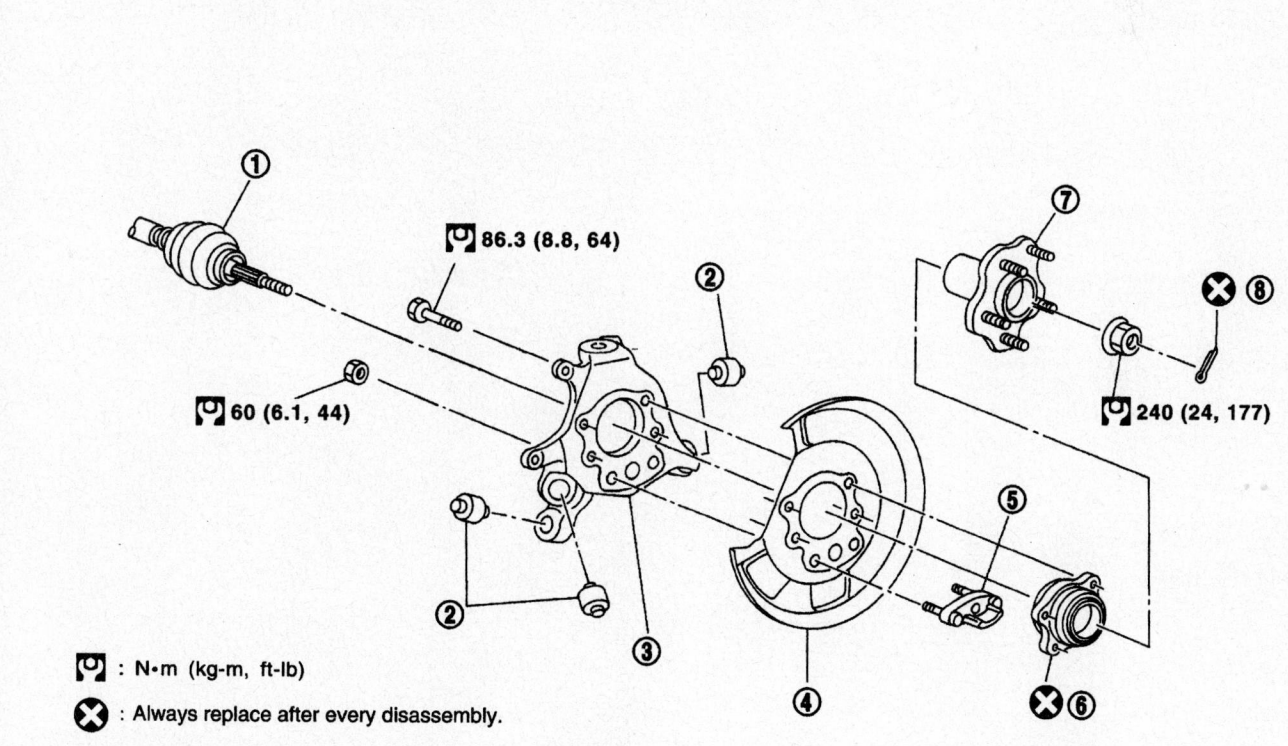

86.3 (8.8, 64)

60 (6.1, 44)

240 (24, 177)

: N•m (kg-m, ft-lb)

: Always replace after every disassembly.

1. Drive shaft	2. Bushing	3. Axle
4. Back plate	5. Anchor block	6. Wheel bearing
7. Wheel hub	8. Cotter pin	

Fig. 255 Exploded view of rear wheel hub mounting

67162-FX35-G230

30. Install the wheel hub/bearing assembly onto the axle, then install the mounting bolts of the wheel hub/bearing assembly and tighten to 64 ft. lbs. (86 Nm).

31. Install the halfshaft to the wheel hub/bearing assembly.

32. Install and tighten the locknut on the halfshaft to 177 ft. lbs. (240 Nm), then install a new cotter pin.

33. Install the wheel sensor on the axle.

34. Install the disc rotor.

35. Install the brake caliper.

36. Install the appropriate wheel.

➡ **Perform the final tightening of the rear suspension fasteners under unladen conditions with the tires on level ground.**

37. Check the wheel alignment.

38. After adjusting wheel alignment, adjust the neutral position of the steering angle sensor.

INFINITI

G35 • G35x

SPECIFICATIONS AND MAINTENANCE CHARTS

ENGINE AND VEHICLE IDENTIFICATION

		Engine							Model Year	
Code	Liters (cc)	Cu. In.	Cyl.	Fuel Sys.	Engine Type	Eng. Mfg.			Code ②	Year
VQ35DE	3.5 (3498)	213	6	MFI	DOHC	Nissan			6	2006
VQ35HR	3.5 (3498)	213	6	MFI	DOHC	Nissan			7	2007
									8	2008

MFI: Multi-port Fuel Injection

DOHC: Double Overhead Camshaft

22140_IG35_C0001

GENERAL ENGINE SPECIFICATIONS

Year	Model	Engine Displacement Liters	Engine ID	Net Horsepower @ rpm	Net Torque @ rpm (ft. lbs.)	Bore x Stroke (in.)	Com-pression Ratio	Oil Pressure @ rpm
2006	G35	3.5	VQ35DE	280@6400	270@4800	3.76X3.21	10.3:1	43@2000
	G35 Coupe	3.5	VQ35DE	306@6800	268@5200	3.76X3.20	10.6:1	43@2000
	G35x	3.5	VQ35DE	298@6200	260@4800	3.76X3.21	10.6:1	43@2000
2007	G35	3.5	VQ35HR	306@6800	268@5200	3.76X3.20	10.6:1	43@2000
	G35 Coupe	3.5	VQ35DE	306@6800	268@5200	3.76X3.20	10.6:1	43@2000
	G35x	3.5	VQ35HR	306@6800	268@5200	3.76X3.20	10.6:1	43@2000
2008	G35	3.5	VQ35HR	306@6800	268@5200	3.76X3.20	10.6:1	43@2000
	G35x	3.5	VQ35HR	306@6800	268@5200	3.76X3.20	10.6:1	43@2000

22140_IG35_C0002

ENGINE TUNE-UP SPECIFICATIONS

Year	Engine Displacement Liters	Engine ID	Spark Plug Gap (in.)	Ignition Timing (deg.) MT	AT	Fuel Pump (psi) ①	Idle Speed (rpm) MT	AT	Valve Clearance Intake	Exhaust
2006	3.5	VQ35DE	0.043	15B	15B	34	600-700	600-700	②	②
2007	3.5	VQ35DE	0.043	15B	15B	34	600-700	600-700	②	②
	3.5	VQ35HR	0.043	15B	15B	34	600-700	600-700	②	②
2008	3.5	VQ35HR	0.043	15B	15B	34	600-700	600-700	②	②

NOTE: The Vehicle Emission Control Information label often reflects specification changes made during production.

The label figures must be used if they differ from those in this chart.

B: Before top dead center

① 43 psi with regulator vacuum hose disconnected

② Hydraulic lash adjuster. See text for procedure.

22140_IG35_C0003

CAPACITIES

Year	Model	Engine Displacement Liters	Engine ID	Engine Oil with Filter	Transmission (pts.) Manual	Transmission (pts.) Auto.	Transfer Case (pts.)	Drive Axle (pts.)	Fuel Tank (gal.)	Cooling System (qts.)
2006	G35	3.5	VQ35DE	5.00	6.25	21.8	2.6	①	20	9.20
	G35 Coupe	3.5	VQ35DE	5.00	6.25	21.8	2.6	A	20	9.20
	G35x	3.5	VQ35DE	5.00	6.25	21.8	2.6	A	20	9.20
2007	G35	3.5	VQ35HR	5.00	6.19	21.8	2.6	①	20	9.20
	G35 Coupe	3.5	VQ35DE	5.00	6.25	21.8	2.6	①	20	9.20
	G35x	3.5	VQ35HR	5.00	6.19	21.8	2.6	A	20	9.20
2008	G35	3.5	VQ35HR	5.00	6.19	21.8	2.6	①	20	9.20
	G35x	3.5	VQ35HR	5.00	6.19	21.8	2.6	A	20	9.20

NOTE: All capacities are approximate. Add fluid gradually and check to be sure a proper fluid level is obtained.

① Rear axle: 3.00

Front Axle 1.38 (AWD)

22140_IG35_C0005

FLUID SPECIFICATIONS

Year	Model	Engine Displacement Liters	Engine ID/VIN	Engine Oil	Manual Trans	Auto. Trans. ①	Drive Axle	Power Steering Fluid	Brake Master Cylinder
2006	G35	3.5	VQ35DE	5W-30	75W-85	Matic J ATF	80W-90	Dexron III	DOT 3
	G35 Coupe	3.5	VQ35DE	5W-30	75W-85	Matic J ATF	80W-90	Dexron III	DOT 3
	G35x	3.5	VQ35DE	5W-30	75W-85	Matic J ATF	80W-90	Dexron III	DOT 3
2007	G35	3.5	VQ35HR	5W-30	75W-85	Matic J ATF	80W-90	Dexron III	DOT 3
	G35 Coupe	3.5	VQ35DE	5W-30	75W-85	Matic J ATF	80W-90	Dexron III	DOT 3
	G35x	3.5	VQ35HR	5W-30	75W-85	Matic J ATF	80W-90	Dexron III	DOT 3
2008	G35	3.5	VQ35HR	5W-30	75W-85	Matic J ATF	80W-90	Dexron III	DOT 3
	G35x	3.5	VQ35HR	5W-30	75W-85	Matic J ATF	80W-90	Dexron III	DOT 3

DOT: Department Of Transpotation

22140_IG35_C0004

VALVE SPECIFICATIONS

Year	Engine Displacement Liters	Engine ID	Seat Angle (deg.)	Face Angle (deg.)	Spring Test Pressure (lbs. @ in.)	Spring Free Height (in.)	Stem-to-Guide Clearance (in.)		Stem Diameter (in.)	
							Intake	Exhaust	Intake	Exhaust
2006	3.5	VQ35DE	45.15-45.45	NA	①	②	0.0008-0.0021	0.0012-0.0025	0.2348-0.2354	0.2344-0.2350
2007	3.5	VQ35DE	45.15-45.45	NA	①	②	0.0008-0.0021	0.0012-0.0025	0.2348-0.2354	0.2344-0.2350
	3.5	VQ35HR	45.15-45.45	NA	97.3 @ 1.094	1.726	0.0008-0.0021	0.0012-0.0025	0.2351-0.2354	0.2344-0.2350
2008	3.5	VQ35HR	45.15-45.45	NA	97.3 @ 1.094	1.726	0.0008-0.0021	0.0012-0.0025	0.2348-0.2354	0.2344-0.2350

NA: Not Available

① Man. Trans. Models: 97.6-110.2 lbs. @ 1.0551 in.

   Auto Trans. Models: 84-95 lbs. @ 1.0709 in.

② Man. Trans. Models: 1.7886 in.

   Auto Trans. Models: 1.8531 in.

22140_IG35_C0006

CAMSHAFT AND BEARING SPECIFICATIONS CHART

All measurements are given in inches.

Year	Engine Displ. Liters	Engine ID/VIN	Journal Dia.	Brg. Oil Clearance	Shaft End-play	Runout	Journal Bore	Lobe Height	
								Intake	Exhaust
2006	3.5	VQ35DE	①	②	0.0045-0.0074	0.0008	NA	③	③
2007	3.5	VQ35DE	①	②	0.0045-0.0074	0.0008	NA	③	③
	3.5	VQ35HR	①	②	0.0045-0.0074	0.001	NA	1.8057-1.8132	1.8061-1.8136
2008	3.5	VQ35HR	①	②	0.0045-0.0074	0.001	NA	1.8057-1.8132	1.8061-1.8136

NA: Not Available

① Nos. 2 through 4: 0.9230-0.9238

   No.1: 1.0211 - 1.0218 in.

② Nos. 2 through 4: 0.0014-0.0030

   No.1: 0.0018 - 0.0034 in.

③ M/T: Exhause: 1.8061-1.8136

   Intake: 1.8057 - 1.8132 in.

   A/T: Exhaust: 1.7663 - 1.7738 in.

   Intake: 1.7663 - 1.7738 in.

22140_IG35_C0008

CRANKSHAFT AND CONNECTING ROD SPECIFICATIONS

All measurements represent standard values and are given in inches.

Year	Engine Displacement Liters	Engine ID	Crankshaft				Connecting Rod		
			Main Brg. Journal Dia.	Main Brg. Oi Clearance	Shaft End-play	Thrust on No.	Journal Diameter	Oil Clearance	Side Clearance
2006	3.5	VQ35DE	2.3603-2.3612	0.0014-0.0018	0.0039-0.0098	NA	2.0460-2.0462	0.0013-0.0023	0.0079-0.0138
2007	3.5	VQ35DE	2.3603-2.3612	0.0014-0.0018	0.0039-0.0098	NA	2.0460-2.0462	0.0013-0.0023	0.0079-0.0138
	3.5	VQ35HR	2.5572	0.0014-0.0018	0.0040-0.0098	NA	2.0460-2.0462	0.0014-0.0018	0.0079-0.0138
2008	3.5	VQ35HR	2.5572	0.0014-0.0018	0.0040-0.0098	NA	2.0460-2.0462	0.0014-0.0023	0.0079-0.0138

NA: Not Available

22140_IG35_C0009

PISTON AND RING SPECIFICATIONS

All measurements are given in inches.

Year	Engine Displacement Liters	Engine ID	Piston Clearance	Ring Gap			Ring Side Clearance		
				Top Compression	Bottom Compression	Oil Control	Top Compression	Bottom Compression	Oil Control
2006	3.5	VQ35DE	0.0004-0.0012	0.0091-0.0130	①	0.0079-0.0197	0.0018-0.0031	0.0012-0.0028	0.0026-0.0053
2007	3.5	VQ35DE	0.0004-0.0012	0.0091-0.0130	①	0.0079-0.0236	0.0018-0.0031	0.0012-0.0028	0.0026-0.0053
	3.5	VQ35HR	0.0004-0.0012	0.0091-0.0130	0.0130-0.0189	0.0067-0.0185	0.0018-0.0031	0.0012-0.0028	NA
2008	3.5	VQ35HR	0.0004-0.0012	0.0091-0.0130	0.0130-0.0189	0.0067-0.0185	0.0018-0.0031	0.0012-0.0028	NA

NA: Not Available

① A/T Models: 0.0130-0.0189
   M/T Models: 0.0091-0.013

22140_IG35_C0007

TORQUE SPECIFICATIONS
All readings in ft. lbs.

Year	Engine Displacement Liters	Engine ID	Cylinder Head Bolts	Main Bearing Bolts	Rod Bearing Bolts	Crankshaft Damper Bolts	Flywheel Bolts	Manifold Intake	Manifold Exhaust	Spark Plugs	Oil Pan Drain Plug
2006	3.5	VQ35DE	①	②	③	④	61-69	⑤	21-23	15-21	22-29
2007	3.5	VQ35DE	①	②	③	④	61-69	⑤	21-23	15-21	22-29
	3.5	VQ35HR	①	②	③	④	61-69	⑤	21-23	15-21	22-29
2008	3.5	VQ35HR	①	②	③	④	61-69	⑤	21-23	15-21	22-29

① Step 1: 72 ft. lbs.
Step 2: Loosen bolts completely
Step 3: 25-33 ft. lbs.
Step 4: Tighten an additional 90 degrees
Step 5: Repeat Step 4

② Step 1: Shift crankshaft to align the bearing beam
Step 2: Tighten all bolts to 24-28 ft. lbs.
Step 3: Tighten an additional 90-95 degrees

③ Step 1: Tighten to 15 ft. lbs.
Step 2: Tighten an additional 90-95 degrees

④ Step 1: 29-36 ft. lbs.
Step 2: Tighten an additional 60-66 degrees

⑤ Step 1: Tighten to 4-7 ft. lbs.
Step 2: Tighten to 20-23 ft. lbs.
Step 3: Tighten, again, to 20-23 ft. lbs.

22140_IG35_C0010

WHEEL ALIGNMENT

Year	Model		Caster Range (+/-Deg.)	Caster Preferred Setting (Deg.)	Camber Range (+/-Deg.)	Camber Preferred Setting (Deg.)	Toe-in (in.)
2006	G35	F	0.75	①	0.75	②	0.04 +/- 0.04
	Coupe	R	—	—	0.50	-1.50	0.11 +/- 0.11
	G35	F	0.75	7.75	0.75	-0.08	0.04 +/- 0.04
	2WD Sedan	R	—	—	0.75	-0.58	0.11 +/- 0.11
	G35	F	0.75	6.67	0.75	-0.25	0.04 +/- 0.04
	AWD Sedan	R	—	—	0.50	-0.58	0.11 +/- 0.11
2007	G35	F	0.75	①	0.75	②	0.04 +/- 0.04
	Coupe	R	—	—	0.75	-1.25	0.12 +/- 0.12
	G35	F	0.75	7.75	0.75	-0.08	0.04 +/- 0.04
	2WD Sedan	R	—	—	0.75	-0.58	0.11 +/- 0.11
	G35	F	0.75	6.67	0.75	-0.25	0.04 +/- 0.04
	AWD Sedan	R	—	—	0.50	-0.58	0.11 +/- 0.11
2008	G35	F	0.75	①	0.75	②	0.04 +/- 0.04
	Coupe	R	—	—	0.75	-1.25	0.12 +/- 0.12
	G35	F	0.75	7.75	0.75	-0.08	0.04 +/- 0.04
	2WD Sedan	R	—	—	0.75	-0.58	0.11 +/- 0.11
	G35	F	0.75	6.67	0.75	-0.25	0.04 +/- 0.04
	AWD Sedan	R	—	—	0.50	-0.58	0.11 +/- 0.11

① 17" Wheel: 7.24
18" Wheel: 7.25
19" Wheel: 8.08

① 17" and 18" Wheel: -0.50
19" Wheel: -0.42

22140_IG35_C0011

TIRE, WHEEL AND BALL JOINT SPECIFICATIONS

Year	Model	OEM Tires Standard	OEM Tires Optional	Tire Pressures (psi) Front	Tire Pressures (psi) Rear	Wheel Size	Ball Joint Inspection	Lug Nut (Ft. Lbs.)
2006	G35 Coupe	①	②	30	30	Std: ③ Opt: 8.0-JJ	④	72-87
	G35 Sedan	P205/65R16	P215/55R17	30	30	Std: 6.5-JJ Opt: 7-JJ	④	72-87
2007	G35 Coupe	①	②	30	30	Std: ③ Opt: 8-JJ	④	72-87
	G35 Sedan	P205/65R16	P215/55R17	30	30	Std: 6.5-JJ Opt: 7-JJ	④	72-87
2008	G35 Coupe	①	② ⑤	30	30	Std: 7.0-JJ Opt: ⑥	④	72-87
	G35 Sedan	P215/55R17	P235/45R18⑦	30	30	Std: 7.0-JJ Opt: 7.5-JJ	④	72-87

OEM: Original Equipment Manufacturer
PSI: Pounds Per Square Inch
STD: Standard
OPT: Optional
NA: Not Available
M/T: Manual Transmission

① Front: P225/50R17. Rear: P235/50R17
② Front: P225/45R18, Rear: P245/45R18
③ Front: 7.5JJ, Rear: 8.0JJ
④ Replace if any measurable movement is found.
⑤ Front: P225/40R19, Rear: P245/40R19 (Standard 6-Spd M/T)
⑥ 18" Wheel: 8.0-JJ;   19" Wheel: Front 8.0-JJ, Rear 8.5-JJ
⑦ 2WD only

22140_IG35_C0012

BRAKE SPECIFICATIONS
All measurements in inches unless noted

Year	Model	Front Brake Disc Original Thickness	Front Brake Disc Minimum Thickness	Front Brake Disc Maximum Run-out	Rear Brake Disc Original Thickness	Rear Brake Disc Minimum Thickness	Rear Brake Disc Maximum Run-out	Minimum Lining Thickness Front	Minimum Lining Thickness Rear	Brake Caliper Bracket Bolts (ft. lbs.)	Brake Caliper Mounting Bolts (ft. lbs.)
2006	G35	1.102	1.024	0.0014	0.630	0.551	0.0022	NA	NA	①	②
2007	G35	1.102 ③	1.024 ④	0.0014	0.630	0.551 ⑤	0.0022	NA	NA	①	②
2008	G35	1.102 ③	1.024 ④	0.0014	0.630	0.551 ⑤	0.0022	NA	NA	①	②

NA: Not Available
① Front: 113-114
  Rear: 53-71
② Front: 17-22
  Rear: 28-36
③ With single piston caliper: 1.260
④ With single piston caliper: 1.181
⑤ With single piston caliper: 0.591

22140_IG35_C0013

SCHEDULED MAINTENANCE INTERVALS
2006-2008 Infiniti— G35

TO BE SERVICED	TYPE OF SERVICE	VEHICLE MILEAGE INTERVAL (x1000)												
		7.5	15	22.5	30	37.5	45	52.5	60	67.5	75	82.5	90	97.5
Engine oil & filter	R	✓	✓	✓	✓	✓	✓	✓	✓	✓	✓	✓	✓	✓
Automatic transaxle fluid ①	S/I		✓		✓		✓		✓		✓		✓	
Brake lines & cables	S/I		✓		✓		✓		✓		✓		✓	
Brake pads & discs	S/I		✓		✓		✓		✓		✓		✓	
Differential gear oil ①	S/I		✓		✓		✓		✓		✓		✓	
Driveshaft boots	S/I		✓		✓		✓		✓		✓		✓	
Manual transaxle oil (G20)	S/I		✓		✓		✓		✓		✓		✓	
In-cabin microfilter	R		✓		✓		✓		✓		✓		✓	
Air cleaner filter ②	R				✓				✓				✓	
Exhaust system	S/I				✓				✓				✓	
Fuel lines	S/I				✓				✓				✓	
Steering gear & linkage, axle & suspension parts	S/I				✓				✓				✓	
Vapor lines	S/I				✓				✓				✓	
Engine coolant ③	R								✓				✓	
Spark plugs (Conventional) ④	R				✓				✓					
Drive belts ⑤	S/I								✓					

R: Replace   S/I: Service or Inspect

① If towing a trailer, using a camper or car-top carrier, or driving on rough or muddy roads, CHANGE oil every 30,000 miles or 24 months.

② If operating in dusty conditions, more frequent maintenance may be required.

③ After 60,000 miles or 48 months, replace coolant every 30,000 miles or 24 months.

④ Platinum-tipped spark plugs should be changed every 105,000 miles.

⑤ After 60,000 miles or 48 months, inspect every 15,000 miles or 12 months. Replace belts if found damaged.

FREQUENT OPERATION MAINTENANCE (SEVERE SERVICE)

If a vehicle is operated under any of the following conditions it is considered severe service:

- Extremely dusty areas.

- 50% or more of the vehicle operation is in 32°C (90°F) or higher temperatures, or constant operation in temperatures below 0°C (32°F).

- Prolonged idling (vehicle operation in stop and go traffic).

- Frequent short running periods (engine does not warm to normal operating temperatures).

- Police, taxi, delivery or trailer towing usage.

Oil & oil filter: change every 3750 miles.

Brake pads & discs: service or inspect every 7500 miles.

Exhaust system: service or inspect every 7500 miles.

Steering gear, linkage, axle & suspension ball joints: service or inspect every 7500 miles.

Steering linkage, ball joints & front suspension ball joints: service or inspect every 7500 miles.

22140_IG35_C0014

PRECAUTIONS

Before servicing any vehicle, please be sure to read all of the following precautions, which deal with personal safety, prevention of component damage, and important points to take into consideration when servicing a motor vehicle:

• Never open, service or drain the radiator or cooling system when the engine is hot; serious burns can occur from the steam and hot coolant.

• Observe all applicable safety precautions when working around fuel. Whenever servicing the fuel system, always work in a well-ventilated area. Do not allow fuel spray or vapors to come in contact with a spark, open flame, or excessive heat (a hot drop light, for example). Keep a dry chemical fire extinguisher near the work area. Always keep fuel in a container specifically designed for fuel storage; also, always properly seal fuel containers to avoid the possibility of fire or explosion. Refer to the additional fuel system precautions later in this section.

• Fuel injection systems often remain pressurized, even after the engine has been turned **OFF**. The fuel system pressure must be relieved before disconnecting any fuel lines. Failure to do so may result in fire and/or personal injury.

• Brake fluid often contains polyglycol ethers and polyglycols. Avoid contact with the eyes and wash your hands thoroughly after handling brake fluid. If you do get brake fluid in your eyes, flush your eyes with clean, running water for 15 minutes. If eye irritation persists, or if you have taken

brake fluid internally, IMMEDIATELY seek medical assistance.

• The EPA warns that prolonged contact with used engine oil may cause a number of skin disorders, including cancer. You should make every effort to minimize your exposure to used engine oil. Protective gloves should be worn when changing oil. Wash your hands and any other exposed skin areas as soon as possible after exposure to used engine oil. Soap and water, or waterless hand cleaner should be used.

• All new vehicles are now equipped with an air bag system, often referred to as a Supplemental Restraint System (SRS) or Supplemental Inflatable Restraint (SIR) system. The system must be disabled before performing service on or around system components, steering column, instrument panel components, wiring and sensors. Failure to follow safety and disabling procedures could result in accidental air bag deployment, possible personal injury and unnecessary system repairs.

• Always wear safety goggles when working with, or around, the air bag system. When carrying a non-deployed air bag, be sure the bag and trim cover are pointed away from your body. When placing a non-deployed air bag on a work surface, always face the bag and trim cover upward, away from the surface. This will reduce the motion of the module if it is accidentally deployed. Refer to the additional air bag system precautions later in this section.

• Clean, high quality brake fluid from a sealed container is essential to the safe and

proper operation of the brake system. You should always buy the correct type of brake fluid for your vehicle. If the brake fluid becomes contaminated, completely flush the system with new fluid. Never reuse any brake fluid. Any brake fluid that is removed from the system should be discarded. Also, do not allow any brake fluid to come in contact with a painted surface; it will damage the paint.

• Never operate the engine without the proper amount and type of engine oil; doing so WILL result in severe engine damage.

• Timing belt maintenance is extremely important. Many models utilize an interference-type, non-freewheeling engine. If the timing belt breaks, the valves in the cylinder head may strike the pistons, causing potentially serious (also time-consuming and expensive) engine damage. Refer to the maintenance interval charts for the recommended replacement interval for the timing belt, and to the timing belt section for belt replacement and inspection.

• Disconnecting the negative battery cable on some vehicles may interfere with the functions of the on-board computer system(s) and may require the computer to undergo a relearning process once the negative battery cable is reconnected.

• When servicing drum brakes, only disassemble and assemble one side at a time, leaving the remaining side intact for reference.

• Only an MVAC-trained, EPA-certified automotive technician should service the air conditioning system or its components.

BRAKES

GENERAL INFORMATION

PRECAUTIONS

• Certain components within the ABS system are not intended to be serviced or repaired individually.

• Do not use rubber hoses or other parts not specifically specified for and ABS system. When using repair kits, replace all parts included in the kit. Partial or incorrect repair may lead to functional problems and require the replacement of components.

• Lubricate rubber parts with clean, fresh brake fluid to ease assembly. Do not

use shop air to clean parts; damage to rubber components may result.

• Use only DOT 3 brake fluid from an unopened container.

• If any hydraulic component or line is removed or replaced, it may be necessary to bleed the entire system.

• A clean repair area is essential. Always clean the reservoir and cap thoroughly before removing the cap. The slightest amount of dirt in the fluid may plug an orifice and impair the system function. Perform repairs after components have been thoroughly cleaned; use only denatured alcohol

ANTI-LOCK BRAKE SYSTEM (ABS)

to clean components. Do not allow ABS components to come into contact with any substance containing mineral oil; this includes used shop rags.

• The Anti-Lock control unit is a microprocessor similar to other computer units in the vehicle. Ensure that the ignition switch is **OFF** before removing or installing controller harnesses. Avoid static electricity discharge at or near the controller.

• If any arc welding is to be done on the vehicle, the control unit should be unplugged before welding operations begin.

BRAKES
BLEEDING THE BRAKE SYSTEM

BLEEDING PROCEDURE

BLEEDING PROCEDURE

1. Before servicing the vehicle, refer to the Precautions Section.

❊❊ CAUTION

While bleeding the brake system, pay attention to the master cylinder fluid level.

2. Turn the ignition switch OFF and disconnect the VDC actuator connectors or negative battery cable.

3. Raise and safely support the vehicle.

➡**Bleed air in the following order, right rear, left front, left rear and right front.**

4. Attach a vinyl tube to the right, rear bleeder valve.

5. Depress the brake pedal fully 4 or 5 times.

6. With the brake pedal depressed, loosen the bleeder valve to let the air out, then tighten it immediately.

7. Repeat steps 3 and 4 until no more air comes out.

8. Tighten the bleeder valve.

9. Fill the master cylinder reservoir.

10. Check that the fluid level in the reservoir tank is within the specified range after air bleeding.

11. Check for brake fluid leakage from the master cylinder mounting face, reservoir tank mounting face and brake tube connections.

BRAKES
FRONT DISC BRAKES

❊❊ CAUTION

Dust and dirt accumulating on brake parts during normal use may contain asbestos fibers from production or aftermarket brake linings. Breathing excessive concentrations of asbestos fibers can cause serious bodily harm. Exercise care when servicing brake parts. Do not sand or grind brake lining unless equipment used is designed to contain the dust residue. Do not clean brake parts with compressed air or by dry brushing. Cleaning should be done by dampening the brake components with a fine mist of water, then wiping the brake components clean with a dampened cloth. Dispose of cloth and all residue containing asbestos fibers in an impermeable container with the appropriate label. Follow practices prescribed by the Occupational Safety and Health Administration (OSHA) and the Environmental Protection Agency (EPA) for the handling, processing, and disposing of dust or debris that may contain asbestos fibers.

BRAKE CALIPER

REMOVAL & INSTALLATION
See Figures 1 through 3.

❊❊ WARNING

Clean any dust from the brake caliper and brake pads with a vacuum dust collector. Never blow with compressed air.

❊❊ CAUTION

Never depress the brake pedal. Brake fluid may splash while removing the brake hose.

❊❊ CAUTION

Never spill or splash any grease and moisture on the brake caliper assembly mounting face, threads, mounting bolts and washers. Wipe out any grease and moisture.

➡**Never reuse the copper washer.**

1. Before servicing the vehicle, refer to the precautions in the beginning of this section.

2. Remove the front wheels.

3. If necessary, drain the brake fluid.

4. Remove both guide pin bolts securing the caliper to the steering knuckle.

5. Loosen and remove the brake hose connector from the caliper.

6. Remove the caliper assembly from the vehicle.

➡**Put matching marks on the wheel hub and bearing assembly and the disc rotor before removing the disc rotor. Never drop disc rotor.**

7. If necessary, remove the rotor.

To install:

➡**See the accompanying illustration for torque values.**

8. Using new copper washers, install the brake line to the brake caliper.

9. Install the caliper to the steering knuckle using the guide pins bolts.

10. Install the wheels and tighten the lug nuts to the proper specification.

11. Bleed the brake system and top off the master cylinder as necessary.

DISC BRAKE PADS

REMOVAL & INSTALLATION

❊❊ WARNING

Clean any dust from the brake caliper and brake pads with a vacuum dust collector. Never blow with compressed air.

❊❊ CAUTION

Never depress the brake pedal. Brake fluid may splash while removing the brake hose.

❊❊ CAUTION

Never spill or splash any grease and moisture on the brake caliper assembly mounting face, threads, mounting bolts and washers. Wipe out any grease and moisture.

❊❊ CAUTION

Do not damage piston boot. Keep rotor clean, from brake fluid.

➡**Never reuse the copper washer.**

1. Before servicing the vehicle, refer to the precautions in the beginning of this section.

2. Remove the wheels.

3. Remove the bottom guide pin from the caliper and swing the caliper cylinder body upward. Support the caliper with a wire.

4. Remove the brake pad retainers and the pads.

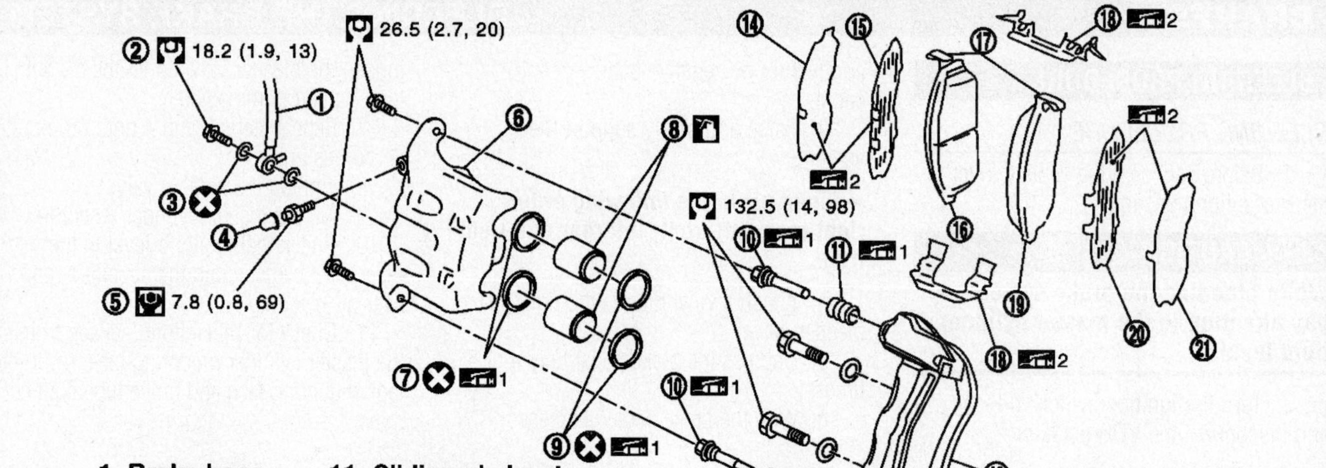

1. Brake hose
2. Union bolt
3. Copper washer
4. Cap
5. Bleeder valve
6. Cylinder body
7. Piston seal
8. Piston
9. Piston boot
10. Sliding pin
11. Sliding pin boot
12. Bushing
13. Torque member
14. Inner shim cover
15. Inner shim
16. Inner pad
17. Pad wear sensor
18. Pad retainer
19. Outer pad
20. Outer shim
21. Outer shim cover

22140_IG35_G0006

Fig. 1 Front caliper and related parts—2006 coupe and sedan, 2007—coupe

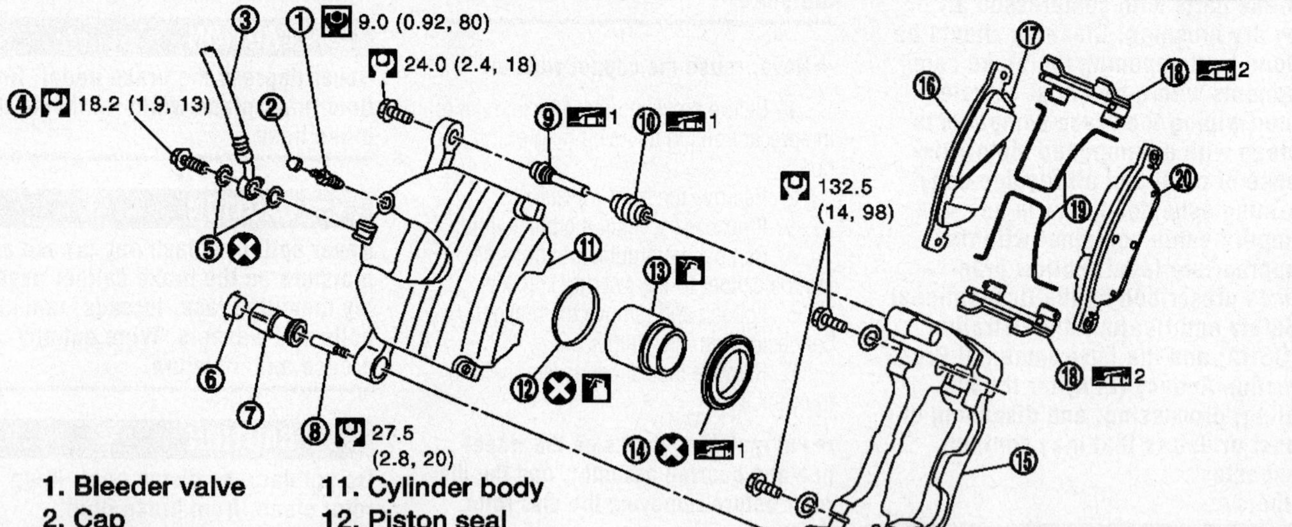

1. Bleeder valve
2. Cap
3. Brake hose
4. Union bolt
5. Copper washer
6. Protector
7. Bushing
8. Location pin
9. Sliding pin
10. Sliding pin boot
11. Cylinder body
12. Piston seal
13. Piston
14. Piston boot
15. Torque member
16. Inner pad
17. Pad wear sensor
18. Pad retainer
19. Pad return spring
20. Outer pad

22140_IG35_G0005

Fig. 2 Front caliper and related parts— single piston type—2007 sedan

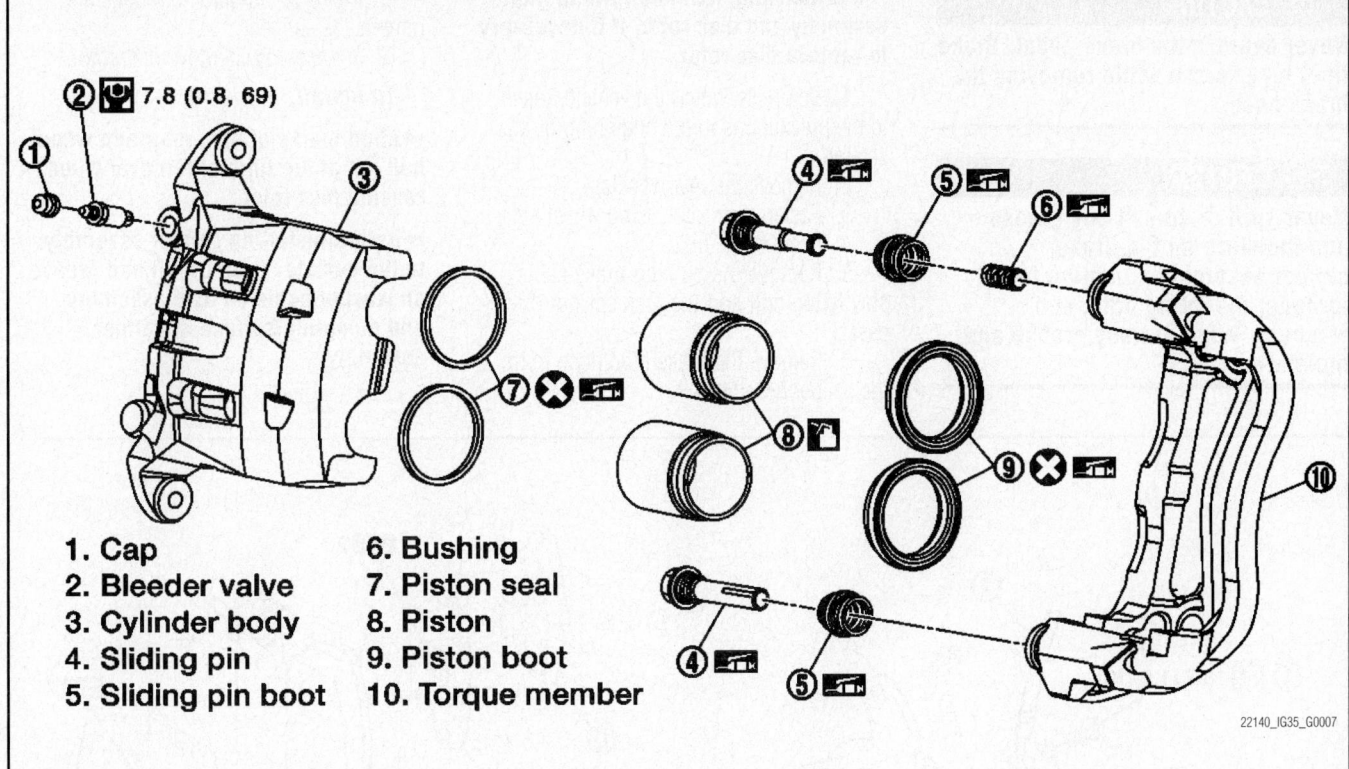

1. Cap
2. Bleeder valve
3. Cylinder body
4. Sliding pin
5. Sliding pin boot
6. Bushing
7. Piston seal
8. Piston
9. Piston boot
10. Torque member

22140_IG35_G0007

Fig. 3 Front caliper and related parts—2008 sedan

To install:

> ✳✳ **CAUTION**
>
> Inner pad and outer pad have pad-return mechanism on upper side of pad retainer. When installing pad to torque member, be sure to install pad return lever to pad wear sensor securely.

> ✳✳ **CAUTION**
>
> When replacing pads with new ones, press in piston until pads can be installed. In this case, carefully monitor brake fluid level in reservoir tank because brake fluid will return to master cylinder reservoir tank.

➡ Single piston brake pads are directional. Make sure the pads are installed in the correct direction.

5. Compress the piston of the disc brake caliper.

6. Install the brake pads and caliper assembly. Refer to appropriate illustration under Brake Caliper for specifications.

7. Apply Copper based brake grease as necessary to the pad retainers before installing it to the torque member if the pad retainers have been removed.

8. Install the wheels.

9. Check the master cylinder and add fluid if necessary.

BRAKES

> ✳✳ **CAUTION**
>
> Dust and dirt accumulating on brake parts during normal use may contain asbestos fibers from production or aftermarket brake linings. Breathing excessive concentrations of asbestos fibers can cause serious bodily harm. Exercise care when servicing brake parts. Do not sand or grind brake lining unless equipment used is designed to contain the dust residue. Do not clean brake parts with compressed air or by dry brushing. Cleaning should be done by dampening the brake components with a fine mist of water, then wiping the brake components clean with a dampened cloth. Dispose of cloth and all residue containing asbestos fibers in an impermeable container with the appropriate label. Follow practices prescribed by the Occupational Safety and Health Administration (OSHA) and the Environmental Protection Agency (EPA) for the handling, processing, and disposing of dust or debris that may contain asbestos fibers.

REAR DISC BRAKES

BRAKE CALIPER

REMOVAL & INSTALLATION
See Figures 4 and 5.

> ✳✳ **WARNING**
>
> Clean any dust from the brake caliper and brake pads with a vacuum dust collector. Never blow with compressed air.

⁂⁑ **CAUTION**

Never depress the brake pedal. Brake fluid may splash while removing the brake hose.

⁂⁑ **CAUTION**

Never spill or splash any grease and moisture on the brake caliper assembly mounting face, threads, mounting bolts and washers. Wipe out any grease and moisture.

➡Put matching marks on wheel hub assembly and disc rotor, if it necessary to remove disc rotor.

1. Before servicing the vehicle, refer to the precautions in the beginning of this section.

2. Remove the rear wheels.

3. Fasten disc rotor using wheel nut

4. Drain brake fluid.

5. Remove the parking brake cable stay fixing bolt and the lock spring, if necessary.

6. Remove the brake fluid hose from the caliper assembly.

7. Remove the guide pin bolts and remove the caliper.

8. If necessary, remove disc rotor.

To install:

➡Align marks of disc rotor and wheel hub put at the time of removal when reusing disc rotor.

➡Before installing caliper assembly to the vehicle, wipe off oil and grease on washer seats on axle assembly and mounting surface of caliper assembly.

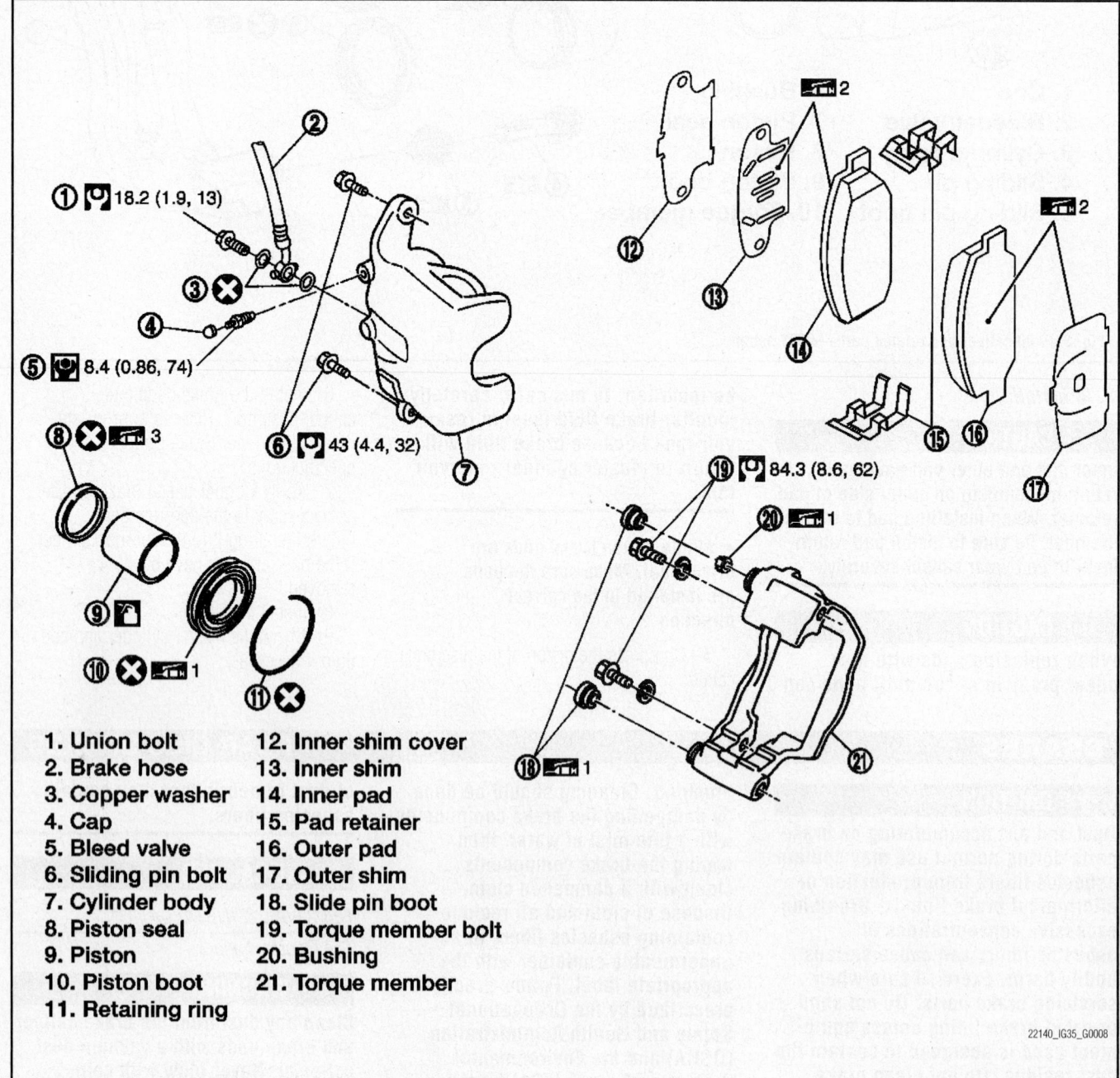

1. Union bolt
2. Brake hose
3. Copper washer
4. Cap
5. Bleed valve
6. Sliding pin bolt
7. Cylinder body
8. Piston seal
9. Piston
10. Piston boot
11. Retaining ring
12. Inner shim cover
13. Inner shim
14. Inner pad
15. Pad retainer
16. Outer pad
17. Outer shim
18. Slide pin boot
19. Torque member bolt
20. Bushing
21. Torque member

22140_IG35_G0008

Fig. 4 Rear brake caliper and related parts—Except 2008 models

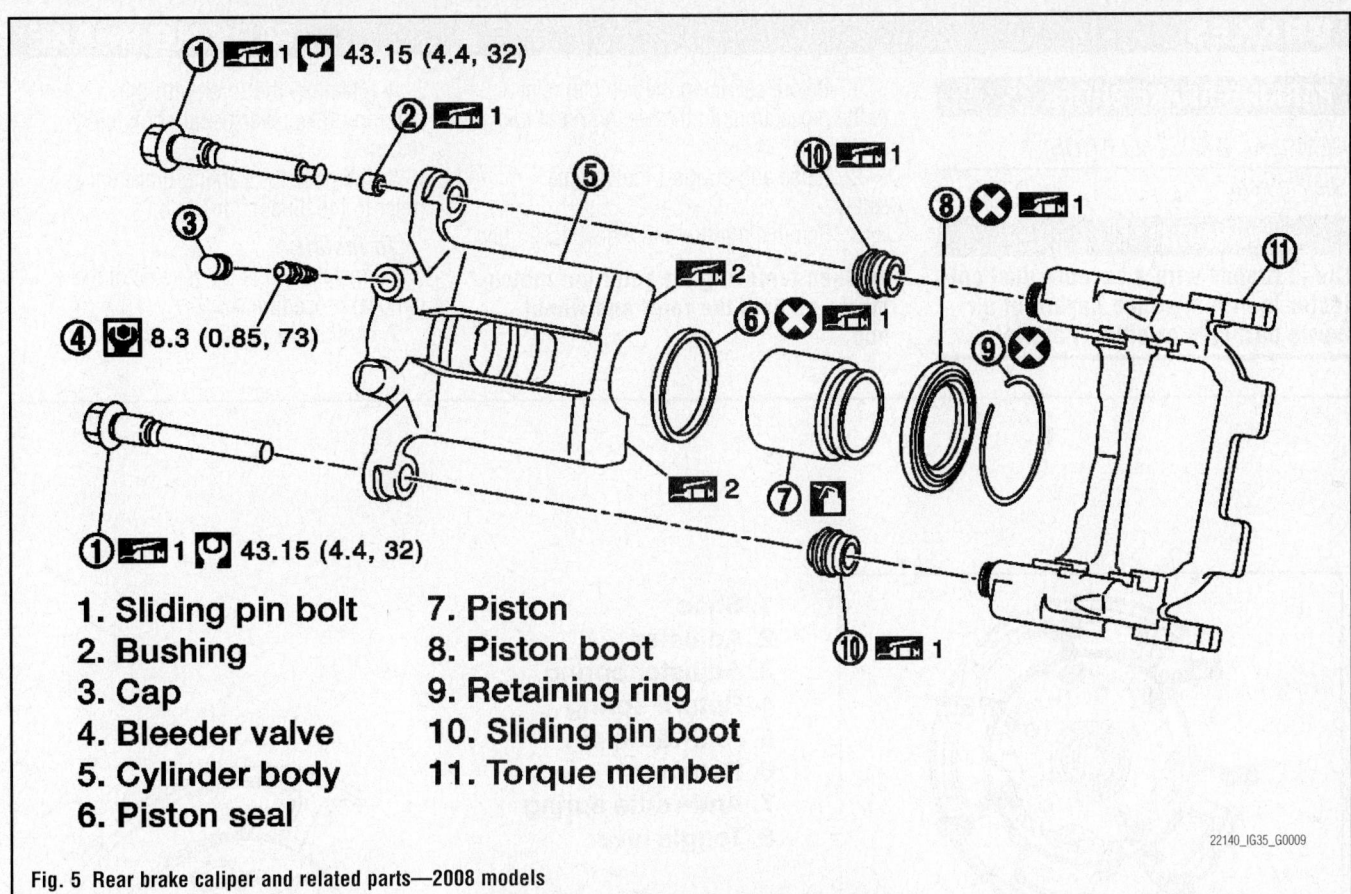

1. **Sliding pin bolt**
2. **Bushing**
3. **Cap**
4. **Bleeder valve**
5. **Cylinder body**
6. **Piston seal**
7. **Piston**
8. **Piston boot**
9. **Retaining ring**
10. **Sliding pin boot**
11. **Torque member**

22140_IG35_G0009

Fig. 5 Rear brake caliper and related parts—2008 models

9. Install the caliper body into position Refer to the accompanying illustrations for torque values and procedures.

10. Reconnect the brake fluid hose and tighten the flare nut to 12–14 ft. lbs. (17–20 Nm).

11. Install the lock spring and the parking brake stay attaching bolt.

12. Bleed the brake system and top off the master cylinder as necessary.

13. Install the wheels.

DISC BRAKE PADS

REMOVAL & INSTALLATION

✳✳ WARNING

Clean any dust from the brake caliper and brake pads with a vacuum dust collector. Never blow with compressed air.

✳✳ CAUTION

Never depress the brake pedal. Brake fluid may splash while removing the brake hose.

✳✳ CAUTION

Never spill or splash any grease and moisture on the brake caliper assembly mounting face, threads, mounting bolts and washers. Wipe out any grease and moisture.

✳✳ CAUTION

Do not damage piston boot. Keep rotor clean, from brake fluid.

➡**Never reuse the copper washer.**

1. Before servicing the vehicle, refer to the precautions in the beginning of this section.

2. Remove the rear wheels.

3. Remove the parking brake cable mounting bolt and lock spring.

4. Disconnect the cable from the caliper.

5. Remove the upper pin bolt.

6. Pivot the caliper body downward.

7. Pull out the pad springs and then remove the pads and shims.

To install:

8. Turn the piston clockwise back into the caliper body. Take care not to damage the piston boot.

9. Coat the pad contact area on the mounting support with grease.

10. Install the pads, shims, and the pad springs.

11. Position the caliper body in the mounting support and tighten the pin bolts to 16–23 ft. lbs. (22–31 Nm).

12. Install the wheels.

13. Check the master cylinder and add fluid if necessary.

BRAKES   PARKING BRAKE

PARKING BRAKE SHOES

REMOVAL & INSTALLATION

See Figure 6.

✳✳ WARNING

Clean brakes with a vacuum dust collector to minimize the hazard of air borne particles or other materials.

1. Before servicing the vehicle, refer to the precautions in the beginning of this section.

2. Raise and support the vehicle safely.

3. Remove the tire and wheel assembly.

➡ **When removing the rotor put match-marks on both the rotor and wheel hub.**

4. Remove the rotor with the parking brake lever/pedal completely returned.

5. Remove the parking brake shoes (refer to the illustration).

To install:

6. Installation is the reverse of the removal procedure.

7. Adjust the parking brake.

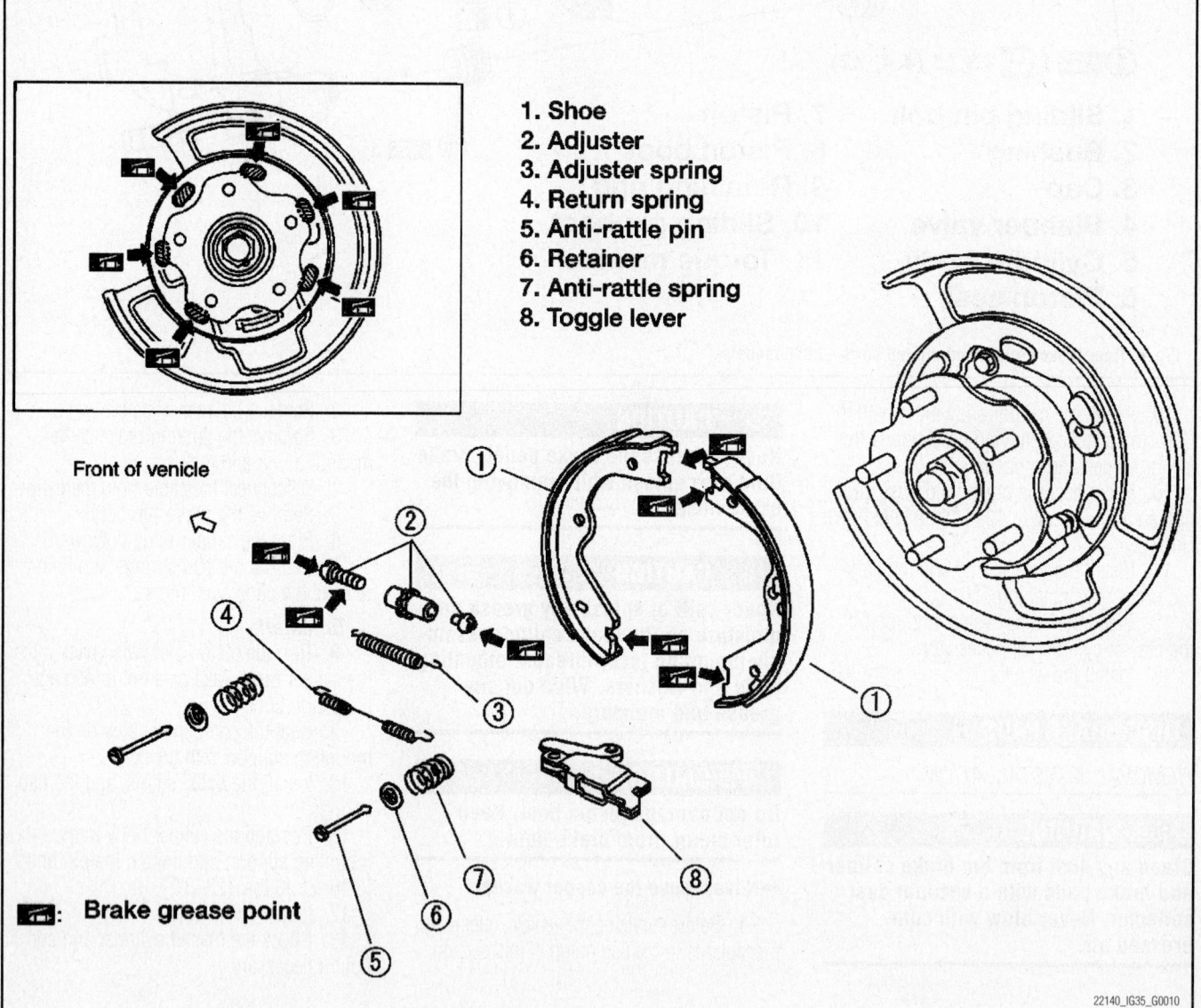

1. Shoe
2. Adjuster
3. Adjuster spring
4. Return spring
5. Anti-rattle pin
6. Retainer
7. Anti-rattle spring
8. Toggle lever

Front of vehicle

▨: Brake grease point

22140_IG35_G0010

Fig. 6 Parking brake shoe and related components

CHASSIS ELECTRICAL   **AIR BAG (SUPPLEMENTAL RESTRAINT SYSTEM)**

GENERAL INFORMATION

✳✳ CAUTION

These vehicles are equipped with an air bag system. The system must be disarmed before performing service on, or around, system components, the steering column, instrument panel components, wiring and sensors. Failure to follow the safety precautions and the disarming procedure could result in accidental air bag deployment, possible injury and unnecessary system repairs.

SERVICE PRECAUTIONS

Disconnect and isolate the battery negative cable before beginning any airbag system component diagnosis, testing, removal, or installation procedures. Allow system capacitor to discharge for two minutes before beginning any component service. This will disable the airbag system. Failure to disable the airbag system may result in accidental airbag deployment, personal injury, or death.

Do not place an intact undeployed airbag face down on a solid surface. The airbag will propel into the air if accidentally deployed and may result in personal injury or death.

When carrying or handling an undeployed airbag, the trim side (face) of the airbag should be pointing towards the body to minimize possibility of injury if accidental deployment occurs. Failure to do this may result in personal injury or death.

Replace airbag system components with OEM replacement parts. Substitute parts may appear interchangeable, but internal differences may result in inferior occupant protection. Failure to do so may result in occupant personal injury or death.

Wear safety glasses, rubber gloves, and long sleeved clothing when cleaning powder residue from vehicle after an airbag deployment. Powder residue emitted from a deployed airbag can cause skin irritation. Flush affected area with cool water if irritation is experienced. If nasal or throat irritation is experienced, exit the vehicle for fresh air until the irritation ceases. If irritation continues, see a physician.

Do not use a replacement airbag that is not in the original packaging. This may result in improper deployment, personal injury, or death.

The factory installed fasteners, screws and bolts used to fasten airbag components have a special coating and are specifically designed for the airbag system. Do not use substitute fasteners. Use only original equipment fasteners listed in the parts catalog when fastener replacement is required.

During, and following, any child restraint anchor service, due to impact event or vehicle repair, carefully inspect all mounting hardware, tether straps, and anchors for proper installation, operation, or damage. If a child restraint anchor is found damaged in any way, the anchor must be replaced. Failure to do this may result in personal injury or death.

Deployed and non-deployed airbags may or may not have live pyrotechnic material within the airbag inflator.

Do not dispose of driver/passenger/curtain airbags or seat belt tensioners unless you are sure of complete deployment. Refer to the Hazardous Substance Control System for proper disposal.

Dispose of deployed airbags and tensioners consistent with state, provincial, local, and federal regulations.

After any airbag component testing or service, do not connect the battery negative cable. Personal injury or death may result if the system test is not performed first.

If the vehicle is equipped with the Occupant Classification System (OCS), do not connect the battery negative cable before performing the OCS Verification Test using the scan tool and the appropriate diagnostic information. Personal injury or death may result if the system test is not performed properly.

Never replace both the Occupant Restraint Controller (ORC) and the Occupant Classification Module (OCM) at the same time. If both require replacement, replace one, then perform the Airbag System test before replacing the other.

Both the ORC and the OCM store Occupant Classification System (OCS) calibration data, which they transfer to one another when one of them is replaced. If both are replaced at the same time, an irreversible fault will be set in both modules and the OCS may malfunction and cause personal injury or death.

If equipped with OCS, the Seat Weight Sensor is a sensitive, calibrated unit and must be handled carefully. Do not drop or

handle roughly. If dropped or damaged, replace with another sensor. Failure to do so may result in occupant injury or death.

If equipped with OCS, the front passenger seat must be handled carefully as well. When removing the seat, be careful when setting on floor not to drop. If dropped, the sensor may be inoperative, could result in occupant injury, or possibly death.

If equipped with OCS, when the passenger front seat is on the floor, no one should sit in the front passenger seat. This uneven force may damage the sensing ability of the seat weight sensors. If sat on and damaged, the sensor may be inoperative, could result in occupant injury, or possibly death.

DISARMING THE SYSTEM

➡ **All Air Bag electrical wiring harnesses and connectors are covered with YELLOW outer insulation. Do not use electrical test equipment on any circuit related to the Air Bag sensors. When installing Air Bag components, always install with the arrow marks facing the front of the vehicle.**

1. Before servicing the vehicle, refer to the precautions in the beginning of this section.

2. Turn the ignition switch to the **OFF** position.

3. Disconnect both battery cables starting with the negative cable first and wait at least 10 minutes after the cables are disconnected. Be sure to insulate the battery terminal ends.

ARMING THE SYSTEM

1. Before servicing the vehicle, refer to the precautions in the beginning of this section.

2. Turn the ignition switch to the **OFF** position.

3. Connect both battery cables starting with the positive cable first.

➡ **The Air Bag or Air Bag system is equipped with a self-diagnostic operation. After turning the ignition key to the ON or START position, the AIR BAG warning lamp will illuminate for 7 seconds. After 7 seconds, the AIR BAG lamp will extinguish if no malfunction is detected. If the AIR BAG lamp does not extinguish after 7 seconds, check the Air Bag self-diagnostic system for a malfunction.**

DRIVETRAIN

AUTOMATIC TRANSMISSION ASSEMBLY

REMOVAL & INSTALLATION

2006 Coupe and Sedan; 2007 Coupe

See Figures 7 through 11.

➡**When removing the Automatic Transmission (A/T) assembly from the engine, first remove the Crankshaft Position Sensor (CKP) from the A/T assembly.**

1. Before servicing the vehicle, refer to the precautions in the beginning of this section.

2. Remove or disconnect the following:
- Negative battery cable
- Engine cover
- A/T fluid level gauge
- Engine under cover
- Exhaust pipe
- Rear driveshaft
- Front driveshaft (AWD), if equipped
- A/T control rod and solenoid valve harness connector
- CKP sensor from the transmission
- Starter motor
- Rear cover plate (2WD)
- Dust cover from converter housing

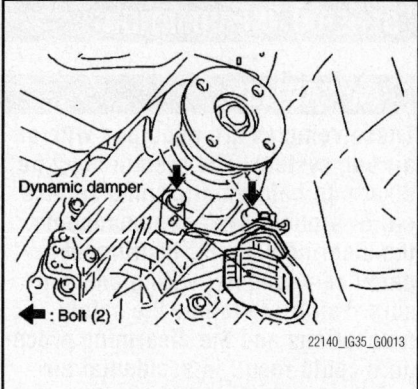

Dynamic damper

◄ : Bolt (2)

22140_IG35_G0013

Fig. 7 Dynamic damper assembly (AWD)

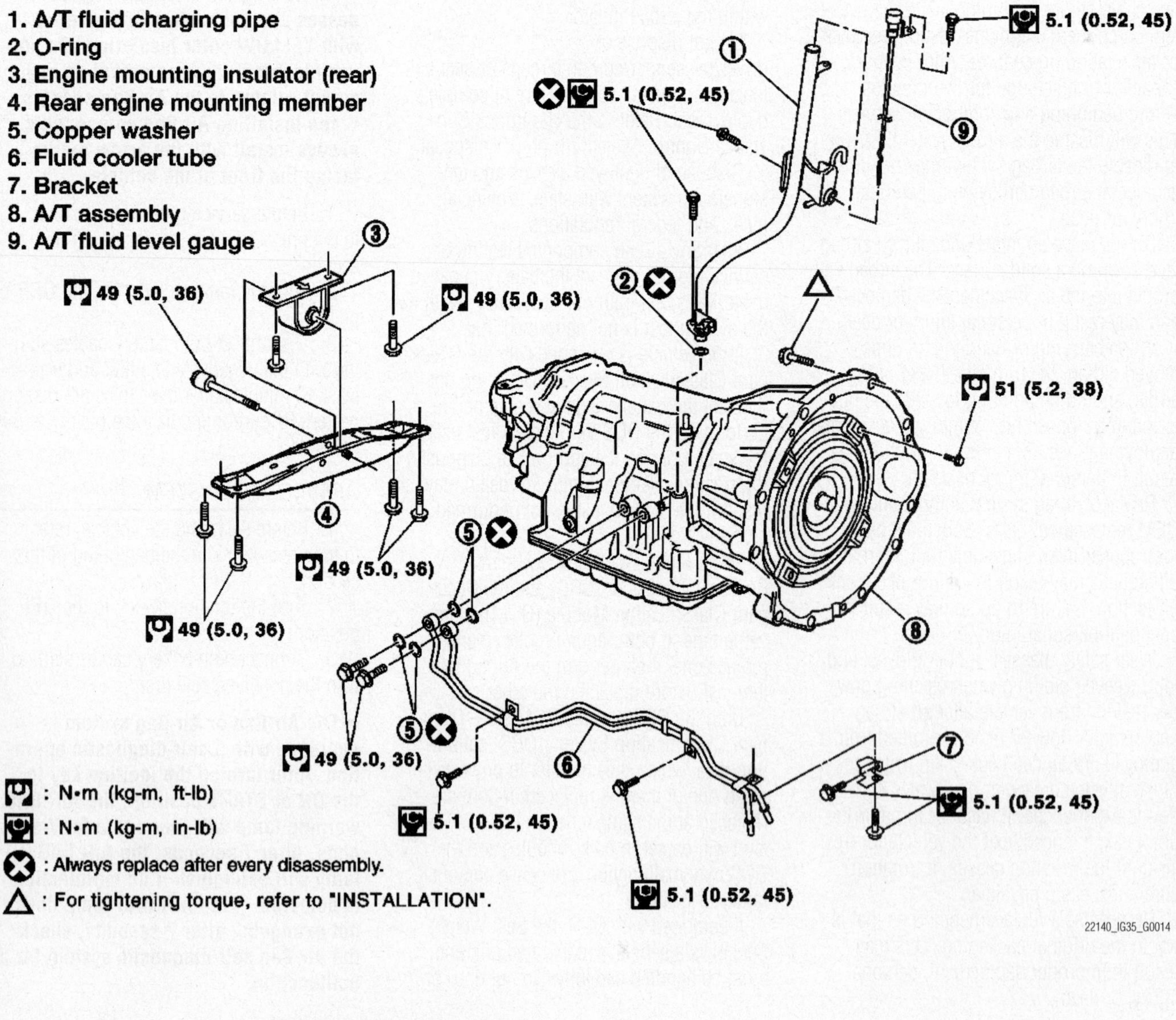

---

1. A/T fluid charging pipe
2. O-ring
3. Engine mounting insulator (rear)
4. Rear engine mounting member
5. Copper washer
6. Fluid cooler tube
7. Bracket
8. A/T assembly
9. A/T fluid level gauge

5.1 (0.52, 45)

✕ 5.1 (0.52, 45)

49 (5.0, 36)

49 (5.0, 36)

51 (5.2, 38)

49 (5.0, 36)

49 (5.0, 36)

49 (5.0, 36)

5.1 (0.52, 45)

5.1 (0.52, 45)

5.1 (0.52, 45)

: N•m (kg-m, ft-lb)

: N•m (kg-m, in-lb)

✕ : Always replace after every disassembly.

⚠ : For tightening torque, refer to "INSTALLATION".

22140_IG35_G0014

Fig. 8 Exploded view of the automatic transmission mounting—2WD

→**When turning the crankshaft, turn it clockwise as viewed from the front of the engine.**

3. Turn the crankshaft, then remove the 4 tightening bolts for the drive plate and torque converter.

4. Remove the dynamic damper (AWD), if equipped.

5. Support the transmission/transfer (AWD) assembly with a suitable jack. Be careful not the let the jack hit the drain plug.

6. Remove or disconnect the following:
• Rear member
• Engine mounting insulator (rear)

• Air breather hose
• A/T assembly harness connector
• A/T fluid charging pipe from A/T assembly
• O-ring from A/T fluid charging pipe
• Fluid cooler tube
• Engine-to-transmission bolts

❄❄ WARNING

Before removal, secure the transmission to the jack and secure the torque converter to prevent it from falling.

• Transmission from the vehicle by carefully lowering it with the jack

7. Remove transfer assembly mounting bolts and separate transfer assembly from transmission (AWD; Refer to Transfer Case Assembly procedures).

To install:

8. Installation is the reverse of the removal procedure, noting the following:

a. Tighten the transmission-to-engine bolts as shown in the illustration.

b. Align the positions of the tightening bolts for the drive plate with those of the torque converter and hand-tighten.

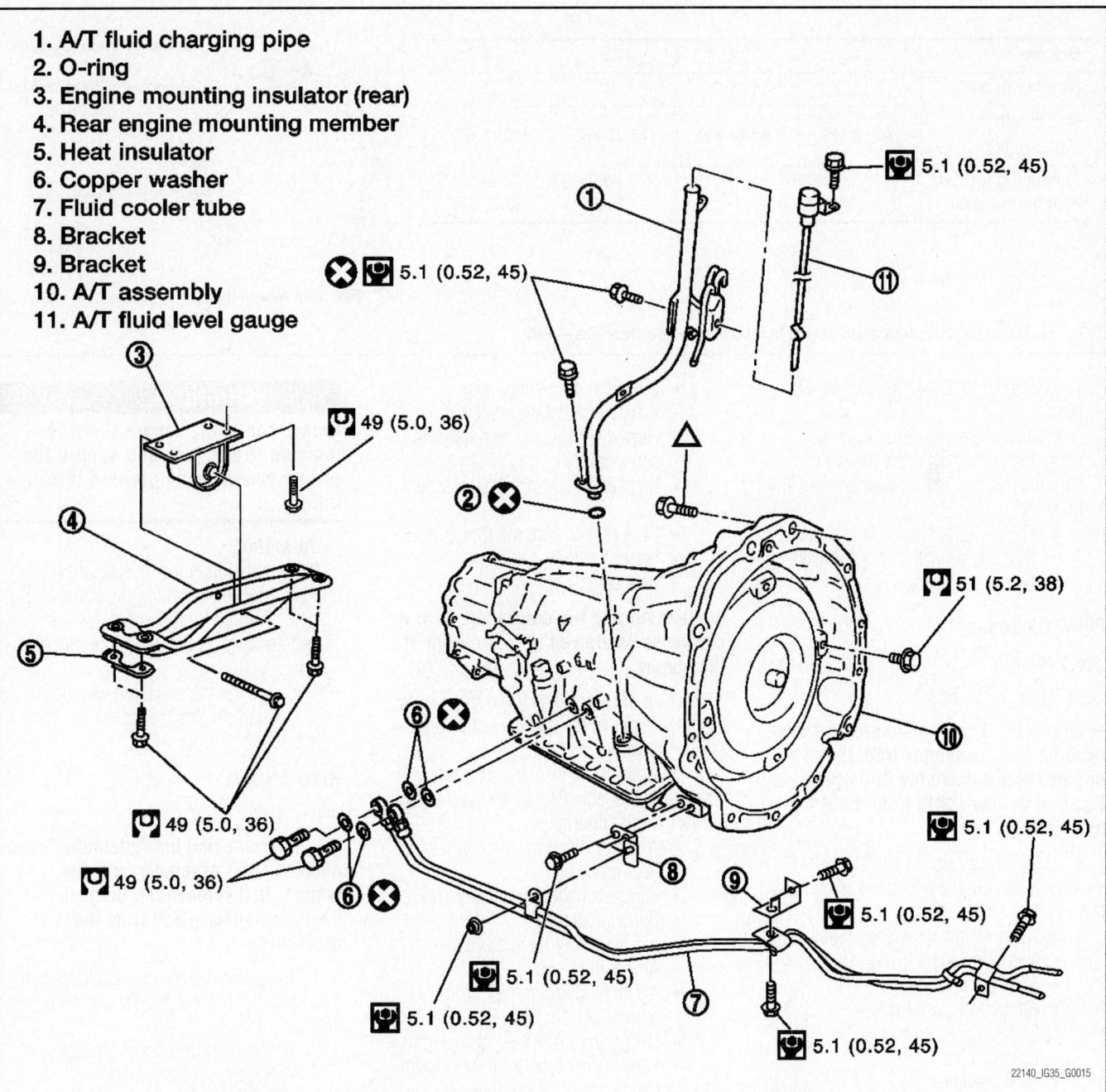

1. A/T fluid charging pipe
2. O-ring
3. Engine mounting insulator (rear)
4. Rear engine mounting member
5. Heat insulator
6. Copper washer
7. Fluid cooler tube
8. Bracket
9. Bracket
10. A/T assembly
11. A/T fluid level gauge

5.1 (0.52, 45)
5.1 (0.52, 45)
49 (5.0, 36)
49 (5.0, 36)
49 (5.0, 36)
51 (5.2, 38)
5.1 (0.52, 45)
5.1 (0.52, 45)
5.1 (0.52, 45)
5.1 (0.52, 45)

Fig. 9 Exploded view of the automatic transmission mounting—AWD

22140_IG35_G0015

Bolt No.	1	2	3	4
Number of bolts	1	5	2	2
Bolt length "ℓ"mm (in)	55 (2.17)	65 (2.56)	50 (2.20)	35 (1.38)
Tightening torque N·m (kg-m, ft-lb)	75 (7.7, 55)		55 (5.6, 41)	47 (4.8, 35)

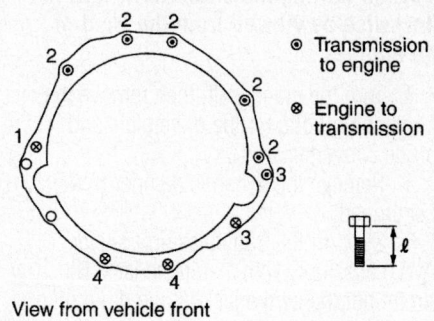

View from vehicle front

09482_INFI_G0411

Fig. 10 Automatic transmission-to-engine bolt tightening specifications—2WD

Bolt No.	1	2	3	4
Number of bolts	1	5	2	1
Bolt length "ℓ"mm (in)	55 (2.17)	65 (2.56)	35 (1.38)	40 (1.57)
Tightening torque N·m (kg-m, ft-lb)	75 (7.7, 55)		47 (4.8, 35)	34 (3.5, 25)

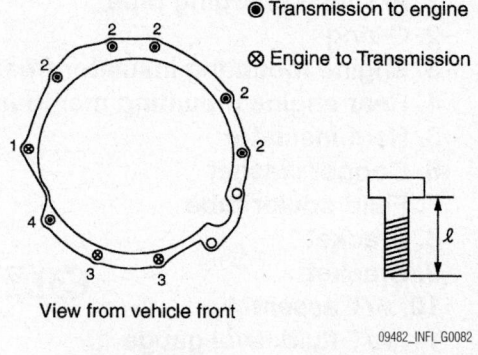

View from vehicle front

09482_INFI_G0082

Fig. 11 Automatic transmission-to-engine bolt tightening specifications—AWD

Then, tighten to 33–42 ft. lbs. (44–58 Nm).

c. After the converter is installed, rotate the crankshaft a few times to be sure the transmission rotates without any binding

d. Fill the transmission with fluid.

e. Start the vehicle, check for leaks and repair if necessary.

2007–08 Sedan

2WD Models

See Figures 12 and 13.

➡When removing the Automatic Transmission (A/T) assembly from the engine, first remove the Crankshaft Position Sensor (CKP) from the A/T assembly.

1. Before servicing the vehicle, refer to the precautions in the beginning of this section.

2. Remove or disconnect the following:
- Negative battery cable
- A/T fluid level gauge
- Air cleaner case (RH)
- Engine under cover
- Exhaust pipe
- Heat insulator
- Rear driveshaft
- Suspension member stay
- Exhaust mounting bracket
- Heated O2 sensors and harness connectors
- Bracket from transmission assembly
- CKP sensor from the transmission
- Starter motor
- Rear cover plate

➡When turning the crankshaft, turn it clockwise as viewed from the front of the engine.

3. Turn the crankshaft, then remove the 4 tightening bolts for the drive plate and torque converter.

4. Support the transmission assembly with a suitable jack. Be careful not the let the jack hit the drain plug.

5. Remove or disconnect the following:
- Rear member
- Engine mounting insulator (rear)
- Dynamic damper
- A/T assembly harness connector
- Air breather hose
- A/T fluid charging pipe from A/T assembly
- O-ring from A/T fluid charging pipe
- Fluid cooler tube from A/T assembly
- Engine-to-transmission bolts

✳✳ WARNING

Before removal, secure the transmission to the jack and secure the torque converter to prevent it from falling.

To install:

6. Installation is the reverse of the removal procedure, noting the following:

a. Tighten the transmission-to-engine bolts as shown in the illustration.

b. Fill the transmission with fluid.

c. Start the vehicle, check for leaks and repair if necessary.

AWD Models

See Figures 14 and 15.

➡When removing the Automatic Transmission (A/T) assembly from the engine, first remove the Crankshaft Position Sensor (CKP) from the A/T assembly.

1. Before servicing the vehicle, refer to the precautions in the beginning of this section.

2. Remove or disconnect the following:
- Negative battery cable
- A/T fluid level gauge

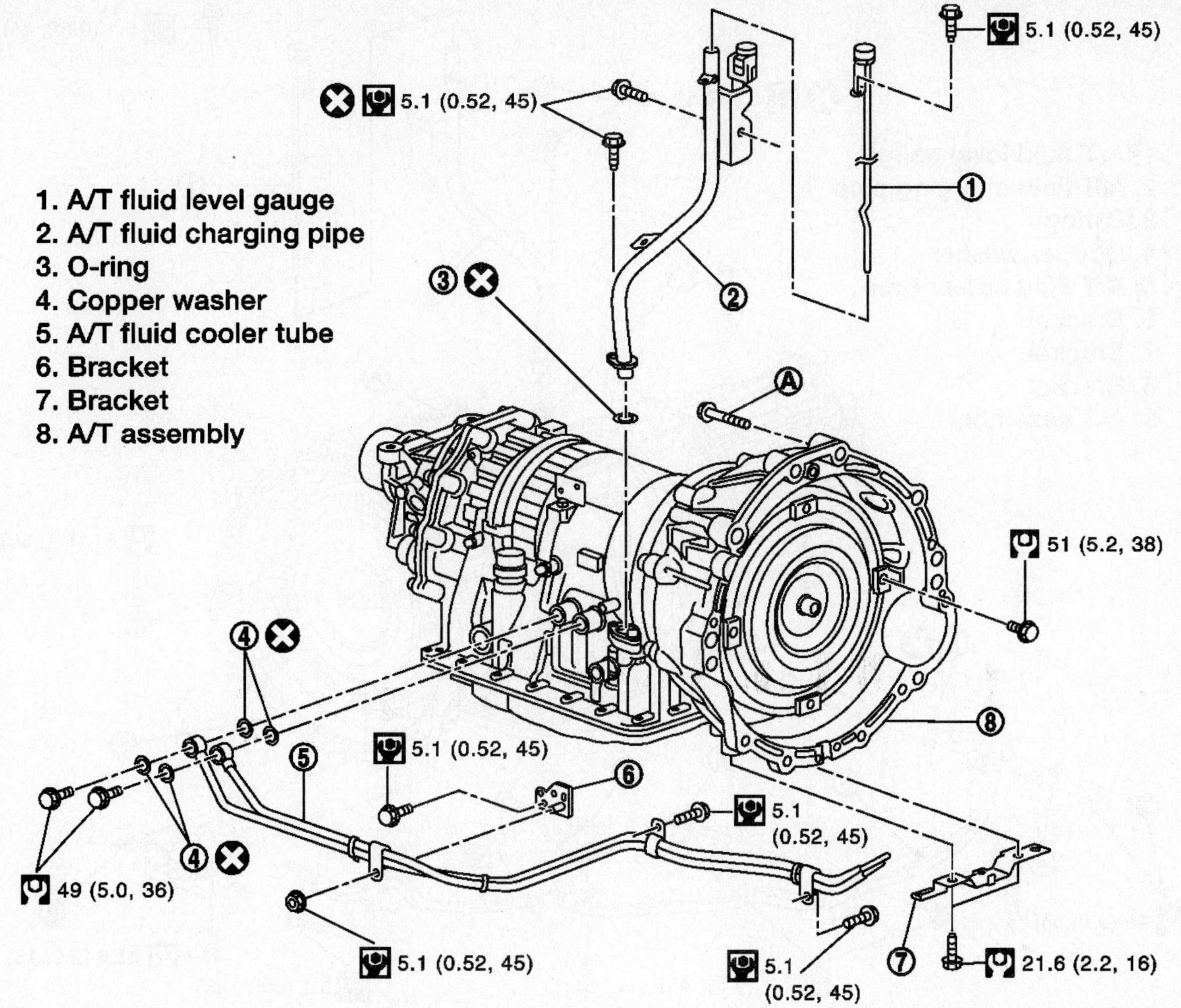

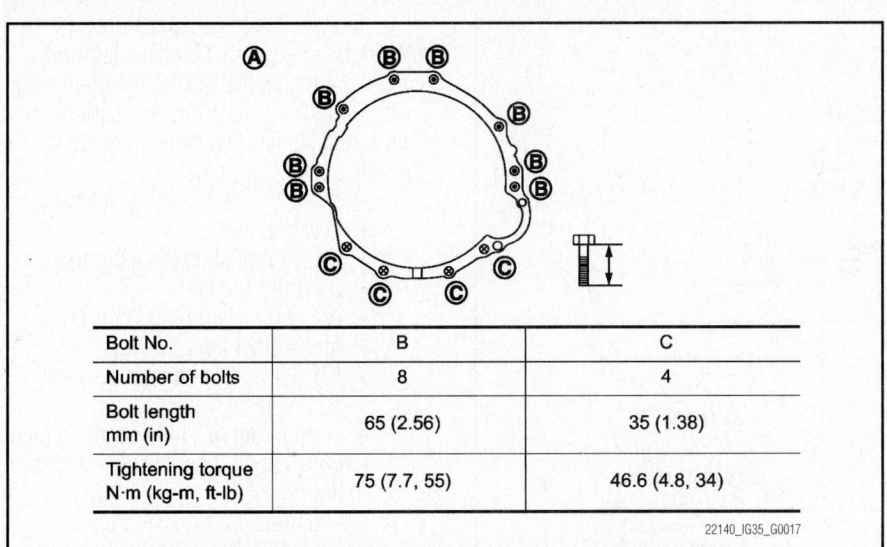

1. A/T fluid level gauge
2. A/T fluid charging pipe
3. O-ring
4. Copper washer
5. A/T fluid cooler tube
6. Bracket
7. Bracket
8. A/T assembly
5.1 (0.52, 45)
5.1 (0.52, 45)
51 (5.2, 38)
5.1 (0.52, 45)
5.1 (0.52, 45)
49 (5.0, 36)
5.1 (0.52, 45)
5.1 (0.52, 45)
21.6 (2.2, 16)
22140_IG35_G0016

Fig. 12 Exploded view of the automatic transmission mounting—2WD

Bolt No.	B	C
Number of bolts	8	4
Bolt length mm (in)	65 (2.56)	35 (1.38)
Tightening torque N·m (kg-m, ft-lb)	75 (7.7, 55)	46.6 (4.8, 34)

22140_IG35_G0017

Fig. 13 Automatic transmission-to-engine bolt tightening specifications—2WD

- Air cleaner case (RH)
- Engine under cover
- Exhaust pipe
- Heat insulator
- Rear driveshaft
- Front cross bar
- Three way catalyst (right bank)
- Front driveshaft
- Control rod
- CKP sensor from the transmission
- Starter motor
- Rear cover plate

➡When turning the crankshaft, turn it clockwise as viewed from the front of the engine.

3. Turn the crankshaft, then remove the 4 tightening bolts for the drive plate and torque converter.

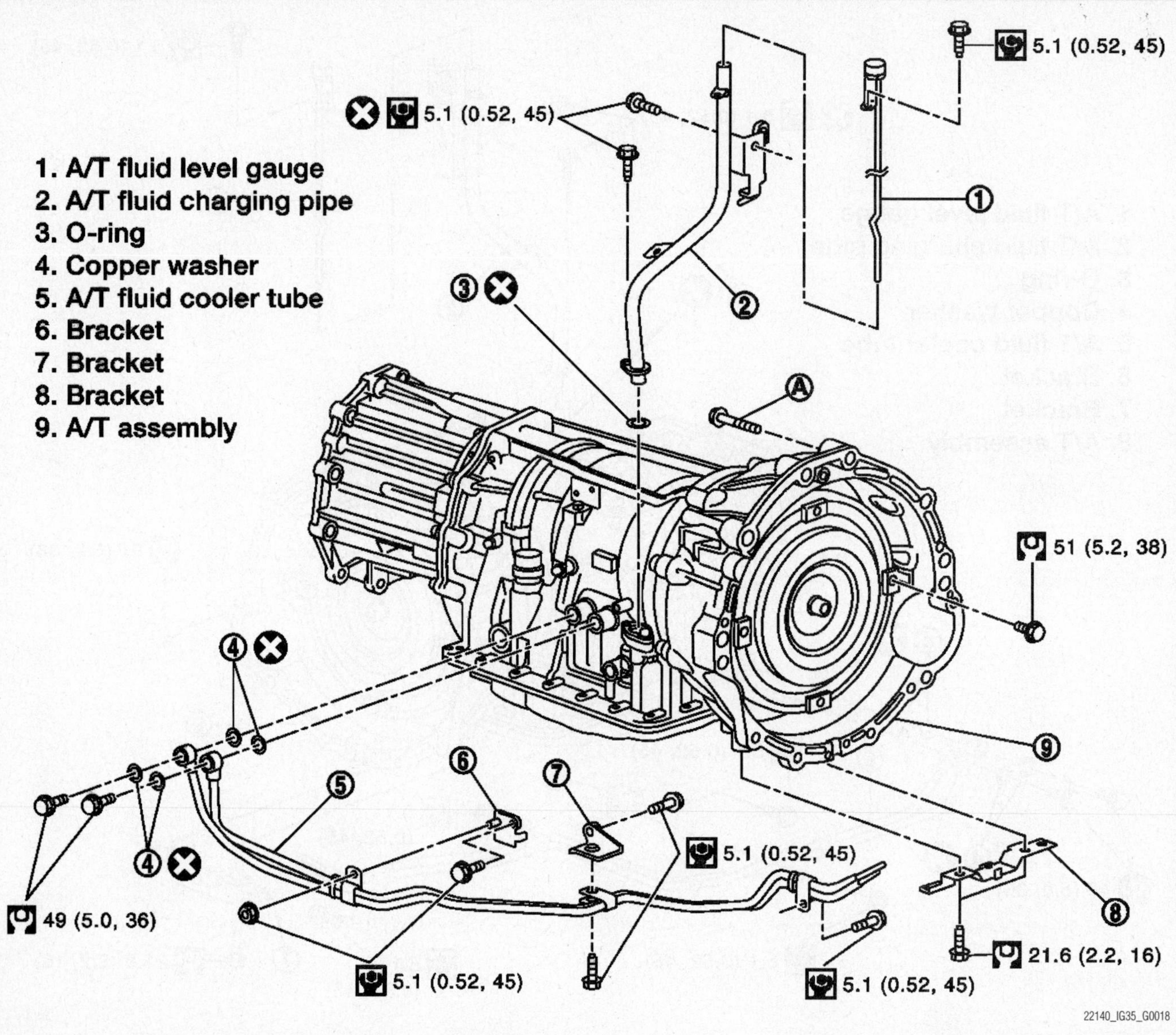

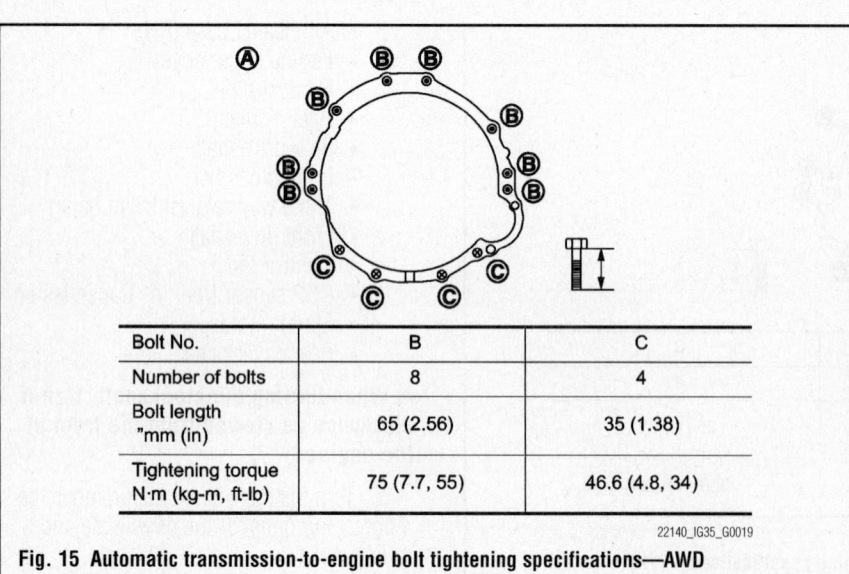

1. A/T fluid level gauge
2. A/T fluid charging pipe
3. O-ring
4. Copper washer
5. A/T fluid cooler tube
6. Bracket
7. Bracket
8. Bracket
9. A/T assembly

5.1 (0.52, 45)

5.1 (0.52, 45)

51 (5.2, 38)

49 (5.0, 36)

5.1 (0.52, 45)

5.1 (0.52, 45)

5.1 (0.52, 45)

21.6 (2.2, 16)

22140_IG35_G0018

Fig. 14 Exploded view of the automatic transmission mounting—AWD

Bolt No.	B	C
Number of bolts	8	4
Bolt length "mm (in)	65 (2.56)	35 (1.38)
Tightening torque N·m (kg-m, ft-lb)	75 (7.7, 55)	46.6 (4.8, 34)

22140_IG35_G0019

Fig. 15 Automatic transmission-to-engine bolt tightening specifications—AWD

4. Support the transmission/transfer (AWD) assembly with a suitable jack. Be careful not the let the jack hit the drain plug.

5. Support the transmission assembly with a suitable jack. Be careful not the let the jack hit the drain plug.

6. Remove or disconnect the following:
- Rear member
- Engine mounting insulator (rear)
- Dynamic damper
- A/T assembly harness connector
- Air breather hose
- A/T fluid charging pipe from A/T assembly
- O-ring from A/T fluid charging pipe
- Fluid cooler tube from A/T assembly
- Engine-to-transmission bolts

✳✳ WARNING

Before removal, secure the transmission to the jack and secure the torque converter to prevent it from falling.

To install:

7. Installation is the reverse of the removal procedure, noting the following:

a. Tighten the transmission-to-engine bolts as shown in the illustration.

b. Align the positions of the tightening bolts for the drive plate with those of the torque converter and hand-tighten. Then, tighten to 33–42 ft. lbs. (44–58 Nm).

c. After the converter is installed, rotate the crankshaft a few times to be sure the transmission rotates without any binding

d. Fill the transmission with fluid.

e. Start the vehicle, check for leaks and repair if necessary.

MANUAL TRANSMISSION ASSEMBLY

REMOVAL & INSTALLATION

2006 Coupe and Sedan; 2007 Coupe

See Figures 16 and 17.

1. Before servicing the vehicle, refer to the precautions in the beginning of this section.

2. Remove or disconnect the following:
- Negative battery cable
- Exhaust mounting bracket
- Catalytic converter and front exhaust tube
- Rear driveshaft
- Control rod mounting bolts and then separate control lever assembly from the control rod assembly.
- Shift lever assembly
- Clutch slave cylinder
- Crankshaft position sensor (CKP)

➡**Handle sensor with care; do not place in an area affected by magnetism. Do not allow iron dust, etc., to get on the sensor's front edge magnetic area. Do not subject it to impact by dropping or hitting. Do not disassemble.**

- Neutral safety switch and back-up lamp switch
- Oxygen sensor, POS, and back-up lamp harness
- Starter motor

3. Support weight of transmission on a suitable transmission jack

➡**Make certain transmission does not rest on switch terminals**

4. Remove rear engine mounting member.

5. Remove engine and transmission mounting bolts.

✳✳ CAUTION

Do not hold control lever housing to prevent bushing of control lever housing from deformation when moving transmission assembly.

6. Lower jack and remove transmission from vehicle.

To install:

✳✳ CAUTION

When installing, be careful to avoid interference between transmission main drive shaft and clutch cover.

➡**If flywheel is removed, align dowel pin with the smallest hole of flywheel.**

7. Installation is the reverse of the removal procedure, observing mounting bolt position and torque values shown in chart.

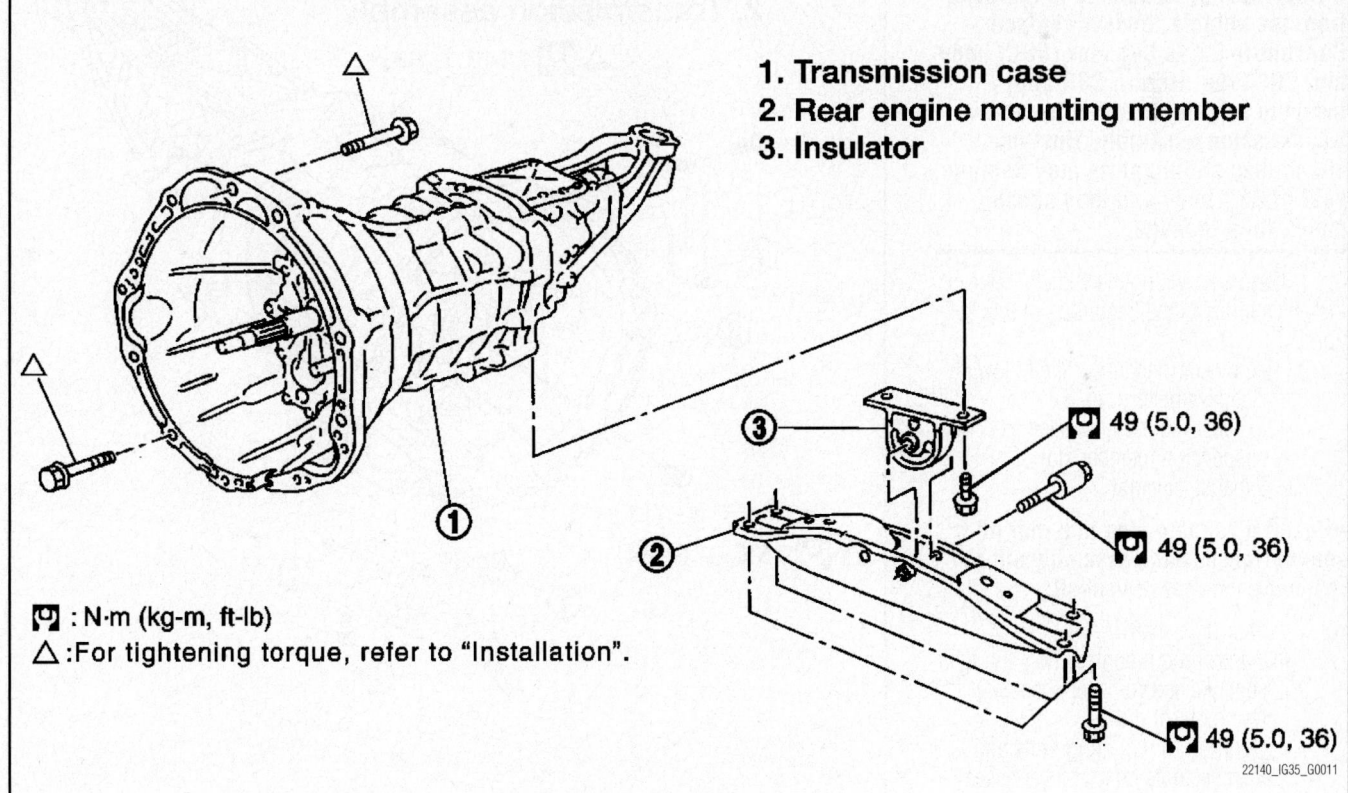

1. **Transmission case**
2. **Rear engine mounting member**
3. **Insulator**

🔧 49 (5.0, 36)

🔧 49 (5.0, 36)

🔧 49 (5.0, 36)

🔧 : N·m (kg-m, ft-lb)
△ :For tightening torque, refer to "Installation".

22140_IG35_G0011

Fig. 16 Manual transmission and mounting member—2006

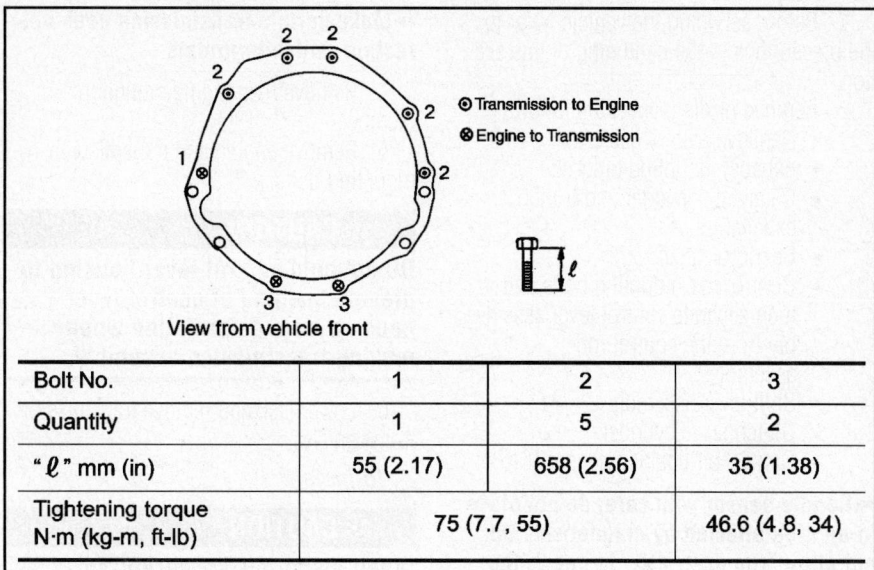

Bolt No.	1	2	3
Quantity	1	5	2
"ℓ" mm (in)	55 (2.17)	658 (2.56)	35 (1.38)
Tightening torque N·m (kg-m, ft-lb)	75 (7.7, 55)		46.6 (4.8, 34)

22140_IG35_G0012

Fig. 17 Bolt position and torque values for manual transmission

⊙ Transmission to Engine
⊗ Engine to Transmission

View from vehicle front

8. After installation, check for oil leakage, oil level and proper operation of shifting mechanism.

2007 Sedan

See Figures 18 and 19.

✱✱ CAUTION

If transmission assembly is removed from the vehicle, always replace Concentric Slave Cylinder (CSC) body and CSC tube. Return CSC body insert to original position to remove transmission assembly. Dust on clutch disc sliding parts may damage seal of CSC body and may cause clutch fluid leakage.

1. Before servicing the vehicle, refer to the precautions in the beginning of this section.

2. Remove or disconnect the following:
- Negative battery cable
- Exhaust mounting bracket
- suspension member stay
- Exhaust assembly

➡Insert a suitable plug into rear oil seal of transmission assembly after removing the rear driveshaft.

- Rear driveshaft
- Control rod mounting bolt and then separate control lever assembly from control rod
- Control rod mounting bolts and then separate control lever assembly from the control rod assembly.
- Shift lever assembly

✱✱ CAUTION

Keep painted surface on the body or other parts free of clutch fluid. If it spills, wipe up immediately and wash the affected area with water. Do not depress clutch pedal during removal procedure.

➡Insert a suitable plug into clutch hose and CSC (Concentric Slave Cylinder) tube after removing clutch tube

- clutch tube, clutch hose and lock plate.
- Crankshaft position sensor (CKP)

➡Handle sensor with care; do not place in an area affected by magnetism. Do not allow iron dust, etc., to get on the sensor's front edge magnetic area. Do not subject it to impact by dropping or hitting. Do not disassemble.

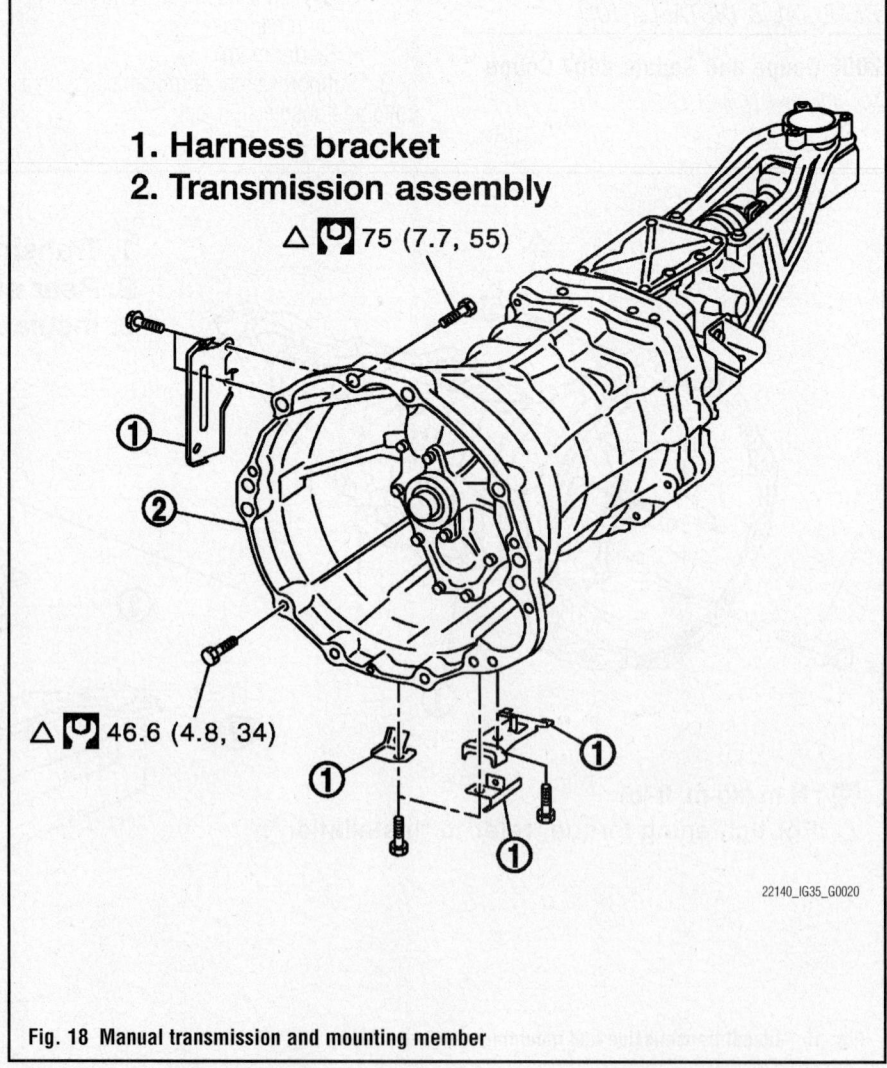

1. Harness bracket
2. Transmission assembly

△ 🔧 75 (7.7, 55)

△ 🔧 46.6 (4.8, 34)

22140_IG35_G0020

Fig. 18 Manual transmission and mounting member

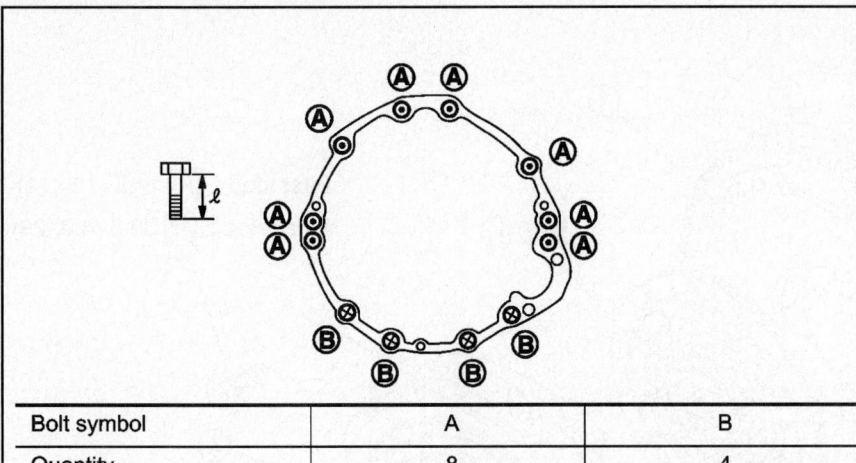

Bolt symbol	A	B
Quantity	8	4
" ℓ " mm (in)	65 (2.56)	35 (1.38)

22140_IG35_G0021

Fig. 19 Bolt position and torque values for manual transmission

- Starter motor
- Rear plate cover
- Park/Neutral Position (PNP) switch harness connector
- Oxygen sensor and connectors
- Harness brackets

3. Support weight of transmission on a suitable transmission jack

➡**Make certain transmission does not rest on switch terminals**

4. Remove rear engine mounting member.

5. Remove engine and transmission mounting bolts.

✸✸ CAUTION

Do not hold control lever housing to prevent bushing of control lever housing from deformation when moving transmission assembly.

✸✸ CAUTION

Secure transmission assembly to a suitable jack while removing it. The transmission assembly must not interfere with the three way catalyst (right bank) and three way catalyst (left bank). The transmission assembly must not interfere with the wire harnesses and clutch hose. Do not hold control lever housing to prevent bushing of control lever housing from deformation when moving transmission assembly.

6. Lower jack and remove transmission from vehicle.

7. Remove CSC body and CSC tube.
8. Remove dynamic damper.

To install:

✸✸ CAUTION

The transmission assembly must not interfere with the three way catalyst (right bank) and three way catalyst (left bank). The transmission assembly must not interfere with the wire harnesses and clutch hose. When installing transmission assembly, be careful not to bring main drive gear into contact with clutch cover. Do not hold control lever housing to prevent bushing of control lever housing from deformation when moving transmission assembly.

➡**If flywheel is removed, align dowel pin with the smallest hole of flywheel.**

9. Installation is the reverse of the removal procedure, observing mounting bolt position and torque values shown in chart.

10. After installation, check for oil leakage, oil level and proper operation of shifting mechanism.

CLUTCH

REMOVAL & INSTALLATION

See Figures 20 through 22.

1. Before servicing the vehicle, refer to the precautions in the beginning of this section.

2. Raise and support the vehicle safely.

3. Negative battery cable
4. Manual transmission assembly
5. Loosen clutch cover mounting bolts and remove clutch cover and clutch disc.

To install:

✸✸ CAUTION

Be sure to apply grease to the points specified. Otherwise, noise, poor disengagement, or damage to the clutch may result. Excessive grease may cause slip or quiver. Wipe off any grease oozing from the parts.

6. Lightly lubricate the transaxle input shaft, input shaft collar, clutch lever assembly and the clutch release bearing with a lithium based grease.

➡**Keep clutch disc and all clutch components clean during installation. Do not allow grease to contact the clutch disc.**

7. Insert an alignment tool into the clutch disc hub. Install the clutch disc and pressure plate on the tool and torque the pressure plate bolts in two passes in the order shown in diagram:
- First pass to 11 ft. lbs. (14.5 Nm)
- Second pass to 29 ft. lbs. (39.6 Nm)

8. Remove the tool.
9. Install or connect the following:
- Throw out bearing in the transmission housing; ensure that the bearing retainer clips are fully engaged
- Transmission
- Negative battery cable

10. If necessary, adjust clutch pedal height and free-play.

BLEEDING

✸✸ CAUTION

Monitor fluid level in reservoir tank to make sure it does not empty

✸✸ CAUTION

Do not spill clutch fluid onto painted surfaces. If it spills, wipe up immediately and wash the affected area with water.

➡**Do not use a vacuum assist or any other type of power bleeder on this system. Use of vacuum assist or power bleeder will not purge all the air from the system.**

1. Fill master cylinder reservoir tank with new clutch fluid.

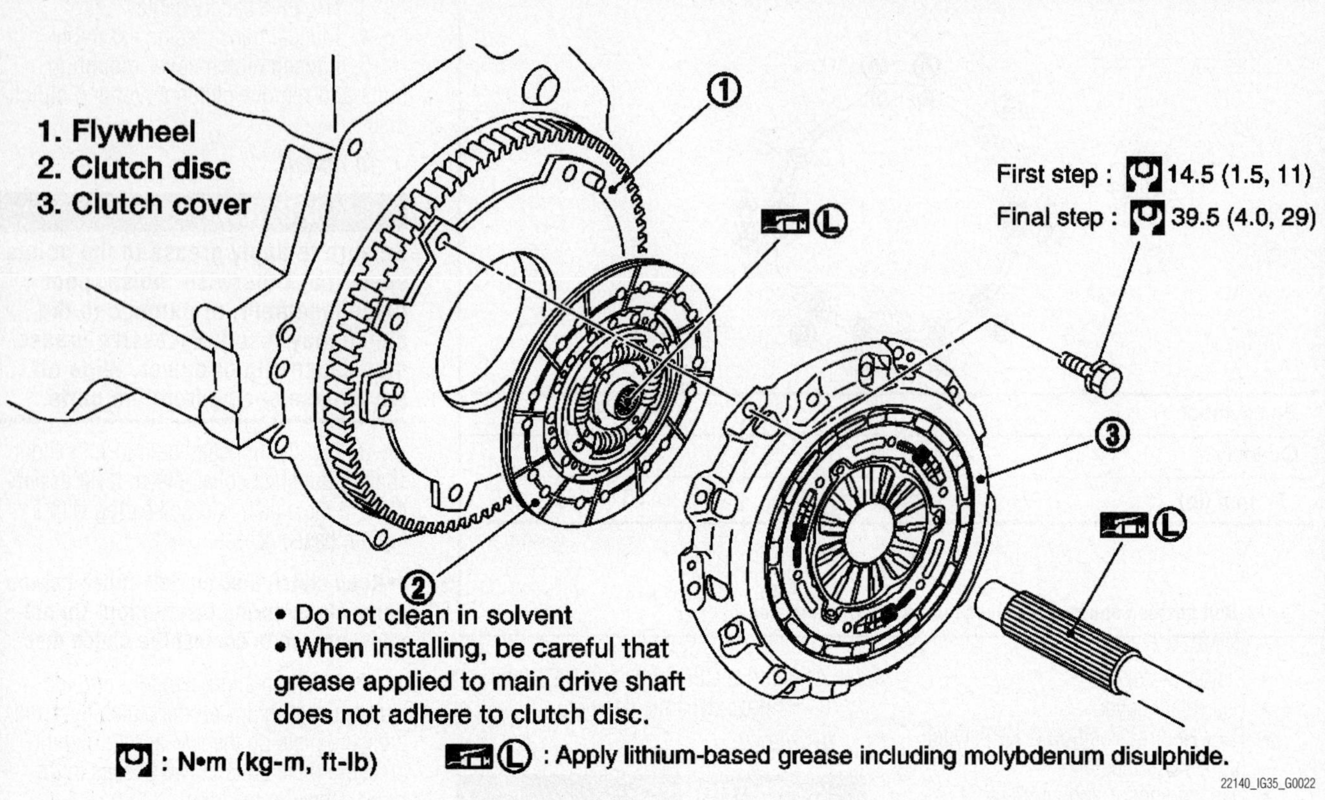

1. Flywheel
2. Clutch disc
3. Clutch cover

First step : 14.5 (1.5, 11)
Final step : 39.5 (4.0, 29)

• Do not clean in solvent
• When installing, be careful that grease applied to main drive shaft does not adhere to clutch disc.

: N•m (kg-m, ft-lb)   : Apply lithium-based grease including molybdenum disulphide.

22140_IG35_G0022

Fig. 20 Clutch disc and cover—2006 Coupe and Sedan, 2007 Coupe

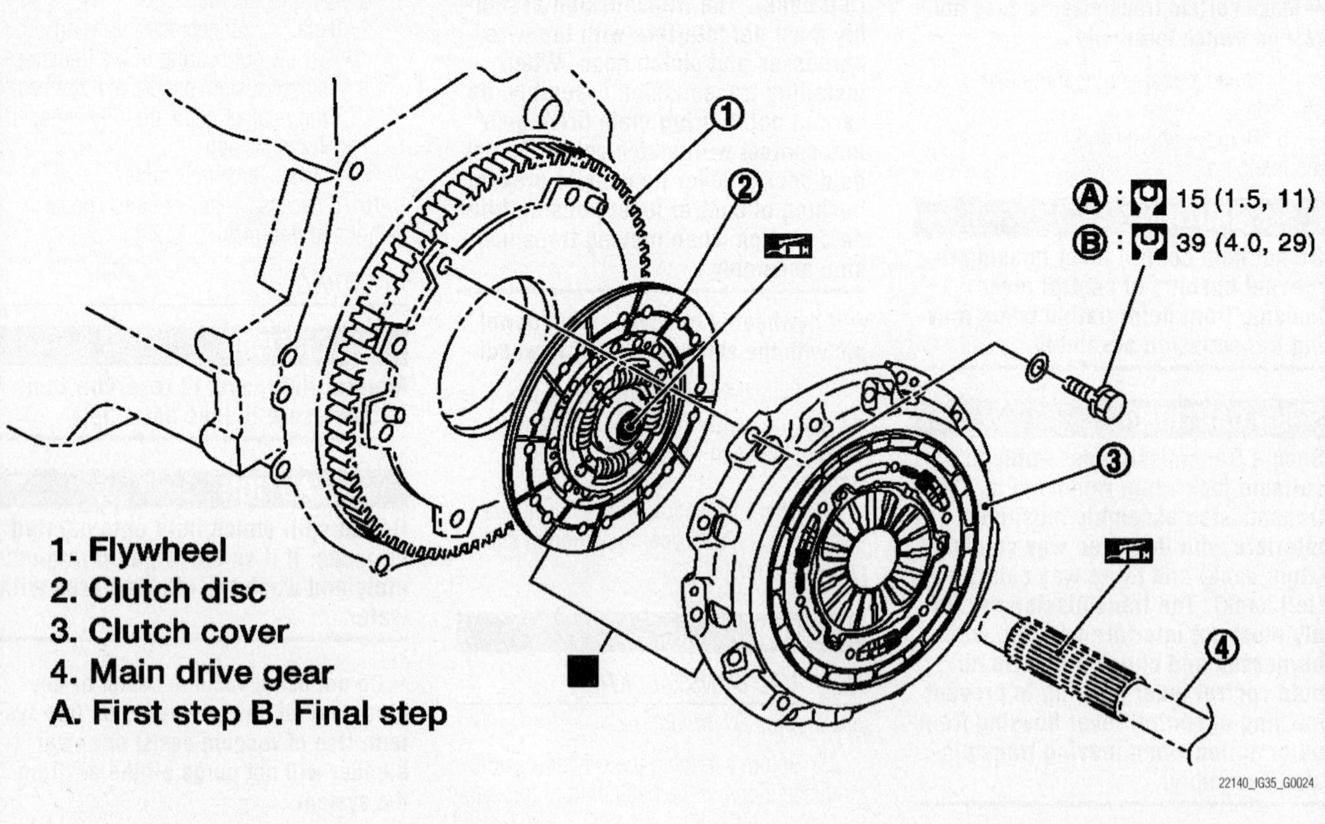

1. Flywheel
2. Clutch disc
3. Clutch cover
4. Main drive gear
A. First step B. Final step

Ⓐ : 15 (1.5, 11)
Ⓑ : 39 (4.0, 29)

22140_IG35_G0024

Fig. 21 Clutch disc and cover—2007 Sedan and 2008 Sedan

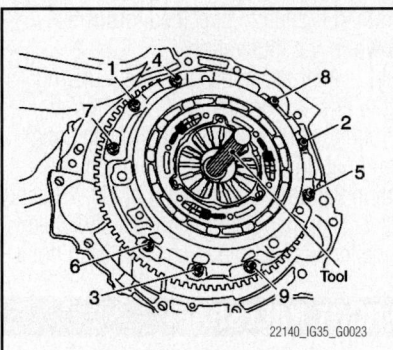

Fig. 22 Tighten clutch cover bolts in the order shown

2. Connect a transparent vinyl hose to air bleeder.

3. Depress clutch pedal quickly and fully a few times and hold it.

4. With the clutch pedal depressed, open air bleeder.

6. Remove rear engine mounting member.

7. Remove transfer assembly mounting bolts and separate transfer assembly from transmission.

> ❋❋ **WARNING**
>
> **Secure transfer assembly and transmission assembly to a jack.**

To install:

8. Installation is the reverse of the removal procedure.

9. Tighten transfer mounting bolts to 27 ft. lbs (37 Nm).

2007–08 Models

See Figure 24.

1. Before servicing the vehicle, refer to the precautions in the beginning of this section.

2. Raise and support the vehicle safely.

3. Remove exhaust system.

4. Remove front and rear drive shaft.

5. Disconnect AWD solenoid harness connector and separate harness from transfer assembly.

6. Remove transfer air breather hose.

7. Remove control rod.

8. Support transfer assembly and transmission assembly with a jack

9. Remove rear engine mounting member.

10. Remove transfer assembly mounting bolts and separate transfer assembly from transmission.

> ❋❋ **WARNING**
>
> **Secure transfer assembly and transmission assembly to a jack.**

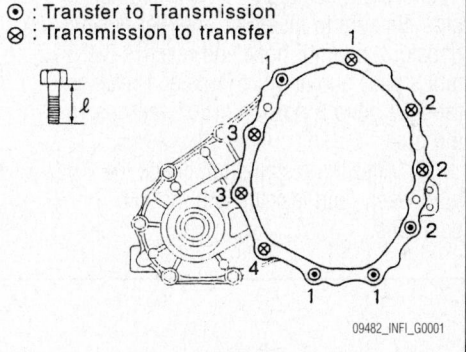

⊙ : Transfer to Transmission
⊗ : Transmission to transfer

Bolt No.	1	2	3	4
Quantity	4	3	2	1
Bolt length " ℓ " mm (in)	75 (2.95)	45 (1.77)	40 (1.57)	30 (1.18)
Tightening torque N·m (kg-m, ft-lb)	37 (3.8, 27)			

09482_INFI_G0001

Fig. 23 Use the accompanying chart for proper bolt placement when installing the transfer case.

5. Close the air bleeder.

6. Release clutch pedal and wait for 5 seconds.

7. Repeat steps 3 to 6 until no bubbles can be observed in brake fluid.

8. Tighten air bleeder.

TRANSFER CASE ASSEMBLY

REMOVAL & INSTALLATION

2006 Models

See Figure 23.

1. Before servicing the vehicle, refer to the precautions in the beginning of this section.

2. Raise and support the vehicle safely.

3. Disconnect transfer assembly harness connector and separate harness from transfer assembly.

4. Remove air breather hose.

5. Support transfer assembly and transmission assembly with a jack.

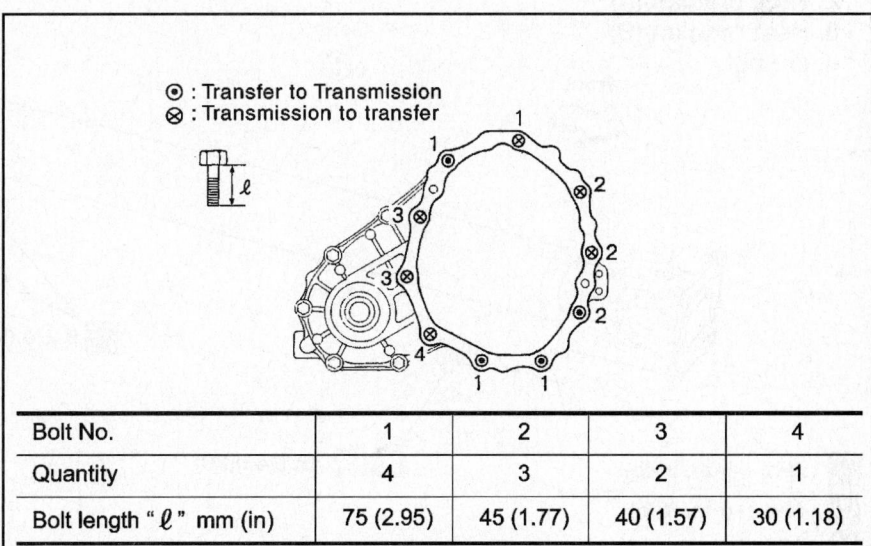

⊙ : Transfer to Transmission
⊗ : Transmission to transfer

Bolt No.	1	2	3	4
Quantity	4	3	2	1
Bolt length " ℓ " mm (in)	75 (2.95)	45 (1.77)	40 (1.57)	30 (1.18)

22140_IG35_G0026

Fig. 24 Use the accompanying chart for proper bolt placement when installing the transfer case.

To install:

11. Installation is the reverse of the removal procedure.

12. Tighten transfer mounting bolts to 27 ft. lbs (37 Nm).

13. When installing transfer air breather hose, make sure there are no pinched or restricted areas on the transfer air breather hose caused by bending or winding. Set transfer air breather hose with paint mark facing upward. Be sure to insert transfer air breather hose into breather tube until hose end reaches the tube's base.

14. Do not deviate from the range (L) of the transfer air breather when installing the transfer air breather hose to the harness bracket of the transfer.

15. Install transfer air breather hose with bracket between the adapter case and the transmission case.

16. Check that transfer breather hose is on the ( ) side when installing the transfer air breather hose to A/T fluid charging pipe. Be sure to insert air breather hose to transfer tube until hose end reaches the tube's base and another hose end reaches the tube bend R portion of A/T fluid charging pipe.

17. After the installation, check the fluid level, fluid leakage and the A/T positions.

FRONT DRIVESHAFT

REMOVAL & INSTALLATION
See Figure 25.

1. Before servicing the vehicle, refer to the precautions in the beginning of this section.

2. Remove exhaust front pipe.

3. Remove engine undercover.

4. Remove the right bank catalytic converter.

5. Put matchmarks on flanges and separate driveshaft from final drive.

➡ **For matchmark, use paint. Do not damage the drive shaft flange and companion flange on the front final drive.**

6. Remove the drive shaft fixing bolts

7. Set the transmission jack at the transfer, remove rear engine mounting bolts, and then lower transmission jack about 0.16 - 0.21 inches (40-50 mm).

8. Remove propeller shaft from the front final drive and transfer case.

To install:

9. To install, reverse removal procedure.

10. Align matching marks to install propeller shaft to final drive companion flange, and then tighten to specified torque as shown in the illustration.

11. After assembly, perform a driving test to check propeller shaft vibration. If vibration occurred, separate propeller shaft from final drive or transfer. Reinstall companion flange after rotating it by 90, 180 and 270 °. Then perform driving test and check propeller shaft vibration again at each point.

FRONT HALFSHAFT

REMOVAL & INSTALLATION

2006 Sedan

AWD Models
See Figure 26.

✳✳ CAUTION

The amount of force need to loosen the front wheel bearing nut is high enough to cause the vehicle to fall off the jack. Remove cotter pin and loosen or tighten this nut with the vehicle on the ground.

1. Before servicing the vehicle, refer to the precautions in the beginning of this section.

2. Remove or disconnect the following:
   • Front wheels
   • Engine undercover

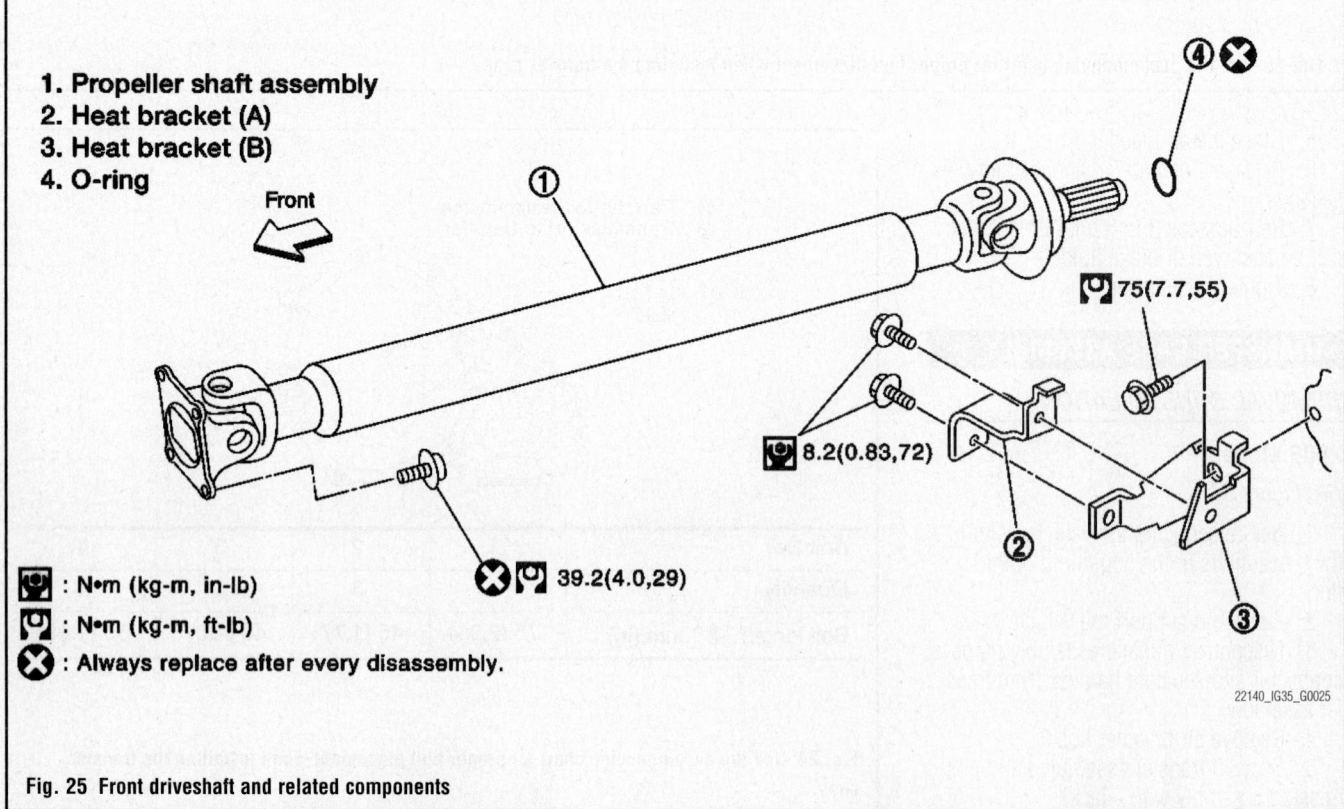

1. **Propeller shaft assembly**
2. **Heat bracket (A)**
3. **Heat bracket (B)**
4. **O-ring**

Front

75 (7.7, 55)

8.2 (0.83, 72)

39.2 (4.0, 29)

: N•m (kg-m, in-lb)

: N•m (kg-m, ft-lb)

✗ : Always replace after every disassembly.

22140_IG35_G0025

Fig. 25 Front driveshaft and related components

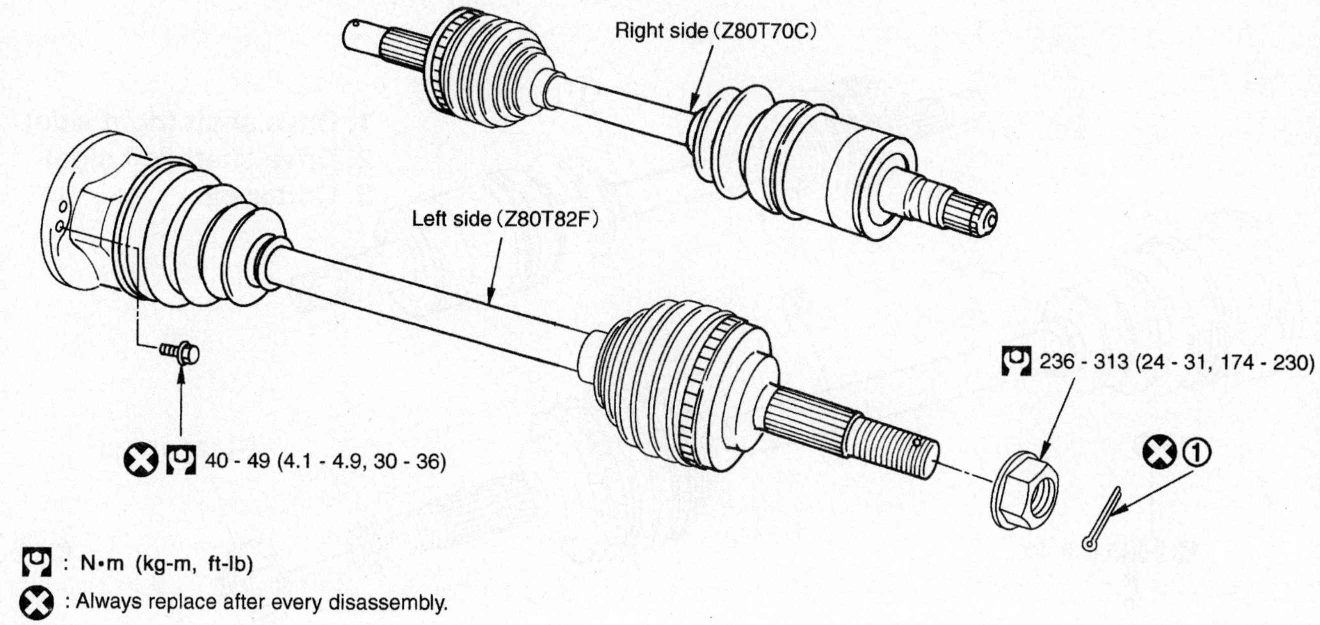

Right side (Z80T70C)

Left side (Z80T82F)

$\otimes$ $\boxed{}$ 40 - 49 (4.1 - 4.9, 30 - 36)

$\boxed{}$ 236 - 313 (24 - 31, 174 - 230)

$\otimes$ ①

$\boxed{}$: N•m (kg-m, ft-lb)

$\otimes$: Always replace after every disassembly.

1.   Cotter pin

09482_INFI_G0091

Fig. 26 Front halfshafts—AWD

- Brake calipers; use wire to support calipers where they will not interfere with work
- Brake rotors
- Antilock Brake wheel sensors

➡**Do not pull on sensor harness**

- Brake hose bracket from steering knuckle
- Front axle nut (nut should be loosened while vehicle is on the ground)
- Tie rod end

➡**Do not damage threads or boot on tie rod end**

- Upper link ball joint

3. Remove halfshaft from wheel hub and bearing assembly

4. On left side, remove bolts securing halfshaft from side shaft and remove halfshaft from vehicle

5. On right side, remove halfshaft from splines in final drive

6. Remove halfshaft from vehicle

To install:

7. Installation is in reverse order of removal. Observe the following:

- Tighten left halfshaft flange to side shaft bolts to 30–36 ft. lbs. (40–49 Nm)

- Tighten axle nuts to 174–230 ft. lbs. (236–313 Nm)
- Tighten upper link ball joint to 40–46 ft. lbs. (54–63.7 Nm)
- Tighten tie rod end ball joint to 22–28 ft. lbs. (29.5–39.2 Nm)

8. **Always** install new cotter pins.

9. Check and perform front end alignment if needed.

2007–08 Sedan

AWD Models

See Figure 27.

1. Before servicing the vehicle, refer to the precautions in the beginning of this section.

2. Remove or disconnect the following:

- Front wheels
- Shock absorber from transverse link

3. On left side, remove bolts securing halfshaft from side shaft and remove halfshaft from vehicle.

4. On right side, using Special Tool: KV40107500, remove halfshaft from splines in final drive.

To install:

5. Installation is in reverse order of removal.

6. Always replace the front final drive oil seal with a new one when installing the drive shaft.

REAR DRIVESHAFT

REMOVAL & INSTALLATION

See Figures 28 through 30.

1. Remove the rear driveshaft as follows:

a. Move the transmission select lever to N position and release the parking brake.

b. Floor rain force

c. Remove the center muffler.

d. Loosen the center bearing mounting bracket fixing nuts.

e. Put matchmarks on flange and rear driveshaft.

➡**For matchmark, use paint. Do not damage the propeller shaft flange and companion flange on the rear final drive.**

f. Remove the driveshaft fixing bolts and nuts.

g. Remove the center bearing mounting bracket fixing nuts.

h. Remove driveshaft from the vehicle.

2. To install, reverse removal procedure.

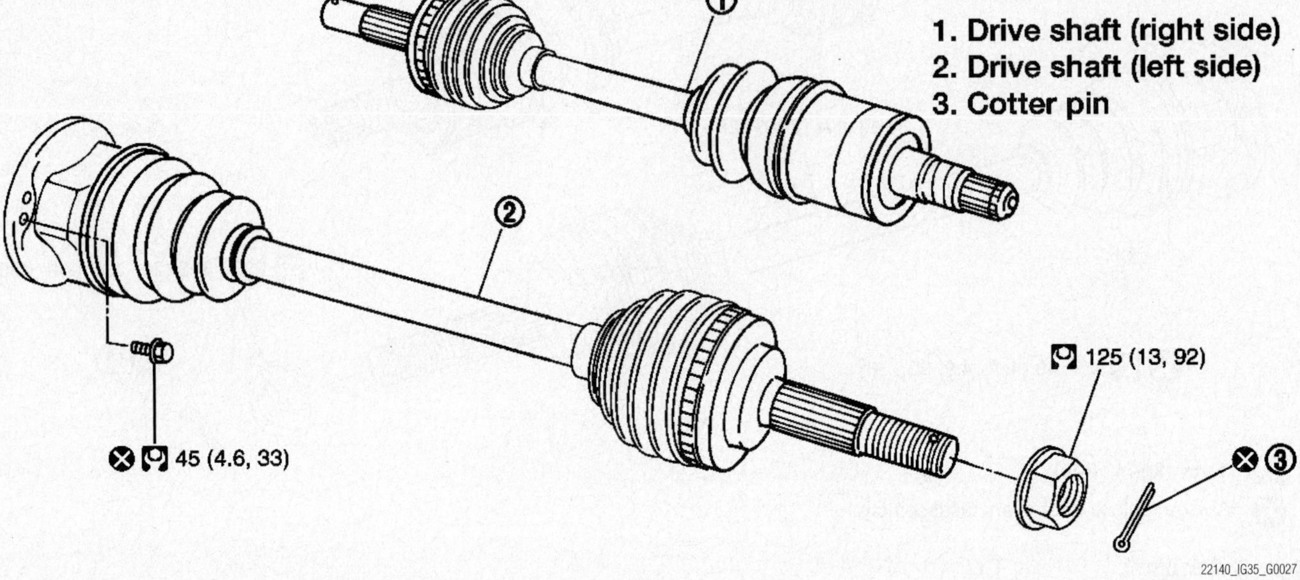

1. Drive shaft (right side)
2. Drive shaft (left side)
3. Cotter pin

125 (13, 92)

45 (4.6, 33)

22140_IG35_G0027

Fig. 27 Front halfshaft and components—AWD

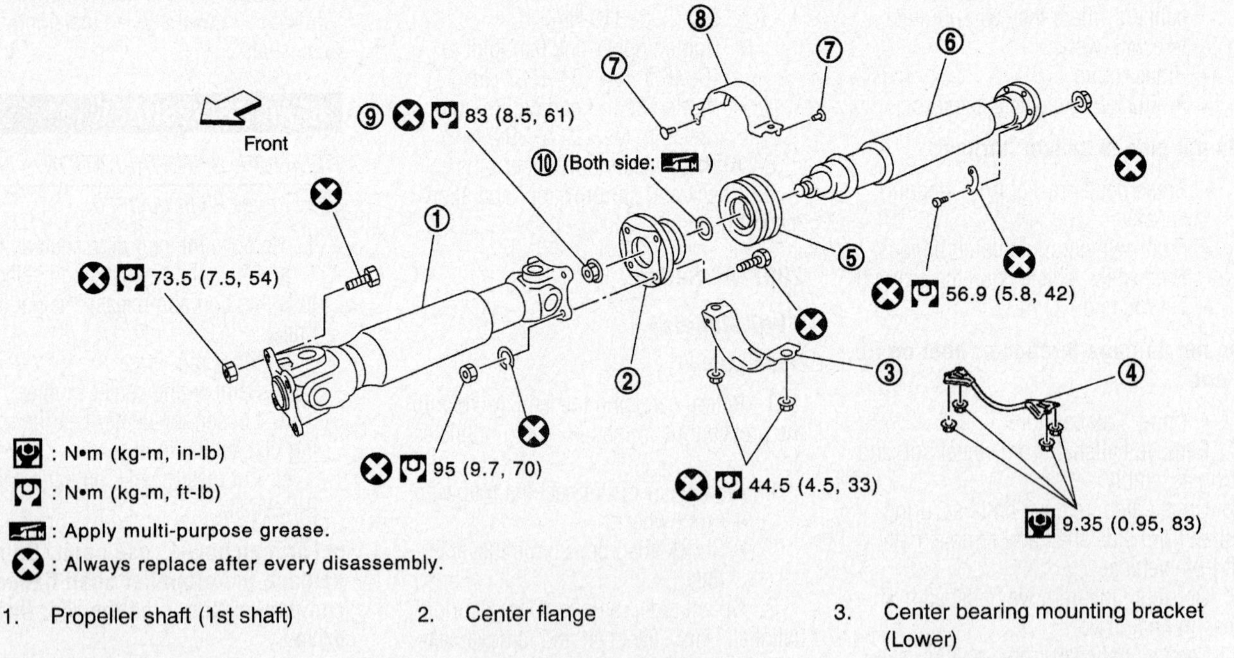

Front

9   83 (8.5, 61)

10 (Both side:)

73.5 (7.5, 54)

56.9 (5.8, 42)

95 (9.7, 70)

44.5 (4.5, 33)

9.35 (0.95, 83)

: N•m (kg-m, in-lb)

: N•m (kg-m, ft-lb)

: Apply multi-purpose grease.

: Always replace after every disassembly.

1. Propeller shaft (1st shaft)	2. Center flange	3. Center bearing mounting bracket (Lower)
4. Floor rain force	5. Center bearing assembly	6. Propeller shaft (2nd shaft)
7. Clip	8. Center bearing mounting bracket (Upper)	9. Lock nut
10. Washer		

09482_INFI_G0003

Fig. 28 Rear driveshaft components—2006 Sedan and 2007 Coupe—AWD

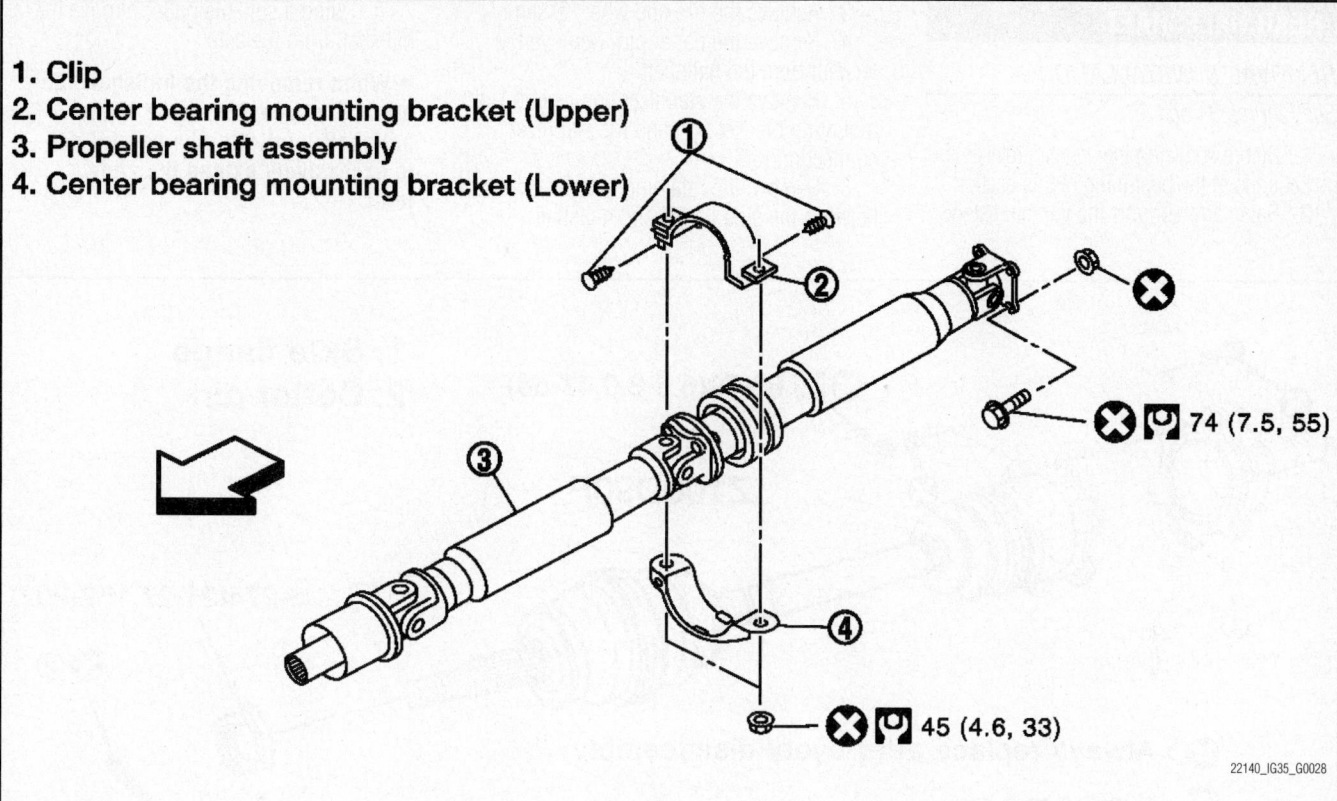

1. Clip
2. Center bearing mounting bracket (Upper)
3. Propeller shaft assembly
4. Center bearing mounting bracket (Lower)

74 (7.5, 55)

45 (4.6, 33)

22140_IG35_G0028

Fig. 29 Rear driveshaft components—2007 and 2008 Sedan

56.9 (5.8, 42)

1. Clip
2. Center bearing mounting bracket (Upper)
3. Propeller shaft assembly
4. Center bearing mounting bracket (Lower)

45 (4.6, 33)

74 (7.5, 55)

22140_IG35_G0029

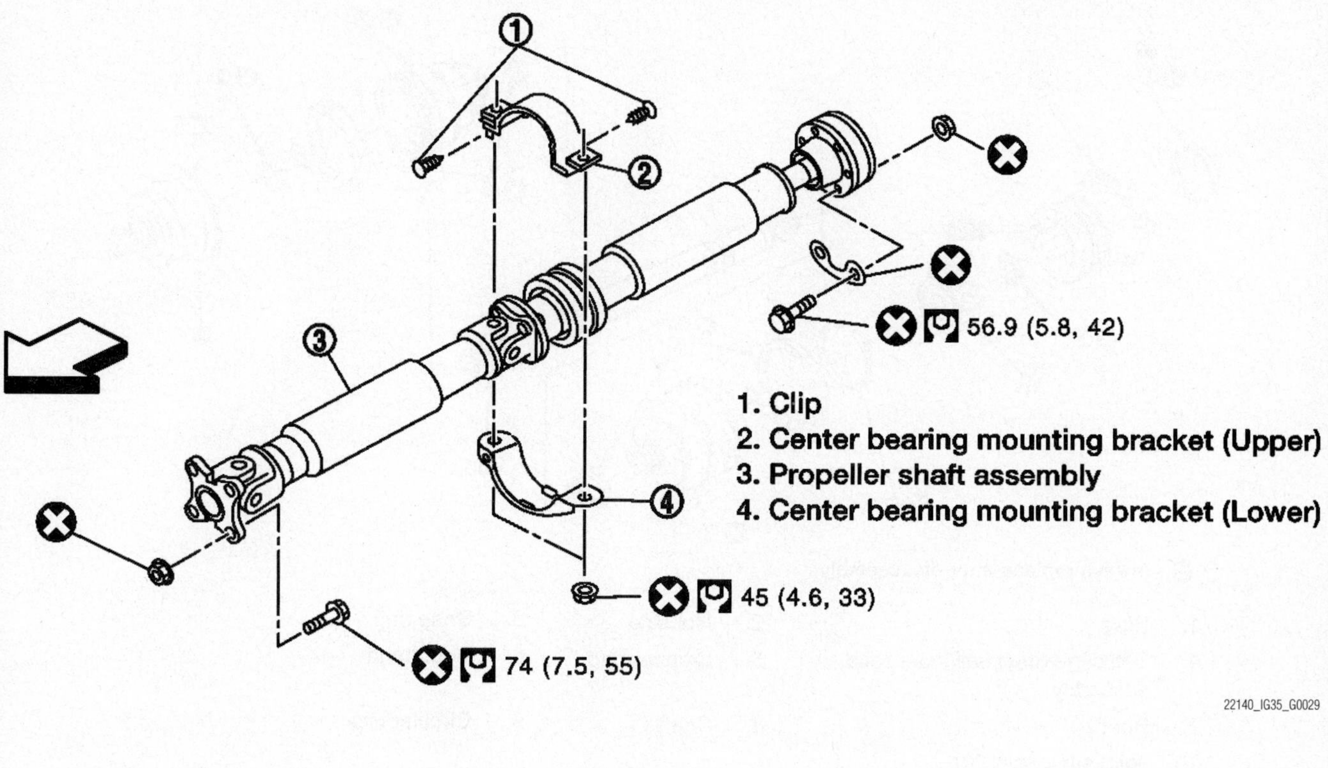

Fig. 30 Rear driveshaft components—2007 and 2008 Sedan—3F80A-1VL107

REAR HALFSHAFT

REMOVAL & INSTALLATION

See Figures 31 and 32.

1. Before servicing the vehicle, refer to the precautions in the beginning of this section.
2. Raise and support the vehicle safely.

3. Remove the tire and wheel assembly.
4. Remove the cotter pin. Remove the locknut from the halfshaft.
5. Remove the stabilizer connecting rod mounting bracket bolt and the stabilizer connecting rod.
6. Remove the retaining nuts and bolts between the side flange and halfshaft.

7. Using a suitable puller remove the halfshaft from the axle.

➡**When removing the halfshaft, do not apply excessive force to the halfshaft joint. Also be careful not to excessively extend the slide joint.**

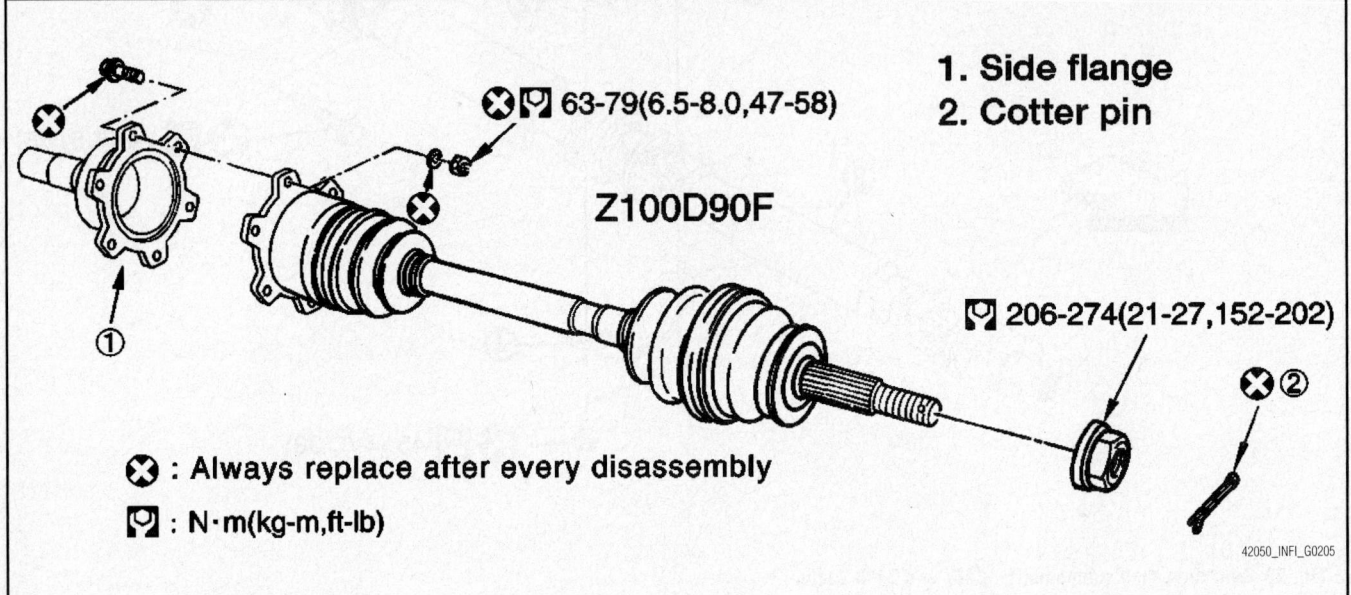

1. Side flange
2. Cotter pin

Z100D90F

❌⌷ 63-79(6.5-8.0,47-58)

⌷ 206-274(21-27,152-202)

❌ : Always replace after every disassembly

⌷ : N·m(kg-m,ft-lb)

42050_INFI_G0205

Fig. 31 Halfshaft and related components

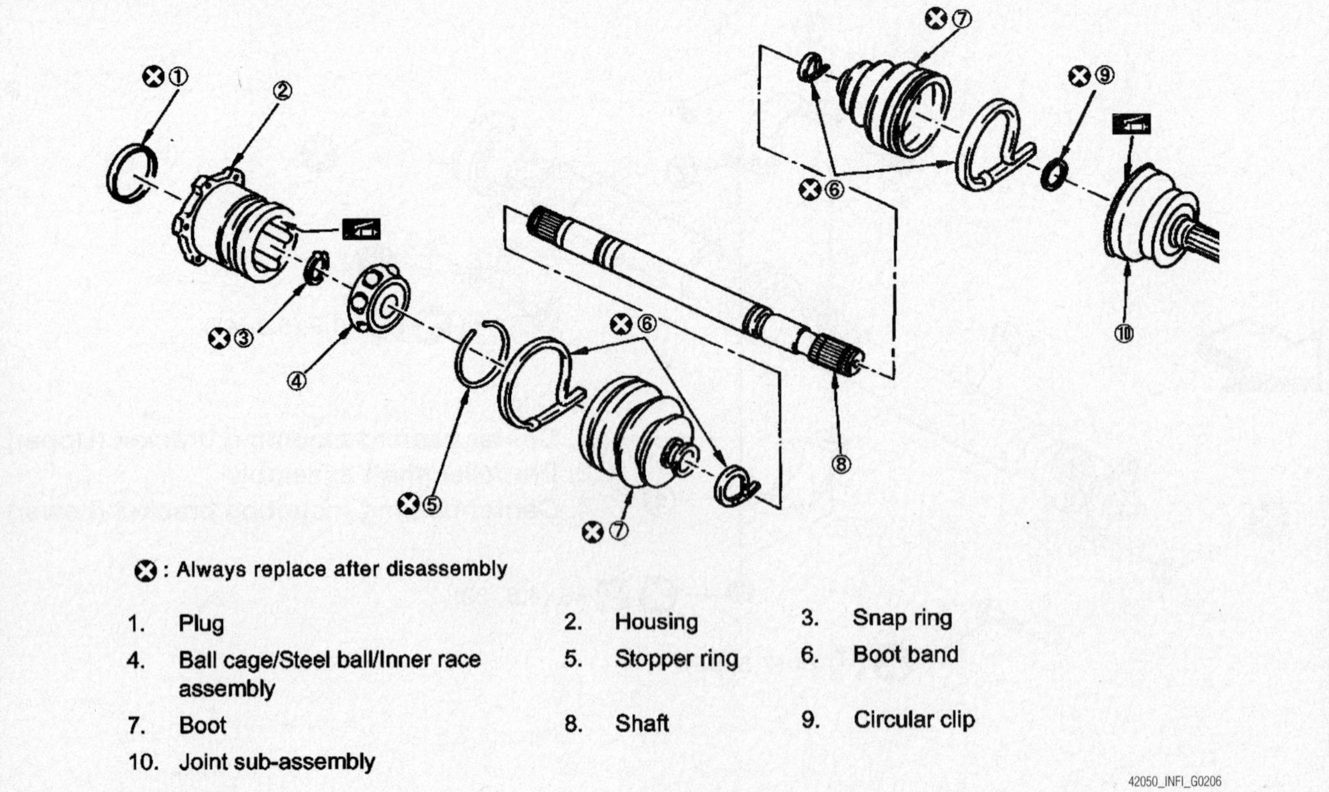

❌ : Always replace after disassembly

1. Plug	2. Housing	3. Snap ring
4. Ball cage/Steel ball/Inner race assembly	5. Stopper ring	6. Boot band
7. Boot	8. Shaft	9. Circular clip
10. Joint sub-assembly		

42050_INFI_G0206

Fig. 32 Halfshaft exploded view

To install:

8. Installation is the reverse of the removal procedure.

REAR PINION SEAL

REMOVAL & INSTALLATION

1. Before servicing the vehicle, refer to the precautions in the beginning of this section.

2. Raise and support the vehicle safely.

3. Remove the rear drive shaft.

4. Remove self-lock nut of companion flange using flange wrench

5. Put matching mark on the end of the mainshaft. The mark should be in line with the mark on the companion flange

6. Remove the companion flange using a puller.

7. Remove the rear oil seal using the puller Special Tool: KV381054S0 (J-34286).

To install:

8. Apply ATF to rear oil seal, install it with drifts: ST30720000 (J-25405) and KV40104830.

9. Align the matching mark of mainshaft with the mark of companion flange, then install the companion flange.

10. Using a flange wrench, install the self-lock nut of companion flange.

11. Reverse remaining removal procedure to complete installation.

12. Check fluid levels.

ENGINE COOLING

THERMOSTAT

REMOVAL & INSTALLATION

2006 Coupe and Sedan; 2007 Coupe

See Figure 33.

1. Before servicing the vehicle, refer to the Precautions Section.

2. Disconnect the negative battery cable.

3. Remove the undercover.

4. Drain the coolant from the radiator and both sides of the cylinder block. Properly dispose the coolant.

➡ **Be sure the engine is cold before draining the radiator. Do not allow coolant to spill on the drive belts.**

5. Remove the air duct and air cleaner case.

6. Remove the front water drain plug on the water pump side of the cylinder block.

7. Disconnect the lower radiator hose and oil cooler hose from the water inlet and thermostat assembly.

8. Remove the water inlet and thermostat assembly.

➡ **Do not disassemble the water inlet and thermostat assembly. Replace them as a unit, if required.**

To install:

9. Installation is the reverse of the removal procedure.

10. Be sure to refill the cooling using the proper grade and type engine coolant.

11. Start the engine and check for leaks.

12. Start the engine and allow it to reach operation temperature. Recheck the coolant level, fill as required.

2007–08 Sedan

See Figure 34.

1. Before servicing the vehicle, refer to the Precautions Section.

2. Disconnect the negative battery cable.

3. Remove air duct and air cleaner case assembly (LH).

4. Remove oil cooler water pipe mounting bolt, and move aside water pipe. (AWD models)

5. Remove engine undercover.

6. Drain engine coolant from radiator drain plug at the bottom of radiator and recycle coolant.

➡ **Be sure the engine is cold before draining the radiator. Do not allow coolant to spill on the drive belts.**

7. Disconnect radiator hose (lower).

8. Disconnect intake valve timing control valve harness connector (LH), and remove intake valve timing control solenoid.

9. Remove water inlet and thermostat assembly

To install:

10. Installation is the reverse of the removal procedure.

11. Be sure to refill the cooling using the proper grade and type engine coolant.

12. Start the engine and check for leaks.

13. Start the engine and allow it to reach operation temperature. Recheck the coolant level, fill as required.

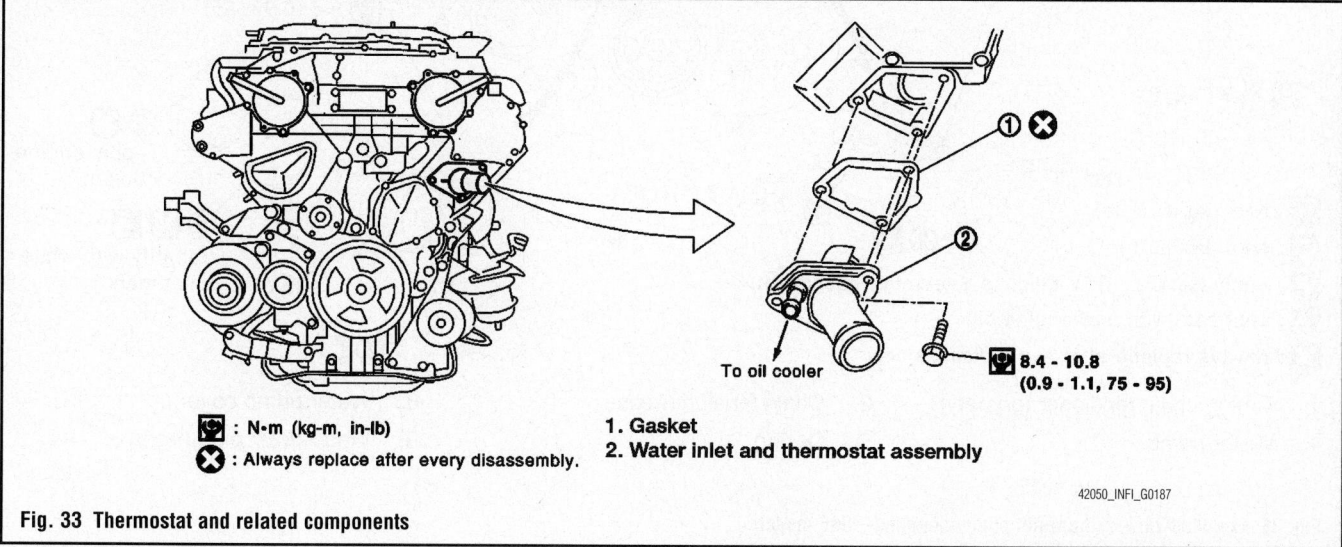

Fig. 33 Thermostat and related components

42050_INFI_G0187

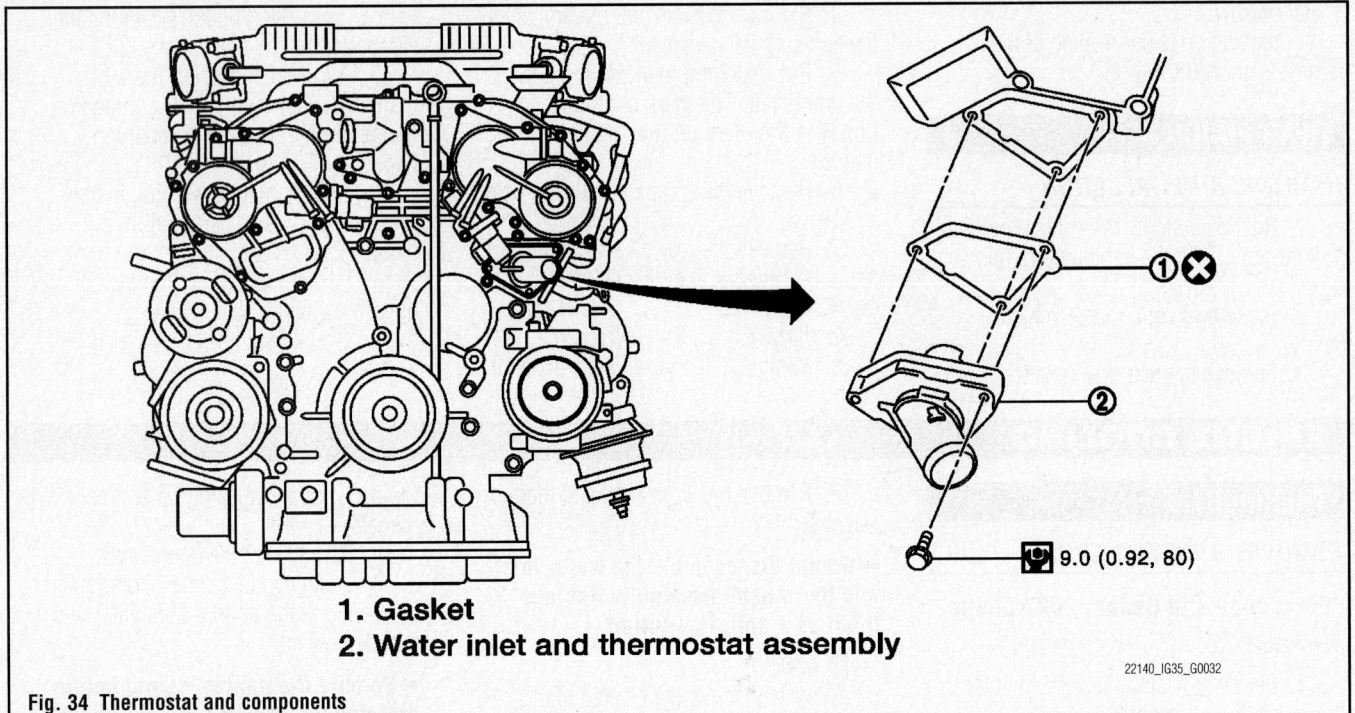

1. Gasket
2. Water inlet and thermostat assembly

22140_IG35_G0032

Fig. 34 Thermostat and components

WATER PUMP

REMOVAL & INSTALLATION

2006 Coupe and Sedan; 2007 Coupe

See Figure 35.

1. Before servicing the vehicle, refer to the precautions in the beginning of this section.
2. Drain the cooling system.
3. Remove or disconnect the following:

- Engine undercover
- Drive belts
- Air duct
- Radiator upper and lower hoses
- Radiator shrouds
- Cooling fan

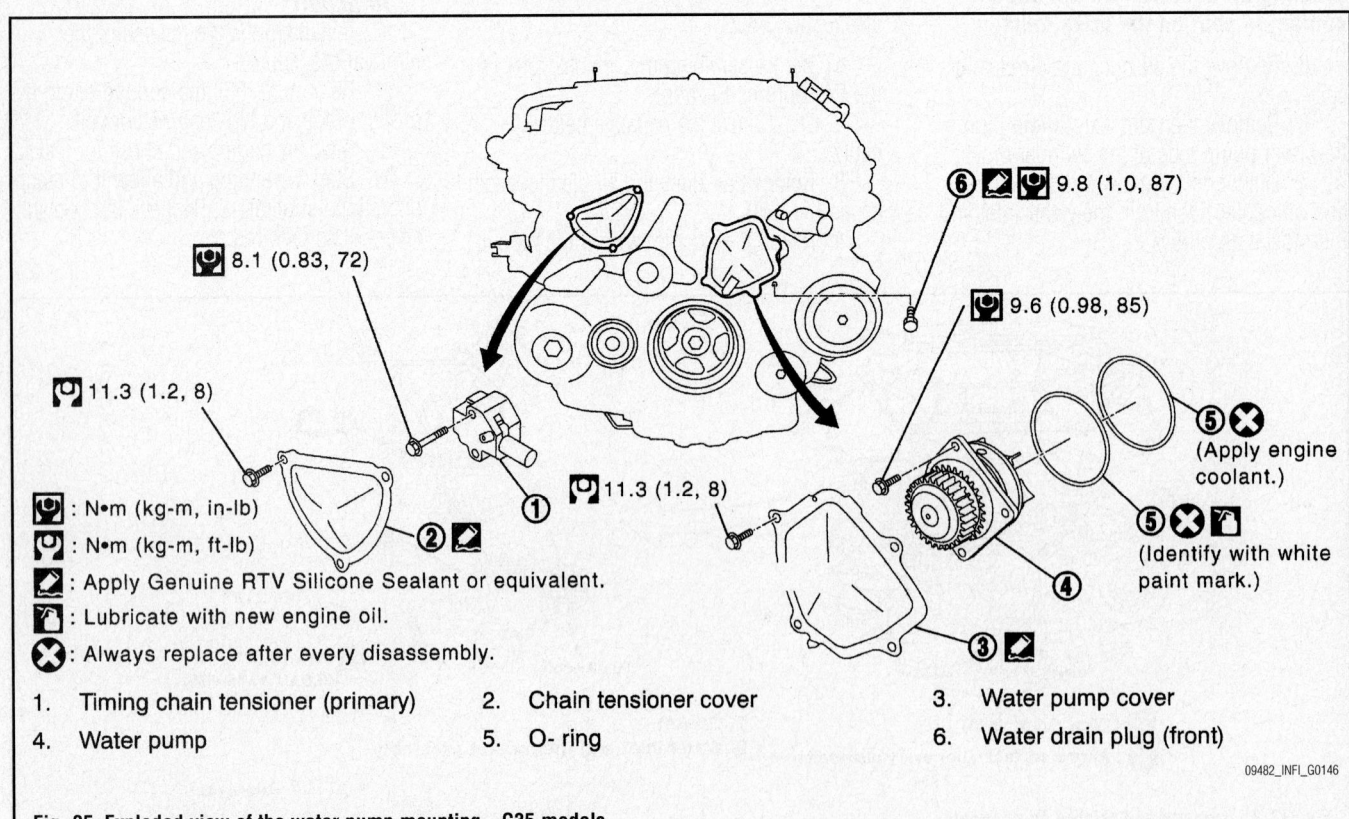

: N•m (kg-m, in-lb)

: N•m (kg-m, ft-lb)

: Apply Genuine RTV Silicone Sealant or equivalent.

: Lubricate with new engine oil.

: Always replace after every disassembly.

1. Timing chain tensioner (primary)
2. Chain tensioner cover
3. Water pump cover
4. Water pump
5. O- ring
6. Water drain plug (front)

09482_INFI_G0146

Fig. 35 Exploded view of the water pump mounting—G35 models

- Water drain plug from the water pump side of the cylinder block
- Timing chain tensioner cover and water pump cover

✳✳ WARNING

Be careful not the drop the mounting bolts inside the chain case.

- Timing chain tensioner
- 3 bolts that secure the water pump

4. Rotate the crankshaft 20° counterclockwise to provide timing chain slack.

5. Put the 2 grade M8 bolts in the 2 M8 threaded holes of the water pump.

6. Tighten each bolt by turning alternately ½ turn until they reach the timing chain rear case. Be sure to turn each bolt ½ turn at a time to prevent damage.

7. Lift the water pump straight out to remove it.

8. When removing the water pump, do not allow the water pump gear to hit the timing chain.

9. Remove and discard the O-rings from the water pump.

10. Clean all traces of liquid gasket from the water pump and covers.

To install:

11. Install or connect the following:
- Water pump with new O-rings. Torque the bolts to 75–95 inch lbs. (8–11 Nm) and rotate the crankshaft pulley to its original position by turning it 20° clockwise.
- Timing chain tensioner. Torque the bolts to 89 inch lbs. (10 Nm). Remove the stopper pin from the timing chain tensioner.

12. Apply a continuous 0.091–0.130 in. (2–3mm) bead of liquid sealant to the mating surfaces of the timing chain tensioner and water pump covers.
- Timing chain tensioner and water pump covers to the engine block. Torque the cover bolts to 89 inch lbs. (10 Nm).
- Water drain plug to the water pump side of the cylinder block
- Cooling fan
- Radiator shrouds
- Radiator upper and lower hoses
- Air duct
- Drive belts
- Engine undercover
- Negative battery cable

13. Fill and bleed the cooling system.

14. Start the vehicle, check for leaks and repair if necessary.

2007–08 Sedan

See Figure 36.

✳✳ CAUTION

When removing water pump assembly, be careful not to get engine coolant on drive belts.

✳✳ CAUTION

Water pump cannot be disassembled and should be replaced as a unit.

✳✳ CAUTION

After installing water pump, connect hose and clamp securely, then check for leaks using the radiator cap tester (commercial service tool) and the radiator cap tester adapter SST: EG17650301 (J33984-A).

1. Before servicing the vehicle, refer to the precautions in the beginning of this section.

2. Remove engine cover.

3. Properly relieve the fuel system pressure.

4. Disconnect the negative battery cable.

5. Remove air duct and air cleaner case assembly (RH and LH).

6. Remove reservoir tank

7. Separate engine harness removing their brackets from front timing chain case.

8. Remove engine undercover.

9. Drain engine oil.

10. Drain the cooling system.

11. Remove cooling fan assembly.

12. Remove radiator hose (upper and lower).

13. Remove front timing chain case

14. Remove timing chain tensioner (primary).

15. Remove water pump as follows:

a. Remove three water pump mounting bolts. Secure a gap between water pump gear and timing chain, by turning crankshaft counterclockwise until timing chain looseness on water pump and the sprocket becomes maximum.

b. Put the 2 grade M8 bolts in the 2 M8 threaded holes of the water pump.

c. Tighten each bolt by turning alternately ½ turn until they reach the timing chain rear case. Be sure to turn each bolt ½ turn at a time to prevent damage.

16. Lift the water pump straight out to remove it.

17. When removing the water pump, do not allow the water pump gear to hit the timing chain.

18. Remove and discard the O-rings from the water pump.

19. Clean all traces of liquid gasket from the water pump and covers.

20. Install or connect the following:
- Water pump with new O-rings. Torque the bolts to 75–95 inch lbs. (8–11 Nm) and rotate the crankshaft

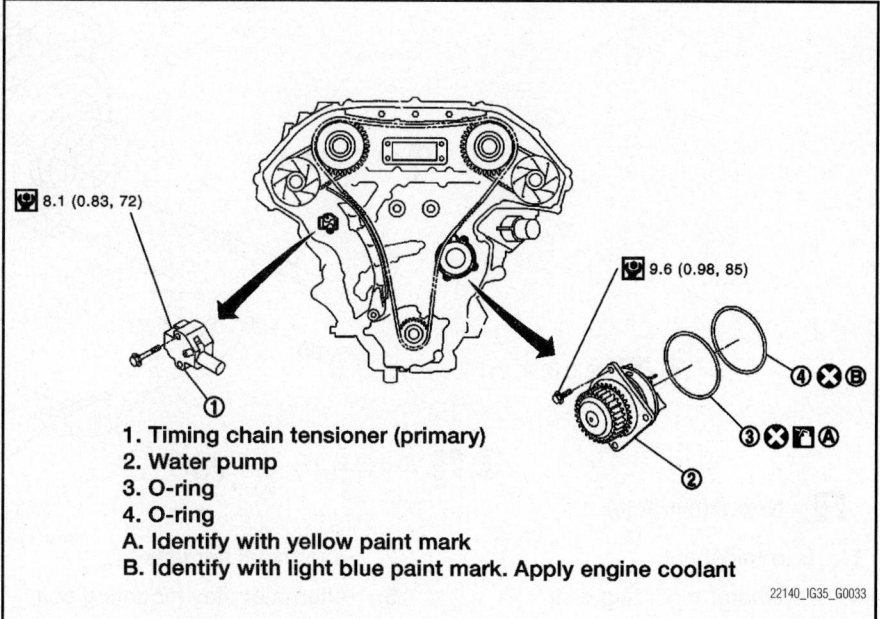

8.1 (0.83, 72)

9.6 (0.98, 85)

1. Timing chain tensioner (primary)
2. Water pump
3. O-ring
4. O-ring
A. Identify with yellow paint mark
B. Identify with light blue paint mark. Apply engine coolant

22140_IG35_G0033

Fig. 36 Water pump and related components

pulley to its original position by turning it 20° clockwise.
- Timing chain tensioner. Torque the bolts to 89 inch lbs. (10 Nm). Remove the stopper pin from the timing chain tensioner.

21. Apply a continuous 0.091–0.130 in. (2–3mm) bead of liquid sealant to the mating surfaces of the timing chain tensioner and water pump covers.

- Timing chain tensioner and water pump covers to the engine block. Torque the cover bolts to 89 inch lbs. (10 Nm).
- Water drain plug to the water pump side of the cylinder block
- Cooling fan

- Radiator shrouds
- Radiator upper and lower hoses
- Air duct
- Drive belts
- Engine undercover
- Negative battery cable

22. Fill and bleed the cooling system.
23. Start the vehicle, check for leaks and repair if necessary.

ENGINE ELECTRICAL

ALTERNATOR

REMOVAL & INSTALLATION

2006 Models

Automatic Transmission

1. Before servicing the vehicle, refer to the precautions in the beginning of this section.
2. Remove or disconnect the following:
   - Negative battery cable
   - Engine undercover
   - Stabilizer bar clamps, then slide stabilizer downward
   - Loosen drive belts
   - Alternator electrical connector

- Oil pressure switch harness connector
- Terminal mounting nut
- Upper and lower alternator mounting bolt
- Both alternator bracket bolts
- Alternator from the vehicle

To install:

3. Installation is the reverse order of removal.

AWD & Manual Transmission

See Figure 37.

1. Before servicing the vehicle, refer to the precautions in the beginning of this section.
2. Remove or disconnect the following:

CHARGING SYSTEM

- Negative battery cable
- Engine undercover
- Radiator fan assembly
- Remove drive belts
- Oil pressure switch harness connector
- Alternator stay mounting bolts and stay
- Alternator mounting bolt
- Alternator electrical connector
- "B" terminal mounting nut
- Harness clip and water hose bracket
- Alternator from the vehicle

To install:

3. Installation is the reverse order of removal.

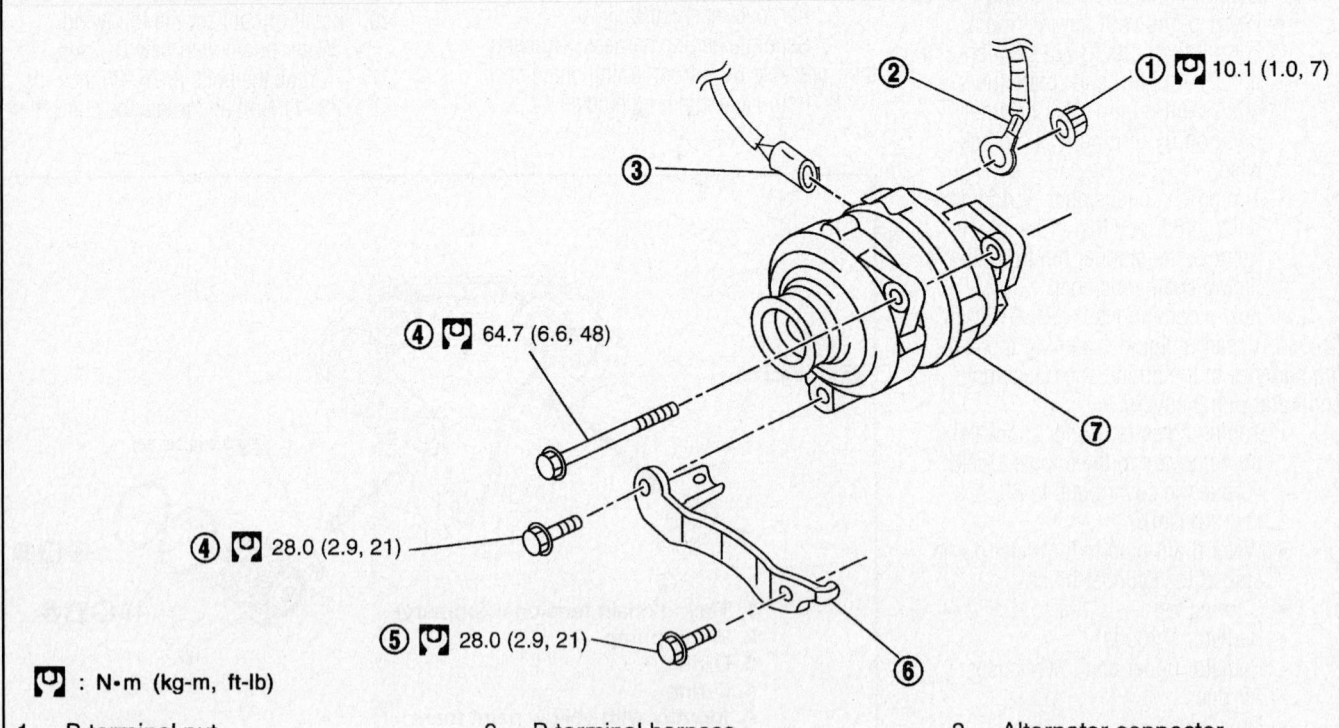

: N•m (kg-m, ft-lb)

1. B terminal nut
2. B terminal harness
3. Alternator connector
4. Alternator mounting bolt
5. Alternator stay mounting bolt
6. Alternator stay
7. Alternator

09482_INFI_G0021

Fig. 37 Alternator mounting and torque specifications

2007–08 Models

2WD Models

See Figure 38.

1. Before servicing the vehicle, refer to the precautions in the beginning of this section.

2. Remove or disconnect the following:
   - Negative battery cable
   - Engine undercover
   - Stabilizer bar clamps, then slide stabilizer downward
   - Radiator cooling fan assembly
   - Drive belt
   - Alternator connector
   - "B" terminal nut
   - Harness bracket bolts
   - Oil pressure switch harness clip
   - Oil pressure switch connector and oil temperature sensor connector

- Alternator mounting bolt and alternator stay mounting bolt, then remove alternator stay
- Alternator mounting bolt
- Alternator assembly downward from the vehicle.

To install:

3. Installation is the reverse order of removal.

AWD Models

See Figure 39.

1. Before servicing the vehicle, refer to the precautions in the beginning of this section.

2. Remove or disconnect the following:
   - Negative battery cable
   - Air cleaner case
   - Power steering oil pressure sensor connector
   - Clip from the harness bracket

and "B" terminal harness from the clip
- Engine undercover
- Drive belt
- Radiator cooling fan assembly
- Alternator mounting bolt and alternator stay mounting bolt, then remove alternator stay
- Alternator mounting bolt
- Alternator connector
- "B" terminal nut
- Power steering oil pump mounting bolts and power steering oil pump hose bracket bolts.
- Move power steering pump and reservoir tank forward
- Remove alternator assembly upward from the vehicle.

To install:

3. Installation is the reverse order of removal.

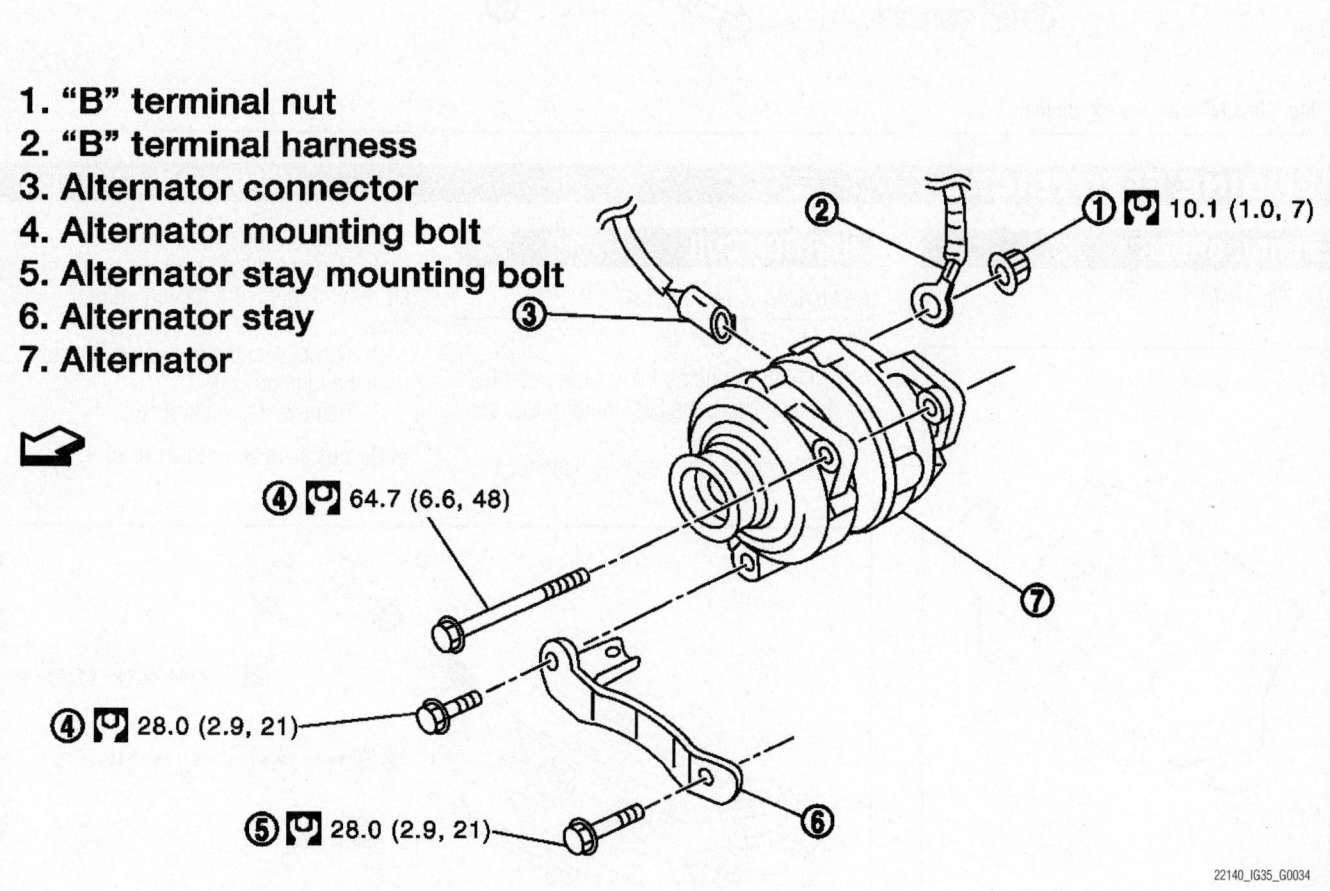

1. "B" terminal nut
2. "B" terminal harness
3. Alternator connector
4. Alternator mounting bolt
5. Alternator stay mounting bolt
6. Alternator stay
7. Alternator

① 10.1 (1.0, 7)

④ 64.7 (6.6, 48)

④ 28.0 (2.9, 21)

⑤ 28.0 (2.9, 21)

22140_IG35_G0034

Fig. 38 Alternator and components

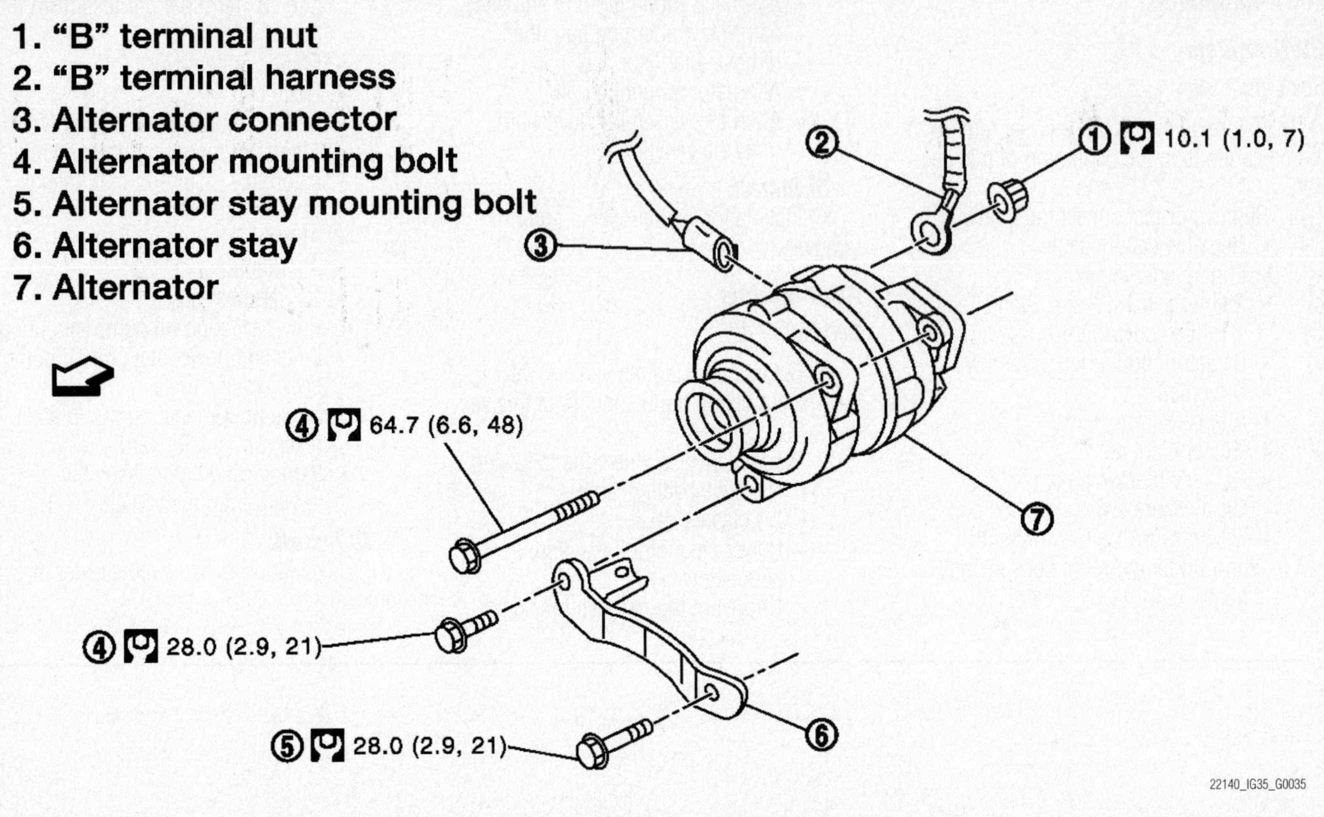

1. "B" terminal nut
2. "B" terminal harness
3. Alternator connector
4. Alternator mounting bolt
5. Alternator stay mounting bolt
6. Alternator stay
7. Alternator

① 10.1 (1.0, 7)

④ 64.7 (6.6, 48)

④ 28.0 (2.9, 21)

⑤ 28.0 (2.9, 21)

22140_IG35_G0035

Fig. 39 Alternator and components

ENGINE ELECTRICAL   **IGNITION SYSTEM**

FIRING ORDER

See Figure 40.

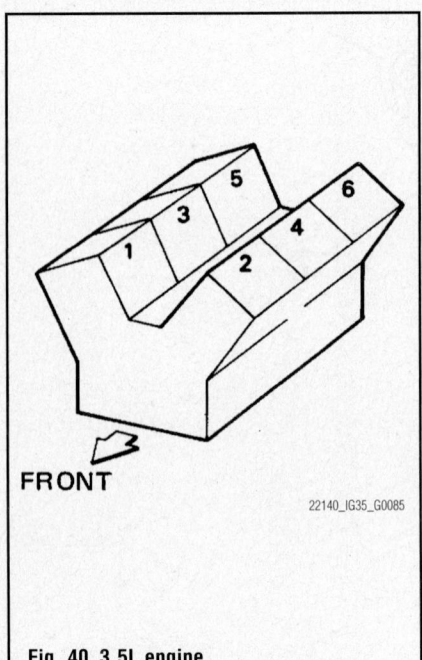

FRONT

22140_IG35_G0085

Fig. 40 3.5L engine
   Firing order: 1–2–3–4–5–6
   Distributorless ignition system

IGNITION COIL

REMOVAL & INSTALLATION
See Figures 41 and 42.

   1. Before servicing the vehicle, refer to the precautions in the beginning of this section.
   2. Disconnect the negative battery cable.

   3. Remove the engine cover.
   4. Remove the air cleaner case and air duct, if removing the left side coils.
   5. Disconnect the harness connector from the ignition coil.
   6. Remove the ignition coil.

➡**Do not drop or shock the coil.**

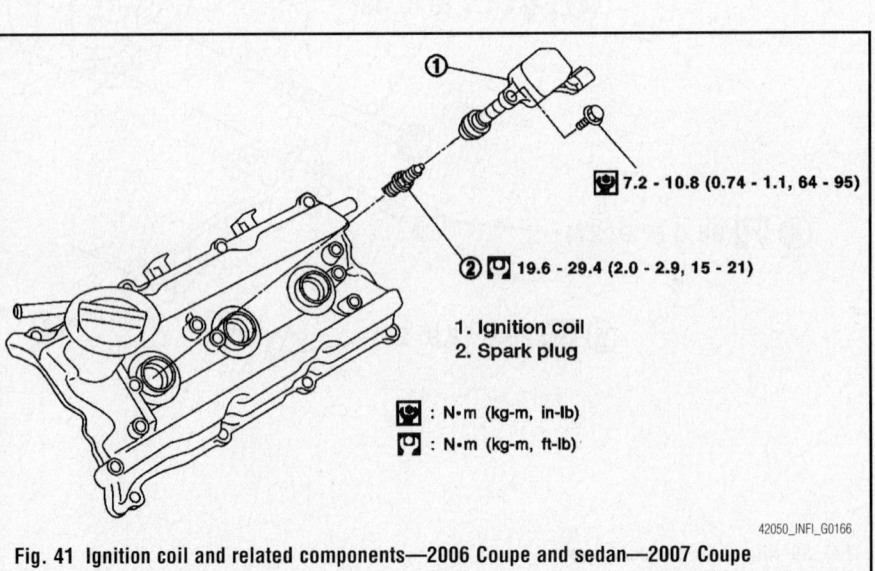

① 7.2 – 10.8 (0.74 - 1.1, 64 - 95)

② 19.6 – 29.4 (2.0 - 2.9, 15 - 21)

1. Ignition coil
2. Spark plug

: N·m (kg-m, in-lb)

: N·m (kg-m, ft-lb)

42050_INFI_G0166

Fig. 41 Ignition coil and related components—2006 Coupe and sedan—2007 Coupe

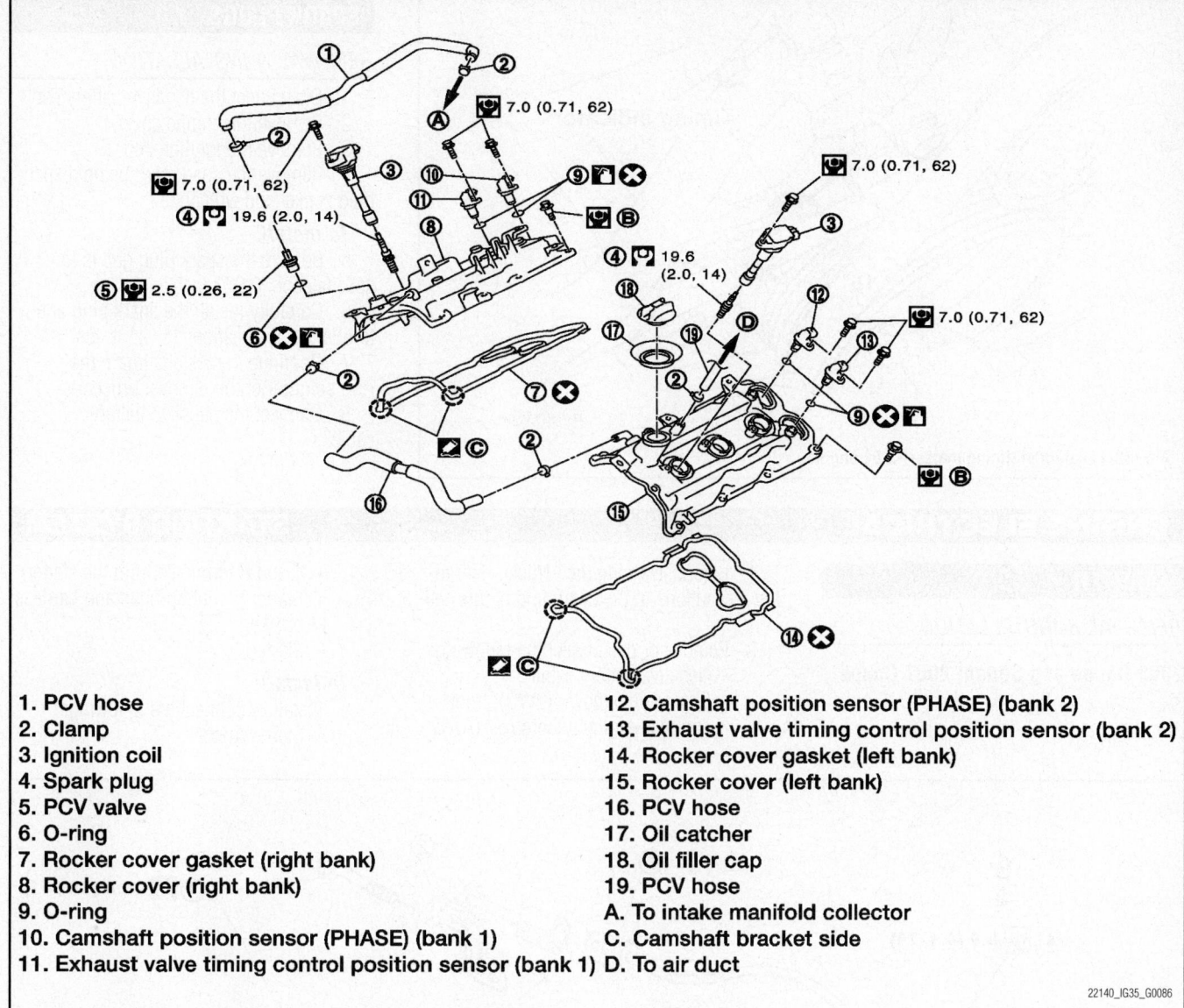

1. PCV hose
2. Clamp
3. Ignition coil
4. Spark plug
5. PCV valve
6. O-ring
7. Rocker cover gasket (right bank)
8. Rocker cover (right bank)
9. O-ring
10. Camshaft position sensor (PHASE) (bank 1)
11. Exhaust valve timing control position sensor (bank 1)
12. Camshaft position sensor (PHASE) (bank 2)
13. Exhaust valve timing control position sensor (bank 2)
14. Rocker cover gasket (left bank)
15. Rocker cover (left bank)
16. PCV hose
17. Oil catcher
18. Oil filler cap
19. PCV hose
A. To intake manifold collector
C. Camshaft bracket side
D. To air duct

22140_IG35_G0086

Fig. 42 Ignition coil and related components—2007 and 2008 sedan

To install:

7. Installation is the reverse of the removal procedure.

8. Tighten the coil retaining bolt to 64–94 inch lbs.

IGNITION TIMING

ADJUSTMENT

See Figure 43.

➡**The engine should be in good mechanical condition and all electrical connectors and vacuum hoses attached before making this adjustment.**

1. Before servicing the vehicle, refer to the precautions in the beginning of this section.

2. Start the engine and let it warm up to normal operating temperature.

3. Open the hood and run the engine under no load at about 2,000 rpm for about 2 minutes.

4. Perform Diagnostic Test Mode II and repair any causes of trouble codes as needed.

5. Run the engine under no load at 2,000 rpm for about 2 minutes. Rev the engine 2 or 3 times and let it idle for 1 minute.

6. Turn **OFF** the engine and disconnect the Throttle Position sensor connector. Remove the No. 1 ignition coil. Connect the coil to the spark plug using a spare piece of high-tension wire so you have a place to connect your timing light. Start the engine.

7. Run the engine under no load at 2,000 rpm for about 2 minutes. Rev the engine 2 or 3 times and let it idle.

8. Check the ignition timing and adjust if needed (Refer to Engine Tune-Up Specifications at beginning of chapter for proper settings; always defer to timing settings on tag under hood if different)

9. Adjustment is made by loosening the screws and turning the Camshaft Position (CMP) sensor until the mark on the crankshaft pulley is pointing at 10° BTDC. Tighten the mounting screws and confirm ignition timing has not changed.

10. Turn the engine **OFF** and connect the TP sensor connector.

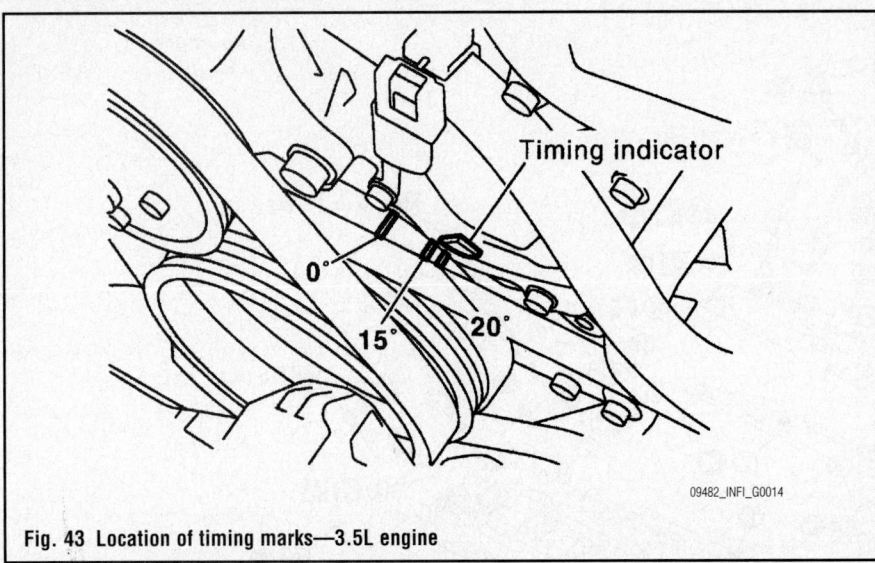

Fig. 43 Location of timing marks—3.5L engine

SPARK PLUGS

REMOVAL & INSTALLATION

1. Disconnect the negative battery cable.
2. Remove the engine cover.
3. Remove the ignition coil.
4. Remove the spark plug using a spark plug socket and wrench.

To install:

5. Be sure the spark plug gap is to specification, 0.043 inch.
6. Carefully install the spark plug and torque to specification, 15–21 ft. lbs.
7. Continue the installation in the reverse order of the removal procedure.
8. Connect the negative battery cable.

ENGINE ELECTRICAL

STARTER

REMOVAL & INSTALLATION

2006 Coupe and Sedan; 2007 Coupe

See Figures 44 and 45.

1. Before servicing the vehicle, refer to the precautions in the beginning of this section.
2. Remove or disconnect the following:
   - Negative battery cable
   - Engine undercover (2WD); front and rear engine undercover (AWD).

STARTING SYSTEM

   - S and B terminals from the starter
   - Starter mounting bolts and harness bracket
   - Starter

To install:

3. Install or connect the following:
   - Starter motor

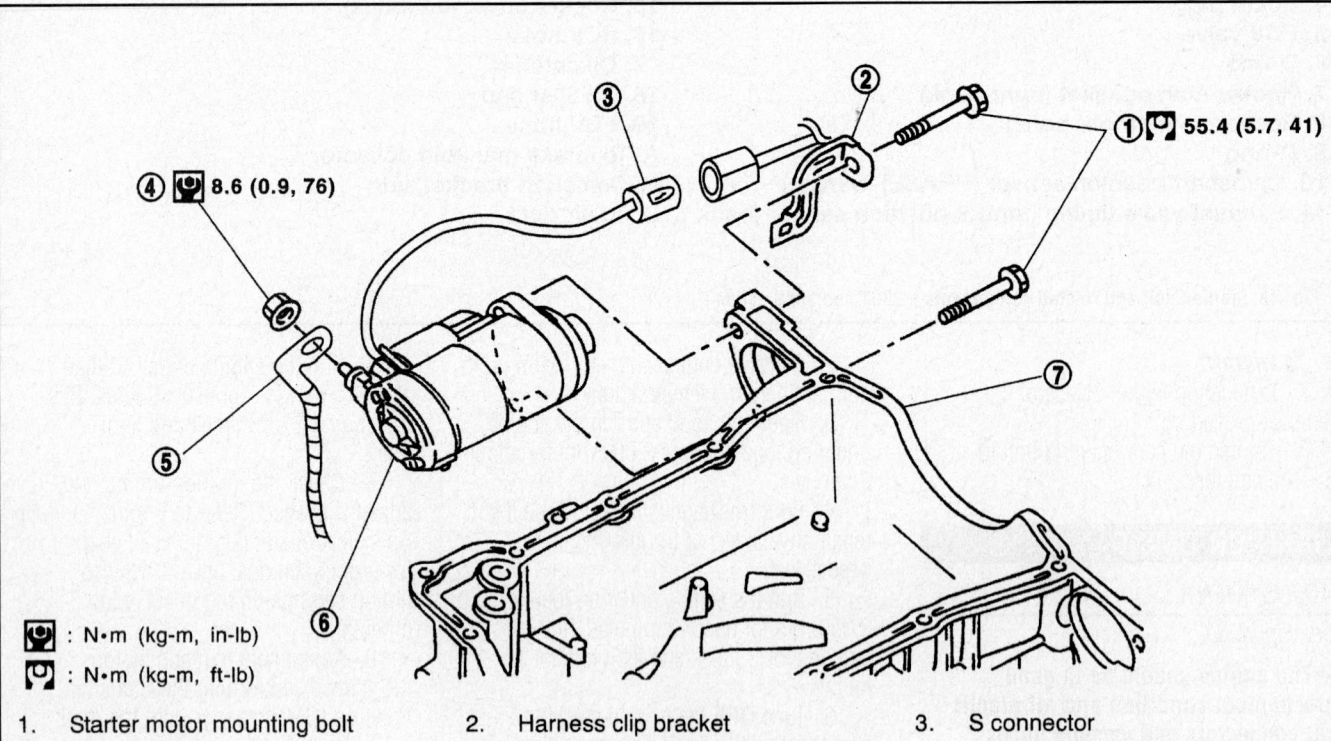

$\boxed{\text{N}}$: N•m (kg-m, in-lb)

$\boxed{\text{N}}$: N•m (kg-m, ft-lb)

1.	Starter motor mounting bolt	2.	Harness clip bracket	3.	S connector
4.	B terminal nut	5.	B terminal harness	6.	Starter motor
7.	Oil pan				

Fig. 44 Starter mounting and torque specifications—2WD

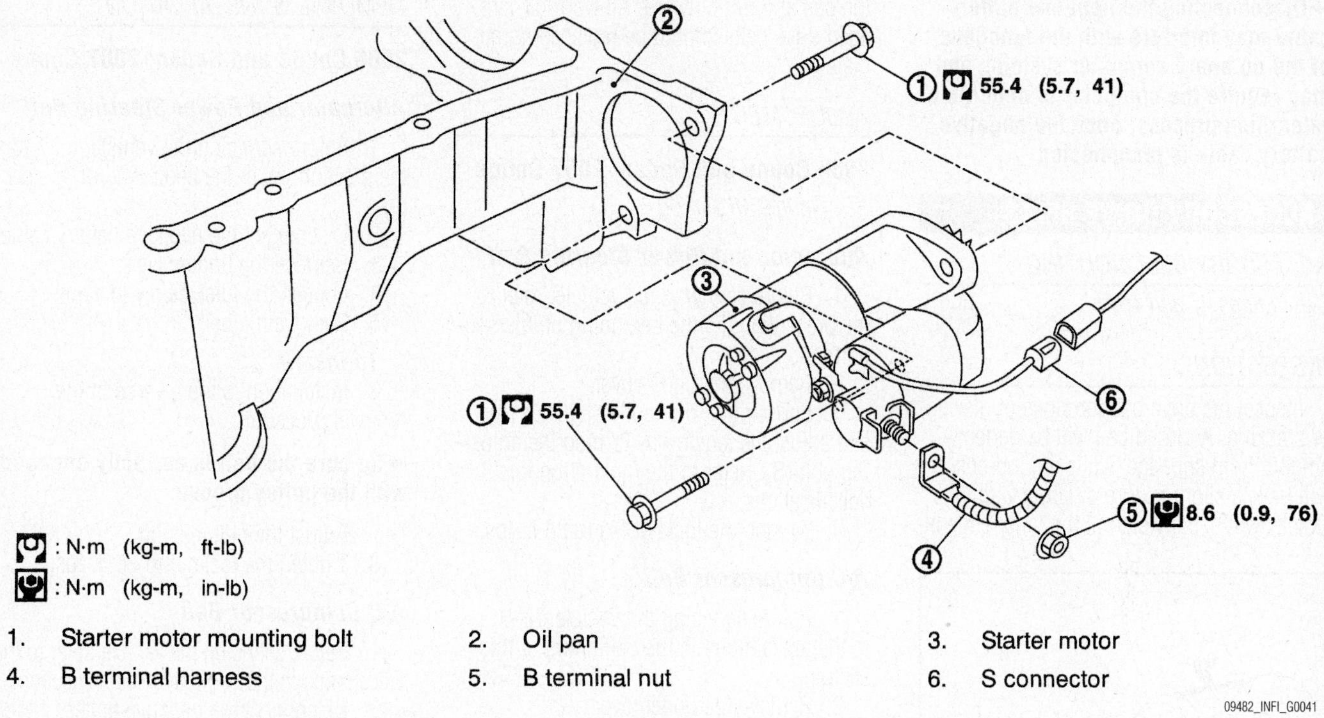

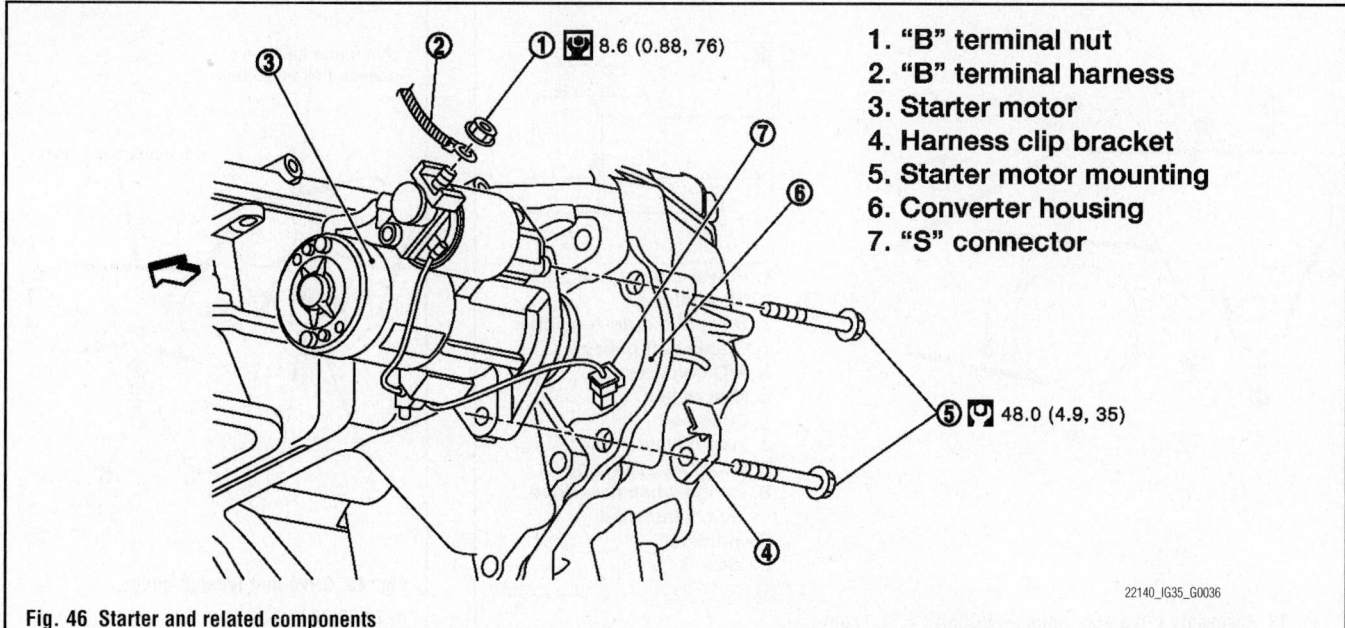

Fig. 45 Starter mounting and torque specifications—AWD

[N·m (kg-m, ft-lb)]
[N·m (kg-m, in-lb)]

1. Starter motor mounting bolt
2. Oil pan
3. Starter motor
4. B terminal harness
5. B terminal nut
6. S connector

- Terminals to the starter
- Engine undercover (2WD); front and rear engine undercover (AWD)
- Negative battery cable

2007–08 Sedan

See Figure 46.

1. *Before servicing the vehicle, refer to the precautions in the beginning of this section.*
2. Remove or disconnect the following:
   - Negative battery cable
   - Engine undercover
   - Exhaust mounting bracket
   - Steering lower joint
   - "B" terminal nut
   - "S" connector
   - Starter motor mounting bolts
   - Starter motor downward from the vehicle

To install:
3. To install, reverse removal procedure.

1. "B" terminal nut
2. "B" terminal harness
3. Starter motor
4. Harness clip bracket
5. Starter motor mounting
6. Converter housing
7. "S" connector

Fig. 46 Starter and related components

ENGINE MECHANICAL

→Disconnecting the negative battery cable may interfere with the functions of the on board computer systems and may require the computer to undergo a relearning process, once the negative battery cable is reconnected.

ACCESSORY DRIVE BELTS

ACCESSORY BELT ROUTING

See Figures 47 and 48.

INSPECTION

Inspect the drive belt for signs of glazing or cracking. A glazed belt will be perfectly smooth from slippage, while a good belt will have a slight texture of fabric visible. Cracks will usually start at the inner edge of

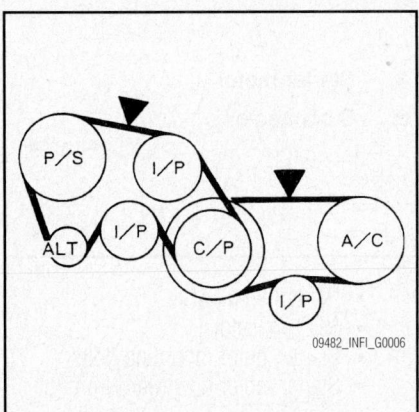

09482_INFI_G0006

Fig. 47 Accessory drive belt routing— VQ35DE 3.5L Engine

the belt and run outward. All worn or damaged drive belts should be replaced immediately.

ADJUSTMENT

2006 Coupe and Sedan; 2007 Coupe

See Figures 49 and 50.

Alternator and Power Steering Belt

1. Before servicing the vehicle, refer to the precautions in the beginning of this section.
2. Remove the undercover.
3. Loosen the idler pulley locknut (A) and adjust the tension by turning the adjusting bolt (B). Refer to the illustration for specifications.
4. Tighten the locknut (A) to 26 ft. lbs.

A/C Compressor Belt

1. Before servicing the vehicle, refer to the precautions in the beginning of this section.
2. Remove the undercover.
3. Loosen the idler pulley locknut (C) and adjust the tension by turning the adjusting bolt (D). Refer to the illustration for specifications.
4. Tighten the locknut (C) to 26 ft. lbs.

2007–08 Sedan

Belt tension is not necessary, as it is automatically adjusted by drive belt auto-tensioner.

REMOVAL & INSTALLATION

2006 Coupe and Sedan; 2007 Coupe

Alternator and Power Steering Belt

1. Before servicing the vehicle, refer to the precautions in the beginning of this section.
2. Disconnect the negative battery cable.
3. Remove the undercover.
4. Loosen the idler pulley locknut.
5. Remove the belt.

To install:

6. Installation is the reverse of the removal procedure.

→Be sure the belt is correctly engaged with the pulley groove.

7. Adjust the belt tension.
8. Tighten the locknut to 26 ft. lbs.

A/C Compressor Belt

1. Before servicing the vehicle, refer to the precautions in the beginning of this section.
2. Disconnect the negative battery cable.
3. Remove the undercover.
4. Loosen the idler pulley locknut.
5. Remove the belt.

To install:

6. Installation is the reverse of the removal procedure.

→Be sure the belt is correctly engaged with the pulley groove.

7. Adjust the belt tension.
8. Tighten the locknut to 26 ft. lbs.

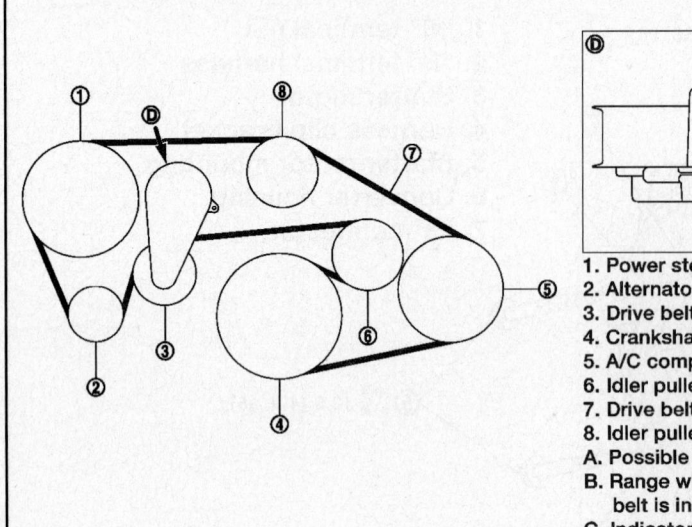

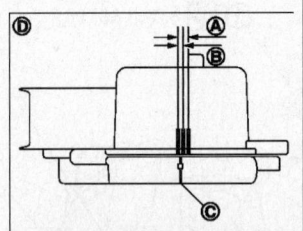

1. Power steering oil pump
2. Alternator
3. Drive belt auto-tensioner
4. Crankshaft pulley
5. A/C compressor
6. Idler pulley
7. Drive belt
8. Idler pulley
A. Possible use range
B. Range when new drive belt is installed
C. Indicator
D. View D

22140_IG35_G0037

Fig. 48 Accessory drive belt routing—VQ35HR 3.5L engine

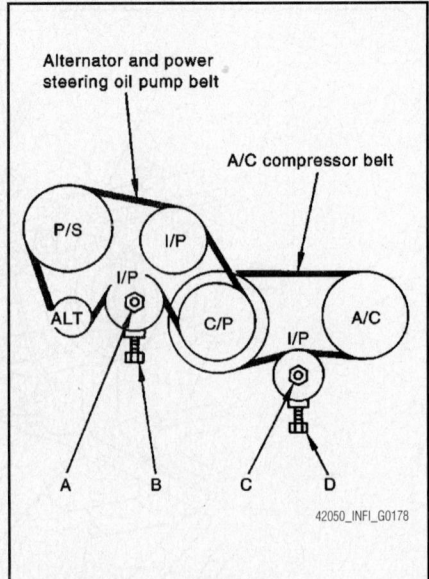

42050_INFI_G0178

Fig. 49 Drive belt tension gauge positioning

Deflection adjustment			Unit: mm (in)	Tension adjustment*			Unit: N (kg, lb)
	Used belt		New belt		Used belt		New belt
	Limit	After adjustment			Limit	After adjustment	
Alternator and power steering oil pump belt	7 (0.28)	4 - 5 (0.16 - 0.20)	3.5 - 4.5 (0.138 - 0.177)		294 (30, 66)	730 - 818 (74.5 - 83.5, 164 - 184)	838 - 926 (85.5 - 94.5, 188 - 208)
A/C compressor belt	12 (0.47)	9 - 10 (0.35 - 0.39)	8 - 9 (0.31 - 0.35)		196 (20, 44)	348 - 436 (35.5 - 44.5, 78 - 98)	470 - 559 (48 - 57, 106 - 126)
Applied pushing force	98 N (10 kg, 22 lb)				—		

*: If belt tension gauge cannot be installed at check points shown, check drive belt tension at different location on the belt.

42050_INFI_G0178A

Fig. 50 Drive belt tension specifications

2007–08 Sedan

1. Remove radiator reservoir tank.
2. Remove engine undercover
3. Remove radiator cooling fan assembly
4. While securely holding the square hole in pulley center of auto tensioner with a spinner handle, move spinner handle counter-clockwise (loosening direction of drive belt).
5. Under the above condition, insert a metallic bar of approximately 0.24 inches (6mm) in diameter [hexagonal wrench through the holding boss to lock the auto-tensioner pulley arm.
6. Remove drive belt.
7. To install, reverse removal procedure.
8. Make sure drive belt is securely installed around all pulleys.
9. Make sure drive belt is correctly engaged with the pulley groove.
10. Check for engine oil and engine coolant is not adhered drive belt and pulley groove.

CAMSHAFT AND VALVE LIFTERS

REMOVAL & INSTALLATION

2006 Coupe and Sedan; 2007 Coupe

See Figures 51 through 54.

1. Before servicing the vehicle, refer to the precautions in the beginning of this section.
2. Drain the engine oil and cooling system.
3. Relieve the fuel system pressure.
4. Remove or disconnect the following:
   - Negative battery cable

- Front timing chain case
- Camshaft sprocket
- Timing chain
- Rear timing chain case
- Camshaft Position (CMP) sensors (PHASE) from the back side of the cylinder heads
- Intake valve timing control solenoid valves from the No. 1 camshaft bracket

➡**Before removal, matchmark the position of the camshafts, brackets and bolts, so they are reinstalled in their original locations.**

- Intake and exhaust camshaft brackets. Loosen the bolts in several stages, in reverse of the sequence shown.
- Camshafts
- Valve lifters if necessary, noting their installed positions

To install:

5. Install or connect the following:
   - Valve lifters, in their original positions

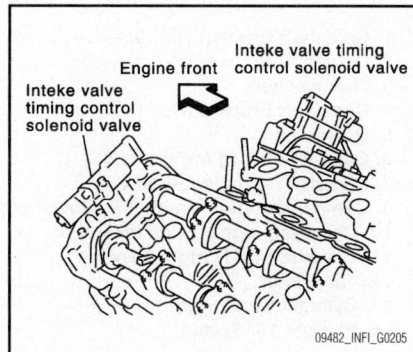

Fig. 51 Location of the intake valve timing control solenoid valves

- Camshafts with the dowel pin attached to its front end face on the exhaust side
- Camshaft brackets, as shown in the illustration

6. Torque the camshaft bracket bolts, in sequence, as follows:
   a. Step 1: Nos. 7–10, then Nos. 1–6 to 17 inch lbs. (2 Nm).
   b. Step 2: Nos. 1–10 to 52 inch lbs. (5.9 Nm).
   c. Step 3: Nos. 1–6 to 80–104 inch lbs. (9–12 Nm).
   d. Step 4: Nos. 7–10 to 74–91 inch lbs.

7. Measure the difference in levels between the front end faces of the No. 1 camshaft bracket and the cylinder head. If the measurement falls out of the range

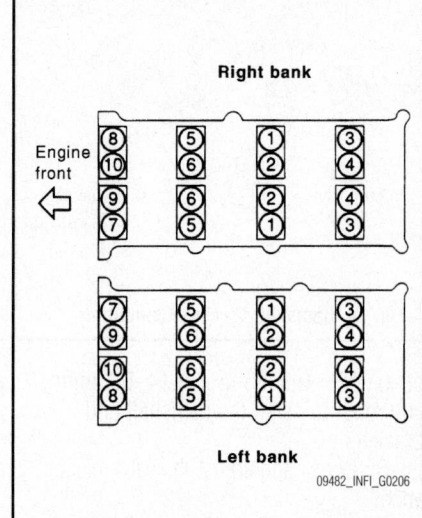

Fig. 52 Intake and exhaust camshaft bracket bolts tightening sequence. Use the reverse sequence for removal

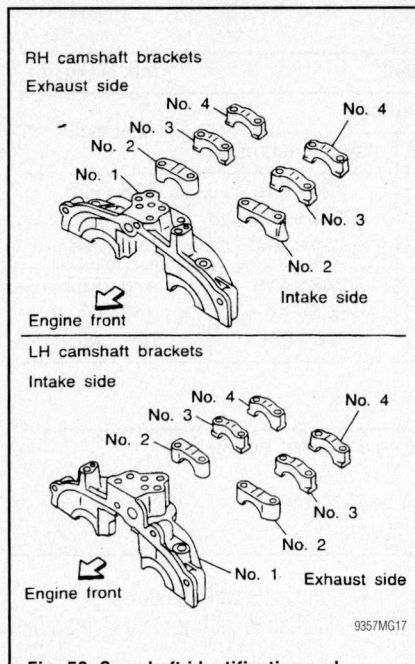

Fig. 53 Camshaft identification and installation

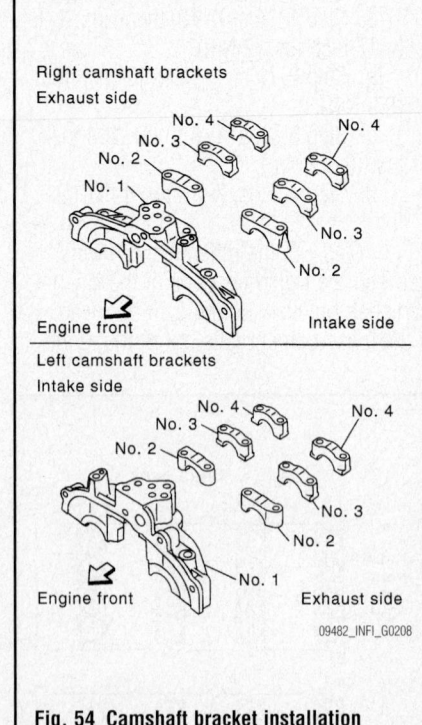

Fig. 54 Camshaft bracket installation

of -0.0055–0.0055 in. (-0.14–0.14mm), you must reinstall the camshaft and brackets.

8. Check and adjust the valve clearance.
   • Intake valve timing control solenoid valves
   • CMP sensors (PHASE)
   • Rear timing chain case, timing

chain, camshaft sprocket and front timing chain case
   • Negative battery cable
9. Fill the engine with oil.
10. Fill and bleed the cooling system.
11. Start the vehicle, check for leaks and repair if necessary.

2007–08 Sedan

See Figures 55 through 60.

1. Before servicing the vehicle, refer to the precautions in the beginning of this section.
2. Remove front timing chain case, camshaft sprocket and timing chain.
3. Remove fuel sub tube.
4. Loosen camshaft sensor bracket bolts.
5. Mark camshafts, camshaft brackets and bolts so they are placed in the same position and direction for installation.

6. Equally loosen camshaft bracket bolts in several steps in reverse order as shown in the illustration.
7. Remove camshaft.
8. Remove valve lifter. Identify installation positions, and store them without mixing them up.
9. Remove timing chain tensioners (secondary) from cylinder head.
10. Remove timing chain tensioners (secondary) with its stopper pin attached.

To install:
11. Install timing chain tensioners (secondary) (1) on both sides of cylinder head.
12. Install timing chain tensioner with its stopper pin (C) attached.
13. Install timing chain tensioner with sliding part facing downward on right-side cylinder head, and with sliding

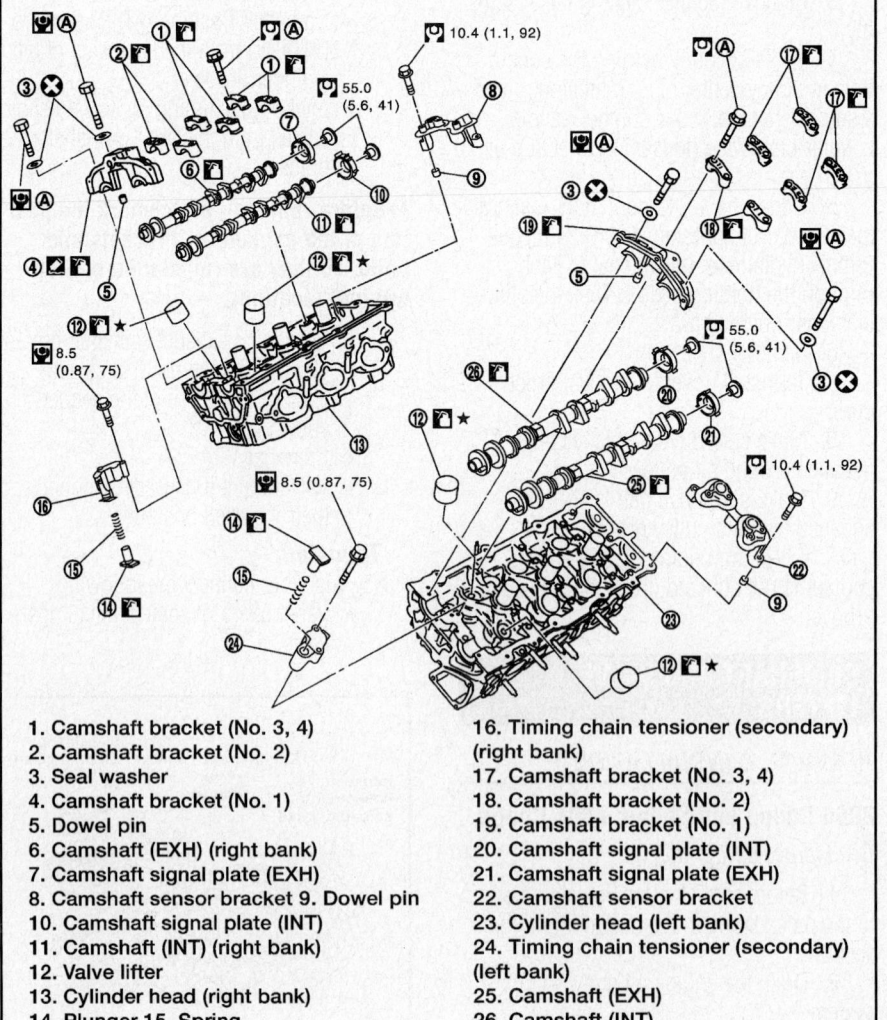

1. Camshaft bracket (No. 3, 4)
2. Camshaft bracket (No. 2)
3. Seal washer
4. Camshaft bracket (No. 1)
5. Dowel pin
6. Camshaft (EXH) (right bank)
7. Camshaft signal plate (EXH)
8. Camshaft sensor bracket 9. Dowel pin
10. Camshaft signal plate (INT)
11. Camshaft (INT) (right bank)
12. Valve lifter
13. Cylinder head (right bank)
14. Plunger 15. Spring

16. Timing chain tensioner (secondary) (right bank)
17. Camshaft bracket (No. 3, 4)
18. Camshaft bracket (No. 2)
19. Camshaft bracket (No. 1)
20. Camshaft signal plate (INT)
21. Camshaft signal plate (EXH)
22. Camshaft sensor bracket
23. Cylinder head (left bank)
24. Timing chain tensioner (secondary) (left bank)
25. Camshaft (EXH)
26. Camshaft (INT)

Fig. 55 Exploded view of camshafts and components

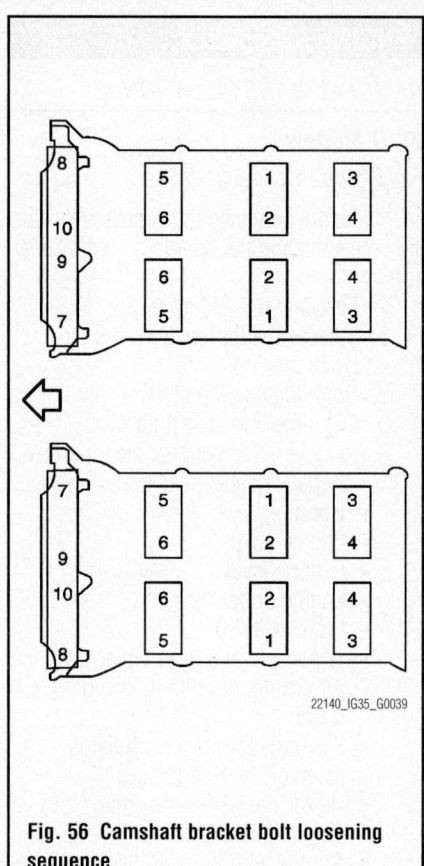

Fig. 56 Camshaft bracket bolt loosening sequence

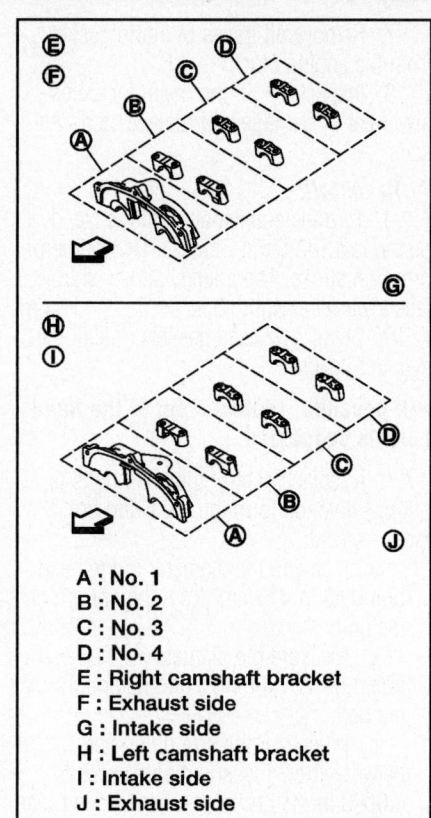

A : No. 1
B : No. 2
C : No. 3
D : No. 4
E : Right camshaft bracket
F : Exhaust side
G : Intake side
H : Left camshaft bracket
I : Intake side
J : Exhaust side

22140_IG35_G0041

Fig. 58 Camshaft bracket installation

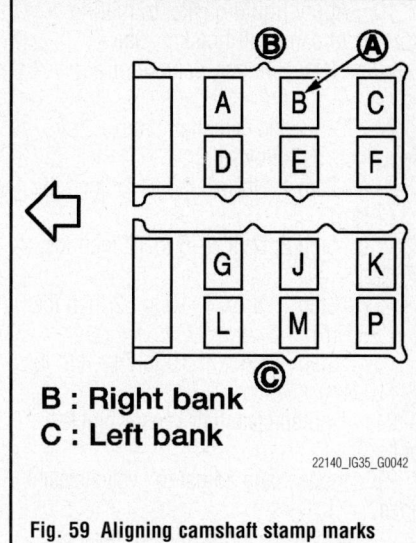

**B : Right bank
C : Left bank**

22140_IG35_G0042

Fig. 59 Aligning camshaft stamp marks

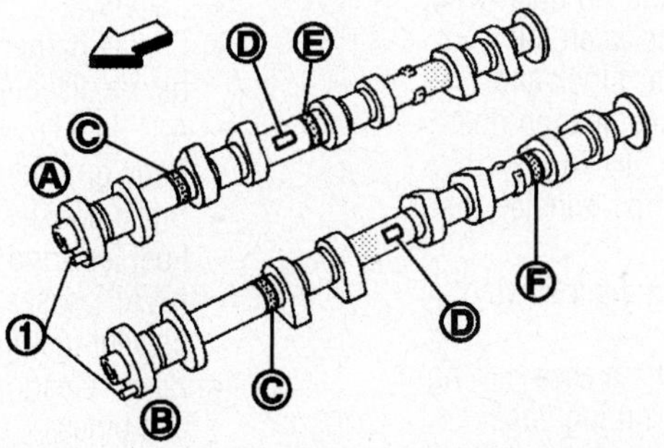

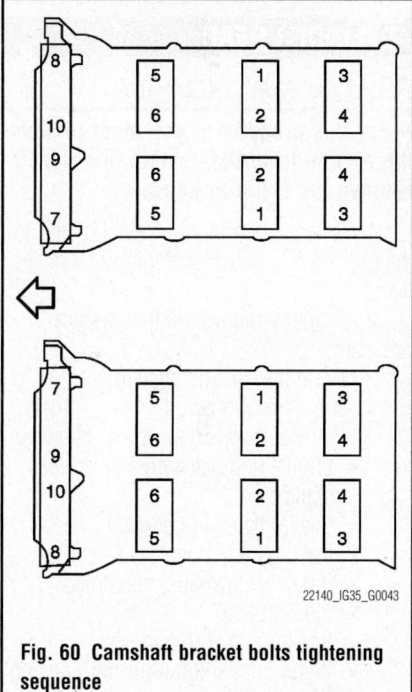

22140_IG35_G0043

Fig. 60 Camshaft bracket bolts tightening sequence

part facing upward on left-side cylinder head.

14. Install valve lifters, in their original positions

15. Follow your identification marks made during removal, or follow the identification marks that are present on new camshafts for proper placement and direction.

16. Install camshafts with the dowel pin attached to its front end face on the exhaust side.

17. Install camshaft brackets.

18. Install camshaft brackets (No. 2 to 4) aligning the stamp marks. There are no identification marks indicating left and right for camshaft bracket (No. 1).

Bank	INT/EXH	Dowel pin (1)	Paint marks			Identification mark (D)
			M1 (E)	M2 (F)	M3 (C)	
RH	EXH (B)	Yes	No	Green	Light blue	1F
	INT (A)	Yes	Green	No	Light blue	1E
LH	INT (A)	Yes	Green	No	Light blue	1G
	EXH (B)	Yes	No	Green	Light blue	1H

22140_IG35_G0040

Fig. 57 Camshaft identification and installation

19. Apply liquid gasket to mating surface of camshaft bracket. Use Genuine RTV Silicone Sealant or equivalent.

20. Torque the camshaft bracket bolts, in sequence, as follows:

   a. Step 1: No's. 7–10 to 17 inch lbs. (2 Nm).

   b. Step 2: No's. 1–6 to 17 inch lbs. (2 Nm).

   c. Step 3: No's. 1–10 to 52 inch lbs. (5.8 Nm).

   d. Step 4: No's. 1–10 to 74– 8 ft. lbs. (10 Nm).

21. Tighten camshaft sensor bracket bolts.

22. Inspect and adjust the valve clearance.

23. To complete installation, reverse remaining removal procedure.

CYLINDER HEAD

REMOVAL & INSTALLATION

➡ **For this procedure, you must remove the engine from the vehicle in order to remove the cylinder head.**

1. Before servicing the vehicle, refer to the precautions in the beginning of this section.

2. Properly relieve the fuel system pressure.

3. Drain the cooling system.

4. Drain the engine oil.

5. Remove or disconnect the following:
- Engine and place on a suitable stand
- Intake manifold collector
- Fuel rail and injector assembly
- Intake and exhaust manifolds
- Ignition coil
- Rocker arm (valve) cover
- Water inlet and thermostat housing
- Water outlet and hoses
- Upper and lower oil pans and strainer
- Front timing chain case, timing chain and rear timing chain case
- Camshaft
- Cylinder head bolts. Loosen in several steps, in the sequence shown.

➡ **A warped or cracked cylinder head could result from removing the bolts in incorrect order.**

- Cylinder heads from the vehicle
- Discard the head gaskets

6. Remove all traces of liquid gasket from the timing chain case and from the water pump covers.

7. Remove all traces of liquid gasket from the engine block.

8. Inspect the timing chain for excessive wear or damage and replace as necessary.

To install:

9. Turn the crankshaft until the No. 1 piston is a Top Dead Center (TDC) on compression stroke. The crankshaft key should face toward the right bank.

10. Using new head gaskets, install the cylinder heads.

➡ **If possible, replacement of the head bolts is suggested.**

11. If replacement of the head bolts is not possible, perform the following bolt measurement:

   a. Measure the diameter of the head bolt 0.43 in. (11mm) from the bottom of the bolt.

   b. Measure the diameter of the head bolt 1.89 in. (48mm) from the bottom of the bolt.

   c. Whenever the size difference between the 2 measurements exceeds 0.0043 in. (0.11mm) the head bolts must be replaced.

12. Install the cylinder head bolts and torque in sequence (Refer to Torque Specifications Chart in beginning of chapter for torque values and procedures)

13. After installing the cylinder head, measure the distance between the front end faces of the cylinder block and cylinder head. If the specification does not fall within 0.555–0.587 in. (14.1–14.9mm), you must reinstall the cylinder head.

14. Install or connect the following:
- Camshaft
- Front timing chain case, timing chain and rear timing chain case
- Oil strainer and upper and lower oil pans
- Water outlet and hoses
- Thermostat housing and water inlet
- Rocker arm (valve) cover
- Ignition coil
- Intake and exhaust manifolds
- Fuel rail and injector assembly
- Intake manifold collector
- Engine into the vehicle
- Negative battery cable

15. Fill the cooling system.

16. Fill the engine with clean oil.

17. Start the vehicle, check for leaks and repair if necessary.

ENGINE ASSEMBLY

REMOVAL & INSTALLATION

2WD Models

See Figures 61 through 64.

1. Before servicing the vehicle, refer to the precautions in the beginning of this section.

2. Evacuate the A/C system.

3. Release the fuel system pressure.

4. Drain the engine oil.

5. Drain the cooling system.

6. Drain the transaxle fluid.

7. Remove or disconnect the following:
- Negative battery cable
- Hood
- Engine cover
- Battery cover
- Engine undercover
- Front wheels
- Wiper arm and cowl top cover
- Air cleaner assembly, including air ducts
- Fan, radiator shroud, radiator, reservoir tank and hoses
- Heater hose from the engine and plug to avoid leaks
- Ground wire from left hand cylinder head
- Positive battery cable from the vehicle and temporarily fasten it on the engine
- Battery
- Engine harness connector; tag before disconnecting
- A/C lines from the A/C compressor
- Body ground cables
- Brake booster vacuum hose
- Fuel feed and Evaporative Emission (EVAP) hoses. Plug the fuel lines to prevent fuel from leaking out.
- Power steering pump reservoir tank and pipes

8. Disconnect the connectors from the passenger compartment as follows:

   a. Remove the passenger side kick plate, dashboard side trim and glove box.

   b. From the engine compartment, detach the connectors from the Transmission Control Module (TCM), and Engine Control Module (ECM).

   c. Unfasten the wire harnesses, pull the harnesses out into the engine compartment, and then temporarily secure them to the engine. Cover all connectors with plastic or similar material to protect them.

9. Remove or disconnect the following:
- Front exhaust pipe from the manifold

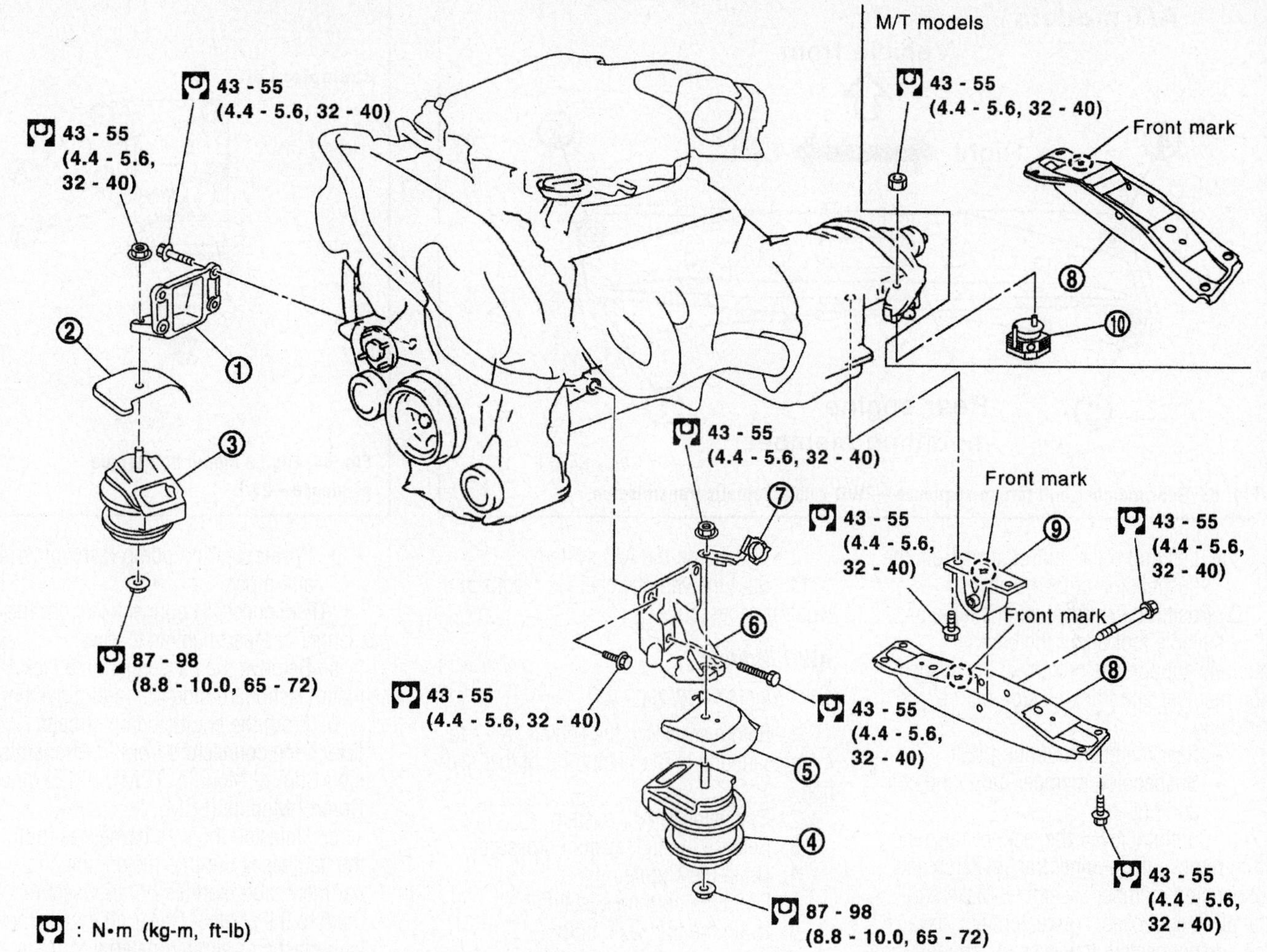

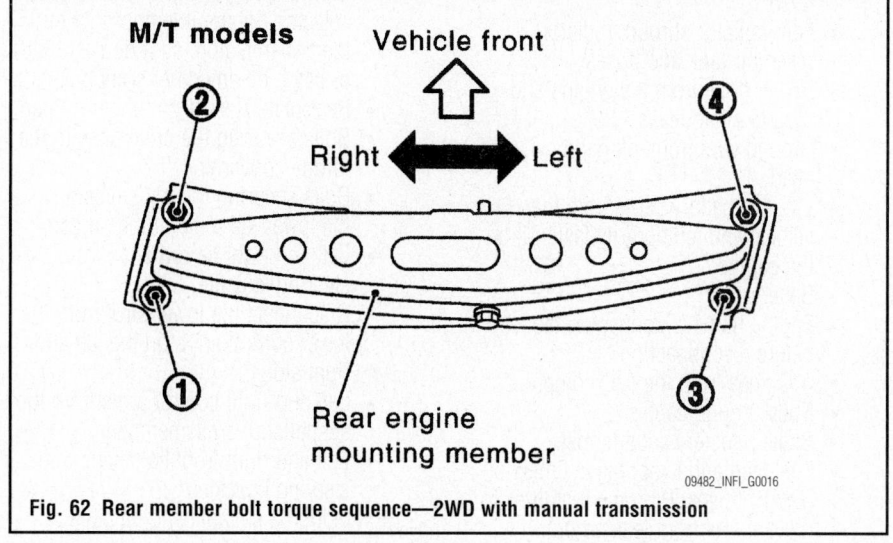

43 - 55 (4.4 - 5.6, 32 - 40)

43 - 55 (4.4 - 5.6, 32 - 40)

43 - 55 (4.4 - 5.6, 32 - 40)

M/T models

43 - 55 (4.4 - 5.6, 32 - 40)

Front mark

43 - 55 (4.4 - 5.6, 32 - 40)

87 - 98 (8.8 - 10.0, 65 - 72)

43 - 55 (4.4 - 5.6, 32 - 40)

Front mark

43 - 55 (4.4 - 5.6, 32 - 40)

43 - 55 (4.4 - 5.6, 32 - 40)

Front mark

43 - 55 (4.4 - 5.6, 32 - 40)

87 - 98 (8.8 - 10.0, 65 - 72)

43 - 55 (4.4 - 5.6, 32 - 40)

: N•m (kg-m, ft-lb)

1. Engine mounting bracket (RH)
2. Heat insulator (RH)
3. Engine mounting insulator (RH)
4. Engine mounting insulator (LH)
5. Heat insulator (LH)
6. Engine mounting bracket (LH)
7. Harness bracket
8. Rear engine mounting member
9. Engine mounting insulator (rear)
10. Dynamic damper

09482_INFI_G0025

Fig. 61 Exploded view of the engine mounts and related components—2WD

M/T models

Vehicle front

Right ⟷ Left

Rear engine mounting member

09482_INFI_G0016

Fig. 62 Rear member bolt torque sequence—2WD with manual transmission

- Steering lower joint and release the steering shaft
- Propeller shaft from the transmission
- Shift control linkage from the gear selector. Secure it temporarily to the transmission, so it doesn't drag or catch on any other components.
- Rear plate cover from upper oil pan
- Bolts securing the drive plate to the torque converter
- Bolts securing the transmission to the lower rear side of the oil pan
- Front stabilizer shaft
- Left and right tie rod ends from the steering knuckle
- Disconnect the lower strut from the lower control arms on the left and right sides

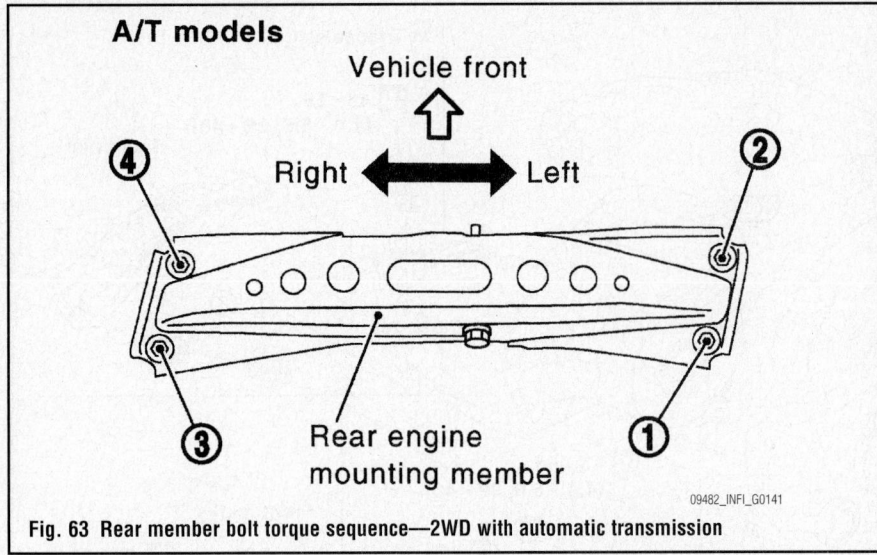

Fig. 63 Rear member bolt torque sequence—2WD with automatic transmission

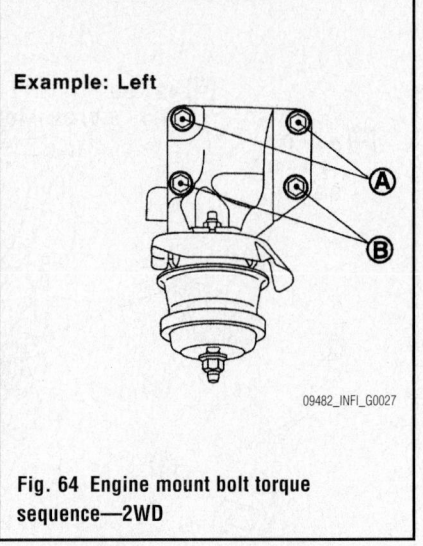

Fig. 64 Engine mount bolt torque sequence—2WD

- Left and right control arms from the suspension crossmember

10. Position a suitable engine table or other suitable tool under the engine. Securely support the bottom of the suspension member and transmission before removing:
  - Rear member mounting bolt
  - Suspension member mounting bolt and nut

11. Carefully lower the jack, or raise the lift to remove the engine, transmission and suspension member assembly. Make sure that all lines, hoses, connectors etc. have been disconnected. If necessary, support the rear of the vehicle at the rear jacking point, as its center of gravity has changed with the engine removed.

12. Remove and components and separate the engine and transmission as necessary.

To install:

13. Installation is the reverse of the removal procedure, noting the following points:
  - Tighten all engine mounts and related components to the specifications shown in the accompanying figure
  - Tighten the rear member mounting bolts in sequence.
  - When installing engine mounting brackets on cylinder block, tighten two upper bolts ("A" in the figure) first, then tighten two lower bolts ("B" in the figure).
  - Make sure to connect all vacuum hoses, lines, and electrical connectors as tagged during removal.
  - Fill the cooling system.
  - Fill the engine with clean oil.
  - Fill the transmission to the proper level.

- Recharge the A/C system.

14. Start the vehicle, check for leaks and repair if necessary.

AWD Models

See Figures 65 through 67.

1. Before servicing the vehicle, refer to the precautions in the beginning of this section.

2. Evacuate the A/C system.

3. Release the fuel system pressure.

4. Drain the engine oil.

5. Drain the cooling system.

6. Drain the transaxle fluid.

7. Remove or disconnect the following:
  - Negative battery cable
  - Hood
  - Engine cover
  - Battery cover
  - Engine undercover
  - Front wheels
  - Wiper arm and cowl top cover
  - Air cleaner assembly, including air ducts
  - Fan, radiator shroud, radiator, reservoir tank and hoses
  - Heater hose from the engine and plug to avoid leaks
  - Ground wire from left hand cylinder head
  - Positive battery cable from the vehicle and temporarily fasten it on the engine
  - Battery
  - Engine harness connector; tag before disconnecting
  - A/C lines from the A/C compressor
  - Body ground cables
  - Brake booster vacuum hose
  - Fuel feed and Evaporative Emission (EVAP) hoses. Plug the fuel lines to prevent fuel from leaking out.

- Power steering pump reservoir tank and pipes

8. Disconnect the connectors from the passenger compartment as follows:
  a. Remove the passenger side kick plate, dashboard side trim and glove box.
  b. From the engine compartment, detach the connectors from the Transmission Control Module (TCM), and Engine Control Module (ECM).
  c. Unfasten the wire harnesses, pull the harnesses out into the engine compartment, and then temporarily secure them to the engine. Cover all connectors with plastic or similar material to protect them.

9. Remove or disconnect the following:
  - Front exhaust pipe from the manifold
  - Steering lower joint and release the steering shaft
  - Propeller shaft from the transmission
  - Shift control linkage from the gear selector. Secure it temporarily to the transmission, so it doesn't drag or catch on any other components.
  - Rear plate cover from upper oil pan
  - Bolts securing the drive plate to the torque converter
  - Bolts securing the transmission to the lower rear side of the oil pan
  - Left and right tie rod ends from the steering knuckle
  - Disconnect the lower strut from the lower control arms on the left and right sides
  - Left and right control arms from the suspension crossmember
  - Left and right front halfshafts from steering knuckles

10. Position a suitable engine table or

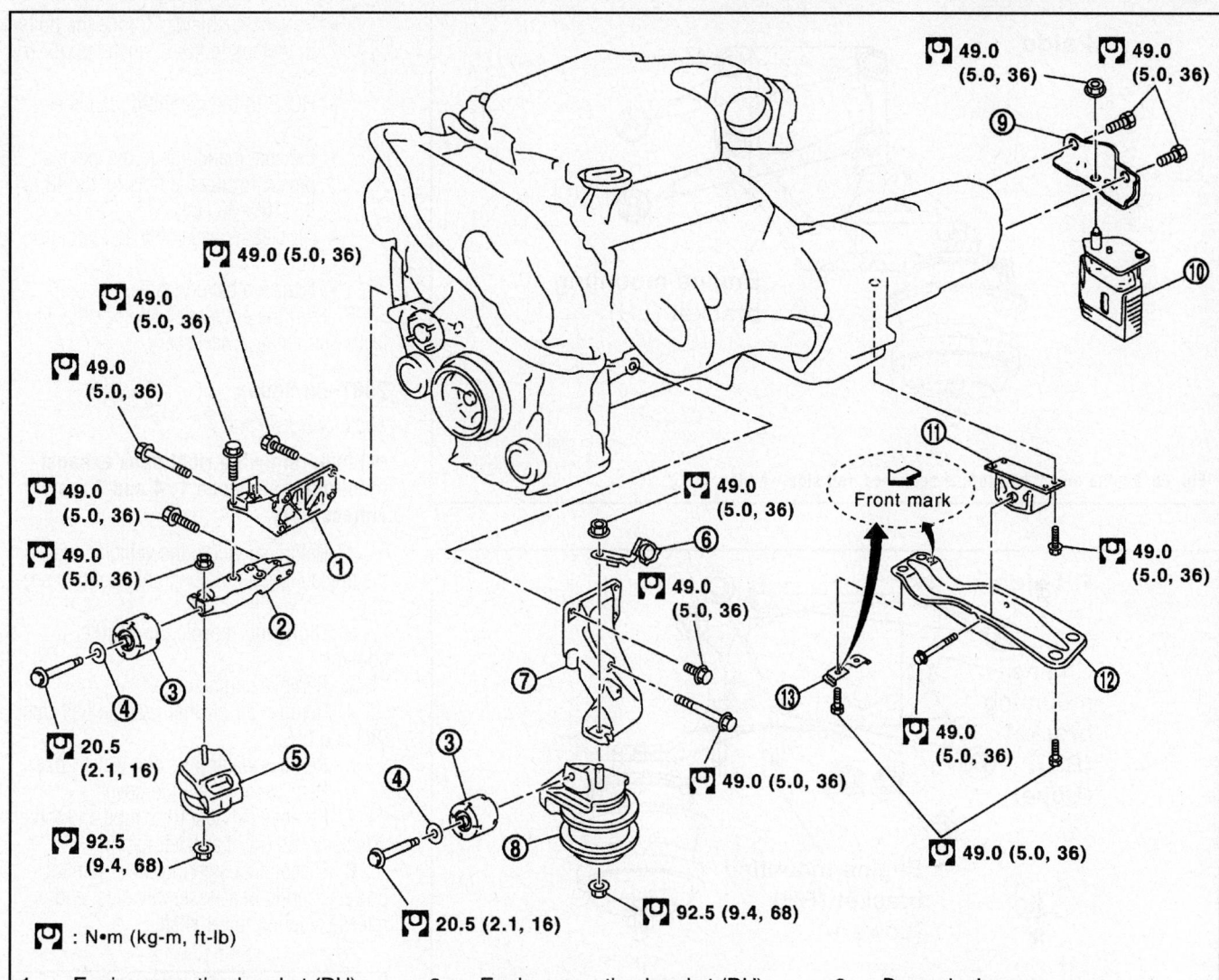

⌷ : N•m (kg-m, ft-lb)

1. Engine mounting bracket (RH) (Upper)
2. Engine mounting bracket (RH) (Lower)
3. Dynamic damper
4. Washer
5. Engine mounting insulator (RH)
6. Harness bracket
7. Engine mounting bracket (LH)
8. Engine mounting insulator (LH)
9. Dynamic damper bracket
10. Dynamic damper
11. Engine mounting insulator (Rear)
12. Rear engine mounting member
13. Heat insulator

09482_INFI_G0026

Fig. 65 Exploded view of the engine mounts and related components—AWD

other suitable tool under the engine. Securely support the bottom of the suspension member and transmission before removing:

• Rear member mounting bolt
• Suspension member mounting bolt and nut

11. Carefully lower the jack, or raise the lift to remove the engine, transmission and suspension member assembly. Make sure that all lines, hoses, connectors etc. have been disconnected. If necessary, support the rear of the vehicle at the rear jacking point, as its center of gravity has changed with the engine removed.

12. Remove and components and separate the engine and transmission as necessary.

To install:

13. Installation is the reverse of the removal procedure, noting the following points:

• Tighten all engine mounts and related components to the specifications shown in the accompanying figure
• When installing left engine mounting bracket on cylinder block, tighten two upper bolts ("A" in the

figure) first, then tighten two lower bolts ("B" in the figure).

• The right engine mounting bracket comprises two pieces. Install right upper engine mounting bracket on cylinder block, tightening two upper bolts ("A" in the figure) first, then tightening two lower bolts ("B" in the figure). Make certain right lower engine mounting bracket is in full contact with the upper mounting bracket and the front final drive assembly before tightening bolts "D" and "E" in sequence.

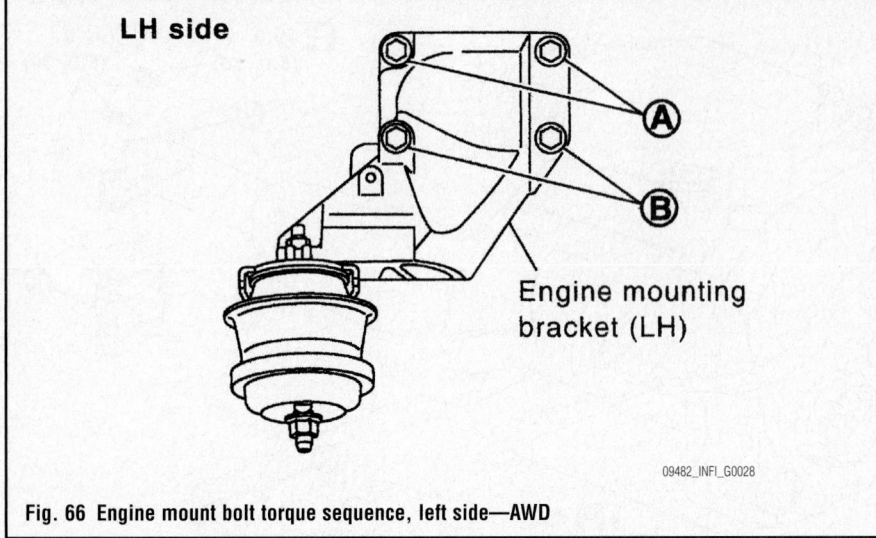

Fig. 66 Engine mount bolt torque sequence, left side—AWD

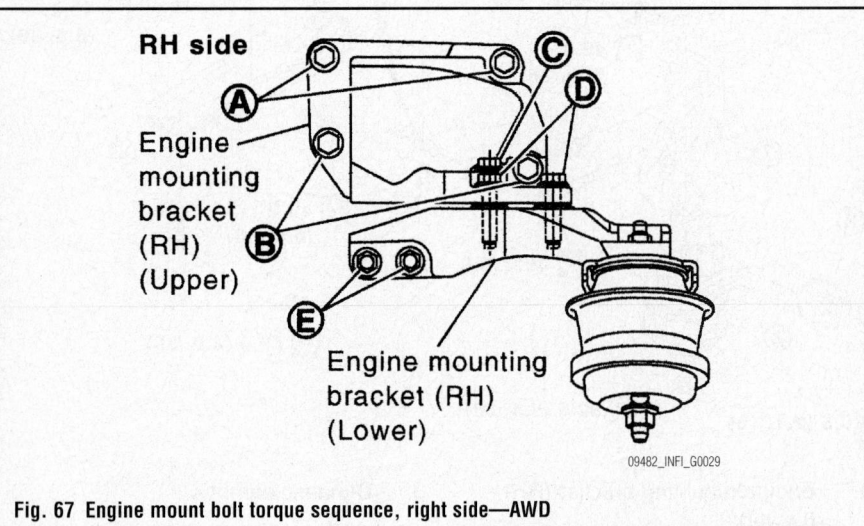

Fig. 67 Engine mount bolt torque sequence, right side—AWD

- Make sure to connect all vacuum hoses, lines, and electrical connectors as tagged during removal.
- Fill the cooling system.
- Fill the engine with clean oil.
- Fill the transmission to the proper level.
- Recharge the A/C system.

14. Start the vehicle, check for leaks and repair if necessary.

EXHAUST MANIFOLD

REMOVAL & INSTALLATION

2006 Coupe and Sedan; 2007 Coupe

See Figures 68 and 69.

1. Before servicing the vehicle, refer to the precautions in the beginning of this section.
2. Remove or disconnect the following:
   - Negative battery cable

➥If necessary, soak the exhaust pipe retaining nuts with penetrating oil to loosen them.

- Engine cover
- Air cleaner assembly and duct
- Front exhaust pipe from the exhaust manifolds
- Heated Oxygen Sensors (HO2S) from the manifold, as necessary
- Protective covers from the manifolds
- Exhaust manifold-to-engine mounting nuts
- Manifold from the engine and discard the gaskets

To install:

3. Clean all gasket mounting surfaces.
4. Install or connect the following:
   - Exhaust manifold with new gaskets. Torque the bolts, in sequence, in 2 steps to 21–23 ft. lbs. (28–32 Nm).

- Protective shields. Torque the bolts in 2 steps to 46–57 inch lbs. (5–6 Nm).
- HO2S to the manifold, as necessary
- Exhaust manifolds to the exhaust pipes. Torque the nuts to 45–48 ft. lbs. (60–66 Nm).
- Air cleaner assembly and engine cover
- Negative battery cable

5. Start the engine, check for exhaust leaks and repair if necessary.

2007–08 Sedan

See Figures 70 and 71.

➥When removing right bank exhaust manifold, only steps 1, 4 and 7 are unnecessary.

1. Before servicing the vehicle, refer to the precautions in the beginning of this section.
2. Drain and recycle the engine coolant.
3. Remove engine cover.
4. Remove air cleaner case and air duct (RH and LH).
5. Remove water pipe and water hose.
6. Remove engine undercover.
7. Remove exhaust front tube and three way catalysts (right and left bank).
8. Disconnect steering lower joint at power steering gear assembly side, and release steering lower shaft.

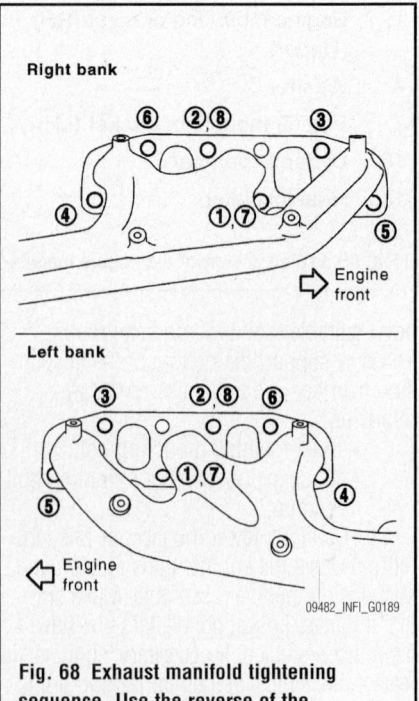

Fig. 68 Exhaust manifold tightening sequence. Use the reverse of the sequence for removal.

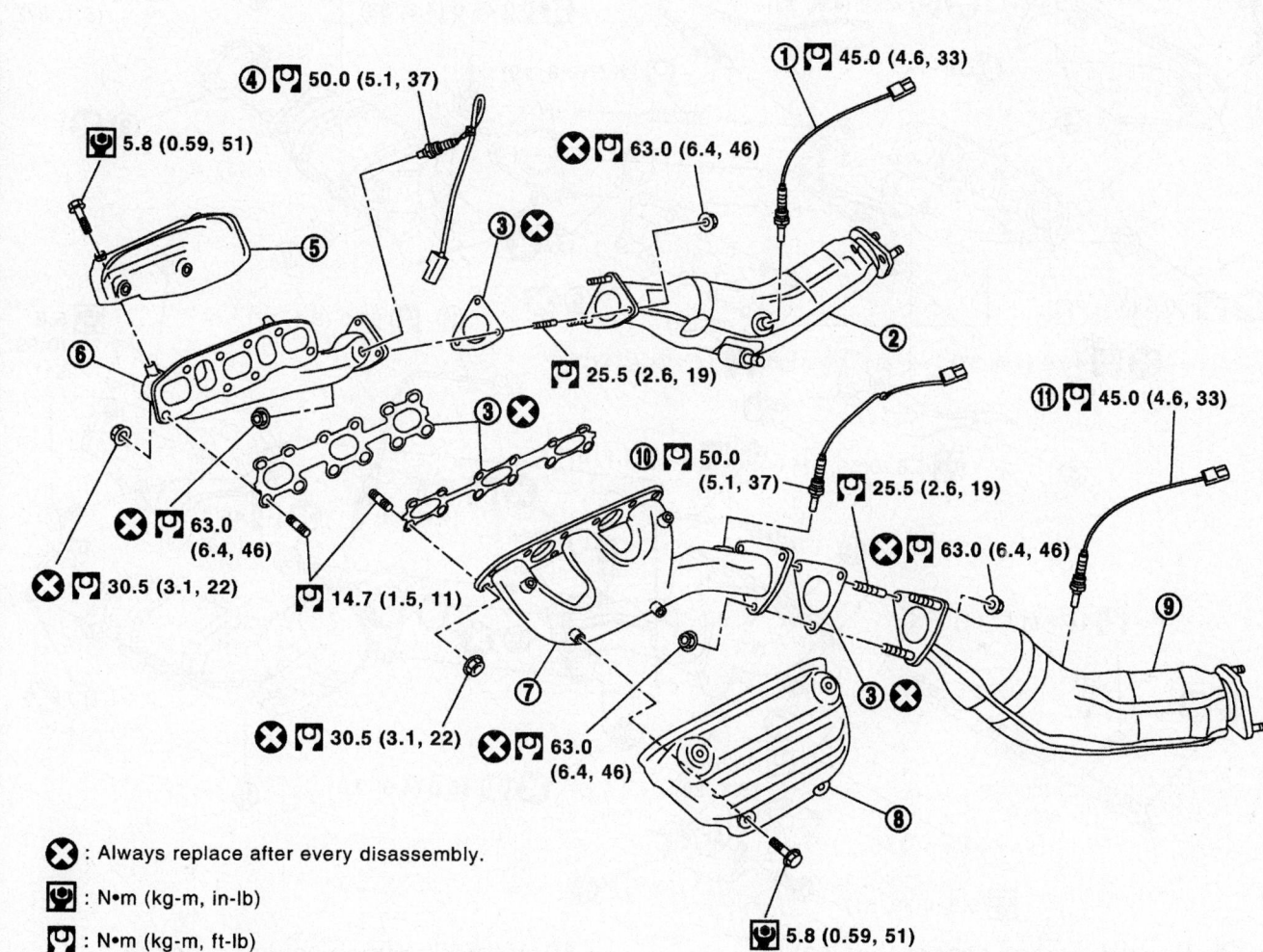

④ ⚙ 50.0 (5.1, 37)

⚙ 5.8 (0.59, 51)

① ⚙ 45.0 (4.6, 33)

❌ ⚙ 63.0 (6.4, 46)

③ ❌

⚙ 25.5 (2.6, 19)

②

⑤

⑥

❌ ⚙ 63.0 (6.4, 46)

❌ ⚙ 30.5 (3.1, 22)

③ ❌

⑩ ⚙ 50.0 (5.1, 37)

⚙ 25.5 (2.6, 19)

⑪ ⚙ 45.0 (4.6, 33)

❌ ⚙ 63.0 (6.4, 46)

⑨

⚙ 14.7 (1.5, 11)

⑦

③ ❌

❌ ⚙ 30.5 (3.1, 22)

❌ ⚙ 63.0 (6.4, 46)

⑧

⚙ 5.8 (0.59, 51)

❌ : Always replace after every disassembly.

⚙ : N•m (kg-m, in-lb)

⚙ : N•m (kg-m, ft-lb)

1.	Heated oxygen sensor 2 (bank 1)	2.	Three way catalyst (right bank)	3.	Gasket
4.	Air fuel ratio sensor 1 (bank 1)	5.	Exhaust manifold cover (right bank)	6.	Exhaust manifold (right bank)
7.	Exhaust manifold (left bank)	8.	Exhaust manifold cover (left bank)	9.	Three way catalyst (left bank)
10.	Air fuel ratio sensor 1 (bank 2)	11.	Heated oxygen sensor 2 (bank 2)		

09482_INFI_G0190

Fig. 69 Exploded view of the exhaust manifold and related components

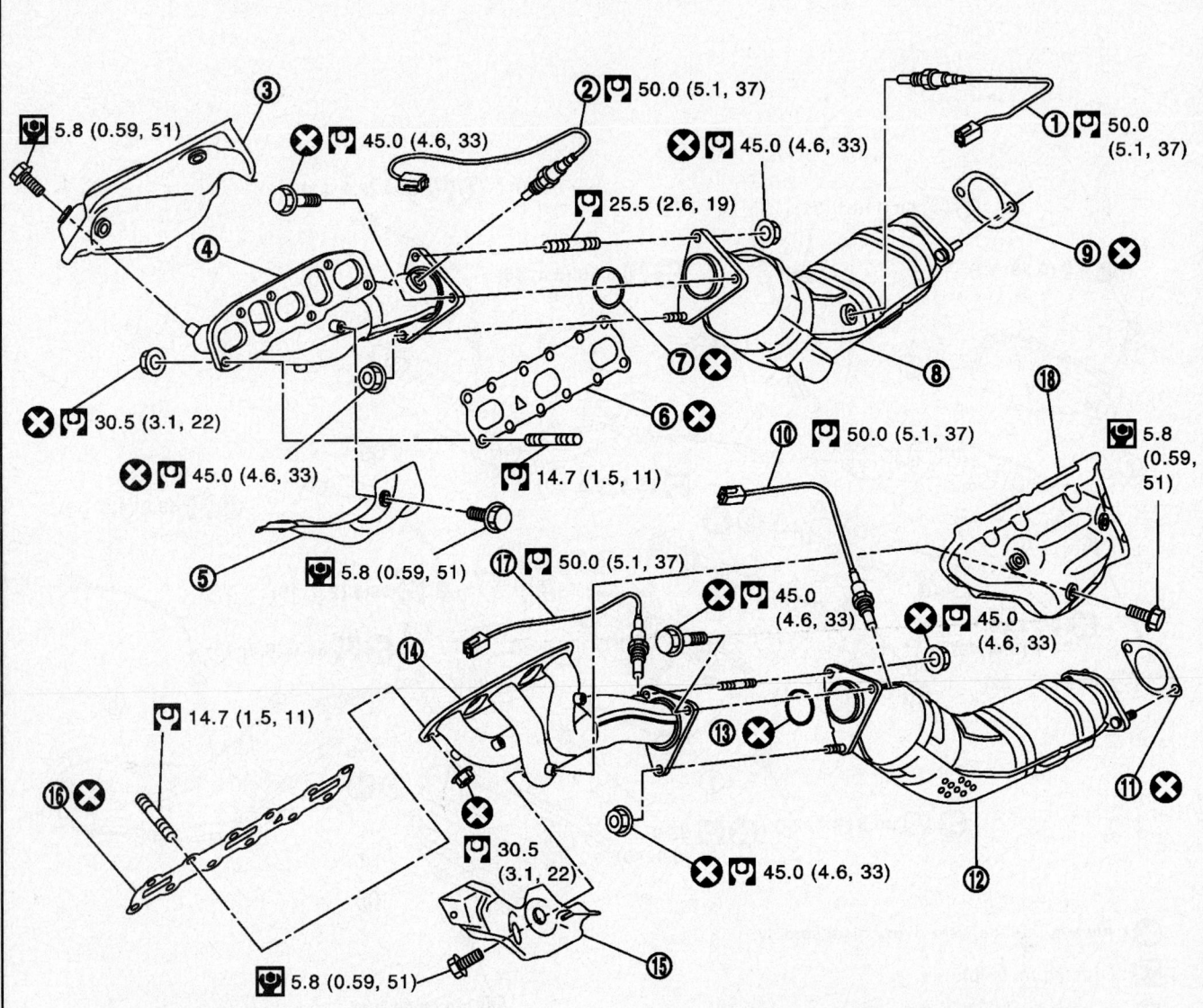

1. Heated oxygen sensor (bank 1)
2. Air fuel ratio sensor (bank 1)
3. Exhaust manifold cover (upper)
4. Exhaust manifold (right bank)
5. Exhaust manifold cover (lower) 6. Gasket
7. Ring gasket
8. Three way catalyst (right bank)
9. Gasket

10. Heated oxygen sensor (bank 2)
11. Gasket
12. Three way catalyst (left bank)
13. Ring gasket
14. Exhaust manifold (left bank)
15. Exhaust manifold cover (lower)
16. Gasket
17. Air fuel ratio sensor (bank 2)
18. Exhaust manifold cover (upper)

22140_IG35_G0088

Fig. 70 Exploded view of the exhaust manifold and related components

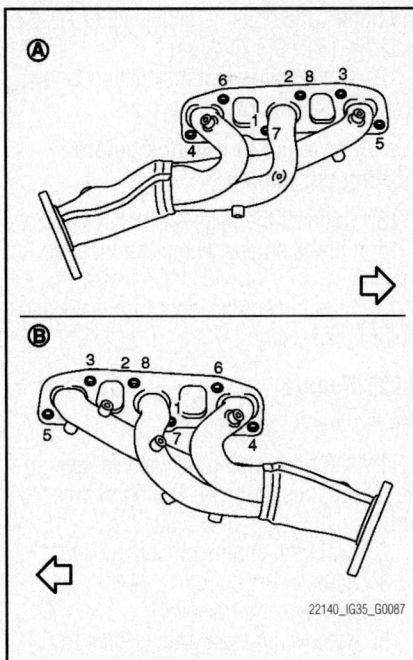

Fig. 71 Exhaust manifold removal and tightening sequence

9. Disconnect air fuel ratio sensor (bank 1 and bank 2) harness connectors and remove harness clip.

10. Remove the HO2S sensor.

11. Remove the upper exhaust manifold cover (right and left bank).

12. Loosen the mounting nuts in the reverse order as shown in the figure to remove exhaust manifold. Disregard No. 7 and 8 in the removal.

13. Remove gaskets and cover openings.

To install:

14. To install, reverse removal procedure.

15. Tighten nuts No. 1 and 2 in two steps. The numerical order No. 7 and 8 shows second step.

INTAKE MANIFOLD

REMOVAL & INSTALLATION

2006 Coupe and Sedan; 2007 Coupe

See Figures 72 and 73.

1. Before servicing the vehicle, refer to the precautions in the beginning of this section.

2. Properly relieve the fuel system pressure.

3. Remove intake manifold collectors (upper and lower).

4. Remove fuel tube and fuel injector assembly.

5. Loosen bolts and nuts in reverse order as shown in the figure to remove intake manifold.

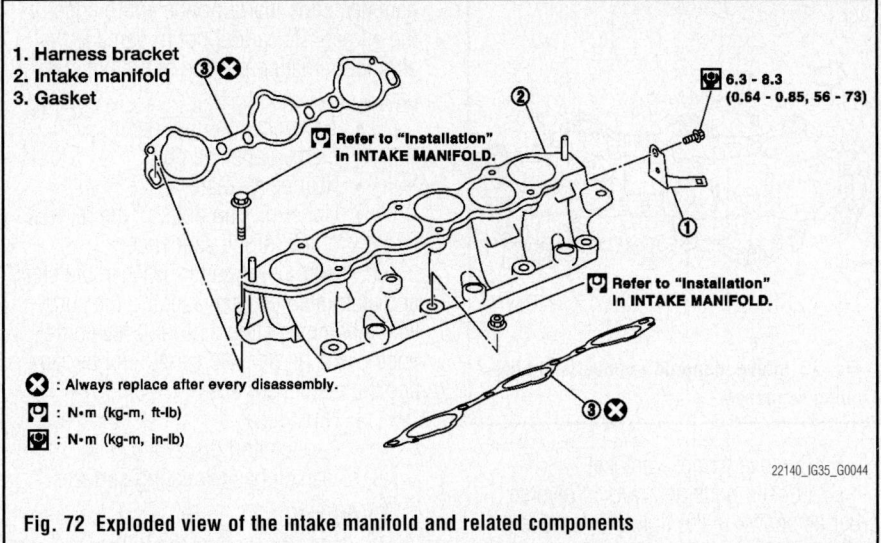

1. Harness bracket
2. Intake manifold
3. Gasket

⊗ : Always replace after every disassembly.
🔧 : N•m (kg-m, ft-lb)
🔧 : N•m (kg-m, in-lb)

Fig. 72 Exploded view of the intake manifold and related components

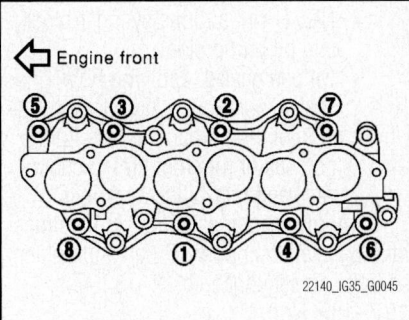

◁ Engine front

Fig. 73 Intake manifold removal and tightening sequence

6. Remove intake manifold gaskets and cover engine openings to avoid entry of foreign materials.

To install:

7. To install, reverse removal procedure.

8. If stud bolts were removed, install them and tighten to 87–104 inch lbs. (10–12 Nm).

9. Tighten all mounting bolts and nuts to the specified torque in two or more steps in sequence:

    a. First step: 4–7 ft. lbs. (5–10 Nm).

    b. Second step: 20–23 ft. lbs. (27–31 Nm).

2007–08 Sedan

See Figures 74 and 75.

1. Before servicing the vehicle, refer to the precautions in the beginning of this section.

2. Properly relieve the fuel system pressure.

3. Remove intake manifold collectors (upper and lower).

4. Remove fuel tube and fuel injector assembly.

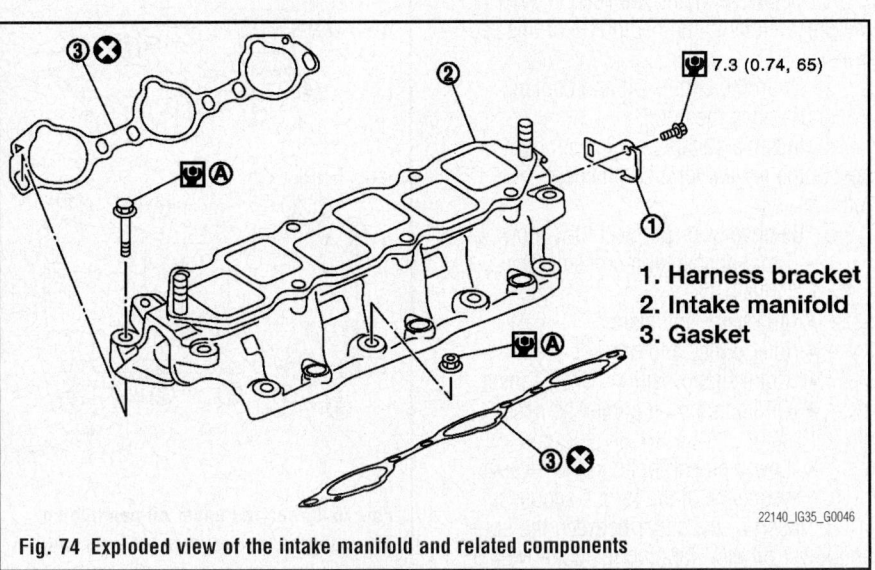

1. **Harness bracket**
2. **Intake manifold**
3. **Gasket**

Fig. 74 Exploded view of the intake manifold and related components

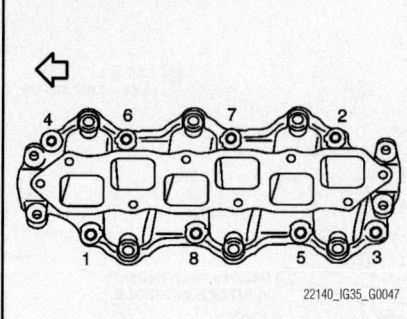

Fig. 75 Intake manifold removal and tightening sequence

5. Remove harness bracket.

6. Loosen bolts and nuts in reverse order as shown in the figure to remove intake manifold.

7. Remove intake manifold gaskets and cover engine openings to avoid entry of foreign materials.

To install:

8. To install, reverse removal procedure.

9. If stud bolts were removed, install them and tighten to 87 inch lbs. (10 Nm).

10. Tighten all mounting bolts and nuts to the specified torque in two or more steps in sequence:

   a. First step: 5 ft. lbs. (7.4 Nm).

   b. Second step: 21 ft. lbs. (30 Nm).

OIL PAN

REMOVAL & INSTALLATION

2006 Models

2WD Models

See Figures 76 through 80.

1. Before servicing the vehicle, refer to the precautions in the beginning of this section.

2. Drain the engine oil and coolant.

3. Remove the hood.

4. Install a suitable engine slinger to secure the engine for crossmember removal.

5. Remove or disconnect the following:

- Front suspension crossmember
- Drive belts
- Alternator and starter
- Idler pulley and bracket
- Crankshaft Position (CKP) sensor
- Oil filter and oil cooler, as necessary
- Lower oil pan bolts in the reverse sequence of the torque sequence

6. Insert a seal cutter between the upper and lower oil pan. Tapping the cutter with a hammer, slide it around the entire edge of the oil pan. Be careful not to damage the aluminum mating surface of the upper oil pan.

- Transmission joint bolts which pierce upper oil pan
- Rear cover plate
- Upper oil pan bolts in the reverse of the torque sequence

7. Insert a seal cutter between the steel and aluminum oil pan. Tapping the cutter with a hammer, slide it around the entire edge of the oil pan. Be careful not to damage the mating surfaces.

- Oil strainer
- O-rings and discard

8. Clean all gasket mating surfaces.

To install:

9. Install or connect the following:

- Oil strainer to the oil pump
- New O-rings to the cylinder block and oil pump side
- Oil pan gasket, applying RTV sealant as shown. Align the protrusion of the oil pan gasket with the notches of the front timing chain case and rear oil seal retainer.

10. Apply sealant as shown in the illustration. Install the upper oil pan and tighten the bolts, in sequence, to 12–13 ft. lbs. (15.7–18.6 Nm).

- Transmission joint bolts

11. For the lower oil pan, apply sealant as shown in the illustrations, then install

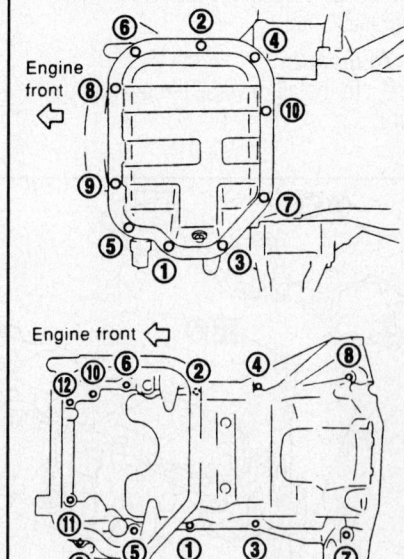

Fig. 76 Lower and upper oil pan tightening sequence (use reverse for loosening)

and tighten the bolts, in sequence to 74–82 inch lbs. (8.3–9.3 Nm).

12. The remainder of installation is the reverse of the removal procedure.

➡**Wait at least 30 minutes before refilling the engine oil.**

13. Connect the negative battery cable

14. Fill the engine with clean oil and coolant.

15. Start the engine, check for leaks and repair if necessary.

AWD Models

See Figures 81 through 86.

1. Before servicing the vehicle, refer to the precautions in the beginning of this section.

2. Drain the engine oil and coolant.

3. Remove the hood.

4. Remove engine cover

5. Remove air hose and air duct to mass air flow sensor side and electronic throttle control actuator side

6. Remove front and rear undercover

7. Install a suitable engine sling to secure the engine for crossmember removal.

8. Remove or disconnect the following:

- Drive belts
- Front suspension crossmember
- Left and right halfshafts
- Side (intermediate) shaft
- Left and right engine mounting brackets and insulators
- Front driveshaft
- Alternator and starter
- Idler pulley and bracket
- Oil filter and oil cooler, as necessary
- Automatic transmission oil cooler
- Crankshaft Position (CKP) sensor

➡**Handle sensor with care; do not allow metal powder to adhere to magnetic tip or place sensor in a location where it might be exposed to magnetism**

- Lower oil pan bolts in the reverse sequence of the torque sequence

9. Insert a seal cutter between the upper and lower oil pan. Tapping the cutter with a hammer, slide it around the entire edge of the oil pan. Be careful not to damage the aluminum mating surface of the upper oil pan.

10. Remove oil strainer

11. Remove transmission joint bolts that pierce upper oil pan

12. Remove rear plate cover

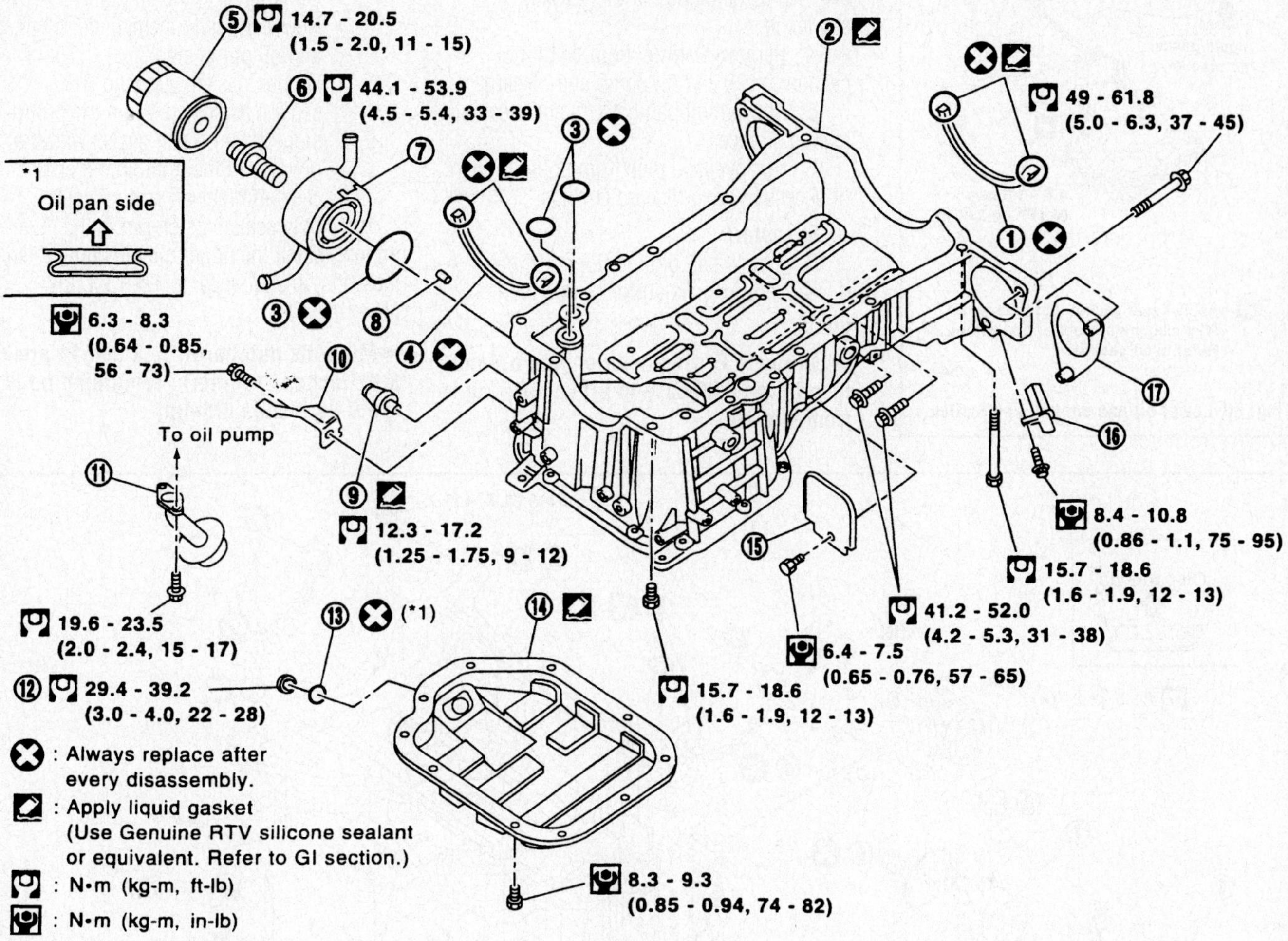

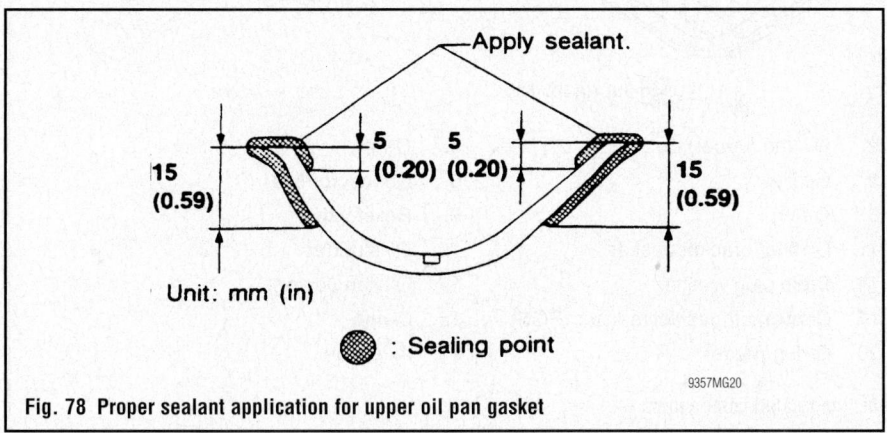

5 ⟦⟧ 14.7 - 20.5
(1.5 - 2.0, 11 - 15)

6 ⟦⟧ 44.1 - 53.9
(4.5 - 5.4, 33 - 39)

*1
Oil pan side

▣ 6.3 - 8.3
(0.64 - 0.85,
56 - 73)

To oil pump

⟦⟧ 49 - 61.8
(5.0 - 6.3, 37 - 45)

9 ▨ ⟦⟧ 12.3 - 17.2
(1.25 - 1.75, 9 - 12)

11

⟦⟧ 19.6 - 23.5
(2.0 - 2.4, 15 - 17)

12 ⟦⟧ 29.4 - 39.2
(3.0 - 4.0, 22 - 28)

13 ⊗ (*1)

14 ▨

▣ 8.4 - 10.8
(0.86 - 1.1, 75 - 95)

⟦⟧ 15.7 - 18.6
(1.6 - 1.9, 12 - 13)

⟦⟧ 41.2 - 52.0
(4.2 - 5.3, 31 - 38)

▣ 6.4 - 7.5
(0.65 - 0.76, 57 - 65)

⟦⟧ 15.7 - 18.6
(1.6 - 1.9, 12 - 13)

▣ 8.3 - 9.3
(0.85 - 0.94, 74 - 82)

⊗ : Always replace after
every disassembly.

▨ : Apply liquid gasket
(Use Genuine RTV silicone sealant
or equivalent. Refer to GI section.)

⟦⟧ : N•m (kg-m, ft-lb)

▣ : N•m (kg-m, in-lb)

1. Oil pan gasket	2. Oil pan (upper)	3. O-ring
4. Oil pan gasket	5. Oil filter	6. Connector bolt
7. Oil cooler	8. Relief valve	9. Oil pressure switch
10. Bracket	11. Oil strainer	12. Drain plug
13. Drain plug washer	14. Oil pan (lower)	15. Rear plate
16. Crankshaft position sensor (POS)	17. Rear cover plate	

9357MG18

Fig. 77 Exploded view of the upper and lower oil pans and related components

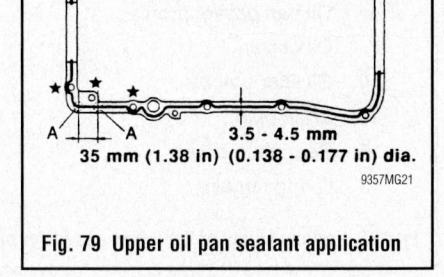

Apply sealant.

5
(0.20)   5
(0.20)

15
(0.59)   15
(0.59)

Unit: mm (in)

⬤ : Sealing point

9357MG20

Fig. 78 Proper sealant application for upper oil pan gasket

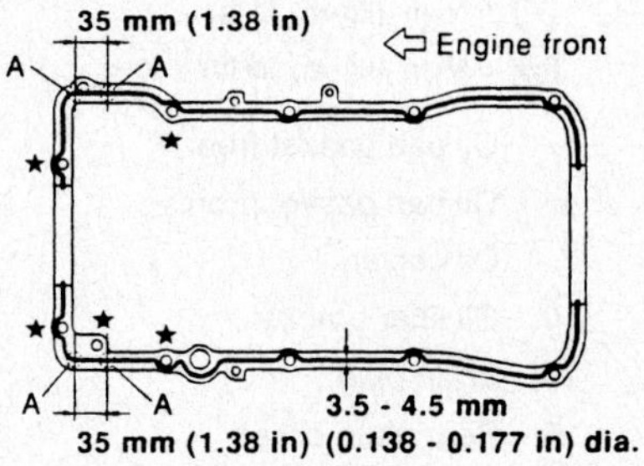

35 mm (1.38 in)   ⟸ Engine front

A   A

★   ★

★   ★

A   A

35 mm (1.38 in)   3.5 - 4.5 mm
(0.138 - 0.177 in) dia.

9357MG21

Fig. 79 Upper oil pan sealant application

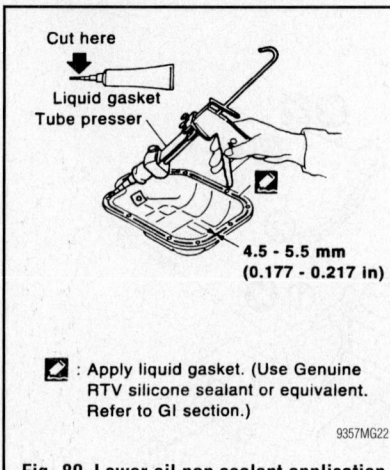

Cut here

Liquid gasket
Tube presser

4.5 - 5.5 mm
(0.177 - 0.217 in)

🖌 : Apply liquid gasket. (Use Genuine
RTV silicone sealant or equivalent.
Refer to GI section.)

9357MG22

Fig. 80 Lower oil pan sealant application

13. Remove upper oil pan bolts in the reverse sequence of the torque sequence

14. Remove O-rings from bottom of cylinder block and oil pump and discard

15. Remove oil pan gaskets and clean all mating surfaces

16. Remove axle pipe from upper oil pan (if necessary) and discard O-rings.

To install:

17. Install axle pipe to upper oil pan, if removed, from axle pipe flange side (left)

➡**Lubricate O-rings with engine oil and insert pipe with care to prevent O-ring from sliding**

18. Install or connect the following:
- Oil strainer to the oil pump
- New O-rings to the cylinder block and oil pump side
- Oil pan gasket, applying RTV sealant as shown. Align the protrusion of the oil pan gasket with the notches of the front timing chain case and rear oil seal retainer.

19. Apply sealant as shown in the illustration. Install the upper oil pan and tighten the bolts, in sequence, to 12–13 ft. lbs. (16–19 Nm).

➡**M8 bolts number 5, 7, 8 and 11 are 3.97 inches (100mm) ; remaining bolts are 0.98 inches (25mm)**

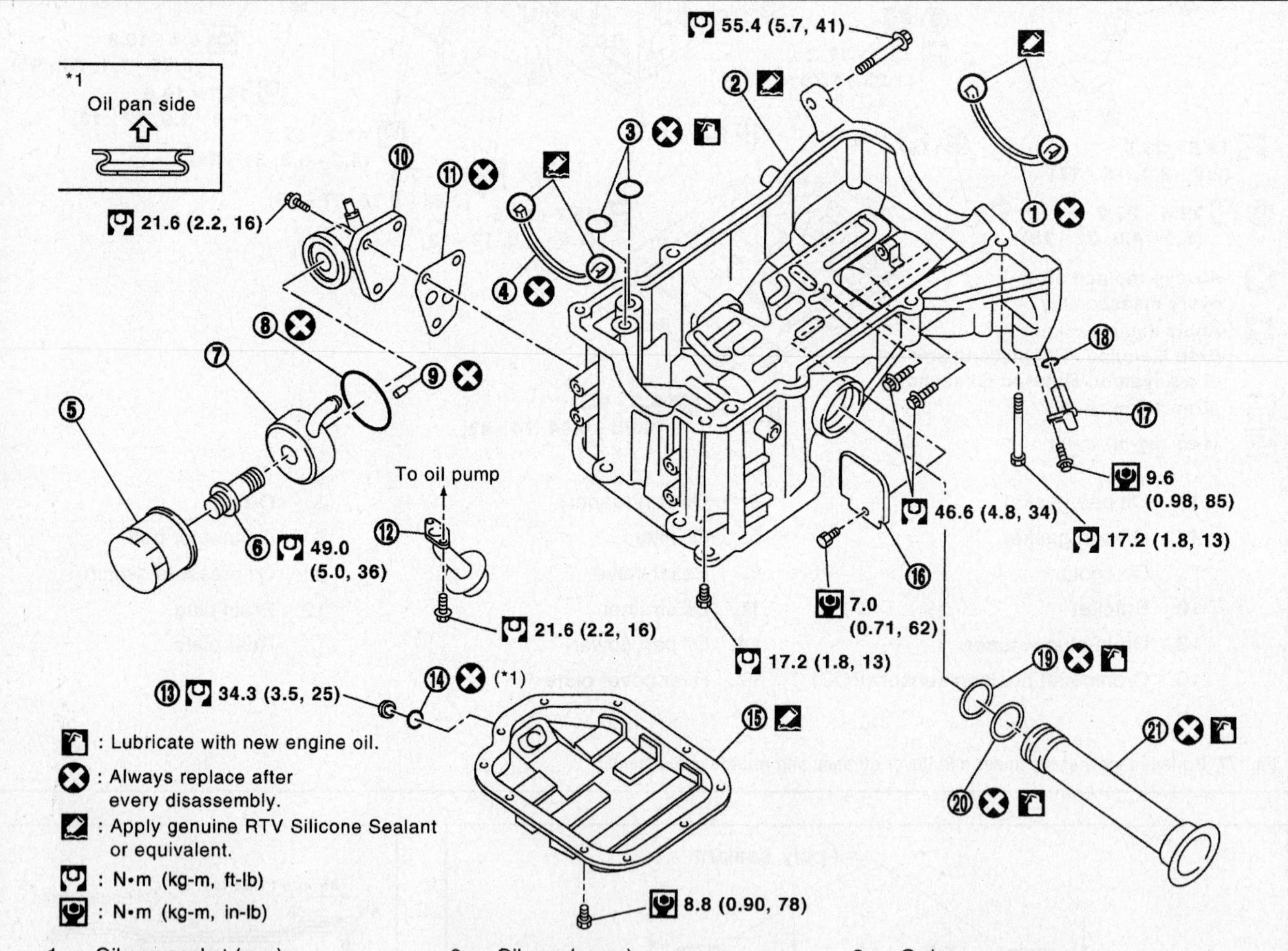

---

*1 Oil pan side

🔧 21.6 (2.2, 16)

🔧 55.4 (5.7, 41)

🔧 9.6 (0.98, 85)

🔧 17.2 (1.8, 13)

🔧 46.6 (4.8, 34)

🔧 7.0 (0.71, 62)

🔧 49.0 (5.0, 36)

🔧 21.6 (2.2, 16)

🔧 17.2 (1.8, 13)

🔧 34.3 (3.5, 25)

🔧 8.8 (0.90, 78)

To oil pump

🖐 : Lubricate with new engine oil.

❌ : Always replace after every disassembly.

🖌 : Apply genuine RTV Silicone Sealant or equivalent.

🔧 : N•m (kg-m, ft-lb)

🔧 : N•m (kg-m, in-lb)

1. Oil pan gasket (rear)	2. Oil pan (upper)	3. O-ring
4. Oil pan gasket (front)	5. Oil filter	6. Connector bolt
7. Oil cooler	8. O-ring	9. Relief valve
10. Oil filter bracket	11. Oil filter bracket gasket	12. Oil strainer
13. Drain plug	14. Drain plug washer	15. Oil pan (lower)
16. Rear plate cover	17. Crankshaft position sensor (POS)	18. O-ring
19. O-ring (small)	20. O-ring (large)	21. Axle pipe

09482_INFI_G0044

Fig. 81 Exploded view of the upper and lower oil pans and related components

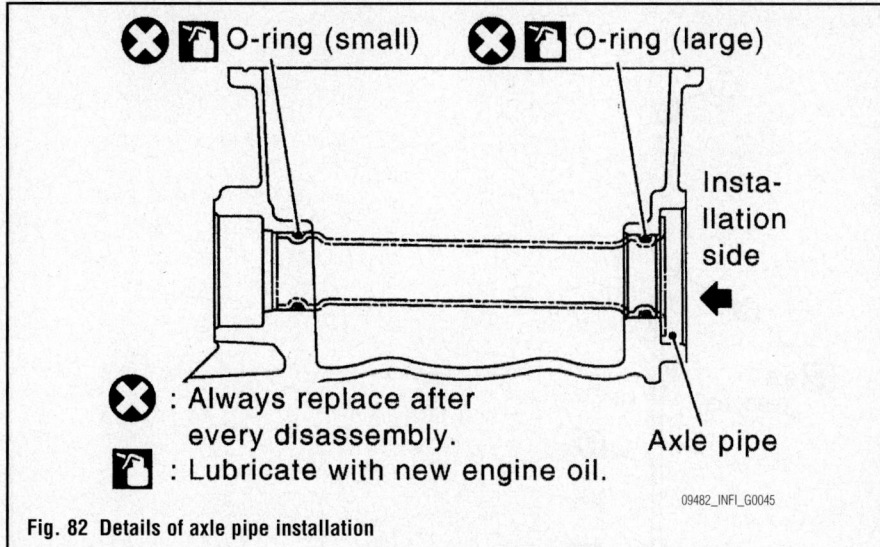

Fig. 82 Details of axle pipe installation

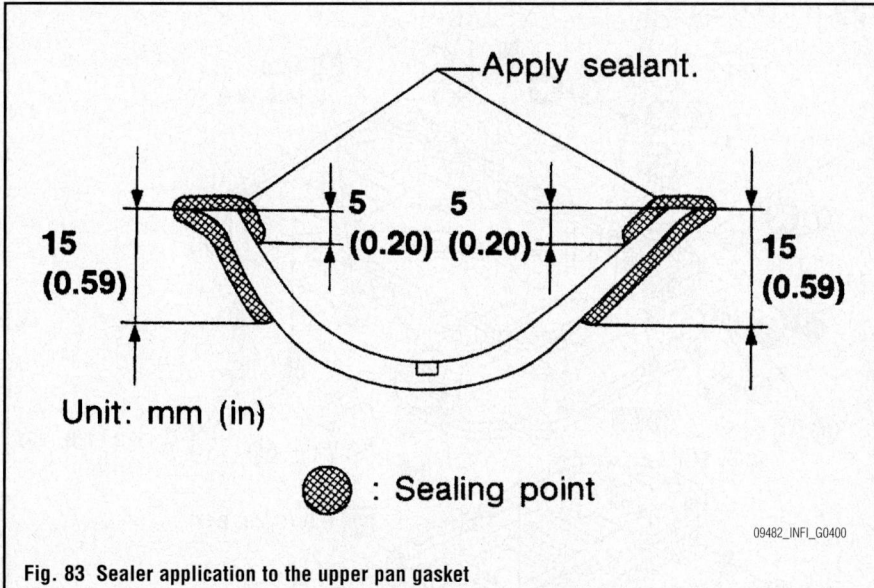

Fig. 83 Sealer application to the upper pan gasket

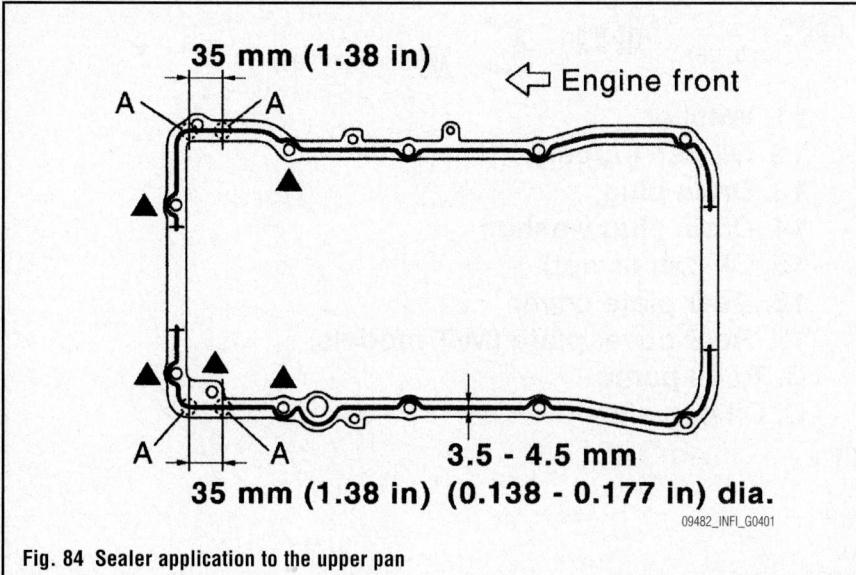

Fig. 84 Sealer application to the upper pan

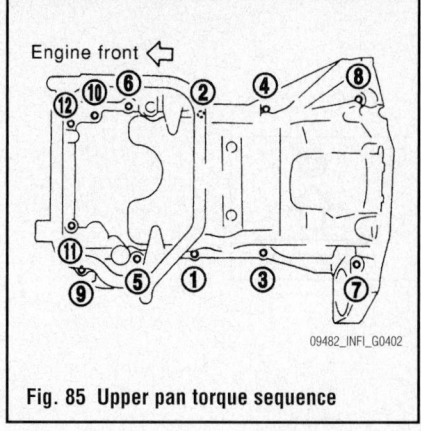

Fig. 85 Upper pan torque sequence

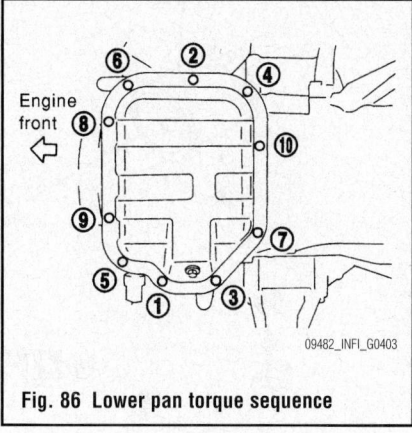

Fig. 86 Lower pan torque sequence

• Transmission joint bolts

20. For the lower oil pan, apply sealant as shown in the illustrations, then install and tighten the bolts, in sequence to 74–82 inch lbs. (8.3–9.3 Nm).

21. The remainder of installation is the reverse of the removal procedure.

➡**Wait at least 30 minutes before refilling the engine oil.**

22. Connect the negative battery cable

23. Fill the engine with clean oil and coolant.

24. Start the engine, check for leaks and repair if necessary.

2007–08 Models

See Figures 87 through 89.

1. Before servicing the vehicle, refer to the precautions in the beginning of this section.

2. Remove oil level gauge, oil pressure switch and oil temperature sensor.

3. Remove oil pan (lower).

4. Remove oil strainer.

5. Loosen mounting bolts in sequence.

6. Insert the seal cutter Special Tool: KV10111100 (J37228) between oil pan (upper) and lower cylinder block. Slide seal

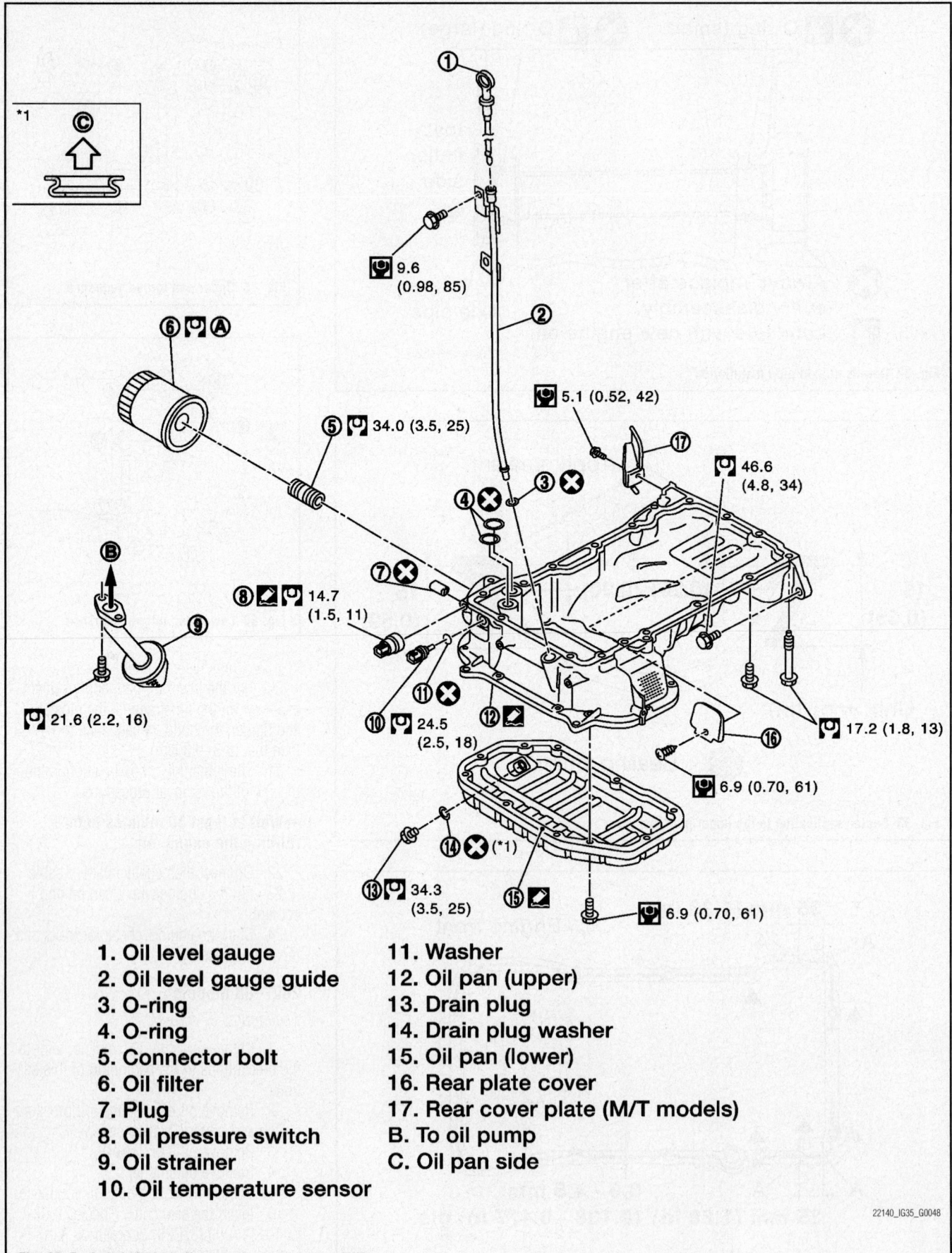

Fig. 87 Exploded view of oil pan and components—2WD

1. Oil level gauge
2. Oil level gauge guide
3. O-ring
4. O-ring
5. Connector bolt
6. Oil filter
7. Plug
8. Oil pressure switch
9. Oil strainer
10. Oil temperature sensor
11. Washer
12. Oil pan (upper)
13. Drain plug
14. Drain plug washer
15. Oil pan (lower)
16. Rear plate cover
17. Rear cover plate (M/T models)
B. To oil pump
C. Oil pan side

22140_IG35_G0048

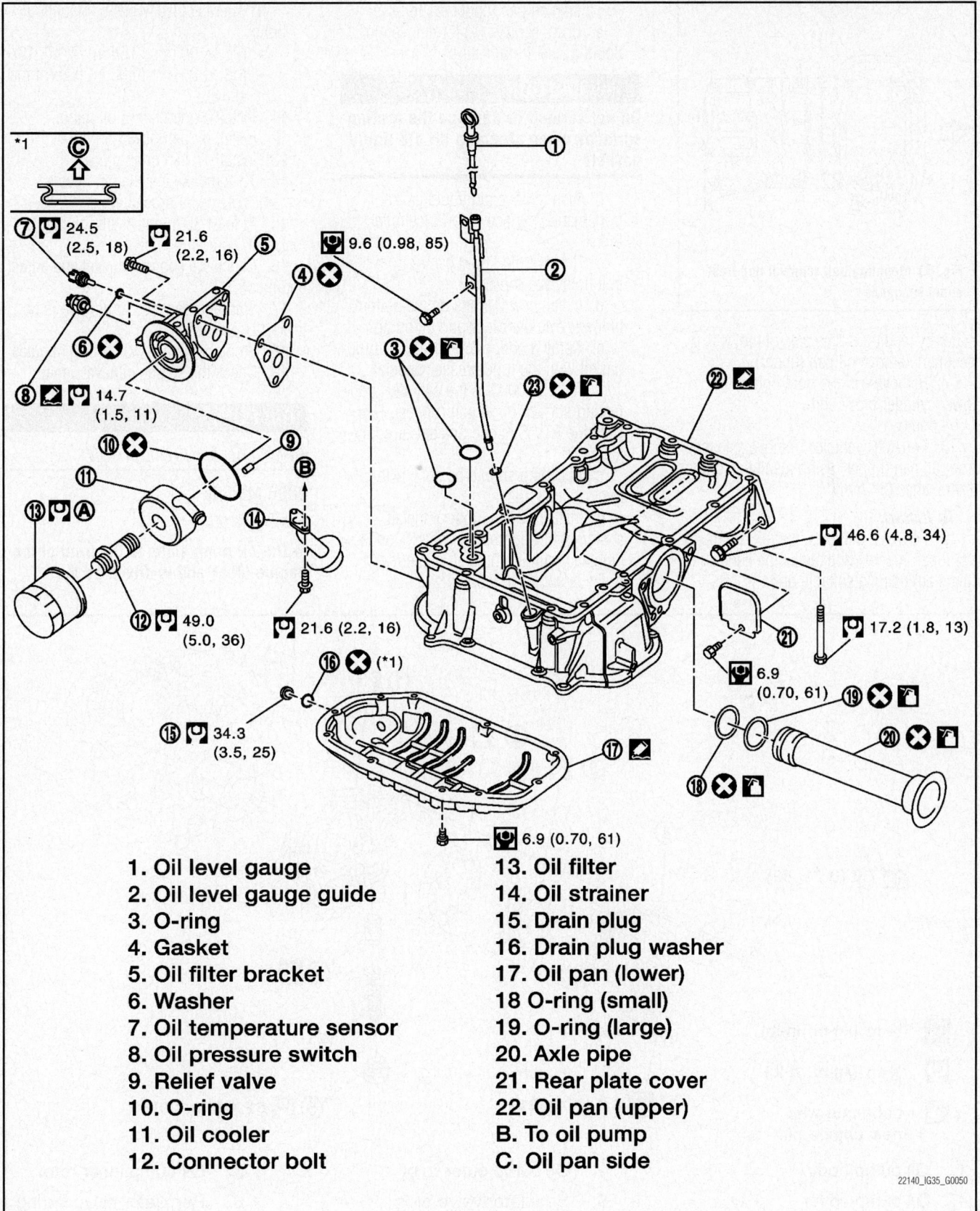

1. Oil level gauge
2. Oil level gauge guide
3. O-ring
4. Gasket
5. Oil filter bracket
6. Washer
7. Oil temperature sensor
8. Oil pressure switch
9. Relief valve
10. O-ring
11. Oil cooler
12. Connector bolt
13. Oil filter
14. Oil strainer
15. Drain plug
16. Drain plug washer
17. Oil pan (lower)
18 O-ring (small)
19. O-ring (large)
20. Axle pipe
21. Rear plate cover
22. Oil pan (upper)
B. To oil pump
C. Oil pan side

22140_IG35_G0050

Fig. 88 Exploded view of oil pan and components—AWD

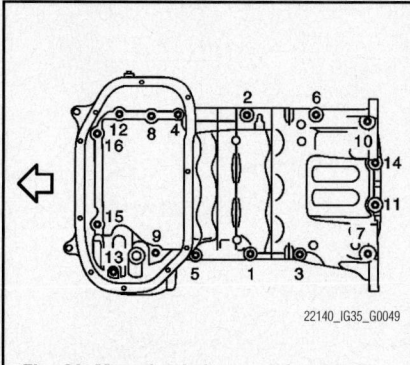

Fig. 89 Mounting bolt removal and tightening sequence

cutter by tapping on the side of tool with a hammer. Remove oil pan (upper).

7. Remove o-rings from bottom of lower cylinder block and oil pump.

8. For AWD vehicles, remove axle pipe from oil pan (upper) using a suitable drift of 1.46 inches (37 mm).

To install:

9. For AWD vehicles, install axle pipe to oil pan (upper) from axle pipe flange side (left side) using a suitable drift.

10. Install oil pan (upper) as follows:
    a. Use a scraper (A) to remove old liquid gasket from mating surfaces

✳✳ CAUTION

Do not scratch or damage the mating surfaces when cleaning off old liquid gasket.

b. Also remove old liquid gasket from mating surface of lower cylinder Block

c. Remove old liquid gasket from the bolt holes and threads

d. Install new O-rings on the bottom of lower cylinder block and oil pump.

e. Apply a continuous bead of liquid gasket with the tube presser Special Tool: WS39930000 to the cylinder block mating surface of oil pan (upper). Use Genuine RTV Silicone Sealant or equivalent.

f. Attaching should be done within 5 minutes after coating.

g. Install oil pan (upper). Install avoiding misalignment of both O-rings. Tighten mounting bolts in numerical order as shown in the figure above.

h. There are two types of mounting bolts:
- M8 × 90 mm (3.54 in): 7, 10, 13
- M8 × 25 mm (0.98 in): Except the above

11. Install oil strainer to oil pump.
12. Install oil pan (lower).
13. Install oil pan drain plug.
14. To complete installation, reverse remaining removal procedure.
15. At least 30 minutes after oil pan is installed, pour engine oil.
16. Check the engine oil level and adjust engine oil.
17. Start engine, and check there is no leak of engine oil.
18. Stop engine and wait for 10 minutes.
19. Check the engine oil level again.

OIL PUMP

REMOVAL & INSTALLATION

2006 Models

See Figure 90.

➡ The oil pump bolts to the front of the engine block and is driven by the

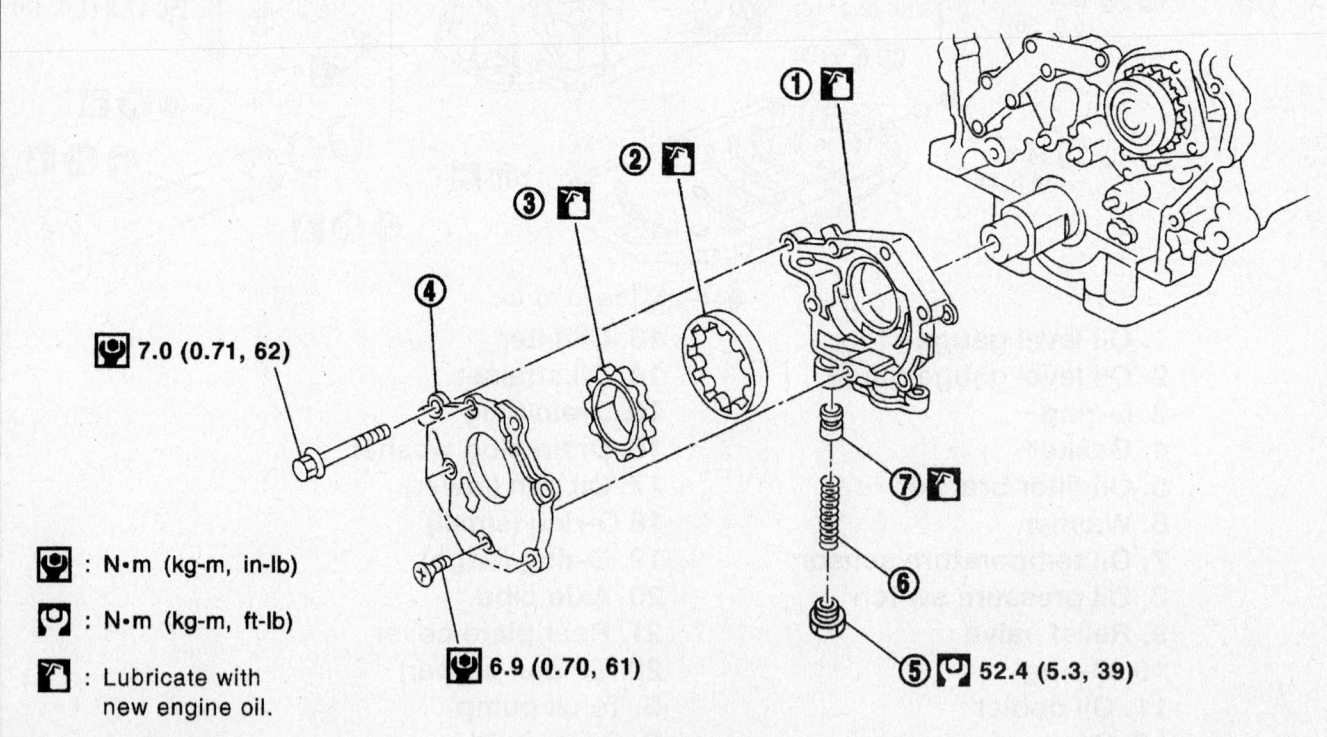

7.0 (0.71, 62)

: N•m (kg-m, in-lb)

: N•m (kg-m, ft-lb)

: Lubricate with new engine oil.

6.9 (0.70, 61)

52.4 (5.3, 39)

1. Oil pump body
2. Oil pump outer rotor
3. Oil pump inner rotor
4. Oil pump cover
5. Regulator valve plug
6. Regulator valve spring
7. Regulator valve

Fig. 90 Exploded view of the oil pump

09482_INFI_G0046

crankshaft. Removal of the timing cover and chains are necessary for oil pump service.

1. Before servicing the vehicle, refer to the precautions in the beginning of this section.

2. Drain the engine oil.

3. Rotate the engine and position it to Top Dead Center (TDC) compression stroke of cylinder No. 1.

4. Remove or disconnect the following:
- Negative battery cable
- Drive belts
- Camshaft Position (CMP) sensor (PHASE) and the Crankshaft Position (CKP) sensor (REF/POS)
- Right front wheel and inner fender cover
- Engine undercover
- Crankshaft pulley
- Front exhaust pipe and its support and support the engine at the left and right side slingers with a suitable hoist
- Engine right side mounting insulator and bracket nuts and bolts
- Center crossmember assembly
- A/C compressor and mounting bracket

- Lower and upper oil pans
- Oil strainer from the oil pump
- Water pump cover and the front cover assembly
- Lower timing chain assembly
- Oil pump

To install:

➡ **When installing the oil pump, be sure to apply engine oil to the gears.**

5. Install or connect the following:
- Oil pump. Refer to illustration for torque values
- Lower timing chain assembly
- Front timing cover and water pump covers
- Oil strainer using a new gasket. Torque the bolts to 12–14 ft. lbs. (16–19 Nm).
- Upper and lower oil pans. Be sure to use new O-rings at the oil pump to upper oil pan mating surface.
- A/C compressor and mounting bracket
- Center crossmember assembly
- Engine right side mounting insulator and bracket and remove the engine support hoist
- Front exhaust pipe and its support

- Crankshaft pulley
- Engine undercover and the right side inner fender cover
- Right front wheel
- CMP sensor (PHASE) and the CKP sensor (REF/POS)
- Engine drive belts and adjust as necessary
- Negative battery cable

6. Fill the engine with clean oil.

7. Start the engine, check the oil pressure, and check for oil leaks.

2007–08 Models

See Figure 91.

1. Before servicing the vehicle, refer to the precautions in the beginning of this section.

2. Remove oil pans (lower) and oil strainer.

3. Remove front timing chain case and timing chain (primary).

4. Remove oil pump assembly.

To install:

5. Before installation, apply new engine oil to the parts as instructed in the figure.

6. To install, reverse removal procedure.

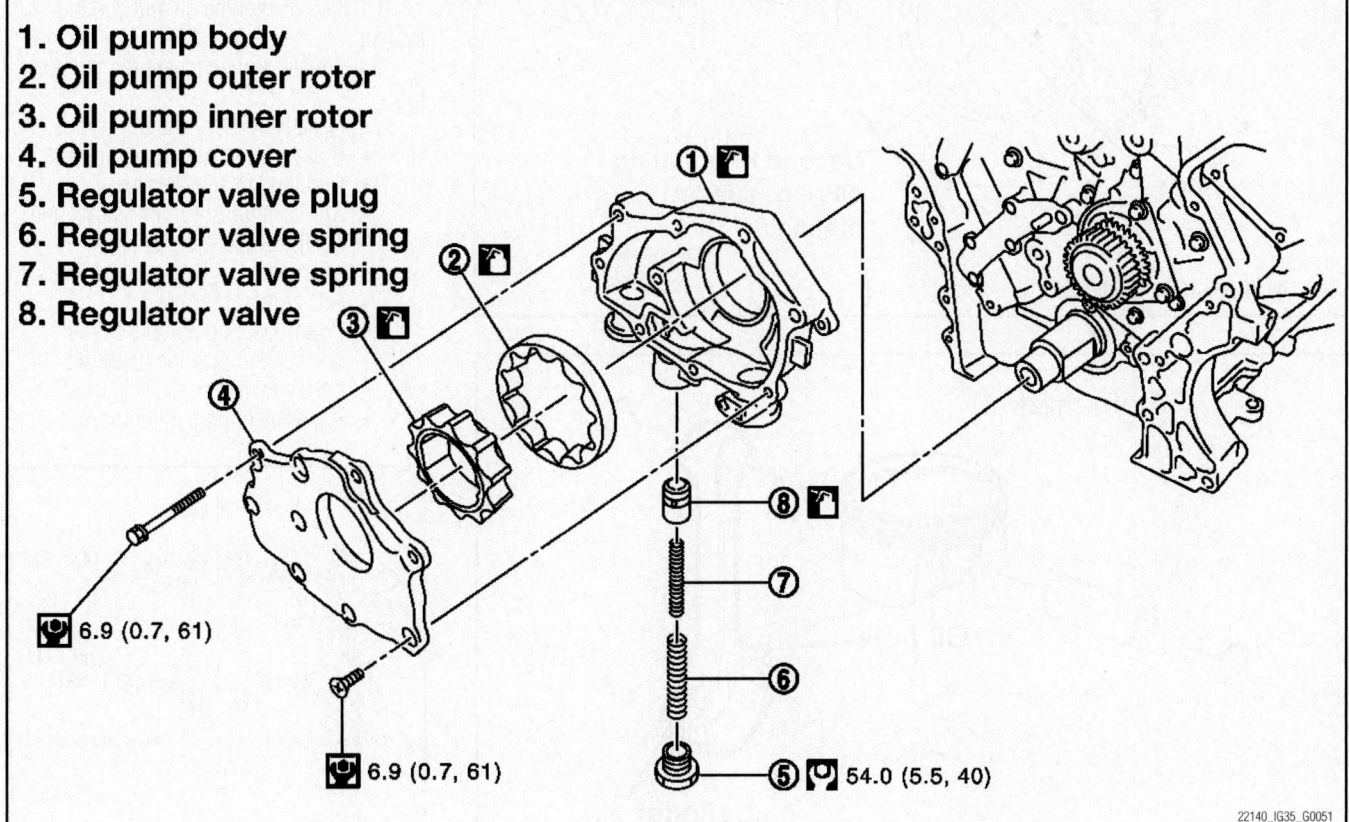

1. Oil pump body
2. Oil pump outer rotor
3. Oil pump inner rotor
4. Oil pump cover
5. Regulator valve plug
6. Regulator valve spring
7. Regulator valve spring
8. Regulator valve

6.9 (0.7, 61)

6.9 (0.7, 61)

54.0 (5.5, 40)

Fig. 91 Exploded view of the oil pump

22140_IG35_G0051

7. When installing, align crankshaft flat faces with oil pump inner rotor flat faces.

8. Check the engine oil level.

9. Start the engine and check there are no leaks.

10. Stop the engine and wait for 10 minutes.

11. Check the engine oil level and adjust the level.

PISTON AND RING

POSITIONING

See Figures 92 and 93.

REAR MAIN SEAL

REMOVAL & INSTALLATION

See Figure 94.

1. Before servicing the vehicle, refer to the precautions in the beginning of this section.

2. Remove or disconnect the following:
- Transmission or transaxle
- Drive plate from the crankshaft

3. Carefully pry the seal out of the retainer without damaging the crankshaft or the seal retainer.

To install:

4. Lubricate the seal with clean engine oil.

5. Install or connect the following:
- Seal into the retainer using the appropriate seal driver

➡Install rear oil seal so that each seal lip is oriented as shown in the figure.

- Driveplate and transmission or transaxle

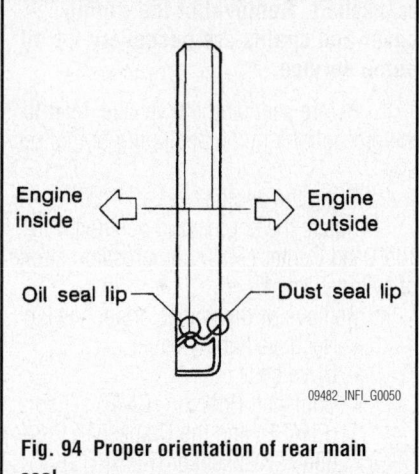

Fig. 94 Proper orientation of rear main seal

TIMING CHAIN, SPROCKETS, FRONT COVER AND SEAL

REMOVAL & INSTALLATION

2006 Coupe and Sedan; 2007 Coupe

See Figures 95 through 103.

1. Before servicing the vehicle, refer to the precautions in the beginning of this section.

2. Properly relieve the fuel system pressure.

3. Drain the engine oil and cooling system.

4. Remove or disconnect the following:
- Negative battery cable
- Right and left side rocker covers
- Cooling fan and radiator
- A/C compressor from bracket and position side with the lines attached
- A/C compressor bracket
- Power steering pump from its bracket and position aside with the lines attached
- Power steering pump bracket

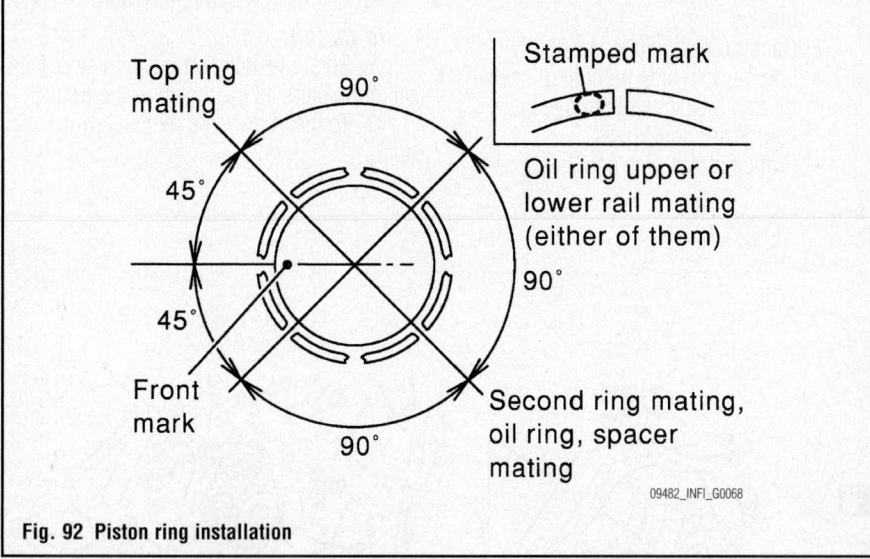

Fig. 92 Piston ring installation

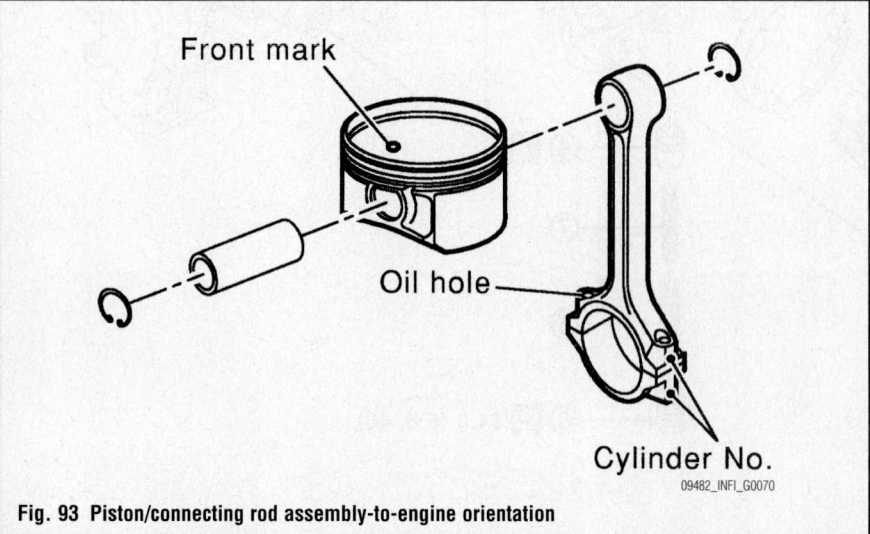

Fig. 93 Piston/connecting rod assembly-to-engine orientation

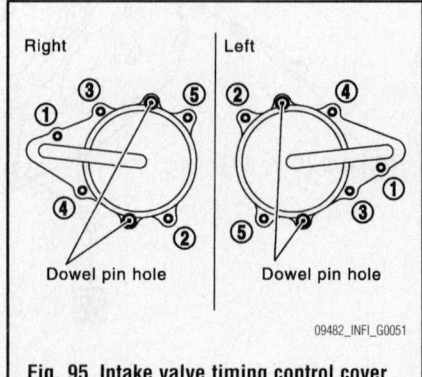

Fig. 95 Intake valve timing control cover tightening sequence—3.5L engine

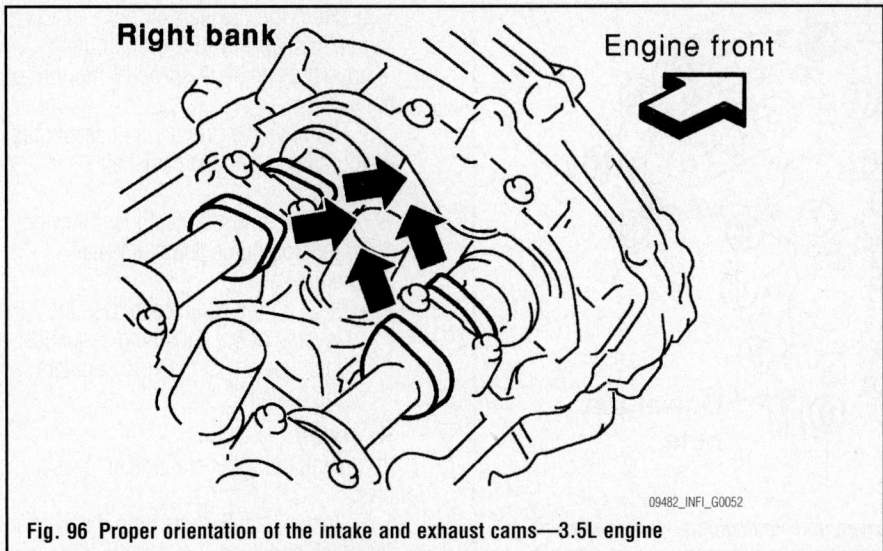

Fig. 96 Proper orientation of the intake and exhaust cams—3.5L engine

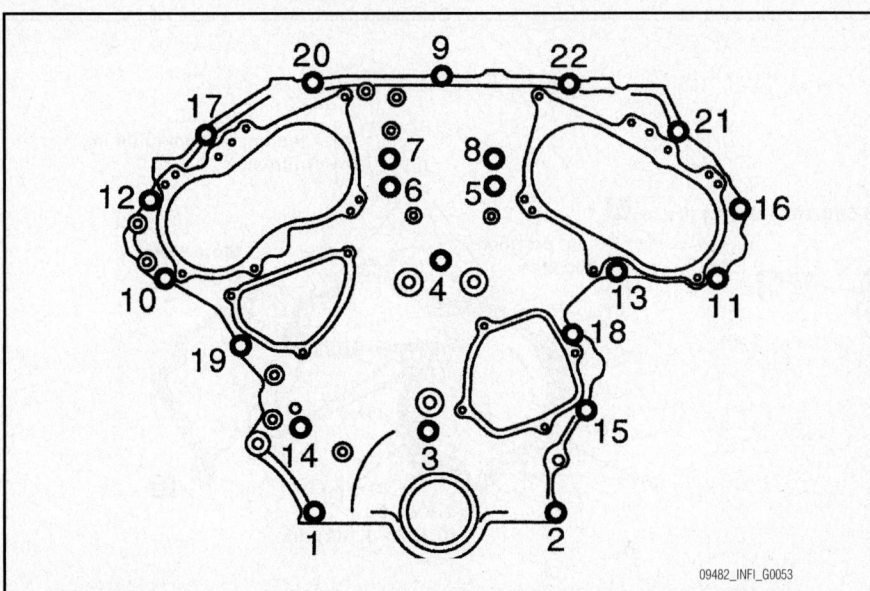

Fig. 97 Front timing chain case torque sequence (loosen in reverse order)—3.5L engine

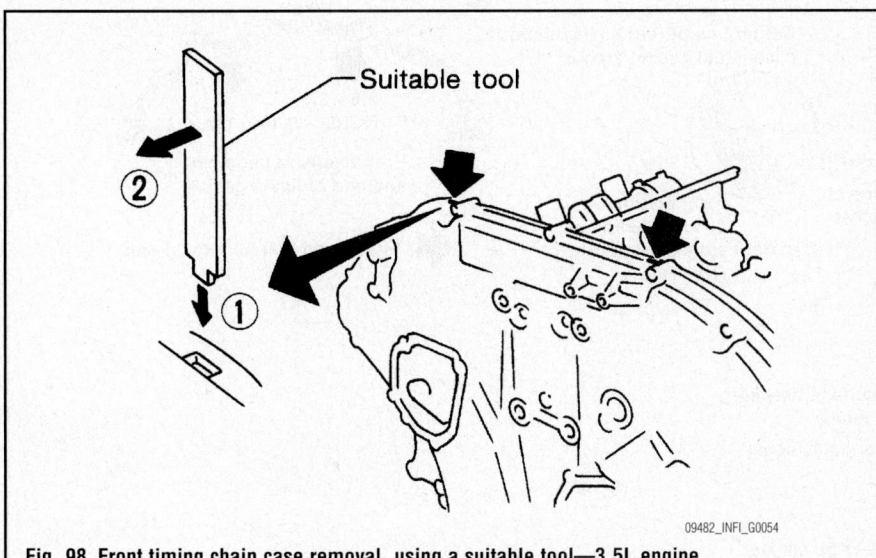

Fig. 98 Front timing chain case removal, using a suitable tool—3.5L engine

• Water bypass hose and cooling fan bracket from the front timing chain case
• Lower and upper oil pans
• Right and left side intake valve timing control covers. Loosen the bolts in the reverse of the tightening sequence. Use a seal cutter to cut the gasket.

5. Set the No. 1 piston to Top Dead Center (TDC) on its compression stroke. Align the timing mark (grooved line) on the crankshaft pulley with the timing indicator on the front cover.

6. Make sure the intake and exhaust cam nose for the No. 1 cylinder is positioned as shown in the illustration. If not, turn the crankshaft on revolution (360°) and align.

7. Remove or disconnect the following:
• Crankshaft pulley using a suitable puller
• Front timing chain case. Loosen the bolts in the reverse of the torque sequence.

✳✳ WARNING

Do not use a screwdriver to pry the case off!

• Timing chain case. Insert the proper tool into the notch at the top of the case as shown. Pry the case off by levering the tool as shown. Use a seat cutter to cut the liquid gasket.
• O-rings from rear timing chain case
• Water pump cover and chain tensioner cover from the case
• Front oil seal from the case using a suitable prytool, being careful not the damage the case
• Internal chain guide, timing chain tensioner and slack guide. Remove the upper chain tensioner by pressing the tensioner in and inserting a 0.098 inch (2.5mm) diameter pin in the pin hole. Once secured remove the bolts and the tensioner.

✳✳ WARNING

After the timing chain is removed, do NOT turn the crankshaft and camshaft separately, or the valve will strike the pistons.

8. Remove the timing chain and camshaft sprocket, as follows:
a. Attach a suitable stopper pin to the right and left camshaft chain tensioners (for secondary timing chains).

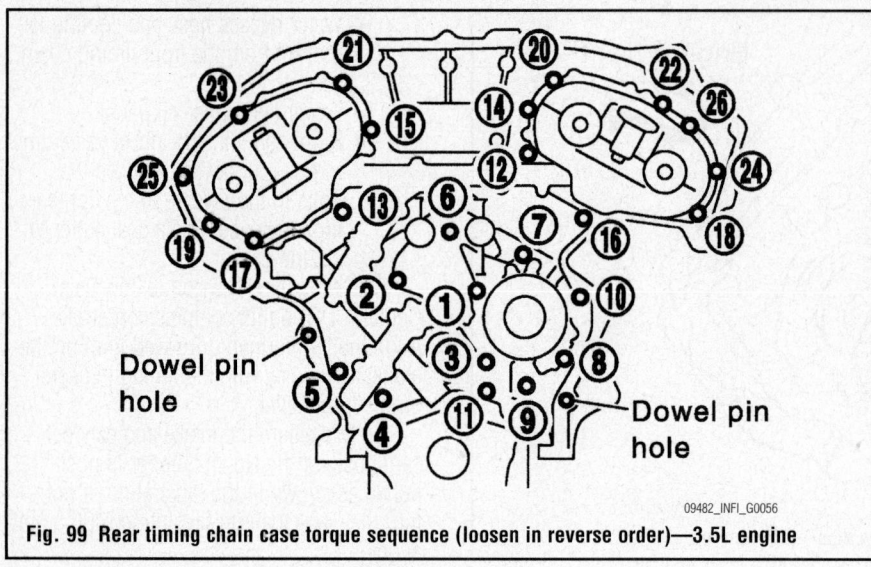

Dowel pin
hole

Dowel pin
hole

09482_INFI_G0056

Fig. 99 Rear timing chain case torque sequence (loosen in reverse order)—3.5L engine

b. Hold the hex part of the camshaft secure with a wrench, then remove the camshaft sprocket mounting bolts.

c. Remove the primary and secondary timing chains with the camshaft sprockets.

9. Remove or disconnect the following:
- Tension guide and crankshaft sprocket
- Rear timing chain case. Use the reverse of the tightening sequence, then use a seal cutter to separate the gasket.

To install:

10. Install the rear timing chain case as follows:

a. Install new O-rings onto the cylinder block and head.

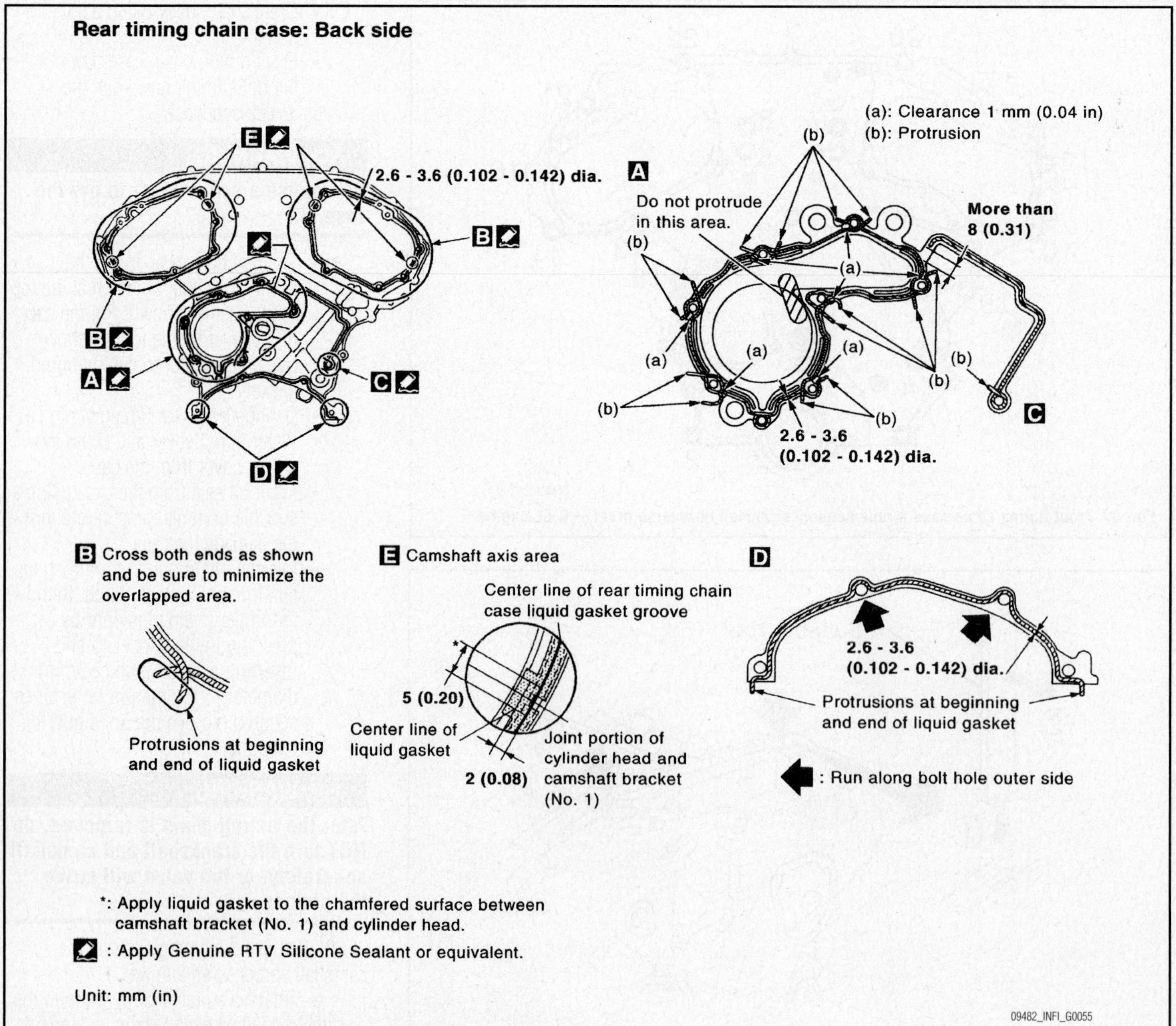

Rear timing chain case: Back side

E✏

2.6 - 3.6 (0.102 - 0.142) dia.

A

B✏

B✏

A✏

C✏

D✏

(a): Clearance 1 mm (0.04 in)
(b): Protrusion

A

(b)

Do not protrude
in this area.

(b)

(a)

(a)

(a)

(a)

(a)

(b)

(b)

More than
8 (0.31)

(b)

(b)

(b)

C

2.6 - 3.6
(0.102 - 0.142) dia.

B Cross both ends as shown and be sure to minimize the overlapped area.

Protrusions at beginning and end of liquid gasket

E Camshaft axis area

Center line of rear timing chain case liquid gasket groove

5 (0.20)

Center line of
liquid gasket

2 (0.08)

Joint portion of cylinder head and camshaft bracket (No. 1)

D

2.6 - 3.6
(0.102 - 0.142) dia.

Protrusions at beginning and end of liquid gasket

◄ : Run along bolt hole outer side

*: Apply liquid gasket to the chamfered surface between camshaft bracket (No. 1) and cylinder head.

✏ : Apply Genuine RTV Silicone Sealant or equivalent.

Unit: mm (in)

09482_INFI_G0055

Fig. 100 Rear timing chain case sealant application—3.5L engine

b. Apply suitable RTV sealant to the back side of the rear timing chain case as shown in the illustration.

c. Install the rear case and tighten the bolts, in sequence, to 9–10 ft. lbs. (11.7–13.7 Nm). There are 2 bolts lengths: Bolts 1, 2, 3, 6–10 are 0.79 in. (20mm) long and the other bolts are 0.63 in. (16mm) long.

11. Set the No. 1 piston to Top Dead Center (TDC) on its compression stroke.

12. Install the crankshaft sprocket, making sure the mating marks on the sprocket face the front of the engine.

13. Push the plunger of the secondary chain tensioner and keep it pressed in with a stopper pin.

14. Install the secondary timing chains and camshaft sprockets, as follows:

a. Align the matchmarks on the secondary timing chain (gold link) with the stamped marks on the intake and exhaust sprockets, then install them.

➡**Matchmarks for the intake sprocket are on the back of the secondary sprocket. There are 2 kinds of marks,** **the right bank uses round marks and the left bank uses oval marks.**

b. Align the dowel pin and pin hole on the camshaft with the groove and dowel pin on the sprocket, then install them.

c. On the intake side, align the pin hold on the small diameter side of the camshaft front end with the dowel pin on the on the back side of the camshaft sprocket, and install them.

d. On the exhaust side, align the dowel pin on the camshaft front end with the pin groove on the camshaft sprocket, then install them.

e. Tighten the camshaft sprocket bolts hand-tight to prevent the dowel pins from dislocating.

15. Install the primary timing chain, as follows:

a. Install the primary chain so the punched mating marks on the cam sprockets are aligned with **matching colored** links on the timing chain, while the notched mating mark on the crankshaft sprocket is aligned with a **different** **color** link on the timing chain. Cam sprocket mating links must be the same color in order to ensure proper cam timing.

b. Use a wrench on the hex portion of the camshaft to secure it in place, tighten the camshaft sprocket mounting bolts to 73–78 ft. lbs. (98–108 Nm).

c. Remove the stopper pins from secondary chain tensioners.

16. Install the internal chain guide, timing chain tensioner, tension guide and slack guide. Do not overtighten the slack guide mounting bolts. It's normal to have a gap under the bolt seats when the mounting bolts are properly tightened to 10–13 ft. lbs. (14–18 Nm).

17. Recheck that all matchmarks are still aligned.

18. Install or connect the following:
- New O-rings on the rear timing chain case
- New front oil seal in to the timing chain cover
- Water pump and chain tensioner covers to the front cover
- Suitable liquid gasket to the back side of the timing chain case
- Dowel pin on the rear timing chain case into the dowel pin hold on the front chain case
- Front timing chain case bolts, in sequence. Tighten the M6 bolts to 9–10 ft. lbs. (11.7–13.7 Nm) and the M8 bolts to 19–23 ft. lbs. (25–31 Nm).

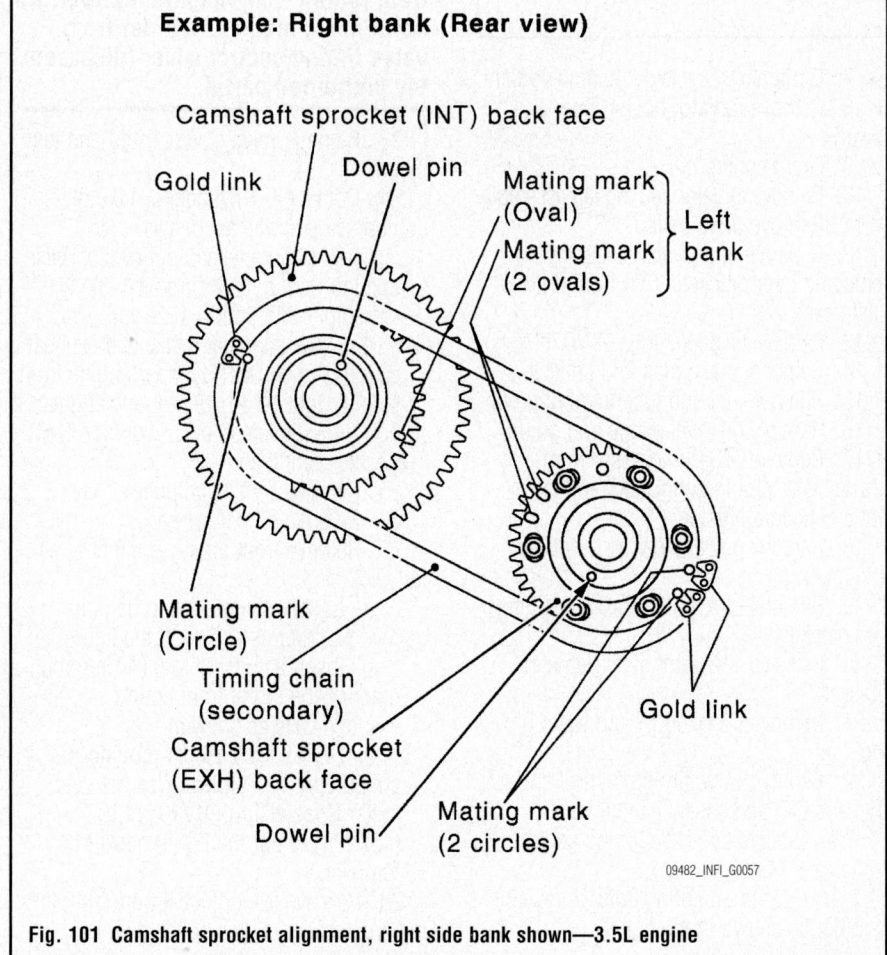

Example: Right bank (Rear view)

Camshaft sprocket (INT) back face
Gold link
Dowel pin
Mating mark (Oval)
Mating mark (2 ovals)
Left bank
Mating mark (Circle)
Timing chain (secondary)
Camshaft sprocket (EXH) back face
Dowel pin
Mating mark (2 circles)
Gold link

09482_INFI_G0057

Fig. 101 Camshaft sprocket alignment, right side bank shown—3.5L engine

Mating mark (pink link)
Mating mark (punched)
Camshaft sprocket
Water pump
Crankshaft sprocket
Mating mark (orange link)
Mating mark (notched)

09482_INFI_G0058

Fig. 102 Proper alignment for the primary timing chain—3.5L engine

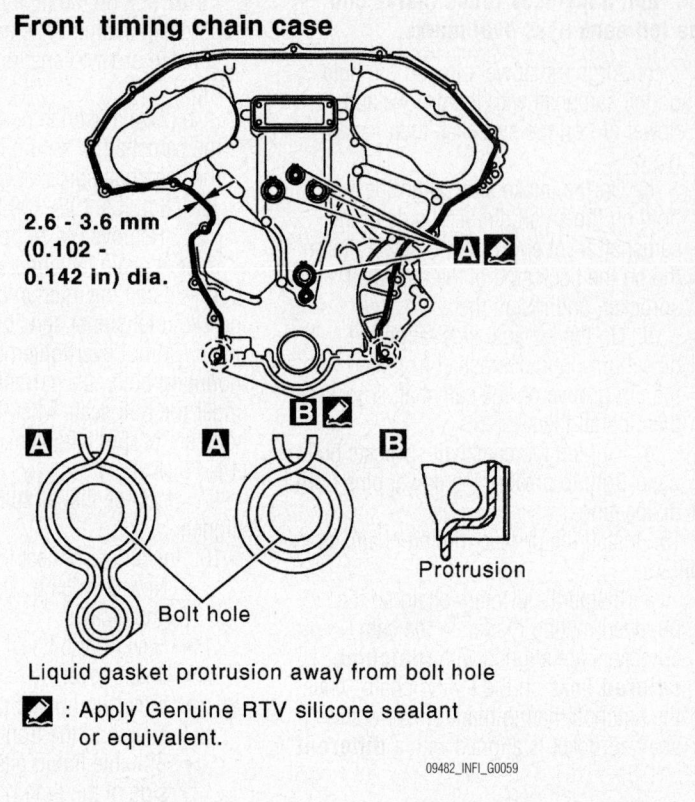

Front timing chain case

2.6 - 3.6 mm
(0.102 -
0.142 in) dia.

Bolt hole

Protrusion

Liquid gasket protrusion away from bolt hole

: Apply Genuine RTV silicone sealant
or equivalent.

09482_INFI_G0059

Fig. 103 Apply proper sealant as shown on the back side of the front timing chain case—3.5L engine

- Right and left intake valve timing control cover
- Crankshaft pulley to the crankshaft and the mounting bolt. Torque the mounting bolt to 29–36 ft. lbs. (39–49 Nm). Torque the crankshaft bolt an additional 60–66° clockwise. This is approximately the angle from 1 hexagon bolt head corner to another.

19. Installation of the remaining components is the reverse of the removal procedure.
20. Fill the engine with clean oil.
21. Fill and bleed the cooling system.
22. Start the engine and check for proper operation.

2007–08 Sedan

See Figures 104 through 112.

1. Before servicing the vehicle, refer to the precautions in the beginning of this section.
2. Properly relieve the fuel system pressure.
3. Remove engine cover.
4. Remove radiator reservoir tank.
5. Remove air duct and air cleaner case assembly (RH and LH).
6. Remove engine undercover with power tool.

7. Drain and recycle the engine coolant.
8. Remove radiator hose (upper and lower).
9. Drain engine oil.
10. Remove radiator cooling fan assembly.
11. Remove drive belts.
12. Separate engine harnesses by removing their brackets from front timing chain case.
13. Remove oil cooler tube (AWD models).
14. Remove intake manifold collector.
15. Remove fuel sub tube mounting bolt.
16. Remove oil level gauge and guide.
17. Remove A/C compressor from bracket with piping connected, and temporarily secure it aside.
18. Remove power steering oil pump and bracket.
19. Remove idler pulley, auto tensioner and bracket.
20. Remove alternator and alternator bracket.
21. Remove water outlet and water piping.
22. Remove valve timing control covers (RH and LH) and gasket as follows:
   a. Disconnect valve timing control harness connector.
   b. Loosen mounting bolts in reverse order as shown in the figure.

✳✳ CAUTION

Shaft is internally jointed with camshaft sprocket (INT) center hole. When removing, keep it horizontal until it is completely disconnected.

   c. Shaft is engaged with intake side camshaft sprocket center hole on inside. Pull straight out so as not to tilt until the joint is disengaged.
   d. The mating surface of magnet retarder may be fitted with the exhaust side camshaft sprocket via the engine oil. Open valve timing control cover carefully.
   e. If the mating surface of magnet retarder is fitted with the camshaft sprocket, open the cover within the range that the load is not applied to the harness. And then, remove it so as to prevent magnet retarder from dropping.

✳✳ CAUTION

Be careful not to damage magnet retarder. When carrying valve timing control cover, face the magnet retarder side up to prevent the cover from falling from magnet retarder. Do not remove magnet retarder from valve timing control cover (disassembly prohibited parts).

23. Remove valve covers (right and left bank).
24. Obtain No.1 cylinder at TDC of its compression stroke as follows:
   a. Rotate crankshaft pulley clockwise to align timing mark (grooved line without color) with timing indicator.
   b. Make sure that intake and exhaust cam noses on No.1 cylinder(engine front side of right bank) are properly aligned. If not, turn crankshaft one revolution (360 degrees) and align.
25. Remove crankshaft pulley.
26. Remove oil pan (lower).
27. Remove front timing chain case as follows:
   a. Loosen the mounting bolts in reverse order as shown in the figure.
   b. Insert a suitable tool (A) into the notch at the top of front timing chain case as shown.
   c. Pry off the case by moving the suitable tool as shown. Use the seal cutter Special Tool: KV10111100 (J37228) to cut the liquid gasket for removal.
28. Remove front oil seal from front timing chain case using a suitable tool.

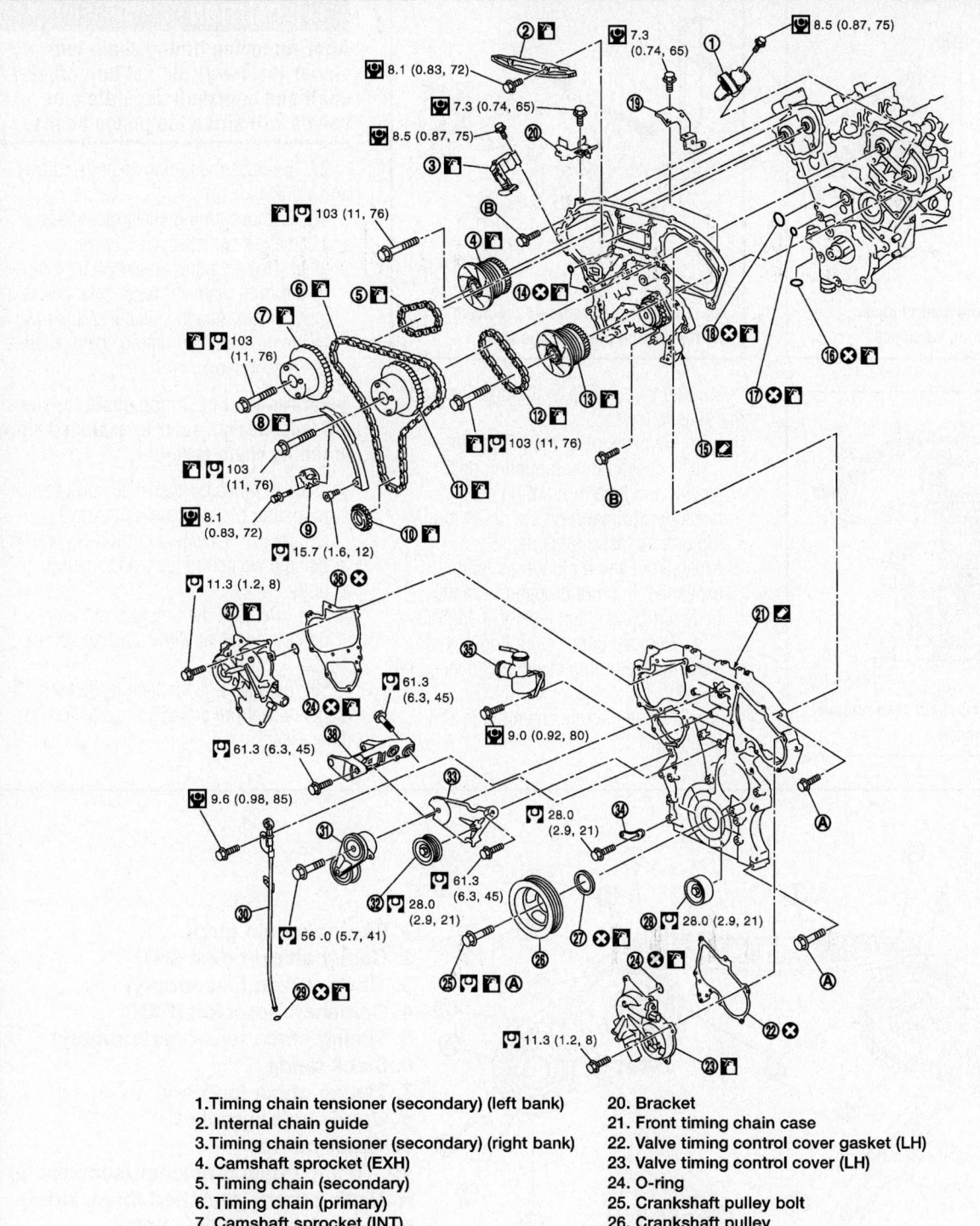

1. Timing chain tensioner (secondary) (left bank)
2. Internal chain guide
3. Timing chain tensioner (secondary) (right bank)
4. Camshaft sprocket (EXH)
5. Timing chain (secondary)
6. Timing chain (primary)
7. Camshaft sprocket (INT)
8. Slack guide
9. Timing chain tensioner (primary)
10. Crankshaft sprocket
11. Camshaft sprocket (INT)
12. Timing chain (secondary)
13. Camshaft sprocket (EXH)
14. O-ring
15. Rear timing chain case
16. O-ring
17. O-ring
18. O-ring
19. Bracket
20. Bracket
21. Front timing chain case
22. Valve timing control cover gasket (LH)
23. Valve timing control cover (LH)
24. O-ring
25. Crankshaft pulley bolt
26. Crankshaft pulley
27. Front oil seal
28. Idler pulley
29. O-ring
30. Oil level gauge guide
31. Drive belt auto-tensioner
32. Idler pulley
33. Idler pulley bracket
34. Alternator bracket
35. Water outlet (front)
36. Valve timing control cover gasket (RH)
37. Valve timing control cover (RH)

22140_IG35_G0097

Fig. 104 Exploded view of timing chain, sprockets, front cover and seal

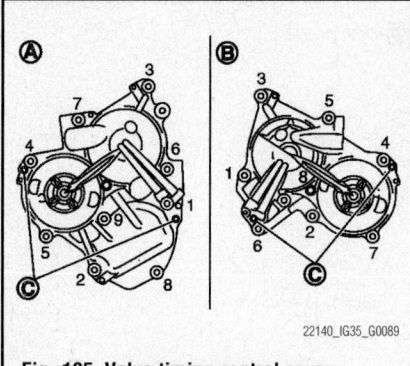

Fig. 105 Valve timing control cover removal and tightening sequence

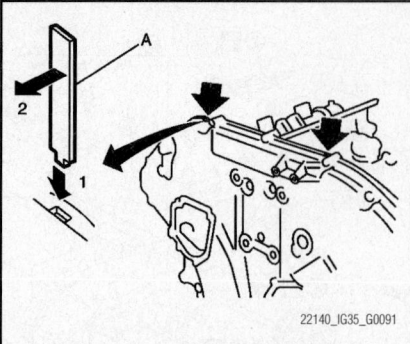

Fig. 107 Inserting tool into the notch at the top of front timing chain case

Front timing chain case removal figure:

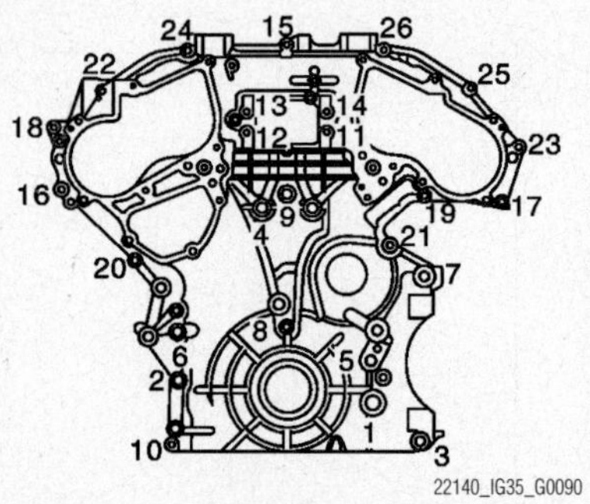

Fig. 106 Front timing chain case removal and tightening sequence

29. Remove timing chain tensioner (primary) as follows:

   a. Remove lower mounting bolt.

   b. Loosen upper mounting bolt slowly, and then turn timing chain tensioner (primary) on the upper mounting bolt so that plunger is fully expanded. Even if plunger is fully expanded, it is not dropped from the body of timing chain tensioner (primary).

   c. Remove upper mounting bolt, and then remove timing chain tensioner (primary).

30. Remove internal chain guide and slack guide.

※※ CAUTION

After removing timing chain tensioner (primary), do not turn crankshaft and camshaft separately or valves will strike the piston heads.

31. Remove the timing chain (primary) and crankshaft sprocket.

32. Remove timing chain (secondary) and camshaft sprockets as follows:

   a. Use an approximately 0.02 inch (0.5 mm)in diameter hard metal pin as a stopper pin. Attach suitable stopper pin to the right and left timing chain tensioners (secondary).

➡**For removal of timing chain tensioners (secondary), refer to exploded view of timing chain assembly.**

   b. Remove the camshaft sprocket mounting bolts (intake and exhaust). Secure the hexagonal portion of camshaft using a wrench to loosen mounting bolts.

   c. Remove the timing chain (secondary) together with camshaft sprockets.

33. Remove timing chain tensioners (secondary) from cylinder head as follows, if necessary:

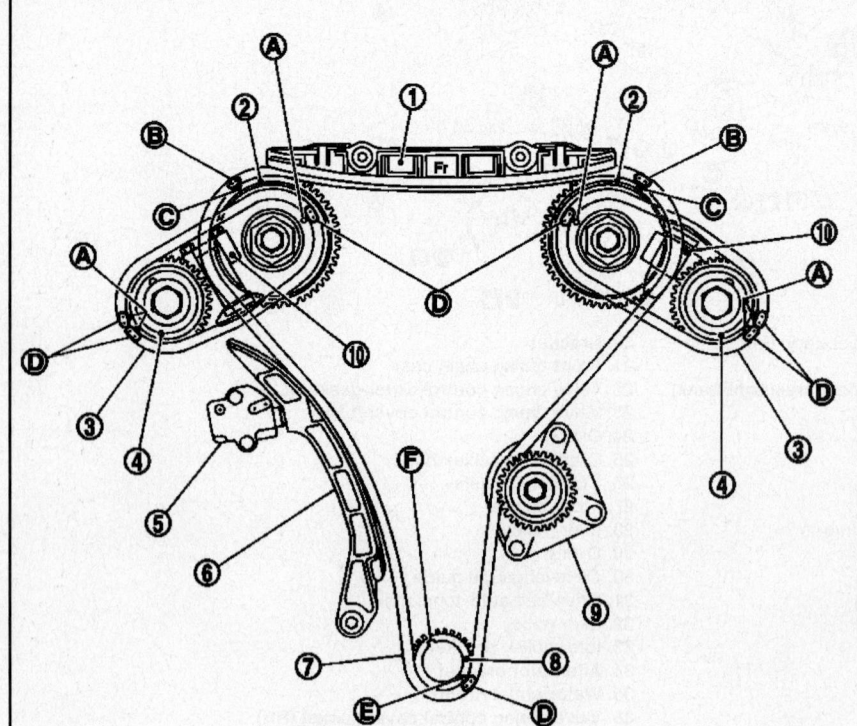

1. Internal chain guide
2. Camshaft sprocket (INT)
3. Timing chain (secondary)
4. Camshaft sprocket (EXH)
5. Timing chain tensioner (primary)
6. Slack guide
7. Timing chain (primary)
8. Crankshaft sprocket
9. Water pump
10. Timing chain tensioner (secondary)
A. Mating mark [punched (back side)]
B. Mating mark (yellow link)
C. Mating mark (punched)
D. Mating mark (orange link)
E. Mating mark (notched)
F. Crankshaft key

Fig. 108 Identifying mating marks on each timing chain and corresponding sprocket, with the components installed

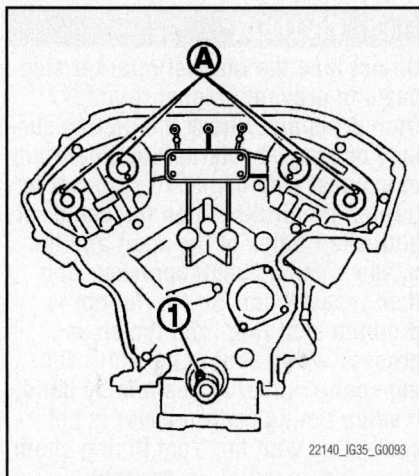

Fig. 109 Dowel pin and crankshaft key positioning

a. Remove the camshaft brackets (No. 1).

b. Remove timing chain tensioners (secondary) with a stopper pin attached.

34. Use a scraper to remove all traces of old liquid gasket from front and rear timing chain cases and oil pan (upper), and liquid gasket mating surfaces. Remove old liquid gasket from bolt hole and thread.

✳✳ CAUTION

Do not let gasket fragments enter oil pan.

To install:

35. The below figure shows the relationship between the mating mark on each timing chain and that on the corre-

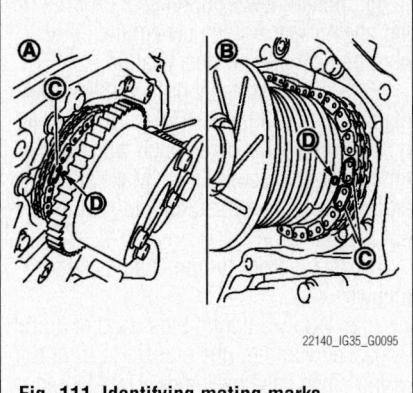

Fig. 111 Identifying mating marks

sponding sprocket, with the components installed.

36. Install timing chain tensioners (secondary) to cylinder head if removed.

37. Make sure that dowel pin (A) and crankshaft key (1) are located as shown in the figure. (No. 1 cylinder at compression TDC).

38. Though camshaft does not stop at the position as shown in the figure, for the placement of cam noses, it is generally accepted camshaft is placed for the same direction of the figure.

39. Install timing chains (secondary) and camshaft sprockets as follows:

a. Push plunger of timing chain tensioner (secondary) and keep it pressed in with a stopper pin.

b. Install timing chains (secondary) and camshaft sprockets.

c. Align the mating marks on timing chain (secondary) (orange link) with the ones on intake and exhaust camshaft sprockets (punched), and install them.

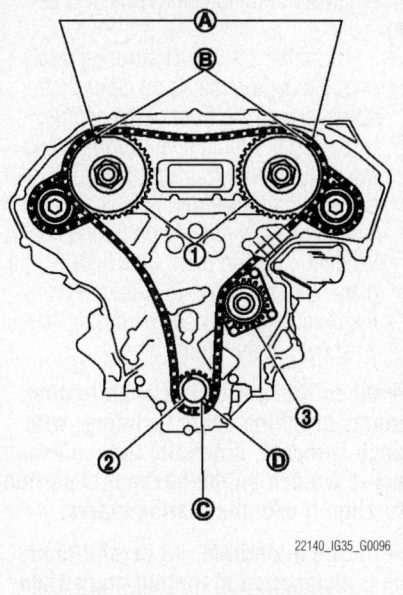

Fig. 112 Installing primary timing chain

➡**Mating marks for camshaft sprockets are on the back side of camshaft sprockets (secondary). There are two types of mating marks, circle and oval types. They should be used for the right and left banks, respectively.**

d. Align dowel pin camshafts with the groove or dowel hole on sprockets, and install them.

e. On the intake side, align dowel pin on camshaft front end with pin groove on the back side of camshaft sprocket, and install them.

f. On the exhaust side, align dowel pin on camshaft front end with pin hole on camshaft sprocket, and install them.

g. In case that positions of each mating mark and each dowel pin are not fit on mating parts, make fine adjustment to the position holding the hexagonal portion on camshaft with wrench or equivalent.

h. Mounting bolts for camshaft sprockets must be tightened in the next step. Tightening them by hand is enough to prevent the dislocation of dowel pins.

i. Make sure the mating marks (punched) (D) on each camshaft sprocket are positioned on the mating marks · (orange link) (C) on timing chain (secondary). A: Intake side. B: Exhaust side.

j. After confirming the mating marks are aligned, tighten camshaft sprocket mounting bolts.

k. Secure camshaft using a wrench at the hexagonal portion to tighten mounting bolts.

l. Pull stopper pins out from timing chain tensioners (secondary).

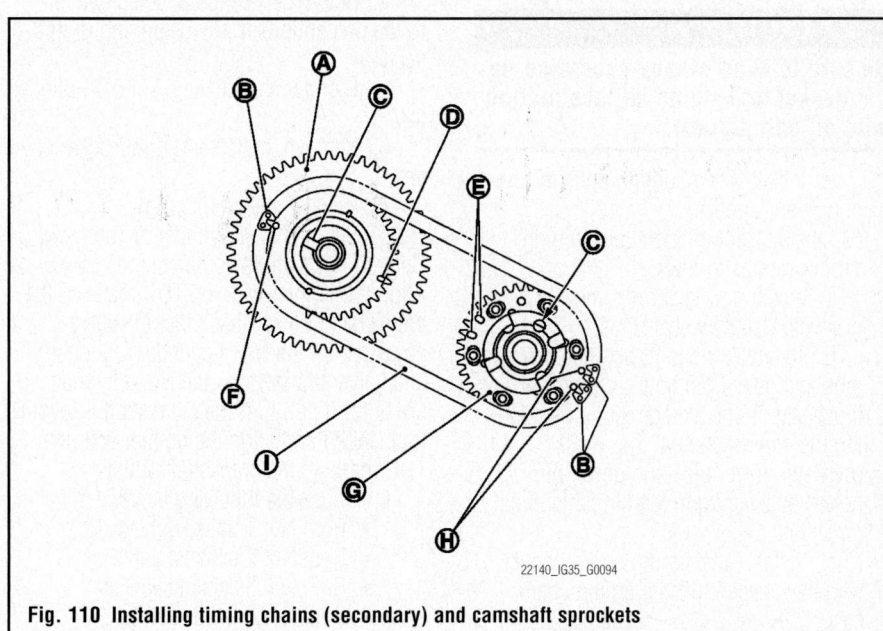

Fig. 110 Installing timing chains (secondary) and camshaft sprockets

40. Install timing chain (primary) as follows:

a. Install crankshaft sprocket. Make sure the mating marks on crankshaft sprocket face the front of the engine.

b. Install timing chain (primary) so the mating mark (punched) (B) on camshaft sprocket (INT) (1) is aligned with the yellow link (A) on timing chain. While the mating mark (notched) (C) on crankshaft sprocket (2) is aligned with the orange link (D) one on timing chain, as shown in the figure.

➡ When it is difficult to align mating marks of timing chain (primary) with each sprocket, gradually turn camshaft using wrench on the hexagonal portion to align it with the mating marks.

➡ During alignment, be careful to prevent dislocation of mating mark alignments of timing chains (secondary).

41. Install internal chain guide, slack guide and timing chain tensioner (primary).

➡ Do not overtighten slack guide mounting bolts. It is normal for a gap to exist under the bolt seats when mounting bolts are tightened to the specification.

42. Install the timing chain tensioner (primary) with the following procedure:

a. Pull plunger stopper tab upwards (or turn lever downward) so as to remove plunger stopper tab from the ratchet of plunger.

b. Push plunger into the inside of the tensioner body.

c. Hold plunger in the fully compressed position by engaging plunger stopper tab with the tip of ratchet.

d. To secure lever, insert stopper pin (E) through hole of lever into tensioner body hole (B). The lever parts and the plunger stopper tab are synchronized.

Therefore, the plunger will be secured under this condition.

e. Install timing chain tensioner (primary). Remove any dirt and foreign materials completely from the back and the mounting surfaces of timing chain tensioner (primary).

f. Pull out stopper pin after installing, and then release plunger

43. Make sure again that the mating marks on sprockets and timing chain have not slipped out of alignment.

44. Install new o-rings on rear timing chain case.

45. Install new front oil seal on front timing chain case. Apply new engine oil to both oil seal lip and dust seal lip.

46. Using a suitable drift with an outer diameter of 2.36 inches (60 mm), press-fit oil seal until it becomes flush with front timing chain case end face. Make sure the garter spring is in position and seal lip is not inverted.

47. Install front timing chain case as follows:

a. Apply a continuous bead of liquid gasket with the tube presser to front timing chain case back side. Use Genuine RTV Silicone Sealant or equivalent.

b. Apply liquid gasket to top surface of oil pan (upper).

c. Assemble front timing chain case.

✳✳ CAUTION

Be careful not to damage front oil seal by interference with front end of crankshaft. Attaching should be done within 5 minutes after liquid gasket application.

d. Install front timing chain case as to fit its dowel pin hole together dowel pin on rear timing chain case.

e. Tighten mounting bolts to the specified torque in numerical order as shown in the previous figure.

- M10 bolts : 1, 2, 3, 4, 5, 6, 7: tighten to 41 ft. lbs. (55 Nm).
- M6 bolts : Except the above: tighten to 9 ft. lbs. (13 Nm).

f. After all bolts are tightened, retighten them to the specified torque in numerical order shown in the figure.

✳✳ CAUTION

Be sure to wipe off any excessive liquid gasket leaking on surface mating with oil pan (upper).

g. Install two mounting bolts in front of oil pan (upper).

48. Install right and left valve timing control covers as follows:

a. Install new seal rings in shaft grooves. Use new seal rings.

b. To check the joint between dowel pins and dowel pin holes, check the looseness in the axle direction by pushing the circumferential looseness (between dowel pins and dowel pin holes) by twisting in the circumferential direction.

c. Install valve timing control cover with new gasket to front timing chain case.

✳✳ CAUTION

Do not face the magnet retarder side down to prevent magnet retarder from dropping. Check the mating surface of magnet retarder and the drum of exhaust side camshaft sprocket for foreign materials. Align the center of both shaft holes of the shaft and the intake side camshaft sprocket, and then insert them. Be careful not to drop the seal ring from the shaft groove. When setting the valve timing control cover in position by hand, if valve timing control cover is not contacting with the front timing chain case, the dowel pin of magnet retarder may not be aligned with the dowel pin holes of cover. In this case, return to step "b".

d. Being careful not to move seal ring from the installation groove, align dowel pins on front timing chain case with holes to install valve timing control covers.

e. Tighten mounting bolts in numerical order.

49. Install oil pan (lower).

50. Install valve covers (right and left banks).

51. Install crankshaft pulley.

VALVE LASH

ADJUSTMENT

See Figures 113 through 115.

➡ Check and adjust the valve clearances while the engine is cold and not running.

1. Before servicing the vehicle, refer to the precautions in the beginning of this section.

2. Remove the intake manifold collector.

3. Remove the left and right rocker covers.

4. Remove the spark plugs.

5. Set the No. 1 cylinder at Top Dead Center (TDC) on its compression stroke. Align the pointer with the TDC mark on the crankshaft pulley. Check that the valve adjusters on the No. 1 cylinder are loose and valve adjusters on the No. 4 cylinder are tight. If not, turn the crankshaft 1 revolution (360°) and align the pointer with the TDC mark on the crankshaft pulley.

6. Check the following valves:
- Both No. 1 intake valves
- Both No. 2 exhaust valves
- Both No. 3 exhaust valves
- Both No. 6 intake valves

7. Using a feeler gauge, measure the clearance between the valve adjuster and the camshaft. Record any valve clearance measurements that are out of specification. Intake valve clearance (cold) is 0.010–0.013 in. (0.26–0.34mm) and exhaust valve clearance (cold) is 0.011–0.015 in. 0.29–0.37mm).

8. Turn the crankshaft 240° and set the No. 3 cylinder to TDC of its compression stroke.

9. Check the following valves:
- Both No. 2 intake valves
- Both No. 3 intake valves
- Both No. 4 exhaust valves
- Both No. 5 exhaust valves

10. Using a feeler gauge, measure the clearance between the valve adjuster and the camshaft. Record any valve clearance measurements that are out of specification. Intake valve clearance (cold) is 0.010–0.013 in. (0.26–0.34mm) and exhaust valve clearance (cold) is 0.011–0.015 in. (0.29–0.37mm).

11. Turn the crankshaft 240° and set the

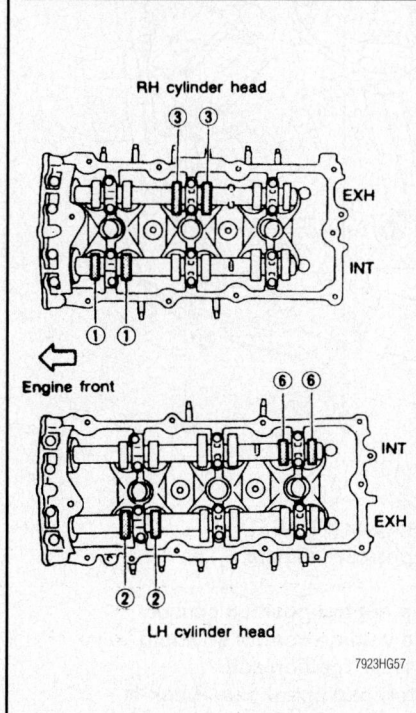

Fig. 113 Valve lash checking sequence at TDC of cylinder No. 1—3.5L engine

No. 5 cylinder to TDC of its compression stroke.

12. Check the following valves:
- Both No. 1 exhaust valves
- Both No. 4 intake valves
- Both No. 5 intake valves
- Both No. 6 exhaust valves

13. Using a feeler gauge, measure the clearance between the valve adjuster and the camshaft. Record any valve clearance measurements that are out of specification. Intake valve clearance (cold) is 0.010–0.013 in. (0.26–0.34mm) and exhaust valve clearance (cold) is 0.011–0.015 inches (0.29–0.37mm).

14. If all the valve clearances are within specification, install the cylinder head cover, spark plugs, and the intake manifold collector.

15. If an adjustment is necessary, adjust the valve clearance while engine is cold by removing the adjusting shim. The adjusting

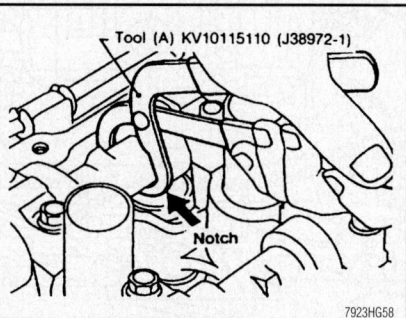

Fig. 114 Install the depressor tool around the camshaft being careful not to damage the surfaces engine shown

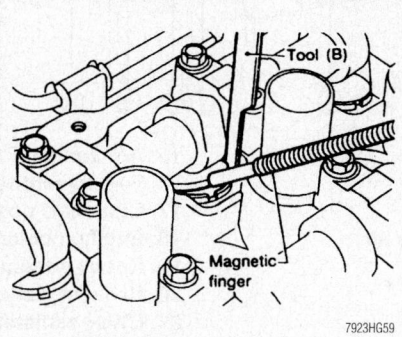

Fig. 115 Use a magnet to remove the shim from the adjuster. Sometimes a shot of compressed air can help lift the shim up

shim can be removed by using the following procedures:

a. Turn the crankshaft so the camshaft lobe of the valve to be adjusted is pointed straight up.

b. Turn the adjuster so the notch is pointed towards the center of the cylinder head; this will facilitate the shim removal process.

c. Using a depressor tool No. KV10115110 push down on the adjuster and insert a keeper tool on the edge of the adjuster to keep the adjuster in the depressed position.

d. Remove the depressor tool and remove the shim with a magnet.

➡ **Compressed air can be blown into the hole of the adjuster to separate the adjusting shim from the adjuster.**

16. Determine the replacement adjusting shim size by using the following procedures and formula:

a. Using a micrometer determine thickness of the removed shim.

b. Calculate the thickness of a new adjusting shim so valve clearance is within the specified values.
- R= thickness of the removed shim
- N= thickness of the new shim
- M= measured valve clearance
- Calculate the Intake Shim as follows: $N = R + M - 0.0118$ in. (0.30mm)
- Calculate the Exhaust Shim as follows: $N = R + M - 0.0130$ in. (0.33mm)

17. Shims are available in 64 sizes from 0.0913–0.1161 in. (2.32–2.95mm) in steps of 0.004 in. (0.01mm). The thickness is stamped on the shim; this side is always installed facing down. Select new shims with thickness as close as possible to calculated valve and install it in the adjuster.

18. Install the new shim onto the adjuster.

19. Depress the adjuster and remove the keeper tool. Remove the depressor tool and recheck the valve clearance. Repeat this procedure for any other valves requiring adjustment.

20. When all valve adjustments are finished, install the cylinder head cover, spark plugs, and the intake manifold collector.

ENGINE PERFORMANCE & EMISSION CONTROL

COMPONENT LOCATIONS

See Figures 116 and 117.

CAMSHAFT POSITION (CMP) SENSOR

LOCATION
See Figure 118.

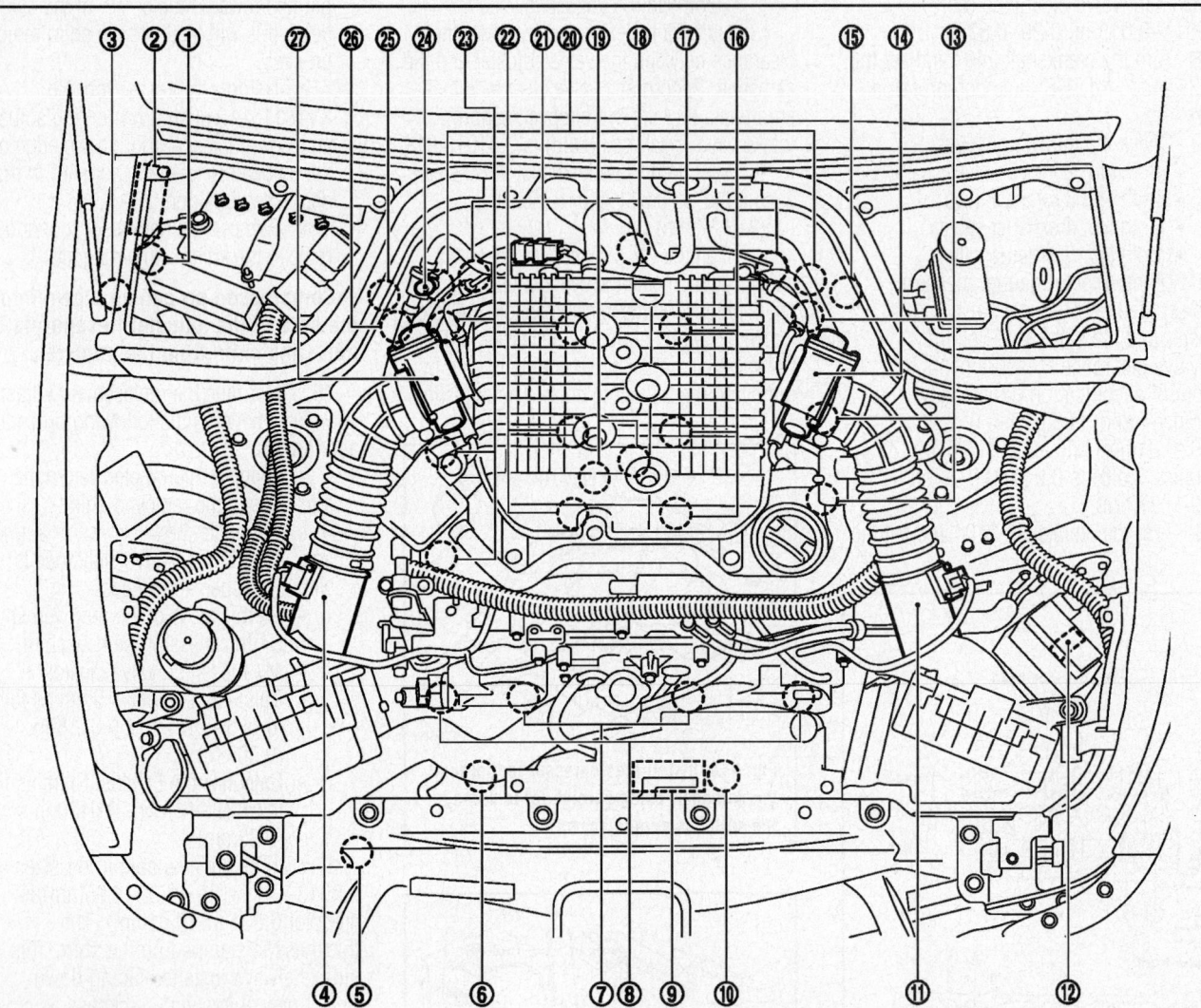

1. Battery current sensor
2. IPDM E/R
3. Cooling fan relay
4. Mass air flow sensor (with intake air temperature sensor) (bank 1)
5. Refrigerant pressure sensor
6. Cooling fan motor-2
7. Intake valve timing control solenoid valve
8. Exhaust valve timing control magnet retarder
9. Cooling fan control module
10. Cooling fan motor-1
11. Mass air flow sensor (with intake air temperature sensor) (bank 2)
12. ICC brake hold relay (ICC models)
13. Ignition coil (with power transistor) and spark plug (bank 2)
14. Electric throttle control actuator (bank 2)
15. A/F sensor 1 (bank 2)
16. Fuel injector (bank 2)
17. Camshaft position sensor (PHASE)
18. Engine coolant temperature sensor
19. Knock sensor
20. Exhaust valve timing control position sensor
21. EVAP canister purge volume control solenoid valve
22. Fuel injector (bank 1) 23. Ignition coil (with power transistor) and spark plug (bank 1)
24. EVAP service port
25. A/F sensor 1 (bank 1)
26. Crankshaft position sensor (POS)
27. Electric throttle control actuator (bank 1)

22140_IG35_G0052

Fig. 116 Engine performance component locations

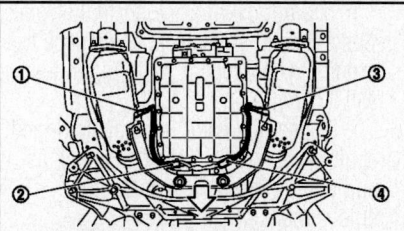

1. Heated oxygen sensor 2 (bank 2)
2. Heated oxygen sensor 2 (bank 2) harness connector
3. Heated oxygen sensor 2 (bank 1)
4. Heated oxygen sensor 2 (bank 1) harness connector

22140_IG35_G0053

Fig. 117 Emission control component locations

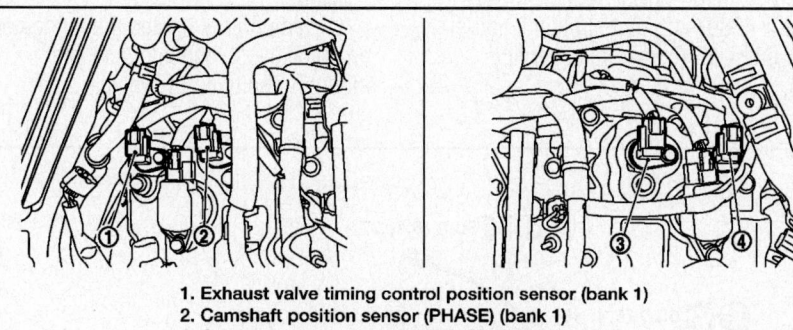

1. Exhaust valve timing control position sensor (bank 1)
2. Camshaft position sensor (PHASE) (bank 1)
3. Camshaft position sensor (PHASE) (bank 2)
4. Exhaust valve timing control position sensor (bank 2)

22140_IG35_G0054

Fig. 118 Location of the camshaft position sensor

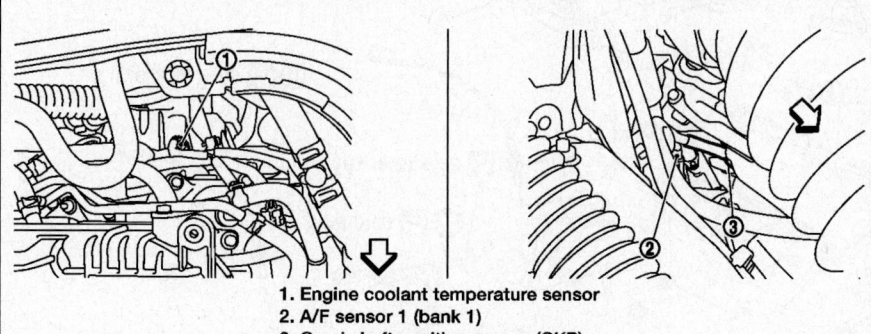

1. Engine coolant temperature sensor
2. A/F sensor 1 (bank 1)
3. Crankshaft position sensor (CKP)

22140_IG35_G0055

Fig. 119 Location of the crankshaft position sensor

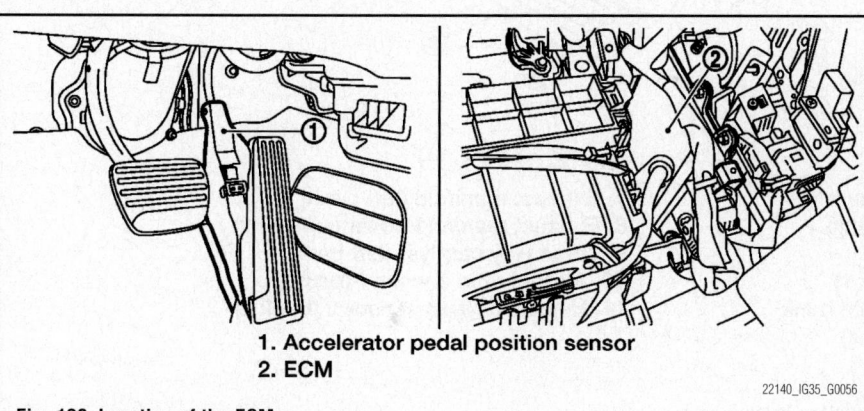

1. Accelerator pedal position sensor
2. ECM

22140_IG35_G0056

Fig. 120 Location of the ECM

REMOVAL & INSTALLATION

To remove the CMP sensor, refer to appropriate illustration.

CRANKSHAFT POSITION (CKP) SENSOR

LOCATION

See Figure 119.

REMOVAL & INSTALLATION

To remove the CKP sensor, refer to appropriate illustration.

ELECTRONIC CONTROL MODULE (ECM)

LOCATION

See Figure 120.

REMOVAL & INSTALLATION

To remove the ECM, refer to appropriate illustration.

ENGINE COOLANT TEMPERATURE (ECT) SENSOR

LOCATION

Refer to illustration under Crankshaft Position (CKP) Sensor to locate the Engine Coolant Temperature Sensor (ECT).

REMOVAL & INSTALLATION

1. Before servicing the vehicle, refer to the Precautions Section.
2. Disconnect the negative battery cable.
3. Drain the cooling system.
4. Remove the sensor electrical connector.
5. Remove the sensor from its mounting.

To install:

6. Installation is the reverse of the removal procedure.

HEATED OXYGEN (HO2S) SENSOR

LOCATION

Refer to Component Locations illustrations for location.

REMOVAL & INSTALLATION

Coupe

See Figure 121.

1. Before servicing the vehicle, refer to the precautions in the beginning of this section.
2. Remove engine cover with power tool.
3. Remove air cleaner case and air duct
4. Remove undercover.
5. Drain engine coolant.
6. Disconnect harness connector and remove heated oxygen sensor 2 on both banks using heated oxygen sensor wrench (SST).

Sedan

See Figure 122.

1. Before servicing the vehicle, refer to the precautions in the beginning of this section.

2. Disconnect each joint and mounting.

3. Refer to illustration and remove necessary components.

❋❋ CAUTION

Be careful not to damage the HO2S.

4. Using HO2S wrench Special Tool: KV10114400 (J-38365), remove the heated oxygen sensor.

To install:

5. To install, reverse removal procedure.

6. Take note of the following:

a. Check for deformation of the grommets.

b. Insert the collar (19 of Components) vertically.

c. Install the collar (5 of Components) with its lower surface horizontal.

d. Temporarily tighten nuts and bolts when installing exhaust pipe assembly. Tighten them to the specified torque when connecting the vehicle rear to the vehicle front.

e. Always replace exhaust tube gaskets with new ones when reassembling.

f. Discard any heated oxygen sensor which has been dropped onto a hard surface such as a concrete floor. Use a new one.

g. Before installing a new heated oxygen sensor, clean exhaust system threads using the heated oxygen sensor thread cleaner, Special Tool: J-43897-18 or J-43897-12, and apply the anti-seize lubricant (commercial service tool).

h. Do not over torque heated oxygen sensor. Doing so may cause damage to heated oxygen sensor, resulting in the "MIL" coming on.

i. If heat insulator is badly deformed, repair or replace it. If deposits such as mud pile up on the heat insulator, remove them.

j. When installing heat insulator avoid large gaps or interference between heat insulator and each exhaust pipe.

k. Remove deposits from the sealing surface of each connection. Connect them securely to avoid gases leakage.

l. Temporarily tighten mounting nuts on the exhaust manifold side and mounting bolts on the vehicle side.

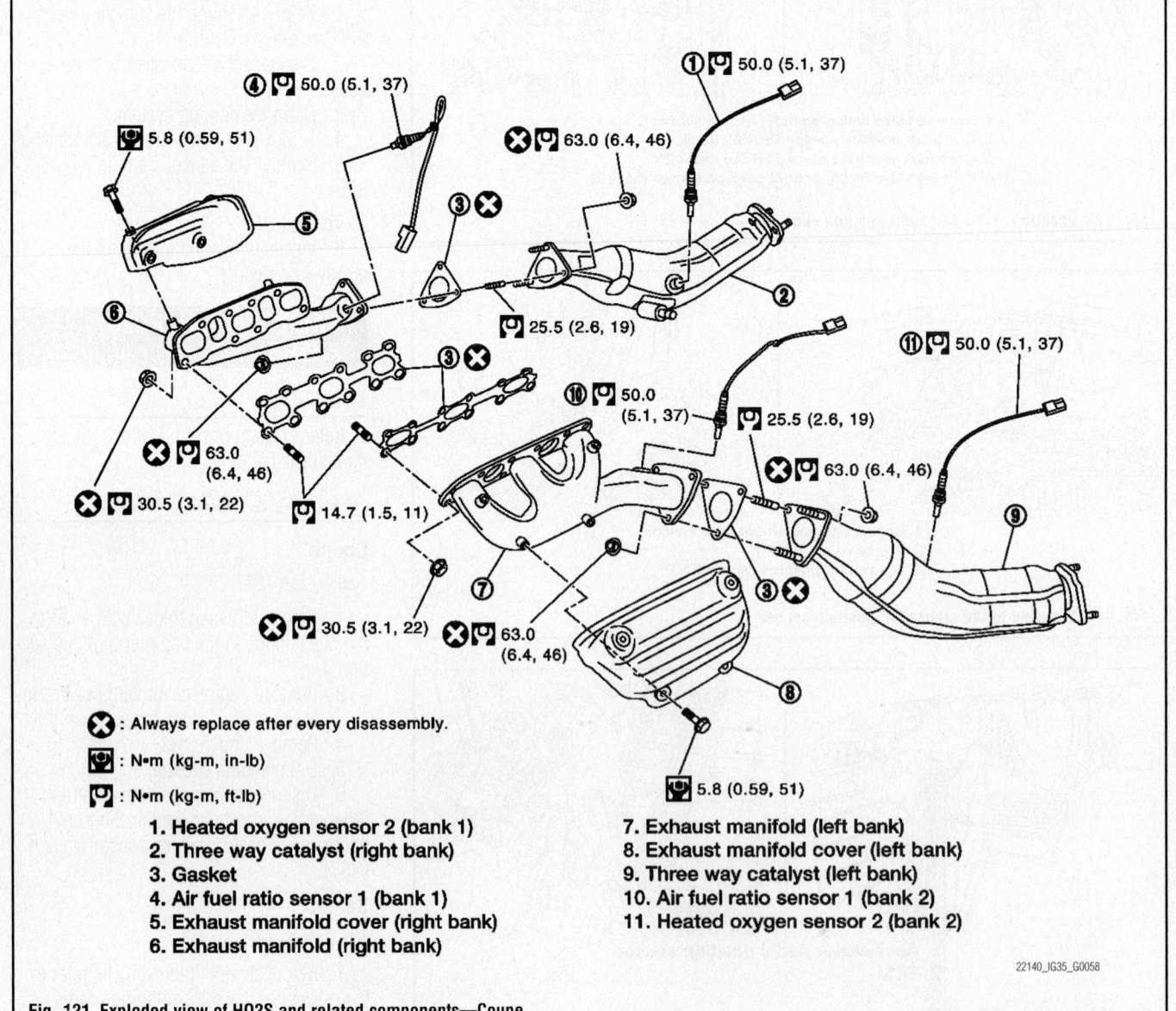

❌ : Always replace after every disassembly.

▣ : N•m (kg-m, in-lb)

▣ : N•m (kg-m, ft-lb)

1. Heated oxygen sensor 2 (bank 1)
2. Three way catalyst (right bank)
3. Gasket
4. Air fuel ratio sensor 1 (bank 1)
5. Exhaust manifold cover (right bank)
6. Exhaust manifold (right bank)
7. Exhaust manifold (left bank)
8. Exhaust manifold cover (left bank)
9. Three way catalyst (left bank)
10. Air fuel ratio sensor 1 (bank 2)
11. Heated oxygen sensor 2 (bank 2)

22140_IG35_G0058

Fig. 121 Exploded view of HO2S and related components—Coupe

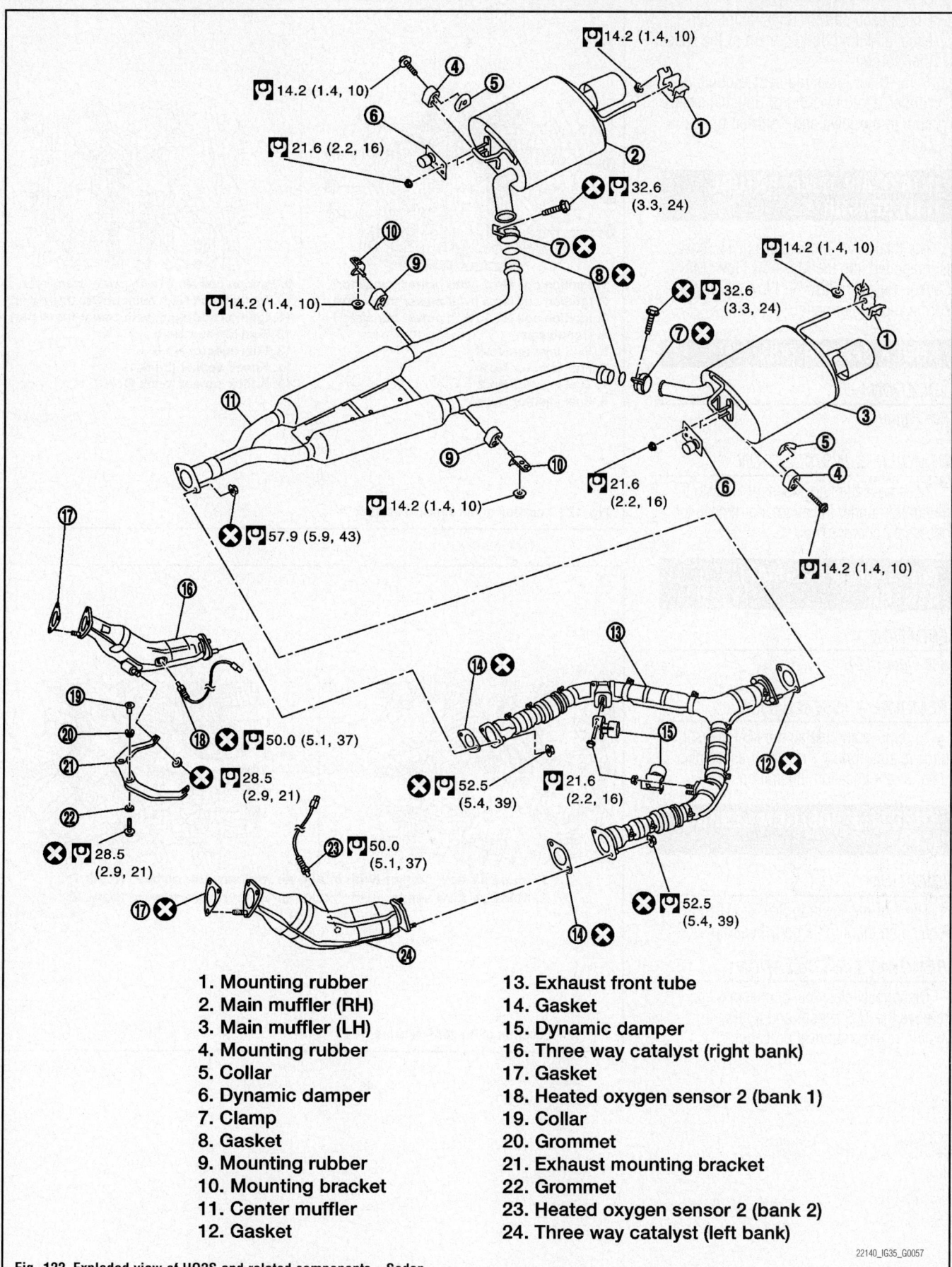

1. Mounting rubber
2. Main muffler (RH)
3. Main muffler (LH)
4. Mounting rubber
5. Collar
6. Dynamic damper
7. Clamp
8. Gasket
9. Mounting rubber
10. Mounting bracket
11. Center muffler
12. Gasket
13. Exhaust front tube
14. Gasket
15. Dynamic damper
16. Three way catalyst (right bank)
17. Gasket
18. Heated oxygen sensor 2 (bank 1)
19. Collar
20. Grommet
21. Exhaust mounting bracket
22. Grommet
23. Heated oxygen sensor 2 (bank 2)
24. Three way catalyst (left bank)

Fig. 122 Exploded view of HO2S and related components—Sedan

22140_IG35_G0057

Check each part for unusual interference, and then tighten them to the specified torque.

m. When installing each mounting rubber, avoid twisting or unusual extension in up/down and right/left directions.

INTAKE AIR TEMPERATURE (IAT) SENSOR

The Intake Air Temperature (IAT) Sensor is integrated into the Mass Air Flow (MAF) Sensor. Refer to Mass Air Flow (MAF) Sensor for more information.

KNOCK SENSOR (KS)

LOCATION

See Figure 123.

REMOVAL & INSTALLATION

To remove the knock sensor, refer to the illustration under Location and remove the necessary components.

MASS AIR FLOW (MAF) SENSOR

LOCATION

See Figure 124.

REMOVAL & INSTALLATION

To remove and install the MAF sensor, remove appropriate air cleaner assemblies. Refer to the Location illustration.

THROTTLE POSITION SENSOR (TPS)

LOCATION

The Throttle Position Sensor (TPS) is located on the throttle body housing.

REMOVAL & INSTALLATION

Disconnect electrical connectors and remove the TPS sensor and bolt(s). To install, reverse removal procedure.

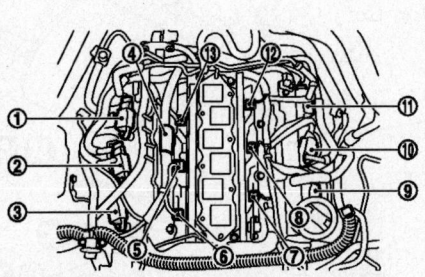
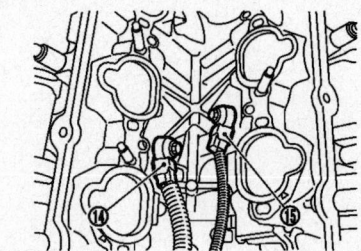

1. Ignition coil No.5 (with power transistor)
2. Ignition coil No.3 (with power transistor)
3. Ignition coil No.1 (with power transistor)
4. Condenser
5. Fuel injector No.3
6. Fuel injector No.1
7. Fuel injector No.2
8. Fuel injector No.4
9. Ignition coil No.2 (with power transistor)
10. Ignition coil No.4 (with power transistor)
11. Ignition coil No.6 (with power transistor)
12. Fuel injector No.6
13. Fuel injector No.5
14. Knock sensor (bank 1)
15. Knock sensor (bank 2)

22140_IG35_G0059

Fig. 123 Location of the knock sensors

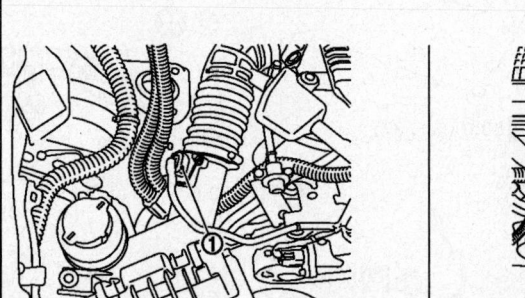
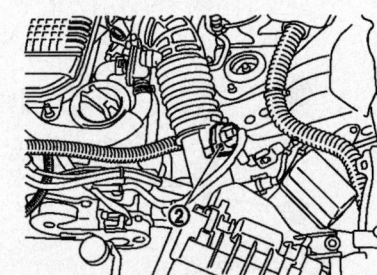

1. Mass air flow sensor (with intake air temperature sensor) (bank 1)
2. Mass air flow sensor (with intake air temperature sensor) (bank 2)

22140_IG35_G0060

Fig. 124 Location of the MAF sensors

FUEL

GASOLINE FUEL INJECTION SYSTEM

FUEL SYSTEM SERVICE PRECAUTIONS

Safety is the most important factor when performing not only fuel system maintenance but any type of maintenance. Failure to conduct maintenance and repairs in a safe manner may result in serious personal injury or death. Maintenance and testing of the vehicle's fuel system components can be accomplished safely and effectively by adhering to the following rules and guidelines.

• To avoid the possibility of fire and personal injury, always disconnect the negative battery cable unless the repair or test procedure requires that battery voltage be applied.

• Always relieve the fuel system pressure prior to disconnecting any fuel system component (injector, fuel rail, pressure regulator, etc.), fitting or fuel line connection. Exercise extreme caution whenever relieving fuel system pressure to avoid exposing skin, face and eyes to fuel spray. Please be advised that fuel under pressure may penetrate the skin or any part of the body that it contacts.

• Always place a shop towel or cloth around the fitting or connection prior to loosening to absorb any excess fuel due to spillage. Ensure that all fuel spillage (should it occur) is quickly removed from engine surfaces. Ensure that all fuel soaked cloths or towels are deposited into a suitable waste container.

• Always keep a dry chemical (Class B) fire extinguisher near the work area.

• Do not allow fuel spray or fuel vapors to come into contact with a spark or open flame.

• Always use a back-up wrench when loosening and tightening fuel line connection fittings. This will prevent unnecessary stress and torsion to fuel line piping.

• Always replace worn fuel fitting O-rings with new Do not substitute fuel hose or equivalent where fuel pipe is installed.

Before servicing the vehicle, make sure to also refer to the precautions in the beginning of this section as well.

RELIEVING FUEL SYSTEM PRESSURE

See Figure 125.

❊❊ CAUTION

Observe all applicable safety precautions when working around fuel.

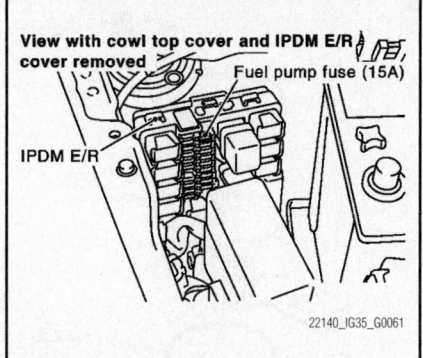

Fig. 125 Location of the fuel pump fuse

Whenever servicing the fuel system, always work in a well ventilated area. Do not allow fuel spray or vapors to come in contact with a spark or open flame. Keep a dry chemical fire extinguisher near the work area. Always keep fuel in a container specifically designed for fuel storage; also, always properly seal fuel containers to avoid the possibility of fire or explosion.

1. Before servicing the vehicle, refer to the precautions in the beginning of this section.

2. Remove fuel pump fuse located in IPDM E/R.

3. Start engine.

4. After engine stalls, crank it two or three times to release all fuel pressure.

5. Turn ignition switch OFF.

6. Reinstall fuel pump fuse after servicing fuel system.

FUEL FILTER

REMOVAL & INSTALLATION
See Figures 126 and 127.

❊❊ CAUTION

Observe all applicable safety precautions when working around fuel. Whenever servicing the fuel system, always work in a well ventilated area. Do not allow fuel spray or vapors to come in contact with a spark or open flame. Keep a dry chemical fire extinguisher near the work area. Always keep fuel in a container specifically designed for fuel storage; also, always properly seal fuel containers to avoid the possibility of fire or explosion.

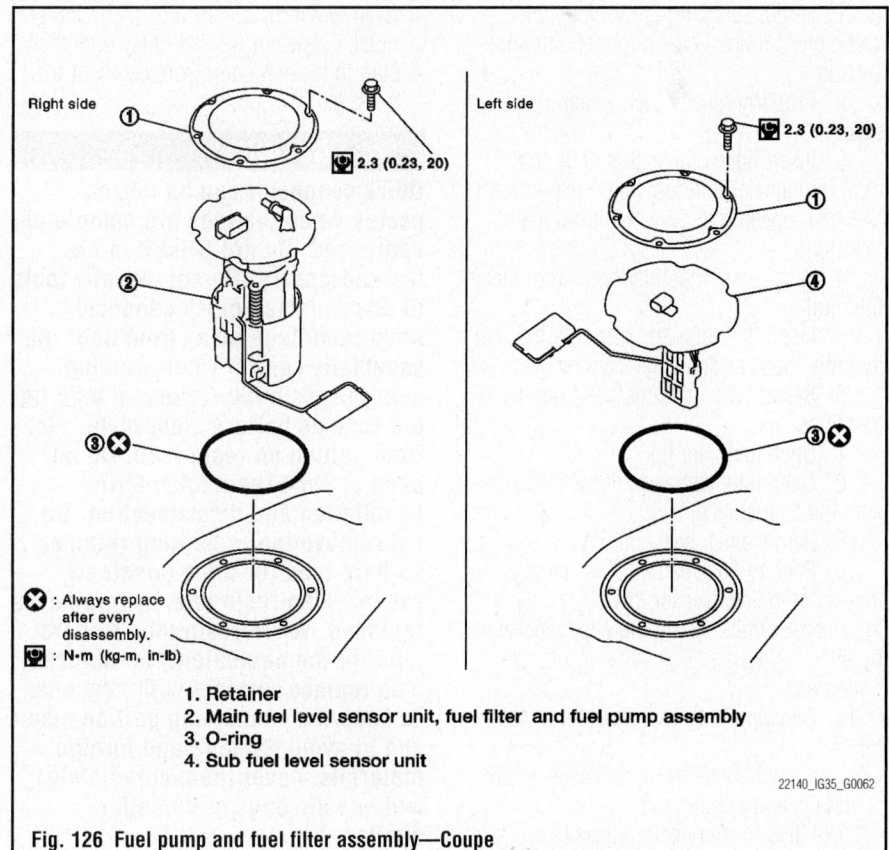

1. Retainer
2. Main fuel level sensor unit, fuel filter and fuel pump assembly
3. O-ring
4. Sub fuel level sensor unit

22140_IG35_G0062

Fig. 126 Fuel pump and fuel filter assembly—Coupe

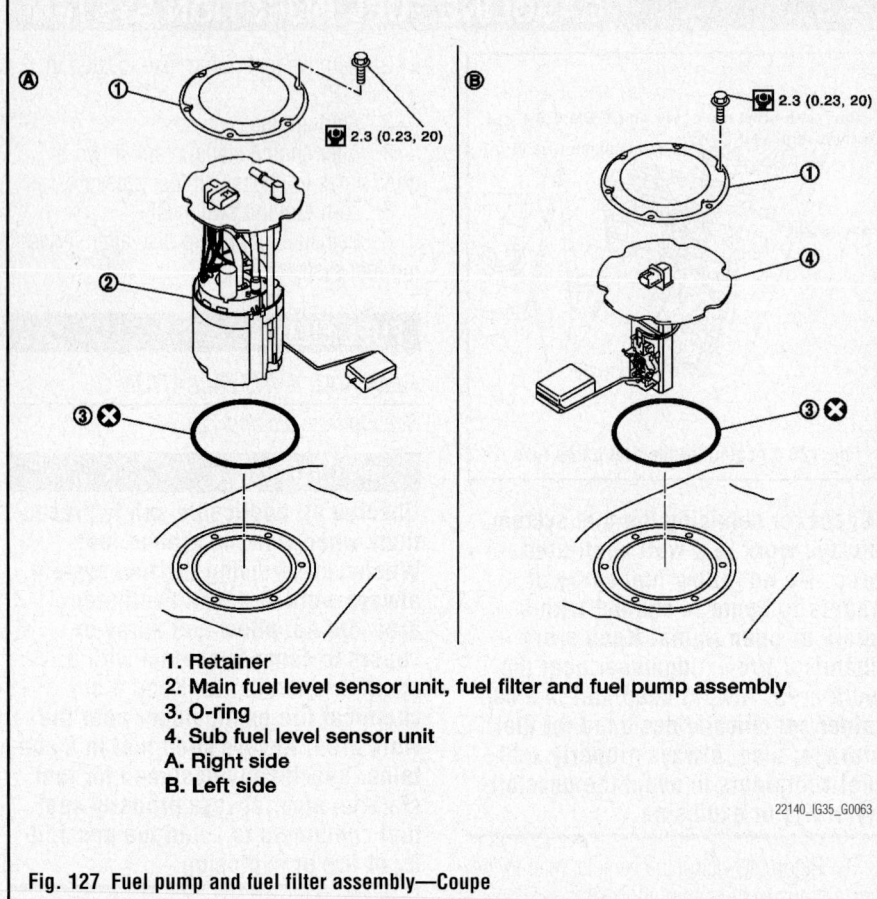

1. Retainer
2. Main fuel level sensor unit, fuel filter and fuel pump assembly
3. O-ring
4. Sub fuel level sensor unit
A. Right side
B. Left side

22140_IG35_G0063

Fig. 127 Fuel pump and fuel filter assembly—Coupe

1. Before servicing the vehicle, refer to the precautions in the beginning of this section.

2. Properly relieve fuel system pressure.

3. Insert hose of less than 0.98 in (25 mm) in diameter into fuel filler tube through fuel filler opening to draw fuel from fuel filler tube.

4. Disconnect fuel filler hose from fuel filler tube.

5. Insert fuel tube into fuel tank through fuel filler hose to draw fuel from fuel tank.

6. Release the fuel pressure from the fuel lines.

7. Open fuel filler lid.

8. Open filler cap and release the pressure inside fuel tank.

9. Remove rear seat cushion.

10. Peel off floor carpet, then remove inspection hole cover for main and sub fuel level sensor units by turning clips clockwise by 90 degrees.

11. Disconnect quick connector as follows:

a. Hold the sides of connector, push in tabs and pull out tube.

b. If quick connector sticks to tube of main fuel level sensor unit, push and pull quick connector several times until they start to move then disconnect them by pulling.

✳✳ CAUTION

Quick connector can be disconnected when the tabs are completely depressed. Do not twist it more than necessary. Do not use any tools to disconnected quick connector. Keep resin tube away from heat. Be especially careful when welding near the resin tube. Prevent acid liquid such as battery electrolyte, etc. from getting on resin tube. Do not bend or twist resin tube during installation and disconnection. Do not remove the remaining retainer on hard tube (or the equivalent) except when resin tube or retainer is replaced. When resin tube or hard tube (or the equivalent) is replaced, also replace retainer with new one. To keep the connecting portion clean and to avoid damage and foreign materials, cover them completely with plastic bags or something similar.

12. Remove main fuel level sensor unit, fuel filter and fuel pump assembly and sub fuel level sensor unit as follows:

a. To remove main fuel level sensor unit, fuel filter and fuel pump assembly:
• Remove retainer
• Raise main fuel level sensor unit, fuel filter and fuel pump assembly, and using snap ring pliers, remove fuel hose connector

b. To remove sub fuel level sensor unit:
• Remove retainer.
• Raise and release sub fuel level sensor unit to remove

To install:

13. To install, reverse removal procedure.

14. Connect quick connector as follows:

a. Check the connection for damage or any foreign materials.

b. Align the connector with the tube, then insert the connector straight into the tube until a click sound is heard.

c. After connecting, check that the connection is secure by following method.
• Pull the tube and the connector to check they are securely connected
• Visually confirm that the two retainer tabs are connected to the connector

15. Turn ignition switch "ON" (with engine stopped), then check connections for leakage by applying fuel pressure to fuel piping.

16. Start engine and let it idle and check there are no fuel leakage at the fuel system connections.

FUEL INJECTORS

REMOVAL & INSTALLATION

1. Before servicing the vehicle, refer to the precautions in the beginning of this section.

2. Relieve the fuel system pressure.

3. Remove or disconnect the following:
• Negative battery cable
• Engine cover, as necessary
• Intake manifold collector
• Vacuum hose from the fuel pressure regulator
• Fuel hoses from fuel rail
• Fuel rail bolts
• Injector harness connectors
• Injectors and the fuel rail as an assembly

- Injector(s) from the fuel rail by pushing them out

4. Remove and discard the fuel injector O-rings

To install:

5. Lubricate the new O-rings with clean engine oil and install the O-rings on the injector(s).

6. Install or connect the following:
- Fuel injectors to the fuel rail
- Fuel rail and injectors as an assembly to the intake manifold

7. Tighten the fuel rail bolts in the following sequence;
   a. Step 1: 7–8 ft. lbs. (9–10 Nm).
   b. Step 2: 15–20 ft. lbs. (21–26 Nm).
- Fuel hoses to the fuel rail
- Vacuum hose to the fuel pressure regulator
- Intake manifold collector
- Engine cover, as necessary
- Negative battery cable

8. Start the vehicle, check for leaks and repair if necessary.

FUEL PUMP

REMOVAL & INSTALLATION

The fuel pump is integrated with the fuel filter. Refer to Fuel Filter in this section for removal and installation procedure.

FUEL TANK

REMOVAL & INSTALLATION

See Figure 128.

1. Before servicing the vehicle, refer to the precautions in the beginning of this section.

2. Relieve the fuel system pressure.

3. Drain fuel from fuel tank if necessary.

4. Remove fuel pump assembly.

5. Remove exhaust front tube, center muffler and main muffler.

6. Remove appropriate drive shaft(s) as necessary.

7. Remove rear parking brake cables.

8. Remove rear suspension assembly.

9. Remove fuel tank protector.

10. Disconnect fuel filler hose, vent hose and EVAP hoses at fuel tank side.

11. Support the lower part of fuel tank with transmission jack.

12. Remove fuel tank mounting bands.

✳✳ CAUTION

Make sure that all connection points have been disconnected. Confirm there is no interference with vehicle.

13. Supporting with hands, descend transmission jack carefully, and remove fuel tank.

14. Remove fuel filler tube, as necessary.

To install:

15. To install, reverse removal procedure.

16. Turn ignition switch "ON" (with engine stopped), and check connections for leakage by applying fuel pressure to fuel piping.

17. Start engine and rev it up and make sure there are no fuel leaks at the fuel system tube and hose connections.

18. After removing/installing rear suspension assembly, make sure to adjust wheel alignment and then, adjust neutral position of steering angle sensor.

IDLE SPEED

ADJUSTMENT

Idle speed is maintained by the Powertrain Control Module (PCM). No adjustment is necessary or possible.

THROTTLE BODY

REMOVAL & INSTALLATION

See Figures 129 and 130.

1. Before servicing the vehicle, refer to the precautions in the beginning of this section.

2. Disconnect the negative battery cable.

3. Remove engine cover.

4. Remove air cleaner case and air duct(s).

5. Drain and recycle coolant.

6. For sedans, disconnect water hoses from electric throttle control actuator. When engine coolant is not drained from radiator, attach plug to water hoses to prevent engine coolant leakage

7. Disconnect harness connector.

8. If necessary, disconnect vacuum hose, PCV hose and EVAP hose from intake manifold collector.

9. Loosen bolts and remove the throttle body.

To install:

10. Install gasket with three protrusions for installation check facing any direction other than upward.

11. To complete installation, reverse remaining removal procedure.

12. Perform Throttle Valve Closed Position Learning procedure:
   a. Make sure that accelerator pedal is fully released.
   b. Turn ignition switch ON.
   c. Turn ignition switch OFF and wait at least 10 seconds. Make sure that throttle valve moves during above 10 seconds by confirming the operating sound.

13. Perform the Idle Air Volume Learning procedure using an ODB II scan tool or equivalent.

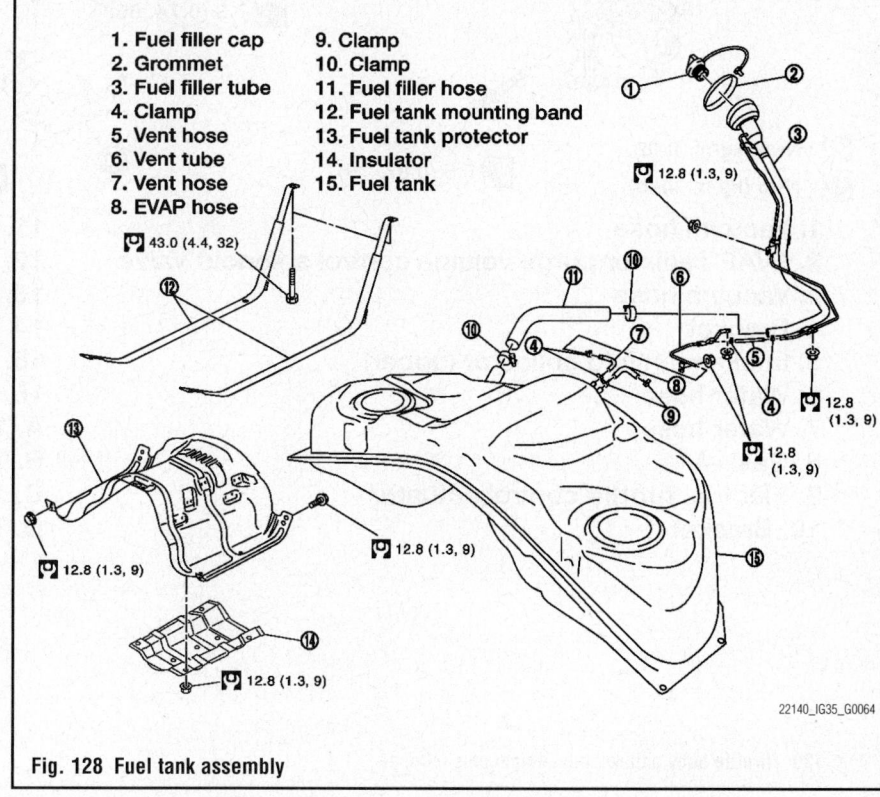

1. Fuel filler cap
2. Grommet
3. Fuel filler tube
4. Clamp
5. Vent hose
6. Vent tube
7. Vent hose
8. EVAP hose
9. Clamp
10. Clamp
11. Fuel filler hose
12. Fuel tank mounting band
13. Fuel tank protector
14. Insulator
15. Fuel tank

43.0 (4.4, 32)

12.8 (1.3, 9)

22140_IG35_G0064

Fig. 128 Fuel tank assembly

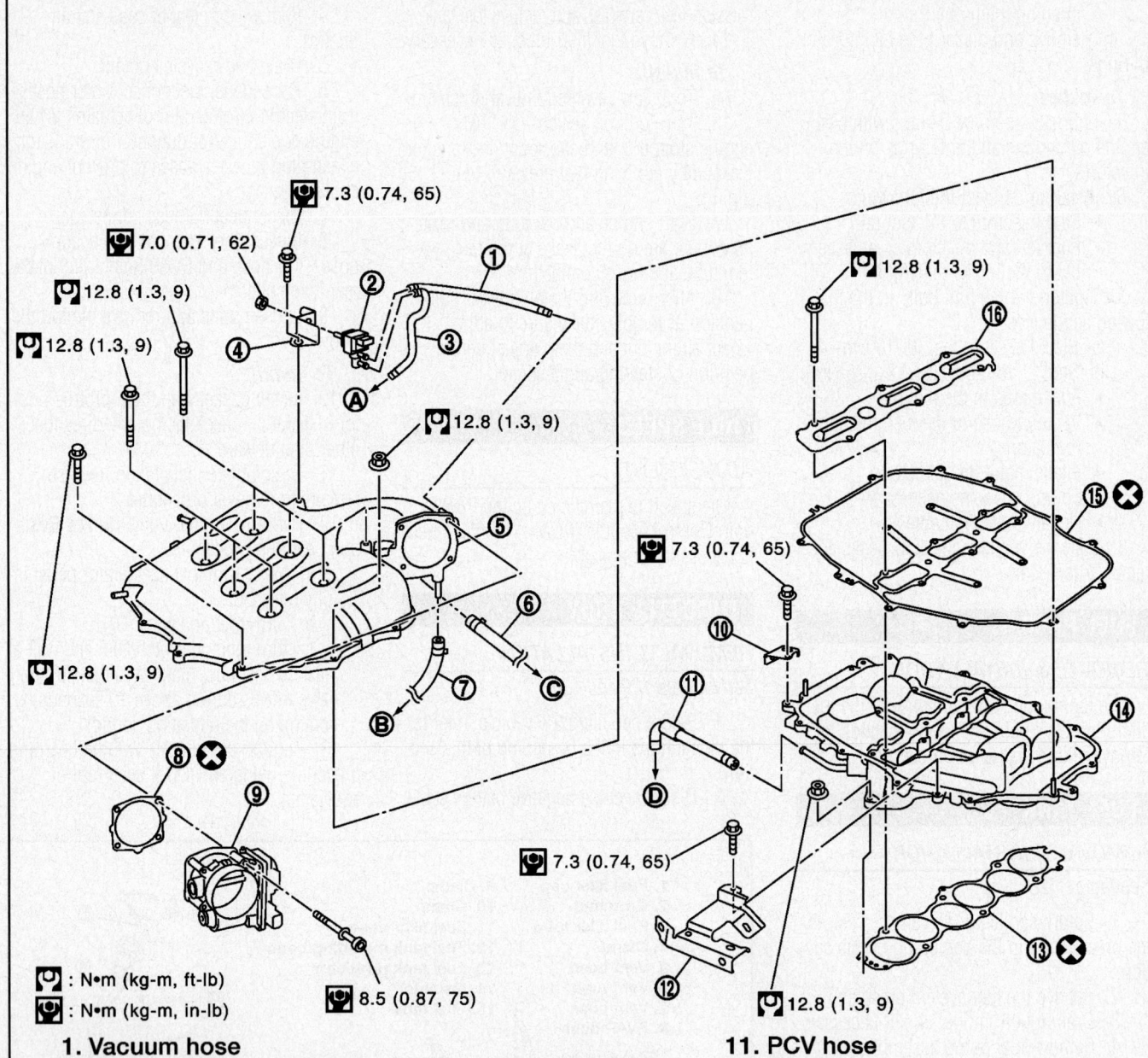

7.3 (0.74, 65)

7.0 (0.71, 62)

12.8 (1.3, 9)

12.8 (1.3, 9)

12.8 (1.3, 9)

12.8 (1.3, 9)

12.8 (1.3, 9)

12.8 (1.3, 9)

7.3 (0.74, 65)

7.3 (0.74, 65)

12.8 (1.3, 9)

8.5 (0.87, 75)

: N•m (kg-m, ft-lb)

: N•m (kg-m, in-lb)

1. Vacuum hose
2. EVAP canister purge volume control solenoid valve
3. Vacuum hose
4. Bracket
5. Intake manifold collector (upper)
6. Water hose
7. Water hose
8. Gasket
9. Electric throttle control actuator
10. Bracket
11. PCV hose
12. Bracket
13. Gasket
14. Intake manifold collector (lower)
15. Gasket
16. Intake manifold collector cover
A. To vacuum pipe
B. To water outlet
C. To heater pipe
D. To PCV valve

22140_IG35_G0065

Fig. 129 Throttle body and related components—Coupe

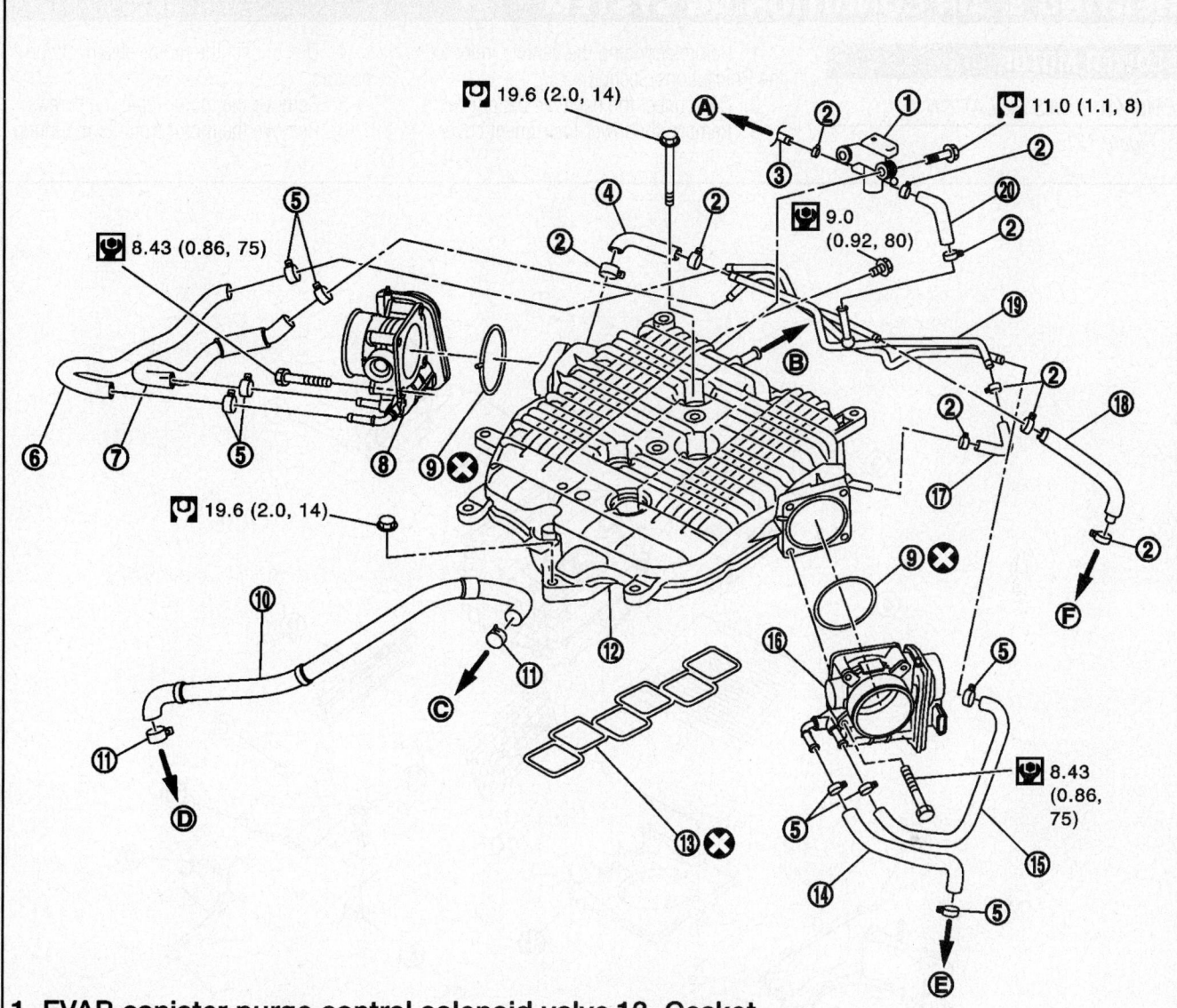

19.6 (2.0, 14)

11.0 (1.1, 8)

9.0 (0.92, 80)

8.43 (0.86, 75)

19.6 (2.0, 14)

8.43 (0.86, 75)

1. EVAP canister purge control solenoid valve
2. Clamp
3. EVAP hose
4. EVAP hose
5. Clamp
6. Water hose
7. Water hose
8. Electric throttle control actuator (bank1)
9. Gasket
10. PCV hose
11. Clamp
12. Intake manifold collector
13. Gasket
14. Water hose
15. Water hose
16. Electric throttle control actuator (bank2)
17. EVAP hose
18. Water hose
19. EVAP tube assembly
20. EVAP hose
A. To vacuum pipe
B. To brake booster
C. To intake manifold collector
D. To PCV valve
E. To heater pipe
F. To water outlet (rear)

22140_IG35_G0066

Fig. 130 Throttle body and related components—Sedan

HEATING & AIR CONDITIONING SYSTEM

BLOWER MOTOR

REMOVAL & INSTALLATION

See Figure 131.

1. Before servicing the vehicle, refer to the Precautions Section.
2. Disconnect the negative battery cable.
3. Remove the lower instrument cover.
4. Disconnect the motor electrical connectors.
5. Remove the motor retaining screws.
6. Remove the motor from its mounting.

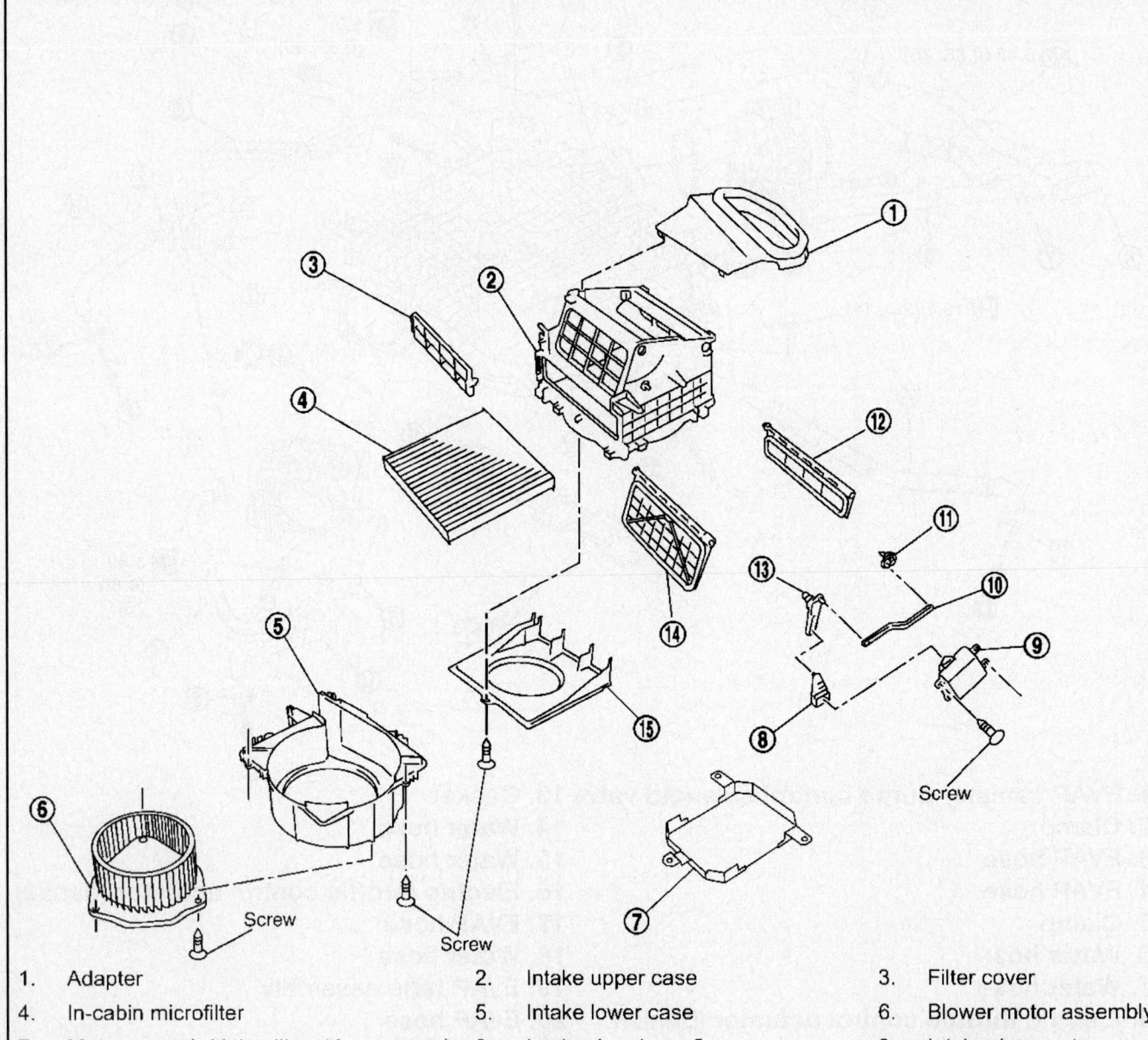

1. Adapter	2. Intake upper case	3. Filter cover
4. In-cabin microfilter	5. Intake lower case	6. Blower motor assembly
7. Motor cover (with intelligent key system)	8. Intake door lever 2	9. Intake door motor
10. Intake door link	11. Intake door lever 3	12. Intake door 2
13. Intake door lever 1	14. Intake door 1	15. Intake bell mouth

22140_IG35_G0500

Fig. 131 Blower motor and related components—2006 models and 2007 coupe

To install:

7. Installation is the reverse of the removal procedure.

HEATER CORE

REMOVAL & INSTALLATION

See Figure 132.

1. Disconnect both battery cables, the negative (() cable first.

✳✳ CAUTION

After disconnecting the battery, wait for a least 3 minutes for the SRS module to deplete its energy before working on the steering column or instrument panel.

2. Drain the cooling system into a clean container for reuse.

3. Use a refrigerant collecting equipment (for HFC-134a) to discharge refrigerant.

4. Remove cowl top cover as follows:
- Remove hood ledge cover.
- Remove both right/left wiper arms.
- Remove cowl top seal rubber.
- Remove clips, cap of cowl top cover and remove cowl top cover (right).
- Remove clips, cap, screws and remove cowl top cover (left).
- Remove washer nozzles and hose from cowl top cover.

5. Disconnect low-pressure flexible hose and high-pressure pipe from evaporator.

✳✳ WARNING

Cap or wrap the joint of the pipe with suitable material such as vinyl tape to avoid the entry of air.

6. Remove air hose and electronic control throttle assembly.

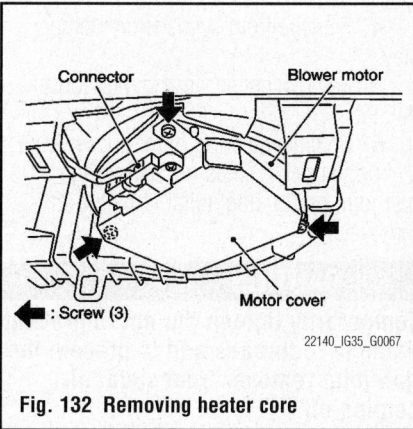

Fig. 132 Removing heater core

7. Disconnect two heater hoses from heater core.

8. Remove instrument panel assembly as follows:
- Remove shift knob (manual transmission)/gear selector knob (automatic transmission)
- Place selector lever in DRIVE position (automatic transmission)
- Insert a thin flat-bladed screwdriver wrapped with tape from behind console boot/console finisher and remove metal clip(s) on back, then remove clips at front (automatic transmission). Then pull up and back to disengage from console.

✳✳ WARNING

Guide pin inserted into automatic transmission device guide can be easily broken; exercise caution during removal.

- Disconnect hazard switch harness connector, and remove console boot/console finisher.

9. Remove or disconnect the following:
- Cluster Lid "C" side finisher screws (left and right) and pawls; remove finisher
- Upper Cluster Lid finisher clips and pawls; remove finisher
- Lower Cluster Lid finisher clips and pawls; remove finisher
- Instrument Panel finisher "B" clips; remove finisher
- Cluster Lid "C" screws; disconnect display/audio unit harness connector and remove unit
- Driver Side Lower Instrument Panel screws and clips; disconnect harness connectors and remove panel
- Hood lock cable and grommet
- Front lower steering column cover
- Upper and lower side steering column covers
- Lighting and turn signal switch
- Wiper and washer switch
- Lower knee protector
- Steering column lock escutcheon
- Cluster Lid "A" bolts; disconnect combination meter and mirror control switch harness connectors and remove cluster lid
- Left and right front kick panels
- Left and right dash side finishers
- Right side lower instrument panel cover
- Glove box assembly
- Left and right Instrument Side Panels
- Center box assembly

- Navigation Control Unit screws and harness connector; remove control unit
- Passenger side Air Bag Module
- Center console
- Display and Amplifier Assembly screws and harness connector; remove assembly
- Left and right front defroster grilles
- Left side ventilator grille
- Center ventilator grille
- Left and right front pillar garnish
- Lower steering column and remove instrument panel bolts and screws; remove instrument panel and pad through passenger-side door

10. Remove Engine Control Module (ECM) with bracket attached.

11. Remove blower assembly mounting bolt and screws; disconnect harness connectors .

12. Remove blower assembly.

✳✳ WARNING

Move assembly toward right side and remove Location of the pin and joint; remove assembly downward

13. Remove clips of vehicle harness from steering member.

14. Remove mounting nuts and bolts, and then remove instrument stays (driver side and passenger side).

15. Remove mounting bolts from heater & cooling unit.

16. Disconnect drain hose.

17. Remove defroster nozzle and ventilator ducts.

18. Remove steering member, and then remove heater & cooling assembly.

19. Remove heater core.

To install:

20. Installation is the reverse order of removal; observe the following:
- Replace air conditioning O-rings with new O-rings
- Exercise caution when reassembling air conditioning piping; the connection points are thin and deform easily.
- Tighten heater & cooling assembly mounting bolts to 61 inch lbs. (6.9 Nm)
- Tighten steering member mounting nut and bolt to 9 ft. lbs. (12 Nm)

21. Refill cooling system and check for leaks.

22. Recharge air conditioning system and check for leaks.

STEERING

POWER STEERING GEAR

REMOVAL & INSTALLATION

2006 Coupe and Sedan; 2007 Coupe

See Figures 133 and 134.

> ❊❊ **CAUTION**
>
> **Spiral cable may snap due to steering operation if steering column is separated from steering gear assembly. Fix steering wheel with a string to avoid turns.**

1. Before servicing the vehicle, refer to the precautions in the beginning of this section.

> ❊❊ **WARNING**
>
> **Do not turn the steering wheel or column with the steering gear removed.**

2. Drain the power steering fluid.
3. Remove or disconnect the following:
   a. Both front wheels.
   b. Engine undercover.
   c. Front sway bar.
   d. Tie rod ends from the steering knuckles.
   e. Pinch bolts from upper and lower sides.
   f. Oil pipings (high pressure side and low pressure side) from steering gear assembly, then drain84 fluid from pipings.

4. Remove power steering solenoid valve harness connector (with EPS).
   a. Loosen bolt on upper yoke of lower joint and remove bolt on lower yoke of joint, then slide lower joint into lower shaft. Separate steering gear assembly from lower shaft.
   b. Tack bolt on upper yoke of lower joint, fix lower joint to lower shaft.
   c. Bolt and remove steering gear assembly, rack mounting bracket and insulator from vehicle.
   d. Bolt from rack mounting bracket insulator.
   e. Bolts attaching the mounting brackets and the steering gear from the vehicle.

To install:

5. Installation is the reverse of the removal procedure, noting the following:
   a. Tighten all fasteners as shown in the illustration.
   b. Fill the power steering reservoir with fluid and bleed the air from the power steering system.
   c. Check the vehicle front end alignment and adjust as necessary.
   d. Bleeding air from the power steering system if necessary.

2007–08 Sedan

2WD Models

See Figure 135.

1. Before servicing the vehicle, refer to the precautions in the beginning of this section.

> ❊❊ **WARNING**
>
> **Do not turn the steering wheel or column with the steering gear removed.**

2. Perform 4WAS front actuator neutral position adjustment.
3. Remove tires
4. Remove front suspension member stay.
5. Remove cotter pin and then loosen the nut
6. Remove steering outer socket from steering knuckle so as not to damage ball joint boot using suitable ball joint remover.

> ❊❊ **CAUTION**
>
> **Temporarily tighten the nut to prevent damage to threads and to prevent the ball joint remover from suddenly coming off.**

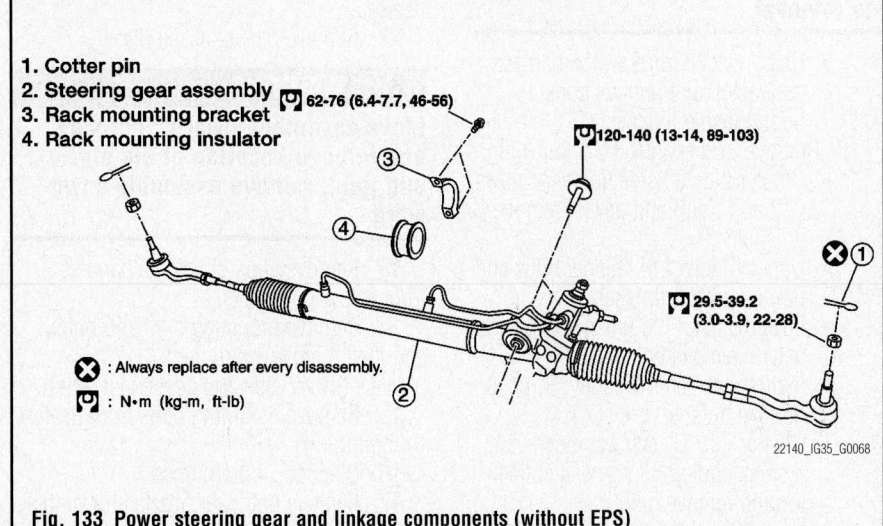

1. Cotter pin
2. Steering gear assembly
3. Rack mounting bracket
4. Rack mounting insulator

62-76 (6.4-7.7, 46-56)
120-140 (13-14, 89-103)
29.5-39.2 (3.0-3.9, 22-28)

❌ : Always replace after every disassembly.

🔧 : N•m (kg-m, ft-lb)

22140_IG35_G0068

Fig. 133 Power steering gear and linkage components (without EPS)

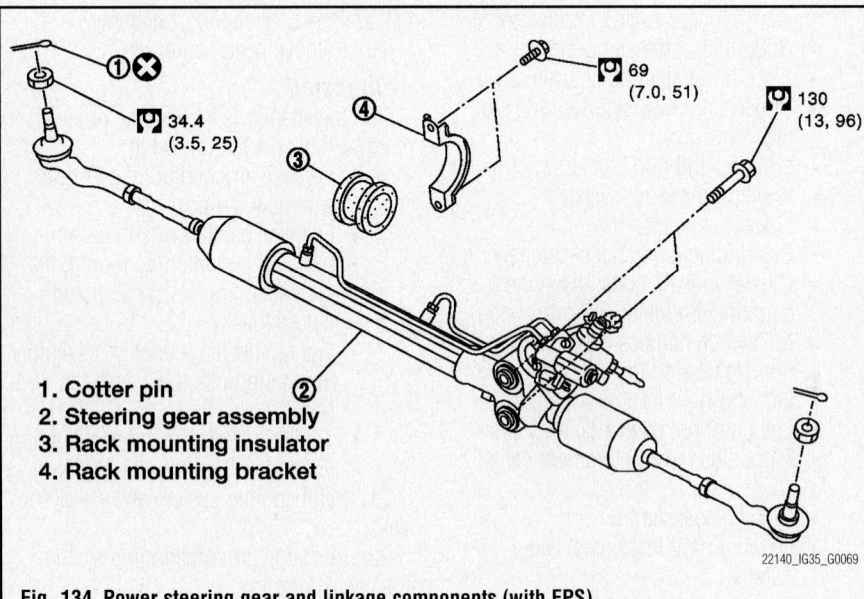

34.4 (3.5, 25)
69 (7.0, 51)
130 (13, 96)

1. Cotter pin
2. Steering gear assembly
3. Rack mounting insulator
4. Rack mounting bracket

22140_IG35_G0069

Fig. 134 Power steering gear and linkage components (with EPS)

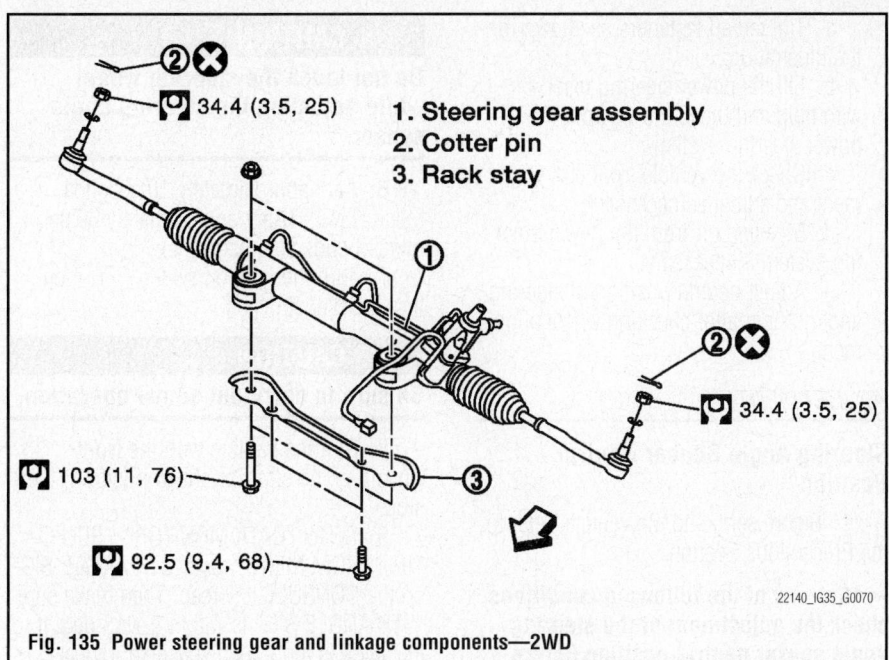

1. Steering gear assembly
2. Cotter pin
3. Rack stay

34.4 (3.5, 25)

34.4 (3.5, 25)

103 (11, 76)

92.5 (9.4, 68)

22140_IG35_G0070

Fig. 135 Power steering gear and linkage components—2WD

7. Remove high pressure piping and low pressure piping of hydraulic piping, and then drain power steering fluid.

8. Remove power steering solenoid valve harness connector.

9. Remove rack stay.

10. Remove lower joint fixing bolt (steering gear side).

11. Separate the lower shaft from the steering gear assembly by sliding the side shaft.

✳✳ CAUTION

Spiral cable may be cut if steering wheel turns while separating steering column assembly and steering

gear assembly. Secure steering wheel using string to avoid turning.

12. Remove steering gear assembly.

To install:

13. Installation is the reverse of the removal procedure, noting the following:

a. Tighten all fasteners as shown in the illustration.

b. Fill the power steering reservoir with fluid and bleed the air from the power steering system.

c. Check the vehicle front end alignment and adjust as necessary.

d. Bleeding air from the power steering system if necessary.

e. Adjust neutral position of steering angle sensor after checking wheel alignment.

f. Perform 4WAS front actuator neutral position adjustment.

AWD Models

See Figure 136.

1. Before servicing the vehicle, refer to the precautions in the beginning of this section.

✳✳ WARNING

Do not turn the steering wheel or column with the steering gear removed.

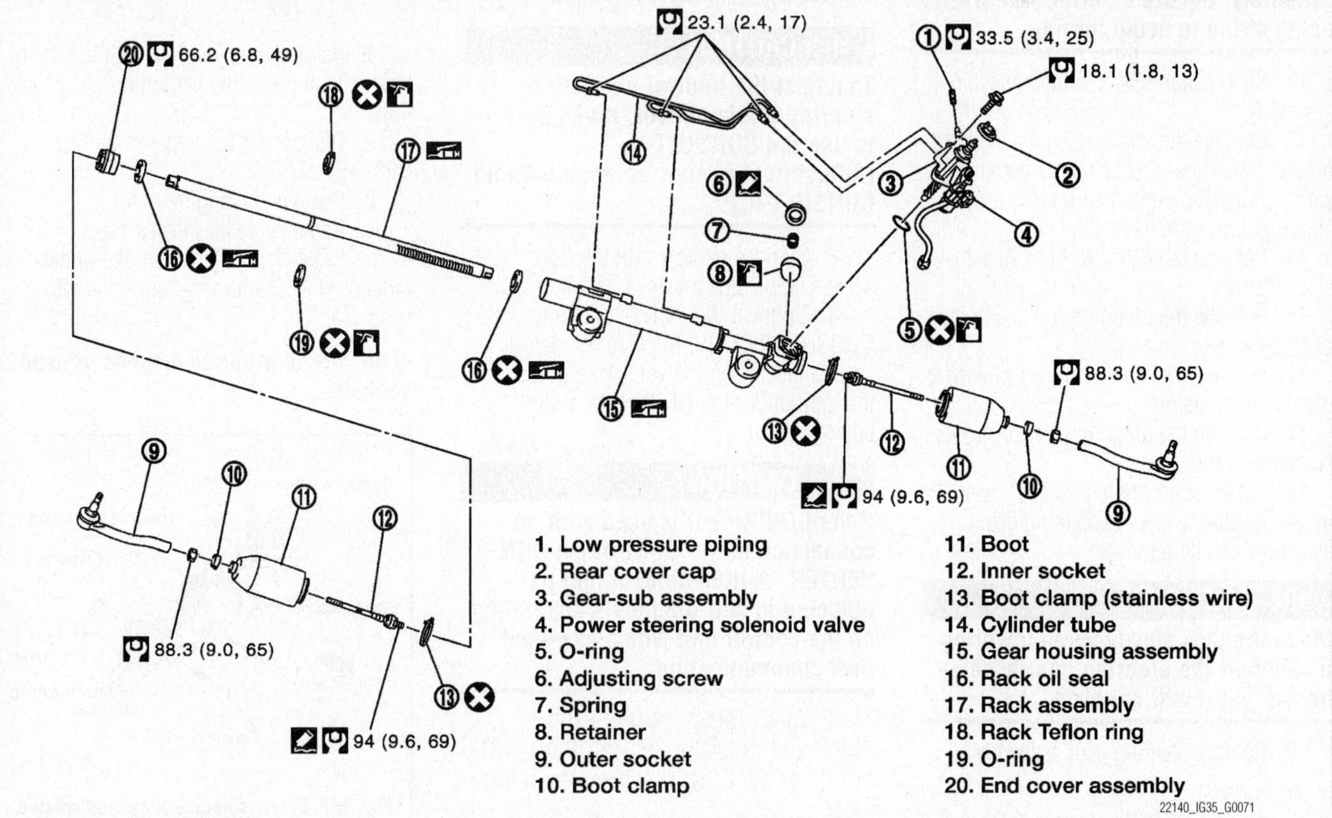

20 66.2 (6.8, 49)

23.1 (2.4, 17)

1 33.5 (3.4, 25)

18.1 (1.8, 13)

18

17

16

6

7

8

19

14

3

2

4

5

15

13

12

11

10

9

88.3 (9.0, 65)

94 (9.6, 69)

9

10

11

12

13

88.3 (9.0, 65)

94 (9.6, 69)

1. Low pressure piping
2. Rear cover cap
3. Gear-sub assembly
4. Power steering solenoid valve
5. O-ring
6. Adjusting screw
7. Spring
8. Retainer
9. Outer socket
10. Boot clamp

11. Boot
12. Inner socket
13. Boot clamp (stainless wire)
14. Cylinder tube
15. Gear housing assembly
16. Rack oil seal
17. Rack assembly
18. Rack Teflon ring
19. O-ring
20. End cover assembly

22140_IG35_G0071

Fig. 136 Power steering gear and linkage components—AWD models

2. Remove tires

3. Remove front stabilizer bar.

4. Remove cotter pin and then loosen the nut

5. Remove steering outer socket from steering knuckle so as not to damage ball joint boot using suitable ball joint remover.

⊛ CAUTION

Temporarily tighten the nut to prevent damage to threads and to prevent the ball joint remover from suddenly coming off.

6. Remove high pressure piping and low pressure piping of hydraulic piping, and then drain power steering fluid.

7. Remove steering hydraulic piping bracket from steering gear assembly.

8. Remove power steering solenoid valve harness connector (steering gear side).

9. Remove rack stay.

10. Remove lower joint fixing bolt (steering gear side).

11. Separate the lower shaft from the steering gear assembly by sliding the side shaft.

⊛ CAUTION

Spiral cable may be cut if steering wheel turns while separating steering column assembly and steering gear assembly. Secure steering wheel using string to avoid turning.

12. Set a suitable jack to transmission assembly.

13. Remove the mounting nuts and bolts on the lower side of shock absorber arm with a power tool, and then remove shock absorber arm from transverse link.

14. Set a suitable jack to front suspension member.

15. Remove the mounting bolts and nuts of steering gear assembly.

16. Remove the mounting nuts of engine mounting insulator.

17. Remove the mounting nuts of front suspension member.

18. Set an appropriate jack and lower it to the position where the steering gear assembly can be removed.

⊛ CAUTION

Move the jack slowly when lowering it. Support the steering gear assembly so that it will not drop.

19. Remove steering gear assembly.

To install:

20. Installation is the reverse of the removal procedure, noting the following:

a. Tighten all fasteners as shown in the illustration.

b. Fill the power steering reservoir with fluid and bleed the air from the power steering system.

c. Check the vehicle front end alignment and adjust as necessary.

d. Bleeding air from the power steering system if necessary.

e. Adjust neutral position of steering angle sensor after checking wheel alignment.

ADJUSTMENT

Steering Angle Sensor Neutral Position

1. Before servicing the vehicle, refer to the Precautions Section.

➡**After any of the following conditions, check the adjustment of the steering angle sensor neutral position before driving the vehicle:**

- Replacing ABS actuator and electric unit (control unit)
- Removing/installing steering angle sensor
- Removing/installing steering components
- Removing/installing suspension components
- Adjusting wheel alignment

⊛ CAUTION

To adjust the neutral position of steering angle sensor, make sure to use the CONSULT-II. (Adjustment cannot be done without CONSULT-II.)

2. Stop the vehicle with the front wheels in the straight-ahead position.

3. Connect the CONSULT-II and CONSULT-II CONVERTER to the data link connector on the vehicle, and turn the ignition switch ON (do not start engine).

⊛ CAUTION

If the CONSULT-II is used with no connection of the CONSULT-II CONVERTER, malfunctions might be detected in self-diagnosis depending on the control unit which carry out CAN communication.

4. Touch "ABS", "WORK SUPPORT" and "ST ANGLE SENSOR ADJUSTMENT" on the CONSULT-II screen in this order.

5. Touch "START".

⊛ CAUTION

Do not touch the steering wheel while adjusting the steering angle sensor.

6. After approximately 10 seconds, touch "END". (After approximately 60 seconds, it ends automatically.)

7. Turn the ignition switch OFF, then turn it ON again.

⊛ CAUTION

Be sure to carry out above operation.

8. Run the vehicle with the front wheels in the straight-ahead position, then stop.

9. Select "DATA MONITOR", "SELECTION FROM MENU", and "STR ANGLE SIG" on the CONSULT-II screen. Then make sure "STR ANGLE SIG" is within 0 ±3.5 deg. If the value is more than specified, repeat steps 3 to 7.

10. Erase memory of the ABS actuator and electric unit (control unit) and the ECM.

11. Turn the ignition switch OFF.

POWER STEERING PUMP

REMOVAL & INSTALLATION

2006 Coupe and Sedan; 2007 Coupe

See Figure 137.

1. Before servicing the vehicle, refer to the precautions in the beginning of this section.

2. Disconnect the negative battery cable.

3. Remove the engine cover.

4. Remove the air cleaner box.

5. Drain the coolant from the upper radiator tank. Remove the upper radiator hose.

➡**Be sure to properly dispose of used coolant.**

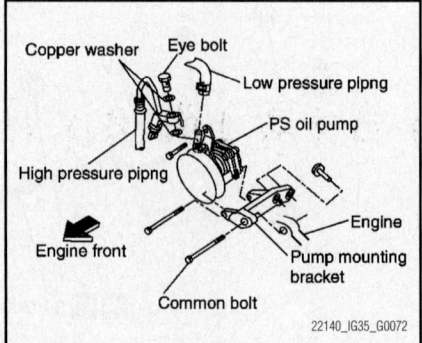

Fig. 137 Power steering pump and related components

6. Remove the radiator shroud.

7. Loosen the idler pulley and remove the belt.

8. Drain the power steering fluid.

9. Disconnect and plug the power steering hose lines.

10. Remove the bolt common to the water pump and power steering pump.

11. Remove the bolt and then the power steering pump.

To install:

12. Installation is the reverse of the removal procedure.

13. Bleed the power steering system.

14. Adjust the belt tension, as required.

2007–08 Sedan

See Figure 138.

1. Before servicing the vehicle, refer to the precautions in the beginning of this section.

2. Disconnect the negative battery cable.

3. Drain power steering fluid from reservoir tank.

4. Remove the right half of the air cleaner and the right half of the air duct.

5. Loosen drive belt.

6. Remove drive belt from oil pump pulley.

7. Remove copper washers and eye bolt (drain fluid from their pipings).

8. Remove suction hose (drain fluid from their pipings).

9. Remove oil pump mounting bolts, and then remove oil pump.

To install:

10. Installation is the reverse of the removal procedure.

11. Bleed the power steering system.

12. Adjust the belt tension, as required.

BLEEDING

1. Before servicing the vehicle, refer to the Precautions Section.

2. Fill the power steering system with the proper grade and type steering fluid.

➡**Do not allow the fluid level in the reservoir tank to go below the MIN level line. Check and add fluid as needed.**

3. Raise and safely support the vehicle.

4. Quickly turn the steering wheel to the full right and left detents and lightly touch the steering stoppers.

➡**Do not hold the steering wheel in the locked position for more than ten seconds.**

5. Repeat this operation until the fluid level no longer decreases.

6. Start the engine.

7. Quickly turn the steering wheel to the full right and left detents and lightly touch the steering stoppers.

➡**Do not hold the steering wheel in the locked position for more than ten seconds.**

8. Check for air bubbles or cloudy fluid. If found, repeat the bleeding procedure.

9. Stop the engine and check the fluid level. Correct as required.

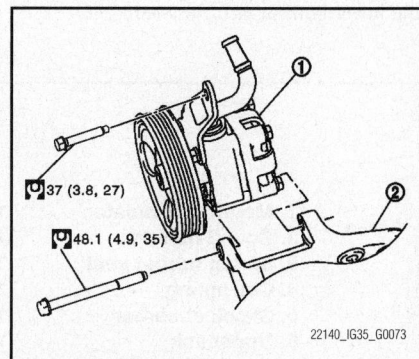

Fig. 138 Power steering pump and related components

SUSPENSION

COIL SPRING

REMOVAL & INSTALLATION

See Figures 139 through 143.

1. Before servicing the vehicle, refer to the precautions in the beginning of this section.

2. Remove the strut and spring assembly.

➡**Make sure piston rod on shock absorber is not damaged when removing components from shock absorber.**

3. Install coil spring compressor tool to shock absorber and fix it in a vise.

❋❋ **CAUTION**

wrap a shop cloth around shock absorber to protect it from damage.

4. Compress coil spring between spring upper seat and spring lower seat (on shock absorber) until coil spring is free.

5. Check that coil spring between spring upper seat and spring lower seat is free and then secure piston rod tip so that piston rod does not turn, and remove piston rod lock nut.

6. Remove mounting insulator, bound bumper, spring upper seat. Then remove coil spring from shock absorber.

7. Gradually release spring compressor and remove coil spring.

❋❋ **CAUTION**

Loosen while making sure coil spring attachment position does not move

8. Check mounting insulator for cracks and rubber parts for wear. Replace them if necessary.

9. Check coil spring for cracks, wear, damage, and replace if necessary.

To install:

➡**Make sure piston rod on shock absorber is not damaged when attaching components to shock absorber.**

10. Install strut attachment (SST) to shock absorber and fix it in a vise.

11. Compress coil spring using a spring compressor and install it onto shock absorber.

FRONT SUSPENSION

12. Install coil spring with large diameter side of 3.94 inches (100 mm) up and small diameter side 3.54 inches (90 mm) down. Identification paint is the 4th winding point from lower side.

❋❋ **CAUTION**

Be sure spring compressor (commercial service tool) is securely attached to coil spring. Compress coil spring.

13. Apply soapy water to bound bumper and insert into mounting insulator. Do not use machine oil.

14. Attach spring upper seat and mounting insulator.

❋❋ **CAUTION**

Make sure coil spring is securely seated in spring mounting groove of spring upper seat.

15. Secure the piston rod tip so that piston rod does not turn, and tighten the specified torque on piston rod lock nut.

16. Gradually release spring compressor and remove coil spring.

✳✳ CAUTION

Loosen spring compressor (commercial service tool) while making sure coil spring attachment position does not move.

17. Remove strut attachment (SST) from shock absorber.
18. Install the strut and spring assembly.

LOWER BALL JOINT

REMOVAL & INSTALLATION

The lower ball joint assembly is part of the lower control arm/transverse link. If replacement of the ball joint is required, the lower control arm needs to be replaced.

LOWER CONTROL ARM

REMOVAL & INSTALLATION

1. Before servicing the vehicle, refer to the precautions in the beginning of this section.
2. Remove or disconnect the following:
   - Negative battery cable
   - Front wheel
   - Engine undercover, if necessary
   - Nuts securing the tension rod to the transverse link (control arm)
   - Nut and separate the ball joint stud from the knuckle
   - Transverse link from the sub-frame

To install:

3. Install or connect the following:
   - Transverse link on the sub-frame. Temporarily install the bolt and nut.
   - Tension rod on the transverse link. Torque the nuts to 87–94 ft. lbs. (118–127 Nm).
   - Nut on the ball joint stud. Torque the nut to 71–88 ft. lbs. (96–120 Nm) for Q45 models or to 59–69 ft. lbs. (75–94 Nm) for G35 models.
   - Front wheel and lower the vehicle to the floor

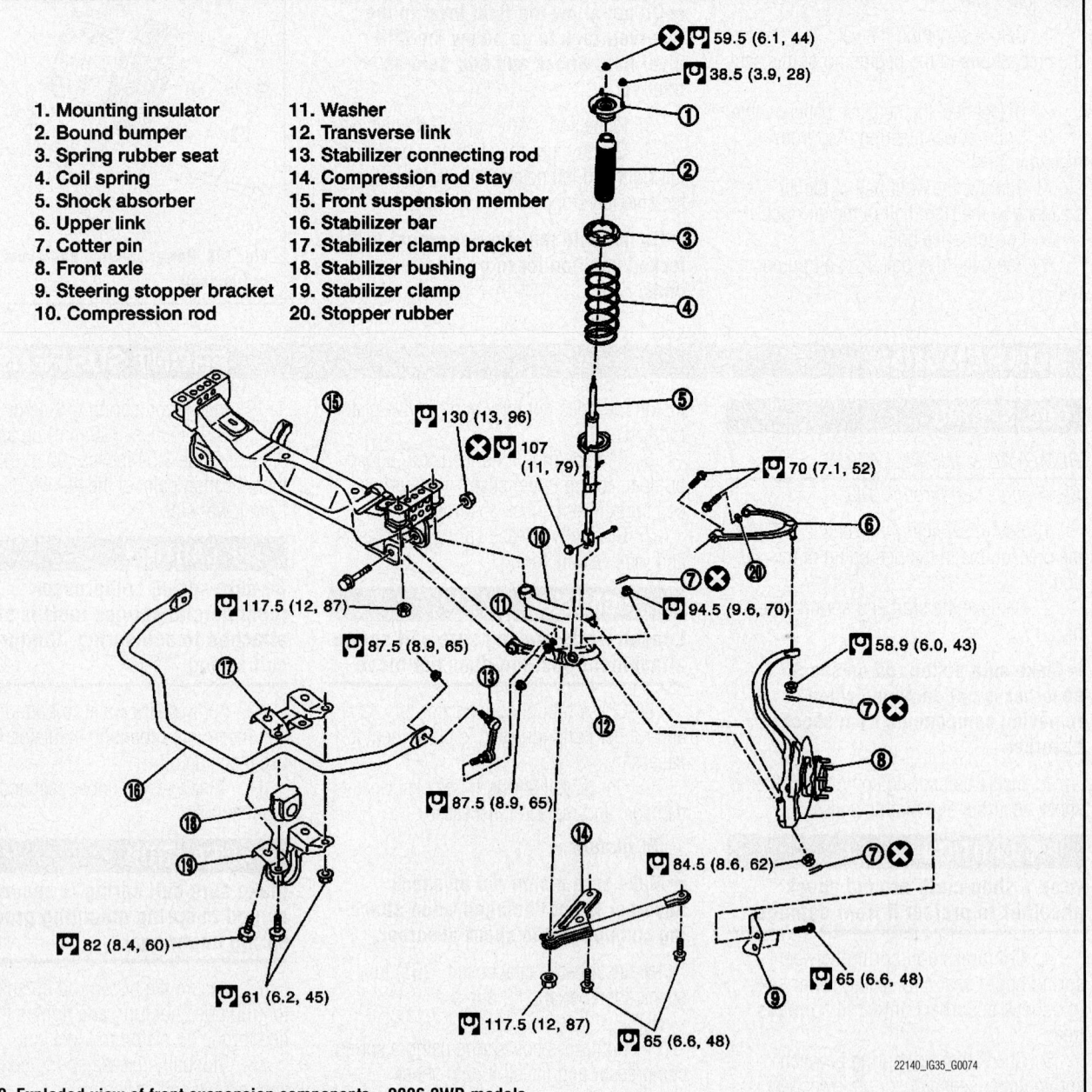

1. Mounting insulator
2. Bound bumper
3. Spring rubber seat
4. Coil spring
5. Shock absorber
6. Upper link
7. Cotter pin
8. Front axle
9. Steering stopper bracket
10. Compression rod
11. Washer
12. Transverse link
13. Stabilizer connecting rod
14. Compression rod stay
15. Front suspension member
16. Stabilizer bar
17. Stabilizer clamp bracket
18. Stabilizer bushing
19. Stabilizer clamp
20. Stopper rubber

Fig. 139 Exploded view of front suspension components—2006 2WD models

22140_IG35_G0074

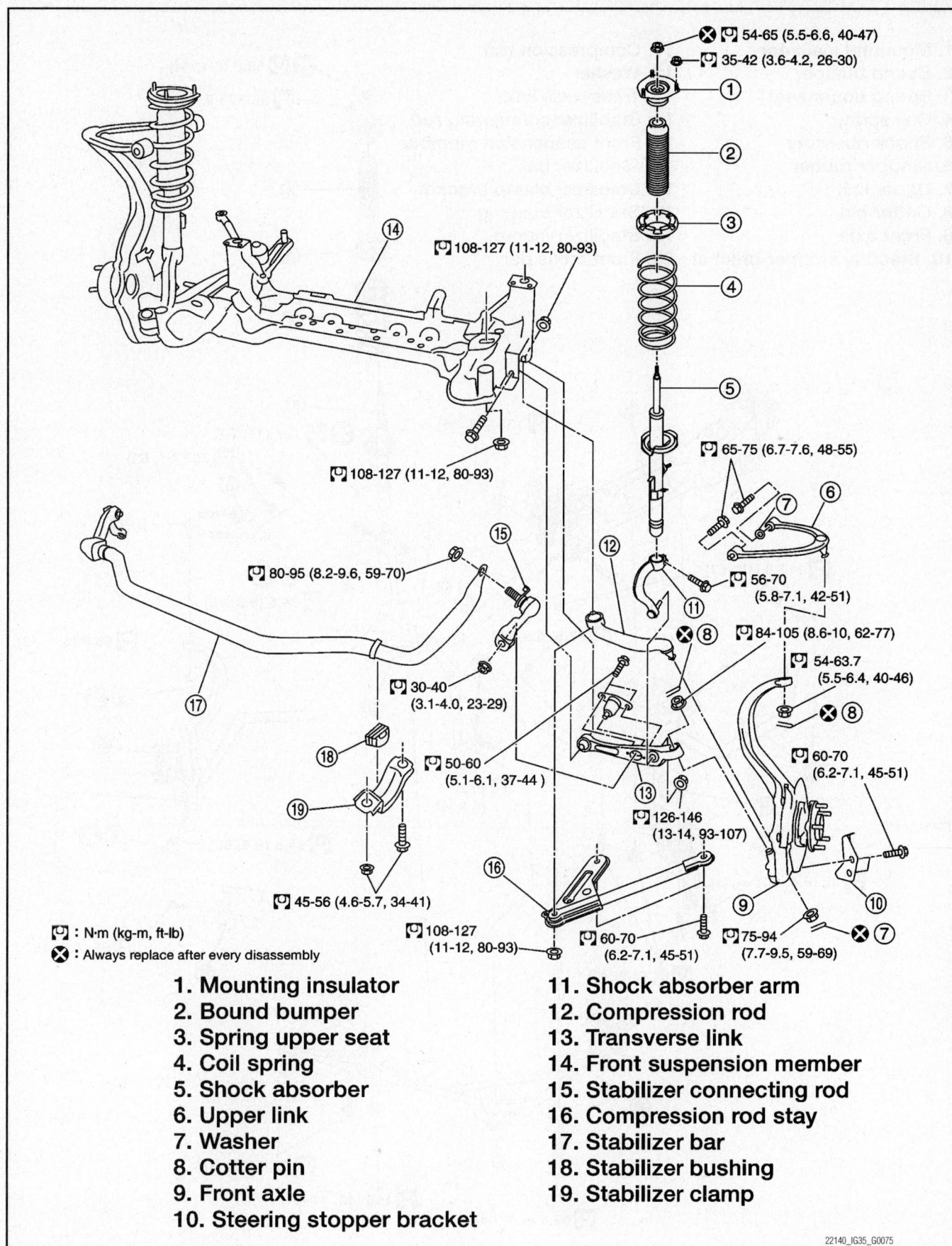

⊡ 54-65 (5.5-6.6, 40-47)
⊡ 35-42 (3.6-4.2, 26-30)
⊡ 108-127 (11-12, 80-93)
⊡ 108-127 (11-12, 80-93)
⊡ 65-75 (6.7-7.6, 48-55)
⊡ 56-70 (5.8-7.1, 42-51)
⊡ 80-95 (8.2-9.6, 59-70)
⊡ 84-105 (8.6-10, 62-77)
⊡ 54-63.7 (5.5-6.4, 40-46)
⊡ 30-40 (3.1-4.0, 23-29)
⊡ 60-70 (6.2-7.1, 45-51)
⊡ 50-60 (5.1-6.1, 37-44)
⊡ 126-146 (13-14, 93-107)
⊡ 45-56 (4.6-5.7, 34-41)
⊡ 108-127 (11-12, 80-93)
⊡ 60-70 (6.2-7.1, 45-51)
⊡ 75-94 (7.7-9.5, 59-69)

⊡ : N·m (kg-m, ft-lb)
⊗ : Always replace after every disassembly

1. Mounting insulator
2. Bound bumper
3. Spring upper seat
4. Coil spring
5. Shock absorber
6. Upper link
7. Washer
8. Cotter pin
9. Front axle
10. Steering stopper bracket
11. Shock absorber arm
12. Compression rod
13. Transverse link
14. Front suspension member
15. Stabilizer connecting rod
16. Compression rod stay
17. Stabilizer bar
18. Stabilizer bushing
19. Stabilizer clamp

22140_IG35_G0075

Fig. 140 Exploded view of front suspension components—2006 AWD models

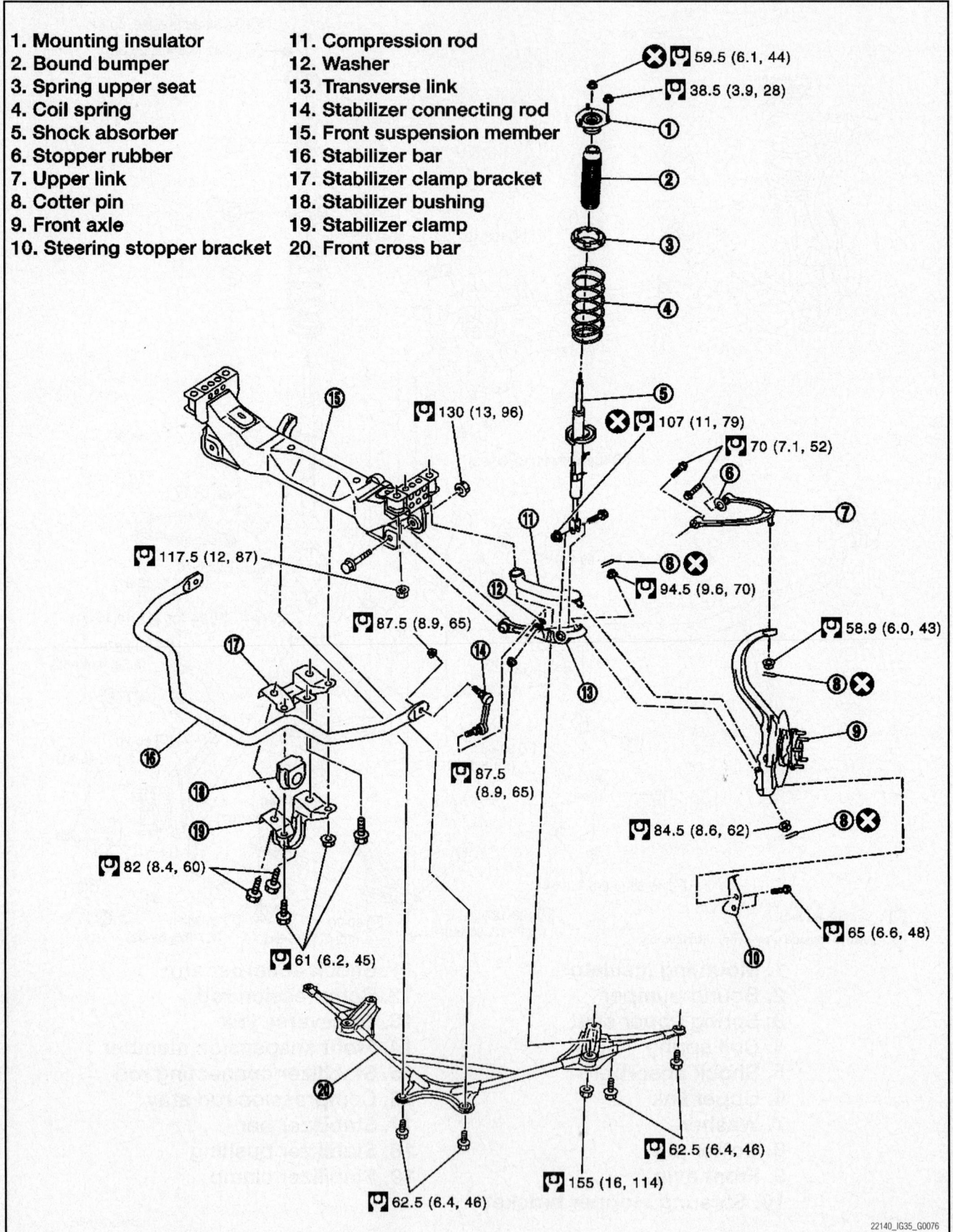

1. Mounting insulator
2. Bound bumper
3. Spring upper seat
4. Coil spring
5. Shock absorber
6. Stopper rubber
7. Upper link
8. Cotter pin
9. Front axle
10. Steering stopper bracket
11. Compression rod
12. Washer
13. Transverse link
14. Stabilizer connecting rod
15. Front suspension member
16. Stabilizer bar
17. Stabilizer clamp bracket
18. Stabilizer bushing
19. Stabilizer clamp
20. Front cross bar

59.5 (6.1, 44)
38.5 (3.9, 28)
130 (13, 96)
107 (11, 79)
70 (7.1, 52)
117.5 (12, 87)
94.5 (9.6, 70)
58.9 (6.0, 43)
87.5 (8.9, 65)
87.5 (8.9, 65)
84.5 (8.6, 62)
82 (8.4, 60)
65 (6.6, 48)
61 (6.2, 45)
62.5 (6.4, 46)
155 (16, 114)
62.5 (6.4, 46)

22140_IG35_G0076

Fig. 141 Exploded view of front suspension components—2007 Coupe

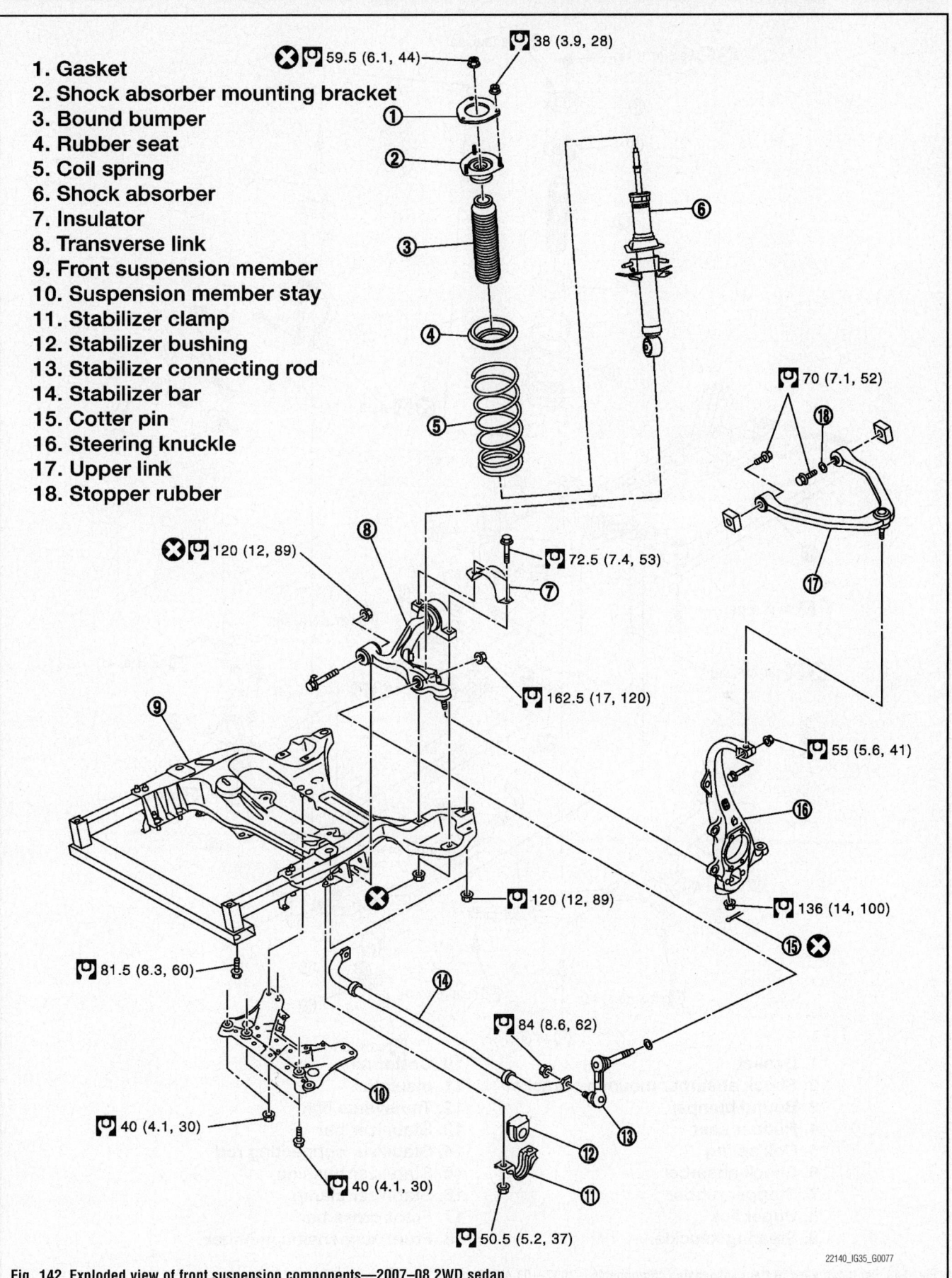

1. Gasket
2. Shock absorber mounting bracket
3. Bound bumper
4. Rubber seat
5. Coil spring
6. Shock absorber
7. Insulator
8. Transverse link
9. Front suspension member
10. Suspension member stay
11. Stabilizer clamp
12. Stabilizer bushing
13. Stabilizer connecting rod
14. Stabilizer bar
15. Cotter pin
16. Steering knuckle
17. Upper link
18. Stopper rubber

38 (3.9, 28)

59.5 (6.1, 44)

70 (7.1, 52)

120 (12, 89)

72.5 (7.4, 53)

162.5 (17, 120)

55 (5.6, 41)

120 (12, 89)

136 (14, 100)

81.5 (8.3, 60)

84 (8.6, 62)

40 (4.1, 30)

40 (4.1, 30)

50.5 (5.2, 37)

22140_IG35_G0077

Fig. 142 Exploded view of front suspension components—2007–08 2WD sedan

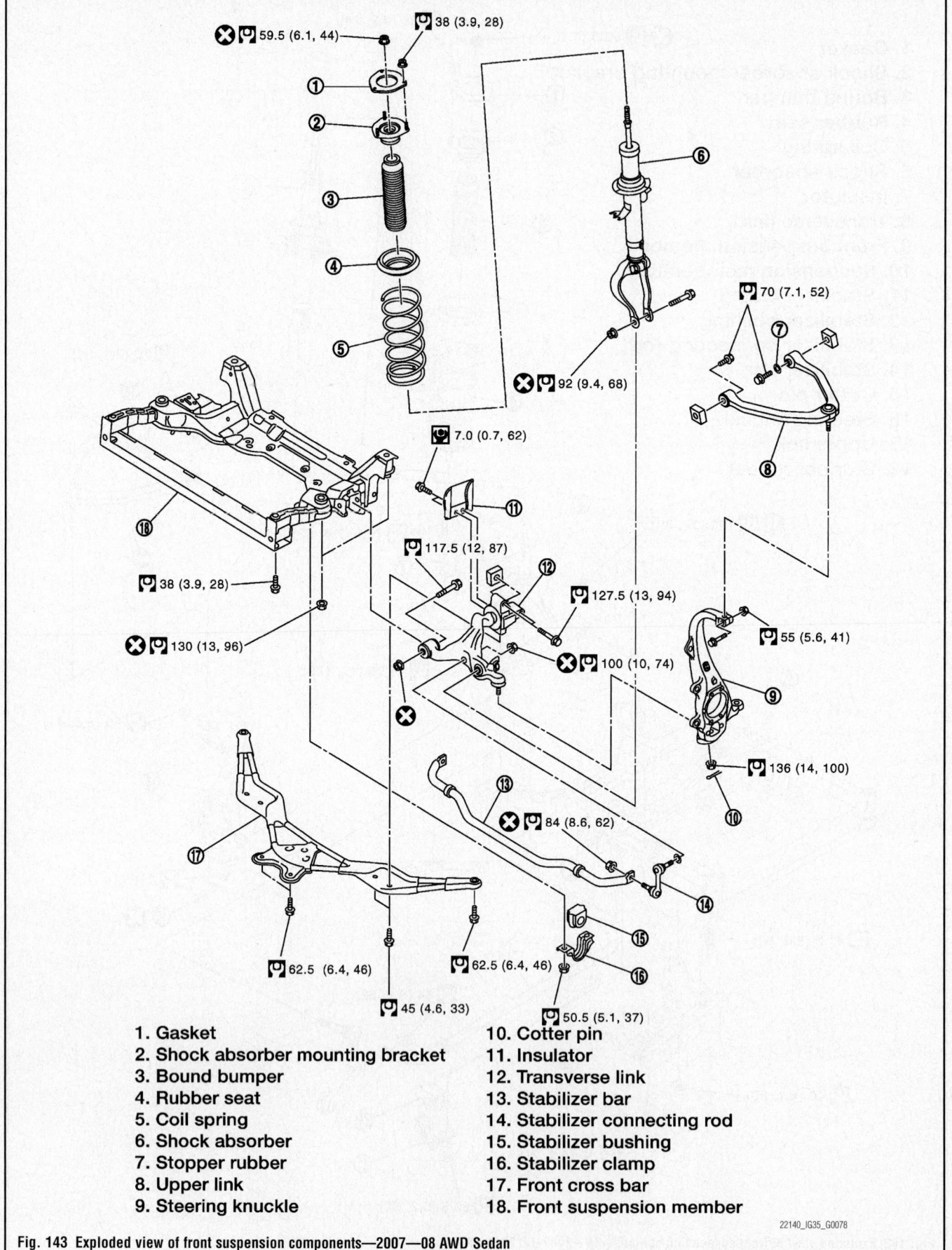

1. Gasket
2. Shock absorber mounting bracket
3. Bound bumper
4. Rubber seat
5. Coil spring
6. Shock absorber
7. Stopper rubber
8. Upper link
9. Steering knuckle
10. Cotter pin
11. Insulator
12. Transverse link
13. Stabilizer bar
14. Stabilizer connecting rod
15. Stabilizer bushing
16. Stabilizer clamp
17. Front cross bar
18. Front suspension member

22140_IG35_G0078

Fig. 143 Exploded view of front suspension components—2007—08 AWD Sedan

- Transverse link mounting bolt. Torque the bolt to 72–87 ft. lbs. (98–118 Nm).
- Engine undercover, if necessary
- Negative battery cable

STABILIZER BAR

REMOVAL & INSTALLATION

2WD Models

See Figure 144.

1. Before servicing the vehicle, refer to the precautions in the beginning of this section.
2. Raise and support the vehicle safely.
3. Remove the wheel.
4. Remove undercover.
5. Remove mounting nut on upper portion of stabilizer connecting rod.
6. Remove fixing bolts and nuts, then remove stabilizer clamp, stabilizer bushing, and stabilizer clamp bracket.
7. Remove stabilizer bar from vehicle.
8. Check stabilizer bar, stabilizer connecting rod, stabilizer bushing, stabilizer clamp and stabilizer clamp bracket for deformation, cracks and damage, and replace if necessary.

To install:

9. Installation is the reverse order of removal.
10. For torque specifications, refer to the appropriate illustrations under Coil Spring.
11. The stabilizer bar uses pillow ball type connecting rod. Position ball joint with case on pillow ball head parallel to stabilizer bar.

AWD Models

1. Before servicing the vehicle, refer to the precautions in the beginning of this section.
2. Raise and support the vehicle safely.
3. Remove tire with power tool.

4. Remove undercover with power tool.
5. Remove mounting nut on upper portion of stabilizer connecting rod with power tool.
6. Remove fixing bolt and nut, then remove stabilizer clamp, stabilizer bushing.
7. Remove stabilizer bar from vehicle.
8. Check stabilizer bar, stabilizer connecting rod, stabilizer bushing and clamp for deformation, cracks and damage, and replace if necessary.

To install:

9. Installation is the reverse order of removal.
10. For torque specifications, refer to the appropriate illustrations under Coil Spring.
11. The stabilizer bar uses pillow ball type connecting rod. Position ball joint with case on pillow ball head parallel to stabilizer bar.

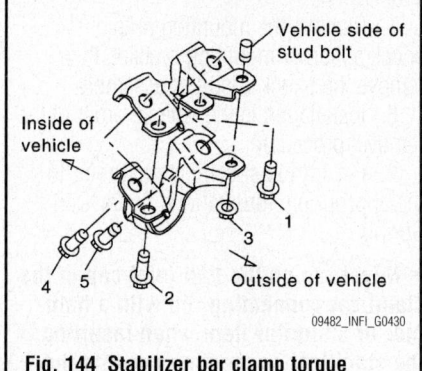

Fig. 144 Stabilizer bar clamp torque sequence—2WD

STEERING KNUCKLE

REMOVAL & INSTALLATION

2006 Coupe and Sedan; 2007 Coupe

See Figures 145 and 146.

➡**This procedure includes wheel hub and bearing removal and installation.**

1. Before servicing the vehicle, refer to the precautions in the beginning of this section.
2. Remove tire and wheel.
3. Remove undercover.
4. Remove front brake caliper. Avoid depressing brake pedal while brake caliper is removed.
5. Remove disc rotor.
6. Remove wheel sensor from steering knuckle. Do not pull on wheel sensor harness

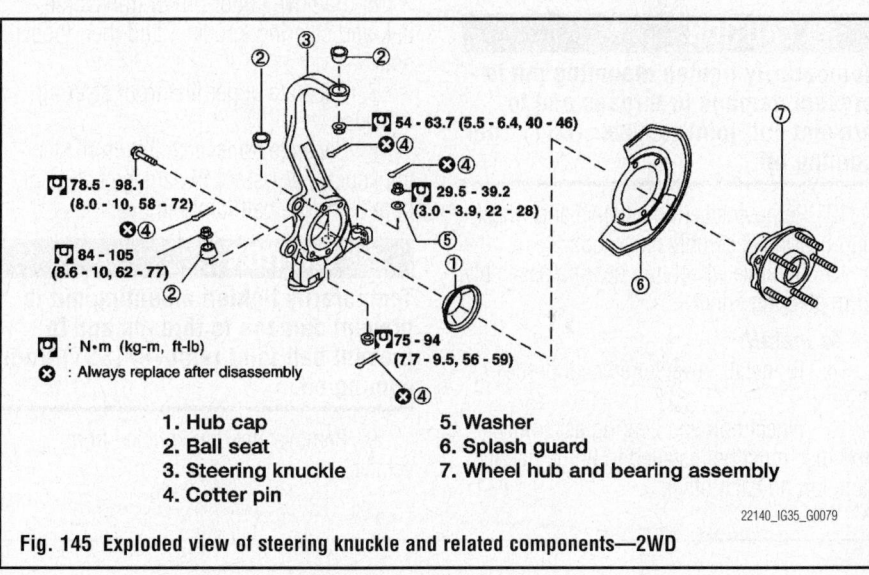

1. Hub cap
2. Ball seat
3. Steering knuckle
4. Cotter pin
5. Washer
6. Splash guard
7. Wheel hub and bearing assembly

Fig. 145 Exploded view of steering knuckle and related components—2WD

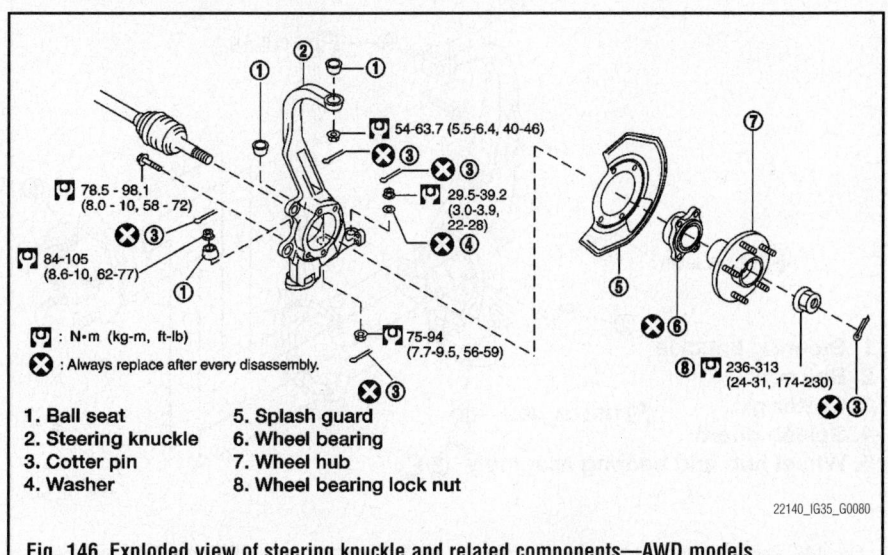

1. Ball seat
2. Steering knuckle
3. Cotter pin
4. Washer
5. Splash guard
6. Wheel bearing
7. Wheel hub
8. Wheel bearing lock nut

Fig. 146 Exploded view of steering knuckle and related components—AWD models

7. Remove brake hose bracket from steering knuckle.

8. Remove cotter pin at steering outer socket, then loosen mounting nut.

9. Use a ball joint remover (SST) to remove steering outer socket from steering knuckle. Be careful not to damage ball joint boot.

> ※※ **CAUTION**
>
> **Temporarily tighten mounting nut to prevent damage to threads and to prevent ball joint remover (SST) from coming off.**

10. After removing upper link, transverse link, compression rod and cotter pin at steering knuckle, loosen mounting nut.

11. Use a ball joint remover (suitable tool) to remove upper link, transverse link and compression rod from steering knuckle. Be careful not to damage ball joint boot.

> ※※ **CAUTION**
>
> **Temporarily tighten mounting nut to prevent damage to threads and to prevent ball joint remover (SST) from coming off.**

12. Remove steering knuckle and wheel hub bearing assembly fixing bolt.

13. Remove wheel hub bearing assembly from steering knuckle.

To install:

14. To install, reverse removal procedure.

15. Wheel hub and bearing assembly and disc must be installed to fit the marked position on each other.

2007–08 Sedan

See Figure 147.

1. Before servicing the vehicle, refer to the precautions in the beginning of this section.

2. Remove wheel hub and bearing assembly and then remove splash guard.

3. Remove brake hose bracket.

4. Remove cotter pin at steering outer socket, then loosen mounting nut.

5. Remove steering outer socket from steering knuckle so as not to damage ball joint boot using the ball joint remover.

> ※※ **CAUTION**
>
> **Temporarily tighten mounting nut to prevent damage to threads and to prevent ball joint remover (SST) from coming off.**

6. Remove cotter pin of transverse link and steering knuckle, and then loosen nut.

7. Separate upper link from steering knuckle.

8. Separate transverse link from steering knuckle so as not to damage ball joint boot using the ball joint remover.

> ※※ **CAUTION**
>
> **Temporarily tighten mounting nut to prevent damage to threads and to prevent ball joint remover (SST) from coming off.**

9. Remove steering knuckle from vehicle.

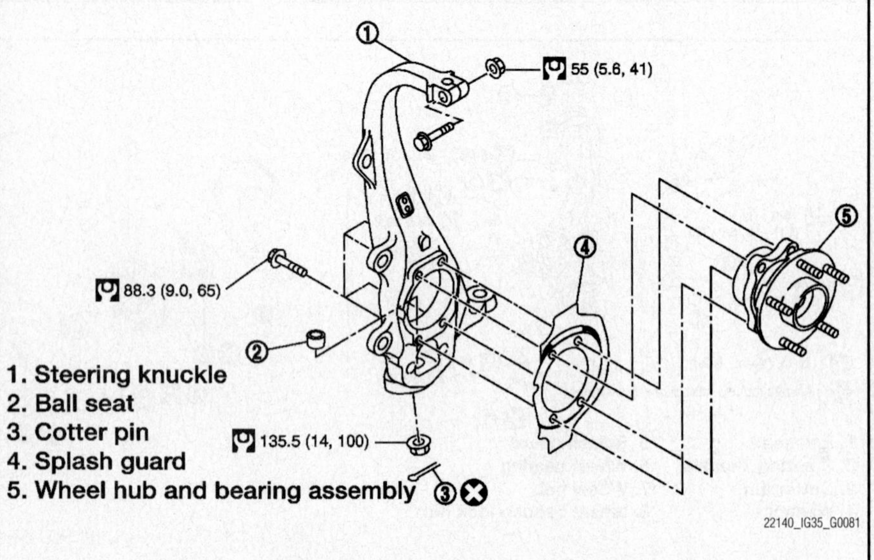

To install:

10. To install, reverse removal procedure.

11. Perform the final tightening of each of parts under unladen conditions, which were removed when removing wheel hub and bearing assembly and steering knuckle.

STRUT & SPRING ASSEMBLY

REMOVAL & INSTALLATION

2006 Coupe and Sedan; 2007 Coupe

1. Before servicing the vehicle, refer to the precautions in the beginning of this section.

2. Remove tire and wheel.

3. Remove undercover.

4. Remove harness of wheel sensor from shock absorber.

5. Remove mounting nuts of brake hose from shock absorber.

6. Remove mounting bolt and nut between shock absorber and transverse link.

7. Remove mounting nuts on mounting insulator with power tool, then remove shock absorber from vehicle.

8. Installation is the reverse order of the removal procedure.

9. For torque specifications, refer to the appropriate illustrations under Coil Spring.

2007–08 Sedan

2WD Models

1. Before servicing the vehicle, refer to the precautions in the beginning of this section.

2. Remove tire and wheel.

3. Remove brake hose bracket.

4. Remove mounting nuts on the lower side of stabilizer connecting rod

5. Remove mounting nuts on upper side of stabilizer connecting rod with power tool, and then remove stabilizer connecting rod from transverse link.

6. Separate upper link from steering knuckle.

7. Remove the mounting nuts of the shock absorber mounting bracket, then remove the shock absorber assembly.

8. Installation is the reverse order of the removal procedure.

9. For torque specifications, refer to the appropriate illustrations under Coil Spring.

➡ Never tap on the ball joint cap of the stabilizer connecting rod with a hammer or a similar item when inserting the stabilizer connecting rod into the transverse link.

55 (5.6, 41)
88.3 (9.0, 65)
135.5 (14, 100)

1. Steering knuckle
2. Ball seat
3. Cotter pin
4. Splash guard
5. Wheel hub and bearing assembly

22140_IG35_G0081

Fig. 147 Exploded view of steering knuckle and related components—2WD and AWD models

→**Perform the final tightening of bolts and nuts at the shock absorber lower side (rubber bushing), under unladen conditions with tires on level ground.**

AWD Models

1. Before servicing the vehicle, refer to the precautions in the beginning of this section.

2. Remove tire and wheel.

3. Remove mounting nuts on the upper side of stabilizer connecting rod with power tool, and then remove stabilizer connecting rod from transverse link.

4. Remove mounting bolts and nuts on the lower side of shock absorber with power tool, and then remove shock absorber from transverse link.

5. Remove drive shaft.

6. Separate upper link from steering knuckle.

7. Remove the mounting nuts of shock absorber mounting bracket, then remove shock absorber assembly.

8. Installation is the reverse order of the removal procedure.

9. For torque specifications, refer to the appropriate illustrations under Coil Spring.

→**Never tap on the ball joint cap of the stabilizer connecting rod with a hammer or a similar item when inserting the stabilizer connecting rod into the transverse link.**

→**Perform the final tightening of bolts and nuts at the shock absorber lower side (rubber bushing), under unladen conditions with tires on level ground.**

UPPER CONTROL ARM

REMOVAL & INSTALLATION

1. Before servicing the vehicle, refer to the precautions in the beginning of this section.

2. Remove tire and wheel.

3. Remove undercover.

4. Remove shock absorber

5. Remove the cotter pin from the upper link ball joint, then loosen the mounting nut.

6. Use a ball joint remover (suitable tool) to remove upper link from steering knuckle. Be careful not to damage ball joint boot.

✳✳ CAUTION

Temporarily tighten the mounting nut to prevent damage to threads and to prevent ball joint remover from coming off.

7. Remove bolts holding upper link to body or steering knuckle.

8. Remove upper link from vehicle.

9. Remove stopper rubber from upper link.

10. Check upper link and bushing for deformation, cracks, or damage. If any non-standard condition is found, replace it.

11. Check boot of ball joint for cracks, or other damage, and also for grease leakage. If any non-standard condition is found, replace it.

To install:

12. Installation is the reverse order of the removal procedure.

13. For torque specifications, refer to the appropriate illustrations under Coil Spring.

→**Perform the final tightening of bolts and nuts at the shock absorber lower side (rubber bushing), under unladen conditions with tires on level ground.**

14. Check wheel alignment.

WHEEL BEARINGS

REMOVAL & INSTALLATION

→**For 2006 Coupe and Sedan and 2007 Coupe models, refer to Steering Knuckle removal and installation procedure.**

2007–08 Sedan Only

2WD Models

→**Refer to illustrations under Steering Knuckle.**

1. Before servicing the vehicle, refer to the precautions in the beginning of this section.

2. Remove the tire and wheel.

3. Remove wheel sensor from steering knuckle. Do not pull on the sensor harness.

4. Remove front brake caliper. Avoid depressing brake pedal while brake caliper is removed.

5. Remove disc rotor.

6. Remove wheel hub and bearing assembly mounting bolts, and then remove splash guard and wheel hub and bearing assembly from steering knuckle.

To install:

7. Installation is the reverse order of the removal procedure.

8. Refer to illustrations under Steering Knuckle for torque specifications.

9. Perform the final tightening of each of parts under unladen conditions, which were removed when removing wheel hub and bearing assembly.

AWD Models

→**Refer to illustrations under Steering Knuckle.**

1. Before servicing the vehicle, refer to the precautions in the beginning of this section.

2. Remove tire and wheel.

3. Remove wheel sensor from steering knuckle. Do not pull on the sensor harness.

4. Remove front brake caliper. Avoid depressing brake pedal while brake caliper is removed.

5. Remove disc rotor.

6. Remove cotter pin, then loosen hub lock nut with power tool.

7. Patch hub lock nut with a piece of wood. Hammer the wood to disengage wheel hub and bearing assembly from drive shaft. Take out the hub lock nut.

✳✳ CAUTION

Never place the drive shaft joint at an extreme angle. Also be careful not to overextend slide joint. Never allow drive shaft to hang down without support for housing (or joint sub-assembly), shaft and the other parts.

→**Use a suitable puller, if wheel hub and bearing assembly and drive shaft cannot be separated even after performing the above procedure.**

8. Remove wheel hub and bearing assembly mounting bolts, and then remove splash guard and wheel hub and bearing assembly from steering knuckle.

To install:

9. Installation is the reverse order of the removal procedure.

10. Refer to illustrations under Steering Knuckle for torque specifications.

11. Perform the final tightening of each of parts under unladen conditions, which were removed when removing wheel hub and bearing assembly.

SUSPENSION

COIL SPRING

REMOVAL & INSTALLATION

See Figures 148 through 150.

➡ **Coil spring is integrated with the rear lower control arm.**

1. Before servicing the vehicle, refer to the precautions in the beginning of this section.

2. Remove the tire and wheel.
3. Set jack under rear lower link.
4. Loosen fixing bolt and nut of rear lower link in side of suspension member,

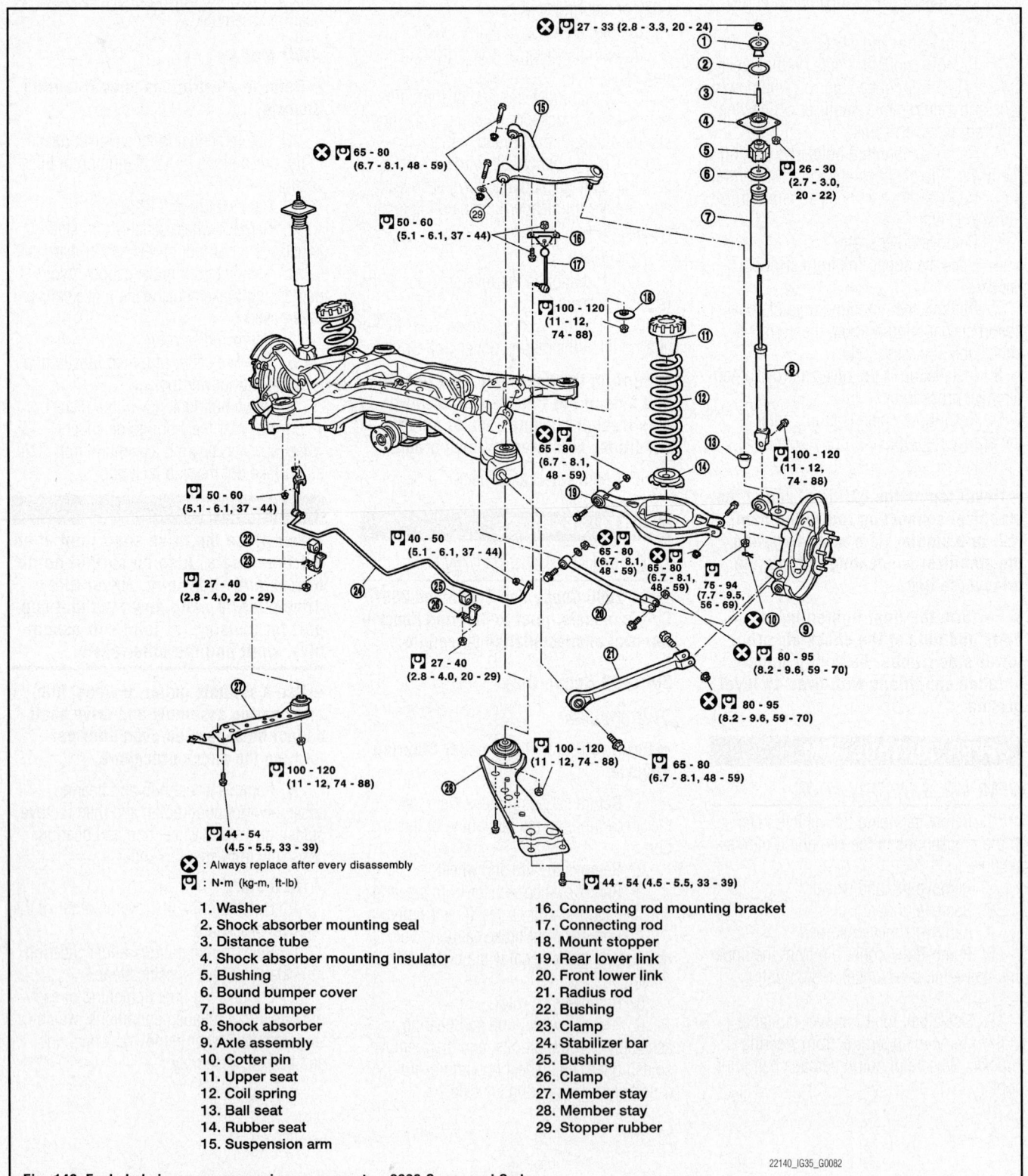

X : Always replace after every disassembly

⊡ : N•m (kg-m, ft-lb)

1. Washer
2. Shock absorber mounting seal
3. Distance tube
4. Shock absorber mounting insulator
5. Bushing
6. Bound bumper cover
7. Bound bumper
8. Shock absorber
9. Axle assembly
10. Cotter pin
11. Upper seat
12. Coil spring
13. Ball seat
14. Rubber seat
15. Suspension arm
16. Connecting rod mounting bracket
17. Connecting rod
18. Mount stopper
19. Rear lower link
20. Front lower link
21. Radius rod
22. Bushing
23. Clamp
24. Stabilizer bar
25. Bushing
26. Clamp
27. Member stay
28. Member stay
29. Stopper rubber

22140_IG35_G0082

Fig. 148 Exploded view rear suspension components—2006 Coupe and Sedan

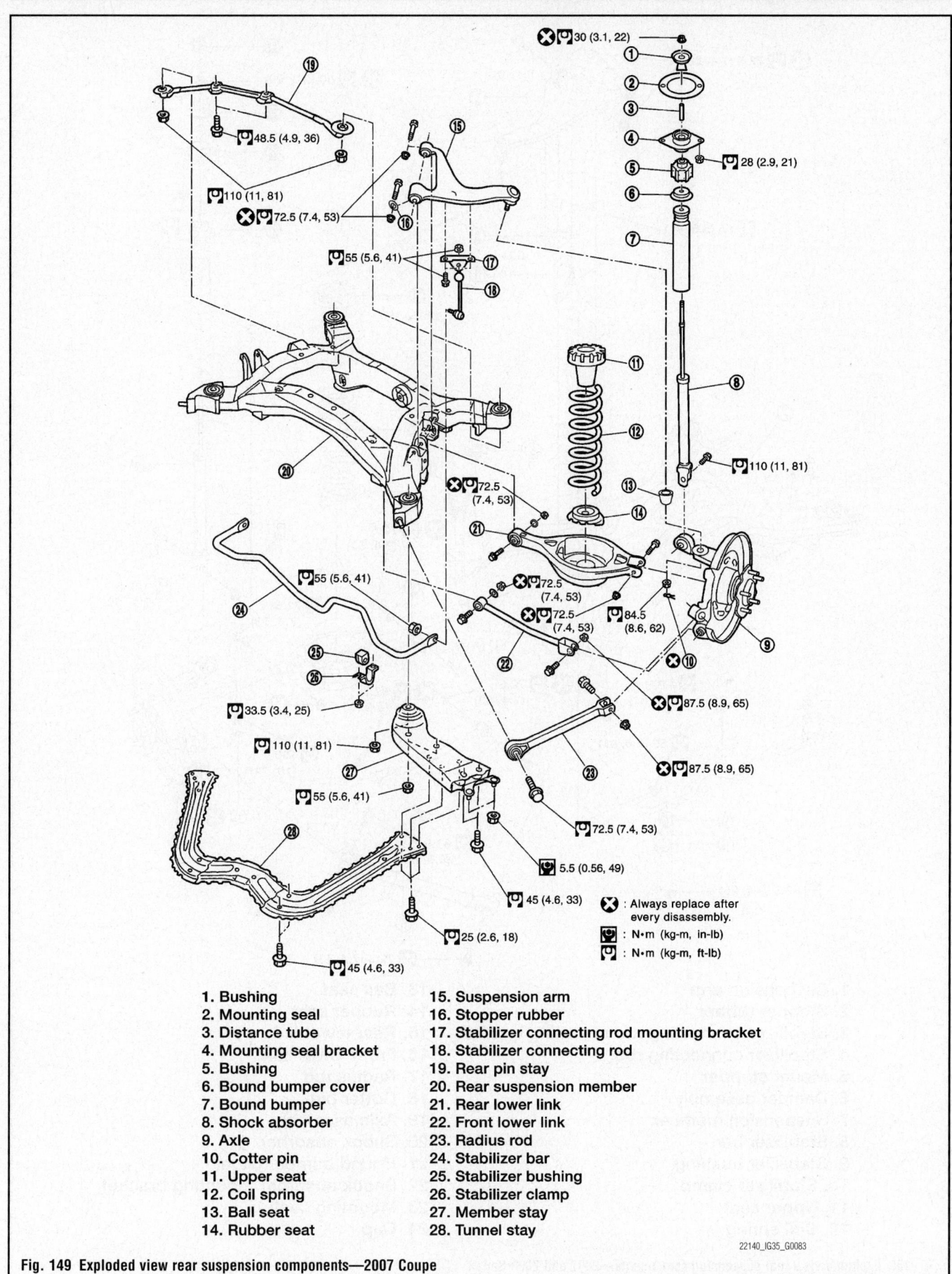

1. Bushing
2. Mounting seal
3. Distance tube
4. Mounting seal bracket
5. Bushing
6. Bound bumper cover
7. Bound bumper
8. Shock absorber
9. Axle
10. Cotter pin
11. Upper seat
12. Coil spring
13. Ball seat
14. Rubber seat
15. Suspension arm
16. Stopper rubber
17. Stabilizer connecting rod mounting bracket
18. Stabilizer connecting rod
19. Rear pin stay
20. Rear suspension member
21. Rear lower link
22. Front lower link
23. Radius rod
24. Stabilizer bar
25. Stabilizer bushing
26. Stabilizer clamp
27. Member stay
28. Tunnel stay

Fig. 149 Exploded view rear suspension components—2007 Coupe

22140_IG35_G0083

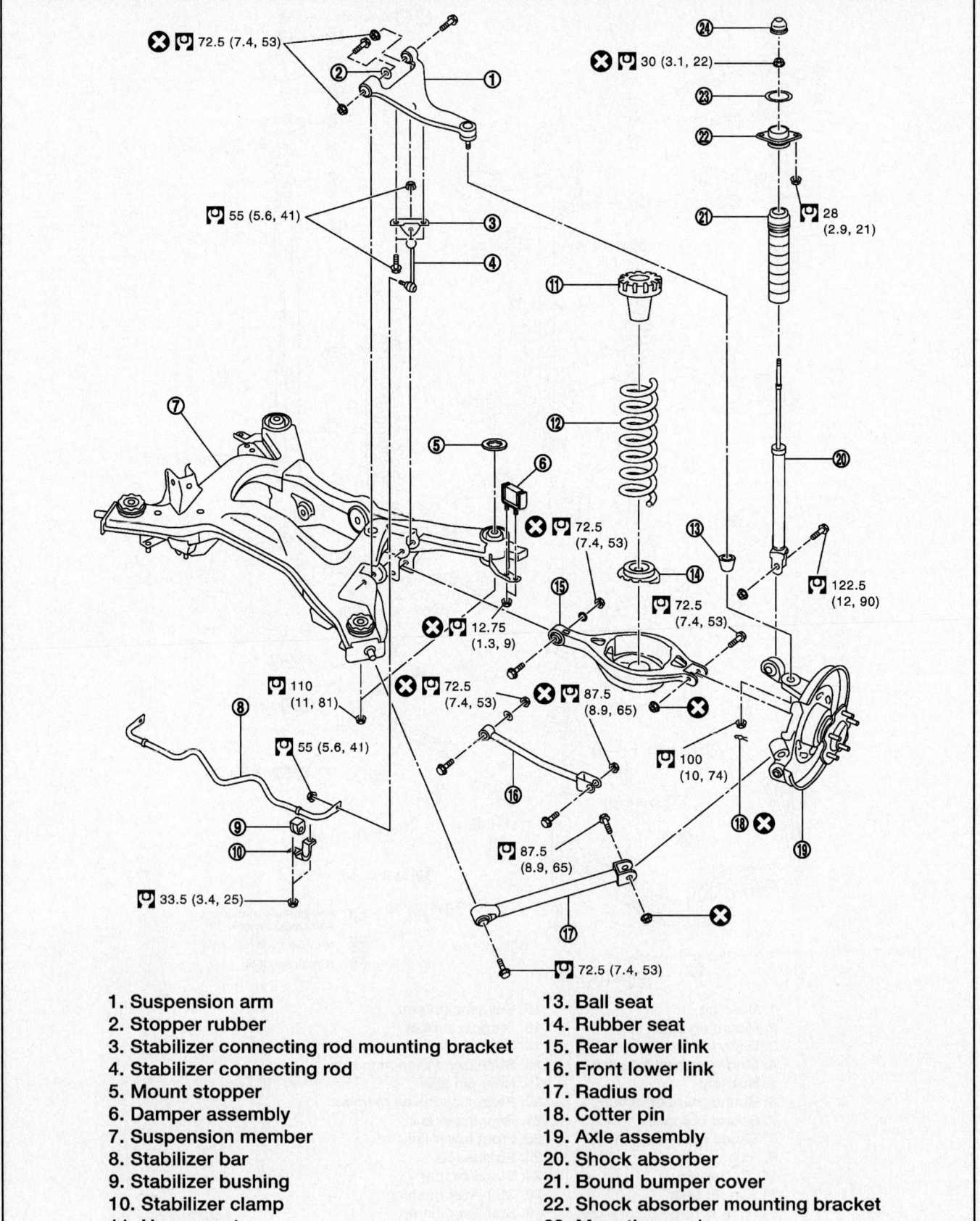

1. Suspension arm
2. Stopper rubber
3. Stabilizer connecting rod mounting bracket
4. Stabilizer connecting rod
5. Mount stopper
6. Damper assembly
7. Suspension member
8. Stabilizer bar
9. Stabilizer bushing
10. Stabilizer clamp
11. Upper seat
12. Coil spring
13. Ball seat
14. Rubber seat
15. Rear lower link
16. Front lower link
17. Radius rod
18. Cotter pin
19. Axle assembly
20. Shock absorber
21. Bound bumper cover
22. Shock absorber mounting bracket
23. Mounting seal
24. Cap

22140_IG35_G0084

Fig. 150 Exploded view rear suspension components—2007 and 2008 Sedan

and then remove fixing bolt and nut in side of axle with power tool.

5. Slowly lower jack, then remove upper seat, coil spring and rubber sheet from rear lower link.

6. Remove fixing bolt and nut in side of suspension member to remove rear lower link.

7. Check rear lower link, bushing and coil spring for deformation, cracks, and damage. Replace rear lower link and coil spring if necessary.

To install:

8. Installation is the reverse order of the removal procedure.

9. Perform final tightening of rear suspension member and axle installation position (rubber bushing) under unladen condition with tires on level ground. Check wheel alignment.

LOWER CONTROL ARM

REMOVAL & INSTALLATION

Refer to Coil Spring for removal and installation procedure. The coil spring is integrated with the rear lower control arm.

SHOCK ABSORBER

REMOVAL & INSTALLATION

1. Before servicing the vehicle, refer to the precautions in the beginning of this section.

2. Remove the tire and wheel.

3. Set jack under rear lower link.

4. Remove the mounting bolt in the lower side of the shock absorber assembly.

5. Remove the mounting seal bracket mounting nuts of the shock absorber upper side with the power tool and remove shock absorber from vehicle.

6. Check shock absorber assembly for deformation, cracks, damage, and replace if necessary.

7. Check piston rod for damage, uneven wear, distortion, and replace if necessary.

8. Check welded and sealed areas for oil leakage, and replace if necessary.

To install:

9. Installation is the reverse order of the removal procedure.

10. Refer to illustrations under Coil Spring for torque specifications.

11. Perform final tightening of shock absorber assembly lower side (rubber bushing) under unladen condition with tires on level ground. Check wheel alignment

STABILIZER BAR

REMOVAL & INSTALLATION

2006 Coupe and Sedan; 2007 Coupe

1. Before servicing the vehicle, refer to the precautions in the beginning of this section.

2. Remove mounting bolts and remove stabilizer connecting rod mount bracket from suspension arm.

3. Remove lower side mounting nut on stabilizer connecting rod and remove stabilizer connecting rod from stabilizer bar.

4. Remove mounting nut on stabilizer clamp and remove stabilizer from vehicle.

5. Check stabilizer bar, stabilizer bushings, stabilizer clamps, stabilizer connecting rod, stabilizer connecting rod mounting bracket for any deformation, crack or damage. Replace if necessary.

To install:

6. Installation is the reverse order of the removal procedure.

7. Refer to illustrations under Coil Spring for torque specifications.

2007–08 Sedan

1. Before servicing the vehicle, refer to the precautions in the beginning of this section.

2. Remove mounting bracket of center muffler and remove mounting rubber of main muffler.

3. Remove lower side mounting nut on stabilizer connecting rod and remove stabilizer connecting rod from stabilizer bar.

4. Remove mounting nut on stabilizer clamp and remove stabilizer from vehicle.

5. Check stabilizer bar, stabilizer bushings, stabilizer clamps, stabilizer connecting rod, stabilizer connecting rod mounting bracket for any deformation, crack or damage. Replace if necessary.

To install:

6. Installation is the reverse order of the removal procedure.

7. Refer to illustrations under Coil Spring for torque specifications.

UPPER CONTROL ARM

REMOVAL & INSTALLATION

2006 Coupe and Sedan; 2007 Coupe

1. Before servicing the vehicle, refer to the precautions in the beginning of this section.

2. Remove the tire and wheel.

3. Remove the mounting bolts and nuts between suspension arm and rear suspension member.

4. Remove cotter pin of suspension arm ball joint, and loosen nut.

5. Use the ball joint remover to remove suspension arm from axle assembly. Be careful not to damage ball joint boot.

❉❉ CAUTION

Temporarily tighten the mounting nut to prevent damage to threads and to prevent ball joint remover from coming off.

6. Remove the suspension arm and the stopper rubber from the vehicle.

7. Check suspension arm and bushing for deformation, cracks or damage. If any non-standard condition is found, replace it.

To install:

8. Installation is the reverse order of the removal procedure.

9. Refer to the illustrations under Coil Spring for torque specifications.

2007–08 Sedan

1. Before servicing the vehicle, refer to the precautions in the beginning of this section.

2. Remove the tire and wheel.

3. Remove brake caliper. Hang bake caliper in a place where it will not interfere with work.

4. Set suitable jack under axle assembly to relieve the coil spring tension.

5. Remove connecting rod mounting bracket from suspension arm

6. Remove drive shaft.

7. Remove the height sensor (with xenon headlamp).

8. Remove cotter pin of suspension arm ball joint, and loosen nut.

9. Remove the mounting bolts and nuts between suspension arm and rear suspension member.

10. Use the ball joint remover to remove suspension arm from axle assembly. Be careful not to damage ball joint boot.

✳ CAUTION

Temporarily tighten the mounting nut to prevent damage to threads and to prevent ball joint remover from coming off.

11. Remove the suspension arm.
12. Check suspension arm and bushing for deformation, cracks or damage. If any non-standard condition is found, replace it.

To install:

13. Installation is the reverse order of the removal procedure.

14. Refer to the illustrations under Coil Spring for torque specifications.

INFINITI

M35 • M45

SPECIFICATIONS AND MAINTENANCE CHARTS

ENGINE AND VEHICLE IDENTIFICATION

Engine							Model Year	
Code ①	Liters (cc)	Cu. In.	Cyl.	Fuel Sys.	Engine Type	Eng. Mfg.	Code ②	Year
VQ35DE	3.5 (3498)	213	6	MFI	DOHC	Nissan	7	2007
VK45DE	4.5 (4494)	274	8	MFI	DOHC	Nissan	8	2008

MFI: Multi-Port Fuel Injection

DOHC: Double Overhead Camshaft

22140_IM35_C0001

GENERAL ENGINE SPECIFICATIONS

Year	Model	Engine Displacement Liters	Engine ID	Net Horsepower @ rpm	Net Torque @ rpm (ft. lbs.)	Bore x Stroke (in.)	Compression Ratio	Oil Pressure @ rpm
2007	M35	3.5	VQ35DE	275@6200	268@4800	3.76X3.20	10.3:1	43@2000
	M45	4.5	VK45DE	325@6400	333@4000	3.66x3.25	10.5:1	43@3200
2008	M35	3.5	VQ35DE	275@6200	268@4800	3.76X3.20	10.3:1	43@2000
	M45	4.5	VK45DE	325@6400	333@4000	3.66x3.25	10.5:1	43@3000

22140_IM35_C0002

ENGINE TUNE-UP SPECIFICATIONS

Year	Engine Displacement Liters	Engine ID	Spark Plug Gap (in.)	Ignition Timing (deg.) MT	Ignition Timing (deg.) AT	Fuel Pump (psi) ①	Idle Speed (rpm) MT	Idle Speed (rpm) AT	Valve Clearance Intake	Valve Clearance Exhaust
2007	3.5	VQ35DE	0.043	—	15 ①	34	—	650±50	②	③
	4.5	VK45DE	0.043	—	12 ①	34	—	650±50	②	③
2008	3.5	VQ35DE	0.043	—	15 ①	34	—	650±50	②	③
	4.5	VK45DE	0.043	—	12 ①	34	—	650±50	②	③

NOTE: The Vehicle Emission Control Information label often reflects specification changes made during production.

The label figures must be used if they differ from those in this chart.

① +/-5° Before top dead center

② Intake 0.010 - 0.013 Cold

③ Exhaust 0.011 - 0.015 Cold

22140_IM35_C0003

CAPACITIES

Year	Model	Engine Displacement Liters	Engine ID	Engine Oil with Filter (qts.)	Transmission (pts.) Manual	Transmission (pts.) Auto.	Transfer Case (pts.)	Drive Axle (pts.)	Fuel Tank (gal.)	Cooling System (qts.)
2007	M35	3.5	VQ35DE	5.00	—	21.7	2.6	①	20	9.3
	M45	4.5	VK45DE	5.75	—	21.7	2.6	①	20	11.0
2008	M35	3.5	VQ35DE	5.00	—	21.7	2.6	①	20	9.3
	M45	4.5	VK45DE	5.75	—	21.7	2.6	①	20	11.0

NOTE: All capacities are approximate. Add fluid gradually and check to be sure a proper fluid level is obtained.

① Front Axle 1.38 (AWD) -Rear Axle 3.00

22140_IM35_C0005

FLUID SPECIFICATIONS

Year	Model	Engine Displ. Liters	Engine Oil	Man. Trans.	Auto. Trans.	Drive Axle Front	Drive Axle Rear	Transfer Case	Power Steering Fluid	Brake Master Cylinder	Cooling System
2007	M35	3.5	①	—	②	GL-5 80 W-90	GL-5 80 W-90	②	NISSAN PSF	DOT 3	③
	M45	4.5	④	—	②	GL-5 80 W-90	GL-5 80 W-90	②	NISSAN PSF	DOT 3	③
2008	M35	3.5	①	—	②	GL-5 80 W-90	GL-5 80 W-90	②	NISSAN PSF	DOT 3	③
	M45	4.5	④	—	②	GL-5 80 W-90	GL-5 80 W-90	②	NISSAN PSF	DOT 3	③

DOT: Department Of Transpotation

① API Certification Mark
   API grade SG/SH, Energy Conserving I & II or
   API grade SJ or SL, Energy Conserving
   ILSAC grade GF-I, GF-II & GF-III
② Genuine NISSAN D ATF (U.S.) or Canada NISSAN ATF

③ NISSAN Long Life Antifreeze/ Coolant or equivalent
④ API Certification Mark
   API grade SJ or SL, Energy Conserving
   ILSAC grade GF-II & GF-III

22140_IM35_C0004

VALVE SPECIFICATIONS

Year	Engine Displacement Liters	Engine ID	Seat Angle (deg.)	Face Angle (deg.)	Spring Test Pressure (lbs. @ in.)	Spring Free Height (in.)	Stem-to-Guide Clearance (in.) Intake	Stem-to-Guide Clearance (in.) Exhaust	Stem Diameter (in.) Intake	Stem Diameter (in.) Exhaust
2007	3.5	VQ35DE	①	②	84 - 95@ 1.0709	1.8531	0.0008- 0.0021	0.0012- 0.0025	0.2348- 0.2354	0.2344- 0.2350
	4.5	VK45DE	①	②	③	1.8247- 1.8444	0.0008- 0.0018	0.0012- 0.0022	1.417- 1.4290	1.228- 1.2400
2008	3.5	VQ35DE	①	②	84 - 95@ 1.0709	1.8531	0.0008- 0.0021	0.0012- 0.0025	0.2348- 0.2354	0.2344- 0.2350
	4.5	VK45DE	①	②	③	1.8247- 1.8444	0.0008- 0.0018	0.0012- 0.0022	1.417- 1.4290	1.228- 1.2400

NA: Not Available

① 45 degrees, 45 minutes
③ 44 degrees, 45 minutes +/-22 minutes
③ 65-74@ 0.0961 - Valve open

22140_IM35_C0006

CAMSHAFT AND BEARING SPECIFICATIONS

All measurements are given in inches.

Year	Engine Displacement Liters	Engine VIN	Journal Diameter	Brg. Oil Clearance	Shaft End-play	Runout	Journal Bore	Lobe Lift	
								Intake	Exhaust
2007	3.5	VQ35DE	①	②	0.0059	0.001	③	1.7663 - 1.7738	1.7663 - 1.7738
	4.5	VK45DE	1.0212- 1.0218	④	0.0045- 0.0074	0.001	1.0236- 1.0244	1.7663 1.7738	0.2168 1.7368
2008	3.5	VQ35DE	①	②	0.0059	0.001	③	1.7663 - 1.7738	1.7663 - 1.7738
	4.5	VK45DE	1.0212- 1.0218	④	0.0045- 0.0074	0.001	1.0236- 1.0244	1.7663 1.7738	0.2173 1.7368

NA: Information not available

Note: (Oil clearance) = (Camshaft bracket inner diameter) – (Camshaft journal diameter).

① No. 1: 1.0211 - 1.0218 inches

No 2,3 and 4 - 0.9230-0.9238 inches

② No. 1: 0.0018 - 0.0033 inches

No 2,3,4 and 5 - 0.0012-0.0027 inches

③ No. 1: 1.0236 - 1.0244 inches

No 2,3 and 4 - 0.9252-0.9260 inches

④ No. 1: 0.0018 - 0.0034 inches

No 2,3,4 and 5 - 0.0014-0.0030 inches

22140_IM35_C0007

CRANKSHAFT AND CONNECTING ROD SPECIFICATIONS

All measurements represent standard values and are given in inches.

Year	Engine Displacement Liters	Engine ID	Crankshaft				Connecting Rod		
			Main Brg. Journal Dia	Main Brg. Oil Clearance	Shaft End-play	Thrust on No.	Journal Diameter	Oil Clearance	Side Clearance
2007	3.5	VQ35DE	2.3603- 2.3612	0.0014- 0.0018	0.0039- 0.0098	3	2.1654- 2.1659	0.0013- 0.0023	0.0079- 0.0138
	4.5	VK45DE	2.5173- 2.5183	①	0.0039- 0.0074	3	2.1654- 2.1659	0.0008- 0.0018	0.0079- 0.0138
2008	3.5	VQ35DE	2.3603- 2.3612	0.0014- 0.0018	0.0039- 0.0098	3	2.1654- 2.1659	0.0013- 0.0023	0.0079- 0.0138
	4.5	VK45DE	2.5173- 2.5183	①	0.0039- 0.0074	3	2.1654- 2.1659	0.0008- 0.0018	0.0079- 0.0138

① Nos. 1 and 5: 0.00004-0.0004 in.

Nos. 2, 3 and 4: 0.0003-0.0007 in.

22140_IM35_C0009

PISTON AND RING SPECIFICATIONS
All measurements are given in inches.

Year	Engine Displacement Liters	Engine ID	Piston Clearance	Ring Gap Top Compression	Ring Gap Bottom Compression	Ring Gap Oil Control	Ring Side Clearance Top Compression	Ring Side Clearance Bottom Compression	Ring Side Clearance Oil Control
2007	3.5	VQ35DE	0.0004-0.0012	0.0091-0.0130	0.0130-0.0189	0.0079-0.0197	0.0018-0.0031	0.0012-0.0028	0.0026-0.0053
	4.5	VK45DE	0.0004-0.0012	0.0087-0.0126	0.0087-0.0126	0.0079-0.0197	0.0018-0.0031	0.0012-0.0028	0.0026-0.0053
2008	3.5	VQ35DE	0.0004-0.0012	0.0091-0.0130	0.0130-0.0189	0.0079-0.0197	0.0018-0.0031	0.0012-0.0028	0.0026-0.0053
	4.5	VK45DE	0.0004-0.0012	0.0087-0.0126	0.0087-0.0126	0.0079-0.0197	0.0018-0.0031	0.0012-0.0028	0.0026-0.0053

22140_IM35_C0008

TORQUE SPECIFICATIONS
All readings in ft. lbs.

Year	Engine Displacement Liters	Engine ID	Cylinder Head Bolts	Main Bearing Bolts	Rod Bearing Bolts	Crankshaft Damper Bolts	Flywheel Bolts	Manifold Intake	Manifold Exhaust	Spark Plugs	Oil Pan Drain Plug
2007	3.5	VQ35DE	①	②	③	33	61-69	④	⑤	15-21	22-29
	4.5	VK45DE	⑥	⑦	⑧	⑨	62-68	⑩ ⑪	⑤	15-21	22-28
2008	3.5	VQ35DE	①	②	③	33	61-69	④	⑤	15-21	22-29
	4.5	VK45DE	⑥	⑦	⑧	⑨	62-68	⑩ ⑪	⑤	15-21	22-28

NA: Not Available

① Step 1: 72 ft. lbs.
  Step 2: Loosen bolts completely
  Step 3: 29 ft. lbs.
  Step 4: Tighten an additional 90 degrees
  Step 5: Again, tighten an additional 90 degrees.
  Step 6: After installing cylinder head, measure
    distance between front end faces of cylinder
    block and cylinder head (left and right banks)
    Standard : 0.555 - 0.587 in (14.1 - 14.9 mm)

② Step 1: 26 ft. lbs.
  Step 2: Tighten an additional 90 degrees

③ Step 1: Tighten to 14 ft. lbs.
  Step 2: Tighten an additional 90 degrees

④ Step 1: Tighten stud bolts to 8 ft. lbs.
  Step 2: Tighten to 5 ft. lbs.
  Step 3: Tighten, again, to 21 ft. lbs.

⑤ Step 1: Tighten stud bolts to 8 ft. lbs.
  Step 2: Tighten to 11 ft. lbs.
  Step 3: Tighten, again, to 22 ft. lbs.

⑥ Step 1: 72 ft.lbs.
  Step 2: Loosen bolts completely
  Step 3: 33 ft. lbs.
  Step 4: Tighten an additional 60 degrees.

⑦ Step 1: M12 bolts: 29 ft. lbs.
  Step 2: M9 bolts: 22 ft. lbs.
  Step 3: M12 bolts: Tighten an additional 40 degrees
  Step 4: M9 bolts: Tighten an additional 30 degrees
  Step 5: M10 side bolts: 36 ft. lbs.

⑧ Step 1: Tighten to 11 ft. lbs.
  Step 2: Tighten an additional 60 degrees

⑨ Step 1: 69 ft. lbs.
  Step 2: Tighten an additional 90 degrees

⑩ Step 1: Tighten Upper Manifold bolts to 9 ft. lbs.

⑪ Step 1: Tighten stud bolts to 8 ft. lbs.
  Step 2: Tighten to 11 ft. lbs.
  Step 3: Tighten, again, to 22 ft. lbs.

22140_IM35_C0010

WHEEL ALIGNMENT

Year	Model		Caster Range (+/-Deg.)	Caster Preferred Setting (Deg.)	Camber Range (+/-Deg.)	Camber Preferred Setting (Deg.)	Toe-in (Deg.)
2007	M35	F	0.75	①	0.75	-0.25	0.04 +/- 0.04
		R	—	—	②	③	0.11 +/- 0.11
	M35	F	0.75	3.83	0.75	-0.25	0.04 +/- 0.04
	AWD	R	—	—	0.50	-0.17	0.11 +/- 0.11
	M45	F	0.75	①	0.75	-0.25	0.04 +/- 0.04
		R	—	—	②	③	0.11 +/- 0.11
	M45	F	0.75	3.83	0.75	-0.25	0.04 +/- 0.04
	AWD	R	—	—	0.50	-0.17	0.11 +/- 0.11
2008	M35	F	0.75	①	0.75	-0.25	0.04 +/- 0.04
		R	—	—	②	③	0.11 +/- 0.11
	M35	F	0.75	3.83	0.75	-0.25	0.04 +/- 0.04
	AWD	R	—	—	0.50	-0.17	0.11 +/- 0.11
	M45	F	0.75	①	0.75	-0.25	0.04 +/- 0.04
		R	—	—	②	③	0.11 +/- 0.11
	M45	F	0.75	3.83	0.75	-0.25	0.04 +/- 0.04
	AWD	R	—	—	0.50	-0.17	0.11 +/- 0.11

Note: Measure wheel alignment under unladen conditions.

"Unladen conditions" means that fuel, engine coolant, and lubricant are full. Spare tire, jack, hand tools and mats are in designated position.

① 18" Wheels 4.50°   ③ Rear 18" Wheels -67°

  19" Wheels 4.58°   Rear 19" Wheels -83°

② Rear 18" Wheels: 40°

  19" Wheels: 50°

22140_IM35_C0011

TIRE, WHEEL AND BALL JOINT SPECIFICATIONS

Year	Model	OEM Tires Standard	OEM Tires Optional	Tire Pressures (psi) Front	Tire Pressures (psi) Rear	Wheel Size	Ball Joint Inspection	Lug Nut (ft. lbs.)
2006	M35	P245/45R18 V	P245/40R19 W	33	33	Std: 8.0-JJ	①	80
	M35 AWD	P245/45R18 V	—	33	33	Opt: 8,5-JJ		
	M45	P245/45R18 V	P245/40R19 W	33	30	Std: 8.0-JJ	①	80
	M45 AWD	P245/45R18 V	—	33	30	Opt: 8,5-JJ		
2007	M35	P245/45R18 V	P245/40R19 W	33	33	Std: 8.0-JJ	①	80
	M35 AWD	P245/45R18 V	—	33	33	Opt: 8,5-JJ		
	M45	P245/45R18 V	P245/40R19 W	33	33	Std: 8.0-JJ	①	80
	M45 AWD	P245/45R18 V	—	33	33	Opt: 8.5-JJ		
2008	M35	P245/45R18 V	P245/40R19 W	33	33	Std: 8.0-JJ	①	80
	M35 AWD	P245/45R18 V	—	33	33	Opt: 8,5-JJ		
	M45	P245/45R18 V	P245/40R19 W	33	33	Std: 8.0-JJ	①	80
	M45 AWD	P245/45R18 V	—	33	33	Opt: 8.5-JJ		

OEM: Original Equipment Manufacturer

PSI: Pounds Per Square Inch

STD: Standard

OPT: Optional

NA: Not Available

① Replace is any measureable movement is found.

22140_IM35_C0012

BRAKE SPECIFICATIONS
All measurements in inches unless noted

| Year | Model | Front Brake Disc | | | Rear Brake Disc | | | Minimum Lining Thickness | | Brake Caliper | |
		Original Thickness	Minimum Thickness	Maximum Run-out	Original Thickness	Minimum Thickness	Maximum Run-out	Front	Rear	Bracket Bolts (ft. lbs.)	Mounting Bolts (ft. lbs.)
2006	M35	1.102	1.024	0.0014	0.631	0.551	0.0022	0.079	0.079	①	②
	M45	1.102	1.024	0.0014	0.631	0.551	0.0022	0.079	0.079	①	②
2007	M35	1.102	1.024	0.0014	0.631	0.551	0.0022	0.079	0.079	①	②
	M45	1.102	1.024	0.0014	0.631	0.551	0.0022	0.079	0.079	①	②
2008	M35	1.102	1.024	0.0014	0.631	0.551	0.0022	0.079	0.079	①	②
	M45	1.102	1.024	0.0014	0.631	0.551	0.0022	0.079	0.079	①	②

① Front: 23-38
   Rear: 32-43

② Front: 17-22
   Rear: 36-28

22140_IM35_C0013

SCHEDULED MAINTENANCE INTERVALS
Infiniti— M35 & M45

TO BE SERVICED	TYPE OF SERVICE	VEHICLE MILEAGE INTERVAL (x1000)												
		7.5	15	22.5	30	37.5	45	52.5	60	67.5	75	82.5	90	97.5
Engine oil & filter	R	✓	✓	✓	✓	✓	✓	✓	✓	✓	✓	✓	✓	✓
Automatic transaxle fluid ①	S/I		✓		✓		✓		✓		✓		✓	
Brake lines & cables	S/I		✓		✓		✓		✓		✓		✓	
Brake pads & discs	S/I		✓		✓		✓		✓		✓		✓	
Differential gear oil	S/I		✓		✓		✓		✓		✓		✓	
Driveshaft boots	S/I		✓		✓		✓		✓		✓		✓	
Active suspension fluid ②	S/I		✓		✓		✓		✓		✓		✓	
In-cabin microfilter	R		✓		✓		✓		✓		✓		✓	
Air cleaner filter ③	R								✓					
Exhaust system	S/I		✓		✓		✓		✓		✓			
Fuel lines	S/I								✓					
Steering gear & linkage, axle & suspension parts	S/I	✓		✓			✓		✓		✓		✓	
Vapor lines	S/I				✓				✓				✓	
Engine coolant ④	R								✓				✓	
Spark plugs(Platinum-tipped type) ⑤	R	Replace every 105,000 miles												
Drive belts ⑥	S/I								✓					

R: Replace   S/I: Service or Inspect

① If towing a trailer, using a camper or car-top carrier, or driving on rough or muddy roads, CHANGE oil every 30,000 miles or 24 months.

② Replace at 60,000 miles (if not previously replaced).

③ If operating in dusty conditions, more frequent maintenance may be required.

④ After 60,000 miles or 48 months, replace coolant every 30,000 miles or 24 months.

⑤ Platinum-tipped spark plugs should be changed every 105,000 miles.

⑥ After 60,000 miles or 48 months, inspect every 15,000 miles or 12 months. Replace belts if found damaged.

FREQUENT OPERATION MAINTENANCE (SEVERE SERVICE)

If a vehicle is operated under any of the following conditions it is considered severe service:

- **Extremely dusty areas.**
- **50% or more of the vehicle operation is in 32°C (90°F) or higher temperatures, or constant operation in temperatures below 0°C (32°F).**
- **Prolonged idling (vehicle operation in stop and go traffic).**
- **Frequent short running periods (engine does not warm to normal operating temperatures).**
- **Police, taxi, delivery or trailer towing usage.**

Oil & oil filter: change every 3750 miles.

Brake pads & discs: service or inspect every 7500 miles.

Driveshaft boots: service or inspect every 7500 miles

Exhaust system: service or inspect every 7500 miles.

Steering gear, linkage, axle & suspension ball joints: service or inspect every 7500 miles.

Steering linkage, ball joints & front suspension ball joints: service or inspect every 7500 miles.

22140_IM35_C0014

PRECAUTIONS

Before servicing any vehicle, please be sure to read all of the following precautions, which deal with personal safety, prevention of component damage, and important points to take into consideration when servicing a motor vehicle:

• Never open, service or drain the radiator or cooling system when the engine is hot; serious burns can occur from the steam and hot coolant.

• Observe all applicable safety precautions when working around fuel. Whenever servicing the fuel system, always work in a well-ventilated area. Do not allow fuel spray or vapors to come in contact with a spark, open flame, or excessive heat (a hot drop light, for example). Keep a dry chemical fire extinguisher near the work area. Always keep fuel in a container specifically designed for fuel storage; also, always properly seal fuel containers to avoid the possibility of fire or explosion. Refer to the additional fuel system precautions later in this section.

• Fuel injection systems often remain pressurized, even after the engine has been turned **OFF**. The fuel system pressure must be relieved before disconnecting any fuel lines. Failure to do so may result in fire and/or personal injury.

• Brake fluid often contains polyglycol ethers and polyglycols. Avoid contact with the eyes and wash your hands thoroughly after handling brake fluid. If you do get brake fluid in your eyes, flush your eyes with clean, running water for 15 minutes. If eye irritation persists, or if you have taken

brake fluid internally, IMMEDIATELY seek medical assistance.

• The EPA warns that prolonged contact with used engine oil may cause a number of skin disorders, including cancer. You should make every effort to minimize your exposure to used engine oil. Protective gloves should be worn when changing oil. Wash your hands and any other exposed skin areas as soon as possible after exposure to used engine oil. Soap and water, or waterless hand cleaner should be used.

• All new vehicles are now equipped with an air bag system, often referred to as a Supplemental Restraint System (SRS) or Supplemental Inflatable Restraint (SIR) system. The system must be disabled before performing service on or around system components, steering column, instrument panel components, wiring and sensors. Failure to follow safety and disabling procedures could result in accidental air bag deployment, possible personal injury and unnecessary system repairs.

• Always wear safety goggles when working with, or around, the air bag system. When carrying a non-deployed air bag, be sure the bag and trim cover are pointed away from your body. When placing a non-deployed air bag on a work surface, always face the bag and trim cover upward, away from the surface. This will reduce the motion of the module if it is accidentally deployed. Refer to the additional air bag system precautions later in this section.

• Clean, high quality brake fluid from a sealed container is essential to the safe and

proper operation of the brake system. You should always buy the correct type of brake fluid for your vehicle. If the brake fluid becomes contaminated, completely flush the system with new fluid. Never reuse any brake fluid. Any brake fluid that is removed from the system should be discarded. Also, do not allow any brake fluid to come in contact with a painted surface; it will damage the paint.

• Never operate the engine without the proper amount and type of engine oil; doing so WILL result in severe engine damage.

• Timing belt maintenance is extremely important. Many models utilize an interference-type, non-freewheeling engine. If the timing belt breaks, the valves in the cylinder head may strike the pistons, causing potentially serious (also time-consuming and expensive) engine damage. Refer to the maintenance interval charts for the recommended replacement interval for the timing belt, and to the timing belt section for belt replacement and inspection.

• Disconnecting the negative battery cable on some vehicles may interfere with the functions of the on-board computer system(s) and may require the computer to undergo a relearning process once the negative battery cable is reconnected.

• When servicing drum brakes, only disassemble and assemble one side at a time, leaving the remaining side intact for reference.

• Only an MVAC-trained, EPA-certified automotive technician should service the air conditioning system or its components.

BRAKES

ANTI-LOCK BRAKE SYSTEM (ABS)

GENERAL INFORMATION

PRECAUTIONS

• Certain components within the ABS system are not intended to be serviced or repaired individually.

• Do not use rubber hoses or other parts not specifically specified for and ABS system. When using repair kits, replace all parts included in the kit. Partial or incorrect repair may lead to functional problems and require the replacement of components.

• Lubricate rubber parts with clean, fresh brake fluid to ease assembly. Do not

use shop air to clean parts; damage to rubber components may result.

• Use only DOT 3 brake fluid from an unopened container.

• If any hydraulic component or line is removed or replaced, it may be necessary to bleed the entire system.

• A clean repair area is essential. Always clean the reservoir and cap thoroughly before removing the cap. The slightest amount of dirt in the fluid may plug an orifice and impair the system function. Perform repairs after components have been thoroughly cleaned; use only denatured alcohol

to clean components. Do not allow ABS components to come into contact with any substance containing mineral oil; this includes used shop rags.

• The Anti-Lock control unit is a microprocessor similar to other computer units in the vehicle. Ensure that the ignition switch is **OFF** before removing or installing controller harnesses. Avoid static electricity discharge at or near the controller.

• If any arc welding is to be done on the vehicle, the control unit should be unplugged before welding operations begin.

BRAKES

FRONT DISC BRAKES

✴✴ CAUTION

Dust and dirt accumulating on brake parts during normal use may contain asbestos fibers from production or aftermarket brake linings. Breathing excessive concentrations of asbestos fibers can cause serious bodily harm. Exercise care when servicing brake parts. Do not sand or grind brake lining unless equipment used is designed to contain the dust residue. Do not clean brake parts with compressed air or by dry brushing. Cleaning should be done by dampening the brake components with a fine mist of water, then wiping the brake components clean with a dampened cloth. Dispose of cloth and all residue containing asbestos fibers in an impermeable container with the appropriate label. Follow practices prescribed by the Occupational Safety and Health Administration (OSHA) and the Environmental Protection Agency (EPA) for the handling, processing, and disposing of dust or debris that may contain asbestos fibers.

BRAKE CALIPER

REMOVAL & INSTALLATION

See Figure 1.

1. Before servicing the vehicle, refer to the precautions in the beginning of this section.

2. Remove the cap from the master cylinder reservoir and extract about ⅓ of the brake fluid from the reservoir to prevent overflow when the caliper piston is compressed.

3. Remove the wheels.

4. Fasten disc rotor using wheel nut to hold rotor in place.

5. Remove union bolts, and then disconnect brake hose from caliper assembly.

6. Remove torque member mounting bolts, and remove brake caliper assembly.

✴✴ WARNING

Do not drop brake pads.

To install:

7. Place an old pad over the caliper piston. Use a C-clamp to compress the piston.

8. Install the union bolts and torque it to 98 ft. lbs. (132 Nm).

9. Install brake hose to brake caliper assembly.

10. Install the wheels and lower the vehicle to the floor.

11. Check and then refill the master cylinder if needed.

DISC BRAKE PADS

REMOVAL & INSTALLATION

✴✴ WARNING

While removing cylinder body, do not depress brake pedal because piston will pop out.

➡It is not necessary to remove bolts on torque member and brake hose except for disassembly or replacement of caliper assembly. In this case, hang cylinder body with a wire so as not to stretch brake hose.

✴✴ WARNING

Do not damage piston boot.

➡If any shim is subject to serious corrosion, replace it with a new one.

➡Always replace shim and shim cover as a set when replacing brake pads.

➡Keep rotor free from brake fluid.

➡Burnish the brake pads and disc rotor mutually contacting surfaces, after refinishing or replacing rotors, after replacing pads, or if a soft pedal occurs at very low mileage.

1. Before servicing the vehicle, refer to the precautions in the beginning of this section.

2. Remove the cap from the master cylinder reservoir and extract about ⅓ of the brake fluid from the reservoir to prevent overflow when the caliper piston is compressed.

3. Remove the wheels.

4. Remove the sliding pin bolt..

5. Pivot the caliper body upward and secure it with a length of wire. Remove the retainers and inner and outer shims and pads.

To install:

6. Place an old pad over the caliper piston. Use a C-clamp to compress the piston.

7. Install the new pads and shims and rotate caliper down onto rotor. Install the pin bolt and torque it to 20 ft. lbs. (26 Nm).

8. Install the wheels and lower the vehicle to the floor.

9. Check and then refill the master cylinder if needed.

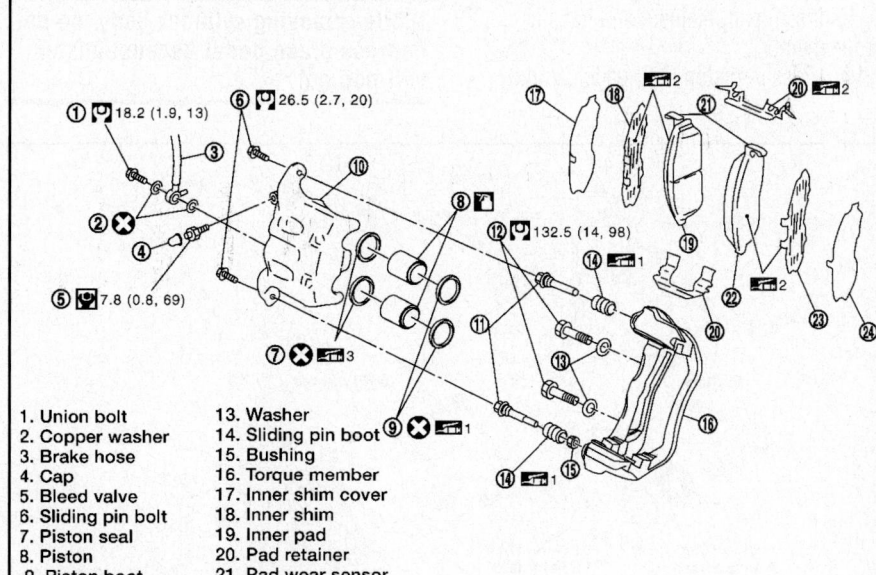

1. Union bolt
2. Copper washer
3. Brake hose
4. Cap
5. Bleed valve
6. Sliding pin bolt
7. Piston seal
8. Piston
9. Piston boot
10. Cylinder body
11. Sliding pin
12. Torque member mounting bolt
13. Washer
14. Sliding pin boot
15. Bushing
16. Torque member
17. Inner shim cover
18. Inner shim
19. Inner pad
20. Pad retainer
21. Pad wear sensor
22. Outer pad
23. Outer shim
24. Outer shim cover

22140_IM35_G0076

Fig. 1 Brake caliper exploded view

BRAKES

REAR DISC BRAKES

※※ CAUTION

Dust and dirt accumulating on brake parts during normal use may contain asbestos fibers from production or aftermarket brake linings. Breathing excessive concentrations of asbestos fibers can cause serious bodily harm. Exercise care when servicing brake parts. Do not sand or grind brake lining unless equipment used is designed to contain the dust residue. Do not clean brake parts with compressed air or by dry brushing. Cleaning should be done by dampening the brake components with a fine mist of water, then wiping the brake components clean with a dampened cloth. Dispose of cloth and all residue containing asbestos fibers in an impermeable container with the appropriate label. Follow practices prescribed by the Occupational Safety and Health Administration (OSHA) and the Environmental Protection Agency (EPA) for the handling, processing, and disposing of dust or debris that may contain asbestos fibers.

BRAKE CALIPER

REMOVAL & INSTALLATION
See Figure 2.

➡ Clean dust on caliper and brake pad with a vacuum dust collector to minimize the hazard of air borne particles or other materials.

※※ WARNING

While removing cylinder body, do not depress brake pedal because piston will pop out.

➡ It is not necessary to remove bolts on torque member and brake hose except for disassembly or replacement of caliper assembly. In this case, hang cylinder body with a wire so as not to stretch brake hose.

※※ WARNING

Do not damage piston boot.

➡ If any shim is subject to serious corrosion, replace it with a new one.

➡ Always replace shim and shim covers as a set when replacing brake pads.

➡ Keep rotor free from brake fluid.

➡ Burnish the brake pads and disc rotor mutually contacting surfaces after refinishing or replacing rotors, after replacing pads, or if a soft pedal occurs at very low mileage.

   1. Before servicing the vehicle, refer to the precautions in the beginning of this section.
   2. Raise and support the vehicle safely.
   3. Remove the tire and wheel assembly.
   4. Fasten disc rotor using wheel nut to hold rotor in place.
   5. Remove the union bolts and torque member bolts, and remove the brake caliper assembly from the vehicle.

➡ Position the assembly to the side. Do not allow the caliper to hang by the brake line. Do not disconnect the brake line from the caliper

※※ WARNING

Do not drop brake pad.

   To install:
   6. Installation is the reverse of the removal procedure.
   7. Apply PBC (Poly Butyl Cuprysil) grease or silicone-based grease to between pad and shim.
   8. Install inner shim, inner shim cover to inner pad, and outer shim to outer pad.
   9. Install pad retainers and pads to torque member.
   10. Press in piston until pads can be

installed, and then install cylinder body to torque member.

➡ In the case of replacing a pad with new one, check a brake fluid level in the reservoir tank because brake fluid returns to master cylinder reservoir tank when pressing piston in.

   11. Install the Torque member mounting bolt and torque it to 62 ft. lbs. (84.3 Nm).
   12. Fasten disc rotor using wheel nut to hold rotor in place.
   13. Remove union bolts, and then disconnect brake hose from caliper assembly.
   14. Remove torque member mounting bolts, and remove brake caliper assembly.

※※ WARNING

Do not drop brake pad.

DISC BRAKE PADS

REMOVAL & INSTALLATION
See Figure 3.

➡ Clean dust on caliper and brake pad with a vacuum dust collector to minimize the hazard of air borne particles or other materials.

※※ WARNING

While removing cylinder body, do not depress brake pedal because piston will pop out.

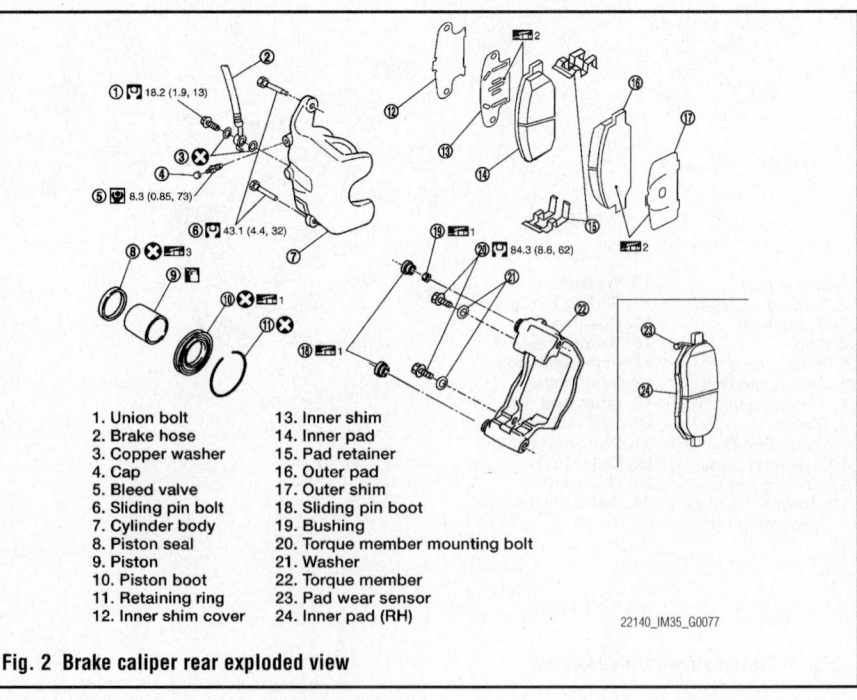

1. Union bolt
2. Brake hose
3. Copper washer
4. Cap
5. Bleed valve
6. Sliding pin bolt
7. Cylinder body
8. Piston seal
9. Piston
10. Piston boot
11. Retaining ring
12. Inner shim cover
13. Inner shim
14. Inner pad
15. Pad retainer
16. Outer pad
17. Outer shim
18. Sliding pin boot
19. Bushing
20. Torque member mounting bolt
21. Washer
22. Torque member
23. Pad wear sensor
24. Inner pad (RH)

22140_IM35_G0077

Fig. 2 Brake caliper rear exploded view

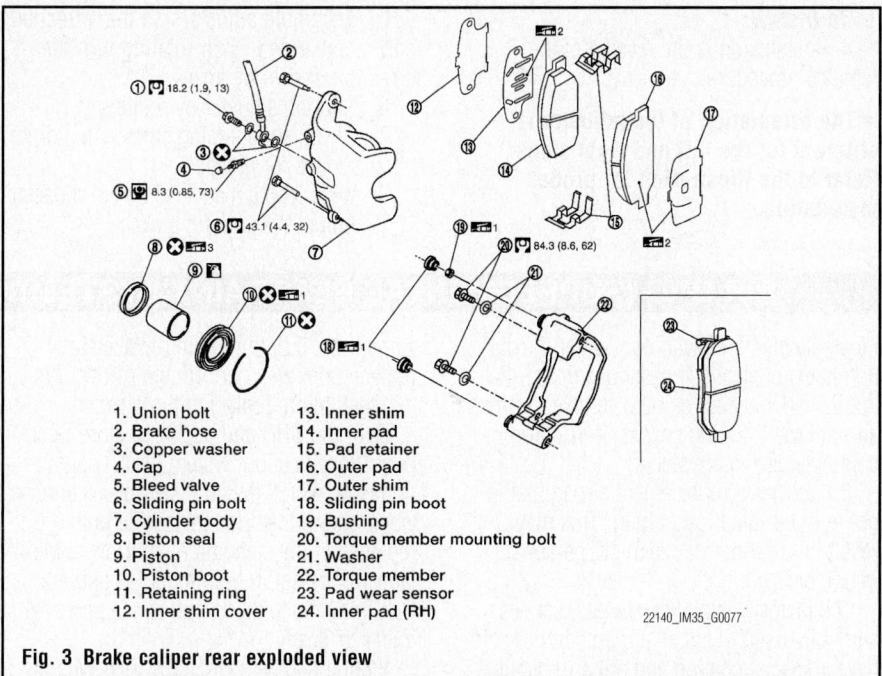

1. Union bolt
2. Brake hose
3. Copper washer
4. Cap
5. Bleed valve
6. Sliding pin bolt
7. Cylinder body
8. Piston seal
9. Piston
10. Piston boot
11. Retaining ring
12. Inner shim cover
13. Inner shim
14. Inner pad
15. Pad retainer
16. Outer pad
17. Outer shim
18. Sliding pin boot
19. Bushing
20. Torque member mounting bolt
21. Washer
22. Torque member
23. Pad wear sensor
24. Inner pad (RH)

22140_IM35_G0077

Fig. 3 Brake caliper rear exploded view

➥ It is not necessary to remove bolts on torque member and brake hose except for disassembly or replacement of caliper assembly. In this case, hang cylinder body with a wire so as not to stretch brake hose.

✳✳ WARNING

Do not damage piston boot.

➥ If any shim is subject to serious corrosion, replace it with a new one.

➥ Always replace shim and shim covers as a set when replacing brake pads.

➥ Keep rotor free from brake fluid.

➥ Burnish the brake pads and disc rotor mutually contacting surfaces after refinishing or replacing rotors, after

replacing pads, or if a soft pedal occurs at very low mileage.

1. Before servicing the vehicle, refer to the precautions in the beginning of this section.

2. Raise and support the vehicle safely.

3. Remove the tire and wheel assembly.

4. Remove the lower pin bolt.

5. Pivot the caliper body upward and secure it with a length of wire. Remove the retainers and inner and outer shims and pads.

To install:

6. Push the piston into the cylinder body.

7. Apply PBC (Poly Butyl Cuprysil) grease or silicone-based grease to between pad and shim.

8. Install inner shim, inner shim cover to inner pad, and outer shim to outer pad.

9. Install pad retainers and pads to torque member.

10. Install the new pads and shims and rotate the caliper down onto rotor. Install the pin bolt and torque it according to the illustration.

11. Install the pin bolts and torque it to 32 ft. lbs. (43.1 Nm).

12. Connect the parking brake cable and install the bracket.

13. Install the wheels.

14. Check rear disc brake for drag.

15. Check and then refill the master cylinder if needed.

BRAKES

PARKING BRAKE SHOES

REMOVAL & INSTALLATION

See Figures 4 and 5.

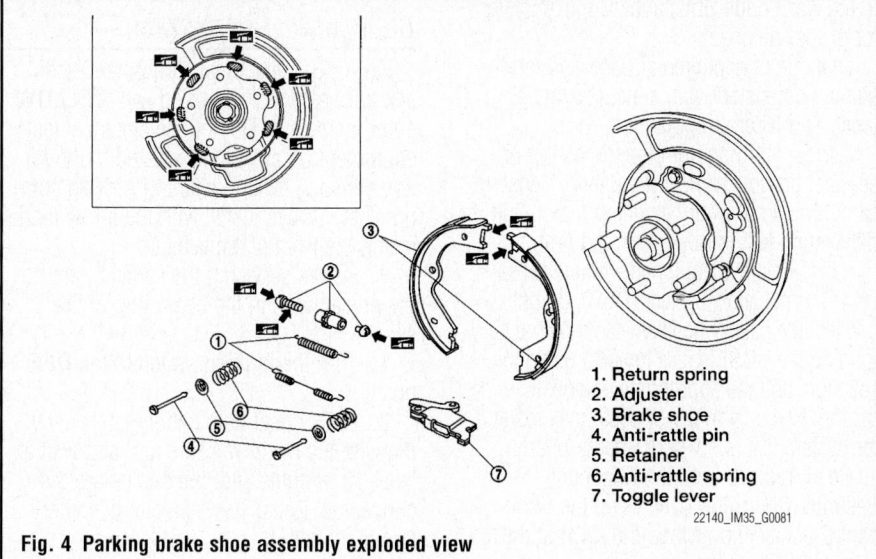

1. Return spring
2. Adjuster
3. Brake shoe
4. Anti-rattle pin
5. Retainer
6. Anti-rattle spring
7. Toggle lever

22140_IM35_G0081

Fig. 4 Parking brake shoe assembly exploded view

➥ Clean brakes with a vacuum dust collector to minimize the hazard of air borne particles or other materials.

PARKING BRAKE

✳✳ CAUTION

Clean dust on disc rotor and back plate using a vacuum dust collector. Do not blow with compressed air.

➥ Put matching marks on both disc rotor and wheel hub when removing disc rotor.

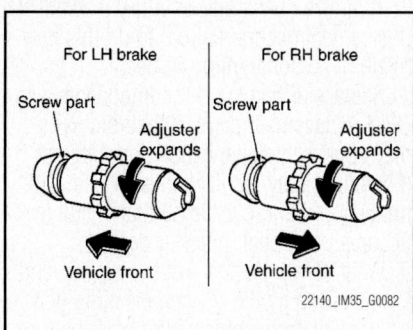

22140_IM35_G0082

Fig. 5 Parking brake shoe adjuster orientation

1. Remove rear tires from vehicle with power tool.

2. Remove disc rotor with parking brake pedal completely in the released position See "Removal and Installation of Brake Caliper Assembly" in this section.

3. Remove the parking brake shoes (refer to the illustration).

To install:

4. Installation is the reverse of the removal procedure.

➡**The orientation of the adjuster is different for the left and right sides. Refer to the illustration for proper installation.**

5. Assemble adjusters so that threaded part is expanded when rotating it in the direction shown by arrow.

6. Shorten adjuster by rotating it.

7. Check shoe sliding surface and drum inner surface for grease.

8. Wipe it off if it adheres on the surfaces.

9. Adjust the parking brake.

CHASSIS ELECTRICAL

AIR BAG (SUPPLEMENTAL RESTRAINT SYSTEM)

GENERAL INFORMATION

✷✷ CAUTION

These vehicles are equipped with an air bag system. The system must be disarmed before performing service on, or around, system components, the steering column, instrument panel components, wiring and sensors. Failure to follow the safety precautions and the disarming procedure could result in accidental air bag deployment, possible injury and unnecessary system repairs.

SERVICE PRECAUTIONS

Disconnect and isolate the battery negative cable before beginning any airbag system component diagnosis, testing, removal, or installation procedures. Allow system capacitor to discharge for two minutes before beginning any component service. This will disable the airbag system. Failure to disable the airbag system may result in accidental airbag deployment, personal injury, or death.

Do not place an intact undeployed airbag face down on a solid surface. The airbag will propel into the air if accidentally deployed and may result in personal injury or death.

When carrying or handling an undeployed airbag, the trim side (face) of the airbag should be pointing towards the body to minimize possibility of injury if accidental deployment occurs. Failure to do this may result in personal injury or death.

Replace airbag system components with OEM replacement parts. Substitute parts may appear interchangeable, but internal differences may result in inferior occupant protection. Failure to do so may result in occupant personal injury or death.

Wear safety glasses, rubber gloves, and long sleeved clothing when cleaning powder residue from vehicle after an airbag deployment. Powder residue emitted from a deployed airbag can cause skin irritation.

Flush affected area with cool water if irritation is experienced. If nasal or throat irritation is experienced, exit the vehicle for fresh air until the irritation ceases. If irritation continues, see a physician.

Do not use a replacement airbag that is not in the original packaging. This may result in improper deployment, personal injury, or death.

The factory installed fasteners, screws and bolts used to fasten airbag components have a special coating and are specifically designed for the airbag system. Do not use substitute fasteners. Use only original equipment fasteners listed in the parts catalog when fastener replacement is required.

During, and following, any child restraint anchor service, due to impact event or vehicle repair, carefully inspect all mounting hardware, tether straps, and anchors for proper installation, operation, or damage. If a child restraint anchor is found damaged in any way, the anchor must be replaced. Failure to do this may result in personal injury or death.

Deployed and non-deployed airbags may or may not have live pyrotechnic material within the airbag inflator.

Do not dispose of driver/passenger/curtain airbags or seat belt tensioners unless you are sure of complete deployment. Refer to the Hazardous Substance Control System for proper disposal.

Dispose of deployed airbags and tensioners consistent with state, provincial, local, and federal regulations.

After any airbag component testing or service, do not connect the battery negative cable. Personal injury or death may result if the system test is not performed first.

If the vehicle is equipped with the Occupant Classification System (OCS), do not connect the battery negative cable before performing the OCS Verification Test using the scan tool and the appropriate diagnostic information. Personal injury or death may result if the system test is not performed properly.

Never replace both the Occupant Restraint Controller (ORC) and the Occupant Classification Module (OCM) at the

same time. If both require replacement, replace one, then perform the Airbag System test before replacing the other.

Both the ORC and the OCM store Occupant Classification System (OCS) calibration data, which they transfer to one another when one of them is replaced. If both are replaced at the same time, an irreversible fault will be set in both modules and the OCS may malfunction and cause personal injury or death.

If equipped with OCS, the Seat Weight Sensor is a sensitive, calibrated unit and must be handled carefully. Do not drop or handle roughly. If dropped or damaged, replace with another sensor. Failure to do so may result in occupant injury or death.

If equipped with OCS, the front passenger seat must be handled carefully as well. When removing the seat, be careful when setting on floor not to drop. If dropped, the sensor may be inoperative, could result in occupant injury, or possibly death.

If equipped with OCS, when the passenger front seat is on the floor, no one should sit in the front passenger seat. This uneven force may damage the sensing ability of the seat weight sensors. If sat on and damaged, the sensor may be inoperative, could result in occupant injury, or possibly death.

DISARMING THE SYSTEM

All Air Bag electrical wiring harnesses and connectors are covered with **YELLOW** outer insulation. Do not use electrical test equipment on any circuit related to the Air Bag sensors. When installing Air Bag components, always install with the arrow marks facing the front of the vehicle.

1. Before servicing the vehicle, refer to the precautions in the beginning of this section.

2. Turn the ignition switch to the **OFF** position.

3. Disconnect both battery cables starting with the negative cable first and wait at least 10 minutes after the cables are disconnected. Be sure to insulate the battery terminal ends.

ARMING THE SYSTEM

ARMING THE SYSTEM

1. Before servicing the vehicle, refer to the precautions in the beginning of this section.
2. Turn the ignition switch to the **OFF** position.

3. Connect both battery cables starting with the positive cable first.
4. The Air Bag or Air Bag system is equipped with a self-diagnostic operation. After turning the ignition key to the **ON** or **START** position, the **AIR BAG** warning lamp will illuminate for 7 seconds. After 7 seconds, the **AIR BAG** lamp will extinguish if no malfunction is detected. If the **AIR BAG** lamp does not extinguish after 7 seconds, check the Air Bag self-diagnostic system for a malfunction.

DRIVETRAIN

AUTOMATIC TRANSMISSION ASSEMBLY

REMOVAL & INSTALLATION

See Figures 6 through 13.

✳✳ WARNING

When removing the Automatic Transmission (A/T) assembly from engine, first remove the crankshaft position sensor (POS) from the A/T assembly. Be careful not to damage sensor edge.

1. Before servicing the vehicle, refer to the precautions section.
2. Drain the transaxle fluid.
3. Remove or disconnect the following:
   - Battery cable from the negative terminal.
   - Engine under cover
   - A/T fluid level gauge
   - Exhaust front tube and center muffler
   - Heat insulator
   - Rear propeller shaft.
   - Rack stay
   - Exhaust mounting bracket
   - Heated oxygen sensor 2 harness connectors (A)
   - Heated oxygen sensor 2 harness (B) from clips (1)
   - Bracket (2) from transmission assembly. M35 models
   - Crankshaft position sensor (POS) from A/T assembly

✳✳ WARNING

Do not subject it to impact by dropping or hitting it. Do not disassemble. Do not allow metal filings, etc., to get on the sensor's front edge magnetic area. Do not place in an area affected by magnetism.

   - Starter motor
   - Rear cover plate 2WD
   - Rear plate cover
   - Tightening bolts for drive plate and torque converter.

✳✳ WARNING

When turning the crankshaft, turn it clockwise as viewed from the front of the engine.

4. Support A/T assembly with a transmission jack.
5. Remove or disconnect the following:
   - Rear engine mounting member
   - Engine mounting insulator (rear)
   - A/T assembly harness connector
   - Air breather hose
   - A/T fluid charging pipe from A/T assembly
   - O-ring from A/T fluid charging pipe
   - Fluid cooler tube from A/T assembly

➡**Plug up openings such as the A/T fluid charging pipe hole, etc.**

   - Bolts from A/T assembly to engine assembly
6. Remove A/T assembly from vehicle.

➡**Secure torque converter to prevent it from dropping.**

7. Secure A/T assembly to a transmission jack.

To install:

8. Check the torque converter distance to transmission case "A" to ensure it is within the reference value limit.
9. Distance "A"
10. M35 models: 0.98 in (25.0 mm) or more
11. M45 models: 0.87 in (22.0 mm) or more
12. Install the bolts from A/T assembly to engine assembly following the procedure below.

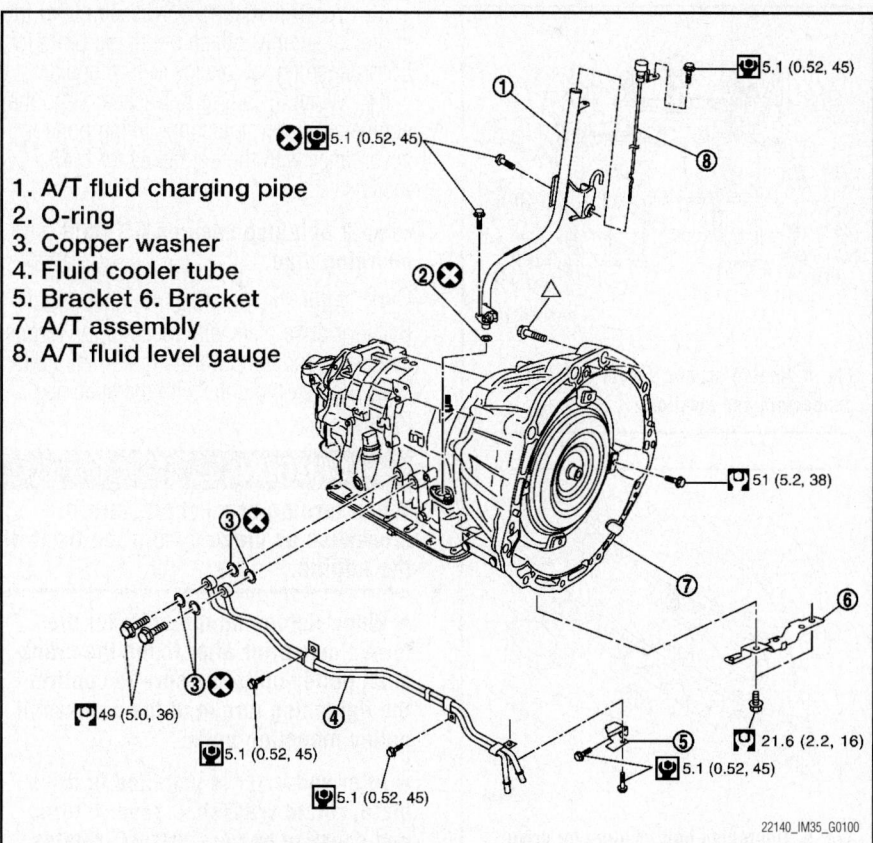

1. A/T fluid charging pipe
2. O-ring
3. Copper washer
4. Fluid cooler tube
5. Bracket 6. Bracket
7. A/T assembly
8. A/T fluid level gauge

5.1 (0.52, 45)
5.1 (0.52, 45)
51 (5.2, 38)
21.6 (2.2, 16)
49 (5.0, 36)
5.1 (0.52, 45)
5.1 (0.52, 45)
5.1 (0.52, 45)

22140_IM35_G0100

Fig. 6 Automatic transmission exploded view M35 models

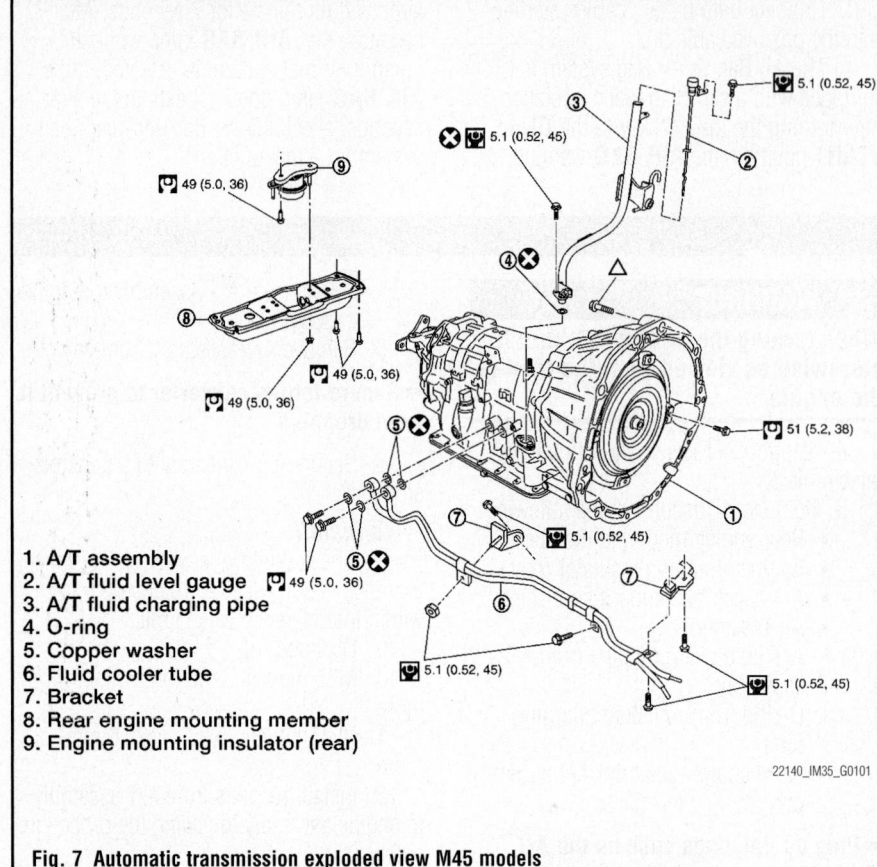

1. A/T assembly
2. A/T fluid level gauge
3. A/T fluid charging pipe
4. O-ring
5. Copper washer
6. Fluid cooler tube
7. Bracket
8. Rear engine mounting member
9. Engine mounting insulator (rear)

Fig. 7 Automatic transmission exploded view M45 models

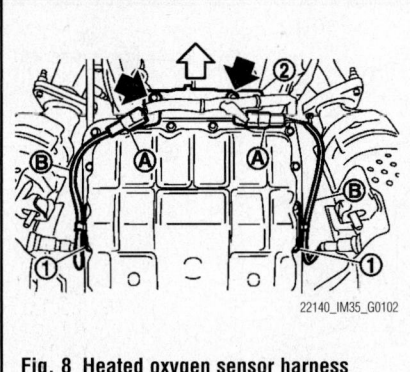

Fig. 8 Heated oxygen sensor harness connectors and locations

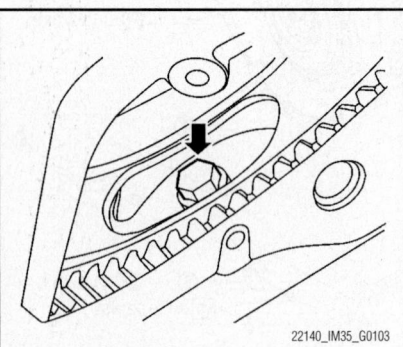

Fig. 9 Tightening bolt location for drive plate to torque converter

13. When installing A/T assembly to the engine assembly, attach the fixing bolts in accordance fig. below for M35 models.

14. When installing A/T assembly to the engine assembly, attach the fixing bolts in accordance with the fig. below for M45 models.

➡ **No.2 bolt also secures A/T fluid charging pipe.**

15. Align the positions of tightening bolts for drive plate with those of the torque converter, and temporarily tighten the bolts. Then, tighten the bolts with the specified torque.

❄❄ CAUTION

When turning crankshaft, turn it clockwise as viewed from the front of the engine.

➡ **When tightening the bolts for the torque converter after fixing the crankshaft pulley bolts, be sure to confirm the tightening torque of the crankshaft pulley mounting bolts.**

➡ **After converter is installed to drive plate, rotate crankshaft several turns and check to be sure that A/T rotates freely without binding.**

16. Install crankshaft position sensor AWD vehicles.

➡ **Remove plugs previously installed into openings such as the A/T fluid charging pipe hole, etc.**

17. Connect or install the following:
- Fluid cooler tube to A/T assembly
- O-ring to A/T fluid charging pipe
- A/T fluid charging pipe to A/T assembly
- Air breather hose
- A/T assembly harness connector
- Engine mounting insulator (rear)
- Rear engine mounting member
- Remove the transmission jack.
- Rear plate cover
- Rear cover plate 2WD
- Starter motor
- Crankshaft Position Sensor (POS) to A/T assembly
- Oxygen sensor bracket (2) from transmission assembly. M35 Models.
- Heated oxygen sensor 2 harness (B) to clips (1)
- Heated oxygen sensor 2 harness connectors (A)
- Exhaust mounting bracket
- Exhaust mounting bracket
- Rack stay
- Rear propeller shaft
- Heat insulator
- Exhaust front tube and center muffler
- A/T fluid level gauge
- Engine under cover
- Battery cable from the negative terminal.

➡ **After completing installation, check A/T fluid leakage, A/T fluid level and A/T position.**

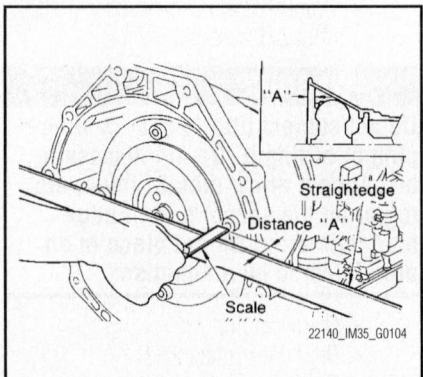

Fig. 10 Checking torque converter distance to transmission case dimensions

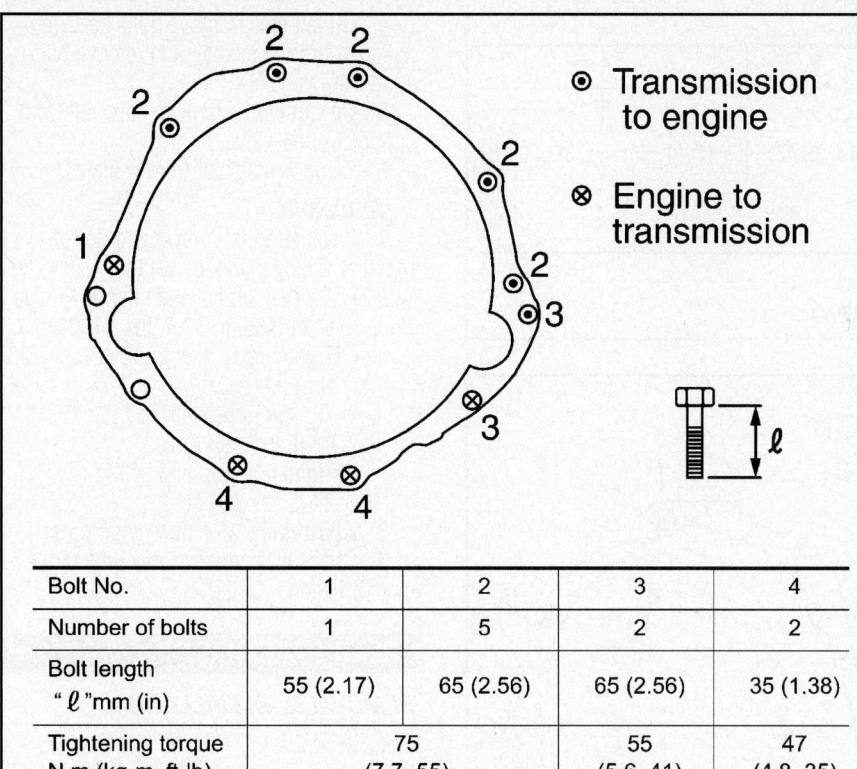

- ⊙ Transmission to engine
- ⊗ Engine to transmission

Bolt No.	1	2	3	4
Number of bolts	1	5	2	2
Bolt length " ℓ "mm (in)	55 (2.17)	65 (2.56)	65 (2.56)	35 (1.38)
Tightening torque N·m (kg-m, ft-lb)	75 (7.7, 55)		55 (5.6, 41)	47 (4.8, 35)

22140_IM35_G0106

Fig. 11 A/T assembly to engine assembly bolt chart—placement, length and torque specifications M35 models

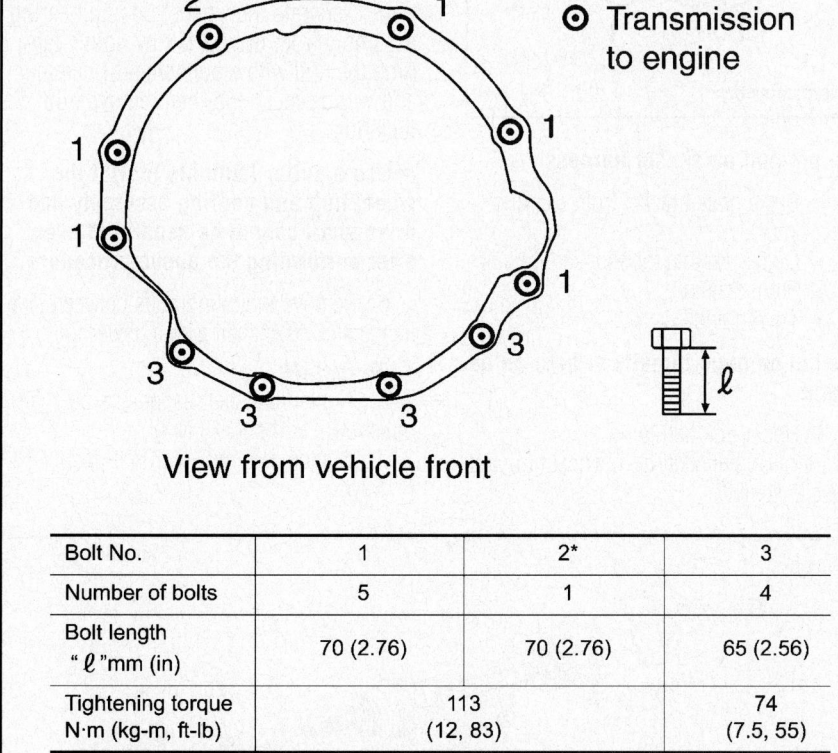

- ⊙ Transmission to engine

View from vehicle front

Bolt No.	1	2*	3
Number of bolts	5	1	4
Bolt length " ℓ "mm (in)	70 (2.76)	70 (2.76)	65 (2.56)
Tightening torque N·m (kg-m, ft-lb)	113 (12, 83)		74 (7.5, 55)

*: No.2 bolt also secures A/T fluid charging pipe.

22140_IM35_G0108

Fig. 12 A/T assembly to engine assembly bolt chart placement, length and torque specifications M45 models

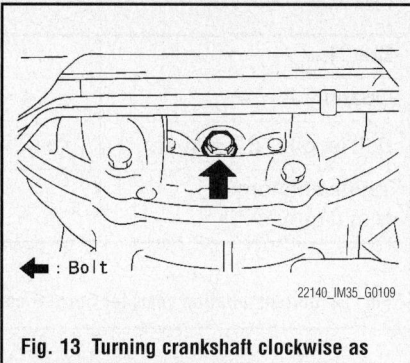

← : Bolt

22140_IM35_G0109

Fig. 13 Turning crankshaft clockwise as viewed from the front of the engine to install torque converter bolts

TRANSFER CASE ASSEMBLY

REMOVAL & INSTALLATION
See Figures 14 and 15.

1. Remove exhaust front tube with power tool.
2. Remove front and rear propeller shaft.
3. Match mark the shaft and companion for reassembly.
4. Disconnect transfer assembly harness connector and separate harness from transfer assembly.
5. Remove air breather hose.
6. Remove control rod.
7. Support transfer assembly and transmission assembly with a jack.
8. Remove rear engine mounting member and engine mounting insulator with power tool.
9. Lower jack to the position where the top transfer mounting bolts can be removed.
10. Remove transfer mounting bolts with power tool and separate transfer from transmission.

✳✳ WARNING
Secure transfer assembly and transmission assembly to a jack.

To install:
11. Install the rear engine mounting member and engine mounting insulator with power tool.
12. Remove the transmission jack.
13. Install the control rod.
14. Install the air breather hose.
15. Install the transfer assembly harness connector.
16. Install the front and rear propeller shafts.
17. Install the exhaust front tube.
18. Tighten the
19. When installing the transfer case to the transmission, install the mounting bolts following the standard below.

Bolt No.	1	2	3	4
Quantity	4	3	2	1
Bolt length "ℓ" mm (in)	75 (2.95)	45 (1.77)	40 (1.57)	30 (1.18)
Tightening torque N·m (kg-m, ft-lb)	37 (3.8, 27)			

22140_IM35_G0121

Fig. 14 Bolt installation chart for transfer case assemblies

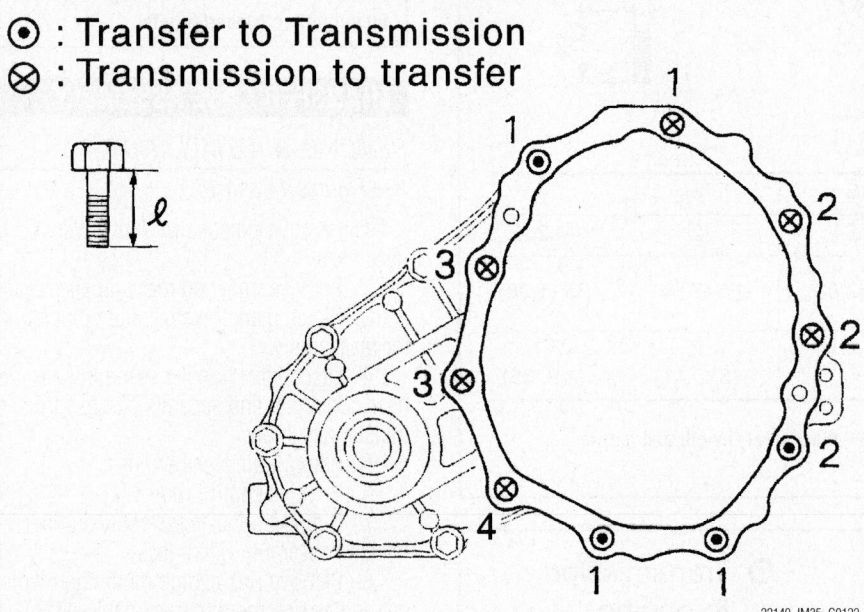

⊙ : Transfer to Transmission
⊗ : Transmission to transfer

22140_IM35_G0122

Fig. 15 Bolt installation pattern for the transfer case to transmission

20. After the installation, check the fluid level, fluid leakage and the A/T positions

FRONT HALFSHAFT

REMOVAL & INSTALLATION

AWD Models

1. Before servicing the vehicle, refer to the precautions section.
2. Remove or disconnect the following:
   - Front wheels
   - Remove wheel sensor from steering knuckle
   - Torque member fixing bolts with power tool. Hang torque member in a place where it will not interfere with work.
   - Brake calipers; use wire to support calipers where they will not interfere with work
   - Brake rotors
   - Antilock Brake wheel sensors

➡**Do not pull on sensor harness**

   - Brake hose bracket from steering knuckle
   - Cotter pin and then loosen the nut.
   - Front axle nut
   - Tie rod end

➡**Do not damage threads or boot on tie rod end**

   - Upper link ball joint
3. Remove halfshaft from wheel hub and bearing assembly

4. On left side, remove bolts securing halfshaft from side shaft and remove halfshaft from vehicle
5. On right side, remove halfshaft from splines in final drive
6. Remove halfshaft from vehicle

To install:
7. Installation is in reverse order of removal. Observe the following:
   - Tighten left halfshaft flange to side shaft bolts to 33 ft. lbs. (45 Nm)
   - Tighten axle nuts to 92 ft. lbs. (125 Nm)
   - Tighten upper link ball joint to 43 ft. lbs. (69 Nm)
   - Tighten tie rod end ball joint to 24 ft. lbs. (34 Nm)
8. **Always** install a new cotter pins.
9. Check and perform front end alignment if needed.

REAR HALFSHAFT

REMOVAL & INSTALLATION

See Figure 16.

1. Remove tires with power tool.
2. Remove cotter pin, then loosen hub lock nut with power tool.
3. Remove stabilizer connecting rod mounting bracket fixing bolt and free stabilizer connecting rod
4. Separate the wheel hub and bearing assembly from drive shaft by lightly tapping the end with a suitable tool hammer and wood block, and then remove hub lock nut.

➡**Use a puller (suitable tool) if the wheel hub and bearing assembly and drive shaft cannot be separated even after performing the above procedure.**

5. Remove mounting bolts between side flange and drive shaft with a power tool.

To install:
6. Tighten halfshaft flange to side shaft bolts to 87 ft. lbs. (118 Nm)
7. Tighten axle nuts to 130 ft. lbs. (175 Nm)

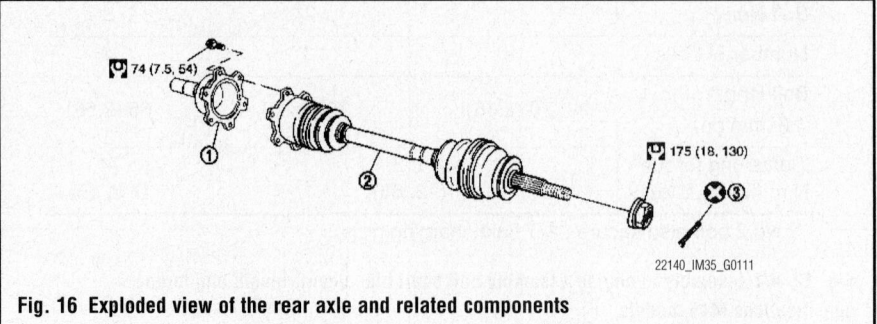

74 (7.5, 54)

175 (18, 130)

22140_IM35_G0111

Fig. 16 Exploded view of the rear axle and related components

CV-BOOTS INSPECTION

See Figure 17.

1. Check boot for cracks, damage, and leakage of grease.
2. Disassemble drive shaft and exchange malfunctioning part if there is a non-standard condition.

REAR PINION SEAL

REMOVAL & INSTALLATION

See Figures 18 through 27.

➡The reuse of collapsible spacer is prohibited in principle. However, it is reusable on a one-time basis only in cases when replacing front oil seal.

➡The diagonally shaded area in the figure shows stamping point for replacement frequency of front oil seal.

➡The following shows if a collapsible spacer replacement is needed before replacing front oil seal.

- Stamp—collapsible spacer replacement
- No stamp—Not required
- "0" or "0" on the far right of stamp—Required
- "01" or "1" on the far right of stamp—Not required

When collapsible spacer replacement is required, disassemble final drive assembly to replace collapsible spacer and front oil seal.

1. Drain gear oil.
2. Make a judgment if a collapsible spacer replacement is required
3. Remove center muffler with a power tool.
4. Remove rear wheel sensor.
5. Match mark the drive shaft prior to removal.
6. Remove drive shaft from final drive. Then suspend it by wire.
7. Install attachment to side flange, on M35 models and then pull out the side flange with the sliding hammer. Tool number A: KV40104100—B: ST36230000 (J- 25840-A)
8. Install attachment to side flange, on M45 models and then pull out the side flange with the sliding hammer. Tool number A: KV40101000—B: ST36230000 (J- 25840-A)

➡Install circle clip in same position it was removed from

9. Match mark the propeller shaft prior to removal.
10. Remove propeller shaft.

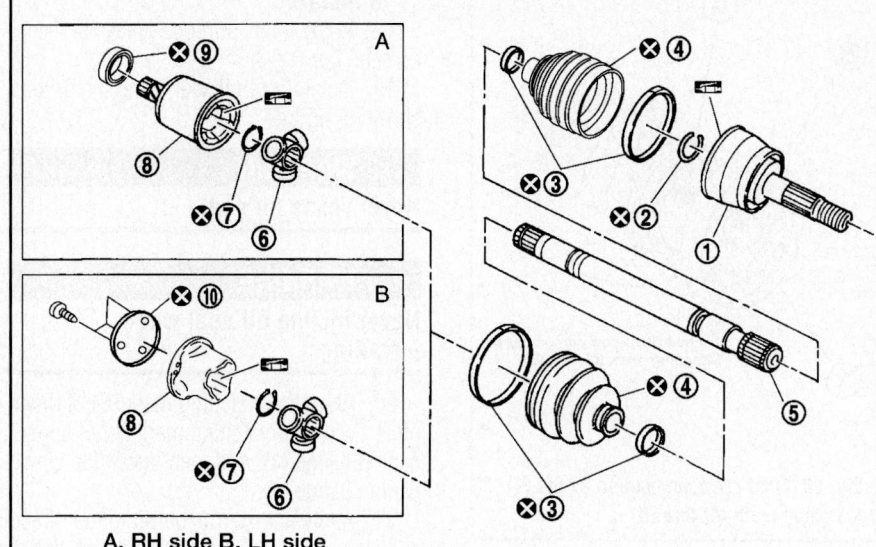

A. RH side B. LH side

1. Joint sub-assembly
2. Circular clip
3. Boot band
4. Boot
5. Shaft
6. Spider assembly
7. Snap ring
8. Housing
9. Dust shield
10. Plug

22140_IM35_G0110

Fig. 17 Rear axle exploded view including related components

11. Measure the total preload with the preload gauge.
12. Tool number A: ST3127S000 (J-25765-A)
13. Record the preload measurement for use on assembly.
14. Put matching mark (B) on the end of the drive pinion. The matching mark (B) should be in line with the matching mark (A) on companion flange (1).

✴✴ WARNING

For matching mark, use paint. Never damage companion flange and drive pinion.

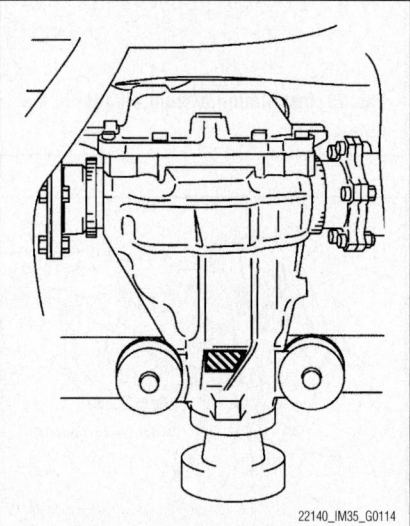

22140_IM35_G0114

Fig. 18 Diagonally shaded area on rear differential showing stamping point for replacement frequency of front oil seal.

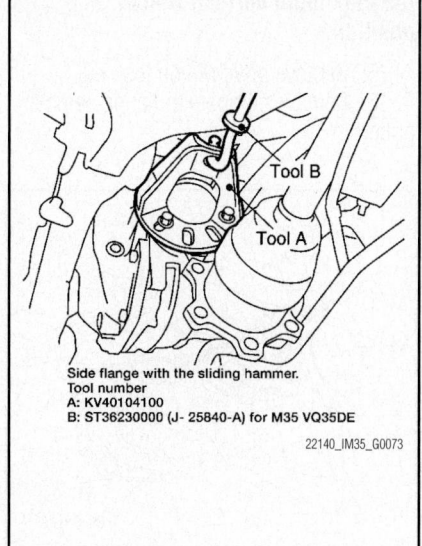

Side flange with the sliding hammer.
Tool number
A: KV40104100
B: ST36230000 (J- 25840-A) for M35 VQ35DE

22140_IM35_G0073

Fig. 19 Special tool used to remove side flange

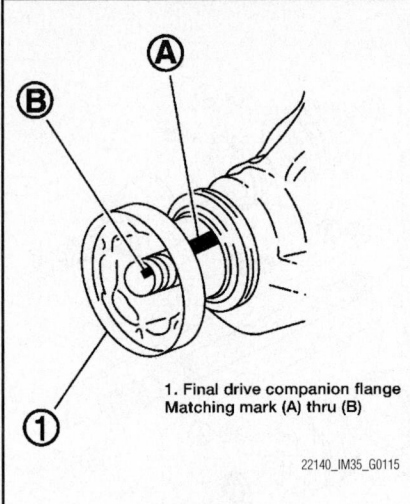

1. Final drive companion flange
Matching mark (A) thru (B)

22140_IM35_G0115

**Fig. 20 Final drive companion flange (1)
Matching mark (A) thru (B)**

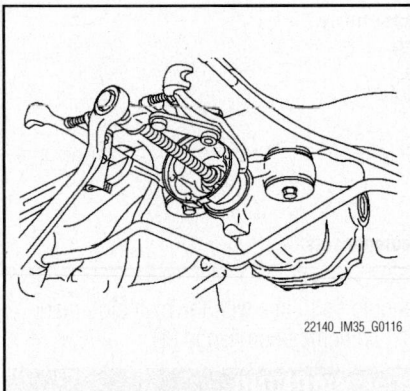

22140_IM35_G0116

**Fig. 21 Differential companion flange
puller**

➡ **The matching mark (A) on the final
drive companion flange (1) indicates
the maximum vertical runout
position.**

15. Remove drive pinion lock nut.
16. Remove companion flange using a
puller.

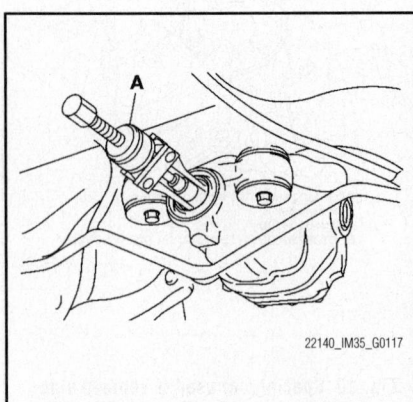

22140_IM35_G0117

Fig. 22 Differential front oil seal puller

17. Remove front oil seal using the puller.

To install:

18. Apply multi-purpose grease to front
oil seal lips.
19. Install front oil seal using the drift as
shown in figure.

> ※※ **WARNING**
> **Never reuse oil seal.**

> ※※ **WARNING**
> **Never incline oil seal when
> installing.**

20. Align the matching mark (B) of drive
pinion with the matching mark (A) of companion flange (1), and then install the companion flange (1).
21. Apply anti-corrosion oil to the thread
and seat of new drive pinion lock nut, and
temporarily tighten drive pinion lock nut to
drive pinion.

> ※※ **WARNING**
> **Never reuse drive pinion lock nut.**

22. Tighten drive pinion lock nut, while
adjust total preload torque with Tool number
A: ST3127S000 (J-25765-A).

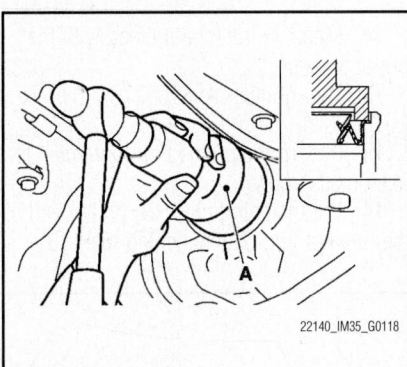

22140_IM35_G0118

**Fig. 23 Installation of front oil seal in the
differential**

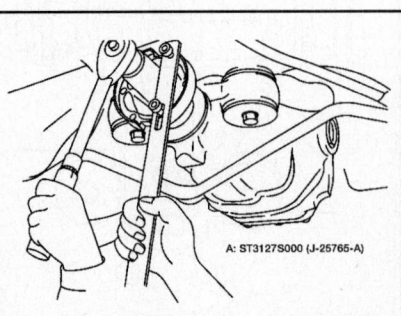

A: ST3127S000 (J-25765-A)

22140_IM35_G0119

**Fig. 24 Tightening drive pinion lock nut
with Tool number A: ST3127S000
(J-25765-A)**

23. Drive pinion lock nut tightening
torque: 109–238 ft. lbs. (147–323 Nm).
24. Total preload torque should equal the
measurement taken during removal plus an
additional 1–3 inch lbs (0.1–0.4 Nm).

> ※※ **WARNING**
> **Adjust to the lower limit of the
> drive pinion lock nut tightening
> torque first.**

➡ **If the preload torque exceeds the
specified value, replace.**

**collapsible spacer and tighten it again
to adjust. Never loosen drive pinion
lock nut to adjust the preload torque.**

25. Make a stamping for identification of
front oil seal replacement frequency.
26. Install propeller shaft.
27. Install side flange with the following
procedure.
28. Attach the protector to side oil seal.

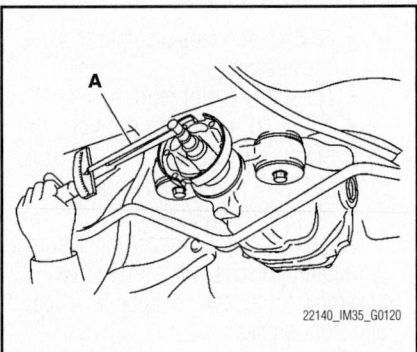

22140_IM35_G0120

**Fig. 25 Tightening drive pinion lock nut to
specification**

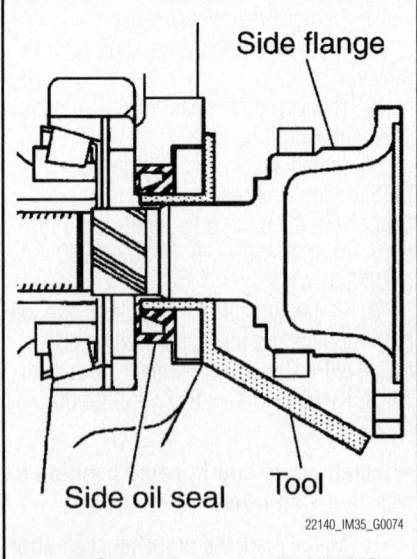

Side flange

Side oil seal   Tool

22140_IM35_G0074

**Fig. 26 Side seal protector KV38107900
(J-39352)**

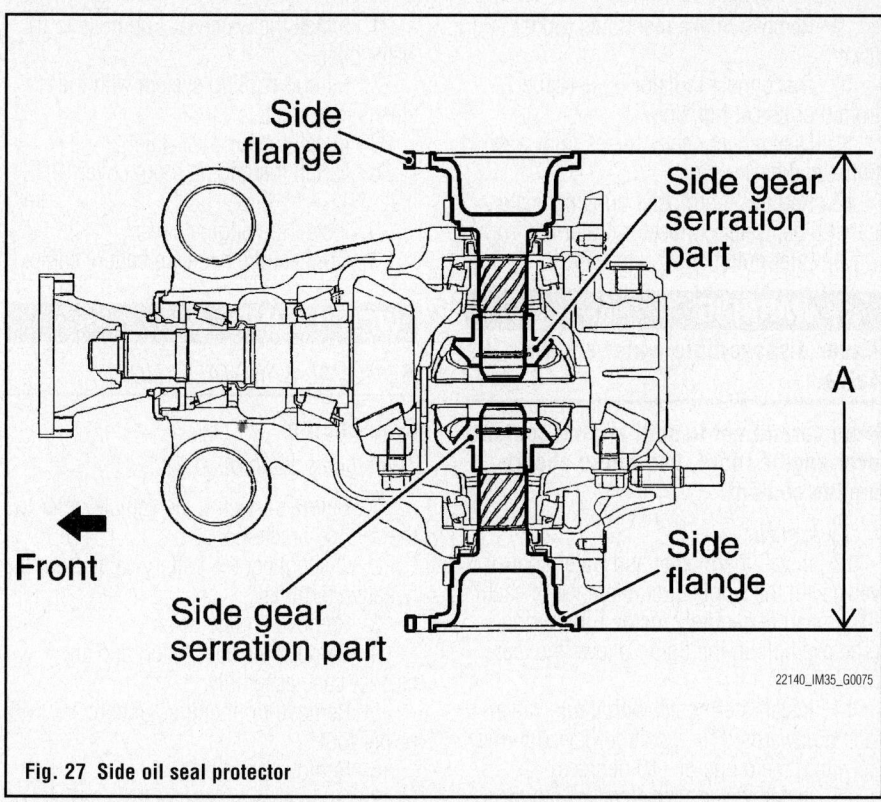

Fig. 27 Side oil seal protector

29. Apply multi-purpose grease to side oil seal lips.

30. Attach the protector to side oil seal. Tool number kv38100200 (j-26233).

31. After the side flange is inserted and the serrated part of side gear has engaged the serrated part of flange, remove the protector.

32. Put a suitable drift on the center of side flange, then drive it until sound changes.

➡**When installation is completed, driving sound of the side flange turns into a sound, which seems to affect the whole final drive.**

33. Confirm that the dimension of the side flange installation measurement (A) in the figure comes into the following.

34. Measurement A: 12.83–12.91 in (326–328 mm)

35. Install drive shaft.

36. Install rear wheel sensor.

37. Install center muffler.

38. Refill gear oil to the final drive and check oil level.

39. Check the final drive for oil leakage

ENGINE COOLING

THERMOSTAT

REMOVAL & INSTALLATION

3.5L Engine

See Figures 28 and 29.

1. Before servicing the vehicle, refer to the service precautions.

2. Disconnect the battery cable from the negative terminal.

3. Remove the engine cover undercover using power tool.

4. Remove the engine room cover (RH and LH).

5. Drain engine coolant from radiator drain plug at the bottom of radiator, and from water drain plug at the front of cylinder block.

✳✳ WARNING

Perform this step when the engine is cold.

➡**Never spill engine coolant on drive belts.**

6. Disconnect radiator hose (lower) and oil cooler water hose from water inlet and thermostat assembly.

7. Remove water inlet and thermostat assembly.

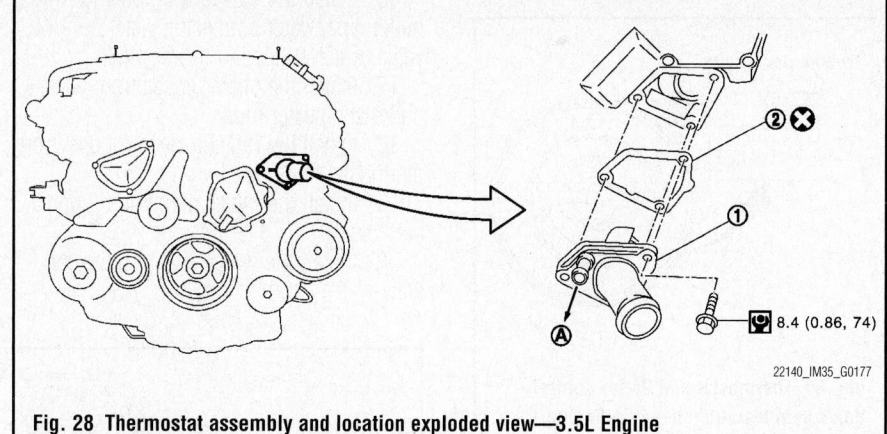

Fig. 28 Thermostat assembly and location exploded view—3.5L Engine

➡**Never disassemble water inlet and thermostat assembly.**

➡**Replace them as a unit, if necessary.**

To install:

8. Install the water inlet and thermostat assembly.

9. Tighten the thermostat bolts to 74 inch lbs.(8.4 Nm).

10. Install the engine room cover (RH and LH).

11. Install the engine cover undercover

12. Fill the cooling system with the specified coolant.

13. Install the negative battery cable.

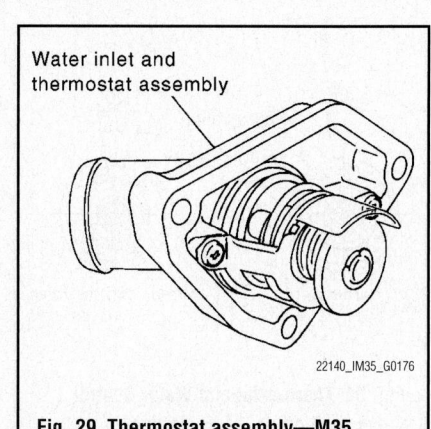

Fig. 29 Thermostat assembly—M35

14. After starting engine check for leaks of engine coolant using the radiator cap tester adapter (commercial service tool) and the radiator cap tester (commercial service tool), let idle for three minutes, then rev engine up to 3,000 rpm under no load to purge air from the high-pressure chamber of chain tensioner.

➡**Engine may produce a rattling noise. This indicates that air still remains in the chamber and is not a matter of concern.**

4.5L Engine

See Figures 30 and 31.

1. Before servicing the vehicle, refer to the service precautions.
2. Disconnect the battery cable from the negative terminal.
3. Remove the engine cover.
4. Remove the engine room cover (RH and LH).
5. Remove the air duct (inlet).
6. Drain engine coolant from drain plugs on radiator and both side of cylinder block.
7. Disconnect water suction hose from water inlet.
8. Remove water inlet and thermostat.

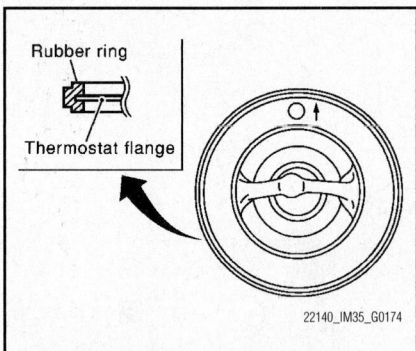

Fig. 30 Thermostat and Water Control Valve seal installation—4.5L Engine

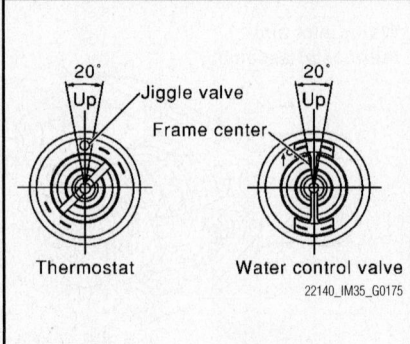

Fig. 31 Thermostat and Water Control Valve positioning—4.5L Engine

9. Remove intake manifolds (upper and lower).
10. Disconnect radiator hose (upper) from thermostat housing.
11. Disconnect heater hoses from water outlet and heater pipe.
12. Remove thermostat housing, water outlet pipe, water connector, water control valve, water outlet and heater pipe.

❄❄ **WARNING**

Never disassemble water control valve.

➡**Be careful not to spill engine coolant over engine room, use rag to absorb engine coolant.**

To install:

13. Install thermostat and water control valve with the whole circumference of each flange part fit securely inside rubber ring. (The example in the figure shows the thermostat).
14. Install thermostat with jiggle valve facing upwards. (The position deviation may be within the range of ±10 degrees).
15. Install water control valve with the up-mark facing up and the frame center part facing upwards. (The position deviation may be within the range of ±10°).
16. Install the thermostat housing, water outlet pipe, water connector, water control valve, water outlet and heater pipe.
17. Install the heater hoses from water outlet and heater pipe.
18. Install the radiator hose (upper) from thermostat housing.
19. Install the intake manifolds (upper and lower).
20. Install the water inlet and thermostat.

21. Install the water suction hose to the water inlet.
22. Fill the cooling system with the specified coolant.
23. Install the air duct (inlet).
24. Install the engine room cover (RH and LH).
25. Install the engine cover.
26. Connect the negative battery cable.

WATER PUMP

REMOVAL & INSTALLATION

3.5L Engine

See Figures 32 through 42.

1. Before servicing the vehicle, refer to the service precautions.
2. Disconnect the battery cable from the negative terminal.
3. Remove engine cover.
4. Remove air duct (inlet) and air cleaner case assembly.
5. Remove front engine undercover with power tool.
6. Remove drive belts.
7. Drain engine coolant from radiator.

❄❄ **WARNING**

Perform this step when the engine is cold.

➡**Never spill engine coolant on drive belts.**

8. Remove water drain plug (front) on water pump side of cylinder block to drain engine coolant from engine inside.
9. Remove chain tensioner cover and water pump cover from front timing chain case.

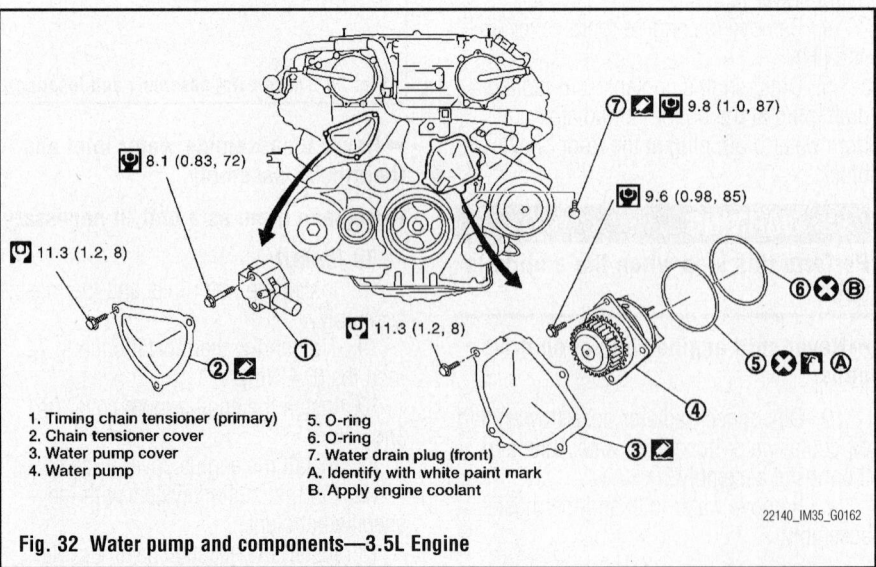

1. Timing chain tensioner (primary)
2. Chain tensioner cover
3. Water pump cover
4. Water pump
5. O-ring
6. O-ring
7. Water drain plug (front)
A. Identify with white paint mark
B. Apply engine coolant

Fig. 32 Water pump and components—3.5L Engine

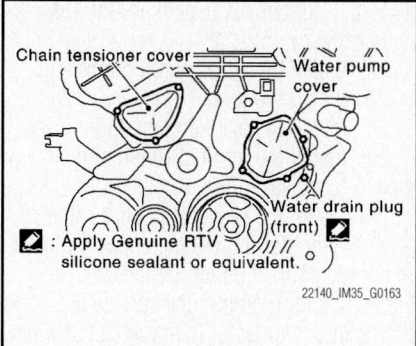

Fig. 33 Drain plug (front) on water pump side of cylinder block—3.5L Engine

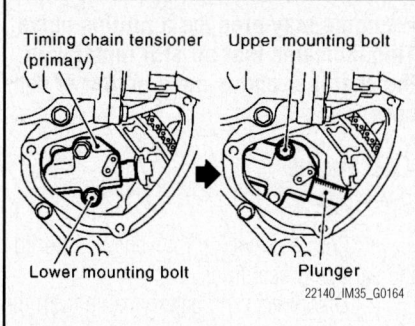

Fig. 34 Removal of the timing chain tensioner (primary)—3.5L Engine

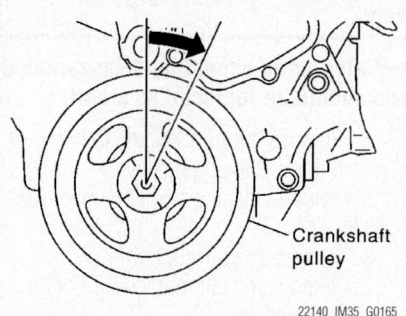

Fig. 35 Crankshaft pulleys clockwise rotation to loosen the timing chain—(primary) side

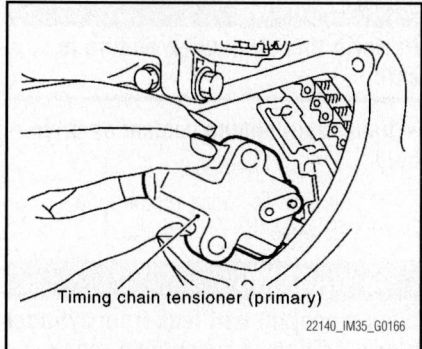

Fig. 36 Removal of the timing chain tensioner (primary)

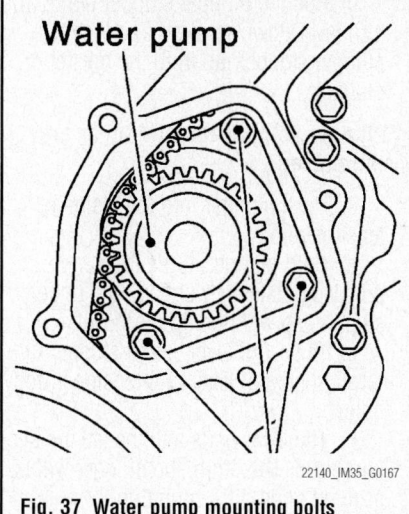

Fig. 37 Water pump mounting bolts

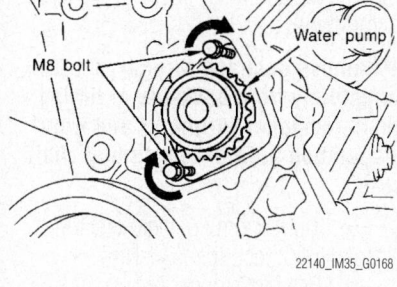

Fig. 38 Water pump removal with M8 bolts

10. Cut the liquid gasket for removal on the pump cover.

11. Remove timing chain tensioner (primary) as follows:

    a. Remove lower mounting bolt.

✳✳ WARNING

Be careful not to drop mounting bolt inside timing chain case.

    b. Loosen upper mounting bolt slowly, and then turn chain tensioner (primary) on the mounting bolt so that plunger is fully expanded.

➡**Even if plunger is fully expanded, it is not dropped from the body of timing chain tensioner (primary).**

    c. Turn crankshaft pulley clockwise so that timing chain on the timing chain tensioner (primary) side is loose.

    d. Remove upper mounting bolt, and then remove timing chain tensioner (primary).

✳✳ WARNING

Be careful not to drop mounting bolt inside timing chain case.

12. Remove water pump as follows:

    a. Remove three water pump mounting bolts. Secure a gap between water pump gear and timing chain, by turning crankshaft pulley counterclockwise until timing chain looseness on water pump sprocket becomes maximum.

    b. Screw M8 bolts [pitch: 0.0492 in (1.25 mm) length: approx. 1.97 in (50 mm)] into water pumps upper and lower mounting bolt holes until they reach timing chain case. Then, alternately tighten each bolt for a half turn, and pull out water pump.

    c. Pull straight out while preventing vane from contacting socket in installation area.

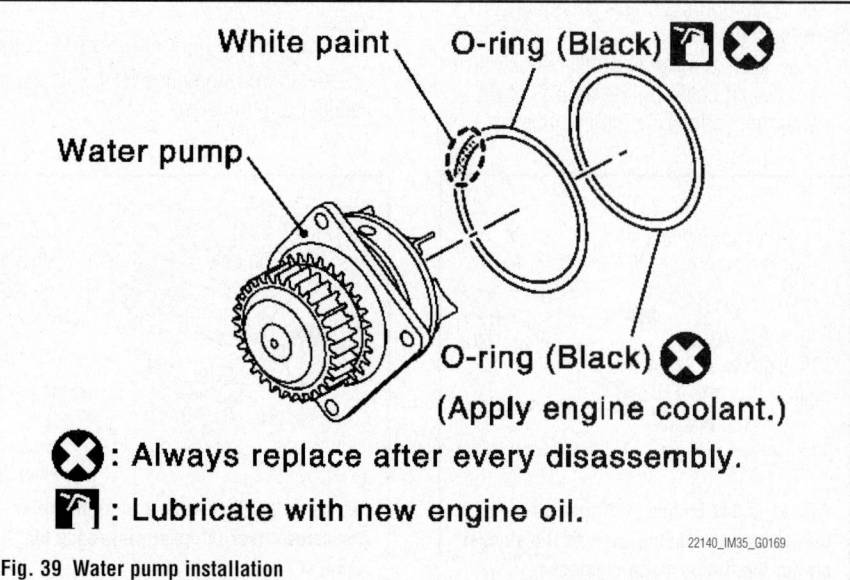

Fig. 39 Water pump installation

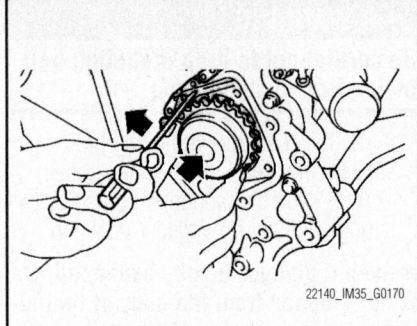

Fig. 40 Installing the timing chain to the water pump

   d. Remove water pump without causing sprocket to contact timing chain.

   e. Remove M8 bolts and O-rings from water pump.

> **※※ WARNING**
>
> **Never disassemble water pump.**

To install:

13. Install new O-rings to water pump.

14. Apply engine oil and engine coolant to O-rings.

15. Locate O-ring with white paint mark to engine front side.

16. Install the water pump.

> **※※ WARNING**
>
> **Never allow cylinder block to nip O-rings when installing water pump.**

17. Torque the bolts to 85 inch lbs. (9.6 Nm)

18. Check timing chain and water pump sprocket are engaged.

19. Insert water pump by tightening mounting bolts alternately and evenly.

20. Install timing chain tensioner (primary) as follows:

   a. Turn crankshaft pulley clockwise so that timing chain on the timing. chain tensioner (primary) side is loose.

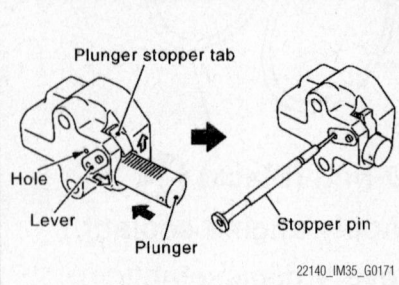

Fig. 41 0.047 inches (1.2 mm) diameter thin screwdriver being used as the stopper pin for the timing chain tensioner

   b. Pull the plunger stopper tab up (or turn lever downward) so as to remove plunger stopper tab from the ratchet of plunger.

➡**Plunger stopper tab and lever are synchronized.**

   c. Push plunger into the inside of tensioner body.

   d. Hold plunger in the fully compressed position by engaging plunger stopper tab with the tip of ratchet.

   e. To secure lever, insert stopper pin through hole of lever into tensioner body hole.

   f. The lever parts and the tab are synchronized. Therefore, the plunger will be secured under this condition.

   g. Figure shows the example of 0.047 inches (1.2 mm) diameter thin screwdriver being used as the stopper pin.

   h. Install timing chain tensioner (primary).

➡**Remove dust and foreign material completely from backside of timing chain tensioner (primary) and from installation area of rear timing chain case**

   i. Tighten timing chain tensioner bolts to 72 inch lbs. (8.1 Nm).

   j. Remove stopper pin.

   k. Check again that timing chain and water pump sprocket are engaged.

21. Install chain tensioner cover and water pump cover as follows:

   a. Before installing, remove all traces of old liquid gasket from mating surface of water pump cover and chain tensioner cover using scraper. Also remove traces of old liquid gasket from the mating surface of front timing chain case.

   b. Apply a continuous 0.091–0.130 in. (2–3mm) of genuine RTV Silicone sealant or equivalent.

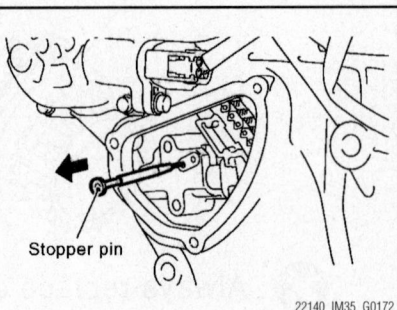

Fig. 42 0.047 inches (1.2 mm) diameter thin screwdriver (stopper pin) ready for removal

   c. Tighten cover bolts to 8 ft. lbs. (11.3 Nm).

22. Install the water pump drain plug.

23. Install the drive belts.

24. Install the front engine undercover.

25. Install the air duct (inlet) and air cleaner case assembly.

26. Install the negative battery cable.

27. After starting engine check for leaks of engine coolant using the radiator cap tester adapter (commercial service tool) and the radiator cap tester (commercial service tool), let idle for three minutes, then rev engine up to 3,000 rpm under no load to purge air from the high-pressure chamber of chain tensioner.

➡**Engine may produce a rattling noise. This indicates that air still remains in the chamber and is not a matter of concern.**

4.5L Engine

See Figure 43.

1. Before servicing the vehicle, refer to the service precautions.

2. Disconnect the battery cable from the negative terminal.

> **※※ WARNING**
>
> **When removing water pump, be careful not to get engine coolant on drive belts.**

➡**Water pump cannot be disassembled and should be replaced as a unit.**

3. Remove the front engine undercover.

4. Remove the engine cover.

5. Remove the engine room cover (RH and LH).

6. Remove the air duct (inlet)

7. Remove the alternator, water pump and A/C compressor belt.

8. Drain engine coolant from drain plugs on radiator and both side of cylinder block.

> **※※ WARNING**
>
> **Perform this step when engine is cold.**

➡**Never spill engine coolant on drive belts.**

9. Remove the water pump pulley.

10. Remove the water pump.

> **※※ WARNING**
>
> **Engine coolant will leak from cylinder block, so have a receptacle ready under vehicle.**

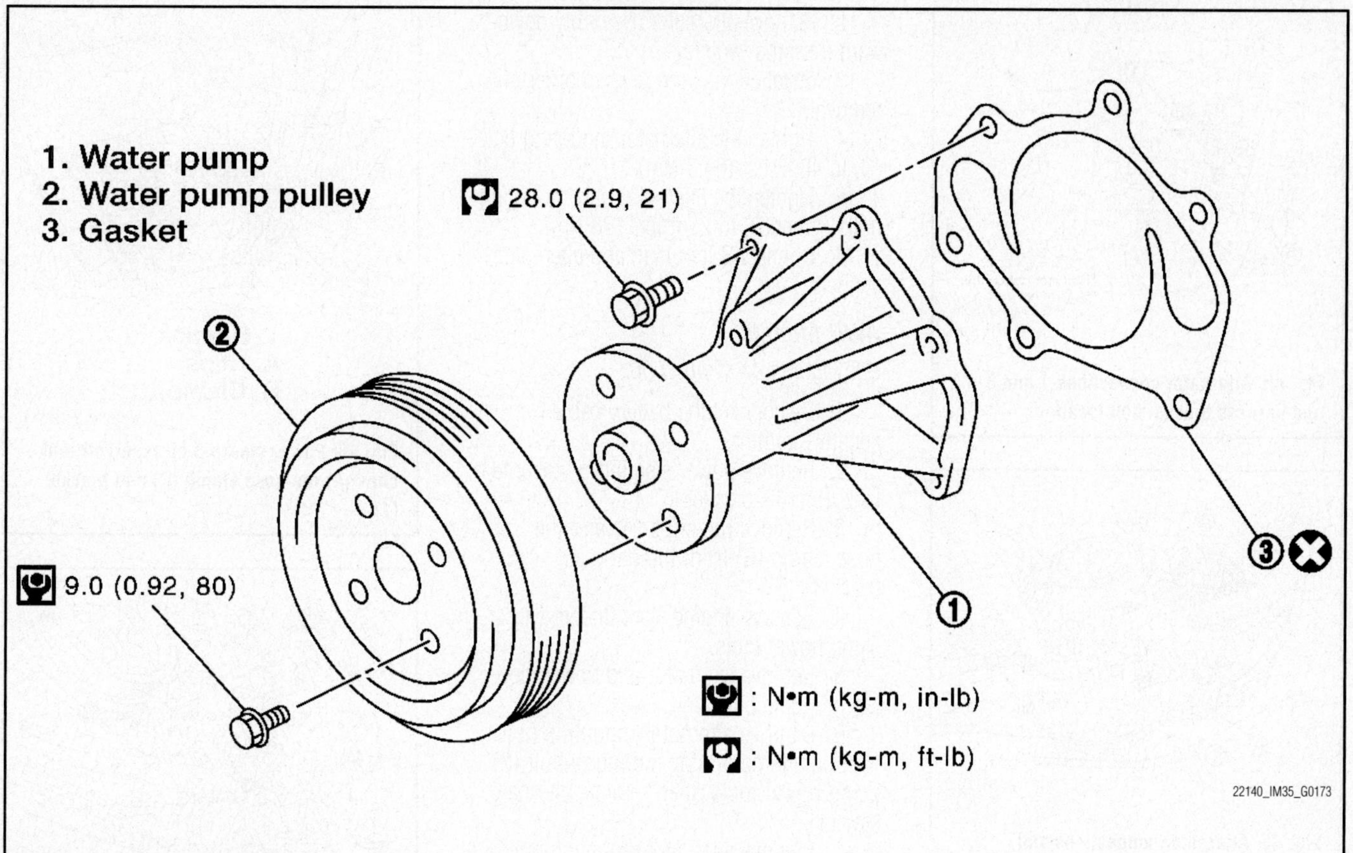

1. Water pump
2. Water pump pulley
3. Gasket

⊞ 28.0 (2.9, 21)

②

①

③ ✖

⊞ 9.0 (0.92, 80)

⊞ : N•m (kg-m, in-lb)

⊞ : N•m (kg-m, ft-lb)

22140_IM35_G0173

Fig. 43 Water pump—4.5L Engine

➡**Never disassemble water pump.**

To install:

11. Installation is the reverse order of removal.

12. Tighten the water pump bolts to 21 ft. lbs. (28 Nm).

13. Tighten the drive belt pulley bolts to 80 inch lbs. (9 Nm).

14. Connect the negative battery cable.

ENGINE ELECTRICAL

ALTERNATOR

REMOVAL & INSTALLATION

3.5L Engine

2WD Models

See Figures 44 through 46.

1. Before servicing the vehicle, refer to the service precautions.

2. Disconnect the battery cable from the negative terminal.

3. Remove engine front undercover, using power tools.

4. Remove alternator and power steering oil pump belt.

5. Disconnect alternator connector (1).

6. Remove "B" terminal nut (2).

7. Remove the harness bracket bolts (A).

8. Remove oil pressure switch harness clip (A) from alternator stay.

CHARGING SYSTEM

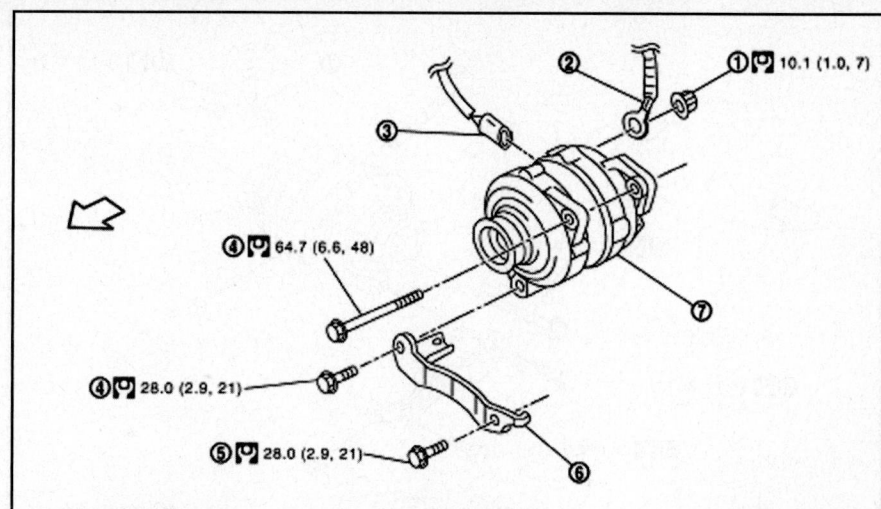

1. "B" terminal nut
2. "B" terminal harness
3. Alternator connector
4. Alternator mounting bolt
5. Alternator stay mounting bolt
6. Alternator stay
7. Alternator

⊞ : N•m (kg-m, ft-lb)   ⇦ : Engine front

22140_IM35_G0008

Fig. 44 Alternator and components exploded view—3.5L Engine 2WD models

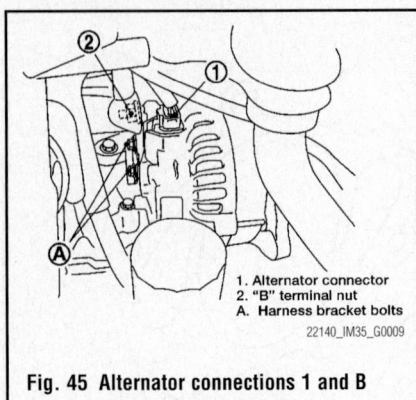

Fig. 45 Alternator connections 1 and B and harness bracket bolt locations

1. Alternator connector
2. "B" terminal nut
A. Harness bracket bolts

22140_IM35_G0009

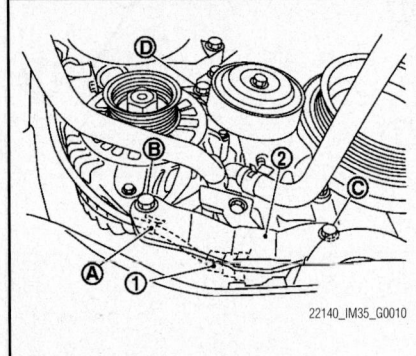

Fig. 46 Alternator connections and bracket bolt locations

22140_IM35_G0010

9. Disconnect oil pressure switch connector (1).

10. Remove alternator mounting bolt (B) and alternator stay mounting bolt (C) using power tools, then remove alternator stay (2).

11. Remove alternator mounting bolt (D), using power tools.

12. Remove alternator assembly downward from the vehicle

13. Installation is the reverse order of removal.

14. Tighten the alternator mounting bolt (C) to 48 ft lbs. (64.7 Nm).

15. Tighten the alternator stay mounting bolts (A and B) to 21 ft lbs. (28 Nm).

16. Connect B terminal; and alternator connector.

AWD Models

See Figures 47 through 50.

1. Disconnect the battery cable from the negative terminal.

2. Remove power steering oil reservoir tank from the bracket.

3. Remove the clips (A) and the hose clamp (B) from the harness bracket (1).

4. Remove engine front undercover, using power tools.

5. Remove alternator and power steering oil pump belt.

6. Remove alternator mounting bolt (A) and alternator stay mounting bolt (B) using power tools, then remove alternator stay (1).

7. Remove alternator mounting bolt (C), using power tools.

8. Pull and turn alternator, and then remove the harness bracket bolts (A).

9. Disconnect alternator connector (1).

10. Remove "B" terminal nut (2).

11. Remove alternator assembly downward from the vehicle.

12. Installation is the reverse order of removal.

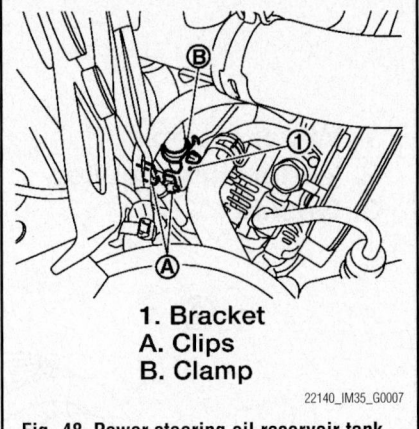

1. Bracket
A. Clips
B. Clamp

22140_IM35_G0007

Fig. 48 Power steering oil reservoir tank clips (A) the hose clamp (B) and bracket (1).

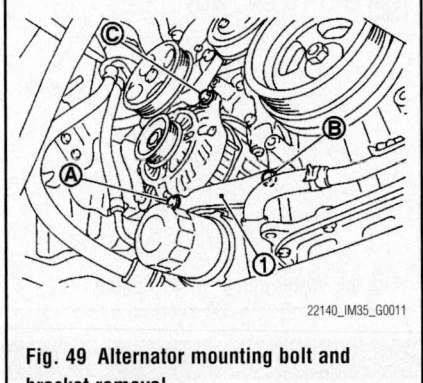

22140_IM35_G0011

Fig. 49 Alternator mounting bolt and bracket removal

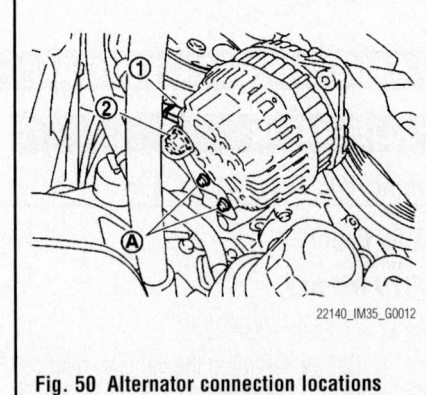

22140_IM35_G0012

Fig. 50 Alternator connection locations

13. Tighten the alternator mounting bolt (C) to 48 ft lbs. (64.7 Nm).

14. Tighten the alternator stay mounting bolts (A and B) to 21 ft lbs. (28 Nm).

15. Connect B terminal; and alternator connector.

4.5L Engine

See Figures 51 through 53.

1. Disconnect the battery cable from the negative terminal.

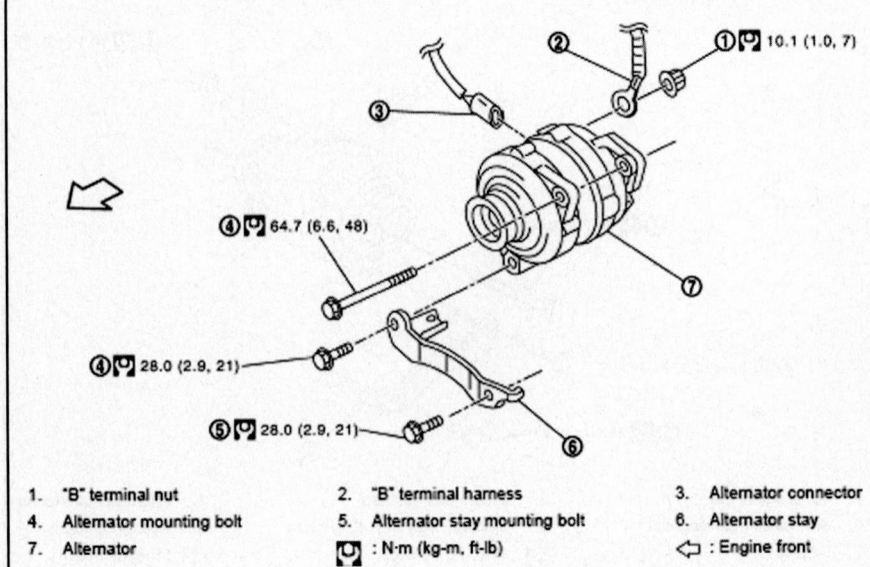

1. "B" terminal nut
2. "B" terminal harness
3. Alternator connector
4. Alternator mounting bolt
5. Alternator stay mounting bolt
6. Alternator stay
7. Alternator
🔧 : N·m (kg-m, ft-lb)
⬅ : Engine front

22140_IM35_G0008

Fig. 47 Alternator and components exploded view—3.5L engine AWD models

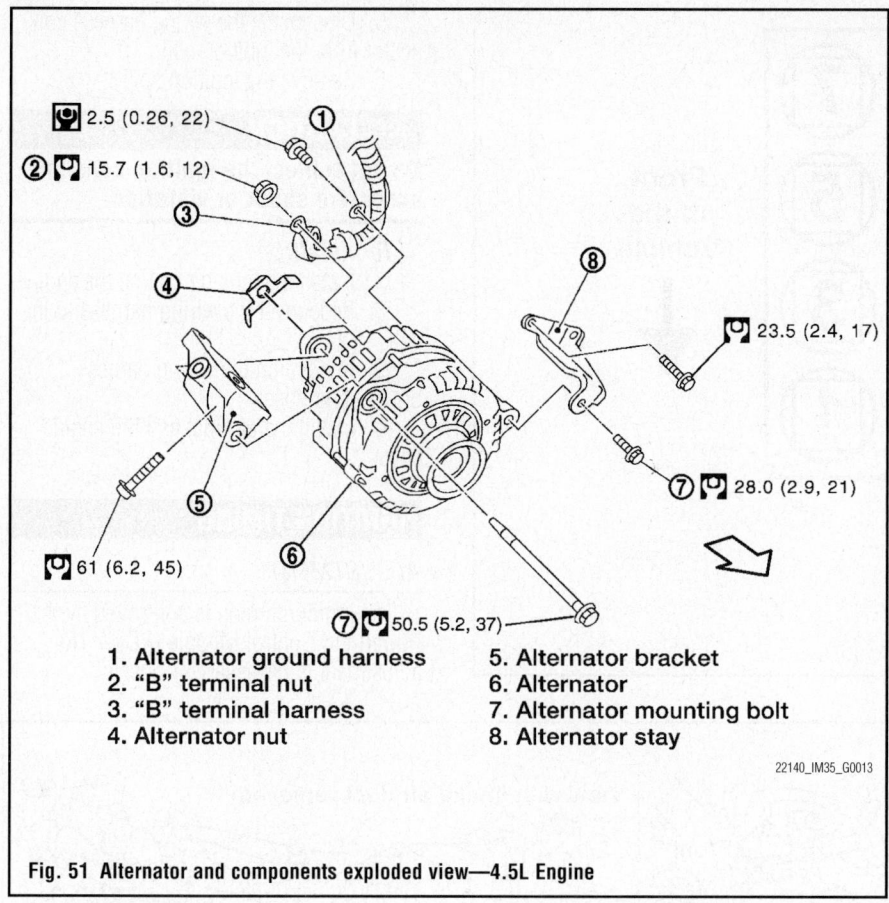

1. Alternator ground harness
2. "B" terminal nut
3. "B" terminal harness
4. Alternator nut
5. Alternator bracket
6. Alternator
7. Alternator mounting bolt
8. Alternator stay

22140_IM35_G0013

Fig. 51 Alternator and components exploded view—4.5L Engine

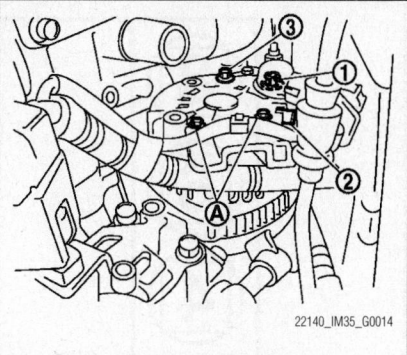

22140_IM35_G0014

Fig. 52 Alternator connections and harness bracket bolt—4.5L Engines

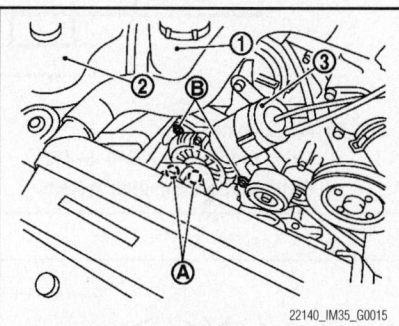

22140_IM35_G0015

Fig. 53 Power steering oil reservoir tank (1) engine coolant reservoir tank (2) vacuum tank (3) harness clips (A) and alternator mounting bolts (B)—4.5L Engine

2. Remove engine front undercover, using power tools.
3. Remove "B" terminal nut (1).
4. Disconnect alternator connector (2).
5. Remove alternator ground harness mounting bolt (3).
6. Remove the harness bracket bolts (A).
7. Remove air intake duct.

8. Remove alternator, water pump and A/C compressor belt.
9. Remove power steering oil reservoir tank (1) from the bracket, engine coolant reservoir tank (2) and vacuum tank (3).
10. Remove the harness clips (A).
11. Remove alternator mounting bolts (B), using power tools.
12. Remove alternator assembly upward.

13. Installation is the reverse order of removal.
14. Tighten the alternator mounting bolt (C) to 48 ft lbs. (64.7Nm).
15. Tighten the alternator stay mounting bolts (A and B) to 21 ft lbs. (28 Nm).
16. Connect B terminal; and alternator connector.

ENGINE ELECTRICAL   **IGNITION SYSTEM**

FIRING ORDER

See Figures 54 and 55.

IGNITION COIL PACK

REMOVAL & INSTALLATION

See Figures 56 and 57.

1. Before servicing the vehicle, refer to the Precautions Section.
2. Remove the engine cover.
3. Remove the air duct (for ignition coil of left bank side).
4. Move aside the wiring harness, wiring harness bracket, and hoses located above ignition coil.

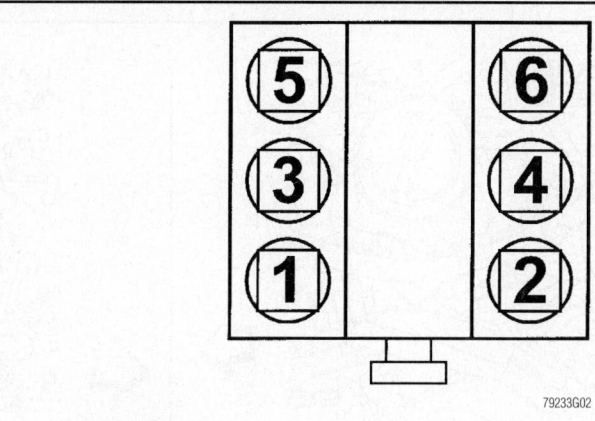

79233G02

Fig. 54 3.5L Engine
Firing order: 1–2–3–4–5–6
Distributor less ignition system

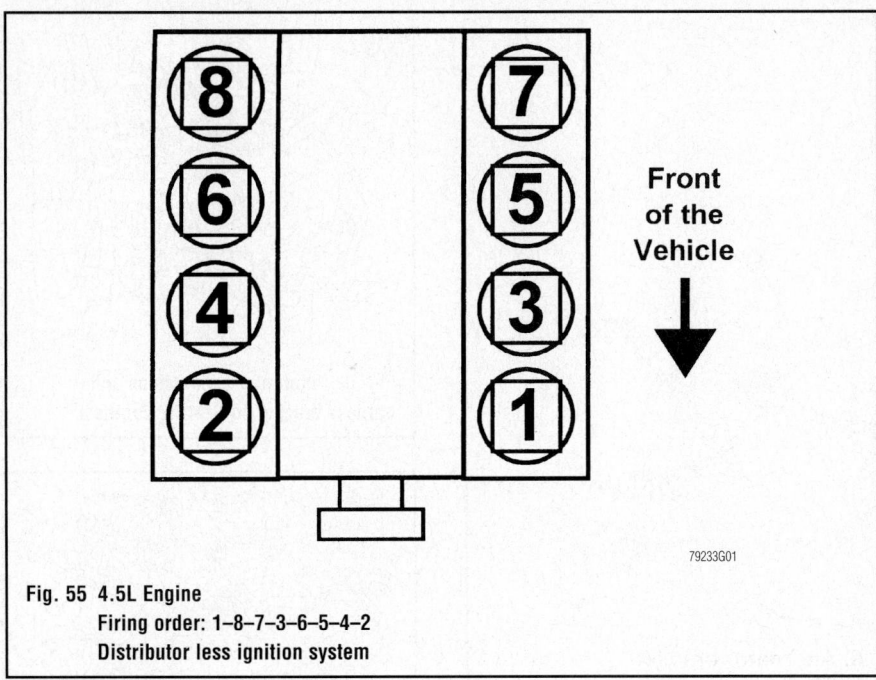

Fig. 55 4.5L Engine
Firing order: 1–8–7–3–6–5–4–2
Distributor less ignition system

5. Disconnect the wiring harness connector from the ignition coil.
6. Remove the ignition coil.

❄❄ CAUTION

Do not subject the ignition coils to excessive shock or vibration.

To install:

7. Install the ignition coil on the engine.
8. Reconnect the wiring harness to the coil.
9. Reposition the wiring harness, bracket and hoses.
10. Install the air duct and the engine cover.

IGNITION TIMING

ADJUSTMENT

The ignition timing is controlled by the Powertrain Control Module (PCM). No adjustment is necessary or possible.

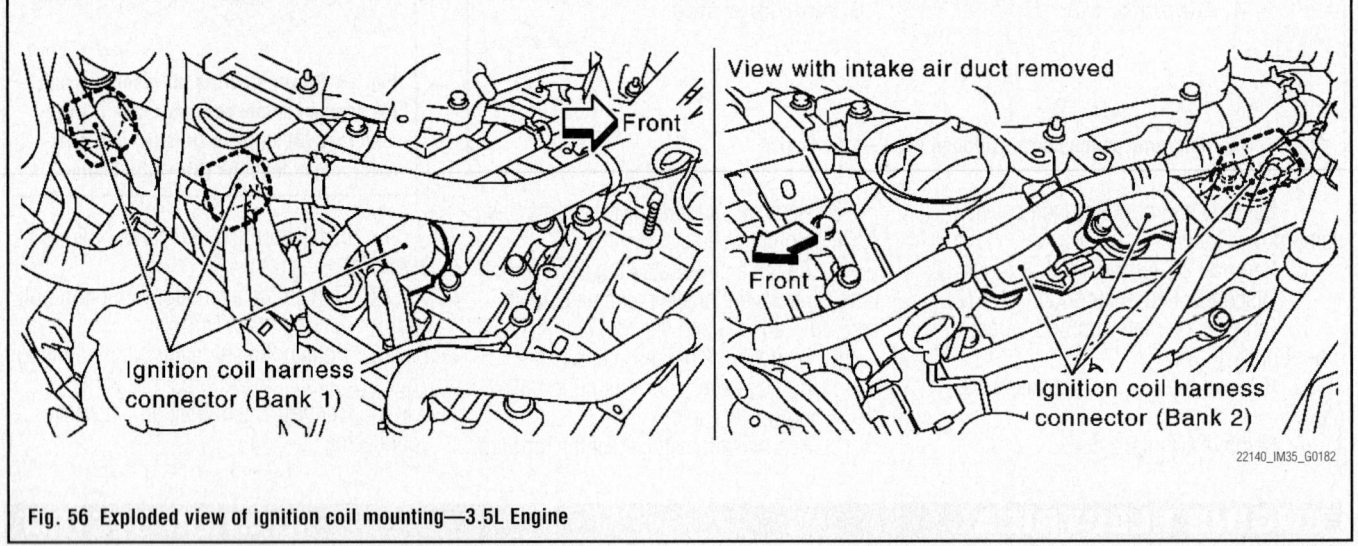

Fig. 56 Exploded view of ignition coil mounting—3.5L Engine

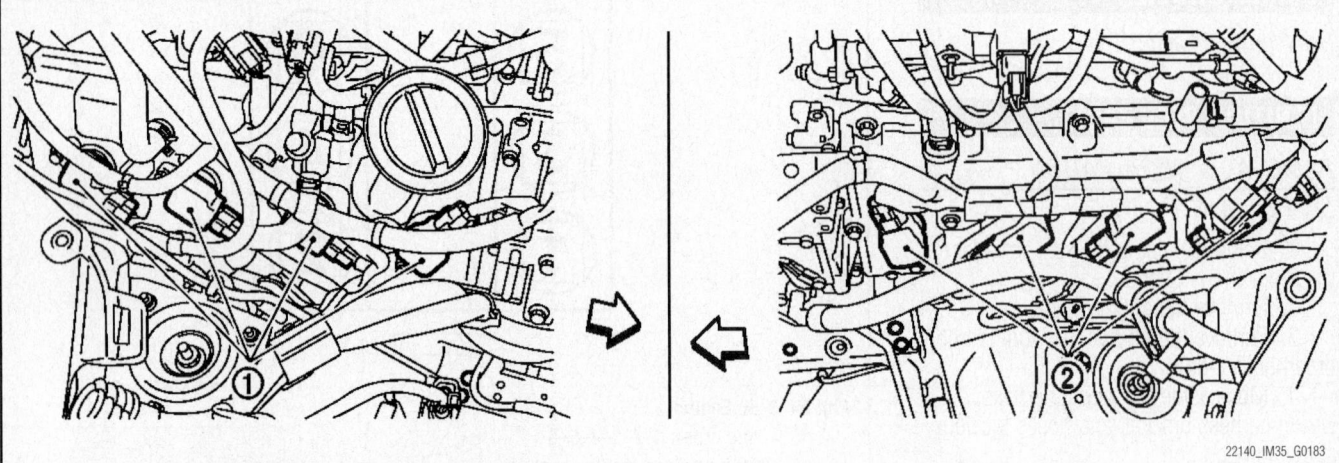

Fig. 57 Exploded view of ignition coil mounting—4.5L Engine

SPARK PLUGS °

REMOVAL & INSTALLATION

1. Disconnect the negative battery cable.
2. Remove the engine cover.
3. Remove the ignition coil retaining bolt.

4. Remove the ignition coil.
5. Remove the spark plug using a spark plug socket and wrench.

To install:

6. Be sure the spark plug gap is to specification (0.043 in).
7. Carefully install the spark

plug and torque to specification, 18 ft. lbs.
8. Install the ignition coil, torque the retaining bolt to 80 inch lbs.
9. Install the engine cover.
10. Connect the negative battery cable.

ENGINE ELECTRICAL

STARTER

REMOVAL & INSTALLATION

3.5L Engine

See Figures 58 and 59.

1. Disconnect the negative battery cable.
2. Remove the engine rear undercover.
3. Remove the starter electrical wires.
4. Remove the starter retaining bolts.
5. On 2WD vehicles, remove the harness clip bracket.
6. Remove the starter from its mounting.

To install:

7. Installation is the reverse of the removal procedure.
8. Tighten the retaining bolts to 41 ft. lbs.
9. Tighten the terminal nut to 87 inch lbs.

4.5L Engine

See Figures 60 and 61.

1. Disconnect the negative battery cable.
2. Remove the front and rear engine undercover.

STARTING SYSTEM

3. Remove the starter electrical wires.
4. Remove the starter retaining bolts.
5. Loosen the automatic transmission fluid cooler tube clip bolts.
6. Remove the starter from its mounting.

To install:

7. Installation is the reverse of the removal procedure.
8. Tighten the retaining bolts to 34 ft. lbs.
9. Tighten the terminal nut to 87 inch lbs.

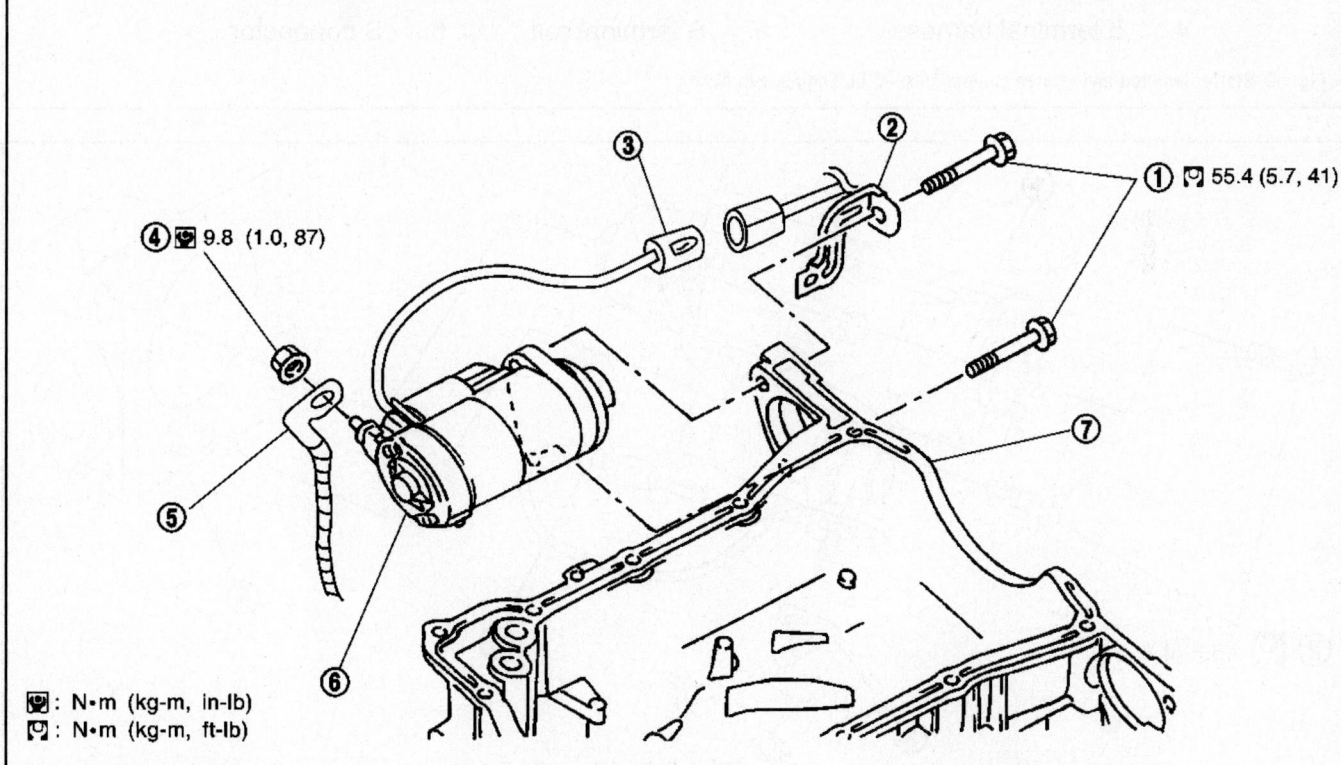

⌷ : N•m (kg-m, in-lb)
⌷ : N•m (kg-m, ft-lb)

1. Starter motor mounting bolt	2. Harness clip bracket	3. S connector
4. B terminal nut	5. B terminal harness	6. Starter motor
7. Oil pan		

42050_FX35_G0108

Fig. 58 Starter location and related components—3.5L Engine with 2WD

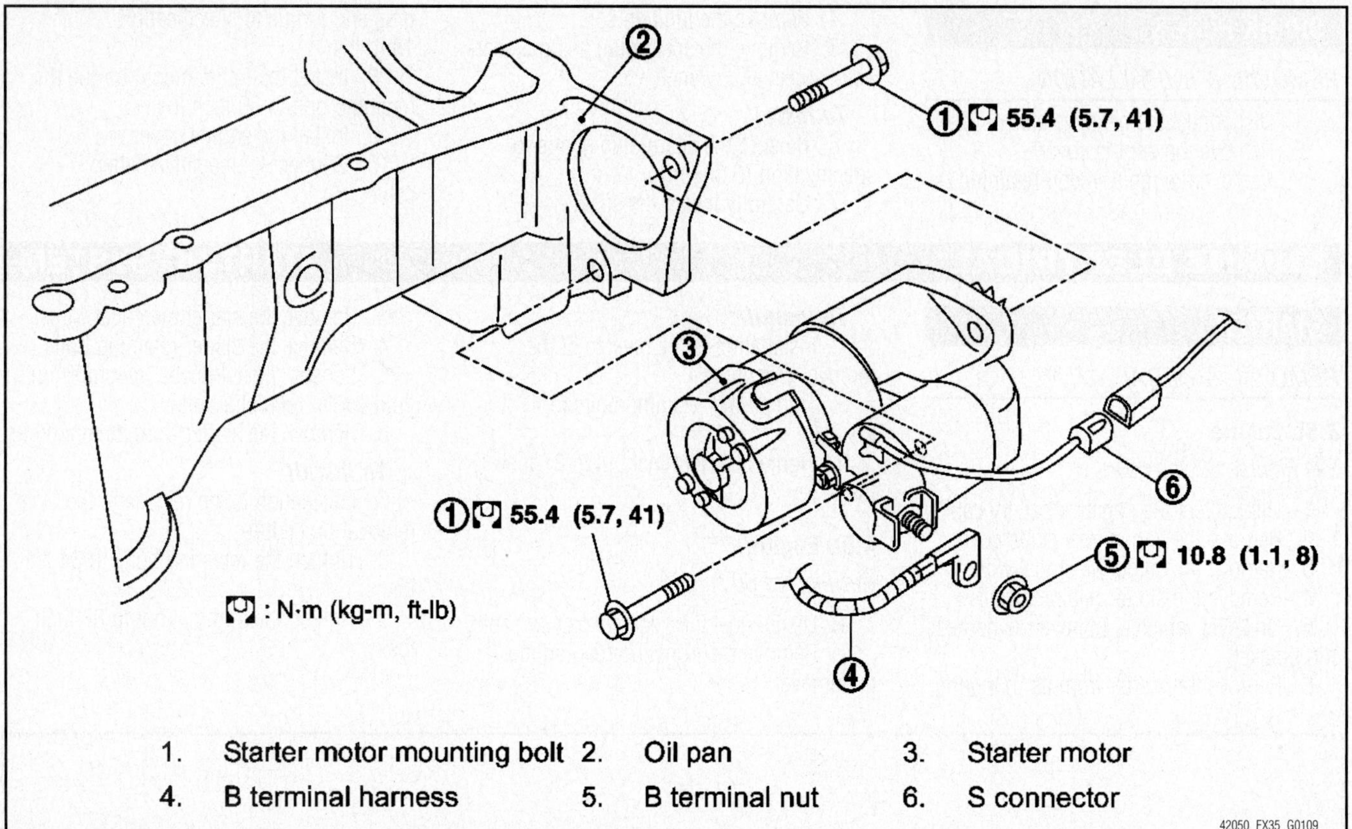

① ⚙ **55.4 (5.7, 41)**

① ⚙ **55.4 (5.7, 41)**

⚙ : N·m (kg-m, ft-lb)

⑥

⑤ ⚙ **10.8 (1.1, 8)**

④

| 1. | Starter motor mounting bolt | 2. | Oil pan | 3. | Starter motor |
| 4. | B terminal harness | 5. | B terminal nut | 6. | S connector |

42050_FX35_G0109

Fig. 59 Starter location and related components—3.5L Engine with AWD

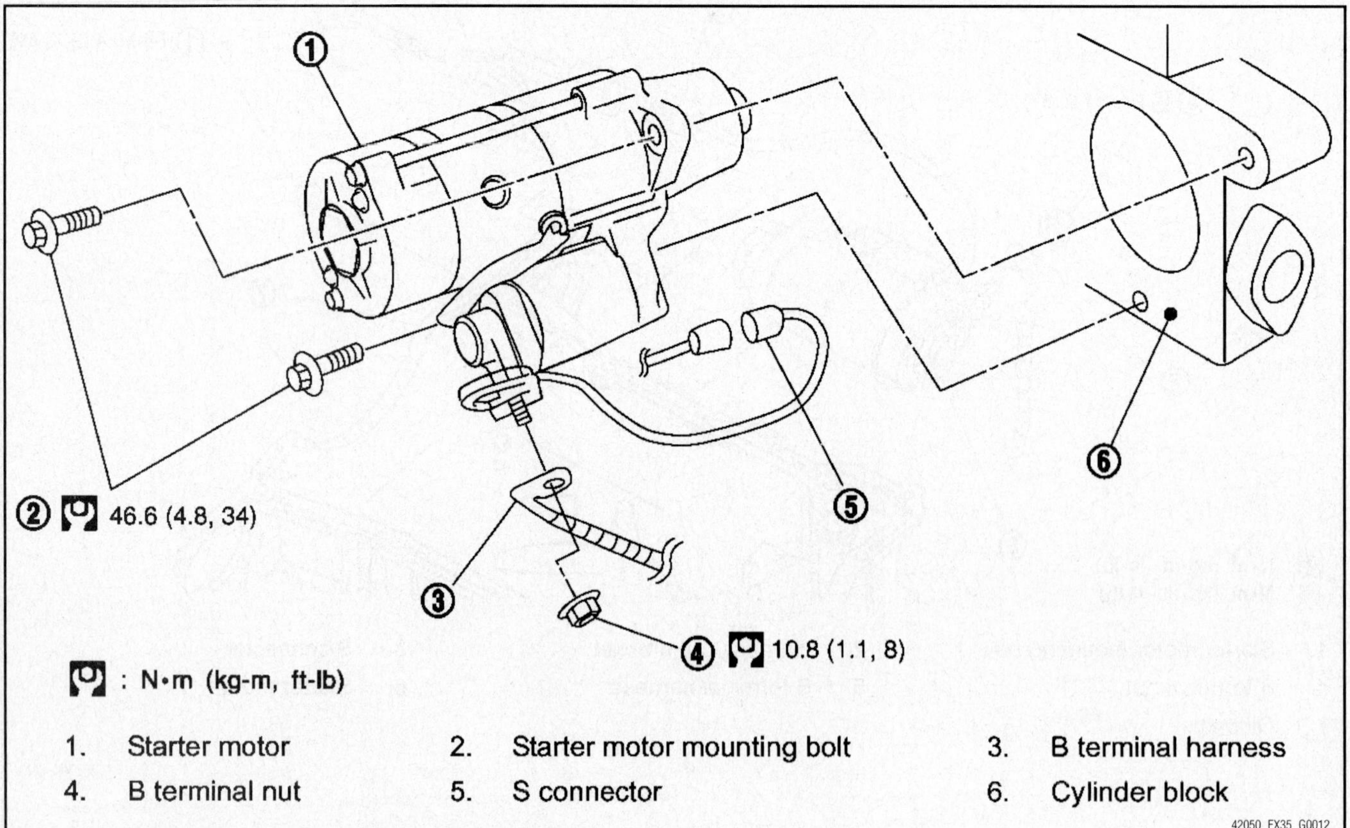

② ⚙ **46.6 (4.8, 34)**

⑥

⑤

③

④ ⚙ **10.8 (1.1, 8)**

⚙ : N·m (kg-m, ft-lb)

| 1. | Starter motor | 2. | Starter motor mounting bolt | 3. | B terminal harness |
| 4. | B terminal nut | 5. | S connector | 6. | Cylinder block |

42050_FX35_G0012

Fig. 60 Starter location and related components—4.5L engine

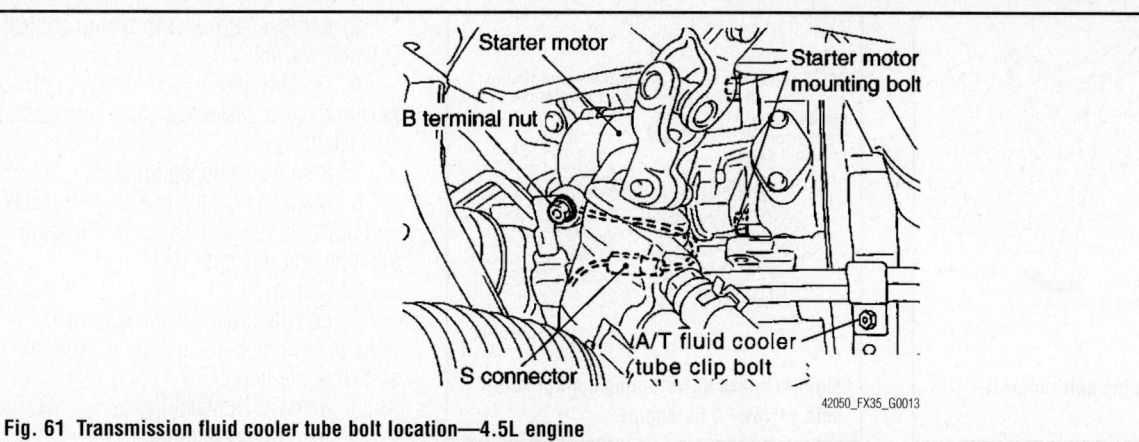

Fig. 61 Transmission fluid cooler tube bolt location—4.5L engine

ENGINE MECHANICAL

➡**Disconnecting the negative battery cable may interfere with the functions of the on board computer systems and may require the computer to undergo a relearning process, once the negative battery cable is reconnected.**

ACCESSORY DRIVE BELTS

REMOVAL & INSTALLATION

3.5L Engine

See Figure 62.

1. Before servicing the vehicle, refer to the service precautions.
2. Disconnect the negative battery cable.
3. Remove the engine undercover.
4. Remove the alternator and power steering belt.
5. Remove the air conditioning compressor belt.

To install:
6. Be sure not to get grease on the belts.
7. Make sure the drive belts are correctly engaged with the pulley groove

8. Tighten idler pulley lock nut to 26 ft. lbs. (34.8 Nm).

2007–08 4.5L Engines

Except Power Steering Belt

See Figure 63.

1. Before servicing the vehicle, refer to the Precautions Section.

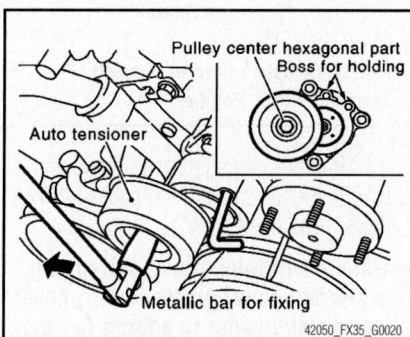

Fig. 63 Alternator, water pump and air conditioning compressor belt removal—4.5L engine

2. Disconnect the negative battery cable.
3. Remove the air duct.
4. With a box wrench, and while securely holding the hexagonal part in the pulley center of the auto tensioner, move the wrench handle in the direction of the arrow shown in the illustration.

➡**Avoid placing your hand in a location where pinching may occur if the holding tool accidentally comes off. Do not loosen the hexagonal part in the center of the drive belt auto tensioner pulley (do not turn it clockwise). If turned clockwise, the complete drive belt auto tensioner must be replaced as a unit including the pulley.**

5. Insert a metallic bar approximately 0.24 inch in diameter through the holding boss to lock the auto tensioner pulley arm.

➡**Leave the auto tensioner pulley arm locked until the belt is reinstalled.**

6. Remove the alternator, water pump and air conditioning compressor belt.

To install:
7. Installation is the reverse of removal.
8. Be sure not to get grease on the belts.
9. Make sure the drive belts are correctly engaged with the pulley groove

Power Steering Belt

See Figure 64.

1. Before servicing the vehicle, refer to the service precautions .
2. Disconnect the negative battery cable.
3. Remove the air duct.
4. Remove the alternator, water pump and air conditioning compressor belt.

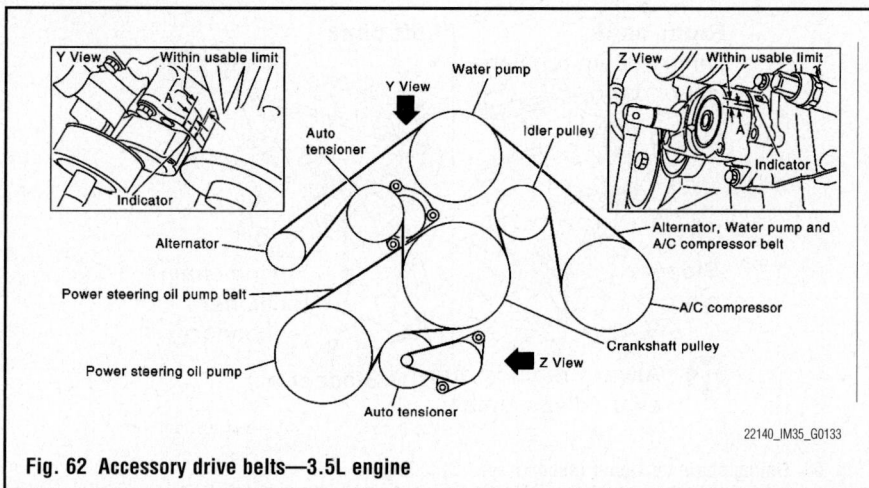

Fig. 62 Accessory drive belts—3.5L engine

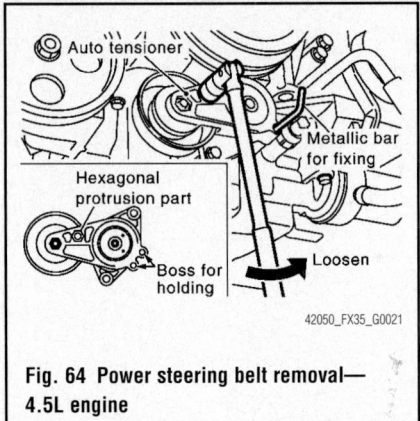

Fig. 64 Power steering belt removal—4.5L engine

5. While securely holding the hexagonal protrusion part of the auto tensioner pulley, with a box wrench move the wrench handle in the direction of the arrow shown in the illustration.

➡ **Avoid placing your hand in a location where pinching may occur if the holding tool accidentally comes off.**

6. Insert a metallic bar approximately 0.24 inch in diameter through the holding boss to lock the auto tensioner pulley arm.

➡ **Leave the auto tensioner pulley arm locked until the belt is reinstalled.**

7. Remove the power steering belt.

To install:

8. Installation is the reverse of removal.
9. Be sure not to get grease on the belts.
10. Make sure the drive belts are correctly engaged with the pulley groove

CAMSHAFT AND VALVE LIFTERS

REMOVAL & INSTALLATION

3.5L Engine

See Figures 65 through 78.

1. Remove front timing chain case, camshaft sprocket, timing chain and rear timing chain case.

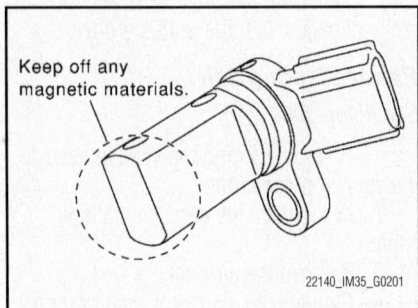

Fig. 65 Camshaft position sensor—3.5L Engine

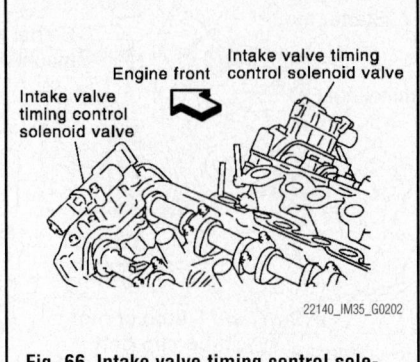

Fig. 66 Intake valve timing control solenoid valves—3.5L Engine

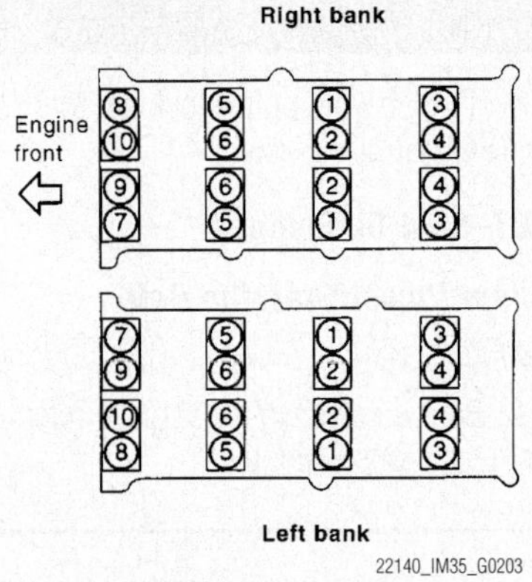

Fig. 67 Camshaft bracket removal sequence—3.5L Engine

2. Remove camshaft position sensor (PHASE) (right bank and left bank) from cylinder head back side.

➡ **Handle carefully to avoid dropping and shocks, never disassemble, never allow metal powder to adhere to magnetic part at sensor tip.**

➡ **Never place sensors in a location where they are exposed to magnetism.**

3. Remove intake valve timing control solenoid valves.
4. Discard intake valve timing control solenoid valve gaskets and use new gaskets for installation.
5. Remove camshaft brackets.
6. Mark camshafts, camshaft brackets and bolts so they are placed in the same position and direction for installation.
7. Equally loosen camshaft bracket bolts in several steps in reverse order as shown in the figure.
8. Remove camshaft.
9. Remove valve lifter.
10. Identify installation positions, and store them without mixing them up.
11. Remove timing chain tensioner (secondary) from cylinder head.
12. Remove timing chain tensioner (secondary) with its stopper pin attached.

➡ **Stopper pin should be attached when timing chain (secondary) is removed.**

To install:

13. Install timing chain tensioners (secondary) on both sides of cylinder head.
14. Install timing chain tensioner with its stopper pin attached.
15. Install timing chain tensioner with sliding part facing downward on right-side cylinder head, and with sliding part facing upward on left-side cylinder head.
16. Install new O-ring as shown in the figure.
17. Install valve lifter in the original position.
18. Install camshafts-Install camshaft with dowel pin attached to its front end face on the exhaust side.
19. Follow your identification marks made during removal, or follow the identification marks that are present on new

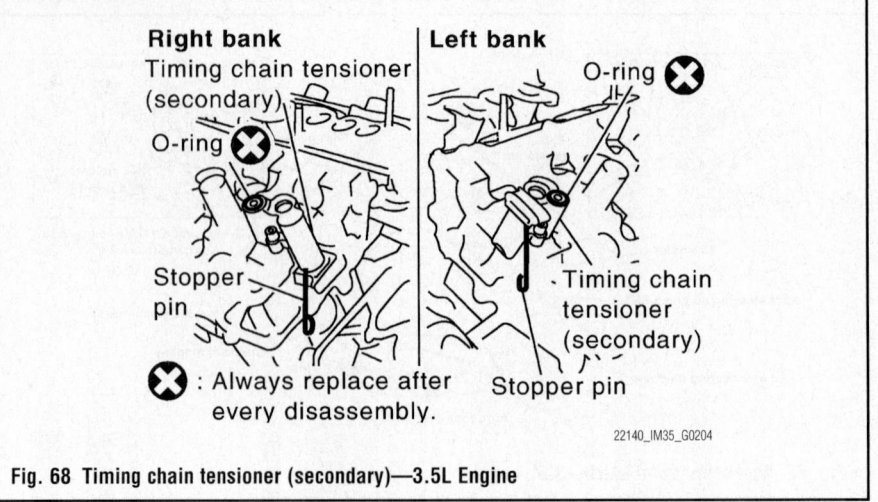

Fig. 68 Timing chain tensioner (secondary)—3.5L Engine

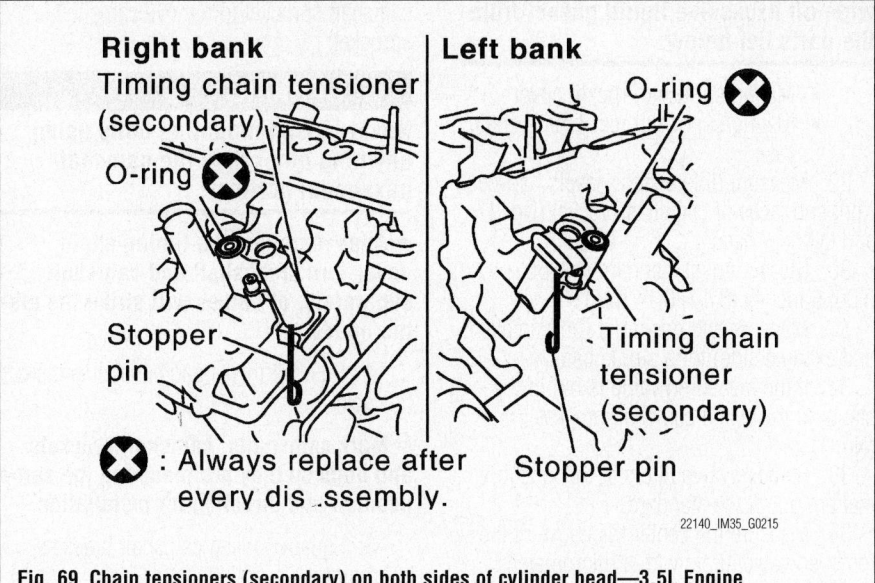

Fig. 69 Chain tensioners (secondary) on both sides of cylinder head—3.5L Engine

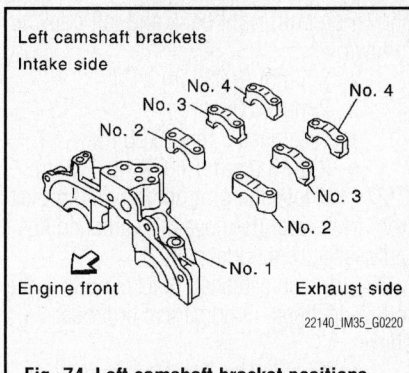

Fig. 74 Left camshaft bracket positions and direction—3.5L Engine

camshafts for proper placement and direction.

20. Install camshaft so that dowel pin hole and dowel pin on front end face are positioned as shown in the figure. (No. 1 cylinder TDC on its compression stroke).

➡**Large and small pin holes are located on front end face of camshaft**

(INT), at intervals of 180 degrees. Face small diameter side pin hole upward (in cylinder head upper face direction).

21. Though camshaft does not stop at the portion as shown in the figure, for the placement of cam nose, it is generally accepted camshaft is placed for the same direction of the figure.

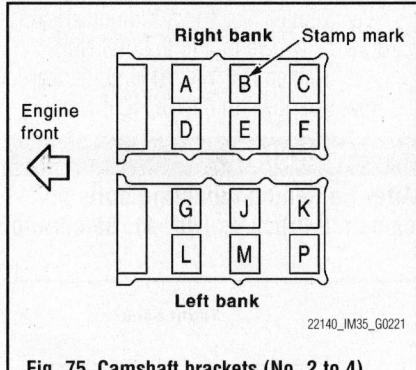

Fig. 75 Camshaft brackets (No. 2 to 4) aligning sequence—3.5L Engine

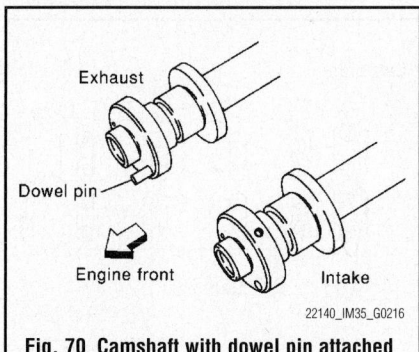

Fig. 70 Camshaft with dowel pin attached to its front end face on the exhaust side—3.5L Engine

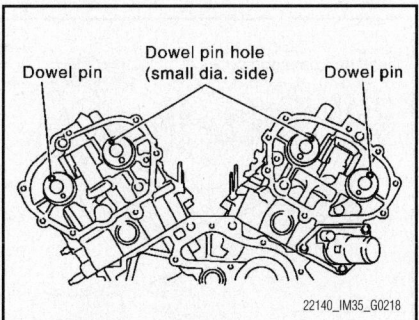

Fig. 72 Dowel pin hole and dowel pin on front end face positioning for TDC on its compression stroke—3.5L Engine

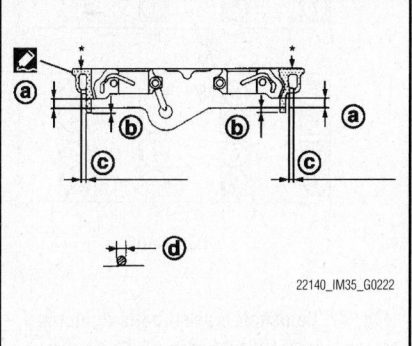

Fig. 76 Gasket application sizing and locations—3.5L Engine

22. Install camshaft brackets.
23. Remove foreign material completely from camshaft bracket backside and from cylinder head installation face.
24. Install camshaft bracket in original position and direction as shown in the figure.
25. Install camshaft brackets (No. 2 to 4) aligning the stamp marks as shown in the figure.

➡**There are no identification marks indicating left and right for camshaft bracket (No. 1).**

26. Apply liquid gasket to mating surface of camshaft bracket (No. 1) as

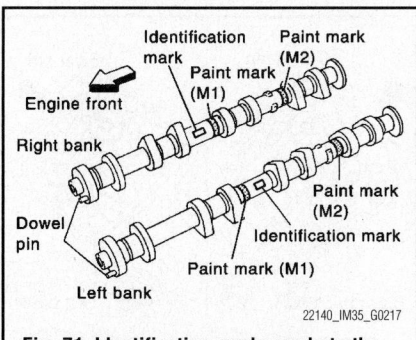

Fig. 71 Identification marks made to the camshaft during removal. for proper orientation on assembly—3.5L Engine

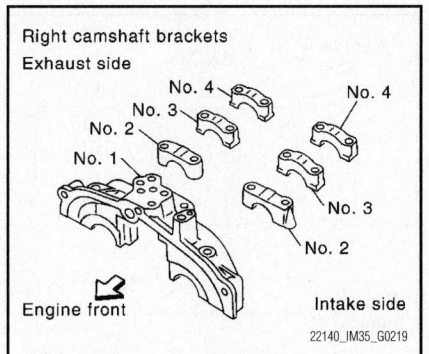

Fig. 73 Right camshaft bracket positions and direction—3.5L Engine

shown on both right bank and left bank as follows:

   a. 8.5 mm (0.335 in).

   b. 2 mm (0.08 in).

   c. Clearance 5 mm (0.20 in).

   d. 2.0 - 3.0 mm (0.079 - 0.118 in).

27. Remove the protruding liquid gasket from front face. (Remove the hardened liquid gasket from surface only).

28. Tighten camshaft bracket bolts in the following steps, in numerical order as shown.

   a. Tighten No. 7 to 10 in numerical order as shown to 1 ft. lb.(1.96 Nm).

   b. Tighten No. 1 to 6 in numerical order as shown to 1 ft. lb.(1.96 Nm).

   c. Tighten No. 1 to 10 in numerical order as shown to 4 ft. lb.(5.88 Nm).

   d. Tighten No. 1 to 10 in numerical order as shown to 8 ft. lb.(10.4 Nm).

✳✳ WARNING

After tightening mounting bolts of camshaft brackets (No. 1), be sure to

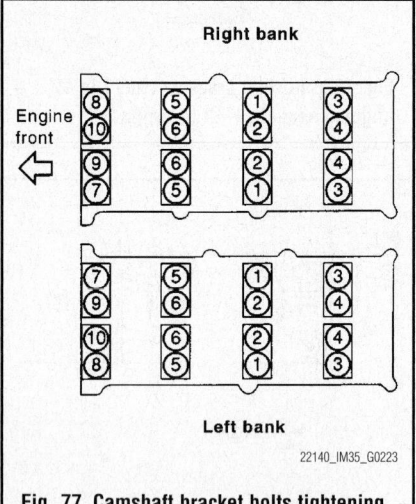

Fig. 77 Camshaft bracket bolts tightening steps, in numerical order—3.5L Engine

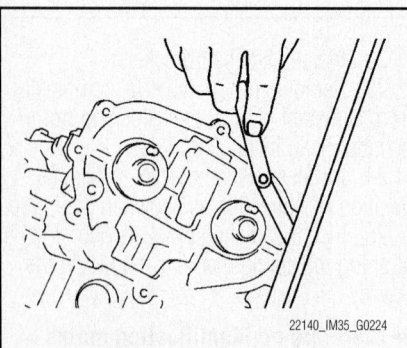

Fig. 78 Difference in levels between the front end faces of the camshaft bracket (No. 1) and the cylinder head—3.5L Engine

wipe off excessive liquid gasket from the parts list below.

- Mating surface of rocker cover
- Mating surface of rear timing chain case

29. Measure difference in levels between front end faces of camshaft bracket (No. 1) and cylinder head.

30. Measurement standard 0.0055–0.0055 inches (0.14–0.14 mm)

31. Measure two positions (both intake and exhaust side) for a single bank.

32. If the measured value is out of the standard, re-install camshaft bracket (No. 1).

33. Remove valve lifters at the locations that are out of the standard.

34. Measure the center thickness of the removed valve lifters with a micrometer.

35. Inspect and adjust the valve clearance.

4.5L Engine

See Figures 79 through 87.

1. Remove engine assembly from vehicle.

2. Remove timing chain.

3. With hexagonal part of camshaft locked with wrench, loosen bolts securing

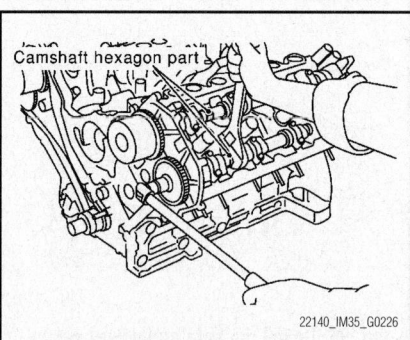

Fig. 79 Hexagonal part of camshaft locked with a wrench while camshaft sprocket is removed—4.5L engine

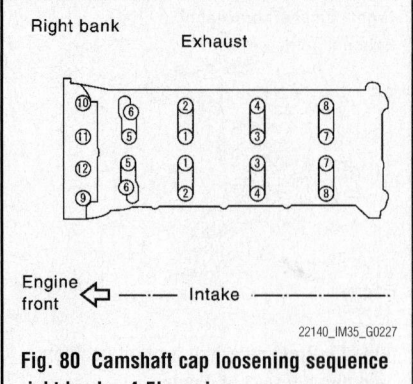

Fig. 80 Camshaft cap loosening sequence right bank—4.5L engine

camshaft sprocket to remove camshaft sprocket.

✳✳ WARNING

Never loosen mounting bolts using anything other than the camshaft hexagonal portion.

➡**After removing the timing chain, never turn crankshaft and camshaft separately, or valves will strike the piston heads.**

4. Remove intake and exhaust camshaft brackets.

➡**Mark camshafts, camshaft brackets and bolts so they are placed in the same position and direction for installation.**

5. Equally loosen camshaft brackets and bolts in several steps in reverse order as shown in the figure.

6. While lightly tapping with a plastic hammer, remove the camshaft bracket (No. 1) and camshaft bracket (No. 6).

➡**The bottom surface of each bracket will be stuck to cylinder the head because of liquid gasket.**

7. Remove camshaft.

8. Remove valve lifter.

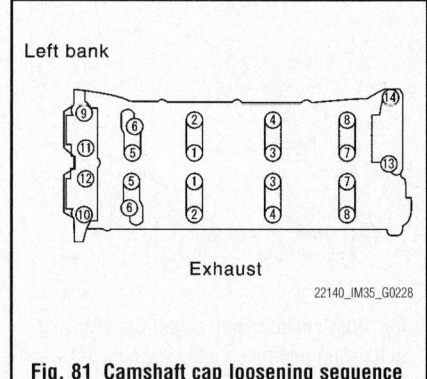

Fig. 81 Camshaft cap loosening sequence left bank—4.5L engine

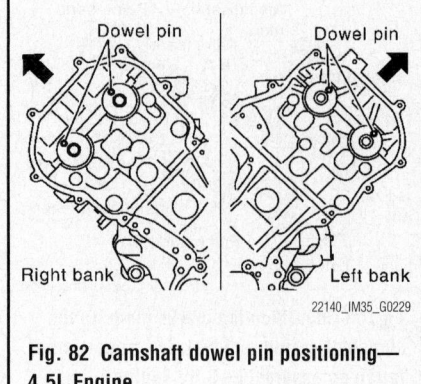

Fig. 82 Camshaft dowel pin positioning—4.5L Engine

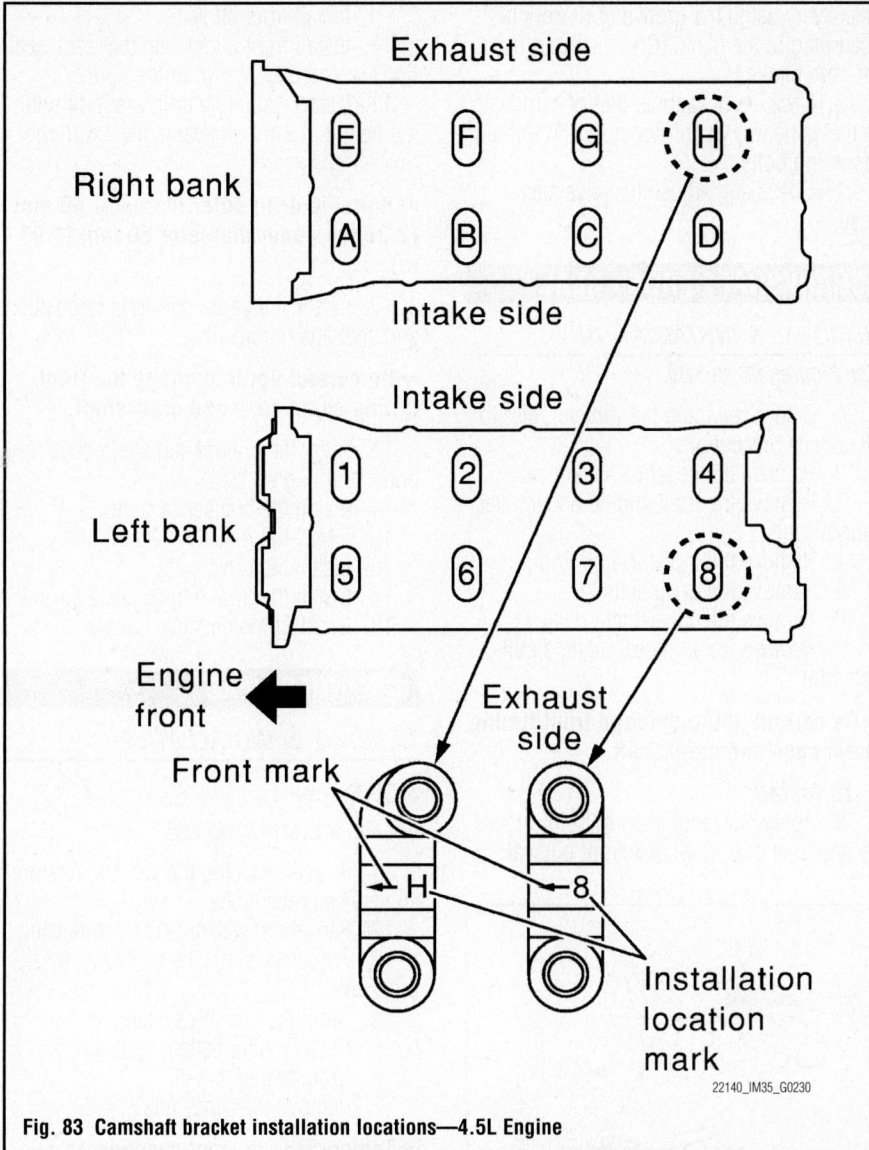

Fig. 83 Camshaft bracket installation locations—4.5L Engine

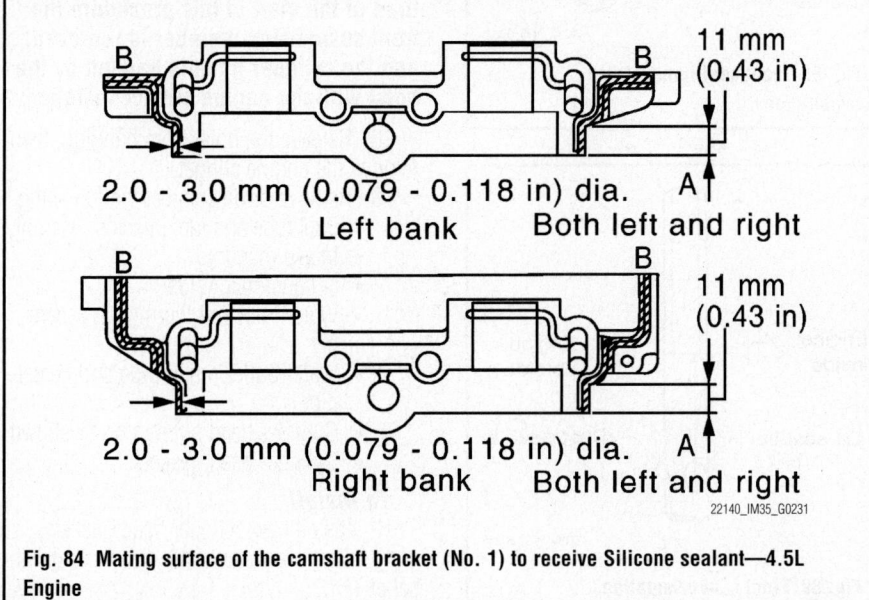

Fig. 84 Mating surface of the camshaft bracket (No. 1) to receive Silicone sealant—4.5L Engine

➡Identify installation positions, and store them without mixing them up.

To install:

9. Install the valve lifter in order they were removed.

10. Install the camshaft.

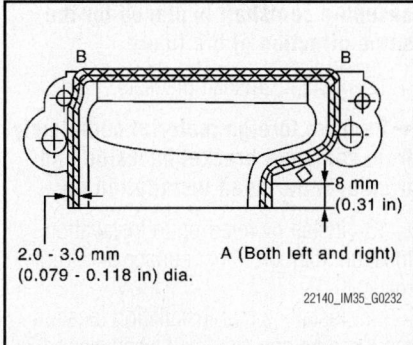

Fig. 85 Mating surface B of the camshaft brackets to receive Silicone sealant—4.5L Engine

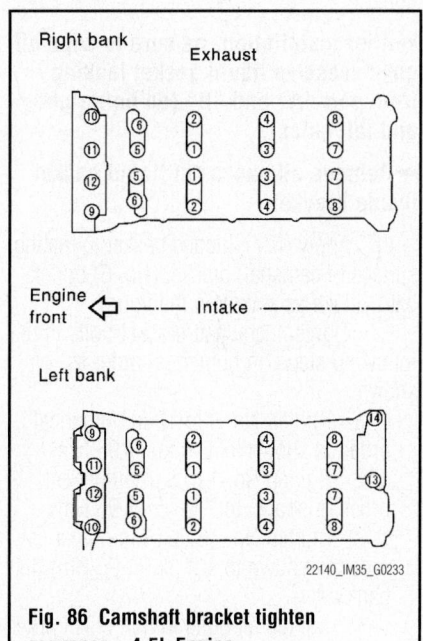

Fig. 86 Camshaft bracket tighten sequence—4.5L Engine

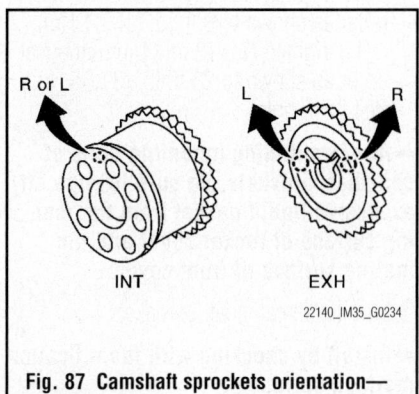

Fig. 87 Camshaft sprockets orientation—4.5L Engine

11. Install camshaft so that dowel pin on front end face are positioned as shown in the figure. (No. 1 cylinder TDC on its compression stroke).

➡**Though camshaft does not stop at the position as shown in the figure, for the placement of cam nose, it is generally accepted camshaft is placed for the same direction of the figure.**

12. Install camshaft brackets.

➡**Remove foreign material completely from camshaft bracket backside and from cylinder head installation face.**

13. Install by referring to installation location mark on upper surface and front mark.

14. Install so that installation location mark can be correctly read when viewed from the side of left exhaust bank.

15. Apply liquid gasket to mating surface of camshaft bracket (No. 1) as shown in the figure.

➡**After installation, be sure to wipe off any excessive liquid gasket leaking from part "A" and "B" (on both right and left sides).**

➡**Remove all excess of liquid gasket inside bracket.**

16. Apply RTV Silicone gasket to mating surface of camshaft bracket (No. 6) on left bank intake as shown in the figure.

17. Tighten camshaft bracket bolts in the following steps, in numerical order as shown.

    a. Tighten No. 9 to 12 in numerical order as shown to 1 ft. lb. (1.96 Nm).

    b. Tighten No. 1 to 8 in numerical order as shown to 1 ft. lb. (1.96 Nm).

    c. Tighten No. 13 to 14 in numerical order as shown to 1 ft. lb. (1.96 Nm) (left bank only).

    d. Tighten all bolts in numerical order as shown to 4 ft. lb. (5.88 Nm).

    e. Tighten No. 1 to 12 in numerical order as shown to 8 ft. lb. (10.41 Nm).

    f. Tighten No. 13 to 14 in numerical order as shown to 23 ft. lb. (31.35 Nm) (left bank only).

➡**After tightening mounting bolts of camshaft brackets, be sure to wipe off excessive liquid gasket from the mating surface of rocker cover and the mating surface of front cover.**

18. Install camshaft sprockets.

➡**Install by checking with identification mark on surface.**

19. Install camshaft sprocket (EXH) by selectively using the groove of dowel pin according to the bank. (Common part used for both banks.)

20. Lock the hexagonal part of camshaft in the same way as for removal, and tighten mounting bolts.

21. Check and adjust the valve clearance.

CRANKSHAFT FRONT SEAL

REMOVAL & INSTALLATION

See Figures 88 and 89.

1. Before servicing the vehicle, refer to all service precautions.
2. Remove the negative battery cable.
3. Remove the front engine undercover (power tool).
4. Remove the radiator 4.5L Engine.
5. Remove the drive belts.
6. Remove the crankshaft pulley.
7. Remove front oil seal using a suitable tool.

➡**Be careful not to damage front timing chain case and crankshaft.**

To install:

8. Apply new engine oil to both oil seal lip and dust seal lip of new front oil seal.

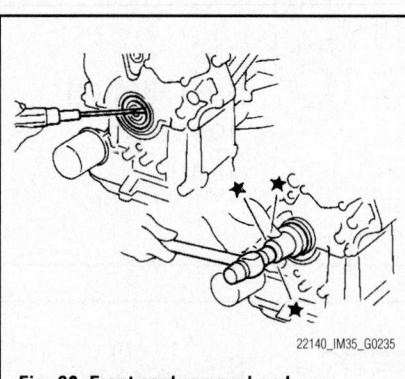

22140_IM35_G0235

Fig. 88 Front seal removal and installation

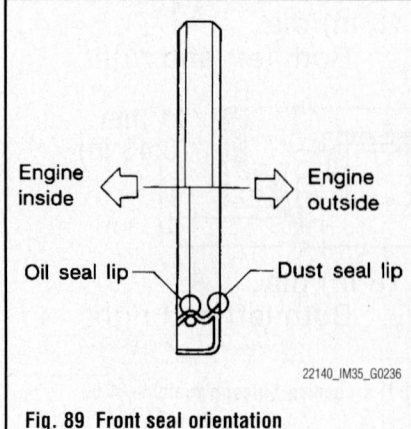

Engine inside ⇐ → **Engine outside**

Oil seal lip ⎯ ⎯ Dust seal lip

22140_IM35_G0236

Fig. 89 Front seal orientation

9. Install front oil seal.
10. Install front oil seal so that each seal lip is oriented as shown in the figure.
11. Using a suitable drift, press-fit until the height of front oil seal is level with the mounting surface.

➡**Suitable drift: outer diameter 60 mm (2.36 in), inner diameter 50 mm (1.97 in).**

12. Check the garter spring is in position and seal lips not inverted.

➡**Be careful not to damage the front timing chain case and crankshaft.**

13. Press-fit straight and avoid causing burrs or tilting oil seal.
14. Install the crankshaft pulley.
15. Install the radiator 4.5L Engine.
16. Install the drive belts.
17. Install the front engine undercover.
18. Connect the negative battery cable.

CYLINDER HEAD

REMOVAL & INSTALLATION

3.5L Engine

See Figures 90 through 96.

1. Before servicing the vehicle, refer to all service precautions.
2. Remove or disconnect the following:
3. Properly relieve the fuel system pressure.
4. Drain the cooling system.
5. Remove both battery cables.
6. Drain the engine oil.
7. Remove the camshafts.

➡**Temporarily fit front suspension member to support the engine. At the time of the start of this procedure the front suspension member is removed, and the cylinder head is hanged by the hoist with the engine slinger installed.**

8. Release the hoist from hanging, then remove the engine slinger.
9. Remove or disconnect the following:

• Fuel tube and fuel injector assembly
• Intake manifold
• Exhaust manifolds
• Water inlet and thermostat assembly
• Water outlet, water pipe and heater pipe
• Cylinder head bolts in order shown
• Cylinder head gaskets

To install:

10. Install new cylinder head gaskets.
11. Turn crankshaft until No. 1 piston is set at TDC.

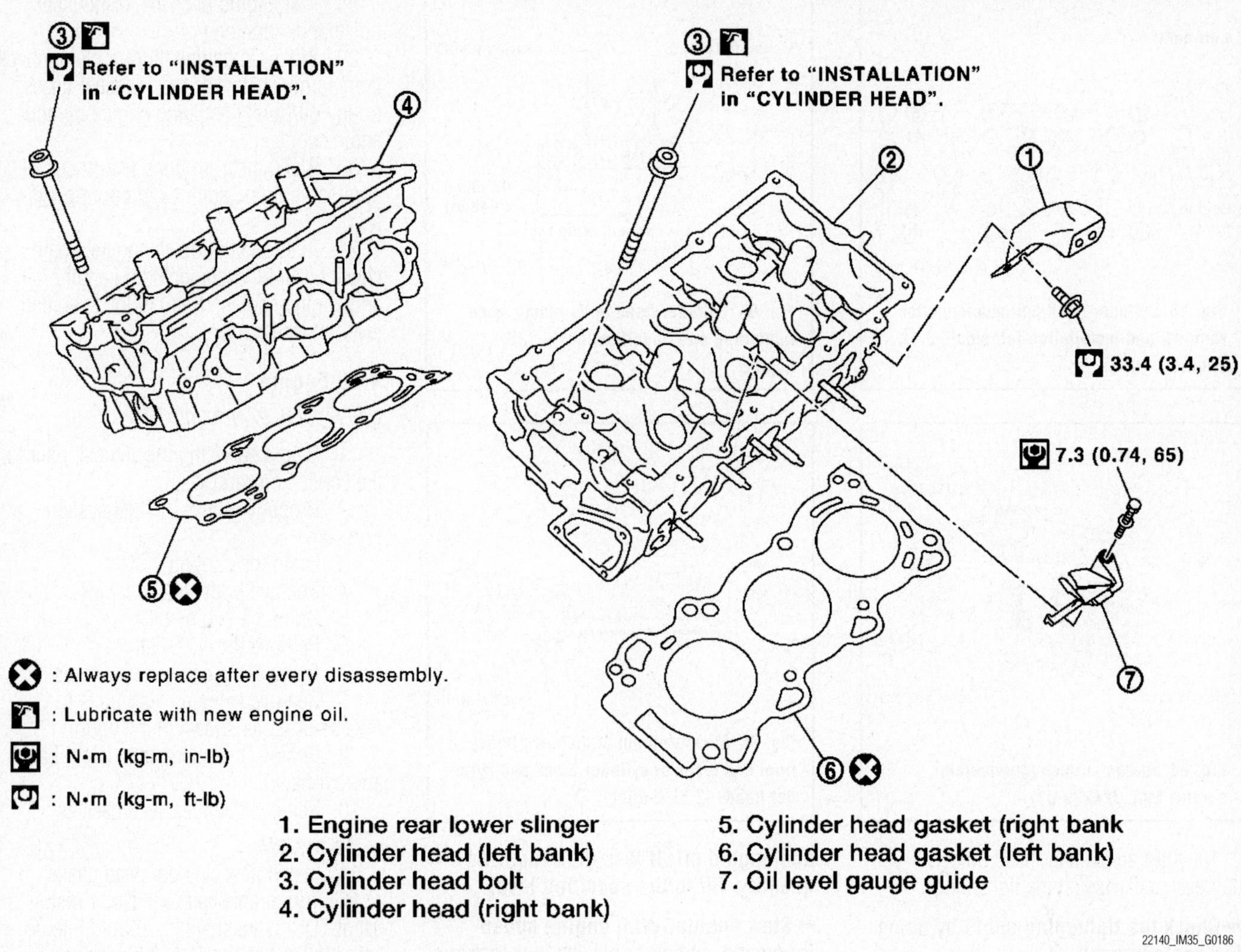

③ Refer to "INSTALLATION" in "CYLINDER HEAD".

③ Refer to "INSTALLATION" in "CYLINDER HEAD".

33.4 (3.4, 25)

7.3 (0.74, 65)

❌ : Always replace after every disassembly.

🛢 : Lubricate with new engine oil.

🔧 : N•m (kg-m, in-lb)

🔧 : N•m (kg-m, ft-lb)

1. Engine rear lower slinger
2. Cylinder head (left bank)
3. Cylinder head bolt
4. Cylinder head (right bank)
5. Cylinder head gasket (right bank)
6. Cylinder head gasket (left bank)
7. Oil level gauge guide

22140_IM35_G0186

Fig. 90 Cylinder head assemblies-exploded view—3.5L Engine

• Crankshaft key should line up with the right bank cylinder center line as shown in the figure.

12. Install cylinder head, follow the steps below to tighten the cylinder head bolts in numerical order as shown in the figure below.

13. Use (commercial service tool: J24239-01) shown in figure below.

➡ **If cylinder head bolts are re-used, check their outer diameters before installation. Refer to "Cylinder Head Bolts Outer Diameter" in the figure below.**

14. Apply new engine oil to threads and seat surfaces of cylinder head bolts.

15. Tighten all cylinder head bolts to 72 ft. lbs. (98.1 Nm).

16. Completely loosen all cylinder head bolts in reverse order of tightening sequence.

17. Tighten all cylinder head bolts to 29 ft. lbs. (39.2 Nm) following tightening sequence.

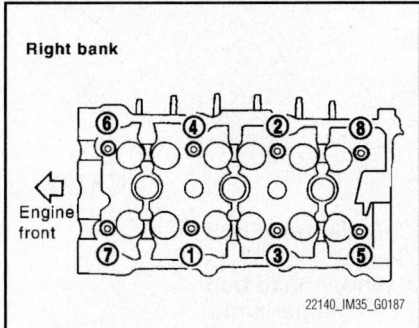

Right bank

Engine front

22140_IM35_G0187

Fig. 91 Cylinder head bolt sequence for removal and installation-right side—3.5L Engine

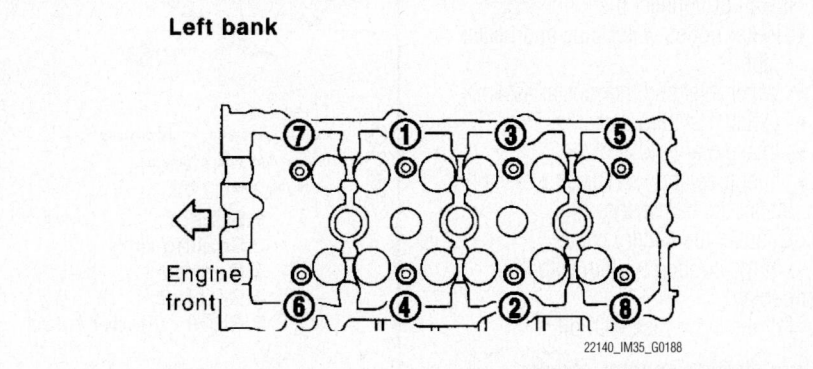

Left bank

Engine front

22140_IM35_G0188

Fig. 92 Cylinder head bolt sequence for removal and installation-left side—3.5L Engine

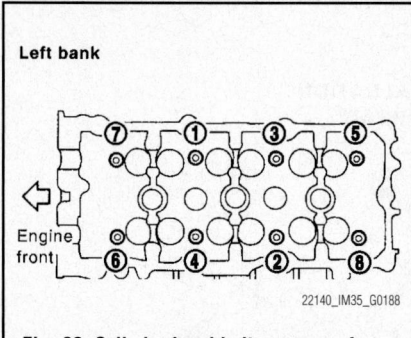

Fig. 93 Cylinder head bolt sequence for removal and installation-left side—3.5L Engine

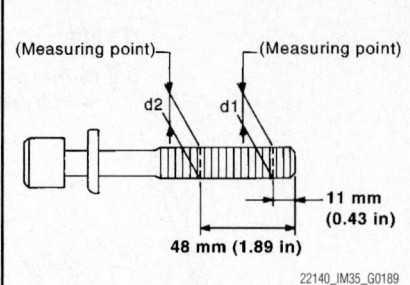

Fig. 95 Cylinder head bolts plastic zone tightening size specifications—3.5L Engine

28. Run engine to check for unusual noise and vibration.

29. Warm up engine thoroughly to check there is no leakage of fuel, exhaust gases, or any oil/fluids including engine oil and engine coolant.

30. Bleed air from lines and hoses of applicable lines, such as in cooling system.

31. After cooling down engine, again check oil/fluid levels including engine oil and engine coolant. Refill to the specified level, if necessary.

4.5L Engine

See Figures 97 through 101.

1. Before servicing the vehicle, refer to the service precautions.

2. Properly relieve the fuel system pressure.

3. Drain the cooling system.

4. Remove both battery cables.

5. Drain the engine oil.

6. Remove the camshafts.

7. Remove the exhaust manifolds.

8. Remove cylinder head bolts in reverse order as shown in the figure

9. Remove cylinder heads (right bank and left bank).

10. Remove cylinder head gaskets.

To install:

11. Install new cylinder head gasket.

12. Turn crankshaft until No. 1 piston is set at TDC, crankshaft key should line up with the left bank cylinder center line as shown in the figure below.

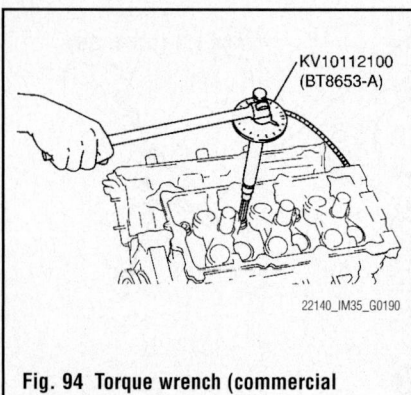

Fig. 94 Torque wrench (commercial service tool: J24239-01)

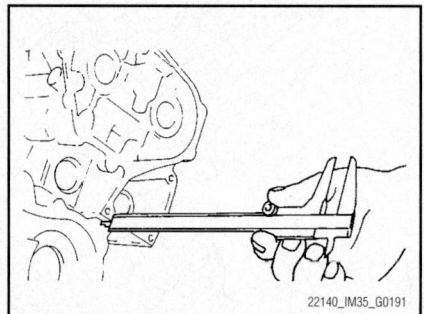

Fig. 96 Measurement of distance between front end faces of cylinder block and cylinder head—3.5L Engine

18. Turn all cylinder head bolts 90 degrees clockwise (angle tightening).

➡**Check the tightening angle by using the angle wrench.**

19. Turn all cylinder head bolts 90 degrees clockwise again (angle tightening) following tightening sequence.

20. After installing cylinder head, measure distance between front end faces of cylinder block and cylinder head (right bank and left bank).

21. Distance must be between 0.555–0.587 inches (14.1–14.9 mm)

22. If measured value is out of the standard, re-install cylinder head.

23. Install or connect the following:
   • Water outlet, water pipe and heater pipe
   • Water inlet and thermostat assembly
   • Exhaust manifolds
   • Intake manifold
   • Fuel tube and fuel injector assembly

24. Install the camshafts.

25. Connect the battery cable.

26. Fill the cooling system with the proper coolant.

27. Fill the crankcase with oil.

➡**Before starting engine, check oil/fluid levels including engine coolant**

and engine oil. If less than required quantity, fill to the specified level

➡**Start engine. With engine speed increased, check again for fuel leakage at connection points.**

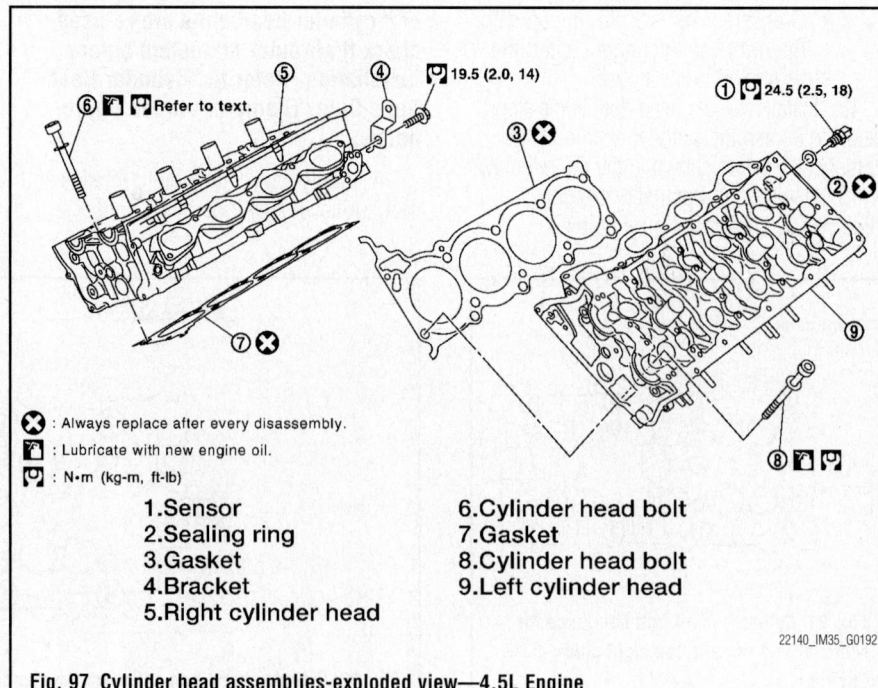

⊗ : Always replace after every disassembly.
⊡ : Lubricate with new engine oil.
⬚ : N•m (kg-m, ft-lb)

1.Sensor	6.Cylinder head bolt
2.Sealing ring	7.Gasket
3.Gasket	8.Cylinder head bolt
4.Bracket	9.Left cylinder head
5.Right cylinder head	

Fig. 97 Cylinder head assemblies-exploded view—4.5L Engine

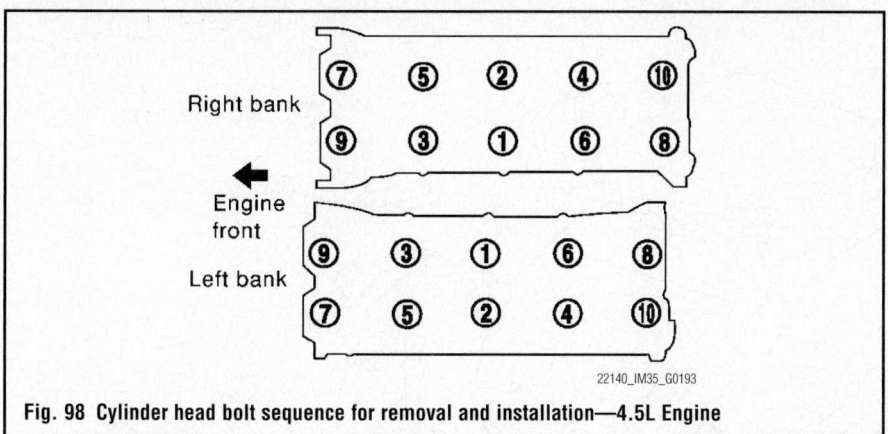

Fig. 98 Cylinder head bolt sequence for removal and installation—4.5L Engine

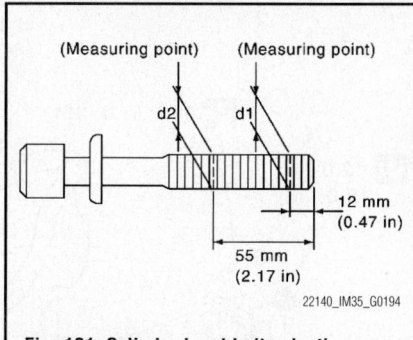

Fig. 101 Cylinder head bolts plastic zone tightening size specifications—4.5L Engine

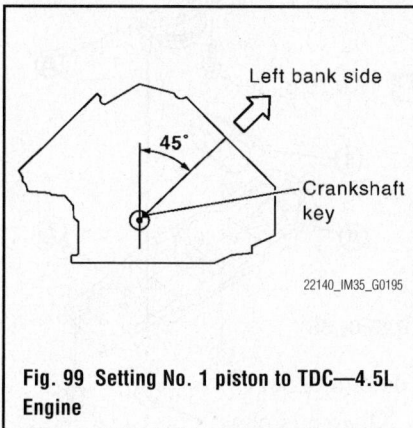

Fig. 99 Setting No. 1 piston to TDC—4.5L Engine

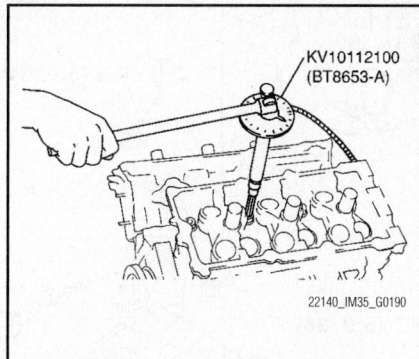

Fig. 100 Torque wrench (commercial service tool: J24239-01)

13. Install cylinder head, follow the steps below to tighten cylinder head bolts in numerical order as shown in the figure below.

14. Use (commercial service tool: J24239-01) shown in figure below.

➡️**If cylinder head bolts are re-used, check their outer diameters before installation. Refer to "Cylinder Head Bolts Outer Diameter" in the figure below.**

15. Apply new engine oil to threads and seat surfaces of cylinder head bolts.

16. Tighten all cylinder head bolts to 72 ft. lbs. (98.1 Nm).

17. Completely loosen all cylinder head bolts in reverse order of tightening sequence.

18. Tighten all cylinder head bolts to 32 ft. lbs. (44 Nm) following tightening sequence.

19. Turn all cylinder head bolts 60 degrees clockwise (angle tightening).

➡️**Check the tightening angle by using the angle wrench.**

20. Turn all cylinder head bolts 60 degrees clockwise again (angle tightening) following tightening sequence.

21. After installing cylinder head, measure distance between front end faces of cylinder block and cylinder head (right bank and left bank).

22. Install the exhaust manifolds.

23. Install the camshafts.

24. Connect the battery cable.

25. Fill the cooling system with the proper coolant.

26. Fill the crankcase with oil.

➡️**Before starting engine, check oil/fluid levels including engine coolant and engine oil. If less than required quantity, fill to the specified level**

➡️**Start engine. With engine speed increased, check again for fuel leakage at connection points.**

27. Run engine to check for unusual noise and vibration.

28. Warm up engine thoroughly to check there is no leakage of fuel, exhaust gases, or any oil/fluids including engine oil and engine coolant.

29. Bleed air from lines and hoses of applicable lines, such as in cooling system.

30. After cooling down engine, again check oil/fluid levels including engine oil and engine coolant. Refill to the specified level, if necessary.

ENGINE ASSEMBLY

REMOVAL & INSTALLATION

3.5L Engine

2WD Models

See Figure 102.

1. Before servicing the vehicle, refer to the service precautions.

2. Evacuate the A/C system.

3. Release the fuel system pressure.

4. Drain the engine oil.

5. Drain the cooling system.

✳️ WARNING

Perform this step when engine is cold.

➡️**Never spill engine coolant on drive belts**

6. Drain the transaxle fluid.

7. Remove or disconnect the following:
- Negative battery cable
- Hood
- Engine room cover(RH and LH)
- Engine cover:
- Front road wheel and tires
- Front and rear engine undercover
- Cowl top cover (RH)
- Air duct and air cleaner case assembly
- Radiator hoses (upper and lower)
- Heater hose from vehicle-side
- Wire bonding (between vehicle to left bank cylinder head)
- A/C piping from A/C compressor
- Brake booster vacuum hose
- Battery positive cable at vehicle side
- Grounding cable
- Fuel feed hose (with damper) and EVAP hose- plug disconnected hoses to prevent fuel leaks
- Reservoir tank of power steering oil pump and piping from vehicle

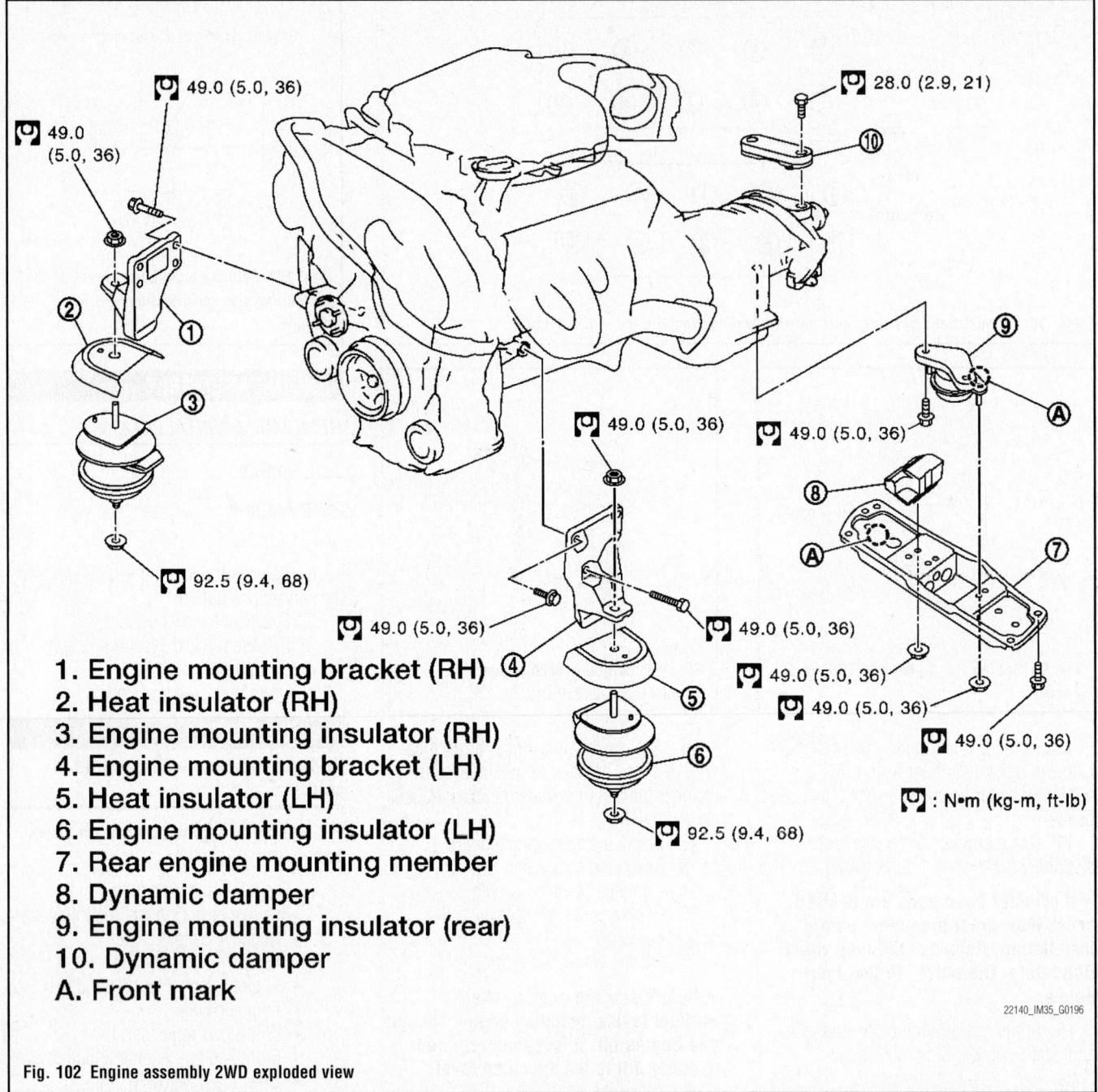

49.0 (5.0, 36)

49.0 (5.0, 36)

28.0 (2.9, 21)

⑩

①

②

③

92.5 (9.4, 68)

49.0 (5.0, 36)

49.0 (5.0, 36)

49.0 (5.0, 36)

49.0 (5.0, 36)

⑨

Ⓐ

⑧

Ⓐ

⑦

④

49.0 (5.0, 36)

⑤

49.0 (5.0, 36)

49.0 (5.0, 36)

⑥

49.0 (5.0, 36)

92.5 (9.4, 68)

: N•m (kg-m, ft-lb)

1. Engine mounting bracket (RH)
2. Heat insulator (RH)
3. Engine mounting insulator (RH)
4. Engine mounting bracket (LH)
5. Heat insulator (LH)
6. Engine mounting insulator (LH)
7. Rear engine mounting member
8. Dynamic damper
9. Engine mounting insulator (rear)
10. Dynamic damper
A. Front mark

22140_IM35_G0196

Fig. 102 Engine assembly 2WD exploded view

- Passenger-side kicking plate, dash side finisher, and glove box.
- Engine room harness connectors at unit sides TCM, ECM and other

8. Disengage intermediate fixing point. Pull out engine room harnesses to engine room side, and temporarily secure them on engine.

9. Remove or disconnect the following:

- A/T fluid cooler hoses and power steering oil pump oil cooler hoses-plug disconnected hoses to prevent leaks
- Heated oxygen sensor harness
- Three way catalyst and exhaust front tube
- Steering lower joint at power steering gear assembly side
- Steering lower shaft

➡**Matchmark the shaft and flange for reassembly.**

- Rear propeller shaft, secure it on the transmission assembly

- Rear plate cover from oil pan (upper)
- Bolts attaching the drive plate to torque converter
- Transmission joint bolts which pierce at oil pan (upper) lower rear
- Front stabilizer at transverse link side
- Lower ends of left and right strut from transverse link
- Steering outer sockets from steering knuckle
- Transverse links mounting bolts at knuckle side

10. Use a manual lift table caddy (commercial service tool) or equivalently rigid

tool such as a transmission jack. Securely support bottom of suspension member and the transmission assembly.

11. Remove rear engine mounting member bolts.

12. Remove front suspension member mounting bolts and nuts.

13. Carefully lower jack, or raise lift to remove the engine, the transmission assembly and front suspension member.

14. When performing work, observe the following caution:
- Confirm there is no interference with the vehicle.
- Check that all connection points have been disconnected.
- Keep in mind the center of vehicle gravity changes. If necessary, use jack(s) to support the vehicle at rear jacking point(s) to prevent it from falling it off the lift.

To install:

15. Carefully raise the jack, or lower the lift to install the engine, the transmission assembly and front suspension member.

16. Install the front suspension member mounting bolts and nuts.

17. Tighten the front suspension member mounting bolts and nuts to 36 ft. lbs. (49. Nm).

18. Install the rear engine mounting member bolts.

19. Tighten the rear engine mounting member bolts and nuts to 36 ft. lbs. (49. Nm).

20. Remove the lift table caddy (commercial service tool) or equivalently rigid tool.

21. Install the transverse links mounting bolts at knuckle side.

22. Tighten the transverse links mounting bolts at knuckle side to 100 ft. lbs. (136. Nm).

23. Install the steering outer sockets from steering knuckle.

24. Tighten the steering outer sockets from steering knuckle to 25 ft. lbs. (34.4. Nm).

25. Install the lower ends of left and right strut from transverse link.

26. Tighten the lower ends of left and right strut from transverse link to 79 ft. lbs. (107. Nm).

27. Install the front stabilizer at transverse link side.

28. Tighten the front stabilizer at transverse link side to 66 ft. lbs. (90. Nm).

29. Install the transmission joint bolts which pierce at oil pan (upper) lower rear side/

30. Tighten the transmission joint bolts

which pierce at oil pan (upper) lower rear side to 38 ft. lbs. (51. Nm).

31. Install the bolts attaching the drive plate to torque converter.

32. Tighten the bolts attaching the drive plate to torque converter to 38 ft. lbs. (51. Nm).

33. Install the rear plate cover from oil pan (upper).

34. Tighten the rear plate cover from oil pan (upper) to 41 ft. lbs. (55. Nm).

35. Install the rear propeller shaft using the matchmarks.

36. Tighten the rear propeller shaft to 55 ft. lbs. (74. Nm).

37. Install the steering lower shaft.

38. Tighten the steering lower shaft to 20 ft. lbs. (26.5. Nm).

39. Install the steering lower joint at power steering gear assembly side.

40. Tighten the steering lower joint at power steering gear assembly side to 20 ft. lbs. (26.5. Nm).

41. Install the three way catalyst and exhaust front tube.

42. Tighten the three way catalyst and exhaust front tube to 43 ft. lbs. (57.9. Nm).

43. Install or connect the following:
- Engine room harnesses to engine room side
- Engine room harness connectors at unit sides TCM, ECM and other
- Passenger-side kicking plate, dash side finisher, and glove box
- Reservoir tank of power steering oil pump and piping
- Fuel feed hose (with damper) and EVAP hose
- Grounding cable
- Battery positive cable at vehicle side
- Brake booster vacuum hose
- A/C piping to A/C compressor
- Wire bonding (between vehicle to left bank cylinder head)
- Heater hose from vehicle-side
- Radiator hoses (upper and lower)
- Air duct and air cleaner case assembly
- Cowl top cover (RH)
- Front and rear engine undercover
- Front road wheel and tires
- Engine cover
- Engine room cover(RH and LH)
- Hood
- Negative battery cable

44. Fill and bleed the cooling system.

45. Fill the engine with clean oil.

46. Fill the transaxle to the proper level.

47. Start the vehicle, check for leaks and repair if necessary.

AWD Models

See Figure 103.

1. Before servicing the vehicle, refer to the service precautions.

2. Evacuate the A/C system.

3. Release the fuel system pressure.

4. Drain the engine oil.

5. Drain the cooling system.

> **✳✳ WARNING**
>
> **Perform this step when engine is cold.**

➡️**Never spill engine coolant on drive belts**

6. Drain the transaxle fluid.

7. Remove or disconnect the following:
- Negative battery cable
- Hood
- Engine room cover(RH and LH)
- Engine cover:
- Front road wheel and tires
- Front and rear engine undercover
- Front cross bar
- Cowl top cover (RH)
- Air duct and air cleaner case assembly
- Radiator hoses (upper and lower)
- Heater hose from vehicle-side
- Wire bonding (between vehicle to left bank cylinder head)
- A/C piping from A/C compressor
- Brake booster vacuum hose
- Battery positive cable at vehicle side
- Grounding cable
- Fuel feed hose (with damper) and EVAP hose- plug disconnected hoses to prevent fuel leaks
- Reservoir tank of power steering oil pump and piping from vehicle
- Passenger-side kicking plate, dash side finisher, and glove box.
- Engine room harness connectors at unit sides TCM, ECM and other

8. Disengage intermediate fixing point. Pull out engine room harnesses to engine room side, and temporarily secure them on engine.

> **✳✳ WARNING**
>
> **When pulling out harnesses, take care not to damage harnesses and connectors.**

9. Remove or disconnect the following:
- A/T fluid cooler hoses and power steering oil pump oil cooler hoses- plug disconnected hoses to prevent leaks
- Heated oxygen sensor harness

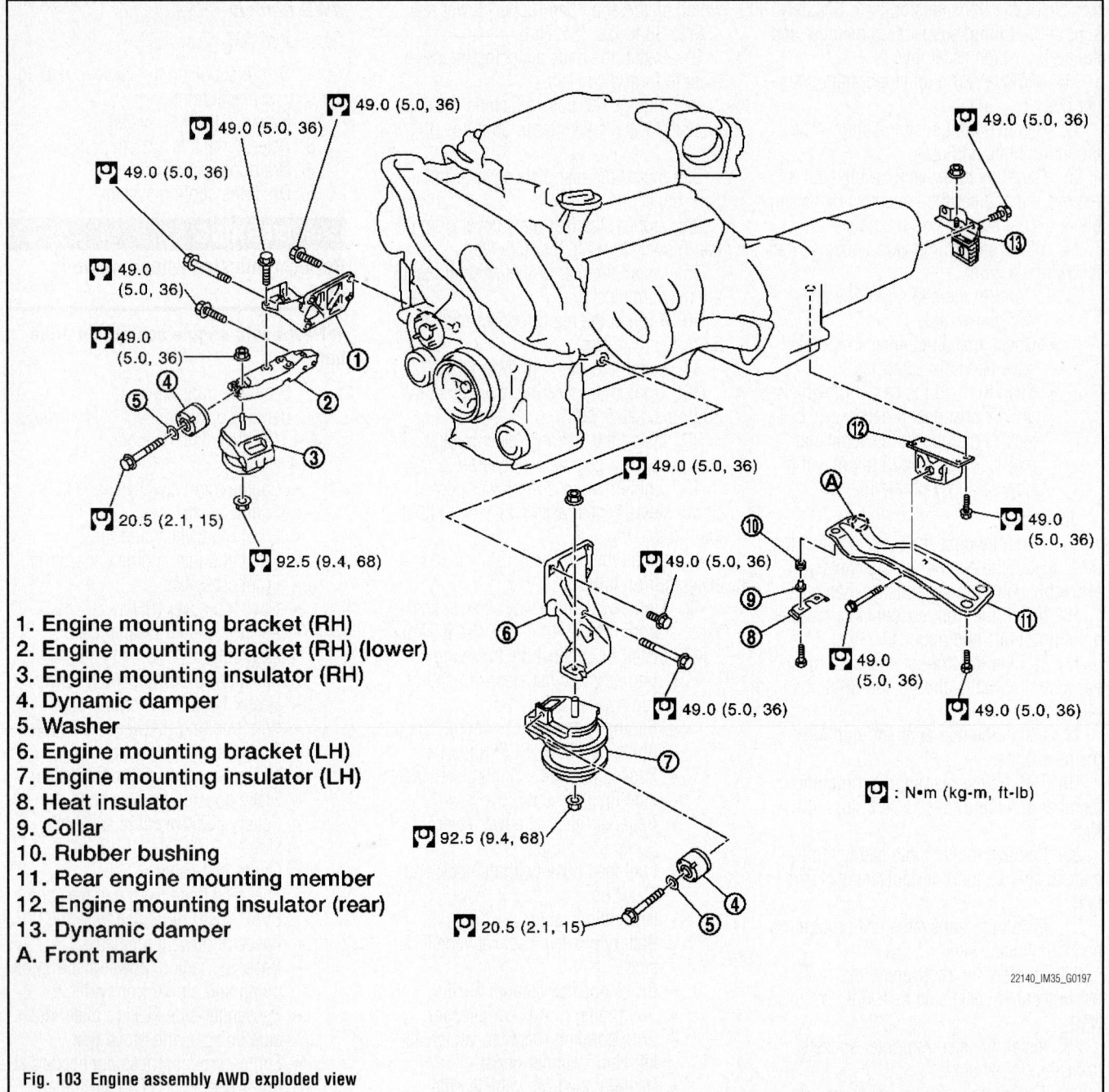

1. Engine mounting bracket (RH)
2. Engine mounting bracket (RH) (lower)
3. Engine mounting insulator (RH)
4. Dynamic damper
5. Washer
6. Engine mounting bracket (LH)
7. Engine mounting insulator (LH)
8. Heat insulator
9. Collar
10. Rubber bushing
11. Rear engine mounting member
12. Engine mounting insulator (rear)
13. Dynamic damper
A. Front mark

Fig. 103 Engine assembly AWD exploded view

- Three way catalyst and exhaust front tube
- Steering lower joint at power steering gear assembly side
- Steering lower shaft

➡Matchmark the shaft and flange for reassembly.

- Rear propeller shaft, secure it on the transmission assembly
- Remove front drive shaft (both side)\
- Harness connector from transmission assembly and transfer assembly

- A/T control rod at control device assembly side
- Rear plate cover from oil pan (upper)
- Bolts attaching the drive plate to torque converter
- Transmission joint bolts which pierce at oil pan (upper) lower rear
- Front stabilizer at transverse link side
- Lower ends of left and right strut from transverse link
- Steering outer sockets from steering knuckle
- Transverse links mounting bolts at knuckle side

10. Use a manual lift table caddy (commercial service tool) or equivalently rigid tool such as a transmission jack. Securely support bottom of suspension member and the transmission assembly.

11. Remove rear engine mounting member bolts.

12. Remove front suspension member mounting bolts and nuts

13. Carefully lower jack, or raise lift to remove the engine, the transmission assembly and front suspension member.

14. When performing work, observe the following caution:

- Confirm there is no interference with the vehicle.
- Check that all connection points have been disconnected.
- Keep in mind the center of vehicle gravity changes. If necessary, use jack(s) to support the vehicle at rear jacking point(s) to prevent it from falling it off the lift

To install:

15. Carefully raise the jack, or lower the lift to install the engine, the transmission assembly and front suspension member.

16. Install the front suspension member mounting bolts and nuts.

17. Tighten the front suspension member mounting bolts and nuts to 36 ft. lbs. (49. Nm).

18. Install the rear engine mounting member bolts.

19. Tighten the rear engine mounting member bolts and nuts to 36 ft. lbs. (49. Nm).

20. Remove the lift table caddy (commercial service tool) or equivalently rigid tool.

21. Install the transverse links mounting bolts at knuckle side.

22. Tighten the transverse links mounting bolts at knuckle side to 100 ft. lbs. (136. Nm).

23. Install the steering outer sockets from steering knuckle.

24. Tighten the steering outer sockets from steering knuckle to 25 ft. lbs. (34.4. Nm).

25. Install the lower ends of left and right strut from transverse link.

26. Tighten the lower ends of left and right strut from transverse link to 79 ft. lbs. (107. Nm).

27. Install the front stabilizer at transverse link side.

28. Tighten the front stabilizer at transverse link side to 66 ft. lbs. (90. Nm).

29. Install the transmission joint bolts, which pierce at oil pan (upper) lower rear side.

30. Tighten the transmission joint bolts which pierce at oil pan (upper) lower rear side to 38 ft. lbs. (51. Nm).

31. Install the bolts attaching the drive plate to torque converter.

32. Tighten the bolts attaching the drive plate to torque converter to 38 ft. lbs. (51. Nm).

33. Install the rear plate cover from oil pan (upper).

34. Install the front drive shaft (both side).

35. Install the harness connector from transmission assembly and transfer assembly.

36. A/T control rod at control device assembly side.

37. Tighten the rear plate cover from oil pan (upper) to 41 ft. lbs. (55. Nm).

38. Install the rear propeller shaft using the matchmarks.

39. Tighten the rear propeller shaft to 55 ft. lbs. (74. Nm).

40. Install the steering lower shaft.

41. Tighten the steering lower shaft to 20 ft. lbs. (26.5. Nm).

42. Install the steering lower joint at power steering gear assembly side.

43. Tighten the steering lower joint at power steering gear assembly side to 20 ft. lbs. (26.5. Nm).

44. Install the three way catalyst and exhaust front tube.

45. Tighten the three way catalyst and exhaust front tube to 43 ft. lbs. (57.9. Nm).

46. Install or connect the following:
- Engine room harnesses to engine room side
- Engine room harness connectors at unit sides TCM, ECM and other
- Passenger-side kicking plate, dash side finisher, and glove box
- Reservoir tank of power steering oil pump and piping
- Fuel feed hose (with damper) and EVAP hose
- Grounding cable
- Battery positive cable at vehicle side
- Brake booster vacuum hose
- A/C piping to A/C compressor
- Wire bonding (between vehicle to left bank cylinder head)
- Heater hose from vehicle-side
- Radiator hoses (upper and lower)
- Air duct and air cleaner case assembly
- Cowl top cover (RH)
- Front cross bar
- Front and rear engine undercover
- Front road wheel and tires
- Engine cover
- Engine room cover(RH and LH)
- Hood
- Negative battery cable

47. Fill and bleed the cooling system.

48. Fill the engine with clean oil.

49. Fill the transaxle to the proper level.

50. Start the vehicle, check for leaks and repair if necessary.

4.5L Engine

2WD Models

See Figure 104.

1. Before servicing the vehicle, refer to the service precautions.

2. Evacuate the A/C system.

3. Release the fuel system pressure.

4. Drain the engine oil.

5. Drain the cooling system.

✳✳ WARNING

Perform this step when engine is cold.

➡**Never spill engine coolant on drive belts.**

6. Drain the transaxle fluid.

7. Remove or disconnect the following:
- Negative battery cable
- Hood
- Engine room cover(RH and LH)
- Engine cover:
- Front road wheel and tires
- Air duct (inlet), air duct and air cleaner case assembly
- Heater hoses
- Wire bonding from exhaust manifold cover to vehicle.
- Vacuum hose between vehicle and engine
- A/C piping from A/C compressor
- Fuel feed hose and EVAP hose
- Ground cable between vehicle and right bank cylinder head
- Reservoir tank of power steering oil pump
- Passenger-side kicking plate, dash side finisher, and glove box
- Engine room harness connectors at unit sides TCM, ECM and other
- Engine room harnesses to engine room side
- A/T fluid cooler hoses and power steering oil pump oil cooler hoses
- Heated oxygen sensor 2 harness
- Exhaust front tube
- Steering lower joint at power steering gear assembly side
- Steering outer sockets from steering knuckle
- A/T control rod at control device assembly side
- Rear plate cover from oil pan
- Bolts fixing drive plate to torque converter
- Transmission joint bolts which pierce at oil pan lower rear side
- Lower ends of left and right strut from transverse link
- Transverse link mounting bolts at knuckle side
- Front stabilizer at transverse link side
- Rear propeller shaft. Matchmark shaft and flange for reassembly

8. Use a manual lift table caddy (commercial service tool) or equivalently rigid

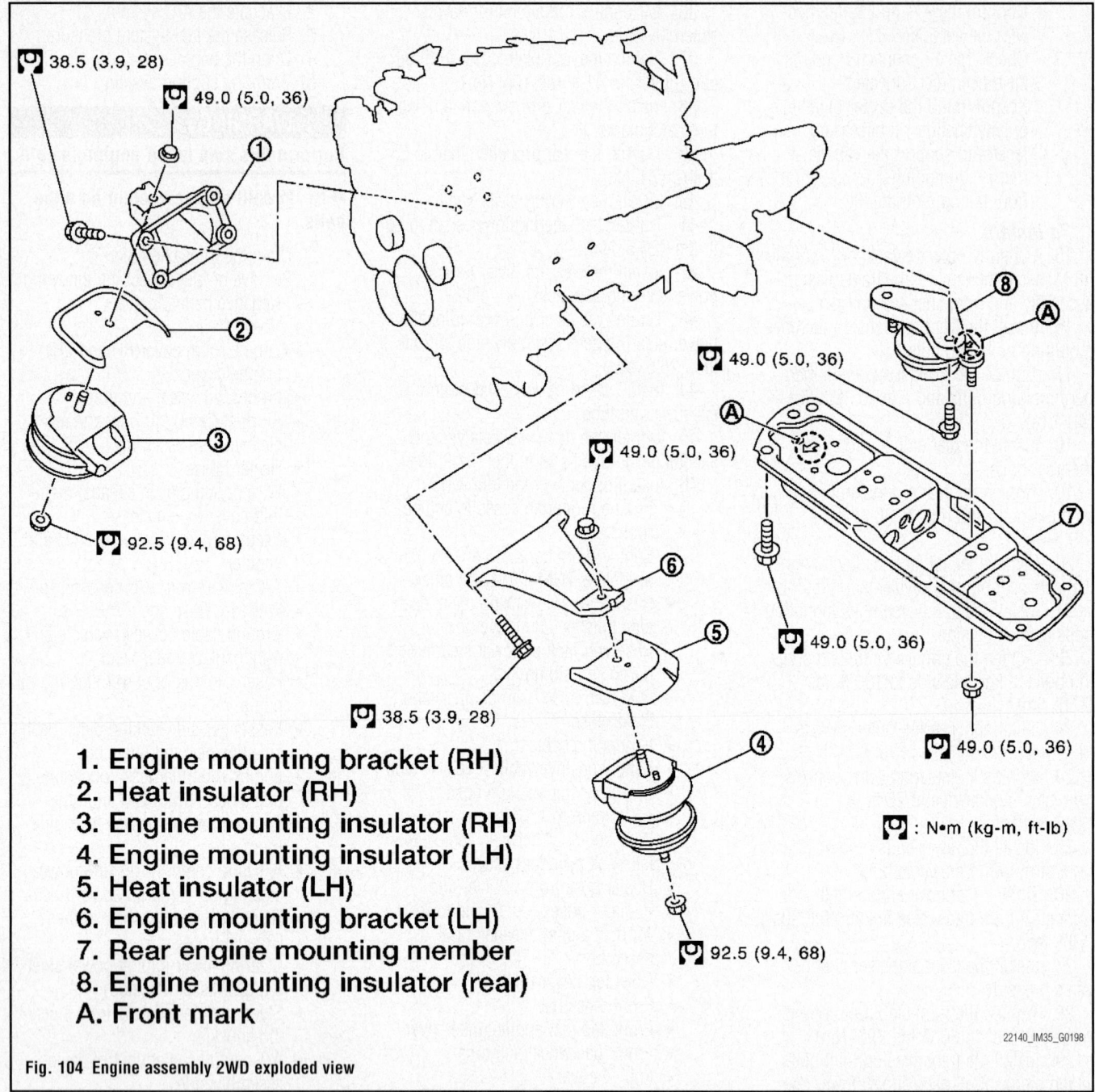

1. Engine mounting bracket (RH)
2. Heat insulator (RH)
3. Engine mounting insulator (RH)
4. Engine mounting insulator (LH)
5. Heat insulator (LH)
6. Engine mounting bracket (LH)
7. Rear engine mounting member
8. Engine mounting insulator (rear)
A. Front mark

38.5 (3.9, 28)
49.0 (5.0, 36)
92.5 (9.4, 68)
49.0 (5.0, 36)
49.0 (5.0, 36)
38.5 (3.9, 28)
49.0 (5.0, 36)
49.0 (5.0, 36)
92.5 (9.4, 68)

: N•m (kg-m, ft-lb)

22140_IM35_G0198

Fig. 104 Engine assembly 2WD exploded view

tool such as a transmission jack. securely support bottom of suspension member and the transmission assembly.

9. Remove rear engine mounting member bolts.

10. Remove front suspension member mounting bolts and nuts.

11. Carefully lower jack, or raise lift to remove the engine, the transmission assembly and front suspension member.

12. When performing work, observe the following caution:
• Confirm there is no interference with the vehicle.

• Check that all connection points have been disconnected.

• Keep in mind the center of vehicle gravity changes. If necessary, use jack(s) to support the vehicle at rear jacking point(s) to prevent it from falling it off the lift.

13. Carefully raise the jack, or lower the lift to install the engine, the transmission assembly and front suspension member.

To install:

14. Install the front suspension member mounting bolts and nuts.

15. Tighten the front suspension mem-

ber mounting bolts and nuts to 36 ft. lbs. (49. Nm).

16. Install the rear engine mounting member bolts.

17. Tighten the rear engine mounting member bolts and nuts to 36 ft. lbs. (49. Nm).

18. Remove the lift table caddy (commercial service tool) or equivalently rigid tool.

19. Install the transverse links mounting bolts at knuckle side.

20. Tighten the transverse links mounting bolts at knuckle side to 100 ft. lbs. (136. Nm).

21. Install the steering outer sockets from steering knuckle.

22. Tighten the steering outer sockets from steering knuckle to 25 ft. lbs. (34.4. Nm).

23. Install the lower ends of left and right strut from transverse link.

24. Tighten the lower ends of left and right strut from transverse link to 79 ft. lbs. (107. Nm).

25. Install the front stabilizer at transverse link side.

26. Tighten the front stabilizer at transverse link side to 66 ft. lbs. (90. Nm).

27. Install the transmission joint bolts, which pierce at oil pan (upper) lower rear side.

28. Tighten the transmission joint bolts which pierce at oil pan (upper) lower rear side to 38 ft. lbs. (51. Nm).

29. Install the bolts attaching the drive plate to torque converter.

30. Tighten the bolts attaching the drive plate to torque converter to 38 ft. lbs. (51. Nm).

31. Install the rear plate cover from oil pan (upper).

32. Install the front drive shaft (both side).

33. Install the harness connector from transmission assembly and transfer assembly.

34. A/T control rod at control device assembly side.

35. Tighten the rear plate cover from oil pan (upper) to 41 ft. lbs. (55. Nm).

36. Install the rear propeller shaft using the matchmarks.

37. Tighten the rear propeller shaft to 55 ft. lbs. (74. Nm).

38. Install the steering lower shaft.

39. Tighten the steering lower shaft to 20 ft. lbs. (26.5. Nm).

40. Install the steering lower joint at power steering gear assembly side.

41. Tighten the steering lower joint at power steering gear assembly side to 20 ft. lbs. (26.5. Nm).

42. Install the three way catalyst and exhaust front tube.

43. Tighten the three way catalyst and exhaust front tube to 43 ft. lbs. (57.9. Nm).

44. Install or connect the following:
- Engine room harnesses to engine room side
- Engine room harness connectors at unit sides TCM, ECM and other
- Passenger-side kicking plate, dash side finisher, and glove box
- Reservoir tank of power steering oil pump and piping
- Fuel feed hose (with damper) and EVAP hose

- Grounding cable
- Battery positive cable at vehicle side
- Brake booster vacuum hose
- A/C piping to A/C compressor
- Wire bonding (between vehicle to left bank cylinder head)
- Heater hose from vehicle-side
- Radiator hoses (upper and lower)
- Air duct and air cleaner case assembly
- Cowl top cover (RH)
- Front cross bar
- Front and rear engine undercover
- Front road wheel and tires
- Engine cover
- Engine room cover(RH and LH)
- Hood
- Negative battery cable

45. Fill and bleed the cooling system.

46. Fill the engine with clean oil.

47. Fill the transaxle to the proper level.

48. Start the vehicle, check for leaks and repair if necessary.

AWD Models

See Figure 105.

1. Before servicing the vehicle, refer to the service precautions.

2. Evacuate the A/C system.

3. Release the fuel system pressure.

4. Drain the engine oil.

5. Drain the cooling system.

> ※※ **WARNING**
>
> **Perform this step when engine is cold.**

➡**Never spill engine coolant on drive belts.**

6. Drain the transaxle fluid.

7. Remove or disconnect the following:
- Negative battery cable
- Hood
- Engine room cover(RH and LH)
- Engine cover:
- Front road wheel and tires
- Front and rear engine undercover
- Air duct (inlet), air duct and air cleaner case assembly
- Front cross bar
- Cowl top cover (RH)
- Air duct and air cleaner case assembly
- Radiator hoses (upper and lower)
- Heater hose from vehicle-side
- Wire bonding (between vehicle to left bank cylinder head)
- A/C piping from A/C compressor
- Brake booster vacuum hose
- Battery positive cable at vehicle side

- Grounding cable
- Fuel feed hose (with damper) and EVAP hose- plug disconnected hoses to prevent fuel leaks
- Reservoir tank of power steering oil pump and piping from vehicle
- Passenger-side kicking plate, dash side finisher, and glove box.
- Engine room harness connectors at unit sides TCM, ECM and other

8. Disengage intermediate fixing point. Pull out engine room harnesses to engine room side, and temporarily secure them on engine.

> ※※ **WARNING**
>
> **When pulling out harnesses, take care not to damage harnesses and connectors.**

9. Remove or disconnect the following:
- A/T fluid cooler hoses and power steering oil pump oil cooler hoses- plug disconnected hoses to prevent leaks
- Heated oxygen sensor harness
- Three way catalyst and exhaust front tube
- Steering lower joint at power steering gear assembly side
- Steering lower shaft

➡**Matchmark the shaft and flange for reassembly.**

- Rear propeller shaft, secure it on the transmission assembly
- Remove front drive shaft (both side)
- Harness connector from transmission assembly and transfer assembly
- A/T control rod at control device assembly side
- Rear plate cover from oil pan (upper)
- Bolts attaching the drive plate to torque converter
- Transmission joint bolts which pierce at oil pan (upper) lower rear
- Front stabilizer at transverse link side
- Lower ends of left and right strut from transverse link
- Steering outer sockets from steering knuckle
- Transverse links mounting bolts at knuckle side

10. Use a manual lift table caddy (commercial service tool) or equivalently rigid tool such as a transmission jack. Securely support bottom of suspension member and the transmission assembly.

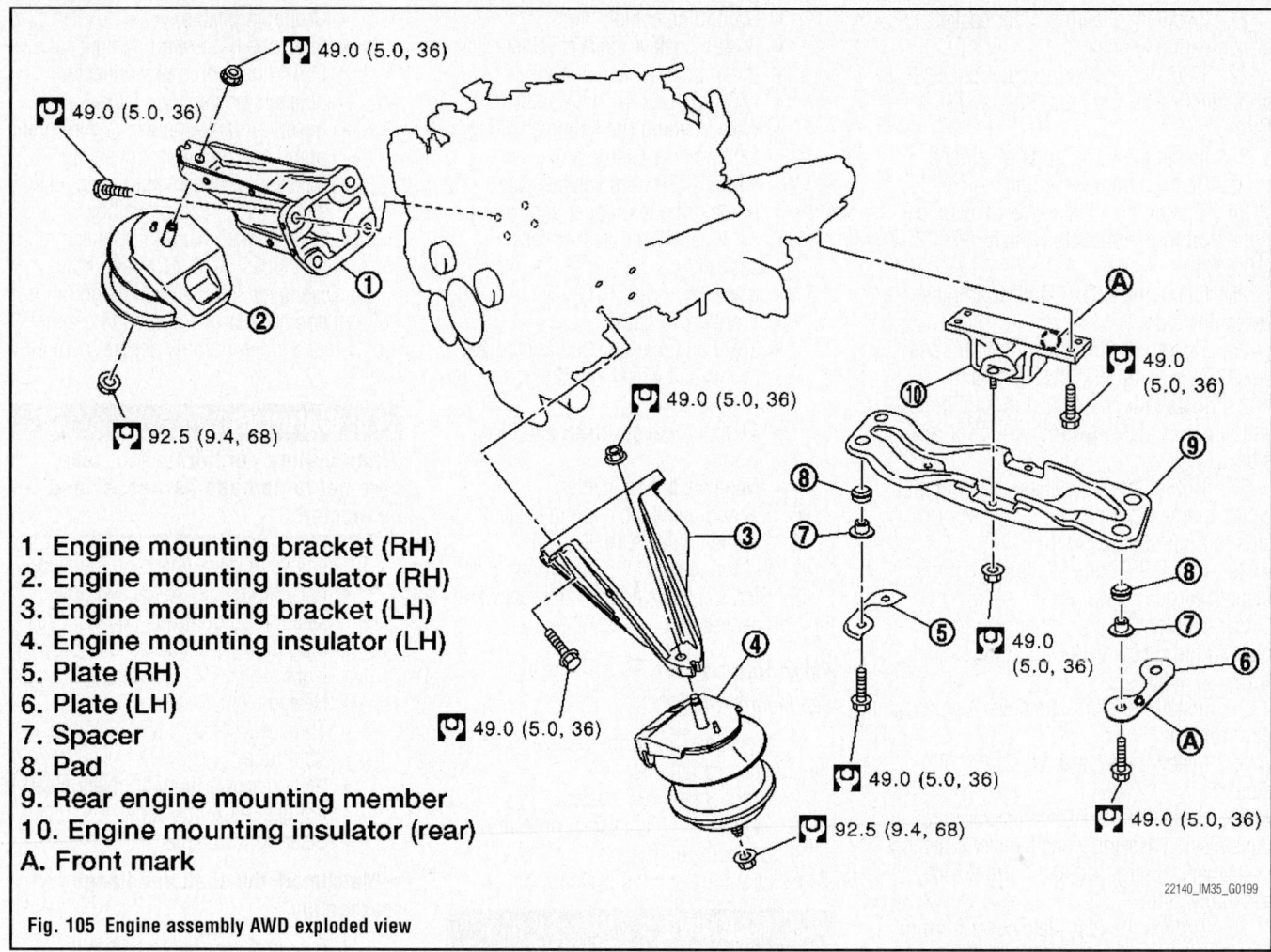

1. Engine mounting bracket (RH)
2. Engine mounting insulator (RH)
3. Engine mounting bracket (LH)
4. Engine mounting insulator (LH)
5. Plate (RH)
6. Plate (LH)
7. Spacer
8. Pad
9. Rear engine mounting member
10. Engine mounting insulator (rear)
A. Front mark

Fig. 105 Engine assembly AWD exploded view

11. Remove rear engine mounting member bolts.

12. Remove front suspension member mounting bolts and nuts

13. Carefully lower jack, or raise lift to remove the engine, the transmission assembly and front suspension member.

14. When performing work, observe the following caution:
- Confirm there is no interference with the vehicle.
- Check that all connection points have been disconnected.
- Keep in mind the center of vehicle gravity changes. If necessary, use jack(s) to support the vehicle at rear jacking point(s) to prevent it from falling it off the lift.

To install:

15. Carefully raise the jack, or lower the lift to install the engine, the transmission assembly and front suspension member.

16. Install the front suspension member mounting bolts and nuts.

17. Tighten the front suspension member mounting bolts and nuts to 36 ft. lbs. (49 Nm).

18. Install the rear engine mounting member bolts.

19. Tighten the rear engine mounting member bolts and nuts to 36 ft. lbs. (49 Nm).

20. Remove the lift table caddy (commercial service tool) or equivalently rigid tool.

21. Install the transverse links mounting bolts at knuckle side.

22. Tighten the transverse links mounting bolts at knuckle side to 100 ft. lbs. (136 Nm).

23. Install the steering outer sockets from steering knuckle.

24. Tighten the steering outer sockets from steering knuckle to 25 ft. lbs. (34.4 Nm).

25. Install the lower ends of left and right strut from transverse link.

26. Tighten the lower ends of left and right strut from transverse link to 79 ft. lbs. (107 Nm).

27. Install the front stabilizer at transverse link side.

28. Tighten the front stabilizer at transverse link side to 66 ft. lbs. (90 Nm).

29. Install the transmission joint bolts which pierce at oil pan (upper) lower rear side/

30. Tighten the transmission joint bolts which pierce at oil pan (upper) lower rear side to 38 ft. lbs. (51 Nm).

31. Install the bolts attaching the drive plate to torque converter.

32. Tighten the bolts attaching the drive plate to torque converter to 38 ft. lbs. (51 Nm).

33. Install the rear plate cover from oil pan (upper).

34. Install the front drive shaft (both side).

35. Install the harness connector to the transmission assembly and transfer assembly.

36. Install the A/T control rod at control device assembly side.

37. Tighten the rear plate cover from oil pan (upper) to 41 ft. lbs. (55 Nm).

38. Install the rear propeller shaft using the matchmarks.

39. Tighten the rear propeller shaft to 55 ft. lbs. (74. Nm)

40. Install the steering lower shaft.

41. Tighten the steering lower shaft to 20 ft. lbs. (26.5. Nm).

42. Install the steering lower joint at power steering gear assembly side.

43. Tighten the steering lower joint at power steering gear assembly side to 20 ft. lbs. (26.5. Nm).

44. Install the three way catalyst and exhaust front tube.

45. Tighten the three way catalyst and exhaust front tube to 43 ft. lbs. (57.9. Nm).

46. Install or connect the following:
- Engine room harnesses to engine room side
- Engine room harness connectors at unit sides TCM, ECM and other connectors

- Passenger-side kicking plate, dash side finisher, and glove box
- Reservoir tank of power steering oil pump and piping
- Fuel feed hose (with damper) and EVAP hose
- Grounding cable
- Battery positive cable at vehicle side
- Brake booster vacuum hose
- A/C piping to A/C compressor
- Wire bonding (between vehicle to left bank cylinder head)
- Heater hose from vehicle-side
- Radiator hoses (upper and lower)
- Air duct and air cleaner case assembly
- Cowl top cover (RH)
- Front cross bar
- Front and rear engine undercover

- Front road wheel and tires
- Engine cover
- Engine room cover(RH and LH)
- Hood
- Negative battery cable

47. Fill and bleed the cooling system.

48. Fill the engine with clean oil.

49. Fill the transaxle to the proper level.

50. Start the vehicle, check for leaks and repair if necessary.

EXHAUST MANIFOLD

REMOVAL & INSTALLATION

3.5L Engine

See Figures 106 through 108.

1. Before servicing the vehicle, refer to the precautions in the beginning of this section.

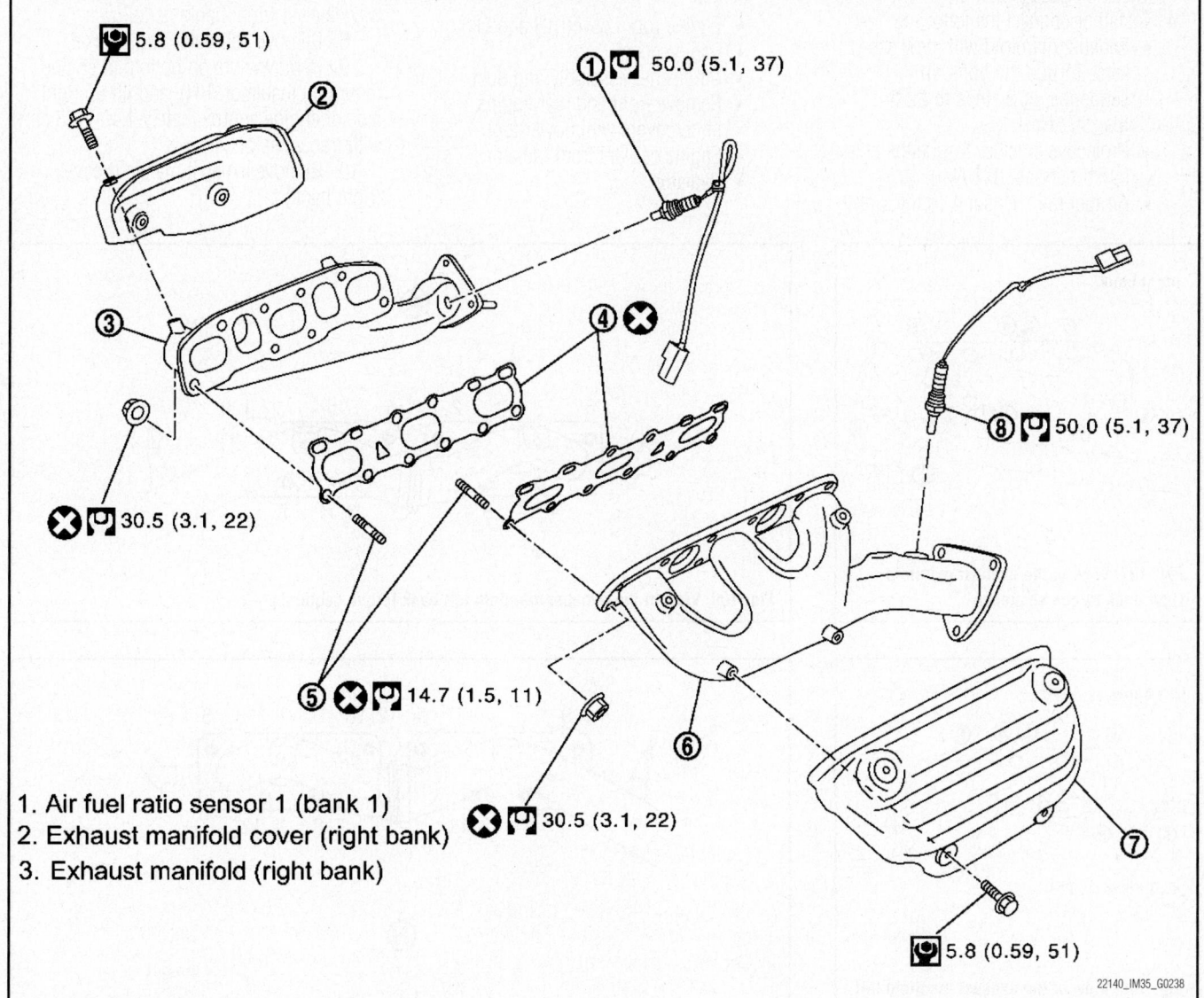

1. Air fuel ratio sensor 1 (bank 1)
2. Exhaust manifold cover (right bank)
3. Exhaust manifold (right bank)

Fig. 106 View of the exhaust manifold and components

2. Remove or disconnect the following:
- Negative battery cable

➡**If necessary, soak the exhaust pipe retaining nuts with penetrating oil to loosen them.**

- Engine room cover(RH and LH)
- Engine cover
- Air cleaner assembly and duct
- Remove exhaust front tube and three way catalysts (right bank and left bank).
- Disconnect air fuel ratio sensor 1 (right bank and left bank) harness connectors and remove harness clip.
- Protective covers from the manifolds
- Manifold from the engine and discard the gaskets

To install:

3. Clean all gasket mounting surfaces.
4. Install or connect the following:
- Exhaust manifold with new gaskets. Torque the bolts, in sequence, in 2 steps to 22 ft. lbs.(30.5Nm).
- Protective shields. Torque the bolts to 51 inch lbs. (5.8 Nm).
- Air fuel ratio sensor 1 (right bank

and left bank) to the manifold, as necessary 37 ft. lbs (50 Nm).
- Exhaust manifolds to the exhaust pipes. Torque the bolts/nuts to 46 ft. lbs. (63 Nm).
- Air cleaner assembly and engine cover
- Engine room cover(RH and LH)
- Negative battery cable

5. Start the engine, check for exhaust leaks and repair if necessary.

4.5L Engine

See Figures 109 and 110.

1. Before servicing the vehicle, refer to the service precautions.
2. Remove or disconnect the following:
- Negative battery cable

➡**If necessary, soak the exhaust pipe retaining nuts with penetrating oil to loosen them.**

- Engine room cover(RH and LH)
- Engine cover
- Air cleaner assembly and duct
- Remove front and rear engine undercovers with power tool.
- Engine coolant from radiator
- Radiator

- Remove exhaust front tube
- Disconnect air fuel ratio sensor 1 (right bank and left bank) harness connectors and remove harness clip
- Exhaust manifold and three way catalyst (left bank)
- A/C piping from A/C compressor
- Steering lower joint to enable steering shaft to move freely
- Starter motor
- Mounting insulator (LH)

3. Raise left side of engine approximately 1.18 in (3 cm).
4. Remove exhaust manifold cover (left bank).
5. Loosen nuts in the reverse order of figure to remove exhaust manifold and three way catalyst (left bank) with power tool.
6. Disregard No. 9 to No. 12 when loosening.
7. Remove exhaust manifold and three way catalyst (right bank) as follows:
8. Remove alternator and bracket.
9. Remove nuts on bottom of engine mounting insulator (RH), and lift up right side of engine approximately 1.18 in (3 cm) with transmission jack.
10. Remove exhaust manifold cover (right bank).

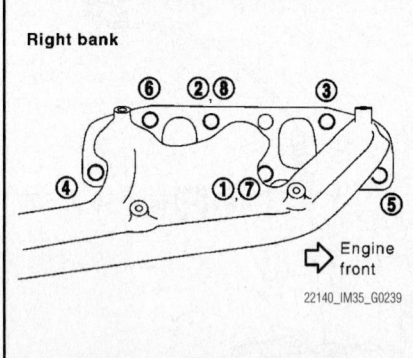

Fig. 107 View of the exhaust manifold right bank torque sequence

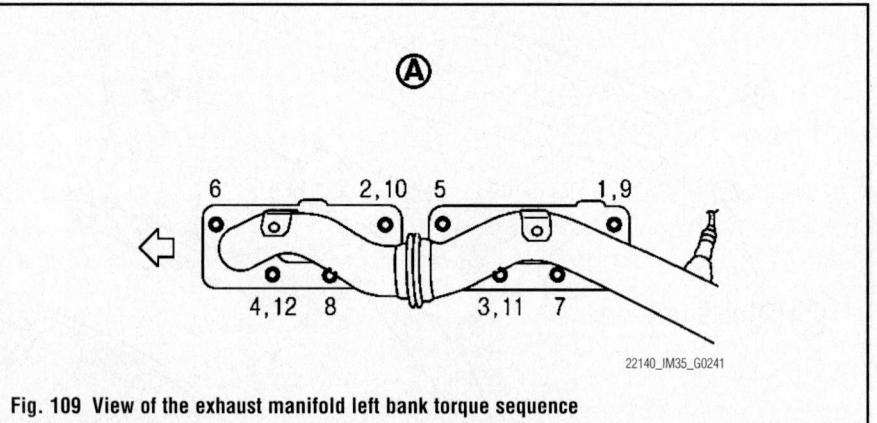

Fig. 109 View of the exhaust manifold left bank torque sequence

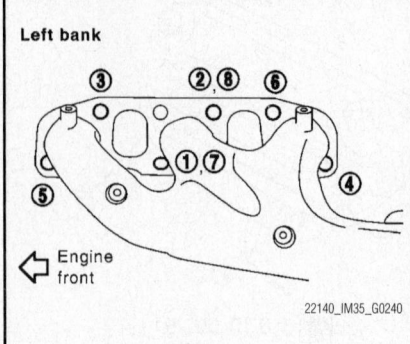

Fig. 108 View of the exhaust manifold left bank torque sequence

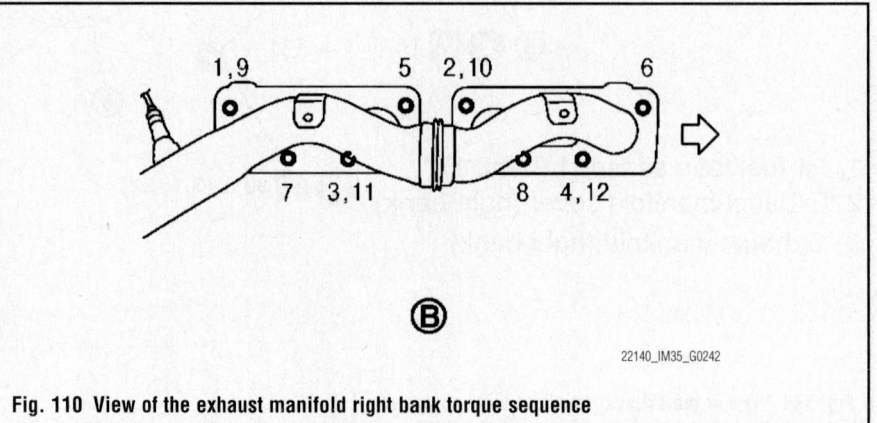

Fig. 110 View of the exhaust manifold right bank torque sequence

11. Loosen nuts in the reverse order of figure to remove exhaust manifold and three way catalyst (right bank) with power tool.

To install:

12. Clean all gasket mounting surfaces.

13. Install or connect the following:
- Right exhaust manifold with new gaskets. Torque the bolts, in sequence, to 21 ft. lbs.(28 Nm)
- Protective shield. Torque the bolts to 51 inch lbs. (5.8 Nm).
- Mounting insulator (RH)
- Alternator and bracket
- Mounting insulator (LH)
- Starter motor
- Steering lower joint
- A/C piping from A/C compressor
- Exhaust manifold and three way catalyst (left bank) Torque the bolts, in sequence, to 21 ft. lbs.(28 Nm)
- Protective shield. Torque the bolts to 51 inch lbs. (5.8 Nm).
- Air fuel ratio sensor 1 (right bank and left bank) to the manifold, as necessary 37 ft. lbs (50 Nm)

❊❊ WARNING

Before installing a new air fuel ratio sensor 1, clean exhaust system threads using oxygen sensor thread cleaner

(commercial service tool: J-43897-18 or J-43897-12), and apply anti-seize lubricant (commercial service tool).

➡ **Never over torque air fuel ratio sensor. Doing so may cause damage to the air fuel ratio sensor 1, resulting in "MIL" coming on.**

- Exhaust manifolds to the exhaust pipes. Torque the bolts/nuts to 39 ft. lbs. (52.5 Nm).
- Radiator
- Air cleaner assembly and engine cover Torque the bolts/nuts to 51 inch lbs. (5.8 Nm).
- Engine coolant
- Engine room cover(RH and LH)
- Negative battery cable

14. Start the engine, check for exhaust leaks and repair if necessary.

INTAKE MANIFOLD

REMOVAL & INSTALLATION

3.5L Engine

See Figures 111 and 112.

1. Before servicing the vehicle, refer to the service precaution.

2. Disconnect the negative battery cable.

3. Release fuel pressure.

4. Remove intake manifold collectors (upper and lower).

5. Remove fuel tube and fuel injector assembly.

6. Loosen mounting bolts and nuts in reverse order as shown in the figure to remove intake manifold with power tool.

7. Remove gaskets.

To install:

8. If stud bolts were removed, install them and tighten to8 ft. lbs. (10.8 Nm).

9. Tighten all mounting bolts and nuts to the specified torque in two or more steps in numerical order shown in the figure.

10. Step 1: Tighten to 5 ft. lbs. (7.4 Nm).

11. Step 2: Tighten to 21 ft. lbs. (29 Nm).

12. Connect the negative battery cable.

4.5L Engine

See Figures 113 through 117.

1. Before servicing the vehicle, refer to the service precaution.

2. Disconnect the negative battery cable.

3. Remove engine room cover (RH and LH).

4. Remove engine cover with power tool.

5. Release fuel pressure.

6. Remove air duct (inlet), air cleaner case and air duct and resonator assembly.

7. Drain engine coolant from radiator.

➡ **Drain engine coolant from radiator when the engine is cold.**

➡ **Never spill engine coolant on drive belts**

8. Remove intake manifold collectors (upper and lower).

9. Remove fuel tube and fuel injector assembly.

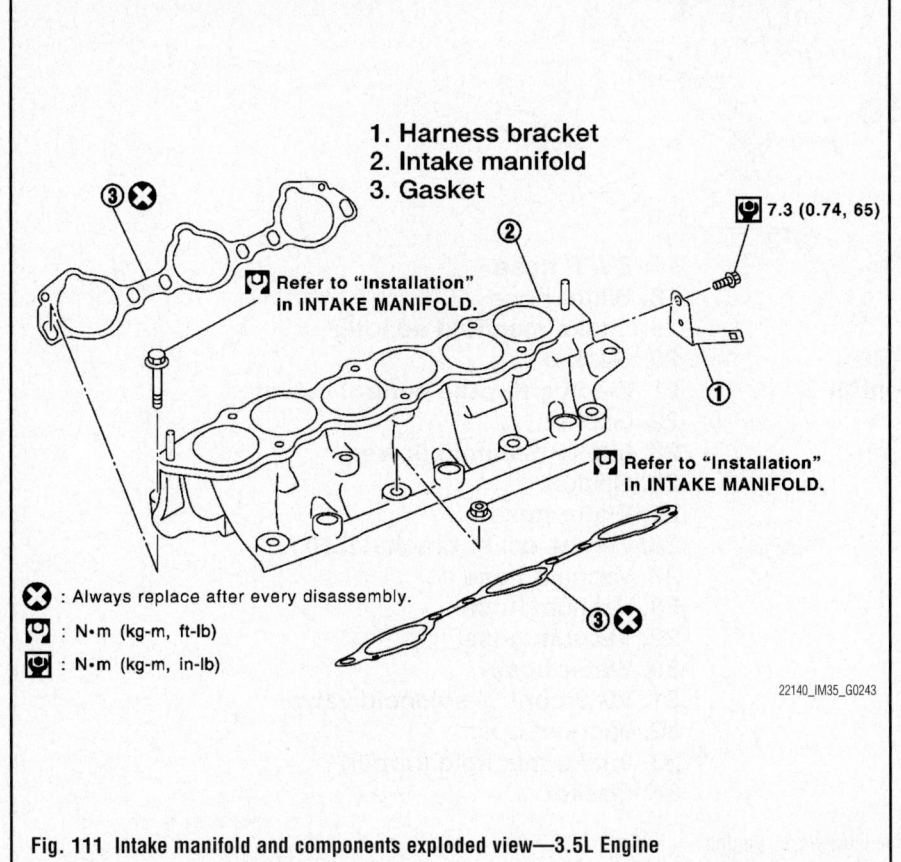

1. Harness bracket
2. Intake manifold
3. Gasket

7.3 (0.74, 65)

Refer to "Installation" in INTAKE MANIFOLD.

Refer to "Installation" in INTAKE MANIFOLD.

❌ : Always replace after every disassembly.

⬚ : N·m (kg-m, ft-lb)

⬚ : N·m (kg-m, in-lb)

22140_IM35_G0243

Fig. 111 Intake manifold and components exploded view—3.5L Engine

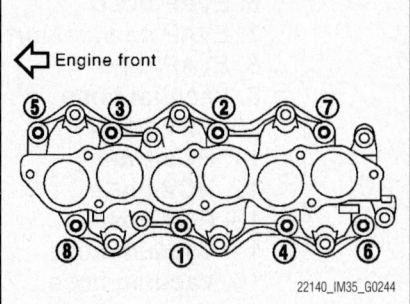

Engine front

22140_IM35_G0244

Fig. 112 Intake manifold and components exploded view—3.5L Engine

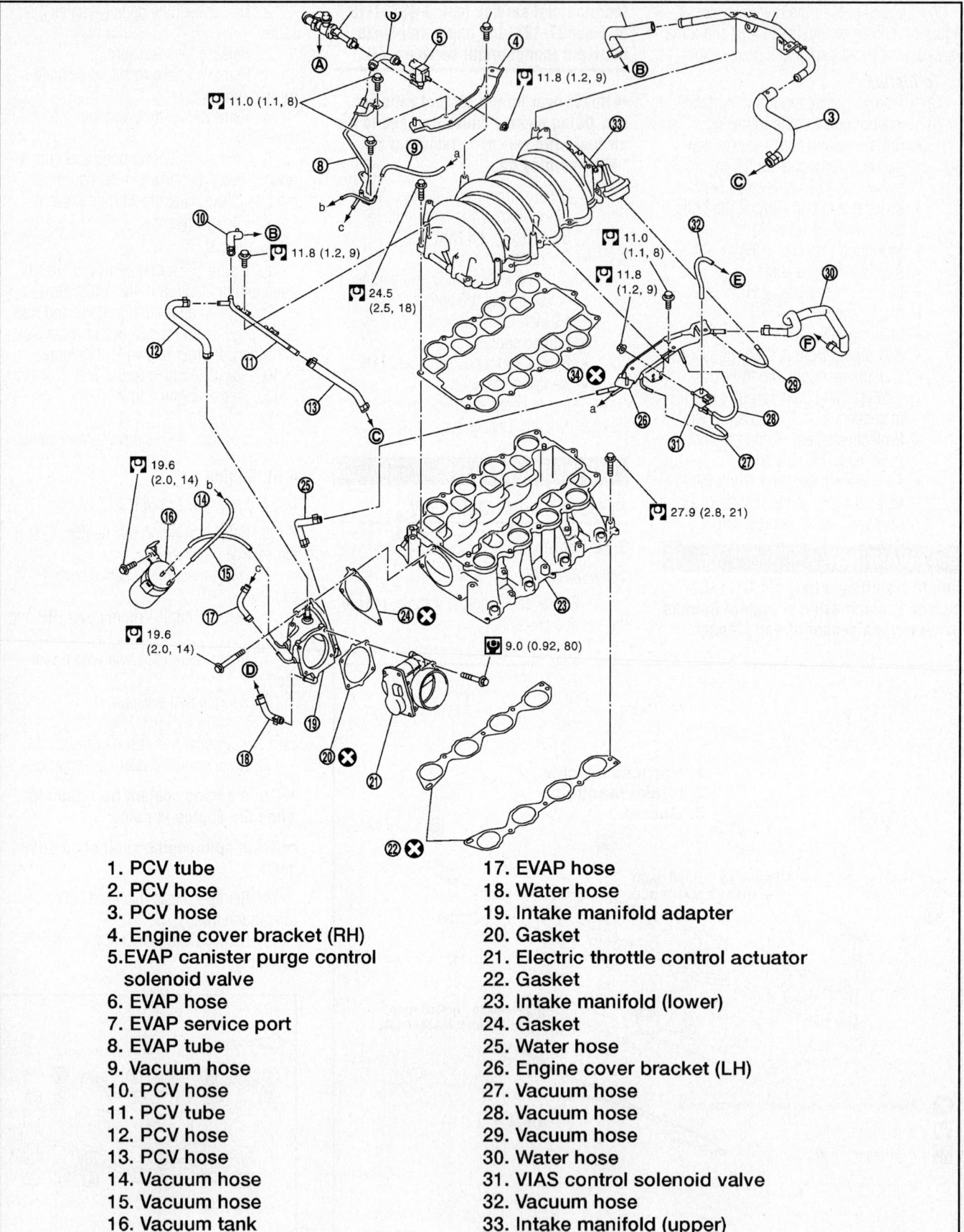

Fig. 113 Intake manifold and components - exploded view—4.5L Engine

1. PCV tube
2. PCV hose
3. PCV hose
4. Engine cover bracket (RH)
5. EVAP canister purge control solenoid valve
6. EVAP hose
7. EVAP service port
8. EVAP tube
9. Vacuum hose
10. PCV hose
11. PCV tube
12. PCV hose
13. PCV hose
14. Vacuum hose
15. Vacuum hose
16. Vacuum tank
17. EVAP hose
18. Water hose
19. Intake manifold adapter
20. Gasket
21. Electric throttle control actuator
22. Gasket
23. Intake manifold (lower)
24. Gasket
25. Water hose
26. Engine cover bracket (LH)
27. Vacuum hose
28. Vacuum hose
29. Vacuum hose
30. Water hose
31. VIAS control solenoid valve
32. Vacuum hose
33. Intake manifold (upper)
34. Gasket

22140_IM35_G0245

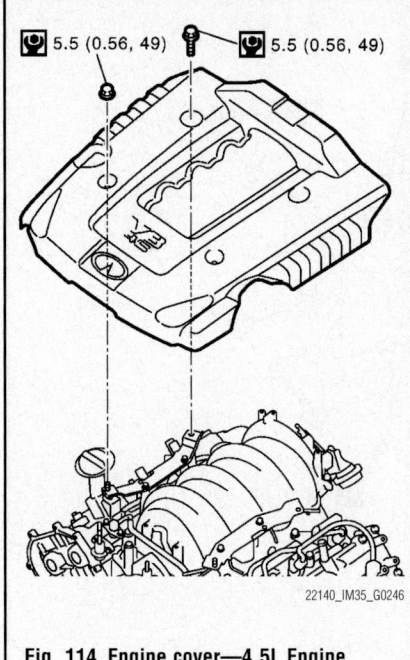

Fig. 114 Engine cover—4.5L Engine

Fig. 115 Fuel feed hose (1) quick connector—4.5L Engine

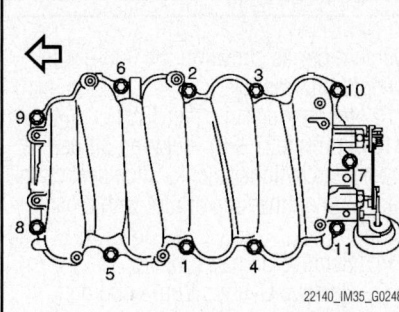

Fig. 116 Upper intake manifold torque sequence—4.5L Engine

10. Disconnect fuel feed hose (1) quick connector on engine side.

11. Remove fuel damper and fuel hose assembly

12. Remove or disconnect harnesses, engine cover bracket (RH and LH), vacuum

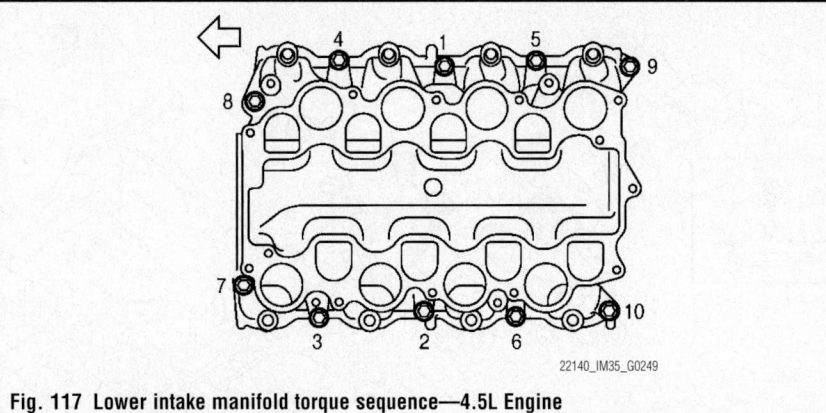

Fig. 117 Lower intake manifold torque sequence—4.5L Engine

hose, EVAP tube and hose and PCV hose and tube from intake manifold (upper).

13. Loosen mounting bolts in reverse order as shown in the figure to remove intake manifold (upper) with power tool.

14. Remove electric throttle control actuator as follows:

15. Disconnect harness connector.

16. Loosen mounting bolts diagonally.

➡**Handle carefully to avoid any shock to electric throttle control actuator, never disassemble.**

17. Remove fuel injector and fuel tube assembly.

18. Disconnect water hoses from intake manifold adaptor.

19. Loosen mounting bolts in reverse order as shown in the figure to remove intake manifold (lower) with power tool.

20. Remove intake manifold adaptor from intake manifold (lower).

21. Remove vacuum tank.

22. Remove intake manifold gaskets.

To install:

23. Install the lower intake manifold and new gaskets.

24. There are two types of mounting bolts. Refer to the following for locating bolts.

- M8 × 90 mm (3.54 in) : 7, 8
- M8 × 35 mm (1.38 in) : Except the above

25. Install the intake manifold adaptor from intake manifold (lower).

26. Tighten the lower intake manifold to 21 ft. lbs. (27.9 Nm).

27. Install the vacuum tank.

28. Install the water hoses to intake manifold adaptor.

29. Install the fuel injector and fuel tube assembly.

30. Install the electric throttle control.

31. Install the intake manifold (upper).

32. Install the harnesses, engine cover bracket (RH and LH), vacuum hose, EVAP

tube and hose and PCV hose and tube to the intake manifold (upper).

33. Install the fuel damper and fuel hose assembly.

34. Install the fuel feed hose (1) quick connector on engine side.

35. Install the fuel tube and fuel injector assembly.

36. Install the intake manifold collectors (upper and lower).

37. Install the air duct (inlet), air cleaner case and air duct and resonator assembly.

38. Fill coolant and run engine to bleed cooling system.

39. Connect the negative battery cable.

40. Check for leaks.

OIL PAN

REMOVAL & INSTALLATION

3.5L Engine

2WD Models

See Figures 118 through 127.

1. Before servicing the vehicle, refer to the service precaution.

2. Disconnect the negative battery cable.

3. Drain engine oil and engine coolant.

➡**Perform this step when the engine is cold, never spill engine oil on drive belts.**

4. Remove or disconnect the following:

- Front and rear engine undercover
- Front tire
- Hood assembly
- Engine room cover (RH and LH).
- Engine cover
- Air duct (inlet).
- Stabilizer clamp, lower stabilizer

5. Install engine slinger to sling engine assembly for positioning.

6. Remove or disconnect the following:

- Front suspension member

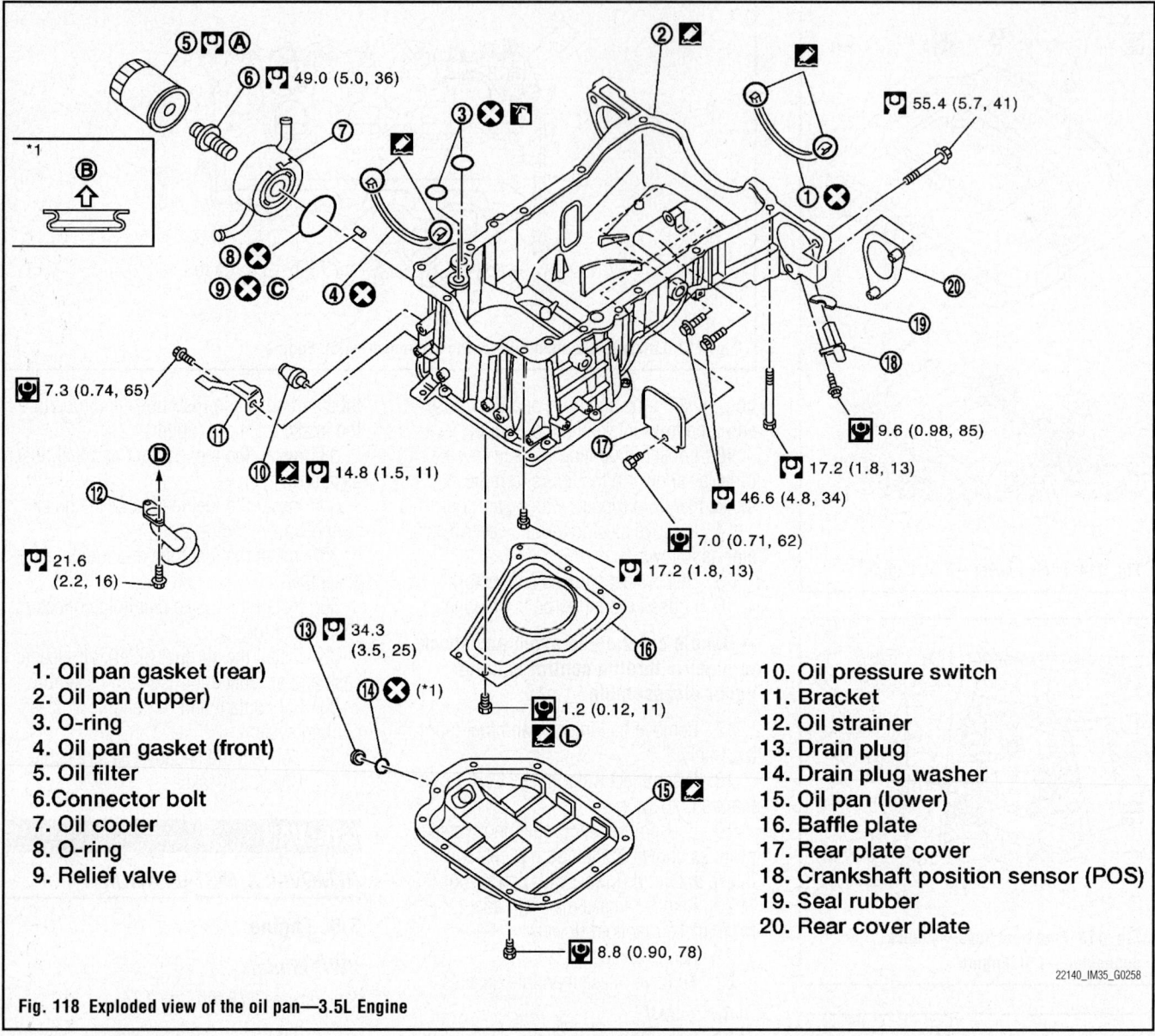

Fig. 118 Exploded view of the oil pan—3.5L Engine

1. Oil pan gasket (rear)
2. Oil pan (upper)
3. O-ring
4. Oil pan gasket (front)
5. Oil filter
6. Connector bolt
7. Oil cooler
8. O-ring
9. Relief valve
10. Oil pressure switch
11. Bracket
12. Oil strainer
13. Drain plug
14. Drain plug washer
15. Oil pan (lower)
16. Baffle plate
17. Rear plate cover
18. Crankshaft position sensor (POS)
19. Seal rubber
20. Rear cover plate

22140_IM35_G0258

• Drive belts.
• Alternator stay
• Starter motor
• Idler pulley and bracket assembly
• Oil cooler water hoses, and remove oil cooler water pipe mounting bolt
• A/T fluid cooler hoses, and remove A/T fluid cooler tube
• Crankshaft position sensor (POS)

➡ **Handle carefully to avoid dropping and shocks, never disassemble, never allow metal powder to adhere to magnetic part at sensor tip, and never place sensors in a location where they are exposed to magnetism.**

• Oil filter
• Oil cooler
7. Remove oil pan (lower) as follows:

a. Loosen mounting bolts in reverse order as shown in the figure to remove.

b. Insert the seal cutter (SST) between oil pan (upper) and oil pan (lower).

➡ **Be careful not to damage the mating surfaces, never insert a screwdriver, this will damage the mating surfaces.**

c. Slide the seal cutter by tapping on the side of tool with a hammer. Remove oil pan (lower).

8. Remove or disconnect the following:
• Baffle plate
• Oil strainer
• Transmission joint bolts, which pierce oil pan (upper)
• Rear cover plate
• Oil pan (upper) transmission joint bolts

9. Loosen mounting bolts in the

reverse order as shown in the figure with power tool to remove.

10. Insert the seal cutter [SST: KV10111100 (J37228)] between oil pan (upper) and cylinder block. Slide seal cutter by tapping on the side of tool with a hammer.

11. Remove oil pan (upper).

12. Remove O-rings from bottom of cylinder block and oil pump

13. Remove oil pan gaskets.

To install:

14. Install oil pan (upper) as follows:

a. Install new oil pan gaskets, apply liquid gasket to oil pan gaskets as shown in the figure.

➡ **Use Genuine RTV Silicone Sealant or equivalent.**

b. To install, align protrusion of oil

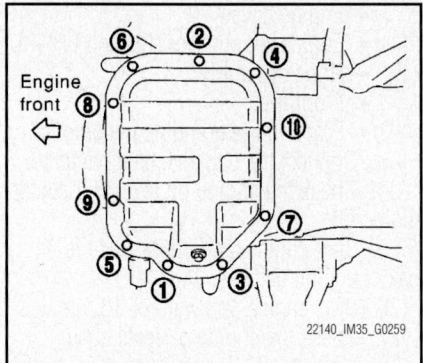

Fig. 119 Oil pan lower bolt sequence 3.5L Engine

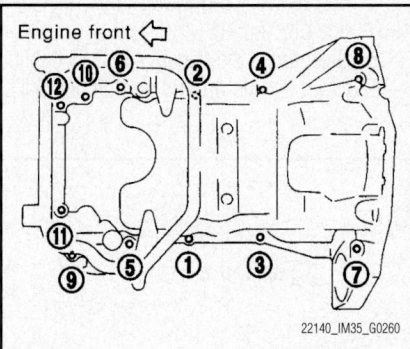

Fig. 120 Oil pan upper bolt sequence 3.5L Engine

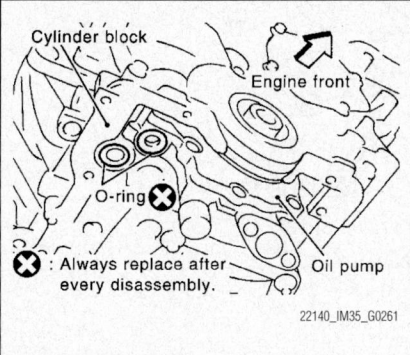

X : Always replace after every disassembly.

Fig. 121 Oil pump and cylinder block O-rings 3.5L Engine

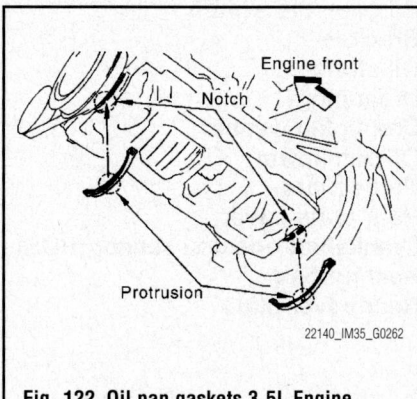

Fig. 122 Oil pan gaskets 3.5L Engine

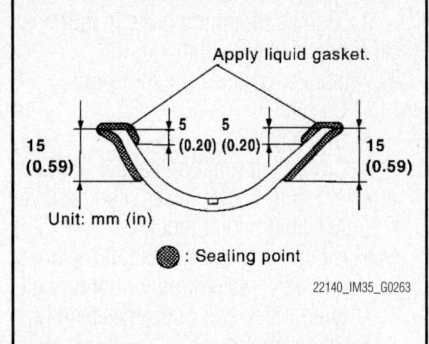

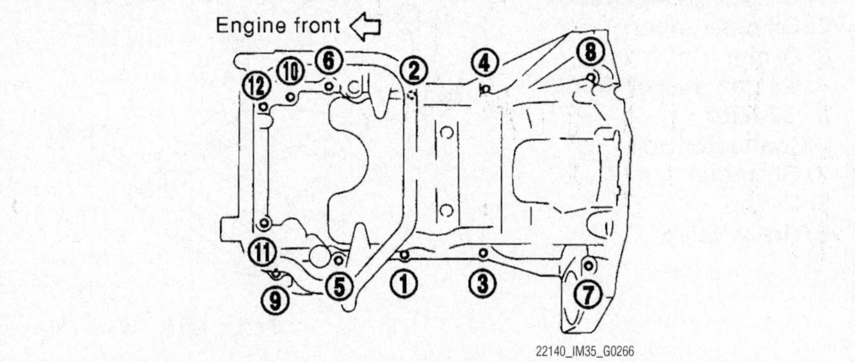

Apply liquid gasket.

Unit: mm (in)

● : Sealing point

Fig. 123 Oil pan gaskets areas to seal with RTV silicone 3.5L Engine

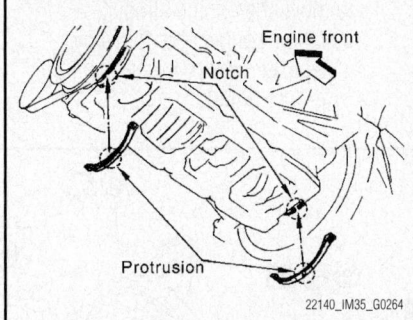

Fig. 124 Alignment of oil pan gasket with notches of front timing chain case and rear oil seal retainer 3.5L Engine

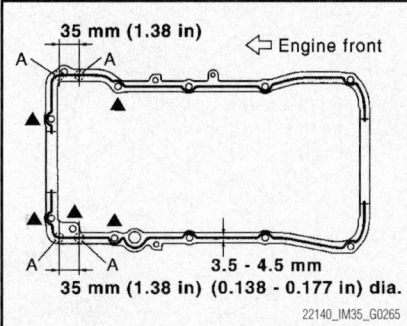

Fig. 125 Cylinder block mating surface of the upper oil pan - silicone application area 3.5L Engine

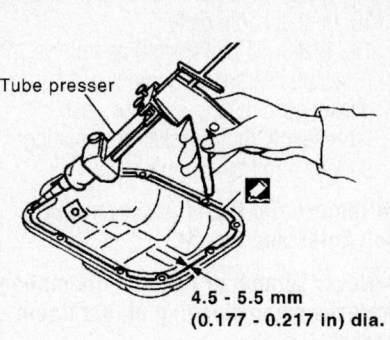

Tube presser

4.5 - 5.5 mm
(0.177 - 0.217 in) dia.

⬚ : Apply Genuine RTV silicone sealant or equivalent. Refer to GI section.

Fig. 127 Lower oil pan silicone application area 3.5L Engine

pan gasket with notches of front timing chain case and rear oil seal retainer.

c. Install oil pan gasket with smaller arc to front timing chain case side.

d. Install new O-rings on the bottom of cylinder block and oil pump.

e. Apply a continuous bead of RTV Silicone to the cylinder block mating surface of oil pan (upper) to a limited portion as shown in the figure.

➡ **Apply a bead of 0.177 to 0.217 inches (4.5 to 5.5 mm) in diameter to area "A," attaching should be done within 5 minutes after coating.**

f. Tighten mounting bolts in numerical order as shown in the figure.

15. There are two types of mounting bolts. Refer to the following for locating bolts.

- 3.94 in (M8 × 100 mm) bolt locations 5, 7, 8, 11
- 0.98 in (M8 × 25 mm) : Except the above

16. Tighten transmission joint bolts.

Fig. 126 Upper oil pan tightening sequence 3.5L Engine

17. Install oil strainer to oil pump.
18. Install baffle plate
19. Install oil pan (lower) as follows:

a. Use scraper to remove old liquid gasket from mating surfaces, also remove old liquid gasket from mating surface of oil pan (upper).

➡**Remove old liquid gasket from the bolt holes and thread.**

➡**Never scratch or damage the mating surfaces when cleaning off old liquid gasket.**

b. Apply a continuous bead of RTV silicone to the oil pan (lower) as shown in the figure.

➡**Attaching should be done within 5 minutes after coating.**

c. Install oil pan (lower).

d. Tighten mounting bolts in numerical order as shown in the figure.
20. Install or connect the following:
- Oil filter
- Oil cooler
- Crankshaft position sensor (POS)
- A/T fluid cooler hoses, and remove A/T fluid cooler tube
- Oil cooler water hoses, and remove oil cooler water pipe mounting bolt
- Idler pulley and bracket assembly
- Starter motor
- Alternator stay
- Drive belts.
- Front suspension member
21. Remove the engine slinger.
22. Install or connect the following:
- Stabilizer clamp, lower stabilizer
- Air duct (inlet).

- Engine cover
- Engine room cover (RH and LH).
- Hood assembly
- Front tire
- Front and rear engine undercover
23. Connect the negative battery cable.
24. Check the engine oil level and adjust engine oil.
25. Start engine, and check there is no leak of engine oil.
26. Stop engine and wait for 10 minutes.
27. Check the engine oil level again

AWD Models

See Figures 128 through 138.

1. Before servicing the vehicle, refer to the service precaution.
2. Disconnect the negative battery cable.
3. Drain engine oil and engine coolant.

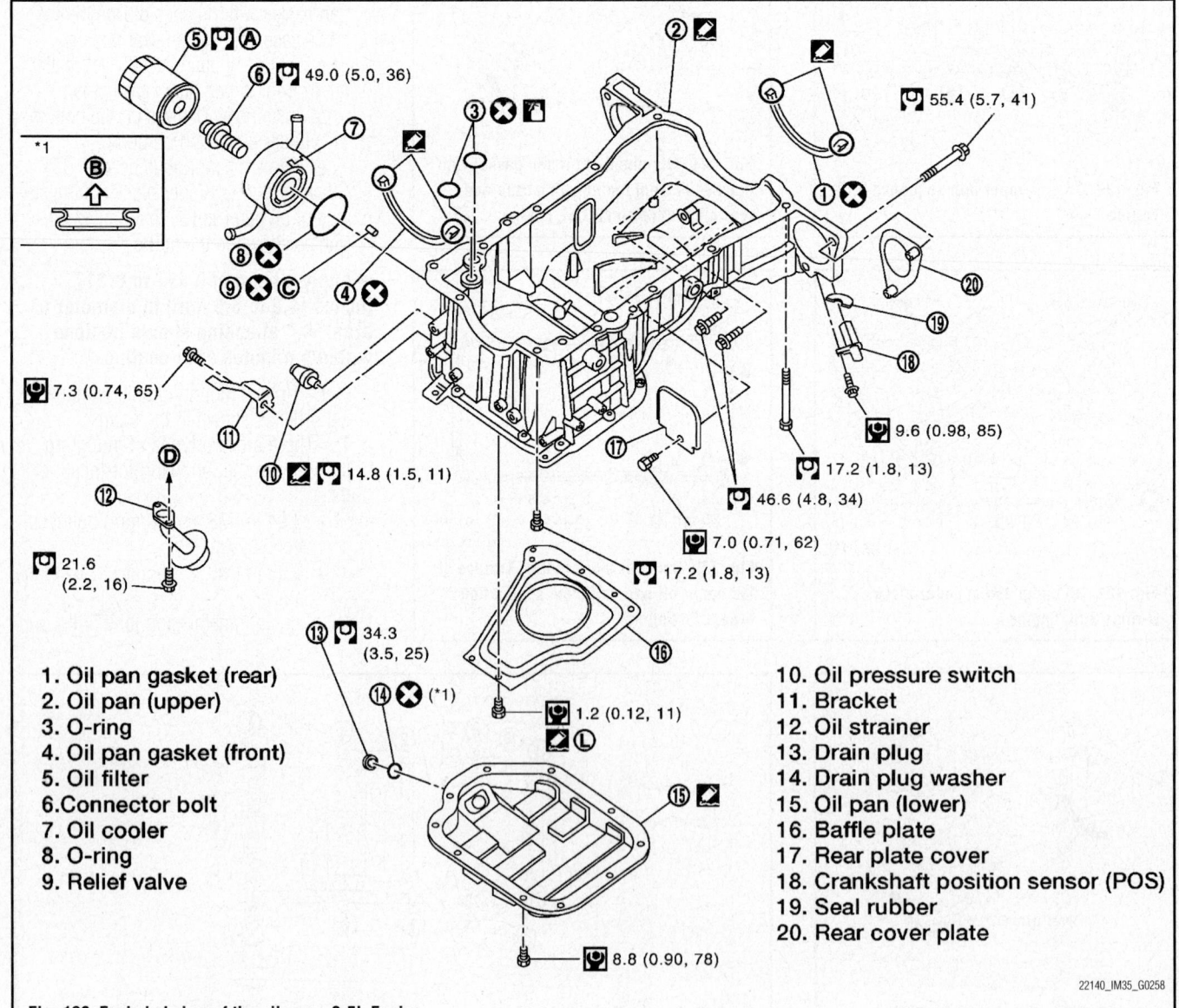

1. Oil pan gasket (rear)
2. Oil pan (upper)
3. O-ring
4. Oil pan gasket (front)
5. Oil filter
6. Connector bolt
7. Oil cooler
8. O-ring
9. Relief valve
10. Oil pressure switch
11. Bracket
12. Oil strainer
13. Drain plug
14. Drain plug washer
15. Oil pan (lower)
16. Baffle plate
17. Rear plate cover
18. Crankshaft position sensor (POS)
19. Seal rubber
20. Rear cover plate

Fig. 128 Exploded view of the oil pan—3.5L Engine

22140_IM35_G0258

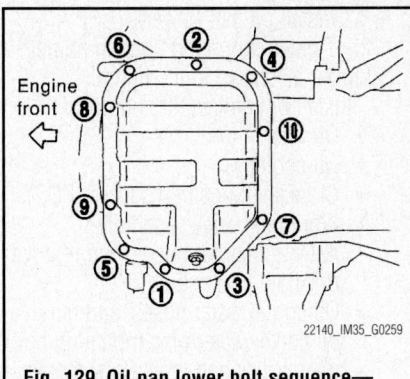

Fig. 129 Oil pan lower bolt sequence—3.5L Engine

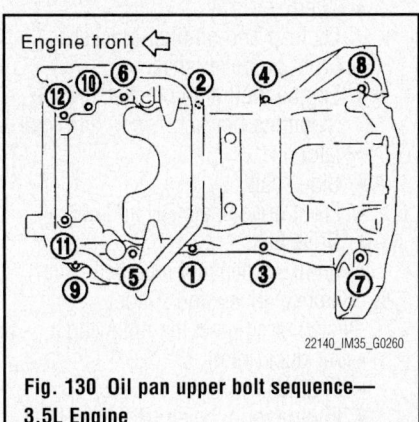

Fig. 130 Oil pan upper bolt sequence—3.5L Engine

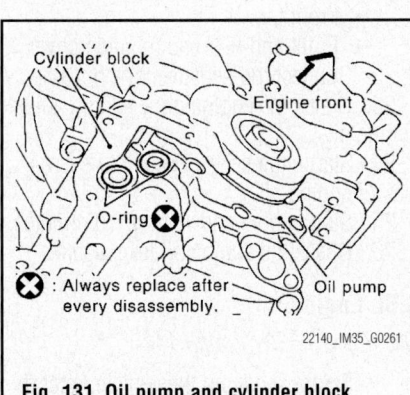

Fig. 131 Oil pump and cylinder block O-rings—3.5L Engine

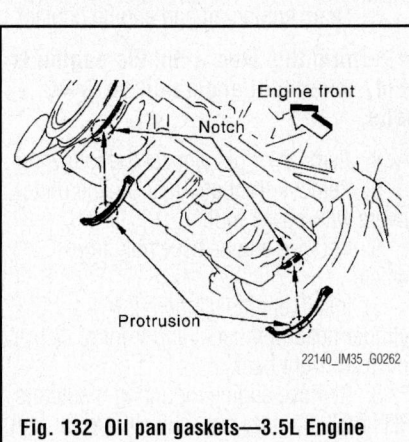

Fig. 132 Oil pan gaskets—3.5L Engine

➡**Perform this step when the engine is cold, never spill engine oil on drive belts.**

4. Remove or disconnect the following:
   • Front and rear engine undercover
   • Front tire
   • Hood assembly
   • Engine room cover (RH and LH).
   • Engine cover
   • Air duct (inlet).
5. Install engine slinger to sling engine assembly for positioning.
6. Remove or disconnect the following:
   • Front suspension member
   • Drive belts.
   • Front drive shaft (RH and LH).
   • Side shaft
   • Engine mounting bracket, engine mounting bracket (lower) and insulator
   • Front propeller shaft
   • Oil filter and oil filter bracket
   • Alternator stay
   • Idler pulley and bracket
   • Starter motor
   • Idler pulley and bracket assembly
   • Oil cooler water hoses, and remove oil cooler water pipe mounting bolt
   • A/T fluid cooler hoses, and remove A/T fluid cooler tube
   • Front final drive assembly.
   • Crankshaft position sensor (POS)

➡**Handle carefully to avoid dropping and shocks, never disassemble, never allow metal powder to adhere to magnetic part at sensor tip, and never place sensors in a location where they are exposed to magnetism.**

7. Remove oil pan (lower) as follows:

a. Loosen mounting bolts in reverse order as shown in the figure to Remove
b. Insert the seal cutter (SST) between oil pan (upper) and oil pan (lower).

➡**Be careful not to damage the mating surfaces, never insert a screwdriver, this will damage the mating surfaces.**

c. Slide the seal cutter by tapping on the side of tool with a hammer. Remove oil pan (lower).
8. Remove or disconnect the following:
   • Baffle plate
   • Oil strainer
   • Transmission joint bolts, which pierce oil pan (upper)
   • Rear cover plate
   • Oil pan (upper) transmission joint bolts
9. Loosen mounting bolts in the reverse order as shown in the figure with power tool to remove.
10. Insert the seal cutter [SST: KV10111100 (J37228)] between oil pan (upper) and cylinder block. Slide seal cutter by tapping on the side of tool with a hammer. Remove oil pan (upper).
11. Remove O-rings from bottom of cylinder block and oil pump
12. Remove oil pan gaskets.

To install:
13. Install oil pan (upper) as follows:
   a. Install new oil pan gaskets, apply liquid gasket to oil pan gaskets as shown in the figure.

➡**Use Genuine RTV Silicone Sealant or equivalent.**

   b. To install, align protrusion of oil pan gasket with notches of front timing chain case and rear oil seal retainer.

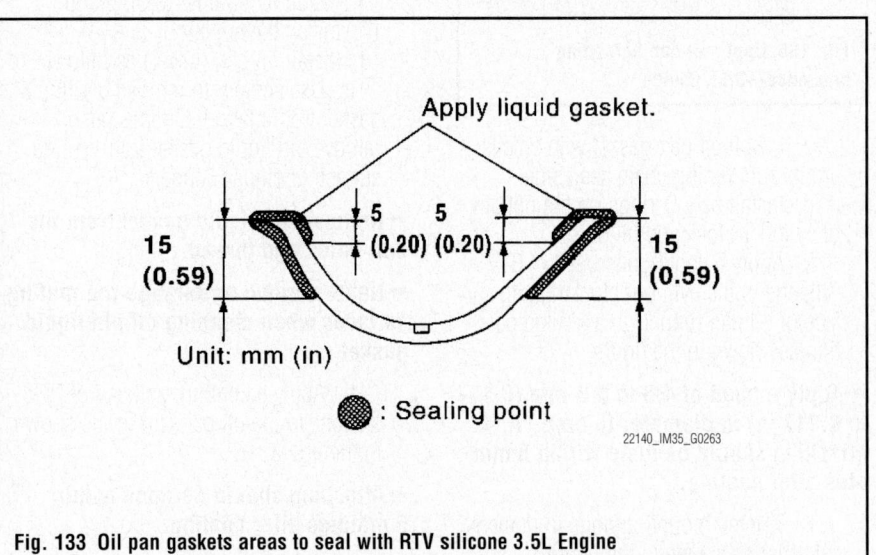

Fig. 133 Oil pan gaskets areas to seal with RTV silicone 3.5L Engine

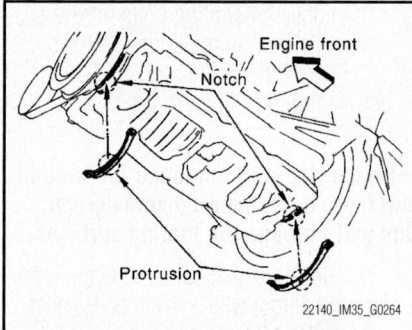

Fig. 134 Alignment of oil pan gasket with notches of front timing chain case and rear oil seal retainer—3.5L Engine

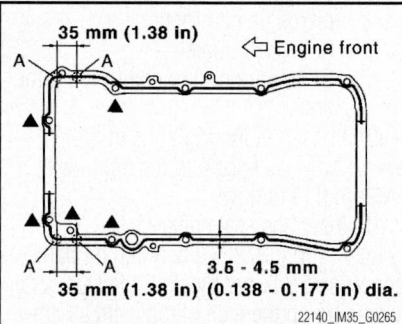

Fig. 135 Cylinder block mating surface of the upper oil pan - silicone application area—3.5L Engine

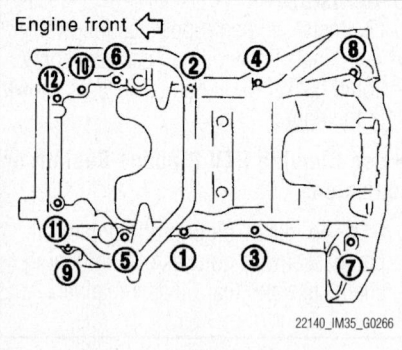

Fig. 136 Upper oil pan tightening sequence—3.5L Engine

c. Install oil pan gasket with smaller arc to front timing chain case side.

d. Install new O-rings on the bottom of cylinder block and oil pump.

e. Apply a continuous bead of RTV Silicone to the cylinder block mating surface of oil pan (upper) to a limited portion as shown in the figure.

➡ **Apply a bead of 4.5 to 5.5 mm (0.177 to 0.217 in) in diameter to area "A," attaching should be done within 5 minutes after coating.**

f. Tighten mounting bolts in numerical order as shown in the figure.

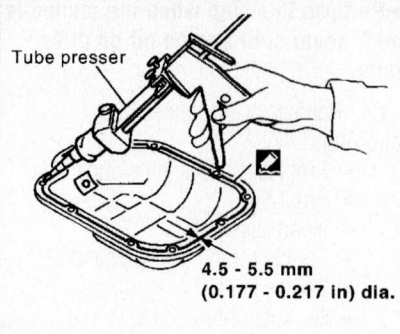

Fig. 137 Lower oil pan - silicone application area—3.5L Engine

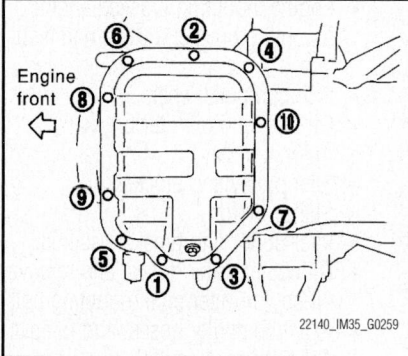

Fig. 138 Oil pan lower bolt sequence—3.5L Engine

14. There are two types of mounting bolts. Refer to the following for locating bolts.

- 3.94 in (M8 × 100 mm) bolt locations 5, 7, 8, 11
- 0.98 in (M8 × 25 mm) : Except the above

15. Tighten transmission joint bolts.
16. Install oil strainer to oil pump.
17. Install baffle plate.
18. Install oil pan (lower) as follows:

a. Use scraper to remove old liquid gasket from mating surfaces, also remove old liquid gasket from mating surface of oil pan (upper).

➡ **Remove old liquid gasket from the bolt holes and thread.**

➡ **Never scratch or damage the mating surfaces when cleaning off old liquid gasket.**

b. Apply a continuous bead of RTV silicone to the oil pan (lower) as shown in the figure.

➡ **Attaching should be done within 5 minutes after coating.**

c. Install oil pan (lower).

d. Tighten mounting bolts in numerical order as shown in the figure.

19. Install or connect the following:
- Oil filter
- Oil cooler
- Crankshaft position sensor (POS)
- Front final drive assembly.
- A/T fluid cooler hoses, and remove A/T fluid cooler tube
- Oil cooler water hoses, and remove oil cooler water pipe mounting bolt
- Idler pulley and bracket assembly
- Starter motor
- idler pulley and bracket
- Alternator stay
- Oil filter and oil filter bracket
- Front propeller shaft
- Engine mounting bracket, engine mounting bracket (lower) and insulator
- Side shaft
- Front drive shaft (RH and LH).
- Drive belts.
- Front suspension member

20. Remove the engine slinger.
21. Install or connect the following:
- Air duct (inlet).
- Engine cover
- Engine room cover (RH and LH).
- Hood assembly
- Front tire
- Front and rear engine undercover.

22. Connect the negative battery cable.
23. Check the engine oil level and adjust engine oil.
24. Start engine, and check there is no leak of engine oil.
25. Stop engine and wait for 10 minutes.
26. Check the engine oil level again.

4.5L Engine

See Figures 139 through 142.

1. Before servicing the vehicle, refer to the service precaution.
2. Disconnect the negative battery cable.
3. Drain engine oil and engine coolant

➡ **Perform this step when the engine is cold, never spill engine oil on drive belts.**

4. Remove or disconnect the following.
5. Remove front and rear engine undercovers with power tool.
6. Remove engine assembly from vehicle.
7. Install engine slingers into front of cylinder head (left bank) and front of cylinder head (right bank).
8. Remove engine mounting insulators (RH and LH) under side nut with power tool.

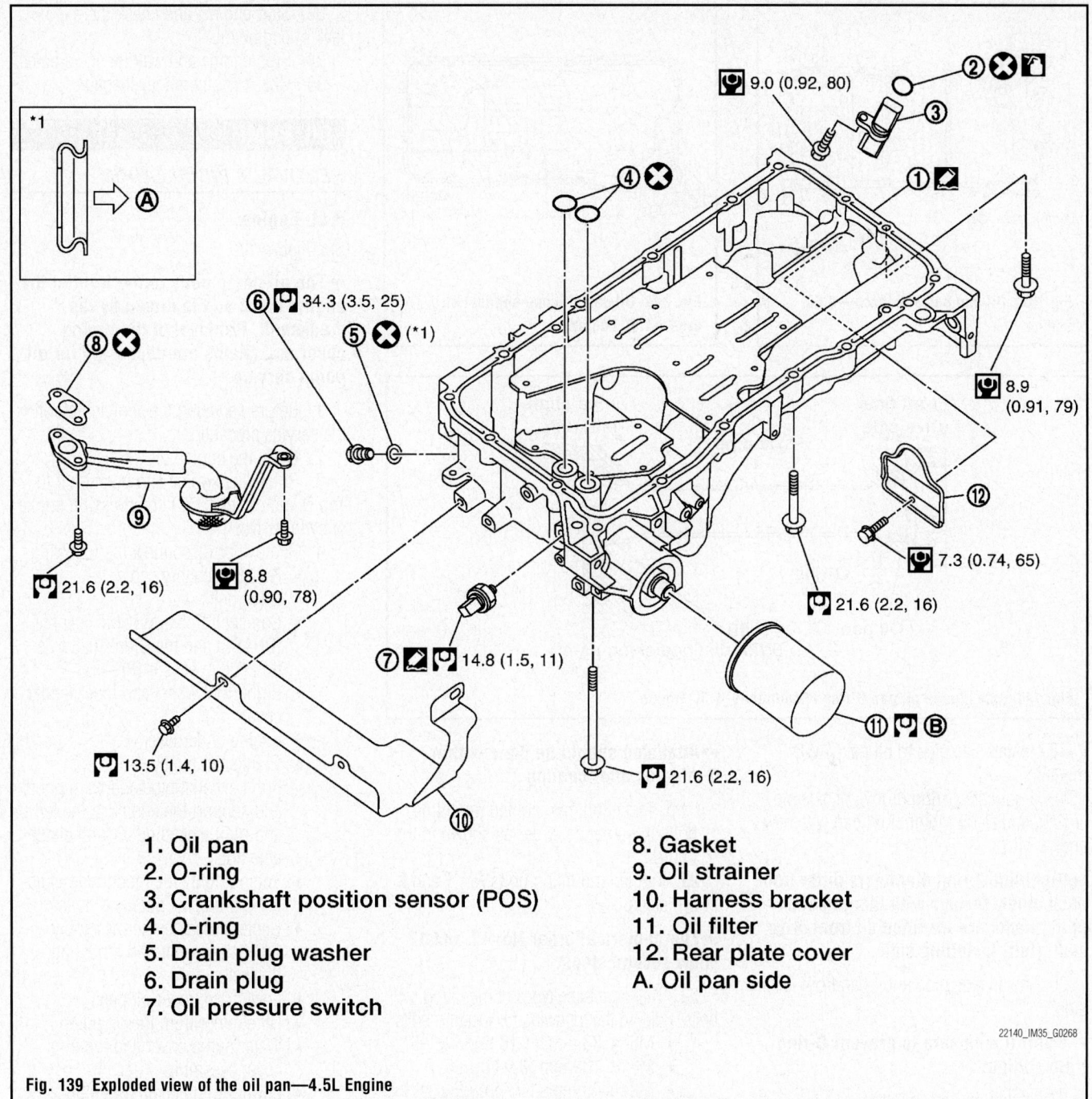

1. Oil pan
2. O-ring
3. Crankshaft position sensor (POS)
4. O-ring
5. Drain plug washer
6. Drain plug
7. Oil pressure switch
8. Gasket
9. Oil strainer
10. Harness bracket
11. Oil filter
12. Rear plate cover
A. Oil pan side

22140_IM35_G0268

Fig. 139 Exploded view of the oil pan—4.5L Engine

9. Lift with hoist and separate engine and transmission assembly from front suspension member.

➡**Avoid damage to and oil/grease smearing or spills onto engine mounting insulator.**

10. Remove harness bracket from oil pan. (2WD models).
11. Remove oil filter.
12. Remove oil pan as the follows:
   a. Remove rear plate cover.
   b. Remove transmission joint bolts which pierce oil pan.

   c. Loosen mounting bolts with power tool in reverse order as shown in the figure.

➡**Disregard the numerical order No. 11 and 17 in removal.**

   d. Insert seal cutter (SST) between oil pan and cylinder block. Slide seal cutter by tapping on the side of seal cutter with hammer.
   e. Remove oil pan.

✳✳ WARNING

Be careful not to damage the mating surfaces, never insert screw-

driver, this will damage the mating surface.

   f. Remove O-rings from bottom of oil pump and front cover.
13. Remove oil pressure switch.
14. As necessary, pull axle pipe from oil pan. (AWD models).
15. Hold the pipes and pull them out to front drive shaft (left) installing side.
16. Remove oil strainer.

To install:
17. Install the oil strainer.

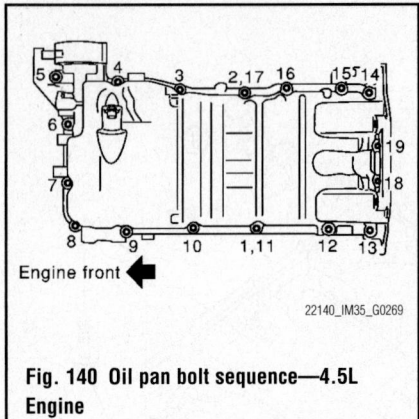

Fig. 140 Oil pan bolt sequence—4.5L Engine

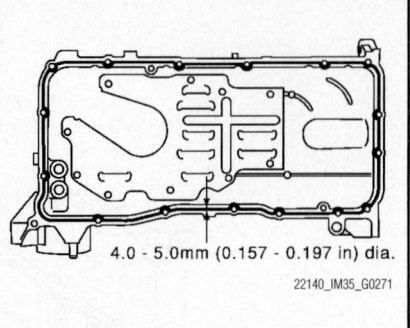

4.0 - 5.0mm (0.157 - 0.197 in) dia.

Fig. 142 Oil pan silicone application area—4.5L Engine

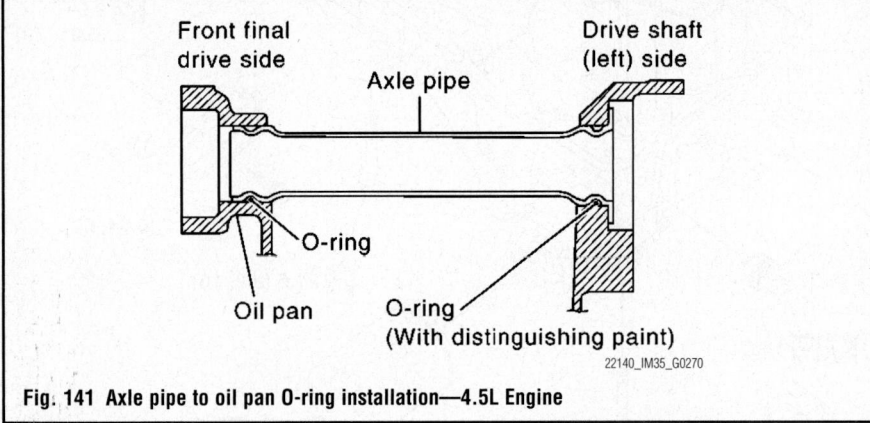

Fig. 141 Axle pipe to oil pan O-ring installation—4.5L Engine

18. Install axle pipe to oil pan (AWD models).

19. Lubricate O-ring groove of axle pip, O-ring, and O-ring joint of oil pan with new engine oil.

➡**Right/left O-ring diameters differ from each other. O-rings with identification paint marks are installed on front drive shaft (left) installing side.**

20. Install axle pipe to oil pan from left side.

➡**Insert it with care to prevent O-ring from sliding.**

21. Install oil pan as follows:
   a. Use scraper to remove old liquid gasket from mating surfaces, remove the old liquid gasket from mating surface of cylinder block, bolt holes and threads.

➡**Never scratch or damage the mating surfaces when cleaning off old liquid gasket.**

   b. Install new O-rings to oil pump and front cover side.
   c. Apply a continuous bead of RTV silicone to the cylinder block mating surfaces and oil pan as shown in the figure.

➡**Attaching should be done within 5 minutes after coating.**

   d. Install oil pan, tighten mounting bolts in numerical order as shown in the figure.

22. Tighten mounting bolts No. 1 and 2 in two steps.

➡**The numerical order No. 11 and 17 show second steps.**

23. There are three types of mounting bolts. Refer to the following for locating bolts.
   • M6 × 30 mm. (1.18 in) : 18, 19
   • M8 × 100 mm (3.94 in) : 5, 9
   • M8 × 45 mm (1.77 in) : Except the above

24. Tighten transmission joint bolts.
25. Install the rear plate cover.
26. Install the oil filter.
27. Install the harness bracket to the oil pan (2WD models).
28. Install the engine mounting insulators (RH and LH) under side nuts.
29. Install the engine assembly into the vehicle.
30. Install the front and rear engine undercovers.
31. Connect the negative battery cable.
32. Check engine oil level and adjust engine oil.

33. Start engine, and check there is no leak of engine oil.
34. Stop engine and wait for 15 minutes.
35. Check engine oil level again.

OIL PUMP

REMOVAL & INSTALLATION

3.5L Engine

See Figure 143.

➡**The oil pump bolts to the front of the engine block and is driven by the crankshaft. Removal of the timing cover and chains are necessary for oil pump service.**

1. Before servicing the vehicle, refer to the service precaution.
2. Drain the engine oil.
3. Rotate the engine and position it to Top Dead Center (TDC) compression stroke of cylinder No. 1.
4. Remove or disconnect the following:
   • Negative battery cable
   • Drive belts
   • Camshaft Position (CMP) sensor (PHASE) and the Crankshaft Position (CKP) sensor (REF/POS)
   • Right front wheel and inner fender cover
   • Engine undercover
   • Crankshaft pulley
   • Front exhaust pipe and its support and support the engine at the left and right side slingers with a suitable hoist
   • Engine right side mounting insulator and bracket nuts and bolts
   • Center crossmember assembly
   • A/C compressor and mounting bracket
   • Lower and upper oil pans
   • Oil strainer from the oil pump
   • Water pump cover and the front cover assembly
   • Lower timing chain assembly
   • Oil pump

To install:

➡**When installing the oil pump, be sure to apply engine oil to the gears.**

5. Install or connect the following:
   • Oil pump. Refer to illustration for torque values
   • Lower timing chain assembly
   • Front timing cover and water pump covers
   • Oil strainer using a new gasket. Torque the bolts to 62 inch. lbs. (7 Nm).

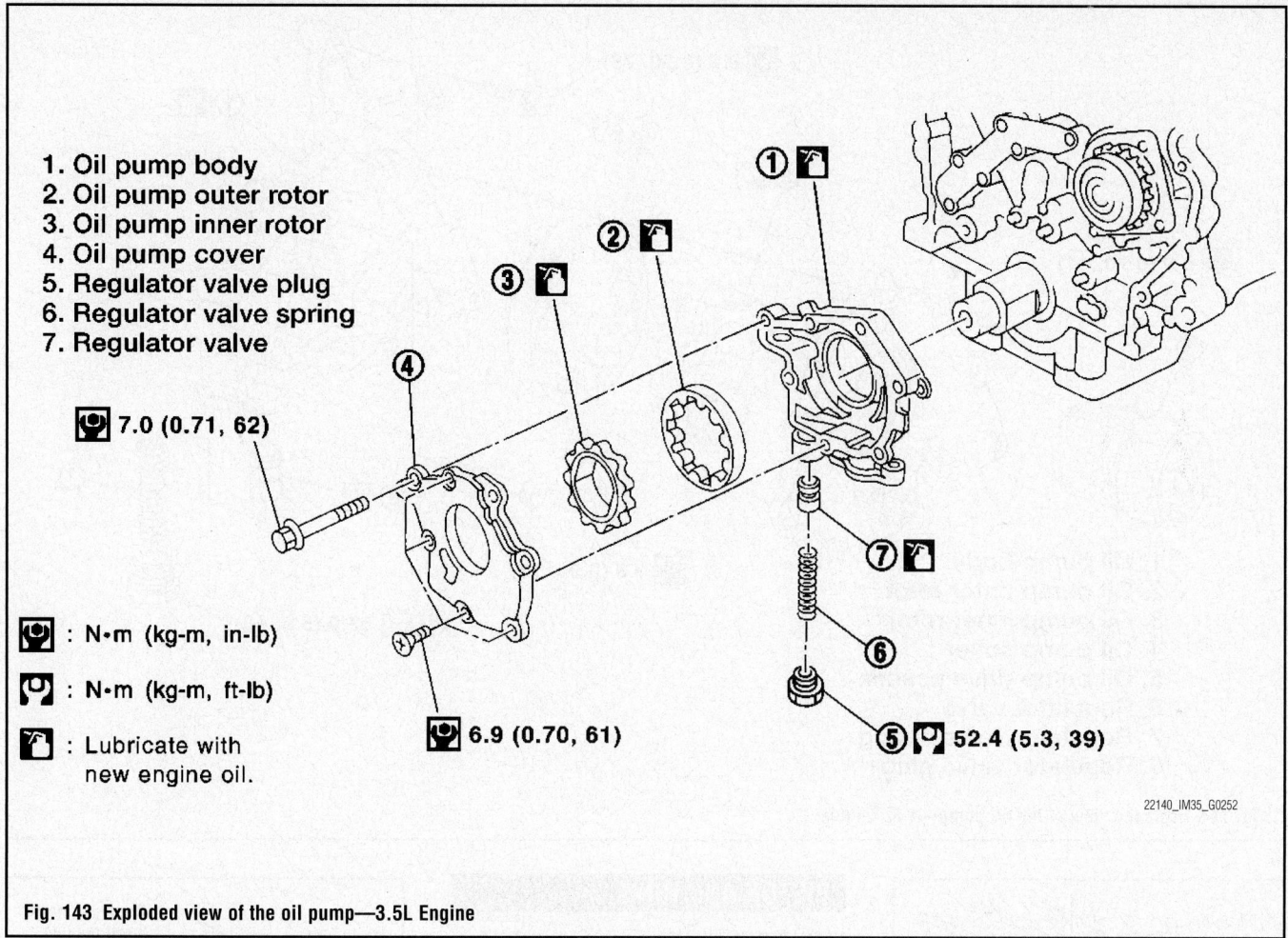

1. Oil pump body
2. Oil pump outer rotor
3. Oil pump inner rotor
4. Oil pump cover
5. Regulator valve plug
6. Regulator valve spring
7. Regulator valve

7.0 (0.71, 62)

: N•m (kg-m, in-lb)

: N•m (kg-m, ft-lb)

: Lubricate with new engine oil.

6.9 (0.70, 61)

52.4 (5.3, 39)

22140_IM35_G0252

Fig. 143 Exploded view of the oil pump—3.5L Engine

- Upper and lower oil pans. Be sure to use new O-rings at the oil pump to upper oil pan mating surface.
- A/C compressor and mounting bracket
- Center crossmember assembly
- Engine right side mounting insulator and bracket and remove the engine support hoist
- Front exhaust pipe and its support
- Crankshaft pulley
- Engine undercover and the right side inner fender cover
- Right front wheel
- CMP sensor (PHASE) and the CKP sensor (REF/POS)
- Engine drive belts and adjust as necessary
- Negative battery cable

6. Fill the engine with clean oil.

7. Start the engine, check the oil pressure, and check for oil leaks.

4.5L Engine

See Figures 144 through 146.

➡**The oil pump is mounted in the cylinder block below the left bank and behind the left timing chain; engine removal is required before servicing. Refer to Engine Assembly Removal and Installation.**

1. Before servicing the vehicle, refer to the precautions in the beginning of this section.

2. Drain the engine oil.

3. Remove or disconnect the following:

- Negative battery cable
- Engine assembly
- Front cover and timing chains
- Oil pump drive spacer.

4. Set bolts in the two bolt holes [M6 × pitch 1.0 mm (0.039 in)] on the front surface. Using suitable puller, pull oil pump drive spacer off from crankshaft.

5. Remove oil pump assembly from the front of the engine

To install:

6. Clean the oil pump mounting surface.

7. Install the oil pump.

8. Tighten the oil pump housing bolts to 78 inch lbs. (8.8 Nm).

9. Tighten the oil pump cover bolts to 51 inch lbs. (6.9 Nm).

10. Install oil pump drive spacer as follows:

    a. Insert oil pump drive spacer according to the directions of crankshaft key and the two flat surfaces of oil pump inner rotor.

➡**If the positional relationship does not allow the insertion, rotate oil pump inner rotor with a finger to allow spacer.**

    b. After confirming that the position of each part is in correct condition to allow for spacer, force fit spacer by lightly tapping with plastic hammer until it contacts and does not go further

11. Install or connect the following:

- Timing chains and front cover
- Negative battery cable

12. Fill the engine with clean oil.

13. Start the vehicle, check for leaks and repair if necessary.

14. Check the engine oil level and adjust engine oil.

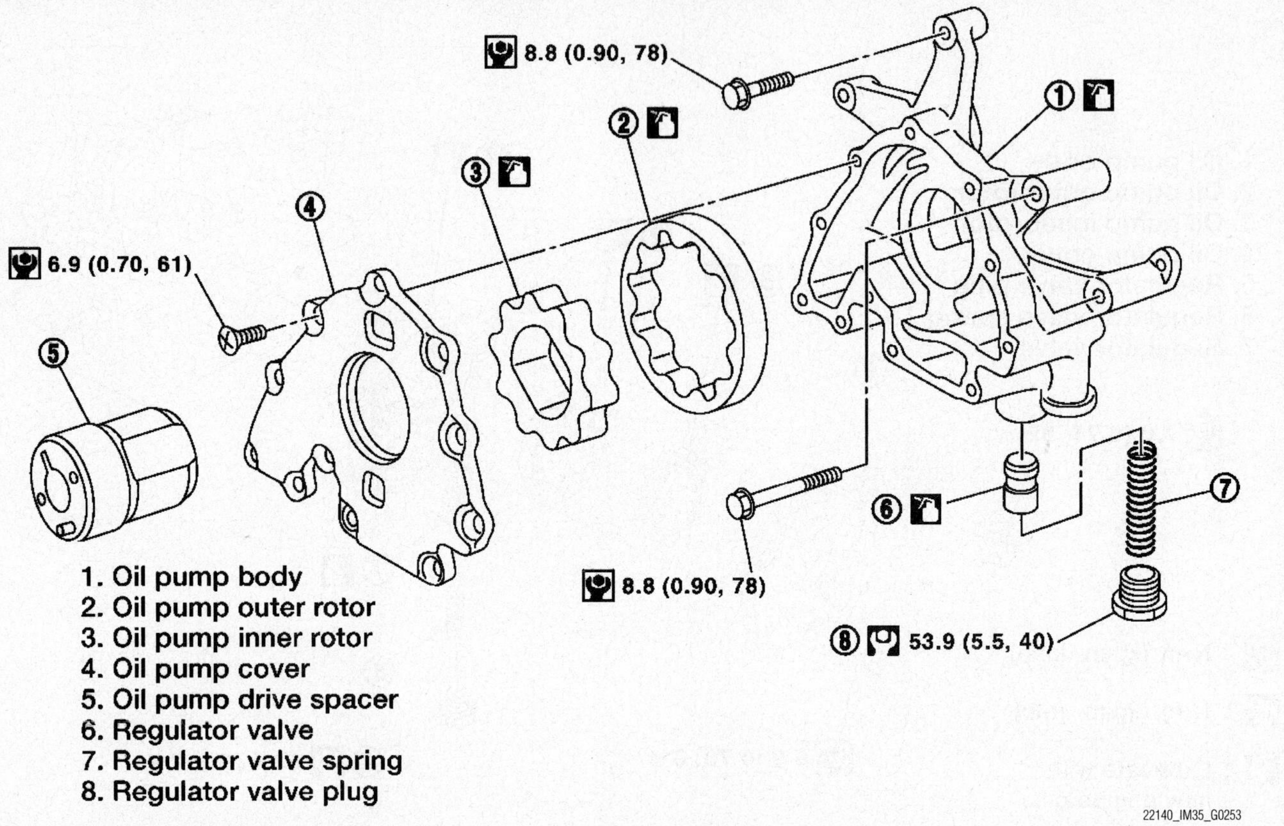

8.8 (0.90, 78)

6.9 (0.70, 61)

8.8 (0.90, 78)

53.9 (5.5, 40)

1. Oil pump body
2. Oil pump outer rotor
3. Oil pump inner rotor
4. Oil pump cover
5. Oil pump drive spacer
6. Regulator valve
7. Regulator valve spring
8. Regulator valve plug

22140_IM35_G0253

Fig. 144 Exploded view of the oil pump—4.5L Engine

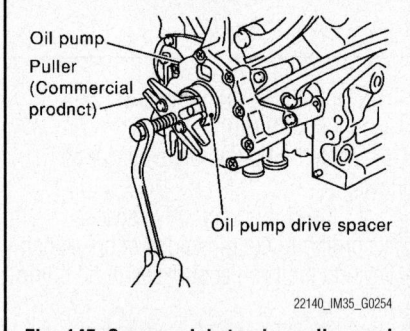

Oil pump

Puller (Commercial prodnct)

Oil pump drive spacer

22140_IM35_G0254

Fig. 145 Commercial steering puller used to pull oil pump drive spacer off from crankshaft—4.5L Engine

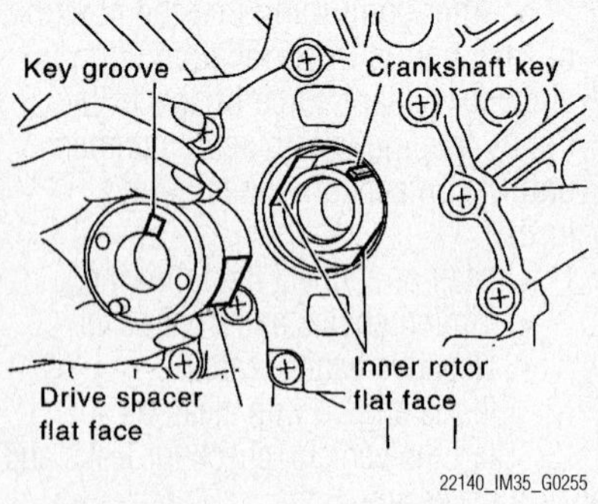

Key groove

Crankshaft key

Drive spacer flat face

Inner rotor flat face

22140_IM35_G0255

Fig. 146 Oil pump drive spacer assembly—4.5L Engine

PISTON AND RING

POSITIONING

3.5L Engine

See Figures 147 and 148.

1. If there is stamped mark on ring, mount it with marked side up.
2. If there is no stamp on ring, no specific orientation is required for installation.
3. Stamped mark:
4. Top ring : —
5. Second ring : R
6. Position each ring with the gap as

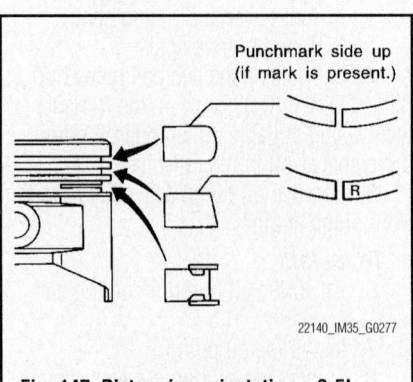

Punchmark side up (if mark is present.)

R

22140_IM35_G0277

Fig. 147 Piston ring orientation—3.5L Engine

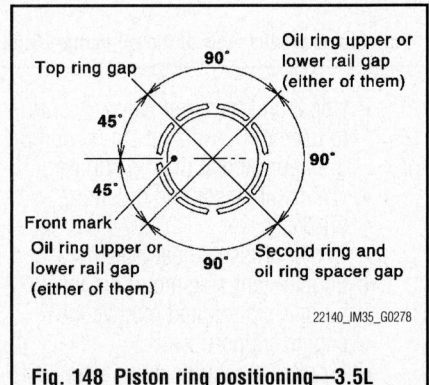

Top ring gap

Oil ring upper or lower rail gap (either of them)

90°

45°

45°

90°

Front mark

Oil ring upper or lower rail gap (either of them)

90°

Second ring and oil ring spacer gap

22140_IM35_G0278

Fig. 148 Piston ring positioning—3.5L Engine

shown in the figure referring to the piston front mark.

4.5L Engine

See Figure 149.

1. Position each ring with the gap as shown in the figure, referring to the piston front mark.
2. Install top ring and second ring with the stamped surface facing upward.
3. Stamped mark are as follows:
   - Top ring : R
   - Second ring : 2 R

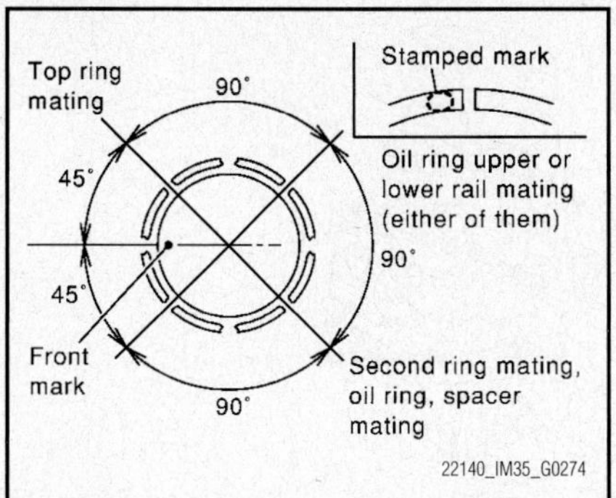

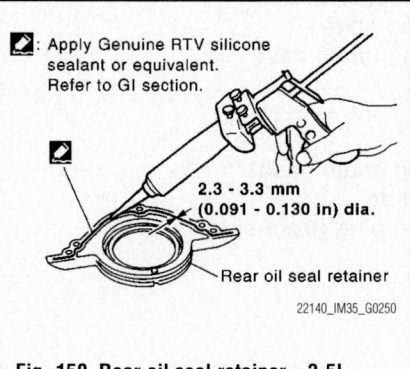

Fig. 149 Piston ring positioning—4.5L Engine

REAR MAIN SEAL

REMOVAL & INSTALLATION

3.5L Engine

See Figure 150.

1. Before servicing the vehicle, refer to the service precaution.
2. Disconnect the negative battery cable.
3. Remove oil pan (upper).
4. Remove transmission assembly.
5. Remove drive plate.
6. Cut away liquid gasket and remove rear oil seal retainer.

✲✲ WARNING

Be careful not to damage mounting surface.

➡ **Regard both rear oil seal and retainer as an assembly.**

7. Remove old liquid gasket on mating surfaces of cylinder block and oil pan (upper) using a scraper.

To install:

8. Apply new engine oil to both oil seal lip and dust seal lip of new rear oil seal retainer.

🔧 : Apply Genuine RTV silicone sealant or equivalent. Refer to GI section.

2.3 - 3.3 mm (0.091 - 0.130 in) dia.

Rear oil seal retainer

22140_IM35_G0250

Fig. 150 Rear oil seal retainer—3.5L Engine

9. Apply a continuous bead of liquid gasket with the tube presser.

➡ **Assembly should be done within 5 minutes after coating.**

10. Install rear oil seal retainer to cylinder block.

➡ **Check the garter spring is in position and seal lips not inverted.**

11. Install the drive plate.
12. Install the transmission assembly.
13. Install the oil pan (upper).
14. Connect the negative battery cable.

4.5L Engine

See Figure 151.

1. Before servicing the vehicle, refer to the service precaution.
2. Disconnect the negative battery cable
3. Remove transmission assembly.
4. Remove drive plate.
5. Remove rear oil seal using suitable tool.

➡ **Be careful not to damage crankshaft and oil seal retainer surface.**

To install:

6. Apply new engine oil to both oil seal lip and dust seal lip of new rear oil seal.
7. Install rear oil seal using suitable tool.
8. Install the drive plate.
9. Install the transmission assembly.
10. Connect the negative battery cable

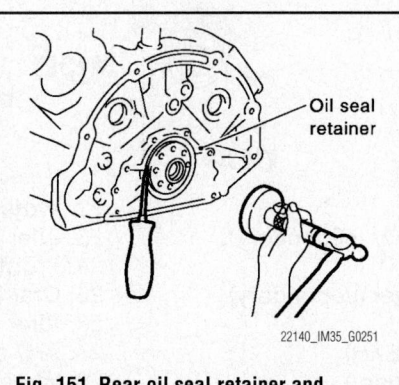

Fig. 151 Rear oil seal retainer and installer—4.5L Engine

TIMING CHAIN, SPROCKETS, FRONT COVER AND SEAL

REMOVAL & INSTALLATION

3.5L Engine

See Figures 152 through 188.

1. Before servicing the vehicle, refer to the precautions in the beginning of this section.

2. Properly relieve the fuel system pressure.
3. Drain the engine oil and cooling system.
4. Remove or disconnect the following:
   - Negative battery cable
   - Engine room cover (RH and LH).
   - Cooling fan and radiator
   - Engine cover with power tool
   - Air duct (inlet) and air cleaner case assembly
   - Front and rear engine undercover with power tool
   - Radiator hose (upper and lower) and A/T fluid cooler hose
   - Engine harnesses removing their brackets from front timing chain case
   - Drive belts
   - Intake manifold collectors (upper and lower)
   - Power steering oil pump from bracket with piping connected, and temporarily secure it aside
   - Power steering oil pump bracket
   - Alternator
   - Water bypass hose, water hose clamp and idler pulley bracket from front timing chain case
   - Intake valve timing control covers
5. Loosen mounting bolts in reverse order as shown in the figure.
6. Shaft is internally jointed with camshaft sprocket (INT) center hole. When removing, keep it horizontal until it is completely disconnected.

➡ **Use the seal cutter (SST: KV10111100 (J37228) to cut liquid gasket for removal.**

7. Remove or disconnect the following:
   - Collared O-ring from front timing chain case (left and right side).
   - Rocker covers (right bank and left bank)
8. Obtain No. 1 cylinder at TDC of its compression stroke as follows:
   a. Rotate crankshaft pulley clockwise to align timing mark (grooved line without color) with timing indicator.
   b. Check that intake and exhaust cam noses on No. 1 cylinder (engine front side of right bank) are located as shown in the figure. If not, turn crankshaft one revolution (360 degrees) and align as shown in the figure.
9. Remove crankshaft pulley as follows:
   a. Remove rear cover plate (2WD models) or starter motor (AWD models) and set ring gear stopper.

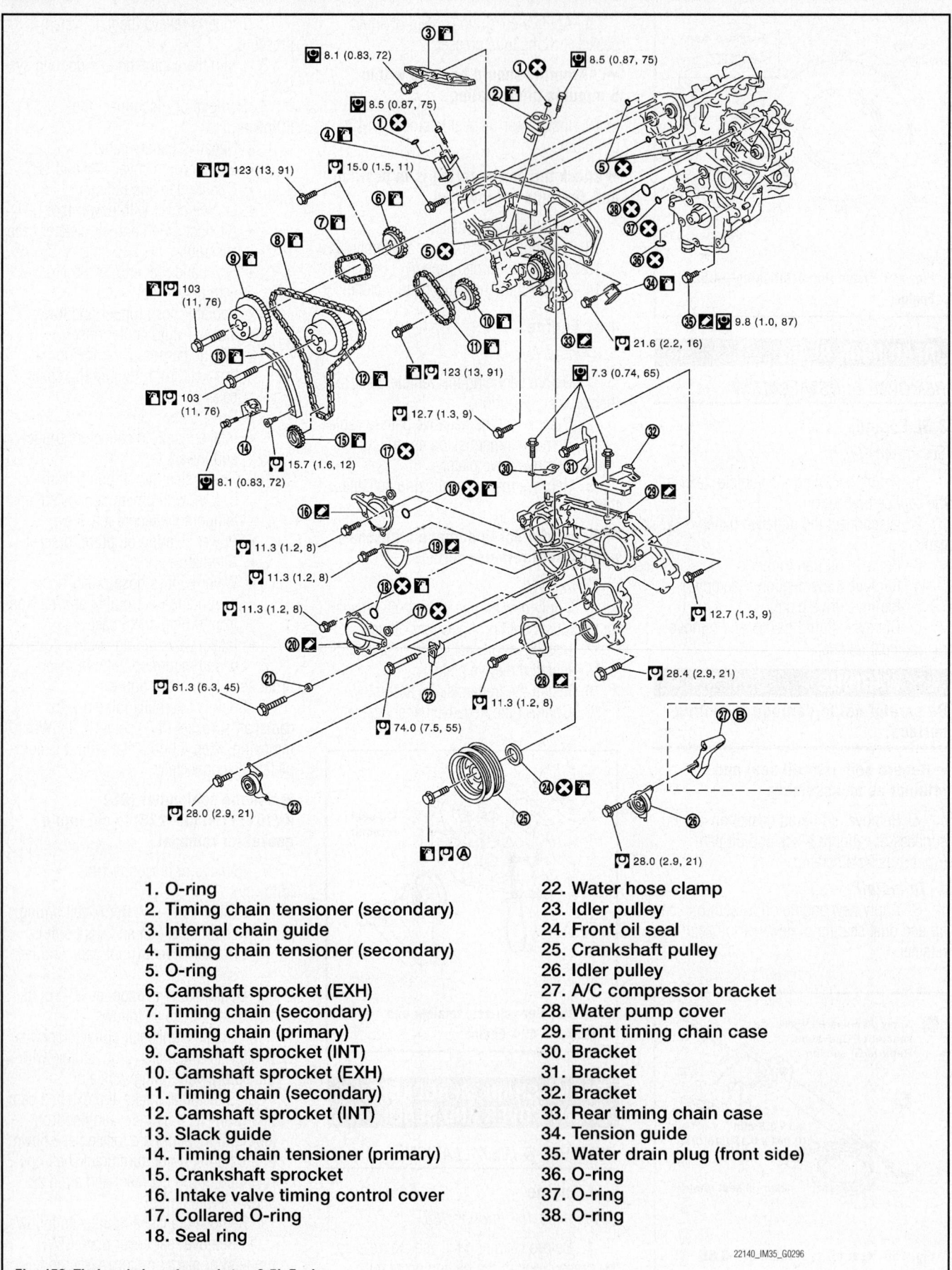

Fig. 152 Timing chain and sprockets—3.5L Engine

1. O-ring
2. Timing chain tensioner (secondary)
3. Internal chain guide
4. Timing chain tensioner (secondary)
5. O-ring
6. Camshaft sprocket (EXH)
7. Timing chain (secondary)
8. Timing chain (primary)
9. Camshaft sprocket (INT)
10. Camshaft sprocket (EXH)
11. Timing chain (secondary)
12. Camshaft sprocket (INT)
13. Slack guide
14. Timing chain tensioner (primary)
15. Crankshaft sprocket
16. Intake valve timing control cover
17. Collared O-ring
18. Seal ring

22. Water hose clamp
23. Idler pulley
24. Front oil seal
25. Crankshaft pulley
26. Idler pulley
27. A/C compressor bracket
28. Water pump cover
29. Front timing chain case
30. Bracket
31. Bracket
32. Bracket
33. Rear timing chain case
34. Tension guide
35. Water drain plug (front side)
36. O-ring
37. O-ring
38. O-ring

22140_IM35_G0296

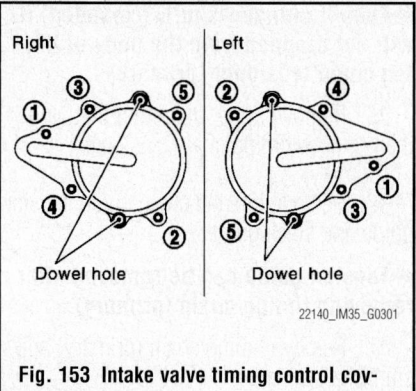

Fig. 153 Intake valve timing control covers—3.5L Engine

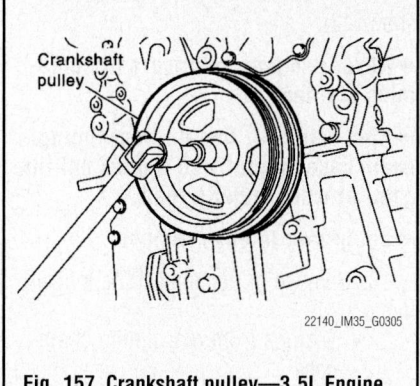

Fig. 157 Crankshaft pulley—3.5L Engine

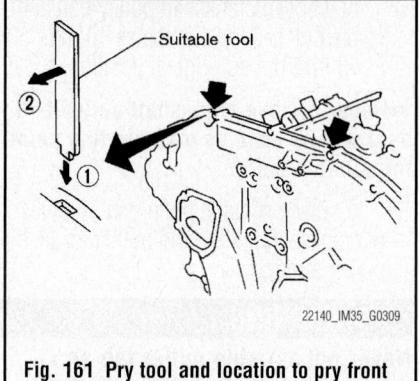

Fig. 161 Pry tool and location to pry front cover from motor—3.5L Engine

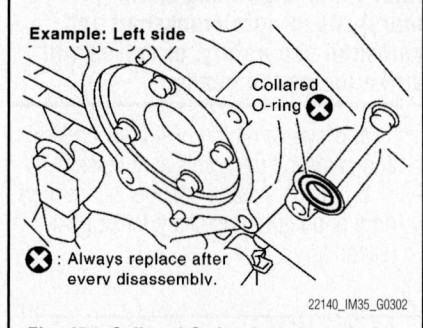

Fig. 154 Collared O-ring from front timing chain case (left and right side) left side shown—3.5L Engine

Fig. 158 Crankshaft pulley removal (upper)—3.5L Engine

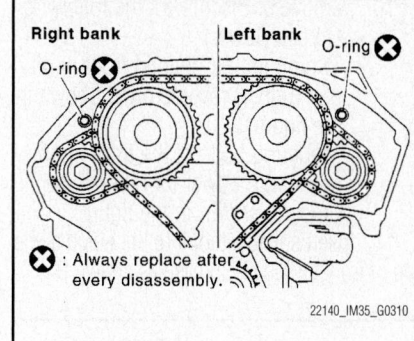

Fig. 162 Oil seal rear timing chain case—3.5L Engine

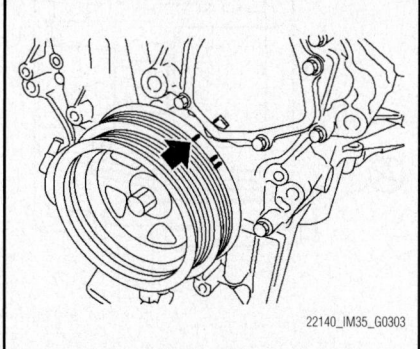

Fig. 155 Crankshaft pulley with grooved timing indicator—3.5L Engine

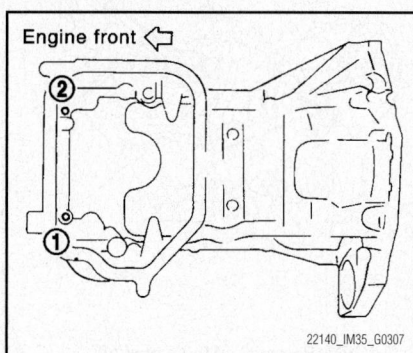

Fig. 159 Mounting bolts of the front oil pan (upper)—3.5L Engine

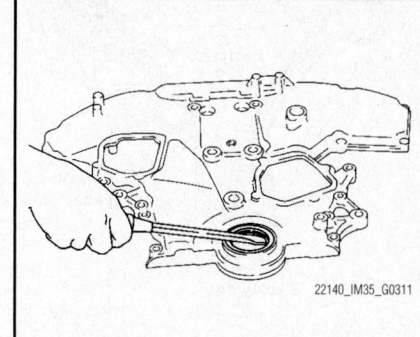

Fig. 163 Front oil seal front timing chain case—3.5L Engine

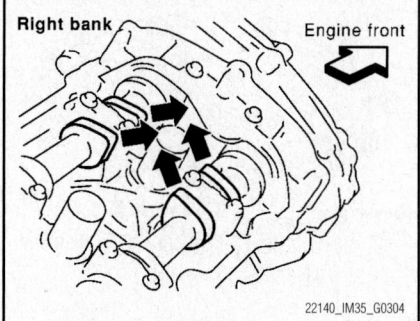

Fig. 156 Intake and exhaust cam noses on No. 1 cylinder (engine front side of right bank) 3.5L Engine

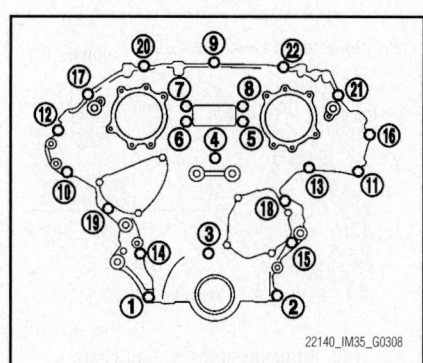

Fig. 160 Front timing chain cover bolt locations key—3.5L Engine

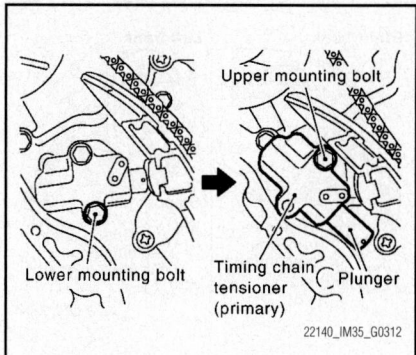

Fig. 164 Primary timing chain tensioner and bolt locations—3.5L Engine

b. Loosen crankshaft pulley bolt and locate bolt seating surface as 10 mm (0.39 in) from its original position.

➡**Never remove crankshaft pulley bolt as it will be used as a supporting point for suitable puller.**

c. Place suitable puller tab on holes of crankshaft pulley, and pull crankshaft pulley through.

❈❈ **WARNING**

Never put suitable puller tab on crankshaft pulley, as this will damage internal damper.

10. Remove or disconnect the following:
  • Oil pan (lower)
  • Two mounting bolts in front of oil pan (upper) reverse order shown in the figure
  • Mounting bolts of the front timing cover with power tool in reverse order as shown in the figure.

11. Insert suitable tool into the notch at the top of front timing chain case as shown (1).

12. Pry off case by moving a tool as shown (2).

➡**Never use a screwdrivers or something similar.**

➡**After removal, handle front timing chain case carefully so it does not tilt, cant, or warp under a load.**

➡**Cut gasket to ease removal.**

13. Remove or disconnect the following:

  • O-rings from rear timing chain case
  • Oil pan gasket (front)
  • Water pump cover
  • Chain tensioner cover from front timing chain case
  • Front oil seal from front timing chain case
  • Timing chain tensioner (primary)
  • Lower mounting bolt.

14. Loosen upper mounting bolt slowly, and then turn timing chain tensioner (primary) on the mounting bolt so that the plunger is fully expanded.

➡**Even if plunger is fully expanded, it will not dropped from the body of timing chain tensioner (primary).**

15. Remove upper mounting bolt, and then remove timing chain tensioner (primary).

16. Remove internal chain guide, tension guide and slack guide.

➡**Tension guide can be removed after removing timing chain (primary).**

17. Remove timing chain (primary) and crankshaft sprocket.

❈❈ **WARNING**

After removing timing chain (primary), never turn crankshaft and camshaft separately, or valves will strike the piston heads.

a. Remove timing chain (secondary) and camshaft sprockets as follows:

b. Attach suitable stopper pin to the right and left timing chain tensioners (secondary).

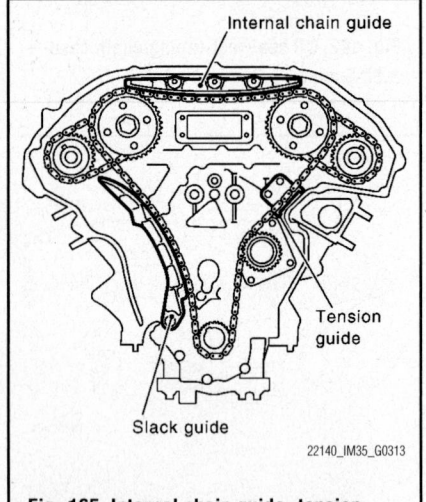

Fig. 165 Internal chain guide, tension guide and slack guide—3.5L Engine

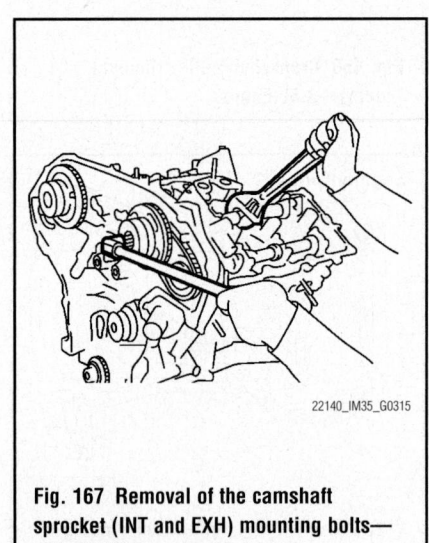

Fig. 167 Removal of the camshaft sprocket (INT and EXH) mounting bolts—3.5L Engine

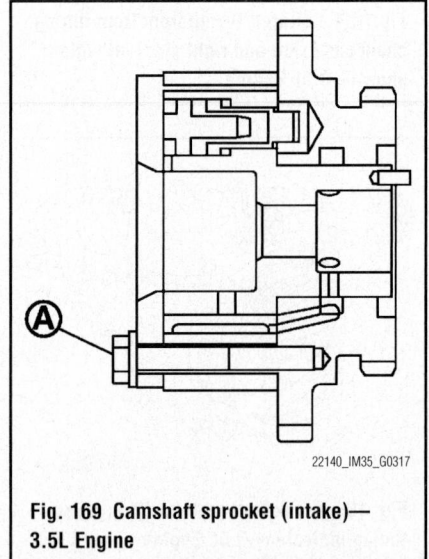

Fig. 169 Camshaft sprocket (intake)—3.5L Engine

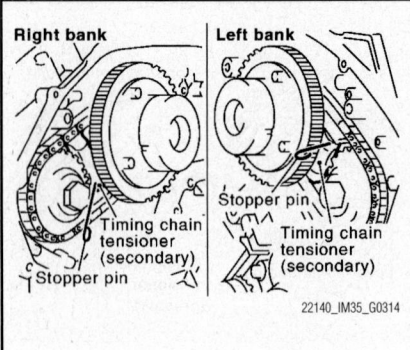

Fig. 166 Removal of timing chain tensioner (secondary)—3.5L Engine

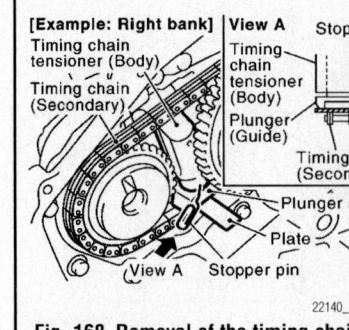

Fig. 168 Removal of the timing chain (secondary) together with camshaft sprockets—3.5L Engine

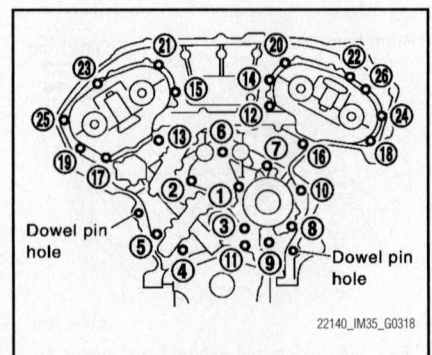

Fig. 170 Rear timing chain case mounting bolts locations—3.5L Engine

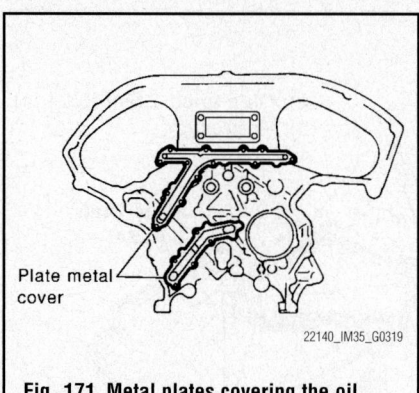

Fig. 171 Metal plates covering the oil passages—3.5L Engine

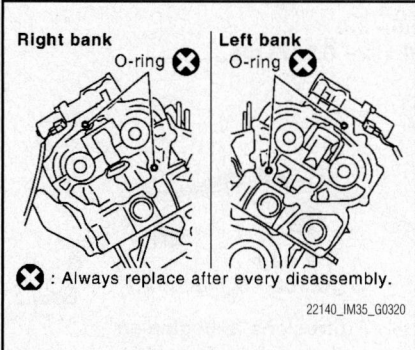

Fig. 172 Passage O-rings in cylinder heads—3.5L Engine

➡Use approximately 0.5 mm (0.020 in) dia. hard metal pin as a stopper pin.

   c. Remove camshaft sprocket (INT and EXH) mounting bolts.

   d. Secure the hexagonal portion of camshaft using a wrench to loosen mounting bolts.

➡Never loosen the mounting bolts with securing anything other than the camshaft hexagonal portion or with tensioning the timing chain.

   e. Remove timing chain (secondary) together with camshaft sprockets.

   f. Turn camshaft slightly to secure slackness of timing chain on timing chain tensioner (secondary) side.

   g. Insert 0.020 inch (0.5 mm) thick metal or resin plate between timing chain and timing chain tensioner plunger (guide). Remove timing chain (secondary) together with camshaft sprockets with timing chain loose from the guide groove.

✳✳ WARNING

Be careful of plunger coming-off when removing timing chain (secondary). This is because the plunger of timing chain tensioner (secondary)

moves during operation, leading it to coming-off of the fixed stopper pin.

➡Camshaft sprocket (INT) is two-for-one structure of primary and secondary sprockets.

✳✳ WARNING

Handle carefully to avoid any shock to camshaft sprocket, Never disassemble. (Never loosen bolts "A" as shown in the figure).

18. Remove rear timing chain case.
19. Remove mounting bolts in reverse order as shown in the figure.
20. Cut gasket to remove the case.

➡Never remove the metal plate covering the oil passage.

➡After removal, handle rear timing chain case carefully so it does not tilt, cant, or warp under a load.

21. Remove O-rings from cylinder head.
22. Remove O-rings from cylinder block.
23. Use a scraper to remove all traces of gasket from front and rear timing chain cases, water pump cover, chain tensioner cover, intake valve timing control covers. and all opposite mating surfaces.

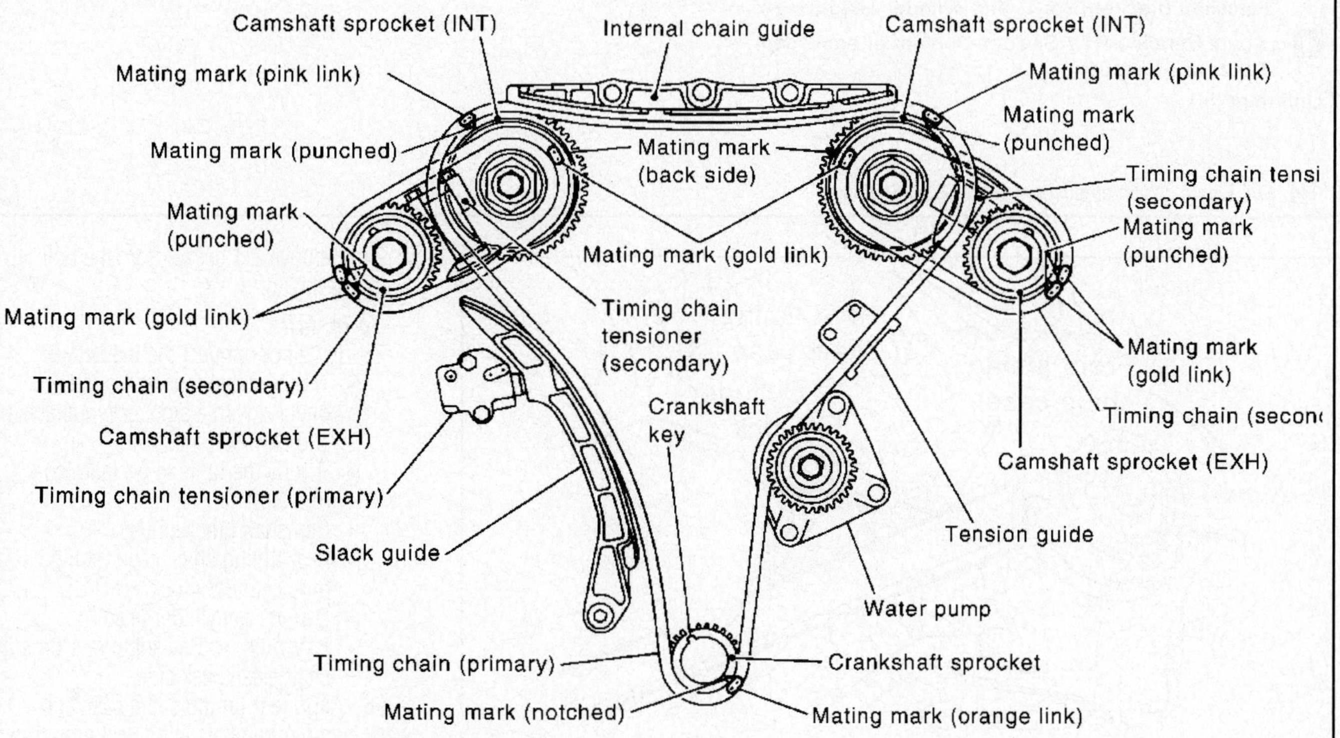

Fig. 173 Timing chain, sprockets guides and tensioners—3.5L Engine

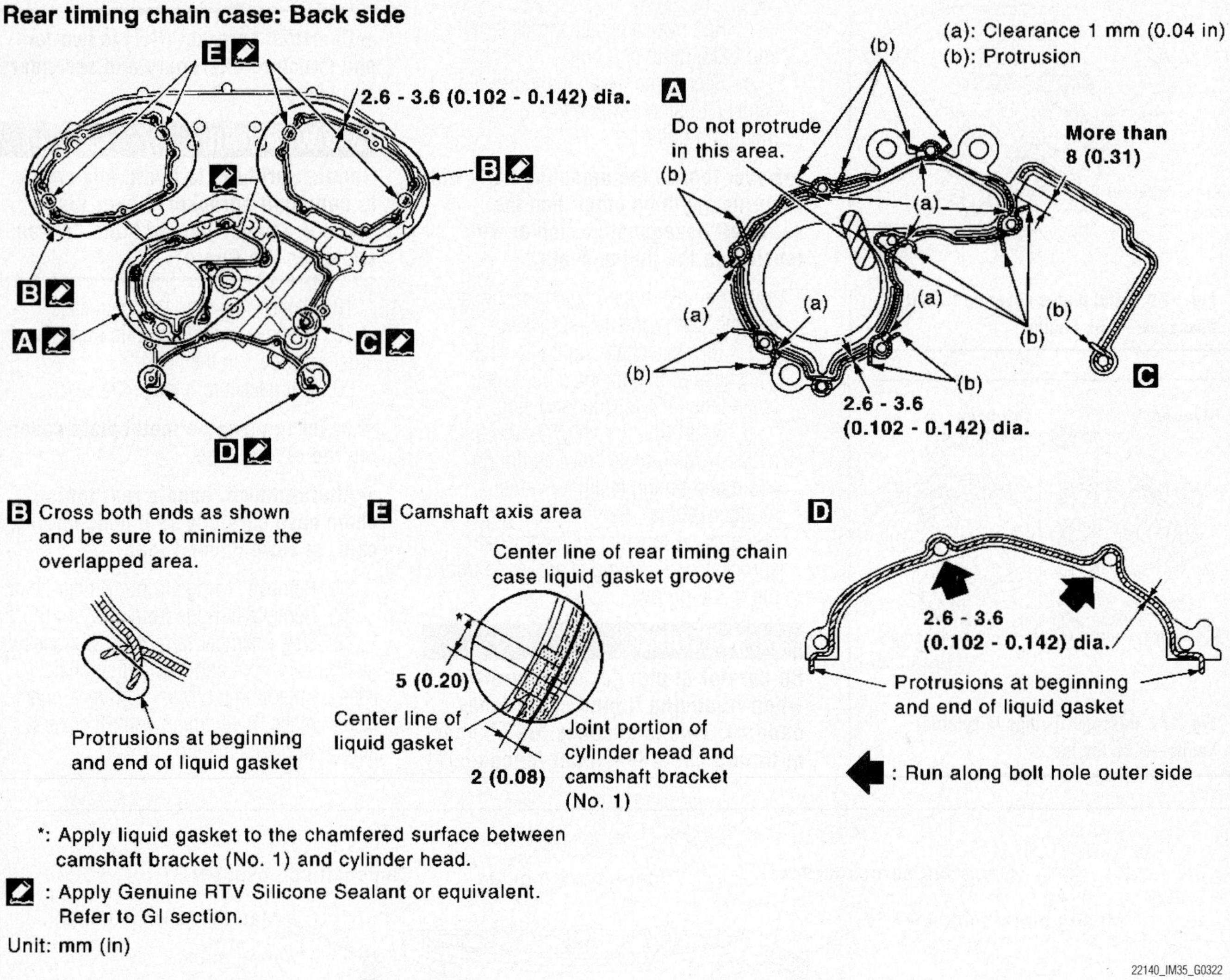

Rear timing chain case: Back side

E✏ 2.6 - 3.6 (0.102 - 0.142) dia.

B✏

B✏ A✏

A✏ C✏

D✏

(a): Clearance 1 mm (0.04 in)
(b): Protrusion

A Do not protrude in this area.
(b)

More than 8 (0.31)

(a)

(b)

C

2.6 - 3.6
(0.102 - 0.142) dia.

B Cross both ends as shown and be sure to minimize the overlapped area.

Protrusions at beginning and end of liquid gasket

E Camshaft axis area

Center line of rear timing chain case liquid gasket groove

Center line of liquid gasket

5 (0.20)

2 (0.08)

Joint portion of cylinder head and camshaft bracket (No. 1)

D

2.6 - 3.6
(0.102 - 0.142) dia.

Protrusions at beginning and end of liquid gasket

◀ : Run along bolt hole outer side

*: Apply liquid gasket to the chamfered surface between camshaft bracket (No. 1) and cylinder head.

✏ : Apply Genuine RTV Silicone Sealant or equivalent. Refer to GI section.

Unit: mm (in)

22140_IM35_G0322

Fig. 174 Liquid gasket application chart—3.5L Engine

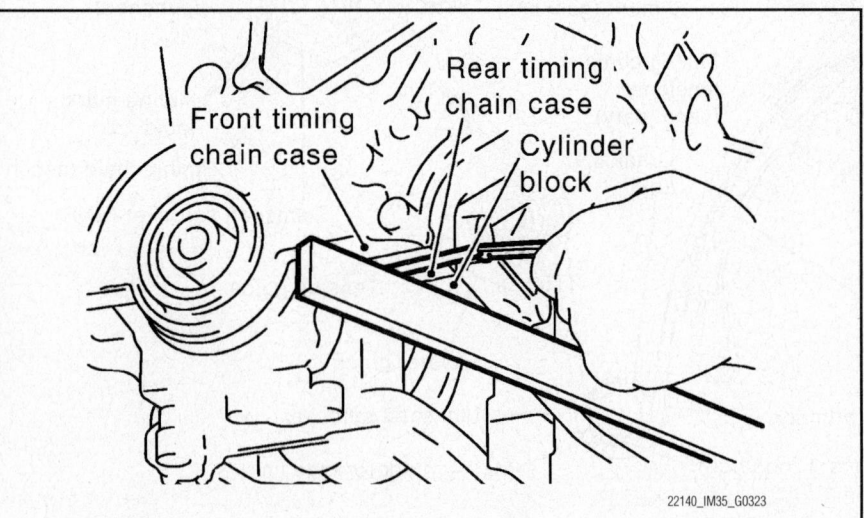

Front timing chain case

Rear timing chain case

Cylinder block

22140_IM35_G0323

Fig. 175 Checking surface height difference between rear timing chain case and cylinder block—3.5L Engine

24. Remove old gasket from the bolt hole and thread.

To install:
25. Install or connect the following:
- Timing chain tensioners (secondary) with a stopper pin attached and new O-rings.
- Torque the bolts to 61 inch lbs. (6.9 Nm)
- Camshaft brackets (No. 1).
- Rear timing chain case O-rings onto cylinder block
- O-rings to cylinder head
- RTV Silicone Sealant to rear timing chain case back side

26. Align rear timing chain case and water pump assembly with dowel pins (right bank and left bank) on cylinder block and install rear timing chain case.

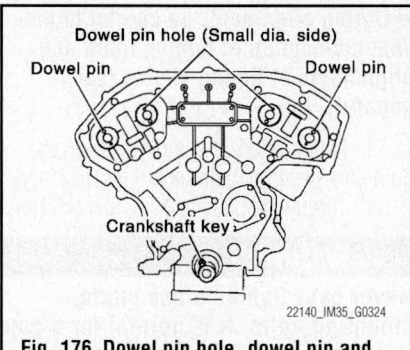

Fig. 176 Dowel pin hole, dowel pin and crankshaft key alignment for No. 1 cylinder at compression TDC—3.5L Engine

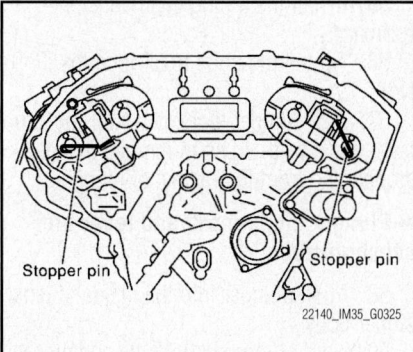

Fig. 177 Locking the plunger of the timing chain tensioner (secondary)—3.5L Engine

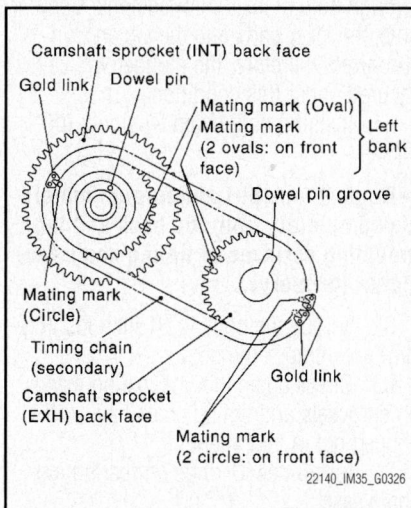

Fig. 178 Timing chains (secondary) and camshaft sprockets, align the mating marks—3.5L Engine

➡**Check O-rings stay in place during installation to cylinder block and cylinder head.**

27. Tighten mounting bolts in numerical order as shown in the figure.

28. There are two types of mounting bolts. Refer to the following for locating bolts.

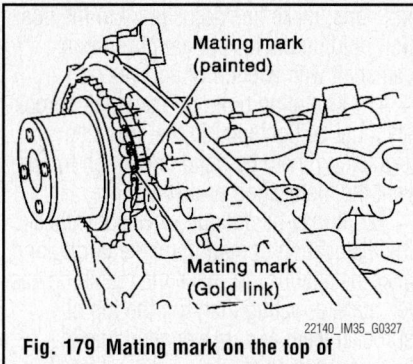

Fig. 179 Mating mark on the top of sprocket teeth to align the camshaft sprockets easily—3.5L Engine

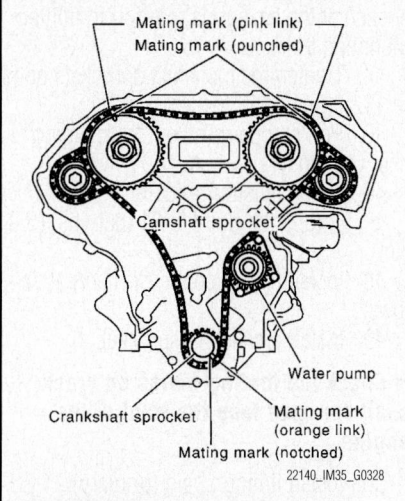

Fig. 180 Mating marks on the top of the camshaft sprockets and the crankshaft sprocket—3.5L Engine

29. Bolt length: 0.79 inches (20 mm); 1, 2, 3, 6, 7, 8, 9, and 10 tighten to 11 ft. lbs. (15 Nm).

30. Bolt length: 0.63 inches (16 mm); 11, 12, 13, 14, 15, 16, 17, 18, 19, 20, 21, 22, 23, 24, 25, 26, tighten to 9 ft. lbs. (12.7 Nm).

31. After installing rear timing chain case, check the surface height difference between the following parts on the oil pan (upper) mounting surface.

32. Rear timing chain case and cylinder block: -0.0094–0.0055 inch (-0.24–0.14 mm).

➡**If the measurement is not within the standard, repeat the installation procedure.**

33. Install water pump with new O-rings.

34. Check that the dowel pin hole, dowel pin and crankshaft key are located as shown in the figure. (No. 1 cylinder at compression TDC)

➡**Though the camshaft does not stop at the position as shown in the figure, to show the placement of cam nose, it**

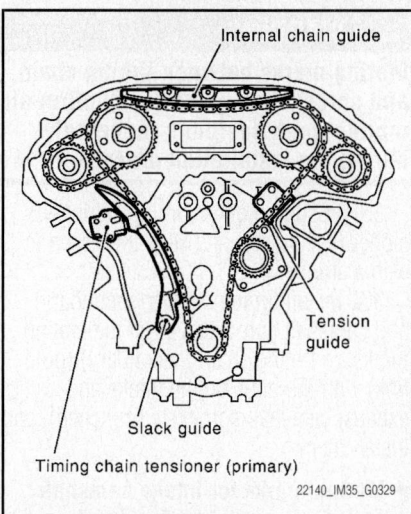

Fig. 181 Internal chain guide, slack guide and timing chain tensioner (primary)—3.5L Engine

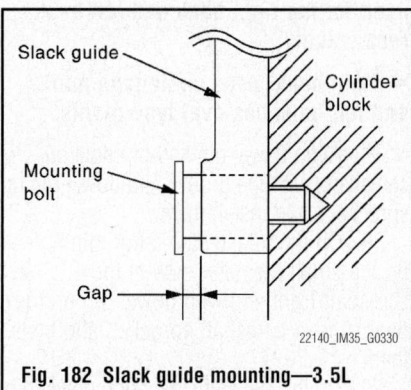

Fig. 182 Slack guide mounting—3.5L Engine

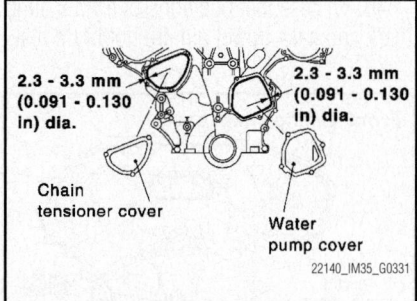

Fig. 183 Water pump cover and chain tensioner cover to front timing chain case sealant guide—3.5L Engine

is generally accepted that the camshaft is placed in the same direction for the figure.

✳✳ WARNING

Hole on small diameter. side must be used for the intake side dowel pin hole, Never misidentify (ignore big diameter. side).

❊❊ **WARNING**

Mating marks between timing chain and sprockets slip easily. Confirm all mating mark positions repeatedly during the installation process.

35. Push plunger of timing chain tensioner (secondary) and keep it pressed in with a stopper pin.

36. Install timing chains (secondary) and camshaft sprockets, align the mating marks on timing chain (secondary) (gold link) with the ones on the intake and exhaust camshaft sprockets (punched), and install them.

➡**Mating marks for intake camshaft sprocket are on the back side of camshaft sprocket (secondary).**

➡**There are two types of mating marks, circle and oval types, they should be used for the right bank and left bank, respectively.**

➡**Right bank, uses circle type marks and left bank use oval type marks.**

37. Align dowel pin and pin hole on camshafts with the groove and dowel pin on sprockets, and install them.

38. On the intake side, align pin hole on the small diameter side of the camshaft front end with dowel pin on the back side of camshaft sprocket, and install them.

39. On the exhaust side, align dowel pin on camshaft front end with pin groove on camshaft sprocket, and install them.

40. In case that positions of each mating mark and each dowel pin are not fit on mat-

ing parts, make fine adjustment to the position holding the hexagonal portion on camshaft with wrench or equivalent.

41. Mounting bolts for camshaft sprockets must be tightened in the next step. Tightening them by hand is enough to prevent the dislocation of dowel pins.

42. It may be difficult to visually check the dislocation of mating marks during and after installation. To make the matching easier, make a mating mark on the top of sprocket teeth and its extended line in advance with paint

43. After confirming the mating marks are aligned, tighten camshaft sprocket mounting bolts, secure camshaft using a wrench at the hexagonal portion to tighten mounting bolts.

44. Tighten the camshaft sprockets bolts to 112 ft. lbs. (152 Nm).

45. Pull stopper pins out from timing chain tensioners (secondary).

46. Install tension guide.

47. Torque the bolts to 61 inch lbs. (6.9 Nm).

48. Install timing chain (primary) as follows:

49. Install crankshaft sprocket.

➡**Check the mating marks on crankshaft sprocket face the front of the engine.**

50. Install timing chain (primary).

51. Install timing chain (primary) so the mating mark (punched) on camshaft sprocket is aligned with the pink link on timing chain, while the mating mark (notched) on crankshaft sprocket is aligned with the orange one on timing chain, as shown in the figure.

52. When it is difficult to align mating marks of timing chain (primary) with each sprocket, gradually turn camshaft using wrench on the hexagonal portion to align it with the mating marks.

➡**During alignment, be careful to prevent dislocation of mating mark and alignments of timing chains (secondary).**

53. Install internal chain guide, slack guide and timing chain tensioner (primary).

54. Torque the bolts to 12 ft. lbs. (16 Nm).

❊❊ **WARNING**

Never over tighten slack guide mounting bolts. It is normal for a gap to exist under the bolt seats when mounting bolts are tightened to the specification.

55. Install the timing chain tensioner (primary).

56. Torque the bolts to 61 inch lbs. (6.9 Nm).

57. Pull plunger stopper tab up (or turn lever downward) so as to remove plunger stopper tab from the ratchet of plunger.

➡**Plunger stopper tab and lever are synchronized.**

58. Push plunger into the inside of tensioner body.

59. Hold plunger in the fully compressed position by engaging plunger stopper tab with the tip of ratchet.

60. To secure lever, insert stopper pin through hole of lever into tensioner body hole, the lever parts and the tab are synchronized, therefore, the plunger will be secured under this condition.

61. Install timing chain tensioner (primary).

➡**Remove any dirt and foreign materials completely from the back and the mounting surfaces of timing chain tensioner (primary).**

62. Torque the bolts to 61 inch lbs. (6.9 Nm) and remove the pins.

63. Check again that the mating marks on sprockets and timing chain have not slipped out of alignment.

64. Install new O-rings on rear timing chain case.

65. Install new front oil seal on front timing chain case.

66. Apply new engine oil to both oil seal lip and dust seal lip.

67. Using a suitable drift [outer diameter: 60 mm (2.36 in)], pressfit oil seal until it becomes flush with front timing chain case end face.

68. Check the garter spring is in position and seal lip is not inverted.

69. Install water pump cover and chain tensioner cover to front timing chain case.

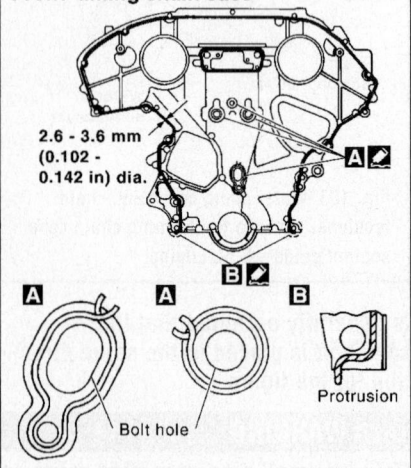

Fig. 184 Liquid gasket application chart for the front timing chain case back side— 3.5L Engine

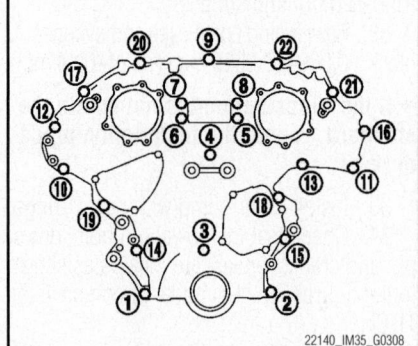

Fig. 185 Front timing chain case bolts location key—3.5L Engine

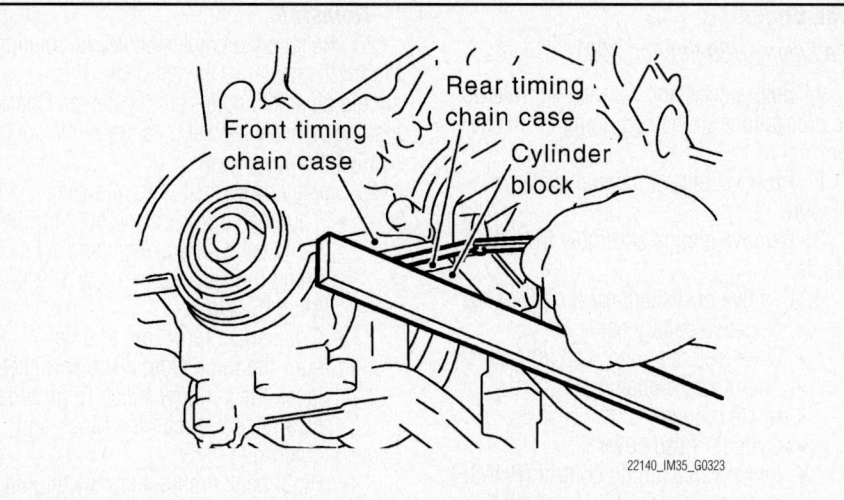

Fig. 186 Surface height difference between rear timing chain case and cylinder block—3.5L Engine

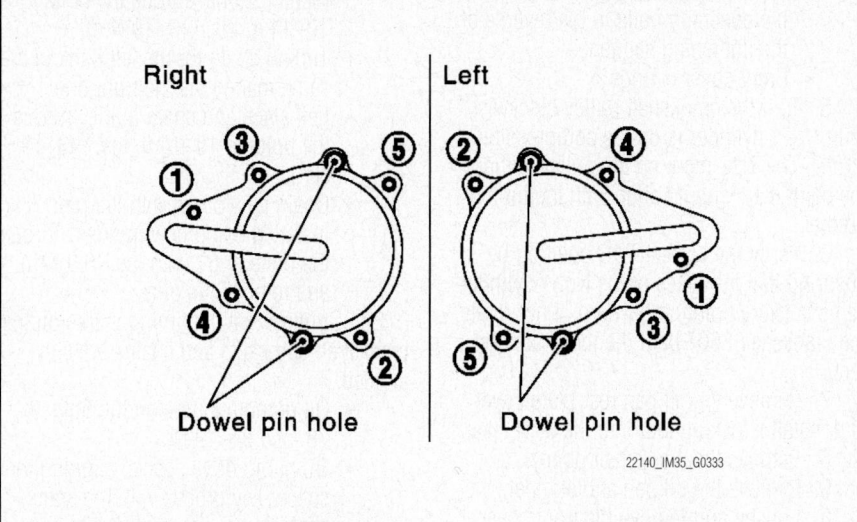

Fig. 187 Intake valve timing control covers tightening sequence—3.5L Engine

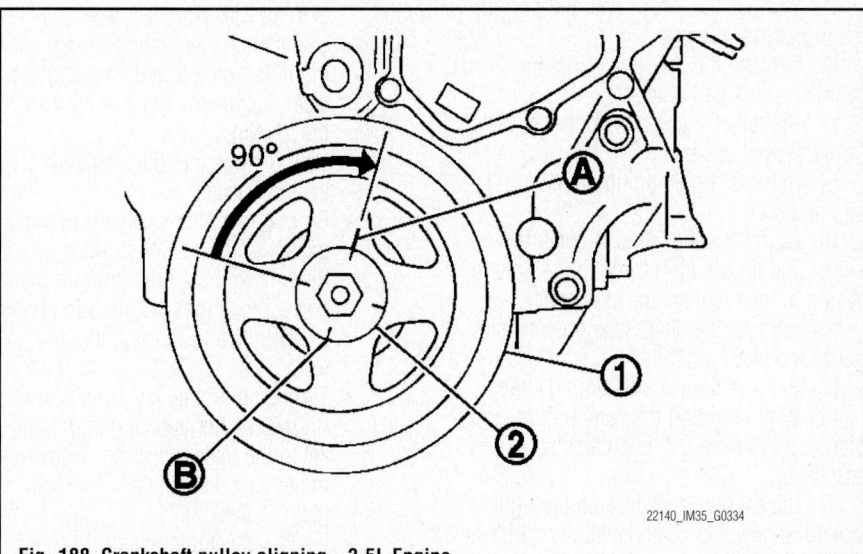

Fig. 188 Crankshaft pulley aligning—3.5L Engine

70. Apply a continuous bead of RTV Silicone) to the front timing chain case as shown in the figure.

71. Install front timing chain case as follows:

72. Apply a continuous bead of RTV Silicone) to the front timing chain case backside as shown in the figure.

73. Install the front timing chain case as to fit its dowel pin holes together with the dowel pin on the rear timing chain case.

74. Tighten mounting bolts to the specified torque in numerical order

75. 21 ft. lbs. (28.4 Nm). For the M8 bolts 1and 2.

76. 9 ft. lbs. (27.7 Nm). For the M6 bolts except the above.

77. After installing front timing chain case, check the surface height difference between the following parts on the oil pan (upper) mounting surface.

78. After installing rear timing chain case, check the surface height difference between the following parts on the oil pan (upper) mounting surface.

79. Rear timing chain case and cylinder block:-0.0055–0.0055 inch (-0.14–0.14 mm).

➡ If the measurement is not within the standard, repeat the installation procedure.

80. Install right and left intake valve timing control covers as follows:

81. Install new seal rings in shaft grooves.

82. Apply a continuous bead of RTV Silicone Sealant to the intake valve timing control covers 0.083–0.122 inch (2.1–3.1 mm) diameter.

83. Install new collared O-rings in front timing chain case oil hole (left and right sides).

84. Being careful not to move seal ring from the installation groove, align dowel pins on front timing chain case with holes to install intake valve timing control covers.

85. Tighten mounting bolts in numerical order as shown in the figure

86. Install oil pans (upper and lower).

87. Install rocker covers (right bank and left bank).

88. Install crankshaft pulley as follows:

89. Hold the crankshaft using the ring gear stopper [SST: KV10117700 (J44716)].

90. Install crankshaft pulley, taking care not to damage front oil seal.

91. When press-fitting crankshaft pulley with plastic hammer, tap on its center portion (not circumference).

92. Tighten crankshaft pulley bolt to 33 ft. lbs. (44.1 Nm).

93. Place a paint mark (A) on crankshaft pulley (1) aligning with the angle mark (B) on crankshaft pulley bolt (2) Tighten the bolt 90 degrees (angle tightening).

94. Rotate crankshaft pulley in normal direction (clockwise when viewed from front) to confirm it turns smoothly.

95. Install or connect the following:
- Rocker covers (right bank and left bank)
- Rear cover plate (2WD models) or starter motor (AWD models)
- Water bypass hose, water hose clamp and idler pulley bracket from front timing chain case
- Alternator
- Power steering oil pump bracket
- Power steering oil pump to bracket with piping
- Intake manifold collectors (upper and lower)
- Drive belts
- Front and rear engine undercover with power tool
- Radiator hose (upper and lower) and A/T fluid cooler hose
- Engine harnesses installing their brackets to the front timing chain case
- Air duct (inlet) and air cleaner case assembly
- Engine cover
- Cooling fan and radiator
- Engine room cover (RH and LH).
- Negative battery cable

96. Before starting engine, check oil/fluid levels including engine coolant and engine oil. If less than required. quantity, fill to the specified level.

97. Turn ignition switch **ON** (with engine stopped). With fuel pressure applied to fuel piping, check for fuel leakage at connection points.

98. Start engine. With engine speed increased, check again for fuel leakage at connection points.

99. Run engine to check for unusual noise and vibration.

100. Warm up engine thoroughly to check there is no leakage of fuel, or any oil/fluids including engine oil and engine coolant.

101. Bleed air from lines and hoses of applicable lines, such as in cooling system.

102. After cooling down engine, again check oil/fluid levels including engine oil and engine coolant. Refill to the specified level, if necessary.

4.5L Engine

See Figures 189 through 209.

1. Before servicing the vehicle, refer to the precautions in the beginning of this section.

2. Properly relieve the fuel system pressure.

3. Remove engine assembly from vehicle.

4. Remove or disconnect the following:
- Negative battery cable
- Drive belt tensioner and idler pulley
- Thermostat housing
- Ignition coils
- Cylinder head cover
- Intake valve timing control (PHASE) sensors for right bank and left bank
- Camshaft Position sensor (CMP)
- Intake valve timing control solenoid from both sides
- Intake valve timing control covers by loosening bolts in the reverse of the tightening sequence
- Front cover o-rings

5. Turn the crankshaft pulley clockwise until No. 1 cylinder is on the compression stroke. The TDC mark on the pulley without the paint mark should align with the timing pointer.

6. Verify the correct TDC position by ensuring that the lobes of the No. 1 cylinder camshafts are pointing outward. If not rotate the crankshaft 360° until the lobes are correct.

7. Remove the oil pan rear plate cover and install a locking tool into the drive gear.

8. Remove the crankshaft pulley.

9. Remove the oil pan and strainer.

10. Loosen and remove the front cover bolts in the reverse of the tightening sequence.

11. Remove the front cover and pry out the front cover oil seal.

12. Remove the 3 O-rings from the cylinder heads and block.

13. Remove the chain tensioner cover from the front cover.

14. Remove the oil pump drive spacer and oil pump.

15. Compress the left side chain tensioner and install a pin through the hole to secure it, then remove the tensioner.

16. Remove the chain tensioner, tension guide and slack guide.

17. Remove the left side timing chain.

18. While holding the camshaft in place, remove the left side camshaft sprockets.

19. Repeat the procedure on the right side to remove the chain tensioner, guides, timing chain and camshaft sprockets.

To install:

20. Be sure the crankshaft key is pointing toward the center of the left bank. This should be a 45°angle from the center. Check that the camshaft dowel pins are in the correct location as shown.

21. Install or connect the following:
- Camshaft sprockets by holding the camshaft with a wrench and tightening the sprocket bolts to 112 ft. lbs. (152 Nm)
- Crankshaft sprockets and be sure the thick side of the sprocket faces the cylinder block to provide clearance between the block and chain
- Right bank timing chain by aligning the marks on the chain with the marks on the sprockets
- Right slack and chain guides. Be sure to install the bolts in the correct locations. Torque the bolts to 10–14 ft. lbs. (13–19 Nm).
- Timing chain for the left bank in the same manner as the right one
- Left slack and chain guide. Torque the bolts to 10–14 ft. lbs. (13–19 Nm).
- Chain tensioners with the pins installed using new gaskets. Torque the bolts to 61 inch lbs. (6.9 Nm) and remove the pins.

22. Confirm that the timing marks on the crankshaft sprockets and chains are still aligned.
- Oil pump and tighten the bolts to 78 inch lbs. (9 Nm).
- Oil pump drive spacer, aligning the spacer key groove with the crankshaft key, and tapping it in with a plastic hammer
- New seal into the front cover until it is flush with the cover face
- Chain tensioner cover after applying liquid gasket to the mating surface. Tighten the bolts to 78 inch lbs. (9 Nm)
- New O-rings into the block and cylinder heads
- Front cover after applying liquid gasket to the mating surface. Be sure to install the bolts in the correct locations. Tighten the bolts in sequence to 78 inch lbs. (9 Nm).
- Timing control cover using new O-rings and after applying liquid gasket to the mating surface. Tighten the bolts in sequence to 97 inch lbs. (11 Nm).
- Intake valve timing control sensor to both sides

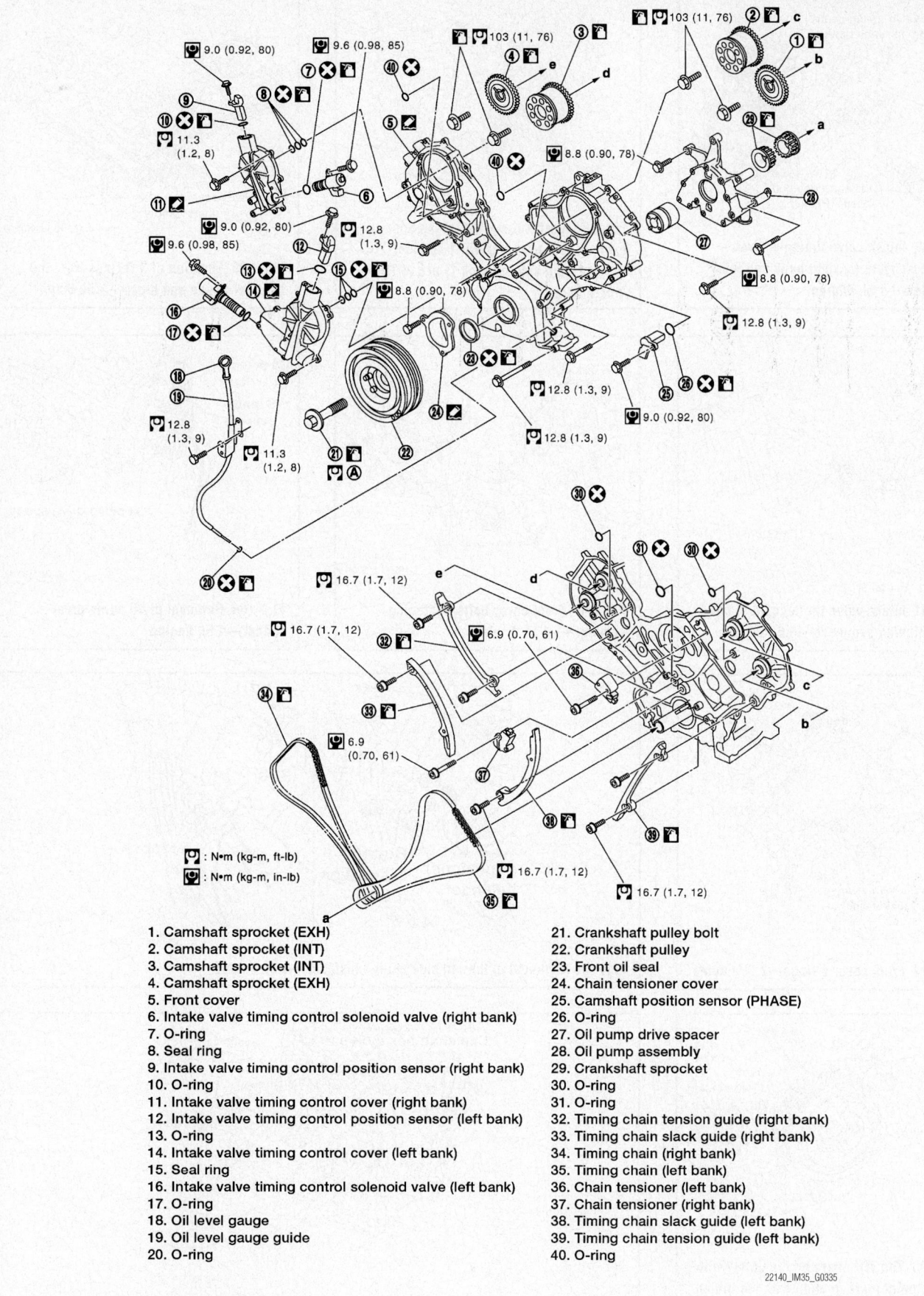

1. Camshaft sprocket (EXH)
2. Camshaft sprocket (INT)
3. Camshaft sprocket (INT)
4. Camshaft sprocket (EXH)
5. Front cover
6. Intake valve timing control solenoid valve (right bank)
7. O-ring
8. Seal ring
9. Intake valve timing control position sensor (right bank)
10. O-ring
11. Intake valve timing control cover (right bank)
12. Intake valve timing control position sensor (left bank)
13. O-ring
14. Intake valve timing control cover (left bank)
15. Seal ring
16. Intake valve timing control solenoid valve (left bank)
17. O-ring
18. Oil level gauge
19. Oil level gauge guide
20. O-ring

21. Crankshaft pulley bolt
22. Crankshaft pulley
23. Front oil seal
24. Chain tensioner cover
25. Camshaft position sensor (PHASE)
26. O-ring
27. Oil pump drive spacer
28. Oil pump assembly
29. Crankshaft sprocket
30. O-ring
31. O-ring
32. Timing chain tension guide (right bank)
33. Timing chain slack guide (right bank)
34. Timing chain (right bank)
35. Timing chain (left bank)
36. Chain tensioner (left bank)
37. Chain tensioner (right bank)
38. Timing chain slack guide (left bank)
39. Timing chain tension guide (left bank)
40. O-ring

22140_IM35_G0335

Fig. 189 Timing chain and sprockets exploded view—4.5L Engine

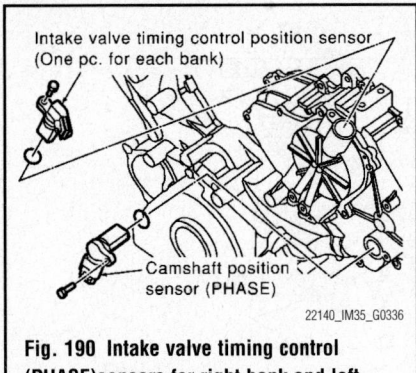

Fig. 190 Intake valve timing control (PHASE)sensors for right bank and left bank view—4.5L Engine

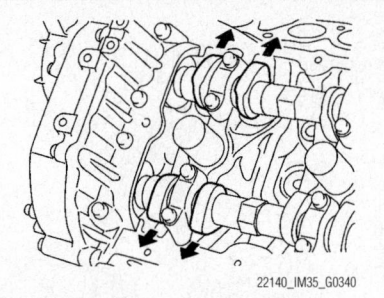

Fig. 194 Verification that intake and exhaust cam noses of No. 1 cylinder (engine front side of bank 1) are at TDC—4.5L Engine

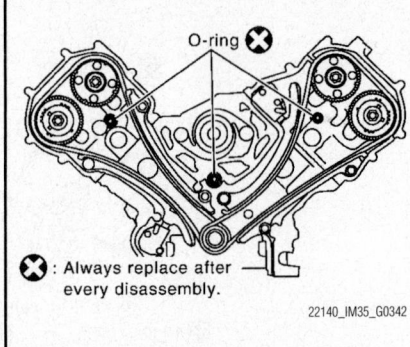

Fig. 198 Location of 3 O-rings from the cylinder heads and block—4.5L Engine

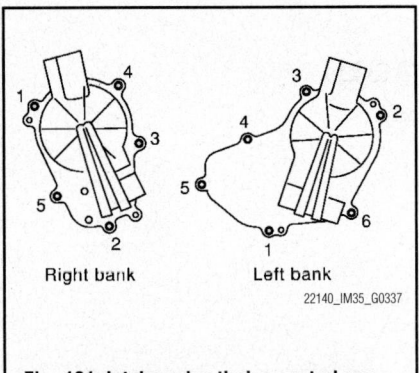

Fig. 191 Intake valve timing control covers tightening sequence—4.5L Engine

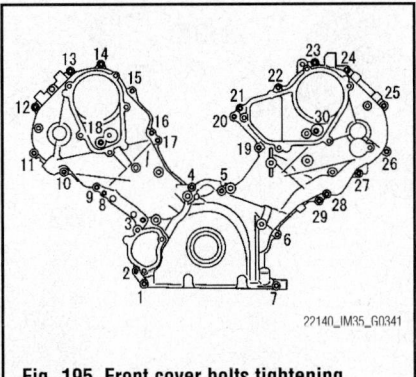

Fig. 195 Front cover bolts tightening sequence—4.5L Engine

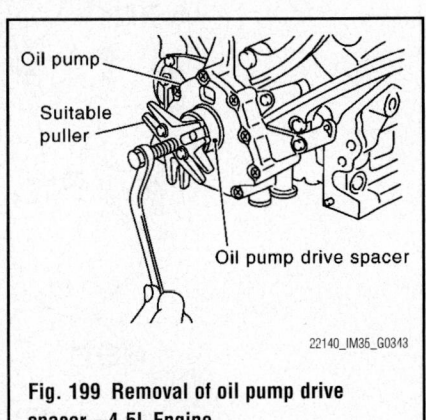

Fig. 199 Removal of oil pump drive spacer—4.5L Engine

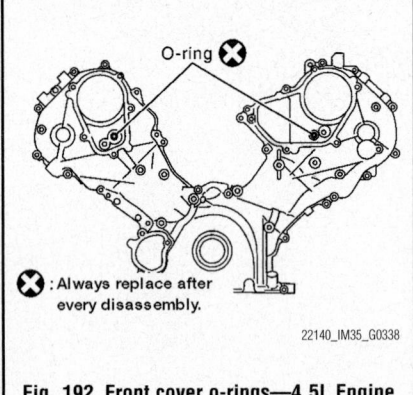

Fig. 192 Front cover o-rings—4.5L Engine

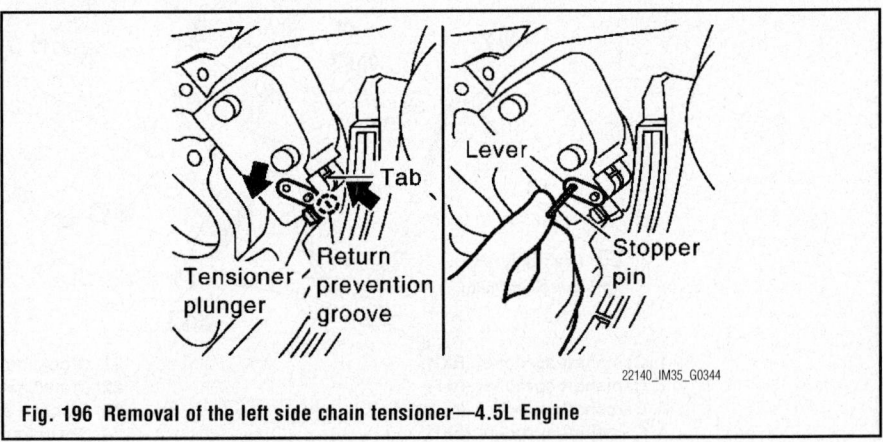

Fig. 196 Removal of the left side chain tensioner—4.5L Engine

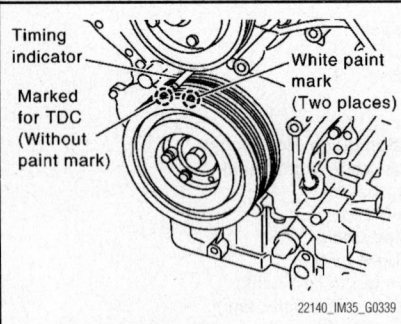
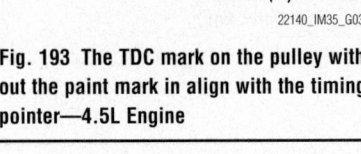

Fig. 193 The TDC mark on the pulley without the paint mark in align with the timing pointer—4.5L Engine

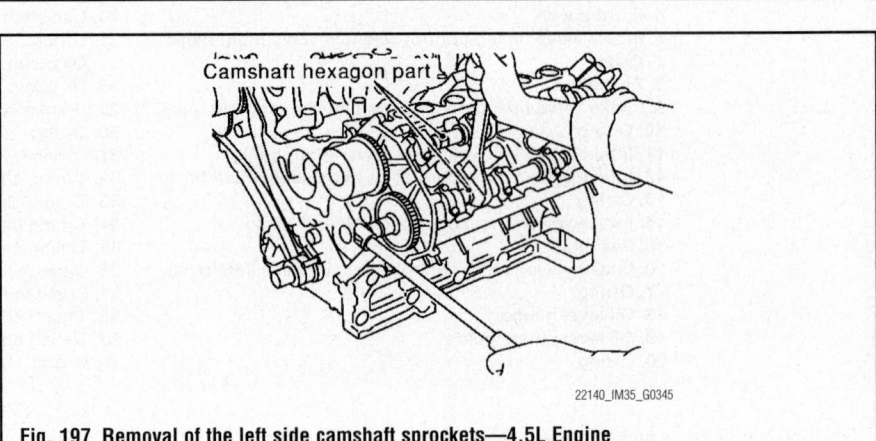

Fig. 197 Removal of the left side camshaft sprockets—4.5L Engine

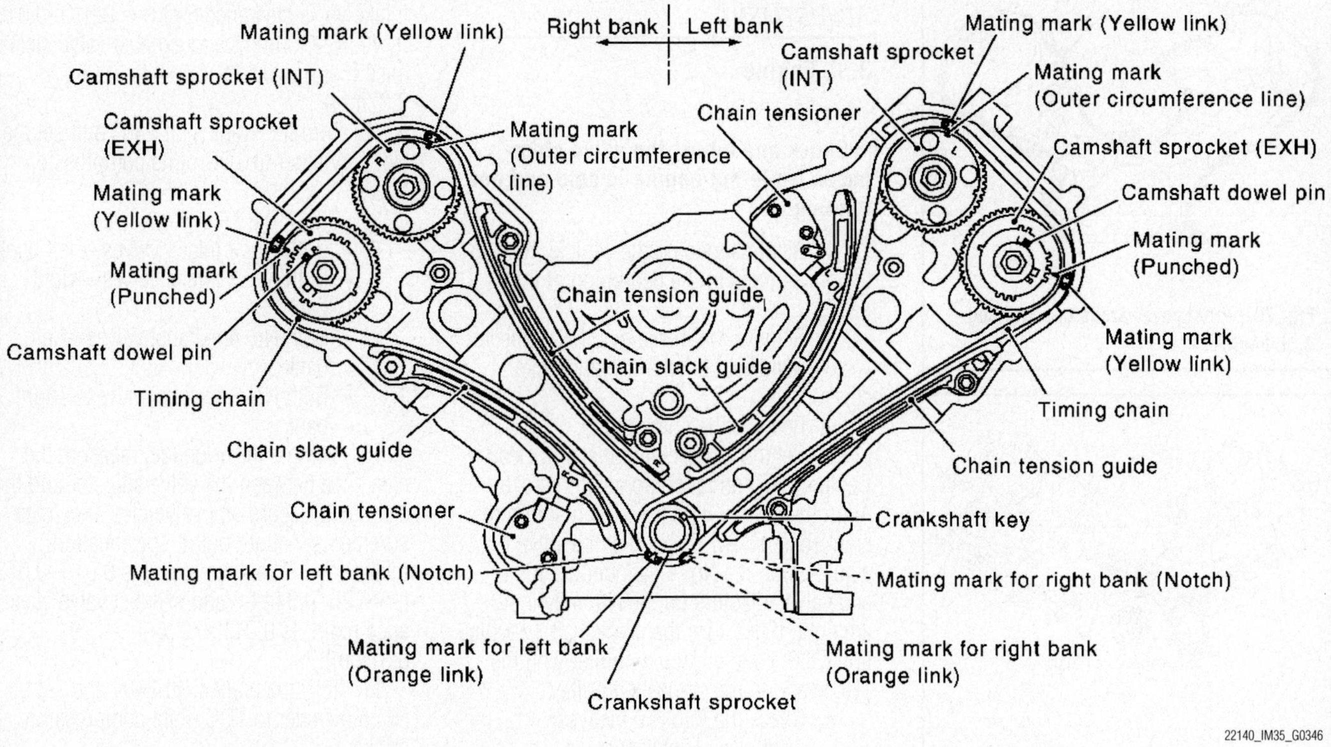

Mating mark (Yellow link)
Right bank
Left bank
Mating mark (Yellow link)
Camshaft sprocket (INT)
Camshaft sprocket
(INT)
Mating mark
(Outer circumference line)
Camshaft sprocket
(EXH)
Mating mark
(Outer circumference
line)
Chain tensioner
Camshaft sprocket (EXH)
Camshaft dowel pin
Mating mark
(Yellow link)
Mating mark
(Punched)
Mating mark
(Punched)
Chain tension guide
Camshaft dowel pin
Mating mark
(Yellow link)
Chain slack guide
Timing chain
Timing chain
Chain slack guide
Chain tension guide
Chain tensioner
Crankshaft key
Mating mark for left bank (Notch)
Mating mark for right bank (Notch)
Mating mark for left bank
(Orange link)
Mating mark for right bank
(Orange link)
Crankshaft sprocket
22140_IM35_G0346

Fig. 200 Exploded view of timing chain and front cover components—4.5L Engine

Dowel pin
Dowel pin
45°
Crankshaft
key
22140_IM35_G0347

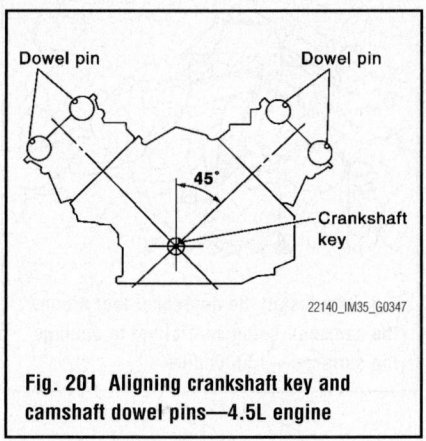

Fig. 201 Aligning crankshaft key and camshaft dowel pins—4.5L engine

Flange
Flange
Crankshaft sprocket
(For left bank)
Crankshaft sprocket
(For right bank)
22140_IM35_G0349

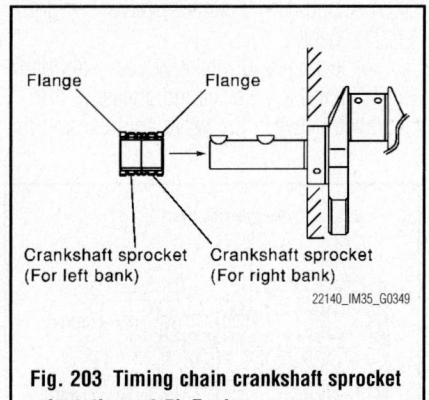

Fig. 203 Timing chain crankshaft sprocket orientation—4.5L Engine

22140_IM35_G0351

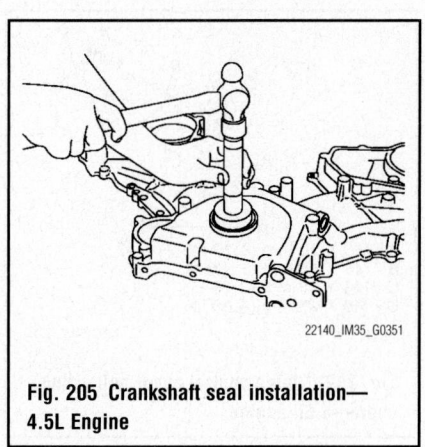

Fig. 205 Crankshaft seal installation—4.5L Engine

R or L
L
R
INT
EXH
22140_IM35_G0348

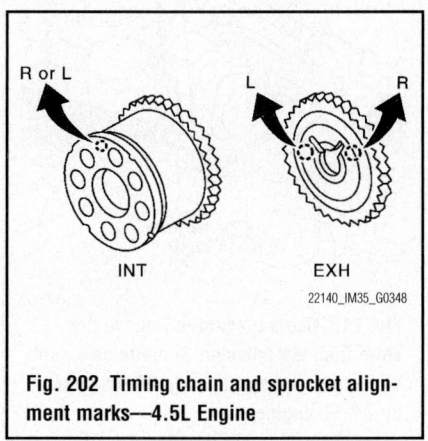

Fig. 202 Timing chain and sprocket alignment marks—4.5L Engine

Key groove
Crankshaft key
Inner rotor
flat face
Drive spacer
flat face
22140_IM35_G0350

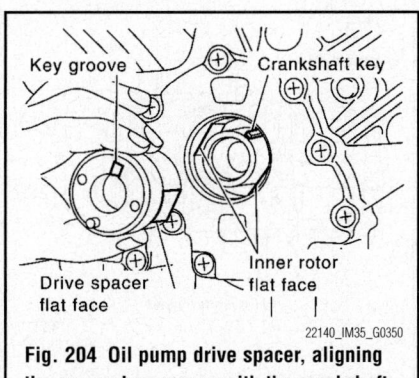

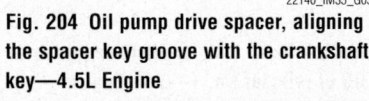

Fig. 204 Oil pump drive spacer, aligning the spacer key groove with the crankshaft key—4.5L Engine

2.6 - 3.1 mm
(0.102 - 0.122 in) dia.
Front cover
Chain tensioner cover
22140_IM35_G0352

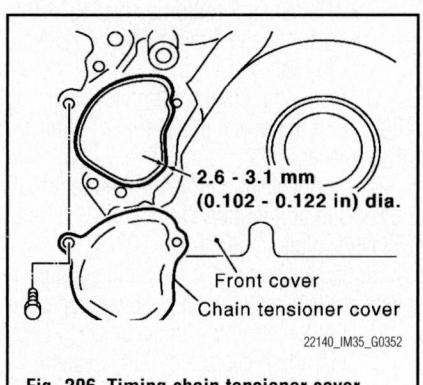

Fig. 206 Timing chain tensioner cover areas to be sealed—4.5L Engine

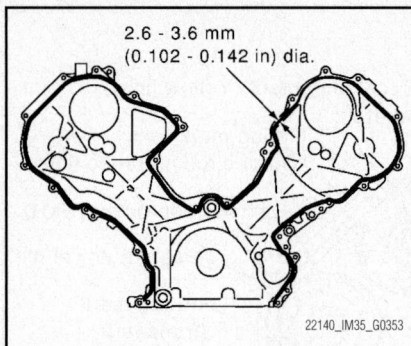

Fig. 207 Front cover areas to be sealed—4.5L Engine

2.6 - 3.6 mm
(0.102 - 0.142 in.) dia.

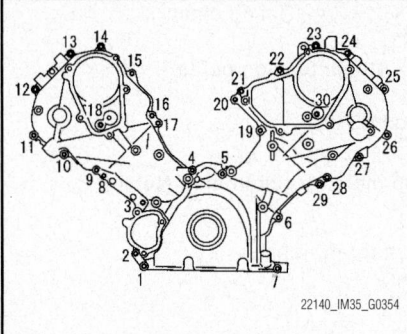

Fig. 208 Timing control cover tightening sequence—4.5L engine

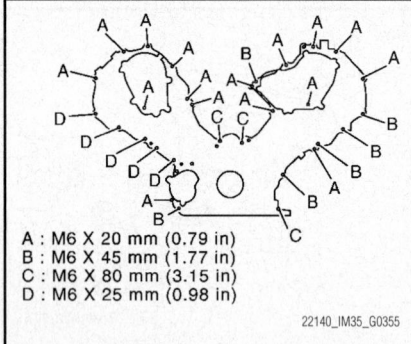

A : M6 X 20 mm (0.79 in)
B : M6 X 45 mm (1.77 in)
C : M6 X 80 mm (3.15 in)
D : M6 X 25 mm (0.98 in)

Fig. 209 Timing control cover bolt sizing chart—4.5L engine

- Camshaft Position sensor (CMP)
- Intake valve timing control solenoid to both sides

23. Install the crankshaft pulley so it aligns with the dowel pin of the oil pump drive spacer.

24. Apply clean engine oil the crankshaft pulley bolt and tighten the bolt to 69 ft. lbs. (93 Nm), plus an additional 90°.

25. Installation of the remaining components is the reverse of the removal procedure.

26. Fill the engine with clean oil.

27. Fill and bleed the cooling system.

28. Start the engine and check for proper operation.

VALVE LASH

ADJUSTMENT

3.5L Engine

See Figures 210 through 212.

➡**Check and adjust the valve clearances while the engine is cold and not running.**

1. Before servicing the vehicle, refer to the precautions in the beginning of this section.

2. Remove the intake manifold collector.

3. Remove the left and right rocker covers.

4. Remove the spark plugs.

5. Set the No. 1 cylinder at Top Dead Center (TDC) on its compression stroke. Align the pointer with the TDC mark on the crankshaft pulley. Check that the valve adjusters on the No. 1 cylinder are loose and valve adjusters on the No. 4 cylinder are tight. If not, turn the crankshaft 1 revolution (360°) and align the pointer with the TDC mark on the crankshaft pulley.

6. Check the following valves:
- Both No. 1 intake valves—right bank
- Both No. 2 exhaust valves—left bank
- Both No. 3 exhaust valves—right bank
- Both No. 6 intake valves—left bank

7. Using a feeler gauge, measure the clearance between the valve adjuster and the

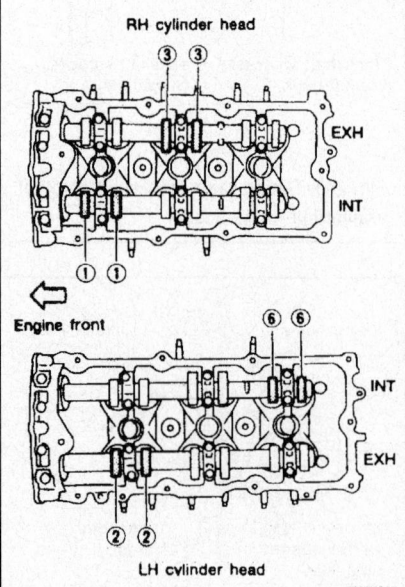

Fig. 210 Valve lash checking sequence at TDC of cylinder No. 1—3.5L Engine

camshaft. Record any valve clearance measurements that are out of specification. Intake valve clearance (cold) is 0.010–0.013 in. (0.26–0.34mm) and exhaust valve clearance (cold) is 0.011–0.015 in. 0.29–0.37mm).

8. Turn the crankshaft 240° and set the No. 3 cylinder to TDC of its compression stroke.

9. Check the following valves:
- Both No. 2 intake valves—left bank
- Both No. 3 intake valves—right bank
- Both No. 4 exhaust valves—left bank
- Both No. 5 exhaust valves—right bank

10. Using a feeler gauge, measure the clearance between the valve adjuster and the camshaft. Record any valve clearance measurements that are out of specification. Intake valve clearance (cold) is 0.010–0.013 in. (0.26–0.34mm) and exhaust valve clearance (cold) is 0.011–0.015 in. (0.29–0.37mm).

11. Turn the crankshaft 240° and set the No. 5 cylinder to TDC of its compression stroke.

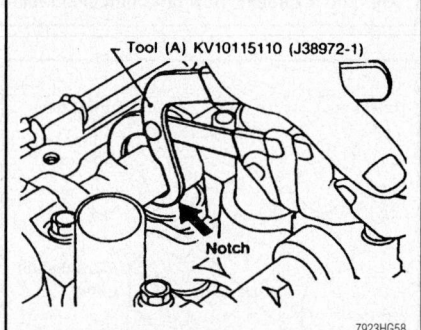

Fig. 211 Install the depressor tool around the camshaft being careful not to damage the surfaces—3.5L Engine

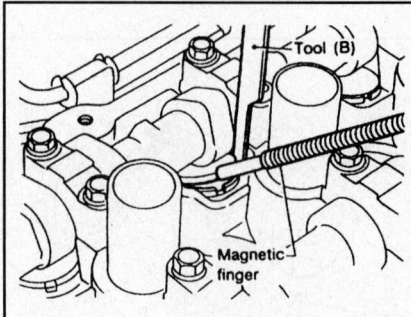

Fig. 212 Use a magnet to remove the shim from the adjuster. Sometimes a shot of compressed air can help lift the shim up—3.5L Engine

12. Check the following valves:
- Both No. 1 exhaust valves—right bank
- Both No. 4 intake valves—left bank
- Both No. 5 intake valves—right bank
- Both No. 6 exhaust valves—left bank

13. Using a feeler gauge, measure the clearance between the valve adjuster and the camshaft. Record any valve clearance measurements that are out of specification. Intake valve clearance (cold) is 0.010–0.013 in. (0.26–0.34mm) and exhaust valve clearance (cold) is 0.011–0.015 inches (0.29–0.37mm).

14. If all the valve clearances are within specification, install the cylinder head cover, spark plugs, and the intake manifold collector.

15. If an adjustment is necessary, adjust the valve clearance while engine is cold by removing the adjusting shim. The adjusting shim can be removed by using the following procedures:

a. Turn the crankshaft so the camshaft lobe of the valve to be adjusted is pointed straight up.

b. Turn the adjuster so the notch is pointed towards the center of the cylinder head; this will facilitate the shim removal process.

c. Using a depressor tool No. KV10115110 push down on the adjuster and insert a keeper tool on the edge of the adjuster to keep the adjuster in the depressed position.

d. Remove the depressor tool and remove the shim with a magnet.

➡**Compressed air can be blown into the hole of the adjuster to separate the adjusting shim from the adjuster.**

16. Determine the replacement adjusting shim size by using the following procedures and formula:

a. Using a micrometer determine thickness of the removed shim.

b. Calculate the thickness of a new adjusting shim so valve clearance is within the specified values.
- $t1$= thickness of the removed shim
- t= thickness of the new shim
- $C1$= measured valve clearance
- $C2$= Standard valve clearance:
- Calculate the Intake Shim as follows: $t = t1 +(c1-C2) - 0.012$ in. (0.30mm)
- Calculate the Exhaust Shim as follows: $N = R + M - 0.013$ in. (0.33mm)

17. Shims are available in 64 sizes from

0.0913–0.1161 in. (2.32–2.95mm) in steps of 0.004 in. (0.01mm). The thickness is stamped on the shim; this side is always installed facing down. Select new shims with thickness as close as possible to calculated valve and install it in the adjuster.

18. Install the new shim onto the adjuster.

19. Depress the adjuster and remove the keeper tool. Remove the depressor tool and recheck the valve clearance. Repeat this procedure for any other valves requiring adjustment.

20. When all valve adjustments are finished, install the cylinder head cover, spark plugs, and the intake manifold collector.

4.5L Engine

See Figure 213.

➡**Check the valve clearances while the engine is warm and not running. Adjustments must be made when the engine is COLD.**

➡**The 4.5L firing order is 1-8-7-3-6-5-4-2. The left bank has cylinders No. 1, 3, 5 and 7 from front to rear and the right bank has cylinders No. 2, 4, 6 and 8 from front to rear.**

1. Before servicing the vehicle, refer to the precautions in the beginning of this section.

2. Remove the engine appearance cover.

3. Remove the left and right rocker covers.

4. Turn the crankshaft clockwise until the TDC mark without the paint mark aligns with the timing pointer.

5. Check that the camshaft lobes on the number one cylinder are pointing outward.

6. Check the following valves:
- Cylinder numbers 1 and 2 intake valves
- Cylinder number 1 exhaust valves

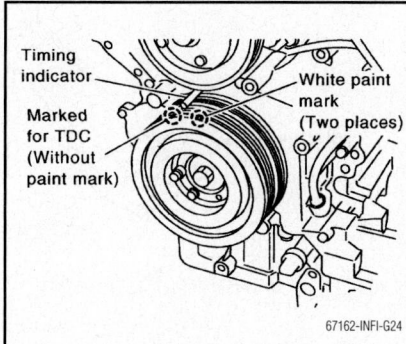

Fig. 213 Identifying No. 1 cylinder TDC mark—4.5L engine

- Cylinder numbers 4 and 5 intake valves
- Cylinder numbers 7 and 8 exhaust valves

7. Using a feeler gauge, measure the clearance between the valve lifter and the camshaft. Record any valve clearance measurements that are out of specification. Intake valve clearance (hot) is 0.012–0.016 in. (0.30–0.41mm) and exhaust valve clearance (hot) is 0.012–0.017 in. 0.30–0.43mm).

8. Turn the crankshaft 270° and set the No. 3 cylinder to TDC of its compression stroke.

9. Check the following valves:
- Cylinder numbers 3 and 4 exhaust valves
- Cylinder numbers 3 and 7 intake valves
- Cylinder numbers 5 and 8 exhaust valves

10. Using a feeler gauge, measure the clearance between the valve lifter and the camshaft. Record any valve clearance measurements that are out of specification. Intake valve clearance (hot) is 0.012–0.016 in. (0.30–0.41mm) and exhaust valve clearance (hot) is 0.012–0.017 in. 0.30–0.43mm).

11. Turn the crankshaft 90° (360° from No. 1 TDC) and set the No. 6 cylinder to TDC of its compression stroke.

12. Check the following valves:
- Cylinder numbers 2 and 6 exhaust valves
- Cylinder number 6 intake valves

13. If all the valve clearances are within specification, install the cylinder head cover and the engine appearance cover.

14. If an adjustment is necessary, adjust the valve clearance while engine is **COLD** by removing the adjusting shim. Refer to Tune Up Specifications for cold clearance measurements.

15. The adjusting shim can be removed by using the following procedures:

a. Turn the crankshaft so the camshaft lobe of the valve to be adjusted is pointed straight up.

b. Using an extra fine screwdriver, turn the round hole of the adjusting shim so it faces toward the center of the cylinder head.

c. Using a depressor tool No. KV10115110 push down on the adjuster and insert a keeper tool on the edge of the adjuster to keep the adjuster in the depressed position.

d. Remove the depressor tool and remove the shim with a magnet.

➡Compressed air can be blown into the hole of the adjuster to separate the adjusting shim from the adjuster.

16. Determine the replacement adjusting shim size by using the following procedures and formula:

a. Using a micrometer determine thickness of the removed shim.

b. Calculate the thickness of a new adjusting shim so valve clearance is within the specified values.

- t1= thickness of the removed shim
- t= thickness of the new shim
- C1= measured valve clearance
- C2= Standard valve clearance
- Calculate the Shims as follows: t = t1 + (C1 - C2) 0.012 in. (0.30mm)
- Calculate the Exhaust Shim as follows: t = t1 + (C1 - C2) 0.013 in. (0.33mm)

17. Shims are available in 64 sizes from 0.0913–0.1161 in. (2.32–2.95mm) in steps of 0.004 in. (0.01mm). The thickness is stamped on the shim; this side is always installed facing down. Select new shims with thickness as close as possible to calculated valve and install it in the adjuster.

18. Install the new shim onto the adjuster.

19. Depress the adjuster and remove the keeper tool. Remove the depressor tool and recheck the valve clearance. Repeat this procedure for any other valves requiring adjustment.

20. When all valve adjustments are finished, install the cylinder head cover and the engine appearance cover.

ENGINE PERFORMANCE & EMISSION CONTROL

ACCELERATOR PEDAL POSITION (APP) SENSOR

LOCATION

See Figure 214.

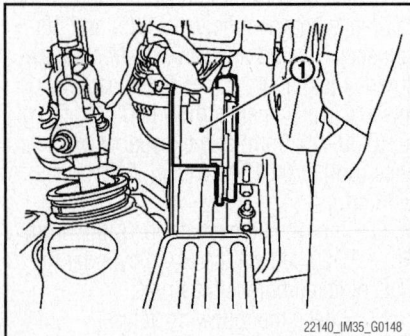

22140_IM35_G0148

Fig. 214 Accelerator Pedal Position (APP) sensor

REMOVAL & INSTALLATION

See Figure 215.

1. Disconnect accelerator pedal position sensor harness connector.
2. Remove front kicking plate and dash side finisher.
3. Remove the cap and the inside mounting nut, and then disassemble the accelerator pedal pad from the floor carpet.
4. Press the pin (1) with long-nose pliers and pull them out in the direction shown by the arrow. Then remove the accelerator pedal pad (2).

> ✳✳ **WARNING**
>
> **Never disengage any part s(the link) other than the pins, remove accelerator pedal stopper cover.**

5. Pull up the floor carpet.
6. Remove mounting nuts of accelerator pedal bracket.

7. Remove accelerator pedal bracket and lever assembly.

> ✳✳ **WARNING**
>
> **Never disassemble accelerator lever. Never remove accelerator pedal position sensor from accelerator lever.**

> ✳✳ **WARNING**
>
> **Avoid impact from dropping etc. during handling.**

> ✳✳ **WARNING**
>
> **Be careful to keep accelerator lever away from water.**

8. Installation is the reverse order of removal.

9. Tighten the base of the assembly to 45 inch lbs. (5 Nm).

10. Tighten the top of the assembly to 45 inch lbs. (5 Nm).

AIR FUEL RATIO (A/F) SENSORS

LOCATION

See Figure 216.

The air fuel ratio (A/F) sensors are placed in the right and left exhaust banks.

REMOVAL & INSTALLATION

1. Using the heated oxygen sensor wrench (SST), remove air fuel ratio sensor 1 (right bank and left bank).

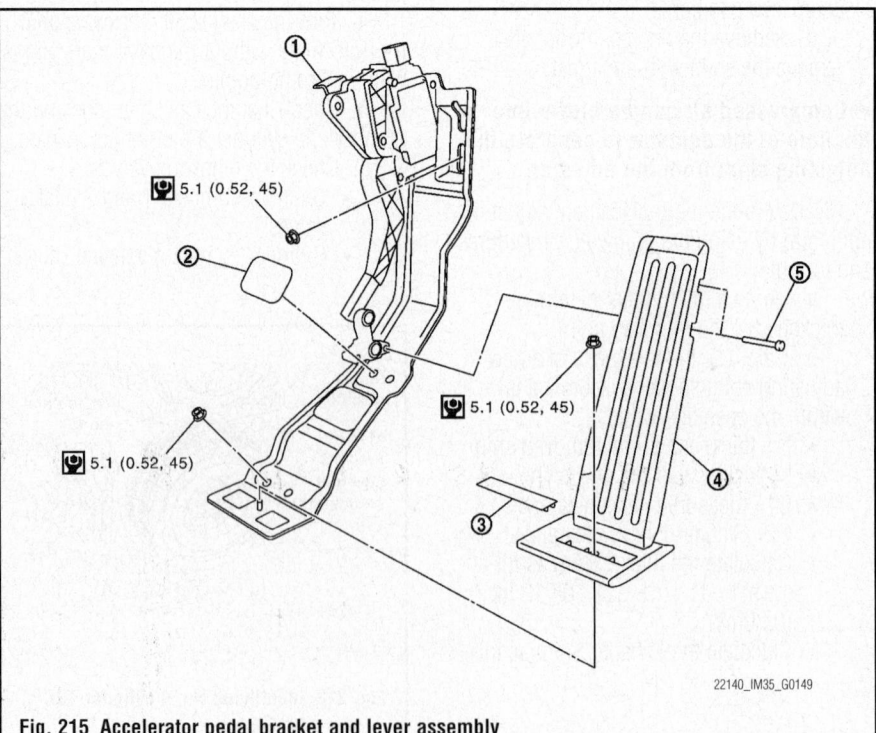

22140_IM35_G0149

Fig. 215 Accelerator pedal bracket and lever assembly

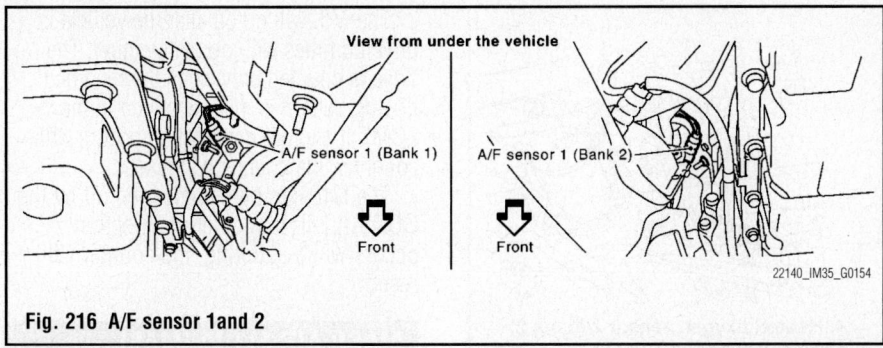

Fig. 216 A/F sensor 1 and 2

CAMSHAFT POSITION (CMP) SENSOR

LOCATION

See Figures 217 and 218.

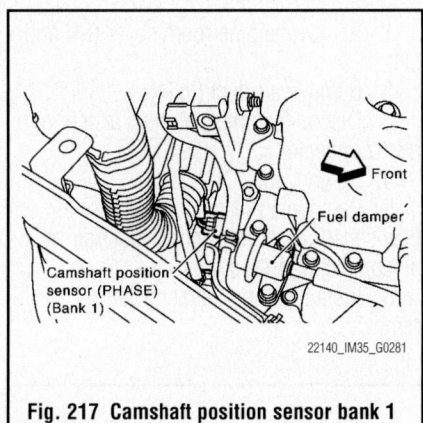

Fig. 217 Camshaft position sensor bank 1

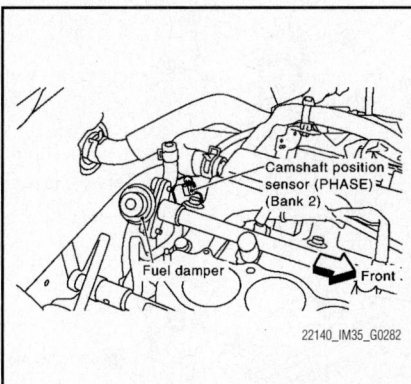

Fig. 218 Camshaft position sensor bank 1

REMOVAL & INSTALLATION

1. Remove the Camshaft Position Sensor (CMP) connector
2. Remove the CMP.

To install:

3. Install the CMP.
4. Install the CMP connector.
5. Tighten the bolt to 7.5 inch lbs. (10 Nm).

CRANKSHAFT POSITION (CKP) SENSOR

LOCATION

The crankshaft position sensor (POS) is located on the A/T converter housing facing the gear teeth (cogs) of the signal plate.

REMOVAL & INSTALLATION

See Figure 219.

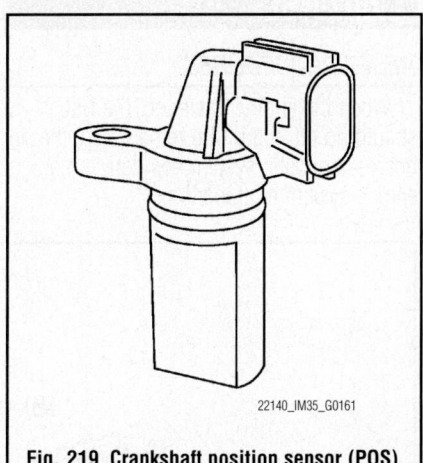

Fig. 219 Crankshaft position sensor (POS)

1. Loosen the fixing bolt of the sensor.
2. Disconnect crankshaft position sensor (POS) harness connector.
3. Remove the sensor.
4. Visually check the sensor for chipping.

ELECTRONIC CONTROL MODULE (ECM)

LOCATION

The ECM is located behind the passenger side instrument lower panel. For this inspection, remove passenger side instrument lower panel.

REMOVAL & INSTALLATION

See Figure 220.

1. Remove ECM harness connector.

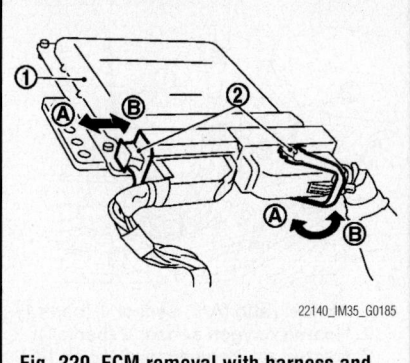

Fig. 220 ECM removal with harness and connector

2. When disconnecting ECM harness connector, loosen (A) it with levers (2) as far as they will go as shown in the figure.

ENGINE COOLANT TEMPERATURE (ECT) SENSOR

LOCATION

See Figure 221.

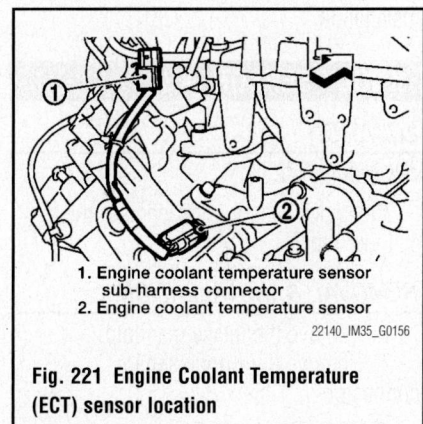

1. Engine coolant temperature sensor sub-harness connector
2. Engine coolant temperature sensor

Fig. 221 Engine Coolant Temperature (ECT) sensor location

REMOVAL & INSTALLATION

1. Disconnect the harness connector.
2. Remove the Engine Coolant Temperature (ECT) sensor.

HEATED OXYGEN (HO2S) SENSOR

LOCATION

See Figure 222.

The Heated oxygen sensor sensors are placed in the right and left exhaust banks.

REMOVAL & INSTALLATION

1. Using the heated oxygen sensor wrench (SST), heated oxygen sensor 1 (right bank and left bank).

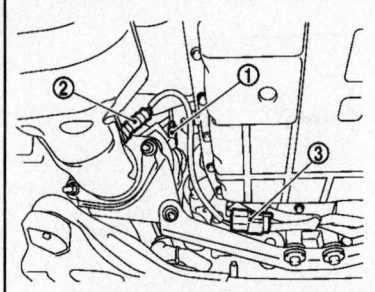
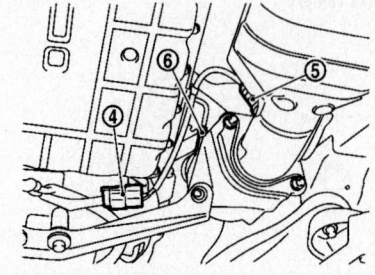

1. Air fuel ratio (A/F) sensor 1 (bank 1)
2. Heated oxygen sensor 2 (bank 1)
3. Heated oxygen sensor 2 (bank 1) harness connector

4. Heated oxygen sensor 2 (bank 2) harness connector
5. Heated oxygen sensor 2 (bank 2)
6. Air fuel ratio (A/F) sensor 1 (bank 2)

22140_IM35_G0159

Fig. 222 Heated oxygen (HO2S2) sensor and Air Fuel (A/F) ratio sensors locations

INTAKE AIR TEMPERATURE (IAT) SENSOR

LOCATION

See Mass Air Flow (MAF) sensor.

REMOVAL & INSTALLATION

See Mass Air Flow Sensor removal and installation.

KNOCK SENSOR (KS)

LOCATION

See Figure 223.

The knock sensor is attached to the cylinder block.

REMOVAL & INSTALLATION

1. Remove the intake manifold.
2. Remove the Knock sensor connector.

To install:

3. Install knock sensor so that connector faces front of the engine.
4. After installing knock sensor, connect harness connector, and lay it out to rear of the engine.

➡**Make sure that knock sensor does not interfere with other parts.**

5. Tighten the knock sensor to 20 ft. lbs. (27 Nm).

MALFUNCTION INDICATOR LIGHT (MIL)

RESET PROCEDURES

When the engine is started, the MIL should go off. If the MIL remains on, the on board diagnostic system has detected an engine system malfunction.

The MIL will go off after the vehicle is driven 3 times with no malfunction. The drive is counted only when the recorded driving pattern is met (as stored in the ECM). If another malfunction occurs while counting, the counter will reset.

The MIL can be commended off by the CONSULT-II, If another malfunction occurs while counting, the counter will reset.

MASS AIR FLOW (MAF) SENSOR

LOCATION

The Mass Air Flow (MAF) sensor is placed in the stream of intake air.

REMOVAL & INSTALLATION

1. Remove engine room cover (RH and LH).
2. Remove air duct (inlet).
3. Disconnect mass air flow sensor harness connector.
4. Disconnect PCV hose.
5. Remove air cleaner case/mass air flow sensor assembly and air duct/air hose disconnecting their joints.
6. Install in the reverse order of removal.

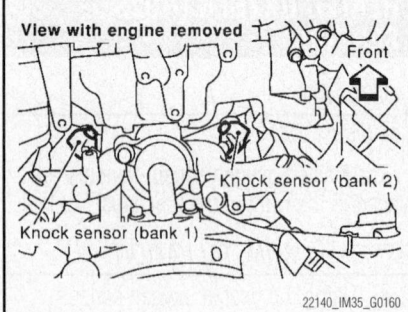

View with engine removed

Knock sensor (bank 2)

Knock sensor (bank 1)

22140_IM35_G0160

Fig. 223 Knock sensor (KS) location bank 1 and bank 2

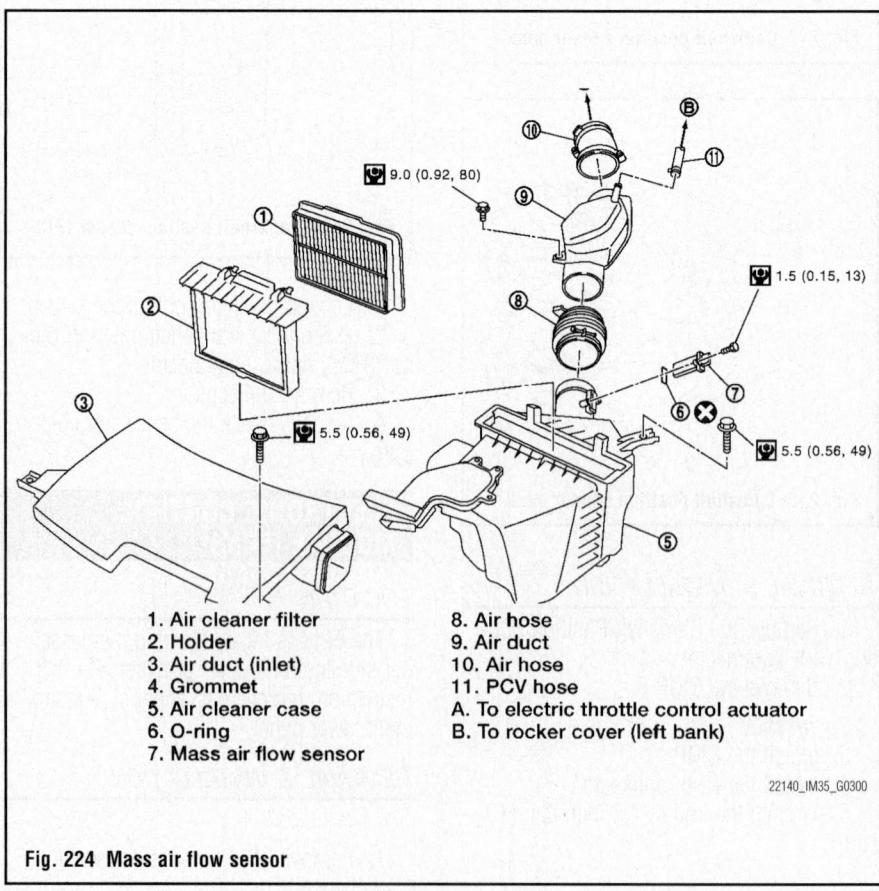

1. Air cleaner filter
2. Holder
3. Air duct (inlet)
4. Grommet
5. Air cleaner case
6. O-ring
7. Mass air flow sensor
8. Air hose
9. Air duct
10. Air hose
11. PCV hose
A. To electric throttle control actuator
B. To rocker cover (left bank)

22140_IM35_G0300

Fig. 224 Mass air flow sensor

THROTTLE POSITION SENSOR (TPS)

LOCATION
See Figure 225.

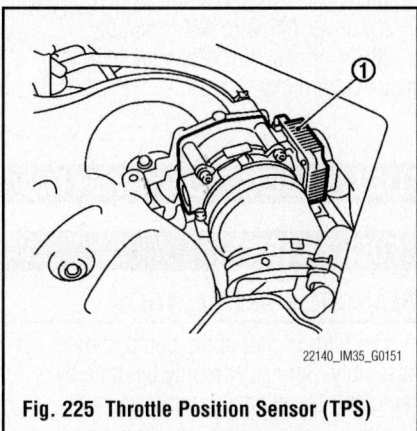

22140_IM35_G0151

Fig. 225 Throttle Position Sensor (TPS)

REMOVAL & INSTALLATION

1. Disconnect electric throttle control actuator harness connector.
2. Remove rubber duct clamp
3. Remove mounting screws.

VEHICLE SPEED SENSOR (VSS)

LOCATION
See Figure 226.

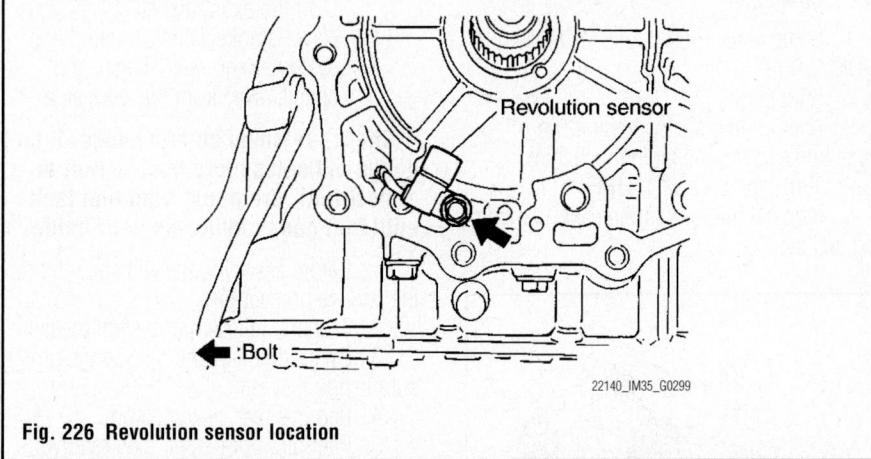

22140_IM35_G0299

Fig. 226 Revolution sensor location

Located in the transmission main transmission housing between the tail shaft section and the rear case.

REMOVAL & INSTALLATION
See Figures 226 through 228.

1. Disconnect the battery cable from the negative terminal.
2. Drain ATF through drain plug.

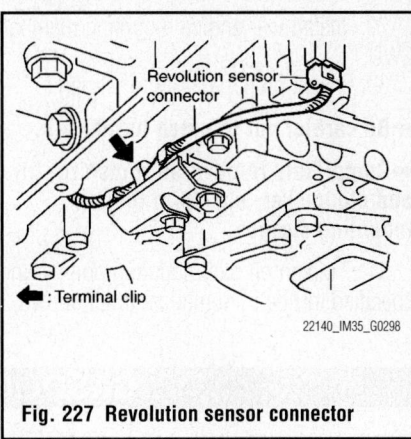

22140_IM35_G0298

Fig. 227 Revolution sensor connector

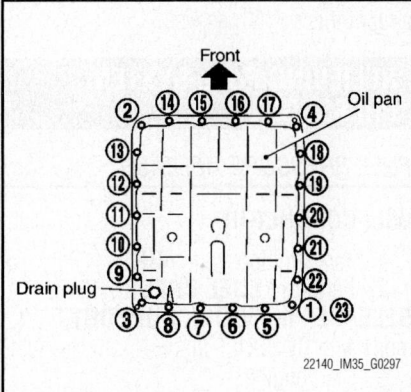

22140_IM35_G0297

Fig. 228 Oil pan mounting bolts tightening sequence

3. Remove exhaust front tube and center muffler with power tool.
4. Remove rear propeller shaft.
5. Remove control rod.
6. Disconnect heated oxygen sensor 2 harness connectors (A).
7. Remove heated oxygen sensor 2 harness (B) from clips (1).
8. Remove bracket (2) from transmission assembly. (3.5L Engine)

9. Remove clips (1).
10. Remove oil pan (2) and oil pan gasket.
11. Remove clips (1) and brackets (2). (4.5L Engine)
12. Remove oil pan (3) and oil pan gasket
13. Remove rear extension assembly (3.5L Engine) or output shaft & companion flange complement (4.5L Engine) according to the following procedures.
14. Remove tightening bolts for rear extension assembly and transmission case
15. Remove rear extension assembly from transmission case. (With needle bearing.)
16. Remove tightening bolts (1) for output shaft & companion flange complement and transmission case 4.5L Engine.
17. Remove output shaft & companion flange complement from transmission case.
18. Disconnect revolution sensor connector.

➡**Be careful not to damage connector**

19. Straighten terminal clip to free revolution sensor harness.
20. Remove revolution sensor from transmission case.

➡**Do not subject it to impact by dropping or hitting it.**

➡**Do not disassemble.**

➡**Do not allow metal filings, etc., to get on the sensor's front edge magnetic area.**

➡**Do not place in an area affected by magnetism.**

To install:

21. Connect revolution sensor connector.
22. Securely fasten revolution sensor harness with clip.
23. Install rear extension assembly (3.5L Engine) or output shaft & companion flange complement (4.5L Engine) according to the following procedures.

➡**Completely remove all moisture, oil and old sealant, etc. from transmission case and rear extension assembly mounting surfaces.**

24. Install rear extension assembly to transmission case. (With needle bearing.)

➡**Insert the tip of parking rod between the parking pole and the parking actuator support when assembling the rear extension assembly.**

25. Tighten rear extension assembly bolts to the specified torque (3.5L Engine).

➡**Do not reuse self-sealing bolts.**

26. Apply recommended sealant (Genuine Anaerobic Liquid Gasket or equivalent.

27. Install output shaft & companion flange complement to transmission case 4.5L Engine.

28. Tighten output shaft & companion flange complement bolts to the specified torque.

➡**Do not reuse self-sealing bolts.**

29. Install rear engine mounting member 2WD.

30. Install oil pan gasket to oil pan

➡**Be careful not to pinch harnesses.**

➡**Completely remove all moisture, oil and old gasket, etc. from oil pan mounting surface.**

31. Tighten oil pan mounting bolts to the specified torque in numerical order shown

in the figure after temporarily tightening them.

32. Tighten oil pan mounting bolts to 70 inch lbs. (7.9 Nm).

33. Install rear propeller shaft.

34. Install exhaust front tube and center muffler

35. Pour ATF into A/T assembly

36. Connect the battery cable to the negative terminal.

FUEL   GASOLINE FUEL INJECTION SYSTEM

FUEL SYSTEM SERVICE PRECAUTIONS

Safety is the most important factor when performing not only fuel system maintenance but any type of maintenance. Failure to conduct maintenance and repairs in a safe manner may result in serious personal injury or death. Maintenance and testing of the vehicle's fuel system components can be accomplished safely and effectively by adhering to the following rules and guidelines.

• To avoid the possibility of fire and personal injury, always disconnect the negative battery cable unless the repair or test procedure requires that battery voltage be applied.

• Always relieve the fuel system pressure prior to disconnecting any fuel system component (injector, fuel rail, pressure regulator, etc.), fitting or fuel line connection. Exercise extreme caution whenever relieving fuel system pressure to avoid exposing skin, face and eyes to fuel spray. Please be advised that fuel under pressure may penetrate the skin or any part of the body that it contacts.

• Always place a shop towel or cloth around the fitting or connection prior to loosening to absorb any excess fuel due to spillage. Ensure that all fuel spillage (should it occur) is quickly removed from engine surfaces. Ensure that all fuel soaked cloths or towels are deposited into a suitable waste container.

• Always keep a dry chemical (Class B) fire extinguisher near the work area.

• Do not allow fuel spray or fuel vapors to come into contact with a spark or open flame.

• Always use a back-up wrench when loosening and tightening fuel line connection fittings. This will prevent unnecessary stress and torsion to fuel line piping.

• Always replace worn fuel fitting O-rings with new Do not substitute fuel hose or equivalent where fuel pipe is installed.

Before servicing the vehicle, make sure to also refer to the precautions in the beginning of this section as well.

RELIEVING FUEL SYSTEM PRESSURE

FUEL PRESSURE RELEASE

With CONSULT-III

1. Turn ignition switch **ON**.
2. Perform **"FUEL PRESSURE RELEASE"** in **"WORK SUPPORT"** mode with CONSULT-III.
3. Start engine.
4. After engine stalls, crank it two or three times to release all fuel pressure.
5. Turn ignition switch **OFF**.

Without CONSULT-III

See Figure 229.

1. Remove fuel pump fuse (1) located in IPDM E/R (2).
2. Start engine.
3. After engine stalls, crank it two or three times to release all fuel pressure.
4. Turn ignition switch **OFF**.
5. Reinstall fuel pump fuse after servicing fuel system.

22140_IM35_G0280

Fig. 229 Fuel pump fuse (1) located in IPDM E/R (2)

FUEL FILTER

REMOVAL & INSTALLATION

The filter is part of the pump and sender assembly. Service can only be done by removing the sender - pump assembly.

FUEL PUMP

REMOVAL & INSTALLATION

See Figure 230.

❊❊ WARNING

When replacing fuel line parts, be sure to observe the following:

• Put a **CAUTION FLAMMABLE** sign in the workshop.
• Be sure to work in a well ventilated area and furnish workshop with a CO2 fire extinguisher.
• Never smoke while servicing fuel system. Keep open flames and sparks away from the work area.

➡**Check fuel level on fuel gauge. If fuel gauge indicates more than ⅞ (full or almost full), drain fuel from fuel tank until fuel gauge indicates ⅞ or below.**

1. Before servicing the vehicle, refer to the service precautions.

2. Properly relieve fuel system pressure.

3. Open filler cap and release the pressure inside fuel tank.

4. Remove rear seat cushion.

5. Peel off floor carpet, then remove inspection hole cover units by turning clips clockwise by 90 degrees.

6. The right side contains the main fuel level sensor unit, fuel filter and fuel pump assembly.

7. Remove or disconnect the following:
• Negative battery cable
• Fuel feed tube
• Fuel pump connector
• Fuel pump assembly
• Fuel pump

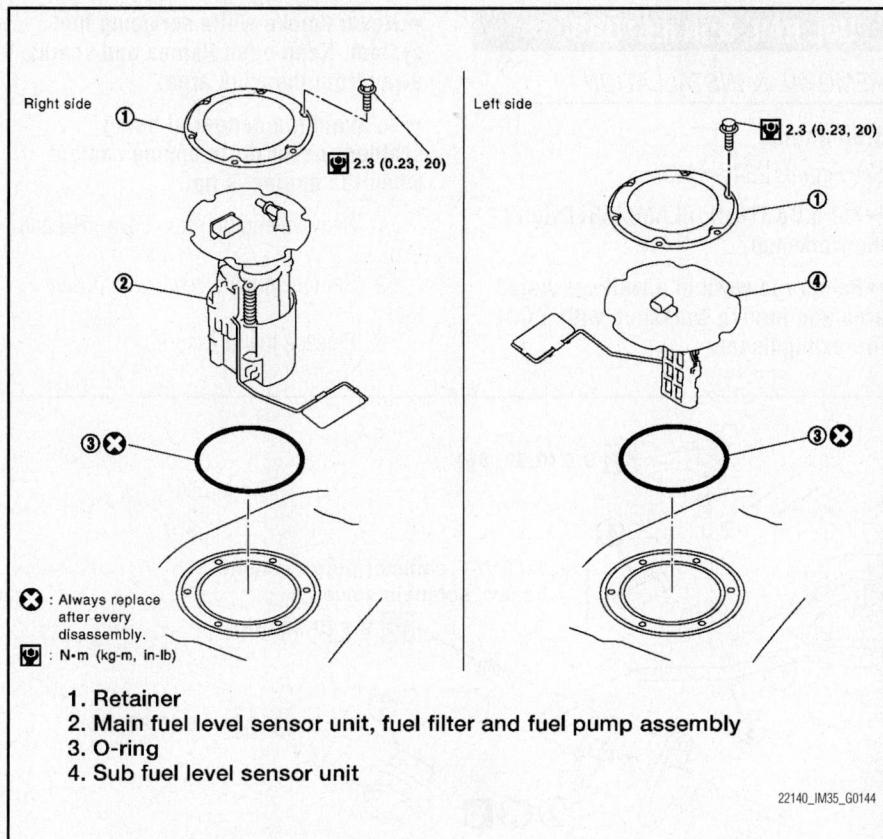

Right side

Left side

2.3 (0.23, 20)

2.3 (0.23, 20)

❌ : Always replace after every disassembly.

N•m (kg-m, in-lb)

1. Retainer
2. Main fuel level sensor unit, fuel filter and fuel pump assembly
3. O-ring
4. Sub fuel level sensor unit

22140_IM35_G0144

Fig. 230 Main fuel level sensor unit, fuel filter and fuel pump assembly

To install:

8. Installation is the reverse of the removal procedure.

9. Tighten the assembly to 20 inch lbs. (2.3 Nm).

10. Start the vehicle, check for leaks and repair if necessary.

FUEL TANK

REMOVAL & INSTALLATION

See Figures 231 through 233.

1. Drain fuel from fuel tank if necessary.

2. Perform work on level place.

3. Remove fuel level sensor and fuel pump units

4. Remove exhaust front tube, center muffler and main muffler.

5. Remove propeller shaft.

6. Remove parking rear brake cables.

7. Remove rear suspension assembly.

8. Remove fuel tank protector.

9. Disconnect fuel filler hose, vent hose and EVAP hoses at fuel tank side.

10. Remove fuel tank mounting bands.

11. Supporting with hands, descend transmission jack carefully, and remove fuel tank.

1. Fuel filler cap
2. Grommet
3. Fuel filler tube
4. Vent hose
5. Vent tube
6. EVAP hose
7. Clamp
8. Fuel filler hose
9. Vent hose
10. Fuel tank mounting band
11. Fuel tank protector
12. Insulator
13. Fuel tank

43.0 (4.4, 32)

5.0 (0.51, 44)

5.0 (0.51, 44)

12.8 (1.3, 9)

12.8 (1.3, 9)

5.0 (0.51, 44)

22140_IM35_G0145

Fig. 231 Fuel tank and components

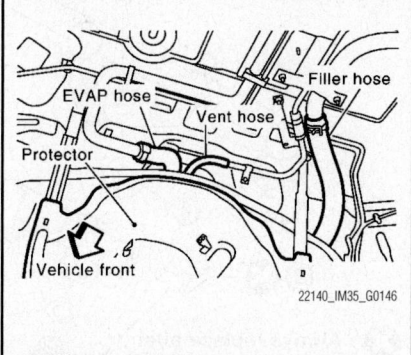

EVAP hose

Filler hose

Vent hose

Protector

Vehicle front

22140_IM35_G0146

Fig. 232 Fuel filler hose, vent hose and EVAP hoses at fuel tank side

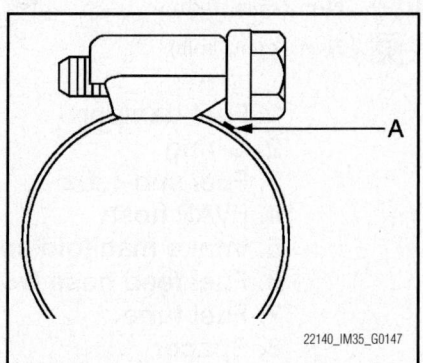

A

22140_IM35_G0147

Fig. 233 Clamp marks to gauge the proper positioning

To install:

➡Surely clamp fuel hoses and insert hose to the length below.

➡Be sure hose clamp is not placed on swelled area of fuel tube.

➡Tighten the clamp band with the top mark (A) until the mark is on the bolt head flange.

12. Fuel filler hose : 1.38 inches (35 mm)
13. The other hoses : 0.98 inches (25 mm)

FUEL RAIL & INJECTORS

REMOVAL & INSTALLATION

3.5L Engine

See Figures 234 and 235.

➡Put a CAUTION: FLAMMABLE sign in the workshop.

➡Be sure to work in a well ventilated area and furnish workshop with a CO2 fire extinguisher.

➡Never smoke while servicing fuel system. Keep open flames and sparks away from the work area.

➡To avoid the danger of being scalded, never drain engine coolant when the engine is hot.

1. Remove engine room cover (RH and LH).
2. Remove engine cover with power tool.
3. Release fuel pressure.

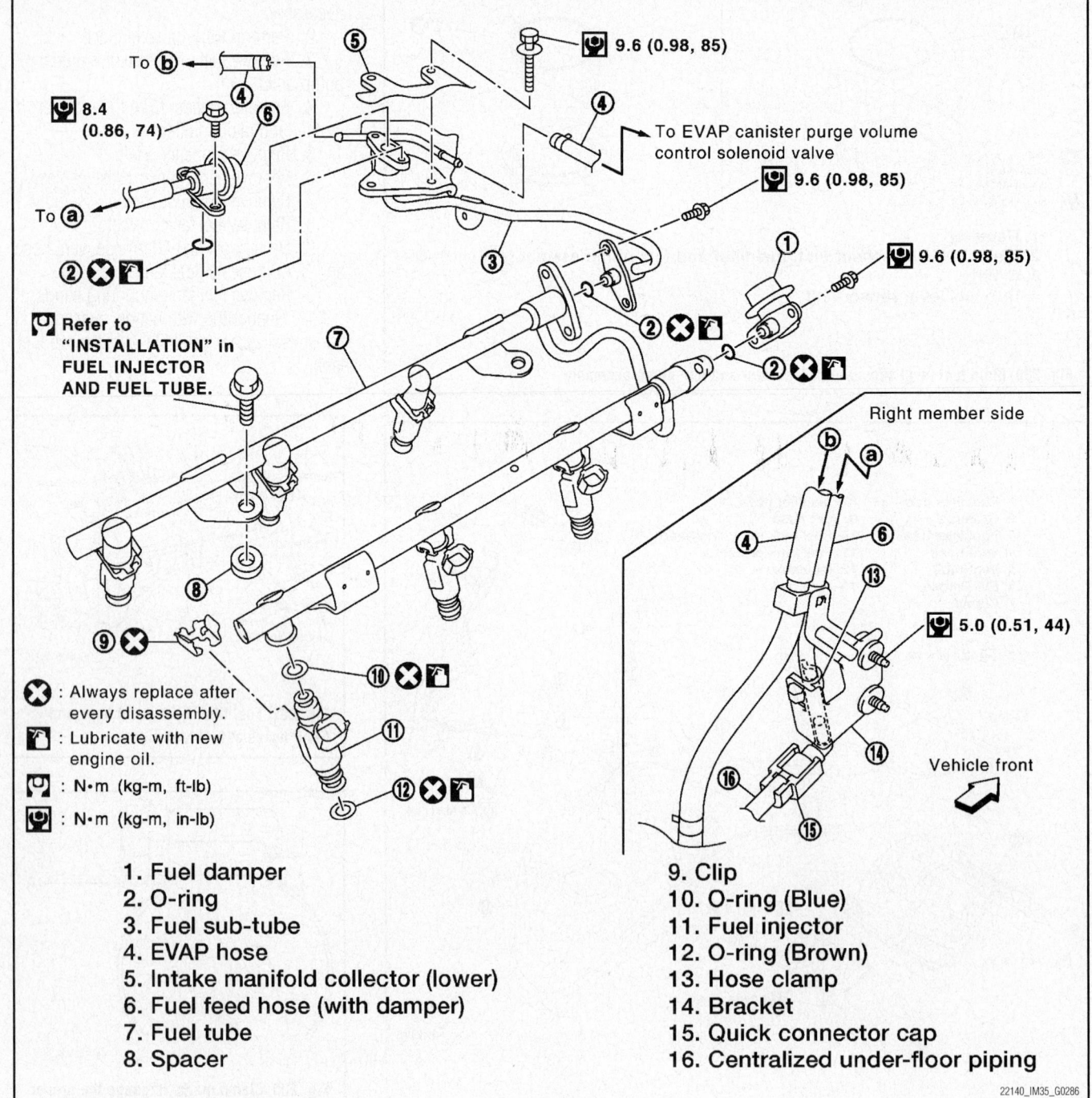

1. Fuel damper
2. O-ring
3. Fuel sub-tube
4. EVAP hose
5. Intake manifold collector (lower)
6. Fuel feed hose (with damper)
7. Fuel tube
8. Spacer
9. Clip
10. O-ring (Blue)
11. Fuel injector
12. O-ring (Brown)
13. Hose clamp
14. Bracket
15. Quick connector cap
16. Centralized under-floor piping

22140_IM35_G0286

Fig. 234 Fuel injectors and fuel rail components

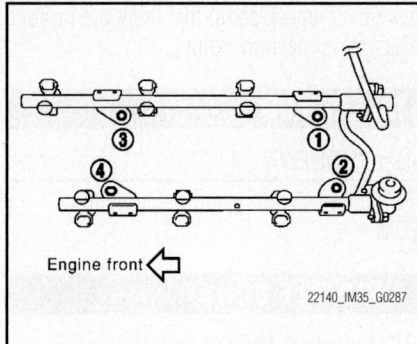

Fig. 235 Numerical order of tightening steps for the fuel rail

4. Drain engine coolant, or when water hoses are disconnected, attach plug to prevent engine coolant leakage.

5. Perform this step when the engine is cold.

6. Remove fuel feed hose (with damper) from fuel sub-tube.

➡There is no fuel return route.

➡While hoses are disconnected, plug them to prevent fuel from draining.

❊❊ WARNING

Never separate damper and hose.

7. When separating fuel feed hose (with damper) and centralized under-floor piping connection, disconnect quick connector as follows.

  a. Remove quick connector cap from quick connector connection on right member side.

  b. Disconnect fuel feed hose (with damper) from bracket hose clamp

8. Remove intake manifold collectors (upper and lower).

9. Disconnect harness connector from fuel injector.

10. Loosen mounting bolts in reverse order as shown in the figure, and remove fuel tube and fuel injector assembly.

11. Remove spacers on intake manifold.

12. Remove fuel sub-tube and fuel damper.

To install:

13. Install fuel damper and fuel sub-tube.

➡When handling new O-rings, be careful of the following caution:

  a. Handle O-ring with bare hands. Never wear gloves.

  b. Lubricate O-ring with new engine oil.

  c. Never clean O-ring with solvent.

  d. Check that O-ring and its mating part are free of foreign material.

  e. When installing O-ring, be careful not to scratch it with tool or fingernails. Also be careful not to twist or stretch O-ring. If O-ring was stretched while it was being attached, never insert it quickly into fuel tube.

  f. Insert new O-ring straight into fuel tube. Never move from center or twist it.

14. Insert fuel damper and fuel sub-tube straight into fuel tube.

15. Tighten mounting bolts evenly in turn.

16. After tightening mounting bolts, check that there is no gap between flange and fuel tube.

17. Install new O-rings to fuel injector, paying attention to the following.

➡Upper and lower O-ring are different, be careful not to confuse them.

18. Fuel tube side o rings are blue.

19. Nozzle side o rings are brown.

20. Install fuel injector to fuel tube as follows:

  a. Insert clip into clip mounting groove on fuel injector.

  b. Check that installation is complete by checking that fuel injector does not rotate or come off.

21. Install spacers on intake manifold.

22. Install fuel tube and fuel injector assembly to intake manifold.

23. Tighten mounting bolts in two steps in numerical order as shown in the figure.

24. Tightening torque is as follows

25. Step 1: 7 ft. lb. (10.1 Nm).

26. Step 2: 17 ft. lb. (23.6 Nm).

27. Connect injector sub-harness.

28. Install intake manifold collectors (upper and lower).

29. Install fuel sub-tube on rear end of intake manifold collector (lower).

30. Connect fuel feed hose (with damper).

31. Connect quick connector between fuel feed hose (with damper) and centralized under-floor piping connection.

32. Remove engine cover with power tool.

33. Remove engine room cover (RH and LH).

34. Fill engine coolant.

35. Turn ignition switch **ON** (with the engine stopped). With fuel pressure applied to fuel piping, check there are no fuel leaks at connection points.

➡Use mirrors for checking at points out of clear sight.

36. Start the engine, with engine speed increased, check again that there are no fuel leaks at connection Points.

4.5L Engine

See Figures 236 and 237.

➡Put a CAUTION: FLAMMABLE sign in the workshop.

➡Be sure to work in a well ventilated area and furnish workshop with a CO2 fire extinguisher.

➡Never smoke while servicing fuel system. Keep open flames and sparks away from the work area.

➡To avoid the danger of being scalded, never drain engine coolant when the engine is hot.

1. Remove engine room cover (RH and LH).

2. Remove engine cover with power tool.

3. Release fuel pressure.

4. Drain engine coolant, or when water hoses are disconnected, attach plug to prevent engine coolant leakage.

5. Disconnect fuel feed hose (1) on engine side as follows: (Perform same procedure for the side of centralized under-floor piping as well.)

6. Remove quick connector cap from quick connector connection.

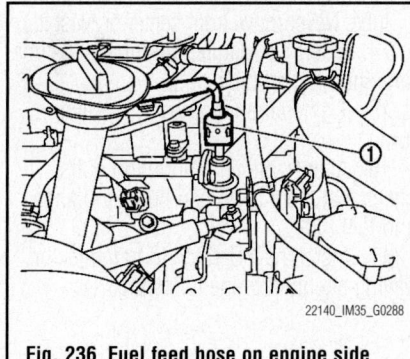

Fig. 236 Fuel feed hose on engine side

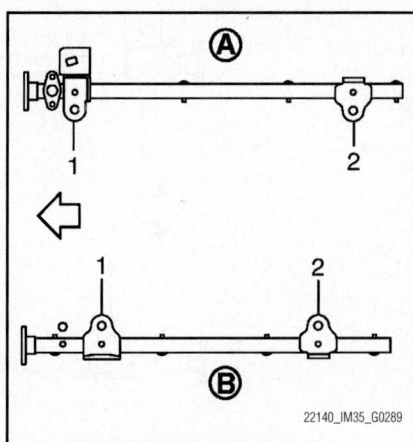

Fig. 237 Fuel tube and fuel injector assembly mounting order

7. Disconnect fuel damper and fuel hose assembly from fuel tubes (RH and LH).

8. Disconnect harness connector from fuel injector.

9. Loosen mounting bolts in reverse order as shown in the figure, and remove fuel tube and fuel injector assembly.

10. Remove spacers on intake manifold (lower).

11. Remove fuel injector from fuel tube.

12. Remove fuel feed damper.

To install:

13. Install fuel damper and fuel sub-tube.

➡**When handling new O-rings, be careful of the following caution:**

a. Handle O-ring with bare hands. Never wear gloves.

b. Lubricate O-ring with new engine oil.

c. Never clean O-ring with solvent.

d. Check that O-ring and its mating part are free of foreign material.

e. When installing O-ring, be careful not to scratch it with tool or fingernails. Also be careful not to twist or stretch O-ring. If O-ring was stretched while it was being attached, never insert it quickly into fuel tube.

f. Insert new O-ring straight into fuel tube. Never move from center or twist it.

14. Insert fuel damper and fuel sub-tube straight into fuel tube.

15. Tighten mounting bolts evenly in turn.

16. After tightening mounting bolts, check that there is no gap between flange and fuel tube.

17. Install new O-rings to fuel injector, paying attention to the following.

➡**Upper and lower O-ring are different, be careful not to confuse them.**

18. Fuel tube side o rings are black.

19. Nozzle side o rings are green.

20. Install fuel injector to fuel tube as follows:

a. Insert clip into clip mounting groove on fuel injector.

b. Check that installation is complete by checking that fuel injector does not rotate or come off.

21. Install spacers on intake manifold.

22. Install fuel tube and fuel injector assembly to intake manifold.

23. Tighten mounting bolts in two steps in numerical order as shown in the figure.

24. Tightening torque is as follows

a. Step 1: 7 ft. lb. (10.1 Nm).

b. Step 2: 17 ft. lb. (23.6 Nm).

25. Connect injector sub-harness.

26. Install intake manifold collectors (upper and lower).

27. Install fuel sub-tube on rear end of intake manifold collector (lower).

28. Connect fuel feed hose (with damper).

29. Connect quick connector between fuel feed hose (with damper) and centralized under-floor piping connection.

30. Remove engine cover with power tool.

31. Remove engine room cover (RH and LH).

32. Fill engine coolant.

33. Turn ignition switch **ON**(with the engine stopped). With fuel pressure applied to fuel piping, check there are no fuel leaks at connection points.

➡**Use mirrors for checking at points out of clear sight.**

34. Start the engine, with engine speed increased, check again that there are no fuel leaks at connection Points.

IDLE SPEED

ADJUSTMENT

The idle speed is controlled by the ECM.

THROTTLE BODY

REMOVAL & INSTALLATION

See Figure 239.

1. Remove engine cover (1) with power tool

2. Disconnect water hoses from intake manifold collector (upper), attach blind plug to prevent engine coolant leakage.

➡**Perform this step when the engine is cold.**

➡**Never spill engine coolant on drive belts.**

3. Remove air cleaner case and air duct.

4. Remove electric throttle control actuator as follows:

a. Disconnect harness connector.

b. Loosen mounting bolts in reverse order as shown in the figure.

➡**Never disassemble.**

To install:

5. Install gasket with positioning no-protrusion surface upward or downward.

6. Tighten in numerical order as shown in the figure.

7. Tighten to 75 inch lbs. (8.5 Nm).

➡**Perform the "Throttle Valve Closed Position Learning" when harness connector of electric throttle control actuator is disconnected**

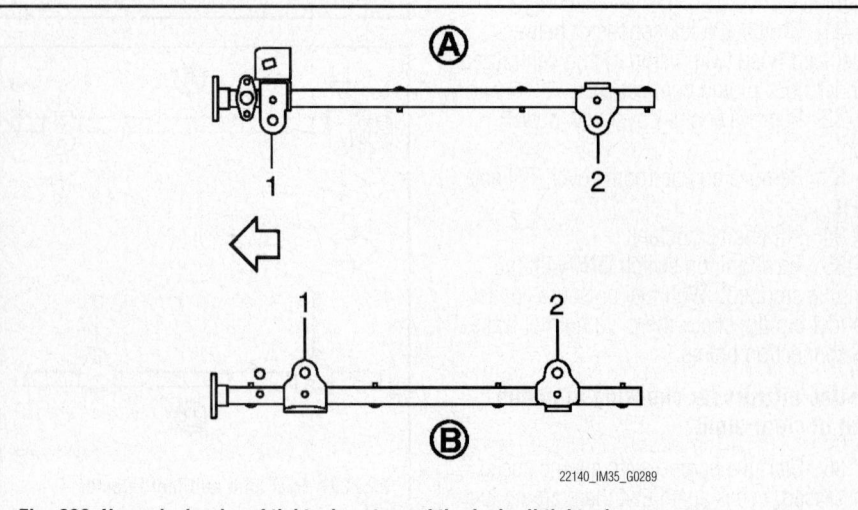

Fig. 238 Numerical order of tightening steps of the fuel rail tightening sequence

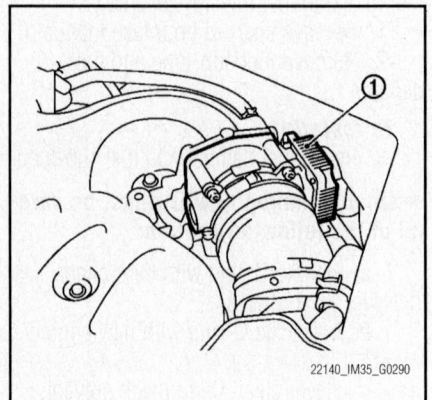

Fig. 239 Throttle body control actuator (1) harness connector

HEATING & AIR CONDITIONING SYSTEM

BLOWER MOTOR

REMOVAL & INSTALLATION

See Figure 240.

1. Remove instrument passenger lower cover.
2. Disconnect blower motor connector.
3. Remove mounting screws, and then remove blower motor.
4. Installation is the reverse order of removal.

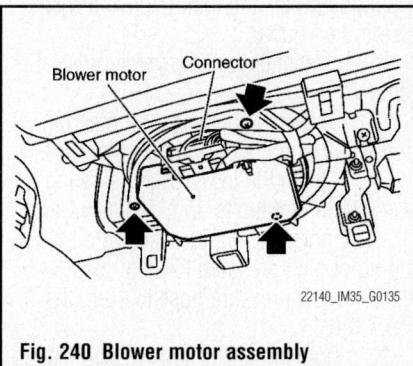

Fig. 240 Blower motor assembly

HEATER CORE

REMOVAL & INSTALLATION

See Figures 241 through 243.

1. Before servicing the vehicle, refer to the service precautions.
2. Disconnect both battery cables, the negative (–) cable first.

✳✳ CAUTION

After disconnecting the battery, wait for a least 3 minutes for the SRS module to deplete its energy before working on the steering column or instrument panel.

3. Recover the air conditioning refrigerant.
4. Drain the engine coolant.
5. Remove the cowl top cover.
6. Remove or disconnect the following:

- Evaporator lines
- Electric throttle control actuator
- Heater hoses
- Instrument panel and pad
- Clips of vehicle harness from steering member
- Driver side lower instrument panel cover
- Instrument stay right and left
- Top instrument cluster cover
- Combination meter
- Inner instrument cluster panel
- Instrument cluster
- Clock
- Automatic transmission shift lever cover
- Cup holder
- Ashtray and center lower panel
- Console upper finish panel
- Audio unit
- Instrument panel lower cover
- Glove box
- Glove box cover
- Center ventilation grille
- Console box assembly
- Front defroster grille
- Both pillar trim panels
- Instrument panel reinforcement bracket
- Instrument panel and pad
- Blower motor, intake door motor and amplifier connectors
- Blower unit
- Vehicle harness clips from steering member
- Defroster nozzle
- Ventilator ducts
- Heating and cooling unit

- Foot duct from heating and cooling unit
- Heater core from heating and cooling unit

To install:

7. Connect or attach the following:

- Heater core to the heating and cooling unit
- Foot duct to the heating and cooling unit
- Ventilator ducts
- Defroster nozzle
- Vehicle harness clips to the steering member
- Blower unit, tighten heater and cooling unit assembly mounting bolts to 61 inch lbs. (6.9 Nm).
- Blower motor, intake door motor and amplifier connector
- Instrument panel and pad
- Both pillar trim panels
- Front defroster grille
- Console box assembly
- Glove box
- Instrument panel lower cover
- Audio unit
- Console upper finish panel
- Center lower panel
- Automatic transmission shift lever cover
- Cup holder
- Instrument cluster
- Cluster panel
- Top instrument cluster cover
- Instrument stay right and left, tighten instrument stay mounting bolts to 9 ft lbs. (12 Nm).
- Driver side lower instrument panel cover
- Clips of vehicle harness from steering member
- Instrument panel and pad
- Heater hoses

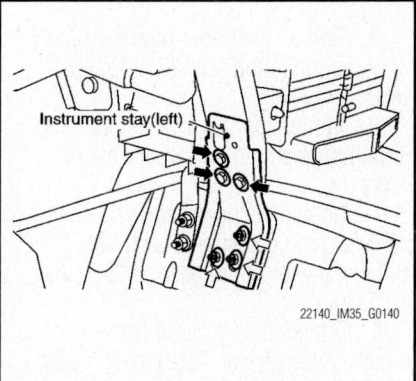

Fig. 241 Left instrument stay

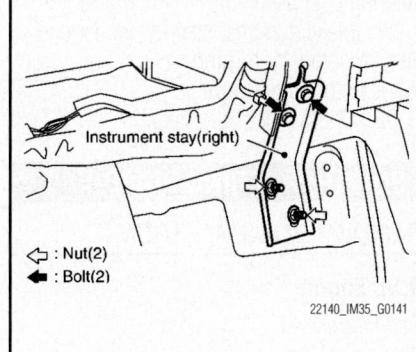

Fig. 242 Right instrument stay

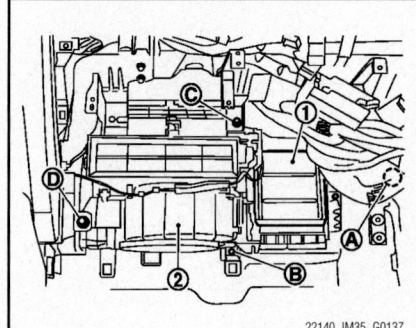

Fig. 243 Blower unit and ECM location and retainer locations

- Electric throttle control actuator
- Evaporator lines

8. Refill the cooling system.

9. Connect both battery cables, the negative (–) cable last.

10. Operate the engine to normal operating temperatures; then, check the climate control operation and check for leaks.

STEERING

POWER RACK & PINION STEERING GEAR

REMOVAL & INSTALLATION

See Figure 244.

1. Set vehicle to the straight-ahead position.

2. Remove tires from vehicle with a power tool.

3. Remove undercover from vehicle with a power tool.

4. Remove lower side fixing bolt of lower joint.

5. Remove cotter pin (1), and then loosen the nut.

6. Remove steering outer socket (2) from steering knuckle (3) so as not to damage ball joint boot (4) using the ball joint remover (suitable tool).

7. Temporarily tighten the nut to prevent damage to threads and to prevent the ball joint remover from suddenly coming off.

8. Remove high and low pressure piping of hydraulic piping, and then drain power steering fluid.

9. Remove steering hydraulic piping bracket from front suspension member.

10. Remove power steering solenoid valve harness connector.

11. Remove rack stay (2WD) or front cross bar (AWD).

12. Remove mounting bolts and nuts of steering gear assembly, and then remove steering gear assembly from vehicle.

To install:

13. Set rack of steering gear in the neutral position.

➡To get the neutral position of rack, turn gear-sub assembly and measure the distance of inner socket, and then measure the intermediate position of the distance.

14. Align rear cover cap projection (A) with the marking position (B) of gear housing assembly.

15. Install slit part of lower joint (C) aligning with the projection (A) of rear cover cap (1). Make sure that the slit part of lower joint (C) is aligned with both the projection (A) of rear cover cap (1) and the marking position (B) of gear housing assembly.

16. Install the steering gear assembly.

17. Install the mounting bolts and nuts.

18. Tighten the mounting bolts and nuts to 60 ft. lbs. (81.5 Nm).

19. Install the rack stay (2WD) or front cross bar (AWD).

20. Tighten the mounting bolt of member stay to 53 ft.lbs.(70 Nm).

21. Tighten the mounting nut of member stay to 96 ft.lbs.(130 Nm).

22. Connect the power steering solenoid valve harness connector.

23. Steering hydraulic piping bracket to the front suspension member.

24. Install the lower side fixing bolt of lower joint.

25. Tighten the lower side fixing bolt of lower joint to 100 ft.lbs.(130 Nm).

26. Install the cotter pin.

27. Install the undercover to the vehicle.

28. Install the tires from vehicle.

29. After installation, bleed air from the steering hydraulic system.

30. Perform final tightening of nuts and bolts on each part under unladen conditions with tires on level ground when removing steering gear assembly.

31. Make sure that steering wheel operates smoothly by turning several times from full left stop to full right stop.

32. Check wheel alignment.

POWER STEERING PUMP

REMOVAL & INSTALLATION

3.5L Engine

See Figure 245.

1. Drain power steering fluid from reservoir tank.

2. Remove undercover from vehicle with a power tool.

3. Remove alternator and power steering oil pump belt.

4. Remove piping of high pressure and low pressure (drain fluid from their piping's).

5. Remove power steering oil pump mounting bolts, and then remove power steering oil pump.

6. Installation is the reverse order of removal.

7. Tighten the upper power steering oil pump mounting bolt 13 ft. lbs. (18 Nm).

8. Tighten the lower power steering oil pump mounting bolts 35 ft. lbs. (48 Nm).

9. Tighten the power steering oil pump piping of high pressure to 25 ft. lbs. (33.5 Nm).and low pressure hose to 15 ft. lbs. (19.7 Nm).

10. Tighten the Eye-joint (assembled to high-pressure side hose) to 27 ft. lbs. (37 Nm)

4.5L Engine

See Figure 246.

1. Drain power steering fluid from reservoir tank.

2. Remove undercover from vehicle with a power tool.

3. Loosen drive belt.

4. Remove drive belt from oil pump pulley.

5. Remove piping of high pressure and low pressure (drain fluid from their piping's).

6. Remove power steering oil pump mounting bolts, and then remove power steering oil pump.

7. Installation is the reverse order of removal.

8. Tighten the upper power steering oil pump mounting bolt 23 ft. lbs. (31 Nm).

9. Tighten the lower power steering oil pump mounting bolt 35 ft. lbs. (48 Nm).

10. Tighten the power steering oil pump piping of high pressure to 24 ft. lbs. (32.9 Nm).and low pressure hose to 15 ft. lbs. (19.7 Nm).

11. Tighten the Eye-joint (assembled to high-pressure side hose) to 41 ft. lbs. (55 Nm).

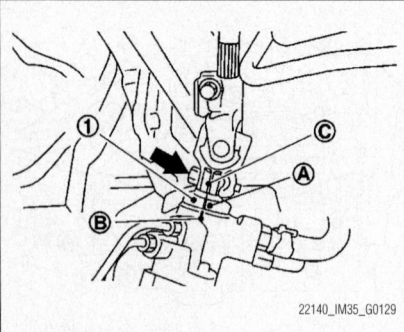

22140_IM35_G0129

Fig. 244 Steering gear in the neutral position

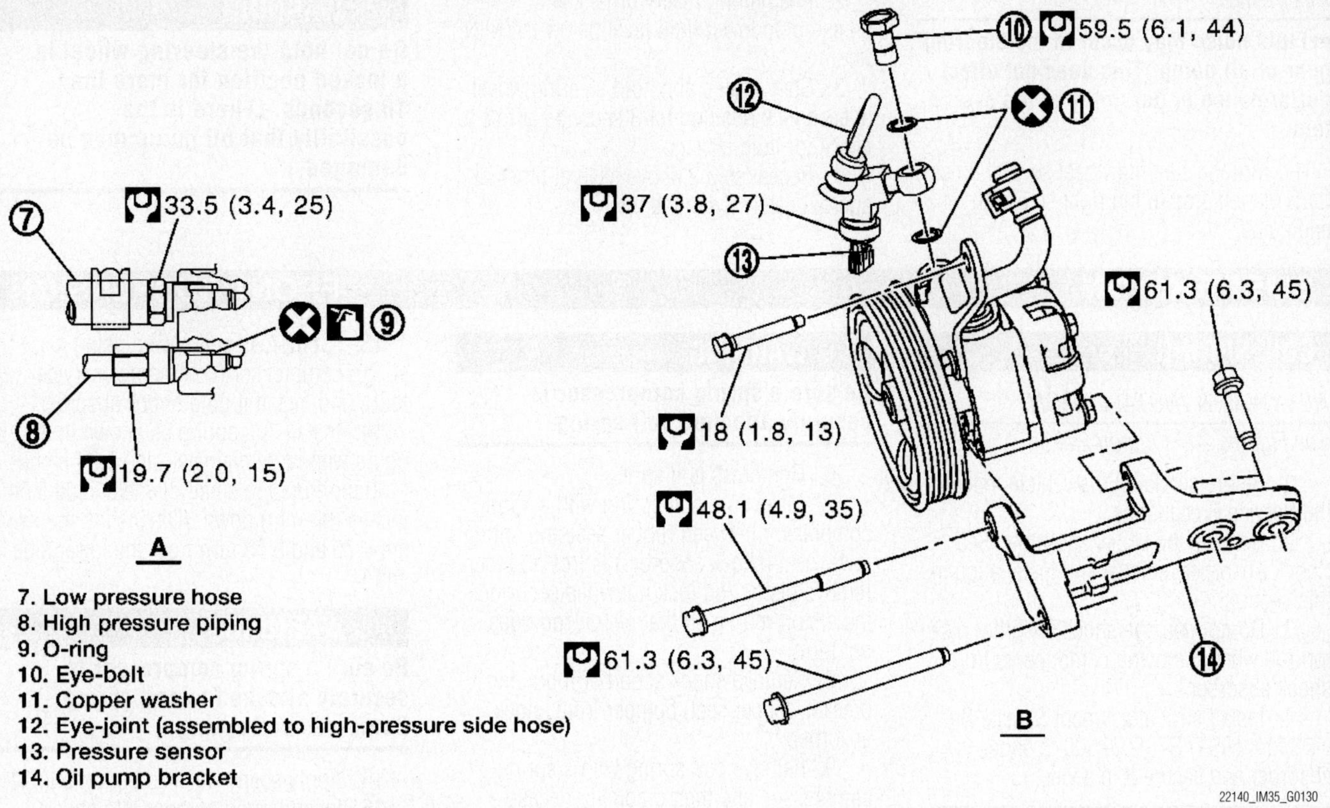

7. Low pressure hose
8. High pressure piping
9. O-ring
10. Eye-bolt
11. Copper washer
12. Eye-joint (assembled to high-pressure side hose)
13. Pressure sensor
14. Oil pump bracket

22140_IM35_G0130

Fig. 245 Steering oil pump and components—M35

7. Low pressure hose
8. Eye-bolt
9. Copper washer
10. Eye-bolt
11. Copper washer
12. Eye-joint (assembled to high-pressure side hose)
13. Pressure sensor
14. Oil pump bracket

22140_IM35_G0131

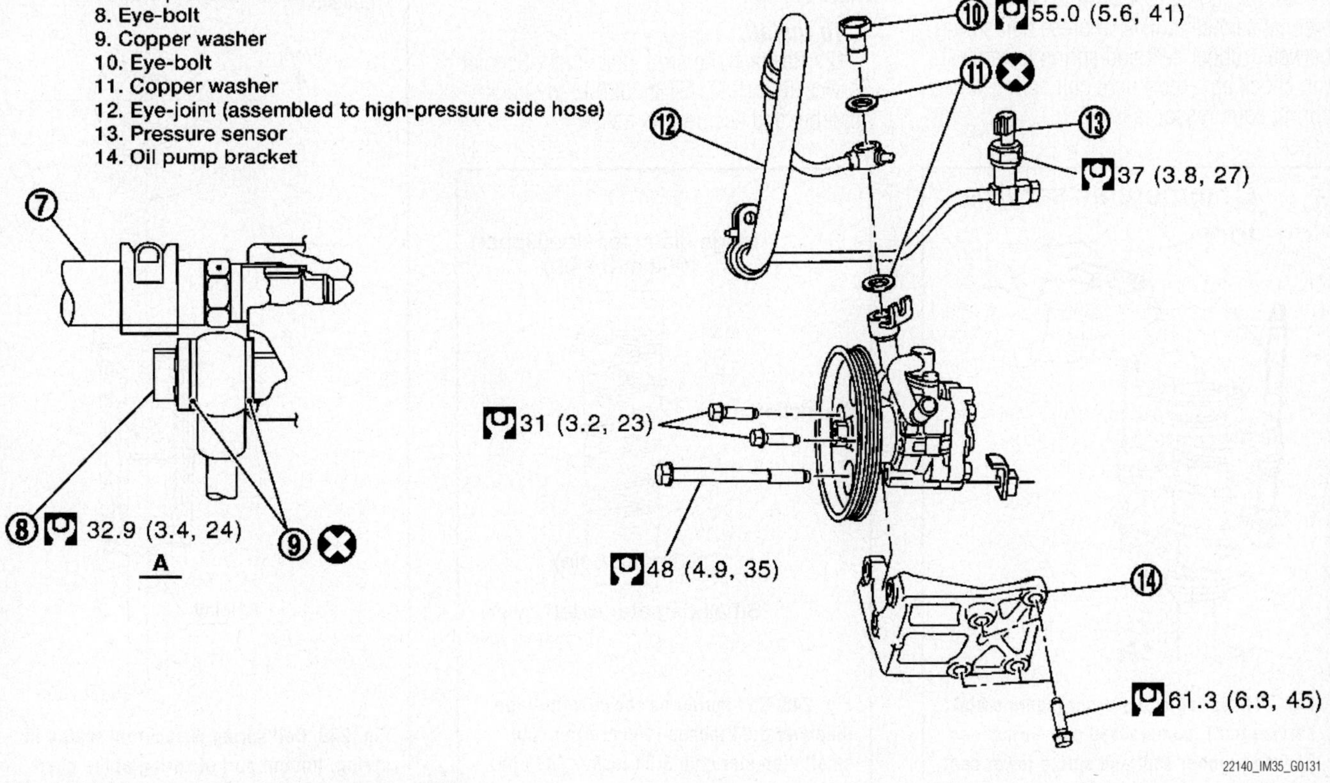

Fig. 246 Steering oil pump and components—M45

BLEEDING

→ **Fluid noise may occur in the steering gear or oil pump. This does not affect performance or durability of the system.**

1. Turn the steering wheel several times from full left stop to full right stop with engine off.

2. Keep filling reservoir tank with fluid so as not to lower fluid level below the MIN line.

3. Start engine and hold steering wheel at each lock position for 3 seconds at idle to check for fluid leakage.

4. Repeat step 2 above several times at approximately 3 second intervals.

✴✴ WARNING
Do not hold the steering wheel in a locked position for more than 10 seconds. (There is the possibility that oil pump may be damaged.)

SUSPENSION

FRONT SUSPENSION

COIL SPRING

REMOVAL & INSTALLATION
See Figures 247 through 249.

1. Before servicing the vehicle, refer to the Service Precautions.

2. Remove the shock assembly. See Shock Absorber removal and installation in this section.

3. Do not damage shock absorber piston rod when removing components from shock absorber.

4. Install strut attachment Special Service Tool (SST) ST35652000 to shock absorber and secure it in a vise.

✴✴ WARNING
When installing the strut attachment to shock absorber, wrap a shop cloth around strut to protect it from damage.

5. Using a spring compressor (commercial service tool), compress coil spring between rubber seat and spring lower seat (on shock absorber) until coil spring with a spring compressor is free.

✴✴ WARNING
Be sure a spring compressor is securely attached coil spring.

6. Compress coil spring

7. Make sure coil spring with a spring compressor between rubber seat and spring lower seat (shock absorber) is free and then remove piston rod lock nut while securing the piston rod tip so that piston rod does not turn.

8. Remove shock absorber mounting bracket, rubber seat, bumper from shock absorber.

9. Remove coil spring with a spring compressor, and then gradually release a spring compressor.

10. Loosen while making sure coil spring attachment position does not move.

11. Remove the strut attachment from shock absorber.

To install:

12. Reattach the strut attachment Special Service Tool (SST) ST35652000 to shock absorber and secure it in a vise.

13. Compress coil spring using a spring compressor (commercial service tool), and install it onto shock absorber.

14. Install coil spring as shown in the figure with large diameter side 3.94 inches (100 mm) up and small diameter side 3.54 inches (90 mm) down. (Distinction marks are 4.75 and 5.75 turn from the lower side end.)

✴✴ WARNING
Be sure a spring compressor is securely attached to coil spring. Compress coil spring.

15. Apply soapy water to bound bumper.

16. Insert bound bumper into shock absorber mounting bracket, and then install it to shock absorber together with rubber seat.

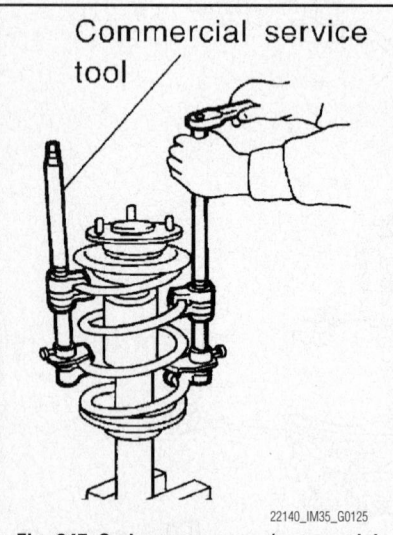

Fig. 247 Spring compressor (commercial service tool), compressed coil spring between rubber seat and spring lower seat (on shock absorber)

Large diameter side(Upper)
100mm(3.94in)

Distinction mark

90mm(3.54in)

Small dameter side(Lower)

22140_IM35_G0126

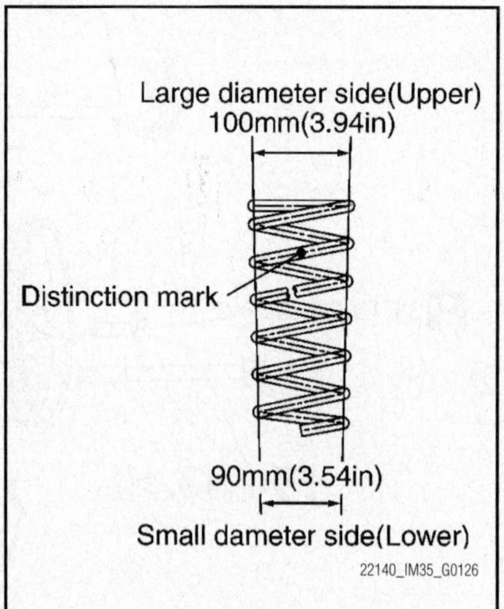

Fig. 248 Coil spring as shown with large diameter 3.94 inches (100 mm) up and small diameter side 3.54 inches (90 mm) down side

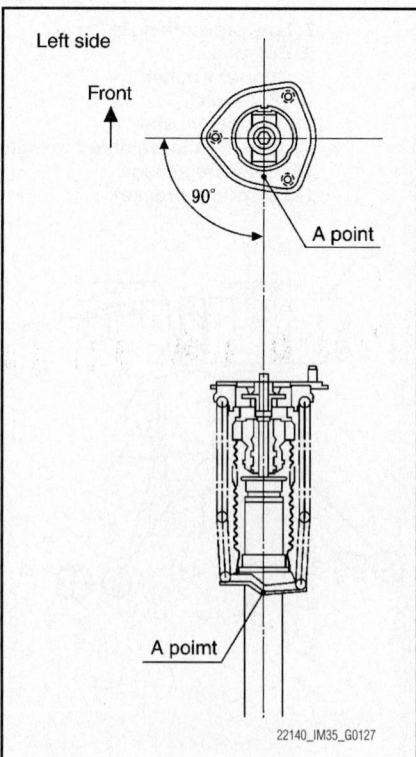

Fig. 249 Coil spring is securely seated in spring, bottom part of spring at the position of A point of spring seat

➡ Do not use machine oil.

17. Install shock absorber mounting bracket as shown in the figure.

18. Be sure coil spring is securely seated in spring mounting groove of rubber seat.

19. The bottom part of spring should be at the position of A point of spring seat.

20. Secure piston rod tip so that piston rod does not turn, then tighten piston rod lock nut with specified torque.

21. Gradually release a spring compressor, and remove coil spring.

❋ WARNING

Loosen while making sure coil spring attachment position does not move.

22. Remove the strut attachment from shock absorber.

23. Install assembly in to vehicle. See Shock Absorber removal and installation in this section.

LOWER BALL JOINT

REMOVAL & INSTALLATION

The lower ball joint is part of the transverse link and is not serviceable as a separate unit. The ball joint and arm must be replaced as an assembly if a malfunction is detected.

SHOCK ABSORBERS

REMOVAL & INSTALLATION

1. Before servicing the vehicle, refer to the Service Precautions.

2. Remove tires from vehicle with a power tool.

3. Remove harness of wheel sensor from shock absorber.

❋ WARNING

Do not pull on wheel sensor harness.

4. Remove brake hose bracket.

5. Remove the mounting nut on the upper side of stabilizer connecting rod with a power tool, and then remove stabilizer connecting rod from transverse link.

6. Remove mounting nut and bolt on the lower side of shock absorber arm with a power tool, and then remove shock absorber arm from transverse link.

7. Remove cotter pin of transverse link and steering knuckle, and then loosen nut.

8. Remove transverse link from steering knuckle so as not to damage ball joint boot using the ball joint remover (suitable tool).

❋ WARNING

Temporarily tighten the nut to prevent damage to threads and to prevent ball joint remover (suitable tool) from suddenly coming off.

9. Remove the mounting bolt on the upper side of shock absorber arm with a power tool, and then remove shock absorber arm from shock absorber.

10. Remove the mounting nuts of shock absorber mounting bracket, then remove shock absorber from vehicle.

To install:

11. Install the shock to the vehicle hand tighten the mounting nuts of shock absorber mounting bracket.

12. Install the mounting bolt on the upper side of shock absorber arm.

13. Install the transverse link to the steering knuckle.

14. Tighten the shock absorber upper nuts to 29 ft lbs. (39 Nm).

15. Tighten the shock absorber lower bolt to 79 ft lbs. (107 Nm).

16. Tighten the transverse link to the steering knuckle nut to 100 ft. lbs. 136 Nm).

17. Install the mounting nut on the upper side of stabilizer connecting rod.

18. Tighten the mounting nut on the upper side of stabilizer connecting rod to 66 ft. lbs. (90 Nm).

19. Perform final tightening of bolt and nut at the shock absorber arm lower side (rubber bushing) under unladen conditions with tires on level ground.

20. Install the brake hose bracket.

21. Install the harness of the wheel sensor to the shock absorber.

22. Install tire and wheel assemblies.

23. Check wheel alignment.

24. Adjust neutral position of steering angle sensor after checking wheel alignment.

STEERING KNUCKLE

REMOVAL & INSTALLATION

1. Before servicing the vehicle, refer to the Service Precautions.

2. Remove tires from vehicle with a power tool.

3. Remove harness of wheel sensor from shock absorber.

❋ WARNING

Do not pull on wheel sensor harness.

4. Remove brake hose bracket.

5. Remove the mounting nut on the upper side of stabilizer connecting rod with a power tool, and then remove stabilizer connecting rod from transverse link.

6. Remove mounting nut and bolt on the lower side of shock absorber arm with a power tool, and then remove shock absorber arm from transverse link.

7. Remove cotter pin of transverse link and steering knuckle, and then loosen nut.

8. Remove transverse link from steering knuckle so as not to damage ball joint boot using the ball joint remover (suitable tool).

❋ WARNING

Temporarily tighten the nut to prevent damage to threads and to prevent ball joint remover (suitable tool) from suddenly coming off.

9. Remove the mounting bolt on the upper side of shock absorber arm with a power tool, and then remove shock absorber arm from shock absorber.

10. Remove the mounting nuts of shock absorber mounting bracket, then remove shock absorber from vehicle.

11. Remove upper link mounting nut and bolt with a power tool, and then remove upper link from steering knuckle.

12. Remove mounting nuts and bolts, and then remove upper link and stopper rubber from vehicle.

To install:

13. Install upper link and stopper rubber to vehicle.

14. Tighten the upper link arm bushing bolts to 52 ft lbs. (70 Nm).

15. Tighten the upper link arm ball joint nut and bolts to 41 ft lbs. (55 Nm).

16. Install the shock to the vehicle hand tighten the mounting nuts of shock absorber mounting bracket.

17. Install the mounting bolt on the upper side of shock absorber arm.

18. Install the transverse link to the steering knuckle.

19. Tighten the shock absorber upper nuts to 29 ft lbs. (39 Nm).

20. Tighten the shock absorber lower bolt to 79 ft lbs. (107 Nm).

21. Tighten the transverse link to the steering knuckle nut to 100 ft. lbs. 136 Nm).

22. Install the mounting nut on the upper side of stabilizer connecting rod.

23. Tighten the mounting nut on the upper side of stabilizer connecting rod to 66 ft. lbs. (90 Nm).

24. Perform final tightening of bolt and nut at the shock absorber arm lower side (rubber bushing) under unladen conditions with tires on level ground.

25. Install the brake hose bracket.

26. Install the harness of the wheel sensor to the shock absorber.

27. Install tire and wheel assemblies.

28. Check wheel alignment.

29. Adjust neutral position of steering angle sensor after checking wheel alignment.

STABILIZER BAR

REMOVAL & INSTALLATION

1. Before servicing the vehicle, refer to the Service Precautions.

2. Remove tires from vehicle with a power tool.

3. Remove undercover with a power tool.

4. Remove the mounting nut on the lower side of stabilizer connecting rod with a power tool, and then remove stabilizer connecting rod from stabilizer bar.

5. If necessary remove the mounting nut on the upper side of stabilizer connecting rod with a power tool, and then remove stabilizer connecting rod from transverse link.

6. Remove the mounting nuts of stabilizer clamp, and then remove stabilizer clamp and stabilizer bushing.

7. Remove stabilizer bar from vehicle.

8. Installation is the reverse order of removal.

9. Tighten the stabilizer bar connecting rods to 66 ft. lbs. (90 Nm).

SUSPENSION

COIL SPRING

REMOVAL & INSTALLATION

See Figures 251 and 252.

1. Before servicing the vehicle, refer to the Service Precautions.

2. Remove tire with a power tool.

3. Set a jack under rear lower link to relieve the coil spring tension.

4. Loosen mounting bolt and nut of rear lower link inside of suspension member, and then remove mounting bolt and nut inside of axle with a power tool.

5. Slowly lower jack, then remove upper seat, coil spring and rubber sheet from rear lower link.

6. Remove mounting bolt and nut inside of suspension member to remove

To install:

7. Make sure that upper seat is attached as shown in the figure.

10. Tighten the stabilizer clamps to 37 ft. lbs. (50 Nm).

UPPER BALL JOINT

REMOVAL & INSTALLATION

The upper ball joint in not serviceable by itself, the upper link must be replaced as an assembly.

WHEEL HUB AND BEARING

REMOVAL & INSTALLATION

1. Before servicing the vehicle, refer to the Service Precautions.

2. Disconnect the negative battery cable.

3. Remove wheel sensor from steering knuckle.

❈❈ WARNING
Do not pull on wheel sensor harness.

4. Remove brake hose brackct.

5. Remove torque member fixing bolts with a power tool. hang torque member in a place where it will not interfere with work.

❈❈ WARNING
Do not depress brake.

6. Remove tires from vehicle with a power tool.

8. Make sure that the projecting parts on upper seat inside is securely fitted on the bracket tabs.

9. Match up rubber seat indentions and rear lower link grooves and attach.

❈❈ WARNING
Make sure spring is not upside down. The top and bottom are indicated by paint color.

10. Perform the final tightening of rear suspension member and axle installation position (rubber bushing) under unladen condition with tires on level ground.

11. Tighten of rear suspension member to 53 ft lbs. (72 Nm).

12. Check wheel alignment.

➡**Adjust neutral position of steering angle sensor after checking the wheel alignment.**

7. Remove rear brake caliper with a power tool. hang it in a place where it will not interfere with work.

8. Put matching mark on disc rotor and the wheel hub and bearing assembly then removing disc rotor.

9. Remove cotter pin, then loosen hub lock nut with a power tool.

10. Separate the wheel hub and bearing assembly from drive shaft by lightly tapping the end with a hammer (suitable tool) and wood block, and then remove hub lock nut.

11. Remove the wheel hub and bearing assembly mounting bolts.

12. Remove the wheel hub and bearing assembly.

To install:

13. Install the wheel hub and bearing assembly while sliding the axle shaft into the center of the bearing.

14. Tighten the hub assembly retaining nuts and bolts to 65 ft. lbs. (88 Nm).

15. Tighten the axle nut to 130 ft. lbs. (175 Nm).

16. Assemble disc rotor by aligning the matching marks made at disassembly.

17. Install rear brake calipers.

18. Tighten the rear caliper bolts to 62 ft lbs. (84.3 Nm).

19. Install wheel assemblies.

20. Connect negative battery cable.

REAR SUSPENSION

CONTROL ARMS/LINKS

REMOVAL & INSTALLATION

1. Before servicing the vehicle, refer to the Service Precautions.

2. Remove tire with a power tool.

3. Remove mounting bolt and nut of front lower link (Rear Suspension) (21) attached to the suspension member, and then remove mounting bolt and nut from axle assembly (10)

To install:

4. Install front lower link (Rear Suspension) (21) to axle assembly.

5. Tighten front lower link to axle assembly to 55 ft. lbs. (88 Nm).

6. Install front lower link (Rear Suspension) (21) to the suspension member.

7. Tighten front lower link to the suspension member to 53 ft. lbs. (72 Nm).

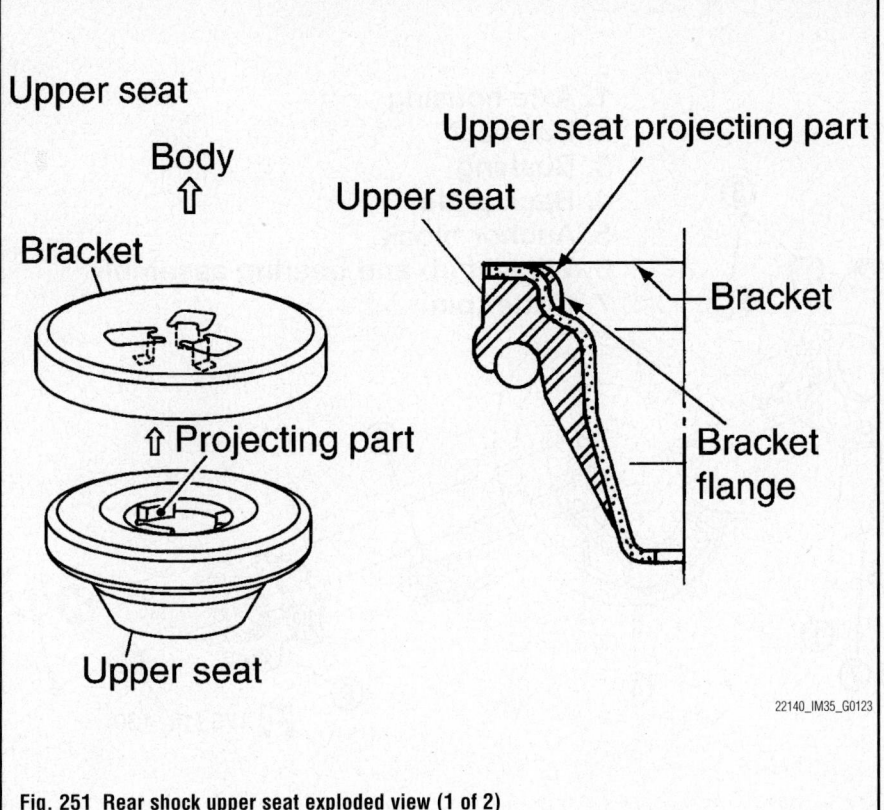

Fig. 251 Rear shock upper seat exploded view (1 of 2)

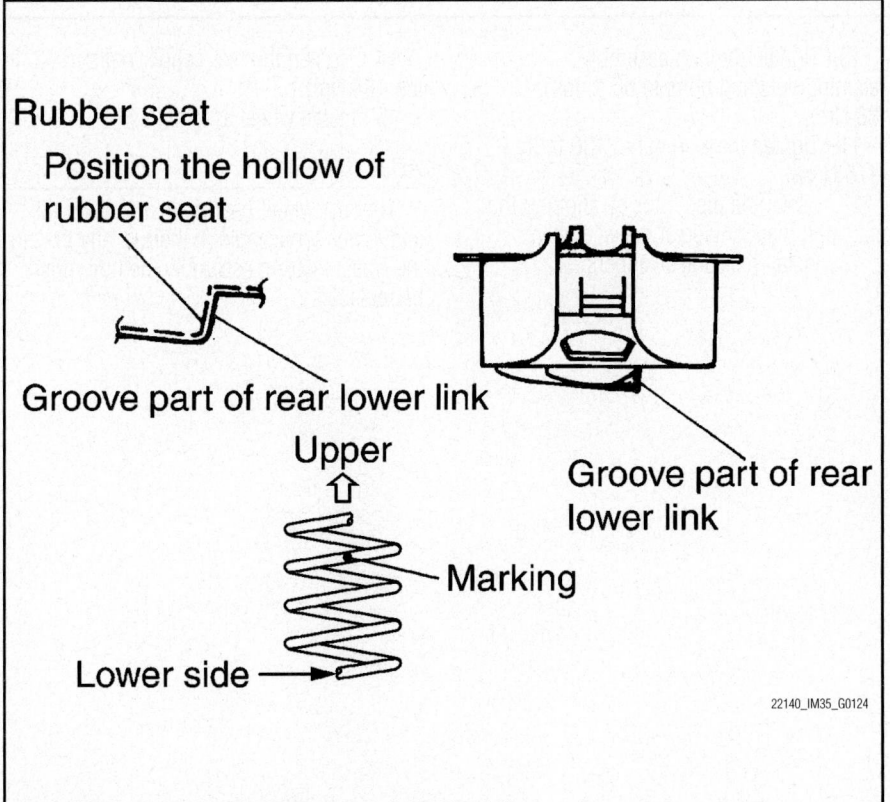

Fig. 252 Rear shock rubber seat exploded view (2 of 2)

SHOCK ABSORBER

REMOVAL & INSTALLATION

1. Before servicing the vehicle, refer to the Service Precautions.
2. Remove tires from vehicle with a power tool.
3. Set a jack under rear lower link to relieve the coil spring tension.
4. Remove shock absorber lower end bolt with a power tool.
5. Gradually lower the jack to remove it from rear lower link.
6. Remove shock absorber assembly upper end nuts with a power tool, and then remove shock absorber assembly from vehicle.

To install:
7. Install shock absorber assembly to vehicle.
8. Install shock absorber assembly upper end nuts.
9. Tighten shock absorber assembly upper end nuts to 21 ft lbs. (28 Nm).
10. Install shock absorber rear lower link
11. Tighten shock absorber assembly rear lower link bolt and nut to 81 ft lbs. (110 Nm).
12. Tighten the rear caliper bolts to 62 ft lbs. (84 Nm).

TESTING

Check for oil leakage, damage and breakage of installation positions.

WHEEL HUB AND BEARING

REMOVAL & INSTALLATION

See Figure 253.

1. Before servicing the vehicle, refer to the Precautions Section.
2. Remove tires from vehicle with a power tool.
3. Remove rear brake caliper with a power tool, hang it in a place where it will not interfere with work.
4. Put matching mark on disc rotor and the wheel hub and bearing assembly then removing disc rotor.
5. Remove cotter pin, then loosen hub lock nut with a power tool.
6. Separate the wheel hub and bearing assembly from drive shaft by lightly tapping the end with a hammer (suitable tool) and wood block, and then remove hub lock nut.

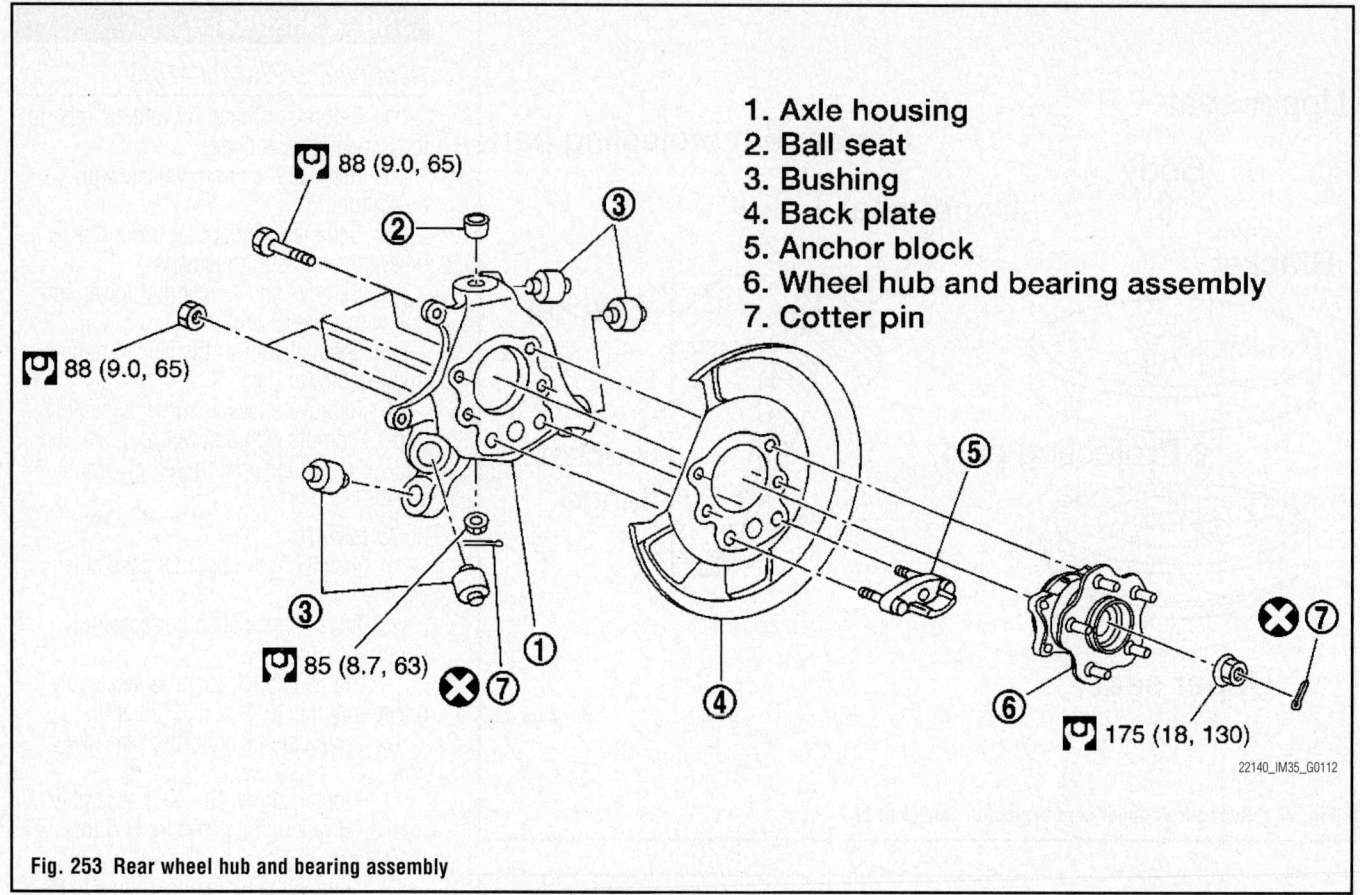

1. Axle housing
2. Ball seat
3. Bushing
4. Back plate
5. Anchor block
6. Wheel hub and bearing assembly
7. Cotter pin

88 (9.0, 65)

88 (9.0, 65)

85 (8.7, 63)

175 (18, 130)

22140_IM35_G0112

Fig. 253 Rear wheel hub and bearing assembly

7. Remove the wheel hub and bearing assembly mounting bolts.

8. Remove the wheel hub and bearing assembly.

To install:

9. Install the wheel hub and bearing assembly while sliding the axle shaft into the center of the bearing.

10. Tighten the hub assembly retaining nuts and bolts to 65 ft. lbs. (88 Nm).

11. Tighten the axle nut to 130 ft. lbs. (175 Nm).

12. Assemble disc rotor by aligning the matching marks made at disassembly.

13. Install rear brake calipers

14. Tighten the rear caliper bolts to 62 ft lbs. (84 Nm).

15. Install wheel assemblies

ADJUSTMENT

The rear wheel bearing is a sealed unit and is not serviceable. If there is any play in the wheel bearing assembly the unit must be replaced.

INFINITI

QX56

SPECIFICATIONS AND MAINTENANCE CHARTS

ENGINE AND VEHICLE IDENTIFICATION

			Engine						
Code ①	Liters (cc)	Cu. In.	Cyl.	Fuel Sys.	Engine	Eng. Mfg.		Code ②	Year
ZH56DE	5.6 (5552)	338.8	8	MFI	DOHC	Nissan		7	2007
								8	2008

MFI: Multi-port Fuel Injection

DOHC: Double Overhead Camshafts

① Located on the timing belt cover

② 10th digit of the Vehicle Identification Number (VIN)

22140_QX56_C0001

GENERAL ENGINE SPECIFICATIONS

Year	Model	Engine Displacement Liters	Engine ID	Net Horsepower @ rpm	Net Torque @ rpm (ft. lbs.)	Bore x Stroke (in.)	Com- pression Ratio	Oil Pressure @ rpm
2007	QX56	5.6	ZH56DE	305@4900	385@3600	3.86X3.62	9.8:1	43@2000
2008	QX56	5.6	ZH56DE	305@4900	385@3600	3.86X3.62	9.8:1	43@2000

22140_QX56_C0002

ENGINE TUNE-UP SPECIFICATIONS

Year	Engine Displacement Liters	Engine ID	Spark Plug Gap (in.)	Ignition Timing	Fuel Pump (psi) ①	Idle Speed ②	Valve Clearance (in.) In.	Valve Clearance (in.) Ex.
2007	5.6	ZH56DE	0.043	15B	51	600-700	0.010-0.013	0.011-0.016
2008	5.6	ZH56DE	0.043	15B	51	600-700	0.010-0.013	0.011-0.016

NOTE: The Vehicle Emission Control Information label often reflects specification changes made during production. The label figures

must be used if they differ from those in this chart.

B: Before top dead center

① System pressure at idle with vacuum hose connected

Should increase when disconnected

② Automatic transmission in Neutral

22140_QX56_C0003

CAPACITIES

Year	Model	Engine Displacement Liters	Engine ID	Engine Oil with Filter (qts.)	Transmission (pts.)	Transfer Case (pts.)	Drive Axle Front (pts.)	Rear (pts.)	Fuel Tank (gal.)	Cooling System (qts.)
2007	QX56	5.6	ZH56DE	6.5	22.5	6.25	3.375	3.75	28.0	15
2008	QX56	5.6	ZH56DE	6.5	22.5	6.25	3.375	3.75	28.0	15

NOTE: All capacities are approximate. Add fluid gradually and check to be sure a proper fluid level is obtained.

22140_QX56_C0004

VALVE SPECIFICATIONS

Year	Engine Displacement Liters	Engine ID	Seat Angle (deg.)	Face Angle (deg.)	Spring Test Pressure (lbs. @ in.)	Spring Installed Height (in.)	Stem-to-Guide Clearance (in.) Intake	Exhaust	Stem Diameter (in.) Intake	Exhaust
2007	5.6	ZH56DE	45.15-45.45	45	37.0@1.457	1.991	0.0008-0.0021	0.0012-0.0025	0.2348-0.2354	0.2344-0.2350
2008	5.6	ZH56DE	45.15-45.45	45	37.0@1.457	1.991	0.0008-0.0021	0.0012-0.0025	0.2348-0.2354	0.2344-0.2350

22140_QX56_C0006

CAMSHAFT SPECIFICATIONS

All measurements are given in inches.

Year	Engine Displ. Liters	Engine ID/VIN	Journal Dia.	Brg. Oil Clearance	Shaft End-play	Runout	Journal Bore	Lobe Height Intake	Exhaust
2007	5.6	ZH56DE	1.0218-1.0224	0.0012-0.0027	0.0045-0.0074	0.0008	1.0236-1.0244	1.7636-1.7738	1.7746-1.7821
2008	5.6	ZH56DE	1.0218-1.0224	0.0012-0.0027	0.0045-0.0074	0.0008	1.0236-1.0244	1.7636-1.7738	1.7746-1.7821

22140_QX56_C0007

CRANKSHAFT AND CONNECTING ROD SPECIFICATIONS

All measurements are given in inches.

Year	Engine Displ. Liters	Engine ID	Crankshaft				Connecting Rod		
			Main Brg. Journal Dia.	Main Brg. Oil Clearance	Shaft End-play	Thrust on No.	Journal Diameter	Oil Clearance	Side Clearance
2007	5.6	ZH56DE	①	②	0.0039-0.0102	3	③	0.0008-0.0015	0.0079-0.0157
2008	5.6	ZH56DE	①	②	0.0039-0.0102	3	③	0.0008-0.0015	0.0079-0.0157

① There are 24 different grades, ranging from 2.5173 - 2.5183.

② No. 1 and 5: 0.00004-0.0004

No. 2, 3 and 4: 0.0003-0.0007

③ Grade 0: 2.1247-2.1250

Grade 1: 2.1245-2.1247

Grade 2: 2.1243-2.1245

22140_QX56_C0005

PISTON AND RING SPECIFICATIONS

All measurements are given in inches.

Year	Engine Displacement Liters	Engine ID	Piston Clearance	Ring Gap			Ring Side Clearance		
				Top Comp.	Bottom Comp.	Oil Control	Top Comp.	Bottom Comp.	Oil Control
2007	5.6	ZH56DE	0.0004-0.0012	0.0091-0.0130	0.0098-0.0157	0.0079-0.0236	0.0014-0.0033	0.0012-0.0028	0.0006-0.0020
2008	5.6	ZH56DE	0.0004-0.0012	0.0091-0.0130	0.0098-0.0157	0.0079-0.0236	0.0014-0.0033	0.0012-0.0028	0.0006-0.0020

22140_QX56_C0008

TORQUE SPECIFICATIONS

All readings in ft. lbs.

Year	Engine Displacement Liters	Engine ID	Cylinder Head Bolts	Main Bearing Bolts	Rod Bearing Bolts	Crankshaft Damper Bolts	Flywheel Bolts	Manifold		Spark Plugs	Oil Pan Drain Plug
								Intake	Exhaust		
2007	5.6	ZH56DE	①	②	③	④	65	6	25	18	25
2008	5.6	ZH56DE	①	②	③	④	65	6	25	18	25

① Step 1: 72 ft. lbs

Step 2: Loosen all bolts completely

Step 3: 33 ft. lbs.

Step 4: +60 degrees

Step 5: +60 degrees

② Step 1: Main Bolts to 29 ft. lbs.

Step 2: Sub-bolts to 22 ft. lbs.

Step 3: Main Bolts +40 degrees

Step 4: Sub-Bolts +30 degrees

Step 5: Side Bolts to 36 ft. lbs.

③ Step 1: 11 ft. lbs.

Step 2: +90 degrees

④ Step 1: 65 ft. lbs.

Step 2: +90 degrees

22140_QX56_C0009

WHEEL ALIGNMENT

Year	Model	Caster Range (+/-Deg.)	Caster Preferred Setting (Deg.)	Camber Range (+/-Deg.)	Camber Preferred Setting (Deg.)	Toe-in (in.)
2007	QX56 ①	0.75	②	0.75	③	0.08+/-0.03
2008	QX56 ①	0.75	②	0.75	③	0.08+/-0.03

① Assumes P275/60R20 tire

② 4x2: +4.00
   4x4: +3.50

③ 4x2: -0.10
   4x4: +0.20

22140_QX56_C0010

TIRE, WHEEL AND BALL JOINT SPECIFICATIONS

Year	Model	OEM Tires Standard	OEM Tires Optional	Tire Pressures (psi) Front	Tire Pressures (psi) Rear	Wheel Size	Ball Joint Inspection	Lugnut Torque (ft. lbs.)
2007	QX56	P275/60R20	None	35	35	18	①	98
2008	QX56	P275/60R20	None	35	35	18	①	98

OEM: Original Equipment Manufacturer

PSI: Pounds Per Square Inch

① Axial play

   Upper: 0

22140_QX56_C0011

BRAKE SPECIFICATIONS

All measurements in inches unless noted

Year	Model		Brake Disc Original Thickness	Brake Disc Minimum Thickness	Brake Disc Maximum Runout	Minimum Pad Thickness	Brake Caliper Bracket Bolts (ft. lbs.)	Brake Caliper Mounting Bolts (ft. lbs.)
2007	QX56	F	1.024	0.965	0.0016	0.039	155	32
		R	0.551	0.472	0.0020	0.039	—	24
2008	QX56	F	1.024	0.965	0.0016	0.039	155	32
		R	0.551	0.472	0.0020	0.039	—	24

22140_QX56_C0012

SCHEDULED MAINTENANCE INTERVALS
NFINITI QX56

TO BE SERVICED	TYPE OF SERVICE	7.5	15	22.5	30	37.5	45	52.5	60
Engine oil & filter	R	✓	✓	✓	✓	✓	✓	✓	✓
Brake lines & cables	S/I		✓		✓		✓		✓
Brake pads and rotors	I	✓	✓	✓	✓	✓	✓	✓	✓
Driveshaft boots & propeller shaft (4x4)	L/I		✓		✓		✓		✓
Transmission, transfer & differential gear oil	I		✓		✓		✓		✓
Air cleaner filter	R				✓				✓
Engine coolant ①	R								✓
Spark plugs (Platinum)	R	Replace every 120,000 miles							
Drive belt(s) ②	S/I								✓
Cabin air filter	R		✓		✓		✓		✓
Exhaust system	I				✓				✓
Fuel lines	S/I				✓				✓
Fuel filter ③									
Steering gear (box) & linkage, axle & suspension parts	I				✓				✓
Vapor lines	S/I				✓				✓

R: Replace   S/I: Service or Inspect   L: Lubricate

① Coolant: After 60,000 miles, inspect every 30,000 miles.

② Drive Belts: After 60,000 miles, inspect every 15,000 miles. Replace belts if damaged.

③ Fuel Filter: Maintenance free item.

FREQUENT OPERATION MAINTENANCE (SEVERE SERVICE)

If a vehicle is operated under any of the following conditions it is considered severe service:

- Extremely dusty areas.

- Rough, muddy, or salt spread roads.

- 50% or more of the vehicle constant operation is in 32°C (90°F) or higher temperatures, or temperatures below 0°C (32°F).

- Prolonged idling (vehicle operation in stop and go traffic).

- Frequent short running periods (engine does not warm to normal operating temperatures).

- Police, taxi, delivery usage or trailer towing usage.

Oil & oil filter: replace every 3750 miles.

Brake pads, discs, drums & linings: service or inspect every 7500 miles.

Driveshaft boots & propeller shaft: service or inspect every 7500 miles.

Exhaust system: service or inspect every 7500 miles.

Steering gear (box) & linkage, (steering damper-4x4), axle & suspension parts: service or inspect every 7500 miles.

Steering linkage ball joints & front suspension ball joints: service or inspect every 7500 miles.

22140_QX56_C0013

PRECAUTIONS

Before servicing any vehicle, please be sure to read all of the following precautions, which deal with personal safety, prevention of component damage, and important points to take into consideration when servicing a motor vehicle:

• Never open, service or drain the radiator or cooling system when the engine is hot; serious burns can occur from the steam and hot coolant.

• Observe all applicable safety precautions when working around fuel. Whenever servicing the fuel system, always work in a well-ventilated area. Do not allow fuel spray or vapors to come in contact with a spark, open flame, or excessive heat (a hot drop light, for example). Keep a dry chemical fire extinguisher near the work area. Always keep fuel in a container specifically designed for fuel storage; also, always properly seal fuel containers to avoid the possibility of fire or explosion. Refer to the additional fuel system precautions later in this section.

• Fuel injection systems often remain pressurized, even after the engine has been turned **OFF**. The fuel system pressure must be relieved before disconnecting any fuel lines. Failure to do so may result in fire and/or personal injury.

• Brake fluid often contains polyglycol ethers and polyglycols. Avoid contact with the eyes and wash your hands thoroughly after handling brake fluid. If you do get brake fluid in your eyes, flush your eyes with clean, running water for 15 minutes. If eye irritation persists, or if you have taken

brake fluid internally, IMMEDIATELY seek medical assistance.

• The EPA warns that prolonged contact with used engine oil may cause a number of skin disorders, including cancer. You should make every effort to minimize your exposure to used engine oil. Protective gloves should be worn when changing oil. Wash your hands and any other exposed skin areas as soon as possible after exposure to used engine oil. Soap and water, or waterless hand cleaner should be used.

• All new vehicles are now equipped with an air bag system, often referred to as a Supplemental Restraint System (SRS) or Supplemental Inflatable Restraint (SIR) system. The system must be disabled before performing service on or around system components, steering column, instrument panel components, wiring and sensors. Failure to follow safety and disabling procedures could result in accidental air bag deployment, possible personal injury and unnecessary system repairs.

• Always wear safety goggles when working with, or around, the air bag system. When carrying a non-deployed air bag, be sure the bag and trim cover are pointed away from your body. When placing a non-deployed air bag on a work surface, always face the bag and trim cover upward, away from the surface. This will reduce the motion of the module if it is accidentally deployed. Refer to the additional air bag system precautions later in this section.

• Clean, high quality brake fluid from a sealed container is essential to the safe and

proper operation of the brake system. You should always buy the correct type of brake fluid for your vehicle. If the brake fluid becomes contaminated, completely flush the system with new fluid. Never reuse any brake fluid. Any brake fluid that is removed from the system should be discarded. Also, do not allow any brake fluid to come in contact with a painted surface; it will damage the paint.

• Never operate the engine without the proper amount and type of engine oil; doing so WILL result in severe engine damage.

• Timing belt maintenance is extremely important. Many models utilize an interference-type, non-freewheeling engine. If the timing belt breaks, the valves in the cylinder head may strike the pistons, causing potentially serious (also time-consuming and expensive) engine damage. Refer to the maintenance interval charts for the recommended replacement interval for the timing belt, and to the timing belt section for belt replacement and inspection.

• Disconnecting the negative battery cable on some vehicles may interfere with the functions of the on-board computer system(s) and may require the computer to undergo a relearning process once the negative battery cable is reconnected.

• When servicing drum brakes, only disassemble and assemble one side at a time, leaving the remaining side intact for reference.

• Only an MVAC-trained, EPA-certified automotive technician should service the air conditioning system or its components.

BRAKES

GENERAL INFORMATION

PRECAUTIONS

• Certain components within the ABS system are not intended to be serviced or repaired individually.

• Do not use rubber hoses or other parts not specifically specified for and ABS system. When using repair kits, replace all parts included in the kit. Partial or incorrect repair may lead to functional problems and require the replacement of components.

• Lubricate rubber parts with clean, fresh brake fluid to ease assembly. Do not

use shop air to clean parts; damage to rubber components may result.

• Use only DOT 3 brake fluid from an unopened container.

• If any hydraulic component or line is removed or replaced, it may be necessary to bleed the entire system.

• A clean repair area is essential. Always clean the reservoir and cap thoroughly before removing the cap. The slightest amount of dirt in the fluid may plug an orifice and impair the system function. Perform repairs after components have been thoroughly cleaned; use only denatured alcohol

ANTI-LOCK BRAKE SYSTEM (ABS)

to clean components. Do not allow ABS components to come into contact with any substance containing mineral oil; this includes used shop rags.

• The Anti-Lock control unit is a microprocessor similar to other computer units in the vehicle. Ensure that the ignition switch is **OFF** before removing or installing controller harnesses. Avoid static electricity discharge at or near the controller.

• If any arc welding is to be done on the vehicle, the control unit should be unplugged before welding operations begin.

BRAKES

BLEEDING THE BRAKE SYSTEM

BLEEDING PROCEDURE

BLEEDING PROCEDURE

➡**Be sure that the master cylinder is full of clean fresh brake fluid before starting the bleeding process. Use only the recommended brake fluid when bleeding the system. Do not allow brake fluid to spill on painted surfaces as damage will occur.**

1. Before servicing the vehicle, refer to the Precautions Section.

2. Disconnect the negative battery cable.
3. Turn the ignition switch OFF. Disconnect the ABS actuator and electric control unit connector.
4. Connect a vinyl tube to the rear right bleed valve. Be sure to have a catch pan handy to catch excess brake fluid.
5. Fully depress the brake pedal four or five times.
6. With the brake pedal depressed, loosen the bleed valve to let air out, then tighten it immediately.

7. Repeat the above steps until all air is removed from the system. Be sure to keep watch on the brake fluid level and replenish, as necessary.
8. Tighten the bleed valve.
9. Repeat the above steps at each wheel, with the master cylinder reservoir tank filled at least half way.
10. Bleed the remaining components in the following order: front left, rear left and front right.

BRAKES

FRONT DISC BRAKES

✳✳ CAUTION

Dust and dirt accumulating on brake parts during normal use may contain asbestos fibers from production or aftermarket brake linings. Breathing excessive concentrations of asbestos fibers can cause serious bodily harm. Exercise care when servicing brake parts. Do not sand or grind brake lining unless equipment used is designed to contain the dust residue. Do not clean brake parts with compressed air or by dry brushing. Cleaning should be done by dampening the brake components with a fine mist of water, then wiping the brake components clean with a dampened cloth. Dispose of cloth and all residue containing asbestos fibers in an impermeable container with the appropriate label. Follow practices prescribed by the Occupational Safety and Health Administration (OSHA) and the Environmental Protection Agency (EPA) for the handling, processing, and disposing of dust or debris that may contain asbestos fibers.

BRAKE CALIPER

REMOVAL & INSTALLATION
See Figure 1.

1. Before servicing the vehicle, refer to the Precautions Section.
2. Drain brake fluid as necessary.
3. Remove or disconnect the following:
   - Wheel
   - Union bolt
   - Caliper-to-torque member slide pins, or remove the caliper and torque member as an assembly
   - Brake caliper

To install:
4. Install or connect the following:
   - Brake caliper, tighten torque member bolts to 155 ft. lbs. (210 Nm); the caliper slide pins to 20 ft. lbs. (27 Nm)
   - Union bolt and tighten to 13 ft. lbs. (18 Nm)

5. Fill the master cylinder and bleed the brake system.
6. Install the wheels.

DISC BRAKE PADS

REMOVAL & INSTALLATION

1. Before servicing the vehicle, refer to the Precautions Section.
2. Remove the wheel.
3. Remove lower sliding pin bolt.
4. Suspend brake caliper with a remove and remove brake pad and shim from torque member.

To install:
5. Push pistons in so that the pad is firmly installed, using a suitable tool.
6. Mount the brake caliper to torque member.
7. Attach pad retainer to torque member.
8. Lubricate lower sliding pin bolt with a thin layer of silicone grease and install. Torque to 20 ft. lbs. (27 Nm).
9. Install the wheel.

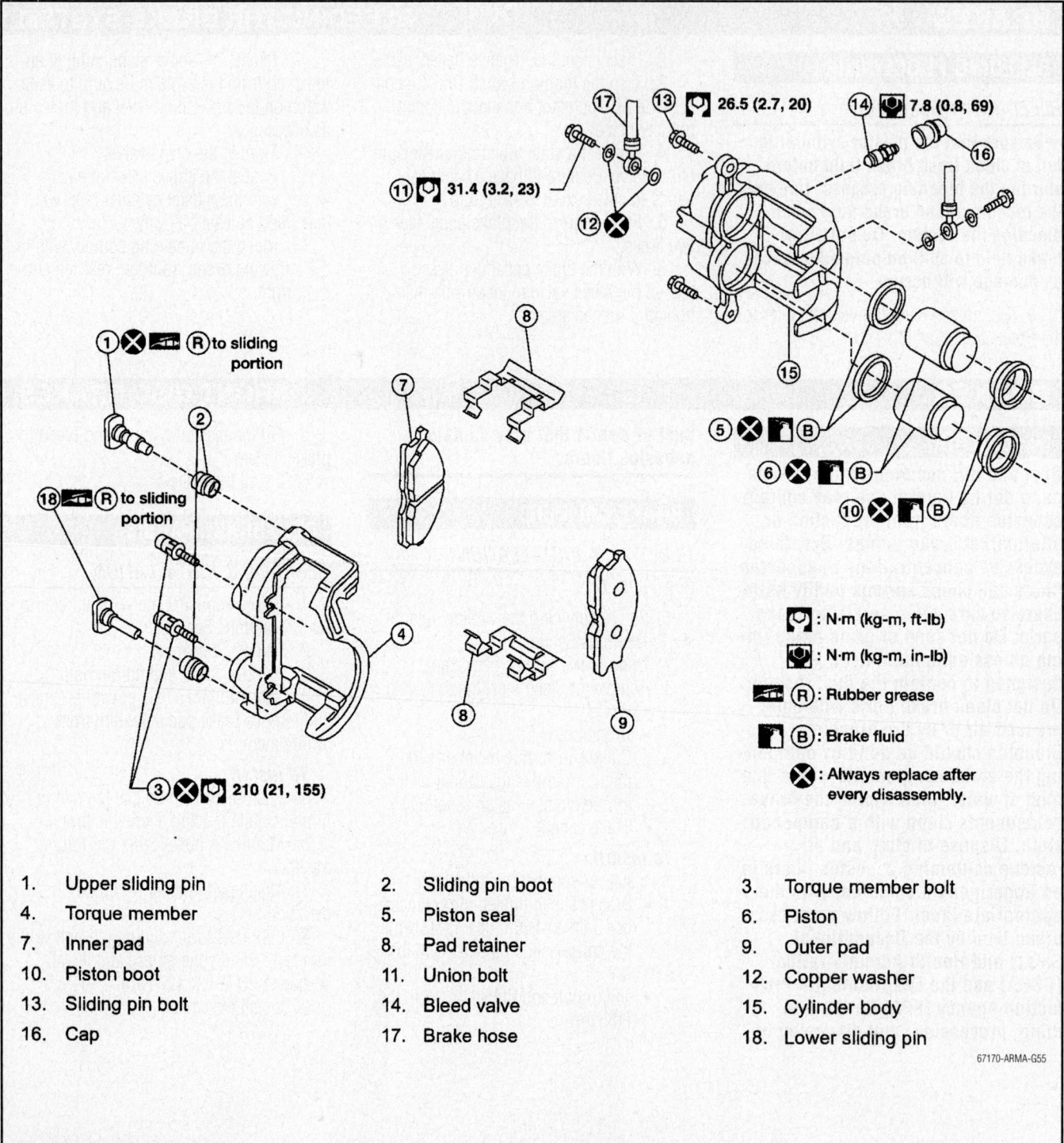

1. Upper sliding pin
4. Torque member
7. Inner pad
10. Piston boot
13. Sliding pin bolt
16. Cap

2. Sliding pin boot
5. Piston seal
8. Pad retainer
11. Union bolt
14. Bleed valve
17. Brake hose

3. Torque member bolt
6. Piston
9. Outer pad
12. Copper washer
15. Cylinder body
18. Lower sliding pin

67170-ARMA-G55

Fig. 1 Front brake components

BRAKES

REAR DISC BRAKES

✳✳ CAUTION

Dust and dirt accumulating on brake parts during normal use may contain asbestos fibers from production or aftermarket brake linings. Breathing excessive concentrations of asbestos fibers can cause serious bodily harm. Exercise care when servicing brake parts. Do not sand or grind brake lining unless equipment used is designed to contain the dust residue.

Do not clean brake parts with compressed air or by dry brushing. Cleaning should be done by dampening the brake components with a fine mist of water, then wiping the brake components clean with a dampened cloth. Dispose of cloth and all residue containing asbestos fibers in an impermeable container with the appropriate label. Follow practices prescribed by the Occupational

Safety and Health Administration (OSHA) and the Environmental Protection Agency (EPA) for the handling, processing, and disposing of dust or debris that may contain asbestos fibers.

BRAKE CALIPER

REMOVAL & INSTALLATION

See Figure 2.

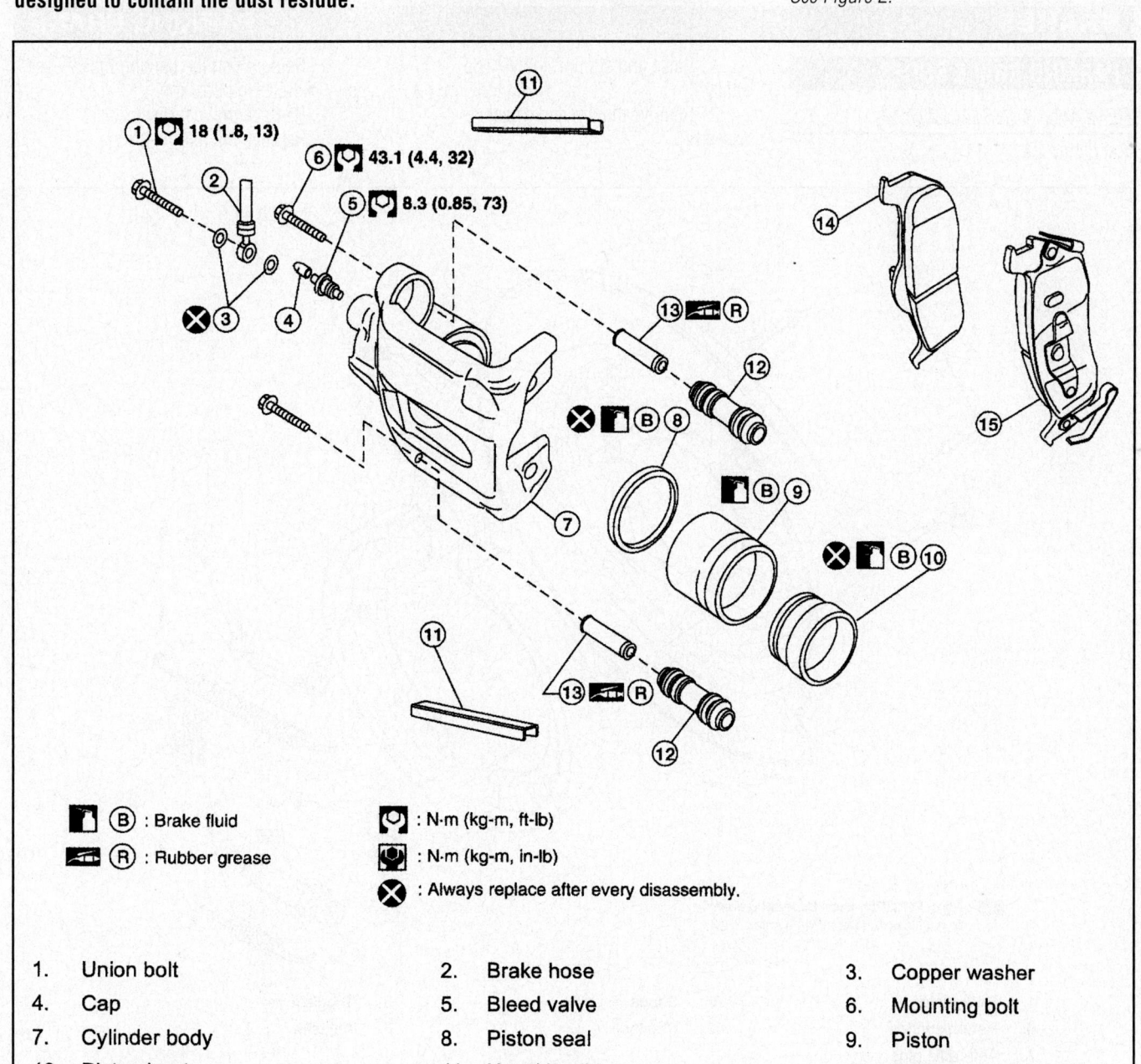

B : Brake fluid

R : Rubber grease

: N·m (kg-m, ft-lb)

: N·m (kg-m, in-lb)

✕ : Always replace after every disassembly.

1. Union bolt	2. Brake hose	3. Copper washer
4. Cap	5. Bleed valve	6. Mounting bolt
7. Cylinder body	8. Piston seal	9. Piston
10. Piston boot	11. Knuckle slide	12. Sliding sleeve boot
13. Sliding sleeve	14. Inner pad	15. Outer pad

67170-ARMA-G56

Fig. 2 Rear brake components

1. Before servicing the vehicle, refer to the Precautions Section.
2. Drain brake fluid as necessary.
3. Remove or disconnect the following:
   - Wheel
   - Union bolt
   - Mounting bolts
   - Brake caliper assembly

To install:

4. Install or connect the following:
   - Brake caliper assembly and tighten mounting bolts to 23ft. lbs. (44 Nm)

- Union bolt and tighten to 13 ft. lbs. (18 Nm)
5. Fill the master cylinder and bleed the brake system.
6. Install the wheels.

DISC BRAKE PADS

REMOVAL & INSTALLATION

1. Before servicing the vehicle, refer to the Precautions Section.
2. Remove the wheel.

3. Remove mounting bolt from the top mount.
4. Swing brake caliper open and remove the brake pads.

To install:

5. Push pistons in so that the pad is firmly installed, using a suitable tool.
6. Install pads to the brake caliper.
7. Install top mounting bolt and tighten to 32 ft. lbs. (44 Nm).
8. Install the wheel.

BRAKES

PARKING BRAKE

PARKING BRAKE SHOES

REMOVAL & INSTALLATION

See Figure 3.

1. Raise and support the vehicle safely.
2. Remove the tire and wheel assembly.

3. Be sure that the parking brake lever is in the released position.
4. Remove the rear disc rotor.
5. Remove the return springs.

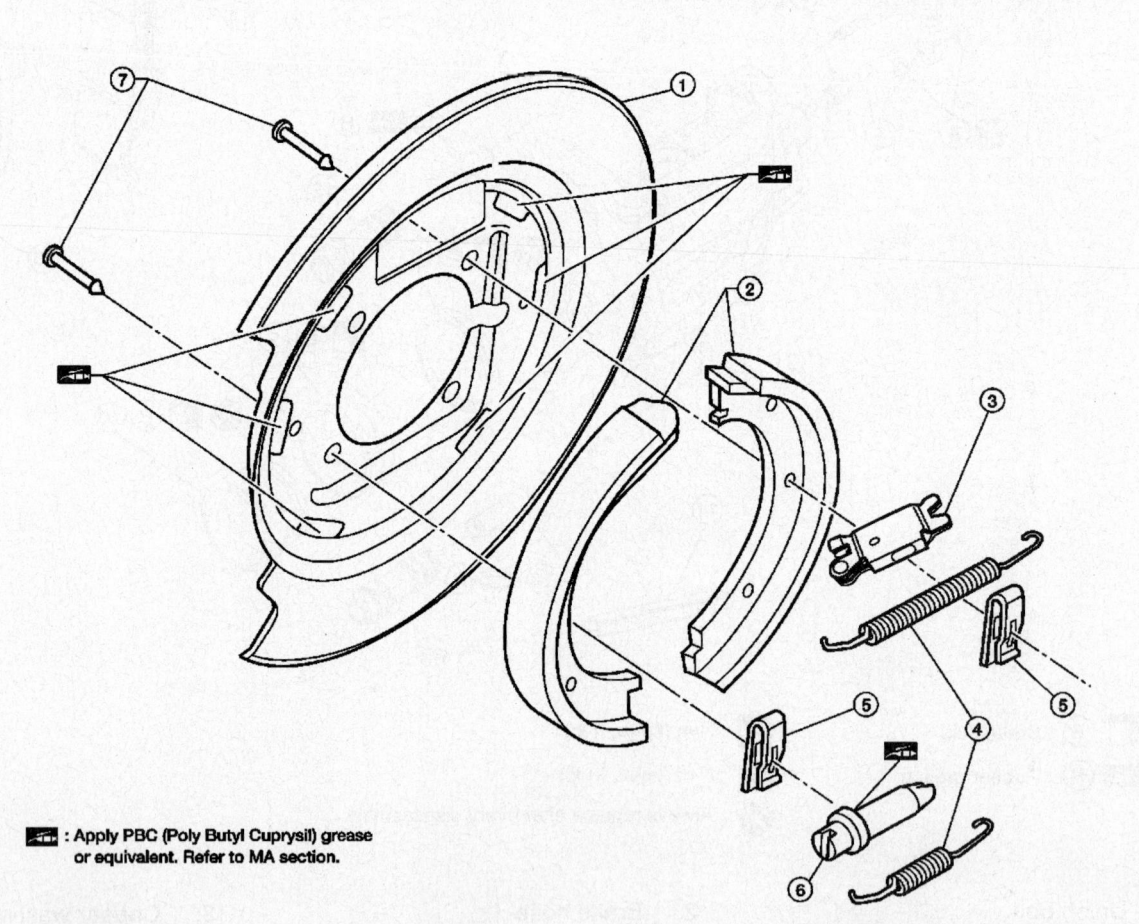

⊞ : Apply PBC (Poly Butyl Cuprysil) grease or equivalent. Refer to MA section.

1.	Back plate	2.	Shoes	3.	Toggle lever
4.	Return springs	5.	Retainers	6.	Adjuster
7.	Anti-rattle pins				

42050_QX56_G0002

Fig. 3 Parking brake shoe and related components

6. Remove the adjuster.

7. Disconnect the parking brake cable from the toggle lever.

8. Remove the retainers.

9. Remove the anti rattle pins and shoes

To install:

10. Apply brake grease to the specified points during reassembly, see illustration for locating points.

11. Install the adjuster so that the threaded part expands when rotating it in the proper direction.

12. Continue the installation in the reverse order of the removal procedure.

13. Adjust the parking brake.

14. Perform the parking brake burnishing operation.

CHASSIS ELECTRICAL

GENERAL INFORMATION

✳✳ CAUTION

These vehicles are equipped with an air bag system. The system must be disarmed before performing service on, or around, system components, the steering column, instrument panel components, wiring and sensors. Failure to follow the safety precautions and the disarming procedure could result in accidental air bag deployment, possible injury and unnecessary system repairs.

SERVICE PRECAUTIONS

Disconnect and isolate the battery negative cable before beginning any airbag system component diagnosis, testing, removal, or installation procedures. Allow system capacitor to discharge for two minutes before beginning any component service. This will disable the airbag system. Failure to disable the airbag system may result in accidental airbag deployment, personal injury, or death.

Do not place an intact undeployed airbag face down on a solid surface. The airbag will propel into the air if accidentally deployed and may result in personal injury or death.

When carrying or handling an undeployed airbag, the trim side (face) of the airbag should be pointing towards the body to minimize possibility of injury if accidental deployment occurs. Failure to do this may result in personal injury or death.

Replace airbag system components with OEM replacement parts. Substitute parts may appear interchangeable, but internal differences may result in inferior occupant protection. Failure to do so may result in occupant personal injury or death.

Wear safety glasses, rubber gloves, and long sleeved clothing when cleaning powder residue from vehicle after an airbag deployment. Powder residue emitted from a deployed airbag can cause skin irritation.

AIR BAG (SUPPLEMENTAL RESTRAINT SYSTEM)

Flush affected area with cool water if irritation is experienced. If nasal or throat irritation is experienced, exit the vehicle for fresh air until the irritation ceases. If irritation continues, see a physician.

Do not use a replacement airbag that is not in the original packaging. This may result in improper deployment, personal injury, or death.

The factory installed fasteners, screws and bolts used to fasten airbag components have a special coating and are specifically designed for the airbag system. Do not use substitute fasteners. Use only original equipment fasteners listed in the parts catalog when fastener replacement is required.

During, and following, any child restraint anchor service, due to impact event or vehicle repair, carefully inspect all mounting hardware, tether straps, and anchors for proper installation, operation, or damage. If a child restraint anchor is found damaged in any way, the anchor must be replaced. Failure to do this may result in personal injury or death.

Deployed and non-deployed airbags may or may not have live pyrotechnic material within the airbag inflator.

Do not dispose of driver/passenger/curtain airbags or seat belt tensioners unless you are sure of complete deployment. Refer to the Hazardous Substance Control System for proper disposal.

Dispose of deployed airbags and tensioners consistent with state, provincial, local, and federal regulations.

After any airbag component testing or service, do not connect the battery negative cable. Personal injury or death may result if the system test is not performed first.

If the vehicle is equipped with the Occupant Classification System (OCS), do not connect the battery negative cable before performing the OCS Verification Test using the scan tool and the appropriate diagnostic information. Personal injury or death may result if the system test is not performed properly.

Never replace both the Occupant Restraint Controller (ORC) and the Occupant Classification Module (OCM) at the same time. If both require replacement,

replace one, then perform the Airbag System test before replacing the other.

Both the ORC and the OCM store Occupant Classification System (OCS) calibration data, which they transfer to one another when one of them is replaced. If both are replaced at the same time, an irreversible fault will be set in both modules and the OCS may malfunction and cause personal injury or death.

If equipped with OCS, the Seat Weight Sensor is a sensitive, calibrated unit and must be handled carefully. Do not drop or handle roughly. If dropped or damaged, replace with another sensor. Failure to do so may result in occupant injury or death.

If equipped with OCS, the front passenger seat must be handled carefully as well. When removing the seat, be careful when setting on floor not to drop. If dropped, the sensor may be inoperative, could result in occupant injury, or possibly death.

If equipped with OCS, when the passenger front seat is on the floor, no one should sit in the front passenger seat. This uneven force may damage the sensing ability of the seat weight sensors. If sat on and damaged, the sensor may be inoperative, could result in occupant injury, or possibly death.

DISARMING THE SYSTEM

1. Before servicing the vehicle, refer to the Precautions Section.

2. Disconnect both battery cables.

3. Wait at least 3 minutes before working on the vehicle. The air bag system is designed to retain enough power to deploy the air bag for a short time after the battery has been disconnected.

4. After repairs are complete, connect the negative battery cable. Turn the ignition switch to the **ON** position and check the air bag warning light blinks for proper operation.

ARMING THE SYSTEM

After repairs are complete, connect the negative battery cable. Turn the ignition switch to the **ON** position and check the air bag warning light blinks for proper operation.

CLOCKSPRING CENTERING

See Figures 4 and 5.

➡ **Before servicing, or working around, the SRS system, turn the ignition switch OFF, disconnect both battery cables and wait at least three minutes. When servicing, or working around, the SRS system do not work directly in front of the air bag module.**

1. Before servicing the vehicle, refer to the Precautions Section.
2. Position the front wheels in the straight ahead position.
3. Disconnect the negative battery cable. Disconnect the positive battery cable.
4. Remove the air bag module.
5. Remove the steering wheel.
6. Remove the upper and lower steering column covers.
7. Remove the wiper washer switch connector. Pinch the tabs at the wiper and washer switch base and slide the switch away from the steering column to remove it.
8. While pressing the tabs, pull the headlight and turn signal switch toward the driver's door and disconnect it from the base.
9. Remove the spiral cable retaining screws, release the clip and remove the spiral cable.

➡ **Do not disassemble the spiral cable. Do not apply lubricant to the spiral cable.**

10. Remove the spiral cable connectors.

➡ **With the steering linkage disconnected, the spiral cable may snap by turning the steering wheel beyond the limited number of turns. The spiral cable can be turned counterclockwise about 2.5 turns from the right end position.**

To install:

11. Installation is the reverse of the removal procedure.
12. Be sure to align the spiral cable correctly when installing the steering wheel. Make sure that the spiral cable is in the neutral position.

➡ **The neutral position is detected by turning to the left 2.6 revolutions from the right end position and ending with the knob at the top. The spiral cable may snap due to steering operation if the cable is installed incorrectly. Also, with the steering linkage disconnected the cable may snap by turning the steering wheel beyond the limited number of turns (2.6 from the neutral position to both the left and right).**

13. If equipped with VDC adjust the steering angle sensor.
14. Use the CONSULT-II tool and perform self diagnosis to ensure no malfunction is detected.

➡ **With the steering linkage disconnected, the spiral cable may snap by turning the steering wheel beyond the limited number of turns. The spiral cable can be turned counterclockwise about 2.5 turns from the right end position.**

15. Tighten the steering wheel retaining nut to 25 ft. lbs.
16. When reinstalling the air bag module, use new bolts. Tighten the bolts to 8 ft. lbs.

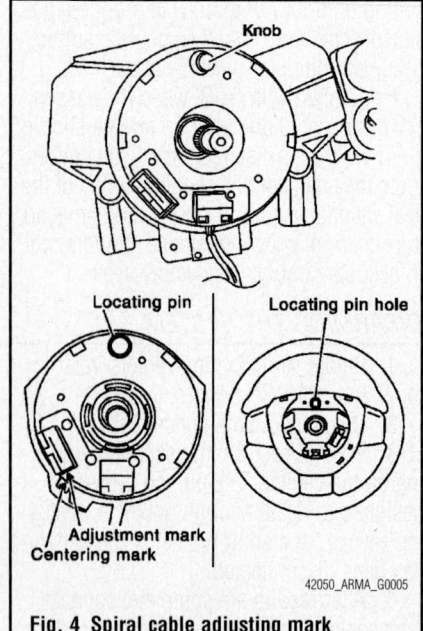

Fig. 4 Spiral cable adjusting mark

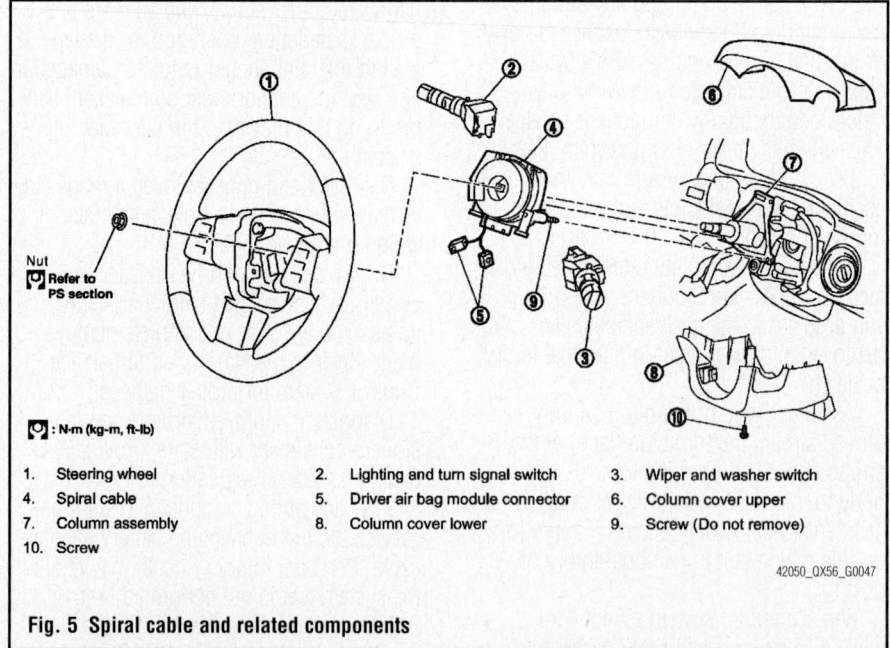

Fig. 5 Spiral cable and related components

DRIVETRAIN

AUTOMATIC TRANSMISSION ASSEMBLY

REMOVAL & INSTALLATION

2WD Models

See Figure 6.

1. Before servicing the vehicle, refer to the Precautions Section.
2. Remove or disconnect the following:
   - Negative battery cable
   - Engine cover
   - Transmission fluid indicator gauge
   - Engine splash guard
   - Exhaust front pipe
   - Center muffler
   - Rear driveshaft
   - Transmission control cable
   - Crankshaft position sensor
   - Transmission cooler tube
   - Dust cover from converter housing
3. Turning crankshaft clockwise, remove the four tightening bolts for drive plate and torque converter.
4. Support the transmission with a suitable jack.
5. Remove or disconnect the following:
   - Transmission cross member
   - Air breather hose
   - Transmission assembly connector
   - Fluid indicator tube from transmission assembly
   - Transmission assembly to engine bolts
   - Transmission assembly from vehicle

To install:

6. Install or connect the following:
   - Transmission assembly into vehicle

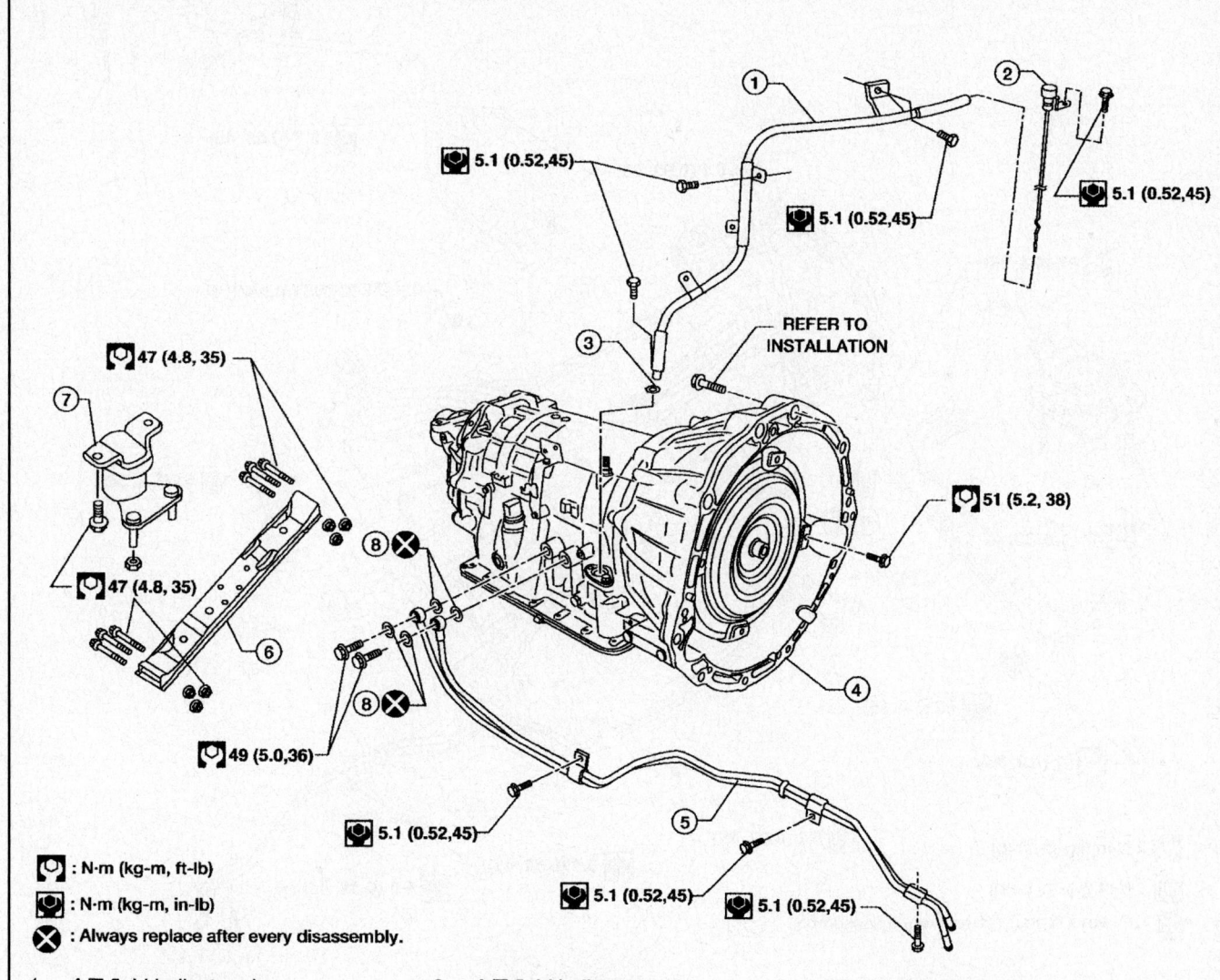

1. A/T fluid indicator pipe	2. A/T fluid indicator	3. O-ring
4. Transmission assembly	5. A/T fluid cooler tube	6. A/T crossmember
7. Insulator	8. Copper washers	

⬚ : N·m (kg-m, ft-lb)

⬚ : N·m (kg-m, in-lb)

✕ : Always replace after every disassembly.

Fig. 6 Transmission and related parts—2WD

67170-ARMA-G39

- Transmission assembly to engine bolts tightening to 83 ft. lbs. (113 Nm)
- Fluid indicator tube to transmission assembly
- Transmission assembly connector
- Air breather hose
- Transmission cross member

7. Turning crankshaft clockwise, install the torque converter to drive plate.

➡**After torque converter is installed, rotate the crankshaft to ensure transmission rotates freely.**

8. Install or connect the following:
- Dust cover for converter housing
- Fluid cooler tube
- Crankshaft position sensor
- Transmission control cable
- Rear driveshaft
- Center muffler
- Exhaust front pipe
- Engine splash guard
- Transmission fluid indicator gauge
- Engine cover
- Negative battery cable

9. Start engine and check for leaks.

4WD Models

See Figure 7.

1. Before servicing the vehicle, refer to the Precautions Section.
2. Remove or disconnect the following:
- Negative battery cable
- Engine cover
- Transmission fluid indicator gauge
- Engine splash guard
- Exhaust front pipe
- Center muffler
- Driveshafts
- Transmission control cable

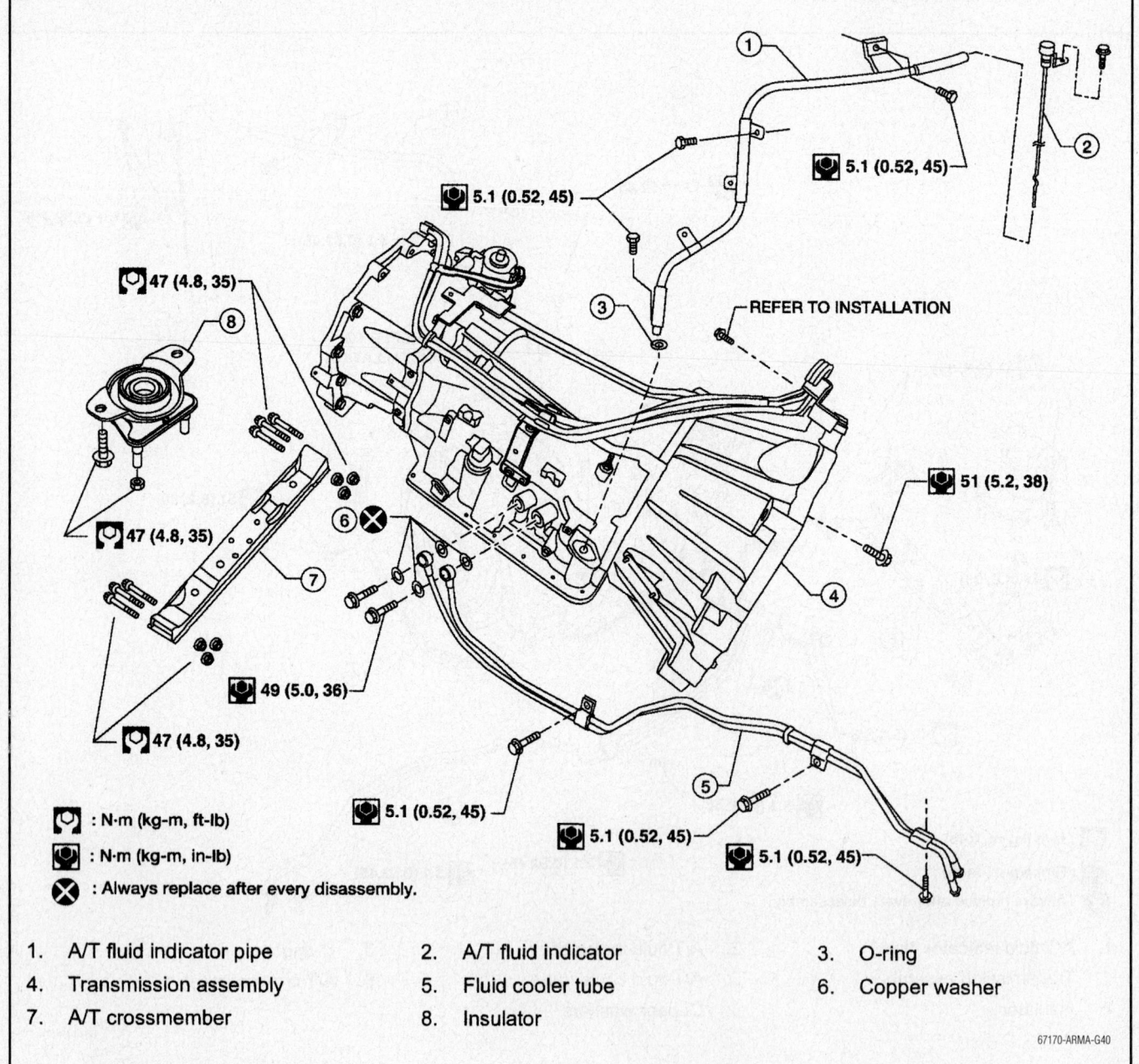

47 (4.8, 35)

47 (4.8, 35)

47 (4.8, 35)

49 (5.0, 36)

5.1 (0.52, 45)

5.1 (0.52, 45)

5.1 (0.52, 45)

5.1 (0.52, 45)

5.1 (0.52, 45)

51 (5.2, 38)

REFER TO INSTALLATION

[symbol] : N·m (kg-m, ft-lb)

[symbol] : N·m (kg-m, in-lb)

[symbol] : Always replace after every disassembly.

1. A/T fluid indicator pipe	2. A/T fluid indicator	3. O-ring
4. Transmission assembly	5. Fluid cooler tube	6. Copper washer
7. A/T crossmember	8. Insulator	

67170-ARMA-G40

Fig. 7 Transmission and related parts—with 4WD

- Crankshaft position sensor
- Fluid cooler tube
- Dust housing for torque converter

3. Turning the crankshaft clockwise, remove the four tightening bolts for drive plate and torque converter.

4. Support the transmission assembly with a suitable jack.

5. Remove transmission cross member.

6. Tilt the transmission slightly to keep clearance between the body and the transmission assembly, then disconnect the air breather hose.

7. Remove or disconnect the following:
- Transmission assembly connector and transfer case connector
- Fluid indicator pipe
- Transmission assembly to engine bolts
- Transmission assembly, with transfer case attached, from vehicle
- Transmission assembly from transfer case

To install:

8. Install or connect the following:
- Transfer case to transmission assembly
- Transmission assembly into vehicle
- Transmission assembly to engine bolts tightening to 83 ft. lbs. (113 Nm)

9. With the transmission slightly tilted to allow clearance between body and transmission, connect the air breather hose.

10. Install the transmission cross member.

11. Turning crankshaft clockwise, install the torque converter to drive plate.

➡**After torque converter is installed, rotate the crankshaft to ensure transmission rotates freely.**

12. Install or connect the following:
- Dust housing for torque converter
- Fluid cooler tube
- Crankshaft position sensor
- Transmission control cable
- Driveshaft
- Center muffler
- Front exhaust pipe
- Engine splash guard
- Transmission fluid indicator gauge
- Engine cover
- Negative battery cable

13. Start engine and check for leaks.

TRANSFER CASE ASSEMBLY

REMOVAL & INSTALLATION

See Figure 8.

1. Before servicing the vehicle, refer to the Precautions Section.

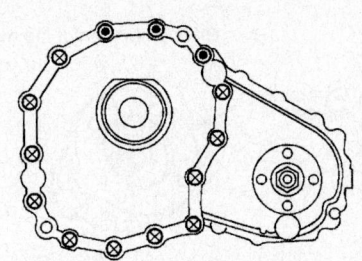

◉ : Transfer → Automatic transmission
⊗ : Automatic transmission → Transfer

67170-ARMA-G41

Fig. 8 Transfer case mounting bolt locations

2. Remove or disconnect the following:
- Transmission splash guard
- Center exhaust pipe and muffler
- Front and rear driveshafts

➡**Plug rear oil seal after removing rear driveshaft.**

- Transmission assembly mounting bolts

3. Support the transmission assembly with a suitable jack and remove the cross-member.

4. Remove or disconnect the following:
- ATP switch, neutral 4LO switch, wait detection switch, transfer motor and transfer control device electrical connectors
- Breather hoses
- Shift actuator from the extension housing
- Transfer case to transmission assembly bolts
- Transfer case assembly

To install:

5. Install or connect the following:
- Transfer case to transmission assembly bolts tightening to 26 ft. lbs. (36 Nm)
- Shift actuator
- Breather hoses
- ATP switch, neutral 4LO switch, wait detection switch, transfer motor and transfer control device electrical connectors
- Support crossmember
- Transmission mounting bolts
- Driveshafts
- Muffler and center exhaust pipe
- Transmission splash guard

FRONT HALFSHAFT

REMOVAL & INSTALLATION

1. Remove wheel and tire using power tool.

2. Remove engine under cover using power tool.

3. Remove wheel sensor harness from mount on knuckle.

➡**Do not pull on wheel sensor harness.**

4. Without disassembling the hydraulic lines, remove brake caliper using power tool. Reposition it aside with wire.

➡**Avoid depressing brake pedal while brake caliper is removed.**

5. Remove coil spring and shock absorber assembly using power tool.

6. Separate upper link ball joint stud from steering knuckle using tool. Support lower link with jack.

7. Remove cotter pin, then remove drive shaft nut.

8. Remove drive shaft mounting bolts from front final drive.

9. Remove drive shaft from wheel hub and bearing assembly.

➡**When removing drive shaft, do not apply an excessive angle to drive shaft joint. Also be careful not to excessively extend slide joint.**

To install:

10. Installation is in the reverse order of removal.

➡**When installing drive shaft onto front final drive, use Tool to prevent damage to the oil seal while inserting drive shaft. Slide drive shaft sliding joint and tap with a hammer to install securely.**

➡**Never reuse the differential side oil seal.**

REAR HALFSHAFT

REMOVAL & INSTALLATION

1. Before servicing the vehicle, refer to the Precautions Section.

2. Remove or disconnect the following:
- Wheel
- Stabilizer bar clamp
- Cotter pin and driveshaft nut
- Bolts from the inside flange of the driveshaft

3. Separate the driveshaft from the wheel hub by lightly tapping the end with suitable hammer and wood block.

4. Remove the halfshaft.

✻✻ CAUTION

Do not excessively extend the slide joint.

To install:

5. Install or connect the following:
- Halfshaft
- Bolts for the inside flange and tighten to 87 ft. lbs. (118 Nm)
- Driveshaft nut and tighten nut to 101 ft. lbs. (137 Nm) and replace cotter pin
- Stabilizer bar clamp
- Wheel

REAR PINION SEAL

REMOVAL & INSTALLATION

See Figures 9 through 12.

1. Before servicing the vehicle, refer to the Precautions Section.
2. Remove the rear driveshaft.

➡**Matchmark driveshaft position.**

3. Measure and record the total preload torque.
4. Matchmark the drive pinion to position 'B' on the companion flange.
5. Remove the drive pinion nut using suitable tool.
6. Remove the companion flange using suitable tool.
7. Remove the rear pinion seal using special tool J-34286.

To install:

8. Press the rear pinion seal into the carrier using suitable tool.
9. Align the matchmark on the companion flange to the drive pinion and install the companion flange.

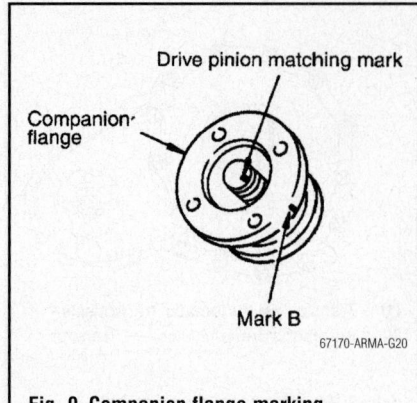

Fig. 9 Companion flange marking

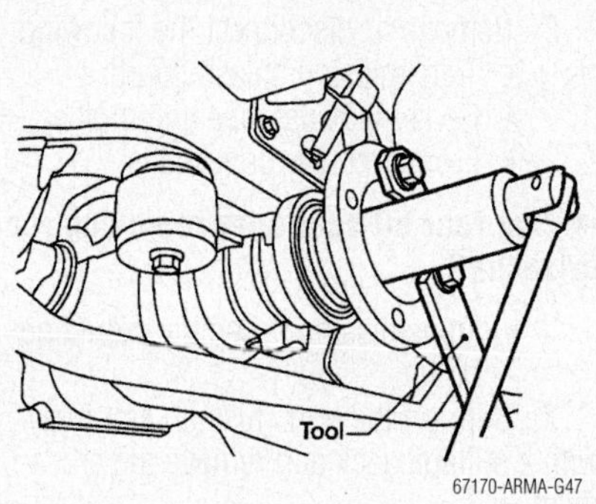

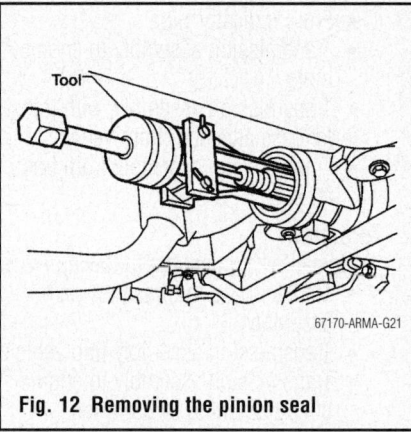

Fig. 10 Loosening the flange nut

10. Lubricate the drive pinion threads and seating surfaces of the drive pinion nut with grease.
11. Using a new drive pinion nut, tighten to 124-274 ft. lbs. (167-372 Nm).

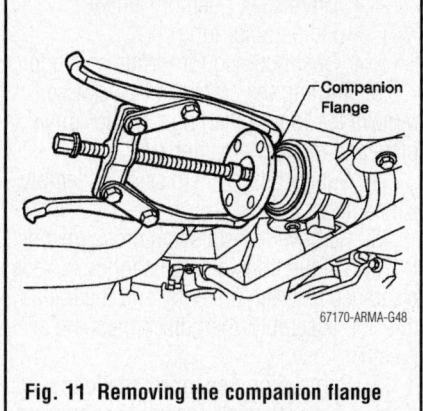

Fig. 11 Removing the companion flange

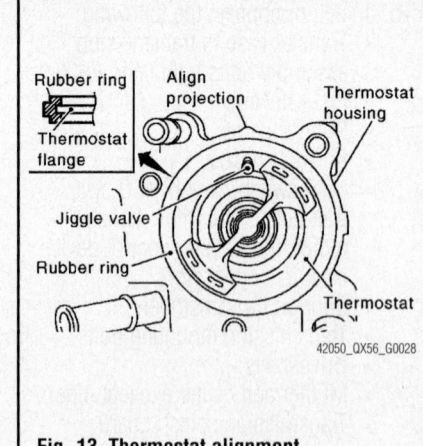

Fig. 12 Removing the pinion seal

➡**Final torque is determined when adjusting total preload using special tool J-25765-A.**

12. Install rear driveshaft using matchmarks.

ENGINE COOLING

THERMOSTAT

REMOVAL & INSTALLATION

See Figures 13 and 14.

➡**Never remove the radiator cap when the engine is hot. Serious burns could occur from high-pressure engine coolant escaping from the radiator.**

1. Be sure the engine is cold.
2. Disconnect the negative battery cable.
3. Remove the air duct and resonator assembly.
4. Remove the engine front undercover.
5. Disconnect the water suction hose from the water inlet.
6. Remove the water inlet and thermostat.

To install:

7. Installation is the reverse of the removal procedure.
8. Be sure to use a new gasket.

Fig. 13 Thermostat alignment

9. Install the thermostat with the whole circumference of each flange part fitting securely inside the rubber ring, as shown in the illustration.

10. Install the thermostat with the jiggle valve facing upward.
11. Be sure to refill the cooling using the proper grade and type engine coolant.
12. Start the engine and check for leaks.
13. Start the engine and allow it to reach operation temperature. Recheck the coolant level, fill as required.

WATER PUMP

REMOVAL & INSTALLATION

See Figure 15.

1. Before servicing the vehicle, refer to the Precautions Section.
2. Drain the cooling system.
3. Remove or disconnect the following:
- Engine splash guard
- Air intake assembly
- Accessory drive belt

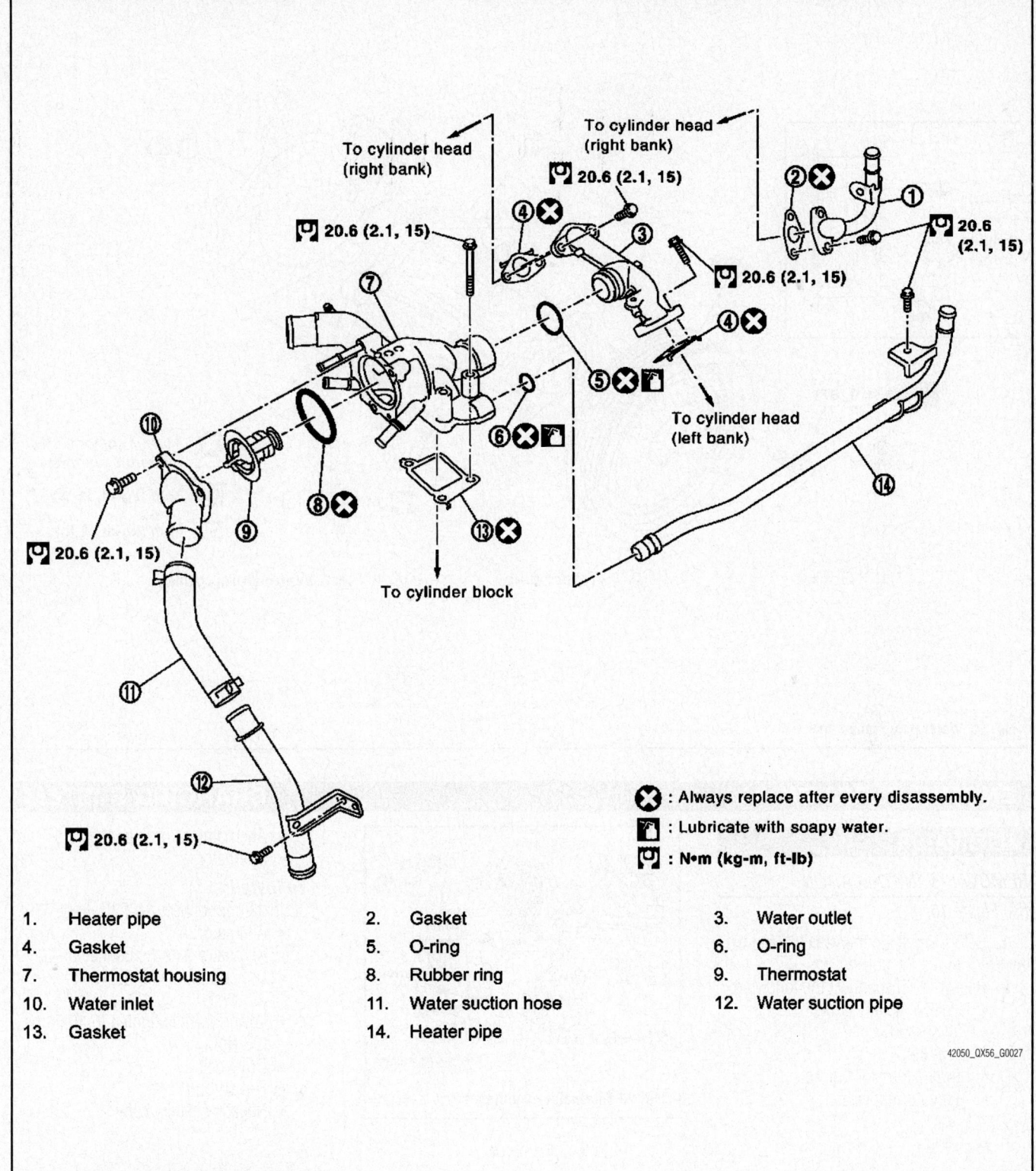

Fig. 14 Thermostat and related components

1.	Heater pipe	2.	Gasket	3.	Water outlet
4.	Gasket	5.	O-ring	6.	O-ring
7.	Thermostat housing	8.	Rubber ring	9.	Thermostat
10.	Water inlet	11.	Water suction hose	12.	Water suction pipe
13.	Gasket	14.	Heater pipe			

42050_QX56_G0027

➡️**Leave tensioner pulley in its fixed position.**

- Water pump pulley
- Water pump

To install:

4. Install or connect the following:

- Water pump with a new gasket. Tighten bolts to 18 ft. lbs. (25 Nm).
- Water pump pulley and tighten bolts to 87 in. lbs. (10 Nm).
- Accessory drive belt

- Air intake assembly
- Engine splash guard

5. Refill the cooling system.

6. Start the engine and check for leaks.

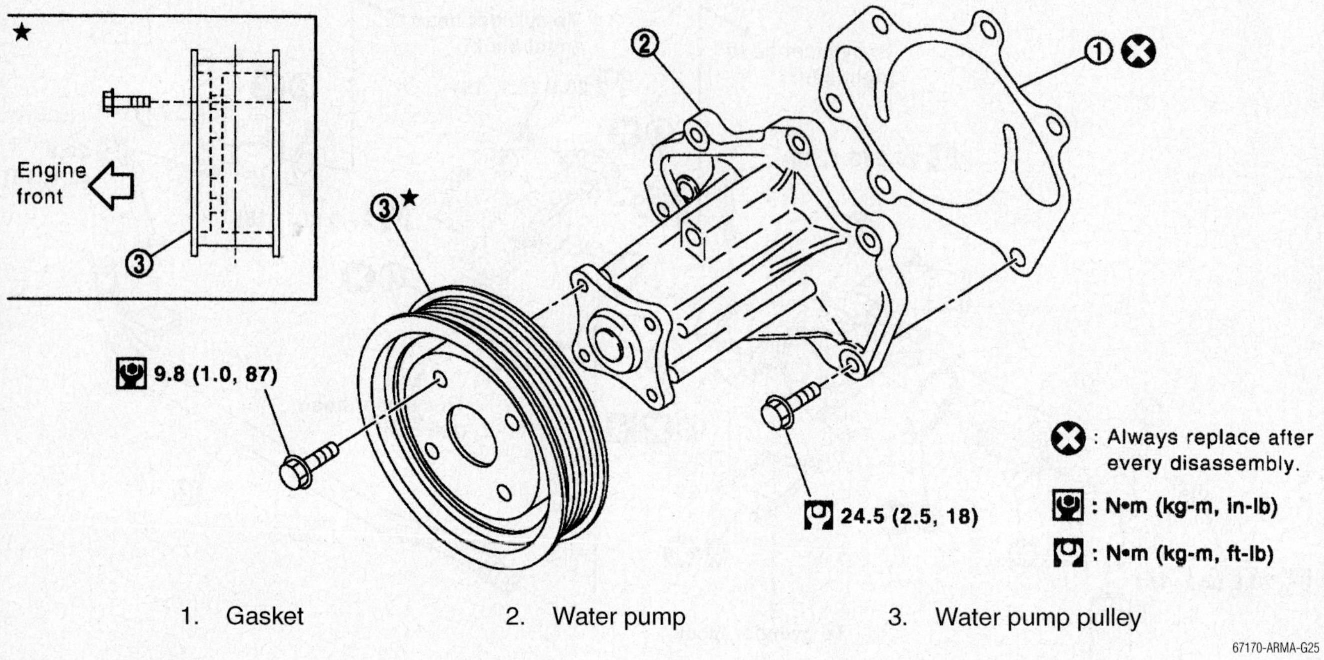

★

Engine front ⟵

③

9.8 (1.0, 87)

②

① ✖

24.5 (2.5, 18)

✖ : Always replace after every disassembly.

⚙ : N•m (kg-m, in-lb)

⚙ : N•m (kg-m, ft-lb)

1. Gasket
2. Water pump
3. Water pump pulley

67170-ARMA-G25

Fig. 15 Water pump mounting

ENGINE ELECTRICAL

ALTERNATOR

REMOVAL & INSTALLATION

See Figure 16.

1. Before servicing the vehicle, refer to the Precautions Section.

2. Remove or disconnect the following:
- Negative battery cable
- Fan shroud
- Drive belt
- Lower alternator bracket
- Alternator upper bolt

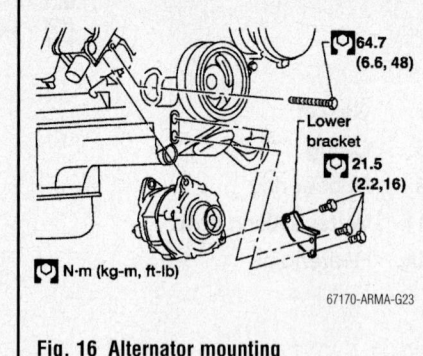

64.7 (6.6, 48)

Lower bracket

21.5 (2.2,16)

N•m (kg-m, ft-lb)

67170-ARMA-G23

Fig. 16 Alternator mounting

CHARGING SYSTEM

- Alternator harness connectors
- Alternator

To install:

3. Install or connect the following:
- Alternator
- Alternator harness connectors
- Upper bolt, tighten to 48 ft. lbs. (65 Nm)
- Lower bracket, tighten to 16 ft. lbs (22 Nm)
- Drive belt
- Fan shroud
- Negative battery cable

ENGINE ELECTRICAL   **IGNITION SYSTEM**

See Figure 17.

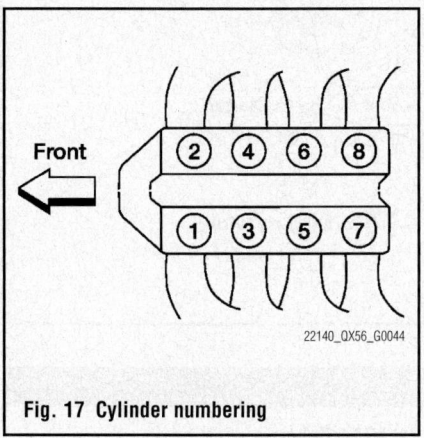

Fig. 17 Cylinder numbering

IGNITION COIL

REMOVAL & INSTALLATION

See Figure 18.

1. Disconnect the negative battery cable.
2. Remove the engine room cover.
3. Disconnect the harness connector from the ignition coil.
4. Remove the ignition coil retaining bolt.
5. Remove the ignition coil.

To install:

6. Install the ignition coil, torque the retaining bolt to 80 inch lbs.
7. Connect the harness coil.
8. Connect the negative battery cable.

IGNITION TIMING

ADJUSTMENT

The ignition timing is controlled by the Powertrain Control Module (PCM). No adjustment is necessary or possible.

SPARK PLUGS

REMOVAL & INSTALLATION

See Figure 19.

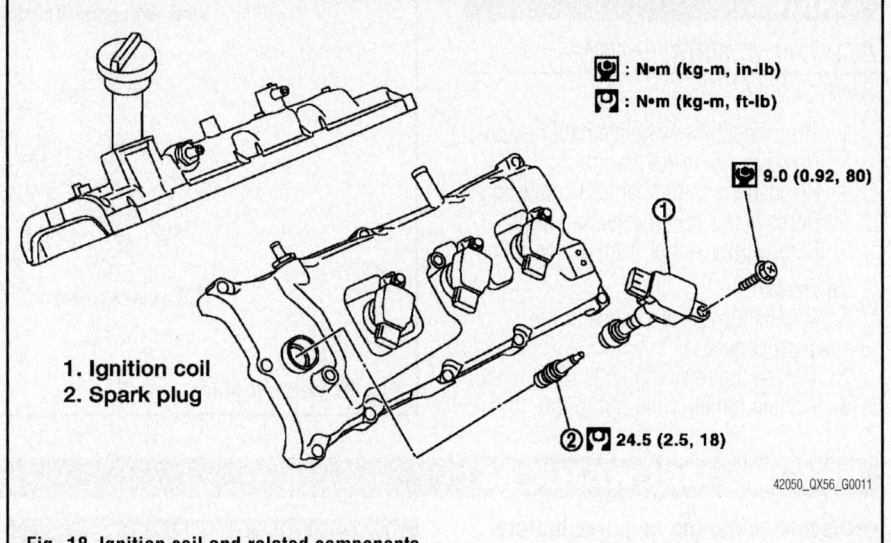

1. Ignition coil
2. Spark plug

Fig. 18 Ignition coil and related components

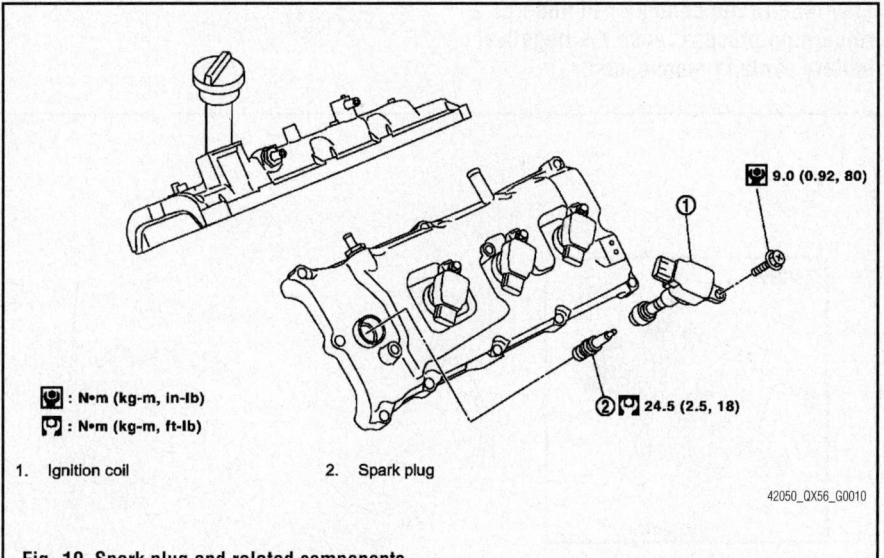

1. Ignition coil
2. Spark plug

Fig. 19 Spark plug and related components

1. Disconnect the negative battery cable.
2. Disconnect the harness connector from the ignition coil.
3. Remove the ignition coil retaining bolt.
4. Remove the ignition coil.
5. Remove the spark plug using a spark plug socket and wrench.

To install:

6. Be sure the spark plug gap is to specification (0.043 in).
7. Carefully install the spark plug and torque to specification, 18 ft. lbs.
8. Install the ignition coil, torque the retaining bolt to 80 inch lbs.
9. Connect the harness coil.
10. Connect the negative battery cable.

ENGINE ELECTRICAL

STARTING SYSTEM

STARTER

REMOVAL & INSTALLATION

See Figure 20.

1. Disconnect the negative battery cable.
2. Remove the intake manifold.
3. Remove the starter harness connectors.
4. Remove the starter retaining bolts.
5. Remove the starter from its mounting.

To install:

6. Installation is the reverse of the removal procedure.
7. Tighten the retaining bolts to 34 ft. lbs.
8. Tighten the terminal nut to 8 ft. lbs.

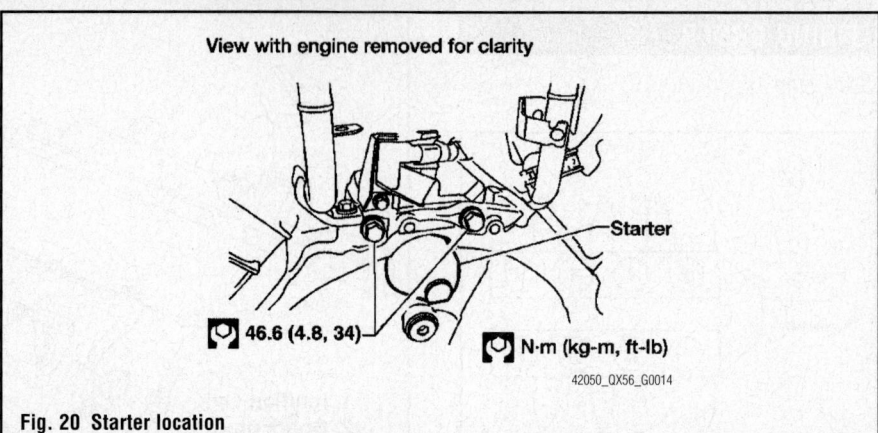

View with engine removed for clarity

46.6 (4.8, 34)   N·m (kg-m, ft-lb)

42050_QX56_G0014

Fig. 20 Starter location

ENGINE MECHANICAL

➡ **Disconnecting the negative battery cable may interfere with the functions of the on board computer systems and may require the computer to undergo a relearning process, once the negative battery cable is reconnected.**

ACCESSORY DRIVE BELTS

ACCESSORY BELT ROUTING

See Figure 21.

INSPECTION

Inspect the drive belt for signs of glazing or cracking. A glazed belt will be perfectly smooth from slippage, while a good belt will have a slight texture of fabric visible.

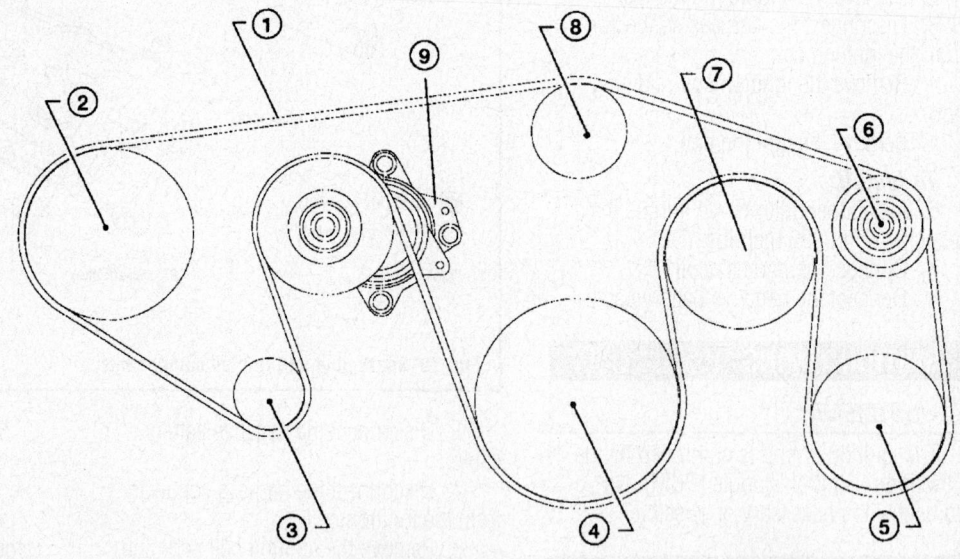

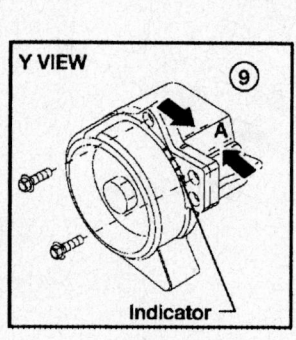

Y VIEW

Indicator

1.	Drive Belt	2.	Power Steering Pump Pulley	3.	Generator pulley
4.	Crankshaft Pulley	5.	A/C Compressor	6.	Idler Pulley
7.	Cooling Fan Pulley	8.	Water Pump Pulley	9.	Drive Belt Tensioner

67162-QX56-G47

Fig. 21 Accessory drive belt routing

Cracks will usually start at the inner edge of the belt and run outward. All worn or damaged drive belts should be replaced immediately.

ADJUSTMENT

See Figure 22.

Drive belt tension is not necessary, as it is automatically adjusted by the auto tensioner.

REMOVAL & INSTALLATION

See Figure 23.

1. Disconnect the negative battery cable.
2. Remove the air duct and resonator assembly.
3. Install special tool J-46535, or equivalent on the auto tensioner pulley bolt and move it upward.

➡**Avoid placing your hand in a location where pinching may occur if the holding tool accidentally comes off.**

4. Remove the drive belt from the vehicle.

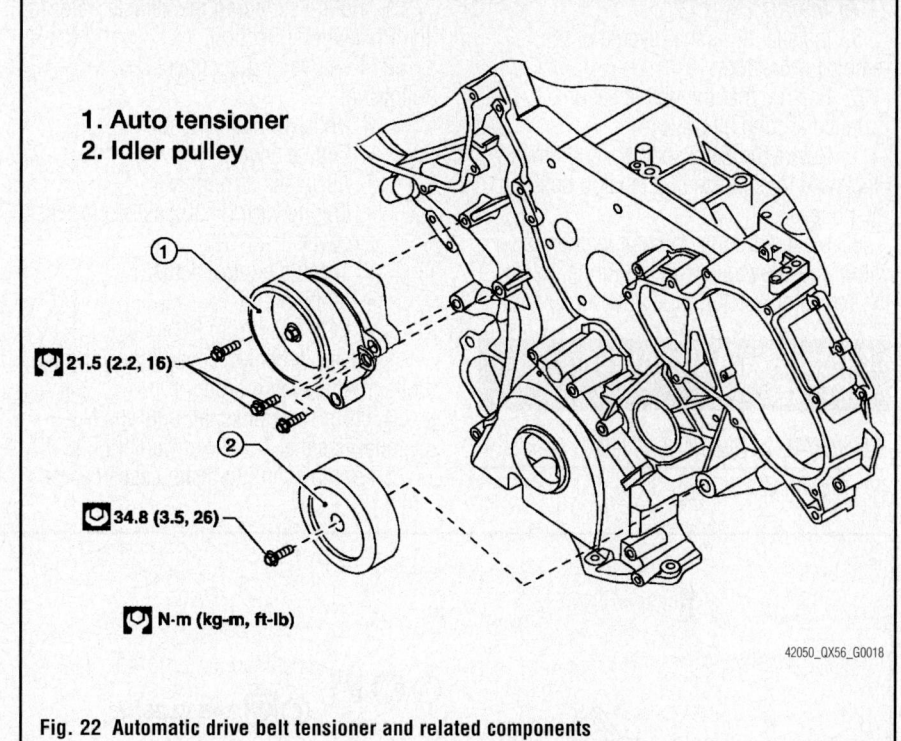

1. Auto tensioner
2. Idler pulley

21.5 (2.2, 16)

34.8 (3.5, 26)

N·m (kg-m, ft-lb)

42050_QX56_G0018

Fig. 22 Automatic drive belt tensioner and related components

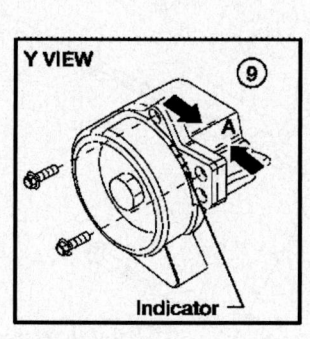

Y VIEW

Indicator

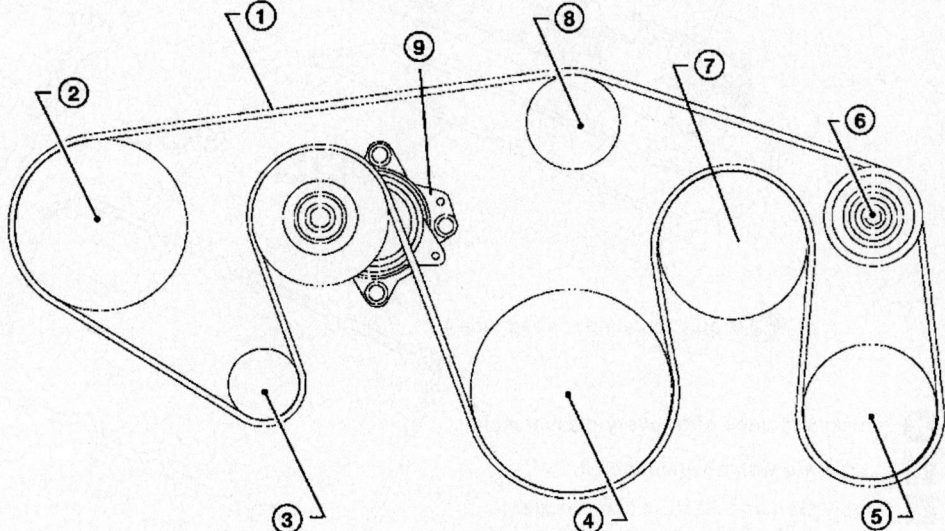

1. Drive belt	2. Power steering oil pump pulley	3. Generator pulley
4. Crankshaft pulley	5. A/C compressor	6. Idler pulley
7. Cooling fan pulley	8. Water pump pulley	9. Drive belt tensioner

42050_QX56_G0019

Fig. 23 Drive belt tensioner indicator (view Y) and related components

To install:

5. Installation is the reverse of the removal procedure.

6. Be sure that the belt is securely installed around all pulleys.

7. Rotate the crankshaft several times clockwise to equalize belt tension between the pulleys.

8. Make sure that the belt tension is within the allowable working range, using the indicator notch on the auto tensioner.

CAMSHAFT AND VALVE LIFTERS

REMOVAL & INSTALLATION

See Figures 24 through 30.

1. Before servicing the vehicle, refer to the Precautions Section.

2. Remove or disconnect the following:

- Negative battery cable
- Engine cover
- Air intake assembly
- Engine wiring harnesses on rocker cover
- Throttle control actuator
- Ignition coil
- PCV hose from PCV valve

3. Remove rocker cover, loosening the bolts in the reverse order.

4. Turn the crankshaft until the No. 1 cylinder is set at Top Dead Center (TDC).

5. Remove timing chain case covers.

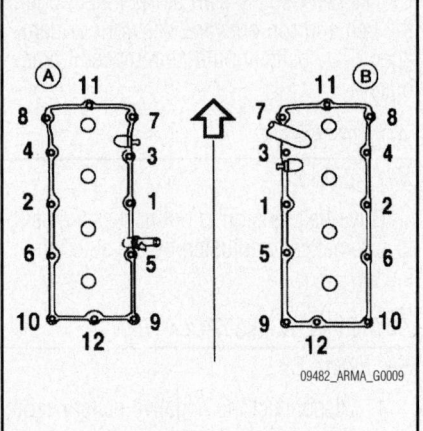

09482_ARMA_G0009

Fig. 25 Rocker cover torque sequence

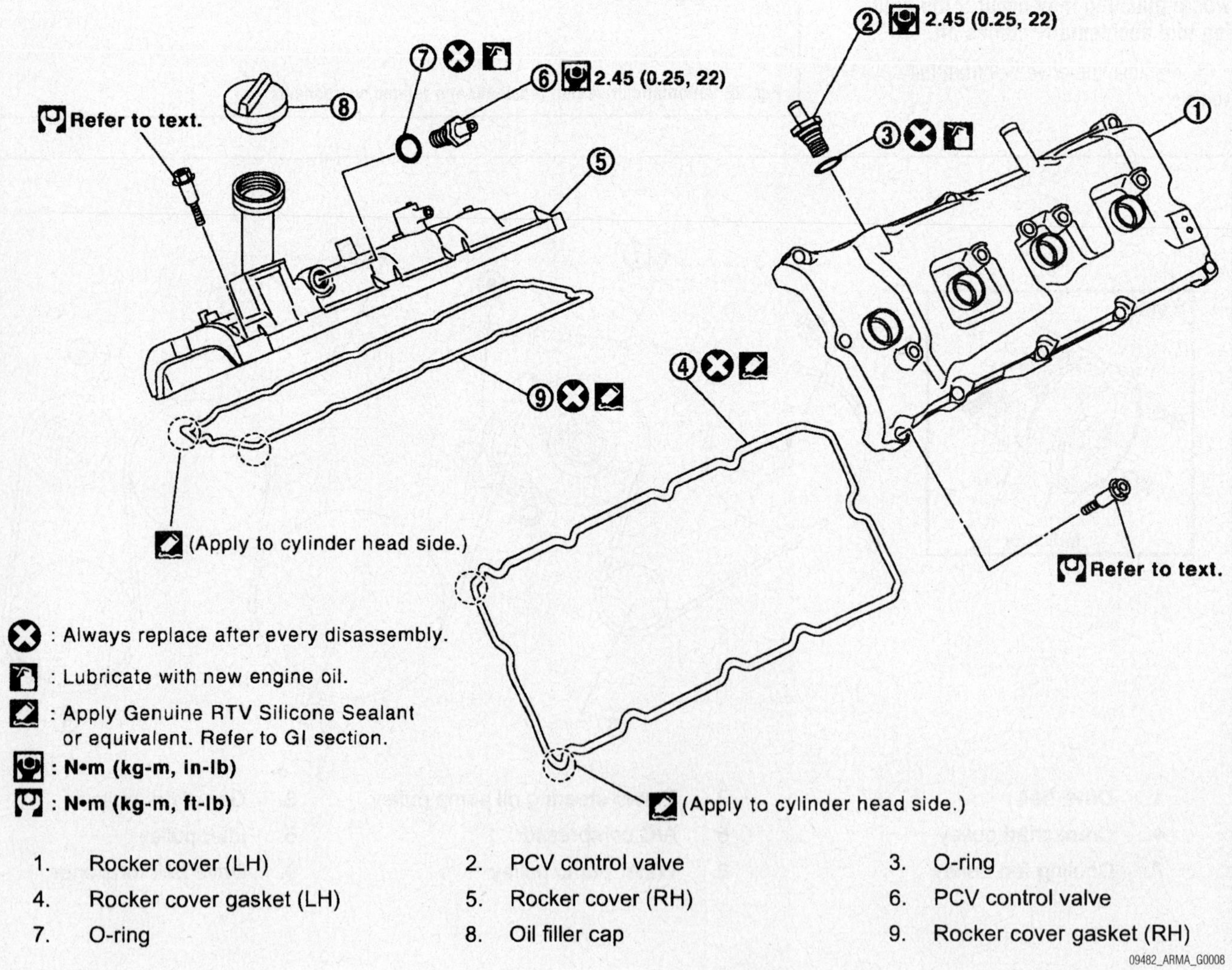

⚿ **Refer to text.**

⑦ ✖ 🖐   ⑥ 🔧 2.45 (0.25, 22)

⑧

② 🔧 2.45 (0.25, 22)

③ ✖ 🖐

⑨ ✖ ✎

④ ✖ ✎

①

✎ (Apply to cylinder head side.)

✎ (Apply to cylinder head side.)

⚿ **Refer to text.**

✖ : Always replace after every disassembly.

🖐 : Lubricate with new engine oil.

✎ : Apply Genuine RTV Silicone Sealant or equivalent. Refer to GI section.

🔧 : N•m (kg-m, in-lb)

⚿ : N•m (kg-m, ft-lb)

1. Rocker cover (LH)	2. PCV control valve	3. O-ring
4. Rocker cover gasket (LH)	5. Rocker cover (RH)	6. PCV control valve
7. O-ring	8. Oil filler cap	9. Rocker cover gasket (RH)

09482_ARMA_G0008

Fig. 24 Exploded view of the rocker cover assembly

Bank	INT EXH	Identification paint (front)	Identification paint (rear)	Identification rib
RH	INT	White	—	Yes.
	EXH	—	Light blue	Yes.
LH	INT	White	—	No.
	EXH	—	Light blue	No.

67162-QX56-G16

Fig. 26 Camshaft installation markings

6. Matchmark the timing chain, aligning with the camshaft sprocket marks.

7. Remove chain tensioner from left bank as follows:

   a. Squeeze end clips and push plunger into tensioner body.

   b. Secure plunger using stopper pin.

   c. Remove the chain tensioner.

8. Remove the chain tensioner from right bank as follows:

   a. Remove the chain tensioner cover using special tool J-37228.

   b. Squeeze end clips and push plunger into tensioner body.

   c. Secure plunger using stopper pin.

   d. Remove the chain tensioner.

9. With camshaft locked with a wrench, loosen bolts to remove camshaft sprocket.

10. Remove front cover bolts.

11. Remove camshaft brackets, removing bolts in reverse order shown in figure.

12. Remove camshaft.

13. Remove valve lifters.

➡**Matchmark the drivetrain components so each part can be reinstalled in its original position.**

To install:

14. Install valve lifters.

15. Install camshaft, refer to table for correct placement.

16. Install camshaft brackets as follows:

   a. Refer to location mark on upper surface of bracket.

   b. Installation mark should be correctly read when viewed from intake side.

17. Install camshaft bracket No. 1 as follows:

   a. Apply liquid gasket to bracket and backside of front cover as shown in figure.

   b. Carefully position and mount camshaft bracket No. 1.

   c. Temporarily tighten front cover bolts.

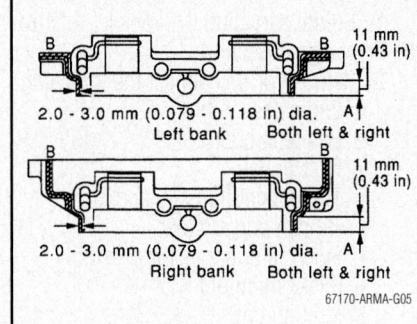

67170-ARMA-G05

Fig. 27 Gasket application for camshaft bracket

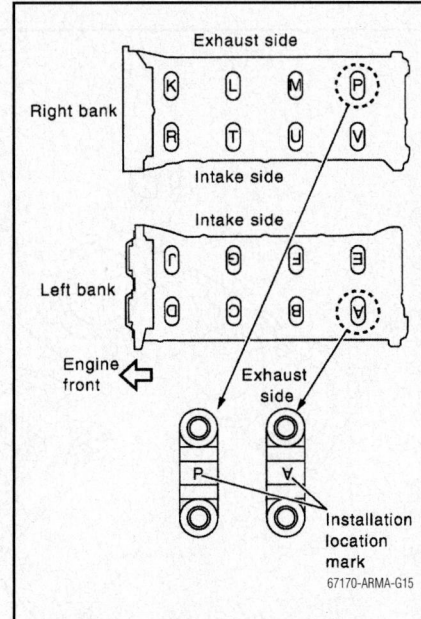

67170-ARMA-G15

Fig. 28 Camshaft bracket installation markings

18. Tighten fixing bolts for camshaft brackets as follows:

   a. Step 1: Bolts 9-12: 17 in. lbs. (1.9 Nm).

   b. Step 2: Bolts 1-8: 17 in. lbs. (1.9 Nm).

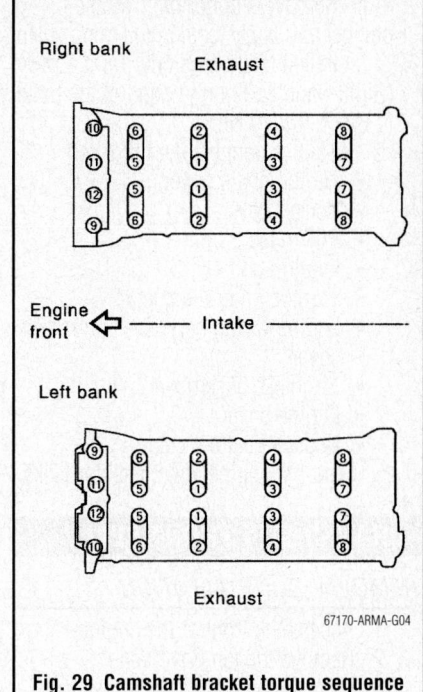

67170-ARMA-G04

Fig. 29 Camshaft bracket torque sequence

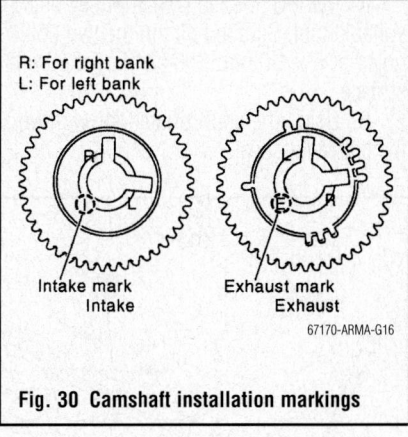

67170-ARMA-G16

Fig. 30 Camshaft installation markings

   c. Step 3: All bolts: 52 in. lbs. (5.9 Nm).

   d. Step 4: All bolts: 92 in. lbs. (10 Nm).

19. Tighten front cover bolts to 8 ft. lbs. (11 Nm).

20. Install camshaft sprocket as follows:

   a. Install camshaft sprocket aligning matchmarks with timing chain. Align camshaft sprocket key groove with dowel pin on camshaft front edge.

   b. Temporarily tighten bolts.

   c. Lock the camshaft with a wrench and tighten the bolts.

21. Install chain tensioner as shown:

   a. Install chain tensioner, compress plunger and hold with stopper pin.

   b. Tighten chain tensioner bolts to 61 in. lbs. (7 Nm).

c. Remove stopper pin, release plunger and apply tension to timing chain.

d. Install chain tensioner front cover (Right-hand bank only) and tighten bolts to 80 in. lbs. (9 Nm).

22. Install or connect the following:
   • Timing chain cover
   • Rocker cover.
   • PCV hose
   • Ignition coil
   • Throttle control actuator
   • Engine wiring harnesses on rocker cover
   • Air intake assembly
   • Engine cover
   • Negative battery cable

23. Start the engine and check for leaks.

CRANKSHAFT FRONT SEAL

REMOVAL & INSTALLATION

1. Access the front of the engine.
2. Remove the fan drive belt.
3. Remove the crankshaft pulley from the crankshaft using tool.
4. Remove the crankshaft pulley using suitable tool. Set the bolts in the two bolt holes [M6 x 1.0 mm (0.04 in)] on the front surface.
5. Using the appropriate tool, remove the crankshaft seal.

To install:

6. Install the front oil seal using suitable tool.

➡ **Do not scratch or make burrs on the circumference of the oil seal.**

7. Install the crankshaft pulley using suitable tool.
8. Install the fan drive belt.
9. For remaining parts, reverse removal procedure.

CYLINDER HEAD

REMOVAL & INSTALLATION

See Figures 31 and 32.

1. Before servicing the vehicle, refer to the Precautions Section.
2. Remove or disconnect the following:
   • Engine assembly
   • Belt tensioner
   • Idler pulley
   • Thermostat housing and hose
   • Oil pan and strainer
   • Fuel rail and injector assembly
   • Intake manifold
   • Ignition coil
   • Rocker cover
   • Crankshaft pulley
   • Front engine cover

   • Oil pump
   • Timing chain
   • Camshaft sprockets
   • Camshafts
   • Cylinder head, removing bolts in reverse order shown in figure

To install:

3. Install the cylinder head with a new gasket. Tighten the bolts in sequence as follows:
   a. Step 1: 72 ft. lbs. (98 Nm).
   b. Step 2: Loosen all bolts completely.
   c. Step 3: 33 ft. lbs. (44 Nm).
   d. Step 4: Plus 60 degrees.
   e. Step 5: Plus 60 degrees.

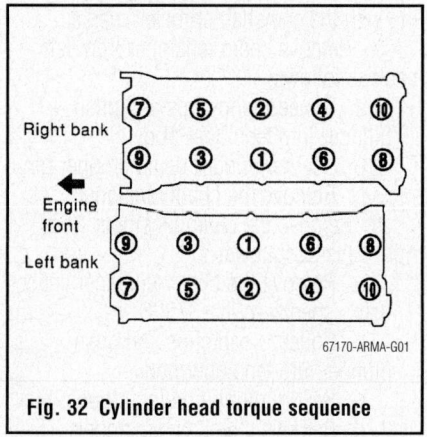

Fig. 32 Cylinder head torque sequence

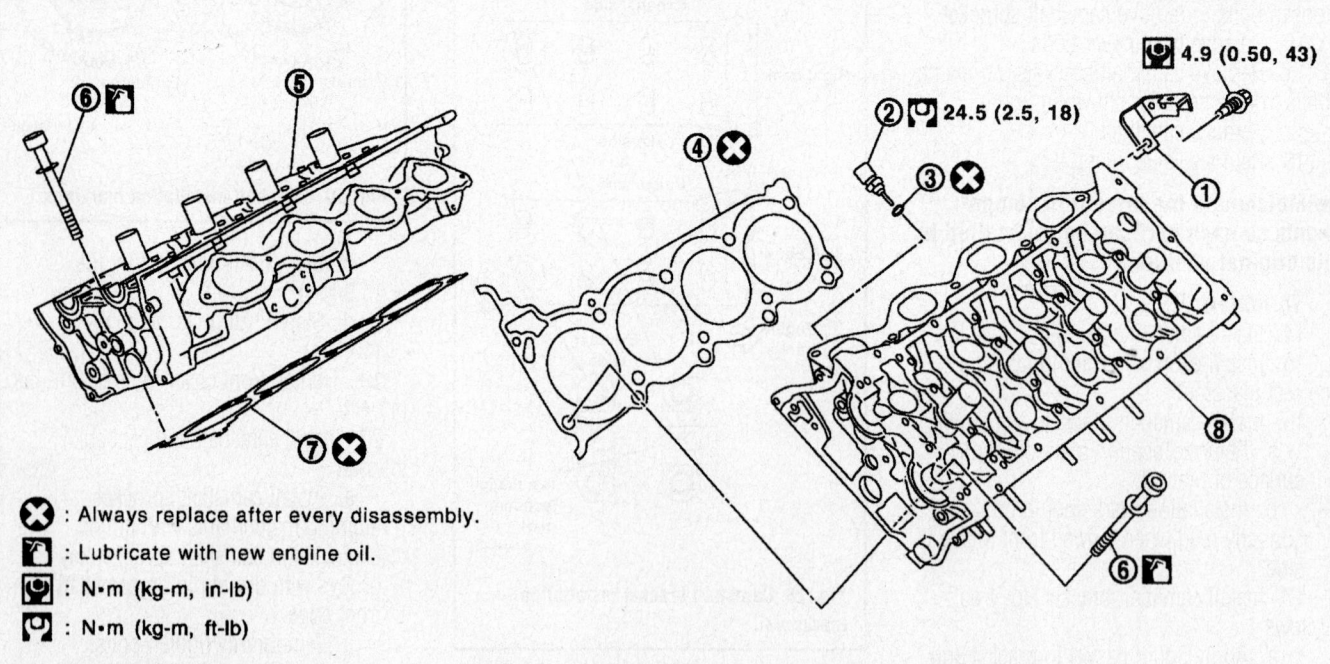

⊗ : Always replace after every disassembly.

🛢 : Lubricate with new engine oil.

🔧 : N•m (kg-m, in-lb)

🔧 : N•m (kg-m, ft-lb)

1. Harness bracket
2. Engine coolant temperature sensor
3. Washer
4. Cylinder head gasket (left bank)
5. Cylinder head (right bank)
6. Cylinder head bolt
7. Cylinder head gasket (right bank)
8. Cylinder head (left bank)

67170-ARMA-G28

Fig. 31 Exploded view of the cylinder head assembly

4. Install or connect the following:
- Camshaft
- Camshaft sprockets
- Timing chain
- Oil pump
- Front engine cover
- Crankshaft pulley
- Rocker cover
- Ignition coil
- Intake manifold
- Fuel tube and injector assembly
- Oil pain and strainer
- Thermostat housing and hose
- Idler pulley
- Belt tensioner
- Engine assembly
5. Start the engine and check for leaks.

ENGINE ASSEMBLY

REMOVAL & INSTALLATION

See Figure 33.

1. Before servicing the vehicle, refer to the Precautions Section.

2. Drain the cooling system.
3. Partially drain the automatic transmission fluid.
4. Relieve the fuel system pressure.
5. Remove or disconnect the following:
- Hood
- Cowl extension
- Engine cover
- Air intake assembly
- Vacuum hose between vehicle and engine
- Radiator hoses
- Radiator
- Drive belts
- Engine fan
- Wiring harness
- ECM
- Power steering reservoir tank and oil pump
- A/C compressor
- Brake booster vacuum line
- EVAP line
- Fuel hose

- Heater hoses
- Exhaust manifolds
- Front final drive assembly
- Automatic transmission dipstick tube assembly
- Automatic transmission

6. Install engine slings onto the left and right cylinder heads and tighten to 33 ft. lbs. (45 Nm).
7. Attach an engine hoist to slings and lift engine out of the vehicle.

To install:

8. Lower engine into the vehicle.
9. Install or connect the following:
- Automatic transmission
- Automatic transmission dipstick tube assembly
- Front final drive assembly
- Exhaust manifolds
- Heater hoses
- Fuel hose
- EVAP line
- Brake booster vacuum line
- A/C compressor

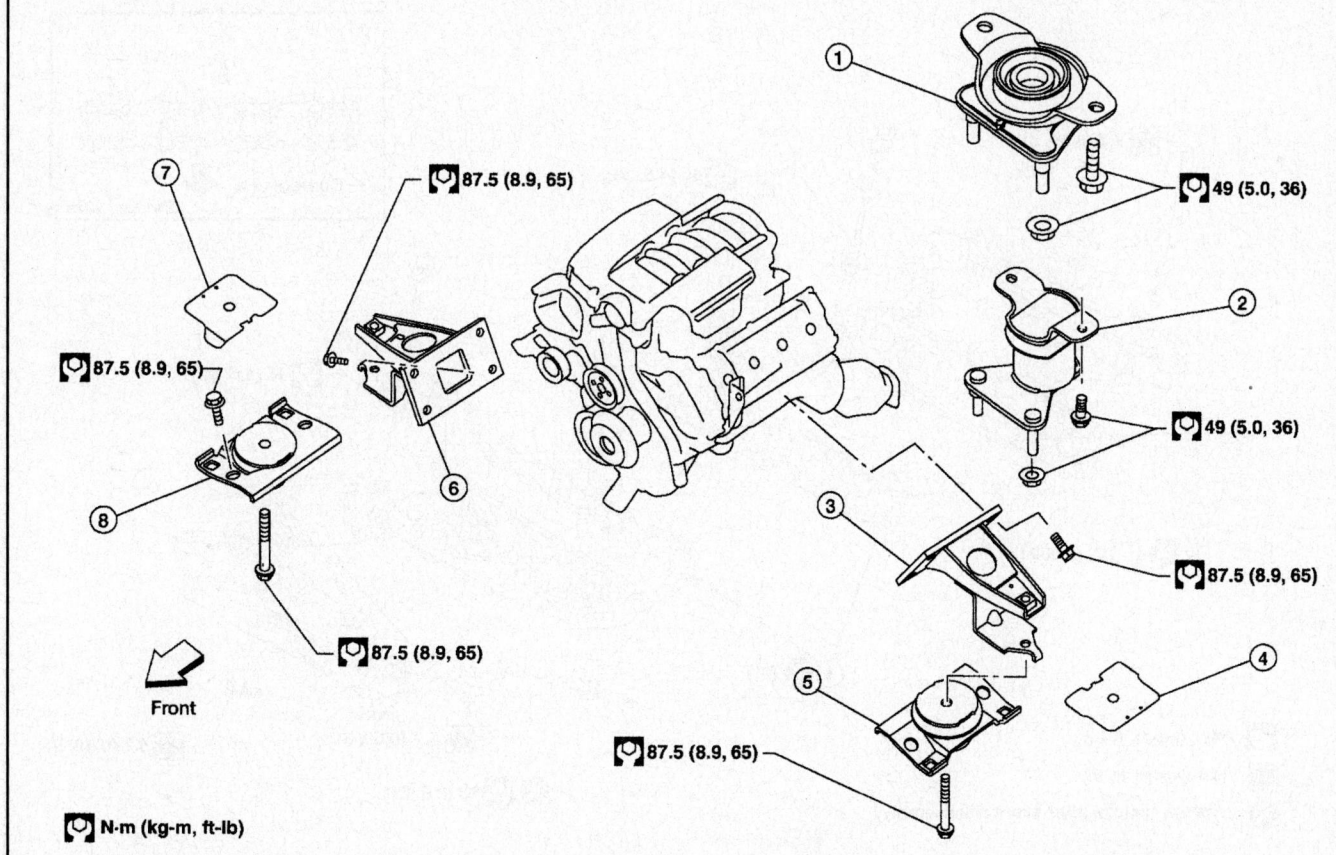

87.5 (8.9, 65)

49 (5.0, 36)

87.5 (8.9, 65)

87.5 (8.9, 65)

49 (5.0, 36)

87.5 (8.9, 65)

87.5 (8.9, 65)

87.5 (8.9, 65)

Front

N·m (kg-m, ft-lb)

1. Rear engine mounting insulator 4x4	2. Rear engine mounting insulator 4x2	3. LH engine mounting bracket
4. LH Heat shield plate	5. LH engine mounting insulator	6. RH engine mounting bracket
7. RH Heat shield plate	8. RH engine mounting insulator	

67170-ARMA-G24

Fig. 33 Engine mounts—QX56

- Power steering reservoir tank and oil pump
- ECM
- Wiring harness
- Engine fan
- Drive belts
- Radiator and radiator hoses
- Vacuum hose between vehicle and engine
- Air intake assembly
- Engine cover
- Cowl extension
- Hood

10. Refill the automatic transmission fluid.
11. Refill the cooling system.
12. Start the engine and check for leaks.

EXHAUST MANIFOLD

REMOVAL & INSTALLATION

See Figures 34 through 36.

1. Before servicing the vehicle, refer to the Precautions Section.
2. Drain the cooling system.

3. Remove or disconnect the following:
- Air intake assembly
- Engine splash guard
- Radiator and radiator hoses
- Accessory drive belt

4. Remove the air fuel ratio sensors as follows:
- Engine cover
- Wiring harness from each sensor

- Sensors, using special tool J-38356
- Front cross bar

5. Remove the left exhaust manifold as follows:
a. Remove the exhaust front tube.
b. Remove the exhaust manifold cover.
c. Loosen the nuts in reverse order shown in figure.
d. Remove studs from position 2, 4, 6, and 8 and remove manifold.

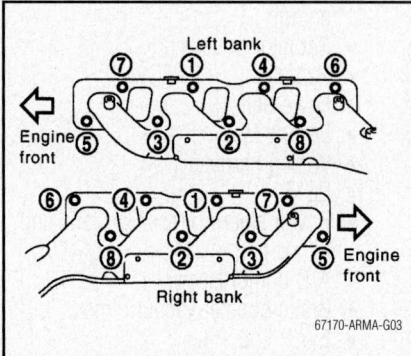

Fig. 35 Exhaust manifold torque sequence

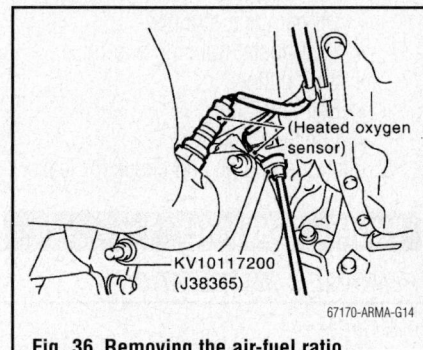

Fig. 36 Removing the air-fuel ratio sensors

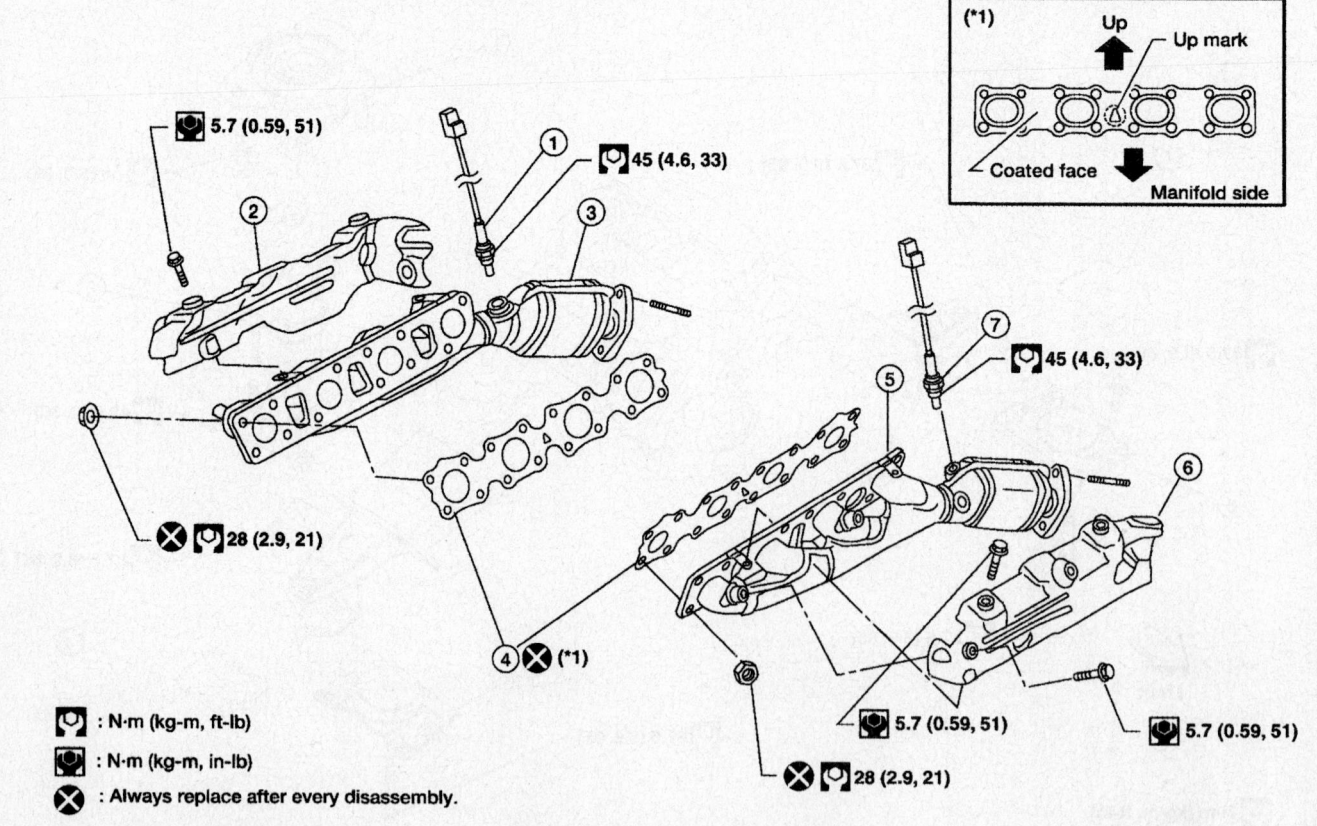

1. Air fuel ratio (A/F) sensor 1 (bank 2)
2. Exhaust manifold cover (right bank)
3. Exhaust manifold (right bank)
4. Gaskets
5. Exhaust manifold (left bank)
6. Exhaust manifold cover (left bank)
7. Air fuel ratio (A/F) sensor 1 (bank 1)

Fig. 34 Exhaust manifolds and related components

6. Remove right exhaust manifold as follows:

    a. Remove the exhaust front tube.

    b. Remove the oil level gauge guide.

    c. Remove the exhaust manifold cover.

    d. Loosen the nuts in reverse order shown in figure.

    e. Remove studs from position 2, 4, 6, and 8 and remove manifold.

To install:

7. Install or connect the following:

- Exhaust manifold gasket with triangle mark facing up and coated (gray) face toward exhaust manifold.
- Exhaust manifold, tightening the nuts as shown in figure
- Exhaust manifold cover
- Oil level gauge guide (right side only)
- Exhaust front tube
- Front cross bar

- Air fuel ratio sensors, with anti-seize lubricant
- Engine cover
- Drive belts
- Radiator and radiator hoses
- Engine splash guard
- Air intake assembly

8. Refill the cooling system.

9. Start engine and check for leaks.

INTAKE MANIFOLD

REMOVAL & INSTALLATION

See Figures 37 and 38.

1. Before servicing the vehicle, refer to the Precautions Section.

2. Drain the cooling system.

3. Relieve the fuel system pressure.

4. Remove or disconnect the following:

- Engine cover
- Air intake assembly

- Fuel supply hose quick connector using special tool J-45488
- Wiring harnesses and brackets from manifold
- Vacuum hoses
- PCV hose and tube

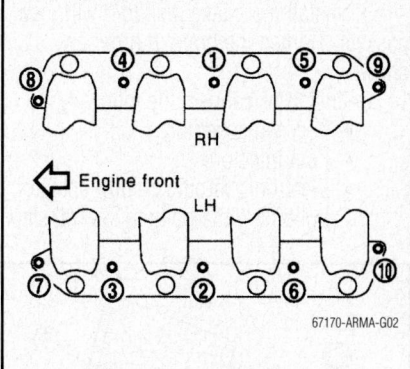

Fig. 38 Intake manifold torque sequence

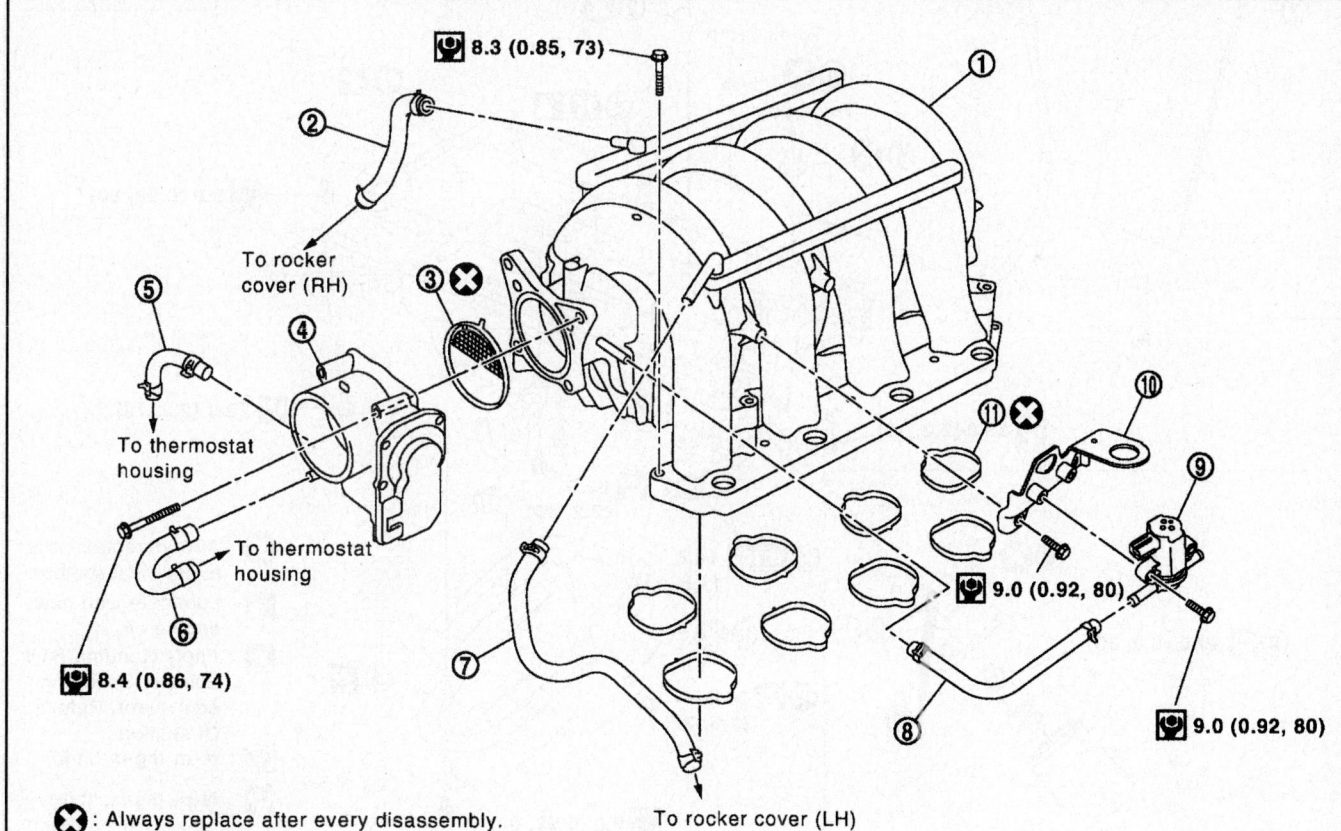

⊗ : Always replace after every disassembly.

🔧 : N•m (kg-m, in-lb)

1.	Intake manifold	2.	PCV hose	3.	Gasket
4.	Electric throttle control actuator	5.	Water hose	6.	Water hose
7.	PCV hose	8.	EVAP hose	9.	EVAP canister purge control solenoid valve
10.	Bracket	11.	Gasket		

67170-ARMA-G29

Fig. 37 Intake manifold and related parts

- Electric throttle control actuator, loosening bolts diagonally
- Fuel injectors
- Fuel rail assembly
- Intake manifold, removing bolts in reverse order shown in figure

To install:

5. Install the intake manifold with new gaskets. Tighten the bolts in order as shown.

6. Install or connect the following:
- Fuel rail assembly
- Fuel injectors
- Electronic throttle control actuator, tightening the bolts in several steps

- PCV hose
- Vacuum hoses
- Wiring harnesses

7. Connect the fuel supply hose as follows:

a. Apply a thin layer of engine oil on the tube from tip end to spool end.

b. Insert tube into quick connector past the white identification mark.

c. Insert tube into quick connector until top spool is completely inside the connector and 2nd level spool is exposed right below the connector.

d. Pull slightly on the quick connector to ensure it is fully engaged.

e. Install quick connector cap on quick connector joint.

8. Install or connect the following:
- Air intake assembly
- Engine cover

9. Refill the cooling system.

10. Start engine and check for leaks.

OIL PAN

REMOVAL & INSTALLATION

See Figures 39 through 41.

1. Before servicing the vehicle, refer to the Precautions Section.

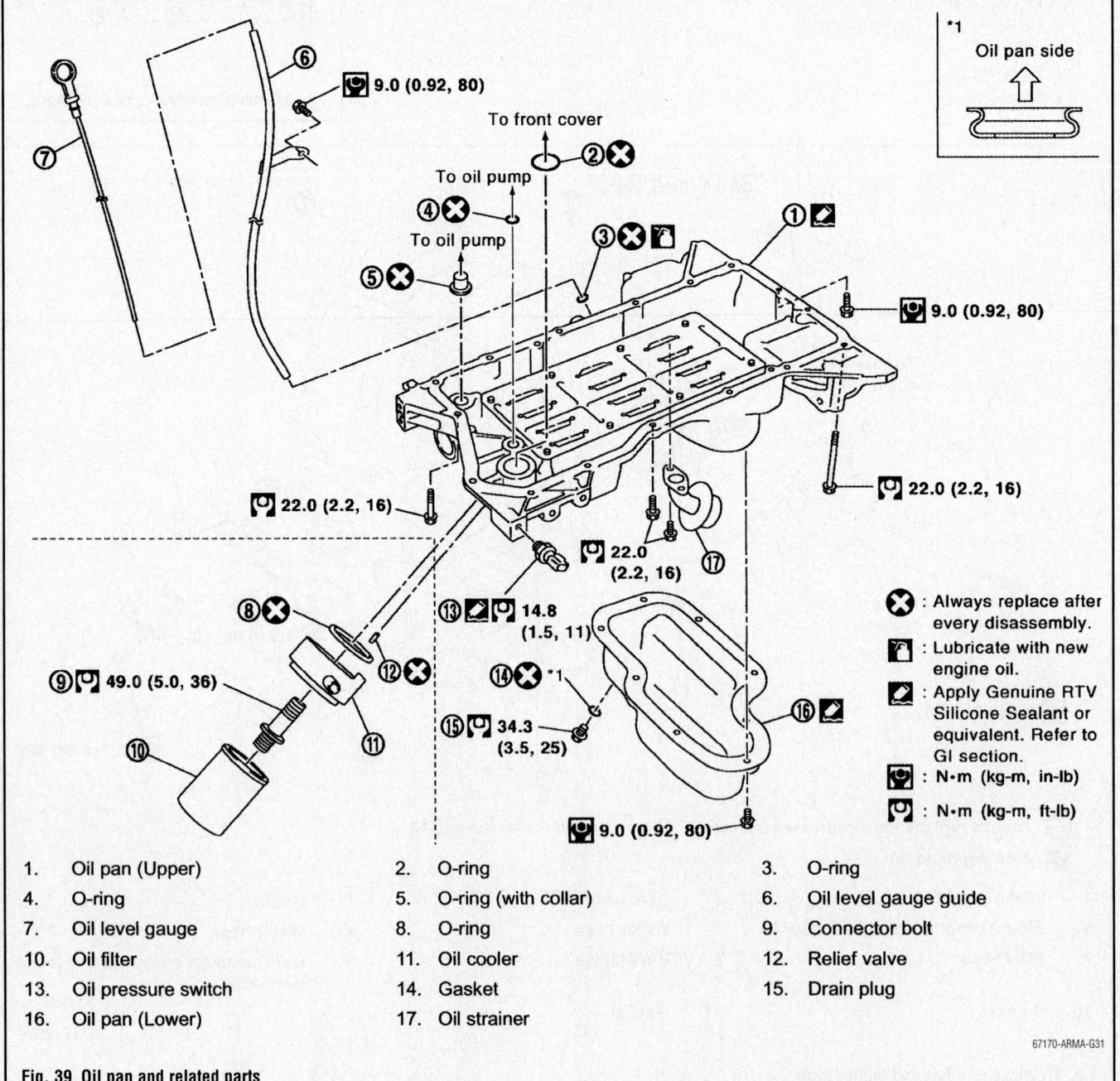

1.	Oil pan (Upper)	2.	O-ring	3.	O-ring
4.	O-ring	5.	O-ring (with collar)	6.	Oil level gauge guide
7.	Oil level gauge	8.	O-ring	9.	Connector bolt
10.	Oil filter	11.	Oil cooler	12.	Relief valve
13.	Oil pressure switch	14.	Gasket	15.	Drain plug
16.	Oil pan (Lower)	17.	Oil strainer		

67170-ARMA-G31

Fig. 39 Oil pan and related parts

2. Remove engine assembly.

3. Remove lower oil pan, loosening bolts in reverse order shown in figure.

4. Remove oil strainer from upper oil pan.

5. Gently pry and remove upper oil pan from engine block.

➡️**Bolts are different sizes and should be kept in the correct order for reinstallation.**

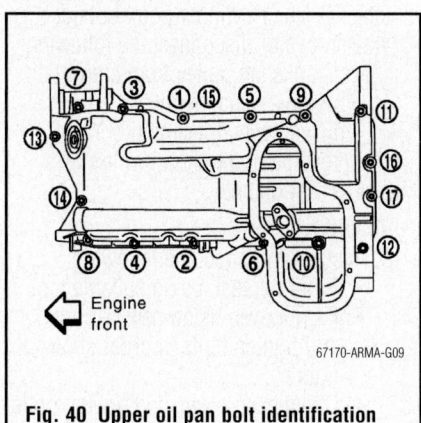

Fig. 40 Upper oil pan bolt identification

To install:

6. Apply liquid gasket to upper oil pan mating surfaces.

7. Install new O-rings to oil pump and front cover side.

8. Tighten upper oil pan bolts to 16 ft. lbs. (22 Nm) in following numerical order:

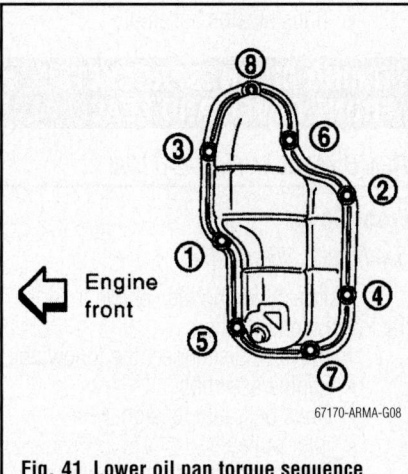

Fig. 41 Lower oil pan torque sequence

9. Install or connect the following:
- Rear plate cover
- Oil strainer to upper oil pan
- Lower oil pan, tightening bolts to 25 ft. lbs. (34 Nm) in order shown in figure

OIL PUMP

REMOVAL & INSTALLATION

See Figure 42.

1. Before servicing the vehicle, refer to the Precautions Section.

2. Remove or disconnect the following:
- Timing chain cover
- Oil pump drive spacer
- Oil pump

To install:

3. Install or connect the following:
- Oil pump
- Oil pump drive spacer
- Timing chain cover

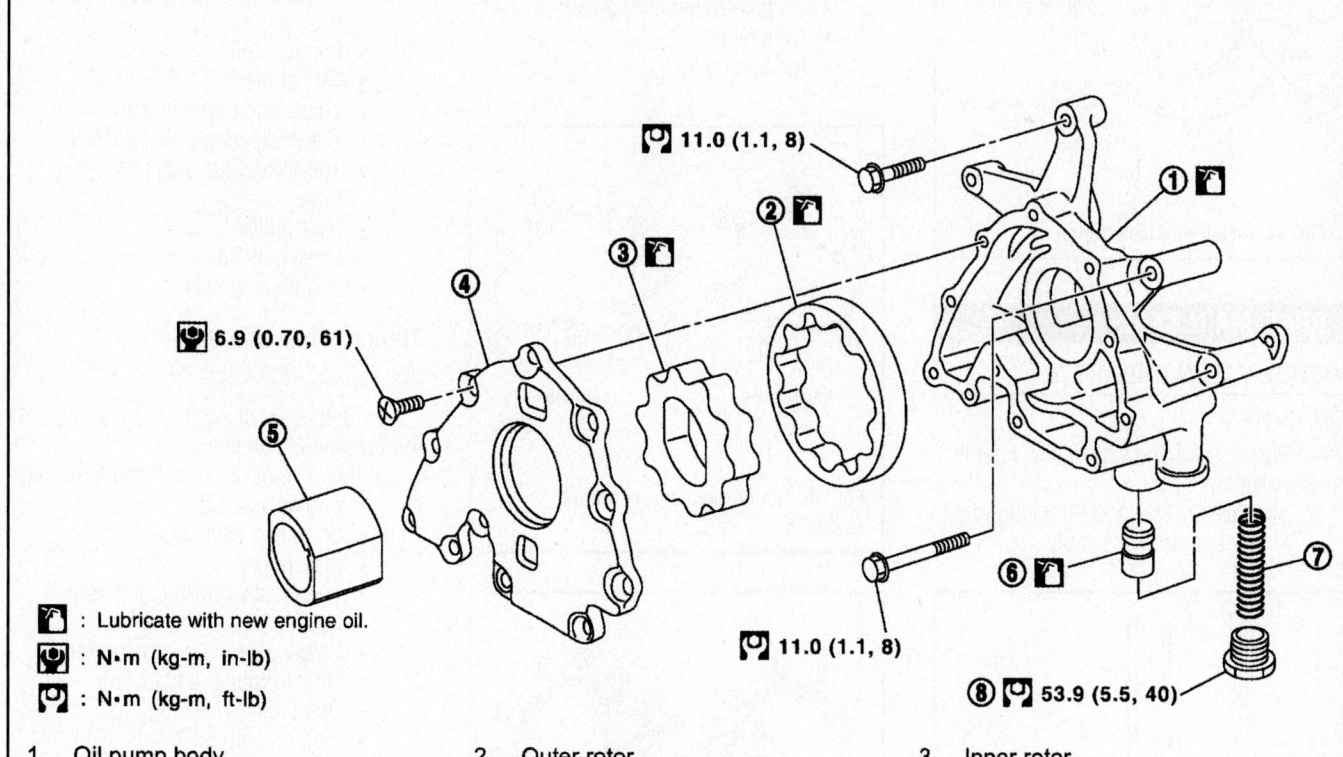

🔩 : Lubricate with new engine oil.

🔧 : N•m (kg-m, in-lb)

🔧 : N•m (kg-m, ft-lb)

1. Oil pump body
4. Oil pump cover
7. Regulator spring

2. Outer rotor
5. Oil pump drive spacer
8. Regulator plug

3. Inner rotor
6. Regulator valve

Fig. 42 Oil pump exploded view

PISTON AND RING

POSITIONING

See Figures 43 and 44.

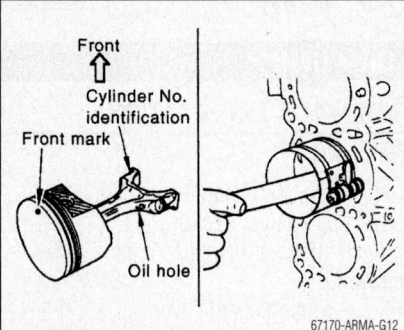

Fig. 43 Piston and rod positioning and identification

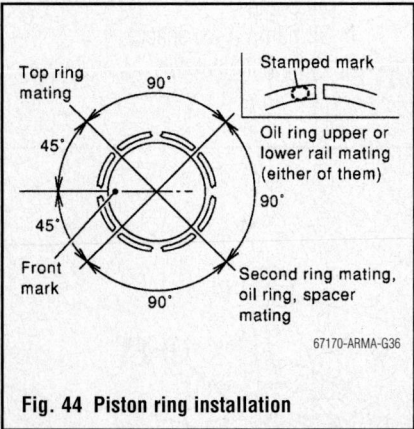

Fig. 44 Piston ring installation

REAR MAIN SEAL

REMOVAL & INSTALLATION

See Figure 45.

1. Before servicing the vehicle, refer to the Precautions Section.
2. Remove or disconnect the following:
   - Transmission assembly

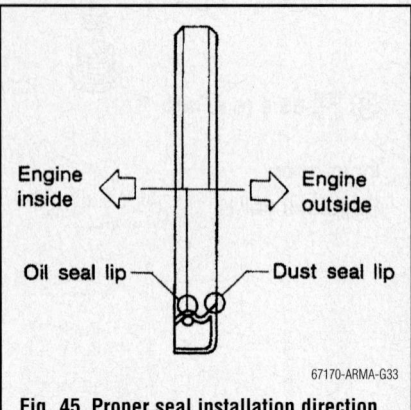

Fig. 45 Proper seal installation direction

- Drive plate
- Engine rear plate
- Rear main seal using suitable tool

To install:

3. Install or connect the following:
   - Rear main seal using suitable tool
   - Engine rear plate
   - Drive plate
   - Transmission assembly

TIMING CHAIN, SPROCKETS, FRONT COVER AND SEAL

REMOVAL & INSTALLATION

Front Cover

See Figures 46 and 47.

1. Before servicing the vehicle, refer to the Precautions Section.
2. Remove or disconnect the following:
   - Engine assembly
   - Drive belt auto tensioner
   - Idler pulley
   - Thermostat housing and water hose
   - Power steering pump bracket
   - Oil pan (upper and lower)
   - Oil strainer
   - Ignition coil

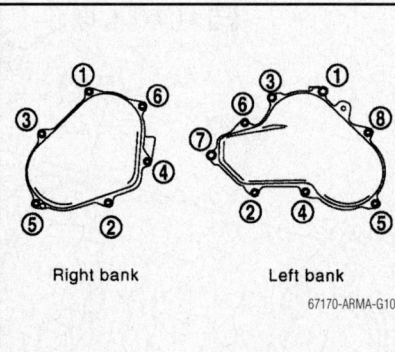

Fig. 46 Timing case covers torque sequence

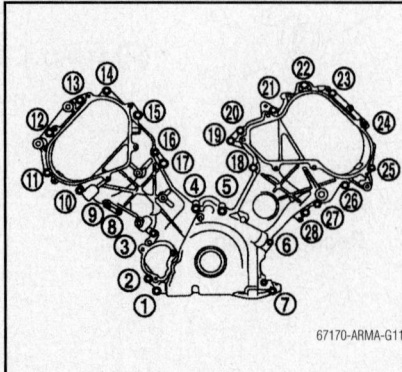

Fig. 47 Front cover torque sequence

- Rocker cover
- Timing chain case cover, loosening bolts in reverse order shown in figure

3. Obtain compression TDC of No. 1 cylinder as follows:

   a. Turn crankshaft pulley to align the TDC identification notch with timing indicator on front cover.

   b. Ensure intake and exhaust cam lobes of No. 1 cylinder point outside.

4. Remove or disconnect the following:
   - Crankshaft pulley from crankshaft using a suitable puller
   - Front cover, loosening bolts in reverse order shown in figure
   - Front oil seal

To install:

5. Install or connect the following:
   - Front oil seal, using suitable tool
   - Front cover, using new O-rings and tighten bolts in order shown in figure
   - Chain case cover, and tighten bolts in order shown in figure
   - Crankshaft pulley and tighten bolt to 69 ft. lbs. (93 Nm) plus 90 degrees
   - Ignition coil
   - Oil strainer
   - Lower and upper oil pan
   - Power steering pump bracket
   - Thermostat housing and water hose
   - Idler pulley
   - Drive belt auto tensioner
   - Engine assembly

Timing Chain and Sprockets

See Figures 48 through 50.

1. Before servicing the vehicle, refer to the Precautions Section.
2. Remove or disconnect the following:
   - Engine assembly
   - Drive belt auto tensioner
   - Idler pulley
   - Thermostat housing and water hose
   - Power steering pump bracket
   - Oil pan (upper and lower)
   - Oil strainer
   - Ignition coil
   - Rocker cover
   - Timing chain case cover, loosening bolts in reverse order shown in figure

3. Obtain compression TDC of No. 1 cylinder as follows:

   a. Turn crankshaft pulley to align the TDC identification notch with timing indicator on front cover.

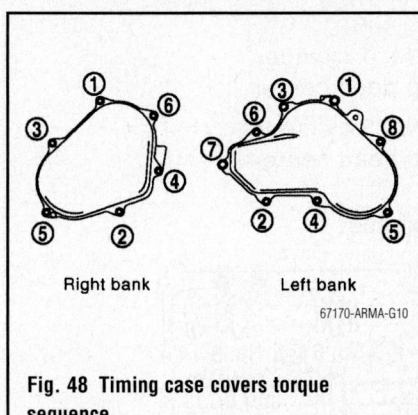

Fig. 48 Timing case covers torque sequence

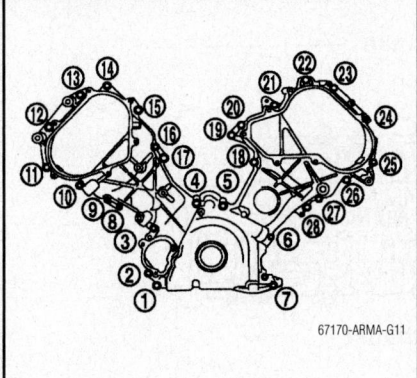

Fig. 49 Front cover torque sequence

b. Ensure intake and exhaust cam lobes of No. 1 cylinder point outside.

4. Remove or disconnect the following:
- Crankshaft pulley from crankshaft using a suitable puller
- Front cover, loosening bolts in reverse order shown in figure
- Front oil seal
- Oil pump drive spacer
- Oil pump

5. Remove the timing chain tensioner as follows:

a. Squeeze the return-proof clip ends using suitable tool and push the plunger into the tensioner body.

b. Secure the plunger using stopper pin.

➡**Stopper pin is made from hard wire approximately 0.04 in (1mm) in diameter.**

c. Remove the bolts and chain tensioner.

6. Remove the following:
- Chain tension guide and slack guide
- Timing chain

7. Using a wrench to hold the hexagon part of the camshaft, loosen the camshaft sprocket bolts.

8. Remove the camshaft sprockets.

To install:

9. Ensure that the crankshaft key and dowel pin of each camshaft are facing the same direction.

10. Install or connect the following:
- Camshaft sprockets and tighten to 112 ft. lbs. (152 Nm).
- Timing chain
- Chain tension guide and slack guide
- Oil pump
- Oil pump drive spacer
- Front oil seal, using suitable tool
- Front cover, using new O-rings and tighten bolts in order shown in figure
- Chain case cover, and tighten bolts in order shown in figure
- Crankshaft pulley and tighten bolt to 69 ft. lbs. (93 Nm) plus 90 degrees
- Ignition coil
- Oil strainer
- Lower and upper oil pan
- Power steering pump bracket
- Thermostat housing and water hose
- Idler pulley
- Drive belt auto tensioner
- Engine assembly

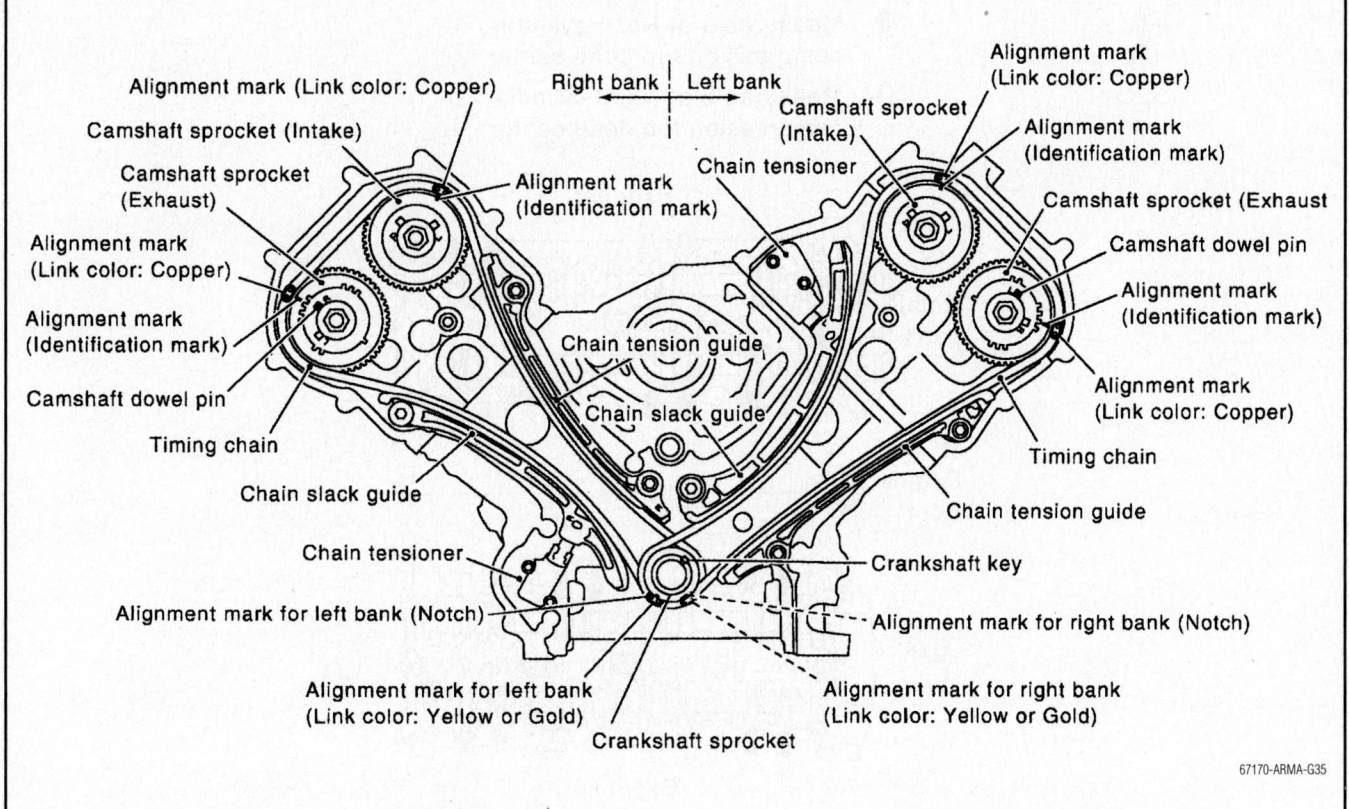

Fig. 50 Timing chain orientation and timing mark alignment

VALVE LASH

ADJUSTMENT

See Figures 51 and 52.

➡**Perform the following inspection after removal, installation or replacement of camshaft or valve-related parts, or if there are unusual engine conditions due to changes in valve clearance over time (starting, idling, and/or noise).**

1. Run engine to operating temperature.
2. Remove or disconnect the following:
   - Engine cover
   - Battery cover
   - Air intake assembly
   - Left and right rocker covers
3. Turn the crankshaft pulley clockwise to Top Dead Center (TDC) identification notch with timing indicator.
4. Ensure that both the intake and exhaust cam noses of the No. 1 cylinder face outside.
5. Measure the valve clearances at locations shown in figure.
6. Turn the crankshaft pulley clockwise 270 degrees from the position of No. 1 cylinder compression to obtain No. 3 cylinder compression TDC.

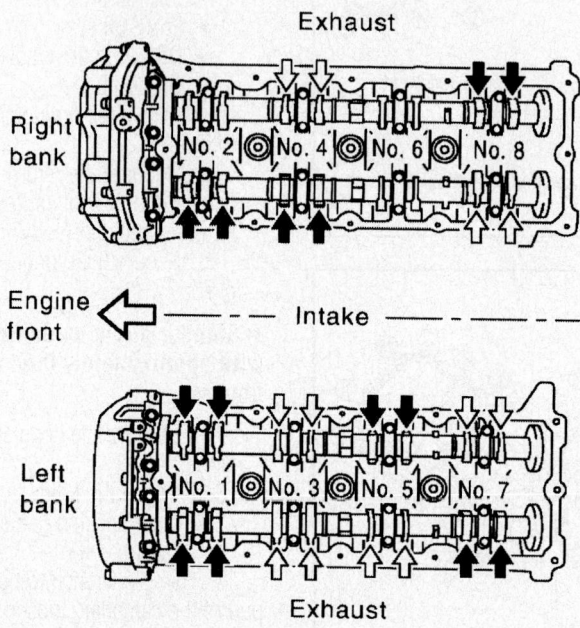

▲ : Measurable at No. 1 cylinder compression top dead center

⇧ : Measurable at No. 3 cylinder compression top dead center

67170-ARMA-G07

Fig. 52 Locations to measure clearance with No. 3 cylinder at TDC

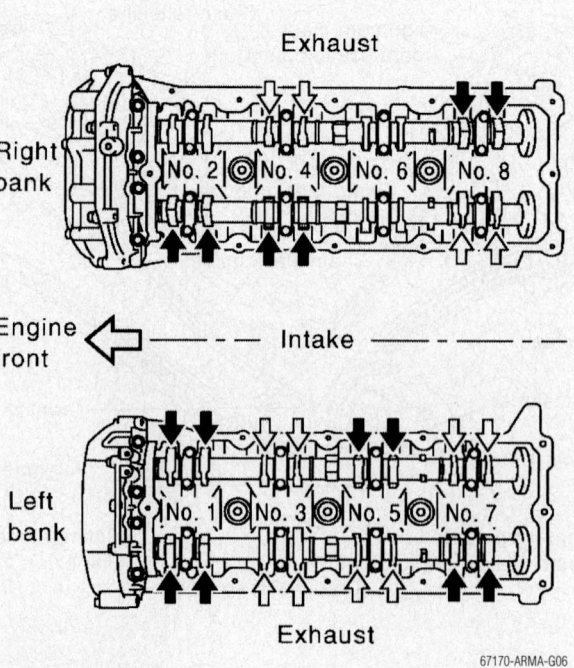

▲ : Measurable at No. 1 cylinder compression top dead center

⇧ : Measurable at No. 3 cylinder compression top dead center

67170-ARMA-G06

Fig. 51 Locations to measure clearance with No. 1 cylinder at TDC

7. Measure the valve clearances at locations shown in the figure.

8. Turn crankshaft pulley clockwise 90 degrees and measure the intake and exhaust valve clearance of No. 6 cylinder and exhaust valve clearance of No. 2 cylinder.

9. To adjust the valves, remove camshaft and valve lifter(s) out of specification.

10. Install replacement valve lifter(s).

11. Install the camshaft.

12. Manually turn the crankshaft pulley several turns.

13. Recheck valve clearances with engine at operating temperature.

ENGINE PERFORMANCE & EMISSION CONTROL

COMPONENT LOCATIONS

See Figures 53 through 56.

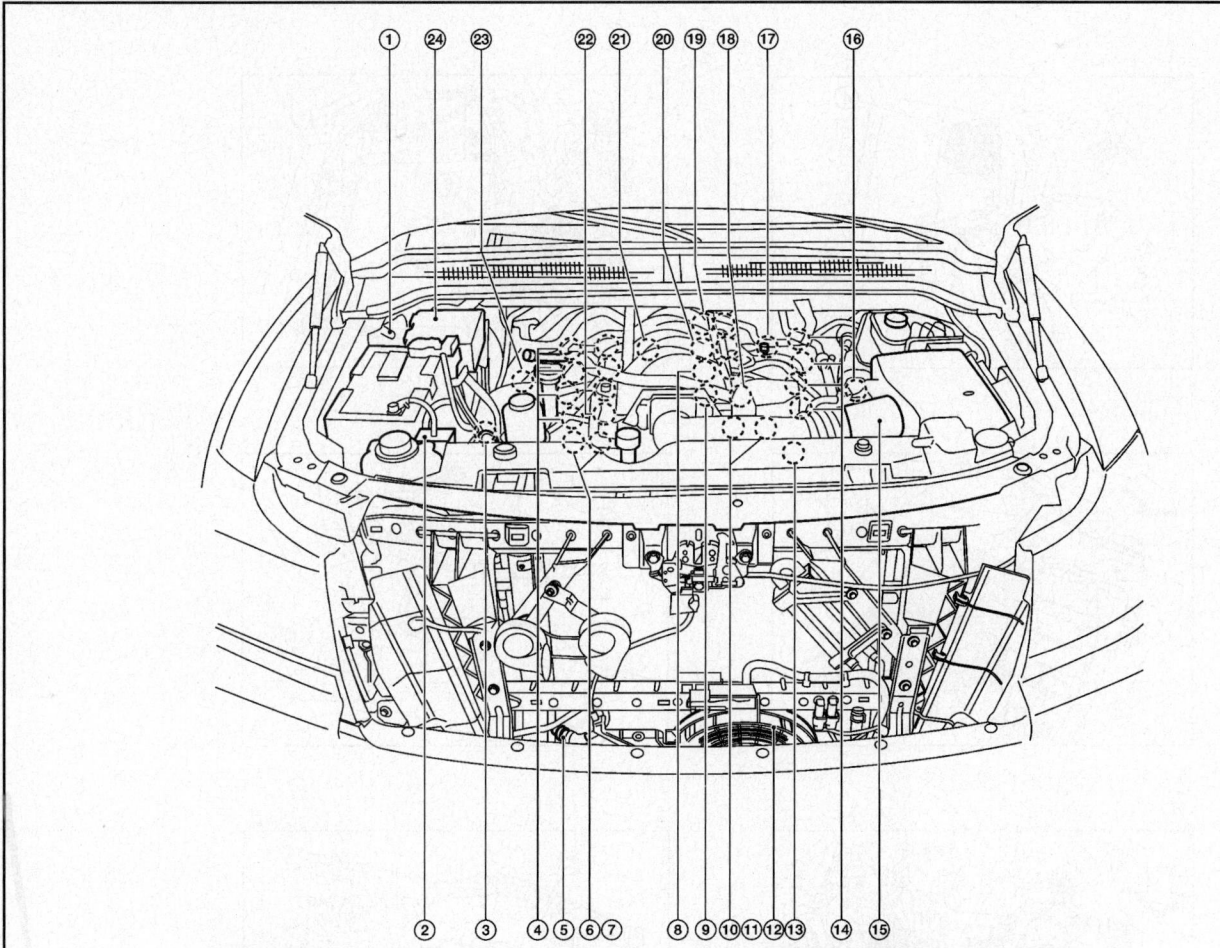

1. ECM
2. Battery current sensor
3. Power steering pressure sensor
4. Ignition coil (with power transistor) and spark plug (bank 2)
5. Refrigerant pressure sensor
6. Intake valve timing control position sensor (bank 2)
7. Intake valve timing control solenoid valve (bank 2)
8. Engine coolant temperature sensor
9. Electric throttle control actuator
10. Intake valve timing control position sensor (bank 1)
11. Intake valve timing control solenoid valve (bank 1)
12. Cooling fan motor
13. Camshaft position sensor (PHASE)
14. Ignition coil (with power transistor) and spark plug (bank 1)
15. Mass air flow sensor (with intake air temperature sensor)
16. A/F sensor 1 (bank 1)
17. EVAP service port
18. Fuel injector (bank 1)
19. Knock sensor (bank 1)
20. EVAP canister purge volume control solenoid valve
21. Knock sensor (bank 2)
22. Fuel injector (bank 2)
23. A/F sensor 1 (bank 2)
24. IPDM E/R

22140_QX56_G0030

Fig. 53 Engine performance components locations (1 of 4)

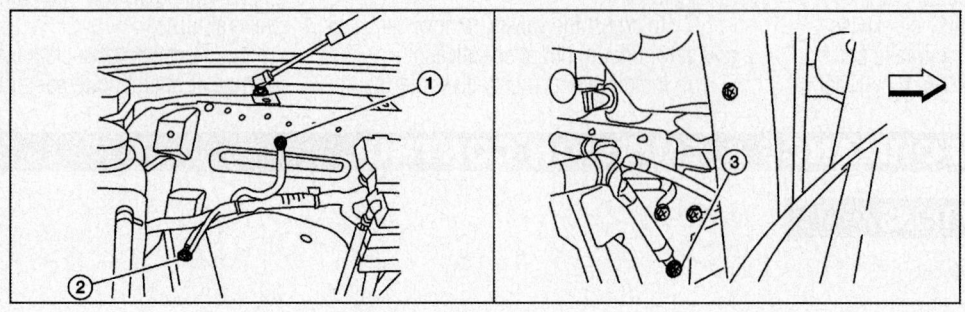

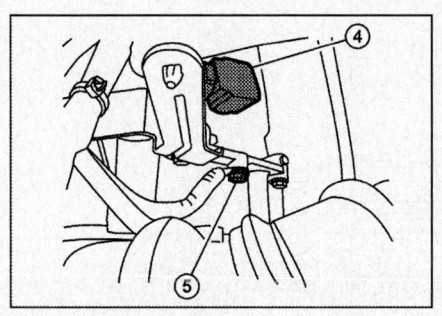
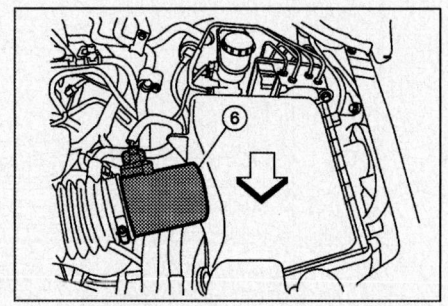

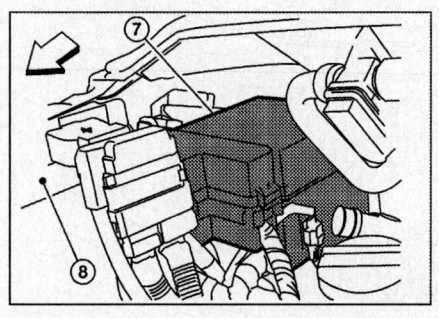
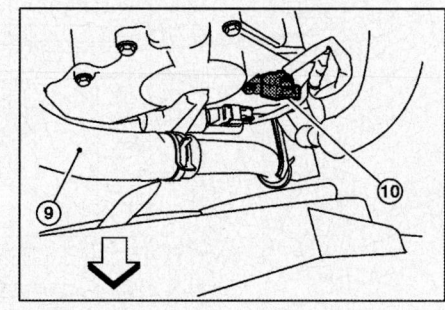

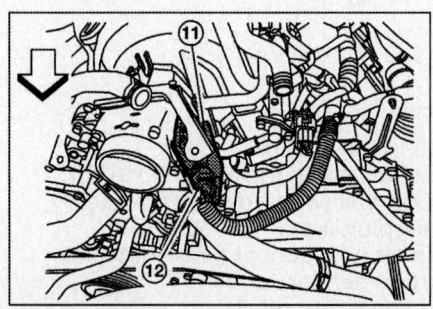
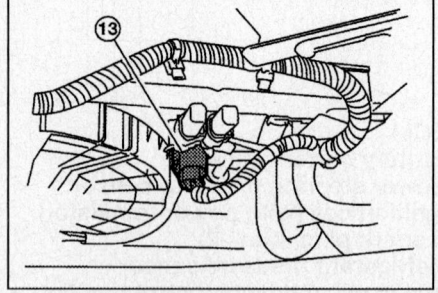

1. Body ground (view with battery removed)
2. Body ground (view with battery removed)
3. Body ground
4. No. 1 ignition coil
5. Engine ground
6. Mass air flow sensor (with intake air temperature sensor)

7. IPDM E/R
8. Battery
9. Radiator hose
10. Camshaft position sensor (PHASE)
11. Electric throttle control actuator (view with intake air duct removed)
12. Cooling fan motor harness connector

22140_QX56_G0031

Fig. 54 Engine performance components locations (2 of 4)

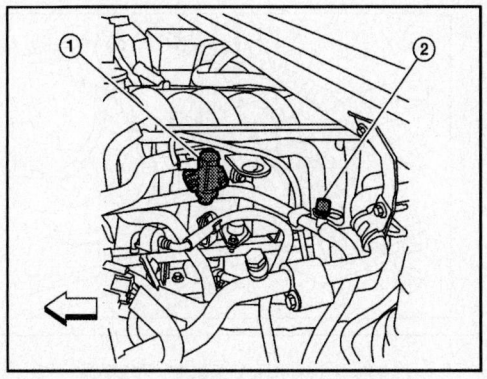
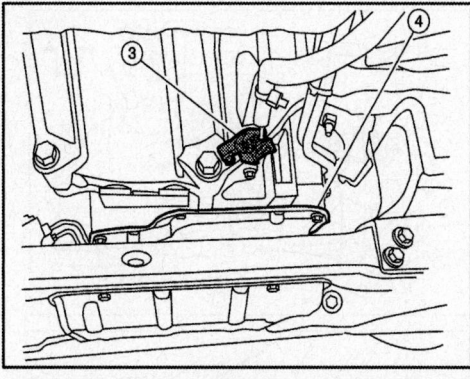
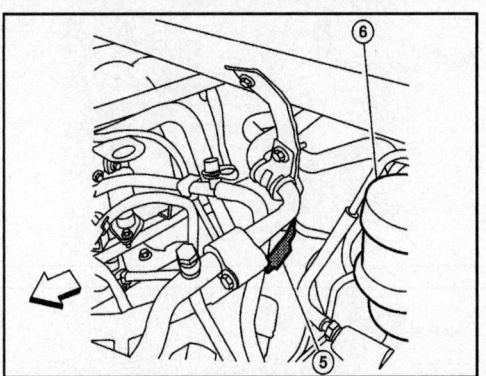
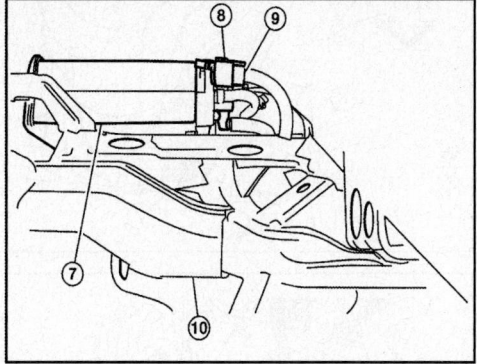
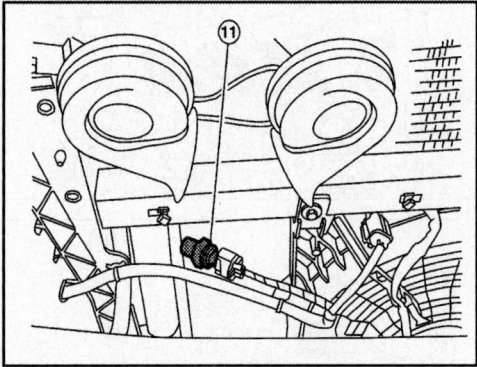
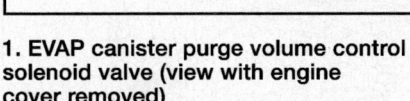

1. EVAP canister purge volume control solenoid valve (view with engine cover removed)
2. EVAP service port (view with engine cover removed)
3. Crankshaft position sensor (POS) (view from under the vehicle)
4. Engine oil pan (view from under the vehicle)
5. Condenser-1 6. Brake fluid reservoir
7. EVAP canister (view with fuel tank removed)
8. EVAP control system pressure sensor (view with fuel tank removed)
9. EVAP canister vent control valve (view with fuel tank removed)

10. Rear suspension member (view with fuel tank removed)
11. Refrigerant pressure sensor (view with front grille removed)
12. Intake valve timing control position sensor (bank 2) (view with engine cover and intake air duct removed)
13. Intake valve timing control position sensor (bank 1) (view with engine cover and intake air duct removed)
14. Intake valve timing control solenoid valve (bank 2) (view with engine cover and intake air duct removed)
15. Drive belt (view with engine cover and intake air duct removed)
16. Radiator hose (view with engine cover and intake air duct removed)
17. Intake valve timing control solenoid valve (bank 1) (view with engine cover and intake air duct removed)

22140_QX56_G0032

Fig. 55 Engine performance components locations (3 of 4)

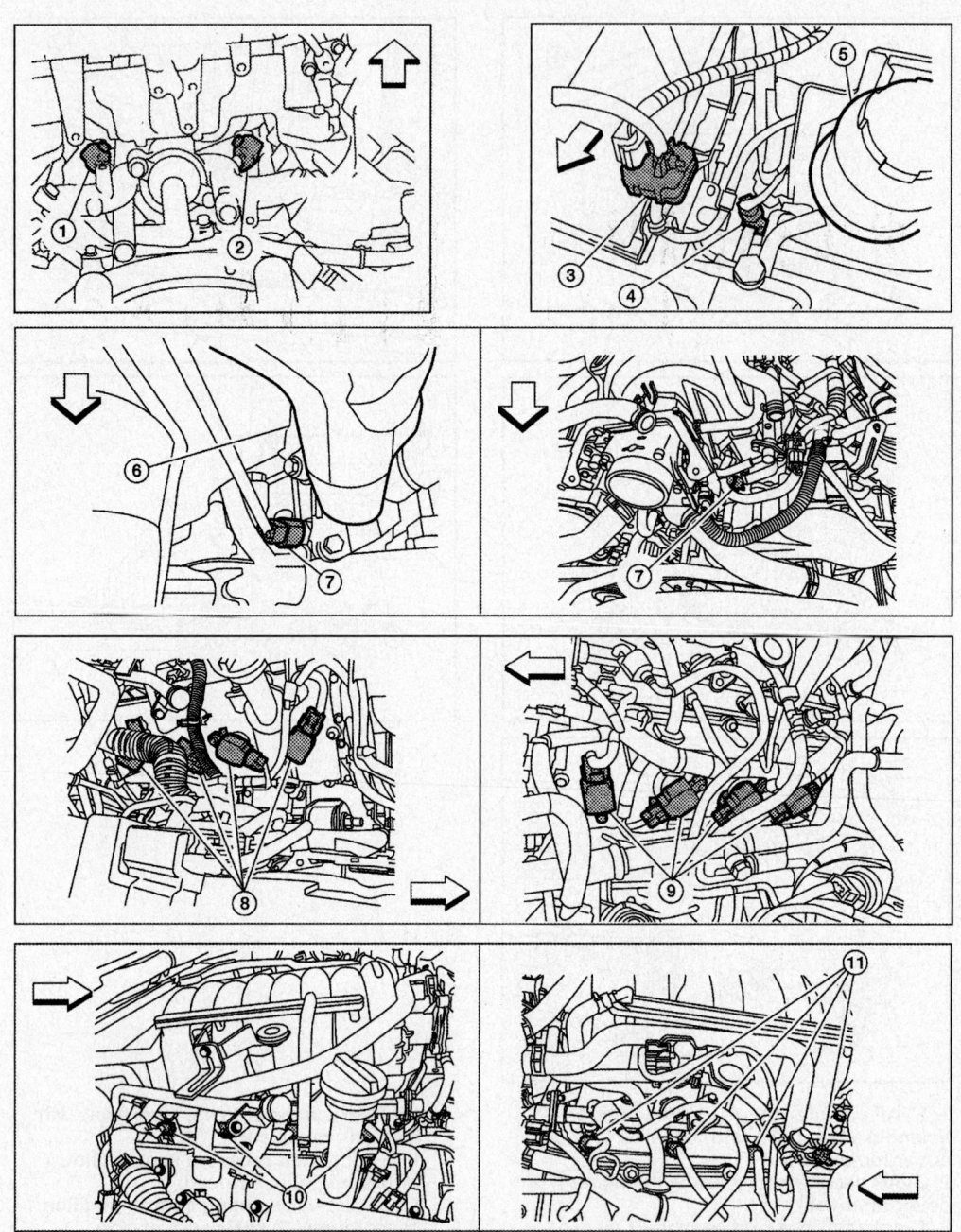

1. Knock sensor (bank 1) (view with engine removed)
2. Knock sensor (bank 2) (view with engine removed)
3. Battery current sensor
4. Power steering pressure sensor
5. Power steering fluid reservoir
6. Intake manifold
7. Engine coolant temperature sensor
8. Ignition coils (with power transistor)
9. Ignition coil (with power transistor)
10. Injector harness connectors (bank 2)
11. Injector harness connectors (bank 1)

22140_QX56_G0033

Fig. 56 Engine performance components locations (4 of 4)

CAMSHAFT POSITION (CMP) SENSOR

LOCATION

Camshaft position sensor is located in the front timing chain cover.

REMOVAL & INSTALLATION

1. Loosen the fixing bolt of the sensor.
2. Disconnect camshaft position sensor (PHASE) harness connector.
3. Remove the sensor.

To install:

4. To install, reverse removal procedure.

CRANKSHAFT POSITION (CKP) SENSOR

LOCATION

See Figure 57.

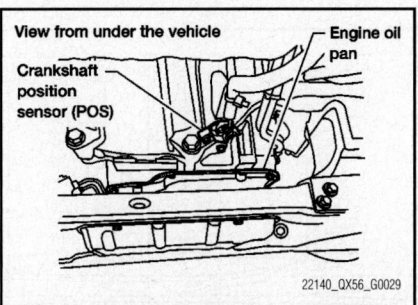

Fig. 57 Crankshaft position sensor location

REMOVAL & INSTALLATION

1. Loosen the fixing bolt of the sensor.
2. Disconnect crankshaft position sensor harness connector.
3. Remove the sensor.

To install:

4. To install, reverse removal procedure.

ELECTRONIC CONTROL MODULE (ECM)

LOCATION

See Figure 58.

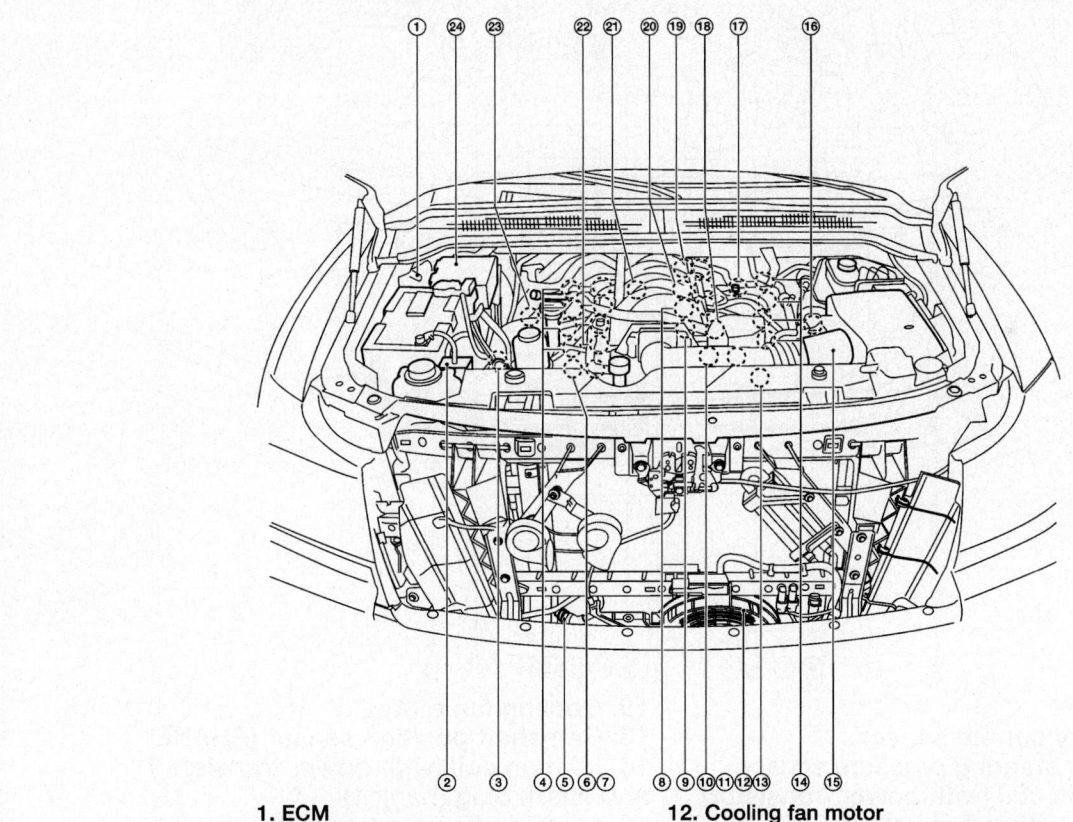

1. ECM
2. Battery current sensor
3. Power steering pressure sensor
4. Ignition coil (with power transistor) and spark plug (bank 2)
5. Refrigerant pressure sensor
6. Intake valve timing control position sensor (bank 2)
7. Intake valve timing control solenoid valve (bank 2)
8. Engine coolant temperature sensor
9. Electric throttle control actuator
10. Intake valve timing control position sensor (bank 1)
11. Intake valve timing control solenoid valve (bank 1)
12. Cooling fan motor
13. Camshaft position sensor (PHASE)
14. Ignition coil (with power transistor) and spark plug (bank 1)
15. Mass air flow sensor (with intake air temperature sensor)
16. A/F sensor 1 (bank 1)
17. EVAP service port
18. Fuel injector (bank 1)
19. Knock sensor (bank 1)
20. EVAP canister purge volume control solenoid valve
21. Knock sensor (bank 2)
22. Fuel injector (bank 2)
23. A/F sensor 1 (bank 2)
24. IPDM E/R

Fig. 58 Engine performance components locations 1

ENGINE COOLANT TEMPERATURE (ECT) SENSOR

LOCATION

See Figure 59.

REMOVAL & INSTALLATION

See Figure 60.

1. Before servicing the vehicle, refer to the Precautions Section.
2. Disconnect the negative battery cable.
3. Drain the cooling system.
4. Remove the sensor electrical connector.
5. Remove the sensor from its mounting.

To install:

6. Installation is the reverse of the removal procedure.

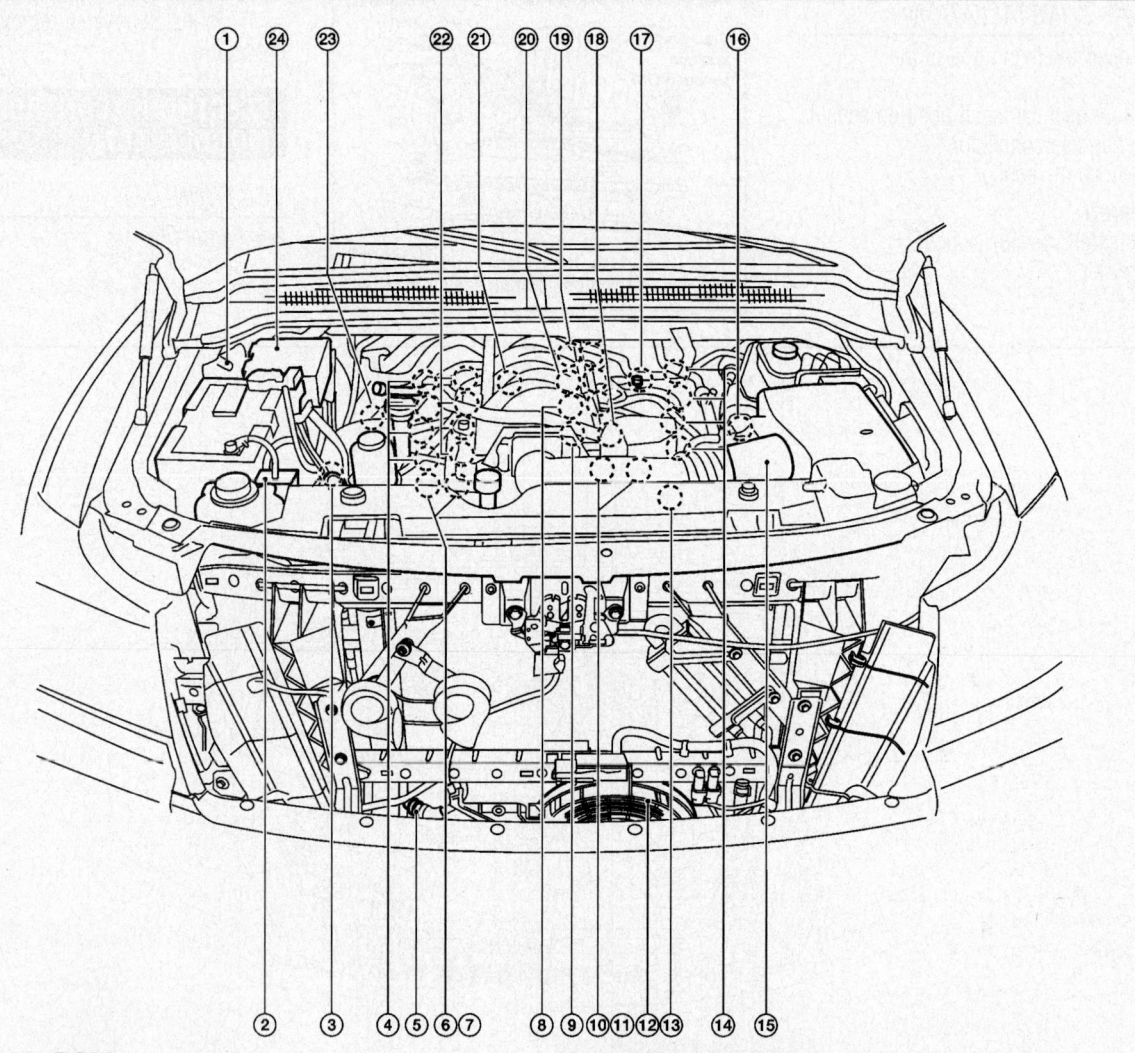

1. ECM
2. Battery current sensor
3. Power steering pressure sensor
4. Ignition coil (with power transistor) and spark plug (bank 2)
5. Refrigerant pressure sensor
6. Intake valve timing control position sensor (bank 2)
7. Intake valve timing control solenoid valve (bank 2)
8. Engine coolant temperature sensor
9. Electric throttle control actuator
10. Intake valve timing control position sensor (bank 1)
11. Intake valve timing control solenoid valve (bank 1)
12. Cooling fan motor
13. Camshaft position sensor (PHASE)
14. Ignition coil (with power transistor) and spark plug (bank 1)
15. Mass air flow sensor (with intake air temperature sensor)
16. A/F sensor 1 (bank 1)
17. EVAP service port
18. Fuel injector (bank 1)
19. Knock sensor (bank 1)
20. EVAP canister purge volume control solenoid valve
21. Knock sensor (bank 2)
22. Fuel injector (bank 2)
23. A/F sensor 1 (bank 2)
24. IPDM E/R

22140_QX56_G0030

Fig. 59 Engine performance components locations

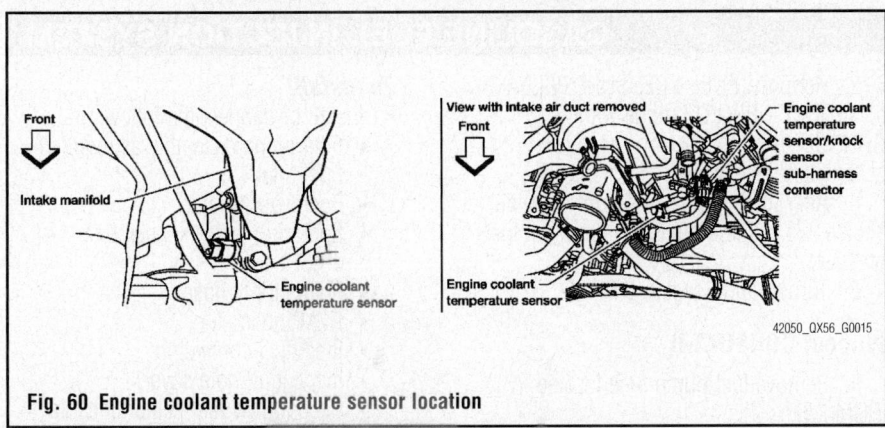

Fig. 60 Engine coolant temperature sensor location

HEATED OXYGEN (HO2S) SENSOR

LOCATION

See Figure 61.

REMOVAL & INSTALLATION

1. Disconnect the oxygen sensor connector.
2. Using the proper tool, remove the oxygen sensor.

To install:

3. To install, reverse removal procedure.

INTAKE AIR TEMPERATURE (IAT) SENSOR

LOCATION

The IAT is located in the Mass air flow sensor.

REMOVAL & INSTALLATION

1. Disconnect sensor connector.
2. Remove retaining screw
3. Remove IAT sensor from air intake duct.

To install:

4. To install, reverse removal procedure.

MASS AIR FLOW (MAF) SENSOR

LOCATION

See Figure 62.

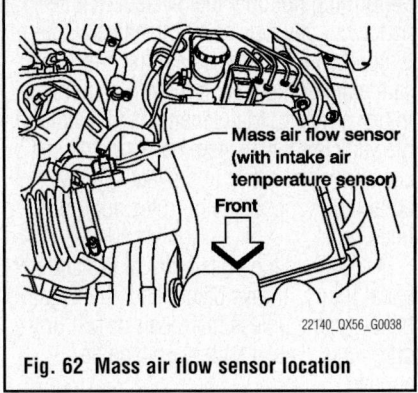

Fig. 62 Mass air flow sensor location

REMOVAL & INSTALLATION

1. Disconnect the sensor connector.
2. Remove the retaining screw.
3. Remove the MAF sensor from air intake duct.

To install:

4. To install, reverse removal procedure.

THROTTLE POSITION SENSOR (TPS)

LOCATION

See Figure 63.

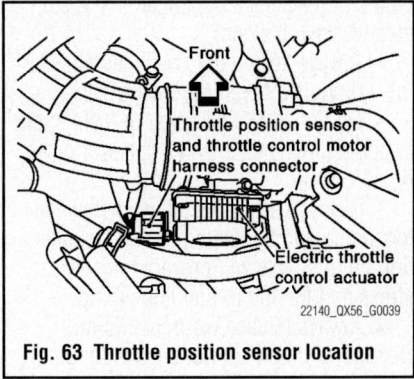

Fig. 63 Throttle position sensor location

REMOVAL & INSTALLATION

1. Disconnect the sensor connector.
2. Remove the retaining screws.
3. Remove the throttle body assembly from intake manifold.

To install:

4. To install, reverse removal procedure.

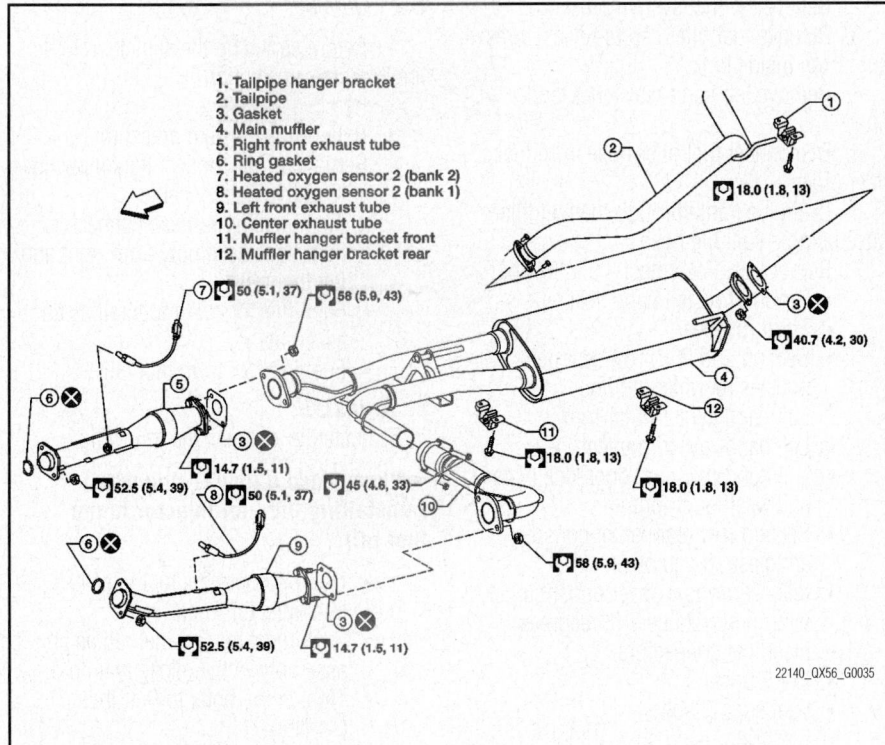

1. Tailpipe hanger bracket
2. Tailpipe
3. Gasket
4. Main muffler
5. Right front exhaust tube
6. Ring gasket
7. Heated oxygen sensor 2 (bank 2)
8. Heated oxygen sensor 2 (bank 1)
9. Left front exhaust tube
10. Center exhaust tube
11. Muffler hanger bracket front
12. Muffler hanger bracket rear

Fig. 61 Heated oxygen sensor location

FUEL   **GASOLINE FUEL INJECTION SYSTEM**

FUEL SYSTEM SERVICE PRECAUTIONS

Safety is the most important factor when performing not only fuel system maintenance but any type of maintenance. Failure to conduct maintenance and repairs in a safe manner may result in serious personal injury or death. Maintenance and testing of the vehicle's fuel system components can be accomplished safely and effectively by adhering to the following rules and guidelines.

• To avoid the possibility of fire and personal injury, always disconnect the negative battery cable unless the repair or test procedure requires that battery voltage be applied.

• Always relieve the fuel system pressure prior to disconnecting any fuel system component (injector, fuel rail, pressure regulator, etc.), fitting or fuel line connection. Exercise extreme caution whenever relieving fuel system pressure to avoid exposing skin, face and eyes to fuel spray. Please be advised that fuel under pressure may penetrate the skin or any part of the body that it contacts.

• Always place a shop towel or cloth around the fitting or connection prior to loosening to absorb any excess fuel due to spillage. Ensure that all fuel spillage (should it occur) is quickly removed from engine surfaces. Ensure that all fuel soaked cloths or towels are deposited into a suitable waste container.

• Always keep a dry chemical (Class B) fire extinguisher near the work area.

• Do not allow fuel spray or fuel vapors to come into contact with a spark or open flame.

• Always use a back-up wrench when loosening and tightening fuel line connection fittings. This will prevent unnecessary stress and torsion to fuel line piping.

• Always replace worn fuel fitting O-rings with new Do not substitute fuel hose or equivalent where fuel pipe is installed.

Before servicing the vehicle, make sure to also refer to the precautions in the beginning of this section as well.

RELIEVING FUEL SYSTEM PRESSURE

With CONSULT-II

1. Turn ignition switch **ON**.

2. Perform "FUEL PRESSURE RELEASE" in "WORK SUPPORT" mode with CONSULT-II.

3. Start the engine.

4. After the engine stalls, turn over the engine two or three times to release all fuel pressure.

5. Turn ignition switch **OFF**.

Without CONSULT-II

1. Remove fuel pump fuse located in IPDM E/R.

2. Start the engine.

3. After the engine stalls, turn over engine two or three times to release all fuel pressure.

4. Turn ignition switch **OFF**.

5. Reinstall fuel pump fuse after servicing fuel system.

FUEL FILTER

REMOVAL & INSTALLATION

See Figures 64 and 65.

➡**The fuel filter is part of the fuel pump assembly.**

1. Before servicing the vehicle, refer to the Precautions Section.

2. Relieve the fuel system pressure.

3. Remove fuel filler cap to release pressure from inside tank.

4. Remove left hand rear inner fender liner.

5. Disconnect fuel filler hose from fuel filler pipe.

6. Drain fuel tank through the fuel filler hose using a suitable hose.

7. Remove or disconnect the following:
• Second row left hand seat
• Third row seat
• Second and third row seat belt buckles mounted on floor
• Left hand center pillar trim
• Left hand rear trim panel
• Left hand rear side door kick plate and weather stripping
• Second row rear center console and base, if equipped
• Inspection hole cover under carpet by turning retainers 90 degrees
• Electrical connectors
• EVAP hose
• Fuel supply hose
• Lock ring using special tool J-46214
• Fuel level sensor
• Fuel filter
• Fuel pump assembly, as required

To install:

8. Install or connect the following:
• Fuel pump assembly as required
• Fuel filter
• Fuel level sensor
• Lock ring using special tool J-46214
• Fuel supply hose
• EVAP hose
• Electrical connectors
• Inspection hole cover
• Second row rear center console and base, if equipped
• Left hand rear side door kick plate and weather stripping
• Left hand rear trim panel
• Left hand center pillar trim
• Second and third row seat belt buckles
• Third row seat
• Second row left hand seat
• Fuel filler hose to fuel filler pipe
• Left hand rear inner fender liner

9. Start the engine and check for leaks.

FUEL INJECTORS

REMOVAL & INSTALLATION

See Figure 66.

1. Before servicing the vehicle, refer to the Precautions Section.

2. Remove engine cover.

3. Relieve fuel system pressure.

4. Remove or disconnect the following:
• Negative battery cable
• Fuel injector harness connectors
• Fuel hose assembly from right and left fuel rails
• Fuel injectors with fuel rail as an assembly
• Fuel injector from fuel rail

To install:

5. Install or connect the following:

➡**Always use a new O-ring when reinstalling the fuel injector to the fuel rail.**

• New clip onto the fuel injector
• Fuel injector to fuel rail
• Fuel injectors and fuel rail as an assembly to the intake manifold. Tighten the bolts to 8 ft. lbs. (11 Nm).
• Fuel hose assembly
• Fuel injector harness connectors
• Negative battery cable
• Engine cover

6. Start engine and check for leaks.

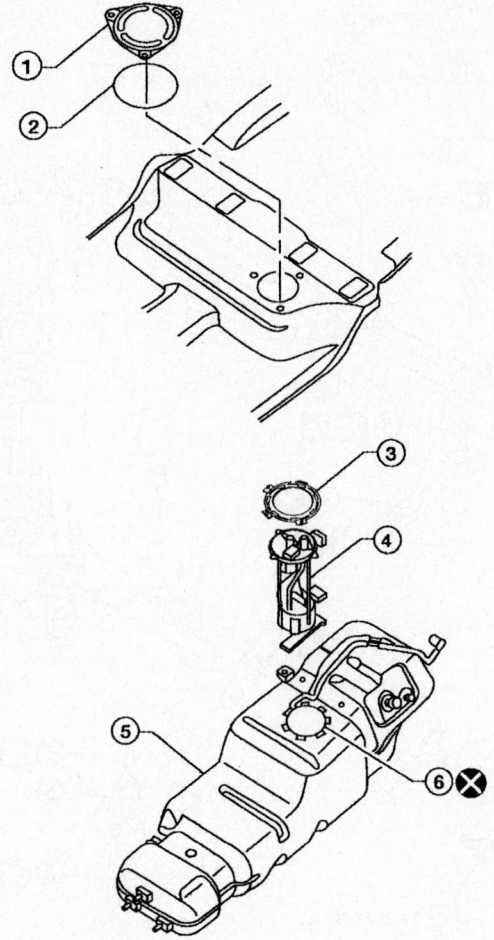

⊗ : Always replace after every disassembly

1. Inspection hole cover
4. Fuel level sensor, fuel filter, and fuel pump assembly

2. Inspection hole cover O-ring
5. Fuel tank

3. Lock ring
6. Fuel level sensor, fuel filter, and fuel pump assembly O-ring

67170-ARMA-G37

Fig. 64 Fuel pump and related parts

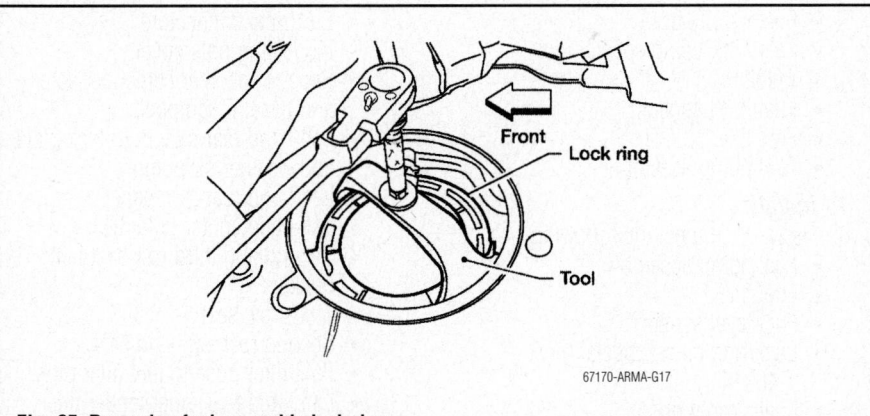

67170-ARMA-G17

Fig. 65 Removing fuel assembly lock ring

FUEL PUMP

REMOVAL & INSTALLATION

See Figures 67 and 68.

1. Before servicing the vehicle, refer to the Precautions Section.
2. Relieve the fuel system pressure.
3. Remove fuel filler cap to release pressure from inside tank.
4. Remove left hand rear inner fender liner.
5. Disconnect fuel filler hose from fuel filler pipe.
6. Drain fuel tank through the fuel filler hose using a suitable hose.

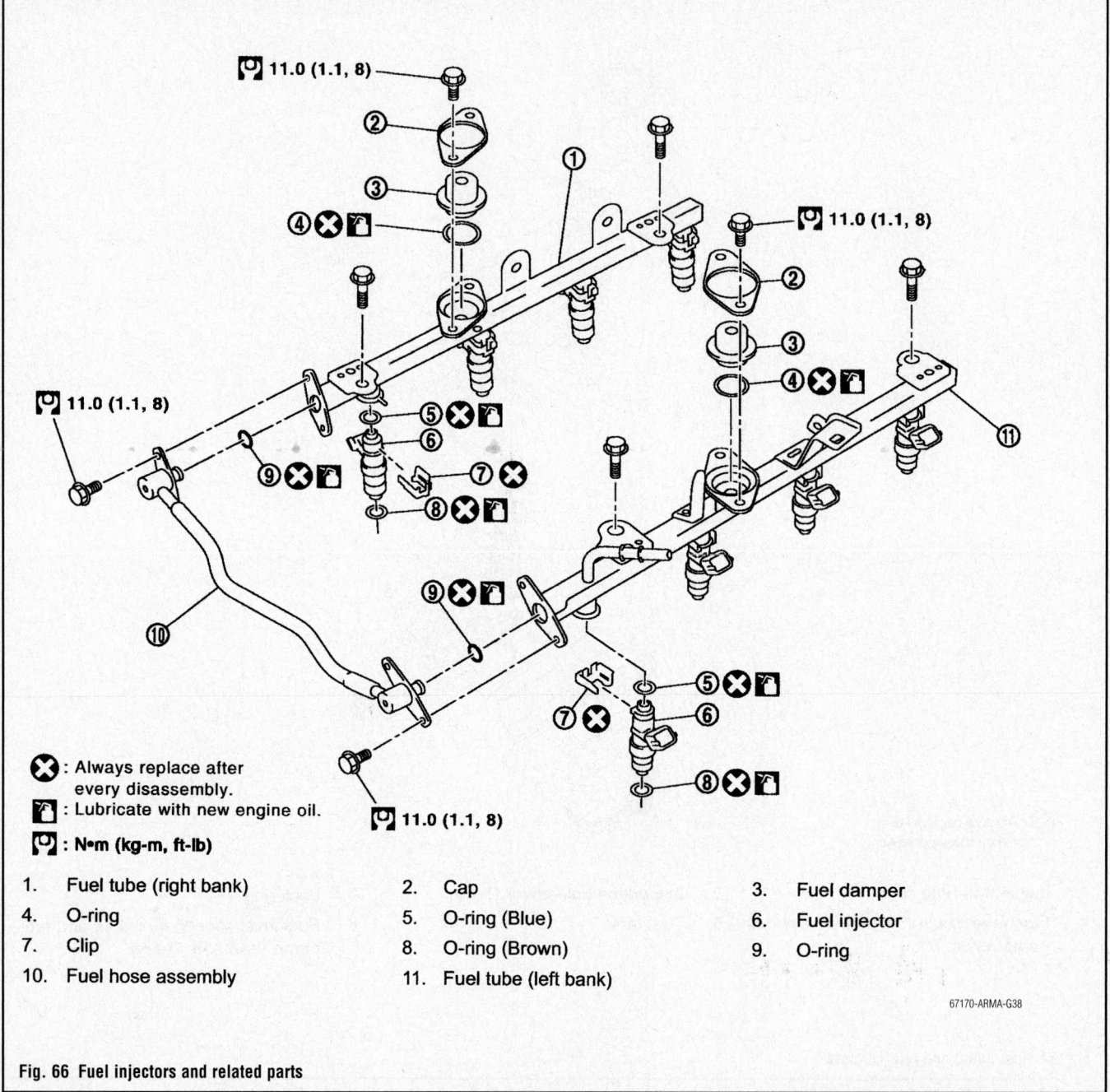

⊗ : Always replace after every disassembly.

🛢 : Lubricate with new engine oil.

🔧 : N•m (kg-m, ft-lb)

1.	Fuel tube (right bank)	2.	Cap	3.	Fuel damper
4.	O-ring	5.	O-ring (Blue)	6.	Fuel injector
7.	Clip	8.	O-ring (Brown)	9.	O-ring
10.	Fuel hose assembly	11.	Fuel tube (left bank)		

67170-ARMA-G38

Fig. 66 Fuel injectors and related parts

7. Remove or disconnect the following:
- Second row left hand seat
- Third row seat
- Second and third row seat belt buckles mounted on floor
- Left hand center pillar trim
- Left hand rear trim panel
- Left hand rear side door kick plate and weather stripping
- Second row rear center console and base, if equipped
- Inspection hole cover under carpet by turning retainers 90 degrees
- Electrical connectors
- EVAP hose
- Fuel supply hose
- Lock ring using special tool J-46214
- Fuel level sensor
- Fuel filter
- Fuel pump assembly

To install:
8. Install or connect the following:
- Fuel pump assembly
- Fuel filter
- Fuel level sensor
- Lock ring using special tool J-46214
- Fuel supply hose
- EVAP hose
- Electrical connectors
- Inspection hole cover
- Second row rear center console and base, if equipped
- Left hand rear side door kick plate and weather stripping
- Left hand rear trim panel
- Left hand center pillar trim
- Second and third row seat belt buckles
- Third row seat
- Second row left hand seat
- Fuel filler hose to fuel filler pipe
- Left hand rear inner fender liner

9. Start the engine and check for leaks.

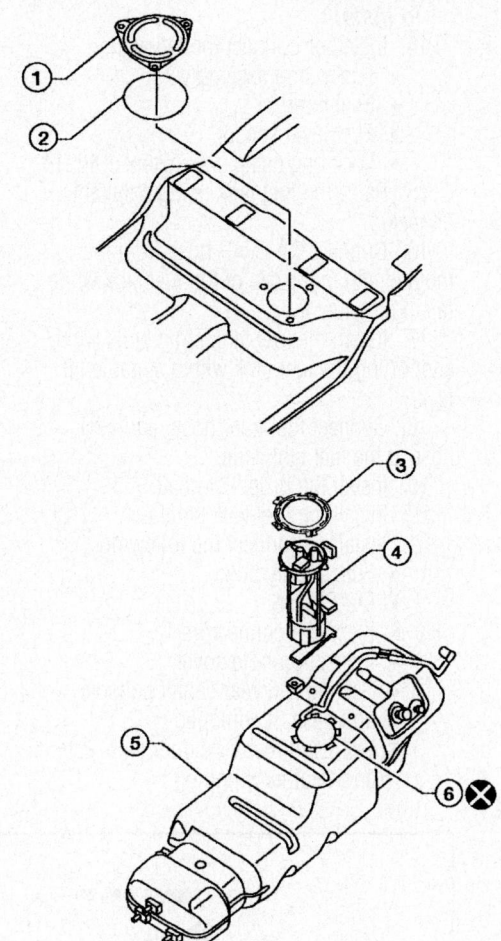

⊗ : **Always replace after every disassembly**

1. Inspection hole cover
2. Inspection hole cover O-ring
3. Lock ring
4. Fuel level sensor, fuel filter, and fuel pump assembly
5. Fuel tank
6. Fuel level sensor, fuel filter, and fuel pump assembly O-ring

67170-ARMA-G37

Fig. 67 Fuel pump and related parts

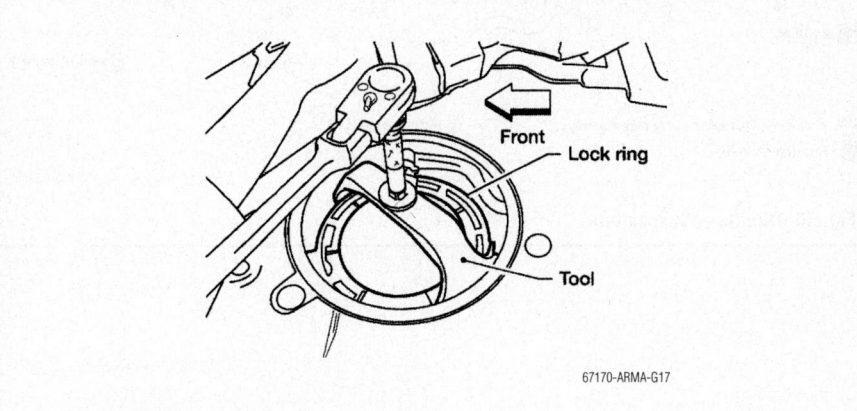

67170-ARMA-G17

Fig. 68 Removing fuel assembly lock ring

FUEL TANK

REMOVAL & INSTALLATION
See Figure 69.

The fuel level sending unit detects a fuel level in the fuel tank and transmits a signal to the combination meter. The combination meter sends the fuel level sensor signal to the ECM through CAN communication line. This component is removed along with the fuel pump.

1. Before servicing the vehicle, refer to the Precautions Section.
2. Relieve the fuel system pressure.
3. Remove fuel filler cap to release pressure from inside tank.

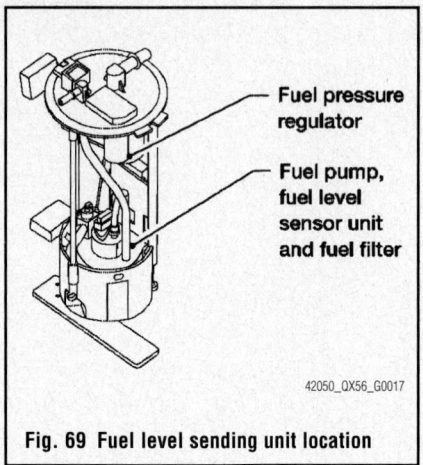

Fuel pressure regulator

Fuel pump, fuel level sensor unit and fuel filter

42050_QX56_G0017

Fig. 69 Fuel level sending unit location

4. Remove left hand rear inner fender liner.

5. Disconnect fuel filler hose from fuel filler pipe.

6. Drain fuel tank through the fuel filler hose using a suitable hose.

7. Remove or disconnect the following:
- Second row left hand seat
- Third row seat
- Second and third row seat belt buckles mounted on floor
- Left hand center pillar trim
- Left hand rear trim panel
- Left hand rear side door kick plate and weather stripping
- Second row rear center console and base, if equipped
- Inspection hole cover under carpet by turning retainers 90 degrees
- Electrical connectors
- EVAP hose
- Fuel supply hose
- Lock ring using special tool J-46214
- Fuel level sensor
- Fuel filter
- Fuel pump assembly

8. Remove the four bolts and remove the fuel tank shield.

9. Remove the propeller shaft.

10. Disconnect fuel filler hose, and vent hose at the fuel tank side.

11. Remove the fuel tank strap bolts while supporting the fuel tank with a suitable lift jack.

12. Disconnect the EVAP hose from the molded clip in the top of the fuel tank while lowering the fuel tank.

13. Lower the fuel tank using a suitable lift jack and remove it.

To install:

14. Install or connect the following:
- Fuel pump assembly
- Fuel filter
- Fuel level sensor
- Lock ring using special tool J-46214

15. Raise the fuel tank using a suitable lift jack.

16. Connect the EVAP hose from the molded clip in the top of the fuel tank while raising the fuel tank.

17. Install the fuel tank strap bolts while supporting the fuel tank with a suitable lift jack.

18. Connect fuel filler hose, and vent hose at the fuel tank side.

19. Install the propeller shaft.

20. Install the fuel tank shield.

21. Install or connect the following:
- Fuel supply hose
- EVAP hose
- Electrical connectors
- Inspection hole cover
- Second row rear center console and base, if equipped
- Left hand rear side door kick plate and weather stripping
- Left hand rear trim panel
- Left hand center pillar trim
- Second and third row seat belt buckles
- Third row seat
- Second row left hand seat
- Fuel filler hose to fuel filler pipe
- Left hand rear inner fender liner

22. Start the engine and check for leaks.

IDLE SPEED

ADJUSTMENT

Idle speed is controlled by the ECM and is not adjustable.

THROTTLE BODY

REMOVAL & INSTALLATION

See Figure 70.

1. Disconnect the sensor connector.
2. Remove retaining screws.
3. Remove the throttle body assembly from intake manifold.

To install:

4. To install, reverse removal procedure.

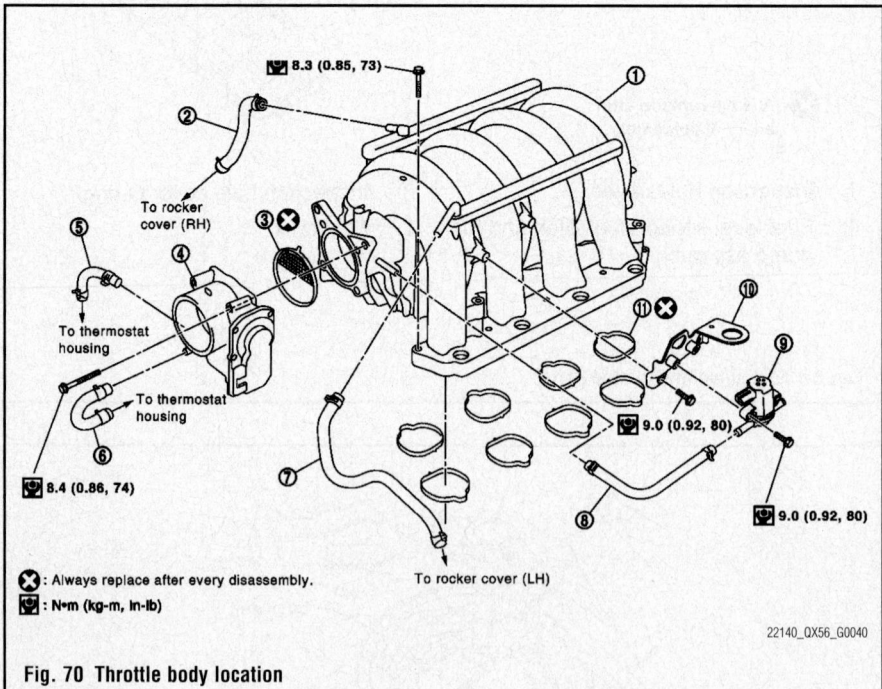

⌐⌐ 8.3 (0.85, 73)

To rocker cover (RH)

To thermostat housing

To thermostat housing

⌐⌐ 8.4 (0.86, 74)

⌐⌐ 9.0 (0.92, 80)

⌐⌐ 9.0 (0.92, 80)

To rocker cover (LH)

✖ : Always replace after every disassembly.

⌐⌐ : N•m (kg-m, in-lb)

Fig. 70 Throttle body location

22140_QX56_G0040

HEATING & AIR CONDITIONING SYSTEM

BLOWER MOTOR

REMOVAL & INSTALLATION

Front

See Figure 71.

➡Before servicing, or working around, the SRS system, turn the ignition switch OFF, disconnect both battery cables and wait at least three minutes. When servicing, or working around, the SRS system do not work directly in front of the air bag module.

1. Before servicing the vehicle, refer to the Precautions Section.
2. Disconnect the negative battery cable.
3. Remove the glove box assembly.
4. Disconnect the front blower motor electrical connector.
5. Remove the blower retaining screws.
6. Remove the blower motor from its mounting.

To install:

7. Installation is the reverse of the removal procedure.

Rear

See Figure 72.

1. Remove the luggage side lower finisher RH.
2. Disconnect the rear blower motor electrical connector.

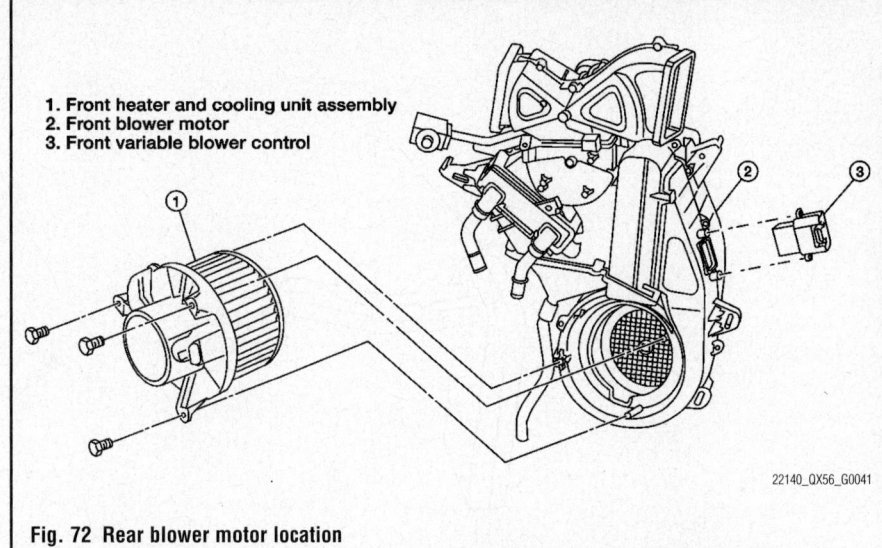

1. Front heater and cooling unit assembly
2. Front blower motor
3. Front variable blower control

22140_QX56_G0041

Fig. 72 Rear blower motor location

3. Remove the three screws and remove the rear blower motor.

To install:

4. Installation is the reverse of the removal procedure.

HEATER CORE

REMOVAL & INSTALLATION

Front

See Figures 73 through 75.

➡Before servicing, or working around, the SRS system, turn the ignition switch OFF, disconnect both battery cables and wait at least three minutes. When servicing, or working around, the SRS system do not work directly in front of the air bag module.

1. Before servicing the vehicle, refer to the Precautions Section.
2. Discharge the refrigerant from the A/C system.
3. Drain the cooling system.
4. Remove or disconnect the following:
   - Negative battery cable
   - Front heater hoses from the front heater core
   - Pressure pipes from the front expansion valve
   - Left-hand instrument panel lower cover
   - Steering column
   - Right-hand instrument panel lower cover
   - Glove box assembly
   - Center console lower covers
   - Center console electrical connectors
   - Center console
5. Remove the power sockets (cigarette lighter) as follows:
   a. Remove inner socket from the ring, while pressing the hook on the ring out from square hole.
   b. Disconnect power socket connector.
   c. Remove ring from power socket finisher while pressing pawls.
6. Remove or disconnect the following:
   - Radio cluster cover
   - Radio unit
   - Display unit cluster lid
   - Display unit and center speaker connectors

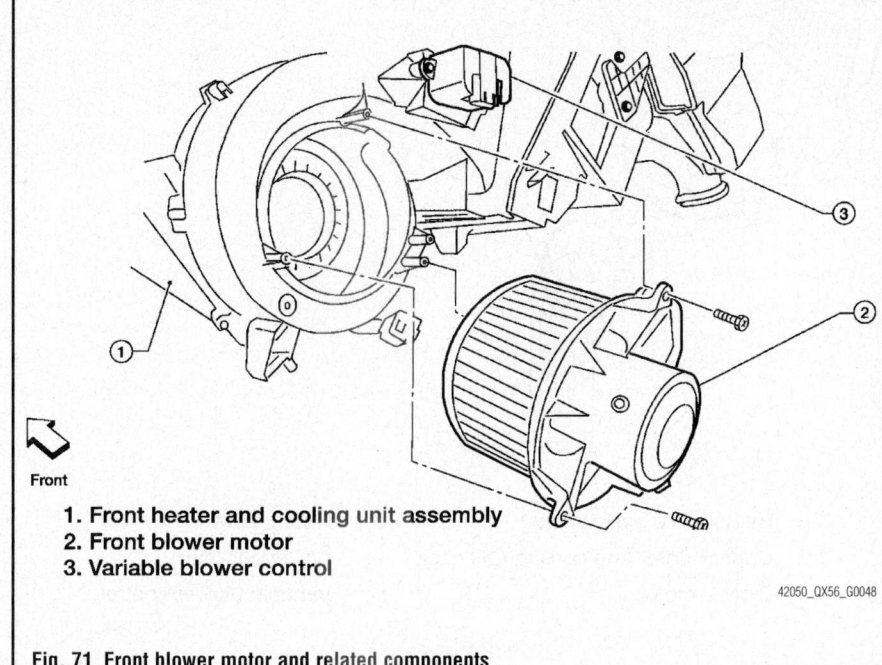

Front

1. Front heater and cooling unit assembly
2. Front blower motor
3. Variable blower control

42050_QX56_G0048

Fig. 71 Front blower motor and related components

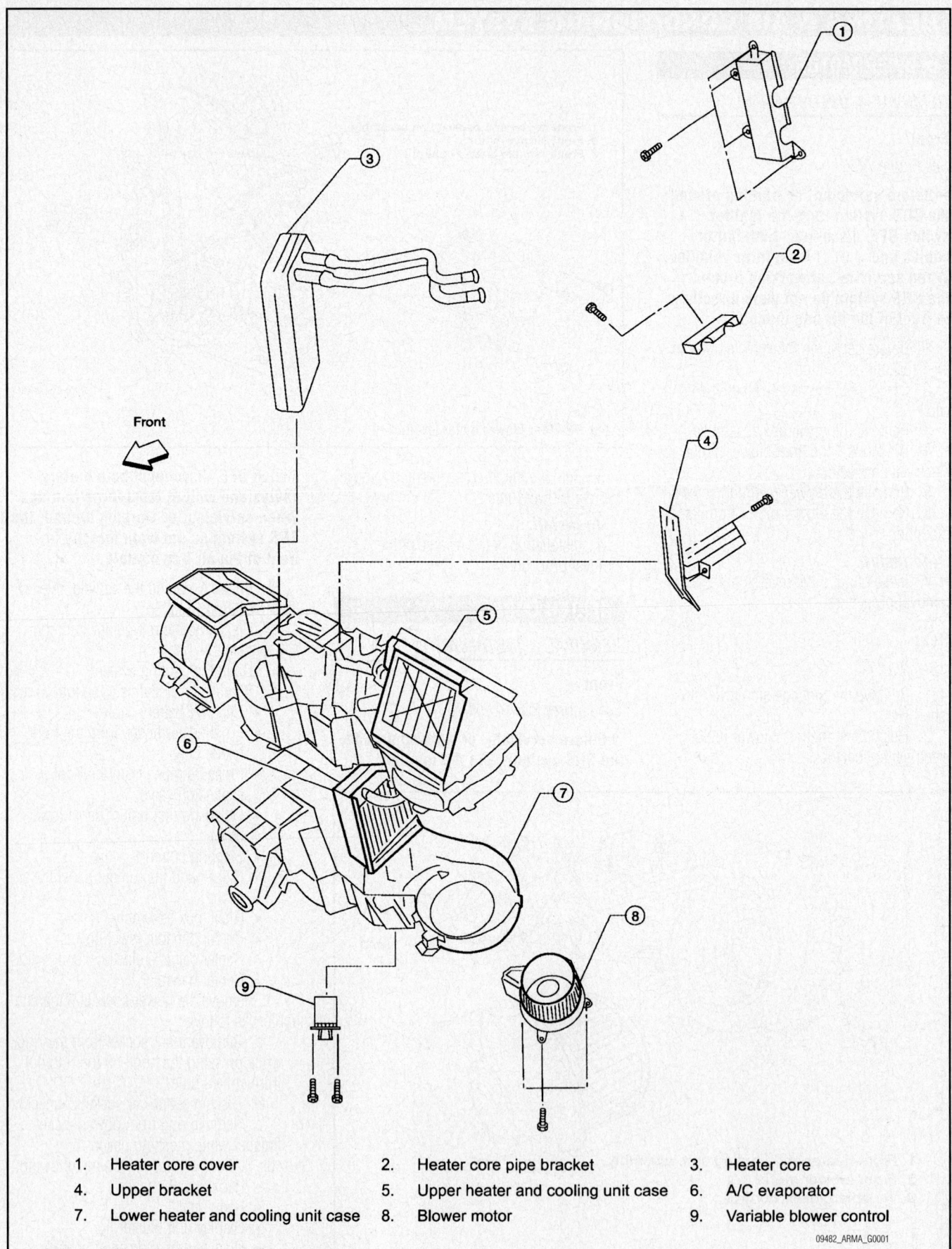

Front

1. Heater core cover
2. Heater core pipe bracket
3. Heater core
4. Upper bracket
5. Upper heater and cooling unit case
6. A/C evaporator
7. Lower heater and cooling unit case
8. Blower motor
9. Variable blower control

09482_ARMA_G0001

Fig. 73 Exploded view front heater core assembly

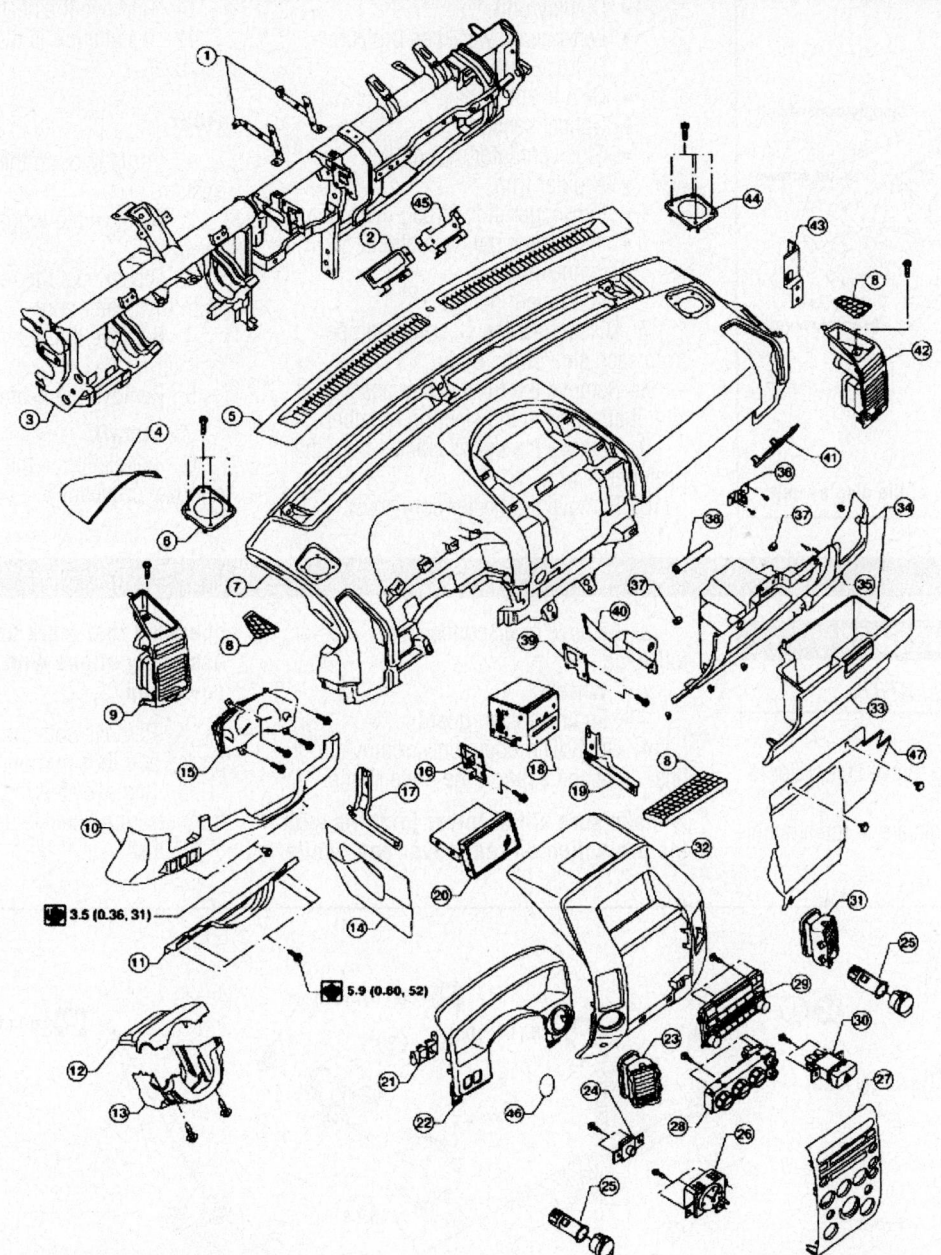

3.5 (0.36, 31)

5.9 (0.60, 52)

1. Steering member brackets	2. Bracket assembly LH	3. Steering member assembly
4. Combination meter cover	5. Defrost grille	6. Instrument panel speaker LH
7. Instrument panel	8. Bin mat	9. Side ventilator assembly LH
10. Instrument lower cover LH	11. Lower knee protector LH	12. Steering column cover upper
13. Steering column cover lower	14. Center console lower cover LH	15. Combination meter
16. Audio unit bracket LH	17. Driver instrument stay	18. Audio unit
19. Passenger instrument stay	20. Display assembly	21. Switch assembly
22. Cluster lid A	23. Cluster lid D ventilator LH	24. Front passenger air bag status light
25. Power socket LH and power socket RH	26. Clock	27. Cluster lid C
28. Front air control	29. Audio control	30. Hazard switch
31. Cluster lid D ventilator RH	32. Cluster lid D	33. Glove box door
34. Glove box assembly	35. Lower instrument panel RH	36. Glove box striker
37. Rubber bumpers	38. Glove box damper	39. Audio unit bracket RH
40. Center console lower cover RH	41. Fuse block cover	42. Side ventilator assembly RH
43. Instrument panel bracket	44. Instrument panel speaker RH	45. Bracket assembly RH
46. Key cylinder escutcheon	47. Instrument lower cover RH	

09482_ARMA_G0002

Fig. 74 Exploded view of the Instrument Panel assembly

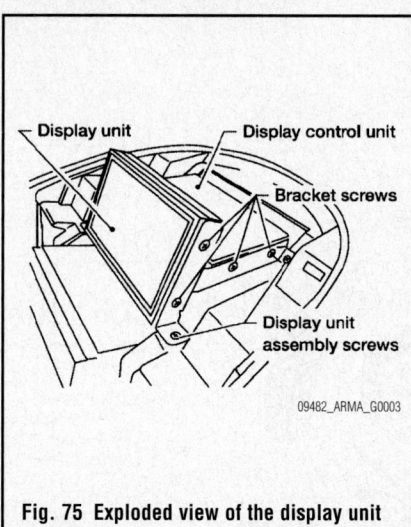

Fig. 75 Exploded view of the display unit

- Display unit
- Left-hand lower knee protector
- Defroster grille
- GPS antenna
- Optical sensor
- Side ventilator assemblies
- A-pillar trim
- Passenger side airbag module
- Instrument panel electrical connectors
- Instrument panel.

7. Disconnect the steering member from each side of the body.

8. Remove the front heater assembly with it attached to the steering member.

9. Remove the upper bracket from the heater assembly.

10. Remove the heater core cover.

11. Remove the heater core.

12. Installation is the reverse orders of removal.

Rear

1. Partially drain the engine cooling system.

2. Remove the luggage side finisher lower RH.

3. Disconnect the rear heater hoses from the heater core.

4. Remove the rear heater core bracket.

5. Remove the heater core.

To install:

6. Installation is the reverse of the removal procedure.

STEERING

POWER STEERING GEAR

REMOVAL & INSTALLATION

See Figure 76.

1. Before servicing the vehicle, refer to the Precautions Section.

2. Ensure the wheels are in the straight-ahead position.

3. Remove or disconnect the following:
- Wheels
- Engine splash guard

4. On 4WD models only, remove front final drive and support the drive shafts.

➡**Make sure slit of lower joint fits with the projection on rear cover cap, while checking that mark on steering gear assembly aligns with mark on rear cover cap.**

5. Remove cotter pin at steering outer socket and loosen mounting nut.

6. Remove steering outer socket from steering knuckle using special tool J-25730-A.

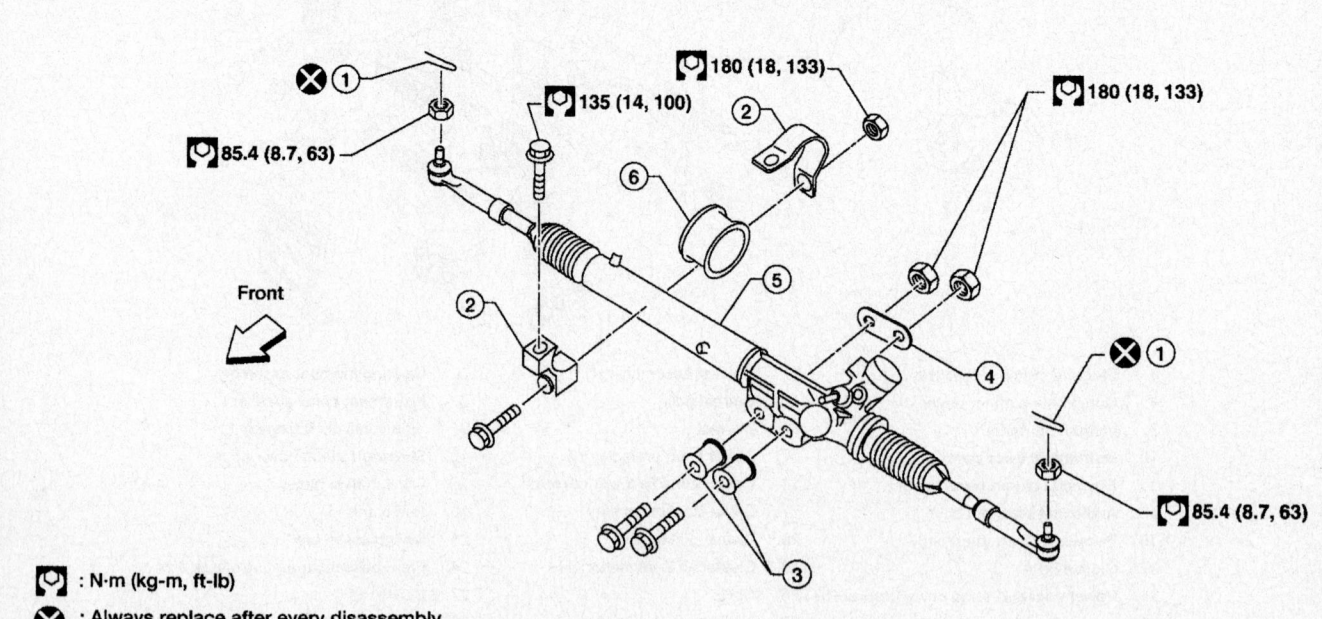

🔧 : N·m (kg-m, ft-lb)

❌ : Always replace after every disassembly.

1. Cotter pin
2. Mounting bracket
3. Bushing
4. Washer
5. Steering gear assembly
6. Mounting insulator

67170-ARMA-G49

Fig. 76 Steering gear assembly

➡**Temporarily tighten mounting nut to prevent damage to threads and to prevent tool from coming off.**

7. On 2WD drive models only, remove stabilizer bar mounting bolts and secure the stabilizer bar.

8. Remove or disconnect the following:
- Oil pipes from steering gear assembly
- Lower joint mounting bolt from lower shaft
- Mounting bolts and nuts from steering gear assembly
- Steering gear assembly

To install:

9. Install or connect the following:
- Steering gear assembly, tighten nuts to 133 ft. lbs. (180 Nm)
- Lower joint mounting bolt
- Oil pipes to steering gear assembly
- Stabilizer bar, 2WD models only
- Steering outer socket to steering knuckle, tighten nut to 63 ft. lbs. (86 Nm)

- Front final drive, 4WD models only
- Engine splash guard
- Wheels

10. Check the wheel alignment and adjust as necessary.

POWER STEERING PUMP

REMOVAL & INSTALLATION

See Figure 77.

1. Before servicing the vehicle, refer to the Precautions Section.
2. Disconnect the negative battery cable.
3. Drain the power steering fluid into a suitable container. Properly discard the used fluid.
4. Remove the engine room cover.
5. Remove the air duct assembly.
6. Remove the power steering reservoir tank.
7. Remove the drive belt.
8. Disconnect the pressure sensor electrical connector.

9. Remove the high pressure and the low pressure lines from the power steering fluid pump.
10. Remove the pump mounting bolts.
11. Remove the pump from the vehicle.

To install:

12. Installation is the reverse of the removal procedure.
13. Bleed the power steering system.

➡**The drive belt tension is automatic and requires no adjustment.**

BLEEDING

1. Before servicing the vehicle, refer to the Precautions Section.
2. Stop the engine.
3. Turn the steering wheel fully to the right and left several times.

➡**Do not allow the fluid level in the reservoir tank to go below the MIN level line. Check and add fluid as needed.**

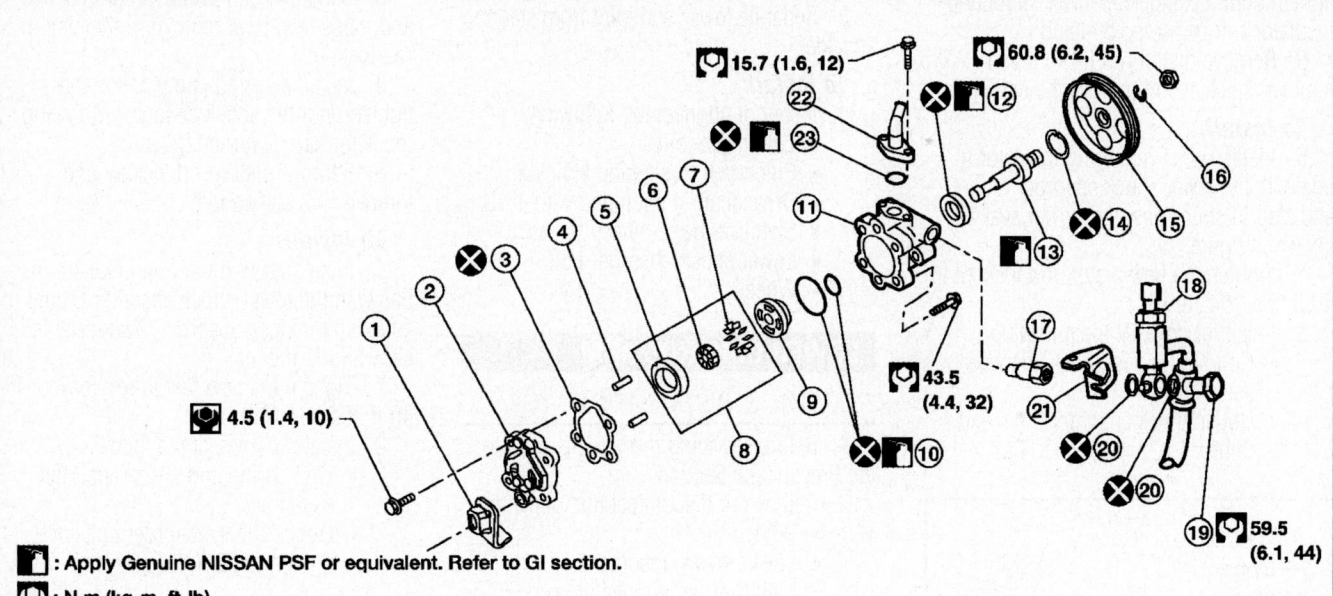

🔧 15.7 (1.6, 12)   🔧 60.8 (6.2, 45)

🔧 4.5 (1.4, 10)

🔧 43.5 (4.4, 32)

🔧 59.5 (6.1, 44)

🛢 : Apply Genuine NISSAN PSF or equivalent. Refer to GI section.

🔧 : N·m (kg-m, ft-lb)

✖ : Always replace after every disassembly.

1. Bracket	2. Rear cover	3. Gasket
4. Lock pin	5. Cam ring	6. Rotor
7. Vane	8. Cartridge	9. Side plate
10. O-ring	11. Body assembly	12. Oil seal
13. Drive shaft assembly	14. Snap ring	15. Pulley
16. Spring washer	17. Flow control valve	18. Pressure sensor
19. Connector bolt	20. Copper washer	21. Bracket
22. Suction pipe	23. O-ring	

42050_QX56_G0043

Fig. 77 Power steering pump—exploded view

4. Run the engine at idle speed. Turn the steering wheel fully to the right and then fully to the left. Hold for about three seconds. Check for fluid leakage.

5. Repeat the above step several times at three second intervals.

➡**Do not hold the steering wheel in the locked position for more than ten seconds.**

6. Check for air bubbles or cloudy fluid. If found, repeat the bleeding procedure.

7. Stop the engine and check the fluid level. Correct as required.

SUSPENSION

COIL SPRING

REMOVAL & INSTALLATION

See Figure 78.

1. Before servicing the vehicle, refer to the Precautions Section.
2. Remove or disconnect the following:
   - Wheel
   - Lower shock absorber bolt
   - Upper shock absorber bolts
   - Coil spring and shock absorber assembly
3. Secure the shock absorber in a vice and loosen (without removing) the piston rod locknut.
4. Install a spring compressor and tighten until the shock absorber mounting insulator can be turned by hand.
5. Remove piston rod locknut and remove shock absorber from the coil spring.

To install:

6. Install upper mounting insulator in line with the lower shock absorber mount and step in shock absorber lower seat as shown in figure.
7. Tighten the new piston rod locknut to 40 ft. lbs. (54 Nm).
8. Install or connect the following:
   - Coil spring and shock absorber assembly
   - Upper shock absorber bolts and tighten to 22 ft. lbs (30 Nm)

- Lower shock absorber bolt and tighten to 99 ft. lbs. (134 Nm)
- Wheel

9. Check wheel alignment and adjust as necessary.

LOWER BALL JOINT

REMOVAL & INSTALLATION

1. Before servicing the vehicle, refer to the Precautions Section.
2. Remove or disconnect the following:
   - Wheel
   - Lower shock absorber bolt
   - Stabilizer bar connecting rod
   - Drive shaft, if equipped with 4WD
   - Pinch bolt from steering knuckle
3. Separate lower ball joint from steering knuckle.

To install:

4. Install or connect the following:
   - Lower ball joint
   - Pinch bolt to steering knuckle
   - Drive shaft, if equipped with 4WD
   - Stabilizer bar connecting rod
   - Lower shock absorber bolt
   - Wheel

LOWER CONTROL ARM

REMOVAL & INSTALLATION

1. Before servicing the vehicle, refer to the Precautions Section.
2. Remove or disconnect the following:
   - Wheel
   - Lower shock absorber bolt
   - Stabilizer bar connecting rod
   - Drive shaft, if equipped with 4WD
   - Pinch bolt from steering knuckle
3. Separate the lower ball joint from the steering knuckle.
4. Remove the following:
   - Lower link adjusting bolts
   - Lower link

To install:

5. Install or connect the following:
   - Lower link and tighten adjusting bolts to 98 ft. lbs. (133 Nm)
   - Lower ball joint
   - Pinch bolt
   - Drive shaft, if equipped with 4WD
   - Stabilizer bar connected rod

FRONT SUSPENSION

- Lower shock absorber bolt
- Wheel

SHOCK ABSORBERS

REMOVAL & INSTALLATION

See Figures 79 and 80.

1. Before servicing the vehicle, refer to the Precautions Section.
2. Remove or disconnect the following:
   - Wheel
   - Lower shock absorber bolt
   - Upper shock absorber bolts
   - Coil spring and shock absorber assembly
3. Secure the shock absorber in a vice and loosen (without removing) the piston rod locknut.
4. Install a spring compressor and tighten until the shock absorber mounting insulator can be turned by hand.
5. Remove piston rod locknut and remove shock absorber.

To install:

6. Install upper mounting insulator in line with the lower shock absorber mount and step in shock absorber lower seat as shown in figure.
7. Tighten the new piston rod locknut to 40 ft. lbs. (54 Nm).
8. Install or connect the following:
   - Coil spring and shock absorber assembly
   - Upper shock absorber bolts and tighten to 22 ft. lbs (30 Nm)

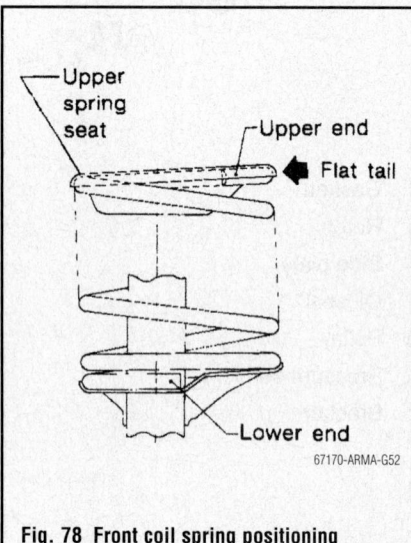

67170-ARMA-G52

Fig. 78 Front coil spring positioning

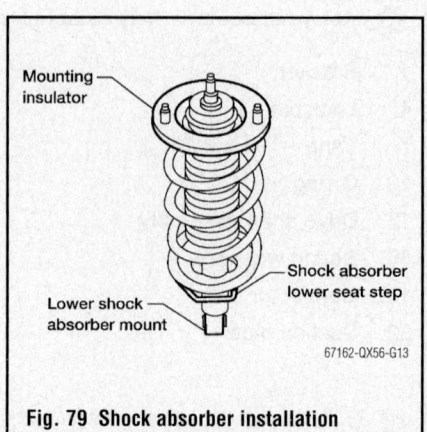

67162-QX56-G13

Fig. 79 Shock absorber installation

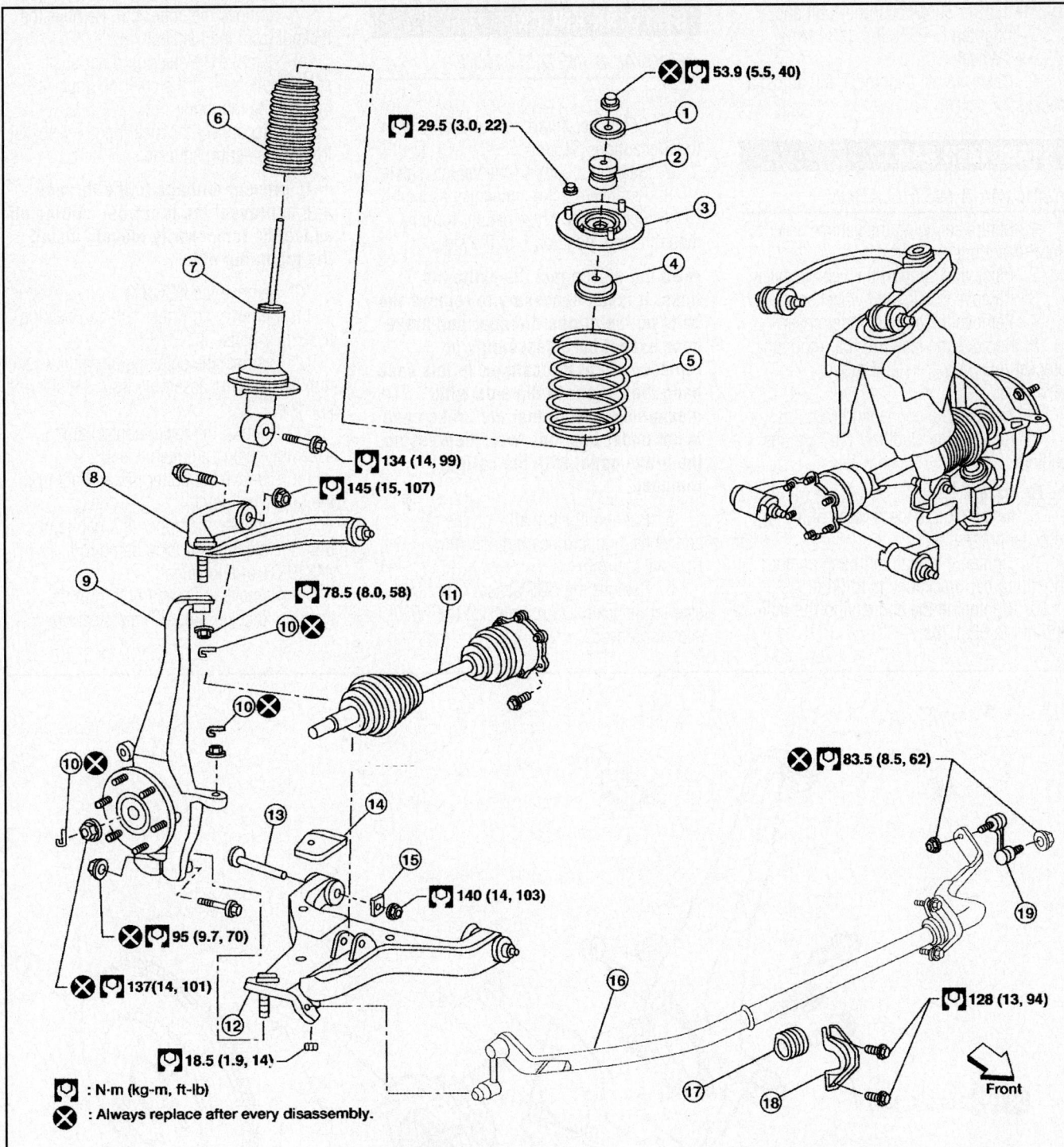

53.9 (5.5, 40)
29.5 (3.0, 22)
134 (14, 99)
145 (15, 107)
78.5 (8.0, 58)
83.5 (8.5, 62)
95 (9.7, 70)
140 (14, 103)
137(14, 101)
128 (13, 94)
18.5 (1.9, 14)

: N·m (kg-m, ft-lb)
: Always replace after every disassembly.

1. Washer	2. Shock absorber bushing	3. Shock absorber mounting insulator
4. Upper seat	5. Coil spring	6. Dust cover
7. Shock absorber	8. Upper link	9. Steering knuckle
10. Cotter pin	11. Drive shaft	12. Lower link
13. Cam bolt	14. Jounce bumper	15. Cam washer
16. Stabilizer bar	17. Stabilizer bar bushing	18. Stabilizer bar mounting bracket
19. Connecting rod		

67170-ARMA-G13

Fig. 80 Front suspension

- Lower shock absorber bolt and tighten to 99 ft. lbs. (134 Nm)
- Wheel

9. Check wheel alignment and adjust as necessary.

STABILIZER BAR

REMOVAL & INSTALLATION

1. Before servicing the vehicle, refer to the Precautions Section.
2. Raise and support the vehicle safely.
3. Remove the tire and wheel assembly.
4. Remove the engine under cover.
5. Remove the stabilizer bar mounting bracket retaining bolts and rubber bushings.
6. Remove the connecting rod nuts.
7. Remove the stabilizer bar from the vehicle.

To install:

8. Installation is the reverse of the removal procedure.
9. Tighten the retaining bushing and mounting bar bracket bolts to 94 ft. lbs.
10. Tightening the connecting rod bolt and nut to 62 ft. lbs.

STEERING KNUCKLE

REMOVAL & INSTALLATION

See Figure 81.

1. Before servicing the vehicle, refer to the Precautions Section.
2. Raise and support the vehicle safely.
3. Remove the tire and wheel assembly.
4. Remove the brake caliper from its mounting and position it to the side.

➡**Do not disconnect the hydraulic lines. It is not necessary to remove the bolts on the torque member and brake hose except for disassembly or replacement of the caliper. In this case hang the caliper to the side with mechanics wire so that the brake hose is not under tension. Avoid depressing the brake pedal with the caliper removed.**

5. Put alignment marks on the rotor and wheel hub and bearing assembly. Remove the rotor.
6. Remove the ABS sensor from the steering knuckle. Do not pull on the ABS sensor harness.

7. Remove the cotter pin. Remove the locknut from the halfshaft.
8. Remove the steering outer shaft socket cotter pin at the steering knuckle. Loosen the mounting nut.
9. Disconnect the steering outer socket from the steering knuckle.

➡**To prevent damage to the threads and to prevent the tool from coming off suddenly, temporarily loosely install the mounting nut.**

10. Remove the halfshaft.
11. Remove the wheel hub and bearing assembly bolts.
12. Remove the splash guard and wheel hub and bearing assembly from the steering knuckle.
13. Support the lower control arm assembly, using a suitable jack.
14. Remove the cotter pin and nut from the upper ball joint.
15. Separate the upper link ball joint from the steering knuckle using tool J-24319-01 or equivalent.
16. Remove the pinch bolt from the steering knuckle. Remove the steering

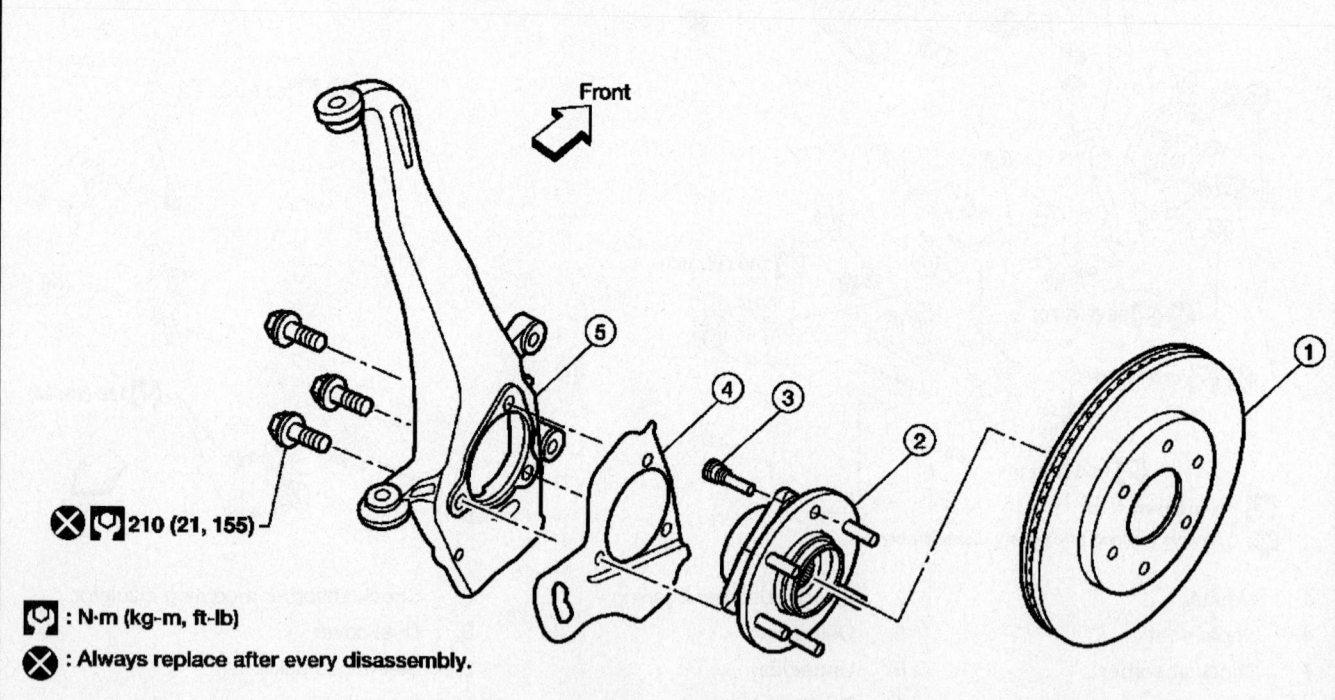

⊗ 🔧 210 (21, 155)

🔧 : N·m (kg-m, ft-lb)

⊗ : Always replace after every disassembly.

1. Disc rotor
2. Wheel hub and bearing assembly
3. Wheel stud
4. Splash guard
5. Steering knuckle

42050_QX56_G0037

Fig. 81 Steering knuckle and related components

knuckle from the lower control arm ball joint.

17. Remove the steering knuckle from the vehicle.

To install:

18. Installation is the reverse of the removal procedure.

19. Be sure to use the alignment marks made during the removal procedure when reinstalling removed components.

20. Check and adjust the front end alignment, as required.

UPPER BALL JOINT

REMOVAL & INSTALLATION

1. Before servicing the vehicle, refer to the Precautions Section.

2. Remove or disconnect the following:
   - Wheel
   - Coil spring and shock absorber assembly
   - Cotter pin and nut from upper ball joint

3. Separate upper ball joint from steering knuckle using special tool J-24319-01.

To install:

4. Install or connect the following:
   - Upper ball joint
   - New cotter pin and tighten nut to 58 ft. lbs. (79 Nm)
   - Coil spring and shock absorber assembly
   - Wheel

UPPER CONTROL ARM

REMOVAL & INSTALLATION

1. Before servicing the vehicle, refer to the Precautions Section.

2. Remove or disconnect the following:
   - Wheel
   - Coil spring and shock absorber assembly
   - Cotter pin and nut from upper ball joint

3. Separate upper ball joint stud from steering knuckle using special tool J-24319-01.

4. Remove the following:
   - Upper control arm mounting bolts
   - Upper control arm

To install:

5. Install or connect the following:
   - Upper control arm and tighten bolts to 107 ft. lbs. (145 Nm)
   - Upper ball joint with new cotter pin and tighten nut to 58 ft. lbs. (79 Nm)
   - Coil spring and shock absorber assembly
   - Wheel

WHEEL BEARINGS

REMOVAL & INSTALLATION

See Figure 82.

1. Before servicing the vehicle, refer to the Precautions Section.

2. Remove or disconnect the following:
   - Wheel
   - Engine splash guard
   - Brake caliper without disconnecting the hydraulic lines, and reposition aside with wire

3. Matchmark the brake rotor to the wheel hub and remove the brake rotor.

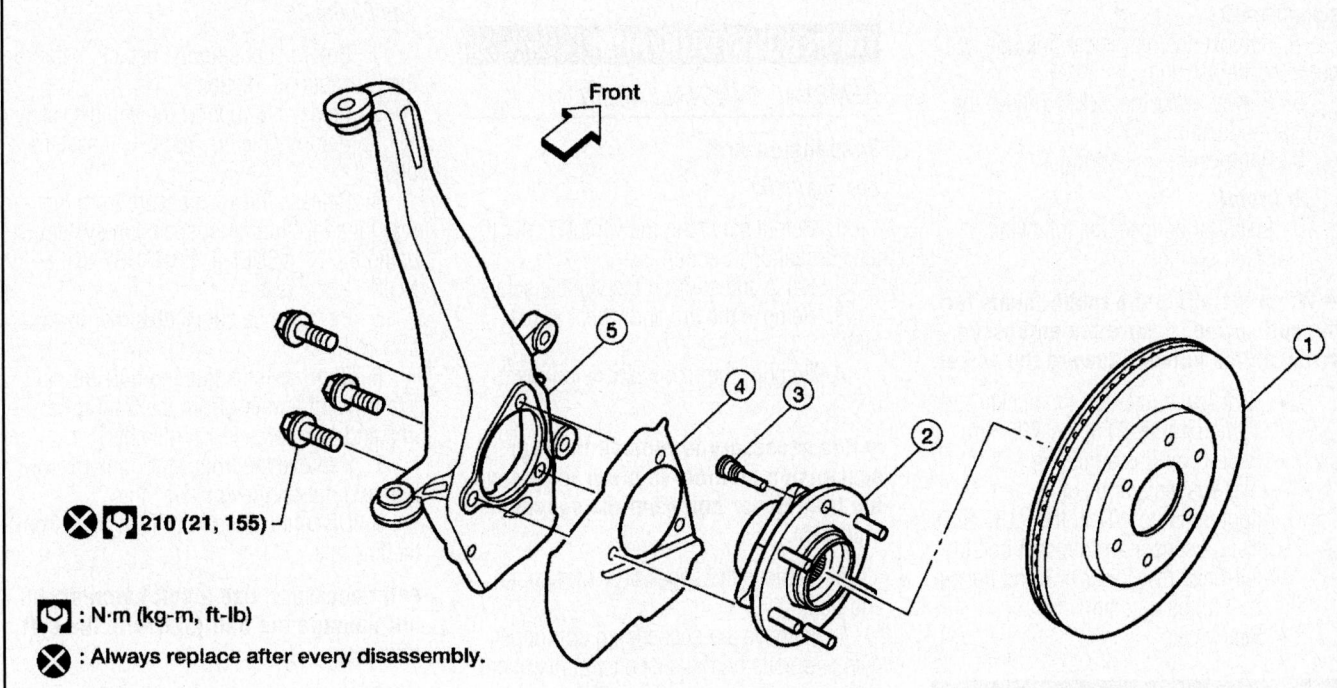

210 (21, 155)

⊡ : N·m (kg-m, ft-lb)

⊗ : Always replace after every disassembly.

| 1. | Disc rotor | 2. | Wheel hub and bearing assembly | 3. | Wheel stud |
| 4. | Splash guard | 5. | Steering knuckle | | |

67170-ARMA-G53

Fig. 82 Front hub/bearing assembly

4. Remove or disconnect the following:
- Cotter pin and locknut from driveshaft
- Driveshaft from wheel hub and bearing assembly
- ABS sensor
- Wheel hub and bearing assembly bolts

- Wheel hub and bearing assembly

To install:

5. Install or connect the following:
- Wheel hub and bearing assembly, using new bolts and tighten to 155 ft. lbs. (210 Nm)
- ABS sensor

- Driveshaft to wheel hub and bearing assembly
- Cotter pin and locknut and tighten to 101 ft. lbs. (137 Nm)
- Brake rotor
- Brake caliper
- Engine splash guard
- Wheel

SUSPENSION

COIL SPRING

REMOVAL & INSTALLATION

1. Before servicing the vehicle, refer to the Precautions Section.
2. Remove the rear wheel.
3. Release the air pressure from the rear load leveling air suspension system using the CONSULT-II "EXHAUST SOLENOID" active test.
4. Remove the height sensor arm bracket bolt from the left-hand rear lower link.
5. Place a suitable jack under the rear lower link and relieve the coil spring tension.
6. Loosen the rear lower link adjusting bolt and nut connected to the rear suspension member.
7. Remove the rear lower link bolt and nut from the knuckle.
8. Slowly lower the jack to relieve the coil spring tension.
9. Remove the coil spring.

To install:

10. Install or connect the following:
- Coil spring

➡**When installing the rubber seats for the coil spring, ensure the embossed arrow points outward toward the wheel.**

- Rear lower link bolt to knuckle and tighten nut to 70 ft. lbs. (95 Nm)
- Rear lower link adjusting bolt to rear suspension member and tighten nut to 101 ft. lbs. (137 Nm)
- Height sensor arm bracket bolt to left-head rear lower link and tighten to 9 ft. lbs. (12 Nm)
- Rear wheel

SHOCK ABSORBER

REMOVAL & INSTALLATION

See Figures 83 and 84.

1. Before servicing the vehicle, refer to the Precautions Section.
2. Remove the rear wheel.
3. Release the air pressure from the rear load leveling air suspension system using the CONSULT-II "EXHAUST SOLENOID" active test.
4. Remove or disconnect the following:
- Rear fender protector
- Rear load leveling air suspension hose from the shock absorber
- Shock absorber upper and lower end bolts
- Shock absorber

To install:

5. Install or connect the following:
- Shock absorber and tighten end bolts to 129 ft. lbs. (175 Nm)
- Rear load leveling air suspension hose
- Rear fender protector
- Rear wheel

UPPER CONTROL ARM

REMOVAL & INSTALLATION

Suspension Arm

See Figure 85.

1. Before servicing the vehicle, refer to the Precautions Section.
2. Raise and support the vehicle safely.
3. Remove the tire and wheel assemblies.
4. Remove the rear suspension member.

➡**It is necessary to remove the rear suspension member in order to remove the front upper bolt from the suspension arm.**

5. Remove the shock absorber upper end bolt.
6. Remove the suspension arm upper nuts and bolts on the suspension member side.
7. Remove the suspension arm pinch bolt and nut on the knuckle side.
8. Disconnect the suspension arm from the knuckle.

➡**If necessary, use a soft hammer. Do not damage the ball joint with the soft hammer.**

9. Remove the suspension arm.

To install:

10. Installation is the reverse of the removal procedure.
11. Perform the final tightening of the nuts and bolts for the links (rubber bushing) with the vehicle in the unladen condition with the tires on level ground.

➡**Unladen condition means that the fuel tank, engine coolant and lubricants are at the full specification and the spare tire, jack, hand tools and mats are in their designated positions.**

12. Check and adjust the alignment, as required.

Upper Link

See Figure 85.

1. Before servicing the vehicle, refer to the Precautions Section.
2. Raise and support the vehicle safely.
3. Remove the tire and wheel assemblies.
4. Release the air pressure from the rear load leveling air suspension system using the CONSULT-II "EXHAUST SOLE-NOID" active test.
5. Remove the shock absorber lower end bolt.
6. Remove the adjusting bolt and nut, and the bolt and nut from the front lower link and rear suspension member.
7. Remove the front lower link pinch bolt and nut on the knuckle side.
8. Disconnect the front lower link from the knuckle.

➡**If necessary, use a soft hammer. Do not damage the ball joint with the soft hammer.**

9. Remove the front lower link.

To install:

10. Installation is the reverse of the removal procedure.
11. Perform the final tightening of the nuts and bolts for the links (rubber bushing) with the vehicle in the unladen condition with the tires on level ground.

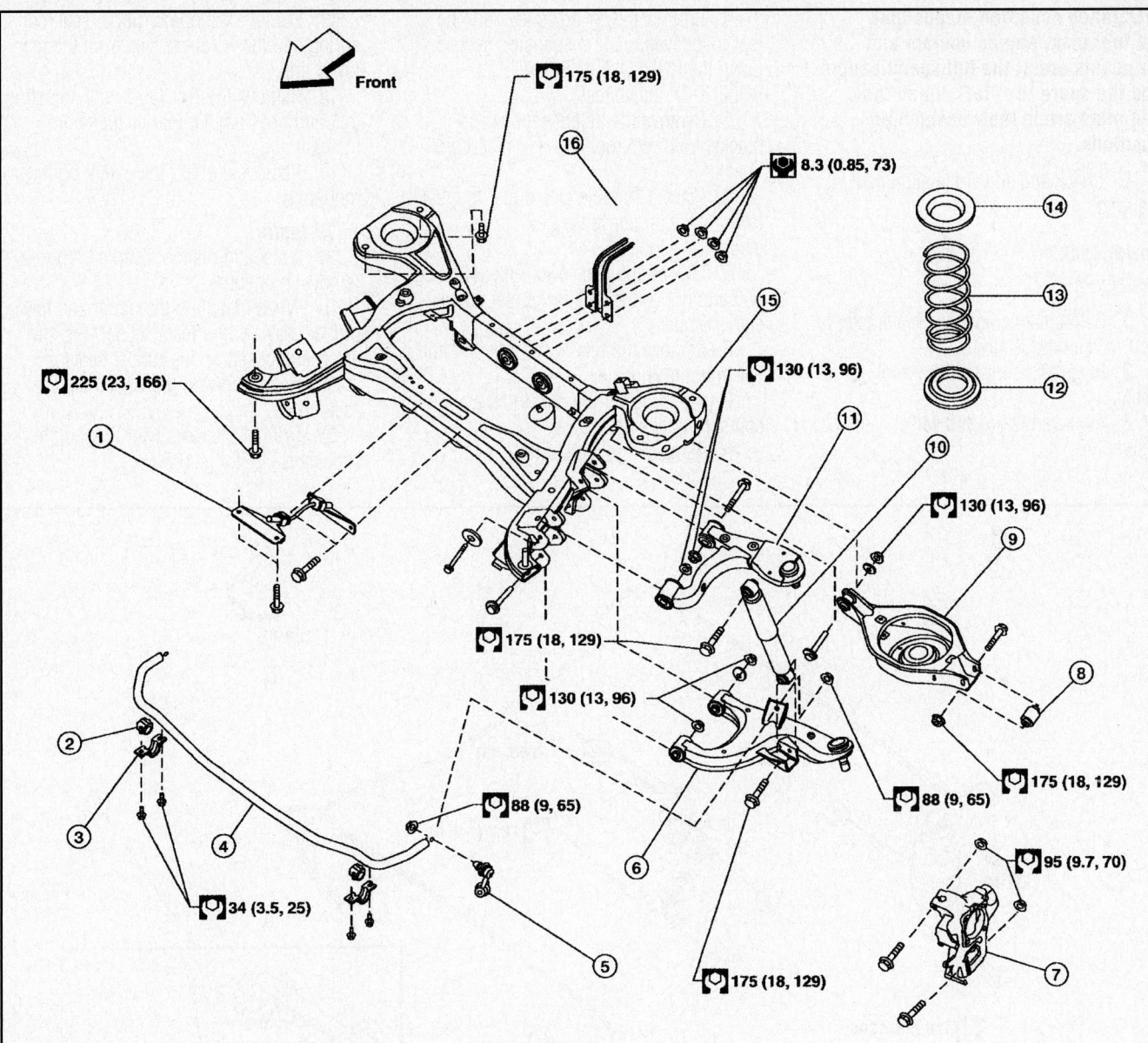

Front

175 (18, 129)

8.3 (0.85, 73)

225 (23, 166)

130 (13, 96)

130 (13, 96)

175 (18, 129)

130 (13, 96)

88 (9, 65)

34 (3.5, 25)

175 (18, 129)

88 (9, 65)

175 (18, 129)

95 (9.7, 70)

N·m (kg-m, in-lb)

N·m (kg-m, ft-lb)

1. Seat belt latch anchor	2. Stabilizer bar bushing	3. Stabilizer bar clamp
4. Stabilizer bar	5. Connecting rod	6. Front lower link
7. Wheel hub and spindle assembly	8. Bushing	9. Rear lower link
10. Shock absorber	11. Suspension arm	12. Lower rubber seat
13. Coil spring	14. Upper rubber seat	15. Rear suspension member
16. Spare tire bracket		

67170-ARMA-G50

Fig. 83 Standard rear suspension

➡**Unladen condition means that the fuel tank, engine coolant and lubricants are at the full specification and the spare tire, jack, hand tools and mats are in their designated positions.**

12. Check and adjust the alignment, as required.

Lower Link

See Figure 85.

1. Before servicing the vehicle, refer to the Precautions Section.

2. Raise and support the vehicle safely.

3. Remove the tire and wheel assemblies.

4. Release the air pressure from the rear load leveling air suspension system using the CONSULT-II "EXHAUST SOLENOID" active test.

5. Remove the height sensor arm bracket bolt from the left-hand rear lower link.

6. Place a suitable jack under the rear lower link and relieve the coil spring tension.

7. Loosen the rear lower link adjusting bolt and nut connected to the rear suspension member.

8. Remove the rear lower link bolt and nut from the knuckle.

9. Slowly lower the jack to relieve the coil spring tension.

10. Remove the coil spring.

11. Remove the upper rubber seat, coil spring and lower rubber seat from the rear lower link.

12. Remove the rear lower link adjusting bolt and nut from the rear suspension member.

13. Remove the rear lower link from its mounting.

To install:

14. Installation is the reverse of the removal procedure.

15. When installing the upper and lower rubber seats for the rear coil springs, the arrow embossed on the rubber seats must point out toward the wheel and tire assembly.

16. Tighten the rear lower link bolt to knuckle to 70 ft. lbs. (95 Nm).

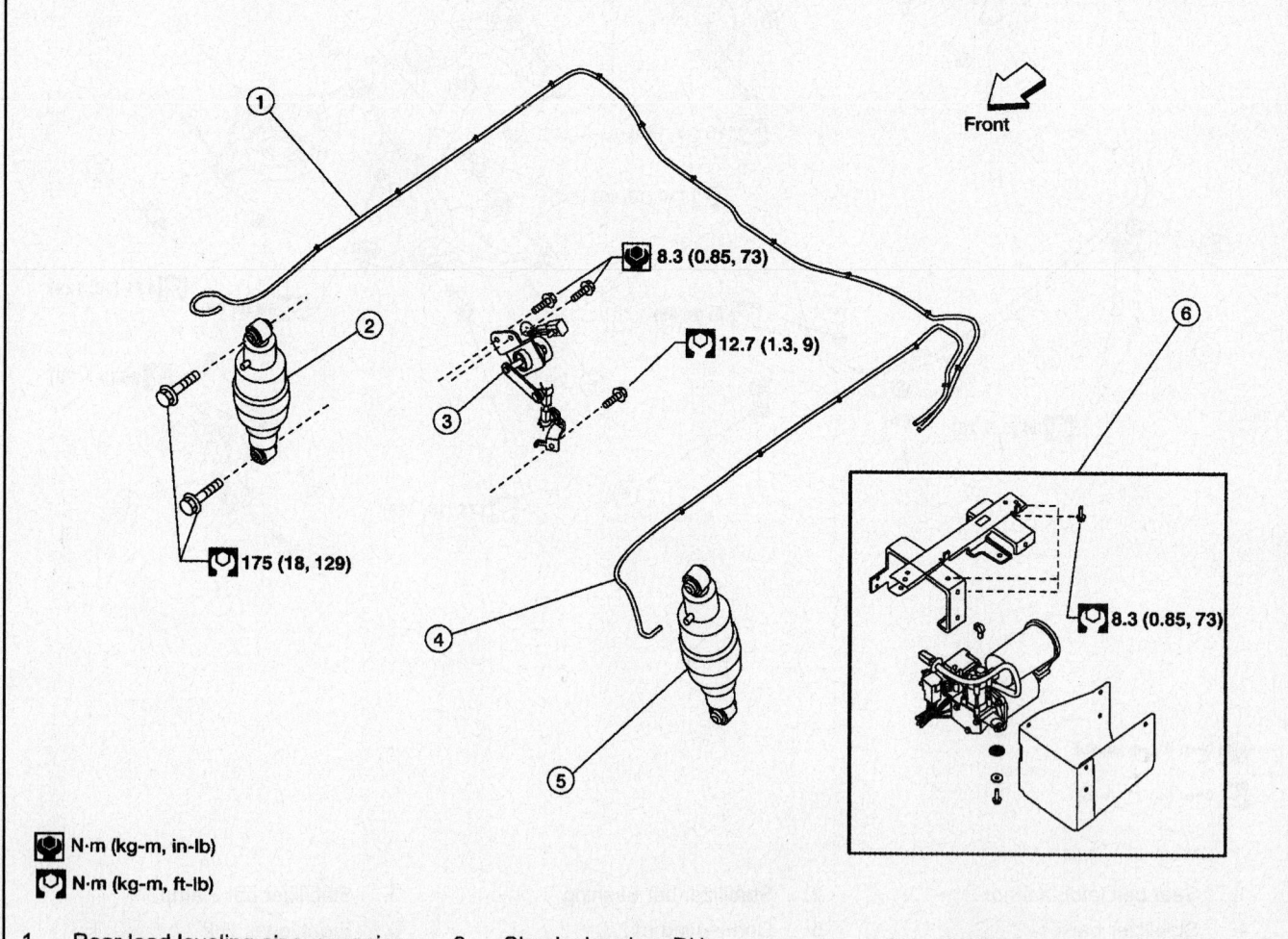

Front

8.3 (0.85, 73)

12.7 (1.3, 9)

175 (18, 129)

8.3 (0.85, 73)

N·m (kg-m, in-lb)

N·m (kg-m, ft-lb)

1. Rear load leveling air suspension hose, RH	2. Shock absorber, RH	3. Height sensor
4. Rear load leveling air suspension hose, LH	5. Shock absorber, LH	6. Rear load leveling air suspension compressor assembly

67170-ARMA-G51

Fig. 84 Rear load leveling air suspension

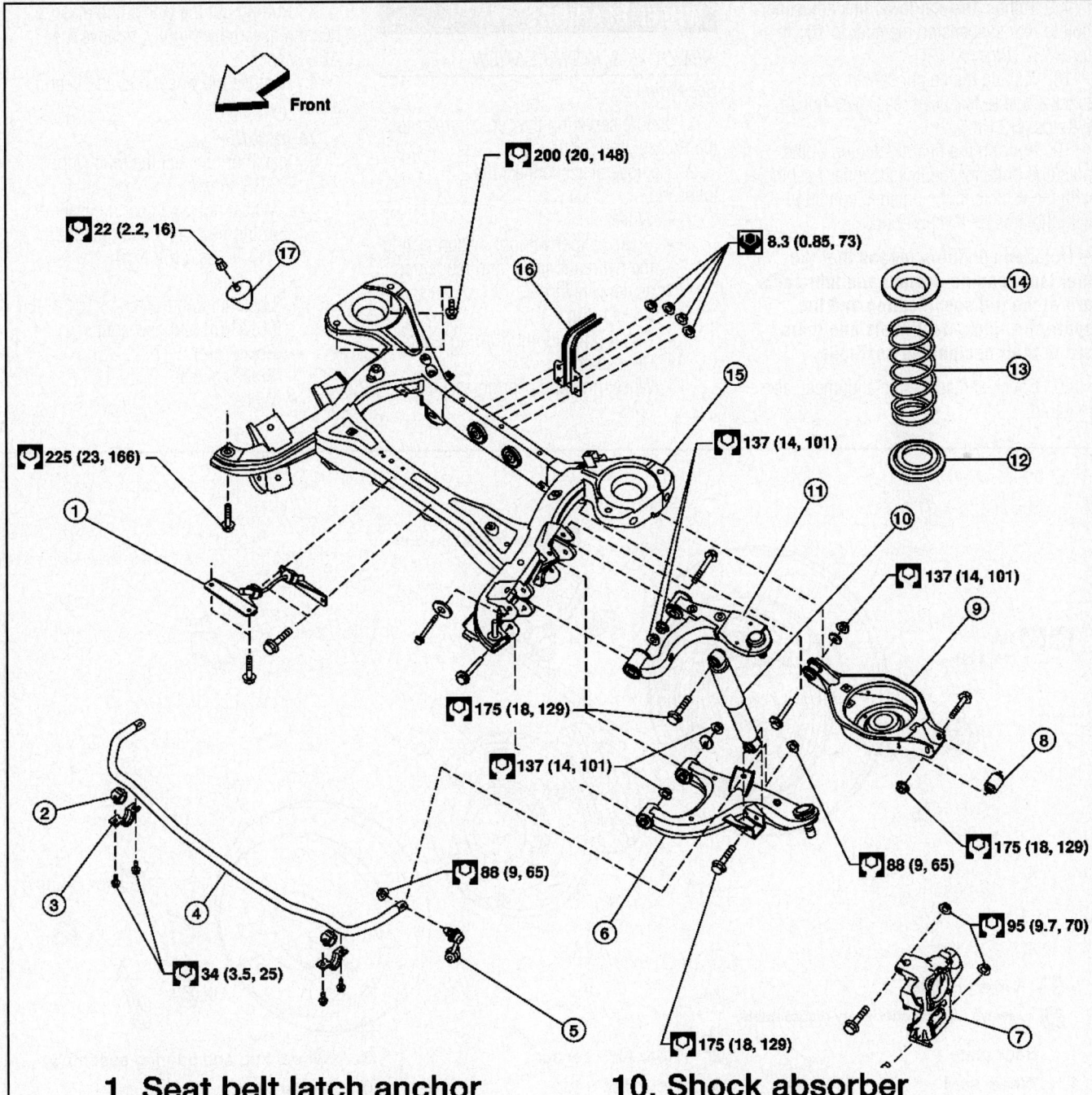

Front

200 (20, 148)

22 (2.2, 16)

8.3 (0.85, 73)

137 (14, 101)

225 (23, 166)

137 (14, 101)

175 (18, 129)

137 (14, 101)

175 (18, 129)

88 (9, 65)

88 (9, 65)

95 (9.7, 70)

34 (3.5, 25)

175 (18, 129)

1. Seat belt latch anchor
2. Stabilizer bar bushing
3. Stabilizer bar clamp
4. Stabilizer bar
5. Connecting rod
6. Front lower link
7. Knuckle
8. Bushing
9. Rear lower link

10. Shock absorber
11. Suspension arm
12. Lower rubber seat
13. Coil spring
14. Upper rubber seat
15. Rear suspension member
16. Spare tire bracket
17. Bound bumper

22140_QX56_G0043

Fig. 85 Rear suspension component locations

17. Tighten the rear lower link adjusting bolt to rear suspension member to 101 ft. lbs. (137 Nm).

18. Tighten the height sensor arm bracket bolt to left-head rear lower link to 9 ft. lbs. (12 Nm).

19. Perform the final tightening of the nuts and bolts for the links (rubber bushing) with the vehicle in the unladen condition with the tires on level ground.

➡**Unladen condition means that the fuel tank, engine coolant and lubricants are at the full specification and the spare tire, jack, hand tools and mats are in their designated positions.**

20. Check and adjust the alignment, as required.

WHEEL BEARINGS

REMOVAL & INSTALLATION

See Figure 86.

1. Before servicing the vehicle, refer to the Precautions Section.

2. Remove or disconnect the following:
- Wheel
- Brake caliper without disconnecting the hydraulic lines, and reposition aside with wire
- Brake rotor
- Cotter pin and nut from driveshaft
- Driveshaft
- Wheel hub and bearing assembly bolts

3. Pulling out the wheel hub and bearing assembly slightly, remove the ABS sensor.

4. Remove the wheel hub and bearing assembly.

To install:

5. Install or connect the following:
- ABS sensor
- Wheel hub and bearing assembly, using new bolts and tighten to 111 ft. lbs. (150 Nm)
- Driveshaft
- Lock nut and tighten to 101 ft. lbs. (137 Nm) and new cotter pin
- Brake rotor
- Brake caliper
- Wheel

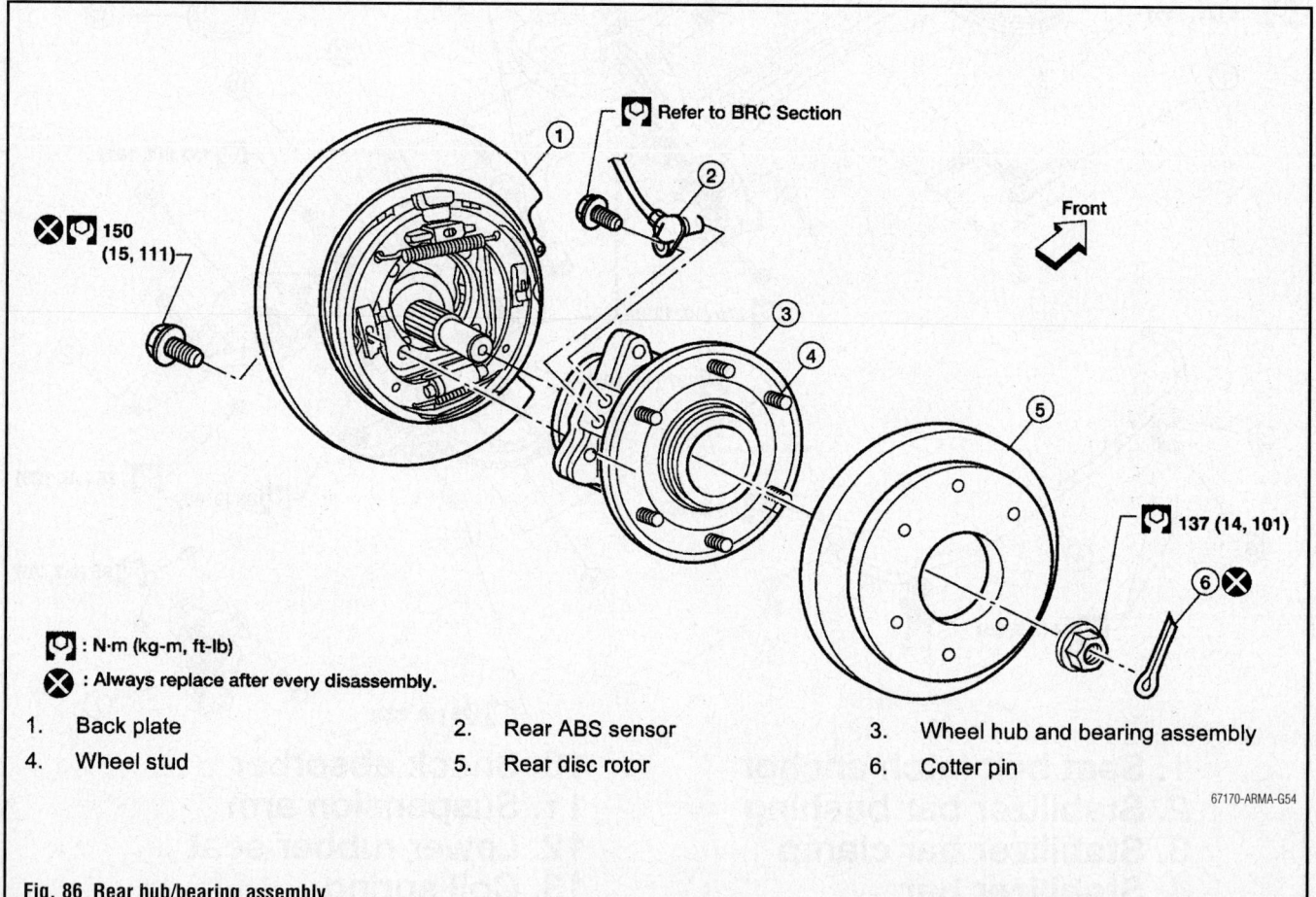

Refer to BRC Section

Front

150 (15, 111)

137 (14, 101)

[⚙] : N·m (kg-m, ft-lb)

[⊗] : Always replace after every disassembly.

1. Back plate
2. Rear ABS sensor
3. Wheel hub and bearing assembly
4. Wheel stud
5. Rear disc rotor
6. Cotter pin

67170-ARMA-G54

Fig. 86 Rear hub/bearing assembly

KIA

Amanti

SPECIFICATIONS AND MAINTENANCE CHARTS

ENGINE AND VEHICLE IDENTIFICATION

Engine							Model Year	
Code ①	Liters (cc)	Cu. In.	Cyl.	Fuel Sys.	Engine Type	Eng. Mfg.	Code ②	Year
4	3.5 (3497)	213.32	6	MPFI	DOHC	KIA	7	2007
5	3.8 (3778)	230.55	6	MPFI	DOHC	KIA	8	2008

MPFI: Multi-Point Fuel Injection

DOHC: Dual Overhead Camshafts

① 8th Digit of VIN

② 10th Digit of VIN

22140_AMAN_C0001

GENERAL ENGINE SPECIFICATIONS

Year	Engine Displacement Liters	Engine ID/VIN	Net Horsepower @ rpm	Net Torque @ rpm (ft. lbs.)	Bore x Stroke (in.)	Com-pression Ratio	Oil Pressure @ rpm
2007	3.8	5	200@5500	220@3500	3.78 x 3.43	10.4:1	18.8@1000
2008	3.8	5	200@5500	220@3500	3.78 x 3.43	10.4:1	18.8@1000

22140_AMAN_C0002

GASOLINE ENGINE TUNE-UP SPECIFICATIONS

Year	Engine Displacement Liters	Engine ID/VIN	Spark Plugs Gap (in.)	Ignition Timing (deg.) MT	AT	Fuel Pump (psi)	Idle Speed (rpm) MT	AT	Valve Clearance In.	Ex.
2007	3.8	5	0.039-0.043	—	10B	54-56	—	700	HYD	HYD
2008	3.8	5	0.039-0.043	—	10B	54-56	—	700	HYD	HYD

HYD: Hydraulic Valve Lifters

B: Before Top Dead Center

22140_AMAN_C0003

CAPACITIES

Year	Model	Engine Displacement Liters	Engine ID/VIN	Engine Oil with Filter	Transmission (pts.) 5–Spd	Transmission (pts.) Auto.	Fuel Tank (gal.)	Cooling System (qts.)
2007	Amanti	3.8	5	5.5	—	23.0	18.4	9.2
2008	Amanti	3.8	5	5.5	—	23.0	18.4	9.2

NOTE: All capacities are approximate. Add fluid gradually and check to be sure a proper fluid level is obtained.

22140_AMAN_C0004

FLUID SPECIFICATIONS

Year	Model	Engine Displ. Liters (VIN)	Engine Oil	Man. Trans.	Auto. Trans.	Drive Axle Front	Drive Axle Rear	Transfer Case	Power Steering Fluid	Brake Master Cylinder	Cooling System
2007	Amanti	3.8 (5)	①	—	②	—	—	—	PSF III	DOT 3 or 4	③
2008	Amanti	3.8 (5)	①	—	②	—	—	—	PSF III	DOT 3 or 4	③

DOT: Department Of Transpotation

① If 5W-20 is not available, 5W-30 can be used

② Diamond ATF SP-III or SK ATF SP III

③ Diamond ATF SP-III or SK ATF SP III

22140_AMAN_C0014

VALVE SPECIFICATIONS

Year	Engine Displacement Liters	Engine ID/VIN	Seat Angle (deg.)	Face Angle (deg.)	Spring Test Pressure (lbs. @ in.)	Spring Free Height (in.)	Stem-to-Guide Clearance (in.) Intake	Stem-to-Guide Clearance (in.) Exhaust	Stem Diameter (in.) Intake	Stem Diameter (in.) Exhaust
2007	3.8	5	45	45	53@1.492	1.7267	0.0008-0.0019	0.0012-0.0021	0.2151-0.2157	0.2149-0.2153
2008	3.8	5	45	45	53@1.492	1.7267	0.0008-0.0019	0.0012-0.0021	0.2151-0.2157	0.2149-0.2153

22140_AMAN_C0007

CAMSHAFT AND BEARING SPECIFICATIONS

All measurements are given in inches unless noted.

| Year | Engine Displacement Liters | Engine ID/VIN | Camshaft | | | Bearing Cap Torque (ft. lbs.) | |
			Cam Height Intake	Cam Height Exhaust	Shaft End-play	Outer	Inner
2007	3.8	5	1.8425	1.8031	0.0008-0.0071	① 7.2 - 8.7	① 7.2 - 8.7
2008	3.8	5	1.8425	1.8031	0.0008-0.0071	① 7.20 - 8.7	① 7.2 - 8.7

① Step 1: 7.2 ft. lbs.
  Step 2: 8.7 ft. lbs.

22140_AMAN_C0005

CRANKSHAFT AND CONNECTING ROD SPECIFICATIONS

All measurements are given in inches.

| Year | Engine Displacement Liters | Engine ID/VIN | Crankshaft | | | | Connecting Rod | | |
			Main Brg. Journal Dia.	Main Brg. Oil Clearance	Shaft End-play	Thrust on No.	Journal Diameter	Oil Clearance	Side Clearance
2007	3.8	5	2.7142-2.7149	0.0008-0.0016	0.0039-0.0110	3	2.2834-2.2842	0.0015-0.0022	0.0039-0.0098
2008	3.8	5	2.7142-2.7149	0.0008-0.0016	0.0039-0.0110	3	2.2834-2.2842	0.0015-0.0022	0.0039-0.0098

22140_AMAN_C0006

PISTON AND RING SPECIFICATIONS

| Year | Engine Displacement Liters | Engine ID/VIN | Piston Clearance | Ring Gap | | | Ring Side Clearance | | |
				Top Compression	Bottom Compression	Oil Control	Top Compression	Bottom Compression	Oil Control
2007	3.8	5	0.0012-0.0020	0.0067-0.0126	0.0126-0.0185	0.0078-0.0275	0.0012-0.0027	0.0012-0.0027	0.0024-0.0059
2008	3.8	5	0.0012-0.0020	0.0067-0.0126	0.0126-0.0185	0.0078-0.0275	0.0012-0.0027	0.0012-0.0027	0.0024-0.0059

NA: Not Available

22140_AMAN_C0008

TORQUE SPECIFICATIONS
All readings in ft. lbs.

Year	Engine Displacement Liters	Engine ID/VIN	Cylinder Head Bolts	Main Bearing Bolts	Rod Bearing Bolts	Crankshaft Damper Bolts	Flywheel Bolts	Manifold Intake	Exhaust	Spark Plugs	Oil Pan Drain Plug
2007	3.8	5	①	②	③	210-224	53-56	20-23	29-33	18-22	25-33
2008	3.8	5	①	②	③	210-224	53-56	20-23	29-33	18-22	25-33

① Step 1: 25 ft. lbs.
Step 2: Plus 92 degrees

② Step 1: 29 ft. lbs.
Step 2: Plus 120 degrees
Step 3: Plus 90 degrees

③ Inner bolts: Step 1: 36 ft. lbs.
Step 2: Plus 90 degrees
Outer bolts: Step 1: 14.46 ft. lbs.
Step 2: Plus 120 degrees
Side bolts: 21.70-23.14 ft. lbs.

④ Step 1: 14.5 ft. lbs.
Step 2: Plus 90 degrees

22140_AMAN_C0009

WHEEL ALIGNMENT

Year	Model		Caster Range (+/-Deg.)	Caster Preferred Setting (Deg.)	Camber Range (+/-Deg.)	Camber Preferred Setting (Deg.)	Toe-in (in.)
2007	Amanti	F	1.0	3.25	0.50	0	0 +/- 0.08
		R	—	—	0.50	-0.50	0.08 +/- 0.08
2008	Amanti	F	1.0	3.25	0.50	0	0 +/- 0.08
		R	—	—	0.50	-0.50	0.08 +/- 0.08

22140_AMAN_C0010

TIRE, WHEEL AND BALL JOINT SPECIFICATIONS

Year	Model	OEM Tires		Tire Pressures (psi)		Wheel Size	Ball Joint Inspection	Lug Nut Torque ①
		Standard	Optional	Front	Rear			
2007	Amanti	P215/65R16	P235/60R16	30	30	6.5Jx16	②	66-81
2008	Amanti	P215/65R16	P235/60R16	30	30	6.5Jx16	②	66-81

OEM: Original Equipment Manufacturer

PSI: Pounds Per Square Inch

① ft. lbs.

② If rotating torque exceeds 1.1 ft. lbs., replace the ball joint assembly

22140_AMAN_C0011

BRAKE SPECIFICATIONS

All measurements in inches unless noted

Year	Model		Brake Disc			Brake Drum Diameter			Minimum Lining Thickness		Brake Caliper	
			Original Thickness	Minimum Thickness	Maximum Run-out	Original Inside Diameter	Max. Wear Limit	Maximum Machine Diameter	Front	Rear	Bracket Bolts (ft. lbs.)	Mounting Bolts (ft. lbs.)
2007	Amanti	F	1.100	1.040	0.002	—	—	—	0.079	—	49-61	16-23
		R	0.390	0.310	0.002	—	—	—	—	0.080	36-43	16-24
2008	Amanti	F	1.100	1.040	0.002	—	—	—	0.079	—	49-61	16-23
		R	0.390	0.310	0.002	—	—	—	—	0.080	36-43	16-24

F: Front

R: Rear

22140_AMAN_C0012

SCHEDULED MAINTENANCE INTERVALS
Kia Amanti 2007 - 08

TO BE SERVICED	TYPE OF SERVICE	VEHICLE MILEAGE INTERVAL (x1000)												
		7.5	15	22.5	30	37.5	45	52.5	60	67.5	75	82.5	90	97.5
Accessory drive belts	S/I			✓			✓			✓			✓	
Air cleaner filter	R		✓		✓		✓		✓		✓		✓	
Air conditioner system	S/I													
Brake lines, hoses and connections	S/I		✓		✓		✓		✓		✓		✓	
Chassis and body fasteners	T	✓	✓	✓	✓	✓	✓	✓	✓	✓	✓	✓	✓	✓
Cooling system hoses and coolant level	S/I		✓		✓		✓		✓		✓		✓	
CV-joint boots	S/I				✓				✓				✓	
Engine coolant	R				✓				✓				✓	
Engine oil and filter	R	✓	✓	✓	✓	✓	✓	✓	✓	✓	✓	✓	✓	✓
Exhaust system heat shields	S/I				✓				✓				✓	
Front and rear brakes	S/I				✓				✓				✓	
Front ball joints	S/I				✓				✓				✓	
Fuel filter	R								✓					
Fuel lines and hoses	S/I				✓				✓				✓	
Locks and hinges	L	✓	✓	✓	✓	✓	✓	✓	✓	✓	✓	✓	✓	✓
Spark plugs	R				✓				✓				✓	
Steering operation and linkage	S/I				✓				✓				✓	
Timing belt	R								✓					

R: Replace   S/I: Service or Inspect   L: Lubricate   T: Tighten

FREQUENT OPERATION MAINTENANCE (SEVERE SERVICE)

If a vehicle is operated under any of the following conditions it is considered severe service

- Towing a trailer or using a camper or car-top carrier

- Repeated short trips of less than 5 miles in temperatures below freezing, or trips of less than 10 miles in any temperature

- Prolonged idling (vehicle operation in stop and go traffic).

- Operating on rough, muddy, unpaved, dusty or salt-covered roads.

- Police, taxi, delivery usage or trailer towing usage.

- Driving in extremely hot (over 90°F) conditions

Oil & oil filter: change every 5000 miles or 5 months, whichever occurs first.

Air cleaner filter: inspect every 15,000 miles or 15 months and replace everything 30,000 miles or 30 months, whichever occur

Fuel system hoses (California models only): replace every 105,000 miles

Emission system hoses (non-CA models): inspect every 55,000 or 55 months, whichever occurs first

Emission system hoses (CA models): inspect every 60,000 miles or 60 months, which occurs first

Front and rear brakes: inspect every 15,000 miles or 15 months, whichever occurs first

Chassis and body fasteners: tighten every 15,000 miles or 15 months, whichever occurs first

Locks and hinges: lubricate every 5000 miles or 5 months, whichever occurs first

22140_AMAN_C0013

PRECAUTIONS

Before servicing any vehicle, please be sure to read all of the following precautions, which deal with personal safety, prevention of component damage, and important points to take into consideration when servicing a motor vehicle:

• Never open, service or drain the radiator or cooling system when the engine is hot; serious burns can occur from the steam and hot coolant.

• Observe all applicable safety precautions when working around fuel. Whenever servicing the fuel system, always work in a well-ventilated area. Do not allow fuel spray or vapors to come in contact with a spark, open flame, or excessive heat (a hot drop light, for example). Keep a dry chemical fire extinguisher near the work area. Always keep fuel in a container specifically designed for fuel storage; also, always properly seal fuel containers to avoid the possibility of fire or explosion. Refer to the additional fuel system precautions later in this section.

• Fuel injection systems often remain pressurized, even after the engine has been turned **OFF**. The fuel system pressure must be relieved before disconnecting any fuel lines. Failure to do so may result in fire and/or personal injury.

• Brake fluid often contains polyglycol ethers and polyglycols. Avoid contact with the eyes and wash your hands thoroughly after handling brake fluid. If you do get brake fluid in your eyes, flush your eyes with clean, running water for 15 minutes. If eye irritation persists, or if you have taken

brake fluid internally, IMMEDIATELY seek medical assistance.

• The EPA warns that prolonged contact with used engine oil may cause a number of skin disorders, including cancer. You should make every effort to minimize your exposure to used engine oil. Protective gloves should be worn when changing oil. Wash your hands and any other exposed skin areas as soon as possible after exposure to used engine oil. Soap and water, or waterless hand cleaner should be used.

• All new vehicles are now equipped with an air bag system, often referred to as a Supplemental Restraint System (SRS) or Supplemental Inflatable Restraint (SIR) system. The system must be disabled before performing service on or around system components, steering column, instrument panel components, wiring and sensors. Failure to follow safety and disabling procedures could result in accidental air bag deployment, possible personal injury and unnecessary system repairs.

• Always wear safety goggles when working with, or around, the air bag system. When carrying a non-deployed air bag, be sure the bag and trim cover are pointed away from your body. When placing a non-deployed air bag on a work surface, always face the bag and trim cover upward, away from the surface. This will reduce the motion of the module if it is accidentally deployed. Refer to the additional air bag system precautions later in this section.

• Clean, high quality brake fluid from a sealed container is essential to the safe and

proper operation of the brake system. You should always buy the correct type of brake fluid for your vehicle. If the brake fluid becomes contaminated, completely flush the system with new fluid. Never reuse any brake fluid. Any brake fluid that is removed from the system should be discarded. Also, do not allow any brake fluid to come in contact with a painted surface; it will damage the paint.

• Never operate the engine without the proper amount and type of engine oil; doing so WILL result in severe engine damage.

• Timing belt maintenance is extremely important. Many models utilize an interference-type, non-freewheeling engine. If the timing belt breaks, the valves in the cylinder head may strike the pistons, causing potentially serious (also time-consuming and expensive) engine damage. Refer to the maintenance interval charts for the recommended replacement interval for the timing belt, and to the timing belt section for belt replacement and inspection.

• Disconnecting the negative battery cable on some vehicles may interfere with the functions of the on-board computer system(s) and may require the computer to undergo a relearning process once the negative battery cable is reconnected.

• When servicing drum brakes, only disassemble and assemble one side at a time, leaving the remaining side intact for reference.

• Only an MVAC-trained, EPA-certified automotive technician should service the air conditioning system or its components.

BRAKES

GENERAL INFORMATION

The ABS system detects wheel revolution while braking and improves handling stability during sudden braking by electrically preventing wheel lockup. Maneuverability is also improved for avoiding obstacles.

BLEEDING THE BRAKE SYSTEM

BLEEDING PROCEDURE
See Figures 1 and 2.

✳✳ WARNING

Clean, high quality brake fluid is essential to the safe and proper operation of the brake system. You should always buy the highest quality brake fluid that is available. If the brake

fluid becomes contaminated, drain and flush the system, then refill the master cylinder with new fluid. Never reuse any brake fluid. Any brake fluid that is removed from the system should be discarded. Also, do not allow any brake fluid to come in contact with a painted surface; it will damage the paint.

✳✳ CAUTION

Brake fluid contains polyglycol ethers and polyglycols. Avoid contact with the eyes and wash your hands thoroughly after handling brake fluid. If you do get brake fluid in your eyes, flush your eyes with clean, running water for 15 minutes. If eye irritation

ANTI-LOCK BRAKE SYSTEM (ABS)

persists, or if you have taken brake fluid internally, IMMEDIATELY seek medical assistance.

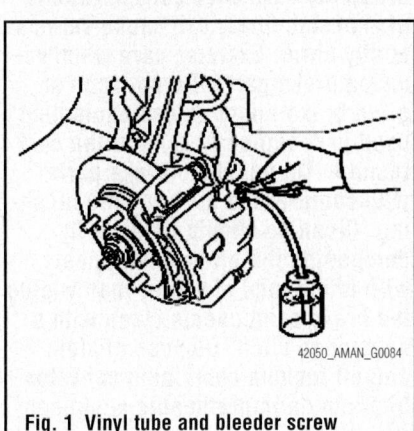

42050_AMAN_G0084

Fig. 1 Vinyl tube and bleeder screw

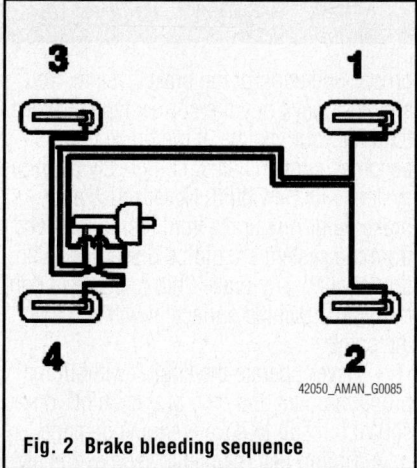

Fig. 2 Brake bleeding sequence

1. Remove the reservoir cap and fill the brake reservoir with brake fluid.

2. Connect a vinyl tube to the wheel cylinder bleeder screw and insert the other end of the tube in a clear container.

3. Slowly depress the brake pedal several times.

4. While depressing the brake pedal fully, loosen the bleeder screw until fluid runs out. Then close the bleeder screw and release the brake pedal.

5. Repeat these steps until there are no more bubbles in the fluid escaping to the clear container.

6. Tighten the bleeder screw to specification: 60–84 inch lbs. (7–9 Nm).

7. Repeat the above procedure for each wheel in the sequence shown in the illustration.

BLEEDING THE ABS SYSTEM

✳✳ CAUTION

Brake fluid contains polyglycol ethers and polyglycols. Avoid contact with the eyes and wash your hands thoroughly after handling brake fluid. If you do get brake fluid in your eyes, flush your eyes with clean, running water for 15 minutes. If eye irritation persists, or if you have taken brake fluid internally, IMMEDIATELY seek medical assistance.

✳✳ WARNING

Clean, high quality brake fluid is essential to the safe and proper operation of the brake system. You should always buy the highest quality brake fluid that is available. If the brake fluid becomes contaminated, drain and flush the system, then refill the master cylinder with new fluid. Never reuse any brake fluid. Any brake fluid that is removed from the system should be discarded. Also, do not allow any brake fluid to come in contact with a painted surface; it will damage the paint.

1. Remove the reservoir cap and fill the brake reservoir with brake fluid.

2. Connect a clear plastic tube to the wheel cylinder bleeder plug and insert the other end of the tube into a clear plastic bottle.

3. Connect the Hi-Scan (Pro) to the Data Link Connector located underneath the dash panel.

4. Select and operate according to the instructions on the Hi-Scan (Pro) screen:

- Select KIA vehicle diagnosis
- Select vehicle name
- Select Anti-Lock Brake system

✳✳ CAUTION

You must obey the maximum operating time of the ABS motor with the Hi-Scan (Pro) to prevent the motor pump from burning.

- Select air bleeding mode
- Press **YES** to operate motor pump and solenoid valve
- Wait 60 seconds before operating the air bleeding, otherwise may damage the motor

5. Pump the brake pedal several times, and then loosen the bleeder screw until fluid starts to run out without bubbles. Then close the bleeder screw.

6. Repeat step 5 until there are no more bubbles in the fluid for each wheel.

7. Tighten the bleeder screw to specification: 60–79 inch lbs. (7–9 Nm).

8. After completion of the repair or correction of the problem, erase the stored fault codes using the clear key on the Hi-Scan (Pro).

9. Disconnect the Hi-Scan (Pro).

10. Fill the brake reservoir with the proper amount of brake fluid.

BRAKES

✳✳ CAUTION

Dust and dirt accumulating on brake parts during normal use may contain asbestos fibers from production or aftermarket brake linings. Breathing excessive concentrations of asbestos fibers can cause serious bodily harm. Exercise care when servicing brake parts. Do not sand or grind brake lining unless equipment used is designed to contain the dust residue. Do not clean brake parts with compressed air or by dry brushing. Cleaning should be done by dampening the brake components with a fine mist of water, then wiping the brake components clean with a dampened cloth. Dispose of cloth and all residue containing asbestos fibers in an impermeable container with the appropriate label. Follow practices prescribed by the Occupational Safety and Health Administration (OSHA) and the Environmental Protection Agency (EPA) for the handling, processing, and disposing of dust or debris that may contain asbestos fibers.

BRAKE CALIPER

REMOVAL & INSTALLATION

1. Before servicing the vehicle, refer to the Precautions Section.

2. Remove or disconnect the following:

- Wheel
- Brake hose at the caliper
- Caliper mounting bolts
- Caliper

FRONT DISC BRAKES

To install:

3. Install or connect the following:

- Caliper. Tighten the mounting bolts to 49–61 ft. lbs. (69–85 Nm).
- Brake line to the caliper. Torque the brake line bolt to 12–14 ft. lbs. (17–20 Nm).

4. Bleed the system.

5. Install the wheel.

DISC BRAKE PADS

REMOVAL & INSTALLATION

1. Before servicing the vehicle, refer to the Precautions Section.

2. Remove or disconnect the following:

- Front wheel
- Caliper mounting bolt
- Suspend the caliper from a wire

- Pads from the caliper support
- Pad retainers, if necessary

To install:
3. Install or connect the following:

- Pad retainers, if removed
- Pads onto the pad retainers

4. Compress the caliper piston using a C-clamp. Rotate the caliper downward and

install the mounting bolt. Tighten to 16–24 ft. lbs. (22–32 Nm).
5. Install the wheel.

BRAKES

❊❊ CAUTION

Dust and dirt accumulating on brake parts during normal use may contain asbestos fibers from production or aftermarket brake linings. Breathing excessive concentrations of asbestos fibers can cause serious bodily harm. Exercise care when servicing brake parts. Do not sand or grind brake lining unless equipment used is designed to contain the dust residue. Do not clean brake parts with compressed air or by dry brushing. Cleaning should be done by dampening the brake components with a fine mist of water, then wiping the brake components clean with a dampened cloth. Dispose of cloth and all residue containing asbestos fibers in an impermeable container with the appropriate label. Follow practices prescribed by the Occupational Safety and Health Administration (OSHA) and the Environmental

Protection Agency (EPA) for the handling, processing, and disposing of dust or debris that may contain asbestos fibers.

BRAKE CALIPER

REMOVAL & INSTALLATION

1. Before servicing the vehicle, refer to the Precautions Section.
2. Release the parking brake.
3. Remove or disconnect the following:
   - Wheel
   - Brake line at the caliper
   - Caliper mounting bolts
   - Caliper

To install:
4. Install or connect the following:
   - Caliper onto its mounting
   - Mounting bolts. Torque the bolts to 16–24 ft. lbs. (22–32 Nm).
   - Brake line to the caliper. Torque the brake line bolt to 12–14 ft. lbs. (17–20 Nm).

REAR DISC BRAKES

5. Bleed the system.
6. Install the wheel.

DISC BRAKE PADS

REMOVAL & INSTALLATION

1. Before servicing the vehicle, refer to the Precautions Section.
2. Remove or disconnect the following:
   - Rear wheel
   - Caliper mounting bolt
   - Suspend the caliper from a wire
   - Pads from the caliper support
   - Pad retainers, if necessary

To install:
3. Install or connect the following:
   - Pad retainers, if removed
   - Pads onto the pad retainers

4. Compress the caliper piston using a C-clamp. Rotate the caliper downward and install the mounting bolt. Tighten to 16–24 ft. lbs. (22–32 Nm).
5. Install the wheel.

CHASSIS ELECTRICAL

GENERAL INFORMATION

❊❊ CAUTION

Some vehicles are equipped with an air bag system. The system must be disarmed before performing service on, or around, system components, the steering column, instrument panel components, wiring and sensors. Failure to follow the safety precautions and the disarming procedure could result in accidental air bag deployment, possible injury and unnecessary system repairs.

SERVICE PRECAUTIONS

Disconnect and isolate the battery negative cable before beginning any airbag system component diagnosis, testing, removal, or installation procedures. Allow system capacitor to discharge for two minutes before beginning any component service. This will disable the airbag system. Failure to disable the airbag system may result in

AIR BAG (SUPPLEMENTAL RESTRAINT SYSTEM)

accidental airbag deployment, personal injury, or death.

Do not place an intact undeployed airbag face down on a solid surface. The airbag will propel into the air if accidentally deployed and may result in personal injury or death.

When carrying or handling an undeployed airbag, the trim side (face) of the airbag should be pointing towards the body to minimize possibility of injury if accidental deployment occurs. Failure to do this may result in personal injury or death.

Replace airbag system components with OEM replacement parts. Substitute parts may appear interchangeable, but internal differences may result in inferior occupant protection. Failure to do so may result in occupant personal injury or death.

Wear safety glasses, rubber gloves, and long sleeved clothing when cleaning powder residue from vehicle after an airbag deployment. Powder residue emitted from a deployed airbag can cause skin irritation. Flush affected area with cool water if irritation is experienced. If nasal or throat irrita-

tion is experienced, exit the vehicle for fresh air until the irritation ceases. If irritation continues, see a physician.

Do not use a replacement airbag that is not in the original packaging. This may result in improper deployment, personal injury, or death.

The factory installed fasteners, screws and bolts used to fasten airbag components have a special coating and are specifically designed for the airbag system. Do not use substitute fasteners. Use only original equipment fasteners listed in the parts catalog when fastener replacement is required.

During, and following, any child restraint anchor service, due to impact event or vehicle repair, carefully inspect all mounting hardware, tether straps, and anchors for proper installation, operation, or damage. If a child restraint anchor is found damaged in any way, the anchor must be replaced. Failure to do this may result in personal injury or death.

Deployed and non-deployed airbags may or may not have live pyrotechnic material within the airbag inflator.

Do not dispose of driver/passenger/curtain airbags or seat belt tensioners unless you are sure of complete deployment. Refer to the Hazardous Substance Control System for proper disposal.

Dispose of deployed airbags and tensioners consistent with state, provincial, local, and federal regulations.

After any airbag component testing or service, do not connect the battery negative cable. Personal injury or death may result if the system test is not performed first.

If the vehicle is equipped with the Occupant Classification System (OCS), do not connect the battery negative cable before performing the OCS Verification Test using the scan tool and the appropriate diagnostic information. Personal injury or death may result if the system test is not performed properly.

Never replace both the Occupant Restraint Controller (ORC) and the Occupant Classification Module (OCM) at the same time. If both require replacement, replace one, then perform the Airbag System test before replacing the other.

Both the ORC and the OCM store Occupant Classification System (OCS) calibration data, which they transfer to one another when one of them is replaced. If both are replaced at the same time, an irreversible fault will be set in both modules and the OCS may malfunction and cause personal injury or death.

If equipped with OCS, the Seat Weight Sensor is a sensitive, calibrated unit and must be handled carefully. Do not drop or handle roughly. If dropped or damaged, replace with another sensor. Failure to do so may result in occupant injury or death.

If equipped with OCS, the front passenger seat must be handled carefully as well. When removing the seat, be careful when setting on floor not to drop. If dropped, the sensor may be inoperative, could result in occupant injury, or possibly death.

If equipped with OCS, when the passenger front seat is on the floor, no one should sit in the front passenger seat. This uneven force may damage the sensing ability of the seat weight sensors. If sat on and damaged, the sensor may be inoperative, could result in occupant injury, or possibly death.

DISARMING THE SYSTEM

1. Before servicing the vehicle, refer to the Precautions Section.
2. Disconnect and isolate the negative battery cable. Wait 3 minutes for the system capacitor to discharge before performing any service.

➡ Wait at least 3 minutes before working on the vehicle. The air bag system is designed to retain enough power to deploy the air bag for a short time after the battery has been disconnected.

ARMING THE SYSTEM

1. After repairs are complete, connect the negative battery cable. Turn the ignition switch to the **ON** position and check that the air bag warning light blinks as it would for normal operation.

CLOCKSPRING CENTERING

The clockspring is under the steering wheel. It ensures a positive connection between the steering column wiring harness and whatever controls are on the steering wheel, and especially the airbag igniter.

1. Before servicing the vehicle, refer to the Precautions Section.
2. Disconnect negative battery cable.
3. Remove ignition key from vehicle
4. Remove the Steering Wheel and Air Bag Module. Refer to Steering Wheel removal and installation.
5. Set the clockspring in the center position. Make certain front wheels are in the straight-ahead position.
6. Make sure the mating mark of the clockspring is properly aligned.

✳✳ CAUTION

If the mating mark is not properly aligned, the steering wheel may not completely rotate during a turn, or the flat cable within the clockspring may be broken causing an obstruction of the normal operation of the SRS and possibly lead to a serious injury to the driver of the vehicle.

DRIVETRAIN

AUTOMATIC TRANSAXLE ASSEMBLY

REMOVAL & INSTALLATION

See Figures 3 through 12.

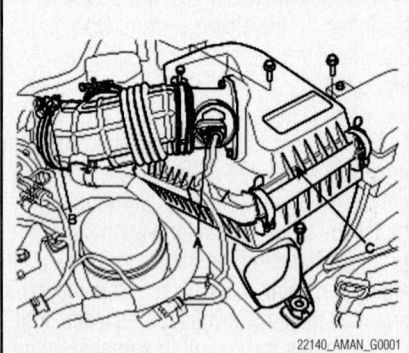

22140_AMAN_G0001

Fig. 3 Remove the air cleaner assembly (C) after disconnecting the air flow sensor connector (A) and loosening the clamp (B)

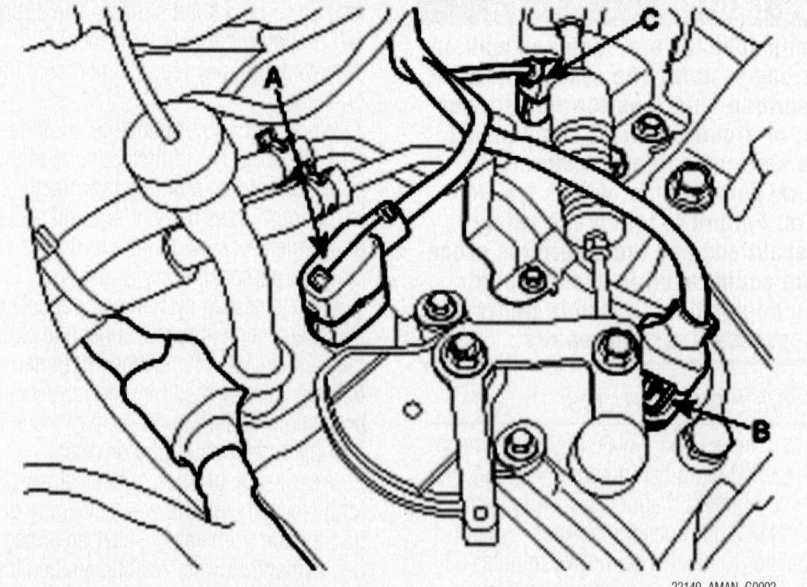

22140_AMAN_G0002

Fig. 4 Disconnect the inhibitor switch connector (A), the solenoid valve connector (B) and the input shaft speed sensor connector (C)

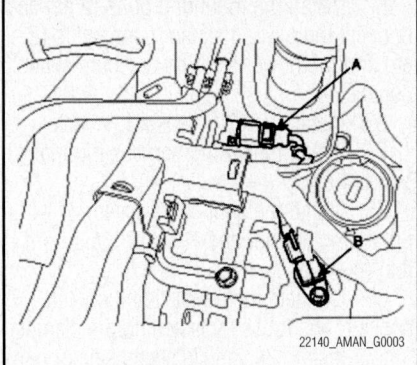

Fig. 5 Disconnect the vehicle speed sensor connector (A) and output shaft speed sensor connector (B)

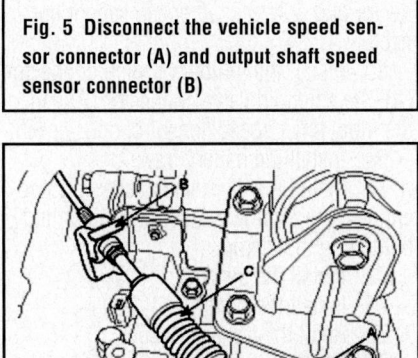

Fig. 6 Remove the shift cable assembly (C) by removing the nut (A) and clip (B)

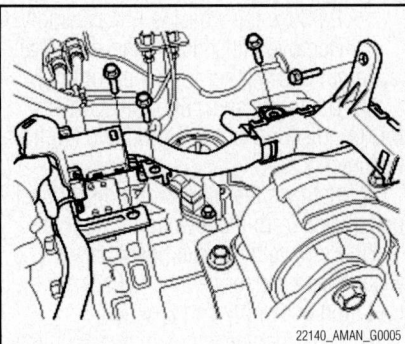

Fig. 7 Remove the wiring harness mounting bolts

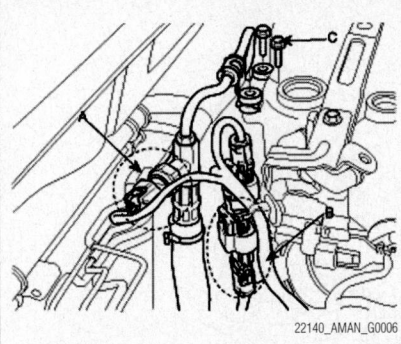

Fig. 8 Disconnect the oxygen sensor connectors (B-2 ea) and the power steering pressure sensor connector (A) and remove the mounting bolts

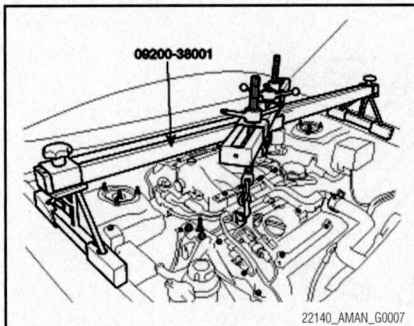

Fig. 9 Using the special tool (09200-38001), hold the engine and transaxle assembly safely

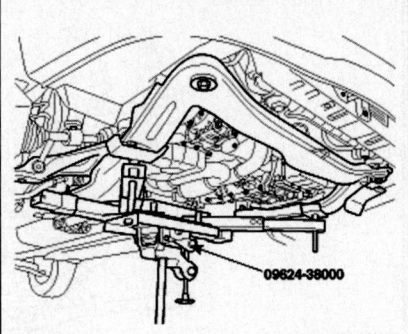

Fig. 10 Remove the mounting bolts from the sub frame by supporting the sub frame by using the special tool(09624-38000)

1. Remove the engine cover.
2. Remove the battery.
3. Remove the air duct.
4. Remove the air cleaner assembly (C) after disconnecting the air flow sensor connector (A) and loosening the clamp (B).
5. Remove the battery tray.
6. Disconnect the inhibitor switch connector (A), the solenoid valve connector (B) and the input shaft speed sensor connector (C).
7. Disconnect the vehicle speed sensor connector (A) and output shaft speed sensor connector (B).
8. Disconnect the transaxle oil cooler hoses from the tubes by loosening the clamps.
9. Remove the shift cable assembly (C) by removing the nut (A) and clip (B).
10. Remove the wiring harness mounting bolts (4 ea).
11. Remove the Crankshaft Position (CKP) sensor connector.
12. Disconnect the oxygen sensor connectors (B-2 ea) and the power steering pressure sensor connector (A) and remove the mounting bolts (C-2 ea).
13. Remove the transaxle mounting bolts (3 ea).
14. Using the special tool(09200-38001), hold the engine and transaxle assembly safely.
15. Remove the transaxle support bracket bolts (4 ea) and nuts (2 ea).
16. Remove the power steering column joint bolt.
17. Remove the front wheels.
18. Remove the undercover.
19. Drain the transaxle fluid by removing the drain plug.

20. Drain power steering fluid through the return tube.
21. Disconnect the power steering pressure tube from the power steering oil pump.
22. Remove the fork from the front lower arm.
23. Disconnect the lower arm, the tie rod end ball joint, the stabilizer bar link from the front knuckle.
24. Remove the front roll stopper mounting bolt.
25. Remove the muffler rubber hanger.
26. Remove the rear roll stopper mounting bolts (4 ea) and the power steering tube mounting bolt.
27. Remove the mounting bolts from the sub frame by supporting the sub frame by using the special tool(09624-38000).
28. Remove the front muffler assembly.
29. Remove the front roll support from the transaxle.
30. Remove the inner shaft bracket bolts (3 ea).
31. Remove halfshafts from transaxle.
32. Remove the left side cover.
33. Remove the starter motor mounting bolts and the two bolts.
34. Remove the cover.
35. Remove the drive plate bolts and the transaxle lower mounting bolts (6 ea).
36. Lifting the vehicle up and lowering the jack slowly, remove the transaxle assembly.

To install:
37. Installation is in the reverse order of removal. Perform the following :
   a. Adjust the shift cable.
   b. Refill the transaxle with fluid.
   c. Refill the radiator with engine coolant.
   d. Bleed air from the cooling system with the heater valve open.

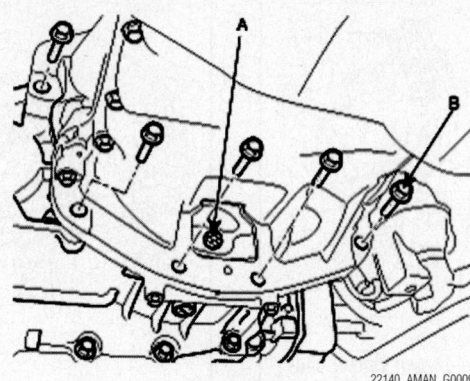

Fig. 11 Tighten the transaxle lower mounting bolts (B, C). Install the drive plate bolts (A) by turning the crankshaft

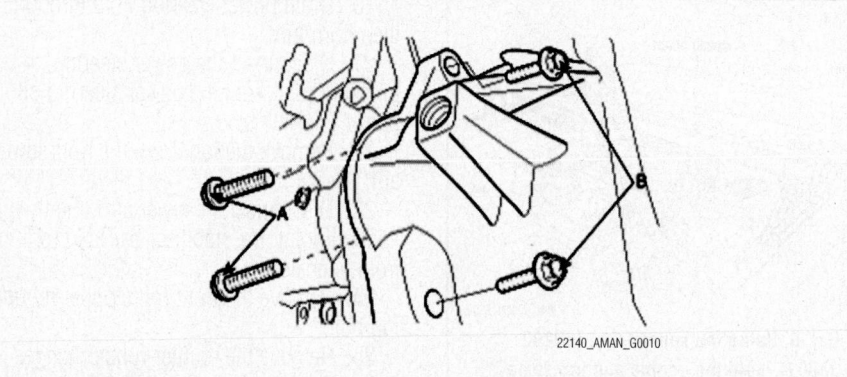

Fig. 12 Install the starter motor mounting bolts (A) and the two bolts (B)

e. Clean the battery posts and cable terminals with sandpaper, assemble them, and apply grease to prevent corrosion.

38. Using a transmission jack, install the transaxle assembly.

39. Tighten the transaxle lower mounting bolts (B-4 ea, C-1 ea). Torque: (B) 29–34 ft. lbs. (40–47 Nm); (C) 58–72 ft. lbs. (80–100 Nm)

40. Install the drive plate bolts (A) by turning the crankshaft. Torque: 33–38 ft. lbs. (46–53 Nm).

41. Install the cover.

42. Install the starter motor mounting bolts (A) and the two bolts (B). Torque: (A) 31–39 ft. lbs. (43–55 Nm) (B) 24–36 ft. lbs. (33–50 Nm).

43. Install the left side cover.

44. After removing the jack, insert the halfshafts.

45. Install the inner shaft bracket bolts (3 ea).

46. Install the front roll support to the transaxle. Torque: 43–57 ft. lbs. (60–80 Nm).

47. Install the front muffler assembly.

48. Install the sub frame supported by the special tool (09624-38000).

49. Tighten the rear roll stopper mounting bolts (4 ea) and the power steering tube mounting bolt.

50. Install the muffler rubber hanger.

51. Tighten the front roll stopper mounting bolt.

52. Install the fork from the front lower arm.

53. Connect the lower arm, the tie rod end ball joint, the stabilizer bar link to the front knuckle.

54. Connect the power steering pressure tube to the power steering oil pump. Torque: 39–47 ft. lbs. (55–65 Nm).

55. Connect the return tube with a clamp.

56. Install the undercover.

57. Install the front wheels and tires.

58. Install the steering column joint bolt.

59. Tighten the transaxle insulator mounting bolt (4 ea) and nuts (2 ea). Torque: 43–58 ft. lbs. (60–80 Nm).

60. Remove the special tool (09200-38001) holding the engine and transaxle assembly.

61. Tighten the transaxle mounting bolts (3 ea). Torque: 47–61 ft. lbs. (65–85 Nm).

62. Install the mounting bolts (2 ea) and connect the oxygen sensor connectors (2 ea) and the power steering pressure sensor connector.

63. Install the CKP sensor connector.

64. Install the wiring harness mounting bolts (4 ea).

65. Install the shift cable assembly by tightening the nut and clip. Torque: 6–9 ft. lbs. (8–12 Nm).

66. Connect the transaxle oil cooler hoses to the tubes by fastening the clamps.

67. Install the vehicle speed sensor connector and output shaft speed sensor connector.

68. Install the inhibitor switch connector (A), the solenoid valve connector (B) and the input shaft speed sensor connector (C).

69. Install the battery tray.

70. Install the air cleaner assembly and connect the air flow sensor connector and tightening the clamp.

71. Install the air duct.

72. Install the battery.

73. Refill the transaxle fluid.

74. Refill the power steering fluid and bleed the air.

75. Install the engine cover.

FRONT HALFSHAFT

REMOVAL & INSTALLATION

See Figure 13.

1. Remove the wheel & tire assembly.

2. Remove the spilt pin and driveshaft castle nut and washer from the front hub.

3. Using the special tool (09568-4A000), disconnect the tie rod end (A) from the knuckle.

4. Remove the 2 bolts and disconnect the knuckle from the lower arm assembly.

5. Remove the brake hose bracket and wheel speed sensor cable bracket from the front strut assembly and knuckle.

6. Using a plastic hammer, disconnect the driveshaft from the axle assembly.

7. Removing the left hand driveshaft (A) from the transaxle by using a pry bar (C) as shown.

8. For the right hand driveshaft, do the following:

a. Remove the stabilizer link from the fork.

b. Remove the fork from the front lower arm.

c. Remove the fork from the front strut assembly.

d. Remove the inner shaft heat over and the heat cover mounting bolts.

e. Remove the inner shaft bracket mounting bolts.

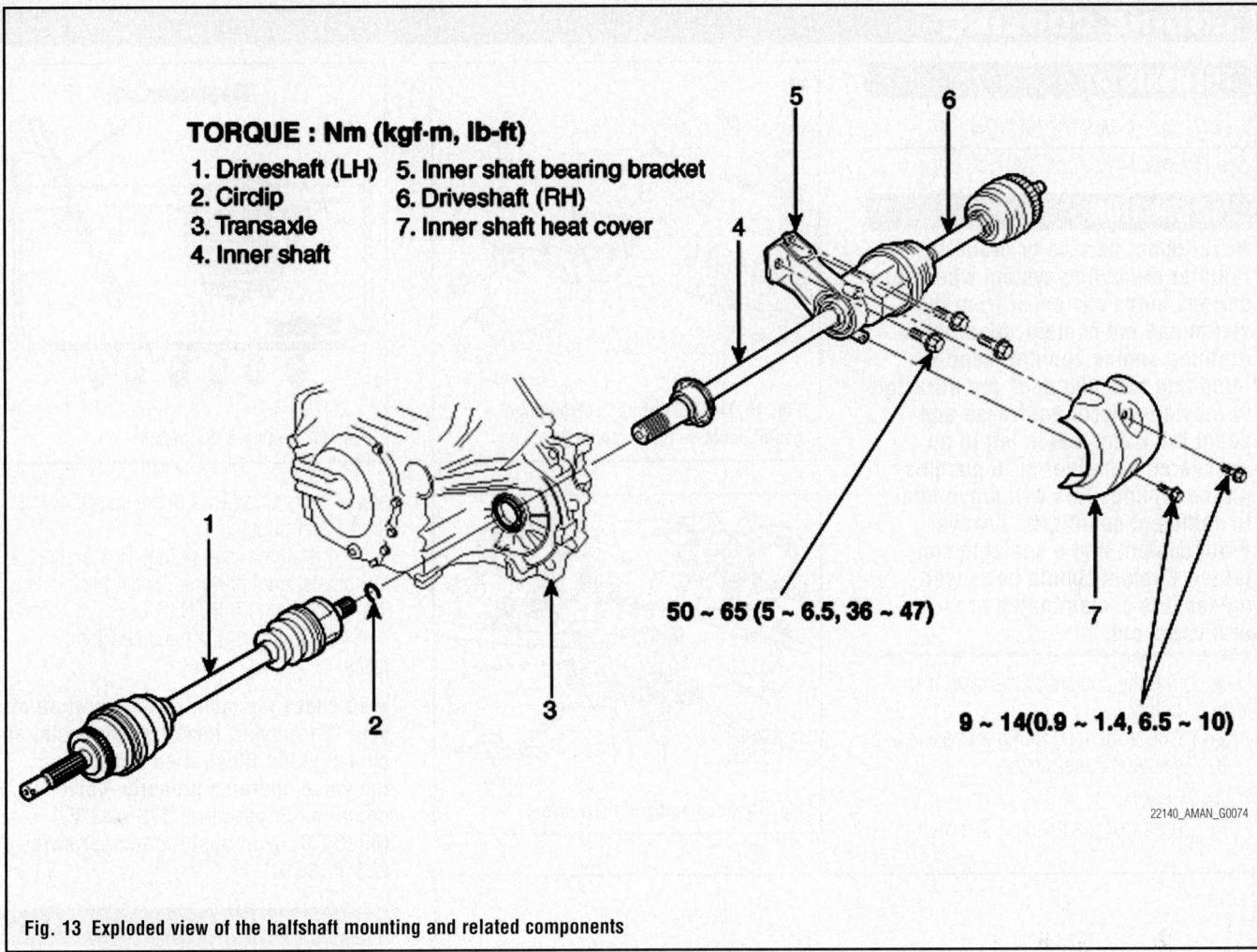

TORQUE : Nm (kgf·m, lb-ft)

1. Driveshaft (LH)
2. Circlip
3. Transaxle
4. Inner shaft
5. Inner shaft bearing bracket
6. Driveshaft (RH)
7. Inner shaft heat cover

50 ~ 65 (5 ~ 6.5, 36 ~ 47)

9 ~ 14(0.9 ~ 1.4, 6.5 ~ 10)

22140_AMAN_G0074

Fig. 13 Exploded view of the halfshaft mounting and related components

f. Remove the front driveshaft assembly with the inner shaft from the transaxle.

> **✳✳ CAUTION**
>
> **Do not try to disconnect the inner shaft from the driveshaft. Because they cannot be disconnected once assembled. Do not reuse the driveshaft which is disassembled from the inner shaft.**

9. Using the special tool (09432-11000), remove the tone wheel.

To install:

> **✳✳ CAUTION**
>
> **Replace the circlip with new ones after removal.**

10. Apply gear oil on the drive shaft splines and the contacting surface of differential case oil seal.

11. After installation, check if the drive shaft cannot be removed.

12. For the right hand driveshaft, do the following:

a. Install the inner shaft bearing bracket mounting bolt (A). Tightening Torque: 36–47 ft. lbs. (50–65 Nm).

b. Install the inner shaft heat cover by installing the heat cover mounting bolts. Tightening Torque: 6.5–10 ft. lbs. (9–14 Nm).

c. Install the fork to the front strut assembly. Tightening Torque: 44–59 ft. lbs. (60–80 Nm).

d. Install the connecting bolt between the fork and the lower arm. Tightening Torque: 101–118 ft. lbs. (140–160 Nm).

e. Install the stabilizer link to the fork. Tightening Torque: 74–88 ft. lbs. (100–120 Nm).

13. Install the drive shaft into the front axle assembly.

14. Install the knuckle in the lower arm assembly and tighten the bolts. Tightening Torque: 74–88 ft. lbs. (100–120 Nm).

15. Install the tie rod end in the knuckle. Tightening Torque: 18–25 ft. lbs. (24–34 Nm)

16. Install the brake hose bracket and wheel speed sensor cable bracket to the front strut assembly and knuckle.

17. After installing the washer with convex surface outward, install the castle nut and the spilt pin. Tightening Torque: 148–207 ft. lbs. (200–280 Nm).

18. Install the wheel & tire assembly.

CV-BOOTS INSPECTION

1. Before servicing the vehicle, refer to the Precautions Section.

2. Check the driveshaft boots for damage and deterioration.
- Raise front of vehicle
- Rotate axle and inspect for cracked or ripped CV boot material on inner and outer CV joints on both sides of vehicle

3. Replace boot if damaged or deteriorated.

ENGINE COOLING

THERMOSTAT

REMOVAL & INSTALLATION

See Figures 14 through 17.

> ❊ **CAUTION**
>
> Never open, service or drain the radiator or cooling system when hot; serious burns can occur from the steam and hot coolant. Also, when draining engine coolant, keep in mind that cats and dogs are attracted to ethylene glycol antifreeze and could drink any that is left in an uncovered container or in puddles on the ground. This will prove fatal in sufficient quantities. Always drain coolant into a sealable container. Coolant should be reused unless it is contaminated or is several years old.

1. Drain the coolant to thermostat level or below.
2. Remove the inlet fitting and gasket.
3. Remove the thermostat.

To install:

4. Check that the flange of the thermo-

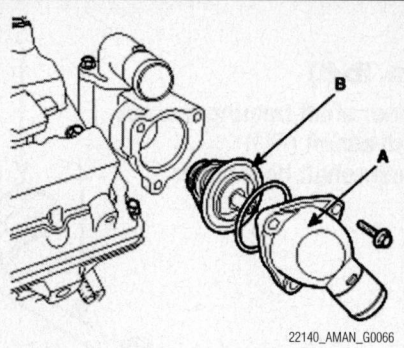

Fig. 15 Thermostat and surrounding components — 3.8L engine

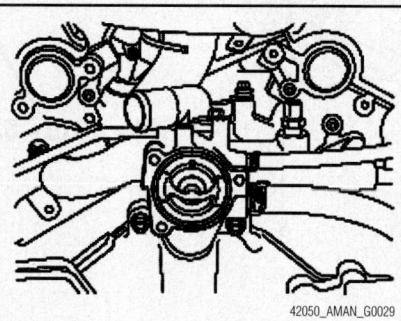

Fig. 16 Positioning of thermostat

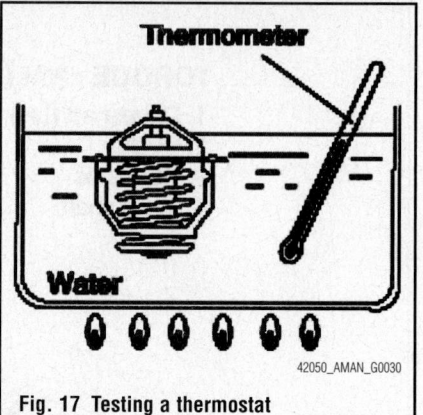

Fig. 17 Testing a thermostat

stat is correctly placed in the socket of the thermostat housing.

5. Install the inlet fitting. Tighten the thermostat inlet fitting bolts to specification: 12–14 ft. lbs. (17–20 Nm).
6. Refill the coolant and check for leaks.

➡ To check the opening temperature of your thermostat, heat the thermostat as shown in the illustration. Check that the valve operates properly. Valve opening temperature: 176–183°F (80–84°C). Full opening temperature: 203°F (95°C).

WATER PUMP

REMOVAL & INSTALLATION

1. Drain the engine coolant.

> ❊ **WARNING**
>
> System is under high pressure when the engine is hot. To avoid danger of releasing scalding engine coolant, remove the cap only when the engine is cool.

2. Remove the accessory drive belt.
3. Remove the 4 bolts and pump pulley.
4. Remove the water pump(A) and gasket.

To install:

5. Installation is the reverse order of the removal.
6. Tightening torque for water pump bolts: 7 ft. lbs. (8 Nm)
7. Refill the engine coolant to the correct level.
8. Start the engine and check for leaks.

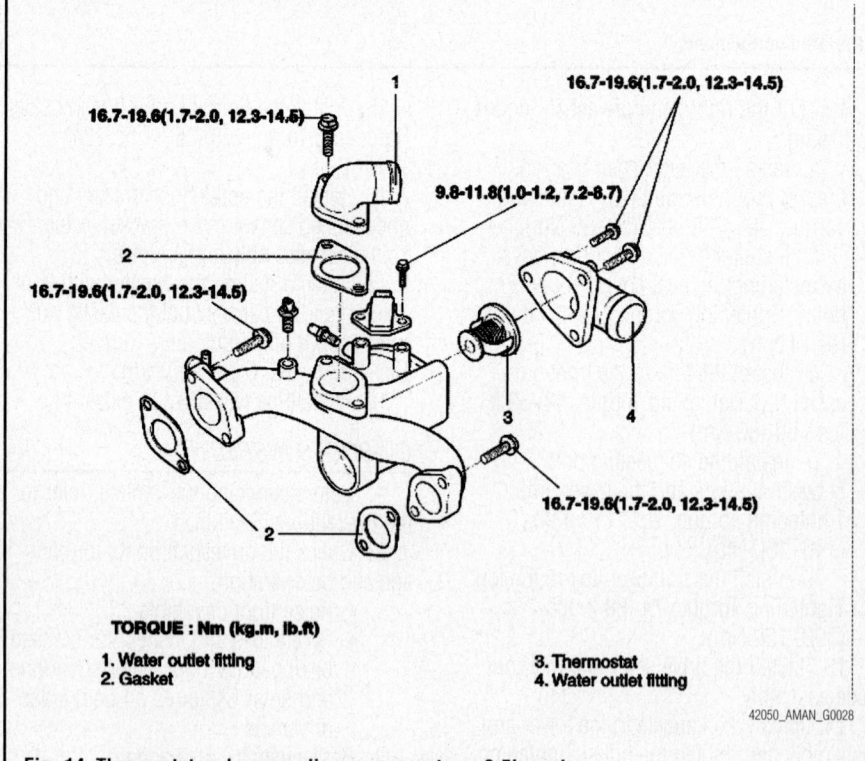

TORQUE : Nm (kg.m, lb.ft)

1. Water outlet fitting
2. Gasket
3. Thermostat
4. Water outlet fitting

Fig. 14 Thermostat and surrounding components — 3.5L engine

ENGINE ELECTRICAL   CHARGING SYSTEM

➥Disconnecting the negative battery cable on some vehicles may interfere with the functions of the on board computer system. The computer may undergo a relearning process once the negative battery cable is reconnected.

ALTERNATOR

REMOVAL & INSTALLATION

1. Before servicing the vehicle, refer to the Precautions Section.

2. Remove or disconnect the following:
- Negative battery cable
- Alternator wiring harness connectors
- Accessory drive belt
- Alternator support bracket
- Alternator

To install:

3. Install the alternator. Tighten the mounting bolts as follows:
   a. Mounting bracket bolts: Tighten to 15–18 ft. lbs. (20–25 Nm)
   b. Support bracket bolt: Tighten to 13–16 ft. lbs. (18–22 Nm)
4. Install or connect the following:
- Accessory drive belt
- Alternator wiring harness connectors
- Negative battery cable

VOLTAGE REGULATOR

REMOVAL & INSTALLATION

The voltage regulator is an integral part of alternator.

ENGINE ELECTRICAL   DISTRIBUTORLESS IGNITION SYSTEM

➥Disconnecting the negative battery cable on some vehicles may interfere with the functions of the on board computer system. The computer may undergo a relearning process once the negative battery cable is reconnected.

FIRING ORDER

See Figure 18.

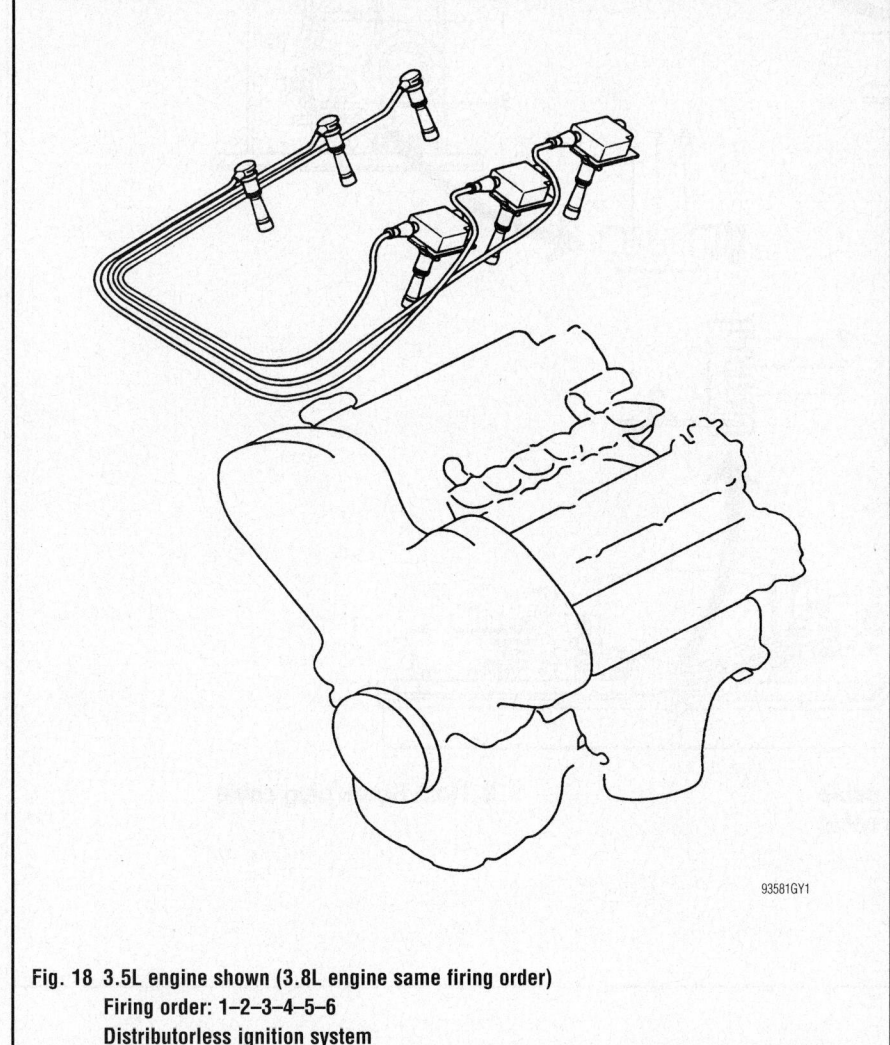

Fig. 18 3.5L engine shown (3.8L engine same firing order)
    Firing order: 1–2–3–4–5–6
    Distributorless ignition system

93581GY1

IGNITION COIL

REMOVAL & INSTALLATION

1. Remove spark plug wires from ignition coil pack.
2. Remove small bolts holding ignition coil in place.
3. Remove ignition coil.

To install:

4. Replace ignition coil and tighten to 80 inch lbs. (9 Nm).
5. Replace spark plug wires.

IGNITION TIMING

ADJUSTMENT

The ignition timing is controlled by the Powertrain Control Module (PCM). No adjustment is necessary.

SPARK PLUGS

REMOVAL & INSTALLATION

See Figure 19.

1. Remove spark plug wire from the spark plug.

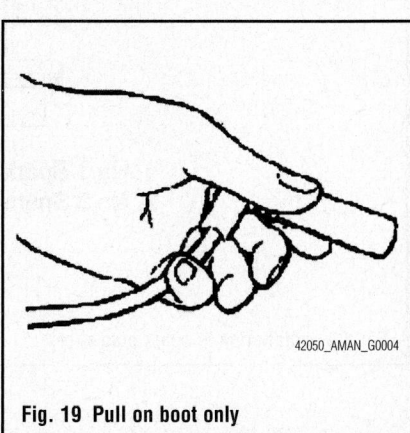

42050_AMAN_G0004

Fig. 19 Pull on boot only

➡**Pull on the spark plug wire boot when removing the spark plug wire, not the wire itself, as it may be damaged.**

2. Clean loose debris away from area of spark plug to keep contaminants from entering engine when spark plug is removed.

3. Remove the spark plug using a spark plug socket and wrench.

To install:

4. Be sure the spark plug gap is set to 0.043 in. (1.09 mm).

5. Carefully install the spark plug and tighten to 15–22 ft. lbs. (20–29 Nm).

SPARK PLUG WIRES

REMOVAL & INSTALLATION

See Figure 20.

1. Remove spark plug wire from the spark plug.

2. Follow the path of the spark plug wire back to the ignition coil removing it from any wire clips.

To install:

3. Replace spark plug wires in correct order and reattached to the wire clips.

➡**Improper arrangement of spark plug wires may induce flashover between the cables, causing misfiring. Therefore, be careful to arrange the spark plug wires properly as shown in the illustration.**

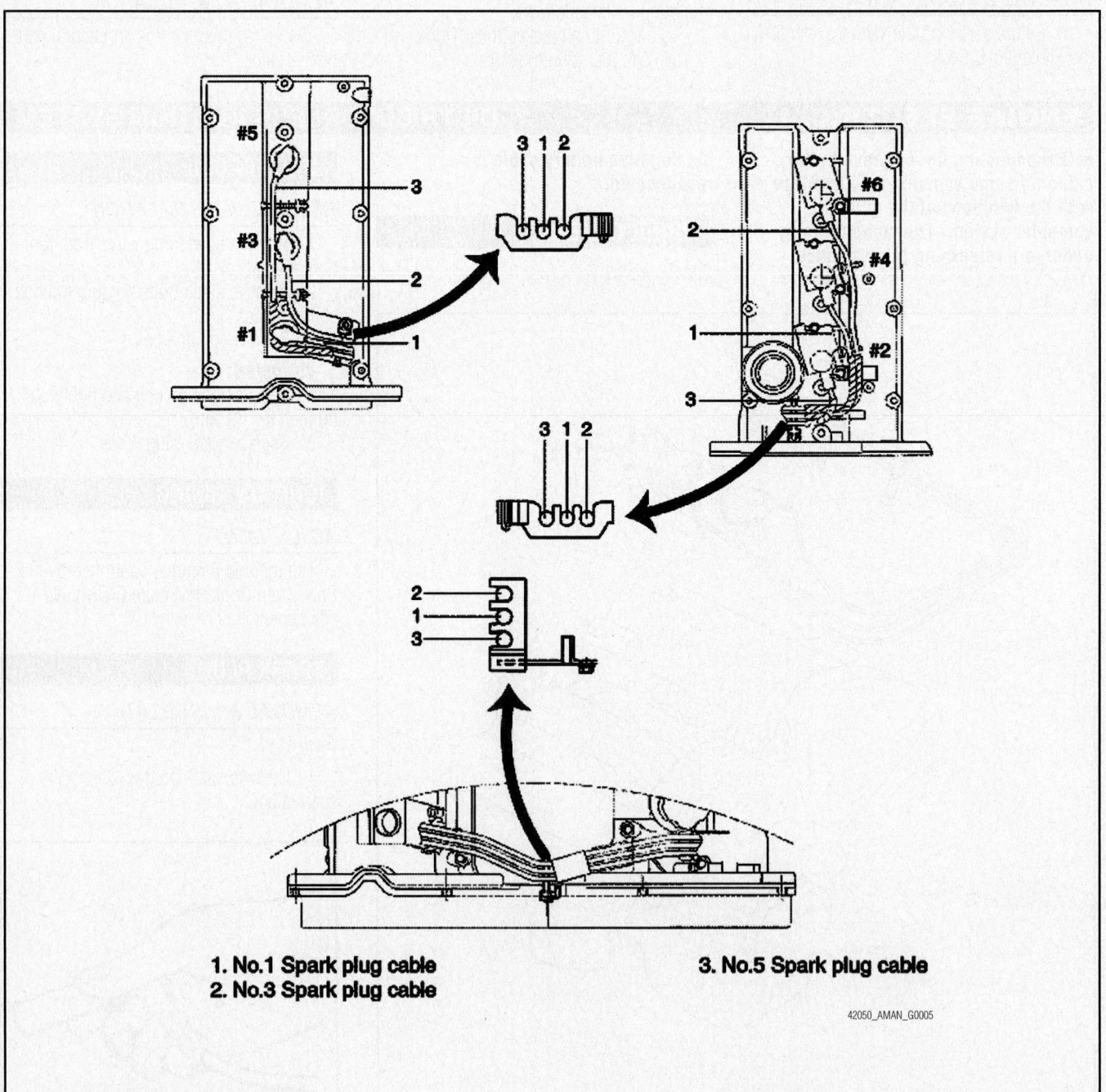

1. No.1 Spark plug cable
2. No.3 Spark plug cable
3. No.5 Spark plug cable

42050_AMAN_G0005

Fig. 20 Installation of spark plug wires

ENGINE ELECTRICAL

→Disconnecting the negative battery cable on some vehicles may interfere with the functions of the on board computer system. The computer may undergo a relearning process once the negative battery cable is reconnected.

ENGINE MECHANICAL

ACCESSORY DRIVE BELTS

ACCESSORY BELT ROUTING

See Figure 21.

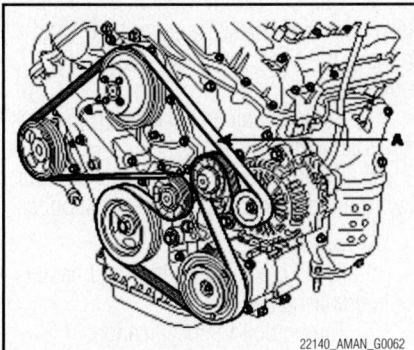

Fig. 21 Accessory belt routing—3.8L engine

INSPECTION

Inspect the drive belt for signs of glazing or cracking. A glazed belt will be perfectly smooth from slippage, while a good belt will have a slight texture of fabric visible. Cracks will usually start at the inner edge of the belt and run outward. All worn or damaged drive belts should be replaced immediately.

ADJUSTMENT

See Figures 22 and 23.

1. Use a tensioner gauge to measure the adjustment of the accessory drive belt.
2. Turn the adjusting bolt (Q) clockwise or counterclockwise until a tension of 22 lbs. (98 N) is measured.

REMOVAL & INSTALLATION

See Figure 24.

1. Release tension from drive belt by turning the belt tension adjusting bolt (Q).
2. Remove drive belt from the vehicle.

STARTER

REMOVAL & INSTALLATION

1. Before servicing the vehicle, refer to the Precautions Section.
2. Remove or disconnect the following:
   • Negative battery cable
   • Starter electrical connectors
   • Starter motor

To install:

3. Install or connect the following:
   • Starter motor. Tighten mounting bolt and nut to 33–40 ft. lbs. (45–55 Nm).
   • Starter electrical connectors. Tighten nut to 20–25 ft. lbs. (27–34 Nm).
   • Speedometer and shift cables
   • Negative battery cable

STARTING SYSTEM

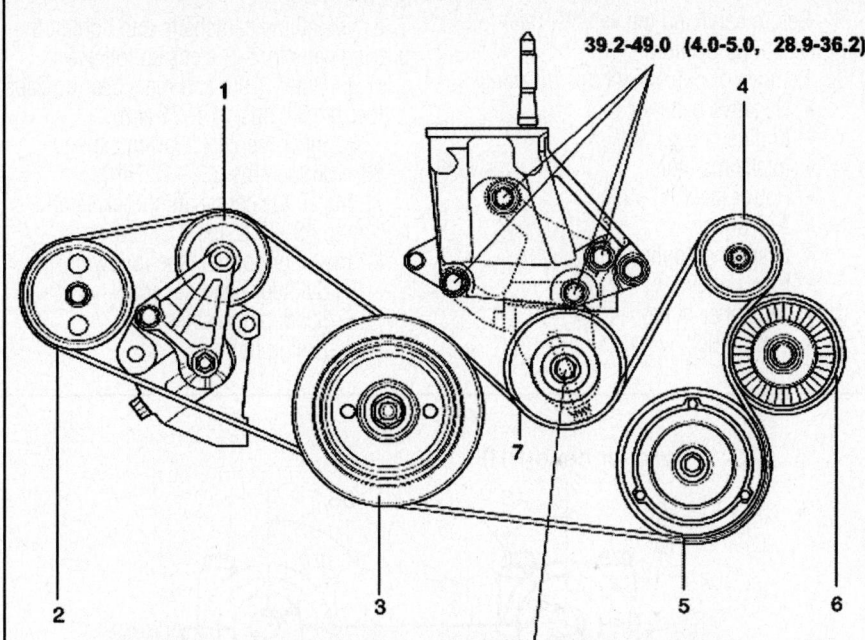

TORQUE : N·m (kg·m, lb·ft)

1. Tension pulley
2. Power steering pulley
3. Crankshaft pulley
4. Generator pulley
5. Air conditioner pulley
6. Idler pulley
7. Tension pulley

Fig. 22 Accessory drive belt routing with component listing

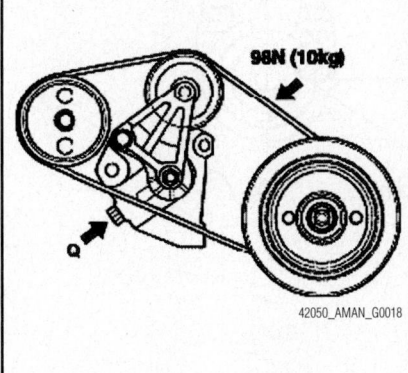

Fig. 23 Belt tension adjusting bolt

Fig. 24 Accessory belt routing for 3.8L engine

To install:

3. Installation is the reverse of the removal procedure.

4. Be sure that the belt is securely installed around all pulleys.

5. Rotate the crankshaft several times clockwise to equalize belt tension between the pulleys.

6. Adjust the drive belt tension.

CAMSHAFT AND VALVE LIFTERS

REMOVAL & INSTALLATION

See Figure 25.

1. Before servicing the vehicle, refer to the Precautions Section.

2. Remove or disconnect the following:
- Negative battery cable
- Engine cover
- Intake manifold
- Power steering pulley
- A/C pulley
- Crankshaft pulley
- Idler pulley
- Tensioner pulley
- Timing belt
- Spark plug cables
- Cylinder head cover
- Camshaft sprockets
- Camshaft bearing caps
- Camshafts

➡**Keep all valvetrain components in order for assembly.**

To install:

3. Rotate the crankshaft to set the No. 1 at Top Dead Center (TDC) of the compression stroke.

4. Ensure the rocker arm is correctly positioned on the lash adjuster and valve.

5. Install the camshaft dowel pins as shown.

6. Install the camshafts and tighten the bearing caps in 2–3 steps as follows:

a. Outer (front and rear) bearing caps to 14–15 ft. lbs. (19–21 Nm).

b. Inner (center) bearing caps to 88–106 inch lbs. (10–12 Nm).

7. Install the camshaft sprockets and tighten to 58–72 ft. lbs. (79–98 Nm).

8. Install or connect the following:
- Cylinder head cover
- Spark plug cables
- Timing belt
- Tensioner pulley
- Idler pulley
- Crankshaft pulley
- A/C pulley
- Power steering pulley
- Intake manifold
- Engine cover
- Negative battery

9. Start the engine and check for leaks.

CRANKSHAFT DAMPER

REMOVAL & INSTALLATION

See Figures 22, 23, 26 and 27.

1. Before servicing the vehicle, refer to the Precautions Section.

2. Remove or disconnect the following:
- Negative battery cable
- Engine cover

3. Using a 16mm wrench, rotate the tensioner arm clockwise (about 14°) and remove the accessory drive belt from the pulleys.

4. Remove bolt and washer from crankshaft damper and pulley.

5. Remove crankshaft damper and pulley from crankshaft using a suitable puller.

To install:

6. Press crankshaft pulley and damper onto crankshaft end.

7. Torque bolt to specification: 130–137 ft. lbs. (177–186 Nm).

8. Install drive accessory belt. Refer to the procedure in this section.

9. Using a 16mm wrench, rotate the tensioner arm counterclockwise (about 14°) and check the tension on the accessory drive belt.

10. Use a tensioner gauge to measure the adjustment of the accessory drive belt.

11. Turn the adjusting bolt (Q) clockwise or counterclockwise until a tension of 22 lb. (98 N) is measured.

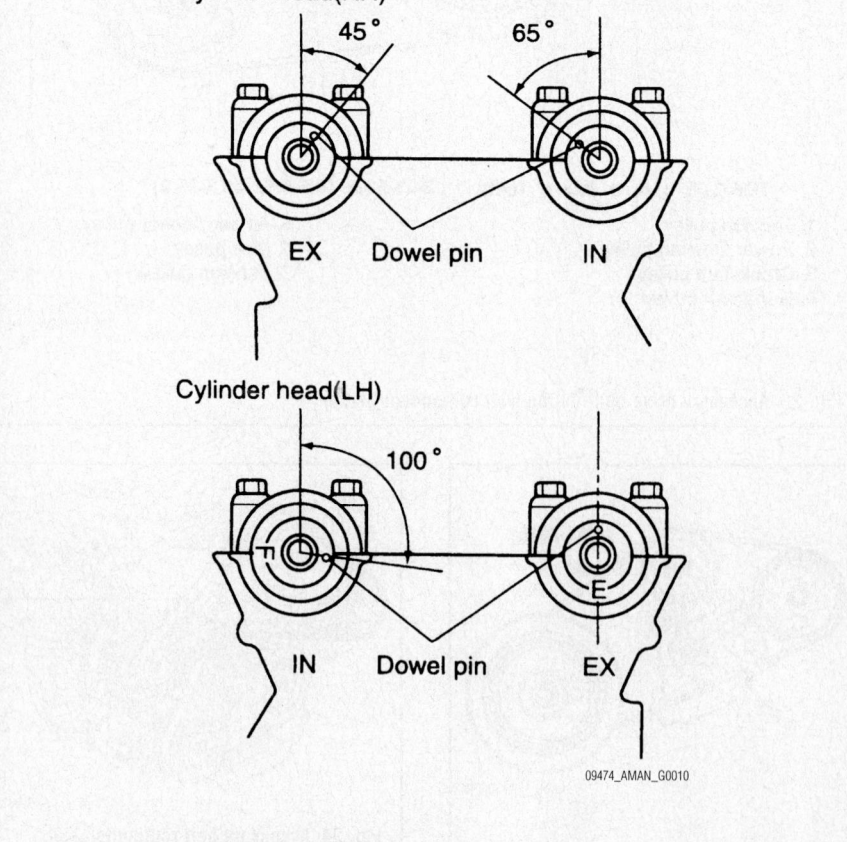

Cylinder head(RH)

45°   65°

EX   Dowel pin   IN

Cylinder head(LH)

100°

IN   Dowel pin   EX

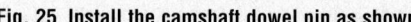
09474_AMAN_G0010

Fig. 25 Install the camshaft dowel pin as shown

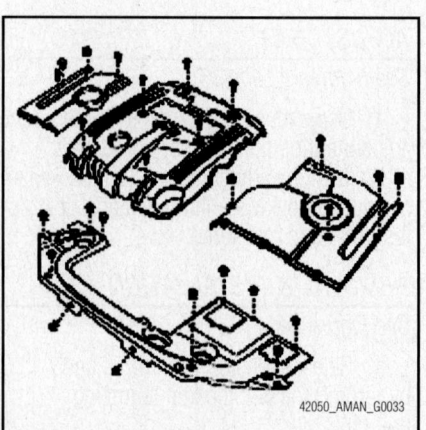
42050_AMAN_G0033

Fig. 26 Removal of engine cover

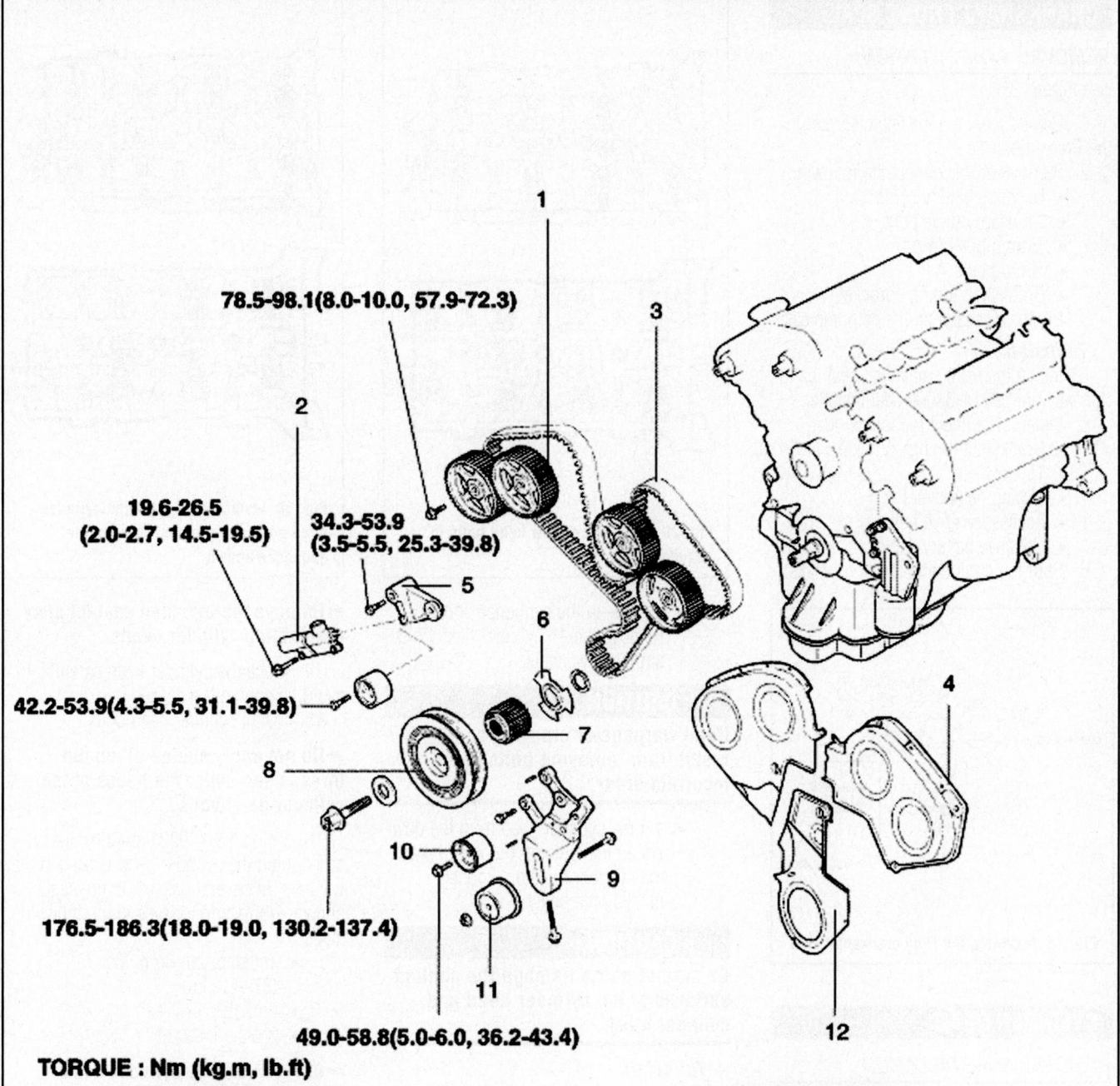

78.5-98.1(8.0-10.0, 57.9-72.3)

19.6-26.5
(2.0-2.7, 14.5-19.5)

34.3-53.9
(3.5-5.5, 25.3-39.8)

42.2-53.9(4.3-5.5, 31.1-39.8)

176.5-186.3(18.0-19.0, 130.2-137.4)

49.0-58.8(5.0-6.0, 36.2-43.4)

TORQUE : Nm (kg.m, lb.ft)

1. Camshaft sprocket
2. Auto tensioner
3. Timing belt
4. Timing belt upper cover
5. Tensioner arm
6. Sensing blade

7. Crankshaft sprocket
8. Crankshaft pulley
9. Engine support lower bracket
10. Timing belt idler bearing
11. Drive belt tensioner pulley
12. Timing belt lower cover

42050_AMAN_G0034

Fig. 27 Crankshaft pulley (with damper) and components

CRANKSHAFT FRONT SEAL

REMOVAL & INSTALLATION

See Figure 28.

1. Before servicing the vehicle, refer to the Precautions Section.
2. Remove or disconnect the following:
   - Negative battery cable
   - Accessory drive belts
   - Timing belt covers
   - Timing belt
   - Crankshaft timing sprocket
3. Pry the oil seal from the oil pump case.

To install:

4. Install the front crankshaft seal using Special Tool 09214-33000 seal installer.
5. Install or connect the following:
   - Crankshaft timing sprocket
   - Timing belt
   - Timing belt covers
   - Accessory drive belts
   - Negative battery cable
6. Start the engine and check for leaks.

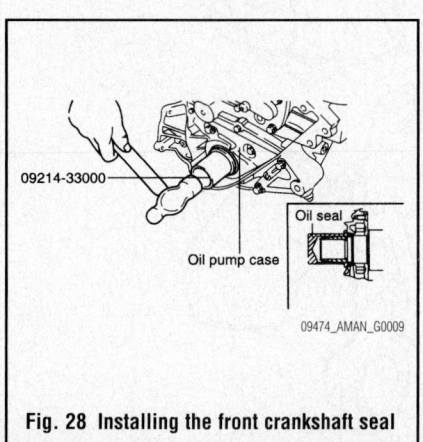

Fig. 28 Installing the front crankshaft seal

CYLINDER HEAD

REMOVAL & INSTALLATION

See Figures 29 through 31.

1. Turn the crankshaft pulley so that the No. 1 piston is at top dead center.

➡**Engine removal is required for this procedure.**

2. Remove exhaust manifold.
3. Remove intake manifold.
4. Remove timing chain.
5. Remove water temperature control assembly.
6. Remove camshaft bearing cap.
7. Remove camshaft assembly.
8. Remove cylinder head bolts, then remove cylinder head as shown:
   - Uniformly loosen and remove the 16 cylinder head bolts, in several

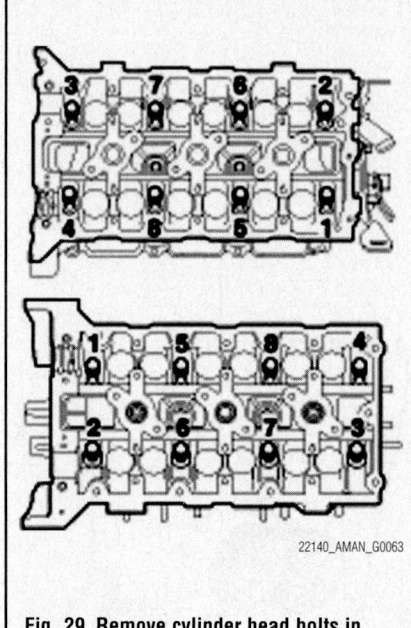

Fig. 29 Remove cylinder head bolts in sequence

passes, in the sequence shown. Remove the 16 cylinder head bolts and plate washers.

❋ CAUTION

Head warpage or cracking could result from removing bolts in an incorrect order.

- Lift the cylinder head from the dowels on the cylinder block and place the cylinder head on wooden blocks on a bench.

❋ CAUTION

Be careful not to damage the contact surfaces of the cylinder head and cylinder block.

To install:

9. Thoroughly clean all parts to be assembled.
10. Always use a new head and manifold gasket.
11. The cylinder head gasket is a metal gasket. Take care not to bend it.
12. Rotate the crankshaft, set the No.1 piston at TDC.
13. Install the cylinder head.

➡**Apply sealant on cylinder block top face before assembling cylinder head gaskets. The part must be assembled within 5 minutes after sealant was applied.**

➡**Be careful of the installation direction.**

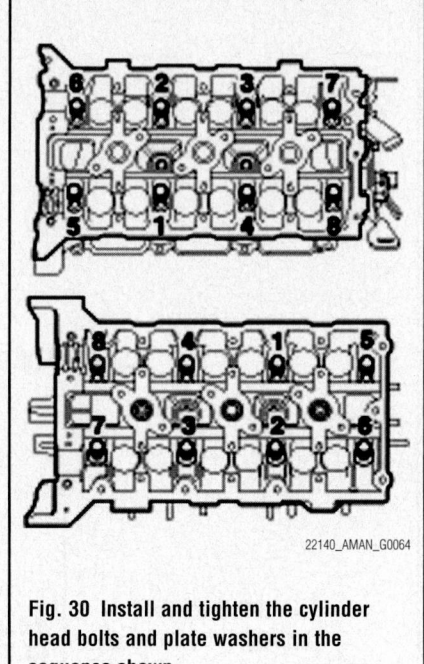

Fig. 30 Install and tighten the cylinder head bolts and plate washers in the sequence shown

➡**Remove the extruded sealant after assembling cylinder heads.**

14. Place the cylinder head carefully to avoid damaging the gasket.
15. Install cylinder head bolts.

➡**Do not apply engine oil on the threads and under the heads of the cylinder head bolts.**

16. Using SST(09221-4A000), install and tighten the cylinder head bolts and plate washers, in several passes, in the sequence shown. Tightening torques in the following steps:
   - 1st step: 28–30 ft. lbs. (37–41 Nm)
   - 2nd step:120°±2°
   - 3rd step: 90°±2°

➡**Always use new cylinder head bolts.**

17. Install the CVVT and camshaft sprocket. Tightening torque: 47.74–56.4ft. lbs. (64.68–76.44 Nm)

➡**Install camshaft-inlet to dowel pin of CVVT assembly. At this time, do not install to oil hole of camshaft-inlet. Hold the hexagonal head wrench portion of the camshaft with a vise, and install the bolt and CVVT assembly. Do not rotate CVVT assembly when camshaft is installed to dowel pin of CVVT assembly.**

18. Install camshafts(A).
   a. Apply a light coat of engine oil on camshaft journals.

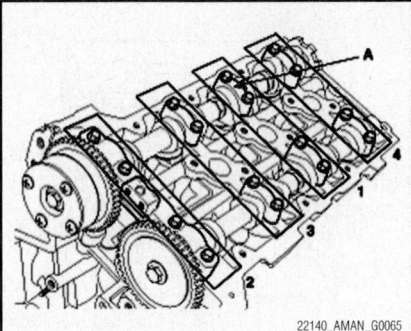

Fig. 31 Install camshaft bearing caps and torque the bolts in sequence

b. Assemble the key groove of camshaft rear side to the same level of head top surface.

c. Be careful the right, left bank, intake, exhaust side before assembling.

19. Install camshaft bearing caps and torque the bolts in the following order:
- 1st step: 4 ft. lbs. (6 Nm)
- 2nd step: 7–9 ft. lbs. (10–12 Nm)

➡ **Be careful to note the right and left bank; intake and exhaust side; and front mark before assembling.**

✳✳ CAUTION

Rotate the crankshaft so as not to contact the valves against the pistons by positioning the pistons 0.3937inch (10mm) from the top of cylinder block.

20. Install water temperature control assembly.
21. Install timing chain.
22. Check and adjust valve clearance.
23. Install the exhaust manifold.
24. Install the intake manifold.

ENGINE ASSEMBLY

REMOVAL & INSTALLATION

See Figures 32 through 34.

1. Before servicing the vehicle, refer to the Precautions Section.
2. Drain the cooling system.
3. Drain the transaxle.
4. Drain the engine oil.
5. Relieve fuel system pressure.
6. Remove or disconnect the following:
- Negative battery cable
- Engine cover
- Battery and battery tray
- Air intake assembly
- Alternator wiring connector and oil pressure switch connector

7. Remove the following engine wiring harnesses:
    a. Crankshaft angle sensor connector
    b. Fuel injector connector
    c. Power steering switch connector
    d. Variable intake motor connector
    e. Accelerator Pedal Position (APP) sensor connector
    f. EGR solenoid connector
    g. Ignition coil harness connector
    h. Oxygen sensor connector
    i. Knock Sensor (KS) connector
    j. Throttle Position Sensor (TPS)
    k. Electronic Throttle System (ETS) motor connector
    l. Limp-home connector
    m. Purge solenoid valve connector
    n. Water temperature sensor connector
8. Remove or disconnect the following:
- Upper and lower radiator hoses
- Ground wire from body frame and dash board panel
- Transmission wiring harnesses
- Ground wire from the transmission assembly
- Heater hoses
- Front and center muffler assembly
- Starter motor wiring harness
- Power steering gear connector
- Steering column intermediate shaft
- Halfshaft from the transmission
- Right side engine under cover
- Accessory drive belt
- A/C compressor and secure to the frame with a wire
- Hood release cable
- Upper radiator support member
- Cooling fan
- Electric fan
- Radiator
- Fuel hoses
9. Support the sub-frame with a suitable jack.
10. Remove the sub-frame mounting bolts.
11. Remove the engine mounting bolt.
12. Remove the transmission mounting bolt.
13. Lift the vehicle up and remove the engine, transmission and sub-frame assembly.

To install:

14. Position the engine and transmission assembly into the vehicle.
15. Install the transmission mounting bolt and tighten to 43–58 ft. lbs. (59–79 Nm).
16. Install the engine mounting bolt and tighten to 43–58 ft. lbs. (59–79 Nm).
17. Install the sub-frame bolts and tighten to 72–87 ft. lbs. (98–118 Nm).

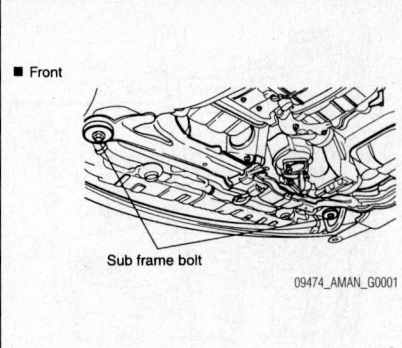

Fig. 32 Location of the front sub-frame mounting bolts

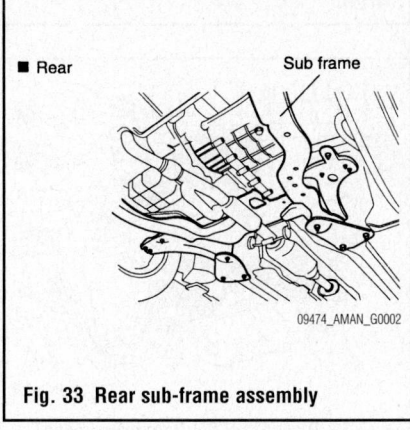

Fig. 33 Rear sub-frame assembly

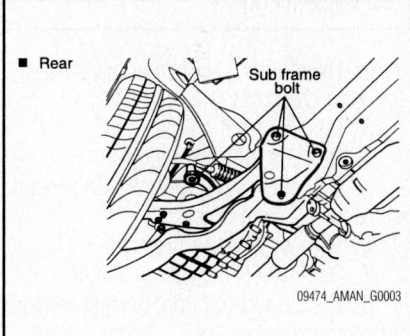

Fig. 34 Location of rear sub-frame mounting bolts

18. The remainder of the installation is the reverse order of removal.
19. Fill the engine with oil to the correct level.
20. Fill the transaxle to the correct level.
21. Fill the cooling system to the proper level.
22. Start the engine and check for leaks.

EXHAUST MANIFOLD

REMOVAL & INSTALLATION

See Figures 35 and 36.

1. Remove under cover.

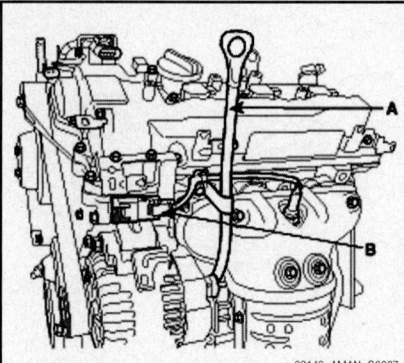

Fig. 35 Remove oil level gauge(A) and disconnect LH front oxygen sensor connector(B).

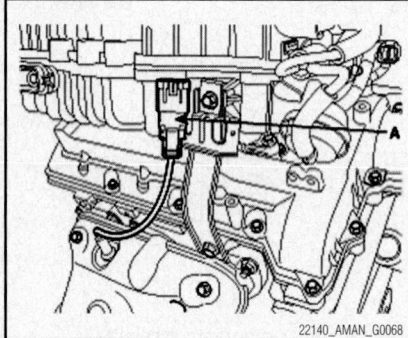

Fig. 36 Disconnect RH front oxygen sensor connector (A).

2. Disconnect LH,RH rear oxygen sensor connector from bracket.
3. Remove front muffler.
4. Remove oil level gauge(A).
5. Disconnect LH front oxygen sensor connector(B) from bracket.
6. Remove LH heat protector.
7. Remove LH exhaust manifold.
8. Disconnect RH front oxygen sensor connector from bracket.
9. Remove RH heat protector.
10. Remove RH exhaust manifold.

To install:

11. Install new gasket and exhaust manifold. Tightening torque: 29–33 ft. lbs. (40–44 Nm)
12. Install heat protectors.
13. Install front muffler.
14. Connect oxygen sensor connectors.
15. Install under cover.

FLEXPLATE

REMOVAL & INSTALLATION

1. Before servicing the vehicle, refer to the Precautions Section.
2. Remove the transmission.
3. Remove the flexplate retaining bolts.

4. Remove the flexplate from the engine.
5. Replace if gears are damaged by starter engagement.

To install:

6. Reverse the removal procedure to install the flexplate.
7. Replace flexplate retaining bolts and tighten in a cross pattern 53–56 ft. lbs. (72–76 Nm).

INTAKE MANIFOLD

REMOVAL & INSTALLATION

See Figure 37.

1. Disconnect AFS and breather hose.
2. Remove air cleaner body and intake hose.
3. Disconnect RH oxygen sensor connector.
4. Disconnect RH injector connector and ignition coil connector.
5. Disconnect PCSV connector, MAP sensor connector and PCSV hose.
6. Disconnect ETC connector and knock sensor connector.
7. Disconnect water hoses from ETC.
8. Disconnect PCV hose.
9. Disconnect brake vacuum hose.
10. Remove surge tank stay.
11. Remove connector bracket from surge tank or connectors(2EA).
12. Remove surge tank.
13. Disconnect breather Pipe assembly.
14. Disconnect LH injector connector.
15. Remove intake manifold and gasket.

To install:

16. Install intake manifold and new gasket on the cylinder head. Tightening torque is done in the following steps:
  * 1: 4 ft. lbs. (6 Nm)
  * 2: 14–17 ft. lbs. (19–24 Nm)
  * 3: Repeat 2nd step twice

➡**Be careful of the installation direction.**

  * a—h: 1st step order
  * 1—8: 2nd step order
17. Install delivery pipe.
18. Connect LH injector connector.
19. Connect breather pipe assembly. Tightening torque: 7–9 ft. lbs. (10–12 Nm).
20. Install surge tank. Tightening torque: 7–9 ft. lbs. (10–12 Nm)
21. Install connector bracket on the surge tank.
22. Install surge tank stay.
23. Connect brake vacuum hose.
24. Connect PCV hose.
25. Connect water hoses to ETC.
26. Connect ETC connector and knock sensor connector.

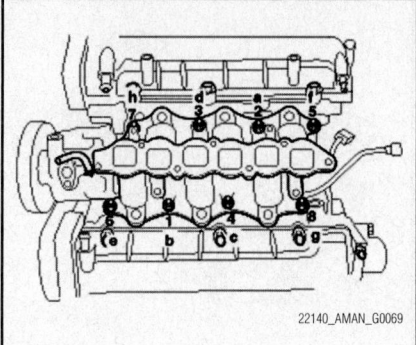

Fig. 37 Intake manifold tightening sequence

27. Connect PCSV connector, MAP sensor connector and PCSV hoe.
28. Connect RH injector connector and ignition coil connector.
29. Connect RH oxygen sensor connector.
30. Install air cleaner upper cover and intake hose.
31. Connect AFS and breather hose.

OIL PAN

REMOVAL & INSTALLATION

See Figures 38 and 39.

1. Before servicing the vehicle, refer to the Precautions Section.
2. Drain the engine oil.
3. Remove or disconnect the following:
  * Negative battery cable
  * Oil pressure switch
  * Oil filter
  * Lower oil pan
  * Upper oil pan

To install:

4. Clean all gasket surfaces of the oil pans and cylinder block.
5. Apply silicone sealant to the groove of the upper oil pan flange.

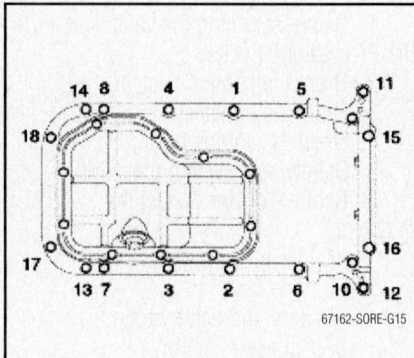

Fig. 38 Upper oil pan tightening sequence

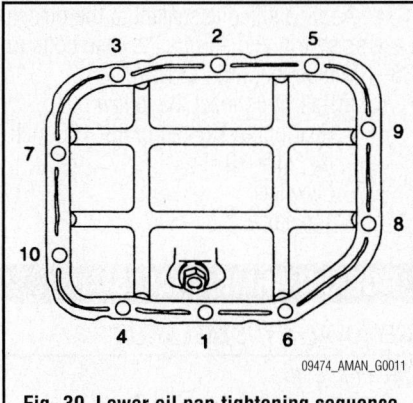

Fig. 39 Lower oil pan tightening sequence

6. Install the upper oil pan and tighten the bolts as follows:

   a. Bolts 1–14: 88–106 inch lbs. (10–12 Nm).

   b. Bolts 15–16: 44–62 inch lbs. (5–7 Nm).

   c. Oil pan-to-transaxle bolts: 22–30 ft. lbs. (30–41 Nm).

7. Apply Loctite<R> 5900 or equivalent to the threaded area of the oil pressure switch. Install the switch and tighten to 70–106 inch lbs. (8–12 Nm).

8. Install the lower oil pan and tighten the bolts in sequence to 88–106 inch lbs. (10–12 Nm).

9. Install or connect the following:
   • Oil filter
   • Negative battery cable

10. Fill the engine with clean oil.

11. Start the vehicle and check for leaks.

OIL PUMP

REMOVAL & INSTALLATION

1. Drain engine oil.

2. Using SST(09215-3C000) remove lower oil pan.

➡**Be careful not to damage the contact surfaces of upper oil pan and lower oil pan.**

3. Remove oil pump chain cover.

4. Remove oil pump chain sprocket.

5. Remove oil pump.

To install:

6. Install oil pump. Tightening torque: 15–17 ft. lbs. (21–23 Nm).

➡**Always use a new O-ring.**

7. Install oil pump sprocket and oil pump chain on the oil pump. Tightening torque: 14–16 ft. lbs. (19–22 Nm).

8. Install oil pump chain cover. Tightening torque: 7–9 ft. lbs. (10–12 Nm).

9. Install lower oil pan as follows:

   a. Clean the sealing face before assembling two parts.

   b. Remove harmful foreign materials on the sealing face before applying sealant.

   c. When applying sealant gasket, sealant must not be protrude into the inside of oil pan.

   d. To prevent leakage of oil, apply sealant gasket to the inner threads of the bolt holes.

10. Install lower oil pan.

11. Uniformly tighten the bolts in several passes. Tightening torque: 7–9 ft. lbs. (10–12 Nm).

➡**After assembly, wait at least 30 minutes before filling the engine with oil.**

MAIN BEARING TORQUE SEQUENCE

See Figures 40 and 41.

1. Install main bearings.

➡**Upper bearings have an oil groove of oil holes; Lower bearings do not.**

2. Align the bearing claw with the claw groove of the cylinder block, push in the 4 upper bearings.

3. Align the bearing claw with the claw groove of the main bearing cap, and push in the 4 lower bearings.

4. Install the 2 thrust bearings (A) under the No.3 journal position of the cylinder block with the oil grooves facing outward.

5. Place crankshaft on the cylinder block.

6. Place main bearing caps on cylinder block.

7. Install main bearing cap bolts.

8. Install and uniformly tighten the bearing cap bolts, in several passes, in the sequence shown.

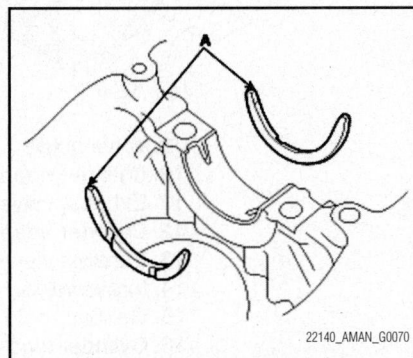

Fig. 40 Install the 2 thrust bearings (A) under the No.3 journal

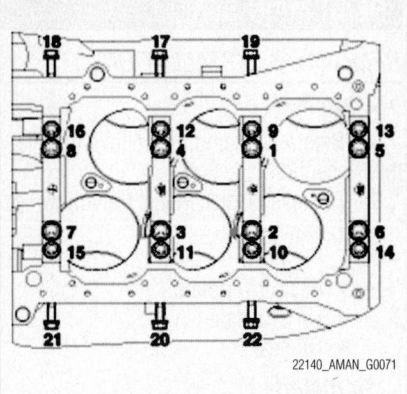

Fig. 41 Install and uniformly tighten the bearing cap bolts, in the sequence shown

9. Tightening torque for main bearing cap bolts:
   • 36 ft. lbs. (49 Nm) + 90° bolts 1 through 8
   • 14 ft. lbs. (20 Nm) + 120° bolts 9 through 16
   • 23 ft. lbs. (30 Nm) bolts 17 through 22

➡**Always use new main bearing cap bolts.**

10. Use SST(09221-4A000), install main bearing cap bolts.

11. Check that the crankshaft turns smoothly.

12. Check crankshaft end play.

PISTON AND RING

POSITIONING

See Figures 42 and 43.

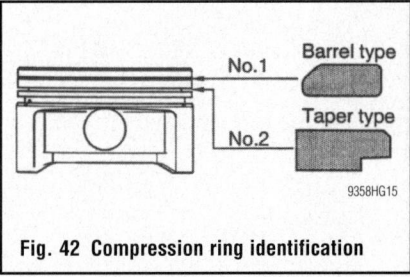

Fig. 42 Compression ring identification

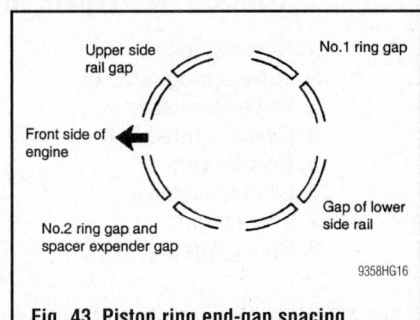

Fig. 43 Piston ring end-gap spacing

REAR MAIN SEAL

REMOVAL & INSTALLATION

See Figure 44.

1. Before servicing the vehicle, refer to the Precautions Section.
2. Remove or disconnect the following:
   - Transaxle
   - Flywheel
   - Rear cover plate
   - Oil seal case
   - Oil seal

To install:

3. Install the oil seal to the oil seal case, using Seal Installer tool 09231-33000.

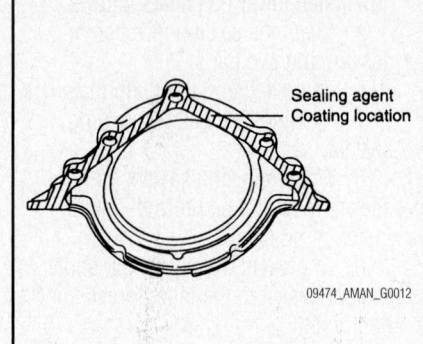

Sealing agent
Coating location
09474_AMAN_G0012

Fig. 44 Rear case sealant application location

4. Apply a silicone sealant to the oil seal case as shown and Torque the case bolts to 88–106 inch lbs. (10–12 Nm).
5. Install or connect the following:
   - Rear plate. Tighten to 88–106 inch lbs. (10–12 Nm).
   - Flywheel
   - Transaxle

ROCKER ARMS/SHAFTS

REMOVAL & INSTALLATION

See Figure 45.

1. Before servicing the vehicle, refer to the Precautions Section.

1
2
3
4
105 - 115 (1050 - 1150, 75 - 82)
5
6
7
8
9
10
11
12
13
14
15
16
TORQUE : N·m (kg·cm, lb·ft)
1. Retainer lock
2. Valve spring retainer
3. Valve stem seal
4. Cylinder head bolt
5. Rocker arm
6. Lash adjuster
7. Vlve spring
8. Spring sheet
9. Valve guide
10. Cylinder head (RH)
11. Exhaust valve seat ring
12. Cylinder head (LH)
13. Exhaust vlave
14. Intake valve
15. Gasket
16. Cylinder block
67162-SORE-G09

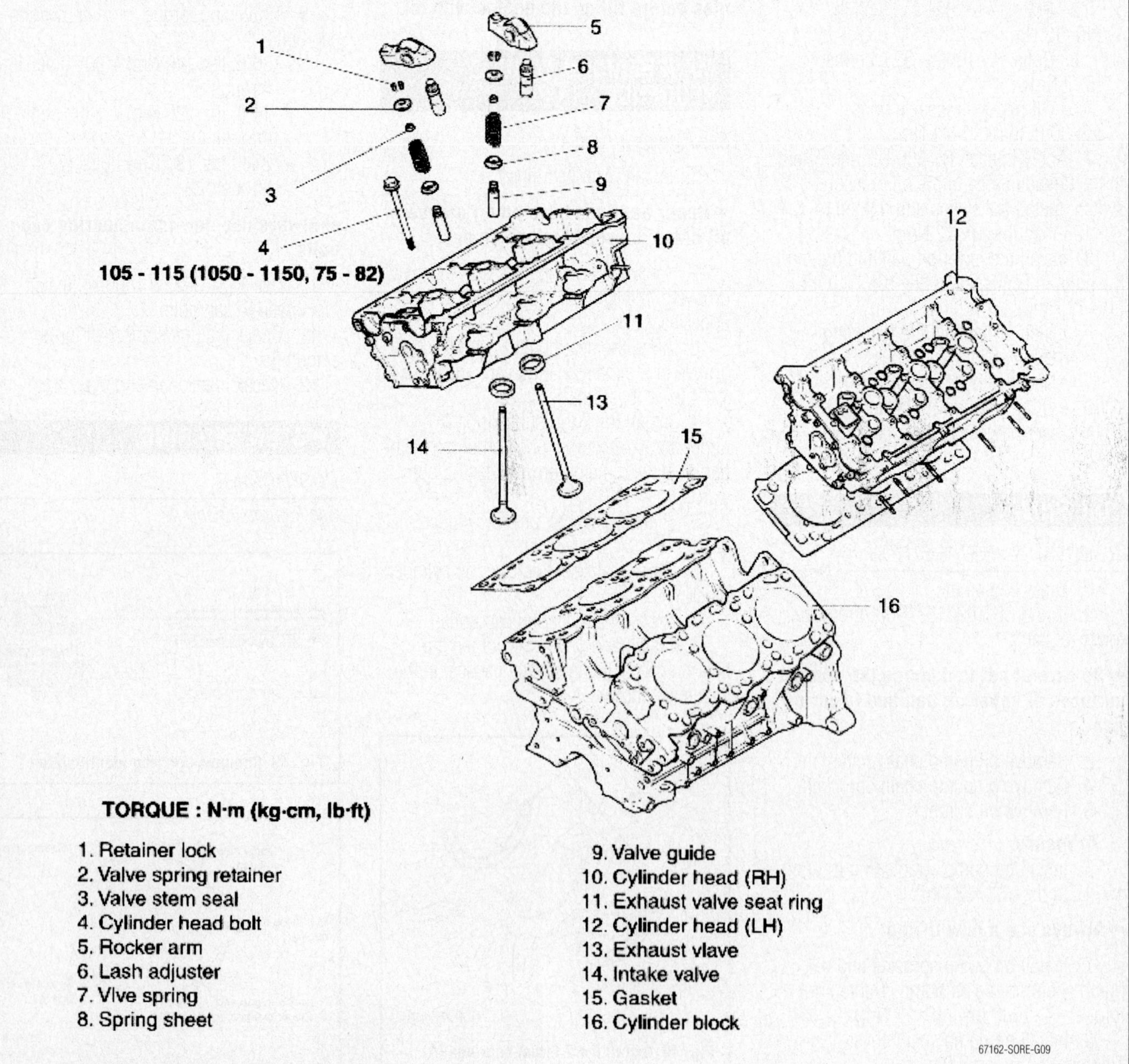

Fig. 45 Exploded view of cylinder head components

2. Drain the cooling system.

3. Relieve the fuel system pressure.

4. Remove or disconnect the following:
   - Negative battery cable
   - Air intake assembly
   - Upper radiator hose
   - Vacuum hose between intake manifold and cylinder head cover
   - Fuel hose
   - Heater hose
   - Intake manifold
   - Spark plug cables
   - Ignition coil
   - Upper and lower timing belt covers
   - Timing belt
   - Camshaft sprockets
   - Exhaust manifold heat shield
   - Exhaust manifold assembly
   - Water pump pulley
   - Cylinder head cover
   - Camshafts
   - Rocker arms and lash adjusters

5. Installation is the reverse of removal.

TIMING CHAIN COVER AND SEAL

REMOVAL & INSTALLATION

See Figures 46 through 60.

1. Disconnect the battery negative cable.

2. Remove the engine cover.

3. Remove the intake air hose and air cleaner assembly.

4. Remove the RH front wheel.

5. Remove the undercover.

6. Remove the side cover.

7. Loosen the drain plug and drain the engine coolant.

8. Drain the engine oil.

9. Loosen the power steering oil cooler return pipe mounting bolt.

10. Remove the surge tank by the following:
   - Disconnect the RH oxygen sensor connector (A) and loosen the power steering hose mounting bolts (B).
   - Disconnect the RH injector connector (A) and ignition coil connector (B).

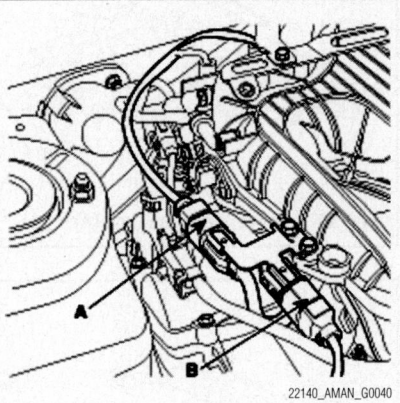

Fig. 47 Disconnect the RH injector connector (A) and ignition coil connector (B)

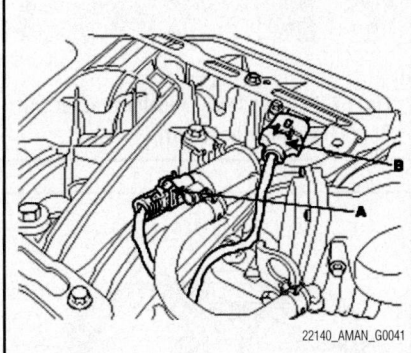

Fig. 48 Disconnect the PCSV connector (A), MAP sensor connector (B) and PCSV hose

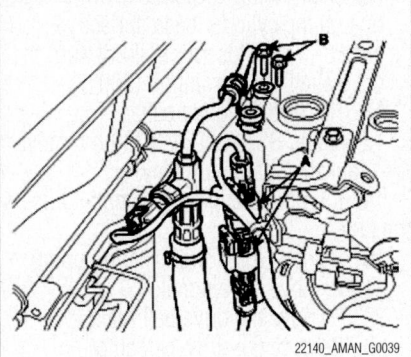

Fig. 46 Disconnect the RH oxygen sensor connector (A) and loosen the power steering hose mounting bolts (B)

- Disconnect the PCSV connector (A), MAP sensor connector (B) and PCSV hose.
- Disconnect the ETC connector (A) and knock sensor connector (B).
- Disconnect the OCV connector (A) and knock sensor connector (B).

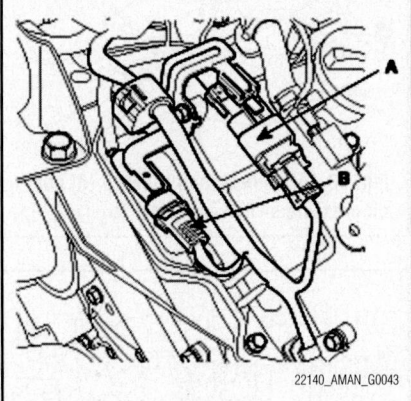

Fig. 50 Disconnect the OCV connector (A) and knock sensor connector (B)

Fig. 51 Disconnect the LH front oxygen sensor connector (A)

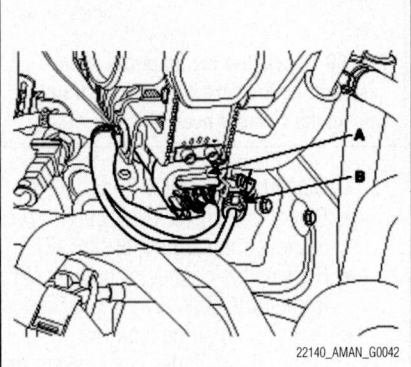

Fig. 49 Disconnect the ETC connector (A) and knock sensor connector (B)

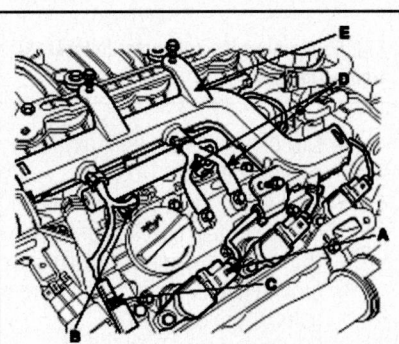

Fig. 52 Disconnect the LH ignition coil connector (A), injector connector (B), condenser connector (C) and ground (D), and remove the wiring harness protector (E)

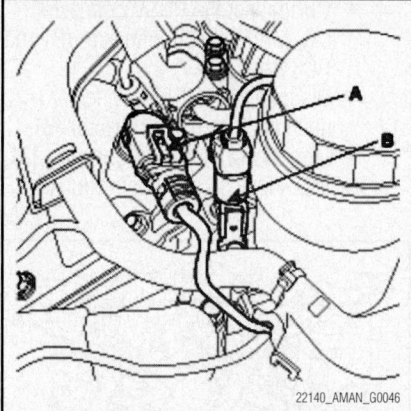

Fig. 53 Disconnect the LH CMPS (A) and oil pressure switch connector (B)

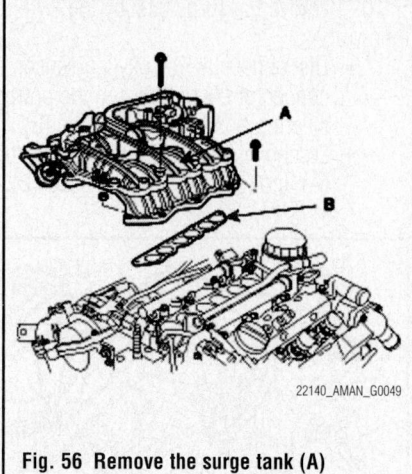

Fig. 56 Remove the surge tank (A)

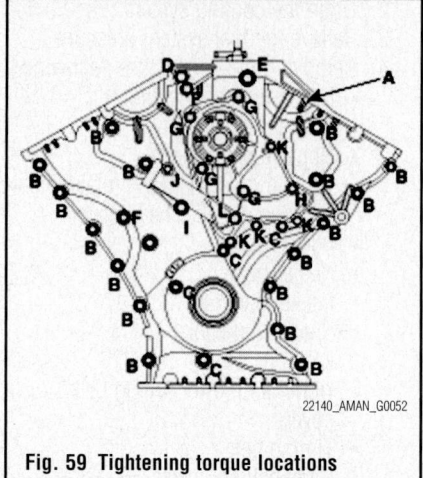

Fig. 59 Tightening torque locations

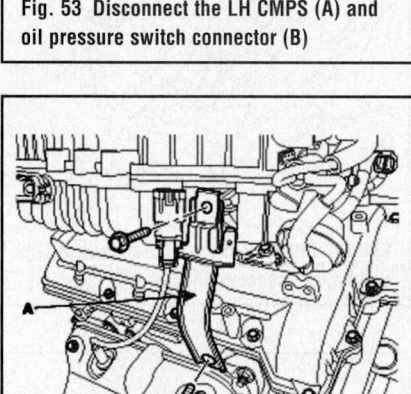

Fig. 54 Remove the surge tank stay (A)

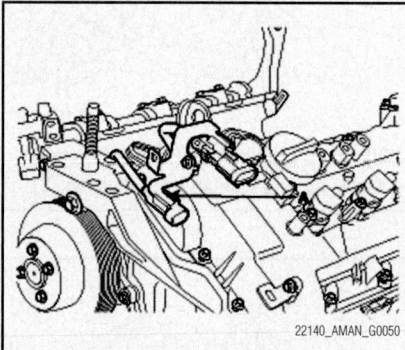

Fig. 57 Remove the connector bracket (A) from LH cylinder head cover

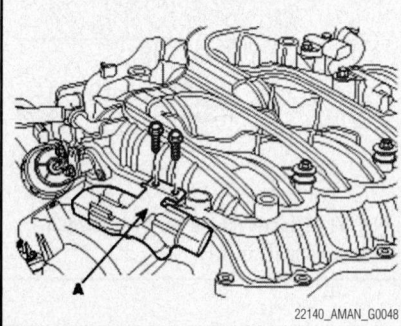

Fig. 55 Remove the connector bracket (A) from surge tank

- Disconnect the LH front oxygen sensor connector (A).
- Disconnect the LH ignition coil connector (A), injector connector (B), condenser connector (C) and ground (D), and remove the wiring harness protector (E).
- Disconnect the LH CMPS (A) and oil pressure switch connector (B)
- Remove the ETC bracket
- Disconnect the water hoses from ETC

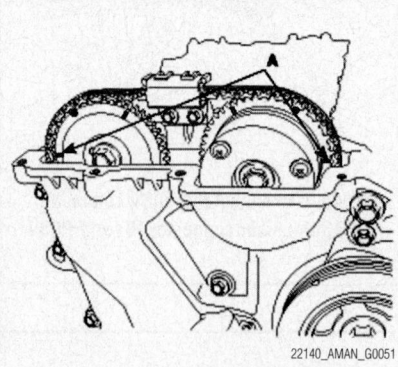

Fig. 58 Check that the mark (A) of the camshaft timing sprockets are in straight line on the cylinder head

- Disconnect the PCV hose
- Disconnect the brake vacuum hose
- Remove the surge tank stay (A)
- Remove the connector bracket (A) from surge tank
- Remove the surge tank (A).

11. Remove the cylinder head covers by the following:
- Remove the connector bracket (A) from LH cylinder head cover

- Disconnect the RH ignition coil connector, condenser connector and remove the wiring bracket.
- Remove the LH, RH ignition coil
- Remove the LH, RH cylinder head cover

12. Using SST (09215-3C000) remove lower oil pan.

➡**Be careful not to damage the contact surfaces of upper oil pan and lower oil pan.**

13. Set a jack to the upper oil pan.
14. Just loosen the transaxle mounting bracket bolts and nuts without removing the transaxle mounting bracket.
15. Remove the engine mounting bracket.
16. Set No.1 cylinder to TDC of compression stroke.
   a. Turn the crankshaft pulley and align its groove with the timing mark "T" of the lower timing chain cover.

➡**Do not rotate engine counterclockwise.**

   b. Check that the mark (A) of the camshaft timing sprockets are in straight line on the cylinder head surface as shown in the illustration. If not, turn the crankshaft one revolution (360°).
17. Remove the drive belt.
18. Using SST (09231-3C300) remove the crankshaft damper pulley.
19. Lift up the engine assembly to using the jack.
20. Remove the power steering pump.
21. Remove the alternator.
22. Remove the drive belt idler.
23. Remove the drive belt auto tensioner.
24. Remove the water pump pulley.
25. Remove the timing chain cover. If necessary remove the water pump first.

To install:

26. Install the timing chain cover by the following:

- The sealant locations on chain cover and on counter parts (cylinder head, cylinder block, and lower oil pan) must be free of engine oil and ETC.
- Before assembling the timing chain cover, the liquid sealant TB1217H should be applied on the gap between cylinder head and cylinder block.
- After applying liquid sealant TB1217H on the timing chain cover, the part must be assembled within 5 minutes after sealant was applied.
- Install the new gasket to the timing chain cover.
- The dowel pins on the cylinder block and holes on the timing chain cover should be used as a reference in order to assemble the timing chain cover to be in exact position.

27. Tightening torques for timing chain cover are as follows:

- B (17 ea): 14–16 ft. lbs. (19–22 Nm)
- C (4 ea): 7–9 ft. lbs. (10–12 Nm)
- D, E: (1 ea): 43–51 ft. lbs. (59–69 Nm)
- F (2 ea): 19 ft. lbs. (26 Nm)
- G (4 ea): 17 ft. lbs. (22 Nm)
- H, I, J, K (1 ea): 7–9 ft. lbs. (10–12 Nm)
- L (1 ea): 16–20 ft. lbs. (22–26 Nm)– New bolt

➡ **The firing and/or blow out test should not be performed within 30 minutes after the timing chain cover was assembled.**

28. Install the water pump pulley. Tightening torque for bolts: 7–9 ft. lbs. (8–10 Nm)

29. Install the drive belt auto tensioner. Tightening torques: large bolt: 60–63 ft. lbs. (81–85 Nm); smaller bolt: 13–16 ft. lbs. (18–22 Nm)

30. Install the drive belt idler. Tightening torque: 39–43 ft. lbs. (53–58 Nm)

31. Install the alternator. Tightening torque: 20–25 ft. lbs. (26–33 Nm)

32. Install the power steering pump.

33. Lower the engine assembly by using the jack.

34. Using SST (09231-3C100), install timing chain cover oil seal.

35. Using SST (09231-3C300) install the crankshaft damper pulley. Tightening torque: 210–224 ft. lbs. (284–304 Nm)

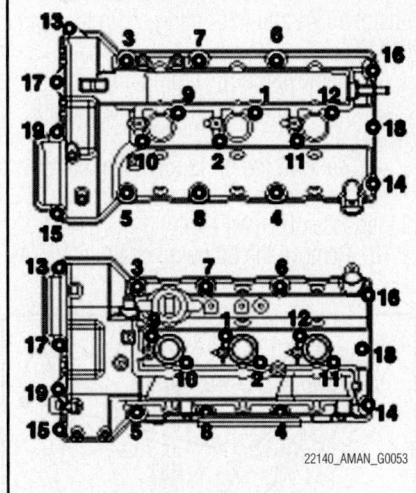

Fig. 60 Install the cylinder head cover bolts as shown

36. Install the drive belt.

37. After putting belt on auto tensioner pulley, release the auto tensioner pulley slowly.

38. Install the cylinder head cover by the following:

- The hardening sealant located on the upper area between timing chain cover and cylinder head should be removed before assembling cylinder head cover.
- After applying sealant (TB1217H), it should be assembled within 5 minutes.

➡ **The firing and/or blow out test should not be performed within 30 minutes after the cylinder head cover was assembled.**

- Install the cylinder head cover bolts as shown. Tightening torque: 7.23–8.68ft. lbs. (9.80–11.76 Nm)
- Install the ignition coil
- Connect the RH ignition coil connector, the condenser connector and install the wiring bracket
- Install the connector bracket to the LH cylinder head cove.

39. Install the surge tank and wiring connectors by the following:

- Install the surge tank. Tightening torque: 7–9 ft. lbs. (10–12 Nm)
- Install the connector bracket to the surge tank. Tightening torque: 5–8 ft. lbs. (7–11 Nm)
- Install the surge tank stay. Tightening torque: 20–23 ft. lbs. (27–31 Nm)
- Connect the brake vacuum hose
- Connect the PCV hose (C)
- Connect the water hoses to the ETC

- Install the ETC bracket
- Connect the LH CMPS and oil pressure switch connector
- Install the wiring harness protector and connect the LH ignition coil connector, injector connector, condenser connector and ground
- Connect the LH front oxygen sensor connector
- Connect the OCV connector and knock sensor connector
- Connect the ETC connector and knock sensor connector
- Connect the PCSV connector, MAP sensor connector and PCSV hose
- Connect the RH injector connector and ignition coil connector
- Connect the RH oxygen sensor connector and tighten the power steering hose mounting bolts

40. Install the engine mounting bracket. Tightening torque: Side bolt: 65–80 ft. lbs. (88–108 Nm); other two bolts and one nut: 43–58 ft. lbs. (59–79 Nm)

41. Install the transaxle mounting bracket bolts and nuts. Tightening torque: 43–58 ft. lbs. (59–79 Nm)

42. Remove the jack from the upper oil pan.

43. Install the lower oil pan by the following:

- Using a gasket scraper, remove all the old packing material from the gasket surfaces
- Before assembling the oil pan, the liquid sealant TB1217H should be applied on oil pan. The part must be assembled within 5 minutes after the sealant was applied.

✳✳ CAUTION

Be sure to do the following:

a. Make clean the sealing face before assembling two parts.

b. Remove harmful foreign matters on the sealing face before applying sealant.

c. When applying sealant gasket, sealant must not be protruded into the inside of oil pan.

d. To prevent leakage of oil, apply sealant gasket to the inner threads of the bolt holes.

- Install the lower oil pan. Tightening torque: 7–9 ft. lbs. (10–12 Nm)

44. Tighten the power steering oil cooler return pipe mounting bolt.

45. Install the side cover and the undercover. Tightening torque: 6–8 ft. lbs. (9–11 Nm)

46. Install the RH front wheel.

47. Install the intake air hose and air cleaner assembly as follows:
- Install the intake air hose and air cleaner body.
- Connect the breather hose to the air cleaner hose.
- Connect the AFS connector.

48. Install the engine cover.
49. Connect the battery negative cable.
50. Refill engine with engine oil.
51. Refill radiator and reservoir tank with engine coolant.
52. Bleed air from the cooling system.

a. Start engine and let it run until it warms up. (until the radiator fan operates 3 or 4 times.)

b. Turn Off the engine. Check the level in the radiator, add coolant if needed. This will allow trapped air to be removed from the cooling system.

c. Put radiator cap on tightly, then run the engine again and check for leaks.

TIMING CHAIN AND SPROCKETS

REMOVAL & INSTALLATION

See Figures 61 through 68.

1. Remove the timing chain cover. Refer to Timing Chain R&I section.

➡ **If necessary remove the water pump (B) first.**

➡ **Be careful not to damage the contact surfaces of cylinder block, cylinder head and timing chain cover.**

2. Before removing the timing chain, mark the RH/LH timing chain with an identification based on the location of the sprocket because the identification mark on the chain for TDC (Top Dead Center) can be erased.
3. Install a set pin after compressing the timing chain tensioner.
4. Remove the RH cam-to-cam guide (A).

5. Remove the RH timing chain auto tensioner (A) and RH timing chain tensioner arm (B).
6. Remove the RH timing chain.
7. Remove the RH timing chain guide (A).
8. Remove the oil pump chain cover.
9. Remove the oil pump chain tensioner assembly (A).
10. Remove the oil pump chain guide (A).

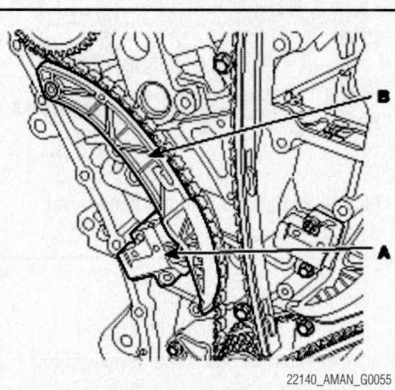

Fig. 62 Remove the RH timing chain auto tensioner (A) and RH timing chain tensioner arm (B)

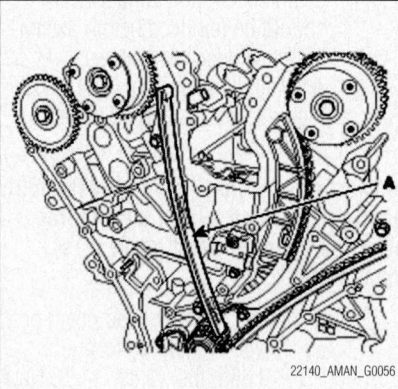

Fig. 63 Remove the RH timing chain guide (A)

11. Remove the oil pump chain sprocket (A) and oil pump chain (B).
12. Remove the crankshaft sprocket (A) (Oil pump & RH camshaft drive).
13. Install a set pin after compressing the LH timing chain tensioner.
14. Remove the LH cam-to-cam guide.
15. Remove the LH timing chain auto tensioner and LH timing chain tensioner arm.
16. Remove the LH timing chain.
17. Remove the LH timing chain guide.
18. Remove the crankshaft sprocket (LH camshaft drive).
19. Remove the tensioner adapter assembly.

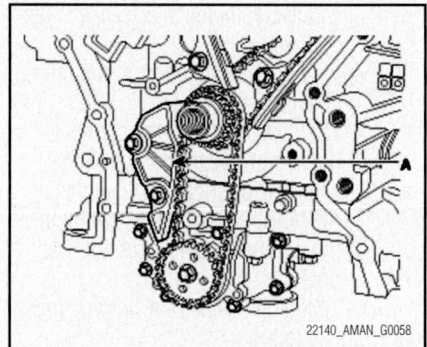

Fig. 65 Remove the oil pump chain guide (A)

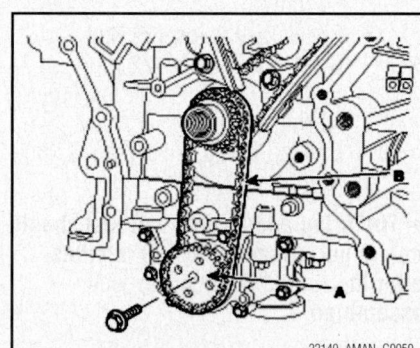

Fig. 66 Remove the oil pump chain sprocket (A) and oil pump chain (B)

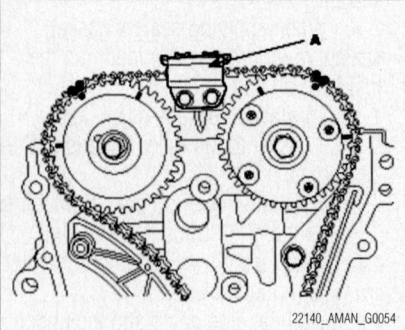

Fig. 61 Remove the RH cam-to-cam guide (A)

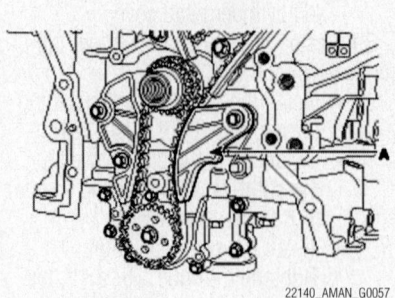

Fig. 64 Remove the oil pump chain tensioner assembly (A)

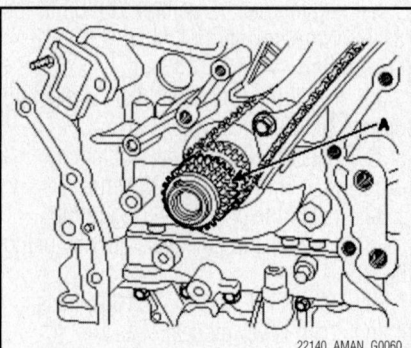

Fig. 67 Remove the crankshaft sprocket (A)

22140_AMAN_G0061

Fig. 68 The key (A) of crankshaft should be aligned with the timing mark (B) of timing chain cover

To install:

20. Check the camshaft sprocket and crankshaft sprocket for abnormal wear, cracks, or damage. Replace as necessary.

21. Inspect the tensioner arm and chain guide for abnormal wear, cracks, or damage. Replace as necessary.

22. Check that the tensioner piston moves smoothly when the ratchet pawl is released with thin rod.

23. Install the jack to the upper oil pan.

24. The key (A) of crankshaft should be aligned with the timing mark (B) of timing chain cover. As a result of this, the piston of No.1 cylinder is placed at the top dead center on compression stroke.

25. Install the tensioner adapter assembly.

26. Install the crankshaft sprocket (LH camshaft drive).

27. Install the LH timing chain guide. Tightening torque: 15–18 ft. lbs. (20–25 Nm)

28. Install LH timing chain. To install the timing chain with no slack between each shaft (cam, crank), follow the procedure below:

- Crankshaft sprocket
- Timing chain guide
- Exhaust camshaft sprocket
- Intake camshaft sprocket

➡️**The timing mark of each sprocket should be matched with timing mark (color link) of the timing chain when installing the timing chain.**

29. Install the LH timing chain tensioner arm. Tightening torque: 14–16 ft. lbs. (19–22 Nm).

30. Install the LH chain tensioner. Tightening torque: 7–9 ft. lbs. (10–12 Nm).

31. Install the LH cam-to-cam guide. Tightening torque: 7–9 ft. lbs. (10–12 Nm).

32. Install the crankshaft sprocket (Oil pump & RH camshaft drive).

33. Install the oil pump chain and oil pump sprocket. Tightening torque: 15 ft. lbs. (20 Nm).

34. Install the RH timing chain guide. Tightening torque: 14–18 ft. lbs. (20–25 Nm)

35. Install the RH timing chain. To install the timing chain with no slack between each shaft (cam, crank), follow the procedure below:

- Crankshaft sprocket
- Intake camshaft sprocket
- Exhaust camshaft sprocket

➡️**The timing mark of each sprocket should be matched with timing mark (color link) of timing chain at installing timing chain.**

36. Install the RH timing chain tensioner arm. Tightening torque: 14–16 ft. lbs. (19–22 Nm)

37. Install the RH timing chain auto tensioner. Tightening torque: 7–9 ft. lbs. (10–12 Nm).

38. Install the RH cam-to-cam guide. Tightening torque: 7–9 ft. lbs. (10–12 Nm).

39. Install the oil pump chain guide. Tightening torque: 7–9 ft. lbs. (10–12 Nm).

40. Install the oil pump chain tensioner assembly. Tightening torque: 7–9 ft. lbs. (10–12 Nm).

41. Pull out the pins of hydraulic tensioners (LH & RH).

42. Install the oil pump chain cover. Tightening torque: 7–9 ft. lbs. (10–12 Nm).

43. After rotating crankshaft 2 revolutions in regular direction (clockwise viewed from front), confirm the timing mark.

➡️**Always turn the crankshaft clockwise.**

44. Install the timing chain cover

VALVE COVERS

REMOVAL & INSTALLATION
See Figures 69 through 71.

1. Remove the engine cover and intake manifold.

2. Disconnect the breather hose and the engine harness.

3. Remove the accessory drive belt.

4. Remove the power steering pulley, air conditioner pulley, crankshaft pulley, idler pulley, and tension pulley.

5. Remove the timing belt cover.

6. Remove the spark plug cables.

7. Loosen the camshaft cover bolts and remove cover.

To install:

8. Install the gasket for the camshaft cover.

9. Clean sealing surface on camshaft cap.

10. Apply sealant to the sealing surface on camshaft cover and cap.

➡️**Be careful of gasket escapement when installing the camshaft cover. You must use the washer when installing the camshaft cover bolts.**

11. Torque the camshaft cover bolts 70–86 inch lbs. (8–10 Nm).

12. Continue the installation in the reverse order of the removal procedure.

VALVE LASH

ADJUSTMENT

This vehicle uses hydraulic valve lash adjusters. Valve lash adjustments are not necessary.

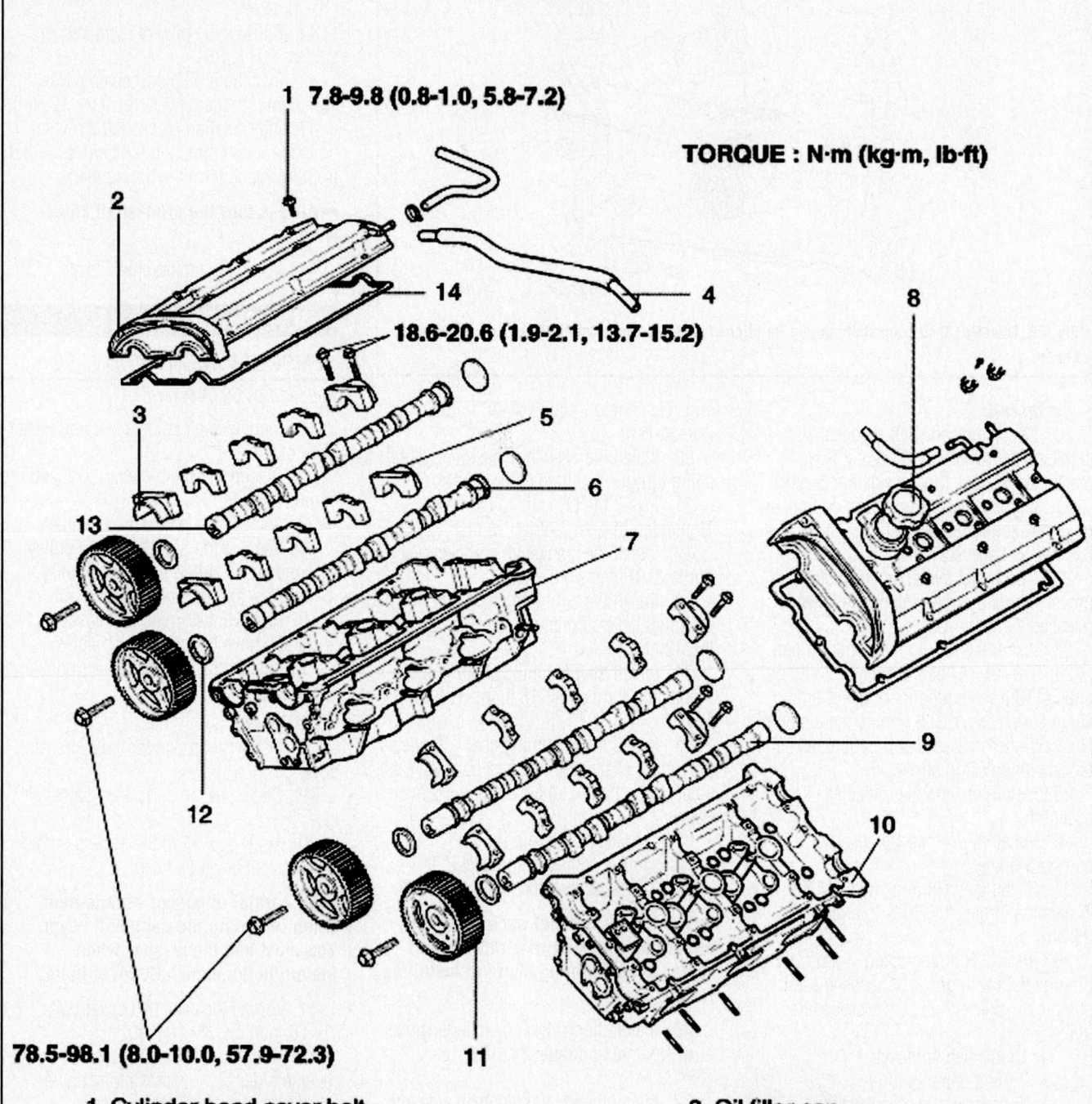

1 7.8-9.8 (0.8-1.0, 5.8-7.2)

TORQUE : N·m (kg·m, lb·ft)

18.6-20.6 (1.9-2.1, 13.7-15.2)

78.5-98.1 (8.0-10.0, 57.9-72.3)

1. Cylinder head cover bolt
2. Cylinder head cover
3. Bearing cap (Front)
4. PCV hose
5. Bearing cap (Rear)
6. Camshaft (IN)
7. Cylinder head (RH)

8. Oil filler cap
9. Camshaft (EX)
10. Cylinder head (LH)
11. Camshaft sprocket
12. Camshaft oil seal
13. Camshaft (EX)
14. Gasket

42050_AMAN_G0020

Fig. 69 Cylinder head components

42050_AMAN_G0021

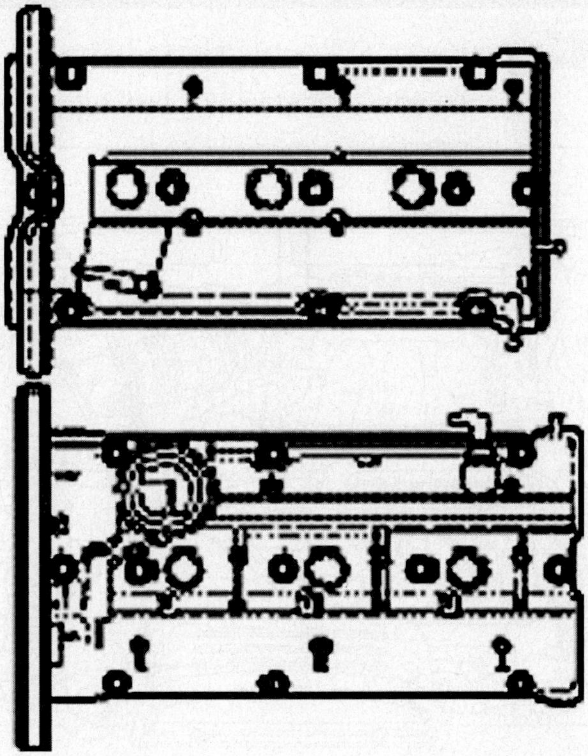

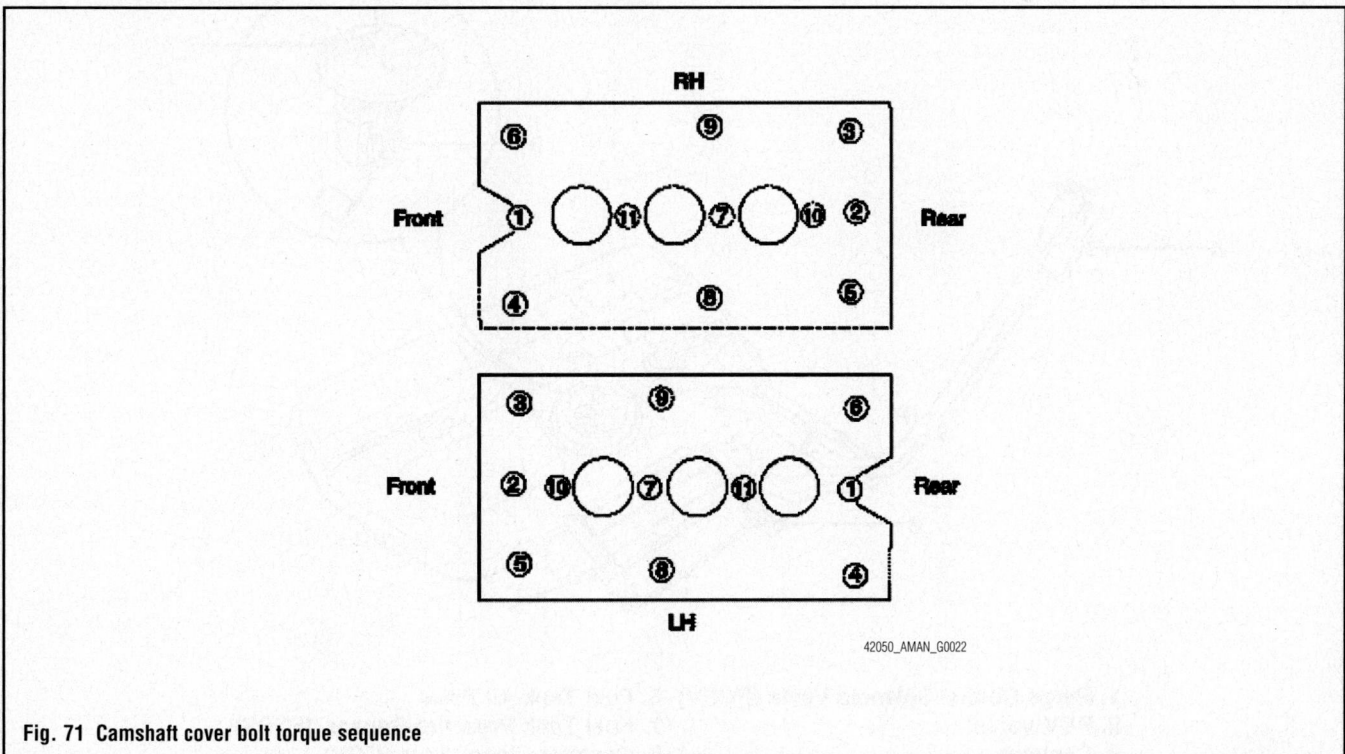

Fig. 70 Location of camshaft cover bolts

42050_AMAN_G0022

Fig. 71 Camshaft cover bolt torque sequence

ENGINE PERFORMANCE & EMISSION CONTROLS

COMPONENT LOCATIONS

See Figure 72.

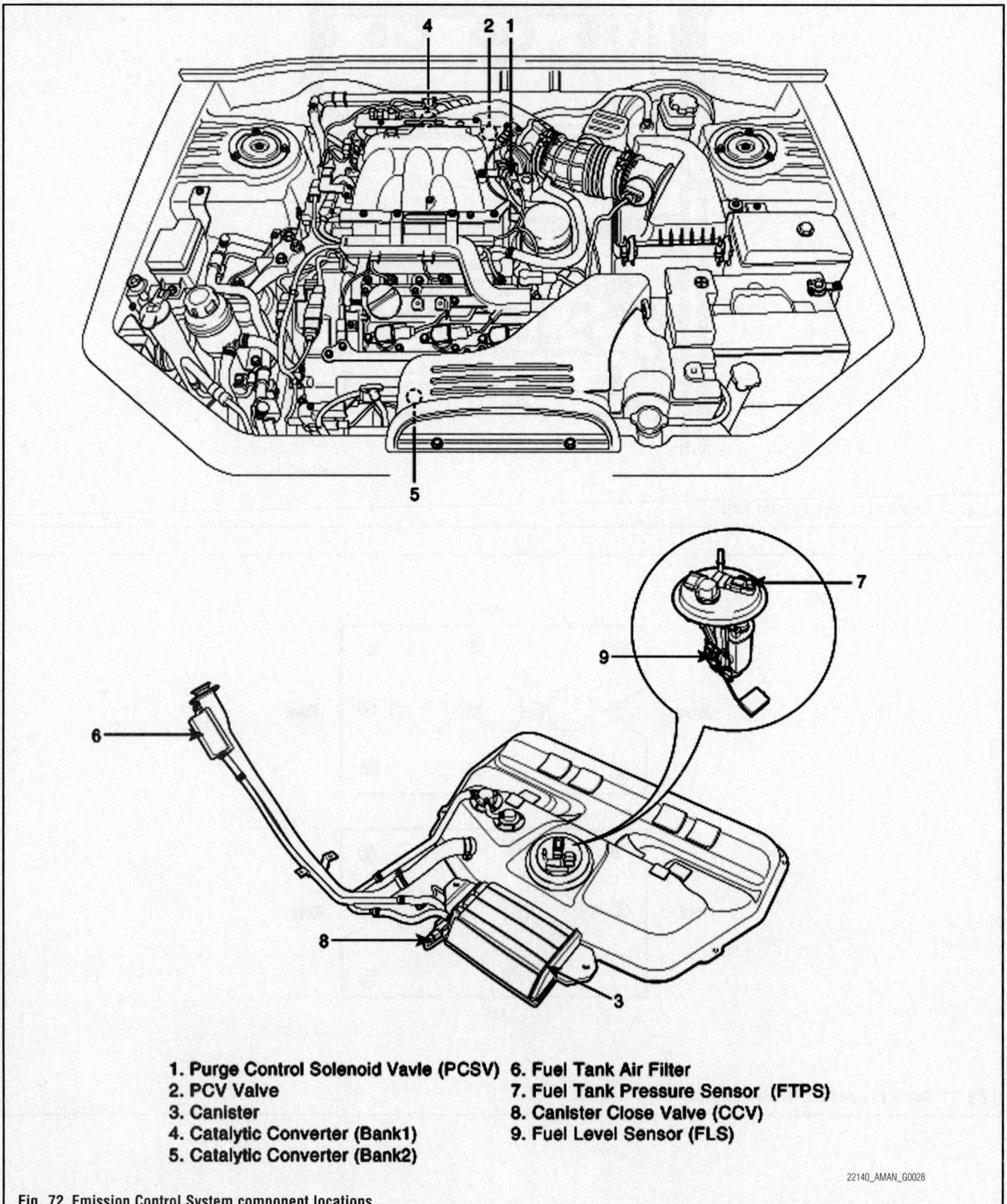

1. Purge Control Solenoid Vavle (PCSV)
2. PCV Valve
3. Canister
4. Catalytic Converter (Bank1)
5. Catalytic Converter (Bank2)
6. Fuel Tank Air Filter
7. Fuel Tank Pressure Sensor (FTPS)
8. Canister Close Valve (CCV)
9. Fuel Level Sensor (FLS)

22140_AMAN_G0028

Fig. 72 Emission Control System component locations

ACCELERATOR PEDAL POSITION (APP) SENSOR

LOCATION

The Accelerator Pedal Position (APP) Sensor is installed on the accelerator pedal module and detects the rotation angle of the accelerator pedal.

CAMSHAFT POSITION (CMP) SENSOR

LOCATION

The two CMPS are installed on engine head cover of bank 1 and 2 and uses a target wheel installed on the camshaft.

REMOVAL & INSTALLATION

1. Disconnect the negative battery cable.
2. Disconnect the connector from the sensor.
3. Remove the bolt that retains the sensor.
4. Remove the sensor.

To install:

5. Installation is the reverse of the removal procedure.

COOLANT TEMPERATURE SENSOR

LOCATION

The Engine Coolant Temperature Sensor (ECTS) is located in the engine coolant passage of the cylinder head for detecting the engine coolant temperature.

REMOVAL & INSTALLATION

See Figure 73.

✴✴ CAUTION

Never open, service or drain the radiator or cooling system when hot; serious burns can occur from the steam and hot coolant. Also, when draining engine coolant, keep in mind that cats and dogs are attracted to ethylene glycol antifreeze and could drink any that is left in an uncovered container or in puddles on the ground. This will prove fatal in sufficient quantities. Always drain coolant into a sealable container. Coolant should be reused unless it is contaminated or is several years old.

1. Drain the engine coolant.
2. Disconnect the ground cable of battery.
3. Remove the electrical connector from the sensor.

---

water temperature connector

42050_AMAN_G0010

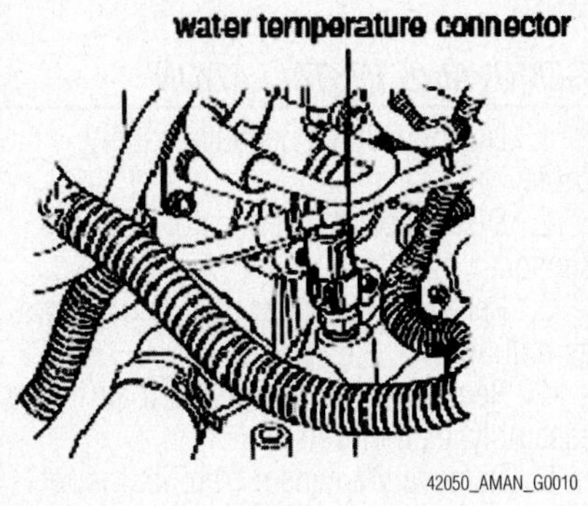

Fig. 73 Location of coolant temperature sensor

4. Remove the coolant temperature sensor.

To install:

5. Apply sealant to sensor threads. Install the sensor and tighten to 15–29 ft. lbs. (20–39 Nm).
6. Connect the coolant sensor to the harness.
7. Connect the ground cable of battery.
8. Refill the coolant.

CRANKSHAFT POSITION (CKP) SENSOR

LOCATION

The Crankshaft Position (CKP) Sensor is installed on transaxle housing.

REMOVAL & INSTALLATION

1. Disconnect the negative battery cable.
2. Disconnect the connector from the sensor.
3. Remove the bolt that retains the sensor in place.
4. Remove the sensor from its mounting.

To install:

5. Installation is the reverse of the removal procedure.
6. Clearance between the sensor and the sensor wheel should be 0.020–0.059 inch.

ENGINE COOLANT TEMPERATURE (ECT) SENSOR

LOCATION

The Engine Coolant Temperature Sensor (ECTS) is located in the engine coolant passage of the cylinder head for detecting the engine coolant temperature.

REMOVAL & INSTALLATION

1. Disconnect the negative battery cable.

---

2. Disconnect the connector from the sensor.
3. Drain the cooling system, as required.
4. Remove the sensor from its mounting.

To install:

5. Installation is the reverse of the removal procedure.

FUEL LEVEL SENDING UNIT

REMOVAL & INSTALLATION

See Figures 74 through 76.

1. Before servicing the vehicle, refer to the Precautions Section.
2. Relieve the fuel system pressure. Gain access to the service cover located in the trunk.
3. Disconnect the fuel pump electrical connector.
4. Start the engine and wait until the fuel in the fuel line is exhausted.
5. After the engine stalls, turn the ignition switch to the **OFF** position and disconnect the negative (–) terminal from the battery.

➡ **Be sure to reduce the fuel pressure before disconnecting the fuel feed hose, otherwise fuel may spill out.**

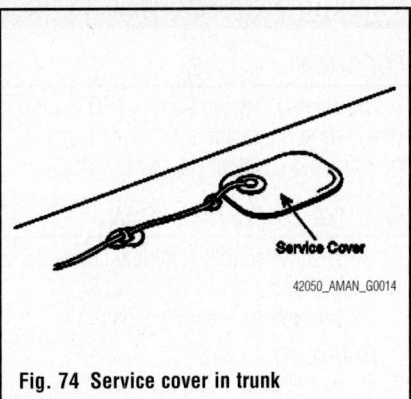

Service Cover

42050_AMAN_G0014

Fig. 74 Service cover in trunk compartment

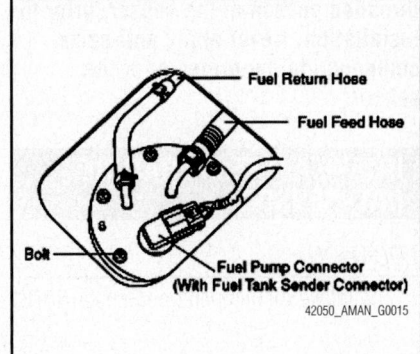

Fuel Return Hose
Fuel Feed Hose
Bolt
Fuel Pump Connector
(With Fuel Tank Sender Connector)

42050_AMAN_G0015

Fig. 75 Fuel pump connector and hoses

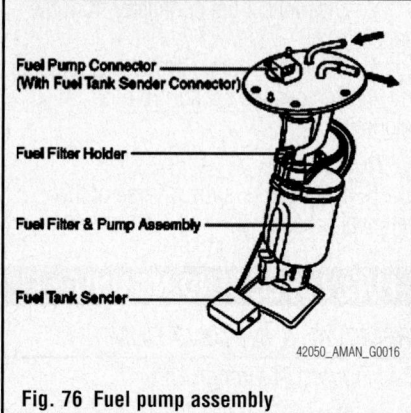

Fuel Pump Connector
(With Fuel Tank Sender Connector)

Fuel Filter Holder

Fuel Filter & Pump Assembly

Fuel Tank Sender

42050_AMAN_G0016

Fig. 76 Fuel pump assembly

6. Disconnect the fuel supply and return hoses from the top of the fuel pump assembly.

7. Remove the 6 mounting bolts and remove the fuel pump assembly from the fuel tank.

8. Check for unrestricted movement of the fuel level sending unit (fuel tank sender).

9. Replace unit if faulty.

To install:

10. Installation is the reverse of removal.

11. Fill the tank with fuel and check for proper fuel pump operation as well as fuel level accuracy.

HEATED OXYGEN (HO2S) SENSOR

LOCATION

The Heated Oxygen Sensor (HO2S) is installed on upstream and downstream of the Manifold Catalyst Converter (MCC).

REMOVAL & INSTALLATION

1. Disconnect the electrical connector from the sensor.

2. Remove the oxygen sensor.

To install:

3. Installation is the reverse of the removal procedure.

➡**Apply anti-seize compound to the threaded portion of the sensor, prior to installation. Never apply anti-seize compound to the protector of the sensor.**

INTAKE AIR TEMPERATURE (IAT) SENSOR

LOCATION

The Intake Air Temperature Sensor (IATS)

is installed inside the Mass Air Flow (MAF) sensor.

REMOVAL & INSTALLATION

1. Disconnect the negative battery cable.

2. Disconnect the connector from the sensor.

3. Remove the sensor retaining screws, as required.

4. Remove the air cleaner and air intake assembly, as required.

5. Remove the sensor from its mounting.

To install:

6. Installation is the reverse of the removal procedure.

KNOCK SENSOR (KS)

LOCATION

The Knock Sensor (KS) consists of two sensors which are installed inside the V-valley of the cylinder block.

REMOVAL & INSTALLATION

1. Disconnect the negative battery cable.

2. Remove the necessary components to gain access to the sensor.

3. Disconnect the sensor connector.

4. Remove the sensor from its mounting.

To install:

5. Installation is the reverse of the removal procedure.

6. Tighten the sensor to 11–18 ft. lbs.

TESTING

➡**If the sensor is suspected of being defective, it should be replaced with a known good component for testing purposes.**

1. Check the sensor torque. It should be 11–18 ft. lbs.

2. If the sensor is still not functioning, replace it with a known good component.

3. Recheck the sensor.

MASS AIR FLOW (MAF) SENSOR

LOCATION

The Mass Air Flow (MAF) Sensor is a located in between the air cleaner and the throttle body.

REMOVAL & INSTALLATION

1. Disconnect the negative battery cable.

2. Disconnect the connector from the sensor.

3. Remove the air cleaner and air intake assembly, as required.

4. Remove the sensor from its mounting.

To install:

5. Installation is the reverse of the removal procedure.

MANIFOLD ABSOLUTE PRESSURE (MAP) SENSOR

LOCATION

The Manifold Absolute Pressure (MAP) Sensor is speed-density type sensor and is installed on the surge tank.

REMOVAL & INSTALLATION

1. Disconnect the negative battery cable.

2. Disconnect the connector from the sensor.

3. Remove the sensor retaining screws.

4. Remove the sensor from its mounting.

To install:

5. Installation is the reverse of the removal procedure.

OIL PRESSURE SENSOR

REMOVAL & INSTALLATION

See Figure 77.

1. Before servicing the vehicle, refer to the Precautions Section.

2. Disconnect the negative battery cable.

3. Raise and safely support the vehicle.

4. Disconnect the oil pressure sensor electrical connector.

5. Position a drain pan under the sensor to catch any spilled engine oil.

6. Remove the sensor from its mounting using a 24 mm deep socket.

To install:

7. Installation is the reverse of the removal procedure.

8. Apply Loctite<R> 5900 or equivalent to the threaded area.

➡**Do not over-tighten the oil pressure sensor.**

9. Torque oil pressure sensor to 72–108 inch lbs. (8–12 Nm).

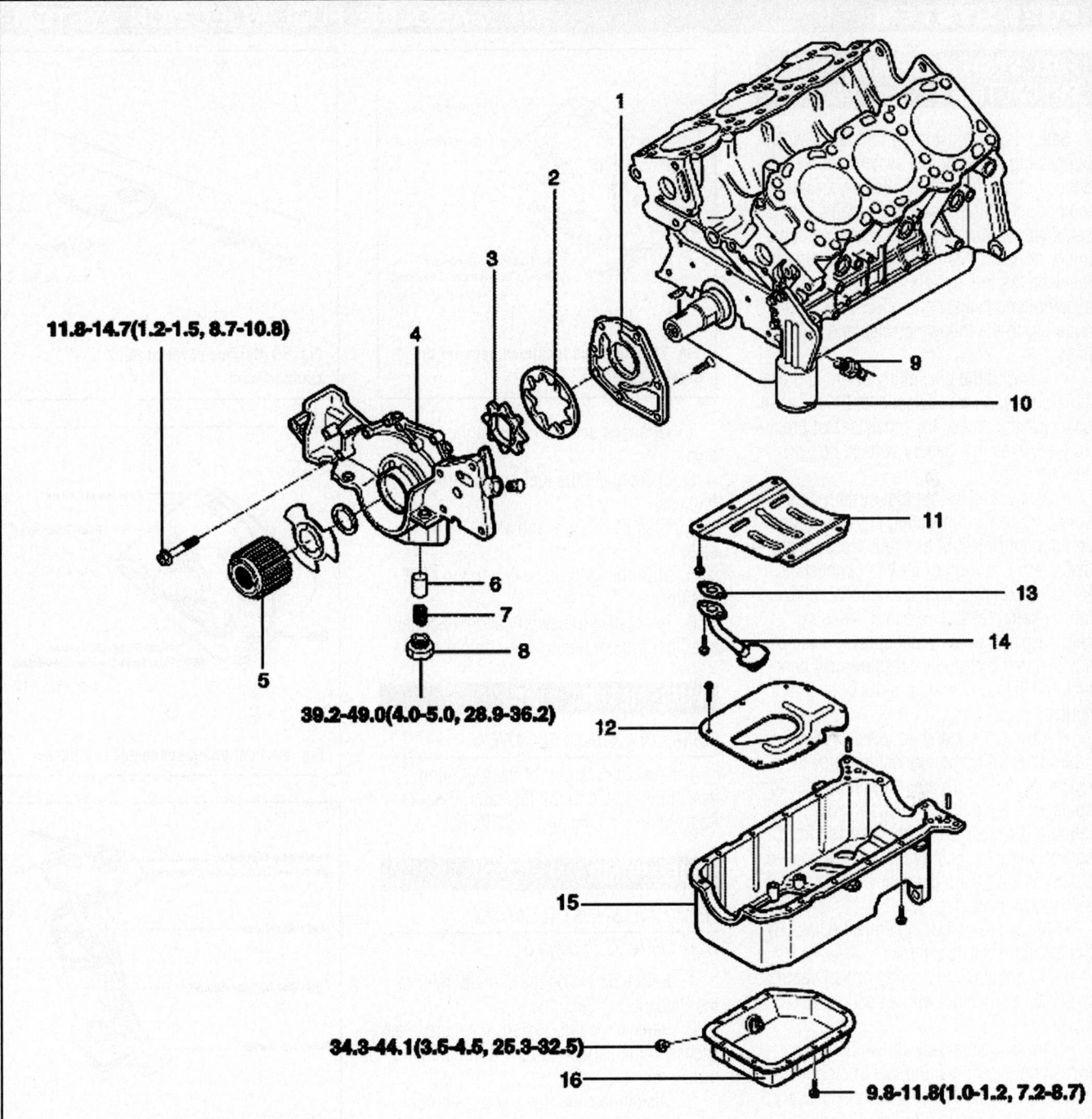

11.8-14.7(1.2-1.5, 8.7-10.8)

39.2-49.0(4.0-5.0, 28.9-36.2)

34.3-44.1(3.5-4.5, 25.3-32.5)

9.8-11.8(1.0-1.2, 7.2-8.7)

TORQUE : Nm (kg·cm, lb.ft)

1. Oil pump cover
2. Oil pump outer rotor
3. Oil pump inner rotor
4. Oil pump case
5. Crankshaft sprocket
6. Relief valve plunger
7. Relief valve spring
8. Relief valve plug

9. Oil pressure switch
10. Oil filter
11. Upper baffle plate
12. Lower baffle plate
13. Gasket
14. Oil screen
15. Upper oil pan
16. Lower oil pan

42050_AMAN_G0011

Fig. 77 Oil sensor location and other lubrication system components

FUEL SYSTEM SERVICE PRECAUTIONS

Safety is the most important factor when performing not only fuel system maintenance but any type of maintenance. Failure to conduct maintenance and repairs in a safe manner may result in serious personal injury or death. Maintenance and testing of the vehicle's fuel system components can be accomplished safely and effectively by adhering to the following rules and guidelines.

• To avoid the possibility of fire and personal injury, always disconnect the negative battery cable unless the repair or test procedure requires that battery voltage be applied.

• Always relieve the fuel system pressure prior to disconnecting any fuel system component (injector, fuel rail, pressure regulator, etc.), fitting or fuel line connection. Exercise extreme caution whenever relieving fuel system pressure to avoid exposing skin, face and eyes to fuel spray. Please be advised that fuel under pressure may penetrate the skin or any part of the body that it contacts.

• Always place a shop towel or cloth around the fitting or connection prior to loosening to absorb any excess fuel due to spillage. Ensure that all fuel spillage (should it occur) is quickly removed from engine surfaces. Ensure that all fuel soaked cloths or towels are deposited into a suitable waste container.

• Always keep a dry chemical (Class B) fire extinguisher near the work area.

• Do not allow fuel spray or fuel vapors to come into contact with a spark or open flame.

• Always use a back-up wrench when loosening and tightening fuel line connection fittings. This will prevent unnecessary stress and torsion to fuel line piping.

• Always replace worn fuel fitting O-rings with new. Do not substitute fuel hose or equivalent where fuel pipe is installed.

Before servicing the vehicle, make sure to also refer to the precautions in the beginning of this section as well.

RELIEVING FUEL SYSTEM PRESSURE

See Figure 78.

1. Before servicing the vehicle, refer to the Precautions Section.

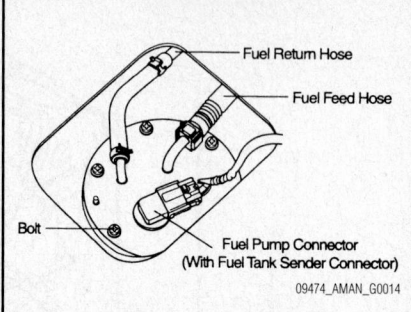

Fig. 78 Fuel hose locations beneath the service cover

2. Open the service cover in the trunk area.
3. Disconnect the fuel pump wiring harness.
4. Start the engine and allow it to stall.
5. Turn the ignition switch to the **OFF** position.
6. Reconnect the electrical connections after fuel system repairs are completed.

FUEL FILTER

REMOVAL & INSTALLATION

The fuel filter is part of the fuel pump assembly located in the fuel tank. Refer to Fuel Pump removal and installation.

FUEL PUMP

REMOVAL & INSTALLATION

See Figures 79 through 81.

1. Before servicing the vehicle, refer to the Precautions Section.
2. Relieve the fuel system pressure. Gain access to the service cover located in the trunk.
3. Disconnect the fuel pump electrical connector.
4. Start the engine and wait until the fuel in the fuel line is exhausted.
5. After the engine stalls, turn the ignition switch to the **OFF** position and disconnect the negative (–) terminal from the battery.

➡**Be sure to reduce the fuel pressure before disconnecting the fuel feed hose, otherwise fuel may spill out.**

6. Disconnect the fuel supply and return hoses from the top of the fuel pump assembly.
7. Remove the 6 mounting bolts and

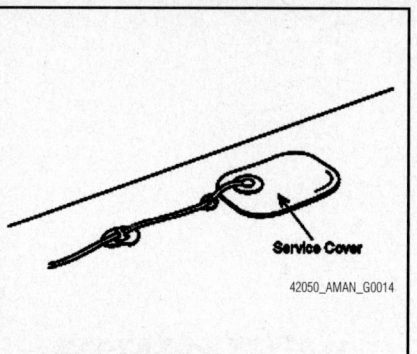

Fig. 79 Service cover in trunk compartment

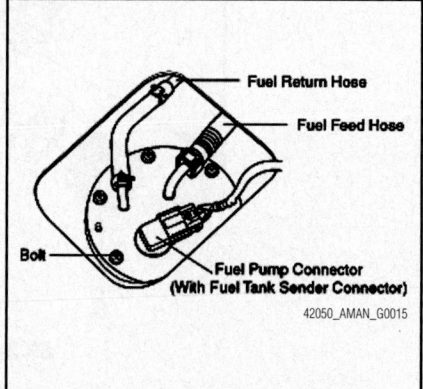

Fig. 80 Fuel pump connector and hoses

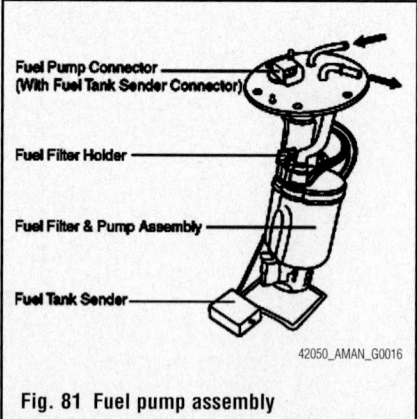

Fig. 81 Fuel pump assembly

remove the fuel pump assembly from the fuel tank.

To install:

8. Installation is the reverse of removal.
9. Fill the tank with fuel and check for proper fuel pump operation.

FUEL PRESSURE REGULATOR

REMOVAL & INSTALLATION

The fuel pressure regulator is built in to the fuel pump.

FUEL RAIL & INJECTORS

REMOVAL & INSTALLATION

1. Before servicing the vehicle, refer to the Precautions Section.
2. Relieve the fuel system pressure.
3. Drain the cooling system.
4. Remove or disconnect the following:
   - Negative battery cable
   - Engine cover
   - Air intake assembly
5. Remove the following engine wiring harnesses:
   a. Crankshaft angle sensor
   b. Camshaft angle sensor
   c. Fuel injector harness

➡**When disconnecting the injector, lift the fuel supply hose and injector assembly upward. Then unscrew the mounting bolt to lift the fuel supply hose. Reinstall the fuel hose and injector assembly after disconnecting the injector harness.**

   d. Power steering switch
   e. Variable intake motor connector
   f. Accelerator position sensor
   g. EGR solenoid
   h. Throttle position sensor
   i. Electronic throttle system (ETS) motor
   j. Limp-home connector
   k. Purge solenoid valve connector
6. Remove or disconnect the following:
   - Vacuum hoses and heater hoses between the intake manifold and cylinder head cover
   - Engine wiring harness bracket
   - EGR valve hose and bracket
   - Surge tank assembly
   - Fuel supply hose
   - Fuel injector assembly
   - Fuel injector
7. Installation is the reverse order of removal. Note the following Torques:
   a. Fuel injector assembly: 84–120 inch lbs. (10–13 Nm).
   b. Surge tank assembly: 11–15 ft. lbs. (15–20 Nm).
   c. EGR valve hose mounting bolts: 12–19 ft. lbs. (17–26 Nm).
   d. EGR pipe fixing bracket bolts: 11–15 ft. lbs. (15–20 Nm).
   e. Surge tank stay mounting bolts: 11–15 ft. lbs. (15–20 Nm).

FUEL TANK

REMOVAL & INSTALLATION

See Figures 82 through 85.

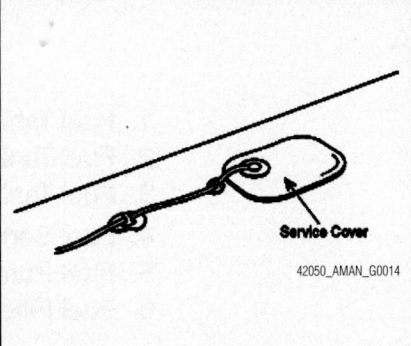

Fig. 82 Service cover in trunk compartment

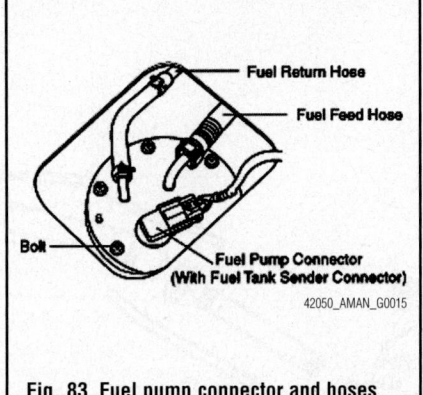

Fig. 83 Fuel pump connector and hoses

✳✳ CAUTION

Observe all applicable safety precautions when working around fuel. Whenever servicing the fuel system, always work in a well ventilated area. Do not allow fuel spray or vapors to come in contact with a spark or open flame. Keep a dry chemical fire extinguisher near the work area. Always keep fuel in a container specifically designed for fuel storage; also, always properly seal fuel containers to avoid the possibility of fire or explosion.

1. Before servicing the vehicle, refer to the Precautions Section.
2. Relieve the fuel system pressure. Gain access to the service cover located in the trunk.
3. Disconnect the fuel pump electrical connector.
4. Start the engine and wait until the fuel in the fuel line is exhausted.
5. After the engine stalls, turn the ignition switch to the **OFF** position and disconnect the negative (–) terminal from the battery.

➡**Be sure to reduce the fuel pressure before disconnecting the fuel feed hose, otherwise fuel may spill out.**

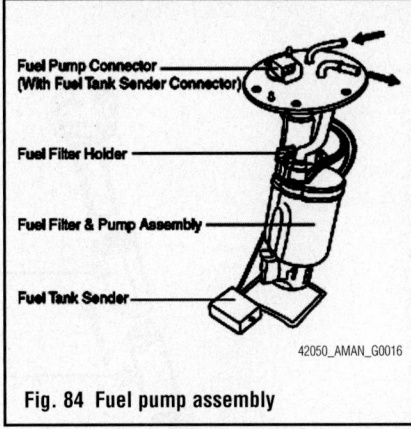

Fig. 84 Fuel pump assembly

6. Disconnect the fuel supply and return hoses from the top of the fuel pump assembly.
7. Remove the 6 mounting bolts and remove the fuel pump assembly from the fuel tank.
8. Raise and safely support the vehicle.
9. Remove the main muffler.
10. Remove the fuel tank cover.
11. Remove the fuel filler hose, the fuel hose (connecting the fuel tank with the canister) and the fuel leveling hose.
12. Remove the fuel tank band mounting bolts and remove the fuel tank.

   To install:
13. Installation is the reverse of removal.
14. Tighten fuel tank mounting bolts to 29–40 ft. lbs. (39–54 Nm).
15. Fill the tank with fuel and check for proper fuel pump operation.

IDLE SPEED

ADJUSTMENT

Idle speed is maintained by the Powertrain Control Module (PCM). No adjustment is necessary or possible.

THROTTLE BODY

REMOVAL & INSTALLATION

See Figure 86.

1. Before servicing the vehicle, refer to the Precautions Section.
2. Remove the engine cover.
3. Disconnect the battery terminal.
4. Disconnect the air flow sensor connector and the vacuum hose from the air intake hose.
5. Remove the air intake hose clamp and bolt.
6. Remove the air intake assembly.
7. Disconnect the following:
   - Crankshaft sensor connector
   - Camshaft sensor connector

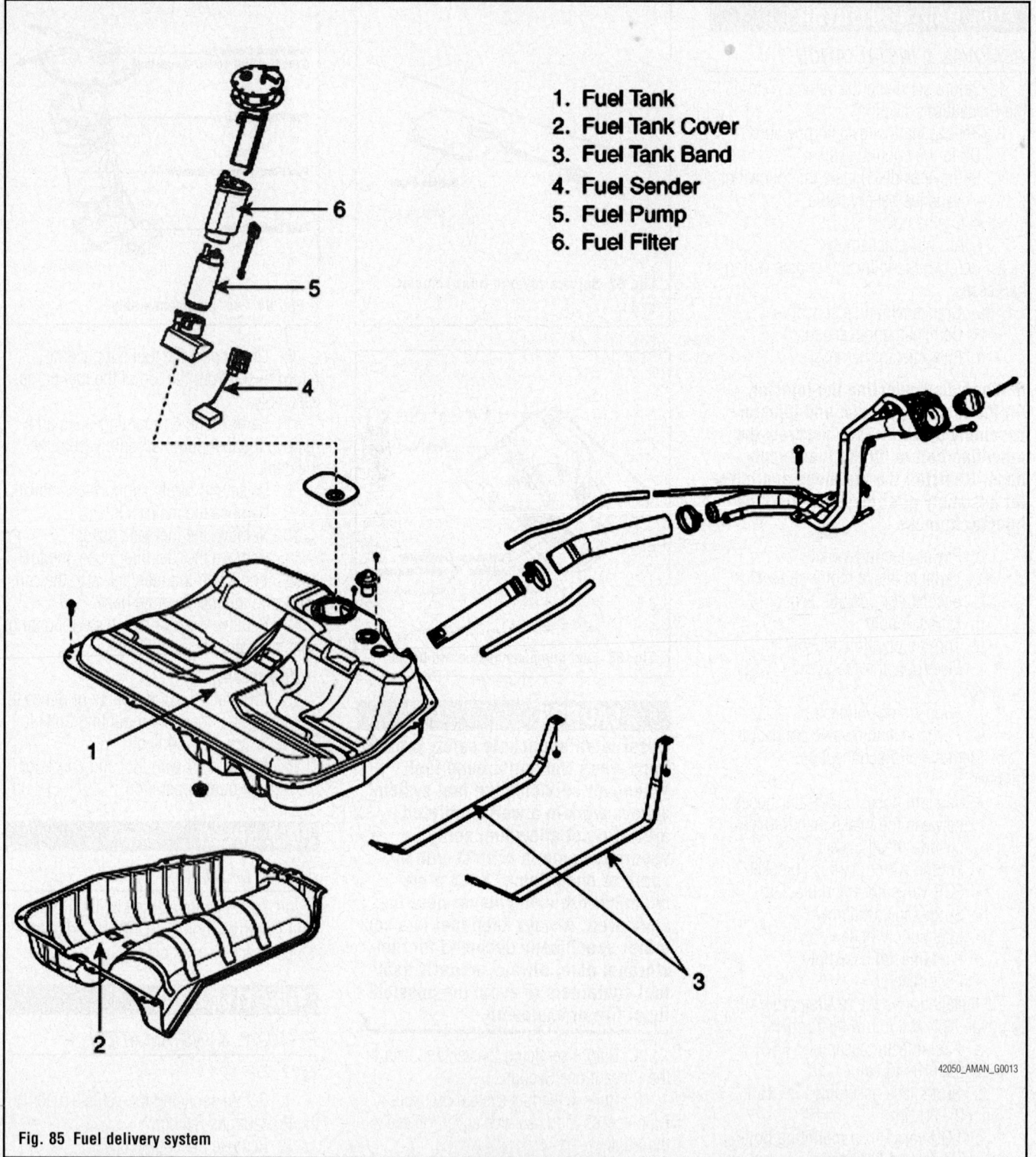

1. Fuel Tank
2. Fuel Tank Cover
3. Fuel Tank Band
4. Fuel Sender
5. Fuel Pump
6. Fuel Filter

42050_AMAN_G0013

Fig. 85 Fuel delivery system

- Injector harness connector
- Power steering switch connector
- Variable intake motor connector
- Accelerator position sensor connector
- EGR solenoid connector.

8. Remove the injector harness bracket bolts and injector connector.

9. Disconnect the throttle position sensor connector, the Electronic Throttle System (ETS) motor connector, the limp-home connector, and the purge solenoid valve connector.

10. Disconnect the vacuum hoses between the cylinder head cover and the intake manifold.

11. Disconnect the fuel pressure regulator vacuum hose, the purge control solenoid valve hose, and the throttle body heater hose.

12. Remove the bracket of the injector harness connector, the crankshaft sensor connector and the camshaft sensor connector.

13. Remove the EGR valve hose, the EGR valve mounting bolts, the EGR pipe bracket bolt, and the surge tank mounting bolts.

14. Remove the throttle body.

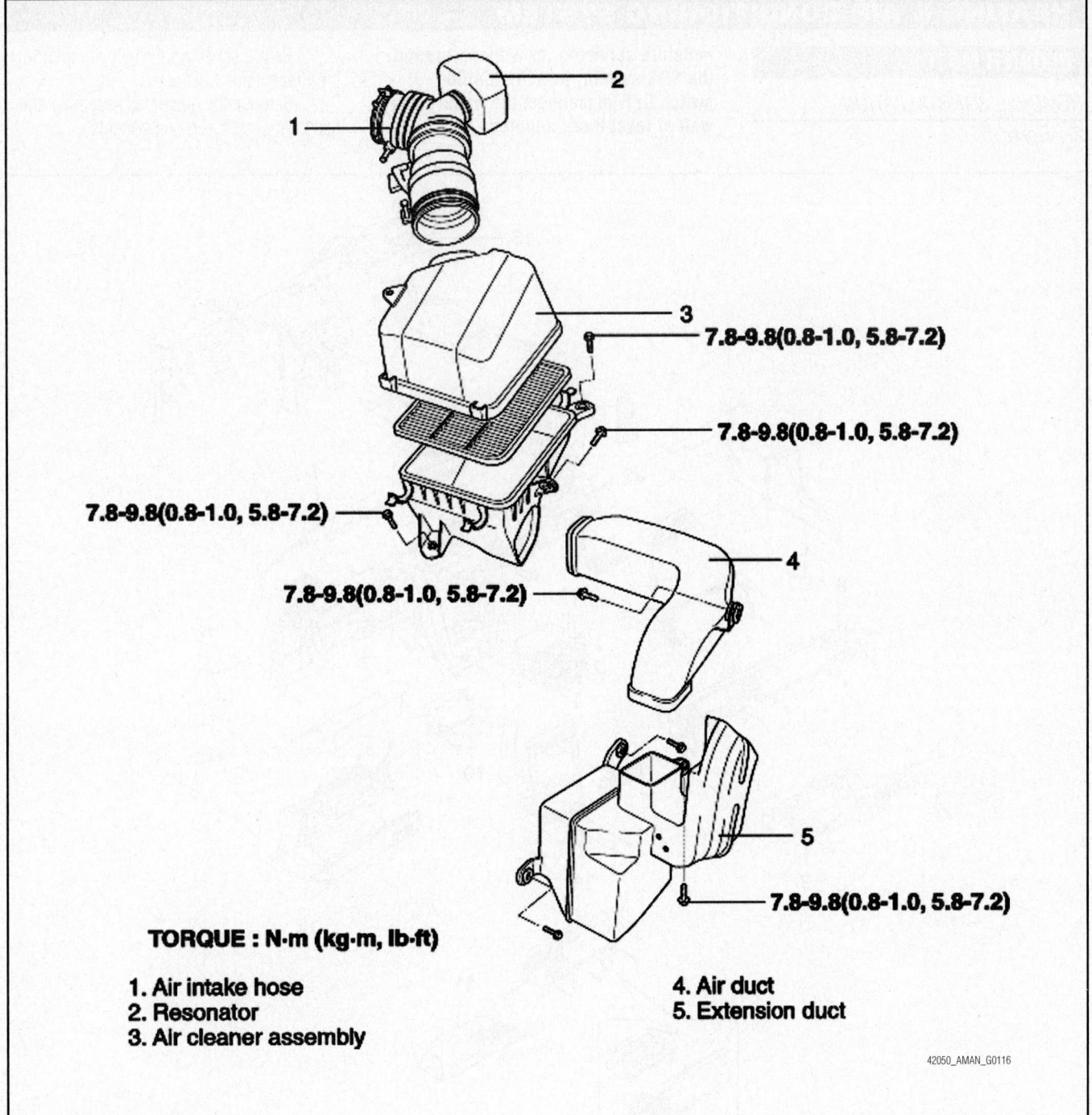

TORQUE : N·m (kg·m, lb·ft)

1. Air intake hose
2. Resonator
3. Air cleaner assembly

4. Air duct
5. Extension duct

42050_AMAN_G0116

Fig. 86 Intake components

To install:

15. Attach throttle body to surge tank assembly with bolts tightened to 15–17 ft. lbs. (20–24 Nm).

16. Install the EGR valve hose, the EGR valve mounting bolts, the EGR pipe fixing bracket bolt and the surge tank stay mounting bolts.

17. Install the bracket of the injector harness connector, the crankshaft angle sensor connector and the camshaft angle sensor connector.

18. Reconnect the vacuum hoses between the cylinder head cover and the intake manifold. Reconnect the fuel pressure regulator vacuum hose, the purge control solenoid valve hose and the throttle body heater hose.

19. Reconnect the throttle position sensor connector, the Electronic Throttle System (ETS) motor connector, the limp-home connector and the purge solenoid valve connector.

20. Reconnect the crankshaft angle sensor connector, the camshaft angle sensor connector, the injector harness connector, the power steering switch connector, the variable intake motor connector, the accelerator position sensor connector, the EGR solenoid connector and the injector connector.

21. Install the air cleaner assembly.

22. Reconnect the air flow sensor connector and the vacuum hose to the air intake hose. Reinstall the air hose clamp.

23. Reconnect the battery terminal.

24. Install the engine cover.

HEATING & AIR CONDITIONING SYSTEM

BLOWER MOTOR

REMOVAL & INSTALLATION

See Figure 87.

➡Before servicing, or working around, the SRS system, turn the ignition switch OFF, disconnect the battery and wait at least three minutes.

1. Before servicing the vehicle, refer to the Precautions Section.
2. Remove the glove box assembly and necessary crash pad components.

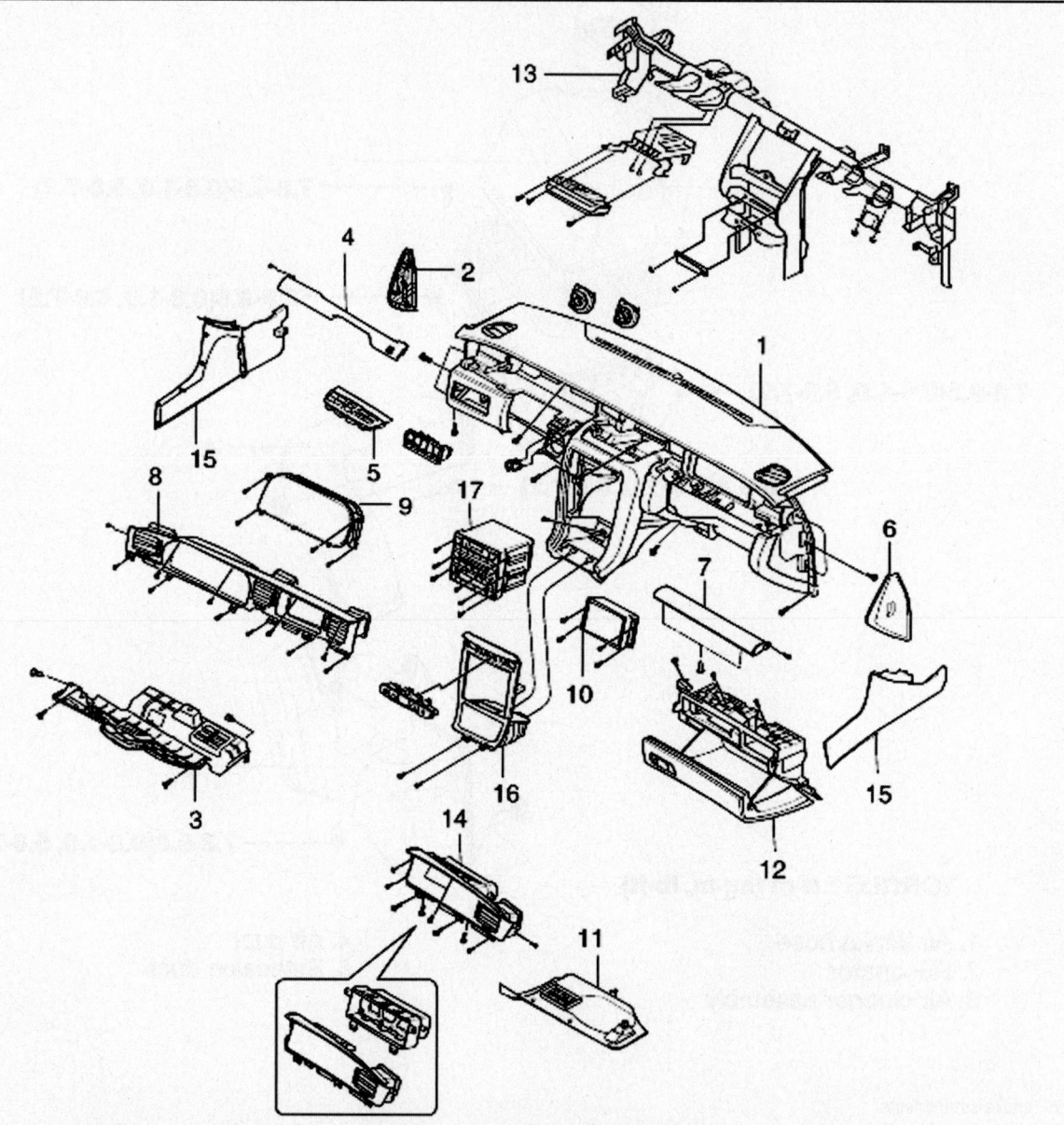

1. Instrument main panel assembly
2. Side mounting cover(LH)
3. Crash pad lower panel(LH)
4. Crash pad plate(LH)
5. Audio keyboard
6. Side mounting cover(RH)
7. Crash pad plate(RH)
8. Cluster facia panel
9. Cluster assembly
10. Audio monitor assembly
11. Under cover(RH)
12. Glove box assembly
13. Cowl cross bar assembly
14. Passenger airbag door & Airbag assembly
15. Side cover
16. Center facia panel
17. Audio & Heater controller assembly

42050_AMAN_G0093

Fig. 87 Crash pad components

3. Disconnect the blower motor electrical connector.

4. Remove the blower retaining screws.

5. Remove the blower motor from its mounting.

To install:

6. Installation is the reverse of the removal procedure.

REMOVAL & INSTALLATION

See Figures 88 and 89.

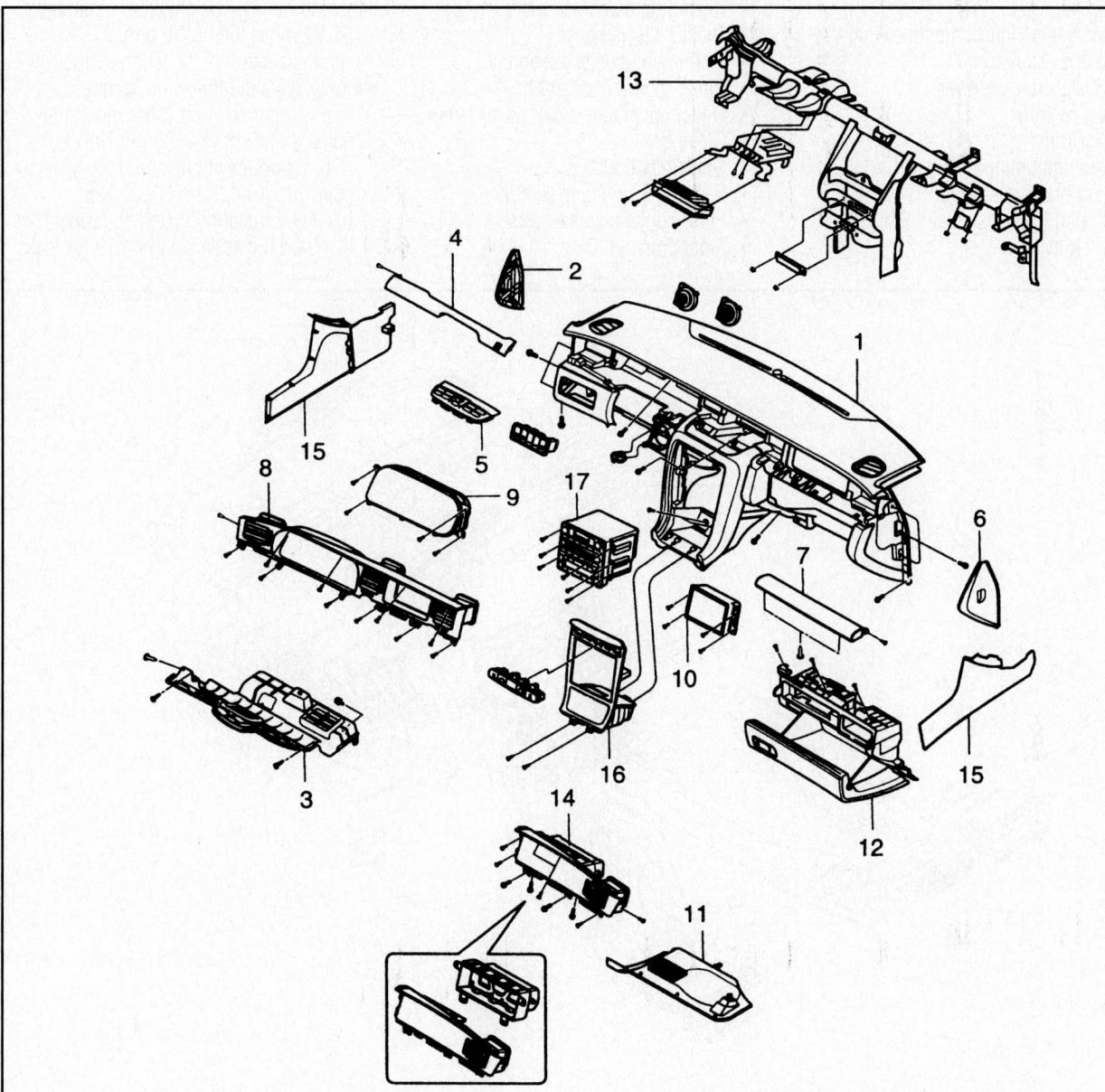

1. Instrument main panel assembly
2. Side mounting cover(LH)
3. Crash pad lower panel(LH)
4. Crash pad plate(LH)
5. Audio keyboard
6. Side mounting cover(RH)
7. Crash pad plate(RH)
8. Cluster facia panel
9. Cluster assembly
10. Audio monitor assembly
11. Under cover(RH)
12. Glove box assembly
13. Cowl cross bar assembly
14. Passenger airbag door & Airbag assembly
15. Side cover
16. Center facia panel
17. Audio & Heater controller assembly

09474_AMAN_G0004

Fig. 88 Exploded view of crash pad components

1. Before servicing the vehicle, refer to the Precautions Section.

2. Drain the engine coolant into a clean container for reuse.

3. Disconnect the negative battery cable.

4. Disconnect the heater hoses from the heater unit.

5. Remove or disconnect the following:
   - Front door scuff trim
   - Hood latch release lever
   - Cowl side trim
   - A-pillar trim
   - Left side mounting trim cover
   - Parking brake release lever
   - Crash pad lower panel and lower panel lamp
   - Steering wheel
   - Steering column shroud
   - Left side crash pad plate
   - Audio keyboard from main instrument panel assembly
   - Right side mounting trim cover
   - Right side crash pad plate
   - Cluster fascia panel
   - Instrument cluster assembly
   - Audio monitor assembly
   - Under cover and under cover lamp
   - Glove box
   - Glove box housing
   - Passenger airbag assembly
   - Floor console side covers
   - Floor console
   - Center fascia panel
   - Radio and heater controller

6. Disconnect the main instrument panel connect and remove the main instrument panel.

7. Remove the cowl crossbar mounting bolts and remove the crossbar assembly.

8. Remove the heater unit.

9. Installation is the reverse order of removal. Note the following Torques:

   a. Upper cowl crossbar mounting bolts: 24–40 ft. lbs. (33–55 Nm).

   b. Lower cowl crossbar mounting bolts: 12–19 ft. lbs. (17–26 Nm).

10. Fill the cooling system to proper level.

11. Start the engine and check for leaks.

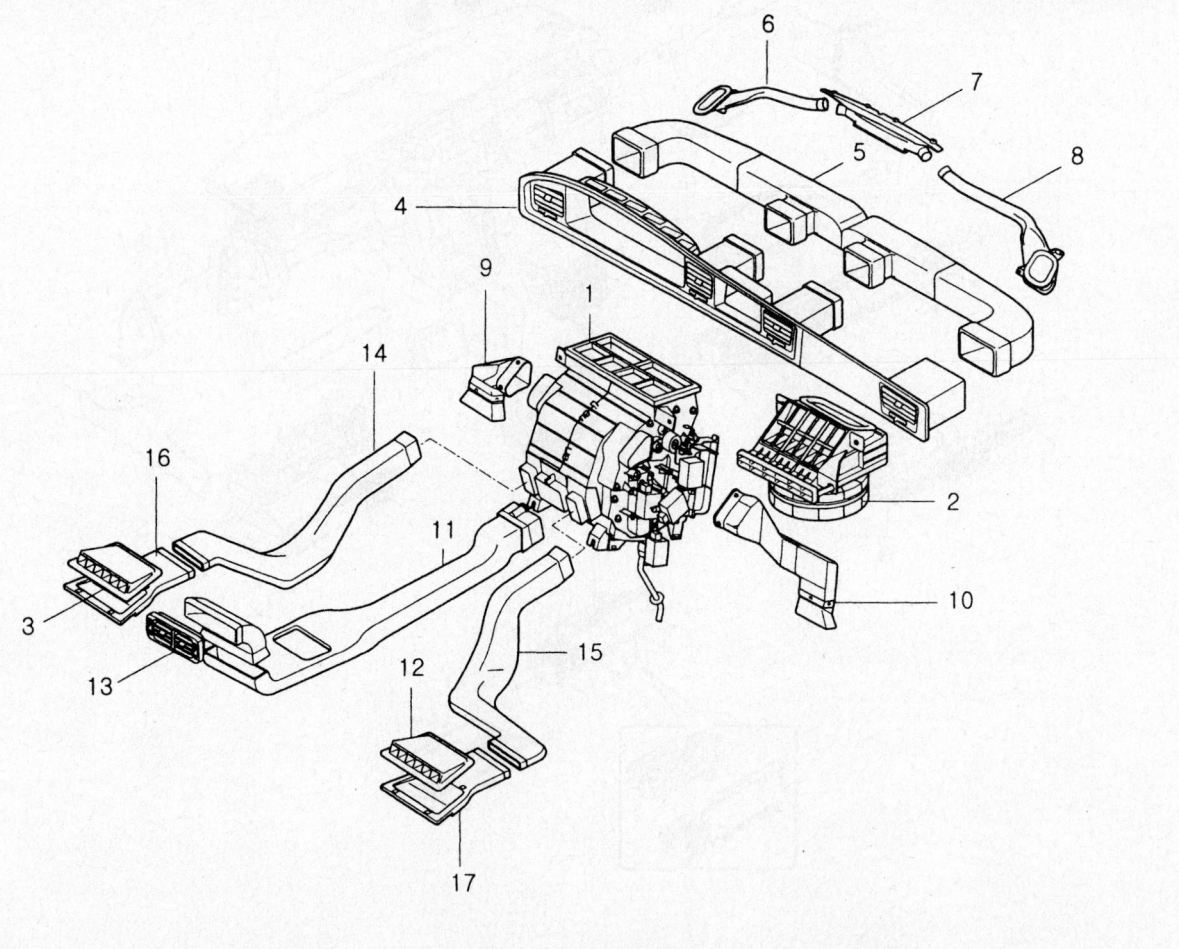

1. Heater unit
2. Blower unit
3. Rear floor duct grill-LH
4. Vent grill
5. Vent duct
6. Dip duct-LH
7. Dip duct center
8. Dip duct-RH
9. Driver side floor duct
10. Passenger side floor duct
11. Rear vent duct
12. Rear floor duct grill-RH
13. Rear vent grill
14. Rear floor duct LH (A)
15. Rear floor duct RH (A)
16. Rear floor duct LH (B)
17. Rear floor duct RH (B)

09474_AMAN_G0005

Fig. 89 Exploded view of heater unit components

STEERING

POWER RACK & PINION STEERING GEAR

REMOVAL & INSTALLATION

See Figures 90 through 92.

1. Before servicing the vehicle, refer to the Precautions Section.
2. Drain the power steering fluid.
3. Remove or disconnect the following:
   • Negative battery cable

• Pressure hose and return tube from reservoir
• Steering joint assembly connecting bolt
• Tie rod end from steering knuckle

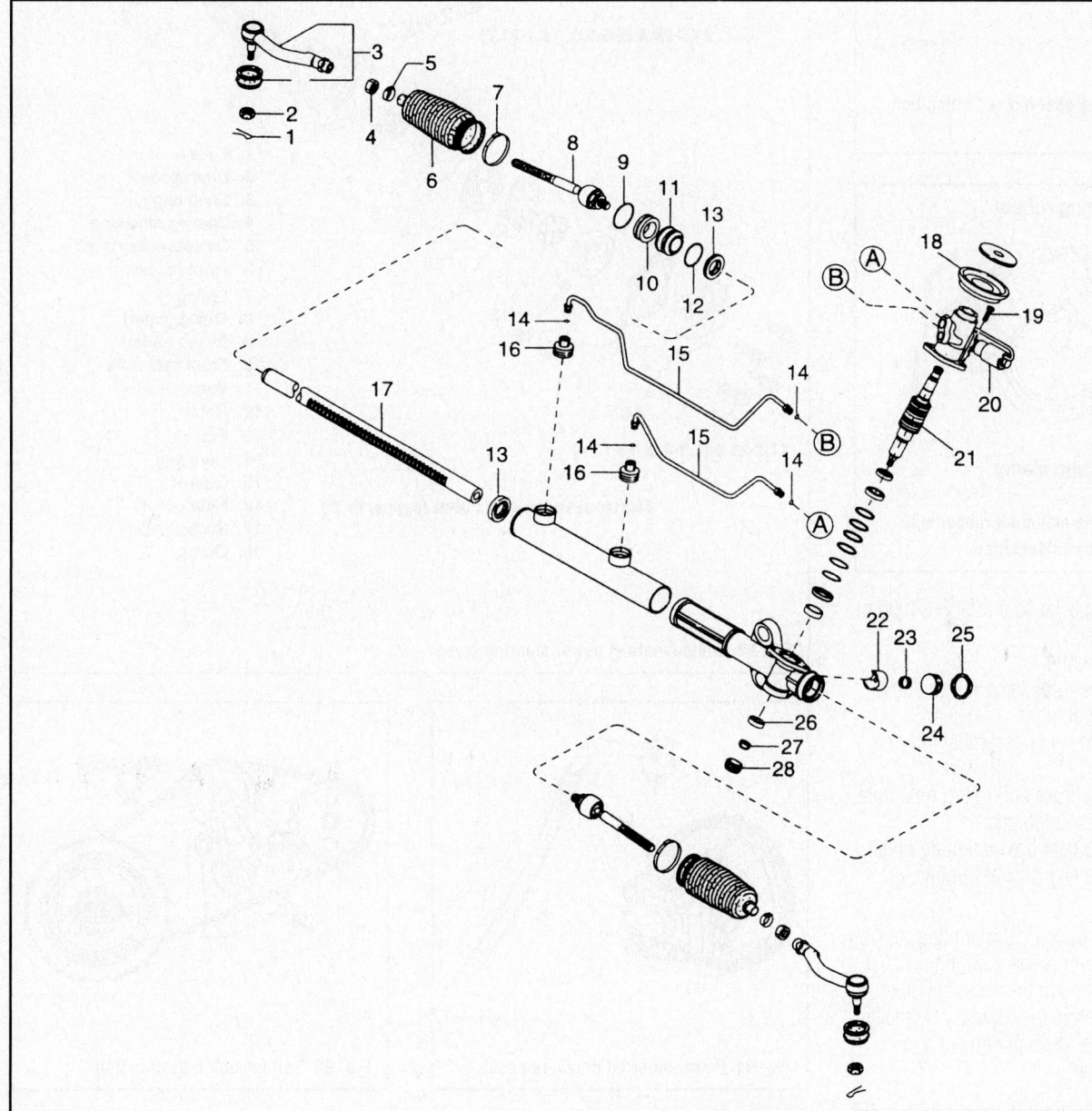

1. Cotter pin	8. Tie rod	15. Pressure pipe	22. Yoke assembly
2. Nut	9. Rack bushing	16. Damper valve assembly	23. Yoke spring
3. Tie rod end assembly	10. Insert bushing	17. Rack bar	24. Yoke cover
4. Lock nut	11. O-ring housing	18. Protector	25. Yoke nut
5. Clip	12. Oil seal	19. Cut off plug	26. Ball bearing
6. Bellows	13. Rack stopper	20. Solenoid valve assembly	27. Self locking nut
7. Steel band	14. Packing	21. Valve assembly	28. Pinion plug

09474_AMAN_G0025

Fig. 90 Exploded view of the rack and pinion assembly

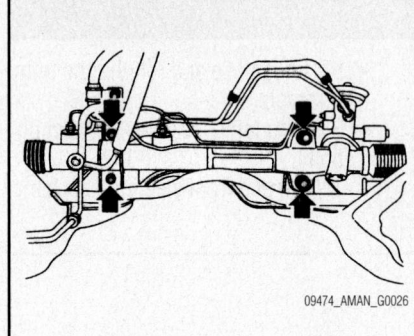

Fig. 91 Steering gear box mounting bolt locations

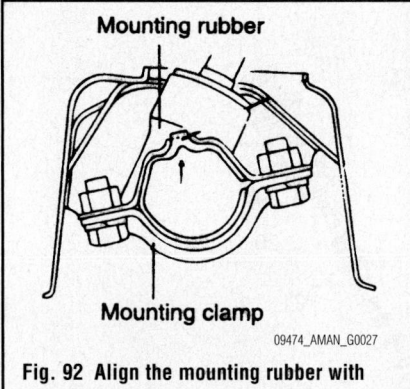

Fig. 92 Align the mounting rubber with the cross-member indentation

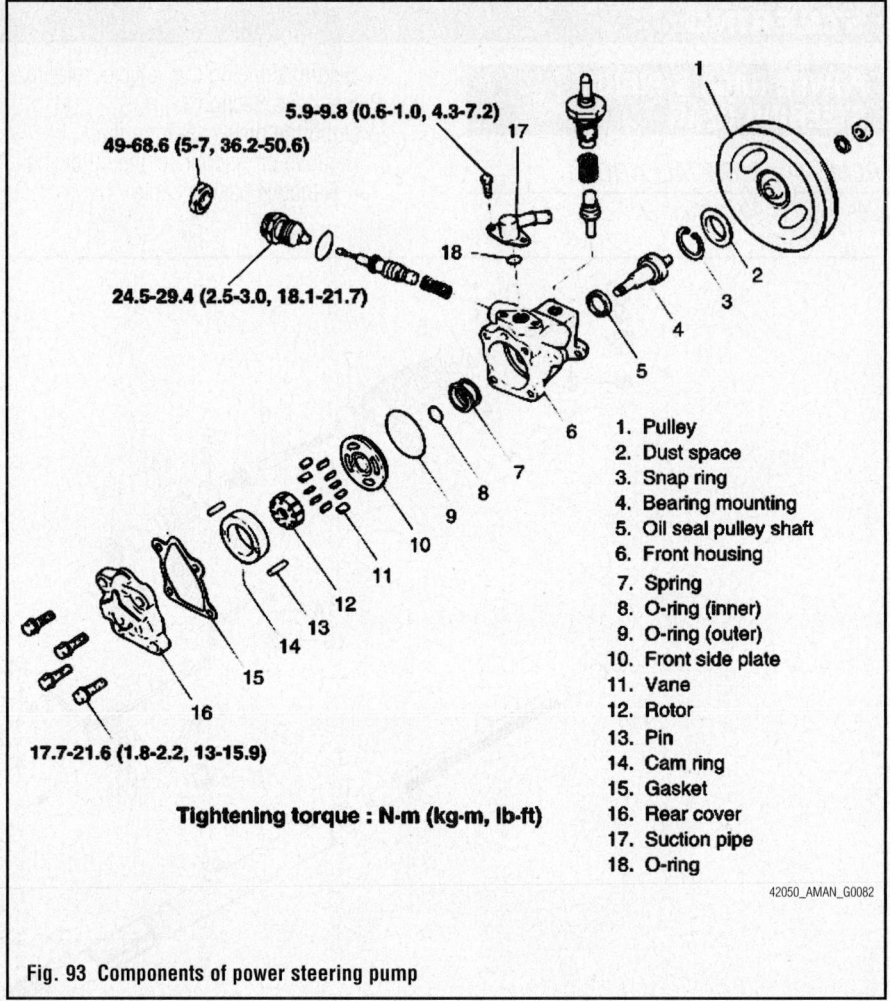

5.9-9.8 (0.6-1.0, 4.3-7.2)

49-68.6 (5-7, 36.2-50.6)

24.5-29.4 (2.5-3.0, 18.1-21.7)

17.7-21.6 (1.8-2.2, 13-15.9)

Tightening torque : N·m (kg·m, lb-ft)

1. Pulley
2. Dust space
3. Snap ring
4. Bearing mounting
5. Oil seal pulley shaft
6. Front housing
7. Spring
8. O-ring (inner)
9. O-ring (outer)
10. Front side plate
11. Vane
12. Rotor
13. Pin
14. Cam ring
15. Gasket
16. Rear cover
17. Suction pipe
18. O-ring

Fig. 93 Components of power steering pump

using Special Tool 09568-34000 or equivalent
• Front muffler
• Connecting bolts of the front and rear roll stopper
• Cross-member assembly mounting bolts
• Pressure hose and return tube from rack and pinion gear
• Steering gear box mounting bolts
• Steering gear box assembly

To install:

4. When installing the mounting rubber, align the projection of the mounting rubber with the indentation in the cross-member.

5. Installation is the reverse of removal.

6. Check the wheel alignment and adjust as necessary.

POWER STEERING PUMP

REMOVAL & INSTALLATION

See Figures 93 through 95.

1. Remove the pressure hose from the power steering pump.

➡**When assembling, use a new O-ring.**

2. Disconnect the suction hose from the suction connector and drain the fluid into a container.

✳✳ CAUTION

When removing the suction hose, cover the alternator with vinyl to keep oil from damaging that component.

3. Remove the drive belt.
4. Loosen the tension adjusting bolt to remove the drive belt.
5. Remove the power steering pump

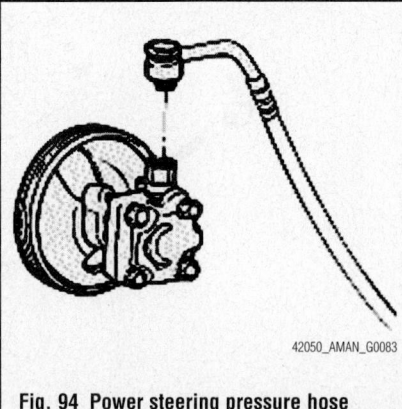

Fig. 94 Power steering pressure hose

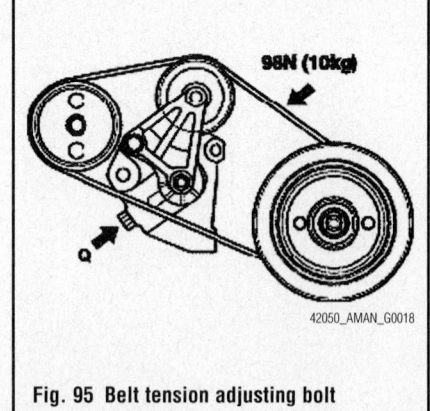

98N (10kg)

Fig. 95 Belt tension adjusting bolt

mounting bolts and disconnect the pressure switch connector.

To install:

6. Install the power steering pump to the power steering pump bracket.

7. Install the suction hose.

8. Install the ribbed V-belt and adjust its tension.

9. Connect the pressure hose to the power steering pump, and the suction hose to the oil reservoir.

→Install the hoses so that they are not twisted and they do not come in contact with any other parts.

10. Replenish the reservoir.
11. Bleed the system.
12. Check the power steering pump pressure.

BLEEDING

1. Before servicing the vehicle, refer to the Precautions Section.

2. With engine off, turn the steering wheel fully to the right and left several times.

→Do not allow the fluid level in the reservoir tank to go below the MIN level line. Check and add fluid as needed.

3. Run the engine at idle speed. Turn the steering wheel fully to the right and then fully to the left. Hold for about three seconds. Check for fluid leakage.

4. Repeat the above step several times at three second intervals.

→Do not hold the steering wheel in the locked position for more than ten seconds.

5. Check for air bubbles or cloudy fluid. If found, repeat the bleeding procedure.

6. Stop the engine and check the fluid level. Fill as required.

SUSPENSION

COIL SPRING

REMOVAL & INSTALLATION

1. Before servicing the vehicle, refer to the Precautions Section.
2. Remove the strut from the vehicle and install a spring compressor.
3. Remove the front cap.
4. Compress the coil spring so that the end of the spring comes away from the spring seat.
5. Remove or disconnect the following:
   • Upper strut mounting nut
   • Insulator
   • Upper spring seat
   • Compressed spring from the strut
   • Spring from the spring compressor

❋ CAUTION

Do not use an impact gun.

To install:
6. Compress the spring and install it on the strut.
7. Install or connect the following:
   • Upper spring seat
   • Insulator
   • Upper strut mount and Torque the nut to 29–36 ft. lbs. (39–49 Nm).
   • Strut to the vehicle
8. Check and/or adjust the wheel alignment.

LOWER BALL JOINT

REMOVAL & INSTALLATION

See Figures 96 through 98.

1. Before servicing the vehicle, refer to the Precautions Section.
2. Remove the wheel and tire.
3. Remove the lower arm connector from the lower arm.
4. Using the special tools(09551-31000, 09545-21100), remove the shock absorber mounting bushing.

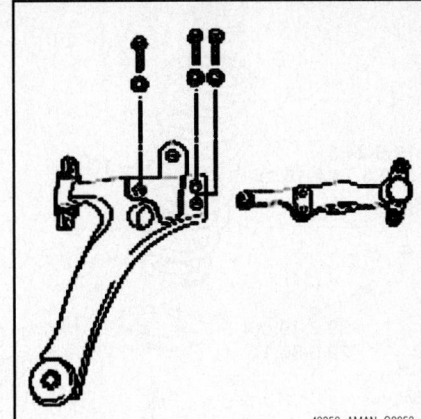

Fig. 96 Remove lower arm connector

42050_AMAN_G0053

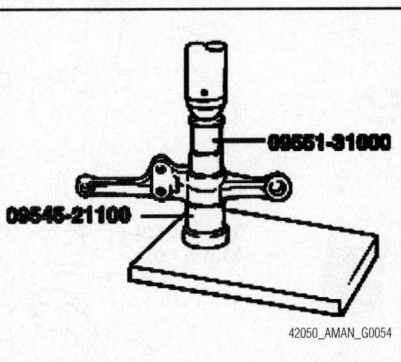

Fig. 97 Removing shock absorber mounting bushing

42050_AMAN_G0054

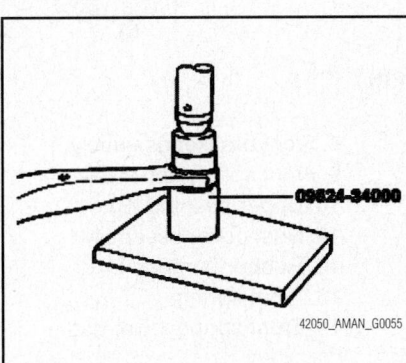

Fig. 98 Remove lower arm bushing

42050_AMAN_G0055

FRONT SUSPENSION

5. Using the special tools (09624-34000), remove the lower arm bushing (G).
6. Check the ball joint for rotating Torque:
   • If a crack is noted in the dust cover, replace the ball joint assembly
   • Measure the lower ball joint for rotating torque
   • Check that it is according to specification: 2–9 inch lbs. (0.2–1 Nm).

LOWER CONTROL ARM

REMOVAL & INSTALLATION

See Figure 99.

1. Before servicing the vehicle, refer to the Precautions Section.
2. Remove the front wheel.
3. Loosen the ball joint nut, but do not remove.
4. Using Special Tool 09455-21000 or equivalent, disconnect the lower arm ball joint from the lower arm connector.
5. Remove the ball joint assembly.
6. Remove the fork to lower arm connector mounting bolt.

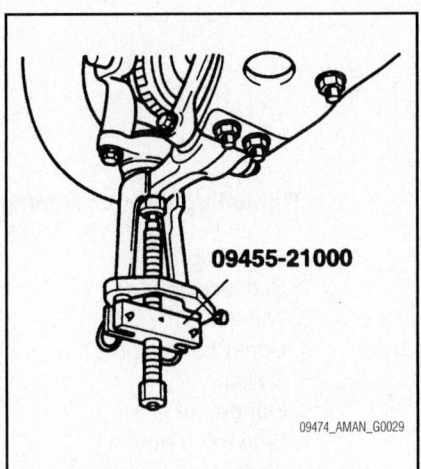

09474_AMAN_G0029

Fig. 99 Disconnect the lower ball joint using Special Tool 094455-21000

7. Remove the stabilizer link from the lower control arm.

8. Remove the lower control arm mounting bolts.

9. Remove the lower control arm.

To install:

10. Install or connect the following:
- Lower control arm and Torque the mounting bolts to 72–87 ft. lbs. (98–118 Nm).
- Tighten the rear bushing bolt to 87–101 ft. lbs. (118–137 Nm).
- Stabilizer bar link
- Fork to lower arm connector mounting bolt and tighten to 72–87 ft. lbs. (98–118 Nm).
- Ball joint assembly
- Front wheel

11. Check and/or adjust the wheel alignment.

MACPHERSON STRUT

REMOVAL & INSTALLATION

See Figure 100.

1. Before servicing the vehicle, refer to the Precautions Section.

2. Remove or disconnect the following:
- Front wheel

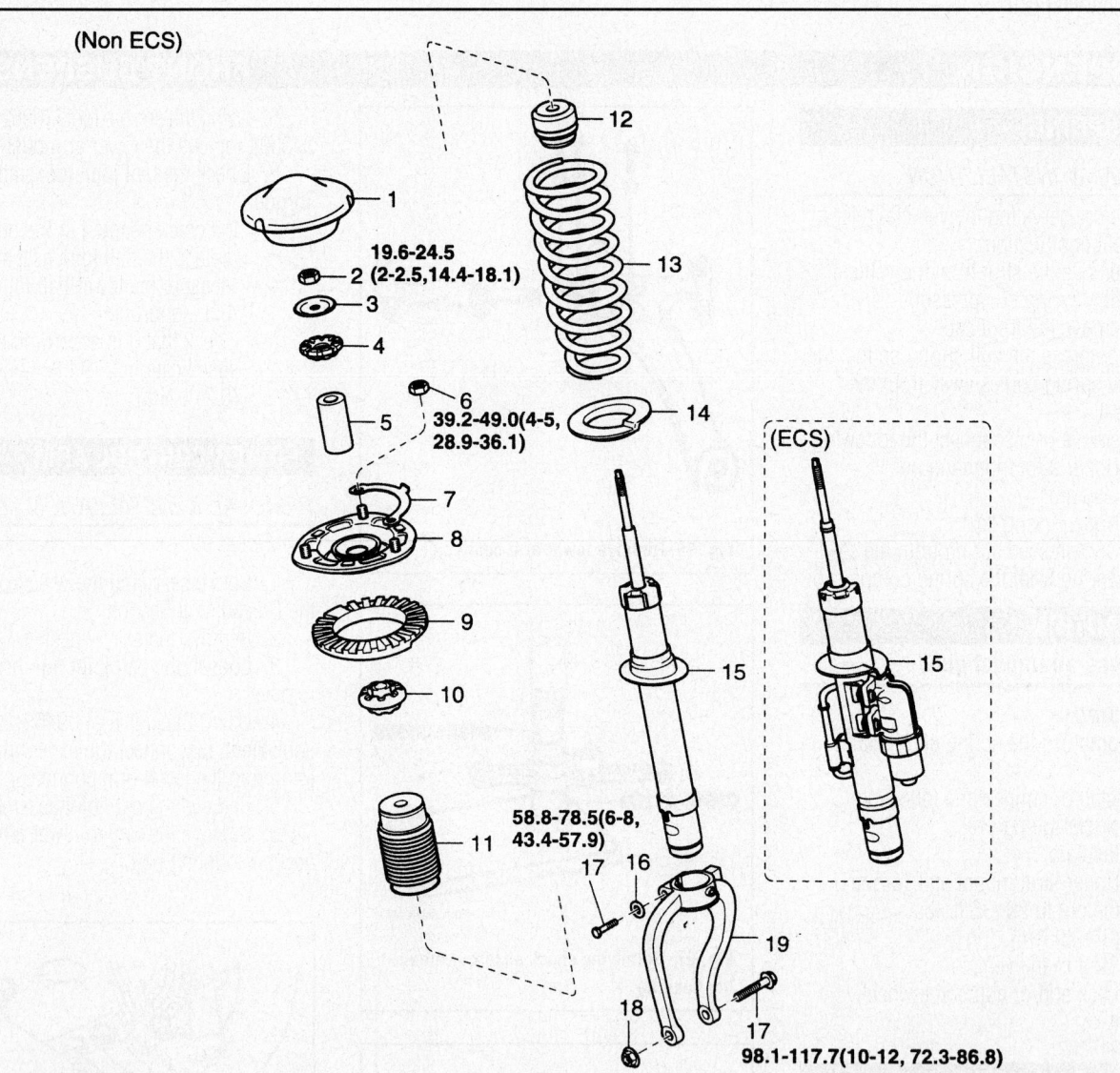

(Non ECS)

19.6-24.5 (2-2.5,14.4-18.1)

39.2-49.0(4-5, 28.9-36.1)

(ECS)

58.8-78.5(6-8, 43.4-57.9)

98.1-117.7(10-12, 72.3-86.8)

Tightening torque : N·m(kg·m,lb·ft)

1. Front cap
2. Self locking nut
3. Washer
4. Upper bushing (A)
5. Collar
6. Flange nut
7. Ring top mounting
8. Front bracket assembly
9. Front spring upper pad
10. Upper bushing (B)
11. Dust cover assembly
12. Rubber bumper
13. Front spring
14. Front spring lower pad
15. Front shock absorber assembly
16. Spring washer
17. Bolt
18. Nut
19. Front fork

09474_AMAN_G0028

Fig. 100 Strut assembly exploded view

- Brake hose bracket from shock absorber mounting fork
- Lower shock absorber mounting fork/lower arm connector mounting bolt
- Mounting fork from the shock absorber
- ECS wiring mounting bolt, if equipped
- Upper strut mounting nuts

3. Push the axle assembly upward and remove the strut assembly.

To install:

4. Install or connect the following:

- Strut assembly and tighten the lower mounting bolt to 72–87 ft. lbs. (98–118 Nm).
- Upper strut mounting nuts and tighten to 29–36 ft. lbs. (39–49 Nm).
- ECS wiring mounting bolt, if equipped
- Brake hose bracket
- Front wheel

5. Check the alignment and adjust as necessary.

OVERHAUL

See Figures 101 through 106.

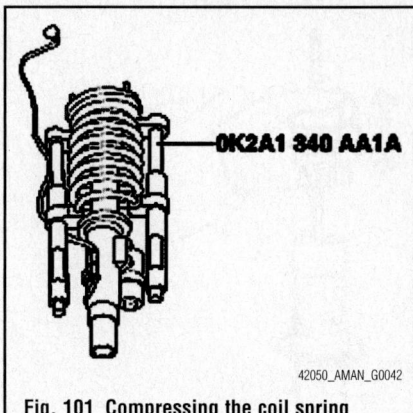

42050_AMAN_G0042

Fig. 101 Compressing the coil spring

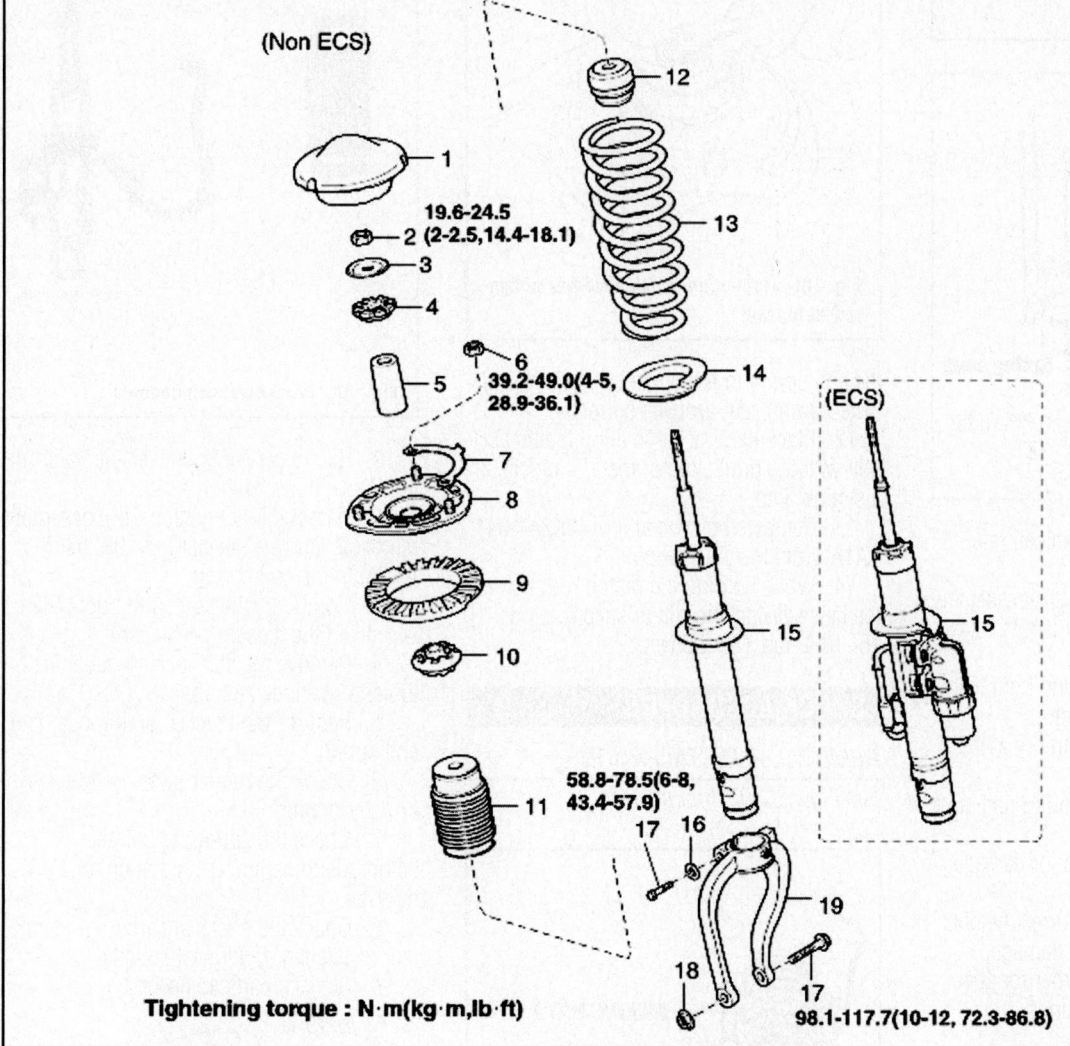

(Non ECS)

19.6-24.5
2 (2-2.5,14.4-18.1)

39.2-49.0(4-5, 28.9-36.1)

58.8-78.5(6-8, 43.4-57.9)

98.1-117.7(10-12, 72.3-86.8)

Tightening torque : N·m(kg·m,lb·ft)

1. Front cap
2. Self locking nut
3. Washer
4. Upper bushing (A)
5. Collar
6. Flange nut
7. Ring top mounting

8. Front bracket assembly
9. Front spring upper pad
10. Upper bushing (B)
11. Dust cover assembly
12. Rubber bumper
13. Front spring
14. Front spring lower pad

15. Front shock absorber assembly
16. Spring washer
17. Bolt
18. Nut
19. Front fork

42050_AMAN_G0048

Fig. 102 Front strut assembly

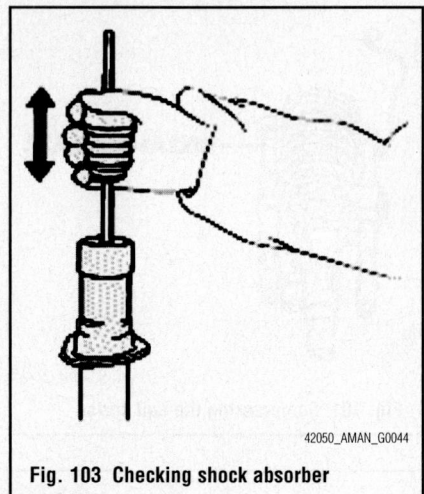

Fig. 103 Checking shock absorber resistance

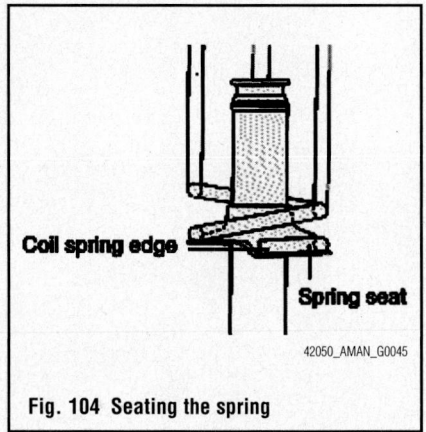

Fig. 104 Seating the spring

1. Before servicing the vehicle, refer to the Precautions Section.

2. With the MacPherson Strut assembly removed, attach a special tool (0K2A1 341 AA1A) to the coil spring.

3. Compress the coil spring until there is only a little tension on the strut.

4. Remove the self-locking nut at the top end of shock absorber.

5. Remove the bracket, spring pad, and coil spring.

6. Check the rubber parts for damage or deterioration.

7. Check the spring for correct height, deformation, deterioration, or damage.

8. Check the shock absorber for abnormal resistance or unusual sounds.

To install:

9. Install the special tool (0K2A1 341 AA1A) and compress the coil spring. After spring is fully compressed, install it on the shock absorber assembly.

10. After seating the dust cover, upper spring pad, bushings, and bracket, tighten the new self-locking nut temporarily.

11. Position the upper and lower ends of

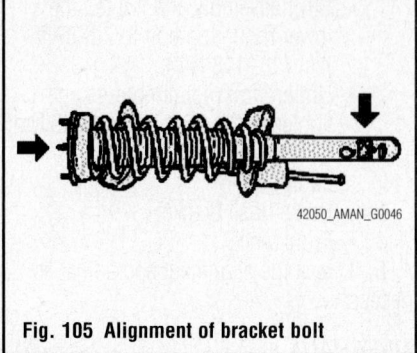

Fig. 105 Alignment of bracket bolt

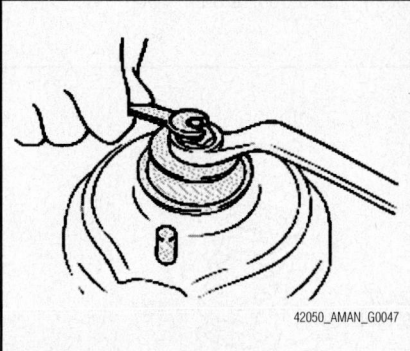

Fig. 106 Tightening shock absorber piston rod-to-bracket

the coil spring in the upper spring pad and lower spring seat grooves correctly.

12. Place the bracket to align the bracket bolt with the projection of the fork bracket in a straight line.

13. Remove the special tool (0K2A1 341 AA1A) from the coil spring.

14. While holding the piston rod, tighten the new self-locking nut to specification: 15–18 ft. lbs. (20–25 Nm).

SHOCK ABSORBERS

REMOVAL & INSTALLATION

See Figures 107 through 110.

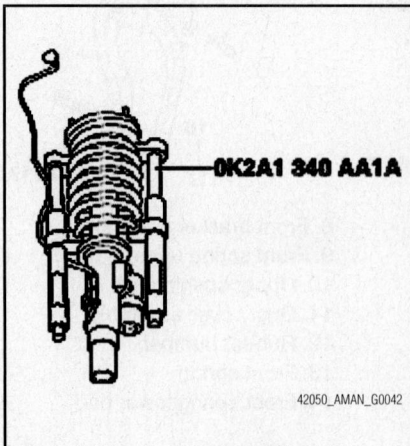

Fig. 107 Compressing the coil spring

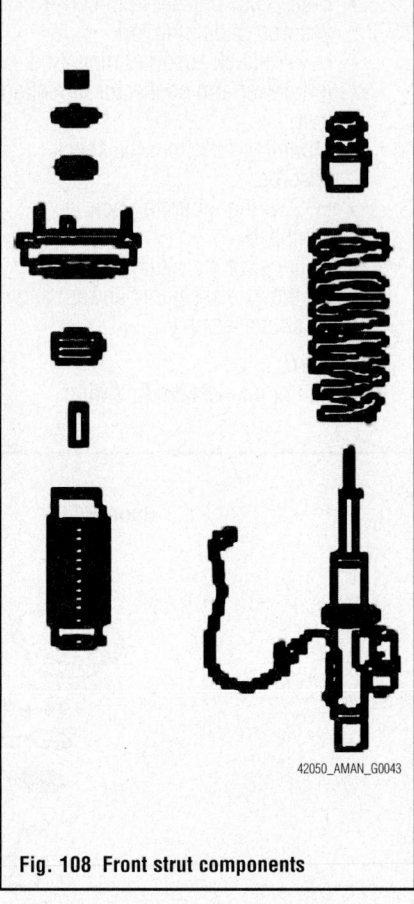

Fig. 108 Front strut components

1. Before servicing the vehicle, refer to the Precautions Section.

2. With the MacPherson Strut assembly removed, attach a special tool (0K2A1 341 AA1A) to the coil spring.

3. Compress the coil spring until there is only a little tension on the strut.

4. Remove the self-locking nut at the top end of shock absorber.

5. Remove the bracket, spring pad, and coil spring.

6. Check the rubber parts for damage or deterioration.

7. Check the spring for correct height, deformation, deterioration, or damage.

8. Check the shock absorber for abnormal resistance or unusual sounds.

9. Replace parts as needed.

To install:

10. Install the special tool (0K2A1 341 AA1A) and compress the coil spring. After spring is fully compressed, install it on the shock absorber assembly.

11. After seating the dust cover, upper spring pad, bushings, and bracket, tighten the new self-locking nut temporarily.

12. Position the upper and lower ends of the coil spring in the upper spring

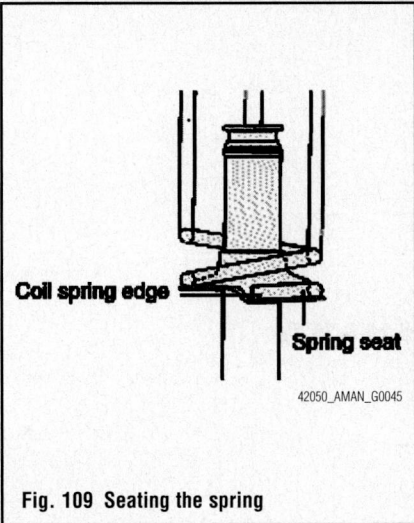

Fig. 109 Seating the spring

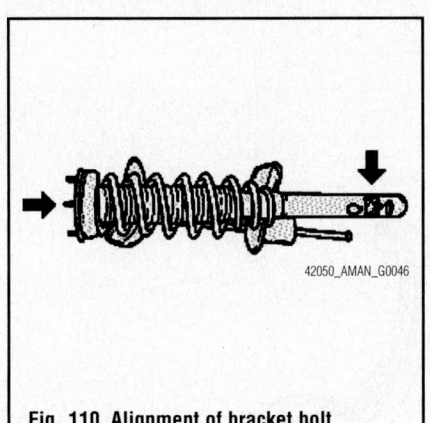

Fig. 110 Alignment of bracket bolt

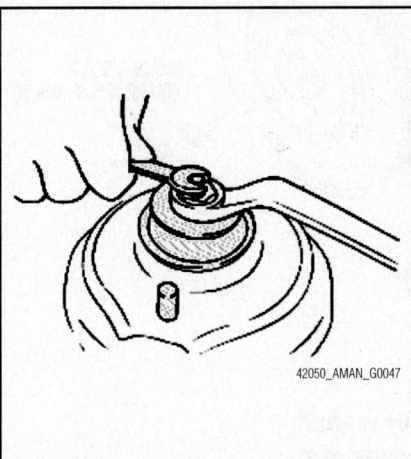

Fig. 111 Tightening shock absorber piston rod to bracket

pad and lower spring seat grooves correctly.

13. Place the bracket to align the bracket bolt with the projection of the fork bracket in a straight line.

14. Remove the special tool (0K2A1 341 AA1A) from the coil spring.

15. While holding the piston rod of the shock absorber, tighten the new self-locking nut to specification: 15–18 ft. lbs. (20–25 Nm).

STEERING KNUCKLE

REMOVAL & INSTALLATION

See Figures 112 through 116.

1. Before servicing the vehicle, refer to the Precautions Section.

2. Raise and support the vehicle safely.

3. Remove the front wheel and tire.

4. Disconnect the wheel speed sensor from the knuckle. Remove the caliper assembly and suspend it with wire.

5. Remove the ABS sensor and wire.

6. Using the special tool (09568-34000), disconnect the tie rod end from the knuckle:

- Be sure to tie the special tool to a nearby part with cord.
- Loosen the nut but do not remove it.

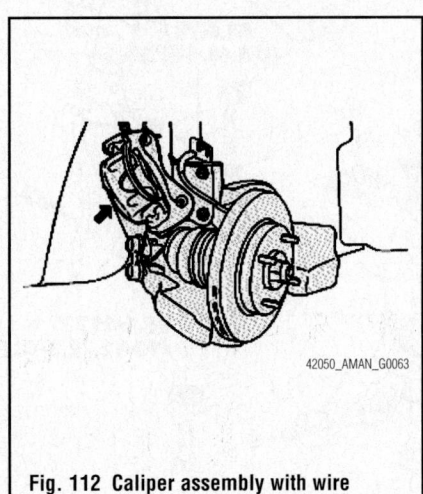

Fig. 112 Caliper assembly with wire

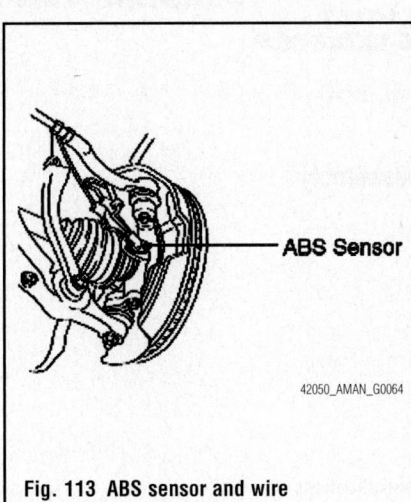

Fig. 113 ABS sensor and wire

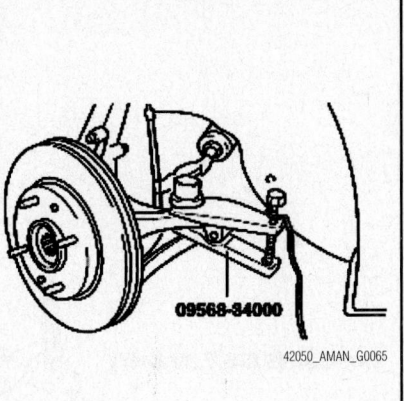

Fig. 114 Disconnecting tie rod end from knuckle

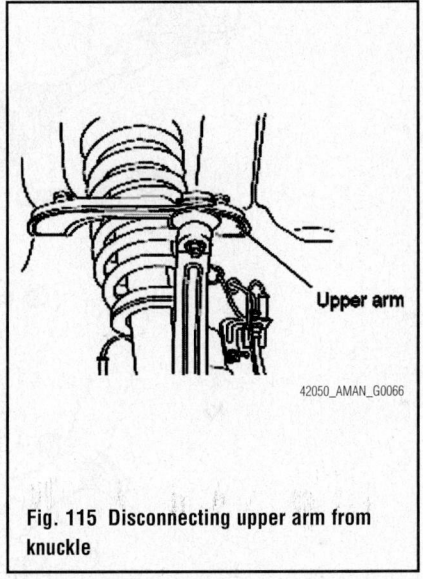

Fig. 115 Disconnecting upper arm from knuckle

7. Remove the 2 bolts and disconnect the ball joint from the knuckle.

8. Remove the brake disc from the knuckle.

9. Remove the split pin and drive shaft castle nut from the front hub.

10. Using a plastic hammer, disconnect the drive shaft from the axle hub.

11. Loosen the upper arm mounting nut, but do not remove it.

12. Using the special tool (09568-34000), disconnect the upper arm from the knuckle

13. Remove the front axle and knuckle together.

To install:

14. Installation is the reverse of the removal procedure.

15. Be careful to follow the Torque specifications in the following illustration.

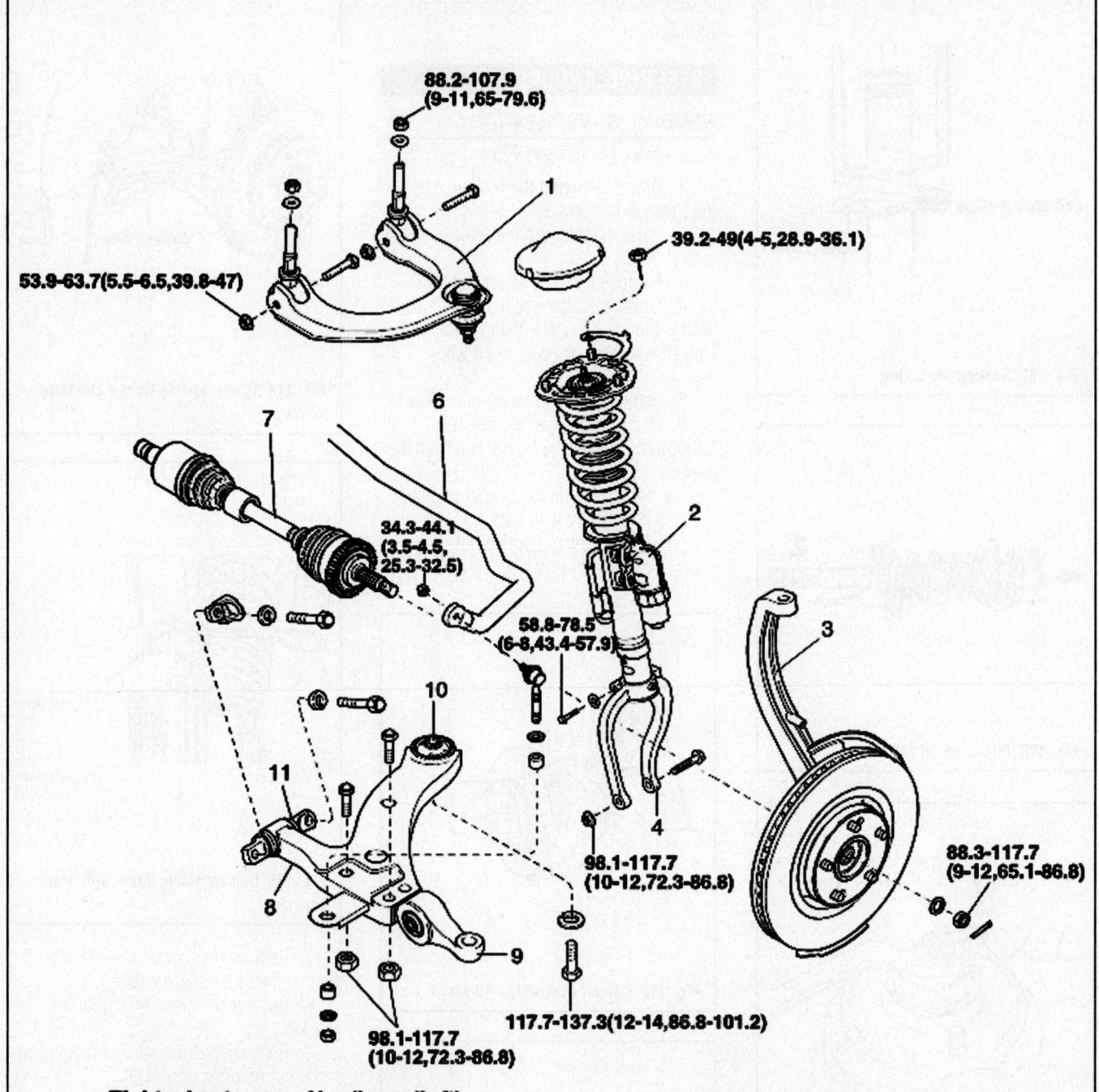

88.2-107.9
(9-11,65-79.6)

39.2-49(4-5,28.9-36.1)

53.9-63.7(5.5-6.5,39.8-47)

34.3-44.1
(3.5-4.5,
25.3-32.5)

58.8-78.5
(6-8,43.4-57.9)

98.1-117.7
(10-12,72.3-86.8)

88.3-117.7
(9-12,65.1-86.8)

117.7-137.3(12-14,86.8-101.2)

98.1-117.7
(10-12,72.3-86.8)

Tightening torque : N·m(kg·m,lb·ft)

1. Upper arm
2. Front shock absorber assembly
3. Knuckle
4. Fork
5. Stabilizer link
6. Stabilizer bar
7. Drive shaft
8. Lower arm
9. Lower arm connector
10. Bushing (G)
11. Bushing (A)

42050_AMAN_G0119

Fig. 116 Front lower arm components with Torque specifications

STABILIZER BAR

REMOVAL & INSTALLATION

See Figures 117 through 123.

1. Before servicing the vehicle, refer to the Precautions Section.
2. Raise and support the vehicle safely.
3. Remove the front wheel and tire.
4. Loosen the ball joint nut, but do not remove it.

Fig. 117 Disconnecting the ball joint

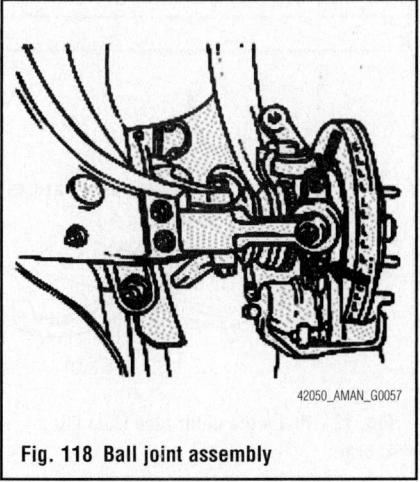

Fig. 118 Ball joint assembly

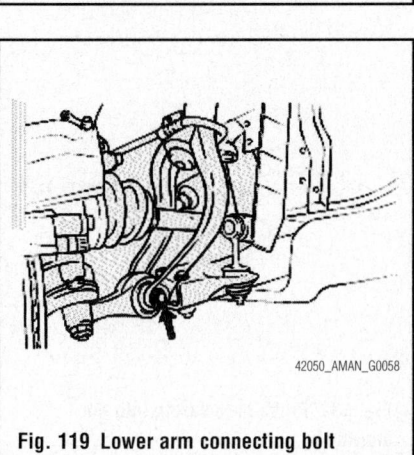

Fig. 119 Lower arm connecting bolt

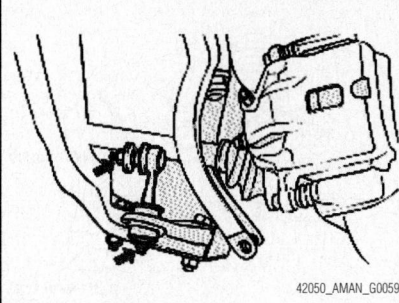

Fig. 120 Disconnecting the stabilizer bar from the lower arm

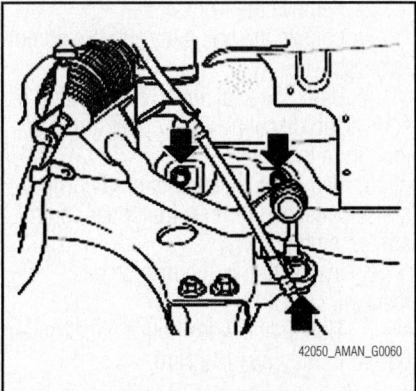

Fig. 121 Stabilizer link mounting bolt

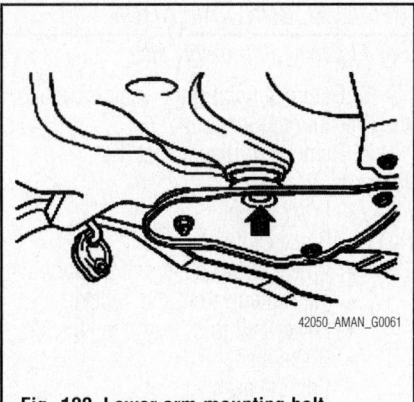

Fig. 122 Lower arm mounting bolt

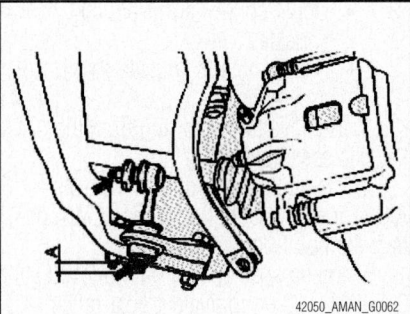

Fig. 123 Standard value for stabilizer link installation

5. Using the special tools (09455-21000), disconnect the lower arm ball joint from the lower arm connector.
6. Remove ball joint assembly.
7. Remove the fork to lower arm connector mounting bolt.
8. Remove the stabilizer link from the lower arm.
9. Remove the two lower arm mounting bolts and the stabilizer link mounting nut.
10. Remove the lower arm mounting bolt.

To install:

11. Installation is the reverse of removal.
12. Install the stabilizer link so that the distance (A) is at the standard value 0.118–0.197 in. (3–5 mm).

UPPER BALL JOINT

REMOVAL & INSTALLATION

See Figures 124 through 127.

1. Before servicing the vehicle, refer to the Precautions Section.
2. Remove the wheel and tire.
3. Loosen the ball joint nut, but do not remove it.
4. Using the special tool (09568-34000), disconnect the upper arm ball joint from the knuckle.

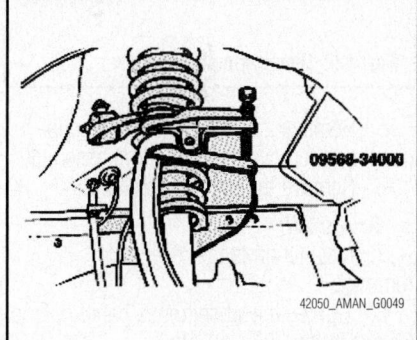

Fig. 124 Disconnecting upper arm ball joint

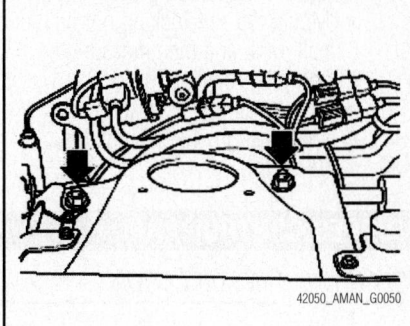

Fig. 125 Upper arm wheelhouse nuts

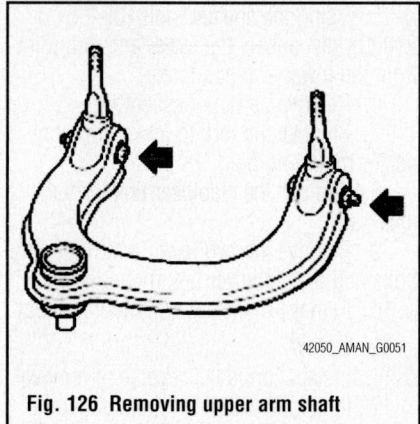

Fig. 126 Removing upper arm shaft

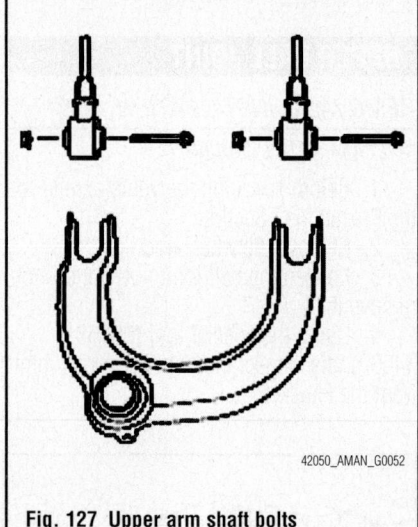

Fig. 127 Upper arm shaft bolts

5. Remove 2 nuts on the wheelhouse panel and remove the upper arm assembly.

6. Remove the upper arm shaft.

To install:

7. Installation is the reverse of removal.

8. Tighten the wheelhouse panel nuts to 65–80 ft. lbs. (88–108 Nm).

9. Check the ball joint for rotating Torque:

- If there is a crack in the dust cover, replace it and add grease.
- Mount the self-locking nut on the ball joint, and then measure the ball joint rotating Torque.
- Check that it is according to specification: 13–22 inch lbs. (2–3 Nm).

UPPER CONTROL ARM

REMOVAL & INSTALLATION

See Figure 128.

1. Before servicing the vehicle, refer to the Precautions Section.

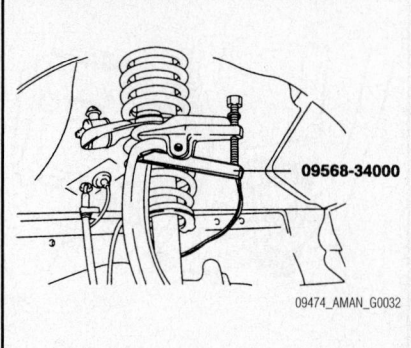

Fig. 128 Disconnecting the upper ball joint from the knuckle

2. Remove the front wheel.

3. Loosen the ball joint nut, but do not remove.

4. Using Special Tool 09568-34000 or equivalent, disconnect the upper arm ball joint from the knuckle.

5. Remove the two nuts on the wheelhouse panel and remove the upper control arm assembly.

6. Installation is the reverse of removal.

7. Tighten the wheelhouse panel nuts to 65–80 ft. lbs. (88–108 Nm).

WHEEL HUB AND BEARING

REMOVAL & INSTALLATION

See Figures 129 through 134.

1. Before servicing the vehicle, refer to the Precautions Section.

2. Remove or disconnect the following:

- Front wheel
- Brake caliper
- Wheel speed sensor, if equipped
- Tie rod end from the knuckle
- Lower ball joint from the knuckle
- Brake disc
- Spindle nut
- Halfshaft from the axle hub, using a plastic hammer if necessary.
- Upper control arm from the knuckle
- Front axle and knuckle as an assembly

3. Remove the snap ring from the knuckle assembly.

4. Using Special Tool 0K-130-331-AA0A or equivalent slide hammer, disconnect the hub from the knuckle.

5. Using special Tool 09455-21000 and 09545-34100 or equivalent gear puller, remove the wheel bearing inner race from the hub.

6. Remove the dust cover.

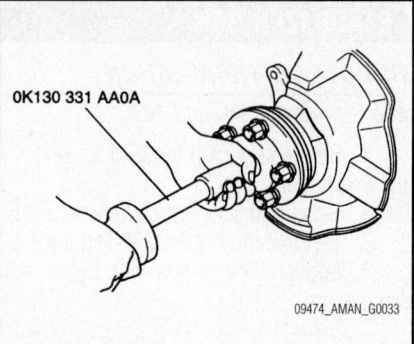

Fig. 129 Disconnect the hub assembly from the knuckle

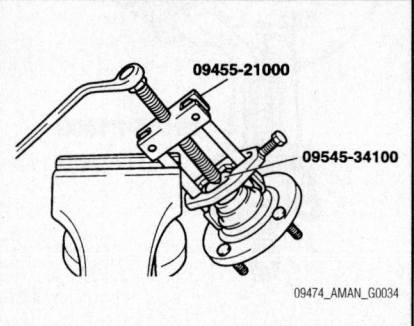

Fig. 130 Remove the wheel bearing inner race from the hub

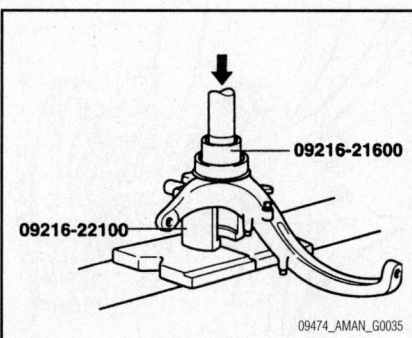

Fig. 131 Press the outer race from the knuckle

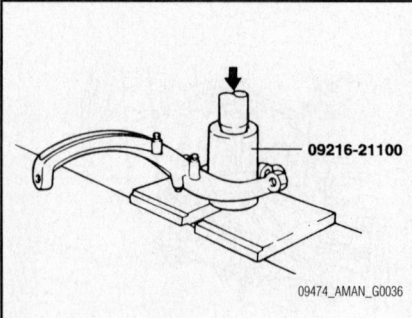

Fig. 132 Press the bearing into the knuckle

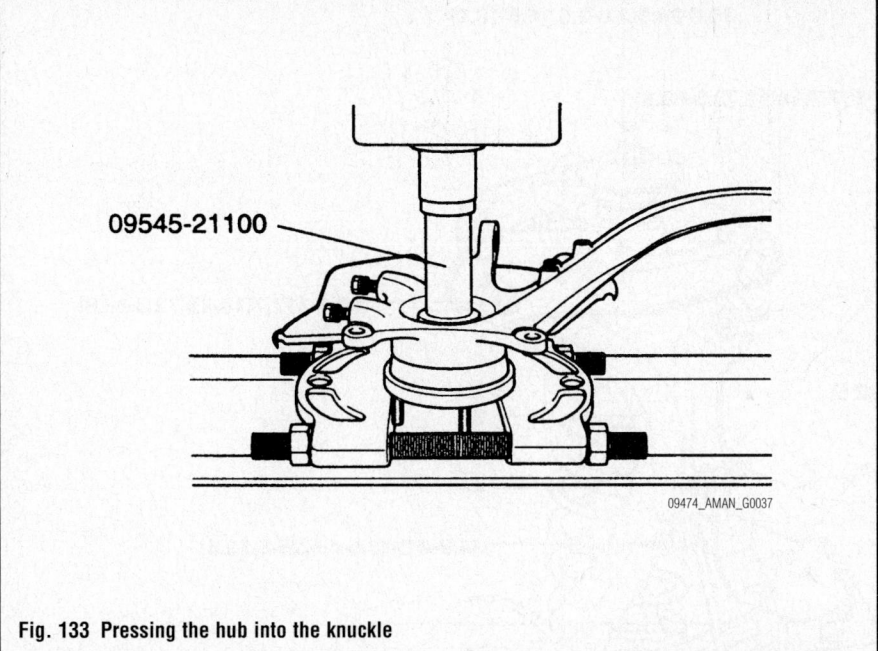

Fig. 133 Pressing the hub into the knuckle

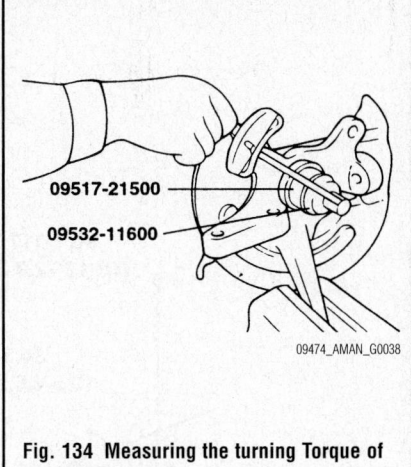

Fig. 134 Measuring the turning Torque of the wheel bearing

7. Using Special Tool 09216-21600 and 09216-22100, remove the wheel bearing outer race from the knuckle.

To install:

8. Apply a thin coat of grease to the knuckle and bearing contact surfaces.

9. Using Special Tool 09216-21100, press the bearing into the knuckle.

10. Install the snap ring into the groove of the knuckle.

11. Install the dust cover.

12. Using Special Tool 09545-21100, press the hub onto the knuckle.

❋❋ WARNING

Do not press against the outer race of the wheel bearing. Damage to the bearing assembly could occur.

13. Tighten the hub to the knuckle using Special Tool 09517-21500 to 148 ft. lbs. (200 Nm).

14. Rotate the hub to the seat the wheel bearing assembly.

15. Measure the wheel bearing turning Torque using Special Tools 0951-21500 and 09532-11600. Torque value should equal 9 inch lbs. (1 Nm) or less.

16. The remainder of the installation is the reverse of removal.

ADJUSTMENT

1. Check the hub for cracks and the splines for wear.

2. Check the snap ring for cracks or damage.

3. Check the knuckle inner surface for scoring and cracks.

SUSPENSION

COIL SPRING

REMOVAL & INSTALLATION

See Figures 135 through 139.

1. Before servicing the vehicle, refer to the Precautions Section.

2. Remove or disconnect the following:
- Rear wheel and tire
- Remove the shock absorber lower mounting bolts.
- Remove the upper arm and rear carrier mounting bolts.
- Remove the shock absorber mounting bracket.
- Remove coil spring and shock absorber as an assembly.

3. With the assembly removed, attach a special tool (0K2A1 341 AA1A) to the coil spring.

4. Compress the coil spring until there is only a little tension on the strut.

5. Remove the self-locking nut at the top end of shock absorber.

6. Remove the shock absorber mounting bracket, dust cover, spring upper pad.

7. Remove the special tool (0K2A1 341 AA1A) and then remove the coil spring.

To install:

8. Install the special tool (0K2A1 341 AA1A) and compress the coil spring. After spring is fully compressed, install it on the shock absorber assembly.

9. Install the dust cover, upper spring pad, shock absorber mounting bracket, and washer

10. Tighten the new self-locking nut temporarily.

11. Position the upper and lower ends of the coil spring in the upper spring pad and lower spring seat grooves correctly.

12. Install so that the upper spring pad is fit in the shock absorber mounting bracket correctly and that the spring urethane tube is located down.

REAR SUSPENSION

13. When the position of the bracket assembly is as shown in the illustration, tighten the new self-locking nut.

14. Remove the special tool.

15. While holding the piston rod, tighten the new self-locking nut to specification: 15–18 ft. lbs. (20–25 Nm).

16. Continue the installation of the coil spring and shock absorber assembly in the reverse order of the removal procedure.

17. Install or connect the following:
- Shock absorber mounting bracket tightening to 72–87 ft. lbs. (98–118 Nm).
- Upper arm and rear carrier mounting bolts to 72–87 ft. lbs. (98–118 Nm).
- Shock absorber lower mounting bolts 58–65 ft. lb. (79–88 Nm).
- Rear wheel and tire tightening lug nuts to 67–82 ft .lbs. (91–112 Nm).

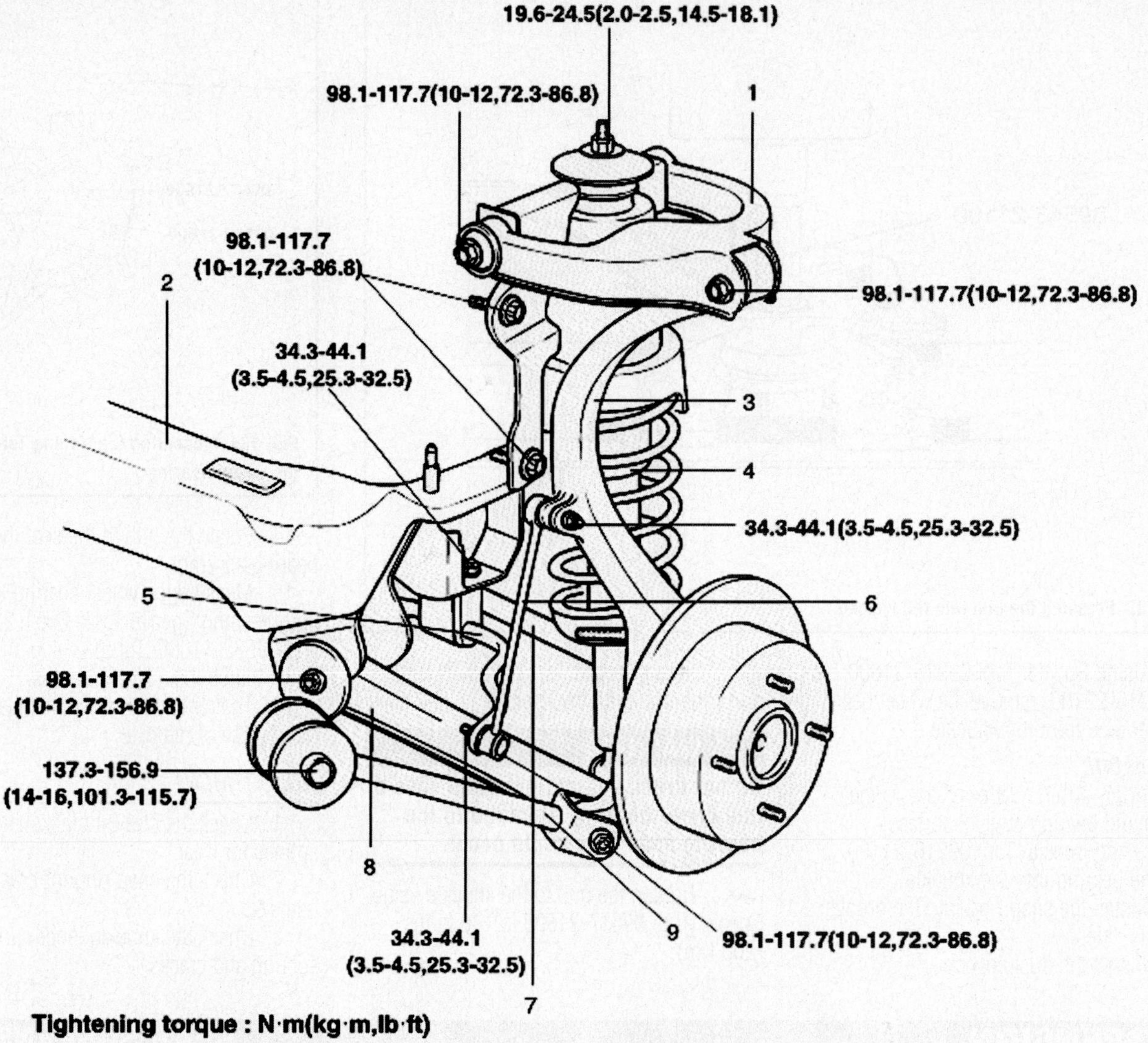

19.6-24.5(2.0-2.5,14.5-18.1)

98.1-117.7(10-12,72.3-86.8)

1

98.1-117.7
(10-12,72.3-86.8)

98.1-117.7(10-12,72.3-86.8)

2

34.3-44.1
(3.5-4.5,25.3-32.5)

3

4

34.3-44.1(3.5-4.5,25.3-32.5)

5

6

98.1-117.7
(10-12,72.3-86.8)

137.3-156.9
(14-16,101.3-115.7)

8

9

98.1-117.7(10-12,72.3-86.8)

34.3-44.1
(3.5-4.5,25.3-32.5)

7

Tightening torque : N·m(kg·m,lb·ft)

1. Upper arm assembly
2. Cross member assembly
3. Rear axle assembly
4. Rear shock absorber complete
5. Stabilizer bar

6. Stabilizer link
7. Assist arm assembly
8. Center arm assembly
9. Trailing arm assembly

42050_AMAN_G0067

Fig. 135 Rear suspension system

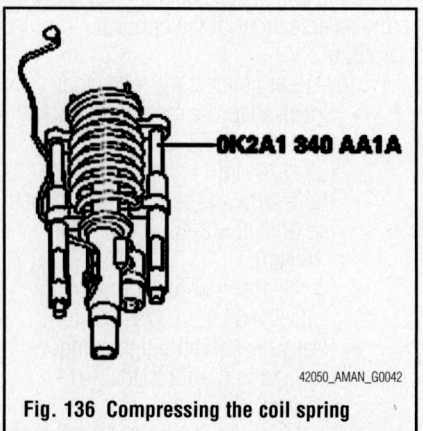

0K2A1 340 AA1A

42050_AMAN_G0042

Fig. 136 Compressing the coil spring

42050_AMAN_G0070

Fig. 137 Removing self-locking nut

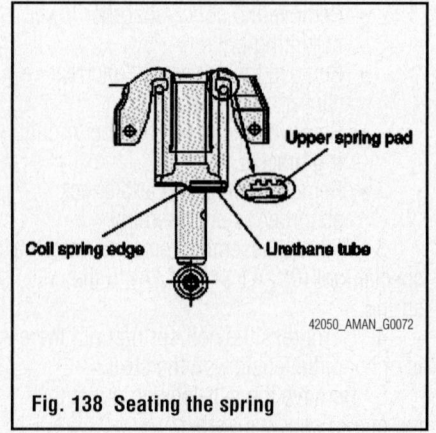

Upper spring pad

Coil spring edge

Urethane tube

42050_AMAN_G0072

Fig. 138 Seating the spring

Bracket assembly
Outside
42050_AMAN_G0073

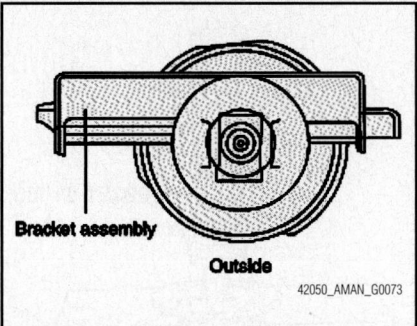

Fig. 139 Bracket assembly position

CONTROL ARMS/LINKS

REMOVAL & INSTALLATION

1. Before servicing the vehicle, refer to the Precautions Section.
2. Remove the rear wheel.
3. Remove upper control arm mounting bolts.
4. Remove the upper arm assembly.
5. Installation is the reverse of removal. Tighten the mounting bolts to specification: 72–87 ft. lbs. (98–118 Nm).

SHOCK ABSORBER

REMOVAL & INSTALLATION

See Figures 140 and 141.

1. Before servicing the vehicle, refer to the Precautions Section.
2. Remove Coil Spring and shock absorber assembly. Refer to Coil Spring removal and installation.
3. Remove the shock absorber lower mounting bolts.

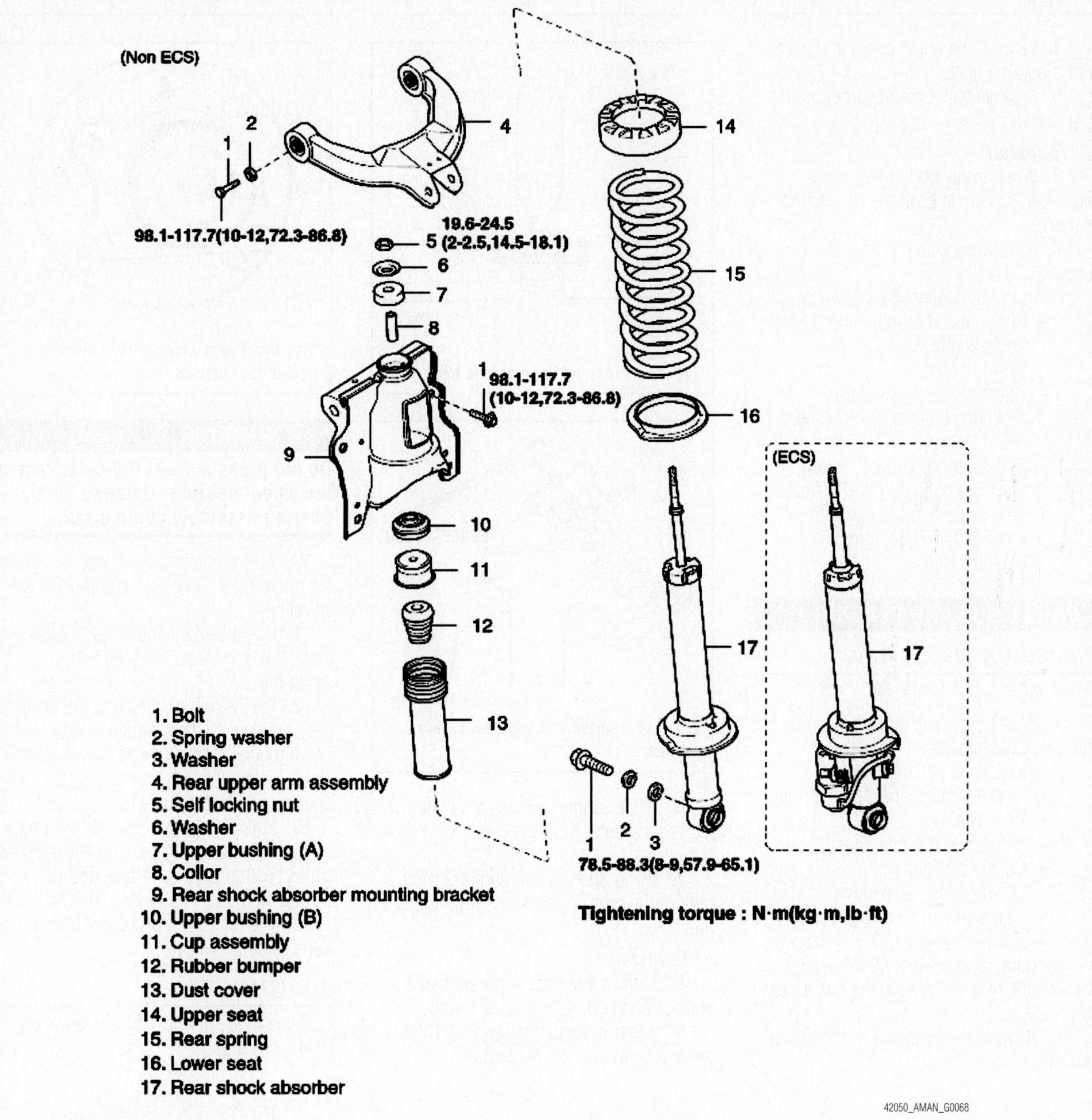

(Non ECS)
98.1-117.7(10-12,72.3-86.8)
19.6-24.5
5 (2-2.5,14.5-18.1)
98.1-117.7
(10-12,72.3-86.8)
(ECS)
78.5-88.3(8-9,57.9-65.1)
Tightening torque : N·m(kg·m,lb·ft)
1. Bolt
2. Spring washer
3. Washer
4. Rear upper arm assembly
5. Self locking nut
6. Washer
7. Upper bushing (A)
8. Collor
9. Rear shock absorber mounting bracket
10. Upper bushing (B)
11. Cup assembly
12. Rubber bumper
13. Dust cover
14. Upper seat
15. Rear spring
16. Lower seat
17. Rear shock absorber
42050_AMAN_G0068

Fig. 140 Rear shock absorber

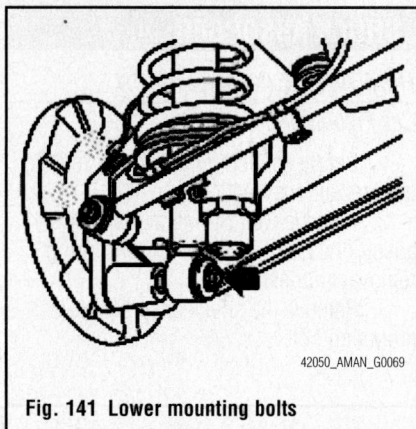

Fig. 141 Lower mounting bolts

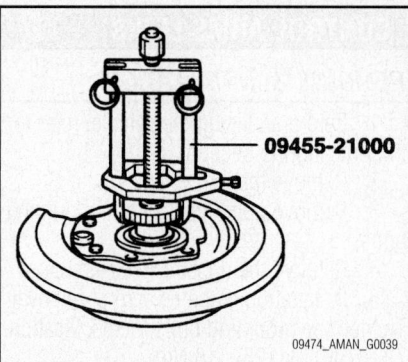

Fig. 142 Removing tone wheel from the hub assembly

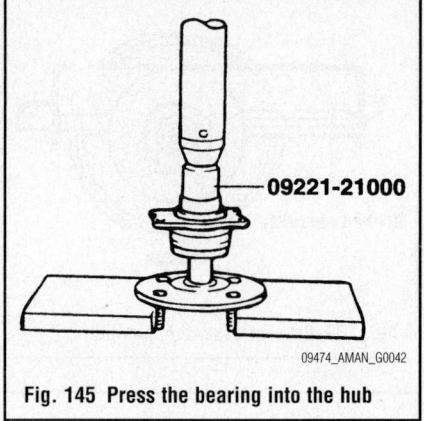

Fig. 145 Press the bearing into the hub

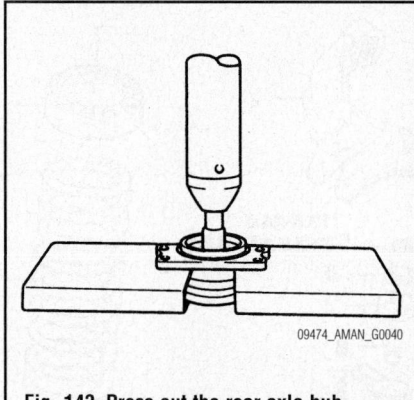

Fig. 143 Press out the rear axle hub

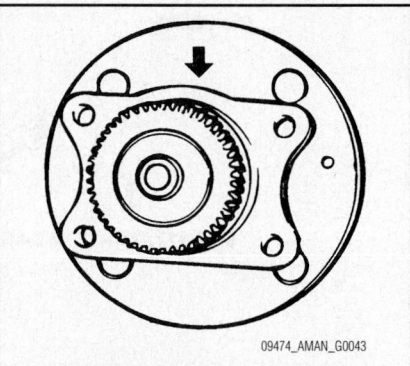

Fig. 146 The rounded area of outer race should face upward

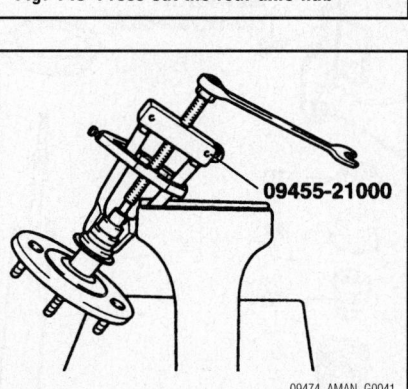

Fig. 144 Remove bearing inner race from the hub

4. Remove the upper arm and rear carrier mounting bolts.

5. Remove the shock absorber mounting bracket.

To install:

6. When installing the rear shock absorber, be sure to clear the connecting surface.

7. Installation is the reverse of removal.

8. Install or connect the following:
- Shock absorber mounting bracket tightening to 72–87 ft. lbs. (98–118 Nm).
- Upper arm and rear carrier mounting bolts to 72–87 ft. lbs. (98–118 Nm).
- Shock absorber lower mounting bolts 58–65 ft. lb. (79–88 Nm).
- Rear wheel and tire tightening lug nuts to 67–82 ft .lbs. (91–112 Nm).

WHEEL HUB AND BEARING

REMOVAL & INSTALLATION

See Figures 142 through 146.

1. Before servicing the vehicle, refer to the Precautions Section.

2. Release the parking brake.

3. Remove or disconnect the following:
- Rear wheel
- Wheel speed sensor, if equipped
- Caliper assembly
- Parking brake assembly
- Rear axle hub mounting bolts
- Hub assembly

4. Using Special Tool 09455-21000, remove the tone wheel from the hub assembly.

5. Remove the flange nut from the hub assembly.

6. While supporting the flange area of the bearing outer race, press out the axle hub.

7. Using Special Tool 09544-21000, remove the bearing inner race from the axle hub.

To install:

8. Apply a thin coat of grease to the knuckle and bearing contact surfaces.

9. Using Special Tool 09221-21000, press the bearing into the hub.

✳✳ WARNING

Do not press against the outer race of the wheel bearing. Damage to the bearing assembly could occur.

10. After tightening the flange nut, stake the nut to meet the concave portion of the spindle.

11. Using Special Tool 09221-21000, press the tone wheel onto the hub assembly.

12. Fix the hub assembly to the brake backing plate so the rounded area of the bearing outer race is placed facing upward.

13. Install the hub mounting bolts and tighten to 59–74 ft. lbs. (100–120 Nm).

14. Rotate the hub to seat the bearing.

15. Using a spring balance, measure the wheel bearing turning Torque. Torque should be 9 inch lbs. (1 Nm) or less.

16. The remainder of the installation is the reverse of removal.

ADJUSTMENT

If bearings are worn, replace with new sealed bearings.

KIA

11

Optima • Rio • Spectra

SPECIFICATIONS AND MAINTENANCE CHARTS

ENGINE AND VEHICLE IDENTIFICATION

Engine							Model Year	
Code ①	Liters (cc)	Cu. In.	Cyl.	Fuel Sys.	Engine Type	Eng. Mfg.	Code ②	Year
3	1.6 (1599)	97.6	4	EGI	DOHC	KIA	6	2006
2	2.0 (1975)	120.5	4	EGI	DOHC	KIA	7	2007
S	2.4 (2359)	144	4	EGI	DOHC	KIA	8	2008
8	2.7 (2656)	163	6	EGI	DOHC	KIA		

EGI: Electronic Gasoline Injection

DOHC: Double Overhead Camshafts

① 8th digit of VIN

② 10th digit of VIN

22140_KIAC_C0001

GENERAL ENGINE SPECIFICATIONS

Year	Model	Engine Displacement Liters	Engine VIN	Net Horsepower @ rpm	Net Torque @ rpm (ft. lbs.)	Bore x Stroke (in.)	Compression Ratio	Oil Pressure @ rpm
2006	Rio	1.6	3	110@6000	107@4500	3.01x3.43	10.0:1	15.6@710-730
	Spectra	2.0	2	138@6000	136@4500	3.23x8.68	10.1:1	43-57@3000
	Optima	2.4	S	149@6000	159@4500	3.41x3.94	10.0:1	43-57@3000
		2.7	8	178@6000	181@4000	3.41x2.95	10.0:1	43-57@3000
2007	Rio	1.6	3	110@6000	107@4500	3.01x3.43	10.0:1	15.6@710-730
	Spectra	2.0	2	138@6000	136@4500	3.23x8.68	10.1:1	43-57@3000
	Optima	2.4	S	149@6000	159@4500	3.41x3.94	10.0:1	43-57@3000
		2.7	8	178@6000	181@4000	3.41x2.95	10.0:1	43-57@3000
2008	Rio	1.6	3	110@6000	107@4500	3.01x3.43	10.0:1	15.6@710-730
	Spectra	2.0	2	138@6000	136@4500	3.23x8.68	10.1:1	43-57@3000
	Optima	2.4	S	149@6000	159@4500	3.41x3.94	10.0:1	43-57@3000
		2.7	8	178@6000	181@4000	3.41x2.95	10.0:1	43-57@3000

22140_KIAC_C0002

ENGINE TUNE-UP SPECIFICATIONS

Year	Engine Displacement Liters	Engine VIN	Spark Plug Gap (in.)	Ignition Timing (deg.) MT	Ignition Timing (deg.) AT	Fuel Pump (psi)	Idle Speed (rpm) MT	Idle Speed (rpm) AT	Valve Clearance Intake	Valve Clearance Exhaust
2006	1.6	3	0.039-0.043	1-10B	1-10B	49.8	710-730	710-730	HYD	HYD
	2.0	2	0.039-0.043	3-7B	3-7B	49.8	750-850	750-850	HYD	HYD
	2.4	S	0.039-0.043	3-7B	3-7B	46-49	700-900	700-900	HYD	HYD
	2.7	8	0.039-0.043	—	7-17B	46-49	—	600-800	HYD	HYD
2007	1.6	3	0.039-0.043	1-10B	1-10B	49.8	710-730	710-730	HYD	HYD
	2.0	2	0.039-0.043	3-7B	3-7B	49.8	750-850	750-850	HYD	HYD
	2.4	S	0.039-0.043	3-7B	3-7B	46-49	700-900	700-900	HYD	HYD
	2.7	8	0.039-0.043	—	7-17B	46-49	—	600-800	HYD	HYD
2008	1.6	3	0.039-0.043	1-10B	1-10B	49.8	710-730	710-730	HYD	HYD
	2.0	2	0.039-0.043	3-7B	3-7B	49.8	750-850	750-850	HYD	HYD
	2.4	S	0.039-0.043	3-7B	3-7B	46-49	700-900	700-900	HYD	HYD
	2.7	8	0.039-0.043	—	7-17B	46-49	—	600-800	HYD	HYD

NOTE: The Vehicle Emission Control Information label often reflects specification changes made during production.

The label figures must be used if they differ from those in this chart.

B: Before top dead center

HYD: Hydraulic

22140_KIAC_C0003

CAPACITIES

Year	Model	Engine Displacement Liters	Engine VIN	Engine Oil with Filter	Transaxle (pts.) Manual	Transaxle (pts.) Auto.	Fuel Tank (gal.)	Cooling System (qts.)
2006	Rio	1.6	3	3.49	4.2	12.9	11.9	6.1
	Spectra	2.0	2	4.23	4.5	13.0	14.5	6.9
	Optima	2.4	S	4.23	4.6	16.4	16.4	7.4
		2.7	8	4.75	—	20.1	17.2	9
2007	Rio	1.6	3	3.49	4.2	12.9	11.9	6.1
	Spectra	2.0	2	4.23	4.5	13.0	14.5	6.9
	Optima	2.4	S	4.23	4.6	16.4	16.4	7.4
		2.7	8	4.75	—	20.1	17.2	9
2008	Rio	1.6	3	3.49	4.2	12.9	11.9	6.1
	Spectra	2.0	2	4.23	4.5	13.0	14.5	6.9
	Optima	2.4	S	4.23	4.6	16.4	16.4	7.4
		2.7	8	4.75	—	20.1	17.2	9

22140_KIAC_C0004

VALVE SPECIFICATIONS

Year	Engine Displacement Liters	Engine VIN	Seat Angle (deg.)	Face Angle (deg.)	Maximum out of Square (in.)	Spring Free Length (in.)	Stem-to-Guide Clearance (in.)		Stem Diameter (in.)	
							Intake	Exhaust	Intake	Exhaust
2006	1.6	3	45-45.5	45-45.5	①	1.732	0.0008-0.0020	0.0014-0.0026	0.2348-0.2354	0.2343-0.2348
	2.0	2	45-45.5	45-45.5	①	1.923	0.0008-0.0020	0.0014-0.0026	0.2348-0.2354	0.2343-0.2348
	2.4	S	44.75-45.1	45.25-45.75	①	1.8677	0.00078-0.0019	0.00118-0.0021	0.2151-0.2157	0.2149-0.2153
	2.7	8	NA	45-45.5	①	1.8425	0.0008-0.0020	0.0014-0.0026	0.2348-0.2354	0.2343-0.2348
2007	1.6	3	45-45.5	45-45.5	①	1.7323	0.0008-0.0020	0.0014-0.0026	0.2348-0.2354	0.2343-0.2348
	2.0	2	45-45.5	45-45.5	①	1.9236	0.0008-0.0020	0.0014-0.0026	0.2348-0.2354	0.2343-0.2348
	2.4	S	44.75-45.1	45.25-45.75	①	1.8677	0.00078-0.0019	0.00118-0.0021	0.2151-0.2157	0.2149-0.2153
	2.7	8	NA	45-45.5	①	1.8425	0.0008-0.0020	0.0014-0.0026	0.2348-0.2354	0.2343-0.2348
2008	1.6	3	45-45.5	45-45.5	①	1.7323	0.0008-0.0020	0.0014-0.0026	0.2348-0.2354	0.2343-0.2348
	2.0	2	NA	NA	②	1.9236	0.0008-0.0019	0.0014-0.0026	0.2348-0.2354	0.2343-0.2348
	2.4	S	44.75-45.1	45-45.5	①	1.804	0.0008-0.0019	0.0020-0.0033	0.2585-0.2591	0.2571-0.2579
	2.7	8	NA	45-45.5	①	1.8425	0.0008-0.0020	0.0014-0.0026	0.2348-0.2354	0.2343-0.2348

NA: Not Available

① Less than 1.5 degrees

② Standard range: 0.0010-0.0023 in.

Maximum value: 0.0080 in.

22140_KIAC_C0005

CRANKSHAFT AND CONNECTING ROD SPECIFICATIONS

All measurements are given in inches.

Year	Engine Displacement Liters	Engine VIN	Crankshaft Main Brg. Journal Dia.	Crankshaft Main Brg. Oil Clearance	Crankshaft Shaft End-play	Crankshaft Thrust on No.	Connecting Rod Journal Diameter	Connecting Rod Oil Clearance	Connecting Rod Side Clearance
2006	1.6	3	1.9665-1.9672	①	0.0020-0.0069	3	1.8898-1.8905	0.0007-0.0014	0.0039-0.0098
	2.0	2	2.2418-2.2426	0.0011-0.0019	0.0024-0.0102	3	1.8898-1.8905	0.0009-0.0017	0.0039-0.0098
	2.4	S	2.0449-2.0456	0.0010-0.0019	0.0027-0.0098	3	2.0079-2.0086	0.0011-0.0018	0.0039-0.0098
	2.7	8	2.4402-2.4409	0.0002-0.0009	0.0028-0.0098	3	2.0079-2.0086	0.0007-0.0014	0.0039-0.0098
2007	1.6	3	1.9665-1.9672	①	0.0020-0.0069	3	1.8898-1.8905	0.0007-0.0014	0.0039-0.0098
	2.0	2	2.2418-2.2426	0.0011-0.0019	0.0024-0.0102	3	1.8898-1.8905	0.0009-0.0017	0.0039-0.0098
	2.4	S	2.0049-2.0456	0.0010-0.0019	0.0027-0.0098	3	2.0079-2.0086	0.0011-0.0018	0.0039-0.0098
	2.7	8	2.4402-2.4409	0.0002-0.0009	0.0028-0.0098	3	2.0079-2.0086	0.0007-0.0014	0.0039-0.0098
2008	1.6	3	1.9665-1.9672	①	0.0020-0.0069	3	1.8898-1.8905	0.0007-0.0014	0.0039-0.0098
	2.0	2	2.2418-2.2426	0.0011-0.0018	0.0023-0.0102	3	1.8898-1.8905	0.0009-0.0017	0.0039-0.0098
	2.4	S	2.0449-2.0456	0.0010-0.0019	0.0027-0.0098	3	2.0079-2.0086	0.0011-0.0018	0.0039-0.0098
	2.7	8	2.4402-2.4409	0.0002-0.0009	0.0028-0.0098	3	2.0079-2.0086	0.0007-0.0014	0.0039-0.0098

① Journal Nos. 1, 2, 4 & 5: 0.0009-0.0016 in.

Journal Nos. 3: 0.0011-0.0018 in.

22140_KIAC_C0007

PISTON AND RING SPECIFICATIONS

All measurements are given in inches.

Year	Engine Displacement Liters	Engine VIN	Piston Clearance	Ring Gap			Ring Side Clearance		
				Top Compression	Bottom Compression	Oil Control	Top Compression	Bottom Compression	Oil Control
2006	1.6	3	0.0008-0.0016	0.0059-0.0118	0.0138-0.0197	0.0079-0.0276	0.0016-0.0033	0.0016-0.0033	0.0031-0.0069
	2.0	2	0.0008-0.0016	0.0091-0.0150	0.0130-0.0189	0.0079-0.0236	0.0016-0.0031	0.0012-0.0028	0.0024-0.0059
	2.4	S	0.0006-0.0014	0.0059-0.0118	0.0118-0.0177	0.0078-0.0275	0.0012-0.0028	0.0012-0.0028	0.0024-0.0059
	2.7	8	0.0008-0.0016	0.0059-0.0118	0.0118-0.0177	0.0078-0.0275	0.0016-0.0031	0.0012-0.0027	0.0024-0.0059
2007	1.6	3	0.0008-0.0016	0.0059-0.0118	0.0138-0.0197	0.0079-0.0276	0.0016-0.0033	0.0016-0.0033	0.0031-0.0069
	2.0	2	0.0008-0.0016	0.0091-0.0150	0.0130-0.0189	0.0079-0.0236	0.0016-0.0031	0.0012-0.0028	0.0024-0.0059
	2.4	S	0.0006-0.0014	0.0059-0.0118	0.0118-0.0177	0.0078-0.0275	0.0012-0.0028	0.0012-0.0028	0.0024-0.0059
	2.7	8	0.0008-0.0016	0.0059-0.0118	0.0118-0.0177	0.0078-0.0275	0.0016-0.0031	0.0012-0.0028	0.0024-0.0059
2008	1.6	3	0.0008-0.0016	0.0059-0.0118	0.0138-0.0197	0.0079-0.0276	0.0016-0.0033	0.0016-0.0033	0.0031-0.0069
	2.0	2	0.0008-0.0016	0.0079-0.0138	0.0146-0.0205	0.0078-0.0236	0.0015-0.0031	0.0012-0.0027	0.0024-0.0059
	2.4	S	0.0006-0.0014	0.0059-0.0118	0.0118-0.0177	0.0078-0.0275	0.0012-0.0028	0.0012-0.0028	0.0024-0.0059
	2.7	8	0.0008-0.0016	0.0059-0.0118	0.0118-0.0177	0.0078-0.0275	0.0016-0.0031	0.0012-0.0027	0.0024-0.0059

22140_KIAC_C0006

TORQUE SPECIFICATIONS
All readings in ft. lbs.

Year	Engine Displacement Liters	Engine VIN	Cylinder Head Bolts	Main Bearing Bolts	Rod Bearing Bolts	Crankshaft Damper Bolts	Flywheel Bolts	Manifold Intake	Manifold Exhaust	Spark Plugs	Oil Pan Drain Plug
2006	1.6	3	①	40-43	23-25	101-109	87-94	11-15	22-29	18-22	29-33
	2.0	2	②	③	36-38	123-130.2	87-94	12-17	31-40	18-22	29-33
	2.4	S	④	⑤	⑥	123-130	87-94	14-20	29-33	15-21	26-32
	2.7	8	⑦	⑧	⑨	123-130.2	87-94	14-17.4	21.7-25.3	15-21	26-33
2007	1.6	3	①	40-43	23-25	101-109	87-94	11-15	22-25	18-22	29-33
	2.0	2	②	③	36-38	123-130.2	87-94	12-17	29-44	18-22	29-33
	2.4	S	④	⑤	⑥	123-130.2	87-94	14-20	29-33	15-21	26-32
	2.7	8	⑦	⑧	⑨	123-130.2	53-56	14-17.5	22-25	15-21	26-33
2008	1.6	3	①	40-43	23-25	101-109	87-94	11-15	22-29	18-22	29-33
	2.0	2	②	③	36-38	116-123	87-94	12-17	31-40	18-22	29-33
	2.4	S	④	⑤	⑥	123-130	87-94	14-20	29-33	15-21	26-32
	2.7	8	⑦	⑧	⑨	123-130	53-56	14-17.5	22-25	15-21	26-33

① Step 1: 22 ft. lbs.
Step 2: plus 90 degrees
Step 3: Loosen fully
Step 4: 22 ft. lbs.
Step 5: Tighten 90 degrees

② M10 bolts; 17-20 ft. lbs. plus 60-65 degrees; plus an additional 6-65 degrees.
M12 bolts: 20.5-33 ft. lbs., plus 60-65 degrees, plus additional 60-65 degrees

③ 20-23 ft. lbs. plus 90 degrees

④ Step 1: 25.3 ft. lbs., plus 90 degrees, plus an additional 90 degrees

⑤ 19.52 ft. lbs., plus 45 degrees

⑥ 14.46 ft. lbs., plus 90 degrees

⑦ Step 1: 18 ft. lbs., plus 60 degrees, plus an additional 45 degrees

⑧ SM10 bolts: 22 ft. lbs. plus 90 degrees.

⑨ 15 ft. lbs.plus 90 degrees

22140_KIAC_C0008

WHEEL ALIGNMENT

Year	Model		Caster Range (+/-Deg.)	Caster Preferred Setting (Deg.)	Camber Range (+/-Deg.)	Camber Preferred Setting (Deg.)	Toe-in (in.)
2006	Rio	F	0.50	0	4.00	+0.5	0.08 +/- 0.08
		R	—	—	1.00	-0.50	0.08 +/- 0.24
	Spectra	F	0.50	+2.60	0.50	0	0 +/- 0.08
		R	—	—	0.50	-0.92	0.16 +/- 0.08
	Optima	F	1.00	+3.15	0.30	0	0.11 +/- 0.12
		R	—	—	0.30	-0.30	0.07 +/- 0.12
2007	Rio	F	0.50	0	4.00	+0.5	0.08 +/- 0.08
		R	—	—	1.00	-0.50	0.08 +/- 0.24
	Spectra	F	0.50	+2.60	0.50	0	0 +/- 0.08
		R	—	—	0.50	-0.92	0.16 +/- 0.08
	Optima	F	1.00	+3.15	0.30	0	0.11 +/- 0.12
		R	—	—	0.30	-0.30	0.07 +/- 0.12
2008	Rio	F	0.50	0	4.00	+0.5	0.08 +/- 0.08
		R	—	—	1.00	-0.50	0.08 +/- 0.24
	Spectra	F	0.50	+2.60	0.50	0	0 +/- 0.08
		R	—	—	0.50	-0.92	0.16 +/- 0.08
	Optima	F	1.00	+3.15	0.30	0	0.11 +/- 0.12
		R	—	—	0.30	-0.30	0.07 +/- 0.12

22140_KIAC_C0009

TIRE, WHEEL AND BALL JOINT SPECIFICATIONS

| Year | Model | OEM Tires | | Tire Pressures (psi) | | Wheel Size | Ball Joint Inspection | Lug Nut Torque (ft. lbs.) |
		Standard	Optional	Front	Rear			
2006	Rio	185/65R14	195/55R14 175/70R14	30	30	Std: 5.0-J Opt: 5.5-J	①	65-79
	Spectra	195/60R15	205/50R16	30	30	6.0-J	①	67-82
	Optima	P205/60HR16	P215/50VR17	30	30	6.5-J	①	65-80
2007	Rio	185/65R14	195/55R15	30	30	Std: 5.0-J Opt: 5.5-J	①	65-79
	Spectra	P195/60R15	P205/50R16	30	30	6.0-J	①	65-80
	Optima	P205/60HR16	P215/50VR17	30	30	6.0-J	①	65-80
2008	Rio	185/65R14	195/55R14 175/70R14	30	30	Std: 5.0-J Opt: 5.5-J	①	65-79
	Spectra	P195/60R15	P205/50R16	30	30	6.0-J	①	67-82
	Optima	P205/60HR16	P215/50VR17	30	30	6.0-J	①	65-80

OEM: Original Equipment Manufacturer

PSI: Pounds Per Square Inch

STD: Standard

OPT: Optional

① Replace if any measureable movent is found.

22140_KIAC_C0010

BRAKE SPECIFICATIONS
All measurements in inches unless noted

| Year | Model | | Brake Disc | | | Brake Drum | | | Minimum Lining Thickness | Brake Caliper | |
			Original Thickness	Minimum Thickness	Maximum Run-out	Original Inside Diameter	Max. Wear Limit	Maximum Machine Diameter		Bracket Bolts (ft. lbs.)	Mounting Bolts (ft. lbs.)
2006	Rio	F	0.870	0.790	0.0012	—	—	—	0.079	58-72	16-23
		R	0.390	0.315	0.0012	8.00	—	8.06	0.079	62-69	16-23
	Spectra	F	1.020	0.945	0.0012	—	—	—	0.079	51-63	16-24
		R	0.390	0.315	0.0012	8.00	—	8.08	0.079	33-49	36-43
	Optima	F	1.020	0.960	0.0016	—	—	—	0.079	58-72	16-23
		R	0.390	0.330	0.0012	—	—	—	0.079	58-72	16-23
2007	Rio	F	0.870	0.790	0.0012	—	—	—	0.079	58-72	16-23
		R	0.390	0.315	0.0012	7.87	—	7.91	0.079	62-69	16-23
	Spectra	F	1.020	0.945	0.0012	—	—	—	0.079	51-63	16-24
		R	0.390	0.315	0.0012	8.00	—	8.08	0.079	51-63	16-24
	Optima	F	1.020	0.960	0.0016	—	—	—	0.079	58-72	16-23
		R	0.390	0.320	0.0039	—	—	—	0.079	51-63	16-24
2008	Rio	F	0.870	0.790	0.0012	—	—	—	0.079	62-69	16-23
		R	0.390	0.315	0.0012	7.87	—	7.91	0.079	62-69	16-23
	Spectra	F	1.020	0.945	0.0012	—	—	—	0.079	51-63	16-24
		R	0.390	0.315	0.0012	8.00	—	8.08	0.079	51-63	16-24
	Optima	F	1.020	0.960	0.0016	—	—	—	0.079	58-72	16-23
		R	0.390	0.330	0.0012	—	—	—	0.079	58-72	16-23

F: Front

R: Rear

22140_KIAC_C0011

SCHEDULED MAINTENANCE INTERVALS
Kia - Spectra, Rio and Optima

TO BE SERVICED	TYPE OF SERVICE	VEHICLE MILEAGE INTERVAL (x1000)												
		7.5	15	22.5	30	37.5	45	52.5	60	67.5	75	82.5	90	97.5
Accessory drive belts	S/I	✓	✓	✓	✓	✓	✓	✓	✓	✓	✓	✓	✓	✓
Air cleaner filter	R			✓			✓			✓			✓	
Air conditioner system	S/I													
Brake lines, hoses and connections	S/I		✓		✓		✓		✓		✓		✓	
Chassis and body fasteners	T											✓		
Cooling system hoses and coolant level	S/I		✓		✓		✓		✓		✓		✓	
CV-joint boots	S/I				✓				✓				✓	
Engine coolant	R				✓				✓				✓	
Engine oil and filter	R	✓	✓	✓	✓	✓	✓	✓	✓	✓	✓	✓	✓	✓
Exhaust system heat shields	S/I				✓				✓				✓	
Front and rear brakes	S/I	✓	✓	✓	✓	✓	✓	✓	✓	✓	✓	✓	✓	✓
Front ball joints	S/I				✓				✓				✓	
Fuel filter	R					✓					✓			
Fuel lines and hoses	S/I	✓	✓	✓	✓	✓	✓	✓	✓	✓	✓	✓	✓	✓
Locks and hinges	L	✓	✓	✓	✓	✓	✓	✓	✓	✓	✓	✓	✓	✓
Spark plugs	R								✓					
Steering operation and linkage	S/I	✓	✓	✓	✓	✓	✓	✓	✓	✓	✓	✓	✓	✓
Timing belt	R								✓					

R: Replace   S/I: Service or Inspect   L: Lubricate   T: Tighten

FREQUENT OPERATION MAINTENANCE (SEVERE SERVICE)

If a vehicle is operated under any of the following conditions it is considered severe service

- Towing a trailer or using a camper or car-top carrier

- Repeated short trips of less than 5 miles in temperatures below freezing, or trips of less than 10 miles in any temperature

- Prolonged idling (vehicle operation in stop and go traffic).

- Operating on rough, muddy, unpaved, dusty or salt-covered roads.

- Police, taxi, delivery usage or trailer towing usage.

- Driving in extremely hot (over 90°F) conditions

Oil & oil filter: change every 5000 miles or 5 months, whichever occurs first.

Air cleaner filter: inspect every 15,000 miles or 15 months and replace everything 30,000 miles or 30 months, whichever occurs fir:

Fuel system hoses (California models only): replace every 105,000 miles

Emission system hoses (non-CA models): inspect every 55,000 or 55 months, whichever occurs first

Emission system hoses (CA models): inspect every 60,000 miles or 60 months, which occurs first

Front and rear brakes: inspect every 15,000 miles or 15 months, whichever occurs first

Chassis and body fasteners: tighten every 15,000 miles or 15 months, whichever occurs first

Locks and hinges: lubricate every 5000 miles or 5 months, whichever occurs first

22140_KIAC_C0012

PRECAUTIONS

Before servicing any vehicle, please be sure to read all of the following precautions, which deal with personal safety, prevention of component damage, and important points to take into consideration when servicing a motor vehicle:

• Never open, service or drain the radiator or cooling system when the engine is hot; serious burns can occur from the steam and hot coolant.

• Observe all applicable safety precautions when working around fuel. Whenever servicing the fuel system, always work in a well-ventilated area. Do not allow fuel spray or vapors to come in contact with a spark, open flame, or excessive heat (a hot drop light, for example). Keep a dry chemical fire extinguisher near the work area. Always keep fuel in a container specifically designed for fuel storage; also, always properly seal fuel containers to avoid the possibility of fire or explosion. Refer to the additional fuel system precautions later in this section.

• Fuel injection systems often remain pressurized, even after the engine has been turned **OFF**. The fuel system pressure must be relieved before disconnecting any fuel lines. Failure to do so may result in fire and/or personal injury.

• Brake fluid often contains polyglycol ethers and polyglycols. Avoid contact with the eyes and wash your hands thoroughly after handling brake fluid. If you do get brake fluid in your eyes, flush your eyes with clean, running water for 15 minutes. If eye irritation persists, or if you have taken

brake fluid internally, IMMEDIATELY seek medical assistance.

• The EPA warns that prolonged contact with used engine oil may cause a number of skin disorders, including cancer. You should make every effort to minimize your exposure to used engine oil. Protective gloves should be worn when changing oil. Wash your hands and any other exposed skin areas as soon as possible after exposure to used engine oil. Soap and water, or waterless hand cleaner should be used.

• All new vehicles are now equipped with an air bag system, often referred to as a Supplemental Restraint System (SRS) or Supplemental Inflatable Restraint (SIR) system. The system must be disabled before performing service on or around system components, steering column, instrument panel components, wiring and sensors. Failure to follow safety and disabling procedures could result in accidental air bag deployment, possible personal injury and unnecessary system repairs.

• Always wear safety goggles when working with, or around, the air bag system. When carrying a non-deployed air bag, be sure the bag and trim cover are pointed away from your body. When placing a non-deployed air bag on a work surface, always face the bag and trim cover upward, away from the surface. This will reduce the motion of the module if it is accidentally deployed. Refer to the additional air bag system precautions later in this section.

• Clean, high quality brake fluid from a sealed container is essential to the safe and

proper operation of the brake system. You should always buy the correct type of brake fluid for your vehicle. If the brake fluid becomes contaminated, completely flush the system with new fluid. Never reuse any brake fluid. Any brake fluid that is removed from the system should be discarded. Also, do not allow any brake fluid to come in contact with a painted surface; it will damage the paint.

• Never operate the engine without the proper amount and type of engine oil; doing so WILL result in severe engine damage.

• Timing belt maintenance is extremely important. Many models utilize an interference-type, non-freewheeling engine. If the timing belt breaks, the valves in the cylinder head may strike the pistons, causing potentially serious (also time-consuming and expensive) engine damage. Refer to the maintenance interval charts for the recommended replacement interval for the timing belt, and to the timing belt section for belt replacement and inspection.

• Disconnecting the negative battery cable on some vehicles may interfere with the functions of the on-board computer system (s) and may require the computer to undergo a relearning process once the negative battery cable is reconnected.

• When servicing drum brakes, only disassemble and assemble one side at a time, leaving the remaining side intact for reference.

• Only an MVAC-trained, EPA-certified automotive technician should service the air conditioning system or its components.

BRAKES

GENERAL INFORMATION

PRECAUTIONS

• Certain components within the ABS system are not intended to be serviced or repaired individually.

• Do not use rubber hoses or other parts not specifically specified for and ABS system. When using repair kits, replace all parts included in the kit. Partial or incorrect repair may lead to functional problems and require the replacement of components.

• Lubricate rubber parts with clean, fresh brake fluid to ease assembly. Do not use shop air to clean parts; damage to rubber components may result.

• Use only DOT 3 brake fluid from an unopened container.

• If any hydraulic component or line is removed or replaced, it may be necessary to bleed the entire system.

• A clean repair area is essential. Always clean the reservoir and cap thoroughly before removing the cap. The slightest amount of dirt in the fluid may plug an orifice and impair the system function. Perform repairs after components have been thoroughly cleaned; use only denatured alcohol to clean components. Do not allow ABS components to come into contact with any substance containing mineral oil; this includes used shop rags.

• The Anti-Lock control unit is a microprocessor similar to other computer units in the vehicle. Ensure that the ignition switch is **OFF** before removing or installing

ANTI-LOCK BRAKE SYSTEM (ABS)

controller harnesses. Avoid static electricity discharge at or near the controller.

• If any arc welding is to be done on the vehicle, the control unit should be unplugged before welding operations begin.

BLEEDING PROCEDURE

BLEEDING THE ABS SYSTEM

The ABS brake system is bled in the usual fashion with no special procedures required. Refer to the bleeding procedure described earlier. Make certain the master cylinder reservoir is filled before the bleeding is begun and check the level frequently.

BRAKES | BLEEDING THE BRAKE SYSTEM

BLEEDING PROCEDURE

BLEEDING PROCEDURE

✳✳ WARNING

When bleeding the brakes, note the following:

- Do not reuse the drained fluid
- Always use Genuine DOT 3 or DOT 4 Brake Fluid. Using a non-Genuine DOT3 or DOT 4 brake fluid can cause corrosion and decrease the life of the system
- Make sure no dirt of other foreign

matter is allowed to contaminate the brake fluid
- Do not spill brake fluid on the vehicle, it may damage the paint; if brake fluid does contact the paint, wash it off immediately with water
- The reservoir on the master cylinder must be at the MAX (upper) level mark at the start of bleeding procedure and checked after bleeding each brake caliper. Add fluid as required

1. Make sure the brake fluid level in the master cylinder fluid reservoir is at the MAX (upper) level line.

2. Have someone slowly pump the brake pedal several times, and then apply steady pressure

3. Loosen the right-rear brake bleed screw to allow air to escape from the system. Then tighten the bleed screw securely

4. Repeat the procedure for each wheel in the sequence shown below until air bubbles no longer appear in the fluid

5. Refill the master cylinder reservoir to the MAX (upper) level line

BRAKES | FRONT DISC BRAKES

✳✳ CAUTION

Dust and dirt accumulating on brake parts during normal use may contain asbestos fibers from production or aftermarket brake linings. Breathing excessive concentrations of asbestos fibers can cause serious bodily harm. Exercise care when servicing brake parts. Do not sand or grind brake lining unless equipment used is designed to contain the dust residue. Do not clean brake parts with compressed air or by dry brushing. Cleaning should be done by dampening the brake components with a fine mist of water, then wiping the brake components clean with a dampened cloth. Dispose of cloth and all residue containing asbestos fibers in an impermeable container with the appropriate label. Follow practices prescribed by the Occupational Safety and Health Administration (OSHA) and the Environmental Protection Agency (EPA) for the handling, processing, and disposing of dust or debris that may contain asbestos fibers.

BRAKE CALIPER

REMOVAL & INSTALLATION

1. Remove guide rod and the caliper up out of the way.
2. Check the hoses and pin boots for damage and deterioration.
3. Remove the pad shims, pad retainers and pads.

To install:
4. Install the pad retainers.

5. Check the foreign material at the pad shims and the back of the pads.

✳✳ CAUTION

Contaminated brake discs or pads reduce stopping ability. Keep grease off the discs and pads.

6. Install the brake pads and pad shims correctly. Install the pad with the wear indicator on the inside.

➡**If you are reusing the pads, always reinstall the brake pads in their original positions to prevent a momentary loss of braking efficiency.**

7. Push in the piston so that the caliper will fit over the pads. Make sure that the piston boot is in position to prevent damaging it when pivoting the caliper down.

➡**Insert the piston in the cylinder using the special tool (09581-11000).**

8. Pivot the caliper down into position. Being careful not to damage the pin boot, install the guide rod bolt.

9. Depress the brake pedal several times to make sure the brakes work, then test-drive.

➡**Engagement of the brake may require a greater pedal stroke immediately after the brake pads have been replaced as a set. Several applications of the brake will restore the normal pedal stroke.**

10. After installation, check for leaks at hose and line joints or connections, and retighten if necessary

DISC BRAKE PADS

REMOVAL & INSTALLATION

1. Remove guide rod and the caliper up out of the way.
2. Check the hoses and pin boots for damage and deterioration.
3. Remove the pad shims, pad retainers and pads.

To install:
4. Install the pad retainers.
5. Check the foreign material at the pad shims and the back of the pads.

✳✳ CAUTION

Contaminated brake discs or pads reduce stopping ability. Keep grease off the discs and pads.

6. Install the brake pads and pad shims correctly. Install the pad with the wear indicator on the inside.

➡**If you are reusing the pads, always reinstall the brake pads in their original positions to prevent a momentary loss of braking efficiency.**

7. Push in the piston so that the caliper will fit over the pads. Make sure that the piston boot is in position to prevent damaging it when pivoting the caliper down.

➡**Insert the piston in the cylinder using the special tool (09581-11000).**

8. Pivot the caliper down into position. Being careful not to damage the pin boot, install the guide rod bolt.

9. Depress the brake pedal several

times to make sure the brakes work, then test-drive.

➡**Engagement of the brake may require a greater pedal stroke immedi-** ately after the brake pads have been **replaced as a set. Several applications of the brake will restore the normal pedal stroke.**

10. After installation, check for leaks at hose and line joints or connections, and retighten if necessary

BRAKES

✳✳ CAUTION

Dust and dirt accumulating on brake parts during normal use may contain asbestos fibers from production or aftermarket brake linings. Breathing excessive concentrations of asbestos fibers can cause serious bodily harm. Exercise care when servicing brake parts. Do not sand or grind brake lining unless equipment used is designed to contain the dust residue. Do not clean brake parts with compressed air or by dry brushing. Cleaning should be done by dampening the brake components with a fine mist of water, then wiping the brake components clean with a dampened cloth. Dispose of cloth and all residue containing asbestos fibers in an impermeable container with the appropriate label. Follow practices prescribed by the Occupational Safety and Health Administration (OSHA) and the Environmental Protection Agency (EPA) for the handling, processing, and disposing of dust or debris that may contain asbestos fibers.

BRAKE CALIPER

REMOVAL & INSTALLATION

1. Raise the vehicle and make sure it is securely supported. Remove the rear wheel.
2. Release the parking brake.
3. Remove the guide rod bolt.
4. Raise the caliper assembly, support it with a wire.
5. Remove the pad shim and pad assembly from caliper bracket.

To install:

6. Install the pad retainers to the caliper.
7. Check the foreign material at the pad shim and the back of the pads.
8. Contaminated brake discs or pads reduce stopping ability. Keep grease off the discs and pads.
9. Install the brake pads and pad shim on the caliper bracket.
10. If you are reusing the pads, always reinstall the brake pads in their original positions to prevent a momentary loss of braking efficiency. Push in the piston using SST (09581-11000) so that the caliper will fit over the pads. Make sure that the piston boot is in position to prevent damaging it when pivoting the caliper down.
11. Pivot caliper down into position. Being careful not to damage the pin boot, install the guide rod bolt and tighten it to the specified torque.
12. Install the brake caliper .
13. After installation, check for leaks at hose and line joints and connections, and retighten if necessary.
14. Depress the brake pedal several times to make sure the brakes work, then test-drive.

➡**Engagement of the brake may require a greater pedal stroke immediately after the brake pads have been replaced as a set. Several applications of the brake pedal will restore the normal pedal stroke.**

DISC BRAKE PADS

REMOVAL & INSTALLATION

1. Raise the vehicle and make sure it is securely supported. Remove the rear wheel.

REAR DISC BRAKES

2. Release the parking brake.
3. Remove the guide rod bolt.
4. Raise the caliper assembly, support it with a wire.
5. Remove the pad shim and pad assembly from caliper bracket.

To install:

6. Install the pad retainers to the caliper.
7. Check the foreign material at the pad shim and the back of the pads.
8. Contaminated brake discs or pads reduce stopping ability. Keep grease off the discs and pads.
9. Install the brake pads and pad shim on the caliper bracket.
10. If you are reusing the pads, always reinstall the brake pads in their original positions to prevent a momentary loss of braking efficiency. Push in the piston using SST (09581-11000) so that the caliper will fit over the pads. Make sure that the piston boot is in position to prevent damaging it when pivoting the caliper down.
11. Pivot caliper down into position. Being careful not to damage the pin boot, install the guide rod bolt and tighten it to the specified torque.
12. Install the brake caliper .
13. After installation, check for leaks at hose and line joints and connections, and retighten if necessary.
14. Depress the brake pedal several times to make sure the brakes work, then test-drive.

➡**Engagement of the brake may require a greater pedal stroke immediately after the brake pads have been replaced as a set. Several applications of the brake pedal will restore the normal pedal stroke.**

BRAKES   **REAR DRUM BRAKES**

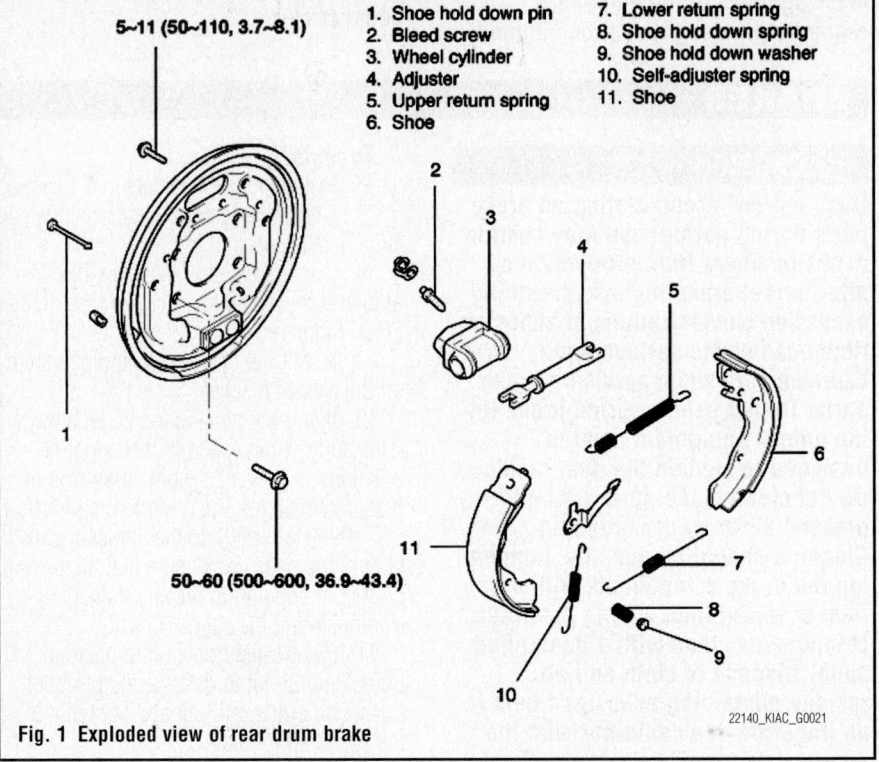

1. Shoe hold down pin
2. Bleed screw
3. Wheel cylinder
4. Adjuster
5. Upper return spring
6. Shoe
7. Lower return spring
8. Shoe hold down spring
9. Shoe hold down washer
10. Self-adjuster spring
11. Shoe

5~11 (50~110, 3.7~8.1)

50~60 (500~600, 36.9~43.4)

22140_KIAC_G0021

Fig. 1 Exploded view of rear drum brake

BRAKE DRUM

REMOVAL & INSTALLATION

Rio & Spectra

See Figure 1.

1. Remove the shoe hole down pins by pushing the shoe hole down washer and turning them.
2. Disengage the upper return spring.
3. Lower the brake shoe assembly, and remove the lower return spring. Make sure not to damage the dust cover on the wheel cylinder.
4. Disconnect the parking brake cable from the parking brake lever.
5. Remove the brake shoe assembly.
6. Remove the upper return spring, self-adjuster lever and self-adjuster spring, and separate the brake shoes.
7. Disconnect the brake line from the wheel cylinder.
8. Remove the bolt and the wheel cylinder from the backing plate.

➡ To install:

➡During installation, note the following:

- Do not spill brake fluid on the vehicle: it may damage the paint; if brake fluid does contact the paint. Wash it off immediately with water.
- To prevent spills, cover the hose joints with rags or shop towels.
- Use only a genuine wheel cylinder special bolt.

9. Apply sealant between the wheel cylinder and backing plate, and install the wheel cylinder.
10. Connect the brake tubes to the wheel cylinder.
11. Connect the parking brake cable to the parking brake lever.
12. Clean the threaded portions of adjuster sleeve and push rod female. Coat the threads of the adjuster assembly with grease. To shorten the clevises, turn the adjuster bolt.
13. Hook the self-adjuster spring to the adjuster lever first, then to the brake shoe.
14. Install the adjuster assembly and upper return spring, noting the installation direction. Be careful not to damage the wheel cylinder dust covers.
15. Install the lower return spring.
16. Apply brake cylinder grease or equivalent rubber grease to the sliding surfaces shown. Wipe off any excess. Don't get grease on the brake linings.
17. Apply brake cylinder grease or equivalent rubber grease to the brake shoe ends and opposite edges of the shoes shown. Wipe off any excess. Don't get grease on the brake linings.
18. Install the brake shoes onto the backing plate. Be careful not to damage the wheel cylinder dust covers.
19. Install the shoe hole down pins and the shoe hole down washers.
20. Hook the upper return spring.
21. Install the brake drum.
22. If the wheel cylinder has been removed, bleed the brake system.
23. Depress the brake pedal several times to set the self-adjusting brake.
24. Adjust the parking brake.

BRAKE SHOES

REMOVAL & INSTALLATION

Rio & Spectra

See Figure 1.

1. Remove the shoe hole down pins by pushing the shoe hole down washer and turning them.
2. Disengage the upper return spring.
3. Lower the brake shoe assembly, and remove the lower return spring. Make sure not to damage the dust cover on the wheel cylinder.
4. Disconnect the parking brake cable from the parking brake lever.
5. Remove the brake shoe assembly.

6. Remove the upper return spring, self-adjuster lever and self-adjuster spring, and separate the brake shoes.

7. Disconnect the brake line from the wheel cylinder.

8. Remove the bolt and the wheel cylinder from the backing plate.

To install:

➡**During installation, note the following:**

- Do not spill brake fluid on the vehicle: it may damage the paint; if brake fluid does contact the paint. Wash it off immediately with water.
- To prevent spills, cover the hose joints with rags or shop towels.
- Use only a genuine wheel cylinder special bolt.

9. Apply sealant between the wheel cylinder and backing plate, and install the wheel cylinder.

10. Connect the brake tubes to the wheel cylinder.

11. Connect the parking brake cable to the parking brake lever.

12. Clean the threaded portions of adjuster sleeve and push rod female. Coat the threads of the adjuster assembly with grease. To shorten the clevises, turn the adjuster bolt.

13. Hook the self-adjuster spring to the adjuster lever first, then to the brake shoe.

14. Install the adjuster assembly and upper return spring, noting the installation direction. Be careful not to damage the wheel cylinder dust covers.

15. Install the lower return spring.

16. Apply brake cylinder grease or equivalent rubber grease to the sliding surfaces shown. Wipe off any excess. Don't get grease on the brake linings.

17. Apply brake cylinder grease or equivalent rubber grease to the brake shoe ends and opposite edges of the shoes shown. Wipe off any excess. Don't get grease on the brake linings.

18. Install the brake shoes onto the backing plate. Be careful not to damage the wheel cylinder dust covers.

19. Install the shoe hole down pins and the shoe hole down washers.

20. Hook the upper return spring.

21. Install the brake drum.

22. If the wheel cylinder has been removed, bleed the brake system.

23. Depress the brake pedal several times to set the self-adjusting brake.

24. Adjust the parking brake.

ADJUSTMENT

1. Depress the brake pedal several times to set the self-adjusting brake.

BRAKES

PARKING BRAKE CABLE

ADJUSTMENT

1. Block the front wheels, then raise the rear of the vehicle and make sure it is securely supported.

2. Make sure the parking brake arm on the rear brake caliper contacts the brake caliper pin on vehicles with rear disc brakes.

3. Pull the parking brake lever up one click.

4. Remove the console.

5. Tighten the adjusting nuts until the parking brakes drag slightly when the rear wheels are turned.

6. Release the parking brake lever fully, and check that parking brakes do not drag when the rear wheels are turned. Readjust if necessary.

7. Make sure that the parking brakes are fully applied when the parking brake lever is pulled up fully.

8. Reinstall the console.

PARKING BRAKE SHOES

REMOVAL & INSTALLATION

1. Remove the console.

2. Loosen the adjusting nut and the parking brake cables.

PARKING BRAKE

3. Disconnect the electrical connector of parking brake switch.

4. Remove the parking brake lever assembly by loosening the bolts.

5. Remove the wheel and tire.

6. Remove the brake shoe.

To install:

7. Install the removed parts in the reverse order of removal.

8. Apply a coating of the specified grease to each sliding parts of the ratchet plate or the ratchet pawl.

9. After installing the cable adjuster, adjust the parking brake lever stroke.

CHASSIS ELECTRICAL

GENERAL INFORMATION

✳✳ CAUTION

These vehicles are equipped with an air bag system. The system must be disarmed before performing service on, or around, system components, the steering column, instrument panel components, wiring and sensors. Failure to follow the safety precautions and the disarming procedure could result in accidental air bag deployment, possible injury and unnecessary system repairs.

AIR BAG (SUPPLEMENTAL RESTRAINT SYSTEM)

SERVICE PRECAUTIONS

Disconnect and isolate the battery negative cable before beginning any airbag system component diagnosis, testing, removal, or installation procedures. Allow system capacitor to discharge for two minutes before beginning any component service. This will disable the airbag system. Failure to disable the airbag system may result in accidental airbag deployment, personal injury, or death.

Do not place an intact undeployed airbag face down on a solid surface. The airbag will propel into the air if accidentally deployed and may result in personal injury or death.

When carrying or handling an undeployed airbag, the trim side (face) of the airbag should be pointing towards the body to minimize possibility of injury if accidental deployment occurs. Failure to do this may result in personal injury or death.

Replace airbag system components with OEM replacement parts. Substitute parts may appear interchangeable, but internal differences may result in inferior occupant protection. Failure to do so may result in occupant personal injury or death.

Wear safety glasses, rubber gloves, and long sleeved clothing when cleaning powder residue from vehicle after an airbag deployment. Powder residue emitted from a deployed airbag can cause skin irritation. Flush affected area with cool water if irritation is experienced. If nasal or throat irritation is experienced, exit the vehicle for fresh air until the irritation ceases. If irritation continues, see a physician.

Do not use a replacement airbag that is not in the original packaging. This may result in improper deployment, personal injury, or death.

The factory installed fasteners, screws and bolts used to fasten airbag components have a special coating and are specifically designed for the airbag system. Do not use substitute fasteners. Use only original equipment fasteners listed in the parts catalog when fastener replacement is required.

During, and following, any child restraint anchor service, due to impact event or vehicle repair, carefully inspect all mounting hardware, tether straps, and anchors for proper installation, operation, or damage. If a child restraint anchor is found damaged in any way, the anchor must be replaced. Failure to do this may result in personal injury or death.

Deployed and non-deployed airbags may or may not have live pyrotechnic material within the airbag inflator.

Do not dispose of driver/passenger/curtain airbags or seat belt tensioners unless you are sure of complete deployment. Refer to the Hazardous Substance Control System for proper disposal.

Dispose of deployed airbags and tensioners consistent with state, provincial, local, and federal regulations.

After any airbag component testing or service, do not connect the battery negative cable. Personal injury or death may result if the system test is not performed first.

If the vehicle is equipped with the Occupant Classification System (OCS), do not connect the battery negative cable before performing the OCS Verification Test using the scan tool and the appropriate diagnostic information. Personal injury or death may result if the system test is not performed properly.

Never replace both the Occupant Restraint Controller (ORC) and the Occupant Classification Module (OCM) at the same time. If both require replacement, replace one, then perform the Airbag System test before replacing the other.

Both the ORC and the OCM store Occupant Classification System (OCS) calibration data, which they transfer to one another when one of them is replaced. If both are replaced at the same time, an irreversible fault will be set in both modules and the OCS may malfunction and cause personal injury or death.

If equipped with OCS, the Seat Weight Sensor is a sensitive, calibrated unit and must be handled carefully. Do not drop or handle roughly. If dropped or damaged, replace with another sensor. Failure to do so may result in occupant injury or death.

If equipped with OCS, the front passenger seat must be handled carefully as well. When removing the seat, be careful when setting on floor not to drop. If dropped, the sensor may be inoperative, could result in occupant injury, or possibly death.

If equipped with OCS, when the passenger front seat is on the floor, no one should sit in the front passenger seat. This uneven force may damage the sensing ability of the seat weight sensors. If sat on and damaged, the sensor may be inoperative, could result in occupant injury, or possibly death.

DISARMING THE SYSTEM

1. Before servicing the vehicle, refer to the Precautions Section.
2. Turn the ignition switch to the **LOCK** position.
3. Disconnect the negative battery cable.
4. Wait 10 minutes for the battery backup power to discharge.

ARMING THE SYSTEM

1. Before servicing the vehicle, refer to the Precautions Section.
2. Connect the negative battery cable.
3. Turn the ignition switch **ON**.
4. Verify that the air bag indicator illuminates for 4–8 seconds, then goes off.

CLOCKSPRING CENTERING

Prior to installing the clock spring, align the mating mark and "NEUTRAL" position indicator of the clock spring, and, after turning the front wheels to the straight-ahead position, install the clock spring to the column switch. If the mating mark of the clock spring is not properly aligned, the steering wheel may not completely rotate during a turn, or the flat cable within the clock spring may be severed, obstructing normal operation of the SRS and possibly leading to serious injury to the vehicle's driver.

DRIVETRAIN

AUTOMATIC TRANSAXLE ASSEMBLY

REMOVAL & INSTALLATION

Optima

See Figures 2 through 12.

1. Drain automatic transaxle fluid.
2. Disconnect the negative terminal from the battery.
3. Remove the engine cover.
4. Disconnect the AFS connector (A) and remove the air cleaner upper body (B).
5. Remove the air cleaner filter out of the air cleaner under body.
6. After disconnecting the PCM connectors (B), remove the air cleaner under body (A).
7. Remove the battery tray.
8. Remove the transaxle oil cooler hoses releasing the clamps.
9. Disconnect the ground wire.
10. Disconnect the connectors related to the transaxle.

a. Disconnect the solenoid valve connector (A).
b. Disconnect the inhibitor switch connector (A).

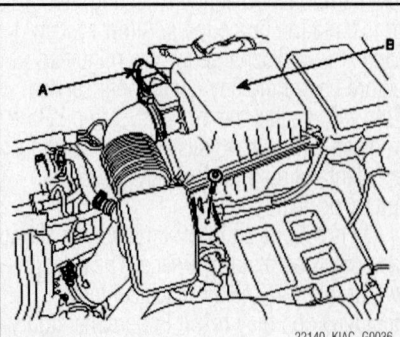

22140_KIAC_G0036

Fig. 2 Disconnect the AFS connector (A) and remove the air cleaner upper body (B)

22140_KIAC_G0037

Fig. 3 Disconnect the PCM connectors (B) and remove the air cleaner under body (A)

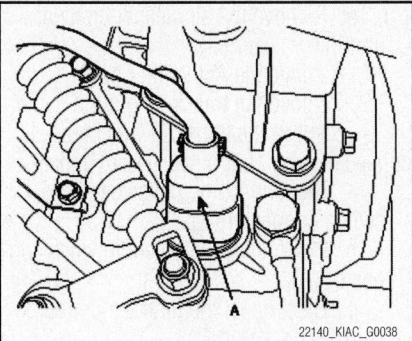

Fig. 4 Disconnect the solenoid valve connector (A)

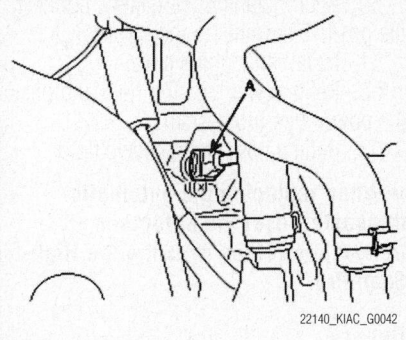

Fig. 8 Disconnect the vehicle speed sensor connector (A).

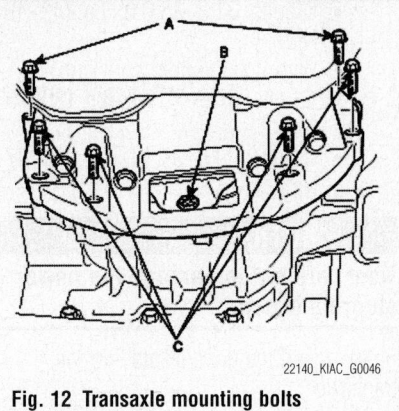

Fig. 12 Transaxle mounting bolts

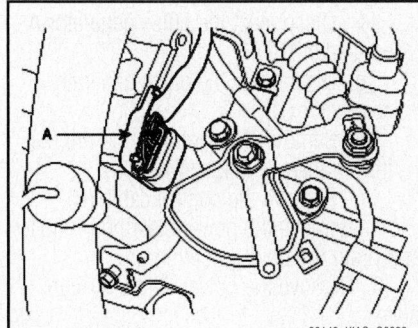

Fig. 5 Disconnect the inhibitor switch connector (A)

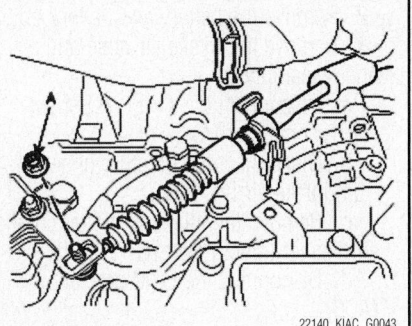

Fig. 9 Disconnect the control cable by removing the nut (A).

c. Disconnect the input speed sensor connector (A).

d. Disconnect the output speed sensor connector (A).

e. Disconnect the vehicle speed sensor connector (A).

11. Disconnect the control cable by removing the nut (A).

12. Raise and safely support the vehicle.

13. Remove the front wheels.

14. Remove the lower cover (A).

15. Drain power steering fluid.

16. Remove the pressure tube from the power steering oil pump.

17. After releasing the power steering hose clamp, disconnect the hose with the tube.

18. Remove the exhaust pipe.

19. Remove the lower arm assembly and tie rod mounting bolts.

20. Supporting the sub frame with a jack, remove the sub frame mounting bolts.

21. After loosening the steering column bolt and lowering a jack, remove the sub frame.

22. Lowering the vehicle down, remove the transaxle upper mounting bolts.

23. Using the SST (09200-38001), support the engine and transaxle assembly safely.

24. Remove the transaxle insulator mounting bolt (A).

25. Remove the starter.

26. Remove the bolts (4EA) which connect the drive plate with the torque converter.

27. Remove the inner shaft mounting bolts and take the drive shafts off the transaxle assembly.

28. After supporting the transaxle with a jack and remove the under mounting bolts.

To install:

29. Lifting up the transaxle slowly with a jack, tighten the under mounting bolts. Toque bolts to the following specifications:

- With 2.4L engine:
- A bolts: 31.3–39.8 ft. lbs. (43–55 Nm)

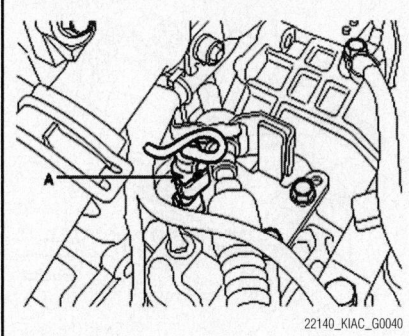

Fig. 6 Disconnect the input speed sensor connector (A)

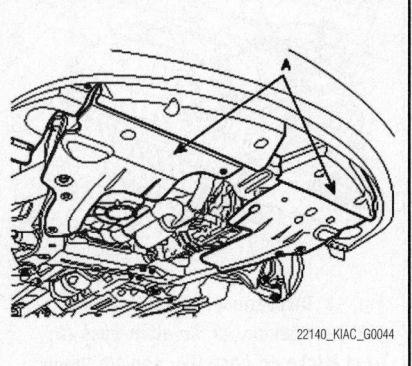

Fig. 10 Remove the lower cover (A).

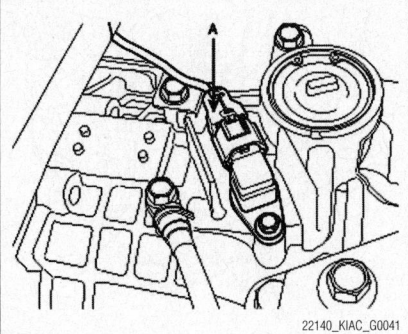

Fig. 7 Disconnect the output speed sensor connector (A).

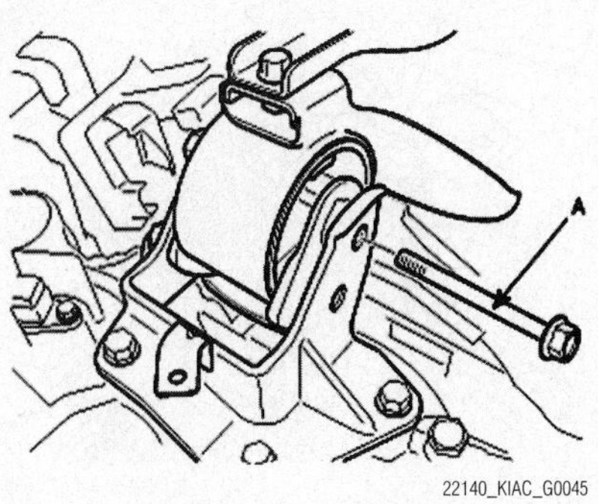

Fig. 11 Remove the transaxle insulator mounting bolt (A).

- C bolts: 31.3–35.6 ft. lbs. (43–49 Nm)
- With 2.7L engine:
- A bolts: 57.9–72.3 ft. lbs. (80–100 Nm)
- C bolts: 29.0–34.1 ft. lbs. (40–47 Nm)

✳✳ CAUTION

Take care not to damage the power steering hoses.

30. Insert the drive shafts into the transaxle.

31. Install the inner shaft mounting bolts (2EA). Torque bolts to 36.2–47.0 ft. lbs. (50–65 Nm).

32. Apply grease on the threaded part of the bolts (4EA) which connect the drive plate with the torque converter. and tighten them. Torque bolts to 33.3–38.3 ft. lbs. (46–53 Nm).

33. Install the starter.

34. Tighten the transaxle insulator mounting bolt. Torque bolt to 47.0–61.5 ft. lbs. (65–85 Nm).

35. Tighten the transaxle upper mounting bolts (4EA). For 2.4L engines, torque bolts to 31.3–39.8 ft. lbs. (43–55 Nm). For 2.7L engines, torque bolts to 47.0–61.5 ft. lbs. (65–85 Nm).

36. Lift the sub frame with a jack.

37. Tighten the steering column bolt.

38. Tighten the sub frame mounting bolts.

39. Install the lower arm assembly and the steering bar tie rod mounting bolts.

40. Install the front exhaust pipe.

41. Clamp the power steering hose.

42. Install the lower cover.

43. Connect the transaxle wire harness connectors.

   a. Connect the solenoid valve connector.

   b. Connect the inhibitor switch connector.

   c. Connect the input speed sensor connector.

   d. Connect the output speed sensor connector.

   e. Connect the vehicle speed sensor connector.

44. Connect the ground wire.

45. Connect the control cable by installing the nut.

46. Clamp the transaxle oil cooler hoses.

47. Install the battery tray.

48. Install the air cleaner under body and connect the PCM connectors.

49. Reassemble the air cleaner filter, install the air cleaner upper body and connect the air flow sensor connector.

50. After installing the battery, connect the positive terminal to the battery.

51. Install the engine cover.

52. Refill power steering fluid and bleed the power steering system.

53. Refill automatic transaxle fluid.

➡**When replacing the automatic transaxle, reset the automatic transaxle's values by using the High-Scan Pro.**

Rio

See Figures 13 through 23.

1. Disconnect the battery terminals and then remove battery.

2. Remove the battery and battery tray.

3. Remove the intake air hose and air cleaner assembly.

   a. Disconnect the Air Flow Sensor (AFS) connector (A).

   b. Disconnect the breather hose (B) from intake air hose (D).

   c. Remove the intake air hose (D) and air cleaner upper cover (C).

   d. Disconnect the ECM connector (A, B).

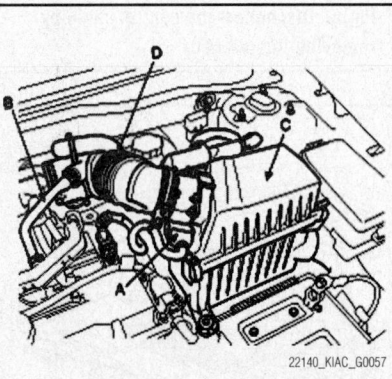

Fig. 13 Disconnect the Air Flow Sensor (AFS) connector (A), breather hose (B) from intake air hose (D), and air cleaner upper cover (C).

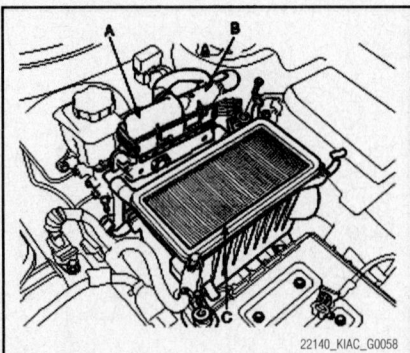

Fig. 14 Disconnect the ECM connector (A, B)

   e. Remove the air cleaner element and lower cover (C).

4. Remove the ATF cooler hose.

5. Remove the transaxle ground cable.

6. Remove the transaxle wire harness connectors and control cable from transaxle.

   a. Disconnect the transaxle range switch connector (A).

   b. Disconnect the solenoid valve connector (B).

   c. Disconnect the fluid temperature sensor connector (C).

   d. Disconnect the vehicle speed sensor connector (A).

   e. Disconnect the band servo switch connector (B).

   f. Disconnect the pulse generator A connector (C).

   g. Disconnect the pulse generator B connector (D).

   h. Remove the control cable nut (A) from transaxle range switch.

   i. Remove the control cable (B).

7. Remove the power steering oil hose and drain the power steering oil.

8. Remove the power steering return hose.

9. Install the special tools (09200-38001), the engine support fixture and the adapter, on the engine assembly.

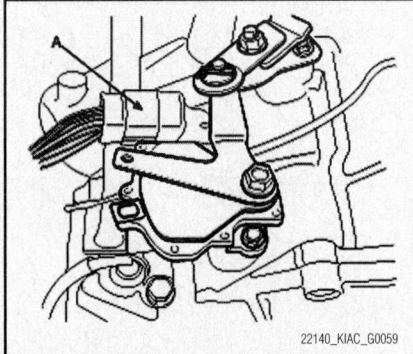

Fig. 15 Disconnect the transaxle range switch connector (A)

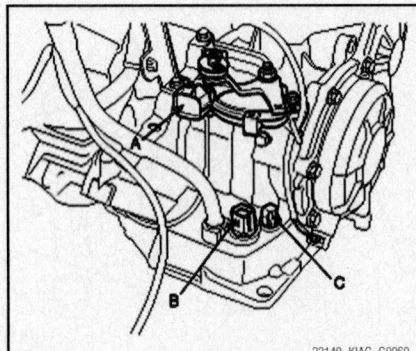

Fig. 16 Disconnect the solenoid valve connector (B)

10. Remove the transaxle mounting support bracket (A).

11. Remove the front roll stopper mounting bolt (A)

12. Remove the rear roll stopper mounting bolt (A)

13. Remove the lower cover.

14. Remove the side cover.

15. Remove the front wheel.

16. Remove the drain plug and drain the automatic transaxle fluid.

17. Separate the tie rod end from the pin and nut.

18. Remove the ABS wheel speed sensor.

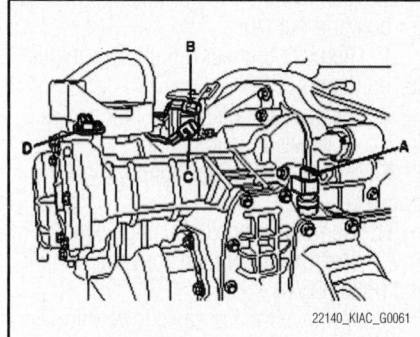

Fig. 17 Disconnect the vehicle speed sensor connector (A)

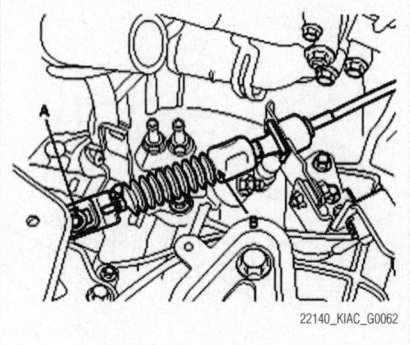

Fig. 18 Remove the control cable nut (A) from transaxle range switch

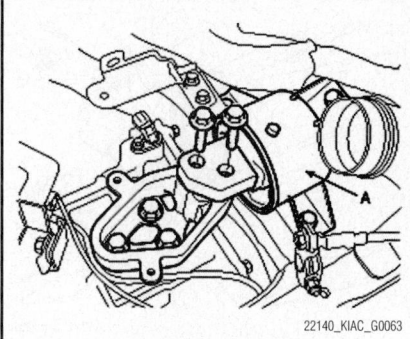

Fig. 19 Remove the transaxle mounting support bracket (A)

19. Remove the front caliper and use a wire to secure the caliper to the vehicle so that it is out of the way.

20. Remove the knuckle mounting bolts.

21. Remove the drive shaft from the transaxle.

22. Remove the steering U-joint mounting bolt.

23. Install the jack for supporting transaxle.

24. Remove the sub-frame bolts and nuts.

25. Remove the start motor mounting bolt.

26. Remove the engine to the torque converter mounting bolts (4EA).

27. Remove the engine to automatic transaxle mounting bolts.

28. Remove the transaxle assembly.

To install:

29. Installation is the reverse of removal.

30. Attach the torque converter on the transaxle side and mount the transaxle assembly onto the engine.

Torque specifications for transaxle installation:

  a. Engine to automatic transaxle mounting bolts. Torque to 43–58 ft. lbs. (60–80 Nm) for 12mm bolts and 32–41 ft. lbs. (43–55 Nm).

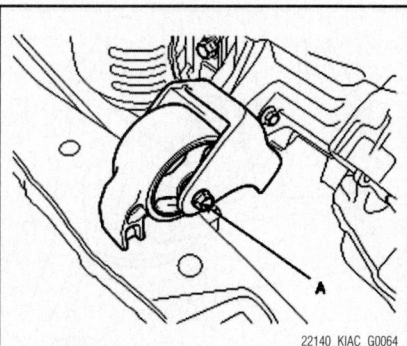

Fig. 20 Remove the front roll stopper mounting bolt (A)

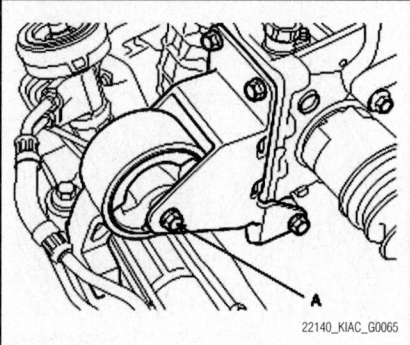

Fig. 21 Remove the rear roll stopper mounting bolt (A)

  b. Engine to the torque converter mounting bolts (4EA). Torque to 33–38 ft. lbs. (46–53 Nm)

  c. Starter motor mounting bolts. Torque to 19–24 ft. lbs. (27–34 Nm)

  d. Sub frame bolt and nut. Torque bolt (A) to 29–40 ft. lbs. (40–55 Nm); nut (B) to 110–126 ft. lbs. (160–180 Nm).

  e. Rear roll stopper insulator bolt. Torque bolt to 36–47 ft. lbs. (50–65 Nm).

  f. Front roll stopper insulator bolt. Torque bolt to 36–47 ft. lbs. (50–65 Nm).

31. During installation, perform the following:

  a. Adjust the shift cable.

  b. Install the transaxle control cable and adjust as follows:

- Move the shift lever and the transaxle range switch to the "N" Position, and install the control cable.

- When connecting the control cable to the transaxle mounting bracket, install the clip until it contacts the control cable.

- Remove any free-play in the control cable by adjusting nut and then

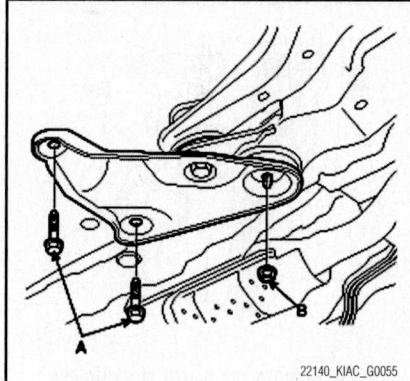

Fig. 22 Sub frame bolts (A) and nut (B)

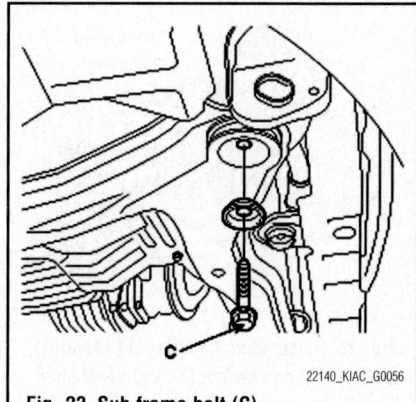

Fig. 23 Sub frame bolt (C)

check to see that the selector lever moves smoothly.

- Check to see that the control cable (A) has been adjusted correctly.

c. Refill the transaxle fluid.

d. Clean the battery posts and cable terminals with sandpaper, assemble them, then apply grease to prevent corrosion.

Spectra

See Figures 24 through 33.

1. Disconnect the positive and negative terminals from the battery.

2. Remove the battery and battery tray.

3. Remove the air cleaner and air intake hose.

   a. Disconnect the air flow sensor connector and breather hose.

   b. Remove the air cleaner upper cover and air intake hose.

   c. Remove the air cleaner lower.

4. Disconnect the ground cable from the transaxle.

5. Remove the power steering eye bolt (A).

6. Disconnect the power steering return hose.

7. Disconnect the solenoid valve (A).

8. Disconnect the transaxle range switch (B).

9. Disconnect the input shaft speed sensor connector (C).

10. Remove the manual control cable (A) from the manual control lever.

11. Disconnect the output shaft speed sensor connector (A).

12. Remove the crank shaft position sensor (A).

13. Remove the oil cooler hose.

14. Remove the starter motor mounting bolt.

15. Remove the steering U-joint mounting bolt.

16. Install the special tools (09200-

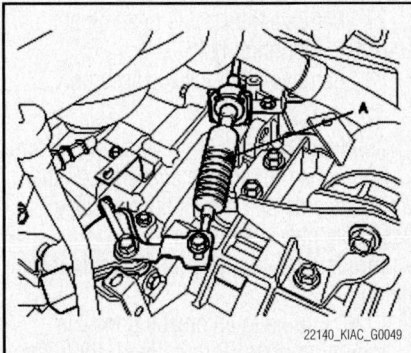

Fig. 26 Remove the manual control cable (A) from the manual control lever

38001), the engine support fixture and the adapter, on the engine assembly.

17. Remove the transaxle mounting insulator bolt (A).

18. Remove the transaxle upper mounting bolts.

19. Lift the vehicle.

20. Remove the front wheels.

21. Loosen the drain plug and then drain the transaxle oil.

22. Separate the tie rod end from the pin and nut.

23. Loosen the stabilizer control link nut.

24. Loosen the lower arm nut.

25. Remove the front muffler.

26. Remove the front roll stopper insulator bolt and nut (A).

27. Remove the rear roll stopper insulator bolt and nut (A).

28. Remove the sub frame bolt and nut.

29. Remove the sub frame.

30. Install the jack for supporting transaxle.

31. Remove the bell housing cover.

32. Remove the torque converter mounting bolts (6EA).

33. Remove the transaxle lower mounting bolt.

34. Remove the transaxle assembly by lifting vehicle.

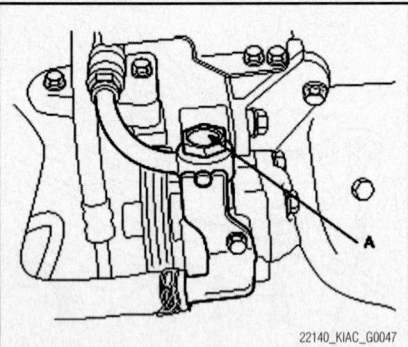

Fig. 24 Remove the power steering eye bolt (A)

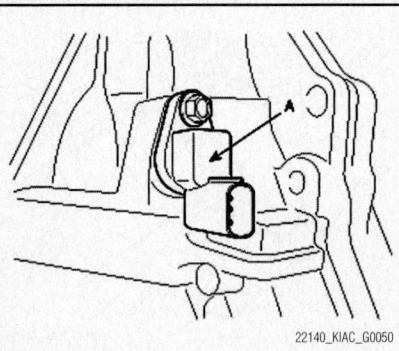

Fig. 27 Disconnect the output shaft speed sensor connector (A)

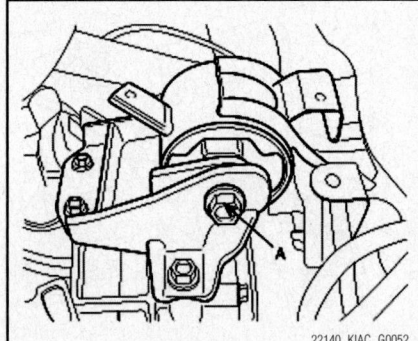

Fig. 29 Remove the transaxle mounting insulator bolt (A)

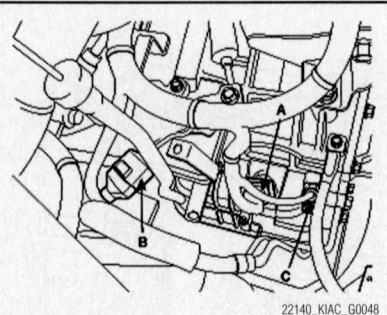

Fig. 25 Disconnect the solenoid valve (A), transaxle range switch (B) and input shaft speed sensor connector (C).

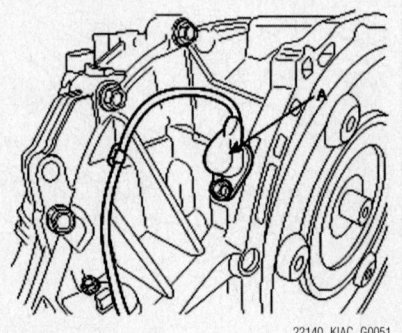

Fig. 28 Remove the crank shaft position sensor (A)

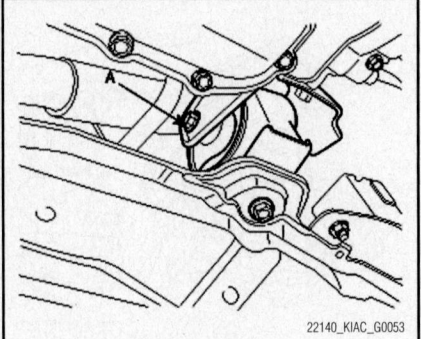

Fig. 30 Remove the front roll stopper insulator bolt and nut (A)

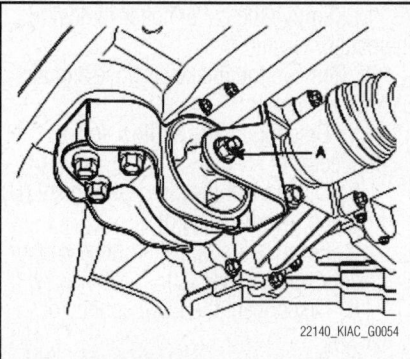

Fig. 31 Remove the rear roll stopper insulator bolt and nut (A)

⁂ **CAUTION**

When removing the transaxle assembly, be careful not to damage any surrounding parts or body components.

To install:

35. Installation is the reverse of removal.
36. Attach the torque converter on the transaxle side and mount the transaxle assembly onto the engine.

Torque specifications for transaxle installation:

a. Sub frame bolt and nut. Torque bolt (A) to 29–40 ft. lbs. (40–55 Nm); nut (B) and bolt (C) to 116–130 ft. lbs. (160–180 Nm).

b. Rear roll stopper insulator bolt and nut. Torque bolt and nut to 47–58 ft. lbs. (65–80 Nm).

c. Front roll stopper insulator bolt and nut. Torque bolt and nut to 65–80 ft. lbs. (90–110 Nm).

d. Lower arm nut. Torque bolt to 43–52 ft. lbs. (60–72 Nm).

e. Stabilizer control link nut. Torque nut to 32–45 ft. lbs. (44–62 Nm).

f. Transaxle mounting insulator bolt (A). Torque bolt and nut to 65–80 ft. lbs. (90–110 Nm).

37. During installation, perform the following:

a. Adjust the shift cable.

b. Install the transaxle control cable and adjust as follows:

• Move the shift lever and the transaxle range switch to the "N" Position, and install the control cable.

• When connecting the control cable to the transaxle mounting bracket, install the clip until it contacts the control cable.

• Remove any free-play in the control cable by adjusting nut and then check to see that the selector lever moves smoothly.

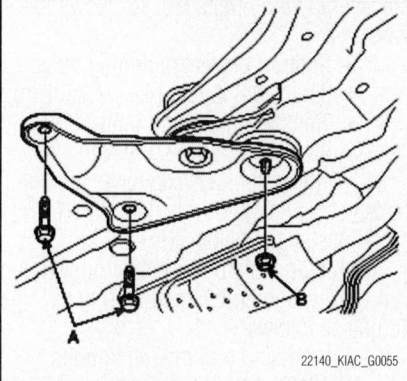

Fig. 32 Sub frame bolts (A) and nut (B)

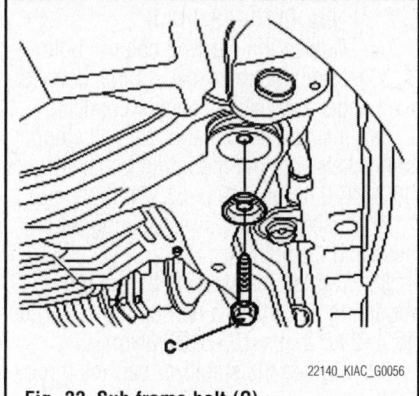

Fig. 33 Sub frame bolt (C).

• Check to see that the control cable (A) has been adjusted correctly.

c. Refill the transaxle fluid.

d. Clean the battery posts and cable terminals with sandpaper, assemble them, then apply grease to prevent corrosion.

MANUAL TRANSAXLE ASSEMBLY

REMOVAL & INSTALLATION

Optima

See Figures 34 through 38.

1. Remove the battery.
2. Remove the engine cover.
3. Disconnect the air flow sensor (AFS) connector (A) and ECU connector (B).
4. Remove the air cleaner assembly.
5. Remove the battery tray.
6. Remove the ground cable from the transaxle.
7. Disconnect the vehicle speed sensor.
8. Disconnect the shift cable assembly (A).
9. Disconnect the backup lamp switch.
10. Disconnect the clip, clutch arm, slave cylinder, and disconnect the hydraulic line.

11. Drain the power steering fluid and disconnect the power steering pressure hose.

12. Install the SST (09200-38001), the engine support fixture and the adapter on the engine and transaxle assembly

13. Remove the under shield cover.

14. Drain the manual transaxle fluid after removing the transaxle drain plug and the oil filler plug.

15. Disconnect the electronic power steering connector.

16. Remove the lower arm ball joint and folk mounting bolts.

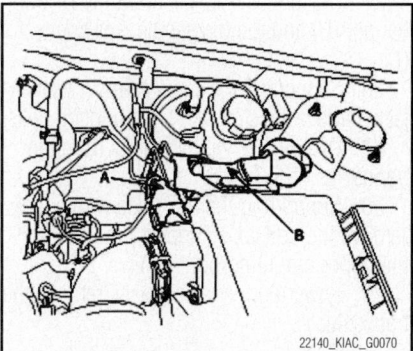

Fig. 34 Disconnect the air flow sensor (AFS) connector (A) and ECU connector (B)

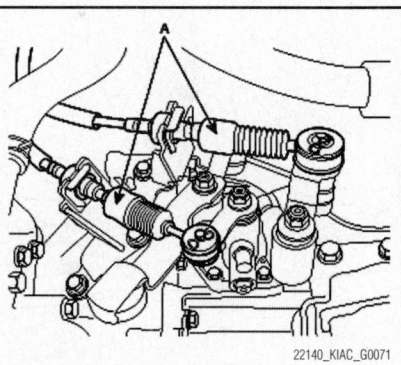

Fig. 35 Disconnect the shift cable assembly (A)

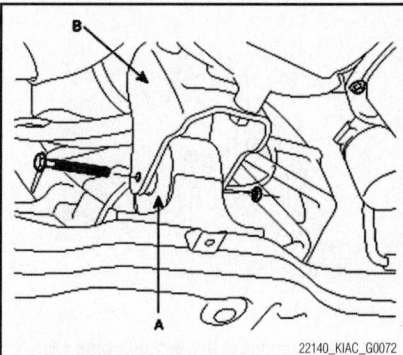

Fig. 36 Disconnect the bracket (B) from the front roll stopper (A)

17. Remove the lower arm ball joint assembly mounting bolts.

18. Remove the stabilizer bar link mounting bolts.

19. After removing a split pin and nut from the steering bar tie rod, disconnect it.

20. Loosen the power steering return hose clamp, disconnect the hose.

21. Disconnect the bracket (B) from the front roll stopper (A).

22. Disconnect the bracket (B) from the rear roll stopper (A).

23. Remove the steering column bolt..

24. Remove the transaxle lower mounting bolts.

25. Disconnect the exhaust pipe (A), the hanger (B) and the oxygen sensor connector (C).

26. Supporting the cross member with a jack, remove the stay with the mounting bolts.

27. Disconnect the halfshafts from the transaxle.

28. Supporting the transaxle with a jack, remove the transaxle support bracket and the upper mounting bolts (B).

29. Lowering the jack slowly, remove the transaxle.

To install:

30. Raising the jack slowly, tighten the transaxle mounting bolts (A) and support

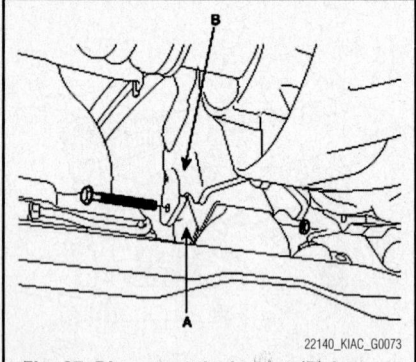

Fig. 37 Disconnect the bracket (B) from the rear roll stopper (A)

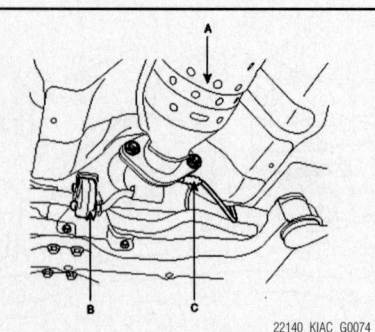

Fig. 38 Disconnect the exhaust pipe (A), the hanger (B) and the oxygen sensor connector (C)

bracket mounting bolts (B). Torque as follows:

- Support bracket mounting bolts: 31.1–39.8 ft. lbs. (42.2–53.9 Nm);
- Transaxle mounting bolts: 43.6–58.2 (59.2–78.9 Nm).

31. Tighten the transaxle lower mounting bolts.

32. Install the halfshafts.

33. Lifting the cross member up with the jack slowly, tighten the mounting bolts. Torque as follows:

- Front and rear mounting bolts: 101.3–115.7 ft. lbs. (137.3–156.9 Nm);
- Stay mounting bolts: 32.5–43.4 ft. lbs. (44.1–58.8 Nm).

34. Tighten the steering column bolt.

35. Assemble the exhaust pipe, connect the hanger and oxygen sensor connector.

36. Install the front and rear roll stoppers to brackets with the mounting bolts. Torque: 36.2–47.0 ft. lbs. (49.0–63.7Nm).

37. Inserting the power steering return hose and clamp.

38. Tighten the steering bar tie rod mounting nut, insert a new split pin. Torque: 17.4–24.6 ft. lbs. (23.5–33.3Nm).

39. Tighten the stabilizer bar link mounting bolt. Torque: 72.3–86.8 ft. lbs. (98.1–117.7Nm).

40. Install the lower arm ball joint assembly to the front axle. Torque: 72.3–86.8 ft. lbs. (98.1–117.7Nm).

41. Connect the electronic power steering connector.

42. Fill the manual transaxle fluid, tighten the oil filler plug. Torque: 21.7–25.3 ft. lbs. (29.4–34.3Nm).

➡ **Manual transaxle fluid: SAE 75W/85**

43. Install the under shield cover.

44. Install the power steering pressure hose, fill the power steering fluid.

45. Install a circlip on the slave cylinder tube and push it till a 'click' sound is heard. Loosen the clamp.

46. Tighten the backup lamp switch.

47. Install the shift cable assembly.

48. Fix the vehicle speed sensor and the ground line.

49. Install the ground cable to the transaxle.

50. Install the battery tray.

51. Install the air cleaner assembly.

52. Install the air flow sensor (AFS) connector and ECU connector.

53. Install the engine cover.

54. Install the battery.

Rio

See Figures 39 through 45.

1. Remove the battery after removing the battery terminal.

2. Remove the intake air hose and air cleaner assembly.

   a. Disconnect the air flow sensor connector (A).

   b. Disconnect the air cleaner hose (D) from the bleeder hose (B).

   c. Remove the intake air hose and air cleaner upper cover (C).

   d. Disconnect the ECM connector (A, B).

   e. Remove the air cleaner element and lower cover (C).

3. Remove the battery tray.

4. Remove the ground cable from transaxle.

5. Remove the transaxle wire harness connectors and control cable from transaxle.

   a. Remove the clutch slave cylinder.

   b. Remove the transaxle control cable (A).

   c. Disconnect the vehicle speed sensor connector (A).

   d. Disconnect the back lamp switch connector (B).

6. Remove the power steering oil hose and drain the power steering oil.

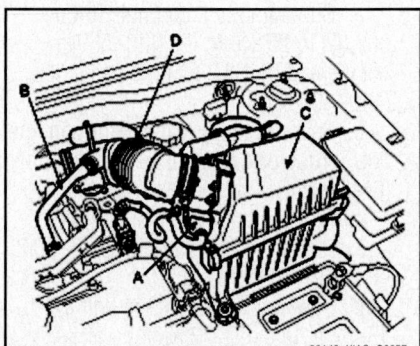

Fig. 39 Disconnect the Air Flow Sensor (AFS) connector (A), breather hose (B) from intake air hose (D), and air cleaner upper cover (C).

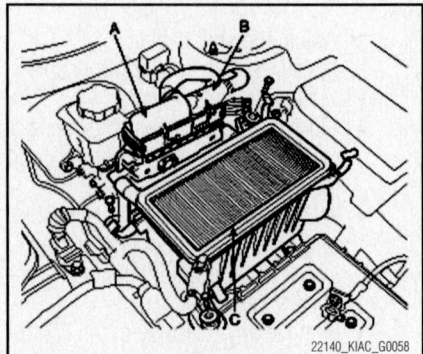

Fig. 40 Disconnect the ECM connector (A, B)

7. Remove the power steering return hose.

8. Install the SST (09200-38001), the engine support fixture and the adapter on the engine and transaxle assembly.

9. Remove the transaxle mounting support bracket (A).

10. Remove the front roll stopper insulator bolt (A).

11. Remove the rear roll stopper mounting bolt (A).

12. Remove the lower cover.

13. Remove the side cover.

14. Remove the front wheels.

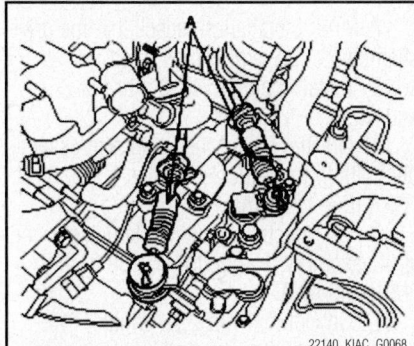

Fig. 41 Remove the transaxle control cable (A)

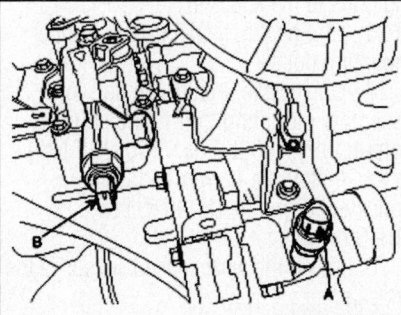

Fig. 42 Disconnect the vehicle speed sensor connector (A) and back lamp switch connector (B)

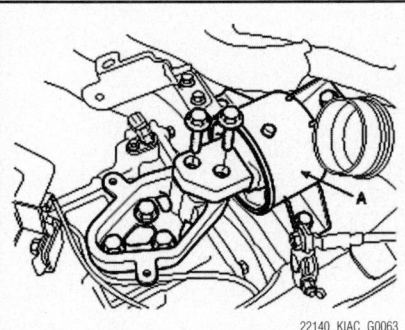

Fig. 43 Remove the transaxle mounting support bracket (A)

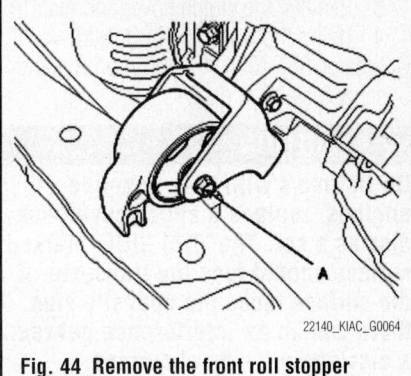

Fig. 44 Remove the front roll stopper mounting bolt (A)

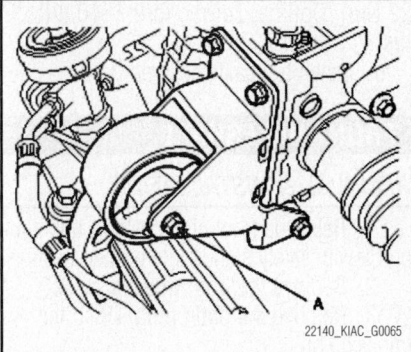

Fig. 45 Remove the rear roll stopper mounting bolt (A)

15. Remove the tie rod end pin and nut, then remove the tie rod end.

16. Remove the ABS wheel speed sensor.

17. Remove the caliper and hang caliper assembly.

18. Remove the knuckle mounting bolts.

19. Remove the drive shaft from transaxle.

20. Remove the steering U-joint mounting bolt.

21. Using a floor jack, support transaxle assembly.

➡ After removing the sub frame mounting bolts, the transaxle assembly may fall downward, and so support them securely with floor jack.

✳✳ CAUTION

Verify that the hoses and connectors are disconnected before removing the transaxle assembly.

22. Remove the sub frame

23. Remove the starter mounting bolts.

24. Remove the transaxle upper mounting bolts from engine.

25. Remove the transaxle lower mounting bolts.

26. Remove the transaxle assembly by lifting vehicle.

27. Installation is in the reverse order of removal.

28. Perform the following :
   a. Adjust the shift cable.
   b. Refill the transaxle with fluid.

29. Use the following torque specifications during installation:
   a. Transaxle lower mounting bolts: Small bolts: 10–16 ft. lbs. (15–20 Nm); Large bolts 32–40 ft. lbs. (43–55 Nm).
   b. Transaxle upper mounting bolts: 43–58 ft. lbs. (60–80 Nm).
   c. Starter motor bolts: 19–24 ft. lbs. (27–34 Nm).
   d. Sub frame nuts: 116–130 ft. lbs. (160–180 Nm) and bolts: 29–40 ft. lbs. (40–55 Nm).
   e. Rear roll stopper insulator bolt: 36–47 ft. lbs. (50–65 Nm).
   f. Front roll stopper insulator bolt: 36–47 ft. lbs. (50–65 Nm).
   g. Slave cylinder mounting bolts: 10–16 ft. lbs. (15–22 Nm).

Spectra

1. Before servicing the vehicle, refer to the Precautions Section.

2. Drain the transaxle oil.

3. Remove or disconnect the following:
   - Negative battery cable
   - Battery heat shield
   - Battery and battery tray
   - Air intake assembly
   - Back-up lamp switch connector
   - Clutch release cylinder
   - Shift cables
   - Vehicle speed sensor
   - Starter mounting bolts
   - Transaxle upper mounting bolts to the engine

4. Support the engine assembly with Special Tool 09200-38001 support fixture.

5. Remove or disconnect the following:
   - Transaxle mounting bracket
   - Front tire
   - Tie rod end
   - Wheel speed sensor
   - Wheel knuckle mounting bolts
   - Caliper assembly
   - Halfshaft
   - Steering U-joint mounting bolt
   - Oxygen sensor
   - Front muffler
   - Power steering hoses
   - Front and rear roll stoppers

6. Using a suitable jack, support the sub-frame assembly.

7. Remove the sub -frame.

8. Using a suitable jack, support the transaxle assembly.

9. Remove the transaxle lower mounting bolts.

10. Lower the transaxle assembly.

11. Installation is the reverse order of removal.

12. Fill the transaxle to the correct level.

CLUTCH

REMOVAL & INSTALLATION

See Figures 46 and 47.

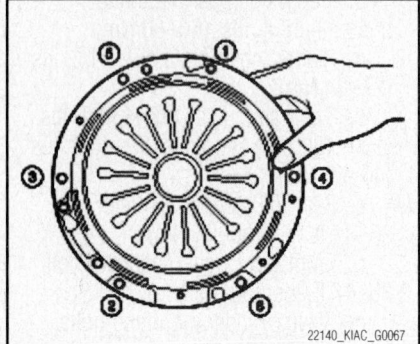

22140_KIAC_G0067

Fig. 46 Tighten clutch cover bolts in sequence—Spectra shown. Rio similar

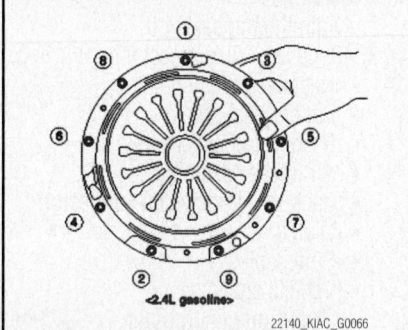

<2.4L gasoline>

22140_KIAC_G0066

Fig. 47 Tighten clutch cover bolts in sequence—Optima with 2.4L engine shown

1. Remove a transaxle assembly.
2. Remove the clutch cover bolts.

3. Remove the clutch cover and disc.

4. Using the SST (09411-11000) on Optima or SST (09411-25000) on Rio and Spectra, install the disc.

❖❖ CAUTION

On Optima's with 2.4L gasoline engines, replace a clutch cover and disc as a set. The 'T/M SIDE' marked surface should face the transaxle. If the surface faces the opposite side, there can be an interference between a disc and a flywheel surface.

5. Tighten the clutch cover. Rio and Spectra: Torque to 10–16 ft. lbs. (15–22 Nm). Optima: Torque to 8.7–10.8 ft. lbs. (11.8–14.7 Nm).

6. Install transaxle assembly.

FRONT HALFSHAFT

REMOVAL & INSTALLATION

1. Raise the front of the vehicle and support it with safety stands in a proper location.

2. Remove the front wheel and tire.

3. Remove the drain plug. Drain the transaxle oil.

4. Remove the split pin, the lock nut and the washer from the front hub.

5. Disconnect the tie rod end ball joint from the knuckle using the Special Tool (09568-34000) after removing the split pin and lock nut.

6. Remove the wheel speed sensor from the knuckle.

7. On Rio, remove the lower arm mounting bolts from the knuckle.

8. On Spectra, remove the strut upper mounting bolts.

9. On Optima, remove the ball joint assembly mounting bolt from the knuckle.

10. On Optima with 2.7L engine:
   a. Remove the heat protector.
   b. Remove the bearing and bracket assembly

11. Using a plastic hammer, disconnect the drive shaft from the axle hub.

12. Push the axle hub outward and separate the drive shaft from the axle hub.

13. Insert a pry bar between the transaxle case and joint case, and separate the drive shaft from the transaxle case.

14. Pull out the drive shaft from the transaxle case.

To install:

15. Apply gear oil on the drive shaft splines and the contacting surface of differential case oil seal.

16. Before installing the drive shaft, set the opening side of the circlip facing downward.

17. After installation, check that the drive shaft cannot be removed by hand.

18. Install the drive shaft into the knuckle.

19. On Rio, install the lower arm assembly bolts to the knuckle. Torque to 72–86 ft. Lbs. (100–120 Nm).

20. On Spectra, install the knuckle in the strut assembly. Torque to 94–108 ft. lbs. (130–150 Nm).

21. On Optima with 2.7L engine:
   a. Install bearing and bracket assembly.
   b. Install heat protector.

22. On Rio, install the lower arm mounting bolts to the knuckle.

23. On Spectra, install the strut upper mounting bolts.

24. On Optima, install the ball joint assembly mounting bolt to the knuckle. Torque to 72–86 ft. Lbs. (100–120 Nm).

25. Install the tie rod end to the knuckle. Torque to 12–25 ft. Lbs. (16–34 Nm).

26. Install the wheel speed sensor to the knuckle.

27. After installing the washer with convex surface outward, install the lock nut and the split pin.

28. Install the wheel and tire

ENGINE COOLING

THERMOSTAT

REMOVAL & INSTALLATION

1. Drain the engine coolant so its level is below thermostat.
2. Remove the water inlet fitting, gasket and thermostat.
3. Installation is the reverse of removal.

WATER PUMP

REMOVAL & INSTALLATION

Optima

2.4L Engine

1. Drain the engine coolant.
2. Remove drive belt.
3. Remove exhaust manifold.
4. Remove water inlet pipe nut.
5. Remove the water pump.

To install:

6. Install the water pump and a new gasket with the 5 bolts. Torque: 14.5–19.5 ft. lbs. (19.6–26.5 Nm).
7. Install water inlet pipe nut. Torque: 14.5–19.5 ft. lbs. (19.6–26.5 Nm).
8. Install exhaust manifold.
9. Install drive belt.
10. Fill with engine coolant.
11. Start engine and check for leaks.
12. Recheck engine coolant level.

2.7L Engine

1. Drain the engine coolant.
2. Remove accessory drive belt.
3. Remove the timing belt.

4. Remove the water pump and gasket.

To install:

5. Install the water pump and a new gasket. Torque: 10.8–15.9 ft. lbs. (14.7–21.6 Nm).
6. Install the timing belt.
7. Install accessory drive belt.
8. Fill with engine coolant.
9. Start engine and check for leaks.
10. Recheck engine coolant level.

Rio

1. Drain the engine coolant.
2. Loosen the water pump pulley bolts.
3. Remove the drive belts.
4. Remove the water pump pulley.
5. Remove the timing belt.
6. Remove the timing belt idler.
7. Remove the water pump.
   a. Remove the 2 bolts and alternator brace.
   b. Remove the 3 bolts and remove the water pump and gasket.

To install:

8. Install the water pump.
   a. Install the water pump and a new gasket with the 3 bolts. Torque: 8.7–10.8 ft. lbs. (11.8–14.7 Nm).
   b. Install the alternator brace with the 2 bolts. Torque: 14.5–19.5 ft. lbs. (19.6–26.5 Nm.
9. Install the timing belt idler.
10. Install the timing belt.
11. Install the water pump pulley.
12. Install the drive belts.

13. Tighten the water pump pulley bolts. Torque: 5.8–7.2 ft. lbs. (7.8–9.8 Nm).
14. Fill with engine coolant.
15. Start engine and check for leaks.
16. Recheck engine coolant level.

Spectra

1. Drain the engine coolant.
2. Remove drive belts.
3. Remove the timing belt.
4. Remove the timing belt idler.
5. Remove the power steering pump and the power steering pump bracket.
6. Remove the water pump.
   a. Remove the 4 bolts and pump pulley.
   b. Remove the 2 bolts, then remove the alternator brace.
   c. Remove the water pump and gasket.

To install:

7. Install the water pump.
   a. Install the water pump and a new gasket with the 3 bolts. Torque: 8.7–10.8 ft. lbs. (11.8–14.7 Nm).
   b. Install the alternator brace with the 2 bolts. Torque: 14.5–19.5 ft. lbs. (19.6–26.5 Nm).
   c. Install the pump pulley.
8. Install the power steering pump and the power steering bracket.
9. Install the timing belt idler.
10. Install the timing belt.
11. Install drive belts.
12. Fill with engine coolant.
13. Start engine and check for leaks.
14. Recheck engine coolant level.

ENGINE ELECTRICAL

ALTERNATOR

REMOVAL & INSTALLATION

Rio

1. Disconnect the battery negative terminal first, then the positive terminal.
2. Temporarily loosen the water pump pulley bolts.
3. Remove the alternator drive belt, after loosening the adjusting bolt and mounting bolt.
4. Remove the power steering pump belt.
5. Remove the water pump pulley.
6. Remove the power steering pump.

7. Remove the power steering pump bracket.
8. Disconnect the alternator connector, and remove the cable from alternator "B" terminal.
9. Remove the adjusting bolt and mounting bolt.
10. Remove the alternator brace.
11. Pull out the through bolt and then remove the alternator.

To install:

12. Installation is the reverse order of removal.
13. Hand tighten the water pump pulley bolts.

CHARGING SYSTEM

14. Adjust the power steering pump belt tension.
15. Adjust the alternator belt tension after installation.
16. Tighten the water pump pulley bolts.

Spectra & Optima

1. Disconnect the battery negative terminal first, then the positive terminal.
2. Disconnect the alternator connector, and remove the cable from alternator "B" terminal.
3. Remove the accessory drive belt.
4. Pull out the through bolt and then remove the alternator.
5. Installation is the reverse of removal.

ENGINE ELECTRICAL　　　　　　　　**IGNITION SYSTEM**

FIRING ORDER

Rio

1.6L Engine

Firing order: 1–3–4–2

Spectra

2.0L Engine

See Figure 48.

Firing order: 1–3–4–2

Optima

2.4L Engine

See Figure 49.

Firing order: 1–3–4–2

2.7L Engine

See Figure 50.

Firing order: 1–2–3–4–5–6

IGNITION COIL

REMOVAL & INSTALLATION

1. Remove the engine cover.
2. Remove the ignition coil.

➡**When removing the ignition coil connector, pull the lock pin and push the clip.**

3. Installation is the reverse of removal.

IGNITION TIMING

ADJUSTMENT

Ignition timing is controlled by the ECM. No adjustment is possible.

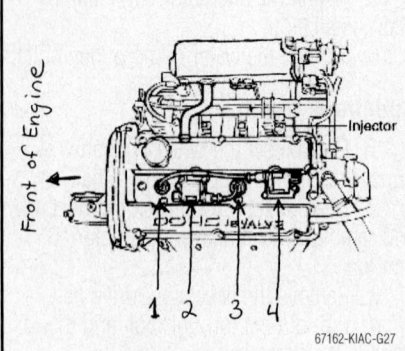

**Fig. 49 2.4L Engine
Firing Order: 1–3–4–2
Distributorless ignition system**

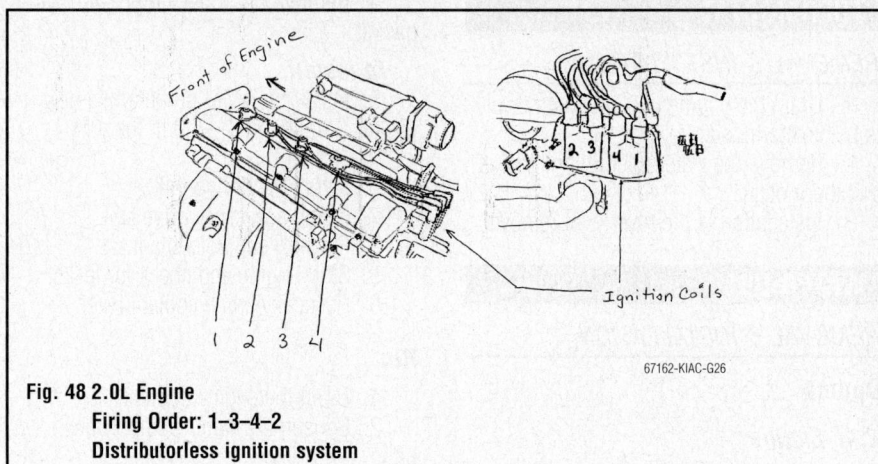

**Fig. 48 2.0L Engine
Firing Order: 1–3–4–2
Distributorless ignition system**

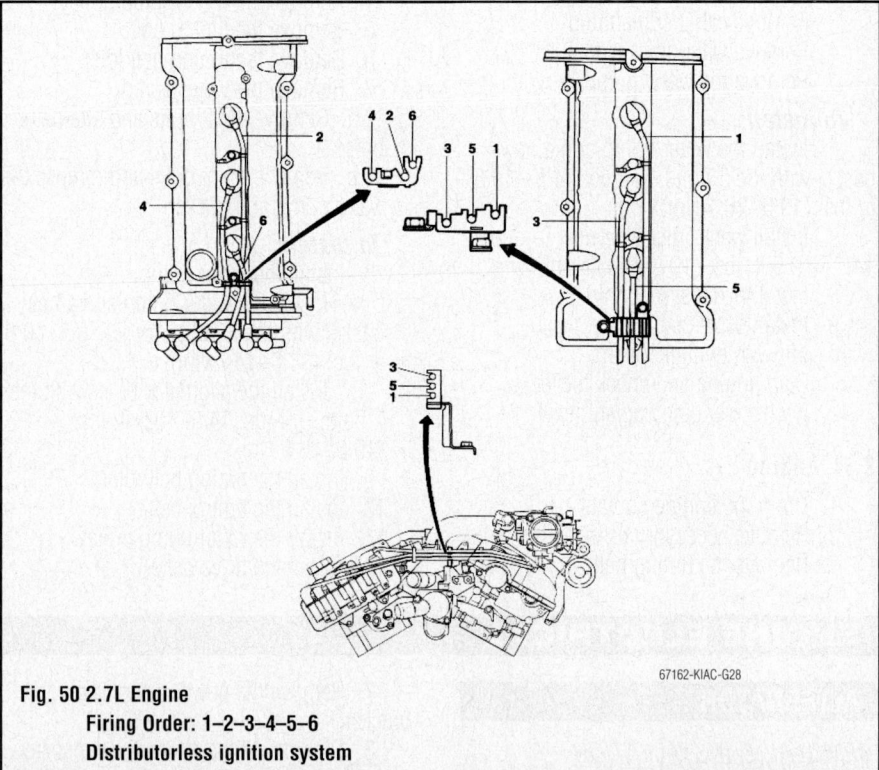

**Fig. 50 2.7L Engine
Firing Order: 1–2–3–4–5–6
Distributorless ignition system**

SPARK PLUGS

REMOVAL & INSTALLATION

1. Disconnect the negative battery cable.

✳✳ WARNING

When disconnecting the cable, only pull on the plug cable boot, never on the wire itself!

2. Detach the spark plug cable and remove the ignition coil.
3. Use spark plug wrench to remove the spark plug (s) from the cylinder head.

✳✳ WARNING

Do not let any dirt or debris get into the engine through the spark plugs holes while when are removed.

To install:

4. Install the spark plug and tighten, to 18–22 ft. lbs. (25–30 Nm)
5. Install the ignition coil (s).
6. Attach the spark plug wire (s) to the ignition coil.
7. Connect the negative battery cable.

ENGINE ELECTRICAL

STARTER

REMOVAL & INSTALLATION

1. Disconnect the battery negative cable.
2. Remove the air cleaner assembly

ENGINE MECHANICAL

→ Disconnecting the negative battery cable may interfere with the functions of the on board computer systems and may require the computer to undergo a relearning process, once the negative battery cable is reconnected.

ACCESSORY DRIVE BELTS

ACCESSORY BELT ROUTING

See Figures 51 through 54.

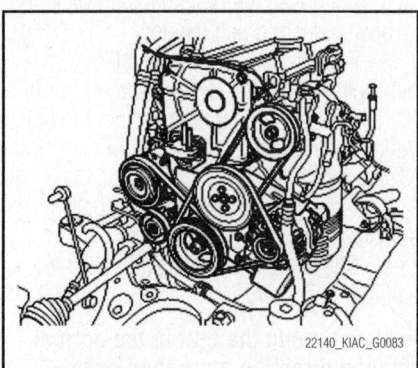

Fig. 51 Accessory drive belt routing—1.6L Engine

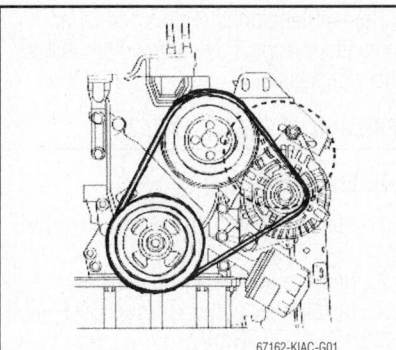

Fig. 52 Accessory drive belt routing—2.0L Engine

INSPECTION

Inspect the drive belt for signs of glazing or cracking. A glazed belt will be perfectly smooth from slippage, while a good belt will have a slight texture of fabric visible. Cracks

3. Remove the shift cable and bracket, if equipped with manual transaxle.
4. Disconnect the starter cable from the B terminal on the solenoid.
5. Disconnect the connector from the S terminal.

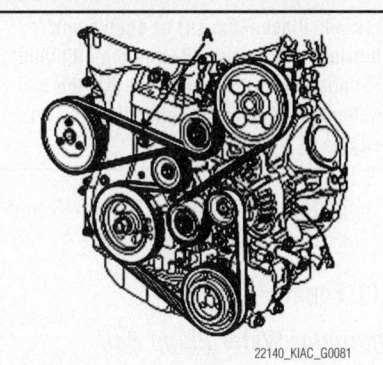

Fig. 53 Accessory drive belt routing—2.4L Engine

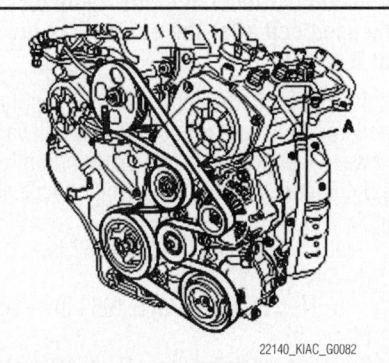

Fig. 54 Accessory drive belt routing—2.7L Engine

will usually start at the inner edge of the belt and run outward. All worn or damaged drive belts should be replaced immediately.

ADJUSTMENT

See Figures 55 through 62.

1.6L Engines

→ Refer to the accompanying illustration for bolt and nut locations.

1. Loosen bolt A and nuts B and C.
2. Turn adjusting bolt D and adjust the belt deflection to within the following specifications (when applying about 22 lbs. of pressure):
   a. New belt: 0.31–0.35 in. (8–9mm)
   b. Used belt: 0.35–0.39 in. (9–10mm)

STARTING SYSTEM

6. Remove the 2 bolts holding the starter, then remove the starter.
7. Installation is the reverse of removal.

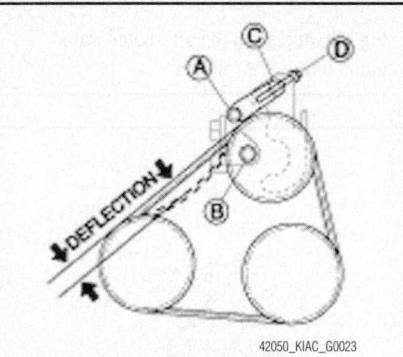

Fig. 55 Belt adjustment—1.6L engines with power steering

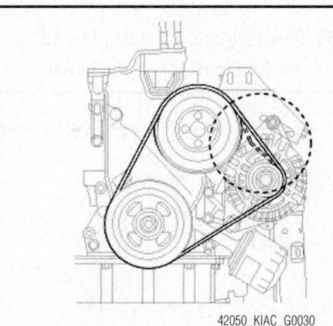

Fig. 56 Measuring the deflection between the alternator and water pump pulleys—2.0L engines

3. Tighten bolt A and nuts B and C, as follows:
   a. Bolt A: 27–39 ft. lbs. (37–53 Nm)

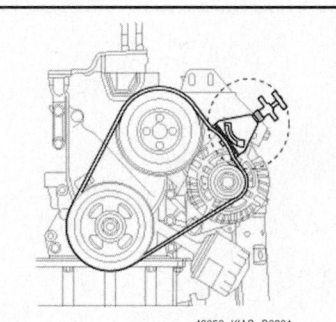

Fig. 57 Using a tension gauge to measure the alternator-water pump belt tension—2.0L engines

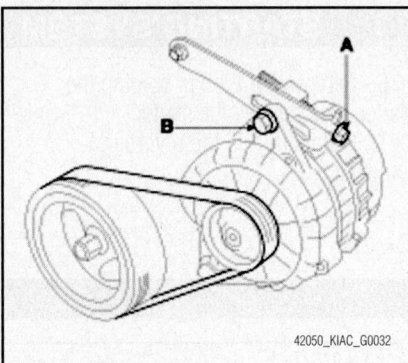

Fig. 58 Adjusting the alternator-water pump belt—2.0L engine

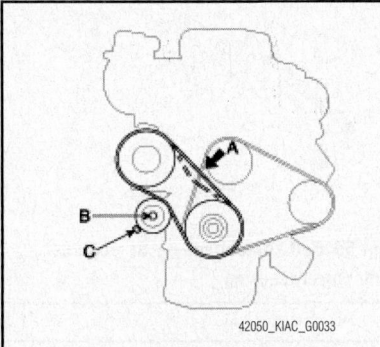

Fig. 59 Checking and adjusting the A/C compressor belt tension—2.0L engine

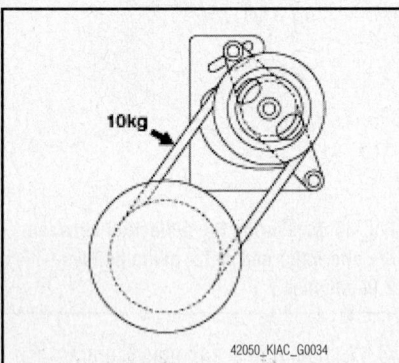

Fig. 60 Checking the power steering belt tension—2.0L engine

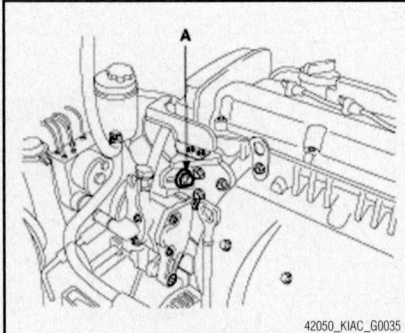

Fig. 61 Loosen the bolt adjusting (A) the power steering belt tension—2.0L engine

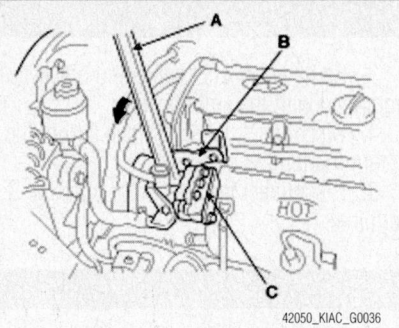

Fig. 62 Place a bar (A) or equivalent, between the bracket (B) and the oil pump (C) and adjust the tension so that the belt deflection is within specifications—2.0L engine

b. Nut B: 14–19 ft. lbs. (19–25 Nm)

c. Nut C: 24–34 ft. lbs. (32–46 Nm)

2.0L Engine

Alternator-Water Pump Belt

➡When using a new belt, first adjust the deflection or tension to the values for the new belt, then readjust the deflection or tension to the value for the used belt after the engine has run for 5 minutes.

1. Using the deflection method, apply moderate pressure (about 22 lbs) halfway between the alternator and water pump and compare with the following specifications.

   a. Used belt: 0.1969–0.2362 in. (5–6mm)

   b. New belt: 0.1575–0.1969 in. (4–5mm)

2. Using a belt tension gauge, measure the drive belt tension and compare with the following specifications:

   a. Alternator belt (new): 86–103 lbs. (383–461 N)

   b. Alternator belt (used): 68–86 lbs. (304–383 N)

   c. If adjustment is necessary, loosen the adjusting bolt (A) and the lock bolt (B), then move the alternator to get the proper tension and retighten the nuts.

3. Recheck the deflection or tension of the belt.

A/C Compressor Belt

1. Operate the A/C one or two times a month, year round and adjust the compressor belt tension from time to time.

2. Using the deflection method, apply moderate pressure (about 22 lbs) halfway between the between the A/C compressor and crankshaft pulley.

   a. New belt: 0.197–0.217 in. (5–5.5mm)

   b. Used belt: 0.236–0.276 in. (6–7mm)

   c. Check after operation: 0.315 in. (8mm)

3. If tension is not within, adjust the belt, as follows:

   a. Loosen the tension mounting bolt (B).

   b. Turn the adjusting bolt (C) to get the proper belt tension, then retighten the mounting bolt (B).

   c. Recheck belt tension.

Power Steering Belt

1. Using the deflection method, apply moderate pressure (about 22 lbs) at the point shown in the accompanying illustration and measure the deflection. It should be within 0.24–0.35 in. (6–9mm).

2. If not within specifications, adjust the belt, as follows:

   a. Loosen the bolt adjusting (A) the power steering belt tension.

3. Place a bar (A) or equivalent, between the bracket (B) and the oil pump (C) and adjust the tension so that the belt deflection is within 0.24–0.35 in. (6–9mm).

4. Tighten the power steering belt adjusting bolt.

5. Check the deflection and adjust it again, if necessary

➡After turning the belt in the normal rotation direction more than once, recheck the belt deflection.

2.4L & 2.7L Engines

The belt tension is maintained by an automatic tensioner. No adjustment is necessary or possible.

REMOVAL & INSTALLATION

2.4L Engines

1. Relieve tension on the tensioner pulley.

2. Remove accessory drive belt

3. Install idler pulley. Torque: 39.7–47.0 ft. lbs. (53.9–63.7Nm).

4. Install accessory drive belt in the following sequence:

   a. Crankshaft pulley

   b. A/C pulley

   c. alternator pulley

   d. idler pulley

   e. P/C pump pulley

   f. idler pulley

   g. Water pump pulley

   h. Tensioner pulley.

5. Rotate auto tensioner arm in the counterclockwise moving auto tensioner pulley bolt with wrench.

6. After putting belt on auto tensioner pulley, release the auto tensioner pulley slowly.

CRANKSHAFT FRONT SEAL

REMOVAL & INSTALLATION

1. Disconnect the negative battery cable.

2. Remove the timing belt covers and belt.

3. Remove the timing belt pulley using a puller.

4. Remove the oil pump bolts and the pump.

5. Wrap a suitable prytool with a rag and work the old seal from the oil pump housing.

To install:

6. Lubricate the seal lip with clean engine oil and push the seal slightly in by hand.

7. Install the seal using a seal installer. Install the seal until it is flush with the oil pump body.

8. Install the timing belt.

9. Connect the negative battery cable.

10. Start the engine and check for leaks.

CYLINDER HEAD

REMOVAL & INSTALLATION

1.6L Engine

See Figures 63 through 78.

1. Disconnect the terminals from battery and remove the battery.
2. Remove the engine cover.
3. Remove the lower cover.
4. Drain the engine coolant.

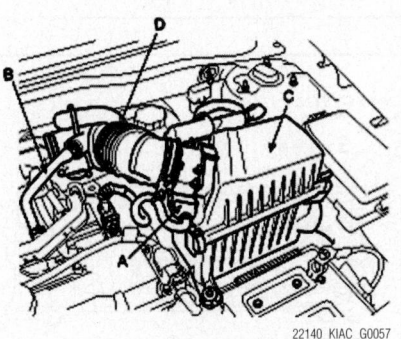

Fig. 63 Disconnect the Air Flow Sensor (AFS) connector (A), breather hose (B) from intake air hose (D), and air cleaner upper cover (C).

5. Remove the intake air hose and air cleaner assembly.

    a. Disconnect the AFS (Air Flow Sensor) connector (A).

    b. Disconnect the breather hose (B) from intake air hose (D).

    c. Remove the intake air hose (D) and air cleaner upper cover (C).

    d. Disconnect the ECM connector (A) and ECM connector (B) (A/T only).

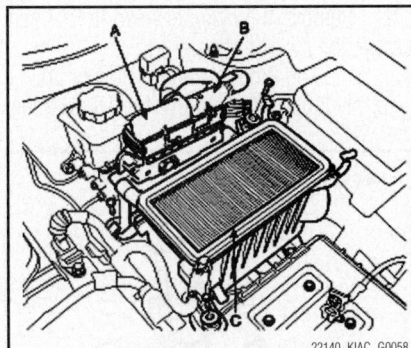

Fig. 64 Disconnect the ECM connector (A) and ECM connector (B) (A/T only)

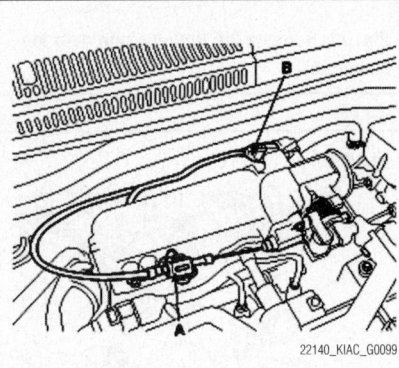

Fig. 65 Remove the accelerator cable (A). Disconnect the Throttle Position Sensor (TPS) connector (B)

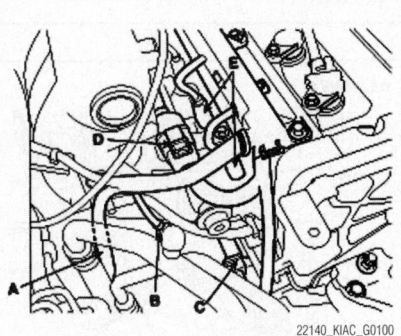

Fig. 66 Disconnect rear oxygen sensor connector (A), A/C compressor switch connector (B), injector connectors (No.3, 4) (D), and injector connectors (No. 1, 2) (E)

    e. Remove the air cleaner element and air cleaner lower cover (C).

6. Remove the battery tray.

7. Remove the upper radiator hose and lower radiator hose.

8. Remove the heater hoses.

9. Remove the fuel hose.

10. Remove the accelerator cable (A) by loosening the lock-nut, then slip the cable end out of the throttle linkage.

11. Disconnect the Throttle Position Sensor (TPS) connector (B).

12. Remove the engine wire harness connectors and wire harness clamps from cylinder head and the intake manifold.

    a. Disconnect the rear oxygen sensor connector (A).

    b. Disconnect the air conditioner compressor switch connector (B).

    c. Disconnect the knock sensor connector (C).

    d. Disconnect the injector connectors (No. 3, 4) (D).

    e. Disconnect the injector connectors (No.1, 2) (E)

    f. Remove the wire harness bracket (A).

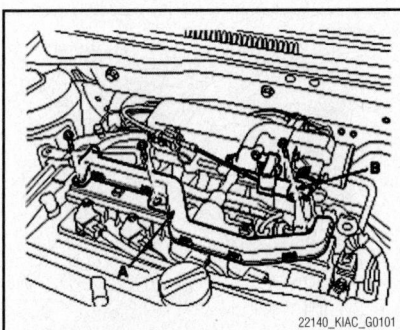

Fig. 67 Remove the wire harness bracket (A). Disconnect the Idle Speed Actuator (ISA) connector (B).

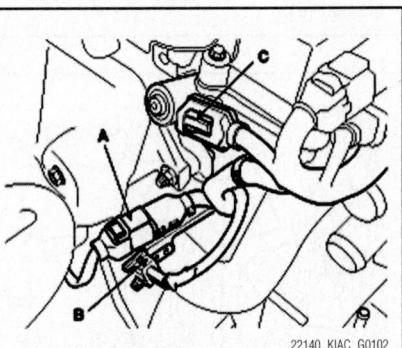

Fig. 68 Disconnect the front oxygen sensor connector (A), Crankshaft Position Sensor (CKP) connector (B), and Oil Control Valve (OCV) connector (C).

g. Disconnect the Idle Speed Actuator (ISA) connector (B).

h. Disconnect the front oxygen sensor connector (A).

i. Disconnect the Crankshaft Position Sensor (CKP) connector (B).

j. Disconnect the Oil Control Valve (OCV) connector (C).

k. Disconnect the ignition coil connector (A).

l. Disconnect the ignition coil condenser connector (B).

m. Disconnect the Camshaft Position Sensor (CMP) connector (C).

n. Disconnect the ground cable (D).

o. Remove the wire harness bracket (E).

13. Disconnect the hose (A) of the Purge Control Solenoid Valve (PCSV) side.

14. Remove the brake booster vacuum hose (B).

15. Remove the power steering pump and suspend the pump with a wire.

16. Remove the ignition coil.

17. Remove the exhaust manifold.

18. Remove the intake manifold.

19. Remove the timing belt.

20. Remove the cylinder head cover.

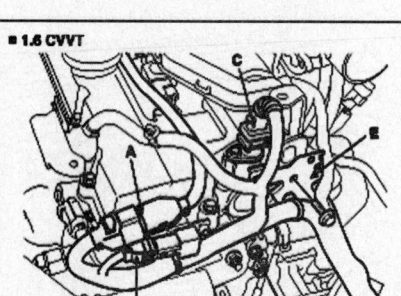

Fig. 69 Disconnect ignition coil connector (A), ignition coil condenser connector (B), Camshaft Position Sensor (CMP) connector (C), ground cable (D) and wire harness bracket (E).

Fig. 70 Disconnect the hose (A) of Purge Control Solenoid Valve (PCVS) and brake booster vacuum hose (B).

21. Remove the camshaft sprocket.

22. Remove the timing chain auto tensioner (A).

23. Remove the camshaft bearing caps and camshafts.

24. Remove the Oil Control Valve (OCV) (A).

25. Remove the Oil Control Valve (OCV) filter (A).

26. Remove the engine mounting support bracket fixing bolts (A).

27. Remove the cylinder head bolts, then remove the cylinder head.

a. Using 8mm hexagon wrench, uniformly loosen and remove the 10 cylinder

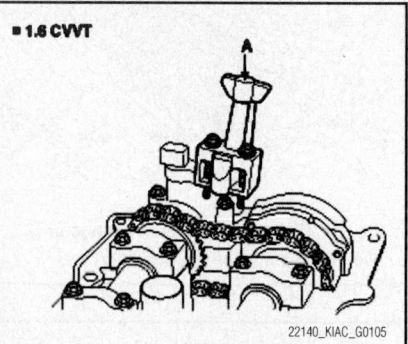

Fig. 71 Remove the timing chain auto tensioner (A)

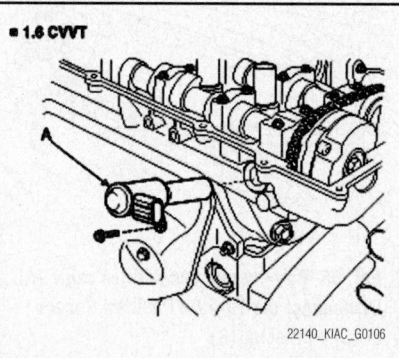

Fig. 72 Remove the Oil Control Valve (OCV) (A)

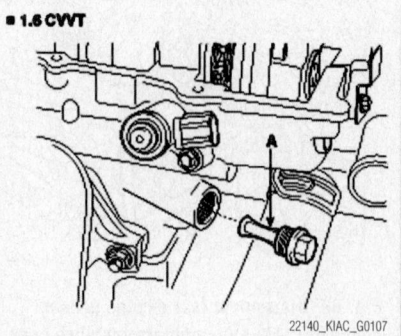

Fig. 73 Remove the Oil Control Valve (OCV) filter (A)

head bolts, in several passes, in the sequence shown.

✲✲ CAUTION

Head warpage or cracking could result from removing bolts in an incorrect order.

b. Lift the cylinder head from the dowels on the cylinder block and replace the cylinder head on wooden blocks on a bench.

To install:

28. Install the cylinder head gasket on the cylinder block.

29. Place the cylinder head carefully in order not to damage the gasket with the bottom part of the end.

30. Install the cylinder head bolts.

a. Apply a light coat if engine oil on the threads and under the heads of the cylinder head bolts.

b. Using 8mm and 10mm hexagon wrench, install and tighten the 10 cylinder head bolts and plate washers, in several passes, in the sequence shown.

Torque: 21.7 ft. lbs. (29.4 Nm) plus 90°; back off all bolts, then re-torque: 21.7 ft. lbs. (29.4 Nm) plus 90°.

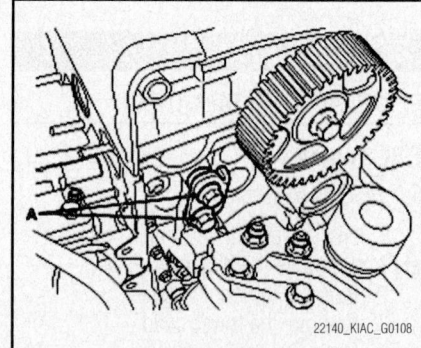

Fig. 74 Remove the engine mounting support bracket fixing bolts (A)

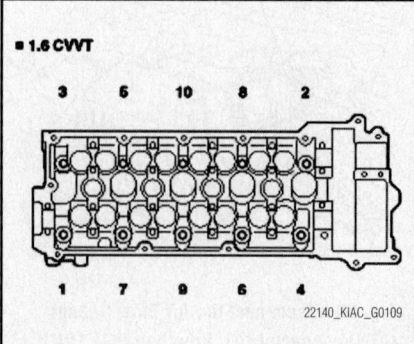

Fig. 75 Remove the cylinder head bolts in the sequence shown

31. Install the engine mounting support bracket fixing bolts.

32. Install the OCV (Oil Control Valve) filter (A). Torque: 29.7–36.9 ft. lbs. (40.2–50.0 Nm).

33. Install the OCV (Oil Control Valve). Torque: 7.2–8.7 ft. lbs. (9.8–11.8 Nm).

❋❋ CAUTION

Do not reuse the OCV (Oil Control Valve) when dropped. Keep clean the OCV (Oil Control Valve). Do not hold the OCV (Oil Control Valve) sleeve during servicing. When the OCV (Oil Control Valve) is installed on the engine, do not move the engine with holding the OCV (Oil Control Valve) yoke.

34. Install the camshafts.

a. Align the camshaft timing chain with the intake timing chain sprocket and exhaust timing chain sprocket as shown.

b. Install the camshaft and bearing caps. Torque: 8.7–10.1 ft. lbs. (11.8–13.7 Nm).

c. Install the timing chain auto tensioner. Torque: 5.8–7.2 ft. lbs. (7.8–9.8 Nm)

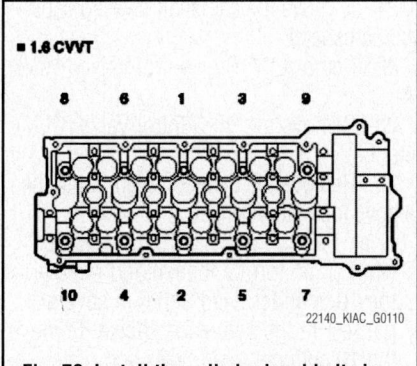

Fig. 76 Install the cylinder head bolts in the sequence shown

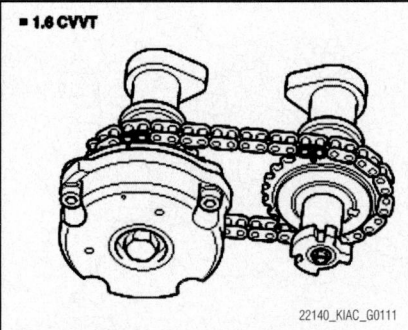

Fig. 77 Align the camshaft timing chain with the intake timing chain sprocket and exhaust timing chain sprocket as shown

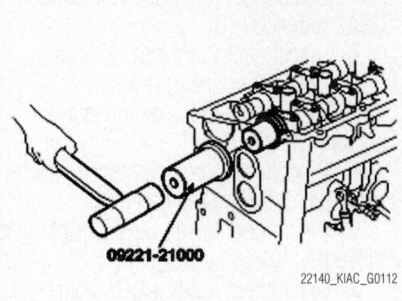

Fig. 78 Using the SST (09221–21000), install the camshaft bearing oil seal.

35. Using the SST (09221–21000), install the camshaft bearing oil seal.

36. Install the camshaft sprocket.

37. Install the cylinder head cover.

a. Install the cylinder head cover gasket in the groove of the cylinder head cover.

➡ **Before installing the cylinder head cover gasket, thoroughly clean the cylinder head cover and the groove. When installing, make sure the cylinder head cover gasket is seated securely in the corners of the recesses with no gap.**

b. Apply liquid gasket to the head cover gasket at the corners of the recess.

c. Install the cylinder head cover with bolts. Uniformly tighten the bolts in several passes. Pre-tighten all bolts: 2.9–3.6 ft. lbs. (3.9–4.9 Nm); then tighten by the specified torque. Torque: 5.8–7.2 ft. lbs. (7.8–9.8 Nm)

38. Install the timing belt.

39. Install the intake manifold.

40. Install the exhaust manifold.

41. Install the ignition coil.

42. Install the power steering pump.

43. Install the brake booster hose.

44. Connect the hose of the PCSV (Purge Control Solenoid Valve) side.

45. Install the engine wire harness connectors and wire harness clamps to the cylinder head and the intake manifold.

a. Install the wire harness bracket.

b. Connect the ground cable.

c. Connect the CMP (Camshaft position sensor) connector.

d. Connect the ignition coil condenser connector.

e. Connect the ignition coil connector.

f. Connect the OCV (Oil Control Valve) connector.

g. Connect the CKP (Crankshaft Position Sensor) connector.

h. Connect the front oxygen sensor connector.

i. Connect the ISA (Idle Speed Actuator) connector.

j. Install the wire harness bracket.

k. Connect the injector connectors (No.1, 2).

l. Connect the injector connectors (No.3, 4).

m. Connect the knock sensor connector.

n. Connect the air conditioner compressor switch connector.

o. Connect the rear oxygen sensor connector.

46. Connect the TPS (Throttle Position Sensor) connector.

47. Install the accelerator cable.

48. Install the fuel hose.

49. Install the heater hoses.

50. Install the upper radiator hose and lower radiator hose.

51. Install the battery tray.

52. Install the intake air hose and air cleaner assembly.

a. Install the air cleaner element and air cleaner lower cover.

b. Connect the ECM connector and ECM connector (A/T only).

c. Install the intake air hose and air cleaner upper cover.

d. Connect the breather hose to intake air hose.

e. Connect the AFS (Air Flow Sensor) connector.

53. Install the lower cover.

54. Install the engine cover.

55. Install the battery and connect the battery terminals.

56. Fill with engine coolant.

57. Start the engine and check for leaks.

58. Recheck engine coolant level and oil level.

2.0L Engine

See Figures 79 through 90.

1. Disconnect the battery terminals.

2. Remove the heat shield and the battery

3. Remove the engine cover.

4. Loosen the radiator drain plug and drain engine coolant.

5. Remove the intake air hose and air cleaner assembly.

a. Disconnect the MAF connector (A).

b. Disconnect the breather hose (B) from air cleaner hose (D).

c. Remove the intake air hose and air cleaner assembly (C).

6. Remove the upper and lower radiator hoses.

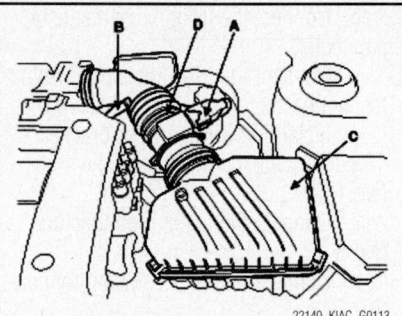

Fig. 79 Disconnect the MAF connector, (A) breather hose (B), remove air cleaner assembly (C) and air cleaner hose (D).

7. Remove the heater hoses.
8. Remove the accelerator cable.
9. Remove the engine wire harness connectors and wire harness clamps from the cylinder head and the intake manifold.

   a. Disconnect Oil control Valve (OCV) connector (A).

   b. Disconnect Oil Temperature Sensor (OTS) connector (B).

   c. Disconnect Engine Coolant Temperature (ECT) sensor connector (C).

   d. Disconnect ignition coil connector (D).

   e. Disconnect Throttle Position Sensor (TPS) connector (A).

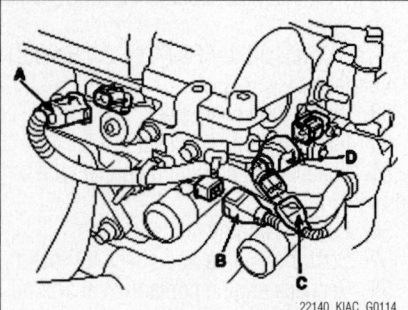

Fig. 80 Remove the engine wire harness connectors and wire harness clamps from the cylinder head and the intake manifold.

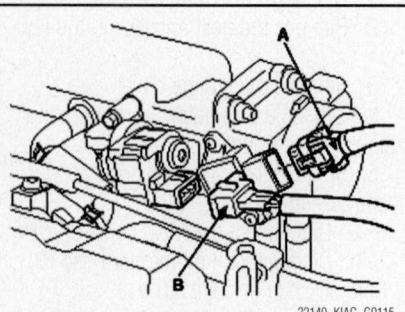

Fig. 81 Disconnect Throttle Position Sensor (TPS) connector (A) and ISA (Idle Speed Actuator) connector (B).

   f. Disconnect Idle Speed Actuator (ISA) connector (B).

   g. Disconnect Camshaft Position Sensor (CMP) connector (A).

   h. Disconnect four fuel injector connectors (B).

   i. Disconnect Knock Sensor connector (C) and the ground cable (D).

   j. Disconnect Purge Control Solenoid Valve (PCSV) connector (E).

   k. Disconnect front heated oxygen sensor connector.

10. Remove the fuel inlet hose from delivery pipe.
11. Remove the PCSV hose.
12. Remove the brake booster vacuum hose.
13. Remove the spark plug cable.
14. Remove the Positive Crankcase Ventilation (PCV) hose (A).
15. Remove the cylinder head cover.
16. Remove the timing belt.
17. Remove the exhaust manifold.
18. Remove the intake manifold.
19. Remove the camshaft sprocket.
20. Remove the timing chain auto tensioner (A).

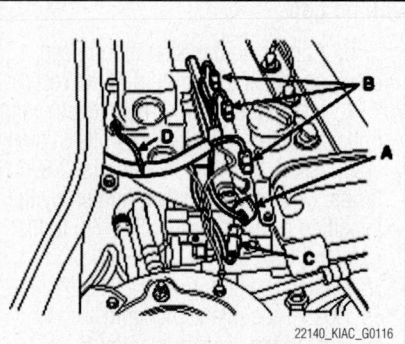

Fig. 82 Disconnect Camshaft Position Sensor (CMP) connector (A), four fuel injector connectors (B), Knock Sensor connector (C) and the ground cable (D).

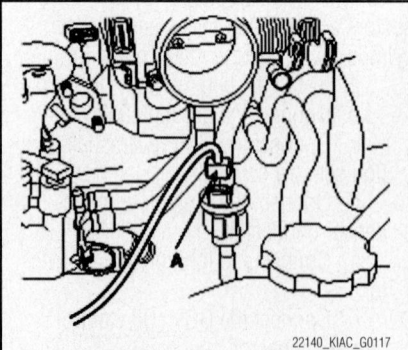

Fig. 83 Disconnect Purge Control Solenoid Valve (PCSV) connector (E)

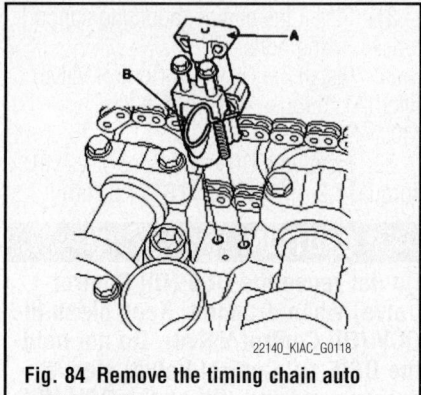

Fig. 84 Remove the timing chain auto tensioner (A)

Fig. 85 Remove the Oil Control Valve (OCV) (A)

21. Remove the camshaft bearing caps and camshafts.
22. Remove the Oil Control Valve (OCV) (A).
23. Remove the Oil Control Valve (OCV) filter (A).
24. Remove the cylinder head bolts, then remove the cylinder head.

   a. Using 8mm and 10mm hexagon wrench, uniformly loosen and remove the 10 cylinder head bolts, in several passes, in the sequence shown. Remove the 10 cylinder head bolts and plate washers.

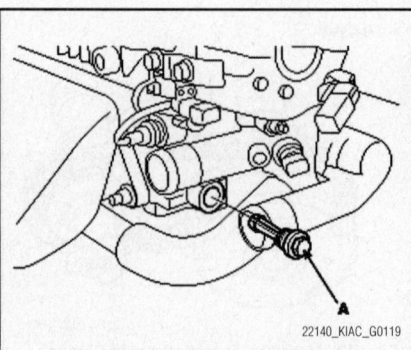

Fig. 86 Remove the OCV (oil control valve) filter (A)

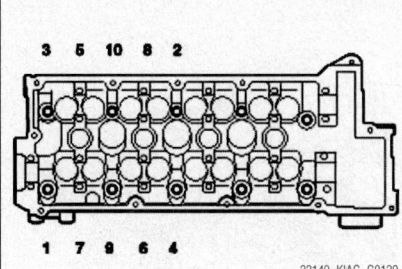

Fig. 87 Remove the cylinder head bolts in the sequence shown

> ❋❋ **CAUTION**
>
> **Head warpage or cracking could result from removing bolts in an incorrect order.**

b. Lift the cylinder head from the dowels on the cylinder block and replace the cylinder head on wooden blocks on a bench.

To install:

25. Install the cylinder head gasket on the cylinder block.

26. Place the cylinder head carefully in order not to damage the gasket with the bottom part of the end.

27. Install the cylinder head bolts.

a. Apply a light coat if engine oil on the threads and under the heads of the cylinder head bolts.

b. Using 8mm and 10mm hexagon wrench, install and tighten the 10 cylinder head bolts and plate washers, in several passes, in the sequence shown.
Torque: M10 bolts: 16.6–19.5 ft. lbs. (22.6–26.5 Nm) plus an additional 60°–65° plus another 60°–65°. M12 bolts: 20.3–23.1 ft. lbs. (27.5–31.4 Nm) plus an additional 60°–65° plus another 60°–65°.

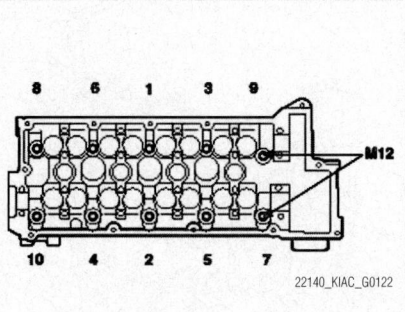

Fig. 88 Install the cylinder head bolts in the sequence shown

28. Install the Oil Control Valve (OCV) filter. Torque: 29.7–36.9 ft. lbs. (40.2–50.0 Nm).

29. Install the Oil Control Valve (OCV). Torque: 7.2–8.7 ft. lbs. (9.8–11.8 Nm).

> ❋❋ **CAUTION**
>
> **Do not reuse the OCV when dropped. Keep clean the OCV. Do not hold the OCV sleeve during servicing. When the OCV is installed on the engine, do not move the engine with holding the OCV yoke.**

30. Install the camshafts.

a. Align the camshaft timing chain with the intake timing chain sprocket and exhaust timing chain sprocket as shown.

b. Install the camshafts and bearing caps. Torque: 10.1–10.8 ft. lbs. (13.7–14.7 Nm).

c. Install the timing chain auto tensioner. Torque: 5.8–7.2 ft. lbs. (7.8–9.8 Nm).

d. Remove the auto tensioner stopper pin.

31. Check and adjust valve clearance.

32. Using the SST (09221-21000), install the camshaft bearing oil seal.

33. Install the camshaft sprocket.

34. Install the timing belt.

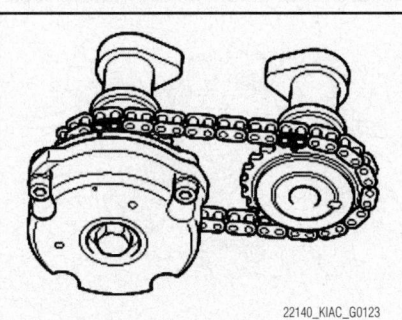

Fig. 89 Align the camshaft timing chain with the intake timing chain sprocket and exhaust timing chain sprocket as shown

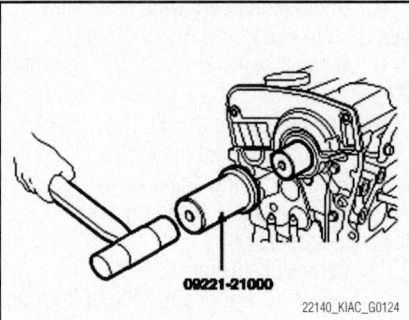

Fig. 90 Using the SST (09221-21000), install the camshaft bearing oil seal

35. Install the cylinder head cover.

a. Install the cylinder head cover gasket in the groove of the cylinder head cover.

b. Apply liquid gasket to the head cover gasket at the corners of the recess.

c. Install the cylinder head cover with the 12 bolts. Uniformly tighten the bolts in several passes. Torque: 5.8–7.2 ft. lbs. (7.8–9.8 Nm).

36. Install the intake manifold.

37. Install the exhaust manifold.

38. Install the Positive Crankcase Ventilation (PCV).

39. Install the spark plug cable.

40. Install the accelerator cable and the auto-cruse cables.

41. Install the bake booster hose.

42. Install the Purge Control Solenoid Valve (PCSV) hose.

43. Install the fuel inlet hose.

44. Install the engine wire harness connectors and wire harness clamps to the cylinder head and the intake manifold.

a. Front heated oxygen sensor connector.

b. Knock sensor connector and the ground cable.

c. Four fuel injector connectors.

d. CMP connector.

e. Purge Control Solenoid Valve (PCSV) connector.

f. Idle Speed Control Actuator (ISCA) connector.

g. Throttle Position Sensor (TPS) connector.

h. Ignition coil connector.

i. Engine Coolant Temperature Sensor (ECTS) sensor connector.

j. Oil temperature sensor connector.

k. Oil Control Valve (OCV) connector.

45. Install the heater hoses.

46. Install the upper and lower radiator hoses.

47. Install the intake air hose and air cleaner assembly.

48. Install the engine cover.

49. Connect the negative terminal to the battery.

50. Fill with engine coolant.

51. Start the engine and check for leaks.

52. Recheck engine coolant level and oil level.

2.4L Engine

See Figures 91 through 104.

1. Disconnect the negative terminal from the battery.

2. Remove engine cover.

3. Remove air duct.

4. Remove the intake air hose and air cleaner assembly.

   a. Disconnect the AFS (B) connector.

   b. Disconnect the breather hose (C) from air cleaner hose.

   c. Disconnect the ECU connector.

   d. Remove the intake air hose and air cleaner (A).

5. Remove front wheels.

6. Remove lower cover.

7. Drain the engine coolant.

8. Remove the upper and lower radiator hoses.

9. Remove the heater hoses.

10. Disconnect A/C switch (A), alternator connector (B), and oil pressure switch (C).

11. Disconnect OCV connector (A) and OTS connector (B) and P/S switch connector (C).

12. Disconnect injector connectors.

13. Disconnect the engine wire harness connectors.

   a. Remove the ETC connector (B), CMP connector (C), knock sensor connector (D).

14. Disconnect ignition coil connectors.

15. Disconnect PCSV connector (A), WTS connector (B), condenser connector (C), and CKP sensor connector (D).

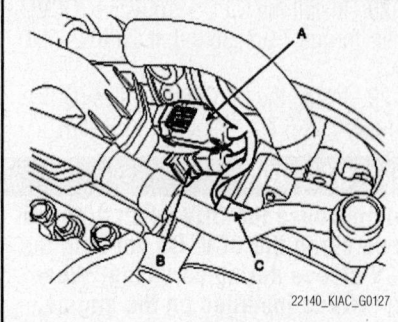

Fig. 93 Disconnect OCV connector (A) and OTS connector (B) and P/S switch connector (C).

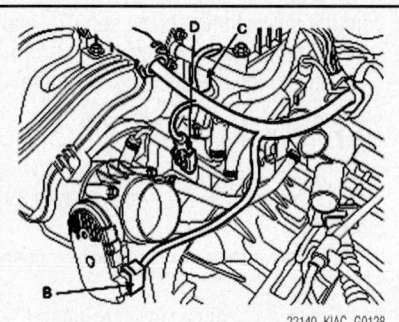

Fig. 94 Remove the ETC connector (B), CMP connector (C), knock sensor connector (D).

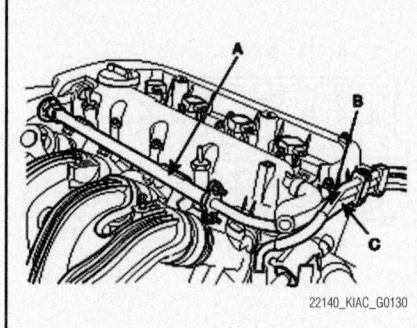

Fig. 96 Remove delivery pipe (A), brake vacuum hose (B), and PCSV hose (C).

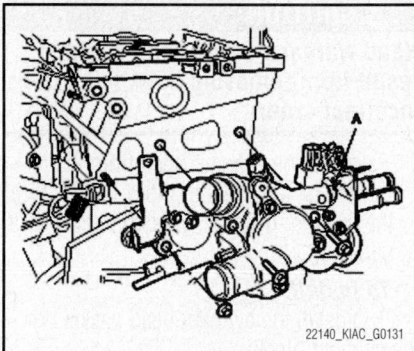

Fig. 97 Remove water temp control assembly (A).

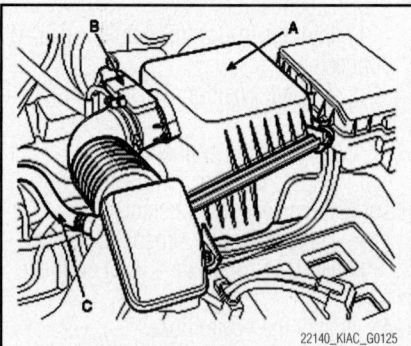

Fig. 91 Disconnect the AFS (B) connector, breather hose (C), and air cleaner (A).

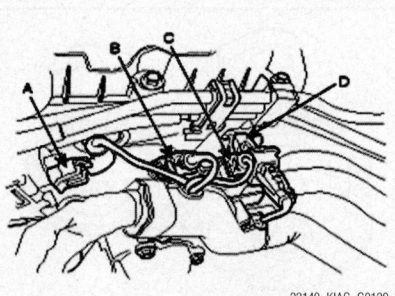

Fig. 95 Disconnect PCSV connector (A), WTS connector (B), condenser connector (C), and CKP sensor connector (D).

Fig. 98 Remove CVVT assembly and camshaft sprocket (A)

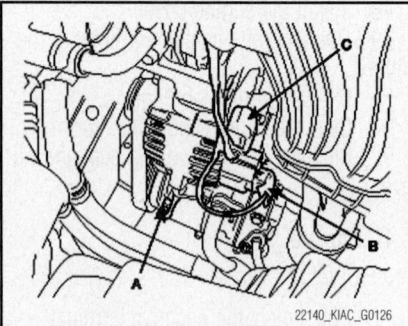

Fig. 92 Disconnect A/C switch (A), alternator connector (B), and oil pressure switch (C).

16. Remove delivery pipe (A), brake vacuum hose (B), and PCSV hose (C).

17. Remove water temp control assembly (A).

18. Remove intake manifold.

19. Remove exhaust manifold.

20. Remove timing chain.

21. Remove CVVT assembly and camshaft sprocket (A).

22. Remove camshaft.

   a. Remove front camshaft bearing cap (A).

   b. Remove camshaft bearing cap (A), in the sequence shown.

   c. Remove camshafts.

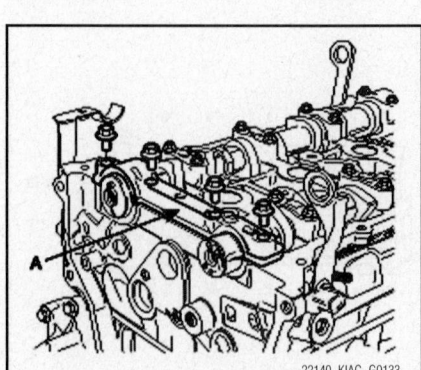

Fig. 99 Remove front camshaft bearing cap (A)

23. Remove OCV (A) and OTS (B).

24. Remove the cylinder head bolts, then remove the cylinder head.

   a. Using triple square wrench, uniformly loosen and remove the 10 cylinder head bolts, in several passes, in the sequence shown. Remove the 10 cylinder head bolts and plate washers.

❋❋ CAUTION

Head warpage or cracking could result from removing bolts in an incorrect order.

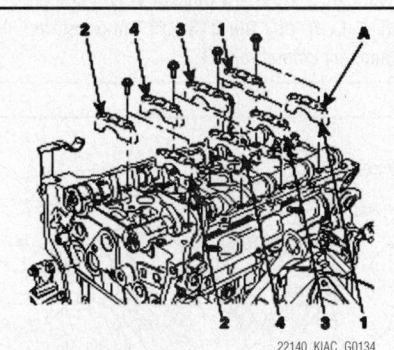

Fig. 100 Remove camshaft bearing cap (A), in the sequence shown

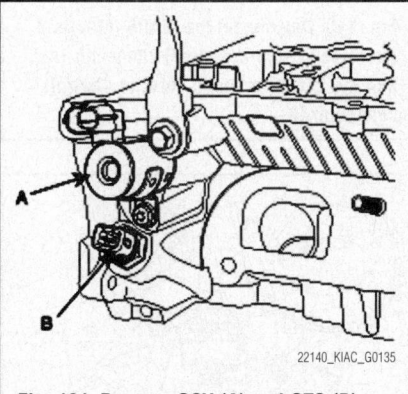

Fig. 101 Remove OCV (A) and OTS (B)

   b. Lift the cylinder head from the dowels on the cylinder block and place the cylinder head on wooden blocks on a bench.

To install:

25. Install the cylinder head gasket on the cylinder block.

➡ **Be careful of the installation direction.**

26. Place the cylinder head carefully in order not to damage the gasket with the bottom part of the end.

27. Install cylinder head bolts.

   a. Apply a light coat if engine oil on the threads and under the heads of the cylinder head bolts.

   b. Install and tighten the 10 cylinder head bolts and plate washers, in several passes, in the sequence shown. Torque: 25.3 ft. lbs. (34.3 Nm) plus 90° plus an additional 90°.

➡ **Always use new cylinder head bolt.**

28. Install OCV and OTS. Torque: OCV: 7.23–8.67 ft. lbs. (9.8–11.76 Nm; OTS: 14.5–17.4 ft. lbs. (19.6–23.52 Nm).

❋❋ CAUTION

Do not reuse the OCV when dropped. Keep the OCV clean. Do not hold the OCV sleeve during servicing. When the OCV is installed on the engine, do not move the engine with holding the OCV yoke.

29. Install the CVVT and camshaft sprocket. Torque: 39.7–47.0 ft. lbs. (53.9–63.7 Nm).

➡ **Hold the hexagonal head wrench portion of the camshaft with a vise, and install the bolt and CVVT assembly.**

30. Install camshafts.

➡ **Apply a light coat of engine oil on camshaft journals.**

31. Install camshaft bearing caps in their proper locations.

Tightening order:
- Group A
- Group B
- Group C.
- Torque: M6 bolts: 8.0–9.4 ft. lbs. (10.8–12.7 Nm; M8 bolts: 20.2–23.1 ft. lbs. (27.4–31.4 Nm).

32. Install timing chain.

33. Check and adjust valve clearance.

34. Install the exhaust manifold.

35. Install the intake manifold.

36. Install water temp control assembly. Torque: Bolt: 10.8–15.9 ft. lbs. (14.7–21.6 Nm; Nut: 14.5–19.5 ft. lbs. (19.6–26.5 Nm).

➡ **Assemble water temp control assembly and water inlet pipe to water pump assembly before nuts for assembling of water inlet pipe to be tightened. Insert after wetting O-ring or inner surface of thermostat housing. Always use a new O-ring.**

37. Install delivery pipe, brake hose, and PCSV hose.

38. Install the PCSV connector, WTS connector, condenser connector, and CKP sensor connector.

39. Install ignition coil connector.

40. Install the engine wire harness connectors.

   a. Install the ETC connector, CMP connector, knock sensor connector.

41. Install the injector connectors.

42. Install the OCV connector, OTS connector, P/S switch connector.

43. Install the A/C switch, alternator connector, and oil pressure switch.

44. Install the upper and lower radiator hoses.

45. Install lower cover.

46. Install the intake air hose and air cleaner assembly.

   a. Install the AFS connector.

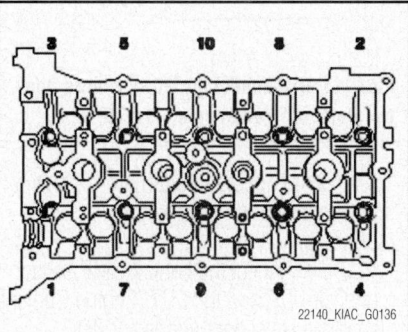

Fig. 102 Remove the cylinder head bolts in the sequence shown

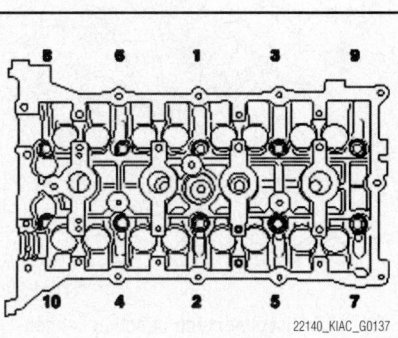

Fig. 103 Install cylinder head bolts in the sequence shown

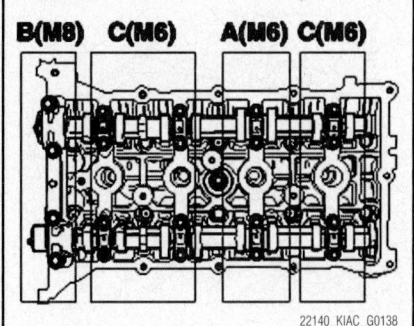

Fig. 104 Install camshaft bearing caps and torque in sequence

b. Install the breather hose from air cleaner hose.

c. Install the ECU connector.

d. Install the intake air hose and air cleaner.

47. Install the engine cover.

48. Install the negative terminal to the battery.

49. Fill with engine coolant.

50. Start the engine and check for leaks.

51. Recheck engine coolant level and oil level.

2.7L Engine

See Figures 105 through 124.

1. Remove the air duct and the battery.

2. Remove the engine cover.

3. Remove the intake air hose and air cleaner assembly.

a. Disconnect the MAF connector (A).

b. Disconnect the breather hose (B) from air cleaner hose.

c. Remove the intake air hose and air cleaner assembly (C).

d. Disconnect the PCM connectors (D).

4. Remove the upper and lower radiator hoses.

5. Remove the fuel inlet hose from the delivery pipe.

6. Disconnect the engine wiring harness connectors.

a. Disconnect the No.1/No.2 knock sensor connectors (A, B), the oil pressure switch connector (C), the ignition coil harness (D) and the No.1 VIS (Variable Induction System) connector (E).

b. Disconnect the bank 1 front/rear O2 sensor connectors (A).

c. Disconnect the injection connectors (A, B, C), the ground lines (D), the condenser connector (E) and the Ignition coil connectors (F).

d. Disconnect the injection harness connector (A), the No.2 VIS (Variable Induction System) connector (B), the No.1/No.2 OCV (Oil Control Valve) connectors (C, D) and the OTS (Oil Temperature Sensor) connector (E).

e. Disconnect the MAPS (Manifold Absolute Pressure Sensor) connector (A), the ETC (Electronic Throttle Control) connector (B) and the PCSV (Purge Control Solenoid Valve) connector (C).

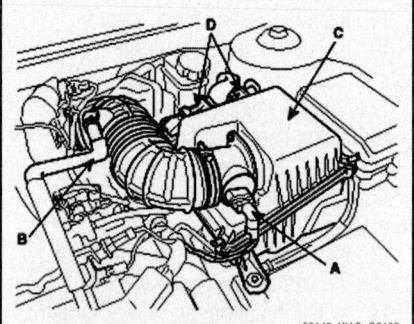

Fig. 105 Disconnect the MAF connector (A), breather hose (B), intake air hose and air cleaner assembly (C), and PCM connectors (D).

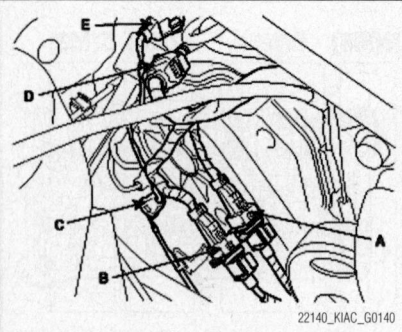

Fig. 106 Disconnect the engine wiring harness connectors

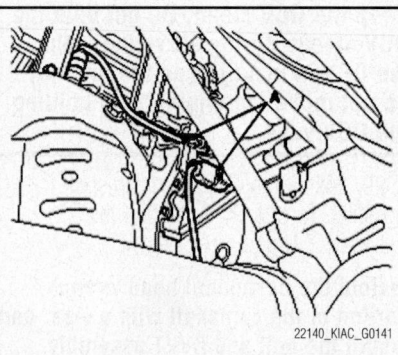

Fig. 107 Disconnect the bank 1 front/rear O2 sensor connectors (A)

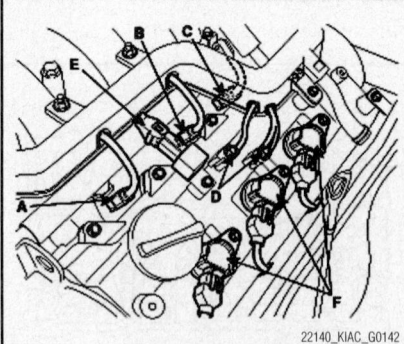

Fig. 108 Disconnect the injection connectors (A, B, C), the ground lines (D), the condenser connector (E) and the Ignition coil connectors (F)

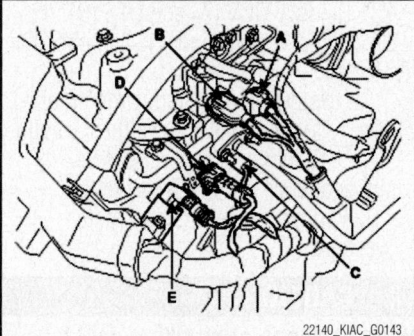

Fig. 109 Disconnect the injection harness connector (A), the No.2 VIS (Variable Induction System) connector (B), the No.1/No.2 OCV (Oil Control Valve) connectors (C, D) and the OTS (Oil Temperature Sensor) connector (E).

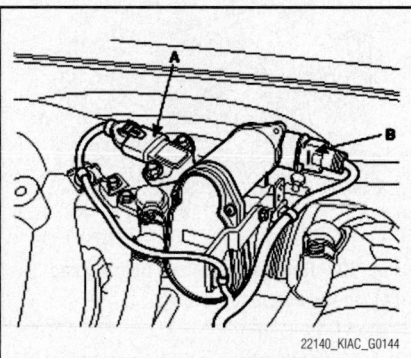

Fig. 110 Disconnect the MAPS (Manifold Absolute Pressure Sensor) connector (A), and the ETC (Electronic Throttle Control) connector (B)

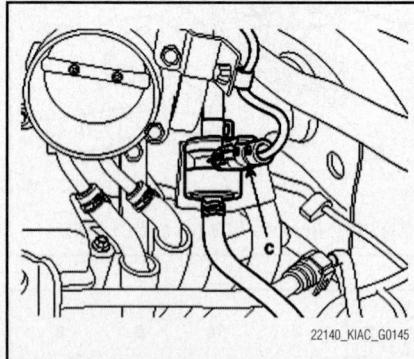

Fig. 111 Disconnect the PCSV (Purge Control Solenoid Valve) connector (C)

f. Disconnect the alternator connector and the air conditioning compressor connector.

g. Disconnect the bank 2 CMP sensor connector (A) and the ECT (Engine Coolant Temperature) sensor connector (B).

h. Disconnect the bank 2 front/rear O2 sensor connectors (A, B) and the CKP sensor connector (C).

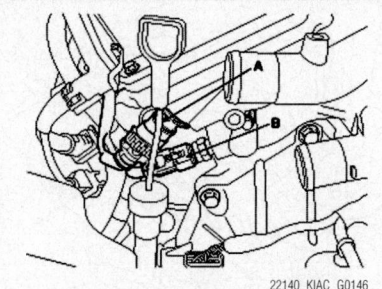

Fig. 112 Disconnect the bank 2 CMP sensor connector (A) and the ECT (Engine Coolant Temperature) sensor connector (B)

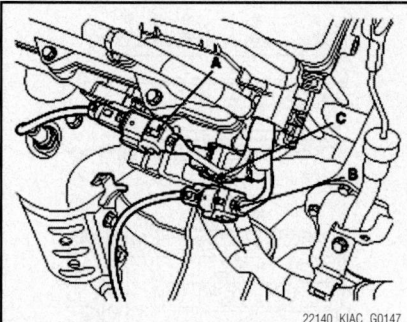

Fig. 113 Disconnect the bank 2 front/rear O2 sensor connectors (A,B) and the CKP sensor connector (C)

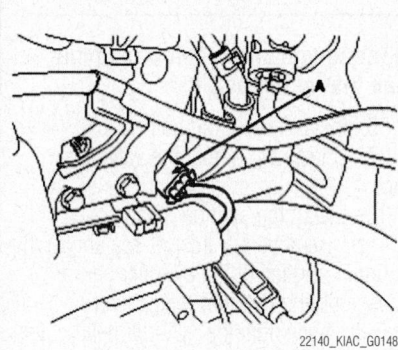

Fig. 114 Disconnect the bank 1 CMP sensor connector (A)

   i. Disconnect the bank 1 CMP sensor connector (A).
   7. Remove the PCV (Purge Control Valve) hose.
   8. Disconnect the brake vacuum hose.
   9. Remove the heater hoses.
   10. Remove the accessory drive belt.
   11. Remove the power steering pump.
   12. Remove the exhaust manifold assembly.
   13. Remove the intake manifold assembly.
   14. Remove the timing belt.
   15. Remove the ignition coils.

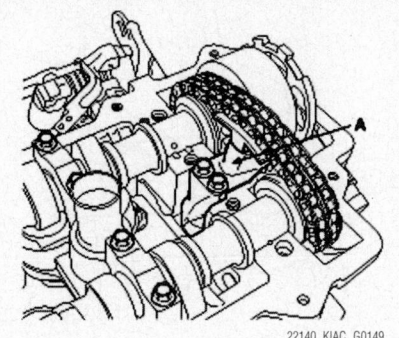

Fig. 115 Remove the timing chain tensioner (A)

   16. Remove the water temp. control assembly.
   17. Remove the cylinder head cover.
   18. Remove the camshaft bearing cap.
   19. Remove the timing chain tensioner (A).
   20. Remove the camshaft.
   21. Remove the bank 1 timing belt rear cover.
   22. Remove the bank 2 timing belt rear cover.
   23. Remove the CKP sensor connector bracket.
   24. Remove the cylinder head assembly.
     a. Remove the bolts in 2–3 steps in the sequence shown.

✳✳ CAUTION

If the bolts are not removed as the order, the deformation of the head assembly can be occurred.

     b. Put the cylinder head assembly on a wooden block after removal from the cylinder block.

   To install:
   25. Install the cylinder head (s) with new head gaskets.

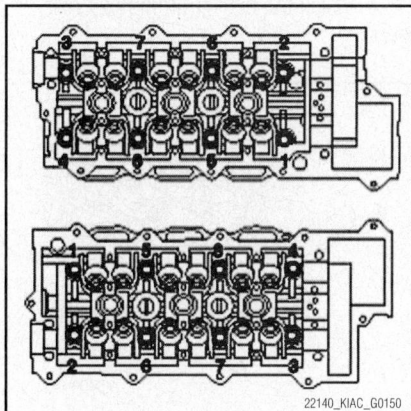

Fig. 116 Remove the bolts in the sequence shown

✳✳ CAUTION

Ensure the LH/RH classification of the cylinder head gasket when installing.

   26. Tighten the cylinder head bolts with the plain washers in several steps in the sequence shown.

➡ **In assembling washers, the marked surface should face upward.**

   27. Before installing the cylinder head bolts, apply engine oil on the thread of the bolts and the surface of the washers. Torque: 18.1 ft. lbs. (24.5 Nm) plus 60° plus an additional 45°.

➡ **Using the SST (09221-4A000), tighten the bolts which need to be tightened with the angular tightening method.**

   28. Install the CVVT assembly and camshaft chain sprocket with the dowel pin in the CVVT installed to the intake camshaft. Ensure that the pin will not be installed in the hole for oil feeding. Torque: 49.2–57.9 ft. lbs. (66.7–78.5 Nm).

➡ **After tightening the CVVT bolts, rotate the CVVT assembly housing counterclockwise by hand to seat the lock pin in the CVVT assembly in good position.**

✳✳ CAUTION

Fix the hexagonal part of the camshaft in a vice when tightening the CVVT bolts. Do not fix the CVVT housing or sprocket in a vice.

   29. Install the camshaft in the cylinder head assembly.
     a. Align the timing mark of the camshaft timing chain.

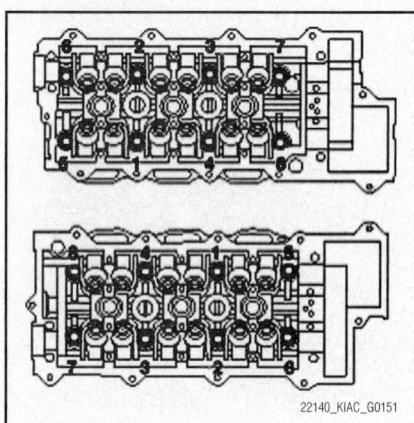

Fig. 117 Install cylinder head bolts in the sequence shown.

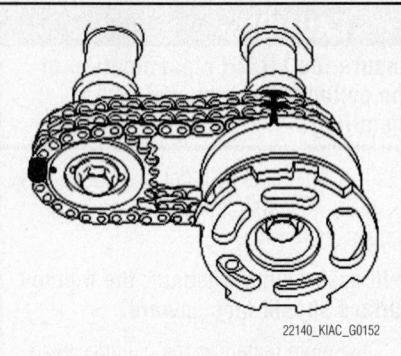

Fig. 118 Left hand camshaft chain timing mark

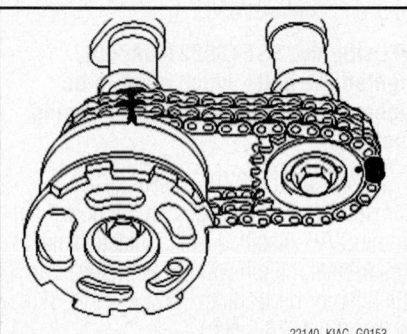

Fig. 119 Right hand camshaft chain timing mark

❈❈ CAUTION

Both timing marks should face upward in reassembly.

30. Install the timing chain tensioner.

a. Insert the set pin by pressing the timing chain tensioner.

b. Install the chain tensioner in the cylinder head assembly.

c. Remove the set pin from the tensioner after installing.

31. Install the camshaft bearing caps. Torque: Bearing cap bolt (A: 6◊38): 8.0–9.4 ft. lbs. (10.8–12.7 Nm). Bearing cap bolt (B: 8◊38): 15.2–18.8 ft. lbs. (20.6–22.5 Nm).

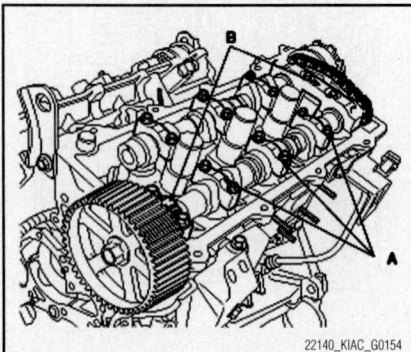

Fig. 120 Install the camshaft bearing caps

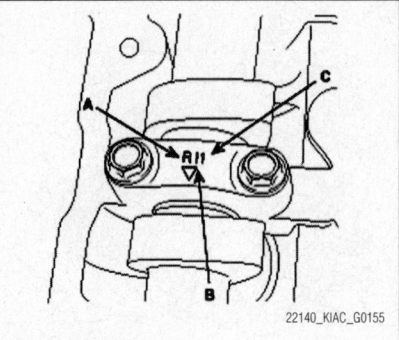

Fig. 121 When installing the bearing caps, check the marks

➡ **When installing the bearing caps, check the marks on them as shown below and install them in its proper position. Note the following designations as shown below:**

- A: (LH/RH HEAD) L (LH), R (RH)
- B: (Intake/Exhaust) I (Intake), E (Exhaust)
- C: (Cap no.): 1, 2, 3

❈❈ CAUTION

When installing the bearing caps, turn the crankshaft to place a piston in the middle of the block because interference between valves and pistons can occur.

32. Using the SST (09214-21000), install the camshaft oil seal.

➡ **Before installing, apply engine oil. The camshaft cap surface should adhere to the cylinder head assembly. Do not press an eccentric load.**

33. Install the CKP sensor connector bracket.

34. Install the bank 2 timing belt rear cover.

35. Install the bank 1 timing belt rear cover.

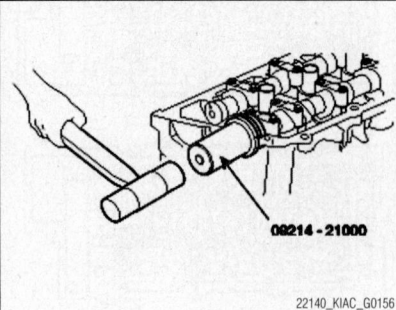

Fig. 122 Using the SST (09214-21000), install the camshaft oil seal

Fig. 123 Install the bank 1 timing belt rear cover

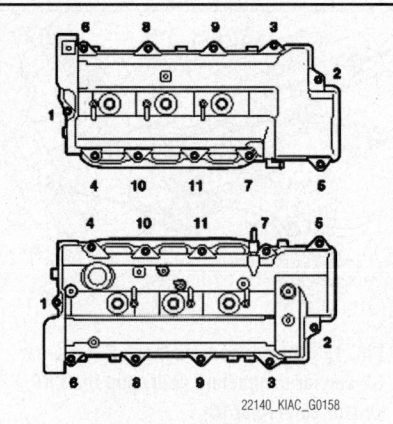

Fig. 124 Tighten the cylinder head cover bolts in sequence

➡ **The length of the bolt B is longer than that of the bolt C.**

36. Install the timing belt.

37. Check and adjust the valve clearance.

38. Install the cylinder head cover.

a. Remove oil, dust or sealant on the upper surface of the cylinder before assembling cylinder head cover.

b. Assemble the cylinder head cover in five minutes after applying liquid gasket (LOCTITE 5900) on the camshaft cap and packing part.

c. Tighten the cylinder head cover bolts in the sequence shown. Torque: 5.8–7.2 ft. lbs. (7.8–9.8 Nm).

➡ **Do not start engine for thirty minutes after assembling the cylinder head cover. Do not reuse the cylinder head cover gasket.**

39. Install the water temp. control assembly.

40. Install the intake manifold assembly.

41. Install the exhaust manifold assembly.

42. Install the power steering pump.

43. Install the accessory drive belt.

44. Install the heater hose.
45. Connect the brake vacuum hose.
46. Install the PCV (Positive Crankcase Ventilation) hose.
47. Connect the engine wiring harness connectors.

a. Connect the bank 1 CMP sensor connector.

b. Connect the bank 2 front/rear O2 sensor connectors and the CKP sensor connector.

c. Connect the bank 2 CMP sensor connector and the WTS (Water Temperature Sensor) connector.

d. Connect the alternator connector and the air conditioning compressor connector.

e. Connect the MAPS (Manifold Absolute Pressure Sensor) connector, the ETC (Electronic Throttle Control) connector and the PCSV (Purge Control Solenoid Valve) connector.

f. Connect the injection harness connector, the No.2 VIS (Variable Induction System) connector, the No.1/No.2 OCV (Oil Control Valve) connectors and the OTS (Oil Temperature Sensor) connector.

g. Connect the injection connectors, the ground lines, the condenser connector and the Ignition coil connectors.

h. Connect the bank 1 front/rear O2 sensor connectors.

i. Connect the No.1/No.2 knock sensor connectors, the oil pressure switch connector, the ignition coil harness and the No.1 VIS (Variable Induction System) connector.

48. Install the fuel inlet hose from the delivery pipe.
49. Install the upper radiator hose and lower radiator hose.
50. Install the intake air hose and air cleaner assembly.

a. Connect the PCM connectors.

b. Install the intake air hose and air cleaner assembly.

c. Connect the breather hose from air cleaner hose.

d. Connect the MAF connector.

51. Install the engine cover.
52. Refill engine coolant.

ENGINE ASSEMBLY

REMOVAL & INSTALLATION

1.6L Engine

See Figures 125 through 138.

1. Disconnect the terminals from battery and remove the battery.
2. Remove the engine cover.

3. Remove the lower cover.
4. Drain the engine coolant.
5. Remove the intake air hose and air cleaner assembly.

a. Disconnect the AFS (Air Flow Sensor) connector (A).

b. Disconnect the breather hose (B) from intake air hose (D).

c. Remove the intake air hose (D) and air cleaner upper cover (C).

d. Disconnect the ECM connector (A) and ECM connector (B) (A/T only).

e. Remove the air cleaner element and air cleaner lower cover (C).

6. Remove the battery tray.
7. Remove the upper radiator hose and lower radiator hose.
8. Remove the heater hoses.
9. Remove the fuel hose.
10. Remove the accelerator cable (A) by loosening the lock-nut, then slip the cable end out of the throttle linkage.
11. Disconnect the TPS (Throttle Position Sensor) connector (B).
12. Remove the engine wire harness connectors and wire harness clamps from cylinder head and the intake manifold.

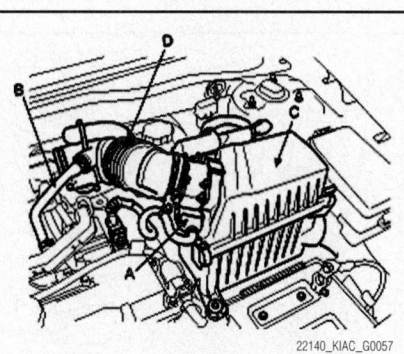

Fig. 125 Disconnect the AFS (Air Flow Sensor) connector (A), breather hose (B) from intake air hose (D), and air cleaner upper cover (C).

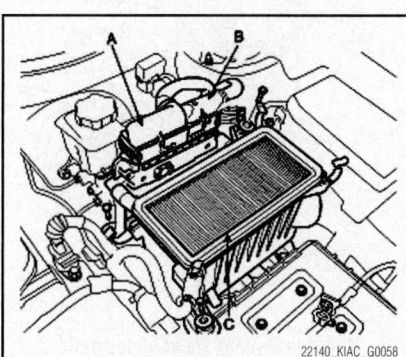

Fig. 126 Disconnect the ECM connector (A) and ECM connector (B) (A/T only)

a. Disconnect the rear oxygen sensor connector (A).

b. Disconnect the air conditioner compressor switch connector (B).

c. Disconnect the knock sensor connector (C).

d. Disconnect the injector connectors (No. 3, 4) (D).

e. Disconnect the injector connectors (No. 1, 2) (E)

f. Remove the wire harness bracket (A).

g. Disconnect the ISA (Idle Speed Actuator) connector (B).

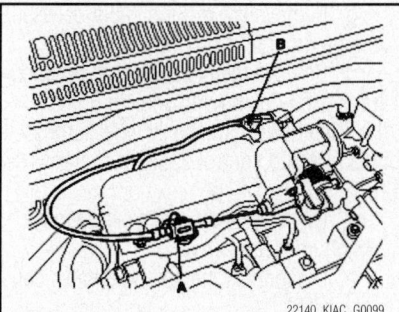

Fig. 127 Remove the accelerator cable (A). Disconnect the TPS (Throttle Position Sensor) connector (B).

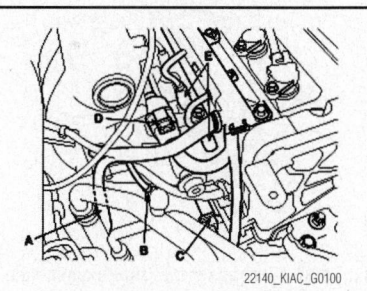

Fig. 128 Disconnect rear oxygen sensor connector (A), A/C compressor switch connector (B), injector connectors (No. 3, 4) (D), and injector connectors (No. 1, 2) (E)

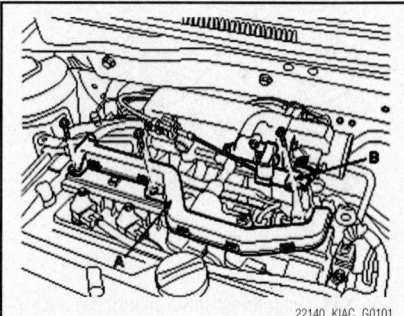

Fig. 129 Remove the wire harness bracket (A). Disconnect the ISA (Idle Speed Actuator) connector (B).

h. Disconnect the front oxygen sensor connector (A).

i. Disconnect the CKP (Crankshaft Position Sensor) connector (B).

j. Disconnect the OCV (Oil Control Valve) connector (C).

k. Disconnect the ignition coil connector (A).

l. Disconnect the ignition coil condenser connector (B).

m. Disconnect the CMP (Camshaft Position Sensor) connector (C).

n. Disconnect the ground cable (D).

o. Remove the wire harness bracket (E).

p. Remove the ground cable between engine mounting and vehicle body.

q. Remove the ground cable between transaxle housing and vehicle body.

13. Remove the transaxle wire harness connectors and control cable from transaxle (A/T).

a. Disconnect the transaxle range switch connector (A).

b. Disconnect the solenoid valve connector (B).

c. Disconnect the ATF oil temperature sensor connector (C).

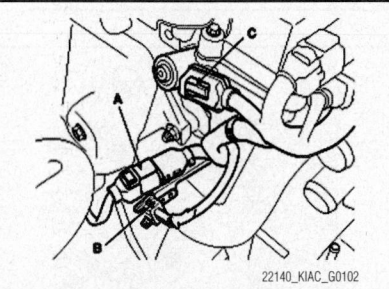

Fig. 130 Disconnect the front oxygen sensor connector (A), CKP (Crankshaft Position Sensor) connector (B), and OCV (Oil Control Valve) connector (C).

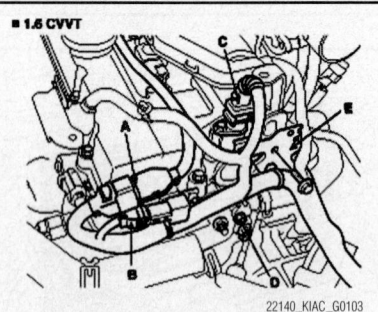

Fig. 131 Disconnect ignition coil connector (A), ignition coil condenser connector (B), CMP (Camshaft Position Sensor) connector (C), ground cable (D) and wire harness bracket (E).

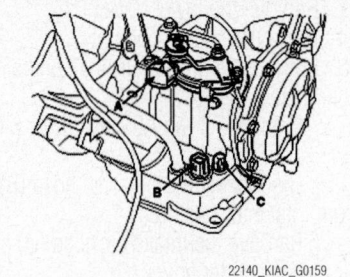

Fig. 132 Disconnect the transaxle range switch connector (A), solenoid valve connector (B), and ATF oil temperature sensor connector (C).

d. Disconnect the vehicle speed sensor connector.

e. Disconnect the band server switch connector.

f. Disconnect the pulse generator connector.

g. Remove the control cable nut (A) from transaxle range switch.

h. Remove the control cable (B).

14. Remove the transaxle wire harness connectors and control cable from transaxle (M/T).

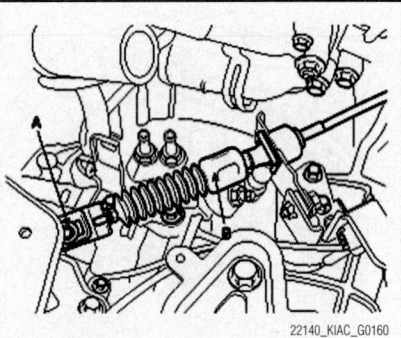

Fig. 133 Remove the control cable nut (A) and control cable (B).

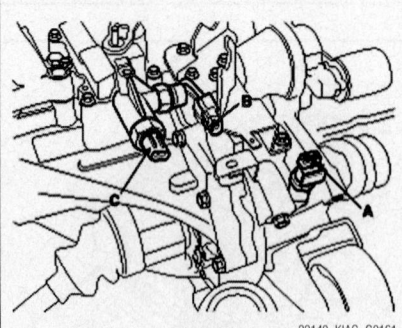

Fig. 134 Disconnect the vehicle speed sensor connector (A), neutral switch connector (B) and back-up lamp switch connector (C).

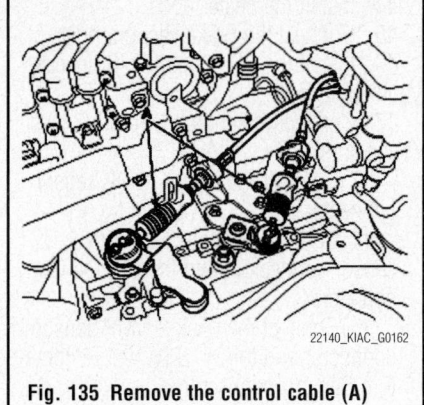

Fig. 135 Remove the control cable (A)

a. Disconnect the vehicle speed sensor connector (A).

b. Disconnect the neutral switch connector (B).

c. Disconnect the back-up lamp switch connector (C).

d. Remove the control cable (A).

15. Move the disconnected wire harnesses to the fuse box side so as to prevent interfering with other parts.

16. Disconnect the hose of the PCSV (Purge Control Solenoid Valve) side.

17. Remove the brake booster vacuum hose.

18. Remove the power steering oil hose and drain the power steering oil.

19. Remove the power steering return hose.

20. Recovering refrigerant and remove the high & low pressure pipe.

21. Remove the nuts (B, D), bolt (C) and engine mounting support bracket (A).

22. Remove the transaxle mounting bracket (A).

23. A/T : Remove the wire harness protector (B) on the transaxle mounting support bracket.

24. M/T : Remove the clutch release cylinder oil hose (B) on the transaxle mounting support bracket.

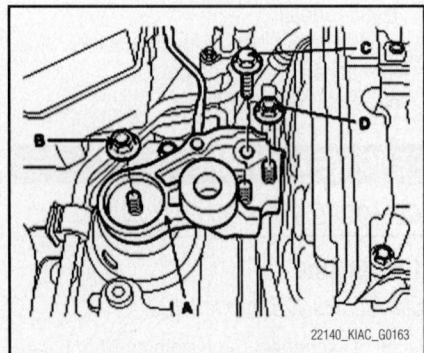

Fig. 136 Remove the nuts (B, D), bolt (C) and engine mounting support bracket (A)

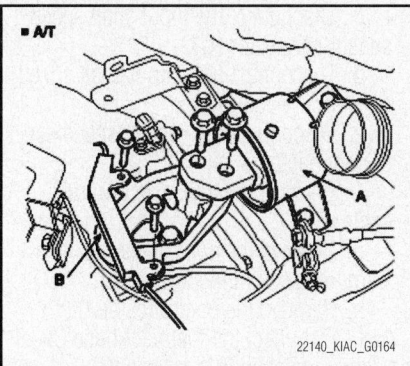

Fig. 137 Remove the A/T transaxle mounting bracket (A)

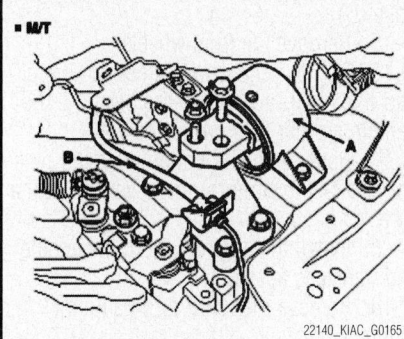

Fig. 138 Remove the M/T transaxle mounting bracket (A)

25. Disconnect the alternator connector and "B" terminal.

26. Remove the front wheels.

27. Remove the ABS wheel speed sensor.

28. Remove the caliper and hang assembly.

29. Remove the knuckle mounting bolts.

30. Remove the steering u-joint mounting bolt.

31. Remove the front muffler heat protector.

32. Remove the front muffler.

33. Using a floor jack, support the engine and transaxle assembly.

✳✳ CAUTION

After removing the sub frame mounting bolt , the engine and transaxle assembly may fall downward, and so support them securely with floor jack. Verify that the hoses and connectors are disconnected before removing the engine and transaxle assembly.

34. Remove the sub frame bolts and nuts.

35. Remove the engine and transaxle assembly by lifting vehicle.

To install:

36. Installation is in the reverse order of removal.

37. Perform the following :

a. Adjust the shift cable.

b. Adjust the throttle cable.

c. Refill the engine with engine oil.

d. Refill the transaxle with fluid.

e. Refill the radiator and reservoir tank with engine coolant.

f. Place the heater control knob on "HOT" position.

g. Bleed air from the cooling system

h. Start engine and let it run until it warms up. (until the radiator fan operates 3 or 4 times.)

i. Turn Off the engine. Check the level in the radiator, add coolant if needed. This will allow trapped air to be removed from the cooling system.

j. Put the radiator cap on tightly, then run the engine again and check for leaks.

k. Clean the battery posts and cable terminals with sandpaper assemble them, then apply grease to prevent corrosion.

l. Inspect for fuel leakage.

m. After assemble the fuel line, turn on the ignition switch (do not operate the starter) so that the fuel pump runs for approximately two seconds and fuel line pressurizes.

n. Repeat this operation two or three times, then check for fuel leakage at any point in the fuel line.

2.0L Engine

See Figures 139 through 148.

1. Disconnect the battery terminals.

2. Remove the heat shield and the battery

3. Remove the engine cover.

4. Loosen the radiator drain plug and drain engine coolant.

5. Remove the intake air hose and air cleaner assembly.

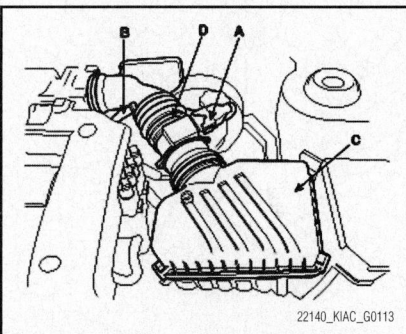

Fig. 139 Disconnect the MAF connector, (A) breather hose (B), remove air cleaner assembly (C) and air cleaner hose (D).

a. Disconnect the MAF connector (A).

b. Disconnect the breather hose (B) from air cleaner hose (D).

c. Remove the intake air hose and air cleaner assembly (C).

6. Remove the upper and lower radiator hoses.

7. Remove the Automatic Transaxle Fluid (ATF) oil cooler hoses, if equipped with A/T.

8. Remove the heater hoses.

9. Remove the accelerator cable.

10. Remove the engine wire harness connectors and wire harness clamps from the cylinder head and the intake manifold.

a. Disconnect OCV (Oil control Valve) connector (A).

b. Disconnect Oil Temperature Sensor (OTS) connector (B).

c. Disconnect Engine Coolant Temperature (ECT) sensor connector (C).

d. Disconnect ignition coil connector (D).

e. Disconnect Throttle Position Sensor (TPS) connector (A).

f. Disconnect Idle Speed Actuator (ISA) connector (B).

g. Disconnect Camshaft Position Sensor (CMP) connector (A).

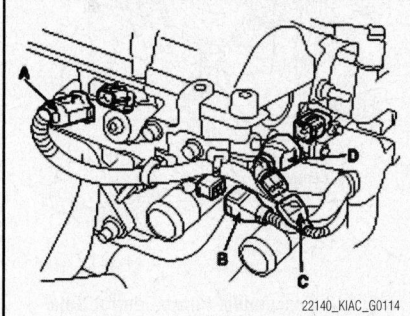

Fig. 140 Remove the engine wire harness connectors and wire harness clamps from the cylinder head and the intake manifold.

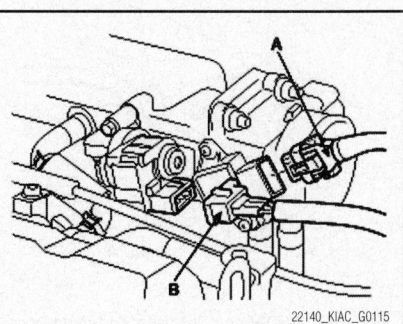

Fig. 141 Disconnect Throttle Position Sensor (TPS) connector (A) and ISA (Idle Speed Actuator) connector (B)

h. Disconnect four fuel injector connectors (B).

i. Disconnect Knock Sensor connector (C) and the ground cable (D).

j. Disconnect Purge Control Solenoid Valve (PCSV) connector (E).

k. Disconnect front heated oxygen sensor connector (A).

l. Disconnect the Crankshaft Angle Position (CKP) sensor connector (B).

m. Disconnect the oil pressure switch connector (C).

11. Remove the transaxle wire harness connectors and control cable from transaxle (A/T).

a. Disconnect the transaxle range switch connector (A).

b. Disconnect the solenoid valve connector (B).

c. Disconnect the input shaft speed sensor connector (C).

d. Disconnect the output shaft speed sensor connector (D).

e. Disconnect the vehicle speed sensor connector (A).

f. Remove the transaxle ground cable.

g. Remove the control cable nut (A) from transaxle range switch.

h. Remove the control cable.

12. Disconnect the fuel inlet hose (A) of the delivery pipe side.

13. Disconnect the hose (B) of the Purge Control Solenoid Valve (PCSV) side.

14. Remove the brake booster vacuum hose (C).

15. Remove the front wheels.

16. Remove the power steering pump and use a wire to secure the pump to the vehicle so that it is out of the way.

17. Remove the air conditioner compressor and fix the compressor to vehicle with a wire.

18. Install the engine jack to the engine and transaxle assembly.

19. Remove the lower arm ball joint mounting bolts.

20. Disconnect the tie-rod from the knuckle.

21. Disconnect the stabilizer bar link from the strut.

22. Remove the front muffler.

23. Remove the steering u-joint mounting bolt.

24. Remove the sub frame bolts.

To install:

25. Installation is in the reverse order of removal.

26. Perform the following:

a. Adjust shift cable.

b. Adjust throttle cable.

c. Refill engine with engine oil.

d. Refill transaxle with fluid.

e. Refill radiator with engine coolant.

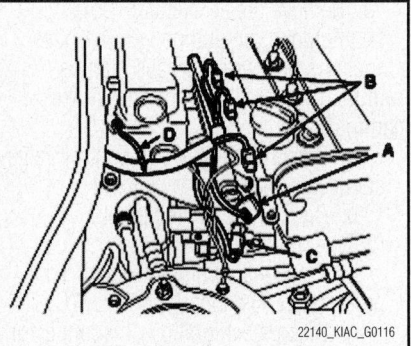

22140_KIAC_G0116

Fig. 142 Disconnect Camshaft Position Sensor (CMP) connector (A), four fuel injector connectors (B), Knock Sensor connector (C) and the ground cable (D).

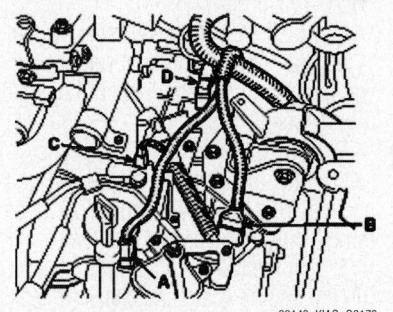

22140_KIAC_G0170

Fig. 145 Disconnect the transaxle range switch connector (A), solenoid valve connector (B), input shaft speed sensor connector (C), and output shaft speed sensor connector (D).

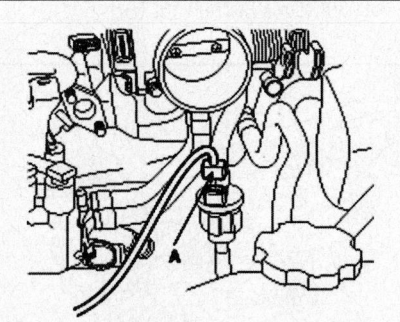

22140_KIAC_G0117

Fig. 143 Disconnect Purge Control Solenoid Valve (PCSV) connector (E)

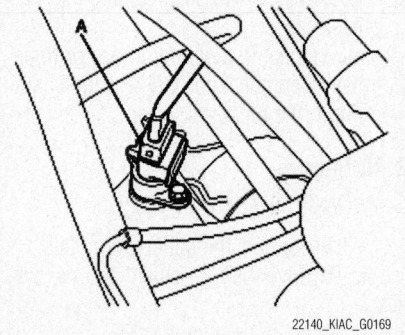

22140_KIAC_G0169

Fig. 146 Disconnect the vehicle speed sensor connector (A).

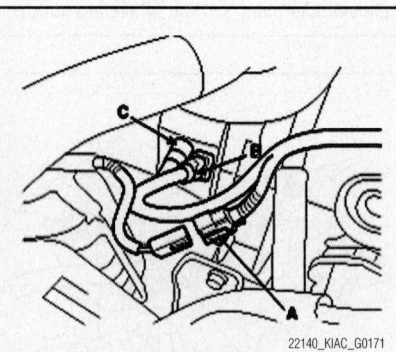

22140_KIAC_G0171

Fig. 144 Disconnect front heated oxygen sensor connector (A), Crankshaft Angle Position (CKP) sensor connector (B), and oil pressure switch connector (C).

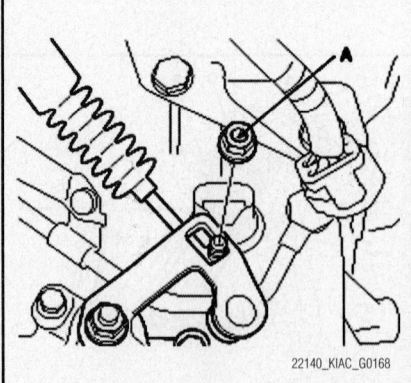

22140_KIAC_G0168

Fig. 147 Remove the control cable nut (A) from transaxle range switch

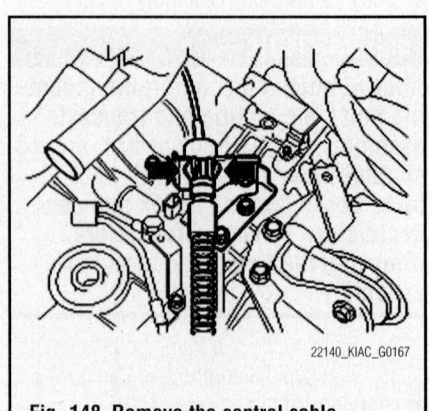

22140_KIAC_G0167

Fig. 148 Remove the control cable

f. Bleed air from the cooling system with the heater valve open.

g. Clean battery posts and cable terminals with sandpaper, assemble them, then apply grease to prevent corrosion.

h. Inspect for fuel leakage.

➡ **After assembling the fuel line, turn on the ignition switch (do not operate the starter) so that the fuel pump runs for approximately two seconds and fuel line pressurizes. Repeat this operation two or three times, then check for fuel leakage at any point in the fuel line.**

2.4L Engine

See Figures 149 through 156.

1. Disconnect the negative terminal from the battery.
2. Remove engine cover.
3. Remove air duct.
4. Remove the intake air hose and air cleaner assembly.

   a. Disconnect the AFS (B) connector.

   b. Disconnect the breather hose (C) from air cleaner hose.

   c. Disconnect the ECU connector.

   d. Remove the intake air hose and air cleaner (A).

5. Remove front wheels.
6. Remove lower cover.
7. Drain the engine coolant.
8. Remove the upper and lower radiator hoses.
9. Remove the heater hoses.
10. Disconnect A/C switch (A), alternator connector (B), and oil pressure switch (C).
11. Disconnect OCV connector (A) and OTS connector (B) and P/S switch connector (C).
12. Disconnect injector connectors.
13. Disconnect the engine wire harness connectors.

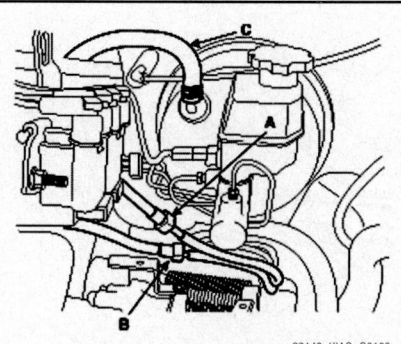

Fig. 149 Disconnect the fuel inlet hose (A), hose (B) of the Purge Control Solenoid Valve (PCSV) side, and brake booster vacuum hose (C).

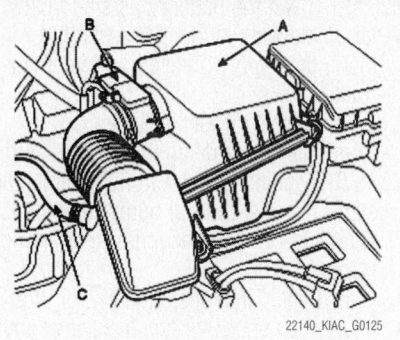

Fig. 150 Disconnect the AFS (B) connector, breather hose (C), and air cleaner (A).

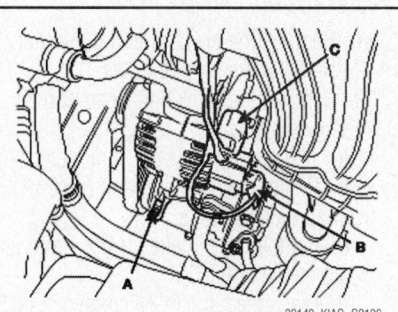

Fig. 151 Disconnect A/C switch (A), alternator connector (B), and oil pressure switch (C).

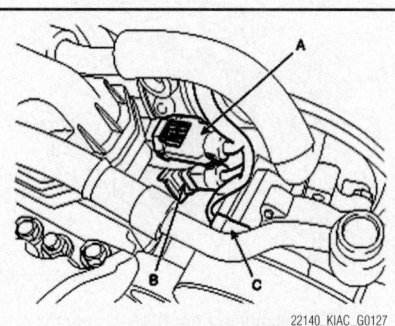

Fig. 152 Disconnect OCV connector (A) and OTS connector (B) and P/S switch connector (C).

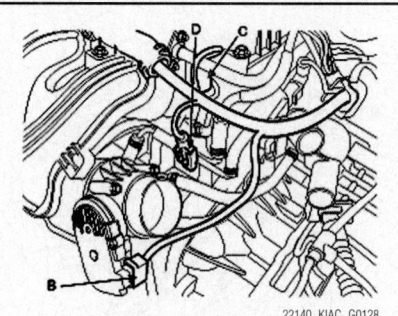

Fig. 153 Remove the ETC connector (B), CMP connector (C), knock sensor connector (D).

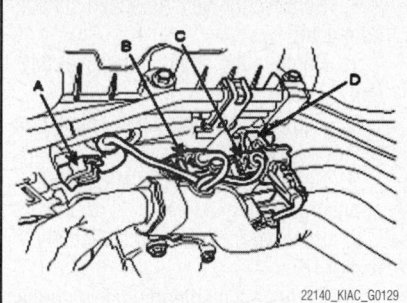

Fig. 154 Disconnect PCSV connector (A), WTS connector (B), condenser connector (C), and CKP sensor connector (D).

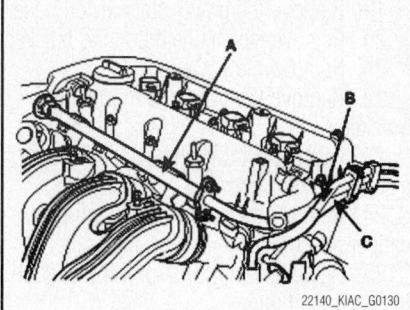

Fig. 155 Remove delivery pipe (A), brake vacuum hose (B), and PCSV hose (C).

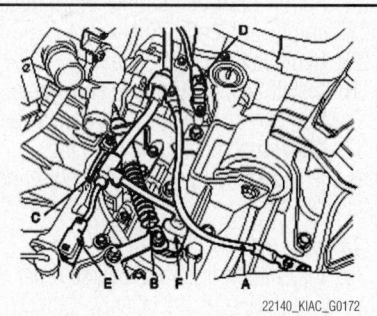

Fig. 156 Remove the ground cable (A), shaft cable assembly (B), input shaft speed connector (C), output shaft speed connector (D), inhibitor switch connector (E), and solenoid valve connector (F).

   a. Remove the ETC connector (B), CMP connector (C), knock sensor connector (D).

14. Disconnect ignition coil connectors.
15. Disconnect PCSV connector (A), WTS connector (B), condenser connector (C), and CKP sensor connector (D).
16. Remove delivery pipe (A), brake vacuum hose (B), and PCSV hose (C).
17. Disconnect the transaxle wire harness connector and control cable.

   a. Remove the ground cable (A) from the transaxle.

b. Remove the vehicle speed sensor connector.

c. Remove the shaft cable assembly (B).

d. Remove the input shaft speed connector (C).

e. Remove the output shaft speed connector (D).

f. Remove the inhibitor switch connector (E).

g. Remove the solenoid valve connector (F).

h. Remove the hose from the A/T oil cooler tube.

18. Remove the steering u-joint mounting bolt.

19. Remove the power steering oil hose.

20. Remove the engine mounting bracket.

21. Remove the A/C pipe.

22. Remove the transaxle mounting bracket.

23. Disconnect the wheel speed sensors from both front knuckles.

24. Remove the caliper and hang the caliper assembly.

25. Remove the front strut lower mounting bolts and nuts.

26. Remove the drive shaft cover, heat protector.

27. Remove the exhaust pipe, hanger, and oxygen sensor.

28. Remove power steering return hose and drain power steering oil.

29. Remove the EPS connector.

30. Install jack and remove sub-frame.

31. Remove drive shaft from transaxle.

32. Separate then engine from the transaxle assembly by loosening mounting bolts.

To install:

33. Installation is in the reverse order of removal.

34. Perform the following:

a. Adjust the shift cable.

b. Adjust the throttle cable.

c. Refill the engine with engine oil.

d. Refill the transaxle with fluid.

e. Refill the radiator with engine coolant.

f. Place the heater control knob on "HOT" position.

g. Bleed air from the cooling system.

h. Start engine and let it run until it warms up (until the radiator fan operates 3 or 4 times).

i. Turn off the engine. Check the level in the radiator, add coolant if needed. This will allow trapped air to be removed from the cooling system.

j. Put the radiator cap on tightly, then run the engine again and check for leaks.

k. Clean the battery posts and cable terminals with sandpaper assemble them, then apply grease to prevent corrosion.

l. Inspect for fuel leakage.

m. After assembling the fuel line, turn on the ignition switch (do not operate the starter) so that the fuel pump runs for approximately two seconds and fuel line pressurizes.

n. Repeat this operation two or three times, then check for fuel leakage at any point in the fuel line.

2.7L Engine

See Figures 157 through 166.

1. Remove the air duct and the battery.

2. Remove the engine cover.

3. Remove the intake air hose and air cleaner assembly.

a. Disconnect the MAF connector (A).

b. Disconnect the breather hose (B) from air cleaner hose.

c. Remove the intake air hose and air cleaner assembly (C).

d. Disconnect the PCM connectors (D).

4. Remove the battery tray while recovering refrigerant.

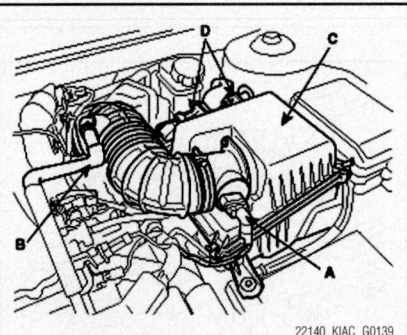

Fig. 157 Disconnect the MAF connector (A), breather hose (B), intake air hose and air cleaner assembly (C), and PCM connectors (D).

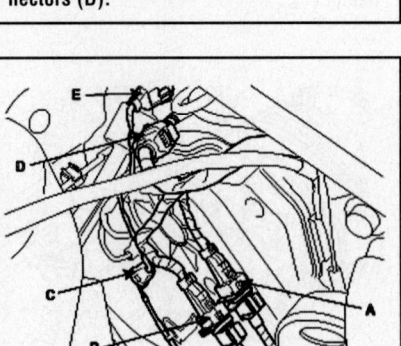

Fig. 158 Disconnect the engine wiring harness connectors

5. Remove the upper and lower radiator hoses.

6. Remove the transaxle oil cooler hoses (A/T vehicles only).

7. Remove the fuel inlet hose from the delivery pipe.

8. Disconnect the engine wiring harness connectors.

a. Disconnect the No.1/No.2 knock sensor connectors (A,B), the oil pressure switch connector (C), the ignition coil harness (D) and the No.1 VIS (Variable Induction System) connector (E).

b. Disconnect the bank 1 front/rear O2 sensor connectors (A).

c. Disconnect the injection connectors (A ,B, C), the ground lines (D), the condenser connector (E) and the Ignition coil connectors (F).

d. Disconnect the injection harness connector (A), the No.2 VIS (Variable Induction System) connector (B), the No.1/No.2 OCV (Oil Control Valve) connectors (C, D) and the OTS (Oil Temperature Sensor) connector (E).

e. Disconnect the MAPS (Manifold Absolute Pressure Sensor) connector (A), the ETC (Electronic Throttle Control)

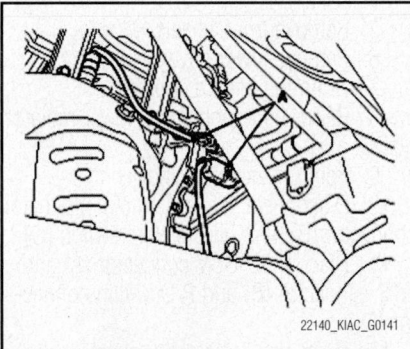

Fig. 159 Disconnect the bank 1 front/rear O2 sensor connectors (A)

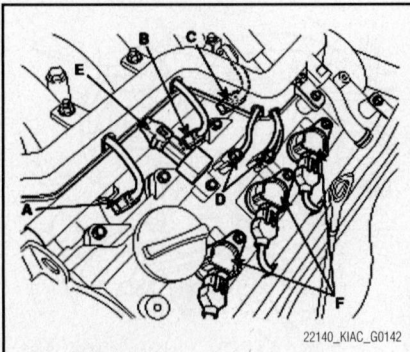

Fig. 160 Disconnect the injection connectors (A, B, C), the ground lines (D), the condenser connector (E) and the Ignition coil connectors (F).

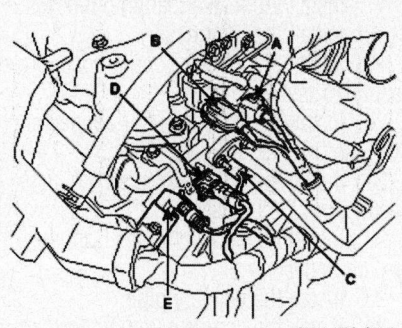

Fig. 161 Disconnect the injection harness connector (A), the No.2 VIS (Variable Induction System) connector (B), the No.1/No.2 OCV (Oil Control Valve) connectors (C, D) and the OTS (Oil Temperature Sensor) connector (E).

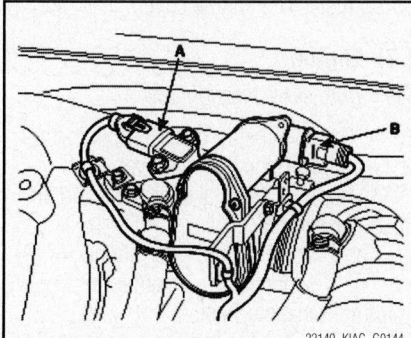

Fig. 162 Disconnect the MAPS (Manifold Absolute Pressure Sensor) connector (A), and the ETC (Electronic Throttle Control) connector (B).

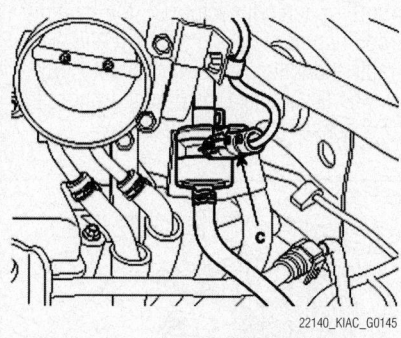

Fig. 163 Disconnect the PCSV (Purge Control Solenoid Valve) connector (C).

connector (B) and the PCSV (Purge Control Solenoid Valve) connector (C).

  f. Disconnect the alternator connector and the air conditioning compressor connector.

  g. Disconnect the bank 2 CMP sensor connector (A) and the ECT (Engine Coolant Temperature) sensor connector (B).

  h. Disconnect the bank 2 front/rear O2 sensor connectors (A, B) and the CKP sensor connector (C).

  i. Disconnect the bank 1 CMP sensor connector (A).

9. Disconnect ground lines from the engine and the transaxle assembly.

10. Disconnect the battery wirings from the engine room fuse & relay box.

11. Remove the heater hoses.

12. Disconnect the brake vacuum hose.

13. Disconnect the transaxle wiring harness connectors.

14. Disconnect the power steering hose.

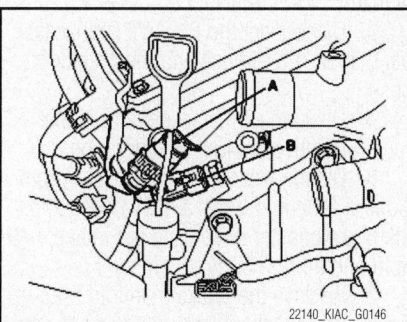

Fig. 164 Disconnect the bank 2 CMP sensor connector (A) and the ECT (Engine Coolant Temperature) sensor connector (B).

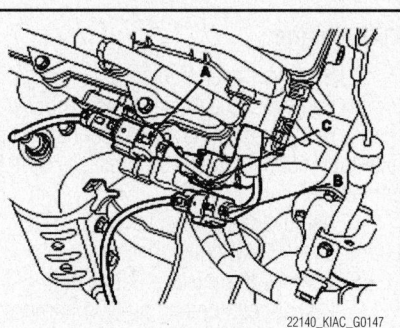

Fig. 165 Disconnect the bank 2 front/rear O2 sensor connectors (A, B) and the CKP sensor connector (C).

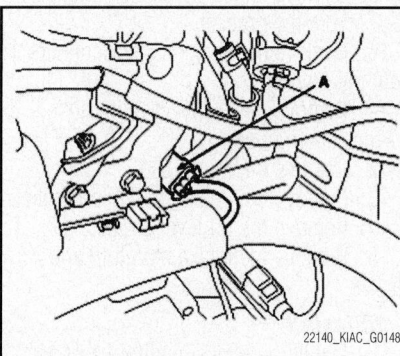

Fig. 166 Disconnect the bank 1 CMP sensor connector (A)

15. Remove the steering column shaft joint bolt.

16. Disconnect the air conditioning compressor hoses.

17. Remove the front wheels and tires.

18. Lifting the vehicle, remove the lower cover.

19. Drain the engine coolant, engine oil and transaxle fluid.

20. Remove the brake caliper.

21. Disconnect the ABS connectors.

22. Disconnect the stabilizer bar link from the struts.

23. Remove the knuckles from the struts.

24. Remove the front muffler.

25. Disconnect the power steering return hose.

26. Remove the engine mounting bracket.

27. Remove the transaxle mounting bracket.

28. Supporting the engine and transaxle assembly with a jack, remove the assembly from the vehicle by loosening the sub frame mounting bolts and lifting up the vehicle slowly.

To install:

29. Installation is in the reverse order of removal.

30. Perform the following :

  a. Adjust the shift cable.

  b. Refill the engine with engine oil.

  c. Refill the transaxle with fluid.

  d. Refill the radiator with engine coolant.

  e. Bleed air from the cooling system with the heater valve open.

  f. Clean the battery posts and cable terminals with "sandpaper" "assemble" them, then apply grease to prevent corrosion.

  g. Inspect for fuel leakage. After assembling the fuel line, turn on the ignition switch (do not operate the starter) so that the fuel pump runs for approximately two seconds and fuel line pressurizes. Repeat this operation two or three times, then check for fuel leakage at any point in the fuel lines.

EXHAUST MANIFOLD

REMOVAL & INSTALLATION

1.6L & 2.0L Engines

1. Remove the engine cover.

2. Disconnect the front oxygen sensor connector.

3. Remove the front muffler.

4. Remove the heat protector.

5. Remove the exhaust manifold and catalytic converter assembly.

6. To install, reverse the removal procedure with new gaskets.

2.4L Engine

1. Remove the oxygen sensor connector.

2. Remove the front muffler.

3. Remove the heat protector.

4. Remove the exhaust manifold bracket.

5. Remove the exhaust manifold and gasket.

To install:

6. Install the exhaust manifold. Torque: 28.92–32.53 ft. lbs. (39.2–44.1 Nm).

7. Install the exhaust manifold bracket

8. Install the heat protector.

9. Install the front muffler.

10. Install the oxygen sensor connector.

2.7L Engine

1. Remove the lower cover.

2. Remove the front muffler.

3. Disconnect the oxygen sensor connectors.

4. Remove the oil level gauge.

5. Remove the heat protector.

6. Remove the exhaust manifold assembly.

To install:

7. Install the exhaust manifold assembly with a new gasket. Torque: 22–25 ft. lbs. (29.4–34.3 Nm).

8. Install the heat protector.

9. Install the front muffler assembly.

10. Connect the oxygen sensor connector.

11. Install the lower cover.

INTAKE MANIFOLD

REMOVAL & INSTALLATION

1.6L Engine

1. Remove the engine cover.

2. Remove the accelerator cable.

3. Disconnect the Throttle Position Sensor (TPS) connector.

4. Disconnect the Idle Speed Actuator (ISA) connector.

5. Disconnect the Positive Crankcase Ventilation (PCV) hose and breather hose.

6. Disconnect the injector connector (No. 3, 4).

7. Disconnect the injector connector (No. 1, 2).

8. Remove the heater hose, Purge Control Solenoid Valve (PCSV) hose, and the brake vacuum hose from throttle body and intake manifold.

9. Disconnect the Purge Control Solenoid Valve (PCSV) and water temperature sensor connector.

10. Remove the fuel delivery pipe.

11. Remove the intake manifold bracket.

12. Remove the intake manifold.

13. Installation is in the reverse order of removal with new gasket.

2.0L Engine

1. Remove the engine cover.

2. Disconnect the Throttle Position Sensor (TPS) connector and the Idle Speed Actuator (ISA) connector.

3. Disconnect the Positive Crankcase Ventilation (PCV) hose and the breather hose.

4. Disconnect the accelerator cable.

5. Remove the fuel delivery pipe.

6. Disconnect the heater hose, Purge Control Solenoid Valve (PCSV) hose and the brake booster hose from the intake manifold and throttle body assembly.

7. Remove the intake manifold bracket.

8. Remove the intake manifold assembly.

9. To install, reverse the removal procedure with new gaskets.

10. Intake manifold assembly: Torque: 11.6–16.6 ft. lbs. (15.7–22.6 Nm).

2.4L Engine

1. Remove the engine cover.

2. Remove the air cleaner assembly.

3. Remove the A/C switch connector, alternator connector and oil pressure switch connector.

4. Remove the OCV connector, OTS connector, P/S switch connector.

5. Remove the injector connector.

6. Disconnect the engine wire harness connectors.

7. Remove the fuel delivery pipe.

8. Remove the coolant hoses from the throttle body.

9. Remove the oil pressure switch connector from the bracket.

10. Remove the knock sensor connector from the bracket.

11. Remove the PCSV vacuum hose, brake vacuum hose.

12. Remove the PCV hose.

13. Remove the intake manifold bracket.

14. Remove the oil level gauge.

15. Remove the intake manifold and gasket.

To install:

16. Install the intake manifold and gasket. Torque: 13.7–20.2 ft. lbs. (18.6–27.44 Nm).

17. Install the intake manifold bracket.

18. Install the PCV hose.

19. Install the PCSV vacuum hose, brake vacuum hose.

20. Install the knock sensor connector from the bracket.

21. Install the oil pressure switch connector from the bracket.

22. Install the coolant hoses from the throttle body.

23. Install the delivery pipe.

24. Install the engine wire harness connectors.

25. Install the injector connector.

26. Install Remove the OCV connector, OTS connector, P/S switch connector.

27. Install the A/C switch connector, alternator connector and oil pressure switch connector.

28. Install the air cleaner assembly.

29. Install the engine cover.

2.7L Engine

See Figures 167 through 180.

1. Remove the air duct and the battery.

2. Remove the engine cover.

3. Remove the intake air hose and air cleaner assembly.

  a. Disconnect the MAF connector (A).

  b. Disconnect the breather hose (B) from air cleaner hose.

  c. Remove the intake air hose and air cleaner assembly (C).

  d. Disconnect the PCM connectors (D).

4. Disconnect the engine wiring harness connectors.

  a. Disconnect the No.1/No.2 knock sensor connectors (A, B), the oil pressure switch connector (C), the ignition coil harness (D) and the No.1 VIS (Variable Induction System) connector (E).

  b. Disconnect the bank 1 front/rear O2 sensor connectors (A).

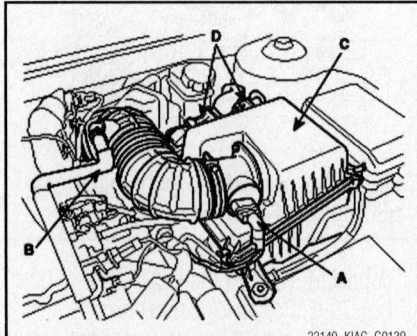

Fig. 167 Disconnect the MAF connector (A), breather hose (B), intake air hose and air cleaner assembly (C), and PCM connectors (D).

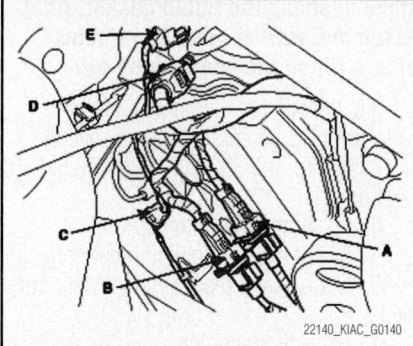

Fig. 168 Disconnect the engine wiring harness connectors.

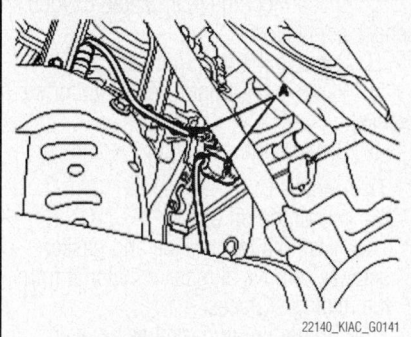

Fig. 169 Disconnect the bank 1 front/rear O2 sensor connectors (A).

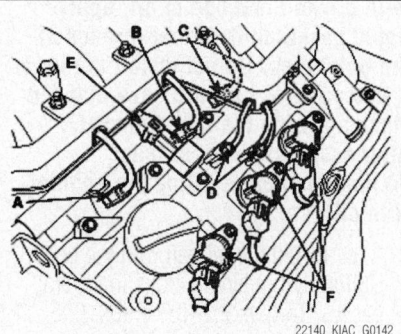

Fig. 170 Disconnect the injection connectors (A ,B, C), the ground lines (D), the condenser connector (E) and the Ignition coil connectors (F).

c. Disconnect the injection connectors (A, B, C), the ground lines (D), the condenser connector (E) and the Ignition coil connectors (F).

d. Disconnect the injection harness connector (A), the No.2 VIS (Variable Induction System) connector (B), the No.1/No.2 OCV (Oil Control Valve) connectors (C, D) and the OTS (Oil Temperature Sensor) connector (E).

e. Disconnect the MAPS (Manifold Absolute Pressure Sensor) connector (A),

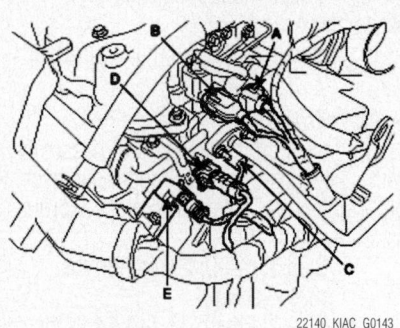

Fig. 171 Disconnect the injection harness connector (A), the No.2 VIS (Variable Induction System) connector (B), the No.1/No.2 OCV (Oil Control Valve) connectors (C, D) and the OTS (Oil Temperature Sensor) connector (E).

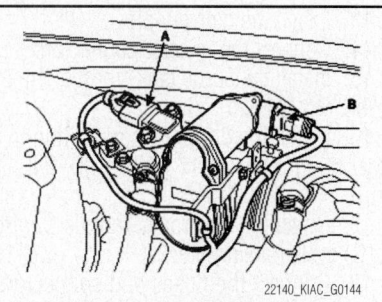

Fig. 172 Disconnect the MAPS (Manifold Absolute Pressure Sensor) connector (A), and the ETC (Electronic Throttle Control) connector (B).

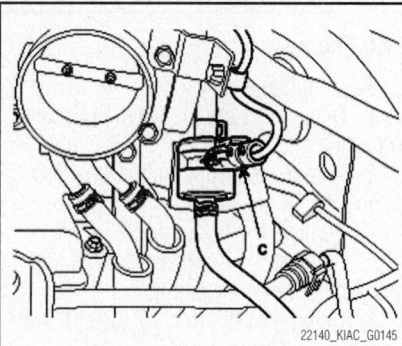

Fig. 173 Disconnect the PCSV (Purge Control Solenoid Valve) connector (C).

the ETC (Electronic Throttle Control) connector (B) and the PCSV (Purge Control Solenoid Valve) connector (C).

f. Disconnect the alternator connector and the air conditioning compressor connector.

g. Disconnect the bank 2 CMP sensor connector (A) and the ECT (Engine Coolant Temperature) sensor connector (B).

h. Disconnect the bank 2 front/rear O2 sensor connectors (A,B) and the CKP sensor connector (C).

i. Disconnect the bank 1 CMP sensor connector (A).

5. Remove the Purge Control Valve (PCV) hose.

6. Remove the Electric Throttle Control (ETC) bracket (A) and the cooling hoses (B).

7. Disconnect the brake vacuum hose.

8. Remove the surge tank mounting bracket.

9. Remove the surge tank (A).

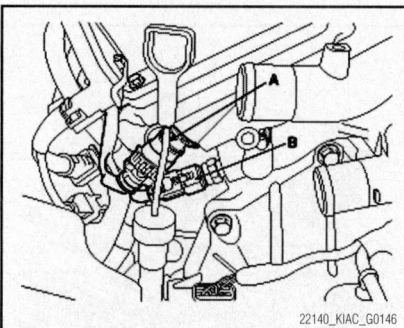

Fig. 174 Disconnect the bank 2 CMP sensor connector (A) and the ECT (Engine Coolant Temperature) sensor connector (B).

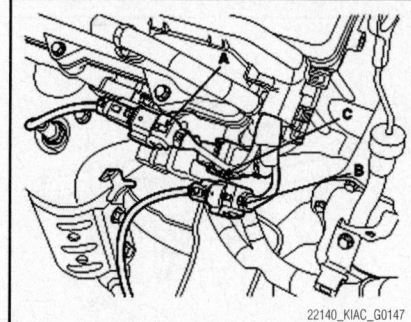

Fig. 175 Disconnect the bank 2 front/rear O2 sensor connectors (A,B) and the CKP sensor connector (C).

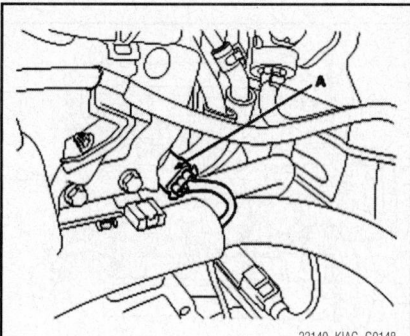

Fig. 176 Disconnect the bank 1 CMP sensor connector (A)

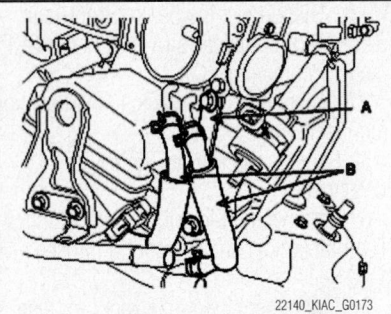

Fig. 177 Remove the Electric Throttle Control (ETC) bracket (A) and the cooling hoses (B)

Fig. 178 Remove the surge tank (A)

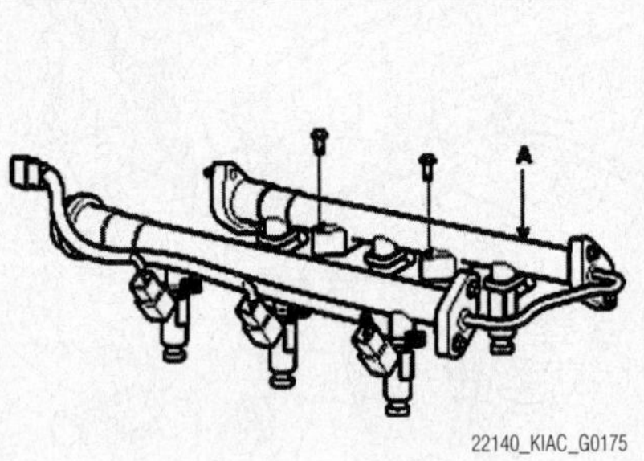

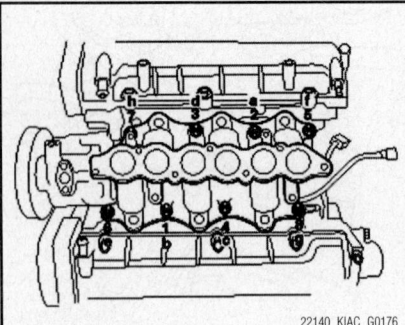

Fig. 179 Remove the fuel delivery pipe assembly (A)

Fig. 180 Install the intake manifold assembly

10. Remove the fuel delivery pipe assembly (A).

11. Remove the intake manifold assembly.

To install:

12. Install the intake manifold assembly with a new gasket to a cylinder head assembly. Tighten the bolts in two steps. Torque as follows:

- Step 1: (a–h): 2.9–4.3 ft. lbs. (3.9–5.9 Nm)
- Step 2: (1–8): 13.7–17.4 ft. lbs. (18.6–23.5 Nm)

✳✳ CAUTION

When installing the gasket on the cylinder head, check the identification marks (LH/RH) to ensure correct installation.

13. Install the delivery pipe.

14. Connect the injector connectors.

15. Install the surge tank. Torque: 14–17 ft. lbs. (19–24 Nm).

16. Install the surge tank mounting bracket. Torque: 14–17 ft. lbs. (19–24 Nm).

17. Install the Electronic Throttle Control (ETC) system fixing bracket.

18. Connect the hoses and connectors.

19. Install the air cleaner assembly.

20. Install the engine cover.

OIL PAN

REMOVAL & INSTALLATION

1.6L Engine

1. Drain the engine oil.

2. Disconnect the rear oxygen sensor connector.

3. Remove the front muffler heat protector

4. Remove the front muffler.

5. Remove the exhaust manifold and catalytic converter assembly.

6. Using the SST (09215-3C000) and remove the oil pan.

To install:

7. Install the oil pan.

a. Using a razor blade and gasket scraper, remove all gasket material from the mating surfaces.

b. Apply liquid gasket as an even bead, centered between the edges of the mating surface. Liquid gasket: MS 721-40A or equivalent

➡To prevent leakage of oil, apply liquid gasket to the inner threads of the bolt holes. Do not install the parts if five minutes or more have elapsed

since applying the liquid gasket. After assembly, wait at least 30 minutes before filling the engine with oil.

c. Install the oil pan with the bolts. Uniformly tighten the bolts in several passes. Torque: 7.2–8.7 ft. lbs. (9.8–11.8 Nm).

8. Install the front muffler.

9. Install the front muffler heat protector

10. Connect the rear oxygen sensor connector.

11. Fill with engine oil.

2.0L Engine

1. Drain engine oil.

2. Disconnect the rear heated oxygen sensor connector.

3. Remove the front muffler.

4. Remove the front muffler mounting bracket.

5. Remove the oil pan.

To install:

6. Install the oil pan.

a. Using a razor blade and gasket scraper, remove all gasket material from the mating surfaces.

b. Apply liquid gasket as an even bead, centered between the edges of the mating surface. Liquid gasket: Loctite NO.5900 or equivalent

➡To prevent leakage of oil, apply liquid gasket to the inner threads of the bolt holes. Do not install the parts if five minutes or more have elapsed since applying the liquid gasket. After assembly, wait at least 30 minutes before filling the engine with oil.

c. Install the oil pan with the bolts. Uniformly tighten the bolts in several passes. Torque: 7.2–8.7 ft. lbs. (9.8–11.8 Nm).

7. Install the front muffler bracket.

8. Install the front muffler.

9. Connect the rear oxygen sensor connector.

10. Fill with engine oil.

OIL PUMP

REMOVAL & INSTALLATION

1.6L Engine

See Figures 181 through 183.

1. Drain the engine oil.

2. Remove the drive belts.

3. Turn the crankshaft pulley, and align its groove with timing mark "T" of the timing belt cover.

4. Remove the timing belt.

5. Remove the timing belt tensioner (A).

6. Remove the oil pan and oil screen.

7. Remove the alternator.

8. Remove the air conditioner compressor tensioner bracket (A).

9. Remove the front case (E).

    a. Remove the screws (B) from the pump housing, then separate the housing and cover (A).

    b. Remove the inner (A) and outer (B) rotors.

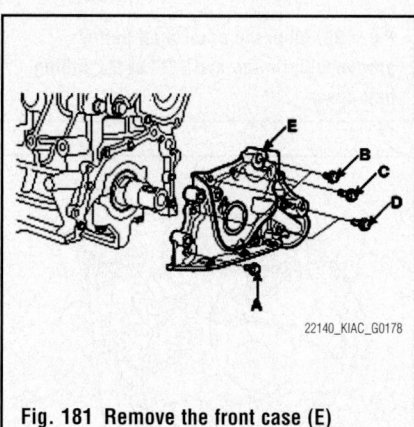

Fig. 181 Remove the front case (E)

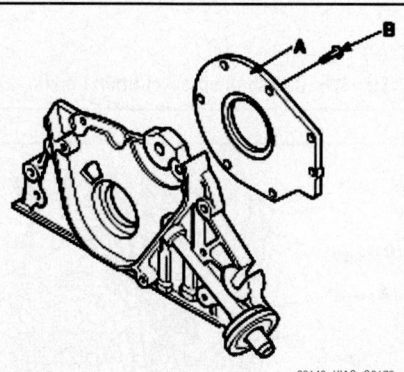

Fig. 182 Remove the screws (B) from the pump housing, then separate the housing and cover (A).

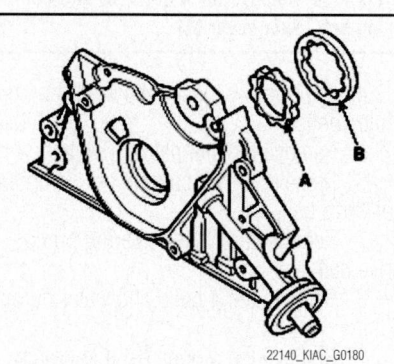

Fig. 183 Remove the inner (A) and outer (B) rotors

To install:

10. Install oil pump.

    a. Place the inner and outer rotors into front case with the marks facing the oil pump cover side.

    b. Install the oil pump cover to front case with the 7 screws. Torque: 4.3–5.1 ft. lbs. (5.9–6.9 Nm).

11. Check that the oil pump turns freely.

12. Install the oil pump on the cylinder block.

    a. Place a new front case gasket on the cylinder block.

    b. Apply engine oil to the lip of the oil pump seal.

    c. Install the oil pump onto the crankshaft. Torque: 13.7–17.4 ft. lbs. (18.6–23.5 Nm).

➡**Bolt lengths:**

    d. A: 1.181 inches (30 mm)

    e. B: 0.866 inches (22 mm)

    f. C: 1.771 inches (45 mm)

    g. D: 2.362 inches (60 mm)

13. Apply a light coat of oil to the seal lip.

14. Using the SST (09214-32000), install the oil seal.

15. Install the air compressor tensioner bracket.

16. Install the alternator.

17. Install the oil screen.

18. Install the oil pan. Torque: 7.2–8.7 ft. lbs. (9.8–11.8 Nm).

➡**Clean the oil pan gasket mating surfaces.**

19. Install the timing tensioner.

20. Install the timing belt.

21. Install the accessory drive belts.

22. Fill the engine oil.

2.0L Engine

See Figures 184 through 187.

1. Drain engine oil.

2. Remove the accessory drive belts.

3. Turn the crankshaft and align the white groove on the crankshaft pulley with the pointer on the lower cover.

4. Remove the timing belt.

5. Remove the timing belt idler.

6. Remove the oil pan and oil screen.

7. Remove the alternator.

8. Remove the air compressor tension bracket (A).

9. Remove the front case (E).

    a. Remove the screws (B) from the pump housing, then separate the housing and cover (A).

    b. Remove the inner (A) and outer (B) rotors.

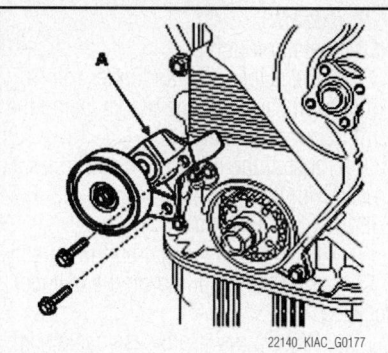

Fig. 184 Remove the air compressor tension bracket (A)

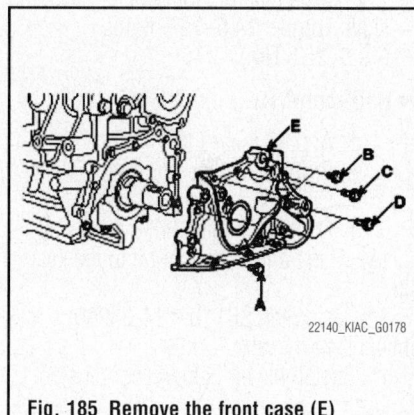

Fig. 185 Remove the front case (E)

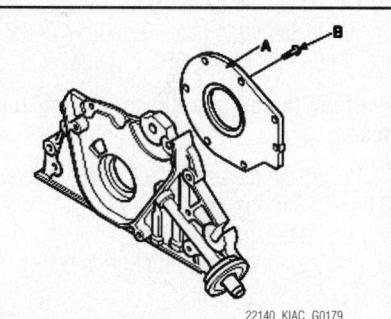

Fig. 186 Remove the screws (B) from the pump housing, then separate the housing and cover (A)

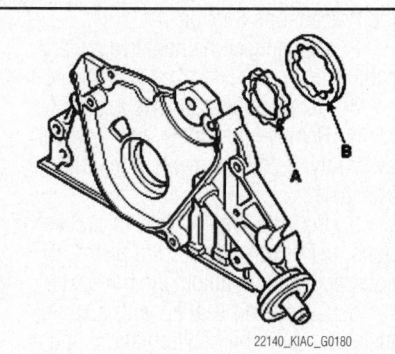

Fig. 187 Remove the inner (A) and outer (B) rotors.

To install:

10. Install oil pump.

    a. Place the inner and outer rotors into front case with the marks facing the oil pump cover side.

    b. Install the oil pump cover to front case with the 7 screws. Torque: 4.3–6.5 ft. lbs. (5.9–8.8 Nm).

11. Check that the oil pump turns freely.

12. Install the oil pump on the cylinder block.

    a. Place a new front case gasket on the cylinder block.

    b. Apply engine oil to the lip of the oil pump seal.

    c. Install the oil pump onto the crankshaft. Torque: 14.5–19.5 ft. lbs. (19.6–26.5 Nm).

➡**Bolt lengths:**

    d. A: 0.98 inches (25 mm)

    e. B: 0.787 inches (20 mm)

    f. C: 1.496 inches (38 mm)

    g. D: 1.771 inches (45 mm)

13. Apply a light coat of oil to the seal lip.

14. Using the SST (09214-33000), install the oil seal.

15. Install the air compressor tension bracket.

16. Install the alternator.

17. Install the oil screen.

18. Install the oil pan. Torque: 7.2–8.7 ft. lbs. (9.8–11.8 Nm).

➡**Clean the oil pan gasket mating surfaces.**

19. Install the timing belt idler. Torque: 31.1–39.8 ft. lbs. (42.2–53.9 Nm).

20. Install the timing belt.

21. Install the accessory drive belt.

22. Fill the engine oil.

REAR MAIN SEAL

REMOVAL & INSTALLATION

1. Before servicing the vehicle, refer to the Precautions Section.

2. Disconnect the negative battery cable.

3. Remove the transaxle assembly.

4. Remove the clutch and flywheel assembly, if equipped with a manual transaxle.

5. Remove the flexplate-to-crankshaft bolts, the flexplate and shim plates, if equipped with an automatic transaxle.

6. Cut the oil seal lip with a knife. Install a rag to the housing and using a screwdriver, carefully pry the oil seal from the oil seal housing. Clean the gasket mounting surfaces.

To install:

7. Clean the oil seal housing. Coat the oil seal and the housing with clean engine oil.

8. Install the oil seal into the housing and tap it evenly into place with a hammer and a large diameter piece of pipe. The seal must be flush with the edge of the rear cover.

9. Install the flywheel assembly or the flexplate, as applicable, and tighten the mounting bolts to 71–76 ft. lbs. (97–102 Nm).

10. Install the clutch assembly, if applicable.

11. Install the transaxle.

12. Connect the negative battery cable.

TIMING BELT, FRONT COVER & SPROCKET

REMOVAL & INSTALLATION

1.6L Engine

See Figures 188 through 203.

Engine removal is not required for this procedure.

1. Remove the engine cover.

2. Remove RH front wheel.

3. Remove 2 bolts (B) and RH side cover (A).

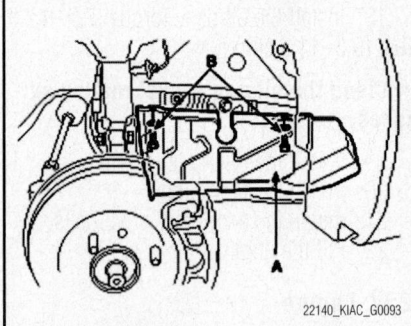

Fig. 188 Remove 2 bolts (B) and RH side cover (A)

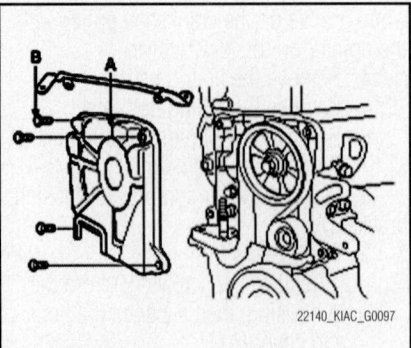

Fig. 189 Remove the 4 bolts (B) and timing belt upper cover (A)

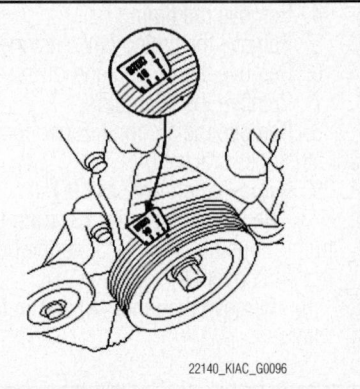

Fig. 190 Align the crankshaft pulley groove with timing mark "T" of the timing belt cover

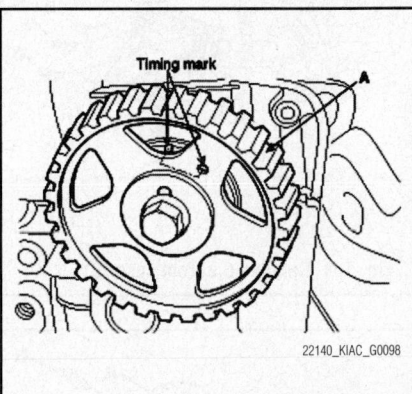

Fig. 191 Camshaft sprocket timing mark

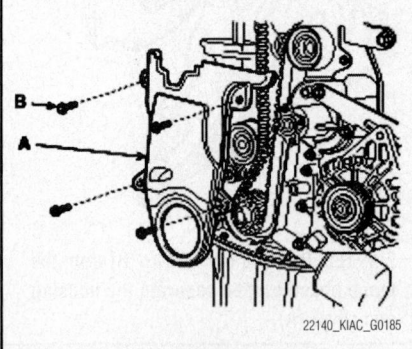

Fig. 192 Remove the 4 bolts (B) and timing belt lower cover (A)

4. Temporarily loosen the water pump pulley bolts.

5. Remove the alternator drive belt.

6. Remove the air conditioner compressor drive belt.

7. Remove the power steering pump drive belt.

8. Remove the 4 bolts and water pump pulley.

9. Remove the 4 bolts (B) and timing belt upper cover (A).

10. Turn the crankshaft pulley, and align its groove with timing mark "T" of the timing

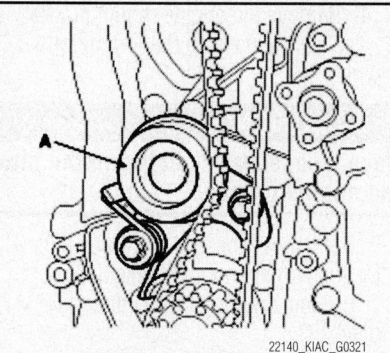

Fig. 193 Remove the timing belt tensioner (A) and timing belt

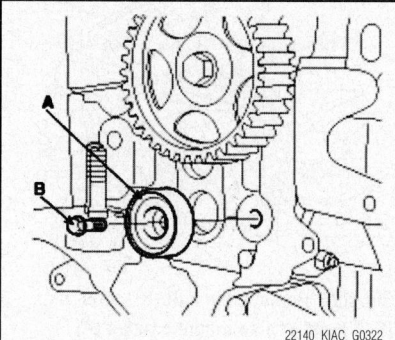

Fig. 194 Remove the bolt (B) and timing belt idler (A)

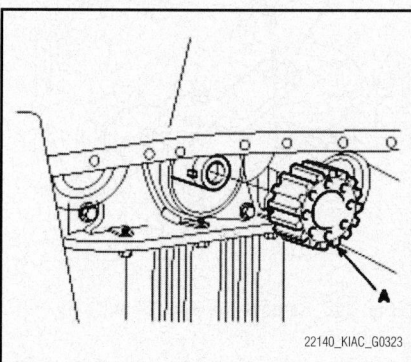

Fig. 195 Remove the crankshaft sprocket (A)

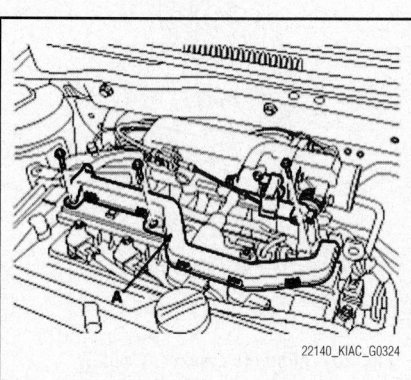

Fig. 196 Remove the wire harness bracket (A)

belt cover. Check that the timing mark of camshaft sprocket is aligned with the timing mark of cylinder head cover. (No. 1 cylinder compression TDC position)

11. Remove the crankshaft pulley bolt and crankshaft pulley.

12. Remove the crankshaft flange.

13. Remove the 4 bolts (B) and timing belt lower cover (A).

14. Remove the timing belt tensioner (A) and timing belt.

15. Remove the bolt (B) and timing belt idler (A).

16. Remove the crankshaft sprocket (A).

17. Remove the cylinder head cover.

   a. Remove the wire harness bracket (A).

   b. Remove the ignition coil.

   c. Remove the PCV (Positive Crankcase Ventilation) hose (A) and the breather hose (B) from the cylinder head cover.

   d. Remove the engine cover bracket (A).

   e. Remove the cylinder head cover bolts (B), the cover (A) and gasket.

18. Remove the camshaft sprocket.

To install:

19. Install the camshaft sprocket and tighten the bolt to the specified torque: 58–72 ft. lbs. (78.5–98.1 Nm).

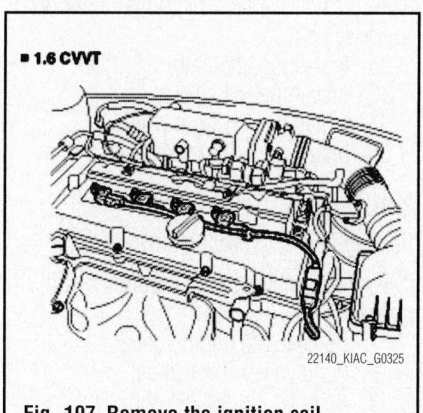

Fig. 197 Remove the ignition coil

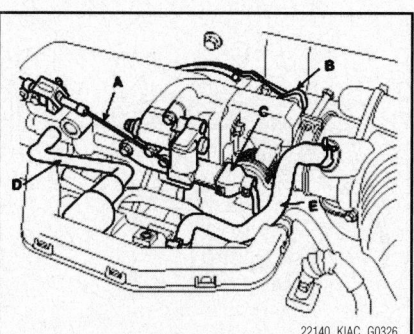

Fig. 198 Remove the PCV (Positive Crankcase Ventilation) hose (A) and the breather hose (B)

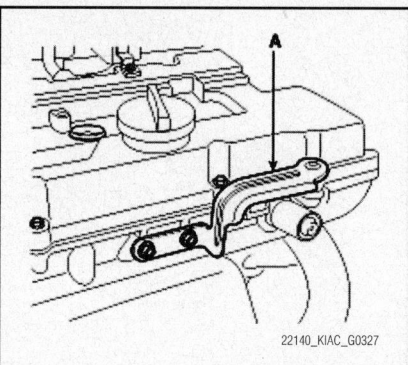

Fig. 199 Remove the engine cover bracket (A)

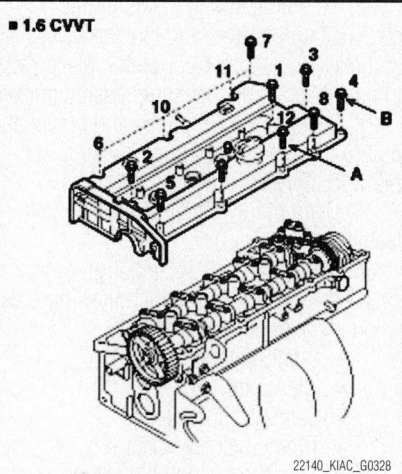

Fig. 200 Remove the cylinder head cover bolts (B), the cover (A) and gasket

20. Install the cylinder head cover.

   a. Install the cylinder head cover and bolts. Torque: 6–7 ft. lbs. (7.8–9.8 Nm).

   b. Install the engine cover bracket.

   c. Install the Positive Crankcase Ventilation (PCV) hose and breather hose to the cylinder head cover.

   d. Install the ignition coil.

21. Install the crankshaft sprocket.

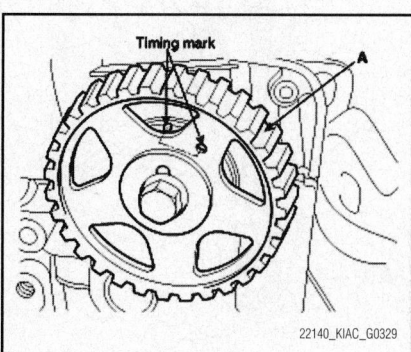

Fig. 201 Align the timing marks of the camshaft sprocket (A)

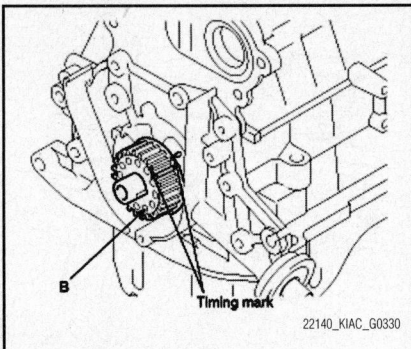

Fig. 202 Align the timing marks of the crankshaft sprocket (B)

22. Align the timing marks of the camshaft sprocket(A) and crankshaft sprocket(B) with the No.1 piston placed at top dead center and its compression stroke.

23. Install the idler pulley and tighten the bolt to the specified torque: 31–40 ft. lbs. (42.2–53.9 Nm).

24. Temporarily install the timing belt tensioner.

25. Install the belt so as not give slack at each center of shaft. Use the following order when installing timing belt:

- Crankshaft sprocket (A)
- Idler pulley (B)
- Camshaft sprocket (C)
- Timing belt tensioner (D)

26. Adjust the timing belt tension.

a. Loosen the tensioner pulley mounting bolt and apply tension to the timing belt.

b. After checking the alignment between each sprocket and each timing belt tooth, tighten the mounting bolts one by one. Torque: 15–20 ft. lbs. (19.6–26.5 Nm).

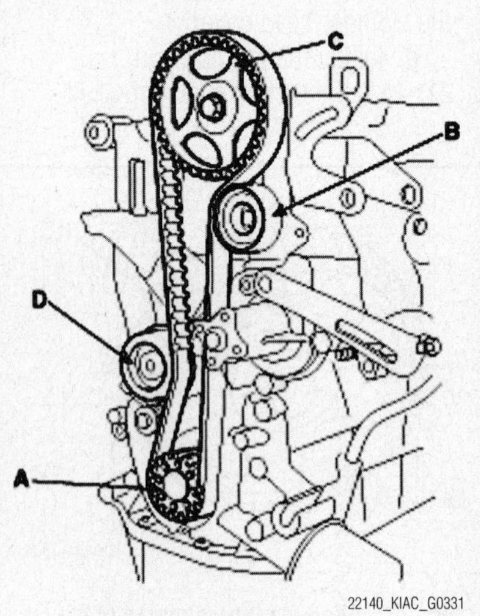

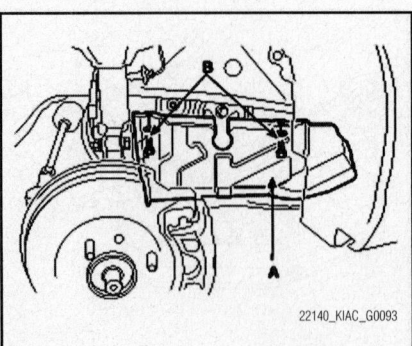

Fig. 203 Install the timing belt

c. Recheck the belt tension.

d. Verify that when the tensioner and the tension side of the timing belt are pushed in horizontally with a moderate force, the timing belt cog end is approximately 1/2 of the tensioner mounting bolt head radius away from the bolt head center.

27. Turn the crankshaft two turns in the operating direction (clockwise) and realign crankshaft sprocket and camshaft sprocket timing mark.

✳✳ CAUTION

Avoid rotating the crankshaft in a counter clockwise direction. Engine damage could occur.

28. Install the timing belt lower cover with 4 bolts. Torque: 5.8–7.2 ft. lbs. (7.8–9.8 Nm).

29. Install the crankshaft flange.

30. Install the crankshaft pulley. Make sure that crankshaft sprocket pin fits the small hole in the pulley. Torque: 101–109 ft. lbs. (137–147 Nm).

31. Install the timing belt upper cover. Torque: 5.8–7.2 ft. lbs. (7.8–9.8 Nm).

32. Install the water pump pulley.

33. Install the power steering pump drive belt.

34. Install the air conditioner compressor drive belt.

35. Install the alternator drive belt.

36. Install the RH side cover.

37. Install the RH front wheel.

38. Install the engine cover with bolts.

2.0L Engine

See Figures 204 through 210.

Engine removal is not required for this procedure.

1. Remove the engine cover.

2. Remove RH front wheel.

3. Remove 2 bolts (B) and RH side cover (A).

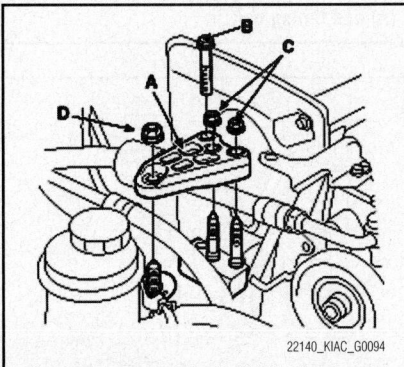

Fig. 204 Remove 2 bolts (B) and RH side cover (A)

4. Remove the engine mount bracket.

a. Set the jack to the engine oil pan.

✳✳ CAUTION

Place wooden block between the jack and engine oil pan.

b. Remove the bolt (B), three nuts (C, D) and engine mount bracket (A).

c. Remove the bolt (B) and stay plate (A).

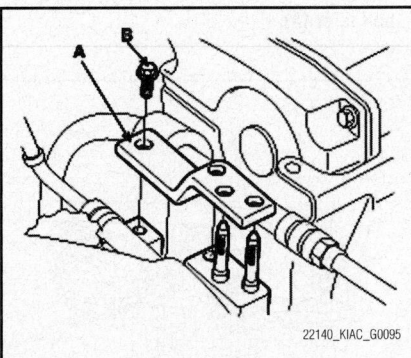

Fig. 205 Remove the bolt (B), three nuts (C, D) and engine mount bracket (A)

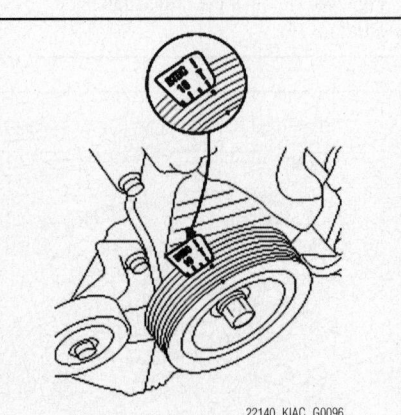

Fig. 206 Remove the bolt (B) and stay plate (A)

Fig. 207 Align the crankshaft pulley groove with timing mark "T" of the timing belt cover.

5. Temporarily loosen the water pump pulley bolts.

6. Remove alternator belt.

7. Remove air compressor belt.

8. Remove power steering belt.

9. Remove four bolts and water pump pulley.

10. Remove the four bolts and timing belt upper cover.

11. Turn the crankshaft pulley, and align its groove with timing mark "T" of the timing belt cover.

12. Remove the crankshaft pulley bolt and crankshaft pulley.

13. Remove the crankshaft flange.

14. Remove the 5 bolts (B) and timing belt lower cover (A).

15. Remove the timing belt tensioner and timing belt.

➡ **If the timing belt is reused, make an arrow indicating the turning direction to make sure that the belt is reinstalled in the same direction as before.**

16. Remove the bolt (B) and timing belt idler (A).

17. Remove the crankshaft sprocket.

18. Remove the cylinder head cover.

   a. Remove the spark plug cable.

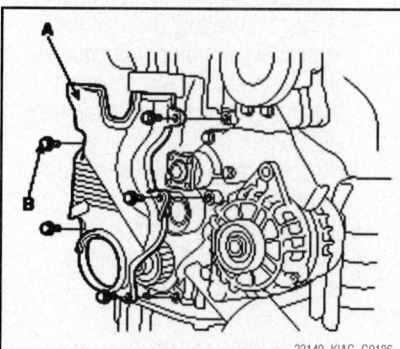

Fig. 208 Remove the 5 bolts (B) and timing belt lower cover (A)

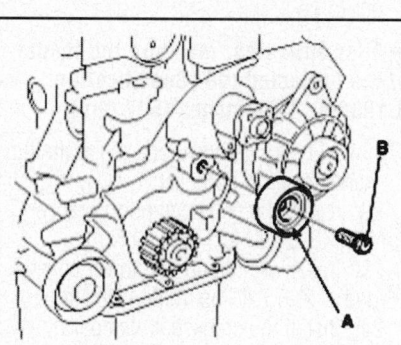

Fig. 209 Remove the bolt (B) and timing belt idler (A)

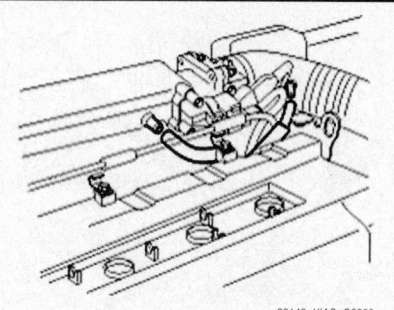

Fig. 210 Remove the Positive Crankcase Ventilation (PCV) hose (A) and breather hose (B)

   b. Remove the accelerator cable and the auto-cruise cable from the cylinder head cover.

   c. Remove the Positive Crankcase Ventilation (PCV) hose (A) and breather hose (B).

   d. Remove the bolts and cylinder head cover.

19. Remove camshaft sprocket.

To install:

20. Install the camshaft sprocket.

21. Install the cylinder head cover.

   a. Install the PVC hose and breather hose.

   b. Install the accelerator cable and the auto-cruise cable on the cylinder head cover.

   c. Install the spark plug cable.

22. Install the timing belt lower cover with 5 bolts. Torque: 5.87.2 ft. lbs. (7.8–9.8 Nm).

23. Install the flange and crankshaft pulley. Torque: 115.7–123.0 ft. lbs. (156.9–166.7 Nm).

➡ **Make sure that crankshaft sprocket pin fits the small hole in the pulley.**

24. Install the timing belt upper cover. Torque: 5.87.2 ft. lbs. (7.8–9.8 Nm).

25. Install the coolant pump pulley.

26. Install power steering belt.

27. Install air compressor bolt.

28. Install alternator belt.

29. Install the engine mount bracket

   a. Install the stay plate. Torque: 31.1–39.8 ft. lbs. (42.2–53.9 Nm).

   b. Install engine mount bracket. Torque: 17mm nut: 50.6–68.7 ft. lbs. (68.6–3.2 Nm); 14mm nuts and bolt: 36.2–47.0 ft. lbs. (49.0–63.7 Nm).

30. Install RH side cover.

31. Install RH front wheel.

32. Install engine cover.

2.7L Engine

See Figures 211 through 219.

1. Remove the engine cover

2. Remove the front right wheel and tire.

3. Remove the right side cover.

4. Remove the drive belt, the idler pulley and the tensioner pulley.

➡ **In removing the drive belt, fix a tool in the auto tensioner pulley bolt and turn the bolt counter clockwise.**

5. Remove the timing belt upper covers (A).

6. Align the groove of the pulley with the timing mark of the timing belt cover by turning the crankshaft pulley clockwise. Check that the timing mark of the camshaft sprocket is aligned with that of the cylinder head cover. (No.1 cylinder piston at TDC)

7. Remove the engine mounting bracket.

   a. Support the engine oil pan with a jack.

✱✱ CAUTION

Put a wooden or rubber block between the jack and the engine oil pan.

   b. Remove the engine mounting bracket (A).

Fig. 211 Remove the timing belt upper covers (A)

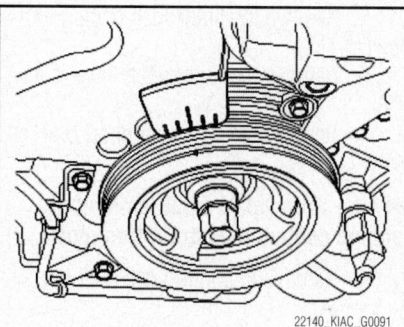

Fig. 212 Align the groove of the pulley with the timing mark of the timing belt cover

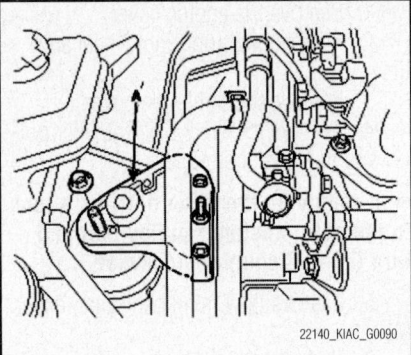

Fig. 213 Remove the engine mounting bracket (A)

Fig. 216 Remove the engine support bracket (A)

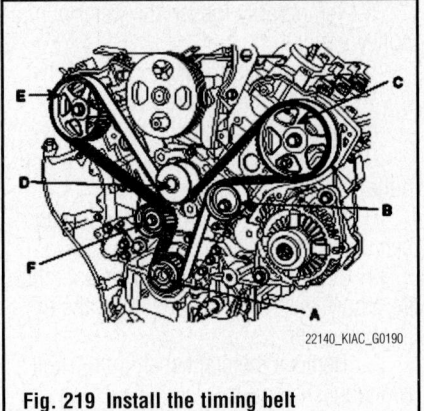

Fig. 219 Install the timing belt

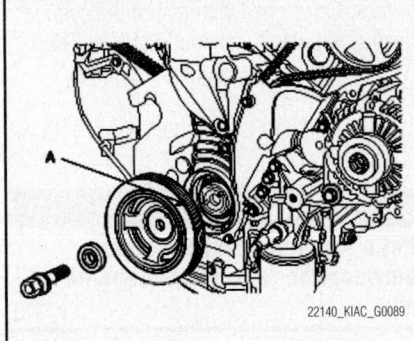

Fig. 214 Remove the crankshaft damper pulley (A)

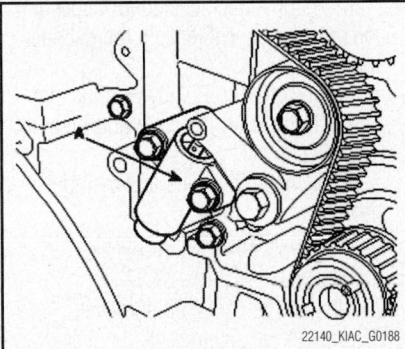

Fig. 217 Remove the timing belt auto tensioner (A)

Fig. 215 Remove timing belt lower cover (A)

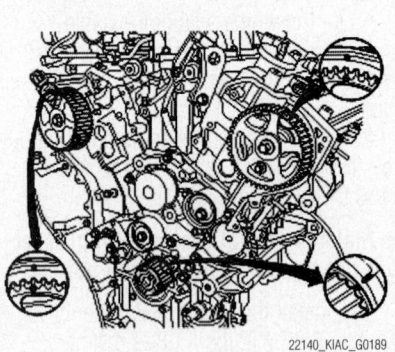

Fig. 218 Ensure the timing marks on the camshaft and the crankshaft sprockets are in the proper positions

8. Remove the crankshaft damper pulley (A).

9. Remove the timing belt lower cover (A).

10. Remove the engine support bracket (A).

➡ **After removal, a small amount of engine coolant may drain from point (B).**

11. Remove the timing belt auto tensioner (A).

12. Remove the timing belt.

➡ **Mark the direction of rotation on the timing belt.**

13. Remove the timing belt tensioner arm assembly and the idler.

14. Remove the crankshaft timing belt sprocket.

To install:

15. Install the crankshaft sprocket.

16. Install the tensioner arm assembly and the idler.

17. Ensure the timing marks on the camshaft and the crankshaft sprockets are in the proper positions.

18. Install the timing belt in the following order:
   a. Crankshaft sprocket (A)
   b. Idler (B)
   c. Bank 2 exhaust cam sprocket (C)
   d. Water pump pulley (D)
   e. Bank 1 exhaust cam sprocket (E)
   f. Tensioner pulley (F).

19. Install the timing belt auto tensioner.
   a. Make the tensioner stand upright for about five minutes before installing.

➡ **When handling the auto-tensioner observe the following:**

- Do not lay down the auto tensioner.
- Do not compress the rod suddenly.
- When reinstalling the auto tensioner, ensure proper orientation.
- Do not press the rod any more when its projection from the body is 2.5mm.
- Keep the auto-tensioner upright at room temperature in winter.

   b. Install the auto tensioner to the front case with the set-pin inserted. Torque: 14.5–19.5 ft. lbs. (19.6–26.5 Nm).

20. Remove the auto tensioner set-pin.

21. Check the tension of the timing belt.
   a. Turn the crankshaft 2 revolutions clockwise, and set the number one cylinder to TDC.

➡ **After 5 minutes, measure the length of the projected rod. Specification: 0.1969–0.2756 inches (5–7 mm).**

   b. Ensure the locations of the timing marks for each sprocket.

22. Install the engine support bracket. Torque: 43–51 ft. lbs. (59–69 Nm).

23. Install the timing belt lower cover. Torque: 7.2–8.7 ft. lbs. (9.8–11.8 Nm).

24. Install the crankshaft damper pulley. Torque: 123–130 ft. lbs. (167–177 Nm).

25. Install the engine mounting bracket. Torque: 47–62 ft. lbs. (64–83 Nm).

26. Install the timing belt upper cover. Torque: 7.2–8.7 ft. lbs. (9.8–11.8 Nm).

27. Install the drive belt tensioner. Torque: 25–40 ft. lbs. (34–54 Nm).

28. Install the drive belt idler and the drive belt. Torque: 25–40 ft. lbs. (34–54 Nm).

29. Install the right side cover.

30. Install the front right wheel and tire.

31. Install the engine cover. Torque: 5.8–8.7 ft. lbs. (7.8–11.8 Nm).

TIMING CHAIN COVER AND SEAL

REMOVAL & INSTALLATION

2.4L Engine

See Figure 220.

1. Remove the engine cover.
2. Remove RH front wheel.
3. Remove RH side cover.
4. Set No.1 cylinder to TDC/compression
5. Remove the engine mount bracket.
6. Remove the accessory drive belt.
7. Remove the idler pulley.
8. Remove the accessory drive belt tensioner.

➡**Tensioner pulley bolt is left handed screw.**

9. Remove the water pump pulley (A).
10. Remove the crankshaft pulley (B).
11. Remove the engine support bracket (C).
12. Disconnect the ignition coil connector.
13. Remove the ignition coils.
14. Remove the PCV hose and breather hose from the cylinder head cover.
15. Remove the cylinder head cover and gasket.
16. Remove the A/C compressor lower bolts.

22140_KIAC_G0191

Fig. 220 Remove the water pump pulley (A), crankshaft pulley (B), and engine support bracket (C).

17. Remove the compressor bracket.
18. Drain the engine oil.
19. Remove the oil pan.

✳✳ CAUTION

Be careful not to damage the contact surfaces of cylinder block and oil pan.

20. Remove the timing chain cover by prying the portions between the cylinder head and cylinder block with a screwdriver.

✳✳ CAUTION

Be careful not to damage the contact surfaces of cylinder block, cylinder head and timing chain cover.

To install:

21. Install timing chain cover.

a. The sealant locations on chain cover and on counter parts (cylinder head, cylinder block, and ladder frame) must be free of engine oil and ETC.

b. Before assembling the timing chain cover, the liquid sealant Loctite 5900 should be applied on the gap between cylinder head and cylinder block.

c. After applying liquid sealant Loctite 5900 on timing chain cover, the part must be assembled within 5 minutes.

d. The dowel pins on the cylinder block and holes on the timing chain cover should be used as a reference in order to assemble the timing chain cover to be in exact position.

e. Torque: M6 bolts: 5.78–7.23 ft. lbs. (7.84–9.8 Nm). M8 bolts: 14–17 ft. lbs. (19–23 Nm).

f. The firing and/or blow out test should not be performed within 30 minutes after the timing chain cover is assembled.

22. Install the oil pan.

a. Using a gasket scraper, remove all the old packing material from the gasket surfaces.

b. Before assembling the oil pan, the liquid sealant Loctite 5900 should be applied on oil pan.

The part must be assembled within 5 minutes after the sealant was applied.

✳✳ CAUTION

When applying sealant gasket, sealant must not be protruded into the inside of oil pan. To prevent leakage of oil, apply sealant gasket to the inner threads of the bolt holes.

c. Install oil pan.

d. Uniformly tighten the bolts in several passes. Torque: M8 bolts: 20–22 ft.

lbs. (26–31 Nm). M6 bolts: 7.23–8.67 ft. lbs. (9.8–11.6 Nm).

e. After assembly, wait at least 30 minutes before filling the engine with oil.

23. Install air compressor bracket. Torque: 14–17 ft. lbs. (20–24 Nm).

24. Install air compressor bolts. Torque: 14–17 ft. lbs. (20–24 Nm).

25. Install cylinder head cover.

a. The hardening sealant located on the upper area between timing chain cover and cylinder head should be removed before assembling cylinder head cover.

b. After applying sealant, it should be assembled within 5 minutes.

c. The firing and/or blow out test should not be performed within 30 minutes after the cylinder head cover was assembled.

d. Torque the cylinder head cover bolts as follows:

- 1st step: 2.89–4.34 ft. lbs. (3.92–5.88 Nm)
- 2st step: 5.78–7.23 ft. lbs. (7.84–9.8 Nm).

✳✳ CAUTION

Do not reuse cylinder head cover gasket.

26. Install ignition coil.

27. Connect ignition coil connector.

28. Install engine support bracket. Torque: M10 bolts: 29–33 ft. lbs. (39–44 Nm); M8 bolts: 14–18 ft. lbs. (20–25 Nm).

29. Install crankshaft pulley. Torque: 123–130 ft. lbs. (167–176 Nm).

30. Install water pump pulley. Torque: 5.78–7.23 ft. lbs. (7.84–9.8 Nm).

31. Install drive belt tensioner and tensioner pulley. Torque: 40–47 ft. lbs. (54–64 Nm).

➡**Tensioner pulley bolt is left-handed screw.**

32. Install idler pulley. Torque: 40–47 ft. lbs. (54–64 Nm).

33. Install the accessory drive belt in the following order:

a. Crankshaft pulley
b. A/C pulley
c. Alternator pulley
d. Idler pulley
e. P/S pump pulley
f. Idler pulley
g. Water pump pulley
h. Tensioner pulley.
i. Rotate auto tensioner arm in the counter—clockwise moving auto tensioner pulley bolt with a wrench.

34. After putting belt on auto tensioner

pulley, release the auto tensioner pulley slowly.

35. Install engine mounting bracket. Torque: 47–61 ft. lbs. (64–83 Nm).
36. Install RH side cover.
37. Install RH front wheel.
38. Install engine cover.

TIMING CHAIN AND SPROCKETS

REMOVAL & INSTALLATION

2.4L Engine

See Figures 221 through 224.

1. Remove the engine cover.
2. Remove RH front wheel.
3. Remove RH side cover.
4. Set No.1 cylinder to TDC/compression
5. Remove the engine mount bracket.
6. Remove the accessory drive belt.
7. Remove the idler pulley.
8. Remove the accessory drive belt tensioner.

➡ **Tensioner pulley bolt is left handed screw.**

9. Remove the water pump pulley (A).
10. Remove the crankshaft pulley (B).
11. Remove the engine support bracket (C).
12. Disconnect the ignition coil connector.
13. Remove the ignition coils.
14. Remove the PCV hose and breather hose from the cylinder head cover.
15. Remove the cylinder head cover and gasket.
16. Remove the A/C compressor lower bolts.
17. Remove the compressor bracket.
18. Drain the engine oil.
19. Remove the oil pan.

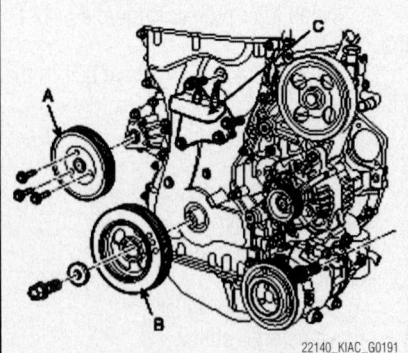

Fig. 221 Remove the water pump pulley (A), crankshaft pulley (B), and engine support bracket (C).

✳✳ CAUTION

Be careful not to damage the contact surfaces of cylinder block and oil pan.

20. Remove the timing chain cover by prying the portions between the cylinder head and cylinder block with a screwdriver.

✳✳ CAUTION

Be careful not to damage the contact surfaces of cylinder block, cylinder head and timing chain cover.

21. The key of crankshaft should be aligned with the mating face of main bearing cap. As a result of this, the piston of No.1 cylinder is placed at the top dead center on compression stroke.
22. Install a set pin after compressing the timing chain tensioner.
23. Remove the timing chain tensioner (A).
24. Remove the timing chain tensioner arm (B).
25. Remove the timing chain.
26. Remove the timing chain guide (A).
27. Remove the timing chain oil jet (A).
28. Remove the crankshaft chain sprocket (B).

To install:

29. Install crankshaft chain sprocket.
30. Install timing chain oil jet. Torque: 5.78–7.23 ft. lbs. (7.84–9.8 Nm).
31. The key of crankshaft should be aligned with the mating surface of main bearing cap. As a result, this places the piston on No.1 cylinder at the top dead center on compression stroke.

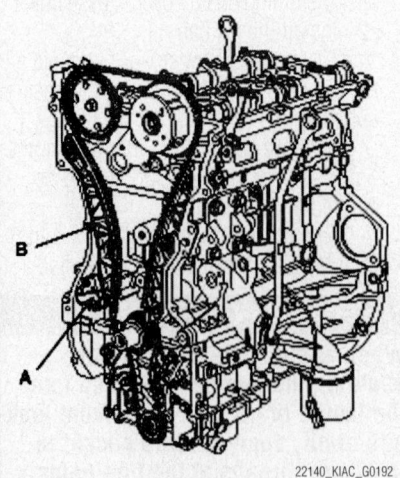

Fig. 222 Remove the timing chain tensioner (A) and timing chain tensioner arm (B)

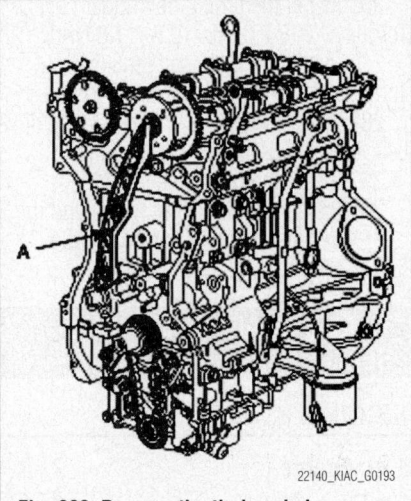

Fig. 223 Remove the timing chain guide (A)

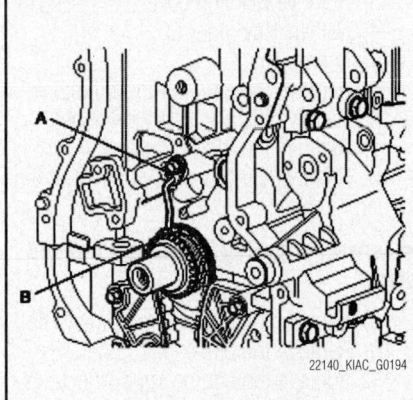

Fig. 224 Remove the timing chain oil jet (A) and crankshaft chain sprocket (B)

32. Install timing chain guide. Torque: 7.23–8.67 ft. lbs. (9.8–11.6 Nm).
33. Install timing chain.

➡ **To install the timing chain with no slack between each shaft (cam, crank), follow the below procedure:**

- Crankshaft sprocket (A)
- Timing chain guide (B)
- Intake camshaft sprocket (C)
- Exhaust camshaft sprocket (D).

The timing mark of each sprockets should be matched with timing mark (color link) of timing chain at installing of the timing chain.

34. Install timing chain tensioner arm. Torque: 7.23–8.67 ft. lbs. (9.8–11.6 Nm).
35. Install timing chain auto tensioner and remove set pin. Torque: 7.23–8.67 ft. lbs. (9.8–11.6 Nm).
36. After rotating crankshaft 2 revolutions in regular direction (clockwise viewed from front), confirm the timing mark.
37. Install timing chain cover.

a. The sealant locations on chain cover and on counter parts (cylinder head, cylinder block, and ladder frame) must be free of engine oil and ETC.

b. Before assembling the timing chain cover, the liquid sealant Loctite 5900 should be applied on the gap between cylinder head and cylinder block.

c. After applying liquid sealant Loctite 5900 on timing chain cover, the part must be assembled within 5 minutes.

d. The dowel pins on the cylinder block and holes on the timing chain cover should be used as a reference in order to assemble the timing chain cover to be in exact position.

e. Torque: M6 bolts: 5.78–7.23 ft. lbs. (7.84–9.8 Nm). M8 bolts: 14–17 ft. lbs. (19–23 Nm).

f. The firing and/or blow out test should not be performed within 30 minutes after the timing chain cover is assembled.

38. Install the oil pan.

a. Using a gasket scraper, remove all the old packing material from the gasket surfaces.

b. Before assembling the oil pan, the liquid sealant Loctite 5900 should be applied on oil pan.

The part must be assembled within 5 minutes after the sealant was applied.

✳✳ CAUTION

When applying sealant gasket, sealant must not be protruded into the inside of oil pan. To prevent leakage of oil, apply sealant gasket to the inner threads of the bolt holes.

c. Install oil pan.

d. Uniformly tighten the bolts in several passes. Torque: M8 bolts: 20–22 ft. lbs. (27–30 Nm). M6 bolts: 7.23–8.67 ft. lbs. (9.8–11.6 Nm).

e. After assembly, wait at least 30 minutes before filling the engine with oil.

39. Install air compressor bracket. Torque: 15–17 ft. lbs. (20–24 Nm).

40. Install air compressor bolts. Torque: 15–17 ft. lbs. (20–24 Nm).

41. Install cylinder head cover.

a. The hardening sealant located on the upper area between timing chain cover and cylinder head should be removed before assembling cylinder head cover.

b. After applying sealant, it should be assembled within 5 minutes.

c. The firing and/or blow out test should not be performed within 30 minutes after the cylinder head cover was assembled.

d. Torque the cylinder head cover bolts as follows:

- 1st step: 2.89–4.34 ft. lbs. (3.92–5.88 Nm)
- 2st step: 5.78–7.23 ft. lbs. (7.84–9.8 Nm).

✳✳ CAUTION

Do not reuse cylinder head cover gasket.

42. Install ignition coil.
43. Connect ignition coil connector.
44. Install engine support bracket. Torque: M10 bolts: 29–33 ft. lbs. (39–44 Nm); M8 bolts: 15–17 ft. lbs. (20–24 Nm).

45. Install crankshaft pulley. Torque: 123–130 ft. lbs. (167–176 Nm).

46. Install water pump pulley. Torque: 5.78–7.23 ft. lbs. (7.84–9.8 Nm).

47. Install drive belt tensioner and tensioner pulley. Torque: 40–47 ft. lbs. (54–64 Nm).

➡**Tensioner pulley bolt is left-handed screw.**

48. Install idler pulley. Torque: 40–47 ft. lbs. (54–64 Nm).

49. Install the accessory drive belt in the following order:

a. Crankshaft pulley
b. A/C pulley
c. Alternator pulley
d. Idler pulley
e. P/S pump pulley
f. Idler pulley
g. Water pump pulley
h. Tensioner pulley.
i. Rotate auto tensioner arm in the counter—clockwise moving auto tensioner pulley bolt with a wrench.

50. After putting belt on auto tensioner pulley, release the auto tensioner pulley slowly.

51. Install engine mounting bracket. Torque: 47–61 ft. lbs. (64–83 Nm).

52. Install RH side cover.
53. Install RH front wheel.
54. Install engine cover.

VALVE LASH

ADJUSTMENT

The valve lash on all engines is kept in adjustment hydraulically. No adjustment is necessary or possible.

ENGINE PERFORMANCE & EMISSION CONTROL

COMPONENT LOCATIONS

1.6L Engine

See Figures 225 and 226.

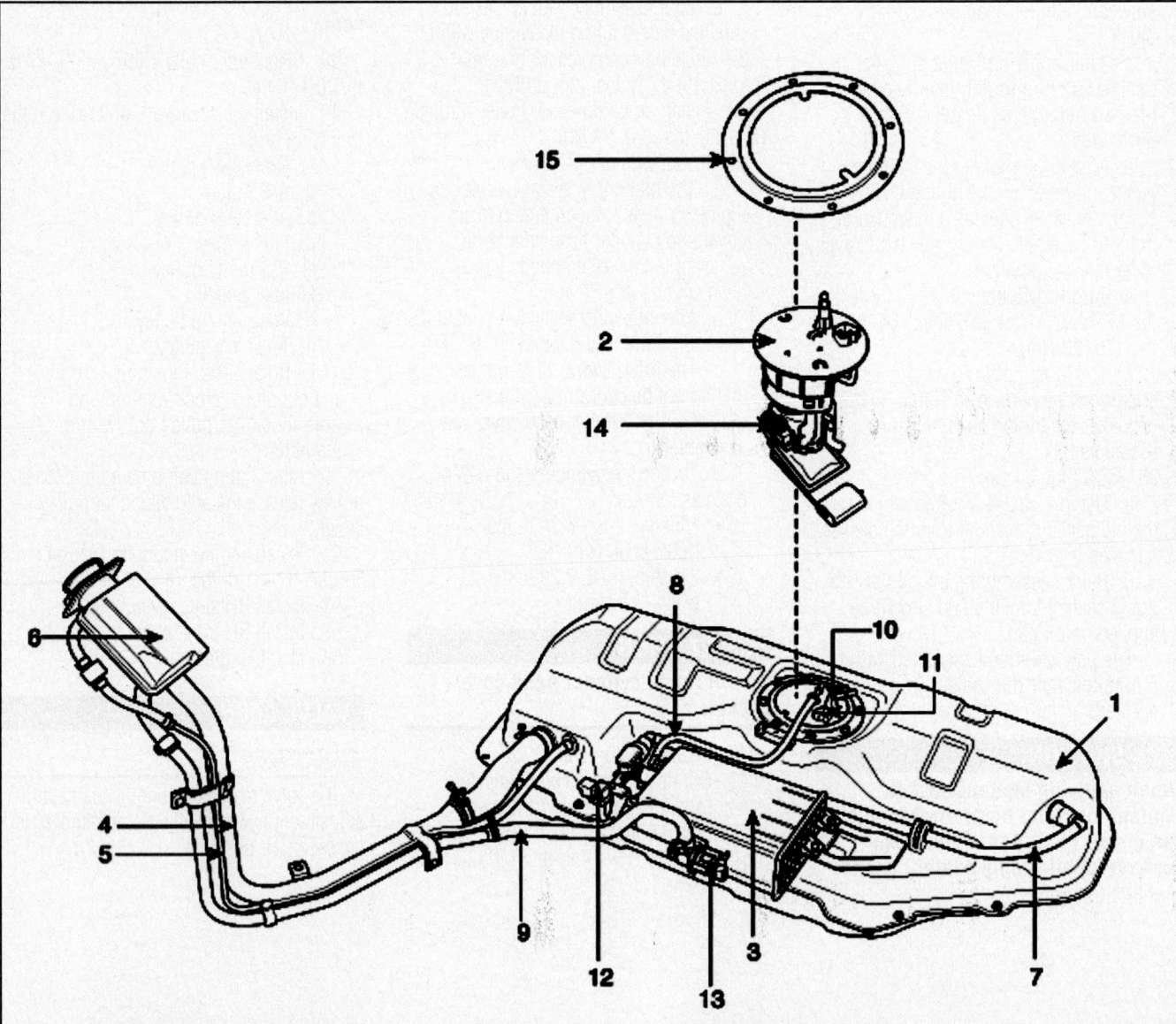

1. Fuel Tank
2. Fuel Pump Assembly
   (including Fuel Filter & Fuel Pressure Regulator)
3. Canister
4. Fuel Filler Pipe
5. Leveling Hose
6. Fuel Tank Air Filter
7. Tube (Canister ↔ Fuel Tank)
8. Tube (Canister ↔ Intake Manifold)

9. Hose (Canister ↔ Fuel Tank Air Filter)
10. Nipple-Fuel Feed Line
11. Fuel Pump Connector
12. Fuel Tank Pressure Sensor (FTPS)
13. Canister Close Valve (CCV)
14. Fuel Level Sensor (FLS)
15. Fuel Pump Locking Ring

22140_KIAC_G0199

Fig. 225 Fuel Delivery System component locations—1.6L engine

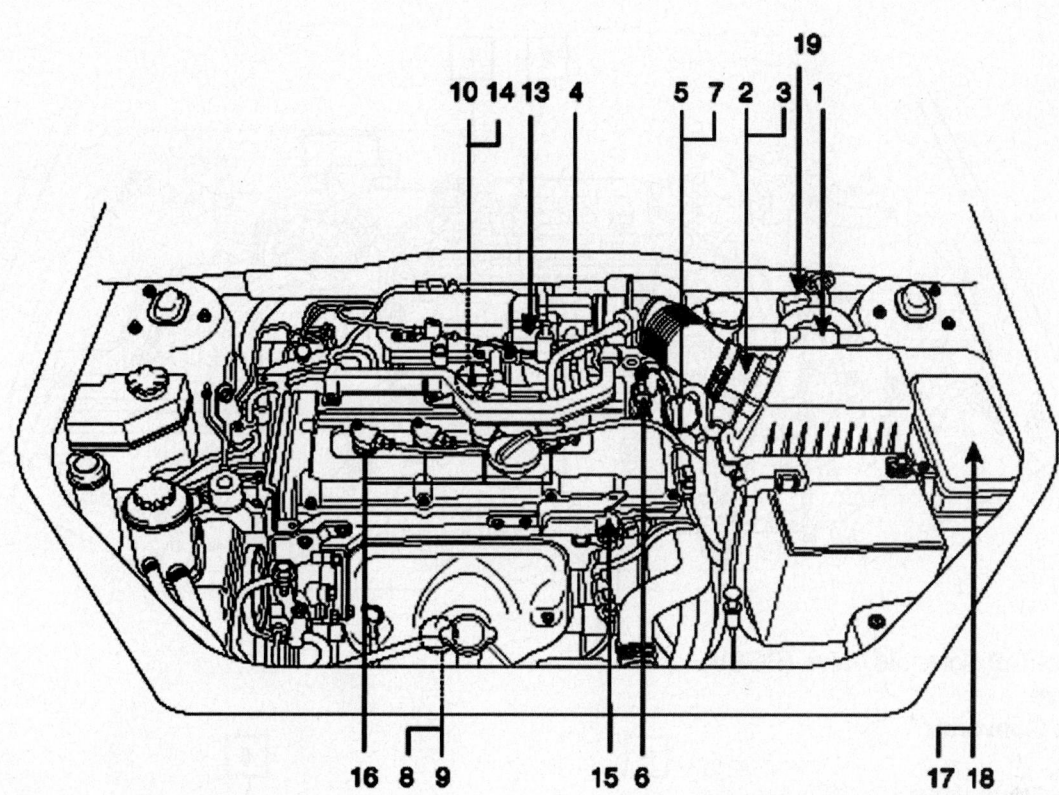

1. ECM (for M/T) / PCM (for A/T)
2. Mass Air Flow Sensor (MAFS)
3. Intake Air Temperature Sensor (IATS)
4. Throttle Position Sensor (TPS)
5. Engine Coolant Temperature Sensor (ECTS)
6. Camshaft Position Sensor (CMPS)
7. Crankshaft Position Sensor (CKPS)
8. Heated Oxygen Sensor (HO2S) [Bank 1/Sensor 1]
9. Heated Oxygen Sensor (HO2S) [Bank 1/Sensor 2]
10. Knock Sensor (KS)
11. Wheel Speed Sensor (WSS)
12. Injector
13. Idle Speed Control Actuator (ISCA)
14. Purge Control Solenoid Valve (PCSV)
15. CVVT Oil Control Valve (OCV)
16. Ignition Coil
17. Main Relay
18. Fuel Pump Relay
19. Multi Purpose Check Connector (20 pin)
20. Data Link Connector (DLC : 16 pin)
21. Fuel Tank Pressure Sensor (FTPS)
22. Canister Close Valve (CCV)
23. Fuel Level Sensor (FLS)

22140_KIAC_G0198

Fig. 226 Control system component locations—1.6L Engine

2.0L Engine

See Figures 227 through 229.

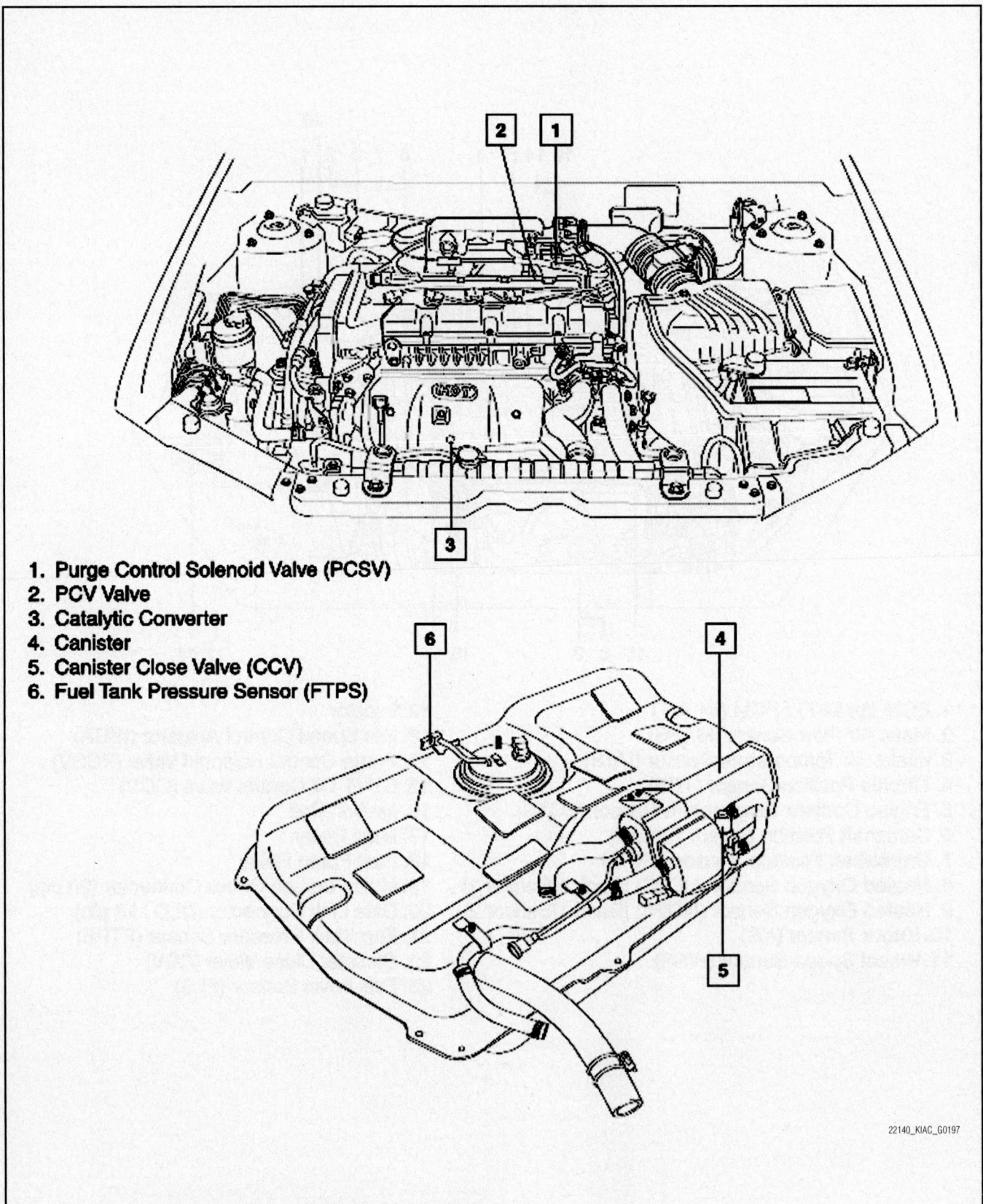

1. Purge Control Solenoid Valve (PCSV)
2. PCV Valve
3. Catalytic Converter
4. Canister
5. Canister Close Valve (CCV)
6. Fuel Tank Pressure Sensor (FTPS)

22140_KIAC_G0197

Fig. 227 Emission control component locations—2.0L Engine

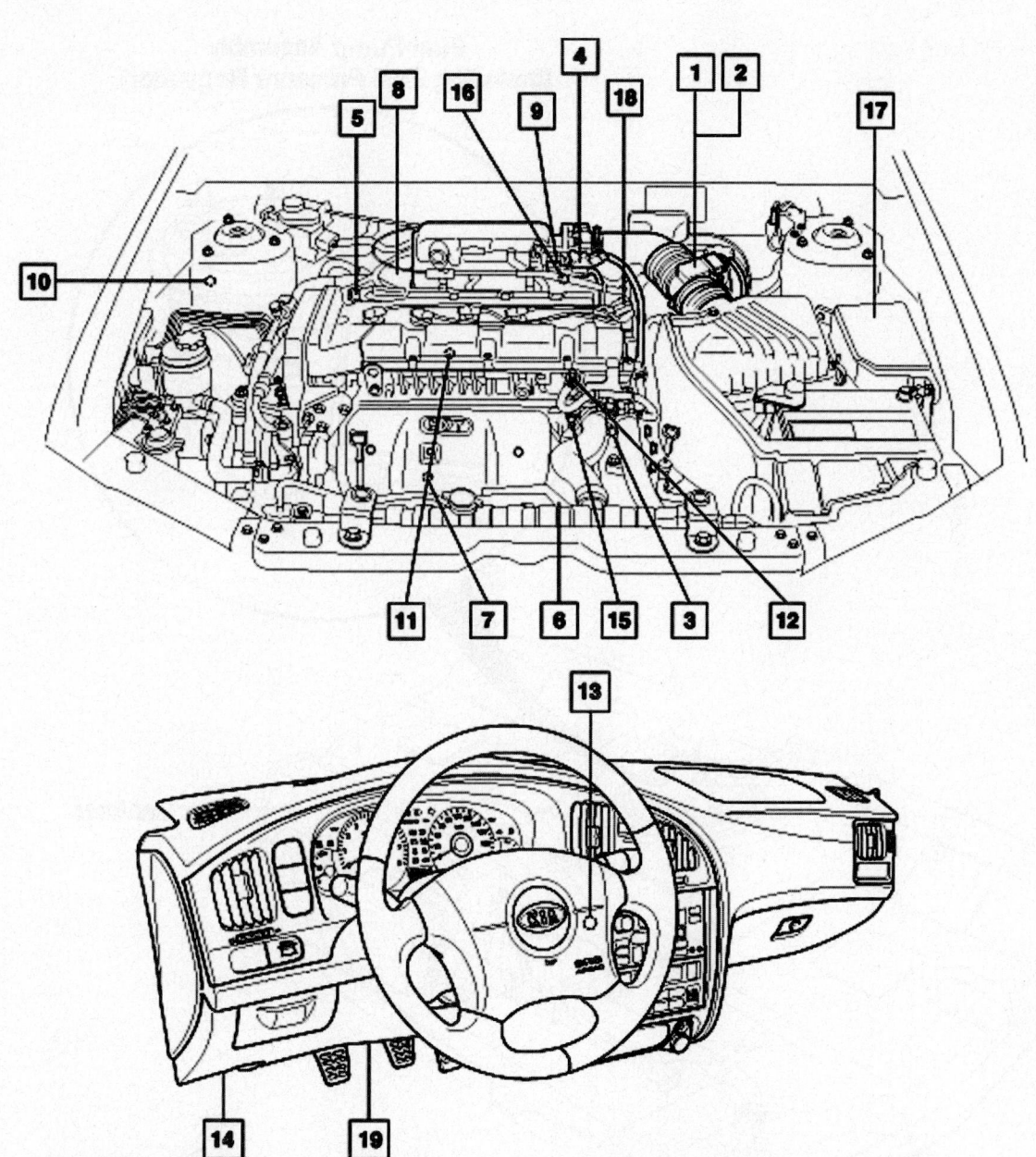

1. Mass Air Flow Sensor (MAFS)
2. Intake Air Temperature Sensor (IATS)
3. Engine Coolant Temperature Sensor (ECTS)
4. Throttle Position Sensor (TPS)
5. Camshaft Position Sensor (CMPS)
6. Crankshaft Position Sensor (CKPS)
7. Heated Oxygen Sensor (HO2S, Sensor 1)
8. Injector
9. Idle Speed Control Actuator (ISCA)
10. Vehicle Speed Sensor (VSS)
11. Knock Sensor
12. CVVT Oil Control Valve (OCV)
13. Ignition Switch
14. ECM
15. CVVT Oil Temperature Sensor (OTS)
16. Purge Control Solenoid Valve (PCSV)
17. Main Relay
18. Ignition Coil
19. DLC (Diagnostic Link Cable)
20. Heated Oxygen Sensor (HO2S, Sensor 2)

22140_KIAC_G0200

Fig. 228 Control system component locations—2.0L Engine

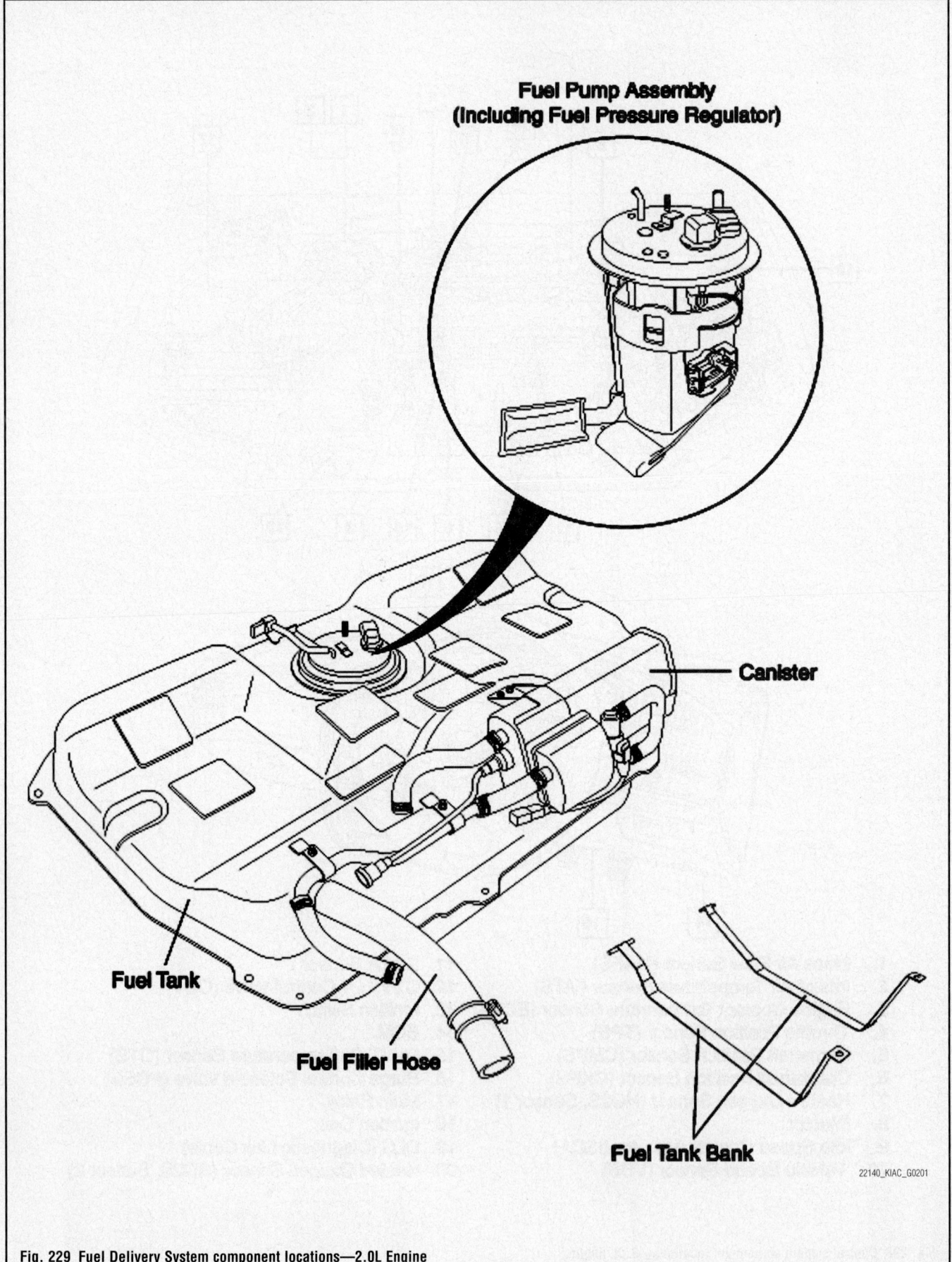

**Fuel Pump Assembly
(Including Fuel Pressure Regulator)**

Canister

Fuel Tank

Fuel Filler Hose

Fuel Tank Bank

22140_KIAC_G0201

Fig. 229 Fuel Delivery System component locations—2.0L Engine

2.4L Engine

See Figures 230 through 232.

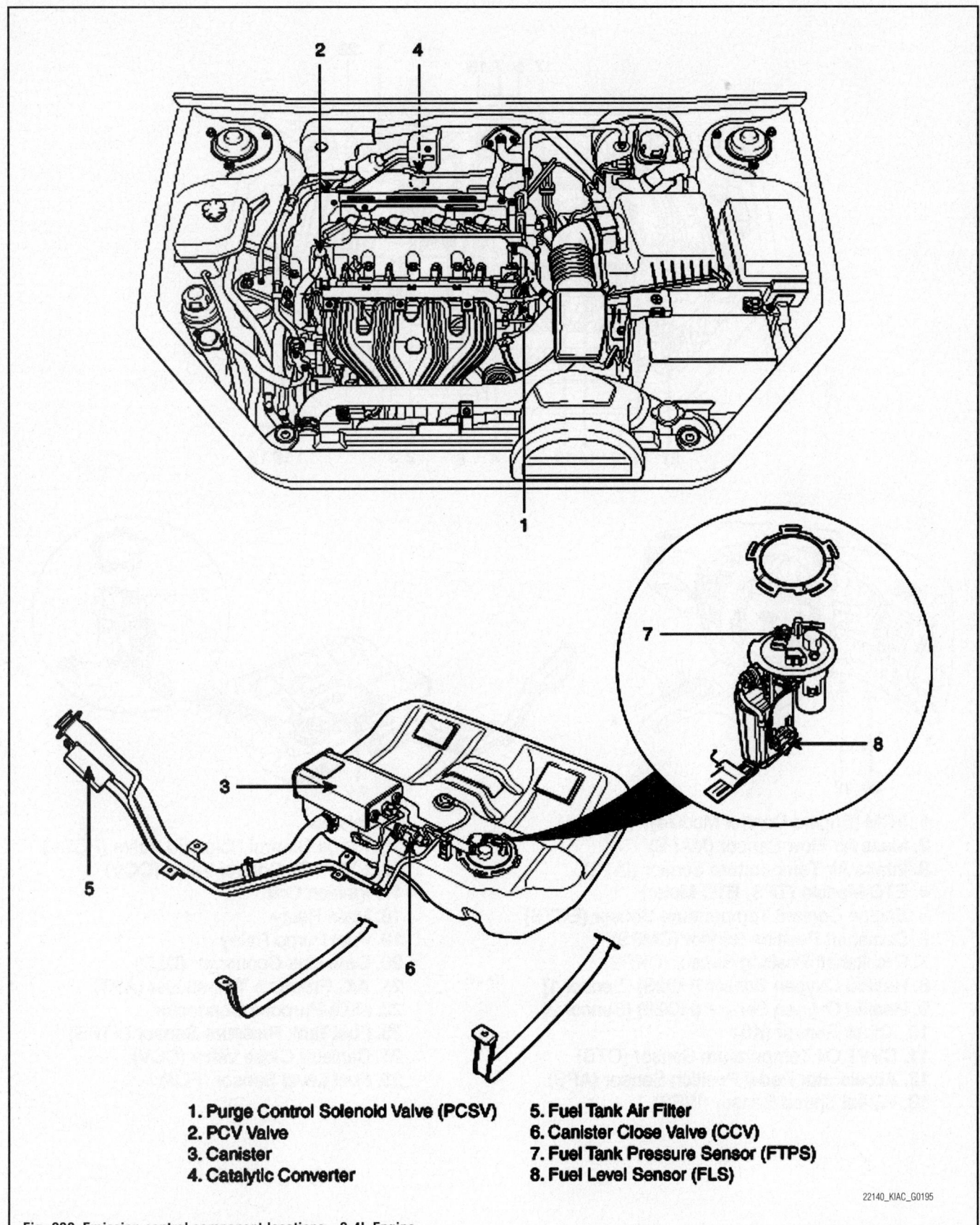

1. Purge Control Solenoid Valve (PCSV)
2. PCV Valve
3. Canister
4. Catalytic Converter
5. Fuel Tank Air Filter
6. Canister Close Valve (CCV)
7. Fuel Tank Pressure Sensor (FTPS)
8. Fuel Level Sensor (FLS)

22140_KIAC_G0195

Fig. 230 Emission control component locations—2.4L Engine

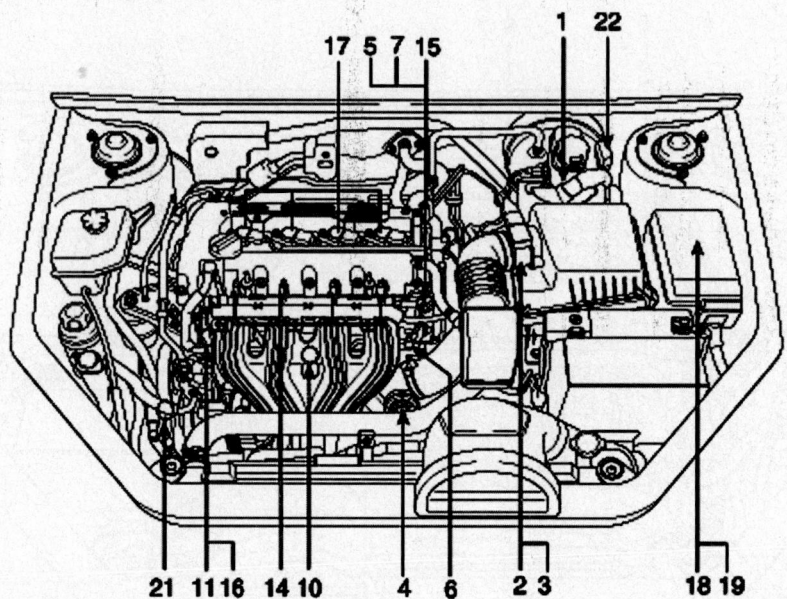

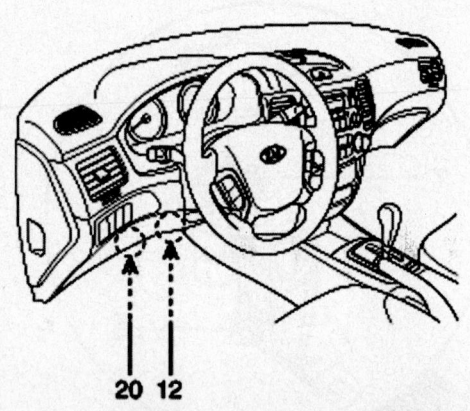

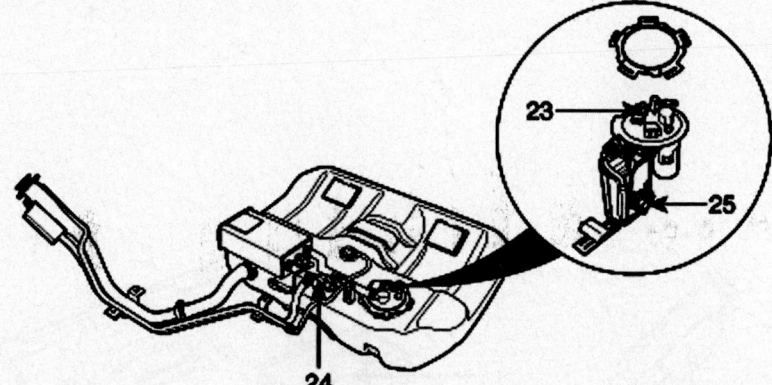

1. ECM (Engine Control Module)
2. Mass Air Flow Sensor (MAFS)
3. Intake Air Temperature Sensor (IATS)
4. ETC Module (TPS, ETC Motor)
5. Engine Coolant Temperature Sensor (ECTS)
6. Camshaft Position Sensor (CMPS)
7. Crankshaft Position Sensor (CKPS)
8. Heated Oxygen Sensor (HO2S) [Sensor 1]
9. Heated Oxygen Sensor (HO2S) [Sensor 2]
10. Knock Sensor (KS)
11. CVVT Oil Temperature Sensor (OTS)
12. Accelerator Pedal Position Sensor (APS)
13. Wheel Speed Sensor (WSS)
14. Injector
15. Purge Control Solenoid Valve (PCSV)
16. CVVT Oil Control Valve (OCV)
17. Ignition Coil
18. Main Relay
19. Fuel Pump Relay
20. Data Link Connector (DLC)
21. A/C Pressure Transducer (APT)
22. Multi-Purpose Connector
23. Fuel Tank Pressure Sensor (FTPS)
24. Canister Close Valve (CCV)
25. Fuel Level Sensor (FLS)

22140_KIAC_G0202

Fig. 231 Control system component locations—2.4L Engine

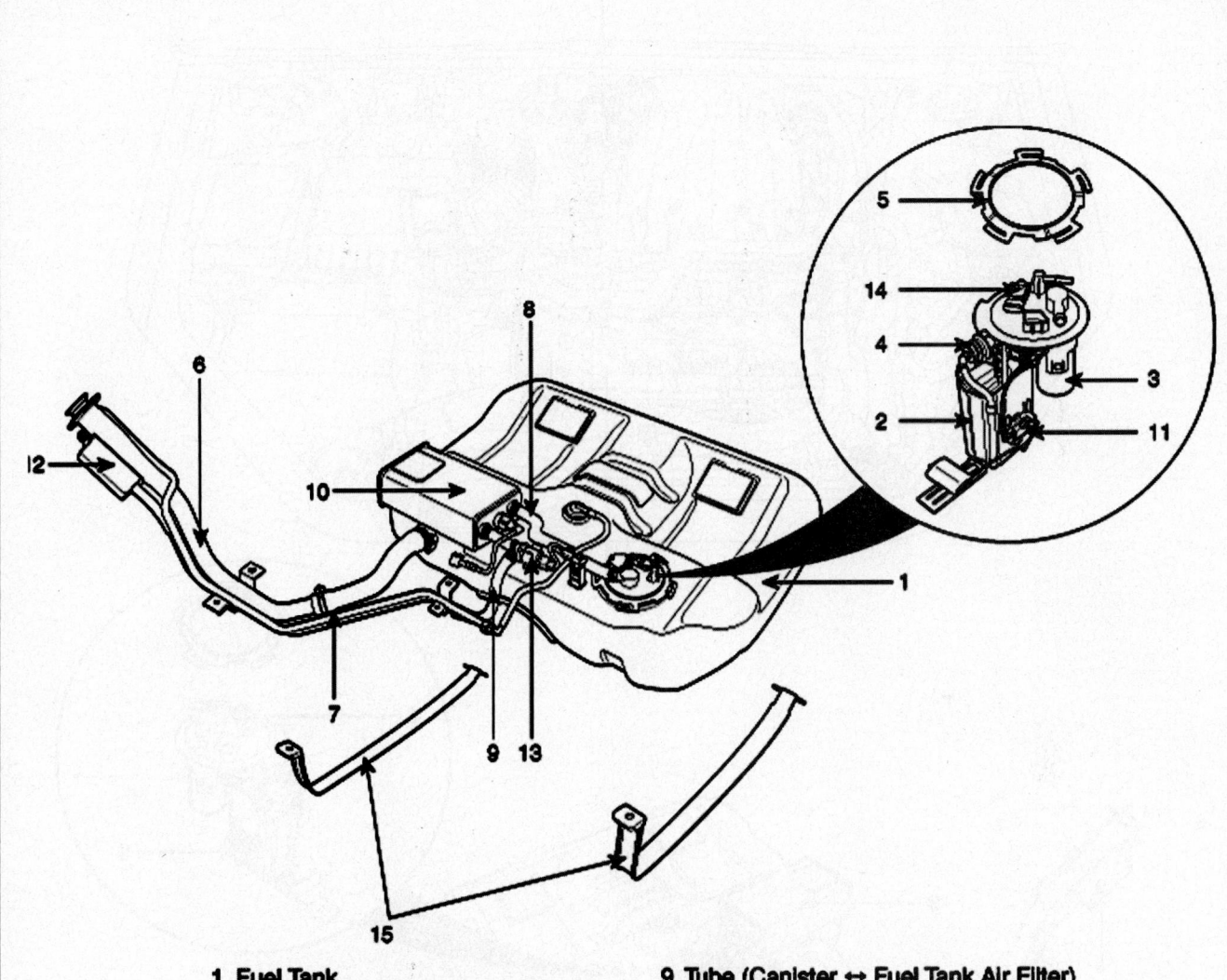

1. Fuel Tank
2. Fuel Pump (Including Fuel Filter)
3. ORVR Valve
4. Fuel Pressure Regulator
5. Locking Ring - Fuel Pump
6. Fuel Filler Pipe
7. Recirculation Pipe
8. Hose (Fuel Tank ↔ Canister)
9. Tube (Canister ↔ Fuel Tank Air Filter)
10. Canister
11. Fuel Level Sensor (FLS)
12. Fuel Tank Air Filter
13. Canister Close Valve (CCV)
14. Fuel Tank Pressure Sensor (FTPS)
15. Fuel Tank Band

22140_KIAC_G0203

Fig. 232 Fuel Delivery System component locations—2.4L Engine

2.7L Engine

See Figures 233 through 235.

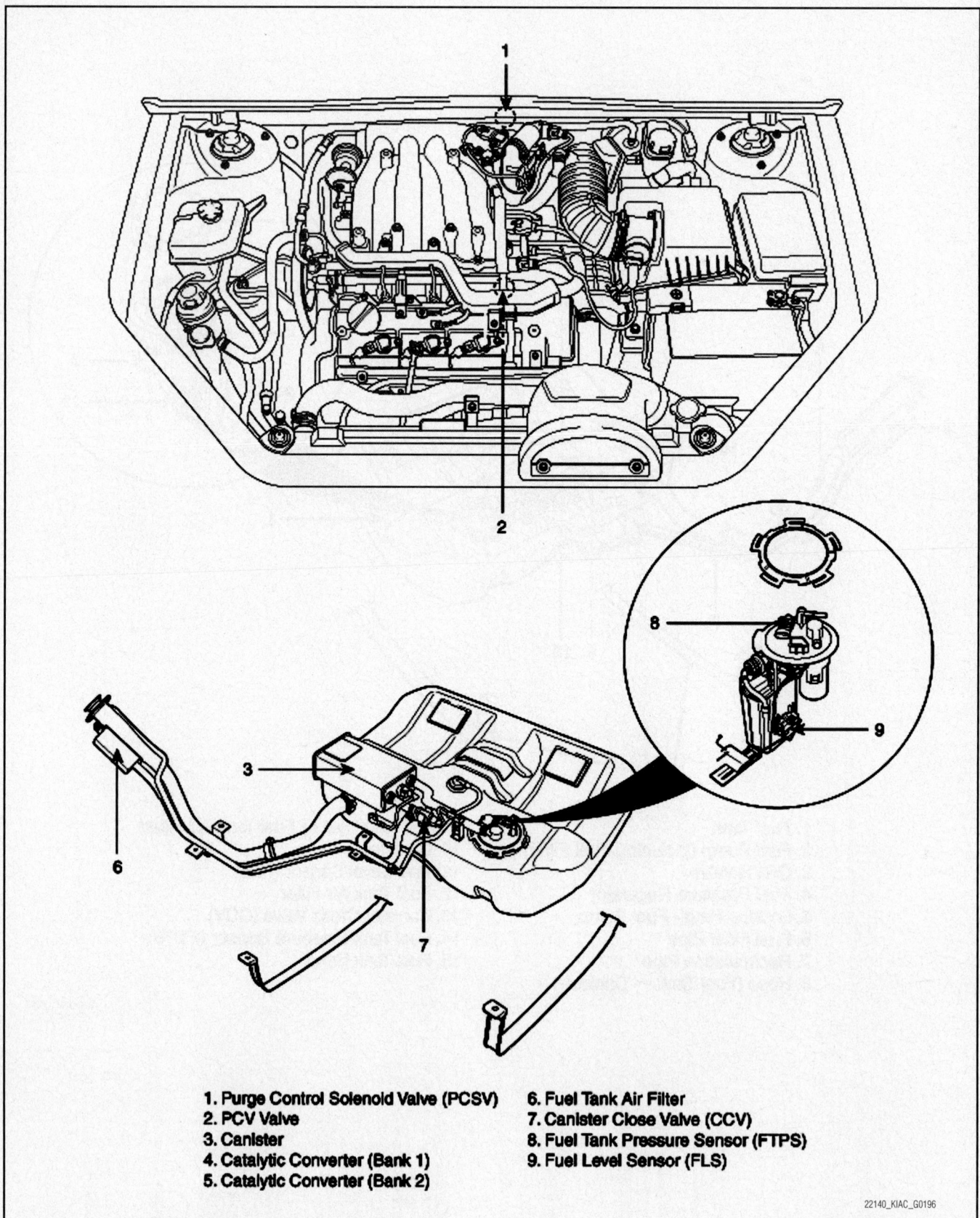

1. Purge Control Solenoid Valve (PCSV)
2. PCV Valve
3. Canister
4. Catalytic Converter (Bank 1)
5. Catalytic Converter (Bank 2)
6. Fuel Tank Air Filter
7. Canister Close Valve (CCV)
8. Fuel Tank Pressure Sensor (FTPS)
9. Fuel Level Sensor (FLS)

22140_KIAC_G0196

Fig. 233 Emission control component locations—2.7L Engine

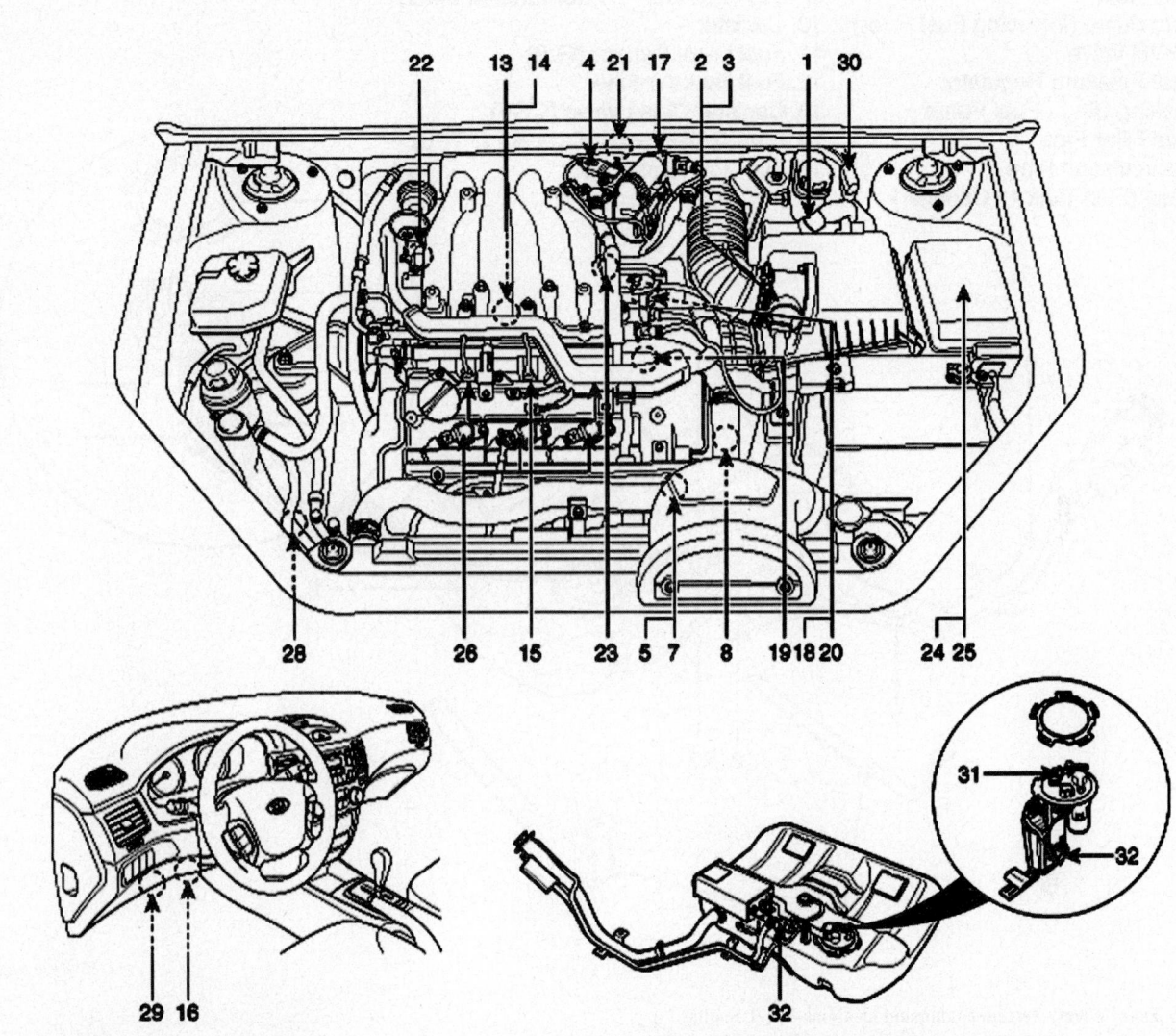

1. PCM (Powertrain Control Module)
2. Mass Air Flow Sensor (MAFS)
3. Intake Air Temperature Sensor (IATS)
4. Manifold Absolute Pressure Sensor (MAPS)
5. Engine Coolant Temperature Sensor (ECTS)
6. Camshaft Position Sensor (CMPS) [Bank 1]
7. Camshaft Position Sensor (CMPS) [Bank 2]
8. Crankshaft Position Sensor (CKPS)
9. Heated Oxygen Sensor (HO2S) [Bank 1 / Sensor 1]
10. Heated Oxygen Sensor (HO2S) [Bank 1 / Sensor 2]
11. Heated Oxygen Sensor (HO2S) [Bank 2 / Sensor 1]
12. Heated Oxygen Sensor (HO2S) [Bank 2 / Sensor 2]
13. Knock Sensor (KS) [Bank 1]
14. Knock Sensor (KS) [Bank 2]
15. Injector
16. Accelerator Position Sensor (APS)

17. ETC Module [Throttle Position Sensor (TPS) + ETC Motor]
18. CVVT Oil Control Valve (OCV) [Bank 1]
19. CVVT Oil Control Valve (OCV) [Bank 2]
20. CVVT Oil Temperature Sensor (OTS)
21. Purge Control Solenoid Valve (PCSV)
22. Variable Intake Solenoid (VIS) Valve #1 (Surge Tank Side)
23. Variable Intake Solenoid (VIS) Valve #2 (Intake Manifold Side)
24. Fuel Pump Relay
25. Main Relay
26. Ignition Coil
27. Wheel Speed Sensor (WSS) [Without ABS/ESC]
28. A/C Pressure Transducer (APT)
29. Data Link Connector (DLC)
30. Multi-Purpose Connector
31. Fuel Tank Pressure Sensor (FTPS)
32. Canister Close Valve (CCV)
33. Fuel Level Sensor (FLS)

22140_KIAC_G0204

Fig. 234 Control system component locations—2.7L Engine

1. Fuel Tank
2. Fuel Pump (Including Fuel Filter)
3. ORVR Valve
4. Fuel Pressure Regulator
5. Locking Ring - Fuel Pump
6. Fuel Filler Pipe
7. Recirculation Pipe
8. Hose (Fuel Tank ↔ Canister)
9. Tube (Canister ↔ Fuel Tank Air Filter)
10. Canister
11. Fuel Level Sensor (FLS)
12. Fuel Tank Air Filter
13. Canister Close Valve (CCV)
14. Fuel Tank Pressure Sensor (FTPS)
15. Fuel Tank Band

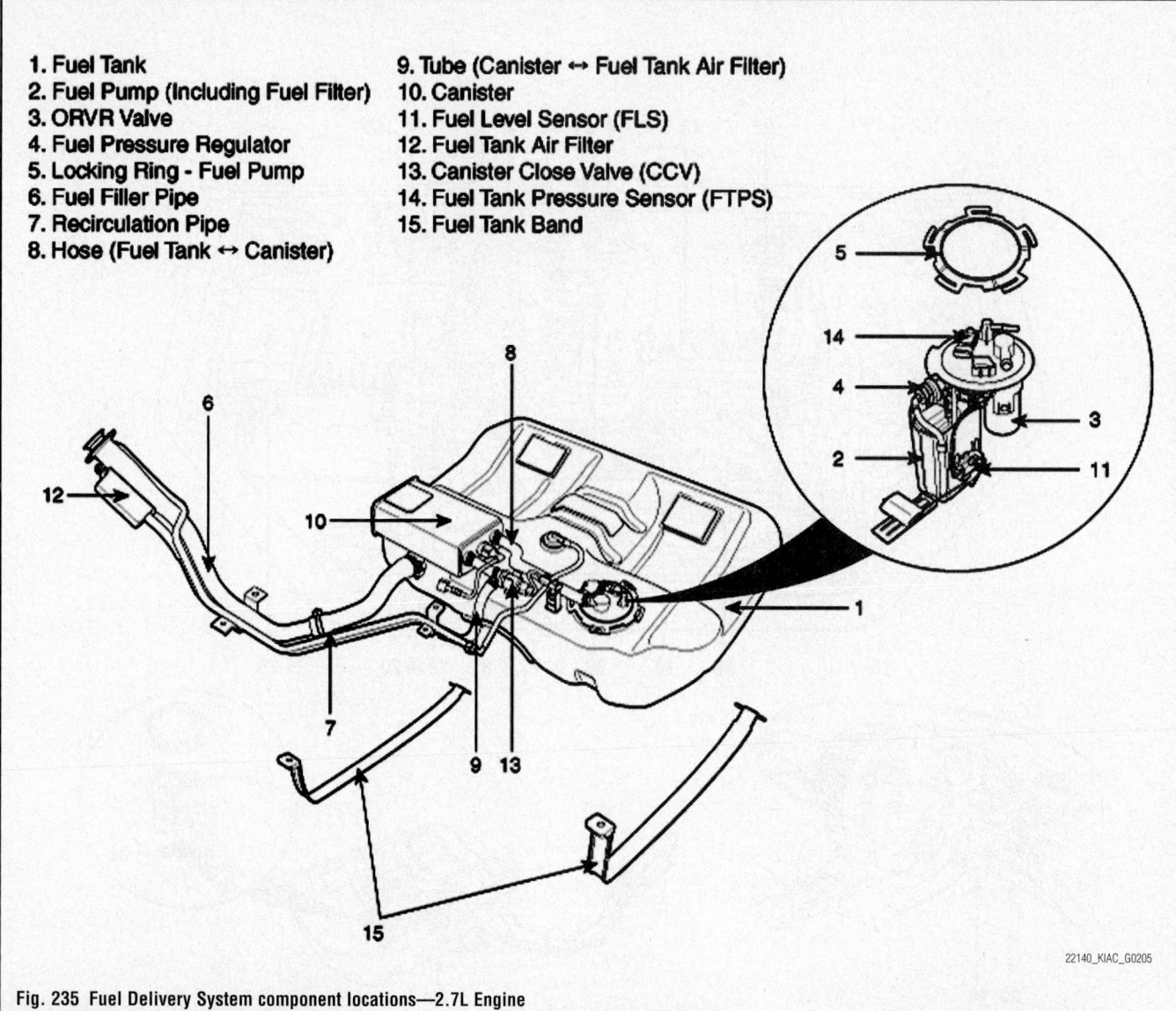

22140_KIAC_G0205

Fig. 235 Fuel Delivery System component locations—2.7L Engine

ELECTRONIC CONTROL MODULE (ECM)

REMOVAL & INSTALLATION

1. Turn the ignition switch off and disconnect the negative battery terminal.
2. Disconnect the ECM connector.
3. Unscrew the mounting bolts behind the air cleaner.
4. Remove the ECM.
5. Installation is reverse of removal.

ENGINE COOLANT TEMPERATURE (ECT) SENSOR

LOCATION

Engine Coolant Temperature Sensor (ECTS) is located in the engine coolant passage of the cylinder head for detecting the engine coolant temperature.

REMOVAL & INSTALLATION

1. Turn ignition switch OFF.
2. Disconnect ECTS connector.
3. Remove the ECTS.
4. Installation is the reverse of removal.

FUEL   **GASOLINE FUEL INJECTION SYSTEM**

FUEL SYSTEM SERVICE PRECAUTIONS

Safety is the most important factor when performing not only fuel system maintenance but any type of maintenance. Failure to conduct maintenance and repairs in a safe manner may result in serious personal injury or death. Maintenance and testing of the vehicle's fuel system components can be accomplished safely and effectively by adhering to the following rules and guidelines.

• To avoid the possibility of fire and personal injury, always disconnect the negative battery cable unless the repair or test procedure requires that battery voltage be applied.

• Always relieve the fuel system pressure prior to disconnecting any fuel system component (injector, fuel rail, pressure regulator, etc.), fitting or fuel line connection. Exercise extreme caution whenever relieving fuel system pressure to avoid exposing skin, face and eyes to fuel spray. Please be advised that fuel under pressure may penetrate the skin or any part of the body that it contacts.

• Always place a shop towel or cloth around the fitting or connection prior to loosening to absorb any excess fuel due to spillage. Ensure that all fuel spillage (should it occur) is quickly removed from engine surfaces. Ensure that all fuel soaked cloths or towels are deposited into a suitable waste container.

• Always keep a dry chemical (Class B) fire extinguisher near the work area.

• Do not allow fuel spray or fuel vapors to come into contact with a spark or open flame.

• Always use a back-up wrench when loosening and tightening fuel line connection fittings. This will prevent unnecessary stress and torsion to fuel line piping.

• Always replace worn fuel fitting O-rings with new Do not substitute fuel hose or equivalent where fuel pipe is installed.

Before servicing the vehicle, make sure to also refer to the precautions in the beginning of this section as well.

RELIEVING FUEL SYSTEM PRESSURE

1. Remove the rear seat cushion
2. For Rio and Spectra models, remove the rear seat cushion.

3. For Optima models, open the trunk.
4. Disconnect the fuel pump connector.
5. Start the engine and wait until fuel in fuel line is exhausted.
6. After engine stalls, turn the ignition switch to OFF position.

FUEL FILTER

REMOVAL & INSTALLATION

The fuel filter is an integral part of the fuel pump assembly.

1. For Rio and Spectra models, remove the rear seat cushion.
2. For Optima models, open the trunk.
3. Remove the service cover
4. Disconnect the fuel pump connector.
5. Start the engine and wait until fuel in fuel line is exhausted.
6. After engine stalls, turn the ignition switch to OFF position.
7. Disconnect the fuel feed line and canister hoses.
8. Unscrew the fuel pump mounting bolts and remove the fuel pump assembly.
9. Install the Fuel Pump according to the reverse order of REMOVAL procedure.

FUEL PUMP

REMOVAL & INSTALLATION

1. For Rio and Spectra models, remove the rear seat cushion.
2. For Optima models, open the trunk.
3. Remove the service cover
4. Disconnect the fuel pump connector.
5. Start the engine and wait until fuel in fuel line is exhausted.
6. After engine stalls, turn the ignition switch to OFF position.
7. Disconnect the fuel feed line and canister hoses.
8. Unscrew the fuel pump mounting bolts and remove the fuel pump assembly.
9. Install the Fuel Pump according to the reverse order of REMOVAL procedure.

FUEL TANK

REMOVAL & INSTALLATION

See Figures 236 through 238.

Rio

1. Remove the rear seat cushion.
2. Remove the service cover
3. Disconnect the fuel pump connector.
4. Start the engine and wait until fuel in fuel line is exhausted.

5. After engine stalls, turn the ignition switch to OFF position.
6. Disconnect the negative terminal from the battery.
7. Disconnect the fuel feed line and canister hose.
8. Raise the vehicle.
9. Remove the center muffler.
10. Support the fuel tank with a jack.
11. Remove the brake hose mounting bolts.
12. Disconnect the fuel filler pipe, the leveling hose and canister hose.
13. Remove the fuel tank mounting bolts and nuts, and then remove the fuel tank.
14. Install the Fuel Tank according to the reverse order to REMOVAL procedure.

Spectra

1. Remove the rear seat cushion.
2. Remove the service cover
3. Disconnect the fuel pump connector.
4. Start the engine and wait until fuel in fuel line is exhausted.
5. After engine stalls, turn the ignition switch to OFF position.
6. Disconnect the negative terminal from the battery.
7. Disconnect the fuel feed line (B) and the Fuel Tank Pressure Sensor (FTPS) connector (C).

➡ **Cover the hose connection with a shop towel to prevent splashing of fuel caused by residual pressure in the fuel line**

8. Lift the vehicle.
9. Remove the main muffler.
10. Disconnect the fuel filler hose, the canister drain hose and the vapor tube.
11. Unfasten two fuel tank band mounting bolts, and then remove the fuel tank from the vehicle.

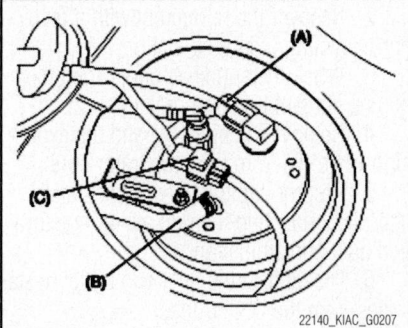

22140_KIAC_G0207

Fig. 236 Disconnect the fuel feed line (B) and the Fuel Tank Pressure Sensor (FTPS) connector (C).

12. Install the Fuel Tank according to the reverse order to REMOVAL procedure.

Optima

1. Open the trunk.
2. Remove the service cover
3. Disconnect the fuel pump connector.
4. Start the engine and wait until fuel in fuel line is exhausted.
5. After engine stalls, turn the ignition switch to OFF position.
6. Disconnect the negative terminal from the battery.
7. Disconnect the fuel feed quick-connector (A), the vapor hose (B), the fuel tank pressure sensor connector (C) and the canister close valve connector (D).
8. Raise the vehicle.
9. Remove the main muffler.
10. Support the fuel tank with a jack and unscrew fuel tank band mounting nuts.

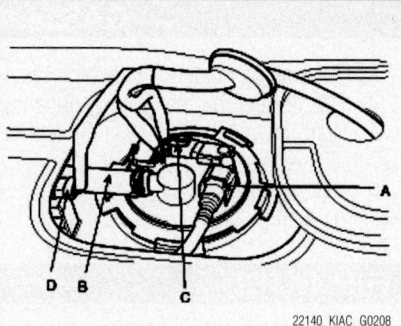

22140_KIAC_G0208

Fig. 237 Disconnect the fuel feed quick-connector (A), the vapor hose (B), the fuel tank pressure sensor connector (C) and the canister close valve connector (D).

11. Disconnect the fuel filler hose.
12. Disconnect the recirculation pipe quick-connector (A), the vapor hose (B) connecting the canister with the fuel tank air filter and the purge tube quick-connector

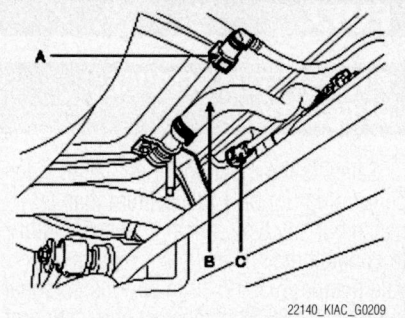

22140_KIAC_G0209

Fig. 238 Disconnect the recirculation pipe quick-connector (A), the vapor hose (B), and purge tube quick-connector (C).

(C) connecting the canister and the intake manifold.

13. Remove the fuel tank from the vehicle with coming down the jack slowly.
14. Install the Fuel Tank according to the reverse order to REMOVAL procedure.

HEATING & AIR CONDITIONING SYSTEM

BLOWER MOTOR

REMOVAL & INSTALLATION

1. Disconnect the negative battery terminal.
2. Disconnect the connector of the blower motor.
3. Remove the blower motor after loosening the mounting screws.
4. Installation is the reverse order of removal.

HEATER CORE

REMOVAL & INSTALLATION

Rio

See Figure 239.

1. Disconnect the negative battery terminal.
2. Recover the refrigerant with a recovery/ recycling/ charging station.
3. When the engine is cool, drain the engine coolant from the radiator.
4. Remove the bolts (A) and the expansion valve (B) from the evaporator core.
5. Plug or cap the lines immediately after disconnecting them to avoid moisture and dust contamination.
6. Disconnect the inlet and outlet heater hoses from the heater unit.
7. Remove the crash pad.
8. Remove the cowl cross bar assembly.
9. Disconnect the electrical connectors from the temperature control actuator, the

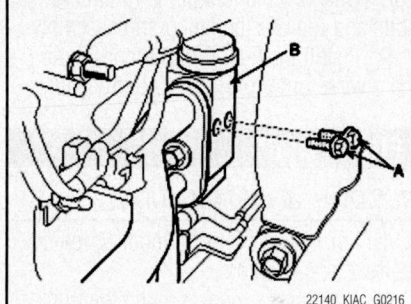

22140_KIAC_G0216

Fig. 239 Remove the bolts (A) and the expansion valve (B) from the evaporator core

mode control actuator and the evaporator temperature sensor.

10. Remove the heater & blower unit after loosening 2 mounting bolts.
11. Remove the blower unit from heater unit after loosening 3 screws.
12. Remove the heater core after removing the cover.

➡**Be careful that the inlet and outlet pipe are not bent during heater core removal, and pull out the heater core.**

To install:

13. Install the heater core in the reverse order of removal.
14. Installation is the reverse order of removal, and note these items :

   a. If you're installing a new evaporator, add refrigerant oil (ND-OIL8).

   b. Replace the O-rings with new ones at each fitting, and apply a thin coat of

refrigerant oil before installing them. Be sure to use the right O-rings for R-134a to avoid leakage.

   c. Immediately after using the oil, replace the cap on the container, and seal it to avoid moisture absorption.

   d. Do not spill the refrigerant oil on the vehicle ; it may damage the paint ; if the refrigerant oil contacts the paint, wash it off immediately.

   e. Apply sealant to the grommets.

   f. Make sure that there is no air leakage.

   g. Charge the system and test its performance.

   h. Do not interchange the inlet and outlet heater hoses and install the hose clamps securely.

   i. Refill the cooling system with engine coolant.

Spectra

1. Disconnect the negative battery terminal.
2. Recover the refrigerant with a recovery/recycling/charging station
3. When the engine is cool, drain the engine coolant from the radiator.
4. Disconnect the inlet and outlet heater hoses from the heater unit.
5. Remove the bolts and the expansion valve from the evaporator core.
6. Plug or cap the lines immediately after disconnecting them to avoid moisture and dust contamination.
7. Remove the crash pad.

8. Disconnect the connectors from the temp. actuator, the mode actuator and the thermistor, then remove the mounting nuts.

9. Remove the heater and evaporator unit after loosening the mounting screws.

10. Remove the self-tapping screws and the upper bracket, the side cover.

➡**Be careful not to bend the inlet and outlet pipes during heater core removal.**

11. Pull out the heater core.

To install:

12. Install the heater core in the reverse order of removal.

13. Install in the reverse order of removal, and note these items :

a. If you're installing a new evaporator, add refrigerant oil (ND-OIL8).

b. Replace the O-rings with new ones at each fitting, and apply a thin coat of refrigerant oil before installing them. Be sure to use the right O-rings for R-134a to avoid leakage.

c. Immediately after using the oil, replace the cap on the container, and seal it to avoid moisture absorption.

d. Do not spill the refrigerant oil on the vehicle ; it may damage the paint ; if the refrigerant oil contacts the paint, wash it off immediately

e. Apply sealant to the grommets.

f. Make sure that there is no air leakage.

g. Charge the system and test its performance.

h. Do not interchange the inlet and outlet heater hoses and install the hose clamps securely.

i. Refill the cooling system with engine coolant.

Optima

See Figures 240 through 242.

1. Disconnect the negative battery terminal.

2. Recover the refrigerant with a recovery/ recycling/ charging station.

3. When the engine is cool, drain the engine coolant from the radiator.

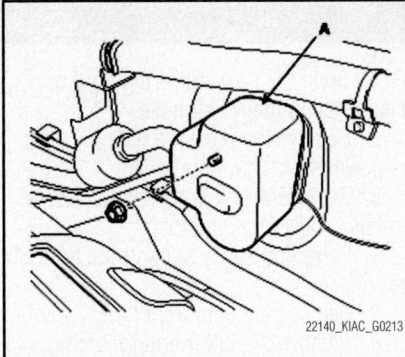

Fig. 240 Remove the expansion valve cover (A)

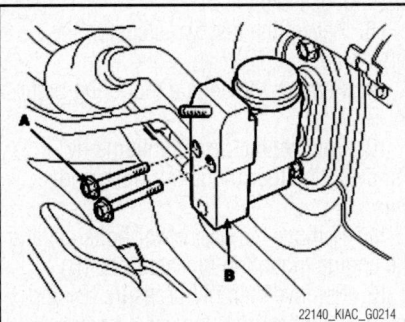

Fig. 241 Remove the bolts (A) and the expansion valve (B) from the evaporator core

4. Remove the expansion valve cover (A).

5. Remove the bolts (A) and the expansion valve (B) from the evaporator core.

6. Plug or cap the lines immediately after disconnecting them to avoid moisture and dust contamination.

7. Disconnect the inlet and outlet heater hoses from the heater unit.

8. Remove the crash pad.

9. Remove the cowl cross bar assembly.

10. Remove the heater & blower unit after loosening 3 mounting bolts.

11. Remove the blower unit from heater unit after loosening 3 screws.

12. Disconnect the heater core cover (A) and remove the heater core (B).

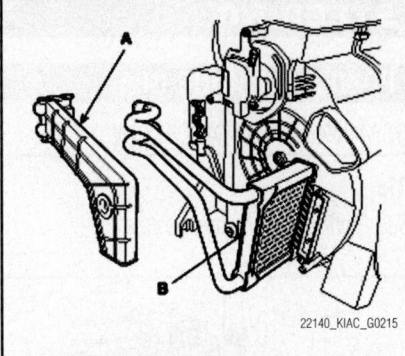

Fig. 242 Disconnect the heater core cover (A) and remove the heater core (B)

13. Disconnect the evaporator cover and remove the evaporator

14. Be careful that the inlet and outlet pipe are not bent during heater core removal, and pull out the heater core.

To install:

15. Install the heater core in the reverse order of removal.

16. Installation is the reverse order of removal, and note these items :

a. If you're installing a new evaporator, add refrigerant oil (ND-OIL8).

b. Replace the O-rings with new ones at each fitting, and apply a thin coat of refrigerant oil before installing them. Be sure to use the right O-rings for R-134a to avoid leakage.

c. Immediately after using the oil, replace the cap on the container, and seal it to avoid moisture absorption.

d. Do not spill the refrigerant oil on the vehicle ; it may damage the paint ; if the refrigerant oil contacts the paint, wash it off immediately.

e. Apply sealant to the grommets.

f. Make sure that there is no air leakage.

g. Charge the system and test its performance.

h. Do not interchange the inlet and outlet heater hoses and install the hose clamps securely.

i. Refill the cooling system with engine coolant.

STEERING

POWER STEERING GEAR

REMOVAL & INSTALLATION

Rio

See Figures 243 through 249.

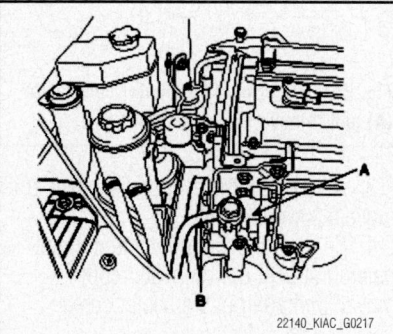

Fig. 243 Remove the pressure pipe (B) from the power steering oil pump (A)

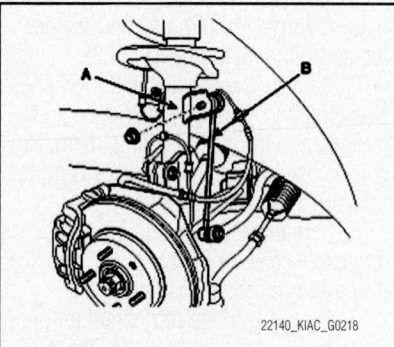

Fig. 244 Remove the stabilizer link (B) from the strut assembly (A)

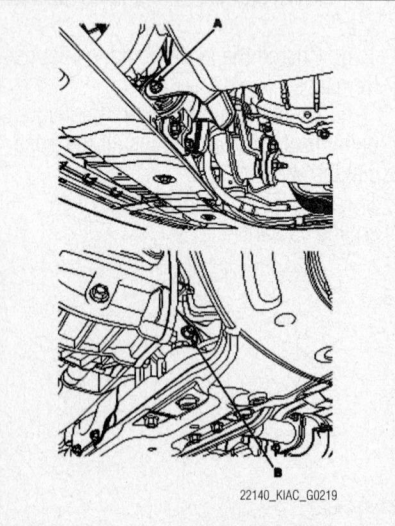

Fig. 245 Remove the engine mounting bolts (A, B)

1. Drain the power steering fluid by disconnecting the return hose.
2. Remove the pressure pipe (B) from the power steering oil pump (A).
3. Remove the steering shaft universal joint assembly mounting bolt.
4. Raise and safely support the front of the vehicle.
5. Remove the both front tires.
6. Remove the lower arm mounting bolts and the tie rod end from the knuckle.
7. Remove the stabilizer link (B) from the strut assembly (A).
8. Repeat the last two steps on the other side.
9. Remove the engine mounting bolts (A, B).
10. Remove the front subframe by removing the four mounting bolts and nuts.
11. Remove the heat cover bolts (A) and the engine mounting bracket bolts (B).
12. Remove the both pressure (B) and return (C) tubes from the valve body housing (A).
13. Remove both pressure and return tubes mounting bracket bolts (A, B).

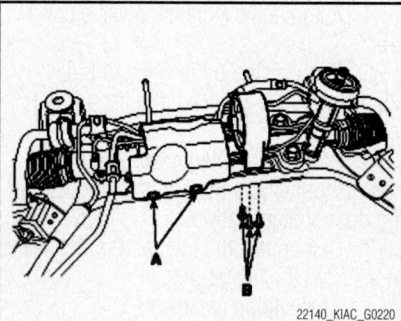

Fig. 246 Remove the heat cover bolts (A) and the engine mounting bracket bolts (B)

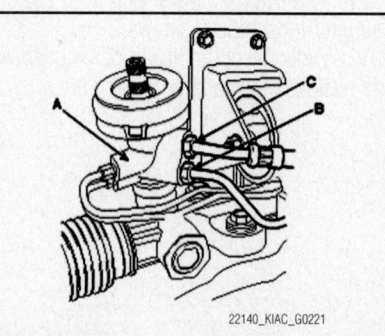

Fig. 247 Remove the both pressure (B) and return (C) tubes from the valve body housing (A)

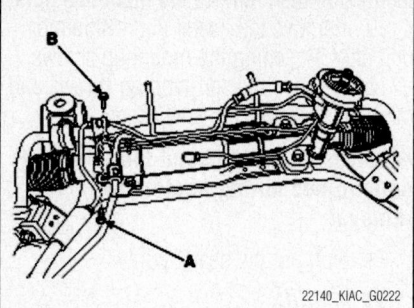

Fig. 248 Remove both pressure and return tubes mounting bracket bolts (A ,B)

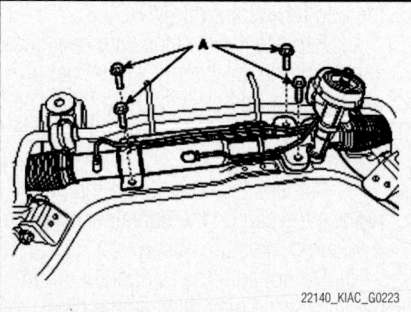

Fig. 249 Remove the power steering gear box (B) mounting bolts (A) from the front subframe

14. Remove the power steering gear box (B) mounting bolts (A) from the front subframe.

To install:

15. Installation is reverse of removal.
16. Use the following torque specifications:
- Engine mounting bracket to subframe: 36–47 ft. lbs. (50–65 Nm).
- Heat cover mounting: 5.8–8.6 ft. lbs. (8–12 Nm).
- Gear box mounting: 65–79 ft. lbs. (90–110 Nm).
- Subframe mounting: 69–86 ft. lbs. (95–120 Nm).
- Engine mounting: 36–47 ft. lbs. (50–65 Nm).
- Pressure pipe to oil pump: 40–47 ft. lbs. (55–65 Nm).

17. Bleed the air in the power steering system.
18. Check wheel alignment

Spectra

See Figures 250 and 251.

1. Drain the power steering fluid.
2. Remove the air intake hose assembly.

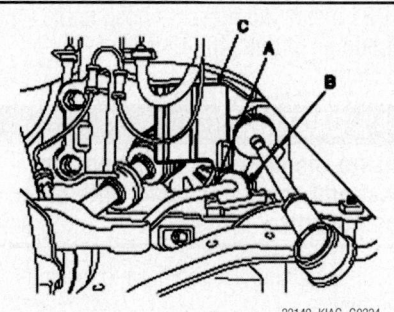

Fig. 250 Remove the dust cover (C) of the stabilizer bar (A) (LH side) mounting bracket (B)

3. Disconnect the pressure tube and the return tube fittings from the gear box.

4. Remove the steering shaft universal joint assembly mounting bolt.

5. Raise and safely support the front of the vehicle.

6. Remove the both front wheels.

7. Disconnect the tie rod from the knuckle by using the special tool (09568-34000).

8. Remove the dust cover (C) of the stabilizer bar (A) (LH side) mounting bracket (B).

9. Remove the mounting bolt and mounting clamp (A) of power steering gear box, and the clamp (B) holding the pressure tube and the return tube.

10. Pull the power steering gear box assembly (A) toward the left side of the vehicle.

To install:

11. Installation is the reverse of removal.

12. Push in the power steering gear box assembly on the left side of the vehicle.

13. Bleed the air in the power steering system.

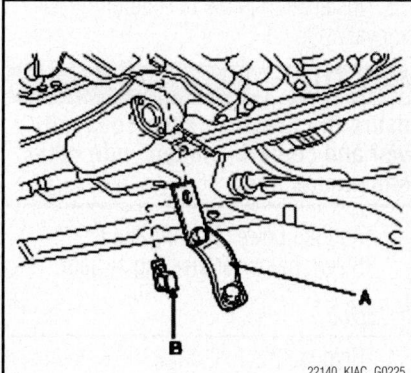

Fig. 251 Remove the mounting bolt and mounting clamp (A) of power steering gear box, and the clamp (B) holding the pressure tube and the return tube

Optima

See Figures 252 through 261.

1. Raise and safely support the front of the vehicle.

2. Remove both front wheels.

3. Drain the power steering fluid by disconnecting the return hose.

4. Disconnect the pressure tube (A) from the power steering pump.

5. Remove the pressure tube bracket bolt (A).

6. Disconnect the stabilizer bar link (A) from the front strut assembly.

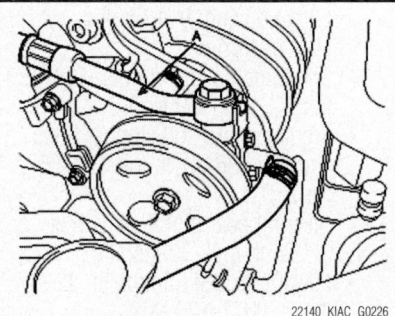

Fig. 252 Disconnect the pressure tube (A) from the power steering pump—2.4L engine

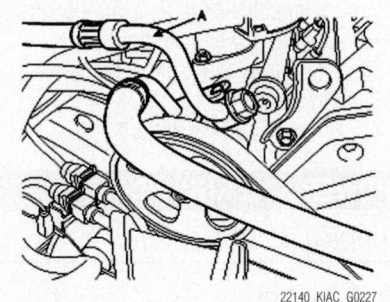

Fig. 253 Disconnect the pressure tube (A) from the power steering pump—2.7L engine

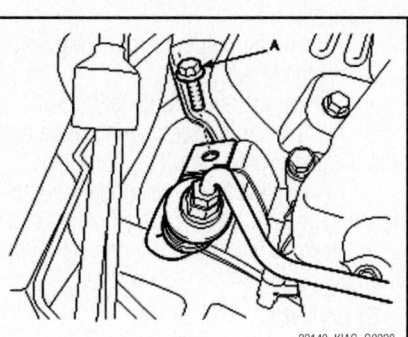

Fig. 254 Remove the pressure tube bracket bolt (A)—2.4L engine

7. Disconnect the tie rod end from the knuckle by using a SST (09568-4A000).

8. Remove the lower arm ball joint bolts (A) on both sides.

9. Remove the bolt connecting steering gear pinion shaft to universal joint.

10. Remove the front and rear roll stopper bolts and nuts (A, B).

11. Remove the sub-frame.

12. Remove the heat protector.

13. Disconnect the pressure tube (A) and return (B) from the valve body housing.

14. Remove the tube bracket bolt.

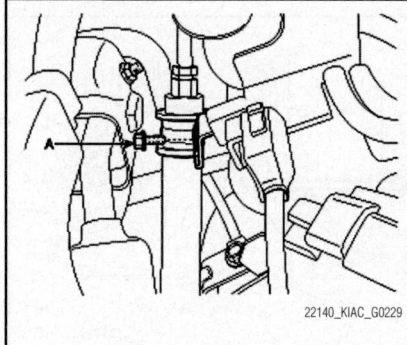

Fig. 255 Remove the pressure tube bracket bolt (A)—2.7L engine

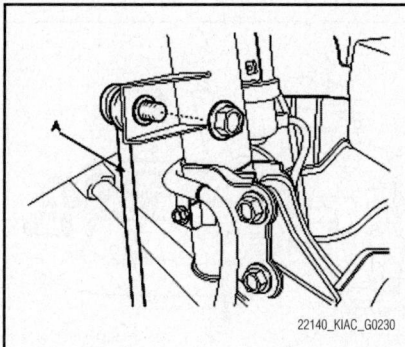

Fig. 256 Disconnect the stabilizer bar link (A) from the front strut assembly

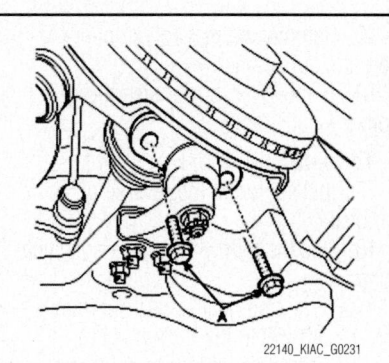

Fig. 257 Remove the lower arm ball joint bolts (A)

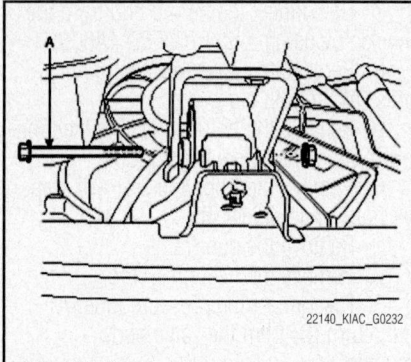

Fig. 258 Remove the front roll stopper bolt and nut (A)

Fig. 259 Remove the rear roll stopper bolt and nut (B)

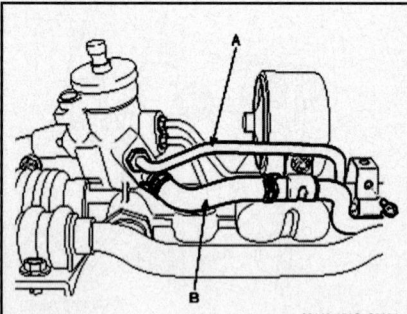

Fig. 260 Disconnect the pressure tube (A) and return (B) from the valve body housing

15. Remove the rear roll stopper (A) from the sub-frame.

16. Remove the power steering gear box from the sub-frame.

To install:

17. Installation is the reverse of removal.

18. Use the following torque specifications:
- Steering gear box to subframe: 43–58 ft. lbs. (60–80 Nm)
- Rear roll stopper to subframe: 40–47 ft. lbs. (55–65 Nm)
- Pressure and return tubes to valve

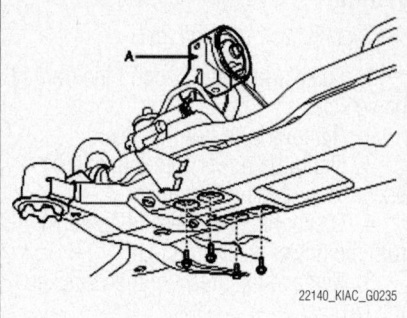

Fig. 261 Remove the rear roll stopper (A) from the sub-frame

body housing: 9–13 ft. lbs. (12–18 Nm)
- Subframe bolts and nuts: 116–130 ft. lbs. (160–180 Nm)
- Front and rear roll stopper bolts and nuts: 36–47 ft. lbs. (50–65 Nm)
- Steering gear pinion shaft bolt: 22–25 ft. lbs. (30–35 Nm)
- Lower arm ball joint bolts: 72–87 ft. lbs. (100–120 Nm)
- Tie rod end castle nut: 17–25 ft. lbs. (24–34 Nm)
- Stabilizer link to front strut assembly: 72–87 ft. lbs. (100–120 Nm)
- Pressure tube to power steering pump: 40–47 ft. lbs. (55–65 Nm)

19. Add the power steering oil.

20. After installation, bleed the power steering system.

21. Adjust the wheel alignment.

POWER STEERING PUMP

REMOVAL & INSTALLATION

Rio & Spectra

1. Remove the pressure hose from the oil pump.

2. Disconnect the suction hose from the suction pipe and drain the fluid into a container.

3. Loosen the oil pump mounting bolts to remove the V belt.

4. Loosen the tension adjusting bolt.

5. Remove the power steering drive belt from the power steering oil pump pulley.

6. Remove the power steering oil pump mounting bolt and the tension adjusting bolt.

7. Remove the power steering oil pump assembly.

To install:

8. After installing the oil pump to the oil pump bracket, install the V belt and tighten the tension adjusting bolt. Torque: 18–24 ft. lbs. (25–33 Nm).

9. Install the pressure hose to the oil pump. Torque: 40–47 ft. lbs. (55–65 Nm).

✸✸ CAUTION

Ensure the pressure hose does not twist and come in contact with other components.

10. Install the suction hose to the suction pipe.

11. Add power steering fluid (PSF-III).

12. Bleed the air in the system.

Optima

See Figure 262.

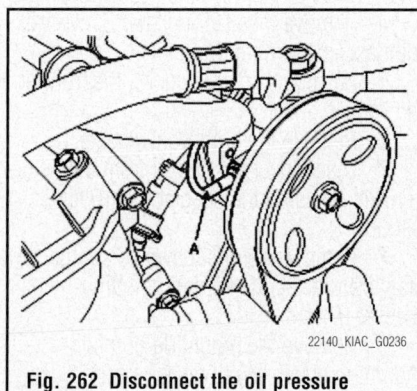

Fig. 262 Disconnect the oil pressure switch (A)

1. Disconnect the oil pressure switch (A).

2. Disconnect the pressure tube and return hose from the power steering pump assembly.

3. Remove the drive belt.

4. Remove the power steering pump assembly from the pump bracket.

To install:

5. Installation is the reverse of removal.

✸✸ CAUTION

Ensure the pressure hose does not twist and come in contact with other components.

6. Add the power steering fluid.

7. Bleed the power steering system.

BLEEDING

1. Remove the fuel pump fuse, then start the engine and wait for the engine to stall. Next, while operating the starting motor intermittently (for 15–20 seconds), turn the steering wheel all the way to the left and then to the right five or six times.

➡During air bleeding, replenish the fluid supply so that the level never falls below the lower position of the filter. If air bleeding is done while the vehicle is idling, the air will be broken up and absorbed into the fluid. Be sure to do the bleeding only while cranking.

2. Reinstall the fuel pump fuse, and start the engine (idling).

3. Turn the steering wheel to the left and the right until there are no air bubbles in the oil reservoir.

✳✳ CAUTION

Do not hold the steering wheel turned all the way to either side for more than ten seconds.

4. Confirm that the fluid is not milky, and that the level is up to the position specified on the level gauge.

5. Confirm that there is little change in the surface of the fluid when the steering wheel is turned left and right.

✳✳ CAUTION

If the surface of the fluid changes considerably, air bleeding should be done again. If the fluid level rises suddenly when the engine is stopped, it indicates that there is still air in the system. If there is air in the system, a jingling noise may be heard from the pump and the control valve may also produce unusual noises. Air in the system will shorten the life of the pump and other parts.

SUSPENSION   FRONT SUSPENSION

LOWER BALL JOINT

REMOVAL & INSTALLATION

Rio

See Figure 263.

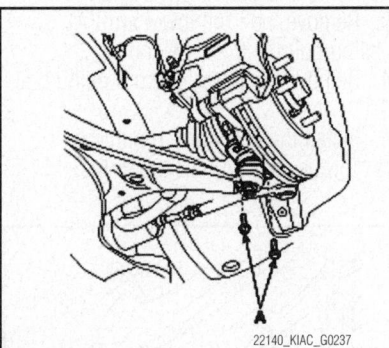

Fig. 263 Remove the lower arm ball joint mounting bolts (A)

1. Raise the front of the vehicle, and make sure it is securely supported.

2. Remove the front wheel and tire.

3. Remove the lower arm ball joint mounting bolts (A).

To install:

4. Install the lower arm ball joint mounting bolts. Torque: 72–86 ft. Lbs. (100–120 Nm).

5. Install the front wheel and tire.

Spectra

See Figures 264 and 265.

1. Remove the front wheel and tire.

2. Remove the split pin, the castle nut and the washer.

3. Loosen the lower arm ball joint nut (A), but do not remove it.

4. Remove the front strut lower mounting bolts (A) from the strut assembly (B).

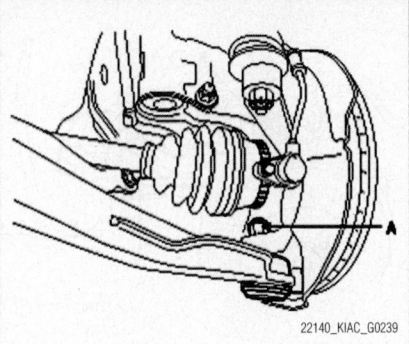

Fig. 264 Loosen the lower arm ball joint nut (A)

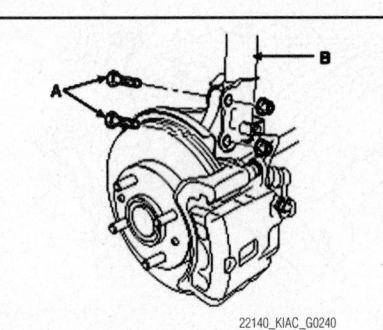

Fig. 265 Remove the front strut lower mounting bolts (A) from the strut assembly (B)

5. Push the axle hub outward to install the Special tool (09568-34000) easily.

6. Using the Special Tool (09568-34000), disconnect the lower arm ball joint from the lower arm.

7. Installation is the reverse of the removal procedure.

LOWER CONTROL ARM

REMOVAL & INSTALLATION

Rio

See Figures 266 and 267.

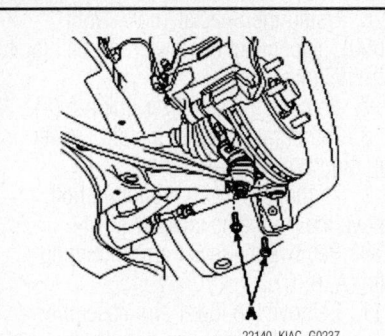

Fig. 266 Remove the lower arm ball joint mounting bolts (A)

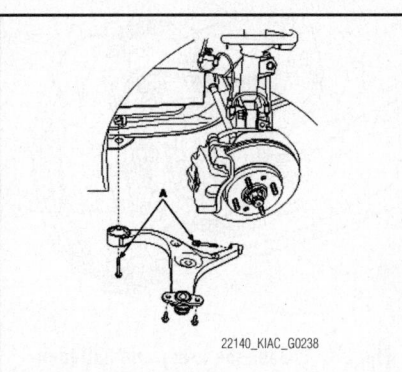

Fig. 267 Remove the lower arm mounting bolts (A)

1. Raise the front of the vehicle, and make sure it is securely supported.

2. Remove the front wheel and tire.

3. Remove the lower arm ball joint mounting bolts (A).

4. Remove the lower arm mounting bolts (A).

5. Remove lower control arm.

To install:

6. Install the lower arm mounting bolts. Torque: A bushing: 72–86 ft. Lbs. (100–120 Nm); G bushing: 72–101 ft. lbs. (100–140 Nm).

7. Install the lower arm ball joint mounting bolts. Torque: 72–86 ft. Lbs. (100–120 Nm).

8. Install the front wheel and tire.

Spectra

See Figures 268 through 274.

1. Remove the front wheel and tire.

2. Remove the split pin, the castle nut and the washer.

3. Loosen the lower arm ball joint nut (A), but do not remove it.

4. Remove the front strut lower mounting bolts (A) from the strut assembly (B).

5. Push the axle hub outward to install the Special tool (09568-34000) easily.

6. Using the Special Tool (09568-34000), disconnect the lower arm ball joint from the lower arm.

7. Remove the stabilizer link nut (A).

8. Temporarily install the strut lower mounting bolt.

9. To the lower arm mounting bolt, remove the right side cover (A).

10. Remove the lower arm mounting bolts (A, B, C).

11. Remove the lower arm assembly after completely removing the nut of lower arm ball joint.

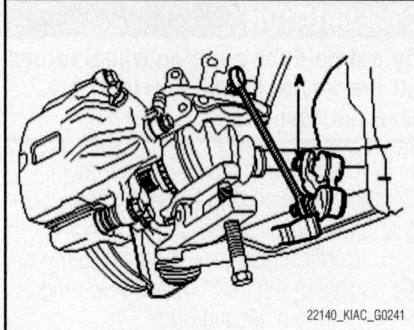

Fig. 270 Remove the stabilizer link nut (A)

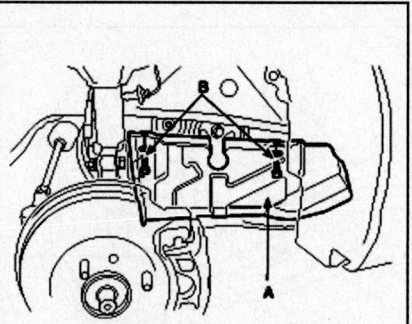

Fig. 271 Remove the right side cover (A)

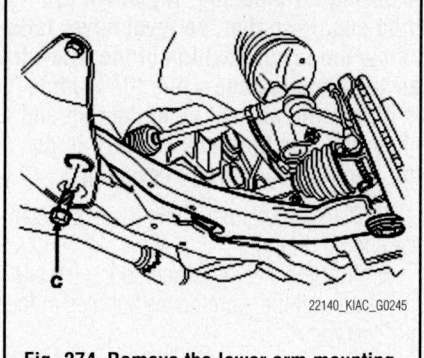

Fig. 274 Remove the lower arm mounting bolt (C)

12. Installation is the reverse of the removal procedure.

Optima

See Figures 275 through 277.

1. Raise the vehicle, and make sure it is securely supported.

2. Remove the front wheel and tire.

3. Remove the front lower arm (A) mounting bolt (B) from the knuckle.

4. Remove the lower arm mounting bolts (A, B).

5. Install the lower arm mounting bolts. Torque: Bolt (A): 72–87 ft. lbs.

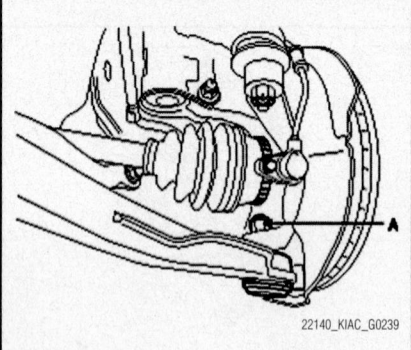

Fig. 268 Loosen the lower arm ball joint nut (A)

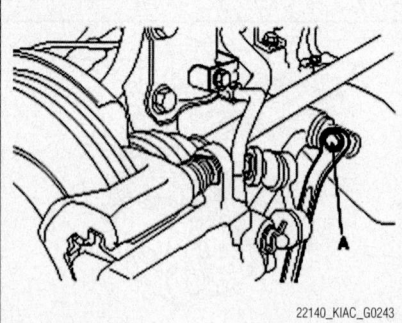

Fig. 272 Remove the lower arm mounting bolt (A)

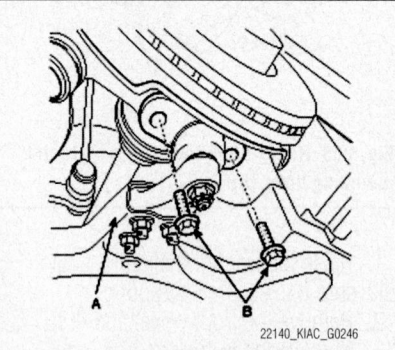

Fig. 275 Remove the front lower arm (A) mounting bolt (B) from the knuckle

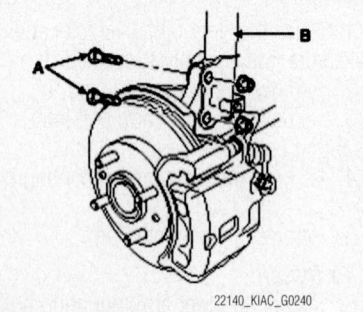

Fig. 269 Remove the front strut lower mounting bolts (A) from the strut assembly (B)

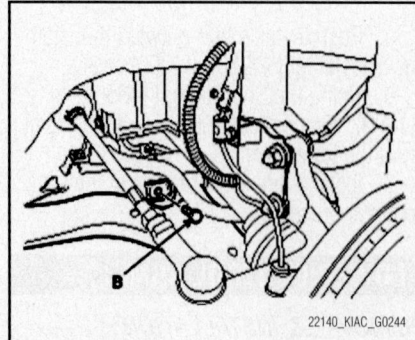

Fig. 273 Remove the lower arm mounting bolt (B)

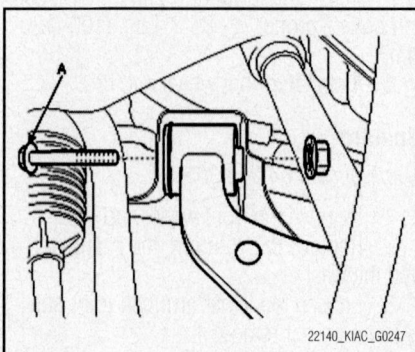

Fig. 276 Remove the lower arm mounting bolts (A)

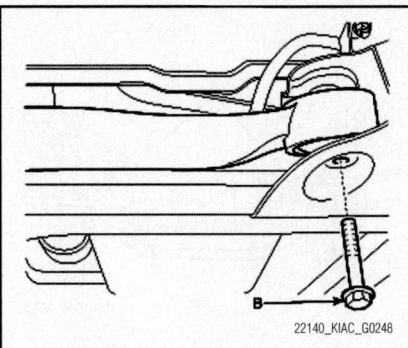

Fig. 277 Remove the lower arm mounting bolts (B)

(98–118 Nm); Bolt (B): 101–116 ft. lbs. (137–157 Nm).

6. Install the front lower arm mounting bolt to the knuckle. Torque: 72–87 ft. lbs. (98–118 Nm).

7. Install the wheel and the tire.

MACPHERSON STRUT

REMOVAL & INSTALLATION

Rio

See Figures 278 through 280.

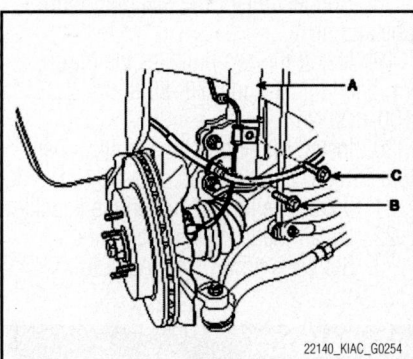

Fig. 278 Remove the brake hose bracket (B) and speed sensor wire mounting bolt (C) from the strut assembly (A)

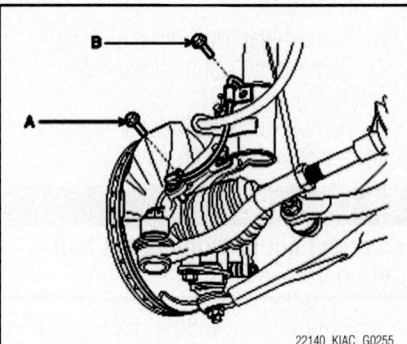

Fig. 279 Remove the speed sensor wire mounting bolt (B) and speed sensor (A)

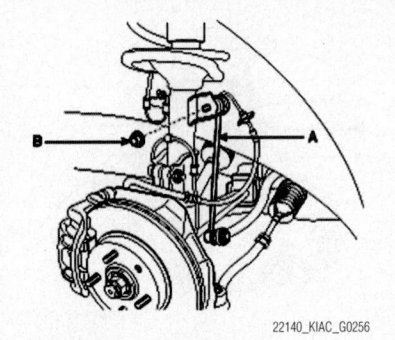

Fig. 280 Remove the nut (B) from the stabilizer bar link (A)

1. Raise the vehicle, and make sure it is securely supported.

2. Remove the front wheel and tire.

3. Remove the brake hose bracket (B) and speed sensor wire mounting bolt (C) from the strut assembly (A).

4. Remove the speed sensor wire mounting bolt (B) and speed sensor (A).

5. Remove the nut (B) from the stabilizer bar link (A).

6. Remove the strut upper mounting nuts.

7. Remove the strut lower mounting bolts and then remove the strut assembly.

To install:

8. Install the strut assembly and lower mounting bolts. Torque: 72–86 ft. lbs. (100–120 Nm)

9. Install the strut upper mounting nuts. Torque: 14–21 ft. lbs. (20–30 Nm).

10. Install the nut on the stabilizer bar link. Torque: 25–32 ft. lbs. (35–45 Nm).

11. Install the speed sensor wire mounting bolt and speed sensor.

12. Install the brake hose bracket and speed sensor wire mounting bolt on the strut assembly.

13. Install the wheel and the tire.

Spectra

See Figures 281 and 282.

1. Raise the vehicle, and make sure it is securely supported.

2. Remove the front wheel and tire.

3. Detach the brake hose bracket (B) from the front strut assembly (A).

4. In case of the vehicles equipped with Anti-lock Brake system, remove the wheel speed sensor (A) from the knuckle (B).

5. Remove the strut upper mounting nuts.

6. Remove the strut lower mounting bolts and then remove the strut assembly.

7. Installation is the reverse of the removal procedure.

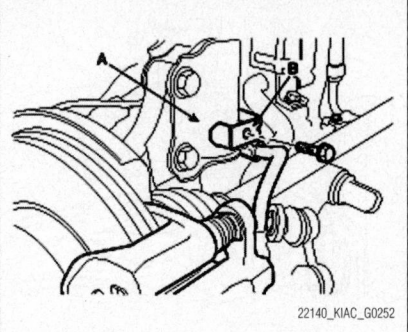

Fig. 281 Detach the brake hose bracket (B) from the front strut assembly (A)

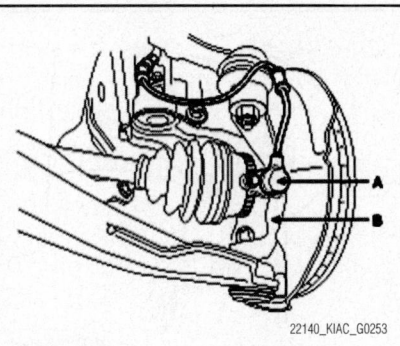

Fig. 282 Remove the wheel speed sensor (A) from the knuckle (B)

Optima

See Figures 283 through 285.

1. Raise the vehicle, and make sure it is securely supported.

2. Remove the front wheel and tire.

3. Remove the brake hose bracket bolt (A, B) from the front strut assembly.

4. Remove the speed sensor (A) and wire (B) bolts from the front knuckle.

5. Remove the front stabilizer link (A) nut (B) from the strut assembly.

6. Remove the strut upper mounting nuts.

7. Remove the front strut assembly from the front knuckle.

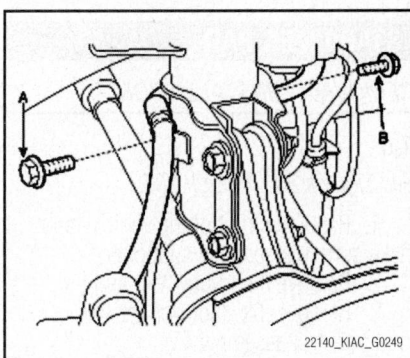

Fig. 283 Remove the brake hose bracket bolt (A, B) from the front strut assembly

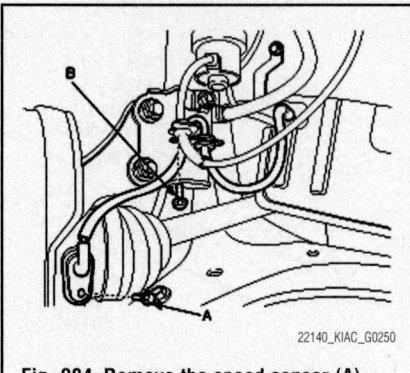

Fig. 284 Remove the speed sensor (A) and wire (B) bolts from the front knuckle

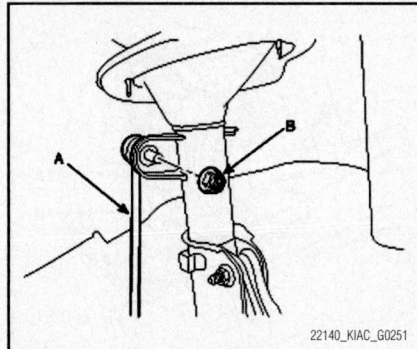

Fig. 285 Remove the front stabilizer link (A) nut (B) from the strut assembly

To install:

8. Install the strut upper mounting nuts. Torque: 33–43 ft. lbs. (44–59 Nm).

9. Install the front strut assembly bolts to the front knuckle. Torque: 101–116 ft. lbs. (137–157 Nm).

10. Install the front stabilizer link nut to the strut assembly. Torque: 72–87 ft. lbs. (98–118 Nm).

11. Install the speed sensor and wire bolts. Torque: 5.1–8.0 ft. lbs. (6.9–10.8 Nm).

12. Install the brake hose bracket bolt to the axle assembly. Torque: 5.1–8.0 ft. lbs. (6.9–10.8 Nm).

13. Install the wheel and the tire.

STABILIZER BAR

REMOVAL & INSTALLATION

Rio

See Figures 286 through 288.

1. Raise the front of the vehicle, and make sure it is securely supported.

2. Remove the front wheel and tire.

3. Remove the stabilizer bar link (B) from the strut assembly (A).

4. Remove the tie rod end from the knuckle by using the special tool (09568-4A000).

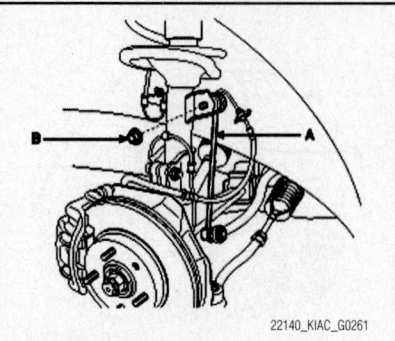

Fig. 286 Remove the stabilizer bar link (B) from the strut assembly (A)

5. Remove the two bolts for lower arm ball joint.

6. Drain power steering oil.

7. Remove the pressure pipe mounting bolt.

8. Disconnect between the return hose and tube.

9. Remove two engine mounting bolts (A,B) and six subframe mounting bolts in order to remove the subframe.

10. Remove both stabilizer brackets (A) and two bushings.

11. Remove the stabilizer bar.

To install:

12. Install the bushing on the stabilizer bar.

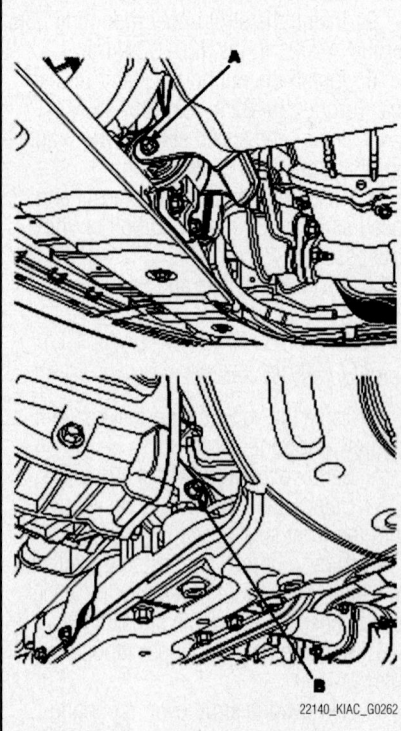

Fig. 287 Remove two engine mounting bolts (A ,B)

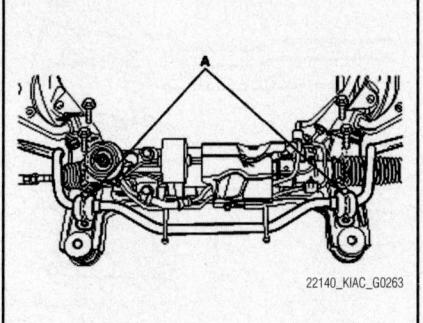

Fig. 288 Remove both stabilizer brackets (A) and two bushings

➡ **Bring clamp of stabilizer bar into contact with bushing.**

13. Install the bracket on the bushing.

14. After tightening the bolts of the bushing bracket temporarily, install the bushing bracket on the opposite side.

15. Install the six subframe mounting bolts. Torque: 68–86 ft. lbs. (95–120 Nm).

16. Install the two engine mounting bolts. Torque: 36–47 ft. lbs. (50–65 Nm).

17. Install the power steering pressure pipe mounting bolt.

18. Connect the power steering return tube and hose.

19. Install the two bolts for the lower arm ball joint. Torque: 72–86 ft. lbs. (100–120 Nm).

20. Install the nut on the stabilizer bar link. Torque: 25–32 ft. lbs. (35–45 Nm).

21. Install the tie rod end on the knuckle.

22. Install the front wheel and tire.

23. Refill the power steering fluid (PSF-3).

✳✳ CAUTION

After installation, bleed the air in the power steering system.

Spectra

See Figure 289.

1. Remove the front wheel and tire.

2. Remove the stabilizer bar link assembly (A).

3. Remove the stabilizer bar link on the opposite side in the same way.

✳✳ CAUTION

Be careful not to damage the ball joint boot.

4. Remove the stabilizer bracket and bushing.

5. Remove the stabilizer bar link on the opposite side in the same way.

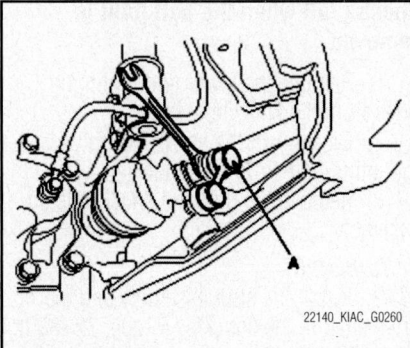

Fig. 289 Remove the stabilizer bar link assembly (A)

6. Remove the stabilizer out of the vehicle's right side.

To install:

7. Install the bushing on the stabilizer bar.

➡ **The distance between the bushing, and the part to which white paint is applied, must continue 10mm outside the vehicle.**

8. Install the bracket on the bushing

9. Align and install the bushing with the white paint on the stabilizer bar.

10. After tightening the bolts of the bushing bracket temporarily, install the bushing bracket on the opposite side.

Optima

See Figures 290 through 293.

1. Remove the connecting bolt (A) between the steering universal joint assembly (B) and the pinion assembly.

2. Raise the vehicle, and make sure it is securely supported.

3. Remove the front wheel and tire.

4. Remove the front stabilizer link from the strut assembly.

5. Remove the brake caliper.

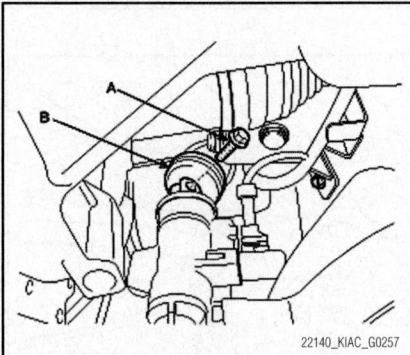

Fig. 290 Remove the connecting bolt (A) between the steering universal joint assembly (B) and the pinion assembly

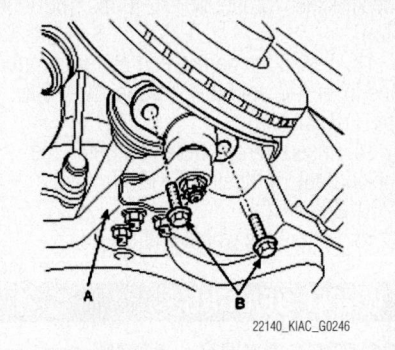

Fig. 291 Remove the lower arm (A) mounting bolt (B)

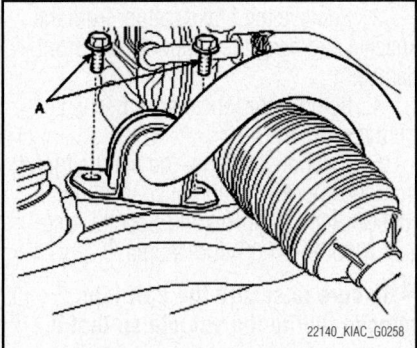

Fig. 292 Remove the stabilizer bar assembly mounting bolts (A)

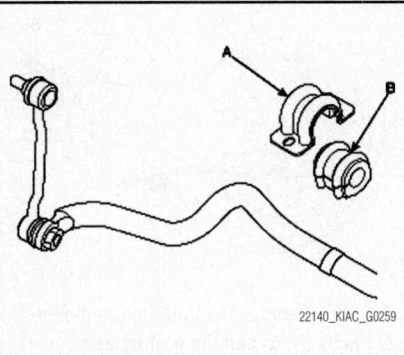

Fig. 293 Remove the brackets (A) and the bushings (B)

6. Remove the ball joint by using the special tool (09568-4A000).

7. Remove the lower arm (A) mounting bolts (B).

8. Remove the engine mounting bolts.

9. Remove the front muffler rubber hanger from the sub-frame.

10. Support the subframe with a jack, and remove the bolts and nuts.

11. After lowering the jack which supports the sub frame, remove both sides of the stabilizer bar assembly mounting bolts (A).

12. Remove the stabilizer bar assembly

through the gap between the body and the rear side of the sub frame.

> ❊❊ **CAUTION**

Be careful not to damage to the power steering related tubes.

13. Remove the brackets (A) and the bushings (B).

To install:

14. Install the bushing on the stabilizer bar.

➡ **Bring clamp of stabilizer bar into contact with bushing.**

15. Install the bracket on the bushing.

16. After tightening the bolts of the bushing bracket temporarily, install the bushing bracket on the opposite side.

17. Install the stabilizer bar bracket mounting bolts to the subframe. Torque: 32.5–39.8 ft. lbs. (44.1–53.9 Nm).

18. After lifting the jack which supports the sub frame, install the four bolts of the sub frame and the eight bolts of the guide bracket. Torque: Bolt (A): 116–130 ft. lbs. (157–177 Nm); Bolt (B): 33–40 ft. lbs. (44–54 Nm).

19. Install the front muffler rubber hanger to the sub-frame.

20. Install the engine mounting bolts. Torque: 36–47 ft. lbs. (49–64 Nm).

21. Install both sides of the lower arm mounting bolts. Torque: 72–87 ft. lbs. (98–118 Nm).

22. Install both sides of the tie rod ends.

23. Install the brake caliper.

24. Install the front stabilizer link to the strut assembly. Torque: 72–87 ft. lbs. (98–118 Nm).

25. Install the wheel and tire.

26. Install the connecting bolt between the steering universal joint assembly and the pinion assembly. Torque: 22–25 ft. lbs. (29–34 Nm).

> ❊❊ **CAUTION**

After installation, if necessary, adjust the alignment of the steering wheel and front tires.

STEERING KNUCKLE

REMOVAL & INSTALLATION

1. Remove the front wheel and tire.

2. Remove the split pin, then remove the locknut and washer from the front hub.

3. Remove the brake caliper from the knuckle and hang the caliper on the front damper.

4. Remove the wheel speed sensor from the knuckle.

5. Disconnect the tie rod end ball joint from the knuckle using the special tool (09568-34000) for Rio and Spectra. Use SST (09568-4A000) for Optima.

➡ **Be sure to secure the ball joint, remove tool to the vehicle so that it doesn't fall when the ball joint is removed.**

6. Remove the lower arm mounting bolts from the knuckle.

7. Disconnect the strut assembly mounting bolts from the knuckle.

8. Remove the hub and knuckle as an assembly.

To install:

9. Install the strut assembly and the drive shaft in the knuckle. Torque: 72–86 ft. lbs. (100–120 Nm).

10. Install the lower arm assembly mounting bolts to the knuckle. Torque: 72–86 ft. lbs. (100–120 Nm).

11. Install the wheel speed sensor to the knuckle.

12. Install the brake caliper assembly to the knuckle. Torque: 58–72 ft. lbs. (80–100 Nm).

13. Install the tie rod end ball joint nut and insert the split pin. Torque: 12–25 ft. lbs. (16–34 Nm).

14. Insert the washer and tighten the locking nut. Torque: 145–188 ft. lbs. (200–260 ➡m).

15. Install the wheel and tire.

WHEEL BEARINGS

REMOVAL & INSTALLATION

1. Remove the front wheel and tire.

2. Remove the split pin, then remove the locknut and washer from the front hub.

3. Remove the brake caliper from the knuckle and hang the caliper on the front damper.

4. Remove the wheel speed sensor from the knuckle.

5. Disconnect the tie rod end ball joint from the knuckle using the special tool (09568-34000) for Rio and Spectra. Use SST (09568-4A000) for Optima.

➡ **Be sure to secure the ball joint, remove tool to the vehicle so that it**

doesn't fall when the ball joint is removed.

6. Remove the lower arm mounting bolts from the knuckle.

7. Disconnect the strut assembly mounting bolts from the knuckle.

8. Remove the hub and knuckle as an assembly.

To install:

9. Install the strut assembly and the drive shaft in the knuckle. Torque: 72–86 ft. lbs. (100–120 Nm).

10. Install the lower arm assembly mounting bolts to the knuckle. Torque: 72–86 ft. lbs. (100–120 Nm).

11. Install the wheel speed sensor to the knuckle.

12. Install the brake caliper assembly to the knuckle. Torque: 58–72 ft. lbs. (80–100 Nm).

13. Install the tie rod end ball joint nut and insert the split pin. Torque: 12–25 ft. lbs. (16–34 Nm).

14. Insert the washer and tighten the locking nut. Torque: 145–188 ft. lbs. (200–260 ➡m).

15. Install the wheel and tire.

SUSPENSION | REAR SUSPENSION

CONTROL ARMS/LINKS

REMOVAL & INSTALLATION

Spectra

Trailing Arm

See Figures 294 and 295.

1. After removing the bolt (C), detach the parking brake cable (B) which is fixed on the rear trailing arm bracket (A).

2. Remove the trailing arm mounting bolts (A, B) and the trailing arm bracket mounting bolts (C, D, E).

3. Remove the trailing arm.

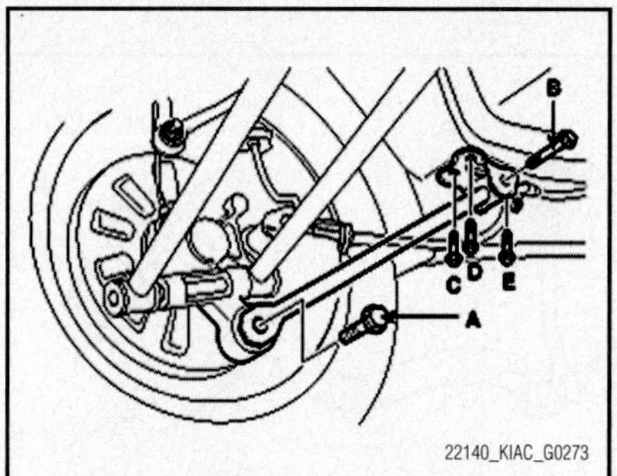

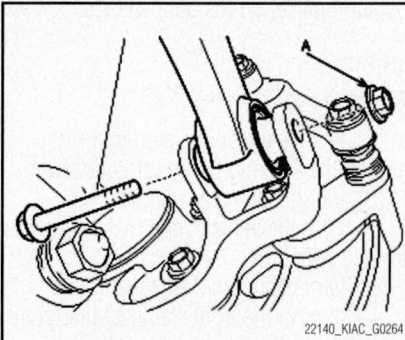

Fig. 295 Remove the trailing arm mounting bolts (A, B) and the trailing arm bracket mounting bolts (C, D, E)

4. Installation is the reverse of the removal procedures.

5. Fully tighten the trailing arm mounting bolts to the specified torque under the unloaded vehicle on the ground.

Optima

Rear Upper Arm

See Figures 296 and 297.

1. Raise the vehicle, and make sure it is securely supported.

2. Remove the rear wheel and tire.

3. Remove the rear upper arm bolt and nut (A) from the knuckle.

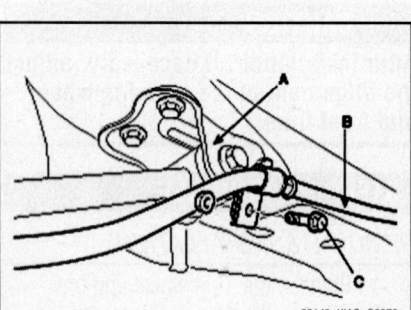

Fig. 294 Remove the bolt (C), detach the parking brake cable (B) on the rear trailing arm bracket (A)

Fig. 296 Remove the rear upper arm bolt and nut (A) from the knuckle

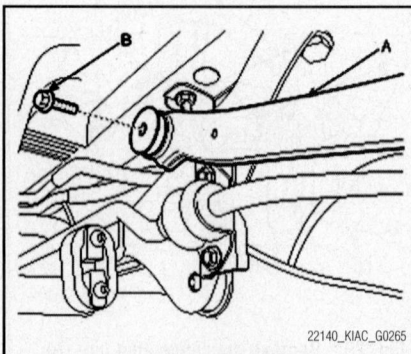

Fig. 297 Remove the rear upper arm (A) bolt (B) from the cross member

4. Remove the rear upper arm (A) bolt (B) from the cross member.

To install:

5. Install the rear upper arm mounting bolt to the cross member. Torque: 72–87 ft. lbs. (98–118 Nm).

6. Install the rear upper arm mounting nut to the knuckle. Torque: 72–87 ft. lbs. (98–118 Nm).

7. Install the wheel and the tire.

Rear Lower Arm

See Figures 298 through 300.

1. Raise the vehicle, and make sure it is securely supported.

2. Remove the rear wheel and tire.

3. Remove the lower arm bolt (B) from the rear knuckle, while supporting the lower arm (A) with a jack.

4. Loosen the lower arm bolt (C) from the cross member.

5. Remove the coil spring (A), the lower seat, and the upper pad.

6. Remove the lower arm (A) mounting bolts (B) from the cross member.

To install:

7. Pre-tighten the lower arm mounting bolts to the cross member.

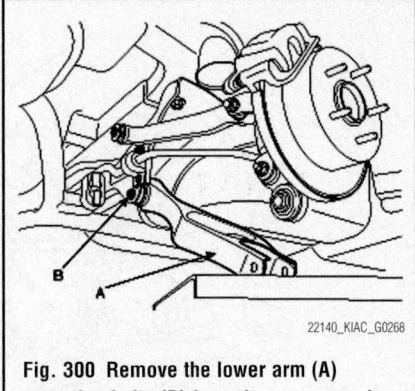

Fig. 300 Remove the lower arm (A) mounting bolts (B) from the cross member

8. Install the coil spring, the lower seat, and the upper pad.

9. Install the lower arm bolt to the rear knuckle and the lower arm bolt to the cross member, while supporting the lower arm with a jack. Torque: 101–116 ft. lbs. (137–157 Nm).

10. Install the wheel and the tire.

Rear Assist Arm

See Figures 301 through 303.

1. Raise the vehicle, and make sure it is securely supported.

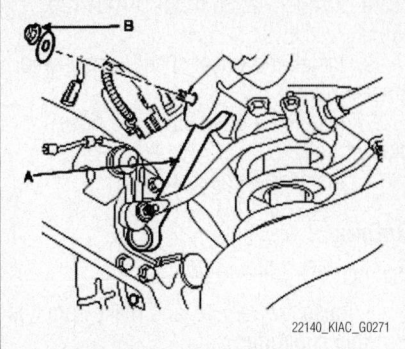

Fig. 303 Remove the rear assist arm (A) mounting nut (B) from the cross member

2. Remove the rear wheel and tire.

3. Remove the brake caliper mounting bolts, and hang the brake caliper assembly with wire.

4. Remove the rear assist arm (A) ball joint self-locking nut (B) and the cotter pin.

5. Remove the rear assist arm ball joint (A) by using the special tool (09568-4A000).

6. Remove the rear assist arm (A) mounting nut (B) from the cross member.

To install:

7. Install the rear assist arm mounting nut to the cross member. Torque: 58–72 ft. lbs. (78–98 Nm).

8. Install the rear assist arm ball joint self-locking nut and the cotter pin. Torque: 33–40 ft. lbs. (44–54 Nm).

9. Install the brake caliper assembly mounting bolts. Torque: 36–43 ft. lbs. (49–59 Nm).

10. Install the wheel and the tire.

COIL SPRING

REMOVAL & INSTALLATION

Rio

1. Remove the wheel and tire.

2. Remove the brake hose bracket.

3. Remove the wheel speed sensor wire bracket.

4. Use a jack at the bottom of the rear torsion axle beam to raise and support, then remove the rear shock absorber lower mounting bolt.

5. Remove the rear coil spring.

To install:

6. Install the upper and lower pads on the coil spring by aligning the grooves on the pads.

7. Place the coil spring with the pads on the torsion axle beam and support it with a jack.

8. Install the rear shock absorber mounting bolt by lifting the rear torsion axle

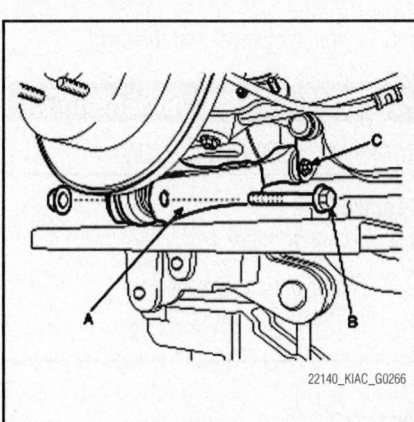

Fig. 298 Remove the lower arm bolt (B)

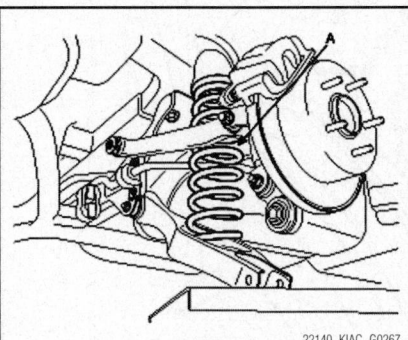

Fig. 299 Remove the coil spring (A), the lower seat, and the upper pad

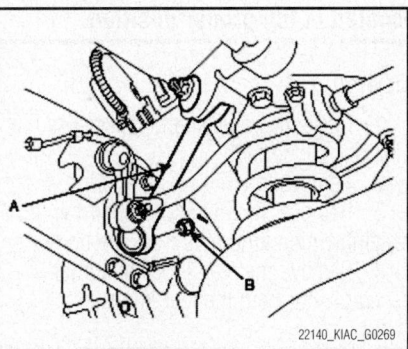

Fig. 301 Remove the rear assist arm (A) ball joint self-locking nut (B) and the cotter pin

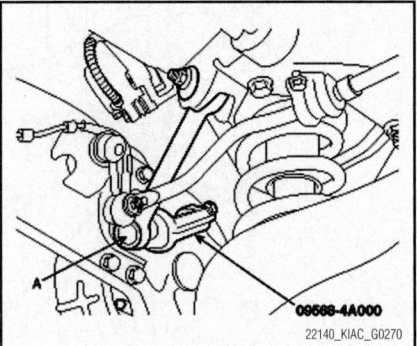

Fig. 302 Remove the rear assist arm ball joint (A) by using the special tool

beam. Torque: 72–86 ft. lbs. (100–120 Nm).

9. Install the wheel speed sensor wire bracket bolt.

10. Install the brake pressure hose bracket bolt.

11. Install the wheel and tire.

Optima

See Figures 304 and 305.

1. Raise the vehicle, and make sure it is securely supported.

2. Remove the rear wheel and tire.

3. Remove the lower arm bolt (B) from the rear knuckle, while supporting the lower arm (A) with a jack.

4. Loosen the lower arm bolt (C) from the cross member.

5. Remove the coil spring (A), the lower seat, and the upper pad.

To install:

6. Install the coil spring, the lower seat, and the upper pad.

7. Install the lower arm bolt to the rear knuckle and the lower arm bolt to the cross member, while supporting the lower arm with a jack. Torque: 101–115.7 ft. lbs. (137–157).

8. Install the wheel and the tire.

SHOCK ABSORBER

REMOVAL & INSTALLATION

Rio

1. Remove the wheel and tire.

2. After supporting the rear torsion axle beam with a jack, remove the rear shock absorber lower mounting bolt. Remove the rear shock absorber.

3. Remove the rear shock absorber mounting bolts.

To install:

4. Tighten the rear shock absorber upper mounting bolt. Torque: 28–43 ft. lbs. (40–60 Nm).

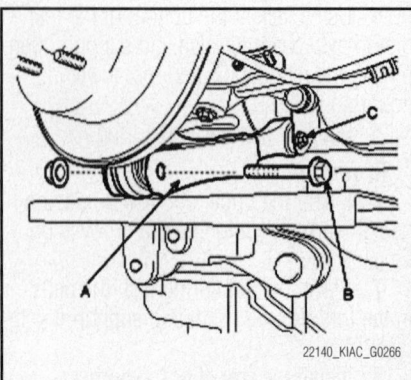

Fig. 304 Remove the lower arm bolt (B)

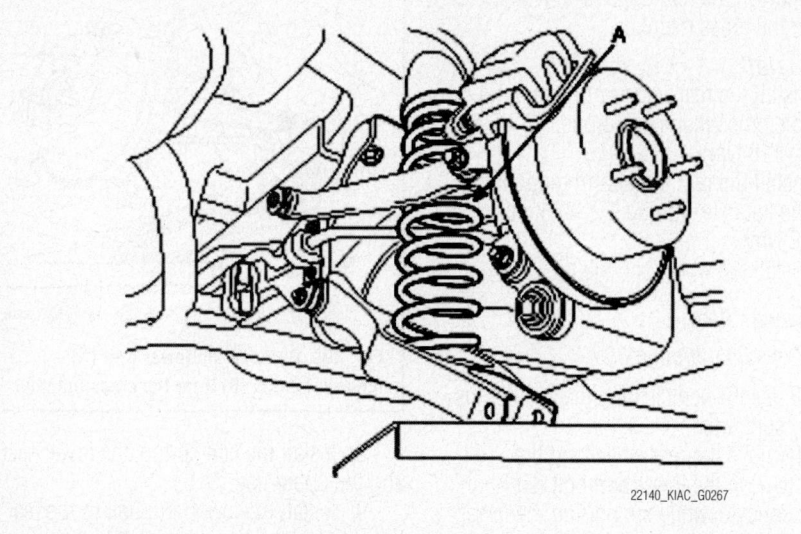

Fig. 305 Remove the coil spring (A), the lower seat, and the upper pad

5. Placing a jack at the bottom of the rear torsion axle beam and raise the vehicle to the proper location.

6. Tighten the rear shock absorber lower mounting bolts. Torque: 72–86 ft. lbs. (100–120 Nm).

✳ CAUTION

Check that the rear coil spring is located in the proper position.

Optima

1. Raise the vehicle, and make sure it is securely supported.

2. Remove the rear wheel and tire.

3. Remove the rear shock absorber assembly mounting nuts from the body.

4. Remove the rear shock absorber assembly nut from the rear knuckle.

5. Remove the shock absorber assembly.

To install:

6. Install the rear shock absorber mounting bolt to the body. Torque: 33–40 ft. lbs. (44–54 Nm).

7. Install the rear shock absorber nut to the knuckle. Torque: 101–116 ft. lbs. (137–157 Nm).

8. Install the wheel and the tire.

STRUT & SPRING ASSEMBLY

REMOVAL & INSTALLATION

Spectra

See Figures 306 and 307.

1. Remove the rear seat.
   a. Raise the rear cushion.

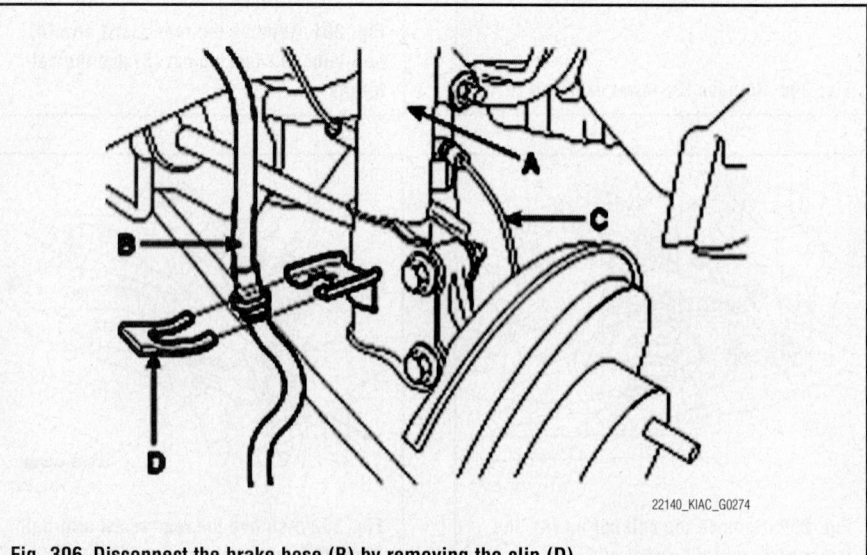

Fig. 306 Disconnect the brake hose (B) by removing the clip (D)

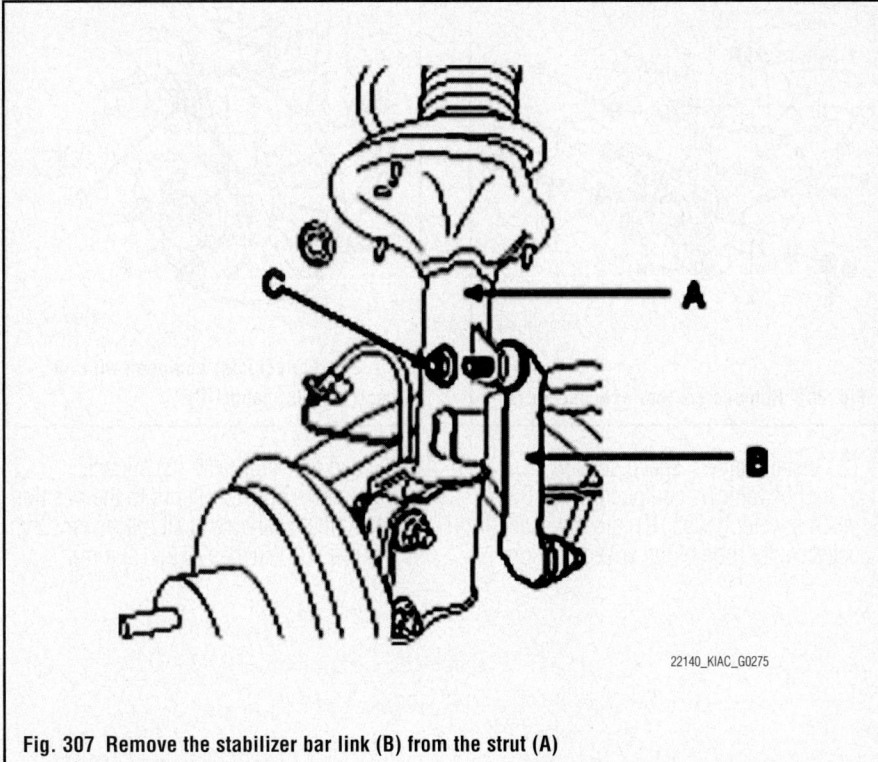

Fig. 307 Remove the stabilizer bar link (B) from the strut (A)

b. Remove the mounting bolts between rear cushion and rear seatback.

c. Remove the mounting bolts to both end parts of rear seatback.

2. Remove the rear strut upper mounting nuts.

3. Raise the vehicle, and make sure it is securely supported.

4. Remove the rear wheel and tire.

5. Disconnect the brake hose (B) and wheel speed sensor wiring (C) from the rear strut (A).

  a. Disconnect the brake hose (B) by removing the clip (D)

  b. Disconnect the wheel speed sensor wiring.

6. After unfastening stabilizer bar link nut (C), remove the stabilizer bar link (B) from the strut (A).

7. Remove the lower strut mounting bolts.

8. Remove the rear strut assembly.

9. Installation is the reverse of the removal procedures.

WHEEL BEARINGS

REMOVAL & INSTALLATION

Rio

See Figure 308.

1. Raise the rear of the vehicle and support it with safety stand.

2. Remove the rear wheel and tire.

3. Remove the wheel speed sensor wire bracket bolt.

4. Remove the parking brake wire bracket bolt.

5. Remove the brake caliper assembly bolt and the brake disk mounting screw.

6. Hang the brake caliper assembly tightly on a proper place with wire.

7. Remove the rear hub bearing assembly bolts (A).

8. Remove the rear hub bearing assembly.

9. The installation is reverse of the removal.

10. Hub bearing assembly: Torque: 36–43 ft. lbs. (50–60 Nm).

Spectra

See Figures 309 through 311.

1. Raise rear of the vehicle.

2. Remove the rear wheel and tire.

3. Remove the brake caliper assembly bolt and the brake disk mounting screw.

4. Hang the brake caliper assembly tightly on a proper place with wire.

5. Remove the brake disc rotor.

6. Remove the bolt (A) and the remove the rear wheel speed sensor (B).

7. Remove the wheel hub dust cap.

8. Remove the wheel bearing nut.

9. Remove the rear wheel hub assembly.

  a. Remove the dust cover.

  b. The rear hub assembly should not be disassembled.

  c. For vehicles equipped with ABS, care must be taken not to scratch or damage the teeth of the rotor.

10. Remove the rear axle carrier (A) by removing bolts, nuts and washers (B, C, D, E).

To install:

11. Tighten the wheel bearing nut.

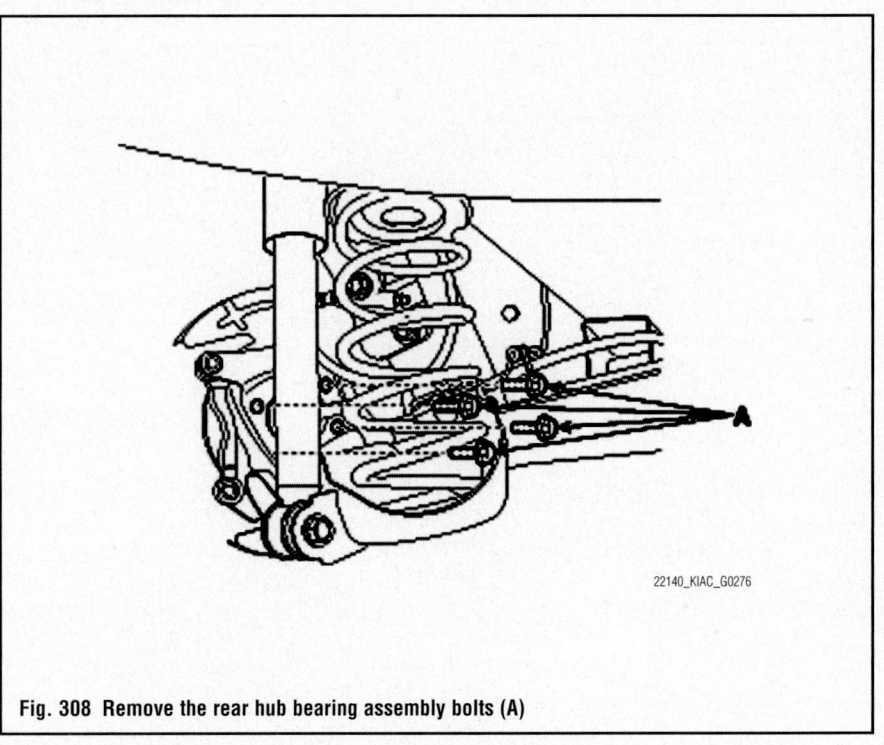

Fig. 308 Remove the rear hub bearing assembly bolts (A)

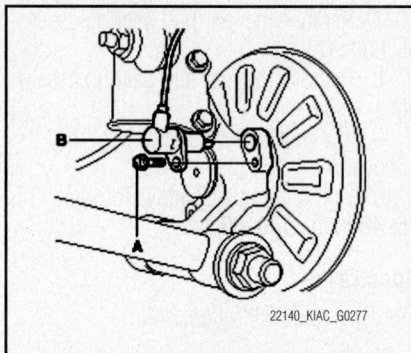

Fig. 309 Remove the bolt (A) and the remove the rear wheel speed sensor (B)

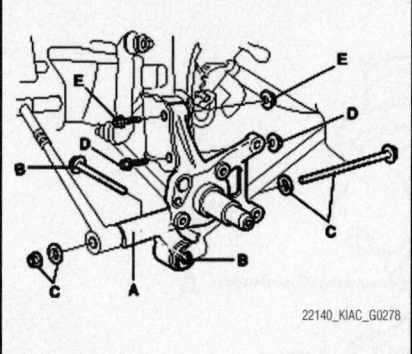

Fig. 310 Remove the rear axle carrier (A)

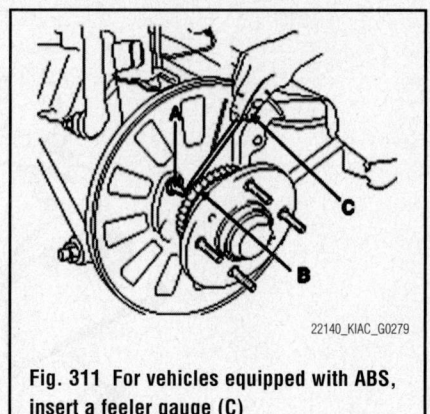

Fig. 311 For vehicles equipped with ABS, insert a feeler gauge (C)

✲✲ CAUTION

Replace the wheel bearing nut with new ones after removal.

12. Install the rear speed sensor.
   a. For vehicles equipped with ABS, insert a feeler gauge (C) into the space between the robe of the speed sensors

(A) and the rotor teeth (B) surface. Tighten the speed sensors to the position where the clearance at all places is within 0.008–0.051 inches (0.2–1.3 mm).

KIA

Rondo

SPECIFICATIONS AND MAINTENANCE CHARTS

ENGINE AND VEHICLE IDENTIFICATION

Engine							Model Year	
Code ①	Liters (cc)	Cu. In.	Cyl.	Fuel Sys.	Engine Type	Eng. Mfg.	Code ②	Year
5	2.4 (2359)	144	4	EGI	DOHC	KIA	7	2007
6	2.7 (2656)	163	6	EGI	DOHC	KIA	8	2008

EGI: Electronic Gasoline Injection

DOHC: Double Overhead Camshafts

22140_ROND_C0001

GENERAL ENGINE SPECIFICATIONS

Year	Model	Engine Displacement Liters	Engine VIN	Net Horsepower @ rpm	Net Torque @ rpm (ft. lbs.)	Bore x Stroke (in.)	Compression Ratio	Oil Pressure @ rpm
2007	Rondo	2.4	5	149@6000	159@4500	3.46x3.82	10.5:1	24@1000
		2.7	6	178@6000	181@4000	3.41x2.95	10.4:1	18@700
2008	Rondo	2.4	5	149@6000	159@4500	3.46x3.82	10.5:1	24@1000
		2.7	6	178@6000	181@4000	3.41x2.95	10.4:1	18@700

22140_ROND_C0002

ENGINE TUNE-UP SPECIFICATIONS

Year	Engine Displacement Liters	Engine VIN	Spark Plug Gap (in.)	Ignition Timing (deg.) MT	AT	Fuel Pump (psi)	Idle Speed (rpm) MT	AT	Valve Clearance Intake	Exhaust
2007	2.4	5	0.039-0.043	NA	①	46-49	NA	700-900	②	③
	2.7	6	0.039-0.043	NA	①	46-49	NA	600-800	②	③
2008	2.4	5	0.039-0.043	NA	①	46-49	NA	700-900	②	③
	2.7	6	0.039-0.043	NA	①	46-49	NA	600-800	②	③

NA: Not available

NOTE: The Vehicle Emission Control Information label often reflects specification changes made during production.

The label figures must be used if they differ from those in this chart.

① Ignition timing is controlled by electronic ignition timing system.

② 0.0039-0.0118 in

③ 0.0097-0.0157 in

22140_ROND_C0003

CAPACITIES

Year	Model	Engine Displacement Liters	Engine VIN	Engine Oil with Filter	Transaxle (pts.) Manual	Auto.	Fuel Tank (gal.)	Cooling System (qts.)
2007	Rondo	2.4	5	4.23	NA	16.4	16.4	7.4
		2.7	6	4.75	NA	20.1	17.2	9.0
2008	Rondo	2.4	5	4.23	NA	16.4	16.4	7.4
		2.7	6	4.75	NA	20.1	17.2	9.0

NA: Not available

22140_ROND_C0004

FLUID SPECIFICATIONS

Year	Model	Engine Displacement Liters (VIN)	Engine Oil	Auto. Trans.	Drive Axle	Power Steering Fluid	Brake Master Cylinder
2007	Rondo	2.4 (5)	①	Diamond ATF SP-III	NA	PSF-4	DOT 3
		2.7 (6)	①	Diamond ATF SP-III	NA	PSF-4	DOT 3
2008	Rondo	2.4 (5)	①	Diamond ATF SP-III	NA	PSF-4	DOT 3
		2.7 (6)	①	Diamond ATF SP-III	NA	PSF-4	DOT 3

① 10W-30 API Classification: SJ or SL

22140_ROND_C0014

VALVE SPECIFICATIONS

Year	Engine Displacement Liters	Engine VIN	Seat Angle (deg.)	Face Angle (deg.)	Maximum out of Square (in.)	Spring Free Length (in.)	Stem-to-Guide Clearance (in.) Intake	Exhaust	Stem Diameter (in.) Intake	Exhaust
2007	2.4	5	44.75-45.1	45.25-45.75	①	1.8677	0.00078-0.0019	0.00118-0.0021	0.2151-0.2157	0.2149-0.2153
	2.7	6	NA	45-45.5	①	1.8425	0.0008-0.0020	0.0014-0.0026	0.2348-0.2354	0.2343-0.2348
2008	2.4	5	44.75-45.1	45-45.5	①	1.804	0.0008-0.0019	0.0020-0.0033	0.2585-0.2591	0.2571-0.2579
	2.7	6	NA	45-45.5	①	1.8425	0.0008-0.0020	0.0014-0.0026	0.2348-0.2354	0.2343-0.2348

NA: Not Available

① Less than 1.5 degrees.

22140_ROND_C0005

CAMSHAFT AND BEARING SPECIFICATIONS

All measurements are given in inches.

Year	Engine Displacement Liters	Engine VIN	Journal Diameter	Brg. Oil Clearance	Shaft End-play	Runout	Journal Bore	Lobe Lift Intake	Lobe Lift Exhaust
2007	2.4	5	①	②	0.0039-0.0086	NA	NA	NA	NA
	2.7	6	1.1009-1.1016	0.0012-0.0022	0.0039-0.0079	NA	NA	NA	NA
2008	2.4	5	①	②	0.0039-0.0086	NA	NA	NA	NA
	2.7	6	1.1009-1.1016	0.0012-0.0022	0.0039-0.0079	NA	NA	NA	NA

NA: Information not available

① Intake No. 1: 1.1811 in.

Intake and Exhaust: No. 2, 3, 4, 5: 0.9449 in.

Exhaust No. 1: 1.5748 in.

② Intake No. 1: 0.00078-0.00224 in.

Intake No. 2, 3, 4, 5: 0.00177-0.00323

Exhaust: 0.00177-0.00323 in.

22140_ROND_C00013

CRANKSHAFT AND CONNECTING ROD SPECIFICATIONS

All measurements are given in inches.

Year	Engine Displacement Liters	Engine VIN	Crankshaft Main Brg. Journal Dia.	Crankshaft Main Brg. Oil Clearance	Crankshaft Shaft End-play	Crankshaft Thrust on No.	Connecting Rod Journal Diameter	Connecting Rod Oil Clearance	Connecting Rod Side Clearance
2007	2.4	5	2.0049-2.0456	0.0010-0.0019	0.0027-0.0098	3	2.0079-2.0086	0.0011-0.0018	0.0039-0.0098
	2.7	6	2.4402-2.4409	0.0002-0.0009	0.0028-0.0098	3	2.0079-2.0086	0.0007-0.0014	0.0039-0.0098
2008	2.4	5	2.0449-2.0456	0.0010-0.0019	0.0027-0.0098	3	2.0079-2.0086	0.0011-0.0018	0.0039-0.0098
	2.7	6	2.4402-2.4409	0.0002-0.0009	0.0028-0.0098	3	2.0079-2.0086	0.0007-0.0014	0.0039-0.0098

22140_ROND_C0007

PISTON AND RING SPECIFICATIONS

All measurements are given in inches.

Year	Engine Displacement Liters	Engine VIN	Piston Clearance	Ring Gap			Ring Side Clearance		
				Top Compression	Bottom Compression	Oil Control	Top Compression	Bottom Compression	Oil Control
2007	2.4	5	0.0006-0.0014	0.0059-0.0118	0.0118-0.0177	0.0078-0.0275	0.0012-0.0028	0.0012-0.0028	0.0024-0.0059
	2.7	6	0.0008-0.0020	0.0059-0.0118	0.0118-0.0177	0.0078-0.0275	0.0016-0.0031	0.0012-0.0028	0.0024-0.0059
2008	2.4	5	0.0006-0.0014	0.0059-0.0118	0.0118-0.0177	0.0078-0.0275	0.0012-0.0028	0.0012-0.0028	0.0024-0.0059
	2.7	6	0.0008-0.0020	0.0059-0.0118	0.0118-0.0177	0.0078-0.0275	0.0016-0.0031	0.0012-0.0027	0.0024-0.0059

22140_ROND_C0006

TORQUE SPECIFICATIONS

All readings in ft. lbs.

Year	Engine Displacement Liters	Engine VIN	Cylinder Head Bolts	Main Bearing Bolts	Rod Bearing Bolts	Crankshaft Damper Bolts	Flywheel Bolts	Manifold		Spark Plugs	Oil Pan Drain Plug
								Intake	Exhaust		
2007	2.4	5	①	②	③	123-130	87-94	14-20	29-33	15-21	26-32
	2.7	6	④	⑤	⑥	123-130	53-56	14-17	22-25	15-21	26-33
2008	2.4	5	①	②	③	123-130	87-94	14-20	29-33	15-21	26-32
	2.7	6	④	⑤	⑥	123-130	53-56	14-17	22-25	15-21	26-33

① Step 1: 25.3 ft. lbs., plus 90 degrees, plus an additional 90 degrees
② 19.52 ft. lbs., plus 45 degrees
③ 14.46 ft. lbs., plus 90 degrees
④ Step 1: 18 ft. lbs., plus 60 degrees, plus an additional 45 degrees

⑤ M10 bolts: 22 ft. lbs., plus 90 degrees
　 M8 bolts: 11.6 ft. lbs., plus 90 degrees
⑥ 15 ft. lbs., plus 90 degrees

22140_ROND_C0008

WHEEL ALIGNMENT

Year	Model		Caster		Camber		Toe-in (in.)
			Range (+/-Deg.)	Preferred Setting (Deg.)	Range (+/-Deg.)	Preferred Setting (Deg.)	
2007	Rondo	F	1.00	+3.15	0.30	0	0.11 +/- 0.12
		R	—	—	0.30	-0.30	0.07 +/- 0.12
2008	Rondo	F	1.00	+3.15	0.30	0	0.11 +/- 0.12
		R	—	—	0.30	-0.30	0.07 +/- 0.12

22140_ROND_C0009

TIRE, WHEEL AND BALL JOINT SPECIFICATIONS

Year	Model	OEM Tires		Tire Pressures (psi)		Wheel Size	Ball Joint Inspection	Lug Nut Torque (ft. lbs.)
		Standard	Optional	Front	Rear			
2007	Rondo	P205/60HR16	P215/50VR17	30	30	6.0-J	①	65-80
2008	Rondo	P205/60HR16	P215/50VR17	30	30	6.0-J	①	65-80

OEM: Original Equipment Manufacturer

PSI: Pounds Per Square Inch

STD: Standard

OPT: Optional

① Replace is any measureable movement is found.

22140_ROND_C0010

BRAKE SPECIFICATIONS
All measurements in inches unless noted

Year	Model		Brake Disc			Brake Drum			Minimum Lining Thickness	Brake Caliper	
			Original Thickness	Minimum Thickness	Maximum Run-out	Original Inside Diameter	Max. Wear Limit	Maximum Machine Diameter		Bracket Bolts (ft. lbs.)	Mounting Bolts (ft. lbs.)
2007	Rondo	F	1.020	0.960	0.0016	NA	NA	NA	0.079	58-72	16-23
		R	0.390	0.320	0.0039	NA	NA	NA	0.079	51-63	16-24
2008	Rondo	F	1.020	0.960	0.0016	NA	NA	NA	0.079	58-72	16-23
		R	0.390	0.330	0.0012	NA	NA	NA	0.079	58-72	16-23

NA: Not available

F: Front

R: Rear

22140_ROND_C0011

SCHEDULED MAINTENANCE INTERVALS
KIA - RONDO

TO BE SERVICED	TYPE OF SERVICE	VEHICLE MILEAGE INTERVAL (x1000)												
		7.5	15	22.5	30	37.5	45	52.5	60	67.5	75	82.5	90	97.5
Accessory drive belts	S/I	✓	✓	✓	✓	✓	✓	✓	✓	✓	✓	✓	✓	✓
Air cleaner filter	R			✓		✓				✓			✓	
Air conditioner system	S/I													
Brake lines, hoses and connections	S/I		✓		✓		✓		✓		✓		✓	
Chassis and body fasteners	T												✓	
Cooling system hoses and coolant level	S/I		✓		✓		✓		✓		✓		✓	
CV-joint boots	S/I			✓					✓				✓	
Engine coolant	R			✓					✓				✓	
Engine oil and filter	R	✓	✓	✓	✓	✓	✓	✓	✓	✓	✓	✓	✓	✓
Exhaust system heat shields	S/I			✓					✓				✓	
Front and rear brakes	S/I	✓	✓	✓	✓	✓	✓	✓	✓	✓	✓	✓	✓	✓
Front ball joints	S/I			✓					✓				✓	
Fuel filter	R					✓					✓			
Fuel lines and hoses	S/I	✓	✓	✓	✓	✓	✓	✓	✓	✓	✓	✓	✓	✓
Locks and hinges	L	✓	✓	✓	✓	✓	✓	✓	✓	✓	✓	✓	✓	✓
Spark plugs	R								✓					
Steering operation and linkage	S/I	✓	✓	✓	✓	✓	✓	✓	✓	✓	✓	✓	✓	✓
Timing belt	R								✓					

R: Replace S/I: Service or Inspect L: Lubricate T: Tighten

FREQUENT OPERATION MAINTENANCE (SEVERE SERVICE)

If a vehicle is operated under any of the following conditions it is considered severe service

- Towing a trailer or using a camper or car-top carrier

- Repeated short trips of less than 5 miles in temperatures below freezing, or trips of less than 10 miles in any temperature

- Prolonged idling (vehicle operation in stop and go traffic).

- Operating on rough, muddy, unpaved, dusty or salt-covered roads.

- Police, taxi, delivery usage or trailer towing usage.

- Driving in extremely hot (over 90°F) conditions

Oil & oil filter: change every 5000 miles or 5 months, whichever occurs first.

Air cleaner filter: inspect every 15,000 miles or 15 months and replace everything 30,000 miles or 30 months, whichever occurs fir:

Fuel system hoses (California models only): replace every 105,000 miles

Emission system hoses (non-CA models): inspect every 55,000 or 55 months, whichever occurs first

Emission system hoses (CA models): inspect every 60,000 miles or 60 months, which occurs first

Front and rear brakes: inspect every 15,000 miles or 15 months, whichever occurs first

Chassis and body fasteners: tighten every 15,000 miles or 15 months, whichever occurs first

Locks and hinges: lubricate every 5000 miles or 5 months, whichever occurs first

22140_ROND_C0012

PRECAUTIONS

Before servicing any vehicle, please be sure to read all of the following precautions, which deal with personal safety, prevention of component damage, and important points to take into consideration when servicing a motor vehicle:

• Never open, service or drain the radiator or cooling system when the engine is hot; serious burns can occur from the steam and hot coolant.

• Observe all applicable safety precautions when working around fuel. Whenever servicing the fuel system, always work in a well-ventilated area. Do not allow fuel spray or vapors to come in contact with a spark, open flame, or excessive heat (a hot drop light, for example). Keep a dry chemical fire extinguisher near the work area. Always keep fuel in a container specifically designed for fuel storage; also, always properly seal fuel containers to avoid the possibility of fire or explosion. Refer to the additional fuel system precautions later in this section.

• Fuel injection systems often remain pressurized, even after the engine has been turned **OFF**. The fuel system pressure must be relieved before disconnecting any fuel lines. Failure to do so may result in fire and/or personal injury.

• Brake fluid often contains polyglycol ethers and polyglycols. Avoid contact with the eyes and wash your hands thoroughly after handling brake fluid. If you do get brake fluid in your eyes, flush your eyes with clean, running water for 15 minutes. If eye irritation persists, or if you have taken brake fluid internally, IMMEDIATELY seek medical assistance.

• The EPA warns that prolonged contact with used engine oil may cause a number of skin disorders, including cancer. You should make every effort to minimize your exposure to used engine oil. Protective gloves should be worn when changing oil. Wash your hands and any other exposed skin areas as soon as possible after exposure to used engine oil. Soap and water, or waterless hand cleaner should be used.

• All new vehicles are now equipped with an air bag system, often referred to as a Supplemental Restraint System (SRS) or Supplemental Inflatable Restraint (SIR) system. The system must be disabled before performing service on or around system components, steering column, instrument panel components, wiring and sensors. Failure to follow safety and disabling procedures could result in accidental air bag deployment, possible personal injury and unnecessary system repairs.

• Always wear safety goggles when working with, or around, the air bag system. When carrying a non-deployed air bag, be sure the bag and trim cover are pointed away from your body. When placing a non-deployed air bag on a work surface, always face the bag and trim cover upward, away from the surface. This will reduce the motion of the module if it is accidentally deployed. Refer to the additional air bag system precautions later in this section.

• Clean, high quality brake fluid from a sealed container is essential to the safe and proper operation of the brake system. You should always buy the correct type of brake fluid for your vehicle. If the brake fluid becomes contaminated, completely flush the system with new fluid. Never reuse any brake fluid. Any brake fluid that is removed from the system should be discarded. Also, do not allow any brake fluid to come in contact with a painted surface; it will damage the paint.

• Never operate the engine without the proper amount and type of engine oil; doing so WILL result in severe engine damage.

• Timing belt maintenance is extremely important. Many models utilize an interference-type, non-freewheeling engine. If the timing belt breaks, the valves in the cylinder head may strike the pistons, causing potentially serious (also time-consuming and expensive) engine damage. Refer to the maintenance interval charts for the recommended replacement interval for the timing belt, and to the timing belt section for belt replacement and inspection.

• Disconnecting the negative battery cable on some vehicles may interfere with the functions of the on-board computer system(s) and may require the computer to undergo a relearning process once the negative battery cable is reconnected.

• When servicing drum brakes, only disassemble and assemble one side at a time, leaving the remaining side intact for reference.

• Only an MVAC-trained, EPA-certified automotive technician should service the air conditioning system or its components.

BRAKES

ANTI-LOCK BRAKE SYSTEM (ABS)

GENERAL INFORMATION

PRECAUTIONS

• Certain components within the ABS system are not intended to be serviced or repaired individually.

• Do not use rubber hoses or other parts not specifically specified for and ABS system. When using repair kits, replace all parts included in the kit. Partial or incorrect repair may lead to functional problems and require the replacement of components.

• Lubricate rubber parts with clean, fresh brake fluid to ease assembly. Do not use shop air to clean parts; damage to rubber components may result.

• Use only DOT 3 brake fluid from an unopened container.

• If any hydraulic component or line is removed or replaced, it may be necessary to bleed the entire system.

• A clean repair area is essential. Always clean the reservoir and cap thoroughly before removing the cap. The slightest amount of dirt in the fluid may plug an orifice and impair the system function. Perform repairs after components have been thoroughly cleaned; use only denatured alcohol to clean components. Do not allow ABS components to come into contact with any substance containing mineral oil; this includes used shop rags.

• The Anti-Lock control unit is a microprocessor similar to other computer units in the vehicle. Ensure that the ignition switch is **OFF** before removing or installing controller harnesses. Avoid static electricity discharge at or near the controller.

• If any arc welding is to be done on the vehicle, the control unit should be unplugged before welding operations begin.

BRAKES

BLEEDING THE BRAKE SYSTEM

BLEEDING PROCEDURE

BLEEDING PROCEDURE

When bleeding the brakes, note the following:

- Do not reuse the drained fluid
- Always use Genuine DOT 3 or DOT 4 Brake Fluid. Using a non-Genuine DOT3 or DOT 4 brake fluid can cause corrosion and decrease the life of the system
- Make sure no dirt of other foreign matter is allowed to contaminate the brake fluid

- Do not spill brake fluid on the vehicle, it may damage the paint; if brake fluid does contact the paint, wash it off immediately with water
- The reservoir on the master cylinder must be at the MAX (upper) level mark at the start of bleeding procedure and checked after bleeding each brake caliper. Add fluid as required

1. Make sure the brake fluid level in the master cylinder fluid reservoir is at the MAX (upper) level line.
2. Have someone slowly pump the brake pedal several times, and then apply steady pressure
3. Loosen the right-rear brake bleed screw to allow air to escape from the system. Then tighten the bleed screw securely
4. Repeat the procedure for each wheel in the sequence shown below until air bubbles no longer appear in the fluid
5. Refill the master cylinder reservoir to the MAX (upper) level line

BRAKES

FRONT DISC BRAKES

✳✳ CAUTION

Dust and dirt accumulating on brake parts during normal use may contain asbestos fibers from production or aftermarket brake linings. Breathing excessive concentrations of asbestos fibers can cause serious bodily harm. Exercise care when servicing brake parts. Do not sand or grind brake lining unless equipment used is designed to contain the dust residue. Do not clean brake parts with compressed air or by dry brushing. Cleaning should be done by dampening the brake components with a fine mist of water, then wiping the brake components clean with a dampened cloth. Dispose of cloth and all residue containing asbestos fibers in an impermeable container with the appropriate label. Follow practices prescribed by the Occupational Safety and Health Administration (OSHA) and the Environmental Protection Agency (EPA) for the handling, processing, and disposing of dust or debris that may contain asbestos fibers.

BRAKE CALIPER

REMOVAL & INSTALLATION

1. Raise and safely support the front of the vehicle.
2. Remove the front wheel and tire.
3. Remove the brake hose bolt and the guide rod bolts from the caliper assembly.
4. Remove the caliper assembly.
5. Installation is the reverse of removal.

DISC BRAKE PADS

REMOVAL & INSTALLATION

1. Raise and safely support the front of the vehicle.
2. Remove the front wheel and tire.
3. Remove the brake hose bolt and the guide rod bolts from the caliper assembly.
4. Remove the caliper assembly.
5. Remove the pads, the pad shims and the pad retainers from the caliper bracket.
6. Installation is the reverse of removal.

BRAKES

REAR DISC BRAKES

✳✳ CAUTION

Dust and dirt accumulating on brake parts during normal use may contain asbestos fibers from production or aftermarket brake linings. Breathing excessive concentrations of asbestos fibers can cause serious bodily harm. Exercise care when servicing brake parts. Do not sand or grind brake lining unless equipment used is designed to contain the dust residue. Do not clean brake parts with compressed air or by dry brushing. Cleaning should be done by dampening the brake components with a fine mist of water, then wiping the brake components clean with a dampened cloth. Dispose of cloth and all residue containing asbestos fibers in an impermeable container with the appropriate label. Follow practices prescribed by the Occupational Safety and Health Administration (OSHA) and the Environmental Protection Agency (EPA) for the handling, processing, and disposing of dust or debris that may contain asbestos fibers.

BRAKE CALIPER

REMOVAL & INSTALLATION

1. Raise and safely support the rear of the vehicle.
2. Remove the rear wheel and tire.
3. Remove the brake hose bolt and the guide rod bolts from the caliper assembly.
4. Remove the caliper assembly.
5. Installation is the reverse of removal.

DISC BRAKE PADS

REMOVAL & INSTALLATION

1. Raise and safely support the rear of the vehicle.
2. Remove the rear wheel and tire.
3. Remove the brake hose bolt and the guide rod bolts from the caliper assembly.
4. Remove the caliper assembly.
5. Remove the pads, the pad shims and the pad retainers from the caliper bracket.
6. Installation is the reverse of removal.

BRACES

PARKING BRAKE

PARKING BRAKE SHOES

REMOVAL & INSTALLATION

See Figure 1.

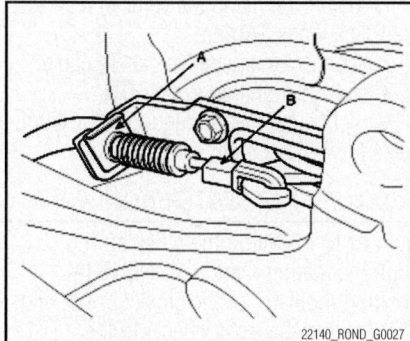

22140_ROND_G0027

Fig. 1 Remove the parking brake cable (B), after removing the clip (A)

1. Raise the vehicle, and make sure it is securely supported.
2. Remove the rear tire and wheel.
3. Remove the brake caliper.
4. Remove the rear brake disc (rotor).
5. Remove the parking brake cable (B), after removing the clip (A).
6. Remove the shoe hold down pin and spring by pressing and rotating the spring.

7. Remove the adjuster assembly and the lower return spring.
8. Remove the upper return spring and the brake shoes.
9. Remove the operating lever assembly.

To install:
10. Install the operating lever assembly.
11. Install the upper return spring and the brake shoes.
12. Install the adjuster assembly and the lower return spring.
13. Install the shoe hold down pin and spring by pressing and rotating the spring.
14. Install the parking brake cable, then install the clip.
15. Install the rear brake disc.
16. Adjust the rear brake shoe clearance.
    a. Remove the plug from the disc.
    b. Rotate the toothed wheel of adjuster by a screw driver until the disc is not moving.
    c. Return it by 5 notches in the opposite direction.
17. Install the brake caliper.
18. Install the tire and wheel.
19. Adjust the parking brake pedal.
20. If the parking brake shoe or the brake disc is replaced with a new one, perform the brake shoe bed-in procedure.

a. While operating the parking brake pedal, drive the vehicle 0.3 miles (500 meters) at the speed of 37 mph (60 kph).
b. Repeat the above procedure more than two times.
c. Parking brake must hold on at 30% grade.

✳✳ CAUTION

After adjusting parking brake, verify the following ;

d. Must be free from malfunction when the parking pedal is operated.
e. Check that all parts move smoothly.
f. The parking brake indicator lamp must be on after the parking pedal is applied and must be off after the pedal is released.

ADJUSTMENT

1. Adjust the rear brake shoe clearance.
   a. Remove the plug from the disc.
   b. Rotate the toothed wheel of adjuster by a screw driver until the disc is not moving.
   c. Return it by 5 notches in the opposite direction.

CHASSIS ELECTRICAL

AIR BAG (SUPPLEMENTAL RESTRAINT SYSTEM)

GENERAL INFORMATION

✳✳ CAUTION

These vehicles are equipped with an air bag system. The system must be disarmed before performing service on, or around, system components, the steering column, instrument panel components, wiring and sensors. Failure to follow the safety precautions and the disarming procedure could result in accidental air bag deployment, possible injury and unnecessary system repairs.

SERVICE PRECAUTIONS

Disconnect and isolate the battery negative cable before beginning any airbag system component diagnosis, testing, removal, or installation procedures. Allow system capacitor to discharge for two minutes before beginning any component service. This will disable the airbag system. Failure

to disable the airbag system may result in accidental airbag deployment, personal injury, or death.

Do not place an intact undeployed airbag face down on a solid surface. The airbag will propel into the air if accidentally deployed and may result in personal injury or death.

When carrying or handling an undeployed airbag, the trim side (face) of the airbag should be pointing towards the body to minimize possibility of injury if accidental deployment occurs. Failure to do this may result in personal injury or death.

Replace airbag system components with OEM replacement parts. Substitute parts may appear interchangeable, but internal differences may result in inferior occupant protection. Failure to do so may result in occupant personal injury or death.

Wear safety glasses, rubber gloves, and long sleeved clothing when cleaning powder residue from vehicle after an airbag deployment. Powder residue emitted from a deployed airbag can cause skin irritation.

Flush affected area with cool water if irritation is experienced. If nasal or throat irritation is experienced, exit the vehicle for fresh air until the irritation ceases. If irritation continues, see a physician.

Do not use a replacement airbag that is not in the original packaging. This may result in improper deployment, personal injury, or death.

The factory installed fasteners, screws and bolts used to fasten airbag components have a special coating and are specifically designed for the airbag system. Do not use substitute fasteners. Use only original equipment fasteners listed in the parts catalog when fastener replacement is required.

During, and following, any child restraint anchor service, due to impact event or vehicle repair, carefully inspect all mounting hardware, tether straps, and anchors for proper installation, operation, or damage. If a child restraint anchor is found damaged in any way, the anchor must be replaced. Failure to do this may result in personal injury or death.

Deployed and non-deployed airbags may or may not have live pyrotechnic material within the airbag inflator.

Do not dispose of driver/passenger/curtain airbags or seat belt tensioners unless you are sure of complete deployment. Refer to the Hazardous Substance Control System for proper disposal.

Dispose of deployed airbags and tensioners consistent with state, provincial, local, and federal regulations.

After any airbag component testing or service, do not connect the battery negative cable. Personal injury or death may result if the system test is not performed first.

If the vehicle is equipped with the Occupant Classification System (OCS), do not connect the battery negative cable before performing the OCS Verification Test using the scan tool and the appropriate diagnostic information. Personal injury or death may result if the system test is not performed properly.

Never replace both the Occupant Restraint Controller (ORC) and the Occupant Classification Module (OCM) at the same time. If both require replacement, replace one, then perform the Airbag System test before replacing the other.

Both the ORC and the OCM store Occupant Classification System (OCS) calibration data, which they transfer to one another when one of them is replaced. If both are replaced at the same time, an irreversible fault will be set in both modules and the OCS may malfunction and cause personal injury or death.

If equipped with OCS, the Seat Weight Sensor is a sensitive, calibrated unit and must be handled carefully. Do not drop or handle roughly. If dropped or damaged, replace with another sensor. Failure to do so may result in occupant injury or death.

If equipped with OCS, the front passenger seat must be handled carefully as well. When removing the seat, be careful when setting on floor not to drop. If dropped, the sensor may be inoperative, could result in occupant injury, or possibly death.

If equipped with OCS, when the passenger front seat is on the floor, no one should sit in the front passenger seat. This uneven force may damage the sensing ability of the seat weight sensors. If sat on and damaged, the sensor may be inoperative, could result in occupant injury, or possibly death.

DISARMING THE SYSTEM

1. Before servicing the vehicle, refer to the Precautions Section.

2. Turn the ignition switch to the **LOCK** position.

3. Disconnect the negative battery cable.

4. Wait 10 minutes for the battery back-up power to discharge.

ARMING THE SYSTEM

1. Before servicing the vehicle, refer to the Precautions Section.

2. Connect the negative battery cable.

3. Turn the ignition switch **ON**.

4. Verify that the air bag indicator illuminates for 4–8 seconds, then goes off.

CLOCKSPRING CENTERING

Prior to installing the clock spring, align the mating mark and "NEUTRAL" position indicator of the clock spring, and, after turning the front wheels to the straight-ahead position, install the clock spring to the column switch. If the mating mark of the clock spring is not properly aligned, the steering wheel may not completely rotate during a turn, or the flat cable within the clock spring may be severed, obstructing normal operation of the SRS and possibly leading to serious injury to the vehicle's driver.

DRIVETRAIN

AUTOMATIC TRANSAXLE ASSEMBLY

REMOVAL & INSTALLATION

See Figures 2 through 12.

1. Remove the battery.
2. Remove the engine cover.
3. Remove the air cleaner upper body (B) after disconnecting the air flow sensor connector (A).

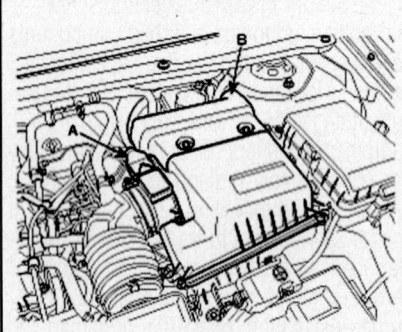

Fig. 2 Remove the air cleaner upper body (B) after disconnecting the air flow sensor connector (A)

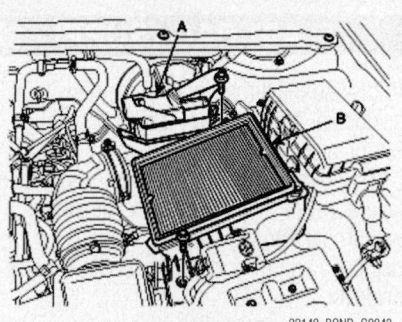

Fig. 3 Remove the PCM connector (A) and air cleaner under body (B)

4. Remove the PCM connector (A) and air cleaner under body (B).
5. Remove the battery tray.
6. Disconnect the ground wire.
7. Disconnect the inhibiter switch connector (A), solenoid valve connector (B) and the input shaft speed sensor connector (C).
8. Disconnect the output shaft speed sensor connector (A), and vehicle speed sensor connector (B).
9. Detach the hoses (A), removing the oil cooler hose clamps.

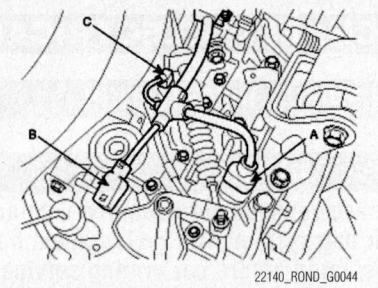

Fig. 4 Disconnect the inhibiter switch connector (A), solenoid valve connector (B) and the input shaft speed sensor connector (C)

10. Remove the control cable assembly by removing the nut (B) and clip (A).
11. Remove the transaxle upper mounting bolts.
12. Disconnect the power steering pressure tube from the power steering oil pump.
13. Using the SST (09200-38001), hold the engine and transaxle assembly safely.
14. Remove the front wheels and Lifting up the vehicle, remove the undercover.

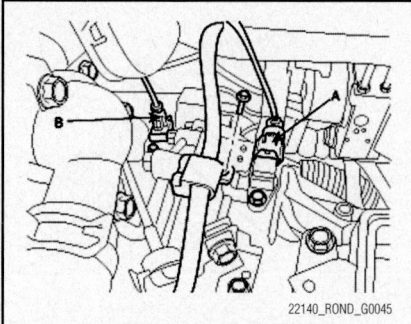

Fig. 5 Disconnect the output shaft speed sensor connector (A), and vehicle speed sensor connector (B)

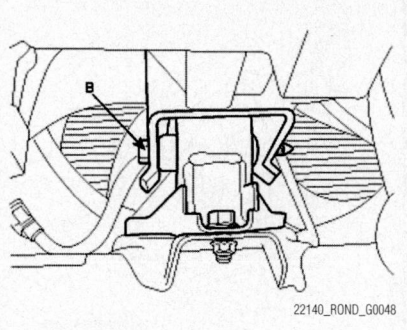

Fig. 8 Remove the front (B) roll stopper mounting bolts

Fig. 11 Remove the drive plate bolts (A)

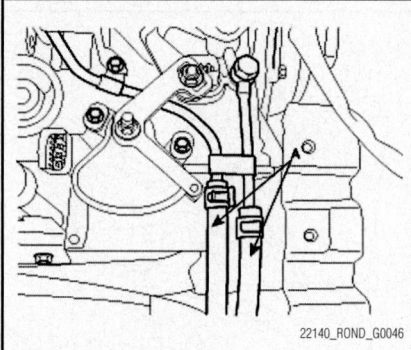

Fig. 6 Detach the hoses (A), removing the oil cooler hose clamps

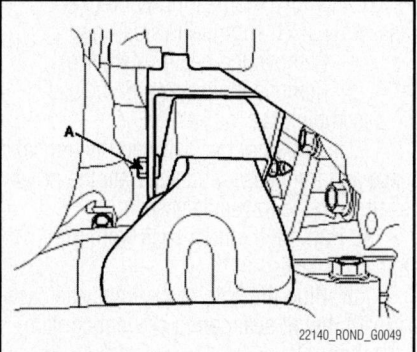

Fig. 9 Remove the rear (A) roll stopper mounting bolts

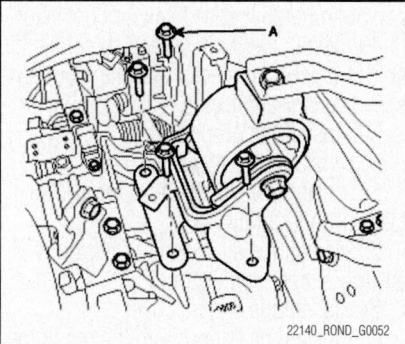

Fig. 12 Remove the transaxle insulator mounting bolts (A)

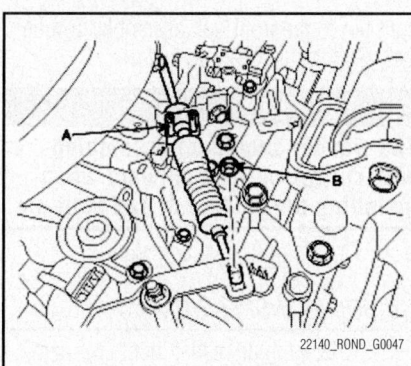

Fig. 7 Remove the control cable assembly by removing the nut (B) and clip (A)

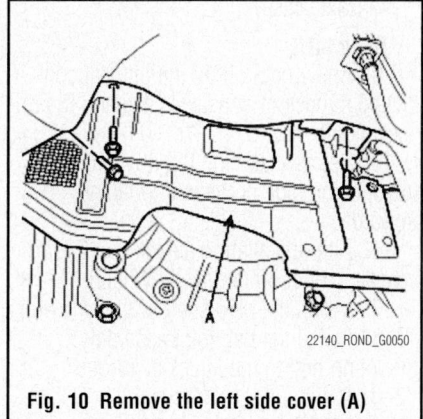

Fig. 10 Remove the left side cover (A)

15. Remove the oil drain plug and drain the fluid.

16. Drain power steering fluid through the return tube by removing the clamp.

17. Remove the lower arm ball joint assembly mounting bolts and tie rod end ball joint.

18. Remove the stabilizer link mounting nuts.

19. Remove the power steering column joint bolt.

20. Remove the front (B) rear (A) roll stopper mounting bolts.

21. Remove the muffler hanger rubber.

22. Supporting the sub frame with a jack and the special tool (09624-38000), remove the mounting bolts.

23. Disconnect the driveshafts from the transaxle.

24. Remove the left side cover (A).

25. Remove the drive plate bolts (A).

➡**Remove the bolts (6 ea) rotating the crankshaft clockwise.**

26. Remove the transaxle insulator mounting bolts (A).

27. Supporting the transaxle with a jack, remove the transaxle lower mounting bolts.

To install:

28. Install the transaxle lower mounting bolts after fitting the transaxle assembly into the engine assembly. Torque: 58–72 ft. lbs. (80–100 Nm).

29. Install the transaxle insulator mounting bracket bolts. Torque: 43–58 ft. lbs. (60–80 Nm).

30. Install the drive plate bolts by rotating the timing gear.
Torque: 33–38 ft. lbs. (46–53 Nm).

➡**Install the bolts 6ea) rotating the crankshaft clockwise.**

31. Install the left side cover.

32. Connect the driveshafts to the transaxle.

33. Supporting the sub frame with a jack and the special tool (09624-38000), install the mounting bolts. Torque: 101–116 ft. lbs. (140–160 Nm).

34. Install the muffler hanger rubber.

35. Install the front rear roll stopper mounting bolts. Torque: 36–47 ft. lbs. (50–65 Nm).

36. Install the steering column joint bolt.

37. Install the stabilizer link mounting nuts.

38. Install the lower arm ball joint assembly mounting bolts and tie rod end ball joint.

39. Connect the power steering return hose by tightening the clamp.
40. Install the oil drain plug.
41. Install the front under cover.
42. Install the front wheels and tires.
43. Remove the special tool (09200-38001).
44. Connect the power steering pressure bolt.
45. Install the transaxle upper mounting bolts. Torque: 31–40 ft. lbs. (43–55 Nm).
46. Install the control cable assembly by installing the nut and clip.
47. Connect the transaxle oil cooler hoses to the tubes by tightening the clamps.
48. Install the vehicle speed sensor connector and output speed sensor connector.
49. Install the inhibiter switch connector, solenoid valve connector and the input shaft speed sensor connector.
50. Install the ground wire.
51. Install the battery tray.
52. Install the PCM connector and air cleaner under body.
53. Install the air cleaner upper body and air flow sensor connector.
54. Install the engine cover.
55. Install the battery with the mounting plate and bolt
56. Refill the automatic transaxle fluid.
57. Refill the power steering fluid and bleed the air.

FRONT HALFSHAFT

REMOVAL & INSTALLATION

See Figures 13 through 15.

1. Raise the vehicle, and make sure it is securely supported.
2. Remove the front wheel and tire.
3. Remove the split pin, the castle nut and washer from the front hub.
4. Remove the ball joint assembly mounting bolt (A) from the knuckle.
5. Using a plastic hammer, disconnect driveshaft from the axle hub.

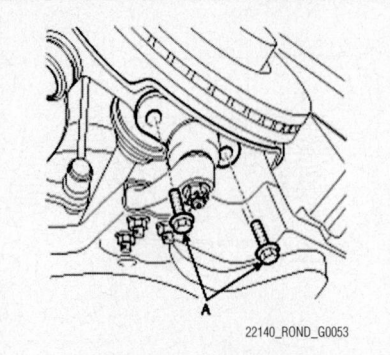

Fig. 13 Remove the ball joint assembly mounting bolt (A) from the knuckle

6. Remove the bearing & bracket assembly. 2.7L engine RH side only.
   a. Remove the heat protector (A).
   b. Remove the bearing & bracket assembly (A).
7. Insert a pry bar between the transaxle case and joint case, and separate the driveshaft from the transaxle case.
8. Pull out the driveshaft from the transaxle case.
   a. Plug the hole of the transaxle case with the oil seal cap to prevent contamination.
   b. Support the driveshaft properly.
   c. Replace the retainer ring whenever the driveshaft is removed from the transaxle case.

To install:

9. Apply gear oil on the oil seal contacting surface of transaxle case and the driveshaft splines.
10. Before installing the driveshaft, set the opening side of the circlip facing upward.
11. After installation, check that the driveshaft cannot be removed by hand.
12. Install the driveshaft to the axle hub.
13. Install the ball joint assembly mounting bolt to the knuckle. Torque: 72–87 ft. lbs. (98–117 Nm).
14. Install the washer, castle nut and new

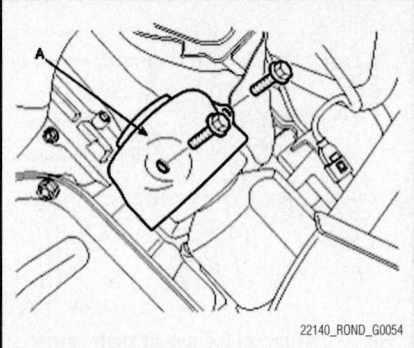

Fig. 14 Remove the heat protector (A)–2.7L engine RH side only

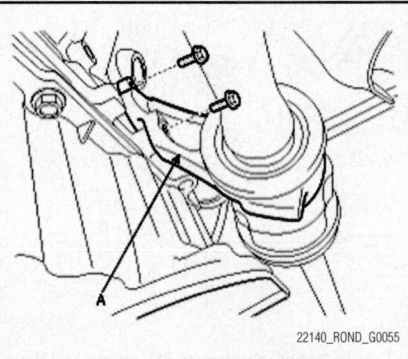

Fig. 15 Remove the bearing & bracket assembly (A)–2.7L engine RH side only

split pin to the front hub assembly. Torque: 145–203 ft. lbs. (196–275 Nm).

✳✳ CAUTION

The washer should be assembled with convex surface outward when installing the castle nut and split pin.

15. Install the wheel and the tire.

CV-BOOTS INSPECTION

1. Check the driveshaft boots for damage and deterioration.
2. Check the boots for grease leak.
3. Replace as needed.

ENGINE COOLING

THERMOSTAT

REMOVAL & INSTALLATION

1. Drain engine coolant so its level is below thermostat.
2. Remove water inlet and thermostat.
3. Installation is the reverse of removal.

WATER PUMP

REMOVAL & INSTALLATION

2.4L Engine

See Figure 16.

1. Drain the engine coolant.
2. Remove drive belt.
3. Remove exhaust manifold.
4. Remove water inlet pipe nut.
5. Remove the water pump (A).
6. Installation is the reverse of removal.

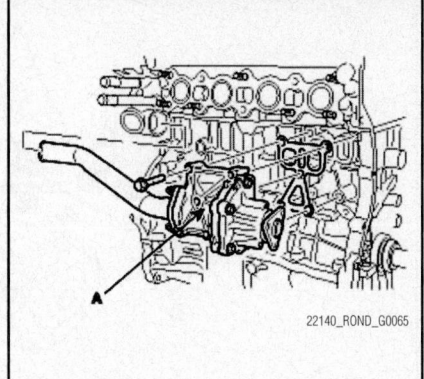

22140_ROND_G0065

Fig. 16 Remove the water pump (A)

2.7L Engine

See Figure 17.

1. Drain the engine coolant.
2. Remove accessory drive belt (A).

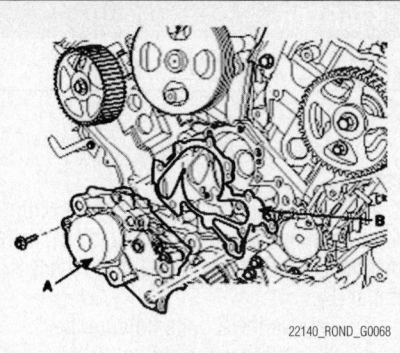

22140_ROND_G0068

Fig. 17 Remove the water pump (A) and gasket (B)

3. Remove the timing belt.
4. Remove the water pump (A) and gasket (B).
5. Installation is the reverse of removal.

ENGINE ELECTRICAL

ALTERNATOR

REMOVAL & INSTALLATION

1. Disconnect the battery negative terminal first, then the positive terminal.

2. Disconnect the alternator connector.
3. Remove the cable from alternator "B" terminal.
4. Remove the drive belt.

CHARGING SYSTEM

5. Pull out the through bolt and remove the alternator
6. Installation is the reverse of removal.

ENGINE ELECTRICAL

FIRING ORDER

2.4L engine firing order: 1–3–4–2
2.7L engine firing order: 1–2–3–4–5–6

IGNITION COIL

REMOVAL & INSTALLATION

1. Remove the engine cover.
2. Remove the ignition coil connector (s).

➡**When removing the ignition coil con-**
nector, pull the lock pin and push the
clip.

3. Remove the ignition coil (s).
4. Installation is the reverse of removal.

IGNITION TIMING

ADJUSTMENT

Ignition timing is controlled by the ECM. No adjustment is possible.

IGNITION SYSTEM

SPARK PLUGS

REMOVAL & INSTALLATION

1. Remove the engine cover.
2. Remove the ignition coil connector (s).

➡**When removing the ignition coil con-**
nector, pull the lock pin and push the
clip.

3. Remove the ignition coil (s).
4. Using a spark plug socket, remove the spark plug (s).
5. Installation is the reverse of removal.

ENGINE ELECTRICAL

STARTER

REMOVAL & INSTALLATION

See Figure 18.

1. Disconnect the battery negative cable.

2. Disconnect the starter cable (A) from the B terminal (B) on the solenoid (C), then disconnect the connector (D) from the S terminal (E).

3. Remove the 2 bolts holding the starter, then remove the starter.

4. Installation is the reverse of removal.

5. Connect the battery negative cable to the battery.

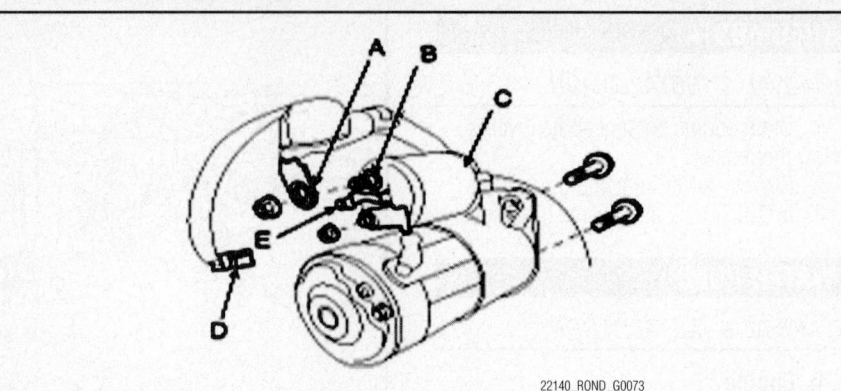

Fig. 18 Disconnect the starter cable A) from the B terminal (B) on the solenoid (C), then disconnect the connector (D) from the S terminal (E)

ENGINE MECHANICAL

➡️**Disconnecting the negative battery cable may interfere with the functions of the on board computer systems and may require the computer to undergo a relearning process, once the negative battery cable is reconnected.**

ACCESSORY DRIVE BELTS

ACCESSORY BELT ROUTING

See Figures 19 and 20.

INSPECTION

Inspect the drive belt for signs of glazing or cracking. A glazed belt will be perfectly smooth from slippage, while a good belt will have a slight texture of fabric visible. Cracks will usually start at the inner edge of the belt and run outward. All worn or dam-

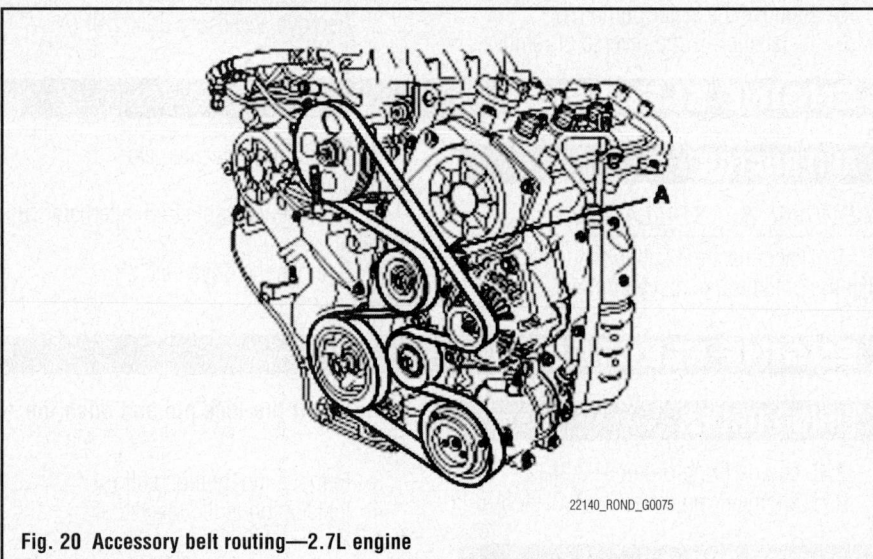

Fig. 20 Accessory belt routing—2.7L engine

aged drive belts should be replaced immediately.

ADJUSTMENT

The belt tension is maintained by an automatic tensioner. No adjustment is necessary or possible.

REMOVAL & INSTALLATION

2.4L Engine

1. Relieve tension on the tensioner pulley.

2. Remove accessory drive belt

To install:

3. Install accessory drive belt in the following sequence:

    a. Crankshaft pulley

    b. A/C pulley

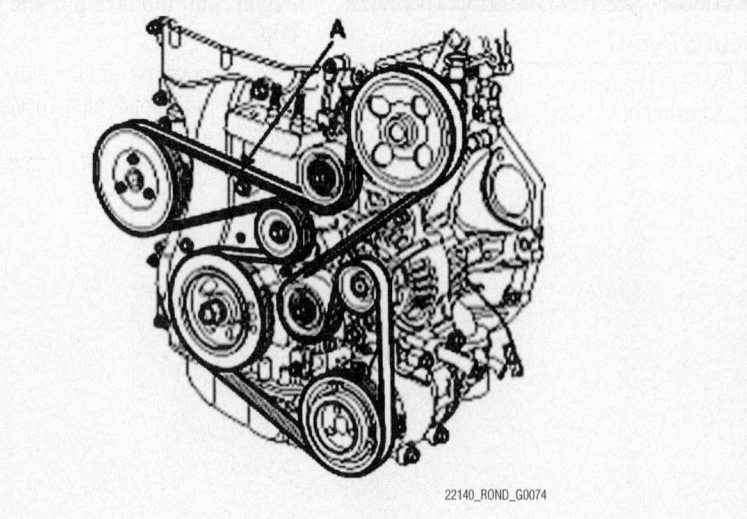

Fig. 19 Accessory belt routing—2.4L engine

c. Alternator pulley
d. Idler pulley
e. P/C pump pulley
f. Idler pulley
g. Water pump pulley
h. Tensioner pulley.

4. Rotate auto tensioner arm in the counterclockwise moving auto tensioner pulley bolt with wrench.

5. After putting belt on auto tensioner pulley, release the auto tensioner pulley slowly.

2.7L Engine

1. Release tension on the accessory belt tensioner.

2. Remove accessory drive belt.

To install:

3. Increase load on the accessory belt tensioner.

4. Install the accessory drive belt.

5. Release the belt tensioner to take up slack in the belt.

BALANCE SHAFT

REMOVAL & INSTALLATION

2.4L Engine

See Figures 21 through 25.

1. Remove the timing chain.

2. Install a set pin after compressing the balance shaft chain tensioner.

3. Remove the balance shaft chain tensioner (A).

4. Remove the balance shaft chain tensioner arm (B).

5. Remove the balance shaft chain guide (C).

6. Remove the balance shaft module (A) and balance shaft chain (B).

To install:

7. The key of crankshaft should be aligned with the mating face of main bearing

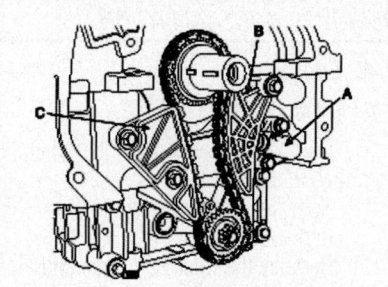

Fig. 21 Remove the balance shaft chain tensioner (A), balance shaft chain tensioner arm (B), and balance shaft chain guide (C).

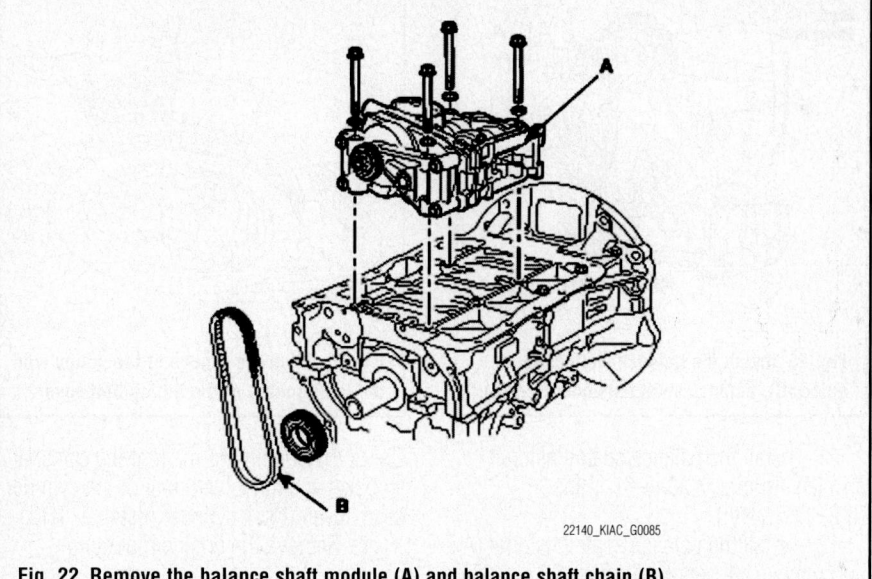

Fig. 22 Remove the balance shaft module (A) and balance shaft chain (B).

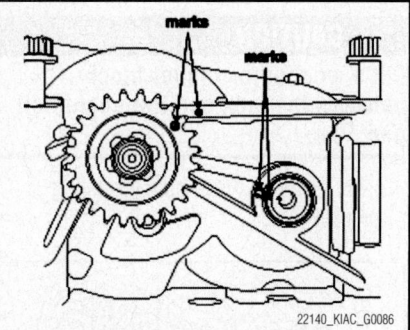

Fig. 23 Timing marks to be visually aligned with centers of adjacent cast timing notches.

cap. As a result of this, the piston of No.1 cylinder is placed at the top dead center on compression stroke.

8. Confirm the balance shaft module timing mark. Timing marks to be visually aligned with centers of adjacent cast timing notches.

9. Install balance shaft module (A) that the timing mark of balance shaft module sprocket should be matched with the timing mark (color link) of balance shaft chain. Torque: 12.3 ft. lbs. (16.66Nm), then an additional 60° plus an additional 60°.

10. Install the balance shaft chain guide (C). Torque: 7.23–8.67 ft. lbs. (9.8–11.76 Nm).

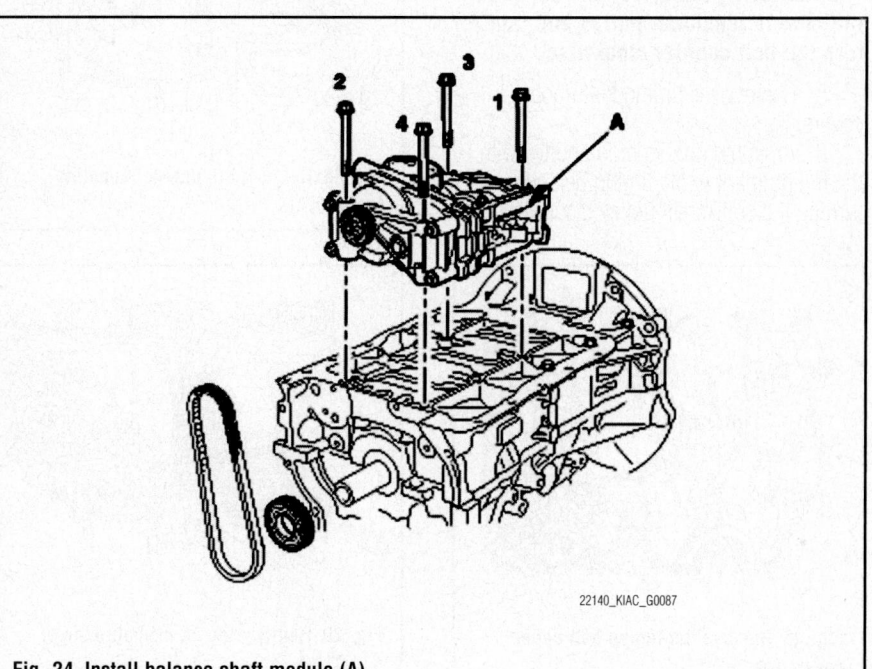

Fig. 24 Install balance shaft module (A)

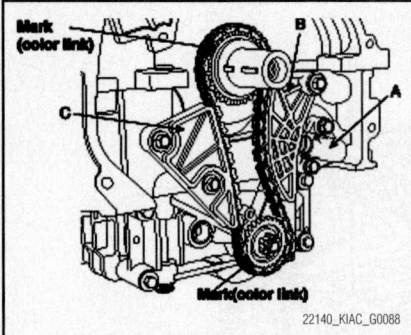

Fig. 25 Install the balance shaft chain guide (C), balance shaft tensioner arm (B),

11. Install the balance shaft tensioner arm (B). Torque: 7.23–8.67 ft. lbs. (9.8–11.76 Nm).

12. Install the balance shaft tensioner (A) and remove the set pin. Torque: 7.23–8.67 ft. lbs. (9.8–11.76 Nm).

13. Confirm the timing marks.

14. Install timing chain.

CRANKSHAFT DAMPER

REMOVAL & INSTALLATION

2.7L Engine

See Figures 26 through 29.

1. Remove the engine cover

2. Remove the front right wheel and tire.

3. Remove the right side cover.

4. Remove the drive belt, the idler pulley and the tensioner pulley.

➡**In removing the drive belt, fix a tool in the auto tensioner pulley bolt and turn the bolt counter clockwise.**

5. Remove the timing belt upper covers (A).

6. Align the groove of the pulley with the timing mark of the timing belt cover by turning the crankshaft pulley clockwise.

Fig. 26 Remove the timing belt upper covers (A)

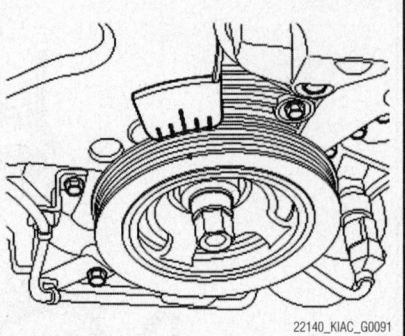

Fig. 27 Align the groove of the pulley with the timing mark of the timing belt cover

Check that the timing mark of the camshaft sprocket is aligned with that of the cylinder head cover. (No.1 cylinder piston at TDC)

7. Remove the engine mounting bracket.

   a. Support the engine oil pan with a jack.

✳✳ CAUTION

Put a wooden or rubber block between the jack and the engine oil pan.

   b. Remove the engine mounting bracket (A).

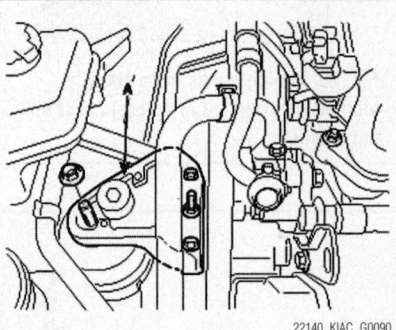

Fig. 28 Remove the engine mounting bracket (A)

Fig. 29 Remove the crankshaft damper pulley (A)

8. Remove the crankshaft damper pulley (A).

To install:

9. Install the crankshaft damper pulley. Torque: 123–130 ft. lbs. (167–177 Nm).

10. Install the engine mounting bracket. Torque: 47–62 ft. lbs. (64–83 Nm).

11. Install the timing belt upper cover. Torque: 7.2–8.7 ft. lbs. (9.8–11.8 Nm).

12. Install the drive belt tensioner. Torque: 25–40 ft. lbs. (34–54 Nm).

13. Install the drive belt idler and the drive belt. Torque: 25–40 ft. lbs. (34–54 Nm).

14. Install the right side cover.

15. Install the front right wheel and tire.

16. Install the engine cover. Torque: 5.8–8.7 ft. lbs. (7.8–11.8 Nm).

CRANKSHAFT FRONT SEAL

REMOVAL & INSTALLATION

1. Disconnect the negative battery cable.

2. Remove the timing belt covers and belt.

3. Remove the timing belt pulley using a puller.

4. Remove the oil pump bolts and the pump.

5. Wrap a suitable prytool with a rag and work the old seal from the oil pump housing.

To install:

6. Lubricate the seal lip with clean engine oil and push the seal slightly in by hand.

7. Install the seal using a seal installer. Install the seal until it is flush with the oil pump body.

8. Install the timing belt.

9. Connect the negative battery cable.

10. Start the engine and check for leaks.

CYLINDER HEAD

REMOVAL & INSTALLATION

2.4L Engine

See Figures 30 through 43.

1. Disconnect the negative terminal from the battery.

2. Remove engine cover.

3. Remove air duct.

4. Remove the intake air hose and air cleaner assembly.

   a. Disconnect the AFS (B) connector.

   b. Disconnect the breather hose (D) and vacuum hose (E) from air cleaner hose (C).

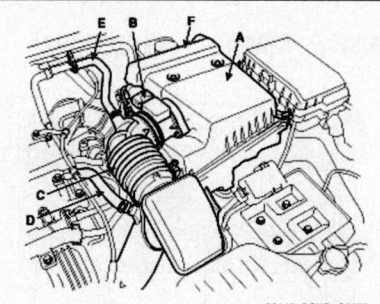

Fig. 30 Disconnect the AFS (B) connector, breather hose (D), and vacuum hose (E) from air cleaner hose (C)

c. Disconnect the PCM connector and cover (F).

d. Remove the intake air hose and air cleaner (A).

5. Remove front wheels.

6. Remove lower cover.

7. Drain the engine coolant.

8. Remove the upper and lower radiator hoses.

9. Remove the heater hoses.

10. Disconnect A/C switch (A), alternator connector (B), and oil pressure switch (C).

11. Disconnect OCV connector (A) and OTS connector (B) and P/S switch connector (C).

12. Disconnect injector connectors.

13. Disconnect the engine wire harness connectors.

a. Remove the ETC connector (B), CMP connector (C), knock sensor connector (D).

14. Disconnect ignition coil connectors.

15. Disconnect PCSV connector (A), WTS connector (B), condenser connector (C), and CKP sensor connector (D).

16. Remove delivery pipe (A), brake vacuum hose (B), and PCSV hose (C).

17. Remove water temp control assembly (A).

18. Remove intake manifold.

19. Remove exhaust manifold.

20. Remove timing chain.

21. Remove CVVT assembly and camshaft sprocket (A).

22. Remove camshaft.

a. Remove front camshaft bearing cap (A).

b. Remove camshaft bearing cap (A), in the sequence shown.

c. Remove camshafts.

23. Remove OCV (A) and OTS (B).

24. Remove the cylinder head bolts, then remove the cylinder head.

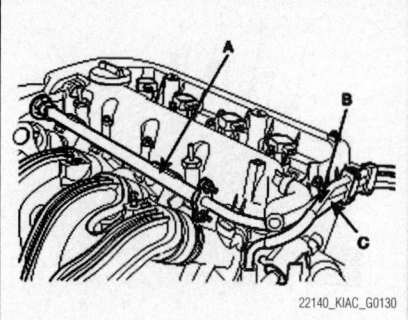

Fig. 35 Remove delivery pipe (A), brake vacuum hose (B), and PCSV hose (C)

Fig. 36 Remove water temp control assembly (A)

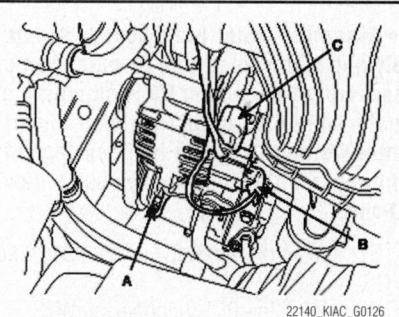

Fig. 31 Disconnect A/C switch (A), alternator connector (B), and oil pressure switch (C)

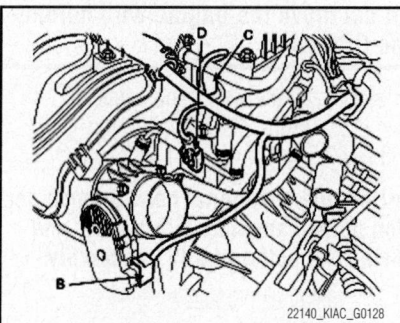

Fig. 33 Remove the ETC connector (B), CMP connector (C), knock sensor connector (D)

Fig. 37 Remove CVVT assembly and camshaft sprocket (A)

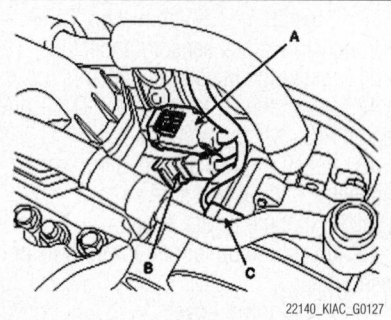

Fig. 32 Disconnect OCV connector (A) and OTS connector (B) and P/S switch connector (C)

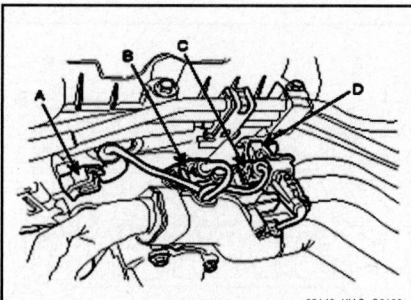

Fig. 34 Disconnect PCSV connector (A), WTS connector (B), condenser connector (C), and CKP sensor connector (D

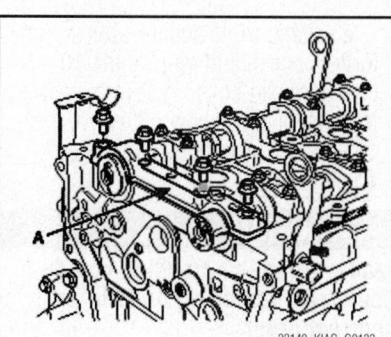

Fig. 38 Remove front camshaft bearing cap (A)

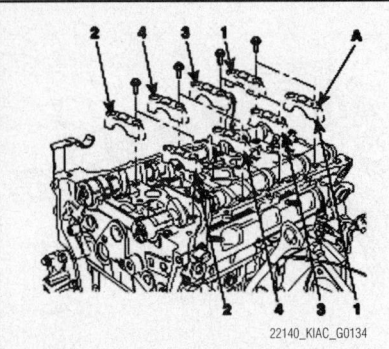

Fig. 39 Remove camshaft bearing cap (A), in the sequence shown

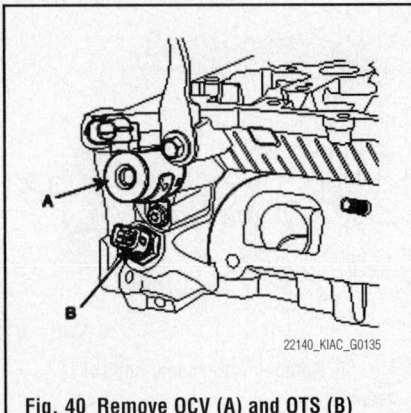

Fig. 40 Remove OCV (A) and OTS (B)

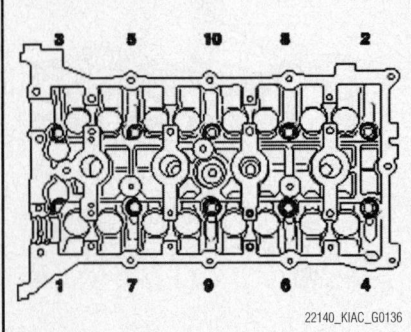

Fig. 41 Remove the cylinder head bolts in the sequence shown

a. Using triple square wrench, uniformly loosen and remove the 10 cylinder head bolts, in several passes, in the sequence shown. Remove the 10 cylinder head bolts and plate washers.

> ✳✳ **CAUTION**
>
> **Head warpage or cracking could result from removing bolts in an incorrect order.**

b. Lift the cylinder head from the dowels on the cylinder block and place the cylinder head on wooden blocks on a bench.

To install:

25. Install the cylinder head gasket on the cylinder block.

➥**Be careful of the installation direction.**

26. Place the cylinder head carefully in order not to damage the gasket with the bottom part of the end.

27. Install cylinder head bolts.

a. Apply a light coat if engine oil on the threads and under the heads of the cylinder head bolts.

b. Install and tighten the 10 cylinder head bolts and plate washers, in several passes, in the sequence shown. Torque: 25.3 ft. lbs. (34.3 Nm) plus 90° plus an additional 90°.

➥**Always use new cylinder head bolt.**

28. Install OCV and OTS. Torque: OCV: 7.23–8.67 ft. lbs. (9.8–11.76 Nm); OTS: 14.5–17.4 ft. lbs. (19.6–23.52 Nm).

> ✳✳ **CAUTION**
>
> **Do not reuse the OCV when dropped. Keep the OCV clean. Do not hold the OCV sleeve during servicing. When the OCV is installed on the engine, do not move the engine with holding the OCV yoke.**

29. Install the CVVT and camshaft sprocket. Torque: 39.7–47.0 ft. lbs. (53.9–63.7 Nm).

➥**Hold the hexagonal head wrench portion of the camshaft with a vise, and install the bolt and CVVT assembly.**

30. Install camshafts.

➥**Apply a light coat of engine oil on camshaft journals.**

31. Install camshaft bearing caps in their proper locations.

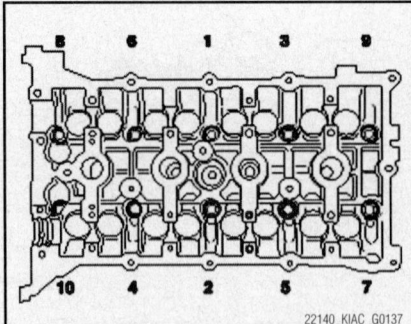

Fig. 42 Install cylinder head bolts in the sequence shown

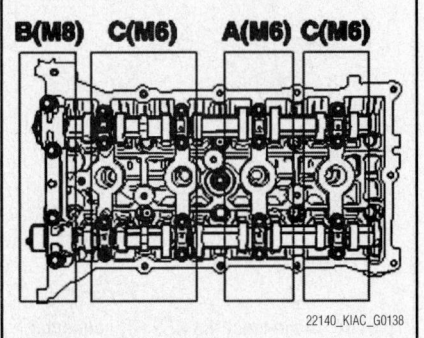

Fig. 43 Install camshaft bearing caps and torque in sequence

Tightening order:
- Group A
- Group B
- Group C.
- Torque: M6 bolts: 8.0–9.4 ft. lbs. (10.8–12.7 Nm); M8 bolts: 20.2–23.1 ft. lbs. (27.4–31.4 Nm).

32. Install timing chain.

33. Check and adjust valve clearance.

34. Install the exhaust manifold.

35. Install the intake manifold.

36. Install water temp control assembly. Torque: Bolt: 10.8–14.5 ft. lbs. (14.7–19.6 Nm); Nut: 13.7–17.4 ft. lbs. (18.6–23.5 Nm).

➥**Assemble water temp control assembly and water inlet pipe to water pump assembly before nuts for assembling of water inlet pipe to be tightened. Insert after wetting O-ring or inner surface of thermostat housing. Always use a new O-ring.**

37. Install delivery pipe, brake hose, and PCSV hose.

38. Install the PCSV connector, WTS connector, condenser connector, and CKP sensor connector.

39. Install ignition coil connector.

40. Install the engine wire harness connectors.

a. Install the ETC connector, CMP connector, knock sensor connector.

41. Install the injector connectors.

42. Install the OCV connector, OTS connector, P/S switch connector.

43. Install the A/C switch, alternator connector, and oil pressure switch.

44. Install the heater hoses.

45. Install the upper and lower radiator hoses.

46. Install lower cover.

47. Install the intake air hose and air cleaner assembly.

a. Install the AFS connector.

b. Install the breather hose and vacuum hose to air cleaner hose.

c. Install the PCM connector and cover.

d. Install the intake air hose and air cleaner.

48. Install the engine cover.

49. Install the negative terminal to the battery.

50. Fill with engine coolant.

51. Start the engine and check for leaks.

52. Recheck engine coolant level and oil level.

2.7L Engine

See Figures 44 through 63.

1. Remove the battery.
2. Remove the air duct.
3. Remove the engine cover.
4. Remove the intake air hose and air cleaner assembly.

   a. Disconnect the MAF connector (A).

   b. Disconnect the breather hose (B) from air cleaner hose.

   c. Remove the intake air hose and air cleaner assembly (C).

   d. Disconnect the PCM connectors (D).

5. Remove the upper and lower radiator hoses.

6. Remove the fuel inlet hose from the delivery pipe.

7. Disconnect the engine wiring harness connectors.

   a. Disconnect the No.1/No.2 knock sensor connectors (A, B), the oil pressure switch connector (C), the ignition coil harness (D) and the No.1 VIS (Variable Induction System) connector (E).

   b. Disconnect the bank 1 front/rear O2 sensor connectors (A).

   c. Disconnect the injection connectors (A B, C), the ground lines (D), the

condenser connector (E) and the Ignition coil connectors (F).

   d. Disconnect the injection harness connector (A), the No.2 VIS (Variable Induction System) connector (B), the No.1/No.2 OCV (Oil Control Valve) connectors (C, D) and the OTS (Oil Temperature Sensor) connector (E).

   e. Disconnect the MAPS (Manifold Absolute Pressure Sensor) connector (A), the ETC (Electronic Throttle Control)

connector (B) and the PCSV (Purge Control Solenoid Valve) connector (C).

   f. Disconnect the alternator connector and the air conditioning compressor connector.

   g. Disconnect the bank 2 CMP sensor connector (A) and the ECT (Engine Coolant Temperature) sensor connector (B).

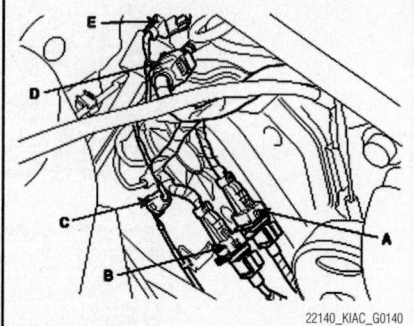

Fig. 45 Disconnect the engine wiring harness connectors

Fig. 46 Disconnect the bank 1 front/rear O2 sensor connectors (A)

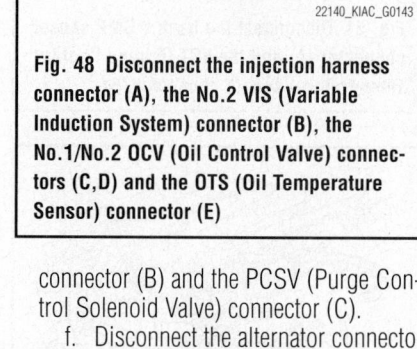

Fig. 48 Disconnect the injection harness connector (A), the No.2 VIS (Variable Induction System) connector (B), the No.1/No.2 OCV (Oil Control Valve) connectors (C,D) and the OTS (Oil Temperature Sensor) connector (E)

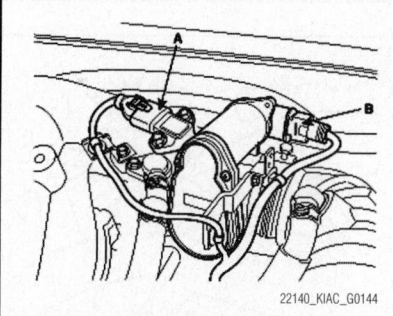

Fig. 49 Disconnect the MAPS (Manifold Absolute Pressure Sensor) connector (A), and the ETC (Electronic Throttle Control) connector (B)

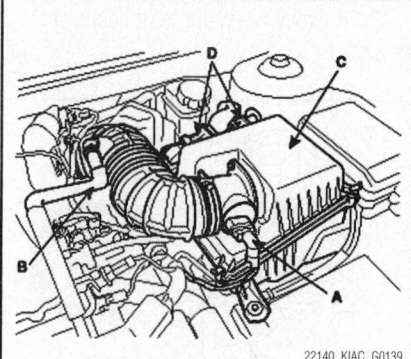

Fig. 44 Disconnect the MAF connector (A), breather hose (B), intake air hose and air cleaner assembly (C), and PCM connectors (D)

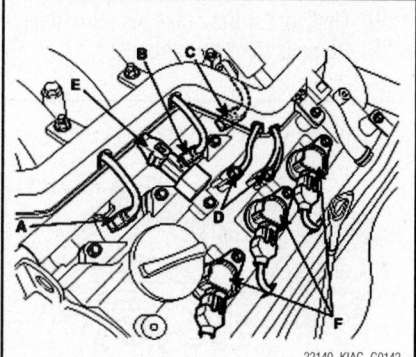

Fig. 47 Disconnect the injection connectors (A, B, C), the ground lines (D), the condenser connector (E) and the ignition coil connectors (F)

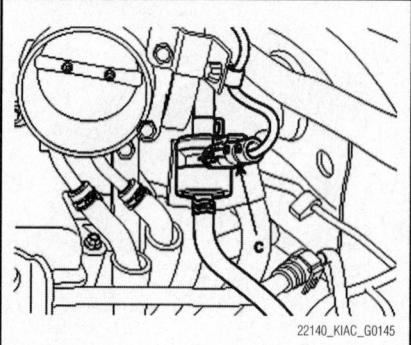

Fig. 50 Disconnect the PCSV (Purge Control Solenoid Valve) connector (C)

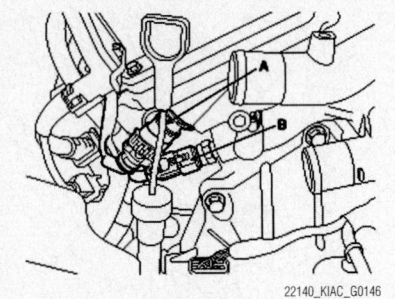

Fig. 51 Disconnect the bank 2 CMP sensor connector (A) and the ECT (Engine Coolant Temperature) sensor connector (B)

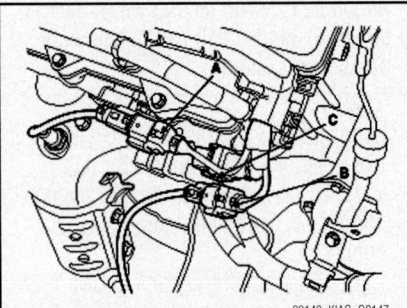

Fig. 52 Disconnect the bank 2 front/rear O2 sensor connectors (A,B) and the CKP sensor connector (C)

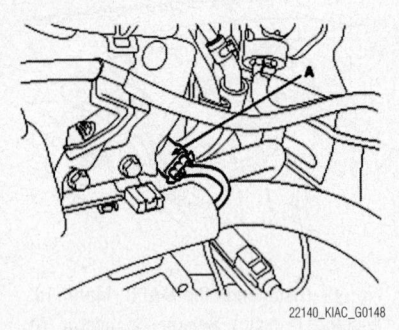

Fig. 53 Disconnect the bank 1 CMP sensor connector (A)

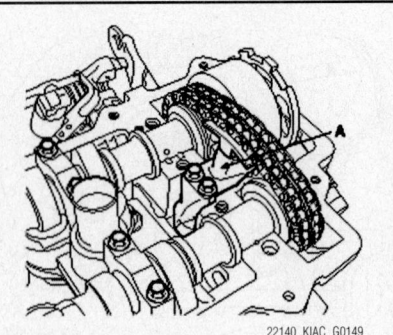

Fig. 54 Remove the timing chain tensioner (A)

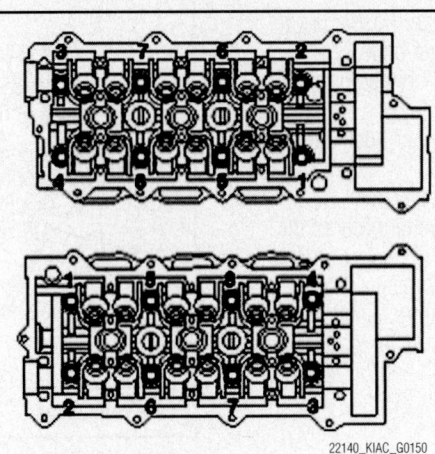

Fig. 55 Remove the bolts in the sequence shown

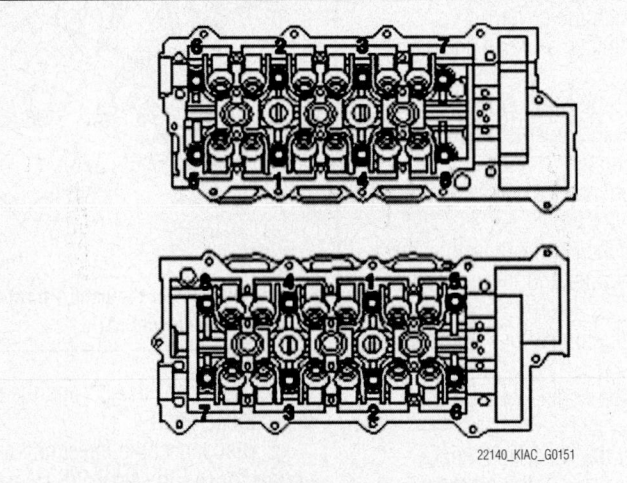

Fig. 56 Install cylinder head bolts in the sequence shown

h. Disconnect the bank 2 front/rear O2 sensor connectors (A,B) and the CKP sensor connector (C).

i. Disconnect the bank 1 CMP sensor connector (A).

8. Remove the PCV (Purge Control Valve) hose.

9. Disconnect the brake vacuum hose.

10. Remove the heater hoses.

11. Remove the accessory drive belt.

12. Remove the power steering pump.

13. Remove the exhaust manifold assembly.

14. Remove the intake manifold assembly.

15. Remove the timing belt.

16. Remove the ignition coils.

17. Remove the water temp. control assembly.

Fig. 57 Left hand camshaft chain timing mark

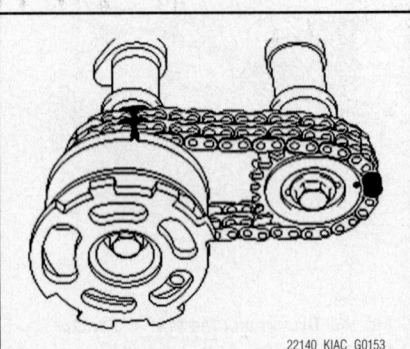

Fig. 58 Right hand camshaft chain timing mark

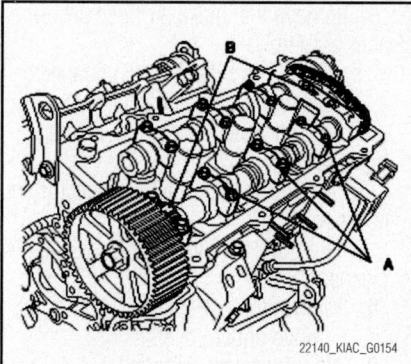

Fig. 59 Install the camshaft bearing caps

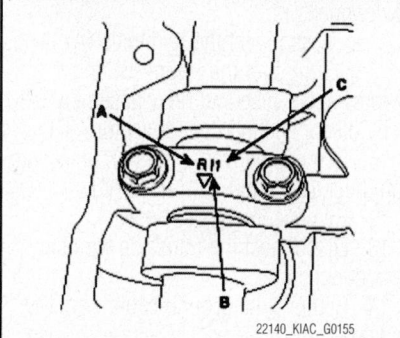

Fig. 60 When installing the bearing caps, check the marks

18. Remove the cylinder head cover.
19. Remove the camshaft bearing cap.
20. Remove the timing chain tensioner (A).
21. Remove the camshaft.
22. Remove the bank 1 timing belt rear cover.
23. Remove the bank 2 timing belt rear cover.
24. Remove the CKP sensor connector bracket.
25. Remove the cylinder head assembly.
    a. Remove the bolts in 2–3 steps in the sequence shown.

✳✳ CAUTION

If the bolts are not removed as the order, the deformation of the head assembly can be occurred.

b. Put the cylinder head assembly on a wooden block after removal from the cylinder block.

To install:
26. Install the cylinder head (s) with new head gaskets.

✳✳ CAUTION

Ensure the LH/RH classification of the cylinder head gasket when installing.

27. Tighten the cylinder head bolts with the plain washers in several steps in the sequence shown.

➡ **In assembling washers, the marked surface should face upward.**

28. Before installing the cylinder head bolts, apply engine oil on the thread of the bolts and the surface of the washers. Torque: 18.1 ft. lbs. (24.5 Nm) plus 60° plus an additional 45°.

➡ **Using the SST (09221-4A000), tighten the bolts which need to be tightened with the angular tightening method.**

29. Install the CVVT assembly and camshaft chain sprocket with the dowel pin in the CVVT installed to the intake camshaft. Ensure that the pin will not be installed in the hole for oil feeding. Torque: 49.2–57.9 ft. lbs. (66.7–78.5 Nm).

➡ **After tightening the CVVT bolts, rotate the CVVT assembly housing counterclockwise by hand to seat the lock pin in the CVVT assembly in good position.**

✳✳ CAUTION

Fix the hexagonal part of the camshaft in a vice when tightening the CVVT bolts. Do not fix the CVVT housing or sprocket in a vice.

30. Install the camshaft in the cylinder head assembly.
    a. Align the timing mark of the camshaft timing chain.

✳✳ CAUTION

Both timing marks should face upward in reassembly.

31. Install the timing chain tensioner.
    a. Insert the set pin by pressing the timing chain tensioner.
    b. Install the chain tensioner in the cylinder head assembly.
    c. Remove the set pin from the tensioner after installing.
32. Install the camshaft bearing caps. Torque: Bearing cap bolt (A: 6 × 38): 8.0–9.4 ft. lbs. (10.8–12.7 Nm). Bearing cap bolt (B: 8 × 38): 15.2–18.8 ft. lbs. (20.6–22.5 Nm).

➡ **When installing the bearing caps, check the marks on them as shown below and install them in its proper position. Note the following designations as shown below:**

- A: (LH/RH HEAD) L (LH), R (RH)
- B: (Intake/Exhaust) I (Intake), E (Exhaust)
- C: (Cap no.): 1, 2, 3

✳✳ CAUTION

When installing the bearing caps, turn the crankshaft to place a piston in the middle of the block because interference between valves and pistons can occur.

33. Using the SST (09214-21000), install the camshaft oil seal.

➡ **Before installing, apply engine oil. The camshaft cap surface should adhere to the cylinder head assembly. Do not press an eccentric load.**

34. Install the CKP sensor connector bracket.
35. Install the bank 2 timing belt rear cover.
36. Install the bank 1 timing belt rear cover.

➡ **The length of the bolt B is longer than that of the bolt C.**

37. Install the timing belt.
38. Check and adjust the valve clearance.

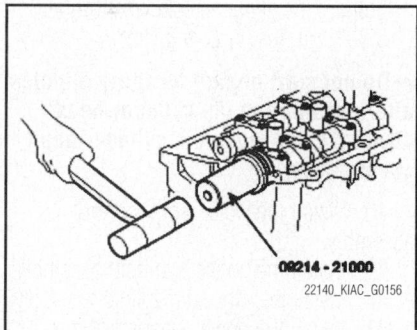

Fig. 61 Using the SST (09214-21000), install the camshaft oil seal

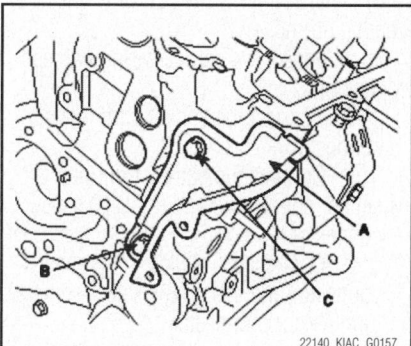

Fig. 62 Install the bank 1 timing belt rear cover

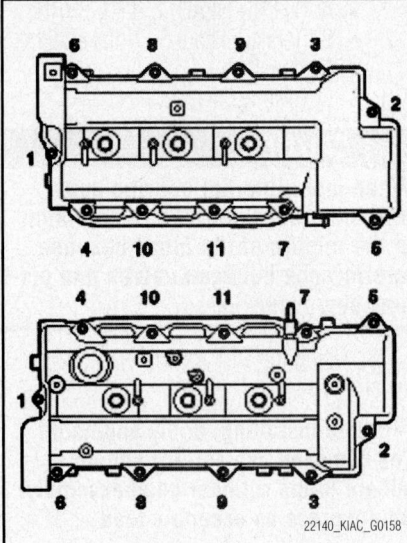

Fig. 63 Tighten the cylinder head cover bolts in sequence

39. Install the cylinder head cover.

a. Remove oil, dust or sealant on the upper surface of the cylinder before assembling cylinder head cover.

b. Assemble the cylinder head cover in five minutes after applying liquid gasket (LOCTITE 5900) on the camshaft cap and packing part.

c. Tighten the cylinder head cover bolts in the sequence shown. Torque: 5.8–7.2 ft. lbs. (7.8–9.8 Nm).

➡ **Do not start engine for thirty minutes after assembling the cylinder head cover. Do not reuse the cylinder head cover gasket.**

40. Install the water temp. control assembly.

41. Install the intake manifold assembly.

42. Install the exhaust manifold assembly.

43. Install the power steering pump.

44. Install the accessory drive belt.

45. Install the heater hose.

46. Connect the brake vacuum hose.

47. Install the PCV (Positive Crankcase Ventilation) hose.

48. Connect the engine wiring harness connectors.

a. Connect the bank 1 CMP sensor connector.

b. Connect the bank 2 front/rear O2 sensor connectors and the CKP sensor connector.

c. Connect the bank 2 CMP sensor connector and the WTS (Water Temperature Sensor) connector.

d. Connect the alternator connector and the air conditioning compressor connector.

e. Connect the MAPS (Manifold Absolute Pressure Sensor) connector, the ETC (Electronic Throttle Control) connector and the PCSV (Purge Control Solenoid Valve) connector.

f. Connect the injection harness connector, the No.2 VIS (Variable Induction System) connector, the No.1/No.2 OCV (Oil Control Valve) connectors and the OTS (Oil Temperature Sensor) connector.

g. Connect the injection connectors, the ground lines, the condenser connector and the Ignition coil connectors.

h. Connect the bank 1 front/rear O2 sensor connectors.

i. Connect the No.1/No.2 knock sensor connectors, the oil pressure switch connector, the ignition coil harness and the No.1 VIS (Variable Induction System) connector.

49. Install the fuel inlet hose from the delivery pipe.

50. Install the upper radiator hose and lower radiator hose.

51. Install the intake air hose and air cleaner assembly.

a. Connect the PCM connectors.

b. Install the intake air hose and air cleaner assembly.

c. Connect the breather hose from air cleaner hose.

d. Connect the MAF connector.

52. Install the engine cover.

53. Refill engine coolant.

ENGINE ASSEMBLY

REMOVAL & INSTALLATION

2.4L Engine

See Figures 64 through 68.

1. Remove the engine cover.
2. Recover A/C refrigerant.
3. Remove the air duct.
4. Remove the battery (A).

5. Remove the intake air hose and air cleaner assembly.

a. Disconnect the AFS (B) connector.

b. Disconnect the breather hose (D) and vacuum hose (E) from air cleaner hose (C).

c. Disconnect the PCM connector and cover (F).

d. Remove the intake air hose and air cleaner (A).

6. Remove the battery tray.

7. Remove under cover.

8. Drain the engine coolant.

9. Remove the upper and lower radiator hoses.

10. Disconnect the engine room junction box connectors.

a. Disconnect the terminals (A).

b. Turn over the levers (B) and remove the fuse and relay assembly (A) by disconnecting the connectors (C).

11. Remove delivery pipe (A), brake vacuum hose (B), and PCSV hose (C).

12. Remove the heater hoses.

13. Disconnect the transaxle wire harness connector.

14. Remove the ground cable from the transaxle.

15. Remove the shaft cable assembly.

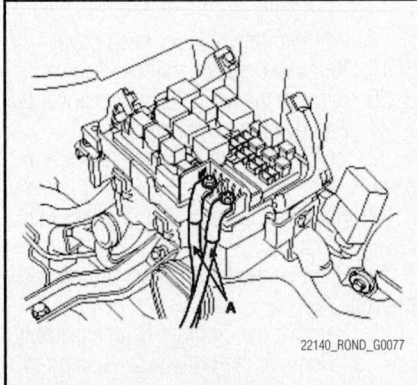

Fig. 65 Disconnect the terminals (A)

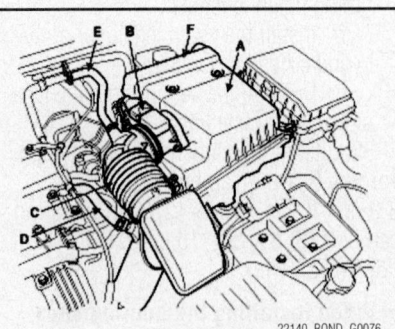

Fig. 64 Disconnect the AFS (B) connector, breather hose (D), and vacuum hose (E) from air cleaner hose (C)

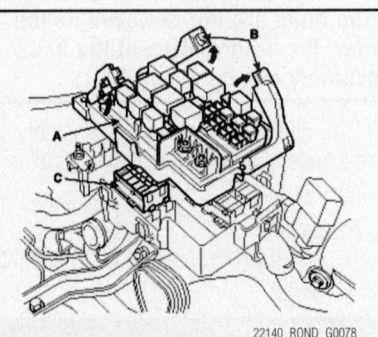

Fig. 66 Turn over the levers (B) and remove the fuse and relay assembly (A) by disconnecting the connectors (C)

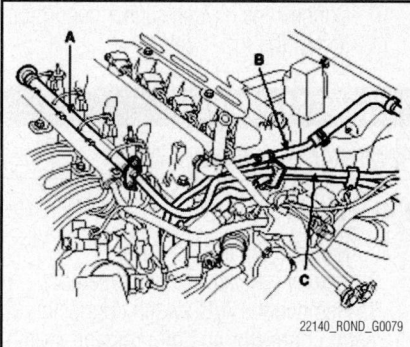

Fig. 67 Remove delivery pipe (A), brake vacuum hose (B), and PCSV hose (C)

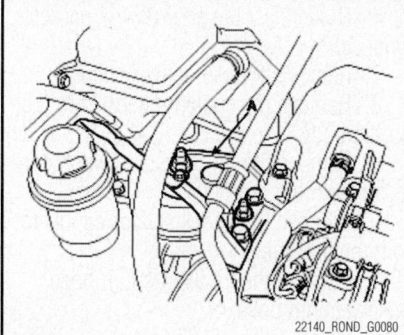

Fig. 68 Remove the engine mounting bracket (A)

16. Remove the steering u-joint mounting bolt.
17. Remove the power steering oil hose.
18. Remove the engine mounting bracket (A).
19. Remove the air conditioning compressor pipe.
20. Remove the transaxle mounting bracket.
21. Remove the front tires.
22. Remove the tie rod end nuts after removing the cotter pins.
23. Remove the stabilizer bar link nuts.
24. Remove the lower arm mounting bolts.
25. Remove the drive shaft cover and the heat protector.
26. Remove the exhaust pipe and the hanger.
27. Remove power steering return hose and drain power steering oil.
28. Install a jack and remove sub-frame with the engine and transaxle assembly.
29. Separate the engine from the transaxle assembly.

To install:
30. Installation is the reverse of removal.
31. Adjust the shift cable.
32. Adjust the throttle cable.
33. Refill the engine with engine oil.

34. Refill the transaxle with fluid.
35. Refill the radiator with engine coolant.
36. Place the heater control knob in "HOT" position.
37. Bleed air from cooling system.
38. Start engine and let it run until it warms up.
39. Turn off the engine. Check the level in the radiator, add coolant if needed.
40. Put the radiator cap on tightly, then run the engine again and check for leaks.
41. Clean the battery posts and cable terminals with sandpaper, then apply grease to prevent corrosion.
42. Inspect for fuel leakage.
43. After assembling the fuel line, turn on the ignition switch (do not operate the starter) so that the fuel pump runs for approximately two seconds and fuel line pressurizes.
44. Repeat this operation two or three times, then check for fuel leakage at any point in the fuel line.

2.7L Engine

See Figures 69 through 74.

1. Remove the battery.
2. Remove the engine cover.
3. Remove the air duct.
4. Remove the undercover.
5. Drain the engine coolant
6. Remove the air cleaner.
   a. Remove the power train control module (PCM) cover (D).
   b. Disconnect the MAF connector (A).
   c. Disconnect the breather hose (B) from air cleaner hose.
   d. Disconnect the Power Train Control Module (PCM) connector (F).
   e. Remove the air cleaner assembly.
7. Remove the battery tray.
8. Disconnect the engine wire harness connectors.
   a. Remove the relay & fuse box cover.

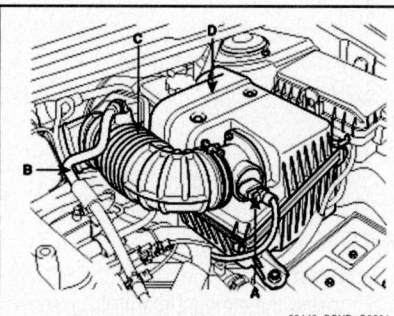

Fig. 69 Remove the Power Train Control Module (PCM) cover (D), MAF connector (A) and breather hose (B)

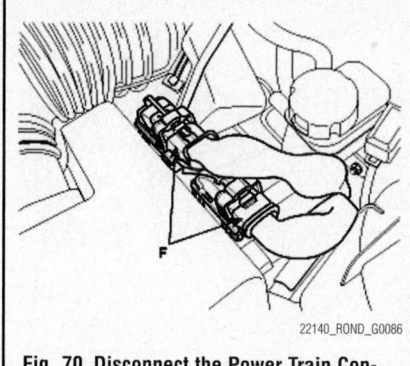

Fig. 70 Disconnect the Power Train Control Module (PCM) connector (F)

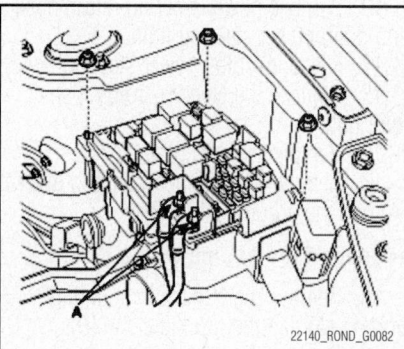

Fig. 71 Disconnect the cable (A) and the relay & fuse box mounting nuts

   b. Disconnect the cable (A) and the relay & fuse box mounting nuts.
   c. Remove the relay and fuses connector by pulling lever (A).
   d. Remove the control connector (B).
   e. Remove the wire harness mounting bolt and the ground cable mounting bolt.
9. Remove the heater hoses.
10. Disconnect the brake vacuum hose.
11. Remover the transaxle control cable.
12. Disconnect the Automatic Transaxle Fluid (ATF) cooler hoses from the radiator.

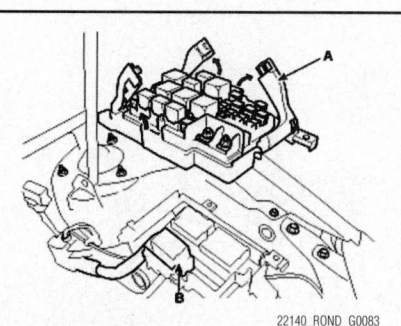

Fig. 72 Remove the relay and fuses connector by pulling lever (A). Remove the control connector (B)

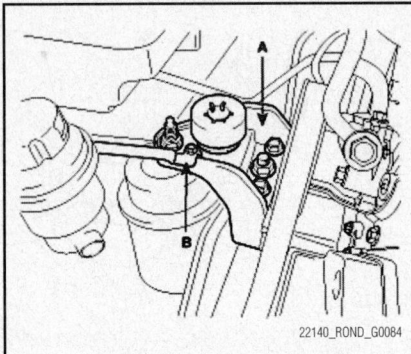

Fig. 73 Remove the engine mounting bracket (A) and the ground cable (B)

13. Remove power steering return hose and drain power steering fluid.

14. Disconnect the power steering hose.

15. Remove the steering column shaft joint bolt.

16. Remove the A/C pipe.

17. Remove the engine mounting bracket (A) and the ground cable (B).

➡**Support the engine with a jack.**

18. Remove the transaxle mounting bracket (A) and the ground cable (B).

19. Remove the tie rod end.

20. Remove the stabilizer bar link nuts.

21. Remove the lower arm mounting bolts.

22. Remove the drive shaft cover and the heat protector.

23. Remove the front muffler.

24. Supporting the engine and transaxle assembly with a jack, remove the assembly from the vehicle by removing the subframe mounting bolts and lifting up the vehicle slowly.

To install:

25. Installation is in the reverse order of removal.

26. Perform the following :

a. Adjust shift cable.

b. Adjust throttle cable.

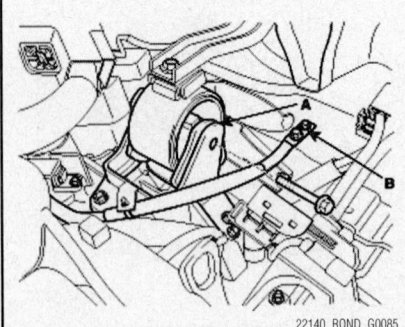

Fig. 74 Remove the transaxle mounting bracket (A) and the ground cable (B)

c. Refill engine with engine oil.

d. Refill transaxle with fluid.

e. Refill radiator with engine coolant.

f. Place the heater control knob in "HOT" position.

g. Bleed air from cooling system.

h. Start engine and let it run until it warms up.

i. Turn off the engine. Check the level in the radiator, add coolant if needed.

j. Put the radiator cap on tightly, then run the engine again and check for leaks.

k. Clean the battery posts and cable terminals with sandpaper, then apply grease to prevent corrosion.

l. Inspect for fuel leakage.

m. After assembling the fuel line, turn on the ignition switch (do not operate the starter) so that the fuel pump runs for approximately two seconds and fuel line pressurizes.

n. Repeat this operation two or three times, then check for fuel leakage at any point in the fuel line.

EXHAUST MANIFOLD

REMOVAL & INSTALLATION

2.4L Engine

1. Remove the oxygen sensor connector.

2. Remove the front muffler.

3. Remove the heat protector.

4. Remove the exhaust manifold bracket.

5. Remove the exhaust manifold and gasket.

To install:

6. Install the exhaust manifold. Torque: 28.92–32.53 ft. lbs. (39.2–44.1 Nm).

7. Install the exhaust manifold bracket

8. Install the heat protector.

9. Install the front muffler.

10. Install the oxygen sensor connector.

2.7L Engine

1. Remove the lower cover.

2. Remove the front muffler.

3. Disconnect the oxygen sensor connectors.

4. Remove the oil level gauge.

5. Remove the heat protector.

6. Remove the exhaust manifold assembly.

To install:

7. Install the exhaust manifold assembly with a new gasket. Torque: 21.7–25.3 ft. lbs. (29.4–34.3 Nm).

8. Install the heat protector.

9. Install the front muffler assembly.

10. Connect the oxygen sensor connector.

11. Install the lower cover.

INTAKE MANIFOLD

REMOVAL & INSTALLATION

2.4L Engine

1. Remove the engine cover.

2. Remove the air cleaner assembly.

3. Remove the A/C switch connector, alternator connector and oil pressure switch connector.

4. Remove the OCV connector, OTS connector, P/S switch connector.

5. Remove the injector connector.

6. Disconnect the engine wire harness connectors.

7. Remove the fuel delivery pipe.

8. Remove the coolant hoses from the throttle body.

9. Remove the oil pressure switch connector from the bracket.

10. Remove the knock sensor connector from the bracket.

11. Remove the PCSV vacuum hose, brake vacuum hose.

12. Remove the PCV hose.

13. Remove the intake manifold bracket.

14. Remove the oil level gauge.

15. Remove the intake manifold and gasket.

To install:

16. Install the intake manifold and gasket. Torque: 13.7–20.2 ft. lbs. (18.6–27.44 Nm).

17. Install the intake manifold bracket.

18. Install the PCV hose.

19. Install the PCSV vacuum hose, brake vacuum hose.

20. Install the knock sensor connector from the bracket.

21. Install the oil pressure switch connector from the bracket.

22. Install the coolant hoses from the throttle body.

23. Install the delivery pipe.

24. Install the engine wire harness connectors.

25. Install the injector connector.

26. Install Remove the OCV connector, OTS connector, P/S switch connector.

27. Install the A/C switch connector, alternator connector and oil pressure switch connector.

28. Install the air cleaner assembly.

29. Install the engine cover.

2.7L Engine

See Figures 75 through 87.

1. Remove the engine cover.

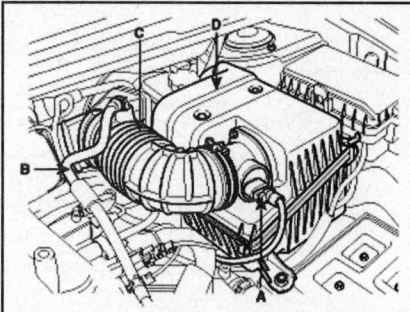

Fig. 75 Remove the Powertrain Control Module (PCM) cover (D), MAF connector (A) and breather hose (B)

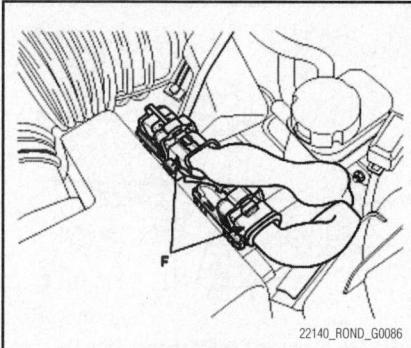

Fig. 76 Disconnect the Powertrain Control Module (PCM) connector (F)

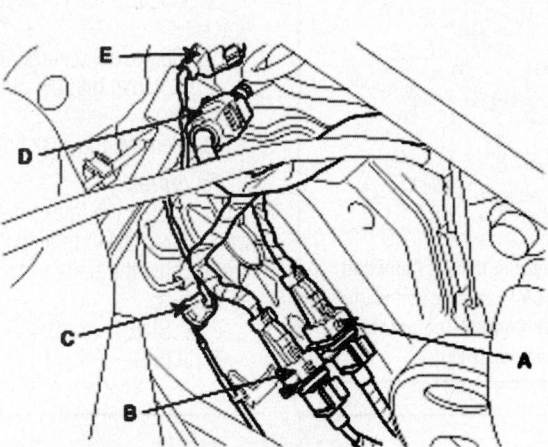

A – No.1 knock sensor
B – No.2 knock sensor
C – Oil pressure Switch connector
D – Ignition coil harness
E – No.1 Variable Induction System (VIS) connector

Fig. 77 Disconnect the No.1/No.2 knock sensor connectors (A, B), the oil pressure switch connector (C), the ignition coil harness (D) and the No.1 Variable Induction System (VIS) connector (E)

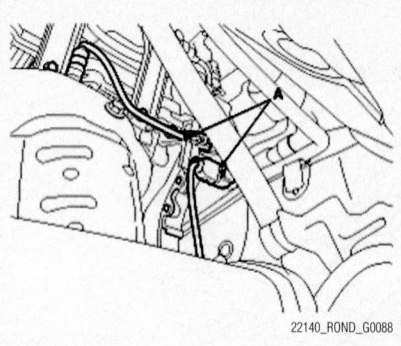

Fig. 78 Disconnect bank 1 front/rear O2 sensor connectors (A)

2. Remove the air cleaner.
  a. Remove the powertrain control module (PCM) cover (D).
  b. Disconnect the MAF connector (A).
  c. Disconnect the breather hose (B) from air cleaner hose.
  d. Disconnect the Powertrain Control Module (PCM) connector (F).
  e. Remove the air cleaner assembly.
3. Disconnect the engine wiring harness connectors.
  a. Disconnect the No.1/No.2 knock sensor connectors (A, B), the oil pressure switch connector (C), the ignition coil harness (D) and the No.1

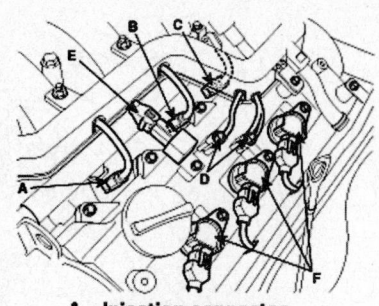

A – Injection connector
B – Injection connector
C – Injection connector
D – Ground lines
E – Condenser connectors
F – Ignition coil connectors

Fig. 79 Disconnect the injection connectors (A, B, C), the ground lines (D), the condenser connector (E) and the Ignition coil connectors (F)

Variable Induction System (VIS) connector (E).
  b. Disconnect bank 1 front/rear O2 sensor connectors (A).
  c. Disconnect the injection connectors (A ,B, C), the ground lines (D), the condenser connector (E) and the Ignition coil connectors (F).
  d. Disconnect the injection harness connector (A), the No.2 Variable Induction System (VIS) connector (B), the No.1/No.2 Oil Control Valve (OCV) connectors (C,D) and the Oil Temperature Sensor (OTS) connector (E).
  e. Disconnect the Manifold Absolute Pressure Sensor (MAP) connector (A).
  f. Disconnect the Electronic Throttle Control (ETC) connector (B) and the Purge Control Solenoid Valve (PCSV) connector (C).
  g. Disconnect the generator connector and the air conditioning compressor connector.
  h. Disconnect the bank 2 Camshaft Position Sensor (CMP) sensor connector (A) and the Engine Coolant Temperature (ECT) sensor connector (B).
  i. Disconnect the bank 2 front/rear O2 sensor connectors (A,B) and the Crankshaft Position Sensor (CKP) connector (C).
  j. Disconnect the bank 1 Camshaft Position Sensor (CMP) connector (A).
4. Remove the Purge Control Valve (PCV) hose.
5. Remove the Electronic Throttle Control (ETC) bracket (A) and the cooling hoses (B).

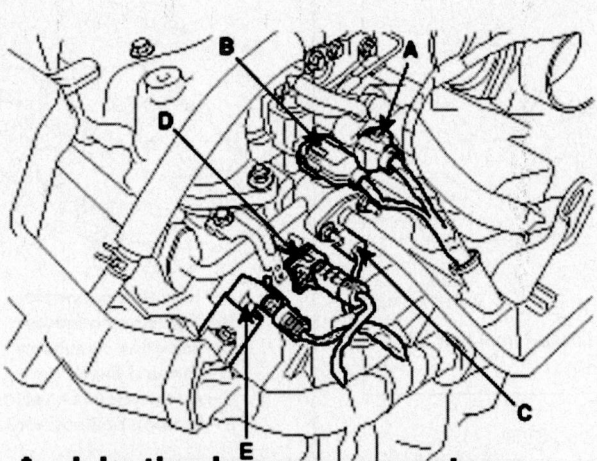

A – Injection harness connector
B – No.2 Variable Induction System
(VIS) connector
C – No.1 Oil Control Valve (OCV) connector
D – No.2 Oil Control Valve (OCV) connector
E – Oil Temperature Sensor (OTS) connector

22140_ROND_G0090

Fig. 80 Disconnect the injection harness connector (A), the No.2 Variable Induction System (VIS) connector (B), the No.1/No.2 Oil Control Valve (OCV) connectors (C,D) and the Oil Temperature Sensor (OTS) connector (E)

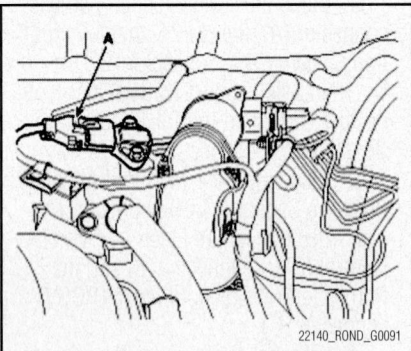

22140_ROND_G0091

Fig. 81 Disconnect the Manifold Absolute Pressure Sensor (MAP) connector (A)

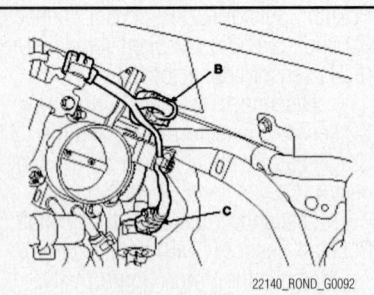

22140_ROND_G0092

Fig. 82 Disconnect the Electronic Throttle Control (ETC) connector (B) and the Purge Control Solenoid Valve (PCSV) connector (C)

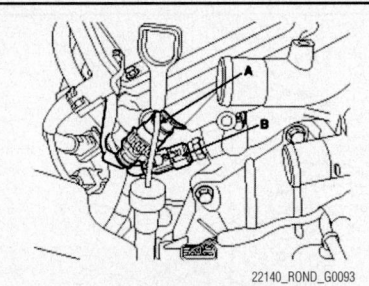

22140_ROND_G0093

Fig. 83 Disconnect the bank 2 Camshaft Position Sensor (CMP) sensor connector (A) and the Engine Coolant Temperature (ECT) sensor connector (B)

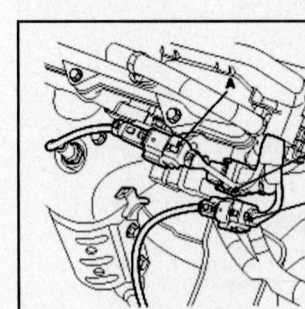

22140_ROND_G0094

Fig. 84 Disconnect the bank 2 front/rear O2 sensor connectors (A,B) and the Crankshaft Position Sensor (CKP) connector (C)

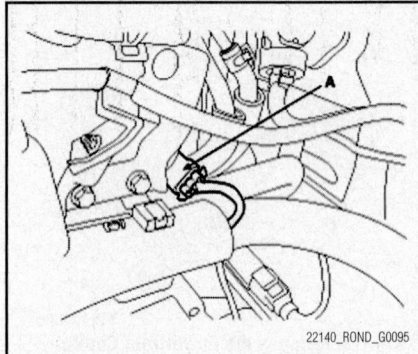

22140_ROND_G0095

Fig. 85 Disconnect the bank 1 Camshaft Position Sensor (CMP) connector (A)

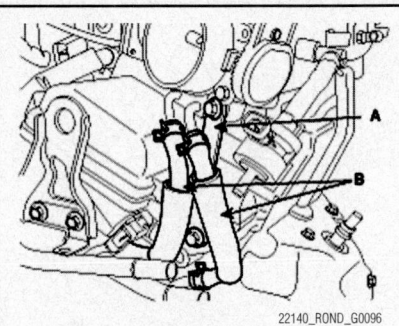

22140_ROND_G0096

Fig. 86 Remove the Electronic Throttle Control (ETC) bracket (A) and the cooling hoses (B)

6. Disconnect the brake vacuum hose.
7. Remove the surge tank mounting bracket.
8. Remove the surge tank.
9. Remove the fuel delivery pipe assembly.
10. Remove the intake manifold assembly.

To install:

11. Install the intake manifold assembly with a new gasket to a cylinder head assembly. Tighten the bolts in two steps. Torque as follows:
- Step 1: (a–h): 2.9–4.3 ft. lbs. (3.9–5.9 Nm)

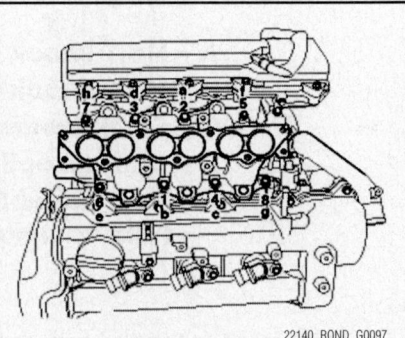

22140_ROND_G0097

Fig. 87 Install the intake manifold assembly in two steps

- Step 2: (1–8): 13.7–17.4 ft. lbs.
(18.6–23.5 Nm)
12. Install the delivery pipe.
13. Connect the LH injector connector.
14. Install the surge tank.
Tightening torque 18.6–23.5 Nm (14–17 ft. lbs.)
15. Install the surge tank mounting bracket.
Tightening torque 18.6–23.5 Nm (2.4kgf.m, 14–17 ft. lbs.)
16. Install the Electronic Throttle Control (ETC) system fixing bracket.
17. Connect all hoses and connectors.
18. Install the air cleaner assembly.
19. Install the engine cover.

OIL PAN

REMOVAL & INSTALLATION

2.7L Engine

See Figures 88 and 89.

1. Drain engine oil.
2. Remove the front right wheel and tire.
3. Remove the front right side cover.
4. Remove the front muffler.
5. Remove the alternator.
6. Remove the timing belt.
7. Remove the oil filter bracket.

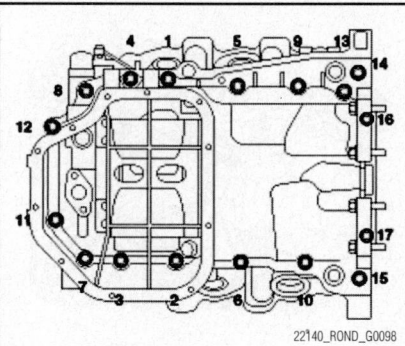

Fig. 88 Tighten the bolts in several steps uniformly

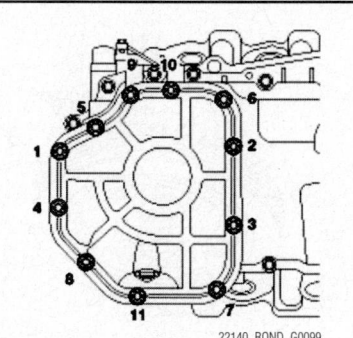

Fig. 89 Tighten the bolts in several steps uniformly

8. Using SST (09215-3C000), remove the lower oil pan.
9. Remove the oil screen.
10. Remove the upper oil pan, using the SST (09215-3C000).

To install:

11. Install the upper oil pan.
12. Tighten the bolts in several steps uniformly. Torque: Bolts 1–15: 13.7–17.4 ft. lbs. (18.6–23.5 Nm). Bolts 16 and 17: 3.6–5.1 ft. lbs. (4.9–6.9 Nm).
13. Install the oil screen. Torque: 10.8–15.9 ft. lbs. (14.7–21.6 Nm).
14. Install the lower oil pan.
15. Tighten the bolts in several steps uniformly. Torque: 7.2–8.7 ft. lbs. (9.8–11.8 Nm).
16. Install the oil filter bracket. Torque: 13.7–17.4 ft. lbs. (18.6–23.5 Nm).
17. Install the timing belt.
18. Install the alternator.
19. Install the front muffler.
20. Install the front right side cover.
21. Install the wheel and tire.
22. Fill with engine coolant.
23. Start engine and check for leaks.
24. Recheck engine coolant level.

OIL PUMP

REMOVAL & INSTALLATION

2.7L Engine

See Figures 88 and 89.

1. Drain engine oil.
2. Remove the front right wheel and tire.
3. Remove the front right side cover.
4. Remove the front muffler.
5. Remove the alternator.
6. Remove the timing belt.
7. Remove the oil filter bracket.
8. Using SST (09215-3C000), remove the lower oil pan.
9. Remove the oil screen.
10. Remove the upper oil pan, using the SST (09215-3C000).
11. Remove the oil pump case.

To install:

12. Install oil pump case. Torque: 13.7–17.4 ft. lbs. (18.6–23.5 Nm).

➡**In the installation of the oil pump, always use a new o-ring.**

13. Using the SST (09214~33000), install the oil pump case oil seal.
14. Install the upper oil pan.
15. Tighten the bolts in several steps uniformly. Torque: Bolts 1–15: 13.7–17.4 ft. lbs. (18.6–23.5 Nm). Bolts 16 and 17: 3.6–5.1 ft. lbs. (4.9–6.9 Nm).

16. Install the oil screen. Torque: 10.8–15.9 ft. lbs. (14.7–21.6 Nm).
17. Install the lower oil pan.
18. Tighten the bolts in several steps uniformly. Torque: 7.2–8.7 ft. lbs. (9.8–11.8 Nm).
19. Install the oil filter bracket. Torque: 13.7–17.4 ft. lbs. (18.6–23.5 Nm).
20. Install the timing belt.
21. Install the alternator.
22. Install the front muffler.
23. Install the front right side cover.
24. Install the wheel and tire.
25. Fill with engine coolant.
26. Start engine and check for leaks.
27. Recheck engine coolant level.

REAR MAIN SEAL

REMOVAL & INSTALLATION

1. Before servicing the vehicle, refer to the Precautions Section.
2. Disconnect the negative battery cable.
3. Remove the transaxle assembly.
4. Remove the flexplate-to-crankshaft bolts, the flexplate and shim plates.
5. Cut the oil seal lip with a knife. Install a rag to the housing and using a screwdriver, carefully pry the oil seal from the oil seal housing. Clean the gasket mounting surfaces.

To install:

6. Clean the oil seal housing. Coat the oil seal and the housing with clean engine oil.
7. Install the oil seal into the housing and tap it evenly into place with a hammer and a large diameter piece of pipe. The seal must be flush with the edge of the rear cover.
8. Install the flexplate and tighten the mounting bolts to 71–76 ft. lbs. (97–102 Nm).
9. Install the transaxle.
10. Connect the negative battery cable.

TIMING BELT FRONT COVER

REMOVAL & INSTALLATION

2.7L Engine

See Figures 90 through 94.

1. Remove the engine cover
2. Remove the front right wheel and tire.
3. Remove the right side cover.
4. Remove the drive belt, the idler pulley and the tensioner pulley.

➡**In removing the drive belt, fix a tool in the auto tensioner pulley bolt and turn the bolt counter clockwise.**

Fig. 90 Remove the timing belt upper covers (A)

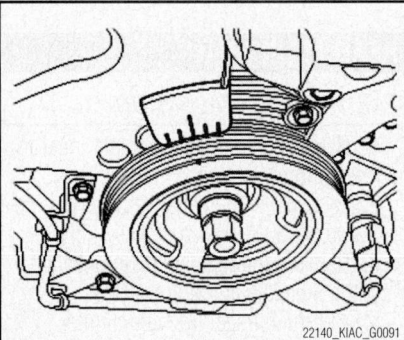

Fig. 91 Align the groove of the pulley with the timing mark of the timing belt cover

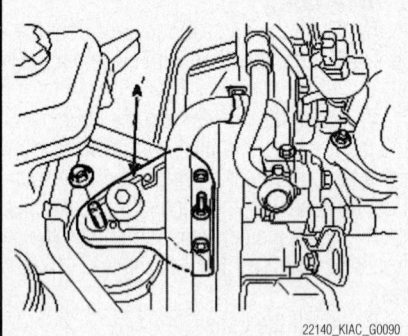

Fig. 92 Remove the engine mounting bracket (A)

Fig. 93 Remove the crankshaft damper pulley (A)

Fig. 94 Remove Timing belt lower cover (A)

5. Remove the timing belt upper covers (A).

6. Align the groove of the pulley with the timing mark of the timing belt cover by turning the crankshaft pulley clockwise. Check that the timing mark of the camshaft sprocket is aligned with that of the cylinder head cover. (No.1 cylinder piston at TDC)

7. Remove the engine mounting bracket.

   a. Support the engine oil pan with a jack.

❈❈ CAUTION

Put a wooden or rubber block between the jack and the engine oil pan.

   b. Remove the engine mounting bracket (A).

8. Remove the crankshaft damper pulley (A).

9. Remove the timing belt lower cover (A)

To install:

10. Install the timing belt lower cover. Torque: 7.2–8.7 ft. lbs. (9.8–11.8 Nm).

11. Install the crankshaft damper pulley. Torque: 123.0–130.2 ft. lbs. (166.7–176.5 Nm).

12. Install the engine mounting bracket. Torque: 47.0–61.5 ft. lbs. (63.7–83.4 Nm).

13. Install the timing belt upper cover. Torque: 7.2–8.7 ft. lbs. (9.8–11.8 Nm).

14. Install the drive belt tensioner. Torque: 25.3–39.8 ft. lbs. (34.3–53.9 Nm).

15. Install the drive belt idler and the drive belt. Torque: 25.3–39.8 ft. lbs. (34.3–53.9 Nm).

16. Install the right side cover.

17. Install the front right wheel and tire.

18. Install the engine cover. Torque: 5.8–8.7 ft. lbs. (7.8–11.8 Nm).

TIMING BELT AND SPROCKETS

REMOVAL & INSTALLATION

2.7L Engine

See Figures 95 through 102.

1. Remove the engine cover

2. Remove the front right wheel and tire.

3. Remove the right side cover.

4. Remove the drive belt, the idler pulley and the tensioner pulley.

➡**In removing the drive belt, fix a tool in the auto tensioner pulley bolt and turn the bolt counter clockwise.**

5. Remove the timing belt upper covers (A).

6. Align the groove of the pulley with the timing mark of the timing belt cover by turning the crankshaft pulley clockwise. Check that the timing mark of the camshaft sprocket is aligned with that of the cylinder head cover. (No.1 cylinder piston at TDC)

7. Remove the engine mounting bracket.

   a. Support the engine oil pan with a jack.

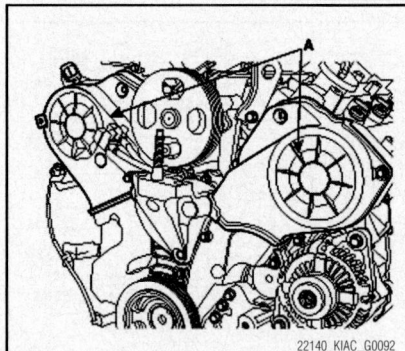

Fig. 95 Remove the timing belt upper covers (A)

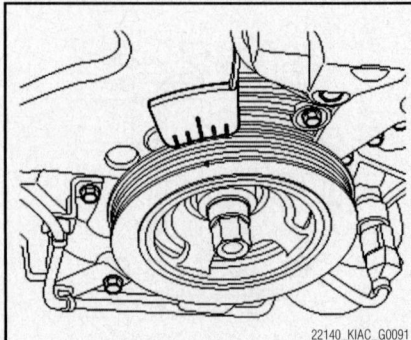

Fig. 96 Align the groove of the pulley with the timing mark of the timing belt cover

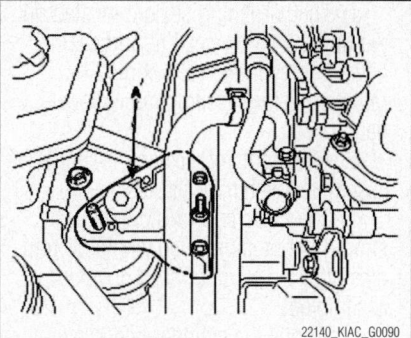

Fig. 97 Remove the engine mounting bracket (A)

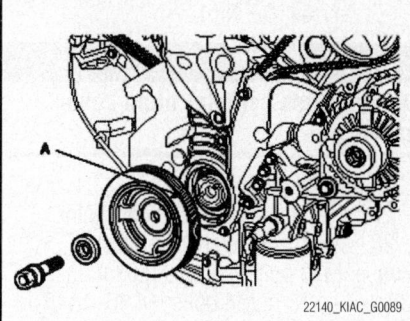

Fig. 98 Remove the crankshaft damper pulley (A)

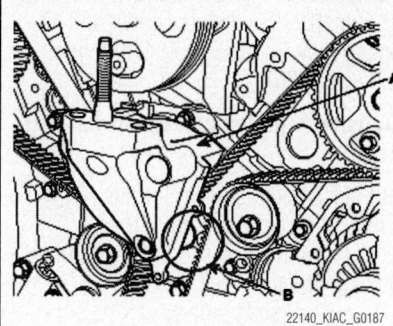

Fig. 99 Remove timing belt lower cover (A)

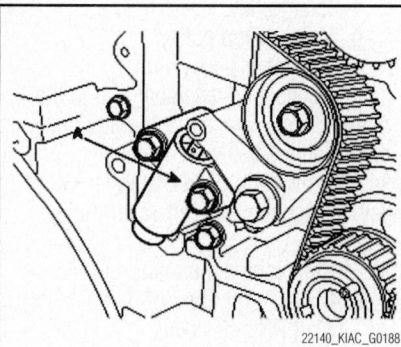

Fig. 100 Remove the timing belt auto tensioner (A)

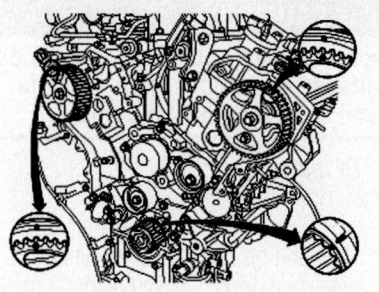

Fig. 101 Ensure the timing marks on the camshaft and the crankshaft sprockets are in the proper positions

✳✳ CAUTION

Put a wooden or rubber block between the jack and the engine oil pan.

   b. Remove the engine mounting bracket (A).

   8. Remove the crankshaft damper pulley (A).

   9. Remove the timing belt lower cover (A).

   10. Remove the engine support bracket (A).

➡**After removal, a small amount of engine coolant may drain from point (B).**

   11. Remove the timing belt auto tensioner (A).

   12. Remove the timing belt.

➡**Mark the direction of rotation on the timing belt.**

   13. Remove the timing belt tensioner arm assembly and the idler.

   14. Remove the crankshaft timing belt sprocket.

To install:

   15. Install the crankshaft sprocket.

   16. Install the tensioner arm assembly and the idler.

   17. Ensure the timing marks on the camshaft and the crankshaft sprockets are in the proper positions.

   18. Install the timing belt in the following order:

     a. Crankshaft sprocket (A)

     b. Idler (B)

     c. Bank 2 exhaust cam sprocket (C)

     d. Water pump pulley (D)

     e. Bank 1 exhaust cam sprocket (E)

     f. Tensioner pulley (F).

   19. Install the timing belt auto tensioner.

   a. Make the tensioner stand upright for about five minutes before installing.

➡**When handling the auto-tensioner observe the following:**

- Do not lay down the auto tensioner.
- Do not compress the rod suddenly.
- When reinstalling the auto tensioner, ensure proper orientation.
- Do not press the rod any more when its projection from the body is 2.5mm.
- Keep the auto-tensioner upright at room temperature in winter.

   b. Install the auto tensioner to the front case with the set-pin inserted. Torque: 14.5–19.5 ft. lbs. (19.6–26.5 Nm).

   20. Remove the auto tensioner set-pin.

   21. Check the tension of the timing belt.

   a. Turn the crankshaft 2 revolutions clockwise, and set the number one cylinder to TDC.

➡**After 5minutes, measure the length of the projected rod. Specification: 0.1969–0.2756 inches (5–7 mm).**

   b. Ensure the locations of the timing marks for each sprocket.

   22. Install the engine support bracket. Torque: 43.4–50.6 ft. lbs. (58.8–68.6 Nm).

   23. Install the timing belt lower cover. Torque: 7.2–8.7 ft. lbs. (9.8–11.8 Nm).

   24. Install the crankshaft damper pulley. Torque: 123.0–130.2 ft. lbs. (166.7–176.5 Nm).

   25. Install the engine mounting bracket. Torque: 47.0–61.5 ft. lbs. (63.7–83.4 Nm).

   26. Install the timing belt upper cover. Torque: 7.2–8.7 ft. lbs. (9.8–11.8 Nm).

   27. Install the drive belt tensioner. Torque: 25.3–39.8 ft. lbs. (34.3–53.9 Nm).

   28. Install the drive belt idler and the drive belt. Torque: 25.3–39.8 ft. lbs. (34.3–53.9 Nm).

   29. Install the right side cover.

   30. Install the front right wheel and tire.

   31. Install the engine cover. Torque: 5.8–8.7 ft. lbs. (7.8–11.8 Nm).

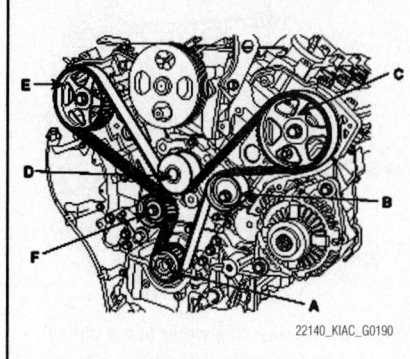

Fig. 102 Install the timing belt

TIMING CHAIN COVER AND SEAL

REMOVAL & INSTALLATION

2.4L Engine

See Figure 103.

1. Remove the engine cover.
2. Remove RH front wheel.
3. Remove RH side cover.
4. Set No.1 cylinder to TDC/compression
5. Remove the engine mount bracket.
6. Remove the accessory drive belt.
7. Remove the idler pulley.
8. Remove the accessory drive belt tensioner.

➡**Tensioner pulley bolt is left - handed screw.**

9. Remove the water pump pulley (A).
10. Remove the crankshaft pulley (B).
11. Remove the engine support bracket (C).
12. Disconnect the ignition coil connector.
13. Remove the ignition coils.
14. Remove the PCV hose and breather hose from the cylinder head cover.
15. Remove the cylinder head cover and gasket.
16. Remove the A/C compressor lower bolts.
17. Remove the compressor bracket.
18. Drain the engine oil.
19. Remove the oil pan.

✳✳ CAUTION

Be careful not to damage the contact surfaces of cylinder block and oil pan.

20. Remove the timing chain cover by prying the portions between the cylinder head and cylinder block with a screwdriver.

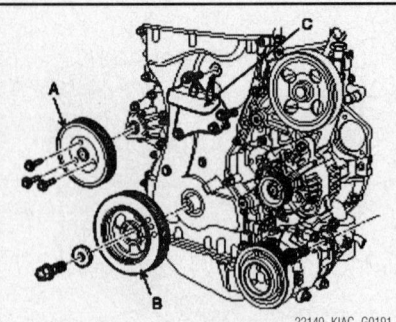

22140_KIAC_G0191

Fig. 103 Remove the water pump pulley (A), crankshaft pulley (B), and engine support bracket (C)

✳✳ CAUTION

Be careful not to damage the contact surfaces of cylinder block, cylinder head and timing chain cover.

To install:

21. Install timing chain cover.
    a. The sealant locations on chain cover and on counter parts (cylinder head, cylinder block, and ladder frame) must be free of engine oil and ETC.
    b. Before assembling the timing chain cover, the liquid sealant Loctite 5900 should be applied on the gap between cylinder head and cylinder block.
    c. After applying liquid sealant Loctite 5900 on timing chain cover, the part must be assembled within 5 minutes.
    d. The dowel pins on the cylinder block and holes on the timing chain cover should be used as a reference in order to assemble the timing chain cover to be in exact position.
    e. Torque: M6 bolts: 5.78–7.23 ft. lbs. (7.84–9.8 Nm). M8 bolts: 13.74–16.63 ft. lbs. (18.62–22.54 Nm).
    f. The firing and/or blow out test should not be performed within 30 minutes after the timing chain cover is assembled.
22. Install the oil pan.
    a. Using a gasket scraper, remove all the old packing material from the gasket surfaces.
    b. Before assembling the oil pan, the liquid sealant Loctite 5900 should be applied on oil pan.
    The part must be assembled within 5 minutes after the sealant was applied.

✳✳ CAUTION

When applying sealant gasket, sealant must not be protruded into the inside of oil pan. To prevent leakage of oil, apply sealant gasket to the inner threads of the bolt holes.

    c. Install oil pan.
    d. Uniformly tighten the bolts in several passes. Torque: M8 bolts: 19.52–22.41 ft. lbs. (26.46–30.38 Nm). M6 bolts: 7.23–8.67 ft. lbs. (9.8–11.6 Nm).
    e. After assembly, wait at least 30 minutes before filling the engine with oil.
23. Install air compressor bracket. Torque: 14.46–17.35 ft. lbs. (19.6–23.52 Nm).
24. Install air compressor bolts. Torque: 14.46–17.35 ft. lbs. (19.6–23.52 Nm).
25. Install cylinder head cover.

    a. The hardening sealant located on the upper area between timing chain cover and cylinder head should be removed before assembling cylinder head cover.
    b. After applying sealant, it should be assembled within 5 minutes.
    c. The firing and/or blow out test should not be performed within 30 minutes after the cylinder head cover was assembled.
    d. Torque the cylinder head cover bolts as follows:
    - 1st step: 2.89–4.34 ft. lbs. (3.92–5.88 Nm)
    - 2st step: 5.78–7.23 ft. lbs. (7.84–9.8 Nm)

✳✳ CAUTION

Do not reuse cylinder head cover gasket.

26. Install ignition coil.
27. Connect ignition coil connector.
28. Install engine support bracket. Torque: M10 bolts: 28.92–32.53 ft. lbs. (39.2–44.1 Nm); M8 bolts: 14.46–18.07 ft. lbs. (19.6–24.5 Nm).
29. Install crankshaft pulley. Torque: 122.9–130.13 ft. lbs. (166.6–176.4 Nm).
30. Install water pump pulley. Torque: 5.78–7.23 ft. lbs. (7.84–9.8 Nm).
31. Install drive belt tensioner and tensioner pulley. Torque: 39.7–47.0 ft. lbs. (53.9–63.7 Nm).

➡**Tensioner pulley bolt is left-handed screw.**

32. Install idler pulley. Torque: 39.7–47.0 ft. lbs. (53.9–63.7 Nm).
33. Install the accessory drive belt in the following order:
    a. Crankshaft pulley
    b. A/C pulley
    c. Alternator pulley
    d. Idler pulley
    e. P/S pump pulley
    f. Idler pulley
    g. Water pump pulley
    h. Tensioner pulley.
    i. Rotate auto tensioner arm in the counter—clockwise moving auto tensioner pulley bolt with a wrench.
34. After putting belt on auto tensioner pulley, release the auto tensioner pulley slowly.
35. Install engine mounting bracket. Torque: 47.0–61.4 ft. lbs. (63.7–83.3 Nm).
36. Install RH side cover.
37. Install RH front wheel.
38. Install engine cover.

TIMING CHAIN AND SPROCKETS

REMOVAL & INSTALLATION

2.4L Engine

See Figures 104 through 107.

1. Remove the engine cover.
2. Remove RH front wheel.
3. Remove RH side cover.
4. Set No.1 cylinder to TDC/compression
5. Remove the engine mount bracket.
6. Remove the accessory drive belt.
7. Remove the idler pulley.
8. Remove the accessory drive belt tensioner.

➡**Tensioner pulley bolt is left - handed screw.**

9. Remove the water pump pulley (A).
10. Remove the crankshaft pulley (B).
11. Remove the engine support bracket (C).
12. Disconnect the ignition coil connector.
13. Remove the ignition coils.
14. Remove the PCV hose and breather hose from the cylinder head cover.
15. Remove the cylinder head cover and gasket.
16. Remove the A/C compressor lower bolts.
17. Remove the compressor bracket.
18. Drain the engine oil.
19. Remove the oil pan.

✳✳ CAUTION

Be careful not to damage the contact surfaces of cylinder block and oil pan.

20. Remove the timing chain cover by prying the portions between the cylinder head and cylinder block with a screwdriver.

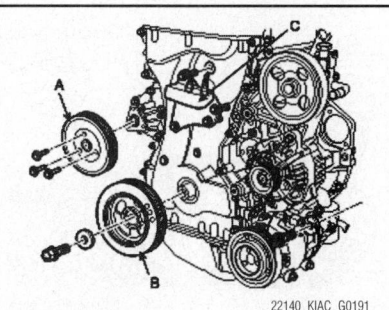

Fig. 104 Remove the water pump pulley (A), crankshaft pulley (B), and engine support bracket (C)

✳✳ CAUTION

Be careful not to damage the contact surfaces of cylinder block, cylinder head and timing chain cover.

21. The key of crankshaft should be aligned with the mating face of main bearing cap. As a result of this, the piston of No.1 cylinder is placed at the top dead center on compression stroke.
22. Install a set pin after compressing the timing chain tensioner.
23. Remove the timing chain tensioner (A).
24. Remove the timing chain tensioner arm (B).

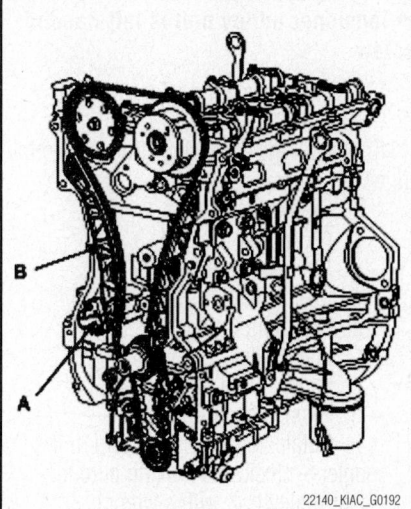

Fig. 105 Remove the timing chain tensioner (A) and timing chain tensioner arm (B)

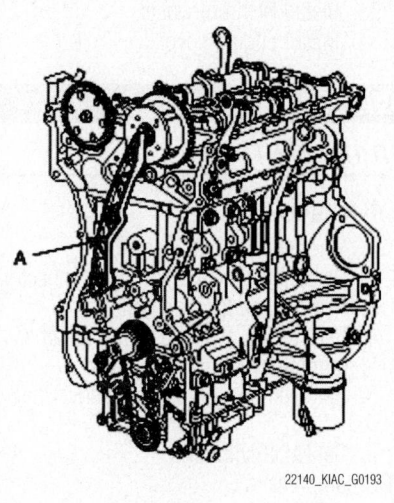

Fig. 106 Remove the timing chain guide (A)

25. Remove the timing chain.
26. Remove the timing chain guide (A).
27. Remove the timing chain oil jet (A).
28. Remove the crankshaft chain sprocket (B).

To install:

29. Install crankshaft chain sprocket.
30. Install timing chain oil jet. Torque: 5.78–7.23 ft. lbs. (7.84–9.8 Nm).
31. The key of crankshaft should be aligned with the mating surface of main bearing cap. As a result, this places the piston on No.1 cylinder at the top dead center on compression stroke.
32. Install timing chain guide. Torque: 7.23–8.67 ft. lbs. (9.8–11.6 Nm).
33. Install timing chain.

➡**To install the timing chain with no slack between each shaft (cam, crank), follow the below procedure:**

- Crankshaft sprocket (A)
- Timing chain guide (B)
- Intake camshaft sprocket (C)
- Exhaust camshaft sprocket (D).

The timing mark of each sprockets should be matched with timing mark (color link) of timing chain at installing of the timing chain.

34. Install timing chain tensioner arm. Torque: 7.23–8.67 ft. lbs. (9.8–11.6 Nm).
35. Install timing chain auto tensioner and remove set pin. Torque: 7.23–8.67 ft. lbs. (9.8–11.6 Nm).
36. After rotating crankshaft 2 revolutions in regular direction (clockwise viewed from front), confirm the timing mark.
37. Install timing chain cover.
   a. The sealant locations on chain cover and on counter parts (cylinder head, cylinder block, and ladder frame) must be free of engine oil and ETC.
   b. Before assembling the timing chain cover, the liquid sealant Loctite 5900 should be applied on the gap between cylinder head and cylinder block.

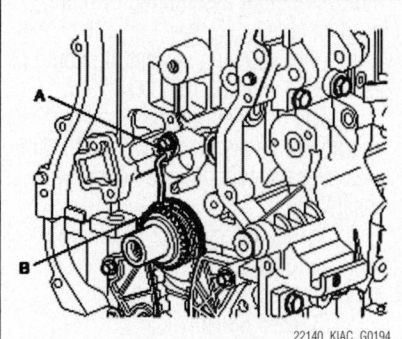

Fig. 107 Remove the timing chain oil jet (A) and crankshaft chain sprocket (B)

c. After applying liquid sealant Loctite 5900 on timing chain cover, the part must be assembled within 5 minutes.

d. The dowel pins on the cylinder block and holes on the timing chain cover should be used as a reference in order to assemble the timing chain cover to be in exact position.

e. Torque: M6 bolts: 5.78–7.23 ft. lbs. (7.84–9.8 Nm). M8 bolts: 13.74–16.63 ft. lbs. (18.62–22.54 Nm).

f. The firing and/or blow out test should not be performed within 30 minutes after the timing chain cover is assembled.

38. Install the oil pan.

a. Using a gasket scraper, remove all the old packing material from the gasket surfaces.

b. Before assembling the oil pan, the liquid sealant Loctite 5900 should be applied on oil pan.

The part must be assembled within 5 minutes after the sealant was applied.

✳✳ CAUTION

When applying sealant gasket, sealant must not be protruded into the inside of oil pan. To prevent leakage of oil, apply sealant gasket to the inner threads of the bolt holes.

c. Install oil pan.

d. Uniformly tighten the bolts in several passes. Torque: M8 bolts: 19.52–22.41 ft. lbs. (26.46–30.38 Nm). M6 bolts: 7.23–8.67 ft. lbs. (9.8–11.6 Nm).

e. After assembly, wait at least 30 minutes before filling the engine with oil.

39. Install air compressor bracket. Torque: 14.46–17.35 ft. lbs. (19.6–23.52 Nm).

40. Install air compressor bolts. Torque: 14.46–17.35 ft. lbs. (19.6–23.52 Nm).

41. Install cylinder head cover.

a. The hardening sealant located on the upper area between timing chain cover and cylinder head should be removed before assembling cylinder head cover.

b. After applying sealant, it should be assembled within 5 minutes.

c. The firing and/or blow out test should not be performed within 30 minutes after the cylinder head cover was assembled.

d. Torque the cylinder head cover bolts as follows:

- 1st step: 2.89–4.34 ft. lbs. (3.92–5.88 Nm)
- 2st step: 5.78–7.23 ft. lbs. (7.84–9.8 Nm).

✳✳ CAUTION

Do not reuse cylinder head cover gasket.

42. Install ignition coil.

43. Connect ignition coil connector.

44. Install engine support bracket. Torque: M10 bolts: 28.92–32.53 ft. lbs. (39.2–44.1 Nm); M8 bolts: 14.46–18.07 ft. lbs. (19.6–24.5 Nm).

45. Install crankshaft pulley. Torque: 122.9–130.13 ft. lbs. (166.6–176.4 Nm).

46. Install water pump pulley. Torque: 5.78–7.23 ft. lbs. (7.84–9.8 Nm).

47. Install drive belt tensioner and tensioner pulley. Torque: 39.7–47.0 ft. lbs. (53.9–63.7 Nm).

➡ **Tensioner pulley bolt is left-handed screw.**

48. Install idler pulley. Torque: 39.7–47.0 ft. lbs. (53.9–63.7 Nm).

49. Install the accessory drive belt in the following order:

a. Crankshaft pulley
b. A/C pulley
c. Alternator pulley
d. Idler pulley
e. P/S pump pulley
f. Idler pulley
g. Water pump pulley
h. Tensioner pulley.
i. Rotate auto tensioner arm in the counter—clockwise moving auto tensioner pulley bolt with a wrench.

50. After putting belt on auto tensioner pulley, release the auto tensioner pulley slowly.

51. Install engine mounting bracket. Torque: 47.0–61.4 ft. lbs. (63.7–83.3 Nm).

52. Install RH side cover.

53. Install RH front wheel.

54. Install engine cover.

VALVE LASH

ADJUSTMENT

2.4L Engine

1. Remove the engine cover.
2. Disconnect the ignition coil connector.
3. Remove the ignition coils.
4. Remove the PCV hose and breather hose from the cylinder head cover.
5. Remove the cylinder head cover and gasket.
6. Set No.1 cylinder to TDC/compression.

a. Turn the crankshaft pulley and align its groove with the timing mark "T" of the lower timing chain cover.

b. Check that the mark of the camshaft timing sprockets are in straight line on the cylinder head surface.

If not, turn the crankshaft one revolution (360°)

7. Inspect the valve clearance.

a. Check only the No. 1 cylinder: TDC/Compression. Measure the valve clearance. Using a thickness gauge, measure the clearance between the tappet and the base circle of camshaft.

b. Record the out-of-specification valve clearance measurements. They will be used later to determine required replacement of adjusting tappet (s). Valve clearance specification with engine coolant temperature of 68°F (20°C) Limit: Intake: 0.0039–0.0118inches (0.10–0.30 mm); Exhaust: 0.0079–0.0157inches (0.20–0.40 mm).

a. Turn the crankshaft pulley one revolution (360°) and align the groove with timing mark "T" of the lower timing chain cover.

b. Check only valves at TDC/compression. Measure the valve clearance.

8. Adjust the intake and exhaust valve clearance.

2.7L Engine

See Figures 108 through 110.

➡ **Inspect and adjust the valve clearance when the engine is cold and cylinder head is installed on the cylinder block.**

1. Remove the engine cover.
2. Remove air cleaner assembly.
3. Remove the surge tank.

a. Disconnect the ignition coil connector and remove the ignition coil.

b. Remove the cylinder head cover.

4. Set the piston of the No.1 cylinder to Top Dead Center (TDC) position.

a. Turn the crankshaft pulley clock-

22140_ROND_G0100

Fig. 108 Check that the timing marks on the camshaft sprocket are in a straight line with the rocker cover mark

wise and align its groove with the timing mark "T" of the timing chain cover.

b. Check that the timing marks on the camshaft sprocket are in a straight line with the rocker cover mark for No. 1 cylinder TDC as shown in the illustration.

➡ **If not, turn the crankshaft one revolution clockwise.**

5. Inspect the intake and the exhaust valve clearance.

a. With No. 1 cylinder at TDC the valves which can be measured its clearance are as shown below.

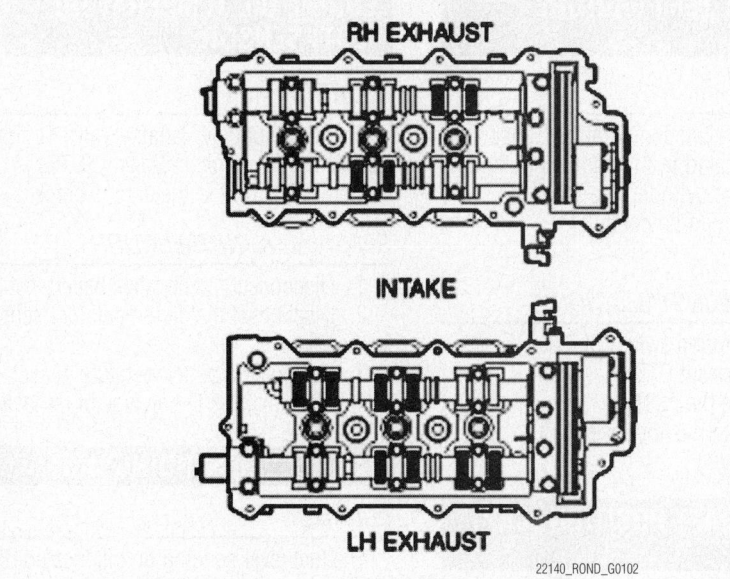

Fig. 110 With the piston of the No.4 cylinder positioning at TDC, the valves which can be measured

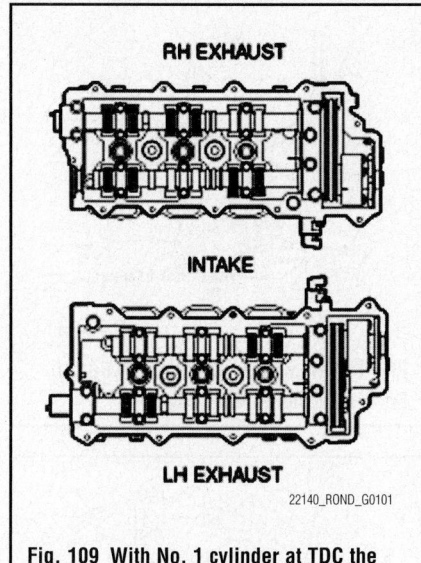

Fig. 109 With No. 1 cylinder at TDC the valves which can be measured

6. Measurement method:

a. Using a thickness gauge, measure the clearance between the tappet and the base circle of camshaft.

b. Record the out-of-specification valve clearance measurements. They will be used later to determine the required adjusting tappet (s) for replacement.

Valve clearance specification with engine coolant temperature of 68°F (20°C) Limit: Intake: 0.0039–0.0118inches (0.10–0.30 mm); Exhaust: 0.0079–0.0157inches (0.20–0.40 mm).

a. Turn the crankshaft pulley one revolution (360°) clockwise and align the groove with the timing mark "T" of the timing chain cover.

b. With the piston of the No.4 cylinder positioning at TDC, the valves which can be measured its clearance are as shown below.

7. Adjust the intake and the exhaust valve clearances.

ENGINE PERFORMANCE & EMISSION CONTROL

ACCELERATOR PEDAL POSITION (APP) SENSOR

LOCATION

Accelerator Position Sensor (APS) is installed on the accelerator pedal module and detects the rotation angle of the accelerator pedal.

REMOVAL & INSTALLATION

1. Turn ignition switch off and disconnect the negative battery cable.
2. Disconnect the accelerator position sensor connector.
3. Unfasten the four mounting nuts
4. Remove the accelerator pedal from the vehicle.
5. Installation is the reverse of removal.

AIR CHARGE TEMPERATURE (ACT) SENSOR

LOCATION

The Intake Air Temperature Sensor (IATS) is located in the engine compartment on the air cleaner housing.

CAMSHAFT POSITION (CMP) SENSOR

LOCATION

2.4L Engine
See Figure 111.

The Camshaft Position Sensor (CMPS) is installed on engine head cover and uses a target wheel installed on the camshaft.

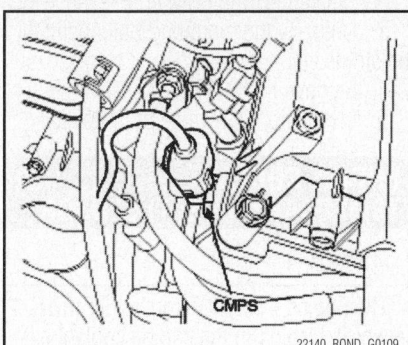

Fig. 111 The CMPS is installed on engine head cover– 2.4L engine.

2.7L Engine

The two Camshaft Position Sensors (CMPS) are installed on engine head cover of bank 1 and 2 and use a target wheel installed on each camshaft.

COOLANT TEMPERATURE SENSOR

LOCATION

Engine Coolant Temperature Sensor (ECTS) is located in the engine coolant passage of the cylinder head for detecting the engine coolant temperature.

REMOVAL & INSTALLATION

1. Turn ignition switch OFF.
2. Disconnect ECTS connector.
3. Remove the ECTS.
4. Installation is the reverse of removal.

CRANKSHAFT POSITION (CKP) SENSOR

LOCATION

The Crankshaft Position Sensor is installed on transaxle housing

ENGINE CONTROL MODULE (ECM)

LOCATION

2.4L Engine

The Engine Control Module (ECM) is installed near the air cleaner.

REMOVAL & INSTALLATION

2.4L Engine

1. Turn the ignition switch off and disconnect the negative battery terminal.
2. Disconnect the ECM connector.
3. Unscrew the mounting bolts behind the air cleaner.
4. Remove the ECM.
5. Installation is reverse of removal.

ENGINE COOLANT TEMPERATURE (ECT) SENSOR

LOCATION

The Engine Coolant Temperature (ECT) sensor is located in the engine coolant passage of the cylinder head for detecting the engine coolant temperature.

REMOVAL & INSTALLATION

1. Turn ignition switch OFF.
2. Disconnect ECTS connector.
3. Remove the ECTS.
4. Installation is the reverse of removal.

ENGINE OIL TEMPERATURE (EOT) SENSOR

LOCATION

The Continuously Variable Valve Timing (CVVT) Oil Temperature Sensor (OTS) is located at the front of the engine block.

REMOVAL & INSTALLATION

1. Disconnect the negative battery cable.
2. Disconnect the oil temperature sensor connector
3. Remove the oil temperature sensor.
4. Installation is the reverse of removal.

FUEL LEVEL SENDING UNIT

LOCATION

The fuel level sending unit is located on the fuel pump in the fuel tank.

REMOVAL & INSTALLATION

1. Disconnect the fuel pump connector.
2. Start the engine and wait until fuel in fuel line is exhausted.
3. After engine stalls, turn the ignition switch to OFF position.
4. Disconnect the fuel feed quick-connector and the vapor hose .
5. Unscrew the fuel pump locking ring with the special service tool (SST No.: 09310-3K000) and remove the fuel pump assembly.
6. Installation is reverse of removal.

HEATED OXYGEN (HO2S) SENSOR

LOCATION

See Figures 112 through 116.

The Heated Oxygen Sensor (HO2S) is installed on upstream and downstream of the Manifold Catalyst Converter (MCC).

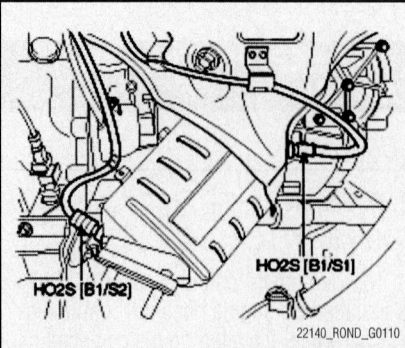

Fig. 112 Heated Oxygen Sensor (HO2S)—2.4L engine

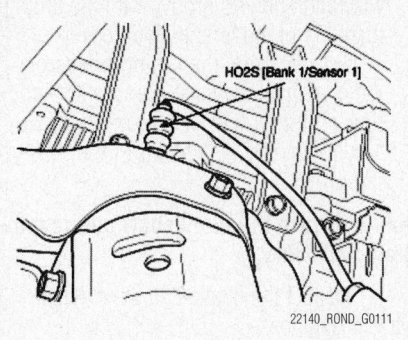

Fig. 113 Heated Oxygen Sensor (HO2S) Bank 1/Sensor 1—2.7L engine

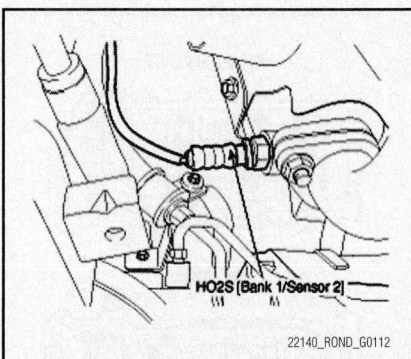

Fig. 114 Heated Oxygen Sensor (HO2S) Bank 1/Sensor 2—2.7L engine

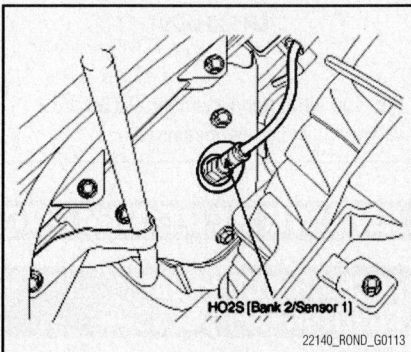

Fig. 115 Heated Oxygen Sensor (HO2S) Bank 2/Sensor 1—2.7L engine

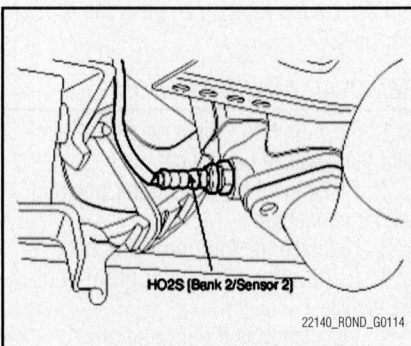

Fig. 116 Heated Oxygen Sensor (HO2S) Bank 2/Sensor 2—2.7L engine

REMOVAL & INSTALLATION

1. Disconnect the negative battery cable.
2. Disconnect the HO2S sensor connector
3. Remove the HO2S sensor.
4. Installation is the reverse of removal.

INTAKE AIR TEMPERATURE (IAT) SENSOR

LOCATION

The Intake Air Temperature Sensor (IATS) is installed inside the Mass Air Flow Sensor (MAFS).

KNOCK SENSOR (KS)

LOCATION

The Knock Sensor (KS) senses engine knocking and is installed on the cylinder block.

MASS AIR FLOW (MAF) SENSOR

LOCATION

The Mass Air Flow (MAF) sensor is located in between the air cleaner and the throttle body.

REMOVAL & INSTALLATION

See Figure 117.

1. Disconnect the negative battery cable.
2. Disconnect and remove the MAFS from the air cleaner assembly.
3. Installation is the reverse of removal.

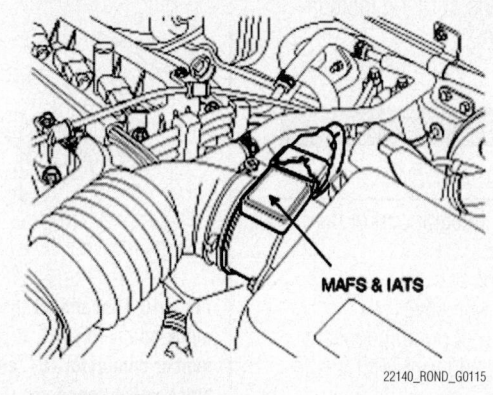

Fig. 117 Disconnect and remove the MAFS from the air cleaner assembly

22140_ROND_G0115

MANIFOLD ABSOLUTE PRESSURE (MAP) SENSOR

LOCATION

The Manifold Absolute Pressure (MAP) sensor is installed on the surge tank.

POWERTRAIN CONTROL MODULE (PCM)

LOCATION

2.7L Engine

The Powertrain Control Module (PCM) is installed on the air cleaner.

REMOVAL & INSTALLATION

2.7L Engine

1. Turn the ignition switch off and disconnect the negative battery terminal.

2. Disconnect the PCM connector.
3. Unscrew the mounting bolts behind the air cleaner.
4. Remove the PCM.
5. Installation is reverse of removal.

VARIABLE CAMSHAFT TIMING OIL CONTROL SOLENOID

LOCATION

The Continuously Variable Valve Timing (CVVT) Oil Control Valve (OCV)is located at the front of the engine block.

REMOVAL & INSTALLATION

1. Disconnect the negative battery cable.
2. Disconnect the oil control valve connector
3. Remove the oil control valve.
4. Installation is the reverse of removal.

FUEL   **GASOLINE FUEL INJECTION SYSTEM**

FUEL SYSTEM SERVICE PRECAUTIONS

Safety is the most important factor when performing not only fuel system maintenance but any type of maintenance. Failure to conduct maintenance and repairs in a safe manner may result in serious personal injury or death. Maintenance and testing of the vehicle's fuel system components can be accomplished safely and effectively by adhering to the following rules and guidelines.

• To avoid the possibility of fire and personal injury, always disconnect the negative battery cable unless the repair or test procedure requires that battery voltage be applied.

• Always relieve the fuel system pressure prior to disconnecting any fuel system com-

ponent (injector, fuel rail, pressure regulator, etc.), fitting or fuel line connection. Exercise extreme caution whenever relieving fuel system pressure to avoid exposing skin, face and eyes to fuel spray. Please be advised that fuel under pressure may penetrate the skin or any part of the body that it contacts.

• Always place a shop towel or cloth around the fitting or connection prior to loosening to absorb any excess fuel due to spillage. Ensure that all fuel spillage (should it occur) is quickly removed from engine surfaces. Ensure that all fuel soaked cloths or towels are deposited into a suitable waste container.

• Always keep a dry chemical (Class B) fire extinguisher near the work area.

• Do not allow fuel spray or fuel vapors to come into contact with a spark or open flame.

• Always use a back-up wrench when loosening and tightening fuel line connection fittings. This will prevent unnecessary stress and torsion to fuel line piping.

• Always replace worn fuel fitting O-rings with new Do not substitute fuel hose or equivalent where fuel pipe is installed.

Before servicing the vehicle, make sure to also refer to the precautions in the beginning of this section as well.

RELIEVING FUEL SYSTEM PRESSURE

1. Remove the third seat.
2. Open the service cover.
3. Disconnect the fuel pump connector.
4. Start the engine and wait until fuel in fuel line is exhausted.

5. After engine stalls, turn the ignition switch to OFF position.

6. Disconnect the negative battery cable.

FUEL FILTER

REMOVAL & INSTALLATION

The fuel filter is an integral part of the fuel pump assembly.

1. Remove the third seat.
2. Remove the service cover
3. Disconnect the fuel pump connector.
4. Start the engine and wait until fuel in fuel line is exhausted.
5. After engine stalls, turn the ignition switch to OFF position.
6. Disconnect the fuel feed line and canister hoses.
7. Unscrew the fuel pump mounting bolts and remove the fuel pump assembly.
8. Install the fuel pump in the reverse order of removal.

FUEL PUMP

REMOVAL & INSTALLATION

1. Remove the third seat.
2. Remove the service cover
3. Disconnect the fuel pump connector.
4. Start the engine and wait until fuel in fuel line is exhausted.
5. After engine stalls, turn the ignition switch to OFF position.
6. Disconnect the fuel feed line and canister hoses.
7. Unscrew the fuel pump mounting bolts and remove the fuel pump assembly.
8. Install the fuel pump in the reverse order of removal.

FUEL TANK

REMOVAL & INSTALLATION

See Figures 118 through 120.

1. Remove the third seat.

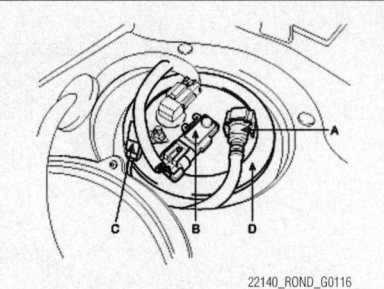

22140_ROND_G0116

Fig. 118 Disconnect the fuel feed tube quick-connector (A), the fuel tank pressure sensor connector (B), and the canister close valve connector (C)

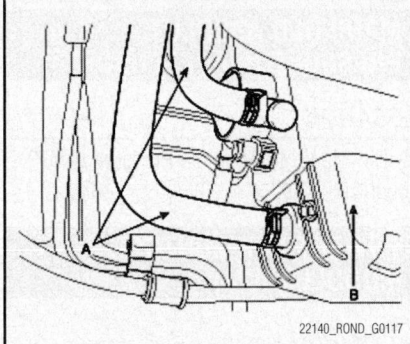

22140_ROND_G0117

Fig. 119 Disconnect the vacuum hoses (A) from the canister (B)

2. Remove the service cover
3. Disconnect the fuel pump connector.
4. Start the engine and wait until fuel in fuel line is exhausted.
5. After engine stalls, turn the ignition switch to OFF position.
6. Disconnect the fuel feed tube quick-connector (A), the fuel tank pressure sensor connector (B), and the canister close valve connector (C).
7. Raise and safely support the vehicle.
8. Remove the muffler assembly.
9. Support the fuel tank with a jack.

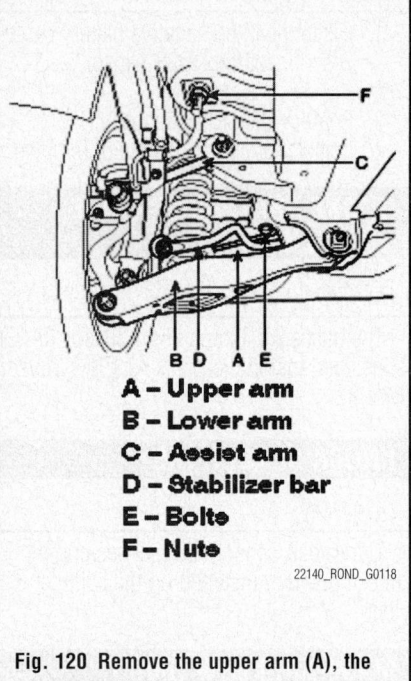

A – Upper arm
B – Lower arm
C – Assist arm
D – Stabilizer bar
E – Bolts
F – Nuts

22140_ROND_G0118

Fig. 120 Remove the upper arm (A), the lower arm (B), the assist arm (C), and the stabilizer bar (D)

10. Disconnect the fuel filler hose, the leveling hose, and the vacuum hose.
11. Disconnect the vacuum hoses (A) from the canister (B).
12. Remove the upper arm (A), the lower arm (B), the assist arm (C), and the stabilizer bar (D) from the fuel tank & suspension member assembly.
13. Remove the mounting bolts (E) and nuts (F), and then remove the fuel tank & suspension member assembly from the vehicle.
14. Installation is reverse of removal.

IDLE SPEED

ADJUSTMENT

The idle speed is controlled by the ECM. No adjustment is possible.

HEATING & AIR CONDITIONING SYSTEM

BLOWER MOTOR

REMOVAL & INSTALLATION

See Figures 121 and 122.

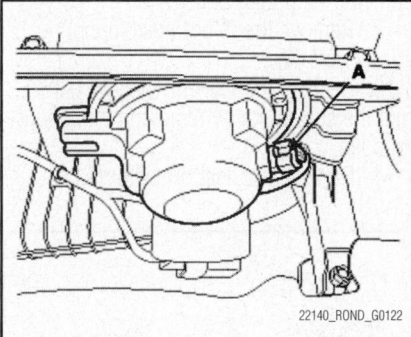

Fig. 121 Disconnect the connector (A) of the blower motor

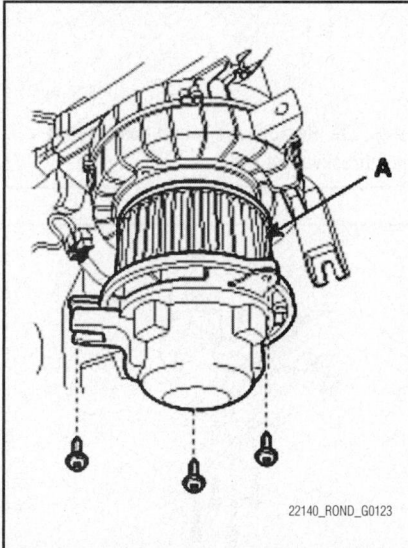

Fig. 122 Remove the blower motor (A) after removing the mounting screws

1. Disconnect the negative battery terminal.
2. Disconnect the connector (A) of the blower motor.
3. Remove the blower motor (A) after removing the mounting screws.
4. Installation is the reverse order of removal.

HEATER CORE

REMOVAL & INSTALLATION

See Figures 123 through 125.

1. Disconnect the negative battery terminal.
2. Recover the refrigerant with a recovery/ recycling/ charging station.
3. When the engine is cool, drain the engine coolant from the radiator.
4. Remove the bolts (A) and the expansion valve (B) from the evaporator core.

➡**Plug or cap the lines immediately after disconnecting them to avoid moisture and dust contamination.**

5. Disconnect the inlet and outlet heater hoses from the heater unit.

✳✳ CAUTION

Engine coolant will run out when the hoses are disconnected; drain it into a clean drip pan. Be sure not to let coolant spill on electrical parts or painted surfaces. If any coolant spills, rinse it off immediately.

6. Remove the crash pad (instrument panel).

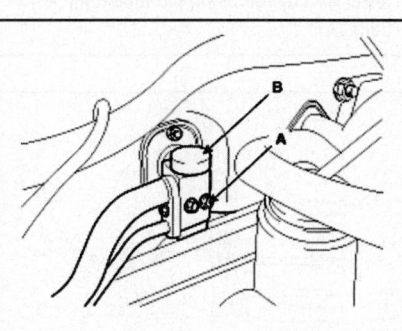

Fig. 123 Remove the bolts (A) and the expansion valve (B) from the evaporator core

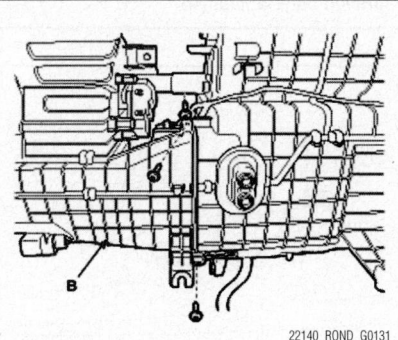

Fig. 124 Remove the blower unit (B) from heater unit

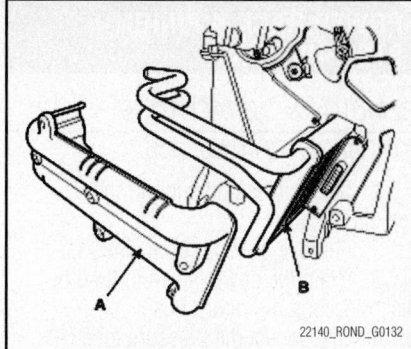

Fig. 125 Remove the heater core (B) after removing the cover (A)

7. Remove the cowl cross bar assembly.
8. Disconnect the connectors from the temperature control actuator, the mode control actuator and the evaporator temperature sensor.
9. Remove the heater & blower unit after removing 3 mounting bolts.
10. Remove the blower unit (B) from heater unit after removing 3 screws.
11. Remove the heater core (B) after removing the cover (A).

To install:

12. Install the heater core in the reverse order of removal.
13. Installation is the reverse order of removal, and note these items :

   a. If you're installing a new evaporator, add refrigerant oil (ND-OIL8).

   b. Replace the O-rings with new ones at each fitting, and apply a thin coat of refrigerant oil before installing them. Be sure to use the right O-rings for R-134a to avoid leakage.

   c. Immediately after using the oil, replace the cap on the container, and seal it to avoid moisture absorption.

   d. Do not spill the refrigerant oil on the vehicle ; it may damage the paint ; if the refrigerant oil contacts the paint, wash it off immediately.

   e. Apply sealant to the grommets.

   f. Make sure that there is no air leakage.

   g. Charge the system and test its performance.

   h. Do not interchange the inlet and outlet heater hoses and install the hose clamps securely.

   i. Refill the cooling system with engine coolant.

STEERING

POWER RACK & PINION STEERING GEAR

REMOVAL & INSTALLATION

See Figures 126 through 135.

1. Safely raise and support the vehicle.
2. Remove the both front wheels.
3. Drain the power steering fluid by disconnecting the return hose.
4. Disconnect the pressure tube (A) from the power steering pump by removing the eye bolt.
5. Disconnect the stabilizer link (A) with the strut assembly by removing the nut.
6. Remove the tie-rod end with the knuckle using a SST (09568-34000).
7. Separate the knuckle with the lower arm by removing the mounting bolts (A).
8. Loosen the bolt and disconnect the steering universal joint assembly with the pinion.
9. Remove the muffler rubber hanger.
10. Remove the front and rear roll stopper through bolts & nuts (A, B).
11. Remove the sub-frame and stay by removing mounting bolts.
12. Remove the return hose mounting bracket bolt (A).
13. Remove the rear roll stopper (A) from the sub-frame by removing the mounting bolts.
14. Remove the heat protector (A).

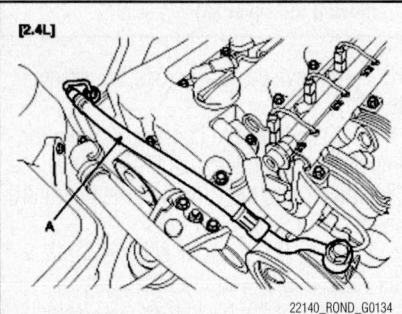

Fig. 126 Disconnect the pressure tube (A) from the power steering pump–2.4L engine

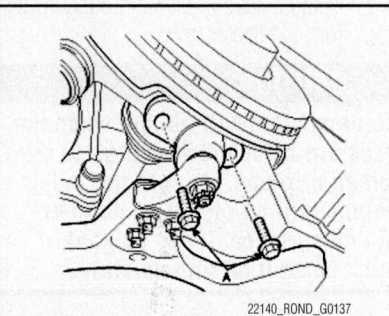

Fig. 129 Separate the knuckle with the lower arm by removing the mounting bolts (A)

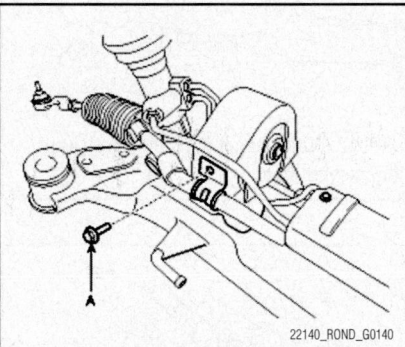

Fig. 132 Remove the return hose mounting bracket bolt (A)

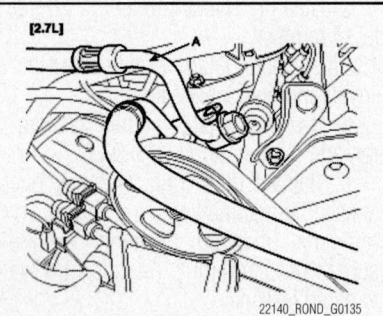

Fig. 127 Disconnect the pressure tube (A) from the power steering pump–2.7L engine

Fig. 130 Remove the front roll stopper through bolts & nuts (A)

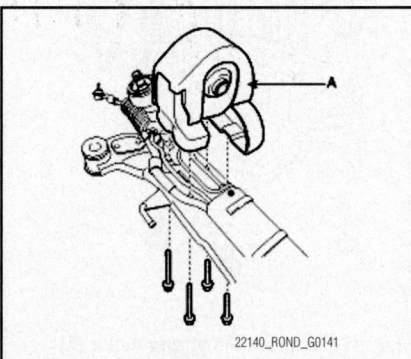

Fig. 133 Remove the rear roll stopper (A) from the sub-frame

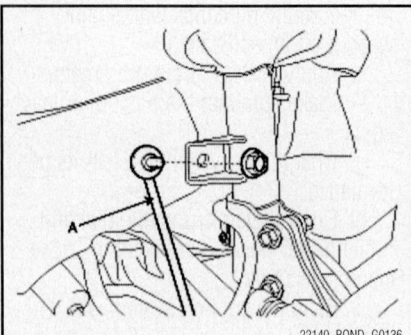

Fig. 128 Disconnect the stabilizer link (A) with the strut assembly

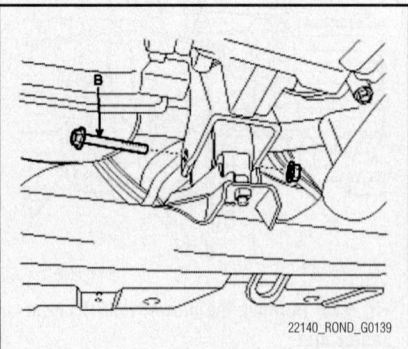

Fig. 131 Remove the rear roll stopper through bolts & nuts (B)

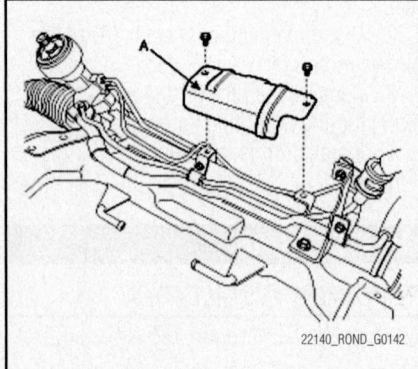

Fig. 134 Remove the heat protector (A)

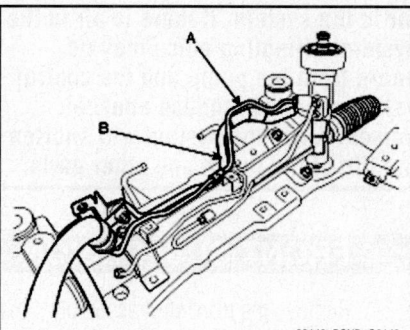

Fig. 135 Remove the pressure (A) and return tube (B) from the steering gear box

15. Remove the pressure (A) and return tube (B) from the steering gear box.

16. Remove the steering gear box from the sub-frame by removing the mounting bolts.

To install:

17. Install the steering gear box to the sub frame by tightening the mounting bolts. Torque: 43–58 ft. lbs. (60–80 Nm).

18. Install the pressure tube and return tube to the steering gear box and tighten the flare nut. Torque: 9–13 ft. lbs. (12–18 Nm).

19. Install the heat protector.

20. Install the rear roll stopper to the sub-frame by tightening the mounting bolts. Torque: 36–47 ft. lbs. (50–65 Nm).

21. Install the return hose mounting bracket to rear roll stopper.

22. Install the sub-frame and stay by tightening the mounting bolts. Torque: Sub-frame mounting bolts: 116–130 ft. Lbs. (160–180 Nm); Sub-frame stay mounting bolts: 33–40 ft. lbs. 45–55 Nm).

23. Install the front and rear roll stopper through bolts and nut. Torque: 36–47 ft. lbs. (50–65 Nm).

24. Connect the return hose.

25. Install the muffler rubber hanger.

26. Connect the lower arm with the knuckle by tightening the bolts. Torque: 72–87 ft. lbs. 100–120 Nm).

27. Connect the tie rod end with the knuckle and then install the castle nut and split pin. Torque: 17–25 ft. lbs. (24–34 Nm).

28. Connect the stabilizer link with the front strut assembly and tighten the nut. Torque: 72–87 t. lbs. (100–120 Nm).

29. Connect the universal joint assembly with the pinion of the steering gear box and tighten the bolt. Torque: 22–25 ft. lbs. (30–35 Nm).

30. Connect the pressure tube to power steering pump by tightening the eye bolt. Torque: 40–47 ft. lbs. (55–65 Nm).

31. Add power steering fluid to reservoir.

32. Bleed the power steering system.

33. Check and adjust front wheel alignment.

POWER STEERING PUMP

REMOVAL & INSTALLATION

See Figures 136 through 139.

1. Remove the accessory drive belt.

2. Disconnect the pressure tube (A) and return hose (B) from the power steering pump.

3. Loosen the mounting bolts and remove the power steering pump.

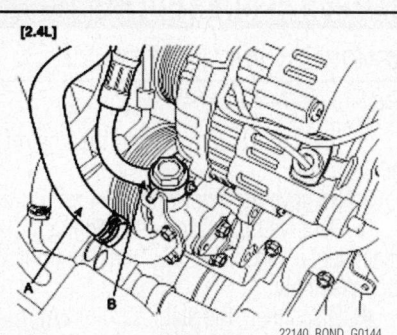

Fig. 136 Disconnect the pressure tube (A) and return hose (B) from the power steering pump–2.4L engine

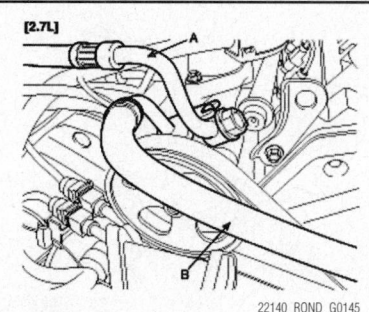

Fig. 137 Disconnect the pressure tube (A) and return hose (B) from the power steering pump–2.7L engine

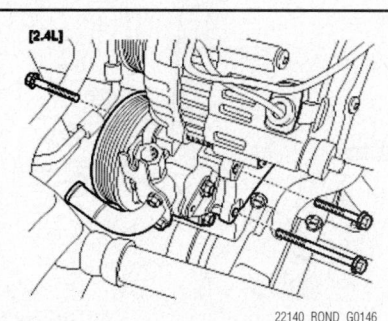

Fig. 138 Loosen the mounting bolts and remove the power steering pump–2.4L engine

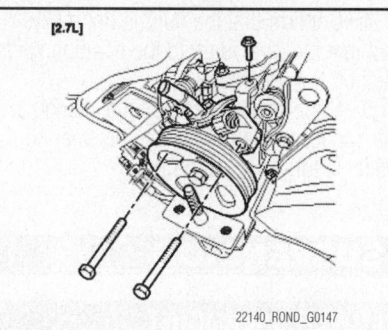

Fig. 139 Loosen the mounting bolts and remove the power steering pump–2.7L engine

To install:

4. Install the power steering pump and tighten the mounting bolts.

   a. 2.4L engine: Torque: 12–19 ft. lbs. (17–26 Nm).

   b. 2.7L engine: Torque: Long bolts: 26–35 ft. lbs. (35–50 Nm); Small bolt: 12–19 ft. lbs. (17–26 Nm).

5. Connect the suction hose with suction pipe.

6. Connect the pressure tube to power steering pump by installing the eye bolt. Torque: 36–47 ft. lbs. (55–65 Nm).

7. Install the accessory drive belt.

8. Add power steering fluid to reservoir.

9. Bleed the power steering system.

10. Check power steering pump relief pressure.

BLEEDING

1. Remove the fuel pump fuse, then start the engine and wait for the engine to stall. Next, while operating the starting motor intermittently (for 15–20 seconds), turn the steering wheel all the way to the left and then to the right five or six times.

➡**During air bleeding, replenish the fluid supply so that the level never falls below the lower position of the filter. If air bleeding is done while the vehicle is idling, the air will be broken up and absorbed into the fluid. Be sure to do the bleeding only while cranking.**

2. Reinstall the fuel pump fuse, and start the engine (idling).

3. Turn the steering wheel to the left and the right until there are no air bubbles in the oil reservoir.

❊❊ CAUTION

Do not hold the steering wheel turned all the way to either side for more than ten seconds.

4. Confirm that the fluid is not milky, and that the level is up to the position specified on the level gauge.

5. Confirm that there is little change in the surface of the fluid when the steering wheel is turned left and right.

❊❊ CAUTION

If the surface of the fluid changes considerably, air bleeding should be done again. If the fluid level rises suddenly when the engine is stopped, it indicates that there is still

air in the system. If there is air in the system, a jingling noise may be heard from the pump and the control valve may also produce unusual noises. Air in the system will shorten the life of the pump and other parts.

SUSPENSION

LOWER BALL JOINT

REMOVAL & INSTALLATION

1. Raise and safely support the front of the vehicle.
2. Remove the front wheel & tire.

❊ CAUTION

Be careful not to damage the hub bolts when removing the front wheel & tire.

3. Remove the split pin and the castle nut from the lower arm ball joint.
4. Separate the lower arm from the lower arm ball joint by using SST (09568-34000).
5. Remove the lower arm from the sub-frame.

To install:

6. Install the front lower arm to the sub-frame. Torque: Bolt: 101–116 ft. lbs. (140–160 Nm); Bolt & nut: 72–87 ft. lbs. (100–120 Nm).
7. Connect the lower arm with the ball joint. Torque: 58–65 ft. lbs. (80–90 Nm).
8. Install the front wheel & tire.

LOWER CONTROL ARM

REMOVAL & INSTALLATION

1. Raise and safely support the front of the vehicle.
2. Remove the front wheel & tire.

❊❊ CAUTION

Be careful not to damage the hub bolts when removing the front wheel & tire.

3. Remove the split pin and the castle nut from the lower arm ball joint.
4. Separate the lower arm from the lower arm ball joint by using SST (09568-34000).
5. Remove the lower arm from the sub-frame.

To install:

6. Install the front lower arm to the sub-frame. Torque: Bolt: 101–116 ft. lbs. (140–160 Nm); Bolt & nut: 72–87 ft. lbs. (100–120 Nm).

7. Connect the lower arm with the ball joint. Torque: 58–65 ft. lbs. (80–90 Nm).
8. Install the front wheel & tire.

MACPHERSON STRUT

REMOVAL & INSTALLATION

See Figures 140 and 141.

1. Raise and safely support the front of the vehicle.
2. Remove the front wheel & tire.
3. Remove the brake hose and the wheel speed sensor bracket from the front strut assembly
4. Disconnect the stabilizer link (A) with the front strut assembly.
5. Disconnect the front strut assembly (A) from the knuckle.

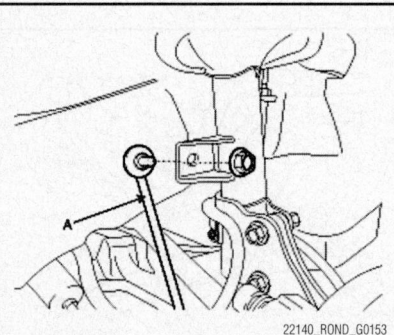

22140_ROND_G0153

Fig. 140 Disconnect the stabilizer link (A) with the front strut assembly

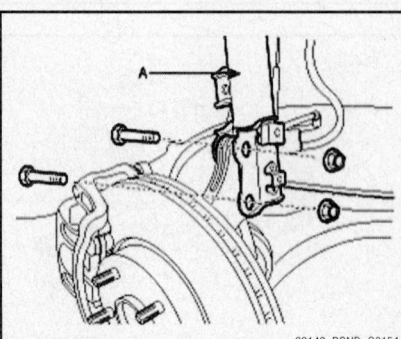

22140_ROND_G0154

Fig. 141 Disconnect the front strut assembly (A) from the knuckle

FRONT SUSPENSION

6. Remove the front strut assembly from the wheel housing panel.

To install:

7. Install the front strut assembly to the wheel housing panel. Torque: 33–43 ft. lbs. (45–60 Nm).
8. Connect the front strut assembly with the knuckle. Torque: 101–116 ft. lbs. (140–160 Nm).
9. Install the stabilizer link to the front strut assembly. Torque: 72–87 ft. lbs. (100–120 Nm).
10. Install the brake hose and wheel speed sensor bracket to front strut assembly.
11. Install the front wheel & tire.

STEERING KNUCKLE

REMOVAL & INSTALLATION

1. Raise the vehicle, and make sure it is securely supported.
2. Remove the front wheel and tire.
3. Remove the wheel speed sensor from the knuckle.
4. Remove the brake caliper mounting bolts, and hang the brake caliper assembly with wire.
5. Remove the split pin, castle nut and washer from the front hub
6. Remove the ball joint assembly mounting bolt from the knuckle.
7. Remove the tie rod end ball joint from the knuckle.
   a. Remove the split pin and castle nut.
   b. Disconnect the ball joint from knuckle by using the special tool (09568-4A000).
8. Remove the brake disc (rotor) from the front hub assembly
9. Remove the strut assembly mounting bolts and nuts.
10. Remove the hub and knuckle assembly.

To install:

11. Install the hub and knuckle assembly to the drive shaft.
12. Install the knuckle to the strut assembly. Torque: 101.3–115.7 ft. lbs. (137.3–156.9 Nm).

13. Install the brake disc (rotor) to the front hub assembly.

14. Install the tie rod end ball joint to the knuckle.

15. Install the nut and split pin. Torque: 17.4–24.6 ft. lbs. (23.5–33.3 Nm).

16. Install the ball joint assembly mounting bolt to the knuckle. Torque: 72.3–86.8 ft. lbs. 98.1–117.7 Nm).

17. Install the washer, castle nut and new split pin to the front hub assembly. Torque: 144.7–202.5 ft. lbs. (196.1–274.6 Nm).

※※ CAUTION

The washer should be assembled with convex surface outward when installing the castle nut and split pin.

18. Install the brake caliper. Torque: 57.9–72.3 ft. lbs. (78.5–98.1 Nm).

19. Install the wheel speed sensor to the knuckle.

20. Install the wheel and the tire

STABILIZER BAR

REMOVAL & INSTALLATION

See Figures 142 through 146.

1. Raise and safely support the front of the vehicle.

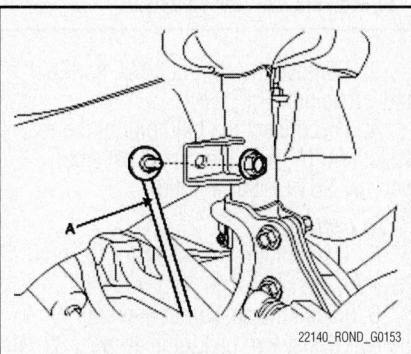

Fig. 142 Disconnect the stabilizer link (A) with the front strut assembly

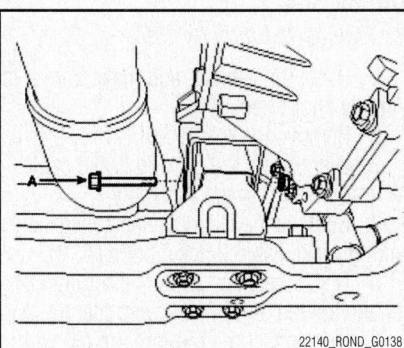

Fig. 143 Remove the front roll stopper through bolts & nuts (A)

2. Remove the front wheel & tire.

3. Disconnect the universal joint assembly from the pinion of the steering gear box.

4. Disconnect the pressure tube from the power steering pump

5. Drain the power steering fluid by disconnecting the return hose.

6. Disconnect the stabilizer link (A) from the front strut assembly

7. Disconnect the tie-rod end from the front knuckle by using SST (09568-34000).

8. Disconnect the front lower arm with the knuckle.

9. Remove the muffler rubber hanger.

10. Remove the front and rear roll stopper bolts & nuts (A, B).

11. Remove the sub-frame and sub-frame stay.

12. Remove the stabilizer bar from the sub-frame.

13. Disconnect the stabilizer link (A) with the stabilizer bar.

14. Remove the bushing and the bracket from the stabilizer bar.

To install:

15. Install the bushing and the bracket to the stabilizer bar.

16. Connect the stabilizer link with stabilizer bar. Torque: 72–87 ft. lbs. (100–120 Nm).

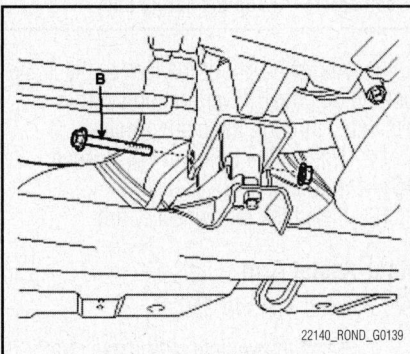

Fig. 144 Remove the rear roll stopper through bolts & nuts (B)

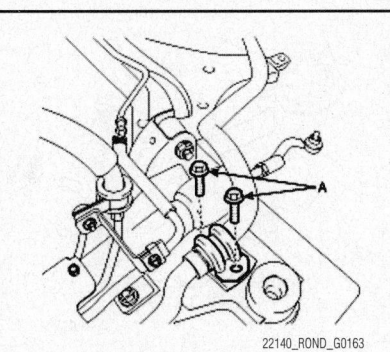

Fig. 145 Remove the stabilizer bar from the sub-frame

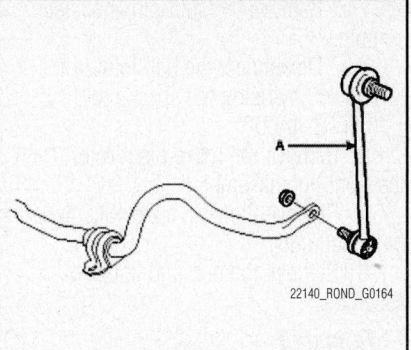

Fig. 146 Disconnect the stabilizer link (A) with the stabilizer bar

17. Install the stabilizer to the sub-frame. Torque: 33–40 ft. lbs. (45–55 Nm).

18. Install the sub-frame and stay. Torque: Sub-frame mounting bolts: 116–130 ft. lbs. (160–180 Nm); Sub-frame stay mounting bolts: 33–40 ft. lbs. (45–55 Nm).

19. Install the front and rear roll stopper through bolts and nuts. Torque: 36–47 ft. lbs. (50–65 Nm).

20. Install the muffler rubber hanger.

21. Connect the lower arm with the knuckle

22. Connect the tie-rod end with the knuckle. Torque: 17–25 ft. lbs. (24–34 Nm).

23. Connect the stabilizer link with front strut assembly. Torque: 72–87 ft. lbs. (100–120 Nm).

24. Connect the power steering return hose.

25. Connect the pressure tube to the power steering pump. Torque: 40–47 ft. lbs. (55–65 Nm).

26. Connect the universal joint assembly with the pinion of the steering gear box. Torque: 22–25 ft. lbs. (30–35 Nm).

27. Install the front wheel & tire.

28. Bleed the power steering system.

WHEEL HUB AND BEARING

REMOVAL & INSTALLATION

1. Raise the vehicle, and make sure it is securely supported.

2. Remove the front wheel and tire.

3. Remove the wheel speed sensor from the knuckle.

4. Remove the brake caliper mounting bolts, and hang the brake caliper assembly with wire.

5. Remove the split pin, castle nut and washer from the front hub

6. Remove the ball joint assembly mounting bolt from the knuckle.

7. Remove the tie rod end ball joint from the knuckle.

a. Remove the split pin and castle nut.

b. Disconnect the ball joint from knuckle by using the special tool (09568-4A000).

8. Remove the brake disc (rotor) from the front hub assembly

9. Remove the strut assembly mounting bolts and nuts.

10. Remove the hub and knuckle assembly.

To install:

11. Install the hub and knuckle assembly to the drive shaft.

12. Install the knuckle to the strut assembly. Torque: 101.3–115.7 ft. lbs. (137.3–156.9 Nm).

13. Install the brake disc (rotor) to the front hub assembly.

14. Install the tie rod end ball joint to the knuckle.

15. Install the nut and split pin. Torque: 17.4–24.6 ft. lbs. (23.5–33.3 Nm).

16. Install the ball joint assembly mounting bolt to the knuckle. Torque: 72.3–86.8 ft. lbs. 98.1–117.7 Nm).

17. Install the washer, castle nut and new split pin to the front hub assembly.

Torque: 144.7–202.5 ft. lbs. (196.1–274.6 Nm).

❋❋ CAUTION

The washer should be assembled with convex surface outward when installing the castle nut and split pin.

18. Install the brake caliper. Torque: 57.9–72.3 ft. lbs. (78.5–98.1 Nm).

19. Install the wheel speed sensor to the knuckle.

20. Install the wheel and the tire

SUSPENSION

CONTROL ARMS/LINKS

REMOVAL & INSTALLATION

Rear Lower Arm

See Figures 147 and 148.

1. Raise the vehicle, and make sure it is securely supported.

2. Remove the rear wheel & tire.

3. Support the lower portion of the rear lower arm (A) with a jack.

4. Temporarily loosen the bolt (B) holding the cross member to the rear lower arm. Do not remove it.

5. Remove the bolt & nut (C) holding the rear lower arm to the carrier assembly.

6. Lower the jack and remove the coil spring (A) and the spring pad.

7. Remove the rear lower arm (C) from the cross member

To install:

8. Connect the rear lower arm with the cross member and then temporarily tighten the bolt.

9. Install the coil spring and support the lower portion of the rear lower arm with a jack.

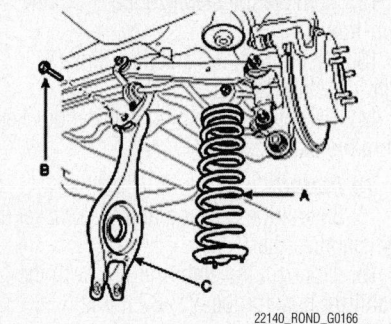

Fig. 148 Lower the jack and remove the coil spring (A) and the spring pad

10. Adjust height of the jack to place the bolt holding rear lower arm and carrier assembly through the mating holes.

11. Tighten the bolt and nut. Torque: 101–116 ft. lbs. (140–160 Nm).

12. Install the rear wheel & tire.

Rear Assist Arm

See Figures 149 and 150.

1. Raise the vehicle, and make sure it is securely supported.

2. Remove the rear wheel & tire.

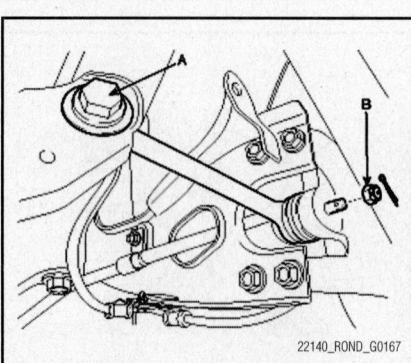

Fig. 149 Remove the cam bolt (A), split pin, and castle nut (B)

REAR SUSPENSION

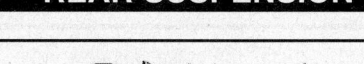
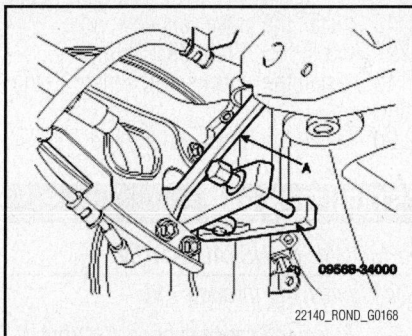

Fig. 150 Disconnect the ball joint of the rear assist arm (A) with the carrier assembly with the SST (09568-34000)

3. Remove the cam bolt (A), split pin, and castle nut (B).

4. Disconnect the ball joint of the rear assist arm (A) with the carrier assembly with the SST (09568-34000).

To install:

5. Install the rear assist Arm. Torque: 58–72 ft. lbs. (80–100 Nm).

6. Install the castle nut and split pin to the rear assist arm ball joint. Torque: 33–40 ft. lbs. 45–55 Nm).

7. Install the rear wheel & tire.

Trailing Arm

See Figures 151 through 155.

1. Raise the vehicle, and make sure it is securely supported.

2. Remove the rear wheel & tire.

3. Disconnect the parking brake cable (A) from the rear brake assembly.

4. Remove the wheel speed sensor and parking brake cable bracket bolts (A, B).

5. Disconnect the rear stabilizer link (B) with the trailing arm by removing the nut (A).

6. Disconnect the trailing arm (B) with the carrier assembly by removing the mounting bolts (A).

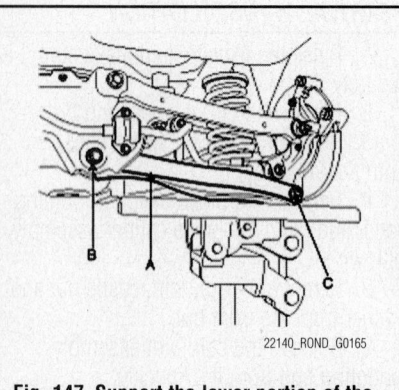

Fig. 147 Support the lower portion of the rear lower arm (A) with a jack

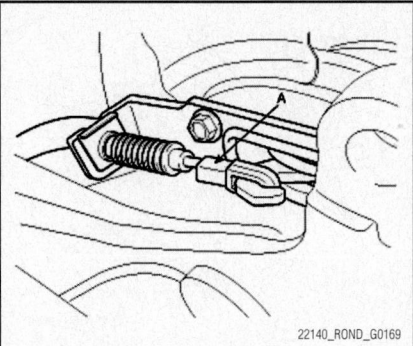

Fig. 151 Disconnect the parking brake cable (A) from the rear brake assembly

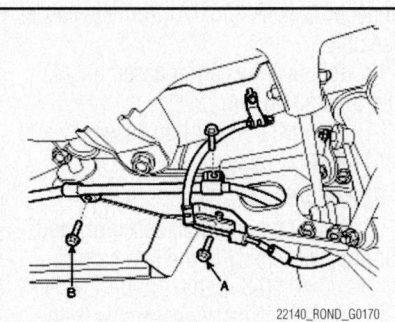

Fig. 152 Remove the wheel speed sensor and parking brake cable bracket bolts (A, B)

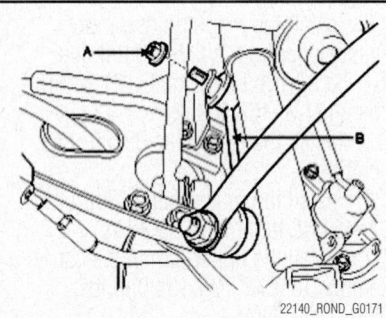

Fig. 153 Disconnect the rear stabilizer link (B) with the trailing arm by removing the nut (A)

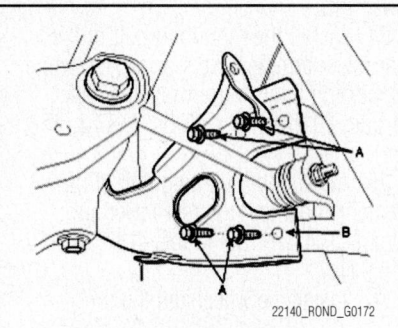

Fig. 154 Disconnect the trailing arm (B) with the carrier assembly by removing the mounting bolts (A)

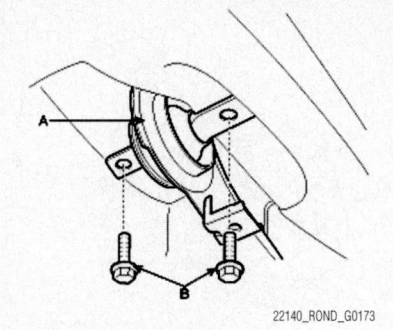

Fig. 155 Remove the rear trailing arm (A) by removing the mounting bolt (B)

7. Remove the rear trailing arm (A) by removing the mounting bolt (B).

To install:

8. Install the rear trailing arm to the body. Torque: 101–116 ft. lbs. (140–160 Nm).

9. Connect the trailing arm with the carrier assembly. Torque: 33–40 ft. lbs. (45–55 Nm).

10. Connect the rear stabilizer link with the trailing arm. Torque: 33–40 ft. lbs. (45–55 Nm).

11. Connect the parking brake cable with the rear brake assembly.

12. Install the wheel speed sensor and parking brake cable bracket.

13. Install the rear wheel & tire.

Upper Arm

See Figure 156.

1. Raise the vehicle, and make sure it is securely supported.

2. Remove the rear wheel & tire.

3. Remove the rear upper arm (C) by removing the mounting bolts & nut (A, B).

To install:

4. Install the rear upper arm between the rear cross member and the carrier assembly. Torque: 80–94 ft. lbs. (110–130 Nm).

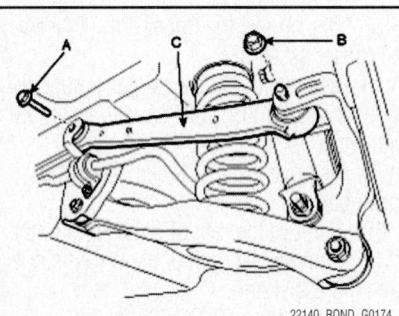

Fig. 156 Remove the rear upper arm (C) by removing the mounting bolts & nut (A, B)

Install the rear upper arm so that the letter "R" can face the rear of vehicle.

5. Install the rear wheel & tire.

SHOCK ABSORBER

REMOVAL & INSTALLATION

See Figure 157.

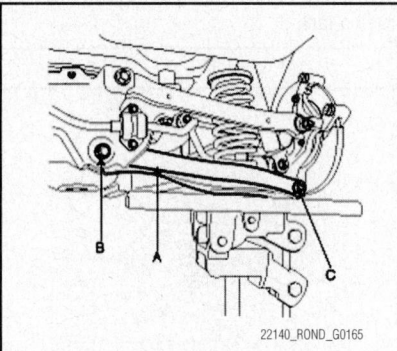

Fig. 157 Support the lower portion of the rear lower arm (A) with a jack

1. Raise the vehicle, and make sure it is securely supported.

2. Remove the rear wheel & tire.

3. Support the lower portion of the rear lower arm (A) with a jack.

4. Remove the bolt & nut holding the rear shock absorber to the carrier assembly.

5. Remove the rear shock absorber from the wheel housing by removing the mounting bolts.

6. Install the rear shock absorber to the wheel housing. Torque: 33–40 ft. lbs. (45–55 Nm).

7. Connect the rear shock absorber to the carrier assembly. Torque: 101–116 ft. lbs. (140–160 Nm).

8. Install the rear wheel & tire.

TESTING

1. Check the rubber parts for wear and deterioration.

2. Compress and extend the piston rod and check that there is no abnormal resistance or unusual sound during operation.

WHEEL HUB AND BEARING

REMOVAL & INSTALLATION

See Figures 158 through 164.

1. Raise the vehicle, and make sure it is securely supported.

2. Remove the rear wheel & tire.

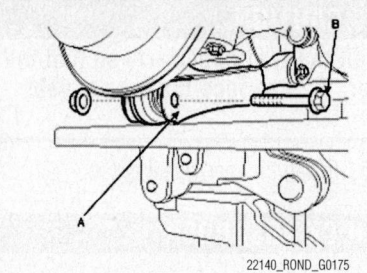

Fig. 158 Remove the mounting bolt (B) of the rear lower arm (A) and the rear carrier, while supporting the lower arm (A) with a jack

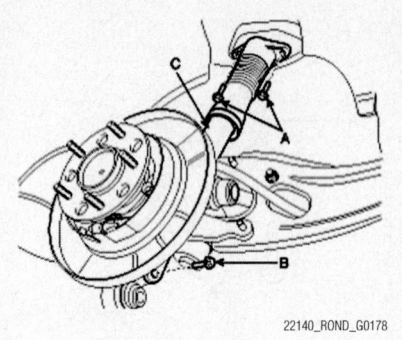

Fig. 161 Remove the rear strut assembly (C)

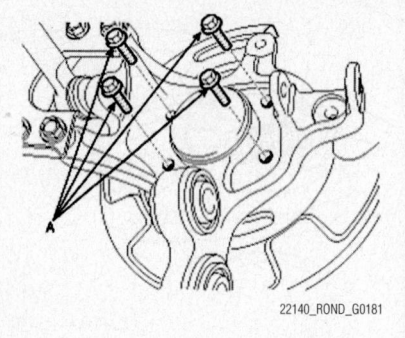

Fig. 164 Remove the rear hub assembly and rear brake assembly

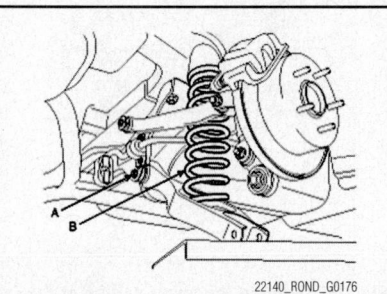

Fig. 159 Remove the mounting nut (A) of the cross member and the rear lower arm, then remove the coil spring (B) by taking down the jack

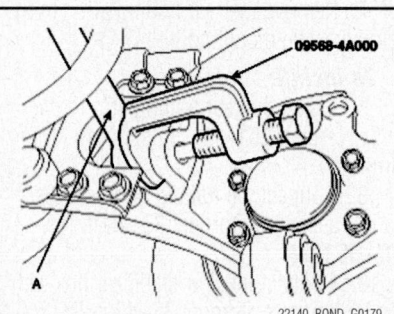

Fig. 162 Using the SST (09568-4A000), disconnect the assist arm (A) from the carrier assembly

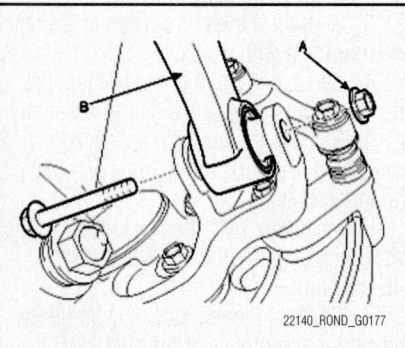

Fig. 160 Disconnect the upper arm (B) from the carrier assembly

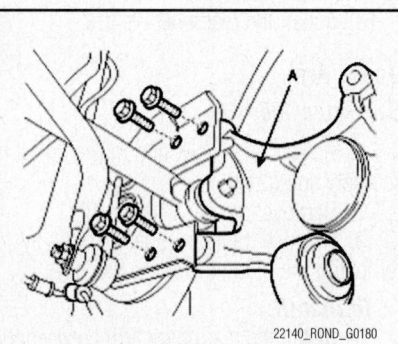

Fig. 163 Remove the carrier assembly (A) from the trailing arm

3. Remove the mounting bolt (B) of the rear lower arm (A) and the rear carrier, while supporting the lower arm (A) with a jack.

4. Remove the mounting nut (A) of the cross member and the rear lower arm, then remove the coil spring (B) by taking down the jack.

5. Remove the wheel speed sensor.

6. Disconnect the parking brake cable.

7. Disconnect the upper arm (B) from the carrier assembly.

8. Remove the brake caliper mounting bolts, and hang the brake caliper assembly with wire.

9. Remove the rear brake disc (rotor) assembly.

10. Remove the rear strut assembly (C).

11. Remove the split pin and castle nut from the assist arm.

12. Using the SST (09568-4A000), dis-

connect the assist arm (A) from the carrier assembly.

13. Remove the carrier assembly (A) from the trailing arm.

14. Remove the rear hub assembly and rear brake assembly.

To install:

15. Install the rear hub assembly and rear brake assembly to the rear carrier. Torque: 43.4–50.6 ft. lbs. (58.8–68.6 Nm).

16. Install the carrier assembly to the trailing arm. Torque: 32.5–39.8 ft. lbs. (44.1–53.9 Nm).

17. Install the assist arm to the carrier assembly. Torque: 32.5–39.8 ft. lbs. (44.1–53.9 Nm).

18. Install the rear strut assembly. Torque: Upper bolts: 32.5–39.8 ft. lbs. (44.1–53.9 Nm); Lower bolt: 101.3–115.7 ft. lbs. (137.3–156.9 Nm).

19. Install the rear brake disc (rotor) assembly.

20. Install the brake caliper. Torque: 36.2–43.4 ft. lbs. (49.0–58.8 Nm).

21. Install the upper arm to the carrier assembly. Torque" 79.6–94.0 ft. lbs. (107.9–127.5 Nm).

22. Install the wheel speed sensor and the parking brake cable.

23. Install the coil spring on the rear lower arm, then slowly jack-up the rear lower arm.

24. Install the mounting bolt of the rear lower arm and the rear carrier, while supporting the lower arm with a jack. Torque: 101.3–115.7 ft. lbs. (137.3–156.9 Nm).

25. Install the mounting bolt of the cross member and the rear lower arm. Torque: 101.3–115.7 ft. lbs. (137.3–156.9 Nm).

26. Install the wheel and the tire

SPECIFICATIONS AND MAINTENANCE CHARTS

ENGINE AND VEHICLE IDENTIFICATION

Engine								Model Year	
Code ①	Liters (cc)	Cu. In.	Cyl.	Fuel Sys.	Engine Type	Eng. Mfg.		Code ②	Year
3	3.8 (3778)	231	6	EGI	DOHC	KIA		6	2006
								7	2007
								8	2008

EGI: Electronic Gasoline Injection

DOHC: Double Overhead Camshafts

22140_SEDO_C0001

GENERAL ENGINE SPECIFICATIONS

Year	Model	Engine Displacement Liters	Engine VIN	Net Horsepower @ rpm	Net Torque @ rpm (ft. lbs.)	Bore x Stroke (in.)	Compression Ratio	Oil Pressure @ rpm
2006	Sedona	3.8	3	250@6000	253@3500	3.78x3.43	10.4:01	18.77@1000
2007	Sedona	3.8	3	250@6000	253@3500	3.78x3.43	10.4:01	18.77@1000
2008	Sedona	3.8	3	250@6000	253@3500	3.78x3.43	10.4:01	18.77@1000

22140_SEDO_C0002

ENGINE TUNE-UP SPECIFICATIONS

Year	Engine Displacement Liters	Engine VIN	Spark Plug Gap (in.)	Ignition Timing (deg.) ① MT	Ignition Timing (deg.) ① AT	Fuel Pump (psi)	Idle Speed (rpm) MT	Idle Speed (rpm) AT	Valve Clearance Intake	Valve Clearance Exhaust
2006	3.8	3	0.0394-0.0433	NA	②	③	NA	550-750	0.0067-0.0090	0.0106-0.0129
2007	3.8	3	0.0394-0.0433	NA	②	③	NA	550-750	0.0067-0.0090	0.0106-0.0129
2008	3.8	3	0.0394-0.0433	NA	②	③	NA	550-750	0.0067-0.0090	0.0106-0.0129

NOTE: The Vehicle Emission Control Information label often reflects specification changes made during production.

The label figures must be used if they differ from those in this chart

NA: Not Available

① Computer controled, no adjustment possible

② 5-15 degrees

③ 54.3-55.8 at idle

22140_SEDO_C0003

CAPACITIES

Year	Model	Engine Displacement Liters	Engine VIN	Engine Oil with Filter (qts.)	Transaxle (pts.) Manual	Transaxle (pts.) Auto.	Fuel Tank (gal.)	Cooling System (qts.)
2006	Sedona	3.8	3	5.5	—	22.0	21.1	9.1
2007	Sedona	3.8	3	5.5	—	22.0	21.1	9.1
2008	Sedona	3.8	3	5.5	—	22.0	21.1	9.1

NOTE: All capacities are approximate. Add fluid gradually and ensure a proper level is obtained.

22140_SEDO_C0004

FLUID SPECIFICATIONS

Year	Model	Engine Displ. Liters (VIN)	Engine Oil	Man. Trans.	Auto. Trans.	Drive Axle Front	Drive Axle Rear	Transfer Case	Power Steering Fluid	Brake Master Cylinder	Cooling System
2006	Sedona	3.8 (3)	①	NA	②	NA	NA	NA	PSF-3	DOT 3	③
2007	Sedona	3.8 (3)	①	NA	②	NA	NA	NA	PSF-3	DOT 3	③
2008	Sedona	3.8 (3)	①	NA	②	NA	NA	NA	PSF-3	DOT 3	③

NA: Not Available

DOT: Department Of Transpotation

① SL (SJ) or above, 5W-20

② Diamond ATF SP-III or SK ATF SP-III

③ Ethylene glycol base for aluminum radiator

22140_SEDO_C0014

VALVE SPECIFICATIONS

Year	Engine Displacement Liters	Engine VIN	Seat Angle (deg.)	Face Angle (deg.)	Maximum out of Square (degrees)	Spring Free Length (in.)	Stem-to-Guide Clearance (in.) Intake	Stem-to-Guide Clearance (in.) Exhaust	Stem Diameter (in.) Intake	Stem Diameter (in.) Exhaust
2006	3.8	3	44.75-45.20	45.25-45.75	NA	1.7267	0.0008-0.0019	0.0012-0.0021	0.215-0.216	0.215-0.216
2007	3.8	3	44.75-45.20	45.25-45.75	NA	1.7267	0.0008-0.0019	0.0012-0.0021	0.215-0.216	0.215-0.216
2008	3.8	3	44.75-45.20	45.25-45.75	NA	1.7267	0.0008-0.0019	0.0012-0.0021	0.215-0.216	0.215-0.216

NA: Not Available

22140_SEDO_C0005

CAMSHAFT AND BEARING SPECIFICATIONS

All measurements are given in inches.

Year	Engine Displacement Liters	Engine VIN	Journal Diameter	Brg. Oil Clearance	Shaft End-play	Runout	Journal Bore	Lobe Lift Intake	Lobe Lift Exhaust
2006	3.8	3	①	②	③	NA	NA	NA	NA
2007	3.8	3	①	②	③	NA	NA	NA	NA
2008	3.8	3	①	②	③	NA	NA	NA	NA

NA: Not Available

① No. 1: 1.1009-1.1016 inches

  No. 2, 3, 4: 0.9430-0.9437 inches

② No. 1: 0.0008-0.0022 inches

  No. 2, 3, 4: 0.0012-0.0026 inches

③ 0.0008-0.0071 inches

22140_SEDO_C0013

CRANKSHAFT AND CONNECTING ROD SPECIFICATIONS

All measurements are given in inches.

Year	Engine Displacement Liters	Engine VIN	Crankshaft Main Brg. Journal Dia.	Crankshaft Main Brg. Oil Clearance	Crankshaft Shaft End-play	Crankshaft Thrust on No.	Connecting Rod Journal Diameter	Connecting Rod Oil Clearance	Connecting Rod Side Clearance
2006	3.8	3	2.7142-2.7149	0.0008-0.0016	0.0039-0.0110	3	2.2834-2.2842	0.0015-0.0022	0.0039-0.0098
2007	3.8	3	2.7142-2.7149	0.0008-0.0016	0.0039-0.0110	3	2.2834-2.2842	0.0015-0.0022	0.0039-0.0098
2008	3.8	3	2.7142-2.7149	0.0008-0.0016	0.0039-0.0110	3	2.2834-2.2842	0.0015-0.0022	0.0039-0.0098

22140_SEDO_C0007

PISTON AND RING SPECIFICATIONS

All measurements are given in inches.

Year	Engine Displacement Liters	Engine VIN	Piston Clearance	Ring Gap Top Compression	Ring Gap Bottom Compression	Ring Gap Oil Control	Ring Side Clearance Top Compression	Ring Side Clearance Bottom Compression	Ring Side Clearance Oil Control
2006	3.8	3	0.0012-0.0020	0.0067-0.0126	0.0126-0.0185	0.0078-0.0275	0.0012-0.0027	0.0012-0.0027	0.0024-0.0059
2007	3.8	3	0.0012-0.0020	0.0067-0.0126	0.0126-0.0185	0.0078-0.0275	0.0012-0.0027	0.0012-0.0027	0.0024-0.0059
2008	3.8	3	0.0012-0.0020	0.0067-0.0126	0.0126-0.0185	0.0078-0.0275	0.0012-0.0027	0.0012-0.0027	0.0024-0.0059

22140_SEDO_C0006

TORQUE SPECIFICATIONS
All readings in ft. lbs.

Year	Engine Displacement Liters	Engine VIN	Cylinder Head Bolts	Main Bearing Bolts	Rod Bearing Bolts	Crankshaft Damper Bolts	Flywheel Bolts	Manifold Intake	Manifold Exhaust	Spark Plugs	Oil Pan Drain Plug
2006	3.8	3	①	②	③	210-224	53-57	14-17	29-33	15-22	25-32
2007	3.8	3	①	②	③	210-224	53-57	14-17	29-33	15-22	25-32
2008	3.8	3	①	②	③	210-224	53-57	14-17	29-33	15-22	25-32

① Step 1: 28.93 ft. lbs.
 Step 2: Plus 120 degrees
 Step 3: Plus 90 degrees

② Bolts 1 thru 8 (inside cap bolts): 36.16 ft. lbs. plus 90 degrees
 Bolts 9 thru 16 (outside cap bolts) 14.46 ft. lbs. plus 120 degrees
 Bolts 17 thru 22 (side cap bolts) 21.7-23.2 ft. lbs.
 See illustration in text for location information

③ Step 1: 14.46 ft. lbs.
 Step 2: Plus 90 degrees

22140_SEDO_C0008

WHEEL ALIGNMENT

Year	Model		Caster Range (+/-Deg.)	Caster Preferred Setting (Deg.)	Camber Range (+/-Deg.)	Camber Preferred Setting (Deg.)	Toe-in (in.)
2006	Sedona	F	—	4° 05' +/- 30'	—	0 +/- 30'	0 +/- .0787
		R	—	—	—	-20 +/- 30'	.1378 +/- .0787
2007	Sedona	F	—	4° 05' +/- 30'	—	0 +/- 30'	0 +/- .0787
		R	—	—	—	-20 +/- 30'	.1378 +/- .0787
2008	Sedona	F	—	4° 05' +/- 30'	—	0 +/- 30'	0 +/- .0787
		R	—	—	—	-20 +/- 30'	.1378 +/- .0787

22140_SEDO_C0009

TIRE, WHEEL AND BALL JOINT SPECIFICATIONS

Year	Model	OEM Tires		Tire Pressure (psi)		Wheel Size	Ball Joint Inspection	Lug Nut Torque (ft. lbs.)
		Standard	Optional	Front	Rear			
2006	Sedona	P225/70R16	P235/60R17	35	35	②	①	65-79
2007	Sedona	P225/70R16	P235/60R17	35	35	②	①	65-79
2008	Sedona	P225/70R16	P235/60R17	35	35	②	①	65-79

OEM: Original Equipment Manufacturer

PSI: Pounds Per Square Inch

① Replace if any measurable movement is found.

② STD: 6.5Jx16

  OPT: 6.5Jx17

22140_SEDO_C0010

BRAKE SPECIFICATIONS

All measurements in inches unless noted

Year	Model		Brake Disc			Brake Drum			Minimum Lining Thickness	Brake Caliper	
			Original Thickness	Minimum Thickness	Maximum Run-out	Original Inside Diameter	Max. Wear Limit	Maximum Machine Diameter		Bracket Bolts (ft. lbs.)	Mounting Bolts (ft. lbs.)
2006	Sedona	F	1.180	1.100	0.0012	—	—	—	—	62-72	16-23
		R	0.470	NA	0.0020	—	—	—	—	36-43	16-23
2007	Sedona	F	1.180	1.100	0.0012	—	—	—	—	62-72	16-23
		R	0.470	NA	0.0020	—	—	—	—	36-43	16-23
2008	Sedona	F	1.180	1.100	0.0012	—	—	—	—	62-72	16-23
		R	0.470	NA	0.0020	—	—	—	—	36-43	16-23

NA: Not Available

F: Front

R: Rear

22140_SEDO_C0011

SCHEDULED MAINTENANCE INTERVALS
Kia—Sedona

TO BE SERVICED	TYPE OF SERVIC	VEHICLE MILEAGE INTERVAL (x1000)																
		7.5	15	22.5	30	37.5	45	52.5	60	67.5	75	82.5	90	97.5	100	105	112.5	120
Accessory drive belts	S/I	✓	✓	✓	✓	✓	✓	✓	✓	✓	✓	✓	✓	✓	✓	✓	✓	✓
Air cleaner element	I/R		✓		✓		✓		✓		✓		✓			✓		✓
Air conditioner system	S/I	Inspect the system operation and refrigerant amount annually.																
Brake lines, hoses and connections	S/I		✓		✓		✓		✓		✓		✓			✓		✓
Chassis and body fasteners	T	✓	✓	✓	✓	✓	✓	✓	✓	✓	✓	✓	✓	✓	✓	✓	✓	✓
Cooling system hoses and coolant level	S/I		✓		✓		✓		✓		✓		✓			✓		✓
CV-joint boots					✓				✓				✓					✓
Engine coolant	R								✓				✓					✓
Engine oil and filter	R	✓	✓	✓	✓	✓	✓	✓	✓	✓	✓	✓	✓	✓	✓	✓	✓	✓
Exhaust system heat shields	S/I				✓				✓				✓					✓
Front and rear brakes	S/I				✓				✓				✓					✓
Front ball joints	S/I				✓				✓				✓					✓
Fuel filter	R					✓					✓						✓	
Fuel tank air filter	R				✓				✓				✓					
Fuel lines and hoses	S/I				✓				✓				✓					✓
Idle speed	A				✓				✓				✓					✓
Locks and hinges	L	✓	✓	✓	✓	✓	✓	✓	✓	✓	✓	✓	✓	✓	✓	✓	✓	✓
Spark plugs	R														✓			
Steering operation and linkage	S/I				✓				✓				✓					✓
Timing belt	R								✓									✓
Valve Clearance	A								✓									✓

R: Replace   S/I: Inspect and service, if needed   L: Lubricate   A: Adjust   T: Tighten

FREQUENT OPERATION MAINTENANCE (SEVERE SERVICE)

If a vehicle is operated under any of the following conditions it is considered severe service

- Towing a trailer or using a camper or car-top carrier.

- Repeated short trips of less than 5 miles in temperatures below freezing, or trips of less than 10 miles in any temperature.

- Extensive idling or low-speed driving for long distances as in heavy commercial use, such as delivery, taxi or police cars.

- Operating on rough, muddy or salt-covered roads.

- Operating on unpaved or dusty roads.

- Driving in extremely hot (over 90°F) conditions.

Engine oil and filter: replace every 5000 miles or 5 months, whichever occurs first.

Air cleaner element: inspect ever 15,000 miles or 15 months and replace every 30,000 miles or 30 months, whichever occurs first.

Fuel system hoses (California models only): replace every 105,000 miles.

Emission system hoses (non-California models): inspect every 55,000 miles or 55 months, whichever occurs first.

Emission system hoses (California models): inspect every 60,000 miles or 60 months, whichever occurs first.

Front and rear disc brakes: inspect every 15,000 miles or 15 months, whichever occurs first.

Chassis and body fasteners: tighten every 15,000 miles or 15 months, whichever occurs first.

Locks and hinges: lubricate every 5000 miles or 5 months, whichever occurs first.

22140_SEDO_C0012

PRECAUTIONS

Before servicing any vehicle, please be sure to read all of the following precautions, which deal with personal safety, prevention of component damage, and important points to take into consideration when servicing a motor vehicle:

• Never open, service or drain the radiator or cooling system when the engine is hot; serious burns can occur from the steam and hot coolant.

• Observe all applicable safety precautions when working around fuel. Whenever servicing the fuel system, always work in a well-ventilated area. Do not allow fuel spray or vapors to come in contact with a spark, open flame, or excessive heat (a hot drop light, for example). Keep a dry chemical fire extinguisher near the work area. Always keep fuel in a container specifically designed for fuel storage; also, always properly seal fuel containers to avoid the possibility of fire or explosion. Refer to the additional fuel system precautions later in this section.

• Fuel injection systems often remain pressurized, even after the engine has been turned **OFF**. The fuel system pressure must be relieved before disconnecting any fuel lines. Failure to do so may result in fire and/or personal injury.

• Brake fluid often contains polyglycol ethers and polyglycols. Avoid contact with the eyes and wash your hands thoroughly after handling brake fluid. If you do get brake fluid in your eyes, flush your eyes with clean, running water for 15 minutes. If eye irritation persists, or if you have taken

brake fluid internally, IMMEDIATELY seek medical assistance.

• The EPA warns that prolonged contact with used engine oil may cause a number of skin disorders, including cancer. You should make every effort to minimize your exposure to used engine oil. Protective gloves should be worn when changing oil. Wash your hands and any other exposed skin areas as soon as possible after exposure to used engine oil. Soap and water, or waterless hand cleaner should be used.

• All new vehicles are now equipped with an air bag system, often referred to as a Supplemental Restraint System (SRS) or Supplemental Inflatable Restraint (SIR) system. The system must be disabled before performing service on or around system components, steering column, instrument panel components, wiring and sensors. Failure to follow safety and disabling procedures could result in accidental air bag deployment, possible personal injury and unnecessary system repairs.

• Always wear safety goggles when working with, or around, the air bag system. When carrying a non-deployed air bag, be sure the bag and trim cover are pointed away from your body. When placing a non-deployed air bag on a work surface, always face the bag and trim cover upward, away from the surface. This will reduce the motion of the module if it is accidentally deployed. Refer to the additional air bag system precautions later in this section.

• Clean, high quality brake fluid from a sealed container is essential to the safe and

proper operation of the brake system. You should always buy the correct type of brake fluid for your vehicle. If the brake fluid becomes contaminated, completely flush the system with new fluid. Never reuse any brake fluid. Any brake fluid that is removed from the system should be discarded. Also, do not allow any brake fluid to come in contact with a painted surface; it will damage the paint.

• Never operate the engine without the proper amount and type of engine oil; doing so WILL result in severe engine damage.

• Timing belt maintenance is extremely important. Many models utilize an interference-type, non-freewheeling engine. If the timing belt breaks, the valves in the cylinder head may strike the pistons, causing potentially serious (also time-consuming and expensive) engine damage. Refer to the maintenance interval charts for the recommended replacement interval for the timing belt, and to the timing belt section for belt replacement and inspection.

• Disconnecting the negative battery cable on some vehicles may interfere with the functions of the on-board computer system(s) and may require the computer to undergo a relearning process once the negative battery cable is reconnected.

• When servicing drum brakes, only disassemble and assemble one side at a time, leaving the remaining side intact for reference.

• Only an MVAC-trained, EPA-certified automotive technician should service the air conditioning system or its components.

BRAKES

ANTI-LOCK BRAKE SYSTEM (ABS)

GENERAL INFORMATION

PRECAUTIONS

• Certain components within the ABS system are not intended to be serviced or repaired individually.

• Do not use rubber hoses or other parts not specifically specified for and ABS system. When using repair kits, replace all parts included in the kit. Partial or incorrect repair may lead to functional problems and require the replacement of components.

• Lubricate rubber parts with clean, fresh brake fluid to ease assembly. Do not

use shop air to clean parts; damage to rubber components may result.

• Use only DOT 3 brake fluid from an unopened container.

• If any hydraulic component or line is removed or replaced, it may be necessary to bleed the entire system.

• A clean repair area is essential. Always clean the reservoir and cap thoroughly before removing the cap. The slightest amount of dirt in the fluid may plug an orifice and impair the system function. Perform repairs after components have been thoroughly cleaned; use only denatured alcohol

to clean components. Do not allow ABS components to come into contact with any substance containing mineral oil; this includes used shop rags.

• The Anti-Lock control unit is a microprocessor similar to other computer units in the vehicle. Ensure that the ignition switch is **OFF** before removing or installing controller harnesses. Avoid static electricity discharge at or near the controller.

• If any arc welding is to be done on the vehicle, the control unit should be unplugged before welding operations begin.

BRAKES — BLEEDING THE BRAKE SYSTEM

BLEEDING PROCEDURE

BLEEDING PROCEDURE

✳ WARNING

When bleeding the brakes, note the following:

- Do not reuse the drained fluid
- Always use Genuine DOT 3 or DOT 4 Brake Fluid. Using a non-Genuine DOT3 or DOT 4 brake fluid can cause corrosion and decrease the life of the system

- Make sure no dirt of other foreign matter is allowed to contaminate the brake fluid
- Do not spill brake fluid on the vehicle, it may damage the paint; if brake fluid does contact the paint, wash it off immediately with water
- The reservoir on the master cylinder must be at the MAX (upper) level mark at the start of bleeding procedure and checked after bleeding each brake caliper. Add fluid as required

1. Make sure the brake fluid level in the master cylinder fluid reservoir is at the MAX (upper) level line.
2. Have someone slowly pump the brake pedal several times, and then apply steady pressure
3. Loosen the right-rear brake bleed screw to allow air to escape from the system. Tighten the bleed screw securely
4. Repeat the procedure for each wheel in the sequence shown below until air bubbles no longer appear in the fluid
5. Refill the master cylinder reservoir to the MAX (upper) level line

BRAKES — FRONT DISC BRAKES

✳ CAUTION

Dust and dirt accumulating on brake parts during normal use may contain asbestos fibers from production or aftermarket brake linings. Breathing excessive concentrations of asbestos fibers can cause serious bodily harm. Exercise care when servicing brake parts. Do not sand or grind brake lining unless equipment used is designed to contain the dust residue. Do not clean brake parts with compressed air or by dry brushing. Cleaning should be done by dampening the brake components with a fine mist of water, then wiping the brake components clean with a dampened cloth. Dispose of cloth and all residue containing asbestos fibers in an impermeable container with the appropriate label. Follow practices prescribed by the Occupational Safety and Health Administration (OSHA) and the Environmental Protection Agency (EPA) for the handling, processing, and disposing of dust or debris that may contain asbestos fibers.

BRAKE CALIPER

REMOVAL & INSTALLATION

1. Raise and safely support the front of the vehicle.
2. Remove the front wheel and tire.
3. Remove the guide rod bolts from the caliper assembly.
4. Remove the caliper assembly.
5. Installation is the reverse of removal.

DISC BRAKE PADS

REMOVAL & INSTALLATION

1. Raise and safely support the front of the vehicle.
2. Remove the front wheel and tire.
3. Remove the guide rod bolts from the caliper assembly.
4. Remove the caliper assembly.
5. Remove the pads, the pad shims and the pad retainers from the caliper bracket.
6. Installation is the reverse of removal.

BRAKES — REAR DISC BRAKES

✳ CAUTION

Dust and dirt accumulating on brake parts during normal use may contain asbestos fibers from production or aftermarket brake linings. Breathing excessive concentrations of asbestos fibers can cause serious bodily harm. Exercise care when servicing brake parts. Do not sand or grind brake lining unless equipment used is designed to contain the dust residue. Do not clean brake parts with compressed air or by dry brushing. Cleaning should be done by dampening the brake components with a fine mist of water, then wiping the brake components clean with a dampened cloth. Dispose of cloth and all residue containing asbestos fibers in an impermeable container with the appropriate label. Follow practices prescribed by the Occupational Safety and Health Administration (OSHA) and the Environmental Protection Agency (EPA) for the handling, processing, and disposing of dust or debris that may contain asbestos fibers.

BRAKE CALIPER

REMOVAL & INSTALLATION

1. Raise and safely support the rear of the vehicle.
2. Remove the rear wheel and tire.
3. Remove the guide rod bolts from the caliper assembly.
4. Remove the caliper assembly.
5. Installation is the reverse of removal.

DISC BRAKE PADS

REMOVAL & INSTALLATION

1. Raise and safely support the rear of the vehicle.
2. Remove the rear wheel and tire.
3. Remove the guide rod bolts from the caliper assembly.
4. Remove the caliper assembly.
5. Remove the pads, the pad shims and the pad retainers from the caliper bracket.
6. Installation is the reverse of removal.

BRACES   PARKING BRAKE

PARKING BRAKE CABLES

ADJUSTMENT

See Figure 1.

1. Adjust the adjusting nut (A) so that the parking pedal stoke is 3.46–3.86 inches (88–98 mm) after full stroke operation of parking pedal more than 3 times.

2. The parking brake adjustment must be completed after adjusting the rear shoe.

3. After adjusting parking brake, notice following:

    a. Must be free clearance between adjusting nut and pin.

    b. Check securely that the brake is not dragging.

PARKING BRAKE SHOES

REMOVAL & INSTALLATION

1. Raise and safely support the rear of the vehicle.

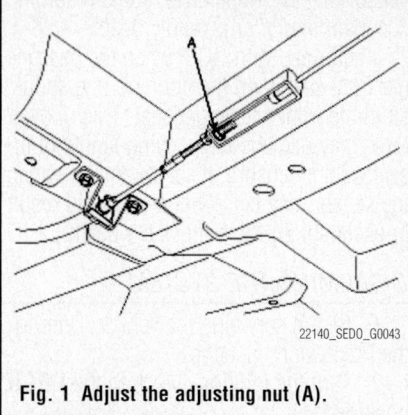

22140_SEDO_G0043

Fig. 1 Adjust the adjusting nut (A).

2. Remove the rear wheel and tire.

3. Remove the dust cap and the rotor.

4. Remove the hub nut and washer and remove the rear hub.

5. Remove the shoe hold down pin and spring by pressing and rotating the spring.

6. Remove the adjuster assembly and the lower return spring.

7. Remove the strut and the upper return spring.

8. Remove the retaining clip from the parking brake cable at the back of the backing plate.

9. Disconnect the parking brake cable from the brake shoe.

To install:

10. Connect the parking brake cable to the parking brake shoe.

11. Install the shoe hold down pin and spring to hold the brake shoe.

12. Install the adjuster assembly and the lower return spring.

13. Install the upper return spring and strut.

14. Grease where is necessary.

15. Install the rear hubs.

16. Install the hub nut.

17. Install the dust cap and the rotor.

18. Install the rear wheels and tires.

19. Tighten the parking brake adjusting nut.

CHASSIS ELECTRICAL   AIR BAG (SUPPLEMENTAL RESTRAINT SYSTEM)

GENERAL INFORMATION

✳✳ CAUTION

These vehicles are equipped with an air bag system. The system must be disarmed before performing service on, or around, system components, the steering column, instrument panel components, wiring and sensors. Failure to follow the safety precautions and the disarming procedure could result in accidental air bag deployment, possible injury and unnecessary system repairs.

SERVICE PRECAUTIONS

Disconnect and isolate the battery negative cable before beginning any airbag system component diagnosis, testing, removal, or installation procedures. Allow system capacitor to discharge for two minutes before beginning any component service. This will disable the airbag system. Failure to disable the airbag system may result in accidental airbag deployment, personal injury, or death.

Do not place an intact undeployed airbag face down on a solid surface. The airbag will propel into the air if accidentally deployed and may result in personal injury or death.

When carrying or handling an undeployed airbag, the trim side (face) of the airbag should be pointing towards the body to minimize possibility of injury if accidental deployment occurs. Failure to do this may result in personal injury or death.

Replace airbag system components with OEM replacement parts. Substitute parts may appear interchangeable, but internal differences may result in inferior occupant protection. Failure to do so may result in occupant personal injury or death.

Wear safety glasses, rubber gloves, and long sleeved clothing when cleaning powder residue from vehicle after an airbag deployment. Powder residue emitted from a deployed airbag can cause skin irritation. Flush affected area with cool water if irritation is experienced. If nasal or throat irritation is experienced, exit the vehicle for fresh air until the irritation ceases. If irritation continues, see a physician.

Do not use a replacement airbag that is not in the original packaging. This may result in improper deployment, personal injury, or death.

The factory installed fasteners, screws and bolts used to fasten airbag components have a special coating and are specifically designed for the airbag system. Do not use substitute fasteners. Use only original equipment fasteners listed in the parts catalog when fastener replacement is required.

During, and following, any child restraint anchor service, due to impact event or vehicle repair, carefully inspect all mounting hardware, tether straps, and anchors for proper installation, operation, or damage. If a child restraint anchor is found damaged in any way, the anchor must be replaced. Failure to do this may result in personal injury or death.

Deployed and non-deployed airbags may or may not have live pyrotechnic material within the airbag inflator.

Do not dispose of driver/passenger/ curtain airbags or seat belt tensioners unless you are sure of complete deployment. Refer to the Hazardous Substance Control System for proper disposal.

Dispose of deployed airbags and tensioners consistent with state, provincial, local, and federal regulations.

After any airbag component testing or service, do not connect the battery negative cable. Personal injury or death may result if the system test is not performed first.

If the vehicle is equipped with the Occupant Classification System (OCS), do not connect the battery negative cable before performing the OCS Verification Test using the scan tool and the appropriate diagnostic information. Personal injury or death may result if the system test is not performed properly.

Never replace both the Occupant

Restraint Controller (ORC) and the Occupant Classification Module (OCM) at the same time. If both require replacement, replace one, then perform the Airbag System test before replacing the other.

Both the ORC and the OCM store Occupant Classification System (OCS) calibration data, which they transfer to one another when one of them is replaced. If both are replaced at the same time, an irreversible fault will be set in both modules and the OCS may malfunction and cause personal injury or death.

If equipped with OCS, the Seat Weight Sensor is a sensitive, calibrated unit and must be handled carefully. Do not drop or handle roughly. If dropped or damaged, replace with another sensor. Failure to do so may result in occupant injury or death.

If equipped with OCS, the front passenger seat must be handled carefully as well.

When removing the seat, be careful when setting on floor not to drop. If dropped, the sensor may be inoperative, could result in occupant injury, or possibly death.

If equipped with OCS, when the passenger front seat is on the floor, no one should sit in the front passenger seat. This uneven force may damage the sensing ability of the seat weight sensors. If sat on and damaged, the sensor may be inoperative, could result in occupant injury, or possibly death.

DISARMING THE SYSTEM

1. Before servicing the vehicle, refer to the Precautions Section.
2. Turn the ignition switch to the **LOCK** position.
3. Disconnect the negative battery cable.
4. Wait 10 minutes for the battery back-up power to discharge.

ARMING THE SYSTEM

1. Before servicing the vehicle, refer to the Precautions Section.
2. Connect the negative battery cable.
3. Turn the ignition switch **ON**.
4. Verify that the air bag indicator illuminates for 4–8 seconds, then goes off.

CLOCKSPRING CENTERING

Prior to installing the clock spring, align the mating mark and "NEUTRAL" position indicator of the clock spring, and, after turning the front wheels to the straight-ahead position, install the clock spring to the column switch. If the mating mark of the clock spring is not properly aligned, the steering wheel may not completely rotate during a turn, or the flat cable within the clock spring may be severed, obstructing normal operation of the SRS and possibly leading to serious injury to the vehicle's driver.

DRIVETRAIN

AUTOMATIC TRANSAXLE

REMOVAL & INSTALLATION

See Figures 2 through 14.

1. Remove the battery.
2. Disconnect the Air Flow Sensor (AFS) connector (A).
3. Remove the air cleaner upper cover by loosening the clips.
4. Remove the air cleaner assembly.
5. Disconnect the air cleaner hose.
6. Remove the battery tray.
7. Disconnect the following transaxle wire harness connectors:
   a. Remove the inhibiter switch connector (A).
   b. Remove the solenoid valve connector (A).
   c. Remove the input speed sensor connector (A).
   d. Remove the output speed sensor connector (A).
   e. Remove the vehicle speed sensor connector (A).
   f. Remove the Crankshaft Position (CKP) sensor connector (A).
8. Remove the shift cable by removing the bolt (A) and clip (B).
9. Disconnect the transaxle oil cooler hoses from the tubes.
10. Remove the transaxle mounting bolts (B, C).
11. Using the SST(09200-38001), hold the engine and transaxle assembly safely.
12. Remove the transaxle insulator mounting bolt (A).
13. Remove the front wheels.
14. Remove the power steering column joint bolt.
15. Raise the vehicle.
16. Remove the undercover.
17. Drain the transaxle oil.
18. Drain the power steering oil through the return tube.
19. Disconnect the power steering pressure tube from the power steering oil pump.

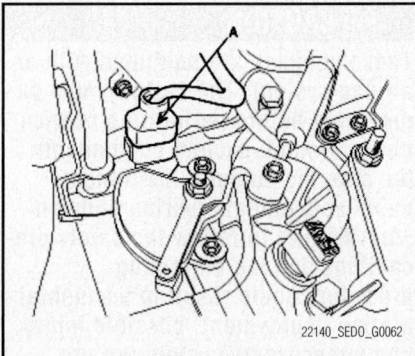

Fig. 4 Remove the solenoid valve connector (A)

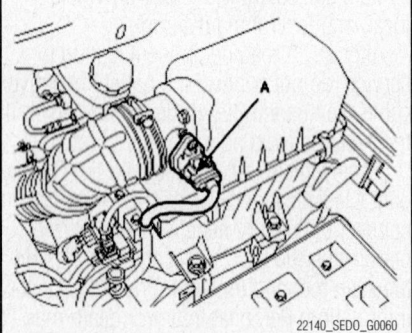

Fig. 2 Disconnect the Air Flow Sensor (AFS) connector (A)

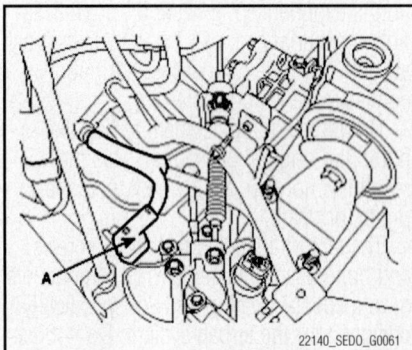

Fig. 3 Remove the inhibiter switch connector (A)

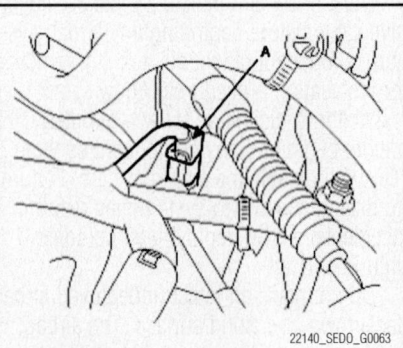

Fig. 5 Remove the put speed sensor connector (A)

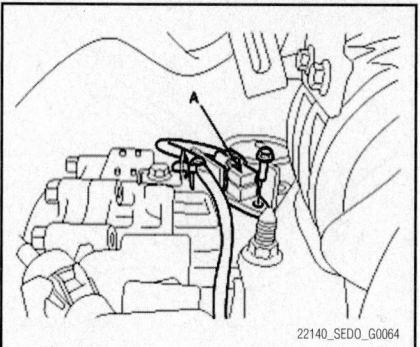

Fig. 6 Remove the output speed sensor connector (A)

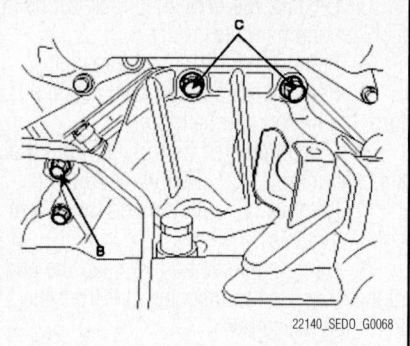

Fig. 10 Remove the transaxle mounting bolts (B, C)

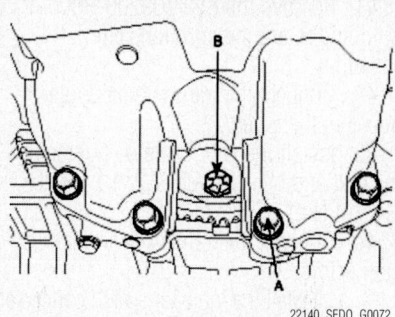

Fig. 14 Remove the transaxle under mounting bolts (A) and the drive plate bolts (B)

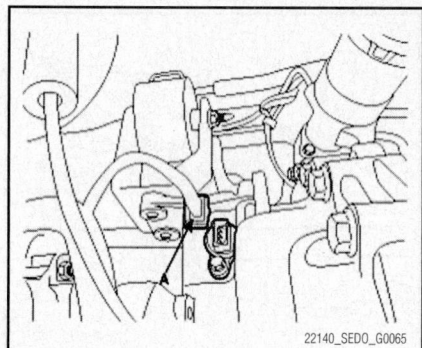

Fig. 7 Remove the vehicle speed sensor connector (A)

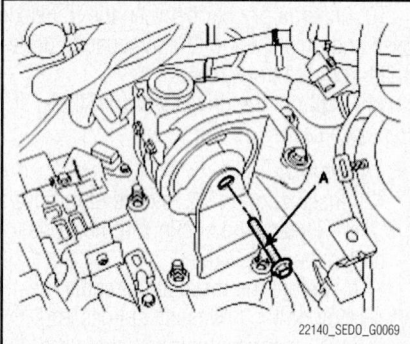

Fig. 11 Remove the transaxle insulator mounting bolt (A)

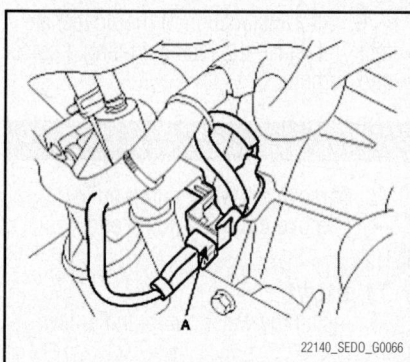

Fig. 8 Remove the Crankshaft Position (CKP) sensor connector (A)

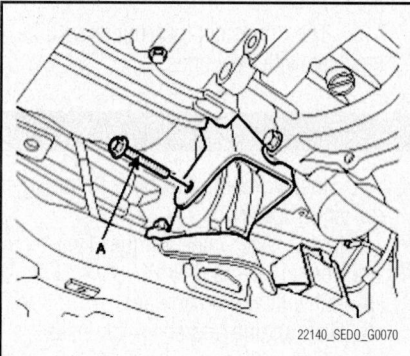

Fig. 12 Remove the roll stopper mounting bolts (A)

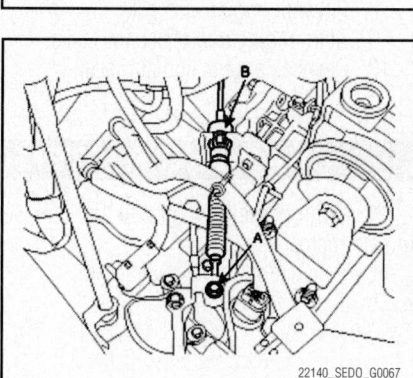

Fig. 9 Remove the shift cable by removing the bolt (A) and clip (B)

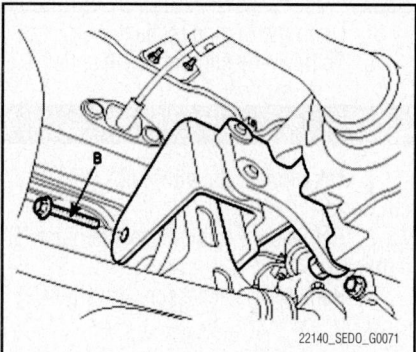

Fig. 13 Remove the roll stopper mounting bolts (B)

20. Disconnect the lower arm, the tie rod end ball joint, the stabilizer bar link from the front knuckle.

21. Remove the roll stopper mounting bolts (A,B).

22. Remove the mounting bolts from the sub frame by supporting the sub frame with a jack.

23. Remove drive shaft from transaxle.

24. Install a jack for supporting the transaxle assembly.

25. Remove the transaxle under mounting bolts (A) and the drive plate bolts (B).

To install:

26. Lowering the vehicle and/or lifting up a jack, install the transaxle assembly.

27. Tighten the transaxle under mounting bolts. Torque: 47–62 ft. lbs. (65–85 Nm).

28. Install the starter motor. Torque: 47–62 ft. lbs. (65–85 Nm).

29. Tighten the transaxle under mounting bolts. Torque: 33–38 ft. lbs. (46–53 Nm).

30. After removing a jack, insert the drive shafts.

31. Install the sub frame.

32. Tighten the roll stopper mounting bolts. Torque: 65–80 ft. lbs. (90–110 Nm).

33. Connect the return tube with a clamp.

34. Connect the lower arm, the tie rod end ball joint and the stabilizer bar link to the front knuckle.

35. Connect the power steering pressure tube to the power steering oil pump.

36. Install the undercover.

37. Install the steering column joint bolt..

38. Install the front wheels and tires.

39. Tighten the transaxle insulator mounting bolt. Torque: 65–80 ft. lbs. (90–110 Nm).

40. Tighten the transaxle mounting bolts. Torque: 47–62 ft. lbs. (65–85 Nm).

41. Remove the SST (09200-38001) holding the engine and transaxle assembly.

42. Connect the transaxle oil cooler hoses to the tubes.

43. Install the shift cable by tightening the bolt and clip. Torque: 7.2–10.1 ft. lbs. (10–14 Nm).

44. Connect the transaxle wire harness connectors.

    a. Install the inhibiter switch connector.

    b. Install the solenoid valve connector.

    c. Install the input speed sensor connector.

    d. Install the output speed sensor connector.

    e. Install the vehicle speed sensor connector.

    f. Install the CKP sensor connector.

45. Install the battery tray

46. Connect the air cleaner hose

47. Install the air cleaner assembly

48. Install the air cleaner upper cover

49. Connect the AFS connector

50. Install the battery

FRONT HALFSHAFT

REMOVAL & INSTALLATION

See Figures 15 and 16.

1. Raise the vehicle and remove the wheel & tire assembly.

2. Remove the drain plug and drain the ATF.

3. Unstake the driveshaft lock nut using a chisel and hammer.

4. Remove the driveshaft lock nut.

5. Remove the split pin and castle nut form the tie rod end ball joint.

6. Disconnect the tie rod end from the knuckle using a SST (09568-4A000).

7. Remove the split pin and lower arm bolt and nut (A).

8. Using a plastic hammer, tap the end of the driveshaft to disconnect it from the front hub assembly.

9. Remove the driveshaft.

    a. Remove the heat protector.

    b. Remove the inner shaft bearing bracket assembly mounting bolts (A).

10. Insert a pry bar between the transaxle case and driveshaft joint and separate driveshaft from the transaxle.

11. Pull out the driveshaft from the transaxle case.

To install:

12. Installation is the reverse of removal.

    a. Replace the circlip with new ones before the installation.

    b. Before the installation, apply the gear oil on the driveshaft splines and contacting surface of differential case oil seal.

    c. After installing the driveshaft joint to the transaxle case, be sure it will not come out.

    d. The driveshaft lock nut should be replaced with new ones.

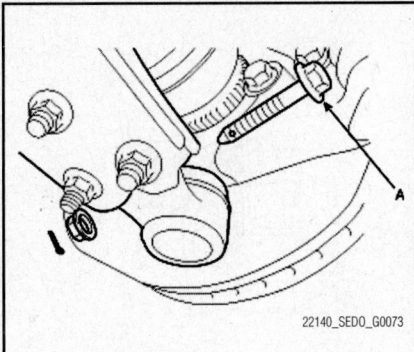

Fig. 15 Remove the split pin and lower arm bolt and nut (A)

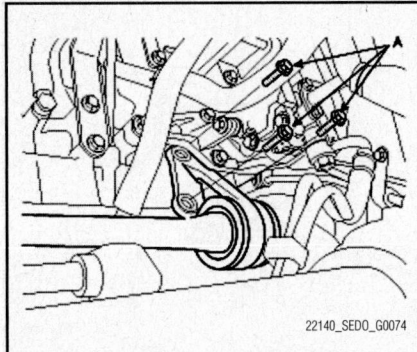

Fig. 16 Remove the inner shaft bearing bracket assembly mounting bolts (A)

    e. After installation of the driveshaft lock nut, stake the lock nut using a chisel and hammer.

ENGINE COOLING

THERMOSTAT

REMOVAL & INSTALLATION

1. Drain engine coolant so its level is below thermostat.

2. Remove water inlet and thermostat.

To install:

3. Place thermostat in thermostat housing.

    a. Install a new thermostat and gasket.

    b. Install the thermostat with the jiggle valve upward.

4. Install water inlet. Torque: 12.3—14.5 ft. lbs. (16.7–19.6 Nm).

5. Fill with engine coolant.

6. Start engine and check for leaks.

WATER PUMP

REMOVAL & INSTALLATION

1. Drain the engine coolant.

2. Remove the serpentine drive belt

3. Remove the water pump pulley

4. Remove the water pump and gasket.

To install:

5. Install the water pump and a new gasket.

6. Install the water pump pulley.

7. Install the serpentine drive belt

8. Fill with engine coolant.

9. Start engine and check for leaks.

10. Recheck engine coolant level.

ENGINE ELECTRICAL

ALTERNATOR

REMOVAL & INSTALLATION

1. Disconnect the battery negative terminal first, then the positive terminal.

2. Disconnect the alternator connector.

3. Remove the cable from alternator "B" terminal.

4. Remove the accessory drive belt.

CHARGING SYSTEM

5. Pull out the through bolt and remove the alternator

6. Installation is the reverse of removal.

ENGINE ELECTRICAL

FIRING ORDER

3.8L engine firing order: 1–2–3–4–5–6

IGNITION COIL

REMOVAL & INSTALLATION

1. Remove the engine cover.
2. Remove the ignition coil connector (s).

➡**When removing the ignition coil connector, pull the lock pin and push the clip.**

ENGINE ELECTRICAL

STARTER

REMOVAL & INSTALLATION

See Figure 17.

1. Disconnect the battery negative cable.
2. Disconnect the starter cable (A) from the B terminal (B) on the solenoid (C), and disconnect the connector (D) from the S terminal (E).
3. Remove the 2 bolts holding the starter, and remove the starter.
4. Installation is the reverse of removal.
5. Connect the battery negative cable to the battery.

ENGINE MECHANICAL

➡**Disconnecting the negative battery cable may interfere with the functions of the on board computer systems and may require the computer to undergo a relearning process, once the negative battery cable is reconnected.**

ACCESSORY DRIVE BELTS

ACCESSORY BELT ROUTING

See Figure 18.

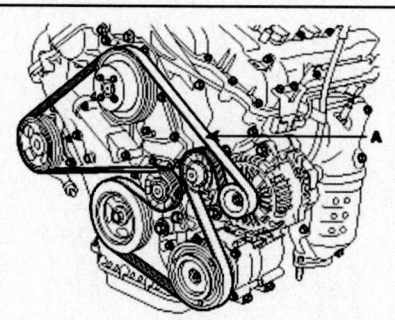

Fig. 18 Accessory belt routing—3.8L engine

22140_SEDO_G0085

3. Remove the ignition coil (s).
4. Installation is the reverse of removal.

IGNITION TIMING

ADJUSTMENT

Ignition timing is controlled by the ECM. No adjustment is possible.

SPARK PLUGS

REMOVAL & INSTALLATION

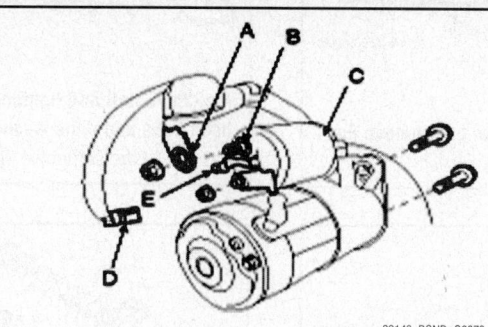

Fig. 17 Disconnect the starter cable (A) from the B terminal (B) on the solenoid (C), and disconnect the connector (D) from the S terminal (E)

22140_ROND_G0073

INSPECTION

Inspect the drive belt for signs of glazing or cracking. A glazed belt will be perfectly smooth from slippage, while a good belt will have a slight texture of fabric visible. Cracks will usually start at the inner edge of the belt and run outward. All worn or damaged drive belts should be replaced immediately.

ADJUSTMENT

The belt tension is maintained by an automatic tensioner. No adjustment is necessary or possible.

REMOVAL & INSTALLATION

1. Relieve tension on the tensioner pulley.
2. Remove accessory drive belt

To install:
3. Install accessory drive belt in the following sequence:
   a. Crankshaft pulley
   b. A/C pulley
   c. Idler pulley

IGNITION SYSTEM

1. Remove the engine cover.
2. Remove the ignition coil connector (s).

➡**When removing the ignition coil connector, pull the lock pin and push the clip.**

3. Remove the ignition coil (s).
4. Using a spark plug socket, remove the spark plug (s).
5. Installation is the reverse of removal.

STARTING SYSTEM

CYLINDER HEAD

   d. Alternator pulley
   e. Water pump pulley
   f. P/S pump pulley
   g. Tensioner pulley.
4. Rotate auto tensioner arm in the counterclockwise direction moving auto tensioner pulley bolt with wrench.
5. After putting belt on auto tensioner pulley, release the auto tensioner pulley slowly.

CYLINDER HEAD

REMOVAL & INSTALLATION

See Figures 19 through 26.

➡**Engine removal is required for this procedure.**

1. Remove exhaust manifold.
2. Remove intake manifold.
3. Remove timing chain.
4. Remove water temperature control assembly.
5. Remove camshaft bearing caps.
6. Remove camshaft assembly.

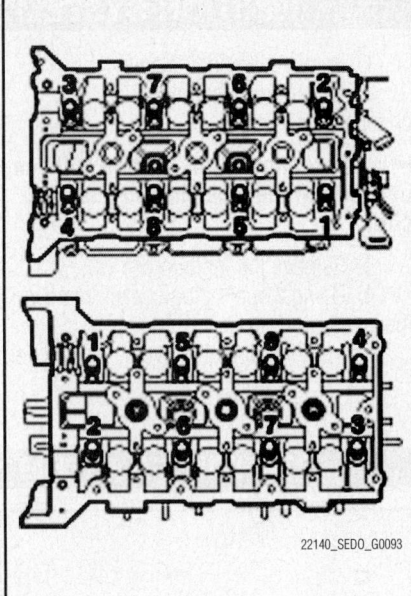

Fig. 19 Uniformly loosen and remove the 16 cylinder head bolts

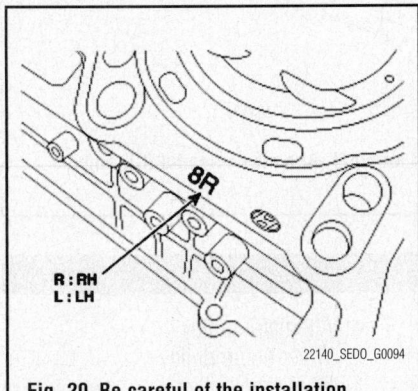

Fig. 20 Be careful of the installation direction

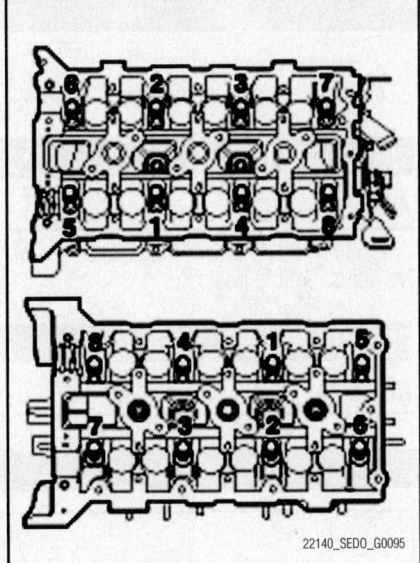

Fig. 21 Install and tighten the cylinder head bolts and plate washers, in several passes, in the sequence shown

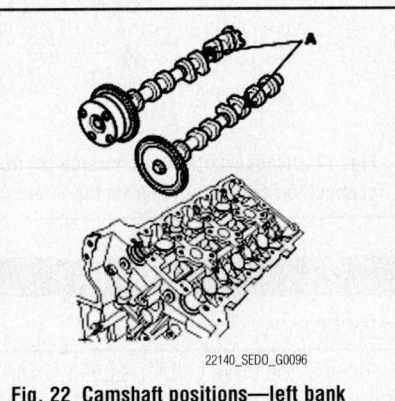

Fig. 22 Camshaft positions—left bank shown

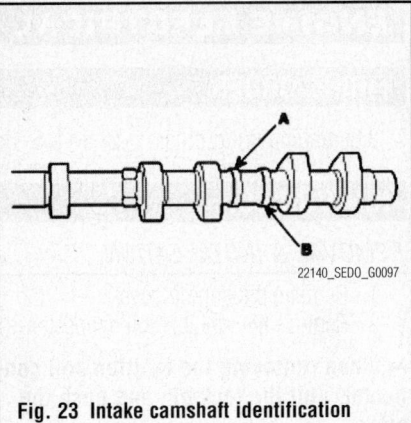

Fig. 23 Intake camshaft identification

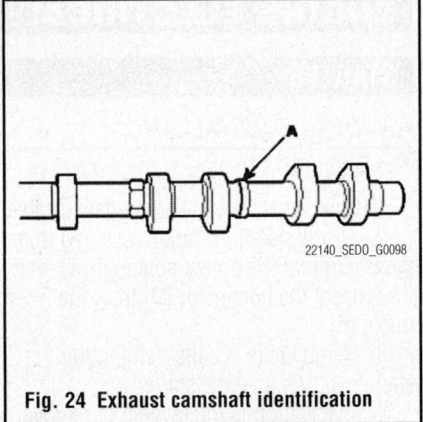

Fig. 24 Exhaust camshaft identification

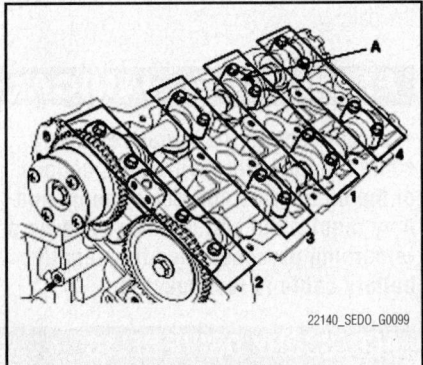

Fig. 25 Assemble camshaft bearing caps in the order shown

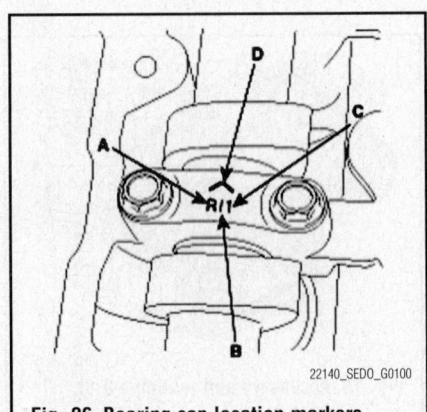

Fig. 26 Bearing cap location markers

7. Remove cylinder head bolts, then remove cylinder head.

a. Uniformly loosen and remove the 16 cylinder head bolts, in several passes, in the sequence shown.

b. Remove the 16 cylinder head bolts and plate washers.

✳✳ CAUTION

Head warpage or cracking could result from removing bolts in an incorrect order.

c. Lift the cylinder head from the dowels on the cylinder block and place the cylinder head on wooden blocks on a bench.

To install:

➡**Thoroughly clean all parts to be assembled. Always use a new head**

and manifold gasket. **The cylinder head gasket is a metal gasket. Take care not to bend it.**

8. If necessary, rotate the crankshaft to set the No.1 piston at TDC.

9. Install the cylinder head.

a. The sealant locations on cylinder head and cylinder block must be free from contamination.

b. Apply sealant on cylinder block top face before assembling cylinder head gaskets. The part must be assembled within 5 minutes after sealant was applied.

c. Apply sealant on cylinder head gaskets after assembling cylinder head gaskets on cylinder block. The part must be assembled within 5 minutes after sealant was applied.

➡**Be careful of the installation direction.**

d. Install the cylinder head.

➡**Remove the extruded sealant after assembling cylinder heads.**

10. Place the cylinder head carefully in order not to damage the gasket with the bottom part of the end.

11. Install cylinder head bolts.

a. Do not apply engine oil on the threads and under the heads of the cylinder head bolts.

b. Using SST(09221-4A000), install and tighten the cylinder head bolts and plate washers, in several passes, in the sequence shown. Torque: 1st step: 27.5–30.4 ft. lbs. (37.3–41.2 Nm); 2nd step: Additional 120° ± 2°; 3rd step: Additional 90° ± 2°.

➡**Always use new cylinder head bolt.**

12. Install the CVVT and camshaft sprocket. Torque: 48–56 ft. lbs. (65–76 Nm).

➡**Install camshaft-inlet to dowel pin of CVVT assembly. Ensure that the camshaft oil inlet is not obstructed. Do not rotate CVVT assembly when camshaft is installed to dowel pin of CVVT assembly.**

13. Install camshafts (A).

➡**Apply a light coat of engine oil on camshaft journals. Assemble the key groove of camshaft rear side to the same level of head top surface. Ensure that the camshaft components are installed in the correct locations.**

14. Install camshaft bearing caps. Assemble camshaft bearing caps in the order shown.

Torque: 1st step: 4.3 ft. lbs. (5.9 Nm); 2nd step: 7.23–8.68 ft. lbs. (9.80–11.76 Nm).

➡**Ensure that the cam bearing caps are installed in the correct location and direction.**

A: L (LH), R (RH) B: I (Intake), None (Exhaust) C: Journal number D: Front mark.

15. Install water temperature control assembly.

16. Install timing chain.

17. Check and adjust valve clearance.

18. Install the exhaust manifold.

19. Install the intake manifold.

ENGINE ASSEMBLY

REMOVAL & INSTALLATION

See Figures 27 through 33.

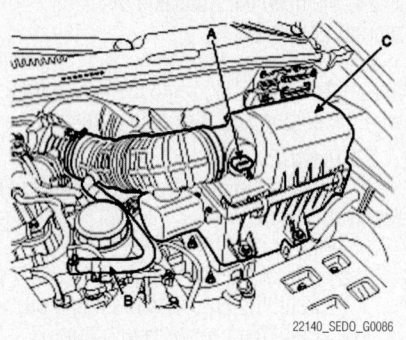

Fig. 27 **Remove the intake air hose and air cleaner assembly**

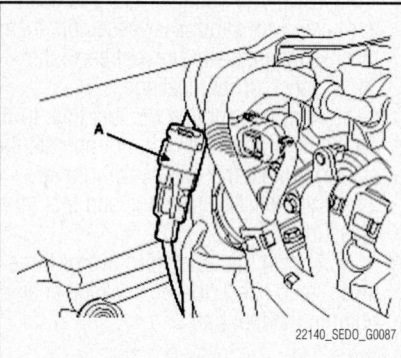

Fig. 28 **Disconnect the RH oxygen sensor connectors (A)**

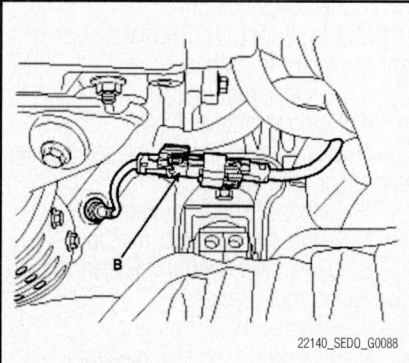

Fig. 29 **Disconnect the LH oxygen sensor connectors (B)**

1. Remove the engine cover.

➡**If your vehicle is equipped with air conditioning, refer to the precautions for information regarding the implications of servicing your A/C system yourself. Only a MVAC-trained, EPA-certified, automotive technician should service the A/C system or its components.**

2. Recover refrigerant, opening the high & low pressure pipe caps and connecting the refrigerant station.

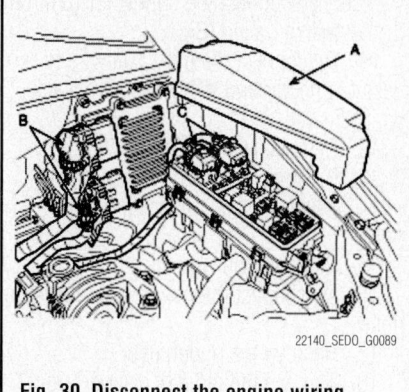

Fig. 30 **Disconnect the engine wiring**

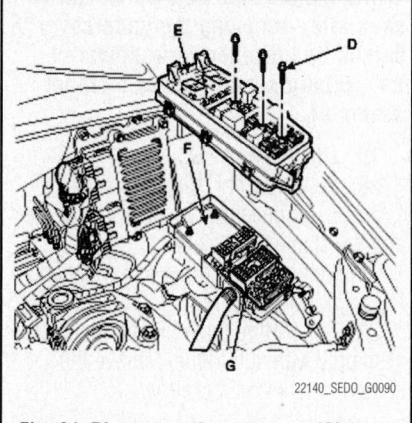

Fig. 31 **Disconnect the connector (G) from the splash shield (B)**

3. Disconnect the terminals from the battery and remove the battery.

4. Remove the intake air hose and air cleaner assembly.

a. Disconnect the MAF connector (A).

b. Disconnect the breather hose (B) from air cleaner hose.

c. Remove the intake air hose and air cleaner assembly (C).

5. Remove the battery tray.

6. Remove the radiator grille upper cover.

7. Disconnect the RH and LH oxygen sensor connectors (A, B).

8. Disconnect the automatic transaxle fluid cooler hoses.

9. Disconnect the A/C high and low pressure from the radiator or the compressor.

10. Disconnect the engine wiring.

a. Remove the engine room fuse and relay box cover (A).

b. Disconnect the PCM connectors (B).

c. Disconnect the FAM connectors (C).

d. Unscrew the FAM mounting bolts (D-3EA) and take the FAM (E) out of the splash shield (F).

e. Disconnect the connector (G) from the splash shield (B).

11. Disconnect the transaxle wire harness connector and remove the transaxle control cable.

12. Remove heater hose.

13. Remove the brake vacuum hose.

14. Remove power steering pump hose.

15. Remove front wheels.

16. Disconnect the power steering return hose.

17. Remove the undercover.

18. Remove the radiator support upper member assembly.

➡ **The bottom side bolt which can be seen after removing the undercover should be loosened for removal of the radiator support upper member assembly.**

19. Drain engine oil.

20. Remove front exhaust pipe.

21. Disconnect the wheel speed sensor and the stabilizer bar link.

22. Remove the caliper assembly and suspend it with a wire.

23. Supporting the engine and transaxle assembly with sub frame, remove the engine mounting bracket(A).

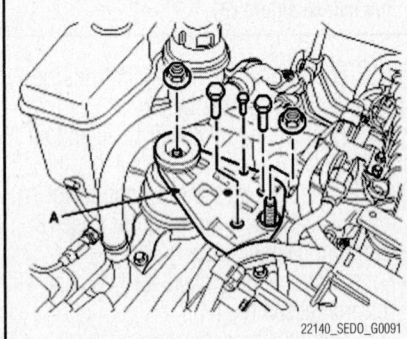

Fig. 32 Remove the engine mounting bracket (A)

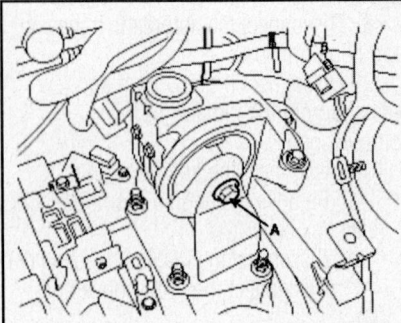

Fig. 33 Remove the transaxle insulator mounting bolt (A)

24. Remove the transaxle insulator mounting bolt (A)

25. Supporting the engine and transaxle assembly with a jack, remove the sub frame with the engine and transaxle assembly.

To install:

26. Installation is in the reverse order of removal.

27. Perform the following :

  a. Adjust the shift cable.

  b. Refill the engine with engine oil.

  c. Refill the transaxle with fluid.

  d. Refill the radiator with engine coolant.

  e. Bleed air from the cooling system with the heater valve open.

  f. Clean the battery posts and cable terminals with sandpaper assemble them, then apply grease to prevent corrosion.

28. Inspect for fuel leakage.

  a. After assembling the fuel line, turn on the ignition switch (do not operate the starter) so that the fuel pump runs for approximately two seconds and fuel line pressurizes.

  b. Repeat this operation two or three times, then check for fuel leakage at any point in the fuel line.

EXHAUST MANIFOLD

REMOVAL & INSTALLATION

1. Remove undercover

2. Disconnect LH, RH rear oxygen sensor connector from bracket.

3. Remove front muffler

4. Remove oil level gauge.

5. Disconnect LH front oxygen sensor connector from bracket.

6. Remove LH heat protector.

7. Remove LH exhaust manifold.

8. Disconnect RH front oxygen sensor connector from bracket.

9. Remove RH heat protector.

10. Remove RH exhaust manifold.

To install:

11. Install new gasket and exhaust manifold. Torque: 29–32 ft. lbs. (39–44 Nm).

12. Install heat protector.

13. Install front muffler. Torque: 29–32 ft. lbs. (39–44 Nm).

14. Connect oxygen sensor connector.

15. Install undercover.

INTAKE MANIFOLD

REMOVAL & INSTALLATION

See Figures 34 through 43.

1. Remove the engine cover

2. Remove the intake air hose and air cleaner assembly.

  a. Disconnect the MAF connector (A).

  b. Disconnect the breather hose (B) from air cleaner hose.

  c. Remove the intake air hose and air cleaner assembly (C).

3. Disconnect RH oxygen sensor connector (A).

4. Disconnect RH injector connector (A) and ignition coil connector (B).

5. Disconnect PCSV connector (A), MAP sensor connector (B) and PCSV hose.

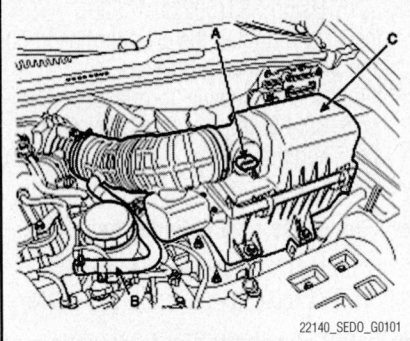

Fig. 34 Remove the intake air hose and air cleaner assembly

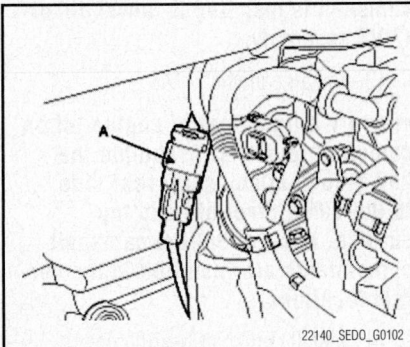

Fig. 35 Disconnect RH oxygen sensor connector (A)

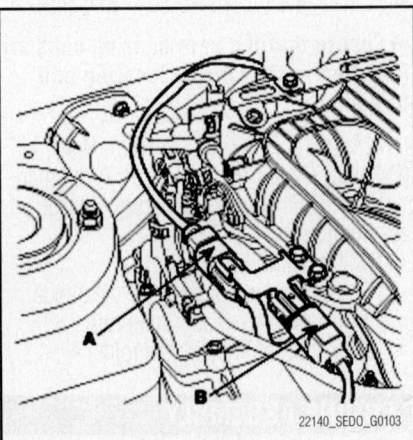

Fig. 36 Disconnect RH injector connector (A) and ignition coil connector (B)

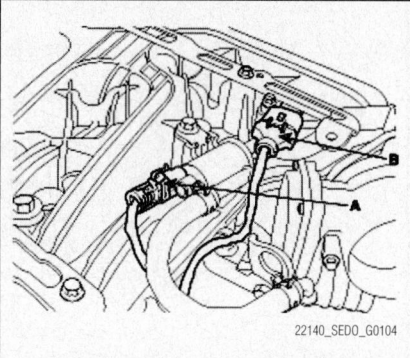

Fig. 37 Disconnect PCSV connector (A), MAP sensor connector (B) and PCSV hose

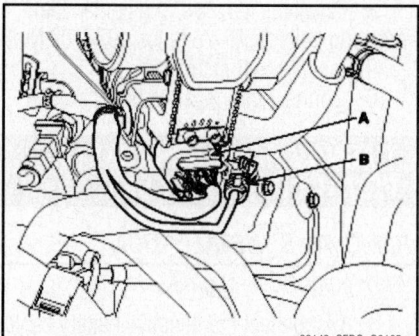

Fig. 38 Disconnect ETC connector (A) and knock sensor connector (B)

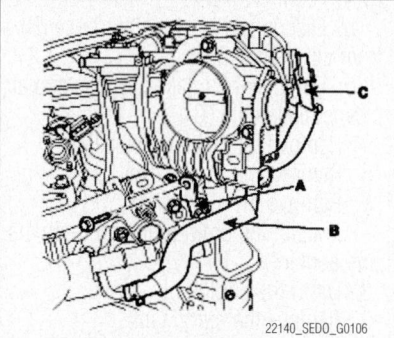

Fig. 39 Remove ETC bracket (A). Disconnect water hose (B) and PCV (C) hose.

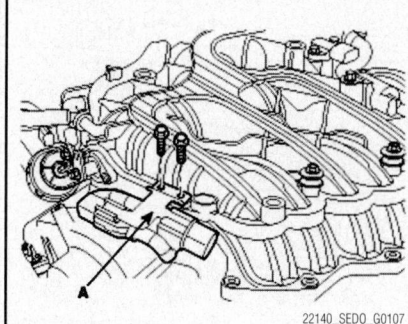

Fig. 40 Remove connector bracket (A) from surge tank

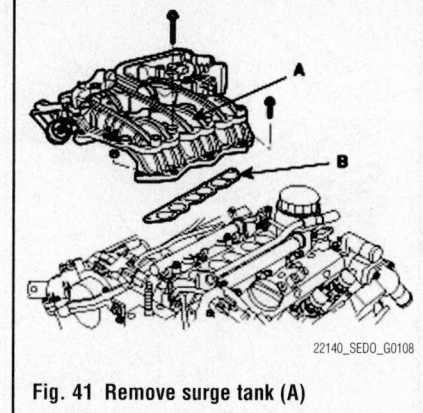

Fig. 41 Remove surge tank (A)

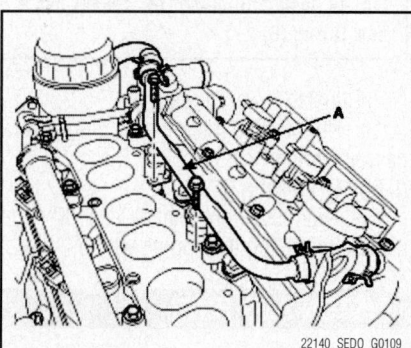

Fig. 42 Disconnect breather pipe assembly (A)

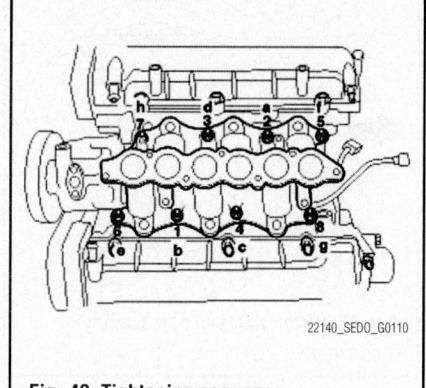

Fig. 43 Tightening sequence

6. Disconnect the Electric Throttle Control (ETC) connector (A) and knock sensor connector (B).

7. Remove the ETC bracket (A).

8. Disconnect the water hoses (B) from ETC.

9. Disconnect the PCV (C) hose.

10. Disconnect the brake vacuum hose.

11. Remove the surge tank stay.

12. Remove the connector bracket (A) from surge tank.

13. Remove surge tank (A).

14. Disconnect the breather pipe assembly (A).

15. Disconnect the LH injector connector.

16. Remove the delivery pipe and intake manifold together.

To install:

17. Install the intake manifold and new gasket on the cylinder head. Torque: 1st Step: 2.9–4.3 ft. lbs. (3.9–5.9 Nm); 2nd Step: 14–17 ft. lbs. (19–24 Nm); 3rd Step: Repeat 2nd step twice or more.

➡ **Be careful of the installation direction.**

- 1st step order: a—h.
- 2nd step order: 1–8.

18. Install delivery pipe.

19. Connect LH injector connector.

20. Connect breather Pipe assembly.

21. Install surge tank.

22. Install connector bracket on the surge tank.

23. Install surge tank stay. Torque: 20–23 ft. lbs. (27–31 Nm).

24. Connect brake vacuum hose.

25. Connect PCV hose.

26. Connect water hoses to ETC.

27. Install ETC bracket. Torque: 12–19 ft. lbs. (16–26 Nm).

28. Connect ETC connector and knock sensor connector.

29. Connect PCSV connector, MAP sensor connector and PCSV hose.

30. Connect RH injector connector and ignition coil connector.

31. Connect RH oxygen sensor connector.

32. Install air cleaner upper cover and intake hose.

33. Connect MAF and breather hose.

OIL PAN

REMOVAL & INSTALLATION

1. Drain engine oil.

2. Using SST(09215-3C000) remove lower oil pan.

To install:

3. Install lower oil pan.

4. Uniformly tighten the bolts in several passes. Torque: 7.23–8.68 ft. lbs. (9.80–11.76 Nm).

5. After assembly, wait at least 30 minutes before filling the engine with oil.

OIL PUMP

REMOVAL & INSTALLATION

See Figures 44 through 47.

1. Drain engine oil.

2. Using SST(09215-3C000) remove lower oil pan.

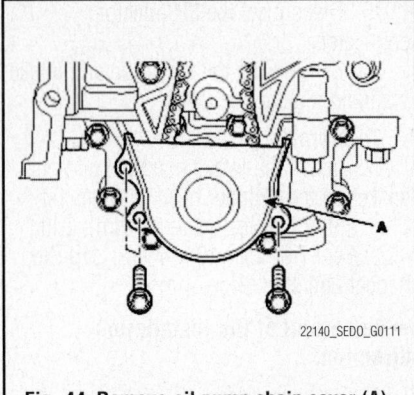

Fig. 44 Remove oil pump chain cover (A)

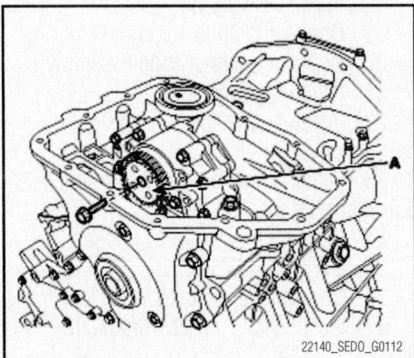

Fig. 45 Remove oil pump chain sprocket (A)

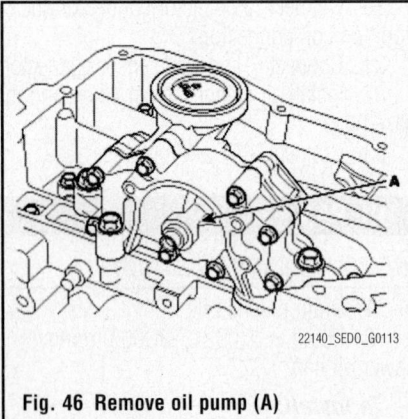

Fig. 46 Remove oil pump (A)

3. Remove oil pump chain cover (A).
4. Remove oil pump chain sprocket (A).
5. Remove oil pump (A).

To install:

6. Install oil pump (A). Tighten to 15–17 ft. lbs. (20–24 Nm).

➡ **Always use a new O-ring (B).**

7. Install oil pump sprocket and oil pump chain on the oil pump. Torque: 14–16 ft. lbs. (19–22 Nm).
8. Install oil pump chain cover. Torque: 7.23–8.68 ft. lbs. (9.80–11.76 Nm).

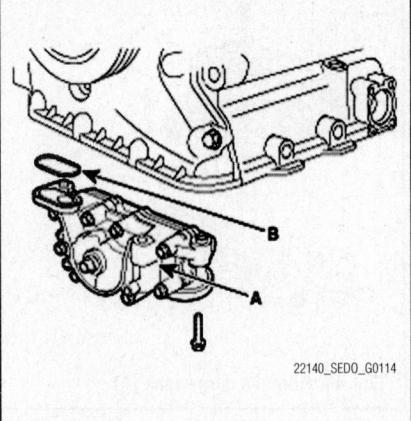

Fig. 47 Install oil pump (A). Always use a new O-ring (B)

9. Install lower oil pan.
10. Uniformly tighten the bolts in several passes. Torque: 7.23–8.68 ft. lbs. (9.80–11.76 Nm).
11. After assembly, wait at least 30 minutes before filling the engine with oil.

PISTON AND RING

POSITIONING

See Figure 48.

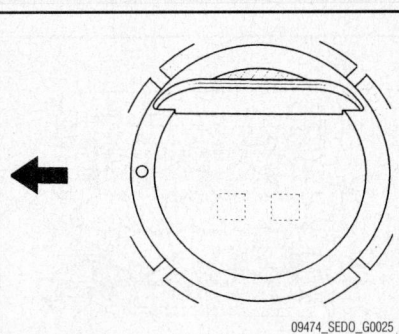

Fig. 48 Piston ring end gap spacing—3.8L Engine

REAR MAIN SEAL

REMOVAL & INSTALLATION

See Figure 49.

1. Disconnect the negative battery cable.
2. Remove the transaxle assembly.
3. Remove the flexplate-to-crankshaft bolts, the flexplate and shim plates.
4. Remove the rear oil seal case (A).
5. Remove the oil seal.

To install:

6. Install a new oil seal.
7. Install rear oil seal case. Torque: 7.23–8.68 ft. lbs. (9.80–11.76 Nm).

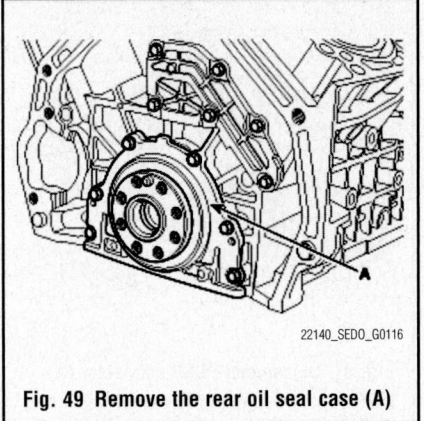

Fig. 49 Remove the rear oil seal case (A)

8. Install the flexplate, and tighten the mounting bolts to 71–76 ft. lbs. (97–102 Nm).
9. Install the transaxle.
10. Connect the negative battery cable.

TIMING CHAIN, SPROCKETS, FRONT COVER AND SEAL

REMOVAL & INSTALLATION

See Figures 50 through 86.

1. Disconnect the negative battery cable.
2. Remove the engine cover
3. Remove the intake air hose and air cleaner assembly.
   a. Disconnect the Mass Air Flow (MAF) connector (A).
   b. Disconnect the breather hose (B) from air cleaner hose.
   c. Remove the intake air hose and air cleaner assembly (C).
4. Remove the RH front wheel.
5. Remove the undercover
6. Remove the side cover.
7. Loosen the drain plug and drain the engine coolant.
8. Drain the engine oil.
9. Remove the surge tank.
   a. Disconnect the RH oxygen sensor connector (A) and loosen the power steering hose mounting bolts (B).

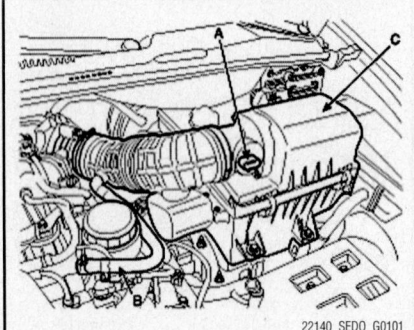

Fig. 50 Remove the intake air hose and air cleaner assembly

b. Disconnect the RH injector connector (A) and ignition coil connector (B).

c. Disconnect the PCSV connector (A), MAP sensor connector (B) and PCSV hose.

d. Disconnect the ETC connector (A) and knock sensor connector (B).

e. Disconnect the Oil Control Valve (OCV) connector (A) and knock sensor connector (B).

f. Disconnect the LH front oxygen sensor connector (A).

g. Disconnect the LH ignition coil connector (A), injector connector (B), condenser connector (C) and ground (D), and remove the wiring harness protector (E).

h. Disconnect the LH Camshaft Position Sensor (CMPS) (A) and oil pressure switch connector (B).

i. Remove the Electric Throttle Control (ETC) bracket (A).

j. Disconnect the water hoses (B) from ETC.

k. Disconnect the PCV hose (C).

l. Disconnect the brake vacuum hose.

m. Remove the surge tank stay (A).

n. Remove the connector bracket (A) from surge tank.

o. Remove the surge tank (A).

10. Remove the cylinder head cover.

a. Remove the connector bracket (A) from LH cylinder head cover.

b. Disconnect the RH ignition coil connector (A), condenser connector (B) and remove the wiring bracket (C).

c. Remove the LH, RH ignition coil.

d. Remove the LH, RH cylinder head covers.

11. Using SST (09215-3C000) remove lower oil pan.

12. Set a jack to the upper oil pan.

13. Remove the coolant reservoir tank.

14. Remove the engine mounting bracket (A).

15. Set No.1 cylinder to TDC/compression.

a. Turn the crankshaft pulley and align its groove with the timing mark "T" of the lower timing chain cover.

➡**Do not rotate the engine counterclockwise.**

b. Check that the mark (A) of the camshaft timing sprockets are in straight line on the cylinder head surface as shown. If not, turn the crankshaft one revolution (360°).

16. Remove the accessory drive belt.

17. Using SST (09231-3C300) remove the crankshaft damper pulley.

18. Lift up the engine assembly to using the jack.

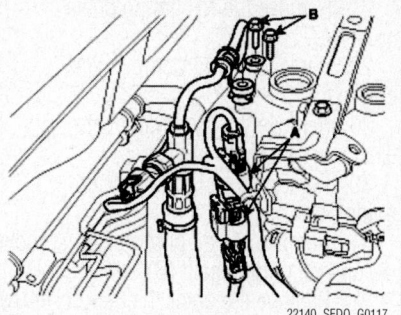
Fig. 51 Disconnect the RH oxygen sensor connector (A) and loosen the power steering hose mounting bolts (B)

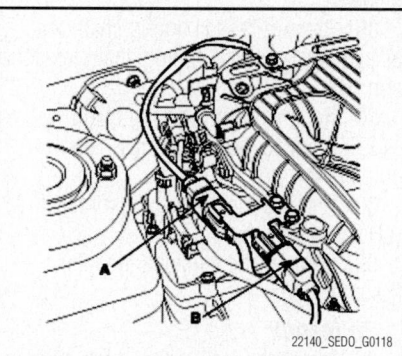
Fig. 52 Disconnect the RH injector connector (A) and ignition coil connector (B)

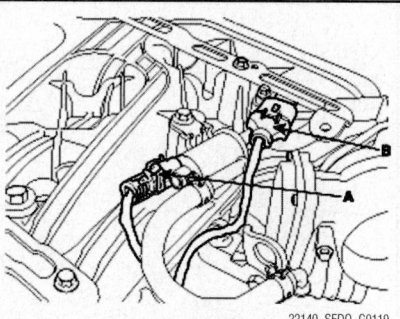
Fig. 53 Disconnect the PCSV connector (A), MAP sensor connector (B) and PCSV hose

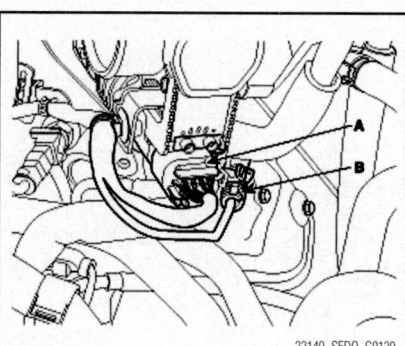
Fig. 54 Disconnect the ETC connector (A) and knock sensor connector (B)

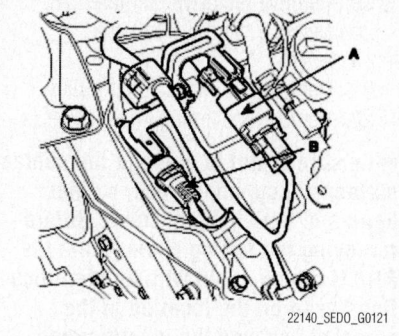
Fig. 55 Disconnect the OCV connector (A) and knock sensor connector (B)

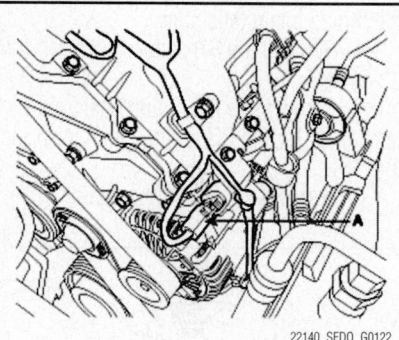
Fig. 56 Disconnect the LH front oxygen sensor connector (A)

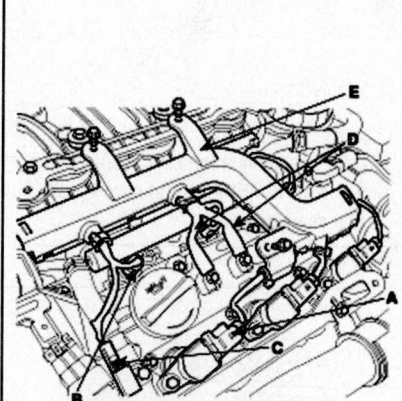
A: LH Ignition Coil Connector
B: Injector Connector
C: Condenser Connector
D: Ground Cable
E: Wiring Harness Protector

Fig. 57 Disconnect the LH ignition coil connector (A), injector connector (B), condenser connector (C) and ground (D), and remove the wiring harness protector (E).

19. Remove the drive belt idler (A).
20. Remove the drive belt auto tensioner (A).
21. Remove the water pump pulley.
22. Remove the timing chain cover (A).

→Be careful not to damage the contact surfaces of cylinder block, cylinder head and timing chain cover. Before removing the timing chain, mark the RH/LH timing chain with an identification based on the location of the sprocket because the identification mark on the chain for TDC (Top Dead Center) can be erased.

23. Install a set pin after compressing the timing chain tensioner.
24. Remove the RH cam-to-cam guide (A).
25. Remove the RH timing chain auto tensioner (A) and RH timing chain tensioner arm (B).
26. Remove the RH timing chain.

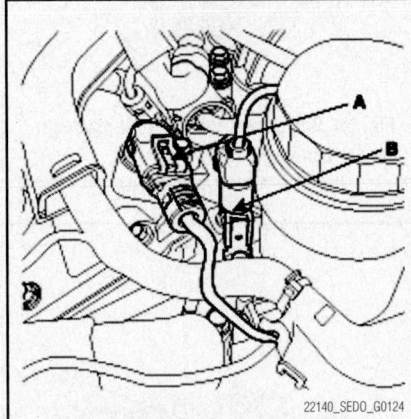

Fig. 58 Disconnect the LH Camshaft Position Sensor (CMPS) (A) and oil pressure switch connector (B)

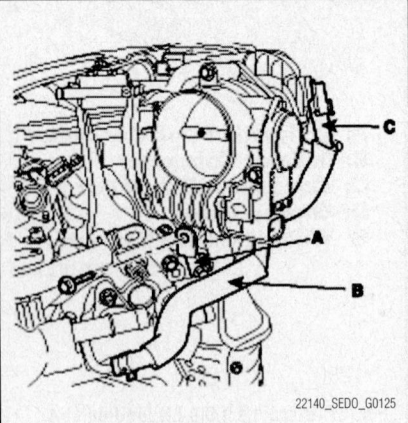

Fig. 59 Remove the Electric Throttle Control (ETC) bracket (A)

27. Remove the RH timing chain guide (A).
28. Remove the oil pump chain cover (A).
29. Remove the oil pump chain tensioner assembly (A).
30. Remove the oil pump chain guide (A).
31. Remove the oil pump chain sprocket (A) and oil pump chain (B).
32. Remove the crankshaft sprocket (A) (Oil pump & RH camshaft drive).
33. Install a set pin after compressing the LH timing chain tensioner.
34. Remove the LH cam-to-cam guide (A).
35. Remove the LH timing chain auto tensioner (A) and LH timing chain tensioner arm (B).
36. Remove the LH timing chain.
37. Remove the LH timing chain guide (A).
38. Remove the crankshaft sprocket (A) (LH camshaft drive).
39. Remove the tensioner adapter assembly (A).

To install:
40. Install the jack to the upper oil pan.
41. The key (A) of crankshaft should be aligned with the timing mark (B) of timing

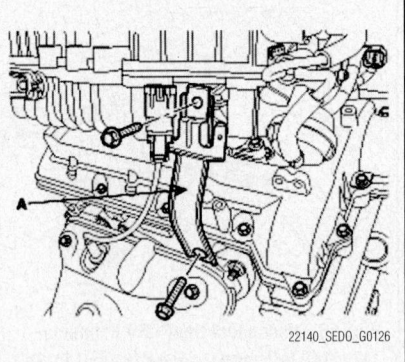

Fig. 60 Remove the surge tank stay (A)

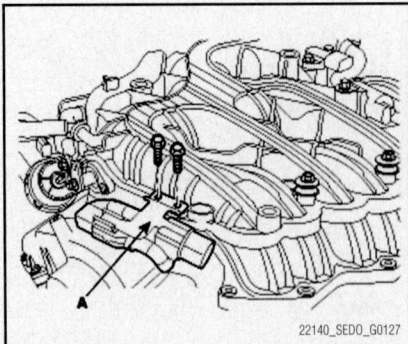

Fig. 61 Remove the connector bracket (A) from surge tank

chain cover. As a result of this, the piston of No.1 cylinder is placed at the top dead center on compression stroke.
42. Install the tensioner adapter assembly.
43. Install the crankshaft sprocket (LH camshaft drive).
44. Install the LH timing chain guide. Torque: 15–18 ft. lbs. (20–25 Nm).
45. Install LH timing chain.

→To install the timing chain with no slack between each shaft (cam, crank), follow the procedure below:

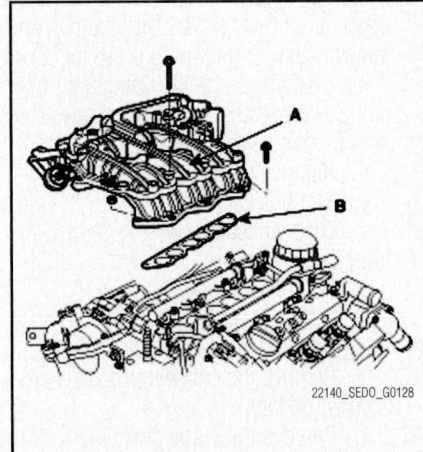

Fig. 62 Remove the surge tank (A)

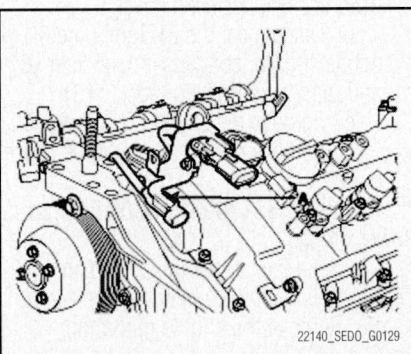

Fig. 63 Remove the connector bracket (A) from LH cylinder head cover

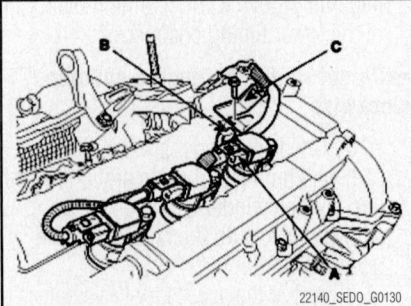

Fig. 64 Disconnect the RH ignition coil connector (A), condenser connector (B) and remove the wiring bracket (C).

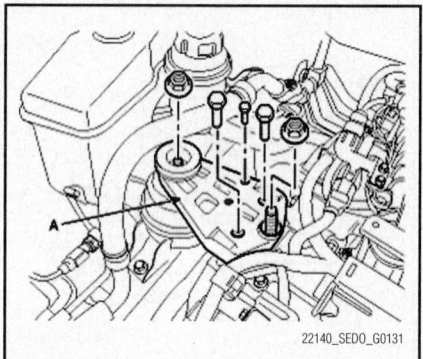

Fig. 65 Remove the engine mounting bracket (A)

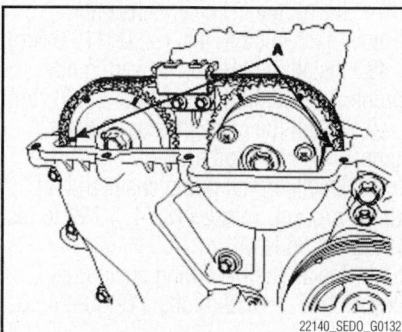

Fig. 66 Check that the mark (A) of the camshaft timing sprockets are in straight line on the cylinder head

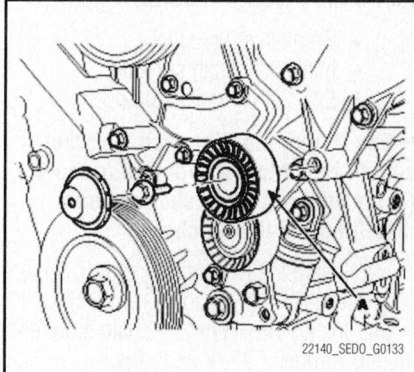

Fig. 67 Remove the drive belt idler (A)

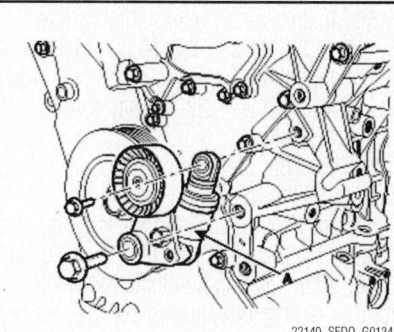

Fig. 68 Remove the drive belt auto tensioner (A)

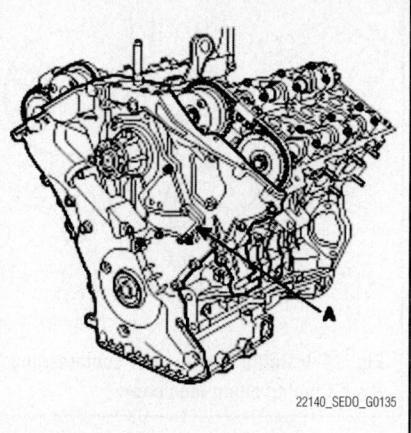

Fig. 69 Remove the timing chain cover (A)

Fig. 70 Install a set pin after compressing the timing chain tensioner

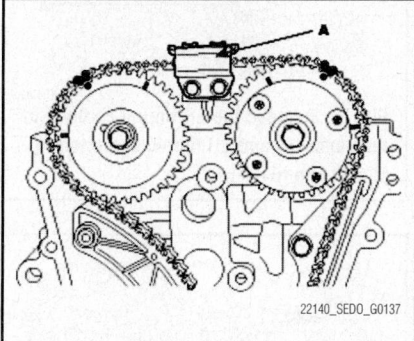

Fig. 71 Remove the RH cam-to-cam guide (A)

- Crankshaft sprocket
- Timing chain guide
- Exhaust camshaft sprocket
- Intake camshaft sprocket.

➥**The timing mark of each sprocket should be matched with**

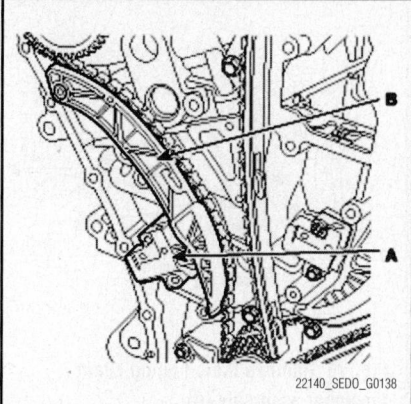

Fig. 72 Remove the RH timing chain auto tensioner (A) and RH timing chain tensioner arm (B).

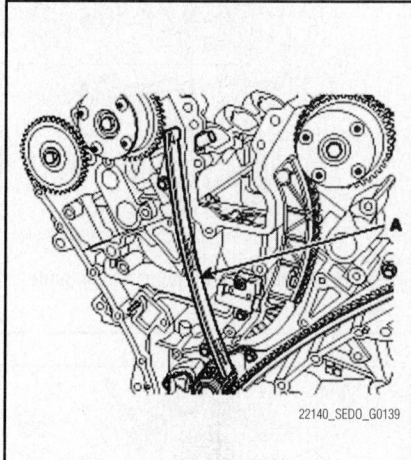

Fig. 73 Remove the RH timing chain guide (A)

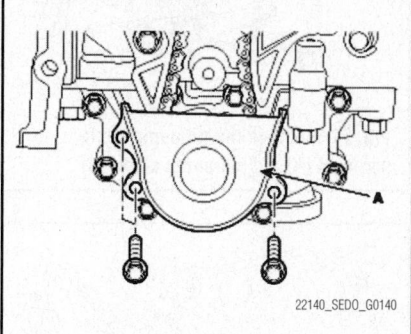

Fig. 74 Remove the oil pump chain cover (A)

timing mark (color link) of timing chain when installing the timing chain.

46. Install the LH timing chain tensioner arm. Torque: 13.74–15.91 ft. lbs. (18.62–21.56 Nm).

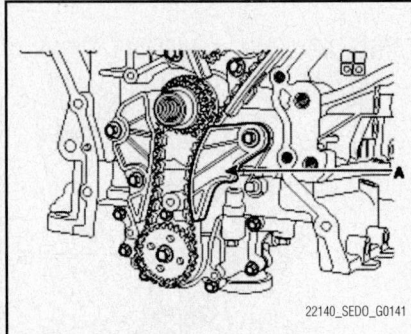

Fig. 75 Remove the oil pump chain tensioner assembly (A)

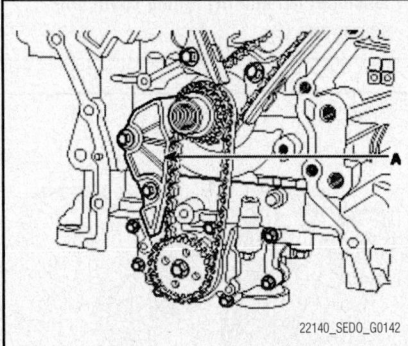

Fig. 76 Remove the oil pump chain guide (A)

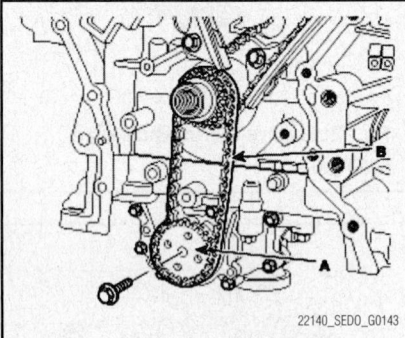

Fig. 77 Remove the oil pump chain sprocket (A) and oil pump chain (B)

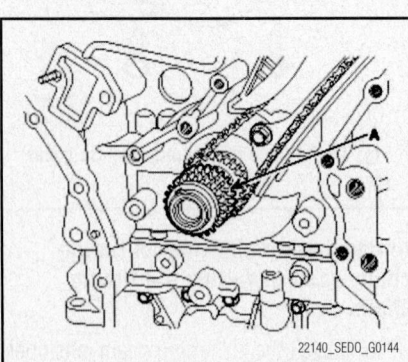

Fig. 78 Remove the crankshaft sprocket (A) (Oil pump & RH camshaft drive)

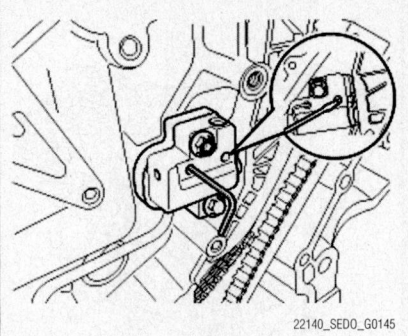

Fig. 79 Install a set pin after compressing the LH timing chain tensioner

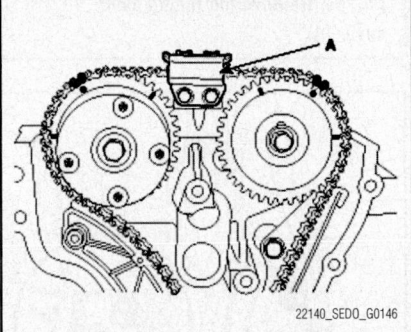

Fig. 80 Remove the LH cam-to-cam guide (A)

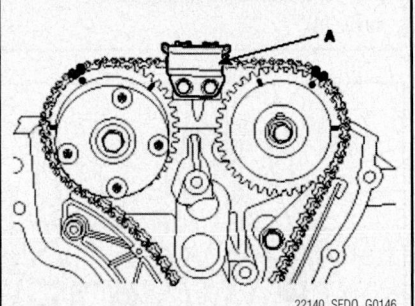

Fig. 81 Remove the LH timing chain auto tensioner (A) and LH timing chain tensioner arm (B)

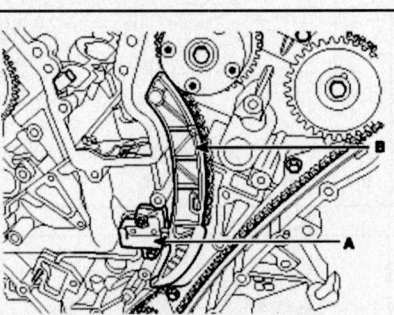

Fig. 82 Remove the LH timing chain guide (A)

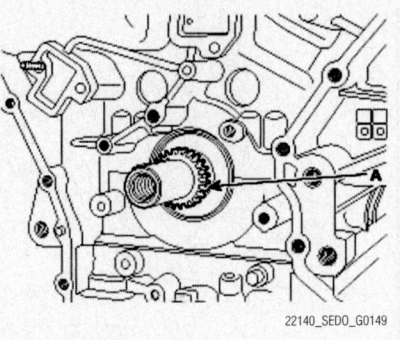

Fig. 83 Remove the crankshaft sprocket (A) (LH camshaft drive)

47. Install the LH chain tensioner. Torque: 7.23–8.68 ft. lbs. (9.80–11.76 Nm).

48. Install the LH cam-to-cam guide. Torque: 7.23–8.68 ft. lbs. (9.80–11.76 Nm).

49. Install the crankshaft sprocket (Oil pump & RH camshaft drive).

50. Install the oil pump chain and oil pump sprocket. Torque: 13.74–15.91 ft. lbs. (18.62–21.56 Nm).

51. Install the RH timing chain guide. Torque: 14.17–18.08 ft. lbs. (19.60–24.50 Nm).

52. Install the RH timing chain.

➡ **To install the timing chain with no slack between each shaft (cam, crank), follow the procedure below:**

- Crankshaft sprocket
- Intake camshaft sprocket
- Exhaust camshaft sprocket

➡ **The timing mark of each sprocket should be matched with timing mark (color link) of timing chain when installing the timing chain.**

53. Install the RH timing chain tensioner arm. Torque: 14–16 ft. lbs. (19–22 Nm).

54. Install the RH timing chain auto tensioner. Torque: 7.23–8.68 ft. lbs. (9.80–11.76 Nm).

55. Install the RH cam-to-cam guide. Torque: 7.23–8.68 ft. lbs. (9.80–11.76 Nm).

56. Install the oil pump chain guide. Torque: 7.23–8.68 ft. lbs. (9.80–11.76 Nm).

57. Install the oil pump chain tensioner assembly. Torque: 7.23–8.68 ft. lbs. (9.80–11.76 Nm).

58. Pull out the pins of hydraulic tensioners (LH & RH).

59. Install the oil pump chain cover. Torque: 7.23–8.68 ft. lbs. (9.80–11.76 Nm).

60. After rotating crankshaft 2 revolutions in regular direction (clockwise viewed from front), confirm the timing mark.

➡ **Always turn the crankshaft clockwise.**

61. Install the timing chain cover.

a. The sealant locations on chain cover and on counter parts (cylinder head, cylinder block, and lower oil pan) must be free of engine oil.

b. Before assembling the timing chain cover, the liquid sealant TB1217H should be applied on the gap between cylinder head and cylinder block. The part must be assembled within 5 minutes after sealant was applied.

c. Install the new gasket to the timing chain cover.

d. Install the timing chain cover.

e. The sealant locations on chain cover and on counter parts (cylinder head, cylinder block, and lower oil pan) must be free of engine oil.

➡**During timing cover installation, care not to take off applied sealant on the timing cover by contact with other parts.**

f. The dowel pins on the cylinder block and holes on the timing chain cover should be used as a reference in order to assemble the timing chain cover to the exact position.

Torque as follows:
- B (17ea): 14–16 ft. lbs. (19–22 Nm).
- C (4ea): 7.23–8.68 ft. lbs. (9.80–11.76 Nm).
- D (1ea): 43–51 ft. lbs. (59–69 Nm).
- E (1ea): 44–51 ft. lbs. (59–69 Nm).
- F (2ea): 19–20 ft. lbs. (25–27 Nm).
- G (4ea): 16–17 ft. lbs. (22–24 Nm).
- H (1ea): 7.23–8.68 ft. lbs. (9.80–11.76 Nm).
- I (1ea): 7.23–8.68 ft. lbs. (9.80–11.76 Nm).
- J (1ea): 7.23–8.68 ft. lbs. (9.80–11.76 Nm).
- K (4ea): 7.23–8.68 ft. lbs. (9.80–11.76 Nm).
- L (1ea): 16–20 ft. lbs. (22–26 Nm).

g. The firing and/or blow out test should not be performed within 30 minutes after the timing chain cover was assembled.

62. Install the water pump pulley. Torque: 5.78–7.23 ft. lbs. (7.84–9.80 Nm).

63. Install the drive belt auto tensioner. Torque: Large Bolt: 60–64 ft. lbs. (81–85 Nm); Small Bolt: 13–16 ft. lbs. (18–22 Nm).

64. Install the drive belt idler. Torque: 39–43 ft. lbs. (53–58 Nm).

65. Lower the engine assembly by using the jack.

66. Using SST (09231-3C100), install timing chain cover oil seal.

67. Using SST (09231-3C300) install the crankshaft damper pulley. Torque: 210–224 ft. lbs. (284–304 Nm).

68. Install the accessory drive belt in the following sequence:
- Crankshaft pulley
- A/C pulley
- Idler pulley
- Alternator pulley
- Water pump pulley
- P/S pump pulley
- Tensioner pulley.

69. Rotate auto tensioner arm in the counterclockwise moving auto tensioner pulley bolt with wrench.

70. After putting belt on auto tensioner pulley, release the auto tensioner pulley slowly.

71. Install the cylinder head cover.

a. The hardening sealant located on the upper area between timing chain cover and cylinder head should be removed before assembling cylinder head cover.

b. After applying sealant (TB1217H), it should be assembled within 5 minutes.

c. The firing and/or blow out test should not be performed within 30 minutes after the cylinder head cover was assembled.

d. Install the cylinder head cover

bolts in sequence. Torque: 7.23–8.68 ft. lbs. (9.80–11.76 Nm).

e. Install the ignition coil.

f. Connect the RH ignition coil connector, the condenser connector and install the wiring bracket.

72. Install the surge tank and wiring connectors.

a. Install the surge tank. Torque: 7.23–8.68 ft. lbs. (9.80–11.76 Nm).

b. Install the connector bracket to the surge tank. Torque: 5.06–7.96 ft. lbs. (6.86–10.78 Nm).

c. Install the surge tank stay. Torque: 20–23 ft. lbs. (27–31 Nm).

d. Connect the brake vacuum hose.

e. Connect the PCV hose.

f. Connect the water hoses to the ETC.

g. Install the ETC bracket.

h. Connect the LH CMPS connector and oil pressure switch connector.

i. Install the wiring harness protector and connect the LH ignition coil connector, injector connector, condenser connector and ground.

j. Connect the LH front oxygen sensor connector.

k. Connect the OCV connector and knock sensor connector.

l. Connect the ETC connector and knock sensor connector.

m. Connect the PCSV connector, MAP sensor connector and PCSV hose.

n. Connect the RH injector connector and ignition coil connector.

o. Connect the RH oxygen sensor connector and tighten the power steering hose mounting bolts.

73. Install the engine mounting bracket. Torque: 65–79 ft. lbs. (88–108 Nm).

74. Install the coolant reservoir tank.

75. Remove the jack from the upper oil pan.

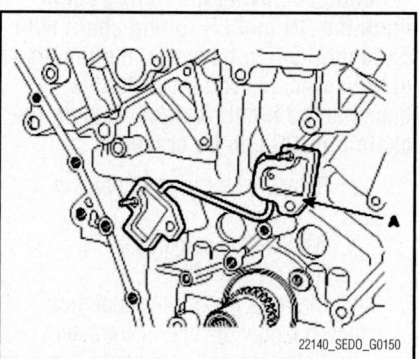

Fig. 84 Remove the tensioner adapter assembly (A)

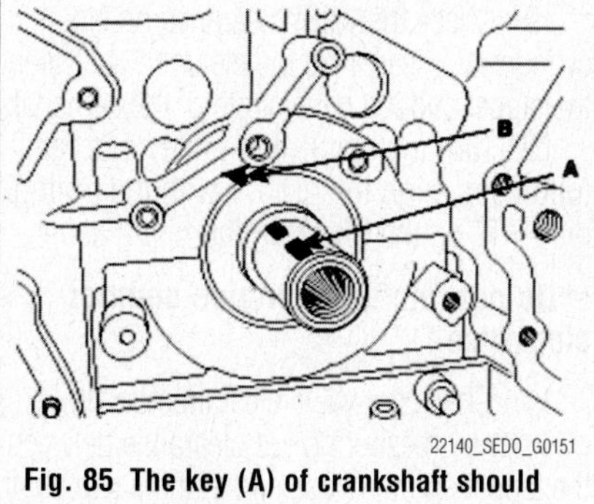

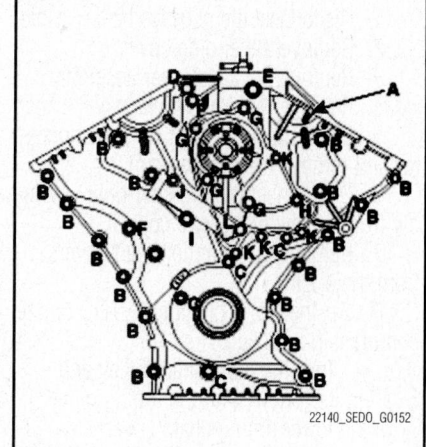

Fig. 85 The key (A) of crankshaft should be aligned with the timing mark (B) of timing chain cover

Fig. 86 Various torque locations

76. Install lower oil pan.

a. Using a gasket scraper, remove all the old packing material from the gasket surfaces.

b. Before assembling the oil pan, the liquid sealant TB1217H should be applied on oil pan.

c. Install the lower oil pan. Torque: 7.23–8.68 ft. lbs. (9.80–11.76 Nm).

77. Install the side cover.

78. Install the undercover.

79. Install the RH front wheel.

80. Install the intake air hose and air cleaner assembly.

a. Install the intake air hose and air cleaner assembly.

b. Connect the breather hose to the air intake hose.

c. Connect the MAF sensor connector.

81. Install the engine cover.

82. Connect the battery negative cable.

➡ **After installation, complete the following:**

- Refill engine with engine oil.
- Refill radiator and reservoir tank with engine coolant.
- Bleed air from the cooling system.
- Start engine and let it run until it warms up (until the radiator fan operates 3 or 4 times.)
- Turn off the engine. Check the level in the radiator, add coolant if needed. This will allow trapped air to be removed from the cooling system.
- Put radiator cap on tightly, then run the engine again and check for leaks.

VALVE LASH

ADJUSTMENT

See Figures 87 and 88.

1. Disconnect the negative battery cable.
2. Remove the engine cover.
3. Remove the air cleaner assembly.
4. Remove the surge tank.
5. Disconnect the ignition coil connector and remove the ignition coil.
6. Disconnect the breather pipe assembly from the cylinder head cover.
7. Remove the cylinder head covers from the engine.
8. Set the No. 1 cylinder to TDC on the compression stroke.

a. Turn the crankshaft pulley and align its groove with the timing mark "T" of the lower timing chain cover.

b. Check that the mark of the camshaft timing sprockets are in straight line posi-

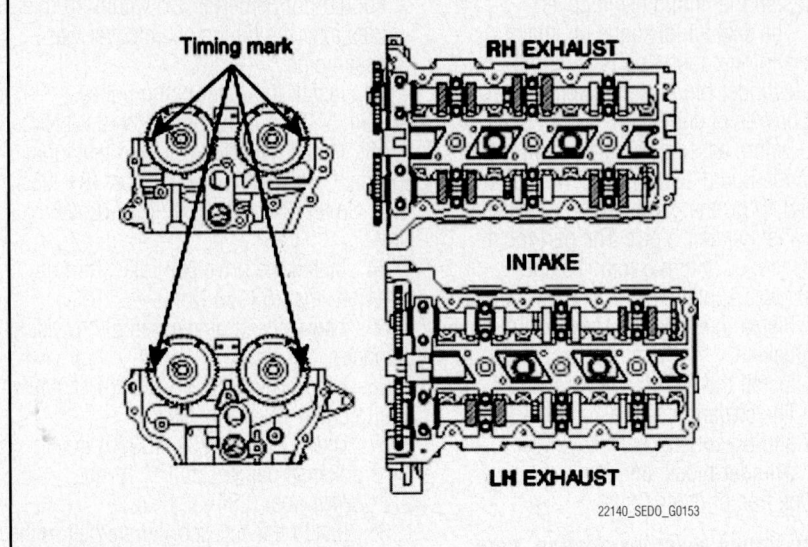

Fig. 87 Valve adjustment No. 1 cylinder

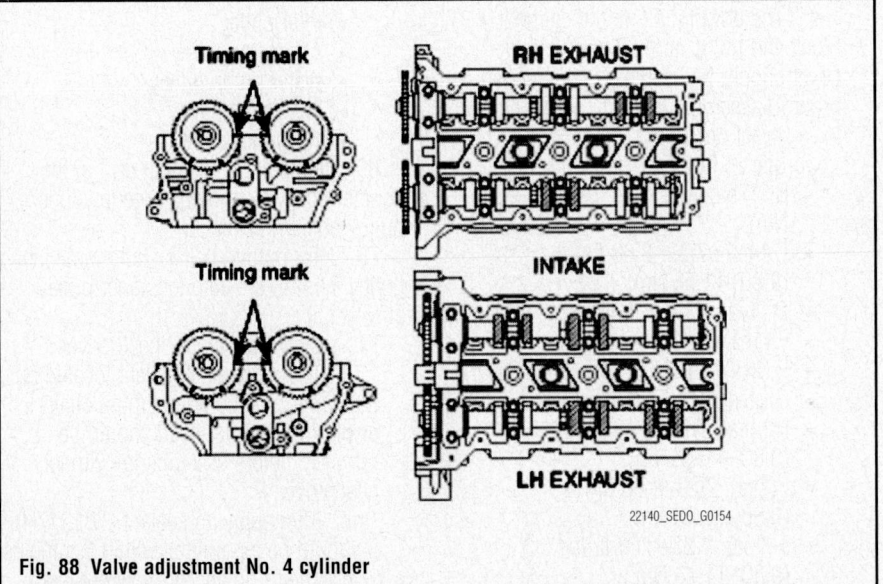

Fig. 88 Valve adjustment No. 4 cylinder

tioning on the cylinder head surface. If not rotate the crankshaft 360 degrees.

➡ **Do not rotate the engine counterclockwise.**

9. Check the valve clearance on No. 1 cylinder by measuring the clearance between the tappet and the base circle of the camshaft.

10. Turn the crankshaft pulley one revolution and align the groove with the timing mark "T" on the lower timing chain cover.

➡ **Do not rotate the engine counterclockwise.**

11. Check the valve clearance on No. 4 cylinder by measuring the clearance between the tappet and the base circle of the camshaft.

12. Adjust the intake and exhaust valve clearance.

a. Set the No. 1 cylinder to TDC on the compression stroke.

b. Remove the timing chain.

➡ **Before removing the timing chain mark the RH and LH timing chain with an identification based on the location of the sprocket. You must do this because the identification mark on the chain for TDC can be erased.**

c. Remove the camshaft bearing caps.

d. Remove the camshafts.

e. Remove the MLA's.

f. Measure the thickness of the removed tappet using a micrometer.

g. Calculate the thickness of the new tappet so that the valve clearance comes within the specified value.

Valve clearance: Engine coolant temperature: 68°F (20°C

- T: Thickness of removed tappet
- A: Measured valve clearance
- N: Thickness of new tappet
- Intake: N = T + [A - 0.0079 inches (0.20 mm)]
- Exhaust: N = T + [A - 0.0118inches (0.30 mm)]
- h. Select a new tappet with a thickness as close as possible to the calculated value.

➡**Shims are available in 41 size increments ranging of 0.0006 inches from 0.118 inches to 0.1417 inches.**

13. Place a new tappet on the cylinder head.

➡**Apply clean engine oil at the** selected tappet on the periphery and top surface.

14. Install the camshafts.
15. Install the bearing caps.
16. Install the timing chain.
17. Turn the crankshaft two turns in the clockwise direction and realign the crankshaft sprocket and camshaft sprocket timing marks.
18. Recheck the valve clearance.

ENGINE PERFORMANCE & EMISSION CONTROL

COMPONENT LOCATIONS

See Figures 89 and 90.

CAMSHAFT POSITION (CMP) SENSOR

LOCATION

The two CMPS are installed on engine head covers of banks 1 and 2.

REMOVAL & INSTALLATION

1. Turn ignition switch OFF.
2. Disconnect the CMPS connector.
3. Remove the bolt and the CMPS.
4. Installation is the reverse of removal.

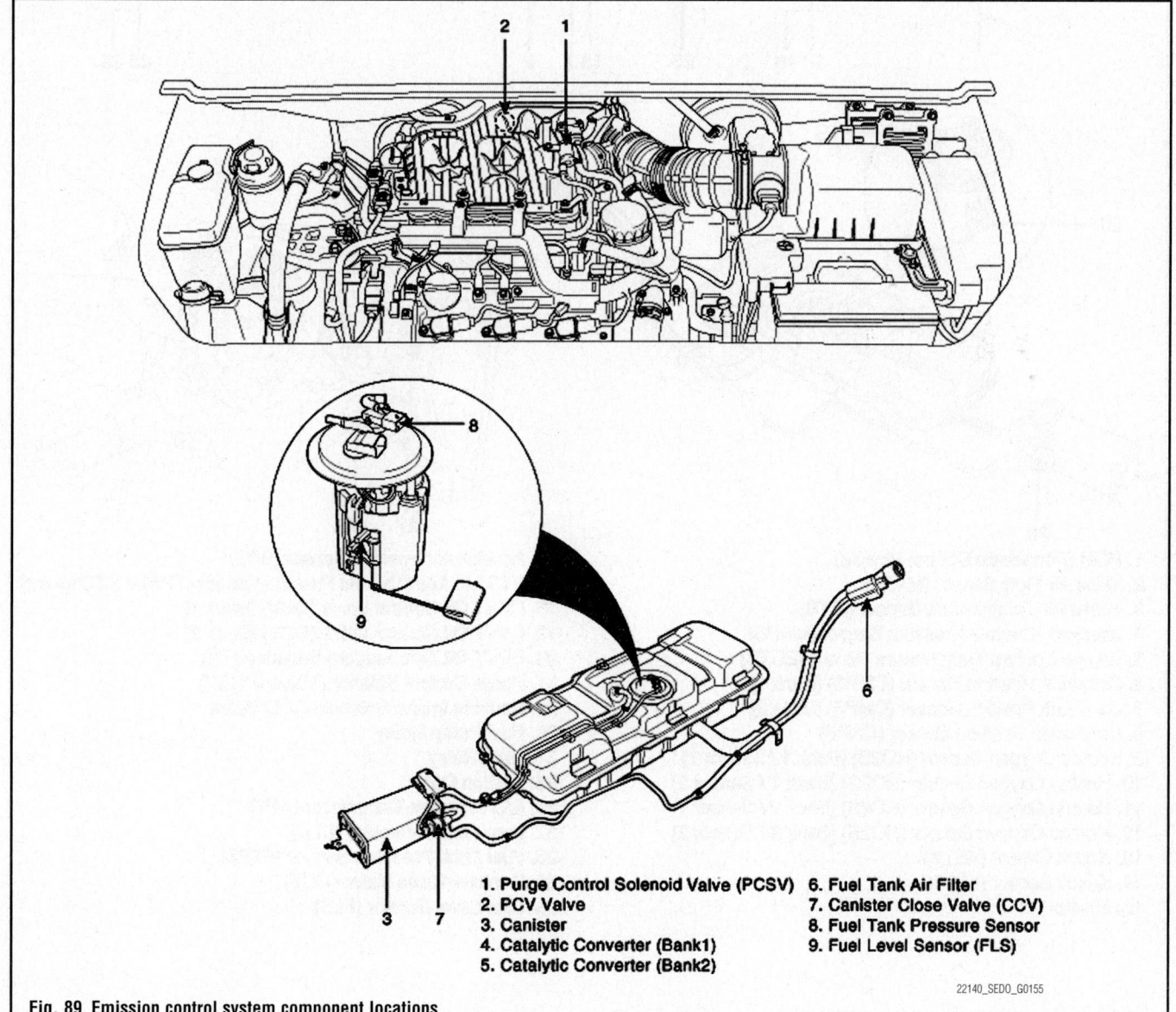

1. **Purge Control Solenoid Valve (PCSV)**
2. **PCV Valve**
3. **Canister**
4. **Catalytic Converter (Bank1)**
5. **Catalytic Converter (Bank2)**
6. **Fuel Tank Air Filter**
7. **Canister Close Valve (CCV)**
8. **Fuel Tank Pressure Sensor**
9. **Fuel Level Sensor (FLS)**

22140_SEDO_G0155

Fig. 89 Emission control system component locations

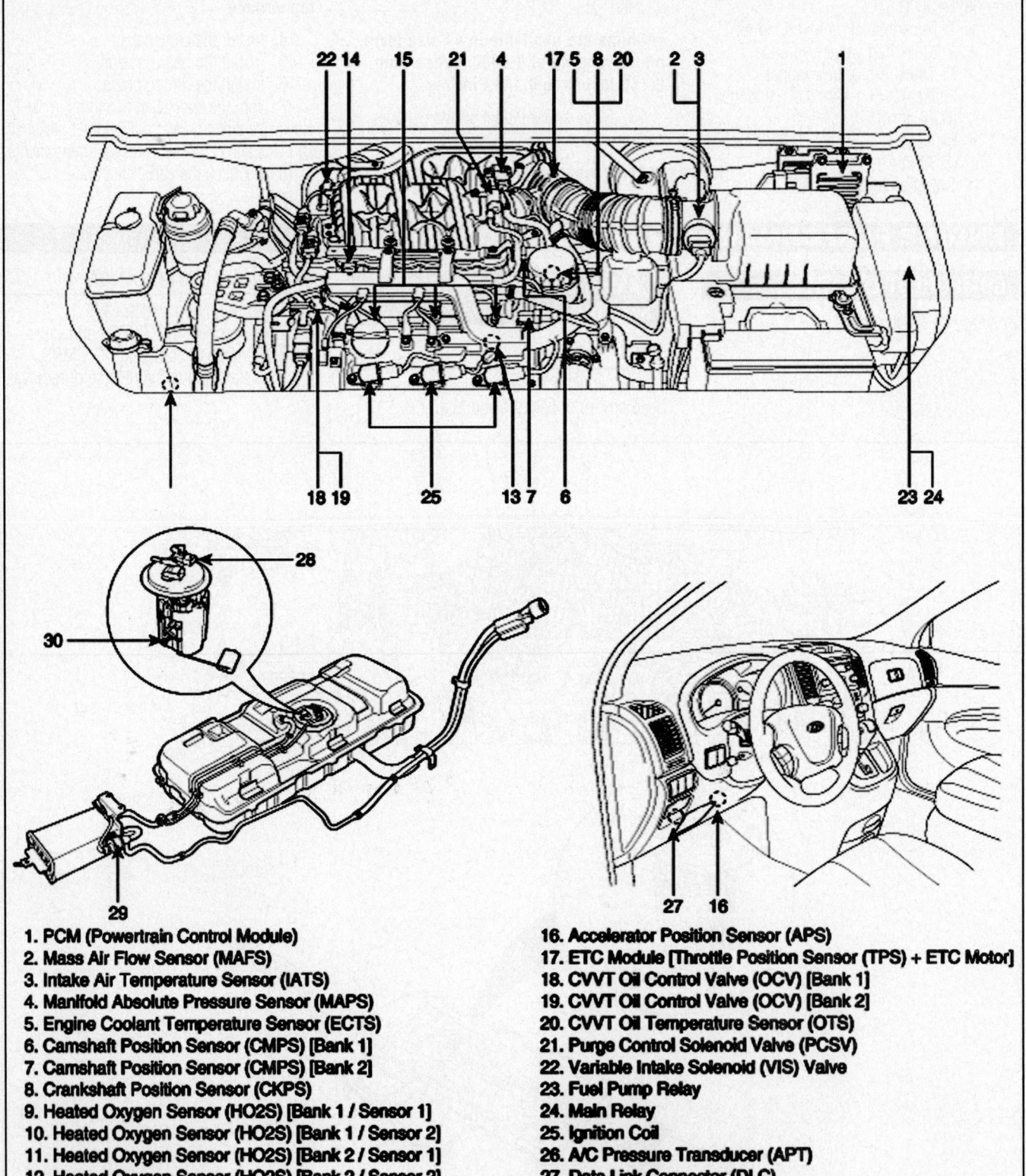

1. PCM (Powertrain Control Module)
2. Mass Air Flow Sensor (MAFS)
3. Intake Air Temperature Sensor (IATS)
4. Manifold Absolute Pressure Sensor (MAPS)
5. Engine Coolant Temperature Sensor (ECTS)
6. Camshaft Position Sensor (CMPS) [Bank 1]
7. Camshaft Position Sensor (CMPS) [Bank 2]
8. Crankshaft Position Sensor (CKPS)
9. Heated Oxygen Sensor (HO2S) [Bank 1 / Sensor 1]
10. Heated Oxygen Sensor (HO2S) [Bank 1 / Sensor 2]
11. Heated Oxygen Sensor (HO2S) [Bank 2 / Sensor 1]
12. Heated Oxygen Sensor (HO2S) [Bank 2 / Sensor 2]
13. Knock Sensor (KS) #1
14. Knock Sensor (KS) #2
15. Injector

16. Accelerator Position Sensor (APS)
17. ETC Module [Throttle Position Sensor (TPS) + ETC Motor]
18. CVVT Oil Control Valve (OCV) [Bank 1]
19. CVVT Oil Control Valve (OCV) [Bank 2]
20. CVVT Oil Temperature Sensor (OTS)
21. Purge Control Solenoid Valve (PCSV)
22. Variable Intake Solenoid (VIS) Valve
23. Fuel Pump Relay
24. Main Relay
25. Ignition Coil
26. A/C Pressure Transducer (APT)
27. Data Link Connector (DLC)
28. Fuel Tank Pressure Sensor (FTPS)
29. Canister Close Valve (CCV)
30. Fuel Level Sensor (FLS)

22140_SEDO_G0156

Fig. 90 Engine control system component locations

CRANKSHAFT POSITION (CKP) SENSOR

LOCATION

The Crankshaft Position Sensor (CKPS) is installed on transaxle housing.

REMOVAL & INSTALLATION

1. Turn ignition switch OFF.
2. Disconnect the CKPS connector.
3. Remove the CKPS.
4. Installation is the reverse of removal.

ENGINE COOLANT TEMPERATURE (ECT) SENSOR

LOCATION

Engine Coolant Temperature Sensor (ECTS) is located in the engine coolant passage of the cylinder head for detecting the engine coolant temperature.

REMOVAL & INSTALLATION

1. Turn ignition switch OFF.
2. Disconnect ECTS connector.
3. Remove the ECTS.
4. Installation is the reverse of removal.

HEATED OXYGEN (HO2S) SENSOR

LOCATION

Heated Oxygen Sensor (HO2S) is installed on upstream and downstream of the Manifold Catalyst Converter (MCC).

REMOVAL & INSTALLATION

1. Disconnect the negative battery cable.
2. Disconnect the HO2S sensor connector
3. Remove the HO2S sensor.
4. Installation is the reverse of removal.

INTAKE AIR TEMPERATURE (IAT) SENSOR

LOCATION

Intake Air Temperature Sensor (IATS) is installed inside the Mass Air Flow Sensor (MAFS).

REMOVAL & INSTALLATION

1. Disconnect the negative battery cable.
2. Disconnect the IAT sensor connector
3. Remove the IAT sensor.
4. Installation is the reverse of removal.

KNOCK SENSOR (KS)

LOCATION

Knock Sensor (KS) senses engine knocking and the two sensors are installed inside the V-valley of the cylinder block.

MANIFOLD ABSOLUTE PRESSURE (MAP) SENSOR

LOCATION

Manifold Absolute Pressure Sensor (MAPS) is installed on the surge tank.

REMOVAL & INSTALLATION

1. Disconnect the negative battery cable.
2. Disconnect the MAPS sensor connector
3. Remove the MAPS sensor.
4. Installation is the reverse of removal.

MASS AIR FLOW (MAF) SENSOR

LOCATION

Mass Air Flow Sensor (MAFS) is located in between the air cleaner and the throttle body.

REMOVAL & INSTALLATION

1. Disconnect the negative battery cable.
2. Disconnect and remove the MAFS from the air cleaner assembly.
3. Installation is the reverse of removal.

POWERTRAIN CONTROL MODULE (PCM)

LOCATION

The Powertrain Control Module (PCM) is installed on the firewall near the air cleaner.

REMOVAL & INSTALLATION

1. Turn the ignition switch off and disconnect the negative battery terminal.
2. Disconnect the PCM connector.
3. Unscrew the mounting bolts behind the air cleaner.
4. Remove the PCM.
5. Installation is reverse of removal.

FUEL   GASOLINE FUEL INJECTION SYSTEM

FUEL SYSTEM SERVICE PRECAUTIONS

Safety is the most important factor when performing not only fuel system maintenance but any type of maintenance. Failure to conduct maintenance and repairs in a safe manner may result in serious personal injury or death. Maintenance and testing of the vehicle's fuel system components can be accomplished safely and effectively by adhering to the following rules and guidelines.

• To avoid the possibility of fire and personal injury, always disconnect the negative battery cable unless the repair or test procedure requires that battery voltage be applied.

• Always relieve the fuel system pressure prior to disconnecting any fuel system component (injector, fuel rail, pressure regulator, etc.), fitting or fuel line connection.

Exercise extreme caution whenever relieving fuel system pressure to avoid exposing skin, face and eyes to fuel spray. Please be advised that fuel under pressure may penetrate the skin or any part of the body that it contacts.

• Always place a shop towel or cloth around the fitting or connection prior to loosening to absorb any excess fuel due to spillage. Ensure that all fuel spillage (should it occur) is quickly removed from engine surfaces. Ensure that all fuel soaked cloths or towels are deposited into a suitable waste container.

• Always keep a dry chemical (Class B) fire extinguisher near the work area.

• Do not allow fuel spray or fuel vapors to come into contact with a spark or open flame.

• Always use a back-up wrench when loosening and tightening fuel line connection fittings. This will prevent

unnecessary stress and torsion to fuel line piping.

• Always replace worn fuel fitting O-rings with new Do not substitute fuel hose or equivalent where fuel pipe is installed.

Before servicing the vehicle, make sure to also refer to the precautions in the beginning of this section as well.

RELIEVING FUEL SYSTEM PRESSURE

1. Remove the third seat.
2. Open the service cover.
3. Disconnect the fuel pump connector.
4. Start the engine and wait until fuel in fuel line is exhausted.
5. After engine stalls, turn the ignition switch to OFF position.
6. Disconnect the negative battery cable.

FUEL FILTER

REMOVAL & INSTALLATION

See Figures 91 through 93.

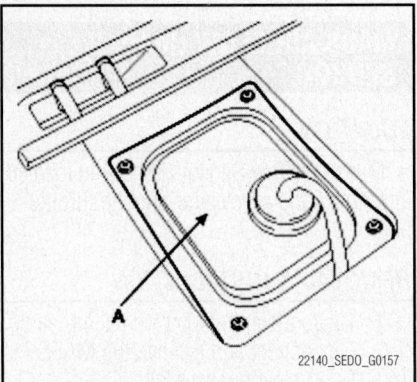

Fig. 91 Remove the service cover (A)

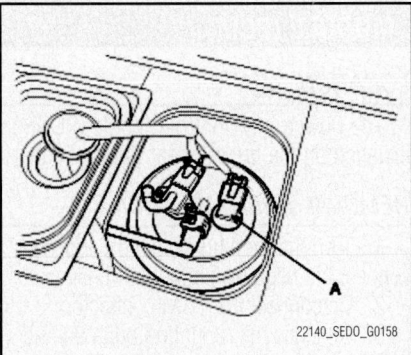

Fig. 92 Disconnect the fuel pump connector (A).

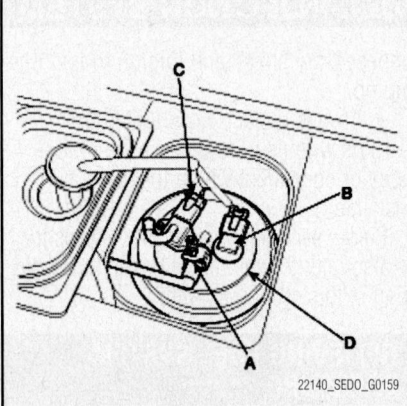

Fig. 93 Disconnect the fuel feed quick connector (A), the fuel pump connector (B) and fuel tank pressure sensor connector (C), then remove the fuel pump cover (D).

1. Remove the 2nd seat (s).
2. Remove the service cover (A).
3. Disconnect the fuel pump connector (A).
4. Start the engine and wait until fuel in fuel line is exhausted.
5. After the engine stalls, turn the ignition switch OFF.
6. Disconnect the fuel feed quick connector (A), the fuel pump connector (B) and fuel tank pressure sensor connector (C), then remove the fuel pump cover (D).
7. Remove the fuel pump from the fuel tank.
8. Installation is the reverse of removal.

FUEL PUMP

REMOVAL & INSTALLATION

See Figures 91 through 93.

1. Remove the 2nd seat (s).
2. Remove the service cover (A).
3. Disconnect the fuel pump connector (A).
4. Start the engine and wait until fuel in fuel line is exhausted.
5. After the engine stalls, turn the ignition switch OFF.
6. Disconnect the fuel feed quick connector (A), the fuel pump connector (B) and fuel tank pressure sensor connector (C), then remove the fuel pump cover (D).
7. Remove the fuel pump from the fuel tank.
8. Installation is the reverse of removal.

FUEL TANK

REMOVAL & INSTALLATION

See Figures 91, 92, 94 through 96.

1. Remove the rear seat(s).
2. Remove the service cover (A).
3. Disconnect the fuel pump connector (A).
4. Start the engine and wait until fuel in fuel line is exhausted.
5. After the engine stalls, turn the ignition switch OFF.
6. Disconnect the fuel pump connector (A) and the fuel tank pressure sensor connector (B).
7. Lift the vehicle and support the fuel tank with a jack.
8. Disconnect the fuel filler hose (A) and the leveling hose (B).
9. Disconnect the fuel feed quick connector (A) near canister.

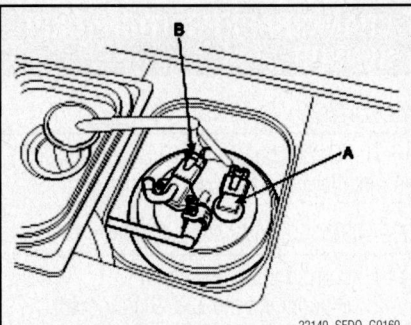

Fig. 94 Disconnect the fuel pump connector (A) and the fuel tank pressure sensor connector (B).

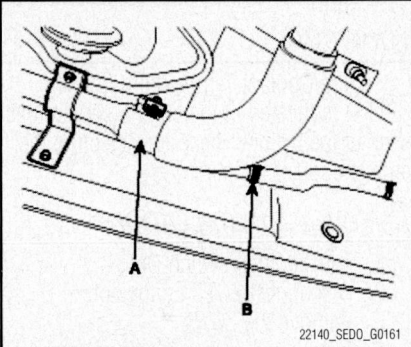

Fig. 95 Disconnect the fuel filler hose (A) and the leveling hose (B).

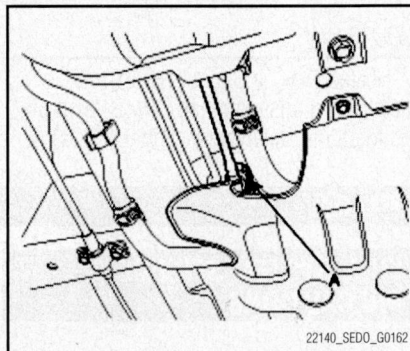

Fig. 96 Disconnect the fuel feed quick connector (A) near canister.

10. Remove the fuel tank bands and remove the fuel tank.
11. Installation is the reverse of removal.

IDLE SPEED

ADJUSTMENT

The idle speed is controlled by the PCM. No adjustment is necessary or possible.

HEATING & AIR CONDITIONING SYSTEM

BLOWER MOTOR

REMOVAL & INSTALLATION

See Figures 97 and 98.

1. Disconnect the negative battery terminal.
2. Disconnect the connector (A) of the blower motor.
3. Remove the blower motor (A) after removing the mounting screws.
4. Installation is the reverse order of removal.

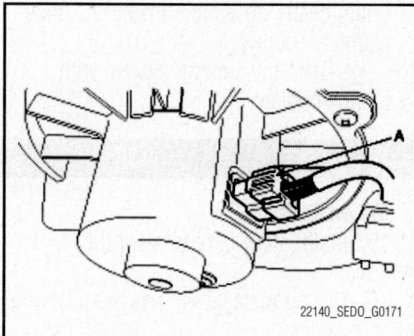

Fig. 97 Disconnect the connector (A) of the blower motor

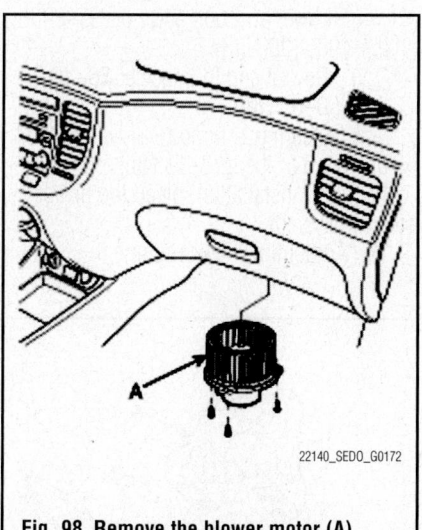

Fig. 98 Remove the blower motor (A) after removing the mounting screws

HEATER CORE

REMOVAL & INSTALLATION

See Figures 99 through 102.

1. Disconnect the negative battery terminal.
2. Recover the refrigerant with a recovery/recycling/charging station.

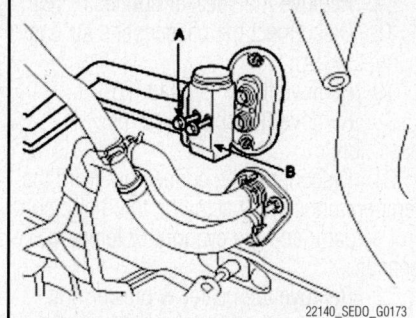

Fig. 99 Remove the bolts (A) and the expansion valve (B) from the evaporator core

3. Drain the engine coolant from the radiator.
4. Remove the bolts (A) and the expansion valve (B) from the evaporator core.

➡ **Plug or cap the lines immediately after disconnecting them to avoid moisture and dust contamination.**

5. Disconnect the inlet and outlet heater hoses from the heater unit.
6. Remove the front seat.
7. Tilt the steering column down.

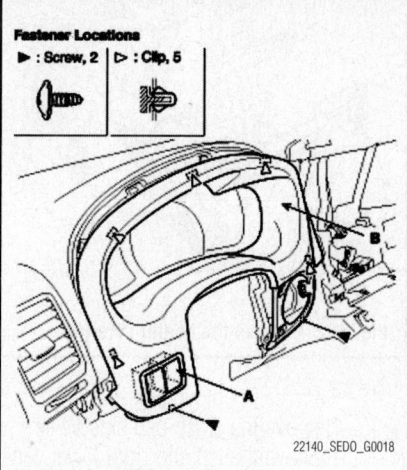

Fig. 100 Disconnect connector (A). Remove the cluster fascia panel (B).

8. Remove screws from the center cluster fascia panel.
9. Disconnect connector (A).
10. Remove the cluster fascia panel (B).
11. Remove mounting screws, disconnect all connectors and remove the audio assembly.

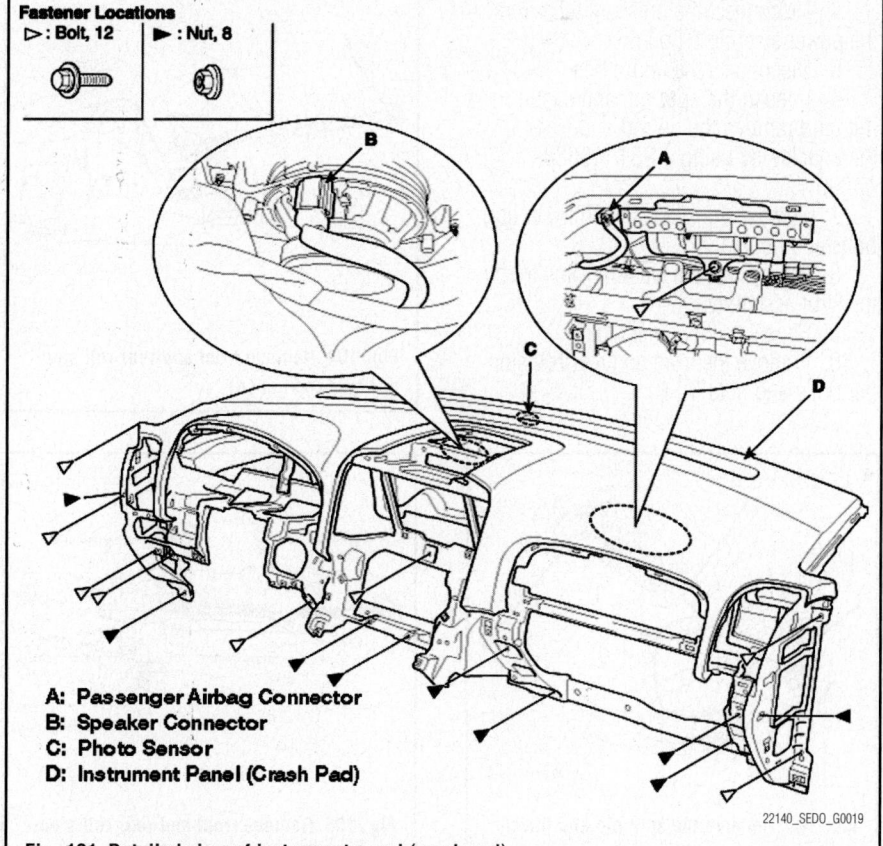

A: Passenger Airbag Connector
B: Speaker Connector
C: Photo Sensor
D: Instrument Panel (Crash Pad)

Fig. 101 Detailed view of instrument panel (crash pad).

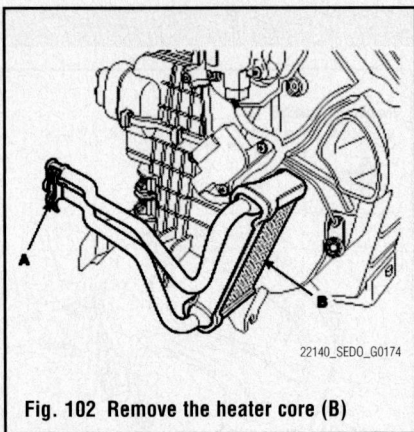

Fig. 102 Remove the heater core (B)

12. Remove the glove box.
13. Remove the crash pad side covers.
14. Remove the left and right lower center covers.

15. Remove the front pillar trim.
16. Remove the photo sensor (C).
17. Remove the speaker connector (B).
18. Disconnect the passenger's air bag connector (A).
19. Remove the crash pad (D).
20. Remove the cowl cross bar assembly.
21. Disconnect the connectors from the temperature control actuator, the mode control actuator and the evaporator temperature sensor.
22. Remove the heater & blower unit.
23. Remove the blower unit from heater unit.
24. Remove the heater core (B).

➡ **Be careful that the inlet and outlet pipe are not bent during heater core removal.**

To install:
25. Installation is the reverse order of removal, and note these items :
   a. If you're installing a new evaporator, add refrigerant oil (ND-OIL8).
   b. Replace the O-rings with new ones at each fitting, and apply a thin coat of refrigerant oil before installing them. Be sure to use the right O-rings for R-134a to avoid leakage.
   c. Apply sealant to the grommets.
   d. Make sure that there is no air leakage.
   e. Charge the system and test its performance.
   f. Do not interchange the inlet and outlet heater hoses and install the hose clamps securely.
   g. Refill the cooling system with engine coolant.

STEERING

POWER STEERING GEAR

REMOVAL & INSTALLATION

See Figures 103 through 106.

1. Remove the both front wheels.
2. Drain the power steering fluid.
3. Remove the bolt connecting steering column to universal joint.
4. Disconnect the pressure tube from the power steering oil pump.
5. Disconnect the return hose.
6. Loosen the split pin and castle nut, and remove the tie rod end from the knuckle by using a SST (09568-4A000).
7. Remove the split pin and lower arm bolts and nut (A).
8. Disconnect the stabilizer link from the strut assembly.
9. Repeat on the other side.
10. Remove the front and rear roll stopper bolts and nuts (A, B).

11. Remove the sub-frame.
12. Remove the rear roll stopper (A) from the sub-frame.
13. Disconnect the pressure and return lines from the valve body housing.
14. Remove the power steering gear box from the sub-frame.

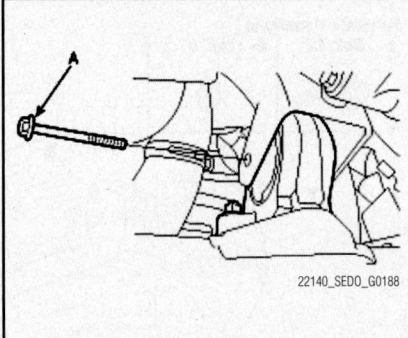

Fig. 104 Remove front and rear roll stopper bolt and nut (A)

To install:
15. Installation is reverse of the removal.
16. During installation, use the following torque values:
   a. Power steering gear box to sub-frame: 65–80 ft. lbs. (90–110 Nm).
   b. Stabilizer link to strut assembly: 72–87 ft. lbs. (100–120 Nm).
   c. Lower arm bolt and nut: 65–87 ft. lbs. (90–120 Nm).
   d. Tie rod end to knuckle: 36–40 ft. lbs. (50–55 Nm).
   e. Steering column to universal joint: 9.4–13.0 ft. lbs. (13–18 Nm).
17. After installation, bleed the power steering system.
18. Adjust the wheel alignment.

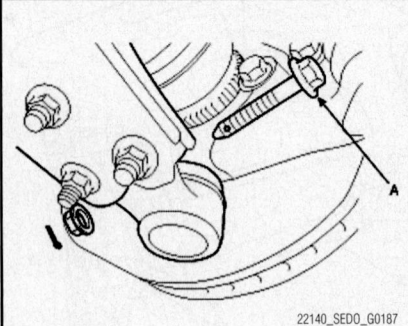

Fig. 103 Remove the split pin and lower arm bolts and nut (A)

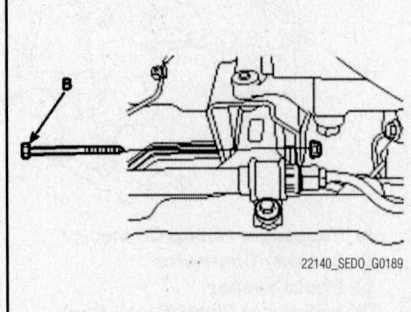

Fig. 105 Remove front and rear roll stopper bolt and nut (B)

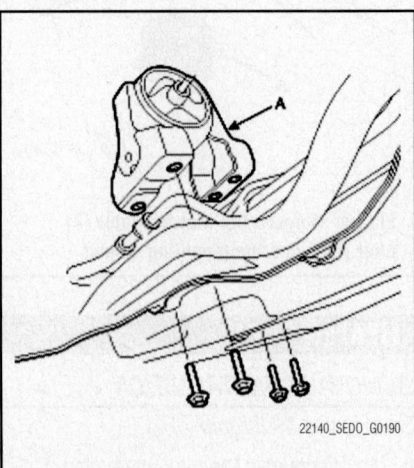

Fig. 106 Remove the rear roll stopper (A) from the sub-frame

POWER STEERING PUMP

REMOVAL & INSTALLATION

1. Disconnect the pressure tube from the power steering pump.
2. Disconnect the suction hose from the suction pipe.
3. Remove the accessory drive belt.
4. Remove the power steering pump assembly from the pump bracket.

To install:

5. Installation is the reverse of removal.

> ✳✳ **CAUTION**
>
> **The pressure tube does not twist and come in contact with other components.**

6. Add power steering fluid
7. Bleed the power steering system.
8. Check the oil pump pressure.

BLEEDING

1. Remove the fuel pump fuse, then start the engine and wait for the engine to stall. Next, while operating the starting motor intermittently (for 15–20 seconds), turn the steering wheel all the way to the left and then to the right five or six times.

➡ **During air bleeding, replenish the fluid supply so that the level never falls below the lower position of the filter. If air bleeding is done while the vehicle is idling, the air will be broken up and absorbed into the fluid. Be sure to do the bleeding only while cranking.**

2. Reinstall the fuel pump fuse, and start the engine (idling).
3. Turn the steering wheel to the left and the right until there are no air bubbles in the oil reservoir.

> ✳✳ **CAUTION**
>
> **Do not hold the steering wheel turned all the way to either side for more than ten seconds.**

4. Confirm that the fluid is not milky, and that the level is up to the position specified on the level gauge.
5. Confirm that there is little change in the surface of the fluid when the steering wheel is turned left and right.

> ✳✳ **CAUTION**
>
> **If the surface of the fluid changes considerably, air bleeding should be done again. If the fluid level rises suddenly when the engine is stopped, it indicates that there is still air in the system. If there is air in the system, a jingling noise may be heard from the pump and the control valve may also produce unusual noises. Air in the system will shorten the life of the pump and other parts.**

SUSPENSION

FRONT SUSPENSION

LOWER BALL JOINT

REMOVAL & INSTALLATION

See Figures 107 and 108.

1. Raise the front of the vehicle, and make sure it is securely supported.
2. Remove the front wheel and tire
3. Remove the front lower arm (A) mounting bolt (B) from the knuckle.
4. Remove the lower arm mounting bolts (A, B).
5. Install the lower arm mounting bolts. Torque: 116–130 ft. lbs. (160–180 Nm).
6. Install the front lower arm ball joint

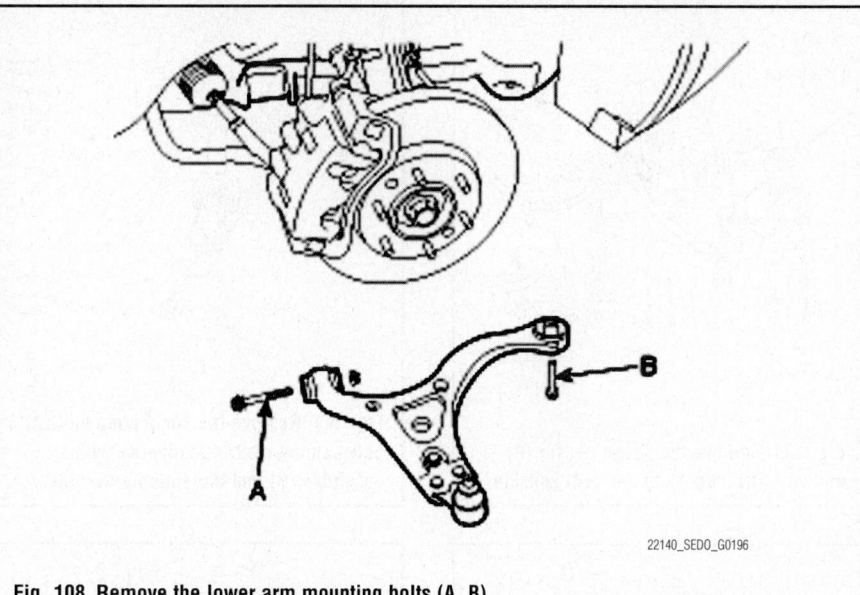

Fig. 108 Remove the lower arm mounting bolts (A, B)

22140_SEDO_G0196

mounting bolt to the knuckle. Torque: 65–88 ft. lbs. (90–120 Nm).

7. Install the wheel and the tire.

LOWER CONTROL ARM

REMOVAL & INSTALLATION

See Figures 107 and 108.

1. Raise the front of the vehicle, and make sure it is securely supported.

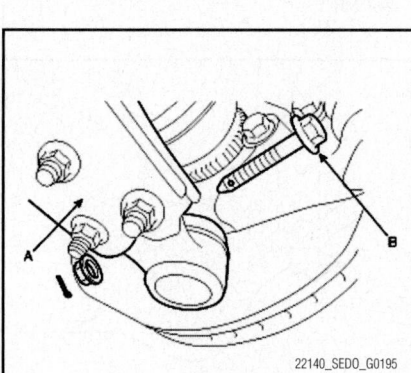

Fig. 107 Remove the front lower arm (A) mounting bolt (B) from the knuckle

22140_SEDO_G0195

2. Remove the front wheel and tire
3. Remove the front lower arm (A) mounting bolt (B) from the knuckle.
4. Remove the lower arm mounting bolts (A, B).
5. Install the lower arm mounting bolts. Torque: 116–130 ft. lbs. (160–180 Nm).
6. Install the front lower arm ball joint mounting bolt to the knuckle. Torque: 65–87 ft. lbs. (90–120 Nm).
7. Install the wheel and the tire.

MACPHERSON STRUT

REMOVAL & INSTALLATION

See Figures 109 and 110.

1. Raise the front of the vehicle, and make sure it is securely supported.
2. Remove the front wheel and tire.
3. Remove the brake hose bracket bolts from the front strut assembly.
4. Remove the speed sensor (A) and wire (B) bolts from the front knuckle.
5. Remove the front stabilizer link nut from the strut assembly.
6. Remove the strut upper mounting nuts.
7. Remove the front strut assembly (A) bolts (B) from the front knuckle.

To install:

8. Install the strut upper mounting nuts. Torque: 33–43 ft. lbs. (45–60 Nm).
9. Install the front strut assembly bolts to the front knuckle. Torque: 72–87 ft. lbs. (100–120 Nm).
10. Install the front stabilizer link nut to the strut assembly. Torque: 72–87 ft. lbs. (100–120 Nm).

11. Install the speed sensor and wire bolts.
12. Install the brake hose bracket bolt to the axle assembly.
13. Install the wheel and the tire.

STABILIZER BAR

REMOVAL & INSTALLATION

See Figures 111 through 118.

1. Remove the connecting bolt (A) between the steering universal joint assembly (B) and the pinion assembly.
2. Raise the front of the vehicle, and make sure it is securely supported.
3. Remove the front wheel and tire.
4. Remove the front stabilizer link (A) nut (B) from the strut assembly.
5. After removing both sides of the tie rod end self-locking nuts and cotter pins, remove the ball joints by using the special tool(09568-4A000).
6. Remove both sides of the lower arm (A) mounting bolts (B).
7. Remove the engine mounting bolts (A, B)

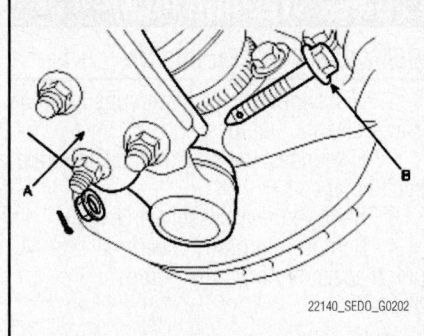

Fig. 113 Remove both sides of the lower arm (A) mounting bolts (B)

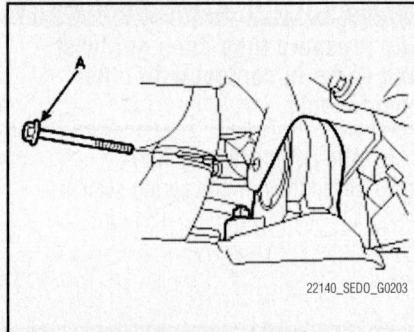

Fig. 114 Remove the engine mounting bolt (A)

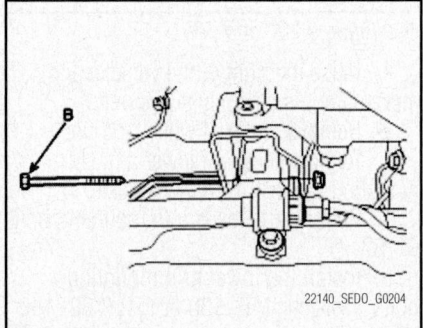

Fig. 115 Remove the engine mounting bolt (B)

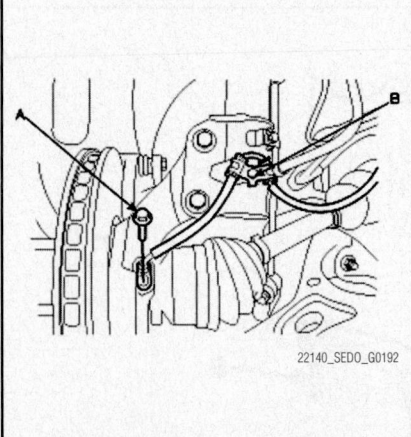

Fig. 109 Remove the speed sensor (A) and wire (B) bolts from the front knuckle

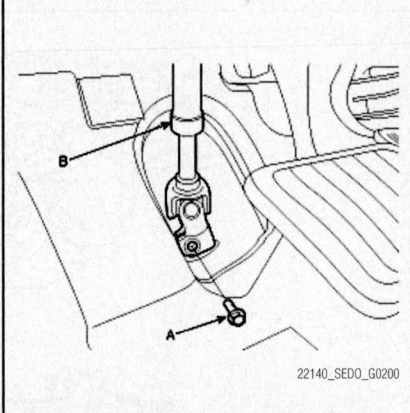

Fig. 111 Remove the connecting bolt (A) between the steering universal joint assembly (B) and the pinion assembly

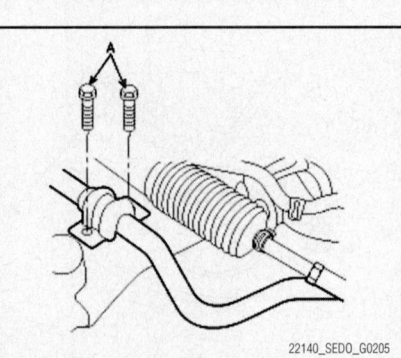

Fig. 116 Remove both sides of the stabilizer bar assembly mounting bolts (A)

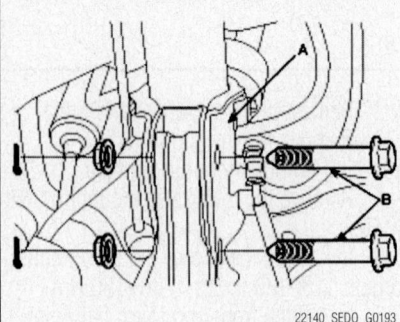

Fig. 110 Remove the front strut assembly (A) bolts (B) from the front knuckle

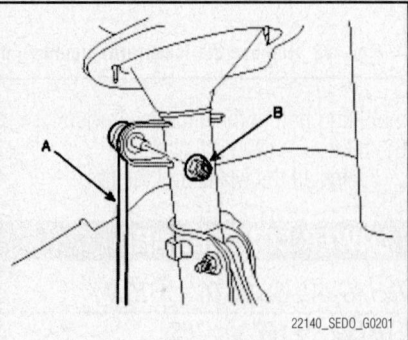

Fig. 112 Remove the front stabilizer link (A) nut (B) from the strut assembly

8. Remove the twelve bolts and nuts of the sub frame by supporting it with a jack.

9. After lowering the jack which supports the sub frame, remove both sides of the stabilizer bar assembly mounting bolts (A).

10. Remove the stabilizer bar assembly through the gap between the body and the rear side of the sub frame.

11. Remove the brackets (A) and the bushings (B).

To install:

12. Install the bushings on the stabilizer bar.

➡ **Bring clamp of stabilizer bar into contact with bushing.**

13. Install the brackets on the bushings.

14. Install the stabilizer bar bracket mounting bolts to the sub frame. Torque: 28–43 ft. lbs. (39–60 Nm).

15. Raise the jack which supports the sub frame and install the four bolts (A) of the sub frame and the eight bolts (B) of the guide bracket. Torque: Bolt (A): 116–130 ft. lbs. (160–180 Nm); Bolt (B): 33–43 ft. lbs. (45–60 Nm).

16. Install the engine mounting bolts. Torque: 47–62 ft. lbs. (65–85 Nm).

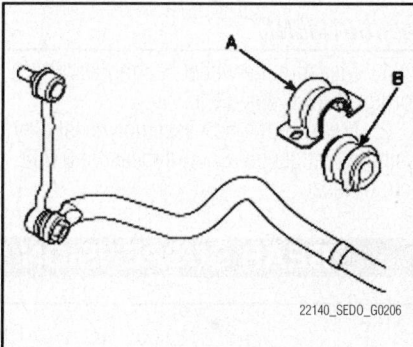

Fig. 117 Remove the brackets (A) and the bushings (B)

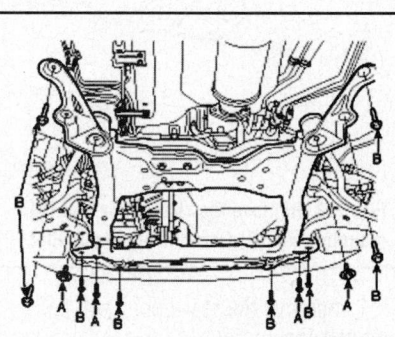

Fig. 118 Install the four bolts (A) of the sub frame and the eight bolts (B) of the guide bracket

17. Install both sides of the lower arm mounting bolts. Torque: 65–87 ft. lbs. (90–120 Nm).

18. Install both sides of the tie rod end self-locking nuts. Insert cotter pin after nut is torqued to specifications. Torque: 43–58 ft. lbs. (65–80 Nm).

19. Install the front stabilizer link nut to the strut assembly. Torque: 72–87 ft. lbs. (100–120 Nm).

20. Install the wheel and the tire.

21. Install the connecting bolt between the steering universal joint assembly and the pinion assembly. Torque: 9–13 ft. lbs. (13–18 Nm).

✳✳ CAUTION

After installation, if necessary, adjust the alignment of the steering wheel and front tires.

STEERING KNUCKLE

REMOVAL & INSTALLATION

See Figures 119 through 121.

1. Raise the vehicle and remove the wheel & tire.

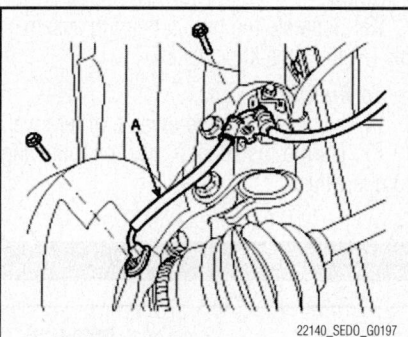

Fig. 119 Remove the wheel speed sensor & wire (A)

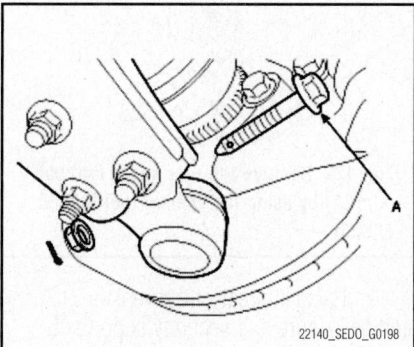

Fig. 120 Remove the split pin and lower arm bolt and nut (A)

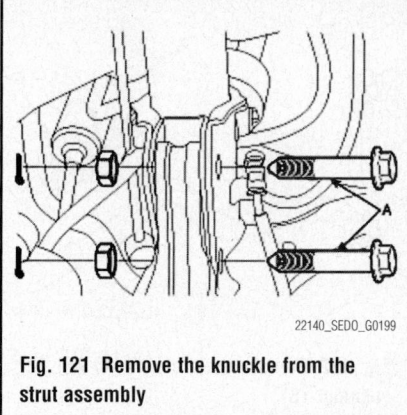

Fig. 121 Remove the knuckle from the strut assembly

2. Remove the wheel speed sensor & wire (A).

3. Disconnect the brake hose from the strut assembly.

4. Unstake the driveshaft lock nut using a chisel and hammer.

5. Remove the driveshaft lock nut.

6. Remove the caliper assembly from the knuckle and suspend it with wire.

7. Remove the split pin and castle nut from the tie rod end ball joint.

8. Disconnect the tie rod end from the knuckle using a SST (09568-4A000).

9. Remove the split pin and lower arm bolt and nut (A).

10. Using a plastic hammer, disconnect the driveshaft from the front hub assembly.

11. Remove the disc brake rotor from the front hub assembly.

12. Remove the knuckle from the strut assembly.

To install:

13. Installation is the reverse of removal.

14. During installation, use the following torque values:

   a. Knuckle to strut assembly: 72–87 ft. lbs. (100–120 Nm).

   b. Lower arm bolt and nut: 65–87 ft. lbs. (90–120 Nm).

   c. Castle nut for tie rod end ball joint: 43–58 ft. lbs. (60–80 Nm).

   d. Brake caliper assembly to knuckle: 61–72 ft. lbs. (85–100 Nm).

   e. Driveshaft lock nut: 177–199 ft. lbs. (245–275 Nm).

➡ **Be sure to stake the driveshaft lock nut.**

WHEEL BEARINGS

REMOVAL & INSTALLATION

See Figures 122 through 125.

1. Raise the vehicle and remove the wheel & tire.

2. Remove the wheel speed sensor & wire (A).

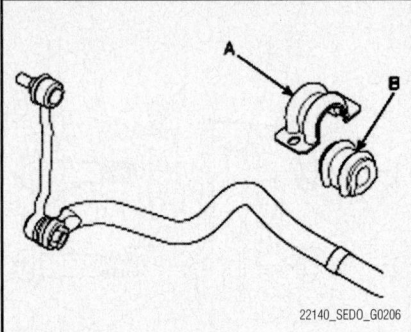

Fig. 122 Remove the brackets (A) and the bushings (B)

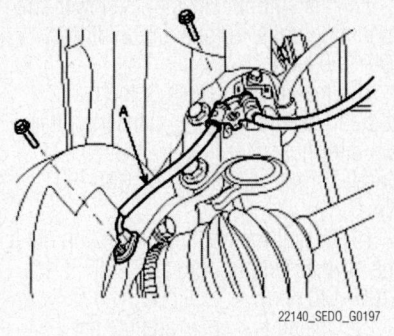

Fig. 124 Remove the wheel speed sensor & wire (A)

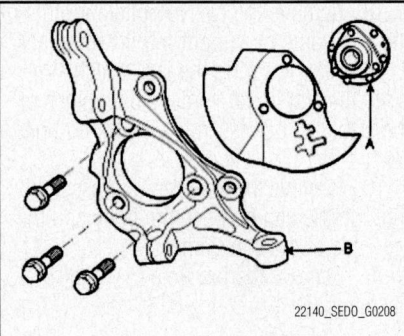

Fig. 125 Separate the hub & bearing assembly (A) from the knuckle (B)

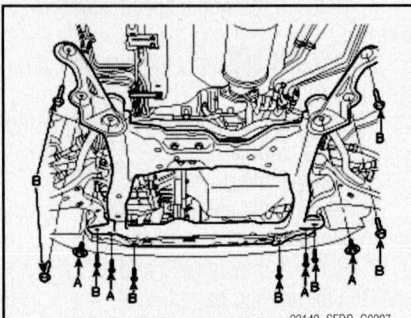

Fig. 123 Install the four bolts (A) of the sub frame and the eight bolts (B) of the guide bracket

3. Disconnect the brake hose from the strut assembly.
4. Unstake the driveshaft lock nut using a chisel and hammer.
5. Remove the driveshaft lock nut.

6. Remove the caliper assembly from the knuckle and suspend it with wire.
7. Remove the split pin and castle nut from the tie rod end ball joint.
8. Disconnect the tie rod end from the knuckle using a SST (09568-4A000).
9. Remove the split pin and lower arm bolt and nut (A).
10. Using a plastic hammer, disconnect the driveshaft from the front hub assembly.
11. Remove the disc brake rotor from the front hub assembly.
12. Remove the knuckle from the strut assembly.
13. Separate the hub & bearing assembly (A) from the knuckle (B).

To install:
14. Installation is the reverse of removal.
15. During installation, use the following torque values:

a. Hub & bearing assembly to knuckle: 116–130 ft. lbs. (160–180 Nm).
b. Knuckle to strut assembly: 72–87 ft. lbs. (100–120 Nm).
c. Lower arm bolt and nut: 65–87 ft. lbs. (90–120 Nm).
d. Castle nut for tie rod end ball joint: 43–58 ft. lbs. (60–80 Nm).
e. Brake caliper assembly to knuckle: 61–72 ft. lbs. (85–100 Nm).
f. Driveshaft lock nut: 177–199 ft. lbs. (245–275 Nm).

➡**Be sure to stake the driveshaft lock nut.**

ADJUSTMENT

1. Measure the wheel bearing starting torque: 1.45 ft. lbs. (1.97 Nm).
2. Measure the hub assembly axial play using a dial gauge. Should measure 0.008 mm or less.

SUSPENSION

REAR SUSPENSION

CONTROL ARMS/LINKS

REMOVAL & INSTALLATION

Rear Upper Arm
See Figures 126 through 128.

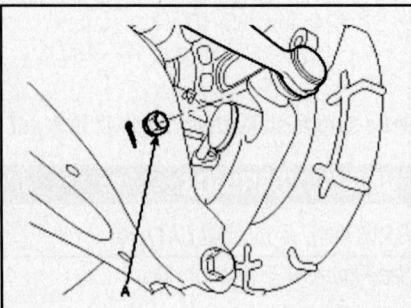

Fig. 126 Remove the rear upper arm ball joint self-locking nut (A) and the cotter pin

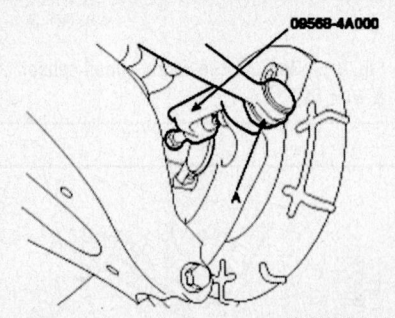

Fig. 127 Remove the rear upper arm ball joint (A) by using the special tool (09568-4A000)

1. Raise the rear of the vehicle, and make sure it is securely supported.
2. Remove the rear wheel and tire.
3. Remove the brake caliper mounting bolts and suspend the brake caliper assembly with wire.

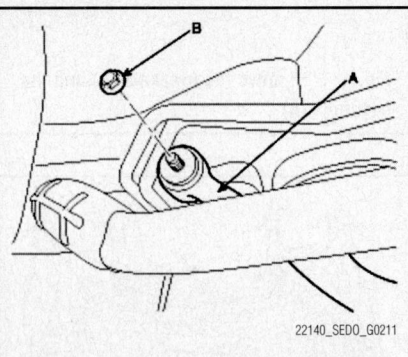

Fig. 128 Remove the rear upper arm (A) mounting nut (B) from the cross member

4. Remove the rear upper arm ball joint self-locking nut (A) and the cotter pin.
5. Remove the rear upper arm ball joint (A) by using the special tool (09568-4A000).

6. Remove the rear upper arm (A) mounting nut (B) from the cross member.

To install:

7. Install the rear upper arm mounting nut to the cross member. Torque: 115.7–130.2 ft. lbs. (160–180 Nm).

8. Install the rear upper arm ball joint self-locking nut and the cotter pin. Torque: 65.1–79.5 ft. lbs. (90–110 Nm).

9. Install the brake caliper mounting bolts. Torque: 36.2–43.4 ft. lbs. (50–60 Nm).

10. Install the wheel and the tire.

Rear Lower Arm

See Figures 129 and 130.

1. Raise the rear of the vehicle, and make sure it is securely supported.

2. Remove the rear wheel and tire.

3. Remove the lower arm bolt (B) from the rear knuckle, while supporting the lower arm (A) with a jack.

4. Loosen the lower arm bolt (C) from the cross member.

5. Remove the spring, the lower seat, and the upper pad.

6. Remove the lower arm (A) mounting bolts (B) from the cross member (C).

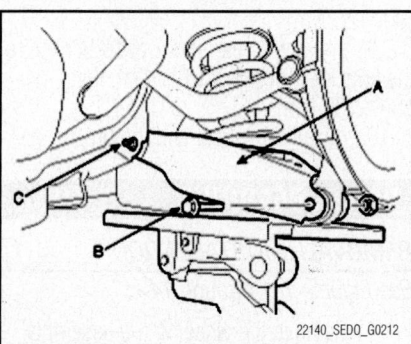

Fig. 129 Remove the lower arm bolt (B) from the rear knuckle, while supporting the lower arm (A) with a jack

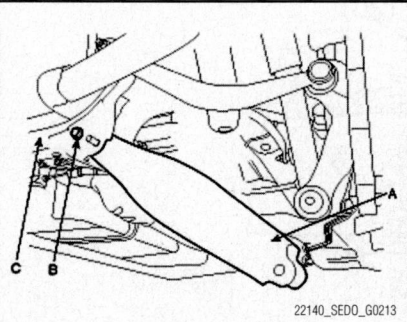

Fig. 130 Remove the lower arm (A) mounting bolts (B) from the cross member (C)

To install:

7. Install the lower arm mounting bolts to the cross member. Torque: 144.7–195.3 ft. lbs. (200–270 Nm).

8. Install the spring, the lower seat, and the upper pad.

9. Install the lower arm bolt to the rear knuckle and the lower arm bolt to the cross member with a specified torque, while supporting the lower arm (A) with a jack. Torque: Lower arm bolt to rear knuckle: 86.8–115.7 ft. lbs. (120–160 Nm); Lower arm bolt to cross-member: 144.7–195.3 ft. lbs. (200–270 Nm).

10. Install the wheel and the tire.

Rear Assist Arm

See Figures 131 through 133.

1. Raise the rear of the vehicle, and make sure it is securely supported.

2. Remove the rear wheel and tire.

3. Remove the brake caliper mounting bolts and suspend the brake caliper assembly with wire.

4. Remove the rear assist arm (A) ball joint self-locking nut (B) and the cotter pin.

5. Remove the rear assist arm ball joint (A) by using the special tool (09568-4A000).

6. Remove the rear assist arm (A) mounting nut (B) from the cross member.

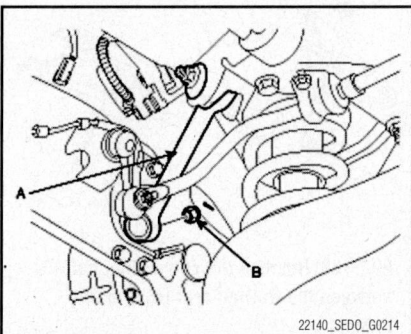

Fig. 131 Remove the rear assist arm (A) ball joint self-locking nut (B) and the cotter pin

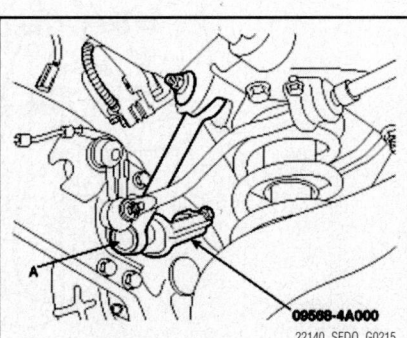

Fig. 132 Remove the rear assist arm ball joint (A) by using the special tool (09568-4A000)

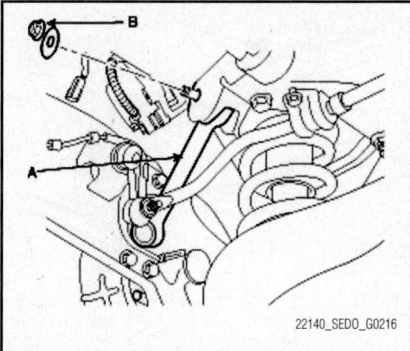

Fig. 133 Remove the rear assist arm (A) mounting nut (B) from the cross member

To install:

7. Install the rear assist arm mounting nut to the cross member. Torque: 65–87 ft. lbs. (90–120 Nm).

8. Install the rear assist arm ball joint self-locking nut and the cotter pin. Torque: 116–130 ft. lbs. (160–180 Nm).

9. Install the brake caliper mounting bolts. Torque: 36–43 ft. lbs. (50–60 Nm).

10. Install the wheel and the tire.

Trailing Arm

See Figures 134 through 139.

1. Raise the rear of the vehicle, and make sure it is securely supported.

2. Remove the rear wheel and tire.

3. Remove the wheel speed sensor wire bracket bolt (A) from the body and the connector.

4. Remove the wheel speed sensor wire bracket bolt (A).

5. Remove the parking brake wire bracket bolt (B).

6. Remove the trailing arm (A) mounting nuts (B) from the knuckle.

7. Remove the stabilizer link (B) nut (C) from the trailing arm (A).

8. Remove the trailing arm bracket (A) mounting bolts (B) from the body.

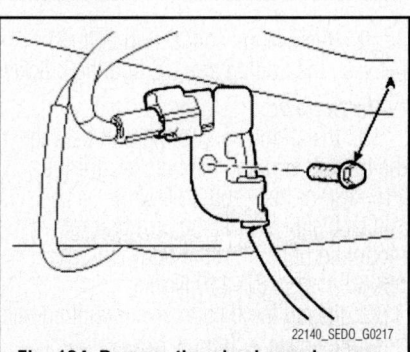

Fig. 134 Remove the wheel speed sensor wire bracket bolt (A) from the body and the connector

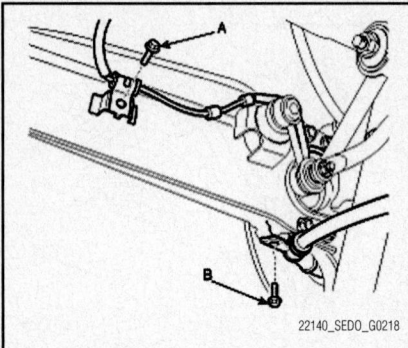

Fig. 135 Remove the wheel speed sensor wire bracket bolt (A)

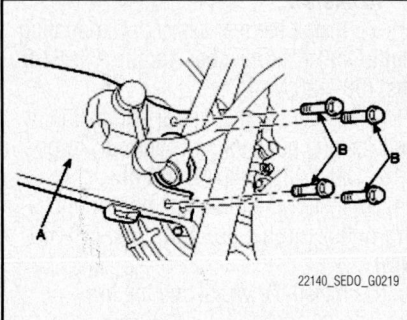

Fig. 136 Remove the trailing arm (A) mounting nuts (B) from the knuckle

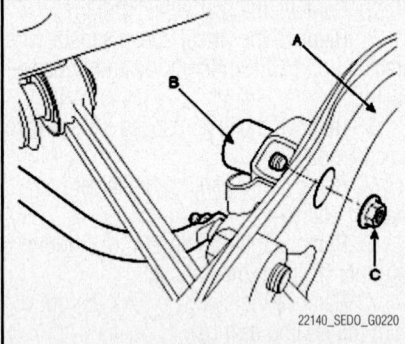

Fig. 137 Remove the stabilizer link (B) nut (C) from the trailing arm (A)

9. Remove the connecting bolt (B) between the trailing arm (A) and the bracket.

To install:

10. Install the connecting bolt between the trailing arm and the bracket. Torque: 116–130 ft. lbs. (160–180 Nm).

11. Install the trailing arm bracket mounting bolts from the body. Torque: 58–80 ft. lbs. (80–110 Nm).

12. Install the trailing arm mounting nuts from the knuckle. Torque: 58–72 ft. lbs. (80–100 Nm).

13. Install the stabilizer link nut to the trailing arm. Torque: 36–47 ft. lbs. (50–65 Nm).

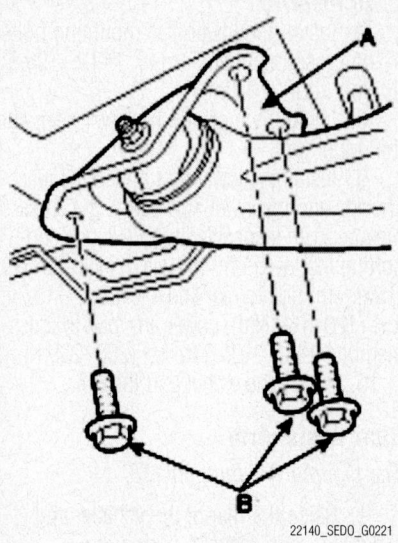

Fig. 138 Remove the trailing arm bracket (A) mounting bolts (B) from the body

14. Install the wheel speed sensor wire bracket bolt and the parking brake wire bracket bolt.

15. Install the wheel speed sensor wire

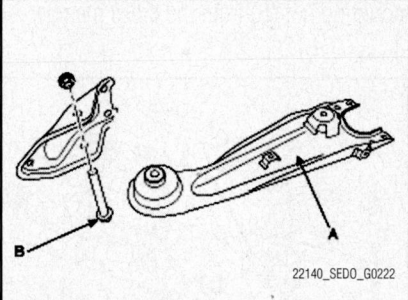

Fig. 139 Remove the connecting bolt (B) between the trailing arm (A) and the bracket

bracket bolt from the body and the connector.

16. Install the wheel and the tire

SHOCK ABSORBER

REMOVAL & INSTALLATION

See Figure 140.

1. Raise the rear of the vehicle, and make sure it is securely supported.

2. Remove the rear wheel and tire.

3. Remove the rear shock absorber assembly mounting bolts (A) and nut (B) from the body.

4. Remove the rear shock absorber assembly nut from the rear knuckle, and remove the shock absorber assembly.

5. Remove the rear shock absorber bracket bolt.

To install:

6. Install the connecting bolt between the rear shock absorber and the bracket. Torque: 116–130 ft. lbs. (160–180 Nm).

7. Install the rear shock absorber to the knuckle temporarily.

8. Install the rear shock absorber bracket mounting bolts and nut. Torque: Bracket bolts: 60–80 ft. lbs. (80–110 Nm); Bracket nut: 65–87 ft. lbs. (90–120 Nm).

9. Install the rear shock absorber nut to the knuckle. Torque: 116–130 ft. lbs. (160–180 Nm).

10. Install the wheel and the tire.

WHEEL BEARINGS

REMOVAL & INSTALLATION

See Figures 141 through 146.

1. Remove the wheel & tire assembly.

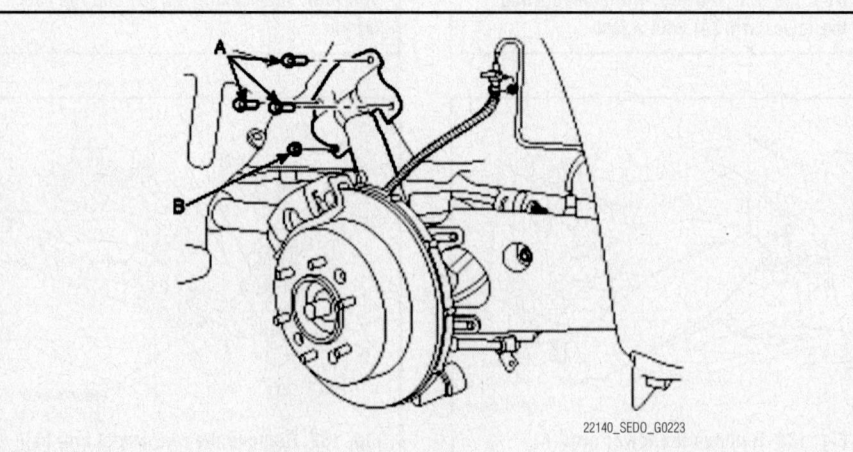

Fig. 140 Remove the rear shock absorber assembly mounting bolts (A) and nut (B) from the body

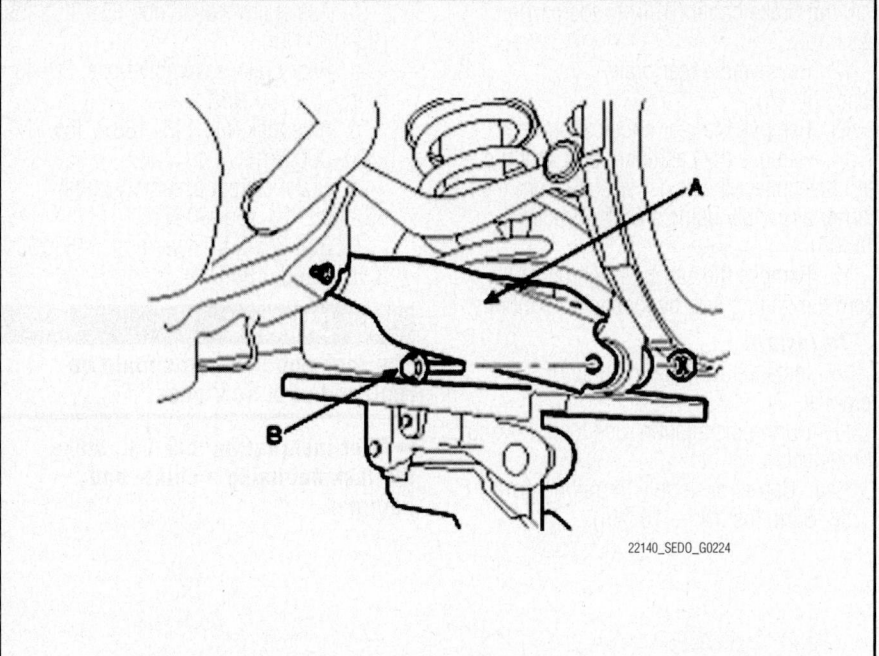

Fig. 141 Support lower part of the lower arm (A) using a jack and remove the bolt & nut (B)

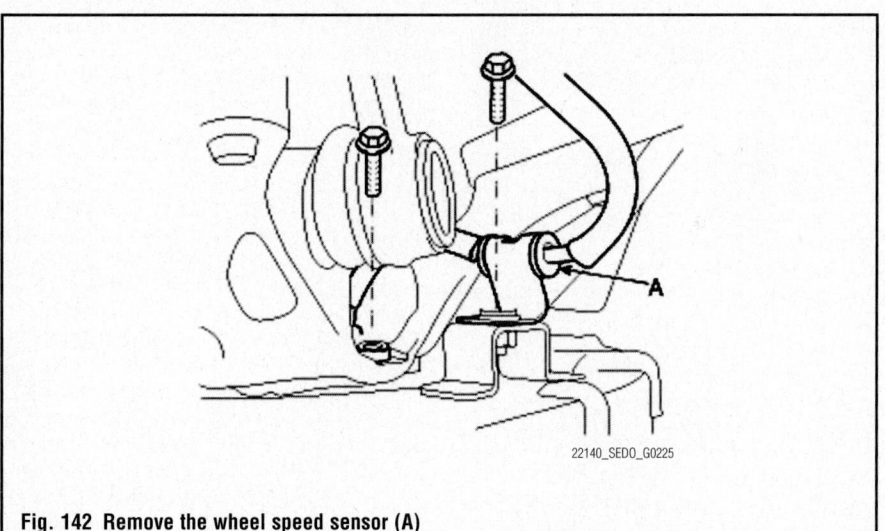

Fig. 142 Remove the wheel speed sensor (A)

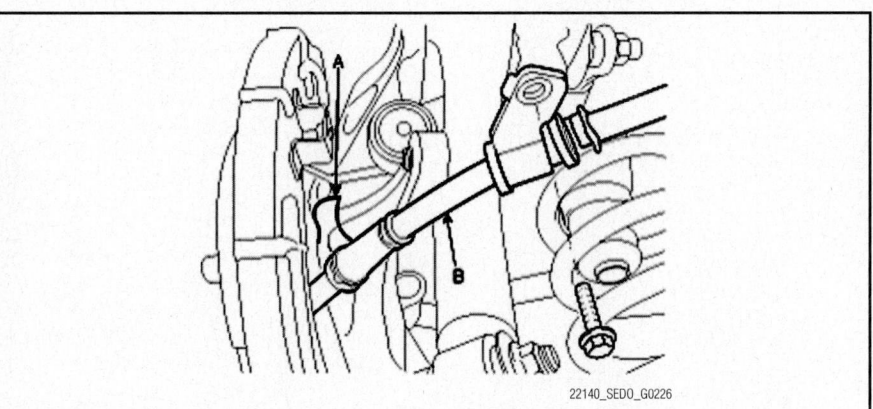

Fig. 143 Remove the clip (A) and bracket mounting bolt and disconnect the parking brake cable (B) from the carrier assembly

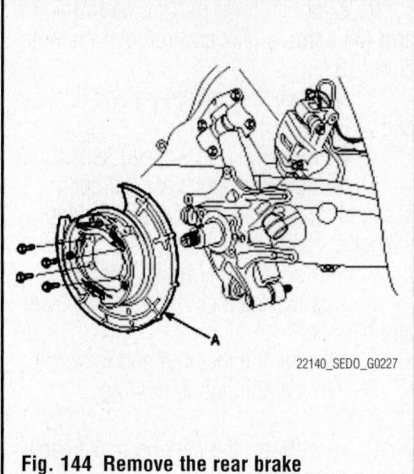

Fig. 144 Remove the rear brake assembly (A)

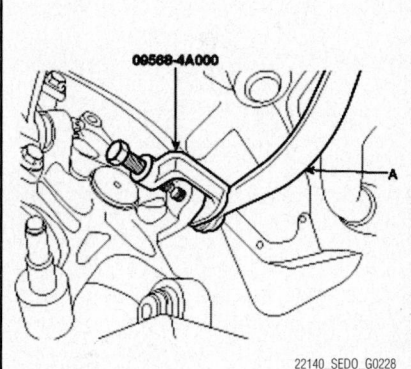

Fig. 145 Remove the castle nut and split pin and disconnect the assist arm (A) from the carrier assembly using a SST (09568-4A000)

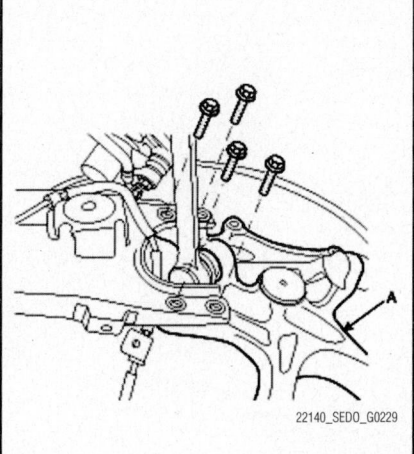

Fig. 146 Remove the carrier assembly (A) from the trailing arm by loosening the bolts

2. Support lower part of the lower arm (A) using a jack and remove the bolt & nut (B).

3. Remove the coil spring and upper pad.

4. Remove the wheel speed sensor (A).

5. Remove the rear brake caliper assembly from the carrier assembly and suspend it with wire.

6. Remove the rear brake rotor disc.

7. Unstake the lock nut using a chisel and hammer.

8. Remove the lock nut and washer.

9. Remove the hub & bearing assembly.

10. Disconnect the parking brake cable from the brake assembly.

11. Remove the clip (A) and bracket mounting bolt and disconnect the parking brake cable (B) from the carrier assembly.

12. Remove the rear brake assembly (A).

13. Remove the rear shock absorber.

14. Remove the castle nut and split pin and disconnect the assist arm (A) from the carrier assembly using a SST (09568-4A000).

15. Remove the carrier assembly (A) from the trailing arm by loosening the bolts.

To install:

16. Installation is the reverse of removal.

17. During installation, use the following torque values:

    a. Carrier assembly to trailing arm: 58–80 ft. lbs. (80–110 Nm).

    b. Assist arm castle nut: 65–72 ft. lbs (90–100 Nm).

    c. Rear brake assembly bolts: 36–43 ft. lbs. (50–60 Nm).

    d. Axle lock nut: 145–188 ft. lbs. (200–260 Nm).

    e. Brake caliper assembly bolts: 36–43 ft. lbs. (50–60 Nm).

    f. Bolt & nut to lower arm: 116–130 ft. lbs. (160–180 Nm).

✳✳ CAUTION

The rear hub lock nut should be replaced with new ones.

➡ **After installation lock nut, stake the lock nut using a chisel and hammer.**

KIA

Sorento

14

SPECIFICATIONS AND MAINTENANCE CHARTS

ENGINE AND VEHICLE IDENTIFICATION

Engine							Model Year	
Code ①	Liters (cc)	Cu. In.	Cyl.	Fuel Sys.	Engine Type	Eng. Mfg.	Code ②	Year
3	3.5 (3497)	213	6	EGI	DOHC	KIA	6	2006
6	3.8 (3778)	231	6	EGI	DOHC	KIA	7	2007
5	3.3 (3342)	204	6	EGI	DOHC	KIA	8	2008

EGI: Electronic Gasoline Injection

DOHC: Double Overhead Camshafts

① 8th digit of VIN

② 10th digit of VIN

22140_SORE_C0001

GENERAL ENGINE SPECIFICATIONS

Year	Model	Engine Displacement Liters	Engine VIN	Net Horsepower @ rpm	Net Torque @ rpm (ft. lbs.)	Bore x Stroke (in.)	Com-pression Ratio	Oil Pressure @ rpm
2006	Sorento	3.5	3	192 @ 5500	217 @ 3000	3.66x3.38	10:01	①
2007	Sorento	3.8	6	262 @ 6000	260 @ 4500	3.78x3.43	10.4:01	18.77@1000
	Sorento	3.3	5	242 @ 6000	228 @ 4500	3.62x3.30	10.4:01	18.77@1000
2008	Sorento	3.8	6	262 @ 6000	260 @ 4500	3.78x3.43	10.4:01	18.77@1000
	Sorento	3.3	5	242 @ 6000	228 @ 4500	3.62x3.30	10.4:01	18.77@1000

① 11.6 psi at idle

22140_SORE_C0002

ENGINE TUNE-UP SPECIFICATIONS

Year	Engine Displacement Liters	Engine VIN	Spark Plug Gap (in.)	Ignition Timing (deg.) MT	Ignition Timing (deg.) AT	Fuel Pump (psi)	Idle Speed (rpm) MT	Idle Speed (rpm) AT	Valve Clearance Intake	Valve Clearance Exhaust
2006	3.5	3	0.039-0.043	①	①	47-48	②	②	HYD	HYD
2007	3.3	5	0.039-0.043	NA	①	54-56	NA	②	③	③
	3.8	6	0.039-0.043	NA	④	54-56	NA	⑤	③	③
2008	3.3	5	0.039-0.043	NA	①	54-56	NA	②	③	③
	3.8	6	0.039-0.043	NA	④	54-56	NA	⑤	③	③

NOTE: The Vehicle Emission Control Information label often reflects specification changes made during production.

   The label figures must be used if they differ from those in this chart

HYD: Hydraulic

① BTDC 10° ± 2°

② 700-900 rpm

③ 0.0008-0.0024 inches

④ BTDC 7° ± 5°

⑤ 550-750 rpm

22140_SORE_C0003

CAPACITIES

Year	Model	Engine Displacement Liters	Engine VIN	Engine Oil with Filter	Transaxle (pts.) Manual	Transaxle (pts.) Auto.	Fuel Tank (gal.)	Cooling System (qts.)
2006	Sorento	3.5	3	4.6	NA	21	21	11.6
2007	Sorento	3.3	5	6.2	NA	23	21	9.4
	Sorento	3.8	6	6.2	NA	23	21	9.4
2008	Sorento	3.3	5	6.2	NA	23	21	9.4
	Sorento	3.8	6	6.2	NA	23	21	9.4

NOTE: All capacities are approximate. Add fluid gradually and ensure a proper level is obtained.

22140_SORE_C0004

FLUID SPECIFICATIONS

Year	Model	Engine Displ. Liters (VIN)	Engine Oil	Man. Trans.	Auto. Trans.	Drive Axle Front	Drive Axle Rear	Transfer Case	Power Steering Fluid	Brake Master Cylinder	Cooling System
2006	Sorento	3.5 (3)	5W30	SAE 75W/85	ATF RED-1	NA	NA	SAE 75W/85	PSF-3	DOT 3	②
2007	Sorento	3.3 (5)	5W20	NA	①	NA	NA	SAE 75W/85	PSF-3	DOT 3	②
	Sorento	3.8 (6)	5W20	NA	①	NA	NA	SAE 75W/85	PSF-3	DOT 3	②
2008	Sorento	3.3 (5)	5W20	NA	①	NA	NA	SAE 75W/85	PSF-3	DOT 3	②
	Sorento	3.8 (6)	5W20	NA	①	NA	NA	SAE 75W/85	PSF-3	DOT 3	②

DOT: Department Of Transpotation

① Diamond ATF SP-3 or SK ATF SP-3

② Ethylene Glycol base for aluminum radiators.

22140_SORE_C0014

VALVE SPECIFICATIONS

Year	Engine Displacement Liters	Engine VIN	Seat Angle (deg.)	Face Angle (deg.)	Maximum out of Square (degrees)	Spring Free Length (in.)	Stem-to-Guide Clearance (in.) Intake	Stem-to-Guide Clearance (in.) Exhaust	Stem Diameter (in.) Intake	Stem Diameter (in.) Exhaust
2006	3.5	3	45	45-45.5	2	1.8268	0.0009-0.0020	0.0020-0.0033	0.258-0.259	0.257-0.258
2007	3.3	5	44.75-45.2	45.25-45.75	Less than 1.5	1.7267	0.00078-0.0019	0.00118-0.0021	0.2151-0.216	0.2149-0.215
	3.8	6	44.75-45.2	45.25-45.75	Less than 1.5	1.7267	0.00078-0.0019	0.00118-0.0021	0.2151-0.216	0.2149-0.215
2008	3.3	5	44.75-45.2	45.25-45.75	Less than 1.5	1.7267	0.00078-0.0019	0.00118-0.0021	0.2151-0.216	0.2149-0.215
	3.8	6	44.75-45.2	45.25-45.75	Less than 1.5	1.7267	0.00078-0.0019	0.00118-0.0021	0.2151-0.216	0.2149-0.215

NA: Not Available

22140_SORE_C0005

CAMSHAFT AND BEARING SPECIFICATIONS

All measurements are given in inches.

Year	Engine Displacement Liters	Engine VIN	Journal Diameter	Brg. Oil Clearance	Shaft End-play	Runout	Journal Bore	Lobe Lift Intake	Lobe Lift Exhaust
2006	3.5	3	1.0220-1.0224	0.0007-0.0024	0.0039-0.0059	NA	NA	NA	NA
2007	3.3	5	①	②	0.0008-0.0071	NA	NA	NA	NA
	3.8	6	①	②	0.0008-0.0071	NA	NA	NA	NA
2008	3.3	5	①	②	0.0008-0.0071	NA	NA	NA	NA
	3.8	6	①	②	0.0008-0.0071	NA	NA	NA	NA

① Intake & Exhaust no. 1: 1.1009-1.1015
  Intake & Exhaust nos. 2, 3, 4: 0.9430-0.9437

② Intake & Exhaust no. 1: 0.0011-0.0022
  Intake & Exhaust nos. 2, 3, 4: 0.0012-0.0026

22140_SORE_C0013

CRANKSHAFT AND CONNECTING ROD SPECIFICATIONS

All measurements are given in inches.

Year	Engine Displacement Liters	Engine VIN	Crankshaft Main Brg. Journal Dia.	Crankshaft Main Brg. Oil Clearance	Crankshaft Shaft End-play	Crankshaft Thrust on No.	Connecting Rod Journal Diameter	Connecting Rod Oil Clearance	Connecting Rod Side Clearance
2006	3.5	3	2.5190-2.5197	0.0009-0.0016	0.002-0.0098	3	2.1650-2.1653	0.0010-0.0017	0.0039-0.0098
2007	3.3	5	2.7142-2.7149	0.0008-0.0016	0.0039-0.0110	3	2.2834-2.2842	0.0015-0.0022	0.0039-0.0098
	3.8	6	2.7142-2.7149	0.0008-0.0016	0.0039-0.0110	3	2.2834-2.2842	0.0015-0.0022	0.0039-0.0098
2008	3.3	5	2.7142-2.7149	0.0008-0.0016	0.0039-0.0110	3	2.2834-2.2842	0.0015-0.0022	0.0039-0.0098
	3.8	6	2.7142-2.7149	0.0008-0.0016	0.0039-0.0110	3	2.2834-2.2842	0.0015-0.0022	0.0039-0.0098

22140_SORE_C0008

PISTON AND RING SPECIFICATIONS

All measurements are given in inches.

Year	Engine Displacement Liters	Engine VIN	Piston Clearance	Ring Gap			Ring Side Clearance		
				Top Compression	Bottom Compression	Oil Control	Top Compression	Bottom Compression	Oil Control
2006	3.5	3	0.0012-0.0020	0.0079-0.0118	0.0177-0.0236	0.0079-0.0276	0.0008-0.0031	0.0016-0.0024	SNUG
2007	3.3	5	0.0012-0.0020	0.0067-0.0126	0.0126-0.0185	0.0078-0.0275	0.0012-0.0027	0.0012-0.0027	0.0024-0.0059
	3.8	6	0.0012-0.0020	0.0067-0.0126	0.0126-0.0185	0.0078-0.0275	0.0012-0.0027	0.0012-0.0027	0.0024-0.0059
2008	3.3	5	0.0012-0.0020	0.0067-0.0126	0.0126-0.0185	0.0078-0.0275	0.0012-0.0027	0.0012-0.0027	0.0024-0.0059
	3.8	6	0.0012-0.0020	0.0067-0.0126	0.0126-0.0185	0.0078-0.0275	0.0012-0.0027	0.0012-0.0027	0.0024-0.0059

22140_SORE_C0007

TORQUE SPECIFICATIONS

All readings in ft. lbs.

Year	Engine Displacement Liters	Engine VIN	Cylinder Head Bolts	Main Bearing Bolts	Rod Bearing Bolts	Crankshaft Damper Bolts	Flywheel Bolts	Manifold		Spark Plugs	Oil Pan Drain Plug
								Intake	Exhaust		
2006	3.5	3	75-82	51-58	①	130-138	53-55	14-17	29-32	15-22	26-32
2007	3.3	5	②	③	④	210-224	53-56	⑤	29-33	15-22	25-33
	3.8	6	②	③	④	210-224	53-56	⑤	29-33	15-22	25-33
2008	3.3	5	②	③	④	210-224	53-56	⑤	29-33	15-22	26-33
	3.8	6	②	③	④	210-224	53-56	⑤	29-33	15-22	26-33

① Step 1: 26 ft. lbs.
Step 2: plus 92 degrees

② Step 1: 29 ft. lbs.
Step 2: plus 120 degrees
Step 3: plus 90 degrees

③ M11 inner bolts:
Step 1: 36 ft. lbs.
Step 2: plus 90 degrees
M8 outer bolts:
Step 1: 15 ft. lbs.
Step 2: plus 120 degrees

④ Step 1: 15 ft. lbs.
Step 2: plus 90 degrees

⑤ Bolts: 20-23 ft. lbs.
Nuts: 14-17 ft. lbs.

22140_SORE_C0006

WHEEL ALIGNMENT

Year	Model		Caster Range (+/-Deg.)	Caster Preferred Setting (Deg.)	Camber Range (+/-Deg.)	Camber Preferred Setting (Deg.)	Toe-in (in.)
2006	Sorento	F	0.50	3.30	0.50	0.39	0.1024
		R	—	—	—	—	—
2007	Sorento	F	0.50	3.89	0.50	0.00	0.079
		R	—	—	—	—	—
2008	Sorento	F	0.50	3.89	0.50	0.00	0.079
		R	—	—	—	—	—

22140_SORE_C0009

TIRE, WHEEL AND BALL JOINT SPECIFICATIONS

Year	Model	OEM Tires Standard	OEM Tires Optional	Tire Pressures (psi) Front	Tire Pressures (psi) Rear	Wheel Size	Ball Joint Inspection	Lug Nut Torque (ft. lbs.)
2006	Sorento	P245/70R16	—	30	30	7JJ	①	65-86
2007	Sorento	P245/70R16	P245/65R17	30	30	7JJ	①	65-86
2008	Sorento	P245/70R16	P245/65R17	30	30	7JJ	①	65-86

OEM: Original Equipment Manufacturer

PSI: Pounds Per Square Inch

STD: Standard

OPT: Optional

① Replace if any measurable movement is found.

22140_SORE_C0010

BRAKE SPECIFICATIONS

All measurements in inches unless noted

Year	Model		Brake Disc Original Thickness	Brake Disc Minimum Thickness	Brake Disc Maximum Run-out	Brake Drum Original Inside Diameter	Brake Drum Max. Wear Limit	Brake Drum Maximum Machine Diameter	Minimum Lining Thickness	Brake Caliper Bracket Bolts (ft. lbs.)	Brake Caliper Mounting Bolts (ft. lbs.)
2006	Sorento	F	1.100	1.020	0.0012	—	—	—	0.079	16-24	47-54
		R	0.787	0.724	—	—	—	—	0.079	16-24	47-54
2007	Sorento	F	1.100	1.020	0.0002	—	—	—	0.079	16-24	47-54
		R	0.752	0.724	—	—	—	—	0.079	16-24	47-54
2008	Sorento	F	1.100	1.020	0.0012	—	—	—	0.079	16-24	47-54
		R	0.752	0.724	—	—	—	—	0.079	16-24	47-54

F: Front

R: Rear

22140_SORE_C0011

SCHEDULED MAINTENANCE INTERVALS
KIA—SORENTO

TO BE SERVICED	OF SERVIC	VEHICLE MILEAGE INTERVAL (x1000)															
		7.5	15	22.5	30	37.5	45	52.5	60	67.5	75	82.5	90	97.5	105	112.5	120
Accessory drive belts	S/I				✓				✓				✓				✓
Air cleaner element	I/R		✓		✓		✓		✓		✓		✓		✓		✓
Air conditioner system	S/I	Inspect the system operation and refrigerant amount annually.															
Brake lines, hoses and connections	S/I		✓		✓		✓		✓		✓		✓		✓		✓
Chassis and body fasteners	T				✓				✓				✓				✓
Cooling system hoses and coolant level	S/I				✓				✓				✓				✓
CV-joint boots	S/I		✓		✓		✓		✓		✓		✓		✓		✓
Engine coolant	R				✓								✓				✓
Engine oil and filter	R	✓	✓	✓	✓	✓	✓	✓	✓	✓	✓	✓	✓	✓	✓	✓	✓
Exhaust system heat shields	S/I				✓				✓				✓				✓
Front and rear brakes	S/I		✓		✓		✓		✓		✓		✓		✓		✓
Front ball joints S/I	S/I				✓				✓				✓				✓
Fuel filter	R								✓								✓
Fuel lines and hoses	S/I				✓				✓				✓				✓
Rear differential fluid	S/I	✓	✓	✓	✓	✓	✓	✓	✓	✓	✓	✓	✓	✓	✓	✓	✓
Front differential fluid (if equipped)	S/I	✓	✓	✓	✓	✓	✓	✓	✓	✓	✓	✓	✓	✓	✓	✓	✓
Automatic transmission fluid	S/I		✓		✓		✓		✓		✓		✓		✓		✓
Manual transmission fluid (if equipped)	S/I	✓	✓	✓	✓	✓	✓	✓	✓	✓	✓	✓	✓	✓	✓	✓	✓
Transfer case oil (if equipped)	S/I	✓	✓	✓	✓	✓	✓	✓	✓	✓	✓	✓	✓	✓	✓	✓	✓
Ignition wires	S/I								✓								✓
Emission hoses	S/I				✓				✓				✓				✓
Driveshaft U-joints	L		✓		✓		✓		✓		✓		✓		✓		✓
Brake fluid	S/I		✓		✓		✓		✓		✓		✓		✓		✓
Clutch fluid (if equipped)	S/I		✓		✓		✓		✓		✓		✓		✓		✓
Idle speed	A				✓				✓				✓				✓
Locks and hinges	L	✓	✓	✓	✓	✓	✓	✓	✓	✓	✓	✓	✓	✓	✓	✓	✓
Spark plugs	R								✓								✓
Steering operation and linkage	S/I				✓				✓				✓				✓
Timing belt	R								✓								✓

R: Replace   S/I: Inspect and service, if needed   L: Lubricate   A: Adjust   T: Tighten

FREQUENT OPERATION MAINTENANCE (SEVERE SERVICE)

If a vehicle is operated under any of the following conditions it is considered severe service

- Towing a trailer or using a camper or car-top carrier.

- Repeated short trips of less than 5 miles in temperatures below freezing, or trips of less than 10 miles in any temperature.

- Extensive idling or low-speed driving for long distances as in heavy commercial use, such as delivery, taxi or police cars.

- Operating on rough, muddy or salt-covered roads.

- Operating on unpaved or dusty roads.

- Driving in extremely hot (over 90°F) conditions.

Engine oil and filter: replace every 3000 miles or 3 months, whichever occurs first.

Air cleaner element: inspect every 10,000 miles or 10 months and replace every 20,000 miles or 20 months, whichever occurs first.

Front differential fluid (if equipped): inspect every 5000 miles and replace every 15,000 miles.

Transfer case fluid (if equipped): inspect every 5000 miles.

Manual transmission fluid (if equipped): inspect every 5000 miles and replace every 60,000 miles.

Emission system hoses: inspect every 30,000 miles or 30 months, whichever occurs first.

Front and rear disc brakes: inspect every 15,000 miles or 15 months, whichever occurs first.

Chassis and body fasteners: tighten every 15,000 miles or 15 months, whichever occurs first.

Locks and hinges: lubricate every 5000 miles or 5 months, whichever occurs first.

PRECAUTIONS

Before servicing any vehicle, please be sure to read all of the following precautions, which deal with personal safety, prevention of component damage, and important points to take into consideration when servicing a motor vehicle:

• Never open, service or drain the radiator or cooling system when the engine is hot; serious burns can occur from the steam and hot coolant.

• Observe all applicable safety precautions when working around fuel. Whenever servicing the fuel system, always work in a well-ventilated area. Do not allow fuel spray or vapors to come in contact with a spark, open flame, or excessive heat (a hot drop light, for example). Keep a dry chemical fire extinguisher near the work area. Always keep fuel in a container specifically designed for fuel storage; also, always properly seal fuel containers to avoid the possibility of fire or explosion. Refer to the additional fuel system precautions later in this section.

• Fuel injection systems often remain pressurized, even after the engine has been turned **OFF**. The fuel system pressure must be relieved before disconnecting any fuel lines. Failure to do so may result in fire and/or personal injury.

• Brake fluid often contains polyglycol ethers and polyglycols. Avoid contact with the eyes and wash your hands thoroughly after handling brake fluid. If you do get brake fluid in your eyes, flush your eyes with clean, running water for 15 minutes. If eye irritation persists, or if you have taken

brake fluid internally, IMMEDIATELY seek medical assistance.

• The EPA warns that prolonged contact with used engine oil may cause a number of skin disorders, including cancer. You should make every effort to minimize your exposure to used engine oil. Protective gloves should be worn when changing oil. Wash your hands and any other exposed skin areas as soon as possible after exposure to used engine oil. Soap and water, or waterless hand cleaner should be used.

• All new vehicles are now equipped with an air bag system, often referred to as a Supplemental Restraint System (SRS) or Supplemental Inflatable Restraint (SIR) system. The system must be disabled before performing service on or around system components, steering column, instrument panel components, wiring and sensors. Failure to follow safety and disabling procedures could result in accidental air bag deployment, possible personal injury and unnecessary system repairs.

• Always wear safety goggles when working with, or around, the air bag system. When carrying a non-deployed air bag, be sure the bag and trim cover are pointed away from your body. When placing a non-deployed air bag on a work surface, always face the bag and trim cover upward, away from the surface. This will reduce the motion of the module if it is accidentally deployed. Refer to the additional air bag system precautions later in this section.

• Clean, high quality brake fluid from a sealed container is essential to the safe and

proper operation of the brake system. You should always buy the correct type of brake fluid for your vehicle. If the brake fluid becomes contaminated, completely flush the system with new fluid. Never reuse any brake fluid. Any brake fluid that is removed from the system should be discarded. Also, do not allow any brake fluid to come in contact with a painted surface; it will damage the paint.

• Never operate the engine without the proper amount and type of engine oil; doing so WILL result in severe engine damage.

• Timing belt maintenance is extremely important. Many models utilize an interference-type, non-freewheeling engine. If the timing belt breaks, the valves in the cylinder head may strike the pistons, causing potentially serious (also time-consuming and expensive) engine damage. Refer to the maintenance interval charts for the recommended replacement interval for the timing belt, and to the timing belt section for belt replacement and inspection.

• Disconnecting the negative battery cable on some vehicles may interfere with the functions of the on-board computer system(s) and may require the computer to undergo a relearning process once the negative battery cable is reconnected.

• When servicing drum brakes, only disassemble and assemble one side at a time, leaving the remaining side intact for reference.

• Only an MVAC-trained, EPA-certified automotive technician should service the air conditioning system or its components.

BRAKES

GENERAL INFORMATION

PRECAUTIONS

• Certain components within the ABS system are not intended to be serviced or repaired individually.

• Do not use rubber hoses or other parts not specifically specified for and ABS system. When using repair kits, replace all parts included in the kit. Partial or incorrect repair may lead to functional problems and require the replacement of components.

• Lubricate rubber parts with clean, fresh brake fluid to ease assembly. Do not

use shop air to clean parts; damage to rubber components may result.

• Use only DOT 3 brake fluid from an unopened container.

• If any hydraulic component or line is removed or replaced, it may be necessary to bleed the entire system.

• A clean repair area is essential. Always clean the reservoir and cap thoroughly before removing the cap. The slightest amount of dirt in the fluid may plug an orifice and impair the system function. Perform repairs after components have been thoroughly cleaned; use only denatured alcohol

ANTI-LOCK BRAKE SYSTEM (ABS)

to clean components. Do not allow ABS components to come into contact with any substance containing mineral oil; this includes used shop rags.

• The Anti-Lock control unit is a microprocessor similar to other computer units in the vehicle. Ensure that the ignition switch is **OFF** before removing or installing controller harnesses. Avoid static electricity discharge at or near the controller.

• If any arc welding is to be done on the vehicle, the control unit should be unplugged before welding operations begin.

BRAKES | BLEEDING THE BRAKE SYSTEM

BLEEDING PROCEDURE

MASTER CYLINDER BLEEDING

→ Immediately wash off any brake fluid that comes into contact with any painted surfaces.

→ Depressing the brake pedal with the reservoir cap removed will cause the fluid to spray.

→ When bleeding, maintain the amount of fluid in the reservoir between the Min. and Max. lines.

1. Fill reservoir with DOT3 brake fluid.
2. Disconnect the brake lines from the master cylinder.

3. Slowly depress the brake pedal and hold it there.
4. Block the outer holes with your fingers, and release the brake pedal.
5. Repeat 3 or 4 times.

BRAKE LINE BLEEDING

1. Connect the vinyl tube to the bleeder plug.
2. Depress the brake pedal several times, then loosen the bleeder plug with the pedal held down.
3. At the point where the fluid stops coming out, tighten the bleeder plug, then release the brake pedal.
4. Repeat until all the air in the fluid has been bled out.

5. Repeat the above procedure to bleed the air out of the brake line for each wheel.
6. Check the fluid level and add fluid if necessary.

BLEEDING THE ABS SYSTEM

The ABS brake system is bled in the usual fashion with no special procedures required. Refer to the bleeding procedure described in this section. Make certain the master cylinder reservoir is filled before the bleeding is begun and check the level frequently.

BRAKES | FRONT DISC BRAKES

⁜ CAUTION

Dust and dirt accumulating on brake parts during normal use may contain asbestos fibers from production or aftermarket brake linings. Breathing excessive concentrations of asbestos fibers can cause serious bodily harm. Exercise care when servicing brake parts. Do not sand or grind brake lining unless equipment used is designed to contain the dust residue. Do not clean brake parts with compressed air or by dry brushing. Cleaning should be done by dampening the brake components with a fine mist of water, then wiping the brake components clean with a dampened cloth. Dispose of cloth and all residue containing asbestos fibers in an impermeable container with the appropriate label. Follow practices prescribed by the Occupational Safety and Health Administration (OSHA) and the Environmental Protection Agency (EPA) for the handling, processing, and disposing of dust or debris that may contain asbestos fibers.

BRAKE CALIPER

REMOVAL & INSTALLATION

1. Remove the front wheel.
2. Disconnect the brake fluid hose.
3. Remove the caliper mounting bolts.
4. Remove the brake caliper.

To install:
5. Install the brake caliper. Tighten the mounting bolts to 16–24 ft. lbs. (22–32 Nm).
6. Install the brake fluid hose. Tighten the hose fitting to 12–14 ft. lbs. (17–20 Nm).
7. Bleed the brake system.
8. Install the front wheel.
9. Before attempting to move the vehicle, pump the brake pedal to seat the pads against the rotors. Make sure the vehicle has a firm brake pedal. Check the level of the brake fluid and add DOT 3 or 4 brake fluid if necessary.

DISC BRAKE PADS

REMOVAL & INSTALLATION

1. Remove the front wheel.
2. Remove the guide pin, lift the caliper assembly up and suspend it with a wire.
3. Remove the following parts from the caliper support:
   - Pad and wear sensor assembly
   - Pad spring
   - Outer shim

To install:
4. Compress the caliper piston into the caliper bore.
5. Install the pad clips.
6. Install the inner and outer pads on each pad clip.
7. Lower the brake caliper carefully so as not to damage the boot.
8. Tighten the guide pin bolt to 16–24 ft. lbs. (22–32 Nm).
9. Install the front wheel.

BRAKES

✳✳ CAUTION

Dust and dirt accumulating on brake parts during normal use may contain asbestos fibers from production or aftermarket brake linings. Breathing excessive concentrations of asbestos fibers can cause serious bodily harm. Exercise care when servicing brake parts. Do not sand or grind brake lining unless equipment used is designed to contain the dust residue. Do not clean brake parts with compressed air or by dry brushing. Cleaning should be done by dampening the brake components with a fine mist of water, then wiping the brake components clean with a dampened cloth. Dispose of cloth and all residue containing asbestos fibers in an impermeable container with the appropriate label. Follow practices prescribed by the Occupational Safety and Health Administration (OSHA) and the Environmental Protection Agency (EPA) for the handling, processing, and disposing of dust or debris that may contain asbestos fibers.

BRAKE CALIPER

REMOVAL & INSTALLATION

1. Remove the rear wheel.
2. Disconnect the brake fluid hose.
3. Remove the caliper guide bolts.
4. Remove the brake caliper.

To install:

5. Install the brake caliper. Tighten the guide bolts to 16–23 ft. lbs. (22–32 Nm).
6. Install the brake fluid hose. Tighten the hose fitting to 12–14 ft. lbs. (17–20 Nm).
7. Bleed the brake system.
8. Install the rear wheel.
9. Before attempting to move the vehicle, pump the brake pedal to seat the pads against the rotors. Make sure the vehicle has a firm brake pedal. Check the level of the brake fluid and add DOT 3 or 4 brake fluid if necessary.

DISC BRAKE PADS

REMOVAL & INSTALLATION

1. Remove the rear wheel.
2. Remove the guide pin bolts, lift the caliper assembly up and suspend it with a wire.
3. Before replacing the brake pads, drain brake fluid from the master cylinder reservoir until it remains half full.
4. Remove the brake pads by turning the piston in the housing assembly using special tool 09581-11000 to compress the piston.
5. Remove the inner and outer pads from the caliper.

To install:

6. Install the inner and outer brake pads, engaging the clips securely onto the caliper assembly.
7. Lower the brake caliper assembly into proper position.
8. Tighten the guide pin bolts to 16–23 ft. lbs. (22–32 Nm).
9. Install the rear wheel.
10. Check the brake fluid level and top off, if necessary.

BRAKES

PARKING BRAKE CABLES

ADJUSTMENT

1. Pull on the parking brake lever with a force of 22 lbs. (10 kg) and count the number of notches. Standard value is 4–6 notches.
2. If the parking brake lever stroke is outside the standard value, adjust it as follows:
   a. Loosen the adjusting nut to release the parking brake cable.
   b. Remove the adjusting hole plug.
   c. Turn the adjuster nut to adjust the parking brake lever stroke to the specification.

3. After adjustment, raise the rear of the vehicle and check that the parking brake does not drag.

PARKING BRAKE SHOES

REMOVAL & INSTALLATION

1. Remove rear disc caliper assembly.
2. Before removing the brake disc, chalk both sides of the screw.

➡**Reduce the shoe gap by turning the adjuster with appropriate tool.**

3. After turning the pin to coincide with hole of spring cap, remove the shoe hold spring.
4. Remove the lower return spring.
5. Remove the parking brake cable mounting nuts.
6. Remove the parking brake shoes.
7. Install the upper return spring and brake shoes
8. Turn the adjuster in clockwise direction and install.
9. Install the lower return spring.
10. Install the shoe hold spring with pliers.
11. Install the disc brake and align the mark while tightening the screw.

CHASSIS ELECTRICAL   AIR BAG (SUPPLEMENTAL RESTRAINT SYSTEM)

GENERAL INFORMATION

✲✲ CAUTION

These vehicles are equipped with an air bag system. The system must be disarmed before performing service on, or around, system components, the steering column, instrument panel components, wiring and sensors. Failure to follow the safety precautions and the disarming procedure could result in accidental air bag deployment, possible injury and unnecessary system repairs.

SERVICE PRECAUTIONS

Disconnect and isolate the battery negative cable before beginning any airbag system component diagnosis, testing, removal, or installation procedures. Allow system capacitor to discharge for two minutes before beginning any component service. This will disable the airbag system. Failure to disable the airbag system may result in accidental airbag deployment, personal injury, or death.

Do not place an intact undeployed airbag face down on a solid surface. The airbag will propel into the air if accidentally deployed and may result in personal injury or death.

When carrying or handling an undeployed airbag, the trim side (face) of the airbag should be pointing towards the body to minimize possibility of injury if accidental deployment occurs. Failure to do this may result in personal injury or death.

Replace airbag system components with OEM replacement parts. Substitute parts may appear interchangeable, but internal differences may result in inferior occupant protection. Failure to do so may result in occupant personal injury or death.

Wear safety glasses, rubber gloves, and long sleeved clothing when cleaning powder residue from vehicle after an airbag deployment. Powder residue emitted from a deployed airbag can cause skin irritation. Flush affected area with cool water if irritation is experienced. If nasal or throat irritation is experienced, exit the vehicle for fresh air until the irritation ceases. If irritation continues, see a physician.

Do not use a replacement airbag that is not in the original packaging. This may result in improper deployment, personal injury, or death.

The factory installed fasteners, screws and bolts used to fasten airbag components have a special coating and are specifically designed for the airbag system. Do not use substitute fasteners. Use only original equipment fasteners listed in the parts catalog when fastener replacement is required.

During, and following, any child restraint anchor service, due to impact event or vehicle repair, carefully inspect all mounting hardware, tether straps, and anchors for proper installation, operation, or damage. If a child restraint anchor is found damaged in any way, the anchor must be replaced. Failure to do this may result in personal injury or death.

Deployed and non-deployed airbags may or may not have live pyrotechnic material within the airbag inflator.

Do not dispose of driver/passenger/curtain airbags or seat belt tensioners unless you are sure of complete deployment. Refer to the Hazardous Substance Control System for proper disposal.

Dispose of deployed airbags and tensioners consistent with state, provincial, local, and federal regulations.

After any airbag component testing or service, do not connect the battery negative cable. Personal injury or death may result if the system test is not performed first.

If the vehicle is equipped with the Occupant Classification System (OCS), do not connect the battery negative cable before performing the OCS Verification Test using the scan tool and the appropriate diagnostic information. Personal injury or death may result if the system test is not performed properly.

Never replace both the Occupant Restraint Controller (ORC) and the Occupant Classification Module (OCM) at the same time. If both require replacement, replace one, then perform the Airbag System test before replacing the other.

Both the ORC and the OCM store Occupant Classification System (OCS) calibration data, which they transfer to one another when one of them is replaced. If both are replaced at the same time, an irreversible

fault will be set in both modules and the OCS may malfunction and cause personal injury or death.

If equipped with OCS, the Seat Weight Sensor is a sensitive, calibrated unit and must be handled carefully. Do not drop or handle roughly. If dropped or damaged, replace with another sensor. Failure to do so may result in occupant injury or death.

If equipped with OCS, the front passenger seat must be handled carefully as well. When removing the seat, be careful when setting on floor not to drop. If dropped, the sensor may be inoperative, could result in occupant injury, or possibly death.

If equipped with OCS, when the passenger front seat is on the floor, no one should sit in the front passenger seat. This uneven force may damage the sensing ability of the seat weight sensors. If sat on and damaged, the sensor may be inoperative, could result in occupant injury, or possibly death.

DISARMING THE SYSTEM

1. Before servicing the vehicle, refer to the Precautions Section.
2. Turn the ignition switch to the **LOCK** position.
3. Disconnect the negative battery cable.
4. Wait 10 minutes for the battery back-up power to discharge.

ARMING THE SYSTEM

1. Before servicing the vehicle, refer to the Precautions Section.
2. Connect the negative battery cable.
3. Turn the ignition switch **ON**.
4. Verify that the air bag indicator illuminates for 4–8 seconds, then goes off.

CLOCKSPRING CENTERING

Prior to installing the clock spring, align the mating mark and "NEUTRAL" position indicator of the clock spring, and, after turning the front wheels to the straight-ahead position, install the clock spring to the column switch. If the mating mark of the clock spring is not properly aligned, the steering wheel may not completely rotate during a turn, or the flat cable within the clock spring may be severed, obstructing normal operation of the SRS and possibly leading to serious injury to the vehicle's driver.

DRIVETRAIN

AUTOMATIC TRANSMISSION ASSEMBLY

REMOVAL & INSTALLATION

2006 Models

1. Disconnect the negative battery cable.
2. Drain the automatic transmission fluid.
3. Remove the control cable.
4. Remove the undercover.
5. Remove the front driveshaft (4WD).
6. Remove the front muffler and the heater protector.
7. Remove the transfer case connector (4WD).
8. Remove the rear driveshaft.
9. Remove the oil cooler pipe.
10. Remove the speed sensor connector.
11. Remove the back-up lamp switch connector.
12. Remove the starter motor.
13. Remove the transmission mounting bolt.
14. Support the transmission on the jack.
15. Remove the transmission with transfer case (4WD vehicle).

To install:

16. Install the transmission and transfer case (if equipped). Tighten the flange bolts to 29–38 ft. lbs. (42–54 Nm).
17. Install the starter motor.
18. Connect the back-up lamp switch connector.
19. Connect the speed sensor connector.
20. Install the oil cooler pipe.
21. Install the rear driveshaft.
22. Connect the transfer case connector. (4WD)
23. Install the front muffler and the heater protector.
24. Install the front driveshaft (4WD).
25. Install the undercover.
26. Install the control cable.
27. Connect the negative battery cable.
28. Fill the transaxle to the correct level with the proper transmission fluid.
29. Start the engine and check for leaks.

2007–08 Models

See Figures 1 through 15.

1. Remove the battery terminal.
2. Remove the engine cover.
3. Remove the O2 sensor connectors (A).
4. Remove the undercover.

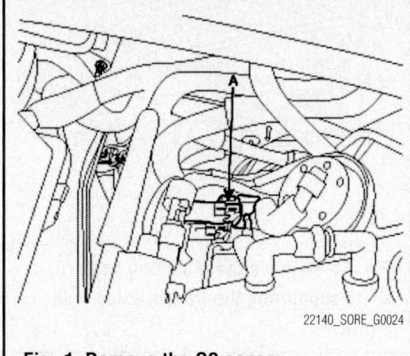

Fig. 1 Remove the O2 sensor connectors (A).

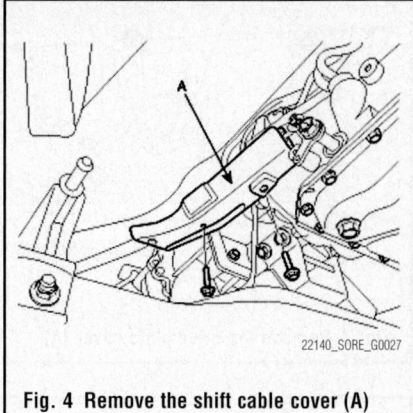

Fig. 4 Remove the shift cable cover (A)

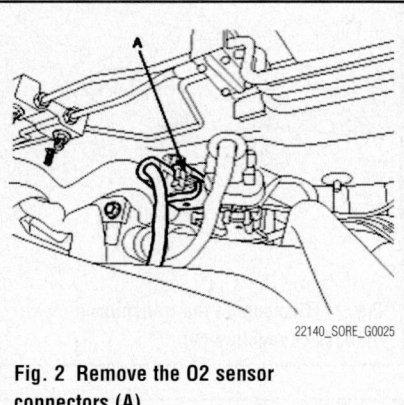

Fig. 2 Remove the O2 sensor connectors (A).

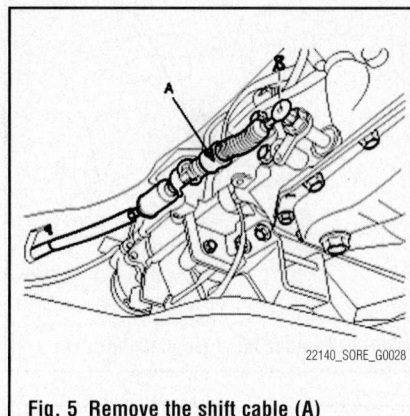

Fig. 5 Remove the shift cable (A)

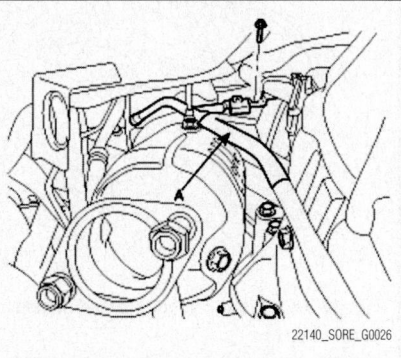

Fig. 3 Remove the transmission oil level gauge (A)

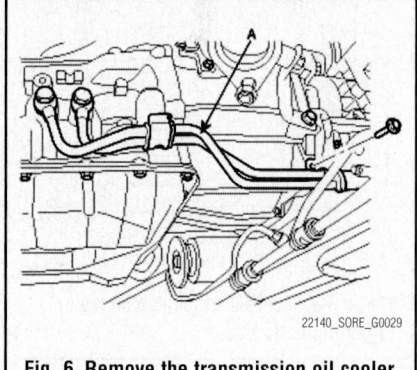

Fig. 6 Remove the transmission oil cooler pipes (A)

5. Drain the automatic transmission fluid.
6. Remove the front propeller shaft. (4WD)
7. Remove the front muffler and the heat protector.
8. Remove the rear propeller shaft.
9. Remove the transmission oil level gauge (A).
10. Remove the shift cable cover (A).
11. Remove the shift cable (A).
12. Remove the transmission oil cooler pipes (A).

13. Remove the drive plate cover (A).
14. Remove the drive plate bolts (A)

➡**Remove the bolts (6ea) while rotating the crankshaft clockwise.**

15. Remove the transmission lower mounting bolts (A).
16. Remove the starter motor mounting bolts (A) and the other bolts (B).
17. Remove the mounting bolts (A) while supporting the transmission with a jack.
18. Lower the jack slightly to simplify

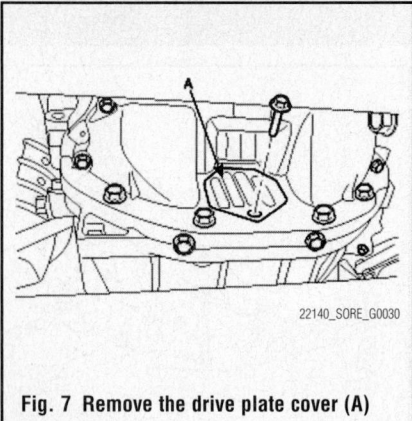

Fig. 7 Remove the drive plate cover (A)

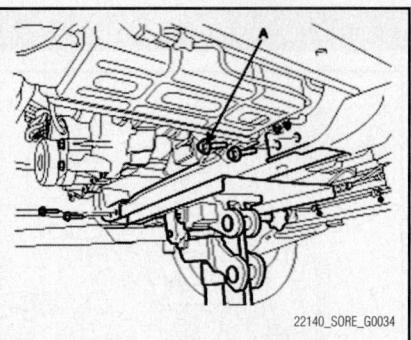

Fig. 11 Remove the mounting bolts (A) while supporting the transmission with a jack

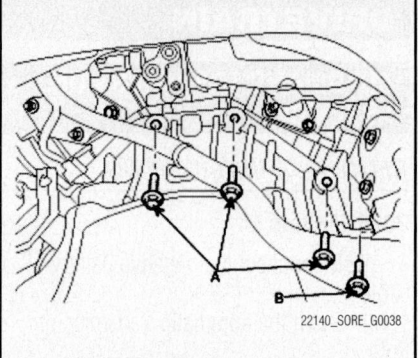

Fig. 15 Remove the transmission upper mounting bolts (A, B)

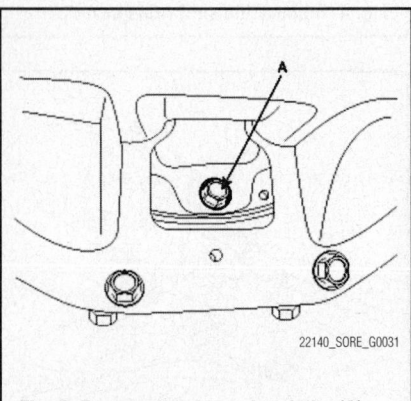

Fig. 8 Remove the drive plate bolts (A)

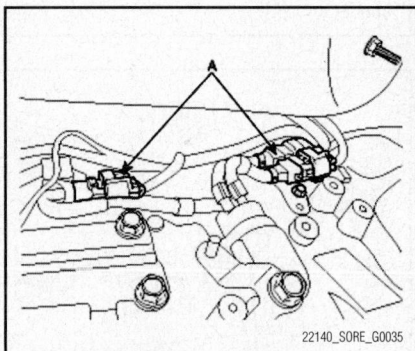
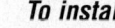

Fig. 12 Disconnect the transmission wire harness connectors (A)

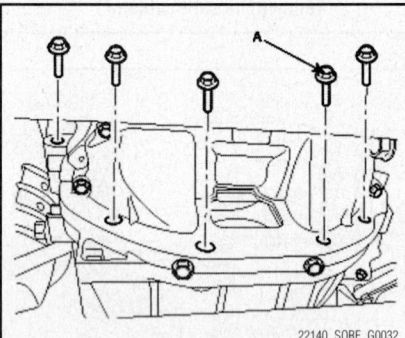

Fig. 9 Remove the transmission lower mounting bolts (A)

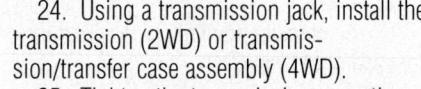

Fig. 13 Remove the vehicle speed sensor connector (A)

Fig. 10 Remove the starter motor mounting bolts (A) and the other bolts (B)

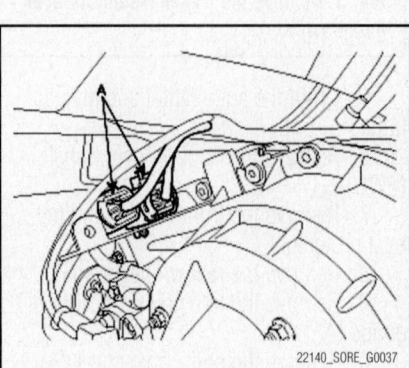
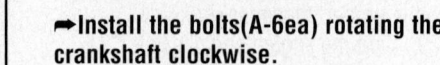

Fig. 14 Remove the transfer case connectors (A) (4WD)

removal of electrical connectors and bolts on the upper part of the transmission.

19. Disconnect the transmission wire harness connectors (A).

20. Remove the vehicle speed sensor connector (A).

21. Remove the transfer case connectors (A) (4WD)

22. Remove the transmission upper mounting bolts (A,B).

23. Remove the transmission (2WD) or transmission/transfer case assembly (4WD).

To install:

24. Using a transmission jack, install the transmission (2WD) or transmission/transfer case assembly (4WD).

25. Tighten the transmission mounting bolts. Torque: A: 21.6–30.3 ft. lbs. (30–42 Nm); B: 57.8–72.3 ft. lbs. (78.4–98 Nm).

26. Install the transfer case connectors. (4WD)

27. Install the vehicle speed sensor connector.

28. Connect the transmission wire harness connectors.

29. Install the crossmember mounting bolts.

30. Install the starter motor mounting bolts and the other bolts. Torque: A: 36.2–47.0 ft. lbs. (50–65 Nm); B: 25.3–33.9 ft. lbs. (34.3–46 Nm).

31. Install the Tighten the transaxle lower mounting bolts.

32. Install the drive plate bolts. Torque: 25.3–30.3 ft. lbs. (34.3–41.1 Nm).

➡ Install the bolts(A-6ea) rotating the crankshaft clockwise.

33. Install the drive plate cover.

34. Install the oil cooler pipes.

35. Install the shift cable.

36. Install the shift cable cover.

37. Install the transmission oil level gauge.

38. Install the rear propeller shaft. Torque: 43.39–50.63 ft. lbs. (58.83–68.64 Nm).

39. Install the front muffler and the heat protector.

40. Install the front propeller shaft. (4WD) Torque: 43.39–50.63 ft. lbs. (58.83–68.64 Nm).

41. Refill the transmission fluid.

42. Install the undercover.

43. Install the O2 sensor connectors.

44. Install the engine cover.

45. Install the battery terminal.

46. Refill the transmission fluid.

MANUAL TRANSMISSION ASSEMBLY

REMOVAL & INSTALLATION

2006 Models

1. Remove the battery terminal.

2. Remove the knob and the control lever.

3. Raise the vehicle.

4. Remove the transmission under cover.

5. Remove the clutch release cylinder.

6. Remove the front propeller shaft (4WD vehicle).

7. Remove the front muffler and the heater protector.

8. Remove the transfer case connector (4WD vehicle).

9. Remove the rear propeller shaft.

10. Support the transmission by the jack.

11. Remove the rear cross member.

12. Remove the transmission with transfer case (4WD vehicle).

13. Installation is the reverse of removal.

CLUTCH

REMOVAL & INSTALLATION

See Figure 16.

2006 Models

1. Disconnect the negative battery cable.

2. Remove the transmission from the vehicle.

3. Insert tool (09411-43000) or equivalent in the clutch disc. This prevents the disc from shifting.

4. Loosen the bolts that attach the clutch pressure plate to the flywheel in a star pattern. Loosen the bolts in succession, one or two turns at a time in order to prevent bending the pressure plate.

To install:

5. Installation is the reverse of the removal procedure.

6. Torque the pressure plate retaining bolts to 11–16 ft. lbs. (15–22 Nm).

Fig. 16 Clutch pressure plate bolt torque sequence

ADJUSTMENTS

1. Measure the clutch pedal height (From the face of the pedal pad to the floorboard) and the clutch pedal free-play (measured at the face of the pedal pad). The standard value is as follows:

   a. Clutch pedal height: 0.29–0.55 inches (7.3–13.9mm)

   b. Clutch pedal free play: 6.45 in (163.8mm)

2. If the clutch pedal free-play is not within the standard value range, adjust as follows :

   a. Turn and adjust the bolt, then secure it by tightening the lock nut.

   b. Turn the push rod to coincide with the standard value and then secure the push rod with the lock nut.

➡ **When adjusting the clutch pedal height or the clutch pedal clevis pin play, be careful not to push the push rod toward the master cylinder.**

3. After completing the adjustments, check that the clutch pedal free play (measured at the face of the pedal pad) falls within the standard value range of 6–13mm (0.2–0.5 in.).

4. If the clutch pedal free play and the distance between the clutch pedal and the floor board when the clutch is disengaged do not meet the standard values, the cause may be either air in the hydraulic system or a faulty master cylinder clutch. Bleed the system or disassemble and inspect the master cylinder or clutch.

BLEEDING

1. With an assistant in the vehicle, raise and safely support the vehicle.

2. Have your assistant pump the clutch pedal three times and hold the pedal to the floor.

3. Open the bleeder valve on the clutch slave cylinder until the air is purged from the cylinder.

4. Tighten the bleeder valve.

5. Have your assistant release the clutch pedal.

6. Fill the clutch master cylinder if below minimum.

7. Repeat Steps 2 through 6 until no air exits from the bleeder valve.

8. Lower the vehicle.

9. Fill the clutch master cylinder fluid reservoir.

TRANSFER CASE ASSEMBLY

REMOVAL & INSTALLATION

1. Drain the transfer case fluid.

2. Disconnect the negative battery cable.

3. Remove the front and rear console mounting screws. Slide the console forward to clear the parking brake handle and set aside. Open the shift boot.

4. Remove the transfer case shift lever locknut and remove the lever knob.

5. Pull the console up to access the wiring connector (s). Unplug the connector (s) and remove the console.

6. Shift the transfer lever to the 4L position.

7. Remove the cover plate.

8. Remove the retaining bolts from the transfer case and lift the shifter lever assembly straight out and properly support the transmission.

9. Matchmark the driveshaft at the flanges and remove the driveshaft.

10. Remove the crossmember bolts and support the transmission on the jack.

11. Disconnect the 4WD light switch connector.

12. Remove the transfer case mounting bolts.

13. Remove the crossmember.

14. Separate the transfer case from the transmission by striking the transfer case with a plastic mallet at the seal area.

15. Lower the transfer case from the vehicle.

To install:

16. Install the transfer case in position with a new gasket. Torque the bolts to 32 ft. lbs. (44 Nm).

17. Install the crossmember.

18. Connect the 4WD light switch connector.

19. Align the matchmarks on the driveshaft to the flanges. Torque the bolts to 27 ft. lbs. (36 Nm) and remove the transmission support

20. Install the retaining bolts to the transfer case and install the shift lever assembly.

21. Remove the transmission support jack.
22. Install the cover plate.
23. Install the floor console unit and connect or install the following:
- Switch wiring connector (s)
- Front console
- Lever knob
- Front console and tie the shift boot draw strings
- Slide the console over the parking brake handle
24. Connect the negative battery cable
25. Fill the transfer case to the proper level
26. Start the vehicle and check for leaks, repair if necessary.

REAR AXLE SHAFT, BEARING & SEAL

REMOVAL & INSTALLATION

1. Disconnect the negative battery cable.
2. Remove the rear wheels.

3. Remove the disc brake and parking brake assembly.
4. Remove the parking brake cable and wheel speed sensor cable.
5. Remove the rear axle shaft mounting bolts.
6. Remove the rear axle shaft.
7. Remove the bearing collar and bearing from the axle.
8. Using a slide hammer, remove the oil seal.

To install:

9. Apply grease to the oil seal lip and using the appropriate seal driver, install the new axle seal into the differential.
10. Install the new wheel bearing and retainer collar to the rear axle shaft.
11. Install the rear axle shaft. Torque the axle shaft mounting bolts to 32–44 ft. lbs. (43–60 Nm).
12. Install the wheel speed sensor and parking brake cables.
13. Install the disc brake and parking brake assembly and the rear wheels.

14. Adjust the parking brake lever.
15. Connect the negative battery cable.

REAR PINION SEAL

REMOVAL & INSTALLATION

1. Drain the gear oil.
2. Remove the driveshaft.
3. Remove the drive pinion.
4. Remove the pinion seal

To install:

5. Install a new pinion seal lightly coated with clean gear oil.
6. Install the drive pinion.
7. Rotate the pinion flange occasionally while tightening the flange nut and make certain that the pinion bearings are seated properly.
8. Install the driveshaft after aligning the matchmarks.
9. Fill the gear oil to the proper level.
10. Start the vehicle and check for leaks, repair if necessary.

ENGINE COOLING

THERMOSTAT

REMOVAL & INSTALLATION

1. Drain the engine coolant.
2. Remove the inlet fitting and gasket.
3. Remove the thermostat.

To install:

4. Install the thermostat with a new gasket, and with the jiggle valve on top.
5. Install the inlet fitting and tighten the bolts to 14 ft. lbs. (20 Nm).
6. Fill the cooling system.
7. Start the engine and check for leaks.

WATER PUMP

REMOVAL & INSTALLATION

3.5L Engine

1. Drain the cooling system.
2. Disconnect the negative battery cable.
3. Remove the drive belt and the water pump pulley.
4. Remove the timing belt cover, timing belt, auto tensioner and idler pulley.
5. Remove the water outlet fitting, the thermostat case and the water pump fitting.
6. Remove the water pump mounting bolts.
7. Remove the water pump.

To install:

8. Clean the gasket surfaces of the water pump body and the cylinder block.
9. Install the new water pump gasket and pump assembly. Tighten the bolts to 12–16 ft. lbs. (17–22 Nm).
10. Install the water pump fitting, the thermostat case and the water outlet fitting.
11. Install the auto tensioner and timing belt. Adjust the timing belt tension, and install the timing belt cover.
12. Install the drive belt, water pump pulley and adjust the auto tensioner.
13. To adjust the drive belt tension, hang the belt on the pulley of the tensioner and install the tensioner.

➡ **If the tensioner is already installed, loosen its mounting bolts to allow belt installation.**

14. Install the drive belt.
15. Torque the tensioner assembly bolt to 33–36 ft. lbs. (45–50 Nm).
16. Refill the coolant.
17. Run the engine and check for leaks.

3.3L & 3.8L Engines

See Figure 17.

1. Drain the engine coolant.
2. Remove the serpentine drive belt(A).

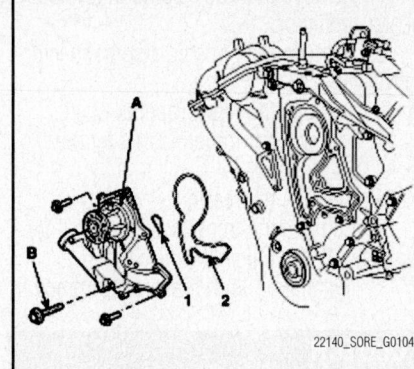

22140_SORE_G0104

Fig. 17 Install the water pump (A) and new gaskets (1, 2)

3. Remove the water pump pulley(A).
4. Remove the cooling fan shroud.
5. Remove the water pump and gasket.

To install:

6. Install the water pump (A) and new gaskets (1, 2). Torque: Large bolt:16–17.36 ft. lbs. (21.56–23.52 Nm); Small bolt: 7.23–8.68 ft. lbs. (9.80–11.76 Nm).
7. Install the water pump pulley. Torque: 5.78–7.23 ft. lbs. (7.84–9.80 Nm).
8. Install the serpentine drive belt.
9. Fill with engine coolant.
10. Start engine and check for leaks.
11. Recheck engine coolant level.

ENGINE ELECTRICAL

ALTERNATOR

REMOVAL & INSTALLATION

1. Disconnect the negative battery cable.
2. Remove the accessory drive belt.

3. Disconnect the wiring to the alternator.
4. Remove the alternator mounting bolts.
5. Remove the alternator.

To install:

6. Install the alternator and insert a support bolt (Do not insert a nut this time).

CHARGING SYSTEM

7. If necessary, insert spacer(s), then insert and tighten nut securely.
8. Connect the alternator electrical connectors.
9. Install the accessory drive belt.
10. Connect the negative battery cable.

ENGINE ELECTRICAL

FIRING ORDER

3.3L, 3.5L and 3.8L engines firing order: 1–2–3–4–5–6

IGNITION COIL

REMOVAL & INSTALLATION

1. Remove the engine cover.
2. Remove the ignition coil connector (s).

➡**When removing the ignition coil connector, pull the lock pin and push the clip.**

3. Remove the ignition coil(s).
4. Installation is the reverse of removal.

IGNITION TIMING

ADJUSTMENT

These vehicles are equipped with a Distributorless Ignition System (DIS). No adjustment is necessary or possible.

SPARK PLUGS

REMOVAL & INSTALLATION

1. Remove the engine cover.

IGNITION SYSTEM

2. Remove the ignition coil connector (s).

➡**When removing the ignition coil connector, pull the lock pin and push the clip.**

3. Remove the ignition coil(s).
4. Using a spark plug socket, remove the spark plug(s).
5. Installation is the reverse of removal.

ENGINE ELECTRICAL

STARTER

REMOVAL & INSTALLATION

3.5L Engine

1. Disconnect the negative battery cable.
2. Disconnect the starter motor electrical connectors.
3. Remove the starter heat shield.
4. Remove the starter motor.

To install:
5. Install the starter motor. Tighten the bolts to 7–9 ft. lbs. (10–12 Nm).
6. Install the starter heat shield.
7. Connect the starter motor electrical connectors. Tighten the battery terminal nut to 3–4 ft. lbs. (4–6 Nm).
8. Connect the negative battery cable.

3.3L & 3.8L Engines

1. Remove the LH exhaust manifold assembly.

STARTING SYSTEM

2. Supporting the engine with a jack, remove the LH side engine mounting bracket.
3. Disconnect the starter cable from the B terminal on the solenoid, and the connector from the S terminal.
4. Remove the remove the starter.
5. Installation is the reverse of removal.

ENGINE MECHANICAL

➡️Disconnecting the negative battery cable may interfere with the functions of the on board computer systems and may require the computer to undergo a relearning process, once the negative battery cable is reconnected.

ACCESSORY DRIVE BELTS

ACCESSORY BELT ROUTING

See Figures 18 and 19.

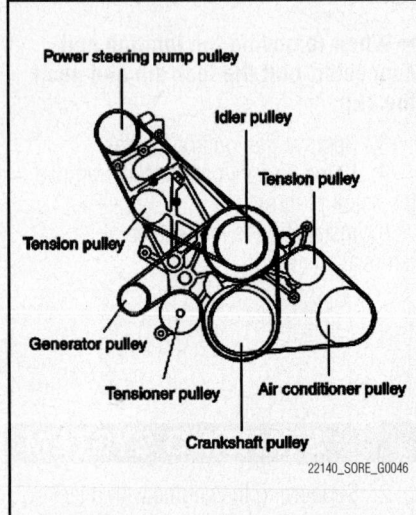

Fig. 18 Accessory belt routing—3.5L engine

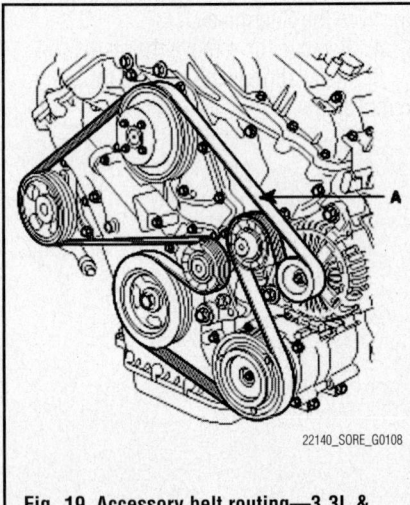

Fig. 19 Accessory belt routing—3.3L & 3.8L engines

INSPECTION

Inspect the drive belt for signs of glazing or cracking. A glazed belt will be perfectly smooth from slippage, while a good belt will have a slight texture of fabric visible.

Cracks will usually start at the inner edge of the belt and run outward. All worn or damaged drive belts should be replaced immediately.

ADJUSTMENT

The belt tension is maintained by an automatic tensioner. No adjustment is necessary or possible.

REMOVAL & INSTALLATION

1. Release tension of the belt tensioner.
2. Remove the serpentine belt.

To install:

3. Hang the belt on the pulley of the tensioner and install the tensioner. (If the tensioner is already installed, loosen its mounting bolts to allow belt installation.) Torque: Tensioner assembly bolt: 33–36 ft. lbs. (45–50 Nm).
4. Install drive belt.
5. Adjust the drive belt tension to specification by turning the adjusting bolt clockwise or counter clockwise. Standard Value: Air - conditioner compressor: 0.28–0.39 inches (7–10 mm); Alternator: 0.39–0.51 inches (10–13 mm); Power steering: 0.31–0.43 inches (8–11 mm).
6. When installing the belt on the pulley, make sure it is centered on the pulley.

CAMSHAFT AND VALVE LIFTERS

REMOVAL & INSTALLATION

3.5L Engine

See Figure 20.

1. Remove the intake manifold.
2. Disconnect the breather hose and the engine harness.
3. Remove the power steering pulley, air conditioner pulley, crankshaft pulley, idler pulley and tensioner pulley.
4. Remove the timing belt cover.
5. Loosen the auto tensioner.
6. Remove the timing belt from the camshaft sprockets.
7. Remove the spark plug cables.
8. Remove the cylinder head covers.
9. Remove the camshaft sprockets.
10. Remove the camshaft bearing caps.
11. Remove the camshafts.
12. Remove the rocker arms and lash adjusters.

➡️Keep all valvetrain components in order for installation.

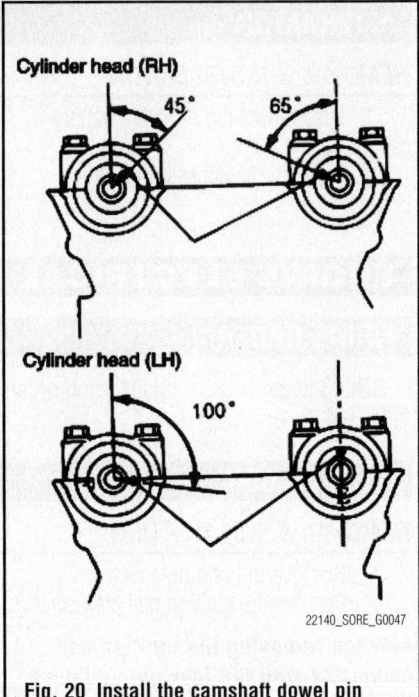

Fig. 20 Install the camshaft dowel pin

To install:

13. Rotate the crankshaft placing the No. 1 cylinder in TDC (Compression stroke) position.
14. Check the position of the rocker arm ensuring it is exactly installed on the lash adjuster and valve.
15. Install the camshaft dowel pin as shown.
16. The left and right banks of the camshafts are different and care should be taken not to confuse them.
Identification marks
- Left bank Intake: I
- Left bank Exhaust: E
- Right bank Intake: J
- Right bank Exhaust: H
17. Confirm the identification mark and the number.

➡️Bearing caps of No.3, No.4, and No.5 have the front mark and arrange the front mark upon the cylinder head while installing the bearing caps.

18. Identification marks:
- Intake: I
- Exhaust: E
19. Tighten the bearing cap bolts using 2 or 3 steps. Torque: Outer (*) 16 each: 13.7–15.2 ft. lbs. (18.6–20.6 Nm); Inner 24 each: 7.2–8.7 ft. lbs. (9.8–11.8 Nm).
20. Install the camshaft sprockets.
21. Install the cylinder head covers.
22. Install the spark plug cables.

23. Install the timing belt on the camshaft sprockets.

24. Adjust the auto tensioner.

25. Install the timing belt cover.

26. Install the power steering pulley, air conditioner pulley, crankshaft pulley, idler pulley and tensioner pulley.

27. Connect the breather hose and the engine harness.

28. Install the intake manifold.

3.3L & 3.8L Engines

See Figures 21 through 36.

1. Remove the engine cover.

2. Remove the engine room resonator (A).

3. Disconnect the MAF sensor connector (A) and the breather hose (B).

4. Remove the air cleaner assembly (C).

5. Disconnect the other breather hose (A), the Purge Control Solenoid Valve (PCSV) hose, the Positive Crankcase Ventilation (PCV) hose (C) and the Electronic Throttle Control (ETC) cooling hoses (D).

6. Remove the wiring over the surge tank.

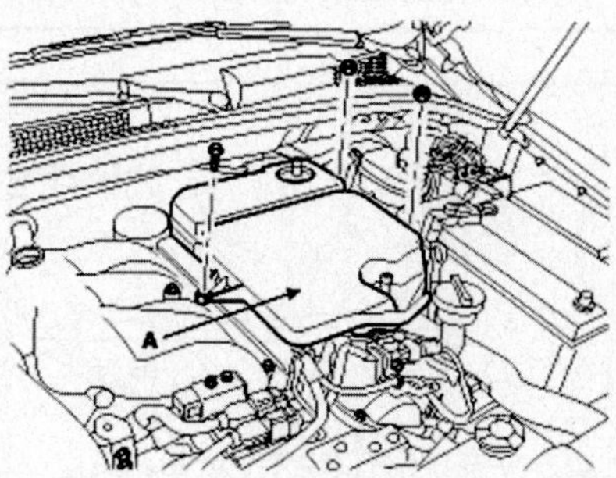

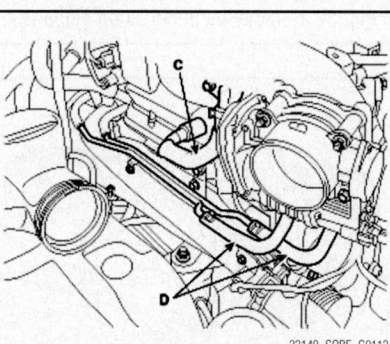

Fig. 21 Remove the engine room resonator (A)

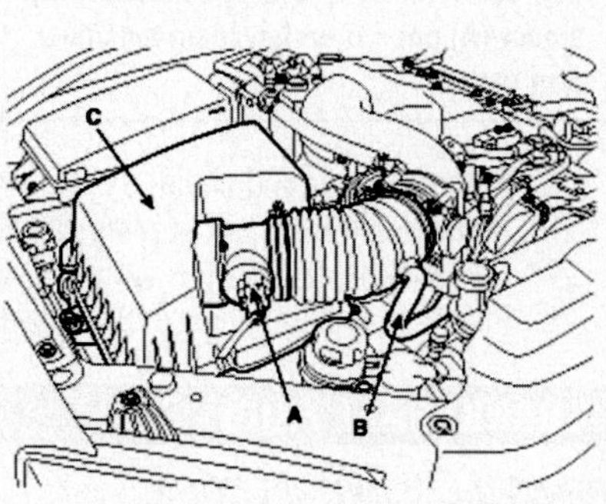

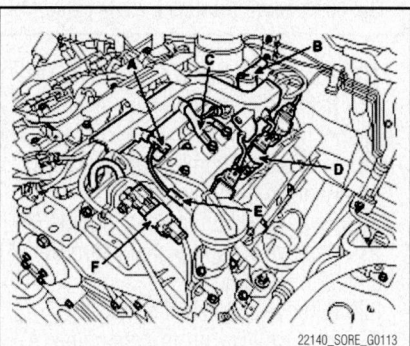

Fig. 22 Disconnect the MAF sensor connector (A) and the breather hose (B). Remove the air cleaner assembly(C).

a. Disconnect the injection harness connector (A).

b. Disconnect the camshaft position sensor (CMP) harness connector (B).

c. Disconnect the ground lines (C).

d. Disconnect the ignition coil harness connector (D).

e. Disconnect the condenser connector (E).

f. Disconnect the oil control valve (OCV) harness connector (F).

7. Remove the surge tank assembly.

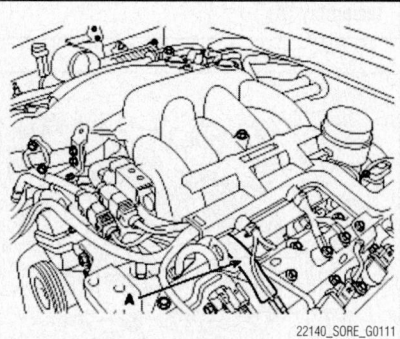

Fig. 23 Disconnect the other breather hose (A)

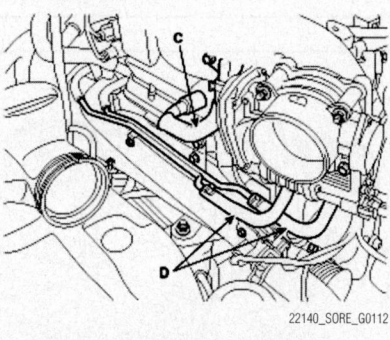

Fig. 24 Disconnect the Positive Crankcase Ventilation (PCV) hose (C) and the Electronic Throttle Control (ETC) cooling hoses (D)

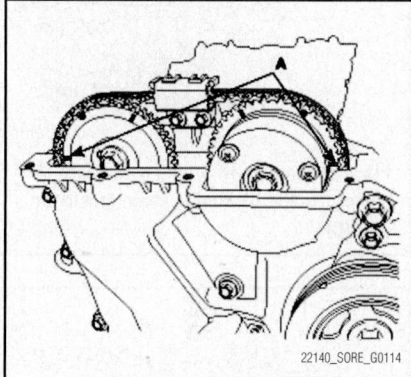

Fig. 25 Remove the wiring over the surge tank.

8. Remove the cylinder head covers and gaskets.

9. Set No.1 cylinder to TDC/compression.

a. Turn the crankshaft pulley and align its groove with the timing mark "T" of the lower timing chain cover.

➡**Do not rotate engine counterclockwise**

b. Check that the mark (A) of the camshaft timing sprockets are in straight line on the cylinder head surface as shown in the illustration. If not, turn the crankshaft one revolution (360°).

10. Remove the lower oil pan.

✳✳ CAUTION

Insert the SST between the oil pan and the ladder frame by tapping it with a plastic hammer in the direction of arrow.

Fig. 26 Check that the mark (A) of the camshaft timing sprockets are in straight line

Fig. 27 Install a set pin after compressing the timing chain tensioner.

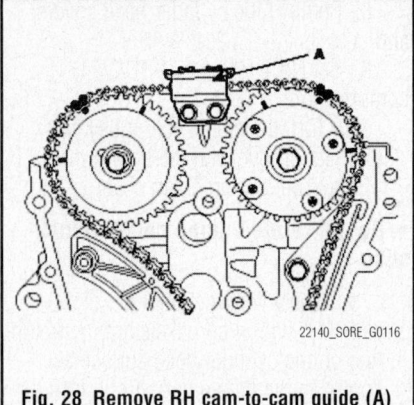

Fig. 28 Remove RH cam-to-cam guide (A)

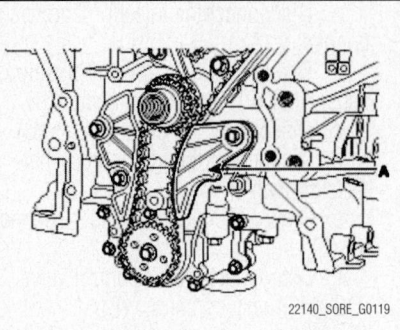

Fig. 31 Remove oil pump chain tensioner assembly (A)

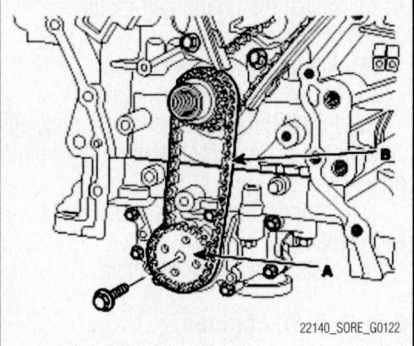

Fig. 34 Remove oil pump chain sprocket (A) and oil pump chain (B)

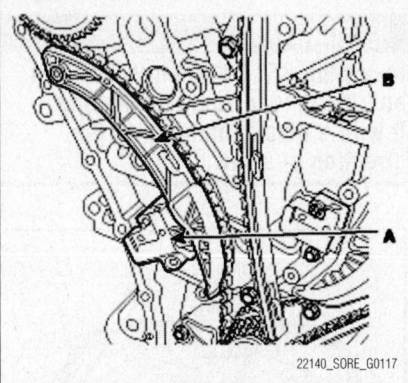

Fig. 29 Remove RH timing chain auto tensioner (A) and RH timing chain tensioner arm (B)

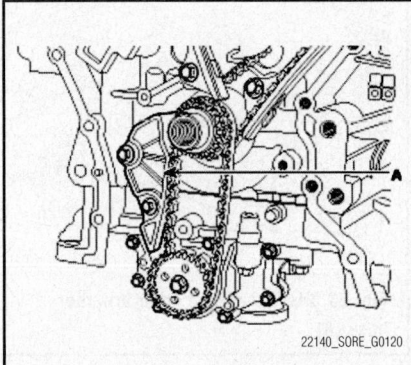

Fig. 32 Remove oil pump chain guide (A)

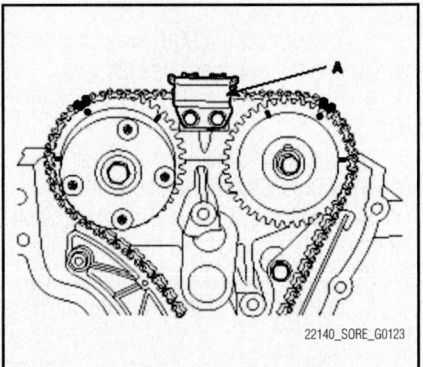

Fig. 35 Remove LH cam-to-cam guide (A)

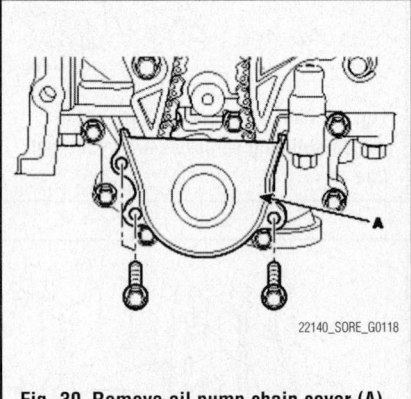

Fig. 30 Remove oil pump chain cover (A)

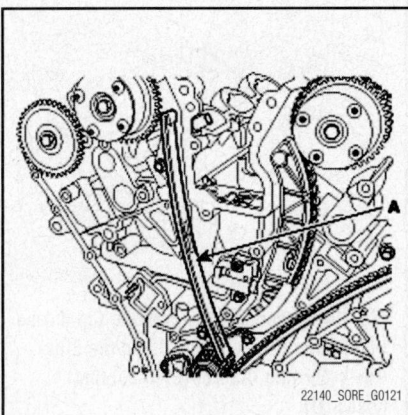

Fig. 33 Remove RH timing chain guide (A)

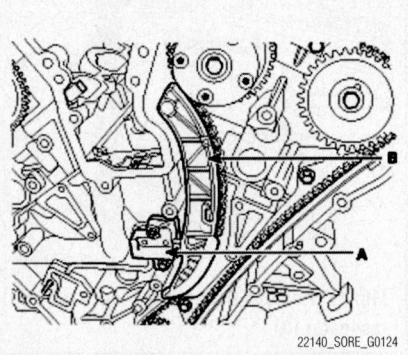

Fig. 36 Remove LH timing chain auto tensioner (A) and LH timing chain tensioner arm (B)

11. Remove the crankshaft damper pulley.
12. Remove the timing chain cover.

➡ **Be careful not to damage the contact surfaces of cylinder block, cylinder head and timing chain cover.**

13. Install a set pin after compressing the timing chain tensioner.
14. Remove RH cam-to-cam guide (A).
15. Remove RH timing chain auto tensioner (A) and RH timing chain tensioner arm (B).

16. Remove oil pump chain cover (A).
17. Remove oil pump chain tensioner assembly (A).
18. Remove oil pump chain guide (A).
19. Remove RH timing chain.
20. Remove RH timing chain guide (A).
21. Remove oil pump chain sprocket (A) and oil pump chain (B).
22. Remove crankshaft sprocket.
23. Remove LH cam-to-cam guide (A).
24. Remove LH timing chain auto tensioner (A) and LH timing chain tensioner arm (B).

25. Remove LH timing chain.
26. Remove the camshaft bearing caps.
27. Remove the camshaft assembly.
28. Installation is the reverse of removal.

CRANKSHAFT FRONT SEAL

REMOVAL & INSTALLATION

3.5L Engine

1. Disconnect the negative battery cable.
2. Remove the accessory drive belts.

3. Remove the idler pulley.

4. Remove the crankshaft pulley.

5. Remove the power steering pump pulley.

6. Remove the belt tensioner pulley.

7. Remove the upper and lower timing belt covers.

8. Remove the timing belt.

9. Remove the timing belt crankshaft sprocket.

10. Remove the Crankshaft Position (CKP) sensor tone ring.

11. Remove the front crankshaft seal.

To install:

12. Install the front crankshaft seal. Use Seal Driver 09214-33000 or similar.

13. Install the CKP sensor tone ring.

14. Install the timing belt crankshaft sprocket.

15. Install the timing belt.

16. Install the upper and lower timing belt covers.

17. Install the belt tensioner pulley.

18. Install the power steering pump pulley.

19. Install the crankshaft pulley.

20. Install the idler pulley.

21. Install the accessory drive belts.

22. Connect the negative battery cable.

CYLINDER HEAD

REMOVAL & INSTALLATION

3.5L Engine

See Figure 37.

1. Drain the cooling system.

2. Relieve the fuel system pressure.

3. Disconnect the negative battery cable.

4. Disconnect the upper radiator hose.

5. Remove the breather hose and air-intake hose.

6. Remove the vacuum hose, fuel hose and coolant hose.

7. Remove the intake manifold.

8. Remove the cables from the spark plugs. The cables should be removed holding the boot portion.

9. Remove the ignition coil.

10. Remove the upper and lower timing belt cover.

11. Remove the timing belt and camshaft sprockets.

12. Remove the heat protector and exhaust manifold assembly.

13. Remove the coolant pump pulley and head cover.

14. Remove the intake and exhaust camshaft.

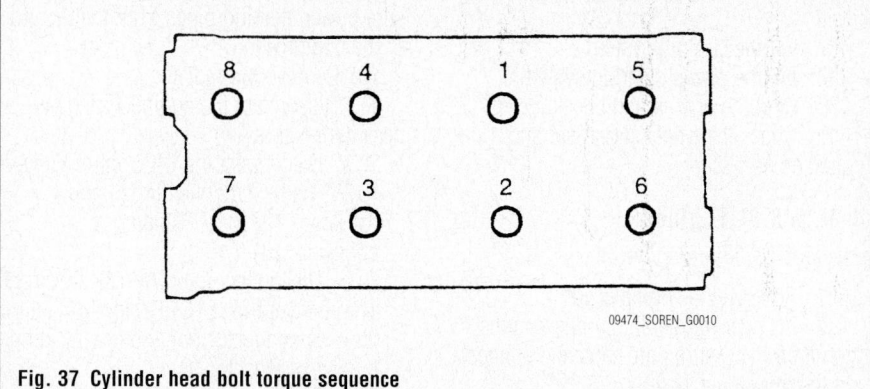

09474_SOREN_G0010

Fig. 37 Cylinder head bolt torque sequence

➡**Keep all valve train components in order for assembly.**

15. Remove the cylinder head assembly. The cylinder head bolts should be removed using the 12 mm socket, in two or three steps. Clean the gasket pieces from the cylinder block top surface and cylinder bottom surface.

To install:

16. Verify the identification marks on the cylinder head gasket. Install the gasket so that the surface of the cylinder head identification mark faces toward the cylinder head.

17. Install the cylinder heads with new gaskets. Do not apply sealant to these surfaces.

18. Tighten the bolts in sequence to 75–82 ft. lbs. (105–115 Nm).

19. Installation is the reverse of the removal procedure.

ENGINE ASSEMBLY

REMOVAL & INSTALLATION

3.5L Engine

1. Remove the battery and air cleaner assembly.

2. Remove the hood from the vehicle.

3. Drain the cooling system. Properly discharge the air conditioning system, as required.

4. Drain the engine oil.

5. Drain the transmission fluid.

6. Relieve fuel system pressure.

7. Remove the upper and lower radiator hoses.

8. Remove the heater hoses.

9. Remove the radiator.

10. Remove the breather hose and air-intake hose.

11. Disconnect the throttle and cruise control cables.

12. Disconnect the oil pressure switch connector.

13. Disconnect the oil pressure sensor connector.

14. Remove the engine ground cable.

15. Disconnect the EVAP canister hose.

16. Disconnect the brake booster vacuum hose.

17. Disconnect the fuel lines.

18. Remove the accessory drive belts.

19. Disconnect the power steering hoses from the pump.

20. Disconnect the alternator wiring harness connectors.

21. If equipped with an automatic transmission, disconnect the oil cooler lines.

22. Disconnect the starter motor wiring harness connectors.

23. Remove the starter motor.

24. Remove the exhaust front pipes from the exhaust manifolds.

25. Disconnect the transmission shift cable or clutch cable, if equipped.

26. Remove the halfshafts.

27. Support the transmission with a jack.

28. Make sure all cable, harness connectors and hoses are disconnected from the engine and transmission.

29. Remove the transmission from the crossmember.

30. Remove the crossmember.

31. Remove the driveshaft.

32. Remove the engine mounting bolts and remove the engine assembly.

To install:

33. Installation is the reverse of removal but please note the following steps:

- Tighten the transaxle mounting bracket bolts to 15–21 ft. lbs. (20–28 Nm)
- Tighten the engine mount bracket bolts to 51–65 ft. lbs. (69–88 Nm)

34. Make sure all cable, harness connectors and hoses are properly connected to the engine and transmission.

35. Fill the engine crankcase to the correct level.

36. Fill the transmission to the correct level.
37. Fill the cooling system.
38. Fill the power steering system.
39. Start the engine and check for leaks.
40. Check the wheel alignment and adjust as necessary.

3.3L & 3.8L Engines

See Figures 38 through 46.

1. Remove the engine cover.
2. Recover refrigerant by opening the high & low pressure pipe caps and connecting the refrigerant station.
3. Remove the undercover.
4. Drain engine oil and engine coolant.
5. Remove the battery.
6. Remove the intake air hose and air cleaner assembly.
   a. Disconnect the MAF connector (A).
   b. Disconnect the breather hose (B) from air cleaner hose.
   c. Remove the intake air hose and air cleaner assembly (C) with the resonator (D).
7. Disconnect the PCM connectors (A).
8. Remove the battery tray.

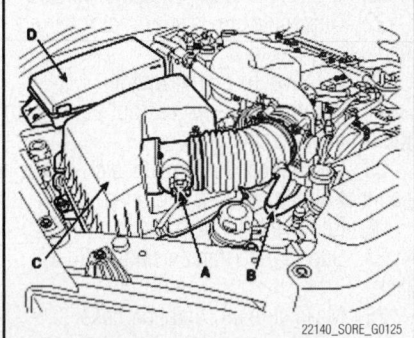

Fig. 38 Disconnect the MAF connector (A), breather hose (B), intake air hose and air cleaner assembly (C) with the resonator (D)

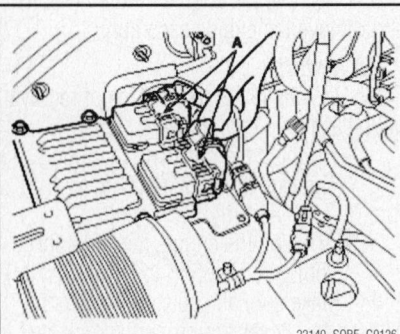

Fig. 39 Disconnect the PCM connectors (A)

9. Disconnect the high and low pressure power steering pipes from the radiator or the compressor.
10. Remove the radiator.
11. Disconnect the engine wiring harness connectors.
   a. Disconnect the Oil Control Valve (OCV) harness connector (A) and the knock sensor (LH) harness connector (B).
   b. Disconnect the MAP (A), Electronic Throttle Control (ETC)(B), ignition coil harness connector (C) and the injection harness connector (D).
   c. Disconnect the battery connector (A), the power steering switch connector (B) and the knock sensor (RH) harness connector (C).
   d. Disconnect the oxygen sensors (A), Camshaft Position Sensor (CMP)(B), Crankshaft Position Sensor (CKP)(C), VIV(D) and the condenser harness connector (E).
   e. Disconnect the harness connectors

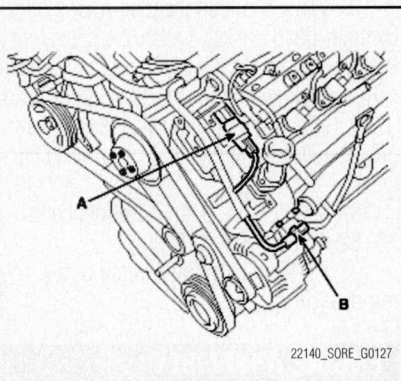

Fig. 40 Disconnect the Oil Control Valve (OCV) harness connector (A) and the knock sensor (LH) harness connector (B)

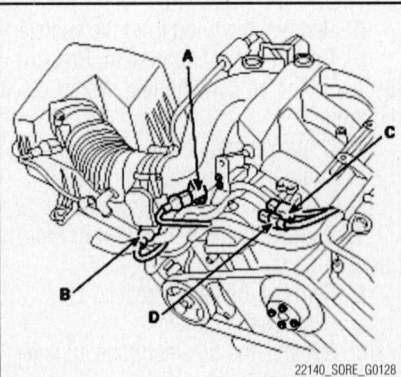

Fig. 41 Disconnect the MAP (A), Electronic Throttle Control (ETC)(B), ignition coil harness connector (C) and the injection harness connector (D)

for the Water Temperature Sensor (WTS)(A), the Oil Temperature Sensor (OTS)(B) and the Purge Control Solenoid Valve (PCSV)(C).

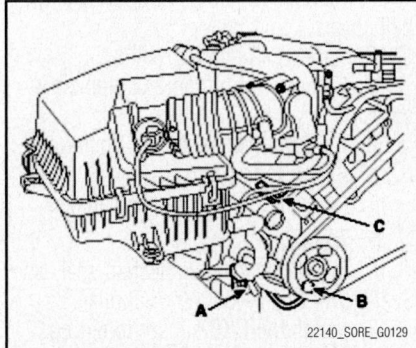

Fig. 42 Disconnect the battery connector (A), the power steering switch connector (B) and the knock sensor (RH) harness connector (C)

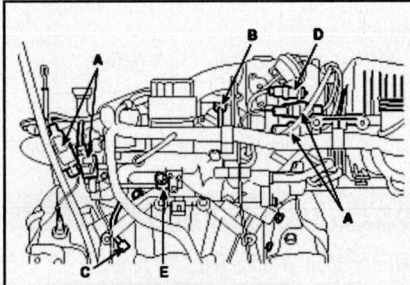

A. Oxygen sensors
B. Camshaft Position Sensor (CMP)
C. Crankshaft Position Sensor (CKP)
D. Variable Induction Sensor (VIS)
E. Condenser harness connector

Fig. 43 Disconnect the oxygen sensors (A), Camshaft Position Sensor (CMP)(B), Crankshaft Position Sensor (CKP)(C), VIV(D) and the condenser harness connector (E)

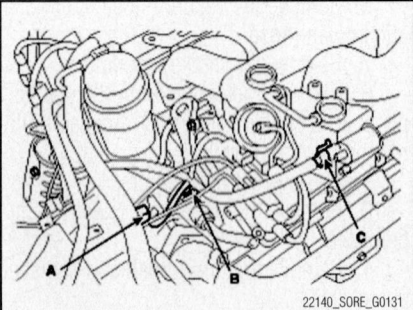

Fig. 44 Disconnect the harness connectors for the Water Temperature Sensor (WTS)(A), the Oil Temperature Sensor (OTS)(B) and the Purge Control Solenoid Valve (PCSV)(C)

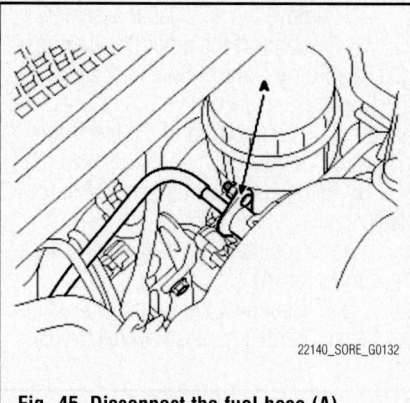

Fig. 45 Disconnect the fuel hose (A).

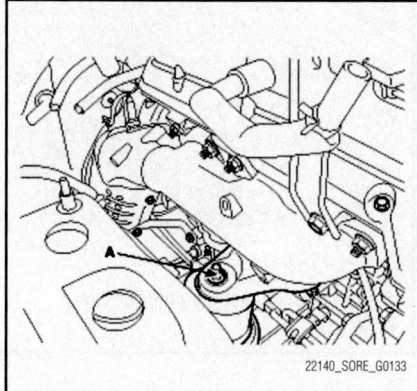

Fig. 46 Remove the engine mounting brackets (A)

12. Disconnect the transmission from the engine assembly.

13. Disconnect the fuel hose (A).

14. Disconnect the front exhaust muffler with the exhaust manifolds.

15. Remove the front wheels.

16. Remove heater hose and disconnect the brake vacuum hose.

17. Remove the exhaust and intake manifold covers.

18. Remove the power steering pump assembly.

19. Remove the hood assembly.

20. Install a jack for supporting the engine assembly.

21. Remove the engine mounting brackets (A).

22. Lift the engine from the vehicle.

To install:

23. Installation is in the reverse order of removal.

24. Perform the following:

a. Adjust the shift cable.

b. Refill the engine with engine oil.

c. Refill the transaxle with fluid.

d. Refill the radiator with engine coolant.

e. Bleed air from the cooling system with the heater valve open.

f. Clean the battery posts and cable terminals with sandpaper assemble them, and apply grease to prevent corrosion.

g. Inspect for fuel leakage.

25. After assembling the fuel line, turn on the ignition switch (do not operate the starter) so that the fuel pump runs for approximately two seconds and fuel line pressurizes.

26. Repeat this operation two or three times, then check for fuel leakage at any point in the fuel line.

EXHAUST MANIFOLD

REMOVAL & INSTALLATION

3.5L Engine

1. Disconnect the negative battery cable.

2. Disconnect the Heated Oxygen (HO$_2$S) sensor connectors.

3. Separate the exhaust Y pipe from the exhaust manifold(s).

4. Remove the heat protector.

5. Remove the exhaust manifold.

6. Remove the exhaust manifold gasket.

To install:

➡**Use only new gaskets and nuts for assembly.**

7. Install the exhaust manifolds with new gaskets. Tighten the fasteners to 20–24 ft. lbs. (27–32 Nm).

8. Install the exhaust manifold heat protectors. Tighten the bolts to 9–11 ft. lbs. (12–15 Nm).

9. Install the exhaust Y pipe.

10. Connect the Heated Oxygen (HO$_2$S) sensor connectors.

11. Connect the negative battery cable.

3.3L & 3.8L Engines

RH Side or Bank 1

See Figures 47 through 49.

1. Remove the engine cover (A).

2. Disconnect the Mass Air Flow (MAF) sensor connector (A) and the breather hose (B).

3. Remove the air cleaner assembly.

4. Remove the RH cooling pipe (A).

5. Remove the RH exhaust manifold heat protector.

❋❋ CAUTION

Handle the heat protector with caution not to be deformed.

6. After removing the undercover,

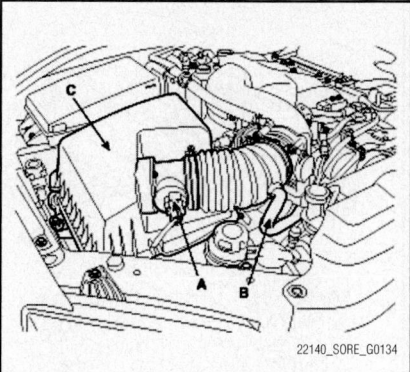

Fig. 47 Disconnect the Mass Air Flow (MAF) sensor connector (A) and the breather hose (B)

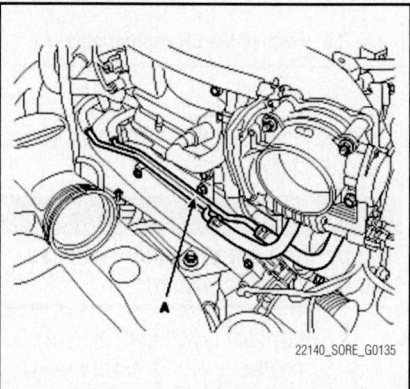

Fig. 48 Remove the RH cooling pipe (A)

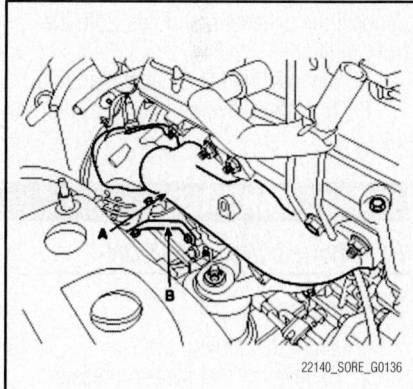

Fig. 49 Remove the RH exhaust manifold (A) and the stay (B)

disconnect the exhaust manifolds from the front muffler.

7. Remove the RH exhaust manifold (A) and the stay (B).

8. To install, reverse the removal procedure.

LH Side or Bank 2

See Figure 50.

1. Remove the engine oil level gauge.

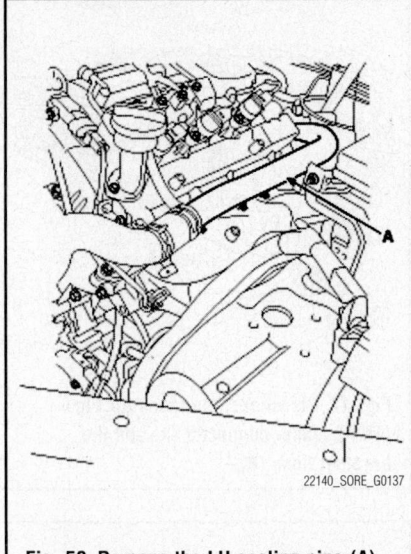

Fig. 50 Remove the LH cooling pipe (A)

2. Remove the battery.
3. Remove the LH exhaust manifold heat protector.

✳✳ CAUTION

Handle the heat protector with caution not to be deformed.

4. Remove the LH cooling pipe (A).
5. Disconnect the oil pressure switch harness connector and the battery ground line.
6. After removing the undercover, disconnect the exhaust manifolds from the front muffler.
7. Remove the LH exhaust manifold.
8. To install, reverse the removal procedure.

INTAKE MANIFOLD

REMOVAL & INSTALLATION

3.5L Engine

1. Drain the cooling system.
2. Relieve the fuel system pressure.
3. Disconnect the negative battery cable.
4. Remove the air intake hose connected to the throttle body.
5. Remove the accelerator and cruise control cables.
6. Remove the engine coolant hose and throttle body.
7. Remove the PCV hose and brake booster vacuum hoses.
8. Disconnect the vacuum hose connections.
9. Remove the surge tank stay.
10. Bleed off the pressure in the fuel pipe line to prevent the fuel from spilling.

11. Disconnect the connector from high pressure hose.
12. Remove the surge tank and gasket.
13. Disconnect the fuel injector harness connector.
14. Remove the delivery pipe with the fuel injector and the pressure regulator.
15. Disconnect the wiring harness of the coolant sensor assembly.
16. Remove the intake manifold.

To install:

17. Installation is the reverse of the removal procedure, while using the following torque values:
- Lower intake manifold nuts, in sequence to: 15–17 ft. lbs. (20–23 Nm)
- Surge tank stay bolts: 11–14 ft. lbs. (15–20 Nm)

3.3L & 3.8L Engine

See Figures 51 through 58.

1. Remove the engine cover.
2. Remove the engine room resonator (A).
3. Disconnect the MAF sensor connector (A) and the breather hose (B).

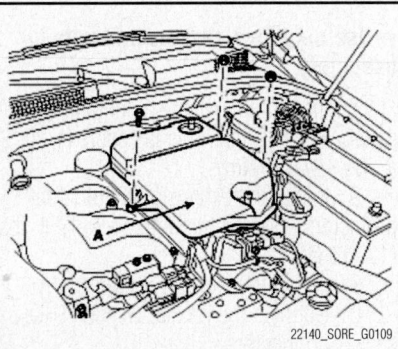

Fig. 51 Remove the engine room resonator (A)

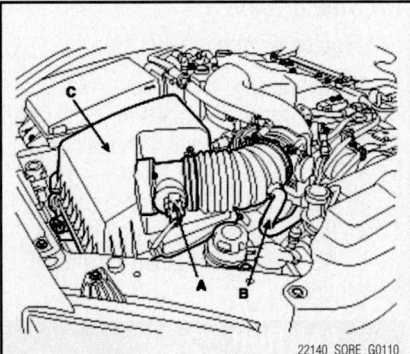

Fig. 52 Disconnect the MAF sensor connector (A) and the breather hose (B). Remove the air cleaner assembly (C).

4. Remove the air cleaner assembly (C).
5. Disconnect the other breather hose (A), the Purge Control Solenoid Valve (PCSV) hose, the Positive Crankcase Ventilation (PCV) hose (C) and the Electronic Throttle Control (ETC) cooling hoses (D).
6. Remove the wiring over the surge tank.
   a. Disconnect the injection harness connector (A).
   b. Disconnect the Camshaft Position Sensor (CMP) harness connector (B).

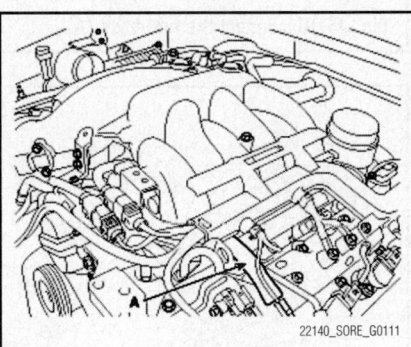

Fig. 53 Disconnect the other breather hose (A)

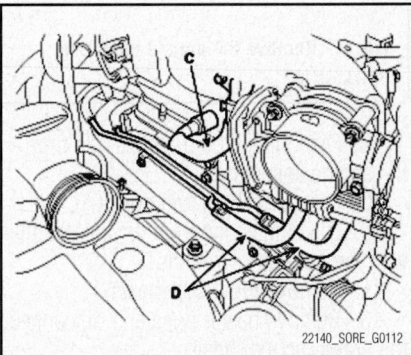

Fig. 54 Disconnect the Positive Crankcase Ventilation (PCV) hose (C) and the Electronic Throttle Control (ETC) cooling hoses (D)

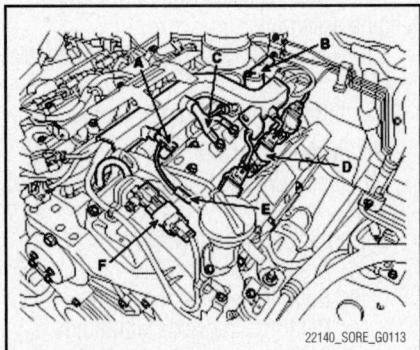

Fig. 55 Remove the wiring over the surge tank.

Fig. 56 Disconnect the Variable Induction System (VIS) solenoid valve connector (G), the injector wiring (H) and ignition coil wiring (I)

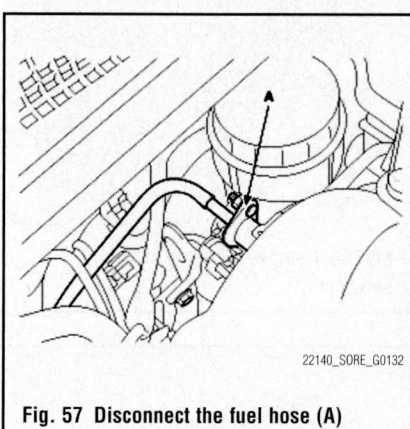

Fig. 57 Disconnect the fuel hose (A)

c. Disconnect the ground lines (C).

d. Disconnect the ignition coil harness connector (D).

e. Disconnect the condenser connector (E).

f. Disconnect the Oil Control Valve (OCV) harness connector (F).

g. Disconnect the Variable Induction System (VIS) solenoid valve connector (G).

h. Disconnect the injector wiring (H) and ignition coil wiring (I).

7. Disconnect the fuel hose tube (A).

8. Remove heater hose.

9. Disconnect the brake vacuum hose.

10. Disconnect the surge tank stay.

11. Remove the surge tank assembly.

12. Disconnect the injector connectors.

13. Disconnect the water hose on intake manifold from the nipple on the chain cover.

14. Remove the delivery pipe and intake manifold as an assembly.

➡**Except such cases as defects of injectors or pipe, do not disassemble a delivery pipe from an intake manifold because it is one of the fuel system**

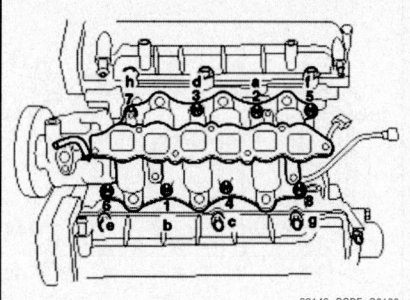

Fig. 58 Intake manifold tightening sequence

parts, or you may have some problems in fuel system.

To install:

15. Install intake manifold and new gasket on the cylinder head. Torque: 1st Step: 2.9–4.3 ft. lbs. (3.9–5.9 Nm); 2nd Step: Bolts: 19.5–23.1 ft. lbs. (26.5–31.4 Nm); Nut: 13.7–17.4 ft. lbs. (18.6–23.5 Nm); 3rd Step: Repeat the 2nd Step twice or more.

a. Use the following tightening sequence:

b. 1st Step order: a—h

c. 2nd Step order: 1–8

16. Connect the water hose on intake manifold to the nipple on the chain cover.

17. Install delivery pipe.

18. Install the surge tank and new gasket on the intake manifold. Torque: Long bolt: 7.23–8.68 ft. lbs. (9.80–11.76 Nm); Short bolt, nut: 13.7–17.4 ft. lbs. (18.6–23.5 Nm).

19. Connect heater hose and the brake vacuum hose.

20. Connect the fuel hose tube.

21. Connect the wiring over the surge tank.

a. Connect the injection harness connector.

b. Connect the Camshaft Position Sensor(CMP) harness connector.

c. Connect the ground lines.

d. Connect the ignition coil harness connector.

e. Connect the condenser connector.

f. Connect the Variable Induction System(VIS) solenoid valve connector.

g. Connect the Oil Control Valve(OCV) harness connector.

22. Connect the other breather hose, the Positive Crankcase Ventilation (PCV) hose and the Electronic Throttle Control(ETC) cooling hoses, ETC connector.

23. Connect the MAF sensor connector and the breather hose.

24. Install the air cleaner assembly.

25. Install the engine room resonator.

26. Install the engine cover.

27. Fill with engine coolant.

OIL PAN

REMOVAL & INSTALLATION

3.5L Engine

See Figures 59 and 60.

1. Drain the engine oil.

2. Disconnect the negative battery cable.

3. Remove the starter motor.

4. Remove the oil pressure switch.

5. Remove the oil filter.

6. Remove the lower oil pan.

7. Remove the upper oil pan.

To install:

8. Apply silicone sealant to the grove of the oil pan flange.

➡**After sealant application do not exceed 15 minutes before installing the oil pan. Be sure sealant does not get inside the oil pan.**

9. Install the upper oil pan and tighten the bolts in sequence as follows:

a. Bolts 1–14: 7–9 ft. lbs. (10–12 Nm).

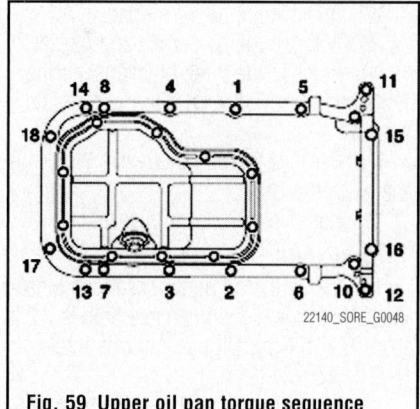

Fig. 59 Upper oil pan torque sequence

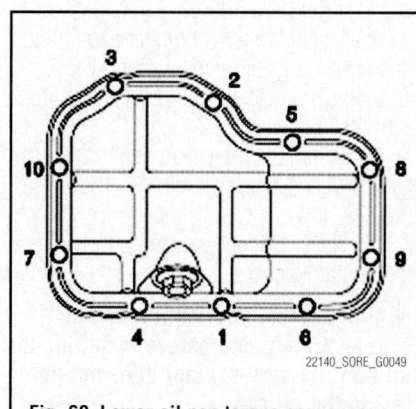

Fig. 60 Lower oil pan torque sequence

b. Bolts 15 and 16: 4–5 ft. lbs. (5–7 Nm).

c. Bolts 17 and 18: Upper oil pan-to-transaxle mounting bolts: 22–30 ft. lbs. (29–41 Nm).

10. Apply sealant to the threads, then install the oil pressure switch and tighten to 6–9 ft. lbs. (8–12 Nm).

11. Install the lower oil pan and tighten the bolts in sequence to 7–9 ft. lbs. (10–12 Nm).

12. Install the oil filter.

13. Install the starter motor.

14. Connect the negative battery cable.

15. Fill the crankcase to the correct level with engine oil.

16. Start the engine and check for leaks.

OIL PUMP

REMOVAL & INSTALLATION

3.5L Engine

See Figures 61 and 62.

1. Drain the engine oil.

2. Disconnect the negative battery cable.

3. Remove the starter motor.

4. Remove the oil pressure switch.

5. Remove the oil filter.

6. Remove the lower oil pan.

7. Remove the upper oil pan.

8. Remove the oil screen and gasket.

9. Remove the three bracket securing bolts and remove the oil filter bracket and gasket.

10. Remove the oil relief valve plug from the oil pump case.

11. Remove the oil pump case.

To install:

12. Install the oil pump case with a new gasket. Torque: Oil pump case bolt: 9–12 ft. lbs. (12–15 Nm); Oil pump cover screw: 6–9 ft. lbs. (8–12 Nm).

13. Install the oil seal into the oil pump case as tightly as possible, using the special tool (09214-33000).

14. Install the relief plunger and spring, and tighten the oil relief valve plug. Torque: Oil relief valve plug: 29–36 ft. lbs. (40–50 Nm).

15. Install the oil screen and a new gasket. Torque: Oil screen bolt: 11–15 ft. lbs. (15–22 Nm).

16. Apply silicone sealant to the grove of the oil pan flange.

➡ **After sealant application do not exceed 15 minutes before installing the oil pan. Be sure sealant does not get inside the oil pan.**

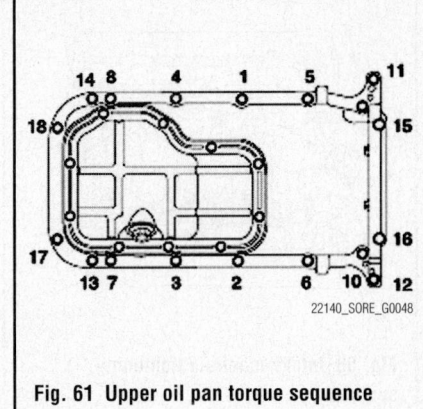

Fig. 61 Upper oil pan torque sequence

22140_SORE_G0048

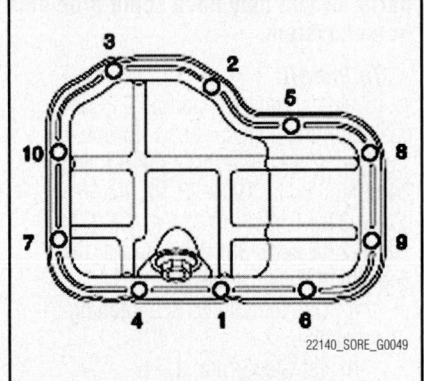

Fig. 62 Lower oil pan torque sequence

22140_SORE_G0049

17. Install the upper oil pan and tighten the bolts in sequence as follows:

a. Bolts 1–14: 7–9 ft. lbs. (10–12 Nm).

b. Bolts 15 and 16: 4–5 ft. lbs. (5–7 Nm).

c. Bolts 17 and 18: Upper oil pan-to-transaxle mounting bolts: 22–30 ft. lbs. (29–41 Nm).

18. Apply sealant to the threads, then install the oil pressure switch and tighten to 6–9 ft. lbs. (8–12 Nm).

19. Install the lower oil pan and tighten the bolts in sequence to 7–9 ft. lbs. (10–12 Nm).

20. Install the oil filter.

21. Install the starter motor.

22. Connect the negative battery cable.

23. Fill the crankcase to the correct level with engine oil.

24. Start the engine and check for leaks.

3.3L & 3.8L Engine

See Figures 63 through 65.

1. Drain engine oil.

2. Remove the front member..

3. Using SST (09215-3C000) remove lower oil pan.

4. Remove oil pump chain cover (A).

5. Remove oil pump chain sprocket (A).

6. Remove oil pump(A).

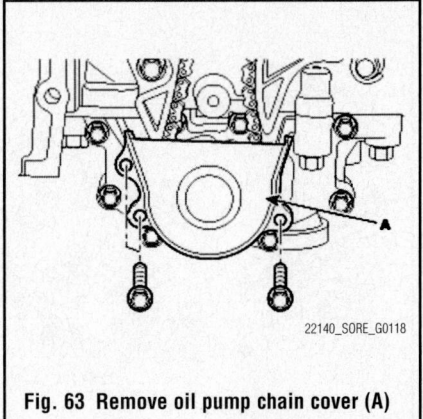

Fig. 63 Remove oil pump chain cover (A)

22140_SORE_G0118

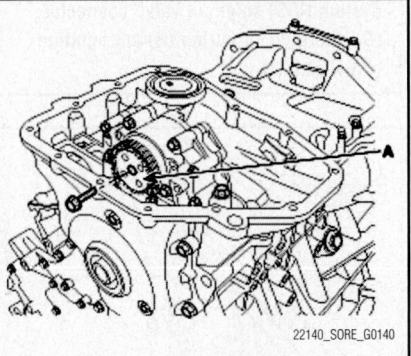

Fig. 64 Remove oil pump chain sprocket (A)

22140_SORE_G0140

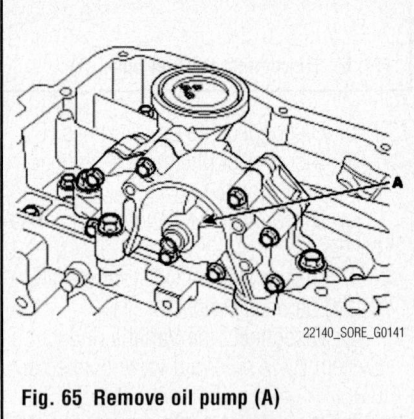

Fig. 65 Remove oil pump (A)

22140_SORE_G0141

To install:

7. Install oil pump. Torque: 14.47–17.36 ft. lbs. (19.60–23.52 Nm).

➡ **Always use a new O-ring.**

8. Install oil pump sprocket and oil pump chain on the oil pump. Torque: 13.74–15.91 ft. lbs. (18.62–21.56 Nm).

9. Install oil pump chain cover.

10. Install lower oil pan. Uniformly tighten the bolts in several passes. Torque: 7.23–8.68 ft. lbs. (9.80–11.76 Nm).

11. Install the front member.

12. After assembly, wait at least 30 minutes before filling the engine with oil.

REAR MAIN SEAL

REMOVAL & INSTALLATION

3.5L Engine

1. Disconnect the negative battery cable.
2. Remove the starter motor.
3. Remove the transmission.
4. Remove the flexplate.
5. Remove the oil seal housing.
6. Remove the oil seal.

To install:

7. Install the oil seal to the seal housing using special tool 09231-33000 or similar seal driver.
8. Apply silicone sealant to the oil seal housing flange.
9. Install the seal housing and tighten the bolts to 7–9 ft. lbs. (10–12 Nm).
10. Install the flexplate. Tighten the bolts to 53–55 ft. lbs. (73–77 Nm).
11. Install the transmission.
12. Install the starter motor.
13. Connect the negative battery cable.

3.3L & 3.8L Engine

See Figures 66 and 67.

1. Disconnect the negative battery cable.

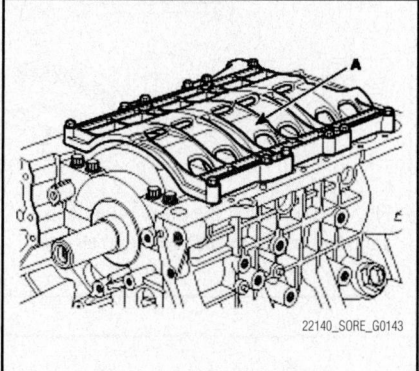

Fig. 66 Remove the baffle plate (A)

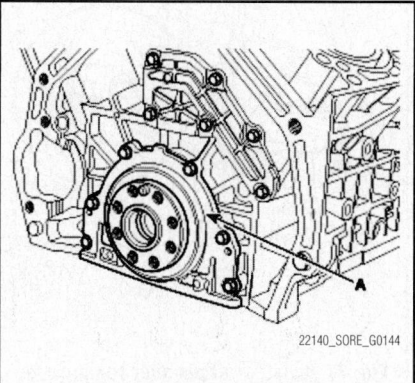

Fig. 67 Remove the rear oil seal case (A)

2. Remove the starter motor.
3. Remove the transmission.
4. Remove the flexplate.
5. Remove the lower and upper oil pans.
6. Remove the baffle plate (A).
7. Remove the rear oil seal case (A).
8. Remove and replace the rear oil seal.
9. Installation is the reverse of removal.

TIMING BELT AND SPROCKETS

REMOVAL & INSTALLATION

3.5L Engine

See Figures 68 through 70.

1. Disconnect the negative battery cable.
2. Rotate the tensioner arm clockwise and remove the belt from the pulley.
3. Remove the power steering pump pulley, idler pulley, tensioner pulley and crankshaft pulley.
4. Remove the upper and lower timing belt covers.
5. Remove the auto tensioner.

➥ **Rotate the crankshaft clockwise and align the timing mark to set the number one piston at TDC on the compression stroke. The timing marks on the camshaft sprocket and cylinder head cover should coincide with one another.**

6. Unbolt the tensioner and remove the timing belt.

➥ **If reusing the timing belt make sure to mark the rotation direction for reassembly.**

7. Remove timing belt sprockets.

To install:

8. Ensure that the engine is set to Top Dead Center (TDC).
9. Install the idler pulley to the idler bracket.

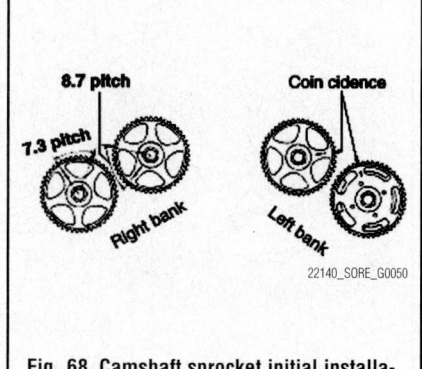

Fig. 68 Camshaft sprocket initial installation position

10. Install the tensioner arm, shaft and plane washer to the cylinder block. Torque the bolts to 25–40 ft. lbs. (35–55 Nm).
11. Install the crankshaft sprocket and align the timing mark. Do not bend the crankshaft sensing blade.
12. Install the camshaft sprocket and adjust the initial installation state according to the illustration.
13. Install the auto tensioner to oil pump case. At this time the auto tensioner set pin should be completely assembled.
14. Align the timing marks of each sprocket and install the timing belt in the following order, crankshaft sprocket, idler pulley, exhaust camshaft sprocket (LH), intake camshaft sprocket (LH), water pump pulley, intake camshaft sprocket (RH), exhaust camshaft sprocket (RH) and tensioner pulley.
15. After installing the timing belt, reconfirm the timing mark.
16. Install the tensioner pulley.
17. Pull the set pin out of the auto tensioner.
18. Rotate the crankshaft 2 revolutions **Clockwise**, then wait 5 minutes for the auto tensioner to adjust.
19. Measure the auto tensioner rod as shown. If the measurement is not

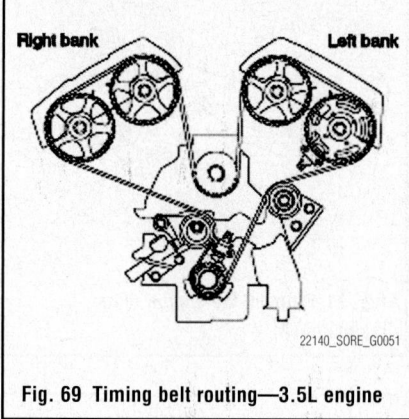

Fig. 69 Timing belt routing—3.5L engine

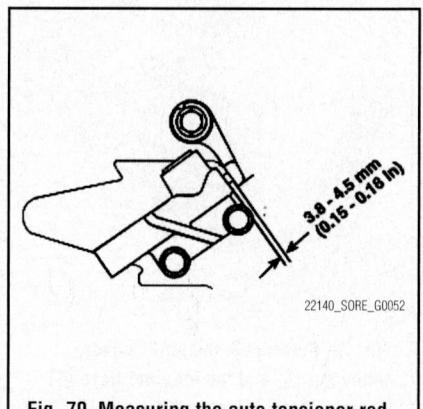

Fig. 70 Measuring the auto tensioner rod

0.15–0.18 inches (3.8–4.5 mm), then repeat the belt tensioning procedure.

20. After the auto tensioner measurement is correct, install the upper and lower timing belt covers.

21. Continue the installation in the reverse order of the removal procedure.

TIMING CHAIN, SPROCKETS, FRONT COVER AND SEAL

REMOVAL & INSTALLATION

3.3L & 3.8L Engine

See Figures 71 through 86.

1. Remove the engine cover.
2. Remove the engine room resonator (A).
3. Disconnect the MAF sensor connector (A) and the breather hose (B).
4. Remove the air cleaner assembly (C).
5. Disconnect the other breather hose (A), the Purge Control Solenoid Valve (PCSV) hose, the Positive Crankcase Ventilation (PCV) hose (C) and the Electronic Throttle Control (ETC) cooling hoses (D).

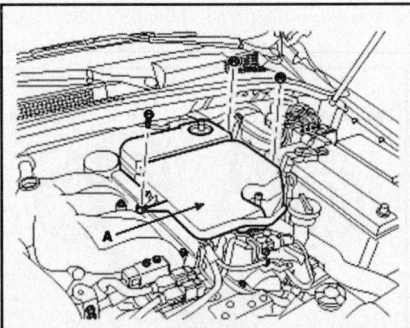

Fig. 71 Remove the engine room resonator (A)

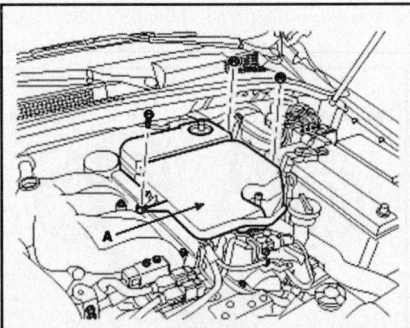

Fig. 72 Disconnect the MAF sensor connector (A) and the breather hose (B). Remove the air cleaner assembly (C)

6. Remove the wiring over the surge tank.
   a. Disconnect the injection harness connector (A).
   b. Disconnect the camshaft position sensor (CMP) harness connector (B).
   c. Disconnect the ground lines (C).
   d. Disconnect the ignition coil harness connector (D).
   e. Disconnect the condenser connector (E).

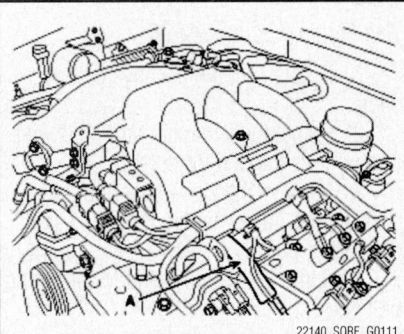

Fig. 73 Disconnect the other breather hose (A)

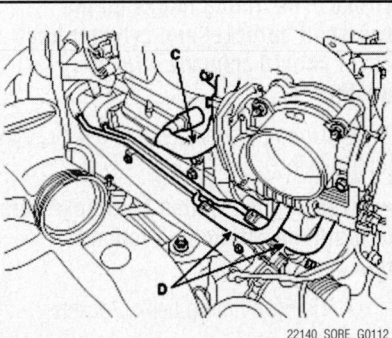

Fig. 74 Disconnect the Positive Crankcase Ventilation (PCV) hose (C) and the Electronic Throttle Control (ETC) cooling hoses (D)

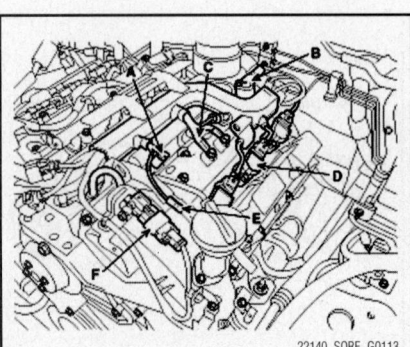

Fig. 75 Remove the wiring over the surge tank.

f. Disconnect the Oil Control Valve (OCV) harness connector (F).

7. Remove the surge tank assembly.
8. Remove the cylinder head covers and gaskets.
9. Set No.1 cylinder to TDC/compression.
   a. Turn the crankshaft pulley and align its groove with the timing mark "T" of the lower timing chain cover.

➡ **Do not rotate engine counterclockwise**

   b. Check that the mark (A) of the camshaft timing sprockets are in straight line on the cylinder head surface as shown in the illustration. If not, turn the crankshaft one revolution (360°).
10. Remove the lower oil pan.

✳✳ CAUTION

Insert the SST between the oil pan and the ladder frame by tapping it

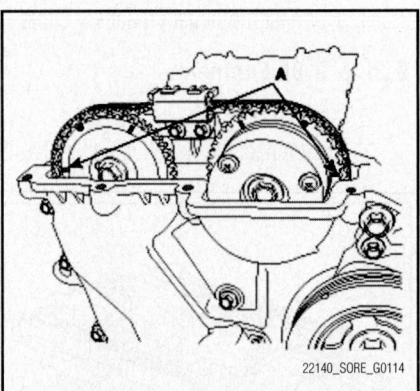

Fig. 76 Check that the mark (A) of the camshaft timing sprockets are in straight line

Fig. 77 Install a set pin after compressing the timing chain tensioner

with a plastic hammer in the direction of arrow.

11. Remove the crankshaft damper pulley.

12. Remove the timing chain cover.

➡**Be careful not to damage the contact surfaces of cylinder block, cylinder head and timing chain cover.**

13. Install a set pin after compressing the timing chain tensioner.

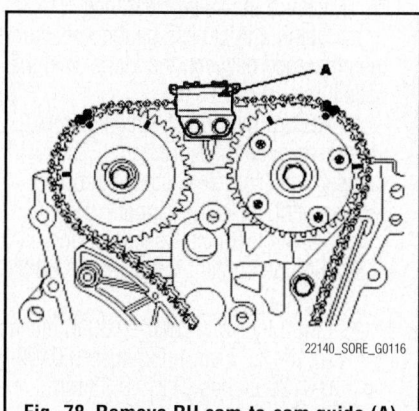

Fig. 78 Remove RH cam-to-cam guide (A)

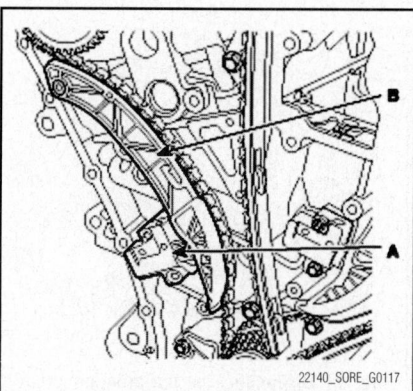

Fig. 79 Remove RH timing chain auto tensioner (A) and RH timing chain tensioner arm (B)

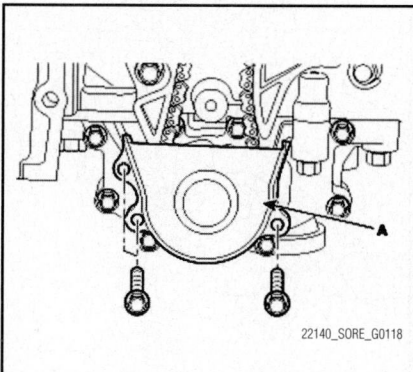

Fig. 80 Remove oil pump chain cover (A)

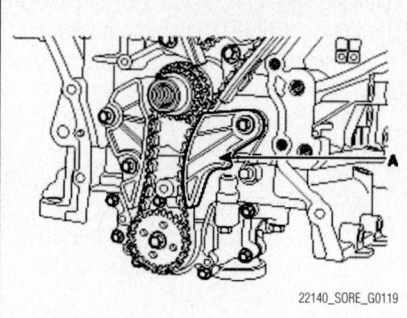

Fig. 81 Remove oil pump chain tensioner assembly (A)

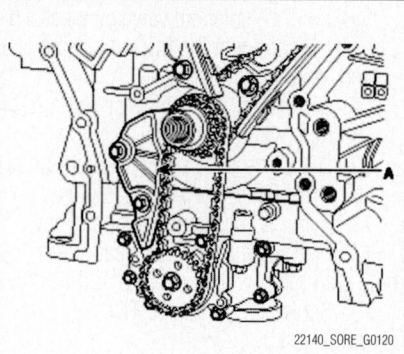

Fig. 82 Remove oil pump chain guide (A)

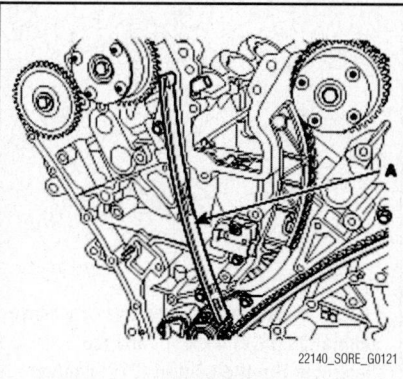

Fig. 83 Remove RH timing chain guide (A)

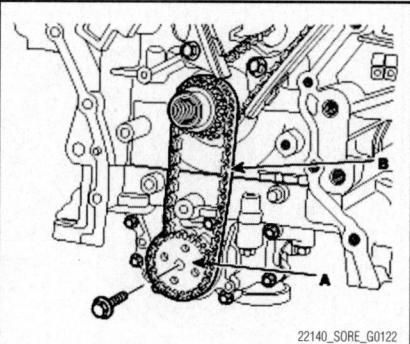

Fig. 84 Remove oil pump chain sprocket (A) and oil pump chain (B)

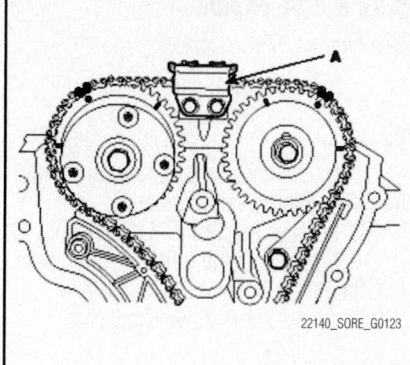

Fig. 85 Remove LH cam-to-cam guide (A)

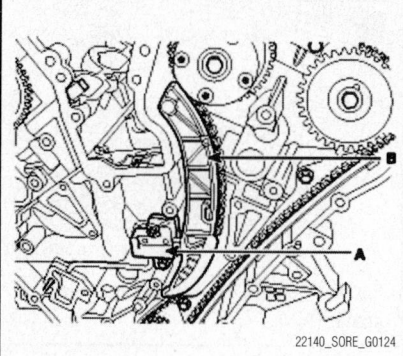

Fig. 86 Remove LH timing chain auto tensioner (A) and LH timing chain tensioner arm (B)

14. Remove RH cam-to-cam guide (A).

15. Remove RH timing chain auto tensioner (A) and RH timing chain tensioner arm (B).

16. Remove oil pump chain cover (A).

17. Remove oil pump chain tensioner assembly (A).

18. Remove oil pump chain guide (A).

19. Remove RH timing chain.

20. Remove RH timing chain guide (A).

21. Remove oil pump chain sprocket (A) and oil pump chain (B).

22. Remove crankshaft sprocket.

23. Remove LH cam-to-cam guide (A).

24. Remove LH timing chain auto tensioner (A) and LH timing chain tensioner arm (B).

25. Remove LH timing chain.

26. Installation is the reverse of removal.

VALVE LASH

ADJUSTMENT

3.5L Engine

This vehicle is equipped with hydraulic valve lifters. No adjustment is necessary.

3.3L & 3.8L Engine

See Figures 87 through 95.

1. Remove the engine cover.
2. Remove the engine room resonator (A).
3. Disconnect the MAF sensor connector (A) and the breather hose (B).
4. Remove the air cleaner assembly (C).
5. Disconnect the other breather hose (A), the Purge Control Solenoid Valve (PCSV) hose, the Positive Crankcase Ventilation (PCV) hose (C) and the Electronic Throttle Control (ETC) cooling hoses (D).
6. Remove the wiring over the surge tank.

    a. Disconnect the injection harness connector (A).

    b. Disconnect the Camshaft Position Sensor (CMP) harness connector (B).

    c. Disconnect the ground lines (C).

    d. Disconnect the ignition coil harness connector (D).

    e. Disconnect the condenser connector (E).

    f. Disconnect the Oil Control Valve (OCV) harness connector (F).

    g. Disconnect the Variable Induction System (VIS) solenoid valve connector (G).

    h. Disconnect the injector wiring (H) and ignition coil wiring (I).

7. Disconnect the fuel hose tube (A).
8. Remove heater hose.
9. Disconnect the brake vacuum hose.
10. Disconnect the surge tank stay.
11. Remove the surge tank assembly.
12. Remove the cylinder head cover bolts and remove the cover(A) and gasket.
13. Set No.1 cylinder to TDC/compression.

    a. Turn the crankshaft pulley and align its groove with the timing mark "T" of the lower timing chain cover.

    b. Check that the mark(A) of the camshaft timing sprockets are in straight line on the cylinder head surface. If not, turn the crankshaft one revolution (360°)

➡ **Do not rotate engine counterclockwise**

14. Inspect the valve clearance.

    a. Check only the valve indicated as shown. (No. 1 cylinder : TDC/Compression).

15. Measure the valve clearance.

    a. Using a thickness gauge, measure the clearance between the tappet and the base circle of camshaft.

    b. Record the out-of-specification valve clearance measurements. They will be used later to determine the required replacement adjusting tappet.

    c. Valve clearance Specification:

- Engine coolant temperature: 68°F (20°C)
- Limit: Intake: 0.0067–0.0090 inches (0.17–0.23 mm); Exhaust: 0.0106–0.0129 inches (0.27–0.33 mm)

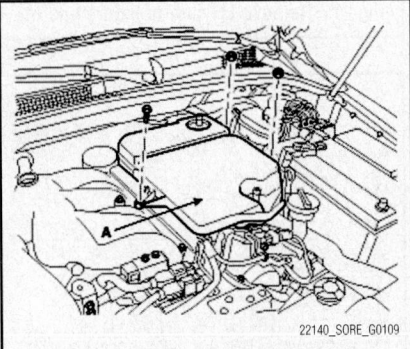

Fig. 87 Remove the engine room resonator (A)

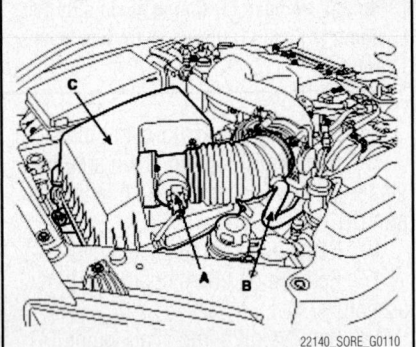

Fig. 88 Disconnect the MAF sensor connector (A) and the breather hose (B). Remove the air cleaner assembly (C).

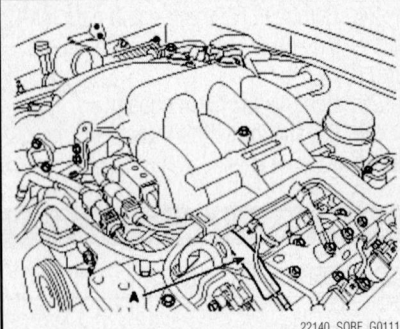

Fig. 89 Disconnect the other breather hose (A)

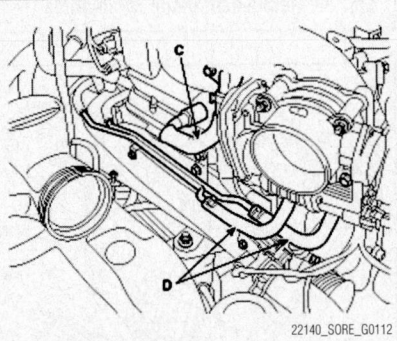

Fig. 90 Disconnect the Positive Crankcase Ventilation (PCV) hose (C) and the Electronic Throttle Control (ETC) cooling hoses (D)

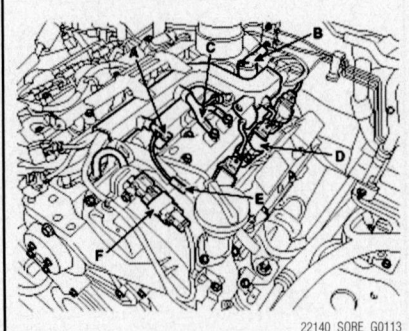

Fig. 91 Remove the wiring over the surge tank

Fig. 92 Disconnect the Variable Induction System (VIS) solenoid valve connector (G), the injector wiring (H) and ignition coil wiring (I)

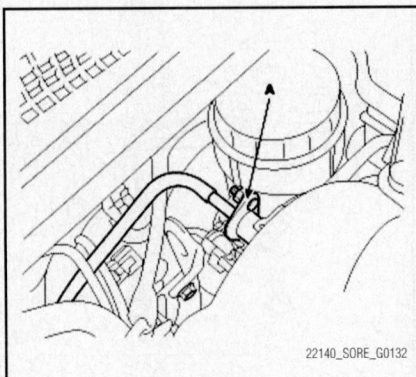

Fig. 93 Disconnect the fuel hose (A)

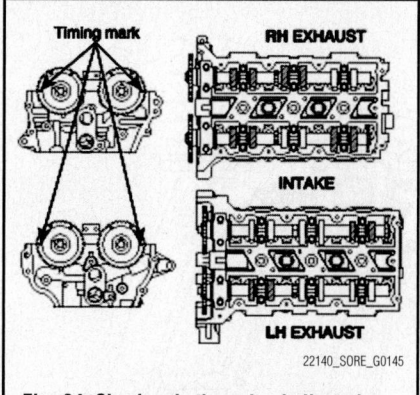

Fig. 94 Check only the valve indicated as shown. No. 1 cylinder TDC/Compression

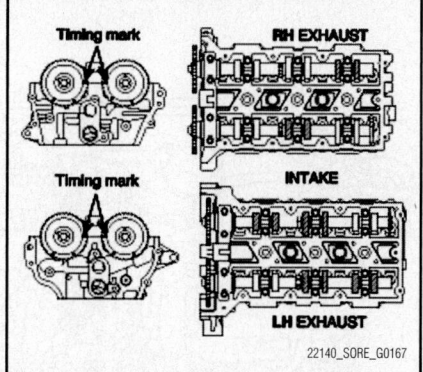

Fig. 95 Check only the valve indicated as shown. No. 4 cylinder TDC/Compression

16. Turn the crankshaft pulley one revolution (360°) and align the groove with timing mark "T" of the lower timing chain cover.

➡**Do not rotate engine counterclockwise**

a. Check only valves indicated as shown. [NO. 4 cylinder : TDC/compression].

17. Measure the valve clearance.

18. Adjust the intake and exhaust valve clearance.

a. Set the No.1 cylinder to the TDC/compression.

b. Mark on the timing chain on the basis of the marking on sprocket and CVVT.

c. Remove the timing chain.

d. Remove the camshaft bearing caps.

e. Remove the camshaft assembly.

f. Remove the Mechanical Lash Adjusters (MLA).

g. Measure the thickness of the removed tappet using a micrometer.

h. Calculate the thickness of a new tappet so that the valve clearance comes within the specified value.

i. Valve clearance:
- Engine coolant temperature: 68° (20°C)
- T: Thickness of removed tappet
- A: Measured valve clearance
- N: Thickness of new tappet

- Intake: N = T + [A − 0.0079 inches (0.20 mm)]
- Exhaust: N = T + [A − 0.0118 inches (0.30 mm)]

j. Select a new tappet with a thickness as close as possible to the calculated value.

➡**Tappets are available in 41 size increments of 0.0006 inches (0.015 mm) from 0.118 inches (3.00 mm) to 0.1417 inches (3.600 mm)**

k. Place a new tappet on the cylinder head.

➡**Appling engine oil at the selected tappet on the periphery and top surface.**

l. Install the intake and exhaust camshaft.

m. Install the bearing caps.

n. Install the timing chain.

o. Turn the crankshaft two turns in the operating direction(clockwise) and realign crankshaft sprocket and camshaft sprocket timing marks.

p. Recheck the valve clearance.

q. Valve clearance Specification:
- Engine coolant temperature: 68°F (20°C)
- Limit: Intake: 0.0067–0.0090 inches (0.17–0.23 mm); Exhaust: 0.0106–0.0129 inches (0.27–0.33 mm)

ENGINE PERFORMANCE & EMISSION CONTROL

COMPONENT LOCATIONS

See Figures 96 through 98.

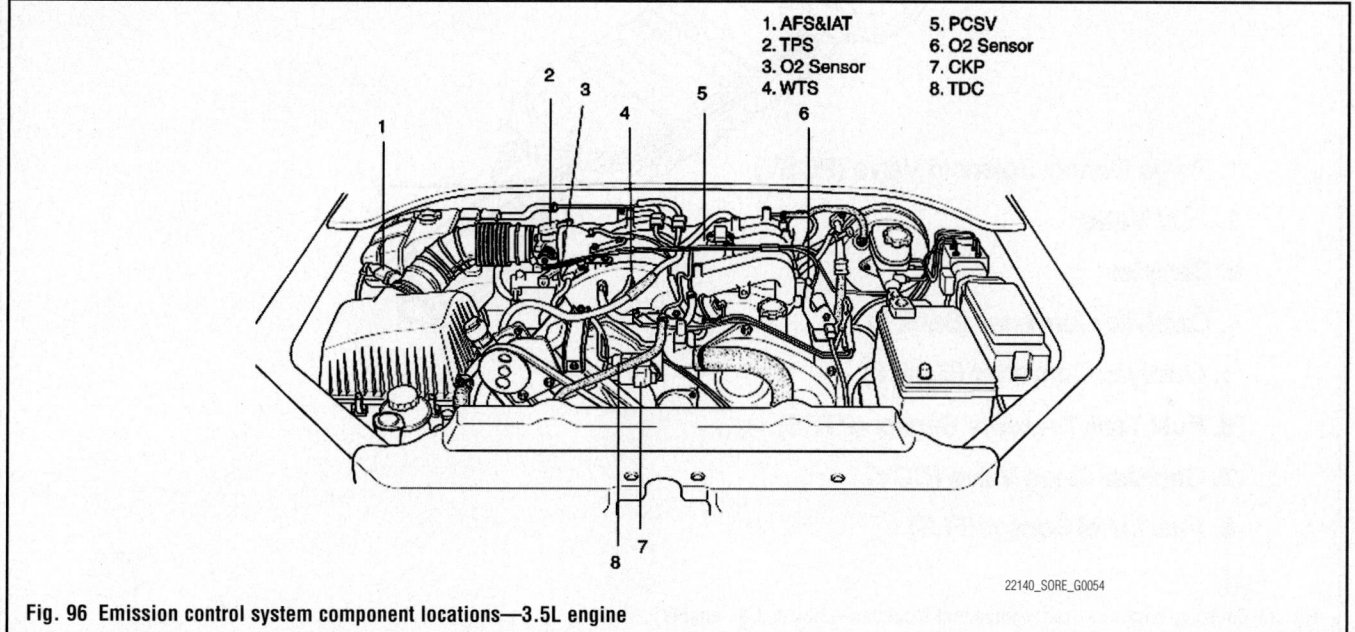

1. AFS&IAT
2. TPS
3. O2 Sensor
4. WTS
5. PCSV
6. O2 Sensor
7. CKP
8. TDC

Fig. 96 Emission control system component locations—3.5L engine

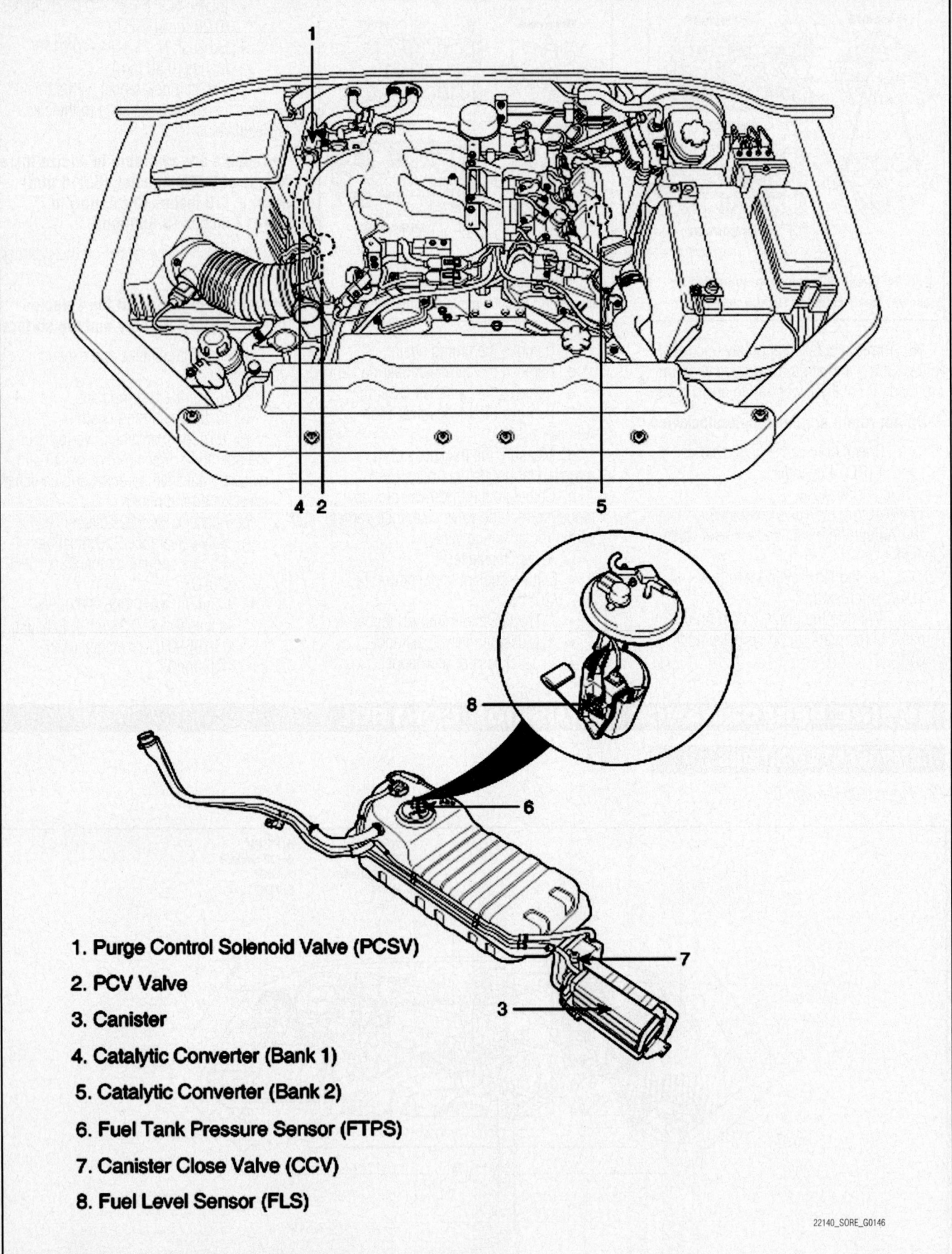

1. Purge Control Solenoid Valve (PCSV)

2. PCV Valve

3. Canister

4. Catalytic Converter (Bank 1)

5. Catalytic Converter (Bank 2)

6. Fuel Tank Pressure Sensor (FTPS)

7. Canister Close Valve (CCV)

8. Fuel Level Sensor (FLS)

22140_SORE_G0146

Fig. 97 Emission control system component locations—3.3L & 3.8L engine

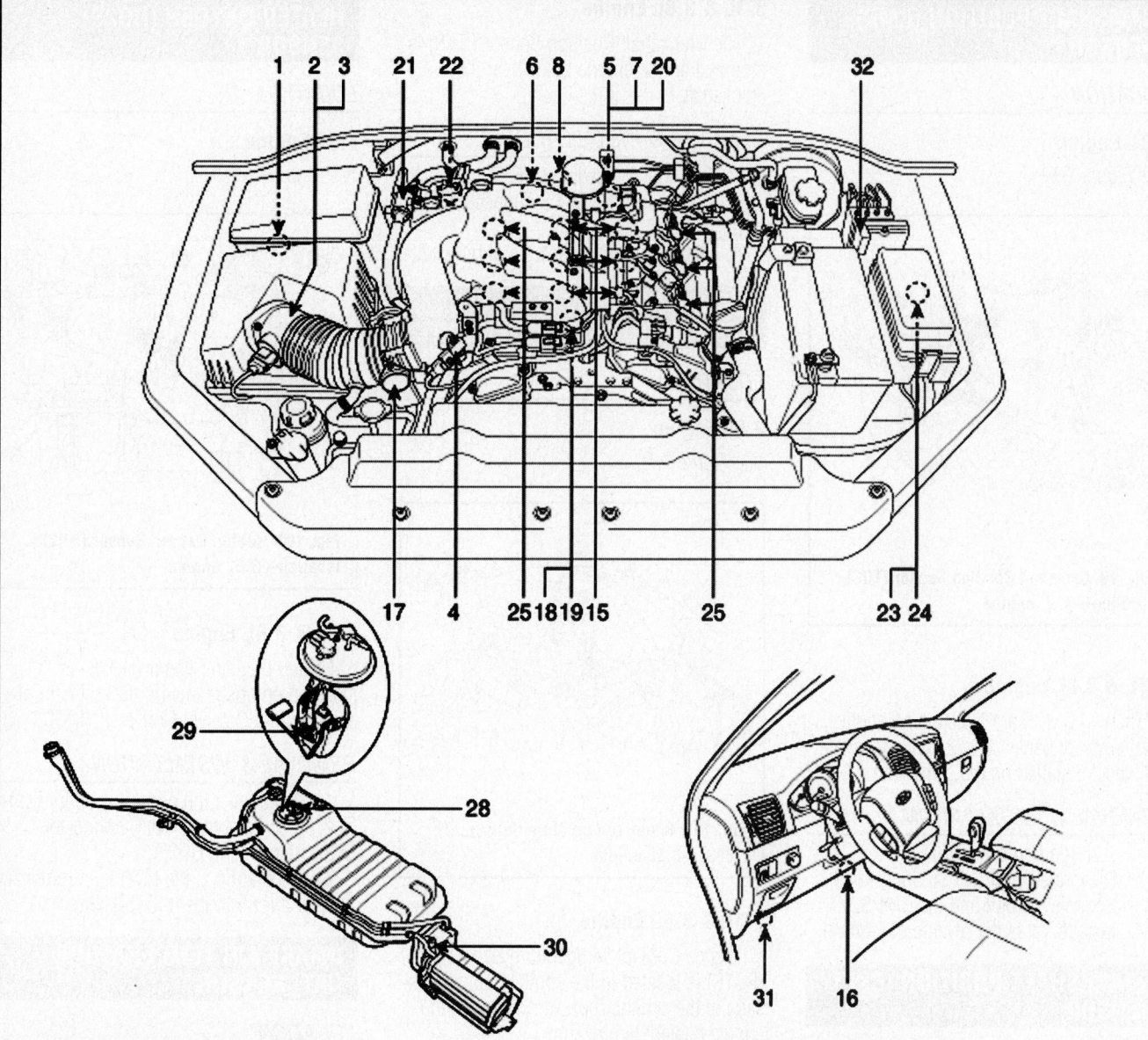

1. ECM (Engine Control Module)
2. Mass Air Flow Sensor (MAFS)
3. Intake Air Temperature Sensor (IATS)
4. Manifold Absolute Pressure Sensor (MAPS)
5. Engine Coolant Temperature Sensor (ECTS)
6. Camshaft Position Sensor (CMPS) [Bank 1]
7. Camshaft Position Sensor (CMPS) [Bank 2]
8. Crankshaft Position Sensor (CKPS)
9. Heated Oxygen Sensor (HO2S) [Bank 1 / Sensor 1]
10. Heated Oxygen Sensor (HO2S) [Bank 1 / Sensor 2]
11. Heated Oxygen Sensor (HO2S) [Bank 2 / Sensor 1]
12. Heated Oxygen Sensor (HO2S) [Bank 2 / Sensor 2]
13. Knock Sensor (KS) [Bank 1]
14. Knock Sensor (KS) [Bank 2]
15. Injector
16. Accelerator Position Sensor (APS)

17. ETC Module [Throttle Position Sensor (TPS) + ETC Motor]
18. CVVT Oil Control Valve (OCV) [Bank 1]
19. CVVT Oil Control Valve (OCV) [Bank 2]
20. CVVT Oil Temperature Sensor (OTS)
21. Purge Control Solenoid Valve (PCSV)
22. Variable Intake Solenoid (VIS) Valve
23. Fuel Pump Relay
24. Main Relay
25. Ignition Coil
26. A/C Pressure Tansducer (APT)
27. Wheel Speed Sensor (WSS)
28. Fuel Tank Pressure Sensor (FTPS)
29. Fuel Level Sensor(FLS)
30. Canister Close Valve (CCV)
31. Data Link Connector (DLC)
32. Multi-Purpose Check Connector

22140_SORE_G0147

Fig. 98 Engine control system component locations—3.3L & 3.8L engine

CAMSHAFT POSITION (CMP) SENSOR

LOCATION

3.5L Engine
See Figure 99.

Fig. 99 Camshaft Position Sensor (TDC) location—3.5L engine

3.3L & 3.8L Engine
The two CMPS are installed on engine head cover of bank 1 and 2 and uses a target wheel installed on the camshaft.

REMOVAL & INSTALLATION
1. Turn ignition switch OFF.
2. Disconnect the CMPS connector.
3. Remove the bolt and the CMPS.
4. Installation is the reverse of removal.

CRANKSHAFT POSITION (CKP) SENSOR

LOCATION

3.5L Engine
See Figure 100.

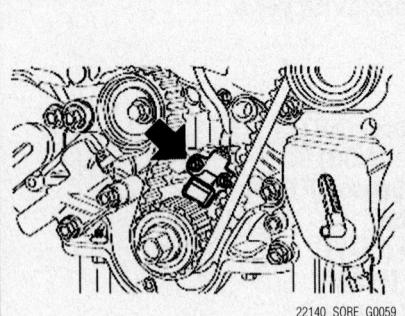

Fig. 100 Crankshaft Position Sensor (CKP) location—3.5L engine

3.3L & 3.8L Engine
The Crankshaft Position Sensor (CKP) is mounted on the engine block near the crankshaft.

REMOVAL & INSTALLATION
1. Turn ignition switch OFF.
2. Disconnect the CKPS connector.
3. Remove the CKPS.
4. Installation is the reverse of removal.

ENGINE COOLANT TEMPERATURE (ECT) SENSOR

LOCATION

3.5L Engine
See Figure 101.

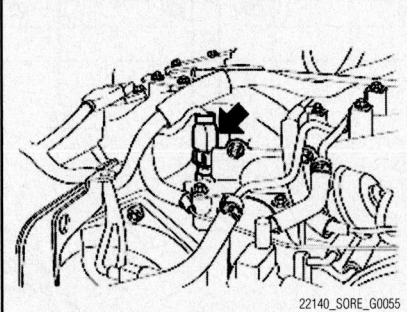

Fig. 101 Water Temperature Sensor (WTS)—3.5L engine

3.3L & 3.8L Engine
Engine Coolant Temperature Sensor (ECTS) is located in the engine coolant passage of the cylinder head for detecting the engine coolant temperature.

REMOVAL & INSTALLATION

3.5L Engine
1. Disconnect the negative battery cable.
2. Remove the water temperature sensor from the intake manifold.

To install:
3. Apply sealant LOCTITE 962T or the equivalent to the threaded portion.
4. Install the water temperature sensor and tighten it to the specified torque. Torque: 14–29 ft. lbs. (20–40 Nm).
5. Securely connect the harness connector.

3.3L & 3.8L Engine
1. Turn ignition switch OFF.
2. Disconnect ECTS connector.
3. Remove the ECTS.
4. Installation is the reverse of removal.

HEATED OXYGEN (HO2S) SENSOR

LOCATION

3.5L Engine
See Figure 102.

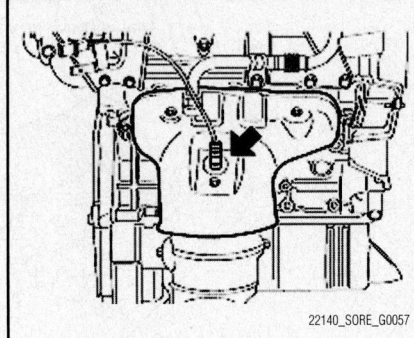

Fig. 102 Heated Oxygen Sensor (HO2S) location—3.5L engine

3.3L & 3.8L Engine
Heated Oxygen Sensor (HO2S) is installed on upstream and downstream of the Manifold Catalyst Converter (MCC).

REMOVAL & INSTALLATION
1. Disconnect the negative battery cable.
2. Disconnect the HO2S connector.
3. Remove the HO2S.
4. Installation is the reverse of removal.
Torque: 29–36 ft. lbs. (40–50 Nm).

INTAKE AIR TEMPERATURE (IAT) SENSOR

LOCATION
Intake Air Temperature Sensor (IATS) is installed inside the Mass Air Flow Sensor (MAFS) and detects the intake air temperature.

REMOVAL & INSTALLATION
1. Disconnect the negative battery cable.
2. Disconnect the IAT sensor connector
3. Remove the IAT sensor.
4. Installation is the reverse of removal.

KNOCK SENSOR (KS)

LOCATION
Knock Sensor (KS) senses engine knocking and the two sensors are installed inside the V-valley of the cylinder block.

REMOVAL & INSTALLATION
1. Turn the ignition off.

2. Disconnect the Knock Sensor (KS) connectors.

3. Remove the KS.

4. Installation is the reverse of removal.

MANIFOLD ABSOLUTE PRESSURE (MAP) SENSOR

LOCATION

3.3L & 3.8L Engine

Manifold Absolute Pressure Sensor (MAPS) is speed-density type sensor and is installed on the surge tank.

REMOVAL & INSTALLATION

1. Disconnect the negative battery cable.

2. Disconnect the MAPS sensor connector

3. Remove the MAPS sensor.

4. Installation is the reverse of removal.

MASS AIR FLOW (MAF) SENSOR

LOCATION

3.5L Engine

The intake air temperature sensor (ATS Sensor), located in the intake air hose, is a resistor-based sensor for detecting the intake air temperature.

3.3L & 3.8L Engine

Mass Air Flow Sensor (MAFS) is a hot-film type sensor and is located in between the air cleaner and the throttle body.

REMOVAL & INSTALLATION

1. Disconnect the negative battery cable.

2. Disconnect and remove the MAFS from the air cleaner assembly.

3. Installation is the reverse of removal.

POWERTRAIN CONTROL MODULE (PCM)

LOCATION

The Powertrain Control Module (PCM) is installed on the firewall near the air cleaner.

REMOVAL & INSTALLATION

1. Turn ignition switch off.

2. Disconnect the battery cable from the battery.

3. Remove the resonator.

4. Disconnect the PCM connectors.

5. Unscrew the PCM mounting bolts and remove the PCM from the bracket.

6. Install a new PCM

THROTTLE POSITION SENSOR (TPS)

LOCATION

3.5L Engine

See Figure 103.

22140_SORE_G0056

Fig. 103 Throttle Position Sensor (TPS) location—3.5L engine

REMOVAL & INSTALLATION

3.5L Engine

1. Disconnect the throttle position sensor connector.

2. Remove the TPS.

3. Installation is the reverse of removal.

4. Torque: 1.1–1.8 ft. lbs. (1.5–2.5 Nm).

FUEL   GASOLINE FUEL INJECTION SYSTEM

FUEL SYSTEM SERVICE PRECAUTIONS

Safety is the most important factor when performing not only fuel system maintenance but any type of maintenance. Failure to conduct maintenance and repairs in a safe manner may result in serious personal injury or death. Maintenance and testing of the vehicle's fuel system components can be accomplished safely and effectively by adhering to the following rules and guidelines.

• To avoid the possibility of fire and personal injury, always disconnect the negative battery cable unless the repair or test procedure requires that battery voltage be applied.

• Always relieve the fuel system pressure prior to disconnecting any fuel system component (injector, fuel rail, pressure regulator, etc.), fitting or fuel line connection. Exercise extreme caution whenever relieving fuel system pressure to avoid exposing

skin, face and eyes to fuel spray. Please be advised that fuel under pressure may penetrate the skin or any part of the body that it contacts.

• Always place a shop towel or cloth around the fitting or connection prior to loosening to absorb any excess fuel due to spillage. Ensure that all fuel spillage (should it occur) is quickly removed from engine surfaces. Ensure that all fuel soaked cloths or towels are deposited into a suitable waste container.

• Always keep a dry chemical (Class B) fire extinguisher near the work area.

• Do not allow fuel spray or fuel vapors to come into contact with a spark or open flame.

• Always use a back-up wrench when loosening and tightening fuel line connection fittings. This will prevent unnecessary stress and torsion to fuel line piping.

• Always replace worn fuel fitting O-rings with new Do not substitute fuel hose or equivalent where fuel pipe is installed.

Before servicing the vehicle, make sure to also refer to the precautions in the beginning of this section as well.

RELIEVING FUEL SYSTEM PRESSURE

1. Remove the rear seat.

2. Disconnect the fuel pump connector located under the center floor carpet..

3. Start the engine and wait until fuel in fuel line is exhausted.

4. After engine stalls, turn the ignition switch to OFF position.

5. Reconnect the fuel pump connector.

6. Disconnect the negative battery cable.

FUEL FILTER

REMOVAL & INSTALLATION

See Fuel Pump Removal & Installation. The fuel filter is located inside the fuel pump.

FUEL PUMP

REMOVAL & INSTALLATION

2006 models

See Figures 104 and 105.

1. Release fuel system pressure.
2. Disconnect the negative battery terminal from the battery.
3. Remove the rear seat.
4. Remove service cover under center floor carpet.
5. Disconnect the fuel pump electrical connector, the EVAP. hose, the main and return hose.
6. Unfasten the assembly fixing-bolt (6 EA).
7. Disconnect the breather hose.
8. Remove the fuel pump.
9. Installation is the reverse of removal.

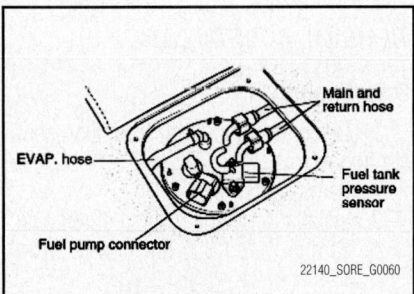

Fig. 104 Disconnect the fuel pump electrical connector, the EVAP. hose, the main and return hose.

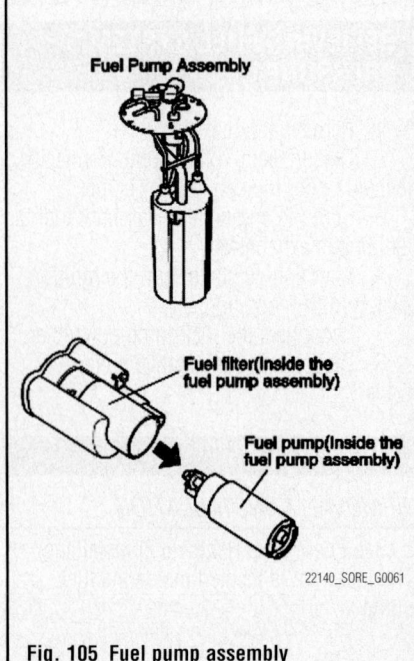

Fig. 105 Fuel pump assembly

2007–08 Models

See Figures 106 through 109.

1. Fold the rear seat cushion.
2. Open the carpet (A).
3. Remove the service cover (B).
4. Disconnect the fuel pump connector (A).

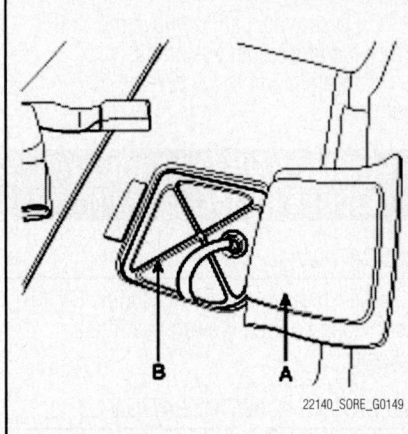

Fig. 106 Open the carpet (A). Remove the service cover (B).

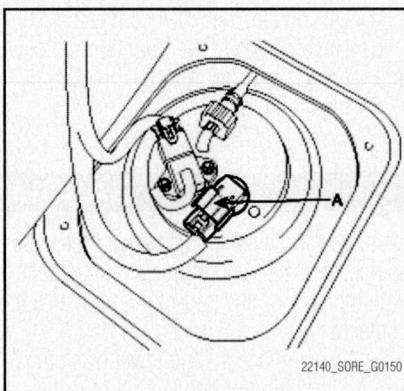

Fig. 107 Disconnect the fuel pump connector (A)

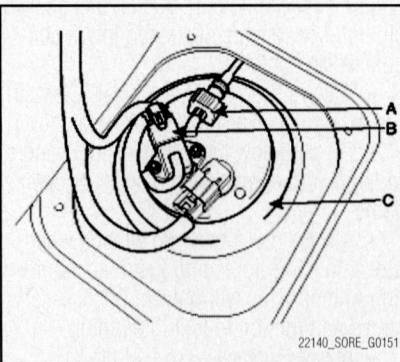

Fig. 108 Disconnect the fuel feed quick-connector (A) and the fuel tank pressure sensor connector (B). Remove the rubber cover (C)

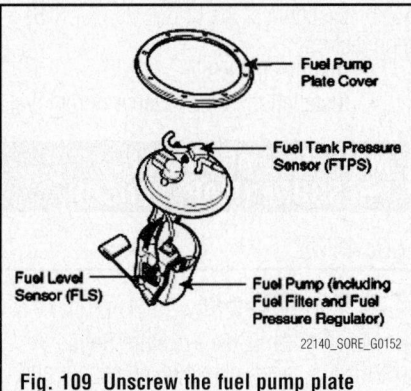

Fig. 109 Unscrew the fuel pump plate cover with the special service tool (SST NO: 09310-2B200) and remove the fuel pump assembly

5. Start the engine and wait until fuel in fuel line is exhausted.
6. After engine stops, turn the ignition switch off.
7. Disconnect the fuel feed quick-connector (A) and the fuel tank pressure sensor connector (B).
8. Remove the rubber cover (C).
9. Unscrew the fuel pump plate cover with the special service tool (SST NO: 09310-2B200) and remove the fuel pump assembly.
10. Install the fuel pump according to the reverse order of "REMOVAL" procedure.

FUEL TANK

REMOVAL & INSTALLATION

2006 models

See Figures 104 and 110.

1. Release fuel system pressure. Refer to "Relieving Fuel System Pressure".
2. Disconnect the negative battery terminal from the battery.
3. Drain the fuel from the fuel tank.
4. Remove the rear seat.
5. Remove the service cover under center floor carpet.
6. Disconnect the fuel pump electrical connector, the EVAP. hose, the main and return hose.
7. Disconnect the breather hose.
8. Disconnect the joint hose.
9. Unfasten the fuel tank-fixing bolts.
10. Remove the fuel tank.
11. Installation is the reverse of removal.

2007–08 Models

See Figures 111 through 117.

1. Fold the rear seat cushion.
2. Open the carpet (A).

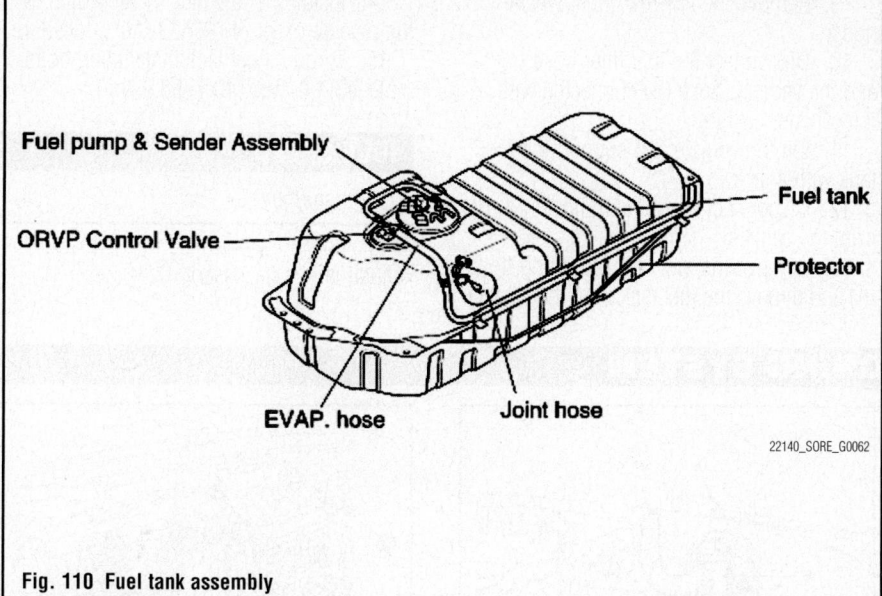

Fuel pump & Sender Assembly

ORVP Control Valve

Fuel tank

Protector

EVAP. hose

Joint hose

22140_SORE_G0062

Fig. 110 Fuel tank assembly

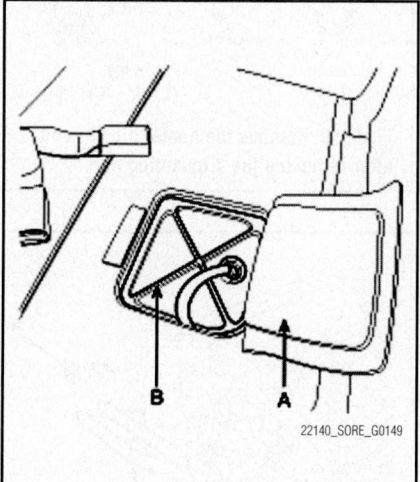

22140_SORE_G0149

Fig. 111 Open the carpet (A). Remove the service cover (B).

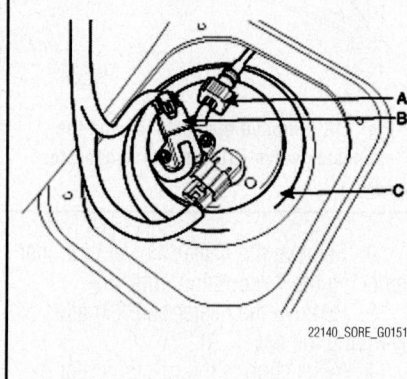

22140_SORE_G0151

Fig. 113 Disconnect the fuel feed quick-connector (A) and the fuel tank pressure sensor connector (B). Remove the rubber cover (C).

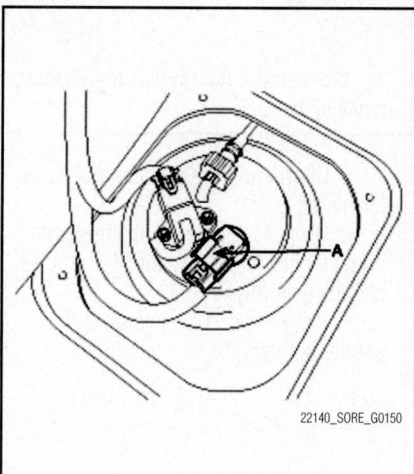

22140_SORE_G0150

Fig. 112 Disconnect the fuel pump connector (A)

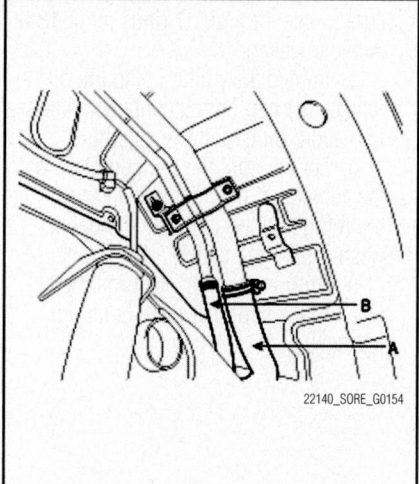

22140_SORE_G0154

Fig. 114 Disconnect the fuel filler hose (A) and the vacuum hose (B).

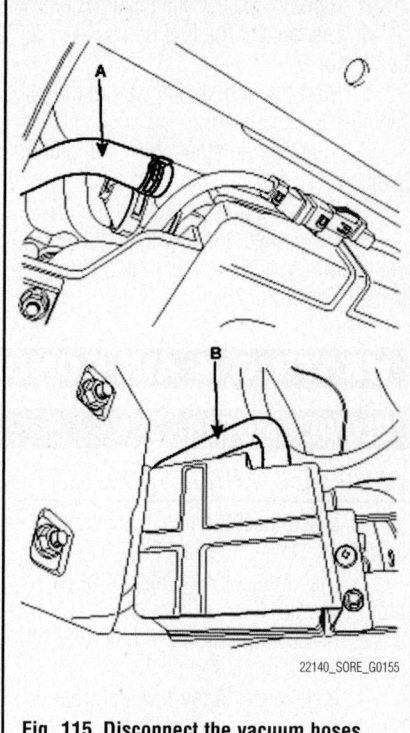

22140_SORE_G0155

Fig. 115 Disconnect the vacuum hoses (A, B) from the canister

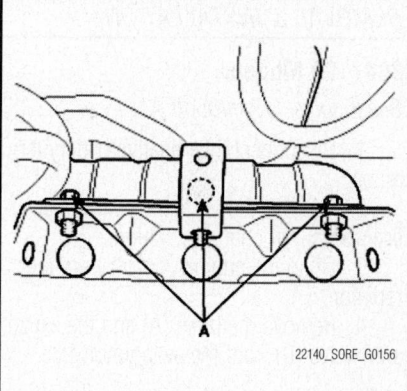

22140_SORE_G0156

Fig. 116 Remove the fuel tank mounting bolts (A)

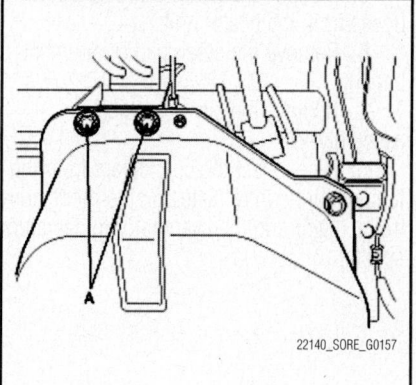

22140_SORE_G0157

Fig. 117 Remove the fuel tank mounting bolts (A)

3. Remove the service cover (B).

4. Disconnect the fuel pump connector (A).

5. Start the engine and wait until fuel in fuel line is exhausted.

6. After engine stops, turn the ignition switch off.

7. Drain the fuel from the fuel tank.

8. Disconnect the fuel feed quick-connector (A) and the fuel tank pressure sensor connector (B).

9. Remove the RH-rear inner wheel house.

10. Disconnect the fuel filler hose (A) and the vacuum hose (B) connected with the canister air filter.

11. Lift the vehicle and support the fuel tank with a jack.

12. Disconnect the vacuum hoses (A, B) from the canister.

13. Remove the fuel tank mounting bolts (A) and remove the fuel tank from the vehicle.

14. Install the fuel tank in according to the reverse order of "REMOVAL" procedure.

15. Torque: Fuel tank installation bolts: 36.2–43.4 ft. lbs. (49.1–58.9 Nm).

IDLE SPEED

ADJUSTMENT

Idle speed is controlled by the PCM. No Adjustment is necessary.

HEATING & AIR CONDITIONING SYSTEM

BLOWER MOTOR

REMOVAL & INSTALLATION

1. Disconnect the negative battery terminal.

2. Disconnect the connector of the blower motor.

3. Remove the blower motor after removing the mounting screws.

4. Installation is the reverse order of removal.

HEATER CORE

REMOVAL & INSTALLATION

2007–08 Models

See Figures 118 through 120.

1. Disconnect the negative battery terminal.

2. Recover the refrigerant with a recovery/ recycling/ charging station.

3. Drain the engine coolant from the radiator.

4. Remove the bolts (A) and the expansion valve (B) from the evaporator core.

➡ Plug or cap the lines immediately after disconnecting them to avoid moisture and dust contamination.

5. Disconnect the inlet and outlet heater hoses from the heater unit.

6. Remove the crash pad (instrument panel).

7. Remove the cowl cross bar assembly.

8. Disconnect the connectors from the temperature control actuator, the mode control actuator and the evaporator temperature sensor.

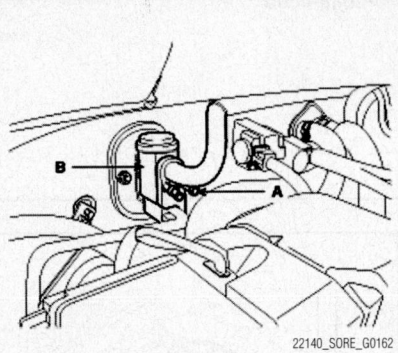

Fig. 118 Remove the bolts (A) and the expansion valve (B) from the evaporator core

9. Remove the heater blower unit after removing the 7 mounting nuts.

10. Remove the heater core (B) after removing the cover (A).

11. Installation is the reverse order of removal, and note these items :

 a. If you're installing a new evaporator, add refrigerant oil (ND-OIL8).

 b. Replace the O-rings with new ones at each fitting, and apply a thin coat of refrigerant oil before installing them. Be sure to use the right O-rings for R-134a to avoid leakage.

 c. Immediately after using the oil, replace the cap on the container, and seal it to avoid moisture absorption.

 d. Do not spill the refrigerant oil on the vehicle ; it may damage the paint ; if the refrigerant oil contacts the paint, wash it off immediately

 e. Apply sealant to the grommets.

 f. Make sure that there is no air leakage.

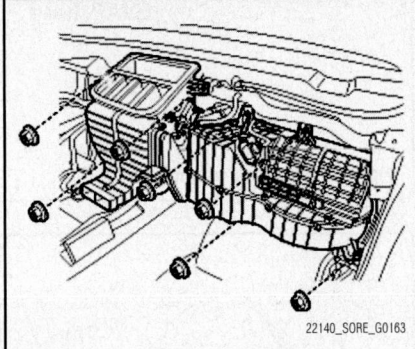

Fig. 119 Remove the heater blower unit after removing the 7 mounting nuts

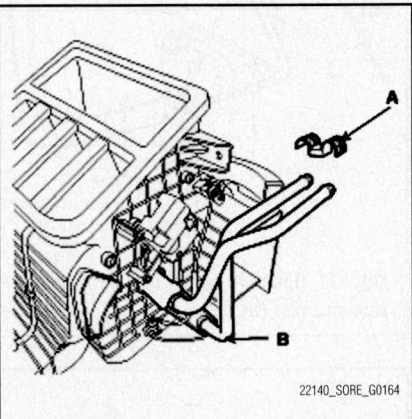

Fig. 120 Remove the heater core (B) after removing the cover (A)

 g. Charge the system and test its performance.

 h. Do not interchange the inlet and outlet heater hoses and install the hose clamps securely.

 i. Refill the cooling system with engine coolant.

STEERING

POWER STEERING GEAR

REMOVAL & INSTALLATION

See Figures 121 and 122.

1. Drain the power steering fluid.
2. Disconnect the pressure tube and return tube.
3. Remove the joint assembly connecting bolt.

Fig. 121 Remove the joint assembly connecting bolt.

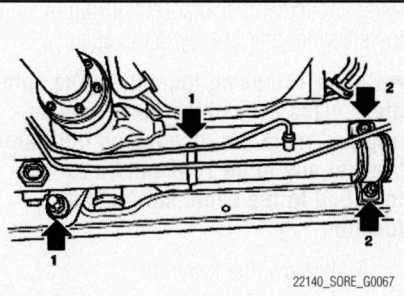

Fig. 122 Remove the steering gear box mounting bolts and remove the steering gear box assembly

4. Using the special tool (0K670-321-019), disconnect the tie rod end from the knuckle arm.
5. Remove the steering gear box mounting bolts and remove the steering gear box assembly together with mounting rubber.
6. Installation is the reverse of removal.

POWER STEERING PUMP

REMOVAL & INSTALLATION

1. Disconnect the pressure hose from the oil pump.
2. Disconnect the suction hose from the suction pipe, and drain the oil.
3. Remove the power steering tension adjusting bolt or flange nut.
4. Separate the belt from the power steering oil pump pulley.
5. Remove the power steering oil pump assembly.

To install:

6. Install the oil pump to the oil pump bracket. Torque: 13–16 ft. lbs. (18–23 Nm).
7. Install the belt and tighten the bolt adjusting tension.
8. Install the suction hose to the oil pump.
9. Install the pressure hose to the oil pump.
10. Add power steering fluid (PSF-III).
11. Air bleed the system.

BLEEDING

1. Raise and safely support the front of the vehicle.
2. Manually turn the oil pump pulley a few times.
3. Turn the steering wheel all the way to the left and to the right five or six times.

4. While operating the starter motor intermittently, turn the steering wheel all the way to the left and right five or six times (for 15 to 20 seconds).

✳✳ CAUTION

During air bleeding, replenish the fluid supply so that the level never falls below the lower position of the filter. If air bleeding is done while engine is running, the air will be broken up and absorbed into the fluid; be sure to do the bleeding only while cranking.

5. Start the engine (idling).
6. Turn the steering wheel to the left and right until there are no air bubbles in the oil reservoir.
7. Confirm that the fluid is not milky, and that the level is up to the specified position on the level gauge.
8. Confirm that there is very little change in the fluid level when the steering wheel is turned left and right.
9. Check whether or not the change in the fluid level is within 0.20 inches (5 mm) when the engine is stopped and when it is running.

✳✳ CAUTION

If the change of the fluid level is 0.20 inches (5 mm) or more, the air has not been completely bled from the system, and thus must be bled completely. If the fluid level rises suddenly after the engine is stopped, the air has not been completely bled. If air bleeding is not complete, there will be abnormal noises from the pump and the flow-control valve, and this condition could cause a lessening of the life of the pump, etc.

LOWER BALL JOINT

REMOVAL & INSTALLATION

See Figures 123 through 126.

1. Raise the front of the vehicle and support it with safety stands.
2. Remove the front wheels.
3. Remove the lower nut of control link of stabilizer bar.
4. Remove the lower nut of shock absorber.
5. Remove the bolts and nuts that joins lower arm and lower arm ball joint.
6. Remove the cotter pin and castle nut from the lower arm ball joint.
7. Remove the lower arm ball joint from the steering knuckle.
8. Remove the steering gear mounting bolts and nuts.
9. Remove the spindle from the front frame crossmember brackets during raising the steering gear box by using suitable bar.

➡Before removing the nuts of the spindles, make note of the numerical setting and mark the location on the frame bracket and plate so it can be

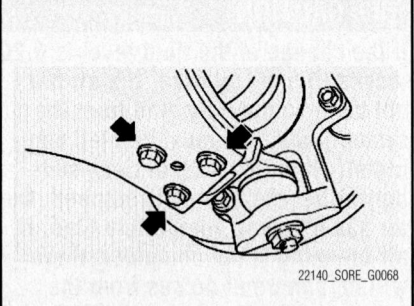

Fig. 123 Remove the bolts and nuts that joins lower arm and lower arm ball joint

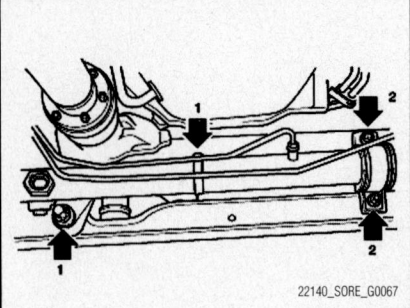

Fig. 124 Remove the steering gear box mounting bolts and nuts

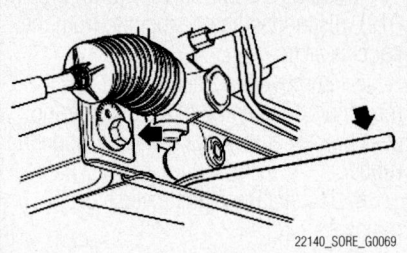

Fig. 125 Remove the spindle from the front frame crossmember brackets during raising the steering gear box by using suitable bar

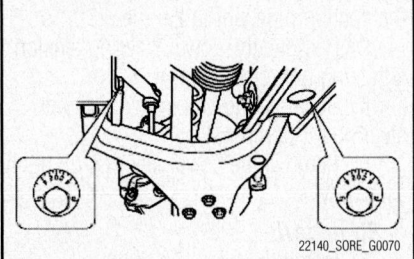

Fig. 126 Make note of the numerical setting and mark the location

re-installed to the same setting and location.

10. Remove the lower arm.
11. Remove and replace ball joints as necessary.

To install:

12. Install the lower arm ball joint to the steering knuckle. Torque: 116–145 ft. lbs. (157–196 Nm).
13. Install a new cotter pin through the castle nut.
14. Position the lower arm to the front frame crossmember brackets.
15. Position the spindle while lifting up the steering gear box by using suitable pry bar.
16. Install the lower arm spindles. Torque: 159–181 ft. lbs. (216–245 Nm).

➡Align the spindle to the numerical setting and marked location on the frame bracket and plate so the same setting and location is maintained.

17. Install the lower nut of the shock absorber. Torque: 88–101 ft. lbs. (122–140 Nm).
18. Install the lower nut of control link of stabilizer bar. Torque: 68–84 ft. lbs. (95–117 Nm).

19. Install the wheels.
20. Remove the safety stands and lower the vehicle.

➡After installation, measure the wheel alignment and adjust if necessary.

LOWER CONTROL ARM

REMOVAL & INSTALLATION

2006 Models

See Figures 123 through 126.

1. Raise the front of the vehicle and support it with safety stands.
2. Remove the front wheels.
3. Remove the lower nut of control link of stabilizer bar.
4. Remove the lower nut of shock absorber.
5. Remove the bolts and nuts that joins lower arm and lower arm ball joint.
6. Remove the cotter pin and castle nut from the lower arm ball joint.
7. Remove the lower arm ball joint from the steering knuckle.
8. Remove the steering gear mounting bolts and nuts.
9. Remove the spindle from the front frame crossmember brackets during raising the steering gear box by using suitable bar.

➡Before removing the nuts of the spindles, make note of the numerical setting and mark the location on the frame bracket and plate so it can be re-installed to the same setting and location.

10. Remove the lower arm.

To install:

11. Install the lower arm ball joint to the steering knuckle. Torque: 116–145 ft. lbs. (157–196 Nm).
12. Install a new cotter pin through the castle nut.
13. Position the lower arm to the front frame crossmember brackets.
14. Position the spindle while lifting up the steering gear box by using suitable pry bar.
15. Install the lower arm spindles. Torque: 159–181 ft. lbs. (216–245 Nm).

➡Align the spindle to the numerical setting and marked location on the frame bracket and plate so the same setting and location is maintained.

16. Install the lower nut of the shock absorber. Torque: 88–101 ft. lbs. (122–140 Nm).

17. Install the lower nut of control link of stabilizer bar. Torque: 68–84 ft. lbs. (95–117 Nm).

18. Install the wheels.

19. Remove the safety stands and lower the vehicle.

➡ **After installation, measure the wheel alignment and adjust if necessary.**

MACPHERSON STRUT

REMOVAL & INSTALLATION

See Figures 127 through 130.

1. Remove the battery.

2. Remove three strut mounting block nuts from the mounting block.

3. Raise the front of the vehicle and support it with safety stands.

4. Remove the front wheels.

5. Remove the bolt on the steering knuckle side that secures the upper arm ball joint.

6. Remove the brake hose bracket and remove the upper arm bolts and nuts.

7. Remove the strut lower nut.

8. Remove the strut assembly from the vehicle.

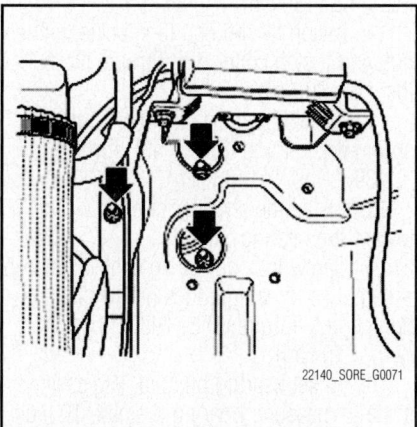

Fig. 127 Remove three strut mounting block nuts from the mounting block

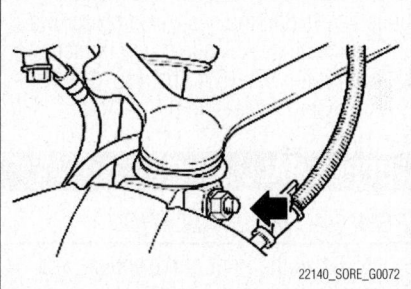

Fig. 128 Remove the bolt on the steering knuckle side that secures the upper arm ball joint

To install:

9. After making sure identification mark on the spring seat. Position the strut assembly into the upper mounting block.

10. Install the mounting block nuts by 3–4 threads only.

11. Insure the front of the vehicle is raised and supported with safety stands.

12. Tighten the lower nut of the strut. Torque: 88–101 ft. lbs. (122–140 Nm).

13. Position the upper arm to the frame brackets, insert the bolts and hand tighten the nuts.

14. Install the upper arm ball joint into the top of the steering knuckle and tighten the side bolt and nut. Torque: 31–39 ft. lbs. (44–55 Nm).

15. Tighten the upper arm bolts and nuts and install brake hose brackets. Torque: 54–68 ft. lbs. (76–95 Nm).

16. Install the front wheels.

17. Lower the vehicle.

18. Tighten the mounting block nuts. Torque: 31–39 ft. lbs. (44–55 Nm).

19. Install the battery mounting bracket and the battery.

20. After installing the front strut assembly, measure the wheel alignment and adjust if necessary.

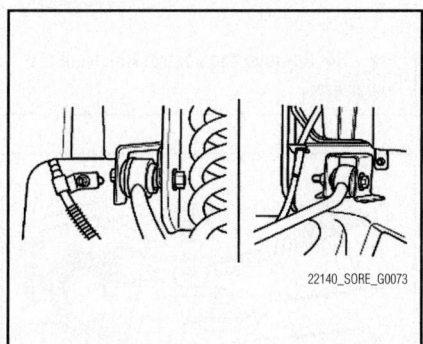

Fig. 129 Remove the brake hose bracket and remove the upper arm bolts and nuts

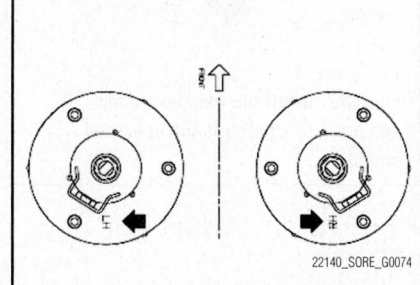

Fig. 130 Identification mark on the spring seat

OVERHAUL

See Figures 131 and 132.

1. Secure the strut in a suitable vise.

2. Loosen the piston rod nut several turns.

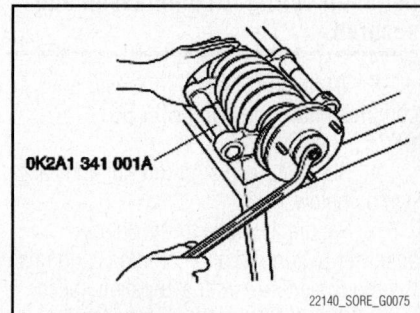

Fig. 131 Compress the coil spring with SST OK2A1-341-001A

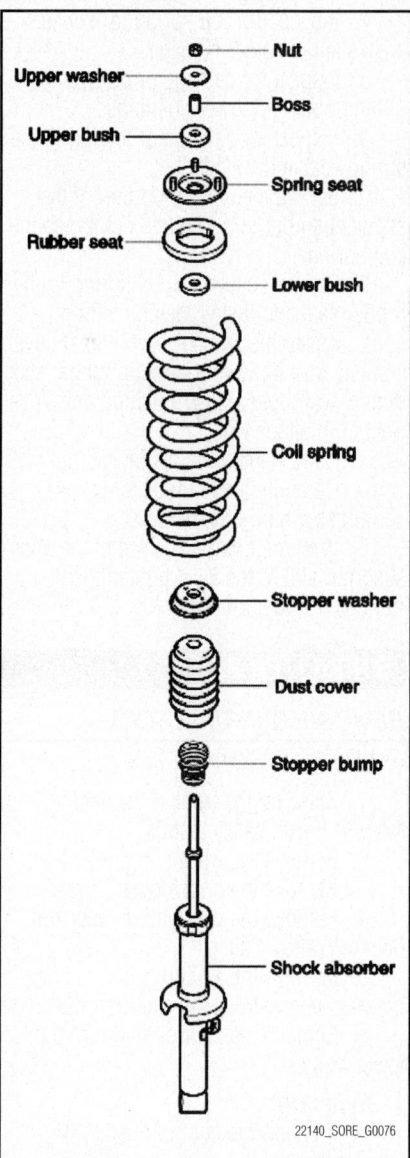

Fig. 132 Exploded view of strut assembly

→Use copperplate in the jaws of the vise to protect the shock absorber bottom bracket.

✳✳ CAUTION

Do not remove the piston rod nut until coil spring is compressed and secured.

3. While still secured in a vise, compress the coil spring with SST OK2A1-341-001A.

4. Remove the piston rod nut and each part as below.

5. Secure a handle to the shock absorber piston rod and compress and raise the rod three times with a constant speed. Inspect for uniform working force and abnormal noise.

6. Inspect the entire shock absorber for signs of oil leakage; replace if required.

7. Inspect the coil spring for stress cracks and/or other damage.

8. Inspect for damage or deterioration of the upper and lower bushings.

9. Inspect for damage or tearing of the spring seat and rubber seat.

10. Secure the bottom portion of the shock absorber in a vise and compress the coil spring.

11. Set the end of the coil spring to the rubber seat and install the coil spring.

12. Assemble stopper bump, dust cover, stopper washer, lower bushing, rubber seat, spring seat, boss, upper bushing and upper washer in sequence.

13. Hand tighten the piston rod nut.

14. Carefully loosen the coil spring compressor and remove it.

15. With the bottom bracket of the shock absorber still in the vice, tighten the piston rod nut. Torque: 54–68 ft. lbs. (76–95 Nm).

STABILIZER BAR

REMOVAL & INSTALLATION

See Figures 133 through 135.

1. Raise up the front of the vehicle and support it with safety stands.

2. Remove the wheels.

3. Remove the undercover.

4. Remove the nuts and damper rubbers of control link.

5. Remove the stabilizer bar bushing brackets and remove the stabilizer bar.

6. Remove the control link from the lower arm.

To install:

7. Position the control links to the lower arm.

8. Loosely tighten the control link nuts.

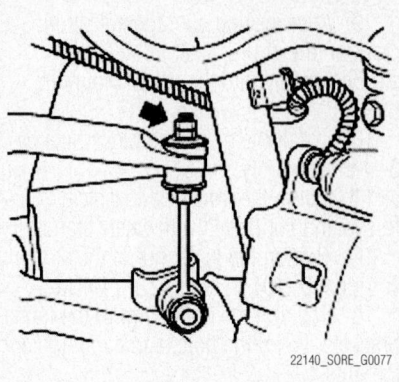

Fig. 133 Remove the nuts and damper rubbers of control link

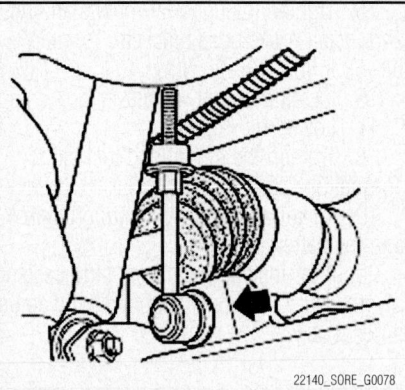

Fig. 134 Remove the control link from the lower arm

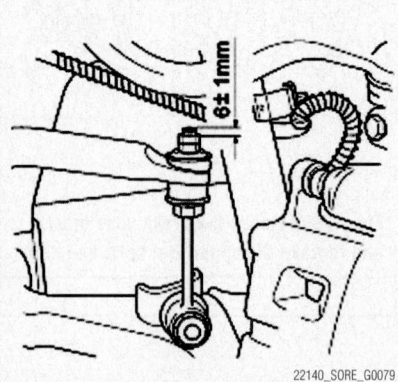

Fig. 135 Install the damper rubber and nut, and tighten to the specified length

9. Install the stabilizer bar on the control link.

10. Align the clamp bushing inside of stabilizer bushing and install bracket. Torque: 31–39 ft. lbs. (44–55 Nm).

11. Install the damper rubber and nut, and tighten to the specified length.

12. Tighten the lower nut of control link. Torque: 68–84 ft. lbs. (95–117 Nm).

STEERING KNUCKLE

REMOVAL & INSTALLATION

1. Remove the vehicle speed sensor.

2. Remove brake caliper from brake rotor. Temporarily tie caliper to vehicle frame with wire.

3. Remove the brake rotor.

4. Using a lock nut wrench (or equivalent), remove lock nut and plain washer (2WD).

5. Remove the upper arm link lock bolt, spring washer and nut.

6. Remove tie rod end from steering knuckle.

7. Remove lower arm from steering knuckle.

8. Remove steering knuckle from vehicle.

To install:

9. Put steering knuckle on the drive shaft end with upper and lower ball joints in mounting holes.

10. Attach lower arm, tighten lock nut, and install cotter pin. Torque: 116–130 ft. lbs. (160–180 Nm).

11. Attach tie rod end to knuckle, tighten nut, and install cotter pin. Torque: 51–57 ft. lbs. (70–80 Nm).

12. Insert upper arm link lock bolt with spring washer and tighten nut. Torque: 32–39 ft. lbs. (44–55 Nm).

13. Install the chamfer of plain washer toward the bearing (2WD)

14. Screw lock nut up against wheel hub assembly and using a lock nut wrench, tighten nut. Torque: 178–198 ft. lbs. (245–275 Nm).

15. To set bearing preload, use spring scale to measure. Bearing preload: 10 inch lbs.

16. Stake the flange of lock nut on the end of drive shaft.

17. Put brake rotor on wheel bearing hub bolts and install the two retaining screws.

18. Install brake caliper and tighten two bolts. Torque: 57–75 ft. lbs. (80–104 Nm).

19. Install wheel and tire.

UPPER BALL JOINT

REMOVAL & INSTALLATION

1. Raise the front of the vehicle and support it with safety stands.

2. Remove the front wheels.

3. Remove the bolt on the steering knuckle side that secures the upper arm ball link.

4. Remove the brake hose bracket.
5. Remove the upper arm.
6. Secure the upper arm in a suitable vise.
7. Inspect for bent, cracked or otherwise damaged upper arm.
8. Inspect for worn or deteriorated upper arm bushing.
9. Inspect for worn or damaged ball link and replace if damaged, deformed or cracked.
10. Replace bushings if worn or deteriorated.
11. Using a standard bearing press, remove the old bushing.

To install:

12. Install the new bushing and press it into the upper arm with a standard bearing press.
13. Position the upper arm to the frame brackets, insert the bolts and hand tighten the nuts.
14. Install the upper arm ball joint into the top of the steering knuckle. Torque: 31–39 ft. lbs. (44–55 Nm).
15. Tighten the upper arm bolts and nuts. Torque: 54–68 ft. lbs. (76–95 Nm).
16. Install brake hose brackets.
17. Install the wheels.

➡**After installation, measure the wheel alignment and adjust if necessary.**

UPPER CONTROL ARM

REMOVAL & INSTALLATION

1. Raise the front of the vehicle and support it with safety stands.
2. Remove the front wheels.
3. Remove the bolt on the steering knuckle side that secures the upper arm ball link.

4. Remove the brake hose bracket.
5. Remove the upper arm.

To install:

6. Position the upper arm to the frame brackets, insert the bolts and hand tighten the nuts.
7. Install the upper arm ball joint into the top of the steering knuckle. Torque: 31–39 ft. lbs. (44–55 Nm).
8. Tighten the upper arm bolts and nuts. Torque: 54–68 ft. lbs. (76–95 Nm).
9. Install brake hose brackets.
10. Install the wheels.

➡**After installation, measure the wheel alignment and adjust if necessary.**

WHEEL BEARINGS

REMOVAL & INSTALLATION

1. Remove the vehicle speed sensor.
2. Remove brake caliper from brake rotor. Temporarily tie caliper to vehicle frame with wire.
3. Remove the brake rotor.
4. Using a lock nut wrench (or equivalent), remove lock nut and plain washer (2WD).
5. Remove the upper arm link lock bolt, spring washer and nut.
6. Remove tie rod end from steering knuckle.
7. Remove lower arm from steering knuckle.
8. Remove steering knuckle from vehicle.
9. Using a screwdriver, pry out oil seal from knuckle (4WD).
10. Press the wheel hub from the knuckle (4WD).

11. Press the knuckle and remove wheel hub (2WD).
12. Inspect bearing for wear or damage.
13. Inspect steering knuckle for wear or damage.

To install:

14. Install the dust cover to the knuckle. Torque: 12–16 ft. lbs. (16–23 Nm).
15. Install new oil seal and install the wheel hub to the knuckle by pressing.
16. Apply grease to the wheel bearing and seal lip.
17. Put steering knuckle on the drive shaft end with upper and lower ball joints in mounting holes.
18. Attach lower arm, tighten lock nut, and install cotter pin. Torque: 116–130 ft. lbs. (160–180 Nm).
19. Attach tie rod end to knuckle, tighten nut, and install cotter pin. Torque: 51–57 ft. lbs. (70–80 Nm).
20. Insert upper arm link lock bolt with spring washer and tighten nut. Torque: 32–39 ft. lbs. (44–55 Nm).
21. Install the chamfer of plain washer toward the bearing (2WD)
22. Screw lock nut up against wheel hub assembly and using a lock nut wrench, tighten nut. Torque: 178–198 ft. lbs. (245–275 Nm).
23. To set bearing preload, use spring scale to measure. Bearing preload: 10 inch lbs.
24. Stake the flange of lock nut on the end of drive shaft.
25. Put brake rotor on wheel bearing hub bolts and install the two retaining screws.
26. Install brake caliper and tighten two bolts. Torque: 57–75 ft. lbs. (80–104 Nm).
27. Install wheel and tire.

COIL SPRING

REMOVAL & INSTALLATION

See Figures 136 and 137.

1. Raise the rear of the vehicle and support it with safety stands.
2. Remove the rear wheels.
3. Raise the rear axle housing with a jack to facilitate removal of the shock absorbers.
4. Remove stabilizer link upper mounting nut.
5. Remove the rear shock absorber lower nut and washer.
6. Remove the shock absorber upper bolt, and remove the shock absorber.
7. Lower the rear axle housing slowly to facilitate removal of the coil spring.
8. Remove the upper rubber seat.

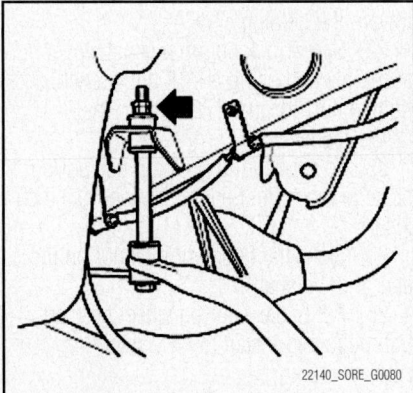

Fig. 136 Remove stabilizer link upper mounting nut

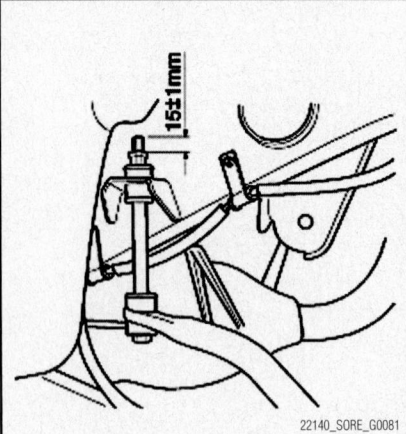

Fig. 137 Install the stabilizer link upper mounting nut to the specified length

To install:

9. Position the upper rubber seat to the coil spring.

> ✲✲ **CAUTION**
>
> **Align the spring end with the groove of the spring pad and fix the spring and the spring pad by adhering the parts with tape.**

10. Slowly raise the rear axle housing while installing the coil spring.
11. Install the shock absorber upper nut. Torque: 88–101 ft. lbs. (122–140 Nm).
12. Install the shock absorber lower bolt. Torque: 88–101 ft. lbs. (122–140 Nm).
13. Install the stabilizer link upper mounting nut to the specified length.
14. Install the wheels
15. Remove the safety stands and lower the vehicle.

CONTROL ARMS/LINKS

REMOVAL & INSTALLATION

Rear Upper Arm

See Figure 138.

Fig. 138 Remove the upper arm bolts and remove the upper arm.

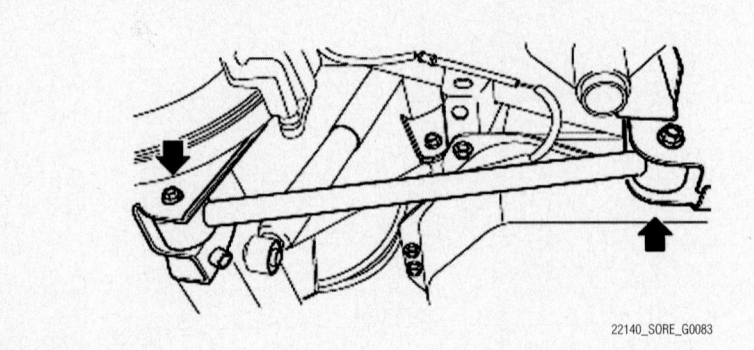

Fig. 139 Remove the lower arm bolts and remove the lower arm

1. Raise the rear of the vehicle and support it with safety stands.
2. Remove the rear wheels.
3. Raise the rear axle housing to facilitate removal of the upper arm.
4. Remove shock absorber lower bolt.
5. Remove the upper arm bolts and remove the upper arm.

To install:

6. Install the upper arm and the bolts. Torque: 88–101 ft. lbs. (122–140 Nm).
7. Install shock absorber lower bolt. Torque: 88–101 ft. lbs. (122–140 Nm).
8. Lower the rear axle housing.
9. Install the wheels.
10. Remove the safety stands and lower the vehicle.

Rear Lower Arm

See Figure 139.

1. Raise the rear of the vehicle and support it with safety stands.
2. Remove the rear wheels.
3. Raise the rear axle housing to facilitate removal of the upper arm.

4. Remove shock absorber lower bolt.

5. Remove wheel speed sensor cable from rear lower arm.

6. Remove the lower arm bolts and remove the lower arm.

To install:

7. Install the lower arm and the bolts. Torque: 101–116 ft. lbs. (137–157 Nm).

8. Install wheel speed sensor cable to the rear lower arm.

9. Install shock absorber lower bolt. Torque: 88–101 ft. lbs. (122–140 Nm).

10. Lower the rear axle housing.

11. Install the wheels.

12. Remove the safety stands and lower the vehicle.

Rear Stabilizer Bar

See Figures 136 and 137.

1. Support the bottom of the rear differential carrier with a jack.

2. Remove the stabilizer link mounting nut.

3. Remove the stabilizer bar bushing bracket.

4. Remove the stabilizer bar.

To install:

5. Install the stabilizer bar.

6. Align the identification mark white paint on stabilizer bar with bushing and install the stabilizer bar bushing bracket. Torque: 13–16 ft. lbs. (19–23 Nm).

7. Install the joint cup and nut and tighten to the specified length.

Lateral Rod

See Figure 140.

1. Raise the rear of the vehicle and support it with safety stands.

2. Remove the rear wheels.

3. Raise the rear axle housing with a jack to facilitate removal of the lateral rod.

4. Remove shock absorber lower bolt.

5. Remove the lateral rod bolts and remove the lateral rod.

To install:

6. Install the lateral rod and the bolts. Torque: 135–155 ft. lbs. (187–215 Nm).

7. Install shock absorber lower bolt. Torque: 88–101 ft. lbs. (122–140 Nm).

8. Lower the rear axle housing.

9. Install the wheels.

10. Remove the safety stands and lower the vehicle.

SHOCK ABSORBER

REMOVAL & INSTALLATION

See Figures 136 and 137.

1. Raise the rear of the vehicle and support it with safety stands.

2. Remove the rear wheels.

3. Raise the rear axle housing with a jack to facilitate removal of the shock absorbers.

4. Remove stabilizer link upper mounting nut.

5. Remove the rear shock absorber lower nut and washer.

6. Remove the shock absorber upper bolt, and remove the shock absorber.

To install:

7. Install the shock absorber upper nut. Torque: 88–101 ft. lbs. (122–140 Nm).

8. Install the shock absorber lower bolt. Torque: 88–101 ft. lbs. (122–140 Nm).

9. Install the stabilizer link upper mounting nut to the specified length.

10. Install the wheels

11. Remove the safety stands and lower the vehicle.

TESTING

1. Compress and expand the shock absorber three to four times and check for uniform working force and abnormal noise.

2. Inspect for gas leakage.

3. Inspect the shock absorber for a worn or deteriorated rubber bushing.

4. Replace the rear shock absorber assembly if a problem is found.

STABILIZER BAR

REMOVAL & INSTALLATION

See Figures 136 and 137.

1. Support the bottom of the rear differential carrier with a jack.

2. Remove the stabilizer link mounting nut.

3. Remove the stabilizer bar bushing bracket.

4. Remove the stabilizer bar.

To install:

5. Install the stabilizer bar.

6. Align the identification mark white paint on stabilizer bar with bushing and install the stabilizer bar bushing bracket. Torque: 13–16 ft. lbs. (19–23 Nm).

7. Install the joint cup and nut and tighten to the specified length.

WHEEL BEARINGS

REMOVAL & INSTALLATION

See Figure 141.

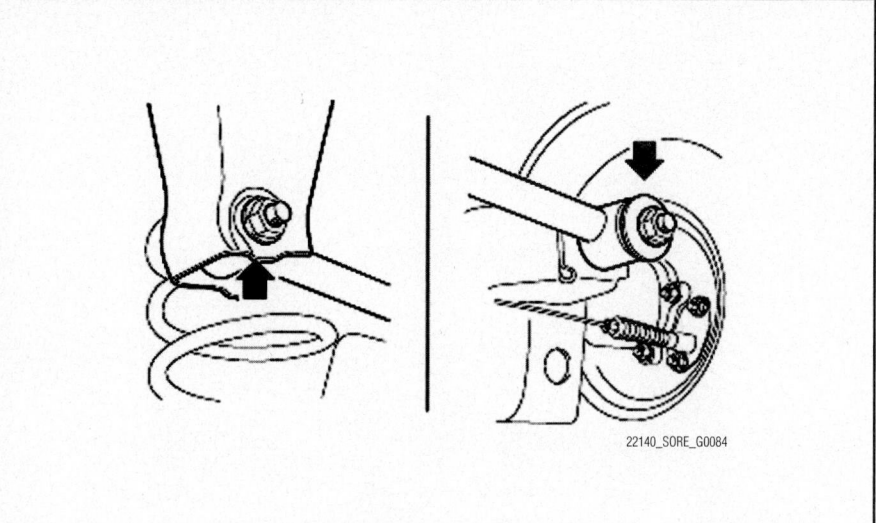

22140_SORE_G0084

Fig. 140 Remove the lateral rod bolts and remove the lateral rod.

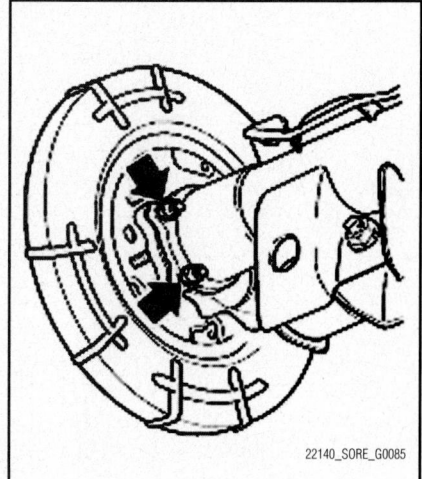

22140_SORE_G0085

Fig. 141 Remove the rear axle shaft mounting nuts

1. Remove the disc brake and parking brake assembly.

2. Remove the parking brake cable and speed sensor cable.

3. Remove the rear axle shaft mounting nuts.

4. Remove the rear axle shaft.

5. Using the special tool (09526-11100), remove the oil seal.

To install:

6. Installation is the reverse of removal.

7. Apply grease to the oil seal lip.

8. Using the special tools (09500-11000, 09532-11500), install the oil seal.

9. After installing the axle shaft, tighten the nuts. Torque: 32–44 ft. lbs. (43–60 Nm).

10. Adjust the parking brake lever stroke.

SPECIFICATIONS AND MAINTENANCE CHARTS

ENGINE AND VEHICLE IDENTIFICATION

Engine								Model Year	
Code ①	Liters (cc)	Cu. In.	Cyl.	Fuel Sys.	Engine Type	Eng. Mfg.		Code ②	Year
2	2.0 (1975)	120.5	4	MFI	DOHC	KIA		7	2007
4	2.0 (1975)	120.5	4	MFI	DOHC	KIA		8	2008
3	2.7 (2656)	162	6	MFI	DOHC	KIA			

MFI: Multi-port Fuel Injection

DOHC: Double Overhead Camshafts

① 8th digit of VIN

② 10th digit of VIN

22140_SPOR_C0001

GENERAL ENGINE SPECIFICATIONS

Year	Model	Engine Displacement Liters	Engine VIN	Net Horsepower @ rpm	Net Torque @ rpm (ft. lbs.)	Bore x Stroke (in.)	Com- pression Ratio	Oil Pressure @ rpm
2007	Sportage	2.0	2, 4	140@6000	136@4500	3.23x3.68	10:01	23 ①
		2.7	3	173@6000	178@4000	3.41x2.95	10:01	7.3 ②
2008	Sportage	2.0	2, 4	140@6000	136@4500	3.23x3.68	10:01	23 ①
		2.7	3	173@6000	178@4000	3.41x2.95	10:01	7.3 ②

① At idle

② Minimum

22140_SPOR_C0002

ENGINE TUNE-UP SPECIFICATIONS

Year	Engine Displacement Liters	Engine VIN	Spark Plug Gap (in.)	Ignition Timing (deg.) MT	Ignition Timing (deg.) AT	Fuel Pump (psi)	Idle Speed (rpm) MT	Idle Speed (rpm) AT	Valve Clearance (in.) Intake	Valve Clearance (in.) Exhaust
2007	2.0	2, 4	0.039-0.043	NA	NA	50	NA	NA	0.0047-0.0110	0.0079-0.0150
	2.7	3	0.039-0.043	NA	NA	50	NA	NA	HYD.	HYD.
2008	2.0	2, 4	0.039-0.043	NA	NA	50	NA	NA	0.0047-0.0110	0.0079-0.0150
	2.7	3	0.039-0.043	NA	NA	50	NA	NA	HYD.	HYD.

B: Before Top Dead Center

HYD: Hydraulic lash adjusters

NA: Information not available

22140_SPOR_C0003

CAPACITIES

Year	Model	Engine Displacement Liters	Engine VIN	Engine Oil with Filter (qts.)	Transmission (pts.) Manual	Transmission (pts.) Auto.	Transfer Case (pts.)	Drive Axle Front (pts.)	Drive Axle Rear (pts.)	Fuel Tank (gal.)	Cooling System (qts.)
2007	Sportage	2.0	2, 4	4.23	4.6	16.4	1.6	①	1.8	15.3	6.4
		2.7	3	4.75	4.6	16.4	1.6	①	1.8	15.3	8.2
2008	Sportage	2.0	2, 4	4.23	4.6	16.4	1.6	①	1.8	15.3	6.4
		2.7	3	4.75	NA	16.4	1.6	①	1.8	15.3	8.2

NA: Not available

NOTE: All capacities are approximate. Add fluid gradually and check to be sure a proper fluid level is obtained.

① Included in transaxle capacity

22140_SPOR_C0004

FLUID SPECIFICATIONS

Year	Model	Engine Displ. Liters (VIN)	Engine Oil	Man. Trans.	Auto. Trans.	Drive Axle Front	Drive Axle Rear	Transfer Case	Power Steering Fluid	Brake Master Cylinder	Cooling System
2007	Sportage	2.0 (2, 4)	5W20	75W/90	①	②	80W/90	80W/90	PSF-3	DOT 3	③
	Sportage	2.7 (3)	5W20	75W/90	①	②	80W/90	80W/90	PSF-3	DOT 3	
2008	Sportage	2.0 (2, 4)	5W20	75W/90	①	②	80W/90	80W/90	PSF-3	DOT 3	③
	Sportage	2.7 (3)	5W20	NA	①	②	80W/90	80W/90	PSF-3	DOT 3	③

DOT: Department Of Transpotation

NA: Not available

① Diamond ATF SP-III or SK ATF SP III

② Included with transaxle

③ Diamond ATF SP-III or SK ATF SP III

22140_SPOR_C0014

VALVE SPECIFICATIONS

Year	Engine Displacement Liters	Engine VIN	Seat Angle (deg.)	Face Angle (deg.)	Spring Test Pressure (lbs. @ in.)	Spring Installed Height (in.)	Stem-to-Guide Clearance (in.) Intake	Stem-to-Guide Clearance (in.) Exhaust	Stem Diameter (in.) Intake	Stem Diameter (in.) Exhaust
2006	2.0	2, 4	45-45.5	45-45.5	①	①	0.0008-0.0020	0.0014-0.0026	0.2348-0.2354	0.2343-0.2348
	2.7	3	45-45.5	45-45.5	48.4 @ 1.3780	1.3780	0.0008-0.0020	0.0012-0.0026	0.2350-0.2354	0.2340-0.2350
2007	2.0	2, 4	45-45.5	45-45.5	①	①	0.0008-0.0020	0.0014-0.0026	0.2348-0.2354	0.2343-0.2348
	2.7	3	45-45.5	45-45.5	48.4 @ 1.3780	1.3780	0.0008-0.0020	0.0012-0.0026	0.2350-0.2354	0.2340-0.2350
2008	2.0	2, 4	45-45.5	45-45.5	①	①	0.0008-0.0020	0.0014-0.0026	0.2348-0.2354	0.2343-0.2348
	2.7	3	45-45.5	45-45.5	48.4 @ 1.3780	1.3780	0.0008-0.0020	0.0012-0.0026	0.2350-0.2354	0.2340-0.2350

① Intake: 41.4 lbs. @ 1.5354 in.

Exhaust: 90.4 lbs. @ 1.2008 in.

Lengths given are installed height.

22140_SPOR_C0006

CAMSHAFT SPECIFICATIONS CHART

All measurements are given in inches.

Year	Engine Displ. Liters	Engine VIN	Journal Dia.	Brg. Oil Clearance	Shaft End-play	Runout	Lobe Height Intake	Exhaust
2006	2.0	2, 4	1.1009-1.1016	0.0008-0.0024	0.0039-0.0079	0.0012	1.7527-1.7605	1.7487-1.7566
	2.7	3	1.0222-1.0228	0.0007-0.0024	0.0039-0.0059	0.0012	1.7303-1.7382	1.7303-1.7382
2008	2.0	2, 4	1.1009-1.1016	0.0008-0.0024	0.0039-0.0079	0.0012	1.7527-1.7605	1.7487-1.7566
	2.7	3	1.0222-1.0228	0.0007-0.0024	0.0039-0.0059	0.0012	1.7303-1.7382	1.7303-1.7382

22140_SPOR_C0007

CRANKSHAFT AND CONNECTING ROD SPECIFICATIONS

All measurements are given in inches.

Year	Engine Displacement Liters	Engine VIN	Crankshaft Main Brg. Journal Dia.	Main Brg. Oil Clearance	Shaft End-play	Thrust on No.	Connecting Rod Journal Diameter	Oil Clearance	Side Clearance
2007	2.0	2, 4	2.2418-2.2426	0.0011-0.0019	0.0024-0.0102	3	1.8898-1.8905	0.0009-0.0017	0.0039-0.0098
	2.7	3	2.4402-2.4409	0.0002-0.0009	0.0028-0.0098	3	1.8891-1.8898	0.0007-0.0014	0.0039-0.0098
2008	2.0	2, 4	2.2418-2.2426	0.0011-0.0019	0.0024-0.0102	3	1.8898-1.8905	0.0009-0.0017	0.0039-0.0098
	2.7	3	2.4402-2.4409	0.0002-0.0009	0.0028-0.0098	3	1.8891-1.8898	0.0007-0.0014	0.0039-0.0098

22140_SPOR_C0005

PISTON AND RING SPECIFICATIONS

All measurements are given in inches.

Year	Engine Displacement Liters	Engine VIN	Piston Clearance	Ring Gap Top Compression	Bottom Compression	Oil Control	Ring Side Clearance Top Compression	Bottom Compression	Oil Control
2007	2.0	2, 4	0.0008-0.0016	0.0091-0.0150	0.0130-0.0189	0.0079-0.0276	0.0016-0.0031	0.0012-0.0028	0.0024-0.0059
	2.7	3	0.0004-0.0012	0.0079-0.0138	0.0146-0.0205	0.0079-0.0276	0.0016-0.0031	0.0012-0.0028	SNUG
2008	2.0	2, 4	0.0008-0.0016	0.0091-0.0150	0.0130-0.0189	0.0079-0.0276	0.0016-0.0031	0.0012-0.0028	0.0024-0.0059
	2.7	3	0.0004-0.0012	0.0079-0.0138	0.0146-0.0205	0.0079-0.0276	0.0016-0.0031	0.0012-0.0028	SNUG

22140_SPOR_C0008

TORQUE SPECIFICATIONS
All readings in ft. lbs.

Year	Engine Displacement Liters	Engine VIN	Cylinder Head Bolts	Main Bearing Bolts	Rod Bearing Bolts	Crankshaft Damper Bolts	Flywheel Bolts	Manifold Intake	Manifold Exhaust	Spark Plugs	Oil Pan Drain Plug
2007	2.0	2, 4	①	②	36-38	123-130	87-94	12-17	31-40	15-22	29-32
	2.7	3	③	④	⑤	130-138	53-56	14-15	22-26	15-22	25-33
2008	2.0	2, 4	①	②	36-38	123-130	87-94	12-17	31-40	15-22	29-32
	2.7	3	③	④	⑤	130-138	53-56	14-15	22-26	15-22	25-33

① 10x99 bolts:
  Step 1: 16.6-19.5 ft. lbs.
  Step 2: + 60-65 degrees
  Step 3: +60-65 degrees
  12x151 bolts:
  Step 1: 20-23 ft. lbs.
  Step 2: + 60-65 degrees
  Step 3: +60-65 degrees

② Step 1: 20-23 ft. lbs.
  Step 2: + 60-65 degrees

③ Step 1: 18 ft. lbs.
  Step 2: +58-62 degrees
  Step 3: +43-47 degrees

④ M10 bolts:
  Step 1: 20-24 ft. lbs
  Step 2: +90-94 degrees
  M8 bolts:
  Step 1: 10-14 ft. lbs
  Step 2: +90-94 degrees

⑤ Step 1: 12-15 ft. lbs.
  Step 2: +90-94 degrees

22140_SPOR_C0009

WHEEL ALIGNMENT

Year	Model		Caster Range (+/-Deg.)	Caster Preferred Setting (Deg.)	Camber Range (+/-Deg.)	Camber Preferred Setting (Deg.)	Toe-in (in.)
2007	Sportage	F	0°30'0	3°36'	0°30'0	0°	0+/-0.79
		R	—	—	0°30'0	0°55'	4.6+3,-1
2008	Sportage	F	0°30'0	3°36'	0°30'0	0°	0+/-0.79
		R	—	—	0°30'0	0°55'	4.6+3,-1

22140_SPOR_C0010

TIRE, WHEEL AND BALL JOINT SPECIFICATIONS

Year	Model	OEM Tires		Tire Pressures (psi)		Wheel Size	Ball Joint Inspection	Lug Nut Torque (ft. lbs.)
		Standard	Optional	Front	Rear			
2007	Sportage	P215/65R16	P235/60R16	30	30	6.5J	①	66-81
2008	Sportage	P215/65R16	P235/60R16	30	30	6.5J	①	66-81

OEM: Original Equipment Manufacturer

PSI: Pounds Per Square Inch

① Replace if any wear shown.

22140_SPOR_C0011

BRAKE SPECIFICATIONS
All measurements in inches unless otherwise noted

Year	Model		Brake Disc			Brake Drum			Minimum Lining Thickness	Caliper Mounting Bolts (ft. lbs.)	Adaptor Plate Bolts (ft. lbs.)
			Original Thickness	Minimum Thickness	Maximum Run-out	Original Inside Diameter	Max. Wear Limit	Maximum Machine Diameter			
2007	Sportage	F	1.020	0.961	0.001	—	—	—	0.079	16-23	—
		R	0.390	0.315	0.001	—	—	—	0.079	16-23	—
2008	Sportage	F	1.020	0.961	0.001	—	—	—	0.079	16-23	—
		R	0.390	0.315	0.001	—	—	—	0.079	16-23	—

22140_SPOR_C0012

SCHEDULED MAINTENANCE INTERVALS
2007-08 Kia Sportage

TO BE SERVICED	TYPE OF SERVICE	7.5	15	22.5	30	37.5	45	52.5	60	67.5	75	82.5	90	97.5	105
Accessory drive belt 2.0L ①	I	✓	✓	✓	✓	✓	✓	✓	✓	✓	✓	✓	✓	✓	✓
Accessory drive belt 2.7L ①	I			✓					✓				✓		
Air cleaner filter	R			✓			✓			✓			✓		
A/C filter	R	Every 10,000 miles													
Automatic transmission fluid	I		✓		✓		✓		✓		✓		✓		✓
Ball joints	S/I				✓				✓				✓		
Battery condition	S/I		✓		✓		✓		✓		✓		✓		✓
Brake/Clutch fluid	I		✓		✓		✓		✓		✓		✓		✓
Brake lines & connections	S/I		✓		✓		✓		✓		✓		✓		✓
Chassis/body fasteners	S/I				✓				✓				✓		
Disc brakes	S/I	✓	✓	✓	✓	✓	✓	✓	✓	✓	✓	✓	✓	✓	✓
Driveshaft	S/I	Clean shaft and retorque bolts every 15,000 miles													
Driveshaft U-joints	L		✓		✓		✓		✓		✓		✓		✓
Engine coolant	R	Replace at 60,000 miles, then every 30,000 mile thereafter													
Engine oil & filter	R	Every 7,500 miles or 12 months													
EVAP canister filter	I				✓				✓				✓		
Exhaust system heat shields	S/I		✓			✓			✓			✓			✓
Fuel filter	R					✓					✓				
Fuel lines & hoses	S/I	✓	✓	✓	✓	✓	✓	✓	✓	✓	✓	✓	✓	✓	✓
Halfshafts and boots	S/I		✓		✓		✓		✓		✓		✓		✓
Locks & hinges	L	✓	✓	✓	✓	✓	✓	✓	✓	✓	✓	✓	✓	✓	✓
Manual transmission fluid	I		✓		✓		✓		✓		✓		✓		✓
Power steering fluid	I	✓	✓	✓	✓	✓	✓	✓	✓	✓	✓	✓	✓	✓	✓
Rear differential fluid ②	S/I	Check every 25,000 miles													
Spark plugs 2.0L platinum	R								✓						
Spark plugs 2.7L iridium	R	Every 100,000 miles or 10 years													
Steering operation & linkage	S/I	✓	✓	✓	✓	✓	✓	✓	✓	✓	✓	✓	✓	✓	✓
Throttle body	I	✓	✓	✓	✓	✓	✓	✓	✓	✓	✓	✓	✓	✓	✓
Timing belt	S/I				✓				✓				✓		
Tires	Rotate	✓	✓	✓	✓	✓	✓	✓	✓	✓	✓	✓	✓	✓	✓
Transfer case fluid ②	S/I	Check every 25,000 miles; replace every 60,000 miles													
Vacuum and crankcase hoses	S/I				✓				✓				✓		
Valve clearance - 2.0L	I	Every 60,000 miles or 4 years													
Water pump	I	Inspect when replacing drive belt													

R: Replace   S/I: Inspect and service, if needed   L: Lubricate

① Replace when excessive cracks occur or tension is not maintained

② Replace fluid whenever submerged in water

22140_SPOR_C0013

SCHEDULED MAINTENANCE INTERVALS
2007-08 Kia Sportage Footnotes

FREQUENT OPERATION MAINTENANCE (SEVERE SERVICE)

If a vehicle is operated under any of the following conditions it is considered severe service

- Towing a trailer or using a camper or car-top carrier

- Repeated short trips of less than 5 miles in temperatures below freezing, or trips of less than 10 miles in any temperature

- Prolonged idling (vehicle operation in stop and go traffic).

- Operating on rough, muddy, unpaved, dusty or salt-covered roads.

- Police, taxi, delivery usage or trailer towing usage.

- Driving in extremely hot (over 90°F) conditions

Oil & oil filter: change every 5000 miles or 5 months, whichever occurs first.

Air cleaner filter: inspect every 15,000 miles or 15 months and replace everything 30,000 miles or 30 months, whichever

Fuel system hoses (California models only): replace every 105,000 miles

Emission system hoses (non-CA models): inspect every 55,000 or 55 months, whichever occurs first

Emission system hoses (CA models): inspect every 60,000 miles or 60 months, which occurs first

Front and rear brakes: inspect every 15,000 miles or 15 months, whichever occurs first

Chassis and body fasteners: tighten every 15,000 miles or 15 months, whichever occurs first

Locks and hinges: lubricate every 5000 miles or 5 months, whichever occurs first

22140_SPOR_C0015

PRECAUTIONS

Before servicing any vehicle, please be sure to read all of the following precautions, which deal with personal safety, prevention of component damage, and important points to take into consideration when servicing a motor vehicle:

• Never open, service or drain the radiator or cooling system when the engine is hot; serious burns can occur from the steam and hot coolant.

• Observe all applicable safety precautions when working around fuel. Whenever servicing the fuel system, always work in a well-ventilated area. Do not allow fuel spray or vapors to come in contact with a spark, open flame, or excessive heat (a hot drop light, for example). Keep a dry chemical fire extinguisher near the work area. Always keep fuel in a container specifically designed for fuel storage; also, always properly seal fuel containers to avoid the possibility of fire or explosion. Refer to the additional fuel system precautions later in this section.

• Fuel injection systems often remain pressurized, even after the engine has been turned **OFF**. The fuel system pressure must be relieved before disconnecting any fuel lines. Failure to do so may result in fire and/or personal injury.

• Brake fluid often contains polyglycol ethers and polyglycols. Avoid contact with the eyes and wash your hands thoroughly after handling brake fluid. If you do get brake fluid in your eyes, flush your eyes with clean, running water for 15 minutes. If eye irritation persists, or if you have taken brake fluid internally, IMMEDIATELY seek medical assistance.

• The EPA warns that prolonged contact with used engine oil may cause a number of skin disorders, including cancer. You should make every effort to minimize your exposure to used engine oil. Protective gloves should be worn when changing oil. Wash your hands and any other exposed skin areas as soon as possible after exposure to used engine oil. Soap and water, or waterless hand cleaner should be used.

• All new vehicles are now equipped with an air bag system, often referred to as a Supplemental Restraint System (SRS) or Supplemental Inflatable Restraint (SIR) system. The system must be disabled before performing service on or around system components, steering column, instrument panel components, wiring and sensors. Failure to follow safety and disabling procedures could result in accidental air bag deployment, possible personal injury and unnecessary system repairs.

• Always wear safety goggles when working with, or around, the air bag system. When carrying a non-deployed air bag, be sure the bag and trim cover are pointed away from your body. When placing a non-deployed air bag on a work surface, always face the bag and trim cover upward, away from the surface. This will reduce the motion of the module if it is accidentally deployed. Refer to the additional air bag system precautions later in this section.

• Clean, high quality brake fluid from a sealed container is essential to the safe and proper operation of the brake system. You should always buy the correct type of brake fluid for your vehicle. If the brake fluid becomes contaminated, completely flush the system with new fluid. Never reuse any brake fluid. Any brake fluid that is removed from the system should be discarded. Also, do not allow any brake fluid to come in contact with a painted surface; it will damage the paint.

• Never operate the engine without the proper amount and type of engine oil; doing so WILL result in severe engine damage.

• Timing belt maintenance is extremely important. Many models utilize an interference-type, non-freewheeling engine. If the timing belt breaks, the valves in the cylinder head may strike the pistons, causing potentially serious (also time-consuming and expensive) engine damage. Refer to the maintenance interval charts for the recommended replacement interval for the timing belt, and to the timing belt section for belt replacement and inspection.

• Disconnecting the negative battery cable on some vehicles may interfere with the functions of the on-board computer system (s) and may require the computer to undergo a relearning process once the negative battery cable is reconnected.

• When servicing drum brakes, only disassemble and assemble one side at a time, leaving the remaining side intact for reference.

• Only an MVAC-trained, EPA-certified automotive technician should service the air conditioning system or its components.

BRAKES

ANTI-LOCK BRAKE SYSTEM (ABS)

GENERAL INFORMATION

ABS limits wheel lockup by automatically modulating the brake pressure during hard braking. ABS can improve steering control during hard braking and reduce stopping distances under most conditions.

The ABS controls all 4 channels (both fronts and rears) independently. The brake force required to engage the ABS function may vary depending on road surface conditions and tire adhesion. A dry paved surface requires a higher brake force than a slippery surface to engage ABS.

You may feel a pedal pulsation (up and down movement of the brake pedal) during an ABS event. In addition, a noise from inside the engine compartment may be heard. This is considered a normal characteristic. The brake pedal effort and pedal feel during normal braking are similar to that of a conventional brake system.

BRAKES   BLEEDING THE BRAKE SYSTEM

BLEEDING PROCEDURE

➡The ABS brake system is bled in the usual fashion with no special procedures required. Refer to the bleeding procedure described in this section. Make certain the master cylinder reservoir is filled before the bleeding is begun and check the level frequently.

1. Before servicing the vehicle, refer to the Precautions Section.

✳✳ WARNING

Do not reuse the drained fluid. Always use Genuine DOT3 or DOT 4

Brake Fluid. Using unapproved brake fluid can cause corrosion and decrease the life of the system. Make sure no dirt of other foreign matter is allowed to contaminate the brake fluid. Do not spill brake fluid on the vehicle, it may damage the paint; if brake fluid does contact the paint, wash it off immediately with water.

➡The reservoir on the master cylinder must be at the MAX (upper) level mark at the start of bleeding procedure and checked after bleeding each brake caliper. Add fluid as required.

2. Have someone slowly pump the brake pedal several times, then apply pressure.

3. Connect a length of clear plastic tube to the bleeder nipple and place the other end in a jar half full of clean brake fluid.

4. Loosen the right-rear brake bleed screw to allow air to escape from the system. Then tighten the bleed screw securely.

5. Repeat the procedure for each wheel in the sequence shown until air bubbles no longer appear in the fluid.

6. Refill the master cylinder reservoir to MAX (upper) level line.

BRAKES   FRONT DISC BRAKES

BRAKE CALIPER

REMOVAL & INSTALLATION

1. Raise and safely support the front of the vehicle.
2. Remove the front wheel.
3. Disconnect the brake fluid hose.
4. Remove the caliper mounting bolts.
5. Remove the brake caliper.

To install:

6. Install the brake caliper. Tighten the mounting bolts to 16–24 ft. lbs. (22–32 Nm).

7. Install the brake fluid hose. Tighten the hose fitting to 12–14 ft. lbs. (17–20 Nm).

8. Bleed the brake system.

9. Install the front wheel.

10. Before attempting to move the vehicle, pump the brake pedal to seat the pads against the rotors. Make sure the vehicle has a firm brake pedal. Check the level of the brake fluid and add DOT 3 brake fluid if necessary.

DISC BRAKE PADS

REMOVAL & INSTALLATION

See Figures 1 through 5.

1. Raise and safely support the front of the vehicle.
2. Remove the front wheel.
3. Remove the guide bolt (B), lift the caliper assembly (A) up and suspend it with a wire.
4. Remove the following parts from the caliper support:
   - Pad shims (A)
   - Pad retainer (B)
   - Pad assembly (C)

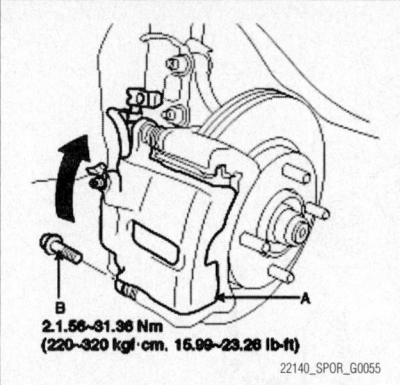

2.1.56~31.36 Nm
(220~320 kgf·cm, 15.99~23.26 lb-ft)

22140_SPOR_G0055

Fig. 1 Remove the guide bolt (B), lift the caliper assembly (A)

To install:

5. Install the pad retainers (A) on the caliper bracket.

6. Check the foreign material at the pad shims and the back of the pads.

➡Contaminated brake discs or pads reduce stopping ability. Keep grease off the discs and pads.

7. Install the brake pads (B) and pad shims (A) correctly. Install the pad with the wear indicator (C) on the inside.

➡If you are reusing the pads, always reinstall the brake pads in their original positions to prevent a momentary loss of braking efficiency.

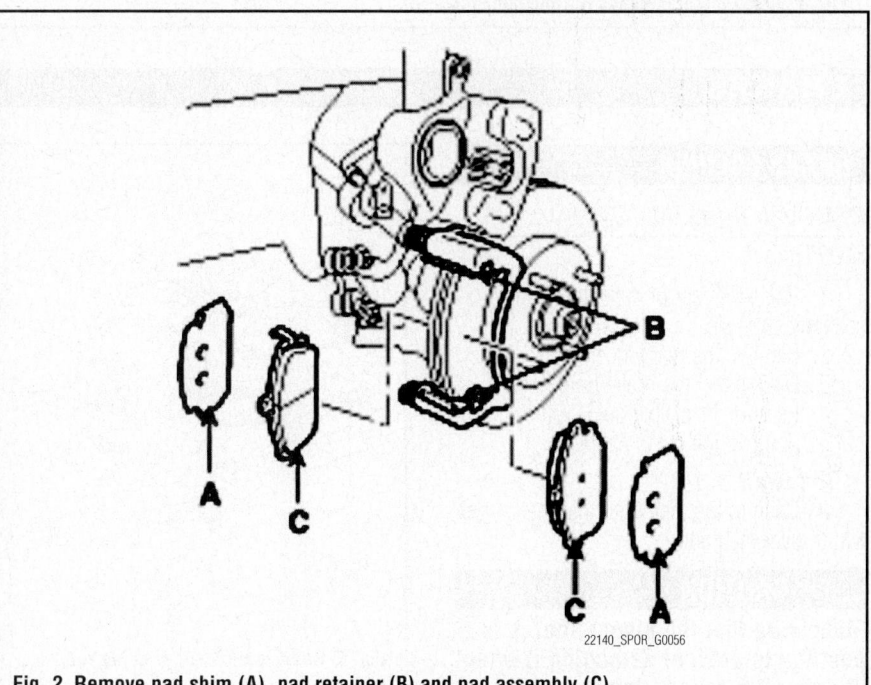

22140_SPOR_G0056

Fig. 2 Remove pad shim (A), pad retainer (B) and pad assembly (C)

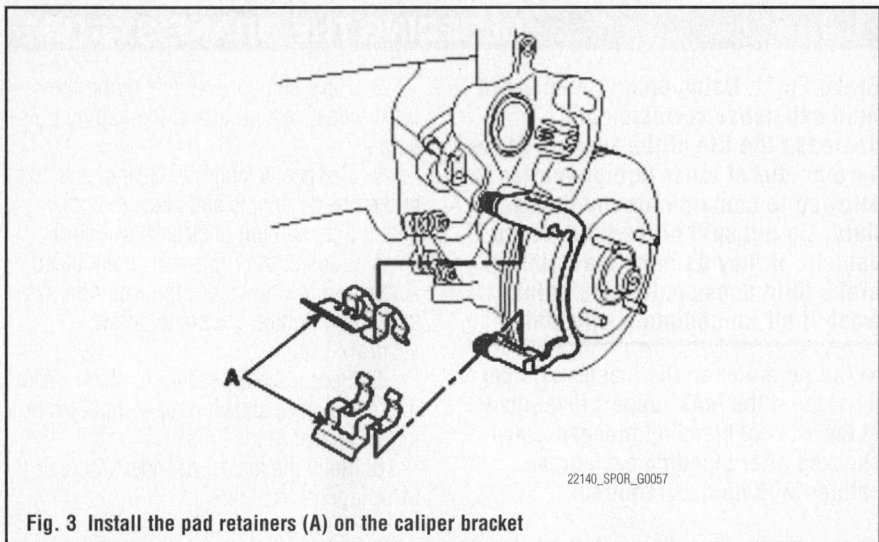

Fig. 3 Install the pad retainers (A) on the caliper bracket

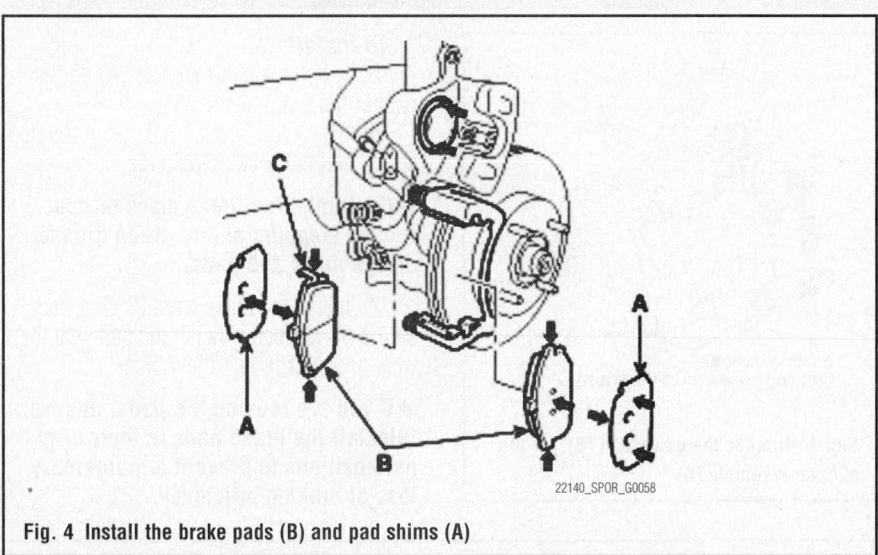

Fig. 4 Install the brake pads (B) and pad shims (A)

BRAKES

BRAKE CALIPER

REMOVAL & INSTALLATION

See Figure 6.

1. Raise the rear of the vehicle and securely support.
2. Remove the rear wheel.
3. Disconnect the brake fluid hose.
4. Remove the caliper guide rod bolts (B).
5. Remove the brake caliper.

To install:

6. Push in the piston so that the caliper will fit over the pads.

✳✳ CAUTION

Make sure that the piston boot is in position to prevent damaging it when pivoting the caliper down.

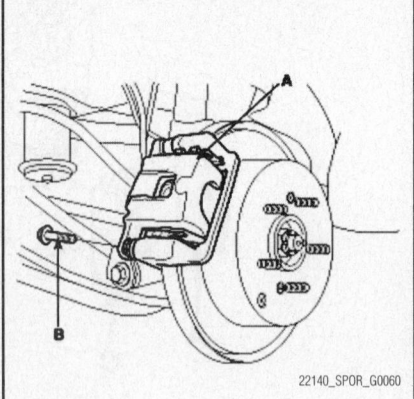

Fig. 6 Remove the caliper guide rod bolts (B)

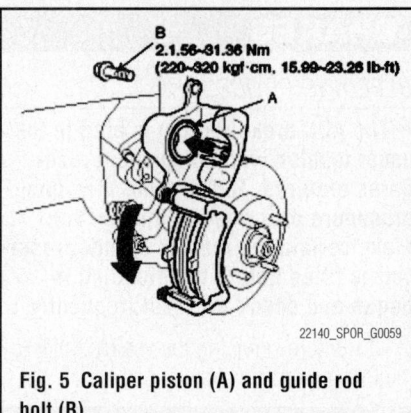

Fig. 5 Caliper piston (A) and guide rod bolt (B)

8. Push in the piston (A) so that the caliper will fit over the pads. Make sure that the piston boot is in position to prevent damaging it when pivoting the caliper down.

9. Pivot the caliper down into position. Being careful not to damage the pin boot, install the guide rod bolt (B). Tighten to 15.99–23.26 ft. lbs. (21.56–31.36 Nm).

10. Depress the brake pedal several times to make sure the brakes work, then test-drive.

➡**Engagement of the brake may require a greater pedal stroke immediately after the brake pads have been replaced as a set. Several applications of the brake will restore the normal pedal stroke.**

11. After installation, check for leaks at hose and line joints or connections, and retighten if necessary.

REAR DISC BRAKES

7. Pivot caliper down into position.
8. Install the guide rod bolt. Tighten to 16–23 ft. lbs. (22–31 Nm).
9. Reconnect the brake fluid line. Tighten the hose fitting to 12–14 ft. lbs. (17–20 Nm).
10. Depress the brake pedal several time to make sure the brakes work, then test-drive.
11. After installation, check for leaks at hose and line joints or connections, and retighten if necessary.

DISC BRAKE PADS

REMOVAL & INSTALLATION

See Figures 6 through 9.

1. Raise the rear of the vehicle and securely support.

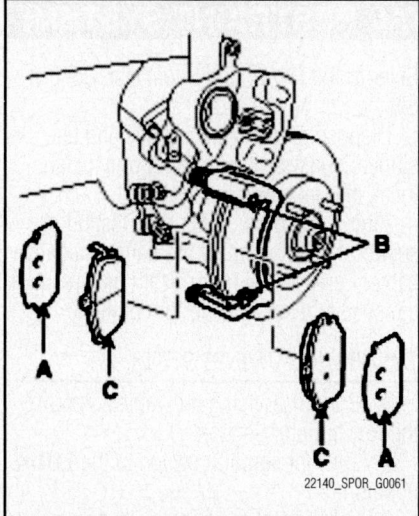

Fig. 7 Remove pad shim (A), pad retainer (B) and pad assembly (C)

Fig. 8 Install pad retainers (A)

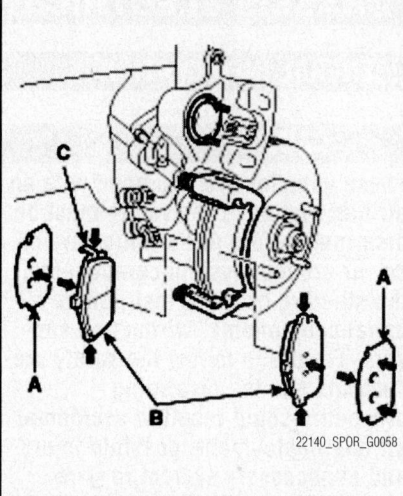

Fig. 9 Install the brake pads (B) and pad shims (A)

2. Remove the rear wheel.
3. Disconnect the brake fluid hose.
4. Remove the caliper guide rod bolts (B).
5. Remove the brake caliper.
6. Remove pad shim (A), pad retainer (B) and pad assembly (C) from the caliper bracket.

To install:

7. Install the pad retainers (A) on caliper bracket.
8. Check the foreign material at the pad shims and the back of the pads.

➡**Contaminated brake discs or pads reduce stopping ability. Keep grease off the discs and pads.**

9. Install the brake pads. Install the pad with the wear indicator on the inside.

➡**If you are reusing the pads, always reinstall the brake pads in their original positions to prevent a momentary loss of braking efficiency.**

10. Push in the piston so that the caliper will fit over the pads.

❋❋ **CAUTION**

Make sure that the piston boot is in position to prevent damaging it when pivoting the caliper down.

11. Pivot caliper down into position.
12. Install the guide rod bolt. Tighten to 15.99–23.26 ft. lbs. (21.56–31.36 Nm).
13. Depress the brake pedal several time to make sure the brakes work, then test-drive.
14. After installation, check for leaks at hose and line joints or connections, and retighten if necessary.

BRAKES

PARKING BRAKE

PARKING BRAKE SHOES

REMOVAL & INSTALLATION

1. Remove rear disc caliper assembly.
2. Before removing the brake disc, chalk both sides of the screw.

➡**Reduce the shoe gap by turning the adjuster with appropriate tool.**

3. After turning the pin to coincide with hole of spring cap, remove the shoe hold spring.

4. Remove the lower return spring.
5. Remove the parking brake cable mounting nuts.
6. Remove the parking brake shoes.

To install:

7. Install the upper return spring and brake shoes
8. Turn the adjuster in clockwise direction and install.
9. Install the lower return spring.
10. Install the shoe hold spring with pliers.

11. Install the disc brake and align the mark while tightening the screw.

ADJUSTMENT

1. Raise the rear of the vehicle, and make sure it is securely supported.
2. Remove the rear wheel and tire.
3. After removing the plug from the disc, rotate the toothed wheel with a screwdriver until the disc does not rotate.
4. Back off the toothed wheel by 5 notches.

CHASSIS ELECTRICAL   AIR BAG (SUPPLEMENTAL RESTRAINT SYSTEM)

GENERAL INFORMATION

�֎ CAUTION

These vehicles are equipped with an air bag system. The system must be disarmed before performing service on, or around, system components, the steering column, instrument panel components, wiring and sensors. Failure to follow the safety precautions and the disarming procedure could result in accidental air bag deployment, possible injury and unnecessary system repairs.

SERVICE PRECAUTIONS

Disconnect and isolate the battery negative cable before beginning any airbag system component diagnosis, testing, removal, or installation procedures. Allow system capacitor to discharge for three minutes before beginning any component service. This will disable the airbag system. Failure to disable the airbag system may result in accidental airbag deployment, personal injury, or death.

Do not place an intact undeployed airbag face down on a solid surface. The airbag will propel into the air if accidentally deployed and may result in personal injury or death.

When carrying or handling an undeployed airbag, the trim side (face) of the airbag should be pointing towards the body to minimize possibility of injury if accidental deployment occurs. Failure to do this may result in personal injury or death.

Replace airbag system components with OEM replacement parts. Substitute parts may appear interchangeable, but internal differences may result in inferior occupant protection. Failure to do so may result in occupant personal injury or death.

Wear safety glasses, rubber gloves, and long sleeved clothing when cleaning powder residue from vehicle after an airbag deployment. Powder residue emitted from a deployed airbag can cause skin irritation. Flush affected area with cool water if irritation is experienced. If nasal or throat irritation is experienced, exit the vehicle for fresh air until the irritation ceases. If irritation continues, see a physician.

Do not use a replacement airbag that is not in the original packaging. This may result in improper deployment, personal injury, or death.

The factory installed fasteners, screws and bolts used to fasten airbag components have a special coating and are specifically designed for the airbag system. Do not use substitute fasteners. Use only original equipment fasteners listed in the parts catalog when fastener replacement is required.

During, and following, any child restraint anchor service, due to impact event or vehicle repair, carefully inspect all mounting hardware, tether straps, and anchors for proper installation, operation, or damage. If a child restraint anchor is found damaged in any way, the anchor must be replaced. Failure to do this may result in personal injury or death.

Deployed and non-deployed airbags may or may not have live pyrotechnic material within the airbag inflator.

Do not dispose of driver/passenger/curtain airbags or seat belt tensioners unless you are sure of complete deployment.

Refer to the Hazardous Substance Control System for proper disposal.

Dispose of deployed airbags and tensioners consistent with state, provincial, local, and federal regulations.

After any airbag component testing or service, do not connect the battery negative cable. Personal injury or death may result if the system test is not performed first.

DISARMING THE SYSTEM

1. Before servicing the vehicle, refer to the Precautions Section.
2. Turn the ignition switch to the **LOCK** position.
3. Disconnect the negative battery cable.
4. Wait 10 minutes for the battery back-up power to discharge.

ARMING THE SYSTEM

1. Before servicing the vehicle, refer to the Precautions Section.
2. Connect the negative battery cable.
3. Turn the ignition switch **ON**.
4. Verify that the air bag indicator illuminates for 4–8 seconds, then goes off.

CLOCKSPRING CENTERING

Prior to installing the clock spring, align the mating mark and "NEUTRAL" position indicator of the clock spring, and, after turning the front wheels to the straight-ahead position, install the clock spring to the column switch. If the mating mark of the clock spring is not properly aligned, the steering wheel may not completely rotate during a turn, or the flat cable within the clock spring may be severed, obstructing normal operation of the SRS and possibly leading to serious injury to the vehicle's driver.

DRIVETRAIN

AUTOMATIC TRANSAXLE ASSEMBLY

REMOVAL & INSTALLATION

See Figures 10 through 23.

1. Remove the air duct.
2. Remove the battery.
3. Remove the battery tray.
4. Remove the air cleaner assembly.
5. Remove the intercooler inlet pipe.
6. Disconnect the connectors relevant to the transaxle.
7. Disconnect the ground wire.
8. Remove the nut (B) which mounts the shift lever cable (A) to the inhibiter switch.
9. Remove the shift lever cable (B) clip (A).
10. Disconnect the ATF cooler hoses (A).
11. Support the engine using SST (09200-38001).
12. Remove the transaxle mounting bracket (A) bolts.
13. Remove the transaxle upper mounting bolts (A)
14. Remove the bolts which mount the transaxle to the front sub frame.
15. Raise the vehicle.
16. Drain the transaxle fluid.
17. Support the transaxle with a jack.
18. Remove the driveshafts.
19. Remove the bolt (A) which mounts the transaxle to the rear sub-frame.

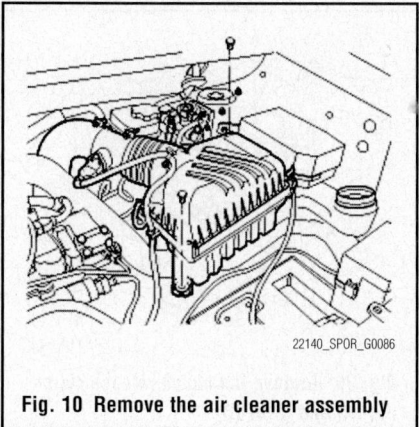

22140_SPOR_G0086

Fig. 10 Remove the air cleaner assembly

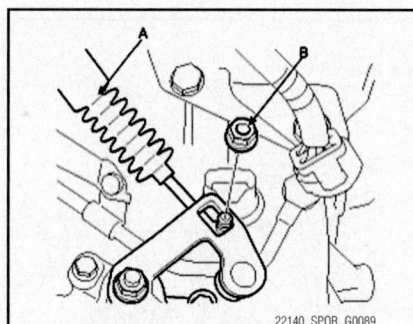

22140_SPOR_G0089

Fig. 13 Remove the nut (B) which mounts the shift lever cable (A) to the inhibiter switch

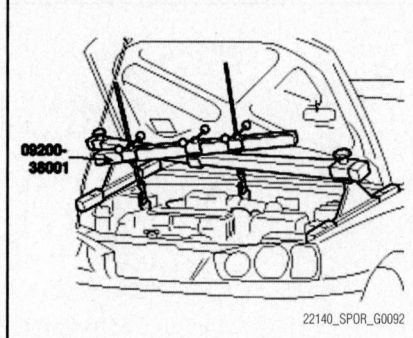

22140_SPOR_G0092

Fig. 16 Support the engine using SST (09200-38001)

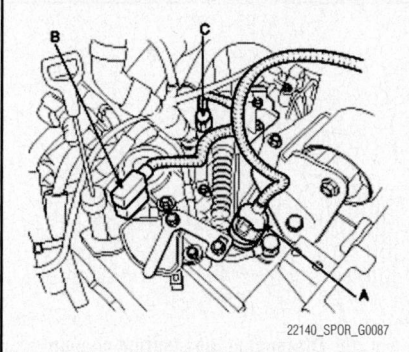

22140_SPOR_G0087

Fig. 11 Disconnect the connectors relevant to the transaxle

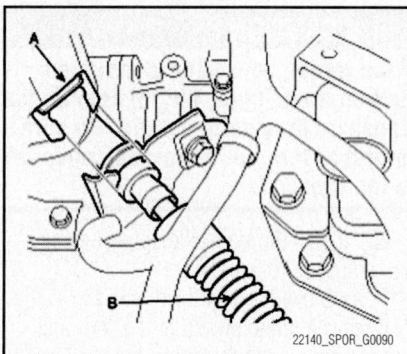

22140_SPOR_G0090

Fig. 14 Remove the shift lever cable (B) clip (A)

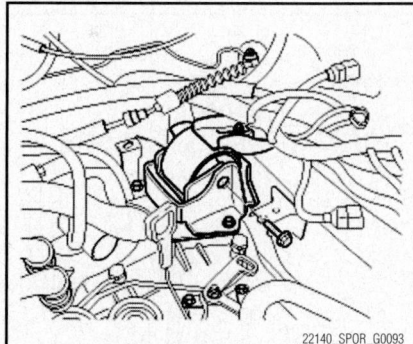

22140_SPOR_G0093

Fig. 17 Remove the transaxle mounting bracket (A) bolts

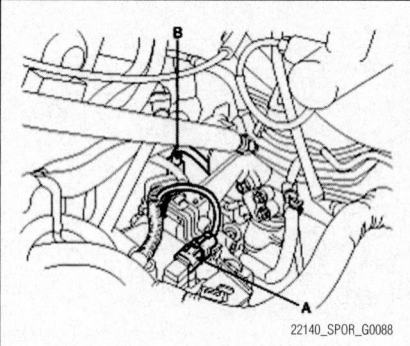

22140_SPOR_G0088

Fig. 12 Disconnect the connectors relevant to the transaxle

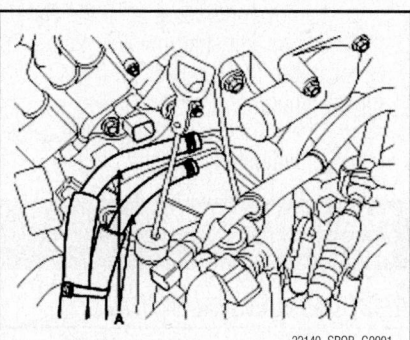

22140_SPOR_G0091

Fig. 15 Disconnect the ATF cooler hoses (A)

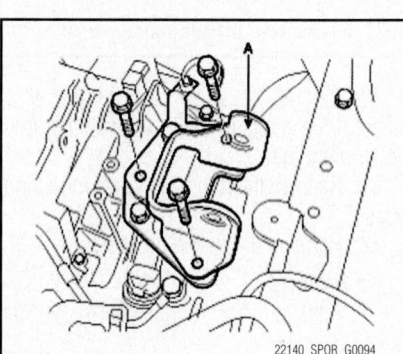

22140_SPOR_G0094

Fig. 18 Remove the transaxle mounting bracket (A) bolts

Fig. 19 Remove the transaxle upper mounting bolts (A)

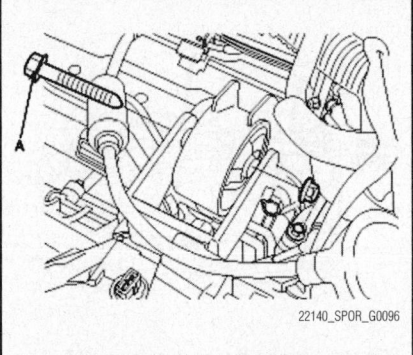

Fig. 20 Remove the bolt (A) which mounts the transaxle to the rear sub-frame

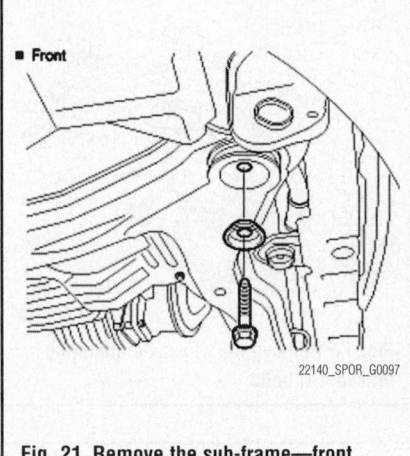

Fig. 21 Remove the sub-frame—front

20. Remove the sub-frame. If it is a 4 wheel drive vehicle (4WD), remove the propeller shaft first

21. Remove the transaxle lower mounting bolts.

22. Remove the transaxle assembly.

To install:

23. Installation is the reverse of removal.

24. Attach the torque converter on the transaxle side and mount the transaxle assembly onto the engine.

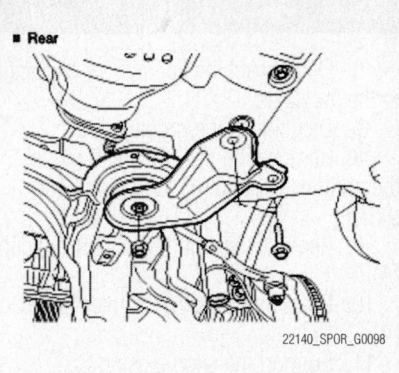

Fig. 22 Remove the sub-frame—RH rear

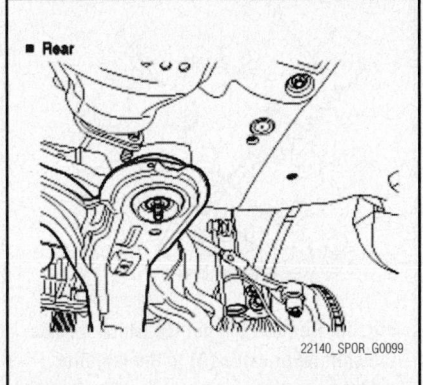

Fig. 23 Remove the sub-frame—LH rear

✳✳ CAUTION

If the torque converter is mounted first on the engine, the oil seal on the transaxle may be damaged. Be sure to first assemble the torque converter to the transaxle.

25. Install the transaxle control cable and adjust as follows.

  a. Move the shift lever and the transaxle range switch to the "N" position, and install the control cable.

  b. When connecting the control cable to the transaxle mounting bracket, install the clip until it contacts the control cable.

  c. Remove any free-play in the control cable by adjusting nut and then check to see that the selector lever moves smoothly.

  d. Check to see that the control cable has been adjusted correctly.

MANUAL TRANSAXLE ASSEMBLY

REMOVAL & INSTALLATION
See Figures 24 through 34.

1. Remove the air duct assembly.
2. Remove the battery and battery tray.

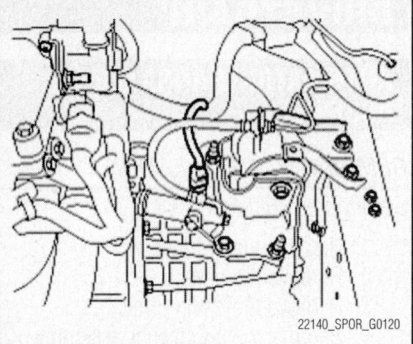

Fig. 24 Remove the back-up lamp connector and the vehicle speed sensor

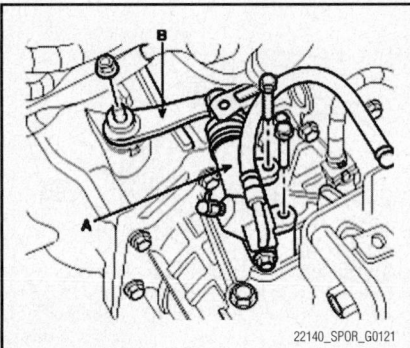

Fig. 25 Remove the clutch release cylinder (A) and lever (B)

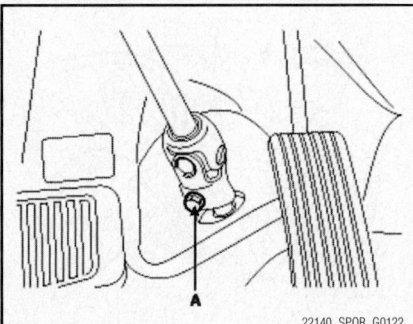

Fig. 26 Disconnect the steering column shaft from the universal joint in the gear box

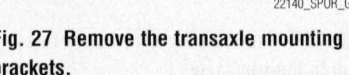

Fig. 27 Remove the transaxle mounting brackets.

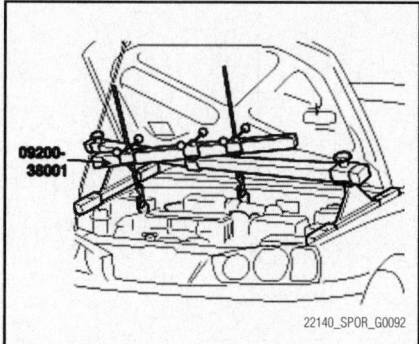

Fig. 28 Using SST (09200-38001), support the engine assembly

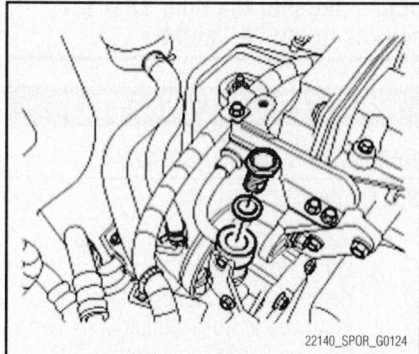

Fig. 29 Disconnect the power steering oil pressure tube from the pump

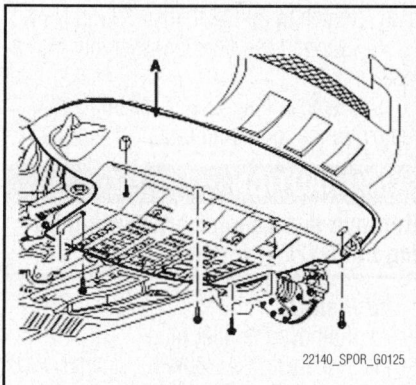

Fig. 30 Remove the undercover (A)

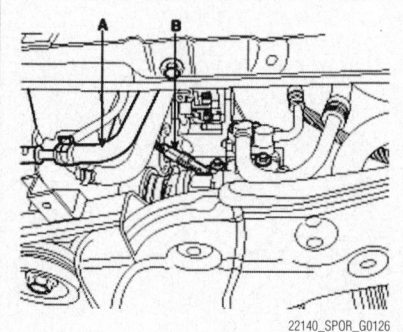

Fig. 31 Disconnect the steering tube (A) and the air conditioner switch (B)

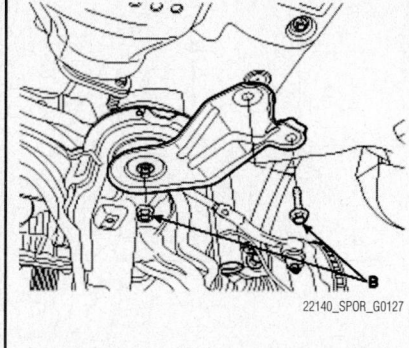

Fig. 32 Remove the cross member mounting brackets

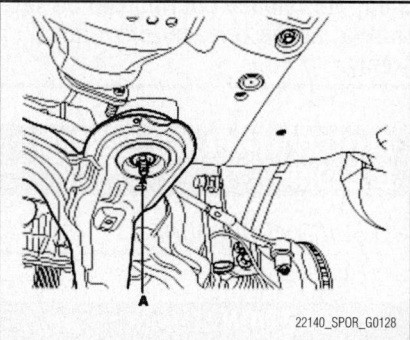

Fig. 33 Remove the cross member mounting brackets

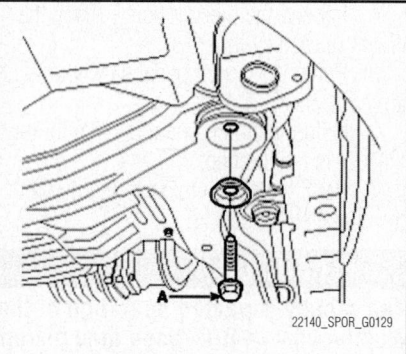

Fig. 34 Remove the cross member mounting brackets

3. Remove the air cleaner assembly and air flow sensor.

4. Remove the back-up lamp connector and the vehicle speed sensor.

5. Disconnect the connectors and the terminals.

6. Remove the clutch release cylinder (A) and lever (B).

7. Separate the transaxle cable from the transaxle assembly.

8. Disconnect the steering column shaft from the universal joint in the gear box.

9. Remove the transaxle clutch housing upper mounting bolts.

10. Remove the transaxle mounting brackets; front, rear, left (A).

11. Using SST (09200-38001), support the engine assembly.

12. Disconnect the power steering oil pressure tube from the pump. Plug after removal to prevent contamination and leaks.

13. Remove the wheel and tire.

14. Disconnect the strut assembly, tie rod and stabilizer bar link from the knuckle.

15. Remove the wheel speed sensor.

16. Remove the brake caliper and suspend it with a wire.

17. Raise the vehicle and remove the undercover (A).

18. Remove the transaxle oil drain plug and drain the fluid.

19. Remove the front muffler.

20. Disconnect the steering tube (A) at the cross member (sub-frame) and the air conditioner switch (B).

21. Install a jack for the removal of a cross member (sub-frame).

22. Remove the cross member.

23. Install a jack for the removal of the transaxle.

24. Remove the front and the rear roll stopper.

25. Remove the engine and the transaxle mounting bolts.

26. Lowering the jack slowly, remove the transaxle.

27. Installation is in the reverse order of removal.

CLUTCH DRIVEN DISC & PRESSURE PLATE

REMOVAL & INSTALLATION
See Figures 35 through 37.

1. Remove the manual transaxle assembly.

2. Insert the special tool (09411-11000) in the clutch disc to prevent the disc from falling.

Fig. 35 Insert the special tool (09411-11000) in the clutch disc to prevent the disc from falling

3. Remove the bolts which attach the clutch cover to the flywheel in a star pattern. Loosen the bolts in succession, one or two turns at a time, to avoid bending the cover flange.

➡ **Do not clean the clutch disc or the release bearing with cleaning solvent.**

4. Remove the release fork shaft and bushing.

To install:

5. Apply multipurpose grease to the spline of the disc.

❋❋ CAUTION

When installing the clutch, apply grease to each part, but be careful not to apply excessive grease. It can cause clutch slippage and shudder.

6. Install the clutch disc assembly to the flywheel using the special tool (09411-11000).

7. Install the clutch cover assembly to the flywheel and temporarily tighten the bolts one or two steps at a time in a star pattern. Tighten to Clutch cover bolt: 11–16 ft. lbs. (15–22 Nm).

8. Align the bearing (A) to the release fork (B) and then install it to the sleeve of the housing.

❋❋ CAUTION

Apply multipurpose grease to the bearing sleeve, contact point of the release fork (B) and the bushing inner surface (C).

9. Install the release lever to the release fork.

10. Install the manual transaxle assembly to the engine.

❋❋ CAUTION

If the transaxle assembly is installed to the engine without performing this

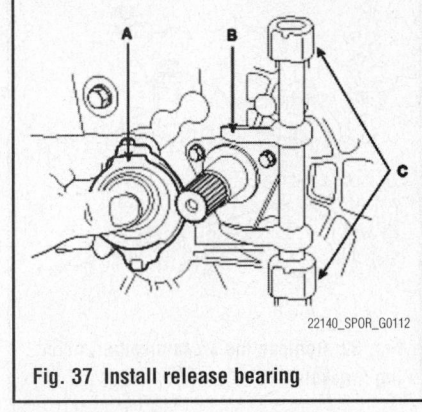

Fig. 37 Install release bearing

step, the release bearing can be separated, as the release fork rotates freely.

CLUTCH HYDRAULIC SYSTEM BLEEDING

BLEEDING PROCEDURE

See Figures 38 and 39.

❋❋ CAUTION

Use only SAE J1703 (DOT 3 or DOT 4). Avoid mixing different brands of fluid.

1. Loosen the bleeder screw (B) at the clutch release cylinder (A).

2. Pump the clutch pedal slowly until all air is expelled.

3. Hold the clutch pedal down until the bleeder is retightened.

4. Refill the clutch master cylinder with the specified fluid.

❋❋ CAUTION

The rapidly-repeated operation of the clutch pedal in B-C range may disrupt the release cylinder's position. During the bleeding operation, press the

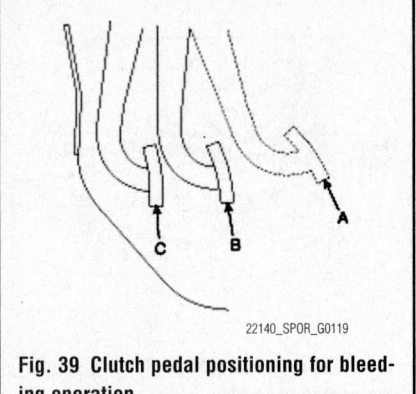

Fig. 39 Clutch pedal positioning for bleeding operation

clutch pedal to the floor after it returns to the "A" point.

TRANSFER CASE ASSEMBLY

REMOVAL & INSTALLATION

See Figure 40.

1. Remove the battery terminal.
2. Lift up the vehicle.
3. Remove the propeller shaft.
4. Remove the front muffler (A).
5. Remove the RH driveshaft.
6. Remove the oil drain plug and drain the fluid.
7. After draining, re-tighten the oil drain plug. Tighten to 29–43 ft. lbs. (39–59 Nm)
8. Support the transfer assembly with a jack.
9. Remove the transfer assembly by removing the mounting bolts.

❋❋ CAUTION

Remove the transfer bracket mounting bolts (2EA) together.

To install:

10. Remove the filler plug.
11. Refill with SAE 80W90. Quantity: 0.8 qt (0.8L)

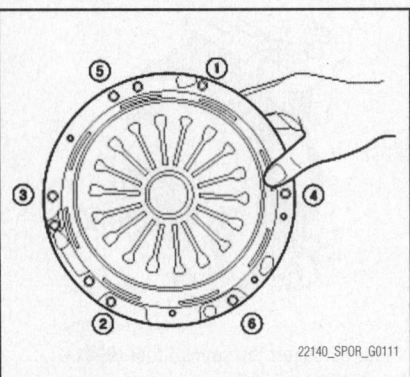

Fig. 36 Star pattern to install clutch cover

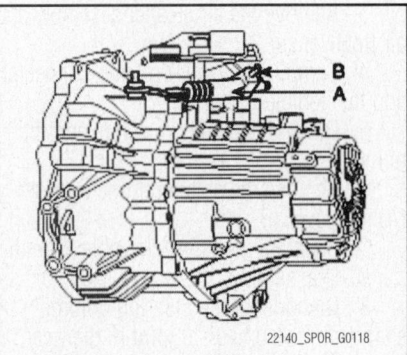

Fig. 38 Loosen the bleeder screw (B) at the clutch release cylinder (A)

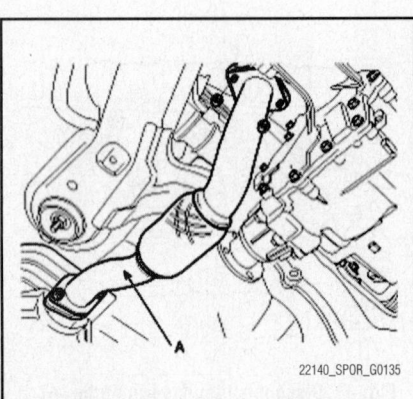

Fig. 40 Remove the front muffler (A)

12. Fix it in proper position with mounting bolts.

FRONT HALFSHAFT

REMOVAL & INSTALLATION

See Figures 41 through 50.

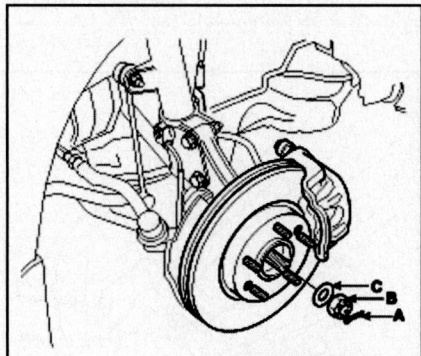

Fig. 41 Remove the split pin (A), the castle nut (B) and the washer (C) from the front hub

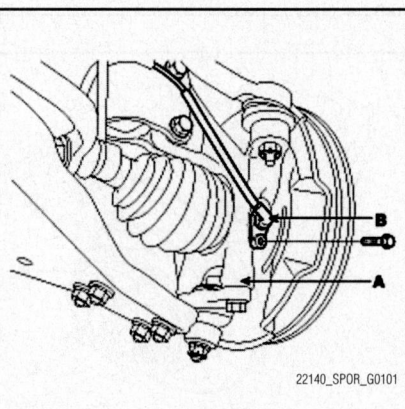

Fig. 42 Remove the wheel speed sensor (B) from the knuckle (A)

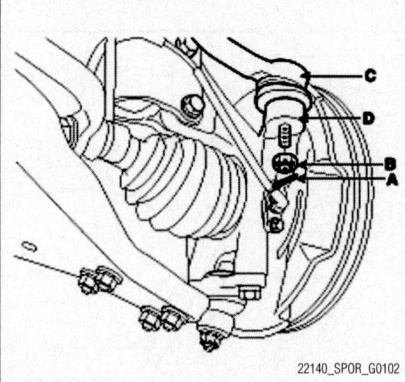

Fig. 43 Remove the split pin (A) and the castle nut (B) from the tie rod end

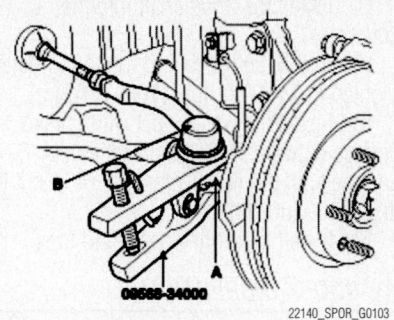

Fig. 44 Disconnect the ball joint (B) from knuckle (A) using the special tool (09568-34000)

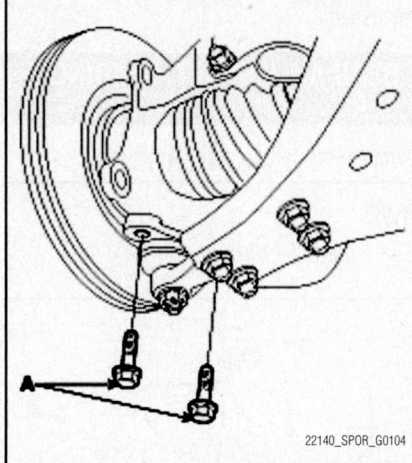

Fig. 45 Remove the lower arm ball joint mounting bolts (A)

1. Raise the front of the vehicle, and make sure it is securely supported.
2. Remove the front wheels and tires.
3. Remove the split pin (A), the castle nut (B) and the washer (C) from the front hub.
4. Remove the wheel speed sensor (B) from the knuckle (A).
5. Remove the split pin (A) and the castle nut (B) from the tie rod end.
6. Disconnect the ball joint (B) from knuckle (A) using the special tool (09568-34000).

✳✳ CAUTION

Apply a few drops of oil to the special tool. (Boot contact part)

7. Remove the lower arm ball joint mounting bolts (A).
8. Using a plastic hammer (A), disconnect the driveshaft (C) from the axle hub (B).
9. Push the axle hub (B) outward and separate the driveshaft (C) from the axle hub (B).

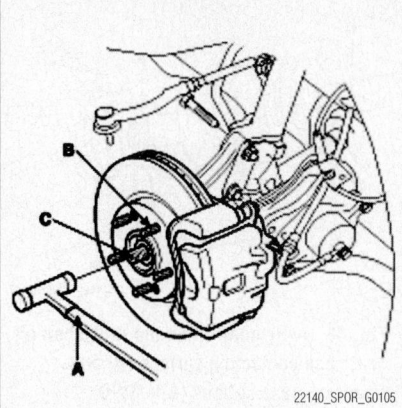

Fig. 46 Using a plastic hammer (A), disconnect the driveshaft (C) from the axle hub (B)

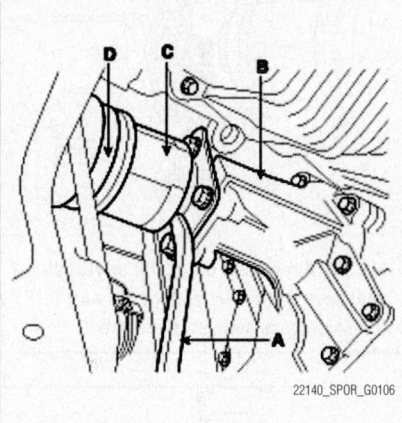

Fig. 47 Separate the driveshaft (D) from the transaxle case

10. Insert a pry bar (A) between the transaxle case (B) and joint case (C), and separate the driveshaft (D) from the transaxle case.

✳✳ CAUTION

Use a pry bar being careful not to damage the transaxle or joint. Do not insert the pry bar too deep, as this may cause damage to the oil seal. Do not pull the driveshaft with excessive force as it may cause components inside the axle shaft joint to dislodge resulting in a torn boot or a damaged bearing. Plug the hole of the transaxle case with the oil seal cap to prevent contamination. Replace the retainer ring whenever the driveshaft is removed from the transaxle case.

To install:
11. Apply gear oil on the driveshaft oil

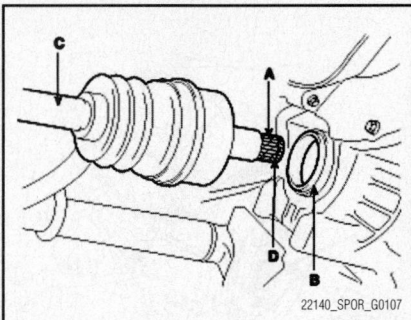

Fig. 48 Apply gear oil on the driveshaft oil seal case contacting surface (B) and transaxle case splines (A)—2WD

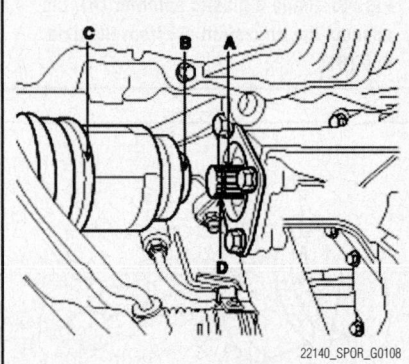

Fig. 49 Apply gear oil on the driveshaft oil seal case contacting surface (B) and transaxle case splines (A)—4WD

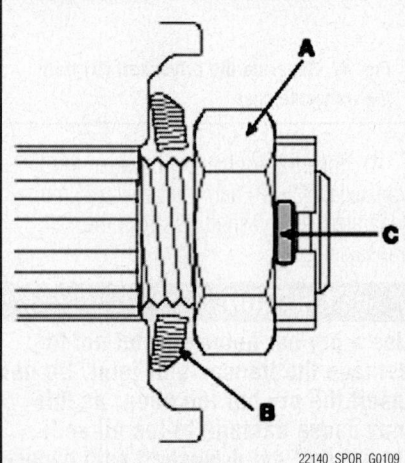

Fig. 50 Install the washer (B) with convex surface outward. Install the castle nut (A) and the split pin (C)

seal case contacting surface (B) and transaxle case splines (A).

12. Before installing the driveshaft (C), set the opening side of the circlip (D) facing downward.

13. After installation, check that the driveshaft cannot be removed by hand.

14. Install the drive shaft into the knuckle.

15. Install the lower arm mounting bolts. Tighten to 74–89 ft. lbs. (100–120 Nm).

16. Install the washer (B) with convex surface outward. Install the castle nut (A) and the split pin (C). Tighten to 148–207 ft. lbs. (200–280 Nm).

17. Install the front wheels and tires.

CV-BOOTS INSPECTION

Inspect the CV-boots for tears, leaks, cracks or any other form of damage and replace if necessary.

Check circlip, snap ring and boot bands for breakage or deformation. Replace as necessary.

REAR AXLE SHAFT, BEARING & SEAL

REMOVAL & INSTALLATION

4WD

See Figures 51 through 55.

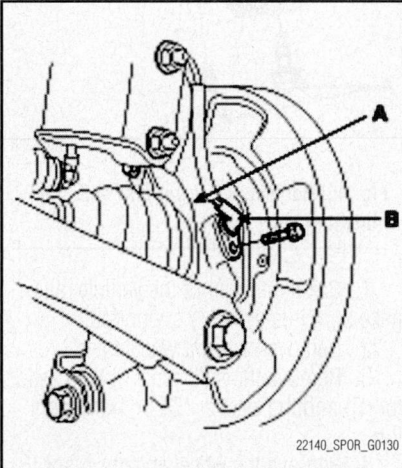

Fig. 51 Remove the wheel speed sensor (B) from the axle carrier (A)

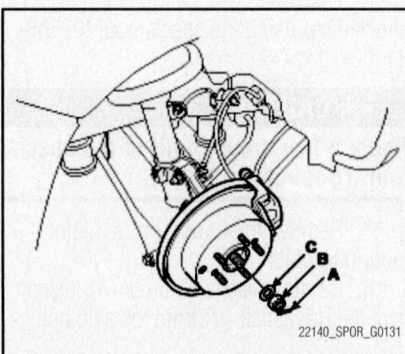

Fig. 52 Remove the split pin (A), remove castle nut (B) and washer (C) from the rear hub

1. Raise the rear of the vehicle, and make sure it is securely supported.

2. Remove the rear wheel and tire.

3. Remove the wheel speed sensor (B) from the axle carrier (A).

4. Remove the split pin (A), remove castle nut (B) and the washer (C) from the rear hub.

5. Remove the trailing arm mounting bolt (B) from the knuckle (A).

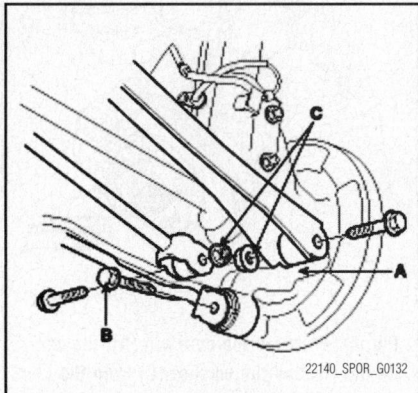

Fig. 53 Remove the trailing arm mounting bolt (B) from the knuckle (A)

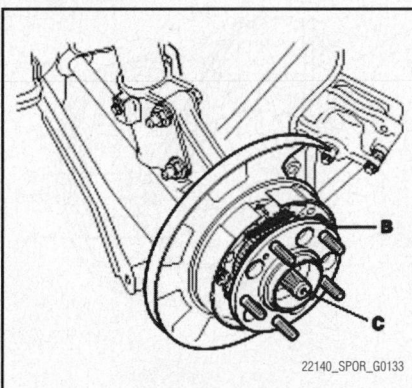

Fig. 54 Push the axle hub (B) outward and separate the driveshaft (C) from the axle hub (B)

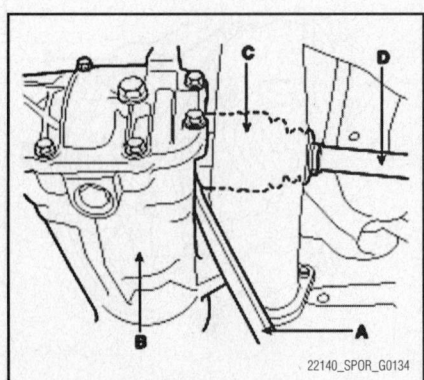

Fig. 55 Insert a pry bar (A) between the differential case (B) and joint case (C)

6. Remove the suspension arm mounting nuts (C).

7. Push the axle hub (B) outward and separate the driveshaft (C) from the axle hub (B).

8. Insert a pry bar (A) between the differential case (B) and joint case (C), and separate the driveshaft (D) from the differential case.

❄❄ CAUTION

Use a pry bar (A) being careful not to damage the transaxle and joint. Do not insert the pry bar (A) too deep, as this may cause damage to the oil seal. Do not pull the driveshaft by

excessive force because it may cause components inside the axle joint kit to dislodge resulting in a torn boot or a damaged bearing. Plug the hole of the transaxle case with the oil seal cap to prevent contamination. Replace the retainer ring whenever the driveshaft is removed from the transaxle case.

To install:

9. Apply gear oil on the driveshaft differential case contacting surface and driveshaft splines.

10. Before installing the driveshaft, set the opening side of the circlip facing downward.

11. After installation, check that the driveshaft cannot be removed by hand.

12. Install the axle into the knuckle.

13. Install the suspension arm mounting nuts and trailing arm mounting bolt to the knuckle. Tighten to
- Suspension arm mounting nuts: 103.8–118.0 ft. lbs. (140–160 Nm)
- Trailing arm mounting bolt: 73.8–88.5 ft. lbs. (100–120 Nm)

14. Install the washer, castle nut and split pin to the rear hub. Tighten to 147.5–206.6 ft. lbs. (200–280 Nm)

15. Install the wheel speed sensor to the knuckle.

16. Install the rear wheel and tire.

ENGINE COOLING

THERMOSTAT

REMOVAL & INSTALLATION

1. Drain the engine coolant so its level is below thermostat.

2. Remove the water inlet fitting, gasket and thermostat.

3. Installation is the reverse of removal.

WATER PUMP

REMOVAL & INSTALLATION

2.0L Engine

See Figures 56 through 59.

1. Drain the engine coolant.

❄❄ WARNING

System is under high pressure when the engine is hot. To avoid danger of releasing scalding engine coolant, remove the cap only when the engine is cool.

2. Remove the drive belts.

3. Remove the timing belt.

4. Remove the timing belt idler.

5. Remove the power steering pump and use a wire to secure the pump to the vehicle so that it is out of the way.

6. Remove the bolts (B, C) and power steering pump bracket (A).

7. Remove the alternator.

8. Remove the water pump.
   a. Remove the 2 bolts (D) and alternator brace (A).

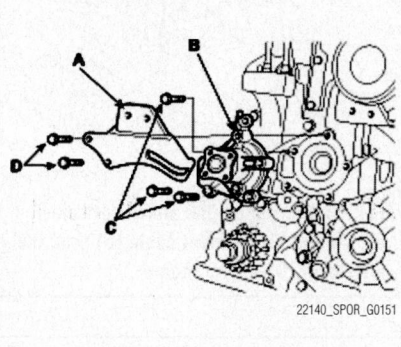

22140_SPOR_G0151

Fig. 57 Remove the water pump

b. Remove the 3 bolts (C) and remove the water pump (B) and gasket.

To install:

9. Install the water pump.
   a. Install the water pump (B) and a new gasket with the 3 bolts (C). Torque to 9–11 ft. lbs. (12–15 Nm).
   b. Install the alternator brace (A) with the 2 bolts (D). Tighten to 15–20 ft. lbs. (20–27 Nm).

10. Install the power steering pump bracket (A) and bolts (B, C). Tighten to Bolts (B): 25–36 ft. lbs. (34–49 Nm); Bolts (C): 11–15 ft. lbs. (15–20 Nm).

11. Install the alternator.

12. Install the power steering pump.

13. Install the timing belt idler.

14. Install the timing belt.

15. Install the water pump pulley.

16. Install the drive belts.

17. Tighten the water pump pulley bolts. Tighten to 7 ft. lbs. (9 Nm).

18. Fill with engine coolant.

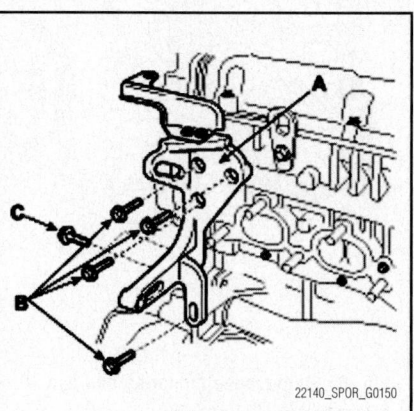

22140_SPOR_G0150

Fig. 56 Remove the bolts (B, C) and power steering pump bracket (A)

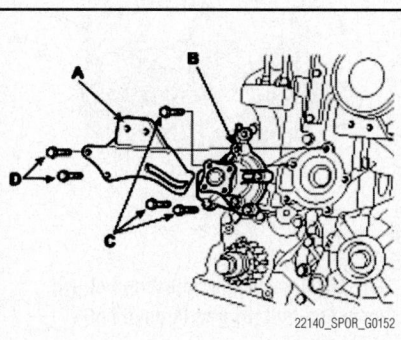

22140_SPOR_G0152

Fig. 58 Torque water pump bolts (A, B, C, D)

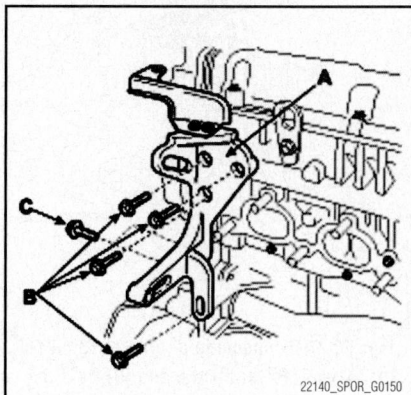

22140_SPOR_G0150

Fig. 59 Install the power steering pump bracket (A) and bolts (B, C)

19. Start engine and check for leaks.
20. Recheck engine coolant level.

2.7L Engine

See Figures 60 and 61.

1. Drain the engine coolant.

2. Remove the drive belt.
3. Remove the timing belt.
4. Remove the timing belt idler.
5. Remove the water pump (A) and gasket (B).

To install:

6. Install the water pump and a new gasket with the 8 bolts. Tighten to 11–16 ft. lbs. (15–22 Nm).

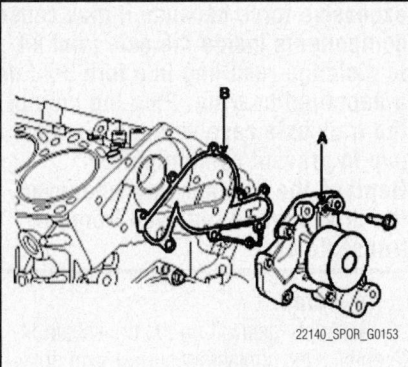

Fig. 60 Remove the water pump (A) and gasket (B)

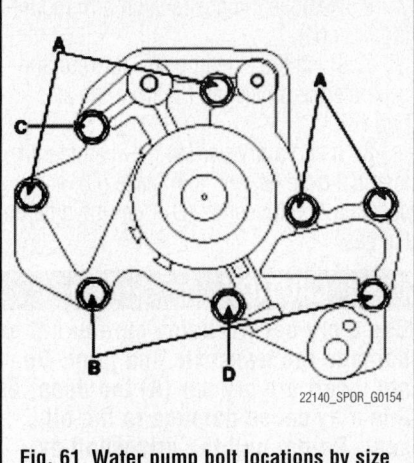

Fig. 61 Water pump bolt locations by size

a. Water pump bolt locations by size:
- A: 4 each 8 x 25
- B: 2 each 8 x 30
- C: 1 each 8 x 32
- D: 1 each 8 x 40

7. Install the timing belt idler.
8. Install the timing belt.
9. Install drive belt.
10. Fill with engine coolant.
11. Start engine and check for leaks.
12. Recheck engine coolant level.

ENGINE ELECTRICAL

ALTERNATOR

REMOVAL & INSTALLATION

See Figures 62 through 65.

1. Disconnect the battery negative terminal first, then the positive terminal.
2. Disconnect the alternator connector (A) and "B" terminal cable (B) from the alternator (C).

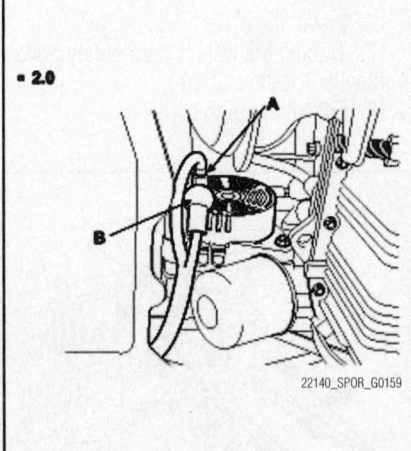

Fig. 62 Disconnect the alternator connector (A) and "B" terminal cable (B) from the alternator (C)—2.0L engine

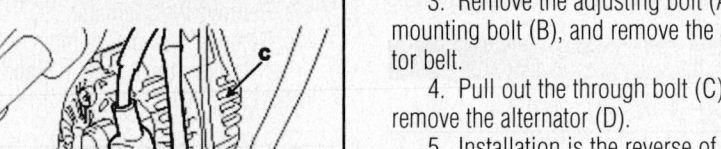

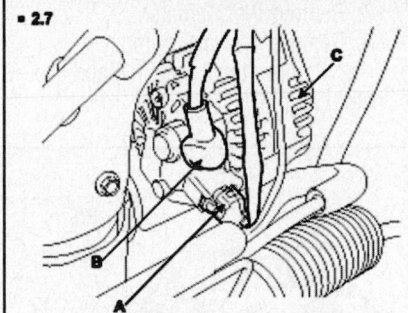

Fig. 63 Disconnect the alternator connector (A) and "B" terminal cable (B) from the alternator (C)—2.7L engine

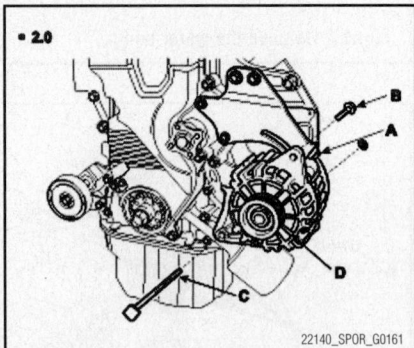

Fig. 64 Remove the adjusting bolt (A), mounting bolt (B) and through bolt (C)—2.0L engine

CHARGING SYSTEM

3. Remove the adjusting bolt (A) and mounting bolt (B), and remove the alternator belt.
4. Pull out the through bolt (C), and remove the alternator (D).
5. Installation is the reverse of removal.
6. Adjust the alternator belt tension after installation.

Fig. 65 Remove the mounting bolt and through bolt—2.7L engine

ENGINE ELECTRICAL | **DISTRIBUTORLESS IGNITION SYSTEM**

Ignition timing is controlled by the electronic control ignition timing system. The standard reference ignition timing data for the engine operating conditions are preprogrammed in the memory of the Engine Control Module (ECM).

The engine operating conditions (speed, load, warm-up condition, etc.) are detected by the various sensors. Based on these sensor signals and the ignition timing data, signals to interrupt the primary current are sent to the ECM. The ignition coil is activated, and timing is controlled.

FIRING ORDER

2.0L engine firing order: 1–3–4–2
2.7L engine firing order: 1–2–3–4–5–6

IGNITION COIL

REMOVAL & INSTALLATION

See Figures 66 and 67.

1. Remove the engine cover.
2. Disconnect the spark plug cable and connector.
3. Remove the ignition coil (A).
4. Installation is the reverse of removal.

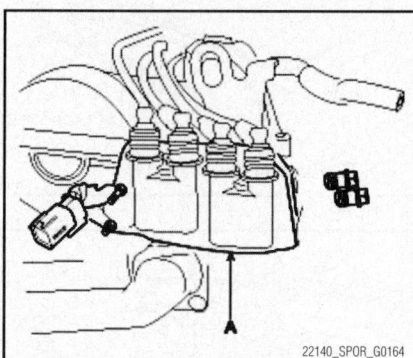

Fig. 66 Remove the ignition coil (A)— 2.0L engine

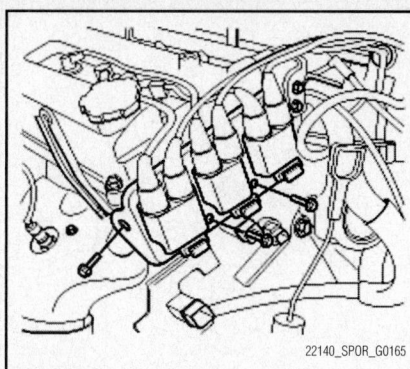

Fig. 67 Remove the ignition coil—2.7L engine

IGNITION TIMING

ADJUSTMENT

These vehicles are equipped with a Distributorless Ignition System (DIS). No adjustment is necessary or possible.

SPARK PLUGS

REMOVAL & INSTALLATION

2.0L Engine

See Figure 68.

1. Remove the spark plug cable (A).

➡ **When removing the spark plug cable, pull on the spark plug cable boot (not the cable), as it may be damaged.**

2. Using a spark plug socket, remove the spark plug (B).

❋❋ CAUTION
Be careful that no contaminates enter through the spark plug holes.

3. Installation is the reverse of removal.

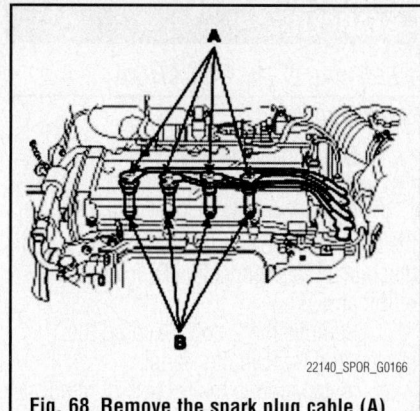

Fig. 68 Remove the spark plug cable (A)

2.7L Engine

See Figure 69.

1. Remove the engine cover.
2. Disconnect the VIS actuator connectors and injector connectors.
3. Remove the accelerator cable.
4. Remove surge tank sub assembly.
5. Remove the spark plug cable.
6. Remove the spark plug.
7. Installation is the reverse of removal.

Fig. 69 Remove surge tank sub assembly

ENGINE ELECTRICAL

STARTER

REMOVAL & INSTALLATION

See Figure 70.

1. Disconnect the battery negative cable.

2. Disconnect the starter cable (A) from the B terminal (B) on the solenoid (C), then disconnect the connector (D) from the "S" terminal (E).

3. Remove the 2 bolts holding the starter, and remove the starter.

4. Installation is the reverse of removal. Torque the nuts and bolts to 20 ft. lbs. (27 Nm).

5. Connect the battery positive cable and negative cable to the battery.

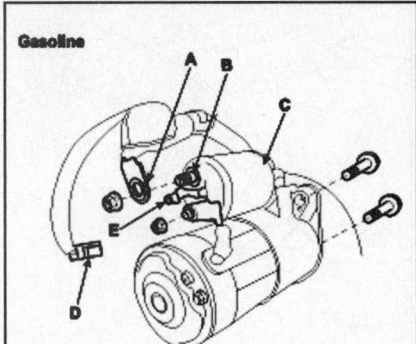

Fig. 70 Disconnect the starter cable (A) from the B terminal (B) on the solenoid (C), then disconnect the connector (D) from the "S" terminal (E)

STARTING SYSTEM

SOLENOID OR RELAY REPLACEMENT

1. Disconnect the negative battery cable.

2. Remove the starter assembly from the vehicle.

3. Remove the nut from the solenoid M-terminal.

4. Remove the solenoid retaining screws and the magnetic switch.

To install:

5. Install the magnetic switch and secure to the drive housing with the two screws.

6. Re-install the nut onto solenoid M-terminal.

7. Reinstall the starter assembly.

8. Reconnect the negative battery cable.

ENGINE MECHANICAL

ACCESSORY DRIVE BELTS

ACCESSORY BELT ROUTING

See Figure 71.

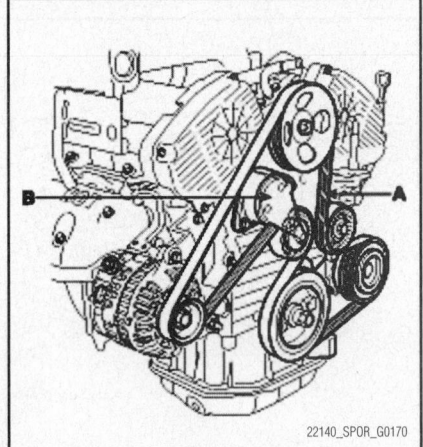

Fig. 71 Serpentine belt routing—2.7L engine

INSPECTION

Inspect the drive belt for signs of glazing or cracking. A glazed belt will be perfectly smooth from slippage, while a good belt will have a slight texture of fabric visible. Cracks will usually start at the inner edge of the belt and run outward. All worn or damaged drive belts should be replaced immediately.

ADJUSTMENT

The belt tension is maintained by an automatic tensioner. No adjustment is necessary or possible.

REMOVAL & INSTALLATION

1. Release tension of the belt tensioner.

2. Remove the serpentine belt.

To install:

3. Hang the belt on the pulley of the tensioner and install the tensioner.

(If the tensioner is already installed, loosen its mounting bolts to allow belt installation.) Tighten to Tensioner assembly bolt: 33–36 ft. lbs. (45–50 Nm).

4. Install drive belt.

5. When installing the belt on the pulley, make sure it is centered on the pulley.

CAMSHAFT AND VALVE LIFTERS

REMOVAL & INSTALLATION

2.0L Engine

See Figures 72 through 89.

1. Remove the air duct (A).

2. Disconnect the terminals from battery.

3. Remove the engine cover.

4. Drain the engine coolant.

5. Remove the intake air hose and air cleaner assembly.

   a. Disconnect the Air Flow Sensor (AFS) connector (A).

   b. Disconnect the breather hose (B) from intake air hose.

   c. Remove the intake air hose and air cleaner assembly (C).

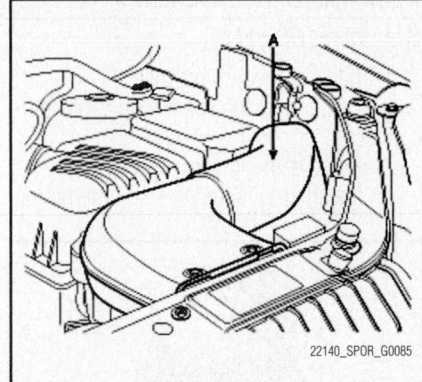

Fig. 72 Remove the air duct (A)

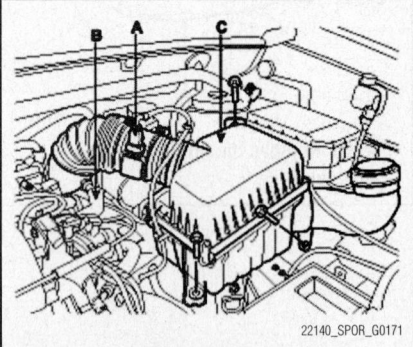

Fig. 73 Disconnect the Air Flow Sensor (AFS) connector (A)

6. Remove the upper radiator hose (A) and lower radiator hose (B).

7. Remove the heater hoses (A).

8. Remove the accelerator cable (A) by loosening the lock-nut, then slip the cable end out of the throttle linkage.

9. Remove the engine wire harness

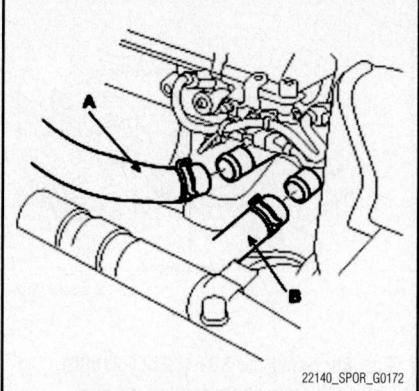

Fig. 74 Remove the upper radiator hose (A) and lower radiator hose (B)

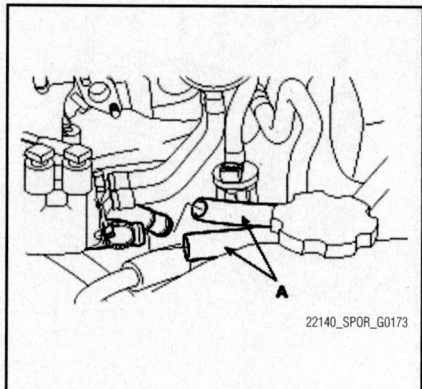

Fig. 75 Remove the heater hoses (A)

Fig. 76 Remove the accelerator cable (A)

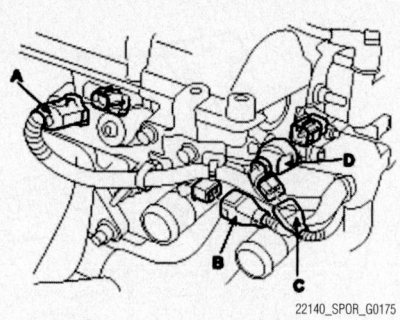

Fig. 77 Remove the engine wire harness connectors

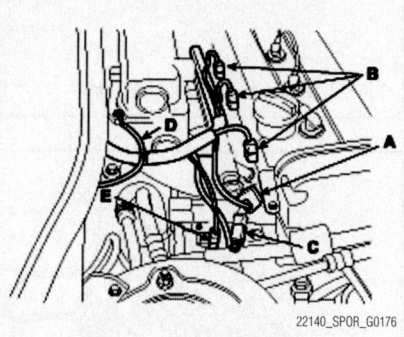

Fig. 78 Disconnect the Camshaft Position Sensor (CMP) connector (A), the four injector connectors (B), the knock sensor connector (C), the ground cables (D) and the air conditioner compressor switch connector (E)

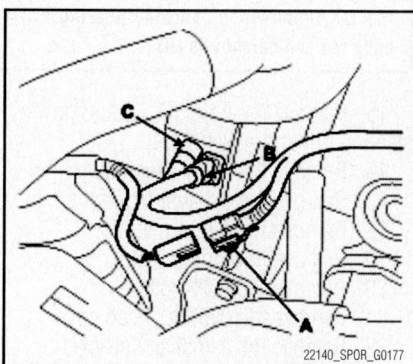

Fig. 79 Disconnect the front heated oxygen sensor connector (A), the Crankshaft Position Sensor (CKP) connector (B) and the oil pressure switch connector (C)

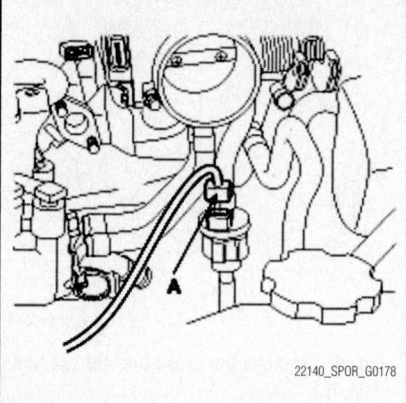

Fig. 80 Disconnect the Purge Control Solenoid Valve (PCSV) connector (A)

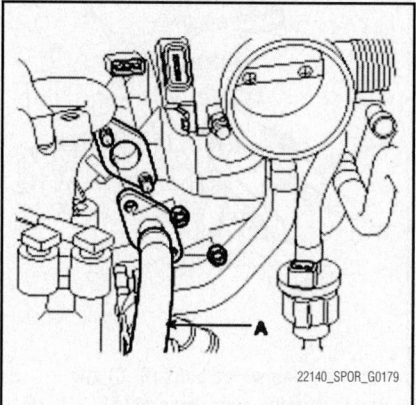

Fig. 81 Disconnect the fuel inlet hose (A) of the delivery pipe side

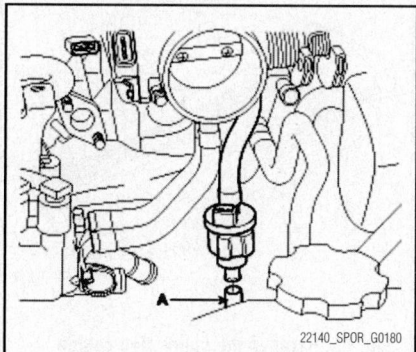

Fig. 82 Disconnect the hose (A) of the PCSV side

connectors and wire harness clamps from cylinder head and the intake manifold.

    a. Disconnect the Oil Control Valve (OCV) connector (A).

    b. Disconnect the oil temperature sensor connector (B).

    c. Disconnect the Engine Coolant Temperature (ECT) sensor connector (C).

    d. Disconnect the ignition coil connector (D).

    e. Disconnect the Throttle Position Sensor (TPS) connector (A).

    f. Disconnect the Idle Speed Actuator (ISA) connector (B).

    g. Disconnect the Camshaft Position Sensor (CMP) connector (A).

    h. Disconnect the four injector connectors (B).

    i. Disconnect the knock sensor connector (C).

    j. Disconnect the ground cables (D) from the intake manifold.

    k. Disconnect the air conditioner compressor switch connector (E).

    l. Disconnect the front heated oxygen sensor connector (A).

    m. Disconnect the Crankshaft Position Sensor (CKP) connector (B).

    n. Disconnect the oil pressure switch connector (C).

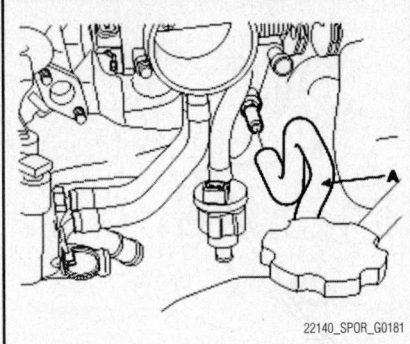

Fig. 83 Remove the brake booster vacuum hose (A)

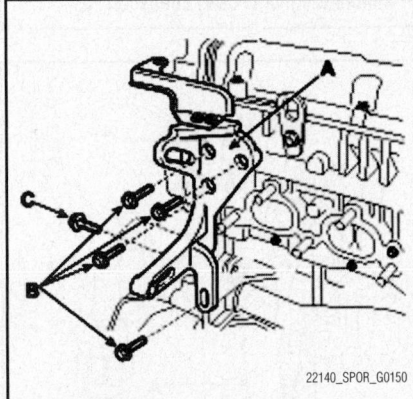

Fig. 84 Remove the bolts (B, C) and power steering pump bracket (A)

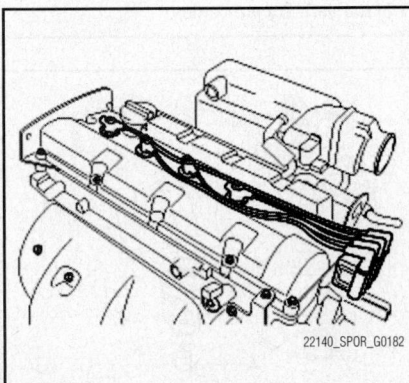

Fig. 85 Remove the spark plug cables

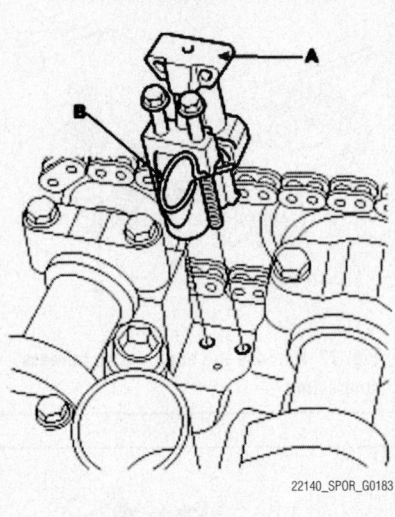

Fig. 86 Remove the auto tensioner

Fig. 87 Remove the camshaft bearing caps (A) and camshafts (B)

15. Remove the bolts (B, C) and power steering pump bracket (A).

16. Remove the spark plug cables.

17. Remove the exhaust manifold.

18. Remove the intake manifold.

19. Remove the timing belt.

20. Remove the PCV hose.

21. Remove the cylinder head cover.

22. Remove the camshaft sprocket.

23. Insert a stopper pin or other device into timing chain auto tensioner and remove the auto tensioner.

24. Remove the camshaft bearing caps (A) and camshafts (B).

To install:

25. Install the camshaft and bearing caps. Tighten to 10 ft. lbs. (14 Nm).

26. Install the timing chain auto tensioner. Tighten to 6 ft. lbs. (8 Nm).

27. Remove the auto tensioner stopper pin.

28. Check and adjust valve clearance.

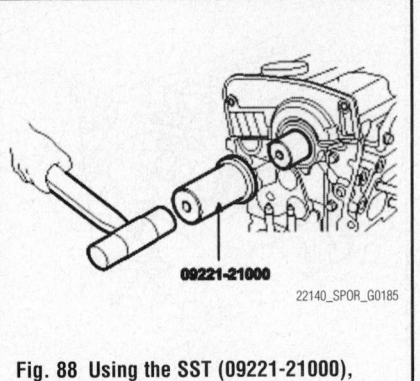

Fig. 88 Using the SST (09221-21000), install the camshaft bearing oil seal

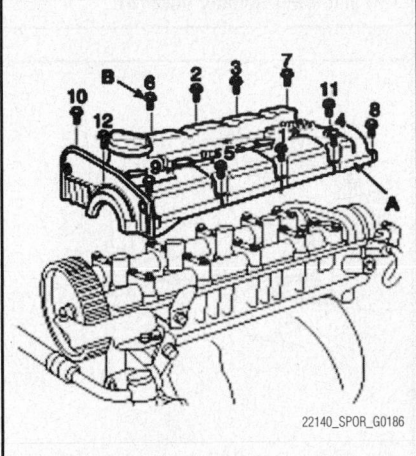

Fig. 89 Install the cylinder head cover (A) in sequence

29. Using the SST (09221-21000), install the camshaft bearing oil seal.

30. Install the camshaft sprocket.

31. Install the cylinder head cover.

a. Install the cylinder head cover gasket in the groove of the cylinder head cover.

➡Before installing the cylinder head cover gasket, thoroughly clean the cylinder head cover and the groove. When installing, make sure the cylinder head cover gasket is seated securely in the corners of the recesses with no gap.

b. Apply liquid gasket to the head cover gasket at the corners of the recess.

➡Use liquid gasket, Loctite No. 5999. Check that the mating surfaces are clean and dry before applying liquid gasket. After assembly, wait at least 30 minutes before filling the engine with oil.

32. Install the cylinder head cover (A) with the 12 bolts (B). Uniformly tighten the

o. Disconnect the Purge Control Solenoid Valve (PCSV) connector (A).

10. Disconnect the fuel inlet hose (A) of the delivery pipe side.

11. Disconnect the hose (A) of the PCSV side.

12. Remove the brake booster vacuum hose (A).

13. Remove the power steering pump drive belt.

14. Remove the power steering pump and use a wire to secure the pump to the vehicle so that it is out of the way.

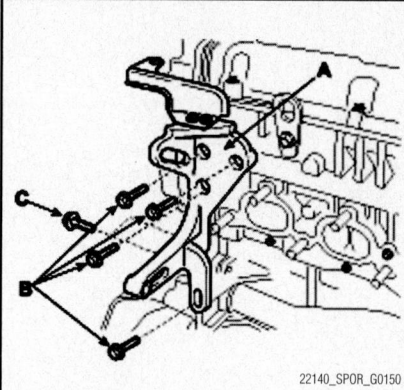

Fig. 90 Torque the power steering bracket bolts

bolts in several passes. Tighten to 6–7 ft. lbs. (8–10 Nm).

33. Install the PCV hose.

34. Install the timing belt.

35. Install the intake manifold.

36. Install the exhaust manifold.

37. Install the spark plug cables.

38. Install the power steering pump bracket and bolts. Tighten to Bolt (B): 25–36 ft. lbs. (34–49 Nm); Bolt (C): 9–11 ft. lbs. (12–15 Nm).

39. Install the power steering pump.

40. Install the accelerator cable.

41. Install the brake booster hose.

42. Connect the hose of the PCSV side.

43. Connect the fuel inlet hose of the delivery pipe side.

44. Install the engine wire harness connectors and wire harness clamps to the cylinder head and the intake manifold.

　a. Connect the PCSV connector.

　b. Connect the front heated oxygen sensor connector.

　c. Connect the CKP connector.

　d. Connect the oil pressure switch connector.

　e. Connect the air conditioner compressor switch connector.

　f. Connect the ground cables to intake manifold.

　g. Connect the knock sensor connector.

　h. Connect the fuel injector connectors.

　i. Connect the CMP connector.

　j. Connect the ISA connector.

　k. Connect the TPS connector.

　l. Connect the ignition coil connector.

　m. Connect the ECT connector.

　n. Connect the oil temperature sensor connector.

　o. Connect the OCV connector.

45. Install the heater hose.

46. Install the upper radiator hose and lower radiator hose.

47. Install the intake air hose and air cleaner assembly.

　a. Install the intake air hose, air cleaner assembly and bolts. Tighten to 6–7 ft. lbs. (8–10 Nm).

　b. Install the breather hose to intake air hose.

　c. Connect the AFS connector.

48. Install the engine cover.

49. Reconnect the battery terminals.

50. Install the air duct.

51. Fill with engine coolant.

52. Start the engine and check for leaks.

53. Recheck engine coolant level and oil level.

2.7L Engine

See Figures 72, 91 through 110.

1. Remove the air duct (A).

2. Remove the cowl grill and wiper motor.

3. Remove the strut bar.

4. Disconnect the negative terminal from the battery.

5. Drain the engine coolant. Remove the radiator cap to speed draining.

6. Remove the engine cover.

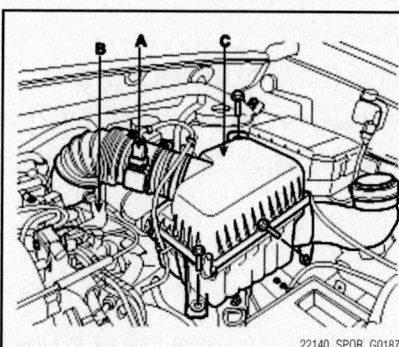

Fig. 91 Remove the intake air hose and air cleaner assembly

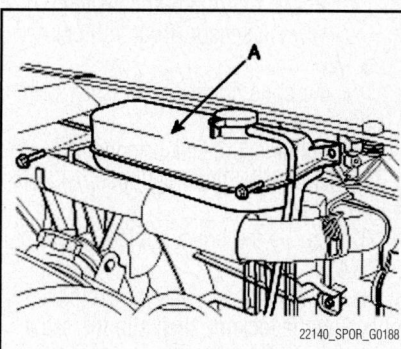

Fig. 92 Remove the coolant reservoir tank (A)

7. Remove the intake air hose and air cleaner assembly.

　a. Disconnect the Air Flow Sensor (AFS) connector (A).

　b. Disconnect the breather hose (B) from air cleaner hose.

　c. Remove the intake air hose and air cleaner assembly (C).

8. Remove the coolant reservoir tank (A).

9. Remove the upper radiator hose (A) and lower radiator hose (B).

10. Remove the heater hoses (A) and throttle body heater hose (B).

11. Remove the engine wire harness connectors and wire harness clamps from the cylinder head and the intake manifold.

　a. Throttle Position Sensor (TPS) connector (A).

　b. Idle Speed Actuator (ISA) connector (B).

　c. Purge Control Solenoid Valve (PCSV) connector (C).

　d. VIS actuator connector (D).

　e. Injector connector (E).

　f. Knock sensor connectors (F).

　g. Camshaft Position Sensor (CMP) connector (G).

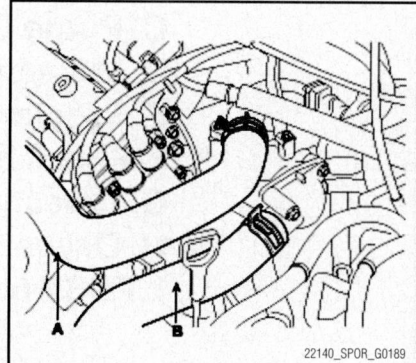

Fig. 93 Remove the upper radiator hose (A) and lower radiator hose (B)

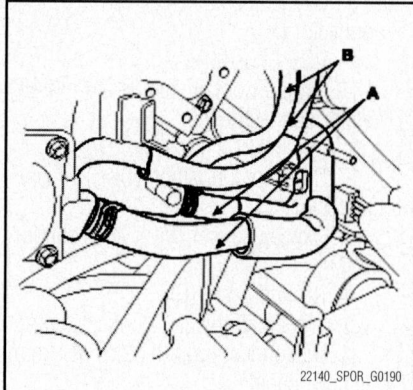

Fig. 94 Remove the heater hoses (A) and throttle body heater hose (B)

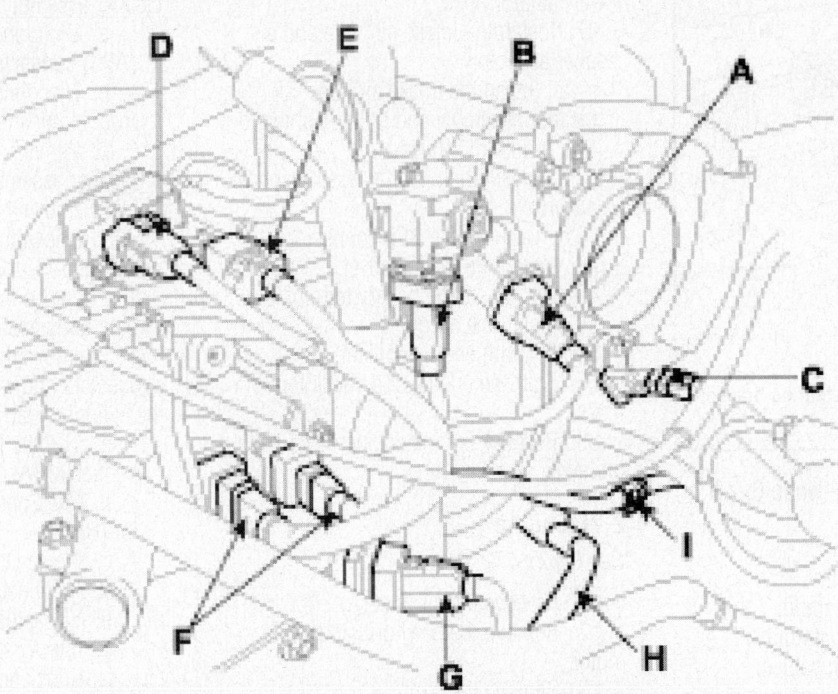

A. **Throttle Position Sensor (TPS) connector**
B. **Idle Speed Actuator (ISA) connector**
C. **Purge Control Solenoid Valve (PCSV) connector**
D. **VIS actuator connector**
E. **Injector connector**
F. **Knock Sensor (KS) connectors**
G. **Camshaft Position Sensor (CMP) connector**
H. **Oxygen sensor (rear 1st) connector**
I. **PCSV hose**

22140_SPOR_G0191

Fig. 95 Remove the engine wire harness connectors and wire harness clamps from the cylinder head and the intake manifold

h. Oxygen sensor (rear 1st) connector (H).

i. PCSV hose (I).

j. Engine Coolant Temperature (ECT) sensor connector (A).

k. Ignition coil connector (B).

l. Crankshaft position (CKP) sensor connector (C).

m. Oxygen sensor (front 2nd) connector (D).

n. Ground cable (E).

o. Three fuel injector connectors (A).

p. Disconnect ground cable (A) from the cowl panel.

q. Intake Air Temperature (IAT) sensor connector (A).

r. VIS actuator connector (B).

s. Oxygen sensor (rear 2nd) connector (C).

t. Power steering pump switch (D).

u. Oxygen sensor (front 1st) connector (A).

v. Air conditioner compressor switch connector (B).

w. Oil pressure sensor connector (C).

12. Remove the fuel inlet hose (A) from delivery pipe.

13. Remove the brake booster vacuum hose (A).

14. Remove the accelerator cable by loosening the locknut, then slip the cable end out of the throttle linkage.

15. Remove the PCV hose (A).

16. Remove the intake manifold.

17. Remove the power steering pump.

18. Remove the exhaust manifold.

19. Remove the timing belt.

20. Remove the spark plug cable.

21. Remove the cylinder head covers (A).

22. Remove the camshaft sprocket.

23. Remove the camshaft bearing caps (A).

24. Remove the camshafts (A).

To install:

25. Install the camshafts.

a. Align the camshaft timing chain with the intake timing chain sprocket and exhaust timing chain sprocket.

b. Install the camshaft.

c. Install the camshaft bearing caps.

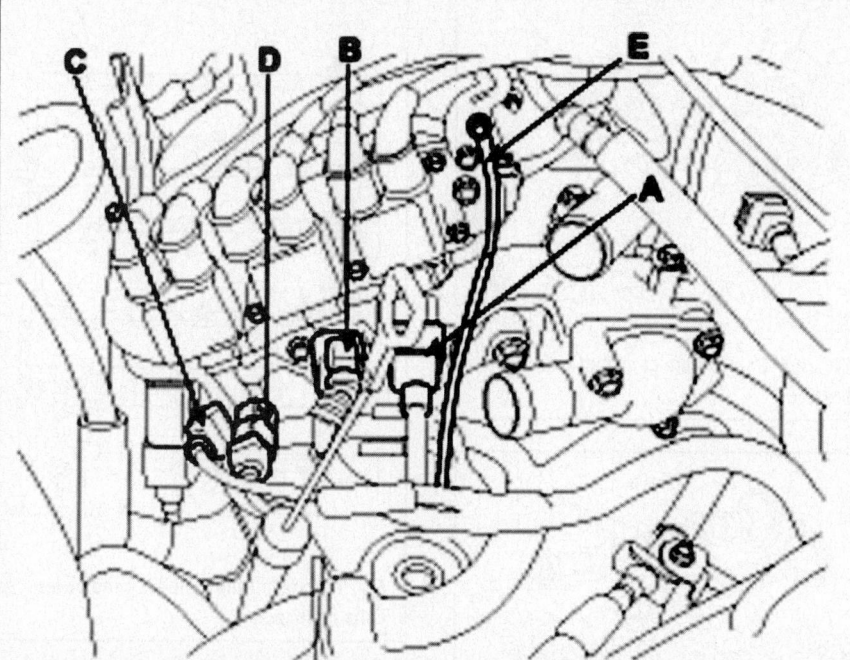

A. Engine Coolant Temperature (ECT) sensor connector
B. Ignition coil connector
C. Crankshaft Position Sensor (CKP) connector
D. Oxygen sensor (front 2nd) connector
E. Ground cable

Fig. 96 Additional engine harness connectors

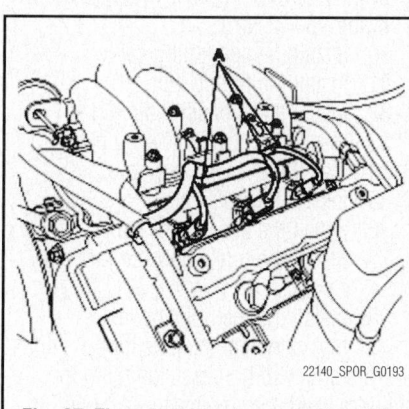

Fig. 97 Three fuel injector connectors (A)

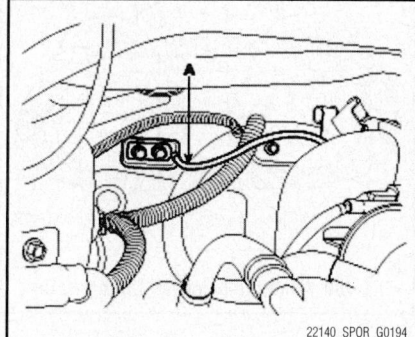

Fig. 98 Disconnect ground cable (A) from the cowl panel

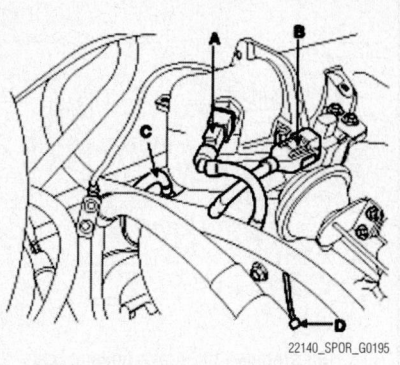

Fig. 99 Intake Air Temperature (IAT) sensor connector (A). VIS actuator connector (B). Oxygen sensor (rear 2nd) connector (C). Power steering pump switch (D).

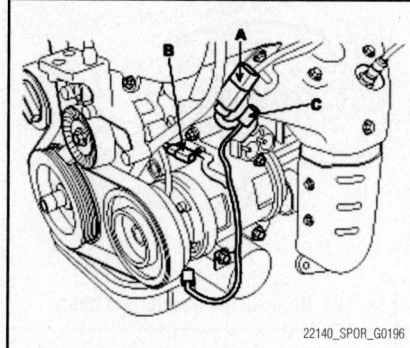

Fig. 100 Oxygen sensor (front 1st) connector (A). Air conditioner compressor switch connector (B). Oil pressure sensor connector (C).

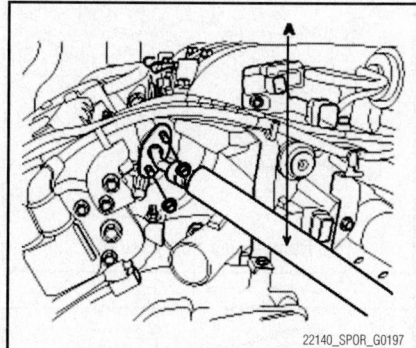

Fig. 101 Remove the fuel inlet hose (A) from delivery pipe

Tighten to M6 (38mm): 7–9 ft. lbs. (10–12 Nm) (Mark7); M6 (50mm): 10–12 ft. lbs. (14–16 Nm) (Mark10)

➡Apply new engine oil to the thrust portion and journal of the camshafts. Apply a light coat of engine oil on the threads and under the heads of the bearing cap bolts.

26. Using the SST (09214-21000), install the camshaft bearing oil seal.

27. Install the camshaft sprocket.
   a. Temporarily install the camshaft sprocket bolts.
   b. Hold the hexagonal head wrench portion of the camshaft with a wrench, and tighten the camshaft sprocket bolts. Tighten to Camshaft sprocket bolt: 65–80 ft. lbs. (90–110 Nm)

28. Install semi-circular packing.
29. Install the cylinder head cover.
   a. Install the cylinder head cover

gasket in the groove of the cylinder head cover.

➡Before installing the head cover gasket, thoroughly clean the head cover gasket and the groove. When installing, make sure the head cover gasket is seated securely in the corners of the recesses with no gap.

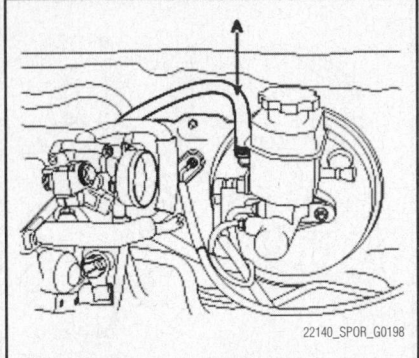

Fig. 102 Remove the brake booster vacuum hose (A)

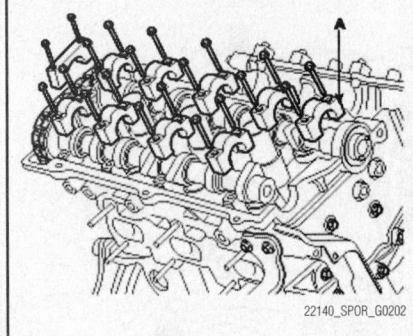

Fig. 106 Remove the camshaft bearing caps (A)

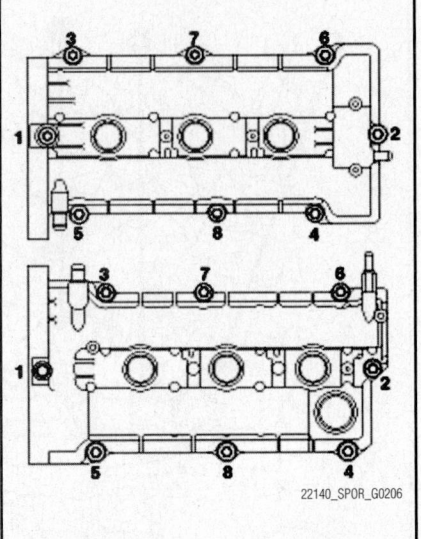

Fig. 110 Install the cylinder head cover bolts in sequence

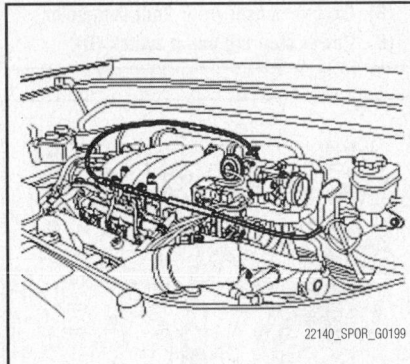

Fig. 103 Remove the accelerator cable

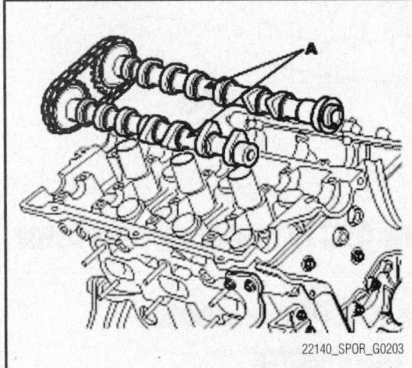

Fig. 107 Remove the camshafts (A)

b. Apply liquid gasket to the head cover gasket at the corners of the recess.

➡**Use liquid gasket, Loctite No.5699. Check that the mating surfaces are clean and dry before applying liquid gasket. After assembly, wait at least 30 minutes before filling the engine with oil.**

c. Install the cylinder head covers with the 16bolts. Uniformly tighten the bolts in several passes. Tighten to 6–7.4 ft. lbs. (8–10 Nm)

30. Install the spark plug cable.
31. Install the timing belt.
32. Install the exhaust manifold.
33. Install the power steering pump.
34. Install the intake manifold.
35. Install the PCV hose.
36. Install the accelerator cable.
37. Install the brake booster vacuum hose.
38. Install the fuel inlet hose.
39. Install the engine wire harness connectors and wire harness clamps to the cylinder head and the intake manifold.

a. Oil pressure sensor connector.
b. Air conditioner compressor switch connector.
c. Oxygen sensor (front 1st) connector.
d. Power steering pump switch.
e. Oxygen sensor (rear 2nd) connector.
f. VIS actuator connector.
g. IAT sensor connector.
h. Connect the ground cable to the cowl panel.
i. Three fuel injector connectors.

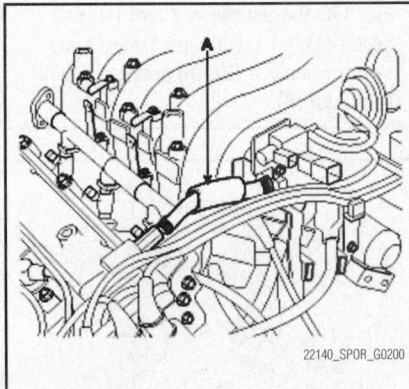

Fig. 104 Remove the PCV hose (A)

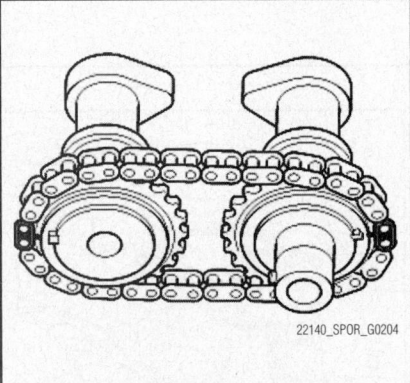

Fig. 108 Align the camshaft timing chain

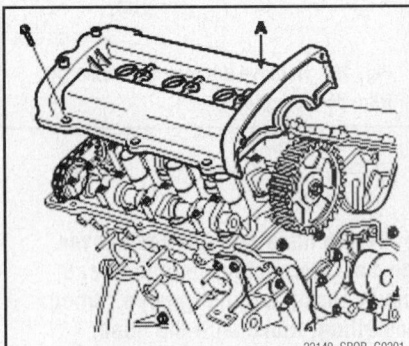

Fig. 105 Remove the cylinder head covers (A)

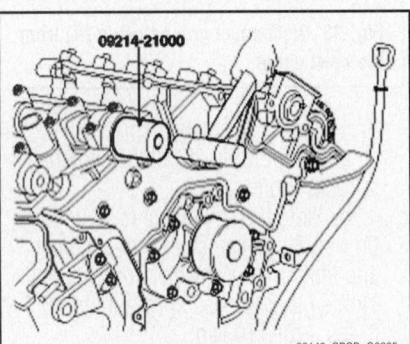

Fig. 109 Using the SST (09214-21000), install the camshaft bearing oil seal

j. Ground cable.

k. Oxygen sensor connector.

l. Crankshaft position sensor connector.

m. Ignition coil connector.

n. ECT sensor connector.

o. PCSV hose.

p. Oxygen sensor (rear 1st) connector.

q. CMP connector.

r. Knock sensor connector.

s. Injector connector.

t. VIS actuator connector.

u. PCSV connector.

v. ISA connector.

w. TPS connector.

40. Install the heater hoses and throttle body heater hose.

41. Install the upper and lower radiator hose.

42. Install the coolant reservoir tank.

43. Install the intake air hose and air cleaner assembly.

    a. Install the intake air hose and air cleaner assembly.

    b. Connect the breather hose from air cleaner hose.

    c. Connect the AFS connector.

44. Install the engine cover.

45. Connect the negative terminal to the battery.

46. Install the strut bar.

47. Install the cowl grill and wiper motor.

48. Install the air duct.

49. Fill with engine coolant.

50. Start the engine and check for leaks.

51. Recheck engine coolant level and oil level.

CRANKSHAFT DAMPER

REMOVAL & INSTALLATION

1. Disconnect the negative battery cable.

2. Remove the accessory drive belts.

3. Remove the crankshaft damper.

To install:

4. Install the crankshaft damper and tighten the bolt.

    a. 2.0L engine: 123–130 ft. lbs. (167–176 Nm).

    b. 2.7L engine: 130–138 ft. lbs. (180–190 Nm).

5. Install and tension the accessory drive belts.

CYLINDER HEAD

REMOVAL & INSTALLATION

2.0L Engine

See Figures 111 through 134.

1. Remove the air duct (A).

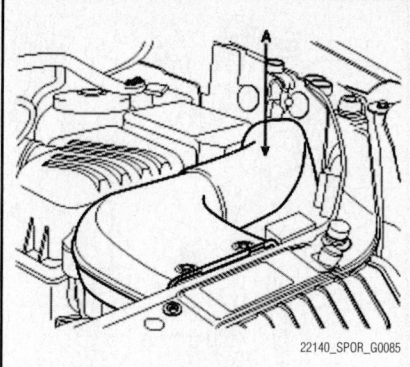

Fig. 111 Remove the air duct (A)

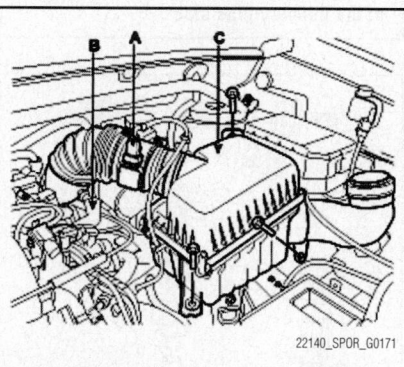

Fig. 112 Disconnect the Air Flow Sensor (AFS) connector (A)

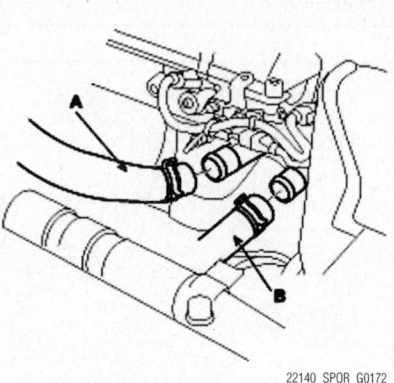

Fig. 113 Remove the upper radiator hose (A) and lower radiator hose (B)

2. Disconnect the terminals from battery.

3. Remove the engine cover.

4. Drain the engine coolant.

5. Remove the intake air hose and air cleaner assembly.

    a. Disconnect the Air Flow Sensor (AFS) connector (A).

    b. Disconnect the breather hose (B) from intake air hose.

    c. Remove the intake air hose and air cleaner assembly (C).

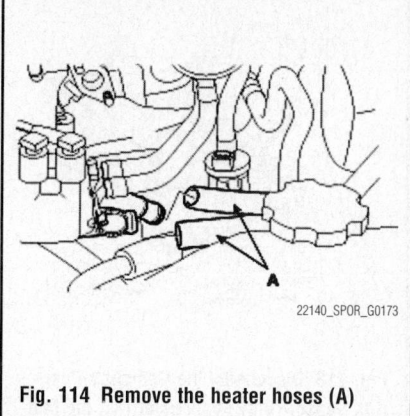

Fig. 114 Remove the heater hoses (A)

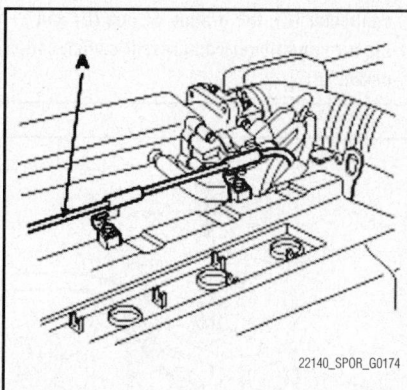

Fig. 115 Remove the accelerator cable (A)

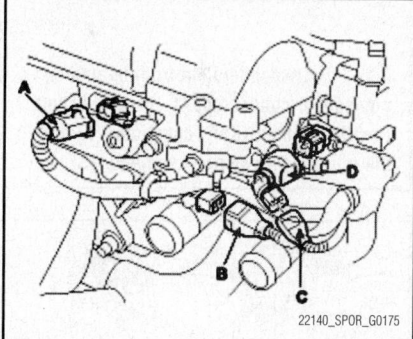

Fig. 116 Remove the engine wire harness connectors

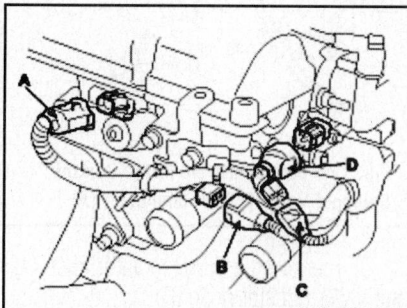

Fig. 117 Disconnect the Throttle Position Sensor (TPS) connector (A) and the Idle Speed Actuator (ISA) connector (B)

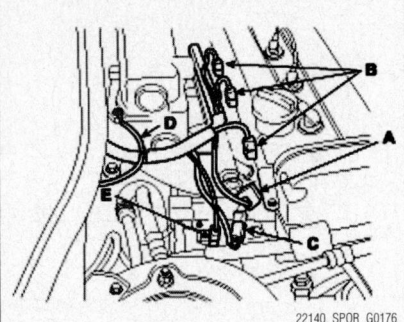

Fig. 118 Disconnect the Camshaft Position Sensor (CMP) connector (A), the four injector connectors (B), the knock sensor connector (C), the ground cables (D) and the air conditioner compressor switch connector (E)

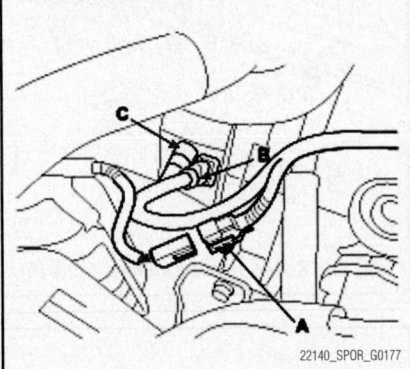

Fig. 119 Disconnect the front heated oxygen sensor connector (A), the Crankshaft Position Sensor (CKP) connector (B) and the oil pressure switch connector (C)

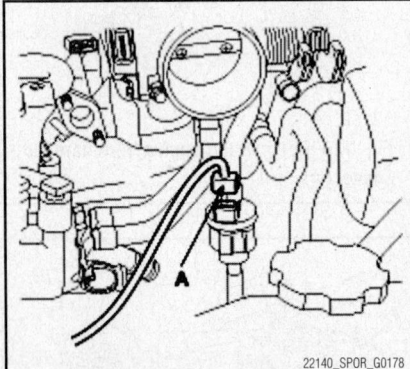

Fig. 120 Disconnect the Purge Control Solenoid Valve (PCSV) connector (A)

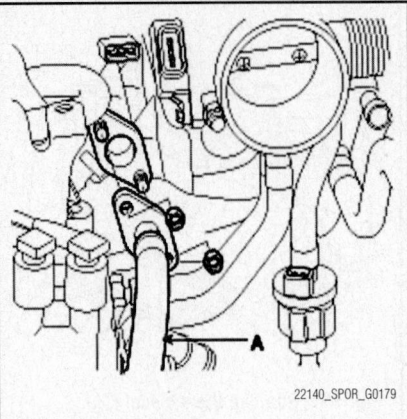

Fig. 121 Disconnect the fuel inlet hose (A) of the delivery pipe side

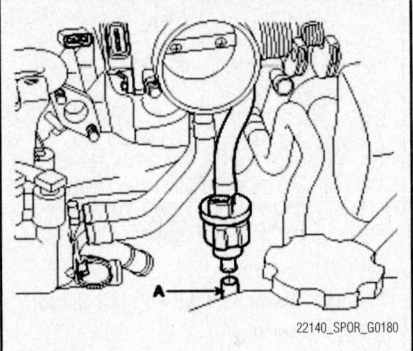

Fig. 122 Disconnect the hose (A) of the PCSV side

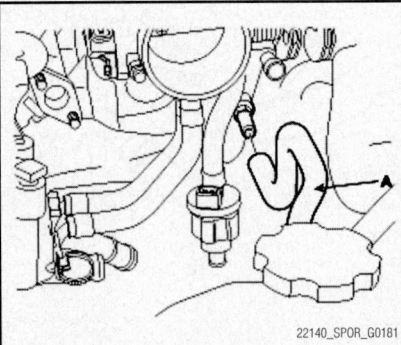

Fig. 123 Remove the brake booster vacuum hose (A)

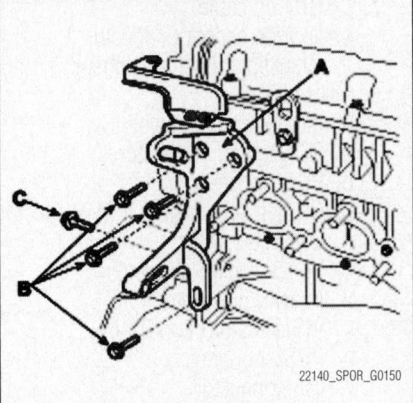

Fig. 124 Remove the bolts (B, C) and power steering pump bracket (A)

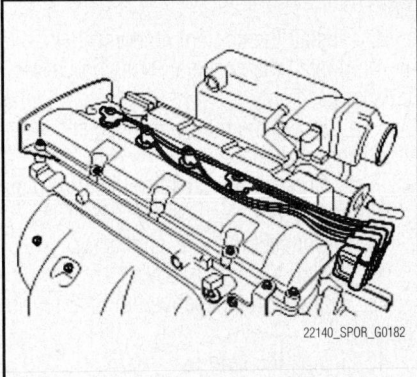

Fig. 125 Remove the spark plug cables

Fig. 126 Remove the auto tensioner

6. Remove the upper radiator hose (A) and lower radiator hose (B).

7. Remove the heater hoses (A).

8. Remove the accelerator cable (A) by loosening the lock-nut, then slip the cable end out of the throttle linkage.

9. Remove the engine wire harness connectors and wire harness clamps from cylinder head and the intake manifold.

   a. Disconnect the Oil Control Valve (OCV) connector (A).

   b. Disconnect the oil temperature sensor connector (B).

   c. Disconnect the Engine Coolant Temperature (ECT) sensor connector (C).

   d. Disconnect the ignition coil connector (D).

   e. Disconnect the Throttle Position Sensor (TPS) connector (A).

   f. Disconnect the Idle Speed Actuator (ISA) connector (B).

   g. Disconnect the Camshaft Position Sensor (CMP) connector (A).

Fig. 127 Remove the camshaft bearing caps (A) and camshafts (B)

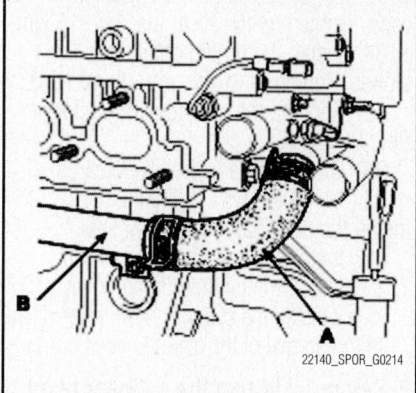

Fig. 130 Remove the water hose (A) from water pipe (B)

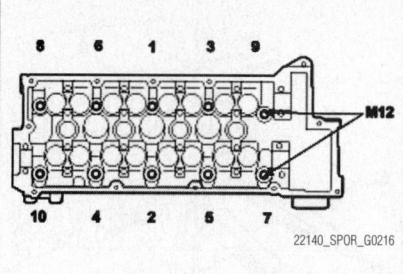

Fig. 132 Install and tighten the 10 cylinder head bolts and plate washers, in several passes, in sequence

Fig. 128 Remove the Oil Control Valve (OCV) (A)

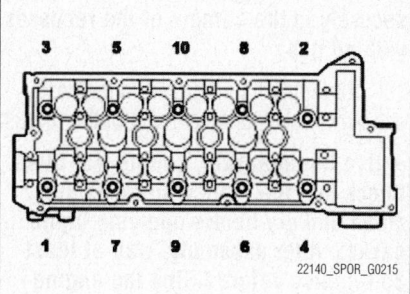

Fig. 131 Remove the 10 cylinder head bolts and plate washers, in several passes, in sequence

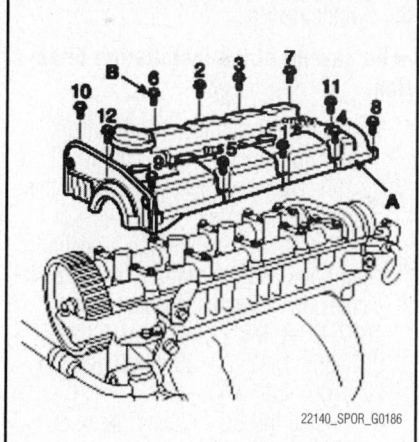

Fig. 133 Using the SST (09221-21000), install the camshaft bearing oil seal

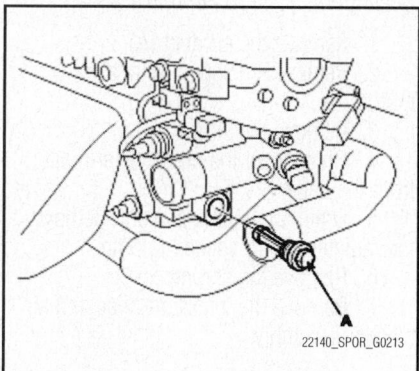

Fig. 129 Remove the OCV filter (A)

h. Disconnect the four injector connectors (B).

i. Disconnect the knock sensor connector (C).

j. Disconnect the ground cables (D) from the intake manifold.

k. Disconnect the air conditioner compressor switch connector (E).

l. Disconnect the front heated oxygen sensor connector (A).

m. Disconnect the Crankshaft Position Sensor (CKP) connector (B).

n. Disconnect the oil pressure switch connector (C).

o. Disconnect the Purge Control Solenoid Valve (PCSV) connector (A).

10. Disconnect the fuel inlet hose (A) of the delivery pipe side.

11. Disconnect the hose (A) of the PCSV side.

12. Remove the brake booster vacuum hose (A).

13. Remove the power steering pump drive belt.

14. Remove the power steering pump and use a wire to secure the pump to the vehicle so that it is out of the way.

15. Remove the bolts (B, C) and power steering pump bracket (A).

16. Remove the spark plug cables.

17. Remove the exhaust manifold.

18. Remove the intake manifold.

19. Remove the timing belt.

20. Remove the PCV hose.

21. Remove the cylinder head cover.

22. Remove the camshaft sprocket.

23. Insert a stopper pin or other device into timing chain auto tensioner and remove the auto tensioner.

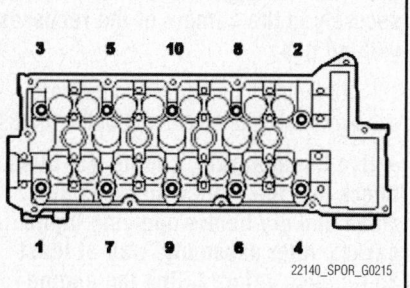

Fig. 134 Install the cylinder head cover (A) in sequence

24. Remove the camshaft bearing caps (A) and camshafts (B).

25. Remove the Oil Control Valve (OCV) (A).

26. Remove the OCV filter (A).

27. Remove the water hose (A) from water pipe (B).

28. Remove the cylinder head bolts, and remove the cylinder head.

a. Uniformly loosen and remove

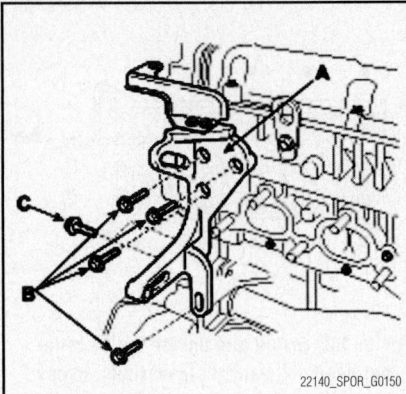

Fig. 135 Torque the power steering bracket bolts

the 10 cylinder head bolts and plate washers, in several passes, in sequence.

> **⁂ CAUTION**
>
> **Head warpage or cracking could result from removing bolts in an incorrect order.**

b. Lift the cylinder head from the dowels on the cylinder block and replace the cylinder head on wooden blocks on a bench.

To install:

29. Install the cylinder head gasket on the cylinder block.

➡ **Be careful of the installation direction.**

30. Install the cylinder head bolts.

a. Apply a light coat if engine oil on the threads and under the heads of the cylinder head bolts.

b. Install and tighten the 10 cylinder head bolts and plate washers, in several passes, in sequence. Tighten to M10: 1677–20 ft. lbs. (23–27 Nm+60°–65° + 60°–65°; M12: 20–23 ft. lbs. (28–31 Nm)+60°–65° +60°–65°

31. Install the OCV filter. Tighten to 30–37 ft. lbs. (40–50 Nm)

➡ **Always use a new OCV filter gasket. Keep the OCV filter clean.**

32. Install the OCV. Tighten to 7–9 ft. lbs. (10–12 Nm)

> **⁂ CAUTION**
>
> **Do not reuse the OCV when dropped. Keep the OCV clean. Do not hold the OCV sleeve during servicing. When the OCV is installed on the engine, be careful not to rotate the engine while holding the yoke.**

33. Install the camshaft and bearing caps. Tighten to 10–11 ft. lbs. (14–15 Nm).

34. Install the timing chain auto tensioner. Tighten to 6–7 ft. lbs. (8–10 Nm).

35. Remove the auto tensioner stopper pin.

36. Check and adjust valve clearance.

37. Using the SST (09221-21000), install the camshaft bearing oil seal.

38. Install the camshaft sprocket.

39. Install the cylinder head cover.

a. Install the cylinder head cover gasket in the groove of the cylinder head cover.

➡ **Before installing the cylinder head cover gasket, thoroughly clean the cylinder head cover and the groove. When installing, make sure the cylinder head cover gasket is seated securely in the corners of the recesses with no gap.**

b. Apply liquid gasket to the head cover gasket at the corners of the recess.

➡ **Use liquid gasket, Loctite No. 5999. Check that the mating surfaces are clean and dry before applying liquid gasket. After assembly, wait at least 30 minutes before filling the engine with oil.**

40. Install the cylinder head cover (A) with the 12 bolts (B). Uniformly tighten the bolts in several passes. Tighten to 6–7 ft. lbs. (8–10 Nm).

41. Install the PCV hose.

42. Install the timing belt.

43. Install the intake manifold.

44. Install the exhaust manifold.

45. Install the spark plug cables.

46. Install the power steering pump bracket and bolts. Tighten to Bolt (B): 25–36 ft. lbs. (34–49 Nm); Bolt (C): 9–11 ft. lbs. (12–15 Nm).

47. Install the power steering pump.

48. Install the accelerator cable.

49. Install the brake booster hose.

50. Connect the hose of the PCSV side.

51. Connect the fuel inlet hose of the delivery pipe side.

52. Install the engine wire harness connectors and wire harness clamps to the cylinder head and the intake manifold.

a. Connect the PCSV connector.

b. Connect the front heated oxygen sensor connector.

c. Connect the CKP connector.

d. Connect the oil pressure switch connector.

e. Connect the air conditioner compressor switch connector.

f. Connect the ground cables to intake manifold.

g. Connect the knock sensor connector.

h. Connect the fuel injector connectors.

i. Connect the CMP connector.

j. Connect the ISA connector.

k. Connect the TPS connector.

l. Connect the ignition coil connector.

m. Connect the ECT connector.

n. Connect the oil temperature sensor connector.

o. Connect the OCV connector.

53. Install the heater hose.

54. Install the upper radiator hose and lower radiator hose.

55. Install the intake air hose and air cleaner assembly.

a. Install the intake air hose, air cleaner assembly and bolts. Tighten to 6–7 ft. lbs. (8–10 Nm).

b. Install the breather hose to intake air hose.

c. Connect the AFS connector.

56. Install the engine cover.

57. Reconnect the battery terminals.

58. Install the air duct.

59. Fill with engine coolant.

60. Start the engine and check for leaks.

61. Recheck engine coolant level and oil level.

2.7L Engine

See Figures 72, 136 through 157.

1. Remove the air duct (A).

2. Remove the cowl grill and wiper motor.

3. Remove the strut bar.

4. Disconnect the negative terminal from the battery.

5. Drain the engine coolant. Remove the radiator cap to speed draining.

6. Remove the engine cover.

7. Remove the intake air hose and air cleaner assembly.

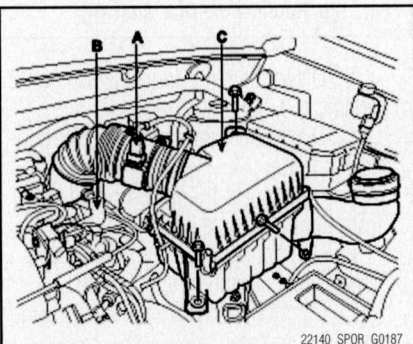

Fig. 136 Remove the intake air hose and air cleaner assembly.

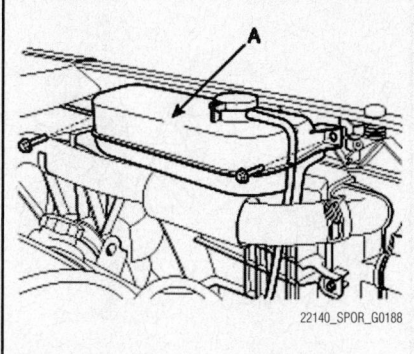

Fig. 137 Remove the coolant reservoir tank (A).

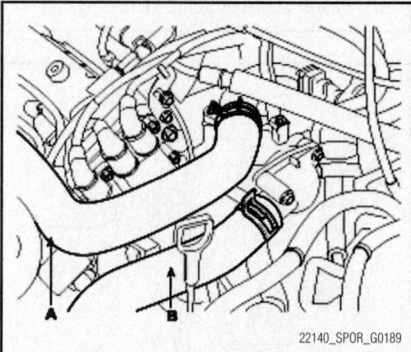

Fig. 138 Remove the upper radiator hose (A) and lower radiator hose (B).

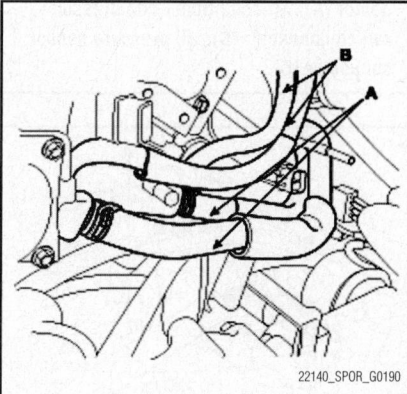

Fig. 139 Remove the heater hoses (A) and throttle body heater hose (B).

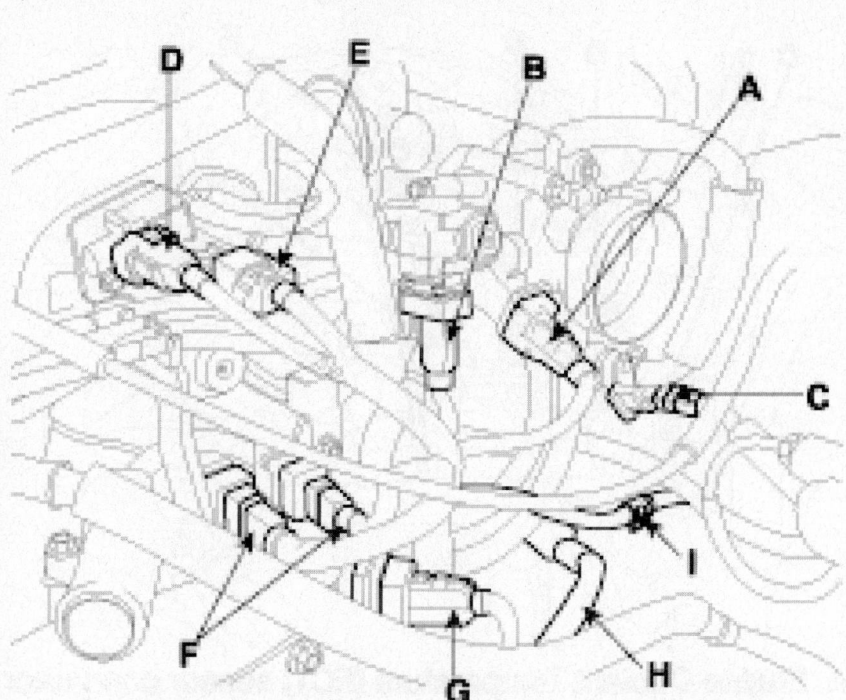

A. Throttle Position Sensor (TPS) connector
B. Idle Speed Actuator (ISA) connector
C. Purge Control Solenoid Valve (PCSV) connector
D. VIS actuator connector
E. Injector connector
F. Knock Sensor (KS) connectors
G. Camshaft Position Sensor (CMP) connector
H. Oxygen sensor (rear 1st) connector
I. PCSV hose

Fig. 140 Remove the engine wire harness connectors and wire harness clamps.

a. Disconnect the Air Flow Sensor (AFS) connector (A).

b. Disconnect the breather hose (B) from air cleaner hose.

c. Remove the intake air hose and air cleaner assembly (C).

8. Remove the coolant reservoir tank (A).

9. Remove the upper radiator hose (A) and lower radiator hose (B).

10. Remove the heater hoses (A) and throttle body heater hose (B).

11. Remove the engine wire harness connectors and wire harness clamps from the cylinder head and the intake manifold.

a. Throttle Position Sensor (TPS) connector (A).

b. Idle Speed Actuator (ISA) connector (B).

c. Purge Control Solenoid Valve (PCSV) connector (C).

d. VIS actuator connector (D).

e. Injector connector (E).

f. Knock sensor connectors (F).

g. Camshaft Position Sensor (CMP) connector (G).

h. Oxygen sensor (rear 1st) connector (H).

i. PCSV hose (I).

j. Engine Coolant Temperature (ECT) sensor connector (A).

k. Ignition coil connector (B).

l. Crankshaft position (CKP) sensor connector (C).

m. Oxygen sensor (front 2nd) connector (D).

n. Ground cable (E).

o. Three fuel injector connectors (A).

p. Disconnect ground cable (A) from the cowl panel.

q. Intake Air Temperature (IAT) sensor connector (A).

r. VIS actuator connector (B).

s. Oxygen sensor (rear 2nd) connector (C).

t. Power steering pump switch (D).

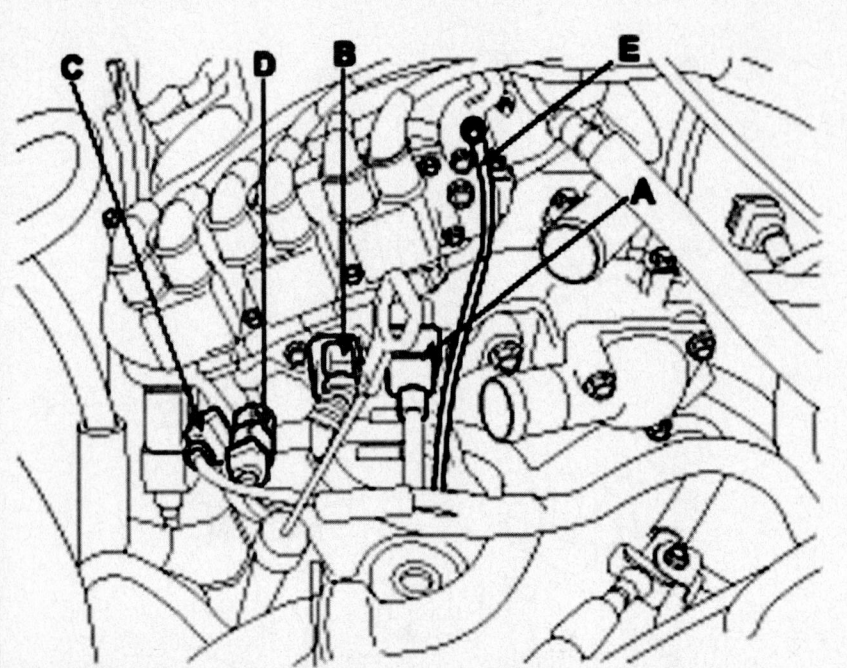

A. Engine Coolant Temperature (ECT) sensor connector
B. Ignition coil connector
C. Crankshaft Position Sensor (CKP) connector
D. Oxygen sensor (front 2nd) connector
E. Ground cable

22140_SPOR_G0192

Fig. 141 Additional engine harness connectors

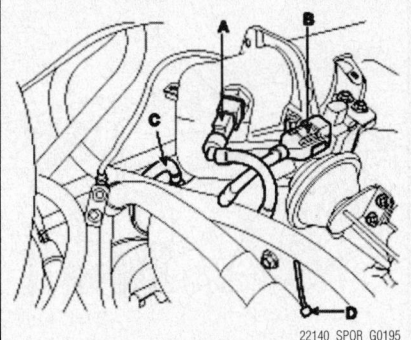

22140_SPOR_G0195

Fig. 144 Intake Air Temperature (IAT) sensor connector (A). VIS actuator connector (B). Oxygen sensor (rear 2nd) connector (C). Power steering pump switch (D).

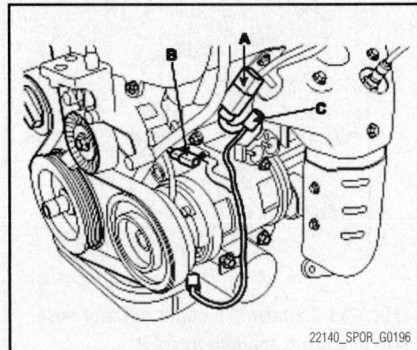

22140_SPOR_G0196

Fig. 145 Oxygen sensor (front 1st) connector (A). Air conditioner compressor switch connector (B). Oil pressure sensor connector (C).

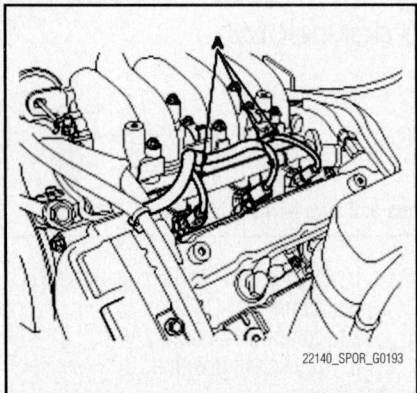

22140_SPOR_G0193

Fig. 142 Three fuel injector connectors (A).

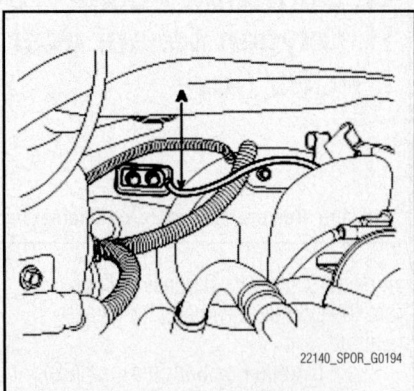

22140_SPOR_G0194

Fig. 143 Disconnect ground cable (A).

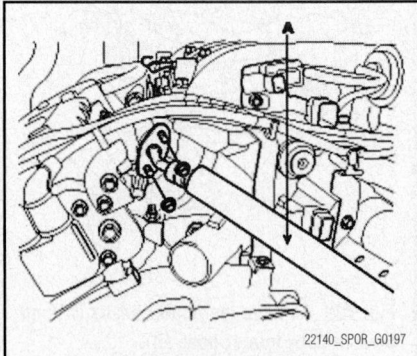

22140_SPOR_G0197

Fig. 146 Remove the fuel inlet hose (A) from delivery pipe.

u. Oxygen sensor (front 1st) connector (A).

v. Air conditioner compressor switch connector (B).

w. Oil pressure sensor connector (C).

12. Remove the fuel inlet hose (A) from delivery pipe.

13. Remove the brake booster vacuum hose (A).

14. Remove the accelerator cable by loosening the locknut, then slip the cable end out of the throttle linkage.

15. Remove the PCV hose (A).

16. Remove the intake manifold.

17. Remove the power steering pump.

18. Remove the exhaust manifold.

19. Remove the timing belt.

20. Remove the spark plug cable.

21. Remove the cylinder head covers (A).

22. Remove the camshaft sprocket.

23. Remove the camshaft bearing caps (A).

24. Remove the camshafts (A).

25. Remove the timing belt rear cover (A).

26. Remove the water temperature control assembly (A) and water pipe.

27. Remove the cylinder head bolts, and remove the cylinder heads.

a. Uniformly loosen and remove the 8 cylinder head bolts and plate washers on

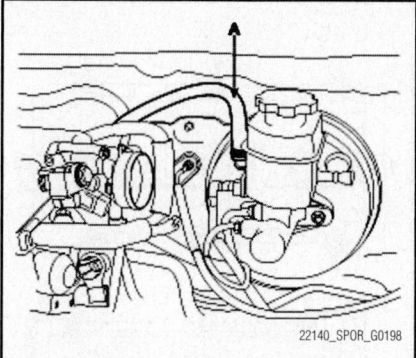

Fig. 147 Remove the brake booster vacuum hose (A).

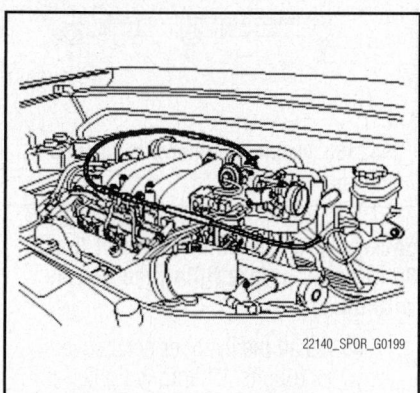

Fig. 148 Remove the accelerator cable

Fig. 149 Remove the PCV hose (A).

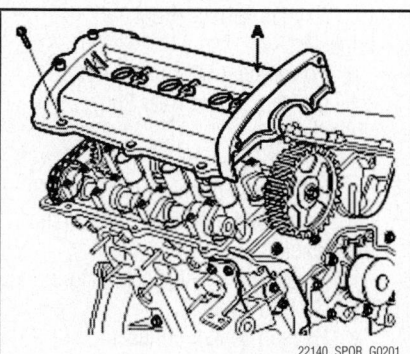

Fig. 150 Remove the cylinder head covers (A).

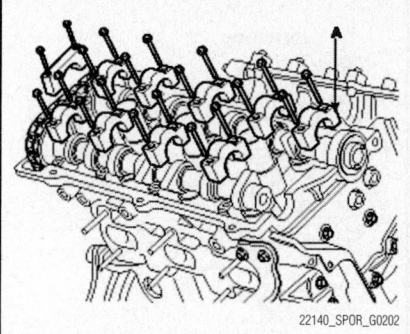

Fig. 151 Remove the camshaft bearing caps (A).

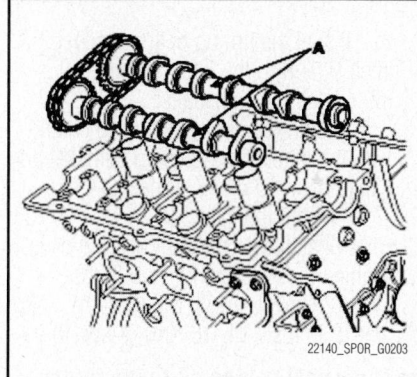

Fig. 152 Remove the camshafts (A).

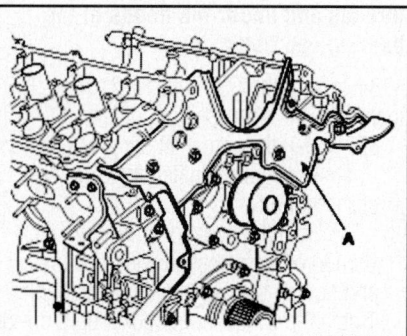

Fig. 153 Remove the timing belt rear cover (A).

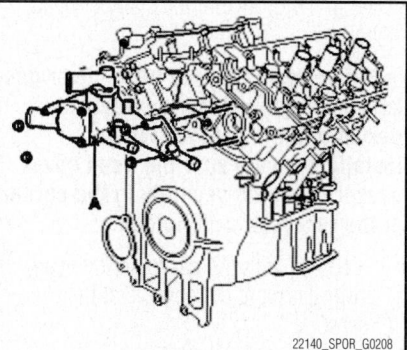

Fig. 154 Remove the water temperature control assembly (A) and water pipe.

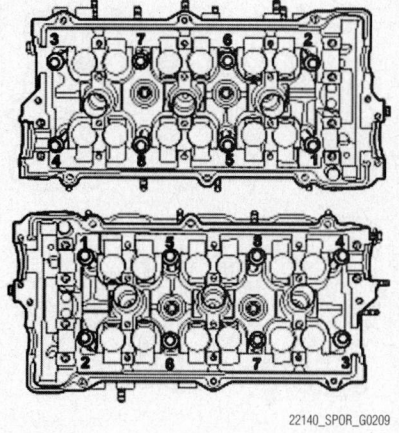

Fig. 155 Remove the 16 cylinder head bolts in several passes and in sequence.

Fig. 156 Install the cylinder head gaskets on the cylinder block.

each cylinder head in several passes and in sequence.

❈❈ CAUTION

Head warpage or cracking could result from removing bolts in an incorrect order.

b. Lift the cylinder head from the dowels on the cylinder block and place the cylinder head on wooden blocks on a bench.

To install:

28. Install the cylinder head gaskets on the cylinder block.

➡**Be careful of the installation direction.**

29. Place the cylinder head quietly in order not to damage the gasket with the bottom part of the end.

30. Install cylinder head bolts.

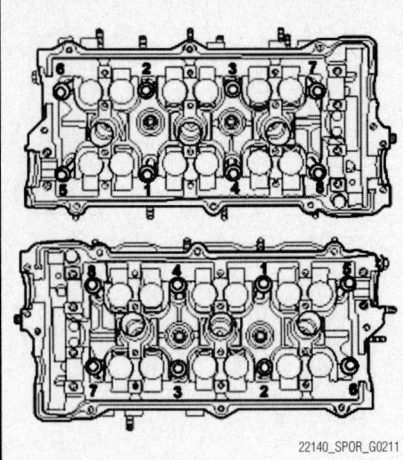

Fig. 157 Install and uniformly tighten the cylinder head bolts on each cylinder head in several passes and in sequence

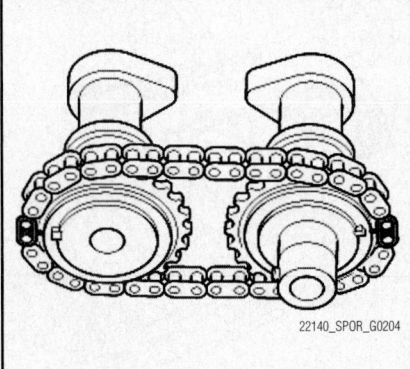

Fig. 158 Align the camshaft timing chain

a. Apply a light coat if engine oil on the threads and under the heads of the cylinder head bolts.

b. Install the plate washer to the cylinder head bolt.

c. Install and uniformly tighten the cylinder head bolts on each cylinder head in several passes and in sequence.

✳✳ CAUTION

If any of the cylinder head bolts does not meet the torque specification, repeat the cylinder head bolt torque tightening sequence.

d. Tighten to 218 ft. lbs. (5 Nm)

e. Retighten the cylinder head bolts by 60° in the same sequential order.

f. Again, retighten the cylinder head bolts by 45° in the same sequential order.

31. Install the water pipe and water temperature control assembly. Tighten to 11–14 ft. lbs. (15–20 Nm)

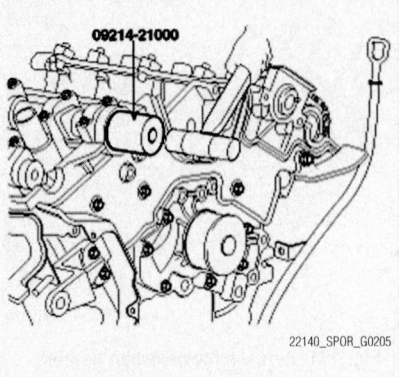

Fig. 159 Using the SST (09214-21000), install the camshaft bearing oil seal.

32. Install the timing belt rear cover. Tighten to 7–9 ft. lbs. (10–12 Nm)

33. Install the camshafts.

a. Align the camshaft timing chain with the intake timing chain sprocket and exhaust timing chain sprocket.

b. Install the camshaft.

c. Install the camshaft bearing caps. Tighten to M6 (38mm): 7–9 ft. lbs. (10–12 Nm) (Mark7); M6 (50mm): 10–12 ft. lbs. (14–16 Nm) (Mark10)

➡**Apply new engine oil to the thrust portion and journal of the camshafts. Apply a light coat of engine oil on the threads and under the heads of the bearing cap bolts.**

34. Using the SST (09214-21000), install the camshaft bearing oil seal.

35. Install the camshaft sprocket.

a. Temporarily install the camshaft sprocket bolts.

b. Hold the hexagonal head wrench portion of the camshaft with a wrench, and tighten the camshaft sprocket bolts. Tighten to Camshaft sprocket bolt: 65–80 ft. lbs. (90–110 Nm)

36. Install semi-circular packing.

37. Install the cylinder head cover.

a. Install the cylinder head cover gasket in the groove of the cylinder head cover.

➡**Before installing the head cover gasket, thoroughly clean the head cover gasket and the groove. When installing, make sure the head cover gasket is seated securely in the corners of the recesses with no gap.**

b. Apply liquid gasket to the head cover gasket at the corners of the recess.

➡**Use liquid gasket, Loctite No.5699. Check that the mating surfaces are clean and dry before applying liquid**

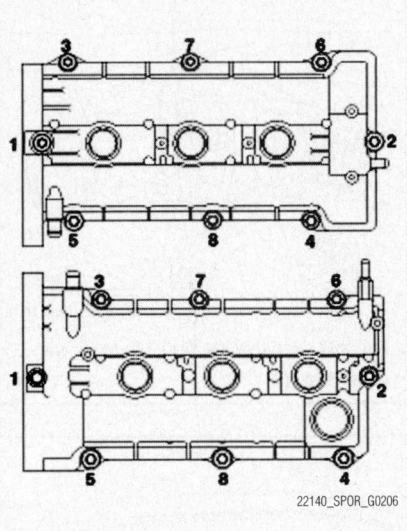

Fig. 160 Install the cylinder head cover bolts in sequence

gasket. After assembly, wait at least 30 minutes before filling the engine with oil.

c. Install the cylinder head covers with the 16bolts. Uniformly tighten the bolts in several passes. Tighten to 6–7.4 ft. lbs. (8–10 Nm)

38. Install the spark plug cable.

39. Install the timing belt.

40. Install the exhaust manifold.

41. Install the power steering pump.

42. Install the intake manifold.

43. Install the PCV hose.

44. Install the accelerator cable.

45. Install the brake booster vacuum hose.

46. Install the fuel inlet hose.

47. Install the engine wire harness connectors and wire harness clamps to the cylinder head and the intake manifold.

a. Oil pressure sensor connector.

b. Air conditioner compressor switch connector.

c. Oxygen sensor (front 1st) connector.

d. Power steering pump switch.

e. Oxygen sensor (rear 2nd) connector.

f. VIS actuator connector.

g. IAT sensor connector.

h. Connect the ground cable to the cowl panel.

i. Three fuel injector connectors.

j. Ground cable.

k. Oxygen sensor connector.

l. Crankshaft position sensor connector.

m. Ignition coil connector.

n. ECT sensor connector.
o. PCSV hose.
p. Oxygen sensor (rear 1st) connector.
q. CMP connector.
r. Knock sensor connector.
s. Injector connector.
t. VIS actuator connector.
u. PCSV connector.
v. ISA connector.
w. TPS connector.
48. Install the heater hoses and throttle body heater hose.
49. Install the upper and lower radiator hose.
50. Install the coolant reservoir tank.
51. Install the intake air hose and air cleaner assembly.
    a. Install the intake air hose and air cleaner assembly.
    b. Connect the breather hose from air cleaner hose.
    c. Connect the AFS connector.
52. Install the engine cover.
53. Connect the negative terminal to the battery.
54. Install the strut bar.
55. Install the cowl grill and wiper motor.
56. Install the air duct.
57. Fill with engine coolant.
58. Start the engine and check for leaks.
59. Recheck engine coolant level and oil level.

ENGINE ASSEMBLY

REMOVAL & INSTALLATION

2.0L Engine

See Figures 161 through 189.

1. Remove the air duct (A).
2. Disconnect the battery terminals and remove the battery.
3. Remove the engine cover.
4. Drain the engine coolant.

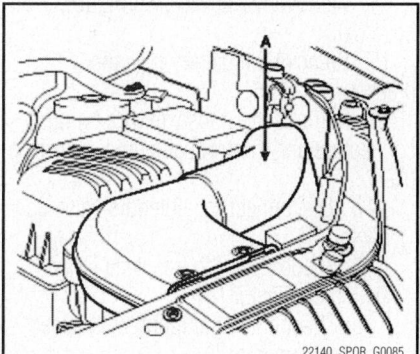

Fig. 161 Remove the air duct (A).

5. Remove the intake air hose and air cleaner assembly.
    a. Disconnect the Air Flow Sensor (AFS) connector (A).
    b. Disconnect the breather hose (B) from intake air hose.
    c. Remove the intake air hose and air cleaner assembly (C).
6. Remove the upper radiator hose (A) and lower radiator hose (B).
7. Remove the heater hoses (A).
8. Remove the accelerator cable (A).

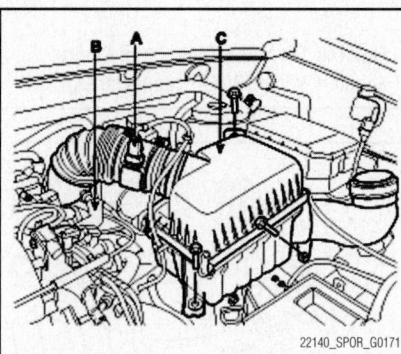

Fig. 162 Disconnect the Air Flow Sensor (AFS) connector (A).

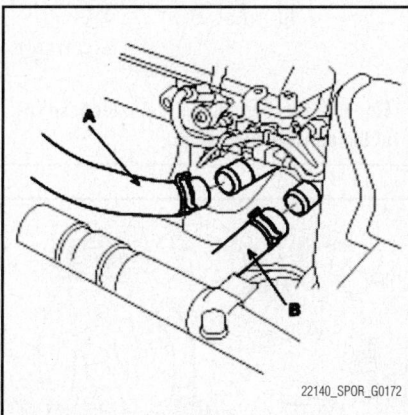

Fig. 163 Remove the upper radiator hose (A) and lower radiator hose (B).

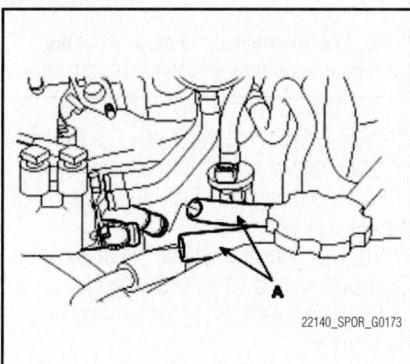

Fig. 164 Remove the heater hoses (A).

9. Remove the engine wire harness connectors and wire harness clamps from the cylinder head and the intake manifold.
    a. Disconnect the OCV (Oil Control Valve) connector (A).
    b. Disconnect the oil temperature sensor connector (B).
    c. Disconnect the Engine Coolant Temperature (ECT) sensor connector (C).

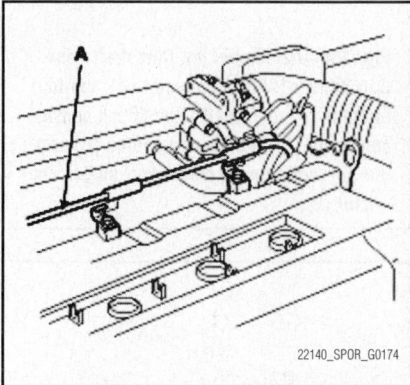

Fig. 165 Remove the accelerator cable (A).

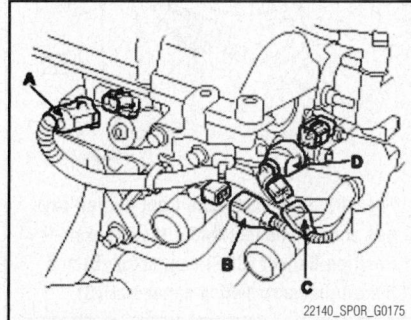

Fig. 166 Disconnect the OCV (Oil Control Valve) connector (A), the oil temperature sensor connector (B), the Engine Coolant Temperature (ECT) sensor connector (C), and the ignition coil connector (D).

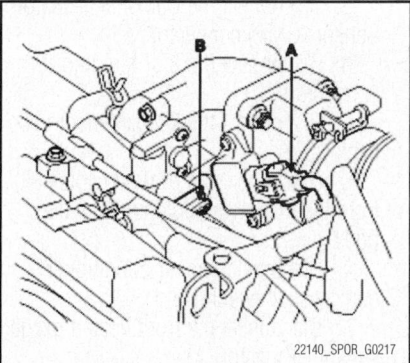

Fig. 167 Disconnect the Throttle Position Sensor (TPS) connector (A) and the Idle Speed Actuator (ISA) connector (B).

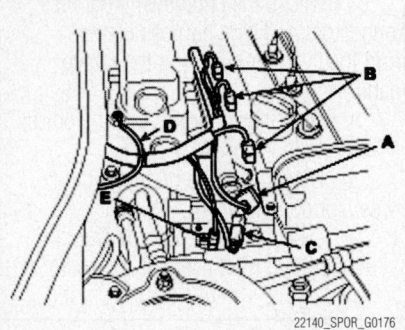

Fig. 168 Disconnect the Camshaft Position Sensor (CMP) connector (A), the four injector connectors (B), the knock sensor connector (C), the ground cables (D) and the air conditioner compressor switch connector (E).

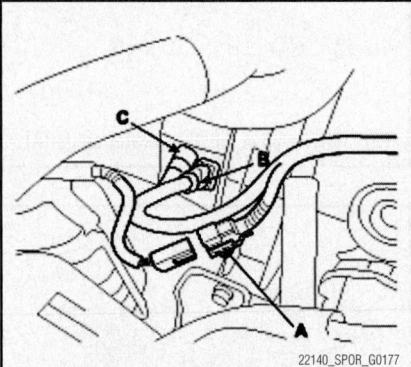

Fig. 169 Disconnect the front heated oxygen sensor connector (A), the Crankshaft Position Sensor (CKP) connector (B) and the oil pressure switch connector (C).

d. Disconnect the ignition coil connector (D).

e. Disconnect the Throttle Position Sensor (TPS) connector (A).

f. Disconnect the Idle Speed Actuator (ISA) connector (B).

g. Disconnect the Camshaft Position Sensor (CMP) connector (A).

h. Disconnect the four fuel injector connectors (B).

i. Disconnect the knock sensor connector (C).

j. Disconnect the ground cables (D) from the intake manifold and vehicle's body.

k. Disconnect the air conditioner compressor switch (E).

l. Disconnect the front heated oxygen sensor connector (A).

m. Disconnect the Crankshaft Position Sensor (CKP) connector (B).

n. Disconnect the oil pressure switch connector (C).

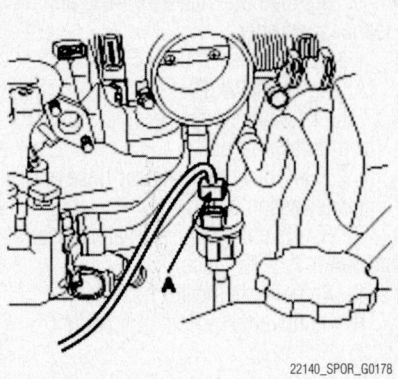

Fig. 170 Disconnect the Purge Control Solenoid Valve (PCSV) connector (A).

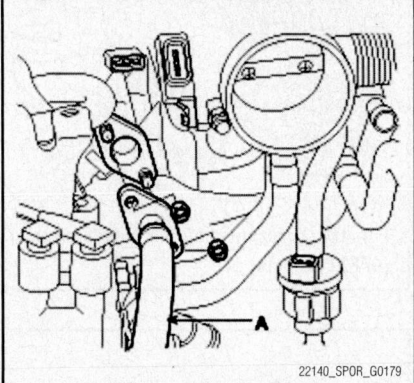

Fig. 171 Disconnect the fuel inlet hose (A) of the delivery pipe side.

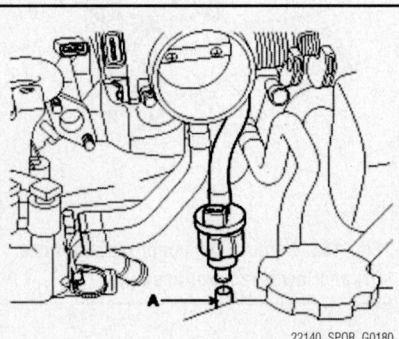

Fig. 172 Disconnect the hose (A) of the Purge Control Solenoid Valve (PCSV) side.

o. Disconnect the Purge Control Solenoid Valve (PCSV) connector (A).

10. Disconnect the fuel inlet hose (A) of the delivery pipe side.

11. Disconnect the hose (A) of the Purge Control Solenoid Valve (PCSV) side.

12. Remove the brake booster vacuum hose (A).

13. Remove the power steering oil hose (A) from the power steering pump.

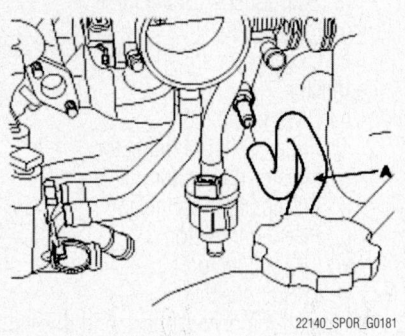

Fig. 173 Remove the brake booster vacuum hose (A).

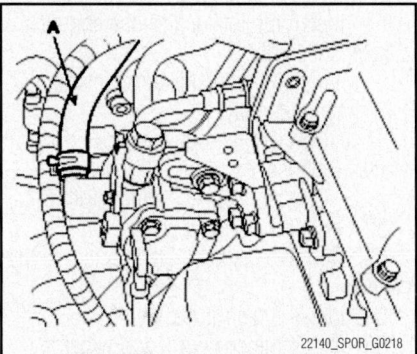

Fig. 174 Remove the power steering oil hose (A) from the power steering pump.

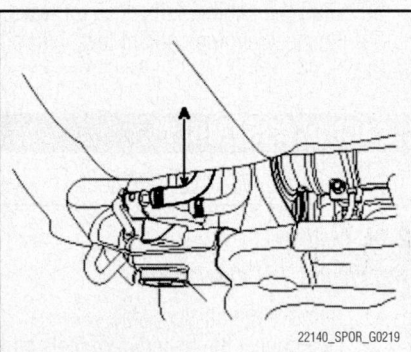

Fig. 175 Remove the power steering lower oil hose (A).

14. Remove the power steering lower oil hose (A).

15. Remove the battery tray and bracket.

16. Remove the transaxle wire harness connectors and control cable from transaxle (A/T).

a. Disconnect the solenoid valve connector (A).

b. Disconnect the transaxle range switch connector (B).

c. Disconnect the input shaft speed sensor connector (C).

d. Disconnect the output shaft speed sensor connector (A).

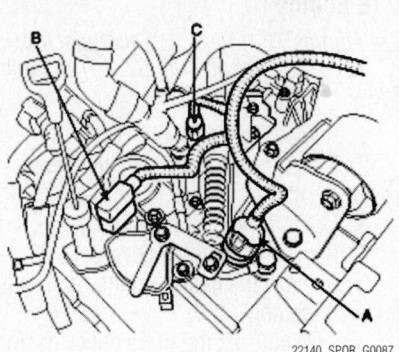

Fig. 176 Remove the transaxle wire harness connectors and control cable from transaxle (A/T).

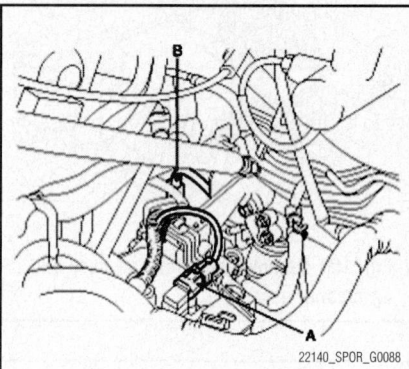

Fig. 177 Disconnect the output shaft speed sensor connector (A) and the vehicle speed sensor connector (B).

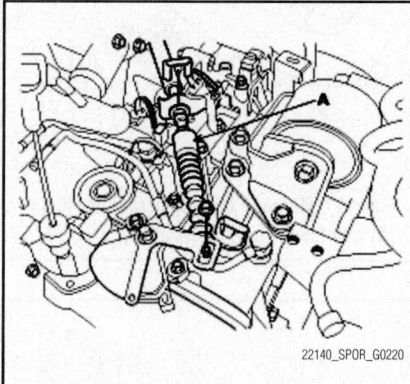

Fig. 178 Remove the control cable (A).

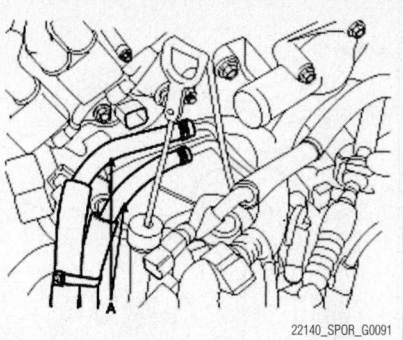

Fig. 179 Remove the transaxle oil cooler hose (A) (A/T).

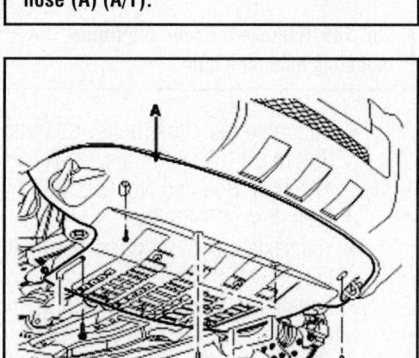

Fig. 180 Remove the undercover (A).

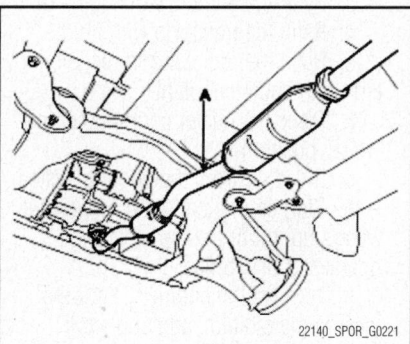

Fig. 181 Remove the front exhaust pipe (A).

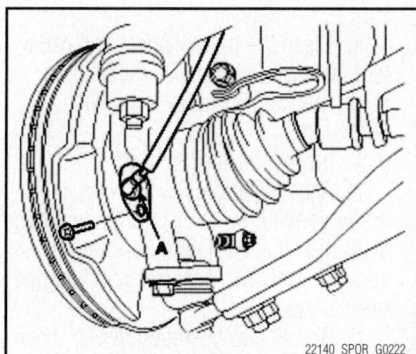

Fig. 182 Disconnect the ABS wheel speed sensor (A) from both front knuckle.

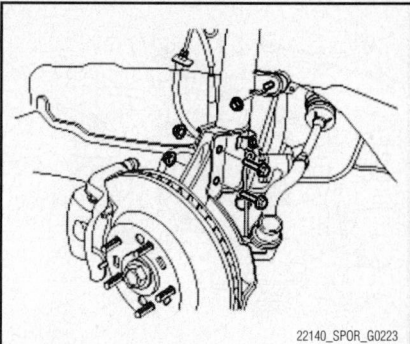

Fig. 183 Remove the front strut lower mounting bolts and nuts.

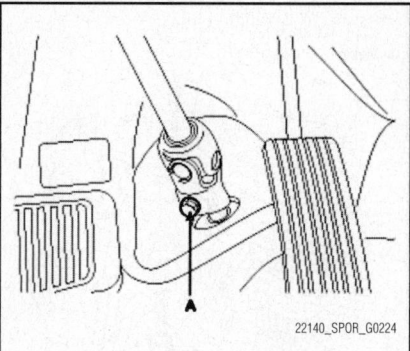

Fig. 184 Remove the steering u-joint mounting bolt (A).

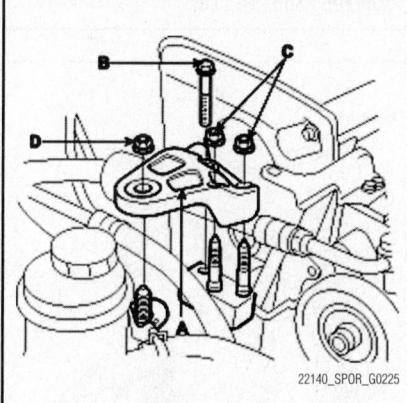

Fig. 185 Remove the engine mounting bracket (A).

    e. Disconnect the vehicle speed sensor connector (B).
    f. Remove the control cable (A).
  17. Remove the transaxle oil cooler hose (A) (A/T).
  18. Remove the undercover (A).
  19. Remove the front exhaust pipe (A).
  20. Disconnect the ABS wheel speed sensor (A) from both front knuckle.

  21. Remove the front strut lower mounting bolts and nuts.
  22. Remove the caliper and hang the caliper assembly.
  23. Remove the steering u-joint mounting bolt (A).
  24. Install the jack for supporting engine and transaxle assembly.
  25. Remove the engine mounting bracket (A).
  26. Remove the transaxle mounting bracket (A).

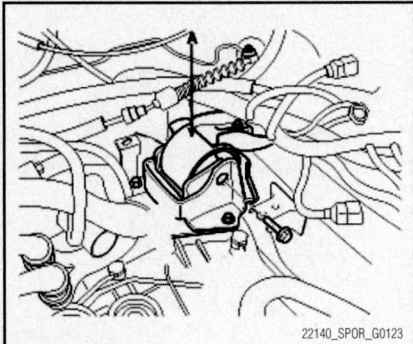

Fig. 186 Remove the transaxle mounting bracket (A).

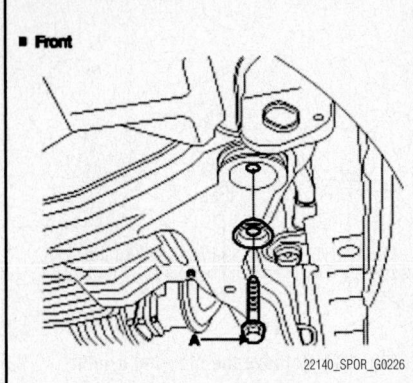

Fig. 187 Remove the front sub frame mounting bolts and nuts.

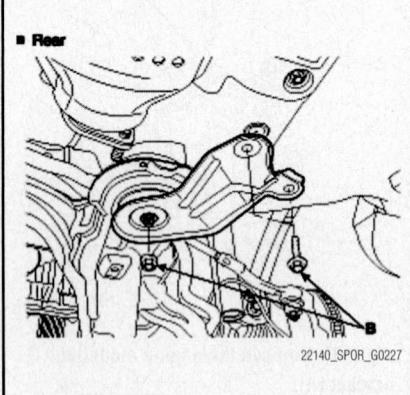

Fig. 188 Remove the rear sub frame mounting bolts and nuts.

27. Remove the sub frame mounting bolts and nuts.

28. Remove the engine and transaxle assembly by lifting vehicle.

To install:

29. Installation is in the reverse order of removal.

30. Use the following torque specifications:

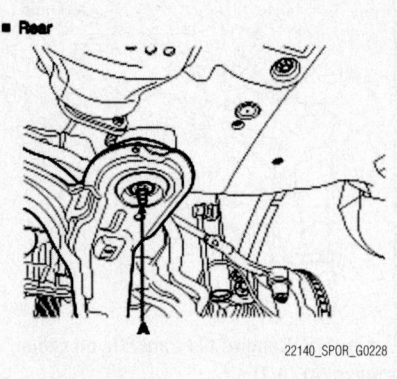

Fig. 189 Remove the rear sub frame mounting bolts and nuts.

- Subframe mounting bolts and nuts: Bolt (A): 116–130 ft. lbs. (157–177 Nm); Bolt and Nut (B): 51–65 ft. lbs. (67–88 Nm)
- Transaxle mounting bolts: 65–80 ft. lbs. (88–108 Nm)
- Engine mounting bracket: Nut (D): 43–58 ft. lbs. (59–79 Nm); Bolt (B) and Nut (C): 36–47 ft. lbs. (49–64 Nm)

31. Perform the following :
   a. Adjust the shift cable.
   b. Adjust the throttle cable.
   c. Refill the engine with engine oil.
   d. Refill the transaxle with fluid.
   e. Refill the radiator and reservoir tank with engine coolant.
   f. Place the heater control knob on "HOT" position.
   g. Bleed air from the cooling system
   h. Start engine and let it run until it warms up. (until the radiator fan operates 3 or 4 times.)
   i. Turn Off the engine. Check the level in the radiator, add coolant if needed. This will allow trapped air to be removed from the cooling system.
   j. Put the radiator cap on tightly, then run the engine again and check for leaks.
   k. Clean the battery posts and cable terminals with sandpaper assemble them, then apply grease to prevent corrosion.
   l. Inspect for fuel leakage.
   m. After assemble the fuel line, turn on the ignition switch (do not operate the starter) so that the fuel pump runs for approximately two seconds and fuel line pressurizes.
   n. Repeat this operation two or three times, then check for fuel leakage at any point in the fuel line.

2.7L Engine

See Figures 72, 11, 12, 182 through 184, 27, 187, 188 through 190, 92, 191 through 208.

1. Remove the air duct (A).
2. Disconnect the negative terminal from the battery.
3. Drain the engine coolant.
4. Remove the engine cover.
5. Remove the intake air hose and air cleaner assembly.
   a. Disconnect the AFS connector (A).

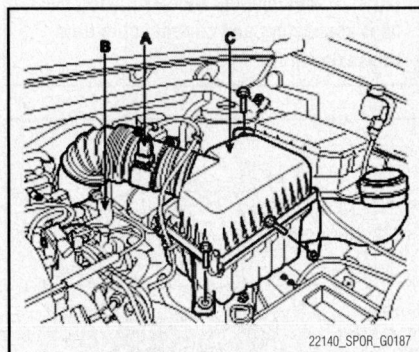

Fig. 190 Remove the intake air hose and air cleaner assembly

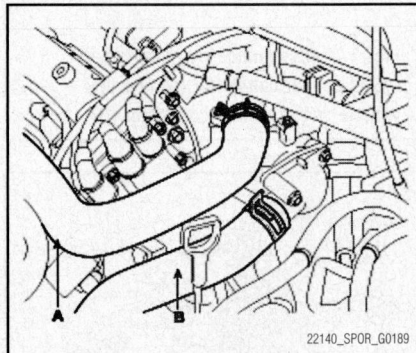

Fig. 191 Remove the upper radiator hose (A) and lower radiator hose (B)

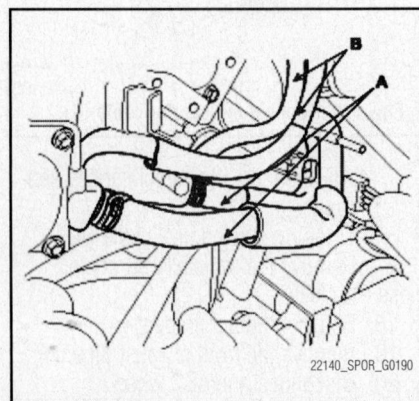

Fig. 192 Remove the heater hoses (A)

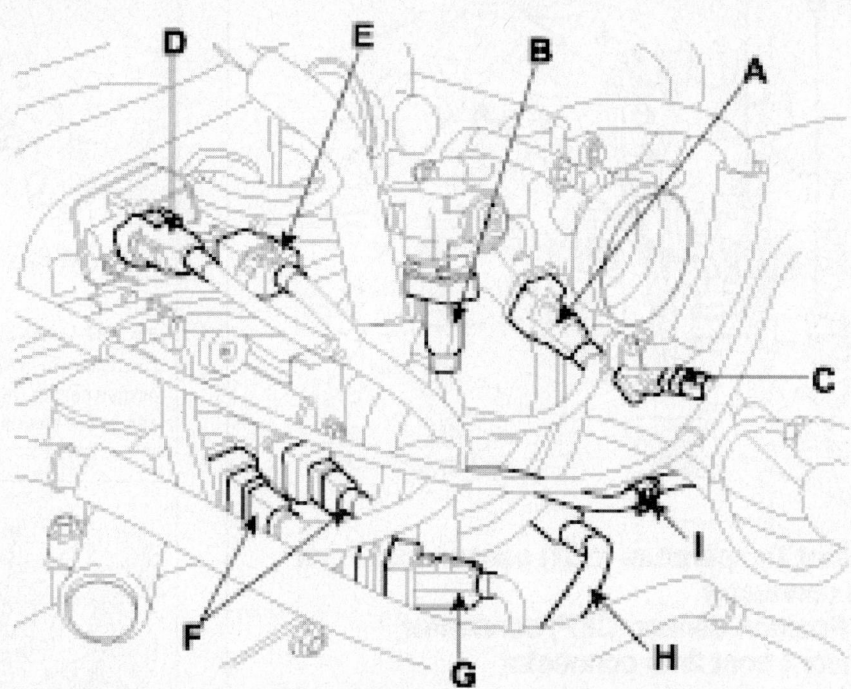

A. Throttle Position Sensor (TPS) connector
B. Idle Speed Actuator (ISA) connector
C. Purge Control Solenoid Valve (PCSV) connector
D. VIS actuator connector
E. Injector connector
F. Knock Sensor (KS) connectors
G. Camshaft Position Sensor (CMP) connector
H. Oxygen sensor (rear 1st) connector
I. PCSV hose

22140_SPOR_G0191

Fig. 193 Remove the engine wire harness connectors

b. Disconnect the breather hose (B) from air cleaner hose.

c. Remove the intake air hose and air cleaner assembly (C).

6. Remove the upper radiator hose (A) and lower radiator hose (B).

7. Remove the heater hoses (A).

8. Remove the engine wire harness connectors and wire harness clamps from the cylinder head and the intake manifold.

a. Throttle Position Sensor (TPS) connector (A).

b. Idle Speed Actuator (ISA) connector (B).

c. Purge Control Solenoid Valve (PCSV) connector (C).

d. VIS actuator connector (D).

e. Injector connector (E).

f. Knock Sensor (KS) connector (F).

g. Camshaft Position Sensor (CMP) connector (G).

h. Oxygen sensor (rear 1st) connector (H).

i. PCSV hose (I).

j. Engine Coolant Temperature (ECT) sensor (A) connector.

k. Ignition coil connector (B).

l. Crankshaft Position (CKP) sensor connector (C).

m. Rear oxygen sensor (front 2nd) connector (D).

n. Ground cable (E).

o. Three fuel injector connectors (A).

p. Disconnect ground cable (A) from the cowl panel.

q. Intake air temperature sensor connector (A).

r. VIS actuator connector (B).

s. Oxygen sensor (rear 2nd) connector (C).

t. Power steering pump switch (D).

9. Disconnect front heated oxygen sensor (LH) connector (A), air compressor switch connector (B) and oil pressure sensor connector (C).

10. Remove the fuel inlet from delivery pipe (A).

11. Remove the brake booster vacuum hose (A).

12. Remove the accelerator cable by

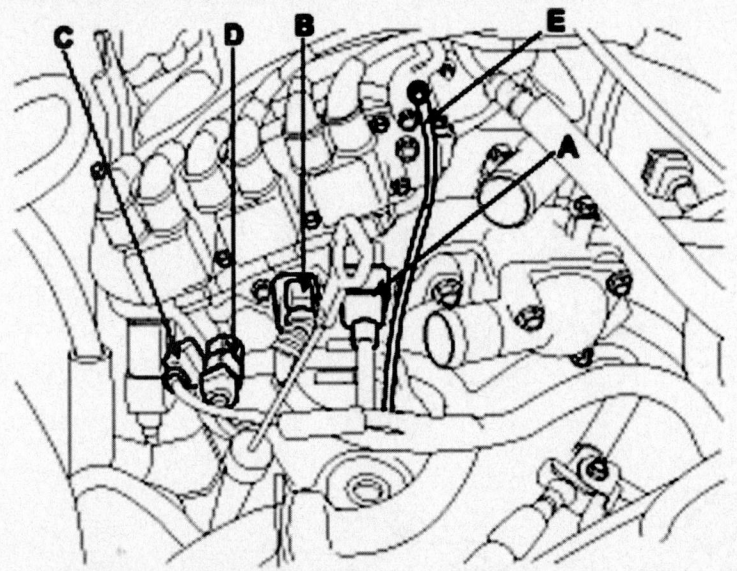

A. Engine Coolant Temperature (ECT) sensor connector
B. Ignition coil connector
C. Crankshaft Position Sensor (CKP) connector
D. Oxygen sensor (front 2nd) connector
E. Ground cable

22140_SPOR_G0192

Fig. 194 Remove additional engine wire harness connectors

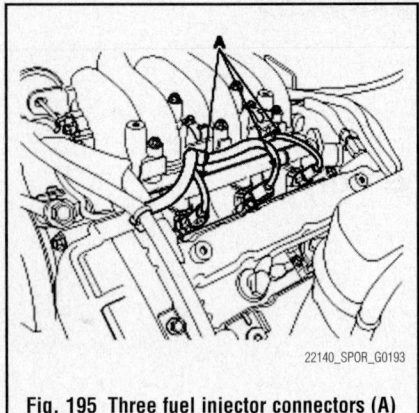

22140_SPOR_G0193

Fig. 195 Three fuel injector connectors (A)

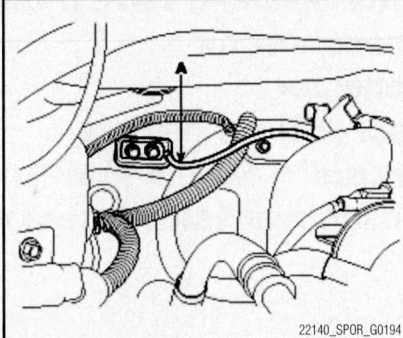

22140_SPOR_G0194

Fig. 196 Disconnect ground cable (A) from the cowl panel

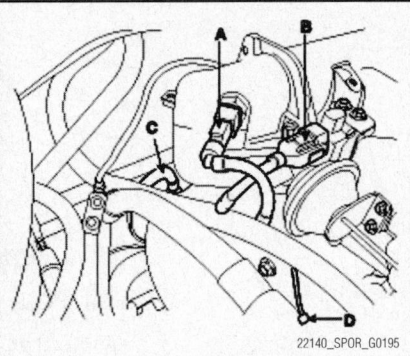

22140_SPOR_G0195

Fig. 197 Intake air temperature sensor connector (A), VIS actuator connector (B), oxygen sensor (rear 2nd) connector (C) and power steering pump switch (D)

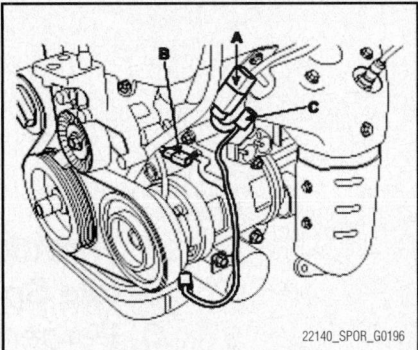

22140_SPOR_G0196

Fig. 198 Disconnect front heated oxygen sensor (LH) connector (A), air compressor switch connector (B) and oil pressure sensor connector (C)

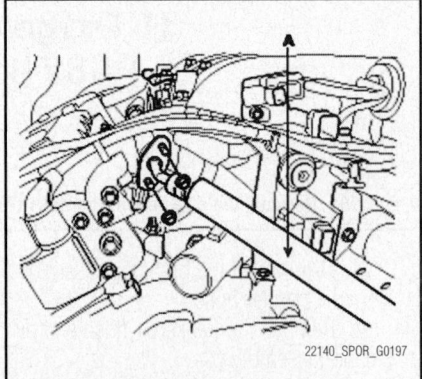

22140_SPOR_G0197

Fig. 199 Remove the fuel inlet from delivery pipe (A)

loosening the locknut, then slip the cable end out of the throttle linkage.

13. Remove the power steering pump hose (A).

14. Remove the battery tray and bracket.

15. Disconnect the transaxle wire harness connector.

    a. Disconnect the inhibitor switch connector (A).

    b. Disconnect the transaxle range connector (B).

    c. Disconnect the input shaft speed connector (C).

    d. Disconnect the output shaft speed connector (A).

    e. Disconnect the vehicle speed sensor connector (B).

16. Remove the control transaxle cable (A).

17. Remove the transaxle oil cooler hoses (A/T) (A).

18. Remove the undercover (A).

19. Remove the front exhaust pipe (A).

20. Disconnect the ABS wheel speed sensor (A) from both front knuckles.

21. Remove the front strut lower mounting bolts and nuts.

22. Remove the caliper and hang the caliper assembly.

23. Remove the steering u-joint mounting bolt (A).

24. Install the jack for supporting engine and transaxle assembly.

25. Remove the engine mounting bracket (A).

26. Remove the transaxle mounting bracket (A).

27. Remove the sub frame mounting bolts and nuts.

28. Remove the engine and transaxle assembly by lifting vehicle.

To install:

29. Installation is in the reverse order of removal.

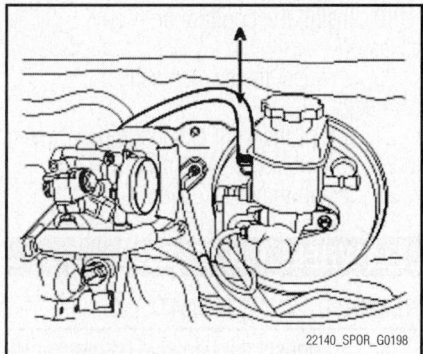

Fig. 200 Remove the brake booster vacuum hose (A)

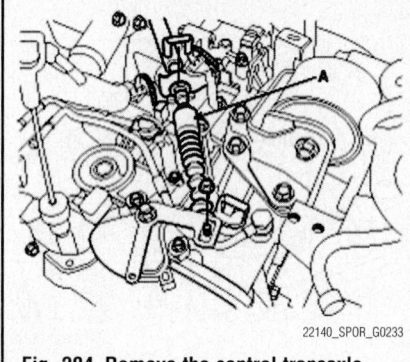

Fig. 204 Remove the control transaxle cable (A)

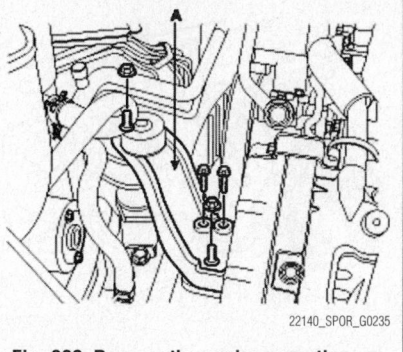

Fig. 208 Remove the engine mounting bracket (A)

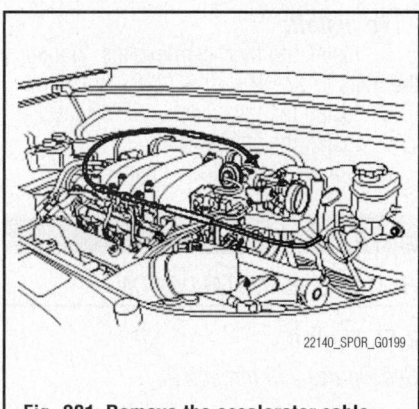

Fig. 201 Remove the accelerator cable

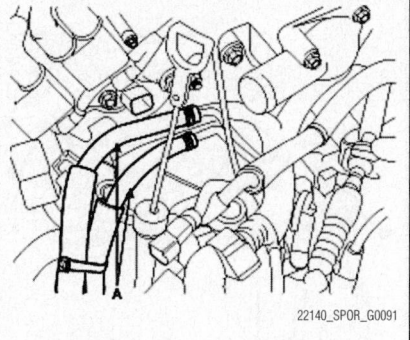

Fig. 205 Remove the transaxle oil cooler hoses (A/T) (A)

30. Perform the following:
   a. Adjust the shift cable.
   b. Adjust the throttle cable.
   c. Refill the engine with engine oil.
   d. Refill the transaxle with fluid.
   e. Refill the radiator with engine coolant.
   f. Bleed air from the cooling system with the heater valve open.
   g. Clean the battery posts and cable terminals with sandpaper assemble them, and apply grease to prevent corrosion.
   h. Inspect for fuel leakage.
   i. After assembling the fuel line, turn on the ignition switch (do not operate the starter) so that the fuel pump runs for approximately two seconds and fuel line pressurizes.
   j. Repeat this operation two or three times, and check for fuel leakage at any point in the fuel line.

EXHAUST MANIFOLD

REMOVAL & INSTALLATION

2.0L Engine

See Figures 209 and 210.

1. Remove the engine cover.

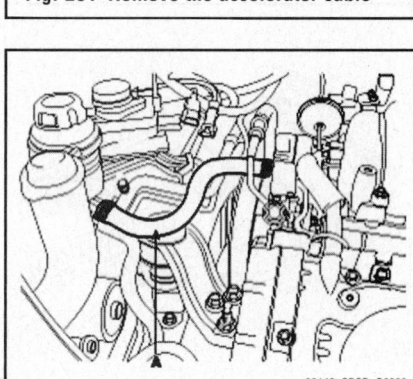

Fig. 202 Remove the power steering pump hose (A)

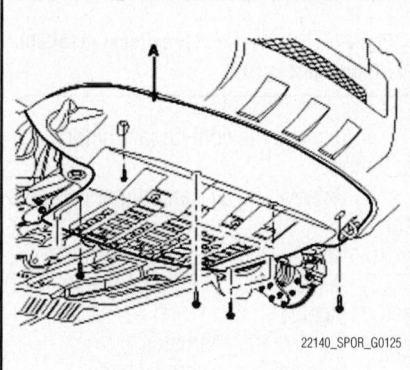

Fig. 206 Remove the undercover (A)

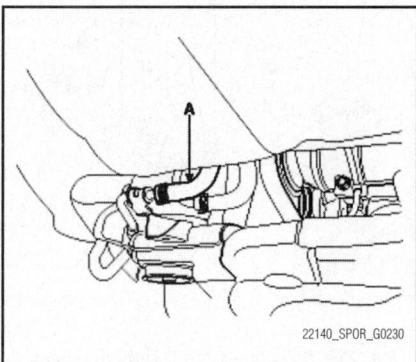

Fig. 203 Remove the power steering pump hose (A)

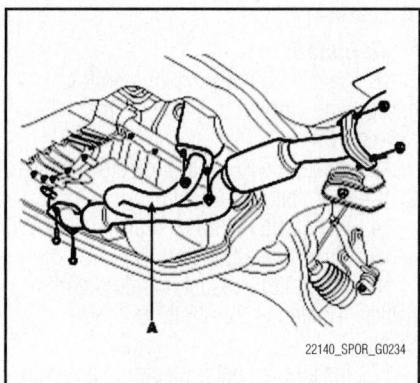

Fig. 207 Remove the front exhaust pipe (A)

Fig. 209 Remove the heat protector

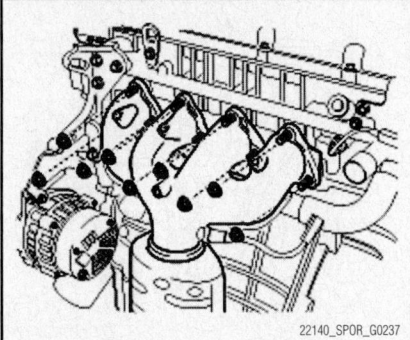

Fig. 210 Remove the exhaust manifold and catalytic converter assembly

2. Disconnect the front oxygen sensor connector.

3. Remove the front muffler.

4. Remove the heat protector.

5. Remove the exhaust manifold and catalytic converter assembly.

6. Installation is in the reverse order of removal

2.7L Engine

See Figures 211 through 214.

1. Remove the undercover (A).

2. Remove the front exhaust pipe (A).

3. Disconnect the oxygen sensor connector.

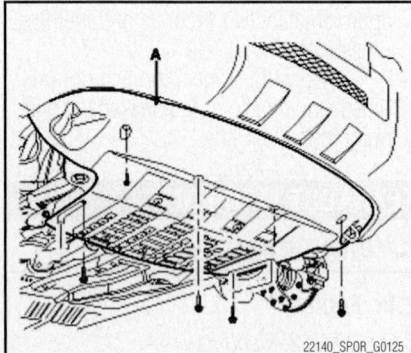

Fig. 211 Remove the undercover (A)

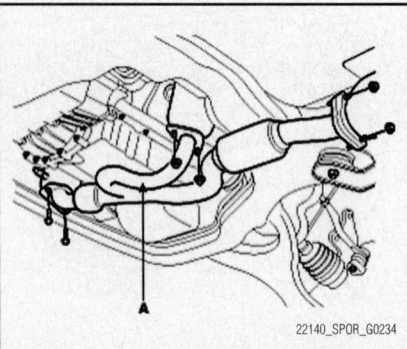

Fig. 212 Remove the front exhaust pipe (A)

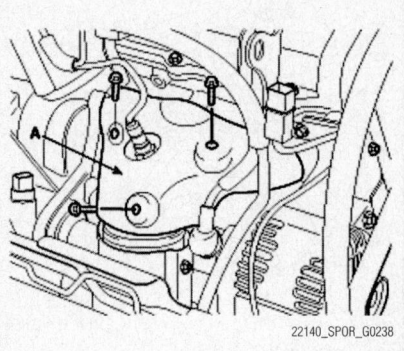

Fig. 213 Remove the LH heat protector (A)

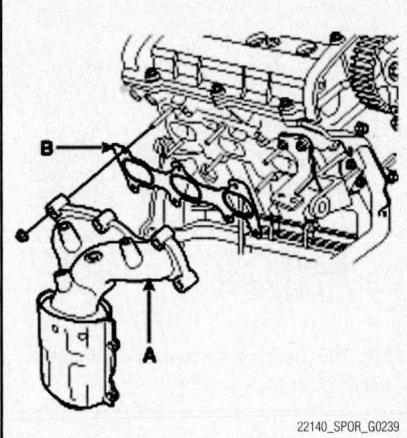

Fig. 214 Remove the LH exhaust manifold (A) and gasket (B)

4. Remove the cooling fan motor assembly.

5. Remove the air conditioner compressor.

6. Remove the LH heat protector (A).

7. Remove the LH exhaust manifold (A) and gasket (B).

8. Remove the alternator.

9. Remove the RH drive shaft.

10. Remove the RH heat protector.

11. Remove the RH exhaust manifold and gasket.

To install:

12. Install the RH exhaust manifold and gasket. Tighten to 22–26 ft. lbs. (30–35 Nm).

13. Install the RH heat protector. Tighten to 12–16 ft. lbs. (17–22 Nm).

14. Install the RH drive shaft.

15. Install the alternator.

16. Install the exhaust manifold and gasket. Tighten to 22–26 ft. lbs. (30–35 Nm).

17. Install the heat protector. Tighten to 9–11 Ft. lbs. (12–15 Nm).

18. Install the air conditioner compressor.

19. Install the cooling fan motor assembly.

20. Connect the oxygen sensor connector.

21. Install the front exhaust pipe. Tighten to 22–30 ft. lbs. (30–40 Nm).

22. Install the undercover.

FLYWHEEL/FLEXPLATE

REMOVAL & INSTALLATION

1. Disconnect the negative battery cable.

2. Remove the starter motor.

3. Remove the transaxle.

4. Remove the flywheel/flexplate.

To install:

5. Install the flywheel/flexplate. Tighten the bolts to 87–94 ft. lbs. (118–128 Nm).

6. Install the transaxle.

7. Install the starter motor.

8. Connect the negative battery cable.

INTAKE MANIFOLD

REMOVAL & INSTALLATION

2.0L Engine

See Figures 215 through 221.

Fig. 215 Remove the engine cover

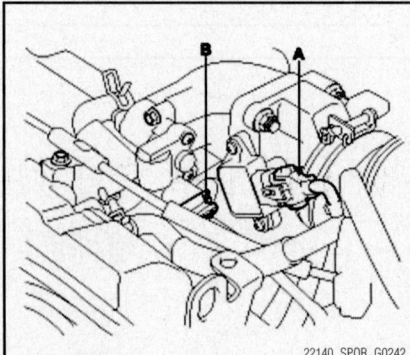

Fig. 216 Disconnect the Throttle Position Sensor (TPS) connector (A) and Idle Speed Actuator (ISA) connector (B)

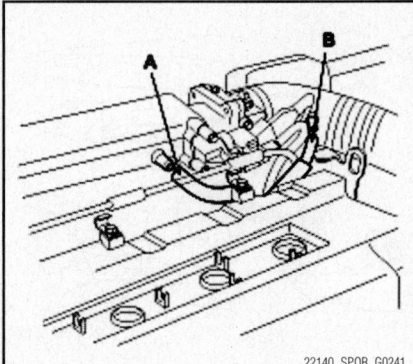

Fig. 217 Disconnect the Positive Crankcase Ventilation (PCV) hose (A) and breather hose (B)

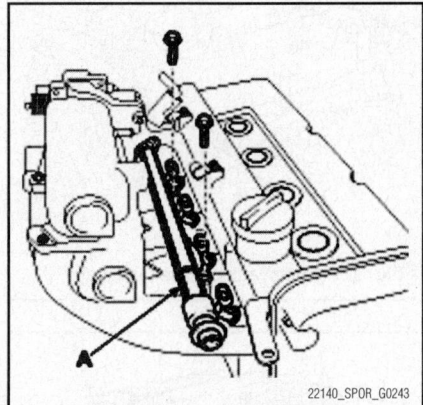

Fig. 218 Remove the fuel delivery pipe (A)

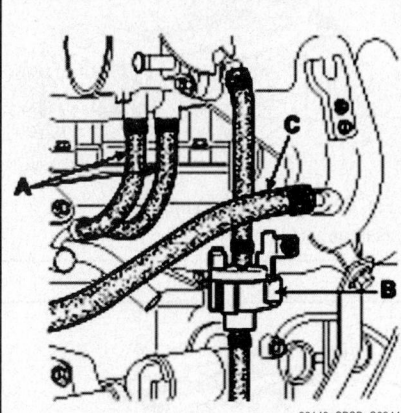

Fig. 219 Remove the heater hose (A), Purge Control Solenoid Valve (PCSV) (B) and the brake vacuum hose (C) from throttle body and intake manifold

1. Remove the engine cover.
2. Disconnect the Throttle Position Sensor (TPS) connector (A) and Idle Speed Actuator (ISA) connector (B).
3. Disconnect the Positive Crankcase

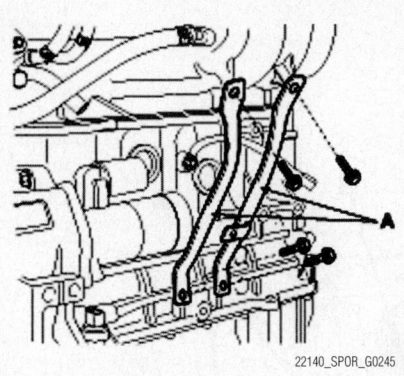

Fig. 220 Remove the intake manifold stay (A)

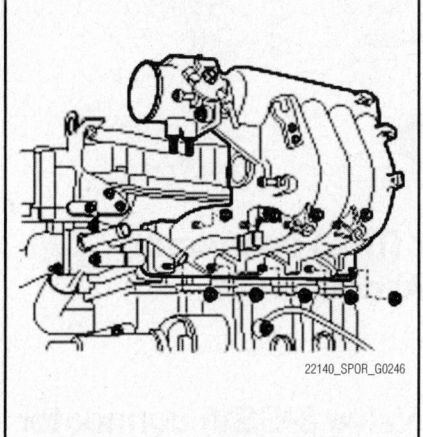

Fig. 221 Remove the intake manifold

Ventilation (PCV) hose (A) and breather hose (B).
4. Remove the accelerator cable.
5. Remove the fuel delivery pipe (A).
6. Remove the heater hose (A), Purge Control Solenoid Valve (PCSV) (B) and the brake vacuum hose (C) from throttle body and intake manifold.
7. Remove the air conditioner compressor.
8. Remove the intake manifold stay (A).
9. Remove the intake manifold.
10. Installation is the reverse order of removal with a new gasket.

2.7L Engine

See Figures 222 through 231.

1. Remove the engine cover.
2. Remove air cleaner hose.
3. Remove surge tank assembly.
   a. Disconnect the accelerator cable.
   b. Disconnect the TPS connector (A).
   c. Disconnect the ISA connector (B).
   d. Disconnect the VIS actuator connector (C).

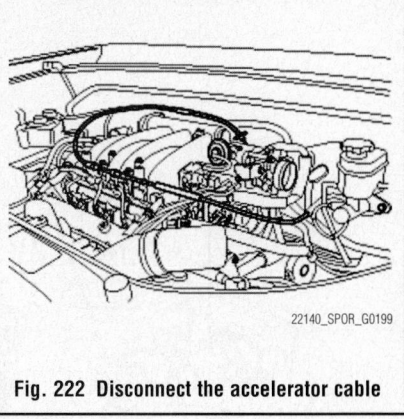

Fig. 222 Disconnect the accelerator cable

e. Disconnect the injector connector (D).
   f. Disconnect the PCSV connector (E).
   g. Disconnect the PCSV hose.
   h. Disconnect the brake booster vacuum hose (A).
   i. Disconnect the PCV hose.
   j. Disconnect the IAT sensor connector (A).
   k. Disconnect the VIS actuator connector (B).
   l. Disconnect the ground cable (A) from the surge tank assembly.
   m. Remove the surge tank stay.
   n. Remove the surge tank assembly (A).
4. Remove the injector assembly (A).
5. Remove the intake manifold (A) and gasket.

To install:

6. Install the intake manifold and gasket. Tighten to 14–15 ft. lbs. (19–21 Nm).
7. Install the injector assembly.
8. Install the surge tank assembly. Tighten to 11–15 ft. lbs. (15–20 Nm).
9. Install the surge tank stay. Tighten to 11–15 ft. lbs. (15–20 Nm).
10. Install the ground cable.
11. Connect the VIS actuator connector.
12. Connect the IAT sensor connector.
13. Connect the PCV hose.
14. Connect the brake booster vacuum hose.
15. Connect the PCSV hose.
16. Connect the PCSV connector.
17. Connect the injector connector.
18. Connector the VIS actuator connector.
19. Connector the ISA connector.
20. Connector the TPS connector.
21. Connector the actuator cable.
22. Install the air cleaner hose.
23. Install the engine cover.

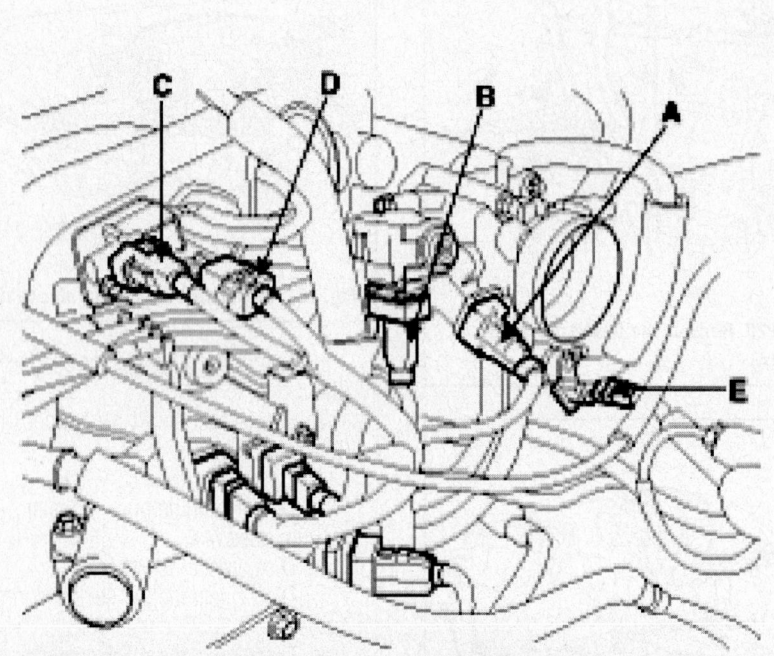

A. Throttle Position Sensor (TPS) connector
B. Idle Speed Actuator (ISA) connector
C. VIS actuator connector
D. Injector connector
E. Purge Control Solenoid Valve (PCSV) connector

22140_SPOR_G0247

Fig. 223 Disconnect the TPS connector (A), the ISA connector (B), the VIS actuator connector (C), the injector connector (D) and the PCSV connector (E)

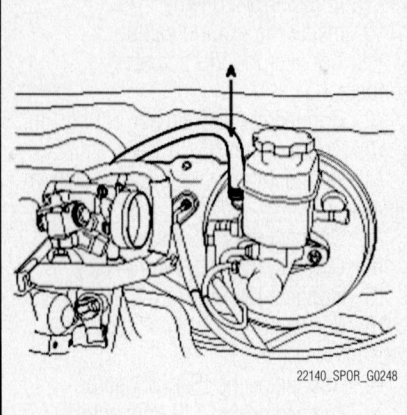

22140_SPOR_G0248

Fig. 224 Disconnect the brake booster vacuum hose (A)

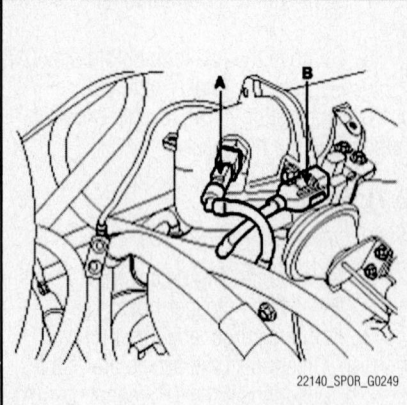

22140_SPOR_G0249

Fig. 225 Disconnect the IAT sensor connector (A) and the VIS actuator connector (B)

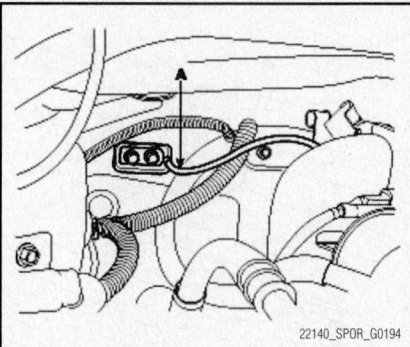

22140_SPOR_G0194

Fig. 226 Disconnect the ground cable (A) from the surge tank assembly

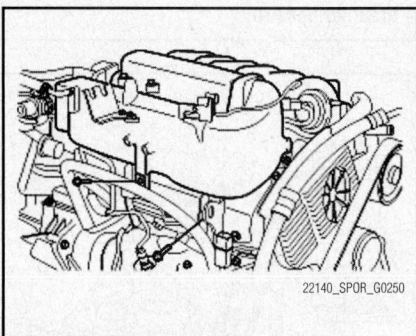

22140_SPOR_G0250

Fig. 227 Remove the surge tank stay

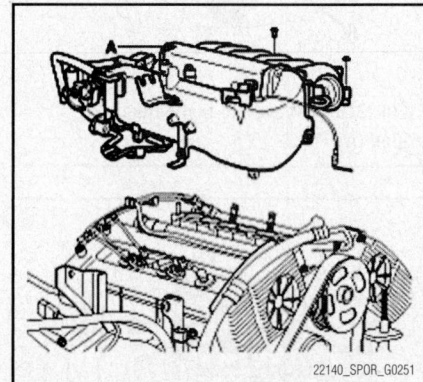

22140_SPOR_G0251

Fig. 228 Remove the surge tank assembly (A)

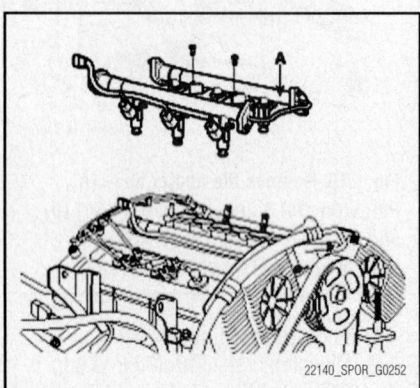

22140_SPOR_G0252

Fig. 229 Remove the injector assembly (A)

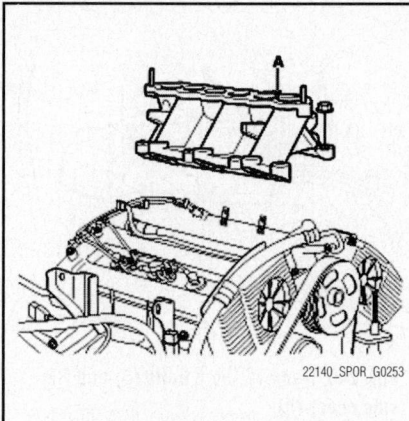

Fig. 230 Remove the intake manifold (A) and gasket

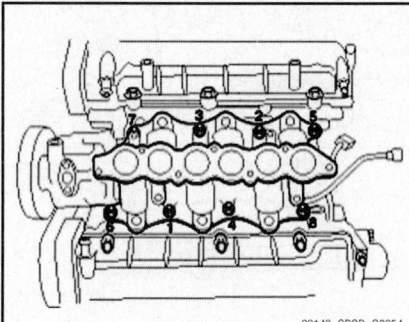

Fig. 231 Install the intake manifold and gasket

OIL PAN

REMOVAL & INSTALLATION

2.0L Engine

See Figures 232 and 233.

1. Drain the engine oil.
2. Disconnect the rear oxygen sensor connector.
3. Remove the front muffler (A).
4. Remove the exhaust manifold.
5. Remove the front muffler bracket (A).
6. Remove the oil pan.

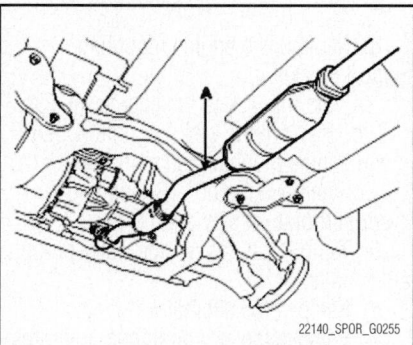

Fig. 232 Remove the front muffler (A)

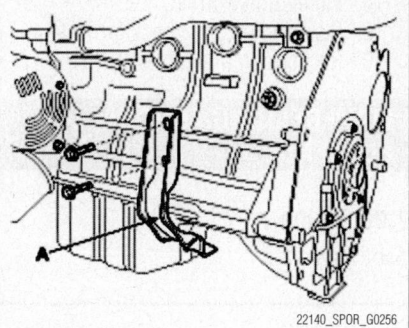

Fig. 233 Remove the front muffler bracket (A)

To install:

7. Install the oil pan.

➡**Check that the mating surfaces are clean and dry before applying liquid gasket.**

a. Apply liquid gasket as an even bead, centered between the edges of the mating surface.

➡**To prevent leakage of oil, apply liquid gasket to the inner threads of the bolt holes. Do not install the parts if five minutes or more have elapsed since applying the liquid gasket. Instead, reapply liquid gasket after removing the residue. After assembly, wait at least 30 minutes before filling the engine with oil.**

b. Install the oil pan with the bolts. Uniformly tighten the bolts in several passes. Tighten to 7–9 ft. lbs. (10–12 Nm)

8. Install the front muffler bracket.
9. Install the exhaust manifold.
10. Install the front muffler.
11. Connect the rear oxygen sensor connector.
12. Fill with engine oil

2.7L Engine

1. Drain engine oil.
2. Remove RH front wheel.
3. Remove RH side cover.
4. Remove the front exhaust pipe.
5. Remove the alternator from engine.
6. Remove the drive belt.
7. Turn the crankshaft and align the white groove on the crankshaft pulley with the pointer on the lower cover.
8. Remove the timing belt.
9. Remove the oil pan and oil screen.

To install:

10. Install the oil pan and oil screen.
11. Install the timing belt.
12. Install the drive belt.

13. Install the alternator.
14. Install the front exhaust pipe.
15. Install the RH front wheel.
16. Fill engine with oil.
17. Start engine and check for leaks.
18. Recheck engine oil level.

OIL PUMP

REMOVAL & INSTALLATION

2.0L Engine

See Figures 234 through 236.

1. Drain the engine oil.
2. Remove the drive belts.
3. Turn the crankshaft pulley, and align its groove with timing mark "T" of the timing belt cover.
4. Remove the timing belt.
5. Remove the bolt (B) and timing belt idler (A).
6. Remove the oil pan and oil screen.
7. Remove the alternator.
8. Remove the air conditioner compressor tensioner bracket (A).

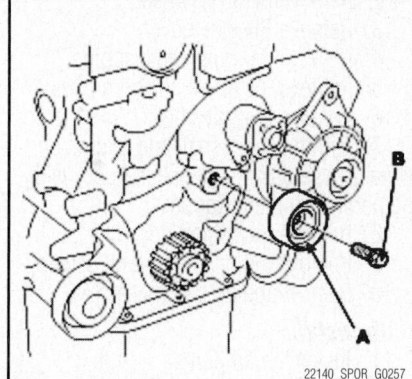

Fig. 234 Remove the bolt (B) and timing belt idler (A)

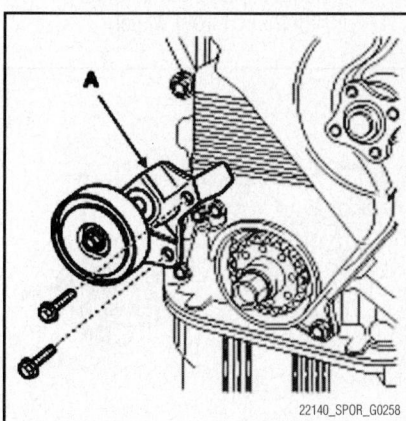

Fig. 235 Remove the air conditioner compressor tensioner bracket (A)

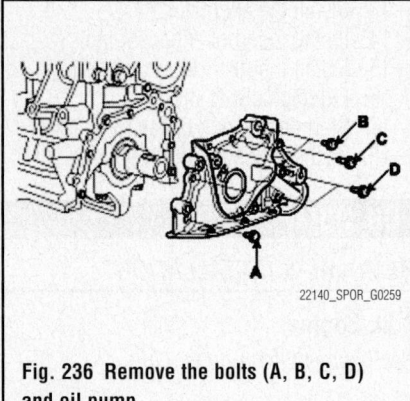

Fig. 236 Remove the bolts (A, B, C, D) and oil pump

9. Remove the bolts (A, B, C, D) and oil pump.

To install:

10. Installation is the reverse of removal.
11. Fill with engine oil.
12. Start the engine and check for leaks.

2.7L Engine

See Figure 237.

1. Drain engine oil.
2. Remove RH front wheel.
3. Remove RH side cover.
4. Remove the front exhaust pipe.
5. Remove the alternator from engine.
6. Remove the drive belt.
7. Turn the crankshaft and align the white groove on the crankshaft pulley with the pointer on the lower cover.
8. Remove the timing belt.
9. Remove the oil pan and oil screen.
10. Remove the oil pump case (A).

To install:

11. Install the oil pump.
12. Install the oil pan and oil screen.
13. Install the timing belt.
14. Install the drive belt.
15. Install the alternator.
16. Install the front exhaust pipe.
17. Install the RH front wheel.

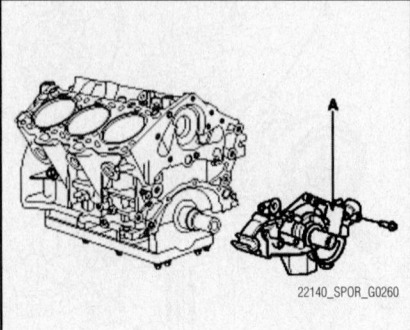

Fig. 237 Remove the oil pump case (A)

18. Fill engine with oil.
19. Start engine and check for leaks.
20. Recheck engine oil level.

MAIN BEARING TORQUE SEQUENCE

2.0L Engine

See Figure 238.

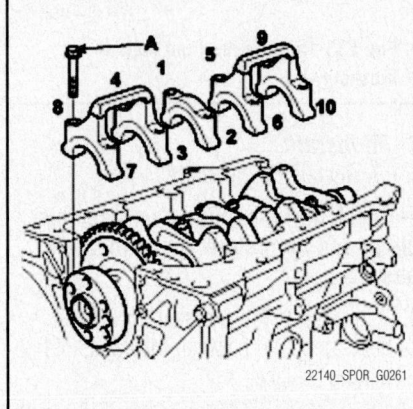

Fig. 238 Main bearing torque sequence— 2.0L engine

2.7L Engine

See Figure 239.

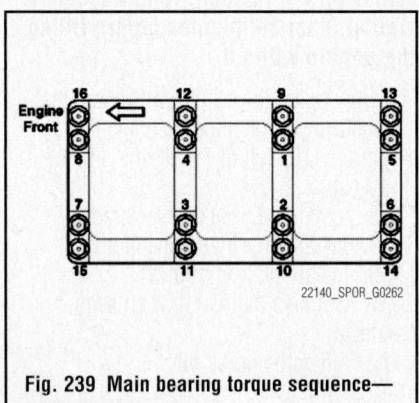

Fig. 239 Main bearing torque sequence— 2.7L engine

TIMING BELT FRONT COVER

REMOVAL & INSTALLATION

2.0L Engine

See Figures 240 through 250.

Engine removal is not required for this procedure.

1. Remove the engine cover.
2. Remove the RH front wheel.
3. Remove the 2 bolts (B) and RH side cover (A).

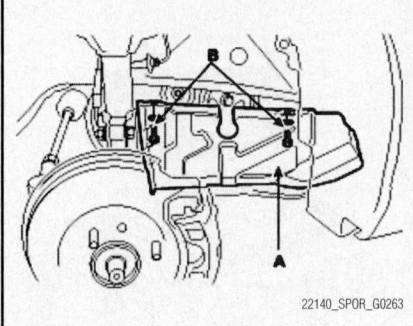

Fig. 240 Remove the 2 bolts (B) and RH side cover (A)

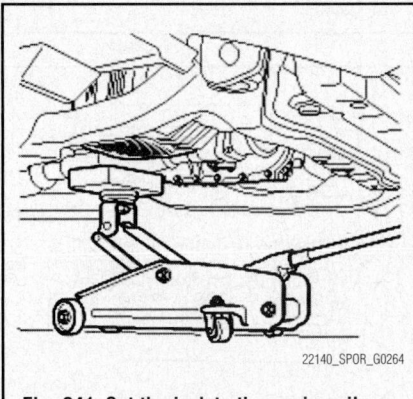

Fig. 241 Set the jack to the engine oil pan

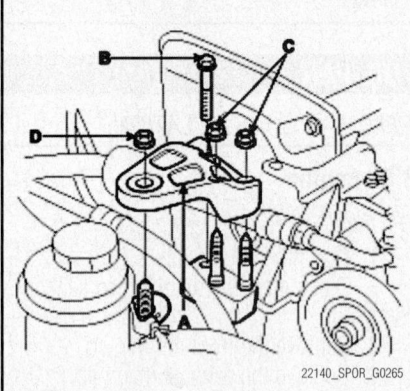

Fig. 242 Remove the bolt (B), nuts (C, D) and engine mounting support bracket (A)

4. Remove the engine mounting support bracket.
   a. Set the jack to the engine oil pan.
   b. Remove the bolt (B), nuts (C, D) and engine mounting support bracket (A).
   c. Remove the bolt (B) and engine support bracket stay plate (A).
5. Temporarily loosen the water pump pulley bolts.
6. Remove the alternator drive belt.
7. Remove the air conditioner compressor drive belt.

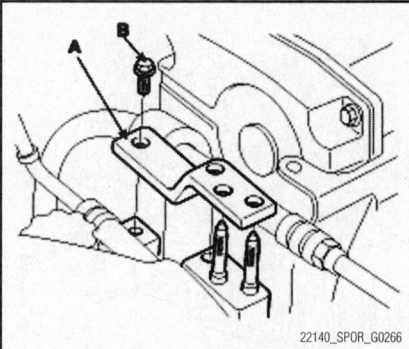

Fig. 243 Remove the bolt (B) and engine support bracket stay plate (A)

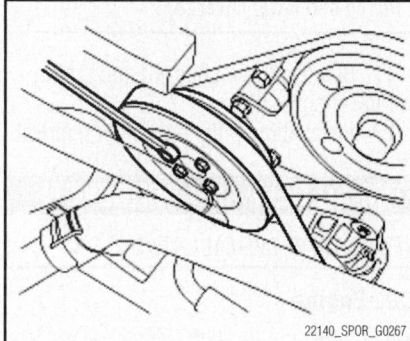

Fig. 244 Temporarily loosen the water pump pulley bolts

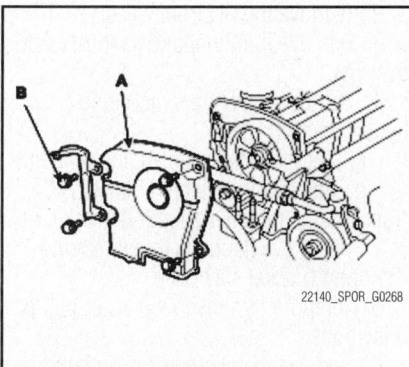

Fig. 245 Remove the 4 bolts (B) and timing belt upper cover (A)

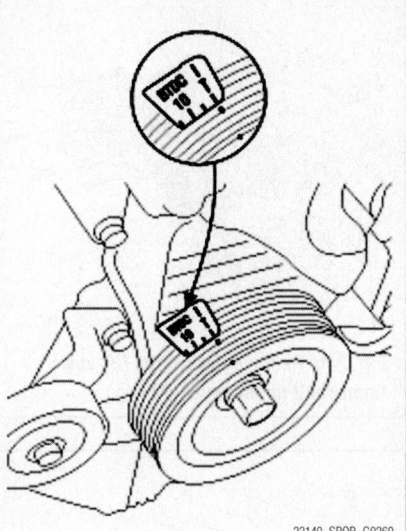

Fig. 246 Turn the crankshaft pulley, and align its groove with timing mark "T" of the timing belt cover

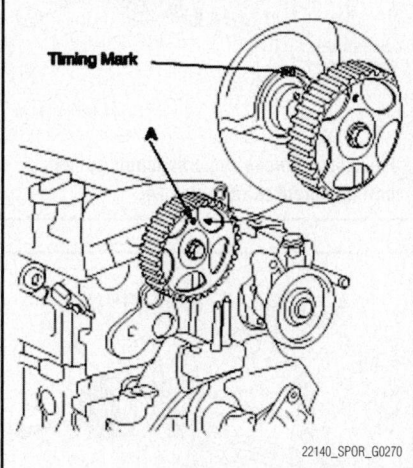

Fig. 247 Check that the timing mark of camshaft sprocket is aligned with the timing mark of cylinder head cover (No. 1 cylinder compression TDC position)

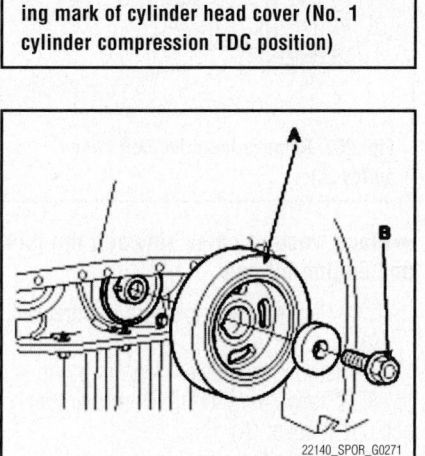

Fig. 248 Remove the crankshaft pulley bolt (B) and crankshaft pulley (A)

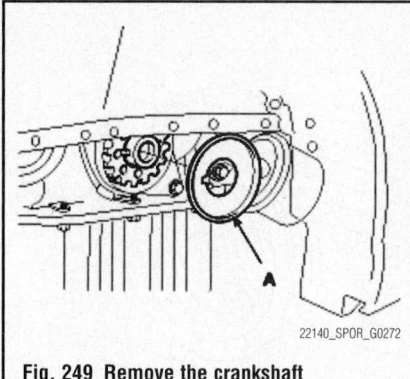

Fig. 249 Remove the crankshaft flange (A).

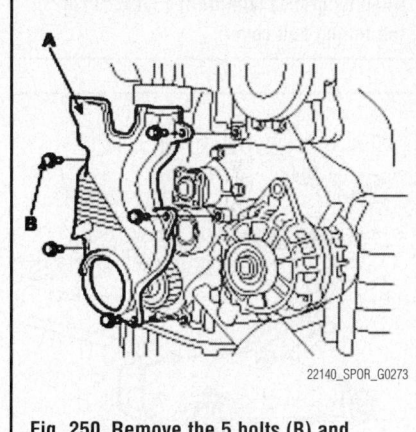

Fig. 250 Remove the 5 bolts (B) and timing belt lower cover (A)

8. Remove the power steering pump drive belt.
9. Remove the 4 bolts and water pump pulley.
10. Remove the 4 bolts (B) and timing belt upper cover (A).
11. Turn the crankshaft pulley, and align its groove with timing mark "T" of the timing belt cover.
12. Check that the timing mark of camshaft sprocket is aligned with the timing mark of cylinder head cover. (No. 1 cylinder compression TDC position)
13. Remove the crankshaft pulley bolt (B) and crankshaft pulley (A).

14. Remove the crankshaft flange (A).
15. Remove the 5 bolts (B) and timing belt lower cover (A)
16. Installation is the reverse of removal.

2.7L Engine

See Figures 251 through 258.

Engine removal is not required for this procedure.
1. Remove the engine cover.
2. Remove RH front wheel.

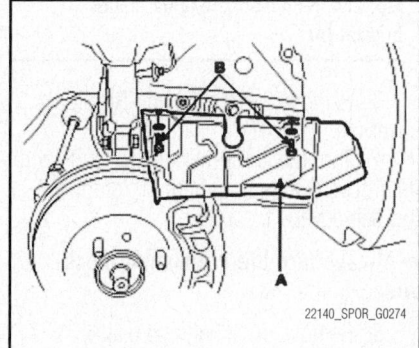

Fig. 251 Remove 2 bolts (B) and RH side cover (A)

Fig. 252 Turn the crankshaft pulley, and align its groove with timing mark "T" of the timing belt cover

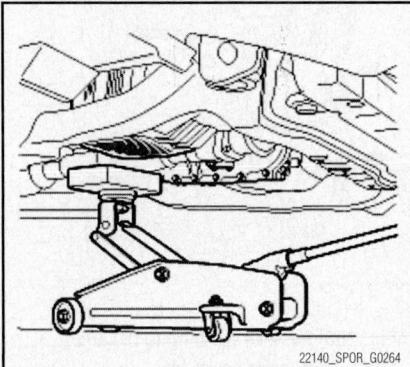

Fig. 253 Set the jack to the engine oil pan

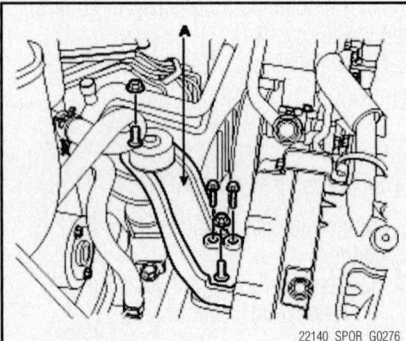

Fig. 254 Remove the engine mount bracket (A)

3. Remove 2 bolts (B) and RH side cover (A).

4. Turn the crankshaft pulley, and align its groove with timing mark "T" of the timing belt cover.

➡Always turn the crankshaft clockwise.

5. Remove drive belt and belt tensioner.

6. Remove the engine mount bracket.
   a. Set the jack to the engine oil pan.

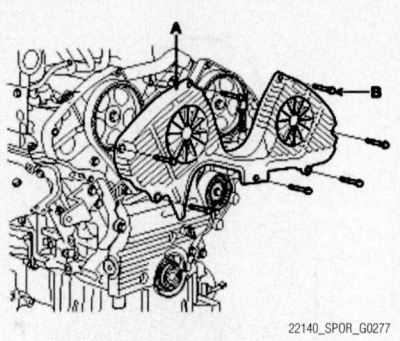

Fig. 255 Remove the 7 bolts (B) and timing belt upper cover (A)

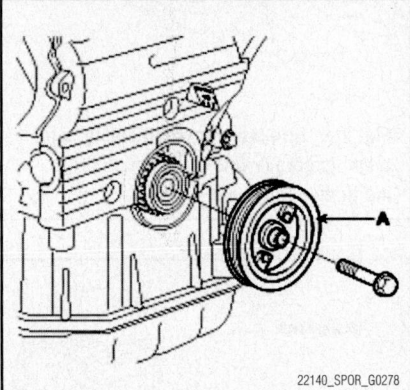

Fig. 256 Remove the crankshaft pulley bolt and crankshaft pulley (A)

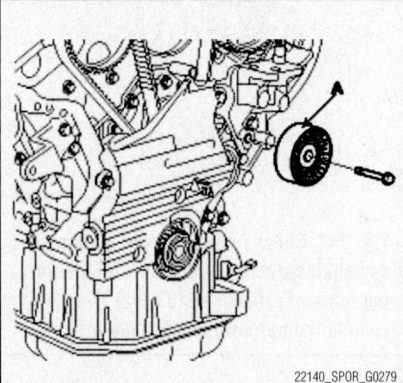

Fig. 257 Remove the drive belt idler pulley (A)

➡Place wooden block between the jack and engine oil pan.

   b. Remove the engine mount bracket (A).

7. Remove the power steering pump.

8. Remove the 7 bolts (B) and timing belt upper cover (A).

9. Remove the crankshaft pulley bolt and crankshaft pulley (A).

10. Remove the drive belt idler pulley (A).

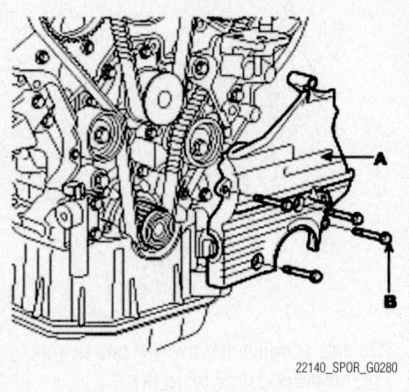

Fig. 258 Remove the 4 bolts (B) and timing belt lower cover (A)

11. Remove the 4 bolts (B) and timing belt lower cover (A).

12. Installation is the reverse of removal.

TIMING BELT AND SPROCKETS

REMOVAL & INSTALLATION

2.0L Engine

See Figures 259 through 277.

Engine removal is not required for this procedure.

1. Remove the engine cover.

2. Remove the RH front wheel.

3. Remove the 2 bolts (B) and RH side cover (A).

4. Remove the engine mounting support bracket.
   a. Set the jack to the engine oil pan.
   b. Remove the bolt (B), nuts (C, D) and engine mounting support bracket (A).
   c. Remove the bolt (B) and engine support bracket stay plate (A).

5. Temporarily loosen the water pump pulley bolts.

6. Remove the alternator drive belt.

7. Remove the air conditioner compressor drive belt.

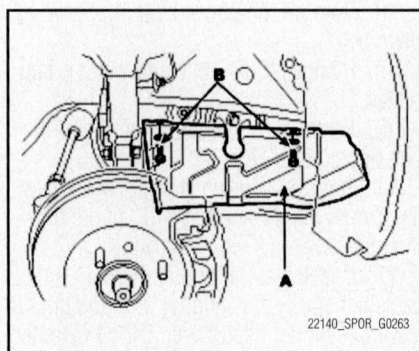

Fig. 259 Remove the 2 bolts (B) and RH side cover (A)

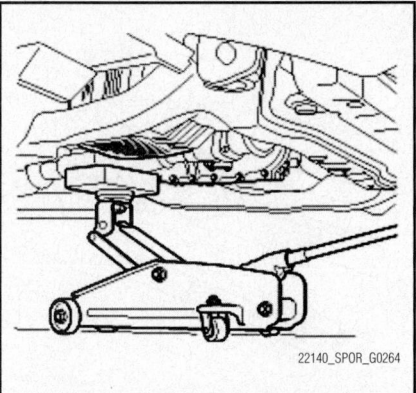

Fig. 260 Set the jack to the engine oil pan

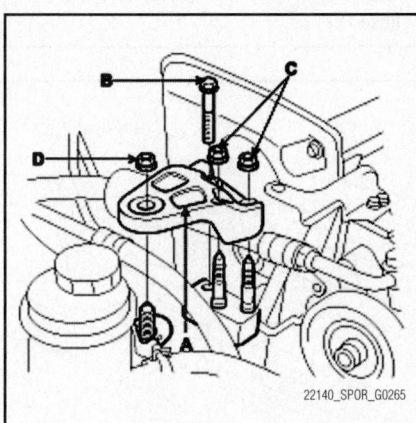

Fig. 261 Remove the bolt (B), nuts (C, D) and engine mounting support bracket (A)

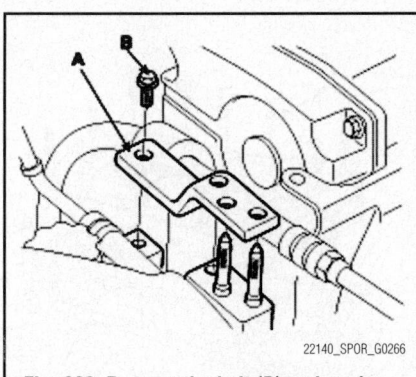

Fig. 262 Remove the bolt (B) and engine support bracket stay plate (A)

8. Remove the power steering pump drive belt.

9. Remove the 4 bolts and water pump pulley.

10. Remove the 4 bolts (B) and timing belt upper cover (A).

11. Turn the crankshaft pulley, and align its groove with timing mark "T" of the timing belt cover.

12. Check that the timing mark of camshaft sprocket is aligned with the timing mark of cylinder head cover. (No.1 cylinder compression TDC position)

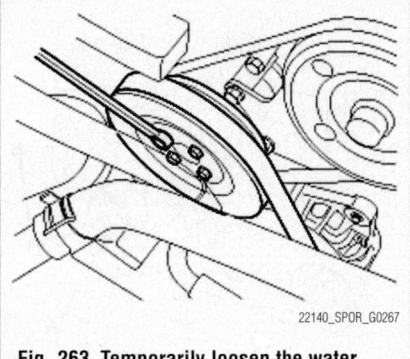

Fig. 263 Temporarily loosen the water pump pulley bolts

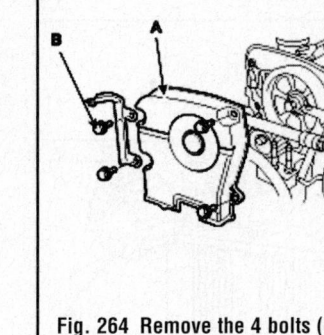

Fig. 264 Remove the 4 bolts (B) and timing belt upper cover (A)

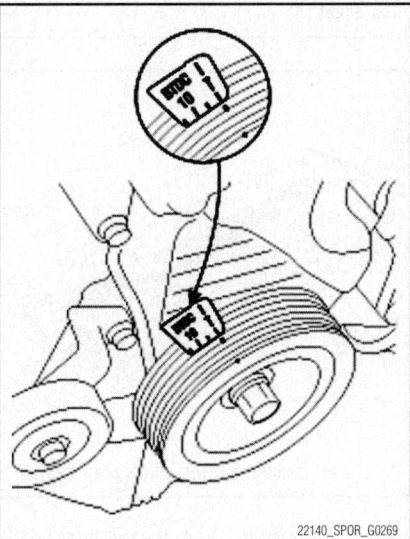

Fig. 265 Turn the crankshaft pulley, and align its groove with timing mark "T" of the timing belt cover

13. Remove the crankshaft pulley bolt (B) and crankshaft pulley (A).

14. Remove the crankshaft flange (A).

15. Remove the 5 bolts (B) and timing belt lower cover (A)

16. Remove the timing belt tensioner (A) and timing belt (B)

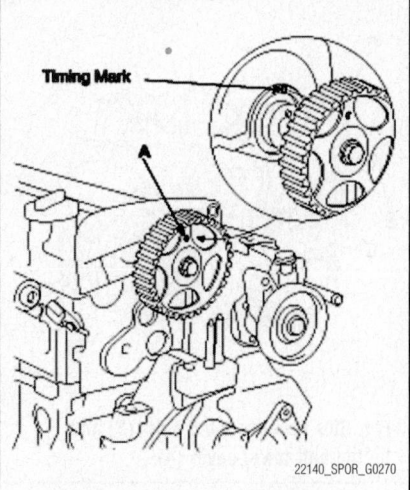

Fig. 266 Check that the timing mark of camshaft sprocket is aligned with the timing mark of cylinder head cover (No.1 cylinder compression TDC position)

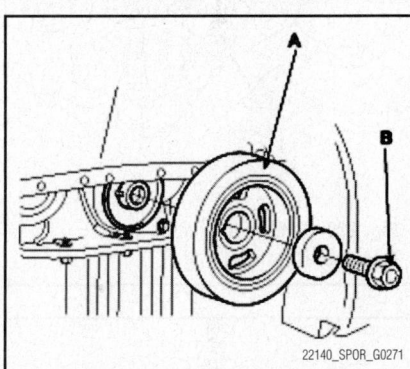

Fig. 267 Remove the crankshaft pulley bolt (B) and crankshaft pulley (A)

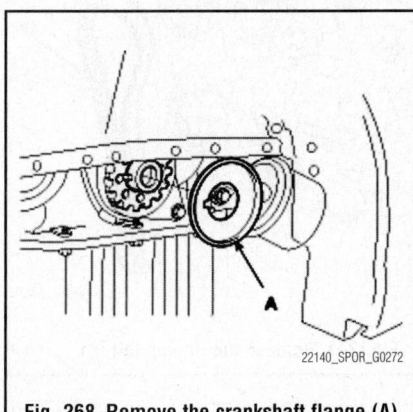

Fig. 268 Remove the crankshaft flange (A)

➡If the timing belt is going to be reused, make an arrow indicating the turning direction to make sure that the belt is reinstalled in the same direction as before.

17. Remove the bolt (B) and timing belt idler (A).

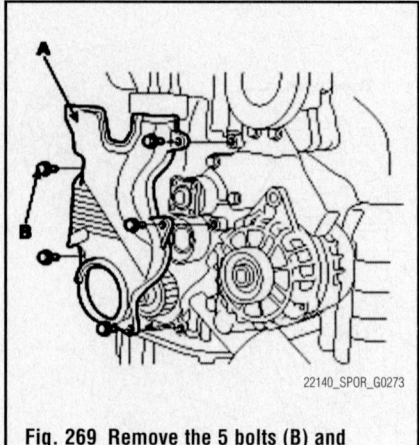

22140_SPOR_G0273

Fig. 269 Remove the 5 bolts (B) and timing belt lower cover (A)

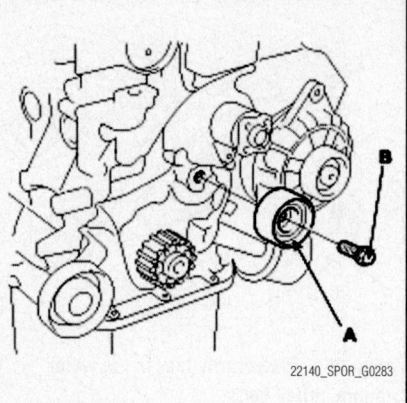

22140_SPOR_G0283

Fig. 272 Remove the bolt (B) and timing belt idler (A)

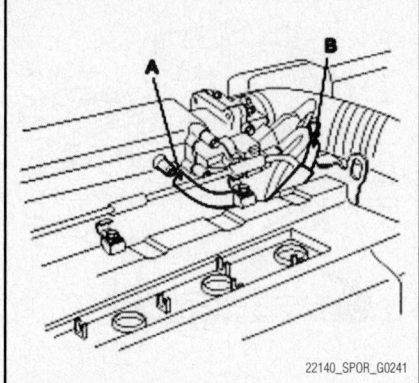

22140_SPOR_G0241

Fig. 275 Remove the Positive Crankcase Ventilation (PCV) hose (A) and breather hose (B)

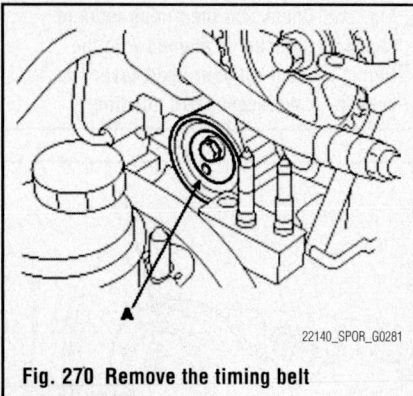

22140_SPOR_G0281

Fig. 270 Remove the timing belt tensioner (A)

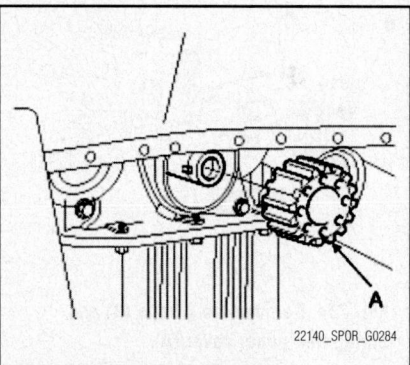

22140_SPOR_G0284

Fig. 273 Remove the crankshaft sprocket (A)

22140_SPOR_G0174

Fig. 276 Remove the accelerator cable (A)

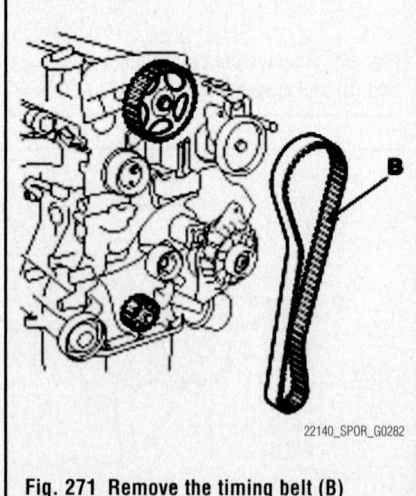

22140_SPOR_G0282

Fig. 271 Remove the timing belt (B)

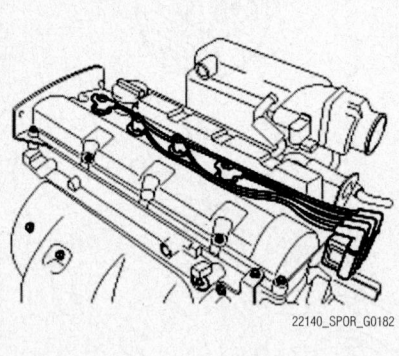

22140_SPOR_G0182

Fig. 274 Disconnect the spark plug cables

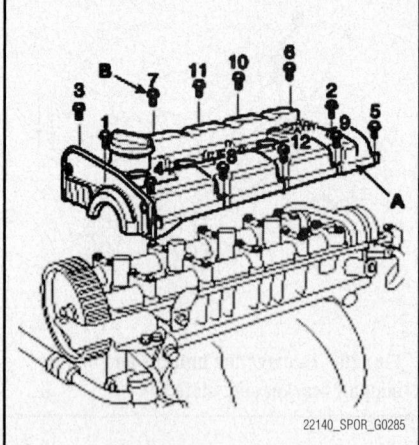

22140_SPOR_G0285

Fig. 277 Remove the cylinder head cover bolts (B) and remove the cover (A) and gasket

18. Remove the crankshaft sprocket (A).
19. Remove the cylinder head cover.
    a. Disconnect the spark plug cables and do not pull on the cable by force.

➡**Pulling on or bending the cables may damage the conductor inside.**

    b. Remove the Positive Crankcase Ventilation (PCV) hose (A) and the

breather hose (B) from the cylinder head cover.
    c. Remove the accelerator cable (A) from the cylinder head cover.
    d. Remove the cylinder head cover bolts (B) and remove the cover (A) and gasket.
20. Remove the camshaft sprocket.
21. Installation is the reverse of removal.

2.7L Engine

See Figures 278 through 289.

Engine removal is not required for this procedure.
    1. Remove the engine cover.
    2. Remove RH front wheel.
    3. Remove 2 bolts (B) and RH side cover (A).
    4. Turn the crankshaft pulley, and align

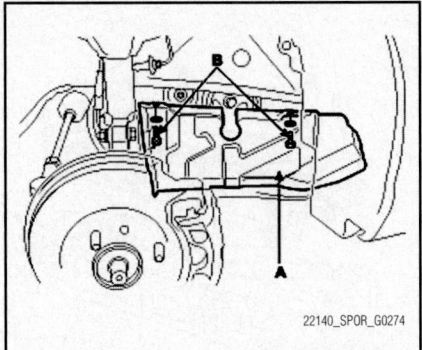

Fig. 278 Remove 2 bolts (B) and RH side cover (A)

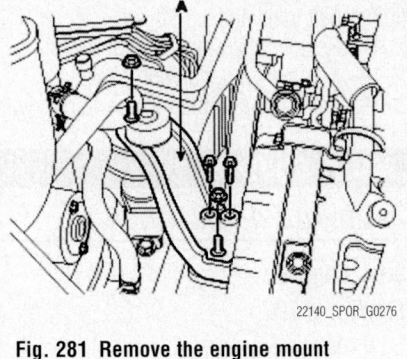

Fig. 281 Remove the engine mount bracket (A)

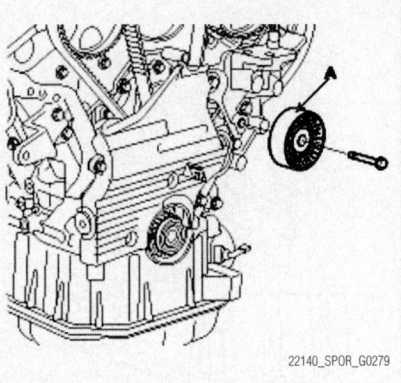

Fig. 284 Remove the drive belt idler pulley (A)

Fig. 279 Turn the crankshaft pulley, and align its groove with timing mark "T" of the timing belt cover

Fig. 282 Remove the 7 bolts (B) and timing belt upper cover (A)

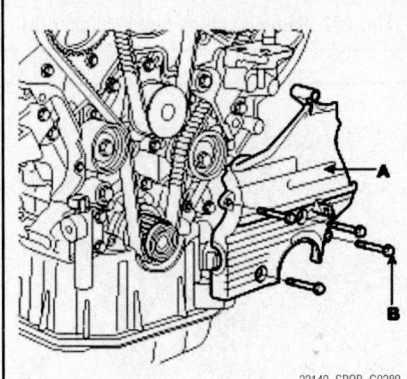

Fig. 285 Remove the 4 bolts (B) and timing belt lower cover (A).

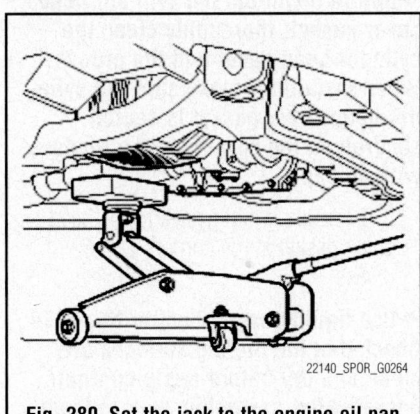

Fig. 280 Set the jack to the engine oil pan

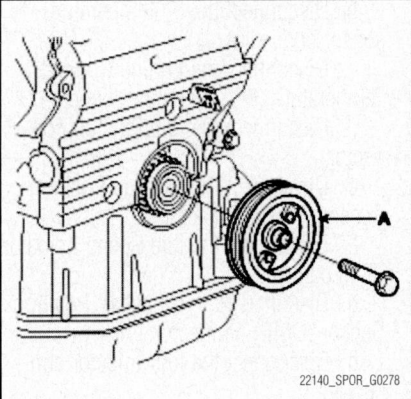

Fig. 283 Remove the crankshaft pulley bolt and crankshaft pulley (A)

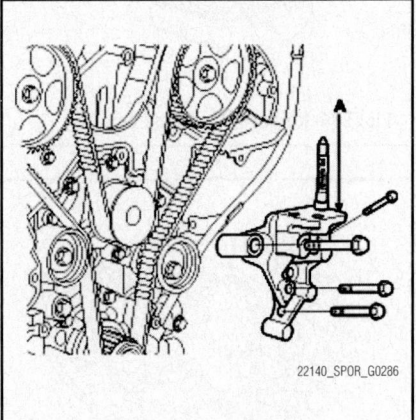

Fig. 286 Remove the engine support bracket (A)

its groove with timing mark "T" of the timing belt cover.

➠**Always turn the crankshaft clockwise.**

5. Remove drive belt and belt tensioner.
6. Remove the engine mount bracket.
  a. Set the jack to the engine oil pan.

➠**Place wooden block between the jack and engine oil pan.**

  b. Remove the engine mount bracket (A).

7. Remove the power steering pump.
8. Remove the 7 bolts (B) and timing belt upper cover (A).
9. Remove the crankshaft pulley bolt and crankshaft pulley (A).
10. Remove the drive belt idler pulley (A).
11. Remove the 4 bolts (B) and timing belt lower cover (A).
12. Remove the engine support bracket (A).
13. Check that timing marks of the camshaft timing pulleys and cylinder head

covers are aligned. If not, turn the crankshaft 1 revolution (360°).
14. Remove timing belt tensioner (A).
15. Remove the timing belt (A).

➠**If the timing belt is reused, make an arrow indicating the turning direction to make sure that the belt is reinstalled in the same direction as before.**

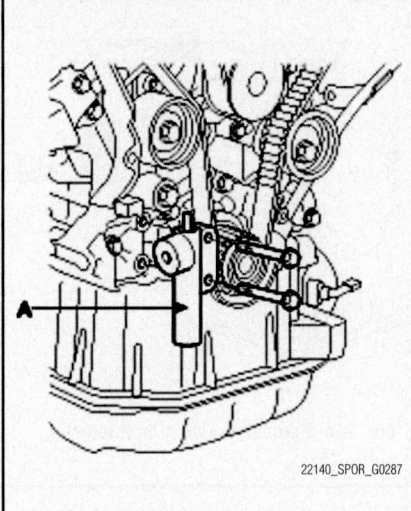

Fig. 287 Remove timing belt tensioner (A)

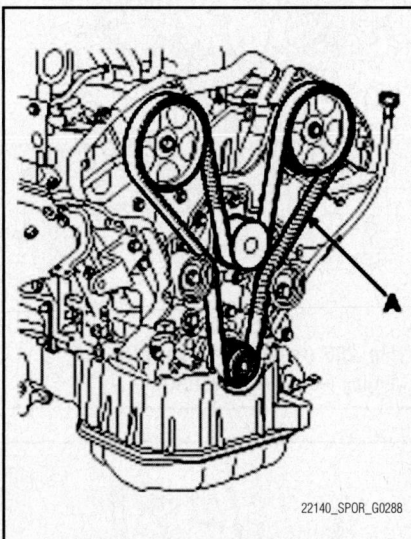

Fig. 288 Remove the timing belt (A)

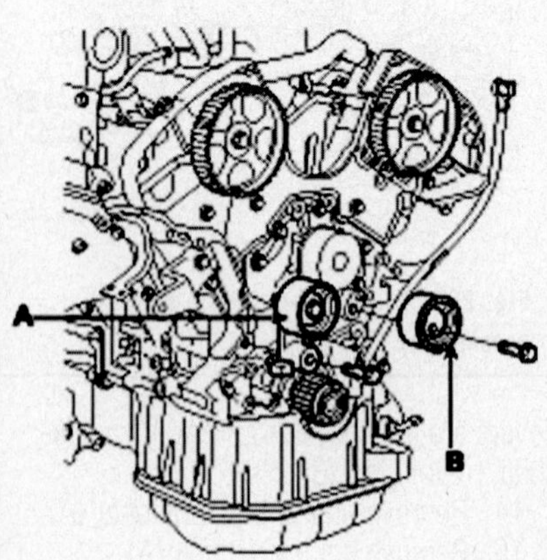

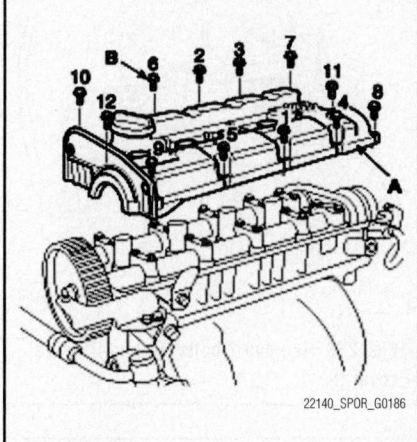

Fig. 289 Remove the tensioner pulley (A) and timing belt idler pulley (B)

16. Remove the tensioner pulley (A) and timing belt idler pulley (B).

17. Remove the crankshaft sprocket.

18. Remove camshaft sprockets.

19. Installation is the reverse of removal.

VALVE COVERS

REMOVAL & INSTALLATION

2.0L Engine

See Figure 290.

1. Remove the air duct.

2. Disconnect the terminals from battery.

3. Remove the engine cover.

4. Remove the intake air hose and air cleaner assembly.

  a. Disconnect the Air Flow Sensor (AFS) connector.

  b. Disconnect the breather hose from intake air hose.

  c. Remove the intake air hose and air cleaner assembly.

5. Remove the accelerator cable by loosening the lock-nut, then slip the cable end out of the throttle linkage.

6. Remove the engine wire harness connectors and wire harness clamps from cylinder head and the intake manifold.

  a. Disconnect the Oil Control Valve (OCV) connector.

  b. Disconnect the oil temperature sensor connector.

  c. Disconnect the Engine Coolant Temperature (ECT) sensor connector.

  d. Disconnect the ignition coil connector.

  e. Disconnect the Throttle Position Sensor (TPS) connector.

  f. Disconnect the Idle Speed Actuator (ISA) connector.

  g. Disconnect the Camshaft Position Sensor (CMP) connector.

  h. Disconnect the four injector connectors.

  i. Disconnect the knock sensor connector.

  j. Disconnect the ground cables from the intake manifold.

  k. Disconnect the air conditioner compressor switch connector.

  l. Disconnect the front heated oxygen sensor connector.

  m. Disconnect the Crankshaft Position Sensor (CKP) connector.

  n. Disconnect the oil pressure switch connector.

  o. Disconnect the Purge Control Solenoid Valve (PCSV) connector.

7. Disconnect the fuel inlet hose of the delivery pipe side.

Fig. 290 Install the cylinder head cover in sequence

8. Disconnect the hose of the PCSV side.

9. Remove the brake booster vacuum hose.

10. Remove the spark plug cables.

11. Remove the PCV hose.

12. Remove the cylinder head cover.

To install:

13. Install the cylinder head cover.

  a. Install the cylinder head cover gasket in the groove of the cylinder head cover.

➡ **Before installing the cylinder head cover gasket, thoroughly clean the cylinder head cover and the groove. When installing, make sure the cylinder head cover gasket is seated securely in the corners of the recesses with no gap.**

  b. Apply liquid gasket to the head cover gasket at the corners of the recess.

➡ **Use liquid gasket, Loctite No. 5999. Check that the mating surfaces are clean and dry before applying liquid gasket. After assembly, wait at least 30 minutes before filling the engine with oil.**

14. Install the cylinder head cover with the 12 bolts. Uniformly tighten the bolts in several passes. Tighten to 6–7 ft. lbs. (8–10 Nm).

15. Install the PCV hose.

16. Install the spark plug cables.

17. Install the accelerator cable.

18. Install the brake booster hose.

19. Connect the hose of the PCSV side.

20. Connect the fuel inlet hose of the delivery pipe side.

21. Install the engine wire harness

connectors and wire harness clamps to the cylinder head and the intake manifold.

   a. Connect the PCSV connector.

   b. Connect the front heated oxygen sensor connector.

   c. Connect the CKP connector.

   d. Connect the oil pressure switch connector.

   e. Connect the air conditioner compressor switch connector.

   f. Connect the ground cables to intake manifold.

   g. Connect the knock sensor connector.

   h. Connect the fuel injector connectors.

   i. Connect the CMP connector.

   j. Connect the ISA connector.

   k. Connect the TPS connector.

   l. Connect the ignition coil connector.

   m. Connect the ECT connector.

   n. Connect the oil temperature sensor connector.

   o. Connect the OCV connector.

22. Install the intake air hose and air cleaner assembly.

   a. Install the intake air hose, air cleaner assembly and bolts. Tighten to 6–7 ft. lbs. (8–10 Nm).

   b. Install the breather hose to intake air hose.

   c. Connect the AFS connector.

23. Install the engine cover.

24. Reconnect the battery terminals.

25. Install the air duct.

2.7L Engine

See Figure 291.

1. Remove the air duct.

2. Remove the cowl grill and wiper motor.

3. Remove the strut bar.

4. Disconnect the negative terminal from the battery.

5. Drain the engine coolant. Remove the radiator cap to speed draining.

6. Remove the engine cover.

7. Remove the intake air hose and air cleaner assembly.

   a. Disconnect the Air Flow Sensor (AFS) connector.

   b. Disconnect the breather hose from air cleaner hose.

   c. Remove the intake air hose and air cleaner assembly.

8. Remove the coolant reservoir tank.

9. Remove the engine wire harness connectors and wire harness clamps from the cylinder head and the intake manifold.

   a. Throttle Position Sensor (TPS) connector.

   b. Idle Speed Actuator (ISA) connector.

   c. Purge Control Solenoid Valve (PCSV) connector.

   d. VIS actuator connector.

   e. Injector connector.

   f. Knock sensor connectors.

   g. Camshaft Position Sensor (CMP) connector.

   h. Oxygen sensor (rear 1st) connector.

   i. PCSV hose.

   j. Engine Coolant Temperature (ECT) sensor connector.

   k. Ignition coil connector.

   l. Crankshaft position (CKP) sensor connector.

   m. Oxygen sensor (front 2nd) connector.

   n. Ground cable.

   o. Three fuel injector connectors.

   p. Disconnect ground cable from the cowl panel.

   q. Intake Air Temperature (IAT) sensor connector.

   r. VIS actuator connector.

   s. Oxygen sensor (rear 2nd) connector.

   t. Power steering pump switch.

   u. Oxygen sensor (front 1st) connector.

   v. Air conditioner compressor switch connector.

   w. Oil pressure sensor connector.

10. Remove the fuel inlet hose from delivery pipe.

11. Remove the brake booster vacuum hose.

12. Remove the accelerator cable by loosening the locknut, then slip the cable end out of the throttle linkage.

13. Remove the PCV hose.

14. Remove the spark plug cable.

15. Remove the cylinder head covers.

To install:

16. Install the cylinder head cover.

   a. Install the cylinder head cover gasket in the groove of the cylinder head cover.

➡**Before installing the head cover gasket, thoroughly clean the head cover gasket and the groove. When installing, make sure the head cover gasket is seated securely in the corners of the recesses with no gap.**

   b. Apply liquid gasket to the head cover gasket at the corners of the recess.

➡**Use liquid gasket, Loctite No. 5699. Check that the mating surfaces are clean and dry before applying liquid**

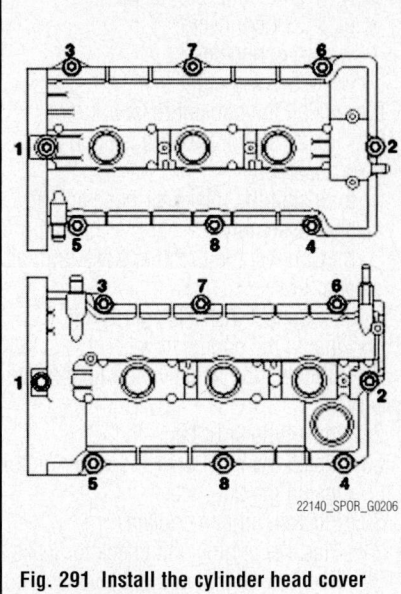

Fig. 291 Install the cylinder head cover bolts in sequence

22140_SPOR_G0206

gasket. After assembly, wait at least 30 minutes before filling the engine with oil.

   c. Install the cylinder head covers with the 16bolts. Uniformly tighten the bolts in several passes. Tighten to 6–7 ft. lbs. (8–10 Nm)

17. Install the spark plug cable.

18. Install the PCV hose.

19. Install the accelerator cable.

20. Install the brake booster vacuum hose.

21. Install the fuel inlet hose.

22. Install the engine wire harness connectors and wire harness clamps to the cylinder head and the intake manifold.

   a. Oil pressure sensor connector.

   b. Air conditioner compressor switch connector.

   c. Oxygen sensor (front 1st) connector.

   d. Power steering pump switch.

   e. Oxygen sensor (rear 2nd) connector.

   f. VIS actuator connector.

   g. IAT sensor connector.

   h. Connect the ground cable to the cowl panel.

   i. Three fuel injector connectors.

   j. Ground cable.

   k. Oxygen sensor connector.

   l. Crankshaft position sensor connector.

   m. Ignition coil connector.

   n. ECT sensor connector.

   o. PCSV hose.

   p. Oxygen sensor (rear 1st) connector.

   q. CMP connector.

   r. Knock sensor connector.

   s. Injector connector.

t. VIS actuator connector.

u. PCSV connector.

v. ISA connector.

w. TPS connector.

23. Install the coolant reservoir tank.

24. Install the intake air hose and air cleaner assembly.

a. Install the intake air hose and air cleaner assembly.

b. Connect the breather hose from air cleaner hose.

c. Connect the AFS connector.

25. Install the engine cover.

26. Connect the negative terminal to the battery.

27. Install the strut bar.

28. Install the cowl grill and wiper motor.

29. Install the air duct.

30. Fill with engine coolant.

31. Start the engine and check for leaks.

32. Recheck engine coolant level and oil level.

VALVE LASH

ADJUSTMENT

2.0L Engine

See Figures 292 through 297.

➡ **Inspect and adjust the valve clearance when the engine is cold. Engine coolant temperature: 68°F ± 9°F (20°C ± 5°C) and cylinder head is installed on cylinder block.**

1. Remove the engine cover.

2. Remove the bolts and timing belt upper cover.

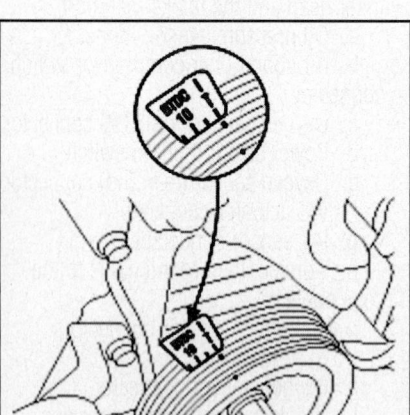

Fig. 292 Turn the crankshaft pulley, and align its groove with timing mark "T" of the timing belt cover

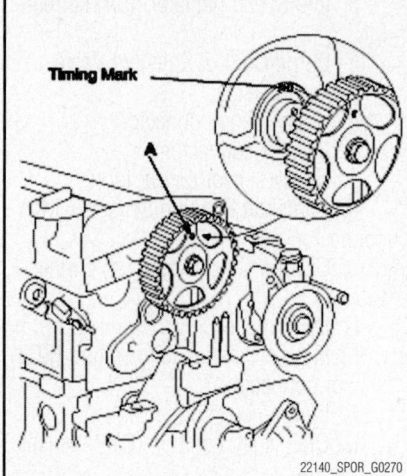

Fig. 293 Check that the timing mark of camshaft sprocket is aligned with the timing mark of cylinder head cover (No. 1 cylinder compression TDC position)

3. Remove the cylinder head cover.

a. Disconnect the spark plug cables.

➡ **Pulling on or bending the cables may damage the conductor inside.**

b. Remove the Positive Crankcase Ventilation (PCV) hose and the breather hose from the cylinder head cover.

c. Remove the accelerator cable from the cylinder head cover.

d. Remove the cylinder head cover bolts and remove the cover and gasket.

4. Set No. 1 cylinder to TDC/compression.

a. Turn the crankshaft pulley and align its groove with the timing mark "T" of the lower timing belt cover.

b. Check that the hole of the camshaft timing pulley (A) is aligned with the timing mark of the bearing cap. If not, turn the crankshaft one revolution (360°).

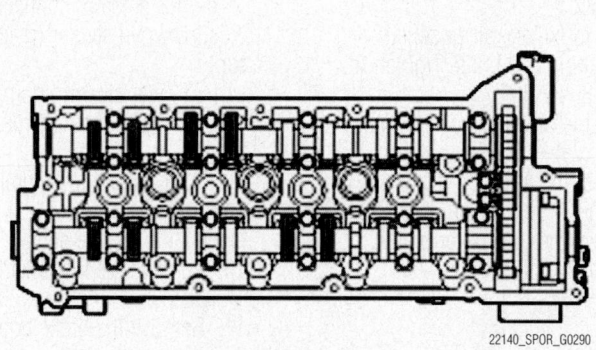

Fig. 294 Check only the valves indicated. No. 1 cylinder: TDC/compression

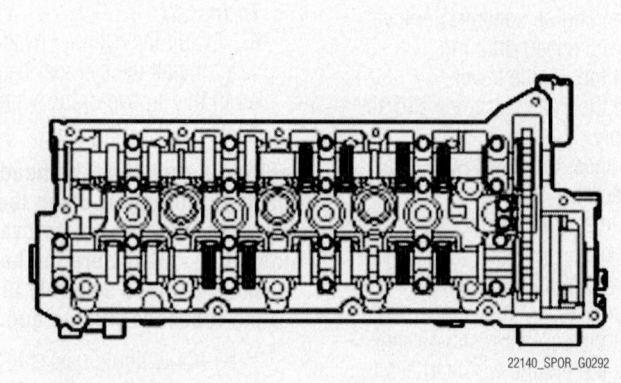

Fig. 295 Check only the valves indicated. No. 4 cylinder: TDC/compression

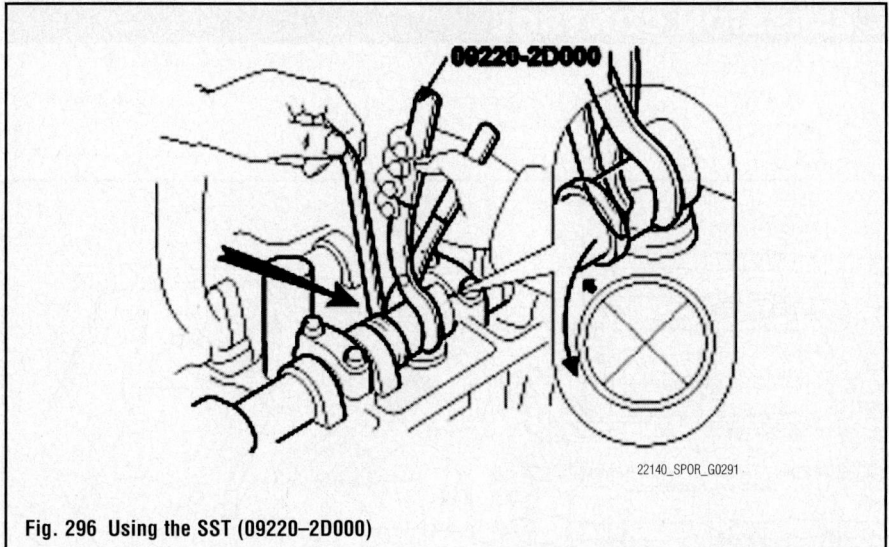

Fig. 296 Using the SST (09220–2D000)

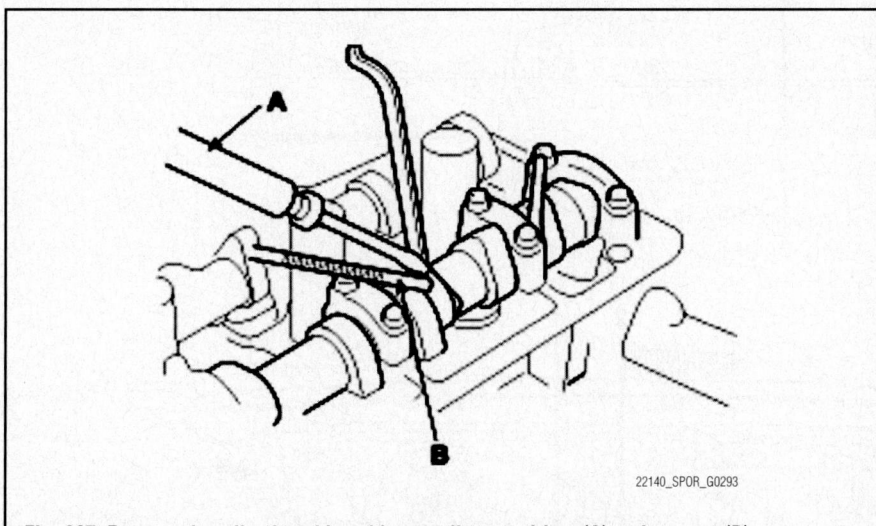

Fig. 297 Remove the adjusting shim with a small screw driver (A) and magnet (B)

5. Inspect the valve clearance.

a. Check only the valves indicated. No. 1 cylinder : TDC/compression. Measure the valve clearance.

b. Using a thickness gauge, measure the clearance between the tappet shim and the base circle of camshaft.

c. Record the out-of-specification valve clearance measurements. They will be used later to determine the required replacement adjusting shim.

d. Valve clearance Specification:

• Engine coolant temperature: 68°F ± 9°F (20°C ± 5°C)

• Intake: 0.0079inches (0.20 mm)
• Exhaust: 0.0110inches (0.28 mm)

e. Limit:

• Intake: 0.0047–0.0110 inches (0.12–0.28mm)
• Exhaust: 0.0079–0.0150 inches (0.20–0.38 mm)

f. Turn the crankshaft pulley one revolution (360°) and align the groove with the timing mark "T" of lower timing belt cover.

g. Check only valves indicated. No. 4 cylinder: TDC/compression. Measure the valve clearance.

6. Adjust the intake and exhaust valve clearance.

a. Turn the crankshaft so that the lobe of the camshaft on the adjusting valve is upward.

b. Using the SST (09220–2D000), press down the valve lifter and place the stopper between the camshaft and valve lifter and remove the special tool.

c. Remove the adjusting shim with a small screw driver (A) and magnet (B).

d. Measure the thickness of the removed shim using a micrometer.

e. Calculate the thickness of a new shim so that the valve clearance comes within the specified value.

f. Valve clearance:

• Engine coolant temperature: 68°F ± 9°F (20°C ± 5°C)
• T: Thickness of removed shim
• A: Measured valve clearance
• N: Thickness of new shim
• Intake: $N = T + [A - 0.0079$ inches $(0.20$ mm$)]$
• Exhaust: $N = T + [A - 0.0110$ inches $(0.28$ mm$)]$

g. Select a new shim with a thickness as close as possible to the calculated value.

➡**Shims are available in 20 size increments of 0.0016 inches (0.04 mm) from 0.0787 inches (2.00 mm) to 0.1087 inches (2.76 mm)**

h. Place a new adjusting shim on the valve lifter.

i. Using SST (09220 - 2D000), press down the valve lifter and remove the stopper.

j. Recheck the valve clearance.

k. Valve clearance:

• Engine coolant temperature: 68°F ± 9°F (20°C ± 5°C)
• Specification:
• Intake: 0.0079 inches (0.20 mm)
• Exhaust: 0.0110 inches (0.28 mm)
• Limit (After adjusting valve clearance)
• Intake: 0.0067–0.0091 inches (0.17–0.23 mm)
• Exhaust: 0.0098–0.0122 inches (0.25–0.31 mm)

ENGINE PERFORMANCE & EMISSION CONTROL

COMPONENT LOCATIONS

See Figures 298 through 300.

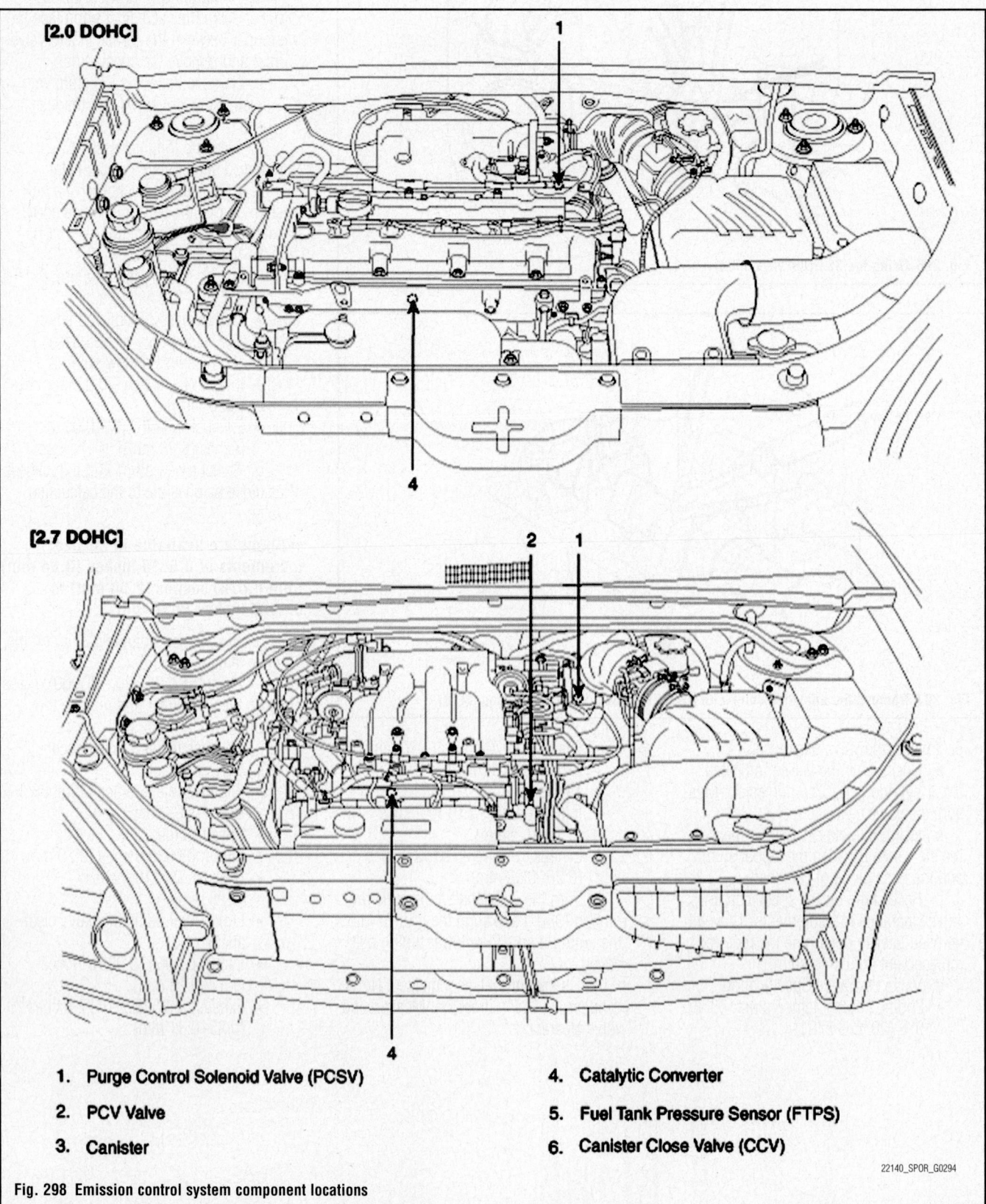

1. Purge Control Solenoid Valve (PCSV)
2. PCV Valve
3. Canister
4. Catalytic Converter
5. Fuel Tank Pressure Sensor (FTPS)
6. Canister Close Valve (CCV)

22140_SPOR_G0294

Fig. 298 Emission control system component locations

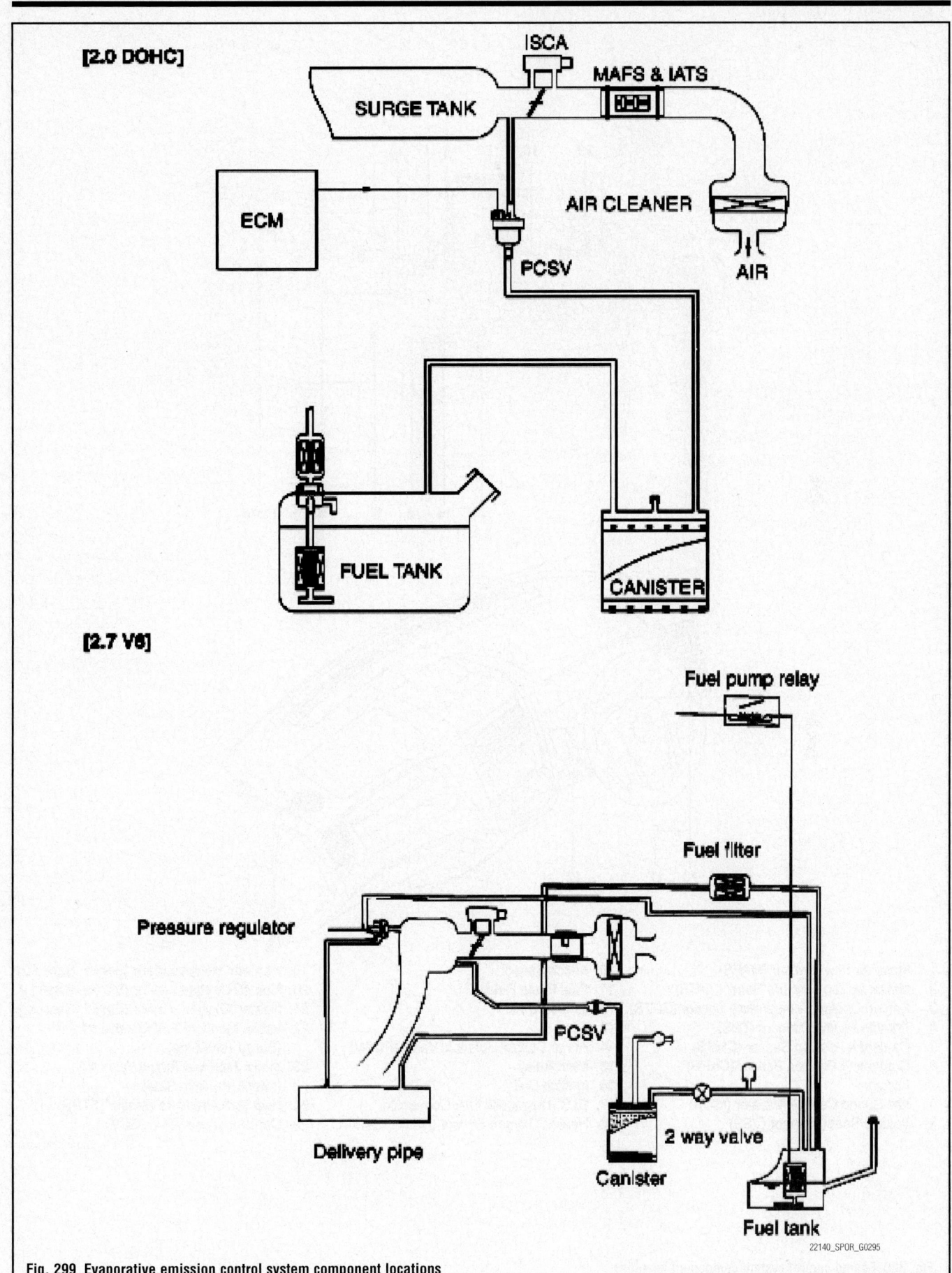

Fig. 299 Evaporative emission control system component locations

22140_SPOR_G0295

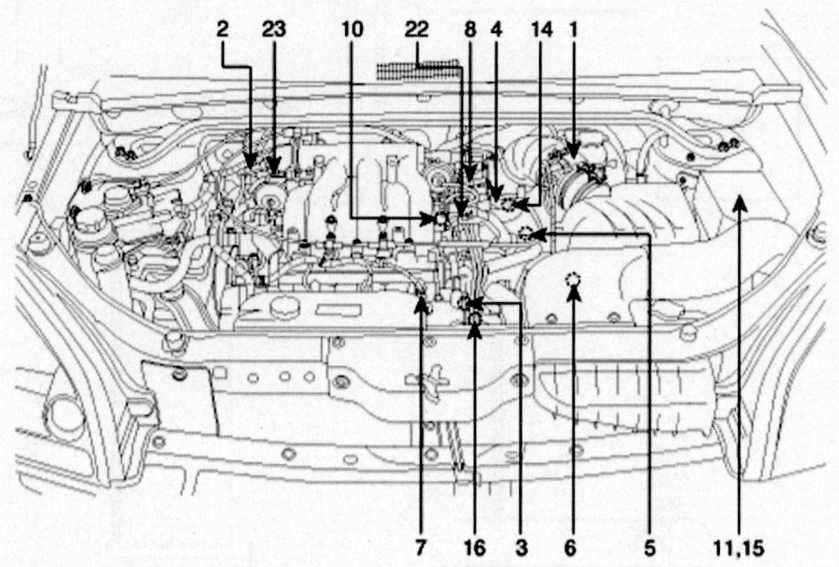

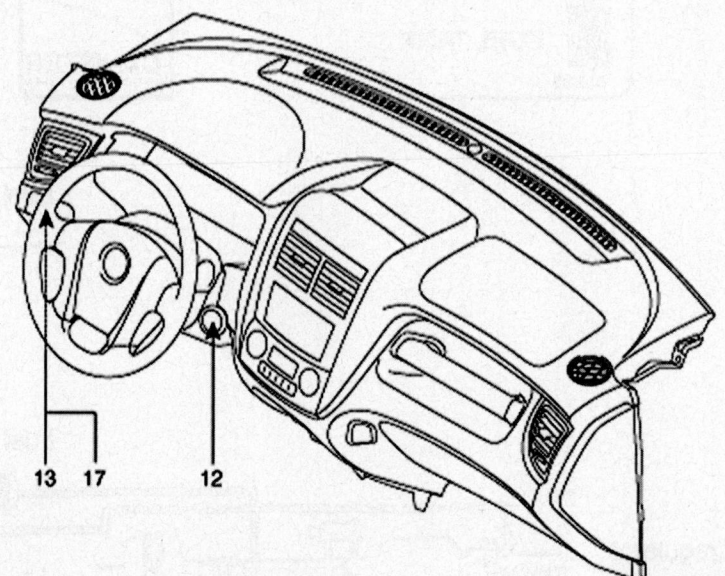

1. **Mass Air Flow Sensor (MAFS)**
2. **Intake Air Temperature Sensor (IATS)**
3. **Engine Coolant Temperature Sensor (ECTS)**
4. **Throttle Position Sensor (TPS)**
5. **Camshaft Position Sensor (CMPS)**
6. **Crankshaft Position Sensor (CKPS)**
7. **Injector**
8. **Idle Speed Control Actuator (ISCA)**
9. **Vehicle Speed Sensor (VSS)**
10. **Knock Sensor**
11. **Fuel Pump Relay**
12. **Ignition Switch**
13. **ECM**
14. **Purge Control Solenoid Valve (PCSV)**
15. **Main Relay**
16. **Ignition Coil**
17. **DLC (Diagnostic Link Connector)**
18. **Heated Oxygen Sensor (Bank1, Sensor1)**
19. **Heated Oxygen seneor (Bank1, Sensor2)**
20. **Heated Oxygen seneor (Bank2, Sensor1)**
21. **Heated Oxygen seneor (Bank2, Sensor2)**
22. **Intake Manifold Tuning Valve #1 (Surge Tank Side)**
23. **Intake Manifold Tuning Valve #1 (Intake Manifold Side)**
24. **Fuel Tank Pressure Sensor (FTPS)**
25. **Canister Close Valve (CCV)**

22140_SPOR_G0296

Fig. 300 Engine control system component locations

CAMSHAFT POSITION (CMP) SENSOR

LOCATION

On 2.0L engines, the camshaft position sensor is located near the top of the engine on the left side. On the 2.7L engine, the camshaft position sensor is located at the front of each cylinder head.

REMOVAL & INSTALLATION

1. Disconnect the negative battery cable.
2. Disconnect the connector from the sensor.
3. Remove the mounting bolts.
4. Remove the sensor from the cylinder head.
5. Installation is the reverse or removal.

CRANKSHAFT POSITION (CKP) SENSOR

REMOVAL & INSTALLATION

1. Disconnect the negative battery cable.
2. Disconnect the connector from the sensor.
3. Remove the mounting bolts.
4. Remove the sensor from the cylinder head.
5. Installation is the reverse or removal.

ELECTRONIC CONTROL MODULE (ECM)

REMOVAL & INSTALLATION

1. Disconnect the negative battery cable.
2. Remove the lower inner trim.
3. Detach the floor mat.
4. Remove the protective cover.
5. Remove the ECM bracket retaining nuts.
6. Remove the clip from the bracket.
7. Disconnect the connectors.
8. Remove the ECM from the vehicle.
9. Installation is the reverse of removal.

➥**When replacing the ECM, be careful not to use the wrong part number as damage to the injection system could occur.**

ENGINE COOLANT TEMPERATURE (ECT) SENSOR

REMOVAL & INSTALLATION

1. Disconnect the negative battery cable.
2. Disconnect the connector from the sensor.
3. Drain the cooling system as required.
4. remove the sensor from its mounting.

5. Installation is the reverse of removal.
6. Refill the cooling system.

HEATED OXYGEN (HO2S) SENSOR

LOCATION

The Heated Oxygen Sensor (HO2S) is positioned in the exhaust pipe ahead of the TWC.

REMOVAL & INSTALLATION

1. Disconnect the electrical connector from the sensor.
2. Remove the oxygen sensor.
3. Installation is the reverse of the removal procedure.

➥**Apply anti-seize compound to the threaded portion of the sensor, prior to installation. Never apply anti-seize compound to the protector of the sensor.**

INTAKE AIR TEMPERATURE (IAT) SENSOR

LOCATION

The Intake Air Temperature (IAT) sensor is installed into the Mass Air Flow Sensor (MAFS).

REMOVAL & INSTALLATION

1. Disconnect the negative battery cable.
2. Disconnect the connector from the sensor.
3. Remove the sensor retaining screws.
4. Remove the air cleaner and air intake assembly, as required.
5. Remove the sensor from its mounting.
6. Installation is the reverse of removal.

KNOCK SENSOR (KS)

LOCATION

See Figure 301.

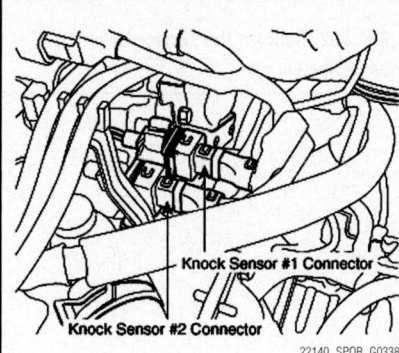

Knock Sensor #1 Connector

Knock Sensor #2 Connector

22140_SPOR_G0338

Fig. 301 Knock sensor location

REMOVAL & INSTALLATION

1. Disconnect the negative battery cable.
2. Remove the intake manifold support bracket.
3. Disconnect the sensor connector.
4. Remove the sensor from its mounting.
5. Installation is the reverse of the removal procedure.

MASS AIR FLOW (MAF) SENSOR

LOCATION

The Mass Air Flow (MAF) sensor is located between the air cleaner assembly and the throttle body.

REMOVAL & INSTALLATION

1. Disconnect the negative battery cable.
2. Disconnect the connector from the sensor.
3. Remove the air cleaner and air intake assembly, as required.
4. Remove the sensor from its mounting.
5. Installation is the reverse of the removal procedure.

MANIFOLD ABSOLUTE PRESSURE (MAP) SENSOR

REMOVAL & INSTALLATION

1. Disconnect the negative battery cable.
2. Disconnect the connector from the sensor.
3. Remove the sensor retaining screws.
4. Remove the sensor from its mounting.
5. Installation is the reverse of the removal procedure.

THROTTLE POSITION SENSOR (TPS)

LOCATION

The Throttle Position Sensor (TPS) is mounted on the throttle body and detects the opening angle of the throttle plate.

REMOVAL & INSTALLATION

1. Turn the ignition switch to the OFF position.
2. Disconnect the negative battery cable.
3. Disconnect the sensor connector.
4. Remove the sensor retaining screws.
5. Remove the sensor from its mounting.
6. Installation is the reverse of the removal procedure.

FUEL SYSTEM — GASOLINE FUEL INJECTION SYSTEM

FUEL SYSTEM SERVICE PRECAUTIONS

Safety is the most important factor when performing not only fuel system maintenance but any type of maintenance. Failure to conduct maintenance and repairs in a safe manner may result in serious personal injury or death. Maintenance and testing of the vehicle's fuel system components can be accomplished safely and effectively by adhering to the following rules and guidelines.

• To avoid the possibility of fire and personal injury, always disconnect the negative battery cable unless the repair or test procedure requires that battery voltage be applied.

• Always relieve the fuel system pressure prior to disconnecting any fuel system component (injector, fuel rail, pressure regulator, etc.), fitting or fuel line connection. Exercise extreme caution whenever relieving fuel system pressure to avoid exposing skin, face and eyes to fuel spray. Please be advised that fuel under pressure may penetrate the skin or any part of the body that it contacts.

• Always place a shop towel or cloth around the fitting or connection prior to loosening to absorb any excess fuel due to spillage. Ensure that all fuel spillage (should it occur) is quickly removed from engine surfaces. Ensure that all fuel soaked cloths or towels are deposited into a suitable waste container.

• Always keep a dry chemical (Class B) fire extinguisher near the work area.

• Do not allow fuel spray or fuel vapors to come into contact with a spark or open flame.

• Always use a back-up wrench when loosening and tightening fuel line connection fittings. This will prevent unnecessary stress and torsion to fuel line piping.

• Always replace worn fuel fitting O-rings with new. Do not substitute fuel hose or equivalent where fuel pipe is installed.

Before servicing the vehicle, make sure to also refer to the precautions in the beginning of this section as well.

RELIEVING FUEL SYSTEM PRESSURE

1. Remove the rear seat.
2. Disconnect the fuel pump connector located under the center floor carpet.
3. Start the engine and wait until fuel in fuel line is exhausted.
4. After engine stalls, turn the ignition switch to OFF position.
5. Reconnect the fuel pump connector.
6. Disconnect the negative battery cable.

FUEL FILTER

REMOVAL & INSTALLATION

See Fuel Pump Removal & Installation. The fuel filter is located inside the fuel pump.

FUEL PUMP

REMOVAL & INSTALLATION

See Figures 302 through 304.

1. Turn ignition switch to "OFF" and disconnect the battery negative terminal.
2. Remove the rear seat cushion.

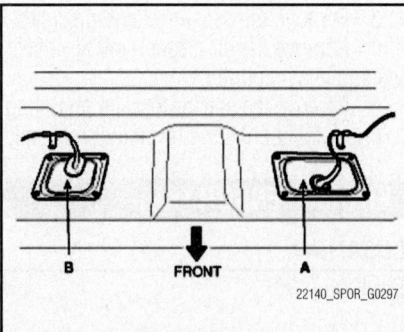

Fig. 302 Remove the service cover (A) under the carpet

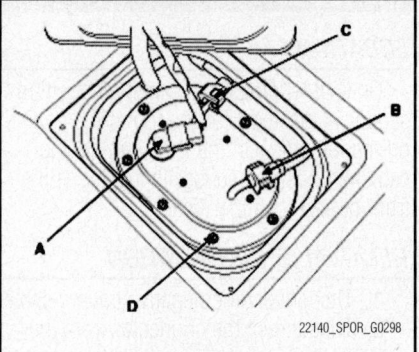

Fig. 303 Disconnect the fuel pump wiring connector (A)

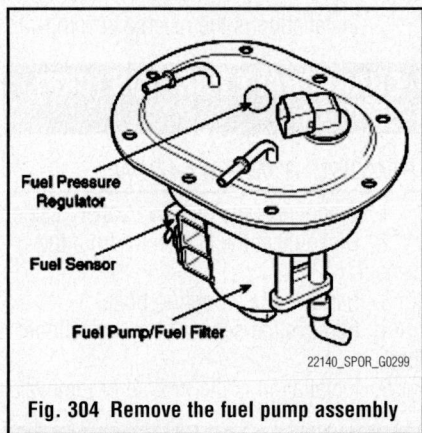

Fig. 304 Remove the fuel pump assembly

3. Remove the service cover (A) under the carpet.
4. Disconnect the fuel pump wiring connector (A).
5. Remove the fuel feed hose (B) and suction hose (C).
6. Remove the fuel pump plate mounting bolts (D).
7. Remove the fuel pump assembly.
8. Installation is the reverse of removal.

FUEL PRESSURE REGULATOR

REMOVAL & INSTALLATION

See Fuel Pump Removal & Installation. The fuel tank pressure sensor is located inside the fuel pump.

HEATING & AIR CONDITIONING SYSTEM

BLOWER MOTOR

REMOVAL & INSTALLATION

See Figure 305.

1. Disconnect the negative battery terminal.

2. Remove the heater unit.

3. Disconnect the connectors from the fresh and recirculation actuator, the blower relay, the blower motor and power mosfet.

4. Remove the self-tapping screws (A), the mounting nut (B), the mounting bolt (C) and the blower unit (D).

➡ **Make sure that there is no air leaking out of the blower and duct joints.**

5. Install in the reverse order of removal.

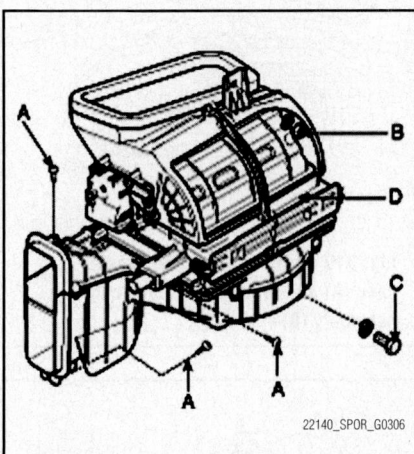

Fig. 305 Remove the self-tapping screws (A), the mounting nut (B), the mounting bolt (C) and the blower unit (D)

HEATER CORE

REMOVAL & INSTALLATION

See Figures 306 through 308.

1. Disconnect the negative battery terminal.

2. Recover the refrigerant with a recovery/recycling/charging station.

3. Remove the bolts (A) and the expansion valve (B) from the evaporator core.

※※ **CAUTION**

Plug or cap the lines immediately after disconnecting them to avoid moisture and dust contamination.

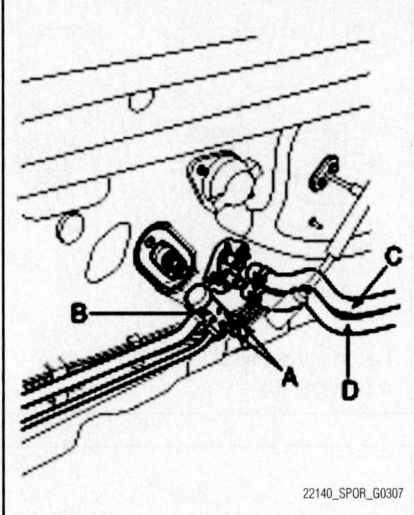

Fig. 306 Remove the bolts (A) and the expansion valve (B) from the evaporator core

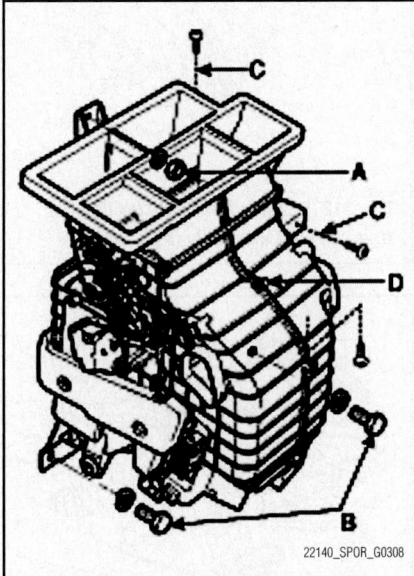

Fig. 307 Remove the mounting nut (A) and the mounting bolts (B)

4. When the engine is cool, disconnect the inlet (D) and outlet (C) heater hoses from the heater unit.

5. Remove the crash pad (instrument panel). Refer to Instrument Panel Removal and Installation.

6. Remove the cross member.

7. Disconnect the connectors from the temperature control actuator, the mode

Fig. 308 Be careful not to bend the inlet and outlet pipes during heater core (A) removal, and pull out the heater core (B)

control actuator and the evaporator temperature sensor, and remove the mounting nut (A) and the mounting bolts (B).

8. Remove the heater and evaporator unit (D) after removing the mounting screws (C).

9. Remove the self-tapping screws and the side bracket.

10. Be careful not to bend the inlet and outlet pipes during heater core (A) removal, and pull out the heater core (B).

11. Install the heater core in the reverse order of removal.

12. Install in the reverse order of removal, and note these items :

a. If you're installing a new evaporator, add refrigerant oil (ND-OIL8).

b. Replace the O-rings with new ones at each fitting, and apply a thin coat of refrigerant oil before installing them. Be sure to use the right O-rings for R-134a to avoid leakage.

c. Apply sealant to the grommets.

d. Make sure that there is no air leakage.

e. Charge the system and test its performance.

f. Do not interchange the inlet and outlet heater hoses and install the hose clamps securely.

g. Refill the cooling system with engine coolant.

STEERING

POWER RACK & PINION STEERING GEAR

REMOVAL & INSTALLATION

See Figures 309 through 318.

1. Disconnect the cover fixing clip (A) on the steering shaft universal joint driver side interior.

2. Remove the noise covers (B).

3. Remove the universal joint and the gear box mounting bolt (A) and disconnect the universal joint (B) from the gear box.

4. Raise the vehicle.

5. Remove the front tires.

6. Remove the engine undercover.

7. After removing the split pin, disconnect the tie rod (A) from the knuckle (B) by using the special tool (09568-34000).

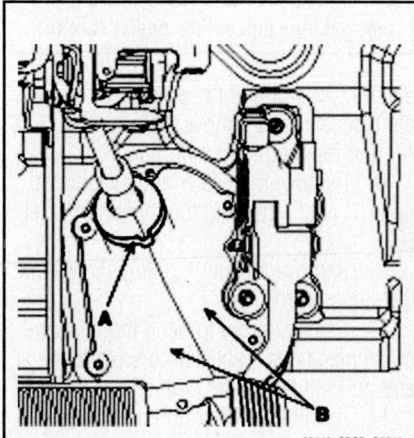

Fig. 309 Disconnect the cover fixing clip (A). Remove the noise covers (B)

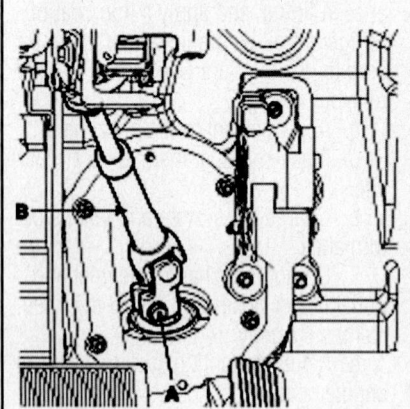

Fig. 310 Disconnect the universal joint from the gear box

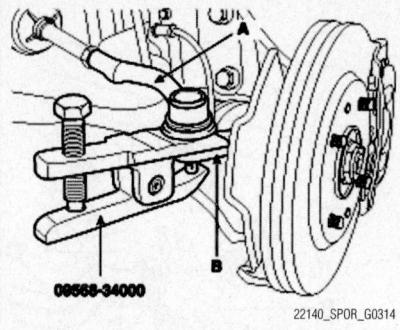

Fig. 311 Disconnect the tie rod (A) from the knuckle (B)

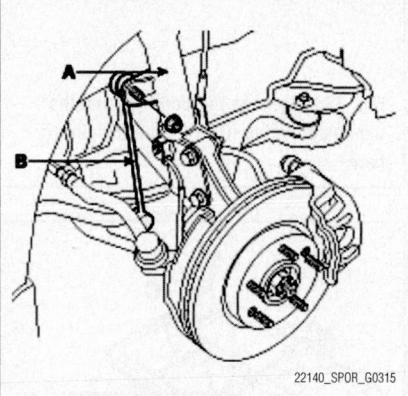

Fig. 312 Remove the stabilizer link (B) from the strut assembly (A)

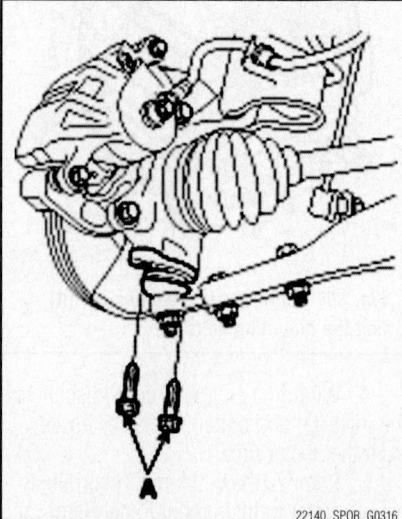

Fig. 313 Remove the two bolts (A) for lower arm ball joint

8. Remove the stabilizer link (B) from the strut assembly (A).

9. Remove the two bolts (A) for lower arm ball joint.

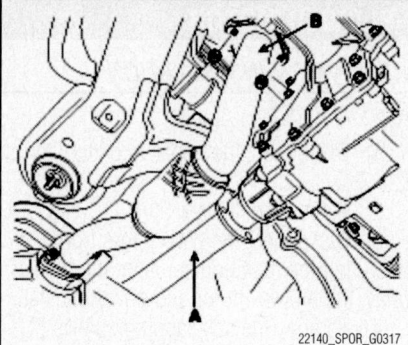

Fig. 314 Remove the propeller shaft (A) (4WD) and the front muffler assembly (B)

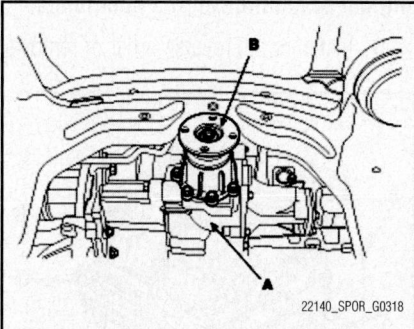

Fig. 315 Drain oil from the transfer case (A) and remove the rear flange assembly (B)

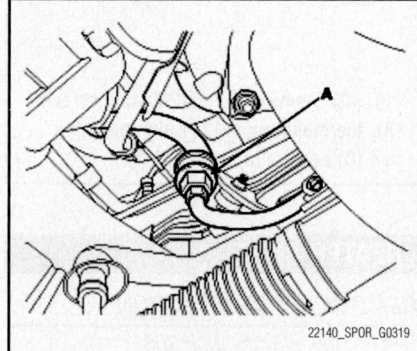

Fig. 316 Remove the connector (A) for the PS pressure tubes

10. Remove the propeller shaft (A) (4WD) and the front muffler assembly (B).

11. Drain oil from the transfer case (A). and remove the rear flange assembly (B).

12. Drain power steering oil.

13. Remove the connector (A) for the PS pressure tubes.

14. Remove two engine mounting bolts (B, C) and six subframe mounting bolts to remove the subframe (A).

15. Remove the power steering gearbox (A) after removing four mounting bolts (B) of the power steering gearbox.

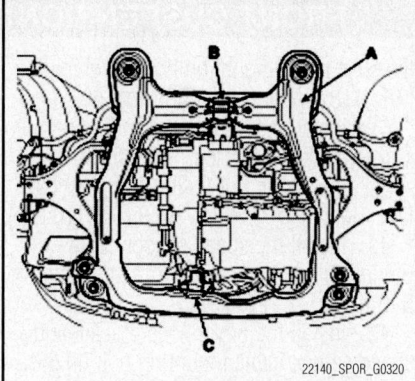

Fig. 317 Remove two engine mounting bolts (B, C) and six subframe mounting bolts to remove the subframe (A)

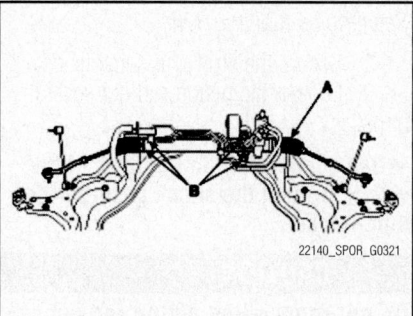

Fig. 318 Remove the power steering gearbox (A) after removing four mounting bolts (B) of the power steering gearbox.

16. Installation is the reverse of removal.

POWER STEERING PUMP

REMOVAL & INSTALLATION

See Figures 319 through 321.

1. Remove the bolt attaching the wiring bracket, and move the wiring to the side.

2. Remove the pressure hose (A) from the oil pump (B), and disconnect the suction hose (C) from the suction connector and drain the fluid into a container.

3. Loosen the tension adjusting bolt (A) on the power steering.

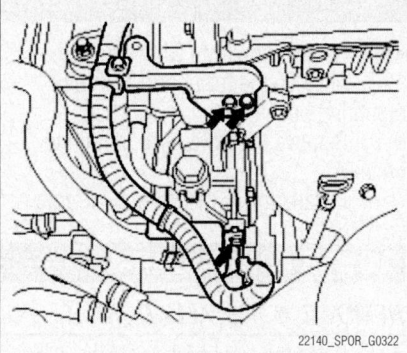

Fig. 319 Remove the bolt attaching the wiring bracket, and move the wiring to the side

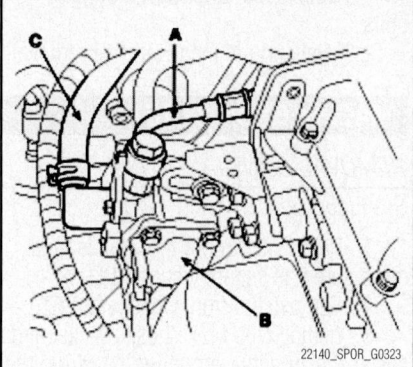

Fig. 320 Remove the pressure hose (A) from the oil pump (B), and disconnect the suction hose (C)

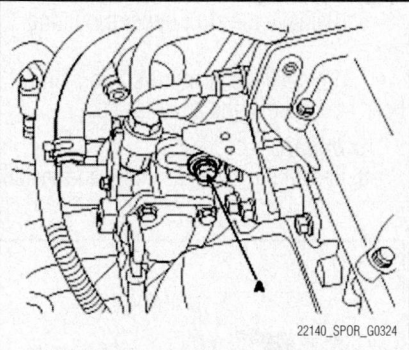

Fig. 321 Loosen the tension adjusting bolt (A) on the power steering pump

4. Remove the "V" belt from the power steering oil pump pulley.

5. Remove the power steering oil pump mounting bolt and the tension adjusting bolt, and remove the steering oil pump assembly.

6. Installation is the reverse of removal.

BLEEDING

1. Fill the power steering fluid reservoir up to the "MAX" position with specified fluid.

2. Raise the front wheels.

3. Disconnect the ignition coil, and while operating the starter motor intermittently (for 15 to 20 seconds), turn the steering wheel all the way to the left and to the right five or six times.

➡️**When bleeding fluid, replenish the fluid so that the level does not fall below the bottom of the filter. Perform air bleeding only while cranking to avoid excessive fluid aeration.**

4. Connect the ignition coil and start the engine (idling).

5. Turn the steering wheel to the left and to the right, until there are no air bubbles in the oil reservoir.

➡️**Do not hold the steering wheel turned all the way to either side for more than ten seconds.**

6. Confirm that the fluid is not milky and that the level is between "MAX" and "MIN" mark on the reservoir.

7. Check that there is a little change in the fluid level when the steering wheel is turned left and right.

➡️**If the fluid level varies 0.2 inches (5 mm) or more, bleed the system again. If the fluid level suddenly rises after stopping the engine, further bleeding is required. Incomplete bleeding will produce a chattering sound in the pump and noise in the flow control valve, and lead to decreased durability of the pump.**

SUSPENSION

CONTROL LINKS

REMOVAL & INSTALLATION

1. Raise the front of the vehicle, and make sure it is securely supported.
2. Remove the front wheel and tire.
3. Remove the stabilizer bar link from the strut assembly.
4. Installation is the reverse of removal.

LOWER BALL JOINT

REMOVAL & INSTALLATION

See Figures 322 and 323.

1. Raise the front of the vehicle, and make sure it is securely supported.

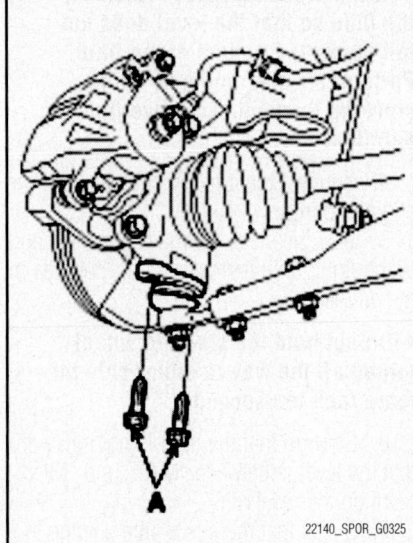

Fig. 322 Remove the lower arm ball joint mounting bolts (A)

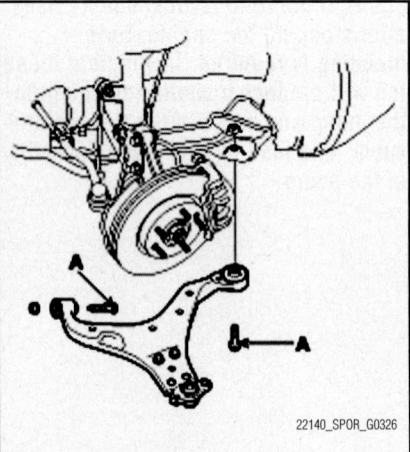

Fig. 323 Remove the lower arm mounting bolts (A)

2. Remove the front wheel and tire.
3. Remove the lower arm ball joint mounting bolts (A).
4. Remove the lower arm mounting bolts (A).
5. Installation is the reverse of removal.

LOWER CONTROL ARM

REMOVAL & INSTALLATION

1. Raise the front of the vehicle, and make sure it is securely supported.
2. Remove the front wheel and tire.
3. Remove the lower arm ball joint mounting bolts.
4. Remove the lower arm mounting bolts.
5. Installation is the reverse of removal.

MACPHERSON STRUT

REMOVAL & INSTALLATION

See Figure 324.

1. Raise the front of the vehicle, and make sure it is securely supported.
2. Remove the front wheel and tire.
3. Remove the brake hose bracket and speed sensor cable-mounting bolt from the strut assembly.
4. Remove the speed sensor cable-mounting bolt and speed sensor.
5. Remove the nut from the stabilizer bar link.
6. Remove the strut upper mounting nuts.
7. Remove the strut lower mounting bolts and remove the strut assembly.

To install:

8. Install the strut assembly and install

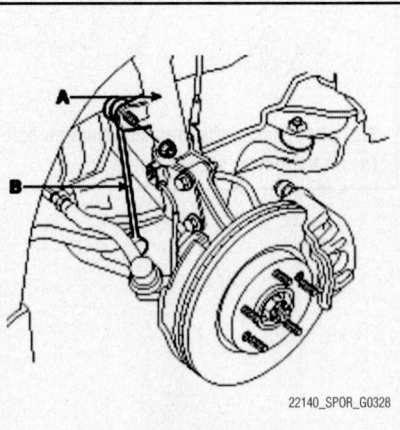

Fig. 324 Remove the stabilizer bar link (B) from the strut assembly (A)

the strut lower mounting bolts. Tighten to 103–118 ft. lbs. (140–160 Nm).
9. Install the strut upper mounting nuts. Tighten to 33–44 ft. lbs. (45–60 Nm).
10. Install the nut on the stabilizer bar link. Tighten to 74–88 ft. lbs. (100–120 Nm).
11. Install the speed sensor cable mounting bolt and speed sensor. Tighten to 5–8 ft. lbs. (7–11 Nm).
12. Install the brake hose bracket and speed sensor cable mounting bolt on the strut assembly. Tighten to 5–8 ft. lbs. (7–11 Nm).
13. Install the front wheel and tire.

OVERHAUL

See Figures 325 and 326.

1. Secure the strut in a suitable vise.
2. Loosen the piston rod nut several turns.

➡**Use copperplate in the jaws of the vise to protect the shock absorber bottom bracket.**

✳✳ CAUTION

Do not remove the piston rod nut until coil spring is compressed and secured.

3. While still secured in a vise, compress the coil spring with SST OK2A1-341-001A.
4. Remove the piston rod nut and each part as below.
5. Secure a handle to the shock absorber piston rod and compress and raise the rod three times with a constant speed. Inspect for uniform working force and abnormal noise.
6. Inspect the entire shock absorber for signs of oil leakage; replace if required.
7. Inspect the coil spring for stress cracks and/or other damage.

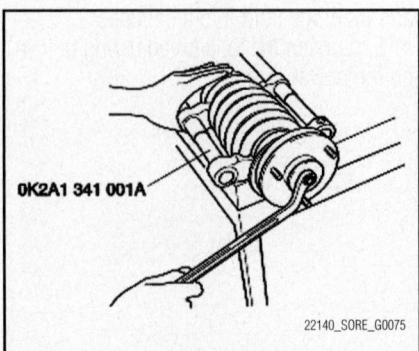

Fig. 325 Compress the coil spring with SST OK2A1-341-001A

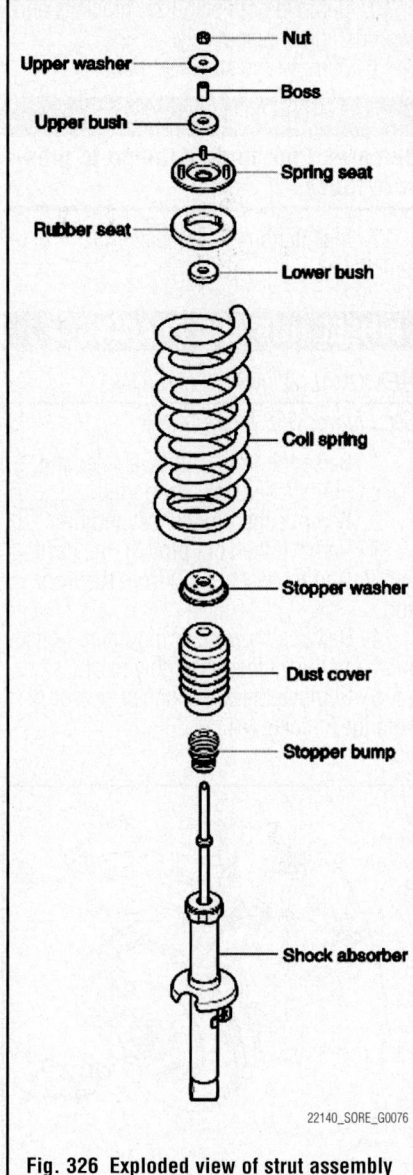

Fig. 326 Exploded view of strut assembly

8. Inspect for damage or deterioration of the upper and lower bushings.

9. Inspect for damage or tearing of the spring seat and rubber seat.

10. Secure the bottom portion of the shock absorber in a vise and compress the coil spring.

11. Set the end of the coil spring to the rubber seat and install the coil spring.

12. Assemble stopper bump, dust cover, stopper washer, lower bushing, rubber seat, spring seat, boss, upper bushing and upper washer in sequence.

13. Hand tighten the piston rod nut.

14. Carefully loosen the coil spring compressor and remove it.

15. With the bottom bracket of the shock absorber still in the vice, tighten the piston rod nut. Tighten to 54–68 ft. lbs. (76–95 Nm).

STEERING KNUCKLE

REMOVAL & INSTALLATION

See Figures 327 through 332.

1. Raise the front of the vehicle, and make sure it is securely supported.

2. Remove the front wheel and tire.

3. Remove the split pin (A), the castle nut (B) and the washer (C) from the front hub.

4. Remove the caliper mounting bolts and hang the caliper assembly to one side.

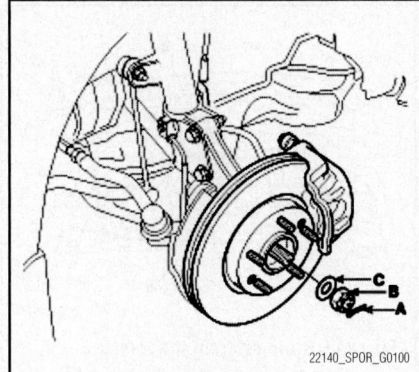

Fig. 327 Remove the split pin (A), the castle nut (B) and the washer (C) from the front hub

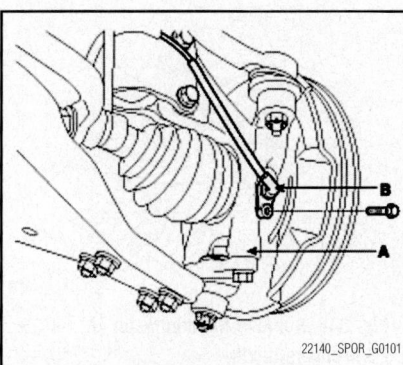

Fig. 328 Remove the wheel speed sensor (B) from the knuckle (A)

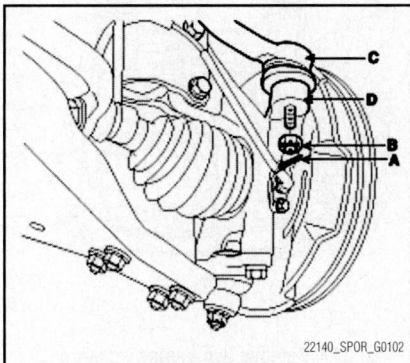

Fig. 329 Disconnect the tie rod end ball joint (C) from the knuckle (D)

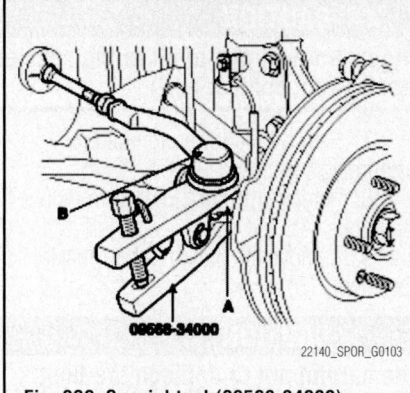

Fig. 330 Special tool (09568-34000)

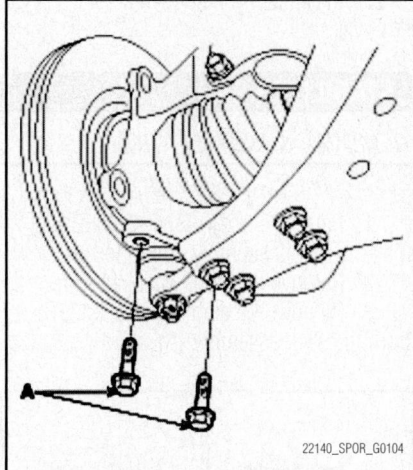

Fig. 331 Remove the lower arm ball joint mounting bolts (A)

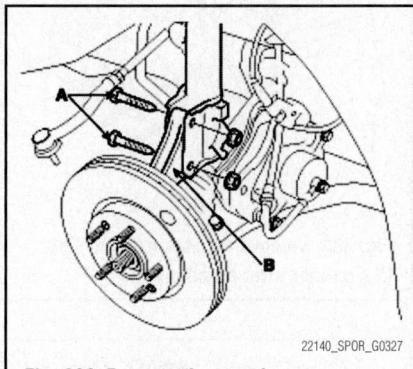

Fig. 332 Remove the strut lower arm mounting bolts (A)

5. Remove the wheel speed sensor (B) from the knuckle (A).

6. Disconnect the tie rod end ball joint (C) from the knuckle (D) using the special tool (09568-34000).

a. Remove the split pin (A).

b. Remove the castle nut (B).

c. Disconnect the ball joint (C) from knuckle (D) using the special tool (09568-34000).

❊ CAUTION

Apply a few drops of oil to the special tool. (Boot contact part)

7. Remove the lower arm ball joint mounting bolts (A).

8. Remove the strut lower arm mounting bolts (A).

9. Remove the hub and the knuckle assembly (B).

❊❊ CAUTION

Be careful not to damage the boot and rotor teeth.

10. Installation is the reverse of removal.

STABILIZER BAR

REMOVAL & INSTALLATION

See Figures 45, 333 through 337.

1. Raise the front of the vehicle, and make sure it is securely supported.

2. Remove the front wheel and tire.

3. Remove the stabilizer bar link (B) from the strut assembly (A).

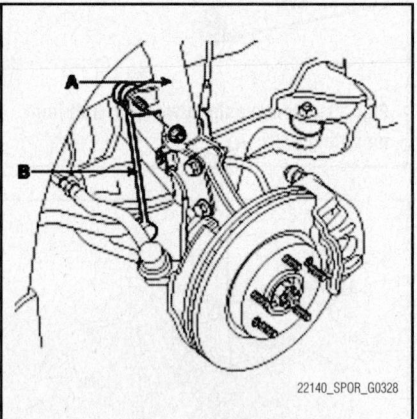

Fig. 333 Remove the stabilizer bar link (B) from the strut assembly (A)

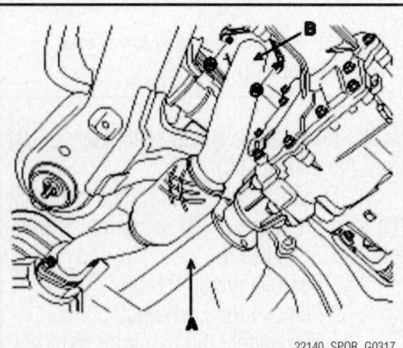

Fig. 334 Remove the propeller shaft (A) (4WD) and the front muffler assembly (B)

4. Remove the two bolts (A) for lower arm ball joint.

5. Remove the propeller shaft (A) (4WD) and the front muffler assembly (B).

6. Drain oil from the transfer case (A) and remove the rear flange assembly (B).

7. Drain power steering oil.

8. Remove the connector (A) for PS pressure tubes.

9. Remove two engine mounting bolts (B, C) and six subframe mounting bolts in order to remove the subframe (A).

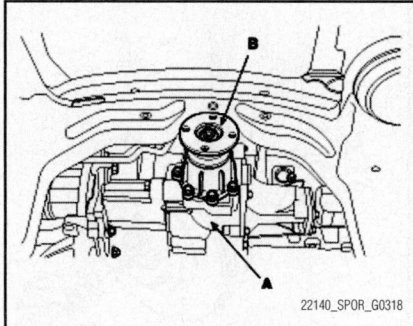

Fig. 335 Drain oil from the transfer case (A) and remove the rear flange assembly (B)

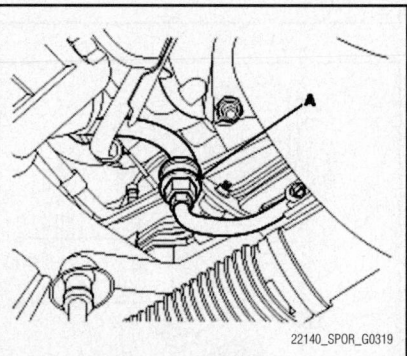

Fig. 336 Remove the connector (A) for the PS pressure tubes

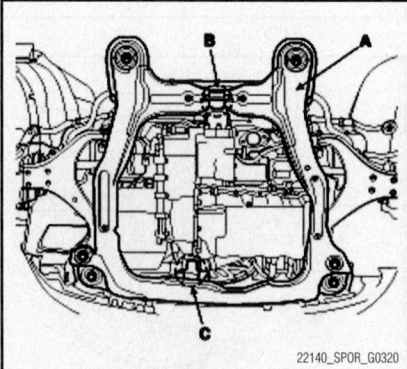

Fig. 337 Remove two engine mounting bolts (B, C) and six subframe mounting bolts to remove the subframe (A)

10. Remove both stabilizer brackets and two bushings respectively.

11. Remove the stabilizer bar.

❊❊ CAUTION

Be careful not to do damage to pressure tubes.

12. Installation is the reverse of removal.

WHEEL HUB AND BEARING

REMOVAL & INSTALLATION

See Figures 338 through 343.

1. Raise the front of the vehicle, and make sure it is securely supported.

2. Remove the front wheel and tire.

3. Remove the split pin (A), the castle nut (B) and the washer (C) from the front hub.

4. Remove the caliper mounting bolts and hang the caliper assembly to one side.

5. Remove the wheel speed sensor (B) from the knuckle (A).

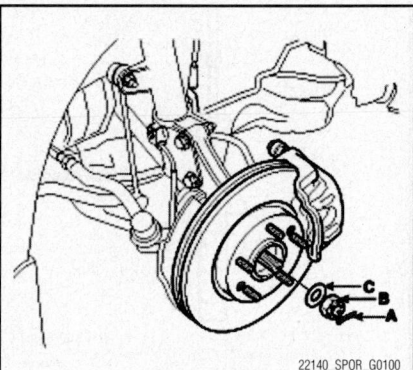

Fig. 338 Remove the split pin (A), the castle nut (B) and the washer (C) from the front hub

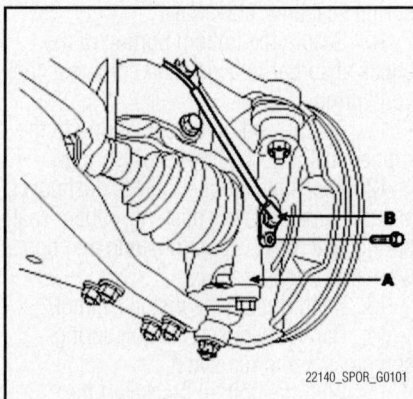

Fig. 339 Remove the wheel speed sensor (B) from the knuckle (A)

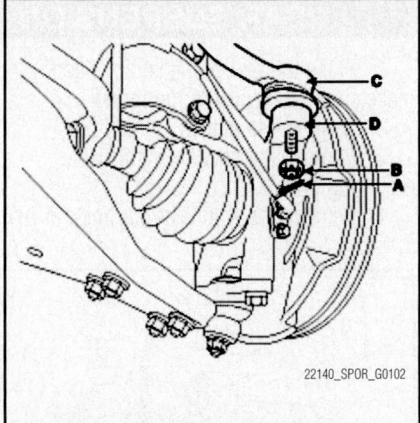

Fig. 340 Disconnect the tie rod end ball joint (C) from the knuckle (D)

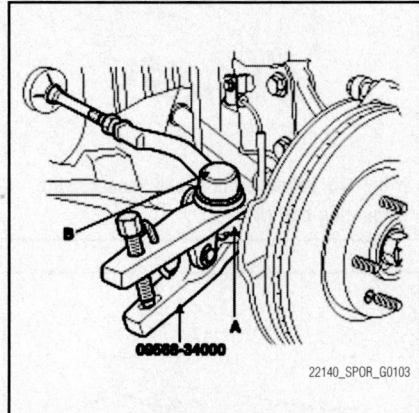

Fig. 341 Special tool (09568-34000)

6. Disconnect the tie rod end ball joint (C) from the knuckle (D) using the special tool (09568-34000).
   a. Remove the split pin (A).
   b. Remove the castle nut (B).
   c. Disconnect the ball joint (C) from knuckle (D) using the special tool (09568-34000).

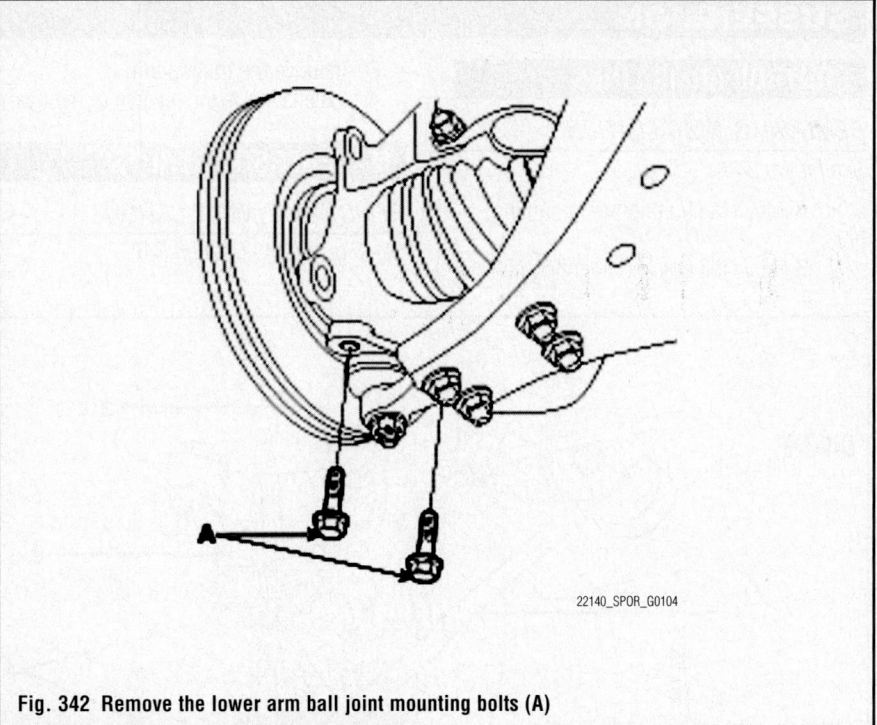

Fig. 342 Remove the lower arm ball joint mounting bolts (A)

> ☀☀ **CAUTION**
>
> **Apply a few drops of oil to the special tool. (Boot contact part)**

7. Remove the lower arm ball joint mounting bolts (A).
8. Remove the strut lower arm mounting bolts (A).
9. Remove the hub and the knuckle assembly (B).

> ☀☀ **CAUTION**
>
> **Be careful not to damage the boot and rotor teeth.**

10. Installation is the reverse of removal.

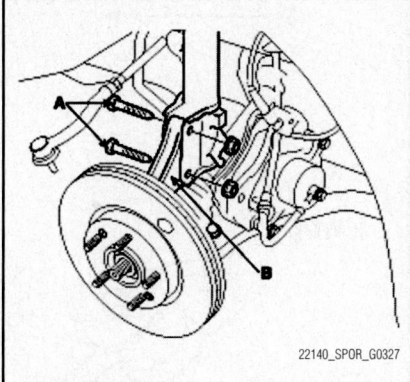

Fig. 343 Remove the strut lower arm mounting bolts (A)

SUSPENSION   REAR SUSPENSION

CONTROL ARMS/LINKS

REMOVAL & INSTALLATION

See Figure 344.

1. Remove the trailing arm mounting bolts.
2. Remove the bracket mounting bolt.

3. Remove the trailing arm.
4. Installation is the reverse of removal.

MACPHERSON STRUTS

REMOVAL & INSTALLATION

See Figures 345 through 347.

1. Raise the rear of the vehicle, and make sure it is securely supported.
2. Remove the rear wheel and tire.
3. Remove the speed sensor cable mounting bolt (A).
4. Remove the stabilizer bar link nut (B).

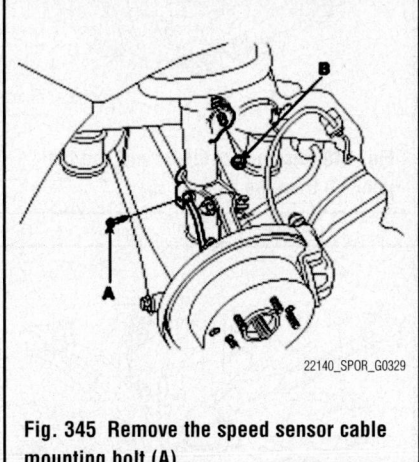

Fig. 345 Remove the speed sensor cable mounting bolt (A)

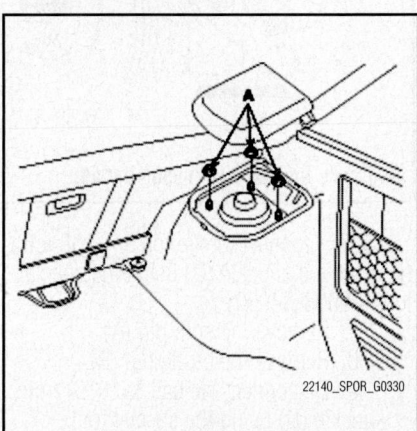

Fig. 346 Remove the strut upper mounting nuts (A)

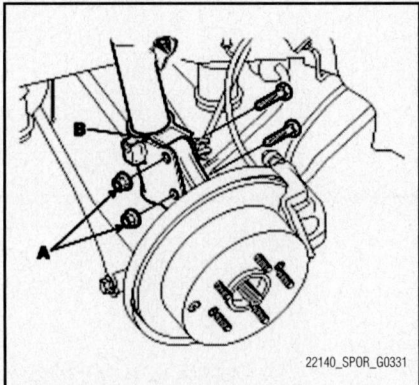

Fig. 347 Remove the strut lower mounting bolts (A) and remove the strut assembly (B).

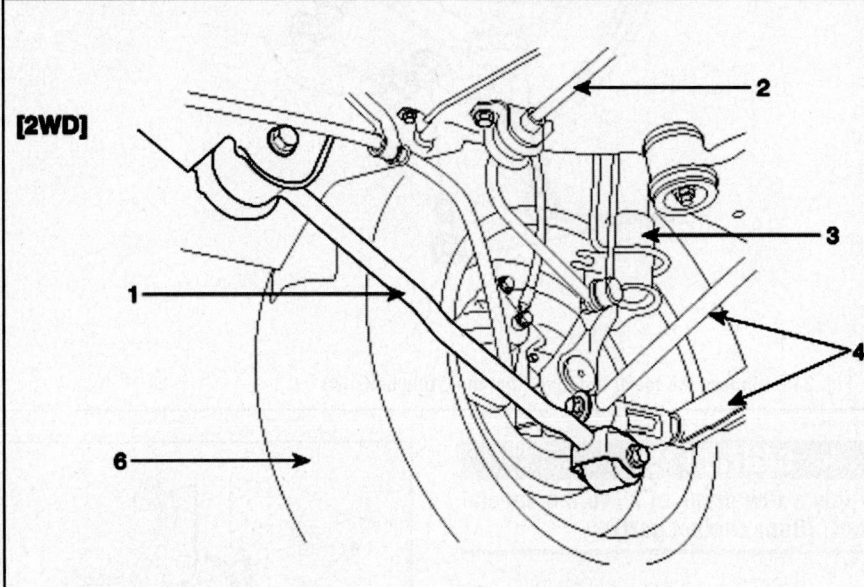

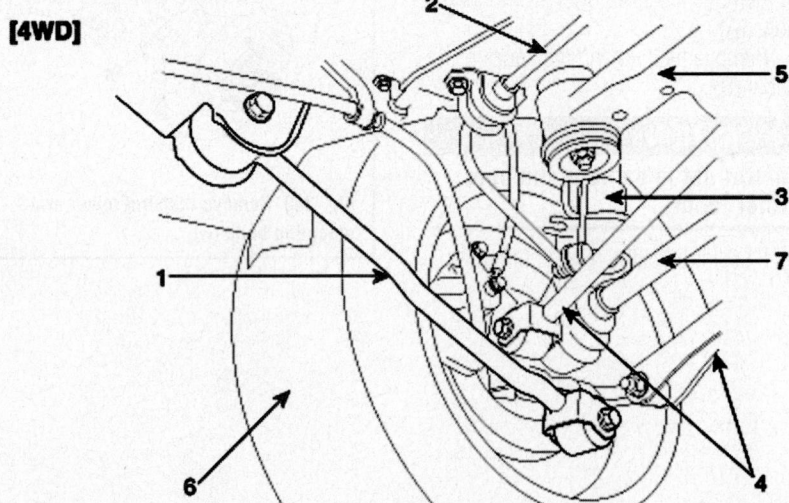

1. Trailing arm
2. Stabilizer bar
3. Strut assembly
4. Suspension arm
5. Cross member
6. Tire
7. Drive shaft

Fig. 344 Trailing arm components of rear suspension

5. Remove the strut upper mounting nuts (A).

6. Remove the strut lower mounting bolts (A) and remove the strut assembly (B).

7. Installation is the reverse of removal.

OVERHAUL

See Figures 348 and 349.

1. Secure the strut in a suitable vise.

2. Loosen the piston rod nut several turns.

➡ **Use copperplate in the jaws of the vise to protect the shock absorber bottom bracket.**

❊❊ CAUTION

Do not remove the piston rod nut until coil spring is compressed and secured.

3. While still secured in a vise, compress the coil spring with SST OK2A1-341-001A.

4. Remove the piston rod nut and each part as below.

5. Secure a handle to the shock absorber piston rod and compress and raise the rod three times with a constant speed. Inspect for uniform working force and abnormal noise.

6. Inspect the entire shock absorber for signs of oil leakage; replace if required.

7. Inspect the coil spring for stress cracks and/or other damage.

8. Inspect for damage or deterioration of the upper and lower bushings.

9. Inspect for damage or tearing of the spring seat and rubber seat.

10. Secure the bottom portion of the shock absorber in a vise and compress the coil spring.

11. Set the end of the coil spring to the rubber seat and install the coil spring.

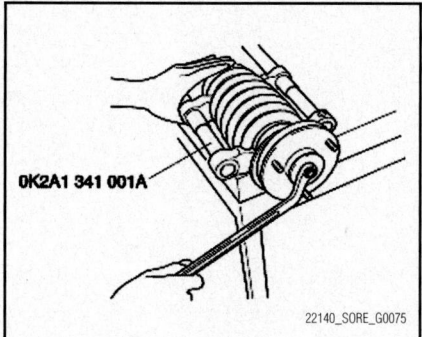

Fig. 348 Compress the coil spring with SST OK2A1-341-001A

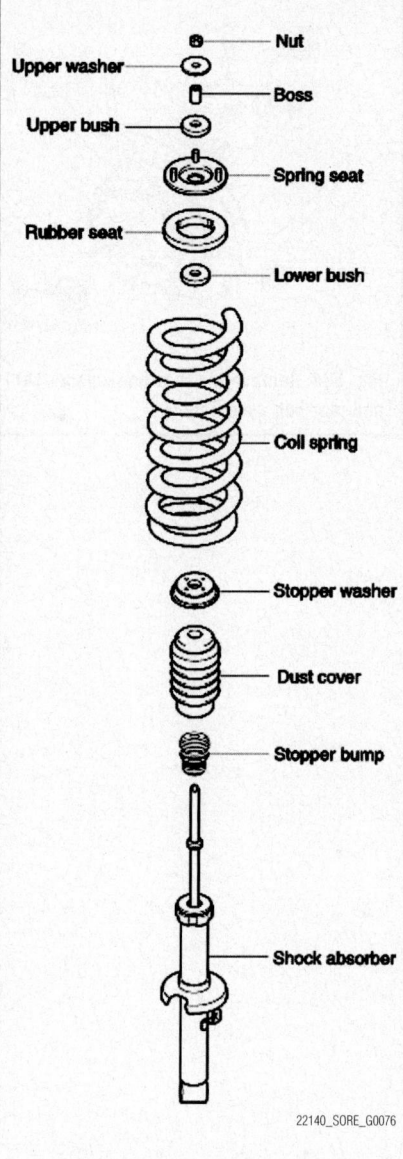

Fig. 349 Exploded view of strut assembly

12. Assemble stopper bump, dust cover, stopper washer, lower bushing, rubber seat, spring seat, boss, upper bushing and upper washer in sequence.

13. Hand tighten the piston rod nut.

14. Carefully loosen the coil spring compressor and remove it.

15. With the bottom bracket of the shock absorber still in the vice, tighten the piston rod nut. Tighten to 54–68 ft. lbs. (76–95 Nm).

WHEEL HUB AND BEARING

REMOVAL & INSTALLATION

See Figures 350 through 354.

1. Raise the rear of the vehicle, and make sure it is securely supported.

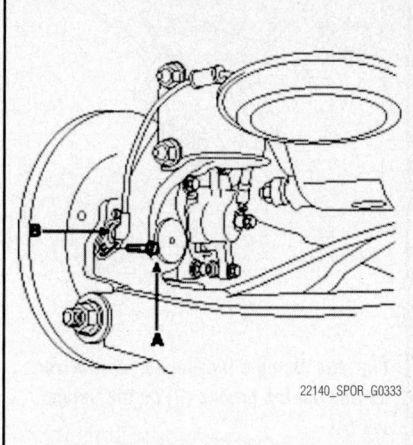

Fig. 350 Remove the wheel speed sensor (B) from the axle carrier (A)

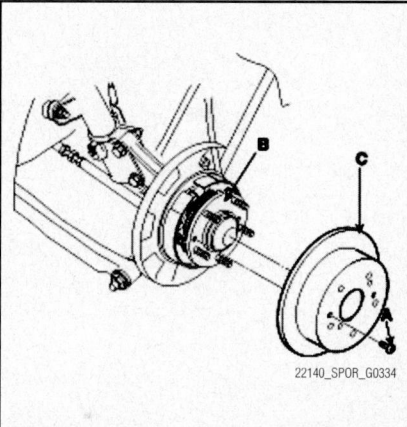

Fig. 351 Remove the brake disc mounting screw (A), and remove the brake disc (C) from the hub (B)

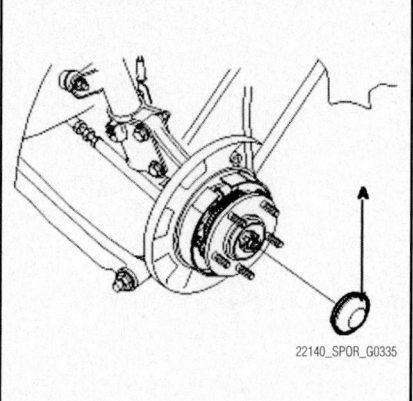

Fig. 352 Using a flat-tipped screwdriver, remove the hub dust cap (A)

2. Remove the rear wheel and tire.

3. Remove the brake caliper mounting bolts and hang the caliper assembly to one side.

4. Remove the wheel speed sensor (B) from the axle carrier (A).

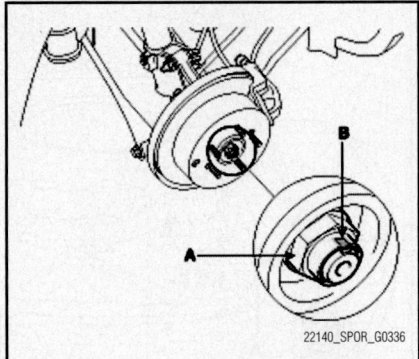

22140_SPOR_G0336

Fig. 353 Using a flat-tipped screwdriver, spread out the groove (B) on the flange nut (A)

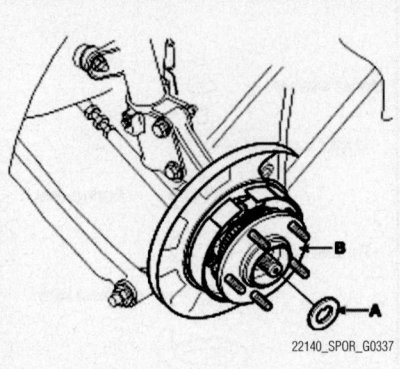

22140_SPOR_G0337

Fig. 354 Remove the rear hub washer (A) and rear hub assembly (B).

5. Remove the brake disc mounting screw (A), and remove the brake disc (C) from the hub (B).

6. Using a flat-tipped screwdriver, remove the hub dust cap (A).

7. Remove the hub bearing flange nut.

a. Using a flat-tipped screwdriver, spread out the groove (B) on the flange nut (A)

b. Remove the hub bearing flange nut (A).

8. Remove the rear hub washer (A) and rear hub assembly (B).

9. Installation is the reverse of removal.

DIAGNOSTIC TROUBLE CODES

OBD II VEHICLE APPLICATIONS

KIA

Amanti
2007-2008
- 3.8L .... VIN 5

Optima
2007-2008
- 2.4L .... VIN 3
- 2.7L .... VIN 4

Rio
2007-2008
- 1.6L .... VIN 3

Rondo
2007-2008
- 2.4L .... VIN 5
- 2.7L .... VIN 6

Sedona
2007-2008
- 3.8L .... VIN 3

Sorento
2007-2008
- 3.3L .... VIN 5
- 3.8L .... VIN 6

Spectra
2007-08
- 2.0L .... VIN 1
- 2.0L .... VIN 2

Sportage
2007-2008
- 2.0L .... VIN 2
- 2.0L .... VIN 4
- 2.7L .... VIN 3

Gas Engine OBD II Trouble Code List (P0xxx Codes)

DTC	Trouble Code Title, Conditions & Possible Causes
DTC: P0011 **2T CCM, MIL:** Yes **Years:** 2007, 2008 **Models:** All **Engines:** 1.6L, 2.0L, 2.4L, 2.7L, 3.5L, 3.8L **Transmissions:** All	**"A" Camshaft Position Timing Over Advanced or System Performance (Bank 1)** Monitor deviation between camshaft position set point and actual value. **Possible Causes:** • Oil leakage • Faulty oil pump • Faulty intake valve control solenoid
DTC: P0012 **2T CCM, MIL:** Yes **Years:** 2007, 2008 **Models:** Amanti, Optima, Rio, Rondo, Sedona, Sorento **Engines:** 1.6L, 2.7L, 3.3L, 3.8L **Transmissions:** All	**"A" Camshaft Position- Timing Over Retarded (Bank 1)** Determines if the phaser is stuck or has a steady error. **Possible Causes:** • Engine oil • OCV • CVVT stuck • Faulty PCM
DTC: P0016 **2T CCM, MIL:** Yes **Years:** 2007, 2008 **Models:** All **Engines:** All **Transmissions:** All	**Crankshaft Position- Camshaft Position Correlation (Bank 1 Sensor "A")** Monitor camshaft position in the full retard condition or during CVVT control. Camshaft switching out of 109 to 141 degrees in full retard position, 70 to 140 degrees CRK during CVVT control. **Possible Causes:** • Abnormal installation of camshaft • Abnormal installation of crankshaft • Abnormal installation of tone wheel
DTC: P0018 **2T CCM, MIL:** Yes **Years:** 2007, 2008 **Models:** Amanti, Optima, Rondo, Sedona, Sorento **Engines:** 2.7L, 3.3L, 3.8L **Transmissions:** All	**Crankshaft Position- Camshaft Position Correlation (Bank 2 Sensor A)** Determines if CAM (B2) target is aligned correctly to the crank. No active faults. **Possible Causes:** • CKPS, CMPS (B2) • CVVT • Timing misalignment • Faulty PCM
DTC: P0021 **2T CCM, MIL:** Yes **Years:** 2007, 2008 **Models:** Amanti, Optima, Rondo, Sedona, Sorento **Engines:** 2.7L, 3.3L, 3.8L **Transmissions:** All	**"A" Camshaft Position- Timing Over Advanced Or System Performance (Bank 2)** Determines if the phaser is moving at an unexpected rate. Cam off set is available. **Possible Causes:** • Excessive phasing • Binding oil pressure (blockage) • Faulty PCM
DTC: P0022 **2T CCM, MIL:** Yes **Years:** 2007, 2008 **Models:** Amanti, Optima, Rondo, Sedona, Sorento **Engines:** 2.7L, 3.3L, 3.8L **Transmissions:** All	**"A" Camshaft Position- Timing Over Retarded (Bank 2)** Determines if the phaser is stuck or has a steady state error. Off sets available. Cam velocity below threshold at 15 CAD/s. **Possible Causes:** • Engine oil • OCV • CVVT stuck • Faulty PCM
DTC: P0030 **2T CCM, MIL:** Yes **Years:** 2007, 2008 **Models:** Amanti, Optima, Rio, Rondo, Sorento, Spectra, Sportage **Engines:** 1.6L, 2.0L, 2.4L, 2.7L, 3.3L, 3.8L **Transmissions:** All	**HO2S-11 Heater Circuit Malfunction** Engine started, engine runtime over 3 minutes, and the PCM determined the resistance of the HO2S heater was more than a calculated amount. **Possible Causes:** • HO2S heater control circuit is open or shorted to ground • HO2S heater control circuit is shorted to system power (B+) • HO2S heater is damaged or has failed • PCM has failed

DTC	Trouble Code Title, Conditions & Possible Causes
DTC: P0031 **2T CCM, MIL: Yes** **Years:** 2007, 2008 **Models:** All **Engines:** All **Transmissions:** All	**O2 Sensor Heater Circuit Low (Bank 1/Sensor 1)** Heater check, low. Open or short circuit. **Possible Causes:** • Open in battery and control circuit • Short to ground in control circuit (pin 48 to 36) • Faulty HO2S heater • Faulty PCM
DTC: P0032 **2T CCM, MIL: Yes** **Years:** 2007, 2008 **Models:** All **Engines:** All **Transmissions:** All	**O2 Sensor Heater Circuit High (Bank 1/Sensor 1)** Heater check, high. Short circuit. **Possible Causes:** • Short to battery in control circuit • Faulty HO2S heater • Faulty PCM
DTC: P0037 **2T CCM, MIL: Yes** **Years:** 2007, 2008 **Models:** All **Engines:** All **Transmissions:** All	**O2 Sensor Heater Circuit Low (Bank 1/Sensor 2)** Heater check, low. Open or short circuit. **Possible Causes:** • Open in battery and control circuit • Short to ground in control circuit (pin 48 to 36) • Faulty HO2S heater • Faulty PCM
DTC: P0038 **2T CCM, MIL: Yes** **Years:** 2007, 2008 **Models:** All **Engines:** All **Transmissions:** All	**O2 Sensor Heater Circuit High (Bank 1/Sensor 2)** Heater check, high. Short circuit. **Possible Causes:** • Short to battery in control circuit • Faulty HO2S heater • Faulty PCM
DTC: P0051 **2T CCM, MIL: Yes** **Years:** 2007, 2008 **Models:** Amanti, Optima, Rondo, Sedona, Sorento, Sportage **Engines:** 2.7L, 3.3L, 3.5L, 3.8L **Transmissions:** All	**HO2S Heater Circuit Low (Bank 2/Sensor 1)** Short circuit to ground on front HO2S heater line. Battery voltage above 10 volts. **Possible Causes:** • Related fuse blown or missing • Open or short to ground in power supply or control harness • Contact resistance in connectors • Faulty HO2S
DTC: P0052 **2T CCM, MIL: Yes** **Years:** 2007, 2008 **Models:** Amanti, Optima, Rondo, Sedona, Sorento, Sportage **Engines:** 2.7L, 3.3L, 3.5L, 3.8L **Transmissions:** All	**HO2S Heater Circuit High (Bank 2/Sensor 1)** Open or short circuit to battery line on front HO2S heater line. Battery voltage above 10 volts. **Possible Causes:** • Open or short to battery in control harness • Contact resistance in connectors • Faulty HO2S
DTC: P0057 **2T CCM, MIL: Yes** **Years:** 2007, 2008 **Models:** Amanti, Optima, Rondo, Sedona, Sorento, Sportage **Engines:** 2.7L, 3.3L, 3.5L, 3.8L **Transmissions:** All	**HO2S Heater Circuit Low (Bank 2/Sensor 2)** Check short circuit to ground on rear HO2S heater line. **Possible Causes:** • Related fuse blown or missing • Open or short to ground in power supply or control harness • Contact resistance in connections • Faulty HO2S
DTC: P0058 **2T CCM, MIL: Yes** **Years:** 2007, 2008 **Models:** Amanti, Optima, Rondo, Sedona, Sorento, Sportage **Engines:** 2.7L, 3.3L, 3.5L, 3.8L **Transmissions:** All	**HO2S Heater Circuit High (Bank 2/Sensor 2)** Check short circuit to ground on rear HO2S heater line. **Possible Causes:** • Open or short to battery in control harness • Contact resistance in connections • Faulty HO2S

DTC	Trouble Code Title, Conditions & Possible Causes
DTC: P0076 **2T CCM, MIL: Yes** **Years:** 2007, 2008 **Models:** All **Engines:** 1.6L, 2.0L, 2.4L, 2.7L, 3.3L, 3.8L **Transmissions:** All	**Intake Valve Control Solenoid Circuit Low (Bank 1)** PCM sets the code if it detects that the intake valve control solenoid control circuit is short to ground. Electrical check. **Possible Causes:** • Faulty ECM/PCM • Short to ground in control circuit • Contact resistance in connectors • Faulty intake valve control solenoid
DTC: P0077 **2T CCM, MIL: Yes** **Years:** 2007, 2008 **Models:** All **Engines:** 1.6L, 2.0L, 2.4L, 2.7L, 3.3L, 3.8L **Transmissions:** All	**Intake Valve Control Solenoid Circuit High (Bank 1)** PCM sets the code if it detects that the OCV control circuit is open or short to battery. Electrical check. **Possible Causes:** • Open or short to battery in control circuit • Contact resistance in connectors • Faulty intake valve control solenoid • Faulty PCM
DTC: P0082 **2T CCM, MIL: Yes** **Years:** 2007, 2008 **Models:** Amanti, Optima, Rondo, Sedona, Sorento **Engines:** 2.7L, 3.3L, 3.8L **Transmissions:** All	**Intake Valve Control Solenoid Circuit Low (Bank 2)** Detects a short to ground or open circuit of VCPD bank 1 intake circuit output. No disabling faults present. Engine running. Enable time delay equal to or greater than 0.5 second. **Possible Causes:** • Poor connection • Open in power circuit • Open or short to ground in control circuit • OCV • Faulty PCM
DTC: P0083 **2T CCM, MIL: Yes** **Years:** 2007, 2008 **Models:** Amanti, Optima, Rondo, Sedona, Sorento **Engines:** 2.7L, 3.3L, 3.8L **Transmissions: All**	**Intake Valve Control Solenoid Circuit High (Bank 2)** Detects a short to battery of VCPD bank 1 intake circuit output. No disabling faults present. Engine running. Enable time delay equal to or greater than 0.5 seconds. **Possible Causes:** • poor connection • Short to battery in control circuit • OCV • Faulty PCM
DTC: P0101 **2T CCM, MIL: Yes** **Years:** 2007, 2008 **Models:** Amanti, Optima, Rondo, Sedona, Sorento, Spectra, Sportage **Engines:** All **Transmissions:** All	**MAF or Volume Airflow Sensor Performance** Engine started, engine running, and the PCM detected too much difference between the Actual MAF sensor signal and a Threshold MAF sensor value in memory during the CCM Rationality test. **Possible Causes:** • MAF sensor signal circuit has high resistance • MAF sensor ground circuit has high resistance • MAF sensor is damaged or has failed (it may be contaminated) • PCM has failed
DTC: P0102 **2T CCM, MIL: Yes** **Years:** 2007, 2008 **Models:** Amanti, Optima, Rondo, Sedona, Sorento, Spectra, Sportage **Engines:** All **Transmissions:** All	**MAF Sensor Circuit Low Input** Key on or engine running and the PCM detected the MAF sensor signal was less than 0.39 volt for more than 1 second. **Possible Causes:** • MAF sensor signal circuit is shorted to ground • MAF sensor power circuit open between sensor and main relay • MAF sensor is damaged or has failed • PCM has failed
DTC: P0103 **2T CCM, MIL: Yes** **Years:** 2007, 2008 **Models:** Amanti, Optima, Rondo, Sedona, Sorento, Spectra, Sportage **Engines:** All **Transmissions:** All	**MAF Sensor Circuit High Input** Key on or engine running and the PCM detected the MAF sensor signal was more than 3.90 volts, condition met for 1 second. **Possible Causes:** • MAF sensor signal circuit shorted to VREF or system power • MAF sensor ground circuit is open between sensor and ground • MAF sensor is damaged or has failed • PCM has failed

DTC	Trouble Code Title, Conditions & Possible Causes
DTC: P0106 **2T CCM, MIL: Yes** **Years:** 2007, 2008 **Models:** Amanti, Optima, Rio, Rondo, Sedona, Sorento, Spectra **Engines:** 1.6L, 2.0L, 2.7L, 3.5L, 3.8L **Transmissions:** All	**Manifold Absolute Pressure Sensor Performance** Engine running and the PCM detected the MAP sensor signal was less than 2.20 volts but more than 0.40 volt during the CCM Rationality test. **Possible Causes:** • Loss of 5-volt supply from the PCM (circuit open or grounded) • MAP sensor signal circuit is open or grounded • MAP sensor is damaged or has failed • PCM has failed
DTC: P0107 **2T CCM, MIL: Yes** **Years:** 2007, 2008 **Models:** Amanti, Optima, Rio, Rondo, Sedona, Sorento, Spectra **Engines:** 1.6L, 2.0L, 2.7L, 3.5L, 3.8L **Transmissions:** All	**Manifold Absolute Pressure Sensor Low Input** Key on or engine running and the PCM detected the MAP sensor signal was less than 0.20 volt, condition met for 1 second. **Possible Causes:** • Loss of 5-volt supply from the PCM (circuit open or grounded) • MAP sensor signal circuit is shorted to ground • MAP sensor is damaged or has failed • PCM has failed
DTC: P0108 **2T CCM, MIL: Yes** **Years:** 2007, 2008 **Models:** Amanti, Optima, Rio, Rondo, Sedona, Sorento, Spectra **Engines:** 1.6L, 2.0L, 2.7L, 3.5L, 3.8L **Transmissions:** All	**Manifold Absolute Pressure Sensor High Input** Key on or engine running and the PCM detected the MAP sensor signal was more than 4.90 volts, condition met for 1 second. **Possible Causes:** • MAP sensor ground circuit open • MAP sensor signal circuit shorted to VREF or system power • MAP sensor is damaged or has failed • PCM has failed
DTC: P0110 **2T CCM, MIL: Yes** **Years:** 2007, 2008 **Models:** Amanti, Optima, Rondo, Sedona, Sorento **Engines:** 2.0L, 2.4L, 2.7L, 3.3L, 3.8L **Transmissions:** All	**Intake Air Temperature Sensor Circuit Malfunction** Engine runtime over 600 seconds and the PCM detected the IAT signal indicated more than 262°F, or that it indicated less than -38°F at any time during the CCM test. **Possible Causes:** • IAT sensor signal circuit open or shorted to ground • IAT sensor signal circuit shorted to VREF or system power • IAT sensor is damaged or has failed • PCM has failed
DTC: P0111 **2T CCM, MIL: Yes** **Years:** 2007, 2008 **Models:** All **Engines:** All **Transmissions:** All	**Intake Air temperature Sensor 1 Circuit Range/Performance** If the sensor is out of specification, a code is set. Output voltage is monitored. Engine coolant is above 167°F. Vehicle speed is above 30MPH for more than 60 seconds. Vehicle speed is below 7MPH for more than 30 seconds. **Possible Causes:** • Poor connection • Faulty IATS • Faulty PCM
DTC: P0112 **2T CCM, MIL: Yes** **Years:** 2007, 2008 **Models:** All **Engines:** All **Transmissions:** All	**Intake Air Temperature Sensor Circuit Low Input** Key on or engine running and the PCM detected the IAT sensor signal was less than 0.2 volt during the CCM test. **Possible Causes:** • IAT sensor signal circuit shorted to ground • IAT sensor is damaged or has failed • PCM has failed
DTC: P0113 **2T CCM, MIL: Yes** **Years:** 2007, 2008 **Models:** All **Engines:** All **Transmissions:** All	**Intake Air Temperature Sensor Circuit High Input** Key on or the engine running and the PCM detected the IAT sensor signal was more than 4.9 volts during the CCM test. **Possible Causes:** • IAT sensor signal circuit shorted to VREF or system power • IAT sensor ground circuit is open • IAT sensor is damaged or has failed • PCM has failed
DTC: P0115 **2T CCM, MIL: Yes** **Years:** 2007, 2008 **Models:** Amanti, Optima, Rio, Rondo, Sedona, Sorento **Engines:** 1.6L, 2.7L, 3.3L, 3.8L **Transmissions:** All	**Engine Coolant Temperature Sensor Circuit Malfunction** Engine runtime over 600 seconds, and the PCM detected the IAT signal indicated more than 280°F, or that it indicated less than -38°F at any time during the CCM test. **Possible Causes:** • ECT sensor signal circuit open or shorted to ground • ECT sensor signal circuit shorted to VREF or system power • ECT sensor is damaged or has failed • PCM has failed

DTC	Trouble Code Title, Conditions & Possible Causes
DTC: P0116 **2T CCM, MIL: Yes** **Years:** 2007, 2008 **Models:** All **Engines:** All **Transmissions:** All	**Engine Coolant Temperature Sensor Range/Performance** Engine running for 20 minutes, and the PCM detected an ECT signal remained at less than a specified value during the CCM test period. **Possible Causes:** • Check for low coolant level or incorrect coolant mixture • Cooling system component failure (thermostat stuck open) • ECT sensor is out of calibration or it is "skewed" • ECT sensor is damaged or has failed
DTC: P0117 **2T CCM, MIL: Yes** **Years:** 2007, 2008 **Models:** All **Engines:** All **Transmissions:** All	**Engine Coolant Temperature Sensor Circuit Low Input** Key on or engine running and the PCM detected the ECT sensor signal was less than 0.20 volt during the CCM test. **Possible Causes:** • ECT sensor signal circuit shorted to ground • ECT sensor is damaged or has failed • PCM has failed
DTC: P0118 **2T CCM, MIL: Yes** **Years:** 2007, 2008 **Models:** All **Engines:** All **Transmissions:** All	**Engine Coolant Temperature Sensor Circuit High Input** Key on or engine running and the PCM detected the ECT sensor signal was more than 4.90 volts during the CCM test. **Possible Causes:** • ECT sensor signal circuit is open between sensor and the PCM • ECT sensor ground circuit is open between sensor and ground • ECT sensor signal circuit is shorted to VREF • ECT sensor is damaged or has failed • PCM has failed
DTC: P0121 **2T CCM, MIL: Yes** **Years:** 2007, 2008 **Models:** Amanti, Optima, Rio, Rondo, Sorento, Spectra, Sportage **Engines:** 1.6L, 2.0L, 2.4L, 2.7L, 3.5L **Transmissions:** All	**Throttle Position Sensor Performance** Engine at idle speed, and the PCM detected the TP sensor signal was outside of the idle speed limit (i.e., 1.20-2.1 volts at idle is the limit). **Possible Causes:** • MAF sensor or TP sensor ground circuit has high resistance • MAF sensor has drifted out of calibration • TP sensor has drifted out of calibration • TP sensor is damaged or has failed
DTC: P0122 **2T CCM, MIL: Yes** **Years:** 2007, 2008 **Models:** All **Engines:** All **Transmissions:** All	**Throttle Position Sensor 'A' Circuit Low Input** Key on or engine running and the PCM detected the TP sensor 'A' signal was less than 0.170-0.200 volt during the CCM test. **Possible Causes:** • TP sensor signal circuit is shorted to ground • TP sensor VREF circuit is open or shorted to ground • TP sensor is damaged or has failed • PCM has failed
DTC: P0123 **2T CCM, MIL: Yes** **Years:** 2007, 2008 **Models:** All **Engines:** All **Transmissions:** All	**Throttle Position Sensor 'A' Circuit High Input** Key on or engine running and the PCM detected the TP sensor signal was more than 4.60-4.80 volts during the CCM test. **Possible Causes:** • TP sensor signal circuit is open • TP sensor ground circuit is open • TP sensor signal circuit is shorted to VREF or system power • TP sensor is damaged or has failed\ • PCM has failed
DTC: P0125 **2T ECT, MIL: Yes** **Years:** 2007, 2008 **Models:** Amanti, Optima, Rondo, Sedona, Sorento, Spectra, Sportage **Engines:** 2.L, 2.4L, 2.7L, 3.3L, 3.5L, 3.8L **Transmissions:** All	**Insufficient Coolant Temperature for Closed Loop** DTC P0116, P0117 and P0118 not set, engine runtime from 6-8 minutes, and the PCM detected the ECT signal did not reach the closed loop temperature of at least 68°F. **Possible Causes:** • Check for low coolant level or incorrect coolant mixture • Cooling system component failure (thermostat stuck open) • ECT sensor is out of calibration ("skewed") or it has failed • PCM has failed

DTC	Trouble Code Title, Conditions & Possible Causes
DTC: P0128 **2T ECT, MIL: Yes** **Years:** 2007, 2008 **Models:** All **Engines:** All **Transmissions:** All	**Thermostat Malfunction** ECT sensor signal less than 140°F at startup, then with the engine running the PCM detected the ECT sensor signal did not reach 170°F with the engine "modeling" temperature (runtime) over 190°F. **Possible Causes:** • ECT sensor is out of calibration or it is "skewed" • Check the operation of the thermostat (it may be stuck open) • Inspect for low coolant level or for an incorrect coolant mixture
DTC: P0130 **2T CCM, MIL: Yes** **Years:** 2007, 2008 **Models:** Amanti, Optima, Rio, Rondo, Sorento, spectra, Sportage **Engines:** 1.6L, 2.0L, 2.4L, 2.7L, 3.5L **Transmissions:** All	**HO2S-11 (Bank 1 Sensor 1) Circuit Malfunction** Engine started, engine running in closed loop at a speed over 3 MPH for 2-3 minutes, and the PCM detected the HO2S signal was more than 1.4 volts, or it was less than 0.02 volt during the CCM test. **Possible Causes:** • HO2S signal circuit open or shorted to ground • HO2S signal circuit shorted to VREF or system power (B+) • HO2S is damaged or has failed • PCM has failed
DTC: P0131 **2T CCM, MIL: Yes** **Years:** 2007, 2008 **Models:** All **Engines:** All **Transmissions:** All	**HO2S-11 (Bank 1 Sensor 1) Circuit Low Input** Engine started, engine running in closed loop at a speed over 3 MPH for 2-3 minutes, and the PCM detected the HO2S signal remained at less than 300 mv during the CCM test. **Possible Causes:** • HO2S signal circuit open or shorted to ground • HO2S signal ground circuit is open • HO2S is contaminated or has failed • PCM has failed
DTC: P0132 **2T CCM, MIL: Yes** **Years:** 2007, 2008 **Models:** All **Engines:** All **Transmissions:** All	**HO2S-11 (Bank 1 Sensor 1) Circuit High Input** Engine started, engine running in closed loop at a speed over 3 MPH for 2-3 minutes, and the PCM detected the front HO2S signal remained fixed at more than 600 mv during the CCM test. **Possible Causes:** • HO2S signal tracking (wet/oily) in connector causing a short • HO2S signal circuit shorted to VREF or system power (B+) • HO2S signal circuit is open, or the ground circuit is open • HO2S heater supply voltage is open or the HO2S has failed • PCM has failed
DTC: P0133 **2T O2S2, MIL: Yes** **Years:** 2007, 2008 **Models:** All **Engines:** All **Transmissions:** All	**HO2S-11 (Bank 1 Sensor 1) Slow Response** Engine started, engine running in closed loop at a speed over 3 MPH for 2-3 minutes, and the PCM detected the average ratio between the HO2S Actual and maximum allowed frequency during 100 Lambda cycles was more than the Threshold value (e.g., 0.66 Hz). **Possible Causes:** • Exhaust leak present in the exhaust manifold or exhaust pipes • Front HO2S failed, or front & rear HO2S connections reversed • HO2S has deteriorated, is contaminated or has failed
DTC: P0134 **2T O2S2, MIL: Yes** **Years:** 2007, 2008 **Models:** All **Engines:** All **Transmissions:** All	**HO2S-11 (Bank 1 Sensor 1) No Activity Detected** Engine started, engine running in closed loop at a speed over 3 MPH for more than 2 minutes, and the PCM detected the HO2S signal stroke rationality was more than the threshold value (0.250 volt) during the Heated Oxygen Sensor Monitor test. **Possible Causes:** • HO2S signal circuit is open or shorted to ground • HO2S has deteriorated, is contaminated or has failed • PCM has failed
DTC: P0135 **2T O2S2 HTR2, MIL: Yes** **Years:** 2007, 2008 **Models:** Rio, Sedona **Engines:** 1.6L, 3.8L **Transmissions:** All	**HO2S-11 (Bank 1 Sensor 1) Heater Circuit Malfunction** Engine started, engine running and PCM detected the HO2S heater current was more than 2 amps, or it was less than 0.25 amps. **Possible Causes:** • HO2S power feed circuit from the Main Relay is open • HO2S heater control circuit is open • HO2S heater element has high resistance, is shorted or open • HO2S heater is damaged or has failed • PCM has failed

DTC	Trouble Code Title, Conditions & Possible Causes
DTC: P0136 **2T CCM, MIL: Yes** **Years:** 2007, 2008 **Models:** Amanti, Optima, Rio, Rondo, Sorento, Spectra, Sportage **Engines:** 1.6L, 2.0L, 2.4L, 2.7L, 3.5L **Transmissions:** All	**HO2S-12 (Bank 1 Sensor 2) Circuit Malfunction** Engine started, engine running in closed loop at a speed over 3 MPH for 2-3 minutes and the PCM detected the HO2S signal was more than 1.4 volts, or that it was less than 0.02 volt during the CCM test. **Possible Causes:** • HO2S signal circuit open or shorted to ground • HO2S signal circuit shorted to VREF or system power (B+) • HO2S is damaged or has failed • PCM has failed
DTC: P0137 **2T CCM, MIL: Yes** **Years:** 2007, 2008 **Models:** All **Engines:** All **Transmissions:** All	**HO2S-12 (Bank 1 Sensor 2) Circuit Low Input** Engine started, engine running in closed loop at a speed over 3 MPH, and the PCM detected the HO2S signal remained at less than 300 mv during the CCM test. **Possible Causes:** • HO2S signal circuit open or shorted to ground • HO2S signal ground circuit is open • HO2S is contaminated or has failed • PCM has failed
DTC: P0138 **2T CCM, MIL: Yes** **Years:** 2007, 2008 **Models:** All **Engines:** All **Transmissions:** All	**HO2S-12 (Bank 1 Sensor 2) Circuit High Input** Engine started, engine running in closed loop at a speed over 3 MPH, and the PCM detected the front HO2S signal remained fixed at more than 600 mv during the CCM test. **Possible Causes:** • HO2S signal tracking (wet/oily) in connector causing a short • HO2S signal circuit shorted to VREF or system power (B+) • HO2S signal circuit is open, or the ground circuit is open • HO2S heater supply voltage is open or the HO2S has failed • PCM has failed
DTC: P0139 **2T O2S2, MIL: Yes** **Years:** 2007, 2008 **Models:** All **Engines:** All **Transmissions:** All	**HO2S-12 (Bank 1 Sensor 2) Slow Response** Engine started, engine running in closed loop at a speed over 3 MPH, and the PCM detected the average ratio between the HO2S Actual and maximum allowed frequency during 100 Lambda cycles was more than the Threshold value (e.g., 0.66 Hz). **Possible Causes:** • Exhaust leak present in the exhaust manifold or exhaust pipes • Front HO2S failed, or front & rear HO2S connections reversed • HO2S has deteriorated, is contaminated or has failed
DTC: P0140 **2T O2S2, MIL: Yes** **Years:** 2007, 2008 **Models:** All **Engines:** All **Transmissions:** All	**HO2S-12 (Bank 1 Sensor 2) No Activity Detected** Engine started, engine running in closed loop at cruise speed at over 3 MPH for more than 2 minutes, and the PCM detected the HO2S signal stroke rationality was more than the threshold value (0.250 volt) during the Oxygen Sensor Monitor test. **Possible Causes:** • HO2S signal circuit is open or shorted to ground • HO2S has deteriorated, is contaminated or has failed • PCM has failed
DTC: P0141 **2T O2S2 HTR2, MIL: Yes** **Years:** 2007, 2008 **Models:** Rio, Sedona **Engines:** 1.6L, 3.8L **Transmissions:** All	**HO2S-12 (Bank 1 Sensor 2) Heater Circuit Malfunction** Engine started, engine running and PCM detected the HO2S heater current was more than 2 amps, or it was less than 0.25 amps. **Possible Causes:** • HO2S power feed circuit from the Main Relay is open • HO2S heater control circuit is open • HO2S heater element has high resistance, is shorted or open • HO2S heater is damaged or has failed • PCM has failed
DTC: P0150 **2T O2S2, MIL: Yes** **Years:** 2007, 2008 **Models:** Amanti, Sorento, Sportage **Engines:** 2.7L, 3.5L **Transmissions:** All	**HO2S-21 (Bank 2 Sensor 1) Slow Response** Engine started, engine running in closed loop at a speed over 3 MPH, and the PCM detected the average ratio between the HO2S Actual and maximum allowed frequency during 100 Lambda cycles was more than the Threshold value (e.g., 0.66 Hz) during the test. **Possible Causes:** • Exhaust leak present in the exhaust manifold or exhaust pipes • Front HO2S failed, or front & rear HO2S connections reversed • HO2S has deteriorated, is contaminated or has failed.

DTC	Trouble Code Title, Conditions & Possible Causes
DTC: P0151 **2T O2S2, MIL: Yes** **Years:** 2007, 2008 **Models:** Amanti, Optima, Rondo, Sedona, Sorento, Sportage **Engines:** 2.7L, 3.3L, 3.5L, 3.8L **Transmissions:** All	**HO2S Circuit Low Voltage (Bank 2 Sensor 1)** Engine started, engine running in closed loop at a speed over 3 MPH, and the PCM detected the average ratio between the HO2S Actual and maximum allowed frequency during 100 Lambda cycles was more than the Threshold value (e.g., 0.66 H z) during the test. **Possible Causes:** • Poor connection • Short to ground in harness • Faulty PCM • HO2S has deteriorated, is contaminated or has failed
DTC: P0152 **2T CCM, MIL: Yes** **Years:** 2007, 2008 **Models:** Amanti, Optima, Rondo, Sorento, Sportage **Engines:** 2.7L, 3.3L, 3.5L, 3.8L **Transmissions:** All	**HO2S-21 (Bank 2 Sensor 1) Circuit Malfunction** Engine started, engine running in closed loop at a speed over 3 MPH for more than 2-3 minutes, and the PCM detected an unexpected voltage condition on the HO2S signal circuit. **Possible Causes:** • HO2S signal circuit open or shorted to ground • HO2S signal circuit is shorted to VREF or to system power (B+) • HO2S is damaged or has failed • PCM has failed
DTC: P0153 **2T O2S, MIL: Yes** **Years:** 2007, 2008 **Models:** Amanti, Optima, Rondo, Sedona, Sorento, Sportage **Engines:** 2.7L, 3.3L, 3.5L, 3.8L **Transmissions:** All	**HO2S-21 (Bank 2 Sensor 1) Slow Response** DTC P0155 not set, engine started, engine running at idle speed in closed loop, and PCM detected the HO2S-12 response time to switch from rich-to-lean or from lean-to-rich was over one second. **Possible Causes:** • HO2S signal circuit is open or shorted to ground • HO2S element is contaminated or it has failed • HO2S heater is damaged or has failed • Intake air leaks, exhaust manifold leaks or PCV system leaks • MAF sensor out of calibration (it may be dirty or contaminated)
DTC: P0154 **2T O2S2, MIL: Yes** **Years:** 2007, 2008 **Models:** Amanti, Optima, Rondo, Sedona, Sorento, Sportage **Engines:** 2.7L, 3.8L **Transmissions:** All	**HO2S-21 (Bank 2 Sensor 1) No Activity Detected** Engine started, engine running in closed loop at a speed over 3 MPH for more than 2 minutes, and the PCM detected the HO2S signal stroke rationality was more than the threshold value (0.250 volt) during the Oxygen Sensor Monitor test. **Possible Causes:** • HO2S signal circuit is open or shorted to ground • HO2S has deteriorated, is contaminated or has failed • PCM has failed
DTC: P0156 **2T CCM, MIL: Yes** **Years:** 2007, 2008 **Models:** Amanti, Sorento, Sportage **Engines:** 2.7L, 3.5L **Transmissions:** All	**HO2S-22 (Bank 2 Sensor 2) Circuit Malfunction** Engine started, engine running in closed loop at a speed over 3 MPH for more than 2 minutes, and the PCM detected an unexpected low voltage condition on the HO2S-22 circuit during the CCM test. **Possible Causes:** • HO2S signal circuit open between the HO2S and the PCM • HO2S signal circuit shorted to ground between HO2S and PCM • HO2S is damaged or has failed • PCM has failed
DTC: P0157 **2T CCM, MIL: Yes** **Years:** 2007, 2008 **Models:** Amanti, Optima, Rondo, Sedona, Sorento, Sportage **Engines:** 2.7L, 3.3L, 3.5L, 3.8L **Transmissions:** All	**HO2S Circuit Low Voltage (Bank 2 Sensor 2)** The signal voltage of the front or rear sensor changes the rear circuit voltage specification when air fuel ratio is rich, a DTC is set. Out of range low failure (ground short open circuit). **Possible Causes:** • Poor connection • Short to ground in HO2S circuit • Faulty HO2S • Faulty PCM
DTC: P0158 **2T CCM, MIL: Yes** **Years:** 2007, 2008 **Models:** Amanti, Optima, Rondo, Sedona, Sorento, Sportage **Engines:** 2.7L, 3.3L, 3.5L, 3.8L **Transmissions:** All	**HO2S Circuit High Voltage (Bank 2 Sensor 2)** The signal voltage is higher than 1.2 volts after open in circuit. Out of range high failure. **Possible Causes:** • Poor connection • Short to battery in HO2S circuit • Faulty HO2S • Faulty PCM

DTC	Trouble Code Title, Conditions & Possible Causes
DTC: P0159 **2T O2S2, MIL: Yes** **Years:** 2007, 2008 **Models:** Amanti, Optima, Rondo, Sedona, Sorento, Sportage **Engines:** 2.7L, 3.3L, 3.5L, 3.8L **Transmissions:** All	**HO2S-22 (Bank 2 Sensor 2) Slow Response** Engine started, engine running in closed loop at a speed over 3 MPH for more than 2 minutes, and the PCM detected the average ratio between the HO2S Actual and maximum allowed frequency during 100 Lambda cycles was more than the Threshold value (e.g., 0.66 Hz) during the Oxygen Sensor Monitor test. **Possible Causes:** • Exhaust leak present in the exhaust manifold or exhaust pipes • Front HO2S failed, or front & rear HO2S connections reversed • HO2S has deteriorated, is contaminated or has failed
DTC: P0160 **2T CCM, MIL: Yes** **Years:** 2007, 2008 **Models:** Amanti, Optima, Rondo, Sedona, Sorento, Sportage **Engines:** 2.7L, 3.3L, 3.5L, 3.8L **Transmissions:** All	**HO2S-22 (Bank 2 Sensor 2) Circuit Malfunction** Engine started, engine running in closed loop at a speed over 3 MPH, and the PCM detected an unexpected high voltage condition on the HO2S circuit during the CCM test. **Possible Causes:** • HO2S signal circuit is shorted to VREF or to the Heater power • HO2S signal circuit is shorted to system power (B+) • HO2S is damaged or has failed • PCM has failed
DTC: P0161 **2T O2S2 HTR2, MIL: Yes** **Years:** 2007, 2008 **Models:** Sedona **Engines:** 3.8L **Transmissions:** All	**HO2S-22 (Bank 2 Sensor 2) Heater Circuit Malfunction** Engine started, engine running and PCM detected the HO2S heater current was more than 2 amps, or it was less than 0.25 amps. **Possible Causes:** • HO2S power feed circuit from the Main Relay is open • HO2S heater control circuit is open • HO2S heater element has high resistance, is shorted or open • HO2S heater is damaged or has failed • PCM has failed
DTC: P0170 **2T FUEL, MIL: Yes** **Years:** 2007, 2008 **Models:** Optima, Rondo, Spectra, Sportage **Engines:** 2.0L, 2.4L, 2.7L **Transmissions:** All	**Fuel System Too Rich or Too Lean (Bank 1)** DTC P0171 and P0172 not set, engine running in closed loop for over 2 minutes, and the PCM detected the amount of rich or lean Fuel Trim correction exceeded the Threshold maximum. **Possible Causes:** • Air leaks present in the exhaust manifold or exhaust pipes • Air being drawn in from leaks in engine gaskets or other seals • Incorrect fuel pressure, or one or more fuel injectors has failed • Front HO2S element is contaminated or has failed • A "fuel control" sensor is out of calibration (BARO, ECT or IAT)
DTC: P0171 **2T FUEL, MIL: Yes** **Years:** 2007, 2008 **Models:** All **Engines:** 1.6L, 2.0L, 2.7L, 3.3L, 3.5L, 3.8L **Transmissions:** All	**Fuel System Too Lean (Bank 1)** Engine running in closed loop at a speed of over 5 MPH for 2-3 minutes, and the PCM detected the Lambda correction value exceeded the "high" Threshold limit, condition met for 200 seconds. **Possible Causes:** • Air leaks in intake manifold, exhaust pipes or exhaust manifold • One or more injectors restricted or pressure regulator has failed • Air is being drawn in from leaks in gaskets or other seals • O2S element is deteriorated or has failed • A "fuel control" sensor is out of calibration (ECT, IAT or MAP)
DTC: P0172 **2T FUEL, MIL: Yes** **Years:** 2007, 2008 **Models:** All **Engines:** 1.6L, 2.0L, 2.4L, 2.7L, 3.3L, 3.5L, 3.8L **Transmissions:** All	**Fuel System Too Rich (Bank 1)** Engine running in closed loop at a speed of over 5 MPH for 2-3 minutes, and the PCM detected the Lambda correction value exceeded the "low" Threshold limit, condition met for 240 seconds. **Possible Causes:** • One or more injectors leaking or pressure regulator is leaking • O2S element is deteriorated or has failed • EVAP vapor recovery system has failed (canister full of fuel) • Base engine fault (i.e., cam timing incorrect, engine oil too high • A "fuel control" sensor is out of calibration (ECT, IAT or MAP)

DTC	Trouble Code Title, Conditions & Possible Causes
DTC: P0173 **2T Fuel, MIL: Yes** **Years:** 2007, 2008 **Models:** Sportage **Engines:** 2.7L **Transmissions:** All	**Fuel Trim Too Rich or Too Lean (Bank 2)** Engine running in closed loop, and the PCM detected the Fuel system was too rich or too lean during two or more consecutive trips. **Possible Causes:** • Base engine "mechanical" fault affecting one or more cylinders • EVAP system component has failed or canister fuel saturated • Exhaust leaks located in front of the HO2S location • Fuel control sensor is out of calibration (i.e., ECT, IAT or MAF) • Fuel delivery system supplying too much fuel during cruise or idle periods (e.g., faulty fuel pump, or faulty pressure regulator) • Fuel injector(s) is leaking or stuck partially open (one or more) • HO2S is contaminated, deteriorated or it has failed
DTC: P0174 **2T FUEL, MIL: Yes** **Years:** 2007, 2008 **Models:** Amanti, Optima, Rondo, Sedona, Sorento, Sportage **Engines:** 2.7L, 3.5L, 3.8L **Transmissions:** All	**Fuel System Too Lean (Bank 2)** Engine running in closed loop at a speed of over 5 MPH for 2-3 minutes, and the PCM detected the Lambda correction value exceeded the "high" Threshold limit, condition met for 200 seconds. **Possible Causes:** • Air leaks in intake manifold, exhaust pipes or exhaust manifold • One or more injectors restricted or pressure regulator has failed • Air is being drawn in from leaks in gaskets or other seals • O2S element is deteriorated or has failed • A "fuel control" sensor is out of calibration (ECT, IAT or MAP)
DTC: P0175 **2T FUEL, MIL: Yes** **Years:** 2007, 2008 **Models:** Amanti, Optima, Rondo, Sedona, Sorento, Sportage **Engines:** 2.7L, 3.3L, 3.5L, 3.8L **Transmissions:** All	**Fuel System Too Rich (Bank 1)** Engine running in closed loop at a speed of over 5 MPH for 2-3 minutes, and the PCM detected the Lambda correction value exceeded the "low" Threshold limit, condition met for 240 seconds. **Possible Causes:** • One or more injectors leaking or pressure regulator is leaking • O2S element is deteriorated or has failed • EVAP vapor recovery system has failed (canister full of fuel) • Base engine fault (i.e., cam timing incorrect, engine oil too high • A "fuel control" sensor is out of calibration (ECT, IAT or MAP)
DTC: P0196 **2T CCM, MIL: Yes** **Years:** 2007, 2008 **Models:** Amanti, Optima, Rondo, Sedona, Sorento, Spectra, Sportage **Engines:** 2.0L, 2.4L, 2.7L, 3.3L, 3.8L **Transmissions:** All	**Engine Oil temperature Sensor Range/Performance** Stuck oil temperature sensor signal or unusual low or high signal. Condition 1 (signal high or low), engine coolant temperature more than 158°F and oil temperature less than 68° F. Condition 2 (signal high or low), engine coolant temperature less than 158°F and oil temperature above 212°F. Condition 3 (stuck signal) engine coolant temperature less than 104°F. **Possible Causes:** • Contact resistance in connectors • faulty OTS
DTC: P0197 **2T CCM, MIL: Yes** **Years:** 2007, 2008 **Models:** Amanti, Optima, Rondo, Sedona, Spectra, Sportage **Engines:** 2.0L, 2.4L, 2.7L, 3.3L, 3.8L **Transmissions:** All	**Engine Oil temperature Sensor Low Input** Signal voltage lower than the possible range of a properly operating OTS. Voltage range check. Engine coolant temperature less than 212°F. Oil temperature above 309°F. **Possible Causes:** • Short circuit to ground • Contact resistance in connectors • faulty OTS
DTC: P0198 **2T CCM, MIL: Yes** **Years:** 2007, 2008 **Models:** Amanti, Optima, Rondo, Sedona, Sorento, Spectra, Sportage **Engines:** 2.0L, 2.4L, 2.7L, 3.3L, 3.8L **Transmissions:** All	**Engine Oil temperature Sensor High Input** Signal voltage higher than the possible range of a properly operating OTS. Voltage range check. Five minutes after engine start if engine coolant temperature less than 14°. Oil temperature minus 33°F. **Possible Causes:** • Open circuit to battery • Contact resistance in connectors • faulty OTS

DTC	Trouble Code Title, Conditions & Possible Causes
DTC: P0201 **2T CCM, MIL: Yes** **Years:** 2007, 2008 **Models:** Amanti, Rio, Sorento **Engines:** 1.6L, 3.5L **Transmissions:** All	**Cylinder 1 Injector Circuit Malfunction** Engine running and the PCM detected the identified fuel injector control circuit signal was more than the upper limit, or that it was less than the lower limit, or that no control signal was present. **Possible Causes:** • Main relay power supply circuit to the injector is open • Fuel injector 1 control circuit is open or shorted to ground • Fuel injector 1 is damaged or has failed • Injector "driver" circuit in the PCM is damaged or has failed
DTC: P0202 **2T CCM, MIL: Yes** **Years:** 2007, 2008 **Models:** Amanti, Rio, Sorento **Engines:** 1.6L, 3.5L **Transmissions:** All	**Cylinder 2 Injector Circuit Malfunction** Engine running and the PCM detected the identified fuel injector control circuit signal was more than the upper limit, or that it was less than the lower limit, or that no control signal was present. **Possible Causes:** • Main relay power supply circuit to the injector is open • Fuel injector 2 control circuit is open or shorted to ground • Fuel injector 2 is damaged or has failed • Injector "driver" circuit in the PCM is damaged or has failed
DTC: P0203 **2T CCM, MIL: Yes** **Years:** 2007, 2008 **Models:** Amanti, Rio, Sorento **Engines:** 1.6L, 3.5L **Transmissions:** All	**Cylinder 3 Injector Circuit Malfunction** Engine running and the PCM detected the identified fuel injector control circuit signal was more than the upper limit, or that it was less than the lower limit, or that no control signal was present. **Possible Causes:** • Main relay power supply circuit to the injector is open • Fuel injector 3 control circuit is open or shorted to ground • Fuel injector 3 is damaged or has failed • Injector "driver" circuit in the PCM is damaged or has failed
DTC: P0204 **2T CCM, MIL: Yes** **Years:** 2007, 2008 **Models:** Amanti, Rio, Sorento **Engines:** 1.6L, 3.5L **Transmissions:** All	**Cylinder 4 Injector Circuit Malfunction** Engine running and the PCM detected the identified fuel injector control circuit signal was more than the upper limit, or that it was less than the lower limit, or that no control signal was present. **Possible Causes:** • Main relay power supply circuit to the injector is open • Fuel injector 4 control circuit is open or shorted to ground • Fuel injector 4 is damaged or has failed • Injector "driver" circuit in the PCM is damaged or has failed
DTC: P0217 **2T CCM, MIL: Yes** **Years:** 2007, 2008 **Models:** Amanti, Optima, Rondo, Sedona, Sorento **Engines:** 2.7L, 3.3L, 3.8L **Transmissions:** All	**Engine Coolant Over Temperature Condition** Engine running and no disabling faults present. Coolant sensor within range. **Possible Causes:** • Poor connection • Lack of engine coolant • Faulty water pump • ECTS • Faulty PCM
DTC: P0222 **2T CCM, MIL: Yes** **Years:** 2007, 2008 **Models:** Amanti, Optima, Rondo, Sedona, Sorento **Engines:** 2.4L, 2.7L, 3.3L, 3.5L, 3.8L **Transmissions:** All	**Throttle/Pedal Position Sensor/Switch "B" Circuit Low Input** The DTC is recorded if the output voltage of the TPS 1 is lower than threshold value (Vtps1 less than or equal to 0.2 volt). TPS 1 low input. **Possible Causes:** • Poor connection • Open or short to ground in TPS circuit • Faulty TPS • Faulty PCM
DTC: P0223 **2T CCM, MIL: Yes** **Years:** 2007, 2008 **Models:** Amanti, Optima, Rondo, Sedona, Sorento **Engines:** 2.4L, 2.7L, 3.3L, 3.5L, 3.8L **Transmissions:** All	**Throttle/Pedal Position Sensor/Switch "B" Circuit High Input** The DTC is recorded if the output voltage of the TPS 1 higher than threshold value (Vtps1 greater than or equal to 4.85 volts, load value, EV less than 70 percent) when TPS 2 (VTPS2 less than or equal to 2.5 volts) is normal. TPS 1 high input. **Possible Causes:** • Poor connection • Open or short to ground in TPS circuit • Faulty TPS • Faulty PCM

DTC	Trouble Code Title, Conditions & Possible Causes
DTC: P0230 **1T CCM, MIL: Yes** **Years:** 2007, 2008 **Models:** All **Engines:** 1.6L, 2.0L, 2.4L, 2.7L, 3.3L, 3.8L **Transmissions:** All	**Fuel Pump Circuit Malfunction** Key on, and then the PCM detected an unexpected voltage condition on the fuel pump circuit through the fuel pump monitoring input. **Possible Causes:** • Fuel pump control circuit is open or shorted to ground • Fuel pump relay power circuit from ignition switch is open • Fuel pump relay is damaged or has failed • PCM has failed
DTC: P0261 **2T CCM, MIL: Yes** **Years:** 2007, 2008 **Models:** All **Engines:** 1.6L, 2.0L, 2.4L, 2.7L, 3.3L, 3.8L **Transmissions:** All	**Fuel Injector 1 Circuit Low Input** Engine running and the PCM detected the Injector 1 signal was in a low signal state (0 volt) with the injector commanded off in the test. **Possible Causes:** • Main relay power supply circuit to the injector is open • Fuel injector 1 control circuit is shorted to ground • Fuel injector 1 is damaged or has failed • Injector "driver" circuit in the PCM is damaged or has failed
DTC: P0262 **2T CCM, MIL: Yes** **Years:** 2007, 2008 **Models:** All **Engines:** 1.6L, 2.0L, 2.4L, 2.7L, 3.3L, 3.8L **Transmissions:** All	**Fuel Injector 1 Circuit High Input** Engine running and the PCM detected the Injector 1 signal was in a high signal state (12 volts) with the injector commanded off in the test. **Possible Causes:** • Fuel injector 1 control circuit is shorted to system power (B+) • Fuel injector 1 is damaged or has failed • Injector 1 "driver" circuit in the PCM is damaged or has failed
DTC: P0264 **2T CCM, MIL: Yes** **Years:** 2007, 2008 **Models:** All **Engines:** 1.6L, 2.0L, 2.4L, 2.7L, 3.3L, 3.8L **Transmissions:** All	**Fuel Injector 2 Circuit Low Input** Engine running and the PCM detected the Injector 2 signal was in a low signal state (0 volt) with the injector commanded off in the test. **Possible Causes:** • Main relay power supply circuit to the injector is open • Fuel injector 2 control circuit is shorted to ground • Fuel injector 2 is damaged or has failed • Injector 2 "driver" circuit in the PCM is damaged or has failed
DTC: P0265 **2T CCM, MIL: Yes** **Years:** 2007, 2008 **Models:** All **Engines:** 1.6L, 2.0L, 2.4L, 2.7L, 3.3L, 3.8L **Transmissions:** All	**Fuel Injector 2 Circuit High Input** Engine running and the PCM detected the Injector 2 signal was in a high signal state (12 volts) with the injector commanded off in the test. **Possible Causes:** • Fuel injector 2 control circuit is shorted to system power (B+) • Fuel injector 2 is damaged or has failed • Injector 2 "driver" circuit in the PCM is damaged or has failed
DTC: P0267 **2T CCM, MIL: Yes** **Years:** 2007, 2008 **Models:** All **Engines:** 1.6L, 2.0L, 2.4L, 2.7L, 3.3L, 3.8L **Transmissions:** All	**Fuel Injector 3 Circuit Low Input** Engine running and the PCM detected the Injector 3 signal was in a low signal state (0 volt) with the injector commanded off in the test. **Possible Causes:** • Main relay power supply circuit to the injector is open • Fuel injector 3 control circuit is shorted to ground • Fuel injector 3 is damaged or has failed • Injector 3 "driver" circuit in the PCM is damaged or has failed
DTC: P0268 **2T CCM, MIL: Yes** **Years:** 2007, 2008 **Models:** All **Engines:** 1.6L, 2.0L, 2.4L, 2.7L, 3.3L, 3.8L **Transmissions:** All	**Fuel Injector 3 Circuit High Input** Engine running and the PCM detected the Injector 3 signal was in a high signal state (12 volts) with the injector commanded off in the test. **Possible Causes:** • Fuel injector 3 control circuit is shorted to system power (B+) • Fuel injector 3 is damaged or has failed • Injector 3 "driver" circuit in the PCM is damaged or has failed
DTC: P0270 **2T CCM, MIL: Yes** **Years:** 2007, 2008 **Models:** All **Engines:** 1.6L, 2.0L, 2.4L, 2.7L, 3.3L, 3.8L **Transmissions:** All	**Fuel Injector 4 Circuit Low Input** Engine running and the PCM detected the Injector 4 signal was in a low signal state (0 volt) with the injector commanded off in the test. **Possible Causes:** • Fuel injector 4 control circuit is shorted to system power (B+) • Fuel injector 4 is damaged or has failed • Injector 4 "driver" circuit in the PCM is damaged or has failed

DTC	Trouble Code Title, Conditions & Possible Causes
DTC: P0271 **2T CCM, MIL:** Yes **Years:** 2007, 2008 **Models:** All **Engines:** 1.6L, 2.0L, 2.4L, 2.7L, 3.3L, 3.8L **Transmissions:** All	**Fuel Injector 4 Circuit High Input** Engine running and the PCM detected the Injector 4 signal was in a high signal state (12 volts) with the injector commanded off in the test. **Possible Causes:** • Fuel injector 4 control circuit is shorted to system power (B+) • Fuel injector 4 is damaged or has failed • Injector 4 "driver" circuit in the PCM is damaged or has failed
DTC: P0273 **2T CCM, MIL:** Yes **Years:** 2007, 2008 **Models:** Amanti, Optima, Rondo, Sedona, Sorento, Sportage **Engines:** 2.7L, 3.3L, 3.8L **Transmissions:** All	**Cylinder 5- Injector Circuit Low** The PCM sets the DTC if the control circuit is shorted to ground. Driver stage check. **Possible Causes:** • Open in power supply harness • Short to ground in control harness • Contact resistance in connectors • Faulty injector
DTC: P0274 **2T CCM, MIL:** Yes **Years:** 2007, 2008 **Models:** Amanti, Optima, Rondo, Sedona, Sorento, Sportage **Engines:** 2.7L, 3.3L, 3.8L **Transmissions:** All	**Cylinder 5- Injector Circuit High** The PCM sets the DTC if the control circuit is open or shorted to battery voltage. Driver stage check. **Possible Causes:** • Open or short to battery control harness • Contact resistance in connectors • Faulty injector
DTC: P0276 **2T CCM, MIL:** Yes **Years:** 2007, 2008 **Models:** Amanti, Optima, Rondo, Sedona, Sorento, Sportage **Engines:** 2.7L, 3.3L, 3.8L **Transmissions:** All	**Cylinder 6- Injector Circuit Low** The PCM sets the DTC if the control circuit is shorted to ground. Driver stage check. **Possible Causes:** • Open in power supply harness • Short to ground in control harness • Contact resistance in connectors • Faulty injector
DTC: P0277 **2T CCM, MIL:** Yes **Years:** 2007, 2008 **Models:** Amanti, Optima, Rondo, Sedona, Sorento, Sportage **Engines:** 2.7L, 33.L, 3.8L **Transmissions:** All	**Cylinder 6- Injector Circuit High** The PCM sets the DTC if the control circuit is open or shorted to battery voltage. Driver stage check. **Possible Causes:** • Open or short to battery control harness • Contact resistance in connectors • Faulty injector
DTC: P0300 **2T MISFIRE, MIL:** Yes **Years:** 2007, 2008 **Models:** All **Engines:** All **Transmissions:** All	**Random Cylinder Misfire Detected** Engine runtime 3 seconds, engine speed change under 1200 RPM, and the PCM detected a random misfire condition in more than one cylinder during the 200 revolution or the 1000 revolution test range. **Note: If the misfire is severe, the MIL will flash on/off on the 1st trip!** **Possible Causes:** • Vehicle driven under low fuel condition (less than 1/8 of a tank) • CKP or CMP sensor signal erratic or out of phase • Base engine problem affecting two or more engine cylinders • Ignition system problem affecting two or more engine cylinders
DTC: P0301 **2T MISFIRE, MIL:** Yes **Years:** 2007, 2008 **Models:** All **Engines:** All **Transmissions:** All	**Cylinder 1 Misfire Detected** Engine runtime 3 seconds, engine speed change under 1200 RPM, and the PCM detected a misfire condition present in one cylinder during the 200 (Catalyst) or 1000 revolution (Emission) test range. **Note: If the misfire is severe, the MIL will flash on/off on the 1st trip!** **Possible Causes:** • Fuel metering (fuel injector dirty) problem affecting Cylinder 1 • Base engine (compression) problem affecting Cylinder 1 • Ignition system (spark plug or plug wire) problem on Cylinder 1

DTC	Trouble Code Title, Conditions & Possible Causes
DTC: P0302 **2T MISFIRE, MIL: Yes** **Years:** 2007, 2008 **Models:** All **Engines:** All **Transmissions:** All	**Cylinder 2 Misfire Detected** Engine runtime 3 seconds, engine speed change under 1200 RPM, and the PCM detected a misfire condition present in one cylinder during the 200 (Catalyst) or 1000 revolution (Emission) test range. **Note: If the misfire is severe, the MIL will flash on/off on the 1st trip!** **Possible Causes:** • Fuel metering (fuel injector dirty) problem affecting Cylinder 2 • Base engine (compression) problem affecting Cylinder 2 • Ignition system (spark plug or plug wire) problem on Cylinder 2
DTC: P0303 **2T MISFIRE, MIL: Yes** **Years:** 2007, 2008 **Models:** All **Engines:** All **Transmissions:** All	**Cylinder 3 Misfire Detected** Engine runtime 3 seconds, engine speed change under 1200 RPM, and the PCM detected a misfire condition present in one cylinder during the 200 (Catalyst) or 1000 revolution (Emission) test range. **Note: If the misfire is severe, the MIL will flash on/off on the 1st trip!** **Possible Causes:** • Fuel metering (fuel injector dirty) problem affecting Cylinder 3 • Base engine (compression) problem affecting Cylinder 3 • Ignition system (spark plug or plug wire) problem on Cylinder 3
DTC: P0304 **2T MISFIRE, MIL: Yes** **Years:** 2007, 2008 **Models:** All **Engines:** All **Transmissions:** All	**Cylinder 4 Misfire Detected** Engine runtime 3 seconds, engine speed change under 1200 RPM, and the PCM detected a misfire condition present in one cylinder during the 200 (Catalyst) or 1000 revolution (Emission) test range. **Note: If the misfire is severe, the MIL will flash on/off on the 1st trip!** **Possible Causes:** • Fuel metering (fuel injector dirty) problem affecting Cylinder 4 • Base engine (compression) problem affecting Cylinder 4 • Ignition system (spark plug or plug wire) problem on Cylinder 4
DTC: P0305 **2T MISFIRE, MIL: Yes** **Years:** 2007, 2008 **Models:** Amanti, Optima, Rondo, Sedona, Sorento, Sportage **Engines:** 2.7L, 3.3L, 3.5L, 3.8L **Transmissions:** All	**Cylinder 5 Misfire Detected** Engine runtime 3 seconds, engine speed change under 1200 RPM, and the PCM detected a misfire condition present in one cylinder during the 200 (Catalyst) or 1000 revolution (Emission) test range. **Note: If the misfire is severe, the MIL will flash on/off on the 1st trip!** **Possible Causes:** • Fuel metering (fuel injector dirty) problem affecting Cylinder 5 • Base engine (compression) problem affecting Cylinder 5 • Ignition system (spark plug or plug wire) problem on Cylinder 5
DTC: P0306 **2T MISFIRE, MIL: Yes** **Years:** 2007, 2008 **Models:** Amanti, Optima, Rondo, Sedona, Sorento, Sportage **Engines:** 2.7L, 3.3L, 3.5L, 3.8L **Transmissions:** All	**Cylinder 6 Misfire Detected** Engine runtime 3 seconds, engine speed change under 1200 RPM, and the PCM detected a misfire condition present in one cylinder during the 200 (Catalyst) or 1000 revolution (Emission) test range. **Note: If the misfire is severe, the MIL will flash on/off on the 1st trip!** **Possible Causes:** • Fuel metering (fuel injector dirty) problem affecting Cylinder 6 • Base engine (compression) problem affecting Cylinder 6 • Ignition system (spark plug or plug wire) problem on Cylinder 6
DTC: P0315 **2T CCM, MIL: Yes** **Years:** 2007, 2008 **Models:** Amanti, Optima, Rondo, Sedona, Sorento, Spectra, Sportage **Engines:** 2.0L, 2.4L, 2.7L, 3.3L, 3.8L **Transmissions:** All	**Segment Time Acquisition Incorrect** A misfire induces a decrease in the engine speed and causes a variation in the segment period. Monitor segment time adaptation. **Possible Causes:** • Improperly installed target wheel • Contact resistance in connectors • Faulty PCM • Faulty CKPS
DTC: P0325 **2T CCM, MIL: Yes** **Years:** 2007, 2008 **Models:** Amanti, Optima, Rondo, Sedona, Sorento, Spectra, Sportage **Engines:** 2.0L, 2.4L, 2.7L, 3.3L, 3.5L, 3.8L **Transmissions:** All	**Knock Sensor Circuit Malfunction (Bank 1)** Engine running at over 1000 RPM for 5 seconds, and the PCM did not detect enough variation in the KS signals (e.g., 0.049 volt). **Possible Causes:** • Knock sensor signal circuit open or shorted to ground • Knock sensor signal circuit shorted to VREF or system power • Knock sensor is damaged or has failed • PCM has failed

DTC	Trouble Code Title, Conditions & Possible Causes
DTC: P0326 **2T CCM, MIL: Yes** **Years:** 2007, 2008 **Models:** Amanti, Optima, Rio, Rondo, Sedona, Sorento **Engines:** 1.6L, 2.7L, 3.3L, 3.8L **Transmissions:** All	**Knock Sensor Circuit Malfunction (Bank 1)** Engine speed from 1000-2200 RPM, ECT sensor signal more than 104°F, engine load more than 2 ms, and the PCM detected the Knock sensor signal was out of range at a calculated engine speed. **Possible Causes:** • Knock sensor signal circuit open or shorted to ground • Knock sensor signal circuit shorted to VREF or system power • Knock sensor is damaged or has failed • PCM has failed
DTC: P0330 **2T CCM, MIL: Yes** **Years:** 2007, 2008 **Models:** Amanti, Optima, Rondo, Sedona, Sorento, Sportage **Engines:** 2.7L, 3.3L, 3.8L **Transmissions:** All	**Knock Sensor Circuit Malfunction (Bank 2)** Engine running at over 1000 RPM for 5 seconds, and the PCM did not detect enough variation in the KS signals (e.g., 0.049 volt). **Possible Causes:** • Knock sensor signal circuit open or shorted to ground • Knock sensor signal circuit shorted to VREF or system power • Knock sensor is damaged or has failed • PCM has failed
DTC: P0331 **2T CCM, MIL: Yes** **Years:** 2007, 2008 **Models:** Optima, Rondo, Sedona, Sorento **Engines:** 2.7L, 3.3L, 3.8L **Transmissions:** All	**Knock Sensor 2 Circuit Range/Performance (Bank 2)** Signal short. Pressure in intake manifold is normal. Engine speed is equal to or less than 1600 RPM. **Possible Causes:** • Poor connection • Short in harness • Faulty knock sensor • Faulty PCM
DTC: P0335 **1T CCM, MIL: Yes** **Years:** 2007, 2008 **Models:** All **Engines:** 1.6L, 2.0L, 2.4L, 2.7L, 3.3L, 3.8L **Transmissions:** All	**Crankshaft Position Sensor Circuit Malfunction** Engine cranking for 5 seconds, and the PCM did not detect any CKP sensor signals, or the vehicle was driven to a speed over 15.5 MPH, and the PCM did not detect any CKP sensor signals for 5 seconds. **Possible Causes:** • CKP sensor signal circuit open or shorted to ground • CKP sensor signal circuit shorted to VREF or system power • CKP sensor is damaged or has failed • PCM has failed
DTC: P0336 **1T CCM, MIL: Yes** **Years:** 2007, 2008 **Models:** Amanti, Optima, Rio, Rondo, Sedona, Sorento **Engines:** 1.6L, 2.4L, 2.7L, 3.3L, 3.8L **Transmissions:** All	**Crankshaft Position Sensor Performance** Engine running and the PCM detected the number of CKP sensor signals counted (between the reference mark gap) did not equal the Actual number of available teeth (i.e., the CKP signals were out or the normal window" of operation with the CMP sensor signals okay). **Possible Causes:** • CKP sensor signal circuit connections loose (intermittent fault) • CKP sensor wiring harness has a connection fault (intermittent) • CKP to Target Wheel "air gap" is incorrect • CKP sensor is damaged or has failed
DTC: P0340 **2T CCM, MIL: Yes** **Years:** 2007, 2008 **Models:** All **Engines:** 1.6L, 2.0L, 2.4L, 2.7L, 3.3L, 3.8L **Transmissions:** All	**Camshaft Position Sensor Circuit Malfunction** Engine speed over 600 RPM, and the PCM detected less than one CMP sensor signal was present, condition met for 1.5 seconds. **Possible Causes:** • CMP sensor signal circuit is open or shorted to ground • CMP sensor signal circuit is shorted to VREF or system power • CMP sensor is damaged or has failed • PCM has failed
DTC: P0341 **2T CCM, MIL: Yes** **Years:** 2007, 2008 **Models:** Amanti, Rio, Optima, Rondo, Sedona, Sorento **Engines:** 1.6L, 2.4L, 2.7L, 3.8L **Transmissions:** All	**Camshaft Position Sensor Circuit Malfunction** No signal or no signal switching is detected. Crankshaft sensor is normal. Battery voltage is between 10 and 16 volts. **Possible Causes:** • Open or short in CMPS circuit • Faulty CMPS • Faulty PCM

DTC	Trouble Code Title, Conditions & Possible Causes
DTC: P0342 **2T CCM, MIL: Yes** **Years:** 2007, 2008 **Models:** Amanti, Rio, Sorento **Engines:** 1.6L, 3.5L **Transmissions:** All	**Camshaft Position Sensor Low Input** Engine speed over 600 RPM, and the PCM detected an unexpected low voltage condition on the CMP sensor signal circuit. **Possible Causes:** • CMP sensor signal circuit is open or shorted to ground • CMP sensor ground circuit is open • CMP sensor is damaged or has failed • PCM has failed
DTC: P0343 **2T CCM, MIL: Yes** **Years:** 2007, 2008 **Models:** Amanti, Rio, Sorento **Engines:** 1.6L, 3.5L **Transmissions:** All	**Camshaft Position Sensor High Input** Engine cranking and the PCM detected the CMP sensor signal was above a threshold value stored in memory during the CCM test. **Possible Causes:** • CMP sensor signal circuit is shorted to VREF or system power • CMP sensor is damaged or has failed • PCM has failed
DTC: P0346 **2T CCM, MIL: Yes** **Years:** 2007, 2008 **Models:** Amanti, Optima, Rondo, Sedona, Sorento **Engines:** 2.7L, 3.8L **Transmissions:** All	**Camshaft Position Sensor "A" Circuit Range/Performance (Bank 2)** Engine running and the PCM detected the CMP sensor signal was above a threshold value stored in memory during the CCM test. **Possible Causes:** • Poor connection • Open or Short in harness • Electrical noise • Target wheel • CMPS • PCM has failed
DTC: P0350 **2T CCM, MIL: Yes** **Years:** 2007, 2008 **Models:** Amanti, Sorento, Sportage **Engines:** 2.7L, 3.5L **Transmissions:** All	**Ignition Coil Primary/Secondary Circuit Malfunction** Engine started, engine running for 4 seconds, and the PCM detected an unexpected voltage condition either the ignition coil primary or on the ignition coil secondary circuit during the CCM test. **Possible Causes:** • Ignition coil primary circuit open or shorted together • Ignition coil secondary components (coil or plug wires) arching • Ignition coil is damaged or has failed • PCM has failed
DTC: P0351 **2T CCM, MIL: Yes** **Years:** 2007, 2008 **Models:** Amanti, Optima, Rondo, Sedona, Sorento, Sportage **Engines:** 2.7L, 3.3L, 3.8L **Transmissions:** All	**Ignition Coil 'A' Circuit Malfunction** Engine started, engine running, and the PCM detected an unexpected voltage condition on the Ignition Coil 'A' primary circuit. **Possible Causes:** • Ignition Coil 'A' primary circuit is open or shorted to ground • Ignition Coil 'A' power circuit is open (test power from I/P fuse) • Ignition Coil 'A' is damaged or has failed • PCM has failed
DTC: P0352 **2T CCM, MIL: Yes** **Years:** 2007, 2008 **Models:** Amanti, Optima, Rondo, Sedona, Sorento, Sportage **Engines:** 2.7L, 3.3L, 3.8L **Engines:** All **Transmissions:** All	**Ignition Coil 'B' Circuit Malfunction** Engine started, engine running, and the PCM detected an unexpected voltage condition on the Ignition Coil 'B' primary circuit. **Possible Causes:** • Ignition Coil 'B' primary circuit is open or shorted to ground • Ignition Coil 'B' power circuit is open (test power from I/P fuse) • Ignition Coil 'B' is damaged or has failed • PCM has failed
DTC: P0353 **2T CCM, MIL: Yes** **Years:** 2007, 2008 **Models:** Amanti, Optima, Rondo, Sedona, Sorento, Sportage **Engines:** 2.7L, 3.3L, 3.8L **Transmissions:** All	**Ignition Coil 'C' Circuit Malfunction** Engine started, engine running, and the PCM detected an unexpected voltage condition on the Ignition Coil 'C' primary circuit. **Possible Causes:** • Ignition Coil 'C' primary circuit is open or shorted to ground • Ignition Coil 'C' power circuit is open (test power from I/P fuse) • Ignition Coil 'C' is damaged or has failed • PCM has failed

DTC	Trouble Code Title, Conditions & Possible Causes
DTC: P0354 **2T CCM, MIL: Yes** **Years:** 2007, 2008 **Models:** Amanti, Optima, Rondo, Sedona, Sorento, Sportage **Engines:** 2.7L, 3.3L, 3.8L **Transmissions:** All	**Ignition Coil 'D' Circuit Malfunction** Engine started, engine running, and the PCM detected an unexpected voltage condition on the Ignition Coil 'D' primary circuit. **Possible Causes:** • Ignition Coil 'D' primary circuit is open or shorted to ground • Ignition Coil 'D' power circuit is open (test power from I/P fuse) • Ignition Coil 'D' is damaged or has failed • PCM has failed
DTC: P0355 **2T CCM, MIL: Yes** **Years:** 2007, 2008 **Models:** Amanti, Rondo, Sedona, Sorento, Sportage **Engines:** 2.7L, 3.3L, 3.8L **Transmissions:** All	**Ignition Coil 'E' Circuit Malfunction** Engine started, engine running, and the PCM detected an unexpected voltage condition on the Ignition Coil 'E' primary circuit. **Possible Causes:** • Ignition Coil 'E' primary circuit is open or shorted to ground • Ignition Coil 'E' power circuit is open (test power from I/P fuse) • Ignition Coil 'E' is damaged or has failed • PCM has failed
DTC: P0356 **2T CCM, MIL: Yes** **Years:** 2007, 2008 **Models:** Amanti, Rondo, Sedona, Sorento, Sportage **Engines:** 2.7L, 3.3L, 3.8L **Transmissions:** All	**Ignition Coil 'F' Circuit Malfunction** Engine started, engine running, and the PCM detected an unexpected voltage condition on the Ignition Coil 'F' primary circuit. **Possible Causes:** • Ignition Coil 'F' primary circuit is open or shorted to ground • Ignition Coil 'F' power circuit is open (test power from I/P fuse) • Ignition Coil 'F' is damaged or has failed • PCM has failed
DTC: P0420 **2T CAT1, MIL: Yes** **Years:** 2007, 2008 **Models:** All **Engines:** All **Transmissions:** All	**Catalyst Efficiency Below Normal (Bank 1)** DTC P0130, P0133, P0134, P0135, P0136, P0139, P0140 and P0141 not set, engine started, engine running in closed loop at a speed of 45-60 MPH for 2-3 minutes, and the PCM detected that the rear HO2S and front HO2S voltage amplitudes were too similar. **Possible Causes:** • Air leaks at the exhaust manifold or in the exhaust pipes • Catalytic converter is damaged or has failed • Front HO2S or rear HO2S is contaminated with fuel or moisture • Front HO2S or rear HO2S is contaminated with fuel or moisture
DTC: P0430 **2T CAT1, MIL: Yes** **Years:** 2007, 2008 **Models:** Amanti, Optima, Rondo, Sedona, Sorento, Sportage **Engines:** 2.7L, 3.3L, 3.5L, 3.8L **Transmissions:** All	**Catalyst Efficiency Below Normal (Bank 2)** DTC P0150, P0153, P0154, P0155, P0156, P0160 and P0161 not set, engine started, engine running in closed loop at 45-60 MPH for 2-3 minutes, and the PCM detected the rear HO2S and front HO2S voltage amplitudes were too similar during the Catalyst Monitor test. **Possible Causes:** • Air leaks at the exhaust manifold or in the exhaust pipes • Catalytic converter is damaged or has failed • Front HO2S or rear HO2S is contaminated with fuel or moisture • Front HO2S or rear HO2S heater is damaged or has failed • Front HO2S or rear HO2S is contaminated with fuel or moisture
DTC: P0441 **2T EVAP, MIL: Yes** **Years:** 2007, 2008 **Models:** Amanti, Optima, Rondo, Sedona, Sorento, Spectra, Sportage **Engines:** 2.0L, 2.4L, 2.7L, 3.3L, 3.5L, 3.8L **Transmissions:** All	**EVAP System Malfunction** ECT sensor signal less than158°F at startup, IAT sensor signal more than 9.05°F, system voltage more than 10.9 volts engine runtime 15-20 minutes at cruise speed, then returned to idle speed, VSS indicating 0 MPH, load value 2.2 ms, canister load factor less than 4.0, fuel tank pressure less than 0.5' Hg, then after the Idle Control system and Fuel Trim stabilized, the PCM detected a continuous purge condition in the EVAP system during the Purge flow test. **Possible Causes:** • Small hoses or cuts present in the EVAP vapor hoses/lines • EVAP canister purge solenoid is damaged or is stuck open • PCM has failed
DTC: P0442 **2T EVAP, MIL: Yes** **Years:** 2007, 2008 **Models:** All **Engines:** All **Transmissions:** All	**EVAP System Small Leak (0.040') Detected** ECT sensor signal less than 158°F at startup, IAT sensor signal more than 9.05°F, system voltage more than 10.9 volts engine runtime 15-20 minutes at cruise speed, then returned to idle speed, VSS indicating 0 MPH, load value 2.2 ms, canister load factor less than 4.0, fuel tank pressure less than 0.5' Hg, then after the Idle Control system and Fuel Trim had stabilized, the PCM detected a fuel vapor leak (as small as 0.040') in the EVAP system during the EVAP Leak Test. **Possible Causes:** • Fuel filler cap damaged, cross-threaded or loosely installed • Small leaks or cuts present in the EVAP vapor hoses/lines • EVAP purge valve is damaged or has failed • PCM has failed

DTC	Trouble Code Title, Conditions & Possible Causes
DTC: P0444 **2T CCM, MIL: Yes** **Years:** 2007, 2008 **Models:** All **Engines:** All **Transmissions:** All	**EVAP Emission System- Purge Control Valve Circuit Open** Engine running. Checking output signals from PCSV every 10 seconds, under detecting condition. **Possible Causes:** • Poor connection • Open or short to ground in harness • PCVS • PCM
DTC: P0445 **2T CCM, MIL: Yes** **Years:** 2007, 2008 **Models:** Amanti, Optima, Rondo, Sedona, Sorento, Spectra, Sportage **Engines:** 2.0L, 2.4L, 2.7L, 3.3L, 3.5L, 3.8L **Transmissions:** All	**EVAP Emission System- Purge Control Valve Circuit Shorted** Engine running. Checking output signals from PCSV every 10 seconds, under detecting condition. **Possible Causes:** • Poor connection • Short to battery in harness • PCVS • PCM
DTC: P0446 **2T CCM, MIL: Yes** **Years:** 2007, 2008 **Models:** Rio **Engines:** 1.6L **Transmissions:** All	**EVAP Emission System- Vent Control Circuit** CCV stuck open. Time after engine start greater than 600 seconds. Idle speed controller activated. Coolant temperature less than 12°F. **Possible Causes:** • Poor connection • CCV • ECM/PCM
DTC: P0447 **2T CCM, MIL: Yes** **Years:** 2007, 2008 **Models:** Amanti, Optima, Rondo, Sedona, Sorento, Spectra, Sportage **Engines:** 2.0L, 2.4L, 2.7L, 3.3L, 3.5L, 3.8L **Transmissions:** All	**EVAP Emission System- Vent Control Circuit Open** Detects a short to ground or open circuit on vent valve output circuit. No disabling faults present. Engine running. **Possible Causes:** • Poor connection • Open or short in power circuit • Open or short in control circuit • CCV • ECM/PCM
DTC: P0448 **2T CCM, MIL: Yes** **Years:** 2007, 2008 **Models:** Amanti, Optima, Rondo, Sedona, Sorento, Spectra, Sportage **Engines:** 2.0L, 2.4L, 2.7L, 3.3L, 3.5L, 3.8L **Transmissions:** All	**EVAP Emission System- Vent Control Circuit Shorted** Detects a short to battery on vent valve output circuit. No disabling faults present. Engine running. **Possible Causes:** • Poor connection • Short to battery in CCV circuit • CCV • ECM/PCM
DTC: P0449 **2T CCM, MIL: Yes** **Years:** 2007, 2008 **Models:** Optima, Rio, Rondo, Spectra, Sportage **Engines:** 1.6L, 2.0L, 2.4L, 2.7L **Transmissions:** All	**EVAP Emission System- Vent Valve/Solenoid Circuit** Circuit continuity check - open. **Possible Causes:** • Poor connection • Open or short to ground in power circuit • CCV • ECM/PCM
DTC: P0451 **2T CCM, MIL: Yes** **Years:** 2007, 2008 **Models:** All **Engines:** All **Transmissions:** All	**EVAP Pressure Sensor Performance** Engine at idle speed with the vehicle speed indicating 0 MPH, then with the EVAP Vent Control solenoid commanded "on", the PCM detected the FTP sensor signal variation was less than 15 mv. **Possible Causes:** • FTP sensor signal or ground circuit has high resistance • FTP sensor is damaged or out of calibration • EVAP canister close valve (CCV) is stuck closed • PCM has failed
DTC: P0452 **2T CCM, MIL: Yes** **Years:** 2007, 2008 **Models:** All **Engines:** All **Transmissions:** All	**EVAP Pressure Sensor Circuit Low Input** Key on or engine running and the PCM detected the Fuel Tank Pressure (FTP) sensor signal was less than 0.14 volt during the test. **Possible Causes:** • FTP sensor signal circuit is shorted to ground • FTP sensor is damaged or has failed • PCM has failed

DTC	Trouble Code Title, Conditions & Possible Causes
DTC: P0453 **2T CCM, MIL:** Yes **Years:** 2007, 2008 **Models:** All **Engines:** All **Transmissions:** All	**EVAP Pressure Sensor Circuit High Input** Key on or engine running and the PCM detected the Fuel Tank Pressure (FTP) sensor signal was more than 4.90 volts during the test. **Possible Causes:** • FTP sensor signal circuit is open • FTP sensor ground circuit is open • FTP sensor is damaged or has failed • PCM has failed
DTC: P0454 **2T CCM, MIL:** Yes **Years:** 2007, 2008 **Models:** Amanti, Optima, Rondo, Sedona, Sorento, Spectra, Sportage **Engines:** 2.0L, 2.4L, 2.7L, 3.3L, 3.5L, 3.8L **Transmissions:** All	**EVAP Emission System- Pressure Sensor Intermittent** The PCM measures pressure stability in the fuel tank, by means of a sensor for a predetermined duration. If fluctuation is larger than predetermined threshold a DTC is set. Sensor signal noise check. **Possible Causes:** • Contact resistance in connectors • Faulty FTPS • Faulty ECM/PCM • Faulty FTPS
DTC: P0455 **2T EVAP, MIL:** Yes **Years:** 2007, 2008 **Models:** All **Engines:** All **Transmissions:** All	**EVAP System Large Leak Detected** ECT input at 38-95°F at startup, engine running at a steady throttle at over 6.2 MPH for over 2 minutes, and the PCM detected the FTP sensor signal indicated more than -15 kPa in the EVAP Leak test. **Possible Causes:** • Fuel filler cap damaged, cross-threaded or loosely installed • Small leaks or cuts present in the EVAP vapor hoses/lines • EVAP purge valve is damaged or has failed • PCM has failed
DTC: P0456 **2T EVAP, MIL:** Yes **Years:** 2007, 2008 **Models:** All **Engines:** All **Transmissions:** All	**EVAP System Very Small Leak (0.020") Detected** ECT input at 38-95°F at startup, vehicle driven at a steady speed of over 6.2 MPH for 2 minutes, and then the PCM detected a very small leak (less than 0.020") in the EVAP system during the Leak test. **Possible Causes:** • Fuel filler is damaged, cross-threaded, loose or missing • Fuel filler pipe is damaged, or a fuel vapor hose is leaking • Rollover valve or ORVR (valve) had failed allowing fuel in lines • Canister close valve clogged or stuck in open or closed position • Purge solenoid valve is damaged or installed improperly • FTP sensor is damaged or has failed • Leaks in the charcoal canister, or at the fuel tank seals
DTC: P0461 **2T CCM, MIL:** Yes **Years:** 2007, 2008 **Models:** Amanti, Optima, Rio, Rondo, Sedona, Sorento, Sportage **Engines:** 1.6L, 2.4L, 2.7L, 3.3L, 3.5L, 3.8L **Transmissions:** All	**Fuel Level Sensor "A" Circuit Range/Performance** Filtered and unfiltered signal of fuel sensor are monitored. **Possible Causes:** • Poor connection • Faulty fuel level sensor • Faulty PCM
DTC: P0462 **2T CCM, MIL:** Yes **Years:** 2007, 2008 **Models:** Amanti, Optima, Rio, Rondo, Sedona, Sorento, Sportage **Engines:** 1.6L, 2.4L, 2.7L, 3.3L, 3.5L, 3.8L **Transmissions:** All	**Fuel Level Sensor Input Low (Sticking)** Key on or engine running system voltage more than 10 volts, and the PCM detected the fuel level sensing unit signal was less than 0.2 volt. **Possible Causes:** • Fuel level sending unit signal circuit shorted to VREF • Fuel level sending unit signal circuit shorted to system power • Fuel level sensing unit is damaged or the fuel tank is damaged • BCM or PCM has failed
DTC: P0463 **2T CCM, MIL:** Yes **Years:** 2007, 2008 **Models:** Amanti, Optima, Rio, Rondo, Sedona, Sorento, Sportage **Engines:** 1.6L, 2.4L, 2.7L, 3.3L, 3.5L, 3.8L **Transmissions:** All	**Fuel Level Sensor Input High (Sticking)** Key on or engine running system voltage more than 10 volts, and the PCM detected the fuel level sensing unit signal was more than 4.5 volts. **Possible Causes:** • Fuel level sending unit signal circuit shorted to VREF • Fuel level sending unit signal circuit shorted to system power • Fuel level sensing unit is damaged or the fuel tank is damaged • BCM or PCM has failed

DTC	Trouble Code Title, Conditions & Possible Causes
DTC: P0464 **2T CCM, MIL: No** **Years:** 2007, 2008 **Models:** Amanti, Optima, Rondo, Sedona, Sorento, Sportage **Engines:** 2.4L, 2.7L, 3.3L, 3.8L **Transmissions:** All	**Fuel Level Sensor "A" Circuit Intermittent** Check signal for fluctuation. The ECM sets the DTC if the fuel level signal is higher than the threshold value (signal fluctuation greater than 50 percent). **Possible Causes:** • Contact resistance in connectors • Short to battery in fuel level (FLS) circuit • Faulty ECM
DTC: P0480 **2T CCM, MIL: Yes** **Years:** 2007, 2008 **Models:** Amanti, Optima, Rondo, Sedona, Sorento **Engines:** 2.7L, 3.3L, 3.8L **Transmissions:** All	**Fan 1 Control Circuit Malfunction** This will detect a short to ground, to battery or open circuit of fan relay output. Fault information provided by an output driver chip. No disabling faults present. Engine running. Enable time delay equal or greater than 0.5 seconds. **Possible Causes:** • Poor connection • Open in power circuit to cooling fan • Open or short in control circuit to PCM • Faulty fan relay • Faulty cooling fan module • Faulty PCM
DTC: P0501 **2T CCM, MIL: Yes** **Years:** 2007, 2008 **Models:** Amanti, Optima, Rio, Rondo, Sedona, Sorento, Spectra, Sportage **Engines:** 1.6L, 2.0L, 2.7L, 3.3L, 3.8L **Transmissions:** All	**Vehicle Speed Sensor Performance** Engine running in gear at high speed and lover for over 1 second, and the PCM did not detect any VSS signals. **Possible Causes:** • VSS signal circuit is open or shorted to ground • VSS signal circuit is shorted to VREF or to system power (B+) • VSS is damaged or has failed • PCM has failed
DTC: P0504 **2T CCM, MIL: Yes** **Years:** 2007, 2008 **Models:** Amanti, Optima, Rondo, Sedona, Sorento **Engines:** 2.4L, 2.7L, 3.3L, 3.8L **Transmissions:** All	**Brake Switch "A"/"B" Correlation (1)** Comparing two brake signals during driving. Case 1: Engine works. Vehicle speed sensor is normal. Case 2: Engine works. Vehicle speed sensor is normal. Vehicle speed is over 20 kph, for at least 1 second. **Possible Causes:** • Poor connection • Open or short • Faulty PCM
DTC: P0506 **2T CCM, MIL: Yes** **Years:** 2007, 2008 **Models:** All **Engines:** All **Transmissions:** All	**Idle Speed Lower Than Expected** Engine running at idle speed while in closed loop, and the PCM detected the Actual idle speed was more than 100 RPM below the Target idle speed during the CCM test. **Possible Causes:** • High resistance between the main relay and IAC valve • High resistance between PCM and IAC valve control circuits • IAC valve is damaged or has failed • The throttle plate is carbon fouled (it may need to be cleaned)
DTC: P0507 **2T CCM, MIL: Yes** **Years:** 2007, 2008 **Models:** All **Engines:** All **Transmissions:** All	**Idle Speed Higher Than Expected** Engine running at idle speed while in closed loop, and the PCM detected the Actual idle speed was more than 200 RPM above the Target idle speed during the CCM test. **Possible Causes:** • High resistance between the main relay and IAC valve • High resistance between PCM and IAC valve control circuits • IAC valve is damaged or has failed • Intake air leak located below the throttle plate assembly
DTC: P0532 **2T CCM, MIL: No** **Years:** 2007, 2008 **Models:** Amanti, Optima, Rio, Rondo, Sedona, Sorento **Engines:** 1.6L, 2.4L, 2.7L, 3.3L, 3.8L **Transmissions:** All	**A/C Refrigerant Pressure Sensor "A" Circuit Low Input** Detects sensor signal short to low voltage. Engine works. Sensor output 0.05 volt. **Possible Causes:** • Poor connection • Open in power circuit • Open or short to ground in signal circuit • Faulty A/C pressure sensor • Faulty PCM

DTC	Trouble Code Title, Conditions & Possible Causes
DTC: P0533 **2T CCM, MIL: No** **Years:** 2007, 2008 **Models:** Amanti, Optima, Rio, Rondo, Sedona, Sorento **Engines:** 1.6L, 2.4L, 2.7L, 3.3L, 3.8L **Transmissions:** All	**A/C Refrigerant Pressure Sensor "A" Circuit High Input** Detects sensor signal short to high voltage. Engine works. Sensor output 4.65 volts. **Possible Causes:** • Poor connection • Open in signal circuit open • Open in ground circuit • Faulty A/C pressure sensor • Faulty PCM
DTC: P0551 **2T CCM, MIL: Yes** **Years:** 2007, 2008 **Models:** Amanti, Rondo, Sorento, Sportage **Engines:** 2.4L, 2.7L, 3.5L **Transmissions:** All	**Power Steering Pressure Sensor/Switch Circuit Range/Performance** If a power steering switch signal is ON when the engine speed is more than 2500 RPM, load value is grater than 55 percent and engine coolant temperature is above 50°F, the DTC will set. Signal of power steering pressure switch is monitored. **Possible Causes:** • Poor connection • Faulty power steering switch • Open or short in power steering switch • Faulty PCM
DTC: P0560 **2T CCM, MIL: Yes** **Years:** 2007, 2008 **Models:** Amanti, Rio, Rondo, Sorento, Spectra, Sportage **Engines:** 1.6L, 2.0L, 2.4L, 2.7L, 3.5L **Transmissions:** All	**Battery Backup Line Circuit Malfunction** Engine runtime over 4 minutes and the PCM did not detect any system voltage on the Battery Backup circuit for 5 seconds. **Possible Causes:** • Battery backup circuit to the PCM is open • Battery backup fuse to the PCM is open or missing • Battery backup circuit to the PCM has high resistance • PCM has failed
DTC: P0561 **2T CCM, MIL: No** **Years:** 2007, 2008 **Models:** Rio **Engines:** 1.6L **Transmissions:** All	**System Voltage Unstable** Engine runtime over 4 minutes, and the PCM detected the system voltage rapidly changed its value by more than 3 volts. **Note: If the Battery Backup circuit is open, the vehicle will not run.** **Possible Causes:** • Charging system problem (charging voltage interrupted) • Backup voltage circuit to the PCM open (intermittent fault) • PCM has failed
DTC: P0562 **2T CCM, MIL: No** **Years:** 2007, 2008 **Models:** All **Engines:** 1.6L, 2.0L, 2.4L, 2.7L, 3.3L, 3.8L **Transmissions:** All	**System Voltage Low Input** Engine runtime over 4 minutes, and the PCM detected the system voltage was less than 8.0 volts, condition met for 5 seconds. **Note: If the Battery Backup circuit is open, the vehicle will not run.** **Possible Causes:** • Charging system problem (charging voltage too low) • Battery backup circuit to the PCM has high resistance • Backup voltage circuit to the PCM open (intermittent fault) • PCM has failed
DTC: P0563 **2T CCM, MIL: No** **Years:** 2007, 2008 **Models:** All **Engines:** 1.6L, 2.0L, 2.4L, 2.7L, 3.3L, 3.8L **Transmissions:** All	**System Voltage High Input** Engine runtime over 4 minutes, and the PCM detected the system voltage was more than 17.0 volts, condition met for 5 seconds. **Note: If the Battery Backup circuit is open, the vehicle will not run.** **Possible Causes:** • Charging system problem (charging voltage too high) • Backup voltage circuit to the PCM open (intermittent fault) • PCM has failed
DTC: P0571 **2T CCM, MIL: Yes** **Years:** 2007, 2008 **Models:** Amanti, Optima, Rondo, Sedona, Sorento **Engines:** 2.7L, 3.3L, 3.8L **Transmissions:** All	**Brake Switch "A" Circuit** PCM detects brake light input signal when the vehicle stops. VSS is normal. Vehicle speed 0 MPH, during one second or more. **Possible Causes:** • Poor connection • Open or short to ground in signal circuit • Faulty PCM
DTC: P0600 **2T CCM, MIL: Yes** **Years:** 2007, 2008 **Models:** Optima, Rondo, Sportage **Engines:** 2.0L, 2.4L **Transmissions:** All	**CAN Communication Bus** CAN message transfer incorrect? **Possible Causes:** • Open or short in CAN line • Contact resistance in connectors • Faulty PCM

DTC	Trouble Code Title, Conditions & Possible Causes
DTC: P0601 **2T PCM, MIL: Yes** **Years:** 2007, 2008 **Models:** Amanti, Optima, Rondo, Sedona, Sorento **Engines:** 2.7L, 3.3L, 3.8L **Transmissions:** All	**PCM or TCM Internal Random Check Sum Error** Key on or engine running and the PCM or the TCM detected that a Random Check Sum Error was present. **Possible Causes:** • Poor terminal contact at the ECM ISC Backup Voltage circuit • PCM or TCM has an internal problem or has failed
DTC: P0602 **2T CCM, MIL: Yes** **Years:** 2007, 2008 **Models:** Amanti, Optima, Rondo, Sedona, Sorento **Engines:** 2.7L, 3.3L, 3.8L **Transmissions:** All	**EEPROM Programming Error** Check internal CPU **Possible Causes:** • Faulty PCM
DTC: P0604 **2T PCM, MIL: Yes** **Years:** 2007, 2008 **Models:** Amanti, Optima, Rondo, Sedona, Sorento **Engines:** 2.7L, 3.3L, 3.8L **Transmissions:** All	**PCM or TCM Internal Random Access Memory Error** Key on or engine running and the PCM or TCM detected an Internal Random Access Memory (RAM) error was present. **Possible Causes:** • Poor terminal contact at the ECM Backup Voltage circuit • PCM or TCM has an internal problem or has failed
DTC: P0605 **1T PCM, MIL: Yes** **Years:** 2007, 2008 **Models:** Optima, Rio, Rondo, Spectra, Sportage **Engines:** 1.6L, 2.0L, 2.4L, 2.7L **Transmissions:** All	**PCM (Internal Controller) ROM Error** Key on for 1 second, and the PCM detected an internal ROM error occurred during the initial Self-Test. **Possible Causes:** • Clear the trouble codes and retest for this trouble code. If the same trouble code resets, the PCM has failed and must be replaced to repair this problem.
DTC: P0606 **2T CCM, MIL: Yes** **Years:** 2007, 2008 **Models:** Amanti, Optima, Rondo, Sedona, Sorento **Engines:** 2.7L, 3.3L, 3.8L **Transmissions:** All	**ECM Processor (ECU-Self Test Failed)** Controller error. No electrical fault of the front HO2S. **Possible Causes:** • Faulty PCM
DTC: P061B **2T CCM, MIL: No** **Years:** 2007, 2008 **Models:** Amanti, Optima, Rondo, Sedona, Sorento **Engines:** 2.7L, 3.3L, 3.8L **Transmissions:** All	**Internal Control Module Torque Calculation Performance** Desired torque error. **Possible Causes:** • Faulty PCM
DTC: P0630 **2T CCM, MIL: Yes** **Years:** 2007, 2008 **Models:** Amanti, Optima, Rio, Rondo, Sedona, Sorento, Sportage **Engines:** All **Transmissions:** All	**VIN Not Programmed Or Incompatible- ECM/PCMECM** PCM internal check. Enable condition, ignition ON. VIN does not exist in boot area. **Possible Causes:** • PCM is new and has not yet been programmed • Faulty PCM
DTC: P0638 **2T CCM, MIL: Yes** **Years:** 2007, 2008 **Models:** Amanti, Optima, Rondo, Sedona, Sorento **Engines:** 2.4L, 2.7L, 3.3L, 3.8L **Transmissions:** All	**Throttle Actuator Control Range/Performance** ETS position control malfunction. Battery voltage more than 5 volts. **Possible Causes:** • Throttle stuck • Open in motor circuit • Faulty motor • Faulty PCM

DTC	Trouble Code Title, Conditions & Possible Causes
DTC: P0641 **2T CCM, MIL: Yes** **Years:** 2007, 2008 **Models:** Amanti, Optima, Rondo, Sedona, Sorento **Engines:** 2.4L, 2.7L, 3.3L, 3.8L **Transmissions:** All	**Sensor Reference Voltage "A" Circuit Open** Sensor reference voltage check. Ignition ON. **Possible Causes:** • Short in sensor power supply line • Faulty PCM
DTC: P0646 **2T CCM, MIL: No** **Years:** 2007, 2008 **Models:** Amanti, Optima, Rio, Rondo, Sedona, Sorento **Engines:** 1.6L, 2.4L, 2.7L, 3.3L, 3.8L **Transmissions:** All	**A/C Clutch Relay Control Circuit Low** Detects circuit short to low voltage. No DTC exists. Engine works. After 0.5 seconds. **Possible Causes:** • Poor connection • Open or short to ground in A/C relay circuit • Faulty A/C relay • Faulty PCM
DTC: P0647 **2T CCM, MIL: No** **Years:** 2007, 2008 **Models:** Amanti, Optima, Rio, Rondo, Sedona, Sorento **Engines:** 1.6L, 2.4L, 2.7L, 3.3L, 3.8L **Transmissions:** All	**A/C Clutch Relay Control Circuit High** Detects circuit short to high voltage. No DTC exists. Engine works. After 0.5 seconds. **Possible Causes:** • Poor connection • Short to power in A/C relay circuit • Faulty A/C relay • Faulty PCM
DTC: P0650 **2T CCM, MIL: No** **Years:** 2007, 2008 **Models:** All **Engines:** 1.6L, 2.0L, 2.4L, 2.7L, 3.3L, 3.8L **Transmissions:** All	**Malfunction Indicator Lamp Circuit Malfunction** Key on or engine running and the PCM detected an unexpected voltage condition on the Malfunction Indicator Lamp (MIL) circuit. **Possible Causes:** • MIL control circuit open • MIL control circuit shorted to ground • MIL "bulb" is damaged or missing • PCM has failed (MIL control "driver" may be open or shorted)
DTC: P0661 **2T CCM, MIL: No** **Years:** 2007, 2008 **Models:** Sorento, Sportage **Engines:** 2.7L, 3.5L **Transmissions:** All	**Intake Manifold Tuning Valve Control Circuit Low (Bank 1) Solenoid Type** DTC is set if the ECM detects that the valve control circuit is shorted to ground. Driver stage check. **Possible Causes:** • Open in power supply harness • Short to ground in control harness • Contact resistance in connectors • Faulty valve
DTC: P0662 **2T CCM, MIL: No** **Years:** 2007, 2008 **Models:** Sorento, Sportage **Engines:** 2.7L, 3.5L **Transmissions:** All	**Intake Manifold Tuning Valve Control Circuit High (Bank 1) Solenoid Type** DTC is set if the ECM detects that the valve control circuit is open or shorted to battery voltage. Driver stage check. **Possible Causes:** • Open or short to battery in control harness • Contact resistance in connectors • Faulty valve
DTC: P0664 **2T CCM, MIL: No** **Years:** 2007, 2008 **Models:** Sportage **Engines:** 2.7L **Transmissions:** All	**Intake Manifold Tuning Valve Control Circuit High (Bank 2) Solenoid Type** DTC is set if the ECM detects that the valve control circuit is shorted to ground. Driver stage check. **Possible Causes:** • Open in power supply harness • Short to ground in control harness • Contact resistance in connectors • Faulty valve
DTC: P0665 **2T CCM, MIL: No** **Years:** 2007, 2008 **Models:** Sportage **Engines:** 2.7L **Transmissions:** All	**Intake Manifold Tuning Valve Control Circuit Low (Bank 1) Solenoid Type** DTC is set if the ECM detects that the valve control circuit is open or shorted to battery voltage. Driver stage check. **Possible Causes:** • Open or short to battery in control harness • Contact resistance in connectors • Faulty valve

DTC	Trouble Code Title, Conditions & Possible Causes
DTC: P0685 **2T CCM, MIL: No** **Years:** 2007, 2008 **Models:** Amanti, Optima, Rondo, Sedona, Sorento **Engines:** 2.7L, 3.3L, 3.8L **Transmissions:** All	**ECM/PCM Power Relay Control Circuit/Open** Engine running. Ignition voltage less than or equal to 11 volts. **Possible Causes:** • poor connection • Open or short to in control circuit • Main relay • PCM
DTC: P0700 **2T CCM, MIL: Yes** **Years:** 2007, 2008 **Models:** Optima, Rio, Rondo, Sedona, Sorento, Spectra, Sportage **Engines:** 1.6L, 2.0L, 2.4L, 2.7L, 3.5L, 3.8L **Transmissions:** A/T	**TCU Request For MIL "ON"** Engine at normal operating temperature. Check for additional DTC's. **Possible Causes:** • poor connection • TCM • PCM/ECM
DTC: P0705 **2T CCM, MIL: Yes** **Years:** 2007, 2008 **Models:** Sorento **Engines:** All **Transmissions:** A/T	**Transmission Range Switch Circuit Malfunction** Engine running with VSS inputs received, and the PCM detected invalid signals or multiple TR switch inputs with the gearshift in drive. **Note: If DTC P1500 is also set, check the Meter Fuse and circuit.** **Possible Causes:** • TR switch signal circuit is open or shorted to ground • TR switch signal circuit is shorted to VREF or system power • TR switch is damaged or has failed • PCM has failed
DTC: P0707 **2T CCM, MIL: Yes** **Years:** 2007, 2008 **Models:** Amanti, Optima, Rio, Rondo, Sedona, Spectra, Sportage **Engines:** All **Transmissions:** A/T	**Transmission Range Switch Circuit Malfunction** Engine running with VSS inputs received, and the PCM did not detect any TR switch signal for 10 seconds during the CCM test. **Possible Causes:** • TR switch signal circuit is open or shorted to ground • TR switch signal circuit is shorted to VREF or system power • TR switch is damaged or has failed • PCM has failed
DTC: P0708 **2T CCM, MIL: Yes** **Years:** 2007, 2008 **Models:** Amanti, Optima, Rio, Rondo, Sedona, Spectra, Sportage **Engines:** All **Transmissions:** A/T	**Transmission Range Switch Circuit Performance** Engine running with VSS inputs received, and the PCM detected two (2) or more TR switch signals simultaneously for over 30 seconds. **Possible Causes:** • TR switch signal circuit is open or shorted to ground • TR switch signal circuit is shorted to VREF or system power • TR switch is damaged or has failed • PCM has failed
DTC: P0712 **2T CCM, MIL: Yes** **Years:** 2007, 2008 **Models:** All **Engines:** All **Transmissions:** A/T	**Transmission Fluid Temperature Low Input** Key on or engine running and the PCM detected the TFT sensor signal indicated less than 0.49 volt 300°F) for more than 1 second. **Note: The TFT sensor signal at 68°F is 4.0 volts, and at 266°F it is 1.5 volts.** **Possible Causes:** • TFT sensor signal circuit shorted to ground (sensor to TCM) • TFT sensor signal circuit shorted to ground (TCM to PCM) • TFT sensor is damaged or has failed • PCM is damaged
DTC: P0713 **2T CCM, MIL: Yes** **Years:** 2007, 2008 **Models:** All **Engines:** All **Transmissions:** A/T	**Transmission Fluid Temperature High Input** Key on or engine running and the PCM detected the TFT sensor signal indicated more than 4.57 volts (-40°F) during the CCM test period. **Note: The TFT sensor signal at 68°F is 4.0 volts, and at 266°F it is 1.5 volts.** **Possible Causes:** • TFT sensor signal circuit is open (sensor circuit to the TCM) • TFT sensor signal circuit is open (TCM circuit to the PCM) • TFT sensor is damaged or has failed • PCM is damaged

DTC	Trouble Code Title, Conditions & Possible Causes
DTC: P0715 **2T CCM, MIL: Yes** **Years:** 2007, 2008 **Models:** Sportage **Engines:** 2.7L **Transmissions:** A/T	**Input/Turbine Speed Sensor Circuit Malfunction** Vehicle driven to a speed of over 15.5 MPH with the engine speed over 1500 RPM in 3rd or 4th gear, and the PCM did not detect any Input/Turbine speed sensor signals for 5 seconds. **Possible Causes:** • Input/Turbine speed sensor signal circuit open or shorted • Input/Turbine speed sensor is damaged or has failed • PCM has failed
DTC: P0720 **2T CCM, MIL: Yes** **Years:** 2007, 2008 **Models:** Sportage **Engines:** 2.7L **Transmissions:** A/T	**Output Shaft Speed Sensor Circuit Malfunction** Vehicle driven in 3rd or 4th gear at a speed of over 20 MPH, and the PCM detected an unexpected voltage condition on the OSS sensor circuit. If this code is generated four (4) times or more, the PCM will lock the transmission into either 2nd or 3rd gear for safety reasons. **Possible Causes:** • Output shaft speed sensor signal circuit open or shorted • Output shaft speed sensor is damaged or has failed • PCM has failed
DTC: P0731 **2T CCM, MIL: Yes** **Years:** 2007, 2008 **Models:** All **Engines:** All **Transmissions:** A/T	**TCM Incorrect First Gear Ratio** Vehicle driven at 12-32 MPH in 3rd gear, shift solenoids 'A', 'B' and 'C', input/turbine speed sensor and TFT sensor inputs all indicating okay, and the PCM detected the 1st gear ratio was too high. **Possible Causes:** • ATF fluid level too low or line pressure low • Control valve stuck or solenoid valve is damaged or has failed • Forward clutch, 3-4 brake band or 1-way clutch No. 1 slippage • PCM has failed
DTC: P0732 **2T CCM, MIL: Yes** **Years:** 2007, 2008 **Models:** All **Engines:** All **Transmissions:** A/T	**TCM Incorrect Second Gear Ratio** Vehicle driven at 17-60 MPH in 2nd gear, shift solenoids 'A', 'B' and 'C', input/turbine speed sensor and TFT sensor inputs all indicating okay, and the PCM detected the 2nd gear ratio was too high. **Possible Causes:** • ATF fluid level too low or line pressure low • Control valve stuck or solenoid valve is damaged or has failed • Forward clutch, 2-4 brake band or 1-way clutch No. 1 slippage • PCM has failed
DTC: P0733 **2T CCM, MIL: Yes** **Years:** 2007, 2008 **Models:** All **Engines:** All **Transmissions:** A/T	**TCM Third Gear Incorrect Ratio** Vehicle driven at 19-32 MPH in 3rd gear, shift solenoids 'A', 'B' and 'C', input/turbine speed sensor and TFT sensor inputs all indicating okay, and the PCM detected the 3rd gear ratio was too high. **Possible Causes:** • ATF fluid level too low or line pressure low • Control valve stuck or solenoid valve is damaged or has failed • Forward clutch, 3-4 brake band or 1-way clutch No. 1 slippage • PCM has failed
DTC: P0734 **2T CCM, MIL: Yes** **Years:** 2007, 2008 **Models:** All **Engines:** All **Transmissions:** A/T	**TCM Fourth Gear Incorrect Ratio** Vehicle driven at 44-65 MPH in 4th gear, shift solenoids 'A', 'B' and 'C', input/turbine speed sensor and TFT sensor inputs all indicating okay, and the PCM detected the 4th gear ratio was too high. **Possible Causes:** • ATF fluid level too low or line pressure low • Control valve stuck or solenoid valve is damaged or has failed • 2-4 brake band or 3-4 clutch slippage • PCM has failed
DTC: P0735 **2T CCM, MIL: Yes** **Years:** 2007, 2008 **Models:** Amanti, Optima, Rondo, Sedona, Sorento **Engines:** 2.7L, 3.3L, 3.5L, 3.8L **Transmissions:** A/T	**TCM Fifth Gear Incorrect Ratio** Vehicle driven at 44-65 MPH in 5th gear, shift solenoids 'A', 'B' and 'C', input/turbine speed sensor and TFT sensor inputs all indicating okay, and the PCM detected the 5th gear ratio was too high. If this code sets 4 times or more, the PCM will lock the gear into 3rd gear. **Possible Causes:** • ATF fluid level too low or line pressure low • Control valve stuck or solenoid valve is damaged or has failed • 2-4 brake band or 3-4 clutch slippage • PCM has failed

DTC	Trouble Code Title, Conditions & Possible Causes
DTC: P0736 **2T CCM, MIL:** Yes **Years:** 2007, 2008 **Models:** Rio, Rondo, Sportage **Engines:** 1.6L, 2.0L, 2.7L, **Transmissions:** A/T	**TCM Reverse Gear Incorrect Ratio** Vehicle driven in Reverse gear, and the PCM detected the output speed sensor signal did not match the input speed sensor signal. If this code sets 4 times, the PCM will lock the gear into 3rd gear. **Possible Causes:** • ATF fluid level too low or line pressure low • Input speed sensor or output speed sensor damaged or failed • UD clutch retainer has failed or DIR planetary carrier has failed • Reverse clutch failure or LR brake line or RED brake line failure
DTC: P0750 **2T CCM, MIL:** Yes **Years:** 2007, 2008 **Models:** Amanti, Optima, Rio, Rondo, Sedona, Spectra, Sportage **Engines:** 1.6L, 2.0L, 2.4L, 2.7L, 3.5L, 3.8L **Transmissions:** A/T	**TCM Shift Solenoid 'A' Circuit Malfunction** Engine running in gear with VSS inputs received, and the PCM detected that the TCC solenoid was always "on" or always "off". **Possible Causes:** • SSA control circuit is open or shorted to ground • SSA is damaged or has failed • PCM or TCM has failed
DTC: P0755 **2T CCM, MIL:** Yes **Years:** 2007, 2008 **Models:** Amanti, Optima, Rio, Rondo, Sedona, Spectra, Sportage **Engines:** 1.6L, 2.0L, 2.4L, 2.7L, 3.5L, 3.8L **Transmissions:** A/T	**TCM Shift Solenoid 'B' Circuit Malfunction** Engine running in gear with VSS inputs received, and the PCM detected that the SSB was always "on" or always "off". **Possible Causes:** • SSB is damaged or has failed • PCM or TCM has failed
DTC: P0760 **2T CCM, MIL:** Yes **Years:** 2007, 2008 **Models:** Amanti, Optima, Rio, Rondo, Sedona, Spectra, Sportage **Engines:** 1.6L, 2.0L, 2.4L, 2.7L, 3.5L, 3.8L **Transmissions:** A/T	**TCM Shift Solenoid 'C' Circuit Malfunction** Engine running in gear, VSS inputs received, and PCM detected an unexpected voltage condition on the SSC circuit during the test. **Possible Causes:** • SSC control circuit open or shorted to ground • SSC control circuit shorted to system power (B+) • SSC is damaged or has failed • PCM or TCM has failed
DTC: P0770 **2T CCM, MIL:** Yes **Years:** 2007, 2008 **Models:** Amanti, Optima, Rondo, Sedona **Engines:** 2.7L, 3.5L, 3.8L **Transmissions:** A/T	**TCM RED Solenoid Circuit Malfunction** Engine running in gear, VSS inputs received, and PCM detected an unexpected voltage condition on the RED solenoid circuit in the test. **Possible Causes:** • RED control circuit open or shorted to ground • RED control circuit shorted to system power (B+) • RED is damaged or has failed • PCM or TCM has failed

Gas Engine OBD II Trouble Code List (P1xxx Codes)

DTC	Trouble Code Title, Conditions & Possible Causes
DTC: P1106 **2T CCM, MIL:** No **Years:** 2007, 2008 **Models:** Amanti, Optima, Rondo, Sedona, Sorento **Engines:** 2.7L, 3.3L, 3.8L **Transmissions:** All	**Manifold Absolute Pressure Sensor Circuit Short- Intermittent High Input** This code detects an intermittent short to high in either the signal circuit or the MAP sensor. **Possible Causes:** • Poor connection • Short to battery in signal circuit • Open in ground circuit • Faulty MAPS • Faulty PCM
DTC: P1107 **2T CCM, MIL:** No **Years:** 2007, 2008 **Models:** Amanti, Optima, Rondo, Sedona, Sorento **Engines:** 2.7L, 3.3L, 3.8L **Transmissions:** All	**Manifold Absolute Pressure Sensor Circuit Short- Intermittent Low Input** This code detects an intermittent short to high in either the signal circuit or the MAP sensor. **Possible Causes:** • Poor connection • Open or short to ground in the power circuit • Open or short to ground in the signal circuit • Faulty MAPS • Faulty PCM

DTC	Trouble Code Title, Conditions & Possible Causes
DTC: P1111 **2T CCM, MIL: No** **Years:** 2007, 2008 **Models:** Amanti, Optima, Rondo, Sedona, Sorento **Engines:** 2.7L, 3.3L, 3.8L **Transmissions:** All	**Intake Air Temperature Sensor Circuit Short- Intermittent High Input** This code detects a continuous short to high in either the signal circuit or the sensor. **Possible Causes:** • Poor connection • Open or short in signal circuit • Open in ground circuit • Faulty IATS • Faulty PCM
DTC: P1112 **2T CCM, MIL: No** **Years:** 2007, 2008 **Models:** Amanti, Optima, Rondo, Sedona, Sorento **Engines:** 2.7L, 3.3L, 3.8L **Transmissions:** All	**Intake Air Temperature Sensor Circuit Short- Intermittent Low Input** This code detects a continuous short to high in either the signal circuit or the sensor. **Possible Causes:** • Poor connection • Short to ground in the signal circuit • Open in ground circuit • Faulty IATS • Faulty PCM
DTC: P1114 **2T CCM, MIL: No** **Years:** 2007, 2008 **Models:** Amanti, Optima, Rondo, Sedona, Sorento **Engines:** 2.7L, 3.3L, 3.8L **Transmissions:** All	**Engine Coolant temperature Sensor Circuit- Intermittent Low Input** This code detects an intermittent short to ground in the signal circuit or the sensor. **Possible Causes:** • Poor connection • Short to ground in signal circuit • Open in ground circuit • Faulty ECTS • Faulty PCM
DTC: P1115 **2T 2 HTR2, MIL: Yes** **Years:** 2007, 2008 **Models:** Amanti, Optima, Rondo, Sedona, Sorento **Engines:** 2.7L, 3.3L, 3.8L **Transmissions:** All	**HO2S-12 (Bank 1 Sensor 2) Heater Circuit Low Input** Engine running, and PCM detected an unexpected voltage condition on the rear HO2S heater control circuit. **Note: Inspect the HO2S connector for oil and water contamination.** **Possible Causes:** • HO2S heater control circuit open between HO2S and the PCM • HO2S heater control circuit is shorted to ground • HO2S heater is damaged or has failed • PCM has failed
DTC: P1295 **2T CCM, MIL: No** **Years:** 2007, 2008 **Models:** Amanti, Optima, Rondo, Sedona, Sorento **Engines:** 2.7L, 3.3L, 3.8L **Transmissions:** All	**Electronic Throttle Control (ETC) System Malfunction- Power Management** This code is set is there is a problem in the power management system. Ignition ON. **Possible Causes:** • TPS malfunction • TPS malfunction plus MAFS malfunction • MAP malfunction plus TPS malfunction • Faulty PCM
DTC: P1505 **2T CCM, MIL: Yes** **Years:** 2007, 2008 **Models:** Rio, Spectra, Sportage **Engines:** 1.6L, 2.0L, 2.7L **Transmissions:** All	**IAC Valve Opening Coil Signal Low** Engine runtime over 5 seconds, and the PCM detected the IAC Valve Opening Coil signal remained in a low state during the test. **Possible Causes:** • IAC valve control signal is open • IAC valve control signal is shorted to ground • IAC valve is damaged or has failed • PCM has failed (IAC "driver" circuit may be open in the PCM)
DTC: P1506 **2T CCM, MIL: Yes** **Years:** 2007, 2008 **Models:** Rio, Spectra, Sportage **Engines:** 1.6L, 2.0L, 2.7L **Transmissions:** All	**IAC Valve Opening Coil Signal High** Engine runtime over 5 seconds, and the PCM detected the IAC Valve Opening Coil signal remained in a high state during the test. **Possible Causes:** • IAC valve control signal is shorted to system power (B+) • IAC valve is damaged or has failed • PCM has failed (IAC "driver" circuit may be shorted in the PCM)

DTC	Trouble Code Title, Conditions & Possible Causes
DTC: P1507 **2T CCM, MIL: Yes** **Years:** 2007, 2008 **Models:** Rio, Spectra, Sportage **Engines:** 1.6L, 2.0L, 2.7L **Transmissions:** All	**IAC Valve Closing Coil Signal Low** Engine runtime over 5 seconds, and the PCM detected the IAC Valve Closing Coil signal remained in a low state during the test. **Possible Causes:** • IAC valve control signal is open • IAC valve control signal is shorted to ground • IAC valve is damaged or has failed • PCM has failed (IAC "driver" circuit may be open in the PCM)
DTC: P1508 **2T CCM, MIL: Yes** **Years:** 2007, 2008 **Models:** Rio, Sportage **Engines:** 1.6L, 2.0L, 2.7L **Transmissions:** All	**IAC Valve Closing Coil Signal High** Engine runtime over 5 seconds, and the PCM detected the IAC Valve Closing Coil signal remained in a high state during the test. **Possible Causes:** • IAC valve control signal is shorted to system power (B+) • IAC valve is damaged or has failed • PCM has failed (IAC "driver" circuit may be shorted in the PCM)
DTC: P1523 **2T CCM, MIL: Yes** **Years:** 2007, 2008 **Models:** Amanti, Optima, Rondo, Sedona, Sorento **Engines:** 2.7L, 3.3L, 3.8L **Transmissions:** All	**VICS Solenoid Valve Circuit Malfunction** Key on or engine running and PCM detected an unexpected voltage condition on the VICS Solenoid valve circuit during the CCM test. **Possible Causes:** • VICS solenoid control circuit is open • VICS solenoid control circuit is shorted to ground • VICS solenoid is damaged or has failed • PCM has failed
DTC: P161B **2T CCM, MIL: Yes** **Years:** 2007, 2008 **Models:** Amanti, Optima, Rondo, Sedona, Sorento **Engines:** 2.7L, 3.3L, 3.8L **Transmissions:** All	**PCM Internal Error- Torque Calculating** This code is set if delivered torque is grossly different from the desired torque. **Possible Causes:** • Intake air leakage • Faulty ETS system • Clogged exhaust system • Faulty PCM
DTC: P2104 **2T CCM, MIL: Yes** **Years:** 2007, 2008 **Models:** Amanti, Optima, Rondo, Sedona, Sorento **Engines:** 2.4L, 2.7L, 3.3L, 3.8L **Transmissions:** All	**Electronic Throttle Control (ETC) System Malfunction- Forced Idle** This code is set if the system is in forced idle mode. Ignition ON. **Possible Causes:** • Faulty AFS • Faulty AFS plus brake • Faulty AFS plus vehicle speed sensor • Faulty AFS plus brake plus vehicle speed sensor • Faulty PCM
DTC: P2105 **2T CCM, MIL: Yes** **Years:** 2007, 2008 **Models:** Amanti, Optima, Rondo, Sedona, Sorento **Engines:** 2.4L, 2.7L, 3.3L, 3.8L **Transmissions:** All	**Electronic Throttle Control (ETC) System Malfunction- Forced Engine Shutdown** This code is set if the system is in forced engine shutdown mode. Ignition ON. **Possible Causes:** • Faulty AFS plus MAPS plus ETS • Faulty PCM
DTC: P2106 **2T CCM, MIL: Yes** **Years:** 2007, 2008 **Models:** Amanti, Optima, Rondo, Sedona, Sorento **Engines:** 2.4L, 2.7L, 3.3L, 3.5L, 3.8L **Transmissions:** All	**Electronic Throttle Control (ETC) System Malfunction- Forced Limited Power** This code is set if the system is in forced limited power mode. Ignition ON. **Possible Causes:** • Faulty APS • Faulty APS + Brake • Faulty APS + vehicle speed sensor • Faulty APS + vehicle speed sensor + brake • Faulty PCM
DTC: P2122 **2T CCM, MIL: Yes** **Years:** 2007, 2008 **Models:** Amanti, Optima, Rondo, Sedona, Sorento **Engines:** 2.4L, 2.7L, 3.3L, 3.5L, 3.8L **Transmissions:** All	**Throttle/Pedal Position Sensor/Switch "D" Circuit Low Input** Accelerator position sensor (APS1) low input. ETS/PCM communication is normal. Output voltage of APS1 is less than 0.2 volt. **Possible Causes:** • Poor connector • Faulty APS1 • Open or short in APS1 circuit • Faulty PCM

DTC	Trouble Code Title, Conditions & Possible Causes
DTC: P2123 **2T CCM, MIL:** Yes **Years:** 2007, 2008 **Models:** Amanti, Optima, Rondo, Sedona, Sorento **Engines:** 2.4L, 2.7L, 3.3L, 3.5L, 3.8L **Transmissions:** All	**Throttle/Pedal Position Sensor/Switch "D" Circuit High Input** Accelerator position sensor (APS1) high input. ETS/PCM communication is normal. Output voltage of APS1 is equal to or greater than 4.9 volts. Output voltage of APS2 is less than 4.1 volts. **Possible Causes:** • Poor connector • Faulty APS1 • Open or short in APS1 circuit • Faulty PCM
DTC: P2127 **2T CCM, MIL:** Yes **Years:** 2007, 2008 **Models:** Amanti, Optima, Rondo, Sedona, Sorento **Engines:** 2.4L, 2.7L, 3.3L, 3.5L, 3.8L **Transmissions:** All	**Throttle/Pedal Position Sensor/Switch "E" Circuit Low Input** Accelerator position sensor (APS2) low input. ETS/PCM communication is normal. Output voltage of APS2 is less than 0.2 volt. **Possible Causes:** • Poor connection • Faulty APS2 • Open or short in APS2 circuit • Faulty PCM
DTC: P2128 **2T CCM, MIL:** Yes **Years:** 2007, 2008 **Models:** Amanti, Optima, Rondo, Sedona, Sorento **Engines:** 2.4L, 2.7L, 3.3L, 3.5L, 3.8L **Transmissions:** All	**Throttle/Pedal Position Sensor/Switch "E" Circuit High Input** Accelerator position sensor (APS2) high input. ETS/PCM communication is normal. Output voltage of APS2 is greater than or equal to 4.9 volts. Output voltage of ASP1 is less than 4.1 volts. **Possible Causes:** • Poor connection • Faulty APS2 • **Open or short in APS2 circuit** • **Faulty PCM**
DTC: P2135 **2T CCM, MIL:** Yes **Years:** 2007, 2008 **Models:** Amanti, Optima, Rondo, Sedona, Sorento **Engines:** 2.4L, 2.7L, 3.3L, 3.5L, 3.8L **Transmissions:** All	**Throttle/Pedal Position Sensor/Switch "A"/"B" Voltage Correlation** Determines if TPS No. 1 disagrees with TPS No. 2. Ignition "ON". **Possible Causes:** • Poor connection • Open or short in TPS circuit • Faulty TPS • Faulty PCM
DTC: P2138 **2T CCM, MIL:** Yes **Years:** 2007, 2008 **Models:** Amanti, Optima, Rondo, Sedona, Sorento **Engines:** 2.4L, 2.7L, 3.3L, 3.8L **Transmissions:** All	**Throttle/Pedal Position Sensor/Switch "D/E" Voltage Correlation** Monitoring abnormal APS. Output voltage of APS1: 0.2 to 4.9 volts. Output voltage of APS2: 0.2 to 4.9 volts. Ignition switch ON. **Possible Causes:** • Poor connection • Faulty APS • Faulty PCM
DTC: P2173 **2T CCM, MIL:** Yes **Years:** 2007, 2008 **Models:** Amanti, Optima, Rondo, Sedona, Sorento **Engines:** 2.7L, 3.3L, 3.8L **Transmissions:** All	**Electronic Throttle Control (ETC) System Malfunction- High Air Flow Detected** The engine airflow measurements are not based on throttle position. They are compared with throttle position based on estimated air flow. If measured air flow is much higher, the throttle body may not be throttling the engine. Engine running. Throttle actuation mode is not off. MAP sensor is not failed. MAF sensor is not failed. IAT sensor is not failed. **Possible Causes:** • Air leakage between TPS and MAFS • Faulty throttle body • Faulty PCM
DTC: P2187 **2T CCM, MIL:** Yes **Years:** 2007, 2008 **Models:** Amanti, Optima, Rio, Rondo, Sedona, Sorento **Engines:** 1.6L, 2.4L, 2.7L, 3.3L, 3.8L **Transmissions:** All	**System Too Lean At Idle (Additive) (Bank 1)** Engine coolant temperature 140°F. Intake air temperature 140°F. System voltage greater than 11 volts. Closed loop active. **Possible Causes:** • Sensors related to fuel trim • Intake system • Fuel pressure • Faulty PCM

DTC	Trouble Code Title, Conditions & Possible Causes
DTC: P2188 **2T CCM, MIL: Yes** **Years:** 2007, 2008 **Models:** Amanti, Optima, Rio, Rondo, Sedona, Sorento **Engines:** 1.6L, 2.4L, 2.7L, 3.3L, 3.8L **Transmissions:** All	**System Too Lean At Idle (Additive) (Bank 2)** Engine coolant temperature 140°F. Intake air temperature 140°F. System voltage greater than 11 volts. Closed loop active. **Possible Causes:** • Sensors related to fuel trim • Intake system • Fuel pressure • Faulty PCM
DTC: P2190 **2T CCM, MIL: Yes** **Years:** 2007, 2008 **Models:** Amanti, Rondo, Sedona, Sorento **Engines:** 2.7L, 3.3L, 3.8L **Transmissions:** All	**System Too Rich At Idle (Additive) (Bank 2)** Engine coolant temperature 140°F. Intake air temperature 140°F. System voltage greater than 11 volts. Closed loop active. **Possible Causes:** • Sensors related to fuel trim • Intake system • Fuel pressure • Faulty PCM
DTC: P2195 **2T CCM, MIL: Yes** **Years:** 2007, 2008 **Models:** Amanti, Rondo, Sedona, Sorento, Spectra **Engines:** 2.0L, 2.7L, 3.3L, 3.8L **Transmissions:** All	**HO2S Signal Stuck Lean (Bank 1 Sensor 1)** Sensor characteristic line shifted to lean. No relevant failure. No misfire detected. Fuel trim control active. **Possible Causes:** • Contact resistance in connectors • Faulty HO2S • Faulty PCM
DTC: P2196 **2T CCM, MIL: Yes** **Years:** 2007, 2008 **Models:** Amanti, Rondo, Sedona, Sorento, Spectra **Engines:** 2.0L, 2.7L, 3.3L, 3.8L **Transmissions:** All	**HO2S Signal Stuck Rich (Bank 1 Sensor 1)** Sensor characteristic line shifted to lean. No relevant failure. No misfire detected. Fuel trim control active. **Possible Causes:** • Contact resistance in connectors • Faulty HO2S • Faulty PCM
DTC: P2197 **2T CCM, MIL: Yes** **Years:** 2007, 2008 **Models:** Amanti, Rondo, Sedona, Sorento **Engines:** 2.7L, 3.3L, 3.8L **Transmissions:** All	**HO2S Signal Stuck Lean (Bank 2 Sensor 1)** Determines if O2 sensor indicates lean exhaust while in power enrichment. Sensor not in cooled status flag. Not in transient conditions status flag. Device control not active. Engine running. Minimum air flow present is equal or greater than 2 g/s. Engine coolant warm (140°F. Above conditions met for at least 1.5L seconds. **Possible Causes:** • Poor connection • Faulty HO2S • Faulty PCM
DTC: P2198 **2T CCM, MIL: Yes** **Years:** 2007, 2008 **Models:** Amanti, Rondo, Sedona, Sorento **Engines:** 2.7L, 3.3L, 3.8L **Transmissions:** All	**HO2S Signal Stuck Rich (Bank 2 Sensor 1)** Determines if O2 sensor indicates rich exhaust while in decal fuel cut off (DFCO). Sensor not in cooled status flag. Not in transient conditions status flag. Device control not active. Engine running. Minimum air flow present is equal or greater than 2 g/s. Ignition voltage equal to or greater than 10 volts. Fuel reduction not active. Engine running long enough (more than 60 seconds). Engine coolant warm (140°F. Above conditions met for at least 1.5L seconds. **Possible Causes:** • Poor connection • Faulty HO2S • Faulty PCM
DTC: P2270 **2T CCM, MIL: Yes** **Years:** 2007, 2008 **Models:** Amanti, Rondo, Sedona, Sorento, Spectra **Engines:** 2.0L, 2.7L, 3.3L, 3.8L **Transmissions:** All	**HO2S Signal Stuck Rich (Bank 1 Sensor 2)** Plausibility check during shift of lambda set point to rich from lean. No fuel cut off. No full load phase. No fuel trim error detected. Delay time to start diagnosis: 13 to 30 seconds. No relevant failure. **Possible Causes:** • Three way catalytic converter (TWC) • Air leakage in exhaust system • Faulty rear HO2S sensor
DTC: P2271 **2T CCM, MIL: Yes** **Years:** 2007, 2008 **Models:** Amanti, Rondo, Sedona, Sorento, Spectra **Engines:** 2.0L, 2.7L, 3.3L, 3.8L **Transmissions:** All	**O2 Signal Stuck Rich (Bank 1/2 Sensor 1)** Plausibility check during shift of lambda set point to rich from lean. No fuel cut off. No full load phase. No fuel trim error detected. Delay time to start diagnosis: 13 to 30 seconds. No relevant failure. **Possible Causes:** • Three way catalytic converter (TWC) • Air leakage in exhaust system • Faulty rear HO2S sensor

DTC	Trouble Code Title, Conditions & Possible Causes
DTC: P2272 **2T CCM, MIL: Yes** **Years:** 2007, 2008 **Models:** Amanti, Rondo, Sedona, Sorento **Engines:** 2.7L, 3.3L, 3.8L **Transmissions:** All	**HO2S Signal Stuck Lean (Bank 2 Sensor 2)** Determines if O2 sensor indicates lean exhaust while in power enrichment mode. Sensor not in cooled status flag. Not in transient conditions status flag. Device control not active. Engine running. Minimum air flow present is equal or greater than 2 g /s. Ignition voltage equal to or greater than 10 volts. Fuel reduction not active. Engine running long enough (more than 60 seconds). Engine coolant warm (140°F. Above conditions met for at least 2.5 seconds. **Possible Causes:** • Poor connection • Faulty HO2S • Faulty PCM
DTC: P2273 **2T CCM, MIL: Yes** **Years:** 2007, 2008 **Models:** Amanti, Rondo, Sedona, Sorento **Engines:** 2.7L, 3.3L, 3.8L **Transmissions:** All	**HO2S Signal Stuck Rich (Bank 2 Sensor 2)** Determines if O2 sensor indicates rich exhaust while in decal fuel cut off (DFCO). Sensor not in cooled status flag. Not in transient conditions status flag. Device control not active. Engine running. Minimum air flow present is equal or greater than 2 g /s. Ignition voltage equal to or greater than 10 volts. Fuel reduction not active. Engine running long enough (more than 60 seconds). Engine coolant warm (140°F. Above conditions met for at least 2.0L seconds. **Possible Causes:** • Poor connection • Faulty HO2S • Faulty PCM
DTC: P2422 **2T CCM, MIL: Yes** **Years:** 2007, 2008 **Models:** Amanti, Rondo, Sedona, Sorento **Engines:** 2.7L, 3.3L, 3.8L **Transmissions:** All	**Evaporative Emission System- Canister Clogging** Ignition voltage 10-16 volts. Barometric pressure 72kpa. Engine run time, one second. **Possible Causes:** • Faulty canister close valve • Clogging of canister air filter • Open in ground harness of FTPS • Faulty PCM
DTC: P2610 **2T CCM, MIL: Yes** **Years:** 2007, 2008 **Models:** Amanti, Optima, Rondo, Sedona, Sorento **Engines:** 2.7L, 3.3L, 3.8L **Transmissions:** All	**ECM/PCM Internal Engine Off Timer Performance** The LPC SPI diagnostic allows the low power counter to count down and simultaneously enables a test timer to run for a calibrated length of time and then compares the lapsed time recorded by the counter to make a pass/fail determination. Engine running. Enough time (10 seconds). Battery voltage 8 volts. No memory failure. **Possible Causes:** • Faulty PCM
DTC: P2A00 **2T CCM, MIL: Yes** **Years:** 2007, 2008 **Models:** Amanti, Optima, Rondo, Sedona, Sorento **Engines:** 2.7L, 3.3L, 3.8L **Transmissions:** All	**O2 Sensor Not ready (Bank 1 Sensor 1)** Detects loss of O2 ready status, which would lead to open loop fueling operation, a default mode. Engine running. Ignition ON. DFCO not present too long (less than 15 seconds). No disabling faults present. All of the above for at least 20 seconds. **Possible Causes:** • Poor connection • Faulty HO2S • Faulty PCM
DTC: P2A03 **2T CCM, MIL: Yes** **Years:** 2007, 2008 **Models:** Amanti, Optima, Rondo, Sedona, Sorento **Engines:** 2.7L, 3.3L, 3.8L **Transmissions:** All	**O2 Sensor Not ready (Bank 1 Sensor 2)** Detects loss of O2 ready status, which would lead to open loop fueling operation, a default mode. Engine running. Ignition ON. DFCO not present too long (less than 15 seconds). No disabling faults present. All of the above for at least 20 seconds. **Possible Causes:** • Poor connection • Faulty HO2S • Faulty PCM

SPECIFICATIONS AND MAINTENANCE CHARTS

ENGINE AND VEHICLE IDENTIFICATION

	Engine						Model Year	
Code ①	Liters (cc)	Cu. In.	Cyl.	Fuel Sys.	Engine Type	Eng. Mfg.	Code ②	Year
VQ30DE	3.5 (3498)	213	6	MFI	DOHC	Nissan	7	2007
VQ35HR	3.5 (3498)	213	6	MFI	DOHC	Nissan	8	2008

MFI: Multi-port Fuel Injection

DOHC: Double Overhead Camshaft

① The Engine Code is stamped on the engine block near the starter.

② 10th position of the Vehicle Identification Number (VIN)

22140_350Z_C0001

GENERAL ENGINE SPECIFICATIONS

Year	Model	Engine Displacement Liters	Engine Series (ID/VIN)	Net Horsepower @ rpm	Net Torque @ rpm (ft. lbs.)	Bore x Stroke (in.)	Compression Ratio	Oil Pressure @ rpm
2007	Maxima	3.5	VQ35DE	255@6000	252@4400	3.76X3.20	10.3:1	43@2000
	350Z	3.5	VQ35HR	306@6000	268@4800	3.76X3.20	10.0:1	43@2000
2008	Maxima	3.5	VQ35DE	255@6000	252@4400	3.76X3.20	10.3:1	43@2000
	350Z	3.5	VQ35HR	306@6000	268@4800	3.76X3.20	10.0:1	43@2000

22140_350Z_C0002

ENGINE TUNE-UP SPECIFICATIONS

Year	Model	Engine Displacement Liters	Engine ID/VIN	Spark Plug Gap (in.)	Ignition Timing (deg.) MT	Ignition Timing (deg.) AT	Fuel Pump (psi) ①	Idle Speed (rpm) MT	Idle Speed (rpm) AT ②	Valve Clearance (in.) Intake ③	Valve Clearance (in.) Exhaust ③
2007	Maxima	3.5	VQ35DE	0.043	NA	NA	51	—	600-700	0.012-0.016	0.012-0.017
	350Z	3.5	VQ35HR	0.043	16B	16B	51	600-700	600-700	0.010-0.013	0.011-0.015
2008	Maxima	3.5	VQ35DE	0.043	NA	NA	51	—	600-700	0.012-0.016	0.012-0.017
	350Z	3.5	VQ35HR	0.043	16B	16B	51	600-700	600-700	0.010-0.013	0.011-0.015

NOTE: The Vehicle Emission Control Information label often reflects specification changes made during production.

The label figures must be used if they differ from those in this chart.

NA: Not Available

B: Before top dead center

① System pressure at idle with vacuum hose connected; should increase to 43 psi when disconnected

② Automatic transmission in neutral

③ Engine cold

22140_350Z_C0003

CAPACITIES

Year	Model	Engine ID/VIN	Engine Displacement Liters	Engine Oil with Filter (qts.)	Transmission (pts.) Man	Transmission (pts.) Auto.	Drive Axle Rear (pts.)	Fuel Tank (gal.)	Cooling System (qts.)
2007	Maxima	VQ35DE	3.5	4.5	—	21.2	—	20.0	11.0
	350Z	VQ35HR	3.5	5.0	6.25	21.3	3.0	20.0	9.25
2008	Maxima	VQ35DE	3.5	4.5	—	21.2	—	20.0	11.0
	350Z	VQ35HR	3.5	5.0	6.25	21.3	3.0	20.0	9.25

NOTE: All capacities are approximate. Add fluid gradually and check to be sure a proper fluid level is obtained.

22140_350Z_C0004

VALVE SPECIFICATIONS

Year	Engine ID/VIN	Engine Displacement Liters	Seat Angle (deg.)	Face Angle (deg.)	Spring Test Pressure (lbs. @ in.)	Spring Installed Height (in.)	Stem-to-Guide Clearance (in.) Intake	Stem-to-Guide Clearance (in.) Exhaust	Stem Diameter (in.) Intake	Stem Diameter (in.) Exhaust
2007	VQ35DE	3.5	45.15-45.45	45	37-42@1.4567	1.466	0.0008-0.0021	0.0016-0.0029	0.2348-0.2354	0.2341-0.2346
	VQ35HR	3.5	45.15-45.45	45	37-42@1.4567	1.466	0.0008-0.0021	0.0012-0.0022	0.2348-0.2354	0.2344-0.2350
2008	VQ35DE	3.5	45.15-45.45	45	37-42@1.4567	1.466	0.0008-0.0021	0.0016-0.0029	0.2348-0.2354	0.2341-0.2346
	VQ35HR	3.5	45.15-45.45	45	37-42@1.4567	1.466	0.0008-0.0021	0.0012-0.0022	0.2348-0.2354	0.2344-0.2350

22140_350Z_C0005

CRANKSHAFT AND CONNECTING ROD SPECIFICATIONS

All measurements are given in inches.

Year	Engine Displacement Liters	Engine ID/VIN	Crankshaft Main Brg. Journal Dia.	Crankshaft Main Brg. Oil Clearance	Crankshaft Shaft End-play	Crankshaft Thrust on No.	Connecting Rod Journal Diameter	Connecting Rod Oil Clearance	Connecting Rod Side Clearance
2007	3.5	VQ35DE	2.3603-2.3612*	0.0014-0.0018	0.0039-0.0098	3	2.0445-2.0462*	0.0013-0.0023	0.0079-0.0138
	3.5	VQ35HR	2.5571-2.5581*	0.0014-0.0018	0.0039-0.0098	3	2.1242-2.1250*	0.0016-0.0021	0.0079-0.0138
2008	3.5	VQ35DE	2.3603-2.3612*	0.0014-0.0018	0.0039-0.0098	3	2.0445-2.0462*	0.0013-0.0023	0.0079-0.0138
	3.5	VQ35HR	2.5571-2.5581*	0.0014-0.0018	0.0039-0.0098	3	2.1242-2.1250*	0.0016-0.0021	0.0079-0.0138

* Based upon grade

22140_350Z_C0008

PISTON AND RING SPECIFICATIONS

All measurements are given in inches.

Year	Engine Displ. Liters	Engine ID/VIN	Piston Clearance	Ring Gap			Ring Side Clearance		
				Top Compression	Bottom Compression	Oil Control	Top Compression	Bottom Compression	Oil Control
2007	3.5	VQ35DE/HR	0.0004- 0.0012	0.0091- 0.0130	0.0130- 0.0189	0.0067- 0.0185	0.0016- 0.0031	0.0012- 0.0028	0.0022- 0.0061
2008	3.5	VQ35DE/HR	0.0004- 0.0012	0.0091- 0.0130	0.0130- 0.0189	0.0067- 0.0185	0.0016- 0.0031	0.0012- 0.0028	0.0022- 0.0061

22140_350Z_C0007

TORQUE SPECIFICATIONS

All readings in ft. lbs.

Year	Engine Displacement Liters	Engine ID/VIN	Cylinder Head Bolts	Main Bearing Bolts	Rod Bearing Bolts	Crankshaft Damper Bolts	Flywheel Bolts	Manifold		Spark Plugs	Oil Drain Plug
								Intake	Exhaust		
2007	3.5	VQ35DE/HR	①	②	③	④	65	⑤	⑥	18	25
2008	3.5	VQ35DE/HR	①	②	③	④	65	⑤	⑥	18	25

① Step 1: 77 ft. lbs.
   Step 2: Loosen bolts completely
   Step 3: 30 ft. lbs.
   Step 4: Tighten an additional 95 degrees
   Step 5: Tighten an additional 95 degrees

② Step 1: 18 ft. lbs.
   Step 2: 26 ft.lbs.
   Step 3: Tighten an additional 90 degrees

③ Step 1: 21 ft. lbs.
   Step 2: Loosen bolts completely
   Step 3: 15 ft.lbs.
   Step 4: Tighten an additional 90 degrees

④ Step 1: 33 ft. lbs.
   Step 2: Tighten an additional 90 degrees

⑤ Step 1: 5 ft. lbs.
   Step 2: 20 ft. lbs.

⑥ Step 1: 11 ft. lbs. in 2 steps

22140_350Z_C0006

WHEEL ALIGNMENT

Year	Model		Caster		Camber		Toe-in (in.)
			Range (+/-Deg.)	Preferred Setting (Deg.)	Range (+/-Deg.)	Preferred Setting (Deg.)	
2007	Maxima	F	0.75	+2.83	0.75	-0.25	0.04 +/- 0.04
		R	—	—	0.50	-0.00	0.11 +/- 0.06
	350Z	F	0.75	+8.17	0.75 / 0.40	-0.58	0.04 +/- 0.04
		R	—	—	0.50	-1.58	0.075 +/- 0.03
2008	Maxima	F	0.75	+2.83	0.75	-0.25	0.04 +/- 0.04
		R	—	—	0.50	-0.00	0.11 +/- 0.06
	350Z	F	0.75	+8.17	0.75 / 0.40	-0.58	0.04 +/- 0.04
		R	—	—	0.50	-1.58	0.075 +/- 0.03

22140_350Z_C0009

TIRE, WHEEL AND BALL JOINT SPECIFICATIONS

| Year | Model | OEM Tires | | Tire Pressures (psi) | | Wheel Size | Lug Nut Torque (ft. lbs.) |
		Standard	Optional	Front	Rear		
2007	Maxima	P225/55R17	245/45R18	33	32	7-JJ/7.5-JJ	80
	350Z Front	225/45R18	245/45R18	35	35	8JJ/8.5JJ	80
	350Z Rear	245/40R18	265/35R19	35	35	9JJ/10JJ	80
2008	Maxima	P225/55R17	245/45R18	33	32	7-JJ/7.5-JJ	80
	350Z Front	225/45R18	245/45R18	35	35	8JJ/8.5JJ	80
	350Z Rear	245/40R18	265/35R19	35	35	9JJ/10JJ	80

OEM: Original Equipment Manufacturer

PSI: Pounds Per Square Inch

22140_350Z_C0010

BRAKE SPECIFICATIONS

All measurements in inches unless noted

| Year | Model | | Brake Disc | | | Minimum Lining Thickness | | Brake Caliper | |
			Original Thickness	Minimum Thickness	Maximum Run-out	Front	Rear	Bracket Bolts (ft. lbs.)	Mounting Bolts (ft. lbs.)
2007	Maxima	F	1.100	0.1.02	0.003	0.079	—	101-129	17-22
		R	0.350	0.310	0.002	—	0.039	62	32
	350Z ①	F	1.102	1.024	0.001	0.079	0.079	113-114	20
		R	0.630	0.551	0.008	0.079	0.079	62	32
	350Z ②	F	1.181	1.118	0.002	0.079	0.079	—	113
		R	0.866	0.795	0.002	0.079	0.079	53-71	—
2008	Maxima	F	1.100	0.1.02	0.003	0.079	—	101-129	17-22
		R	0.350	0.310	0.002	—	0.039	62	32
	350Z ①	F	1.102	1.024	0.001	0.079	0.079	113-114	20
		R	0.630	0.551	0.008	0.079	0.079	62	32
	350Z ②	F	1.181	1.118	0.002	0.079	0.079	—	113
		R	0.866	0.795	0.002	0.079	0.079	53-71	—

NA: Not Available

① Other than Brembo

② Brembo

22140_350Z_C0011

SCHEDULED MAINTENANCE INTERVALS
NISSAN—Maxima & 350Z

TO BE SERVICED	TYPE OF SERVICE	VEHICLE MILEAGE INTERVAL (x1000)												
		7.5	15	22.5	30	37.5	45	52.5	60	67.5	75	82.5	90	97.5
Engine oil & filter	R	✓	✓	✓	✓	✓	✓	✓	✓	✓	✓	✓	✓	✓
Brake lines & cables	S/I		✓		✓		✓		✓		✓		✓	
Brake pads & discs	S/I		✓		✓		✓		✓		✓		✓	
Driveshaft boots	S/I		✓		✓		✓		✓		✓		✓	
Exhaust system	S/I				✓				✓				✓	
Transmission or transaxle fluid	S/I		✓		✓		✓		✓		✓		✓	
Air cleaner filter	R				✓				✓				✓	
Spark plugs (except platinum)	R				✓				✓				✓	
Spark plugs (platinum tip)	R								✓					
Steering gear & linkage, axle & suspension parts	S/I				✓				✓				✓	
Engine coolant	R								✓					
Drive belts	S/I								✓					
Fuel lines	S/I								✓					
Vapor lines	S/I								✓					

R: Replace   S/I: Service or Inspect

FREQUENT OPERATION MAINTENANCE (SEVERE SERVICE)

If a vehicle is operated under any of the following conditions it is considered severe service:

- Extremely dusty areas.
- 50% or more of the vehicle operation is in 32°C (90°F) or higher temperatures, or constant operation in temperatures below 0°C (32°F).
- Prolonged idling (vehicle operation in stop and go traffic).
- Frequent short running periods (engine does not warm to normal operating temperatures).
- Police, taxi, delivery usage or trailer towing usage.

Oil & oil filter: change every 3750 miles.

Brake pads & discs: service or inspect every 7500 miles.

Driveshaft boots: service or inspect every 7500 miles.

Exhaust system: service or inspect every 7500 miles.

Steering gear & linkage, axle & suspension parts: service or inspect every 7500 miles.

Steering linkage ball joints & front suspension ball joints: service or inspect every 7500 miles.

Air cleaner filter: service or inspect every 15,000 miles.

22140_350Z_C0012

PRECAUTIONS

Before servicing any vehicle, please be sure to read all of the following precautions, which deal with personal safety, prevention of component damage, and important points to take into consideration when servicing a motor vehicle:

• Never open, service or drain the radiator or cooling system when the engine is hot; serious burns can occur from the steam and hot coolant.

• Observe all applicable safety precautions when working around fuel. Whenever servicing the fuel system, always work in a well-ventilated area. Do not allow fuel spray or vapors to come in contact with a spark, open flame, or excessive heat (a hot drop light, for example). Keep a dry chemical fire extinguisher near the work area. Always keep fuel in a container specifically designed for fuel storage; also, always properly seal fuel containers to avoid the possibility of fire or explosion. Refer to the additional fuel system precautions later in this section.

• Fuel injection systems often remain pressurized, even after the engine has been turned **OFF**. The fuel system pressure must be relieved before disconnecting any fuel lines. Failure to do so may result in fire and/or personal injury.

• Brake fluid often contains polyglycol ethers and polyglycols. Avoid contact with the eyes and wash your hands thoroughly after handling brake fluid. If you do get brake fluid in your eyes, flush your eyes with clean, running water for 15 minutes. If eye irritation persists, or if you have taken brake fluid internally, IMMEDIATELY seek medical assistance.

• The EPA warns that prolonged contact with used engine oil may cause a number of skin disorders, including cancer. You should make every effort to minimize your exposure to used engine oil. Protective gloves should be worn when changing oil. Wash your hands and any other exposed skin areas as soon as possible after exposure to used engine oil. Soap and water, or waterless hand cleaner should be used.

• All new vehicles are now equipped with an air bag system, often referred to as a Supplemental Restraint System (SRS) or Supplemental Inflatable Restraint (SIR) system. The system must be disabled before performing service on or around system components, steering column, instrument panel components, wiring and sensors. Failure to follow safety and disabling procedures could result in accidental air bag deployment, possible personal injury and unnecessary system repairs.

• Always wear safety goggles when working with, or around, the air bag system. When carrying a non-deployed air bag, be sure the bag and trim cover are pointed away from your body. When placing a non-deployed air bag on a work surface, always face the bag and trim cover upward, away from the surface. This will reduce the motion of the module if it is accidentally deployed. Refer to the additional air bag system precautions later in this section.

• Clean, high quality brake fluid from a sealed container is essential to the safe and proper operation of the brake system. You should always buy the correct type of brake fluid for your vehicle. If the brake fluid becomes contaminated, completely flush the system with new fluid. Never reuse any brake fluid. Any brake fluid that is removed from the system should be discarded. Also, do not allow any brake fluid to come in contact with a painted surface; it will damage the paint.

• Never operate the engine without the proper amount and type of engine oil; doing so WILL result in severe engine damage.

• Timing belt maintenance is extremely important. Many models utilize an interference-type, non-freewheeling engine. If the timing belt breaks, the valves in the cylinder head may strike the pistons, causing potentially serious (also time-consuming and expensive) engine damage. Refer to the maintenance interval charts for the recommended replacement interval for the timing belt, and to the timing belt section for belt replacement and inspection.

• Disconnecting the negative battery cable on some vehicles may interfere with the functions of the on-board computer system(s) and may require the computer to undergo a relearning process once the negative battery cable is reconnected.

• When servicing drum brakes, only disassemble and assemble one side at a time, leaving the remaining side intact for reference.

• Only an MVAC-trained, EPA-certified automotive technician should service the air conditioning system or its components.

BRAKES

BLEEDING PROCEDURE

BLEEDING PROCEDURE

✳✳ CAUTION

Be careful not to splash brake fluid on painted areas; it may cause paint damage. If brake fluid is splashed on painted areas, wash it away with water immediately. All hoses must be free from excessive bending, twisting and pulling.

✳ CAUTION

Pay attention to the following:

• **Carefully monitor brake fluid level at master cylinder during bleeding operation.**

• **If master cylinder is suspected to have air inside, bleed air from master cylinder first.**

• **Fill reservoir with new brake fluid DOT 3. Make sure it is full at all times while bleeding air out of system.**

• **Place a container under master cylinder to avoid spillage of brake fluid.**

• **For models with ABS, turn ignition switch OFF and disconnect ABS actuator connector or battery cable.**

1. Bleed the air from the brake system in the following order:

BLEEDING THE BRAKE SYSTEM

• Right rear brake, left front brake, left rear brake, light front brake

2. Connect a transparent vinyl tube to air bleeder valve.

3. Fully depress brake pedal several times.

4. With brake pedal depressed, open air bleeder valve to release air.

5. Close air bleeder valve.

6. Release brake pedal slowly.

7. Repeat steps 2 through 5 until clear brake fluid comes out of air bleeder valve.

8. Tighten air bleeder valve to 61–78 inch lbs (7–9 Nm).

BRAKES   **FRONT DISC BRAKES**

BRAKE CALIPER

REMOVAL & INSTALLATION

Maxima

1. Before servicing the vehicle, refer to the Precautions Section.
2. Remove or disconnect the following:
   - Front wheels
   - Brake fluid hose
   - Pin bolts
   - Caliper assembly from the vehicle

To install:]
3. Use a large C-clamp to press the caliper piston back into the caliper.
4. Install or connect the following:
   - New pads, new shims and pad retainers
   - Brake caliper and torque the pin bolts to 23 ft. lbs. (31 Nm)
   - Brake line to the caliper, using new copper washers, and torque the connecting bolt to 12–14 ft. lbs. (17–20 Nm).
   - Wheels
5. Bleed the brake system and top off the master cylinder as necessary.

350Z Without Brembo caliper

See Figure 1.

1. Before servicing the vehicle, refer to the Precautions Section.
2. Remove or disconnect the following:
   - Front wheels
   - Brake fluid hose
   - Sliding pin bolts
   - Caliper assembly from the vehicle

To install:
3. Use a large C-clamp to press the caliper piston back into the caliper.
4. Install or connect the following:
   - New pads, new shims and pad retainers
   - Brake caliper and torque the sliding pin bolts to17–22 ft. lbs. (22–31 Nm).
   - Brake line to the caliper, using new copper washers, and torque the connecting bolt to 12–14 ft. lbs. (17–20 Nm).
   - Wheels
5. Bleed the brake system and top off the master cylinder as necessary.

350Z With Brembo caliper

1. Before servicing the vehicle, refer to the Precautions Section.
2. Remove tires from vehicle.
3. Fasten disc rotor using wheel nut.
4. Drain brake fluid gradually (from bleed valve while depressing brake pedal).
5. Remove union bolt, and then remove brake hose from caliper assembly.
6. Remove torque member mounting bolts (from torque member), and remove caliper assembly (from vehicle with a power tool).

❋❋ **WARNING**

Do not drop brake pads.

7. Remove disc rotor.

➡**Put matching marks on both disc rotor and wheel hub when removing disc rotor.**

To install:
8. Install disc rotor.

➡**Align the matching marks of disc rotor and wheel hub, which were marked at the time of removal when reusing disc rotor.**

9. Install caliper assembly to vehicle, and tighten torque member mounting bolts to 113 ft. lbs. (154 Nm).

➡**Before installing torque member to vehicle, wipe oil and grease on washer seats on steering knuckle and mounting surface of torque member.**

10. Install a projection of brake hose metal fitting by aligning with protrusions on cylinder body, and tighten union bolt to 13 ft. lbs. (18 Nm).

❋❋ **WARNING**

Refill with new brake fluid "DOT 3". Never reuse drained brake fluid.

11. Bleed the brake system and top off the master cylinder as necessary.
12. Install tires to vehicle.

DISC BRAKE PADS

REMOVAL & INSTALLATION

Maxima

1. Before servicing the vehicle, refer to the Precautions Section.
2. Remove or disconnect the following:
   - Wheels
   - Bottom guide pin from the caliper and swing the caliper cylinder body up
   - Brake pad retainers and the pads

To install:
3. Compress the piston of the disc brake caliper.
4. Install or connect the following:
   - Brake pads, retainers, and caliper assembly. Torque the guide pin to 16–23 ft. lbs. (22–31 Nm).
   - Wheels
5. Check the master cylinder and add fluid if necessary.

350Z Without Brembo Calipers

1. Before servicing the vehicle, refer to the Precautions Section.
2. Remove or disconnect the following:
   - Front wheels
   - Sliding pin bolts
   - Rotate the caliper up and remove the brake pads

To install:
3. Install or connect the following:
   - New pads, new shims and pad retainers
   - Brake caliper
   - Wheels
4. Bleed the brake system and top off the master cylinder as necessary.

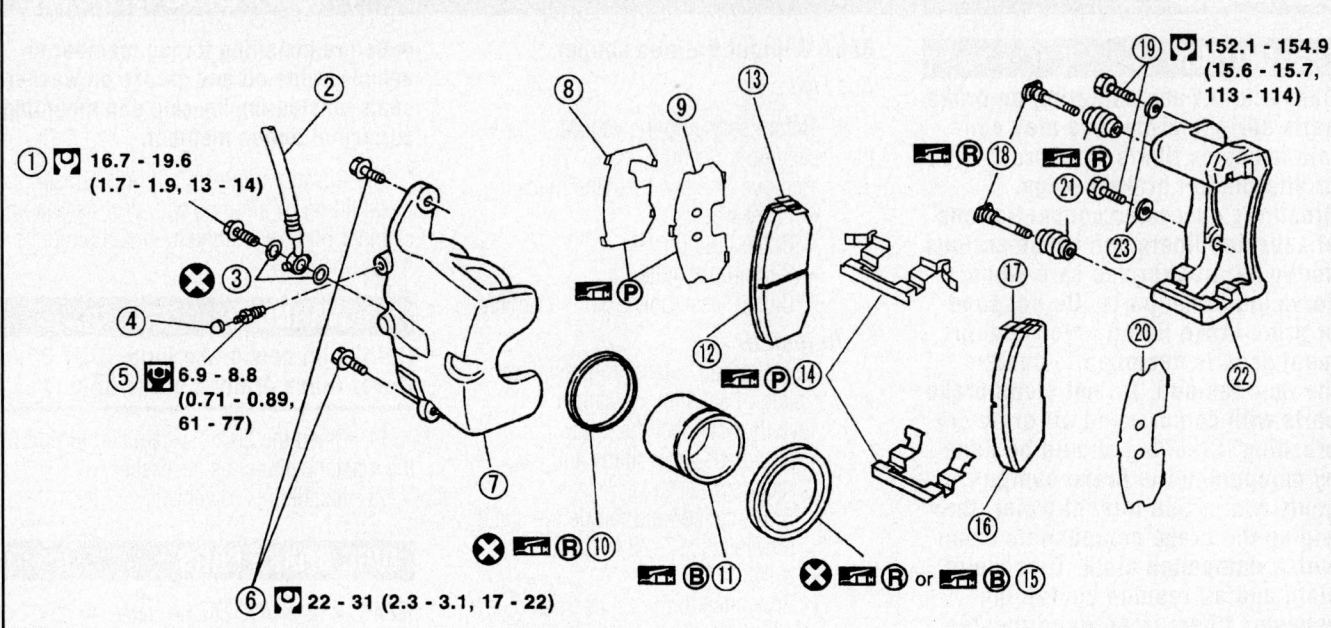

Ⓟ PBC (Poly Butyl Cuprysil)
gress or silicone-based grease

Ⓡ : Rubber Gress

Ⓑ : Brake fluid

🔧 : N•m (kg-m, ft-lb)

🔧 : N•m (kg-m, in-lb)

❌ : Always replace after every disassembly.

1.	Union bolt	2.	Brake hose	3.	Copper washer
4.	Cap	5.	Bleed valve	6.	Sliding pin bolt
7.	Cylinder body	8.	Inner shim cover	9.	Inner shim
10.	Piston seal	11.	Piston	12.	Inner pad
13.	Pad wear sensor	14.	Pad retainer	15.	Piston boot
16.	Outer pad	17.	Pad wear sensor	18.	Sliding pin bolt
19.	Torque member bolts	20.	Outer shim	21.	Slide pin boot
22.	Torque member	23.	Washer		

09482_ZMAX_G0040

Fig. 1 Front brake caliper assembly—350Z with CLZ25VD caliper

350Z With Brembo Caliper

1. Remove tires from vehicle with a power tool.

2. Remove the clip from the pad pin.

3. Remove the pad pin while holding down the cross spring, then remove the cross spring from the caliper.

4. Using pliers, remove the pad from the caliper.

To install:

5. Insert the piston to the position where the pad is attached.

6. Attach pad.

➡**The side of the shim with the larger cutouts should be on the entry side of the disc rotor spin.**

7. Insert the upper pad pin from the inner cylinder side, then insert firmly to the outer cylinder side through the hole in the top of the pad.

8. Place the top of the cross spring over the top pad pin, press in the cross spring, push the lower pad pin from the inner cylinder side to the outer cylinder side, and secure the cross spring.

9. Insert the clip in the small hole at the end of the pad pin.

➡**If the clip is not fully attached, the pad pin or the pad could fall out while the vehicle is in motion.**

10. Install tires to vehicle.

BRAKES

REAR DISC BRAKES

BRAKE CALIPER

REMOVAL & INSTALLATION

Maxima

1. Before servicing the vehicle, refer to the Precautions Section.
2. Remove or disconnect the following:
   - Rear wheels
   - Parking brake cable and the lock spring
   - Brake fluid hose from the caliper
   - Caliper pin bolts and remove the caliper

To install:

3. Turn the piston clockwise back into the caliper body. Remove some brake fluid from the master cylinder, if necessary. Take care not to damage the piston boot.
4. Coat the pad contact area on the mounting support with a silicone based grease.
5. Install or connect the following:
   - New pads, shims and the pad springs
   - Caliper body into position and torque the caliper pin bolts to 32 ft. lbs. (48 Nm)
   - Brake fluid hose, using new copper washers, and tighten the flare nut to 12–14 ft. lbs. (17–20 Nm)
   - Lock spring and the parking brake cable

6. Bleed the brake system and top off the master cylinder as necessary.
7. Replace the wheels.

350Z Without Brembo Caliper

1. Before servicing the vehicle, refer to the Precautions Section.
2. Remove or disconnect the following:
   - Rear wheels
   - Brake fluid hose from the caliper
   - Caliper torque member bolts and remove the caliper

To install:

3. Install or connect the following:
   - New pads, shims and the pad springs
   - Caliper body into position and torque the caliper torque member bolts to 62 ft. lbs. (84.3 Nm)
   - Brake fluid hose, using new copper washers, and tighten the flare nut to 12–14 ft. lbs. (17–20 Nm)
4. Bleed the brake system and top off the master cylinder as necessary.
5. Replace the wheels.

350Z With Brembo Caliper

See Figure 2.

1. Before servicing the vehicle, refer to the Precautions Section.
2. Remove or disconnect the following:
   - Rear wheels
   - Caliper pin bolts and remove the caliper

To install:

3. Install or connect the following:
   - New pads, shims and the pad springs
   - Caliper body into position and torque the caliper mounting bolts to 62 ft. lbs. (84.3 Nm)

DISC BRAKE PADS

REMOVAL & INSTALLATION

Maxima

1. Before servicing the vehicle, refer to the Precautions Section.
2. Remove or disconnect the following:
   - Rear wheels
   - Parking brake cable bracket bolt
   - Pin bolts and lift off the caliper body
   - Pad springs by pulling them out
   Pads and shims

To install:

3. Turn the piston clockwise back into the caliper body. Take care not to damage the piston boot.

4. Coat the pad contact area on the mounting support with a silicone based grease.
5. Install or connect the following:
   - Pads, shims, and the pad springs
   - Caliper body into position and torque the caliper pin bolts to 32 ft. lbs. (48 Nm)
   - Wheels
6. Check the master cylinder and add fluid if necessary.

350Z Without Brembo Calipers

1. Before servicing the vehicle, refer to the Precautions Section.
2. Remove or disconnect the following:
   - Rear wheels
   - Top sliding pin bolt
   - Rotate the caliper up and remove the brake pads

To install:

3. Install or connect the following:
   - New pads, new shims and pad retainers
   - Brake caliper Tighten sliding bolt to 32 ft. lbs. (43 Nm)
   - Wheels
4. Bleed the brake system and top off the master cylinder as necessary.

350Z With Brembo Caliper

1. Remove tires from vehicle with a power tool.
2. Remove the clip from the pad pin.
3. Remove the pad pin while holding down the cross spring, then remove the cross spring from the caliper.
4. Using pliers, remove the pad from the caliper.

To install:

5. Insert the piston to the position where the pad is attached.
6. Attach pad and shim cover.
7. Attach the pad with wear sensor to the outer side.
8. Insert the upper pad pin from the outer cylinder side, then insert firmly to the inner cylinder side through the hole in the top of the pad.
9. Place the top of the cross spring over the top pad pin, press in the cross spring, push the lower pad pin from the outer cylinder side to the inner cylinder side, and secure the cross spring.
10. Insert the clip in the small hole at the end of the pad pin.

➡ **If the clip is not fully attached, the pad pin or the pad could fall out while the vehicle is in motion.**

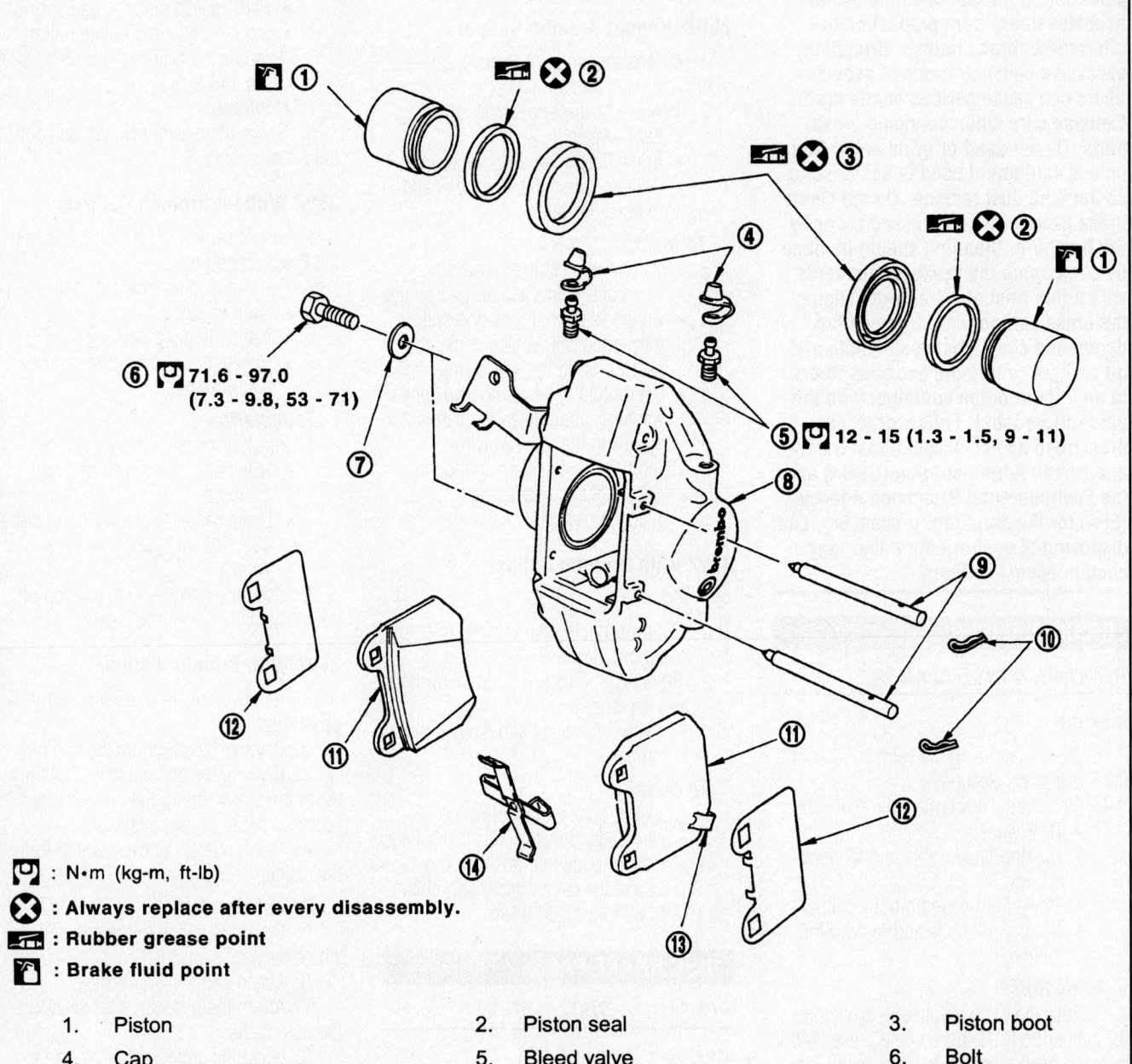

⊡ : N•m (kg-m, ft-lb)

⊗ : **Always replace after every disassembly.**

⬛ : **Rubber grease point**

⬛ : **Brake fluid point**

1.	Piston	2.	Piston seal	3.	Piston boot
4.	Cap	5.	Bleed valve	6.	Bolt
7.	Washer	8.	Caliper	9.	Pad pins
10.	Clips	11.	Brake pad	12.	Shim cover
13.	Pad wear sensor	14.	Cross spring		

09482_ZMAX_G0041

Fig. 2 Rear brake caliper assembly—350Z with OPB13VB/Brembo caliper

BRAKES

PARKING BRAKE

PARKING BRAKE SHOES

REMOVAL & INSTALLATION

See Figure 3.

1. Remove the wheel and tire.
2. Remove the brake rotor with the parking brake lever completely disengaged.
3. If the brake rotor cannot be removed, remove as follows:

   a. Secure the brake rotor with the wheel nut and remove the adjuster hole plug.

   b. Insert a flat-bladed screwdriver through the plug opening and rotate

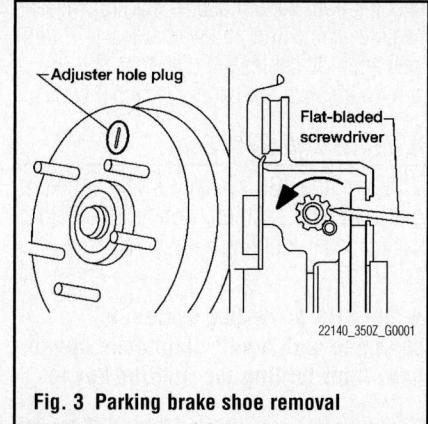

Fig. 3 Parking brake shoe removal

the star wheel on the adjuster assembly in the direction as shown to retract the parking brake shoes.

   c. Remove the parking brake shoe springs using a suitable tool.

   d. Remove the parking brake shoes and adjuster.

To install:

4. Installation is in the reverse order of removal noting the following:

   a. Apply brake grease to the brake shoe contact area.

CHASSIS ELECTRICAL

AIR BAG (SUPPLEMENTAL RESTRAINT SYSTEM)

GENERAL INFORMATION

✳✳ CAUTION

These vehicles are equipped with an air bag system. The system must be disarmed before performing service on, or around, system components, the steering column, instrument panel components, wiring and sensors. Failure to follow the safety precautions and the disarming procedure could result in accidental air bag deployment, possible injury and unnecessary system repairs.

SERVICE PRECAUTIONS

Disconnect and isolate the battery negative cable before beginning any airbag system component diagnosis, testing, removal, or installation procedures. Allow system capacitor to discharge for two minutes before beginning any component service. This will disable the airbag system. Failure to disable the airbag system may result in accidental airbag deployment, personal injury, or death.

Do not place an intact undeployed airbag face down on a solid surface. The airbag will propel into the air if accidentally deployed and may result in personal injury or death.

When carrying or handling an undeployed airbag, the trim side (face) of the airbag should be pointing towards the body to minimize possibility of injury if accidental deployment occurs. Failure to do this may result in personal injury or death.

Replace airbag system components with OEM replacement parts. Substitute parts may appear interchangeable, but internal differences may result in inferior occupant protection. Failure to do so may result in occupant personal injury or death.

Wear safety glasses, rubber gloves, and long sleeved clothing when cleaning powder residue from vehicle after an airbag deployment. Powder residue emitted from a deployed airbag can cause skin irritation. Flush affected area with cool water if irritation is experienced. If nasal or throat irritation is experienced, exit the vehicle for fresh air until the irritation ceases. If irritation continues, see a physician.

Do not use a replacement airbag that is not in the original packaging. This may result in improper deployment, personal injury, or death.

The factory installed fasteners, screws and bolts used to fasten airbag components have a special coating and are specifically designed for the airbag system. Do not use substitute fasteners. Use only original equipment fasteners listed in the parts catalog when fastener replacement is required.

During, and following, any child restraint anchor service, due to impact event or vehicle repair, carefully inspect all mounting hardware, tether straps, and anchors for proper installation, operation, or damage. If a child restraint anchor is found damaged in any way, the anchor must be replaced. Failure to do this may result in personal injury or death.

Deployed and non-deployed airbags may or may not have live pyrotechnic material within the airbag inflator. Do not dispose of driver/passenger/curtain airbags or seat belt tensioners unless you are sure of complete deployment. Refer to the Hazardous Substance Control System for proper disposal.

Dispose of deployed airbags and tensioners consistent with state, provincial, local, and federal regulations.

After any airbag component testing or service, do not connect the battery negative cable. Personal injury or death may result if the system test is not performed first.

If the vehicle is equipped with the Occupant Classification System (OCS), do not connect the battery negative cable before performing the OCS Verification Test using the scan tool and the appropriate diagnostic information. Personal injury or death may result if the system test is not performed properly.

Never replace both the Occupant Restraint Controller (ORC) and the Occupant Classification Module (OCM) at the same time. If both require replacement, replace one, then perform the Airbag System test before replacing the other.

Both the ORC and the OCM store Occupant Classification System (OCS) calibration data, which they transfer to one another when one of them is replaced. If both are replaced at the same time, an irreversible fault will be set in both modules and the OCS may malfunction and cause personal injury or death.

If equipped with OCS, the Seat Weight Sensor is a sensitive, calibrated unit and must be handled carefully. Do not drop or handle roughly. If dropped or damaged, replace with another sensor. Failure to do so may result in occupant injury or death.

If equipped with OCS, the front passenger seat must be handled carefully as well. When removing the seat, be careful when setting on floor not to drop. If dropped, the sensor may be inoperative, could result in occupant injury, or possibly death.

If equipped with OCS, when the passenger front seat is on the floor, no one should sit in the front passenger seat. This uneven force may damage the sensing ability of the seat weight sensors. If sat on and damaged, the sensor may be inoperative, could result in occupant injury, or possibly death.

DISARMING THE SYSTEM

➡**All SRS electrical wiring harnesses and connectors are covered with YELLOW outer insulation. Do not use electrical test equipment on any circuit related to the SRS (air bag) sensors. When installing SRS components, always install with the arrow marks facing the front of the vehicle.**

To disarm the SRS system turn the ignition switch to **OFF** position. Then, disconnect the both battery cables starting with the negative cable first and wait at least 10 minutes after the cables are disconnected. Be sure to insulate the battery terminal ends.

ARMING THE SYSTEM

To arm the SRS system turn the ignition switch to **OFF** position. Connect the both battery cables starting with the positive cable first.

➡**The SRS or air bag system is equipped with a self-diagnostic operation. After turning the ignition key to** the ON or START position, the AIR BAG warning lamp will illuminate for 7 seconds. After 7 seconds, the AIR BAG lamp will extinguish if no malfunction is detected. If the AIR BAG lamp does not extinguish after 7 seconds, check the SRS self-diagnostic system for a malfunction.

CLOCKSPRING CENTERING

Align spiral cable correctly when installing steering wheel. Make sure that the spiral cable is in the neutral position. The neutral position is detected by turning left 2.5 revolutions from the right end position and ending with the knob at the top.

DRIVETRAIN

AUTOMATIC TRANSAXLE ASSEMBLY

REMOVAL & INSTALLATION

Maxima CVT

See Figure 4.

1. Remove the battery, tray and bracket.
2. Remove the air cleaner and air duct assembly.
3. Remove the grille top cover.
4. Remove the hood ledge and engine cover.
5. Disconnect the following:
   a. CVT unit harness connector.
   b. Secondary speed sensor connector
   c. Ground strap connector
   d. Ground cable nut
   e. Remove the harness from the transaxle.
6. Remove the CVT fluid charging pipe.
7. Remove the starter motor from the transaxle.
8. Disconnect the control cable from the transaxle.
9. Remove the bolt from the rear gusset

10. Drain the CVT fluid.
11. Disconnect the CVT fluid cooler hoses from the CVT assembly.
12. Install engine slingers to the rear of both cylinder heads and support the engine using suitable engine support.
13. Remove the upper transaxle to engine bolts.
14. Remove the front exhaust.
15. Remove the crankshaft position sensor from the engine
16. Disconnect the drive shafts.
17. Remove the front suspension member.
18. Support transaxle using a suitable jack.
19. Remove the rear cover plate.
20. Remove the four drive plate to torque converter nuts.

➡**Rotate the crankshaft clockwise as viewed from front of engine for access to drive plate to torque converter nuts. Remove the lower engine to transaxle bolts. Lower the transaxle while supporting it with a jack.**

To install:

21. Installation is in the reverse order of removal.

➡ **Note the following:**

- When replacing an engine or transmission you must make sure any dowels are installed correctly during re-assembly.
- Improper alignment caused by missing dowels may cause vibration, oil leaks or breakage of drive train components.
- Do not reuse O-ring and copper washers.
- When turning crankshaft, turn it clockwise as viewed from the front of the engine.
- When tightening the tightening nuts for the torque converter after installing the crankshaft pulley bolts, be sure to confirm the tightening torque of the crankshaft pulley bolts.
- After converter is installed to drive plate, rotate crankshaft several turns and check to be sure that transaxle rotates freely without binding.
- When installing the drive plate to torque converter nuts, tighten them temporarily. Then tighten the nuts to the specified torque of 39 ft. lbs. (51 Nm)
- After completing installation, check for fluid leakage, fluid level, and the positions of CVT.

AUTOMATIC TRANSMISSION ASSEMBLY

REMOVAL & INSTALLATION

350Z

See Figures 5 and 6.

1. Before servicing the vehicle, refer to the Precautions Section.
2. Drain the fluid from the transmission.
3. Remove or disconnect the following:
   - Negative battery cable
   - Engine undercover
   - Strut tower bar
   - Front under vehicle cross bar
   - Catalytic converter and front exhaust pipe
   - Drive shaft
   - Control rod
   - Transmission harness connector

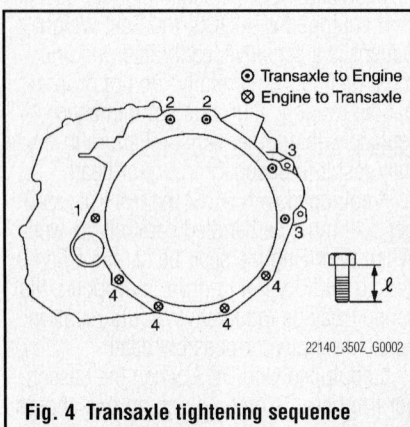

⊚ Transaxle to Engine
⊗ Engine to Transaxle

22140_350Z_G0002

Fig. 4 Transaxle tightening sequence

Bolt No.	1	2	3	4
Number of bolts	1	5	2	2
Bolt length "ℓ" mm (in)	55 (2.17)	65 (2.56)	56 (2.20)	35 (1.38)
Tightening torque N·m (kg-m, ft-lb)	70 - 80 (7.2 - 8.1, 52 - 59)		49.0 - 61.8 (5.0 - 6.3, 37 - 45)	41.2 - 52.0 (4.2 - 5.3, 31 - 38)

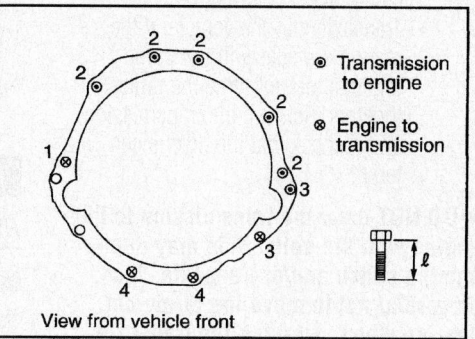

View from vehicle front

⊙ Transmission to engine

⊗ Engine to transmission

67162-NISS-G49

Fig. 5 Automatic transaxle bolt torque specifications and locations—350Z

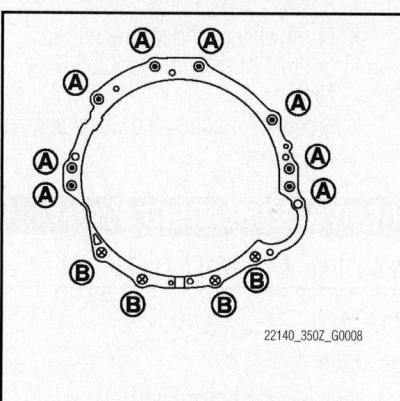

22140_350Z_G0008

Fig. 6 Transmission tightening sequence

- Crankshaft position sensor
- Cooler lines and cap openings
- Fluid charging pip, plug hole
- Air breather hose
- Starter
- Torque converter dust cover
- Torque converter bolts

4. Place a transmission jack under the transmission.

5. Remove the rear engine crossmember.

6. Remove the transmission mounting bolts and the transmission.

To install:

➡ **The transmission mounting bolts are different lengths and require special torque specifications. Use care when installing and tightening these bolts.**

7. Installation is the reverse of the removal procedure noting the following:

a. See illustration and tighten the mounting bolts A to 55 ft. lbs. (75 Nm) and bolts B to 34 ft. lbs. (46.6 Nm)

b. Tighten the rear engine crossmember bolts to 36 ft. lbs. (49 Nm).

c. Tighten the torque converter bolts to 33—42 ft. lbs. (44–58 Nm).

8. Fill the transmission with clean fluid.

9. Start the vehicle, check for leaks and repair if necessary.

MANUAL TRANSMISSION ASSEMBLY

REMOVAL & INSTALLATION

350Z

See Figure 6.

1. Before servicing the vehicle, refer to the Precautions Section.

2. Drain the fluid from the transmission.

3. Remove or disconnect the following:
   - Negative battery cable
   - Engine undercover
   - Strut tower bar
   - Front under vehicle cross bar
   - Catalytic converter and front exhaust pipe
   - Drive shaft
   - Shift lever assembly from the control rod
   - Shifter console and boot
   - Shift lever from shift housing
   - Clutch slave cylinder
   - Crankshaft position sensor
   - Back-up light and neutral safety switch
   - Wire harnesses from transmission
   - Starter
   - Transmission cover plate

4. Place a transmission jack under the transmission.

5. Remove the rear engine crossmember.

6. Remove the transmission mounting bolts and the transmission.

To install:

➡ **The transmission mounting bolts are different lengths and require special**

torque specifications. Use care when installing and tightening these bolts.

7. Installation is the reverse of the removal procedure noting the following:

a. See illustration and tighten the mounting bolts A to 55 ft. lbs. (75 Nm) and bolts B to 34 ft. lbs. (46.6 Nm)

b. Tighten the rear engine crossmember bolts to 36 ft. lbs. (49 Nm).

8. Fill the transmission with clean fluid.

9. Start the vehicle, check for leaks and repair if necessary.

CLUTCH

REMOVAL & INSTALLATION

350Z

1. Before servicing the vehicle, refer to the Precautions Section.

2. Remove or disconnect the following:
   - Transmission assembly

3. Insert a clutch disc centering tool into the clutch disc hub for support.
   - Pressure plate bolts evenly in reverse order of the tightening sequence, a little at a time to prevent distortion
   - Clutch assembly
   - Throw-out bearing from the clutch lever

To install:

4. Apply a light coating of chassis lube to the clutch disc spleens, input shaft and pilot bearing. Use a disc centering tool to aid installation.

5. Install or connect the following:
   - Disc and pressure plate

6. Torque the pressure plate bolts in a crisscross pattern in the following 2 steps:

a. Step 1: 11 ft. lbs. (115 Nm).

b. Step 2: 29 ft. lbs. (340 Nm).

7. Install or connect the following:
   - New throw-out bearing in the

clutch release lever. Remove the clutch disc centering tool.

- Transaxle into the vehicle. If the mating surfaces will not come together, do not force the units together. Remove the transmission and recheck that the disc is centered.

➡**DO NOT draw the transmission to the engine with the bolts. This may damage the clutch and/or transaxle. Also, be careful not to move the throw-out bearing when installing the transaxle.**

8. After the transmission is installed, connect the clutch cable and check operation before complete reassembly.

9. Adjust the clutch pedal as necessary.

BLEEDING

See Figure 7.

Bleeding is required to remove air trapped in the hydraulic system. The bleed screw is located on the clutch slave (operating) cylinder.

Some models are also equipped with a clutch damper mechanism. The clutch damper mechanism is bled in exactly the same manner as the operating cylinder. It should be bled along with the operating cylinder.

1. Before servicing the vehicle, refer to the Precautions Section.

2. Remove the bleed screw dust cap.

3. Attach a transparent vinyl tube to the bleed screw, immersing the free end in a clean container of clean brake fluid.

4. Fill the master cylinder with the proper fluid.

5. Open the bleed screw about ¾ turn.

6. Depress the clutch pedal quickly. Hold it down. Have an assistant tighten the bleed screw. Allow the pedal to return slowly.

7. Repeat the above procedure until no more air bubbles are seen in the fluid container.

8. Remove the bleed tube.

9. Replace the dust cap and refill the master cylinder.

10. Bleed the clutch damper, if equipped.

FRONT HALFSHAFT

REMOVAL & INSTALLATION

Maxima

See Figure 8.

1. Before servicing the vehicle, refer to the precautions in the beginning of this section.

2. Raise and support the front of the vehicle safely and remove the wheels.

3. Remove or disconnect the following:
   - Anti-Lock Brake (ABS) wheel sensor and move it out of the way
   - Brake hose from the strut
   - Wheel bearing locknut
   - Bolts attaching the steering knuckle to the strut. Matchmark the bolts before removal.

➡**Cover axle boots with waste cloth so as not to damage them when removing halfshaft.**

   - Halfshaft from the knuckle by slightly tapping it
   - Bolts attaching the support bearing to the support bearing bracket
   a. Remove the right halfshaft from the vehicle.
   b. Insert a flat bladed tool into the transaxle where the right halfshaft was, place the end of the tool on the halfshaft, then, drive the left shaft from the pinion side gear.

To install:

4. Install or connect the following:
   - New circlip to the halfshaft, then insert the halfshaft into the transaxle

5. With the serration's aligned remove the alignment tool.

6. Push the halfshaft fully into the transaxle to seat the circlip. Try to pull the halfshaft from the transaxle by hand to verify that the circlip is properly seated.
   - Support bearing and torque the bolts to 10–14 ft. lbs. (13–19 Nm)
   - Halfshaft into the steering knuckle and install the hub locknut, do not tighten the hub nut
   - Steering knuckle to the strut
   - Strut mounting bolts to the matchmarks and torque the bolts to 103–117 ft. lbs. (140–159 Nm)
   - Brake hose to the strut
   - ABS wheel sensor and torque the bolt to 13–17 ft. lbs. (18–24 Nm)
   - Front wheels and torque hub locknut to 174–231 ft. lbs. (235–314 Nm)

7. Check and/or adjust the wheel alignment as necessary.

REAR AXLE BEARING & SEAL

REMOVAL & INSTALLATION

Maxima

See Figure 9.

1. Remove the brake rotor.

2. Remove the rear ABS sensor, then move it away from the wheel hub assembly.

3. Remove the wheel hub bolts from knuckle and remove the wheel hub assembly.

To install:

4. Installation is in the reverse order of removal.

REAR HALFSHAFT

REMOVAL & INSTALLATION

350Z

1. Before servicing the vehicle, refer to the Precautions Section.

2. Raise and support the rear of the vehicle safely and remove the wheels.

3. Remove or disconnect the following:
   - Cotter pin and axle nut
   - Stabilizer bar connecting rod

4. Remove the nuts and bolts between the side flange and the drive shaft.

5. Use a suitable puller and remove the drive shaft from the axle.

To install:

6. Install or connect the following:
   - New seal into the transaxle and install halfshaft

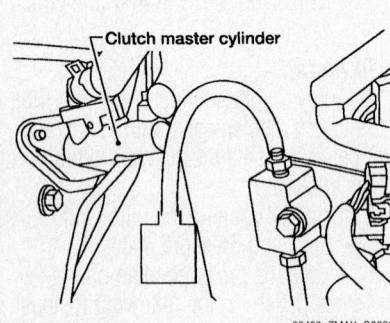

Fig. 7 Clutch system bleeding from the air bleeder valve on the bleed connector— Maxima

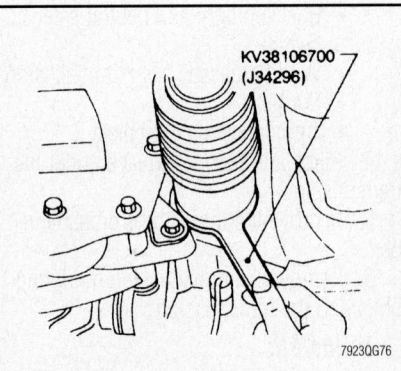

Fig. 8 Left halfshaft alignment tool—Maxima

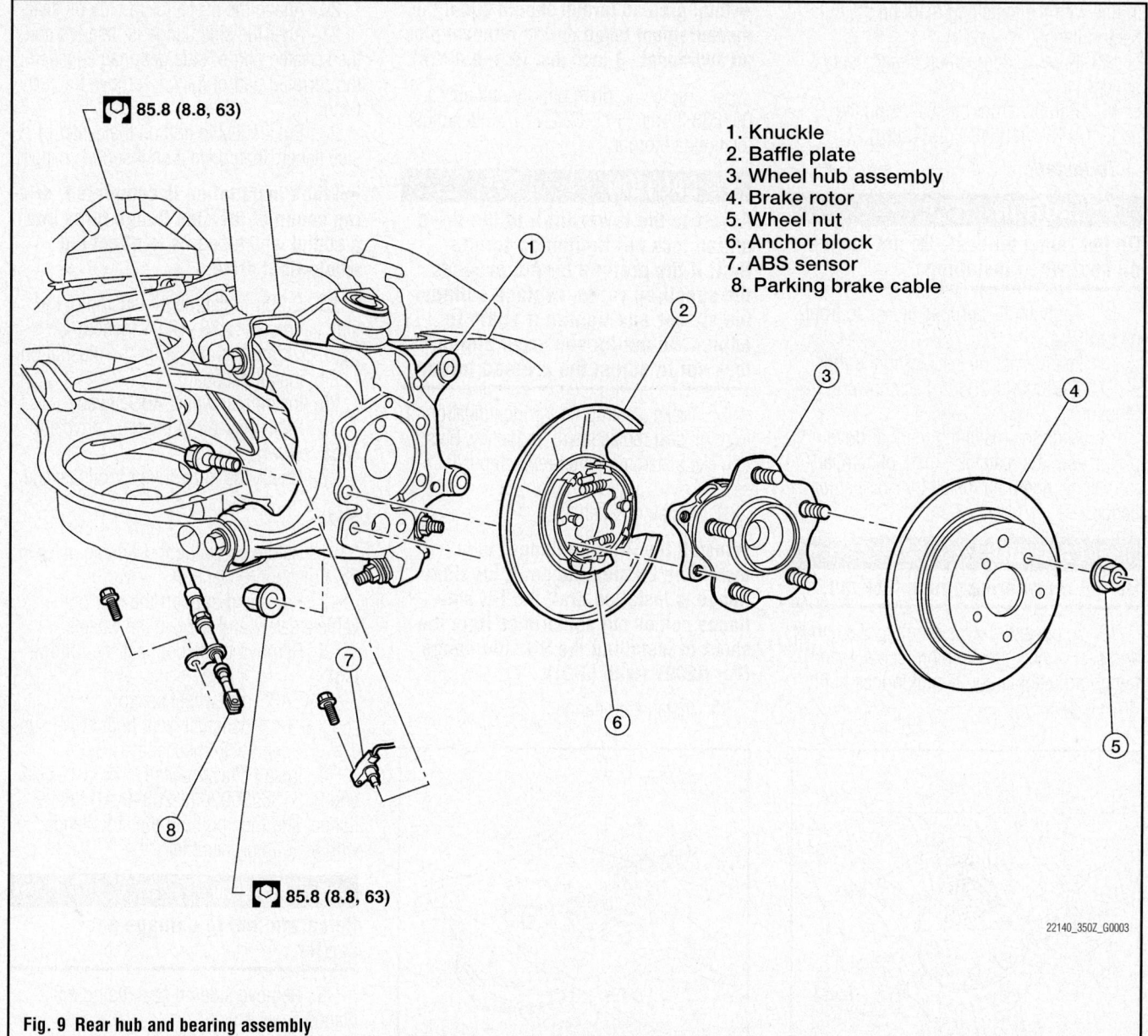

85.8 (8.8, 63)

1. Knuckle
2. Baffle plate
3. Wheel hub assembly
4. Brake rotor
5. Wheel nut
6. Anchor block
7. ABS sensor
8. Parking brake cable

85.8 (8.8, 63)

22140_350Z_G0003

Fig. 9 Rear hub and bearing assembly

- Tighten the side flange bolts to 47–58 ft. lbs. (63–79 Nm)
- Stabilizer bar connecting rod
- Axle nut and new cotter pin. Tighten the axle nut to 152–202 ft. lbs. (206–274 Nm)
- Rear wheels

REAR PINION SEAL

REMOVAL & INSTALLATION

350Z

Front

See Figures 10 through 14.

1. Before servicing the vehicle, refer to the Precautions Section.

➡**Verify the identification stamp for the replacement frequency, installed on the lower part of gear carrier. Use this to determine the replacement of collapsible spacer when replacing front oil seal. If it is necessary to replace the collapsible spacer, remove final drive assembly and disassemble to replace front oil seal and collapsible spacer.**

2. Drain gear oil.
3. Raise and support the rear of the vehicle safely and remove the wheels.
4. Remove or disconnect the following:
   - ABS rear wheel sensor
   - Rear halfshaft, suspend with mechanics wire
5. Install attachments (A: KV40104100 and B: ST36230000 (J-25840-A)) to side

flange, and then pull out the side flange with the sliding hammer.

6. Remove drive shaft.
7. Measure the total preload with the preload gauge tool (ST3127S000 (J-25765-A)) and record the measurement.
8. Remove drive pinion lock nut using the flange wrench KV40104000.

✳✳ CAUTION

For matching mark, use paint. Do not damage drive pinion.

➡ **The matching mark "A" on the final drive companion flange indicates the maximum vertical runout position.**

9. Put matching mark on the end of the drive pinion. The matching mark should be

in line with the matching mark on the companion flange.

10. Remove companion flange using a puller.

11. Remove front oil seal using the puller (A: KV381054S0 (J-34286))

To install:

✳✳ CAUTION

Do not reuse oil seal. Do not incline oil seal when installing.

12. Apply multi-purpose grease to front oil seal lips.

13. Install front oil seal using the drift (A:ST30720000 (J-25405)) as shown in the illustration.

14. Align the matching mark of drive pinion with the matching mark of companion flange, and then install the companion flange.

✳✳ CAUTION

Do not reuse drive pinion lock nut.

15. Apply anti-corrosion oil to the thread and seat of new drive pinion lock nut, and temporarily tighten drive pinion lock nut to drive pinion.

➡**Total preload torque should equal the measurement taken during removal plus an additional −3 inch lbs. (0.1–0.4 Nm).**

16. Tighten to drive pinion lock nut to: 09–238 ft. lbs. (147–323 Nm), while adjust total reload torque.

✳✳ CAUTION

Adjust to the lower limit of the drive pinion lock nut tightening torque first. If the preload torque exceeds the specified value, replace collapsible spacer and tighten it again to adjust. Do not loosen drive pinion lock nut to adjust the preload torque.

17. Make a stamping for identification of front oil seal replacement frequency. Be sure to make a stamping after replacing front oil seal.

18. Install drive shaft

➡**Install the RH side flange, then install the LH side flange. If LH side flange is installed first, the RH side flange comes out sometimes from the shock of installing the RH side flange (For R200V (with LSD)).**

19. Install side flange.

20. Attach the protector to side oil seal.

21. After the side flange is inserted and the serrated part of side gear has engaged the serrated part of flange, remove the protector.

22. Put a suitable drift on the center of side flange, then drive it until sound changes.

➡**When installation is completed, driving sound of the side flange turns into a sound which seems to affect the whole final drive.**

23. Confirm that the dimension of the side flange is: 12.83–12.91 inches (326–328 mm) as shown in the illustration.

24. Install halfshaft

25. Install rear wheel ABS sensor.

26. Refill gear oil to the final drive and check oil level.

27. Check the final drive for oil leakage.

Side

1. Before servicing the vehicle, refer to the Precautions Section.

2. Raise and support the rear of the vehicle safely and remove the wheels.

3. Remove or disconnect the following:

- ABS rear wheel sensor
- Rear halfshaft from final drive, suspend with mechanics wire

4. Install attachments (A: KV40104100 and B: ST36230000 (J-25840-A)) to side flange, and then pull out the side flange with the sliding hammer.

✳✳ CAUTION

Be careful not to damage gear carrier.

5. Remove side oil seal, using a flat-bladed screwdriver.

To install:

6. Apply multi-purpose grease to side oil seal lips.

7. Install side oil seal until it becomes flush with the case end, using the drift (KV38100200 (J-26233)).

✳✳ CAUTION

Do not reuse oil seal. When installing, do not incline oil seal.

➡**Install the RH side flange, then install the LH side flange. If LH side flange is installed first, the RH side flange comes out sometimes from the shock of installing the RH side flange (For R200V (with LSD)).**

8. Install side flange.

9. Attach the protector to side oil seal.

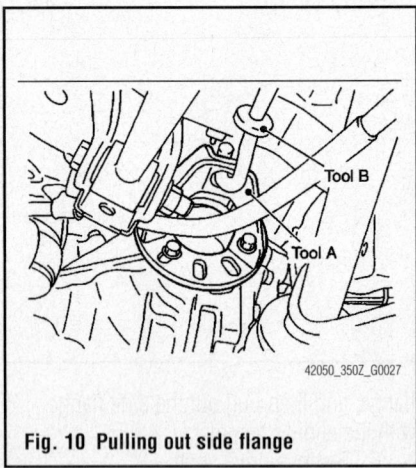

Fig. 10 Pulling out side flange

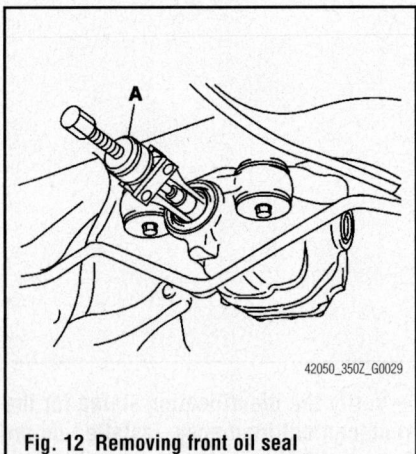

Fig. 12 Removing front oil seal

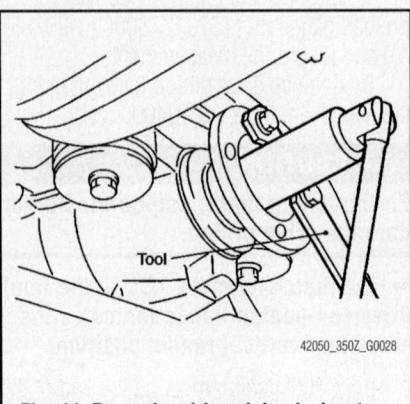

Fig. 11 Removing drive pinion lock nut

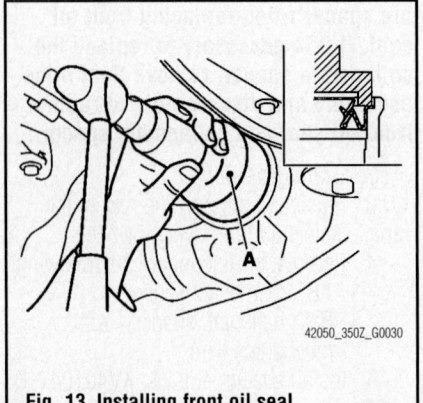

Fig. 13 Installing front oil seal

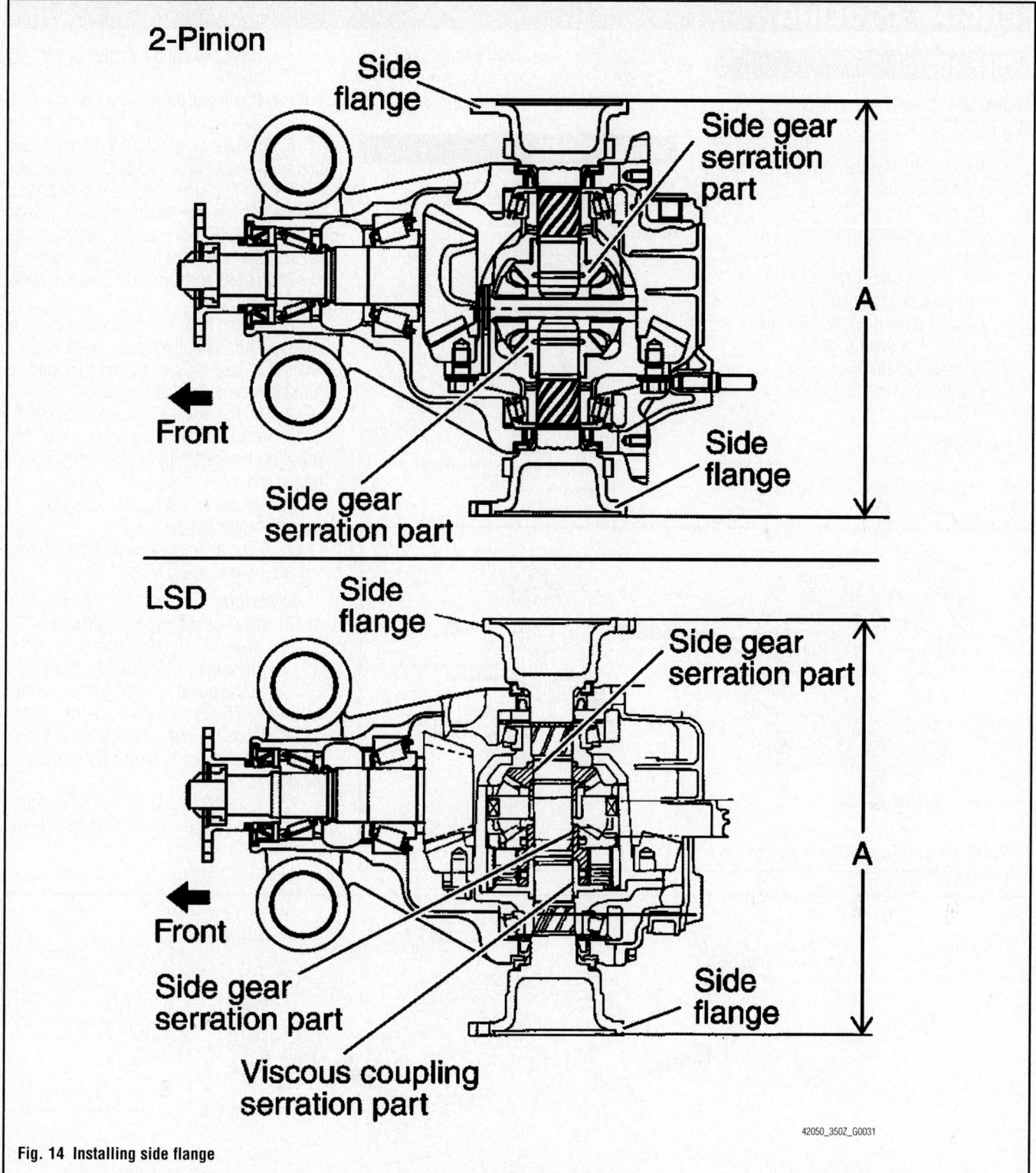

2-Pinion

Side flange

Side gear serration part

Front

Side flange

Side gear serration part

LSD

Side flange

Side gear serration part

Front

Side gear serration part

Side flange

Viscous coupling serration part

42050_350Z_G0031

Fig. 14 Installing side flange

10. After the side flange is inserted and the serrated part of side gear has engaged the serrated part of flange, remove the protector.

11. Put a suitable drift on the center of side flange, then drive it until sound changes.

➡️**When installation is completed, driving sound of the side flange turns into a sound which seems to affect the whole final drive.**

12. Confirm that the dimension of the side flange is: 12.83–12.91 inches (326–328 mm).

13. Install halfshaft

14. Install rear wheel ABS sensor.

15. Refill gear oil to the final drive and check oil level.

16. Check the final drive for oil leakage.

ENGINE COOLING

THERMOSTAT

REMOVAL & INSTALLATION

See Figure 15.

1. Before servicing the vehicle, refer to the Precautions Section.
2. Drain the cooling system.
3. Remove or disconnect the following:
   - Negative battery cable
   - If equipped, engine cover
   - Water drain plug on water pump side of cylinder block
   - Lower radiator hose
   - Water inlet and thermostat assembly

To install:

4. To install, reverse removal procedure
5. Install thermostat with jiggle valve facing upward and tighten bolts to 75 to 99 inch lbs. (8.4 to 11.2 Nm)
6. Run engine and check for leaks

WATER PUMP

REMOVAL & INSTALLATION

Maxima

See Figures 16 and 17.

1. Before servicing the vehicle, refer to the Precautions Section.
2. Drain the cooling system.
3. Position a jack under the oil pan for support. Be sure to place a block of wood on the jack for protection to the engine parts.
4. Remove or disconnect the following:
   - Negative battery cable
   - Right side engine mount and bracket
   - Drive belts and the idler pulley bracket
   - Chain tensioner cover and the water pump cover

5. Push the timing chain tensioner sleeve and apply a stopper pin so it does not return.
   - Timing chain tensioner assembly
   - 3 bolts that secure the water pump

6. Rotate the crankshaft 20 degrees counterclockwise to provide timing chain slack.
7. Put M8 bolts in 2 M8 threaded holes of the water pump.
8. Tighten each bolt by turning alternately ½ turn until they reach the timing chain rear case. Be sure to turn each bolt ½ turn at a time to prevent damage.
9. Lift up the water pump and remove it.
10. When removing the water pump, do not allow the water pump gear to hit the timing chain.
11. Remove and discard the O-rings from the water pump.
12. Clean all traces of liquid gasket from the water pump and covers.

To install:

13. Install or connect the following:
   - Water pump using new O-rings to the engine block. Torque the 3 water pump mounting bolts evenly to 75–95 inch lbs. (8.5–10.7 Nm).
14. Rotate the crankshaft pulley to its original position by turning it 20 degrees clockwise.
   - Timing chain tensioner and torque the bolts to 75–89 inch lbs. (9–10 Nm)

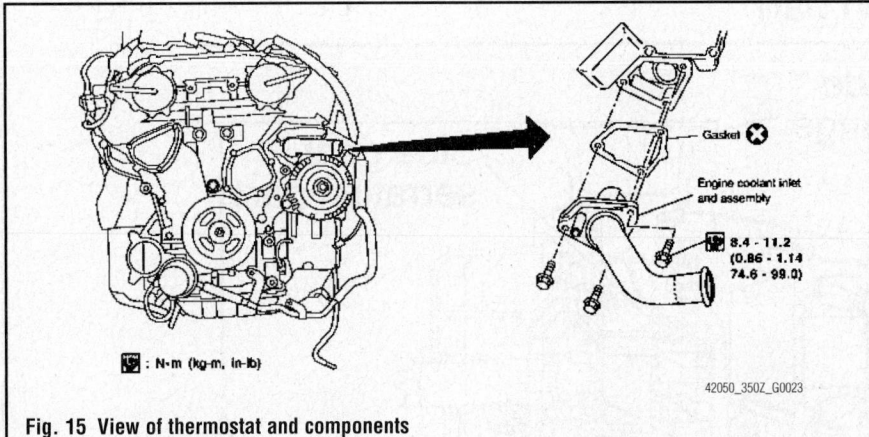

: N·m (kg-m, in-lb)

42050_350Z_G0023

Fig. 15 View of thermostat and components

Gasket ⊗

Engine coolant inlet and assembly

8.4 - 11.2 (0.86 - 1.14 74.6 - 99.0)

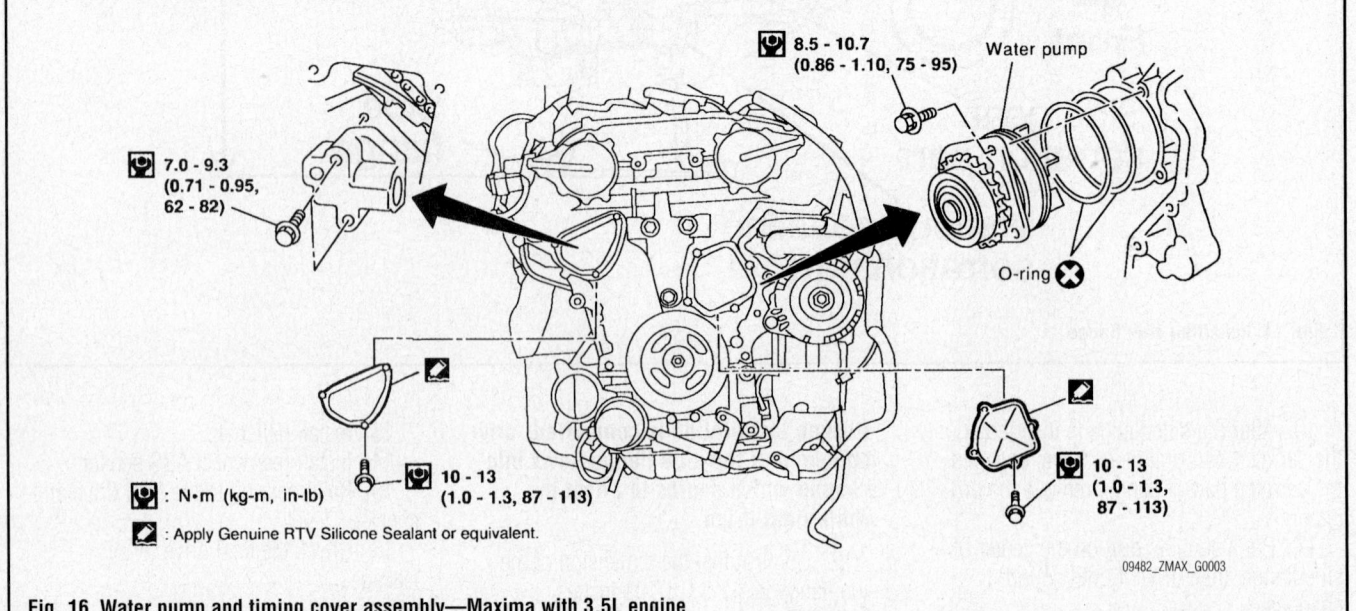

8.5 - 10.7 (0.86 - 1.10, 75 - 95) Water pump

7.0 - 9.3 (0.71 - 0.95, 62 - 82)

O-ring ⊗

10 - 13 (1.0 - 1.3, 87 - 113)

10 - 13 (1.0 - 1.3, 87 - 113)

: N·m (kg-m, in-lb)

: Apply Genuine RTV Silicone Sealant or equivalent.

09482_ZMAX_G0003

Fig. 16 Water pump and timing cover assembly—Maxima with 3.5L engine

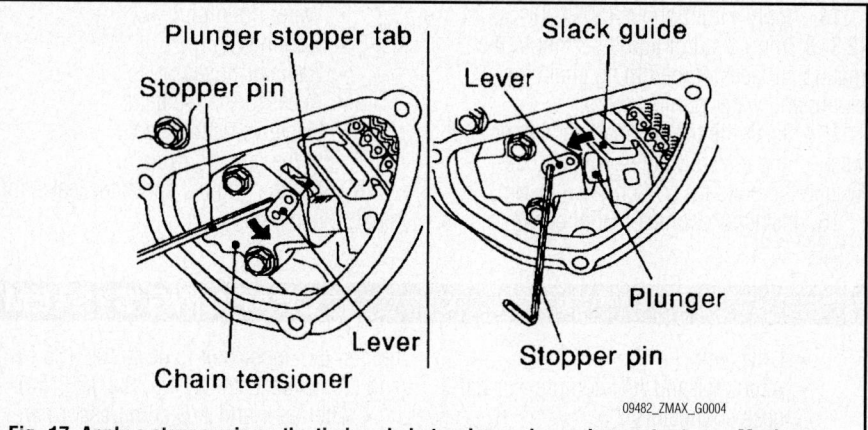

Fig. 17 Apply a stopper pin so the timing chain tensioner sleeve does not return—Maxima with 3.5L engine

15. Remove the stopper pin from the timing chain tensioner.

16. Apply a continuous 0.091–0.130 in. (2.3–3.3mm) bead of liquid sealant to the mating surfaces of the timing chain tensioner and water pump covers.

- Timing chain tensioner and water pump covers to the engine block. Torque the bolts to 87–113 inch lbs. (10–13 Nm).
- Drive belts and the idler pulley bracket
- Right side engine mounting bracket and the engine mount
- Negative battery cable

17. Remove the jack from under the engine and install the drain plugs to the cylinder block.

18. Fill the cooling system.

19. Start the engine, check for leaks and repair if necessary.

350Z

See Figure 18.

1. Before servicing the vehicle, refer to the Precautions Section.

2. Drain the cooling system.

3. Remove or disconnect the following:
- Negative battery cable

- Accessory drive belts
- Radiator hoses
- Cooling fan
- Water drain plug on water pump side of block
- Timing chain tensioner cover
- Water pump cover
- Primary timing chain tensioner
- Water pump mounting bolts

4. Turn the crankshaft pulley counter-clockwise until the timing chain slack on the water pump pulley is at maximum.

5. Place M8 bolts in the upper and lower M8 threaded holes of the water pump.

6. Tighten each bolt by turning alternately ½ turn until they reach the timing chain rear case. Be sure to turn each bolt ½ turn at a time to prevent damage.

7. Lift up the water pump and remove it.

8. When removing the water pump, do not allow the water pump gear to hit the timing chain.

9. Remove and discard the O-rings from the water pump.

10. Clean all traces of liquid gasket from the water pump and covers.

To install:

11. Install the water pump using new O-rings to the engine block. Lubricate the inner O-ring with clean engine oil and the outer O-ring with engine coolant. Ensure

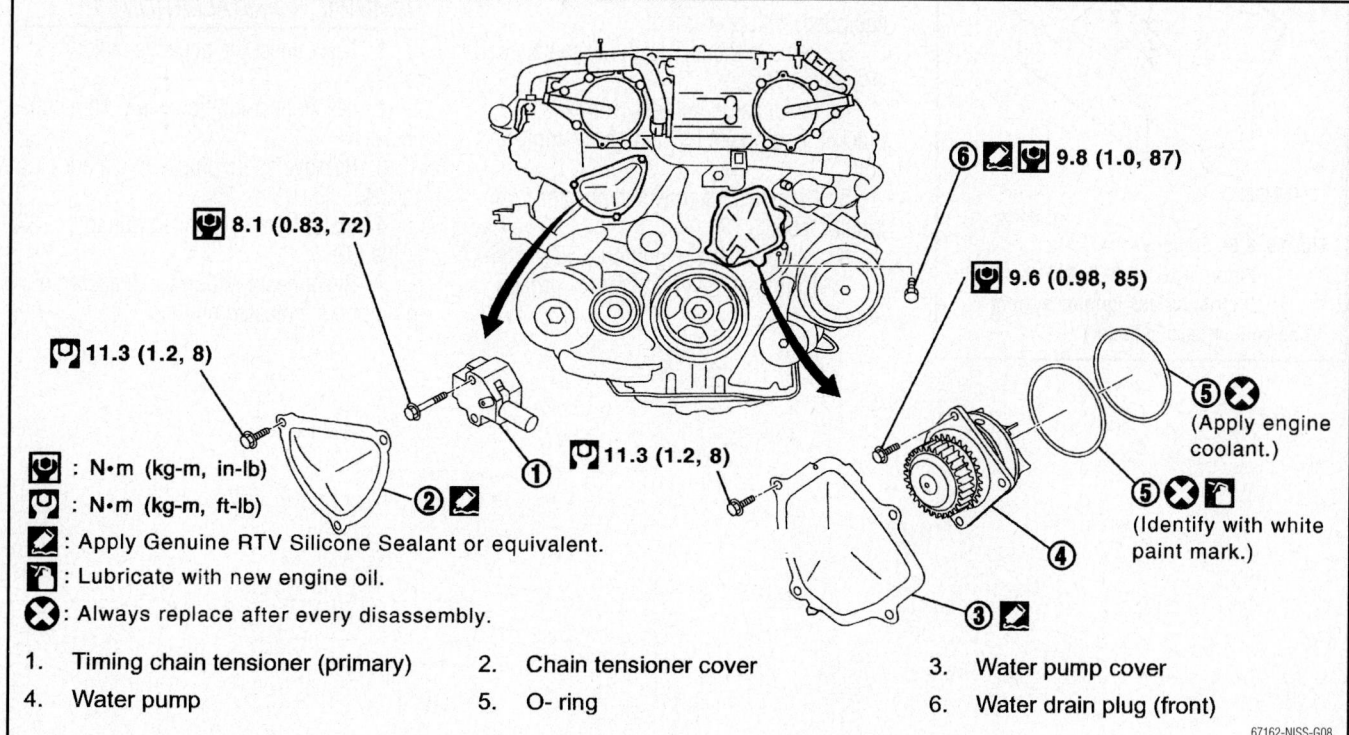

Fig. 18 Exploded view of water pump mounting—350Z with 3.5L engine

1. Timing chain tensioner (primary)
2. Chain tensioner cover
3. Water pump cover
4. Water pump
5. O-ring
6. Water drain plug (front)

the water pump sprocket and timing chain are engaged. Torque the 3 water pump mounting bolts evenly to 85 inch lbs. (10 Nm).

12. Rotate the crankshaft pulley clockwise so the timing chain on the tensioner side is loose.

13. Install the primary timing chain tensioner.

14. Apply a continuous 0.091–0.130 in. (2.3–3.3mm) bead of liquid sealant to the mating surfaces of the timing chain tensioner and water pump covers.

15. Install the timing chain tensioner and water pump covers to the engine block. Torque the bolts to 97 inch lbs. (11 Nm).

16. Install or connect the following:

- Water drain plug
- Cooling fan
- Radiator hoses
- Accessory drive belts
- Negative battery cable

17. Fill the cooling system.

18. Start the engine, check for leaks and repair if necessary.

ENGINE ELECTRICAL

ALTERNATOR

REMOVAL & INSTALLATION

1. Before servicing the vehicle, refer to the Precautions Section.

2. Remove or disconnect the following:
- Negative battery cable
- Right side engine undercover and side inspection cover
- Radiator on Maxima
- Radiator fan on 350Z

- Drive belt
- Alternator and A/C compressor harness connectors
- Upper and lower alternator bolts
- Alternator

Install or connect the following:
- Alternator
- Upper and lower alternator bolts. On Maxima tighten the upper bolt to 12–15 ft. lbs. (16–20 Nm) and the lower bolt to 32–38 ft. lbs. (44–52 Nm). On 350Z,

CHARGING SYSTEM

tighten the upper bolt to 48 ft. lbs. (65 Nm) and the lower bolts to 21 ft. lbs. (28 Nm).

- Alternator and A/C compressor harness connectors
- Drive belt
- Radiator on Maxima
- Radiator fan on 350Z
- Right side engine undercover and side inspection cover
- Negative battery cable

ENGINE ELECTRICAL

FIRING ORDER

See Figure 20.

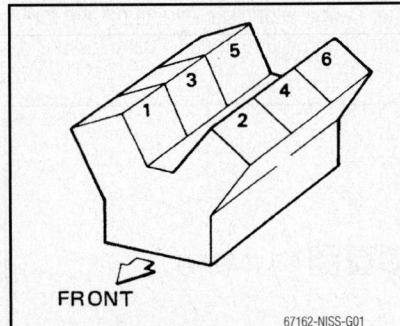

FRONT

67162-NISS-G01

Fig. 19 3.5L Engines
Firing order: 1-2-3-4-5-6
Distributorless ignition system
(one coil on each cylinder)

IGNITION COIL

REMOVAL & INSTALLATION

See Figures 20 and 21.

1. Disconnect the negative battery cable.

2. Remove the engine cover (if equipped) with power tool.

3. Remove the air cleaner assembly and air intake tubes.

4. Move aside harness, harness bracket, and hoses located above ignition coil.

5. Disconnect harness connector from ignition coil.

6. Remove the six ignition coils.

7. To install, reverse removal procedure.

IGNITION SYSTEM

IGNITION TIMING

ADJUSTMENT

No timing adjustment is necessary.

SPARK PLUGS

REMOVAL & INSTALLATION

1. Disconnect the negative battery cable.

2. Disconnect ignition wires from coil packs.

3. Remove spark plugs with spark plug socket.

4. Install spark plugs. Torque to 14–22 ft. lbs. (Nm)

5. Reconnect ignition wires according to numbers indicated on them.

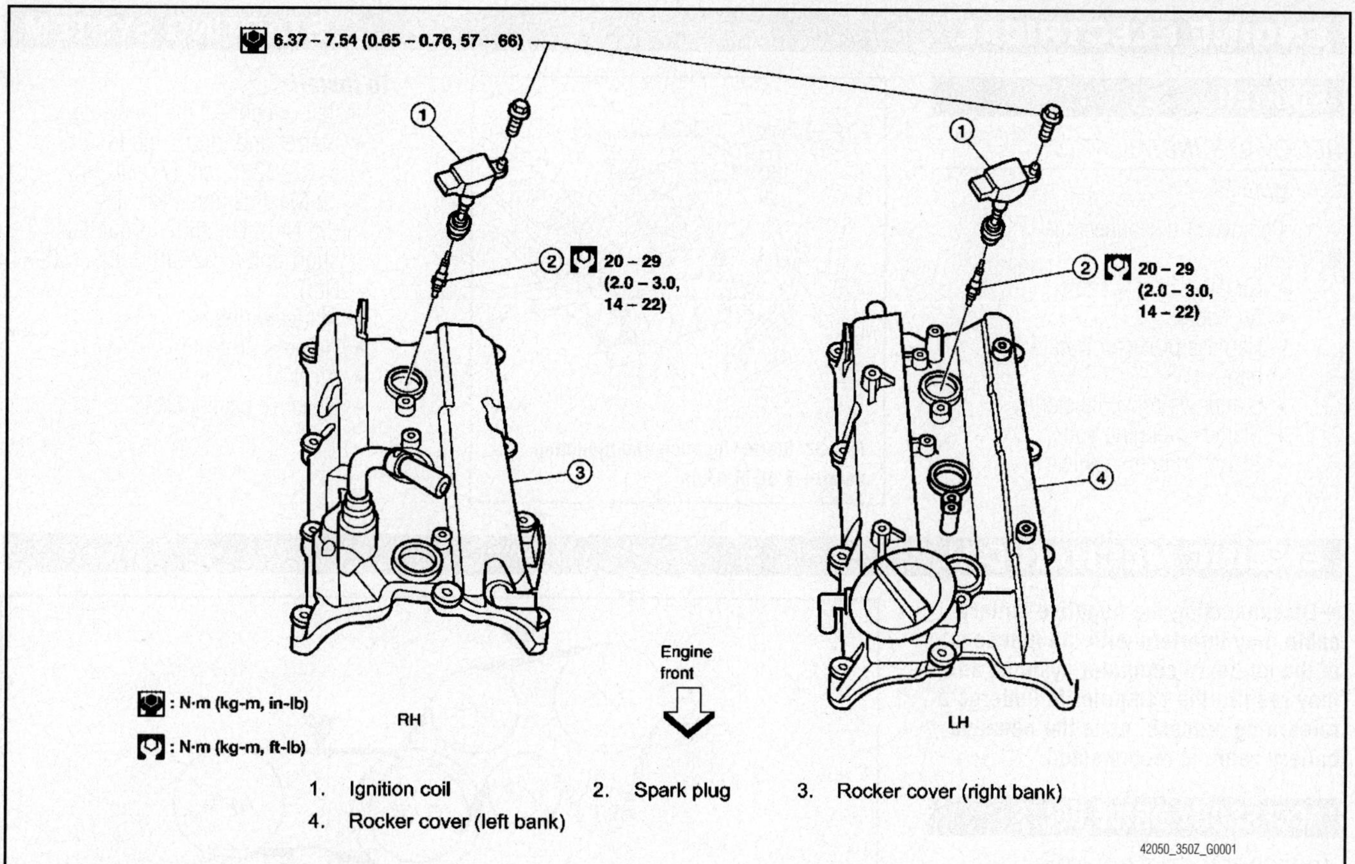

6.37 – 7.54 (0.65 – 0.76, 57 – 66)

① Ignition coil

② 20 – 29 (2.0 – 3.0, 14 – 22)

③ Rocker cover (right bank)

④ Rocker cover (left bank)

Engine front

: N·m (kg-m, in-lb)

: N·m (kg-m, ft-lb)

RH

LH

1. Ignition coil
2. Spark plug
3. Rocker cover (right bank)
4. Rocker cover (left bank)

42050_350Z_G0001

Fig. 20 Removing ignition coils–Maxima

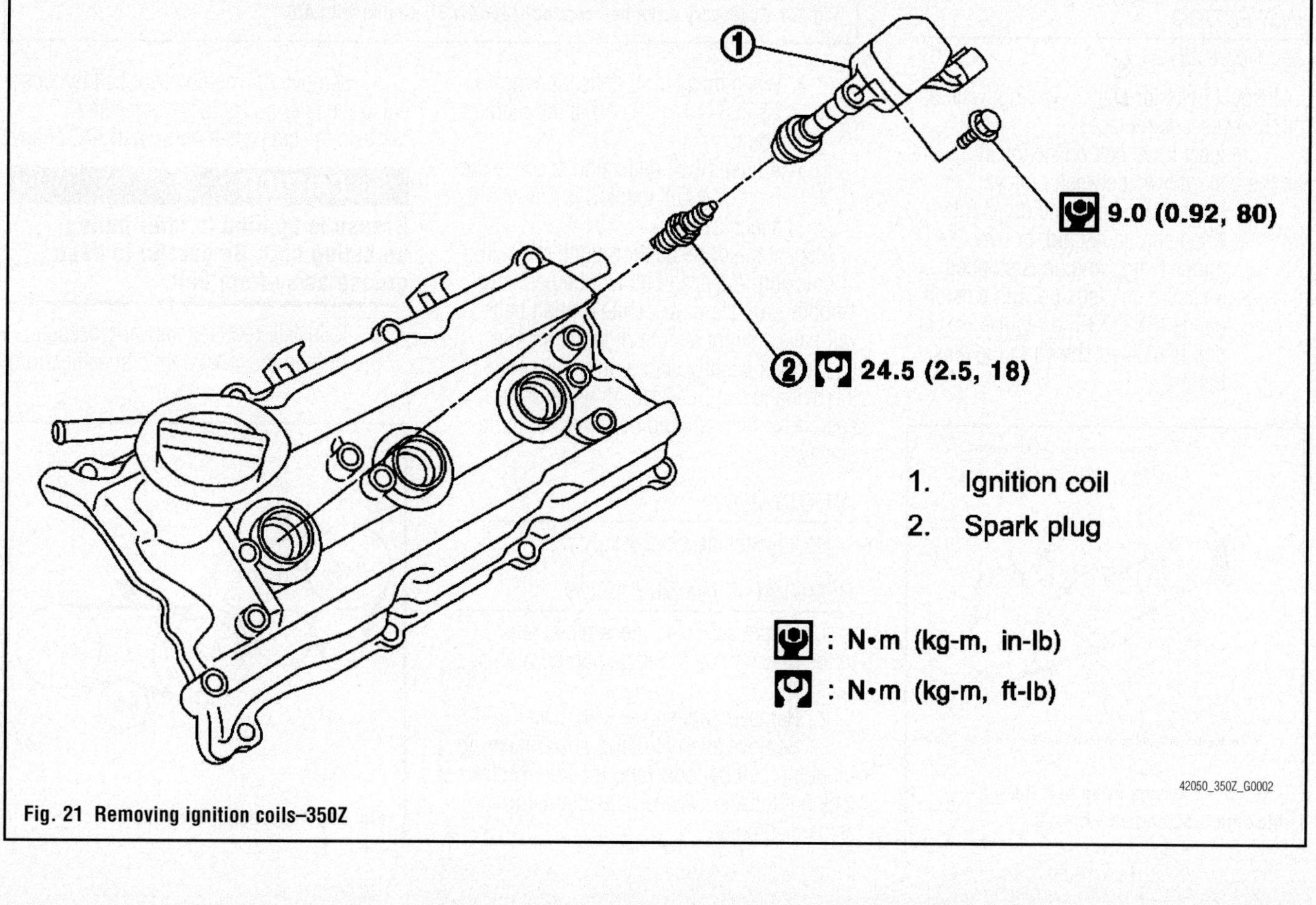

① Ignition coil

9.0 (0.92, 80)

② 24.5 (2.5, 18)

1. Ignition coil
2. Spark plug

: N·m (kg-m, in-lb)

: N·m (kg-m, ft-lb)

42050_350Z_G0002

Fig. 21 Removing ignition coils–350Z

ENGINE ELECTRICAL

STARTER

REMOVAL & INSTALLATION

See Figure 22.

1. Remove or disconnect the following:
   - Negative battery cable
   - Air duct
   - Harness protector from the harness
   - Starter wiring at the starter
   - Starter-to-engine bolts
   - Starter from the vehicle

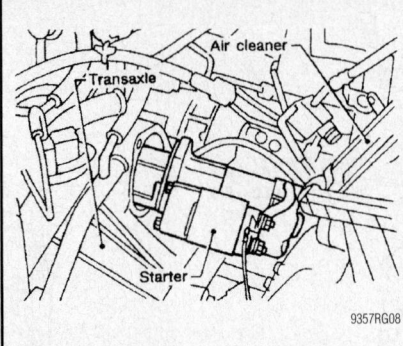

Fig. 22 Starter location and mounting detail—3.5L Maxima

STARTING SYSTEM

To install:

2. Install or connect the following:
   - Starter and torque the long bolt to 57–72 ft. lbs. (77–98 Nm) on Maxima and 41 ft. lbs. (55 Nm). On 350Z torque the short bolt to 22–30 ft. lbs. (30–41 Nm)
   - Starter wiring
   - Harness protector
   - Air duct
   - Negative battery cable

ENGINE MECHANICAL

➡ **Disconnecting the negative battery cable may interfere with the functions of the on board computer systems and may require the computer to undergo a relearning process, once the negative battery cable is reconnected.**

ACCESSORY DRIVE BELTS

ACCESSORY BELT ROUTING

See Figures 23 and 24.

INSPECTION

See Figure 25.

Inspect belts for cracks, fraying, wear and oil. If necessary, replace.

1. Inspect drive belt deflection at a point on the belt midway between pulleys.
   - Inspection should be done only when engine is cold, or over 30 minutes after engine is stopped.
   - Measure belt tension with tension gauge (BT3373-F or equivalent) at points marked shown in the illustration.

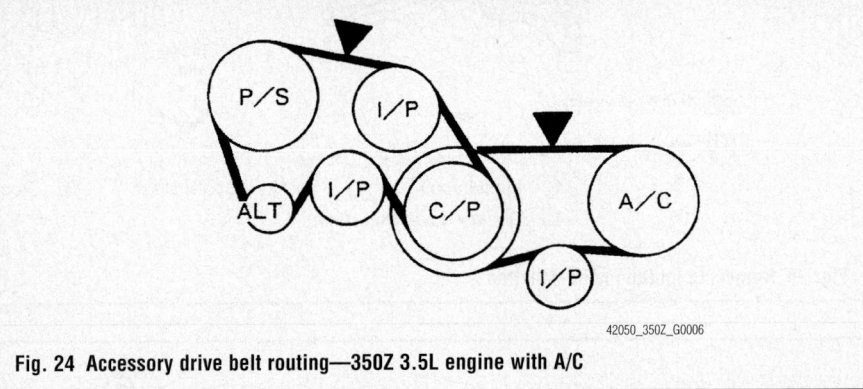

Fig. 24 Accessory drive belt routing—350Z 3.5L engine with A/C

- When measuring deflection, apply 22 lbs. (10 kg, 98 N) at the marked point.
- Adjust if belt deflection exceeds the limit or if belt tension is not within specifications.

Inspect the drive belt for signs of glazing or cracking. A glazed belt will be perfectly smooth from slippage, while a good belt will have a slight texture of fabric visible. Cracks will usually start at the inner edge of the belt and run outward. All worn or damaged drive belts should be replaced immediately.

ADJUSTMENT

No adjustment is necessary.

REMOVAL & INSTALLATION

1. Before servicing the vehicle, refer to the precautions in the beginning of this section.
2. Remove undercover with power tool.
3. Remove alternator and power steering oil pump belt by loosening the idler pulley lock nut located between the alternator and crankshaft pulley.

4. Remove A/C compressor belt by loosening the idler pulley lock nut located between the crankshaft pulley and A/C pulley.

❊❊ CAUTION

Grease is applied to idler pulley adjusting bolt. Be careful to keep grease away from belt.

5. To install, reverse removal procedure.
6. Tighten the following to specification:

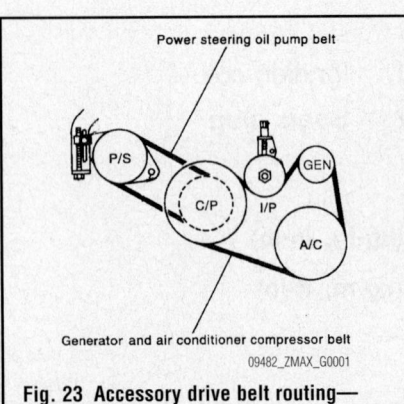

Fig. 23 Accessory drive belt routing—Maxima 3.5L engine with A/C

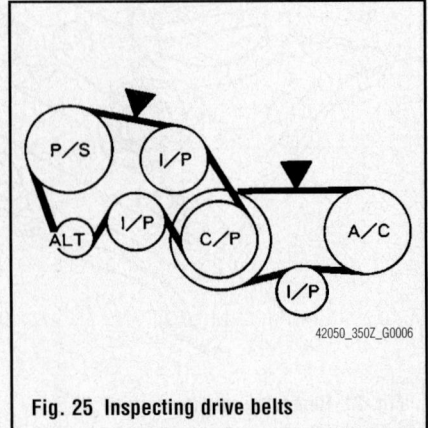

Fig. 25 Inspecting drive belts

- Alternator and power steering oil pump belt idler pulley lock nut: 24–28 ft. lbs. (31–38 Nm).
- A/C compressor belt idler pulley lock nut: 24–28 ft. lbs. (31–38 Nm).

CAMSHAFT AND VALVE LIFTERS

REMOVAL & INSTALLATION

Maxima

See Figures 26 through 29.

1. Before servicing the vehicle, refer to the Precautions Section.
2. Relieve the fuel system pressure.
3. Drain the engine oil.
4. Drain the cooling system.
5. Remove or disconnect the following:
   - Negative battery cable
   - Left side rocker cover

➥**Before detaching any hoses or connectors, note the locations for reassembly.**

- Air duct to intake manifold hose, collector hose, blow-by hose and vacuum hoses
- Fuel hoses and detach the harness connections
- Remove the engine coolant reservoir.
- Canister purge hoses
- Water hoses from the cylinder head and intake manifold
- All 6 ignition coils from the spark plugs
- Spark plugs
- Bolts that secure the Exhaust Gas Recirculation (EGR) tube
- EGR tube
- Intake manifold collector supports and the collector
- Bolts that secure the fuel tube and the fuel tube

- Bolts that secure the intake manifold to the engine block and the manifold. Loosen the bolts in the reverse sequence of the tightening procedure.
- Left-hand and right-hand rocker covers from the cylinder head
- Engine undercovers
- Right front wheel and engine side covers
- Drive belts and idler pulley
- Power steering oil pump belt and the power steering oil pump assembly
- Camshaft Position (CMP) sensor (PHASE) and Crankshaft Position (CKP) sensors (REF)/(POS)

6. Set the No. 1 piston to Top Dead Center (TDC) of compression stroke by rotating the crankshaft.
   - Ring gear cover access plate. Loosen the crankshaft pulley bolt while securing the ring gear so the crankshaft cannot rotate
   - Crankshaft pulley, using a suitable puller
   - Air conditioning compressor and bracket
   - Front exhaust pipe and its support

7. Hang the engine at the right and left side engine slingers with a suitable hoist.
8. Support the transaxle with jack.
   - Right side engine mounting, mounting bracket and nuts
   - Center crossmember assembly
   - Steel (lower) oil pan bolts in the reverse of the installation sequence

9. Insert a seal cutter between the steel and aluminum oil pan.
10. Tapping the cutter with a hammer, slide it around the entire edge of the oil pan. Be careful not to damage the aluminum mating surface of the upper oil pan.
    - Lower oil pan and the oil strainer
    - Aluminum (upper) oil pan bolts in the reverse of the installation sequence

- Transaxle bolts that secure the oil pan

11. Insert a seal cutter between the aluminum oil pan and the engine block.
12. Tapping the cutter with a hammer, slide it around the entire edge of the oil pan. Be careful not to damage the mating surfaces of the oil pan or engine block.
    - Oil pan from the vehicle
    - Water pump cover and the bolts that secure the front timing chain case cover
    - Timing chain case cover, using the seal cutter
    - Internal timing chain guide and the upper chain guide
    - Timing chain tensioner and slack · side chain guide
    - Left and right intake camshaft sprockets first. Be sure to hold the flats of the camshafts while removing the sprocket bolts.
    - Lower timing chain assembly. Be sure to note the aligning marks of the chain before removal.

13. Insert a suitable stopper pin for the left and right camshaft tensioners.
    - Left and right exhaust camshaft sprocket bolts. Be sure to hold the flats of the camshafts while removing the sprocket bolts.
    - Upper timing chain assembly. Be sure to note the aligning marks of the chain before removal.
    - Lower timing chain guide
    - Crankshaft sprocket
    - Bolts that secure the rear timing chain case. The bolts must be loosened in sequence.
    - Rear timing case cover, using the seal cutter

➥**Remove the O-rings from the front of the engine block.**

- Camshaft bearing caps in several steps. The bearing caps MUST be loosened in sequence.

➥**Keep all bearing caps and camshafts in proper order for installation.**

- Left-hand and right-hand camshaft tensioners from the cylinder head
- Camshafts from the cylinder heads
- Lifter assembly from the bore. Be sure to note the locations from where each lifter came.

14. Check the diameter of the valve lifter and the valve lifter guide bore.
15. The diameter of the lifter should be 1.3378–1.3382 in. (33.980–3.0990mm) and the diameter of the bore should be 1.3386–1.3392 in. (34.000–34.016mm).

Bank	INT/EXH	Dowel pin (1)	Paint marks			Identification mark (C)
			M1 (E)	M2 (F)	M3 (D)	
RH	EXH (B)	Yes	No	Green	Light blue	1F
	INT (A)	Yes	Grenn	No	Light blue	1E
LH	INT (A)	Yes	Green	No	Light blue	1G
	EXH (B)	Yes	No	Green	Light blue	1H

22140_350Z_G0009

Fig. 26 Camshaft identification marks legend

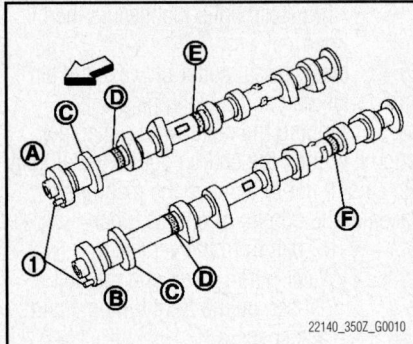

Fig. 27 Camshaft identification marks position

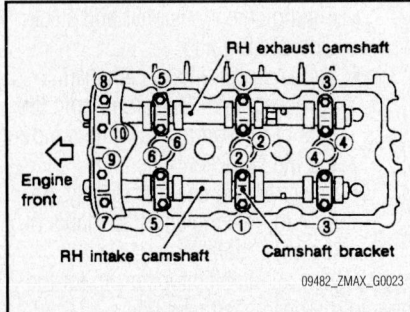

Fig. 28 Right cylinder head camshaft bearing cap tightening sequence—3.5L engine

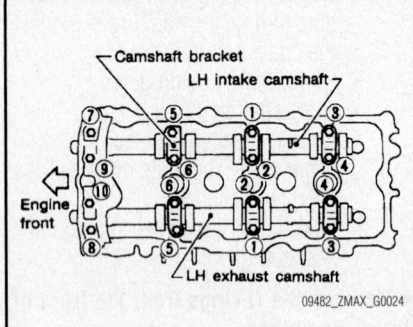

Fig. 29 Left cylinder head camshaft bearing cap tightening sequence—3.5L engine

16. Remove all traces of liquid gasket from the timing chain case and from the water pump covers.

17. Remove all traces of liquid gasket from the engine block.

18. Inspect the camshafts for excessive wear or damage and replace as necessary.

To install:

➡ **Before installing the camshaft brackets, apply RTV sealant to the mating surface of the No. 1 journal head.**

19. Lubricate the valve lifters with clean engine oil and install the lifters into the bore from which they were removed.

20. Turn the crankshaft clockwise until the No. 1 piston is set 240 degrees before TDC on compression stroke.

21. Install or connect the following:
- Camshaft tensioners on both sides of the cylinder heads and torque the bolts to 75–96 inch lbs. (8.4–10.8 Nm)

➡ **The camshafts can be identified by the paint marks on the camshaft. The left cylinder head camshafts have a YELLOW paint mark and the right cylinder head camshafts have a WHITE paint mark.**

- Exhaust and intake camshafts and install the bearing caps. Before installing the No. 1 bearing cap, apply liquid gasket to the corners of the cap.

➡ **When installing the camshafts, position the camshaft keys at the 12 o'clock position in respect to the cylinder head angle.**

22. Torque the camshaft bearing caps as follows:
   a. Bolts No. 7–10: 17 inch lbs. (2 Nm).
   b. Bolts No. 1–6: 17 inch lbs. (2 Nm).
   c. Bolts No. 1–10: 52 inch lbs. (6 Nm).
   d. Bolts No. 1–10: 81–104 inch lbs. (9–11 Nm).

23. Install new O-rings to the front of the engine block.

24. Apply sealant to the hatched portion of the of the rear timing chain case.

25. Align the rear timing chain case with the dowel pins and install onto the cylinder heads and engine block.

26. Torque the rear timing chain case mounting bolts in sequence to 105–121 inch lbs. (11.8–13.7 Nm).

27. Install the crankshaft sprocket with the mating mark facing out.

28. Rotate the crankshaft clockwise and position the crankshaft to TDC of compression stroke and align the dowels of the camshaft sprockets to the 12 o'clock position in respect to the cylinder head.

29. Install the lower chain guide on the dowel pin with the front mark on the guide facing upward.

30. On a workbench, align the marks on the intake and exhaust camshaft sprockets with the marks of the chain.

31. Put the exhaust camshaft sprockets onto the dowel pin and torque the bolts to 88–95 ft. lbs. (119–128 Nm). Be sure to secure the camshafts while tightening the bolts.

32. Install or connect the following:
- Timing chains, sprockets and related components
- Transaxle bolts that secure the oil pan
- Oil pan strainer and torque the bolts to 12–14 ft. lbs. (16–19 Nm)

33. Apply a 0.177–0.217 in. (4.5–5.5mm) continuous bead of liquid gasket to the lower oil pan mating surface and install the oil pan. Torque the bolts in sequence to 57–66 inch lbs. (6.4–7.5 Nm).
- Center crossmember assembly
- Right side engine mounting bracket and mount assembly

34. Remove the engine slinger assembly.
- Front exhaust pipe and its support
- Air conditioning compressor and bracket
- Crankshaft pulley to the crankshaft and install the mounting bolt. Torque the bolt to 14–22 ft. lbs. (20–29 Nm). Torque the crankshaft bolt an additional 60–66 degrees clockwise. This is about the angle from one hexagon bolt head corner to another
- CMP sensor, PHASE and CKP sensors
- Power steering pump
- Idler pulley and all belts
- Engine side and under covers
- Right front wheel
- Intake manifold
- Rocker covers
- Fuel tube
- Intake manifold support and collector
- EGR tube
- Spark plugs and ignition coils
- Coolant hoses
- Canister purge hoses
- Fuel feed and return lines
- Vacuum hoses
- Negative battery cable

35. Fill the cooling system.
36. Fill the engine with clean oil.
37. Start the vehicle, check for leaks and repair if necessary.

350Z

See Figures 30 through 33.

1. Before servicing the vehicle, refer to the Precautions Section.
2. Relieve the fuel system pressure.
3. Drain the engine oil.
4. Drain the cooling system.
5. Disconnect the negative battery cable.
6. Remove the timing chain case, camshaft sprockets, timing chain and rear timing chain case.

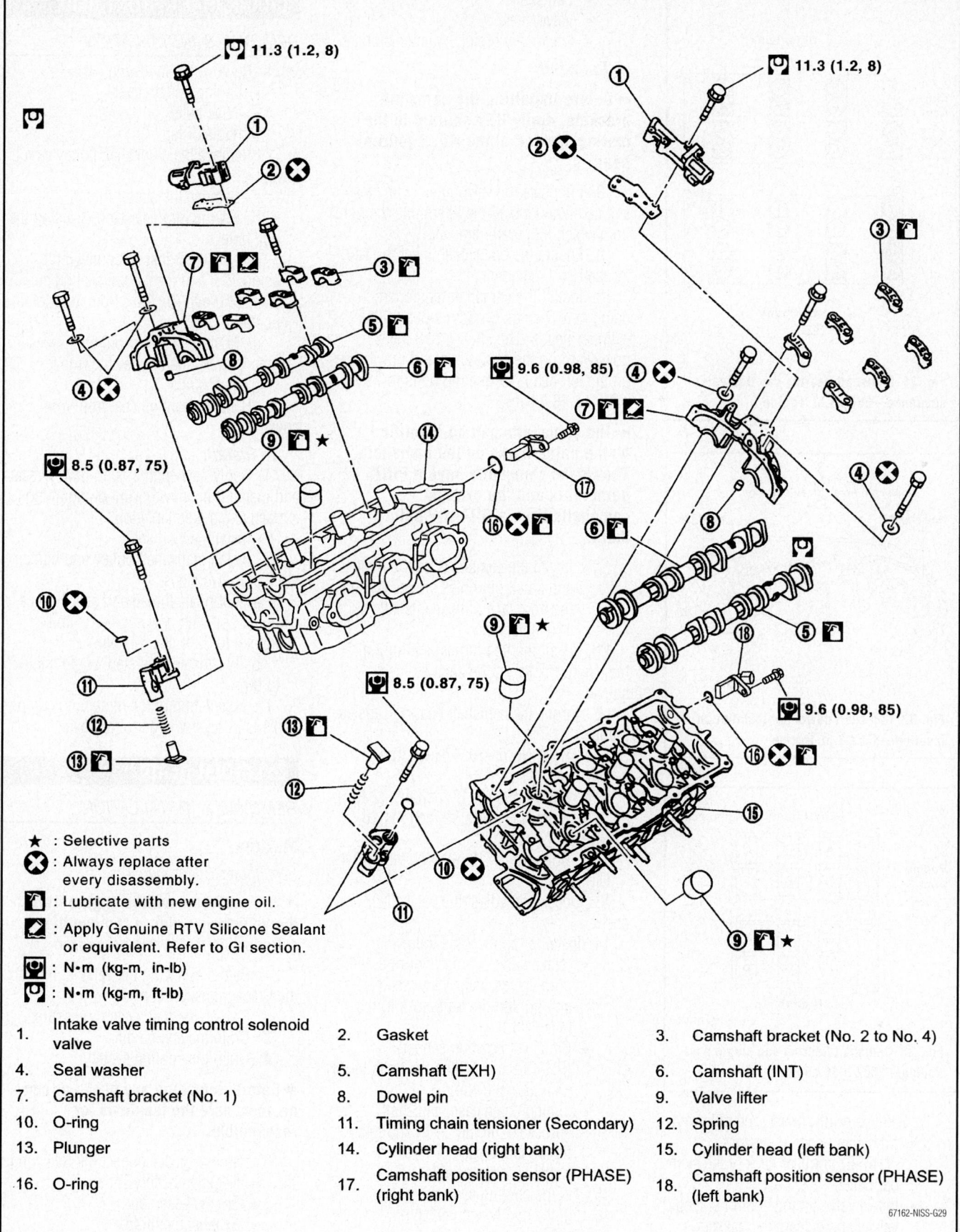

★ : Selective parts

⊗ : Always replace after every disassembly.

▨ : Lubricate with new engine oil.

▨ : Apply Genuine RTV Silicone Sealant or equivalent. Refer to GI section.

▨ : N•m (kg-m, in-lb)

▨ : N•m (kg-m, ft-lb)

1. Intake valve timing control solenoid valve	2. Gasket	3. Camshaft bracket (No. 2 to No. 4)
4. Seal washer	5. Camshaft (EXH)	6. Camshaft (INT)
7. Camshaft bracket (No. 1)	8. Dowel pin	9. Valve lifter
10. O-ring	11. Timing chain tensioner (Secondary)	12. Spring
13. Plunger	14. Cylinder head (right bank)	15. Cylinder head (left bank)
16. O-ring	17. Camshaft position sensor (PHASE) (right bank)	18. Camshaft position sensor (PHASE) (left bank)

67162-NISS-G29

Fig. 30 Exploded view of camshaft assemblies—350Z 3.5L engine

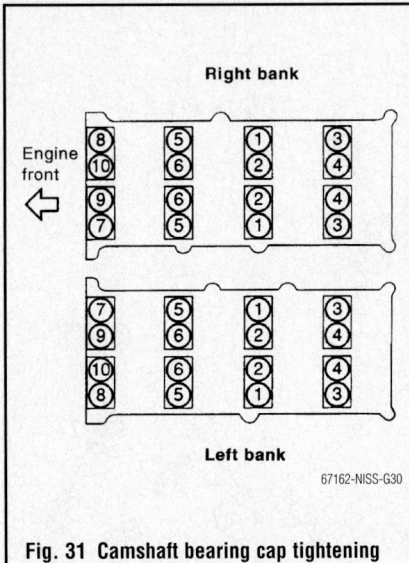

Fig. 31 Camshaft bearing cap tightening sequence—350Z 3.5L engine

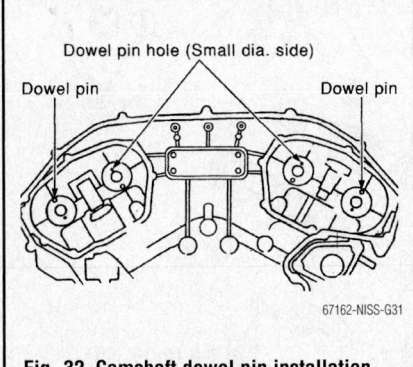

Fig. 32 Camshaft dowel pin installation location—350Z 3.5L engine

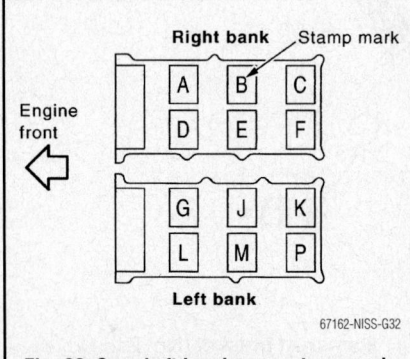

Fig. 33 Camshaft bearing cap stamp positioning—350Z 3.5L engine

7. Remove or disconnect the following:
  • Camshaft position sensors (PHASE) from the back of the cylinder heads
  • Intake valve timing control solenoid valves and discard the gaskets
  • Camshaft bearing caps in the reverse of the tightening sequence

  • Camshafts
  • Valve lifters
  • Secondary timing chain tensioners

To install:

➡ **Before installing the camshaft brackets, apply RTV sealant to the mating surface of the No. 1 journal head.**

8. Lubricate the valve lifters with clean engine oil and install the lifters into the bore from which they were removed.

9. Ensure the crankshaft is set to TDC for the No. 1 cylinder.

10. Install the camshaft tensioners using new O-rings on both sides of the cylinder heads. The sliding part faces downward on the right head and upward on the left head. Torque the bolts to 75 inch lbs. (8.4 Nm).

➡ **The camshafts can be identified by the paint marks on the camshaft. The intake camshafts have a PINK paint mark and the exhaust camshafts have a ORANGE paint mark.**

  • Install the camshafts so the large and small pin holes are located on the front face of the camshafts at 180° intervals.

11. Install the bearing caps aligning the stamp marks on the caps as shown.

12. Torque the camshaft bearing caps as follows:
  a. Bolts No. 7–10: 17 inch lbs. (2 Nm).
  b. Bolts No. 1–6: 17 inch lbs. (2 Nm).
  c. Bolts No. 1–10: 52 inch lbs. (6 Nm).
  d. Bolts No. 1–10: 92 inch lbs. (10.4 Nm).

13. Check and adjust the valve clearance.

14. Install or connect the following:
  • Intake valve timing control solenoid valves using new gaskets and tighten the bolts to 8 ft. lbs. (11.3 Nm)
  • Camshaft position sensors (PHASE) and tighten the bolts to 85 inch lbs. (9.6 Nm)
  • Timing chain case, camshaft sprockets, timing chain and rear timing chain case
  • Negative battery cable

15. Fill the cooling system.
16. Fill the engine with clean oil.
17. Start the vehicle, check for leaks and repair if necessary.

CRANKSHAFT FRONT SEAL

REMOVAL & INSTALLATION

1. Remove the following parts:
  • Engine under cover
  • Drive belts
  • Radiator fan

2. Remove the crankshaft pulley as follows:
  a. Remove the starter motor.
  b. Set the ring gear stopper using the bolt hole.
  c. Loosen crankshaft pulley bolt using Tool and locate bolt seating surface at 10 mm (0.39 in) from its original position.
  d. Position a pulley puller at recess hole of crankshaft pulley to remove crankshaft pulley.

3. Remove front oil seal from front cover.

To install:

4. Apply new engine oil to new oil seal and install it flush with front of mounting surface using a suitable tool.

5. Install new oil seal

6. Install crankshaft pulley and tighten the bolt in two steps.
  a. Lubricate thread and seat surface of the bolt with new engine oil and tighten to 32 ft. lbs. (44 Nm).
  b. For the second step, angle tighten to 90°.

7. Installation of the remaining components is in reverse order of removal.

CYLINDER HEAD

REMOVAL & INSTALLATION

Maxima

See Figures 34 through 42.

➡ **You must remove the engine from the vehicle in order to remove the cylinder head, for this procedure.**

1. Before servicing the vehicle, refer to the Precautions Section.
2. Relieve the fuel system pressure.
3. Drain the engine oil.
4. Drain the cooling system.

➡ **Before detaching any hoses or connectors, note the locations for reassembly.**

5. Remove or disconnect the following:
  • Negative battery cable
  • Engine assembly
  • Exhaust manifold

6. Place the engine on a suitable workstand.

- Oil pan
- Timing chain
- Intake manifold
- Water outlet
- Rear timing chain case bolts, in the sequence shown
- Rear timing chain case
- O-rings from the cylinder head and block
- Intake valve timing control solenoid valves

➡**For installation purposes, matchmark the camshaft brackets before removing them.**

- Intake and exhaust camshafts and brackets. Loosen the bracket bolts in several steps, in the sequence shown.
- Right and left side cam chain tensioner from the cylinder head
- Cylinder head bolts. Loosen in several steps, in the sequence shown.

➡**A warped or cracked cylinder head could result from removing the bolts in incorrect order.**

- Cylinder heads from the vehicle
- Discard the head gaskets
7. Remove all traces of liquid gasket

from the timing chain case and from the water pump covers.

8. Remove all traces of liquid gasket from the engine block.

9. Inspect the timing chain for excessive wear or damage and replace as necessary.

To install:

10. Turn the crankshaft until the No. 1 piston is a Top Dead Center (TDC) on compression stroke. The crankshaft key should face toward the right bank.

11. Using new head gaskets, install the cylinder heads.

➡**If possible, replacement of the head bolts is suggested.**

12. If replacement of the head bolts is not possible, perform the following bolt measurement:

a. Measure the diameter of the head bolt 0.43 in. (11mm) from the bottom of the bolt.

b. Measure the diameter of the head bolt 1.89 in. (48mm) from the bottom of the bolt.

c. Whenever the size difference between the 2 measurements exceeds 0.0043 in. (0.11mm) the head bolts must be replaced.

13. Install the cylinder head bolts and torque in sequence as follows:

a. Step 1: 72 ft. lbs. (98 Nm).
b. Step 2: Completely loosen all bolts.
c. Step 3: 26–32 ft. lbs. (34–44 Nm).
d. Step 4: plus 90–95 degrees clockwise.
e. Step 5: plus 90–95 degrees clockwise.

14. Install or connect the following:

- Camshafts and related components
- Intake valve timing control solenoid valves
- New O-rings to the front of the engine block and cylinder head

15. Apply sealant to the hatched portion of the of the rear timing chain case.

16. Align the rear timing chain case with the dowel pins and install onto the cylinder heads and engine block.

17. Torque the rear timing chain case mounting bolts in sequence to 105–121 inch lbs. (11.8–13.7 Nm).

18. Install or connect the following:

- Water outlet
- Intake manifold
- Timing chain
- Oil pan
- Exhaust manifold
- Engine assembly into the vehicle
- Negative battery cable

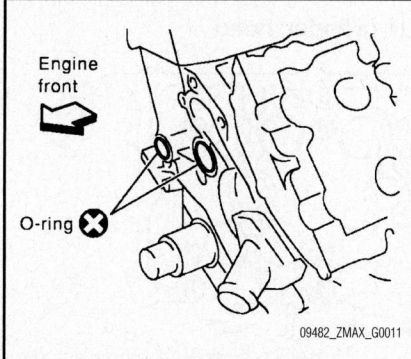

Fig. 34 Loosen the rear timing chain case bolts in sequence—Maxima 3.5L engine

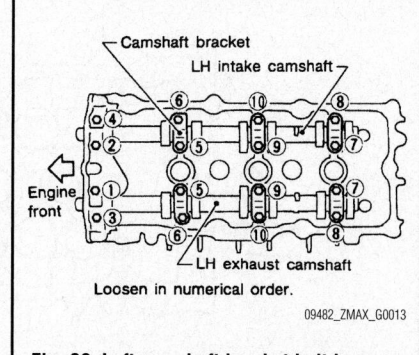

Fig. 36 Remove the O-rings from the engine block—Maxima 3.5L engine

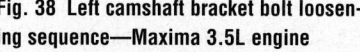

Fig. 38 Left camshaft bracket bolt loosening sequence—Maxima 3.5L engine

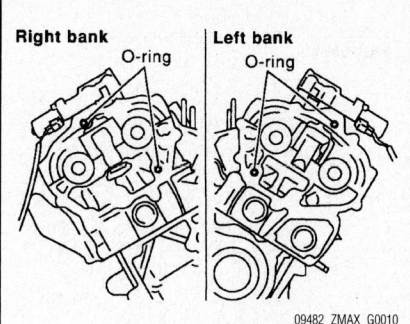

Fig. 35 Remove the O-rings from the cylinder head—Maxima 3.5L engine

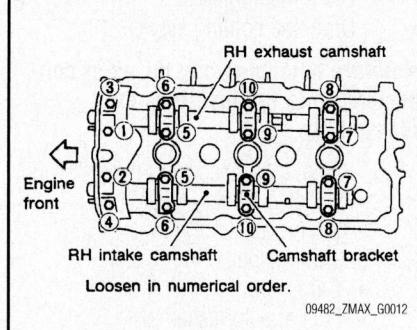

Fig. 37 Right camshaft bracket bolt loosening sequence—Maxima 3.5L engine

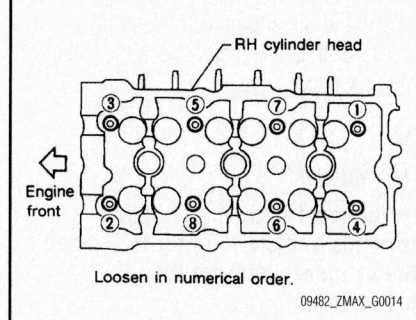

Fig. 39 Right cylinder head bolt loosening sequence—Maxima 3.5L engine

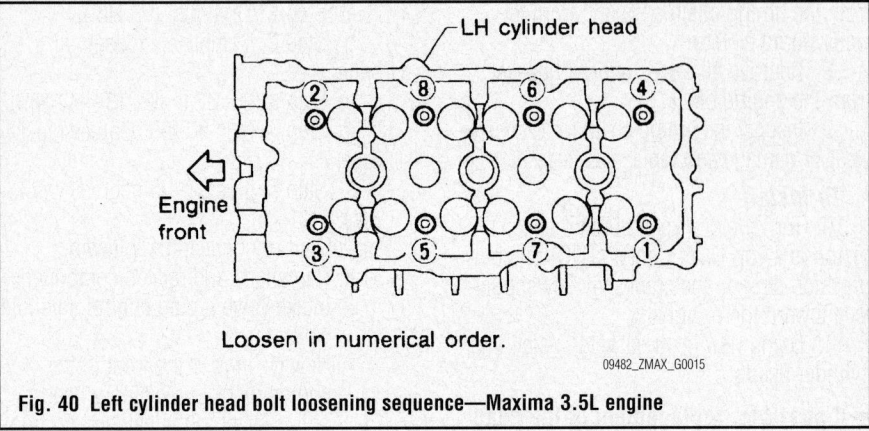

Fig. 40 Left cylinder head bolt loosening sequence—Maxima 3.5L engine

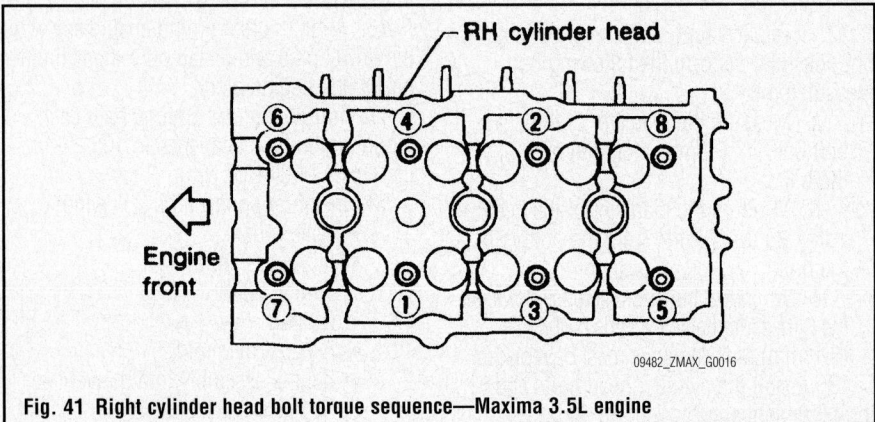

Fig. 41 Right cylinder head bolt torque sequence—Maxima 3.5L engine

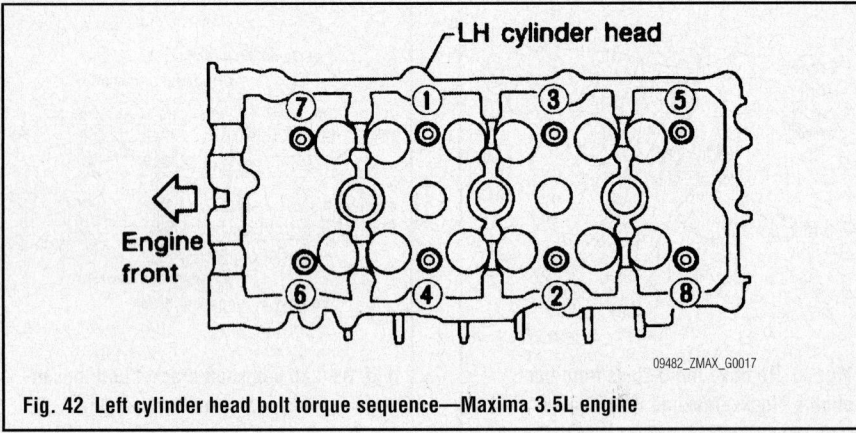

Fig. 42 Left cylinder head bolt torque sequence—Maxima 3.5L engine

19. Fill the cooling system.
20. Fill the engine with clean oil.
21. Start the vehicle, check for leaks and repair if necessary.

350Z

See Figure 43.

➡ **You must remove the engine from the vehicle in order to remove the cylinder head, for this procedure.**

1. Before servicing the vehicle, refer to the Precautions Section.
2. Relieve the fuel system pressure.
3. Drain the engine oil.
4. Drain the cooling system.

➡ **Before detaching any hoses or connectors, note the locations for reassembly.**

5. Remove or disconnect the following:
- Negative battery cable
- Engine assembly
- Timing chain
- Camshafts
- Fuel injector assembly
- Intake manifold
- Exhaust manifold
- Thermostat housing

- Cylinder head bolts in the reverse of the tightening sequence
- Cylinder head gaskets

To install:

6. Turn the crankshaft until the No. 1 piston is a Top Dead Center (TDC) on compression stroke. The crankshaft key should face toward the right bank.
7. Using new head gaskets, install the cylinder heads.

➡ **If possible, replacement of the head bolts is suggested.**

8. If replacement of the head bolts is not possible, perform the following bolt measurement:
   a. Measure the diameter of the head bolt 0.43 in. (11mm) from the bottom of the bolt.
   b. Measure the diameter of the head bolt 1.89 in. (48mm) from the bottom of the bolt.
   c. Whenever the size difference between the 2 measurements exceeds 0.0043 in. (0.11mm) the head bolts must be replaced.
9. Install the cylinder head bolts and torque in sequence as follows:
   a. Step 1: 72 ft. lbs. (98 Nm).
   b. Step 2: Completely loosen all bolts.
   c. Step 3: 26–32 ft. lbs. (34–44 Nm).
   d. Step 4: plus 90–95 degrees clockwise.
   e. Step 5: plus 90–95 degrees clockwise.
10. Install or connect the following:

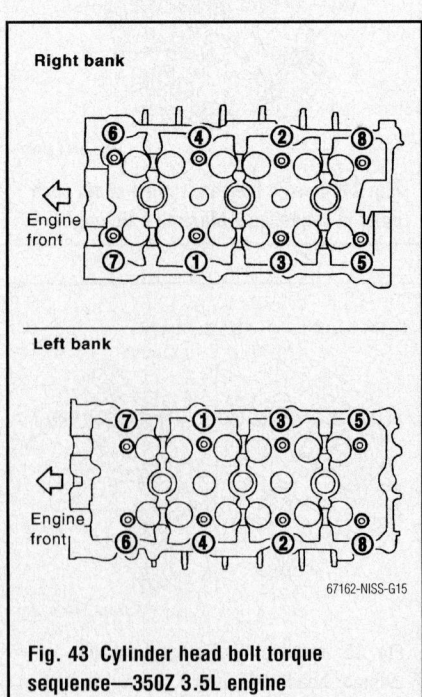

Fig. 43 Cylinder head bolt torque sequence—350Z 3.5L engine

- Thermostat housing
- Exhaust manifold
- Intake manifold
- Fuel injector assembly
- Camshafts
- Timing chain
- Engine assembly
- Negative battery cable

11. Fill the cooling system.

12. Fill the engine with clean oil.

13. Start the vehicle, check for leaks and repair if necessary.

ENGINE ASSEMBLY

REMOVAL & INSTALLATION

Maxima

See Figures 44 through 46.

It is recommended the engine and transaxle be removed as a single unit. If need be, the units may be separated after removal.

➡**The engine and transaxle assembly must be removed from the underside of the vehicle.**

1. Before servicing the vehicle, refer to the Precautions Section.

2. Release the fuel system pressure.

3. Drain the cooling system.

4. Drain the engine oil.

5. Drain the automatic transaxle.

6. Remove or disconnect the following:
- Negative battery cable
- Hood
- Engine under cover
- All vacuum hoses, fuel lines, wires and connectors; tag before disconnecting

7. If necessary, remove the wiper motor and linkage as follows:
- Operate wiper motor one full cycle, then turn off
- Remove the wiper arm nut covers, then the nuts

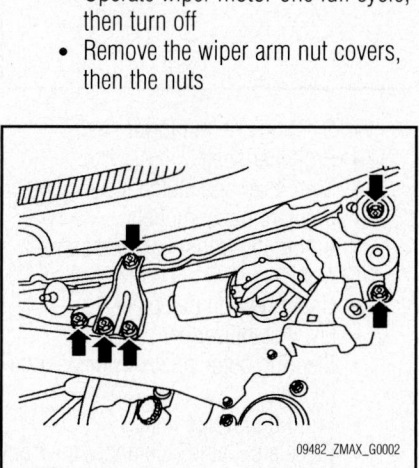

Fig. 44 Removal of motor and bracket assembly—Maxima

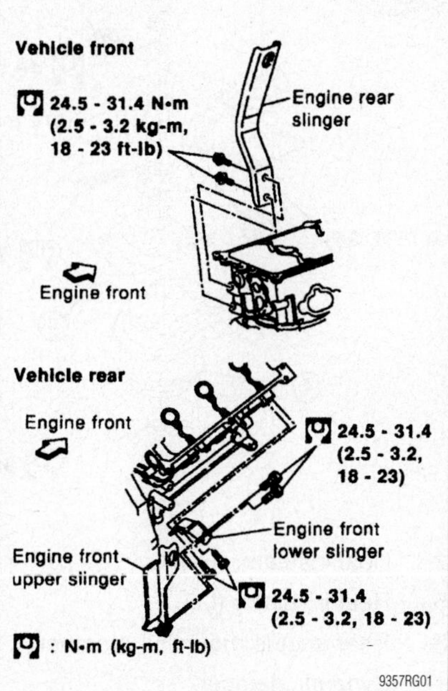

- Remove the wiper arms
- Remove fender covers
- Remove cowl top clips, then partially lift cowl top for access
- Disconnect windshield washer tube
- Remove cowl top seal and weatherstrip seal
- Remove cowl top cover
- Disconnect wiper motor connector
- Remove bracket and wiper motor assembly

8. Remove or disconnect the following:
- Front exhaust pipe from the manifold
- Ball joints from the steering knuckle
- Halfshafts
- Radiator and fans
- Drive belts
- Alternator

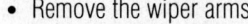

Vehicle front

24.5 - 31.4 N•m (2.5 - 3.2 kg-m, 18 - 23 ft-lb) — Engine rear slinger

Engine front

Vehicle rear

Engine front

24.5 - 31.4 (2.5 - 3.2, 18 - 23)

Engine front lower slinger

Engine front upper slinger

24.5 - 31.4 (2.5 - 3.2, 18 - 23)

: N•m (kg-m, ft-lb)

9357RG01

Fig. 45 Installation of engine slingers to lift the engine

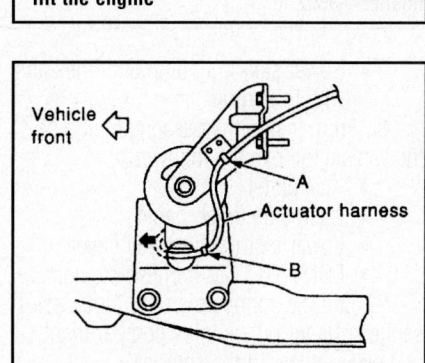

Vehicle front

A — Actuator harness

B

9357RG02

Fig. 46 For electronically controlled engine mounts, the proper length from A to B is 6.69 in. (170mm)

- A/C compressor. Position it aside with the lines attached. Do NOT disconnect the refrigerant lines.
- Power steering pump and position aside with the lines attached. Do NOT disconnect the fluid lines.

9. Place a suitable jack under the transaxle. Install engine slingers and a suitable engine hoist. Raise the engine for access to the left side engine mount.
- Left side engine mount
- Control and support rods from the transaxle, manual transaxle only
- Control cable from the transaxle, automatic transaxle only
- Right side engine mount
- Center member, then carefully and slowly lower the transmission jack

10. Lower the engine/transaxle assembly onto an engine stand.

➡**When lowering the engine out, guide it carefully to avoid hitting any other components.**

To install:

11. Installation is the reverse of the removal procedure, noting the following points:

a. If equipped with electronically controlled engine mounts, install them to the specifications shown in the accompanying figure.

b. Make sure to connect all vacuum hoses, lines, and electrical connectors as tagged during removal.

c. Fill the cooling system.

d. Fill the engine with clean oil.

e. Start the vehicle, check for leaks and repair if necessary.

350Z

See Figures 47 and 48.

It is recommended the engine and transaxle be removed as a single unit. If need be, the units may be separated after removal.

➡**The engine and transaxle assembly must be removed from the underside of the vehicle.**

1. Before servicing the vehicle, refer to the Precautions Section.

2. Release the fuel system pressure.

3. Drain the cooling system.

4. Drain the engine oil.

5. Drain the automatic transaxle, if equipped.

6. Discharge and recover the A/C refrigerant

7. Remove or disconnect the following:
- Negative battery cable
- Hood

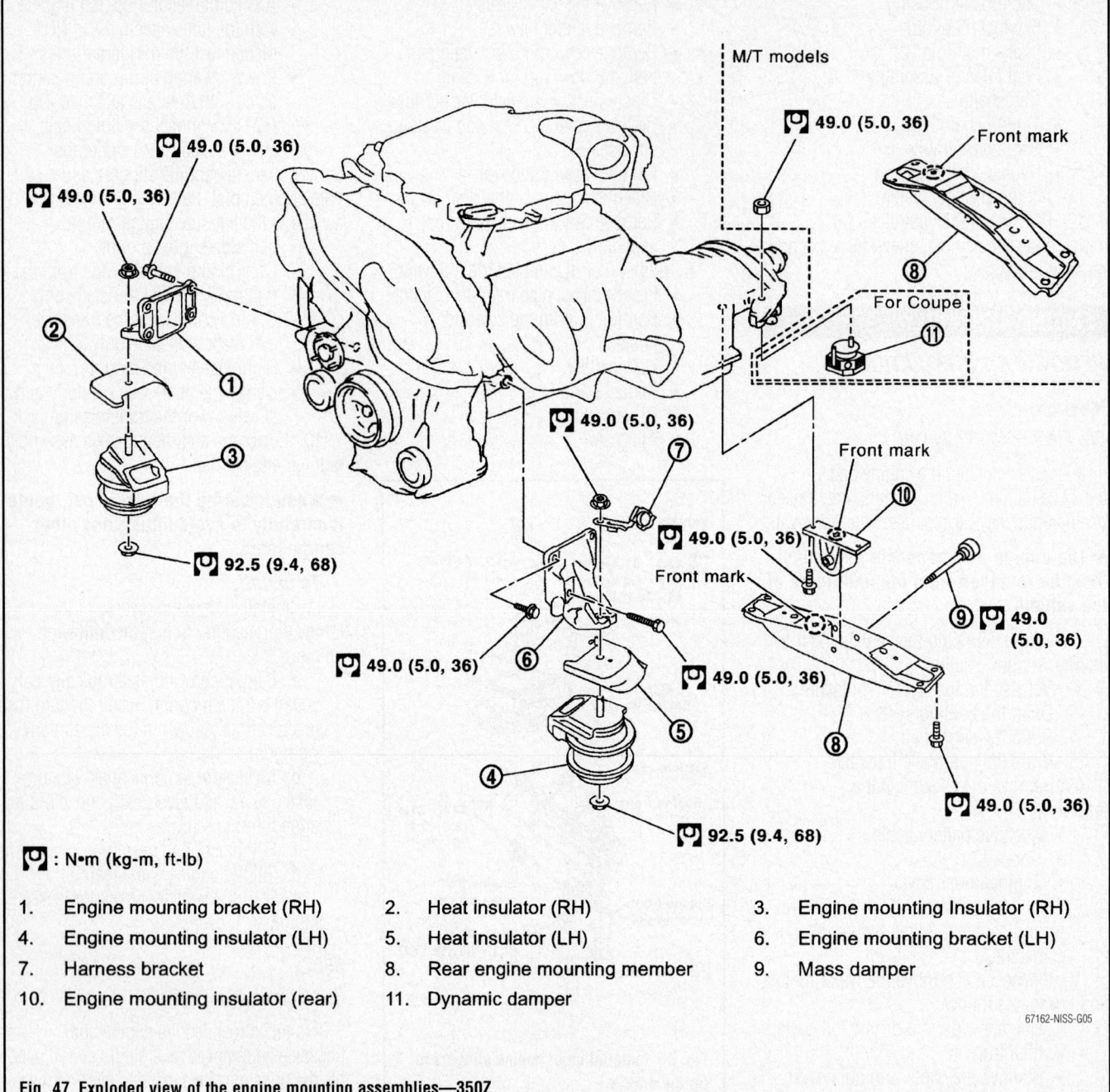

Fig. 47 Exploded view of the engine mounting assemblies—350Z

🔧 : N•m (kg-m, ft-lb)

1. Engine mounting bracket (RH)
2. Heat insulator (RH)
3. Engine mounting Insulator (RH)
4. Engine mounting insulator (LH)
5. Heat insulator (LH)
6. Engine mounting bracket (LH)
7. Harness bracket
8. Rear engine mounting member
9. Mass damper
10. Engine mounting insulator (rear)
11. Dynamic damper

67162-NISS-G05

- Strut tower bar
- Engine under cover
- Wiper arms and cowl top
- All vacuum hoses, fuel lines, wires and connectors; tag before disconnecting
- Front wheels
- Air cleaner case and duct
- Cooling fan, reservoir and hoses
- Heater hoses
- Battery ground at cylinder head
- Battery positive cable harness
- A/C lines from compressor
- 2 body ground cables
- Fuel feed and EVAP hoses

- Power steering pump and lines and wire to engine

8. From inside the passenger side of the vehicle remove the following:
  - Kick panel
  - Dash side finish panel
  - Lower instrument panel cover
  - ECM and TCM harness connectors

9. Pull the connectors out of the passenger side into the engine compartment and secure them to the engine.

10. Remove or disconnect the following:
  - Front exhaust pipe from the manifold

- Steering column lower shaft
- Propeller shaft
- Shift lever and clutch slave cylinder on man. trans. models
- Automatic transmission control rod
- Upper rear oil pan plate
- Front stabilizer bar
- Steering outer socket from steering knuckle
- Front transverse link
- Place a suitable jack under the front suspension member and transmission
- Rear engine crossmember bolts

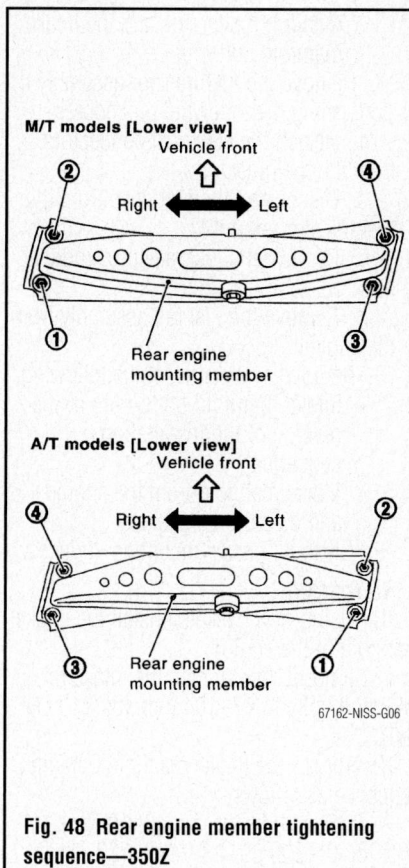

Fig. 48 Rear engine member tightening sequence—350Z

- Front suspension member bolts and nuts, then carefully and slowly lower the jack

➡**When lowering the engine out, guide it carefully to avoid hitting any other components bolts and nuts.**

To install:

11. Installation is the reverse of the removal procedure, noting the following points:

a. Tighten the engine mounts, brackets and mounting members to the specified torque as shown in the illustration. When tightening the engine brackets, tighten the upper bolts first.

b. Tighten the rear engine mounting brackets in the sequence shown.

c. Tighten the suspension member bolts to 65 ft. lbs. (87.5 Nm).

d. Tighten the strut tower bar bolts 33 ft. lbs. (45 Nm).

e. Make sure to connect all vacuum hoses, lines, and electrical connectors as tagged during removal.

f. Fill the cooling system.

g. Fill the engine with clean oil.

h. Start the vehicle, check for leaks and repair if necessary.

EXHAUST MANIFOLD

REMOVAL & INSTALLATION

Maxima

1. Before servicing the vehicle, refer to the Precautions Section.
2. Remove or disconnect the following:
- Negative battery cable
- Exhaust manifolds from the exhaust pipes
- Protective covers from the manifolds
- Heated Oxygen Sensor (HO2S) from the manifold, if equipped
- Exhaust manifold-to-engine mounting nuts
- Manifolds from the engine and discard the gaskets

To install:

3. Clean all gasket mounting surfaces. Install new gaskets.

4. Install or connect the following:
- Exhaust manifold and torque the nuts in steps to 21–24 ft. lbs. (29–33 Nm)
- Protective shields and torque the bolts in steps to 46–57 inch lbs. (5–7 Nm)
- Exhaust manifolds to the exhaust pipes and torque the nuts to 32–37 ft. lbs. (43–50 Nm)
- HO2S sensor to the manifold and torque the fastener to 30–44 ft. lbs. (40–60 Nm), if equipped
- Negative battery cable

5. Start the engine and check for exhaust leaks.

350Z

See Figures 49 and 50.

1. Before servicing the vehicle, refer to the Precautions Section.
2. Drain the engine coolant
3. Remove or disconnect the following:

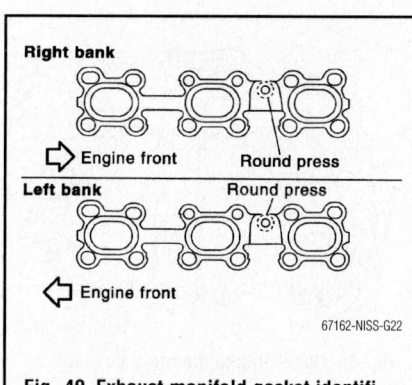

Fig. 49 Exhaust manifold gasket identification—350Z 3.5L engine

- Negative battery cable
- Strut tower bar
- Engine cover
- Air cleaner case and duct
- Heated oxygen sensor No. 2 connectors and sensors
- Exhaust mounting bracket between transmission and catalytic converters
- Catalytic converters
- Heated oxygen sensor No. 1 connectors and sensors
- Water and heater pipe on both sides
Heat shield
- Exhaust manifold bolts in reverse of the tightening sequence
- Manifold gaskets

To install:

4. Clean all gasket mounting surfaces. Install new gaskets noting the correct placement.

5. Installation is the reverse of the removal procedure noting the following:
- Exhaust manifold and torque the nuts in sequence to 22 ft. lbs. (30 Nm)
- Heat shields and torque the bolts in steps to 51 inch lbs. (6 Nm)
- Exhaust manifolds to the exhaust pipes and torque the nuts to 46 ft. lbs. (63 Nm)
- Oxygen sensors and torque the fastener to 33 ft. lbs. (45 Nm)

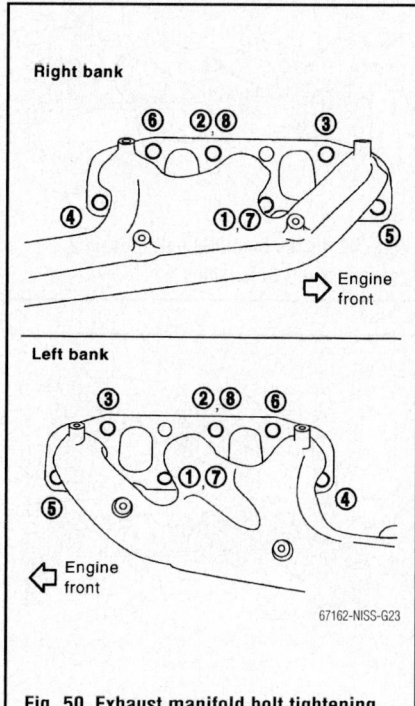

Fig. 50 Exhaust manifold bolt tightening sequence—350Z 3.5L engine

INTAKE MANIFOLD

REMOVAL & INSTALLATION

See Figures 51 through 57.

1. Before servicing the vehicle, refer to the Precautions Section.
2. Drain the cooling system.
3. Release the fuel system pressure.
4. Disconnect the negative battery cable.
5. On 350Z remove or disconnect the following:
   - Strut tower bar
   - Engine cover

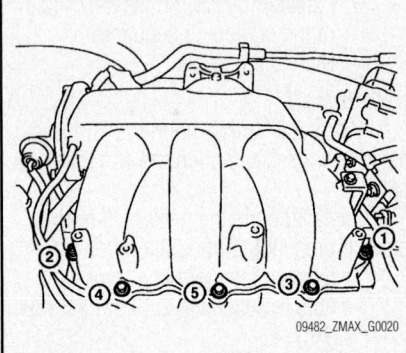

Fig. 51 Upper intake manifold collector loosening sequence—Maxima 3.5L engine

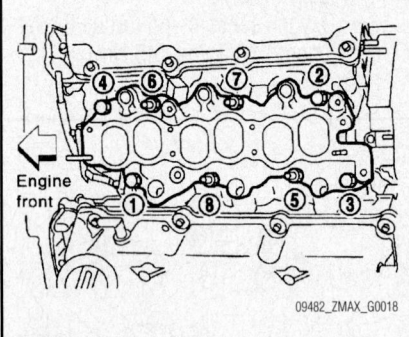

Fig. 52 Intake manifold bolt loosening sequence—3.5L engines

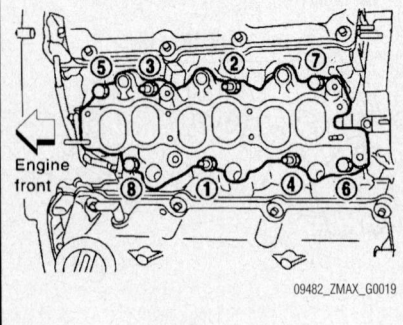

Fig. 53 Intake manifold torque sequence—3.5L engines

   - Air cleaner case and duct
   - Electronic throttle control actuator bolt in the sequence shown
   - Fuel injector and fuel tube assembly
   - Vacuum and water hoses from intake manifold collector
   - EVAP solenoid valve bracket
   - Upper intake manifold collector bolts in reverse of the tightening sequence
   - PCV hose
   - Lower intake manifold collector bolts in reverse of the tightening sequence
6. On Maxima remove or disconnect the following:
   - Throttle body coolant hoses
   - Electrical connectors from the Throttle Actuator
   - Canister purge hose and blow-by hose
   - Intake manifold collector support brackets
   - Right side electrical connectors from the ignition coils
   - Electrical connector from the crank angle sensor and the power transistor, if necessary
   - Intake manifold collector-to-intake

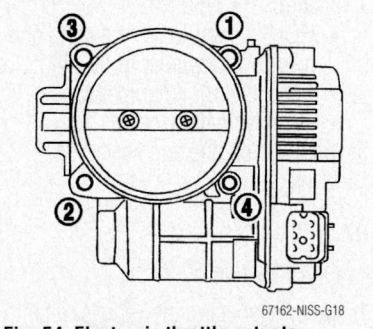

Fig. 54 Electronic throttle actuator removal and installation sequence—350Z 3.5L engine

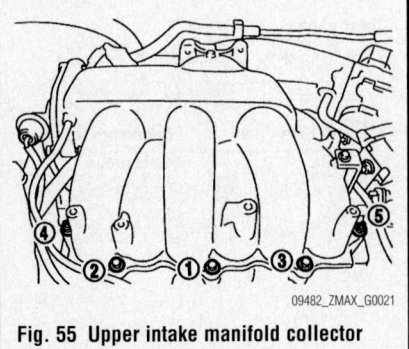

Fig. 55 Upper intake manifold collector tightening sequence—Maxima 3.5L engine

manifold bolts/nuts and the intake manifold collector

7. Remove the fuel injector assembly by performing the following procedures:
   a. Detach the electrical connectors from the fuel injectors.
   b. Disconnect the fuel lines from the fuel injector assembly.
   c. Remove the fuel rail-to-cylinder head bolts.
   d. Remove the fuel rail assembly from the engine.
8. Remove or disconnect the following:
   - Intake manifold bolts/nuts in the reverse of the installation sequence
   - Intake manifold from the engine and discard the gaskets
9. Clean all gasket mounting surfaces.

To install:

10. Using new gaskets, install the intake manifold to the engine.
11. If necessary, tighten the intake manifold stud bolts to 87–104 inch lbs. (10–12 Nm).
12. Torque the intake manifold bolts in sequence as follows:
   a. Step 1:4–7 ft. lbs. (5–10 Nm).
   b. Step 2: 20–23 ft. lbs. (26–31 Nm).

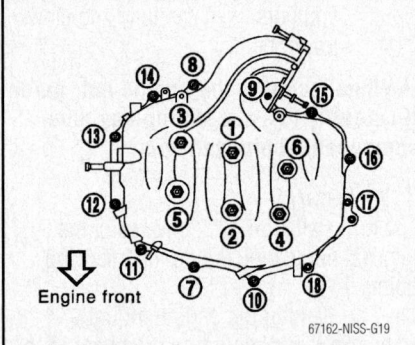

Fig. 56 Upper intake manifold collector tightening sequence—350Z 3.5L engine

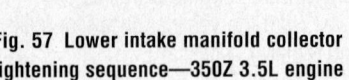

Fig. 57 Lower intake manifold collector tightening sequence—350Z 3.5L engine

c. Step 3: Tighten again to 20–23 ft. lbs. (26–31 Nm).

13. On 350Z install or connect the following:

- Lower intake manifold collector bolts in reverse of the tightening sequence and tighten to 10 ft. lbs. (14 Nm).
- PCV hose
- Upper intake manifold collector bolts in reverse of the tightening sequence and tighten to 10 ft. lbs. (14 Nm).
- EVAP solenoid valve bracket
- Vacuum and water hoses from intake manifold collector
- Fuel injector and fuel tube assembly
- Electronic throttle control actuator bolts in the sequence shown and tighten to 64–85 inch lbs. (7–10 Nm)
- Air cleaner case and duct
- Engine cover
- Strut tower bar and tighten to 24 ft. lbs. (32 Nm).

14. On Maxima install or connect the following:

15. Install the fuel injector assembly by performing the following procedures:

a. Install the fuel rail assembly to the engine.

b. Install the fuel rail-to-cylinder head bolts and tighten the bolts to 15–20 ft. lbs. (21–26 Nm) in 2 progressive steps.

c. Connect the fuel lines to the fuel injector assembly.

d. Connect the electrical connectors to the fuel injectors.

e. Install the intake manifold collector. Torque the fasteners to 13–15 ft. lbs. (17.5–21.5 Nm).

16. Install or connect the following:

- Crank angle sensor and transmitter electrical connectors
- Right side ignition coil electrical connectors
- Intake manifold collector support brackets
- Canister purge and blow by hoses
- Throttle body, EGR valve and intake manifold collector hoses
- IAC valve and fuel pressure regulator hoses
- TP sensor electrical connector
- Throttle body coolant hose

17. On all models, connect the negative battery cable.

18. Fill the cooling system.

19. Start the vehicle, check for leaks and repair if necessary.

OIL PAN

REMOVAL & INSTALLATION

Maxima

See Figures 58 and 59.

1. Before servicing the vehicle, refer to the Precautions Section.
2. Drain the engine oil
3. Remove or disconnect the following:
   - Negative battery cable
   - Engine undercovers
   - Lower oil pan bolts in the reverse of the installation sequence
4. Insert a seal cutter between the steel and aluminum oil pan.
5. Tapping the cutter with a hammer, slide it around the entire edge of the oil pan. Be careful not to damage the aluminum mating surface of the upper oil pan.
   - Lower oil pan and the oil strainer
   - Front exhaust pipe and its support
6. Hang the engine at the right and left side engine slingers with a suitable hoist.
7. Position a suitable jack under the transmission/transaxle.
   - Crankshaft Position (CKP) sensors (REFERENCE and POSITION) from the oil pan

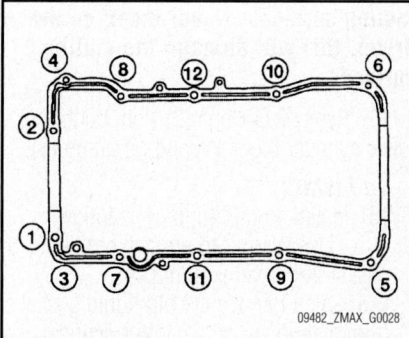

Fig. 58 Aluminum (upper) oil pan bolt loosening sequence—3.5L engine

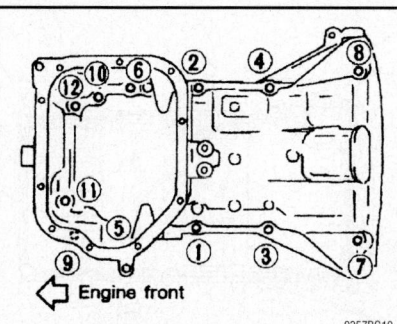

Fig. 59 To prevent pan warpage, tighten the aluminum oil pan bolts in the sequence shown—3.5L engine

- Front and rear engine mounting nuts and bolts
- Center crossmember assembly
- Engine drive belts
- A/C compressor and mounting bracket
- Rear cover plate and the lower transaxle bolts
- Aluminum (upper) oil pan bolts in the reverse of the installation sequence

8. Insert a seal cutter between the aluminum oil pan and the engine block.
9. Tapping the cutter with a hammer, slide it around the entire edge of the oil pan. Be careful not to damage the mating surfaces of the oil pan or engine block.
10. Remove or disconnect the following:
    - Oil pan assembly
    - Bolts that secure the baffle plate and the baffle plate
    - O-rings from the cylinder block and oil pump body

To install:

11. Install or connect the following:
    - Baffle plate to the oil pan and torque the bolts to 22–27 inch lbs. (2.5–3.1 Nm) and apply sealant to the front and rear seal of the oil pan
    - New O-rings to the cylinder block and the oil pump body

12. Apply a 4.5–5.5mm wide continuous bead of liquid gasket to the upper oil pan mating surface and install the oil pan. Torque the bolts in sequence to 16 ft. lbs. (22 Nm).

13. Install or connect the following:
    - Oil pan strainer and torque the bolts to 12–14 ft. lbs. (16–19 Nm)
    - Rear cover plate and lower transaxle bolts
    - A/C compressor and bracket
    - Drive belts
    - Center crossmember
    - Front and rear engine mount hardware
    - CKP sensors
    - Front exhaust tube and support
    - Oil strainer
    - Lower oil pan and torque the bolts, in sequence, to 80 inch lbs. (9 Nm)
    - Engine under covers
    - Negative battery cable

14. After waiting approximately 30 minutes, fill the engine with clean oil.

15. Start the vehicle, check for leaks and repair if necessary.

350Z

See Figures 60 through 63.

➡ **When removing oil pan (lower) only, remove engine assembly is not necessary. Perform step 1, 2 and 10.**

1. Drain engine oil.
2. Remove undercover with power tool

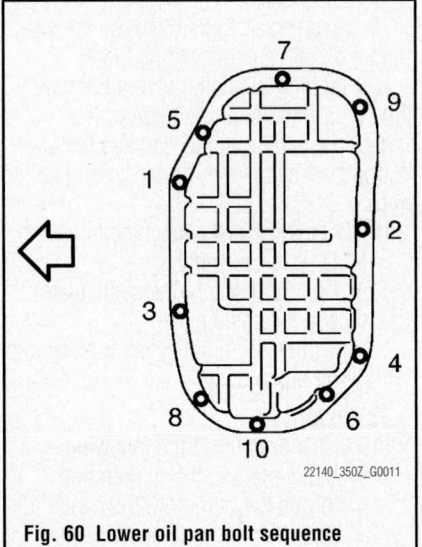

Fig. 60 Lower oil pan bolt sequence

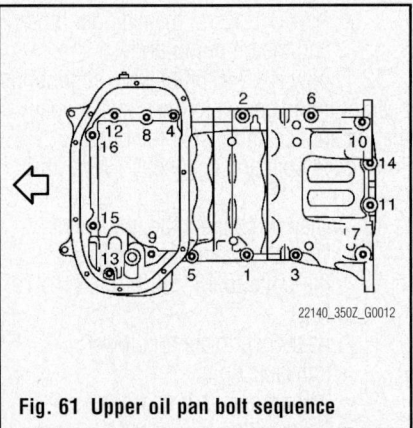

Fig. 61 Upper oil pan bolt sequence

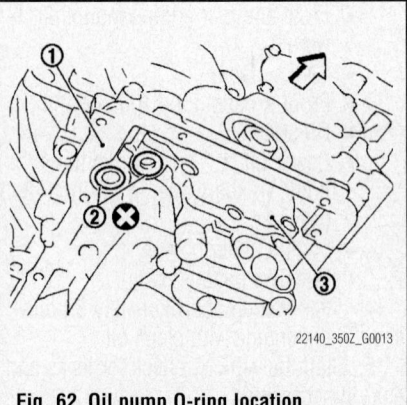

Fig. 62 Oil pump O-ring location

3. Remove engine assembly from the vehicle, and separate front suspension member and transmission from engine.
4. Remove alternator.
5. Remove starter motor.
6. Remove idler pulley and bracket assembly.
7. Remove oil filter, as necessary.
8. Remove oil temperature sensor, as necessary.
9. Remove oil pan (lower) as follows:
10. Loosen mounting bolts with power tool in reverse order as shown in the figure to remove.
    a. Insert seal cutter (SST) between oil pan (upper) and oil pan (lower).

➡ **Be careful not to damage the mating surfaces. Never insert screwdriver, this will damage the mating surface.**

11. Remove oil strainer.
12. Remove rear cover plate.
13. Loosen mounting bolts with power tool in reverse order as shown in the figure to remove oil pan (upper).
14. Insert seal cutter (SST: KV10111100 (J37228) between oil pan (upper) and cylinder block. Slide seal cutter by tapping on the side of tool with hammer. Remove oil pan (upper).

➡ **Be careful not to damage the mating surfaces. Never insert screwdriver, this will damage the mating surface.**

15. Remove O-rings (2) from bottom of lower cylinder block (1) and oil pump (3).

To install:

16. Install oil pan (upper) as follows:
    a. Use scraper to remove old liquid gasket from mating surfaces.
    b. Also remove the old liquid gasket from mating surface of lower cylinder block.
    c. Remove old liquid gasket from the bolt holes and threads.

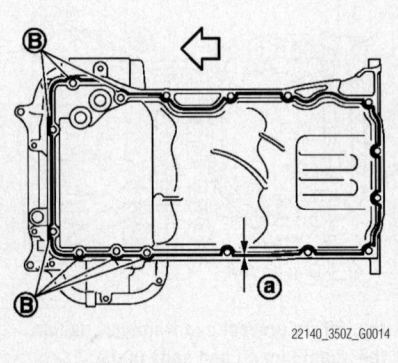

Fig. 63 Sealant location

d. Install new O-rings (2) on the bottom of lower cylinder block (1) and oil pump (3).
    e. Apply a continuous bead of liquid gasket with tube presser (commercial service tool) to the lower cylinder block mating surface of oil pan (upper) to a limited portion as shown in the figure.

➡ **For bolt holes (B) (7 locations), apply liquid gasket outside the holes. Attaching should be done within 5 minutes after coating.**

f. Install oil pan (upper).
    g. Tighten mounting bolts in numerical order as shown in the figure.

➡ **There are two types of mounting bolts. Install the bolts to where they were removed from.**

h. Tighten transmission joint bolts.
17. Install oil strainer to oil pump. Apply locking sealant to the thread of mounting bolts.
18. Install oil pan (lower) as follows:
    a. Use scraper to remove old liquid gasket from mating surfaces.
    b. Also remove the old liquid gasket from mating surface of upper oil pan.
    c. Remove old liquid gasket from the bolt holes and threads.
    d. Apply a continuous bead of liquid gasket with tube presser (commercial service tool) to the oil pan (lower).
    e. Install oil pan (lower). Tighten mounting bolts in numerical order as shown in the figure.
19. Install oil pan drain plug.
20. Install in the reverse order of removal after this step.

OIL PUMP

REMOVAL & INSTALLATION

See Figure 64.

1. Before servicing the vehicle, refer to the Precautions Section.
2. Drain the engine oil.
3. Remove or disconnect the following:
   • Negative battery cable
   • Drive belts
   • Camshaft Position (CMP) sensor (PHASE) and the Crankshaft Position (CKP) sensor (REF)/(POS)
   • Engine lower covers
   • Crankshaft pulley
   • Front exhaust tube and support
   • Right side mounting insulator and bracket
   • Center member
   • A/C compressor and move it aside

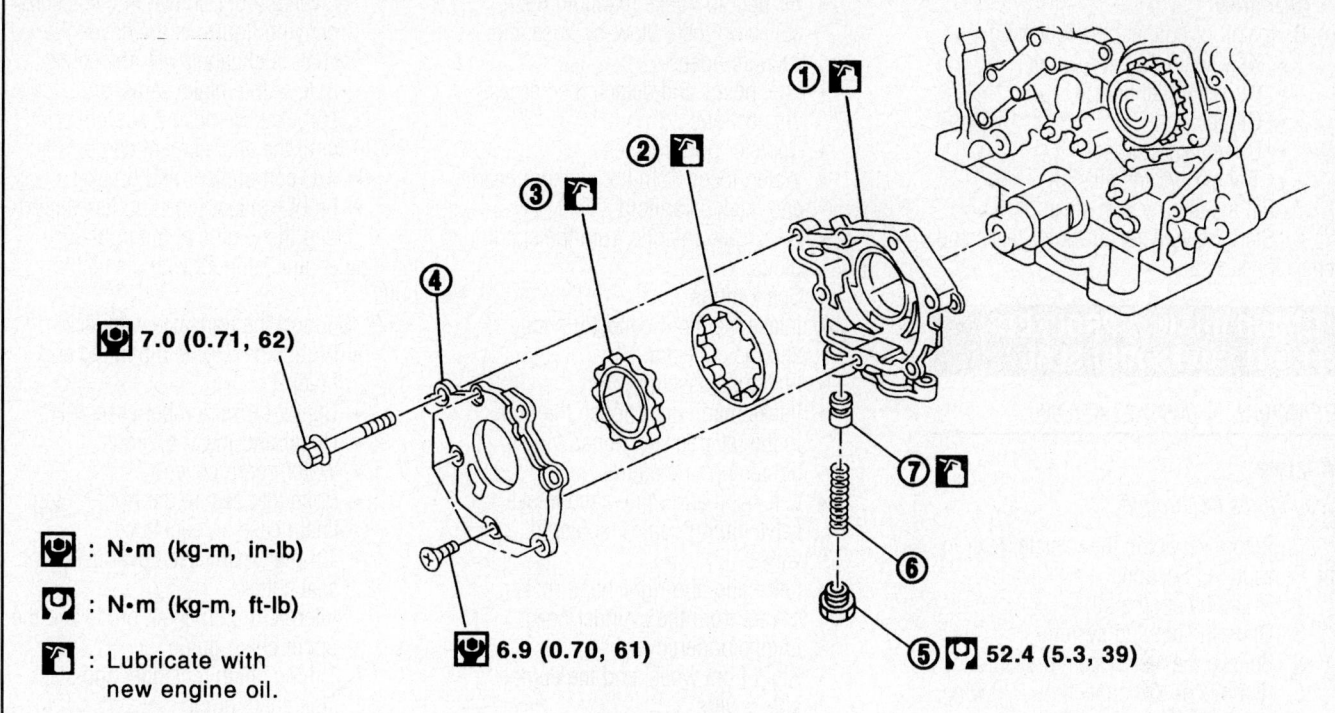

7.0 (0.71, 62)

: N•m (kg-m, in-lb)

: N•m (kg-m, ft-lb)

: Lubricate with new engine oil.

6.9 (0.70, 61)

52.4 (5.3, 39)

1. Oil pump body	2. Oil pump outer rotor	3. Oil pump inner rotor
4. Oil pump cover	5. Regulator valve plug	6. Regulator valve spring
7. Regulator valve		

09482_ZMAX_G0029

Fig. 64 Exploded view of the oil pump assembly–3.5L engine

- Oil pans
- Water pump cover
- Front cover
- Timing chain
- Oil pump assembly
4. Clean all mating surfaces.

To install:
5. Install or connect the following:
- Oil pump
- Timing chain
- Front cover and torque the long bolt to 62 inch lbs. (7 Nm) and the short bolt to 61 inch lbs. (6.5 Nm)
- Water pump cover
- Oil pans
- A/C compressor
- Center member
- Right side mounting insulator and bracket
- Front exhaust tube and support
- Crankshaft pulley
- CMP and CKP sensors
- Engine lower covers
- Drive belts
- Negative battery cable
6. Fill the engine with clean oil.
7. Start the vehicle, check for leaks and repair if necessary.

PISTON AND RING

POSITIONING
See Figure 65.

REAR MAIN SEAL

REMOVAL & INSTALLATION

1. Before servicing the vehicle, refer to the precautions in the beginning of this section.

2. Drain the engine oil.
3. Remove or disconnect the following:
- Transaxle or transmission
- Driveplate/flywheel
- Oil pan
- Oil seal retainer
4. Tap the oil seal out of the retainer with a hammer and drift.
5. Clean all mating surfaces of any residual liquid gasket.

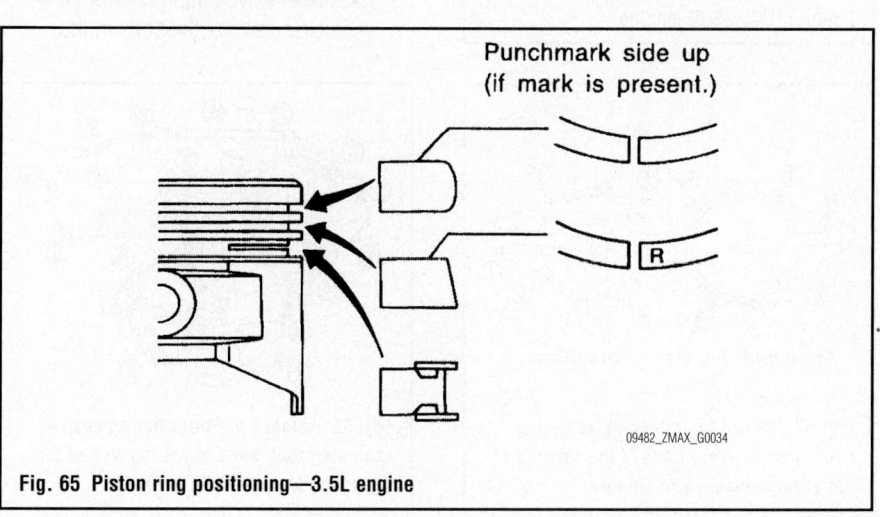

Punchmark side up (if mark is present.)

09482_ZMAX_G0034

Fig. 65 Piston ring positioning—3.5L engine

To install:

6. Install or connect the following:
- New seal into the retainer
- Oil seal retainer
- Oil pan
- Driveplate/flywheel
- Transaxle/transmission

7. Fill the engine with clean oil.

8. Start the vehicle, check for leaks and repair if necessary.

TIMING CHAIN, SPROCKETS, FRONT COVER AND SEAL

REMOVAL & INSTALLATION

Maxima

See Figures 66 through 73.

1. Before servicing the vehicle, refer to the Precautions Section.

2. Drain the engine oil.

3. Drain the cooling system.

4. Relieve the fuel system pressure.

5. Remove or disconnect the following:
- Negative battery cable
- Left side rocker cover ornament

➡**Before detaching any hoses or connectors, note the locations for reassembly.**

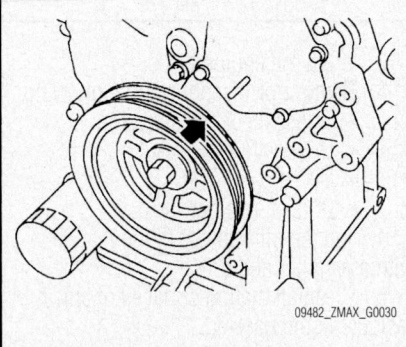

Fig. 66 Set the No. 1 piston to Top Dead Center (TDC)—3.5L engine

- Air duct to intake manifold hose, collector hose, blow-by hose and vacuum hoses
- Fuel hoses and detach the harness connections
- Canister purge hoses
- Water hoses from the cylinder head and intake manifold
- All 6 ignition coils from the spark plugs
- Spark plugs
- Intake manifold collector supports and the collector
- Fuel tube assembly
- Intake manifold. Loosen the bolts in the reverse sequence of the tightening procedure.
- Left-hand and right-hand intake valve timing control solenoid valves
- Left-hand and right-hand rocker covers from the cylinder head
- Engine undercovers
- Right front wheel and the engine side covers
- Drive belts and the idler pulley
- Power steering oil pump belt and the power steering oil pump assembly
- Camshaft Position (CMP) sensor (PHASE) and Crankshaft Position (CKP) sensors (REF)/(POS)

6. Set the No. 1 piston to Top Dead Center (TDC) of compression stroke by rotating the crankshaft.

7. Loosen the crankshaft pulley bolt while securing the ring gear so the crankshaft cannot rotate.
- Ring gear cover access plate

➡**Use care not to damage the ring gear teeth.**

- Crankshaft pulley using a suitable puller
- Intake valve timing control valve cover. Loosen the bolts in the

reverse order shown in the accompanying figure. In the cover, the shaft is engaged with the center hole of the intake camshaft sprocket. Remove it straight out until the engagement comes off.
- A/C compressor and bracket
- Front exhaust pipe and its support

8. Hang the engine at the right and left side engine slingers with a suitable hoist.

9. Support the transaxle with jack.
- Right side engine mounting and bracket
- Center crossmember assembly
- Upper and lower oil pans
- Water pump cover
- Bolts that secure the front timing chain case, in sequence
- Timing chain case cover using a seal cutter
- Internal timing chain guide and the upper chain guide
- Timing chain tensioner and slack side chain guide
- Left and right intake camshaft sprockets first. Be sure to hold the flats of the camshafts while removing the sprocket bolts.
- Lower timing chain assembly. Be sure to note the aligning marks of the chain before removal.

10. Insert a suitable stopper pin for the left and right camshaft tensioners.
- Left and right exhaust camshaft sprocket bolts. Be sure to hold the flats of the camshafts while removing the sprocket bolts.
- Upper timing chain assembly. Be sure to note the aligning marks of the chain before removal.
- Lower timing chain guide
- Crankshaft sprocket
- All traces of liquid gasket from the front timing chain case and from the water pump

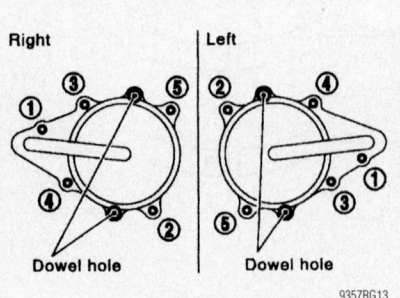

Fig. 67 Loosen the intake valve timing control valve cover bolts in the reverse of the order shown—3.5L engine

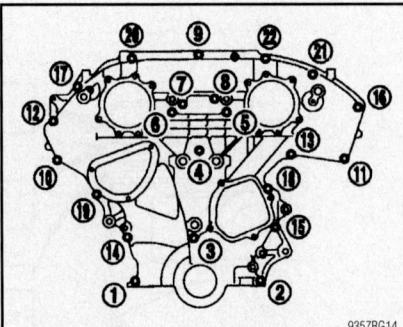

Fig. 68 Remove the front timing chain case mounting bolts in the reverse of the sequence shown—3.5L engines

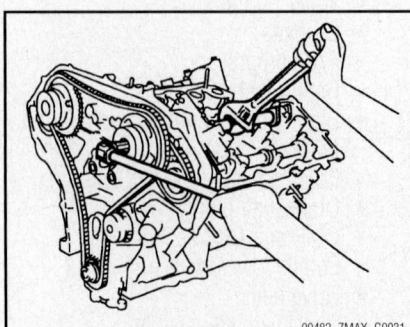

Fig. 69 Hold the camshaft with a wrench while removing the sprocket bolts—3.5L engine

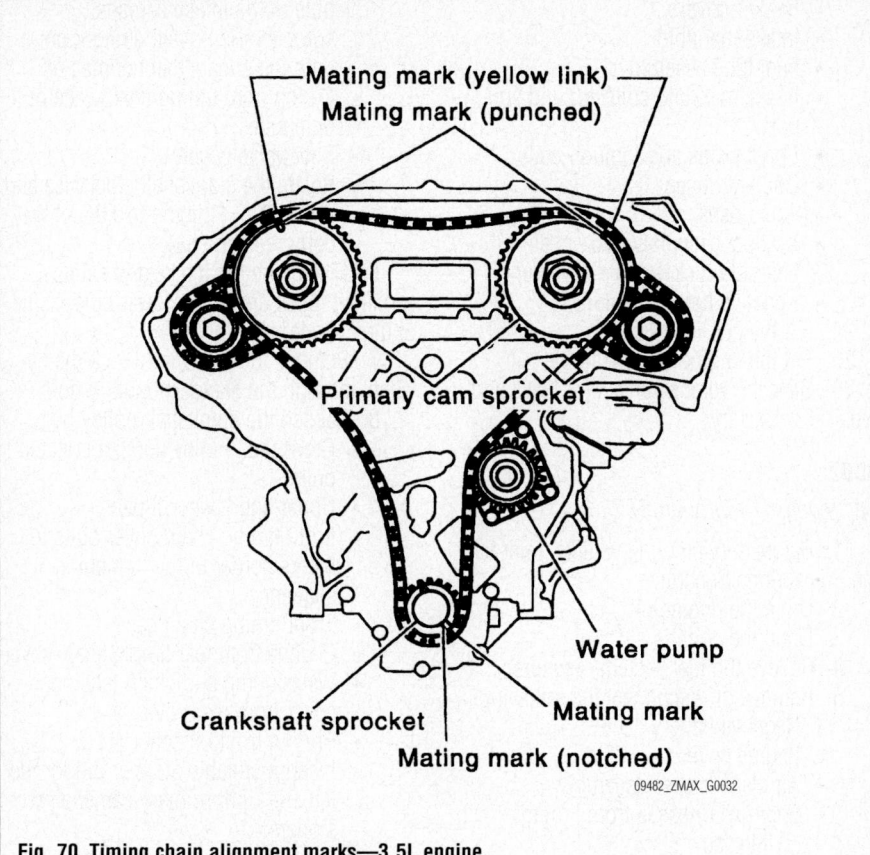

Fig. 70 Timing chain alignment marks—3.5L engine

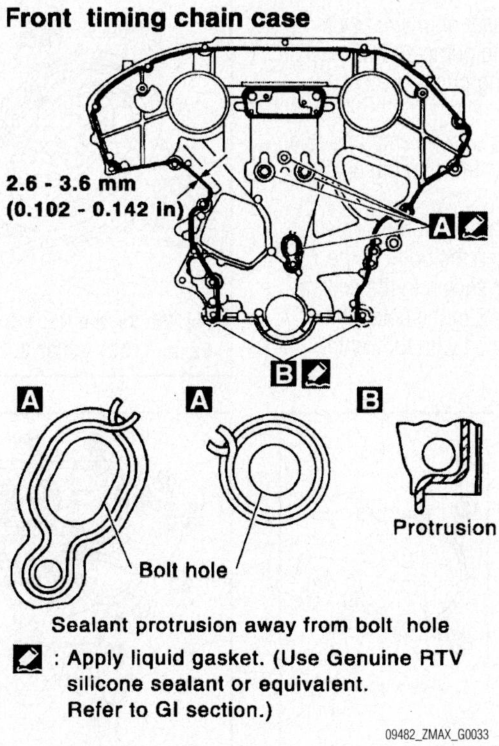

Front timing chain case

2.6 - 3.6 mm
(0.102 - 0.142 in)

Sealant protrusion away from bolt hole

▨ : Apply liquid gasket. (Use Genuine RTV silicone sealant or equivalent. Refer to GI section.)

Fig. 71 Application of liquid gasket to the front timing case—3.5L engine

11. Inspect the timing chain for excessive wear or damage and replace as necessary.

To install:

12. Install or connect the following:
- Crankshaft sprocket with the mating mark facing out

13. Position the crankshaft to TDC of compression stroke and align the dowels of the camshaft sprockets to the 12 o'clock position in respect to the cylinder head.
- Lower timing chain guide. The front mark on the guide should face upwards.

14. On a work bench, align the marks on the intake and exhaust camshaft sprockets with the marks of the chain.
- Exhaust camshaft sprockets onto the dowel pin and torque the mounting bolts to 76 ft. lbs. (102 Nm). Be sure to secure the camshafts while tightening the bolts.
- Timing chains and sprockets to the intake camshafts. Be sure to align the timing chain and sprocket mating marks.
- Left and right camshaft tensioner stopper pins

15. Align the mating mark on the crankshaft with the matchmark (gold link) on the lower timing chain.

16. Install the lower timing chain to the water pump sprocket.

17. Working counterclockwise, install the lower timing chain camshaft sprockets. Be sure to align the sprocket marks with the blue links of the timing chain during installation.

18. Install or connect the following:
- Intake sprocket and torque the bolts to 76 ft. lbs. (102 Nm). Be sure to secure the camshafts while tightening the bolts.
- Internal timing chain guide, upper timing chain guide, lower timing chain tensioner and slack side timing chain guide.

19. Torque the tensioner mounting bolt to 75–96 inch lbs. (8.4–10.8 Nm) and the guide bolts to 108–168 inch lbs. (13–19 Nm).

20. Apply a 0.102–0.142 in. (2.6–3.6 mm) continuous bead of liquid gasket to all necessary areas as shown on the front timing cover.
- Timing cover evenly and gently. Be sure to align the dowel pin holes.

21. Torque the mounting bolts in sequence as follows:
  a. Bolts No. 1 and 2: 19–23 ft. lbs. (26–31 Nm).

b. Bolts No. 3–20: 105–121 inch lbs. (11.8–13.7 Nm).

➡ **Leave the bolts unattended for 30 minutes or more after tightening. This will allow the liquid gasket to cure sufficiently.**

22. Apply a 0.091–0.130 in. (2.3–3.3mm) continuous bead of liquid gasket to the water pump cover and install the cover. Torque the bolts to 84–108 inch lbs. (10–13 Nm).

23. Install or connect the following:
- Oil pans
- Center crossmember
- Right side engine mount and bracket
- Front exhaust pipe and remove the transaxle support
- A/C compressor and bracket
- Crankshaft pulley
- Ring gear access cover plate
- CMP sensor and CKP sensors
- Power steering pump
- Idler pulley and drive belts
- Engine side cover and right front wheel
- Engine under covers
- Intake valve timing control solenoid valves with new covers. Tighten the cover bolts in the proper sequence.

- Rocker covers
- Intake manifold
- Fuel tube assembly
- Intake manifold collector and support
- Spark plugs and ignition coils
- Coolant hoses
- Fuel hoses
- Air duct assembly and hoses
- Left side rocker cover ornament
- Negative battery cable

24. Fill the cooling system.
25. Fill the engine with clean oil.
26. Start the vehicle, check for leaks and repair if necessary.

350Z

See Figures 74 through 81.

1. Before servicing the vehicle, refer to the Precautions Section.
2. Drain the engine oil.
3. Drain the cooling system.
4. Relieve the fuel system pressure.
5. Remove or disconnect the following:
- Negative battery cable
- Engine cover
- Air cleaner case assembly
- Engine harnesses from timing chain case
- Upper and lower intake manifold collectors
- Cooling fan
- Drive belts
- A/C compressor and wire aside
- Power steering pump and wire aside
- Power steering pump bracket
- Alternator
- Water bypass hose, clamp and idler pulley bracket from timing chain case
- Intake valve timing control valve covers. Loosen the bolts in the reverse order shown in the accompanying figure. In the cover, the shaft is engaged with the center

hole of the intake camshaft sprocket. Remove it straight out until the engagement comes off.
- O-ring from timing chain case on both sides
- Both valve covers
- Rotate the crankshaft clockwise and set the No. 1 piston to TDC of the compression stroke

6. Make sure the intake and exhaust camshaft lobes are facing toward the inside of the cylinder head

7. Remove the starter and lock the flywheel through the starter mounting hole.

8. Loosen the crankshaft pulley bolt.
- Crankshaft pulley using a suitable puller
- Upper and lower oil pans
- Front timing chain cover bolts in reverse order of the tightening sequence
- Front timing chain case
- O-rings from rear timing chain cover
- Water pump and chain tensioner cover from rear cover
- Pry out front oil seal
- Insert a suitable stopper pin for the left and right primary camshaft tensioners
- Primary chain tensioners

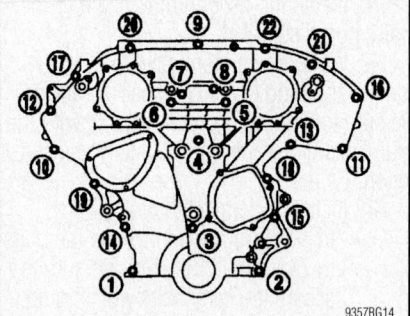

Fig. 72 Tighten the front timing chain case bolts according to the sequence shown—3.5L engines

9357RG14

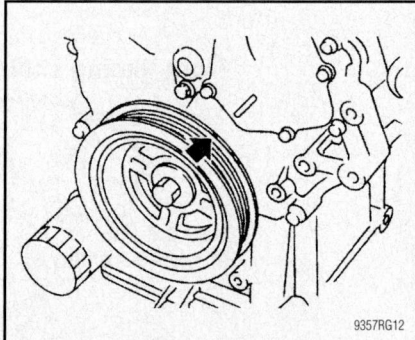

Fig. 75 Set the No. 1 piston to Top Dead Center (TDC)—350Z 3.5L engine

9357RG12

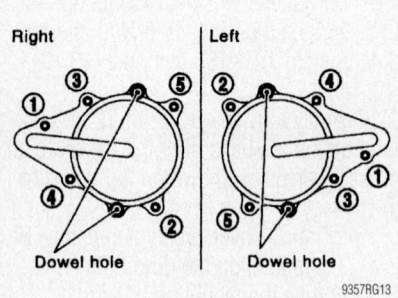

Fig. 73 Tighten the intake valve timing control valve cover bolts in sequence—3.5L engine

9357RG13

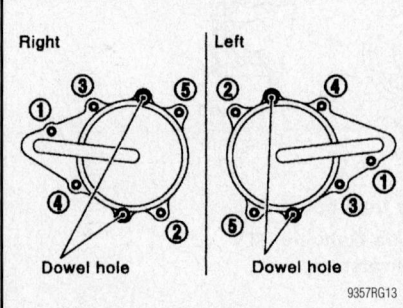

Fig. 74 Loosen the intake valve timing control valve cover bolts in the reverse of the order shown—350Z 3.5L engine

9357RG13

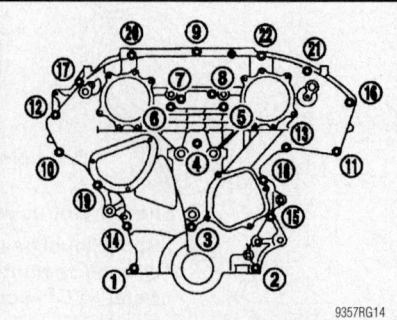

Fig. 76 Remove the front timing chain case mounting bolts in the reverse of the sequence shown—350Z 3.5L engine

9357RG14

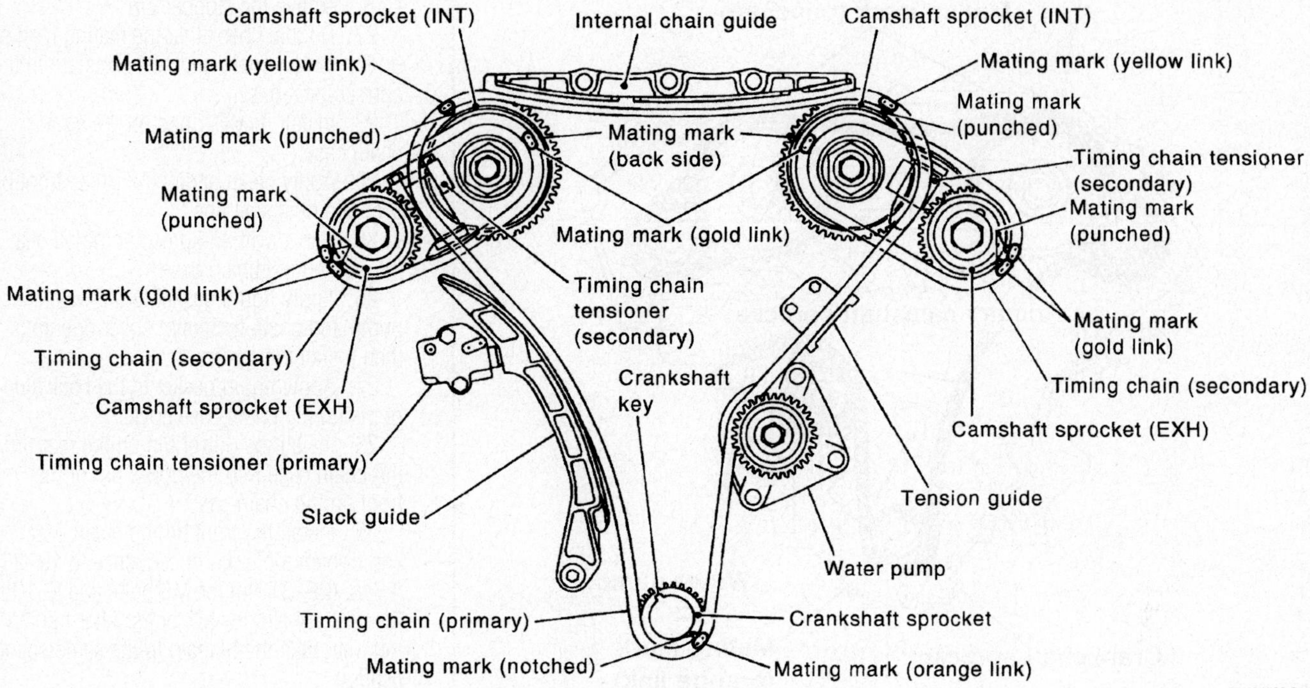

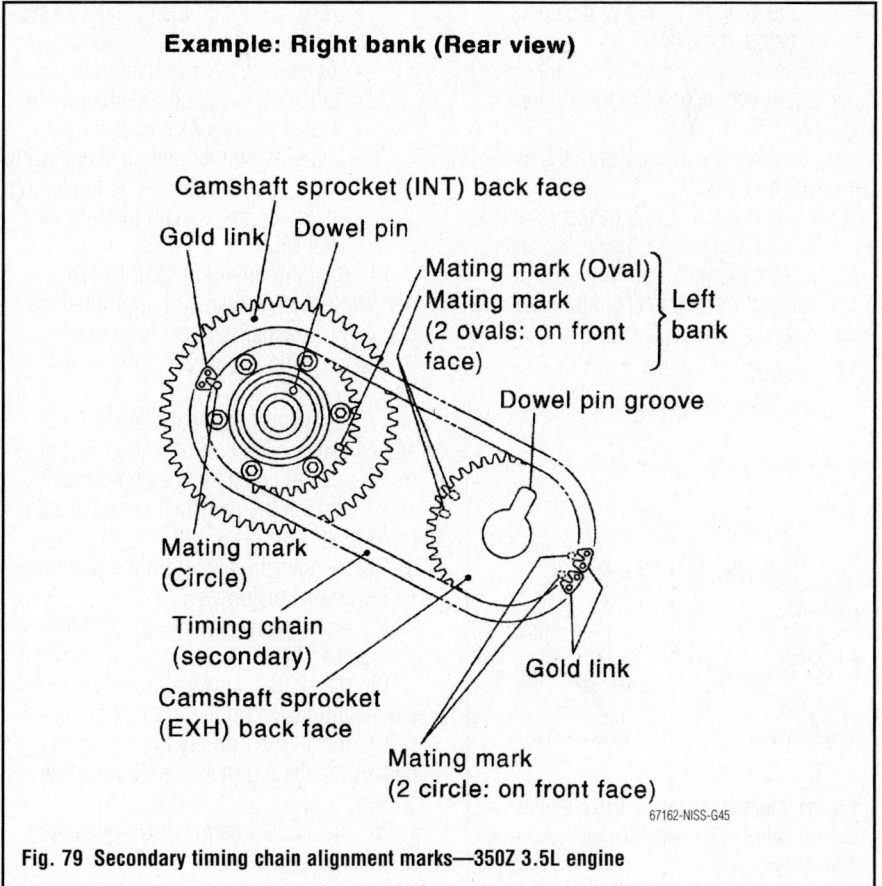

Fig. 77 Timing chain alignment marks—350Z 3.5L engine

- Timing chain guide, slack guide and tension guide
- Primary timing chain and crankshaft sprocket
- Insert a suitable stopper pin for the left and right secondary camshaft tensioners
- Secondary chain tensioners
- Left and right intake camshaft sprocket bolts first, then exhaust sprocket bolts. Be sure to hold the flats of the camshafts while removing the sprocket bolts.
- Lower timing chain assembly with camshaft sprockets. Be sure to note

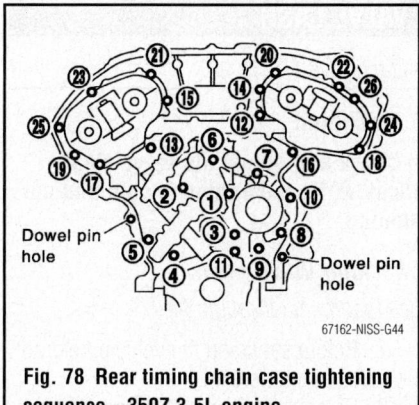

Fig. 78 Rear timing chain case tightening sequence—350Z 3.5L engine

Example: Right bank (Rear view)

Fig. 79 Secondary timing chain alignment marks—350Z 3.5L engine

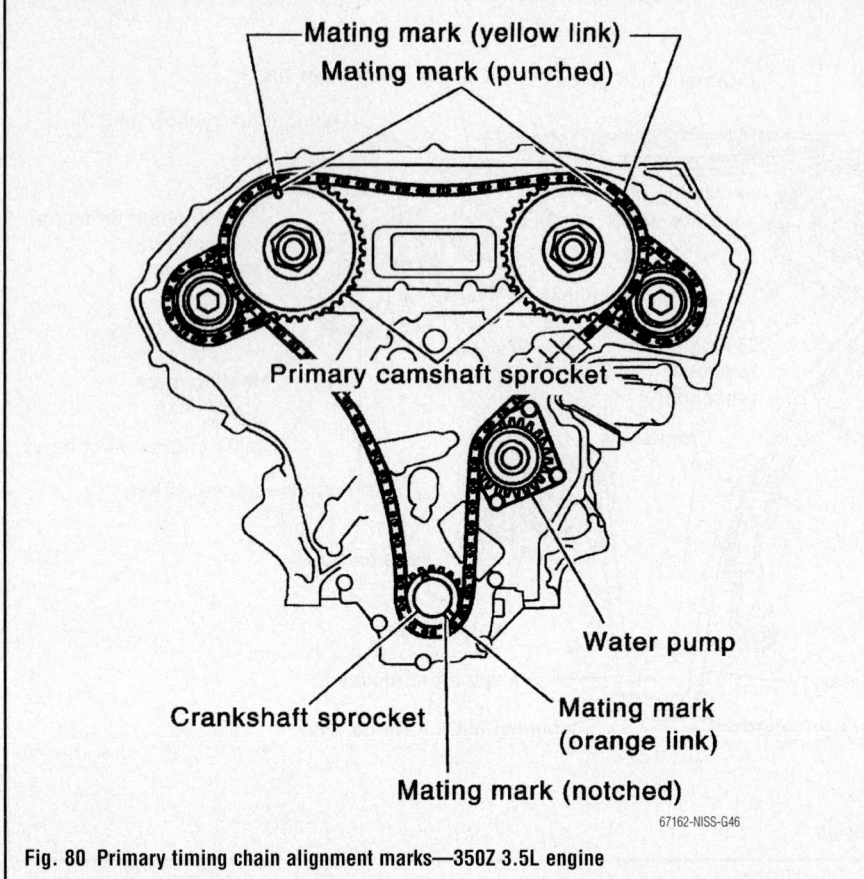

Fig. 80 Primary timing chain alignment marks—350Z 3.5L engine

the aligning marks of the chain before removal.

9. Remove the rear timing chain case bolts in the reverse order of the tightening sequence.

10. Remove the O-rings from the cylinder heads and block.

- All traces of liquid gasket from the front and timing chain case and from the water pump

11. Inspect the timing chain for excessive wear or damage and replace as necessary.

To install:

12. Install or connect the following:

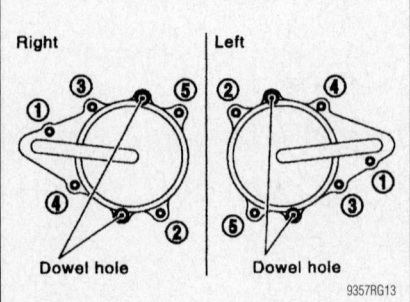

Fig. 81 Tighten the intake valve timing control valve cover bolts in sequence—3.5L engine

- New O-rings to the cylinder heads and block
- Apply a bead of sealant to the back side of the rear timing chain case
- Install the rear timing chain case and tighten the bolts in sequence to 10 ft. lbs. (14 Nm). After tightening, retighten them again to 10 ft. lbs. (14 Nm).

13. Position the crankshaft to TDC of compression stroke and align the dowels of the camshaft sprockets to the 12 o'clock position in respect to the cylinder head.

14. Install the secondary timing chain and camshaft sprockets aligning the timing marks and timing chain links as shown.

15. Tighten the camshaft sprocket bolts to 76 ft. lbs. (103 Nm).

16. Remove the pins from the secondary timing chain tensioners.

17. Install the primary timing chain tension guide.

18. Install the crankshaft sprocket with the mating marks on the front side.

19. Install the primary timing chain and so the mating marks are aligned as shown.

20. Install the internal chain guide and slack guide.

21. Install the primary timing chain tensioner and tighten the bolt to 72 inch lbs. (8 Nm). Remove the stopper pin

22. Double check that the mating marks on the timing chain and sprockets are in the correct locations.

23. Install new O-rings on the timing chain case.

24. Apply clean engine oil to the front oil seal and dust seal lips.

25. Use a drift and press fit the oil seal into the timing chain case.

26. Apply liquid gasket to the water pump and chain tensioner cover openings, then install the covers.

27. Apply liquid gasket to the back side of the timing chain case cover.

28. Install the dowel pin on the rear timing chain case into the dowel pin of the front timing chain case.

29. Install the front timing chain case and tighten the bolts in sequence to 19–23 ft. lbs. (26–31 Nm for M8 bolts and 9–10 ft. lbs. (12–14 Nm for M6 bolts. After tightening, retighten them again to the same specification.

30. Install seal rings in the timing control cover shaft grooves and apply liquid gasket to the covers.

31. Install new O-rings in the timing chain case oil holes.

32. Install the timing control covers and tighten the bolt in sequence to 72 inch lbs. (8 Nm).

33. Install or connect the following:

- Upper and lower oil pans
- Valve covers
- Crankshaft pulley and tighten the bolt to 29–36 ft. lbs. (40–49 Nm), plus an additional 60°.

34. The remainder of the installation is the reverse of the removal procedure.

35. Fill the cooling system.

36. Fill the engine with clean oil.

37. Start the vehicle, check for leaks and repair if necessary.

VALVE LASH

ADJUSTMENT

3.5L Engine

➡Check and adjust the valve clearances while the engine is cold and not running.

Checking Valve Lash

See Figures 82 through 84.

1. Before servicing the vehicle, refer to the Precautions Section.

2. Remove or disconnect the following:

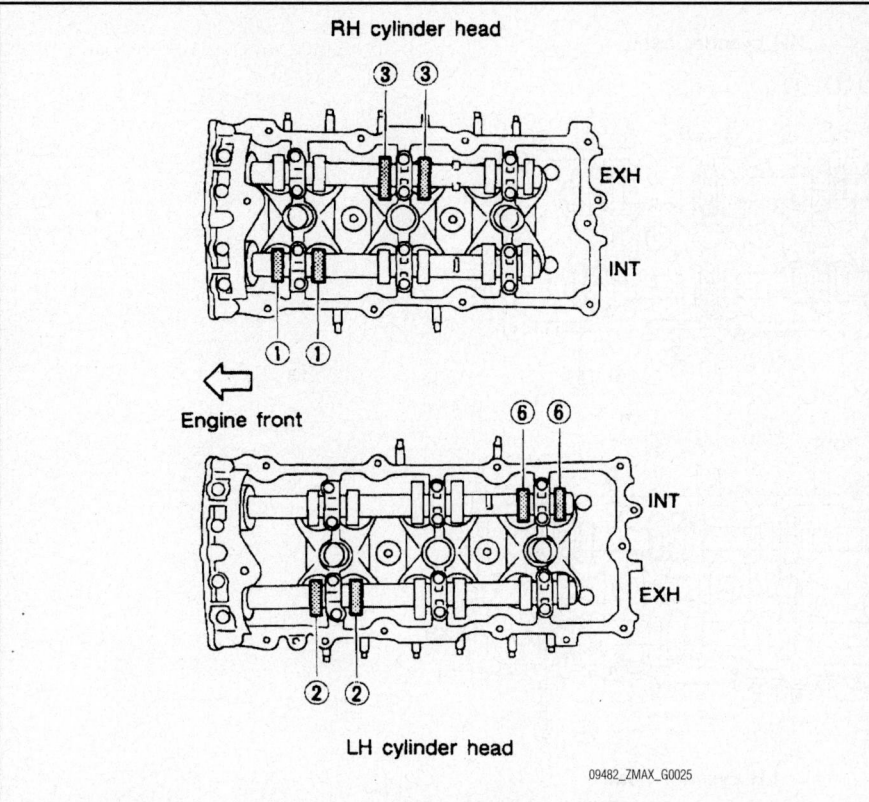

Fig. 82 Measure the valves indicated while the No. 1 piston is at TDC on the compression stroke—3.5L engine

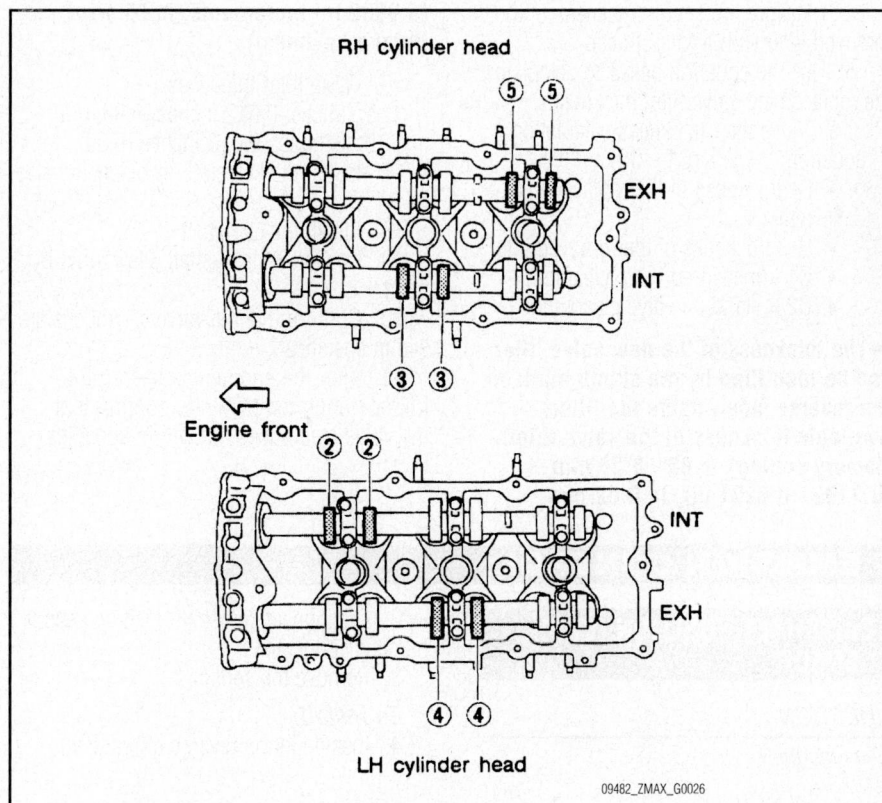

Fig. 83 Measure the valves indicated while the No. 3 piston is at TDC on the compression stroke—3.5L engine

- Intake manifold collector
- Left and right rocker covers
- Spark plugs

3. Set the No. 1 cylinder at Top Dead Center (TDC) on its compression stroke. Align the pointer with the TDC mark on the crankshaft pulley. Check that the valve lifters on the No. 1 cylinder are loose and valve lifters on the No. 4 cylinder are tight. If not, turn the crankshaft 1 revolution (360 degrees) and align the pointer with the TDC mark on the crankshaft pulley.

4. Check the following valves:
- Both No. 1 intake valves
- Both No. 2 exhaust valves
- Both No. 3 exhaust valves
- Both No. 6 intake valves

5. Using a feeler gauge, measure the clearance between the valve lifter and the camshaft. Record any valve clearance measurements that are out of specification. Intake valve clearance (cold) is 0.010–0.013 in. (0.26–0.34mm) and exhaust valve clearance (cold) is 0.011–0.015 in. (0.29–0.37mm).

6. Turn the crankshaft 240 degrees and set the No. 3 cylinder to TDC of its compression stroke.

7. Check the following valves:
- Both No. 2 intake valves
- Both No. 3 intake valves
- Both No. 4 exhaust valves
- Both No. 5 exhaust valves

8. Using a feeler gauge, measure the clearance between the valve lifter and the camshaft. Record any valve clearance measurements that are out of specification. Intake valve clearance (cold) is 0.010–0.013 in. (0.26–0.34mm) and exhaust valve clearance (cold) is 0.011–0.015 in. (0.29–0.37mm).

9. Turn the crankshaft 240 degrees and set the No. 5 cylinder to TDC of its compression stroke.

10. Check the following valves:
- Both No. 1 exhaust valves
- Both No. 4 intake valves
- Both No. 5 intake valves
- Both No. 6 exhaust valves

11. Using a feeler gauge, measure the clearance between the valve lifter and the camshaft. Record any valve clearance measurements that are out of specification. Intake valve clearance (cold) is 0.010–0.013 in. (0.26–0.34mm) and exhaust valve clearance (cold) is 0.011–0.015 in. (0.29–0.37mm).

12. If all the valve clearances are within specification, install the cylinder head cover, spark plugs and the intake manifold collector.

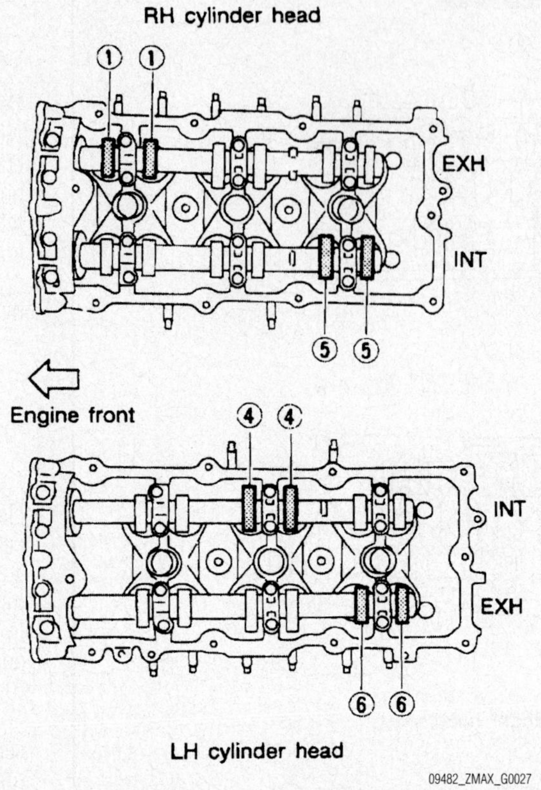

Fig. 84 Measure the valves indicated while the No. 5 piston is at TDC on compression—3.5L engine

Adjusting Valve Lash

1. Before servicing the vehicle, refer to the Precautions Section.

→ **Perform adjustment by selecting the correct head thickness of the valve lifter (adjusting shims are not used). The specified valve lifter thickness is the dimension at normal temperatures. Ignore dimensional differences caused by temperature. Use specifications for hot engine condition to confirm valve clearances.**

2. Remove the camshaft.
3. Remove the valve lifter that was measured as being outside the standard specifications.

4. Measure the center thickness of the removed lifter with a Micrometer.
5. Use the equation below to calculate the replacement valve lifter thickness.
   a. Valve lifter thickness calculation equation: $t = t1 + (C1 - C2)$
   - t = thickness of the replacement lifter
   - $t1$ = thickness of the removed lifter
   - $C1$ = measured valve clearance
   - $C2$ = standard valve clearance

→ **The thickness of the new valve lifter can be identified by the stamp mark on the reverse side (inside the lifter). Available thickness of the valve lifter (factory setting): 7.88 - 8.36 mm (0.3102 - 0.3291 in), in 0.02 mm (0.0008 in) increments, in 25 sizes (intake / exhaust).**

6. Value lifter thickness:
   a. Intake: 0.012 inches (0.30 mm)
   b. Exhaust: 0.013 inches (0.33 mm)
7. Install the selected replacement valve lifter.
8. Install the camshaft.
9. Rotate the crankshaft a few turns by hand.
10. Confirm that the valve clearances are within specification.
11. After the engine has been run to full operating temperature, confirm that the valve clearances are within specification.

ENGINE PERFORMANCE & EMISSION CONTROL

COMPONENT LOCATIONS

3.5L (VQ35DH) V6 Engine
See Figures 85 through 90.

3.5L (VQ35DE) V6 Engine
See Figures 91 through 98.

CAMSHAFT POSITION (CMP) SENSOR

LOCATION
See Figure 99.

REMOVAL & INSTALLATION

1. Loosen the fixing bolt of the sensor.

2. Disconnect camshaft position sensor harness connector.
3. Remove the sensor.

To install:
4. Installation is reverse of removal.

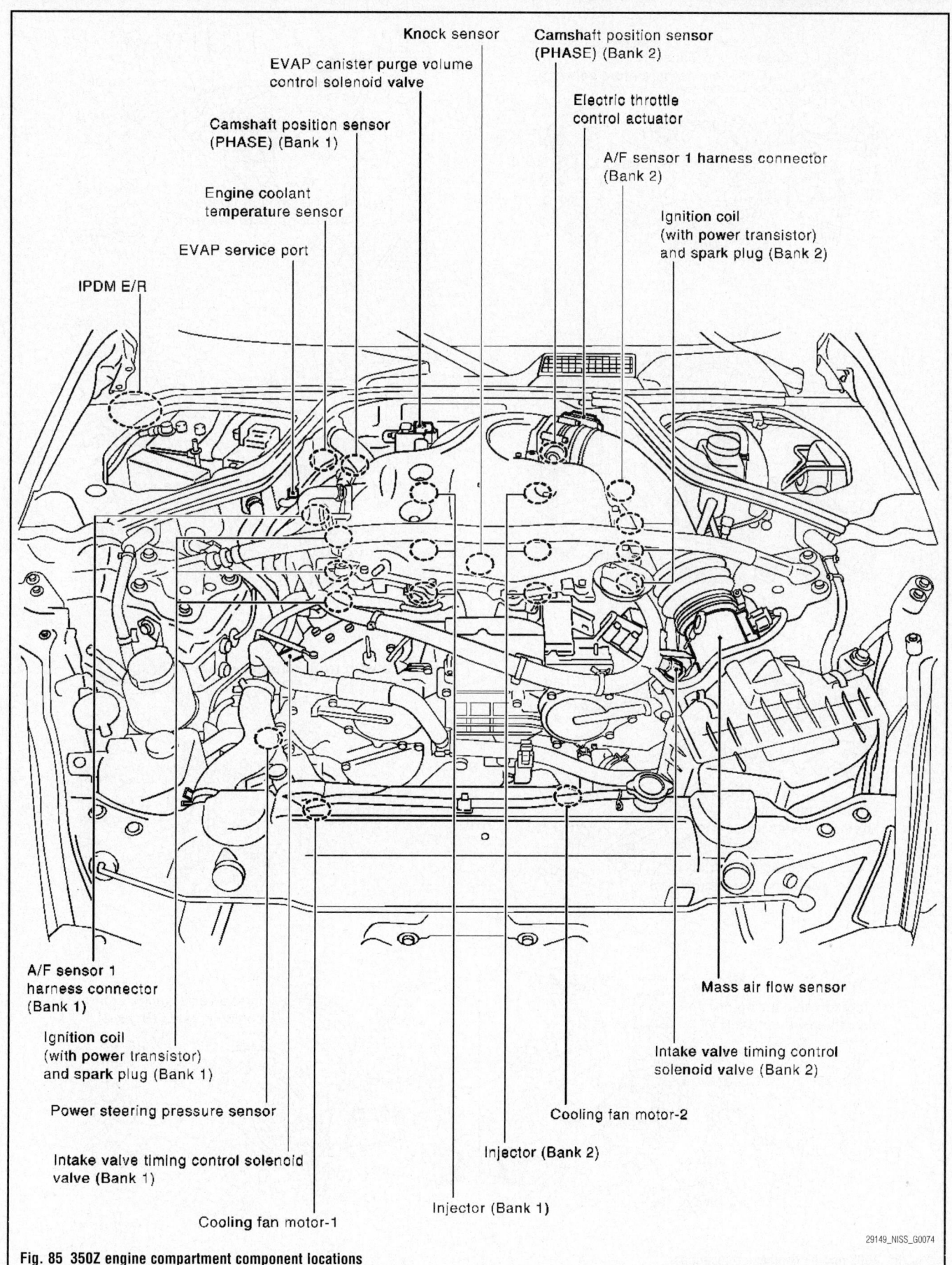

Knock sensor

Camshaft position sensor
(PHASE) (Bank 2)

EVAP canister purge volume
control solenoid valve

Electric throttle
control actuator

Camshaft position sensor
(PHASE) (Bank 1)

A/F sensor 1 harness connector
(Bank 2)

Engine coolant
temperature sensor

Ignition coil
(with power transistor)
and spark plug (Bank 2)

EVAP service port

IPDM E/R

A/F sensor 1
harness connector
(Bank 1)

Mass air flow sensor

Ignition coil
(with power transistor)
and spark plug (Bank 1)

Intake valve timing control
solenoid valve (Bank 2)

Power steering pressure sensor

Cooling fan motor-2

Intake valve timing control solenoid
valve (Bank 1)

Injector (Bank 2)

Cooling fan motor-1

Injector (Bank 1)

29149_NISS_G0074

Fig. 85 350Z engine compartment component locations

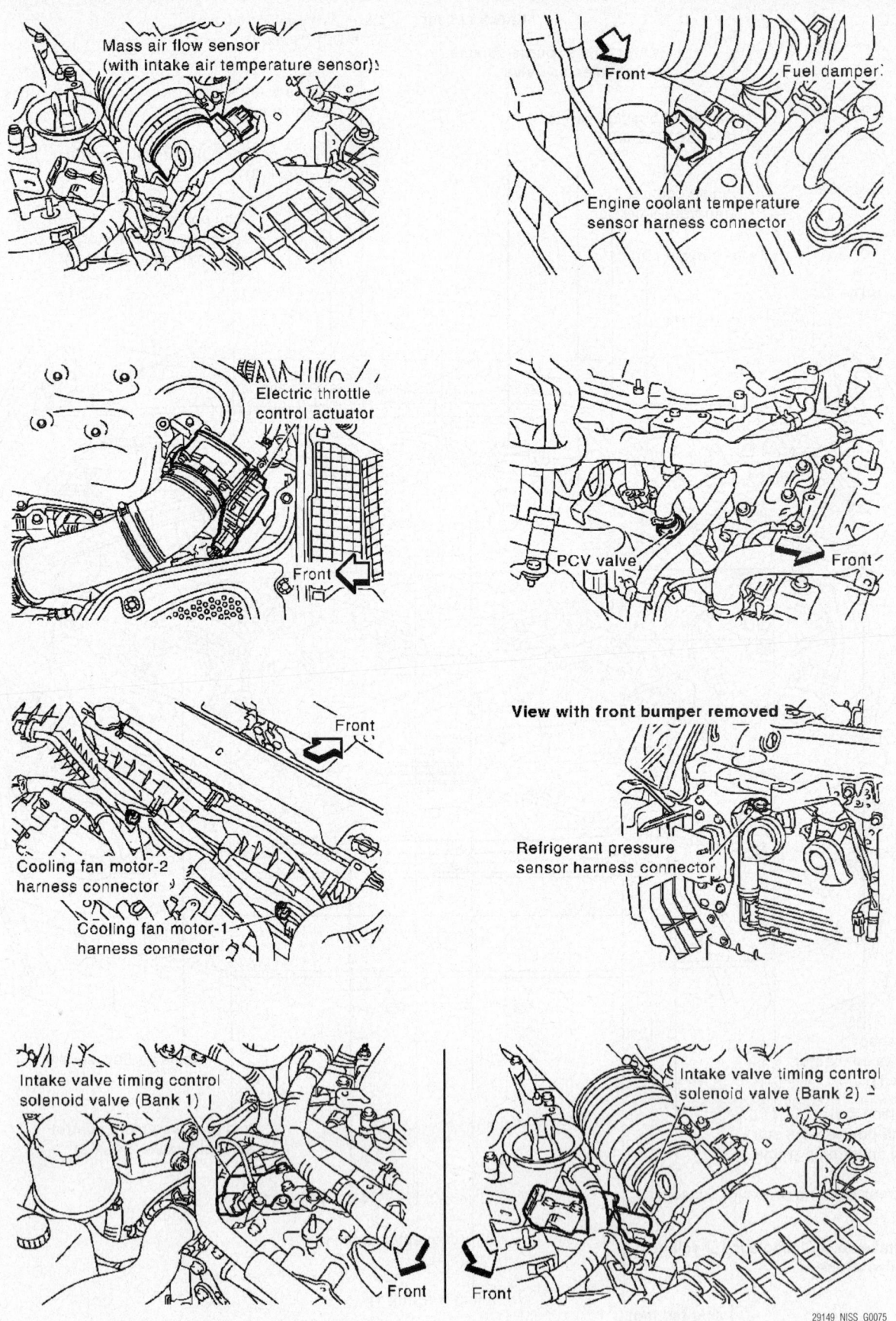

Fig. 86 350Z engine component locations

29149_NISS_G0075

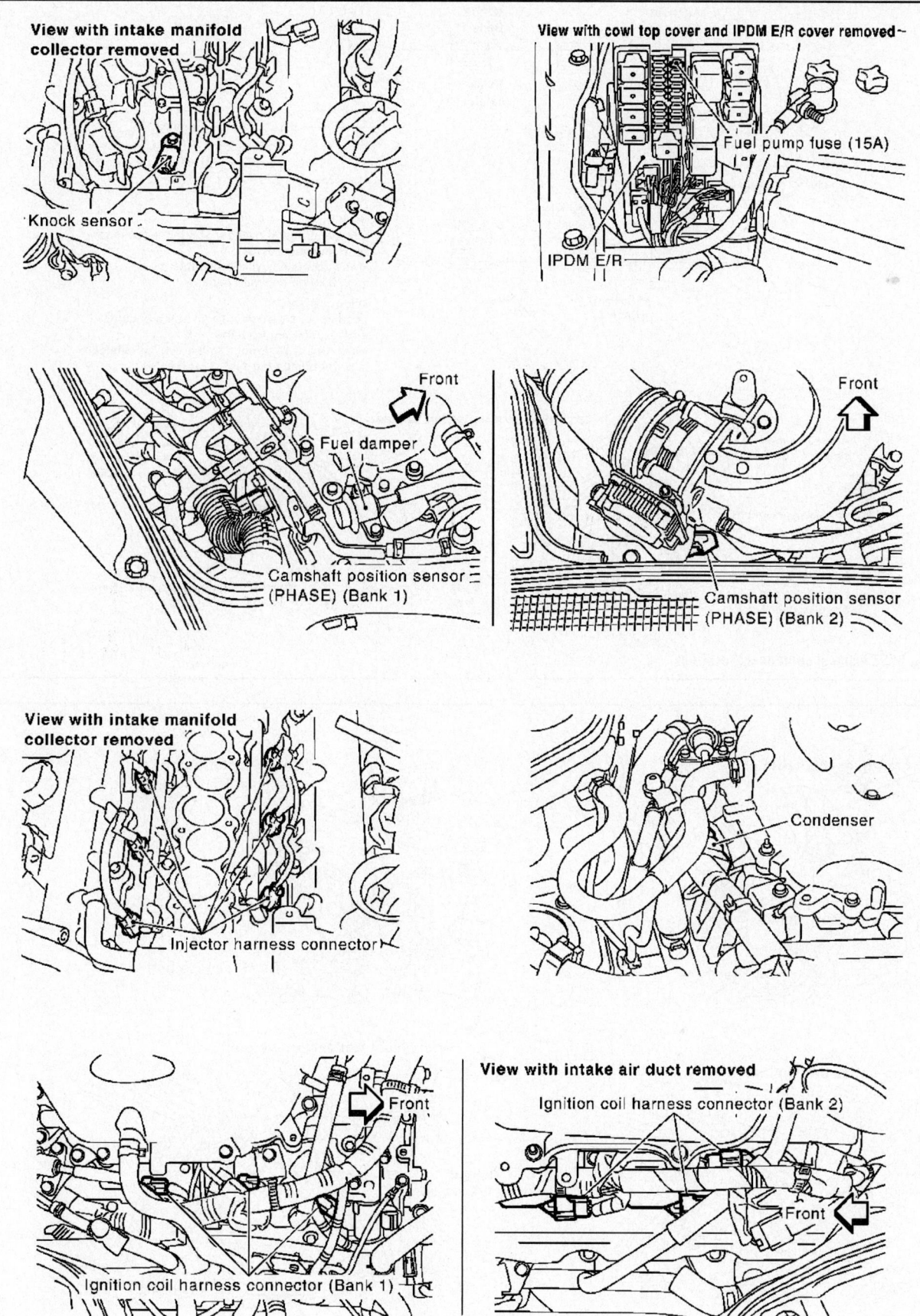

View with intake manifold collector removed

Knock sensor

View with cowl top cover and IPDM E/R cover removed

Fuel pump fuse (15A)

IPDM E/R

Front

Fuel damper

Camshaft position sensor (PHASE) (Bank 1)

Front

Camshaft position sensor (PHASE) (Bank 2)

View with intake manifold collector removed

Injector harness connector

Condenser

Front

Ignition coil harness connector (Bank 1)

View with intake air duct removed

Ignition coil harness connector (Bank 2)

Front

29149_NISS_G0076

Fig. 87 350Z engine component locations

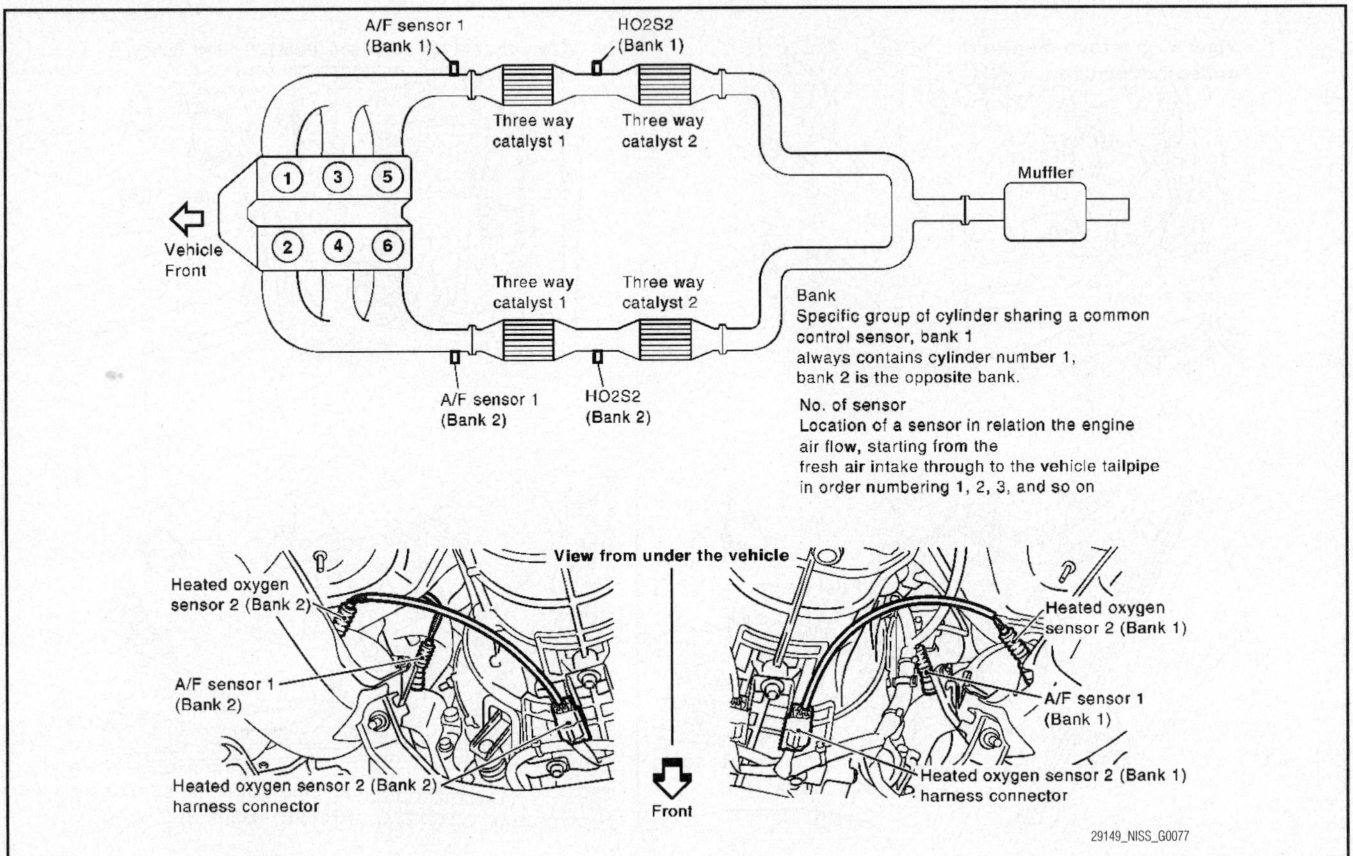

A/F sensor 1 (Bank 1) HO2S2 (Bank 1)

Three way catalyst 1 Three way catalyst 2

1 3 5

2 4 6

Vehicle Front

Muffler

Three way catalyst 1 Three way catalyst 2

A/F sensor 1 (Bank 2) HO2S2 (Bank 2)

Bank
Specific group of cylinder sharing a common control sensor, bank 1 always contains cylinder number 1, bank 2 is the opposite bank.

No. of sensor
Location of a sensor in relation the engine air flow, starting from the fresh air intake through to the vehicle tailpipe in order numbering 1, 2, 3, and so on

View from under the vehicle

Heated oxygen sensor 2 (Bank 2)

A/F sensor 1 (Bank 2)

Heated oxygen sensor 2 (Bank 2) harness connector

Front

Heated oxygen sensor 2 (Bank 1)

A/F sensor 1 (Bank 1)

Heated oxygen sensor 2 (Bank 1) harness connector

29149_NISS_G0077

Fig. 88 350Z exhaust component locations

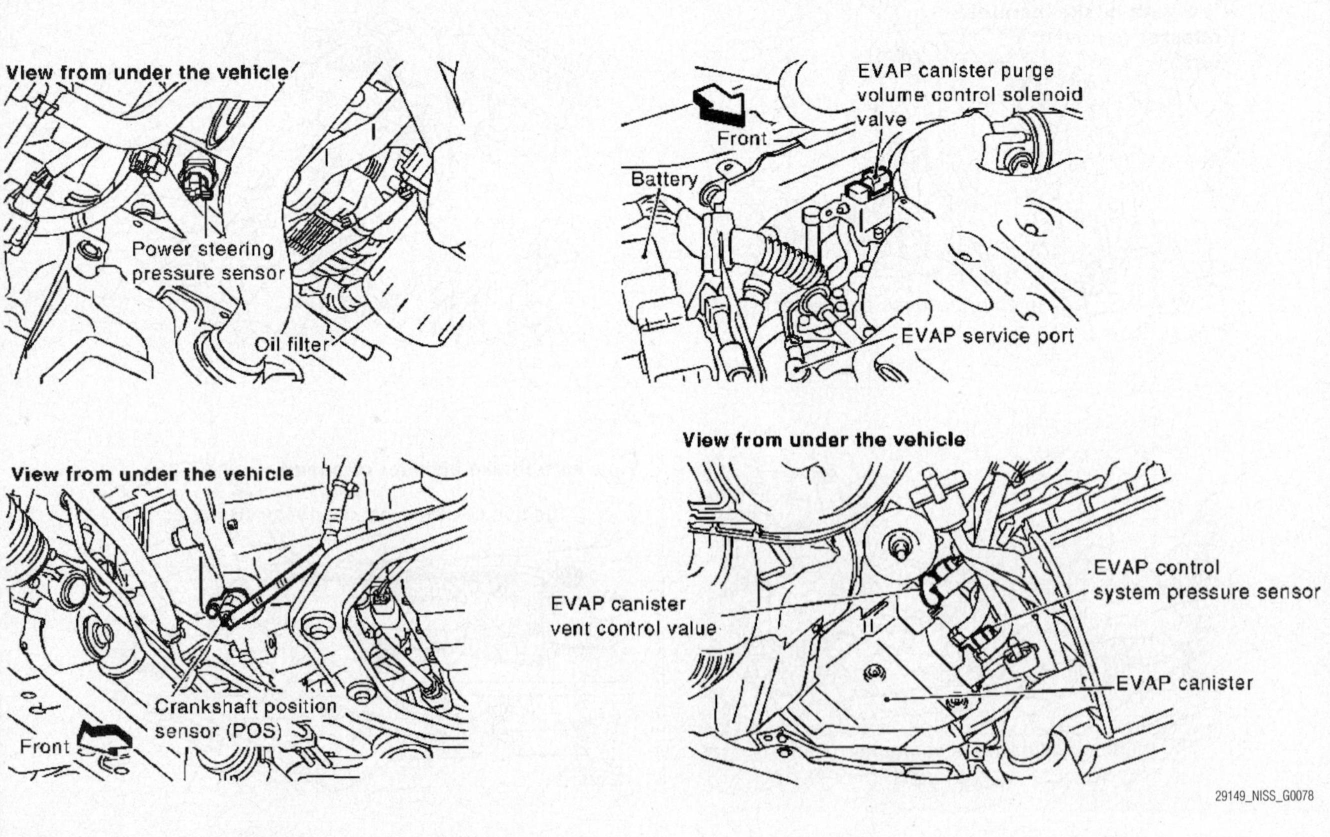

View from under the vehicle

Power steering pressure sensor

Oil filter

Front

EVAP canister purge volume control solenoid valve

Battery

EVAP service port

View from under the vehicle

Crankshaft position sensor (POS)

Front

View from under the vehicle

EVAP canister vent control value

EVAP control system pressure sensor

EVAP canister

29149_NISS_G0078

Fig. 89 350Z component locations

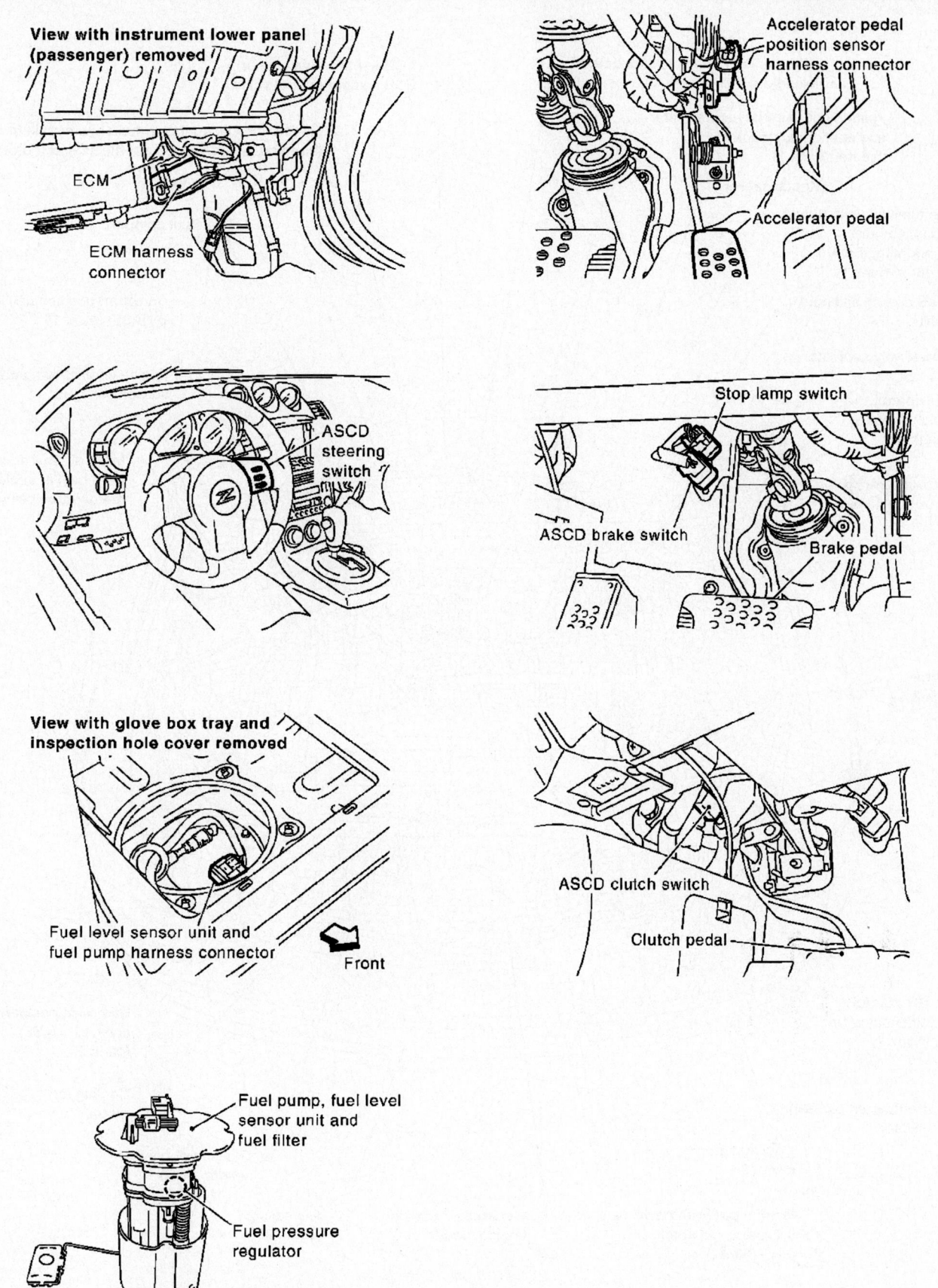

View with instrument lower panel (passenger) removed

ECM

ECM harness connector

Accelerator pedal position sensor harness connector

Accelerator pedal

ASCD steering switch

Stop lamp switch

ASCD brake switch

Brake pedal

View with glove box tray and inspection hole cover removed

Fuel level sensor unit and fuel pump harness connector

Front

ASCD clutch switch

Clutch pedal

Fuel pump, fuel level sensor unit and fuel filter

Fuel pressure regulator

29149_NISS_G0079

Fig. 90 350Z component locations

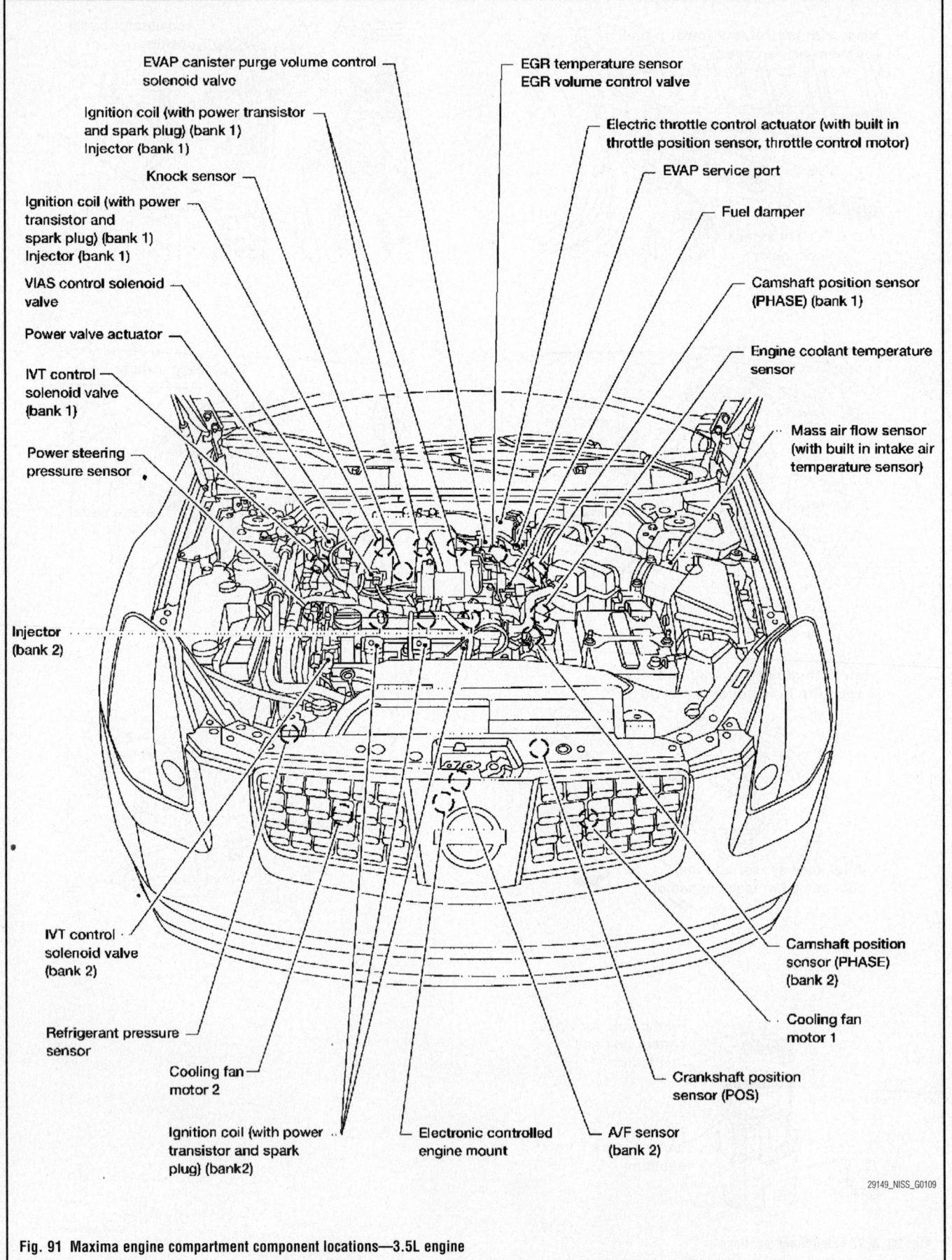

EVAP canister purge volume control
solenoid valve

Ignition coil (with power transistor
and spark plug) (bank 1)
Injector (bank 1)

Knock sensor

Ignition coil (with power
transistor and
spark plug) (bank 1)
Injector (bank 1)

VIAS control solenoid
valve

Power valve actuator

IVT control
solenoid valve
(bank 1)

Power steering
pressure sensor

Injector
(bank 2)

IVT control
solenoid valve
(bank 2)

Refrigerant pressure
sensor

Cooling fan
motor 2

Ignition coil (with power
transistor and spark
plug) (bank2)

Electronic controlled
engine mount

EGR temperature sensor
EGR volume control valve

Electric throttle control actuator (with built in
throttle position sensor, throttle control motor)

EVAP service port

Fuel damper

Camshaft position sensor
(PHASE) (bank 1)

Engine coolant temperature
sensor

Mass air flow sensor
(with built in intake air
temperature sensor)

Camshaft position
sensor (PHASE)
(bank 2)

Cooling fan
motor 1

Crankshaft position
sensor (POS)

A/F sensor
(bank 2)

29149_NISS_G0109

Fig. 91 Maxima engine compartment component locations—3.5L engine

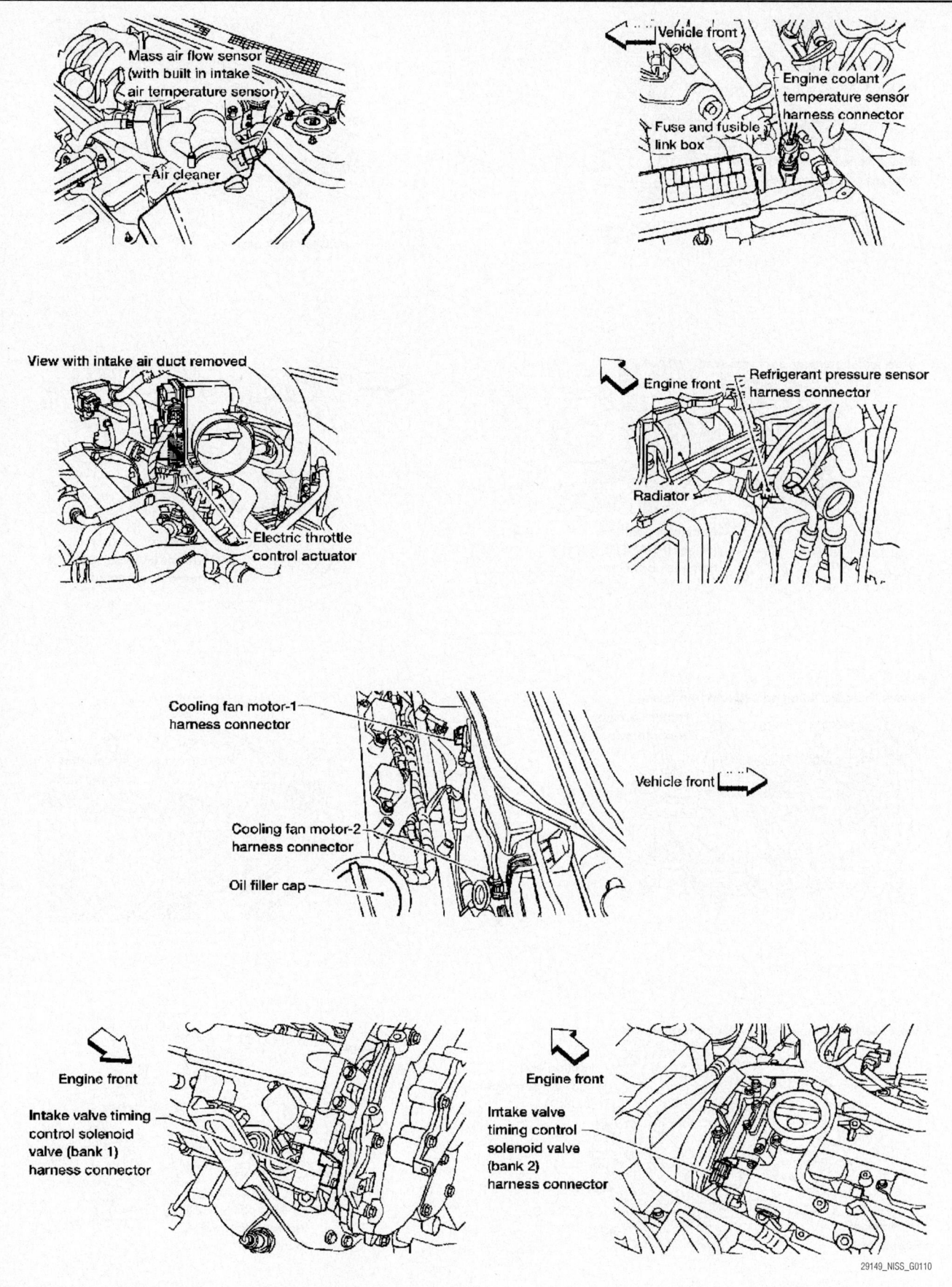

Fig. 92 Maxima engine compartment component locations—3.5L engine

29149_NISS_G0110

View with intake manifold collector removed

Engine front

Knock sensor
harness connector

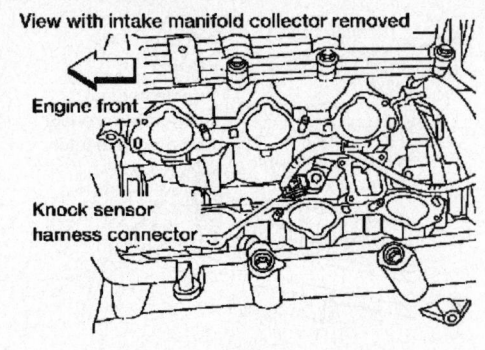

IPDM E/R

Front

Fuel pump fuse
(15A)

Washer tank cap

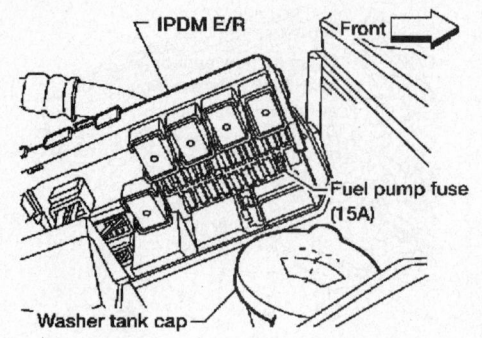

View with intake air duct removed

Camshaft position
sensor (PHASE) (bank 1)
harness connector

Front

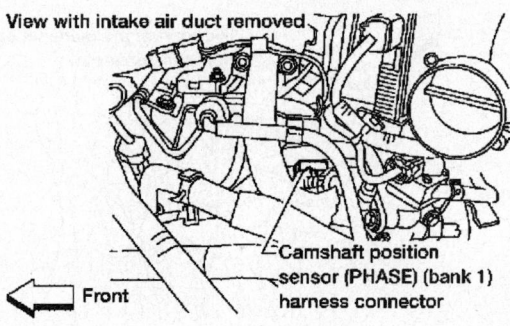

Front

Camshaft position
sensor (PHASE) (bank 2)
harness connector

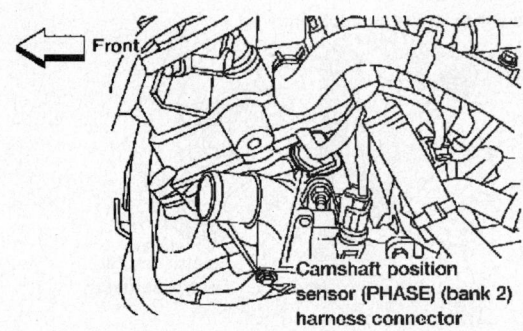

View with intake manifold collector removed

Injector harness
connectors

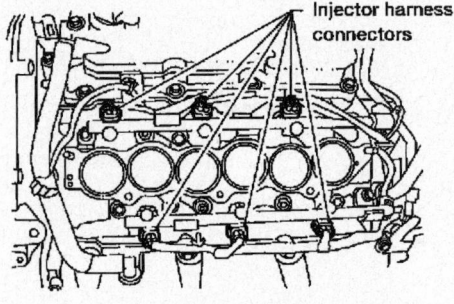

Engine front

VIAS control
solenoid valve

Power valve
actuator

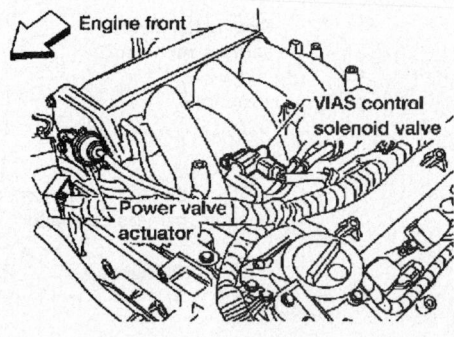

View with intake manifold collector removed

Ignition coil
harness connector
(bank 1)

Engine front

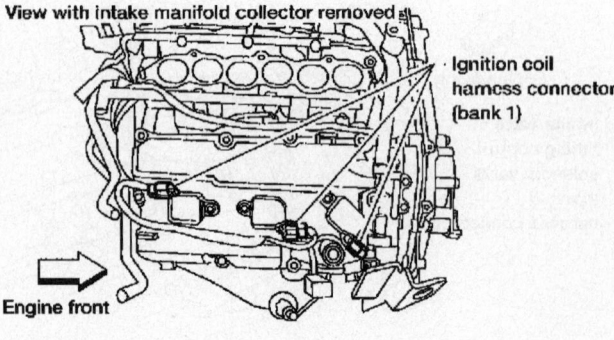

Engine front

Ignition coil
harness connector
(bank 2)

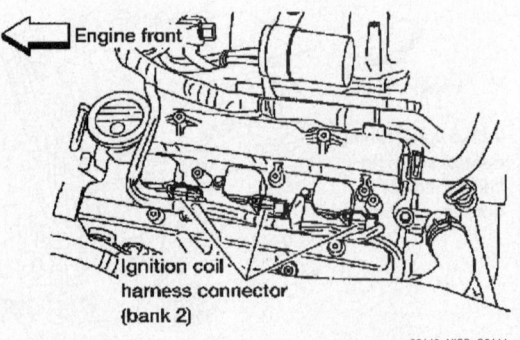

29149_NISS_G0111

Fig. 93 Maxima engine component locations—3.5L engine

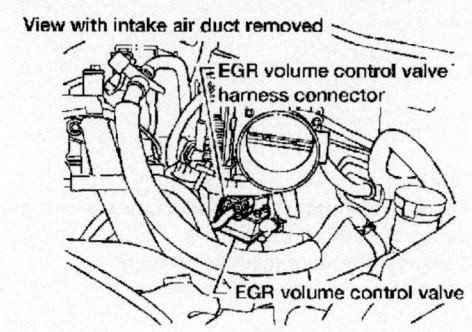

View with intake air duct removed

EGR volume control valve harness connector

EGR volume control valve

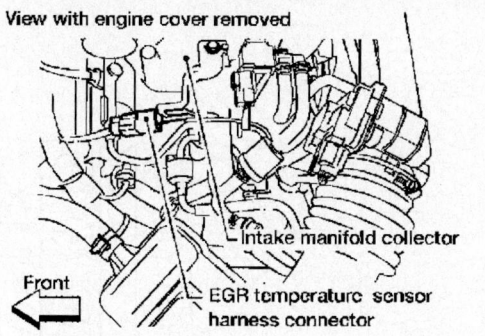

View with engine cover removed

Front

Intake manifold collector

EGR temperature sensor harness connector

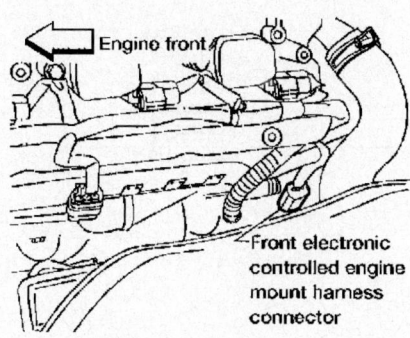

Engine front

Front electronic controlled engine mount harness connector

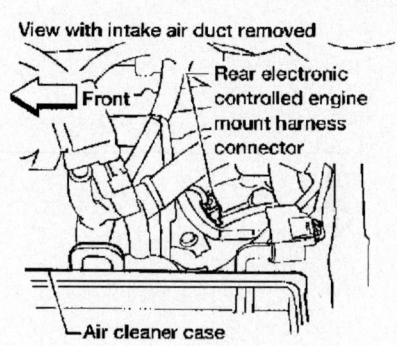

View with intake air duct removed

Front

Rear electronic controlled engine mount harness connector

Air cleaner case

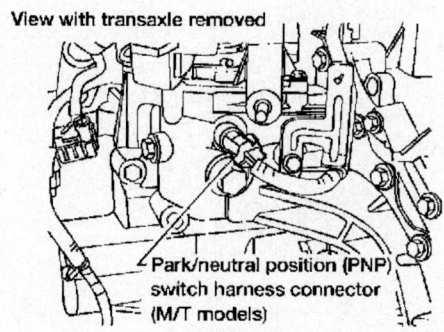

View with transaxle removed

Park/neutral position (PNP) switch harness connector (M/T models)

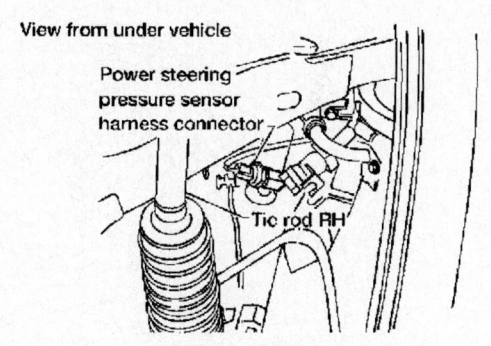

View from under vehicle

Power steering pressure sensor harness connector

Tie rod RH

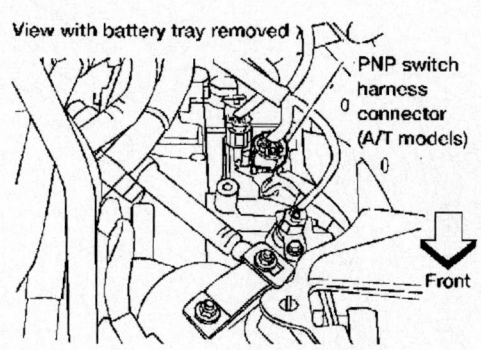

View with battery tray removed

PNP switch harness connector (A/T models)

Front

29149_NISS_G0112

Fig. 94 Maxima component locations—3.5L engine

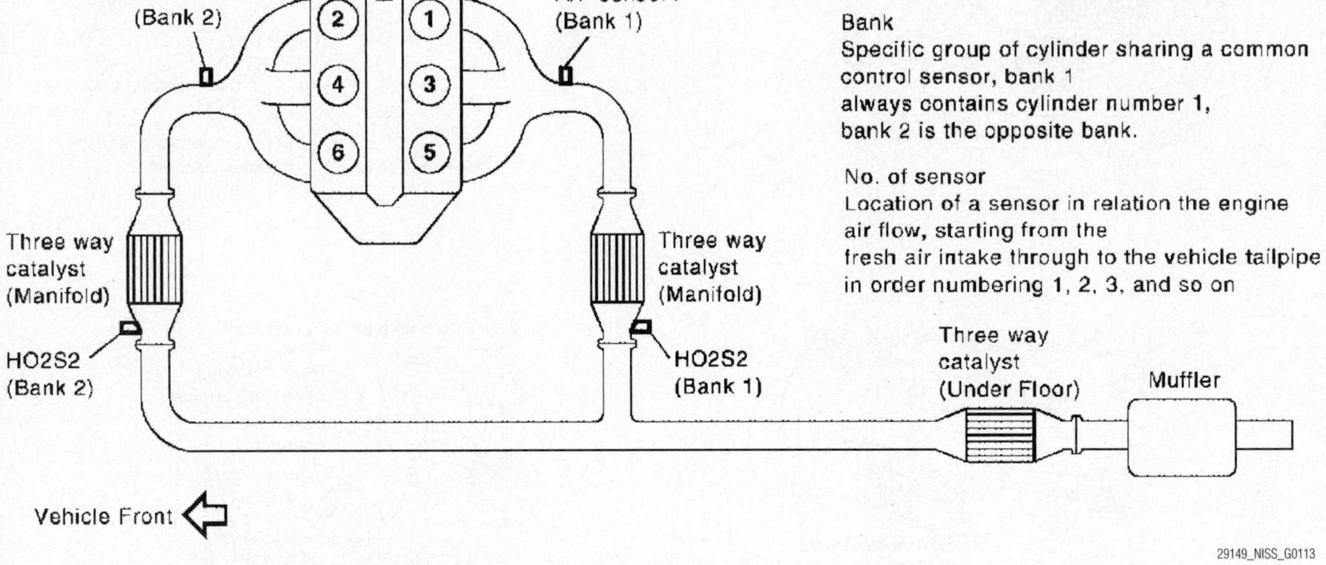

A/F sensor1
(Bank 2)

A/F sensor1
(Bank 1)

2 1
4 3
6 5

Three way
catalyst
(Manifold)

Three way
catalyst
(Manifold)

HO2S2
(Bank 2)

HO2S2
(Bank 1)

Three way
catalyst
(Under Floor)

Muffler

Vehicle Front

Bank
Specific group of cylinder sharing a common
control sensor, bank 1
always contains cylinder number 1,
bank 2 is the opposite bank.

No. of sensor
Location of a sensor in relation the engine
air flow, starting from the
fresh air intake through to the vehicle tailpipe
in order numbering 1, 2, 3, and so on

29149_NISS_G0113

Fig. 95 Maxima component locations—3.5L engine

Engine front

A/F sensor 1
(bank 1)

HO2S2
(bank 1)
harness
connector
(M/T models)

Oil pan

HO2S2
(bank 2)
harness
connector
(M/T models)

Starter

Engine front

A/F sensor 1
(bank 2)

HO2S2 (bank 1)
harness connector
(A/T models)

HO2S2 (bank 2)
harness connector
(A/T models)

Oil pan

29149_NISS_G0114

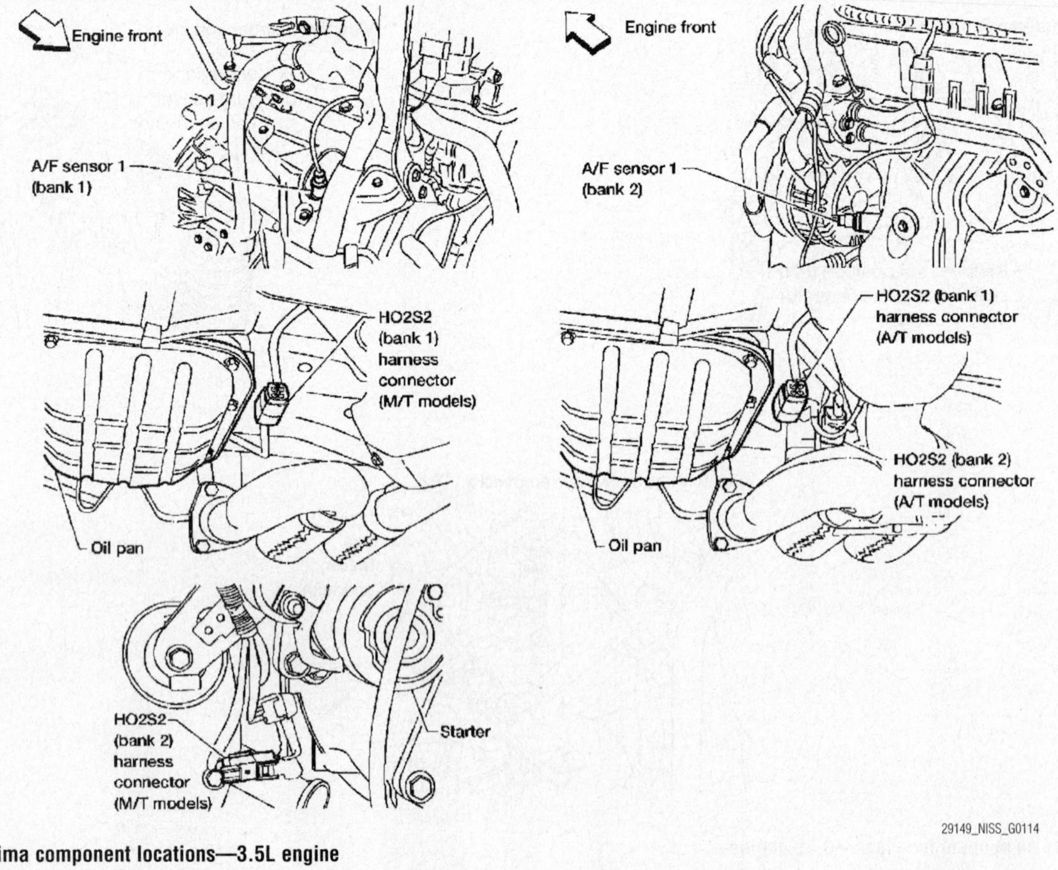

Fig. 96 Maxima component locations—3.5L engine

View from under the vehicle
with rear crossmember removed

EVAP control system
pressure sensor

Water separator

EVAP canister vent
control valve

EVAP canister

View with engine cover removed

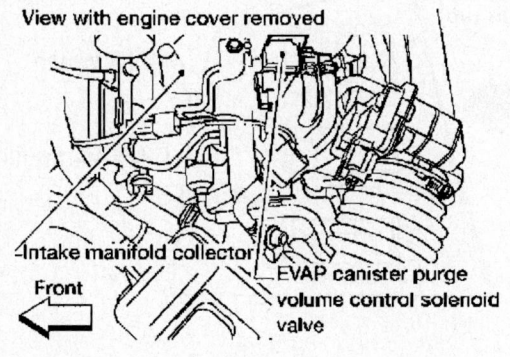

Intake manifold collector

EVAP canister purge
volume control solenoid
valve

Front

View with engine cover removed

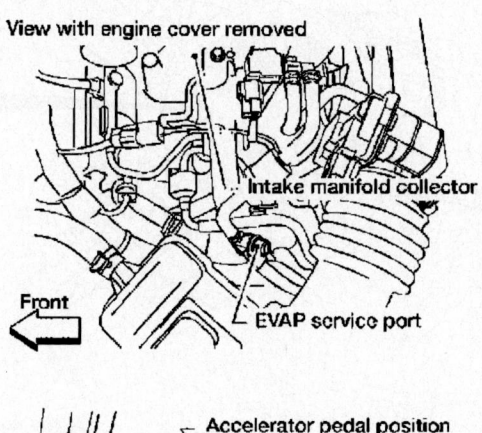

Intake manifold collector

EVAP service port

Front

View with glove box removed

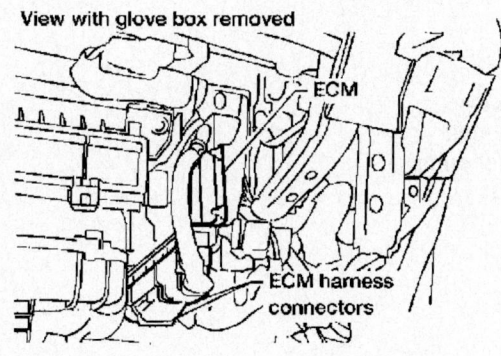

ECM

ECM harness
connectors

Accelerator pedal position
sensor harness connector

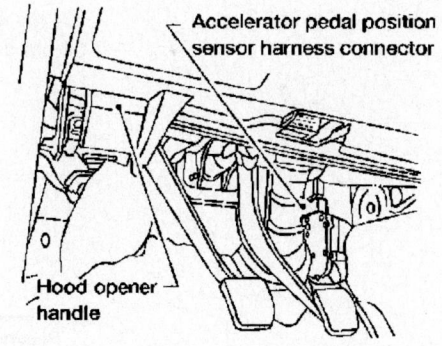

Hood opener
handle

View from under the vehicle

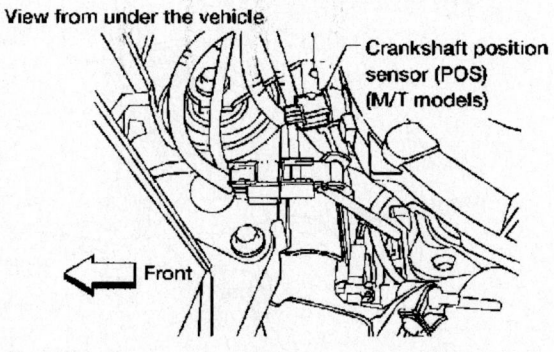

Crankshaft position
sensor (POS) (M/T models)

Front

View from under the vehicle

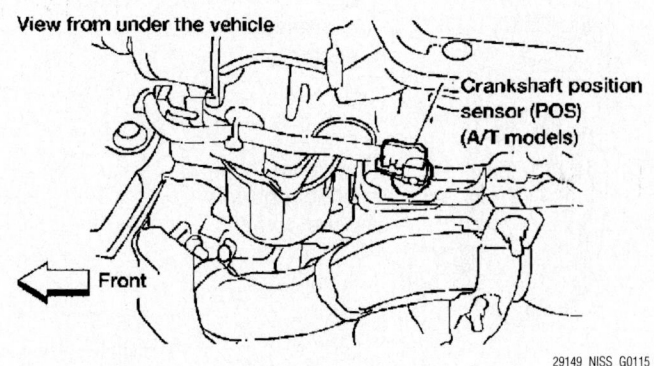

Crankshaft position
sensor (POS)
(A/T models)

Front

29149_NISS_G0115

Fig. 97 Maxima component locations—3.5L engine

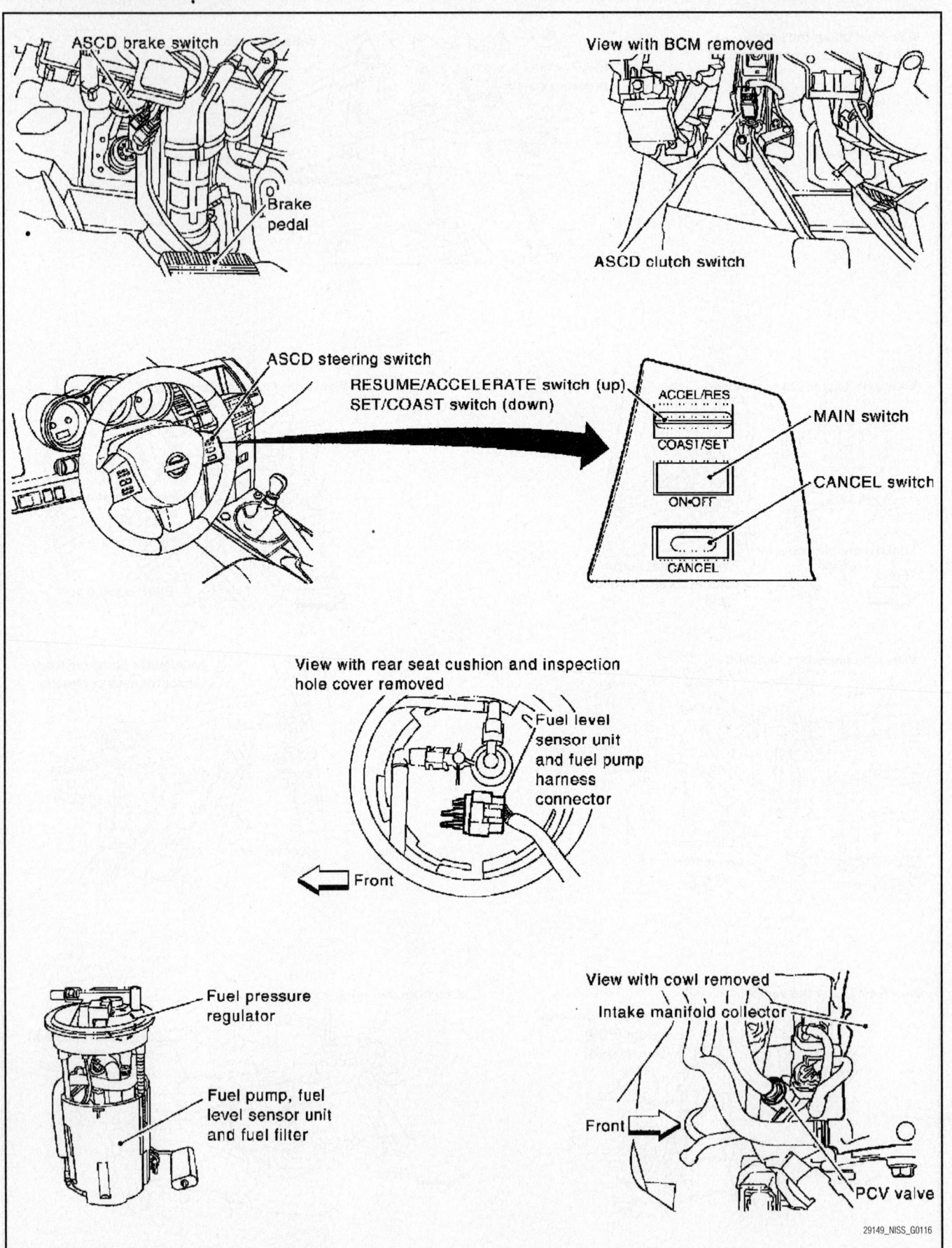

Fig. 98 Maxima component locations—3.5L engine

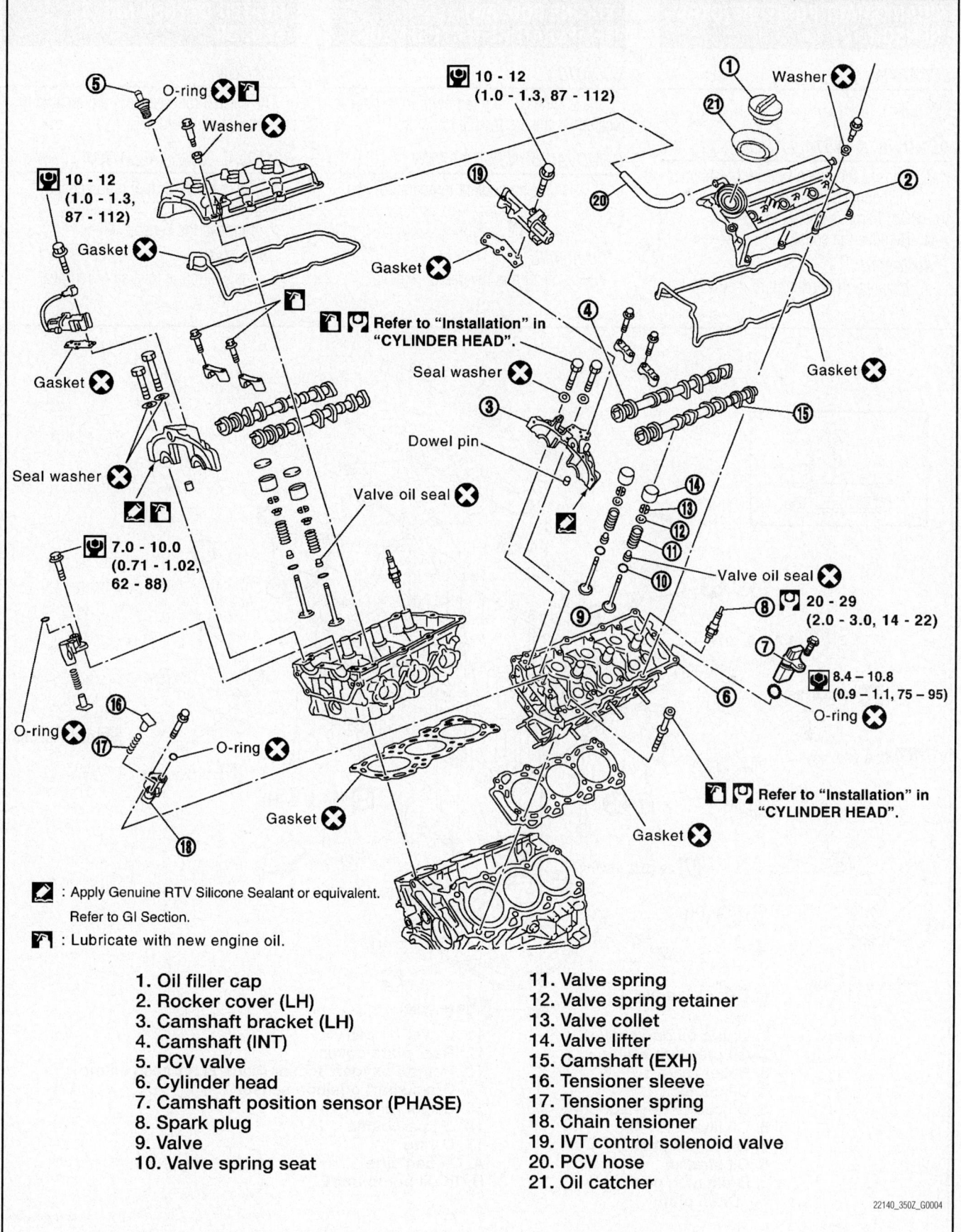

10 - 12 (1.0 - 1.3, 87 - 112)

10 - 12 (1.0 - 1.3, 87 - 112)

O-ring ✕ 🖐

Washer ✕

Washer ✕

Gasket ✕

Gasket ✕

Refer to "Installation" in "CYLINDER HEAD".

Seal washer ✕

Gasket ✕

Gasket ✕

Dowel pin

Seal washer ✕

Valve oil seal ✕

Valve oil seal ✕

7.0 - 10.0 (0.71 - 1.02, 62 - 88)

20 - 29 (2.0 - 3.0, 14 - 22)

8.4 - 10.8 (0.9 - 1.1, 75 - 95)

O-ring ✕

O-ring ✕

O-ring ✕

Refer to "Installation" in "CYLINDER HEAD".

Gasket ✕

Gasket ✕

📝 : Apply Genuine RTV Silicone Sealant or equivalent. Refer to GI Section.

🖐 : Lubricate with new engine oil.

1. Oil filler cap
2. Rocker cover (LH)
3. Camshaft bracket (LH)
4. Camshaft (INT)
5. PCV valve
6. Cylinder head
7. Camshaft position sensor (PHASE)
8. Spark plug
9. Valve
10. Valve spring seat
11. Valve spring
12. Valve spring retainer
13. Valve collet
14. Valve lifter
15. Camshaft (EXH)
16. Tensioner sleeve
17. Tensioner spring
18. Chain tensioner
19. IVT control solenoid valve
20. PCV hose
21. Oil catcher

22140_350Z_G0004

Fig. 99 Camshaft position sensor location

CRANKSHAFT POSITION (CKP) SENSOR

LOCATION

See Figure 100.

REMOVAL & INSTALLATION

1. Loosen the fixing bolt of the sensor.
2. Disconnect crankshaft position sensor (POS) harness connector.
3. Remove the sensor.

To install:

4. Installation is reverse of removal.

ENGINE COOLANT TEMPERATURE (ECT) SENSOR

LOCATION

The engine coolant temperature sensor is located on the thermostat housing.

REMOVAL & INSTALLATION

1. Disconnect coolant temperature sensor harness connector.
2. Remove the sensor.

To install:

3. Installation is reverse of removal.

HEATED OXYGEN (HO2S) SENSOR

LOCATION

The air fuel ratio sensors are located in the exhaust manifold.

REMOVAL & INSTALLATION

1. Disconnect air fuel ratio sensors harness connector.
2. Remove the sensor.

To install:

3. Installation is reverse of removal.

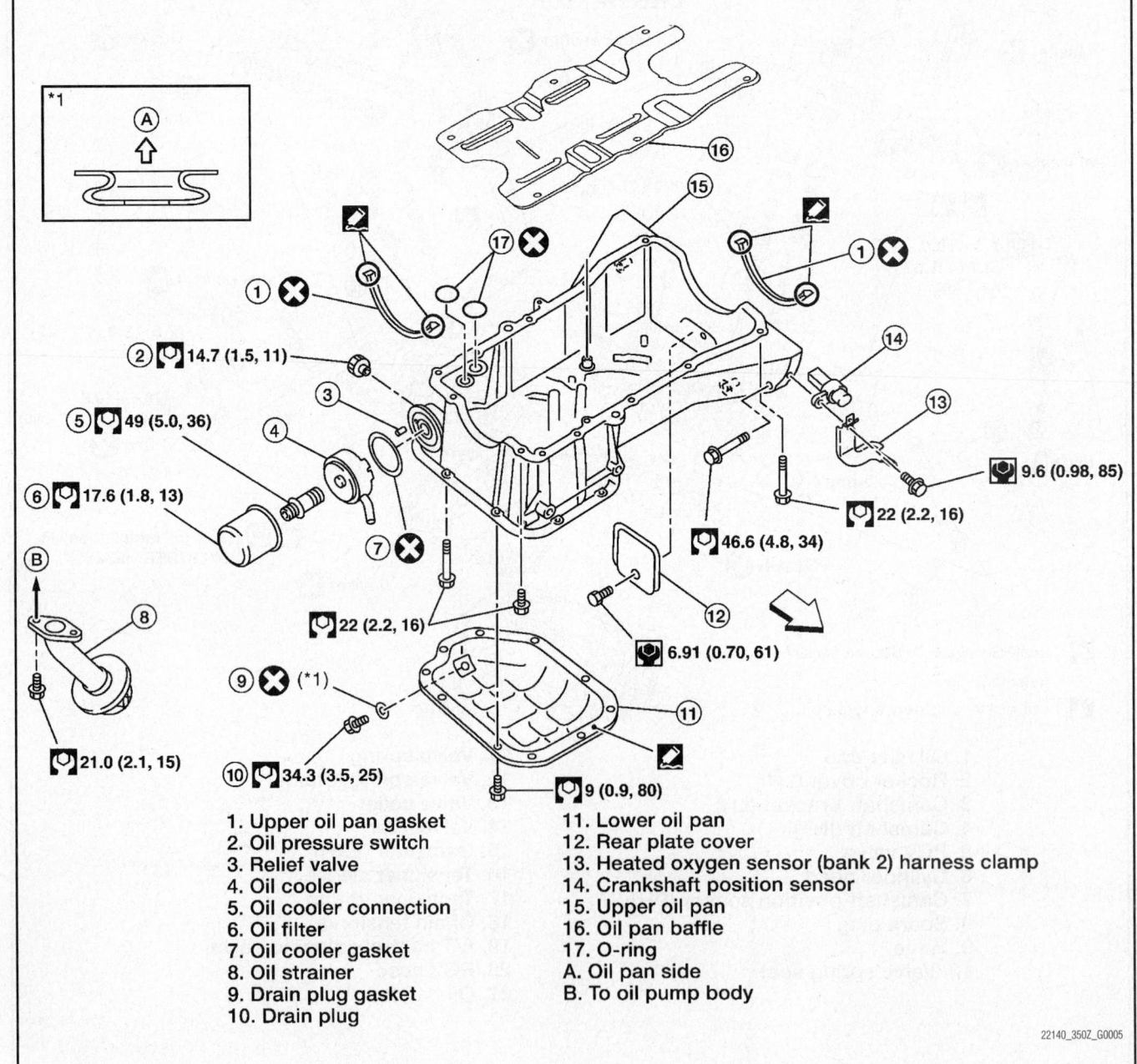

1. Upper oil pan gasket
2. Oil pressure switch
3. Relief valve
4. Oil cooler
5. Oil cooler connection
6. Oil filter
7. Oil cooler gasket
8. Oil strainer
9. Drain plug gasket
10. Drain plug
11. Lower oil pan
12. Rear plate cover
13. Heated oxygen sensor (bank 2) harness clamp
14. Crankshaft position sensor
15. Upper oil pan
16. Oil pan baffle
17. O-ring
A. Oil pan side
B. To oil pump body

22140_350Z_G0005

Fig. 100 Crankshaft position sensor location

INTAKE AIR TEMPERATURE (IAT) SENSOR

LOCATION

The intake air temperature sensor is located on the mass air flow sensor.

REMOVAL & INSTALLATION

1. Disconnect the harness connector from the mass air flow sensor.

2. Disconnect the tube clamp at the electric throttle control actuator and at the fresh air intake tube.

3. Remove air cleaner to electric throttle control actuator tube, air cleaner case (upper) with the mass air flow sensor attached.

4. Remove mass air flow sensor from air cleaner case (upper), as necessary.

5. Remove resonator in the fender, lifting left fender protector, as necessary.

To install:

6. Installation is reverse of removal.

KNOCK SENSOR (KS)

LOCATION

See Figure 101.

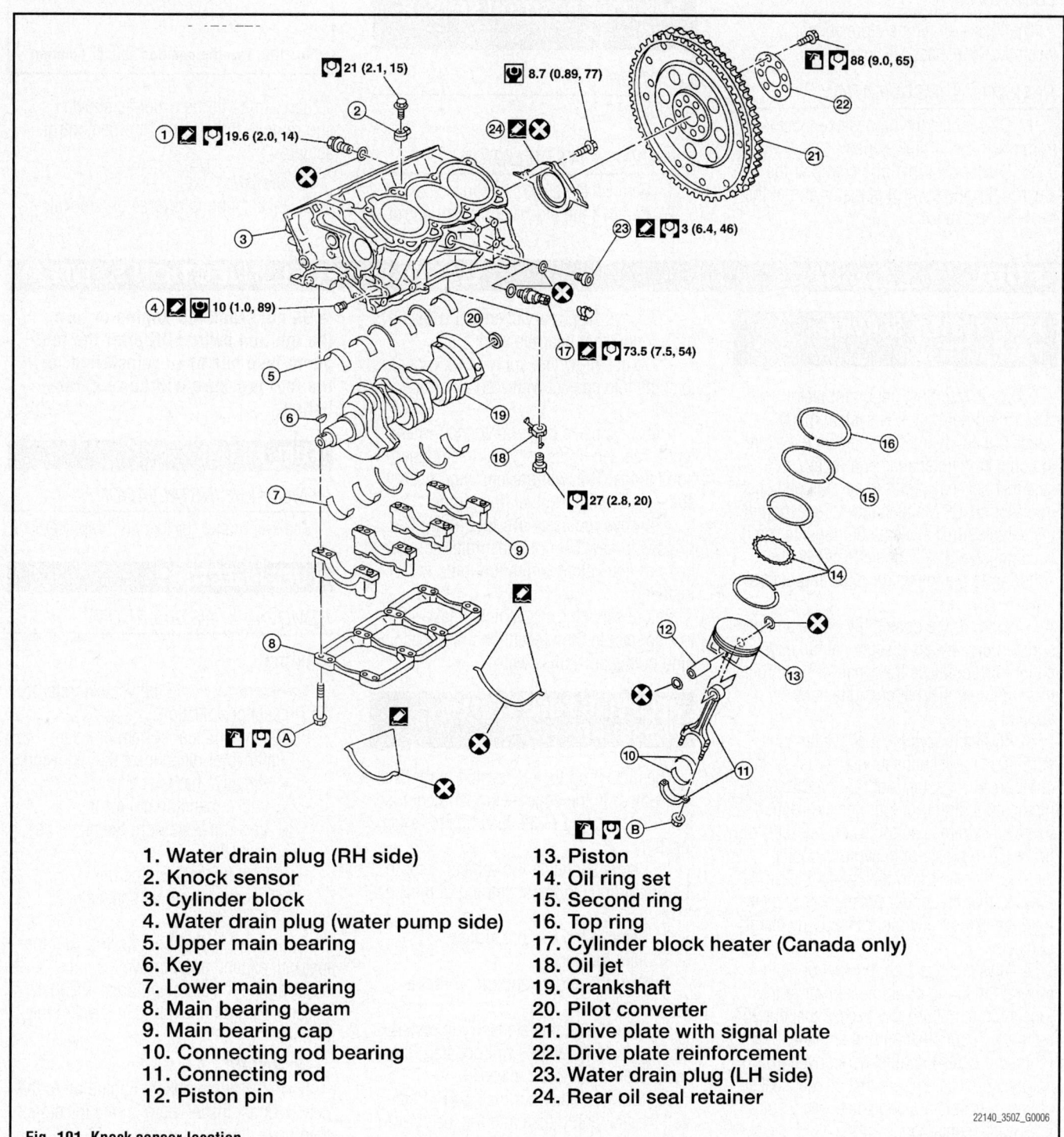

1. Water drain plug (RH side)
2. Knock sensor
3. Cylinder block
4. Water drain plug (water pump side)
5. Upper main bearing
6. Key
7. Lower main bearing
8. Main bearing beam
9. Main bearing cap
10. Connecting rod bearing
11. Connecting rod
12. Piston pin
13. Piston
14. Oil ring set
15. Second ring
16. Top ring
17. Cylinder block heater (Canada only)
18. Oil jet
19. Crankshaft
20. Pilot converter
21. Drive plate with signal plate
22. Drive plate reinforcement
23. Water drain plug (LH side)
24. Rear oil seal retainer

Fig. 101 Knock sensor location

22140_350Z_G0006

REMOVAL & INSTALLATION

1. Remove the engine assembly.
2. Install the engine on the engine stand.
3. Remove the knock sensor.

To install:

4. Installation is reverse of removal.

MASS AIR FLOW (MAF) SENSOR

LOCATION

The intake air temperature sensor is located on the mass air flow sensor.

REMOVAL & INSTALLATION

1. Disconnect the harness connector from the mass air flow sensor.
2. Disconnect the tube clamp at the electric throttle control actuator and at the fresh air intake tube.

3. Remove air cleaner to electric throttle control actuator tube, air cleaner case (upper) with the mass air flow sensor attached.
4. Remove mass air flow sensor from air cleaner case (upper), as necessary.
5. Remove resonator in the fender, lifting left fender protector, as necessary.

To install:

6. Installation is reverse of removal.

THROTTLE POSITION SENSOR (TPS)

LOCATION

See Figure 102.

REMOVAL & INSTALLATION

1. Remove the electrical connector.
2. Remove the electric throttle control

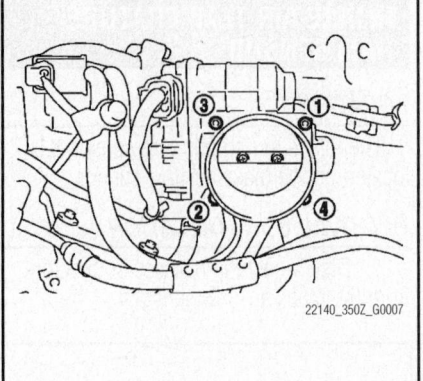

Fig. 102 Throttle position sensor location

actuator bolts in the order as shown and remove the electric throttle control actuator.

To install:

3. Installation is reverse of removal.

FUEL

GASOLINE FUEL INJECTION SYSTEM

FUEL SYSTEM SERVICE PRECAUTIONS

Safety is the most important factor when performing not only fuel system maintenance but any type of maintenance. Failure to conduct maintenance and repairs in a safe manner may result in serious personal injury or death. Maintenance and testing of the vehicle's fuel system components can be accomplished safely and effectively by adhering to the following rules and guidelines.

• To avoid the possibility of fire and personal injury, always disconnect the negative battery cable unless the repair or test procedure requires that battery voltage be applied.

• Always relieve the fuel system pressure prior to disconnecting any fuel system component (injector, fuel rail, pressure regulator, etc.), fitting or fuel line connection. Exercise extreme caution whenever relieving fuel system pressure to avoid exposing skin, face and eyes to fuel spray. Please be advised that fuel under pressure may penetrate the skin or any part of the body that it contacts.

• Always place a shop towel or cloth around the fitting or connection prior to loosening to absorb any excess fuel due to spillage. Ensure that all fuel spillage (should it occur) is quickly removed from engine surfaces. Ensure that all fuel soaked cloths or towels are deposited into a suitable waste container.

• Always keep a dry chemical (Class B) fire extinguisher near the work area.

• Do not allow fuel spray or fuel vapors to come into contact with a spark or open flame.

• Always use a back-up wrench when loosening and tightening fuel line connection fittings. This will prevent unnecessary stress and torsion to fuel line piping.

• Always replace worn fuel fitting O-rings with new Do not substitute fuel hose or equivalent where fuel pipe is installed.

Before servicing the vehicle, make sure to also refer to the precautions in the beginning of this section as well.

RELIEVING FUEL SYSTEM PRESSURE

The fuel pump fuse is located in the dash fuse box or in the engine compartment fuse box. Check the lid of the fuse box for exact location.

1. Before servicing the vehicle, refer to the precautions in the beginning of this section.
2. Remove the fuel pump fuse.
3. Start the engine.
4. Start the engine and run until the engine stalls.
5. After the engine stalls, try to restart the engine; if the engine will not start, the fuel pressure has been released.
6. Turn the ignition switch **OFF**. Reinstall the fuel pump fuse into the fuse block.

➡ Do not crank the engine or turn the ignition switch ON after the fuel pump fuse has been reinstalled, or the fuel pressure will be re-established.

FUEL FILTER

REMOVAL & INSTALLATION

See Fuel pump for fuel filter replacement.

FUEL INJECTORS

REMOVAL & INSTALLATION

Maxima

1. Before servicing the vehicle, refer to the Precautions Section.
2. Relieve the fuel system pressure
3. Remove or disconnect the following:
   • Negative battery cable
   • Intake manifold collector
   • Vacuum hose from the pressure regulator
   • Fuel hoses from the rail
   • Injector electrical connectors
   • Fuel rail bolts
4. To remove the fuel injector from the fuel rail, expand and remove the clips securing the injectors and press the fuel injector out from the fuel rail. Discard the O-rings.

To install:

5. Apply a thin coat of engine oil to the new O-rings, install them on the injectors, then press the injector into the fuel rail.

6. Install or connect the following:
- New injector retaining clips
- New injector gaskets onto the manifold
- Fuel rail assembly to the engine
- Fuel rail-to-cylinder head bolts and torque the bolts to 84–96 inch lbs. (9.3–10.8 Nm). Then tighten them again to 16–19 ft. lbs. (21–26 Nm).
- Fuel lines to the rail assembly
- Vacuum hose to the fuel pressure regulator
- Electrical connectors to the fuel injectors
- Intake manifold collector
- Negative battery cable.

7. Start the engine and check for leaks.

350Z

See Figure 103.

1. Before servicing the vehicle, refer to the Precautions Section.
2. Relieve the fuel system pressure.
3. Drain coolant.
4. Remove or disconnect the following:
- Negative battery cable
- Engine cover
- Remove intake manifold collector.
- Fuel feed hose and damper
- Under vehicle fuel line quick connectors
- Upper and lower intake manifold collectors
- Fuel injector harness connectors
- Fuel rail mounting bolts in reverse of the tightening sequence

5. To remove the fuel injector from the fuel rail, expand and remove the clips securing the injectors and press the fuel injector out from the fuel rail. Discard the O-rings.

To install:

➡**Upper and lower O-rings are different. The Black rings go on the fuel tube**

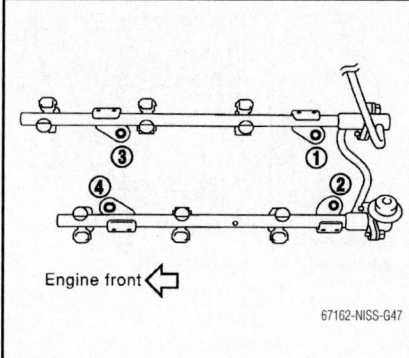

Engine front ←

67162-NISS-G47

Fig. 103 Fuel rail tightening sequence–350Z

side and the Green rings go on the nozzle side.

6. Apply a thin coat of engine oil to the new O-rings, install them on the injectors, then press the injector into the fuel rail.
7. Install or connect the following:
- New injector retaining clips
- Fuel rail mounting bolts in sequence and tighten to 8 ft. lbs. (11 Nm), then to 16–19 ft. lbs. (21–27 Nm).
- Fuel injector harness connectors
- Upper and lower intake manifold collectors
- Under vehicle fuel line quick connectors
- Fuel feed hose and damper
- Engine cover
- Negative battery cable

8. Start the engine and check for leaks.

FUEL PUMP

REMOVAL & INSTALLATION

Maxima

1. Before servicing the vehicle, refer to the precautions in the beginning of this section.
2. Relieve the fuel system pressure
3. Remove or disconnect the following:
- Negative battery cable
- Rear seat or open the access panel in the trunk
- Fuel gauge electrical connector and pump electrical connector
- Fuel outlet and the return hoses
- Fuel tank, if necessary

4. On some models you need to remove the fuel pump assembly-to-fuel tank bolts and lift the fuel pump assembly from the fuel tank.
5. On other models you need to remove the locking ring and raise the fuel pump from the tank. Disconnect the feed tube while raising the pump.
6. Discard the O-ring. Plug the fuel tank opening with a clean rag to prevent dirt from entering the system.

➡**When removing or installing the fuel pump assembly, be careful not to damage or deform it and always install a new O-ring.**

To install:
7. Remove the rag
8. Install or connect the following:
- Fuel pump assembly into the fuel tank using a new O-ring

- Fuel pump assembly-to-fuel tank bolts and torque the bolts to 17–22 inch lbs. (2.0–2.5 Nm)
- Locking ring assembly and tighten
- Fuel tank assembly, if removed
- Fuel lines and the electrical connectors. Always use new clamps when reconnecting fuel line hoses

➡**When installing the upper plate, be sure to align the mark with the center marks on the fuel tank.**

- Negative battery cable
9. Start the engine, check for fuel leaks and repair if necessary.
10. Install the fuel pump access cover.

➡**On some models, the Check Engine Light will stay ON after installation is completed. The memory code in the control unit must be erased. This code is stored for an open fuel pump circuit, this is caused when the fuel pressure is released. To erase the code, disconnect the battery cable for 10 seconds, then reconnect after installation of fuel pump.**

350Z

1. Before servicing the vehicle, refer to the Precautions Section.

➡**If the fuel tank is more than three quarter full, some fuel will have to be drained before removing the fuel pump/fuel filter assembly.**

2. Relieve the fuel system pressure
3. Remove or disconnect the following:
- Negative battery cable
- Rear floor box
- Fuel pump inspection cover
- Harness connector and fuel feed tube
- Fuel feed tube quick connector
- Retainer
- Raise the fuel pump assembly and remove the fuel hose connector
- Remove the fuel pump assembly

4. Reverse the removal procedure to install.

FUEL TANK

REMOVAL & INSTALLATION

Maxima

1. Disconnect the battery negative terminal.
2. Check the fuel level with the vehicle on a level surface. If the fuel gauge indicates more than the level as shown (7/8 full), drain the fuel from the fuel tank until

the fuel gauge indicates a level at or below as shown (7/8 full).

3. Open the fuel door and unscrew the fuel filler cap to release the pressure inside the fuel tank.

4. Release the fuel pressure from the fuel lines.

5. Disconnect the battery negative terminal.

6. Remove rear seat bottom.

7. Reposition the rear floor carpet out of the way to remove the fuel pump inspection hole cover. Turn the four retainers 90° in a clockwise direction and remove the fuel pump inspection hole cover.

8. Disconnect the fuel level sensor unit, fuel filter, and fuel pump assembly electrical connector; and the fuel feed hose from the fuel level sensor unit, fuel filter, and fuel pump assembly.

9. Disconnect the quick connectors.

10. Remove the center exhaust tube, with mufflers.

11. Disconnect the three parking brake cable mounting brackets on each cable and position the cables out of the way.

12. Remove the fuel tank protector.

13. Disconnect the fuel filler hose, recirculation hose and EVAP canister hose at the fuel tank.

14. Disconnect the fuel tank mounting straps while supporting the fuel tank.

15. Remove the fuel tank.

To install:

16. Installation is in the reverse order of removal.

➡ **Before tightening the fuel tank mounting straps, temporarily install the filler hose, recirculation hose, and signal hose. Tighten mounting straps to 32 ft. lbs. (43 Nm).**

350Z

1. Perform steps 2 to 7 of "REMOVAL" in " FUEL LEVEL SENSOR UNIT, FUEL FILTER AND FUEL PUMP ASSEMBLY" on main and sub fuel level sensor units.

2. Remove tunnel stay.

3. Remove exhaust front tube, center muffler and main muffler.

4. Remove propeller shaft.

5. Remove parking rear brake cables.

6. Remove rear suspension assembly.

7. Remove fuel tank protector.

8. Disconnect fuel filler hose, vent hose and EVAP hoses at fuel tank side.

9. Remove fuel tank mounting band bolts while supporting fuel tank.

10. Supporting with hands, descend transmission jack carefully, and remove fuel tank.

11. Remove fuel filler tube, if necessary.

To install:

12. Installation is reverse of removal.

IDLE SPEED

ADJUSTMENT

Idle speed is not adjustable

THROTTLE BODY

REMOVAL & INSTALLATION

See Figure 104.

1. To remove and install throttle body, refer to illustration. Tighten all bolts in sequence to:
   - First pass: 79–95 inch lbs. (9–11 Nm)
   - Second pass: 13–16 ft. lbs. (18–22 Nm)

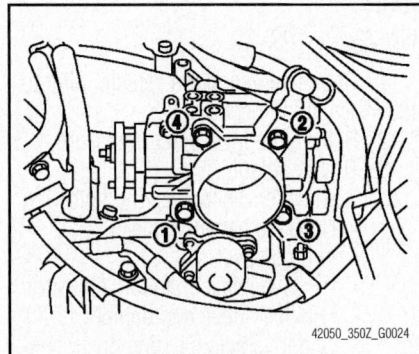

42050_350Z_G0024

Fig. 104 Throttle body tightening sequence

HEATING & AIR CONDITIONING SYSTEM

BLOWER MOTOR

REMOVAL & INSTALLATION

Maxima

See Figure 105.

1. Before servicing the vehicle, refer to the Precautions Section.

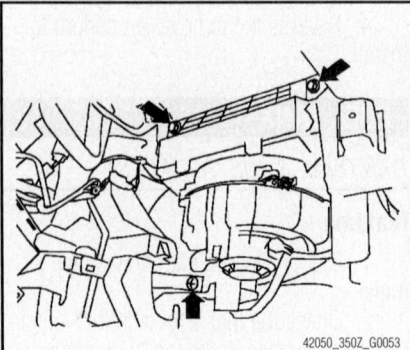

42050_350Z_G0053

Fig. 105 Removing blower unit

2. Disconnect or remove the following:
   - Negative battery cable.
   - Glove box assembly
   - Center console side cover
   - ECM
   - Blower motor, intake door motor, fan control amp connector
   - Main harness from the top of the blower unit
   - Two bolts and one screw from the blower unit
   - Blower unit

3. To install, reverse removal procedure.

350Z

See Figures 106 through 108.

1. Before servicing the vehicle, refer to the Precautions Section.

2. Disconnect or remove the following:
   - Negative battery cable.
   - Lower instrument passenger panel
   - ECM with bracket
   - Intake door motor connector and blower motor connector

✳✳ CAUTION

Move blower unit rightward and remove location pin (1 part) and joint. Then remove blower unit downward.

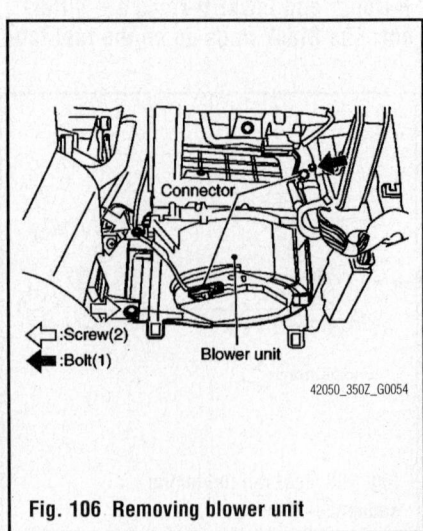

42050_350Z_G0054

Fig. 106 Removing blower unit

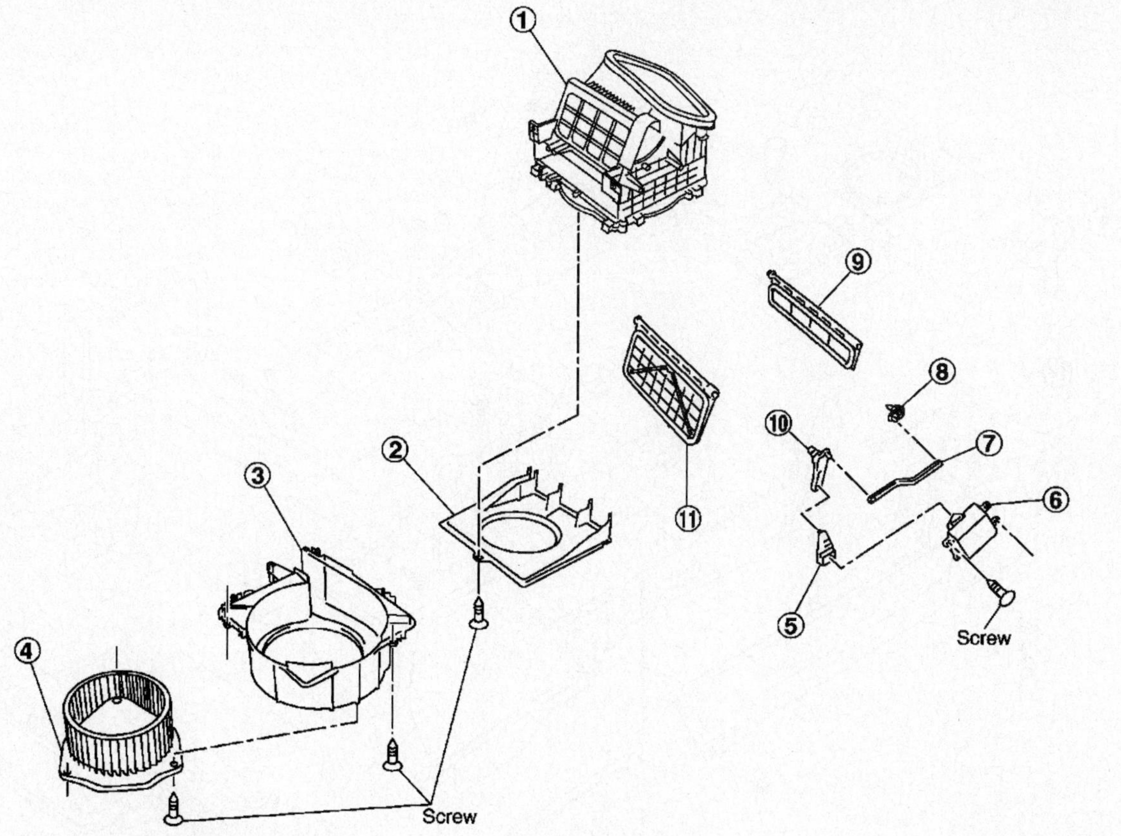

Fig. 107 Exploded view of blower motor and components

1.	Intake upper case	2.	Intake bell mouth	3.	Intake lower case
4.	Blower motor assembly	5.	Intake door lever 2	6.	Intake door motor
7.	Intake door link	8.	Intake door lever 3	9.	Intake door 2
10.	Intake door lever 1	11.	Intake door 1		

42050_350Z_G0055

- Mounting bolts and screws then blower unit
- Blower motor mounting screws then blower motor

3. To install, reverse removal procedure.

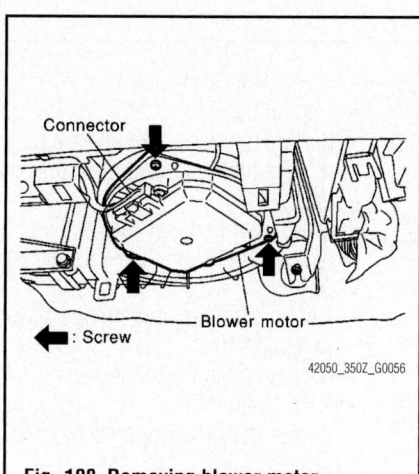

Fig. 108 Removing blower motor

HEATER CORE

REMOVAL & INSTALLATION

Maxima

See Figures 109 through 112.

1. Before servicing the vehicle, refer to the Precautions Section.
2. Disconnect the negative battery terminal.
3. Disconnect battery positive terminal.

✳✳ CAUTION

After disconnecting the negative battery cable, wait for at least 3 minutes for the SRS modules to deplete its energy.

4. Discharge the refrigerant from the A/C system.
5. Drain the coolant from the cooling system.

6. Disconnect the heater hoses from the heater core pipes.
7. Disconnect the refrigerant lines from the evaporator.
8. Remove the instrument panel by removing or disconnecting the following:
- Fuse block cover
- Lower driver instrument panel screws
- Lower driver instrument panel
- Aspirator tube and in vehicle temperature sensor
- Electrical harness connectors
- Right side instrument lower cover
- Glove box pins and remove glove box door
- Glove box housing screws
- Trunk cancel switch harness
- Glove box lamp and harness from glove box housing
- Glove box housing
- Both rear seat cushions and both rear seatbacks

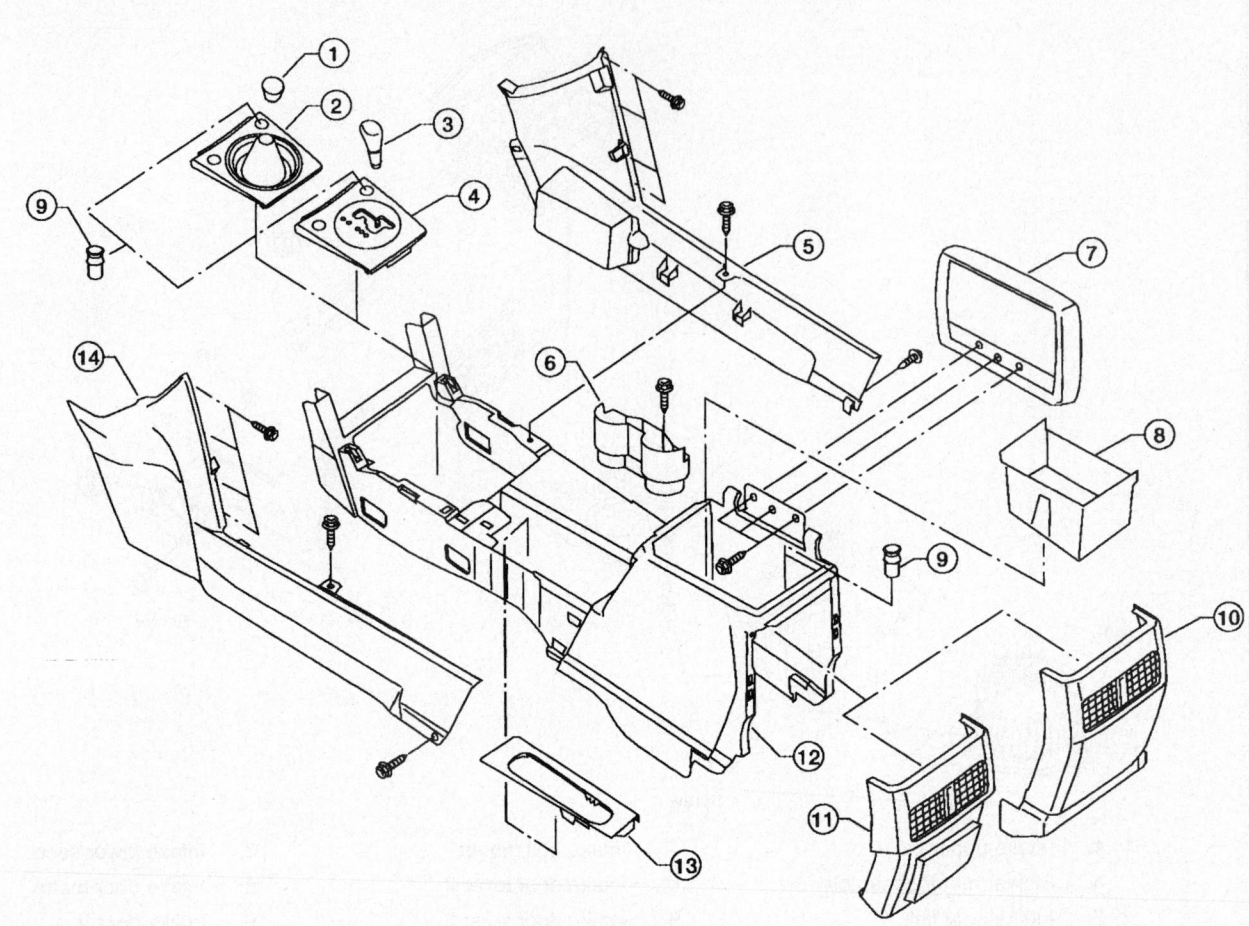

1. M/T shift knob
2. M/t console finisher
3. A/T shift knob
4. A/T console finisher
5. Console cover, RH
6. Cup holder insert
7. Console lid assembly
8. Console rear pocket
9. Power socket assembly
10. Rear finisher assembly, 5 seat model
11. Rear finisher assembly, 4 seat model
12. Front center console
13. Parking brake lever finisher
14. Console cover, LH

09482_ZMAX_G0005

Fig. 109 Exploded view of the center console assembly Maxima

- Screws and pass through assembly
- Clips and right lower side cover and left lower side cover
- Screws, disconnect harness and remove rear console assembly
- A/T or M/T finisher
- Cluster lid C screws
- Pull cluster lid C towards rear of vehicle to release clips
- Cluster lid C electrical connectors
- Cluster lid D screws
- Pull cluster lid D toward rear of vehicle to release clips
- Cluster lid D electrical connectors

✻✻ WARNING

To avoid damage, eject map DVD-ROM disk before removing the NAVI control unit.

✻✻ WARNING

Cover NAVI control unit with a cloth and avoid contact of NAVI control unit with brackets that may cause scratches or damage to NAVI control unit.

- Front center console
- Control device
- Pull up parking brake lever and remove parking brake lever finisher
- Move front seats forward and remove front center console assembly
- NAVI control unit mounting screws (if equipped)
- NAVI control unit connectors
- NAVI control unit
- Center stack mounting screws and center stack

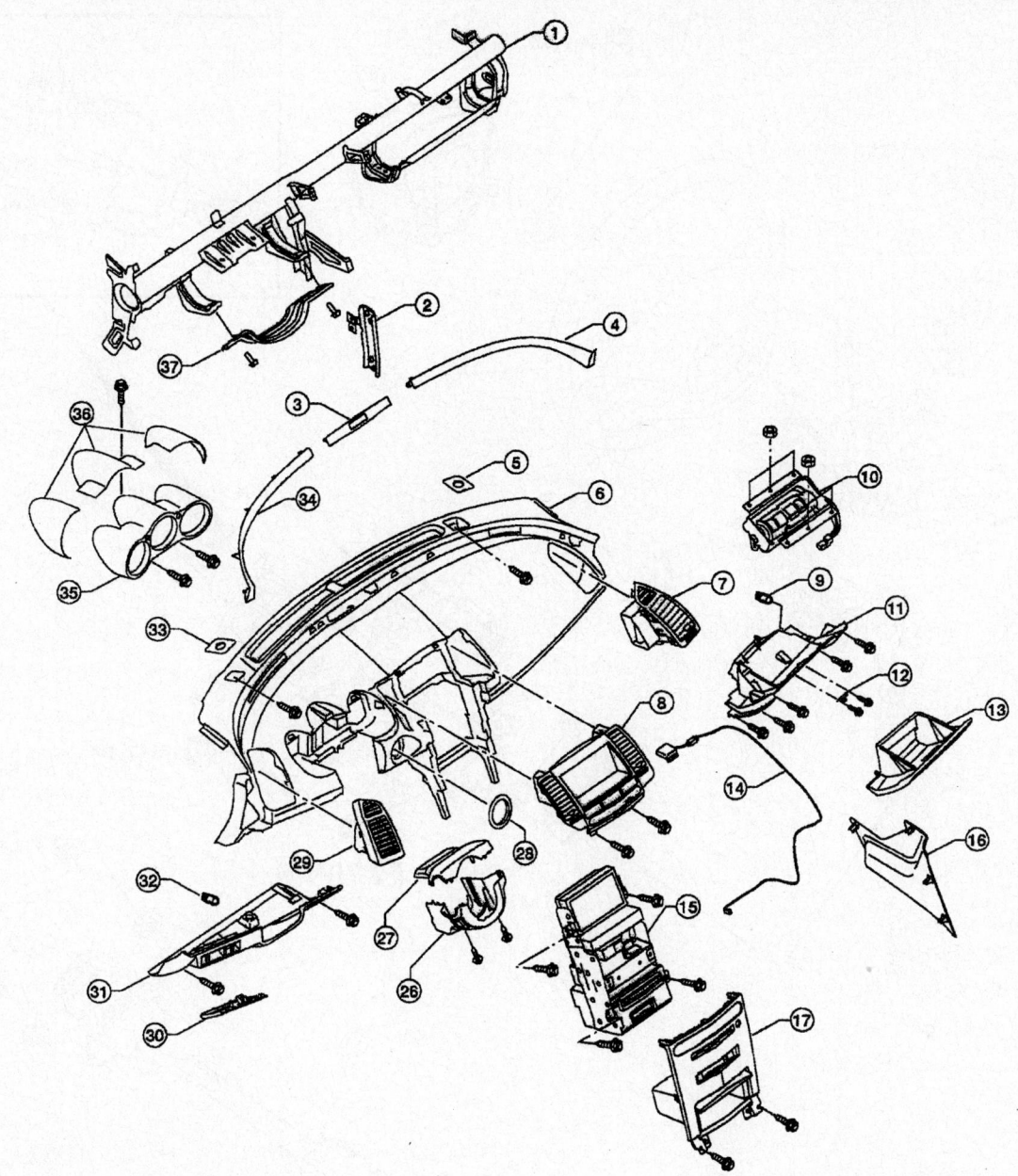

1.	Steering member assembly	2.	Instrument stay, driver	3.	Security light finisher/Passenger air bag off indicator
4.	Instrument panel finisher, RH	5.	Instrument mask RH, sun sensor	6.	Instrument panel
7.	Side ventilator assembly, RH	8.	Cluster lid D	9.	Glove box bulb
10.	Front passenger air bag module	11.	Instrument passenger lower panel	12.	Glove box striker
13.	Glove box assembly	14.	GPS antenna assembly	15.	Center stack
16.	Instrument lower cover, RH	17.	Cluster lid C	18.	M/T shift knob
19.	M/T finisher	20.	A/T shift knob	21.	A/T finisher
22.	Center console assembly	23.	Parking brake lever finisher	24.	Rear upper console assembly
25.	Rear console assembly	26.	Steering column cover, lower	27.	Steering column cover, upper
28.	Steering lock escutcheon	29.	Side ventilator assembly, LH	30.	Fuse block cover
31.	Lower driver instrument panel	32.	Lower driver instrument panel bulb	33.	Instrument mask LH, optical sensor
34.	Instrument panel finisher, LH	35.	Combination meter assembly	36.	Combination meter covers
37.	Lower knee protector, LH				

09482_ZMAX_G0006

Fig. 110 Exploded view of the instrument panel assembly Maxima

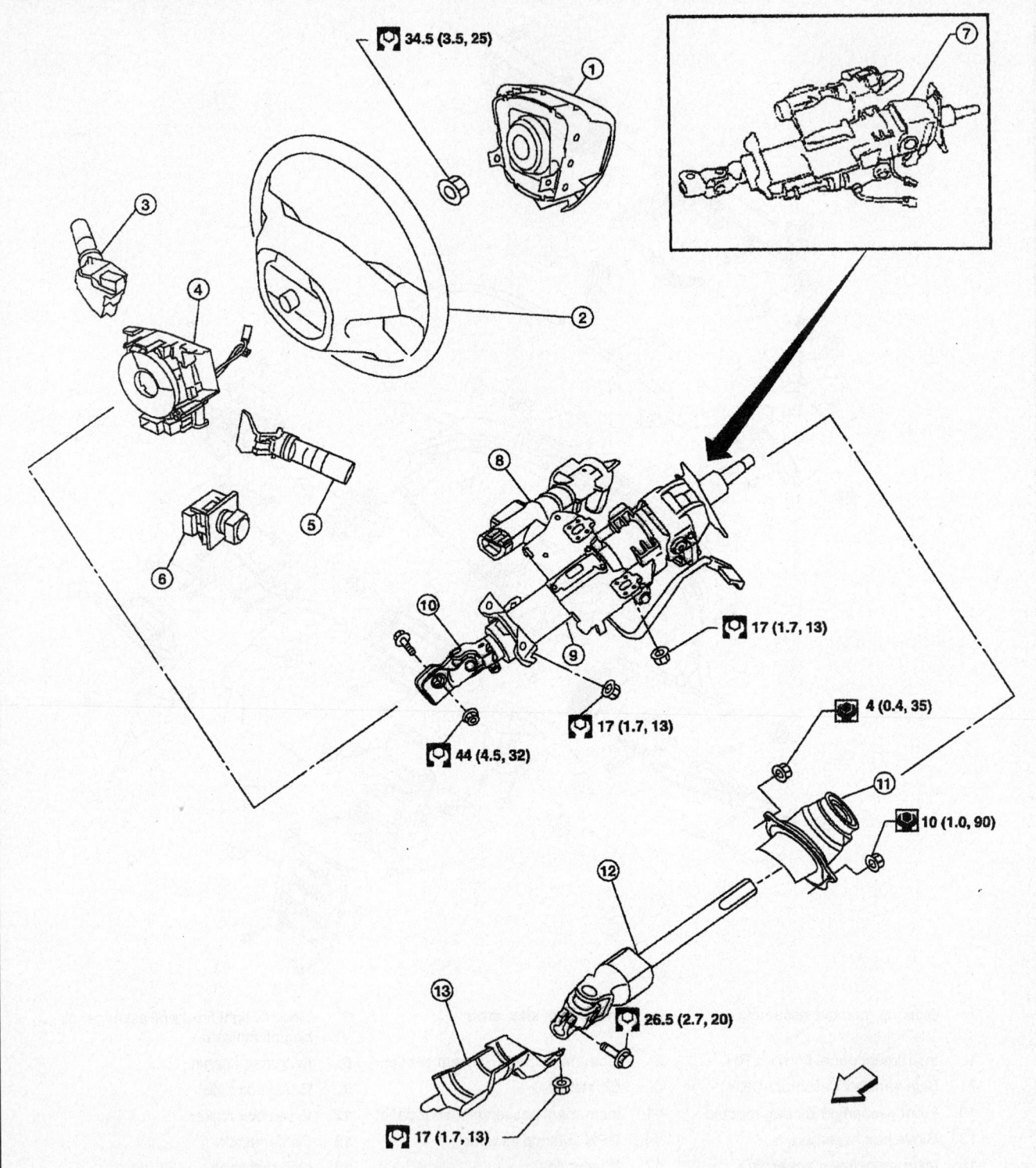

34.5 (3.5, 25)

17 (1.7, 13)

17 (1.7, 13)

44 (4.5, 32)

4 (0.4, 35)

10 (1.0, 90)

26.5 (2.7, 20)

17 (1.7, 13)

1. Driver air bag module
2. Steering wheel
3. Head lamp switch
4. Spiral cable
5. Combination switch
6. ADP steering switch
7. Steering column (electric tilt/telescopic type)
8. Ignition switch
9. Steering column (manual tilt/telescope type)
10. Upper joint
11. Hole cover
12. Lower Joint and shaft assembly
13. Shaft lower cover
⇐ Front

09482_ZMAX_G0007

Fig. 111 Exploded view of the steering wheel and steering column assemblies Maxima

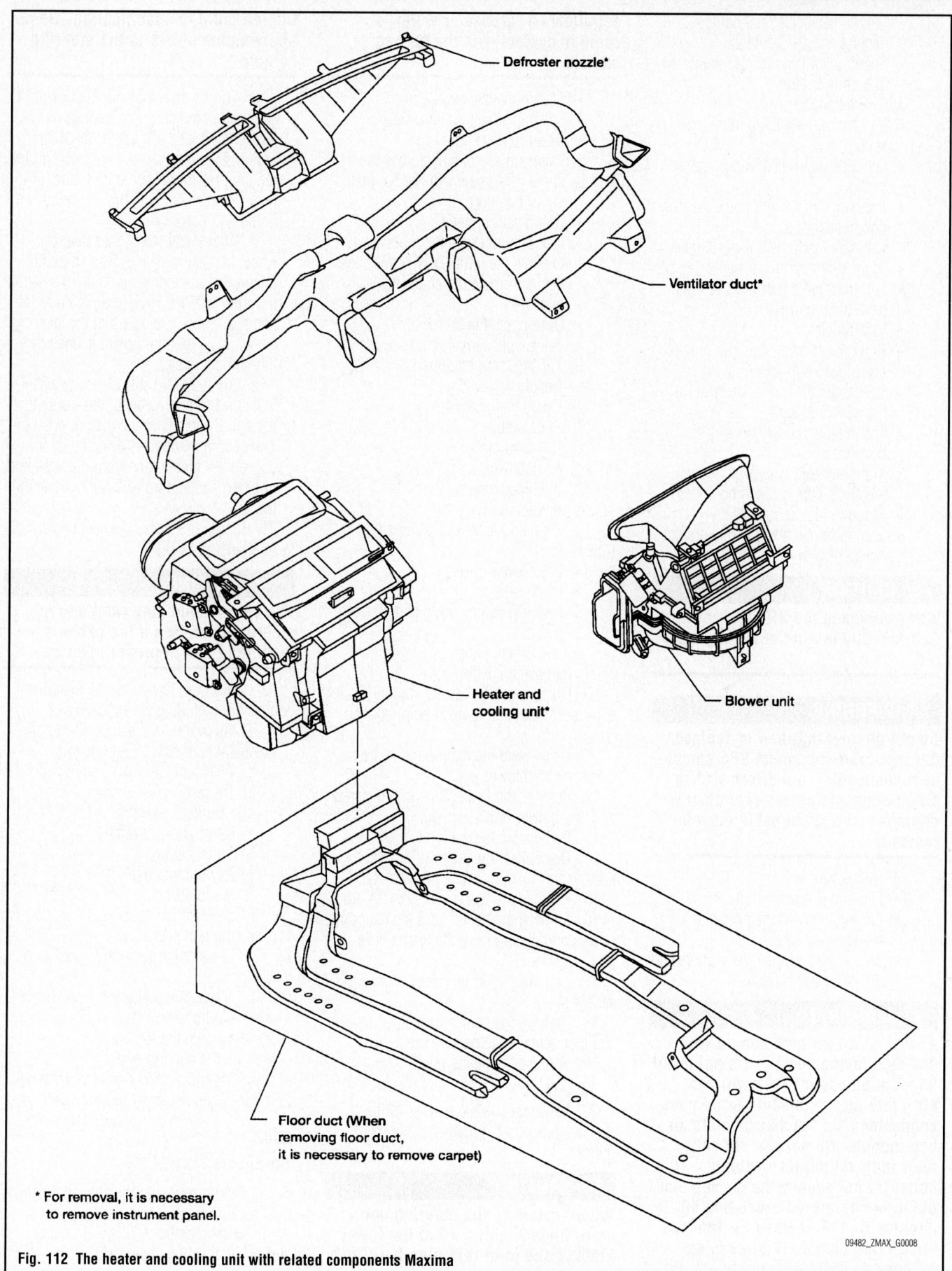

Defroster nozzle*

Ventilator duct*

Heater and
cooling unit*

Blower unit

Floor duct (When
removing floor duct,
it is necessary to remove carpet)

* For removal, it is necessary
to remove instrument panel.

09482_ZMAX_G0008

Fig. 112 The heater and cooling unit with related components Maxima

- Center stack electrical harness connectors and GPS antenna
- Security light flasher/passenger air bag off indicator
- GPS antenna
- Right side instrument mask, sun sensor
- Left side instrument mask, optical sensor
- Partially remove and place the front door welts aside
- Left side instrument panel finisher
- Right side instrument panel finisher
- Left front pillar garnish and right front pillar garnish
- Kicking plate
- Right lower dash side trim and left lower dash side trim
- Passenger air bag to steering member assembly bolt
- Passenger side air bag electrical connector
- Tilt motor and telescopic motor electrical connector (if equipped)
- Steering lock escutcheon
- Set the front wheels in the straight-ahead position.

❊ CAUTION

When servicing the SRS, do not work from directly in front of air bag module.

❊❊ WARNING

Do not attempt to repair or replace damaged direct-connect SRS component connectors. If a driver air bag direct-connect harness connector is damaged, the spiral cable must be replaced.

- Bolt covers
- Left and right side bolts
- Lift the driver air bag module from the steering wheel
- The air bag harness connectors and driver air bag module

❊ WARNING

Always place air bag module with pad side facing upward. Do not insert any foreign objects (screwdriver, etc.) into air bag module or harness connectors. Do not disassemble air bag module. Do not use old bolts after removal; replace with new bolts. Do not expose the air bag module to temperatures exceeding 90 degrees C (194 degrees F). Replace the air bag module if it has been dropped or sustained an impact. Do

not allow oil, grease or water to come in contact with the air bag module.

- Steering wheel center nut
- Steering wheel using a steering wheel removal tool
- Place a piece of tape across the spiral cable so it will not be rotated out of position
- Combination meter
- Left side instrument panel screws to allow the ignition switch to clear the instrument panel during removal
- Disconnect the following:
  a. Telescopic sensor, if equipped
  b. Tilt sensor, if equipped
  c. Headlamp switch
  d. Combination switch
  e. Spiral cable
  f. Key in reminder
  g. Immobilizer
  h. Illumination lamp
  i. Ignition switch
  j. Column harness clips, position aside
  - Shaft lower cover
  - Pinch bolt.
  - Steering column nuts and steering column
  - Instrument panel

9. Remove the ECM.
10. Disconnect the blower motor, intake door motor, and fan control amp. connector.
11. Disconnect the main harness from the top of the blower unit.
12. Remove the two bolts and one screw from the blower unit as shown.
13. Remove the blower unit.
14. Disconnect the mode door motor and the air mix door motor connectors.
15. Remove the heater and cooling unit.
16. Remove the heater core pipe support screws and then remove the heater core pipe support.
17. Remove the air mix door motor (passenger side).
18. Remove the heater core cover screws and then remove the heater core cover.
19. Remove the heater core.

To install:

20. Installation is the reverse of the removal procedure, noting the following points:

❊❊ CAUTION

When installing the steering column, finger-tighten all of the lower bracket and joint retaining bolt; then

tighten them to specification. Do not apply undue stress to the steering column.

a. For the steering column, be sure to align slit of the coupling joint with projection on dust cover. Insert the joint until both surfaces make contact. Torque the joint retaining bolt to 32 ft. lbs. (44 Nm). Torque the lower bracket bolts to 13 ft. lbs. (17 Nm).

b. After installation, turn steering wheel to make sure it moves smoothly. Ensure the number of turns are the same from the straight-forward position to left and right locks. Be sure that the steering wheel is in a neutral position when driving straight ahead.

c. Align spiral cable correctly when installing steering wheel. Make sure that the spiral cable is in the neutral position. The neutral position is detected by turning left 2.5 revolutions from the right end position and ending with the knob at the top.

21. Torque the steering wheel center nut to 25 ft. lbs. (34 Nm).

❊❊ WARNING

The spiral cable may snap due to steering operation if the cable is installed in an improper position.

22. Make sure to connect all electrical connectors including the following:
   a. Tilt motor and telescopic motor electrical connector, if equipped.
   b. Telescopic sensor, if equipped
   c. Tilt sensor, if equipped
   d. Headlamp switch
   e. Combination switch
   f. Spiral cable
   g. Key in reminder
   h. Immobilizer
   i. Illumination lamp
   j. Ignition switch
   k. Column harness clips, position aside
   l. Make sure that any/all vacuum lines and hoses are reconnected.
23. Recharge the A/C system.
24. Fill the cooling system.
25. Reconnect battery positive terminal first; the negative battery terminal second.

350Z

See Figures 113 and 114.

1. Before servicing the vehicle, refer to the Precautions Section.
2. Discharge the A/C system using approved recycling equipment.

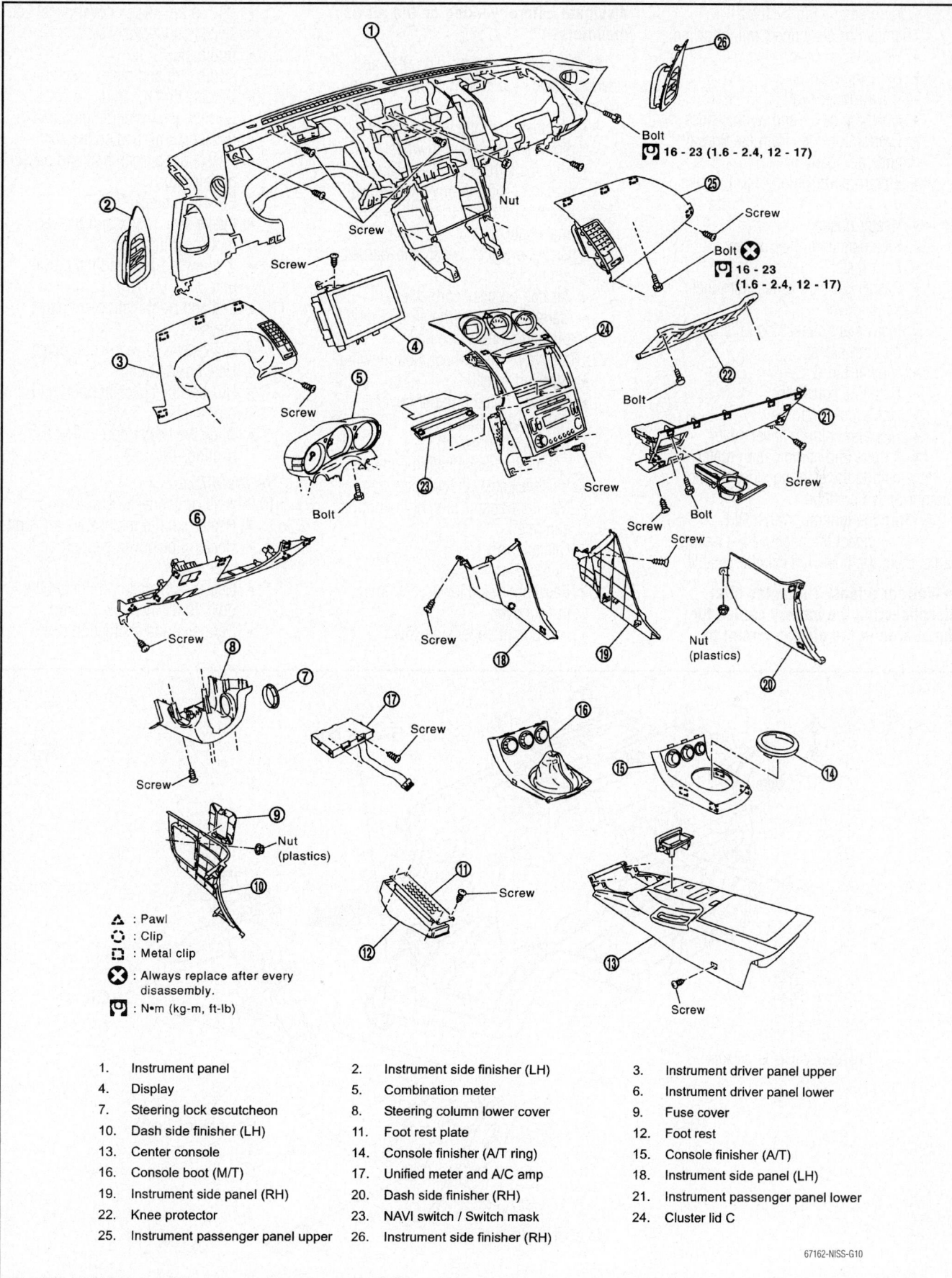

Bolt
🔧 16 - 23 (1.6 - 2.4, 12 - 17)

Bolt ✖
🔧 16 - 23
(1.6 - 2.4, 12 - 17)

△ : Pawl
○ : Clip
▢ : Metal clip
✖ : Always replace after every disassembly.
🔧 : N•m (kg-m, ft-lb)

1.	Instrument panel	2.	Instrument side finisher (LH)	3.	Instrument driver panel upper
4.	Display	5.	Combination meter	6.	Instrument driver panel lower
7.	Steering lock escutcheon	8.	Steering column lower cover	9.	Fuse cover
10.	Dash side finisher (LH)	11.	Foot rest plate	12.	Foot rest
13.	Center console	14.	Console finisher (A/T ring)	15.	Console finisher (A/T)
16.	Console boot (M/T)	17.	Unified meter and A/C amp	18.	Instrument side panel (LH)
19.	Instrument side panel (RH)	20.	Dash side finisher (RH)	21.	Instrument passenger panel lower
22.	Knee protector	23.	NAVI switch / Switch mask	24.	Cluster lid C
25.	Instrument passenger panel upper	26.	Instrument side finisher (RH)		

67162-NISS-G10

Fig. 113 Exploded view of the instrument panel assembly 350Z

3. Drain the cooling system.

4. Remove or disconnect the following:
- Hood ledge cover
- Both wiper arms
- Cowl rubber seal
- Cowl top cover and washer hose
- Evaporator lines from the firewall and cap openings
- Electronic throttle control assembly
- Heater hoses
- Kick panels on both sides
- Foot rests
- Passenger side lower instrument panel cover
- Instrument panel side finish panels on both sides
- Cluster Lid C
- Data link connector
- Hood lock cable
- Steering column lower cover
- 4 bolts and combination meter

5. Position the steering wheel in the straight-ahead position.

6. Turn the ignition switch OFF.

7. Disconnect the negative (–) battery cable; then, the positive (+) battery cable.

➡ **Wait for a least 3 minutes after disconnecting the battery cables for the charge in the air bag circuit to dissipate before working on the air bag module(s).**

8. Remove the driver's side SRS and steering wheel by performing the following procedure:
- Remove both the left and right side lids from the steering wheel.
- Using a tamper resistant Torx® wrench (T50), remove the special bolts from both sides of the steering wheel.
- Steering wheel switch sub-harness connector
- Air bag harness connector
- Carefully remove the air bag module and store it face up.

9. Remove or disconnect the following:
- Steering wheel
- Steering column upper cover
- Spiral cable connector
- Combination switch
- Automatic transmission console finisher panel, if equipped
- Manual transmission shift knob, if equipped
- Center console
- Cup holder
- Passenger side lower instrument panel cover
- Instrument panel side cover
- Navigation switch cover panel and switch connector
- Audio cluster lid
- Audio unit and meter assembly
- Display unit
- Garnish panels and side finishers
- Passenger air bag connector
- Passenger air bag bolt and passenger air bag
- ECM and bracket
- Intake door motor and blower motor connectors
- 2 screws, 1 bolt and the blower unit
- Left and right instrument panel stays
- Defroster and ventilation ducts
- Heating–A/C unit
- Heater pipe cover, support and grommet
- Slide the heater core out of the heating–A/C unit

To install:
10. Install or connect the following:
- Heater core to the heater–A/C unit
- Heater pipe cover, support and grommet
- Heater–A/C unit and tighten the bolts to 61 inch lbs. (7 Nm)
- Defroster and ventilation ducts

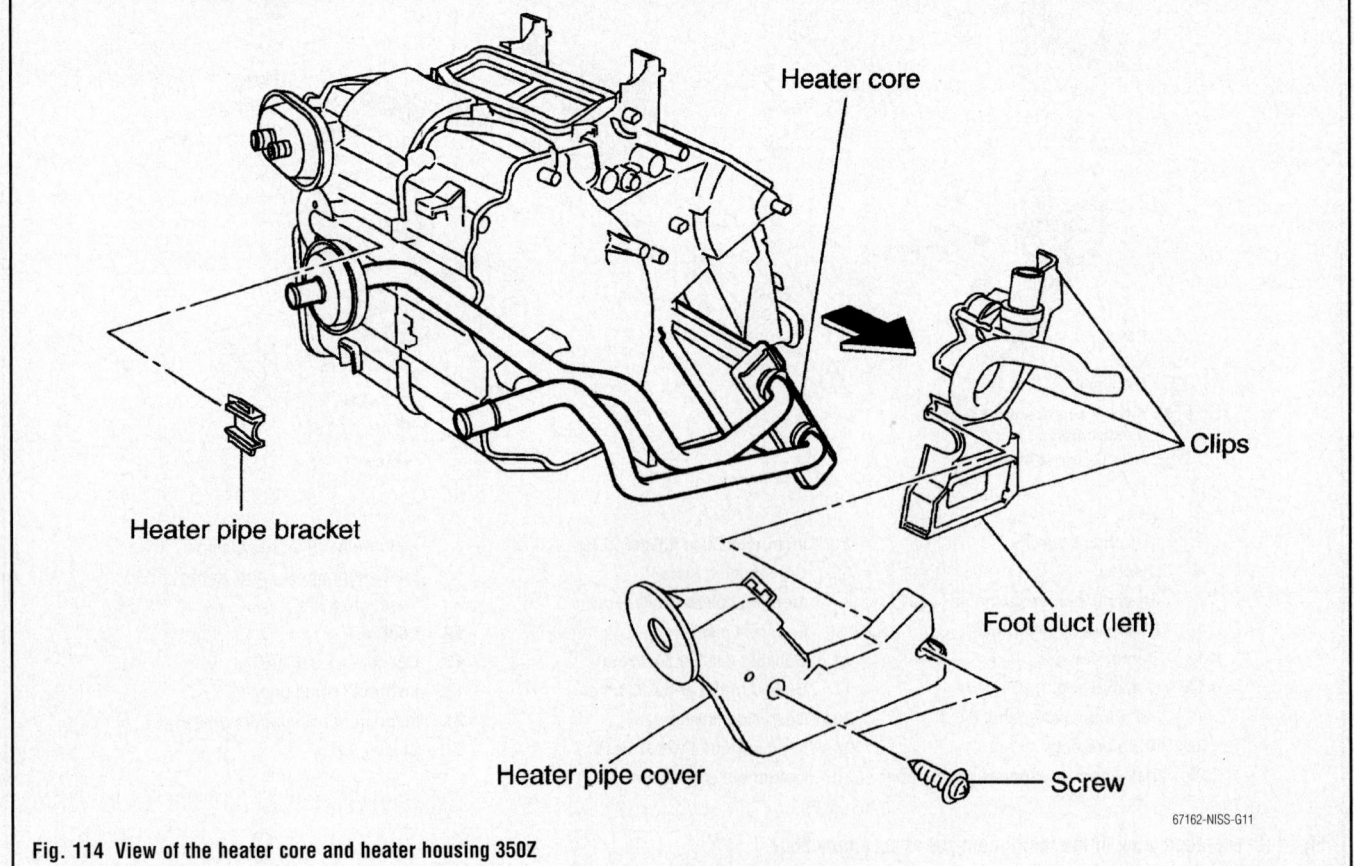

67162-NISS-G11

Fig. 114 View of the heater core and heater housing 350Z

- Left and right instrument panel stays
- 2 screws, 1 bolt and the blower unit
- Intake door motor and blower motor connectors
- ECM and bracket
- Passenger air bag and tighten the bolt to 15–21 ft. lbs. (20–29 Nm)
- Passenger air bag connector
- Garnish panels and side finishers
- Display unit
- Audio unit and meter assembly
- Audio cluster lid
- Navigation switch cover panel and switch connector
- Instrument panel side cover
- Passenger side lower instrument panel cover
- Cup holder
- Center console
- Manual transmission shift knob, if equipped

- Automatic transmission console finisher panel, if equipped
- Combination switch
- Spiral cable connector
- Steering column upper cover
- Steering wheel and tighten the bolt to 22–28 ft. lbs. (30–39 Nm)
- Driver air bag module and tighten the bolts to 83 inch lbs. (9.4 Nm)
- Air bag harness connector
- Steering wheel switch sub-harness connector
- Both the left and right side lids to steering wheel
- 4 bolts and combination meter
- Steering column lower cover
- Hood lock cable
- Data link connector
- Cluster lid C
- Instrument panel side finish panels on both sides

- Passenger side lower instrument panel cover
- Foot rests
- Kick panels on both sides
- Heater hoses
- Electronic throttle control assembly
- Evaporator lines to the firewall using new O-rings
- Cowl top cover and washer hose
- Cowl rubber seal
- Both wiper arms
- Hood ledge cover

11. Refill the cooling system.

12. Connect the positive (+) battery cable; then, the negative (−) battery cable.

13. Evacuate, charge and leak test the air conditioning system refrigerant.

14. Operate the engine to normal operating temperatures; then, check the climate control operation and check for leaks.

STEERING

POWER STEERING GEAR

REMOVAL & INSTALLATION

Maxima

➡Spiral cable may snap due to steering operation if steering column is separated from steering gear assembly. Therefore fix steering wheel with a string to avoid turns.

1. Remove the two front tires with power tool.

2. Remove cotter pin and nut. Discard cotter pin.

3. Disconnect the outer tie-rod ends using Tool.

4. Disconnect the outer stabilizer bar ends with power tool.

5. Remove the stabilizer bracket rear bolts and loosen the front bolts with power tool.

6. Remove the lower joint pinch bolt.

7. Disconnect the power steering high and low pressure lines from the power steering gear.

8. Reposition the stabilizer bar up and out of the way.

9. Remove the two gear housing mounting bolts.

➡**Do not remove the gear housing mounting bracket from the gear housing.**

10. Remove the power steering gear and linkage assembly.

To install:

11. Installation is in the reverse order of removal.

a. Tighten mounting bolts to 90 ft. lbs. (122.5 Nm)

350Z

➡Spiral cable may snap due to steering operation if steering column is separated from steering gear assembly. Therefore fix steering wheel with a string to avoid turns.

1. Set wheels in the straight-ahead position.

2. Remove undercover and tires from vehicle with power tool.

3. Remove front crossbar.

4. Confirm slit of lower joints fits with the projection on rear cover cap, furthermore marking position on steering gear assembly nearly fits with the projection on rear cover cap.

5. Remove cotter pin at steering outer socket, then loosen mounting nut.

6. Use a ball joint remover (SST) to remove steering outer socket from steering knuckle.

7. Remove oil pipes (high pressure side and low pressure side) from steering gear assembly, then drain fluid from pipes.

8. Loosen bolt on upper yoke of lower joint and remove bolt on lower yoke of joint, then slide lower joint into lower shaft. Separate steering gear assembly from lower shaft.

9. Tack bolt on upper yoke of lower joint, fix lower joint to lower shaft.

10. Remove the fixing bolt and remove steering gear assembly, rack mounting bracket and insulator from vehicle.

To install:

11. Installation is reverse of removal.

a. Tighten mounting bracket fixing bolts to 51 ft. lbs. (69 Nm)

b. Tighten mounting bolts to 96 ft. lbs. (130 Nm)

c. Tighten yolk shaft bolts to 18 to 21 ft. lbs. (23.5 to 29.4 Nm)

POWER STEERING PUMP

REMOVAL & INSTALLATION

Maxima

1. Loosen adjustment screw and oil pump through bolt, then remove belt.

2. Remove oil pump union bolts and hose.

3. Remove oil pump bracket bolts.

4. Remove oil pump.

To install:

5. Installation is reverse of removal.

350Z

1. Remove engine cover.

2. Remove air cleaner box.

3. Drain water from radiator upper tank, then remove radiator upper hose.

4. Remove radiator fan shroud.

5. Loosen idler pulley, then remove belt.

6. Drain power steering fluid from reservoir tank.

7. Remove piping of high pressure and low pressure (drain fluid)

8. Remove bolt common to water pump and power steering pump.

9. Remove bolt then remove power steering pump.

To install:

10. Installation is reverse of removal.

BLEEDING

1. Stop engine, and then turn steering wheel fully to right and left several times.

2. Do not allow steering fluid reservoir tank to go below the low-level line. Check tank frequenter and add fluid as needed.

3. Run engine at idle speed. Turn steering wheel fully to the right and then fully to the left, and keep for about 3 seconds. Then check whether a fluid leakage has occurred.

4. Repeat the 2nd procedure several times at about three seconds interval Check generation of air bubbles and cloud in fluid.

5. If air bubbles and the cloud don't fade, stop engine, hold air bleeding until air bubbles and the cloud fade. Perform the 2nd and the 3rd procedures again.

6. Stop engine, check fluid level.

SUSPENSION

FRONT SUSPENSION

COIL SPRING

REMOVAL & INSTALLATION

Maxima

See Figure 115.

1. Before servicing the vehicle, refer to the Precautions Section.
2. Raise and safely support the vehicle.
3. Remove or disconnect the following:
   - Wheel
4. Matchmark the position of the strut-to-steering knuckle location.
   - Brake hose from the strut
   - Anti-Lock Brake (ABS) wheel sensor and move it out of the way
   - Bolts attaching the steering knuckle to the strut. Matchmark the bolts before removal.
5. Open the hood and remove the strut attaching nuts while holding the strut.

❋❋ CAUTION

Do not remove the center locknut from the strut assembly until the strut is safely compressed.

   - Strut from the vehicle
6. Place the strut assembly in a vise with a holding tool or in a spring compressor.
7. Loosen the piston rod locknut.

❋❋ CAUTION

Do not remove the piston rod locknut, the spring is under tension and can cause serious personal injury.

8. Compress the spring with the spring compressor, then remove the piston rod locknut.

➡ **Before removing the strut from the coil spring, note the positioning of the strut in relationship to the coil spring for reassembly.**

9. Remove or disconnect the following:
   - Strut mounting insulator bracket, strut mounting bearing, upper

spring seat and the upper spring rubber seat
   - Strut, leaving the coil spring compressed
   - Piston boot and rebound bumper from the strut

To install:

10. Install or connect the following:
   - Rebound bumper and the boot to the strut piston
   - Strut into the coil spring, be sure the strut and spring are properly positioned with matchmarks
   - Upper spring rubber seat, upper spring seat, strut mounting bearing and the strut mounting insulator bracket. Be sure that the cutout on the upper spring seat is facing outside of the vehicle.
   - Piston rod locknut. Remove the tool and torque the piston rod locknut to 44–65 ft. lbs. (59–88 Nm).
   - Strut into the strut tower
   - New attaching nuts and torque to 32–38 ft. lbs. (43–51 Nm)
   - Bolts attaching the steering knuckle to the strut and align the matchmarks and torque to 103–117 ft. lbs. (140–159 Nm)
   - ABS wheel sensor and torque to 13–17 ft. lbs. (18–24 Nm)
   - Brake hose to the strut
   - Front wheels
11. Lower the vehicle.
12. Check and/or adjust the wheel alignment as necessary.

350Z

See Figure 116.

1. Before servicing the vehicle, refer to the Precautions Section.
2. Raise and safely support the vehicle.
3. Remove or disconnect the following:
   - Wheel
   - Engine undercover
   - Anti-Lock Brake (ABS) wheel sensor and move it out of the way
   - Brake hose from the strut
   - Strut-to-transverse link nut and bolt

4. Open the hood and remove the strut attaching nuts while holding the strut.

❋❋ CAUTION

Do not remove the center locknut from the strut assembly until the strut is safely compressed.

   - Strut from the vehicle
5. Place the strut assembly in a vise with a holding tool or in a spring compressor.
6. Loosen the piston rod locknut.

❋❋ CAUTION

Do not remove the piston rod locknut, the spring is under tension and can cause serious personal injury.

7. Compress the spring with the spring compressor, then remove the piston rod locknut

➡ **Before removing the strut from the coil spring, note the positioning of the strut in relationship to the coil spring for reassembly.**

8. Remove or disconnect the following:
   - Strut mounting insulator bracket, strut mounting bearing, upper spring seat and the upper spring rubber seat
   - Strut, leaving the coil spring compressed
   - Piston boot and rebound bumper from the strut

To install:

9. Install or connect the following:
   - Rebound bumper and the boot to the strut piston
   - Strut into the coil spring, be sure the strut and spring are properly positioned
   - Upper spring rubber seat, upper spring seat, strut mounting bearing and the strut mounting insulator bracket. Be sure that the cutout on the upper spring seat is facing the outside of the vehicle.

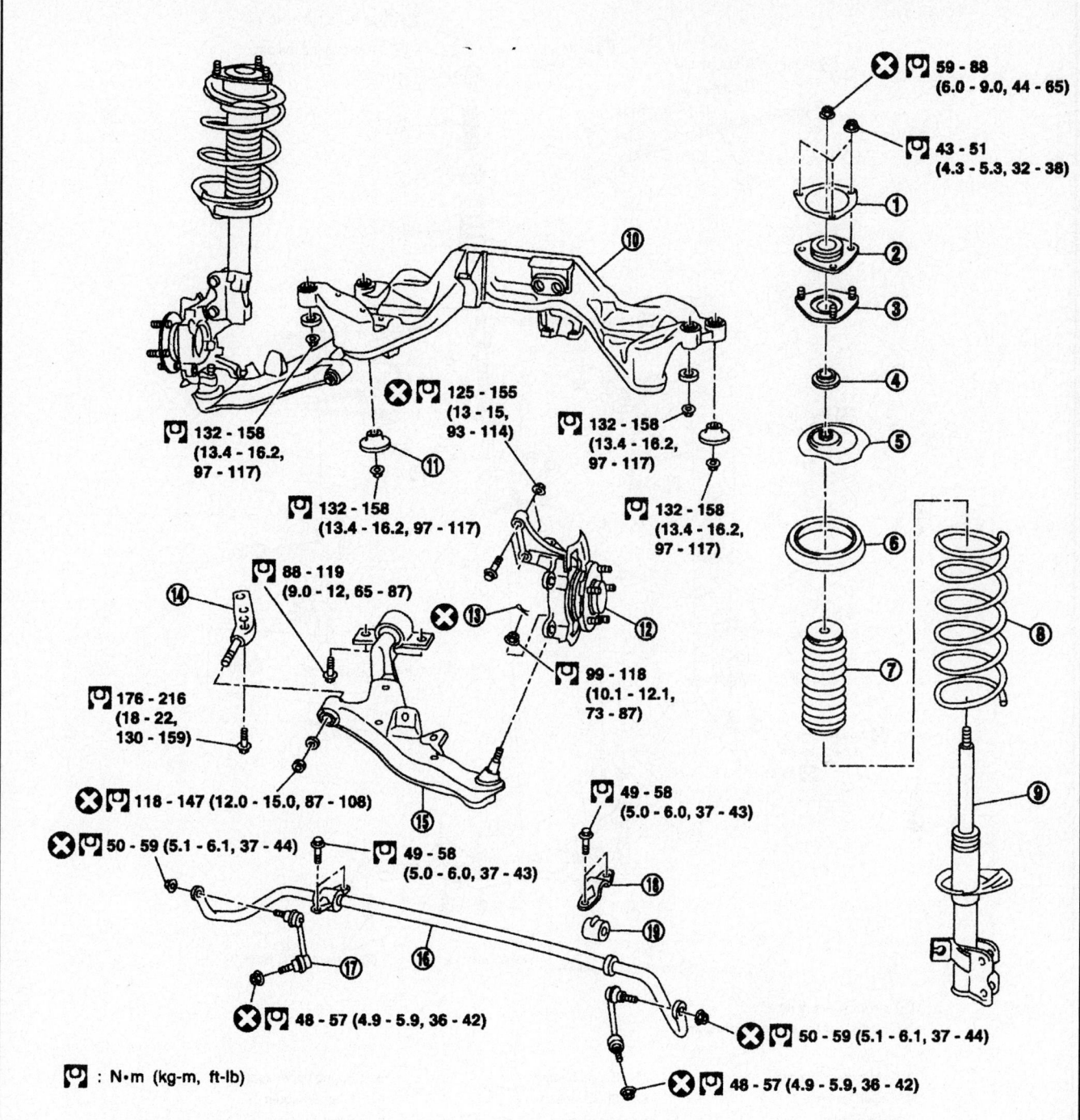

59 - 88
(6.0 - 9.0, 44 - 65)

43 - 51
(4.3 - 5.3, 32 - 38)

125 - 155
(13 - 15,
93 - 114)

132 - 158
(13.4 - 16.2,
97 - 117)

132 - 158
(13.4 - 16.2,
97 - 117)

132 - 158
(13.4 - 16.2, 97 - 117)

132 - 158
(13.4 - 16.2,
97 - 117)

88 - 119
(9.0 - 12, 65 - 87)

176 - 216
(18 - 22,
130 - 159)

99 - 118
(10.1 - 12.1,
73 - 87)

118 - 147 (12.0 - 15.0, 87 - 108)

49 - 58
(5.0 - 6.0, 37 - 43)

50 - 59 (5.1 - 6.1, 37 - 44)

49 - 58
(5.0 - 6.0, 37 - 43)

48 - 57 (4.9 - 5.9, 36 - 42)

50 - 59 (5.1 - 6.1, 37 - 44)

48 - 57 (4.9 - 5.9, 36 - 42)

: N•m (kg-m, ft-lb)

1. Strut spacer
2. Strut mount insulator
3. Strut mount bracket
4. Strut mount bearing
5. Spring upper seat
6. Spring rubber seat
7. Bound bumper rubber

8. Coil spring
9. Shock absorber
10. Suspension member
11. Rebound stopper
12. Wheel hub and steering knuckle
13. Cotter pin

14. Bush link pin
15. Transverse link
16. Stabilizer
17. Connecting rod
18. Stabilizer clamp
19. Bushing

9357RG17

Fig. 115 Exploded view of the front suspension Maxima shown

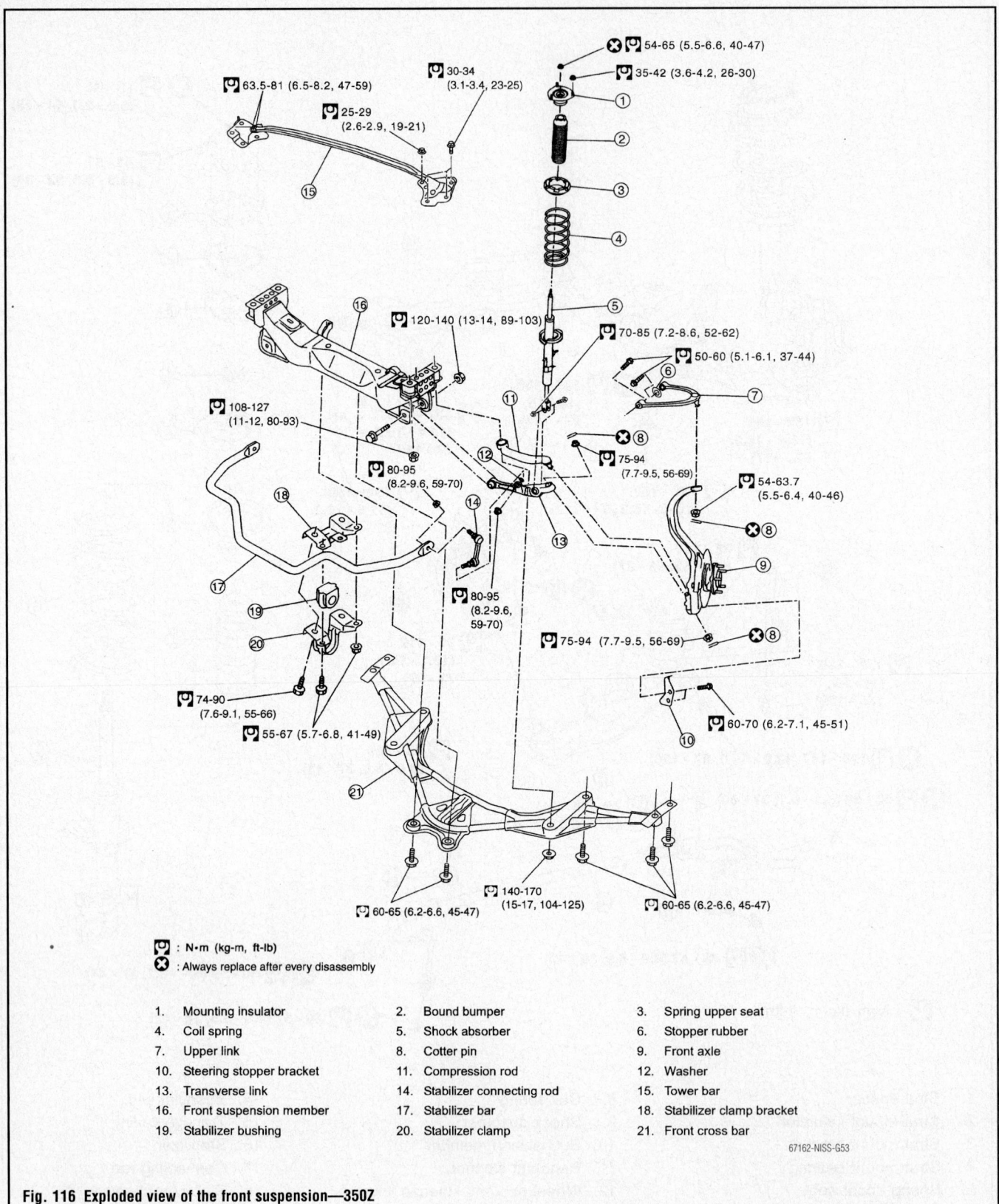

63.5-81 (6.5-8.2, 47-59)
25-29 (2.6-2.9, 19-21)
30-34 (3.1-3.4, 23-25)
54-65 (5.5-6.6, 40-47)
35-42 (3.6-4.2, 26-30)
120-140 (13-14, 89-103)
70-85 (7.2-8.6, 52-62)
50-60 (5.1-6.1, 37-44)
108-127 (11-12, 80-93)
75-94 (7.7-9.5, 56-69)
80-95 (8.2-9.6, 59-70)
54-63.7 (5.5-6.4, 40-46)
80-95 (8.2-9.6, 59-70)
75-94 (7.7-9.5, 56-69)
74-90 (7.6-9.1, 55-66)
55-67 (5.7-6.8, 41-49)
60-70 (6.2-7.1, 45-51)
60-65 (6.2-6.6, 45-47)
140-170 (15-17, 104-125)
60-65 (6.2-6.6, 45-47)

: N•m (kg-m, ft-lb)

: Always replace after every disassembly

1. Mounting insulator	2. Bound bumper	3. Spring upper seat
4. Coil spring	5. Shock absorber	6. Stopper rubber
7. Upper link	8. Cotter pin	9. Front axle
10. Steering stopper bracket	11. Compression rod	12. Washer
13. Transverse link	14. Stabilizer connecting rod	15. Tower bar
16. Front suspension member	17. Stabilizer bar	18. Stabilizer clamp bracket
19. Stabilizer bushing	20. Stabilizer clamp	21. Front cross bar

67162-NISS-G53

Fig. 116 Exploded view of the front suspension—350Z

- Piston rod locknut. Remove the tool and torque the piston rod locknut to 40–47 ft. lbs. (54–65 Nm).
- Strut into the strut tower

- New attaching nuts and torque to 26–30 ft. lbs. (35–42 Nm).
- Bolts attaching the strut to the transverse link and torque to 52–62 ft. lbs. (70–80 Nm)

- ABS wheel sensor
- Brake hose to the strut
- Engine undercover
- Front wheels
10. Lower the vehicle.

11. Check and/or adjust the wheel alignment as necessary.

CONTROL LINKS

REMOVAL & INSTALLATION

1. Remove the wheel and tire.
2. Disconnect the connecting rod end at the stabilizer bar using power tool.

➡**Prevent the stabilizer connecting rod from turning by inserting a hex wrench into the end of the ball stud, then remove nut.**

To install:

3. Installation is in the reverse order of removal.
   a. Tighten nuts to 66 ft. lbs. (89 Nm)

LOWER BALL JOINT

REMOVAL & INSTALLATION

The ball joint is an integral part of the lower control arm. If the ball joint is defective the control arm must be replaced.

LOWER CONTROL ARM

REMOVAL & INSTALLATION

See Traverse Link

STABILIZER BAR

REMOVAL & INSTALLATION

Maxima

1. Before servicing the vehicle, refer to the precautions in the beginning of this section.
2. Raise and safely support the vehicle.
3. Remove or disconnect the following:
   - Undercover
   - Front wheels
   - Front exhaust tube

➡**Prevent the stabilizer connecting rod from turning by inserting a hex wrench into the end of the ball stud, then remove nut.**

   - Connecting rod end at the stabilizer bar using power tool
   - Rear engine mount insulator using power tool
   - Left-hand and right-hand member pin stays using power tool, then remove the rear suspension member mounting nuts

➡**Support rear of suspension member, then lower rear of suspension member**

to gain access to stabilizer bar mounting brackets.

   - Remove the two stabilizer bar brackets from the front suspension member using power tool.
   - Front stabilizer bar and bushings as necessary

To install:

4. To install, reverse removal procedure.
   a. Tighten bracket bolts to 37 ft. lbs. (50 Nm).
5. Tighten rod end nuts to 66 ft. lbs. (89 nm).

350Z

See Figure 117.

1. Before servicing the vehicle, refer to the precautions in the beginning of this section.
2. Raise and safely support the vehicle.
3. Remove or disconnect the following:
   - Undercover
   - Mounting nut on upper portion of stabilizer connecting rod with power tool
   - Fixing bolts and nuts, then remove stabilizer clamp, stabilizer bushing, and stabilizer clamp bracket
   - Stabilizer bar from vehicle

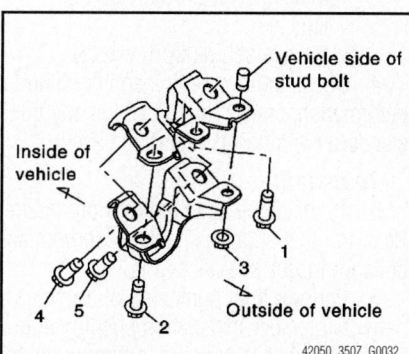

Fig. 117 Stabilizer bracket and stabilizer clamp tightening sequence

To install:

4. To install, reverse removal procedure.
5. When installing stabilizer bracket and stabilizer clamp, refer to illustration and perform the following:
   - (1) fully tighten
   - (2) temporarily tighten
   - (3) temporarily tighten
   - (2) fully tighten
   - (3) fully tighten
   - (4), (5) temporarily tighten then fully tighten

STEERING KNUCKLE

REMOVAL & INSTALLATION

To remove and install steering knuckle, refer to wheel bearings

TOWER BAR

REMOVAL & INSTALLATION

350Z

See Figure 118.

1. Fix center bolt, and then loosen nut in the right and left side.
2. Loosen center bolt 660 degrees (Or turn bolt 1.7 times) to place the black mark of center bolt above.
3. Remove tower bar fixing bolts and nuts, and remove tower bar from vehicle with power tool.

To install:

4. Install tower bar

➡ **If it is hard to install tower bar, install it turning center bolt.**

5. Tighten center bolt 660 degrees, (Or turn bolt 1.7 times) to place the black mark of center bolt above.

➡ **The space between tower bar and engine collector should be**

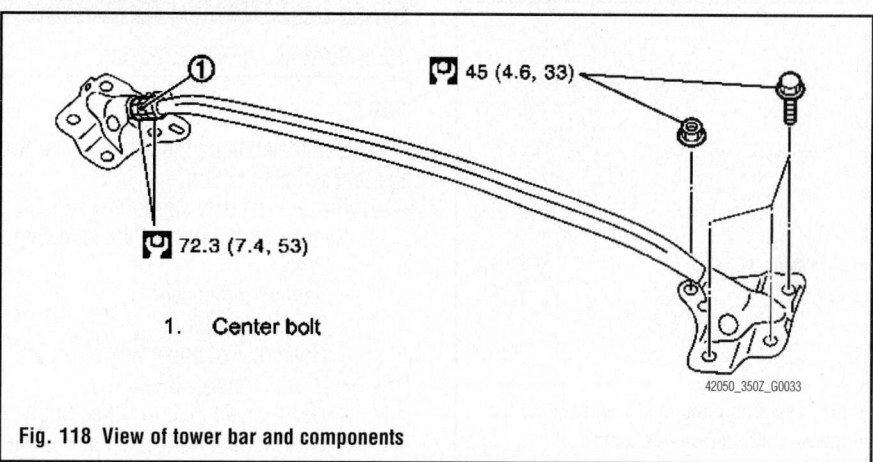

1. Center bolt

Fig. 118 View of tower bar and components

between 0.394 in (10.0 mm) and 0.669 in (17.0 mm).

6. Tighten both nut of the right and left side of center bolt.

TRANSVERSE LINK

REMOVAL & INSTALLATION

Maxima

See Figure 119.

1. Before servicing the vehicle, refer to the precautions in the beginning of this section.
2. Raise and safely support the vehicle.
3. Remove or disconnect the following:
   - Front wheels
   - Steering knuckle and the lower ball joint
   - Stabilizer bar from the lower control arm
   - Bolts attaching the link bushing pin to the chassis
   - Nut attaching the link to the control arm and the link, if necessary
   - Bolts attaching the compression rod bushing clamp
   - Lower control arm/traverse link

To install:

4. Install or connect the following:
   - Lower control arm and the compression rod bushing clamp into the vehicle
   - Link bushing pin, if removed from the control arm
5. Tighten all bolts and nuts until they are snug enough to support the weight of the vehicle but not fully tight, the bolts should be torqued to specification with the vehicle on the floor.

➡Always use a new nut when installing the ball joint to the control arm.

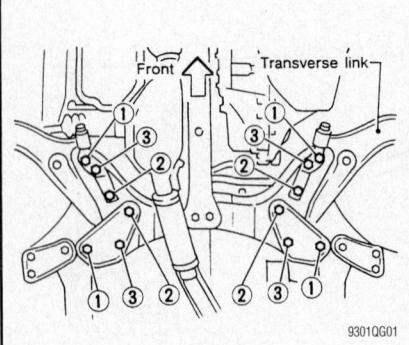

Fig. 119 Bolt tightening sequence for the lower control arms—Maxima

---

- ABS wheel sensor and torque the bolt to 13–17 ft. lbs. (18–24 Nm)
- Front wheels
6. Torque the bolts attaching the compression rod bushing clamp and the link bushing pin, in the proper sequence to 87–108 ft. lbs. (118–147 Nm).
7. If the link bushing pin was removed from the control arm torque the attaching nut to 87–108 ft. lbs. (118–147 Nm).
8. Torque the sway bar attaching nut to 30–35 ft. lbs. (41–47 Nm).
9. Check the vehicle alignment.

350Z

1. Before servicing the vehicle, refer to the precautions in the beginning of this section.
2. Raise and safely support the vehicle.
3. Remove or disconnect the following:
   - Front wheels
   - mounting nut and washer on lower portion of stabilizer connecting rod with power tool
   - mounting nut between transverse link and shock absorber on lower position
   - mounting nut between transverse link and front suspension member with power tool
   - transverse link from steering knuckle
   - transverse link from vehicle
4. Check transverse link and bushing for deformation, cracks, or damage. If any non-standard condition is found, replace it.

To install:

5. To install, reverse removal procedure. Refer to illustration in shocks absorbers and coils for torque specifications.
6. Perform final tightening of front suspension member installation position and shock absorber lower side (rubber bushing) under unladen condition with tires on ground. Check wheel alignment.

UPPER BALL JOINT

REMOVAL & INSTALLATION

350Z

1. Before servicing the vehicle, refer to the Precautions Section.
2. Raise and safely support the vehicle.
3. Remove or disconnect the following:
   - Wheel
   - Engine undercover
   - Shock absorber
   - Cotter pin of upper link ball joint, loosen mounting nut
   - Using a ball joint remover, upper steering link from steering knuckle

---

✴✴ CAUTION

Tighten temporarily mounting nut to prevent damage to threads and to prevent ball joint remover from coming off.

   - Bolts holding upper link to body
   - Upper link

To install:

4. To install, reverse removal procedure. Refer to illustration in shock absorbers and coil spring for torque specifications.

UPPER CONTROL ARM

REMOVAL & INSTALLATION

See upper Ball Joint

WHEEL BEARINGS

REMOVAL & INSTALLATION

Maxima

See Figures 120 and 121.

➡Whenever the hub or bearing assembly is removed, the wheel bearing assembly must be replaced. Never reuse the old bearing assembly.

1. Before servicing the vehicle, refer to the precautions in the beginning of this section.
2. Remove or disconnect the following:
   - Front axle shaft nut and shaft
   - Lower ball joint from knuckle
   - Strut bolts from knuckle
   - Steering link from knuckle
   - Hub with the inner race from the steering knuckle, using a shop press and a suitable tool
   - Bearing inner race from the hub, using a shop press and a suitable tool
   - Outer grease seal
   - Inner grease seal from the steering knuckle, using a prybar

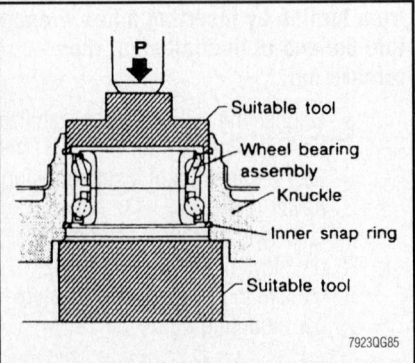

Fig. 120 Typical method of installing the wheel bearing

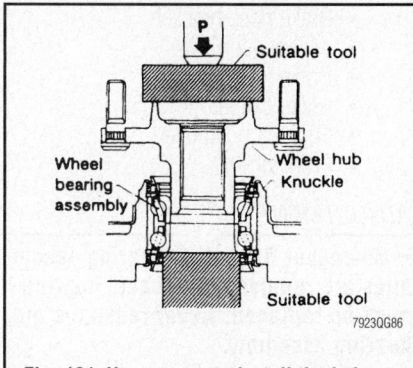

Fig. 121 Use a press to install the hub into the knuckle assembly

- Inner and outer snaprings from the steering knuckle, using snapring pliers
- Sealed bearing assembly from the steering knuckle, using a shop press and a suitable tool

3. Inspect the hub, steering knuckle and snaprings for cracks and/or wear; if necessary, replace the damaged part(s).

To install:

4. Install or connect the following:
- Inner snapring in the steering knuckle groove
- New wheel bearing assembly into the steering knuckle, using a shop press and a suitable tool, until it seats, using a maximum pressure of 3 tons (2722 kg)
- Outer snapring

5. Pack the new grease seal lips with multi-purpose grease.
- New outer grease seal into the steering knuckle, using a shop press and a suitable tool
- Hub into the steering knuckle, using a shop press and a suitable tool, until it seats, using a maximum pressure of 5.5 tons (4990 kg); be careful not to damage the grease seal

6. To check the bearing operation, perform the following procedures:
   a. Increase the press pressure to 3.5–5.0 tons (3175–4536 kg).
   b. Spin the steering knuckle, several turns, in both directions.

   c. Be sure the wheel bearings operate smoothly.
7. If the wheel bearings do not operate smoothly, replace the wheel bearing assembly.
8. Install the knuckle assembly.
9. Install the halfshaft into the hub. Torque the locknut to 174–231 ft. lbs. (235–314 Nm).
10. Install the wheel assembly and lower the vehicle.
11. Road test the vehicle and verify proper operation.

350Z

See Figure 122.

➡ **If the wheel bearing is damaged, the steering knuckle and bearing must be replaced as an assembly.**

1. Before servicing the vehicle, refer to the Precautions Section.
2. Remove or disconnect the following:
- Front wheel
- Engine undercover
- Brake caliper and wire aside

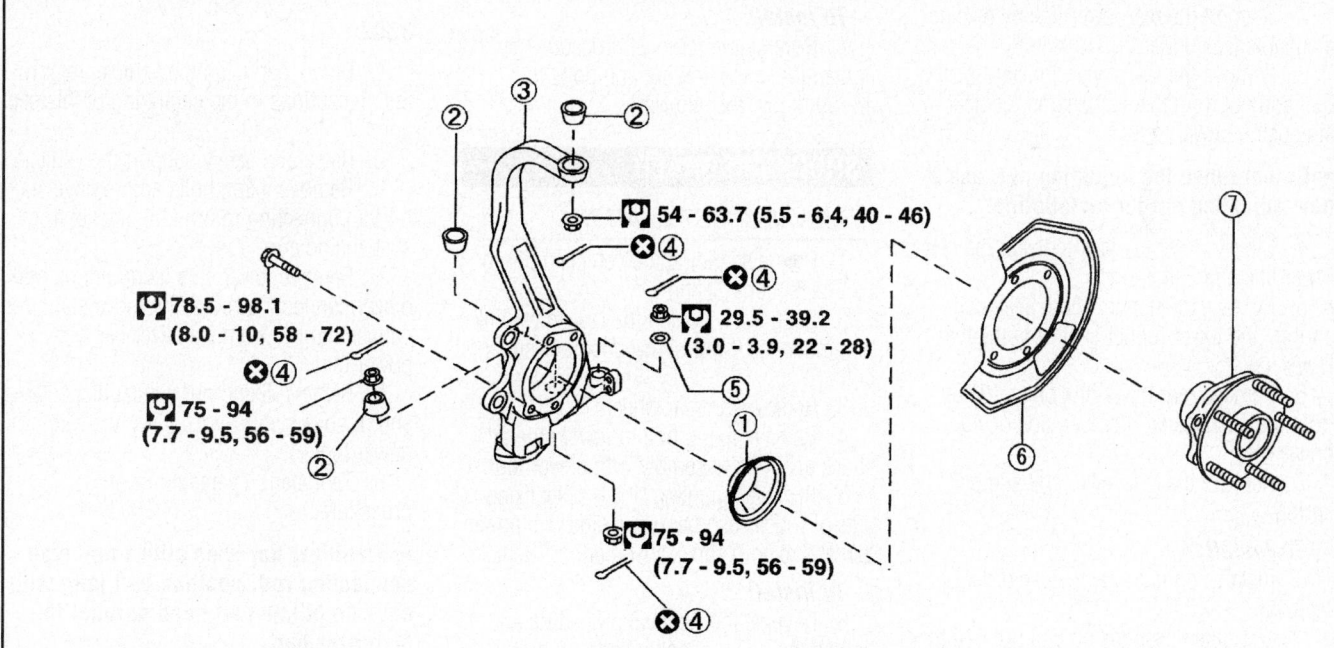

1. Hub cap
2. Ball seat
3. Steering knuckle
4. Cotter pin
5. Washer
6. Splash guard
7. Wheel hub and bearing assembly

Fig. 122 Exploded view of the front steering knuckle and wheel bearing assembly—350Z

- Brake rotor
- ABS sensor
- Brake hose bracket
- Loosen steering outer socket nut
- Separate outer socket from steering knuckle
- Upper link from knuckle
- Transverse link from knuckle
- Compression rod from knuckle
- Loosen steering knuckle nut
- Knuckle and hub assembly from the vehicle
- Separate the wheel hub from the knuckle

To install:

3. Install the wheel hub to the knuckle.

4. Install or connect the following:
- Knuckle and hub assembly
- Tighten the steering knuckle/hub nut to 58–72 ft. lbs. (79–98 Nm)
- Compression rod to knuckle and tighten the nut to 56–69 ft. lbs. (75–94 Nm)
- Transverse link to knuckle and tighten the nut to 56–69 ft. lbs. (75–94 Nm)
- Upper link to knuckle and tighten the nut to 40–46 ft. lbs. (54–64 Nm)
- Install outer socket to steering knuckle
- Steering outer socket nut

- Brake hose bracket
- ABS sensor
- Brake rotor
- Brake caliper
- Engine undercover
- Front wheel

ADJUSTMENT

➡ **Whenever the hub or bearing assemblies are removed, the wheel bearing must be replaced. Never reuse the old bearing assembly.**

The wheel bearings are sealed and are not adjustable. If defective, replacement is the only option.

SUSPENSION

COIL SPRING

REMOVAL & INSTALLATION

Maxima

1. Loosen the rear lower link bolt and nut from the suspension member side.
2. Support the rear lower link by placing a suitable jack under the knuckle.
3. Remove the rear lower link adjusting bolt and nut from the suspension member side using power tool.

➡ **Do not reuse the adjusting nut, use a new adjusting nut for installation.**

4. Slowly lower the jack to lower the rear lower link and coil spring.
5. Remove the upper rubber seat, coil spring, and lower rubber seat from the rear lower link.
6. Remove rear lower link bolt and nut from the suspension member side using power tool.
7. Remove the rear lower link and coil spring.

To install:

8. Installation is in the reverse order of removal.
   a. Tighten suspension end nut/bolt to 73 ft. lbs. (100 Nm)
9. Tighten knuckle end bolt/nut to 78 ft. lbs. (107 Nm)

350Z

See Figure 123.

1. Before servicing the vehicle, refer to the Precautions Section.
2. Place a jack under the rear lower link.
3. Remove the rear wheel.
4. Loosen the lower link nut and bolt on the suspension member side, then remove the bolt and nut on the axle side.
5. Lower the jack slowly and remove the upper seat, coil spring and rubber sheet from the lower link.
6. Remove the lower link nut and bolt on the axle side to remove the lower link.

To install:

7. Reverse the removal procedure and tighten the lower link nut and bolts to 48–59 ft. lbs. (65–80 Nm).

SHOCK ABSORBER

REMOVAL & INSTALLATION

1. Before servicing the vehicle, refer to the Precautions Section.
2. Set a floor jack on the rear lower link to remove the lower shock absorber nut and bolt.
3. Remove the rear wheel.
4. Remove fixing bolt in lower side of shock absorber assembly with power tool.
5. Remove mounting seal bracket fixing nuts of shock absorber upper side with power tool and remove shock absorber from vehicle.

To install:

6. Reverse the removal procedure and tighten the upper shock nuts to 20 ft. lbs. (24 Nm) for the Maxima; 21 ft. lbs. (28 Nm) for the 350Z, and the lower bolt to 74–88 ft. lbs. (100–120 Nm).

STABILIZER BAR

REMOVAL & INSTALLATION

Maxima

1. Before servicing the vehicle, refer to the precautions in the beginning of this section.

REAR SUSPENSION

2. Raise and safely support the vehicle.
3. Disconnect the stabilizer bar ends from the connecting rods using power tool.
4. Remove the stabilizer bar clamps and bushings using power tool.
5. Remove the stabilizer bar.
6. To install, reverse removal procedure.

350Z

1. Before servicing the vehicle, refer to the precautions in the beginning of this section.
2. Raise and safely support the vehicle.
3. Remove fixing bolts and remove stabilizer connecting rod mount bracket from suspension arm.
4. Remove lower side fixing nut on stabilizer connecting rod and remove stabilizer connecting rod from stabilizer bar with power tool.
5. Remove fixing nut on stabilizer clamp and remove stabilizer from vehicle with power tool.
6. To install, reverse removal procedure.

➡ **Stabilizer bar uses pillow ball type connecting rod, position ball joint with case on pillow ball head parallel to stabilizer bar.**

UPPER CONTROL ARM

REMOVAL & INSTALLATION

Maxima

1. Remove the rear suspension assembly.
2. Remove the connecting rod mounting bracket from suspension arm using power tools.

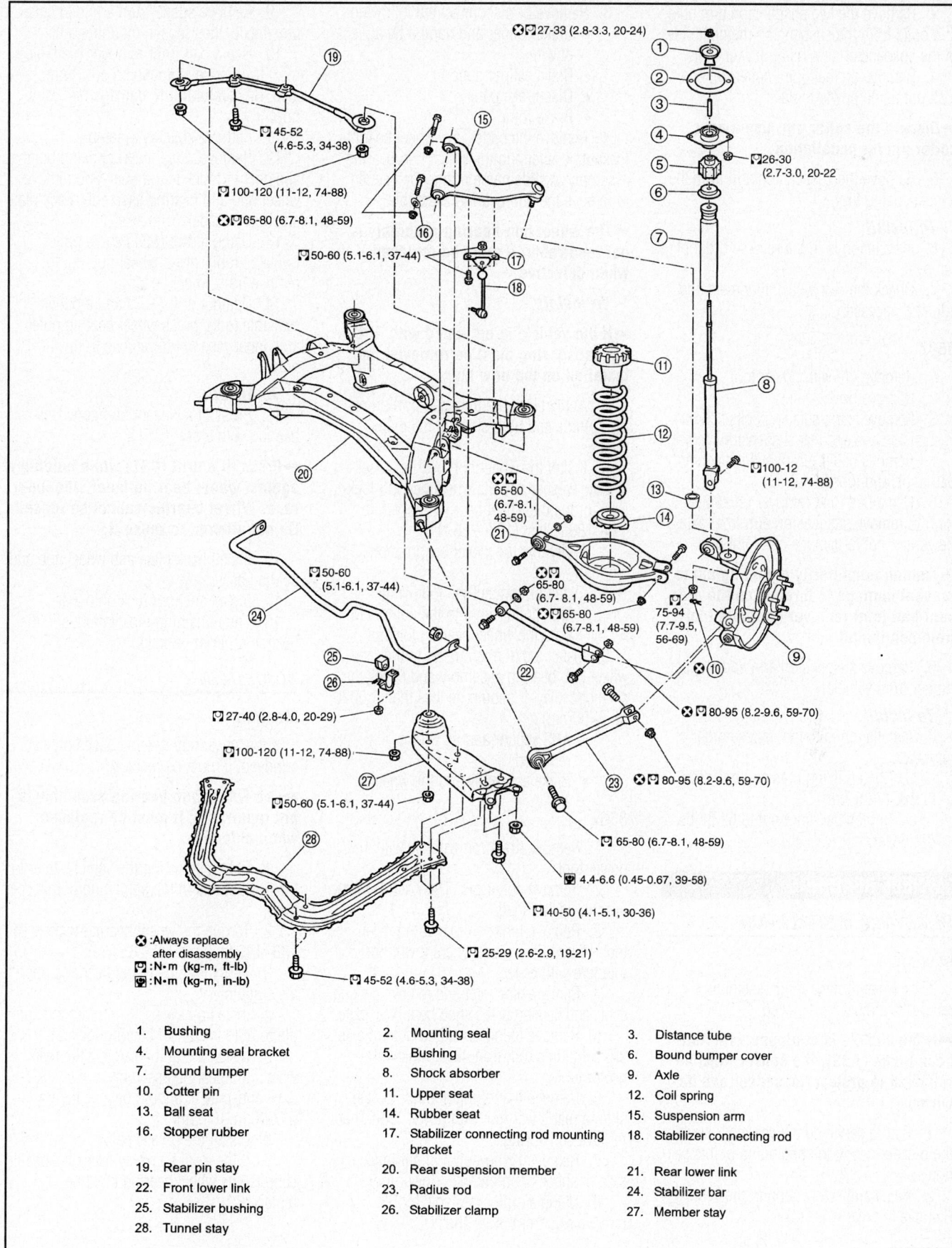

27-33 (2.8-3.3, 20-24)
45-52 (4.6-5.3, 34-38)
100-120 (11-12, 74-88)
65-80 (6.7-8.1, 48-59)
50-60 (5.1-6.1, 37-44)
26-30 (2.7-3.0, 20-22
100-12 (11-12, 74-88)
65-80 (6.7-8.1, 48-59)
50-60 (5.1-6.1, 37-44)
65-80 (6.7-8.1, 48-59)
65-80 (6.7-8.1, 48-59)
75-94 (7.7-9.5, 56-69)
80-95 (8.2-9.6, 59-70)
80-95 (8.2-9.6, 59-70)
27-40 (2.8-4.0, 20-29)
100-120 (11-12, 74-88)
65-80 (6.7-8.1, 48-59)
50-60 (5.1-6.1, 37-44)
4.4-6.6 (0.45-0.67, 39-58)
40-50 (4.1-5.1, 30-36)
25-29 (2.6-2.9, 19-21)
45-52 (4.6-5.3, 34-38)

:Always replace after disassembly
:N•m (kg-m, ft-lb)
:N•m (kg-m, in-lb)

1.	Bushing	2.	Mounting seal	3.	Distance tube
4.	Mounting seal bracket	5.	Bushing	6.	Bound bumper cover
7.	Bound bumper	8.	Shock absorber	9.	Axle
10.	Cotter pin	11.	Upper seat	12.	Coil spring
13.	Ball seat	14.	Rubber seat	15.	Suspension arm
16.	Stopper rubber	17.	Stabilizer connecting rod mounting bracket	18.	Stabilizer connecting rod
19.	Rear pin stay	20.	Rear suspension member	21.	Rear lower link
22.	Front lower link	23.	Radius rod	24.	Stabilizer bar
25.	Stabilizer bushing	26.	Stabilizer clamp	27.	Member stay
28.	Tunnel stay				

Fig. 123 Exploded view of the rear suspension—350Z

67162-NISS-G54

3. Remove the two suspension arm nuts and bolts from the suspension member side of the suspension arm using power tools.

4. Remove the ball joint cotter pin and lock nut using power tools.

➡ **Discard the cotter pin, use a new cotter pin for installation.**

5. Remove the suspension arm from the knuckle using tool.

To install:

6. Installation is in the reverse order of removal.

7. Check the rear wheel alignment and adjust if necessary.

350Z

1. Remove tire with power tool.
2. Remove drive shaft.
3. Remove fixing nuts and bolts between suspension arm and rear suspension member.
4. Remove cotter pin of suspension arm ball joint, and loosen nut.
5. Use a ball joint remover (suitable tool) to remove suspension arm from axle. Be careful not to damage ball joint boot.

➡ **Tighten temporarily mounting nut to prevent damage to threads and to prevent ball joint remover (suitable tool) from coming off.**

6. Remove suspension arm and stopper rubber from vehicle.

To install:

7. Installation is in the reverse order of removal.

    a. Tighten fixing nuts and bolts to 53 ft. lbs. (72.5 Nm)

    b. Tighten ball joint nut to 62 ft. lbs. (84.5 Nm)

WHEEL BEARINGS

REMOVAL & INSTALLATION

Maxima

If the wheel hub bearing assembly is removed, it must be replaced.

➡ **If the vehicle is equipped with Anti-Lock Brake (ABS), the sensor must be removed to protect the sensor and its wiring.**

1. Before servicing the vehicle, refer to the precautions in the beginning of this section.

2. Raise and safely support the vehicle. Remove the rear wheel(s).

3. Remove or disconnect the following:
- Brake caliper and hang it by a piece of wire
- Brake caliper support
- Disc brake pads
- Brake rotor

4. Remove the cotter pin, wheel bearing locknut, washer and the wheel hub bearing assembly. A slide hammer may be needed to remove the hub bearing assembly.

➡ **The wheel hub bearing assembly is not repairable; it must be replaced when defective.**

To install:

➡ **If the vehicle is equipped with ABS, the sensor ring must be removed and installed on the new hub.**

5. Apply oil to the threaded portion of the spindle and both sides of the plain washer.

6. Install the wheel hub bearing assembly, the washer and the wheel bearing locknut. Torque the wheel bearing locknut to 138–188 ft. lbs. (187–255 Nm).

7. Verify that the wheel bearings operate smoothly.

8. Install or connect the following:
- New cotter pin into the spindle to hold the wheel bearing locknut

9. Install a dial micrometer to the rear wheel hub bearing assembly and check the axial end-play. It should be less than 0.0020 in. (0.05mm).
- ABS sensor and its wiring, if removed
- Brake assembly and the wheels

350Z

1. Remove tires from vehicle with power tool.
2. Remove cotter pin. Then remove lock nut from drive shaft.
3. Remove brake caliper with power tool. Hang it in a place where it will not interfere with work.
4. Remove disc rotor and remove parking cable and parking brake shoe from back plate.
5. Remove fixing bolts and nuts in axle side of radius rod, front lower link with power tool.
6. Remove fixing bolt and nut in axle side of rear lower link with power tool. Then remove coil spring.
7. Remove fixing bolt and nut in axle side of shock absorber with power tool.
8. Using a puller (suitable tool), remove axle from drive shaft.

9. Remove suspension arm and cotter pin at axle, then loosen mounting nut.

10. Use a ball joint remover (suitable tool) to remove suspension arm from axle. Be careful not to damage ball joint boot.

11. Remove knuckle assembly

12. Remove wheel bearing fixing bolts and anchor block fixing nuts, and remove wheel hub and bearing assembly, back plate and anchor block from axle.

13. Using a drift (SST) and a puller (suitable tool), press wheel hub out to remove from wheel bearing.

14. Using a drift (SST) and a puller (suitable tool), press wheel bearing outer side inner race out to remove from wheel hub.

To install:

15. Press fit a wheel hub into wheel bearing with a drift (SST).

➡ **Press fit a drift (SST) while holding it against wheel bearing inner side inner race. Wheel bearing cannot be reused. Do not attempt to reuse it.**

16. Install back plate and wheel hub and bearing assembly.

17. Install anchor block onto axle.

18. The remaining installation is in the reverse order of removal.

ADJUSTMENT

Maxima

If the wheel hub bearing assembly is removed, it must be replaced.

➡ **The wheel hub bearing assembly is not repairable; it must be replaced when defective.**

1. Before servicing the vehicle, refer to the precautions in the beginning of this section.

2. Torque the wheel bearing locknut to 138–188 ft. lbs. (187–255 Nm).

3. Verify that the wheel bearings operate smoothly.

4. Install a new cotter pin into the spindle to hold the wheel bearing locknut.

5. Install a dial indicator to the rear wheel hub bearing assembly and check the axial end-play; it should be less than 0.0020 in. (0.05mm).

6. Install the grease cap.

7. If the axial end-play exceeds specifications, the wheel bearing must be replaced.

SPECIFICATIONS AND MAINTENANCE CHARTS

ENGINE AND VEHICLE IDENTIFICATION

Engine								Model Year	
Code ①	Liters (cc)	Cu. In.	Cyl.	Fuel Sys.	Engine Type	Eng. Mfg.		Code ②	Year
MR20DE	2.0 (1997)	122	4	MFI	DOHC	Nissan		7	2007
QR25DE	2.5 (2488)	152	4	MFI	DOHC	Nissan		8	2008
VQ35DE	3.5 (3498)	213	6	MFI	DOHC	Nissan			

MFI: Multi-port Fuel Injection

DOHC: Double Overhead Camshaft

① The Engine Code is stamped on the engine block near the starter.

② 10th position of the Vehicle Identification Number (VIN)

22140_ALTI_C0001

GENERAL ENGINE SPECIFICATIONS

Year	Model	Engine Displacement Liters	Engine Series (ID/VIN)	Net Horsepower @ rpm	Net Torque @ rpm (ft. lbs.)	Bore x Stroke (in.)	Compression Ratio	Oil Pressure @ rpm
2007	Altima	2.5	QR25DE	175@5600	180@3900	3.50X3.94	9.5:1	43@2000
	Altima	3.5	VQ35DE	270@6000	258@4400	3.76X3.20	10.3:1	43@2000
	Altima Hybrid	2.5	QR25DE	158@6200	162@2800	3.50X3.94	9.5:1	43@2000
	Sentra	2.0	MR20DE	140@5100	147@4800	3.31X3.55	10.2:1	29@2000
	Sentra SE-R	2.5	QR25DE	150@5600	154@5600	3.50x3.94	9.5:1	60@3000
2008	Altima	2.5	QR25DE	175@5600	180@3900	3.50X3.94	9.5:1	43@2000
	Altima	3.5	VQ35DE	270@6000	258@4400	3.76X3.20	10.3:1	43@2000
	Altima Hybrid	2.5	QR25DE	158@6200	162@2800	3.50X3.94	9.5:1	43@2000
	Sentra	2.0	MR20DE	140@5100	147@4800	3.31X3.55	10.2:1	29@2000
	Sentra SE-R	2.5	QR25DE	150@5600	154@5600	3.50x3.94	9.5:1	60@3000

22140_ALTI_C0002

ENGINE TUNE-UP SPECIFICATIONS

Year	Engine Displacement Liters	Engine ID/VIN	Spark Plug Gap (in.)	Ignition Timing (deg.) MT	Ignition Timing (deg.) AT	Fuel Pump (psi) ①	Idle Speed (rpm) MT	Idle Speed (rpm) AT ②	Valve Clearance Intake ③	Valve Clearance Exhaust ③
2007	2.0	MR20DE	0.043	6B	6B	51	625-725	650-750	0.010-0.013	0.011-0.015
	2.5	QR25DE	0.043	15B	15B	51	650-750	650-750	0.009-0.013	0.010-0.013
	3.5	VQ35DE	0.043	12B	12B	51	550-650	550-650	0.010-0.013	0.011-0.015
2008	2.0	MR20DE	0.043	6B	6B	51	625-725	650-750	0.010-0.013	0.011-0.015
	2.5	QR25DE	0.043	15B	15B	51	650-750	650-750	0.009-0.013	0.010-0.013
	3.5	VQ35DE	0.043	12B	12B	51	550-650	550-650	0.010-0.013	0.011-0.015

B: Before top dead center

① At idle

② Automatic transmission in neutral

③ Engine cold

22140_ALTI_C0003

CAPACITIES

Year	Model	Engine ID/VIN	Engine Displacement Liters	Engine Oil with Filter (qts.)	Transmission (pts.) 5-Spd	Transmission (pts.) Auto.	Drive Axle Rear (pts.)	Fuel Tank (gal.)	Cooling System (qts.)
2007	Altima	QR25DE	2.5	4.5	7.2	17.5	—	20.0	8.0
	Altima	VQ35DE	3.5	4.5	3.6	10.8	—	20.0	8.5
	Altima Hybrid	QR25DE	2.5	4.8	—	8.75	—	20.0	①
	Sentra	MR20DE	2.0	4	4.25	17.5	—	14.5	8.0
	Sentra SE-R	QR25DE	2.5	4.2	4.9	18	—	13.2	6.4
2008	Altima	QR25DE	2.5	4.5	7.2	17.5	—	20.0	8.0
	Altima	VQ35DE	3.5	4.5	3.6	10.8	—	20.0	8.5
	Altima Hybrid	QR25DE	2.5	4.8	—	8.75	—	20.0	①
	Sentra	MR20DE	2.0	4	4.25	17.5	—	14.5	8.0
	Sentra SE-R	QR25DE	2.5	4.2	4.9	18	—	13.2	6.4

① Engine coolant, 8 qts.: Inverter coolant, 3 qts.

22140_ALTI_C0004

FLUID SPECIFICATIONS

Year	Model	Engine Displ. Liters	Engine Oil	Man. Trans.	Auto. Trans.	Drive Axle Front	Drive Axle Rear	Transfer Case	Power Steering Fluid	Brake Master Cylinder	Cooling System
2007	Sentra	2.0	5W-30	75W-80	Nissan NS-2	—	—	—	—	DOT 3	N-LL
	Sentra SE-R	2.5	5W-30	75W-80	Nissan NS-2	—	—	—	—	DOT 3	N-LL
	Altima	2.5	5W-30	75W-80	Nissan NS-2	—	—	—	Dexron IV	DOT 3	N-LL
	Altima	3.5	5W-30	75W-80	Nissan NS-2	—	—	—	Dexron IV	DOT 3	N-LL
2008	Sentra	2.0	5W-30	75W-80	Nissan NS-2	—	—	—	—	DOT 3	N-LL
	Sentra SE-R	2.5	5W-30	75W-80	Nissan NS-2	—	—	—	—	DOT 3	N-LL
	Altima	2.5	5W-30	75W-80	Nissan NS-2	—	—	—	Dexron IV	DOT 3	N-LL
	Altima	3.5	5W-30	75W-80	Nissan NS-2	—	—	—	Dexron IV	DOT 3	N-LL

N-LL: Nissan Long Life coolant

22140_ALTI_C0005

VALVE SPECIFICATIONS

Year	Engine ID/VIN	Engine Displacement Liters	Seat Angle (deg.)	Face Angle (deg.)	Spring Test Pressure (lbs. @ in.)	Spring Installed Height (in.)	Stem-to-Guide Clearance (in.) Intake	Stem-to-Guide Clearance (in.) Exhaust	Stem Diameter (in.) Intake	Stem Diameter (in.) Exhaust
2007	MR20DE	2.0	45.15-45.45	—	34-39@ 1.39	1.390	0.0008-0.0021	0.0012-0.0025	0.2152-0.2157	0.2148-0.2154
	QR25DE	2.5	45.15-45.45	—	34-39@ 1.39	1.390	0.0008-0.0021	0.0012-0.0025	0.2348-0.2354	0.2344-0.2350
	VQ35DE	3.5	45.15-45.45	—	37-42@ 1.457	1.457	0.0008-0.0021	0.0016-0.0028	0.2348-0.2354	0.2344-0.2350
2008	MR20DE	2.0	45.15-45.45	—	34-39@ 1.39	1.390	0.0008-0.0021	0.0012-0.0025	0.2152-0.2157	0.2148-0.2154
	QR25DE	2.5	45.15-45.45	—	34-39@ 1.39	1.390	0.0008-0.0021	0.0012-0.0025	0.2348-0.2354	0.2344-0.2350
	VQ35DE	3.5	45.15-45.45	—	37-42@ 1.457	1.457	0.0008-0.0021	0.0016-0.0028	0.2348-0.2354	0.2344-0.2350

22140_ALTI_C0006

CAMSHAFT SPECIFICATIONS

All measurements in inches unless noted

Year	Engine Displacement Liters	Engine Code/ID	Journal Dia.	Brg. Oil Clearance	Shaft End-play	Circle Runout	Lobe Height Intake	Lobe Height Exhaust
2007	2.0	MR20DE	①	② 0.0018	0.0030-0.0060	0.0008	1.7560-1.7635	1.6997-1.7072
	2.5	QR25DE	③	⑤ 0.0034	0.0045-0.0074	0.0016	1.7644-1.7718	1.7313-1.7388
	3.5	VQ35DE	④	⑤	0.0045-0.0074	0.0008	1.7904-1.7978	1.7904-1.7978
2008	2.0	MR20DE	①	② 0.0018	0.0030-0.0060	0.0008	1.7560-1.7635	1.6997-1.7072
	2.5	QR25DE	③	⑤ 0.0034	0.0045-0.0074	0.0016	1.7644-1.7718	1.7313-1.7388
	3.5	VQ35DE	④	⑤	0.0045-0.0074	0.0008	1.7904-1.7978	1.7904-1.7978

NA: Not Available

① No. 1: 1.0098-1.1006
All others: 0.9823-0.9831

② No. 1: 0.0018-0.0034
All others: 0.0012-0.0028

③ No. 1: 1.0998-1.1006
All others: 0.9926-0.9234

④ No. 1: 1.0211-1.0218
All others: 0.9230-0.9238

⑤ No. 1: 0.0018-0.0034
All others: 0.0014-0.0030

22140_ALTI_C0007

CRANKSHAFT AND CONNECTING ROD SPECIFICATIONS

All measurements are given in inches.

Year	Engine Displacement Liters	Engine ID/VIN	Crankshaft Main Brg. Journal Dia.	Crankshaft Main Brg. Oil Clearance	Crankshaft Shaft End-play	Thrust on No.	Connecting Rod Journal Diameter	Connecting Rod Oil Clearance	Connecting Rod Side Clearance
2007	2.0	MR20DE	2.0457-2.0464	①	0.0039-0.1020	3	1.7305-1.7311	0.0015-0.0019	0.0079-0.0138
	2.5	QR25DE	2.1636-2.1645	②	0.0039-0.0102	3	1.7699-1.7706	0.0014-0.0018	0.0039-0.0102
	3.5	VQ35DE	2.3603-2.3612	0.0014-0.0018	0.0039-0.0098	3	2.0457-2.0460	0.0010-0.0018	0.0039-0.0098
2008	2.0	MR20DE	2.0457-2.0464	①	0.0039-0.1020	3	1.7305-1.7311	0.0015-0.0019	0.0079-0.0138
	2.5	QR25DE	2.1636-2.1645	②	0.0039-0.0102	3	1.7699-1.7706	0.0014-0.0018	0.0039-0.0102
	3.5	VQ35DE	2.3603-2.3612	0.0014-0.0018	0.0039-0.0098	3	2.0457-2.0460	0.0010-0.0018	0.0039-0.0098

① Nos. 1, 4, 5 : 0.0009-0.0013
Nos. 2, 3 : 0.0005-0.0009

② Nos. 1, 3, 5 : 0.0005-0.0009
Nos. 2, 4 : 0.0007-0.0011

22140_ALTI_C0008

PISTON AND RING SPECIFICATIONS

All measurements are given in inches.

Year	Engine Displacement Liters	Engine ID/VIN	Piston Clearance	Ring Gap Top Compression	Ring Gap Bottom Compression	Ring Gap Oil Control	Ring Side Clearance Top Compression	Ring Side Clearance Bottom Compression	Ring Side Clearance Oil Control
2007	2.0	MR20DE	0.0004-0.0012	0.0080-0.0120	0.0200-0.0260	0.0060-0.0180	0.0020-0.0030	0.0010-0.0030	0.0010-0.0070-
	2.5	QR25DE	0.0004-0.0012	0.0083-0.0122	0.0146-0.0205	0.0079-0.0177	0.0016-0.0031	0.0012-0.0028	0.0018-0.0049
	3.5	VQ35DE	0.0004-0.0012	0.0091-0.0130	0.0091-0.0130	0.0079-0.0177	0.0018-0.0031	0.0012-0.0028	0.0026-0.0049
2008	2.0	MR20DE	0.0004-0.0012	0.0080-0.0120	0.0200-0.0260	0.0060-0.0180	0.0020-0.0030	0.0010-0.0030	0.0010-0.007
	2.5	QR25DE	0.0004-0.0012	0.0083-0.0122	0.0146-0.0205	0.0079-0.0177	0.0016-0.0031	0.0012-0.0028	0.0018-0.0049
	3.5	VQ35DE	0.0004-0.0012	0.0091-0.0130	0.0091-0.0130	0.0079-0.0177	0.0018-0.0031	0.0012-0.0028	0.0026-0.0049

22140_ALTI_C0009

TORQUE SPECIFICATIONS

All readings in ft. lbs.

Year	Engine Displacement Liters	Engine ID/VIN	Cylinder Head Bolts	Main Bearing Bolts	Rod Bearing Bolts	Crankshaft Damper Bolts	Flywheel Bolts	Manifold Intake	Manifold Exhaust	Spark Plugs	Oil Drain Plug
2007	2.0	MR20DE	①	②	③	④	80	20	25	14	25
	2.5	QR25DE	⑤	⑥	⑦	④	76-83	13-15	29-32	18	25
	3.5	VQ35DE	⑧	⑨	⑩	⑪	61-69	⑫	21-24	18	25
2008	2.0	MR20DE	①	②	③	④	80	20	25	14	25
	2.5	QR25DE	⑤	⑥	⑦	④	76-83	13-15	29-32	18	25
	3.5	VQ35DE	⑧	⑨	⑩	⑪	61-69	⑫	21-24	18	25

① Step 1: Tighten all in sequence to 30 ft. lbs.

Step 2: Tighten an additional 100 degrees

Step 3: Loosen all bolts in reverse order

Step 4: Repeat steps 1 and 2

Step 5: Tighten an additional 100 degrees

② Step 1: 25 ft. lbs.

Step 2: Tighten an additional 60 degrees

③ Step 1: 20 ft. lbs.

Step 2: loosen bolts

Step 3: 14 ft. lbs.

④ Step 1: 29-36 ft. lbs.

Step 2: 60-66 degrees

⑤ Step 1: 72 ft. lbs.

Step 2: Loosen completely, then retorque to 26-32 ft. lbs.

Step 3: Turn each bolt, in sequence, an additional 75-80 degrees

Step 4: Turn each bolt, in sequence, an additional 75-80 degrees

⑥ Tighten bolts 11-22 to 19 ft. lbs.

Step 2: Tighten bolts 1-10 to 29 ft. lbs.

Step 3: Tighten bolts 1-10 an additional 60 degrees

⑦ Step 1: Tighten bolts to 20 ft. lbs.

Step 2: Loosen all bolts

Step 3: tighten bolts to 14 ft. lbs.

Step 4: Tighten bolts an additional 90 degrees

⑧ Step 1: Tighten to 72 ft. lbs.

Step 2: Loosen bolts completely

Step 3: 26-32 ft. lbs.

Step 4: Tighten an additional 90-95 degrees

Step 5: Repeat Step 4

⑨ Step 1: Shift crankshaft to align bearing beam

Step 2: Tighten all bolts to 24-28 ft. lbs.

Step 3: Tighten an additional 90-95 degrees

⑩ Step 1: Tighten to 15 ft. lbs.

Step 2: Tighten an additional 90-95 degrees

⑪ Step 1: Tighten to 32 ft. lbs. an additional 90 degrees

⑫ Step 1: Lower manifold bolts; step 1 65 inch lbs.

Step 2: 19 ft. lbs.

Upper maifold: 8 ft. lbs.

22140_ALTI_C0010

WHEEL ALIGNMENT

Year	Model		Caster Range (+/-Deg.)	Caster Preferred Setting (Deg.)	Camber Range (+/-Deg.)	Camber Preferred Setting (Deg.)	Toe-in (in.)
2007	Altima	F	0.75	+5.08	①	②	0.04 +/- 0.04
		R	—	—	0.30	-1.25	0.09 +/- 0.06
	Altima Hybrid	F	0.75	+5.00	0.95	-0.40	0.04 +/- 0.04
		R			0.050	-0.70	0.11 +/- 0.06
	Sentra	F	0.75	+4.92	0.75	0.17	0.00 +/- 0.04
		R	—	—	0.50	-1.50	0.0197 +/- 0.56
2008	Altima	F	0.75	+5.08	①	②	0.04 +/- 0.04
		R	—	—	0.50	-0.62	0.09 +/- 0.06
	Altima Hybrid	F	0.75	+5.00	0.95	-0.40	0.04 +/- 0.04
		R			0.050	-0.70	0.11 +/- 0.06
	Sentra	F	0.75	+4.92	0.75	0.17	0.00 +/- 0.04
		R	—	—	0.50	-1.50	0.0197 +/- 0.56

① Minus 0.25 degrees, plus 0.75 degrees

② Left, -.050 degrees; Right -0.75 degrees

22140_ALTI_C0011

TIRE, WHEEL AND BALL JOINT SPECIFICATIONS

Year	Model	OEM Tires Standard	OEM Tires Optional	Tire Pressures (psi) Front	Tire Pressures (psi) Rear	Wheel Size	Lug Nut Torque (ft. lbs.)
2007	Altima 2.5	P215/60TR16	None	29	29	6.5-JJ	80
	Altima Hybrid	P215/60TR16	None	35	35	6.5-JJ	83
	Altima 3.5 SL	P215/60TR16	None	33	30	7-JJ	80
	Altima 3.5 SE	P215/55/VR17	None	35	35	7.5-JJ	80
	Sentra	P205/60HR15	P205/55HR16	33	30	6.5-JJ	80
	Sentra SE-R	P215/45ZR17	None	33	33	7-JJ	80
2008	Altima	P215/60TR16	None	33	33	6.5-JJ	80
	Altima Hybrid	P215/60TR16	None	35	35	6.5-JJ	83
	Altima 3.5 SL	P215/60/TR16	None	33	33	7-JJ	80
	Altima SE	P215/55/VR17	None	33	33	7.5-JJ	80
	Sentra	P205/60HR15	P205/55HR16	33	30	6.5-JJ	80
	Sentra SE-R	P215/45ZR17	None	33	33	7-JJ	80

OEM: Original Equipment Manufacturer

PSI: Pounds Per Square Inch

22140_ALTI_C0012

BRAKE SPECIFICATIONS

All measurements in inches unless noted

Year	Model		Original Thickness	Minimum Thickness	Maximum Run-out	Original Inside Diameter	Max. Wear Limit	Maximum Machine Diameter	Front	Rear	Bracket Bolts (ft. lbs.)	Mounting Bolts (ft. lbs.)
				Brake Disc		Brake Drum Diameter			Minimum Lining Thickness		Brake Caliper	
2007	Altima	F	1.024	0.945	0.001	—	—	—	0.079	—	53-72	16-23
		R	0.354	0.315	0.001	—	—	—	—	0.039	98	20
	Altima Hybrid	F	1.024	0.945	0.001	—	—	—	0.079	—	98	20
		R	0.354	0.315	0.001	—	—	—	—	0.039	62	32
	Sentra	F	0.945	0.866	0.001		—	—	0.079	—	122	20
		R	—	—	—	9.000	9.079	—	—	0.059	—	—
	Sentra SE-R	F	1.181	1.118	0.002	—	—	—	0.079	—	112	—
		R	0.350	0.315	0.003	—	—	—	—	0.059	28-38	16-23
2008	Altima	F	1.020	0.866	0.003	—	—	—	0.079	—	53-72	16-23
		R	0.354	0.315	0.003	—	—	—	—	0.039	98	20
	Altima Hybrid	F	1.024	0.945	0.001	—	—	—	0.079	—	98	20
		R	0.354	0.315	0.001	—	—	—	—	0.039	62	32
	Sentra	F	0.945	0.866	0.001		—	—	0.079	—	122	20
		R	—	—	—	9.000	9.079	—	—	0.059	—	—
	Sentra SE-R	F	1.181	1.118	0.002	—	—	—	0.079	—	112	—
		R	0.350	0.315	0.003	—	—	—	—	0.059	28-38	16-23

22140_ALTI_C0013

SCHEDULED MAINTENANCE INTERVALS
Nissan—Altima & Sentra

TO BE SERVICED	TYPE OF SERVICE	VEHICLE MILEAGE INTERVAL (x1000)												
		7.5	15	22.5	30	37.5	45	52.5	60	67.5	75	82.5	90	97.5
Engine oil & filter	R	✓	✓	✓	✓	✓	✓	✓	✓	✓	✓	✓	✓	✓
Brake lines & cables	S/I		✓		✓		✓		✓		✓		✓	
Brake pads, discs, drums & linings	S/I		✓		✓		✓		✓		✓		✓	
Driveshaft boots	S/I		✓		✓		✓		✓		✓		✓	
Exhaust system	S/I				✓				✓				✓	
Transaxle fluid	S/I		✓		✓		✓		✓		✓		✓	
Air cleaner filter	R				✓				✓				✓	
Spark plugs (except platinum)	R				✓				✓				✓	
Spark plugs (iridium and platinum)	R	Replace every 105,000 miles												
Steering gear & linkage, axle & suspension parts	S/I				✓				✓				✓	
Engine coolant	R	Replace every 60,000 miles, then every 30,000 miles												
Inverter coolant	R	Replace every 60,000 miles, then every 30,000 miles												
Drive belts	S/I								✓					
Fuel lines	S/I								✓					
Vapor lines	S/I								✓					
Cabin microfilter	R		✓		✓		✓		✓		✓		✓	
Valve adjustment	S/I	As needed												

R: Replace   S/I: Service or Inspect

FREQUENT OPERATION MAINTENANCE (SEVERE SERVICE)

If a vehicle is operated under any of the following conditions it is considered severe service:

- Extremely dusty areas.

- 50% or more of the vehicle operation is in 32°C (90°F) or higher temperatures, or constant operation in temperatures below 0°C (32°F).

- Prolonged idling (vehicle operation in stop and go traffic).

- Frequent short running periods (engine does not warm to normal operating temperatures).

- Police, taxi, delivery usage or trailer towing usage.

Oil & oil filter: change every 3750 miles.

Brake pads & discs: service or inspect every 7500 miles.

Driveshaft boots: service or inspect every 7500 miles.

Exhaust system: service or inspect every 7500 miles.

Steering gear & linkage, axle & suspension parts: service or inspect every 7500 miles.

Steering linkage ball joints & front suspension ball joints: service or inspect every 7500 miles.

Air cleaner filter: service or inspect every 15,000 miles.

22140_ALTI_C0014

PRECAUTIONS

Before servicing any vehicle, please be sure to read all of the following precautions, which deal with personal safety, prevention of component damage, and important points to take into consideration when servicing a motor vehicle:

- Never open, service or drain the radiator or cooling system when the engine is hot; serious burns can occur from the steam and hot coolant.

- Observe all applicable safety precautions when working around fuel. Whenever servicing the fuel system, always work in a well-ventilated area. Do not allow fuel spray or vapors to come in contact with a spark, open flame, or excessive heat (a hot drop light, for example). Keep a dry chemical fire extinguisher near the work area. Always keep fuel in a container specifically designed for fuel storage; also, always properly seal fuel containers to avoid the possibility of fire or explosion. Refer to the additional fuel system precautions later in this section.

- Fuel injection systems often remain pressurized, even after the engine has been turned **OFF**. The fuel system pressure must be relieved before disconnecting any fuel lines. Failure to do so may result in fire and/or personal injury.

- Brake fluid often contains polyglycol ethers and polyglycols. Avoid contact with the eyes and wash your hands thoroughly after handling brake fluid. If you do get brake fluid in your eyes, flush your eyes with clean, running water for 15 minutes. If eye irritation persists, or if you have taken brake fluid internally, IMMEDIATELY seek medical assistance.

- The EPA warns that prolonged contact with used engine oil may cause a number of skin disorders, including cancer. You should make every effort to minimize your exposure to used engine oil. Protective gloves should be worn when changing oil. Wash your hands and any other exposed skin areas as soon as possible after exposure to used engine oil. Soap and water, or waterless hand cleaner should be used.

- All new vehicles are now equipped with an air bag system, often referred to as a Supplemental Restraint System (SRS) or Supplemental Inflatable Restraint (SIR) system. The system must be disabled before performing service on or around system components, steering column, instrument panel components, wiring and sensors. Failure to follow safety and disabling procedures could result in accidental air bag deployment, possible personal injury and unnecessary system repairs.

- Always wear safety goggles when working with, or around, the air bag system. When carrying a non-deployed air bag, be sure the bag and trim cover are pointed away from your body. When placing a non-deployed air bag on a work surface, always face the bag and trim cover upward, away from the surface. This will reduce the motion of the module if it is accidentally deployed. Refer to the additional air bag system precautions later in this section.

- Clean, high quality brake fluid from a sealed container is essential to the safe and proper operation of the brake system. You should always buy the correct type of brake fluid for your vehicle. If the brake fluid becomes contaminated, completely flush the system with new fluid. Never reuse any brake fluid. Any brake fluid that is removed from the system should be discarded. Also, do not allow any brake fluid to come in contact with a painted surface; it will damage the paint.

- Never operate the engine without the proper amount and type of engine oil; doing so WILL result in severe engine damage.

- Timing belt maintenance is extremely important. Many models utilize an interference-type, non-freewheeling engine. If the timing belt breaks, the valves in the cylinder head may strike the pistons, causing potentially serious (also time-consuming and expensive) engine damage. Refer to the maintenance interval charts for the recommended replacement interval for the timing belt, and to the timing belt section for belt replacement and inspection.

- Disconnecting the negative battery cable on some vehicles may interfere with the functions of the on-board computer system(s) and may require the computer to undergo a relearning process once the negative battery cable is reconnected.

- When servicing drum brakes, only disassemble and assemble one side at a time, leaving the remaining side intact for reference.

- Only an MVAC-trained, EPA-certified automotive technician should service the air conditioning system or its components.

BRAKES

GENERAL INFORMATION

PRECAUTIONS

- Certain components within the ABS system are not intended to be serviced or repaired individually.

- Do not use rubber hoses or other parts not specifically specified for and ABS system. When using repair kits, replace all parts included in the kit. Partial or incorrect repair may lead to functional problems and require the replacement of components.

- Lubricate rubber parts with clean, fresh brake fluid to ease assembly. Do not use shop air to clean parts; damage to rubber components may result.

- Use only DOT 3 brake fluid from an unopened container.

- If any hydraulic component or line is removed or replaced, it may be necessary to bleed the entire system.

- A clean repair area is essential. Always clean the reservoir and cap thoroughly before removing the cap. The slightest amount of dirt in the fluid may plug an orifice and impair the system function. Perform repairs after components have been thoroughly cleaned; use only denatured alcohol

ANTI-LOCK BRAKE SYSTEM (ABS)

to clean components. Do not allow ABS components to come into contact with any substance containing mineral oil; this includes used shop rags.

- The Anti-Lock control unit is a microprocessor similar to other computer units in the vehicle. Ensure that the ignition switch is **OFF** before removing or installing controller harnesses. Avoid static electricity discharge at or near the controller.

- If any arc welding is to be done on the vehicle, the control unit should be unplugged before welding operations begin.

BRAKES   BLEEDING THE BRAKE SYSTEM

BLEEDING PROCEDURE

BLEEDING PROCEDURE

❋❋ CAUTION

Be careful not to splash brake fluid on painted areas; it may cause paint damage. If brake fluid is splashed on painted areas, wash it away with water immediately. All hoses must

be free from excessive bending, twisting and pulling.

1. While bleeding, pay attention to master cylinder fluid level.
2. Disconnect ABS actuator and the hydraulic electric unit or disconnect the negative (-) battery cable.
3. Connect a vinyl tube to the rear right bleed valve.
4. Fully depress brake pedal 4 to 5 times.

5. With brake pedal depressed, loosen bleed valve to let the air out, and then tighten it immediately.
6. Repeat until no more air comes out.
7. Tighten bleed to 61–78 inch lbs (7–9 Nm).
8. Following these steps, with master cylinder reservoir filled at least half way, bleed the remaining brake cylinders in order: front left, rear left, and front right.

BRAKES   FRONT DISC BRAKES

❋❋ CAUTION

Dust and dirt accumulating on brake parts during normal use may contain asbestos fibers from production or aftermarket brake linings. Breathing excessive concentrations of asbestos fibers can cause serious bodily harm. Exercise care when servicing brake parts. Do not sand or grind brake lining unless equipment used is designed to contain the dust residue.

Do not clean brake parts with compressed air or by dry brushing. Cleaning should be done by dampening the brake components with a fine mist of water, then wiping the brake components clean with a dampened cloth. Dispose of cloth and all residue containing asbestos fibers in an impermeable container with the appropriate label. Follow practices prescribed by the Occupational

Safety and Health Administration (OSHA) and the Environmental Protection Agency (EPA) for the handling, processing, and disposing of dust or debris that may contain asbestos fibers.

BRAKE CALIPER

REMOVAL & INSTALLATION
See Figures 1 and 2.

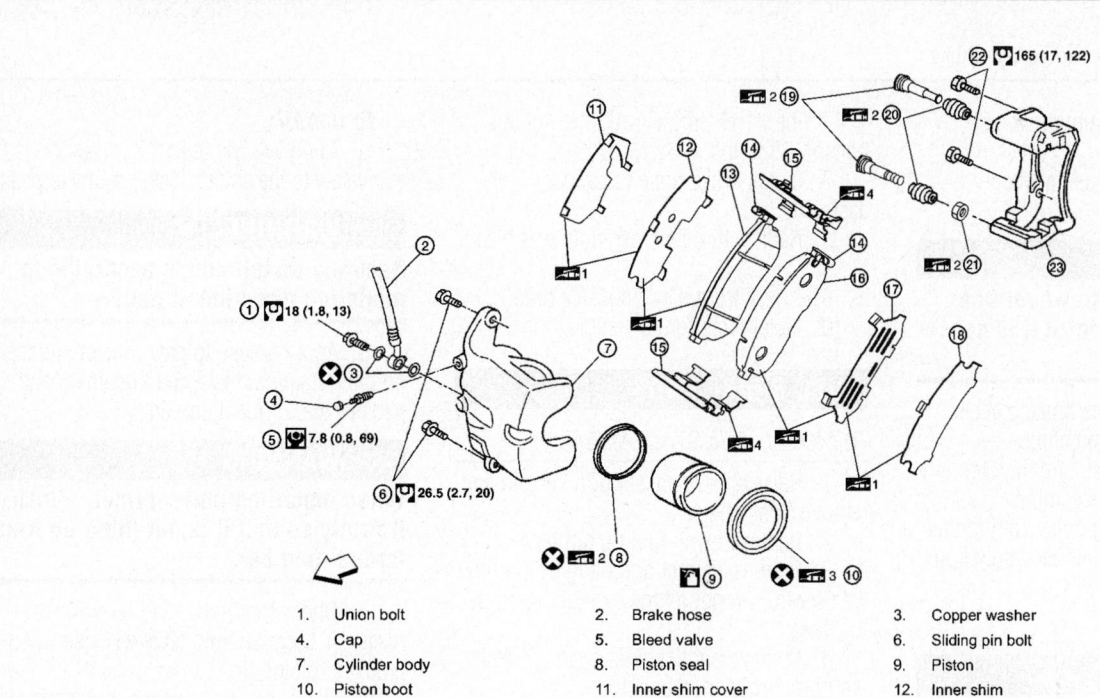

1. Union bolt
2. Brake hose
3. Copper washer
4. Cap
5. Bleed valve
6. Sliding pin bolt
7. Cylinder body
8. Piston seal
9. Piston
10. Piston boot
11. Inner shim cover
12. Inner shim
13. Inner pad
14. Pad wear sensor
15. Pad retainer
16. Outer pad
17. Outer shim
18. Outer shim cover
19. Sliding pin
20. Sliding pin boot
21. Bushing
22. Torque member mounting bolt
23. Torque member
⇐ : Front

▨ : Brake fluid
▨ 1: M-77 grease
▨ 2: Rubber grease
▨ 3: Polyglycol ether based lubricant
▨ 4: M-7439 grease

22140_ALTI_G0001

Fig. 1 Front brake caliper assembly, Sentra

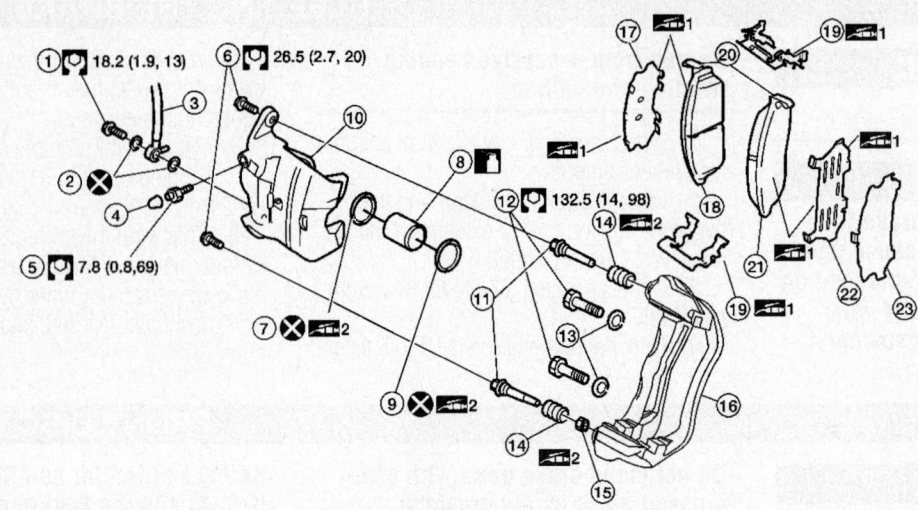

1. Union bolt
2. Copper washer
3. Brake hose
4. Cap
5. Bleed valve
6. Sliding pin bolt
7. Piston seal
8. Piston
9. Piston boot
10. Cylinder body
11. Sliding pin
12. Torque member mounting bolt
13. Washer
14. Sliding pin boot
15. Bushing
16. Torque member
17. Inner multilayered shim
18. Inner pad
19. Pad retainer
20. Pad wear sensor
21. Outer pad
22. Outer shim
23. Outer shim cover

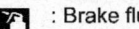 1: Molykote M-77 grease   2: Rubber grease   : Brake fluid

22140_ALTI_G0041

Fig. 2 Front brake caliper assembly, Altima

1. Raise and safely support vehicle and remove tires.
2. Secure disc rotor using wheel nuts.

> ✳✳ **WARNING**
> **Put matching marks on wheel hub assembly and disc rotor if it is necessary to remove rotor.**

3. Attach a tube to the brake bleeder and drain brake fluid from caliper.
4. Remove union bolt, and then remove brake hose from caliper assembly.
5. Remove mounting bolts from torque member and remove caliper assembly from vehicle.

To install:

> ✳✳ **WARNING**
> **Before installing torque member, wipe oil and grease from mounting surface of steering knuckle and torque member.**

6. Install torque member. On Sentra, tighten torque member mounting bolts to

122 ft. lbs. (165 Nm). On Altima, tighten bolts to 98 ft. lbs. (132 Nm).
7. Install brake hose to caliper assembly.
8. Refill with new brake fluid and bleed air.
9. Check front disc brake for drag.
10. Install tires to the vehicle.

DISC BRAKE PADS

REMOVAL & INSTALLATION

1. Raise and safely support vehicle and remove front wheels.
2. Remove lower sliding pin bolt.
3. Remove caliper and hang from body with a wire. Do not let the cylinder hang by the hose.
4. Remove pads, shims and pad retainers from torque member.

> ✳✳ **WARNING**
> **When removing pad retainer from torque member, lift pad retainer in the direction shown by arrow, so as not to deform it.**

To install:

5. Apply Molykote M-77 grease or equivalent to the shims. Install shims to pads.

> ✳✳ **WARNING**
> **Securely install shims according to mounting direction of pads.**

6. Apply grease to pad contact surface on pad retainers. Install pad retainers and pads to the torque member.

> ✳✳ **WARNING**
> **When installing pad retainer, attach it firmly so that it is not lifted up from torque member.**

7. Check the brake fluid level in the reservoir because fluid level will rise when pressing piston in.
8. Use a disc brake piston tool (commercial service tool) to easily press to piston in.
9. Install cylinder body to torque member.
10. Install lower sliding pin bolt (lower side), and tighten it to 20 ft. lbs. (26 Nm).
11. Pump the brake pedal and check the master cylinder reservoir. Add fluid if necessary.

BRAKES

REAR DISC BRAKES

✳✳ CAUTION

Dust and dirt accumulating on brake parts during normal use may contain asbestos fibers from production or aftermarket brake linings. Breathing excessive concentrations of asbestos fibers can cause serious bodily harm. Exercise care when servicing brake parts. Do not sand or grind brake lining unless equipment used is designed to contain the dust residue. Do not clean brake parts with compressed air or by dry brushing. Cleaning should be done by dampening the brake components with a fine mist of water, then wiping the brake components clean with a dampened cloth. Dispose of cloth and all residue containing asbestos fibers in an impermeable container with the appropriate label. Follow practices prescribed by the Occupational Safety and Health Administration (OSHA) and the Environmental Protection Agency (EPA) for the handling, processing, and disposing of dust or debris that may contain asbestos fibers.

BRAKE CALIPER

REMOVAL & INSTALLATION

Altima

See Figure 3.

✳✳ WARNING

While removing cylinder body, do not depress brake pedal because piston will pop out.

1. Raise and safely support the vehicle and remove the rear wheels.
2. Fasten disc rotor in place using wheel nut.
3. Open the bleeder to drain brake fluid from the caliper.
4. Remove union bolt and then disconnect brake hose from caliper assembly.
5. Remove torque member bolts, and remove brake caliper assembly. Do not drop brake pads.
6. Put match marks on wheel hub assembly and disc rotor and remove disc rotor if necessary.

To install:

7. Install disc rotor with match marks aligned.

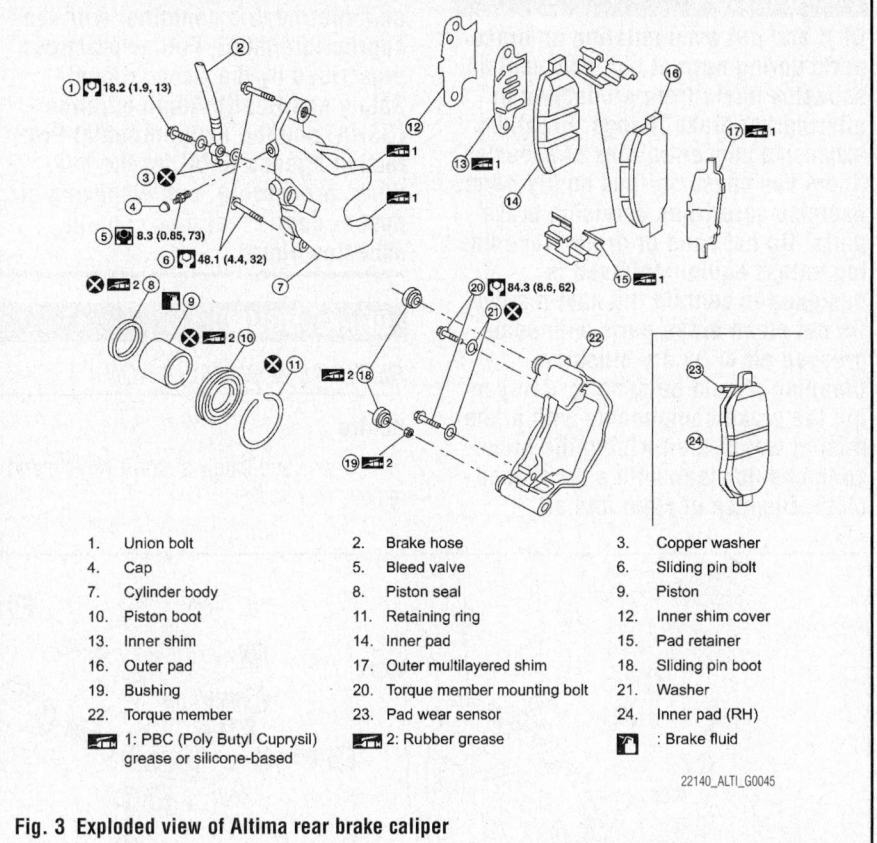

1. Union bolt
2. Brake hose
3. Copper washer
4. Cap
5. Bleed valve
6. Sliding pin bolt
7. Cylinder body
8. Piston seal
9. Piston
10. Piston boot
11. Retaining ring
12. Inner shim cover
13. Inner shim
14. Inner pad
15. Pad retainer
16. Outer pad
17. Outer multilayered shim
18. Sliding pin boot
19. Bushing
20. Torque member mounting bolt
21. Washer
22. Torque member
23. Pad wear sensor
24. Inner pad (RH)

🔧 1: PBC (Poly Butyl Cuprysil) grease or silicone-based
🔧 2: Rubber grease
🛢 : Brake fluid

22140_ALTI_G0045

Fig. 3 Exploded view of Altima rear brake caliper

8. Before installing caliper assembly, clean oil and moisture from all mounting surfaces of caliper assembly and threads, bolts and washers.
9. Install brake caliper assembly vehicle and tighten torque member mounting bolts to 62 ft. lbs. (84 Nm).
10. Install brake hose with new sealing washers and tighten union bolt to 13 ft. lbs. (18 Nm).
11. Refill with new brake fluid and bleed the system.
12. Check rear disc brake for drag, adjust parking brake as necessary.

DISC BRAKE PADS

REMOVAL & INSTALLATION

Altima

➡️It is not necessary to remove bolts on torque member and brake hose except when removing caliper assembly. In this case, hang caliper cylinder from body with a wire so as not to stretch brake hose. If any shim is subject to serious corrosion, replace it with a new one. Always replace shim and shim covers as a set when replacing brake pads.

1. Raise and safely support the vehicle and remove the rear wheels.
2. Remove upper sliding pin bolt and rotate caliper body down.
3. Remove pads, pad retainers, shims and shim cover from torque member.

To install:

4. Apply Molykote M-77 grease or equivalent between shim cover and shim. Install inner shim, inner shim cover to inner pad, and outer multilayered shim to outer pad.
5. Apply grease between pad retainer and pad. Install pad retainers and pads to torque member.
6. Check brake fluid level in the reservoir because fluid level will rise when pressing piston in.
7. Press in piston until pads can be installed, then install caliper to torque member.
8. Install upper sliding pin bolt and tighten to 32 ft. lbs. (48 Nm).
9. Check rear disc brake for drag.

BRAKES

REAR DRUM BRAKES

✳✳ CAUTION

Dust and dirt accumulating on brake parts during normal use may contain asbestos fibers from production or aftermarket brake linings. Breathing excessive concentrations of asbestos fibers can cause serious bodily harm. Exercise care when servicing brake parts. Do not sand or grind brake lining unless equipment used is designed to contain the dust residue. Do not clean brake parts with compressed air or by dry brushing. Cleaning should be done by dampening the brake components with a fine mist of water, then wiping the brake components clean with a dampened cloth. Dispose of cloth and all residue containing asbestos fibers in an impermeable container with the appropriate label. Follow practices prescribed by the Occupational Safety and Health Administration (OSHA) and the Environmental Protection Agency (EPA) for the handling, processing, and disposing of dust or debris that may contain asbestos fibers.

BRAKE DRUM

REMOVAL & INSTALLATION

Sentra

1. Raise and safely support vehicle and remove rear wheel.

2. Slide the drum off the studs. If the drum is difficult to remove:
   - Press up on adjuster lever with a wire or thin rod through adjuster plug hole on the back plate.
   - Turn adjuster with a flat bladed screwdriver to retract the brake shoes.

3. After installing brake drum, adjust brake shoes.

BRAKE SHOES

REMOVAL & INSTALLATION

Sentra

See Figure 4.

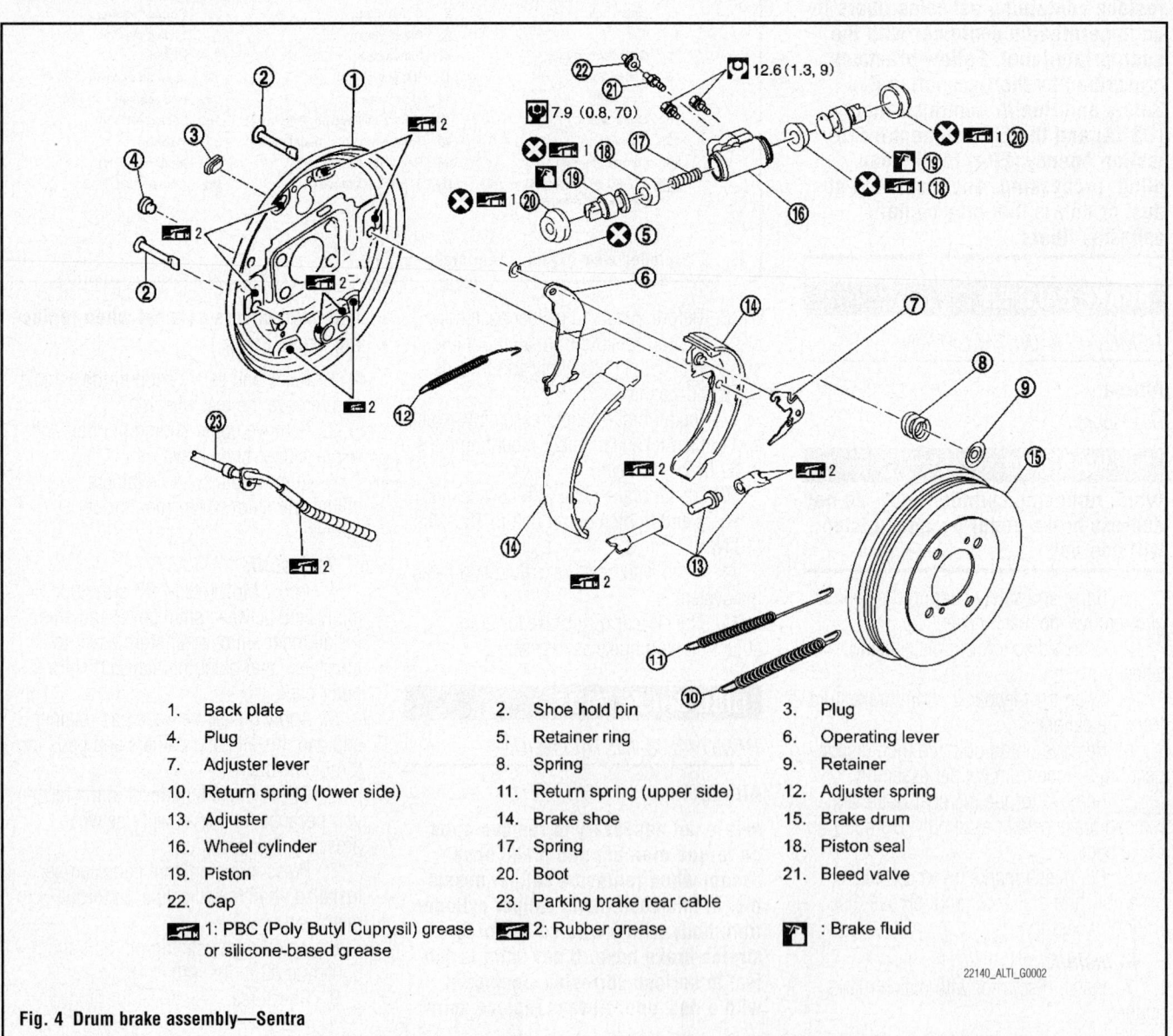

12.6 (1.3, 9)

7.9 (0.8, 70)

1.	Back plate	2.	Shoe hold pin	3.	Plug
4.	Plug	5.	Retainer ring	6.	Operating lever
7.	Adjuster lever	8.	Spring	9.	Retainer
10.	Return spring (lower side)	11.	Return spring (upper side)	12.	Adjuster spring
13.	Adjuster	14.	Brake shoe	15.	Brake drum
16.	Wheel cylinder	17.	Spring	18.	Piston seal
19.	Piston	20.	Boot	21.	Bleed valve
22.	Cap	23.	Parking brake rear cable		

🔧 1: PBC (Poly Butyl Cuprysil) grease or silicone-based grease   🔧 2: Rubber grease   🛢 : Brake fluid

22140_ALTI_G0002

Fig. 4 Drum brake assembly—Sentra

1. Raise and safely support vehicle and remove rear wheels.

2. With the parking brake lever released, remove the brake drum.

3. While pushing and rotating the retainer, pull out shoe hold pin and remove shoe assembly.

✳✳ WARNING

Do not damage the wheel cylinder boot.

4. Remove the parking brake cable from the operating lever. Do not bend the parking brake cable.

5. Disassemble the shoe assembly (shoe, springs, adjuster, adjuster lever).

6. Remove retainer ring to separate operating lever from brake shoe.

To install:

7. If parking brake operating lever is removed, install operating lever to brake shoe. Install and crimp retainer ring.

8. Apply small amount of brake grease to brake shoe sliding surfaces (the worn areas) on the back plate.

9. Apply brake grease to adjuster screw and assemble the screw. (Screws are different for right and left wheels.)

10. Assemble the shoe, adjuster, adjuster lever and springs to the shoe assembly.

11. Connect the parking brake rear cable to the operating lever.

12. Install the shoe assembly. After assembly, be sure that each part is installed properly.

13. Install the brake drum.

14. Depress brake pedal several times.

15. Adjust clearance of brake shoe. See "Parking Brake Adjustment."

16. Install wheels.

BRAKES

PARKING BRAKE SHOES

REMOVAL & INSTALLATION

Altima

See Figure 5.

PARKING BRAKE

1. Before servicing the vehicle, refer to the Precautions Section.

2. Raise and safely support vehicle and remove rear wheels.

3. Release the parking brake lever and remove rear rotors. If rotor cannot be

removed, secure the rotor in place with wheel nuts and remove adjuster hole plug. Pry adjuster up to retract and loosen brake shoes.

4. Remove anti-rattle pins, retainers, anti-rattle springs, and return springs.

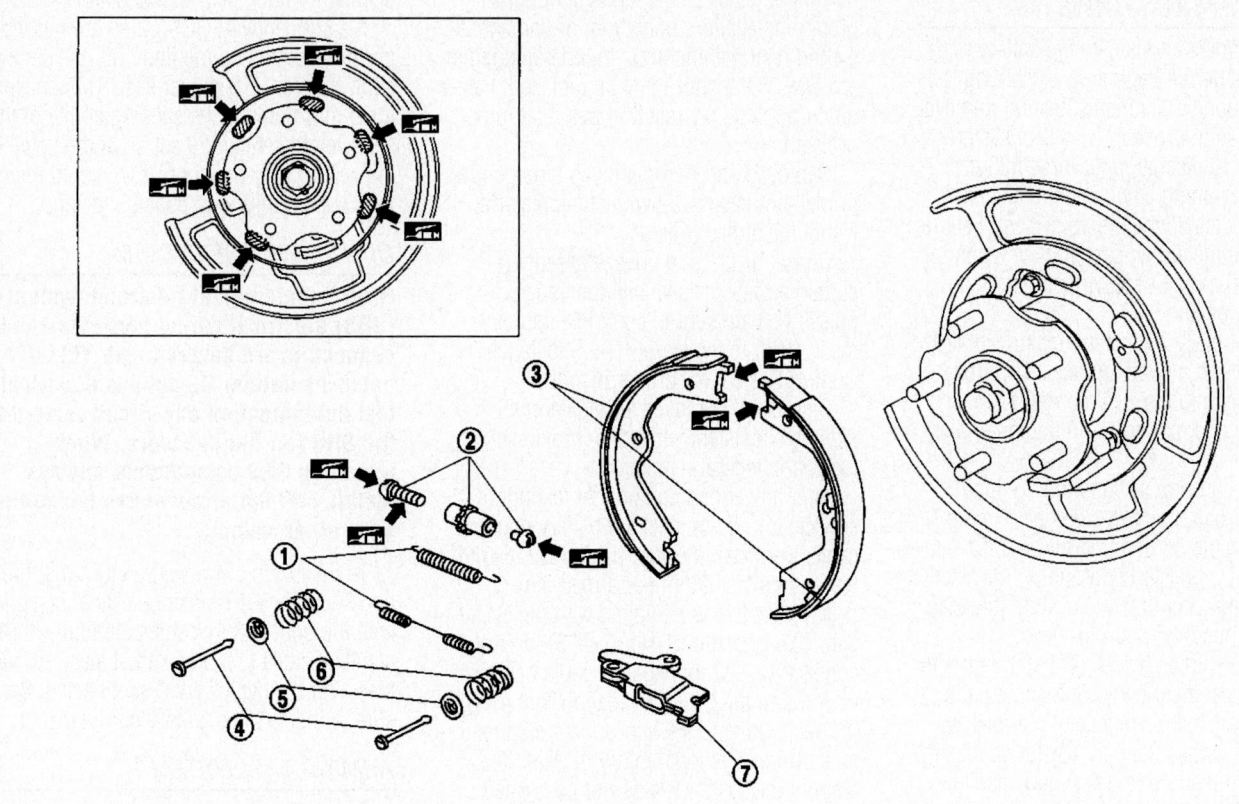

1. Return spring
2. Adjuster
3. Brake shoe
4. Anti-rattle pin
5. Retainer
6. Anti-rattle spring
7. Toggle lever

◢◣ : PBC (Poly Butyl Cuprysil) grease or silicone-based grease

22140_ALTI_G0042

Fig. 5 Altima parking brake assembly

5. Remove parking brake shoes, adjuster assembly, and toggle lever.

6. After cleaning the backplate, carefully apply small dabs of PBC (Poly Butyl Cuprysil) grease or equivalent to the areas shown in the illustration.

7. Lightly grease and assemble adjuster. Shorten adjuster all the way.

8. Clean and dry the drum inner surface.

9. Assemble the brakes, springs and adjusters to the backplate.

10. Install brake rotor and adjust the parking brake.

11. Drive the vehicle at 25 mph (40 kph) for about ten seconds with the parking brake engaged to break in the new shoes.

CHASSIS ELECTRICAL

AIR BAG (SUPPLEMENTAL RESTRAINT SYSTEM)

GENERAL INFORMATION

✷✷ CAUTION

These vehicles are equipped with an air bag system. The system must be disarmed before performing service on, or around, system components, the steering column, instrument panel components, wiring and sensors. Failure to follow the safety precautions and the disarming procedure could result in accidental air bag deployment, possible injury and unnecessary system repairs.

SERVICE PRECAUTIONS

Disconnect and isolate the battery negative cable before beginning any airbag system component diagnosis, testing, removal, or installation procedures. Allow system capacitor to discharge for three minutes before beginning any component service. This will disable the airbag system. Failure to disable the airbag system may result in accidental airbag deployment, personal injury, or death.

Do not place an intact undeployed airbag face down on a solid surface. The airbag will propel into the air if accidentally deployed and may result in personal injury or death.

When carrying or handling an undeployed airbag, the trim side (face) of the airbag should be pointing towards the body to minimize possibility of injury if accidental deployment occurs. Failure to do this may result in personal injury or death.

Replace airbag system components with OEM replacement parts. Substitute parts may appear interchangeable, but internal differences may result in inferior occupant protection. Failure to do so may result in occupant personal injury or death.

Wear safety glasses, rubber gloves, and long sleeved clothing when cleaning powder residue from vehicle after an airbag deployment. Powder residue emitted from a deployed airbag can cause skin irritation. Flush affected area with cool water if irritation is experienced. If nasal or throat irritation is experienced, exit the vehicle for fresh air until the irritation ceases. If irritation continues, see a physician.

Do not use a replacement airbag that is not in the original packaging. This may result in improper deployment, personal injury, or death.

The factory installed fasteners, screws and bolts used to fasten airbag components have a special coating and are specifically designed for the airbag system. Do not use substitute fasteners. Use only original equipment fasteners listed in the parts catalog when fastener replacement is required.

During, and following, any child restraint anchor service, due to impact event or vehicle repair, carefully inspect all mounting hardware, tether straps, and anchors for proper installation, operation, or damage. If a child restraint anchor is found damaged in any way, the anchor must be replaced. Failure to do this may result in personal injury or death.

Deployed and non-deployed airbags may or may not have live pyrotechnic material within the airbag inflator.

Do not dispose of driver/passenger/curtain airbags or seat belt tensioners unless you are sure of complete deployment. Refer to the Hazardous Substance Control System for proper disposal.

Dispose of deployed airbags and tensioners consistent with state, provincial, local, and federal regulations.

After any airbag component testing or service, do not connect the battery negative cable. Personal injury or death may result if the system test is not performed first.

If the vehicle is equipped with the Occupant Classification System (OCS), do not connect the battery negative cable before performing the OCS Verification Test using the scan tool and the appropriate diagnostic information. Personal injury or death may result if the system test is not performed properly.

Never replace both the Occupant Restraint Controller (ORC) and the Occupant Classification Module (OCM) at the same time. If both require replacement, replace one, then perform the Airbag System test before replacing the other.

Both the ORC and the OCM store Occupant Classification System (OCS) calibration data, which they transfer to one another when one of them is replaced. If both are replaced at the same time, an irreversible fault will be set in both modules and the OCS may malfunction and cause personal injury or death.

If equipped with OCS, the Seat Weight Sensor is a sensitive, calibrated unit and must be handled carefully. Do not drop or handle roughly. If dropped or damaged, replace with another sensor. Failure to do so may result in occupant injury or death.

If equipped with OCS, the front passenger seat must be handled carefully as well. When removing the seat, be careful when setting on floor not to drop. If dropped, the sensor may be inoperative, could result in occupant injury, or possibly death.

If equipped with OCS, when the passenger front seat is on the floor, no one should sit in the front passenger seat. This uneven force may damage the sensing ability of the seat weight sensors. If sat on and damaged, the sensor may be inoperative, could result in occupant injury, or possibly death.

DISARMING THE SYSTEM

➡ **All Supplemental Restraint System (SRS) electrical wiring harnesses and connectors are covered with YELLOW outer insulation. Do not use electrical test equipment on any circuit related to the SRS (air bag) sensors. When installing SRS components, always install with the arrow marks facing the front of the vehicle.**

To disarm the SRS system turn the ignition switch to **OFF** position. Then, disconnect the both battery cables starting with the negative cable first and wait at least 10 minutes after the cables are disconnected. Be sure to insulate the battery terminal ends.

ARMING THE SYSTEM

To arm the Supplemental Restraint System (SRS) system turn the ignition switch to **OFF** position. Connect the both battery cables starting with the positive cable first.

➡ **The SRS or air bag system is equipped with a self-diagnostic operation. After turning the ignition key to the ON or START position, the AIR BAG** warning lamp will illuminate for 7 seconds. After 7 seconds, the AIR BAG lamp will extinguish if no malfunction is detected. If the AIR BAG lamp does not extinguish after 7 seconds, check the SRS self-diagnostic system for a malfunction.

CLOCKSPRING CENTERING

1. Slowly turn the clockspring (spiral spring) clockwise till it stops.

2. Turn it counterclockwise about 2.0 turns, then stop turning at the point where the alignment arrows are directly across from each other.

3. Rotate the clockspring slightly as needed so the locating pin is positioned at the top.

➡A new clockspring comes in the neutral position with a stopper clip in place.

DRIVETRAIN

AUTOMATIC TRANSAXLE ASSEMBLY

REMOVAL & INSTALLATION

See Engine Removal & Installation.

MANUAL TRANSAXLE ASSEMBLY

REMOVAL & INSTALLATION

See Engine Removal & Installation.

CLUTCH

REMOVAL & INSTALLATION

Altima
See Figure 6.

✳✳ WARNING
The Concentric Slave Cylinder (CSC) must be replaced whenever transaxle is separated from engine. Return CSC insert to original position to remove transaxle assembly. Clutch dust may damage seal and cause fluid leakage.

✳✳ WARNING
Do not operate CSC/clutch pedal with transaxle and engine separated because piston and stopper will fall off.

1. Before servicing the vehicle, refer to the Precautions Section.

2. Remove the engine and transaxle from the vehicle and separate them. Note the different bolt lengths for reassembly. Remove the CSC.

3. Loosen clutch cover bolts evenly, then remove clutch cover and clutch disc.

To install:
4. Install a new CSC and torque the bolts to 15 ft. lbs. (20 Nm).

5. Clean clutch disc and input shaft splines to remove grease and dust.

6. Apply recommended grease to clutch disc and input shaft splines.

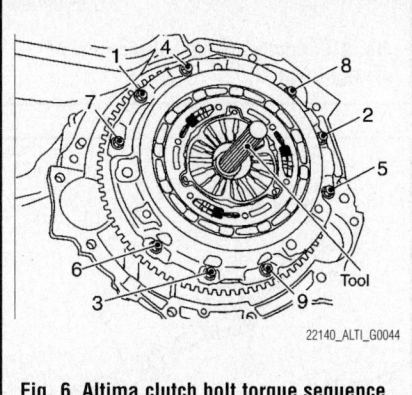

Fig. 6 Altima clutch bolt torque sequence

✳✳ WARNING
Be sure to apply grease only to the points specified. Excess grease may cause slip or shudder. If it adheres to release bearing seal, it will cause clutch fluid leakage.

7. Install clutch disc using a clutch alignment tool.

8. Clean and install clutch cover and hand-tighten bolts.

9. Tighten clutch cover bolts to 11 ft. lbs. (15 Nm) in the order shown, then again to 29 ft. lbs. (39 Nm).

10. Install a new CSC and attach transmission to engine.

Sentra
See Figure 7.

✳✳ WARNING
The Concentric Slave Cylinder (CSC) must be replaced whenever transaxle is separated from engine. Return CSC insert to original position to remove transaxle assembly. Clutch dust may damage seal and cause fluid leakage.

✳✳ WARNING
Do not operate CSC/clutch pedal with transaxle and engine separated because piston and stopper will fall off.

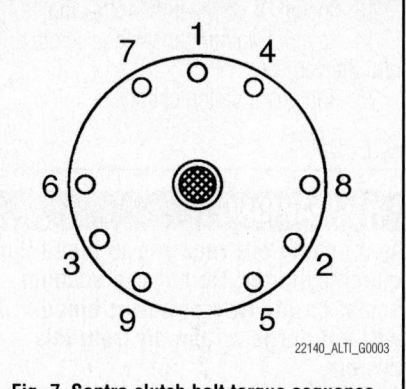

Fig. 7 Sentra clutch bolt torque sequence

➡Any time the transaxle is removed, replace the Concentric Slave Cylinder (CSC).

1. Before servicing the vehicle, refer to the Precautions Section.

2. Remove the engine and transaxle from the vehicle as an assembly. See Engine Removal & Installation.

3. Separate the engine from the transaxle.

4. Insert a clutch disc centering tool into the clutch disc hub for support.

5. Loosen the pressure plate bolts evenly in reverse order of the tightening sequence, a little at a time to prevent distortion

6. Remove clutch assembly.

7. Remove CSC from clutch housing.

To install:
8. Install a new CSC and torque the bolts to 15 ft. lbs. (20 Nm).

9. Apply a light coating of chassis lube to the splines on the transaxle input shaft.

✳✳ WARNING
Keep grease off CSC to prevent fluid leakage.

10. Fit the clutch disc in place using the centering tool.

11. Install the pressure plate and all bolts finger tight.

12. Torque the pressure plate bolts in a crisscross pattern, first to 11 ft. lbs. (15 Nm), then to 19 ft. lbs. (25 Nm).

- Remove the clutch disc centering tool.
- Attach transaxle to engine. Carefully fit input shaft through clutch disc to avoid distorting/damaging clutch cover springs. Torque bolts to 46 ft. lbs. (62 Nm).

➡**DO NOT draw the transaxle to the engine with the bolts. This may damage the clutch and/or transaxle.**

13. Install or connect the following:
14. Install Engine/Transaxle assembly into the vehicle.
15. Bleed the clutch cylinders.

BLEEDING

⚙ WARNING

Two people are required to bleed the clutch cylinder. Do not use vacuum assist or any type of power bleeder. It will not purge all the air from this system.

1. Fill master cylinder reservoir with new brake fluid.
2. Connect a clear tube to the bleeding connector on the Concentric Slave Cylinder (CSC).
3. Push and release the clutch pedal slowly and fully 15 times, waiting 3 seconds between each cycle.
4. Push in the lock pin on the bleeding connector and hold it in.

⚙ WARNING

Hold the lock pin in to prevent the bleeding connector from separating when fluid pressure is applied.

5. Slide the bleeding connector out ³⁄₁₆ (5mm), then press the clutch pedal and hold it down.
6. Push the bleeding connector back in and release the clutch pedal.
7. Repeat until no bubbles are observed in the fluid flow.

FRONT HALFSHAFT

REMOVAL & INSTALLATION

See Figures 8 through 10.

1. Raise and safely support the vehicle and remove the wheel.
2. Remove wheel speed sensor from steering knuckle.

❋❋ WARNING

Do not pull on sensor wiring harness.

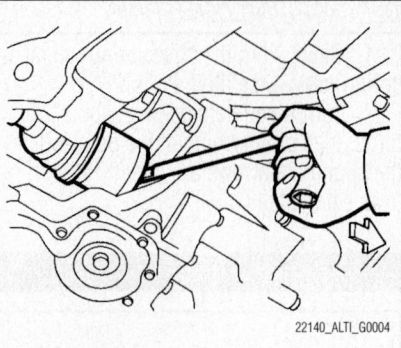
Fig. 8 Carefully pry the halfshaft out of the transaxle.

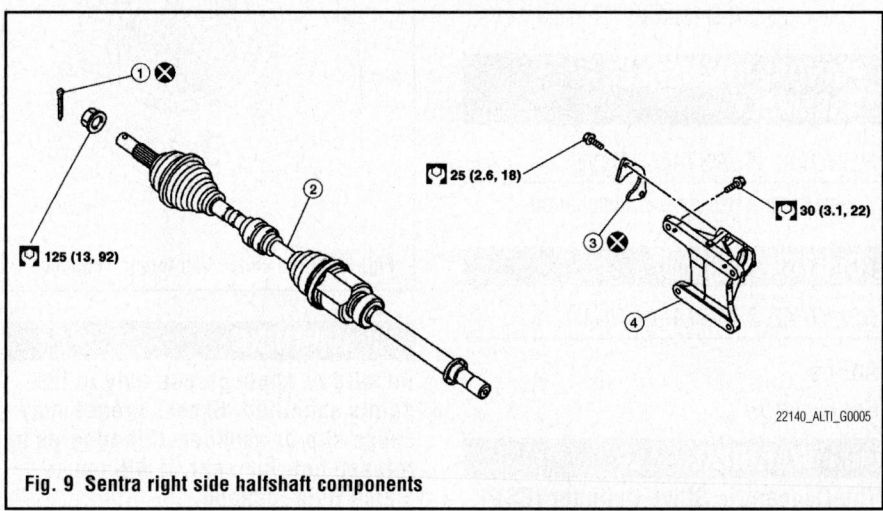
Fig. 9 Sentra right side halfshaft components

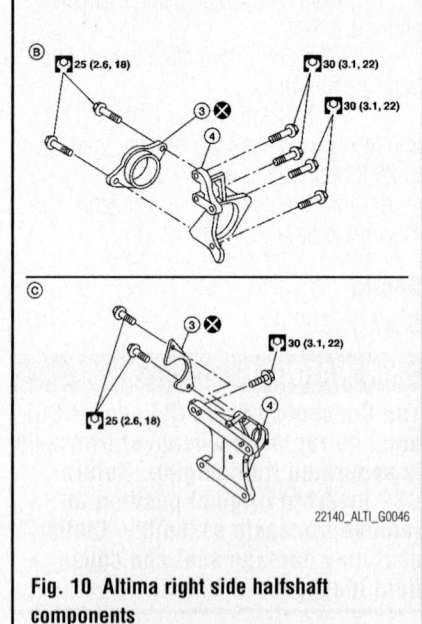
Fig. 10 Altima right side halfshaft components

3. Remove cotter pin, then loosen hub lock nut. Temporarily leave the nut installed to prevent damage to threads.
4. Remove nuts and bolts securing steering knuckle to strut assembly.

Alternately, remove the bolt to separate the lower ball joint from the steering knuckle.

5. Separate the halfshaft from the wheel hub and bearing assembly by lightly tapping the end of the shaft using a hammer and brass drift or wood block, then remove hub lock nut.

➡**NOTE: Use a suitable puller if hub and shaft cannot be separated using the above procedure.**

6. Remove the halfshaft from the wheel hub and bearing assembly.

❋❋ WARNING

Do not apply an excessive angle to halfshaft joint when removing from the wheel hub and bearing assembly. Do not excessively extend inner joint or allow the shaft to hang by the joint. Support the entire halfshaft.

7. To remove the right side halfshaft, remove the bolts and plate from the support bearing bracket.
8. Carefully pry off halfshaft from transaxle assembly.
9. Installation is the reverse of removal. Torque the castle nut to 92 ft. lbs. (125 Nm) and install a new cotter pin. Torque the steering knuckle bolts to 110 ft. lbs. (150 Nm). Torque the ball joint pinch bolt to 41 ft. lbs. (55 Nm).

CV-JOINTS OVERHAUL

Inner Joint

See Figures 11 and 12.

1. Mount halfshaft in a vise.

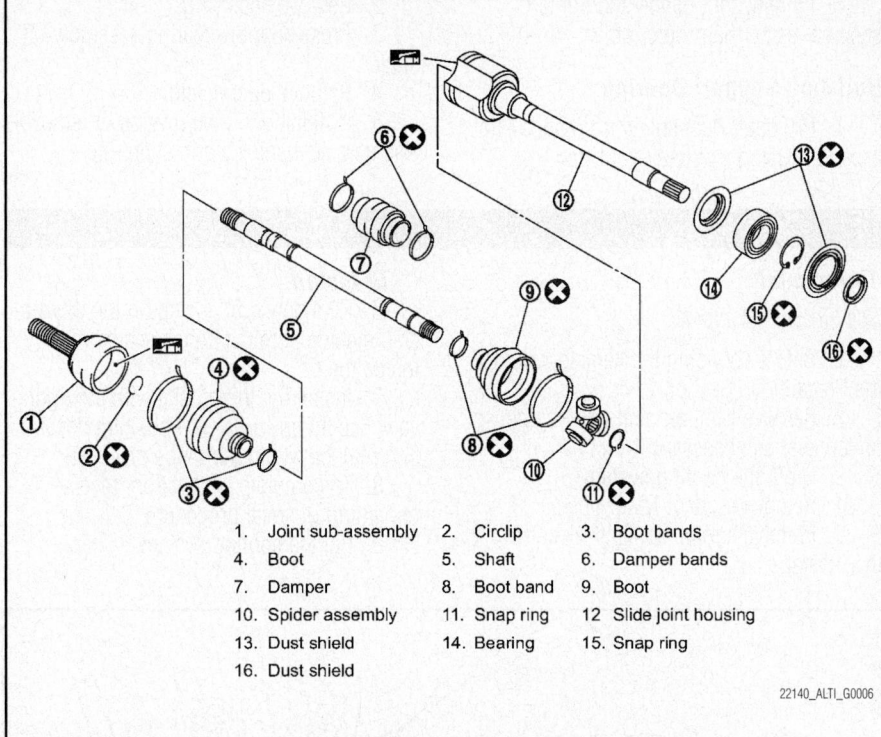

1. Joint sub-assembly
2. Circlip
3. Boot bands
4. Boot
5. Shaft
6. Damper bands
7. Damper
8. Boot band
9. Boot
10. Spider assembly
11. Snap ring
12. Slide joint housing
13. Dust shield
14. Bearing
15. Snap ring
16. Dust shield

22140_ALTI_G0006

Fig. 11 Halfshaft components: inner joint can be rebuilt. Right side halfshaft shown.

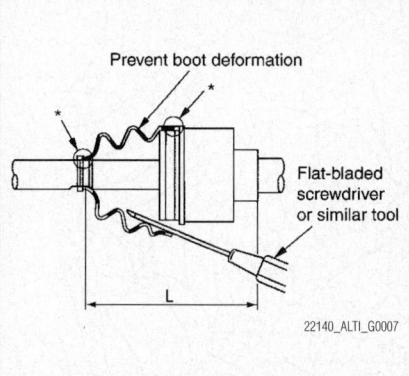

Prevent boot deformation

Flat-bladed screwdriver or similar tool

22140_ALTI_G0007

Fig. 12 When assembling inner CV-joint, make sure boot length "L" is correct

✳✳ WARNING

When mounting shaft in a vise, always use copper or aluminum plates between vise and shaft.

2. To remove inner joint:
   - Remove boot bands and slide the boot back.
   - Put matching marks on slide joint housing and shaft, then pull out shaft from slide joint housing.
   - Put matching marks on spider assembly and shaft.
   - Remove snap ring, then remove spider assembly from shaft.
   - Remove boot from shaft.

To install:

3. Cover halfshaft spline with tape to prevent damage to boot during installation.

4. Install new boot and new small boot band on shaft. Do not reuse boot or boot band.

5. Remove protective tape from half-shaft spline.

6. Align matching mark on spider assembly with matching mark on shaft and install spider assembly with chamfer facing shaft.

7. Install new snapring. Do not reuse old snapring.

8. Apply new ball joint grease to spider assembly and sliding surface.

9. Install the slide joint housing onto the spider assembly and pack with the same grease.

10. Make sure there is no grease on the outside of the joint housing and fit the boot securely into the grooves on the housing.

✳✳ WARNING

Boot may break if boot installation length is not correct.

11. Make sure boot installation length "L" is correct.
   - Altima: L = 7.5 in. (190 mm)
   - Sentra with CVT: L = 7 in. (178 mm)
   - Sentra with manual transmission: L = 6.6 in. (168 mm)

12. Insert a flat-bladed screwdriver into

the large end of boot to bleed air from boot to prevent boot deformation.

13. Secure large end of boot with new boot band. Do not reuse old boot band.

Outer Joint

See Figure 13.

1. Remove boot bands and slide the boot back.

2. Screw a slide hammer 1.2 in (30 mm) or more into threaded part of joint. Pull joint straight off of shaft.

✳✳ WARNING

If joint assembly cannot be removed after five or more unsuccessful attempts, replace the entire halfshaft assembly.

3. Remove circlip from shaft.

4. Remove boot from shaft.

To install:

5. Cover halfshaft spline with tape to prevent damage to the boot.

6. Install a new small boot band on shaft, then the new boot. Do not reuse old boot or boot band.

7. Remove protective tape.

8. Install a new circlip to shaft, making sure it fits securely into the groove. Do not reuse old circlip.

9. Thread castle nut onto joint and fit joint onto shaft.

10. Drive joint home over the circlip with a mallet.

11. Pack the joint with new CV-joint grease. Pack the remainder of the grease into boot.

12. Make sure there is no grease on the outside of the joint housing and fit the boot securely into the grooves on the housing.

✳✳ WARNING

Boot may break if boot installation length is not correct.

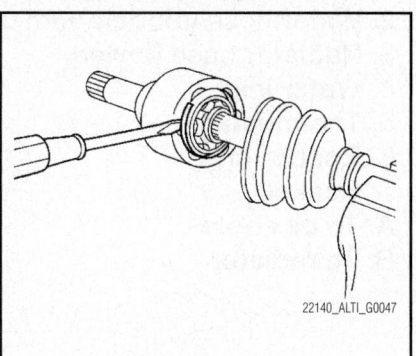

22140_ALTI_G0047

Fig. 13 Altima outer CV-joint snap ring removal

13. Make sure boot installation length from groove on the joint to the groove on the shaft is 5.6 in. (142 mm). Insert a flat-bladed screwdriver into the large end of boot to bleed air from boot to prevent boot deformation.

14. Fit new boot bands and tighten securely using the proper tool.

Halfshaft Support Bearing

1. Pry outer dust shield from slide joint assembly using a suitable tool.

2. Remove snap ring.
3. Press support bearing assembly off shaft.
4. Remove dust shield.
5. Installation is the reverse of removal. Use new dust shields and snap ring.

ENGINE COOLING

THERMOSTAT

REMOVAL & INSTALLATION

2.0 and 2.5L Engines

➡There are two thermostats on these engines. The lower one is referred to as a "thermostat," while the upper one is referred to as a "water control valve."

Thermostat

See Figures 14 and 15.

1. Before servicing the vehicle, refer to the Precautions Section.
2. Remove front air duct and engine undercover as necessary.
3. Drain the cooling system.
4. Remove radiator lower hose.
5. Remove engine coolant inlet and thermostat.

To install:

6. Fit a new rubber ring on the thermostat, making sure the flange seats properly inside the ring.
7. Install the thermostat with the jiggle valve facing upwards. The position deviation may be within the range of +/- 10 °.
8. To complete installation, reverse remaining removal procedure
9. Fill the cooling system.

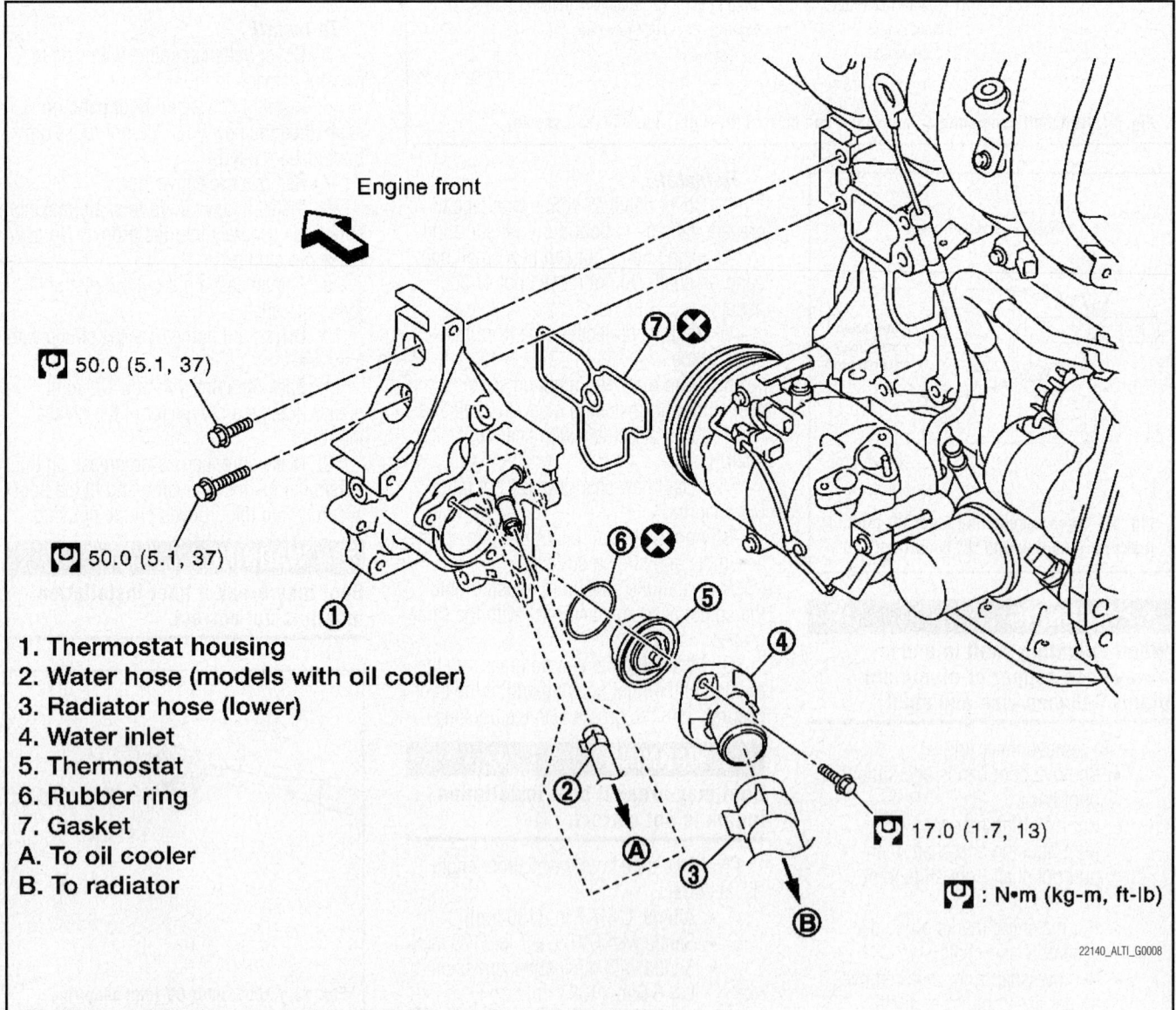

Engine front

50.0 (5.1, 37)

50.0 (5.1, 37)

17.0 (1.7, 13)

1. Thermostat housing
2. Water hose (models with oil cooler)
3. Radiator hose (lower)
4. Water inlet
5. Thermostat
6. Rubber ring
7. Gasket
A. To oil cooler
B. To radiator

: N•m (kg-m, ft-lb)

22140_ALTI_G0008

Fig. 14 Thermostat and components–2.0L engine with manual transmission shown

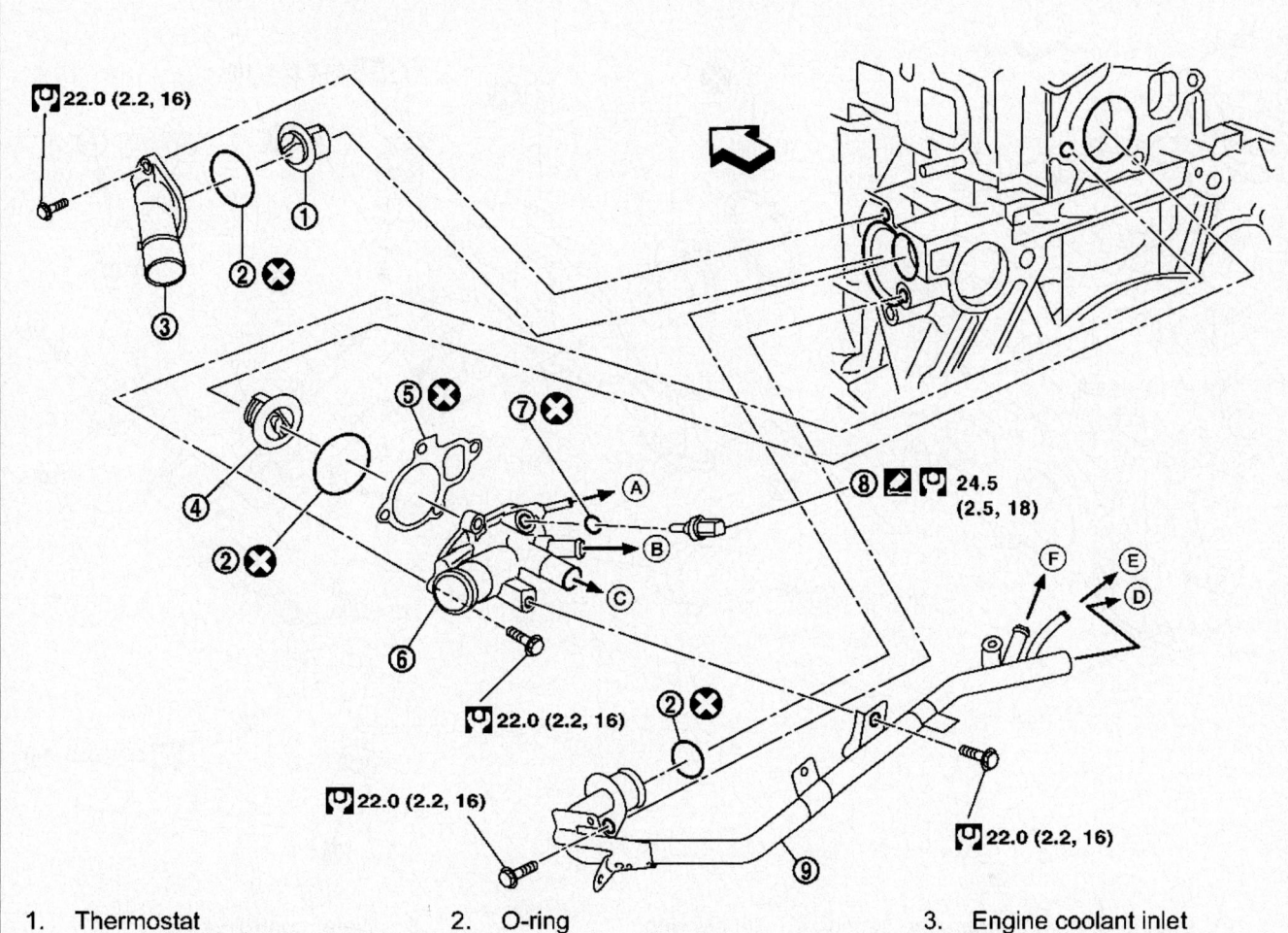

1. Thermostat
4. Water control valve
7. Copper washer
A. To electronic throttle control
D. To heater

2. O-ring
5. Gasket
8. Engine coolant temperature sensor
B. To oil cooler
E. To electronic throttle control

3. Engine coolant inlet
6. Engine coolant outlet
9. Heater pipe
C. To heater
F. To oil cooler

22140_ALTI_G0049

Fig. 15 Thermostat (1) and water valve (4) housings–2.5L engine

10. After installation, run engine for a few minutes and check for leaks.

Water Control Valve

See Figure 16.

✳✳ WARNING

Never remove the radiator cap when the engine is hot. Serious burns could occur from high pressure coolant escaping from the radiator.

1. Before servicing the vehicle, refer to the Precautions Section.
2. Drain the cooling system.

3. Remove the air duct and air cleaner.
4. Remove the upper radiator hose, heater hoses and throttle body hoses.
5. Remove the water outlet and remove the water control valve from the cylinder head.
6. To install, fit a new rubber ring on the water control valve.
7. Fit the valve into the cylinder head with arrow on outer face pointing up. The valve frame should be vertical.
8. Fit a new gasket and install the outlet onto the cylinder head.
9. Reconnect the hoses using new

O-rings, gaskets and copper washers as needed.
10. Fill the cooling system.
11. After installation, run engine for a few minutes and check for leaks.

3.5L Engine

See Figure 17.

1. Remove engine undercover.
2. Drain coolant from radiator.
3. Remove drive belts.
4. Remove water drain plug on water pump side of the engine.
5. Disconnect lower radiator hose.

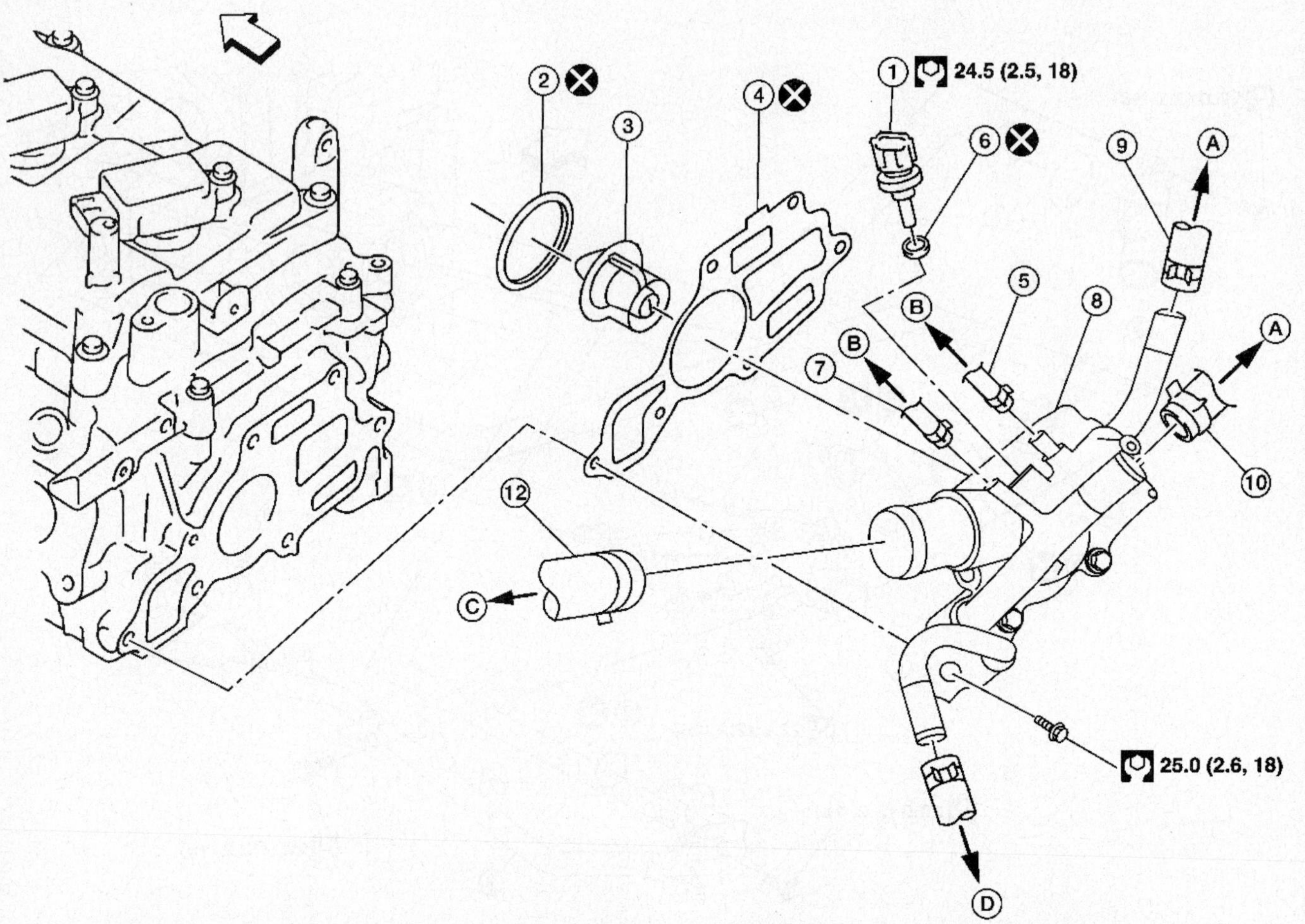

1. Engine coolant temperature sensor
2. Rubber ring
3. Water control valve
4. Gasket
5. Water hose
6. Gasket
7. Water hose
8. Water outlet
9. Heater hose
10. Heater hose
11. Water hose (CVT fluid cooler)
12. Radiator hose (upper)
⇐ Front
A. To heater
B. To electric throttle control actuator
C. To radiator
D. To CVT fluid cooler

22140_ALTI_G0009

Fig. 16 Exploded view of water valve—2.0L engine with CVT shown

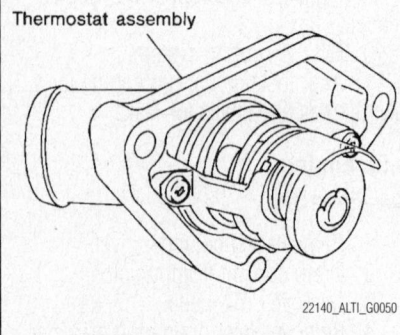

Fig. 17 Do not remove the thermostat from the housing: 3.5L engine

6. Remove engine coolant inlet and thermostat assembly.

➡**Do not remove thermostat from housing. Replace them as a unit if necessary.**

To install:

7. Make sure the thermostat jiggle valve faces up and fit the thermostat/coolant inlet housing into place with a new gasket. Torque the bolts to 87 in. lbs. (10 Nm).

8. Reconnect the hose and install the drain plug.

9. Refill the cooling system and check for leaks.

10. Install remaining components and run the engine to check for leaks again.

WATER PUMP

REMOVAL & INSTALLATION

2.0L Engine

1. Before servicing the vehicle, refer to the precautions in the beginning of this section.

2. Drain the cooling system.

3. To remove the drive belt tensioner:
   • Working underneath the vehicle,

place a wrench on the tensioner idler pulley nut.

• Push the wrench clockwise and insert a short screwdriver in the hole that appears to the right of the tensioner to hold the tensioner in place.

• Remove the drive belt

4. Loosen the water pump bolts and remove the pump.

5. Remove all traces of gasket material from sealing surfaces.

To install:

6. Install a new gasket:

7. Install the water pump and torque the bolts to 18 ft. lbs. (25 Nm).

8. Install the drive belt and hold the tensioner with a wrench to remove the screwdriver.

9. Fill the cooling system.

10. Start the vehicle, check for leaks and repair if necessary.

2.5L Engine

1. Before servicing the vehicle, refer to the precautions in the beginning of this section.

2. Drain the cooling system.

3. Remove the engine undercover.

4. A special tool is needed to remove the drive belt tensioner. To remove the tensioner:

• Working underneath the vehicle, place tool J-46535 on the tensioner idler pulley and push clockwise.

❊❊ WARNING

Do not loosen the belt tensioner pulley bolt or turn the tensioner counterclockwise. If the bolt is loosened, the tensioner assembly must be replaced.

• Insert a short screwdriver into the tensioner retaining boss to lock the tensioner in place

• Remove the drive belt

5. Remove the coolant reservoir.

6. Remove the Intelligent Power Distribution Module (IPDM) by unlocking the pawls and unplugging the connector.

7. Raise and safely support the vehicle and remove the right front wheel.

8. Remove the inner fender.

9. Remove the ground strap.

➡**The alternator and exhaust system may interfere with removal of the water pipe.**

10. Loosen the water pump bolts and remove the pump.

11. Remove all traces of gasket material from sealing surfaces.

To install:

12. Install a new gasket:

13. Install the water pump and torque the bolts to 20 ft. lbs. (28 Nm).

14. Install the drive belt and hold the tensioner with the special tool to remove the screwdriver.

15. Reinstall the remaining components.

16. Fill the cooling system.

17. Start the engine, check for leaks and repair if necessary.

3.5L Engine

See Figures 18 through 20.

➡**The water pump is driven by the timing chain, but only the chain tensioner must be removed to remove the pump.**

1. Drain the radiator and remove the coolant reservoir.

2. Raise and safely support the vehicle and remove the right wheel and the splash shield.

3. Remove drive belts.

4. Remove idler pulley, then the power steering and generator adjusting bars.

5. Support engine and remove the front engine mount and bracket.

6. Remove water drain plug on water pump side of cylinder block.

7. Remove chain tensioner cover and water pump cover.

8. Remove the timing chain tensioner assembly.

• Pull the lever down to release the plunger stopper tab.

• Insert a pin into the tensioner body hole to hold the lever and keep the plunger stopper tab released.

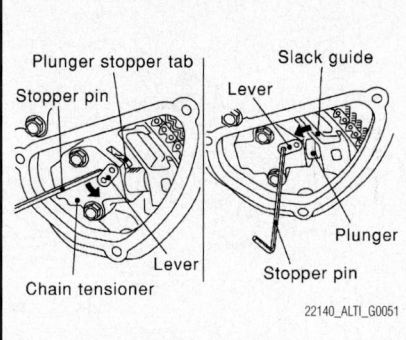

Fig. 19 Move the chain tensioner latching lever and insert a stopper pin

A drill bit or 1.2 mm Allen wrench can be used for a stopper pin.

• Push the plunger into the tensioner body by pressing the slack guide.

• Keep the slack guide pressed and push the stopper pin deeper into the chain tensioner body.

• Make a gap between water pump gear and timing chain by turning the crankshaft pulley approximately 20° clockwise.

• Remove chain tensioner.

❊❊ WARNING

Be careful not to drop the bolts inside the engine.

9. Remove the three water pump bolts. Make a gap between water pump gear and timing chain by turning crankshaft pulley counterclockwise until timing chain loosens on water pump sprocket.

10. Two M8 × 1.25 mm bolts, 2 in. (50 mm) long, are used to extract the pump from the engine. Screw the bolts into the water pump's upper and lower bolt holes

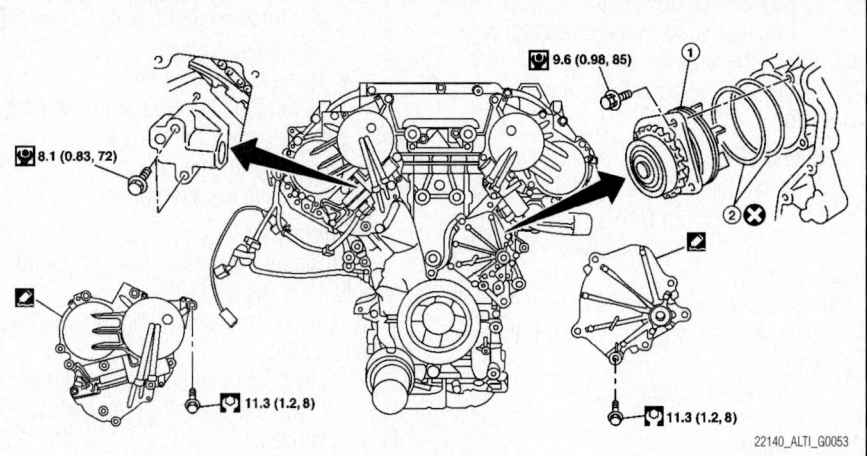

Fig. 18 Altima 3.5L engine: the water pump is on one side of the engine and the timing chain tensioner is on the other

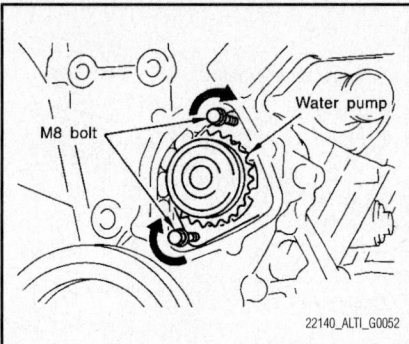

Fig. 20 Two bolts are used to extract the water pump from its housing.

until they reach the timing chain case. Alternately tighten each bolt a half turn to extract the water pump.

➡ **Pull the pump straight out to prevent the pump vanes from contacting housing or timing chain.**

11. Remove the M8 bolts and two O-rings from the pump.

To install:

12. Install two new O-rings to water pump.

13. Apply engine oil and coolant to the O-rings. Locate the O-ring with the white paint mark towards the front of the engine.

14. Carefully install the water pump to avoid damaging or displacing the O-rings.

15. Make sure the timing chain and water pump sprocket are engaged.

16. Gradually tighten the bolts alternately and evenly. Torque to 85 in. lbs. (10 Nm).

17. Turn the crankshaft pulley approximately 20° clockwise so the timing chain is loose on the chain tensioner side.

18. Make sure the chain tensioner and timing chain mounting area are clean, then apply oil to the tensioner oil hole.

19. Install the timing chain tensioner and torque the bolts to 75 in. lbs. (9 Nm).

20. Remove the stopper pin.

21. Clean the sealing surfaces of the chain tensioner cover and water pump cover. Apply RTV sealant and install the covers.

22. Install water drain plug into the cylinder block

23. Install idler pulley and torque bolts to 21 ft. lbs. (34 Nm).

24. Install the remaining components.

25. Refill the engine with coolant.

26. After starting the engine, let it idle for three minutes, then rev to 3,000 rpm under no load to purge air from the high-pressure chamber of the chain tensioner. The engine may produce a rattling noise.

This indicates air still remains in the chamber and is not a matter of concern.

ENGINE ELECTRICAL

ALTERNATOR

REMOVAL & INSTALLATION

2.0L Engine

1. Before servicing the vehicle, refer to the precautions in the beginning of this section.

2. Disconnect the negative battery cable.

3. Working underneath the vehicle, place a wrench on the tensioner idler pulley nut.

4. Push the wrench clockwise and insert a short screwdriver in the hole that appears to the right of the tensioner to hold the tensioner in place.

5. Remove the drive belt.

6. Unplug the alternator connector and disconnect the large B+ cable.

7. Remove the bolts to remove the alternator.

To install:

8. Install the alternator and torque the bolts to 18 ft. lbs. (25 Nm).

9. Connect the wiring. Tighten the B+ cable carefully.

10. Install the drive belt and hold the tensioner with the wrench to remove the screwdriver.

11. Connect the battery cable.

2.5L Engine

1. Before servicing the vehicle, refer to the Precautions Section.

2. Disconnect negative battery cable.

3. Remove engine undercover.

4. A special tool is needed to remove the drive belt tensioner. To remove the tensioner:

- Working underneath the vehicle, place tool J-46535 on the tensioner idler pulley and push clockwise.

✳✳ WARNING

Do not loosen the belt tensioner pulley bolt or turn the tensioner counterclockwise. If the bolt is loosened, the tensioner assembly must be replaced.

- Insert a short screwdriver into the tensioner retaining boss to lock the tensioner in place
- Remove the drive belt

5. Unplug the connector from the alternator and disconnect the B+ cable.

6. Remove the alternator mounting bolts and remove the alternator.

To install:

7. Install the alternator and torque the bolts to 18 ft. lbs. (25 Nm)

8. Connect the wiring. Tighten the B+ cable carefully.

9. Install the drive belt and hold the tensioner with the special tool to remove the screwdriver.

10. Install the engine undercover.

11. Connect the negative battery cable.

CHARGING SYSTEM

3.5L Engine

➡ **The A/C compressor must be removed to remove the alternator.**

1. Disconnect the negative battery cable.

2. Drain engine coolant.

3. Remove engine room cover.

4. Raise and safely support the vehicle and remove the right front wheel.

5. Remove front/right-side engine undercover.

6. Remove air cleaner and duct assembly.

7. Remove battery tray.

8. Remove cooling fan assembly.

9. Remove the drive belt.

10. Remove the A/C compressor.

- Evacuate A/C system.
- Disconnect the high-pressure flexible hose and low-pressure flexible hose from the compressor. Cap or wrap the hoses and compressor openings with vinyl tape to entry of any contaminants or moisture.
- Disconnect the clamp and reposition the power steering pipe out of the way.
- Disconnect the compressor wiring.
- Remove the right hand compressor bolts.
- Remove the front compressor bolts and remove the compressor.

11. Remove idler pulleys.

12. Disconnect the oil pressure switch.

13. Disconnect the alternator wiring.

14. Remove the bolt and nuts and slide the alternator out.

To install:

15. Fit the alternator into place and install the bolts. Torque to 15 ft. lbs. (20 Nm).
16. Connect the alternator wiring.
17. Connect the oil pressure switch.
18. Install idler pulleys.
19. Install the A/C compressor
   - Install the compressor and torque the bolts to 45 ft. lbs. (61 Nm).
   - Install new O-rings on the hoses and lubricate them with A/C oil.
   - Connect the hoses
20. Install the drive belt.
21. Install cooling fan assembly.
22. Install battery tray and battery.
23. Refill the cooling system and run the engine to check for leaks.
24. Recharge the A/C system and check for leaks.
25. Install remaining components.

ENGINE ELECTRICAL

FIRING ORDER

See Figure 21.

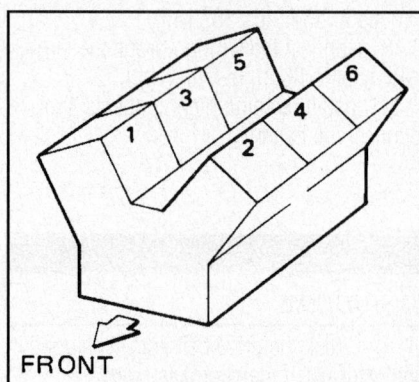

Fig. 21 3.5L Engines
Firing order: 1–2–3–4–5–6
Distributorless ignition system
(one coil on each cylinder)

IGNITION COIL

REMOVAL & INSTALLATION

2.0L and 3.5L Engines

1. Remove the intake manifold.
2. Disconnect coil wiring.
3. Remove bolt to remove ignition coil.
4. Installation is reverse of removal.

2.5L Engine

1. Disconnect the ignition coil wiring as required.
2. Remove ignition coil.
3. Installation is the reverse of removal. Make sure coil seals the opening in the rocker cover.

IGNITION TIMING

ADJUSTMENT

2.0L Engine

See Figures 22 and 23.

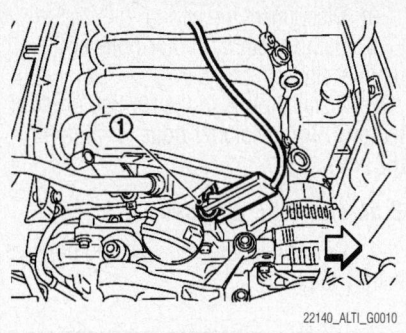

Fig. 22 Some vehicles have a timing loop for connecting a standard timing light

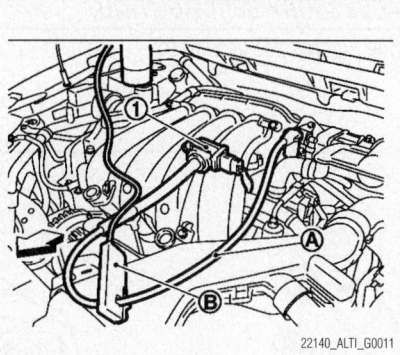

Fig. 23 Add an extension wire between the No. 4 ignition coil and spark plug to connect a timing light

1. There are two different ways to connect a standard timing light. If equipped with a loop wire as shown, simply attach the pick-up clamp and check ignition timing at the crankshaft pulley.
2. If no loop wire is installed:
   - Remove No. 4 ignition coil
   - Make up a high-tension extension wire using a suitable sparkplug wire.
   - Connect No. 4 ignition coil to No. 4 spark plug with high-tension extension wire and attach timing light clamp to this wire.

IGNITION SYSTEM

3. Check ignition timing at the crankshaft pulley. Timing should be 6° ± 5° BTDC (in Neutral), but there is no adjustment.

2.5L and 3.5L Engines

See Figure 24.

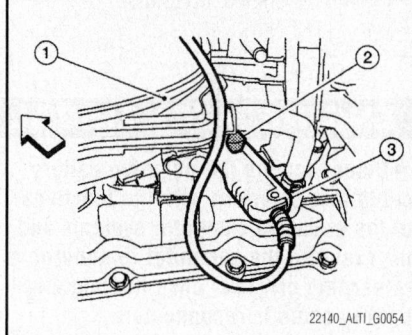

Fig. 24 On 3.5L engine, a timing light pickup can be connected to the No.1 ignition coil on the rear cylinder bank below the intake manifold—3.5L engine

1. Slide the wiring harness protector of ignition coil No.1 back (2) to reach the wires.
2. Attach timing light to the ignition coil No.1 wires as shown in the figure.
3. On 3.5L engine, ignition timing should be 18± 5° BTDC at idle. There is no adjustment.
4. On 2.5L engine, ignition timing should be 15± 5° BTDC at idle. There is no adjustment.

SPARK PLUGS

REMOVAL & INSTALLATION

1. On 2.0L and 3.5L engines, remove intake manifold
2. Remove ignition coils
3. Remove spark plugs.
4. Installation is the reverse of removal.

ENGINE ELECTRICAL

STARTING SYSTEM

STARTER

REMOVAL & INSTALLATION

Manual Transaxle

1. Disconnect the battery negative terminal.

2. On 2.0L engine, remove air inlet duct.

3. Remove "S" terminal nut.

4. Remove "B" terminal nut.

5. Remove starter motor bolts.

6. Remove starter motor.

7. Installation is in the reverse order of removal. Torque starter bolts:
- 2.0L and 3.5L engines: 46 ft. lbs. (62 Nm).
- 2.5L engine, 83 ft. lbs. (112 Nm).

✳✳ WARNING

Tighten "B" terminal nut carefully.

Automatic Transaxle

2.0L Engine

1. Disconnect the battery negative terminal.

2. Remove air inlet duct.

3. Disconnect the wiring.

4. Remove starter motor bolts and remove starter motor.

5. Installation is in the reverse order of removal. Torque starter bolts to 46 ft. lbs. (62 Nm).

2.5L and 3.5L Engines

1. Disconnect the negative and positive battery terminal.

2. Remove the air cleaner assembly and air ducts.

3. On 3.5L engine, remove the following:
- ECM
- CVT control unit
- IPDM/ER (fuel/relay box)

4. Remove the battery tray.

5. Disconnect the starter wiring.

6. Remove the two starter bolts and remove the starter.

To install:

7. Fit starter into place and install bolts. Torque to 45 ft. lbs. (62 Nm).

8. Connect the wiring, torque the large cable nut to 86 in. lbs. (10 Nm).

9. Install remaining components and connect the battery.

ENGINE MECHANICAL

➡Disconnecting the negative battery cable may interfere with the functions of the on board computer systems and may require the computer to undergo a relearning process, once the negative battery cable is reconnected.

ACCESSORY DRIVE BELTS

ACCESSORY BELT ROUTING

See Figures 25 through 27.

INSPECTION

1. Check the drive belt auto-tensioner indicator when the engine is cold.

2. Make sure that the indicator (notch on fixed side) of drive belt auto-tensioner points to range A.

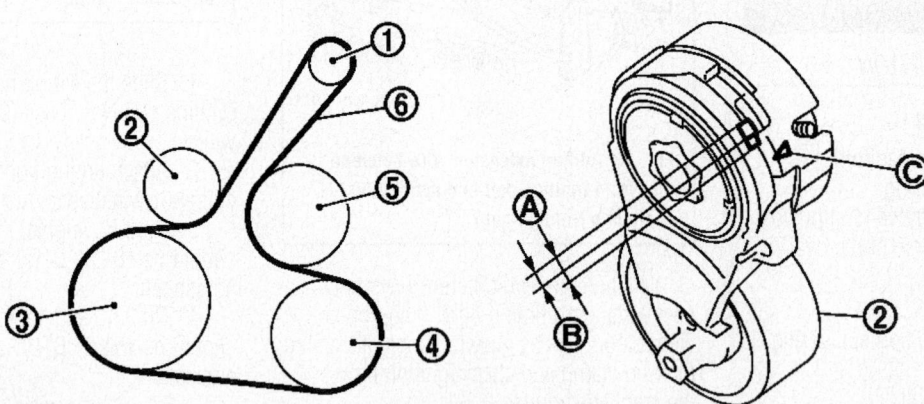

1.	Alternator	2.	Drive belt auto–tensioner	3.	Crankshaft pulley
4.	A/C compressor (models with A/C) Idler pulley (models without A/C)	5.	Water pump	6.	Drive belt
A.	Possible use range	B.	Range when new drive belt is installed	C.	Indicator

22140_ALTI_G0012

Fig. 25 Accessory drive belt and automatic tensioner: 2.0L engine

Drive belt auto-tensioner
View A
ALT
W/P
P/S
C/P
A/C
Indicator (notch)
Engine
front
A
Range of possible use
View A
22140_ALTI_G0037

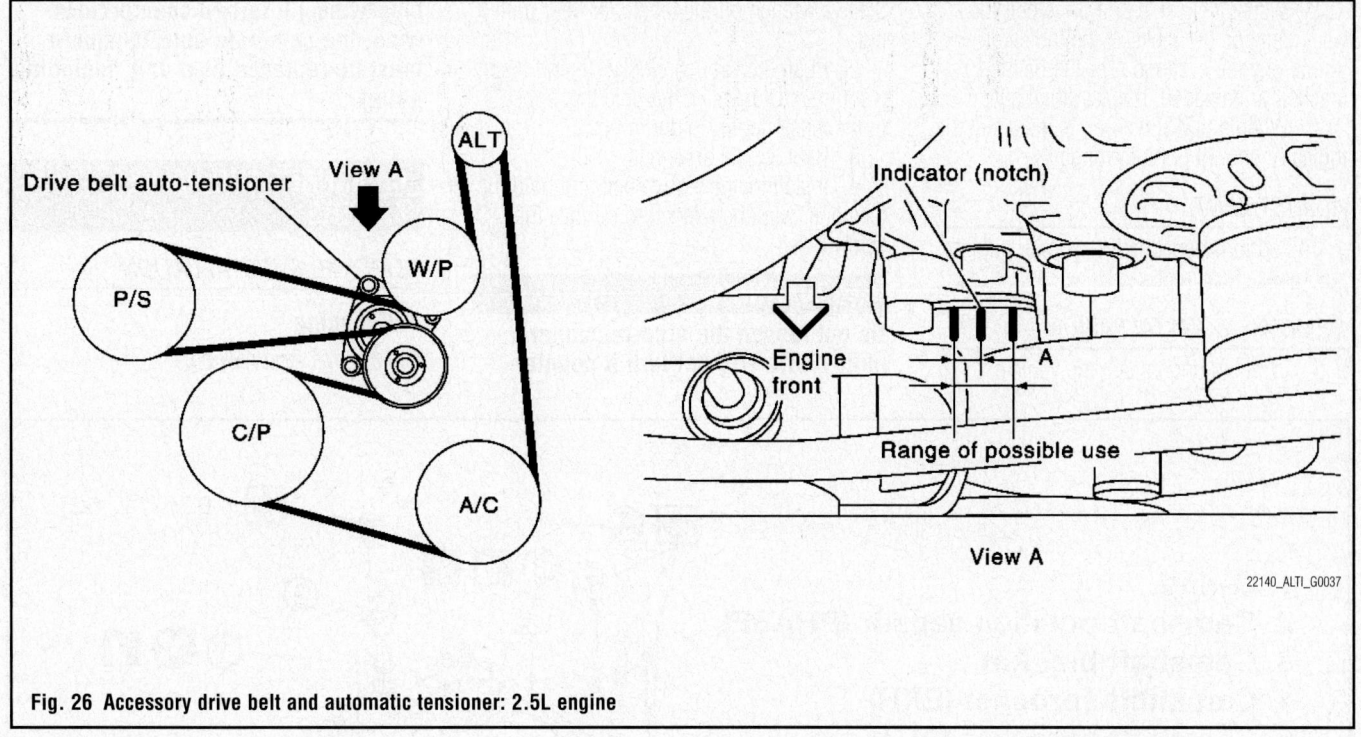

Fig. 26 Accessory drive belt and automatic tensioner: 2.5L engine

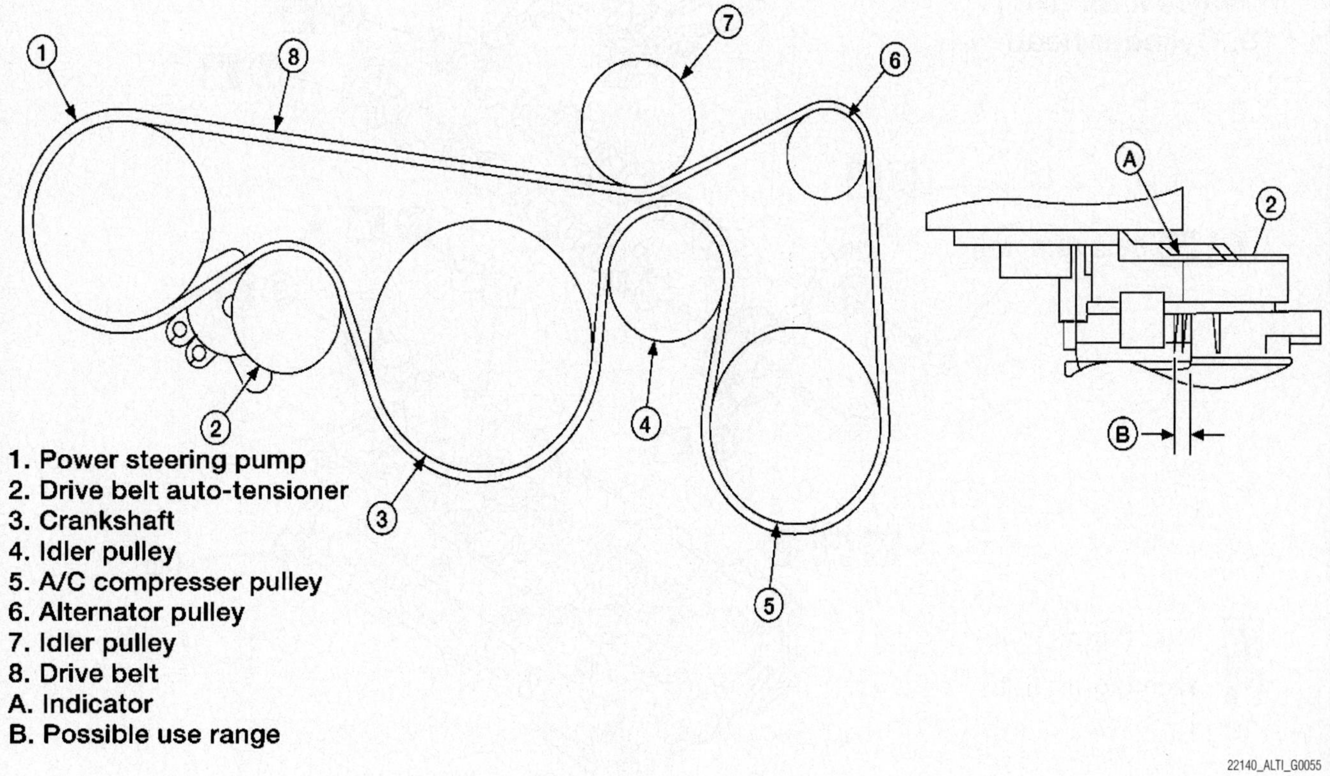

1
8
7
6
A
2
2
4
3
5
B
1. Power steering pump
2. Drive belt auto-tensioner
3. Crankshaft
4. Idler pulley
5. A/C compresser pulley
6. Alternator pulley
7. Idler pulley
8. Drive belt
A. Indicator
B. Possible use range
22140_ALTI_G0055

Fig. 27 Accessory drive belt and automatic tensioner: 3.5L engine

3. Visually check entire drive belt for wear, damage or cracks. If the indicator points outside of range A or the belt is cracked or damaged, replace drive belt.

4. When a new drive belt is installed, the indicator should point to range B.

ADJUSTMENT

The spring-loaded belt tensioner adjusts automatically as required.

REMOVAL & INSTALLATION

1. Working underneath the vehicle, place a wrench on the tensioner idler pulley nut.

2. Push the wrench clockwise and insert a short screwdriver in the hole that appears in the tensioner to hold it in place.

3. Remove the drive belt.

4. Install the new drive belt and hold the tensioner with the wrench to remove the screwdriver.

❄❄ WARNING

Do not loosen the auto-tensioner pulley bolt. (Do not turn it counter-

clockwise.) If turned counterclockwise, the complete auto-tensioner must be replaced as a unit, including pulley.

CAMSHAFT AND VALVE LIFTERS

REMOVAL & INSTALLATION

2.0L Engine
See Figures 28 through 31.

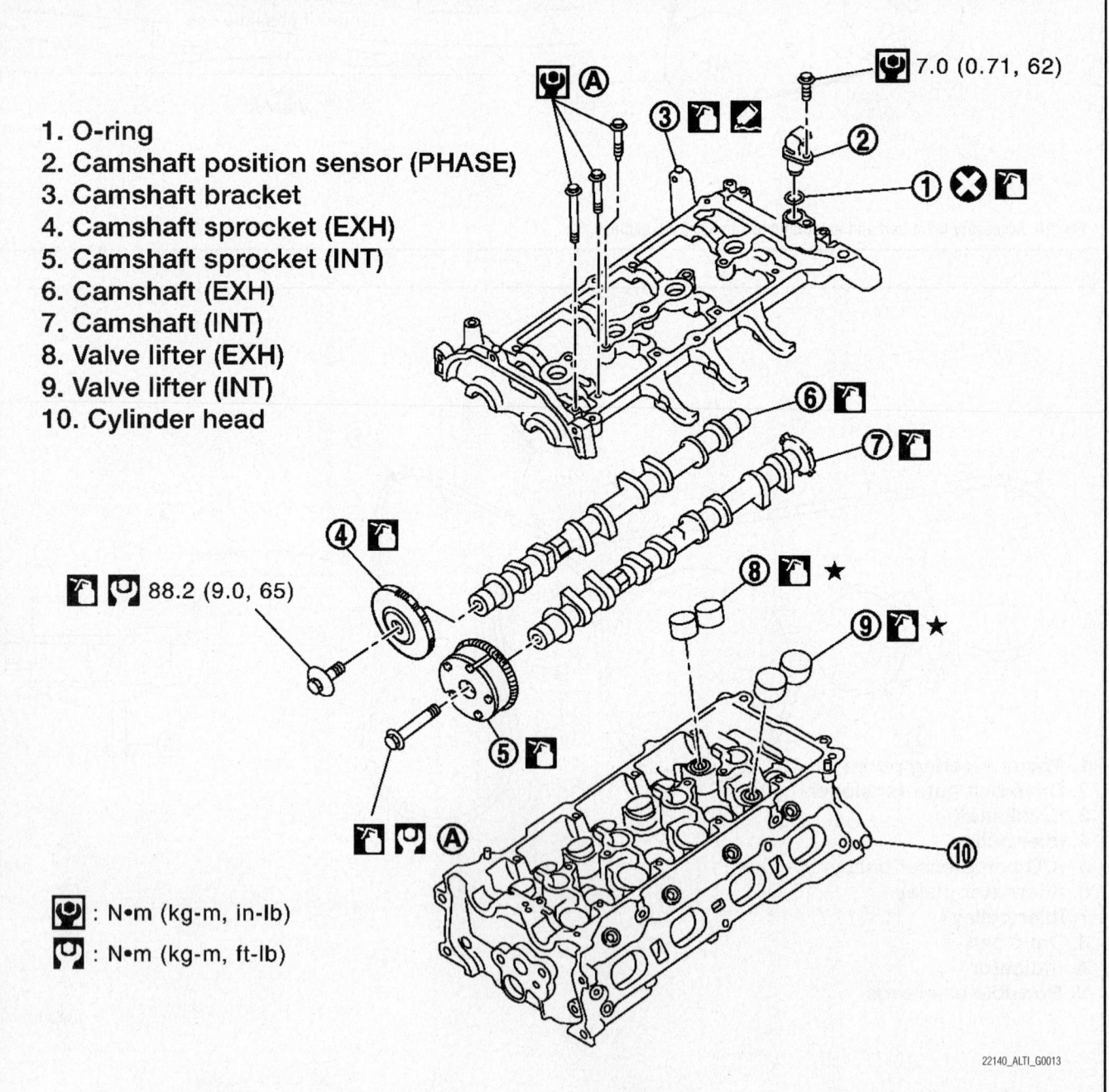

1. O-ring
2. Camshaft position sensor (PHASE)
3. Camshaft bracket
4. Camshaft sprocket (EXH)
5. Camshaft sprocket (INT)
6. Camshaft (EXH)
7. Camshaft (INT)
8. Valve lifter (EXH)
9. Valve lifter (INT)
10. Cylinder head

7.0 (0.71, 62)

88.2 (9.0, 65)

: N•m (kg-m, in-lb)

: N•m (kg-m, ft-lb)

22140_ALTI_G0013

Fig. 28 Exploded view of camshaft components—2.0L engine

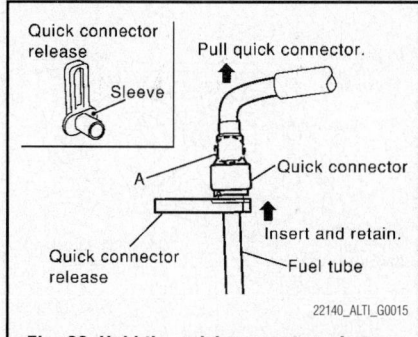

Fig. 29 Hold the quick connector release tool up while pulling fuel line straight off. Don't twist the fuel line.

Fig. 30 Align match marks on intake camshaft sprocket and camshaft bracket

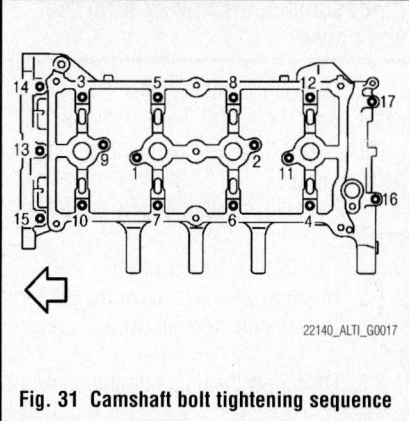

Fig. 31 Camshaft bolt tightening sequence

✳✳ CAUTION

Be sure to work in a well ventilated area and keep a CO2 fire extinguisher handy. Do not smoke while servicing fuel system. Keep open flames and sparks away from the work area.

1. Release the fuel pressure.
   - Remove fuel pump fuse.
   - Start engine.
   - After engine stalls, crank it two or three times to release all fuel pressure.
   - Turn ignition switch OFF.
2. Disconnect negative battery cable.
3. Remove right front wheel.
4. Remove inner front fender.
5. Drain engine coolant.
6. Remove the intake manifold:
7. Remove ignition coils, spark plugs and rocker cover.
8. Remove fuel tube and fuel injector assembly
   - Remove quick connector cap from quick connector connection.
   - Disconnect fuel feed hose from hose clamp.
   - With the sleeve side of quick connector release facing quick connector, install quick connector release tool onto fuel tube.
   - Insert quick connector release into quick connector until sleeve contacts and goes no further. Hold quick connector release on that position. Inserting quick connector release hard will not disconnect quick connector. Hold quick connector release where it contacts and goes no further.
   - Draw and pull out quick connector straight from fuel tube.

➥ Valves are adjusted by replacing the lifters. Now is the time to measure and record valve clearances.

9. Remove the rocker cover.
10. Remove the timing cover, timing chain and related parts.
    - Remove camshaft position sensor from camshaft bracket.

✳✳ WARNING

Handle camshaft sensor carefully to avoid dropping and shocks. Never disassemble. Never allow metal powder to adhere to magnetic part at sensor tip. Never place sensor in a location where it is exposed to magnetism.

11. Align the match marks (A) on the intake camshaft sprocket (2) and the camshaft bracket (1) as shown. This prevents the knock pin of the camshaft from engaging with the incorrect pin hole when installing the camshaft sprocket.
12. Hold the camshaft with a wrench and loosen camshaft sprocket bolts to remove camshaft sprocket.

✳✳ WARNING

Never rotate crankshaft or camshaft while timing chain is removed. It causes interference between valve and piston. Never loosen the sprocket bolts without holding the camshaft securely with a wrench.

13. Loosen bolts in reverse of tightening order.
14. Cut liquid gasket by prying at the right front and left rear corners of the camshaft bracket, then remove the camshaft bracket.

✳✳ WARNING

Be careful not to damage the mating surface. A more adhesive liquid gasket is applied compared to previous types when shipped, so it should not be forced off the position not specified.

15. Carefully lift out the camshafts.

To install:

16. Install valve lifters in their original positions.
17. Install camshafts. Position the dowel pins at the sprocket end at the 12 o'clock position.
18. Carefully clean the camshaft bracket and apply a bead of RTV sealer around the outer edges and spark plug holes.
19. Fit the camshaft bracket into position and install the bolts. The long bolts go into holes 13, 14 and 15.
20. Tighten all bolts in numerical order in three steps. Do not over tighten the bolts.
    - Step 1: 17 in. lb. (2 Nm)
    - Step 2: 52 in. lb. (6 Nm)
    - Step 3: 84 in. lb. (9.5 Nm)
21. Install the intake camshaft sprocket, making sure to align the match marks. Hold the camshaft with a wrench and tighten the bolt to 26 ft. lbs. (35 Nm) plus 67 °.

✳✳ WARNING

Use an angle-measuring wrench.

22. Install the exhaust camshaft sprocket, hold the camshaft with a wrench and torque the bolt to 65 ft. lbs. (88 Nm).
23. Install timing chain and related parts.
24. Inspect and adjust valve clearance.
25. Installation of the remaining components is in the reverse order of removal.

2.5L Engine
See Figures 32 through 37.

✳✳ CAUTION

Be sure to work in a well ventilated area and keep a CO2 fire extinguisher handy. Do not smoke while servicing fuel system. Keep open

flames and sparks away from the work area.

1. Disconnect the negative battery cable.

2. Support the engine using a suitable hoist or jack.

3. Remove the right engine mount and brackets.

4. Remove the rocker cover.

5. Remove the power steering reservoir.

6. Remove the coolant overflow reservoir.

7. Disconnect variable timing control solenoid.

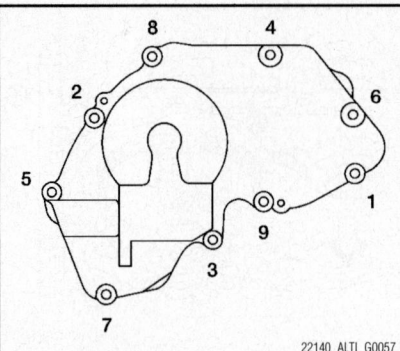

Fig. 33 Valve timing control cover bolt tightening sequence—2.5L engine

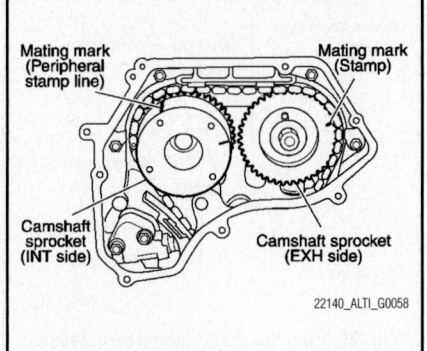

Fig. 34 Align cam timing marks to find TDC

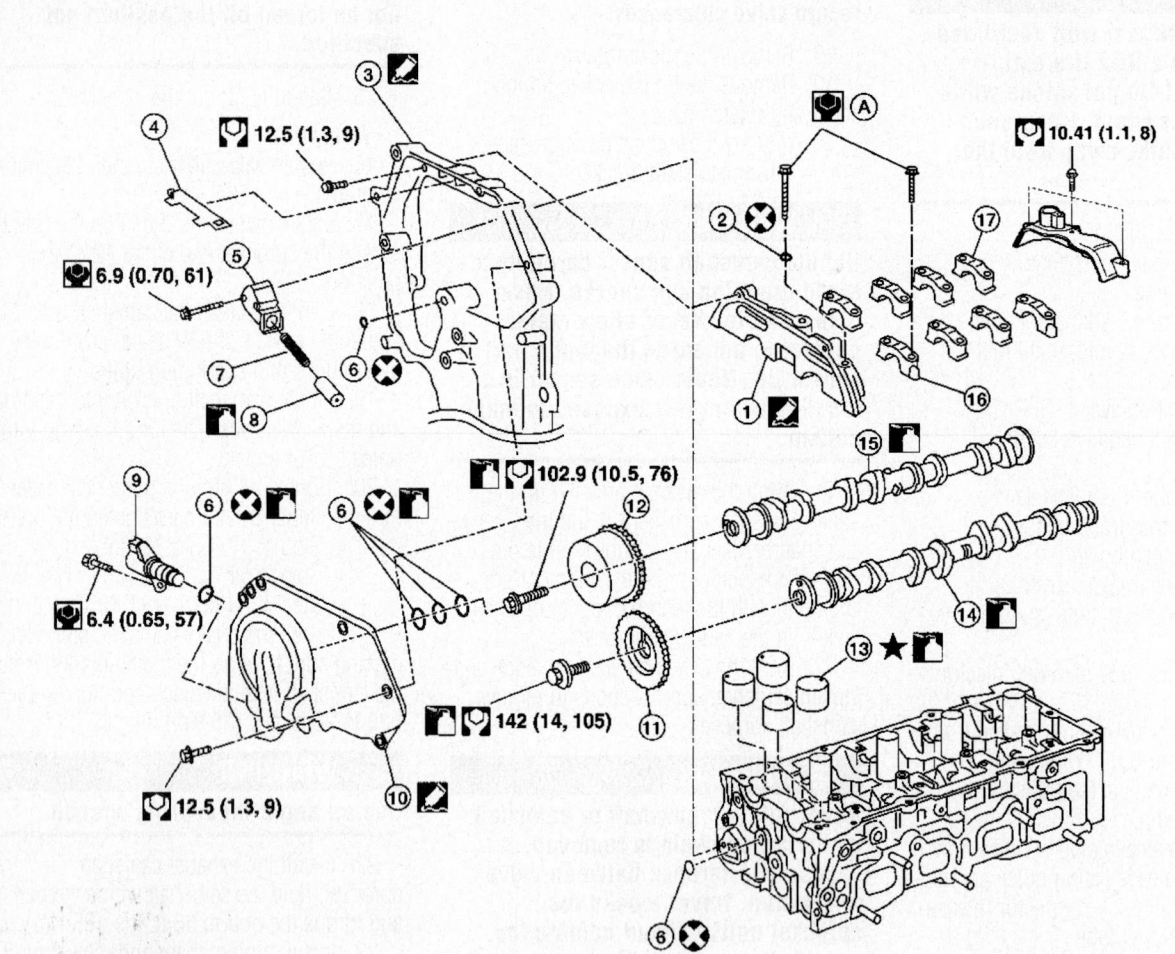

1.	Camshaft bracket (No.1)	2.	Washer	3.	Front cover (partial view)
4.	Chain guide	5.	Chain tensioner	6.	O-ring(s)
7.	Chain tensioner spring	8.	Chain tensioner plunger	9.	IVT control solenoid valve
10.	IVT control cover	11.	Camshaft sprocket (EXH)	12.	Camshaft sprocket (INT)
13.	Valve lifter	14.	Camshaft (EXH)	15.	Camshaft (INT)
16.	Camshaft bracket (No. 2)	17.	Camshaft bracket (No. 5)	A.	Follow installation procedure

Fig. 32 Exploded view of camshaft components—2.5L engine

8. Loosen the camshaft timing control cover bolts in reverse order of the tightening sequence.

9. Set the No.1 cylinder at TDC on its compression stroke with the following procedure:

- Open the splash cover under the engine.
- Rotate crankshaft pulley clockwise to align timing mark for TDC with timing indicator on front cover.
- Make sure the mating marks on camshaft sprockets are lined up with the yellow links in the timing chain as shown. If not, rotate crankshaft pulley one more turn to line up the mating marks to the yellow links.

10. Pull the timing chain guide out between the camshaft sprockets through front cover.

11. Remove camshaft sprockets with the following procedure.

✳✳ WARNING

Do not rotate the crankshaft or camshaft while the timing chain is removed. This is an interference engine.

➡ **Chain tension holding work is not necessary. Crankshaft sprocket and timing chain do not disconnect while front cover is attached.**

- Make sure marks on camshaft sprockets are aligned with the yellow links in the timing chain
- Paint an indelible mating mark on the sprocket and timing chain link plate.
- Push in the chain tensioner plunger and insert a 0.020 in. (0.5 mm) pin into the hole on tensioner body to hold the chain tensioner.
- Remove the timing chain tensioner.

12. Hold the hexagonal part of camshaft with a wrench and loosen the camshaft sprocket mounting bolts to remove the camshaft sprockets.

13. Loosen the camshaft bearing cap bolts in reverse of the order and remove the caps. A rubber mallet may be needed.

14. Remove the camshafts.

15. When removing the valve lifters, keep them in order so they can be installed in the same locations.

To install:

16. Install the valve lifters.

17. Lubricate and carefully fit the camshafts into place. The back of the intake camshaft has the signal plate for the

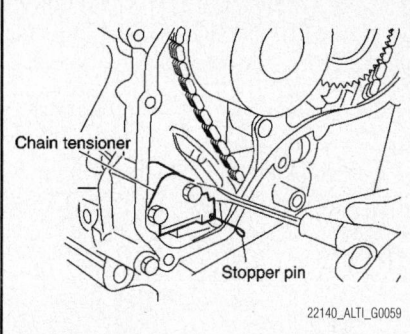

Fig. 35 Push the chain tensioner plunger in and insert a pin to keep it retracted

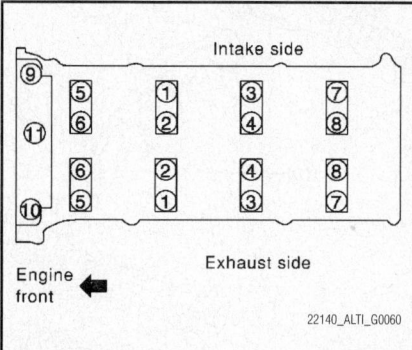

Fig. 36 Camshaft bearing cap bolt tightening order—2.5L engine

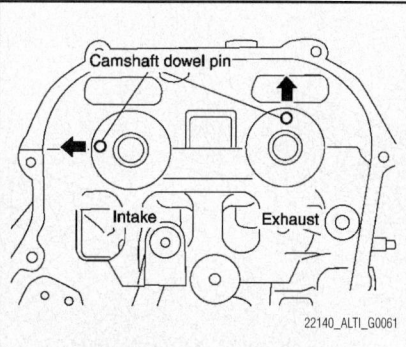

Fig. 37 Make sure camshaft dowel pins are positioned as shown.

camshaft position sensor. Make sure the dowel pins on the front side are positioned as shown.

18. Install the camshaft bearing caps so the numbers or letters can be read from the exhaust side of the cylinder head. Caps 2, 3, 4 and 5 go on the intake camshaft, front to rear of engine. Caps A, B, C and D are front-to-rear exhaust camshaft caps. Make the bolts only finger tight at this time.

19. Make sure all sealing surfaces are clean and apply RTV sealant to the bracket where it contacts the cylinder head and the front cover. The sealant bead should be outside the bolt holes.

20. Position the camshaft bracket near the mounting position and install it without disturbing the sealant.

21. Tighten all camshaft bracket and bearing cap bolts in four steps in the order shown.

- Step 1, bolts 9–11: 17 in. lbs. (2 Nm)
- Step 2, bolts 1–8: 17 in. lbs. (2 Nm)
- Step 3, bolts 1–11 52 in. lbs. (6 Nm)
- Step 4, bolts 1–11: 92 in. lbs. (10 Nm)

22. Wipe off any excess sealant.

23. Install camshaft sprockets, making sure to line up the mating marks on each camshaft sprocket with the ones painted on the timing chain during removal.

➡ **Before installation of chain tensioner, it is possible to re-match the marks on timing chain with the ones on each sprocket.**

24. Install chain tensioner and check again to make sure that mating marks have not slipped.

25. Remove the stopper pin from the tensioner and check the timing marks again.

26. Install chain guide.

27. Install camshaft timing control cover with the following procedure.

- Install control solenoid valve to intake valve timing control cover.
- Install new O-ring to front cover side.

28. Apply RTV sealant to the cover. The bead should be inside the bolt holes.

- Install the control cover and tighten the bolts in numerical order to 9 ft. lbs. (12 Nm).

29. Check and adjust valve clearances.

30. Installation of the remaining components is in the reverse order.

3.5L Engine

See Figures 38 through 41.

✳✳ CAUTION

Be sure to work in a well ventilated area and keep a CO2 fire extinguisher handy. Do not smoke while servicing fuel system. Keep open flames and sparks away from the work area.

1. Remove the timing chains.

2. Remove the fuel rail and injectors.

3. Remove camshaft position sensor brackets.

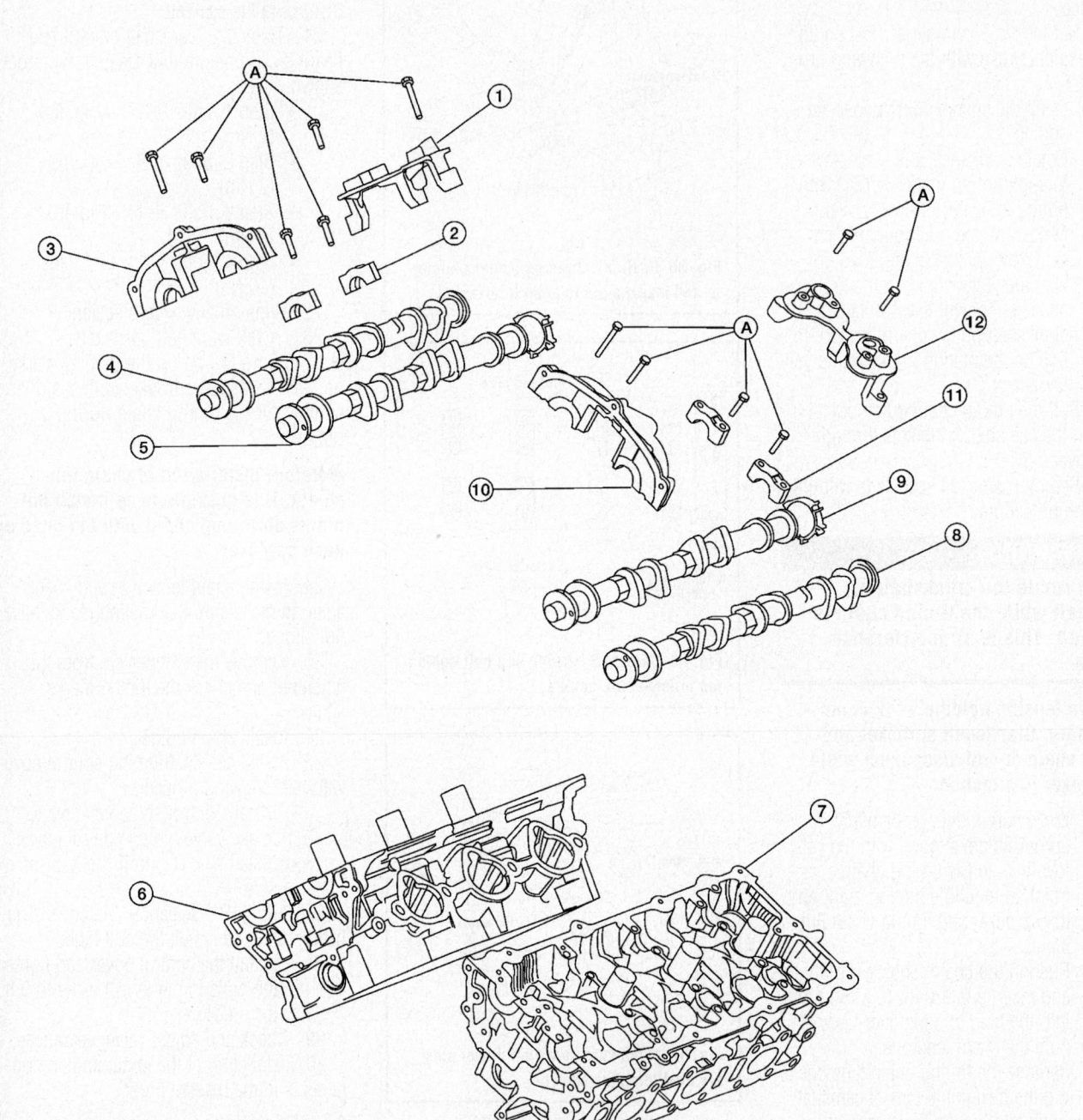

1. Camshaft position sensor bracket (RH) 2. Camshaft brackets 3. No. 1 camshaft bracket (RH)
4. Camshaft (EXH) RH 5. Camshaft (INT) RH 6. Cylinder head (RH)
7. Cylinder head (LH) 8. Camshaft (EXH) LH 9. Camshaft (INT) LH
10. No. 1 camshaft bracket (LH) 11. Camshaft brackets 12. Camshaft position sensor bracket (LH)
A. Follow installation procedure

22140_ALTI_G0062

Fig. 38 Exploded view of camshaft assembly—3.5L engine

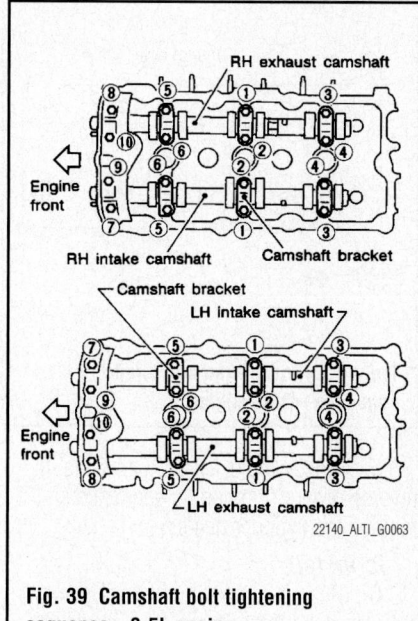

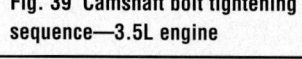

Fig. 39 Camshaft bolt tightening sequence—3.5L engine

4. Mark the camshaft bearing caps so they can be returned to their original positions.

5. Loosen the bolts in reverse of tightening order in several steps.

6. Remove the bolts, bearing caps and camshafts. Mark all parts so they can be installed in their original locations, including the camshafts.

7. Remove valve lifters and mark them so they can be returned to their original locations.

8. Remove secondary timing chain tensioner from cylinder head with its stopper pin still inserted.

To install:

9. Before installation, clean off any old RTV sealant and clean the bolt holes.

Make sure the crankshaft is set at TDC on the No. 1 piston's compression stroke. The crankshaft key should line up with the right bank cylinder center line.

10. Install camshaft chain tensioners on both sides of cylinder head.

11. Install valve lifters.

12. Install exhaust and intake camshafts and fit bearing caps into place. Make the bolts only finger tight.

➡**The intake camshaft has a drill mark on sprocket mounting flange.**

13. Before installing the front bracket, make sure the sealing surfaces are clean. Apply RTV sealant and fit the front bracket into place.

14. Carefully torque all camshaft bolts in the correct sequence in several steps.
- Step 1: tighten bolts 7–10 in order,

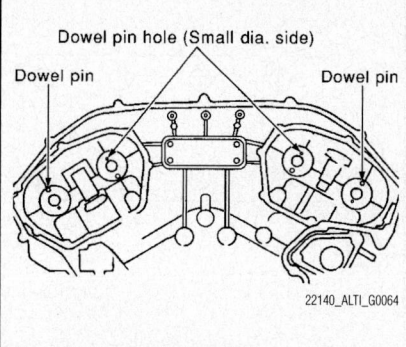

Fig. 40 Camshaft dowel pin positions for camshaft installation

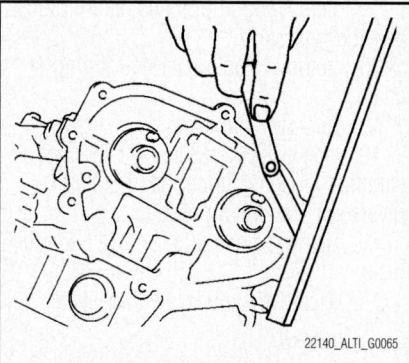

Fig. 41 Measure alignment of camshaft bracket and cylinder head.

then bolts 1–6 in order to 17 in. lb. (2 Nm)
- Step 2: tighten all bolts in order to 52 in. lb. (6 Nm)
- Step 3: tighten bolts 1–6 in order to 80–104 in. lb. 9–12 Nm)

15. Measure the alignment of the front camshaft bracket and cylinder head. If the parts are more than 0.006 in. (0.14 mm) out of alignment, re-install camshaft and camshaft bracket.

16. Install the valve timing control solenoid valves with new gaskets.

17. If necessary, install camshaft position sensor.

18. Install the fuel rail and injectors.

19. Install the timing chains.

CRANKSHAFT FRONT SEAL

REMOVAL & INSTALLATION
See Figure 42.

2.0L Engine

1. Remove front timing cover.
2. Carefully pry out old seal.
3. Using a suitable tool, press-fit the new seal until it is flush with the front of the cover. Make sure the seal is straight and not curled.

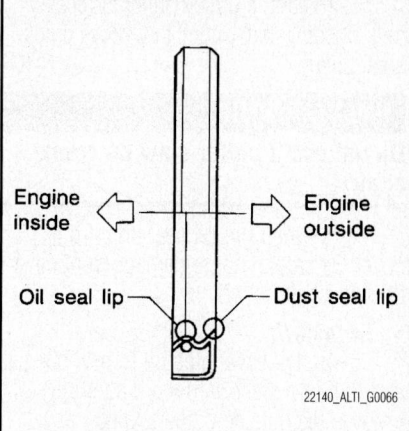

Fig. 42 Front and rear main seals are installed with the longer lip towards the outside of the engine—all engines

2.5L Engine

1. Raise and safely support the vehicle and remove the right front wheel.
2. Remove the engine under cover.
3. Remove the drive belts.
4. Remove the crankshaft pulley.
5. Carefully pry the front oil seal from the front cover. Be careful not to scratch front cover or crankshaft.

To install:

6. Apply new engine oil to new oil seal and fit it into place with the longer lip towards the outside of the engine.

7. Press the seal straight in using an appropriately sized drift. Make sure the garter spring in the oil seal is in position and seal lip is not inverted.

8. Install crankshaft pulley and toque the bolts to 31 ft. lbs. plus an additional 60 °.

9. Installation of the remaining components is in reverse order of removal.

3.5L Engine

1. Remove the drive belts.
2. Remove the radiator fan assembly.
3. Remove the crankshaft pulley as follows:
- Remove the starter.
- Lock the ring gear using a locking tool attached to the starter bolt hole.

❄❄ WARNING

Do not damage the ring gear teeth, or the signal plate teeth behind the ring gear when installing the locking tool.

4. Loosen crankshaft pulley bolt using pulley holder but don't remove the bolt yet.

5. Position a pulley puller at recess hole of crankshaft pulley to remove crankshaft pulley.

⁂ WARNING

Do not use a puller claw on crankshaft

6. Pry the front oil seal from front cover. Be careful not to damage front cover or crankshaft.

To install:

7. Apply new engine oil to new oil seal and fit it into place with the longer lip towards the outside of the engine.

8. Press the seal straight in using an appropriately sized drift. Make sure the garter spring in the oil seal is in position and seal lip is not inverted.

9. Install crankshaft pulley:
- Fit the pulley onto the crankshaft.
- Lubricate the bolt threads and seat surface with new engine oil.
- Torque the bolt to 32 ft. lbs. (44 Nm)
- Tighten the bolt an additional 90 °.

10. Remove locking tool attached to the starter bolt hole.

11. Installation of the remaining components is in reverse order of removal.

CYLINDER HEAD

REMOVAL & INSTALLATION

2.0 Engine

See Figure 43.

⁂ CAUTION

Be sure to work in a well ventilated area and keep a CO2 fire extinguisher handy. Do not smoke while servicing fuel system. Keep open flames and sparks away from the work area.

1. Release the fuel pressure.
- Remove fuel pump fuse.
- Start engine.
- After engine stalls, crank it two or three times to release all fuel pressure.
- Turn ignition switch OFF.
2. Disconnect negative battery cable.
3. Drain engine coolant and engine oil.
4. Remove right front wheel and inner front fender.
5. Remove drive belt.
6. Remove intake and exhaust manifolds.
7. Remove fuel tube and fuel injector assembly
8. Remove water outlet assembly.

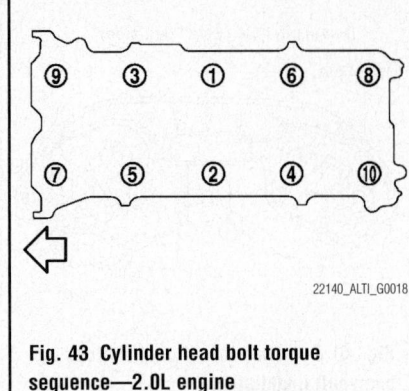

Fig. 43 Cylinder head bolt torque sequence—2.0L engine

9. Remove ignition coils, spark plugs and rocker cover.
10. Remove front cover and timing chain.
11. Remove camshafts.
12. Use a TORX socket size E18 and carefully loosen cylinder head bolts in reverse of installation order.
13. Remove bolts and remove cylinder head.
14. Remove cylinder head gasket.

To install:

➡ **Use new cylinder head bolts.**

15. Clean the bolt hole threads and all sealing surfaces.
16. Install a new cylinder head gasket.
17. Apply new engine oil to threads and seating surface of bolts.
18. Install cylinder head and tighten bolts in numerical order in the following sequence.
- Step a: 30 ft. lb. (40 Nm)
- Step b: 100° clockwise
- Step c: Loosen in the reverse order of tightening sequence.
- Step d: 30 ft. lb. (40 Nm)
- Step e: 100° clockwise
- Step f: 100° clockwise

⁂ WARNING

Check and confirm the tightening angle by using an angle torque wrench or protractor. Never judge by visual inspection without the tool.

19. Install the remaining components in the reverse order of removal.

2.5L Engine

See Figure 44.

1. Remove the engine and transaxle assembly.
2. Remove the timing chain.
3. Remove the camshafts.
4. Remove the spark plugs.

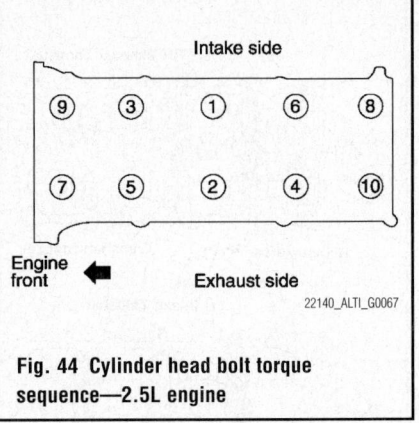

Fig. 44 Cylinder head bolt torque sequence—2.5L engine

5. Loosen the cylinder head bolts in reverse order of tightening sequence and remove the cylinder head.

To install:

6. Clean all gasket surfaces. Use new cylinder head bolts.
7. Fit the new head gasket into place.
8. Fit the cylinder head into place. Lubricate the bolts with new engine oil and install them finger tight.
9. Tighten the head bolts in sequence in several steps:
- Step 1: torque to 72 ft-lb (98 Nm)
- Step 2: Loosen all bolts in the reverse order of tightening.
- Step 3: torque to 29 ft-lb (39 Nm)
- Step 4: tighten an additional 75°
- Step 5: tighten an additional 75°

10. Install remaining components and install engine.

3.5L Engine

See Figures 45 and 46.

1. Remove the intake and exhaust camshafts.
2. Remove the coolant outlet assembly
3. Remove the timing chain case bolts from the cylinder heads.
4. Loosen the cylinder head bolts gradually in three stages in the reverse order of tightening sequence.
5. Remove the cylinder head.

To install:

➡ **New cylinder head bolts are required.**

6. Clean all sealing surfaces and bolt holes.
7. Make sure cylinder No. 1 is at TDC.
8. Fit new cylinder head gaskets in place.
9. Fit the cylinder head into place and lubricate the bolt threads with new engine oil.

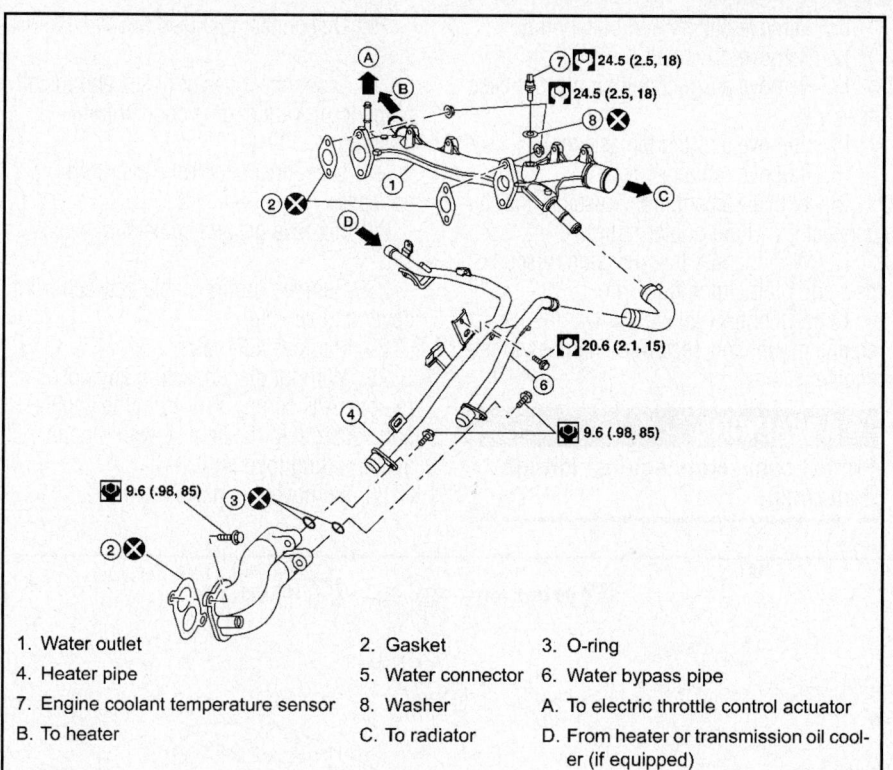

1. Water outlet
2. Gasket
3. O-ring
4. Heater pipe
5. Water connector
6. Water bypass pipe
7. Engine coolant temperature sensor
8. Washer
A. To electric throttle control actuator
B. To heater
C. To radiator
D. From heater or transmission oil cooler (if equipped)

22140_ALTI_G0068

Fig. 45 Coolant outlet assembly—3.5L engine

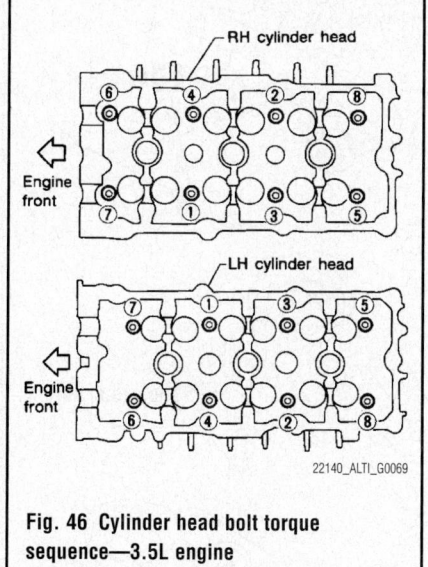

22140_ALTI_G0069

Fig. 46 Cylinder head bolt torque sequence—3.5L engine

10. Install the cylinder head bolts and torque in numerical sequence in the following steps.
- Step 1: torque to 72 ft-lb (98.1 Nm)
- Step 2: Loosen in the reverse order of tightening
- Step 3: torque to 29 ft-lb (33 Nm)
- Step 4: tighten an additional 90°
- Step 5: tighten an additional 90°
11. Install remaining components.

ENGINE ASSEMBLY

REMOVAL & INSTALLATION

Sentra

See Figures 47 through 49.

➡The engine and transmission are lowered together as a unit. This procedure requires a lift, engine hoist and a transmission jack or similar tools. Be prepared for the rearward shift of the vehicle's center of gravity as the engine is removed.

1. Remove engine undercover
2. Remove hood assembly.
3. Remove windshield wipers, cowl top cover and cowl top extension assembly.
4. Drain engine coolant.
5. Remove front wheels and inner fender.
6. Remove front exhaust pipe.
7. Remove halfshaft
8. Remove transaxle joint bolts that pierce upper oil pan.
9. Remove rear torque rod.
10. Release fuel pressure.

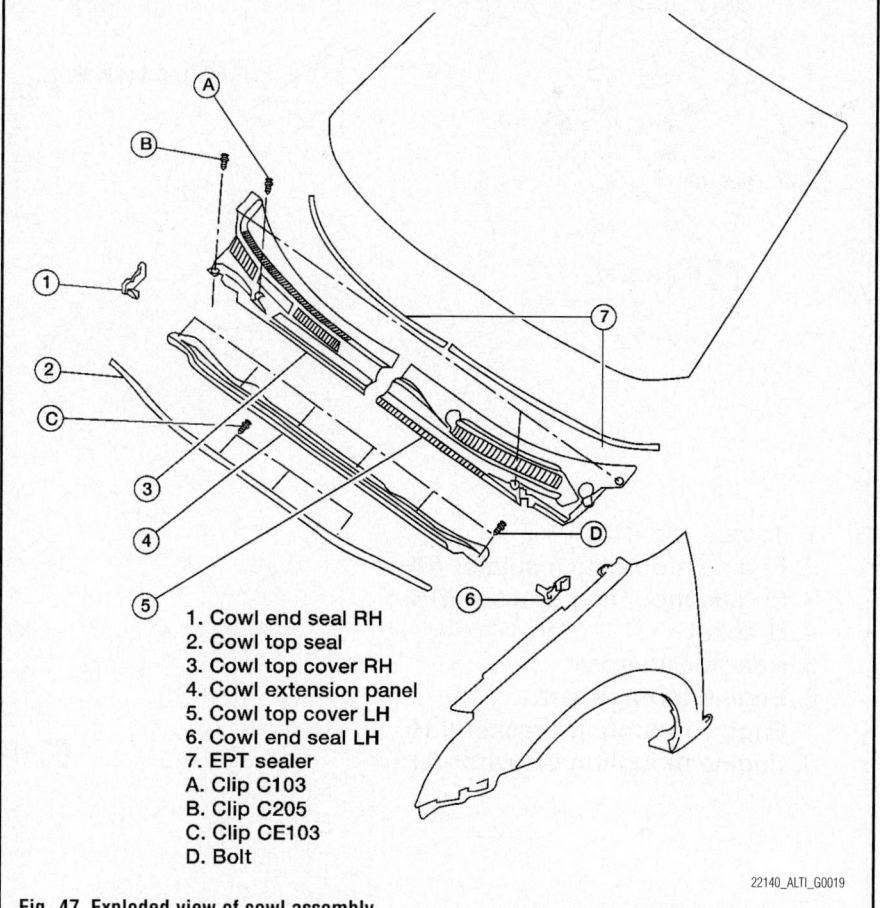

1. Cowl end seal RH
2. Cowl top seal
3. Cowl top cover RH
4. Cowl extension panel
5. Cowl top cover LH
6. Cowl end seal LH
7. EPT sealer
A. Clip C103
B. Clip C205
C. Clip CE103
D. Bolt

22140_ALTI_G0019

Fig. 47 Exploded view of cowl assembly

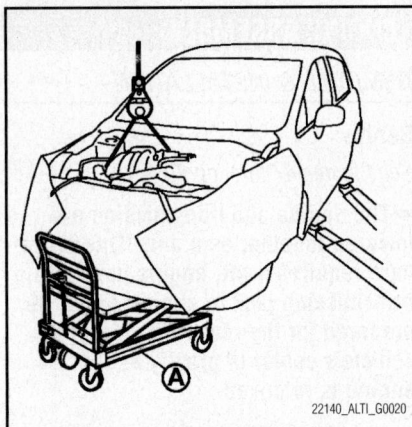

Fig. 48 The engine and transmission are lowered from the vehicle as a unit.

22140_ALTI_G0020

11. Remove battery and battery tray.

12. Remove drive belt.

13. Remove air duct and air cleaner case assembly.

14. Remove cooling fan assembly.

15. Remove radiator hoses.

16. With automatic transmission, disconnect CVT fluid cooler hoses.

17. With manual transmission, disconnect and plug clutch fluid line.

18. Disconnect all wiring near the left engine mount and secure the harness to the engine.

✳✳ WARNING
Protect connectors against foreign materials.

19. Disconnect fuel feed hose at engine side.

20. Disconnect heater hoses and install plugs to prevent engine coolant from draining.

21. Disconnect control cable from transaxle

22. Remove ground cable at transaxle side.

23. Remove ground cable between front cover and vehicle.

24. Remove alternator.

25. Without disconnecting any hoses, remove A/C compressor from the engine and secure it to the vehicle with rope to avoid putting load on it.

26. Remove the intake manifold.

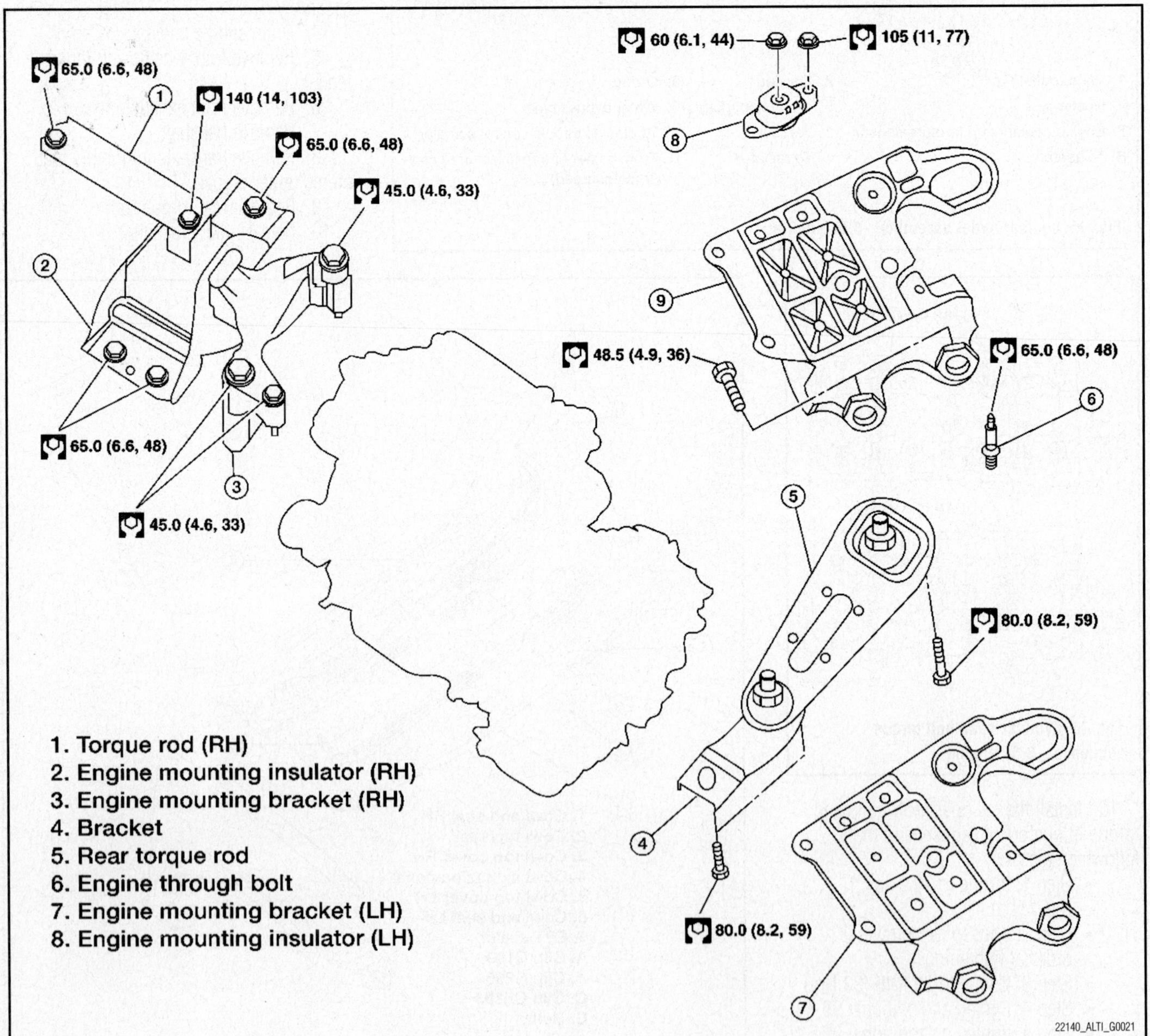

65.0 (6.6, 48)

140 (14, 103)

65.0 (6.6, 48)

45.0 (4.6, 33)

65.0 (6.6, 48)

45.0 (4.6, 33)

60 (6.1, 44)

105 (11, 77)

48.5 (4.9, 36)

65.0 (6.6, 48)

80.0 (8.2, 59)

80.0 (8.2, 59)

1. Torque rod (RH)
2. Engine mounting insulator (RH)
3. Engine mounting bracket (RH)
4. Bracket
5. Rear torque rod
6. Engine through bolt
7. Engine mounting bracket (LH)
8. Engine mounting insulator (LH)

22140_ALTI_G0021

Fig. 49 Engine mounts and bolt torques

27. Install engine sling to cylinder head front left side and rear right side.

28. Support engine and transaxle assembly with an engine hoist and secure the engine in appropriate position.

29. Use a transmission jack or equivalent. Secure the engine and transaxle to the jack and simultaneously adjust hoist tension.

✳✳ WARNING

The vehicle's center of gravity will shift rearward. Support the vehicle with a jack or suitable tool at the rear.

30. Remove right engine mount and torque rod and engine bracket.

31. Remove the large nut from the center of the left engine mount.

32. Slowly begin to lower the engine and the transaxle assembly from the vehicle. Check for any parts, wiring or hoses that are still connected.

To install:

➡**Do not allow engine oil to get on engine mounts. Make sure that each mount is seated properly, and tighten nuts and bolts in order.**

33. Slowly raise the engine/transaxle assembly into the vehicle and loosely engage mounts.

34. When all mounts are engaged, loosely install the large nut in the center of the left engine mount.

35. Install right engine mount and bracket. Install the torque rod.

36. When all mounts are loosely assembled, tighten the rear bolt on the right engine mount, then the two front bolts.

37. Tighten the remaining mounts.

38. Install the intake manifold.

39. Install A/C compressor and alternator.

40. Install ground cables.

41. Connect transaxle linkage and manual transmission clutch fluid line.

42. Connect heater hoses.

43. Connect fuel feed hose.

44. Connect all wiring.

45. With automatic transmission, connect CVT fluid cooler hoses.

46. Install radiator hoses.

47. Install cooling fan assembly.

48. Install air duct and air cleaner assembly.

49. Install drive belt.

50. Install battery and battery tray.

51. Install rear torque rod.

52. Install transaxle joint bolts that pierce upper oil pan.

53. Install halfshafts

54. Install exhaust front tube.

55. Install inner front fenders.

56. Install windshield wipers, cowl top cover and cowl top extension assembly.

57. Install hood assembly.

58. Install engine undercover

59. Before starting engine, fill fluids to the specified levels. Bleed the hydraulic clutch (if equipped).

60. Turn ignition switch "ON" (with engine stopped) and check for fuel leaks.

61. Start engine and check immediately for fuel leakage at connection points.

62. Run engine to check for other fluid leaks or unusual noise and vibration.

63. Bleed air from cooling system.

64. After cooling down engine, again check oil/fluid levels.

Altima

➡**The engine and transmission are lowered together as a unit. This procedure requires a lift, engine hoist and a transmission jack or similar tools. Be prepared for the rearward shift of the vehicle's center of gravity as the engine is removed.**

1. Release fuel system pressure, then disconnect both battery cables.

2. Remove engine undercover

3. Remove hood assembly.

4. Remove windshield wipers, cowl top cover and cowl top extension assembly.

5. Remove front wheels and inner fenders.

6. Drain engine oil, coolant, power steering fluid and transmission fluid.

7. Remove air intake duct and air cleaner case assembly with mass air flow sensor.

8. Remove battery tray.

9. With 3.5L engine, remove CVT control unit if equipped.

10. Remove strut brace.

11. Disconnect/remove vacuum hoses as necessary.

12. Disconnect and set aside the IPDM/ER (fuse/relay box) and remove the bracket.

13. Remove upper and lower radiator hoses (engine side)

14. With automatic transaxle, remove cooler lines and plug the openings.

15. With manual transaxle, disconnect the clutch hydraulic line and cap the openings.

16. Disconnect fuel lines

17. Remove the engine coolant reservoir tank.

18. Disconnect the heater hoses.

19. With 3.5L engine, remove the radiator fan assembly.

20. With 3.5L engine, properly discharge and recover the A/C refrigerant and remove the A/C compressor. Cover the openings to keep dirt and moisture out.

21. Remove flywheel cover plate.

22. Remove the torque converter bolts (CVT models).

23. Remove front exhaust pipe.

24. Remove the left and right halfshafts.

25. Disassemble front suspension as required and remove the front subframe.

26. On 2.5L engine, dismount the A/C compressor with piping connected and secure with wire to the radiator support.

27. Dismount the power steering pump with piping connected and position it aside with wire.

28. Disconnect the clutch operating cylinder fluid line

29. Disconnect the transaxle shift controls.

30. Install engine sling into left front of cylinder head and right rear. On 2.5L engine, use generator bracket bolt holes for the front sling.

31. Support engine and transaxle assembly with engine lifting equipment from the top with the vehicle raised on a hoist.

32. Remove right engine mount and bracket.

33. Remove left transaxle mount through-bolts.

34. Make sure all wires, hoses and other necessary items are disconnected or removed and lower the engine and transaxle assembly from the engine compartment.

35. Separate engine and transaxle.

To install:

36. Assemble engine and transaxle.

37. Lift assembly into vehicle and install left transaxle mount through-bolts and right engine mount and bracket.

38. Connect the transaxle shift controls.

39. Connect the clutch operating cylinder fluid line

40. Install the power steering pump.

41. Install the A/C compressor.

42. Install the front subframe and reassemble front suspension.

43. Install the halfshafts.

44. Install exhaust pipe.

45. Install the torque converter bolts (CVT models) and torque to 80 ft. lbs. (108 Nm).

46. Install flywheel cover plate.

47. Connect the heater hoses and radiator hoses.

48. Refit the engine coolant reservoir tank.

49. Connect fuel lines

50. Install the IPDM/ER (fuse/relay box) and bracket and connect all wiring.

51. Install strut brace.

52. Install battery tray and battery.

53. Refill all fluids.

54. Install inner fenders.

55. Install cowl top cover and cowl top extension assembly. Install windshield wipers.

56. Install hood assembly.

57. Before starting engine, check once more to make sure everything is properly installed and connected. Turn ignition switch ON (with engine stopped) and check for fuel leaks at connection points.

58. Start engine. With engine speed increased, check again for fuel leakage at connection points.

59. Run engine to check for unusual noise and vibration.

60. Warm up engine thoroughly to make sure there is no leakage of fuel, exhaust gas, or any oils/fluids including engine oil and engine coolant.

61. Bleed air from passages in lines and hoses, such as in cooling system.

62. After cooling down engine, again check oils/fluids including engine oil and engine coolant. Refill to specified level as necessary.

EXHAUST MANIFOLD

REMOVAL & INSTALLATION

2.0L Engine

See Figure 50.

1. Disconnect front exhaust pipe.

✳✳ CAUTION

Be careful not to cut your hand on heat insulator edge.

• Disconnect heated oxygen sensor.

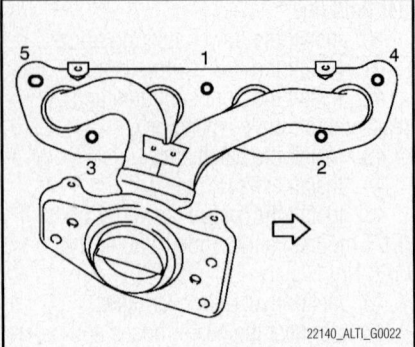

22140_ALTI_G0022

Fig. 50 Exhaust manifold bolt tightening sequence

• Remove oxygen sensor.
• Disconnect each joint and mounting rubber.

2. Remove exhaust manifold heat shield.

3. Remove the A/F sensor.

4. Remove exhaust manifold stay bolt.

5. Loosen nuts in reverse order and remove exhaust manifold.

To install:

6. Install a new gasket.

7. Fit manifold into place and loosely install all nuts. The upper end of the stay is marked "up."

8. Torque nuts in sequence in two steps to 25 ft. lbs. (33 Nm).

9. To connect exhaust pipe:

• Securely insert new seal bearing into exhaust manifold. Be careful not to damage seal bearing surface when installing.
• Install spring and tighten nut. Be careful that the stud bolt nut does not interfere with the flanged area.
• Make sure the spring sits properly on the flange surface by aligning it to the locator dimples.

10. Install the A/F and oxygen sensors. Do not over tighten or the MIL may turn on.

2.5L Engine

See Figure 51.

1. Remove the engine undercover.

2. Disconnect the air fuel ratio (A/F) sensor, unhook the harness from the bracket and remove the sensor.

3. Remove the lower exhaust manifold covers.

4. Remove the front exhaust pipe.

5. Remove the upper exhaust manifold cover.

6. Loosen the manifold nuts and remove the exhaust manifold and three way catalyst. Discard the gasket.

To install:

7. Install a new gasket and fit the manifold/catalyst into place.

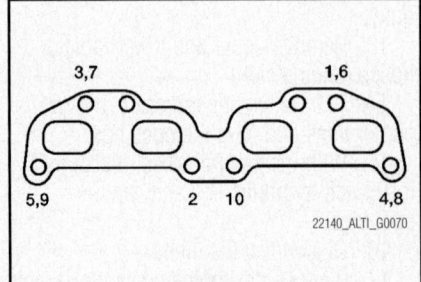

22140_ALTI_G0070

Fig. 51 Exhaust manifold bolt torque sequence—2.5L engine

8. Install the nuts and torque in sequence to 11 ft. lbs. (15 Nm).

9. Install the exhaust pipe and torque the bolts to 11 ft. lbs. (15 Nm).

10. Install the A/F sensor and reconnect the wiring.

3.5L Engine

See Figure 52.

➡ **This procedure requires removing the front subframe. When removing the front and rear engine mount bolts and nuts, lift the engine up slightly for safety.**

1. Raise and safely support the vehicle and remove the front wheel.

2. Remove the engine undercover.

3. Remove the inner wheel well splash shields.

4. Remove the radiator and cooling fan assembly.

5. Remove the front exhaust pipe.

6. Remove the front suspension subframe.

• Remove nut on lower portion of stabilizer bar connecting rod from lower suspension arm.
• Remove suspension arm from subframe and swing it outward.
• Remove front exhaust pipe.
• Support engine or transmission with a jack.
• Remove steering gear bolts. Remove steering gear and power steering tube bracket from suspension member. Hang steering gear.

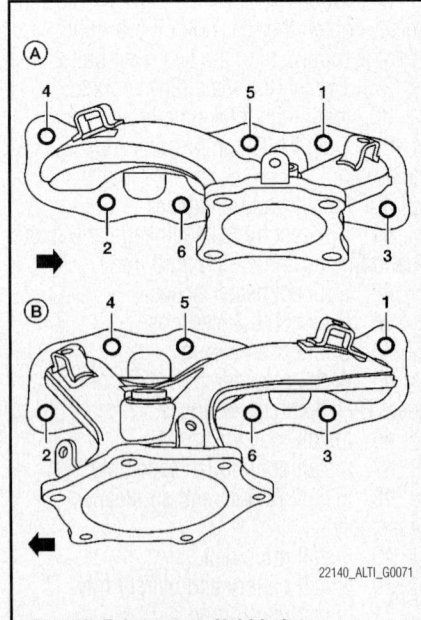

22140_ALTI_G0071

Fig. 52 Exhaust manifold bolt torque sequence—3.5L engine

- Set a jack under subframe and remove the nuts.
- Slowly lower jack to remove subframe from vehicle.

7. Remove the left and right catalyst supports.

8. Disconnect and remove the oxygen sensors and air ratio (A/F) sensors.

9. Remove exhaust manifold and three way catalyst heat shields.

10. Remove the catalysts by loosening the bolts first and then removing the nuts and through bolts.

11. To remove the exhaust manifolds, loosen the nuts in reverse of the installation order.

To install:

12. Install a new manifold gasket and fit the manifold into place.

13. Torque the nuts in order to 23 ft. lbs. (31 Nm)

14. If removed, install the catalysts using new gaskets and torque the bolts to 52 ft. lbs. (70 Nm).

15. Install the sensors and heat shields.

16. Raise the subframe into place and start all fasteners.

- Torque the large subframe bolts to 107 ft. lbs. (145 Nm).
- Torque the suspension arm bolts to 114 ft. lbs. (155 Nm).
- Reconnect steering gear and stabilizer bar.

17. Install remaining components and run the engine to check for leaks.

INTAKE MANIFOLD

REMOVAL & INSTALLATION

2.0L Engine

See Figure 53.

1. Drain engine coolant or disconnect water hoses from electronic throttle control actuator and plug the hoses.

2. Disconnect wiring from electronic throttle control actuator.

❊❊ WARNING

Handle carefully to avoid any shock to electric throttle control actuator. Never disassemble.

3. Remove dipstick and cover the opening to avoid entry of foreign materials.

4. Loosen and install intake manifold bolts in reverse order.

5. Remove manifold and cover intake ports.

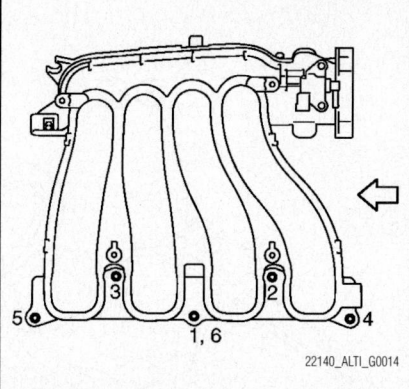

Fig. 53 Intake manifold bolt tightening sequence 2.0L engine

6. Make sure all sealing surfaces are clean and fit a new manifold gasket into place.

7. Install manifold and loosely install all nuts and bolts.

8. Torque the manifold-to-cylinder head bolts in sequence to 20 ft. lbs. (27 Nm).

9. Connect all hoses and wiring.

10. Check/refill coolant level.

11. Run the engine to check for leaks.

2.5L Engine

1. Release the fuel pressure.

2. Drain coolant.

3. Disconnect the Mass Air Flow (MAF) sensor.

4. Remove air cleaner and air duct assembly.

5. Remove windshield wipers and cowl.

6. Disconnect the following components from the intake manifold:

- PCV hose
- EVAP hose and EVAP canister purge volume control solenoid
- Electric throttle control actuator
- Brake booster vacuum hose

7. Disconnect the fuel line quick connector on the engine side.

8. Disconnect throttle control actuator coolant hoses.

9. Remove the bolts and nuts and remove the intake manifold assembly. Cover engine openings to avoid entry of foreign materials.

To install:

10. Install a new manifold gasket and fit the manifold into place. Working from the center towards the ends, torque the nuts and bolts to 14 ft. lbs. (20 Nm), then repeat the torque sequence.

11. Reconnect all wiring, hoses and vacuum lines.

12. Refill the cooling system and run the engine to check for leaks.

13. Reinstall the cowl and wiper arms.

3.5L Engine

1. Remove the upper intake manifold.

- Remove air cleaner case lid, mass air flow sensor, and air intake tube as one assembly.
- Partially drain the coolant.
- Disconnect fuel tube quick connector at the fuel rail by pressing in the tabs on the sides and pulling straight out.
- Disconnect the power brake booster vacuum hose, the coolant hoses from the electric throttle control actuator, the swirl control vacuum lines, the fuel injectors electrical connectors, and the PCV hose.

➡ **Cover any engine openings to avoid the entry of any foreign material.**

- Disconnect the electric throttle control actuator electrical connector and coolant hoses.
- Remove the wiper blades and cowl.
- Remove the upper intake manifold.

OIL PAN

REMOVAL & INSTALLATION

2.0L Engine

See Figures 54 through 56.

1. Drain engine oil.

2. Remove engine and transaxle assembly.

3. Remove oil filter.

4. Remove lower oil pan bolts in reverse order.

5. After removing the pan, clean off the sealant being careful not to damage the mating surfaces.

6. Remove the flywheel.

7. Remove the front cover, timing chain, and the oil pump drive chain

8. Remove oil pan (lower) bolts in reverse order as shown.

To install:

❊❊ WARNING

The rear oil seal should be installed within 5 minutes after installing upper oil pan. Always replace rear oil seal with new one. Never touch oil seal lip.

9. Carefully scrape old liquid gasket from mating surfaces on oil pan and cylinder block.

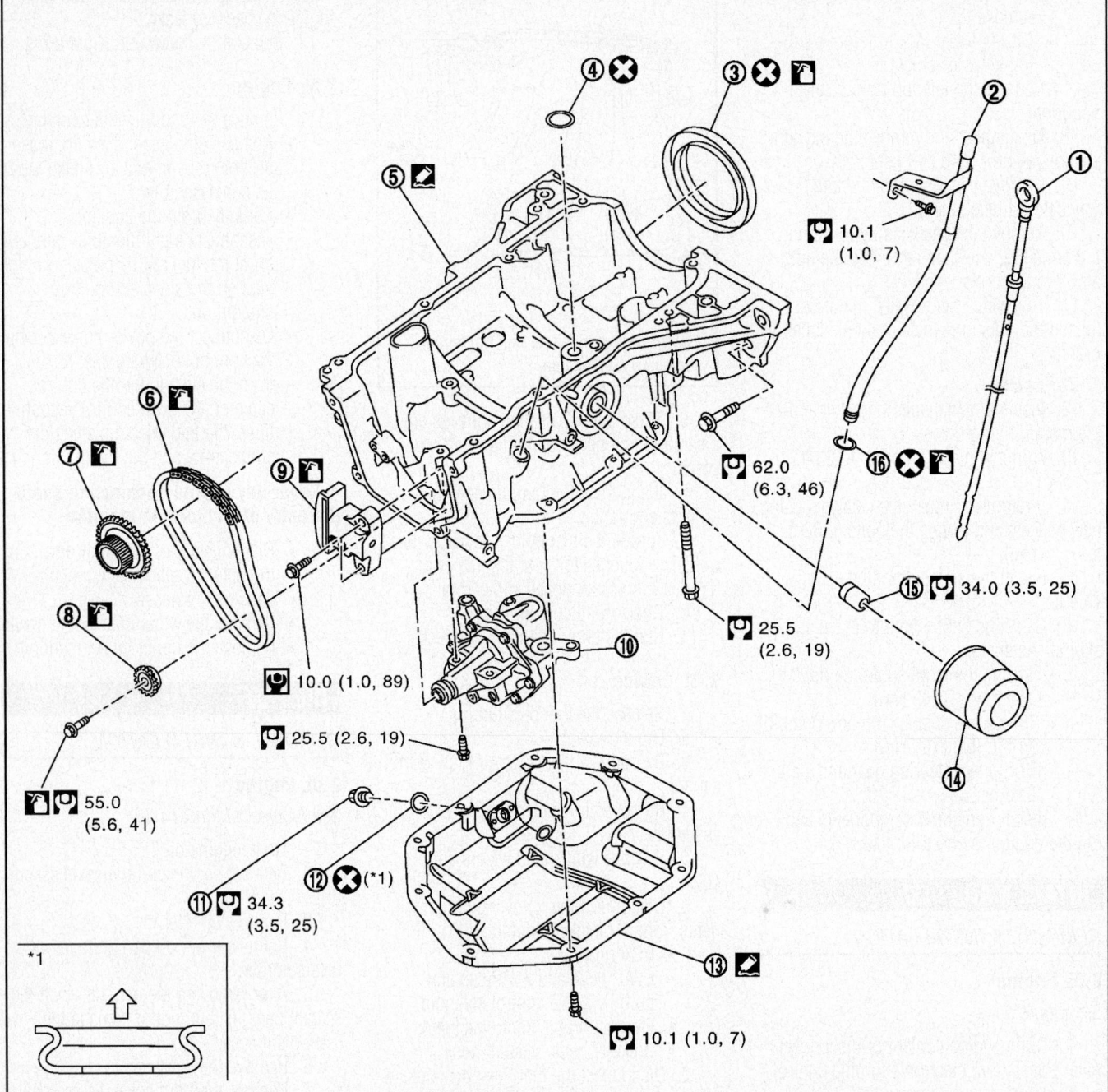

1. Oil level gauge
2. Oil level gauge guide
3. Rear oil seal
4. O-ring
5. Oil pan (upper) and balancer and oil pump assembly
6. Balancer and oil pump drive chain
7. Crankshaft sprocket
8. Balancer sprocket
9. Timing chain tensioner (for balancer and oil pump)
10. Balancer and oil pump
11. Drain plug
12. Drain plug washer
13. Oil pan (lower)
14. Oil filter
15. Connector bolt
16. O-ring
⇦ : Oil pan side

22140_ALTI_G0023

Fig. 54 Oil pan assembly—2.0L engine

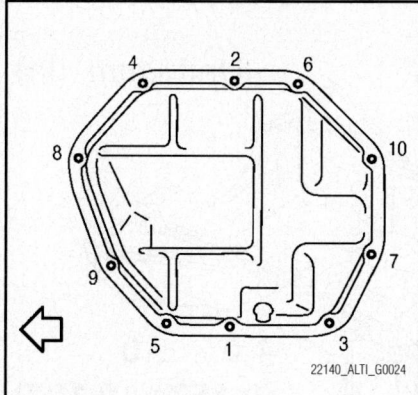

Fig. 55 Lower oil pan bolt torque sequence

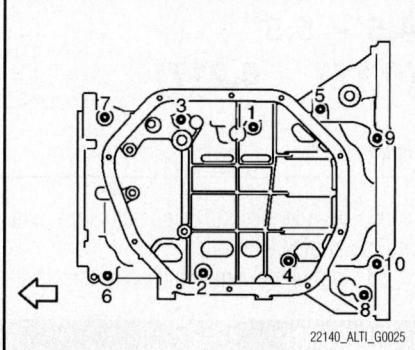

Fig. 56 Upper oil pan bolt torque sequence

10. Remove old liquid gasket from the bolt holes and threads.

11. Apply RTV liquid gasket to the oil pan. The bead should be outside the four center bolt holes and inside the four corner bolt holes.

12. Install new oil filter passage O-ring on the cylinder block. Make sure it's properly aligned.

13. Fit the pan into place on the block and tighten bolts in numerical order to 19 ft. lbs. (25 Nm).

14. Install rear oil seal with the following procedure.

- Wipe off liquid gasket protruding from the block/pan assembly.
- Apply engine oil to entire outside area of rear oil seal.
- Press-fit the rear oil seal using a press tool with outer diameter 4.53 in. (115mm) and inner diameter 3.54 in. (90 mm) Press-fit straight until seal is flush with engine block, making sure that rear oil seal does not curl or tilt.

15. Install oil pump sprocket, oil pump drive chain and other related parts if removed.

16. Use a scraper (A) to remove old liquid gasket from mating surfaces on lower oil pan.

17. Remove old liquid gasket from the bolt holes and threads.

18. Apply a bead of RTV sealant, staying inside all the bolt holes.

19. Tighten bolts in numerical order to 7 ft. lbs. (10 Nm).

20. Install oil filter, hand tight only.

21. Installation of the remaining components is in the reverse order of removal.

2.5L Engine

See Figure 57.

1. Raise and safely support the vehicle and drain the oil.

2. Remove the front exhaust pipe.

3. Remove power steering cooler hose bracket from suspension member.

4. Remove the front subframe for clearance to remove the oil pan.

- Remove nut on lower portion of stabilizer bar connecting rod from lower suspension arm.
- Remove suspension arm from subframe and swing it outward.
- Remove front exhaust pipe.
- Support engine or transmission with a jack.
- Remove steering gear bolts. Remove steering gear and power steering tube bracket from suspension member. Hang steering gear.
- Set a jack under subframe and remove the nuts.

5. Remove the lower oil pan bolts and remove the pan.

6. Remove the oil strainer.

7. Remove rear plate cover and four engine-to transaxle bolts.

8. Loosen the upper oil pan bolts and remove the upper oil pan. Note the different bolt lengths.

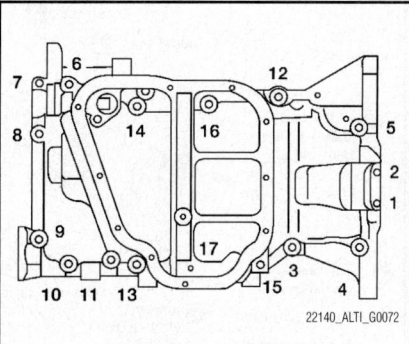

Fig. 57 Upper oil pan bolt torque sequence—2.5L engine

To install:

9. Carefully clean all sealing surfaces and apply a bead of RTV sealant to the upper oil pan. Install new O-rings and fit the pan into place. Start the bolts finger tight.

10. Torque the oil pan bolts in sequence to 16 ft. lbs. (22 Nm).

11. Torque the pan-to-transaxle bolts to 31 ft. lbs. (46 Nm).

12. Install the strainer and torque the bolts

13. Apply RTV to the lower oil pan to 16 ft. lbs. (22 Nm).

14. Install the subframe:

- Raise the subframe into place and start all fasteners.
- Torque the large subframe bolts to 107 ft. lbs. (145 Nm).
- Torque the suspension arm bolts to 114 ft. lbs. (155 Nm).
- Reconnect steering gear and stabilizer bar.

15. Install the remaining components and refill the engine with oil.

16. Run the engine to check for leaks.

3.5L Engine

See Figures 58 through 60.

✳✳ WARNING

When removing the front and rear engine through bolts and nuts, lift the engine up slightly for safety. When removing the upper oil pan from the engine, first remove the crankshaft position sensor. Be careful not to damage sensor edges or signal plate teeth.

1. Disconnect the battery negative cable.

2. Raise and safely support the vehicle and remove the front wheels.

3. Remove the engine undercover and the right inner fender splash shield.

4. Drain the oil and remove the oil dipstick.

5. Drain the engine coolant.

6. Remove the A/C drive belt.

7. Remove the front exhaust pipe.

8. Remove coolant pipe bolts.

9. Remove the A/C compressor with piping attached and position it out of the way securely with wire. Do not pull on or crimp the A/C lines and hoses.

10. Disconnect the coolant lines from the engine oil cooler and plug them to prevent coolant loss.

11. Remove the oil filter and engine oil cooler from the upper oil pan.

12. Remove the oil pressure switch/sensor, and the crankshaft position sensor (POS) from the upper oil pan.

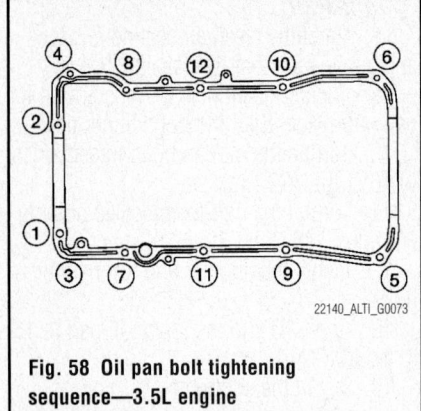

Fig. 58 Oil pan bolt tightening sequence—3.5L engine

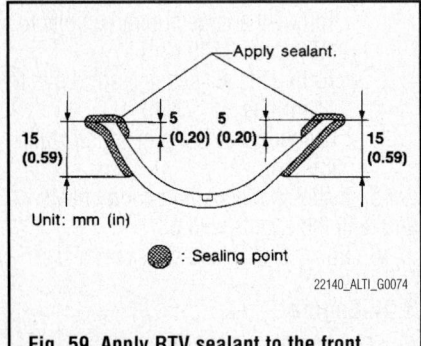

Fig. 59 Apply RTV sealant to the front cover oil pan gasket and rear oil strainer gasket—3.5L engine

13. Remove the right halfshaft.

14. Disconnect the oxygen sensors and remove the two catalytic converters from the exhaust manifolds.

15. Remove the rear plate cover from the upper oil pan.

16. Loosen the lower oil pan bolts and remove the lower oil pan. Be careful not to damage the mating surfaces.

17. If removing the upper oil pan, remove the Crankshaft Position Sensor (CKP).

18. Remove the four upper oil pan-to-transaxle bolts.

19. Loosen the upper oil pan bolts in reverse of the tightening sequence.

20. Insert an appropriate size tool into the notch at the right rear of the upper oil pan and carefully pry the pan off.

21. Remove the O-ring seals from the bottom of the cylinder block and oil pump housing. Discard the O-rings.

22. Remove front cover gasket and rear oil seal retainer gasket.

23. Remove the oil strainer.

To install:

24. Carefully clean all sealing surfaces and bolt holes. Take care to not scratch or damage the sealing surfaces when cleaning off the old sealant.

25. Install oil strainer.

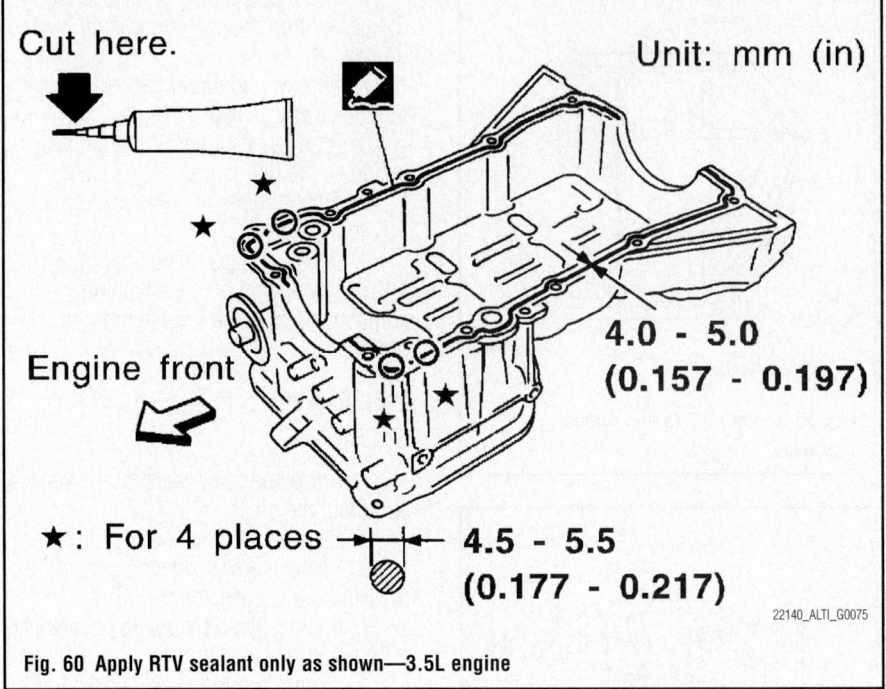

Fig. 60 Apply RTV sealant only as shown—3.5L engine

26. Apply RTV sealant to the front cover gasket and the rear oil seal retainer gasket and install them.

27. Carefully apply RTV sealant to the upper oil pan only as shown in the illustration. Note the bead of sealant is inside most of the bolt holes.

28. Install new O-rings on the cylinder block and oil pump body.

29. Carefully install the upper oil pan and start all the bolts. Torque the bolts in order to 13 ft. lbs. (17 Nm).

30. Install the four upper oil pan-to-transaxle bolts and torque to 13 ft. lb. (17 Nm).

31. Apply RTV sealant to the lower oil pan and install the pan. Torque the bolts in a crossing pattern to 78 in. lbs. (9 Nm).

32. Wait at least 30 minutes before refilling the engine with oil.

33. Install the remaining components.

OIL PUMP

REMOVAL & INSTALLATION

Sentra

See Oil Pan Removal and Installation

Altima

See Figures 61 and 62.

1. Remove the timing chain.
2. Remove oil pump assembly.
3. Installation is in the reverse of removal.

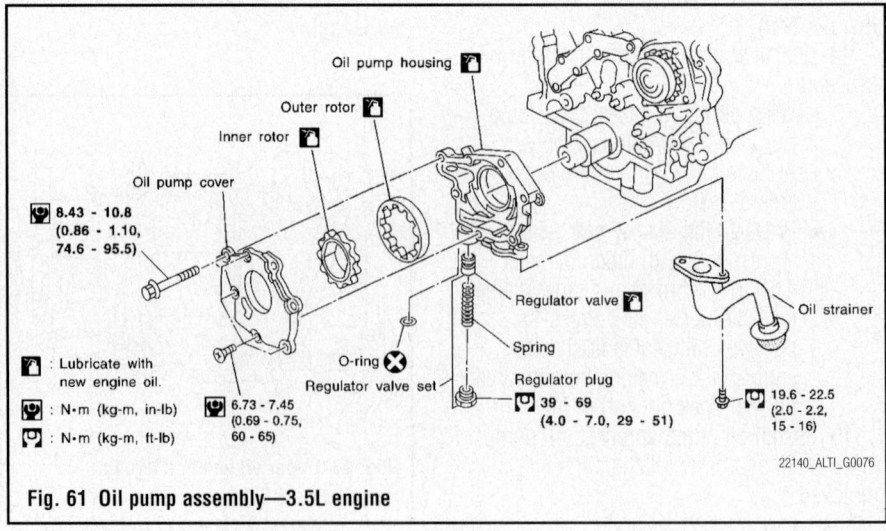

Fig. 61 Oil pump assembly—3.5L engine

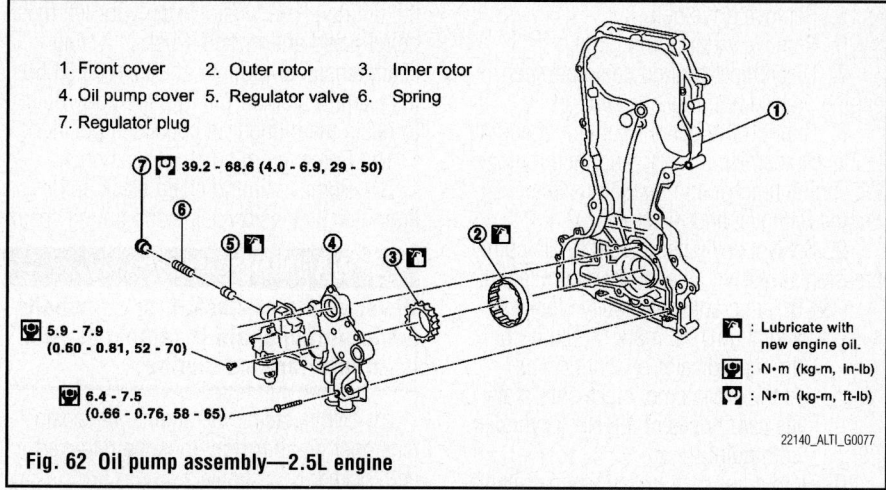

1. Front cover
2. Outer rotor
3. Inner rotor
4. Oil pump cover
5. Regulator valve
6. Spring
7. Regulator plug

⑦ 39.2 - 68.6 (4.0 - 6.9, 29 - 50)

5.9 - 7.9 (0.60 - 0.81, 52 - 70)

6.4 - 7.5 (0.66 - 0.76, 58 - 65)

Lubricate with new engine oil.

N•m (kg-m, in-lb)

N•m (kg-m, ft-lb)

22140_ALTI_G0077

Fig. 62 Oil pump assembly—2.5L engine

PISTON AND RING

POSITIONING

See Figures 63.

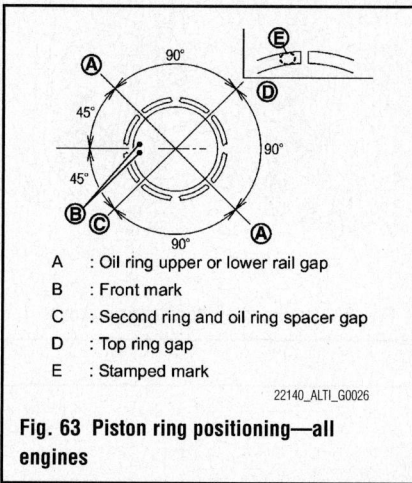

A : Oil ring upper or lower rail gap
B : Front mark
C : Second ring and oil ring spacer gap
D : Top ring gap
E : Stamped mark

22140_ALTI_G0026

Fig. 63 Piston ring positioning—all engines

REAR MAIN SEAL

REMOVAL & INSTALLATION

See Figure 64.

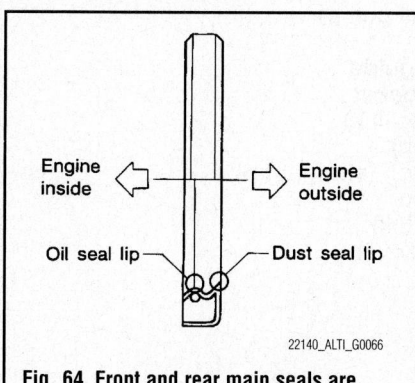

Engine inside ⇦

⇨ Engine outside

Oil seal lip

Dust seal lip

22140_ALTI_G0066

Fig. 64 Front and rear main seals are installed with the longer lip towards the outside of the engine—all engines

1. Remove engine and transaxle assembly and separate them.
2. Remove clutch and flywheel or drive plate.
3. On 3.5L engine, remove seal retainer.
4. On other engines, carefully pry rear

oil seal out with a suitable tool. Be careful to not damage sealing surfaces on the engine block or crankshaft.

To install:

5. Apply RTV sealant to entire outside area of new rear oil seal.
6. Fit the seal into place with the longer lip towards the outside of the engine.
7. Press the seal flush with the engine block or seal retainer using a suitable press tool. Make sure the seal is not tilted or curled. Wipe away any excess sealant.
8. On 3.5L engine, apply RTV sealant to the seal retainer and install it to the engine. Torque the bolts to 78 in. lbs. (9 Nm).

TIMING CHAIN, SPROCKETS, FRONT COVER AND SEAL

REMOVAL & INSTALLATION

2.0L Engine

See Figures 65 through 69.

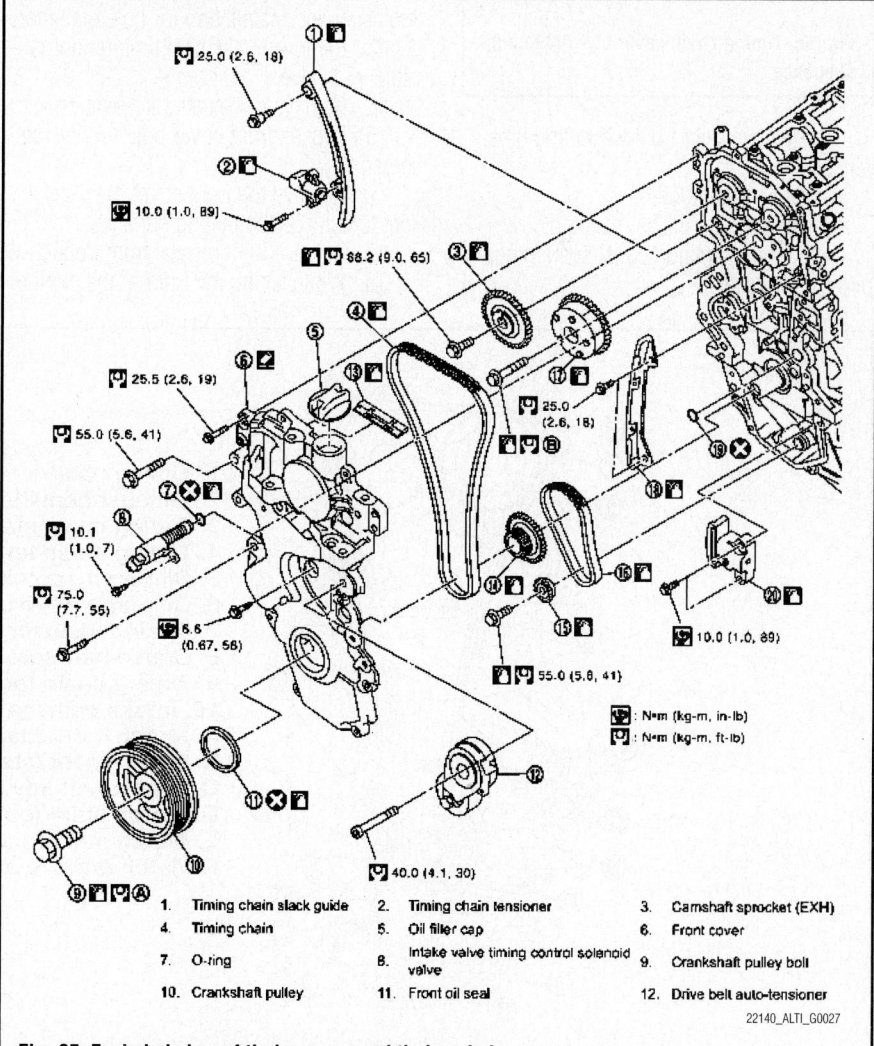

① 25.0 (2.6, 18)

② 10.0 (1.0, 89)

88.2 (9.0, 65)

25.5 (2.6, 19)

55.0 (5.6, 41)

10.1 (1.0, 7)

75.0 (7.7, 55)

6.6 (0.67, 58)

25.0 (2.6, 18)

10.0 (1.0, 89)

55.0 (5.6, 41)

40.0 (4.1, 30)

N•m (kg-m, in-lb)

N•m (kg-m, ft-lb)

1. Timing chain slack guide
2. Timing chain tensioner
3. Camshaft sprocket (EXH)
4. Timing chain
5. Oil filler cap
6. Front cover
7. O-ring
8. Intake valve timing control solenoid valve
9. Crankshaft pulley bolt
10. Crankshaft pulley
11. Front oil seal
12. Drive belt auto-tensioner

22140_ALTI_G0027

Fig. 65 Exploded view of timing cover and timing chain

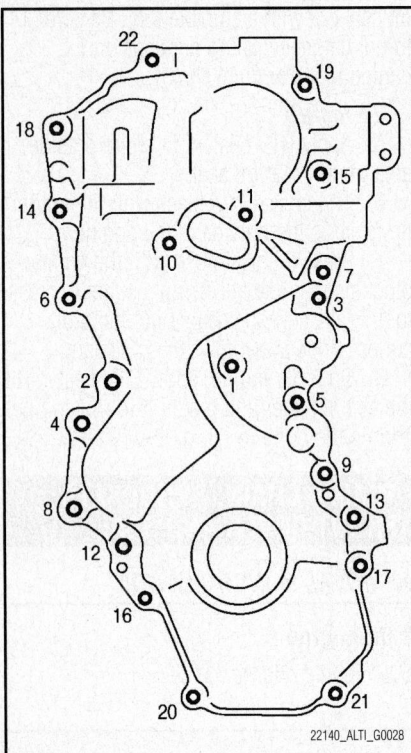

Fig. 66 Timing chain cover bolt tightening sequence.

1. Remove right front wheel and inner fender
2. Drain engine oil.
3. Remove the intake manifold:
4. Remove ignition coils, spark plugs and rocker cover.

5. Remove drive belt.
6. Remove water pump pulley.
7. Disconnect ground cable between engine bracket and radiator support.
8. Support the bottom surface of engine using a transmission jack, and then remove the engine bracket and insulator. (See Engine Removal and Installation)
9. Set No.1 cylinder at TDC on its compression stroke with the following procedure:
   - Rotate crankshaft pulley clockwise and align TDC mark (no paint) to timing indicator on front cover.
   - At the same time, make sure that the cam noses of the No.1 cylinder both pointing up.
10. Hold crankshaft pulley using suitable tool and loosen crankshaft pulley bolt. Do not remove the bolt yet.
11. Attach a pulley puller in the threaded holes on crankshaft pulley and remove the pulley.
12. If removing the crankshaft sprocket, oil pump sprocket or other related parts, remove the oil lower pan. See Oil Pan Removal and Installation for bolt sequence.
13. Remove intake cam timing control solenoid valve.
14. Remove drive belt auto-tensioner.
15. Loosen front cover bolts in reverse order.
16. Remove the front cover. Be careful not to damage the mating surfaces.
17. Remove front oil seal from front cover.
18. While facing the front of the engine,

the timing chain tensioner on your left must be retracted and locked. Push in timing chain tensioner plunger and insert a 0.060 in (1.5mm) stopper pin into the body hole to retain the plunger in collapsed position.
19. Remove timing chain tensioner.
20. Remove timing chain slack guide, timing chain tension guide and timing chain.

❋❋ WARNING
Never rotate crankshaft or camshafts while timing chain is removed. This is an interference engine.

21. While facing the engine, the chain tensioner on your right must be retracted and locked.
   - Fully lift up lever A and push the slack guide B into the chain tensioner (1).

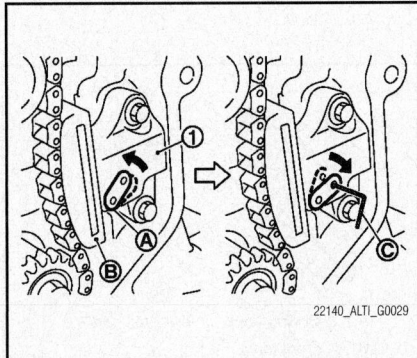

Fig. 67 Timing chain tensioner removal

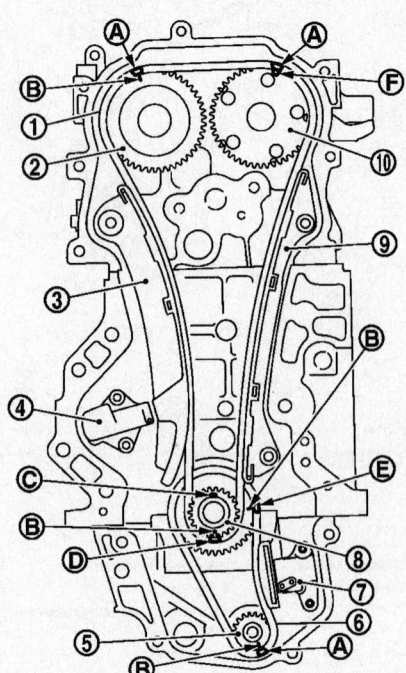

1. Timing chain
2. Exhaust camshaft sprocket
3. Timing chain slack guide
4. Timing chain tensioner
5. Oil pump sprocket
6. Oil pump drive chain
7. Chain tensioner (for oil pump)
8. Crankshaft sprocket
9. Timing chain tension guide
10. Intake camshaft sprocket
A. Match mark (dark blue link)
B. Match mark (stamping)
C. Crankshaft key position (straight up)
D. Match mark (gold link)
E. Match mark (orange link)
F. Match mark (outer groove)

Fig. 68 Timing chain assembly match marks.

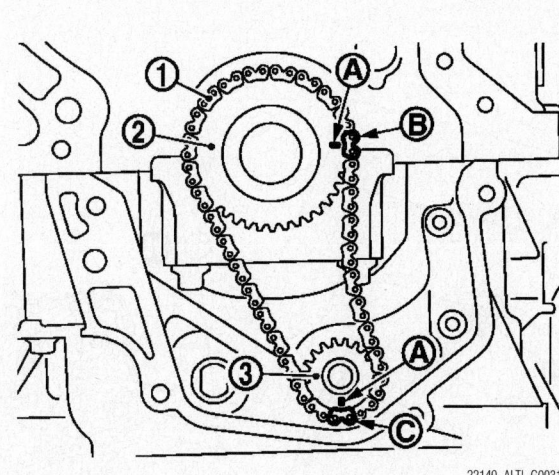

Fig. 69 The oil pump must be indexed to the crankshaft by aligning the marks on the chain and sprockets.

- Matching the hole on lever with the hole on tensioner body, insert a 0.040 in. (1.0mm) stopper pin (C) to secure slack guide.
22. Remove chain tensioner.
23. Remove crankshaft sprocket, oil pump sprocket and oil pump drive chain as a set.
24. Remove timing chain tension guide (front cover side) from front cover if necessary.

To install:

➡**The figure shows the relationship between the match mark on each timing chain and on the corresponding sprocket, with the components installed. Make sure the crankshaft key points straight up. There are two outer grooves on the intake camshaft sprocket. The wider one is a match mark.**

25. If the timing chain tension guide (front cover side) is removed, install it to the front cover.
26. Align the match marks and install crankshaft sprocket, oil pump sprocket and oil pump drive chain. If these marks are not aligned, rotate the oil pump as needed.
27. Make sure the oil pump chain tensioner plunger is compressed and locked a stopper pin, then install it.
28. Pull out the tensioner stopper pin and check the match mark alignment again.
29. Install the timing chain with all match marks aligned.
30. Install the timing chain tension guide and the timing chain slack guide.

31. With the plunger retracted and pinned, install timing chain tensioner.
32. Pull out the stopper pin and check all match mark alignments again.
33. Temporarily install the crankshaft pulley and bolt and rotate the crankshaft two full turns. Check all match mark alignments again. Remove pulley.
34. Apply new engine oil to new front oil seal joint surface. Fit the seal into the front cover with the longer lip towards the outside of the engine.
35. Using a suitable tool, press-fit front oil seal until it is flush with front end surface of front cover. Make sure the seal is straight and not curled.
36. Install new O-ring to cylinder block.
37. Apply RTV sealant to the front cover. Don't forget the opening in the upper center (bolts 10 and 11).
38. Make sure O-ring on cylinder block is correctly installed.
39. Install front cover and bolts. Be careful not to damage front oil seal on front end of crankshaft.
40. Tighten bolts in numerical order in two steps to specified torque.
41. Wipe off any excess liquid gasket.
42. Install crankshaft pulley and apply new oil to the pulley bolt.
43. Hold crankshaft pulley using the proper tool.
44. Tighten crankshaft pulley bolt in four steps.

- 51 ft. lbs. (67 Nm)
- 0 ft. lbs. (0 Nm)
- 22 ft. lbs. (29 Nm)
- 60 °

45. Make sure crankshaft rotates clockwise smoothly.
46. Installation of the remaining components is in the reverse order of removal.

2.5L Engine

See Figures 70 through 74.

1. Raise and safely support the vehicle.
2. Remove the oil pan.
3. Remove the alternator.
4. Disconnect variable timing control solenoid harness connector.
5. Remove the coolant overflow reservoir.
6. Position the engine compartment fuse and relay box aside.
7. Remove the right engine mount and bracket.
8. Remove the camshaft timing control (IVT) cover using.
9. Pull the chain guide (between camshaft sprockets) out through front cover.
10. Set the No.1 cylinder at TDC on the compression stroke. Make sure the timing marks on the camshaft sprockets are aligned. See Camshaft Removal and Installation.
11. Hold the crankshaft pulley and loosen but do not remove the crankshaft pulley bolt.
12. Remove the crankshaft pulley with a bolt-on puller. Do not use a claw type puller.
13. Remove the timing cover bolts in reverse of the tightening sequence.
14. Push in the chain tensioner plunger and secure it with a stopper pin in the hole on the tensioner body. Remove chain tensioner.
15. Remove the timing chain.
16. If necessary, hold the hexagonal part of the camshaft with a wrench and loosen the camshaft sprocket bolt and remove the camshaft sprocket.

✻✻ WARNING

Do not rotate the crankshaft or camshafts while the timing chain is removed. This is an interference engine.

17. To remove the balance shaft chain, compress the chain guide tensioner and remove the tensioner, guide, timing chain, and oil pump drive spacer.

➡**The balance shaft bolts must be replaced once removed.**

18. Secure the left balancer shaft with a wrench and loosen the sprocket bolt.
19. Remove balancer unit timing chain, balancer unit sprocket and crankshaft sprocket.

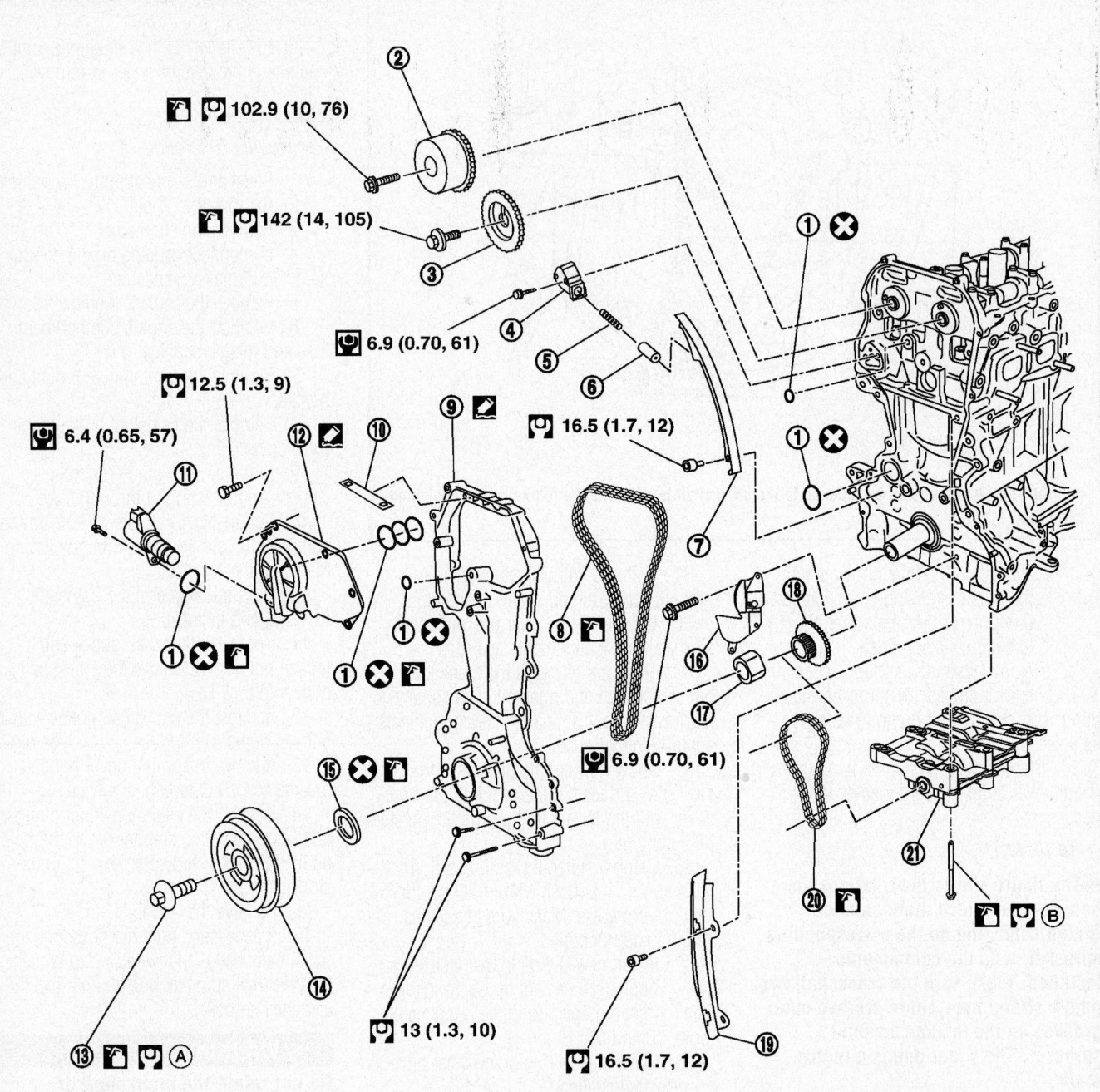

1. O-rings
2. Camshaft sprocket (INT)
3. Camshaft sprocket (EXH)
4. Chain tensioner
5. Spring
6. Chain tensioner plunger
7. Timing chain slack guide
8. Timing chain
9. Front cover
10. Chain guide
11. IVT solenoid valve
12. IVT cover
13. Crankshaft pulley bolt
14. Crankshaft pulley
15. Front oil seal
16. Balancer unit timing chain tensioner
17. Oil pump drive spacer
18. Crankshaft sprocket
19. Timing chain tension guide
20. Balancer unit timing chain
21. Balancer unit
A. Follow installation procedure
B. Follow installation procedure

22140_ALTI_G0083

Fig. 70 Timing chain and balance shaft chain components—2.5L engine

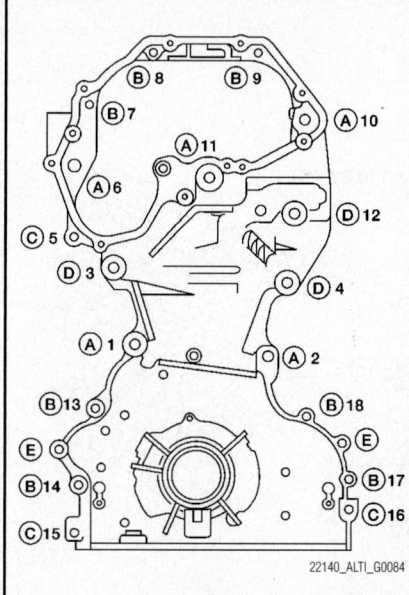

Fig. 71 Timing chain cover bolt torque sequence—2.5L engine

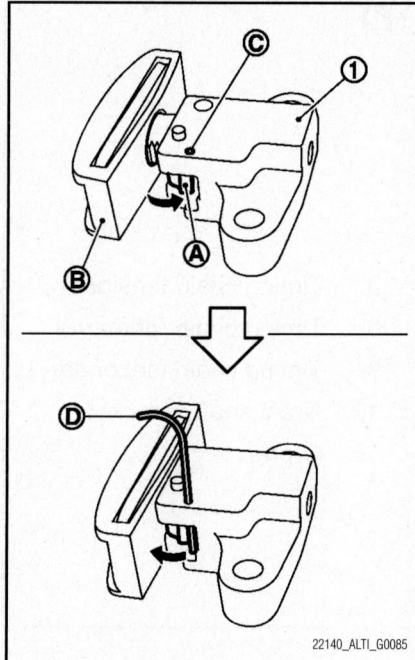

Fig. 72 Release tab A to compress chain guide tensioner B and insert pin D in hole C to lock it in place—2.5L engine

To install:

➡**There may be two color variations of the link marks (link colors) on the timing chain. There are 26 links between the gold/yellow marks on the timing chain; and 64 links between the camshaft sprocket gold/yellow link and the crankshaft sprocket orange/blue link on the timing chain side without the tensioner.**

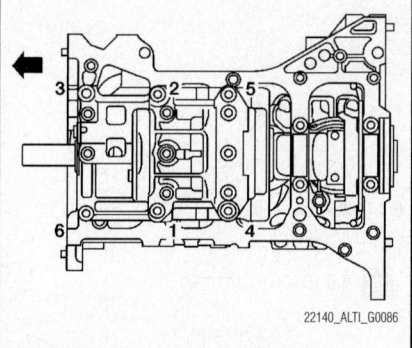

Fig. 73 Balance shaft assembly bolt torque sequence—2.5L engine

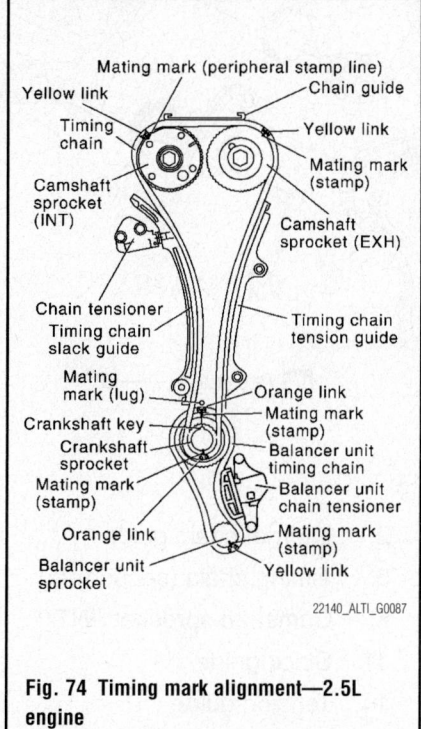

Fig. 74 Timing mark alignment—2.5L engine

20. Make sure the crankshaft key points straight up.

21. Use new bolts for the balance shaft assembly and oil the threads before installation. Install the balancer unit and tighten the bolts in numerical in the following steps:
- Step 1: Bolts 1-5, 31 ft-lb (42 Nm): Bolt 6, 27 ft-lb (36 Nm)
- Step 2: Bolts 1-5, 120 °: Bolt 6, 90 °
- Step 3: Loosen in reverse order or tightening sequence
- Step 4: Bolts 1-5, 31 ft-lb (42 Nm): Bolt 6, 27 ft-lb (36 Nm)
- Step 5: Bolts 1-5, 120 °: Bolt 6, 90 °

22. Install the crankshaft sprocket and timing chain for the balancer unit. Make sure the crankshaft sprocket mark is at the top to align with the mark on the block. The orange or blue link on the chain aligns with this same mark, while the gold or yellow link aligns with the mark on the balance shaft sprocket.

23. Install balancer timing chain tensioner and remove the stopper pin. Make sure the timing marks are still aligned.

24. Install the camshaft timing chain and related parts.

25. After installing timing chain tensioner, remove the stopper pin and make sure that the tensioner moves freely.

26. Rotate the crankshaft two full turns and make sure all timing marks still line up.

27. Install a new front oil seal

28. Carefully clean all sealing surfaces and apply RTV sealant to the timing cover. Install the cover and torque bolts A and D to 36 ft. lbs. (49 Nm) and bolts B and C to 9 ft. lbs. (13 Nm).

29. Carefully clean all sealing surfaces and apply RTV sealant to the rocker cover. Install the cover

30. Install IVT solenoid valve to IVT cover with a new O-ring.

31. Apply RTV sealant to the IVT cover and install it. Torque the bolts in order to 9 ft. lbs. (13 Nm).

32. Install crankshaft pulley and torque bolt to 31 ft. lbs. (42 Nm) plus 60 °.

33. Install remaining components.

3.5L Engine

See Figures 75 through 79.

1. Remove the engine cover.

2. Relieve fuel system pressure.

3. Remove the air cleaner case and mass air flow sensor.

4. Remove the engine coolant reservoir.

5. Remove the windshield wipers and cowl.

6. Remove the IPDM E/R (fuse and relay box) and position aside.

7. Raise and safely support the vehicle and remove the front wheels

8. Remove the engine undercover and right inner fender.

9. Remove the drive belt, idler pulley and drive belt auto-tensioner.

10. Properly discharge and recover the air conditioning refrigerant and remove the A/C compressor.

11. Remove engine oil cooler pipe bolts.

12. Remove the power steering pump and reservoir with lines attached and position them aside.

13. Remove the alternator.

14. Disconnect the engine harness and position aside.

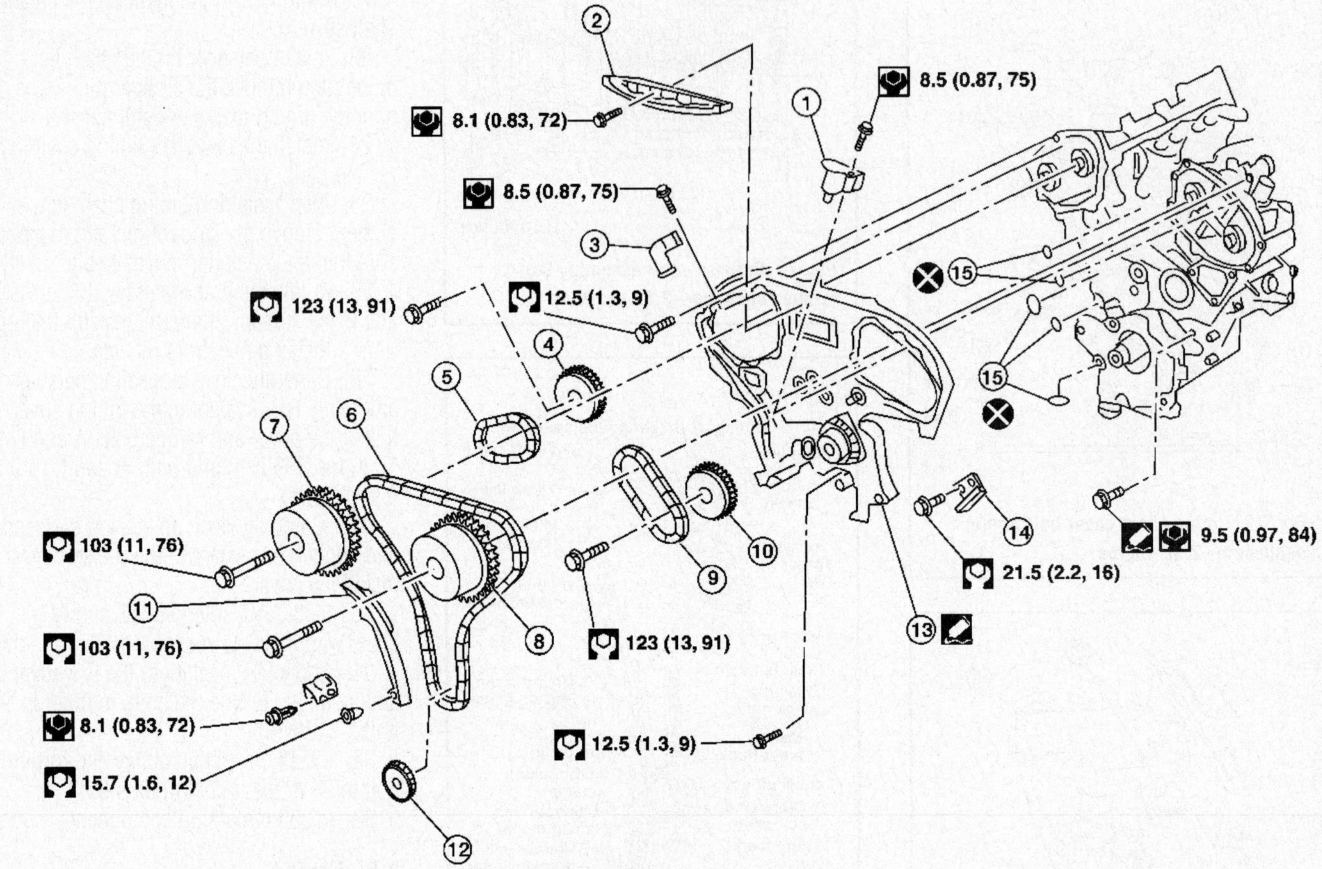

8.5 (0.87, 75)
8.1 (0.83, 72)
8.5 (0.87, 75)
123 (13, 91)
12.5 (1.3, 9)
103 (11, 76)
103 (11, 76)
8.1 (0.83, 72)
15.7 (1.6, 12)
123 (13, 91)
12.5 (1.3, 9)
9.5 (0.97, 84)
21.5 (2.2, 16)

1. Timing chain tensioner	2. Internal chain guide	3. Timing chain tensioner
4. Camshaft sprocket (EXH)	5. Timing chain (secondary)	6. Timing chain (primary)
7. Camshaft sprocket (INT)	8. Camshaft sprocket (INT)	9. Timing chain (secondary)
10. Camshaft sprocket (EXH)	11. Slack guide	12. Crankshaft sprocket
13. Rear timing chain case	14. Tension guide	15. O-ring

22140_ALTI_G0078

Fig. 75 Timing chain assembly—3.5L engine

15. Remove the A/C low-pressure flexible hose. Cap the openings to keep out dirt and moisture.
16. Support the engine and remove the right engine mount and bracket.
17. Remove the water pump cover.
18. Remove the IVT covers.

➡The shaft in the cover is inserted into the center hole of the intake camshaft sprocket. Remove the cover by pulling straight out until the cover disengages from the camshaft sprocket.

19. Remove the starter.
20. Remove the upper intake manifold.
21. Remove the ignition coils and spark plugs. Note the coil's locations for installation.
22. Remove the rocker covers.
23. Set the engine to TDC of No. 1 cylinder.
24. Lock the drive plate or flywheel in place with a locking tool attached to the starter bolt hole. Take care to not damage the ring gear teeth or the signal plate teeth behind the ring gear.
25. Loosen crankshaft pulley bolt but do not remove it.
26. Position a puller on the crankshaft pulley. Make sure the puller grabs the

6
20
7
9
8
22
5
17
21
2
12
10
11
4
13
19
15
18
14
3
2
1

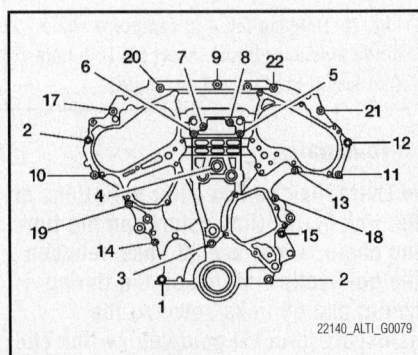

22140_ALTI_G0079

Fig. 76 Timing chain cover bolt torque sequence—3.5L engine

inside of the pulley close to the bolt, not the outside diameter of the pulley. Remove the crankshaft pulley.

27. Remove the oil pans. Temporarily reinstall the lower oil pan.

28. Support front of engine with a suitable jack.

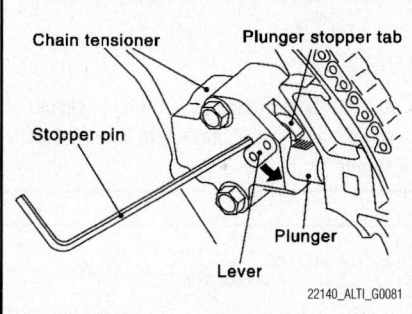

Fig. 77 Pin the lever down to unlatch the stopper tab, push the plunger in, then raise the lever and pin the plunger in place.

29. Remove the front timing chain cover bolts in the reverse of installation order.

30. Make sure the engine is still at TDC and make paint marks on the timing chain and sprockets to indicate the correct position of the components for installation.

31. Remove the internal chain guide.

32. To remove the timing chain tensioner"

- Push the lever down to release plunger latch.
- Partially insert a pin to hold the lever down.
- Push the plunger in and hold it.
- Remove the locking pin, raise the lever and reinsert the pin all the way.
- Remove the timing chain tensioner.

33. Remove the primary timing chain and sprockets:

34. To remove the secondary chain:
- Paint alignment marks on the chains and sprockets.
- Hold the camshaft with a wrench and loosen the sprocket bolt.

- Insert a stopper pin into the chain tensioners.
- Rotate camshaft slightly to slacken timing chain. Make sure the chain comes out of the guide groove in the tensioner.
- Insert a thin plate between the chain and tensioner.
- Remove cam sprockets and secondary timing chain together.

✳✳ WARNING

Even with the stopper pin inserted, the plunger can come out of the tensioner when timing chain is removed. Use caution during removal.

➡**The intake camshaft sprocket is an assembly of primary and secondary sprockets. Do not disassemble.**

To install:
35. Replace the front main seal.

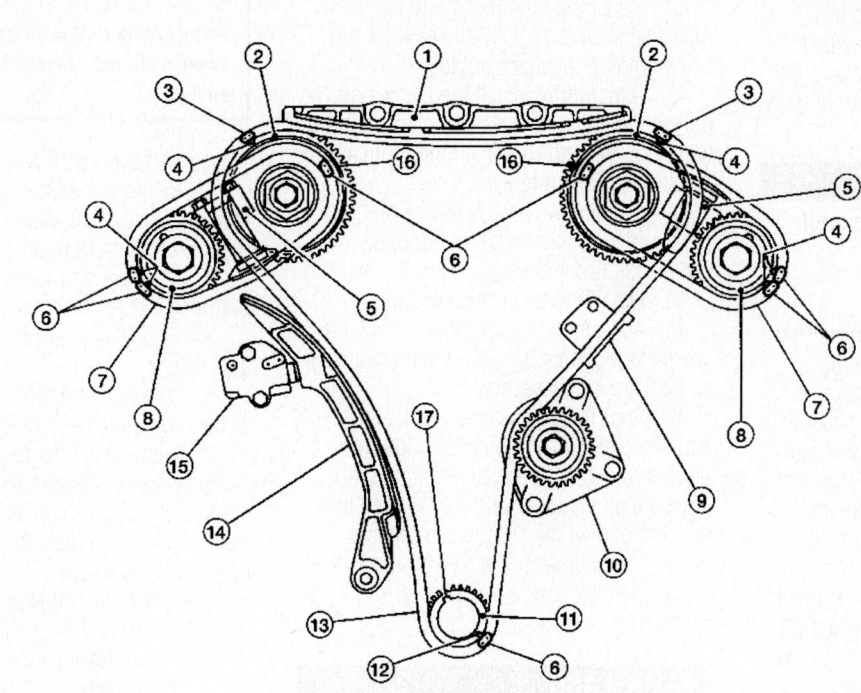

1.	Internal chain guide	2.	Camshaft sprocket (intake)	3.	Mating mark (pink link)
4.	Mating mark (punched)	5.	Secondary timing chain tensioner	6.	Mating mark (orange link)
7.	Secondary timing chain	8.	Camshaft sprocket (exhaust)	9.	Tension guide
10.	Water pump	11.	Crankshaft sprocket	12.	Mating mark (notched)
13.	Primary timing chain	14.	Slack guide	15.	Primary timing chain tensioner
16.	Mating mark (back side)	17.	Crankshaft key		

Fig. 78 Timing chains installed—3.5L engine

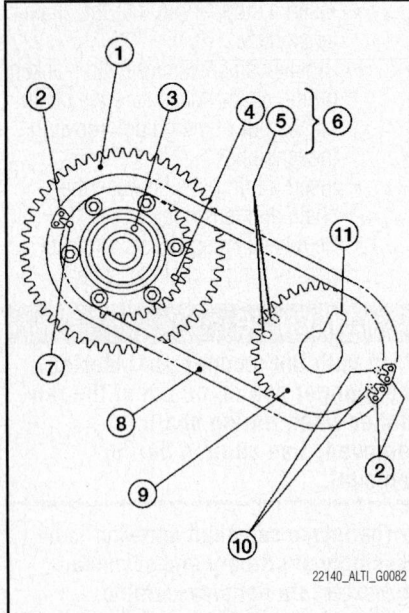

Fig. 79 Secondary timing chain alignment marks—3.5L engine

36. If removed, install the timing chain tension guide.

37. Make sure the crankshaft is at TDC of No. 1 piston. Make sure the camshaft dowel pin holes are both at the top and the crankshaft key is aligned with the right bank of cylinders.

❊❊ WARNING

The dowel pin holes are the small diameter holes for camshaft sprocket dowel pins. Do not misidentify.

38. To install the secondary timing chains:

- Compress the secondary chain tensioner and secure it with a stopper pin. Install the tensioner.
- On a flat surface, assemble the camshaft sprockets and timing chains with all the match marks aligned. Align marks 4, 5, 7 and 10 on the secondary timing chain (8) (orange link) with the ones on the intake and exhaust sprockets (stamped). Matching marks for the intake sprocket are on the back side of the secondary sprocket.

➡There are two types of matching marks, round (7 and 10) for the left cylinder bank and oval (4 and 5) for the right.

- Fit the chain and sprocket assemblies into place on the camshafts, making sure the dowel pins align properly. Install the sprocket bolts

finger tight just to hold the sprockets in place.
- Check chain and sprocket alignments again.
- Hold each camshaft and torque the sprocket bolt to 76 ft. lbs. (103 Nm).
- Remove the tensioner stopper pins and check timing mark alignments again.

39. Install the crankshaft sprocket. Make sure the mating marks on the crankshaft sprocket face the front of the engine.

40. Install the primary timing chain. Make sure the punched mark on the camshaft sprocket is aligned with the pink link on the timing chain, while the notched mark on the crankshaft sprocket is aligned with the orange one on the timing chain. During alignment, be careful to avoid disturbing the secondary timing chain alignment.

41. Install the chain guide and slack guide. Do not overtighten the slack guide bolts. It is normal for a gap to exist under the bolt seats.

42. Install the timing chain tensioners. After installation, pull out the stopper pin while holding the slack guide.

43. Temporarily install the crankshaft pulley and bolt and rotate the crankshaft two full turns. Reconfirm that all timing marks are still properly aligned.

44. Install new O-rings on the rear timing chain case and install the front timing chain case (front cover).

45. Install remaining components.

46. Activate the fuel system. Check for any leaks when the system is repressurized and correct as necessary.

47. Start the engine and check all systems for leaks. Rev engine to 3,000 rpm under no load to purge air from the high-pressure oil chamber of the chain tensioners. The engine may produce a rattling noise. This indicates that air still remains in the chamber and is not a matter of concern.

VALVE LASH

ADJUSTMENT

➡**Valves are adjusted by changing lifters. There are 26 different thicknesses of Nissan valve lifter available in sizes ranging from 3.00 to 3.50 mm (0.1181 to 0.1378 in) in steps of 0.02 mm (0.0008 in.).**

2.0 and 2.5L Engines

See Figures 80 through 82.

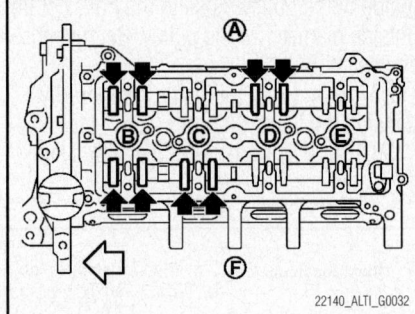

Fig. 80 With engine at TDC No. 1, check valve clearance at these cam lobes—2.0L engine

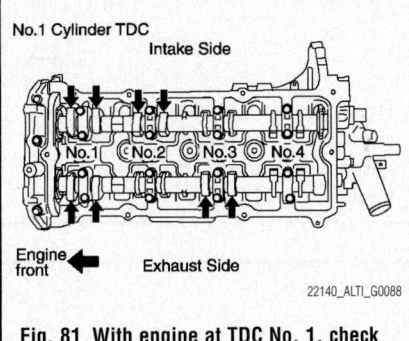

Fig. 81 With engine at TDC No. 1, check valve clearance at these cam lobes—2.5L engine

1. Remove the rocker cover.

2. Rotate crankshaft to TDC on No.1 cylinder. Align TDC mark (no paint) to timing indicator on front cover.

3. Measure and record valve clearance at the cam lobes indicated.

4. Rotate the crankshaft one full turn to TDC No. 4.

5. Measure and record valve clearance at the remaining cam lobes.

6. If adjustment is required, remove the camshafts and measure the valve lifter thickness with a micrometer. Use the equation below to calculate the replacement valve lifter thickness.

- Valve lifter thickness calculation: $t = t1 + (C1 - C2)$
- t = Valve lifter thickness to be replaced
- $t1$ = Removed valve lifter thickness
- $C1$ = Measured valve clearance
- $C2$ = Specified valve clearance

7. Thickness of new valve lifters can be identified by a stamp mark inside the lifter body.

8. Install the selected valve lifters.

9. Install the camshafts and check valve clearance again.

10. Install timing chain and related parts.

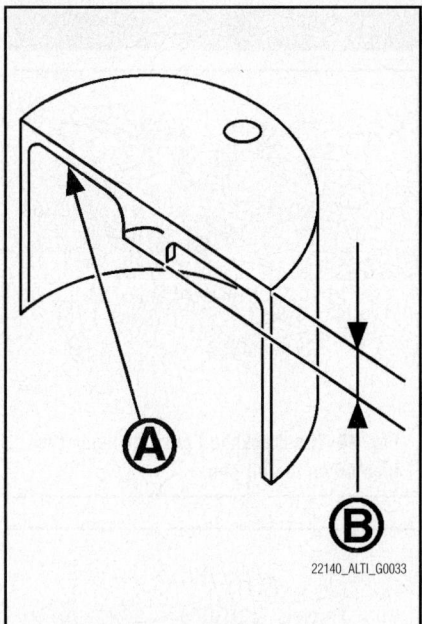

Fig. 82 Valve lifter thickness (B) is marked inside the body (A). Stamp mark "302" indicates a thickness of 3.02 mm (0.1189 in.).

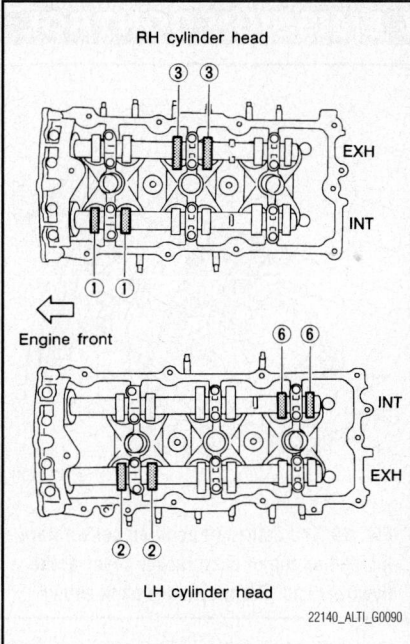

Fig. 83 With engine at TDC No. 1, check valve clearance at these cam lobes—3.5L engine

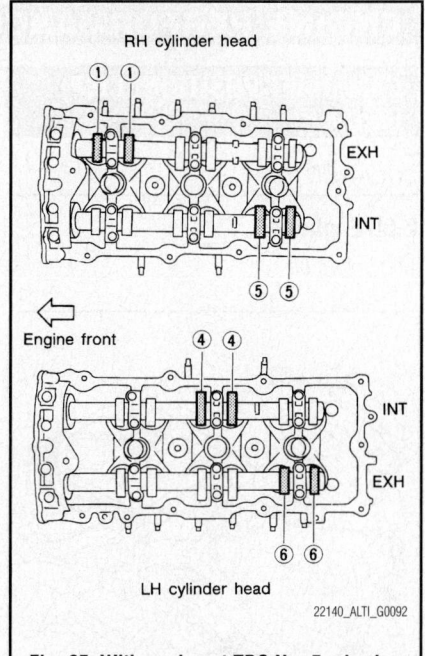

Fig. 85 With engine at TDC No. 5, check valve clearance at these cam lobes—3.5L engine

11. Manually rotate crankshaft pulley a few rotations and check valve clearance again.

12. Installation of the remaining components is the reverse of removal.

3.5L Engine

See Figures 82 through 85.

1. Remove the rocker covers.

2. Rotate crankshaft to TDC on No.1 cylinder. Align TDC mark (no paint) to timing indicator on front cover.

3. Measure and record valve clearance at the cam lobes indicated.

4. Rotate the crankshaft 240 ° to TDC No. 3.

5. Measure and record valve clearance at the indicated cam lobes.

6. Rotate the crankshaft 240 ° to TDC No. 5.

7. Measure and record valve clearance at the indicated cam lobes.

8. If adjustment is required, remove the camshafts and measure the valve lifter

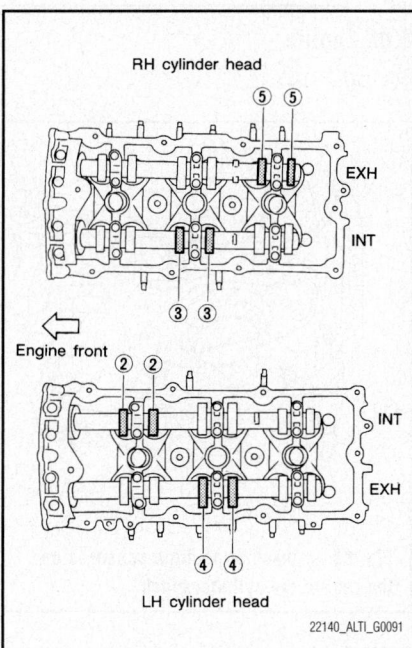

Fig. 84 With engine at TDC No. 3, check valve clearance at these cam lobes—3.5L engine

thickness with a micrometer. Use the equation below to calculate the replacement valve lifter thickness.

- Valve lifter thickness calculation: $t = t1 + (C1 - C2)$
- t = Valve lifter thickness to be replaced
- $t1$ = Removed valve lifter thickness
- $C1$ = Measured valve clearance
- $C2$ = Specified valve clearance

9. Thickness of new valve lifters can be identified by stamp mark inside the lifter body.

10. Install the selected valve lifters.

11. Install the camshafts and check valve clearance again.

12. Install timing chain and related parts.

13. Manually rotate crankshaft pulley a few rotations and check valve clearance again.

14. Installation of the remaining components is the reverse of removal.

ENGINE PERFORMANCE & EMISSION CONTROL

CAMSHAFT POSITION (CMP) SENSOR

LOCATION

2.0L Engine
See Figure 86.

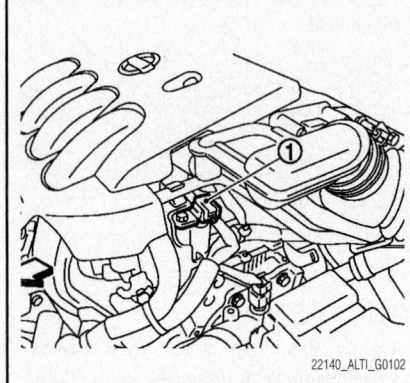

22140_ALTI_G0102

Fig. 86 The camshaft position sensor is on top of the engine near the ignition coil.

2.5L Engine
See Figure 87.

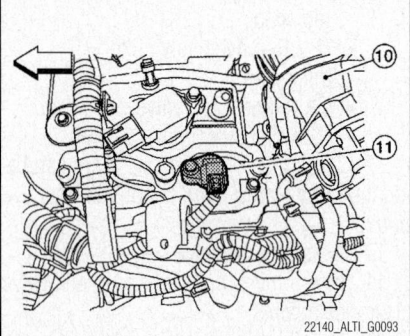

22140_ALTI_G0093

Fig. 87 The cam position sensor is located on top of the engine at the flywheel end.

3.5L Engine
See Figure 88.

REMOVAL & INSTALLATION

1. Unplug the connector.
2. Remove the bolt to remove the sensor.
3. During installation, use a new O-ring and make sure the sensor is clean and free of metal filings on the magnet end.

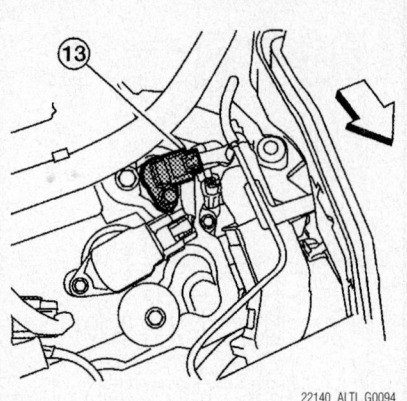

22140_ALTI_G0094

Fig. 88 The camshaft position sensors are located on top of each rocker cover at the flywheel end. Front cylinder bank shown.

CRANKSHAFT POSITION (CKP) SENSOR

LOCATION

2.0L Engine
See Figure 89.

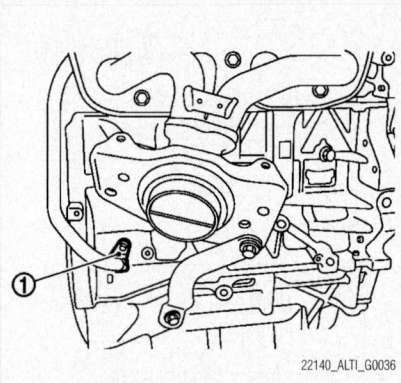

22140_ALTI_G0036

Fig. 89 Crankshaft position sensor is on the rear of the cylinder block.

2.5L Engine
See Figure 90.

3.5L Engine
See Figure 91.

REMOVAL & INSTALLATION

1. Unplug the connector.
2. Remove the bolt to remove the sensor.
3. During installation, use a new O-ring and make sure the sensor is clean and free of metal filings on the magnet end.

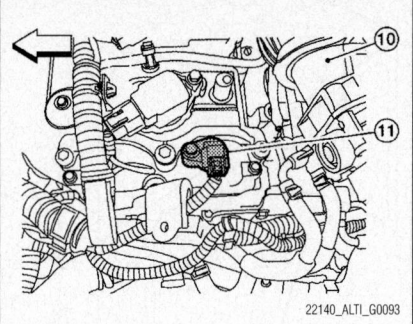

22140_ALTI_G0093

Fig. 90 The crankshaft position sensor is located on the oil pan.

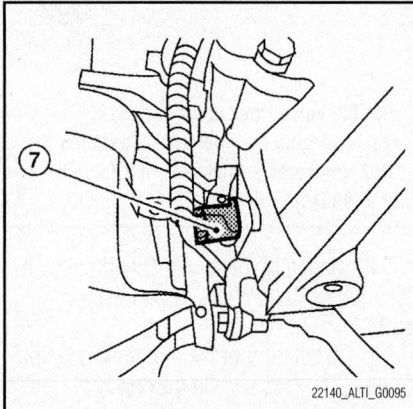

22140_ALTI_G0095

Fig. 91 The crankshaft position sensor is on the flywheel end of the block below the front cylinder head. Manual transmission model shown; automatic similar.

ELECTRONIC CONTROL MODULE (ECM)

LOCATION

Sentra
See Figure 92.

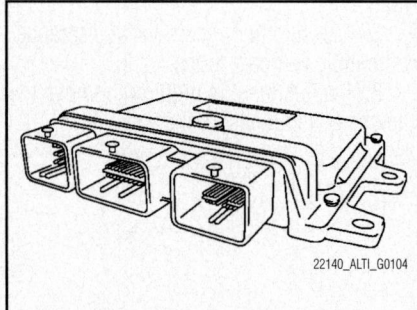

22140_ALTI_G0104

Fig. 92 Sentra ECM is mounted in the engine compartment.

Altima

See Figure 93.

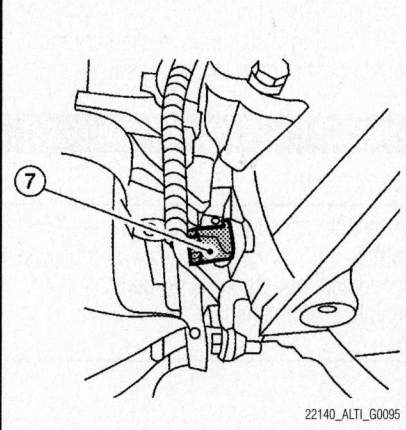

Fig. 93 The ECM is next to the battery—Altima

REMOVAL & INSTALLATION

1. Remove the connector cover and remove the ECM.
2. Unplug the connector.
3. When installing, take care not to bend the pins in the connector.

ENGINE COOLANT TEMPERATURE (ECT) SENSOR

LOCATION

2.0 and 2.5L Engines

See Figure 94.

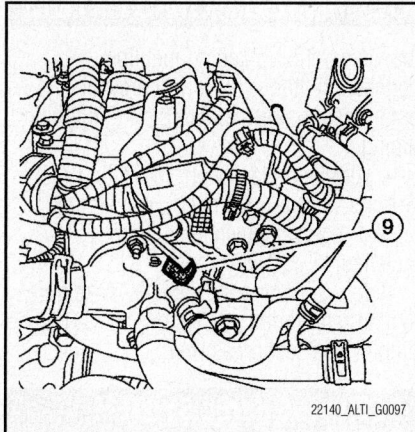

Fig. 94 The ECT is in the water outlet assembly near the upper radiator hose connection.

3.5L Engine

See Figure 95.

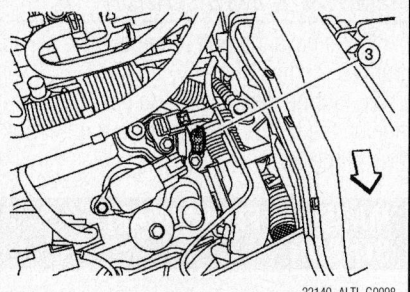

Fig. 95 The ECT is in the water outlet assembly between the cylinder heads near the front head.

REMOVAL & INSTALLATION

1. Unplug the connector.
2. Unscrew the sensor.
3. During installation, use a new O-ring and make sure the sensor is clean.

HEATED OXYGEN (HO2S) SENSOR

LOCATION

2.0L and 2.5L Engine

See Figure 96.

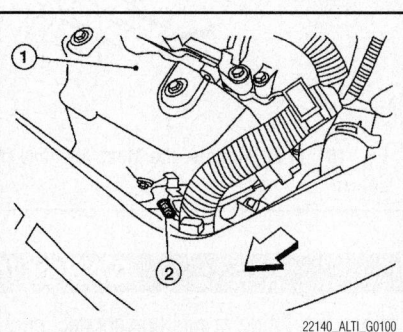

Fig. 96 The primary oxygen sensor (air/fuel ratio sensor) is in the intake manifold below the heat shield. 2.5L engine shown.

3.5L Engine

See Figure 97.

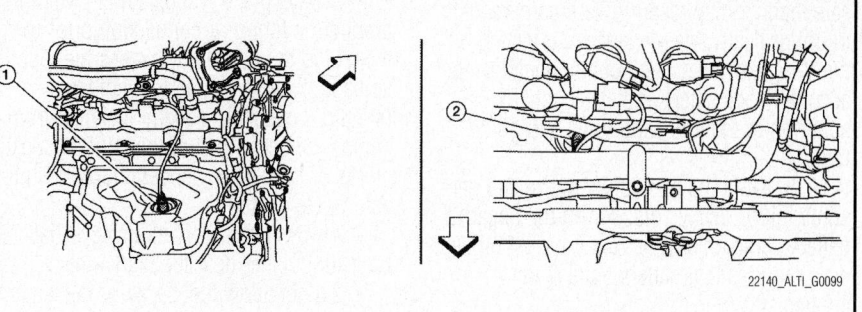

Fig. 97 The primary oxygen sensors (air/fuel ratio sensors) are in the exhaust manifolds. Arrows indicate front of vehicle.

3.5L Engine

See Figure 97.

REMOVAL & INSTALLATION

1. Make sure the exhaust manifold is cool.
2. Unplug the connector and remove the sensor with an oxygen sensor wrench or socket.
3. When installing, lightly coat the threads with anti-seize and torque to 23 ft. lbs. (31 Nm). Do not over tighten or the Malfunction Indicator Light (MIL) may turn on.

INTAKE AIR TEMPERATURE (IAT) SENSOR

LOCATION

The intake air temperature sensor is part of the Mass Air Flow (MAF) sensor.

KNOCK SENSOR (KS)

LOCATION

2.0L Engine

See Figure 98.

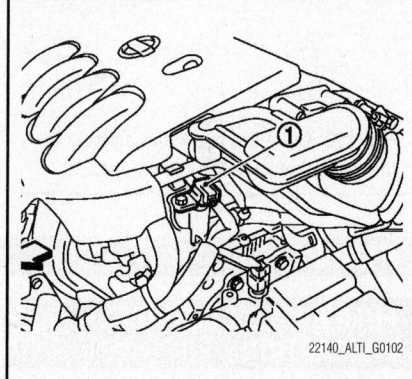

Fig. 98 Knock sensor is mounted on the engine block below the intake manifold.

2.5L Engine

See Figure 99.

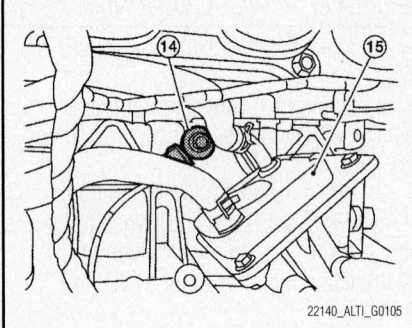

22140_ALTI_G0105

Fig. 99 The knock sensor is on the front of the engine block near the oil cooler.

3.5L Engine

See Figure 100.

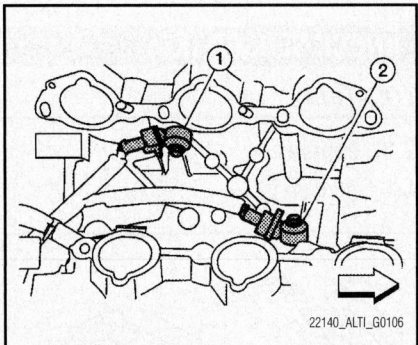

22140_ALTI_G0106

Fig. 100 There are two knock sensors in the valley between the cylinder heads.

REMOVAL & INSTALLATION

1. Remove the bolt to remove the sensor from the engine block.
2. During installation, do not over tighten the bolt or the sensor will not function properly.

MASS AIR FLOW (MAF) SENSOR

LOCATION

See Figure 101.

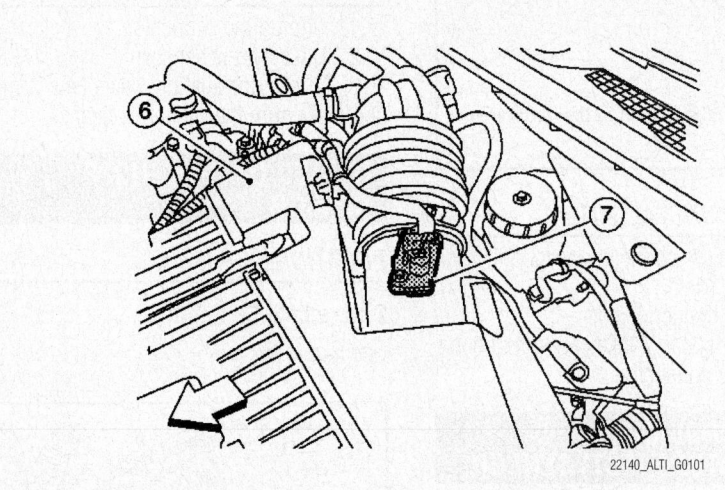

22140_ALTI_G0101

Fig. 101 In all models, the Mass Air Flow (MAF) sensor is in the air filter housing. 2.5L engine shown

REMOVAL & INSTALLATION

1. Unplug the connector.
2. Remove the screws to remove the sensor.
3. During installation, make sure the sensor seats firmly in the housing.

VEHICLE SPEED SENSOR (VSS)

LOCATION

Vehicle speed is calculated by the ABS control unit using inputs from all four wheel speed sensors. There is no dedicated vehicle speed sensor.

FUEL

FUEL SYSTEM SERVICE PRECAUTIONS

Safety is the most important factor when performing not only fuel system maintenance but any type of maintenance. Failure to conduct maintenance and repairs in a safe manner may result in serious personal injury or death. Maintenance and testing of the vehicle's fuel system components can be accomplished safely and effectively by adhering to the following rules and guidelines.

• To avoid the possibility of fire and personal injury, always disconnect the negative battery cable unless the repair or test procedure requires that battery voltage be applied.

• Always relieve the fuel system pressure prior to disconnecting any fuel system component (injector, fuel rail, pressure regulator, etc.), fitting or fuel line connection. Exercise extreme caution whenever relieving fuel system pressure to avoid exposing skin, face and eyes to fuel spray. Please be advised that fuel under pressure may penetrate the skin or any part of the body that it contacts.

• Always place a shop towel or cloth around the fitting or connection prior to loosening to absorb any excess fuel due to spillage. Ensure that all fuel spillage (should it occur) is quickly installed from engine surfaces. Ensure that all fuel soaked cloths or towels are deposited into a suitable waste container.

• Always keep a dry chemical (Class B) fire extinguisher near the work area.

• Do not allow fuel spray or fuel vapors to come into contact with a spark or open flame.

• Always use a back-up wrench when

GASOLINE FUEL INJECTION SYSTEM

loosening and tightening fuel line connection fittings. This will prevent unnecessary stress and torsion to fuel line piping.

• Always replace worn fuel fitting O-rings with new Do not substitute fuel hose or equivalent where fuel pipe is installed.

Before servicing the vehicle, make sure to also refer to the precautions in the beginning of this section as well.

RELIEVING FUEL SYSTEM PRESSURE

1. Remove the fuel pump fuse.
2. Start the engine.
3. After the engine stalls, crank it three times to release all remaining fuel pressure.
4. Turn ignition switch OFF.

FUEL FILTER

REMOVAL & INSTALLATION

➥The fuel filter is part of the fuel pump assembly and not serviced separately. The fuel pump is inside the fuel tank and can be accessed through an open under the rear seat.

FUEL INJECTORS

REMOVAL & INSTALLATION

Altima

See Figures 102 through 104.

1. Remove the engine cover.
2. Release the fuel system pressure.
3. On 2.5L engine, remove the front air duct.

4. On 3.5L engine, remove air cleaner case, mass air flow sensor and air intake tube as an assembly.

✳✳ WARNING

Prepare a container and cloth for catching any spilled fuel. This operation should be performed in a place that is free from any open flames.

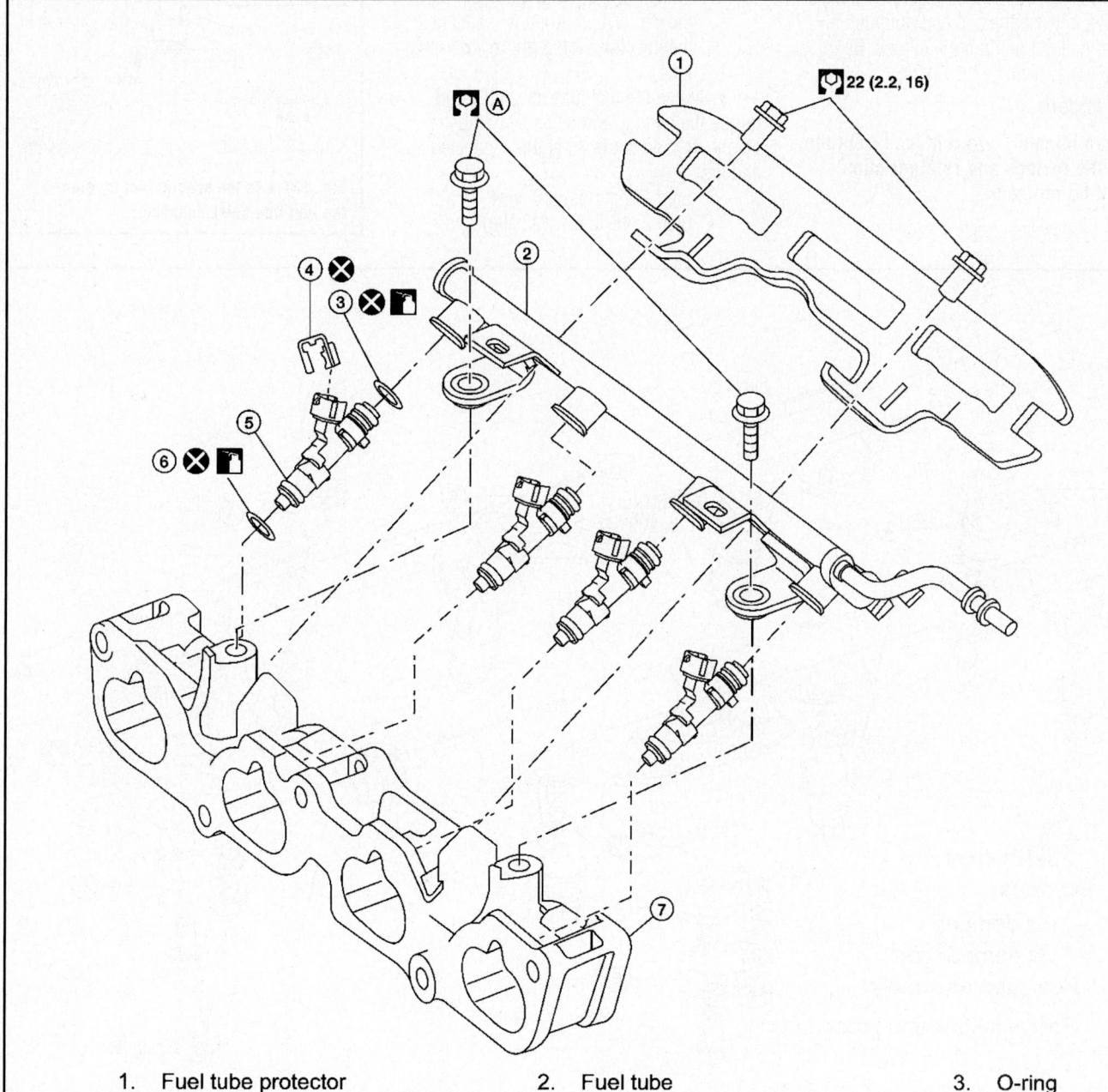

1.	Fuel tube protector	2.	Fuel tube	3.	O-ring
4.	Clip	5.	Fuel injector	6.	O-ring
7.	Intake manifold adapter	A.	Follow installation procedure		

22140_ALTI_G0107

Fig. 102 Fuel injector assembly—2.5L Engine

5. Disconnect the fuel hose quick connector at the fuel tube side. While hoses are disconnected, seal their openings with vinyl bag or similar material to prevent foreign material from entering them.

6. Remove the intake manifold.

7. Disconnect the injector wiring harness at the front of the engine and remove it from the bracket.

8. Remove the bolts and remove the fuel tube and fuel injectors as an assembly.

9. To remove the fuel injectors from the fuel tube, release the clip and pull fuel injector straight out of the fuel tube. Be careful not to damage the nozzle.

To install:

➡ **When injectors are removed from the tube, the O-rings and retainer clips should be replaced.**

10. Fit new O-rings to the injector.

11. Fit a new clip onto the injector. Make sure the tab opposite the electrical connector fits into the notch of the clip.

12. Insert the fuel injector into fuel tube.

- On 2.5L engine, the flange on the tube is longer on one side; make sure it fits into the notch on the clip. When properly installed, the injector should not rotate in the tube.

- On 3.5L engine, make sure the tab on the tube fits into the notch on the clip. When properly installed, the injector should not rotate in the tube.

13. Install the fuel tube assembly and torque the bolts in two steps. On 3.5L engine, tighten the bolts at the connected end first.

- Step 1: 7 ft. lbs. (10 Nm)
- Step 2: 16 ft. lbs. (22 Nm)

14. Install the fuel tube protector bracket.

15. Connect the fuel hose quick connector and pressurize the fuel system to check for leaks. The pump will run for two

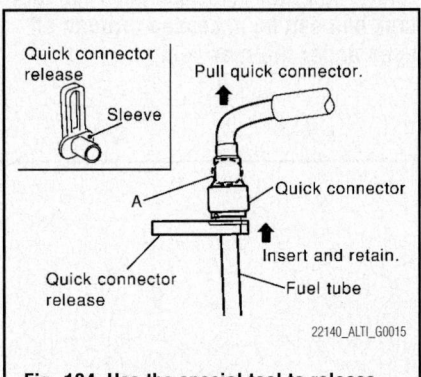

Fig. 104 Use the special tool to release the fuel line quick connector.

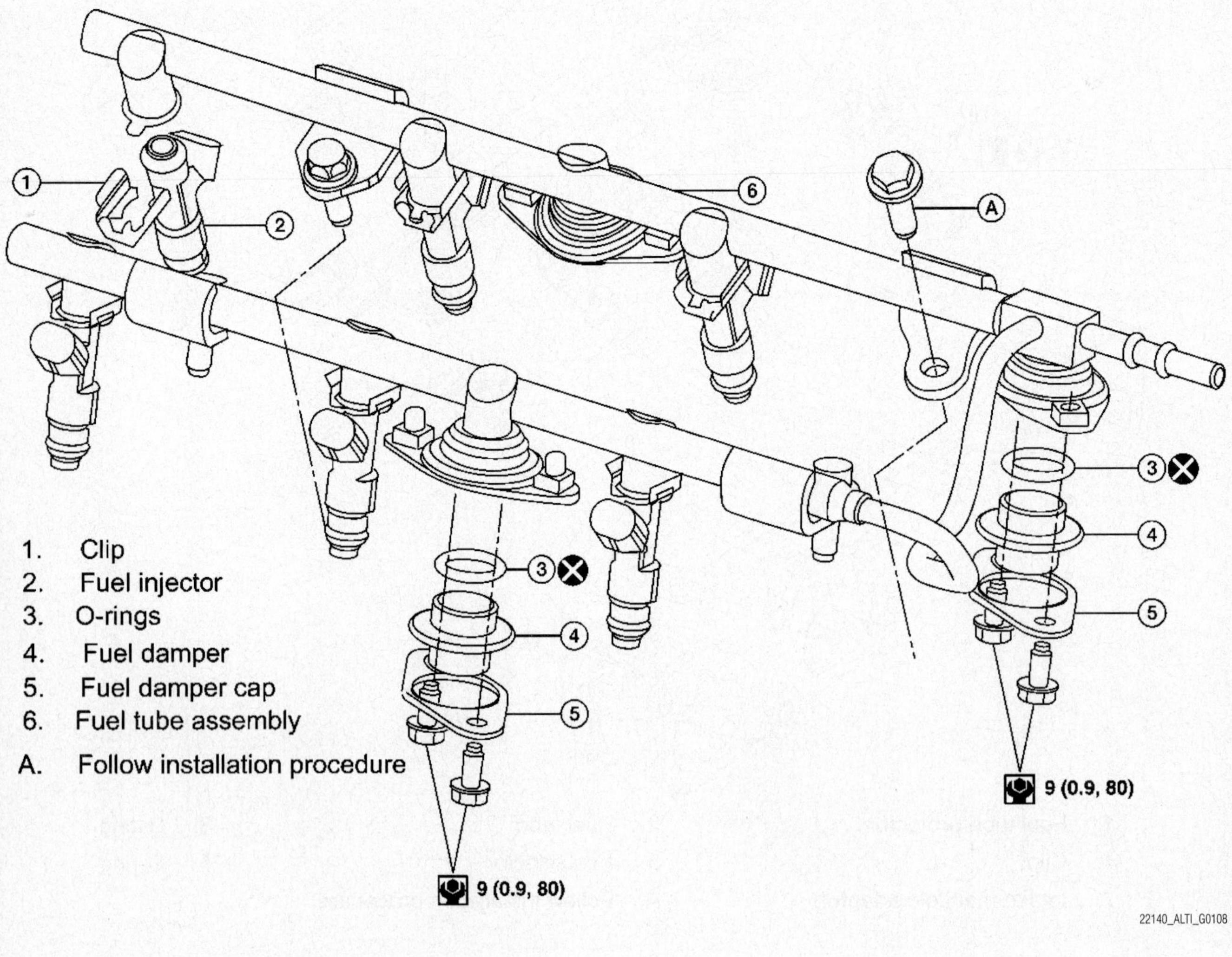

1. Clip
2. Fuel injector
3. O-rings
4. Fuel damper
5. Fuel damper cap
6. Fuel tube assembly

A. Follow installation procedure

9 (0.9, 80)

Fig. 103 Fuel injector assembly—3.5L Engine

seconds each time the ignition switch is turned ON.

16. Install the intake manifold.

FUEL PUMP

REMOVAL & INSTALLATION

See Figures 105 and 106.

➡ **Fuel will be spilled when removing fuel pump assembly if the tank is full. If the fuel gauge indicates more than (⅞ full), drain at least 3 ⅛ gallons (12L) from the fuel tank.**

1. Open fuel door and unscrew the fuel filler cap to release the pressure inside the fuel tank.

2. Release fuel system pressure.

3. Remove rear seat bottom.

4. Turn the three cover retainers 90° in counterclockwise and remove the fuel pump inspection hole cover.

5. Disconnect the wiring and fuel hose quick connectors. To remove the quick connector, hold the sides of the connector, push in the tabs and pull the tube straight out. The tube can be removed only when the tabs are completely depressed. Do not twist or use any tools.

6. To keep the connectors clean and to avoid damage, cover them completely with plastic bags or something similar. Do not insert plugs to prevent damage to O-ring.

7. Remove the locking ring using the correct tool and carefully lift the pump/filter/sending unit out of the tank. Take care not to bend the float arm.

➡ **On Altima, the lock ring and lock ring seal must be replaced.**

To install:

8. Carefully fit the fuel pump assembly into the tank. On Altima, fit a new lock ring and seal.

9. Reconnect the hoses and wiring.

10. Connect the battery and turn the ignition switch ON three or four times to run the pump, pressurize the system and check for leaks.

FUEL TANK

REMOVAL & INSTALLATION

See Figures 107 and 108.

➡ **Fuel will be spilled when removing fuel pump assembly if the tank is full.**

If the fuel gauge indicates more than (⅞ full), drain at least 3 ⅛ gallons (12L) from the fuel tank.

1. Check fuel level with the vehicle on a level surface. If the fuel gauge indicates more than (⅞ full), drain at least 3 ⅛ gallons (12L) from the fuel tank .

2. Siphon fuel from fuel tank if necessary.

3. Open fuel door and unscrew the fuel filler cap to release the pressure inside the fuel tank.

4. Release the fuel pressure.

5. Remove rear seat bottom.

6. Turn the three retainers 90° in a counterclockwise direction and remove the fuel pump inspection hole cover.

7. Disconnect wiring and fuel feed hoses. To keep the connectors clean and to avoid damage, cover them completely with plastic bags or something similar. Do not insert plugs to prevent damage to O-ring.

8. On Sentra, remove center exhaust pipe.

9. On Altima, disconnect the parking brake cables.

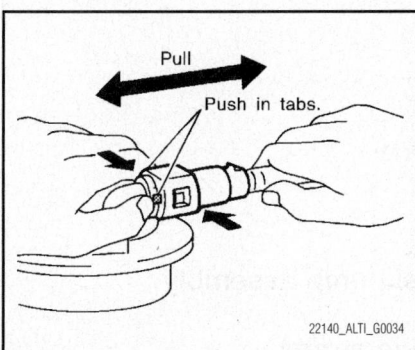

Fig. 105 **Disconnecting the fuel line quick connector**

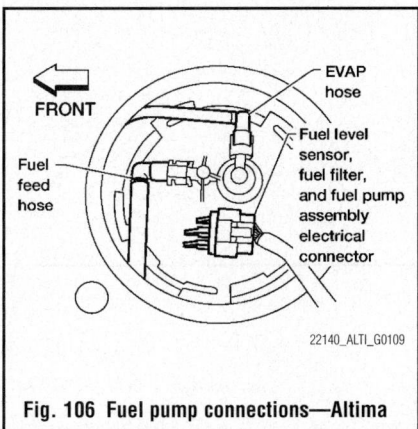

Fig. 106 **Fuel pump connections—Altima**

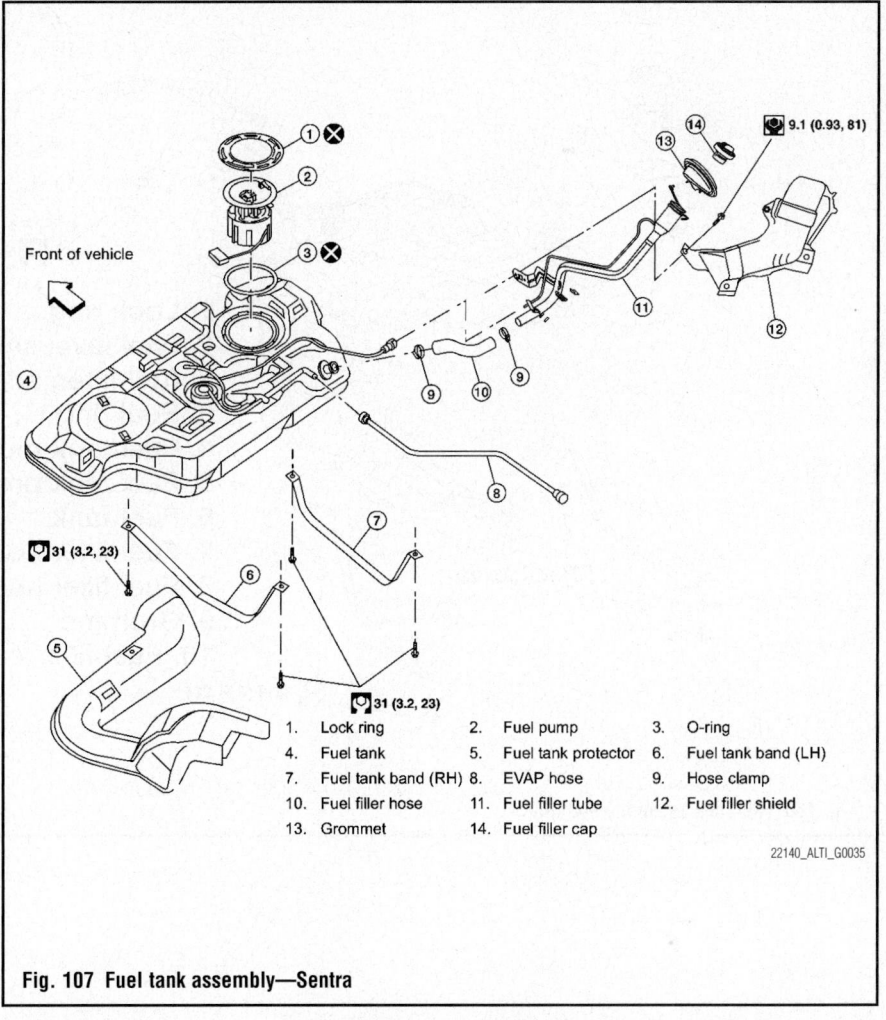

1. Lock ring
2. Fuel pump
3. O-ring
4. Fuel tank
5. Fuel tank protector
6. Fuel tank band (LH)
7. Fuel tank band (RH)
8. EVAP hose
9. Hose clamp
10. Fuel filler hose
11. Fuel filler tube
12. Fuel filler shield
13. Grommet
14. Fuel filler cap

Fig. 107 **Fuel tank assembly—Sentra**

10. Remove fuel tank protector.

11. Loosen fuel filler hose clamp and remove fuel filler hose.

✴✵ WARNING

Do not remove fuel filler hose from fuel filler tube. Mark components for alignment.

12. On Sentra, remove vent hose and EVAP hose at rear of fuel tank.

13. Support center of fuel tank with transmission jack.

14. Remove fuel tank bands. If they are not marked "R" and "L," mark them now.

15. Lower transmission jack carefully to remove fuel tank while supporting it by hand.

✴✵ WARNING

Fuel tank may be in an unstable position because of the shape of fuel tank bottom. Be sure to support tank securely.

To install:

16. Secure tank on a transmission jack and carefully raise it into position

17. Fit the fuel tank bands (marked "R" and "L") and tighten the bolts to 23 ft. lbs. (31 Nm).

18. Connect the filler and vent hoses and tighten the clamps.

19. Connect remaining tubes, wiring and fuel lines.

20. With some fuel in the tank, turn the ignition switch ON to pressurize the system and check for leaks.

21. Install remaining components.

THROTTLE BODY

REMOVAL & INSTALLATION

1. Remove engine cover.

2. Remove air intake ducts and air filter housing as required.

3. Disconnect throttle wiring.

4. Disconnect coolant hoses and plug them immediately to prevent coolant leaks.

5. Remove the bolts to remove throttle assembly.

6. Installation is the reverse of removal. Torque the bolts to 7 ft. lbs. (10 Nm). Do not over tighten these bolts or the throttle may not work properly.

7. A throttle relearn procedure is required.

- Make sure that accelerator pedal is fully released.
- Turn ignition switch ON.
- Turn ignition switch OFF. During the next 10 seconds, you should hear the throttle moving as the ECM learns the "closed throttle" position.

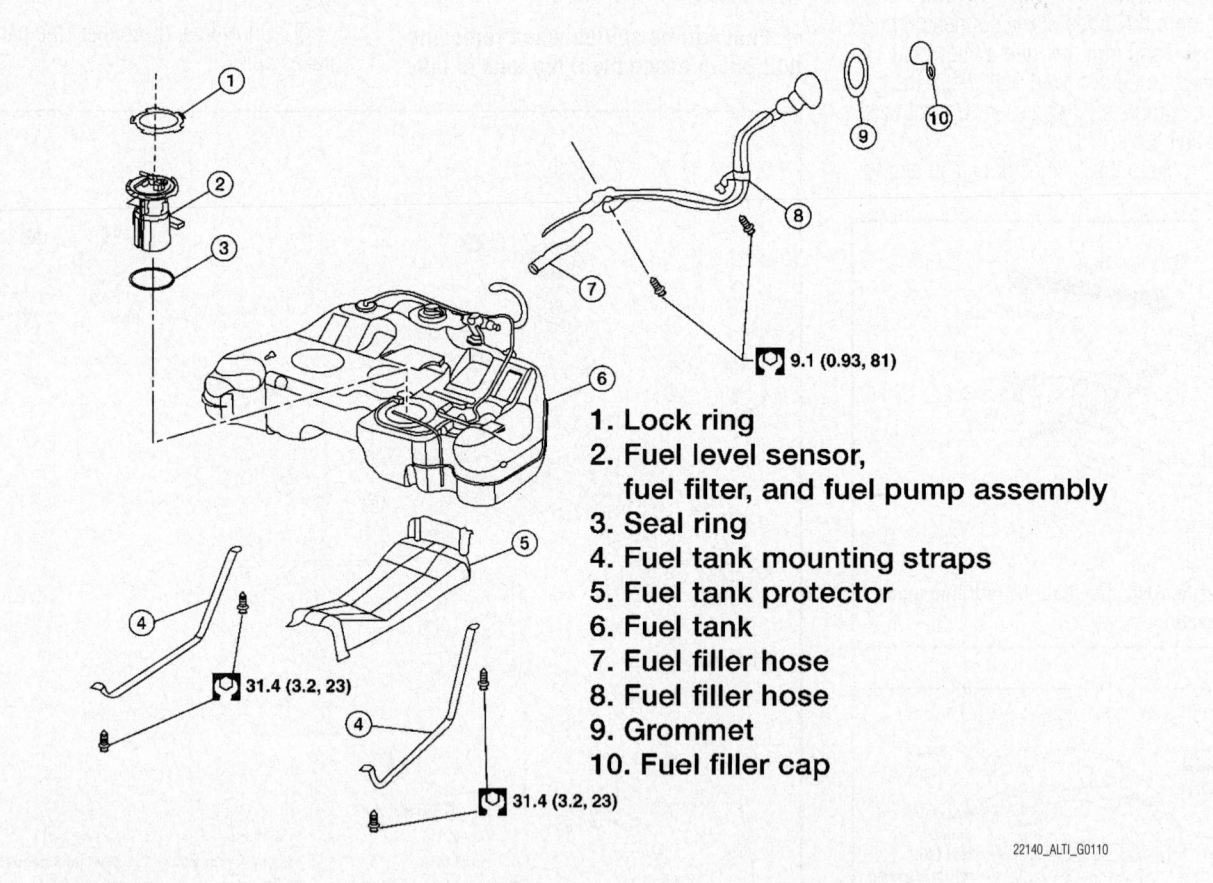

9.1 (0.93, 81)

1. Lock ring
2. Fuel level sensor, fuel filter, and fuel pump assembly
3. Seal ring
4. Fuel tank mounting straps
5. Fuel tank protector
6. Fuel tank
7. Fuel filler hose
8. Fuel filler hose
9. Grommet
10. Fuel filler cap

31.4 (3.2, 23)

31.4 (3.2, 23)

22140_ALTI_G0110

Fig. 108 Fuel tank assembly—Altima

HEATING & AIR CONDITIONING SYSTEM

BLOWER MOTOR

REMOVAL & INSTALLATION

Sentra

1. Remove the instrument panel. See Heater Core Removal & Installation
2. Disconnect the wiring and remove the blower motor.
3. Installation is the reverse of removal.

Altima

1. Remove the glove box.
2. Disconnect the blower motor wiring.
3. Remove the three blower motor screws and remove the blower motor.
4. Installation is the reverse of removal.

HEATER CORE

REMOVAL & INSTALLATION

Sentra

See Figure 109.

➡**Heater core removal requires removing the dashboard.**

1. Before servicing the vehicle, refer to the Precautions Section.
2. Position the steering wheel in the straight-ahead position.
3. Turn the ignition switch OFF.
4. Disconnect the negative (() battery cable; then, the positive (+) battery cable.

➡**Wait at least 10 minutes after disconnecting the battery cables for the charge in the air bag circuit to dissipate before working on the air bag module(s).**

5. Properly discharge and recover the refrigerant from the A/C system.
6. Drain the cooling system.
7. Install the steering column by removing the following:
   - lower dash cover
   - airbag module: store it face up out of the way
   - steering wheel
   - steering column covers
   - combination switch and spiral cable
   - disconnect all wiring
   - install the bolt to disconnect the upper end of the intermediate shaft from the power steering motor.

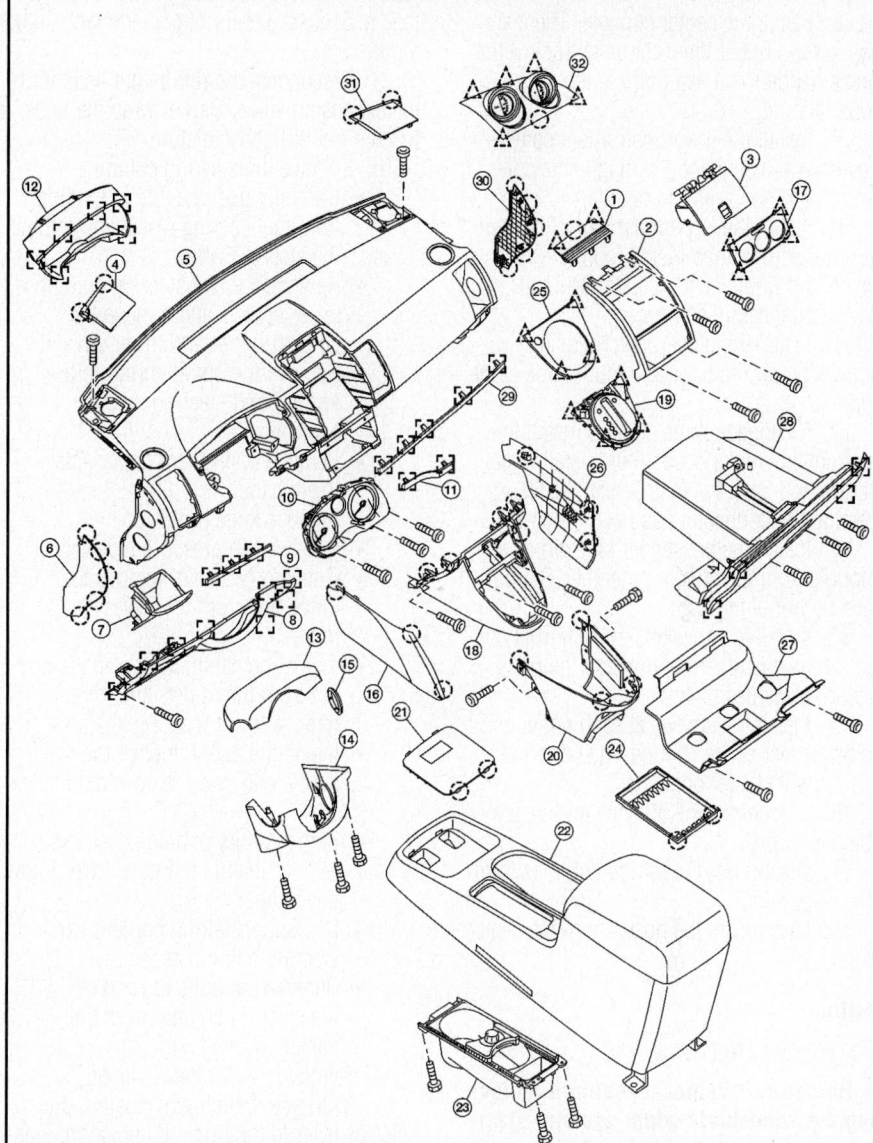

1. Cluster lid C upper mask
2. Cluster lid C
3. Cluster lid C storage bin
4. Speaker grille (LH)
5. Instrument panel
6. Instrument side mask (LH)
7. Fuse block lid storage bin
8. Instrument lower finisher
9. Instrument panel trim (LH)
10. Combination meter
11. Instrument panel trim center
12. Cluster lid A
13. Steering column cover upper
14. Steering column cover lower
15. Steering lock escutcheon
16. Instrument lower cover (LH)
17. Controller finisher
18. Instrument upper cover (center)
19. CVT finisher
20. Instrument lower cover (center)
21. Center console mat
22. Center console
23. Center console cup holder
24. Center console tray
25. MT finisher
26. Instrument lower cover (RH)
27. Glove box lower finisher
28. Glove box assembly
29. Instrument panel trim (RH)
30. Instrument side mask (RH)
31. Speaker grille (RH)

22140_ALTI_G0038

Fig. 109 Exploded view of Sentra dashboard

- install the nuts to lower the power steering assembly from the vehicle.

8. Install the center console. There are two screws under the front panel (below the brake handle) and two more at the back (open the lid).

9. Install the lower instrument panel covers and disengage the diagnostic connector and hood release handle.

10. Install the uppermost part of the center cluster trim, then install four screws to install the whole center cluster. Unplug wiring connectors as needed.

11. Install the glove box. There are two screws below the door and four inside at the top.

12. Disengage three clips to install the instrument cluster cover, then install three screws to install the instrument cluster. Unplug the connectors as necessary.

13. Install the passenger side airbag module from the steering member. Store it face up out of the way.

14. Install the speakers at each front corner of the dashboard and install the bolts securing the dashboard.

15. Install the screws at each upper and lower corner of the dashboard and in the instrument cluster opening.

16. Disconnect the antenna and install the dashboard.

17. Disconnect the refrigerant lines from the evaporator.

18. Disconnect the hoses from the heater core.

Altima

See Figures 110 through 112.

➡ **Heater core removal requires removing the windshield wiper system, steering column and dashboard.**

1. Before servicing the vehicle, refer to the Precautions Section.

2. Position the steering wheel in the straight-ahead position.

3. Turn the ignition switch OFF.

4. Disconnect the negative (()) battery cable; then, the positive (+) battery cable.

➡ **Wait at least 10 minutes after disconnecting the battery cables for the charge in the air bag circuit to dissipate before working on the air bag module(s).**

5. Properly discharge and recover the refrigerant from the A/C system.

6. Drain the cooling system.

7. With 3.5L engine, remove the strut tower brace, the front cowl and the windshield wiper system.

8. Disconnect the heater hoses from the heater core pipes. Cap or wrap the pipe joint with a suitable material such as vinyl tape to avoid the entry of contaminants into the system.

9. Disconnect the refrigerant lines from the expansion valve. Cap or wrap the lines to keep out dirt and moisture.

10. Remove the steering column:
- Carefully peel the airbag module from the steering wheel and remove the steering wheel.
- Remove the two steering column cover screws and remove the column upper and lower covers.
- Remove the spiral spring (clock spring) and steering column switches.
- Remove instrument driver side lower panel.
- Remove knee protector.
- Remove tire pressure receiver.
- Outside the cabin, remove the lower bolt from the lower shaft joint.
- Inside the cabin, remove bolts and nut from the upper joint, then remove lower steering shaft.
- Inside the cabin, loosen the herbie clip, then remove hole cover seal.
- Remove nuts of hole cover and remove clamp and hole cover from dash panel.
- Disconnect wiring harness from steering column assembly.
- Remove the bolts to remove the steering column from the vehicle.

11. Remove the A-pillar finishers.

12. With automatic transmission, put selector lever in the drive (D) position, then remove selector lever knob. With manual transmission, put the shift lever in neutral, then remove shift lever knob.

13. Remove the center cluster lid D and remove the storage bin or CD changer (if equipped).

14. Remove both instrument side masks

15. Open the fuse block cover, remove the instrument lower cover screw (A), then remove the instrument lower cover (1).

16. Disconnect the in-vehicle sensor, VDC switch and trunk lid release switch. Disconnect the aspirator tube.

17. Remove the two screws at the top of the center console.

18. Remove both center console side panels and remove the console side screws.

19. Remove the screws on each side of the console at the rear and remove the center console.

20. Open the glove box door and remove three screws above the opening.

21. Close the door and remove the two screws below it to remove the glove box.

22. Remove the instrument cluster lid (above the cluster itself).

23. Remove the two screws, then remove the instruments. Carefully disconnect the wiring.

24. Remove the center ventilator grilles.

25. Remove the cluster lid and disconnect wiring as required.

26. Remove both tweeter speaker grilles and disconnect the harness connectors.

27. Remove the passenger air bag bolt.

28. Remove the remaining instrument panel screws.

29. Disconnect the audio harness connector located near the RH A-pillar.

30. Lift the instrument panel high enough in order to disconnect all the necessary harness connectors, then remove the instrument panel.

31. Disconnect the drain hose.

32. Remove the heater and cooling unit assembly attached to the steering member as one assembly from the vehicle.

33. Remove the blower unit from the heater and cooling unit and steering member assembly.

34. Remove the heater and cooling case from the steering member.

35. Remove the duct from the drive side of the case and remove the heater core.

To install:

36. Assemble the blower unit and heater/cooling case and install into the vehicle.

37. Connect the drain hose.

38. Fit the instrument panel into place and begin connecting the wiring.

39. Install the remaining instrument panel screws.

40. Remove the passenger air bag.

41. Install the speakers and the center ventilator grilles.

42. Install the instruments and the instrument cluster lid (above the cluster itself).

43. Install the glove box.

44. Install the center console, center cluster and the storage bin or CD changer (if equipped).

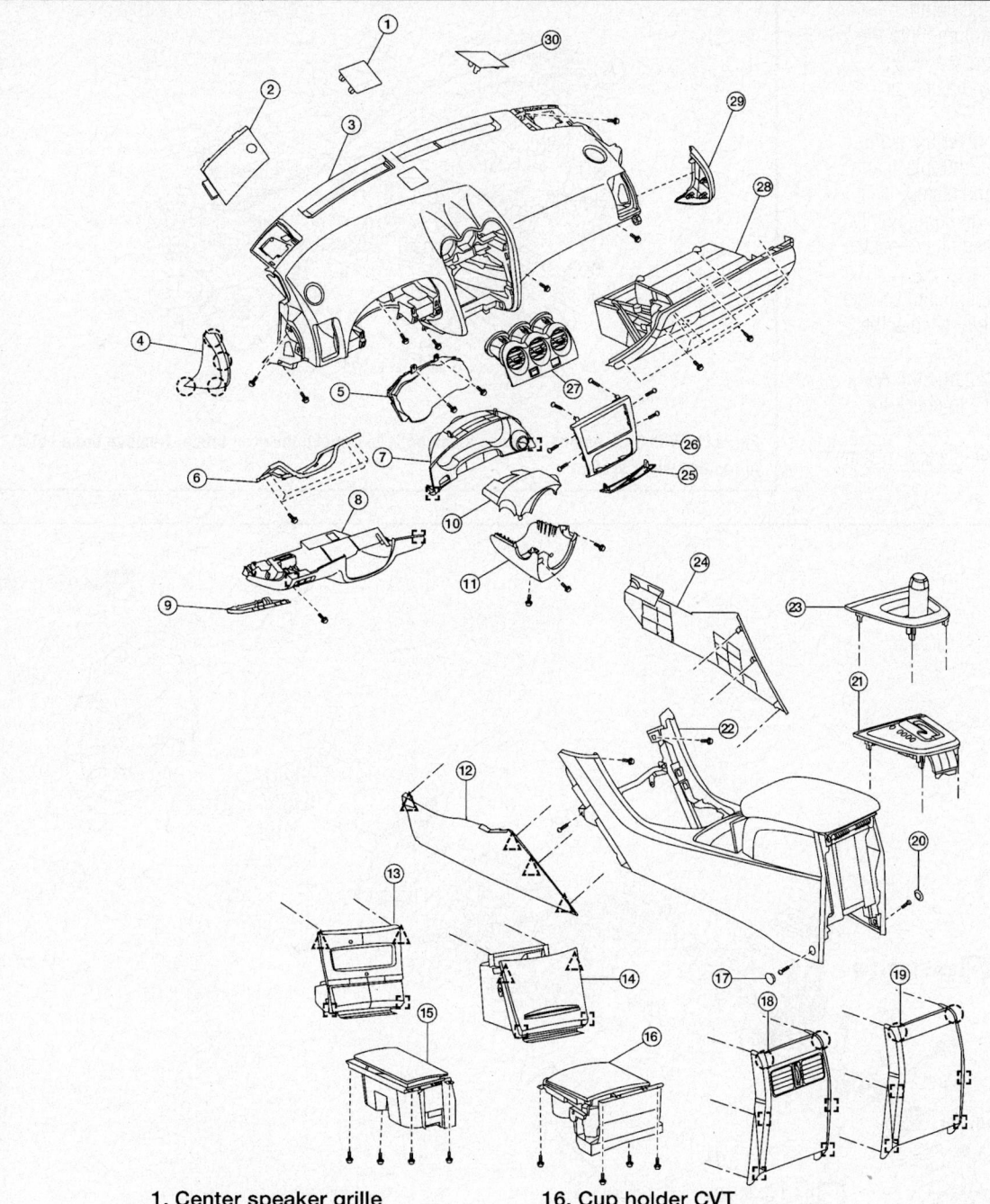

1. Center speaker grille
2. Tweeter speaker grille (LH)
3. Instrument panel
4. Instrument side mask (LH)
5. Combination meter
6. Lower knee protector (LH)
7. Cluster lid A
8. Instrument lower cover (LH)
9. Fuse block cover
10. Steering column cover upper
11. Steering column cover lower
12. Console side finisher (LH)
13. CD changer finisher (if equipped)
14. Storage bin (if equipped)
15. Cup holder M/T
16. Cup holder CVT
17. Console screw cover (LH)
18. Console rear finisher (if equipped with rear duct)
19. Console rear finisher (without rear duct)
20. Console screw cover (RH)
21. CVT finisher
22. Console
23. M/T finisher
24. Console side finisher (RH)
25. Cluster lid D
26. Cluster lid C
27. Center ventilator grilles
28. Glove box assembly
29. Instrument side mask (RH)
30. Tweeter speaker grille (RH)

22140_ALTI_G0111

Fig. 110 Exploded view of instrument panel—Altima

45. Install the steering column:
   - Fit the steering column into the vehicle and secure the bolts.
   - Connect the wiring as far as possible.
   - Install the lower steering shaft.
   - Fit the hole cover into place and install the seal and clamp.

46. Install the remaining components.

47. Connect the refrigerant and heater hoses.

48. Make sure no one is in the vehicle when connecting the battery in case the airbag deploys.

49. Refill all fluids, recharge the A/C system and run the engine to check for leaks or other problems.

50. Install the windshield wiper system and cowl.

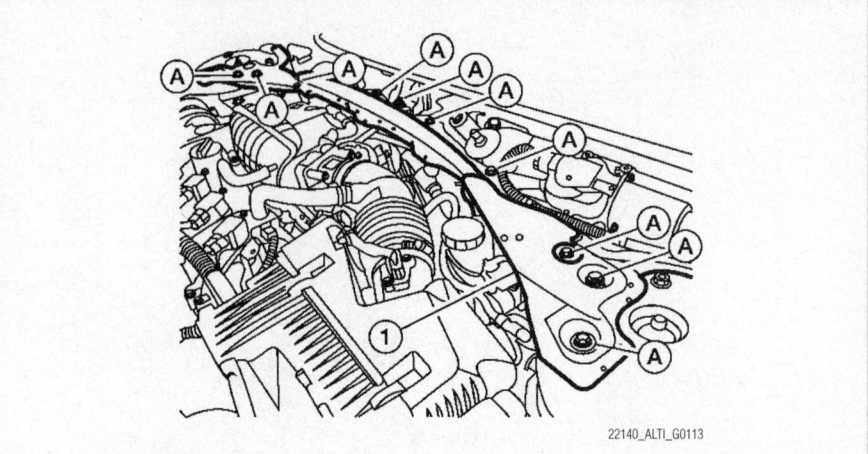

22140_ALTI_G0113

Fig. 111 With 3.5L engine, remove these bolts to remove the strut brace. Remove three bolts to remove wiper system.

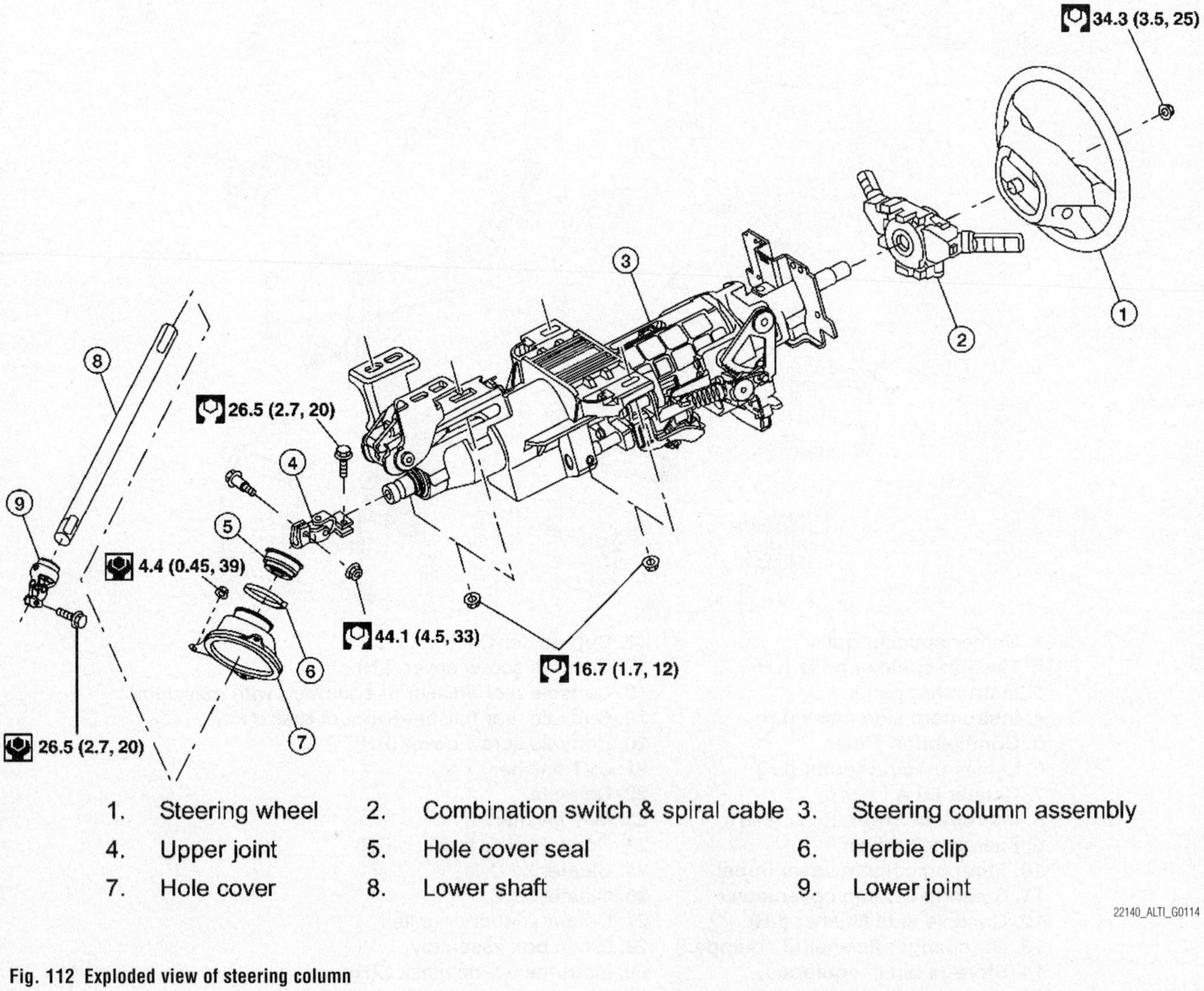

34.3 (3.5, 25)

26.5 (2.7, 20)

4.4 (0.45, 39)

44.1 (4.5, 33)

16.7 (1.7, 12)

26.5 (2.7, 20)

1.	Steering wheel	2.	Combination switch & spiral cable	3.	Steering column assembly
4.	Upper joint	5.	Hole cover seal	6.	Herbie clip
7.	Hole cover	8.	Lower shaft	9.	Lower joint

22140_ALTI_G0114

Fig. 112 Exploded view of steering column

STEERING

POWER STEERING GEAR

REMOVAL & INSTALLATION

Sentra

See Figure 113.

➡**This procedure involves removing the front suspension subframe.**

1. Set the front wheels to the straight-ahead position.

2. Remove the lower intermediate steering shaft clamp bolt.

3. Raise and safely support the vehicle and remove the front wheels.

4. Remove the cotter pins and loosen (but do not remove) the tie rod end-to-steering knuckle nuts.

5. Use a ball joint press to disengage the tie rod ends from the knuckle. Be careful not to damage the boot.

6. Remove the nuts and disconnect the tie rod ends.

7. To remove the subframe:

- Remove the nut on the lower stabilizer bar connecting rod and disconnect the rod from stabilizer bar.

- Remove rear engine mount torque rod.

- Set a jack under the front subframe, then remove the subframe bolts.

- Gradually lower the subframe far enough to remove the steering gear.

8. Remove the bolts and nuts to remove the steering gear assembly.

To install:

9. Fit the steering gear into place, install the nuts and bolts and torque to 105 ft. lbs. (142 Nm).

10. Carefully raise front subframe into place and start all the bolts.

11. Torque the front subframe bolts to 69 ft. lbs. (94 Nm) and the rear bolts to 103 ft. lbs. (140 Nm).

12. Install rear engine mount torque rod.

13. Connect tie rod to steering knuckle and torque nut to 25 ft. lbs. (34 Nm) and tighten as needed to install new cotter pins.

14. Install stabilizer bar links. Torque nuts to 30 ft. lbs. (41 Nm).

15. Install remaining components and check wheel alignment.

Altima

See Figures 114 and 115.

1. Raise and safely support the vehicle and remove the front wheels.

2. Remove the engine undercover.

3. Remove the bolt from the lower steering shaft joint.

4. At the tie rod ends, remove the cotter pin and loosen the nut. Do not remove it yet.

5. Use a ball joint separator to disengage the tie rod end from the steering knuckle. Be careful not to damage the boot.

6. Remove the nut to disconnect the tie rod from the steering knuckle.

7. Disconnect the hydraulic lines and drain the power steering fluid.

8. Remove the hydraulic piping bracket from front suspension member.

9. Disconnect the wiring from the steering gear.

10. Remove the bolts and nuts and remove the steering gear assembly from the vehicle.

To install:

11. Make sure the steering rack is in the neutral position

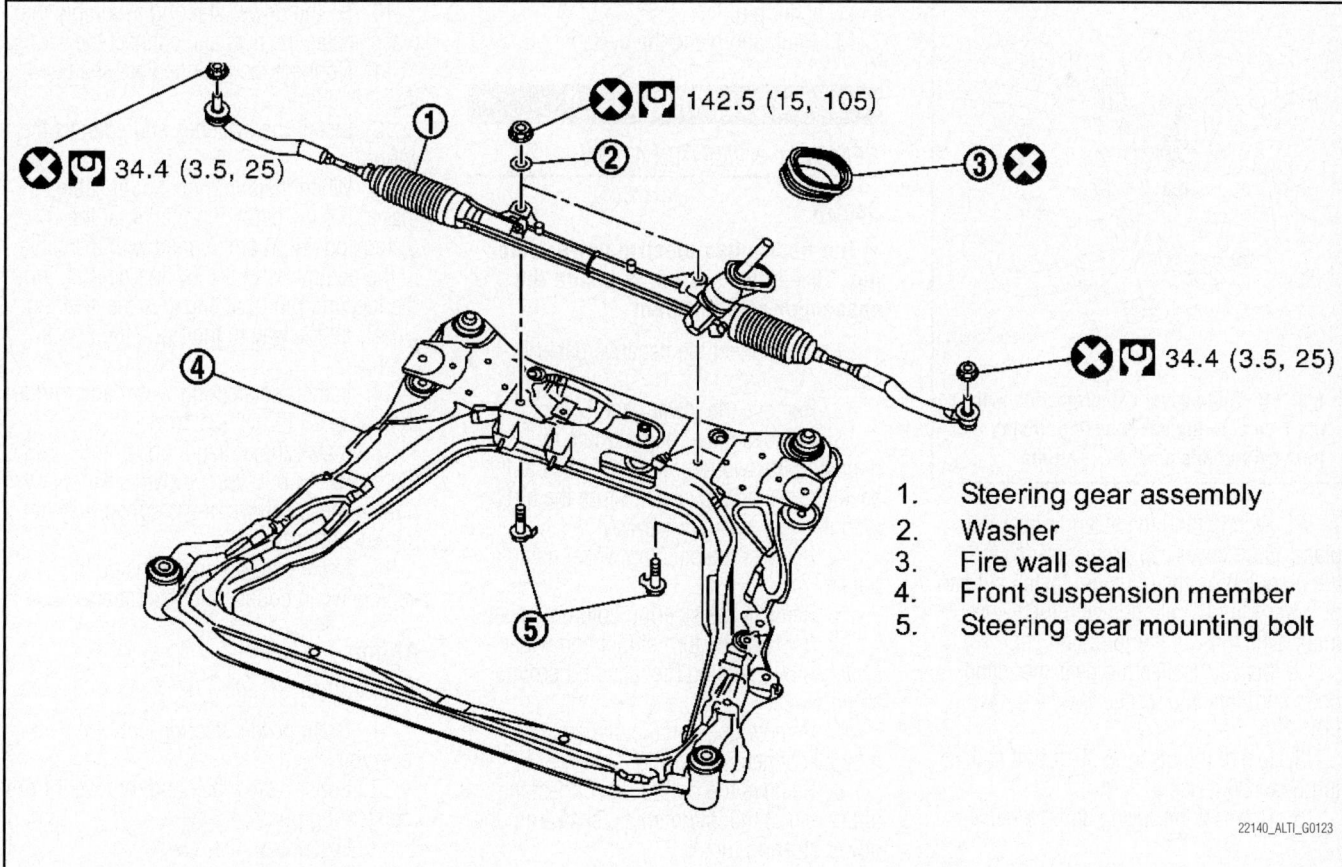

1. Steering gear assembly
2. Washer
3. Fire wall seal
4. Front suspension member
5. Steering gear mounting bolt

Fig. 113 Sentra steering gear removal

22140_ALTI_G0123

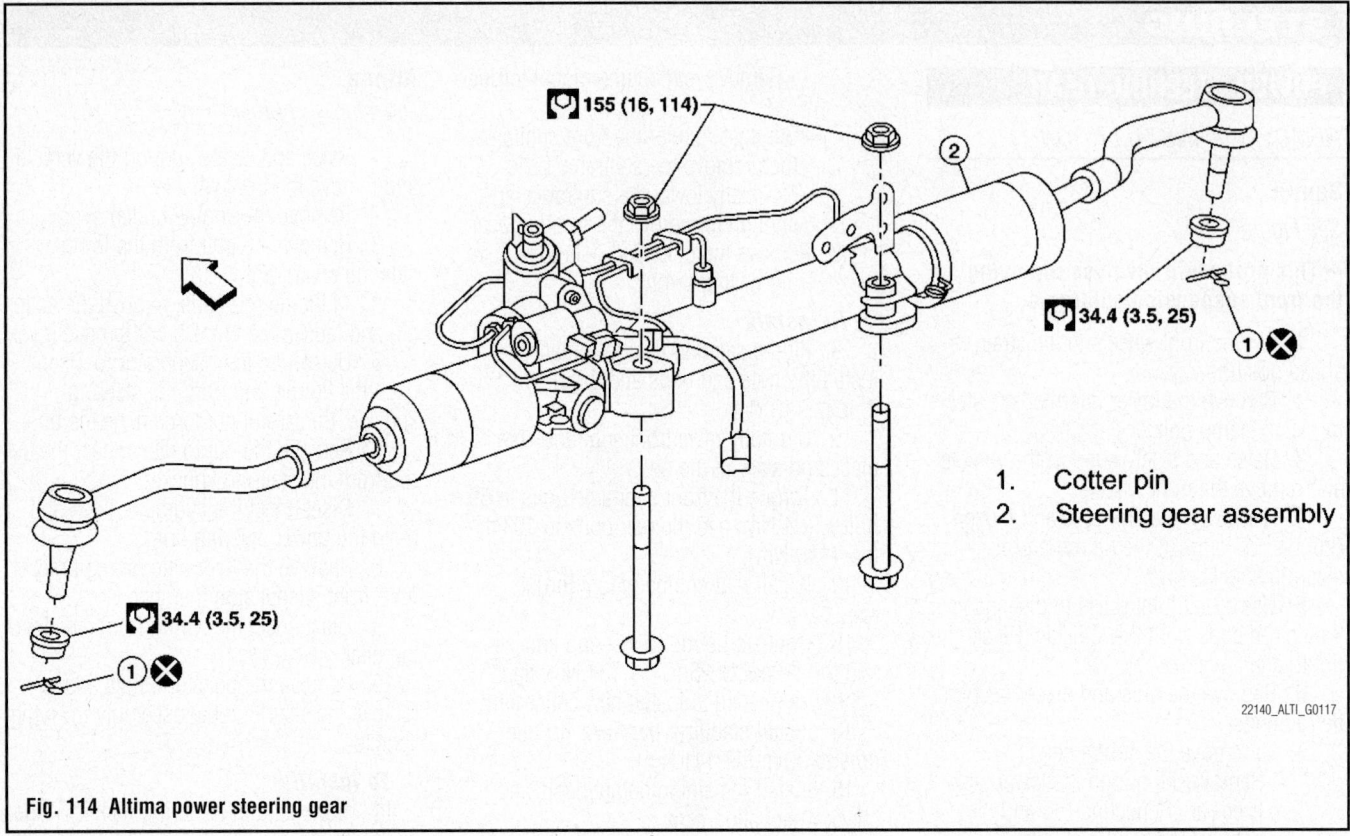

155 (16, 114)

34.4 (3.5, 25)

34.4 (3.5, 25)

1. Cotter pin
2. Steering gear assembly

22140_ALTI_G0117

Fig. 114 Altima power steering gear

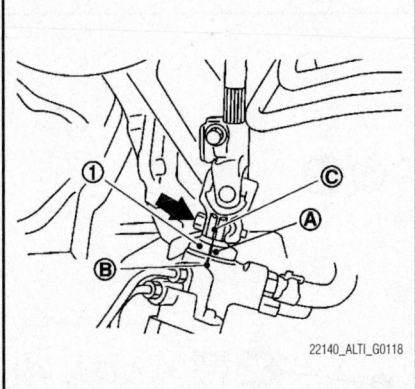

22140_ALTI_G0118

Fig. 115 Align cover cap projection with mark on steering gear housing. Install joint split in this position.—Altima

12. When fitting the steering gear into place, align cover cap projection (A) with the mark (B) on the housing. Install slit part of lower joint (C) aligned with the same mark. Don't install the joint bolt yet.

13. Install the steering gear mounting bolts and nuts and torque to 114 ft. lbs. (155 Nm).

14. Install the steering shaft bolt and torque to 20 ft. lbs. (27 Nm).

15. Connect the wiring and hydraulic lines.

16. Attach the tie rod ends to the steering knuckles. Torque the nuts to 25 ft. lbs.

(34 Nm), then tighten as required to install a new cotter pin.

17. Refill and bleed the system.

POWER STEERING PUMP

REMOVAL & INSTALLATION

Sentra

➡ **The Sentra has electric power steering. The entire system is inside the passenger compartment.**

1. Disconnect the negative battery cable.

2. Remove the lower dash cover.

3. Remove the airbag module by removing the two tamper-proof bolts at the back of the steering wheel. Store the airbag module face up out of the way.

4. Remove the steering wheel using a puller.

5. Remove the steering column covers.

6. Remove the turn signal and wiper switches by pressing the tabs and pulling straight out.

7. Remove the spiral cable and carefully disconnect the wiring.

8. Remove the bolt to disconnect the upper end of the intermediate shaft from the power steering motor.

9. Remove the nuts to lower the power steering assembly from the vehicle.

To install:

10. Fit the power steering assembly into place, install the nuts and connect the wiring.

11. Connect the intermediate shaft and install the bolt.

12. Install the switches and connect the wiring.

13. When installing the spiral cable, make sure the centering marks (at the 7 o'clock position) are aligned with the cable in the neutral position. To find neutral, turn the locating pin (that engages the steering wheel) all the way to the right, then back 2.6 turns.

14. Install the steering wheel and torque the nut to 25 ft. lbs. (31 Nm).

15. Install the airbag module and torque the screws to 8 ft. lbs. (10 Nm). Do not over tighten these screws or the airbag may not deploy properly.

16. Make sure no one is inside the vehicle when connecting the battery.

Altima

See Figures 116 and 117.

1. Drain power steering fluid from reservoir.

2. Loosen drive belt and remove it from the steering pump pulley.

3. Disconnect the hoses.

4. Remove the bolts and remove power steering pump.

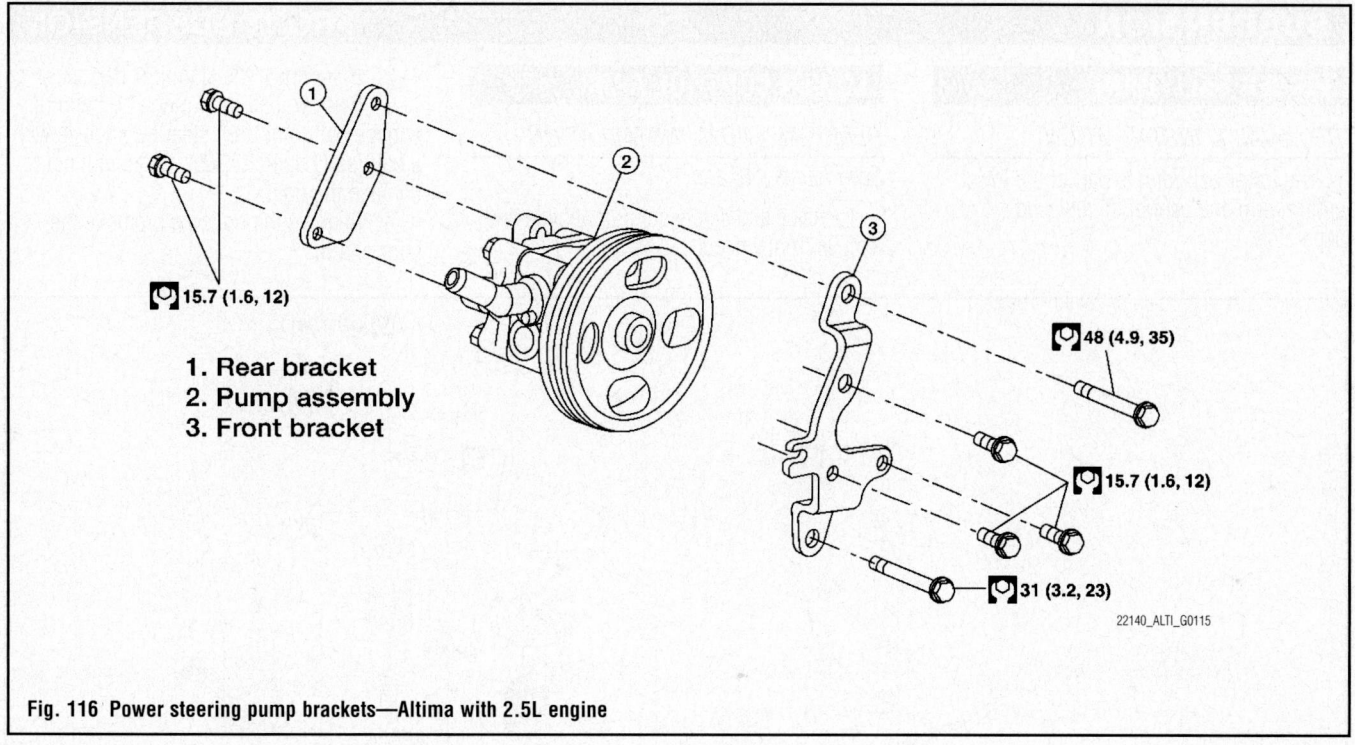

15.7 (1.6, 12)

1. Rear bracket
2. Pump assembly
3. Front bracket

48 (4.9, 35)

15.7 (1.6, 12)

31 (3.2, 23)

22140_ALTI_G0115

Fig. 116 Power steering pump brackets—Altima with 2.5L engine

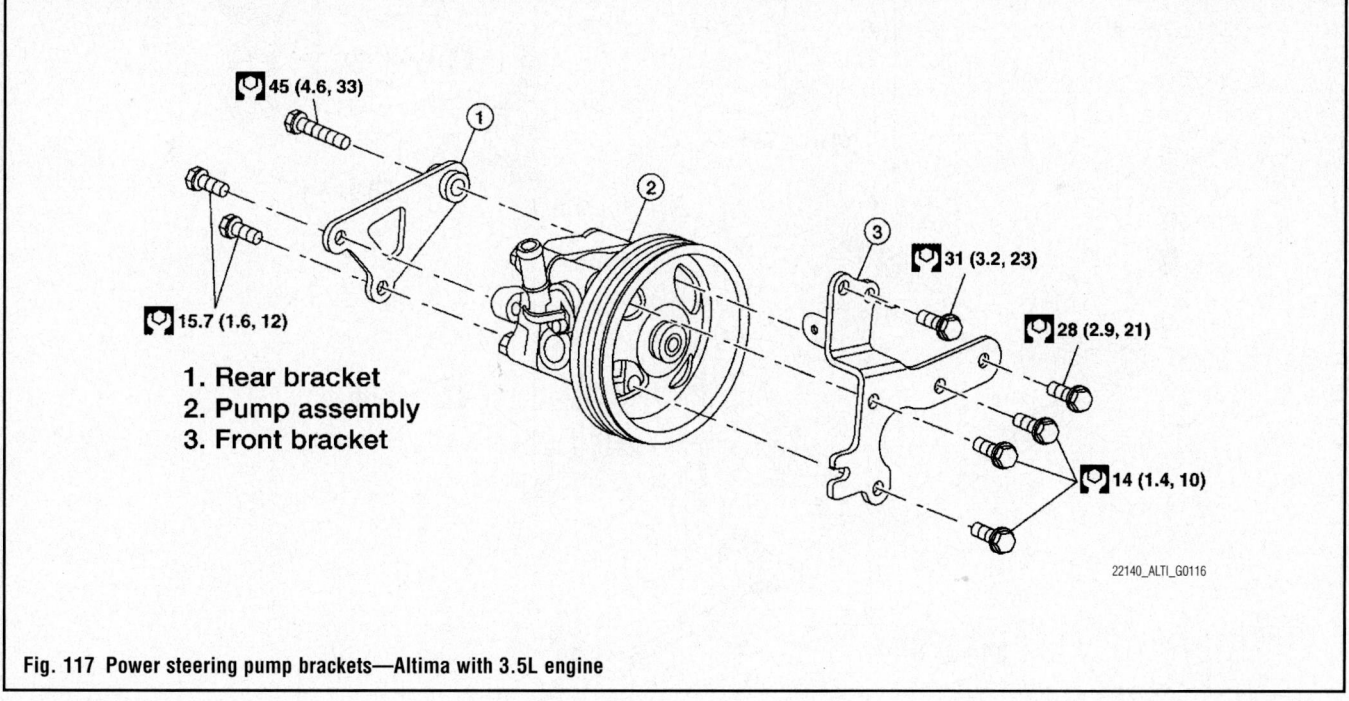

45 (4.6, 33)

15.7 (1.6, 12)

1. Rear bracket
2. Pump assembly
3. Front bracket

31 (3.2, 23)

28 (2.9, 21)

14 (1.4, 10)

22140_ALTI_G0116

Fig. 117 Power steering pump brackets—Altima with 3.5L engine

To install:

5. Installation is the reverse of removal. Do not over tighten the hose connections as this can damage O-ring, washer and connector.

6. Refill the reservoir and bleed the system.

BLEEDING

1. Run engine until the fluid temperature reaches 50 to 80° C (122 to 176°F) in reservoir tank, and keep engine speed idle.

2. Turn steering wheel several times from full left stop to full right stop.

3. Hold steering wheel at each lock position for five seconds and carefully, check for fluid leakage.

✻✻ WARNING

Do not hold the steering wheel in a locked position for more than 10 seconds. There is the possibility that oil pump may be damaged.

4. If fluid leakage at connections is noticed, then loosen flare nut and then retighten. Do not overtighten connector as this can damage O-ring, washer and connector.

5. Check steering gear boots for accumulation of fluid indicating leaks in the steering gear.

6. Check steering fluid level.

SUSPENSION

FRONT SUSPENSION

LOWER BALL JOINT

REMOVAL & INSTALLATION

The lower ball joint is part of the lower control arm and cannot be replaced separately.

LOWER CONTROL ARM

REMOVAL AND & INSTALLATION

See Figures 118 and 119.

1. Raise and safely support vehicle and remove front wheels.

2. Disconnect the stabilizer bar.

3. Remove ball joint pinch bolt and disengage ball joint from steering knuckle with a ball joint separator tool. Be careful not to damage the boot.

4. Remove the bolts and remove the control arm.

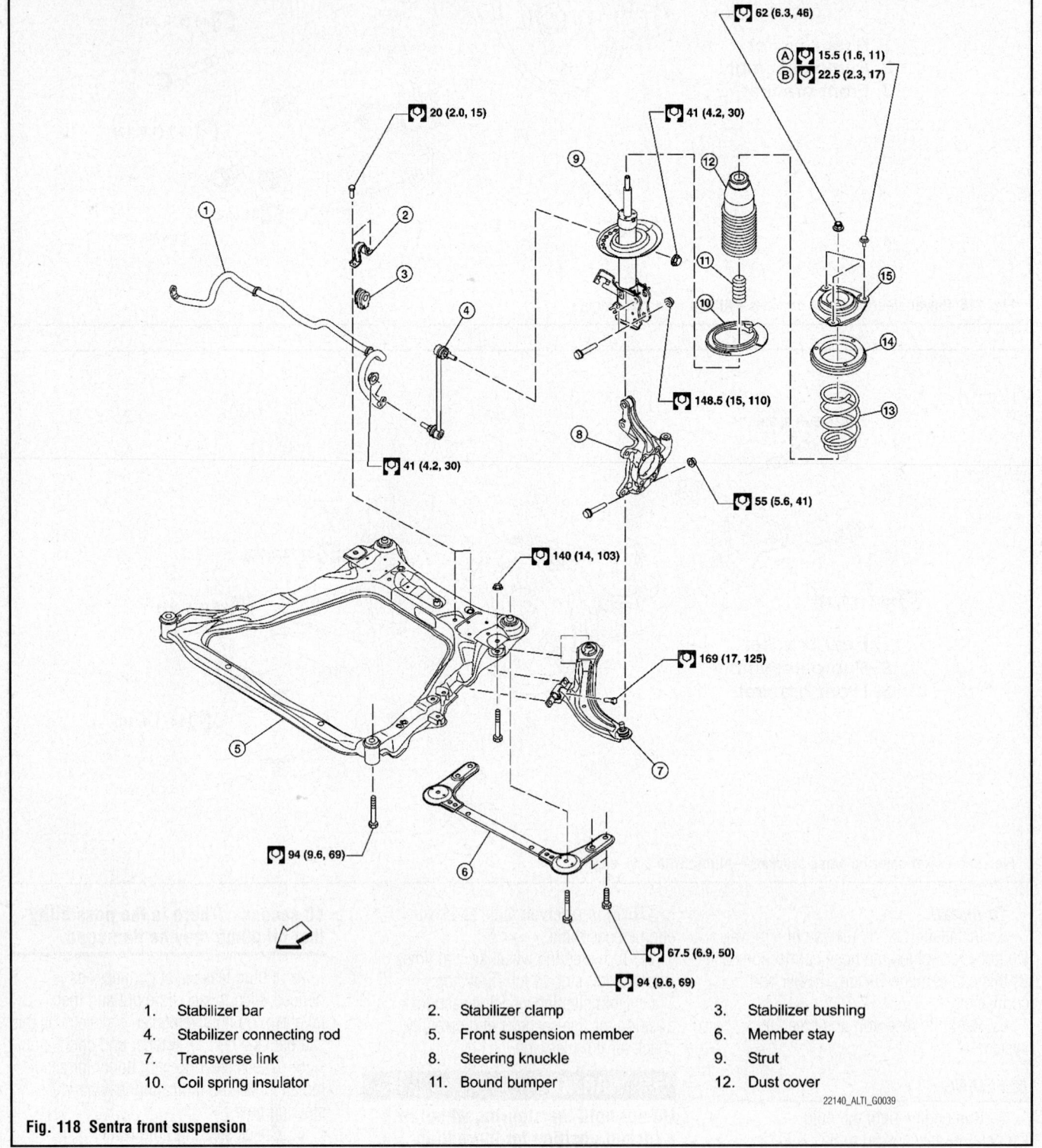

1. Stabilizer bar
2. Stabilizer clamp
3. Stabilizer bushing
4. Stabilizer connecting rod
5. Front suspension member
6. Member stay
7. Transverse link
8. Steering knuckle
9. Strut
10. Coil spring insulator
11. Bound bumper
12. Dust cover

22140_ALTI_G0039

Fig. 118 Sentra front suspension

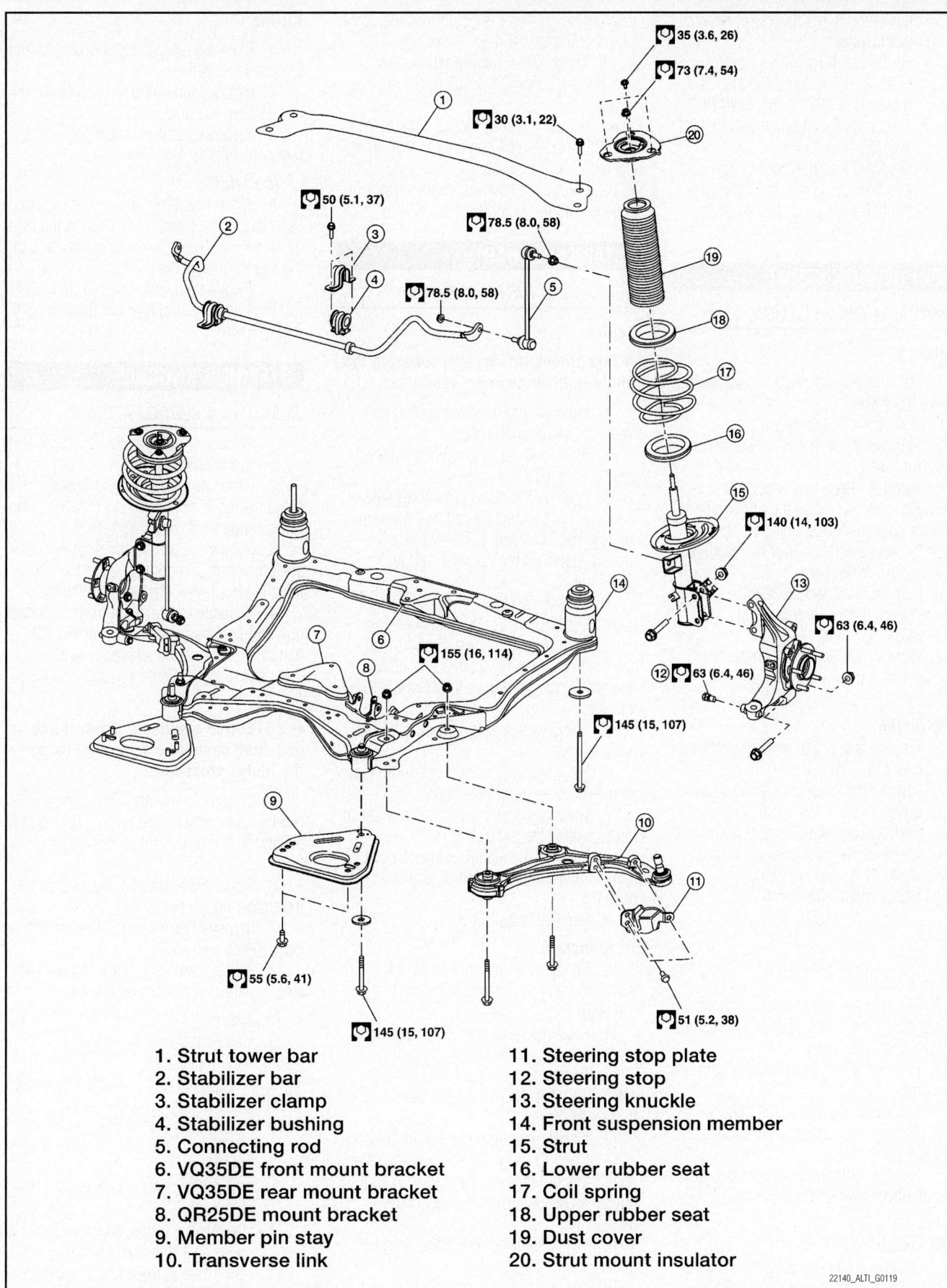

1. Strut tower bar
2. Stabilizer bar
3. Stabilizer clamp
4. Stabilizer bushing
5. Connecting rod
6. VQ35DE front mount bracket
7. VQ35DE rear mount bracket
8. QR25DE mount bracket
9. Member pin stay
10. Transverse link
11. Steering stop plate
12. Steering stop
13. Steering knuckle
14. Front suspension member
15. Strut
16. Lower rubber seat
17. Coil spring
18. Upper rubber seat
19. Dust cover
20. Strut mount insulator

22140_ALTI_G0119

Fig. 119 Altima front suspension

5. Installation is the reverse of removal, noting the following:

- On Sentra, torque front control arm bolts to 125 ft. lbs. (169 Nm) and rear bolt to 103 Ft. lbs. (140 Nm). Torque ball joint pinch bolt to 41 ft. lbs. (55 Nm).
- On Altima, torque control arm nuts and bolts to 114 ft. lbs. (155 Nm) the ball joint pinch bolt to 38 ft. lbs. (51 Nm).

MACPHERSON STRUT

REMOVAL & INSTALLATION

Sentra

1. Raise and safely support vehicle and remove front wheels.
2. Remove cowl top panel.
3. Remove wheel speed sensor wiring from strut assembly.
4. Remove brake hose lock plate.
5. Remove the nut on the upper side of stabilizer connecting rod using power tool, and then disconnect stabilizer connecting rod from strut assembly.
6. Remove nuts and bolts and remove steering knuckle from strut assembly.
7. Remove the upper strut mounting insulator bolts, then remove strut assembly.

To install:

8. Fit strut into upper mount and install bolts finger tight.
9. Attach strut to steering knuckle and torque bolts to 110 ft. lbs. (120 Nm).
10. Perform final tightening of upper mount with tires on level ground. Torque nuts/bolts to 11 ft. lbs. (15 Nm).
11. Check wheel alignment.

Altima

1. Raise and safely support vehicle and remove front wheels.
2. Remove the brake caliper without disconnecting the brake hose and hang it from the body with wire.
3. Remove wheel speed sensor harness from strut.
4. Remove brake hose lock plate.
5. Remove steering knuckle-to-strut bolts and nuts.
6. Remove bolt on strut tower bar, then bolts on strut tower and remove strut from vehicle.

To install:

7. Be sure arrows on strut mount insulator and spring upper seat point towards the outside of the vehicle. Also be sure notch in strut spacer is positioned the same way, then install strut.
8. Assemble upper mounting plate with its notch facing the outside of the vehicle.
9. Torque the strut mount bolts to 26 ft. lbs. (35 Nm) and the center nut to 54 ft. lbs. (73 Nm).
10. Torque the steering knuckle-to-strut bolts to 103 ft. lbs. (140 Nm).

STABILIZER BAR

REMOVAL & INSTALLATION

Sentra

➡This procedure involves lowering the whole front suspension subframe.

1. Separate intermediate shaft from steering gear pinion shaft.
2. Raise and safely support the vehicle and remove the front wheels.
3. Remove the nut on the lower side of stabilizer bar connecting rod and disconnect the rod from stabilizer bar. If necessary remove stabilizer bar connecting rod upper nut.
4. Loosen tie rod joint nut and disengage tie rod from steering knuckle using a ball joint separator. Be careful not to damage the boot. Remove the nut and disconnect the tie rod joint from the steering knuckle.
5. Remove rear engine mount torque rod.
6. Set a jack under the front subframe, then remove the subframe bolts.
7. Gradually lower subframe in order to remove stabilizer bar bolts.
8. Remove the stabilizer clamp bolts, then remove stabilizer clamps and stabilizer bushing.
9. Remove stabilizer bar.

To install:

10. Fit stabilizer bar into place and install the clamps. Torque bolts to 15 ft. lbs. (20 Nm).
11. Carefully raise front subframe into place and start all the bolts.
12. Torque the rear subframe bolts to 103 ft. lbs. (140 Nm) and front bolts to 69 ft. lbs. (94 Nm).
13. Install rear engine mount torque rod.
14. Connect tie rod to steering knuckle and torque nut to 25 ft. lbs. (34 Nm) and tighten as needed to install new cotter pins.
15. Install stabilizer bar links. Torque nuts to 30 ft. lbs. (41 Nm).
16. Install remaining components and check wheel alignment.

Altima

1. Raise and safely support vehicle and remove front wheels.
2. Remove mounting nuts on upper end of stabilizer bar links.
3. Remove stabilizer clamp bolts and remove stabilizer from the vehicle.

To install:

4. When installing, make sure the open bolt hole in the clamps are towards the front of the vehicle. Make sure the cut surface of the bushing faces the rear.
5. Torque the clamp bolts to 37 ft. lbs. (50 Nm) and the stabilizer bar links to 58 ft. lbs. (79 Nm).

STEERING KNUCKLE

REMOVAL & INSTALLATION

1. Raise and safely support vehicle and remove front wheels.
2. Remove brake rotor and brake caliper without disconnecting hydraulic line. Hang caliper from body with wire.
3. Remove cotter pin, then loosen half-shaft lock nut. Temporarily leave the nut installed to prevent damage to threads.
4. Separate the halfshaft from the wheel hub and bearing assembly by tapping the end of the shaft using a hammer and brass drift or wood block, then remove hub lock nut.

➡NOTE: Use a suitable puller if hub and shaft cannot be separated using the above procedure.

5. Loosen tie rod nut and disengage tie rod from steering knuckle with a ball joint separator. Be careful not to damage the boot.
6. Remove the nut and separate tie rod from steering knuckle.
7. Remove lower ball joint pinch bolt from steering knuckle.
8. Remove nuts and bolts and remove steering knuckle from strut assembly.

To install:

9. Fit steering knuckle to strut and install bolts finger tight.
10. Fit halfshaft into hub and install nut finger tight.
11. Fit ball joint into steering knuckle.
- On Sentra, torque bolt to 41 ft. lbs. (55 Nm).
- On Altima, torque bolt to 46 ft. lbs. (63 Nm).
- On Sentra, torque knuckle/strut bolts to 110 ft. lbs. (120 Nm).
- On Altima, torque knuckle/strut bolts to 103 ft. lbs. (140 Nm).

12. Fit tie rod end into steering knuckle. Torque nut to 25 ft. lbs. (35 Nm) and tighten as needed to install a new cotter pin.

13. Torque halfshaft nut to 92 ft. lbs. (125 Nm) and install a new cotter pin.

14. Install remaining components and check wheel alignment.

SUSPENSION

COIL SPRING

REMOVAL & INSTALLATION

Sentra

See Figure 120.

1. Set jack under rear suspension beam.

2. Remove both lower shock absorber bolts.

3. Carefully lower the suspension and remove the spring.

4. Installation is the reverse of removal. Torque upper shock absorber bolts to 50 ft. lbs. (88 Nm) and lower bolts to 92 ft. lbs. (125 Nm).

WHEEL BEARINGS

REMOVAL & INSTALLATION

1. Remove the steering knuckle.

2. Remove the bolts to remove the hub and bearing assembly from the steering knuckle.

Altima

See Figures 121 and 122.

1. Raise and safely support the vehicle and remove the rear wheels.

2. Loosen the rear lower link bolt and nut from the suspension member side.

3. Support the rear lower link with a jack.

4. Remove the rear lower link adjusting bolt and nut from the suspension member side.

➡**Do not reuse the adjusting nut, use a new adjusting nut for installation.**

5. Slowly lower the jack to lower the rear lower link and coil spring.

3. Installation is the reverse of removal.

• On Sentra, torque the hub assembly bolts to 65 ft. lbs. (88 Nm).

• On Altima, torque the hub assembly bolts to 46 ft. lbs. (60 Nm).

REAR SUSPENSION

6. Remove the upper rubber seat, coil spring, and lower rubber seat from the rear lower link.

To install:

7. Fit the spring seats into place.

• Check that the projecting part inside the upper rubber seat and the bracket flange are properly engaged.

• Check that the projection part outside the upper rubber seat is toward the front of the vehicle.

• Position the hollow of the lower rubber seat with the groove part of the rear lower link.

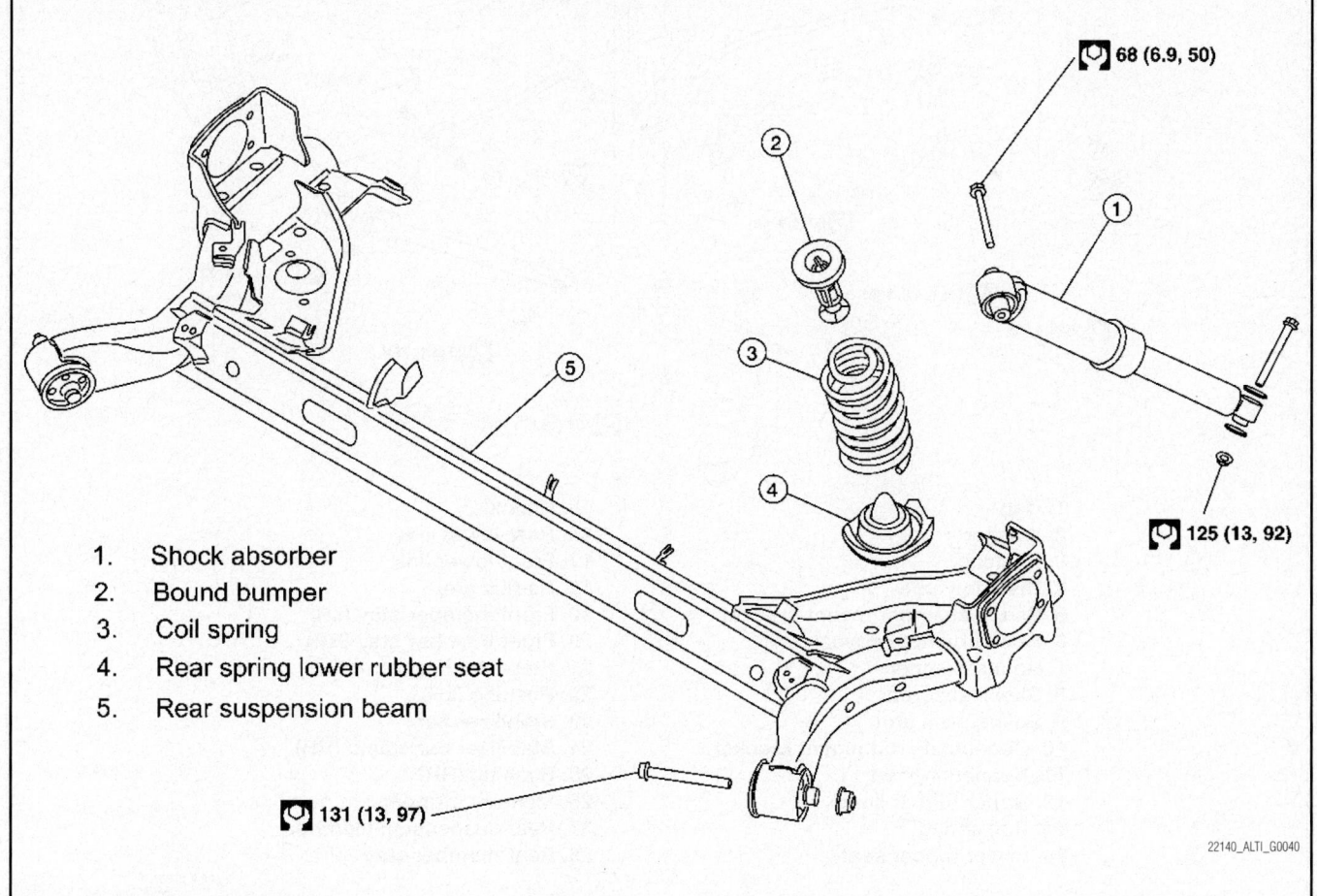

1. Shock absorber
2. Bound bumper
3. Coil spring
4. Rear spring lower rubber seat
5. Rear suspension beam

68 (6.9, 50)

125 (13, 92)

131 (13, 97)

22140_ALTI_G0040

Fig. 120 Sentra rear suspension

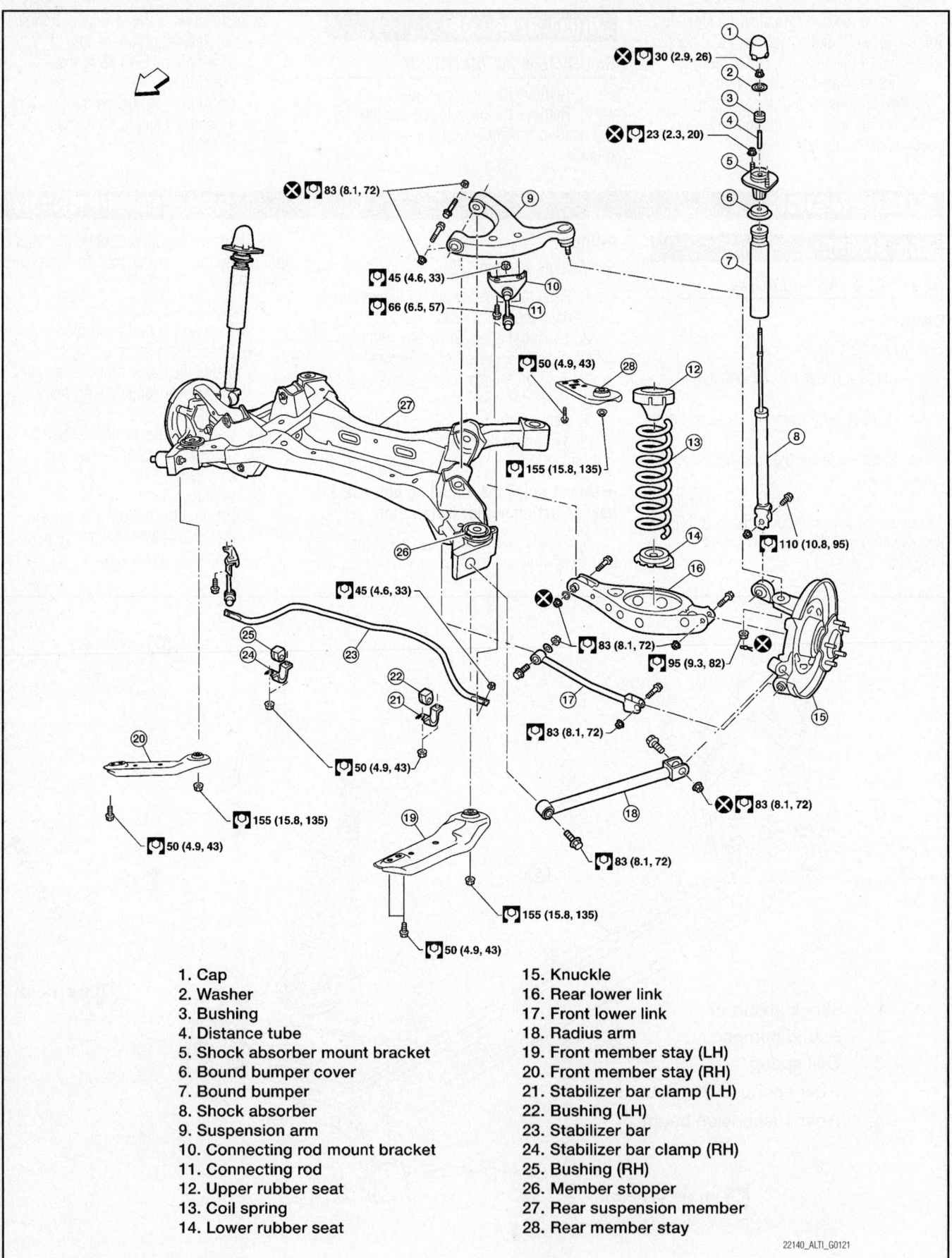

1. Cap
2. Washer
3. Bushing
4. Distance tube
5. Shock absorber mount bracket
6. Bound bumper cover
7. Bound bumper
8. Shock absorber
9. Suspension arm
10. Connecting rod mount bracket
11. Connecting rod
12. Upper rubber seat
13. Coil spring
14. Lower rubber seat
15. Knuckle
16. Rear lower link
17. Front lower link
18. Radius arm
19. Front member stay (LH)
20. Front member stay (RH)
21. Stabilizer bar clamp (LH)
22. Bushing (LH)
23. Stabilizer bar
24. Stabilizer bar clamp (RH)
25. Bushing (RH)
26. Member stopper
27. Rear suspension member
28. Rear member stay

22140_ALTI_G0121

Fig. 121 Altima Coupe rear suspension

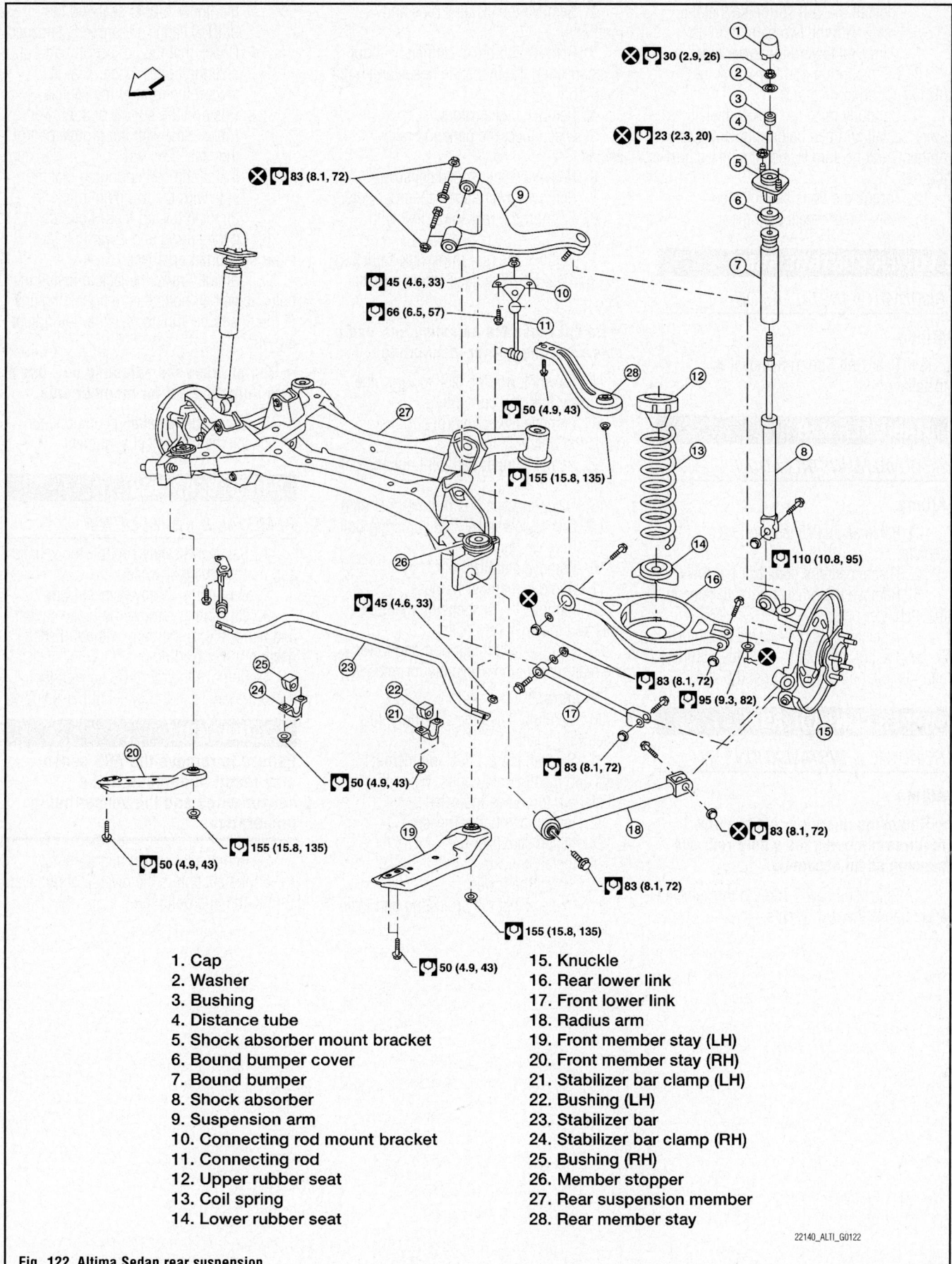

1. Cap
2. Washer
3. Bushing
4. Distance tube
5. Shock absorber mount bracket
6. Bound bumper cover
7. Bound bumper
8. Shock absorber
9. Suspension arm
10. Connecting rod mount bracket
11. Connecting rod
12. Upper rubber seat
13. Coil spring
14. Lower rubber seat
15. Knuckle
16. Rear lower link
17. Front lower link
18. Radius arm
19. Front member stay (LH)
20. Front member stay (RH)
21. Stabilizer bar clamp (LH)
22. Bushing (LH)
23. Stabilizer bar
24. Stabilizer bar clamp (RH)
25. Bushing (RH)
26. Member stopper
27. Rear suspension member
28. Rear member stay

22140_ALTI_G0122

Fig. 122 Altima Sedan rear suspension

- Install the coil spring so that the side with the two paint mark is directed toward the lower side.

8. Fit the spring and lower link into place supported on a jack.

9. Carefully raise the jack to install the bolts. Install all bolts before tightening any of them, and be sure to install a new adjusting nut.

10. Torque the bolts as indicated.

11. Adjust rear wheel alignment.

LOWER CONTROL ARM

REMOVAL & INSTALLATION

Altima

See Rear Coil Spring Removal & Installation

STABILIZER BAR

REMOVAL & INSTALLATION

Altima

1. Raise and safely support the vehicle.

2. Disconnect the stabilizer bar links.

3. Remove the bracket bolts to remove the stabilizer bar.

4. Installation is the reverse of removal. Torque the bracket nuts to 43 ft. lbs. (50 Nm) and the link bolts to 57 ft. lbs. (66 Nm).

UPPER CONTROL ARM

REMOVAL & INSTALLATION

Altima

➡ **Removing the upper control arm requires removing the whole rear suspension as an assembly.**

1. Raise and safely support the vehicle and remove the rear wheels.

2. Remove the exhaust pipe and muffler(s).

3. Remove the brake calipers without disconnecting the hydraulic hoses and hang them from the body

4. Remove brake rotors.

5. Disconnect the parking brake cables.

6. Remove wheel speed sensors.

7. Remove lower shock absorber nuts.

8. Support the rear lower link with jack.

9. Remove the rear lower link adjusting bolt and nut from the suspension member side.

➡ **Do not reuse the adjusting nut, use a new adjusting nut for installation.**

10. Slowly lower the jack to lower the rear lower link and coil spring.

11. Remove lower control arm link adjusting bolt and arm.

12. Remove upper ball joint nut and cotter pin.

13. Disconnect radius rod member side.

14. Disconnect lower link adjusting bolt.

15. Remove knuckle.

16. Remove stabilizer bar.

17. Remove member harness.

18. Remove the suspension member nuts and member stay bolts.

19. Use the jack to support and lower the rear suspension assembly for removal.

To install:

20. Carefully raise the assembly into place with a jack

21. Start all the suspension member nuts and member stay bolts. When they're all started, torque as indicated.

22. Install member harness.

23. Install stabilizer bar.

24. Install knuckle.

25. Install the lower.

- Check that the projecting part inside

the upper rubber seat and the bracket flange are properly engaged.

- Check that the projection part outside the upper rubber seat is toward the front of the vehicle.

- Position the hollow of the lower rubber seat with the groove part of the rear lower link.

- Install the coil spring so that the side with the two paint mark is directed toward the lower side.

26. Fit the spring and lower link into place supported on a jack.

27. Carefully raise the jack to install the bolts. Install all bolts before tightening any of them, and be sure to install a new adjusting nut.

➡ **Do not reuse the adjusting nut, use a new adjusting nut for member side.**

28. Install the remaining components and check the rear wheel alignment

WHEEL BEARINGS

REMOVAL & INSTALLATION

1. Raise and safely support the vehicle and remove the rear wheel.

2. On Sentra, remove brake drum.

3. On Altima, remove the brake caliper and hang it from the body without disconnecting the brake line.

4. Remove the rear ABS sensor, then move it away from the wheel hub assembly.

❄❄ WARNING

Failure to remove the ABS sensor may result in damage to the sensor wires and the sensor being inoperative.

5. Remove the wheel hub assembly.

6. Installation is the reverse of removal. Torque the hub bolts to 65 ft. lbs. (88 Nm).

SPECIFICATIONS AND MAINTENANCE CHARTS

ENGINE AND VEHICLE IDENTIFICATION

			Engine					Model Year	
Code	Liters (cc)	Cu. In.	Cyl.	Fuel Sys.	Engine	Eng. Mfg.		Code ①	Year
VK56DE	5.6 (5552)	338.8	8	MFI	DOHC	Nissan		6	2006
MFI: Multi-port Fuel Injection								7	2007
DOHC: Double Overhead Camshafts								8	2008

① 10th digit of the Vehicle Identification Number (VIN)

22140_ARMA_C0001

GENERAL ENGINE SPECIFICATIONS

Year	Model	Engine Displacement Liters	Engine ID	Net Horsepower @ rpm	Net Torque @ rpm (ft. lbs.)	Bore x Stroke (in.)	Com-pression Ratio	Oil Pressure @ rpm
2006	Armada	5.6	VK56DE	305@4900	385@3600	3.86X3.62	9.8:1	43@2000
2007	Armada	5.6	VK56DE	317@4900	385@3600	3.86X3.62	9.8:1	43@2000
2008	Armada	5.6	VK56DE	317@4900	385@3600	3.86X3.62	9.8:1	43@2000

22140_ARMA_C0002

ENGINE TUNE-UP SPECIFICATIONS

Year	Engine Displacement Liters	Engine ID	Spark Plug Gap (in.)	Ignition Timing	Fuel Pump (psi) ①	Idle Speed ②	Valve Clearance (in.) In.	Ex.
2006	5.6	VK56DE	0.043	15B	51	600-700	③	④
2007	5.6	VK56DE	0.043	15B	51	600-700	③	④
2008	5.6	VK56DE	0.043	15B	51	600-700	③	④

NOTE: The Vehicle Emission Control Information label often reflects specification changes made during production. The label figures must be used if they differ from those in this chart.

B: Before top dead center

① System pressure at idle with vacuum hose connected
Should increase to 43 psi when disconnected

② Automatic transmission in Neutral

③ 0.010-0.013 cold
0.012-0.016 hot

④ 0.011-0.015 cold
0.012-0.017 hot

22140_ARMA_C0003

CAPACITIES

Year	Model	Engine Displacement Liters	Engine ID	Engine Oil with Filter (qts.)	Transmission (pts.)	Transfer Case (pts.)	Drive Axle Front (pts.)	Drive Axle Rear (pts.)	Fuel Tank (gal.)	Cooling System (qts.)
2006	Armada	5.6	VK56DE	6.5	22.5	6.25	3.75	3.23	28.0	15
2007	Armada	5.6	VK56DE	6.5	22.5	6.25	3.75	3.23	28.0	15
2008	Armada	5.6	VK56DE	6.5	22.5	6.25	3.75	3.23	28.0	15

NOTE: All capacities are approximate. Add fluid gradually and check to be sure a proper fluid level is obtained.

22140_ARMA_C0004

FLUID SPECIFICATIONS

Year	Model	Engine Displ. Liters (VIN)	Engine Oil	Auto. Trans.	Drive Axle Front	Drive Axle Rear	Transfer Case	Power Steering Fluid	Brake Master Cylinder	Cooling System
2006	Armada	VK56DE	①	②	GL-5 ③	GL-5 80W-90	④	NISSAN PSF	DOT 3	⑤
	Armada	VK56DE	①	②	GL-5 ③	GL-5 80W-90	④	NISSAN PSF	DOT 3	⑤
2007	Armada	VK56DE	①	②	GL-5 ③	GL-5 80W-90	④	NISSAN PSF	DOT 3	⑤
	Armada	VK56DE	①	②	GL-5 ③	GL-5 80W-90	④	NISSAN PSF	DOT 3	⑤
2008	Armada	VK56DE	①	②	GL-5 ③	GL-5 80W-90	④	NISSAN PSF	DOT 3	⑤
	Armada	VK56DE	①	②	GL-5 ③	GL-5 80W-90	④	NISSAN PSF	DOT 3	⑤

DOT: Department Of Transpotation

① API Certification Mark #1

   API grade SG/SH, Energy Conserving I & II or

   API grade SJ or SL, Energy Conserving

   ILSAC grade GF-I, GF-II & GF-III

② Genuine NISSAN Matic J ATF #2 (Continental U.S. and Alaska) or Canada NISSAN Automatic Transmission Fluid

③ Synthetic 75W-90 Gear Oil (Part No. 999MP-DF200P) or equivalent

④ Genuine NISSAN Matic D ATF (Continental U.S and Alaska) or Canada NISSAN Automatic Transmission Fluid *7

⑤ NISSAN Long Life Antifreeze/ Coolant or equivalent

22140_ARMA_C0005

VALVE SPECIFICATIONS

Year	Engine Displacement Liters	Engine ID	Seat Angle (deg.)	Face Angle (deg.)	Spring Test Pressure (lbs. @ in.)	Spring Installed Height (in.)	Stem-to-Guide Clearance (in.) Intake	Stem-to-Guide Clearance (in.) Exhaust	Stem Diameter (in.) Intake	Stem Diameter (in.) Exhaust
2006	5.6	VK56DE	45.15-45.45	45	37.0@1.457	1.991	0.0008-0.0021	0.0012-0.0025	0.2348-0.2354	0.2344-0.2350
2007	5.6	VK56DE	45.15-45.45	45	37.0@1.457	1.991	0.0008-0.0021	0.0012-0.0025	0.2348-0.2354	0.2344-0.2350
2008	5.6	VK56DE	45.15-45.45	45	37.0@1.457	1.991	0.0008-0.0021	0.0012-0.0025	0.2348-0.2354	0.2344-0.2350

22140_ARMA_C0007

CAMSHAFT SPECIFICATIONS

All measurements are given in inches.

Year	Engine Displ. Liters	Engine ID/VIN	Journal Dia.	Brg. Oil Clearance	Shaft End-play	Runout	Journal Bore	Lobe Height Intake	Lobe Height Exhaust
2006	5.6	VK56DE	1.0218-1.0224	0.0012-0.0027	0.0045-0.0074	0.0008	1.0217-1.0244	1.7663-1.7738	1.7746-1.7821
2007	5.6	VK56DE	1.0217-1.0224	0.0012-0.0028	0.0045-0.0074	0.0008	1.0217-1.0244	1.7663-1.7738	1.7746-1.7821
2008	5.6	VK56DE	1.0217-1.0224	0.0012-0.0028	0.0045-0.0074	0.0008	1.0217-1.0244	1.7663-1.7738	1.7746-1.7821

22140_ARMA_C0008

CRANKSHAFT AND CONNECTING ROD SPECIFICATIONS

All measurements are given in inches.

Year	Engine Displ. Liters	Engine ID	Crankshaft Main Brg. Journal Dia.	Main Brg. Oil Clearance	Shaft End-play	Thrust on No.	Connecting Rod Journal Diameter	Oil Clearance	Side Clearance
2006	5.6	VK56DE	①	②	0.0118	3	③	0.0008-0.0015	0.0079-0.0157
2007	5.6	VK56DE	①	②	0.0118	3	③	0.0008-0.0015	0.0079-0.0157
2008	5.6	VK56DE	①	②	0.0118	3	③	0.0008-0.0015	0.0079-0.0157

① There are 24 different grades, ranging from (2.5182) to (2.174)

② No. 1 and 5: 0.00004-0.0004

No. 2, 3 and 4: 0.0003-0.0007

③ There are 3 different ranges from (2.1247) to (2.150)

22140_ARMA_C0006

PISTON AND RING SPECIFICATIONS

All measurements are given in inches.

Year	Engine Displacement Liters	Engine ID	Piston Clearance	Ring Gap Top Comp.	Ring Gap Bottom Comp.	Ring Gap Oil Control	Ring Side Clearance Top Comp.	Ring Side Clearance Bottom Comp.	Ring Side Clearance Oil Control
2006	5.6	VK56DE	0.0004-0.0012	0.0091-0.0130	0.0098-0.0157	0.0079-0.0236	0.0014-0.0033	0.0012-0.0028	0.0006-0.0020
2007	5.6	VK56DE	0.0004-0.0012	0.0091-0.0130	0.0098-0.0157	0.0079-0.0236	0.0014-0.0033	0.0012-0.0028	0.0006-0.0020
2008	5.6	VK56DE	0.0004-0.0012	0.0091-0.0130	0.0098-0.0157	0.0079-0.0236	0.0014-0.0033	0.0012-0.0028	0.0006-0.0020

22140_ARMA_C0009

TORQUE SPECIFICATIONS
All readings in ft. lbs.

Year	Engine Displacement Liters	Engine ID	Cylinder Head Bolts	Main Bearing Bolts	Rod Bearing Bolts	Crankshaft Damper Bolts	Flywheel Bolts	Manifold Intake	Manifold Exhaust	Spark Plugs	Oil Pan Drain Plug
2006	5.6	VK56DE	①	②	③	④	65	6	25	18	25
2007	5.6	VK56DE	①	②	③	④	65	6	25	18	25
2008	5.6	VK56DE	①	②	③	④	65	6	25	18	25

① Step 1: 72 ft. lbs

   Step 2: Loosen all bolts completely

   Step 3: 33 ft. lbs.

   Step 4: +60 degrees

   Step 5: +60 degrees

② Step 1: Main Bolts to 29 ft. lbs.

   Step 2: Sub-bolts to 22 ft. lbs.

   Step 3: Main Bolts +40 degrees

   Step 4: Sub-Bolts +30 degrees

   Step 5: Side Bolts to 36 ft. lbs.

③ Step 1: 11 ft. lbs.

   Step 2: +90 degrees

④ Step 1: 69 ft. lbs.

   Step 2: +90 degrees

22140_ARMA_C0010

WHEEL ALIGNMENT

Year	Model	Caster Range (+/-Deg.)	Caster Preferred Setting (Deg.)	Camber Range (+/-Deg.)	Camber Preferred Setting (Deg.)	Toe-in (in.)
2006	Armada	0.75	①	0.75	②	0.08+/-0.11
2007	Armada	0.75	①	0.75	②	0.11+/-0.04
2008	Armada	0.75	①	0.75	②	0.11+/-0.04

① 4x2: +3.47

   4x4: +3.05

② 4x2: -0.10

   4x4: +0.20

22140_ARMA_C0011

TIRE, WHEEL AND BALL JOINT SPECIFICATIONS

Year	Model	OEM Tires Standard	OEM Tires Optional	Tire Pressures (psi) Front	Tire Pressures (psi) Rear	Wheel Size	Ball Joint Inspection	Lug Nut Torque (ft. lbs.)
2006	Armada	P265/70R18	—	35	35	18x8J	①	98
2007	Armada	P265/70R18	P275/60R20	35	35	18x8J	①	98
2008	Armada	P265/70R18	P275/60R20	35	35	18x8J	①	98

OEM: Original Equipment Manufacturer

PSI: Pounds Per Square Inch

① Axial play: upper 0

22140_ARMA_C0012

BRAKE SPECIFICATIONS
All measurements in inches unless noted

Year	Model		Brake Disc Original Thickness	Brake Disc Minimum Thickness	Brake Disc Maximum Runout	Minimum Pad Thickness	Brake Caliper Bracket Bolts (ft. lbs.)	Brake Caliper Mounting Bolts (ft. lbs.)
2006	Armada	F	1.024	0.965	0.001	0.039	63	32
		R	0.478	0.472	0.0020	0.039	55	32
2007	Armada	F	1.024	0.965	0.0016	0.039	63	32
		R	0.478	0.472	0.0020	0.039	55	32
2008	Armada	F	1.024	0.965	0.0016	0.039	63	32
		R	0.478	0.472	0.0020	0.039	55	32

22140_ARMA_C0013

SCHEDULED MAINTENANCE INTERVALS
Nissan Armada 2006 - 2008

TO BE SERVICED	TYPE OF SERVICE	7.5	15	22.5	30	37.5	45	52.5	60
Engine oil & filter	R	✓	✓	✓	✓	✓	✓	✓	✓
Brake lines & cables	S/I		✓		✓		✓		✓
Brake pads and rotors	I	✓	✓	✓	✓	✓	✓	✓	✓
Driveshaft boots & propeller shaft (4x4)	L/I		✓		✓		✓		✓
Transmission, transfer & differential gear oil	I		✓		✓		✓		✓
Air cleaner filter	R					✓			✓
Engine coolant ①	R								✓
Spark plugs (Platinum)	R				Replace every 105,000 miles				
Drive belt(s) ②	S/I								✓
Cabin air filter	R		✓		✓		✓		✓
Exhaust system	I				✓				✓
Fuel lines	S/I				✓				✓
Fuel filter ③									
Steering gear (box) & linkage, axle & suspension parts	I				✓				✓
Vapor lines	S/I				✓				✓

R: Replace   S/I: Service or Inspect   L: Lubricate

① Coolant: After 60,000 miles, inspect every 30,000 miles.

② Drive Belts: After 60,000 miles, inspect every 15,000 miles. Replace belts if damaged.

③ Fuel Filter: Maintenance free item.

FREQUENT OPERATION MAINTENANCE (SEVERE SERVICE)

If a vehicle is operated under any of the following conditions it is considered severe service:

- Extremely dusty areas.

- Rough, muddy, or salt spread roads.

- 50% or more of the vehicle constant operation is in 32°C (90°F) or higher temperatures, or temperatures below 0°C (32°F).

- Prolonged idling (vehicle operation in stop and go traffic).

- Frequent short running periods (engine does not warm to normal operating temperatures).

- Police, taxi, delivery usage or trailer towing usage.

Oil & oil filter: replace every 3750 miles.

Brake pads, discs, drums & linings: service or inspect every 7500 miles.

Driveshaft boots & propeller shaft: service or inspect every 7500 miles.

Exhaust system: service or inspect every 7500 miles.

Steering gear (box) & linkage, (steering damper-4x4), axle & suspension parts: service or inspect every 7500 miles.

Steering linkage ball joints & front suspension ball joints: service or inspect every 7500 miles.

22140_ARMA_C0014

PRECAUTIONS

Before servicing any vehicle, please be sure to read all of the following precautions, which deal with personal safety, prevention of component damage, and important points to take into consideration when servicing a motor vehicle:

• Never open, service or drain the radiator or cooling system when the engine is hot; serious burns can occur from the steam and hot coolant.

• Observe all applicable safety precautions when working around fuel. Whenever servicing the fuel system, always work in a well-ventilated area. Do not allow fuel spray or vapors to come in contact with a spark, open flame, or excessive heat (a hot drop light, for example). Keep a dry chemical fire extinguisher near the work area. Always keep fuel in a container specifically designed for fuel storage; also, always properly seal fuel containers to avoid the possibility of fire or explosion. Refer to the additional fuel system precautions later in this section.

• Fuel injection systems often remain pressurized, even after the engine has been turned **OFF**. The fuel system pressure must be relieved before disconnecting any fuel lines. Failure to do so may result in fire and/or personal injury.

• Brake fluid often contains polyglycol ethers and polyglycols. Avoid contact with the eyes and wash your hands thoroughly after handling brake fluid. If you do get brake fluid in your eyes, flush your eyes with clean, running water for 15 minutes. If eye irritation persists, or if you have taken

brake fluid internally, IMMEDIATELY seek medical assistance.

• The EPA warns that prolonged contact with used engine oil may cause a number of skin disorders, including cancer. You should make every effort to minimize your exposure to used engine oil. Protective gloves should be worn when changing oil. Wash your hands and any other exposed skin areas as soon as possible after exposure to used engine oil. Soap and water, or waterless hand cleaner should be used.

• All new vehicles are now equipped with an air bag system, often referred to as a Supplemental Restraint System (SRS) or Supplemental Inflatable Restraint (SIR) system. The system must be disabled before performing service on or around system components, steering column, instrument panel components, wiring and sensors. Failure to follow safety and disabling procedures could result in accidental air bag deployment, possible personal injury and unnecessary system repairs.

• Always wear safety goggles when working with, or around, the air bag system. When carrying a non-deployed air bag, be sure the bag and trim cover are pointed away from your body. When placing a non-deployed air bag on a work surface, always face the bag and trim cover upward, away from the surface. This will reduce the motion of the module if it is accidentally deployed. Refer to the additional air bag system precautions later in this section.

• Clean, high quality brake fluid from a sealed container is essential to the safe and

proper operation of the brake system. You should always buy the correct type of brake fluid for your vehicle. If the brake fluid becomes contaminated, completely flush the system with new fluid. Never reuse any brake fluid. Any brake fluid that is removed from the system should be discarded. Also, do not allow any brake fluid to come in contact with a painted surface; it will damage the paint.

• Never operate the engine without the proper amount and type of engine oil; doing so WILL result in severe engine damage.

• Timing belt maintenance is extremely important. Many models utilize an interference-type, non-freewheeling engine. If the timing belt breaks, the valves in the cylinder head may strike the pistons, causing potentially serious (also time-consuming and expensive) engine damage. Refer to the maintenance interval charts for the recommended replacement interval for the timing belt, and to the timing belt section for belt replacement and inspection.

• Disconnecting the negative battery cable on some vehicles may interfere with the functions of the on-board computer system(s) and may require the computer to undergo a relearning process once the negative battery cable is reconnected.

• When servicing drum brakes, only disassemble and assemble one side at a time, leaving the remaining side intact for reference.

• Only an MVAC-trained, EPA-certified automotive technician should service the air conditioning system or its components.

BRAKES

GENERAL INFORMATION

PRECAUTIONS

• Certain components within the ABS system are not intended to be serviced or repaired individually.

• Do not use rubber hoses or other parts not specifically specified for and ABS system. When using repair kits, replace all parts included in the kit. Partial or incorrect repair may lead to functional problems and require the replacement of components.

• Lubricate rubber parts with clean,

fresh brake fluid to ease assembly. Do not use shop air to clean parts; damage to rubber components may result.

• Use only DOT 3 brake fluid from an unopened container.

• If any hydraulic component or line is removed or replaced, it may be necessary to bleed the entire system.

• A clean repair area is essential. Always clean the reservoir and cap thoroughly before removing the cap. The slightest amount of dirt in the fluid may plug an orifice and impair the system function. Perform repairs after components have been thor-

ANTI-LOCK BRAKE SYSTEM (ABS)

oughly cleaned; use only denatured alcohol to clean components. Do not allow ABS components to come into contact with any substance containing mineral oil; this includes used shop rags.

• The Anti-Lock control unit is a microprocessor similar to other computer units in the vehicle. Ensure that the ignition switch is **OFF** before removing or installing controller harnesses. Avoid static electricity discharge at or near the controller.

• If any arc welding is to be done on the vehicle, the control unit should be unplugged before welding operations begin.

BRAKES

BLEEDING THE BRAKE SYSTEM

BLEEDING PROCEDURE

BLEEDING PROCEDURE

➡Be sure that the master cylinder is full of clean fresh brake fluid before starting the bleeding process. Use only the recommended brake fluid when bleeding the system. Do not allow brake fluid to spill on painted surfaces as damage will occur.

1. Before servicing the vehicle, refer to the Precautions Section.

2. Disconnect the negative battery cable.

3. Turn the ignition switch **OFF** Disconnect the ABS actuator and electric control unit connector.

4. Connect a vinyl tube to the rear right bleed valve. Be sure to have a catch pan handy to catch excess brake fluid.

5. Fully depress the brake pedal four or five times.

6. With the brake pedal depressed, loosen the bleed valve to let air out, then tighten it immediately.

7. Repeat the above steps until all air is removed from the system. Be sure to keep watch on the brake fluid level and replenish, as necessary.

8. Tighten the bleed valve.

9. Repeat the above steps at each wheel, with the master cylinder reservoir tank filled at least half way.

10. Bleed the remaining components in the following order: front left, rear left and front right.

BRAKES

FRONT DISC BRAKES

✳ CAUTION

Dust and dirt accumulating on brake parts during normal use may contain asbestos fibers from production or aftermarket brake linings. Breathing excessive concentrations of asbestos fibers can cause serious bodily harm. Exercise care when servicing brake parts. Do not sand or grind brake lining unless equipment used is designed to contain the dust residue. Do not clean brake parts with compressed air or by dry brushing. Cleaning should be done by dampening the brake components with a fine mist of water, then wiping the brake components clean with a dampened cloth. Dispose of cloth and all residue containing asbestos fibers in an impermeable container with the appropriate label. Follow practices prescribed by the Occupational Safety and Health Administration (OSHA) and the Environmental Protection Agency (EPA) for the handling, processing, and disposing of dust or debris that may contain asbestos fibers.

BRAKE CALIPER

REMOVAL & INSTALLATION

See Figure 1.

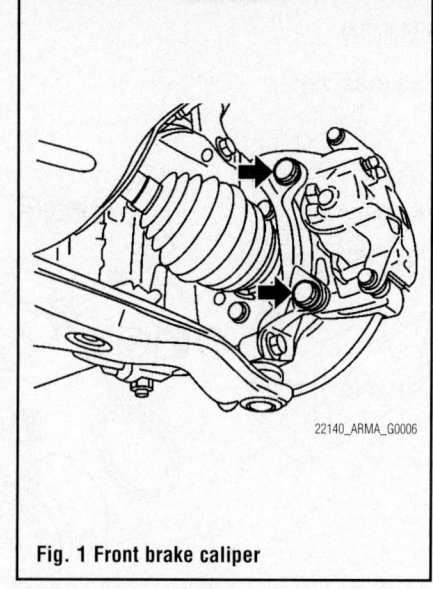

22140_ARMA_G0006

Fig. 1 Front brake caliper

1. Before servicing the vehicle, refer to the Precautions Section.

2. Drain brake fluid as necessary.

3. Remove or disconnect the following:

- Wheel
- Union bolt
- Caliper-to-torque member slide pins, or remove the caliper and torque member as an assembly.
- Brake caliper

To install:

4. Install or connect the following:

- Brake caliper, tighten torque member bolts to 155 ft. lbs. (210 Nm); the caliper slide pins to 20 ft. lbs. (27 Nm)
- Union bolt and tighten to 13 ft. lbs. (18 Nm)

5. Fill the master cylinder and bleed the brake system.

6. Install the wheels.

DISC BRAKE PADS

REMOVAL & INSTALLATION

1. Before servicing the vehicle, refer to the Precautions Section.

2. Remove the wheel.

3. Remove lower sliding pin bolt.

4. Suspend brake caliper with a remove and remove brake pad and shim from torque member.

To install:

5. Push pistons in so that the pad is firmly installed, using a suitable tool.

6. Mount the brake caliper to torque member.

7. Attach pad retainer to torque member.

8. Lubricate lower sliding pin bolt with a thin layer of silicone grease and install. Torque to 24 ft. lbs. (32 Nm).

9. Install the wheels.

BRAKES

REAR DISC BRAKES

※※ CAUTION

Dust and dirt accumulating on brake parts during normal use may contain asbestos fibers from production or aftermarket brake linings. Breathing excessive concentrations of asbestos fibers can cause serious bodily harm. Exercise care when servicing brake parts. Do not sand or grind brake lining unless equipment used is designed to contain the dust residue.

Do not clean brake parts with compressed air or by dry brushing. Cleaning should be done by dampening the brake components with a fine mist of water, then wiping the brake components clean with a dampened cloth. Dispose of cloth and all residue containing asbestos fibers in an impermeable container with the appropriate label. Follow practices prescribed by the Occupational

Safety and Health Administration (OSHA) and the Environmental Protection Agency (EPA) for the handling, processing, and disposing of dust or debris that may contain asbestos fibers.

BRAKE CALIPER

REMOVAL & INSTALLATION

See Figure 2.

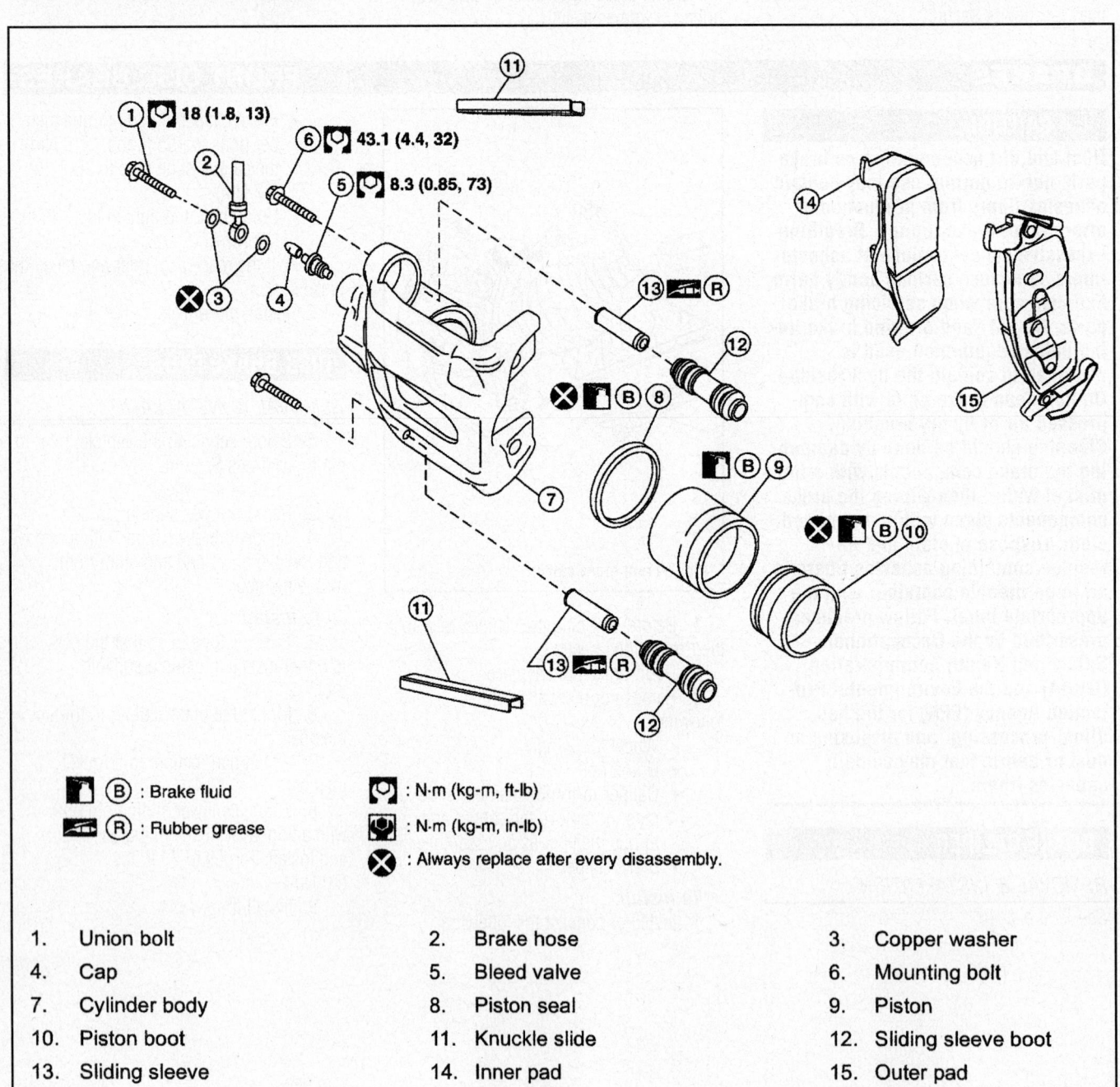

⬛ Ⓑ : Brake fluid

▰ Ⓡ : Rubber grease

Ⓧ : Always replace after every disassembly.

⬚ : N·m (kg-m, ft-lb)

⬚ : N·m (kg-m, in-lb)

1. Union bolt	2. Brake hose	3. Copper washer
4. Cap	5. Bleed valve	6. Mounting bolt
7. Cylinder body	8. Piston seal	9. Piston
10. Piston boot	11. Knuckle slide	12. Sliding sleeve boot
13. Sliding sleeve	14. Inner pad	15. Outer pad

67170-ARMA-G56

Fig. 2 Rear brake components

1. Before servicing the vehicle, refer to the Precautions Section.
2. Drain brake fluid as necessary.
3. Remove or disconnect the following:
   - Wheel
   - Union bolt
   - Mounting bolts
   - Brake caliper assembly

To install:
4. Install or connect the following:
   - Brake caliper assembly and

tighten mounting bolts to 32ft. lbs. (48 Nm).
5. Fill the master cylinder and bleed the brake system.
6. Install the wheels.

DISC BRAKE PADS

REMOVAL & INSTALLATION

1. Before servicing the vehicle, refer to the Precautions Section.
2. Remove the wheel.

3. Remove mounting bolt from the top mount.
4. Swing brake caliper open and remove the brake pads.

To install:
5. Push pistons in so that the pad is firmly installed, using a suitable tool.
6. Install pads to the brake caliper.
7. Install top mounting bolt and tighten to 32 ft. lbs. (48 Nm).
8. Install the wheel.

BRAKES

PARKING BRAKE

PARKING BRAKE CABLES

ADJUSTMENT

See Figure 3.

1. Disconnect the negative battery cable.
2. Remove the lower instrument panel, driver's side.
3. Partially engage the parking brake pedal to access the adjusting nut.

4. Insert a deep socket wrench to rotate the adjusting nut and loosen the cable sufficiently.
5. Disengage the parking brake pedal.
6. Raise and support the vehicle safely.

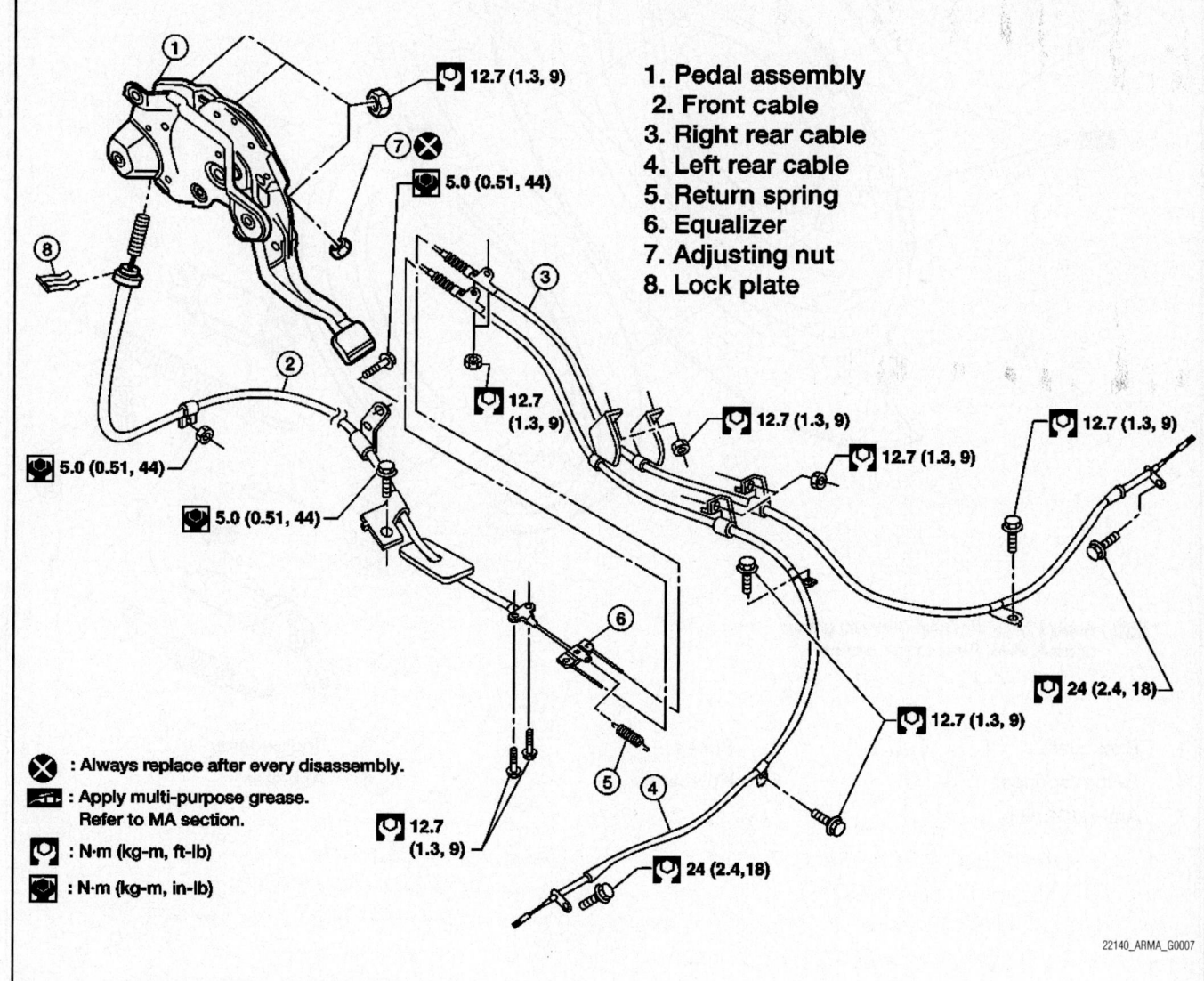

1. **Pedal assembly**
2. **Front cable**
3. **Right rear cable**
4. **Left rear cable**
5. **Return spring**
6. **Equalizer**
7. **Adjusting nut**
8. **Lock plate**

⊗ : Always replace after every disassembly.

: Apply multi-purpose grease.
Refer to MA section.

: N·m (kg-m, ft-lb)

: N·m (kg-m, in-lb)

Fig. 3 Parking brake assembly and related components

22140_ARMA_G0007

7. Remove the tire and wheel assembly.

8. Remove the rear rotors. Measure the inner diameter at the widest point using tool J-21177A or equivalent.

9. Transfer the recorded measurement less 0.6 mm to the parking brake shoes and adjust accordingly.

10. Using wheel nuts, secure the rotor to the hub to prevent it from tilting.

11. Rotate the rotor to make sure that there is no drag.

12. To adjust the cable operate the pedal ten or more times with a force of 110 lb.

13. Rotate the adjusting nut with a deep socket to adjust the pedal stroke to specification. Specification is 3–4 notches with a force of 44.1 lb.

14. With the parking brake pedal com-

pletely disengaged, make sure there is no drag on the parking brake.

15. Reassemble and reinstall any removed components.

PARKING BRAKE SHOES

REMOVAL & INSTALLATION

See Figure 4.

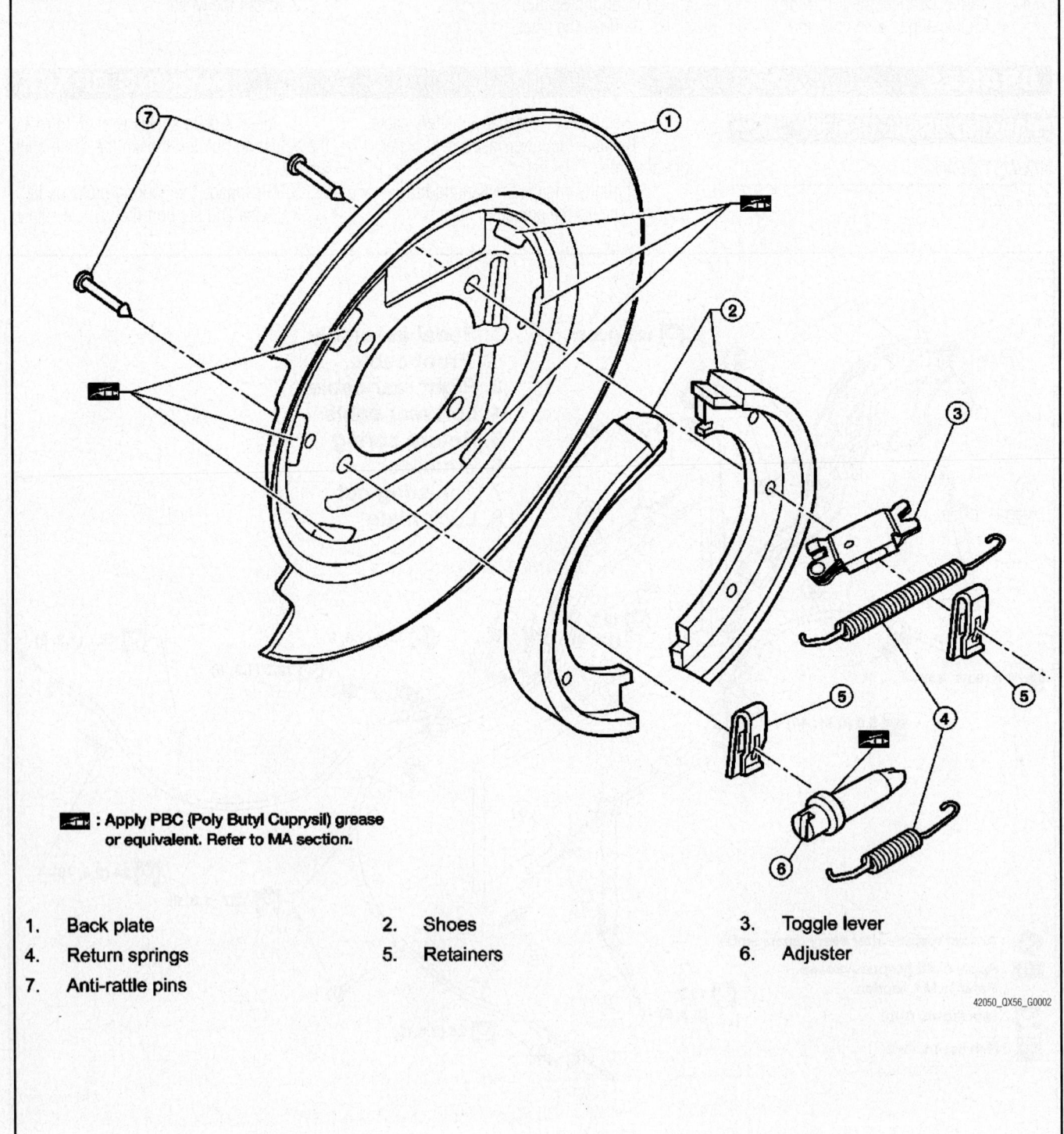

: Apply PBC (Poly Butyl Cuprysil) grease or equivalent. Refer to MA section.

1. Back plate	2. Shoes	3. Toggle lever
4. Return springs	5. Retainers	6. Adjuster
7. Anti-rattle pins		

42050_QX56_G0002

Fig. 4 Parking brake shoe and related components

1. Raise and support the vehicle safely.

2. Remove the tire and wheel assembly.

3. Be sure that the parking brake lever is in the released position.

4. Remove the rear disc rotor.

5. Remove the rear driveshaft.

6. Disconnect the ABS sensor at the harness connector. Remove the ABS sensor wire from the grommet mounts.

7. Remove the wheel hub and bearing assembly.

➡ **Withdraw the ABS sensor harness through the backing plate when removing the wheel and hub bearing assembly.**

8. Remove the return springs. Remove the adjuster. Remove the retainers. Remove the anti rattle pins and shoes.

9. Disconnect the parking brake cable from the toggle lever.

10. Remove the backing plate.

To install:

11. Apply brake grease to the specified points during reassembly, see illustration for locating points.

12. Install the adjuster so that the threaded part expands when rotating it in the proper direction.

13. Continue the installation in the reverse order of the removal procedure.

14. Adjust the parking brake.

15. Perform the parking brake burnishing operation.

CHASSIS ELECTRICAL

AIR BAG (SUPPLEMENTAL RESTRAINT SYSTEM)

GENERAL INFORMATION

✳✳ CAUTION

These vehicles are equipped with an air bag system. The system must be disarmed before performing service on, or around, system components, the steering column, instrument panel components, wiring and sensors. Failure to follow the safety precautions and the disarming procedure could result in accidental air bag deployment, possible injury and unnecessary system repairs.

SERVICE PRECAUTIONS

Disconnect and isolate the battery negative cable before beginning any airbag system component diagnosis, testing, removal, or installation procedures. Allow system capacitor to discharge for two minutes before beginning any component service. This will disable the airbag system. Failure to disable the airbag system may result in accidental airbag deployment, personal injury, or death.

Do not place an intact undeployed airbag face down on a solid surface. The airbag will propel into the air if accidentally deployed and may result in personal injury or death.

When carrying or handling an undeployed airbag, the trim side (face) of the airbag should be pointing towards the body to minimize possibility of injury if accidental deployment occurs. Failure to do this may result in personal injury or death.

Replace airbag system components with OEM replacement parts. Substitute parts may appear interchangeable, but internal differences may result in inferior occupant protection. Failure to do so may result in occupant personal injury or death.

Wear safety glasses, rubber gloves, and long sleeved clothing when cleaning powder residue from vehicle after an airbag deployment. Powder residue emitted from a deployed airbag can cause skin irritation. Flush affected area with cool water if irritation is experienced. If nasal or throat irritation is experienced, exit the vehicle for fresh air until the irritation ceases. If irritation continues, see a physician.

Do not use a replacement airbag that is not in the original packaging. This may result in improper deployment, personal injury, or death.

The factory installed fasteners, screws and bolts used to fasten airbag components have a special coating and are specifically designed for the airbag system. Do not use substitute fasteners. Use only original equipment fasteners listed in the parts catalog when fastener replacement is required.

During, and following, any child restraint anchor service, due to impact event or vehicle repair, carefully inspect all mounting hardware, tether straps, and anchors for proper installation, operation, or damage. If a child restraint anchor is found damaged in any way, the anchor must be replaced. Failure to do this may result in personal injury or death.

Deployed and non-deployed airbags may or may not have live pyrotechnic material within the airbag inflator.

Do not dispose of driver/passenger/curtain airbags or seat belt tensioners unless you are sure of complete deployment. Refer to the Hazardous Substance Control System for proper disposal.

Dispose of deployed airbags and tensioners consistent with state, provincial, local, and federal regulations.

After any airbag component testing or service, do not connect the battery negative cable. Personal injury or death may result if the system test is not performed first.

If the vehicle is equipped with the Occupant Classification System (OCS), do not connect the battery negative cable before performing the OCS Verification Test using the scan tool and the appropriate diagnostic information. Personal injury or death may result if the system test is not performed properly.

Never replace both the Occupant Restraint Controller (ORC) and the Occupant Classification Module (OCM) at the same time. If both require replacement, replace one, then perform the Airbag System test before replacing the other.

Both the ORC and the OCM store Occupant Classification System (OCS) calibration data, which they transfer to one another when one of them is replaced. If both are replaced at the same time, an irreversible fault will be set in both modules and the OCS may malfunction and cause personal injury or death.

If equipped with OCS, the Seat Weight Sensor is a sensitive, calibrated unit and must be handled carefully. Do not drop or handle roughly. If dropped or damaged, replace with another sensor. Failure to do so may result in occupant injury or death.

If equipped with OCS, the front passenger seat must be handled carefully as well. When removing the seat, be careful when setting on floor not to drop. If dropped, the sensor may be inoperative, could result in occupant injury, or possibly death.

If equipped with OCS, when the passenger front seat is on the floor, no one should sit in the front passenger seat. This uneven force may damage the sensing ability of the seat weight sensors. If sat on and damaged, the sensor may be inoperative, could result in occupant injury, or possibly death.

DISARMING THE SYSTEM

1. Before servicing the vehicle, refer to the Precautions Section.

2. Disconnect both battery cables.

3. Wait at least 3 minutes before working on the vehicle. The air bag system is designed to retain enough power to deploy

the air bag for a short time after the battery has been disconnected.

4. After repairs are complete, connect the negative battery cable. Turn the ignition switch to the **ON** position and check the air bag warning light blinks for proper operation.

ARMING THE SYSTEM

After repairs are complete, connect the negative battery cable. Turn the ignition switch to the **ON** position and check the air bag warning light blinks for proper operation.

CLOCKSPRING CENTERING

See Figures 5 and 6.

※※ CAUTION

Before servicing the Supplemental Restraint System (SRS), turn ignition switch OFF, disconnect both battery cables and wait at least 3 minutes.

※※ CAUTION

When servicing the SRS, do not work from directly in front of air bag module.

1. Remove the steering wheel.

2. Remove the column cover upper and lower.

3. Remove wiper washer switch connector, then pinch the tabs at wiper and washer switch base and slide switch away from steering column to remove.

4. While pressing tabs, pull lighting with the turn signal switch toward driver door, disconnect the assembly from the base.

5. Remove the screws, release the clip, and remove the spiral cable.

➡ **Do not disassemble spiral cable.**

➡ **Do not apply lubricant to the spiral cable.**

6. Remove the spiral cable connectors.

※※ WARNING

With the steering linkage disconnected, the spiral cable may snap by turning the steering wheel beyond the limited number of turns. The spiral cable can be turned counterclockwise about 2.5 turns from the neutral position

To install:

7. Align spiral cable correctly when installing steering wheel.

8. Make sure that the spiral cable is in the neutral position.

➡ **The neutral position is detected by turning left 2.6 revolutions from the right end position and ending with the knob at the top.**

9. Connect the spiral cable connectors.

10. While pressing the tabs, push the turn signal switch toward driver door connect the assembly to the base.

11. Remove the screws, release the clip, and remove the spiral cable.

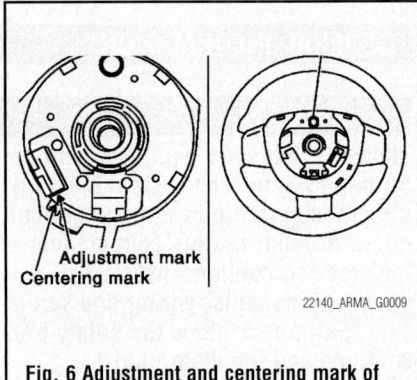

Fig. 6 Adjustment and centering mark of the spiral cable

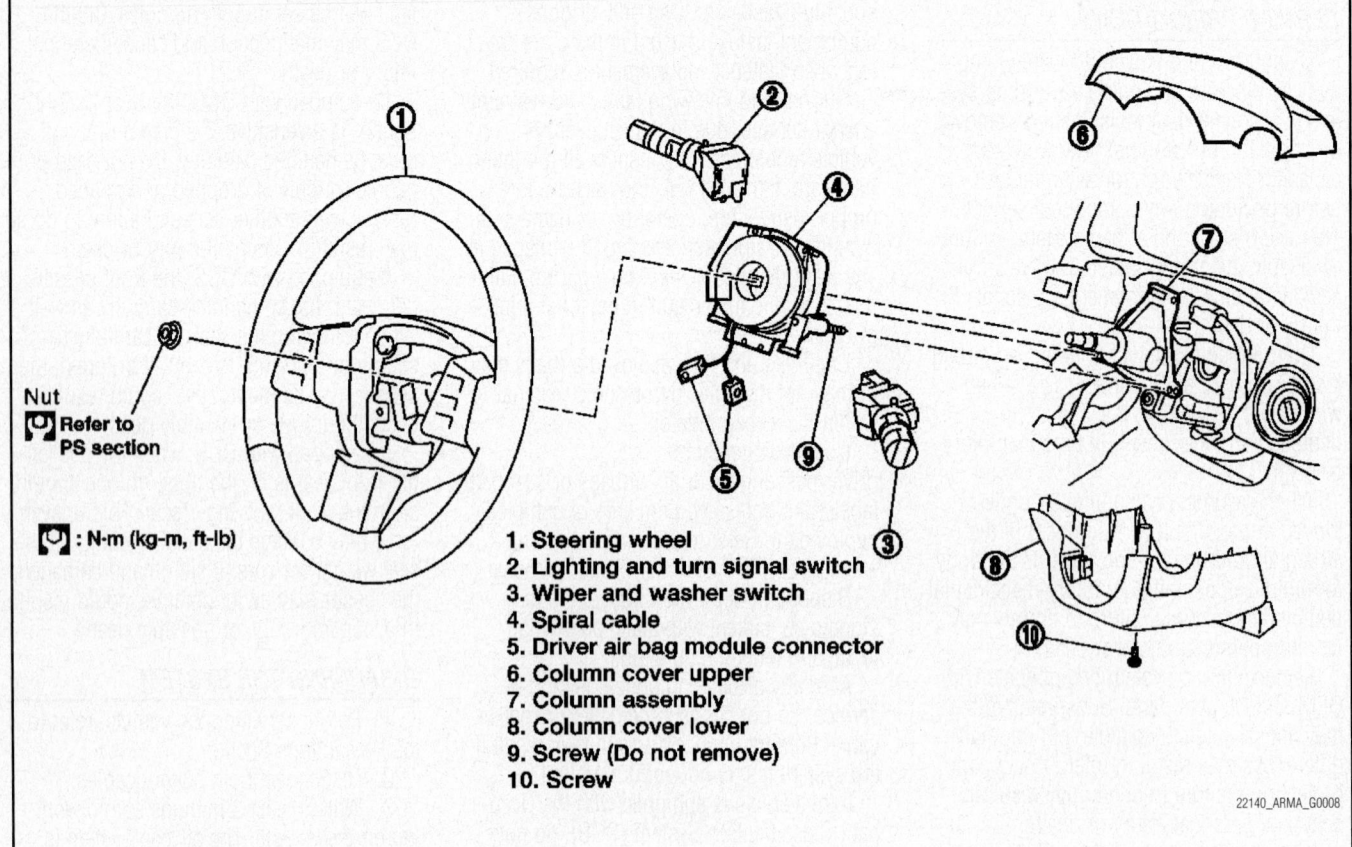

Nut
🔧 Refer to PS section

🔧 : N-m (kg-m, ft-lb)

1. Steering wheel
2. Lighting and turn signal switch
3. Wiper and washer switch
4. Spiral cable
5. Driver air bag module connector
6. Column cover upper
7. Column assembly
8. Column cover lower
9. Screw (Do not remove)
10. Screw

22140_ARMA_G0008

Fig. 5 Driver air bag module assembly and related components

DRIVETRAIN

Before starting diagnosis of the vehicle, understand the symptoms well. Perform correct and systematic operations.

• Check for the correct installation status prior to removal or disassembly. When matching marks are required, be certain they do not interfere with the function of the parts they are applied to.

• Overhaul should be done in a clean work area, a dust proof area is recommended.

• Before disassembly, completely remove sand and mud from the exterior of the unit, preventing them from entering into the unit during disassembly or assembly.

• Always use shop paper for cleaning the inside of components.

• Avoid using cotton gloves or a shop cloth to prevent the entering of lint.

• Check appearance of the disassembled parts for damage, deformation, and abnor-

mal wear. Replace them with new ones if necessary.

• Gaskets, seals and O-rings should be replaced any time the unit is disassembled.

• Clean and flush the parts sufficiently and blow them dry.

• Be careful not to damage sliding surfaces and mating surfaces.

• When applying sealant, remove the old sealant from the mating surface; then remove any moisture, oil, and foreign materials from the application and mating surfaces.

• In principle, tighten nuts or bolts gradually in several steps working diagonally from inside to outside. If a tightening sequence is specified, observe it.

• During assembly, observe the specified tightening torque.

• Add new differential gear oil, petroleum jelly, or multi-purpose grease, as specified.

AUTOMATIC TRANSMISSION ASSEMBLY

REMOVAL & INSTALLATION

2WD

See Figures 7 and 8.

1. Before servicing the vehicle, refer to the Precautions Section.
2. Remove or disconnect the following:
   • Negative battery cable
   • Engine cover
   • Transmission fluid indicator gauge
   • Engine splash guard
   • Exhaust front pipe
   • Center muffler
   • Rear driveshaft
   • Transmission control cable
   • Crankshaft position sensor
   • Transmission cooler tube
   • Dust cover from converter housing

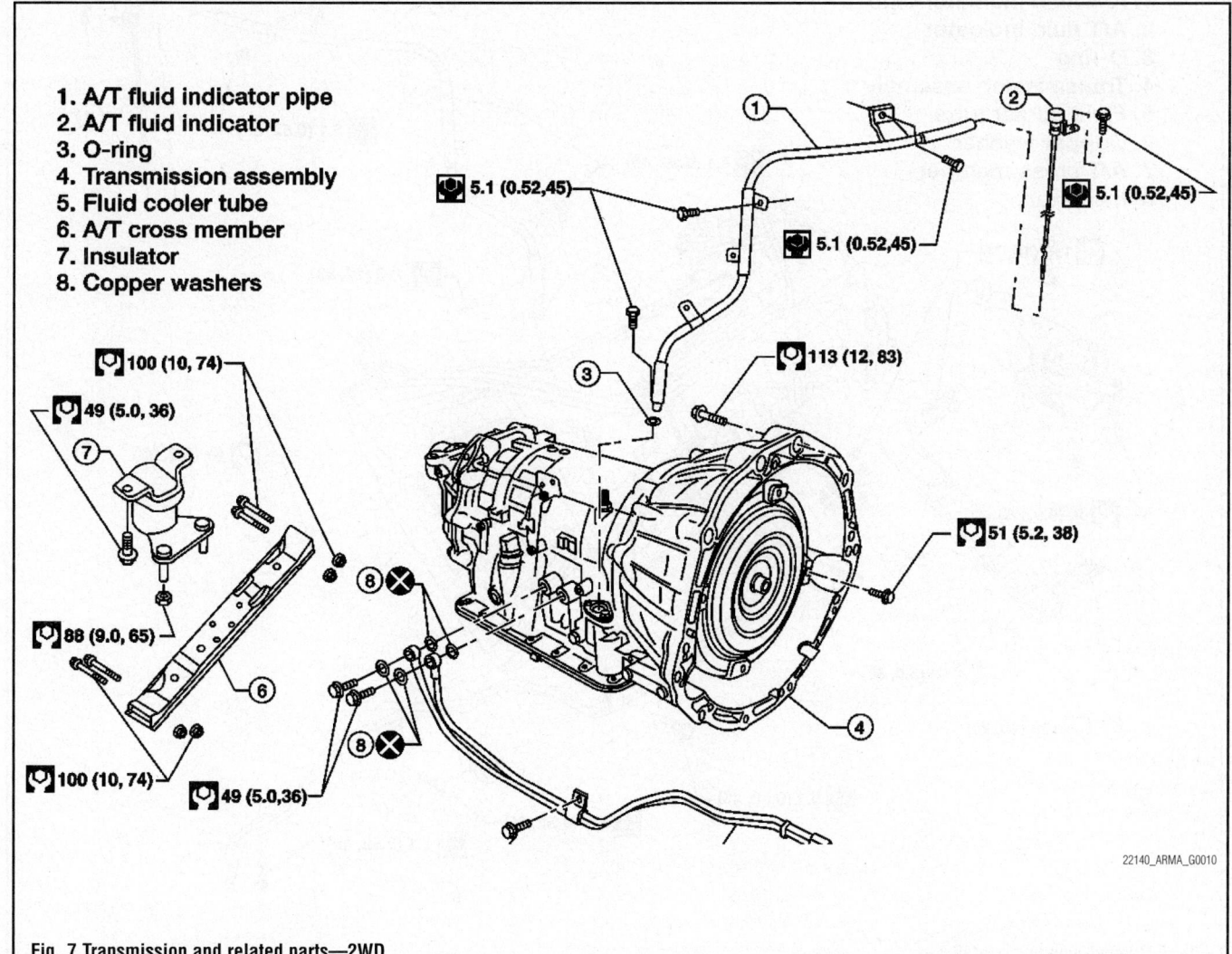

1. A/T fluid indicator pipe
2. A/T fluid indicator
3. O-ring
4. Transmission assembly
5. Fluid cooler tube
6. A/T cross member
7. Insulator
8. Copper washers

5.1 (0.52,45)

5.1 (0.52,45)

5.1 (0.52,45)

113 (12, 83)

100 (10, 74)

49 (5.0, 36)

51 (5.2, 38)

88 (9.0, 65)

100 (10, 74)

49 (5.0,36)

22140_ARMA_G0010

Fig. 7 Transmission and related parts—2WD

3. Turning crankshaft clockwise, remove the four tightening bolts for drive plate and torque converter
4. Support the transmission with a suitable jack.
5. Remove or disconnect the following:

2*
3
⊙A/T to engine
1
4
5
9
6
View from vehicle rear
8
7
22140_ARMA_G0011

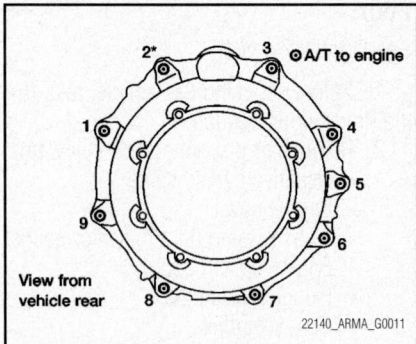

Fig. 8 Transmission to the engine attachment bolt order

- Transmission cross member
- Air breather hose
- Transmission assembly connector
- Fluid indicator tube from transmission assembly
- Transmission assembly to engine bolts
- Transmission assembly from vehicle

To install:

6. Install or connect the following:
   - Transmission assembly into vehicle
   - Transmission assembly to engine bolts tightening to 83 ft. lbs. (113 Nm)
   - Fluid indicator tube to transmission assembly
   - Transmission assembly connector
   - Air breather hose
   - Transmission cross member
7. Turning crankshaft clockwise, install the torque converter to drive plate.

➡After torque converter is installed, rotate the crankshaft to ensure transmission rotates freely.

8. Install or connect the following:
   - Dust cover for converter housing
   - Fluid cooler tube
   - Crankshaft position sensor
   - Transmission control cable
   - Rear driveshaft
   - Center muffler
   - Exhaust front pipe
   - Engine splash guard
   - Transmission fluid indicator gauge
   - Engine cover
   - Negative battery cable
9. Start engine and check for leaks.

4WD

See Figures 9 and 10.

1. Before servicing the vehicle, refer to the Precautions Section.

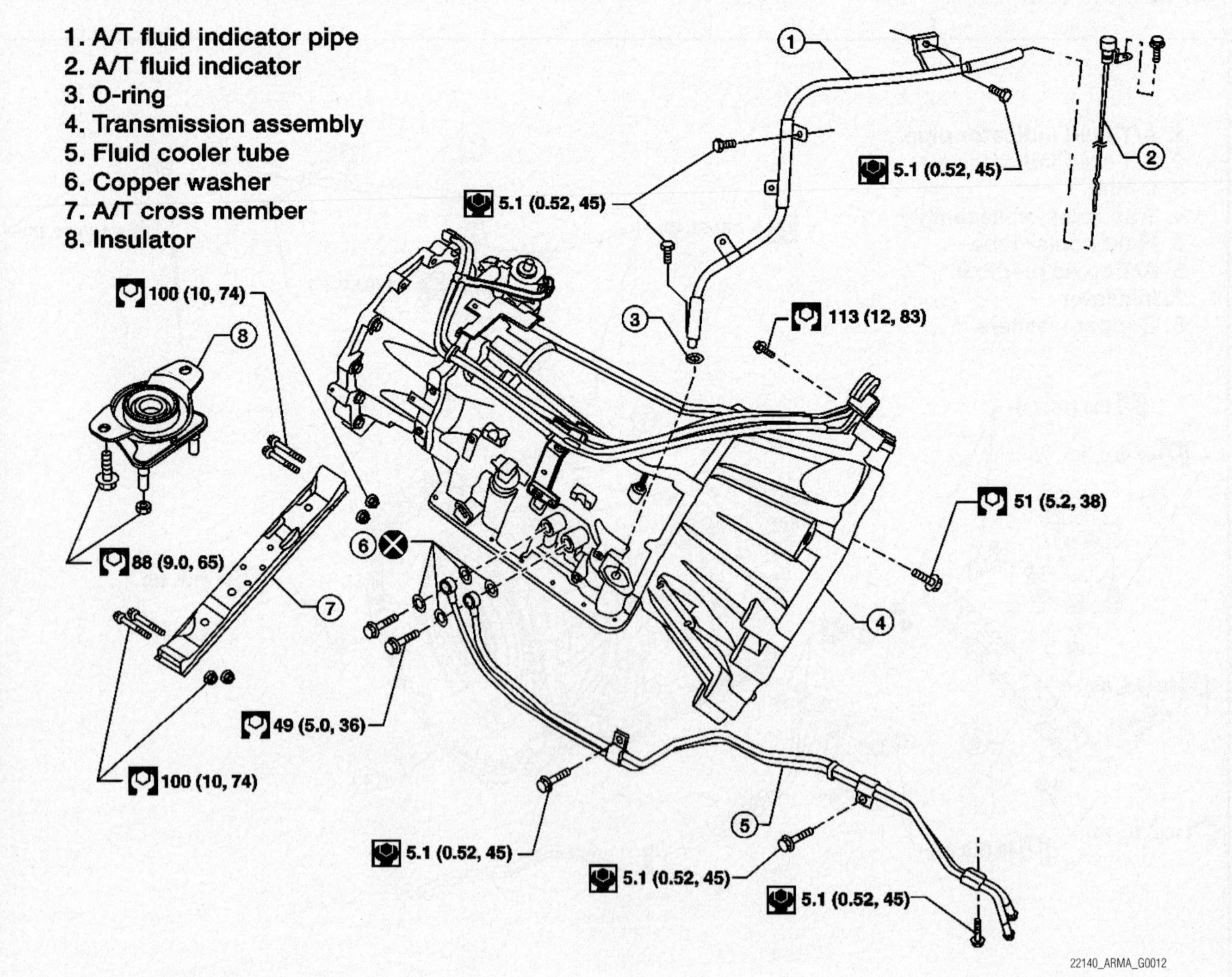

1. A/T fluid indicator pipe
2. A/T fluid indicator
3. O-ring
4. Transmission assembly
5. Fluid cooler tube
6. Copper washer
7. A/T cross member
8. Insulator
5.1 (0.52, 45)
5.1 (0.52, 45)
100 (10, 74)
113 (12, 83)
51 (5.2, 38)
88 (9.0, 65)
49 (5.0, 36)
100 (10, 74)
5.1 (0.52, 45)
5.1 (0.52, 45)
5.1 (0.52, 45)
22140_ARMA_G0012

Fig. 9 Transmission and related parts

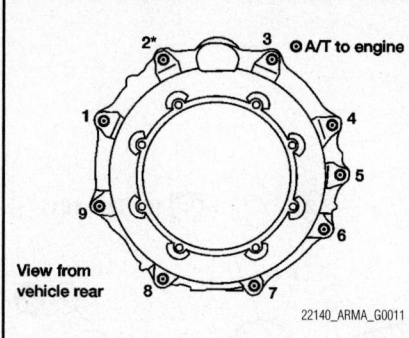

Fig. 10 Transmission-to-engine bolt tightening sequence

2. Remove or disconnect the following:
- Negative battery cable
- Engine cover
- Transmission fluid indicator gauge
- Engine splash guard
- Exhaust front pipe
- Center muffler
- Driveshaft
- Transmission control cable
- Crankshaft position sensor
- Fluid cooler tube
- Dust housing for torque converter

3. Turning the crankshaft clockwise, remove the four tightening bolts for drive plate and torque converter.

4. Support the transmission assembly with a suitable jack.

5. Remove transmission cross member.

6. Tilt the transmission slightly to keep clearance between the body and the transmission assembly, then disconnect the air breather hose.

7. Remove or disconnect the following:
- Transmission assembly connector and transfer case connector
- Fluid indicator pipe
- Transmission assembly to engine bolts
- Transmission assembly, with transfer case attached, from vehicle
- Transmission assembly from transfer case

To install:

8. Install or connect the following:
- Transfer case to transmission assembly
- Transmission assembly into vehicle
- Transmission assembly to engine bolts tightening to 83 ft. lbs. (113 Nm)

9. With the transmission slightly tilted to allow clearance between body and transmission, connect the air breather hose.

10. Install the transmission cross member.

11. Turning crankshaft clockwise, install the torque converter to drive plate.

➡**After torque converter is installed, rotate the crankshaft to ensure transmission rotates freely.**

12. Install or connect the following:
- Dust housing for torque converter
- Fluid cooler tube
- Crankshaft position sensor
- Transmission control cable
- Driveshaft
- Center muffler
- Front exhaust pipe
- Engine splash guard
- Transmission fluid indicator gauge
- Engine cover
- Negative battery cable

13. Start engine and check for leaks.

TRANSFER CASE ASSEMBLY

REMOVAL & INSTALLATION

See Figure 11.

1. Before servicing the vehicle, refer to the Precautions Section.

2. Remove or disconnect the following:
- Transmission splash guard
- Center exhaust pipe and muffler
- Front and rear driveshafts

➡**Plug rear oil seal after removing rear driveshaft.**

- Transmission assembly mounting bolts

3. Support the transmission assembly with a suitable jack and remove the cross-member.

4. Remove or disconnect the following:
- ATP switch, neutral 4LO switch, wait detection switch, transfer motor and transfer control device electrical connectors

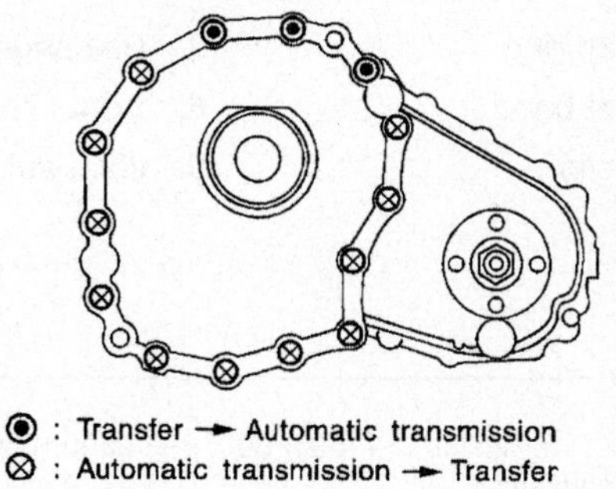

Fig. 11 Transfer case mounting bolt locations

◉ : Transfer ➝ Automatic transmission
⊗ : Automatic transmission ➝ Transfer

67170-ARMA-G41

- Breather hoses
- Shift actuator from the extension housing
- Transfer case to transmission assembly bolts
- Transfer case assembly

To install:

5. Install or connect the following:
- Transfer case to transmission assembly bolts tightening to 38 ft. lbs. (52 Nm)
- Shift actuator
- Breather hoses
- ATP switch, neutral 4LO switch, wait detection switch, transfer motor and transfer control device electrical connectors
- Support crossmember
- Transmission mounting bolts
- Driveshafts
- Muffler and center exhaust pipe
- Transmission splash guard

FRONT HALFSHAFT

REMOVAL & INSTALLATION

See Figure 12.

1. Before servicing the vehicle, refer to the Precautions Section.

2. Remove or disconnect the following:
- Wheel
- Engine splash guard
- Cotter pin and halfshaft nut
- Halfshaft from front differential
- Halfshaft from hub and bearing assembly

To install:

3. Install or connect the following:
- Halfshaft into hub
- Halfshaft into front differential
- Halfshaft nut and tighten to 101 ft. lbs. and replace cotter pin
- Engine splash guard
- Wheel

CV-JOINTS OVERHAUL

Inner

See Figure 13.

1. Before servicing the vehicle, refer to the Precautions Section.

2. Remove the halfshaft from the vehicle.

3. Mount halfshaft in a vise.

4. Remove the dust boot bands.

5. Remove the stopper ring with a flat-bladed screwdriver or suitable tool.

6. Remove the snap ring.

7. Disassemble the cage, ball and inner race assembly and dust boot for cleaning and inspection.

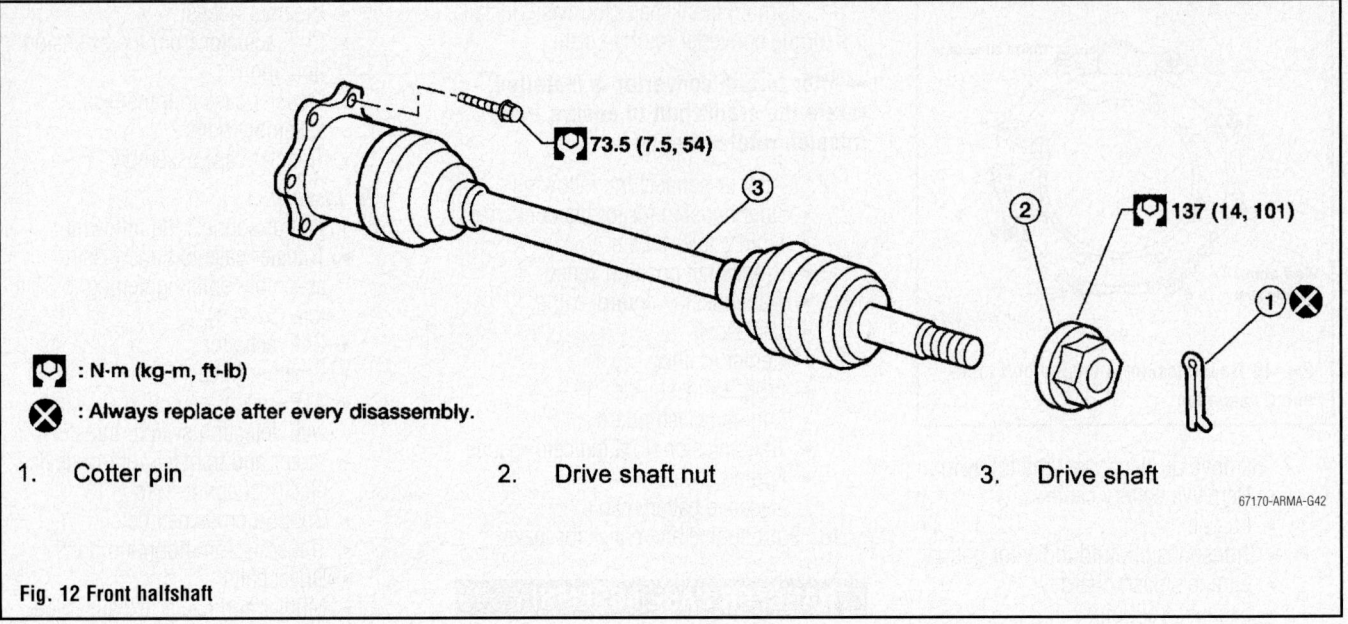

73.5 (7.5, 54)

137 (14, 101)

🔧 : N·m (kg-m, ft-lb)

✖ : Always replace after every disassembly.

1. Cotter pin
2. Drive shaft nut
3. Drive shaft

67170-ARMA-G42

Fig. 12 Front halfshaft

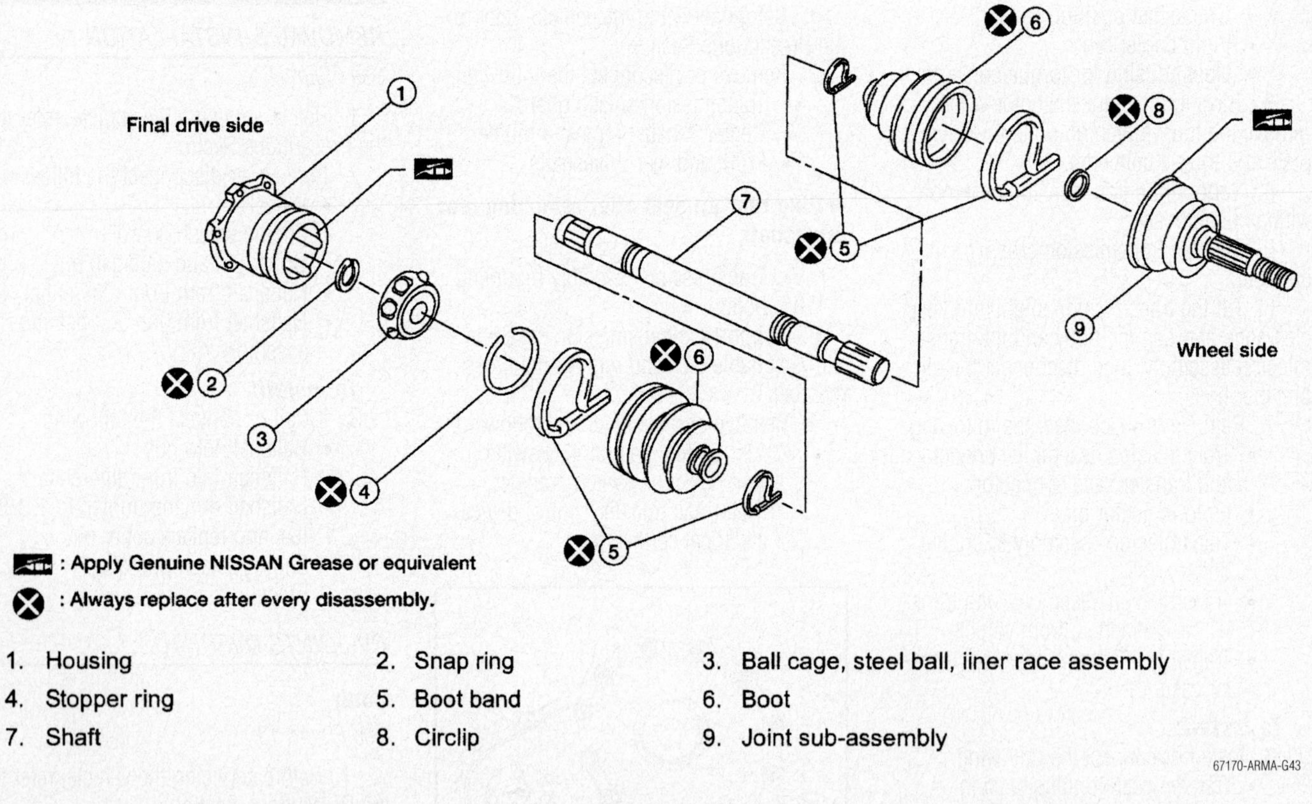

Final drive side

Wheel side

🛢 : Apply Genuine NISSAN Grease or equivalent

✖ : Always replace after every disassembly.

1. Housing
2. Snap ring
3. Ball cage, steel ball, iiner race assembly
4. Stopper ring
5. Boot band
6. Boot
7. Shaft
8. Circlip
9. Joint sub-assembly

67170-ARMA-G43

Fig. 13 Front halfshaft—exploded view

To install:

➡ Discard old dust boot, dust boot bands and snap ring and use new ones for assembly.

8. Wrap the serrated part of the halfshaft with tape.

9. Install new dust boot and band onto halfshaft.

10. Remove tape from serrated part of halfshaft.

11. Install the cage, ball and inner race assembly.

12. Install new snap ring.

13. Insert 4.50-5.3 oz of genuine NISSAN grease or equivalent onto the housing and install onto halfshaft.

14. Install the stopper ring onto the housing.

15. Install the dust boot into the grooves on joint sub-assembly.

16. Secure the big and small ends of the dust boot using new boot bands.

Outer

See Figure 14.

1. Before servicing the vehicle, refer to the Precautions Section.

2. Remove the halfshaft from the vehicle.

3. Mount halfshaft in a vise.

4. Remove the dust boot bands and dust boot from joint sub-assembly.

5. Insert a suitable puller into the threaded part of the halfshaft. Pull the joint sub-assembly off of the halfshaft as shown in figure.

6. Remove dust boot and circlip from halfshaft for cleaning and inspection.

To install:

➡**Discard old dust boot, boot bands and circlip and use new ones for assembly.**

7. Insert genuine NISSAN grease or equivalent into the joint sub-assembly until grease oozes from the ball groove and serration hole.

8. Wrap the serrated part of the halfshaft with tape.

9. Install new dust boot and band onto halfshaft.

10. Remove tape from serrated part of the halfshaft.

11. Press-fit the new circlip to the halfshaft.

12. Insert 5.1-5.8 oz of genuine NISSAN

grease or equivalent into the joint sub-assembly and large end of boot.

13. Install the dust boot into the grooves on the joint sub-assembly.

14. Secure the big and small ends of the dust boot using new boot bands.

FRONT PINION SEAL

REMOVAL & INSTALLATION

See Figures 15 through 18.

1. Before servicing the vehicle, refer to the Precautions Section.

2. Remove or disconnect the following:
- Front driveshaft
- Halfshafts

3. Measure and record the pinion bearing preload using special tool J-25765-A.

4. Loosen the pinion nut while holding

Fig. 15 Removing the companion flange using Special Tool J-44195

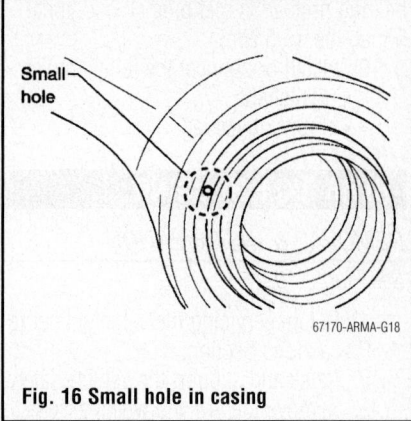

Fig. 16 Small hole in casing

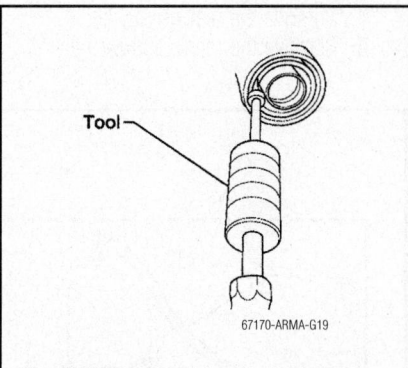

Fig. 17 Removing pinion seal using special tool SP8P

Fig. 18 Front pinion seal installation

the companion flange using special tool J-44195.

5. Remove the companion flange using a suitable tool.

6. Using a punch or drill, place a small hole in the case.

7. Remove the seal using special tool SP8P or equivalent.

To install:

8. Press front seal into carrier using a suitable tool.

9. Install companion flange and new pinion nut. Tighten pinion nut until there is no endplay and until recorded pinion,

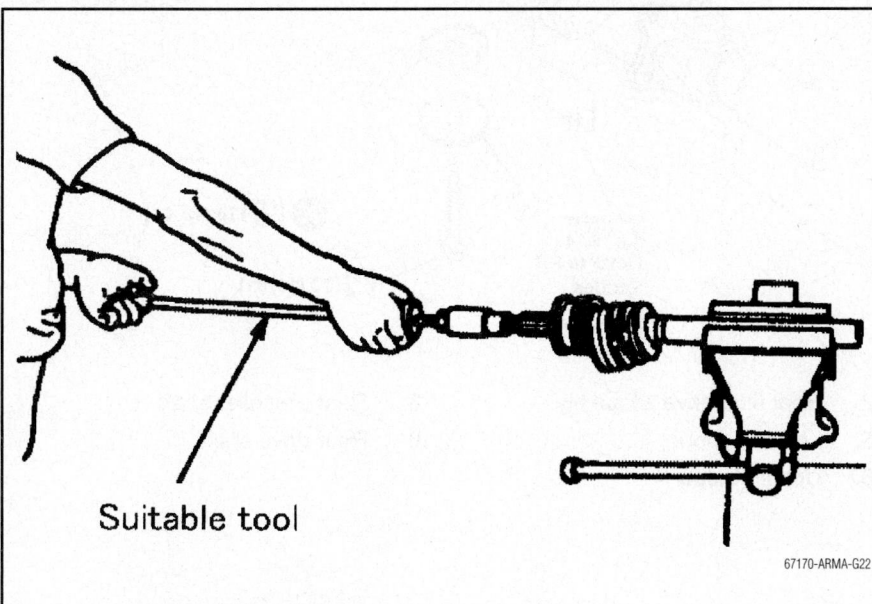

Fig. 14 Using a suitable puller to remove joint sub-assembly.

bearing preload is met plus an additional 5 inch lbs. (0.5 Nm).

10. Install or connect the following:
- Halfshafts
- Front driveshaft

REAR AXLE HOUSING

REMOVAL & INSTALLATION

See Figure 19.

1. Before servicing the vehicle, refer to the Precautions Section.
2. Raise and support the vehicle safely.
3. Remove the tire and wheel assemblies.
4. Remove the rear driveshaft.
5. Remove the rear stabilizer bar.

6. Remove the spare tire.
7. Disconnect the rear halfshaft from the rear axle assembly. Position it aside using mechanics wire or equivalent.
8. Disconnect the breather hose from the axle cover.
9. Position a suitable jack under the assembly.

➡**Do not position the jack under the aluminum cover.**

10. Remove the rear axle assembly retaining bolts and nuts.
11. Carefully remove the assembly from the vehicle.

To install:

12. Installation is the reverse of the removal procedure.

REAR AXLE SHAFT, BEARING & SEAL

REMOVAL & INSTALLATION

See removal of Rear Axle Stub Shaft and Seal

REAR AXLE STUB SHAFT AND SEAL

REMOVAL & INSTALLATION

See Figure 20.

1. Remove the side flange using tools. KV40104100 and ST36230000 (J-25840-A).
2. Remove the side oil seal using suitable tool.

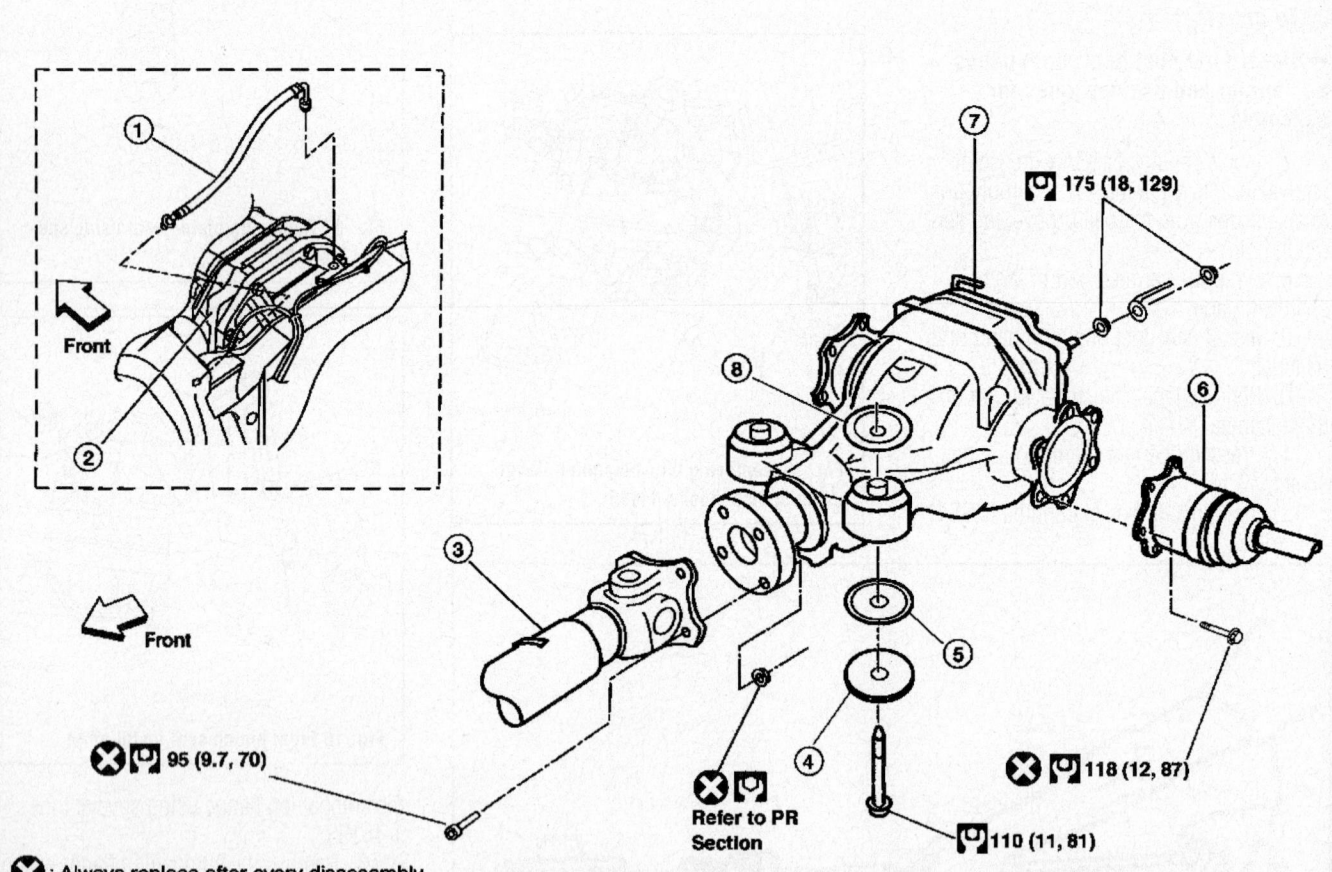

175 (18, 129)

95 (9.7, 70)

Refer to PR Section

118 (12, 87)

110 (11, 81)

❌ : Always replace after every disassembly.

🔧 : N·m (kg - m, lft - lb)

1. Breather hose	2. Rear final drive assembly	3. Rear propeller shaft
4. Washer	5. Lower stopper	6. Rear drive shaft
7. Rear final drive assembly	8. Upper stopper	

42050_QX56_G0036

Fig. 19 Rear axle assembly and related components

To install:

3. Apply multi-purpose grease to the lips of the new front oil seal. Then drive the new front oil seal in evenly until it becomes flush with the gear carrier using ToolST15310000

➡ Do not reuse oil seal.

➡ Apply multi-purpose grease to the lips of the new oil seal.

4. Install the side flange using tool KV38107900 (J-39352)

5. Install the tool to the side oil seal.

6. Drive in the side flange using suitable tool.

➡ Installation is completed when the driving sound of the side flange turns into a sound, which seems to affect the whole rear final drive assembly.

REAR HALFSHAFT

REMOVAL & INSTALLATION

1. Before servicing the vehicle, refer to the Precautions Section.

2. Remove or disconnect the following:
   - Wheel
   - Stabilizer bar clamp

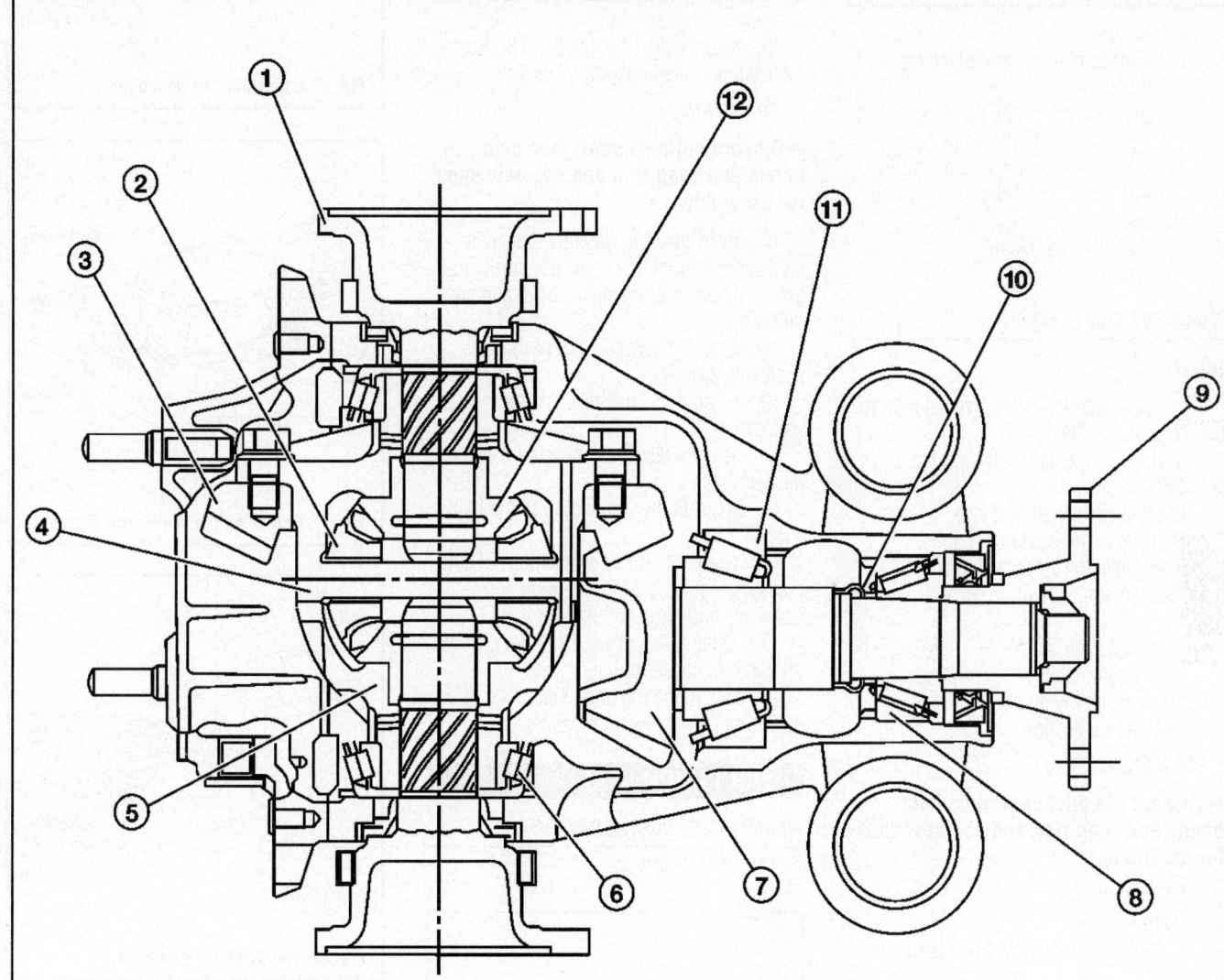

1. **Side flange**
2. **Pinion mate gear**
3. **Drive gear**
4. **Pinion mate shaft**
5. **Differential case**
6. **Side bearing**
7. **Drive pinion**
8. **Drive pinion front bearing**
9. **Companion flange**
10. **Collapsible spacer**
11. **Drive pinion rear bearing**
12. **Side gear**

22140_ARMA_G0013

Fig. 20 Rear axle assembly and related components

- Cotter pin and driveshaft nut
- Bolts from the inside flange of the driveshaft

3. Separate the driveshaft from the wheel hub by lightly tapping the end with suitable hammer and wood block.

4. Remove the halfshaft.

✳✳ CAUTION

Do not excessively extend the slide joint.

To install:

5. Install or connect the following:
- Halfshaft
- Bolts for the inside flange and tighten to 87 ft. lbs. (118 Nm)
- Driveshaft nut and tighten nut to 101 ft. lbs. (137 Nm) and replace cotter pin
- Stabilizer bar clamp
- Wheel

CV-JOINT OVERHAUL

Inner

1. Before servicing the vehicle, refer to the Precautions Section.

2. Remove the halfshaft from the vehicle.

3. Mount halfshaft in a vise.

4. Remove the dust boot bands.

5. Remove the stopper ring with a flat-bladed screwdriver or suitable tool.

6. Remove the snap ring.

7. Disassemble the cage, ball and inner race assembly and dust boot for cleaning and inspection.

To install:

➡ Discard old dust boot, dust boot bands and snap ring and use new ones for assembly.

8. Wrap the serrated part of the half-shaft with tape.

9. Install new dust boot and band onto halfshaft.

10. Remove tape from serrated part of halfshaft.

11. Install the cage, ball and inner race assembly.

12. Install new snap ring.

13. Insert 4.50-5.3 oz of genuine NISSAN grease or equivalent onto the housing and install onto halfshaft.

14. Install the stopper ring onto the housing.

15. Install the dust boot into the grooves on joint sub-assembly.

16. Secure the big and small ends of the dust boot using new boot bands.

Outer

1. Before servicing the vehicle, refer to the Precautions Section.

2. Remove the halfshaft from the vehicle.

3. Mount halfshaft in a vise.

4. Remove the dust boot bands and dust boot from joint sub-assembly.

5. Insert a suitable puller into the threaded part of the halfshaft. Pull the joint sub-assembly off of the halfshaft as shown in figure.

6. Remove dust boot and circlip from halfshaft for cleaning and inspection.

To install:

➡ Discard old dust boot, dust boot bands and snap ring and use new ones for assembly.

7. Insert genuine NISSAN grease or equivalent into the joint sub-assembly until grease oozes from the ball groove and serration hole.

8. Wrap the serrated part of the half-shaft with tape.

9. Install new dust boot and band onto halfshaft.

10. Remove tape from serrated part of the halfshaft.

11. Press-fit the new circlip to the half-shaft.

12. Insert 5.1-5.8 oz of genuine NISSAN grease or equivalent into the joint sub-assembly and large end of boot.

13. Install the dust boot into the grooves on the joint sub-assembly.

14. Secure the big and small ends of the dust boot using new boot bands.

REAR PINION SEAL

REMOVAL & INSTALLATION

See Figures 21 through 24.

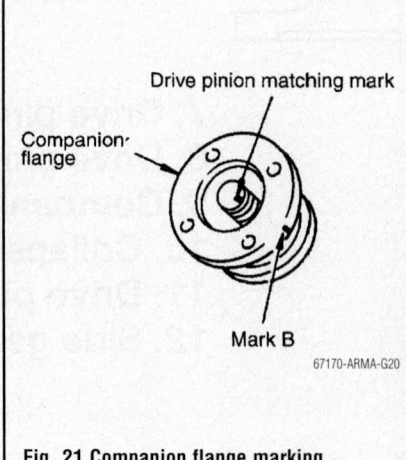

Fig. 21 Companion flange marking

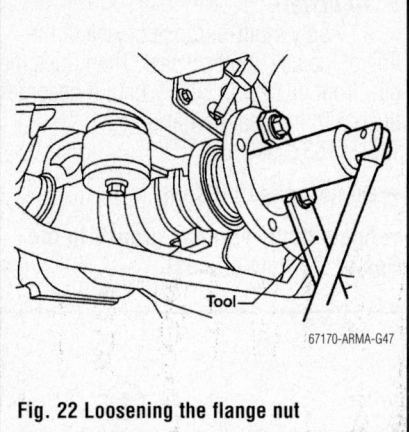

Fig. 22 Loosening the flange nut

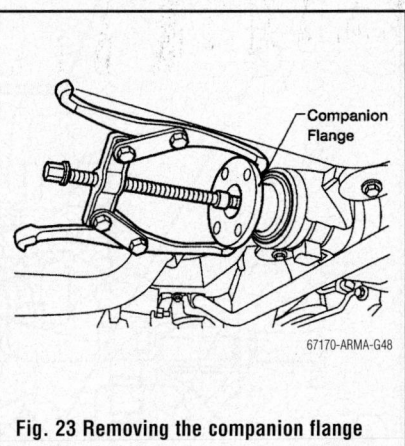

Fig. 23 Removing the companion flange

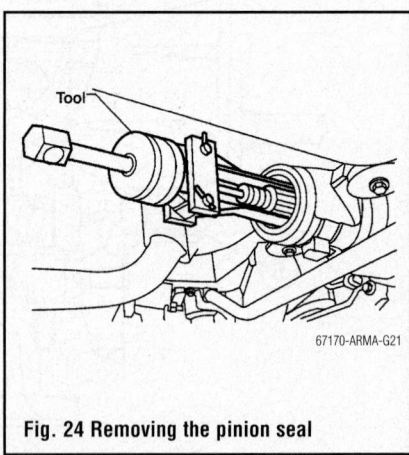

Fig. 24 Removing the pinion seal

1. Before servicing the vehicle, refer to the Precautions Section.

2. Remove the rear driveshaft.

3. Measure and record the total pre-load.

4. Matchmark the drive pinion to position 'B' on the companion flange.

5. Remove the drive pinion nut using suitable tool.

6. Remove the companion flange using suitable tool.

7. Remove the rear pinion seal using special tool J-34286.

To install:

8. Press the rear pinion seal into the carrier using suitable tool.

9. Align the matchmark on the companion flange to the drive pinion and install the companion flange.

10. Lubricate the drive pinion threads and seating surfaces of the drive pinion nut with grease.

11. Using a new drive pinion nut, tighten to 124-274 ft. lbs. (167-372 Nm).

→**Final torque is determined when adjusting total preload using special tool J-25765-A.**

12. Install rear driveshaft.

ENGINE COOLING

THERMOSTAT

REMOVAL & INSTALLATION

See Figures 25 and 26.

→**Never remove the radiator cap when the engine is hot. Serious burns could occur from high-pressure engine coolant escaping from the radiator.**

1. Be sure the engine is cold.
2. Disconnect the negative battery cable.

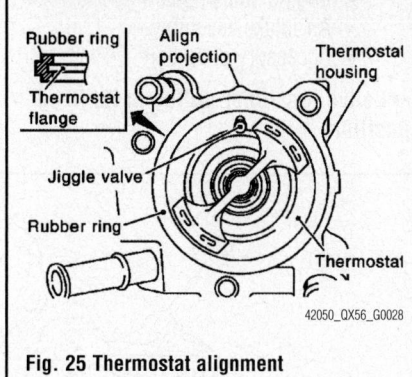

Fig. 25 Thermostat alignment

3. Remove the air duct and resonator assembly.
4. Remove the engine front undercover.
5. Disconnect the water suction hose from the water inlet.
6. Remove the water inlet and thermostat.

To install:

7. Installation is the reverse of the removal procedure.
8. Be sure to use a new gasket.
9. Install the thermostat with the whole

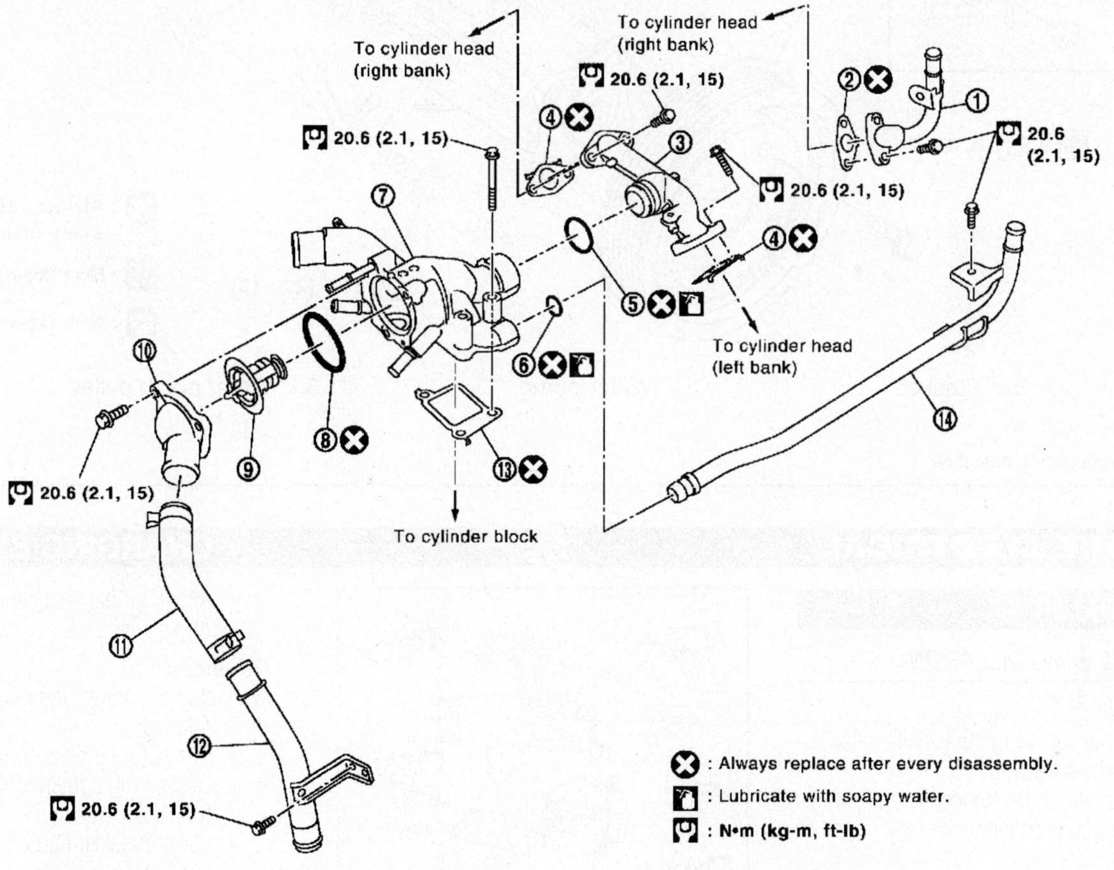

1. Heater pipe	2. Gasket	3. Water outlet
4. Gasket	5. O-ring	6. O-ring
7. Thermostat housing	8. Rubber ring	9. Thermostat
10. Water inlet	11. Water suction hose	12. Water suction pipe
13. Gasket	14. Heater pipe	

⊗ : Always replace after every disassembly.

🛢 : Lubricate with soapy water.

🔧 : N•m (kg-m, ft-lb)

Fig. 26 Thermostat and related components

42050_QX56_G0027

circumference of each flange part fitting securely inside the rubber ring, as shown in the illustration.

10. Install the thermostat with the jiggle valve facing upward.

11. Be sure to refill the cooling using the proper grade and type engine coolant.

12. Start the engine and check for leaks.

13. Start the engine and allow it to reach operation temperature. Recheck the coolant level, fill as required.

WATER PUMP

REMOVAL & INSTALLATION

See Figure 27.

1. Before servicing the vehicle, refer to the Precautions Section.
2. Drain the cooling system.
3. Remove or disconnect the following:
   - Engine splash guard
   - Air intake assembly
   - Accessory drive belt

➡**Leave tensioner pulley in its fixed position.**

- Water pump pulley
- Water pump

To install:

4. Install or connect the following:
   - Water pump with a new gasket. Tighten bolts to 18 ft. lbs. (25 Nm).
   - Water pump pulley and tighten bolts to 87 in. lbs. (10 Nm).
   - Accessory drive belt
   - Air intake assembly
   - Engine splash guard
5. Refill the cooling system.
6. Start the engine and check for leaks.

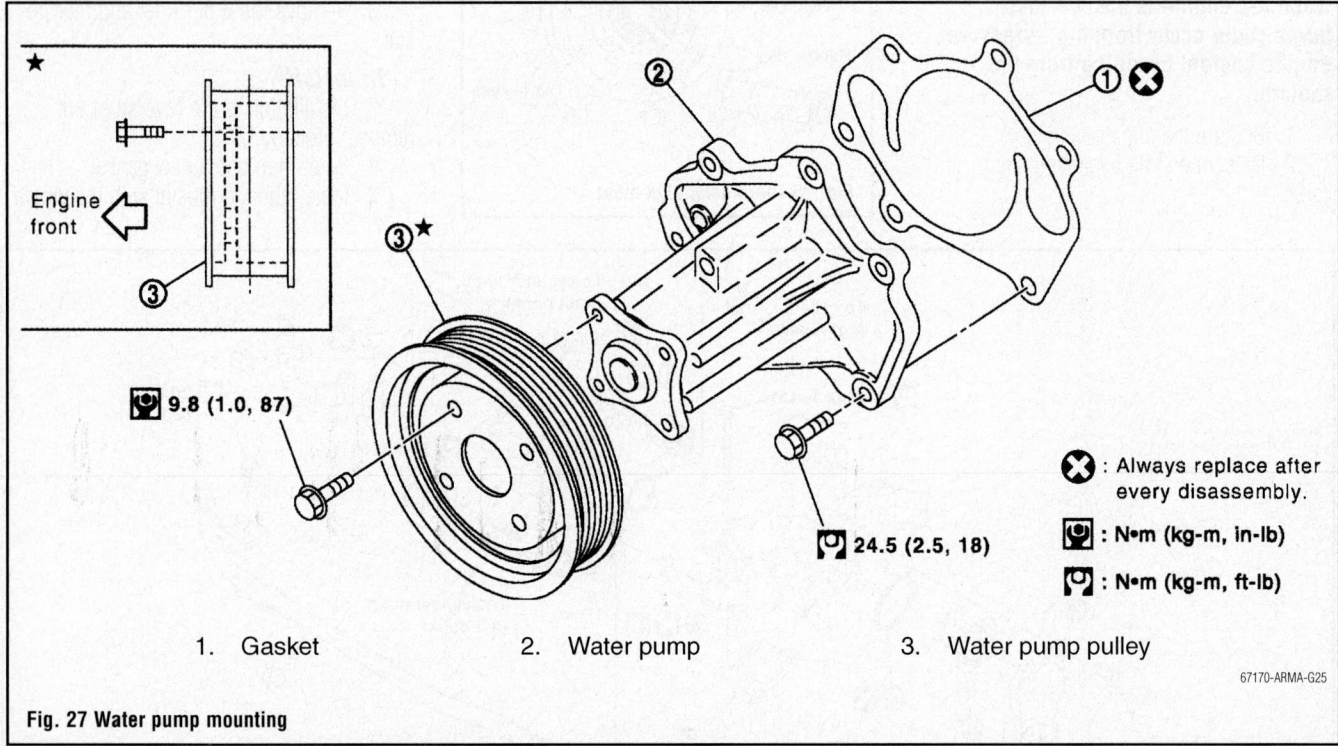

9.8 (1.0, 87)

24.5 (2.5, 18)

: Always replace after every disassembly.

: N•m (kg-m, in-lb)

: N•m (kg-m, ft-lb)

1. Gasket   2. Water pump   3. Water pump pulley

67170-ARMA-G25

Fig. 27 Water pump mounting

ENGINE ELECTRICAL

ALTERNATOR

REMOVAL & INSTALLATION

See Figure 28.

1. Before servicing the vehicle, refer to the Precautions Section.
2. Remove or disconnect the following:
   - Negative battery cable
   - Fan shroud
   - Drive belt
   - Lower alternator bracket
   - Alternator upper bolt

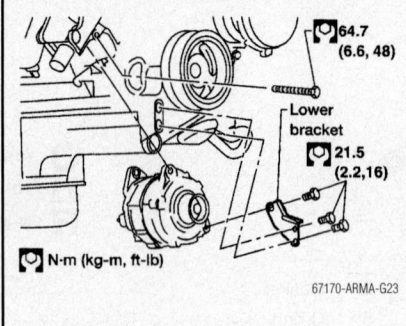

64.7 (6.6, 48)

Lower bracket

21.5 (2.2,16)

N•m (kg-m, ft-lb)

67170-ARMA-G23

Fig. 28 Alternator mounting

CHARGING SYSTEM

- Alternator harness connectors
- Alternator

To install:

3. Install or connect the following:
   - Alternator
   - Alternator harness connectors
   - Upper bolt, tighten to 48 ft. lbs. (65 Nm)
   - Lower bracket, tighten to 16 ft. lbs (22 Nm)
   - Drive belt
   - Fan shroud
   - Negative battery cable

FIRING ORDER

See Figure 29.

Firing order: 1–8–7–3–6–5–4–2

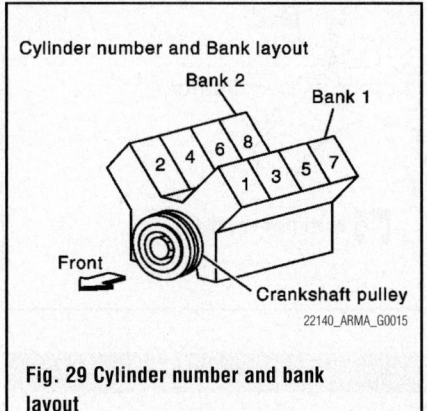

Fig. 29 Cylinder number and bank layout

IGNITION COIL

REMOVAL & INSTALLATION

See Figure 30.

1. Disconnect the negative battery cable.
2. Remove the engine room cover.
3. Disconnect the harness connector from the ignition coil.
4. Remove the ignition coil retaining bolt.
5. Remove the ignition coil.

To install:

6. Install the ignition coil, torque the retaining bolt to 80 inch lbs.
7. Connect the harness coil.
8. Connect the negative battery cable.

IGNITION TIMING

ADJUSTMENT

The ignition timing is controlled by the Powertrain Control Module (PCM). No adjustment is necessary or possible.

SPARK PLUGS

REMOVAL & INSTALLATION

See Figure 31.

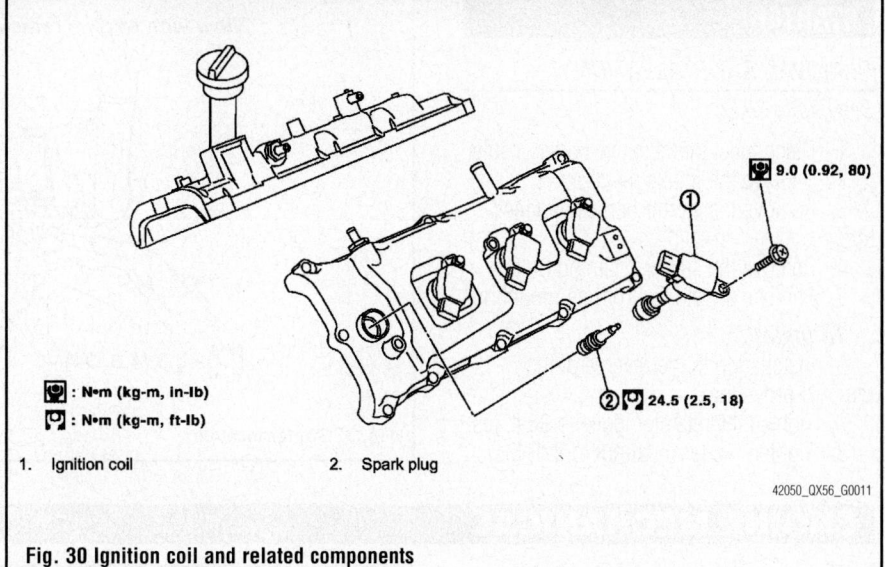

1. Ignition coil   2. Spark plug

Fig. 30 Ignition coil and related components

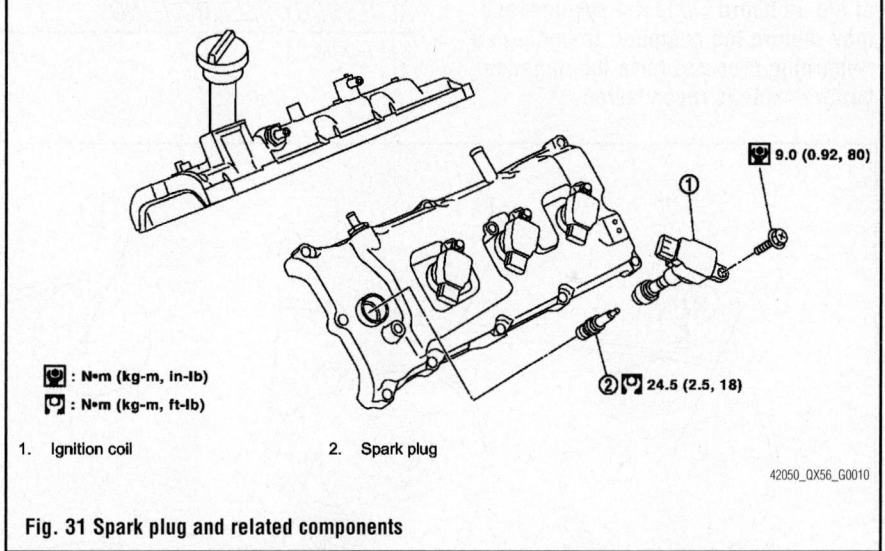

1. Ignition coil   2. Spark plug

Fig. 31 Spark plug and related components

1. Disconnect the negative battery cable.
2. Disconnect the harness connector from the ignition coil.
3. Remove the ignition coil retaining bolt.
4. Remove the ignition coil.
5. Remove the spark plug using a spark plug socket and wrench.

To install:

6. Be sure the spark plug gap is to specification (0.043 in).
7. Carefully install the spark plug and torque to specification, 18 ft. lbs.
8. Install the ignition coil, torque the retaining bolt to 80 inch lbs.
9. Connect the harness coil.
10. Connect the negative battery cable.

ENGINE ELECTRICAL

STARTING SYSTEM

STARTER

REMOVAL & INSTALLATION

See Figure 32.

1. Disconnect the negative battery cable.
2. Remove the intake manifold.
3. Remove the starter harness connectors.
4. Remove the starter retaining bolts.
5. Remove the starter from its mounting.

To install:

6. Installation is the reverse of the removal procedure.
7. Tighten the retaining bolts to 34 ft. lbs.
8. Tighten the terminal nut to 8 ft. lbs.

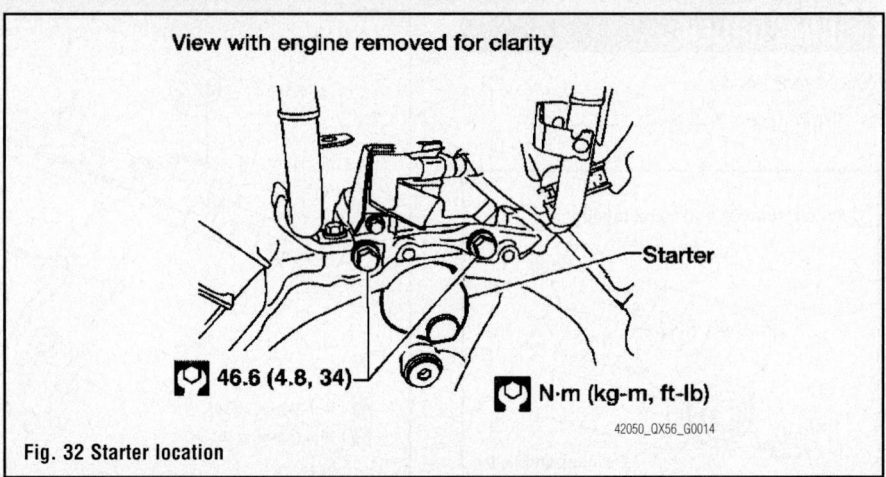

View with engine removed for clarity

Starter

46.6 (4.8, 34)   N·m (kg-m, ft-lb)

42050_QX56_G0014

Fig. 32 Starter location

ENGINE MECHANICAL

➡ Disconnecting the negative battery cable may interfere with the functions of the on board computer systems and may require the computer to undergo a relearning process, once the negative battery cable is reconnected.

ACCESSORY DRIVE BELTS

ACCESSORY BELT ROUTING

See Figure 33.

INSPECTION

Inspect the drive belt for signs of glazing or cracking. A glazed belt will be perfectly smooth from slippage, while a good belt will have a slight texture of fabric visible.

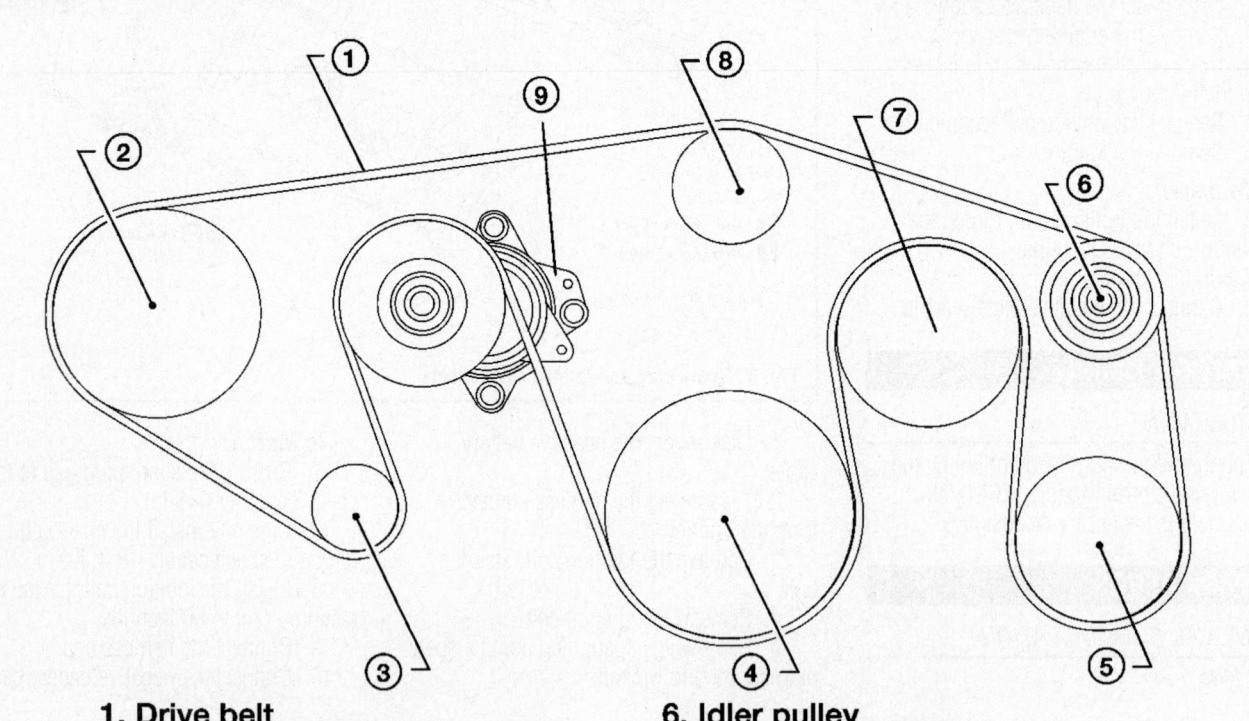

1. Drive belt
2. Power steering pump pulley
3. Generator pulley
4. Crankshaft pulley
5. A/C compressor
6. Idler pulley
7. Cooling fan pulley
8. Water pump pulley
9. Drive belt auto tensioner

22140_ARMA_G0018

Fig. 33 Accessory belt routing

Cracks will usually start at the inner edge of the belt and run outward. All worn or damaged drive belts should be replaced immediately.

ADJUSTMENT

Drive belt tension is not necessary, as it is automatically adjusted by the auto tensioner.

REMOVAL & INSTALLATION

See Figure 34.

1. Disconnect the negative battery cable.
2. Remove the air duct and resonator assembly.
3. Install special tool J-46535, or equivalent on the auto tensioner pulley bolt and move it upward.

➡ **Avoid placing your hand in a location where pinching may occur if the holding tool accidentally comes off.**

4. Remove the drive belt from the vehicle.

To install:

5. Installation is the reverse of the removal procedure.
6. Be sure that the belt is securely installed around all pulleys.
7. Rotate the crankshaft several times clockwise to equalize belt tension between the pulleys.
8. Make sure that the belt tension is within the allowable working range, using the indicator notch on the auto tensioner.

CAMSHAFT AND VALVE LIFTERS

REMOVAL & INSTALLATION

See Figures 35 through 41.

1. Before servicing the vehicle, refer to the Precautions Section.
2. Remove or disconnect the following:
   - Negative battery cable
   - Engine cover
   - Air intake assembly
   - Engine wiring harnesses on rocker cover
   - Throttle control actuator
   - Ignition coil
   - PCV hose from PCV valve

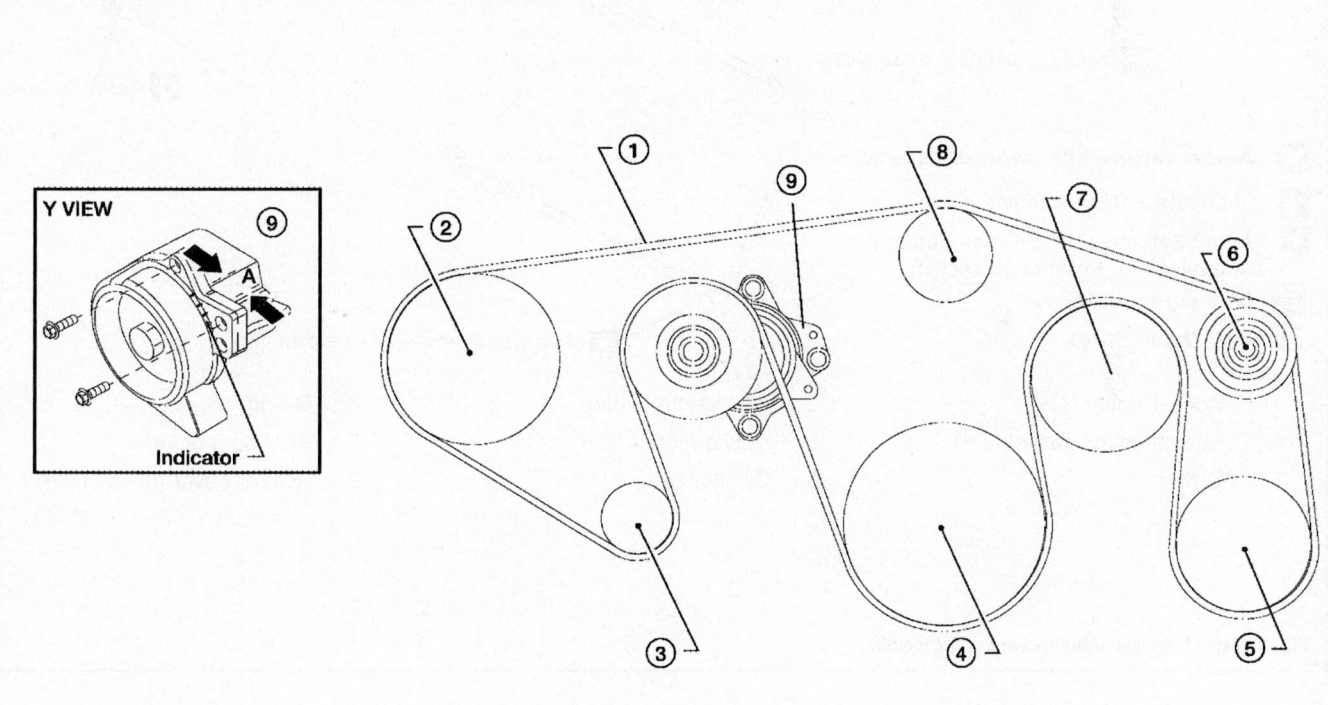

1.	Drive belt	2.	Power steering oil pump pulley	3.	Generator pulley
4.	Crankshaft pulley	5.	A/C compressor	6.	Idler pulley
7.	Cooling fan pulley	8.	Water pump pulley	9.	Drive belt tensioner

42050_QX56_G0019

Fig. 34 Drive belt tensioner indicator and related components

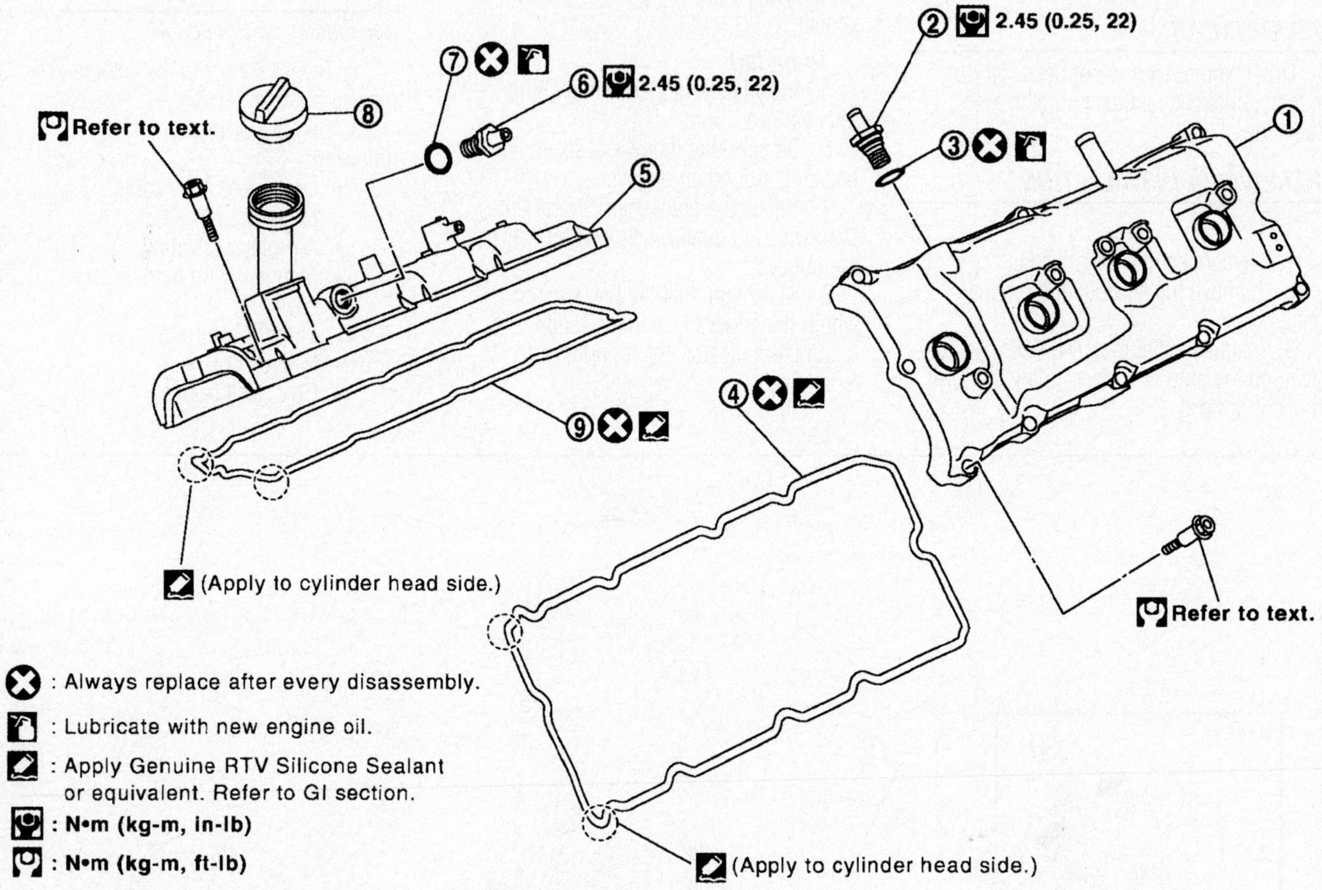

Refer to text.

⑧

⑦ ✗ 🛢

⑥ 🔧 **2.45 (0.25, 22)**

② 🔧 **2.45 (0.25, 22)**

③ ✗ 🛢

①

⑤

④ ✗ ✎

⑨ ✗ ✎

(Apply to cylinder head side.)

Refer to text.

(Apply to cylinder head side.)

✗ : Always replace after every disassembly.

🛢 : Lubricate with new engine oil.

✎ : Apply Genuine RTV Silicone Sealant or equivalent. Refer to GI section.

🔧 : N•m (kg-m, in-lb)

🔧 : N•m (kg-m, ft-lb)

1. Rocker cover (LH)	2. PCV control valve	3. O-ring
4. Rocker cover gasket (LH)	5. Rocker cover (RH)	6. PCV control valve
7. O-ring	8. Oil filler cap	9. Rocker cover gasket (RH)

09482_ARMA_G0008

Fig. 35 Exploded view of the rocker cover assembly

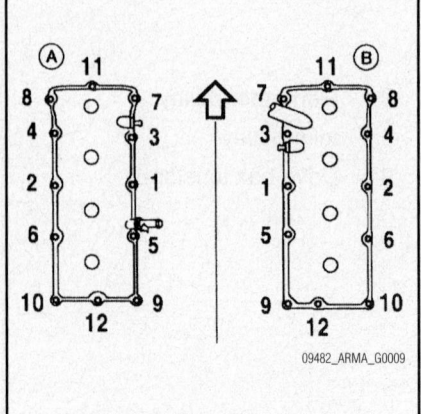

09482_ARMA_G0009

Fig. 36 Rocker cover torque sequence

Bank	INT EXH	Identification paint (front)	Identification paint (rear)	Identification rib
RH	INT	White	—	Yes.
	EXH	—	Light blue	Yes.
LH	INT	White	—	No.
	EXH	—	Light blue	No.

67162-QX56-G16

Fig. 37 Camshaft installation markings

3. Remove rocker cover, loosening the bolts in the reverse order.

4. Turn the crankshaft until the No. 1 cylinder is set at Top Dead Center (TDC).

5. Remove timing chain case covers.

6. Matchmark the timing chain, aligning with the camshaft sprocket marks.

7. Remove chain tensioner from left bank as follows:

   a. Squeeze end clips and push plunger into tensioner body.

   b. Secure plunger using stopper pin.

   c. Remove the chain tensioner.

8. Remove the chain tensioner from right bank as follows:

   a. Remove the chain tensioner cover using special tool J-37228.

   b. Squeeze end clips and push plunger into tensioner body.

   c. Secure plunger using stopper pin.

   d. Remove the chain tensioner.

9. With camshaft locked with a wrench, loosen bolts to remove camshaft sprocket.

10. Remove front cover bolts.

11. Remove camshaft brackets, removing bolts in reverse order shown in figure.

12. Remove camshaft.

13. Remove valve lifters.

➡**Matchmark the drivetrain components so each part can be reinstalled in its original position.**

To install:

14. Install valve lifters.

15. Install camshaft, refer to table for correct placement.

16. Install camshaft brackets as follows:

   a. Refer to location mark on upper surface of bracket.

   b. Installation mark should be correctly read when viewed from intake side.

17. Install camshaft bracket no. 1 as follows:

   a. Apply liquid gasket to bracket and backside of front cover as shown in figure.

   b. Carefully position and mount camshaft bracket #1.

   c. Temporarily tighten front cover bolts

18. Tighten fixing bolts for camshaft brackets as follows:

   a. Step 1: Bolts 9–12: 17 in. lbs. (1.9 Nm)

   b. Step 2: Bolts 1–8: 17 in. lbs. (1.9 Nm)

   c. Step 3: All bolts: 52 in. lbs. (5.9 Nm)

   d. Step 4: All bolts: 92 in. lbs. (10 Nm)

19. Tighten front cover bolts to 8 ft. lbs. (11 Nm)

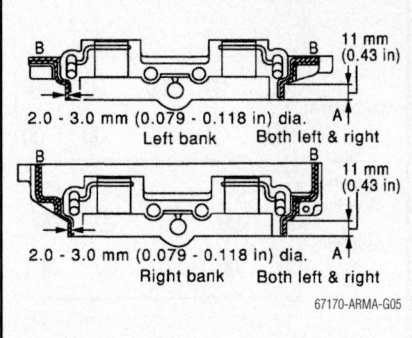

Fig. 38 Gasket application for camshaft bracket

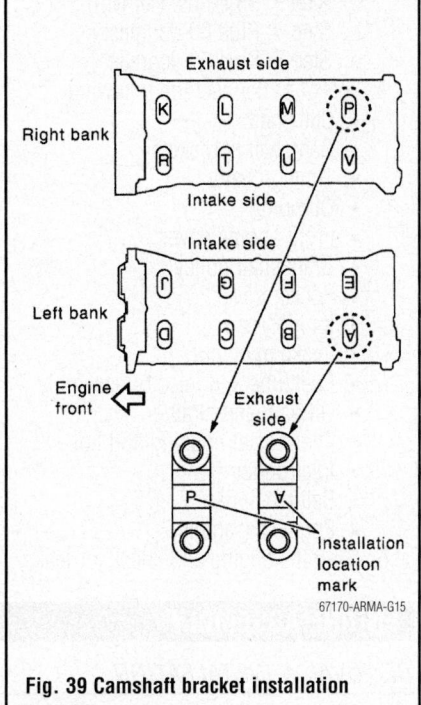

Fig. 39 Camshaft bracket installation markings

20. Install camshaft sprocket as follows:

   a. Install camshaft sprocket aligning matchmarks with timing chain. Align camshaft sprocket key groove with dowel pin on camshaft front edge.

   b. Temporarily tighten bolts.

   c. Lock the camshaft with a wrench and tighten the bolts.

21. Install chain tensioner as shown:

   a. Install chain tensioner, compress plunger and hold with stopper pin.

   b. Tighten chain tensioner bolts to 61 in. lbs. (7 Nm)

   c. Remove stopper pin, release plunger and apply tension to timing chain.

   d. Install chain tensioner front cover (Right-hand bank only) and tighten bolts to 80 in. lbs. (9 Nm).

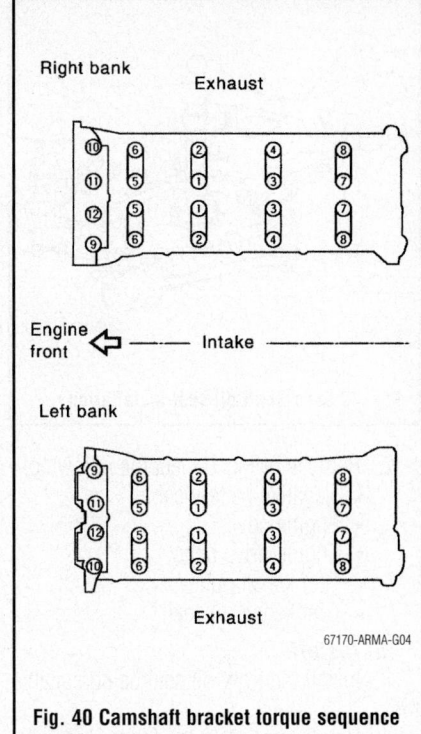

Fig. 40 Camshaft bracket torque sequence

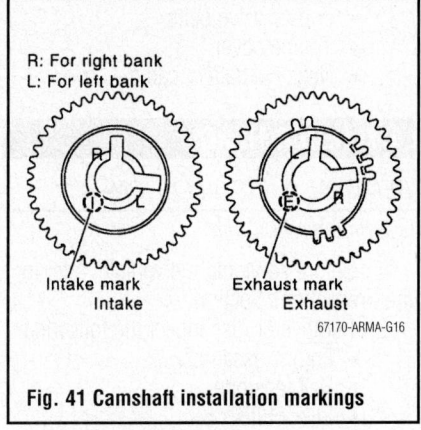

Fig. 41 Camshaft installation markings

22. Install or connect the following:
- Timing chain cover
- Rocker cover.
- PCV hose
- Ignition coil
- Throttle control actuator
- Engine wiring harnesses on rocker cover
- Air intake assembly
- Engine cover
- Negative battery cable

23. Start the engine and check for leaks.

CRANKSHAFT FRONT SEAL

REMOVAL & INSTALLATION

See Figure 42.

1. Before servicing the vehicle, refer to the Precautions Section.

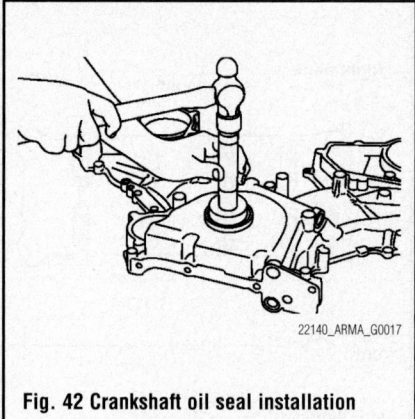

Fig. 42 Crankshaft oil seal installation

22140_ARMA_G0017

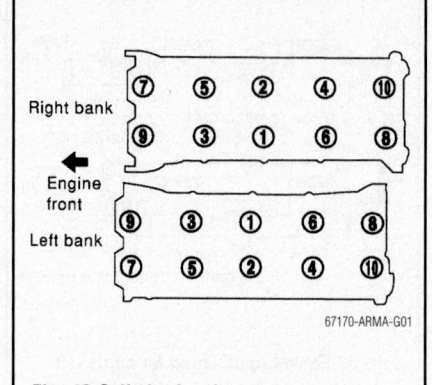

Fig. 43 Cylinder head torque sequence

67170-ARMA-G01

2. Remove or disconnect the following:
   - Negative battery cable
   - Engine cover
   - Engine drive belts
   - Crankshaft pulley
   - Crankshaft oil seal

To install:

3. Install the front oil seal using suitable tool.

4. Install or connect the following:
   - Crankshaft pulley
   - Engine drive belts
   - Engine cover
   - Negative battery cable

CYLINDER HEAD

REMOVAL & INSTALLATION

See Figures 43.

1. Before servicing the vehicle, refer to the Precautions Section.
2. Remove or disconnect the following:
   - Engine assembly
   - Belt tensioner
   - Idler pulley
   - Thermostat housing and hose
   - Oil pan and strainer
   - Fuel rail and injector assembly
   - Intake manifold
   - Ignition coil
   - Rocker cover
   - Crankshaft pulley
   - Front engine cover
   - Oil pump
   - Timing chain
   - Camshaft sprockets
   - Camshafts
   - Cylinder head, removing bolts in reverse order shown in figure

To install:

3. Install the cylinder head with a new gasket. Tighten the bolts in sequence as follows:
   a. Step 1: 72 ft. lbs. (98 Nm)
   b. Step 2: Loosen all bolts completely

c. Step 3: 33 ft. lbs. (44 Nm)
   d. Step 4: Plus 60 degrees
   e. Step 5: Plus 60 degrees

4. Install or connect the following:
   - Camshaft
   - Camshaft sprockets
   - Timing chain
   - Oil pump
   - Front engine cover
   - Crankshaft pulley
   - Rocker cover
   - Ignition coil
   - Intake manifold
   - Fuel tube and injector assembly
   - Oil pain and strainer
   - Thermostat housing and hose
   - Idler pulley
   - Belt tensioner
   - Engine assembly

5. Start the engine and check for leaks

ENGINE ASSEMBLY

REMOVAL & INSTALLATION

1. Before servicing the vehicle, refer to the Precautions Section.
2. Drain the cooling system.
3. Partially drain the automatic transmission fluid.
4. Relieve the fuel system pressure.
5. Remove or disconnect the following:
   - Hood
   - Cowl extension
   - Engine cover
   - Air intake assembly
   - Vacuum hose between vehicle and engine
   - Radiator hoses
   - Radiator
   - Drive belts
   - Engine fan
   - Wiring harness
   - ECM
   - Power steering reservoir tank and oil pump
   - A/C compressor

   - Brake booster vacuum line
   - EVAP line
   - Fuel hose
   - Heater hoses
   - Exhaust manifolds
   - Front final drive assembly
   - Automatic transmission dipstick tube assembly
   - Automatic transmission

6. Install engine slings onto the left and right cylinder heads and tighten to 33 ft. lbs. (45 Nm).
7. Attach an engine hoist to slings and lift engine out of the vehicle

To install:

8. Lower engine into the vehicle
9. Install or connect the following:
   - Automatic transmission
   - Automatic transmission dipstick tube assembly
   - Front final drive assembly
   - Exhaust manifolds
   - Heater hoses
   - Fuel hose
   - EVAP line
   - Brake booster vacuum line
   - A/C compressor
   - Power steering reservoir tank and oil pump
   - ECM
   - Wiring harness
   - Engine fan
   - Drive belts
   - Radiator and radiator hoses
   - Vacuum hose between vehicle and engine
   - Air intake assembly
   - Engine cover
   - Cowl extension
   - Hood

10. Refill the automatic transmission fluid.
11. Refill the cooling system.
12. Start the engine and check for leaks.

EXHAUST MANIFOLD

REMOVAL & INSTALLATION

See Figures 44 through 46.

1. Before servicing the vehicle, refer to the Precautions Section.
2. Drain the cooling system.
3. Remove or disconnect the following:
   - Air intake assembly
   - Engine splash guard
   - Radiator and radiator hoses
   - Accessory drive belt
4. Remove the air fuel ratio sensors as follows:
   - Engine cover
   - Wiring harness from each sensor

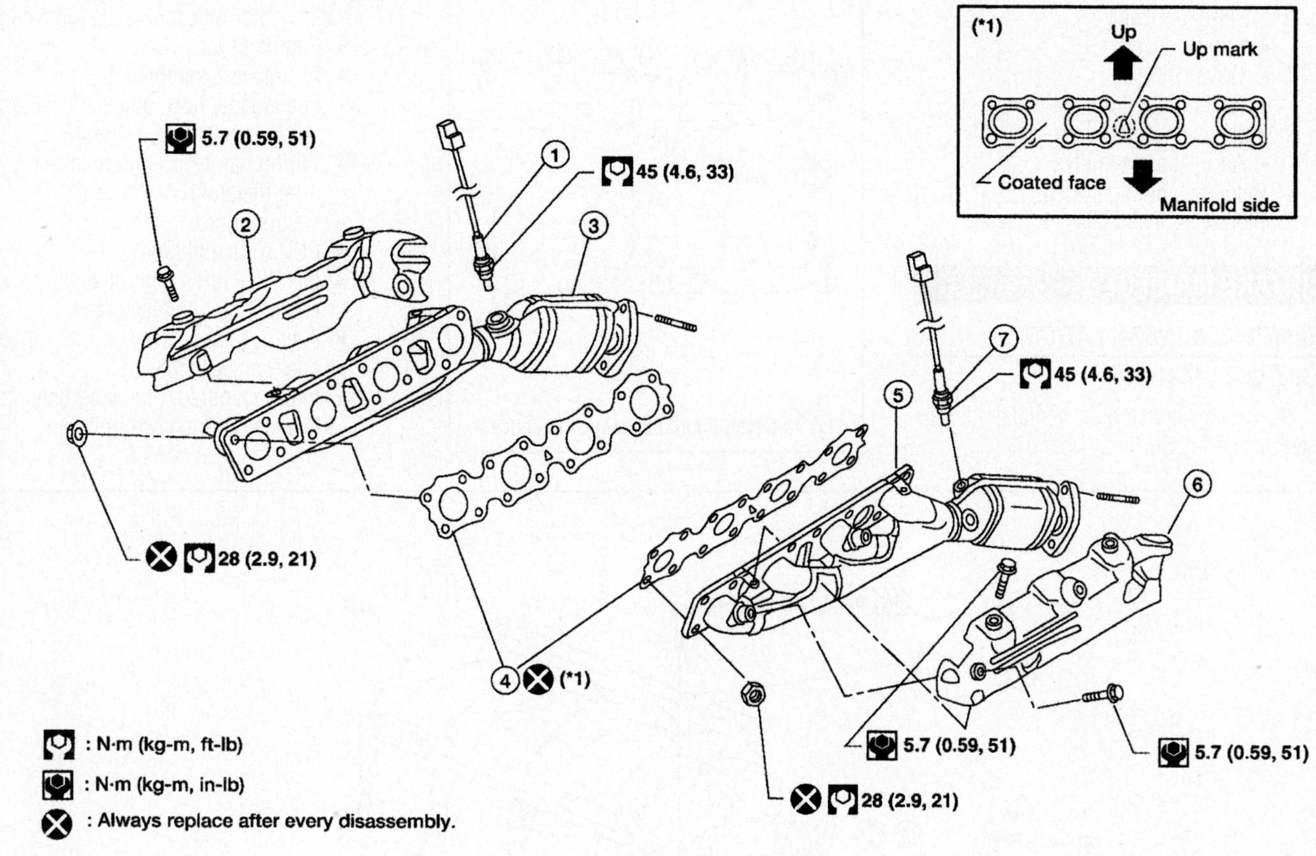

(*1)

Up — Up mark

Coated face

Manifold side

5.7 (0.59, 51)

45 (4.6, 33)

45 (4.6, 33)

28 (2.9, 21)

28 (2.9, 21)

5.7 (0.59, 51)

5.7 (0.59, 51)

(*1)

: N·m (kg-m, ft-lb)

: N·m (kg-m, in-lb)

: Always replace after every disassembly.

1. Air fuel ratio (A/F) sensor 1 (bank 2)
2. Exhaust manifold cover (right bank)
3. Exhaust manifold (right bank)
4. Gaskets
5. Exhaust manifold (left bank)
6. Exhaust manifold cover (left bank)
7. Air fuel ratio (A/F) sensor 1 (bank 1)

67170-ARMA-G30

Fig. 44 Exhaust manifolds and related components

- Sensors, using special tool J-38356
- Front cross bar
5. Remove the left exhaust manifold as follows:
    a. Remove the exhaust front tube.
    b. Remove the exhaust manifold cover.
    c. Loosen the nuts in reverse order shown in figure.
    d. Remove studs from position 2, 4, 6, and 8 and remove manifold.
6. Remove right exhaust manifold as follows:
    a. Remove the exhaust front tube.
    b. Remove the oil level gauge guide.
    c. Remove the exhaust manifold cover.
    d. Loosen the nuts in reverse order shown in figure.
    e. Remove studs from position 2, 4, 6, and 8 and remove manifold.

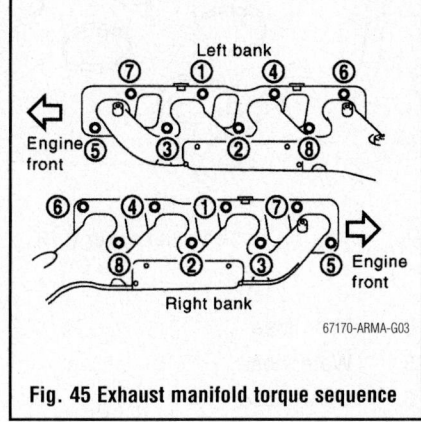

Left bank

Engine front

Right bank

Engine front

67170-ARMA-G03

Fig. 45 Exhaust manifold torque sequence

To install:

7. Install or connect the following:
    - Exhaust manifold gasket with triangle mark facing up and coated (gray) face toward exhaust manifold.

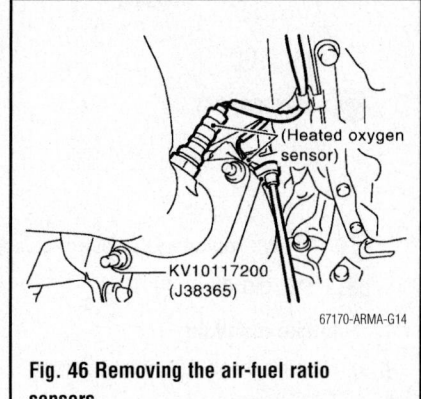

(Heated oxygen sensor)

KV10117200
(J38365)

67170-ARMA-G14

Fig. 46 Removing the air-fuel ratio sensors

- Exhaust manifold, tightening the nuts as shown in figure
- Exhaust manifold cover
- Oil level gauge guide (right side only)
- Exhaust front tube

- Front cross bar
- Air fuel ratio sensors, with anti-seize lubricant
- Engine cover
- Drive belts
- Radiator and radiator hoses
- Engine splash guard
- Air intake assembly

8. Refill the cooling system
9. Start engine and check for leaks.

INTAKE MANIFOLD

REMOVAL & INSTALLATION

See Figures 47 and 48.

1. Before servicing the vehicle, refer to the Precautions Section.

RH
Engine front
LH
67170-ARMA-G02

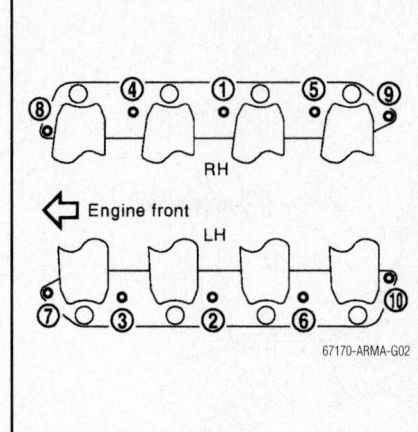

Fig. 48 Intake manifold torque sequence

2. Drain the cooling system.
3. Relieve the fuel system pressure.
4. Remove or disconnect the following:
   - Engine cover
   - Air intake assembly
   - Fuel supply hose quick connector using special tool J-45488
   - Wiring harnesses and brackets from manifold
   - Vacuum hoses
   - PCV hose and tube
   - Electric throttle control actuator, loosening bolts diagonally
   - Fuel injectors
   - Fuel rail assembly
   - Intake manifold, removing bolts in reverse order shown in figure

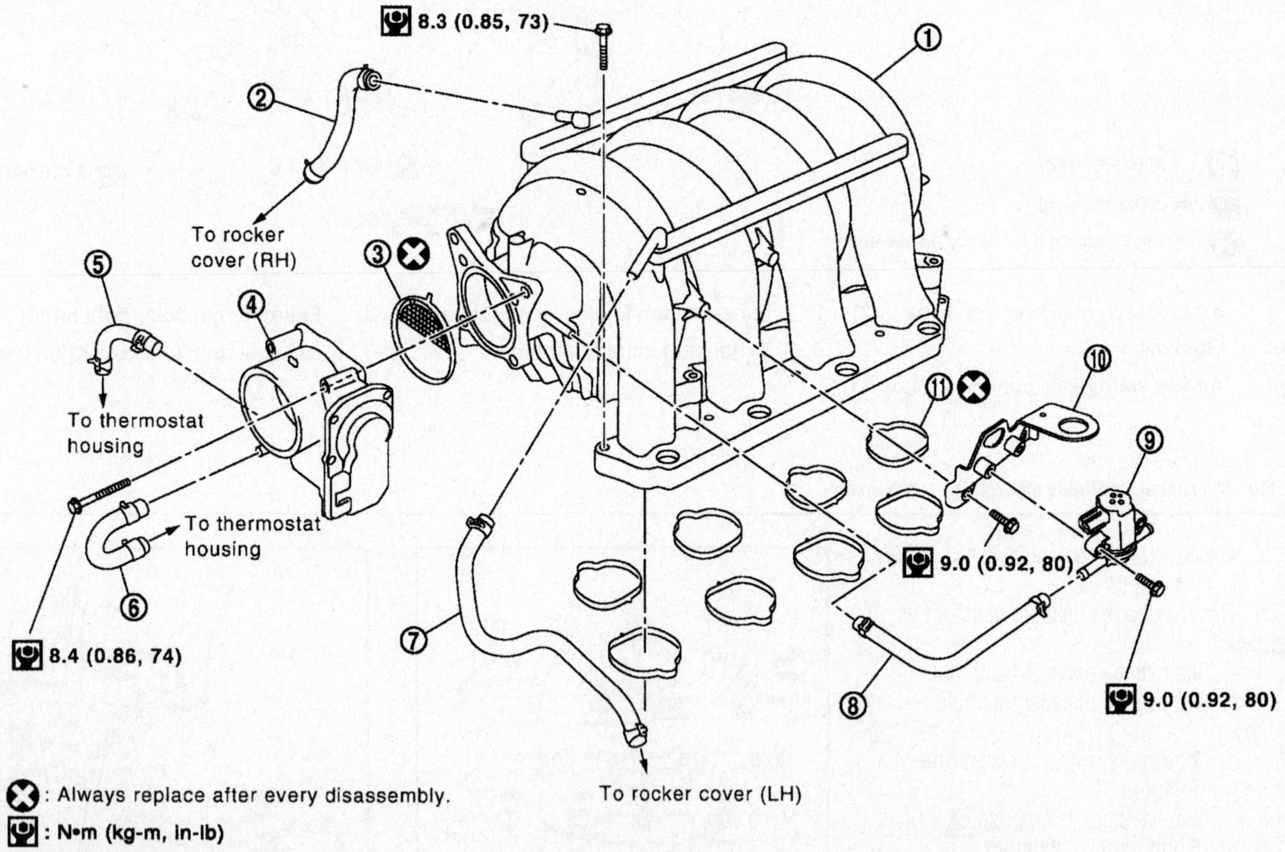

8.3 (0.85, 73)
To rocker cover (RH)
To thermostat housing
To thermostat housing
8.4 (0.86, 74)
9.0 (0.92, 80)
9.0 (0.92, 80)
To rocker cover (LH)
: Always replace after every disassembly.
: N•m (kg-m, in-lb)

1.	Intake manifold	2.	PCV hose	3.	Gasket
4.	Electric throttle control actuator	5.	Water hose	6.	Water hose
7.	PCV hose	8.	EVAP hose	9.	EVAP canister purge control solenoid valve
10.	Bracket	11.	Gasket		

67170-ARMA-G29

Fig. 47 Intake manifold and related parts

To install:

5. Install the intake manifold with new gaskets. Tighten the bolts in order as shown.

6. Install or connect the following:
- Fuel rail assembly
- Fuel injectors
- Electronic throttle control actuator, tightening the bolts in several steps
- PCV hose
- Vacuum hoses
- Wiring harnesses

7. Connect the fuel supply hose as follows:

a. Apply a thin layer of engine oil on the tube from tip end to spool end.

b. Insert tube into quick connector past the white identification mark

c. Insert tube into quick connector until top spool is completely inside the connector and 2nd level spool is exposed right below the connector.

d. Pull slightly on the quick connector to ensure it is fully engaged.

e. Install quick connector cap on quick connector joint.

8. Install or connect the following:
- Air intake assembly
- Engine cover

9. Refill the cooling system.

10. Start engine and check for leaks.

OIL PAN

REMOVAL & INSTALLATION

See Figures 49 through 51.

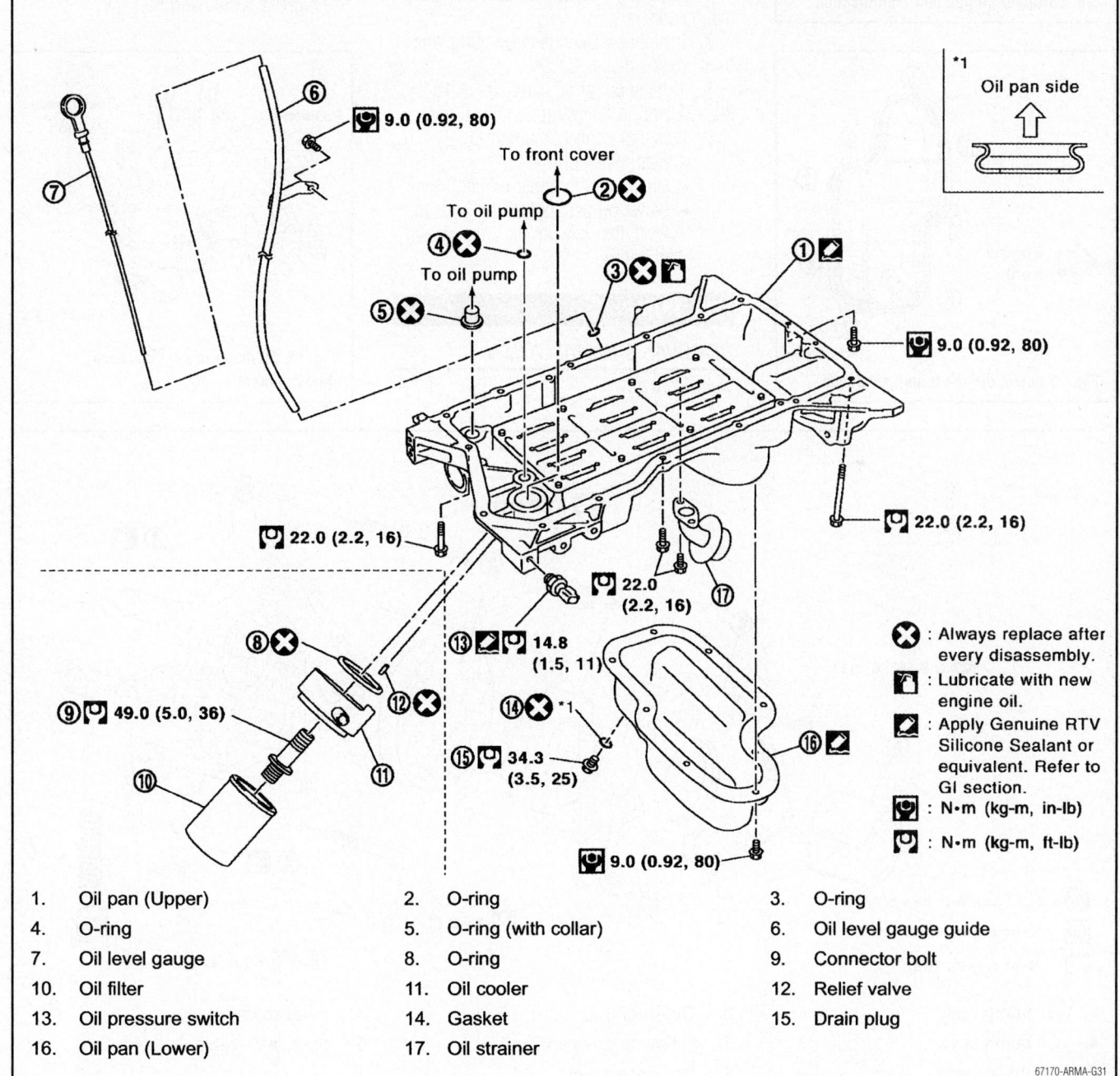

Fig. 49 Oil pan and related parts

No.	Part	No.	Part	No.	Part
1.	Oil pan (Upper)	2.	O-ring	3.	O-ring
4.	O-ring	5.	O-ring (with collar)	6.	Oil level gauge guide
7.	Oil level gauge	8.	O-ring	9.	Connector bolt
10.	Oil filter	11.	Oil cooler	12.	Relief valve
13.	Oil pressure switch	14.	Gasket	15.	Drain plug
16.	Oil pan (Lower)	17.	Oil strainer		

67170-ARMA-G31

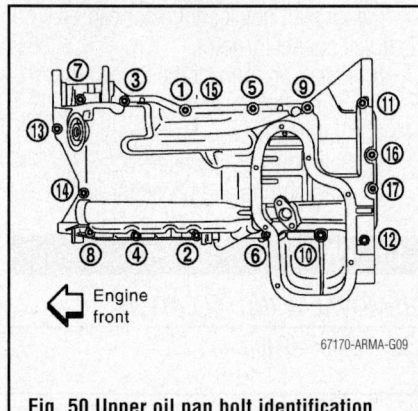

Fig. 50 Upper oil pan bolt identification

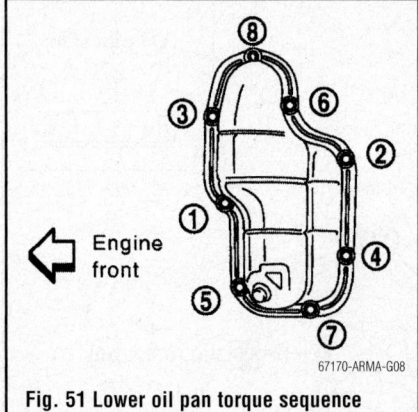

Fig. 51 Lower oil pan torque sequence

1. Before servicing the vehicle, refer to the Precautions Section.

2. Remove engine assembly.

3. Remove lower oil pan, loosening bolts in reverse order shown in figure.

4. Remove oil strainer from upper oil pan.

5. Gently pry and remove upper oil pan from engine block.

➡**Bolts are different sizes and should be kept in the correct order for reinstallation.**

To install:

6. Apply liquid gasket to upper oil pan mating surfaces.

7. Install new O-rings to oil pump and front cover side.

8. Tighten upper oil pan bolts to 16 ft. lbs. (22 Nm) in following numerical order:

9. Install or connect the following:
- Rear plate cover
- Oil strainer to upper oil pan
- Lower oil pan, tightening bolts to 25 ft. lbs. (34 Nm) in order shown in figure

OIL PUMP

REMOVAL & INSTALLATION

See Figure 52.

1. Before servicing the vehicle, refer to the Precautions Section.

2. Remove or disconnect the following:
- Timing chain cover
- Oil pump drive spacer
- Oil pump

To install:

3. Install or connect the following:
- Oil pump
- Oil pump drive spacer
- Timing chain cover

INSPECTION

See Figures 53 through 58.

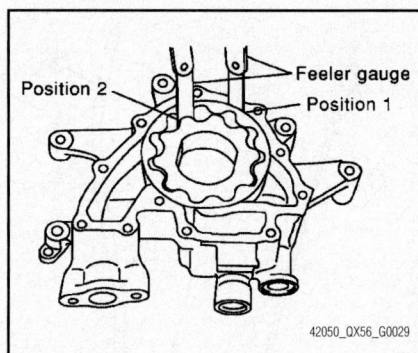

Fig. 53 Oil pump radial clearance measurement

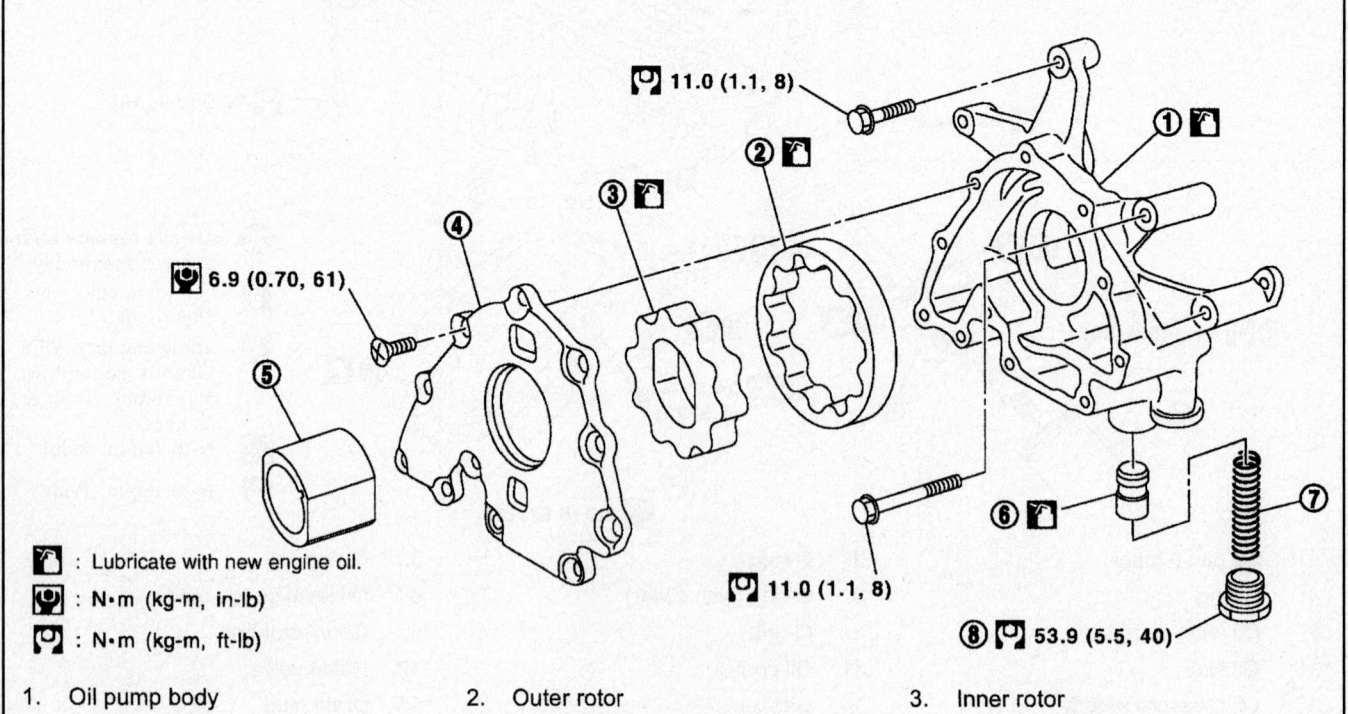

⬛ : Lubricate with new engine oil.

⬛ : N•m (kg-m, in-lb)

⬛ : N•m (kg-m, ft-lb)

1. Oil pump body
4. Oil pump cover
7. Regulator spring

2. Outer rotor
5. Oil pump drive spacer
8. Regulator plug

3. Inner rotor
6. Regulator valve

Fig. 52 Oil pump exploded view

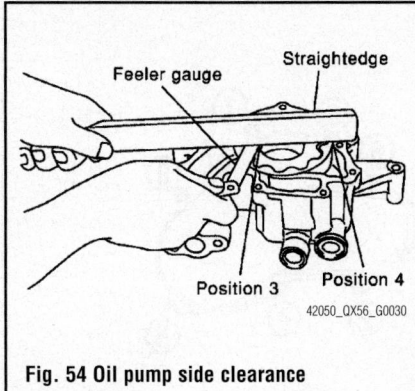

Fig. 54 Oil pump side clearance measurement

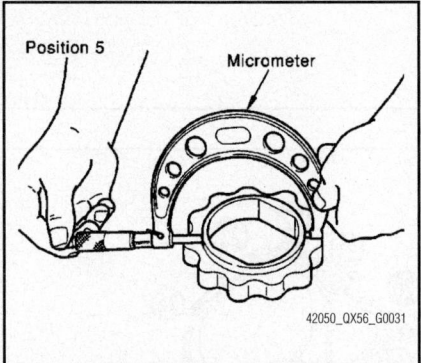

Fig. 55 Oil pump inner rotor and pump body clearance measurement

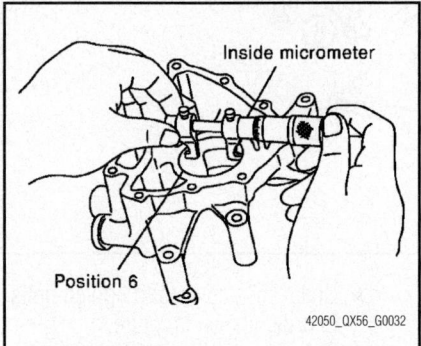

Fig. 56 Oil pump inner diameter clearance measurement

1. Remove the oil pump from the engine.

2. Remove the oil pump cover.

3. Remove the inner and outer rotors from the oil pump body.

4. Remove the regulator plug, regulator spring and regulator valve.

5. Measure the radial clearance using a feeler gauge. Body to outer rotor (position 1) should be 0.0045–0.0079 inch. Inner rotor to outer tip (position 2) should be 0.0071 inch.

6. Measure the side clearance using a feeler gauge. Body to inner rotor (position 3) should be 0.0012–0.0028 inch. Body to

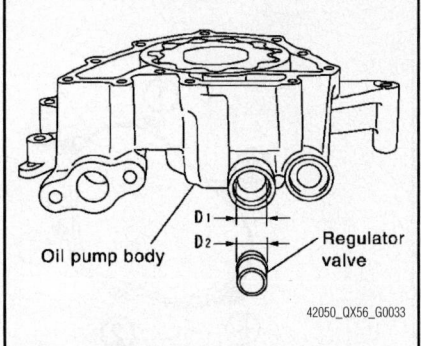

Fig. 57 Oil pump regulator valve clearance measurement

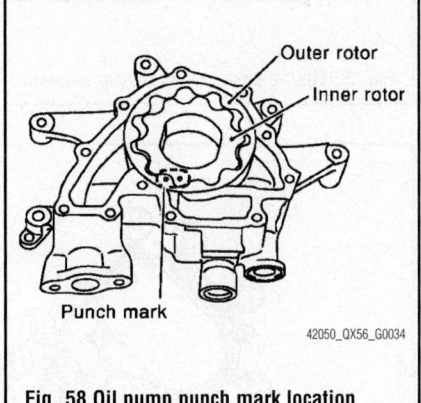

Fig. 58 Oil pump punch mark location

outer rotor (position 4) should be 0.0012–0.0035 inch.

7. Calculate the clearance between the inner rotor and the oil pump body as follows.

8. Measure the outer diameter of the protruded portion of the inner rotor (position 5) using a feeler gauge. Measure the inner diameter of the oil pump body to the brazed portion (position 6) using a feeler gauge. Calculate the clearance using the following formula. Clearance=Inner diameter of oil pump body-Outer diameter of inner rotor. Inner rotor to brazed portion of housing clearance specification is 0.0018–0.0036 inch.

9. Check the regulator valve to oil pump cover clearance using the following formula. Clearance=D1 (valve hole diameter)- D2 (outer diameter of valve). Regulator valve to oil pump specification should be 0.0016–0.0038 inch.

➡**Coat the valve with clean engine oil. Check that it falls smoothly into the regulator valve hole by its own weight.**

10. Assemble the oil pump in the reverse order.

11. Install the inner and outer rotor with the punched marks on the oil pump cover side.

PISTON AND RING

POSITIONING

See Figures 59 and 60.

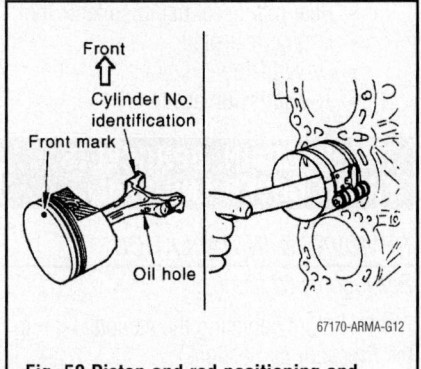

Fig. 59 Piston and rod positioning and identification

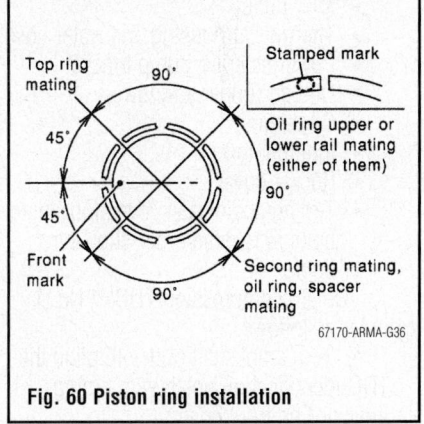

Fig. 60 Piston ring installation

REAR MAIN SEAL

REMOVAL & INSTALLATION

See Figure 61.

1. Before servicing the vehicle, refer to the Precautions Section.

2. Remove or disconnect the following:
   • Transmission assembly

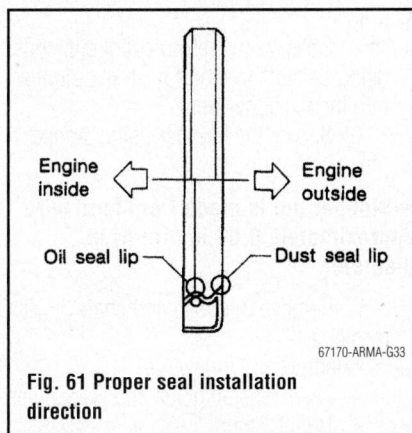

Fig. 61 Proper seal installation direction

- Drive plate
- Engine rear plate
- Rear main seal using suitable tool

To install:

3. Install or connect the following:
- Rear main seal using suitable tool
- Engine rear plate
- Drive plate
- Transmission assembly

TIMING CHAIN, SPROCKETS, FRONT COVER AND SEAL

REMOVAL & INSTALLATION

See Figures 62 and 63.

1. Before servicing the vehicle, refer to the Precautions Section.
2. Remove or disconnect the following:
- Engine assembly
- Drive belt auto tensioner
- Idler pulley
- Thermostat housing and water hose
- Power steering pump bracket
- Oil pan (upper and lower)
- Oil strainer
- Ignition coil
- Rocker cover
- Timing chain case cover, loosening bolts in reverse order shown in figure
3. Obtain compression TDC of No. 1 cylinder as follows:
   a. Turn crankshaft pulley to align the TDC identification notch with timing indicator on front cover.
   b. Ensure intake and exhaust cam lobes of No. 1 cylinder point outside.
4. Remove or disconnect the following:
- Crankshaft pulley from crankshaft using a suitable puller
- Front cover, loosening bolts in reverse order shown in figure
- Front oil seal
- Oil pump drive spacer
- Oil pump
5. Remove the timing chain tensioner as follows:
   a. Squeeze the return-proof clip ends using suitable tool and push the plunger into the tensioner body.
   b. Secure the plunger using stopper pin.

➡ **Stopper pin is made from hard wire approximately 0.04 in (1mm) in diameter.**

   c. Remove the bolts and chain tensioner.
6. Remove the following:
- Chain tension guide and slack guide
- Timing chain

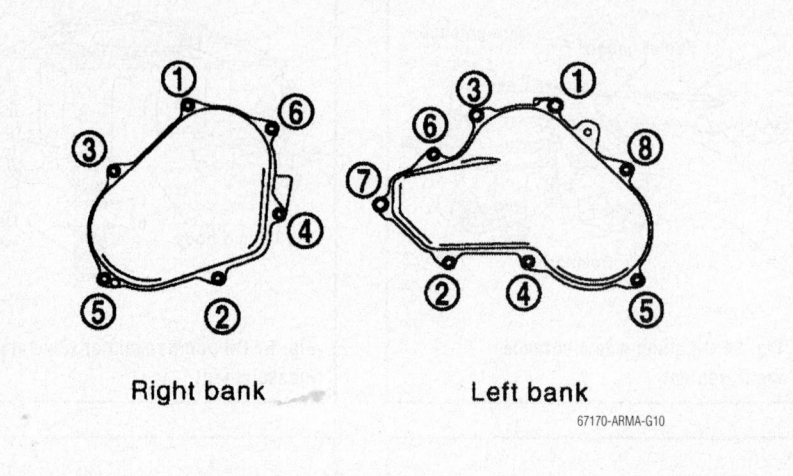

Right bank   **Left bank**

67170-ARMA-G10

Fig. 62 Timing case covers torque sequence

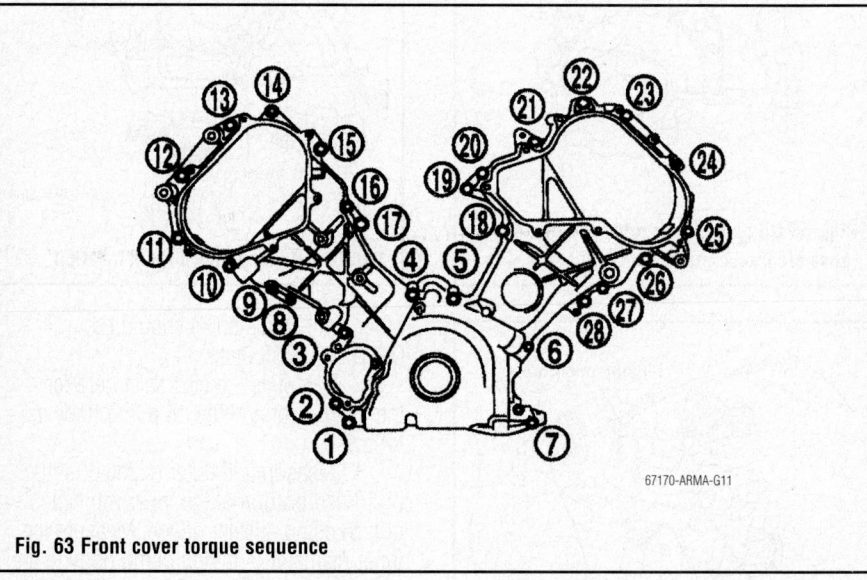

67170-ARMA-G11

Fig. 63 Front cover torque sequence

7. Using a wrench to hold the hexagon part of the camshaft, loosen the camshaft sprocket bolts.
8. Remove the camshaft sprockets.

To install:

9. Ensure that the crankshaft key and dowel pin of each camshaft are facing the same direction.
10. Install or connect the following:
- Camshaft sprockets and tighten to 112 ft. lbs. (152 Nm).
- Timing chain
- Chain tension guide and slack guide
- Oil pump
- Oil pump drive spacer
- Front oil seal, using suitable tool
- Front cover, using new O-rings and tighten bolts in order shown in figure

- Chain case cover, and tighten bolts in order shown in figure
- Crankshaft pulley and tighten bolt to 69 ft. lbs. (93 Nm) plus 90 degrees
- Ignition coil
- Oil strainer
- Lower and upper oil pan
- Power steering pump bracket
- Thermostat housing and water hose
- Idler pulley
- Drive belt auto tensioner
- Engine assembly

VALVE LASH

ADJUSTMENT

See Figures 64 and 65.

➡ **Perform the following inspection after removal, installation or replace-**

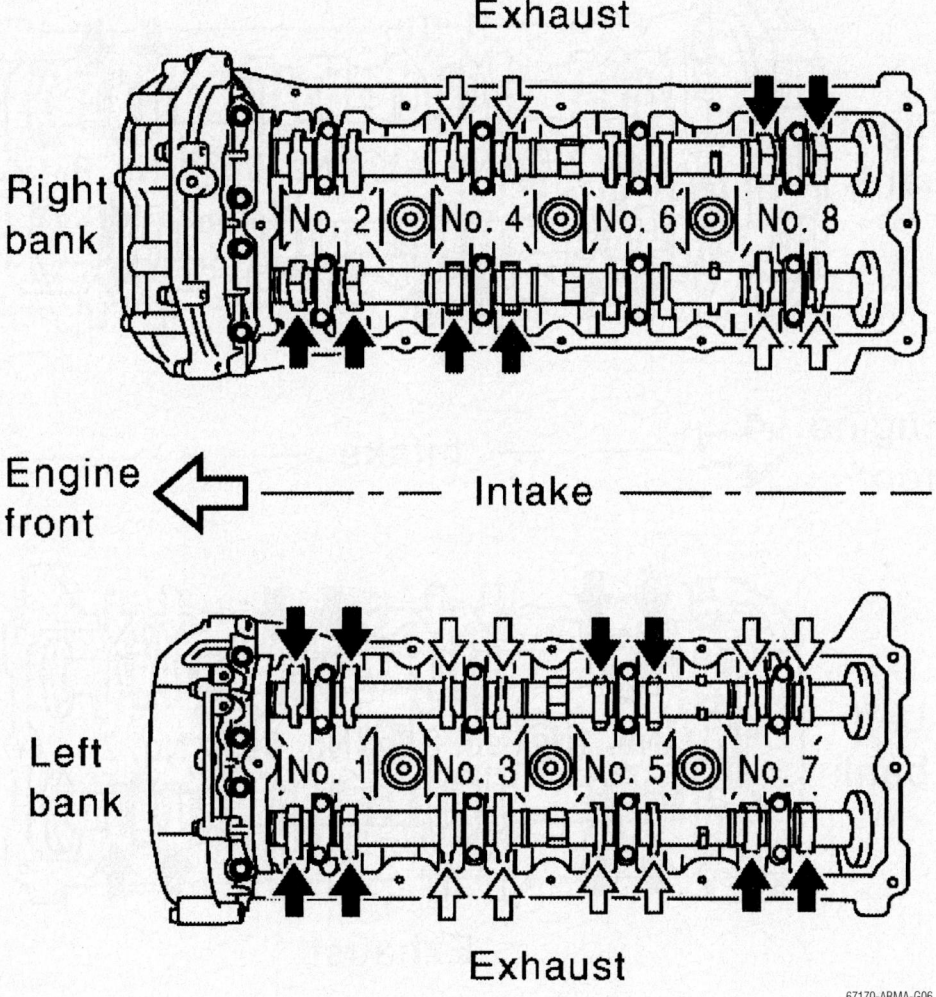

↑ : Measurable at No. 1 cylinder compression top dead center

⇧ : Measurable at No. 3 cylinder compression top dead center

Exhaust

Right bank

No. 2   No. 4   No. 6   No. 8

Engine front ⟵ — · — Intake — · —

Left bank

No. 1   No. 3   No. 5   No. 7

Exhaust

67170-ARMA-G06

Fig. 64 Locations to measure clearance with No. 1 cylinder at TDC

ment of camshaft or valve-related parts, or if there are unusual engine conditions due to changes in valve clearance over time (starting, idling, and/or noise).

1. Run engine to operating temperature.
2. Remove or disconnect the following:

- Engine cover
- Battery cover
- Air intake assembly
- Left and right rocker covers

3. Turn the crankshaft pulley clockwise to Top Dead Center (TDC) identification notch with timing indicator.

4. Ensure that both the intake and exhaust cam noses of the No. 1 cylinder face outside.

5. Measure the valve clearances at locations shown in figure.

6. Turn the crankshaft pulley clockwise 270 degrees from the position of No. 1

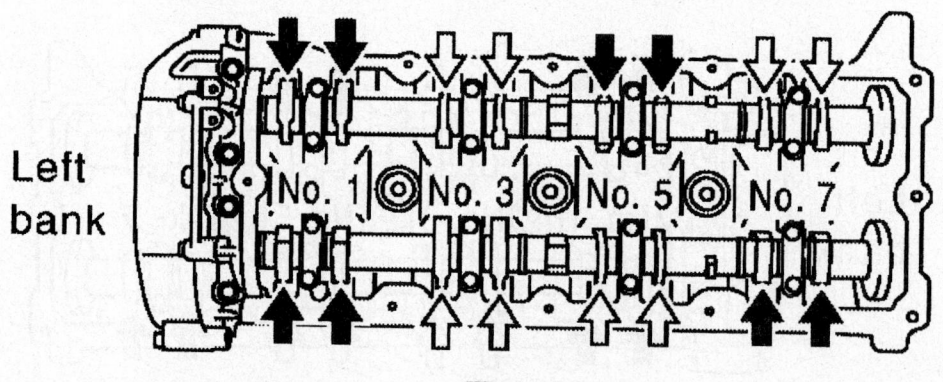

▲ : Measurable at No. 1 cylinder compression top dead center

⇧ : Measurable at No. 3 cylinder compression top dead center

Fig. 65 Locations to measure clearance with No. 3 cylinder at TDC

cylinder compression to obtain No. 3 cylinder compression TDC.

7. Measure the valve clearances at locations shown in the figure.

8. Turn crankshaft pulley clockwise 90 degrees and measure the intake and exhaust valve clearance of No. 6 cylinder and exhaust valve clearance of No. 2 cylinder.

9. To adjust the valves, remove camshaft and valve lifter(s) out of specification.

10. Install replacement valve lifter(s).

11. Install the camshaft.

12. Manually turn the crankshaft pulley several turns.

13. Recheck valve clearances with engine at operating temperature.

ENGINE PERFORMANCE & EMISSION CONTROL

COMPONENT LOCATIONS

See Figure 66.

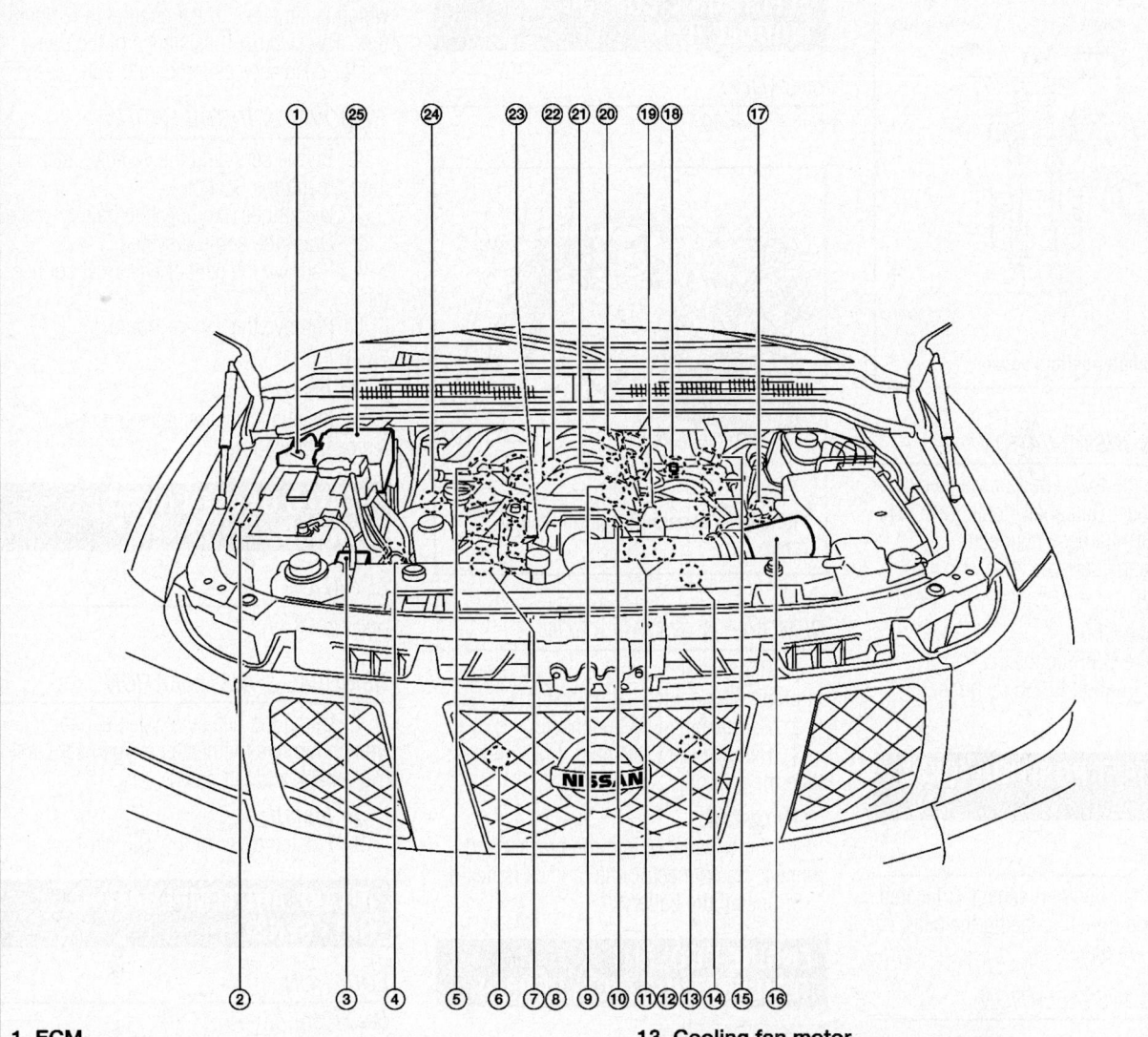

1. ECM
2. Dropping resistor (FFV models only)
3. Battery current sensor
4. Power steering pressure sensor
5. Ignition coil (with power transistor) and spark plug (bank 2)
6. Refrigerant pressure sensor
7. Intake valve timing control position sensor (bank 2)
8. Intake valve timing control solenoid valve (bank 2)
9. Engine coolant temperature sensor
10. Electric throttle control actuator
11. Intake valve timing control position sensor (bank 1)
12. Intake valve timing control solenoid valve (bank 1)
13. Cooling fan motor
14. Camshaft position sensor (PHASE)
15. Ignition coil (with power transistor) and spark plug (bank 1)
16. Mass air flow sensor (with intake air temperature sensor)
17. A/F sensor 1 (bank 1)
18. EVAP service port
19. Fuel injector (bank 1)
20. Knock sensor (bank 1)
21. EVAP canister purge volume control solenoid valve
22. Knock sensor (bank 2)
23. Fuel injector (bank 2)
24. A/F sensor 1 (bank 2)
25. IPDM E/R

22140_ARMA_G0019

Fig. 66 Component locations

CAMSHAFT POSITION (CMP) SENSOR

LOCATION

See Figure 67.

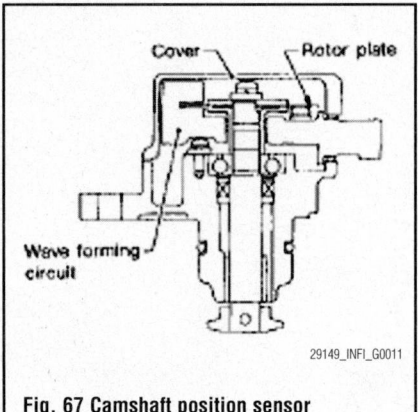

Fig. 67 Camshaft position sensor

REMOVAL & INSTALLATION

1. Loosen the fixing bolt of the sensor.
2. Disconnect Camshaft Position Sensor (CMP) (PHASE) harness connector.
3. Remove the sensor.

To install:
4. Install the CMP.
5. Install the CMP connector.
6. Tighten the bolt to 7.5 inch lbs. (10 Nm).

CRANKSHAFT POSITION (CKP) SENSOR

LOCATION

The crankshaft position sensor is located on the transaxle housing, facing the gear teeth of the drive plate.

REMOVAL & INSTALLATION

See Figure 68.

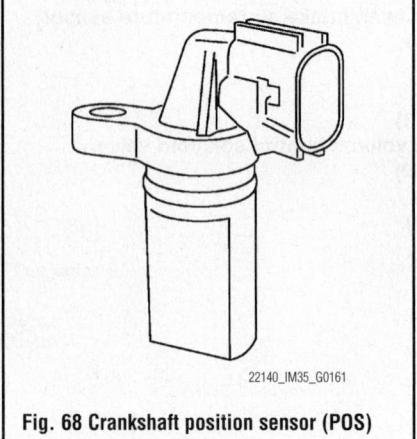

Fig. 68 Crankshaft position sensor (POS)

1. Loosen the fixing bolt of the sensor.
2. Disconnect crankshaft position sensor (POS) harness connector.
3. Remove the sensor.
4. Visually check the sensor for chipping.

ELECTRONIC CONTROL MODULE (ECM)

LOCATION

See Figure 69.

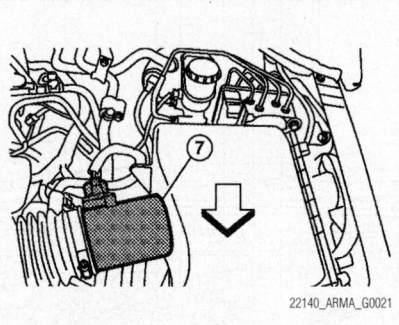

Fig. 69 The Electronic Control Module (ECM)

REMOVAL & INSTALLATION

1. Removal the negative battery cable and then the positive battery cable.
2. Remove the ECM connector.
3. Remove the ECM hold down and slide the unit out.

To install:
4. Slide the ECM into the bracket and connect the connector until a click is heard.
5. Install the battery.

ENGINE COOLANT TEMPERATURE (ECT) SENSOR

LOCATION

See Figure 70.

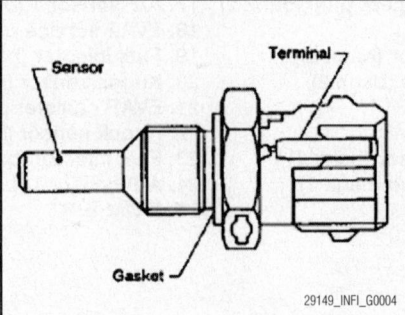

Fig. 70 The engine coolant temperature sensor

OPERATION

The engine coolant temperature sensor is used to detect the engine coolant temperature. The sensor modifies a voltage signal from the ECM. The modified signal returns to the ECM as the engine coolant temperature input. The sensor uses a thermistor, which is sensitive to the change in temperature. The electrical resistance of the thermistor decreases as temperature increases.

REMOVAL & INSTALLATION

1. Before servicing the vehicle, refer to the Precautions Section.
2. Disconnect the negative battery cable.
3. Drain the cooling system.
4. Remove the sensor electrical connector.
5. Remove the sensor from its mounting.

To install:
6. Installation is the reverse of the removal procedure.

HEATED OXYGEN (HO2S) SENSOR

LOCATION

See Figure 71.

REMOVAL & INSTALLATION

1. Using the heated oxygen sensor wrench remove the heated oxygen sensor 1 (right bank and left bank).

To install:
2. Tighten to 37 ft. lbs. (50 Nm).

INTAKE AIR TEMPERATURE (IAT) SENSOR

LOCATION

See Figure 72.

The intake air temperature sensor is mounted to the intake air duct.

REMOVAL & INSTALLATION

See Mass Air Flow (MAF) sensor removal and installation.

KNOCK SENSOR (KS)

LOCATION

The knock sensor is attached to the cylinder block.

REMOVAL & INSTALLATION

1. Remove the intake manifold
2. Remove the knock sensor connector.

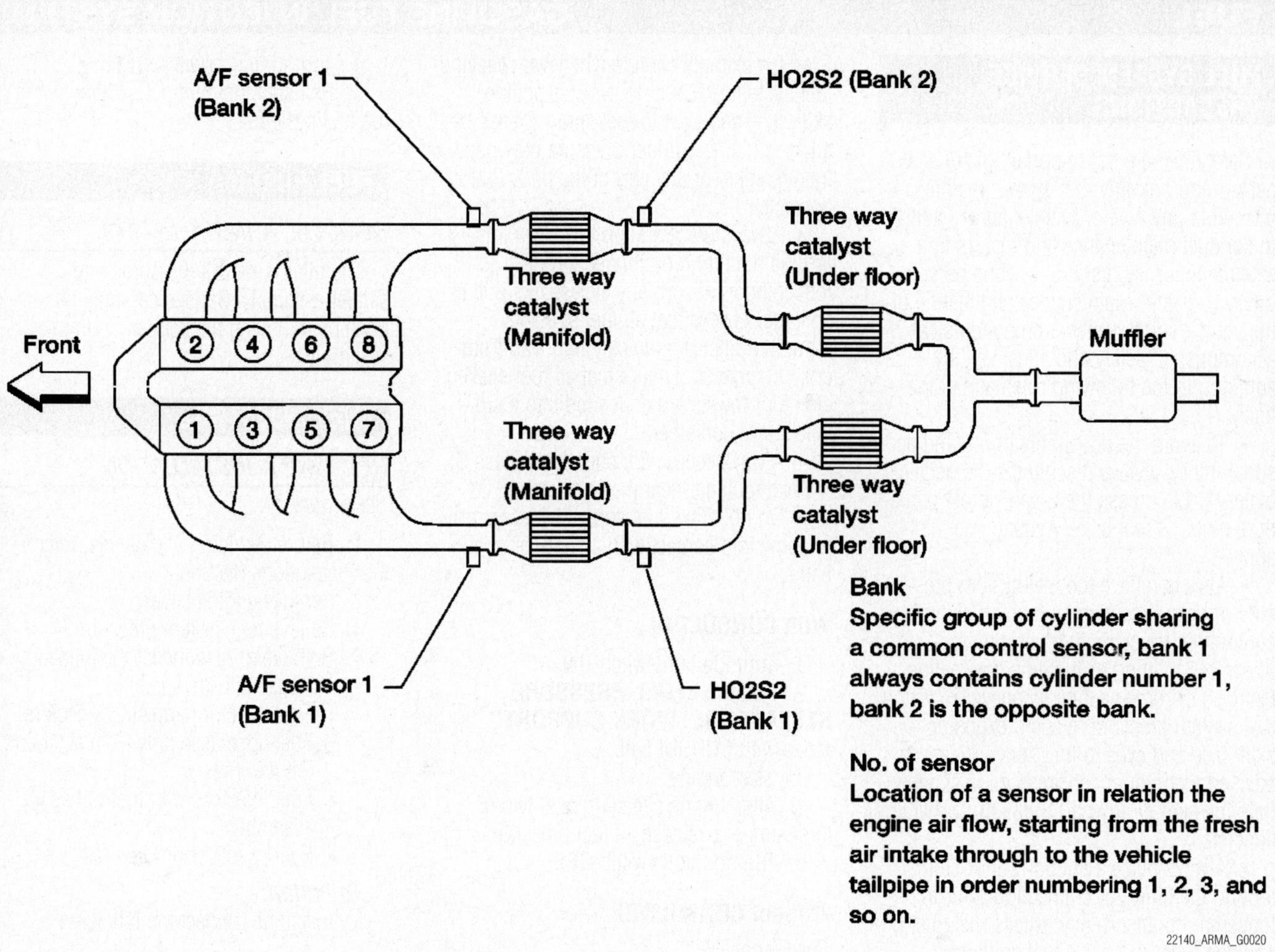

A/F sensor 1
(Bank 2)

HO2S2 (Bank 2)

Three way
catalyst
(Manifold)

Three way
catalyst
(Under floor)

Front

② ④ ⑥ ⑧

① ③ ⑤ ⑦

Muffler

Three way
catalyst
(Manifold)

Three way
catalyst
(Under floor)

A/F sensor 1
(Bank 1)

HO2S2
(Bank 1)

Bank
Specific group of cylinder sharing
a common control sensor, bank 1
always contains cylinder number 1,
bank 2 is the opposite bank.

No. of sensor
Location of a sensor in relation the
engine air flow, starting from the fresh
air intake through to the vehicle
tailpipe in order numbering 1, 2, 3, and
so on.

22140_ARMA_G0020

Fig. 71 Heated oxygen (HO2S) sensor locations

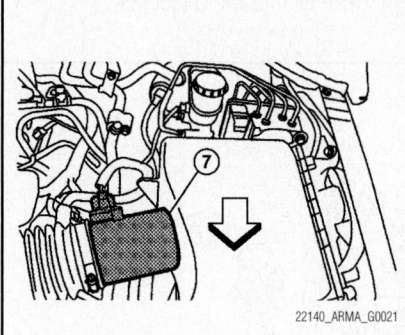

22140_ARMA_G0021

**Fig. 72 Mass Air Flow /Intake Air Temper-
ature (MAF/IAT) Sensor**

To install:

3. Install knock sensor so that connector
faces front of the engine.

4. After installing knock sensor, connect
harness connector, and lay it out to rear of
the engine.

➡**Make sure that knock sensor does
not interfere with other parts.**

5. Tighten the knock sensor to 20 ft. lbs.
(27 Nm).

MASS AIR FLOW (MAF) SENSOR

LOCATION

The mass air flow sensor is located in
the intake air duct.

REMOVAL & INSTALLATION

1. Remove engine room cover (RH
and LH).
2. Remove air duct (inlet).
3. Disconnect mass air flow sensor harness connector.
4. Disconnect PCV hose.
5. Remove air cleaner case/mass air
flow sensor assembly and air duct/air hose
disconnecting their joints.

6. Install in the reverse order of
removal.

THROTTLE POSITION SENSOR (TPS)

LOCATION

The Throttle Position Sensor (TPS) is
located on the throttle housing.

REMOVAL & INSTALLATION

1. Disconnect electric throttle control
actuator harness connector.
2. Remove rubber duct clamp.
3. Remove mounting screws.

VEHICLE SPEED SENSOR (VSS)

LOCATION

The vehicle speed sensor is with ABS
actuator electric unit (control unit)

FUEL   **GASOLINE FUEL INJECTION SYSTEM**

FUEL SYSTEM SERVICE PRECAUTIONS

Safety is the most important factor when performing not only fuel system maintenance but any type of maintenance. Failure to conduct maintenance and repairs in a safe manner may result in serious personal injury or death. Maintenance and testing of the vehicle's fuel system components can be accomplished safely and effectively by adhering to the following rules and guidelines.

• To avoid the possibility of fire and personal injury, always disconnect the negative battery cable unless the repair or test procedure requires that battery voltage be applied.

• Always relieve the fuel system pressure prior to disconnecting any fuel system component (injector, fuel rail, pressure regulator, etc.), fitting or fuel line connection. Exercise extreme caution whenever relieving fuel system pressure to avoid exposing skin, face and eyes to fuel spray. Please be advised that fuel under pressure may penetrate the skin or any part of the body that it contacts.

• Always place a shop towel or cloth around the fitting or connection prior to loosening to absorb any excess fuel due to spillage. Ensure that all fuel spillage (should it occur) is quickly removed from engine surfaces. Ensure that all fuel soaked cloths or towels are deposited into a suitable waste container.

• Always keep a dry chemical (Class B) fire extinguisher near the work area.

• Do not allow fuel spray or fuel vapors to come into contact with a spark or open flame.

• Always use a back-up wrench when loosening and tightening fuel line connection fittings. This will prevent unnecessary stress and torsion to fuel line piping.

• Always replace worn fuel fitting O-rings with new Do not substitute fuel hose or equivalent where fuel pipe is installed.

Before servicing the vehicle, make sure to also refer to the precautions in the beginning of this section as well.

RELIEVING FUEL SYSTEM PRESSURE

• Always relieve the fuel system pressure prior to disconnecting any fuel system component (injector, fuel rail, pressure regulator, etc.), fitting or fuel line connection.

Exercise extreme caution whenever relieving fuel system pressure to avoid exposing skin, face and eyes to fuel spray. Please be advised that fuel under pressure may penetrate the skin or any part of the body that it contacts.

• Always place a shop towel or cloth around the fitting or connection prior to loosening to absorb any excess fuel due to spillage. Ensure that all fuel spillage (should it occur) is quickly removed from engine surfaces. Ensure that all fuel soaked cloths or towels are deposited into a suitable waste container.

• Always keep a dry chemical (Class B) fire extinguisher near the work area.

• Do not allow fuel spray or fuel vapors to come into contact with a spark or open flame.

With CONSULT-III

1. Turn ignition switch **ON**.
2. Perform **"FUEL PRESSURE RELEASE"** in **"WORK SUPPORT"** mode with CONSULT-III.
3. Start engine.
4. After engine stalls, crank it two or three times to release all fuel pressure.
5. Turn ignition switch **OFF**.

Without CONSULT-III

See Figure 73.

1. Remove fuel pump fuse (1) located in IPDM E/R (2).
2. Start engine.
3. After engine stalls, crank it two or three times to release all fuel pressure.

4. Turn ignition switch **OFF**.
5. Reinstall fuel pump fuse after servicing fuel system.

FUEL FILTER

REMOVAL & INSTALLATION

The filter is part of the pump and sender assembly. Service can only be done by removing the sender - pump assembly.

FUEL INJECTORS

REMOVAL & INSTALLATION

See Figure 74.

1. Before servicing the vehicle, refer to the Precautions Section.
2. Remove engine cover
3. Relieve fuel system pressure.
4. Remove or disconnect the following:
   • Negative battery cable
   • Fuel injector harness connectors
   • Fuel hose assembly from right and left fuel rails
   • Fuel injectors with fuel rail as an assembly
   • Fuel injector from fuel rail

To install:

5. Install or connect the following:

➡**Always use a new O-ring when reinstalling the fuel injector to the fuel rail.**

   • New clip onto the fuel injector
   • Fuel injector to fuel rail

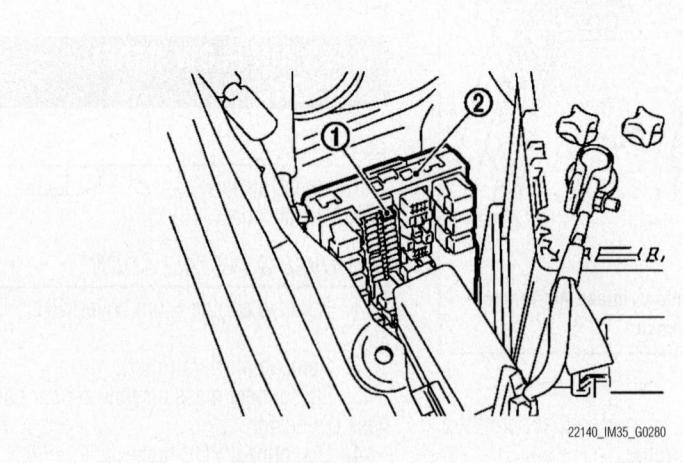

22140_IM35_G0280

Fig. 73 Fuel pump fuse (1) located in IPDM E/R (2)

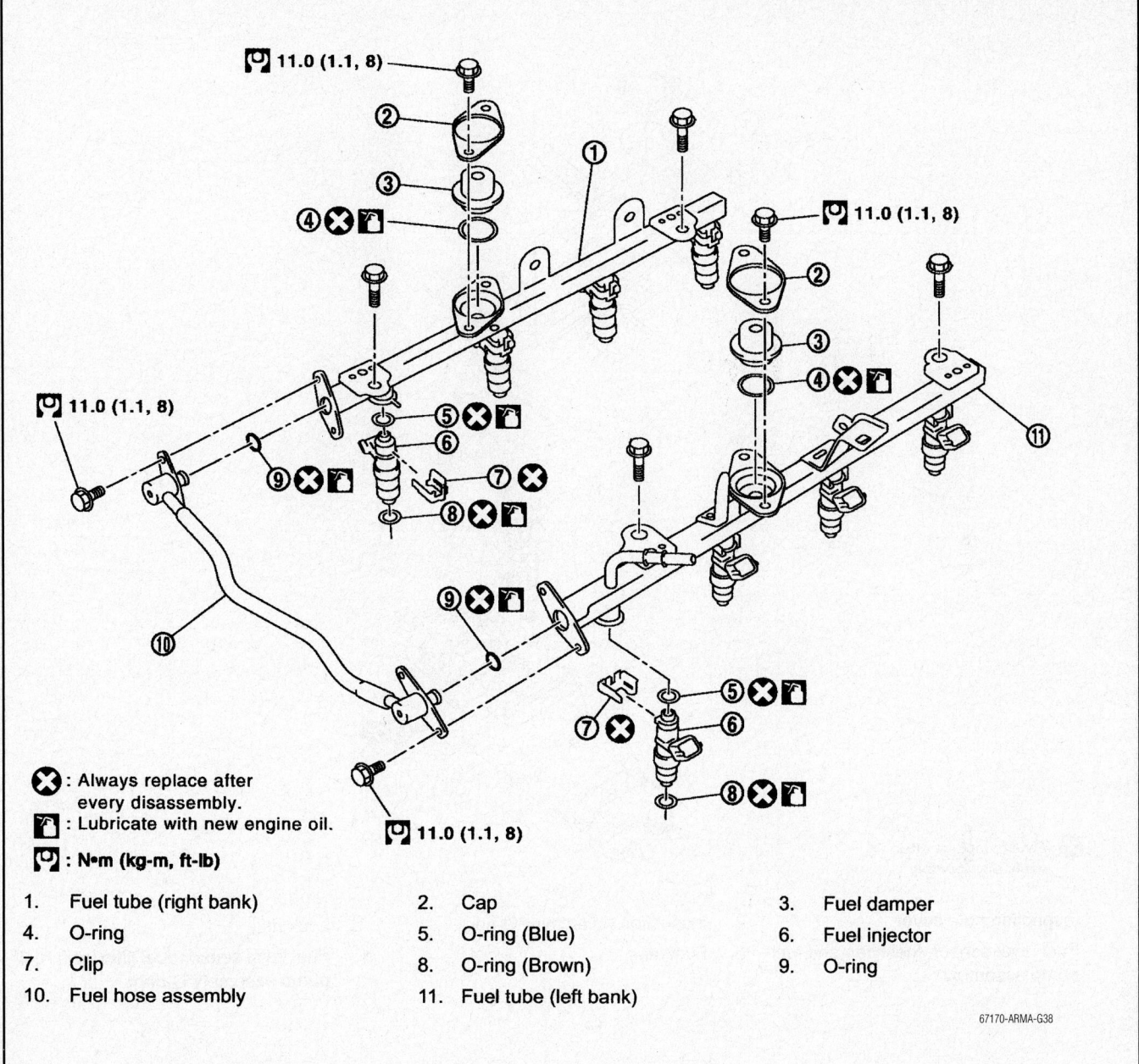

⊗ : Always replace after every disassembly.

🛢 : Lubricate with new engine oil.

🔧 : N•m (kg-m, ft-lb)

1. Fuel tube (right bank)
2. Cap
3. Fuel damper
4. O-ring
5. O-ring (Blue)
6. Fuel injector
7. Clip
8. O-ring (Brown)
9. O-ring
10. Fuel hose assembly
11. Fuel tube (left bank)

67170-ARMA-G38

Fig. 74 Fuel injectors and related parts

- Fuel injectors and fuel rail as an assembly to the intake manifold. Tighten the bolts to 8 ft. lbs. (11 Nm).
- Fuel hose assembly
- Fuel injector harness connectors
- Negative battery cable
- Engine cover
6. Start engine and check for leaks.

FUEL PUMP

REMOVAL & INSTALLATION

See Figures 75 and 76.

➡**The fuel filter is part of the fuel pump assembly.**

1. Before servicing the vehicle, refer to the Precautions Section.
2. Relieve the fuel system pressure.
3. Remove fuel filler cap to release pressure from inside tank.
4. Remove left hand rear inner fender liner.
5. Disconnect fuel filler hose from fuel filler pipe.
6. Drain fuel tank through the fuel filler hose using a suitable hose.
7. Remove or disconnect the following:

- Second row left hand seat
- Third row seat
- Second and third row seat belt buckles mounted on floor
- Left hand center pillar trim
- Left hand rear trim panel
- Left hand rear side door kick plate and weather stripping
- Second row rear center console and base, if equipped
- Inspection hole cover under carpet by turning retainers 90 degrees
- Electrical connectors
- EVAP hose

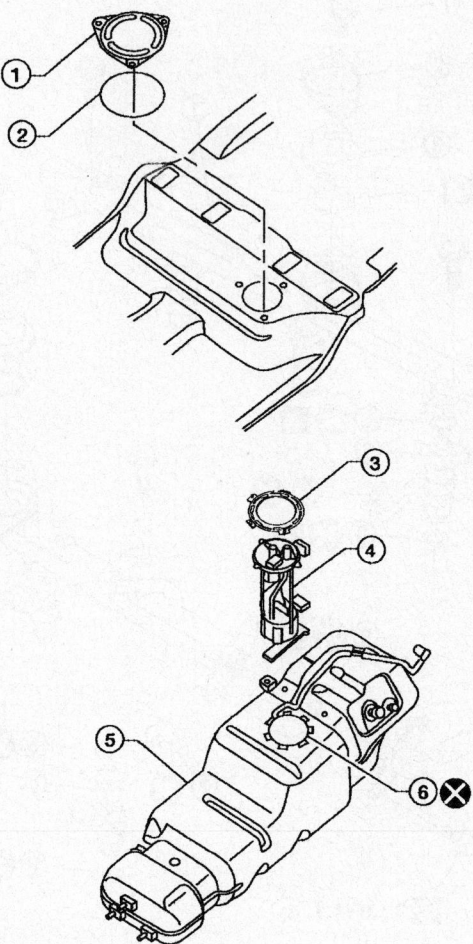

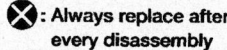

⊗ : Always replace after every disassembly

1. Inspection hole cover
2. Inspection hole cover O-ring
3. Lock ring
4. Fuel level sensor, fuel filter, and fuel pump assembly
5. Fuel tank
6. Fuel level sensor, fuel filter, and fuel pump assembly O-ring

67170-ARMA-G37

Fig. 75 Fuel pump and related parts

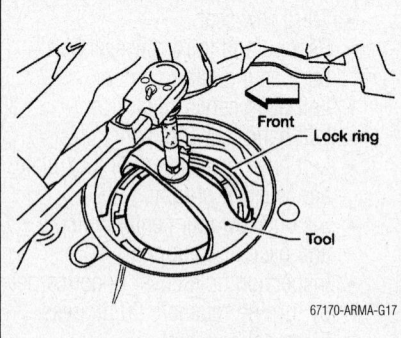

67170-ARMA-G17

Fig. 76 Removing fuel assembly lock ring

- Fuel supply hose
- Lock ring using special tool J-46214
- Fuel level sensor
- Fuel filter
- Fuel pump assembly, as required

To install:
8. Install or connect the following:
  - Fuel pump assembly as required
  - Fuel filter
  - Fuel level sensor
  - Lock ring using special tool J-46214
  - Fuel supply hose
  - EVAP hose
  - Electrical connectors

- Inspection hole cover
- Second row rear center console and base, if equipped
- Left hand rear side door kick plate and weather stripping
- Left hand rear trim panel
- Left hand center pillar trim
- Second and third row seat belt buckles
- Third row seat
- Second row left hand seat
- Fuel filler hose to fuel filler pipe
- Left hand rear inner fender liner
9. Start the engine and check for leaks.

FUEL TANK

REMOVAL & INSTALLATION

See Figure 77.

1. Before servicing the vehicle, refer to the Precautions Section.
2. Relieve the fuel system pressure.
3. Open the fuel filler lid.

4. Open the filler cap and release the pressure inside the fuel tank.
5. Remove the LH rear wheel and tire.

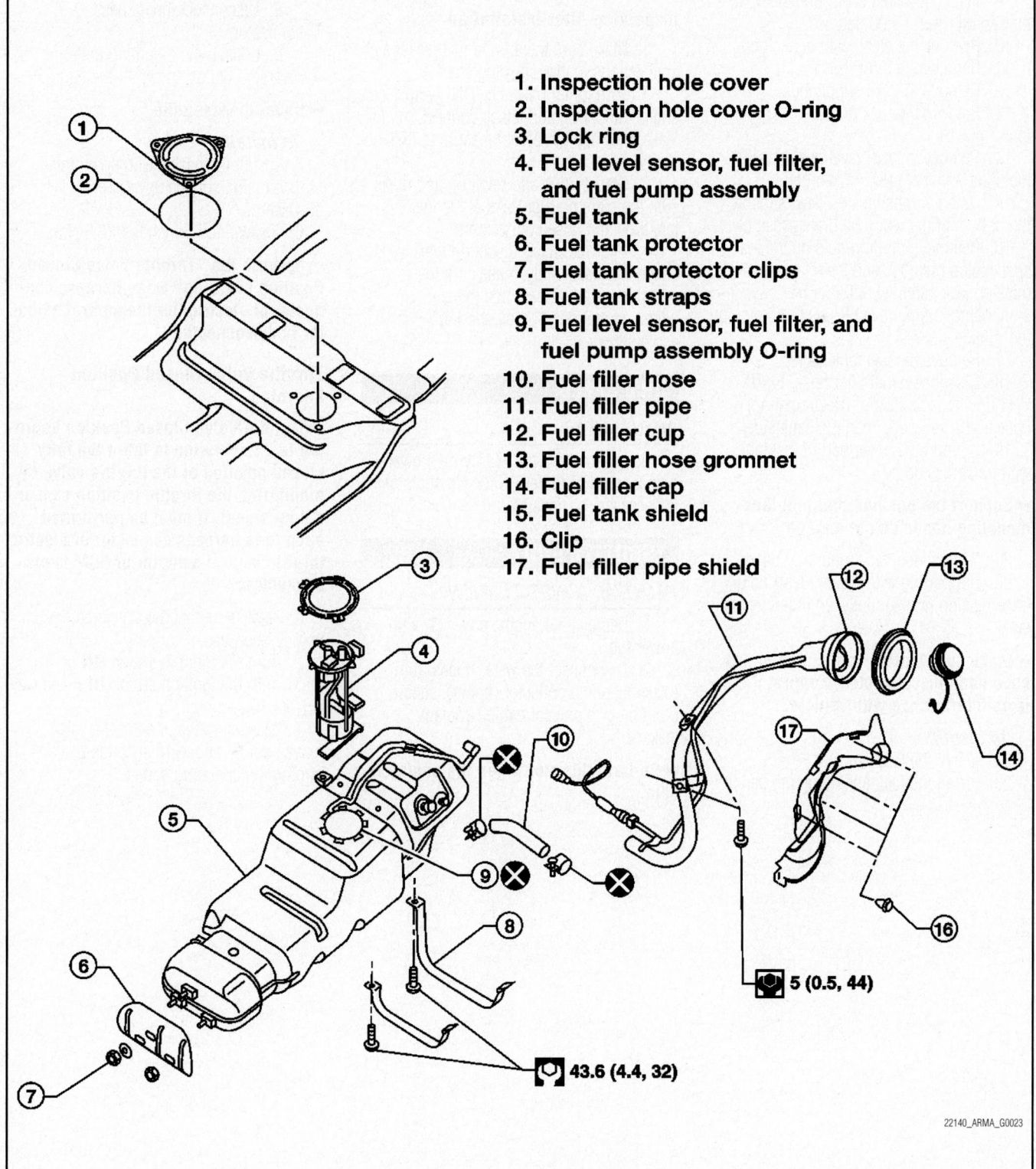

1. Inspection hole cover
2. Inspection hole cover O-ring
3. Lock ring
4. Fuel level sensor, fuel filter, and fuel pump assembly
5. Fuel tank
6. Fuel tank protector
7. Fuel tank protector clips
8. Fuel tank straps
9. Fuel level sensor, fuel filter, and fuel pump assembly O-ring
10. Fuel filler hose
11. Fuel filler pipe
12. Fuel filler cup
13. Fuel filler hose grommet
14. Fuel filler cap
15. Fuel tank shield
16. Clip
17. Fuel filler pipe shield

5 (0.5, 44)

43.6 (4.4, 32)

22140_ARMA_G0023

Fig. 77 Fuel tank and components

6. Remove the four clips and remove the fuel filler pipe shield.

7. Left hand rear side door kick plate and weather stripping.

8. Second row rear center console and base, if equipped.

9. Inspection hole cover under carpet by turning retainers 90 degrees.

10. Remove the electrical connectors.

11. Remove the EVAP hose.

12. Remove the fuel supply hose

13. Remove the lock ring using special tool J-46214

14. Remove the fuel level sensor unit, fuel filter and fuel pump assembly.

15. Using a transmission jack, support the bottom of the rear final drive assembly.

16. Remove mounting nuts on the rear suspension member, and lower the transmission jack carefully until the rear suspension member is removed from stud bolts on the vehicle.

17. Remove the fuel tank protector.

18. Disconnect fuel filler hose, EVAP/Vent line hose and EVAP (Recirculation) hose at the other end than fuel tank side.

19. Support the lower part of fuel tank with transmission jack.

➡ **Support the position that fuel tank mounting bands do not engage.**

20. Remove fuel tank mounting bands.

21. Supporting with hands, descend transmission jack carefully, and remove fuel tank.

➡ **Make sure that all connection points have been disconnected, confirm there is no interference with vehicle.**

To install:

22. Install the fuel tank.

23. Install the fuel tank mounting bands.

24. Tighten the fuel tank mounting bands to 36 ft. lbs. (37 Nm).

25. Install the fuel level sensor unit, fuel filter and fuel pump assembly.

26. Add ¼ tank of fuel for checks outlined below.

Inspection After Installation

1. Make sure there is no fuel leakage at connections in the following steps.

2. Turn ignition switch "ON" (with engine stopped), and check connections for leakage by applying fuel pressure to fuel piping.

3. Start engine and rev it up and make sure there are no fuel leaks at the fuel system tube and hose connections.

4. After removing/installing rear suspension assembly, make sure to adjust wheel alignment and then, adjust neutral position of steering angle sensor.

IDLE SPEED

ADJUSTMENT

Idle speed is maintained by the Powertrain Control Module (PCM). No adjustment is necessary or possible.

THROTTLE BODY

REMOVAL & INSTALLATION

1. Remove the engine cover (1) with power tool

2. Disconnect the water hoses from intake manifold collector (upper), attach blind plug to prevent engine coolant leakage.

➡ **Perform this step when the engine is cold.**

➡ **Never spill the engine coolant on drive belts.**

3. Remove the air cleaner case and air duct.

4. Remove the electric throttle control actuator as follows:

   a. Disconnect the harness connector.

   b. Loosen the mounting bolts in reverse order as shown in the figure.

➡ **Never disassemble.**

To install:

5. Install the gasket with positioning no-protrusion surface upward or downward.

6. Tighten to 75 inch lbs. (8.5 Nm).

➡ **Perform the "Throttle Valve Closed Position Learning" when harness connector of electric throttle control actuator is disconnected**

Throttle Valve Closed Position Learning

➡ **Throttle Valve Closed Position Learning is an operation to learn the fully closed position of the throttle valve by monitoring the throttle position sensor output signal. It must be performed each time harness connector of electric throttle control actuator or ECM is disconnected.**

1. Make sure that the accelerator pedal is fully released.

2. Turn the ignition switch **ON**.

3. Turn the ignition switch **OFF** and wait at least 10 seconds.

4. Make sure that the throttle valve moves during above 10 seconds by confirming the operating sound.

HEATING & AIR CONDITIONING SYSTEM

BLOWER MOTOR

REMOVAL & INSTALLATION

See Figure 78.

➡ **Before servicing, or working around, the SRS system, turn the ignition switch OFF, disconnect both battery cables and wait at least three minutes. When servicing, or working around, the SRS system do not work directly in front of the air bag module.**

1. Before servicing the vehicle, refer to the Precautions Section.
2. Disconnect the negative battery cable.
3. Remove the glove box assembly.
4. Disconnect the front blower motor electrical connector.

5. Remove the blower retaining screws.
6. Remove the blower motor from its mounting.

To install:

7. Installation is the reverse of the removal procedure.

HEATER CORE

REMOVAL & INSTALLATION

See Figures 79 through 82.

➡ **Before servicing, or working around, the SRS system, turn the ignition switch OFF, disconnect both battery cables and wait at least three minutes. When servicing, or working around, the SRS system do not work directly in front of the air bag module.**

1. Before servicing the vehicle, refer to the Precautions Section.
2. Discharge the refrigerant from the A/C system.
3. Drain the cooling system.
4. Remove or disconnect the following:
   - Negative battery cable
   - Wiper arms
   - Cowl top seal
   - Cowl top extension brackets
   - Wiper motor and connecting rod linkage
   - Windshield washer hose
   - Cowl top extension
   - Exhaust system
   - Front heater hoses from front heater core
   - Pressure pipes from the front expansion valve

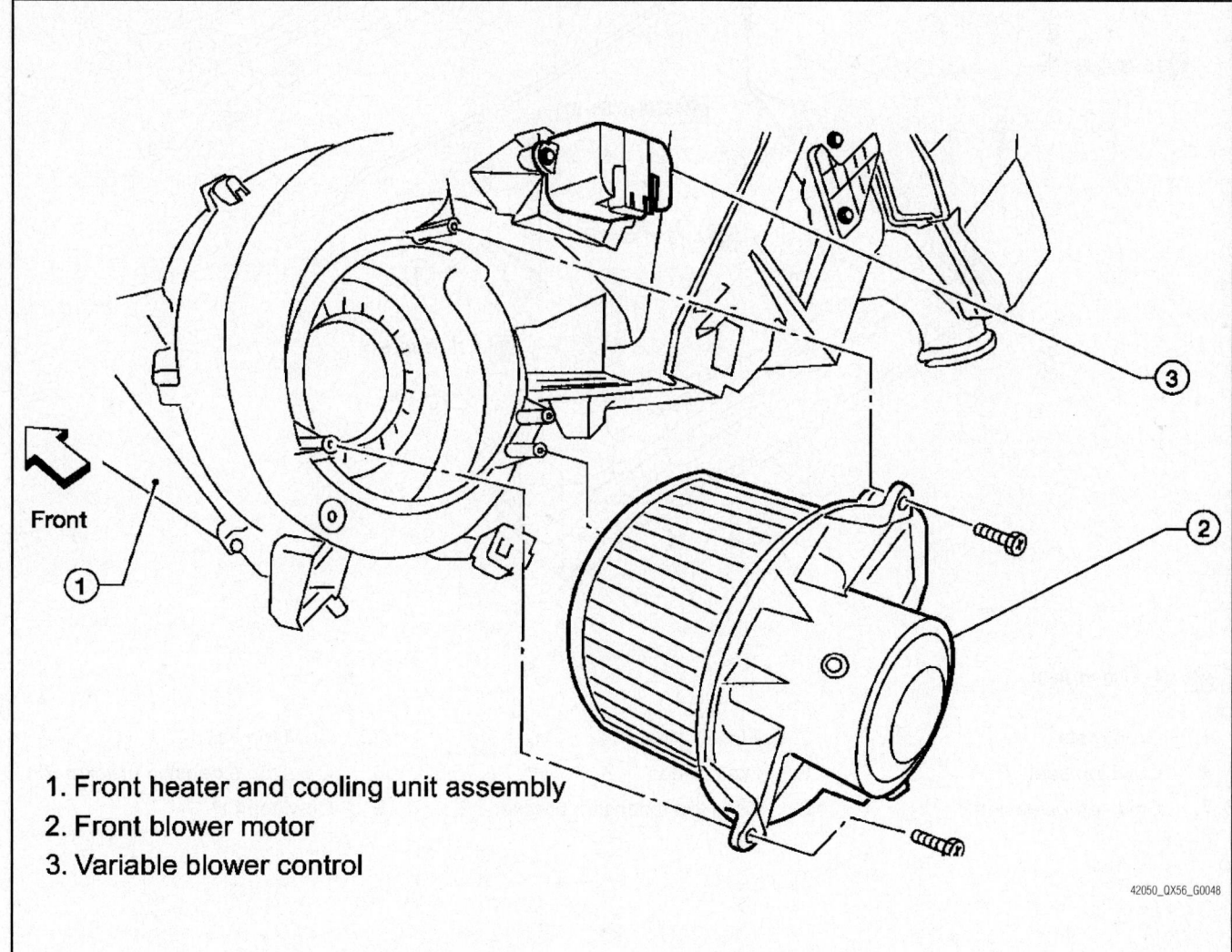

1. Front heater and cooling unit assembly
2. Front blower motor
3. Variable blower control

42050_QX56_G0048

Fig. 78 Front blower motor and related components

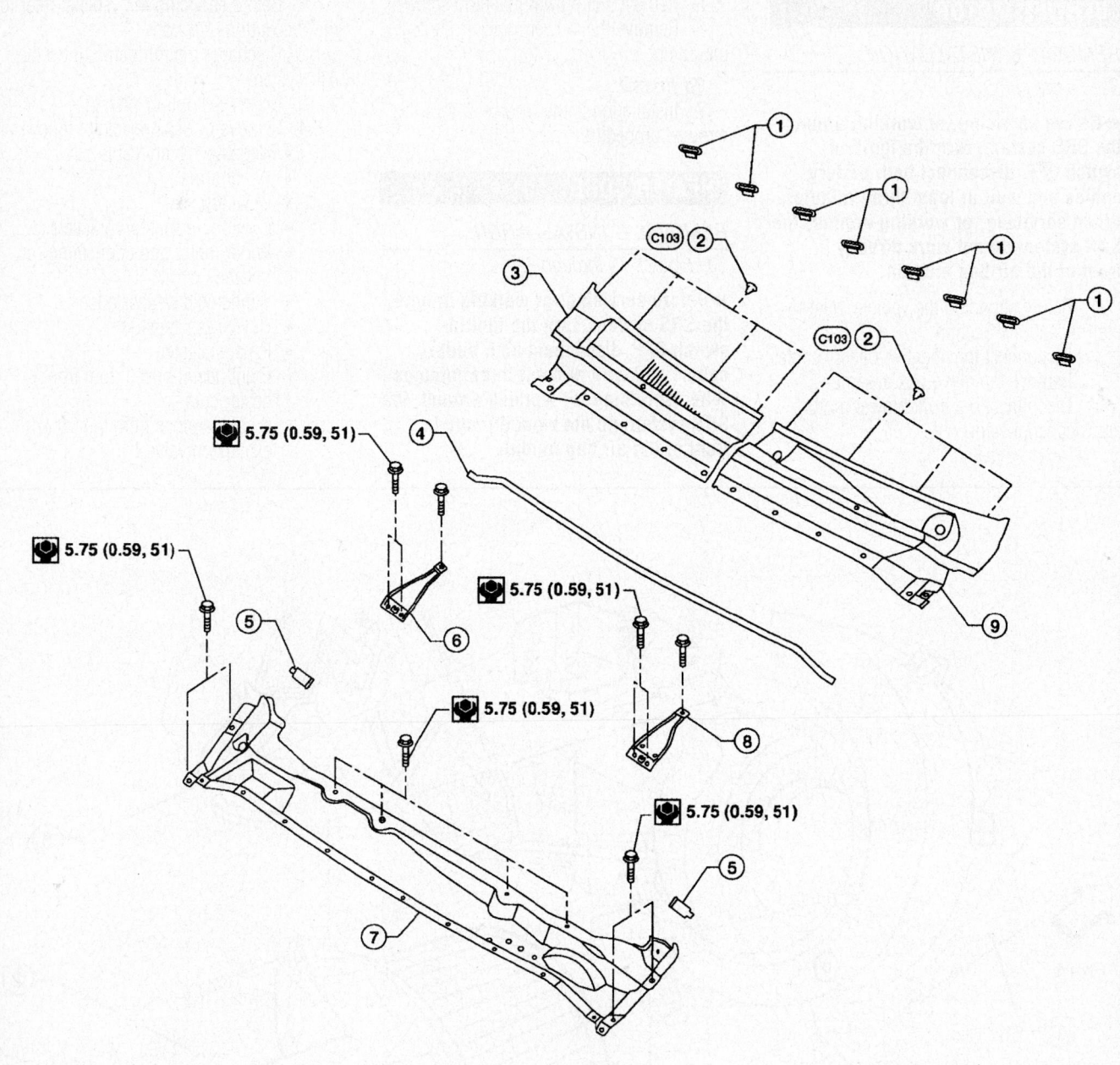

5.75 (0.59, 51)

: N·m (kg-m, in-lb)

1. Grommets
2. Plastic clips
3. Cowl top RH
4. Cowl top seal
5. Drain tubes
6. Cowl top extension bracket RH
7. Cowl top extension
8. Cowl top extension bracket LH
9. Cowl top LH

09482_ARMA_G0004

Fig. 79 Exploded view of the cowl top

- Right-hand lower instrument panel
- Transmission shifter
- Center console
- Steering column
- Left-hand lower instrument panel
- Gauge cluster front and top cover
- Instrument gauge cluster
- Defroster grille
- Optical sensor

- Side ventilator assemblies
- A-pillar trim
- Instrument panel electrical connections
- Instrument panel

5. Disconnect the steering member from each side of the vehicle body.

6. Remove the front heater and cooling unit assembly with it attached to the steering member.

7. Remove the heater unit assembly from the steering member.

8. Remove the upper bracket from the heater assembly.

9. Remove the heater core cover.

10. Remove the heater core.

11. Installation is the reverse orders of removal.

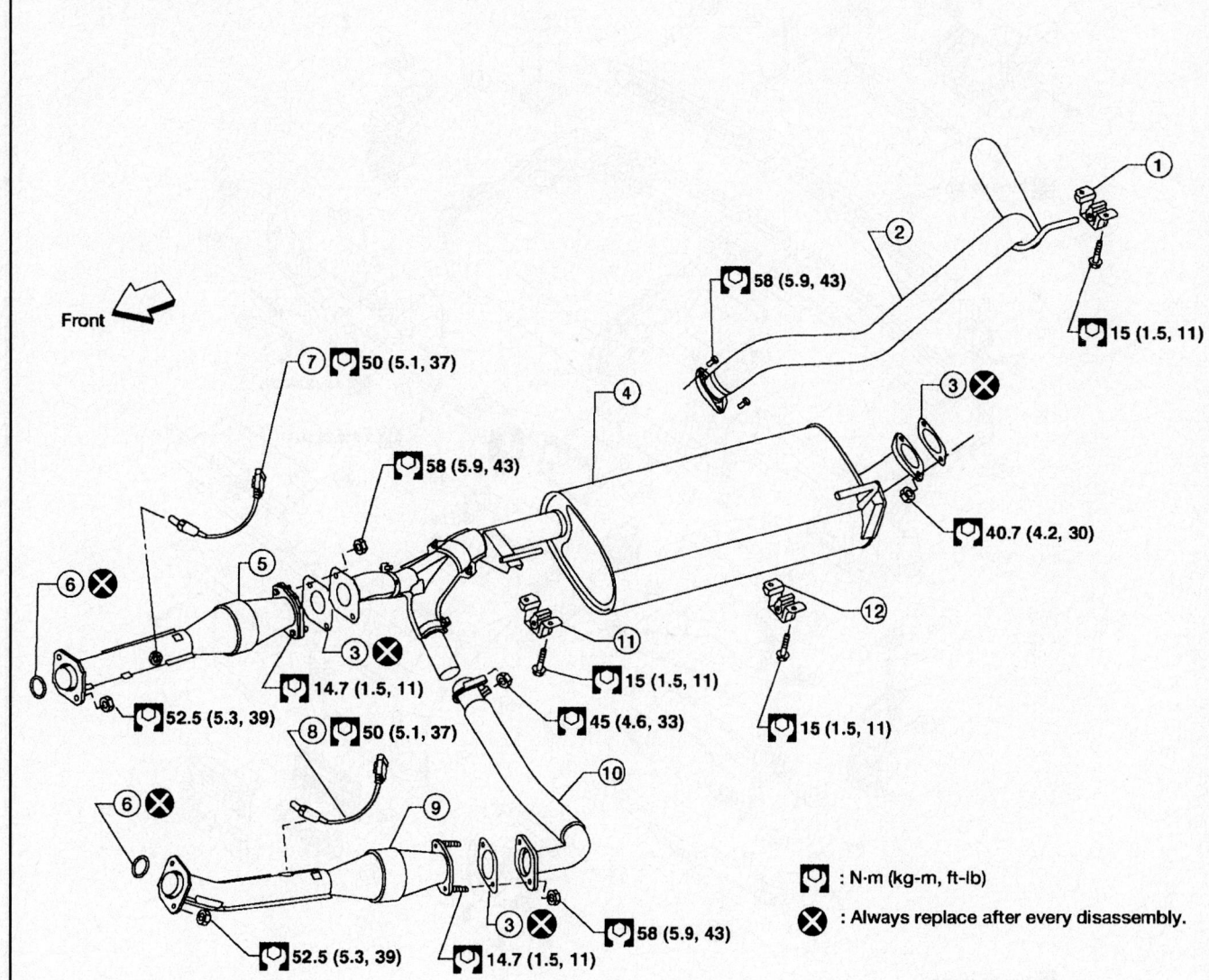

1. Tailpipe hanger bracket
2. Tailpipe
3. Gasket
4. Main muffler
5. Right front exhaust tube
6. Ring gasket
7. Heated oxygen sensor 2 (bank 2)
8. Heated oxygen sensor 2 (bank 1)
9. Left front exhaust tube
10. Center exhaust tube
11. Muffler hanger bracket front
12. Muffler hanger bracket rear

09482_ARMA_G0005

Fig. 80 Exploded view of the exhaust system

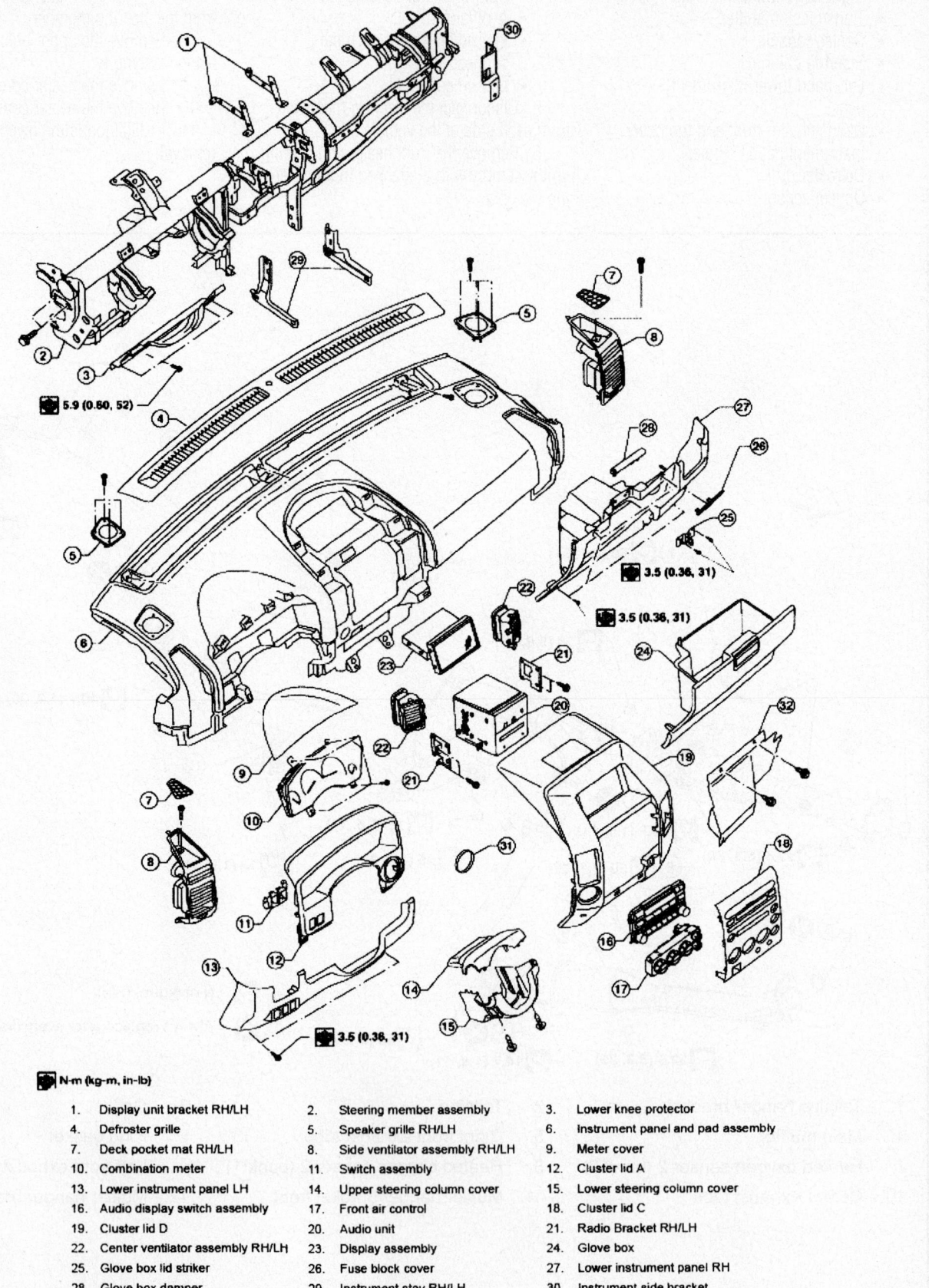

N-m (kg-m, in-lb)

1. Display unit bracket RH/LH
2. Steering member assembly
3. Lower knee protector
4. Defroster grille
5. Speaker grille RH/LH
6. Instrument panel and pad assembly
7. Deck pocket mat RH/LH
8. Side ventilator assembly RH/LH
9. Meter cover
10. Combination meter
11. Switch assembly
12. Cluster lid A
13. Lower instrument panel LH
14. Upper steering column cover
15. Lower steering column cover
16. Audio display switch assembly
17. Front air control
18. Cluster lid C
19. Cluster lid D
20. Audio unit
21. Radio Bracket RH/LH
22. Center ventilator assembly RH/LH
23. Display assembly
24. Glove box
25. Glove box lid striker
26. Fuse block cover
27. Lower instrument panel RH
28. Glove box damper
29. Instrument stay RH/LH
30. Instrument side bracket
31. Key cylinder escutcheon
32. Lower instrument panel RH

09482_ARMA_G0006

Fig. 81 Exploded view of the instrument panel

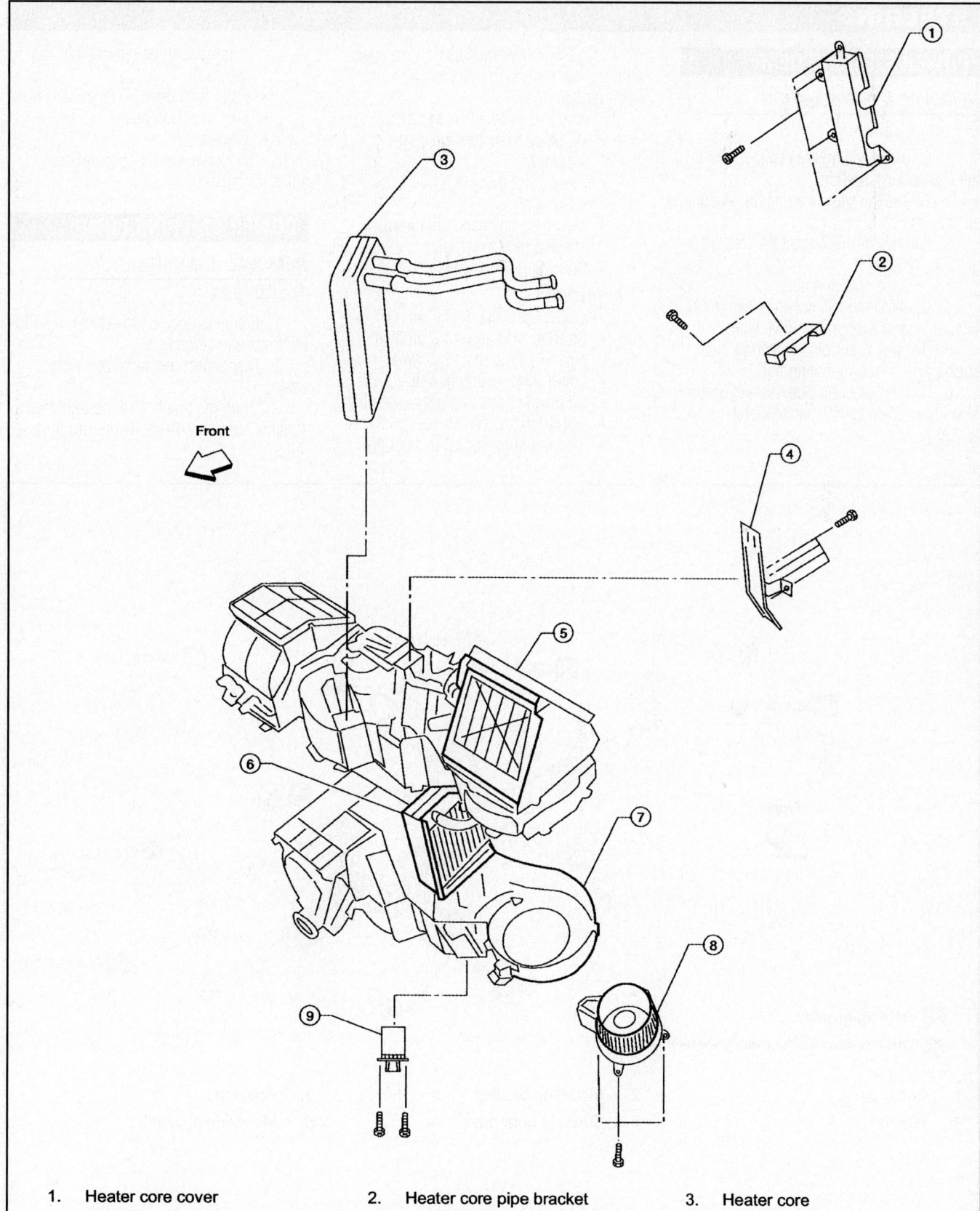

Front

1.	Heater core cover	2.	Heater core pipe bracket	3.	Heater core
4.	Upper bracket	5.	Upper heater and cooling unit case	6.	A/C evaporator
7.	Lower heater and cooling unit case	8.	Blower motor	9.	Variable blower control

67170-ARMA-G26

Fig. 82 Front heater/AC assembly

STEERING

POWER STEERING GEAR

REMOVAL & INSTALLATION

See Figure 83.

1. Before servicing the vehicle, refer to the Precautions Section.
2. Ensure the wheels are in the straight-ahead position.
3. Remove or disconnect the following:
   • Wheels
   • Engine splash guard
4. On 4WD models only, remove front final drive and support the drive shafts.
5. Remove cotter pin at steering outer socket and loosen mounting nut.
6. Remove steering outer socket from steering knuckle using special tool J-25730-A.

7. On 2WD drive models only, remove stabilizer bar mounting bolts and secure the stabilizer bar.
8. Remove or disconnect the following:
   • Oil pipes from steering gear assembly
   • Lower joint mounting bolt from lower shaft
   • Mounting bolts and nuts from steering gear assembly
   • Steering gear assembly

To install:

9. Install or connect the following:
   • Steering gear assembly, tighten nuts to 133 ft. lbs. (180 Nm)
   • Lower joint mounting bolt
   • Oil pipes to steering gear assembly
   • Stabilizer bar, 2WD models only
   • Steering outer socket to steering

knuckle, tighten nut to 63 ft. lbs. (86 Nm)
   • Front final drive, 4WD models only
   • Engine splash guard
   • Wheels
10. Check the wheel alignment and adjust as necessary.

POWER STEERING PUMP

REMOVAL & INSTALLATION

See Figure 84.

1. Before servicing the vehicle, refer to the Precautions Section.
2. Disconnect the negative battery cable.
3. Drain the power steering fluid into a suitable container. Properly discard the used fluid.

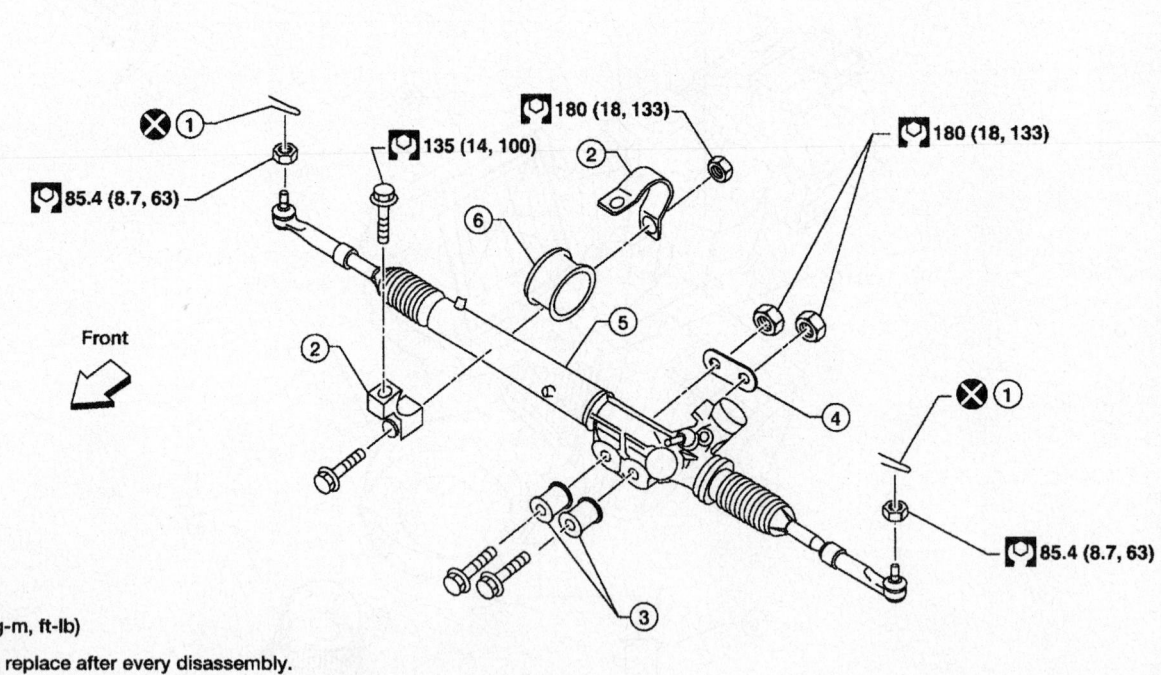

[symbol] : N·m (kg-m, ft-lb)

[symbol] : Always replace after every disassembly.

1. Cotter pin
4. Washer
2. Mounting bracket
5. Steering gear assembly
3. Bushing
6. Mounting insulator

67170-ARMA-G49

Fig. 83 Steering gear assembly

4. Remove the engine room cover.

5. Remove the air duct assembly.

6. Remove the power steering reservoir tank.

7. Remove the drive belt.

8. Disconnect the pressure sensor electrical connector.

9. Remove the high pressure and the low pressure lines from the power steering fluid pump.

10. Remove the pump mounting bolts.

11. Remove the pump from the vehicle.

To install:

12. Installation is the reverse of the removal procedure.

13. Bleed the power steering system.

➡**The drive belt tension is automatic and requires no adjustment.**

BLEEDING

1. Before servicing the vehicle, refer to the Precautions Section.

2. Stop the engine.

3. Turn the steering wheel fully to the right and left several times.

➡**Do not allow the fluid level in the reservoir tank to go below the MIN level line. Check and add fluid as needed.**

4. Run the engine at idle speed. Turn the steering wheel fully to the right and then fully to the left. Hold for about three seconds. Check for fluid leakage.

5. Repeat the above step several times at three second intervals.

➡**Do not hold the steering wheel in the locked position for more than ten seconds.**

6. Check for air bubbles or cloudy fluid. If found, repeat the bleeding procedure.

7. Stop the engine and check the fluid level. Correct as required.

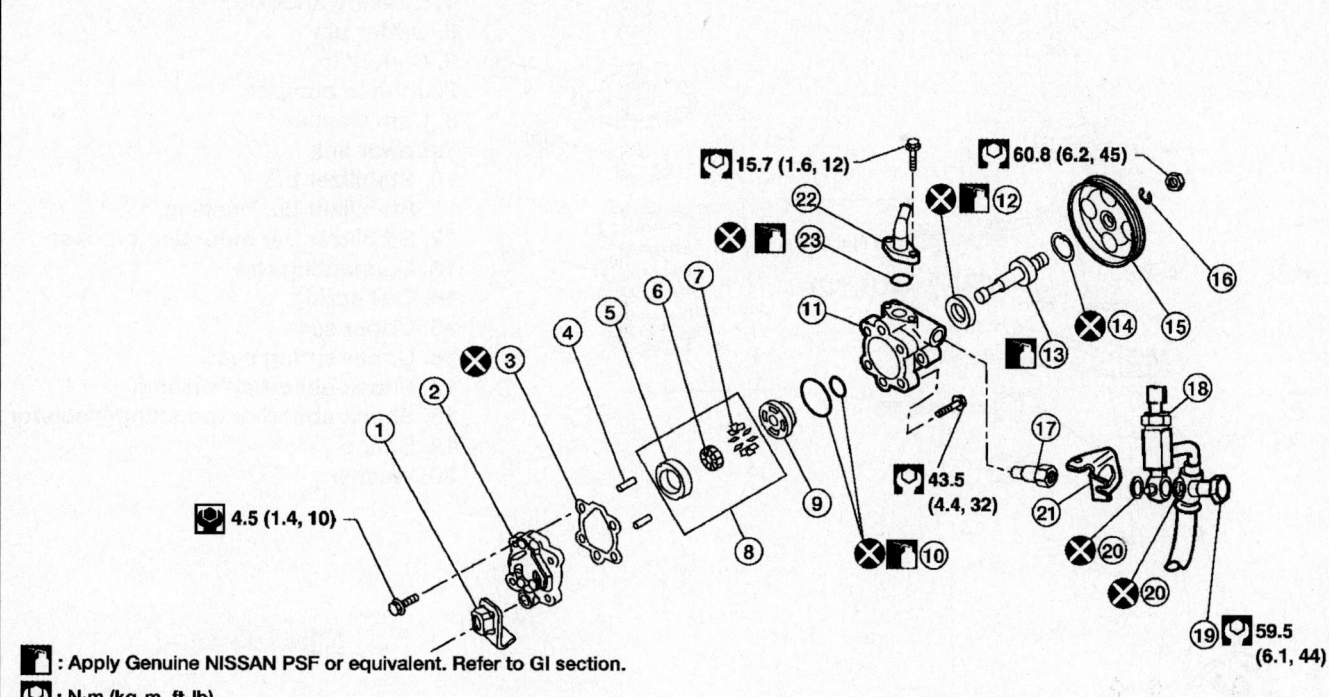

: Apply Genuine NISSAN PSF or equivalent. Refer to GI section.

: N·m (kg-m, ft-lb)

: Always replace after every disassembly.

1.	Bracket	2.	Rear cover	3.	Gasket
4.	Lock pin	5.	Cam ring	6.	Rotor
7.	Vane	8.	Cartridge	9.	Side plate
10.	O-ring	11.	Body assembly	12.	Oil seal
13.	Drive shaft assembly	14.	Snap ring	15.	Pulley
16.	Spring washer	17.	Flow control valve	18.	Pressure sensor
19.	Connector bolt	20.	Copper washer	21.	Bracket
22.	Suction pipe	23.	O-ring		

42050_QX56_G0043

Fig. 84 Power steering pump—exploded view

See Figure 85.

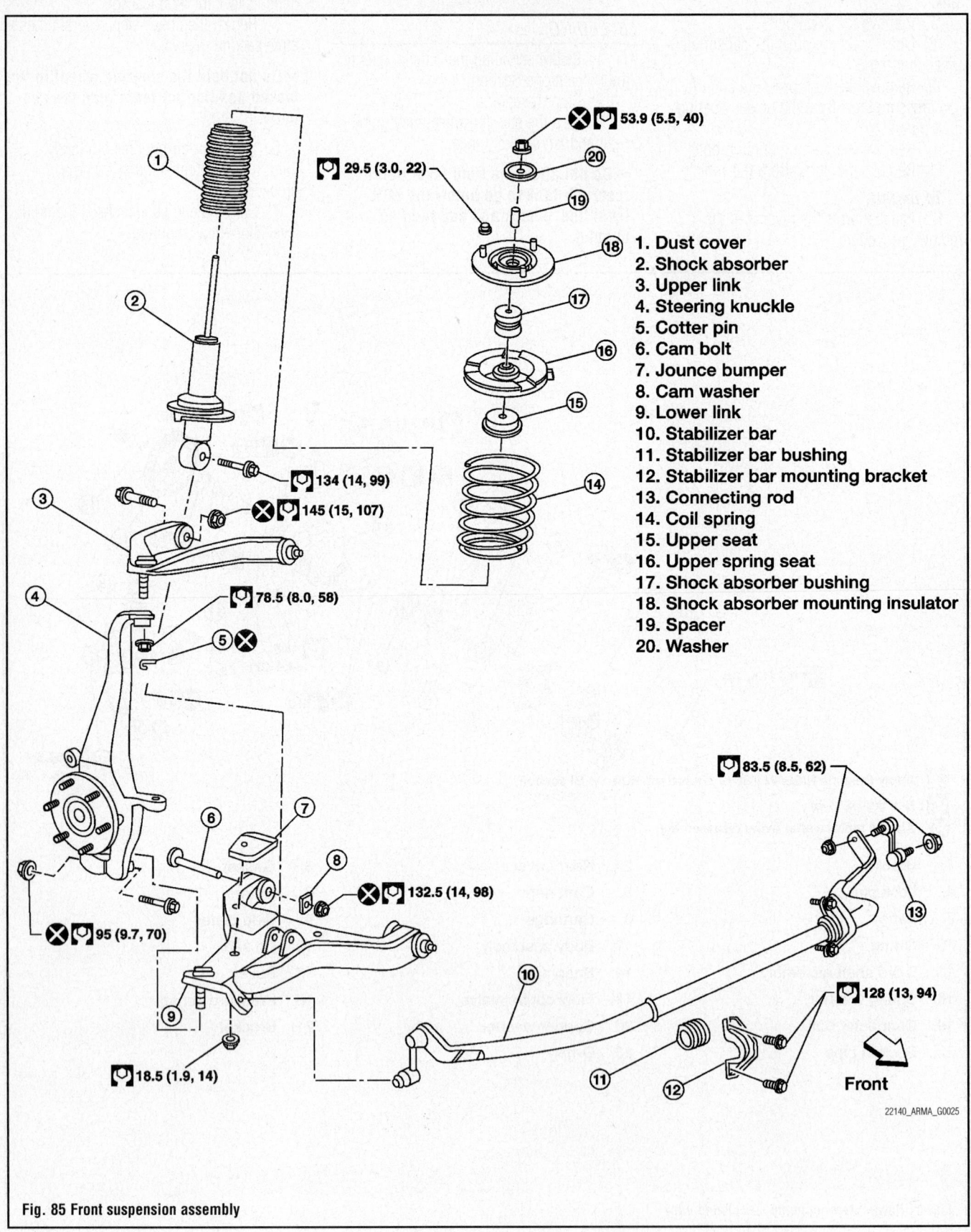

1. Dust cover
2. Shock absorber
3. Upper link
4. Steering knuckle
5. Cotter pin
6. Cam bolt
7. Jounce bumper
8. Cam washer
9. Lower link
10. Stabilizer bar
11. Stabilizer bar bushing
12. Stabilizer bar mounting bracket
13. Connecting rod
14. Coil spring
15. Upper seat
16. Upper spring seat
17. Shock absorber bushing
18. Shock absorber mounting insulator
19. Spacer
20. Washer

53.9 (5.5, 40)
29.5 (3.0, 22)
134 (14, 99)
145 (15, 107)
78.5 (8.0, 58)
83.5 (8.5, 62)
132.5 (14, 98)
95 (9.7, 70)
128 (13, 94)
18.5 (1.9, 14)

Front

22140_ARMA_G0025

Fig. 85 Front suspension assembly

COIL SPRING

REMOVAL & INSTALLATION

See Figure 86.

1. Before servicing the vehicle, refer to the Precautions Section.
2. Remove or disconnect the following:
   - Wheel
   - Lower shock absorber bolt
   - Upper shock absorber bolts
   - Coil spring and shock absorber assembly
3. Secure the shock absorber in a vice and loosen (without removing) the piston rod locknut.
4. Install a spring compressor and tighten until the shock absorber mounting insulator can be turned by hand.
5. Remove piston rod locknut and remove shock absorber from the coil spring.

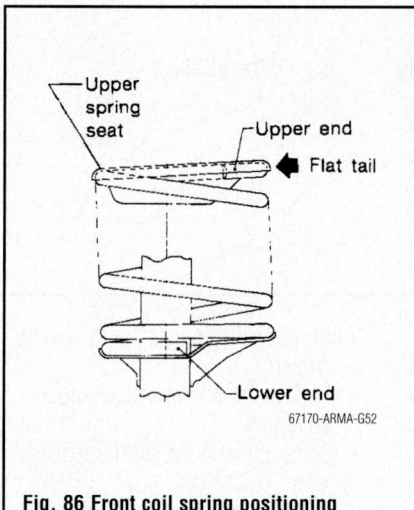

Fig. 86 Front coil spring positioning

Upper spring seat
Upper end
Flat tail
Lower end
67170-ARMA-G52

To install:

6. Install upper mounting insulator in line with the lower shock absorber mount and step in shock absorber lower seat as shown in figure.
7. Tighten the new piston rod locknut to 40 ft. lbs. (54 Nm).
8. Install or connect the following:
   - Coil spring and shock absorber assembly
   - Upper shock absorber bolts and tighten to 22 ft. lbs (30 Nm)
   - Lower shock absorber bolt and tighten to 99 ft. lbs. (134 Nm)
   - Wheel
9. Check wheel alignment and adjust as necessary.

CONTROL LINKS

REMOVAL & INSTALLATION

Lower Link

1. Before servicing the vehicle, refer to the Precautions Section.
2. Remove or disconnect the following:
   - Wheel
   - Lower shock absorber bolt
   - Stabilizer bar connecting rod
   - Drive shaft, if equipped with 4WD
   - Pinch bolt from steering knuckle
3. Separate the lower ball joint from the steering knuckle.
4. Remove the following:
   - Lower link adjusting bolts
   - Lower link

To install:

5. Install or connect the following:
   - Lower link and tighten adjusting bolts to 98 ft. lbs. (133 Nm)
   - Lower ball joint
   - Pinch bolt
   - Drive shaft, if equipped with 4WD
   - Stabilizer bar connected rod
   - Lower shock absorber bolt
   - Wheel

Upper Link

1. Before servicing the vehicle, refer to the Precautions Section.
2. Remove or disconnect the following:
   - Wheel
   - Coil spring and shock absorber assembly
   - Cotter pin and nut from upper ball joint
3. Separate upper ball joint stud from steering knuckle using special tool J-24319-01.
4. Remove the following:
   - Upper control arm mounting bolts
   - Upper control arm

To install:

5. Install or connect the following:
   - Upper control arm and tighten bolts to 107 ft. lbs. (145 Nm)
   - Upper ball joint with new cotter pin and tighten nut to 58 ft. lbs. (79 Nm)
   - Coil spring and shock absorber assembly
   - Wheel

LOWER BALL JOINT

REMOVAL & INSTALLATION

1. Before servicing the vehicle, refer to the Precautions Section.

2. Remove or disconnect the following:
   - Wheel
   - Lower shock absorber bolt
   - Stabilizer bar connecting rod
   - Drive shaft, if equipped with 4WD
   - Pinch bolt from steering knuckle
3. Separate lower ball joint from steering knuckle

To install:

4. Install or connect the following:
   - Lower ball joint
   - Pinch bolt to steering knuckle
   - Drive shaft, if equipped with 4WD
   - Stabilizer bar connecting rod
   - Lower shock absorber bolt
   - Wheel

STABILIZER BAR

REMOVAL & INSTALLATION

1. Before servicing the vehicle, refer to the Precautions Section.
2. Raise and support the vehicle safely.
3. Remove the tire and wheel assembly.
4. Remove the engine under cover.
5. Remove the stabilizer bar mounting bracket retaining bolts and rubber bushings.
6. Remove the connecting rod nuts.
7. Remove the stabilizer bar from the vehicle.

To install:

8. Installation is the reverse of the removal procedure.
9. Tighten the retaining bushing and mounting bar bracket bolts to 94 ft. lbs.
10. Tightening the connecting rod bolt and nut to 62 ft. lbs.

STEERING KNUCKLE

REMOVAL & INSTALLATION

See Figure 87.

1. Before servicing the vehicle, refer to the Precautions Section.
2. Raise and support the vehicle safely.
3. Remove the tire and wheel assembly.
4. Remove the brake caliper from its mounting and position it to the side.

➡**Do not disconnect the hydraulic lines. It is not necessary to remove the bolts on the torque member and brake hose except for disassembly or replacement of the caliper. In this case, hang the caliper to the side with mechanics wire so that the brake hose is not under tension. Avoid depressing the brake pedal with the caliper removed.**

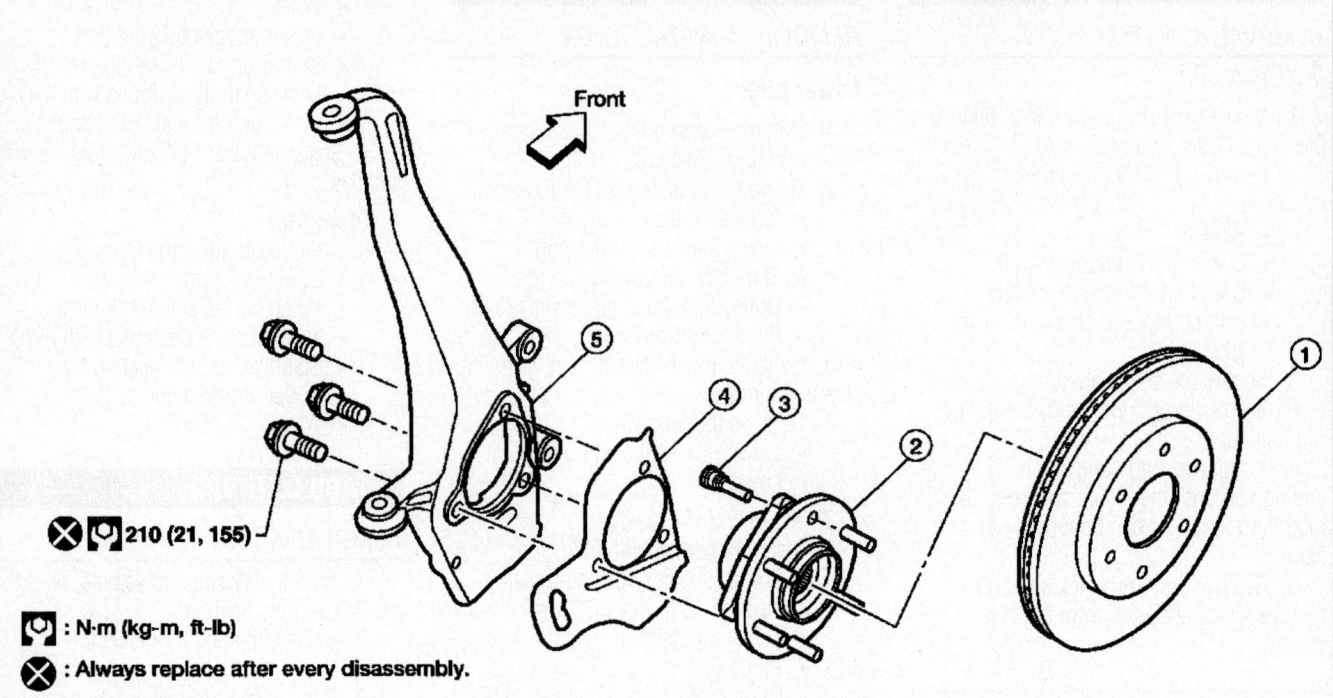

Front

$\otimes$ $\boxdot$ **210 (21, 155)**

$\boxdot$: N·m (kg-m, ft-lb)

$\otimes$: Always replace after every disassembly.

1. Disc rotor
4. Splash guard
2. Wheel hub and bearing assembly
5. Steering knuckle
3. Wheel stud

42050_QX56_G0037

Fig. 87 Steering knuckle and related components

5. Put alignment marks on the rotor and wheel hub and bearing assembly. Remove the rotor.

6. Remove the ABS sensor from the steering knuckle. Do not pull on the ABS sensor harness.

7. Remove the cotter pin. Remove the locknut from the halfshaft.

8. Remove the steering outer shaft socket cotter pin at the steering knuckle. Loosen the mounting nut.

9. Disconnect the steering outer socket from the steering knuckle.

➡ **To prevent damage to the threads and to prevent the tool from coming off suddenly, temporarily loosely install the mounting nut.**

10. Remove the halfshaft.
11. Remove the wheel hub and bearing assembly bolts.
12. Remove the splash guard and wheel hub and bearing assembly from the steering knuckle.
13. Support the lower control arm assembly, using a suitable jack.

14. Remove the cotter pin and nut from the upper ball joint.

15. Separate the upper link ball joint from the steering knuckle using tool J-24319-01 or equivalent.

16. Remove the pinch bolt from the steering knuckle. Remove the steering knuckle from the lower control arm ball joint.

17. Remove the steering knuckle from the vehicle.

To install:

18. Installation is the reverse of the removal procedure.

19. Be sure to use the alignment marks made during the removal procedure when reinstalling removed components.

20. Check and adjust the front end alignment, as required.

UPPER BALL JOINT

REMOVAL & INSTALLATION

1. Before servicing the vehicle, refer to the Precautions Section.

2. Remove or disconnect the following:
   • Wheel
   • Coil spring and shock absorber assembly
   • Cotter pin and nut from upper ball joint

3. Separate upper ball joint from steering knuckle using special tool J-24319-01

To install:

4. Install or connect the following:
   • Upper ball joint
   • New cotter pin and tighten nut to 58 ft. lbs. (79 Nm)
   • Coil spring and shock absorber assembly
   • Wheel

HUB AND BEARING

REMOVAL & INSTALLATION

See Figure 88.

1. Before servicing the vehicle, refer to the Precautions Section.

2. Remove or disconnect the following:
   • Wheel

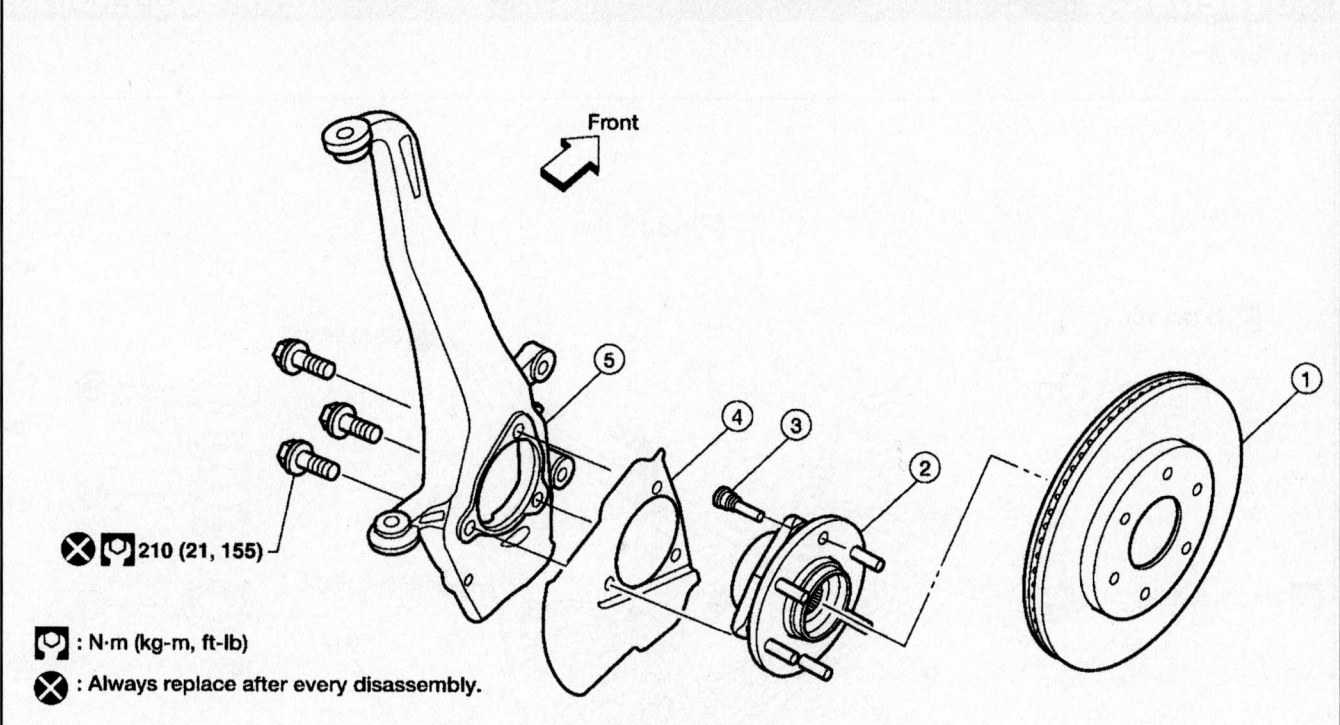

Front

210 (21, 155)

: N·m (kg-m, ft-lb)

: Always replace after every disassembly.

1. Disc rotor
2. Wheel hub and bearing assembly
3. Wheel stud
4. Splash guard
5. Steering knuckle

67170-ARMA-G53

Fig. 88 Front hub/bearing assembly

- Engine splash guard
- Brake caliper without disconnecting the hydraulic lines, and reposition aside with wire

3. Matchmark the brake rotor to the wheel hub and remove the brake rotor.

4. Remove or disconnect the following:
  - Cotter pin and locknut from driveshaft
  - Driveshaft from wheel hub and bearing assembly
  - ABS sensor

- Wheel hub and bearing assembly bolts
- Wheel hub and bearing assembly

To install:

5. Install or connect the following:
  - Wheel hub and bearing assembly, using new bolts and tighten to 155 ft. lbs. (210 Nm)
  - ABS sensor
  - Driveshaft to wheel hub and bearing assembly
  - Cotter pin and locknut and tighten to 101 ft. lbs. (137 Nm)

- Brake rotor
- Brake caliper
- Engine splash guard
- Wheel

ADJUSTMENT

The front wheel bearings are part of a unitized hub and are not adjustable. Move the wheel hub in the axial direction by hand. Make sure that there is no looseness of the wheel bearing. Axial end play is 0.002 inch or less.

See Figure 89.

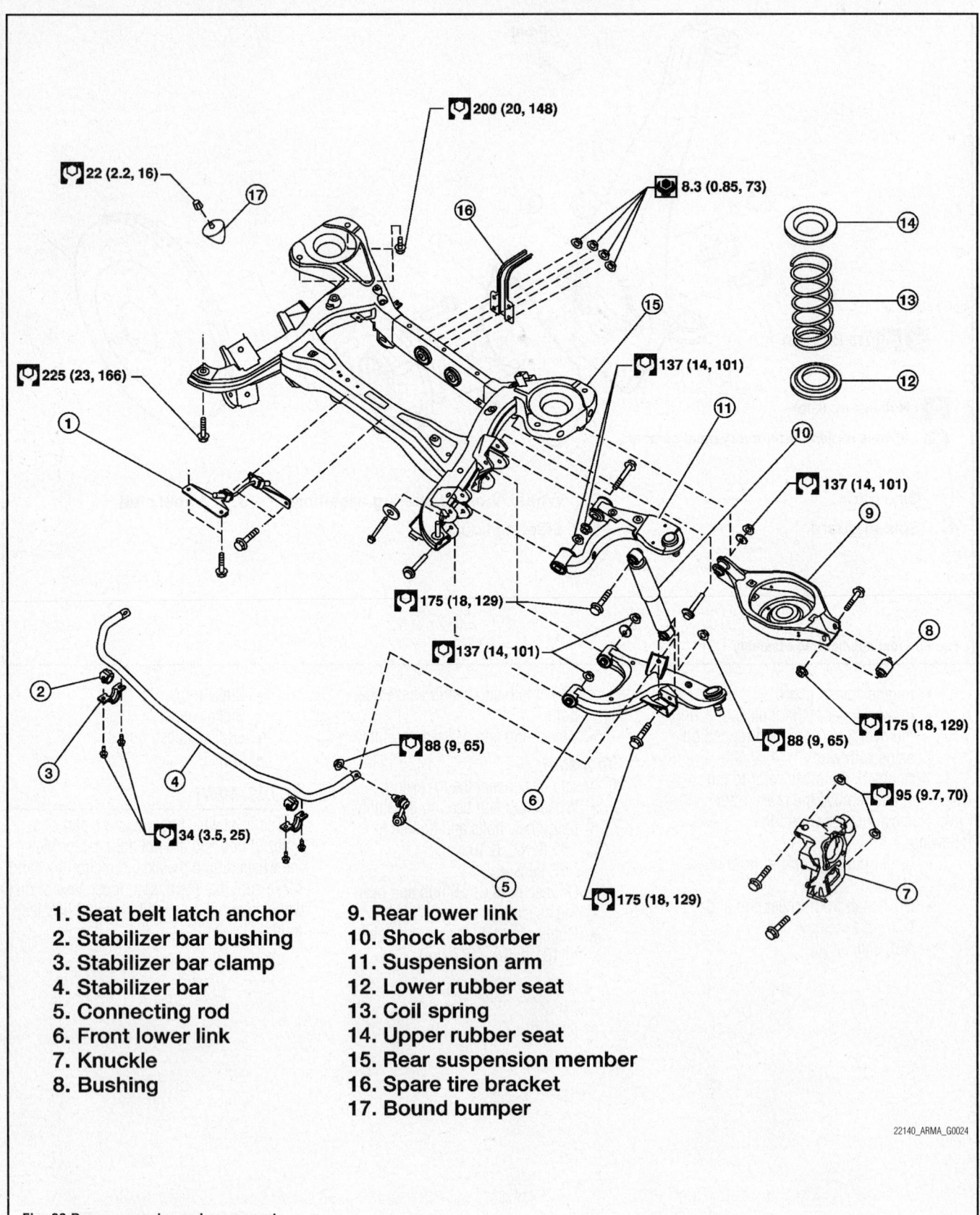

1. Seat belt latch anchor
2. Stabilizer bar bushing
3. Stabilizer bar clamp
4. Stabilizer bar
5. Connecting rod
6. Front lower link
7. Knuckle
8. Bushing
9. Rear lower link
10. Shock absorber
11. Suspension arm
12. Lower rubber seat
13. Coil spring
14. Upper rubber seat
15. Rear suspension member
16. Spare tire bracket
17. Bound bumper

22140_ARMA_G0024

Fig. 89 Rear suspension and components

COIL SPRING

REMOVAL & INSTALLATION

1. Before servicing the vehicle, refer to the Precautions Section.
2. Remove the rear wheel.
3. Release the air pressure from the rear load leveling air suspension system using the CONSULT-II "EXHAUST SOLE-NOID" active test.
4. Remove the height sensor arm bracket bolt from the left-hand rear lower link.
5. Place a suitable jack under the rear lower link and relieve the coil spring tension.
6. Loosen the rear lower link adjusting bolt and nut connected to the rear suspension member.
7. Remove the rear lower link bolt and nut from the knuckle.
8. Slowly lower the jack to relieve the coil spring tension.
9. Remove the coil spring.

To install:
10. Install or connect the following:
   - Coil spring

➡ **When installing the rubber seats for the coil spring, ensure the embossed arrow points outward toward the wheel.**

   - Rear lower link bolt to knuckle and tighten nut to 70 ft. lbs. (95 Nm)
   - Rear lower link adjusting bolt to rear suspension member and tighten nut to 101 ft. lbs. (137 Nm)
   - Height sensor arm bracket bolt to left-head rear lower link and tighten to 9 ft. lbs. (12 Nm)
   - Rear wheel

LOWER CONTROL ARM (LINK)

REMOVAL & INSTALLATION

1. Before servicing the vehicle, refer to the Precautions Section.
2. Raise and support the vehicle safely.
3. Remove the tire and wheel assemblies.
4. Release the air pressure from the rear load leveling air suspension system using the CONSULT-II "EXHAUST SOLE-NOID" active test.
5. Remove the height sensor arm bracket bolt from the left-hand rear lower link.
6. Place a suitable jack under the rear lower link and relieve the coil spring tension.

7. Loosen the rear lower link adjusting bolt and nut connected to the rear suspension member.
8. Remove the rear lower link bolt and nut from the knuckle.
9. Slowly lower the jack to relieve the coil spring tension.
10. Remove the coil spring.
11. Remove the upper rubber seat, coil spring and lower rubber seat from the rear lower link.
12. Remove the rear lower link adjusting bolt and nut from the rear suspension member.
13. Remove the rear lower link from its mounting.

To install:
14. Installation is the reverse of the removal procedure.
15. When installing the upper and lower rubber seats for the rear coil springs, the arrow embossed on the rubber seats must point out toward the wheel and tire assembly.
16. Tighten the rear lower link bolt to knuckle to 70 ft. lbs. (95 Nm)
17. Tighten the rear lower link adjusting bolt to rear suspension member to 101 ft. lbs. (137 Nm)
18. Tighten the height sensor arm bracket bolt to left-head rear lower link to 9 ft. lbs. (12 Nm)
19. Perform the final tightening of the nuts and bolts for the links (rubber bushing) with the vehicle in the unladen condition with the tires on level ground.

➡ **Unladen condition means that the fuel tank, engine coolant and lubricants are at the full specification and the spare tire, jack, hand tools and mats are in their designated positions.**

20. Check and adjust the alignment, as required.

STABILIZER BAR

REMOVAL & INSTALLATION

1. Before servicing the vehicle, refer to the Precautions Section.
2. Raise and support the vehicle safely.
3. Remove the tire and wheel assemblies, as required.
4. Disconnect the stabilizer bar ends from the connecting rods.
5. Remove the stabilizer bar clamps. Remove the stabilizer bar bushings.
6. Remove the stabilizer bar from its mounting.

To install:
7. Installation is the reverse of the removal procedure.

8. Install the stabilizer bar with the ball joint properly aligned.
9. Install the stabilizer bar bushing and clamp so they are positioned inside of the sideslip prevention clamp on the stabilizer bar.
10. Perform the final tightening of the nuts and bolts for the links (rubber bushing) with the vehicle in the unladen condition with the tires on level ground.

➡ **Unladen condition means that the fuel tank, engine coolant and lubricants are at the full specification and the spare tire, jack, hand tools and mats are in their designated positions.**

11. Check and adjust the alignment, as required.

SHOCK ABSORBER

REMOVAL & INSTALLATION

See Figures 90 and 91.

1. Before servicing the vehicle, refer to the Precautions Section.
2. Remove the rear wheel.
3. Release the air pressure from the rear load leveling air suspension system using the CONSULT-II "EXHAUST SOLE-NOID" active test.
4. Remove or disconnect the following:
   - Rear fender protector
   - Rear load leveling air suspension hose from the shock absorber
   - Shock absorber upper and lower end bolts
   - Shock absorber

To install:
5. Install or connect the following:
   - Shock absorber and tighten end bolts to 129 ft. lbs. (175 Nm)
   - Rear load leveling air suspension hose
   - Rear fender protector
   - Rear wheel

SUSPENSION ARM

REMOVAL & INSTALLATION

1. Before servicing the vehicle, refer to the Precautions Section.
2. Raise and support the vehicle safely.
3. Remove the tire and wheel assemblies.
4. Remove the rear suspension member.

➡ **It is necessary to remove the rear suspension member in order to remove the front upper bolt from the suspension arm.**

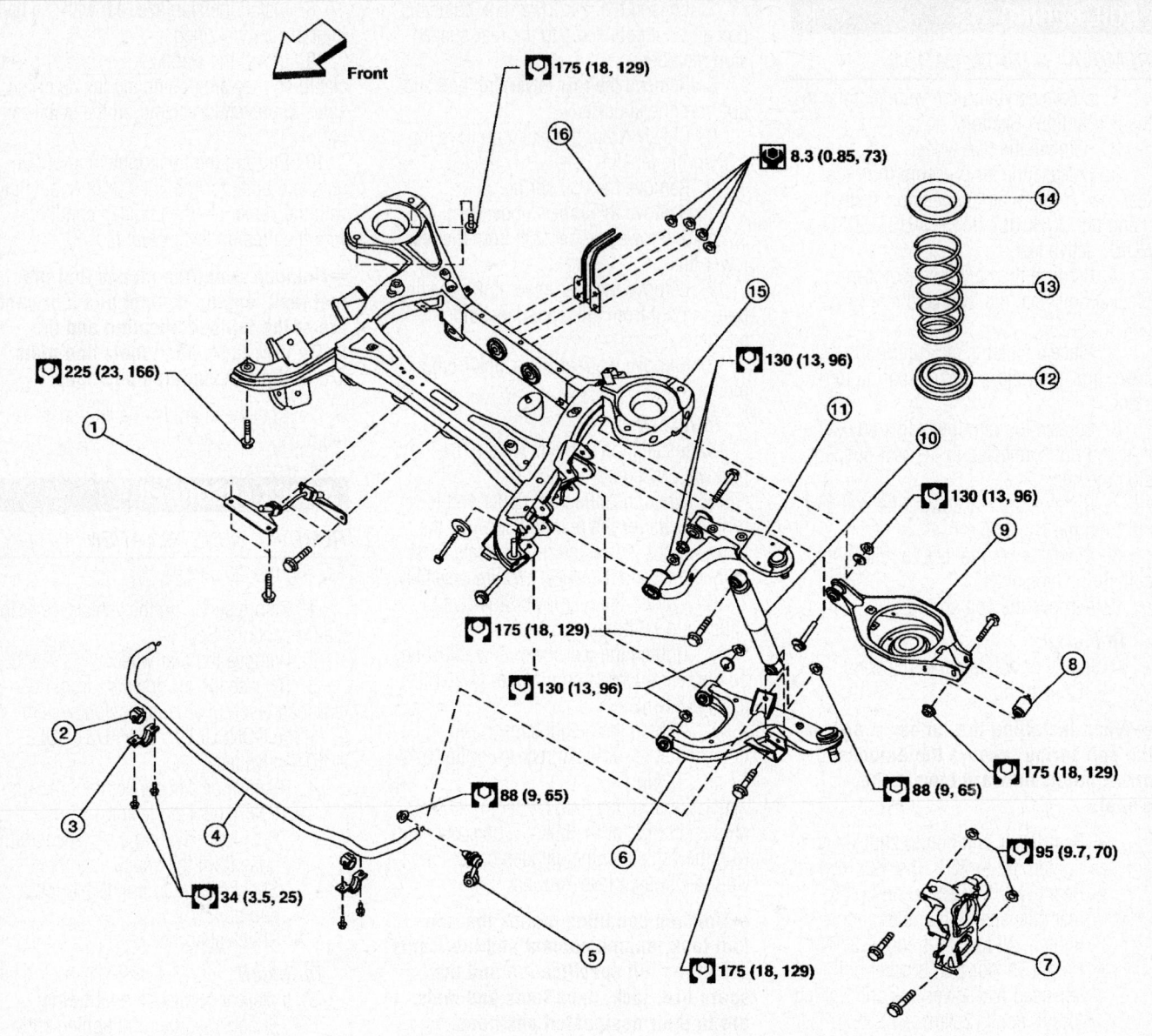

Front

175 (18, 129)

16

8.3 (0.85, 73)

14

13

12

225 (23, 166)

130 (13, 96)

15

11

10

130 (13, 96)

9

1

175 (18, 129)

8

130 (13, 96)

88 (9, 65)

175 (18, 129)

2

88 (9, 65)

3

4

34 (3.5, 25)

95 (9.7, 70)

6

5

175 (18, 129)

7

N·m (kg-m, in-lb)

N·m (kg-m, ft-lb)

1. Seat belt latch anchor	2. Stabilizer bar bushing	3. Stabilizer bar clamp
4. Stabilizer bar	5. Connecting rod	6. Front lower link
7. Wheel hub and spindle assembly	8. Bushing	9. Rear lower link
10. Shock absorber	11. Suspension arm	12. Lower rubber seat
13. Coil spring	14. Upper rubber seat	15. Rear suspension member
16. Spare tire bracket		

67170-ARMA-G50

Fig. 90 Standard rear suspension

5. Remove the shock absorber upper end bolt.

6. Remove the suspension arm upper nuts and bolts on the suspension member side.

7. Remove the suspension arm pinch bolt and nut on the knuckle side.

8. Disconnect the suspension arm from the knuckle.

➡️**If necessary, use a soft hammer. Do not damage the ball joint with the soft hammer.**

9. Remove the suspension arm.

To install:

10. Installation is the reverse of the removal procedure.

11. Perform the final tightening of the nuts and bolts for the links (rubber bushing) with the vehicle in the unladen condition with the tires on level ground.

➡️**Unladen condition means that the fuel tank, engine coolant and lubricants are at the full specification and the spare tire, jack, hand tools and mats are in their designated positions.**

12. Check and adjust the alignment, as required.

UPPER CONTROL ARM

REMOVAL & INSTALLATION

1. Before servicing the vehicle, refer to the Precautions Section.
2. Raise and support the vehicle safely.
3. Remove the tire and wheel assemblies.
4. Release the air pressure from the rear load leveling air suspension system using the CONSULT-II "EXHAUST SOLENOID" active test.

5. Remove the shock absorber lower end bolt.

6. Remove the adjusting bolt and nut, and the bolt and nut from the front lower link and rear suspension member.

7. Remove the front lower link pinch bolt and nut on the knuckle side.

8. Disconnect the front lower link from the knuckle.

➡️**If necessary, use a soft hammer. Do not damage the ball joint with the soft hammer.**

9. Remove the front lower link.

To install:

10. Installation is the reverse of the removal procedure.

11. Perform the final tightening of the nuts and bolts for the links (rubber bushing) with the vehicle in the unladen condition with the tires on level ground.

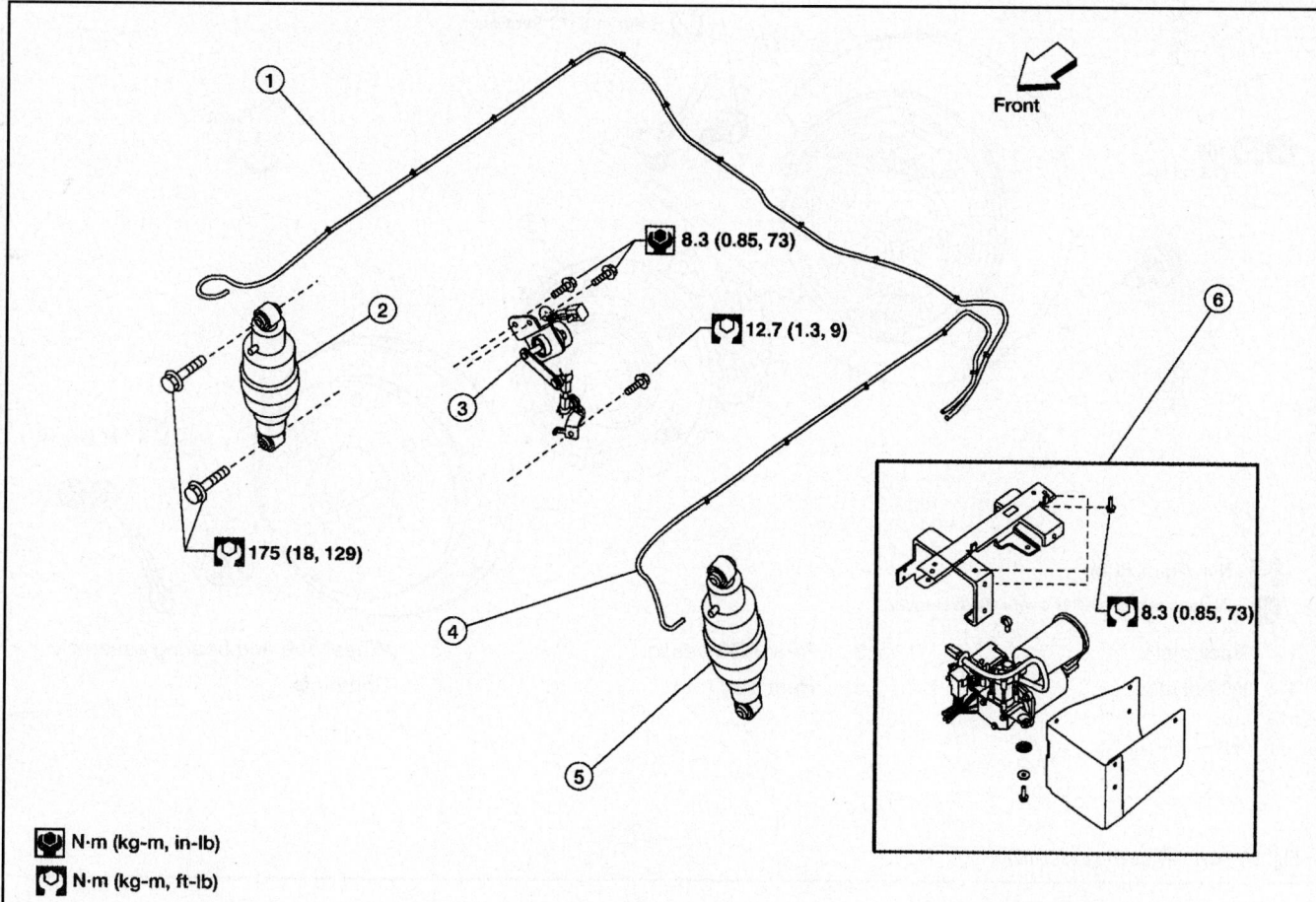

N·m (kg-m, in-lb)

N·m (kg-m, ft-lb)

1. Rear load leveling air suspension hose, RH
2. Shock absorber, RH
3. Height sensor
4. Rear load leveling air suspension hose, LH
5. Shock absorber, LH
6. Rear load leveling air suspension compressor assembly

67170-ARMA-G51

Fig. 91 Rear load leveling air suspension

➡**Unladen condition means that the fuel tank, engine coolant and lubricants are at the full specification and the spare tire, jack, hand tools and mats are in their designated positions.**

12. Check and adjust the alignment, as required.

WHEEL HUB AND BEARINGS

REMOVAL & INSTALLATION

See Figure 92.

1. Before servicing the vehicle, refer to the Precautions Section.
2. Remove or disconnect the following:
   • Wheel

• Brake caliper without disconnecting the hydraulic lines, and reposition aside with wire
• Brake rotor
• Cotter pin and nut from driveshaft
• Driveshaft
• Wheel hub and bearing assembly bolts

3. Pulling out the wheel hub and bearing assembly slightly, remove the ABS sensor.
4. Remove the wheel hub and bearing assembly.

To install:

5. Install or connect the following:
   • ABS sensor

• Wheel hub and bearing assembly, using new bolts and tighten to 111 ft. lbs. (150 Nm)
• Driveshaft
• Lock nut and tighten to 101 ft. lbs. (137 Nm) and new cotter pin
• Brake rotor
• Brake caliper
• Wheel

ADJUSTMENT

The rear wheel bearings are part of a unitized hub and are not adjustable. Move the wheel hub in the axial direction by hand. Make sure that there is no looseness of the wheel bearing. Axial endplay is 0.002 inch or less.

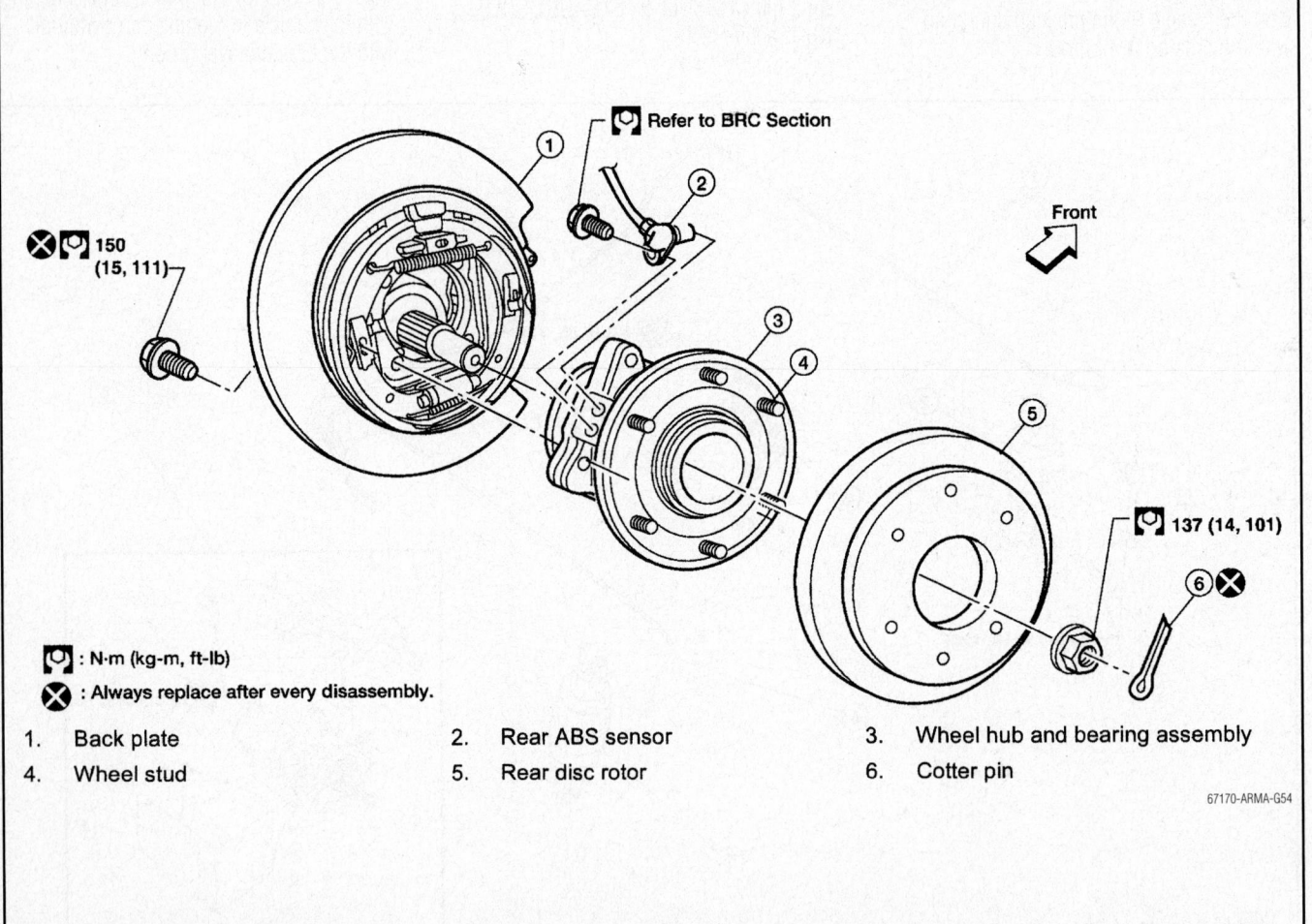

1. Back plate
4. Wheel stud
2. Rear ABS sensor
5. Rear disc rotor
3. Wheel hub and bearing assembly
6. Cotter pin

67170-ARMA-G54

Fig. 92 Rear hub/bearing assembly

NISSAN

Frontier • Xterra

SPECIFICATIONS AND MAINTENANCE CHARTS

ENGINE AND VEHICLE IDENTIFICATION

			Engine				Model Year	
ID/Code ①	Liters (cc)	Cu. In.	Cyl.	Fuel Sys.	Engine Type	Eng. Mfg.	Code ②	Year
QR25DE	2.5 (2488)	152	4	MFI	DOHC	Nissan	7	2007
VQ40DE	4.0 (3954)	241	6	MFI	DOHC	Nissan	8	2008

MFI: Multi-port Fuel Injection

SOHC: Single Overhead Camshaft

DOHC: Double Overhead Camshafts

① 4th digit of the Vehicle Identification Number (VIN)

② 10th digit of the Vehicle Identification Number (VIN)

22140_FRON_C0001

GENERAL ENGINE SPECIFICATIONS

Year	Model	Engine Displacement Liters	Engine ID	Net Horsepower @ rpm	Net Torque @ rpm (ft. lbs.)	Bore x Stroke (in.)	Compression Ratio	Oil Pressure @ rpm
2007	Frontier	2.5	QR25DE	154@5200	173@4400	3.50X3.94	9.5:1	43@2000
	Frontier	4.0	VQ40DE	265@5600	284@4000	3.76X3.62	9.7:1	43@2000
	Xterra	4.0	VQ40DE	265@5600	284@4000	3.76X3.62	9.7:1	43@2000
2008	Frontier	2.5	QR25DE	154@5200	173@4400	3.50X3.94	9.5:1	43@2000
	Frontier	4.0	VQ40DE	265@5600	284@4000	3.76X3.62	9.7:1	43@2000
	Xterra	4.0	VQ40DE	265@5600	284@4000	3.76X3.62	9.7:1	43@2000

22140_FRON_C0002

ENGINE TUNE-UP SPECIFICATIONS

Year	Engine Displ. Liters	Engine ID	Spark Plug Gap (in.)	Ignition Timing (deg.) MT	Ignition Timing (deg.) AT	Fuel Pump (psi)	Idle Speed (rpm) MT	Idle Speed (rpm) AT ①	Valve Clearance (in.) In.	Valve Clearance (in.) Ex.
2007	2.5	QR25DE	0.043	10-20B	10-20B	51 ②	575-675	650-750	③	④
	4.0	VQ40DE	0.043	10-20B	10-20B	51 ②	575-675	650-750	⑤	⑥
2008	2.5	QR25DE	0.043	10-20B	10-20B	51 ②	575-675	650-750	③	④
	4.0	VQ40DE	0.043	10-20B	10-20B	51 ②	575-675	650-750	⑤	⑥

NOTE: The Vehicle Emission Control Information label often reflects specification changes made during production. The label figures must be used if they differ from those in this chart.

B: Before top dead center

HYD: Hydraulic

① Automatic transmission in Neutral

② At idle

③ 0.009-0.013 cold
0.012-0.016 hot

④ 0.010-0.013 cold
0.012-0.017 hot

⑤ 0.010-0.013 cold
0.012-0.016 hot

⑥ 0.011-0.015 cold
0.012-0.017 hot

22140_FRON_C0003

CAPACITIES

Year	Model	Engine Displacement Liters	Engine ID	Engine Oil with Filter (qts.)	Transmission (pts.) Manual	Transmission (pts.) Auto.	Transfer Case (pts.)	Drive Axle Front (pts.)	Drive Axle Rear (pts.)	Fuel Tank (gal.)	Cooling System (qts.)
2007	Frontier	2.5	QR25DE	5.15	6.15	21.50	—	—	3.3	21.2	10.0
	Frontier	4.0	VQ40DE	5.30	①	21.50	2.1	1.75	②	21.2	11.0
	Xterra	4.0	VQ40DE	5.10	①	21.50	2.1	1.75	②	21.2	11.0
2008	Frontier	2.5	QR25DE	5.15	6.15	21.50	—	—	3.3	21.2	10.0
	Frontier	4.0	VQ40DE	5.15	①	21.50	2.1	1.75	②	21.2	11.0
	Xterra	4.0	VQ40DE	5.50	①	21.50	2.1	1.75	②	21.2	11.0

NOTE: All capacities are approximate. Add fluid gradually and check to be sure a proper fluid level is obtained.

① 2WD: 8.3; 4WD: 8.9

② C200: 3.3; M226: 4.25

22140_FRON_C0004

VALVE SPECIFICATIONS

Year	Engine Displacement Liters	Engine ID	Seat Angle (deg.)	Face Angle (deg.)	Spring Test Pressure (lbs. @ in.)	Spring Installed Height (in.)	Stem-to-Guide Clearance (in.) Intake	Stem-to-Guide Clearance (in.) Exhaust	Stem Diameter (in.) Intake	Stem Diameter (in.) Exhaust
2007	2.5	QR25DE	①	②	③	1.390	0.0008-0.0021	0.0012-0.0025	0.2348-0.2354	0.2344-0.2350
	4.0	VQ40DE	①	②	④	1.456	0.0008-0.0021	0.0012-0.0025	0.2348-0.2354	0.2344-0.2350
2008	2.5	QR25DE	①	②	③	1.390	0.0008-0.0021	0.0012-0.0025	0.2348-0.2354	0.2344-0.2350
	4.0	VQ40DE	①	②	④	1.456	0.0008-0.0021	0.0012-0.0025	0.2348-0.2354	0.2344-0.2350

① 44 degrees 22 minutes to 45 degrees 8 minutes

② 45 degrees 15 minutes to 45 degrees 45 minutes

③ Installation: 37-42@1.457

Valve open INT: 79-89@0.996

Valve open EXH: 71-81@1.053

④ Installation: 37-42@1.457

Valve open: 84-95@1.071

22140_FRON_C0005

CAMSHAFT SPECIFICATIONS CHART

All measurements are given in inches.

Year	Engine Displ. Liters	Engine VIN	Journal Dia.	Brg. Oil Clearance	Shaft End-play	Runout	Lobe Height Intake	Lobe Height Exhaust
2007	2.5	QR25DE	①	0.0018-0.0034	0.0045-0.0074	0.0008	1.7722-1.7797	1.7313-1.7388
	4.0	VQ40DE	②	③	0.0045-0.0074	0.0010	1.7900-1.7921	1.7746-1.7821
2008	2.5	QR25DE	①	0.0018-0.0034	0.0045-0.0074	0.0008	1.7722-1.7797	1.7313-1.7388
	4.0	VQ40DE	②	③	0.0045-0.0074	0.0010	1.7900-1.7921	1.7746-1.7821

① No.1: 1.0998-1.1006. No's. 2, 3, 4, 5: 0.9226-0.9234

② No.1: 1.0211-1.0218. No's. 2, 3, 4: 0.9230-0.9238

③ No.1: 1.0018-0.0034. No's. 2, 3, 4: 0.0014-0.0030

22140_FRON_C0006

CRANKSHAFT AND CONNECTING ROD SPECIFICATIONS

All measurements are given in inches.

Year	Engine Displacement Liters	Engine ID	Crankshaft Main Brg. Journal Dia.	Crankshaft Main Brg. Oil Clearance	Crankshaft Shaft End-play	Crankshaft Thrust on No.	Connecting Rod Journal Diameter	Connecting Rod Oil Clearance	Connecting Rod Side Clearance
2007	2.5	QR25DE	①	②	0.0039-0.0102	NA	NA	0.0015-0.0022	0.0079-0.0138
	4.0	VQ40DE	③	0.0014-0.0018	0.0039-0.0098	NA	2.2441-2.2446	0.0013-0.0023	0.0079-0.0138
2008	2.5	QR25DE	①	②	0.0039-0.0102	NA	NA	0.0015-0.0022	0.0079-0.0138
	4.0	VQ40DE	③	0.0014-0.0018	0.0039-0.0098	NA	2.2441-2.2446	0.0013-0.0023	0.0079-0.0138

NA: Not Available

① There are 24 different grades, ranging from A (2.1645) to 7 (2.1636)

② No. 1, 3, 5: 0.0011-0.0017

No's. 2, 4: 0.0016-0.0022

③ There are 24 different grades, ranging from A (2.7549) to 7 (2.7540)

22140_FRON_C0007

PISTON AND RING SPECIFICATIONS

All measurements are given in inches.

Year	Engine Displacement Liters	Engine ID	Piston Clearance	Ring Gap			Ring Side Clearance		
				Top Comp.	Bottom Comp.	Oil Control	Top Comp.	Bottom Comp.	Oil Control
2007	2.5	QR25DE	0.0004-0.0012	0.0083-0.0122	0.0146 0.0205	0.0079-0.0236	0.0016-0.0031	0.0012-0.0028	0.0026-0.0053
	4.0	VQ40DE	0.0004-0.0012	0.0091-0.0130	0.0130-0.0189	0.0079-0.0197	0.0018-0.0031	0.0012-0.0028	0.0026-0.0053
2008	2.5	QR25DE	0.0004-0.0012	0.0083-0.0122	0.0146 0.0205	0.0079-0.0236	0.0016-0.0031	0.0012-0.0028	0.0026-0.0053
	4.0	VQ40DE	0.0004-0.0012	0.0091-0.0130	0.0130-0.0189	0.0079-0.0197	0.0018-0.0031	0.0012-0.0028	0.0026-0.0053

22140_FRON_C0008

TORQUE SPECIFICATIONS

All readings in ft. lbs.

Year	Engine Displacement Liters	Engine ID	Cylinder Head Bolts	Main Bearing Bolts	Rod Bearing Bolts	Crankshaft Damper Bolts	Flywheel Bolts	Manifold		Spark Plugs	Oil Pan Drain Plug
								Intake	Exhaust		
2007	2.5	QR25DE	①	②	③	④	80	⑤	⑥	14-22	25
	4.0	VQ40DE	⑦	⑧	14	⑨	65	⑩	⑪	14-22	25
2008	2.5	QR25DE	①	②	③	④	80	⑤	⑥	14-22	25
	4.0	VQ40DE	⑦	⑧	14	⑨	65	⑩	⑪	14-22	25

① Step 1: 37 ft. lbs.

Step 2: Plus 60 degrees clockwise

Step 3: loosen completely to 0 ft. lbs.

Step 4: 29 ft. lbs.

Step 5: Plus 75 degrees clockwise

Step 6: Plus 75 degrees clockwise

② Step 1: bolts 11-22 19 ft. lbs.

Step 2: bolts 1-10 29 ft. lbs.

Step 3: bolts 1-10 Plus 60-65 degrees

③ Step 1: 20 ft. lbs.

Step 2: loosen to 0 ft. lbs.

Step 3: 14 ft. lbs.

Step 4: Plus 85-95 degrees

④ Step 1: 31 ft. lbs.

Step 2: Plus 60 degrees

⑤ 14 ft. lbs.

⑥ Stud bolt: 11 ft. lbs.

Nuts: 22 ft. lbs.

⑦ Step 1: 72 ft. lbs.

Step 2: loosen completely to 0 ft. lbs.

Step 3: 29 ft. lbs.

Step 4: Plus 90 degrees clockwise

Step 5: Plus 90 degrees clockwise

⑧ Bolts: 17-24 (M8) 16 ft. lbs.

Install rear main seal

Bolts: 1-16 (M10) 26 ft. lbs.

Bolts: 1-16 (M10) Plus 90 degrees clockwise

⑨ Step 1: 33 ft. lbs.

Step 2: Plus 84-90 degrees clockwise

⑩ Intake manifold collector:

Bolts and nuts: 8 ft. lbs.

Stud bolts: 61 inch lbs.

⑪ Intake manifold:

Bolts and nuts: 5 ft. lbs. and

than to 21 ft. lbs.

Studs: 8 ft. lbs.

22140_FRON_C0009

WHEEL ALIGNMENT

| Year | Model | | Caster | | Camber | | Toe-in |
			Range (+/-Deg.)	Preferred Setting (Deg.)	Range (+/-Deg.)	Preferred Setting (Deg.)	(in.)
2007	Frontier	2WD	①	①	②	②	③
		4WD	④	④	⑤	⑤	③
	Xterra	2WD	①	①	②	②	⑥
		4WD	④	④	⑤	⑤	⑥
2008	Frontier	2WD	①	①	②	②	③
		4WD	④	④	⑤	⑤	③
	Xterra	2WD	①	①	②	②	⑥
		4WD	④	④	⑤	⑤	⑥

NOTE: On 2007-2008 vehicles, fuel, coolant and engine oil must be full. Spare tire, jack, hand tools and mats must be in place.

Some 2007-2008 vehicles may be equipped with non adjustable lower link bolts and washers. In order to adjust caster and camber on these vehicles, first replace these bolts with adjustable cam bolts and washers.

① Minimum: 2 degrees 15' (2.25 degrees)

  Nominal: 3 degrees 0' (3.00 degrees)

  Maximum: 3 degrees 45' (3.75 degrees)

② Minimum: 2 degrees 0' (2.00 degrees)

  Nominal: 0 degrees 15' (0.25 degrees)

  Maximum: 1 degrees 0' (1.00 degrees)

③ Minimum: 0.08 inches

  Nominal 0.12 inches

  Maximum 0.16 inches

④ Minimum: 0 degrees 15' (-0.25 degrees)

  Nominal: 2 degrees 45' (2.75 degrees)

  Maximum: 3 degrees 30' (3.50 degrees)

⑤ Minimum: 0 degrees 15' (-0.25 degrees)

  Nominal: 0 degrees 30' (0.50 degrees)

  Maximum: 1 degrees 15' (1.25 degrees)

⑥ Minimum: 0.12 inches

  Nominal 0.16 inches

  Maximum 0.20 inches

22140_FRON_C0010

TIRE, WHEEL AND BALL JOINT SPECIFICATIONS

Year	Model	OEM Tires Standard	OEM Tires Optional	Tire Pressures (psi) Front	Tire Pressures (psi) Rear	Wheel Size	Ball Joint Inspection	Lug Nut Torque (ft. lbs.)
2007	Frontier XE	P235/75R15	None	①	①	7J	②	98
	Frontier SE	P265/70R15	None	①	①	7J	②	98
	Frontier NISMO off road	P265/75R16	None	①	①	7J	②	98
	Frontier LE	P265/65R17	None	①	①	7.5J	②	98
	Xterra S	P265/70R16	None	①	①	7J	②	98
	Xterra S-O/R	P265/75R16	None	①	①	7J	②	98
	Xterra SE	P255/65R17	None	①	①	7.5J	②	98
2008	Frontier XE	P235/75R15	None	①	①	7J	②	98
	Frontier SE (King Cab)	P265/70R16	None	①	①	7J	②	98
	Frontier SE (Club Cab)	P265/65R17	None	①	①	7.5J	②	98
	Frontier NISMO off road	P265/75R16	None	①	①	7J	②	98
	Frontier LE	P265/65R17	None	①	①	7.5J	②	98
	Xterra X	P265/70R16	None	①	①	7J	②	98
	Xterra S	P265/70R16	None	①	①	7J	②	98
	Xterra S-O/R	P265/75R16	None	①	①	7J	②	98
	Xterra SE	P255/65R17	None	①	①	7.5J	②	98

OEM: Original Equipment Manufacturer

PSI: Pounds Per Square Inch

① See placard on vehicle

② Replace if any measurable movement is found.

22140_FRON_C0011

BRAKE SPECIFICATIONS

All measurements in inches unless noted

Year	Model	Brake Disc Original Thickness	Brake Disc Minimum Thickness	Brake Disc Maximum Runout	Brake Drum Diameter Original Inside Diameter	Brake Drum Diameter Max. Wear Limit	Brake Drum Diameter Maximum Machine Diameter	Minimum Lining Thickness Front	Minimum Lining Thickness Rear	Brake Caliper Bracket Bolts (ft. lbs.)	Brake Caliper Mounting Bolts (ft. lbs.)
2007	Frontier	①	②	0.002	③	④	—	0.079	0.079	⑤	⑥
	Xterra	1.102	1.024	0.002	⑦	④	—	0.079	0.079	⑤	⑥
2008	Frontier	①	②	0.002	③	④	—	0.079	0.079	⑤	⑥
	Xterra	1.102	1.024	0.002	⑦	④	—	0.079	0.079	⑤	⑥

① 4-cyl.: 0.710
6-cyl.: 1.100

② 4-cyl.: 0.630
6-cyl.: 1.024

③ Rear disc brakes: 0.710

④ Rear disc brakes: 0.630

⑤ Front: 136 ft. lbs.
Rear: 76 ft. lbs.

⑥ Front: 32 ft. lbs.
Rear: 19 ft. lbs.

⑦ Rear disc brakes: 0.709

22140_FRON_C0012

SCHEDULED MAINTENANCE INTERVALS
Nissan Frontier and Xterra

TO BE SERVICED	TYPE OF SERVICE	VEHICLE MILEAGE INTERVAL (x1000)												
		7.5	15	22.5	30	37.5	45	52.5	60	67.5	75	82.5	90	97.5
Engine oil & filter	R	✓	✓	✓	✓	✓	✓	✓	✓	✓	✓	✓	✓	✓
Brake lines & cables	S/I		✓		✓		✓		✓		✓		✓	
Brake pads, discs, drums & linings	S/I		✓		✓		✓		✓		✓		✓	
Driveshaft boots & propeller shaft	S/I				✓				✓				✓	
Front wheel bearings (4X2)	S/I				✓				✓				✓	
Front wheel bearings (4X4)	S/I				✓				✓				✓	
Automatic & manual transmission, transfer & differential gear oil ①	S/I		✓		✓		✓		✓		✓		✓	
Air cleaner filter	R				✓				✓				✓	
Engine coolant	R								✓				✓	
Spark plugs (platinum)	R	replace every 105,000 miles												
Drive belt(s)	S/I				✓				✓				✓	
Exhaust system	S/I				✓				✓				✓	
Fuel lines	S/I				✓				✓				✓	
Steering gear (box) & linkage, axle & suspension parts	S/I				✓				✓				✓	
Vapor lines	S/I				✓				✓				✓	
Tires (rotate)	S/I	✓	✓	✓	✓	✓	✓	✓	✓	✓	✓	✓	✓	✓
Timing belt ②	R													

R: Replace   S/I: Service or Inspect

① Differential (w/limited-slip differential) oil: replace oil every 30,000 miles, 2007-2008 vehicles.

② Timing belt: replace at 105,000 miles.

FREQUENT OPERATION MAINTENANCE (SEVERE SERVICE)

If a vehicle is operated under any of the following conditions it is considered severe service:

- Extremely dusty areas.

- 50% or more of the vehicle operation is in 32°C (90°F) or higher temperatures, or constant operation in temperatures below 0°C (32°F).

- Prolonged idling (vehicle operation in stop and go traffic).

- Frequent short running periods (engine does not warm to normal operating temperatures).

- Police, taxi, delivery usage or trailer towing usage.

Oil & oil filter: replace every 3750 miles.

Brake pads, discs, drums & linings: service or inspect every 7500 miles.

Driveshaft boots & propeller shaft: service or inspect every 7500 miles.

Exhaust system: service or inspect every 7500 miles.

Steering gear (box) & linkage, (steering damper-4X4), axle & suspension parts: service or inspect every 7500 miles.

Steering linkage ball joints & front suspension ball joints: service or inspect every 7500 miles.

22140_FRON_C0013

PRECAUTIONS

Before servicing any vehicle, please be sure to read all of the following precautions, which deal with personal safety, prevention of component damage, and important points to take into consideration when servicing a motor vehicle:

• Never open, service or drain the radiator or cooling system when the engine is hot; serious burns can occur from the steam and hot coolant.

• Observe all applicable safety precautions when working around fuel. Whenever servicing the fuel system, always work in a well-ventilated area. Do not allow fuel spray or vapors to come in contact with a spark, open flame, or excessive heat (a hot drop light, for example). Keep a dry chemical fire extinguisher near the work area. Always keep fuel in a container specifically designed for fuel storage; also, always properly seal fuel containers to avoid the possibility of fire or explosion. Refer to the additional fuel system precautions later in this section.

• Fuel injection systems often remain pressurized, even after the engine has been turned **OFF**. The fuel system pressure must be relieved before disconnecting any fuel lines. Failure to do so may result in fire and/or personal injury.

• Brake fluid often contains polyglycol ethers and polyglycols. Avoid contact with the eyes and wash your hands thoroughly after handling brake fluid. If you do get brake fluid in your eyes, flush your eyes with clean, running water for 15 minutes. If eye irritation persists, or if you have taken brake fluid internally, IMMEDIATELY seek medical assistance.

• The EPA warns that prolonged contact with used engine oil may cause a number of skin disorders, including cancer. You should make every effort to minimize your exposure to used engine oil. Protective gloves should be worn when changing oil. Wash your hands and any other exposed skin areas as soon as possible after exposure to used engine oil. Soap and water, or waterless hand cleaner should be used.

• All new vehicles are now equipped with an air bag system, often referred to as a Supplemental Restraint System (SRS) or Supplemental Inflatable Restraint (SIR) system. The system must be disabled before performing service on or around system components, steering column, instrument panel components, wiring and sensors. Failure to follow safety and disabling procedures could result in accidental air bag deployment, possible personal injury and unnecessary system repairs.

• Always wear safety goggles when working with, or around, the air bag system. When carrying a non-deployed air bag, be sure the bag and trim cover are pointed away from your body. When placing a non-deployed air bag on a work surface, always face the bag and trim cover upward, away from the surface. This will reduce the motion of the module if it is accidentally deployed. Refer to the additional air bag system precautions later in this section.

• Clean, high quality brake fluid from a sealed container is essential to the safe and proper operation of the brake system. You should always buy the correct type of brake fluid for your vehicle. If the brake fluid becomes contaminated, completely flush the system with new fluid. Never reuse any brake fluid. Any brake fluid that is removed from the system should be discarded. Also, do not allow any brake fluid to come in contact with a painted surface; it will damage the paint.

• Never operate the engine without the proper amount and type of engine oil; doing so WILL result in severe engine damage.

• Timing belt maintenance is extremely important. Many models utilize an interference-type, non-freewheeling engine. If the timing belt breaks, the valves in the cylinder head may strike the pistons, causing potentially serious (also time-consuming and expensive) engine damage. Refer to the maintenance interval charts for the recommended replacement interval for the timing belt, and to the timing belt section for belt replacement and inspection.

• Disconnecting the negative battery cable on some vehicles may interfere with the functions of the on-board computer system(s) and may require the computer to undergo a relearning process once the negative battery cable is reconnected.

• When servicing drum brakes, only disassemble and assemble one side at a time, leaving the remaining side intact for reference.

• Only an MVAC-trained, EPA-certified automotive technician should service the air conditioning system or its components.

BRAKES

GENERAL INFORMATION

The ABS system detects wheel revolution while braking and improves handling stability during sudden braking by electrically preventing wheel lockup. Maneuverability is also improved for avoiding obstacles.

PRECAUTIONS

• Certain components within the ABS system are not intended to be serviced or repaired individually.

• Do not use rubber hoses or other parts not specifically specified for and ABS system. When using repair kits, replace all parts included in the kit. Partial or incorrect repair may lead to functional problems and require the replacement of components.

• Lubricate rubber parts with clean, fresh brake fluid to ease assembly. Do not use shop air to clean parts; damage to rubber components may result.

• Use only DOT 3 brake fluid from an unopened container.

• If any hydraulic component or line is removed or replaced, it may be necessary to bleed the entire system.

• A clean repair area is essential. Always clean the reservoir and cap thoroughly before removing the cap. The slightest amount of dirt in the fluid may plug an orifice and impair the system function. Perform

ANTI-LOCK BRAKE SYSTEM (ABS)

repairs after components have been thoroughly cleaned; use only denatured alcohol to clean components. Do not allow ABS components to come into contact with any substance containing mineral oil; this includes used shop rags.

• The Anti-Lock control unit is a microprocessor similar to other computer units in the vehicle. Ensure that the ignition switch is **OFF** before removing or installing controller harnesses. Avoid static electricity discharge at or near the controller.

• If any arc welding is to be done on the vehicle, the control unit should be unplugged before welding operations begin.

BRAKES   BLEEDING THE BRAKE SYSTEM

BLEEDING PROCEDURE

BLEEDING PROCEDURE

1. Before servicing the vehicle, refer to the Precautions Section.

✳✳ CAUTION

While bleeding the brake system, pay attention to the master cylinder fluid level.

2. Disconnect the negative battery cable.

3. Raise and safely support the vehicle.

4. Attach a vinyl tube to the right, rear bleeder valve.

5. Depress the brake pedal fully 4 or 5 times.

6. With the brake pedal depressed, loosen the bleeder valve to let the air out, then tighten it immediately.

7. Repeat steps 3 and 4 until no more air comes out.

8. Tighten the bleeder valve.

9. Fill the master cylinder reservoir.

10. Repeat the above steps for the left front, left rear, and the right front calipers, in that order.

BRAKES   FRONT DISC BRAKES

✳✳ CAUTION

Dust and dirt accumulating on brake parts during normal use may contain asbestos fibers from production or aftermarket brake linings. Breathing excessive concentrations of asbestos fibers can cause serious bodily harm. Exercise care when servicing brake parts. Do not sand or grind brake lining unless equipment used is designed to contain the dust residue. Do not clean brake parts with compressed air or by dry brushing. Cleaning should be done by dampening the brake components with a fine mist of water, then wiping the brake components clean with a dampened cloth. Dispose of cloth and all residue containing asbestos fibers in an impermeable container with the appropriate label. Follow practices prescribed by the Occupational Safety and Health Administration (OSHA) and the Environmental Protection Agency (EPA) for the handling, processing, and disposing of dust or debris that may contain asbestos fibers.

BRAKE CALIPER

REMOVAL & INSTALLATION

See Figure 1.

1. Before servicing the vehicle, refer to the Precautions Section.

2. Drain the brake fluid, as necessary.

3. Raise the vehicle and support safely.

4. Remove the tire and wheel assembly.

5. Remove the bolt attaching the brake hose to the caliper. Plug the brake hose to prevent brake fluid loss.

6. Remove the caliper support mounting bolts and lift the caliper assembly from the knuckle.

To install

7. Position the caliper assembly onto the knuckle and install the bolts. Make sure the rotor fits between the brake pads. Torque the bolts to specification.

8. Using new copper washers, connect the brake hose to the caliper.

Torque the brake hose attaching bolt to specification.

9. Bleed the brake system.

10. Apply the brake pedal and inspect the system. Ensure proper operation and no leakage.

11. Install tire and wheel assembly. Lower the vehicle and road test.

DISC BRAKE PADS

REMOVAL & INSTALLATION

1. Before servicing the vehicle, refer to the Precautions Section.

2. Drain the brake fluid, as necessary.

3. Raise the vehicle and support safely.

4. Remove the bottom pin from the caliper and swing the caliper cylinder body upward; support the caliper with a wire.

5. Remove the brake pad retainers, shims and the pads.

To install:

6. Compress the piston of the disc brake caliper.

7. Install the brake pads and caliper assembly.

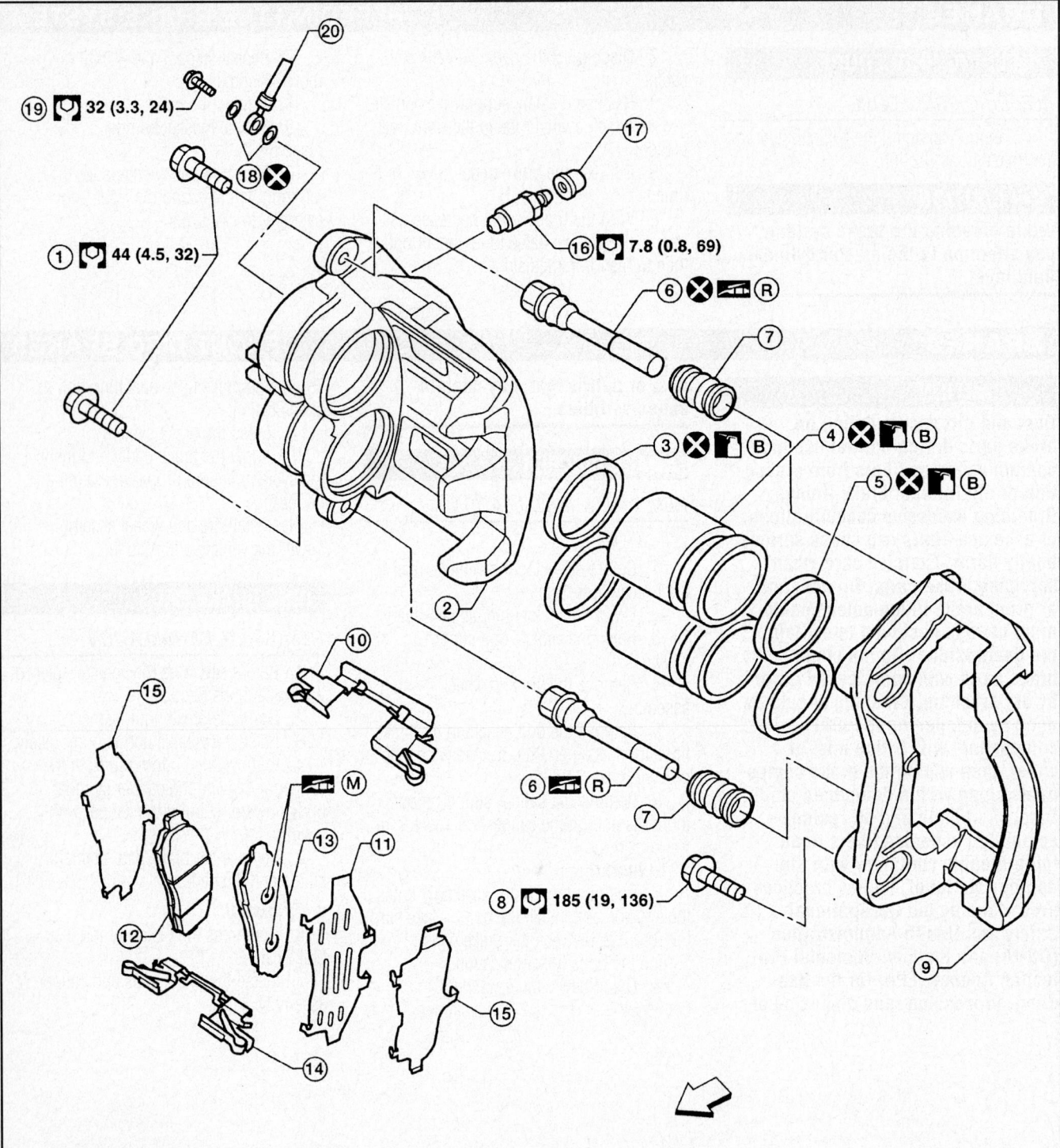

1. Sliding pin bolt
2. Cylinder body
3. Piston seal
4. Piston
5. Piston boot
6. Sliding pin
7. Sliding pin boot
8. Torque member bolt
9. Torque member
10. Pad retainer
11. Inner shim
12. Inner brake pad
13. Outer brake pad
14. Pad retainer
15. Outer shim
16. Bleed valve
17. Cap
18. Copper washers
19. Union bolt
20. Brake hose

Fig. 1 Front disc brake and related components

09482_FRON_G0121

BRAKES

REAR DISC BRAKES

BRAKE CALIPER

REMOVAL & INSTALLATION

See Figures 2 and 3.

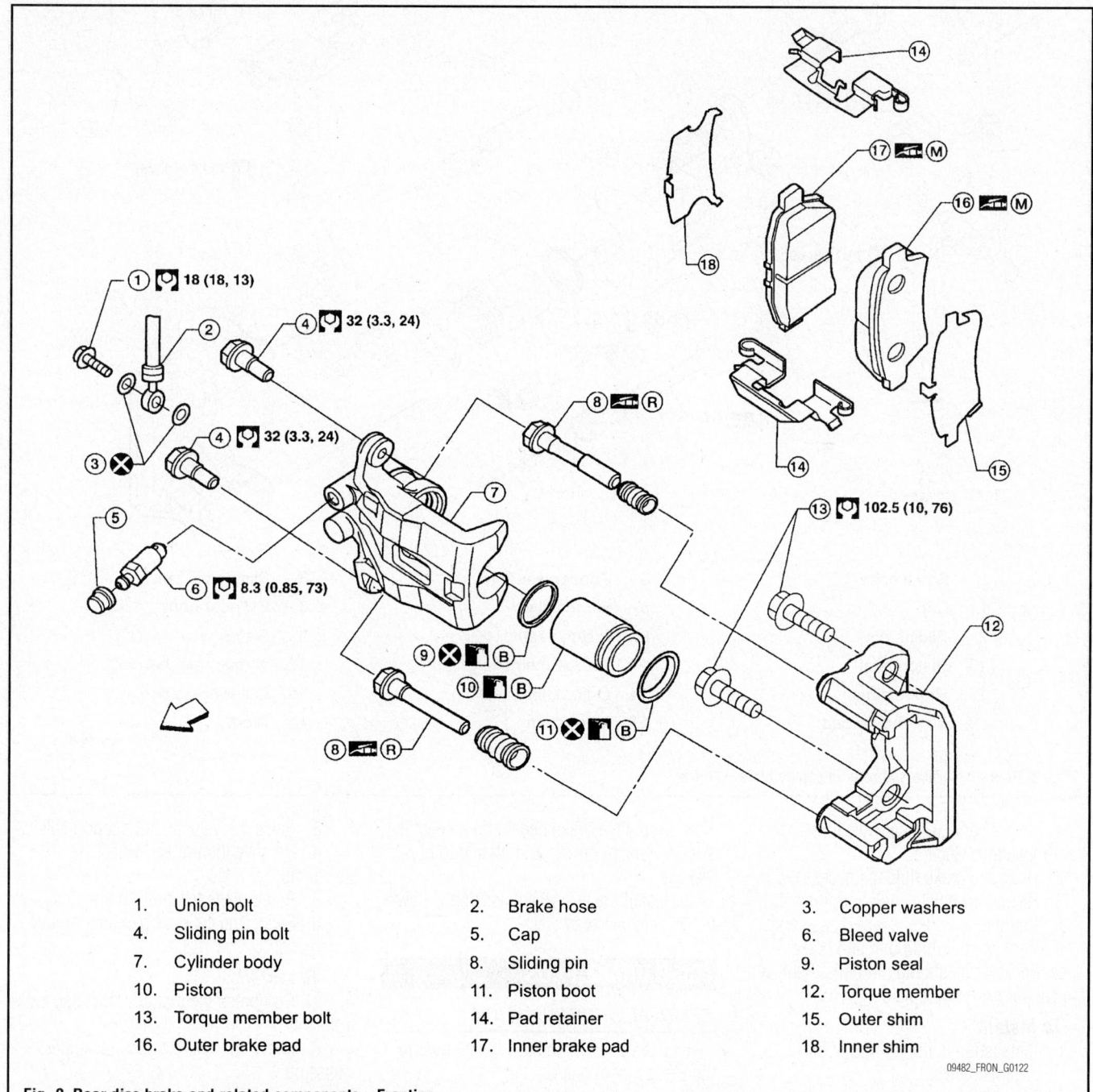

1. Union bolt	2. Brake hose	3. Copper washers
4. Sliding pin bolt	5. Cap	6. Bleed valve
7. Cylinder body	8. Sliding pin	9. Piston seal
10. Piston	11. Piston boot	12. Torque member
13. Torque member bolt	14. Pad retainer	15. Outer shim
16. Outer brake pad	17. Inner brake pad	18. Inner shim

09482_FRON_G0122

Fig. 2 Rear disc brake and related components—Frontier

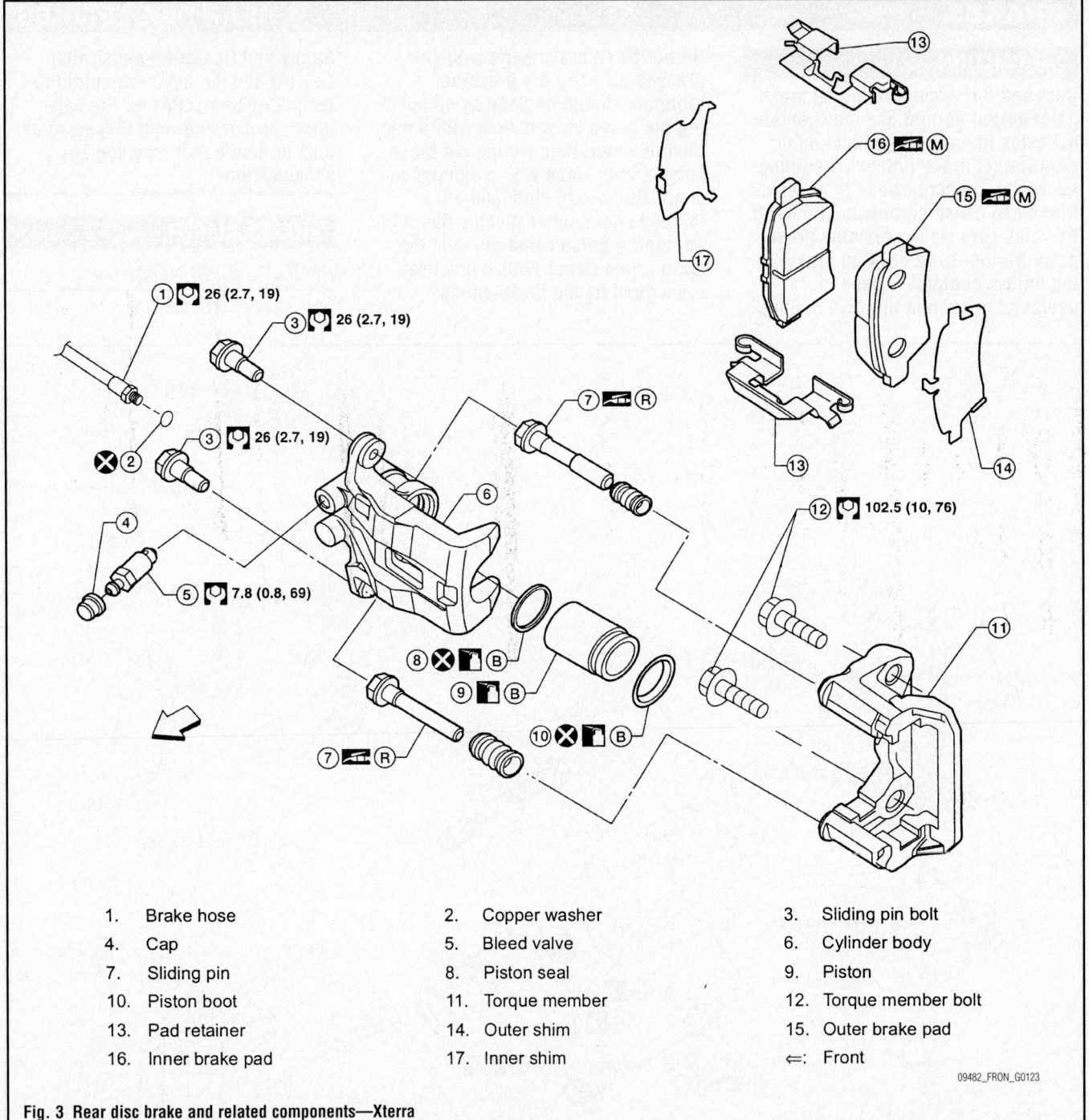

1. Brake hose
2. Copper washer
3. Sliding pin bolt
4. Cap
5. Bleed valve
6. Cylinder body
7. Sliding pin
8. Piston seal
9. Piston
10. Piston boot
11. Torque member
12. Torque member bolt
13. Pad retainer
14. Outer shim
15. Outer brake pad
16. Inner brake pad
17. Inner shim
⇐: Front

09482_FRON_G0123

Fig. 3 Rear disc brake and related components—Xterra

1. Before servicing the vehicle, refer to the Precautions Section.
2. Drain the brake fluid, as necessary.
3. Raise the vehicle and support safely.
4. Remove the tire and wheel assembly.
5. Remove the union bolt and brake hose. Remove the sliding pin bolts. Remove the caliper from the vehicle.

To install:

6. Installation is the reverse of the removal procedure.
7. Bleed the brake system.

8. Apply the brake pedal and inspect the system. Ensure proper operation and no leakage.
9. Install tire and wheel assembly. Lower the vehicle and road test.

DISC BRAKE PADS

REMOVAL & INSTALLATION

1. Before servicing the vehicle, refer to the Precautions Section.
2. Drain the brake fluid, as necessary.

3. Raise the vehicle and support safely.
4. Remove the tire and wheel assembly.
5. Remove the top bolt from the caliper.
6. Swing the caliper open and remove the pads.

To install:

7. Compress the piston of the disc brake caliper.
8. Install the brake pads and caliper assembly.

BRAKES   **PARKING BRAKE**

PARKING BRAKE SHOES

REMOVAL & INSTALLATION

See Figures 4 and 5.

1. Before servicing the vehicle, refer to the Precautions Section.

➡ **Clean the brakes with a vacuum dust collector to minimize the hazard of airborne particles or other materials.**

➡ **Remove the disc rotor with the parking brake completely disengaged.**

2. Raise and safely support the vehicle.

3. Release the parking brake.
4. Remove the rear wheels.
5. Remove the rotor.
6. Remove the return springs.
7. Remove the adjuster.
8. Remove the, retainers, anti-rattle pins and shoes.

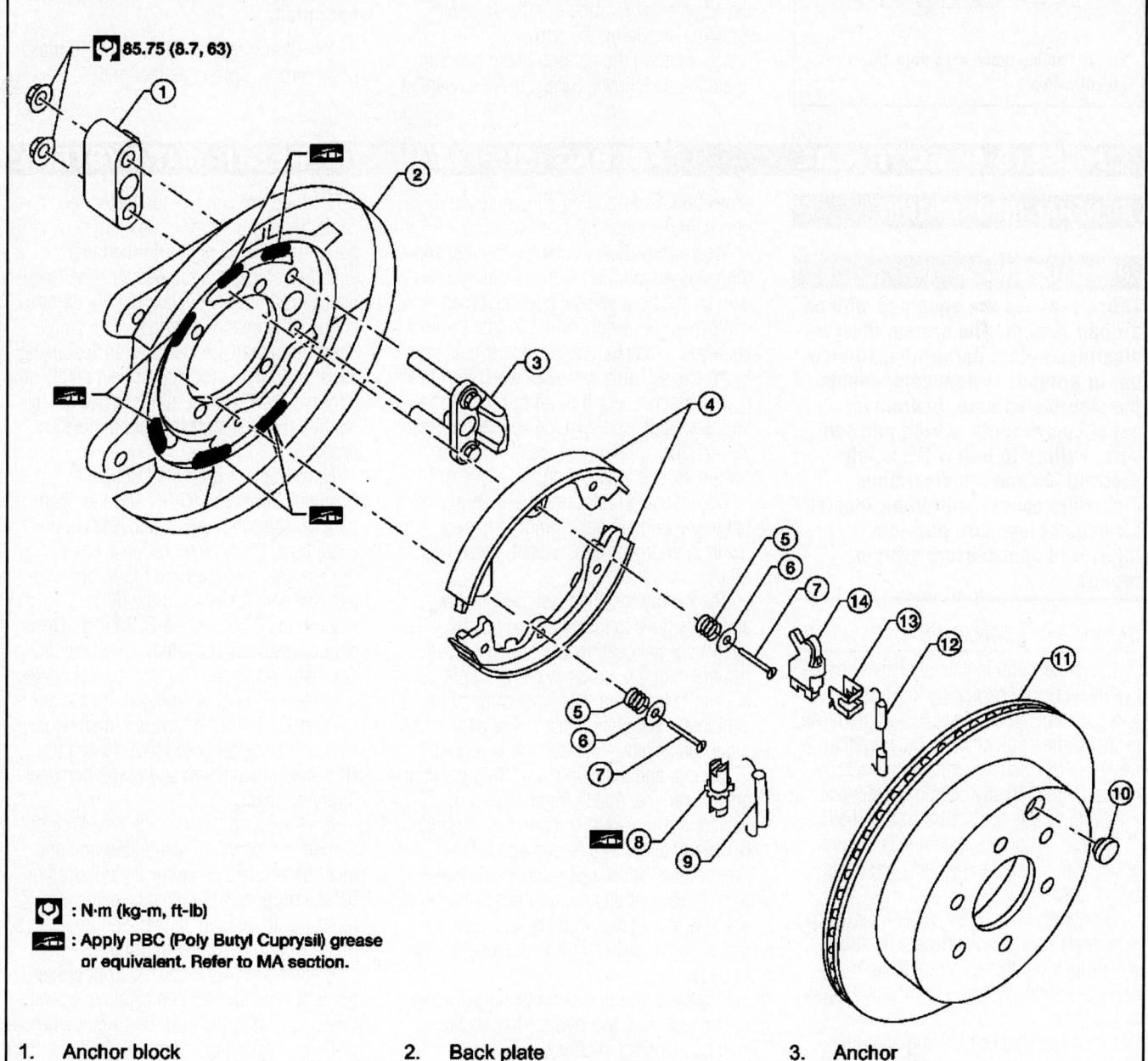

85.75 (8.7, 63)

○ : N·m (kg-m, ft-lb)

: Apply PBC (Poly Butyl Cuprysil) grease or equivalent. Refer to MA section.

1. Anchor block	2. Back plate	3. Anchor
4. Shoes	5. Shoe hold-down spring	6. Retainer
7. Shoe hold-down pin	8. Adjuster	9. Rear return spring
10. Adjuster access plug	11. Disc rotor	12. Front return spring
13. Pin retainer	14. Toggle lever	

42050_FRON_G0055

Fig. 4 Parking brake and related components

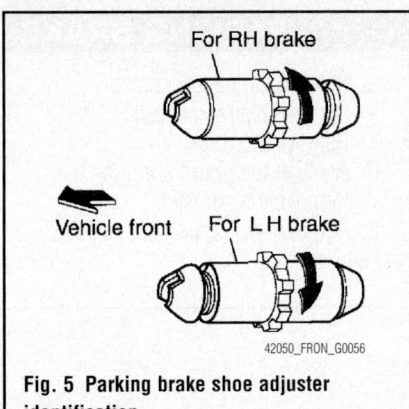

Fig. 5 Parking brake shoe adjuster identification

9. Remove the pin retainer. Disconnect the parking brake cable from the toggle lever.

10. Remove the back plate.

To install:

11. Installation is the reverse of the removal procedure.

12. Assemble the adjuster so that the threaded part expands when rotating it in the direction shown by the arrow. Shorten the adjuster by rotating it in the opposite direction shown by the arrow.

13. Perform the parking brake break-in operation as follows: Safely, drive forward at approximately 40 km/h (25 mph) with the parking brake set with a force of approx. 200 N (45 lbs.) for about 30 seconds.

14. After the break-in operation, check the pedal stroke of parking brake. Readjust if necessary.

➡ **To prevent lining from getting too hot, allow a cool off period of approximately 5 minutes after every break-in operation.**

15. Check and adjust the parking brake pedal stroke. Correct as required.

CHASSIS ELECTRICAL

AIR BAG (SUPPLEMENTAL RESTRAINT SYSTEM)

GENERAL INFORMATION

✳✳ CAUTION

These vehicles are equipped with an air bag system. The system must be disarmed before performing service on, or around, system components, the steering column, instrument panel components, wiring and sensors. Failure to follow the safety precautions and the disarming procedure could result in accidental air bag deployment, possible injury and unnecessary system repairs.

SERVICE PRECAUTIONS

Disconnect and isolate the battery negative cable before beginning any airbag system component diagnosis, testing, removal, or installation procedures. Allow system capacitor to discharge for two minutes before beginning any component service. This will disable the airbag system. Failure to disable the airbag system may result in accidental airbag deployment, personal injury, or death.

Do not place an intact undeployed airbag face down on a solid surface. The airbag will propel into the air if accidentally deployed and may result in personal injury or death.

When carrying or handling an undeployed airbag, the trim side (face) of the airbag should be pointing towards the body to minimize possibility of injury if accidental deployment occurs. Failure to do this may result in personal injury or death.

Replace airbag system components with OEM replacement parts. Substitute parts may appear interchangeable, but internal differences may result in inferior occupant protection. Failure to do so may result in occupant personal injury or death.

Wear safety glasses, rubber gloves, and long sleeved clothing when cleaning powder residue from vehicle after an airbag deployment. Powder residue emitted from a deployed airbag can cause skin irritation. Flush affected area with cool water if irritation is experienced. If nasal or throat irritation is experienced, exit the vehicle for fresh air until the irritation ceases. If irritation continues, see a physician.

Do not use a replacement airbag that is not in the original packaging. This may result in improper deployment, personal injury, or death.

The factory installed fasteners, screws and bolts used to fasten airbag components have a special coating and are specifically designed for the airbag system. Do not use substitute fasteners. Use only original equipment fasteners listed in the parts catalog when fastener replacement is required.

During, and following, any child restraint anchor service, due to impact event or vehicle repair, carefully inspect all mounting hardware, tether straps, and anchors for proper installation, operation, or damage. If a child restraint anchor is found damaged in any way, the anchor must be replaced. Failure to do this may result in personal injury or death.

Deployed and non-deployed airbags may or may not have live pyrotechnic material within the airbag inflator.

Do not dispose of driver/passenger/curtain airbags or seat belt tensioners unless you are sure of complete deployment. Refer to the Hazardous Substance Control System for proper disposal.

Dispose of deployed airbags and tensioners consistent with state, provincial, local, and federal regulations.

After any airbag component testing or service, do not connect the battery negative cable. Personal injury or death may result if the system test is not performed first.

If the vehicle is equipped with the Occupant Classification System (OCS), do not connect the battery negative cable before performing the OCS Verification Test using the scan tool and the appropriate diagnostic information. Personal injury or death may result if the system test is not performed properly.

Never replace both the Occupant Restraint Controller (ORC) and the Occupant Classification Module (OCM) at the same time. If both require replacement, replace one, then perform the Airbag System test before replacing the other.

Both the ORC and the OCM store Occupant Classification System (OCS) calibration data, which they transfer to one another when one of them is replaced. If both are replaced at the same time, an irreversible fault will be set in both modules and the OCS may malfunction and cause personal injury or death.

If equipped with OCS, the Seat Weight Sensor is a sensitive, calibrated unit and must be handled carefully. Do not drop or handle roughly. If dropped or damaged, replace with another sensor. Failure to do so may result in occupant injury or death.

If equipped with OCS, the front passenger seat must be handled carefully as well. When removing the seat, be careful when setting on floor not to drop. If dropped, the sensor may be inoperative, could result in occupant injury, or possibly death.

If equipped with OCS, when the passenger front seat is on the floor, no one should sit in the front passenger seat. This uneven force may damage the sensing ability of the seat weight sensors. If sat on and damaged, the sensor may be inoperative, could result in occupant injury, or possibly death.

DISARMING THE SYSTEM

DISARMING THE SYSTEM

To disarm the **SRS** system turn the ignition switch to the **OFF** position. Then, disconnect both battery cables starting with the negative cable first and wait at least 3 minutes after the cables are disconnected.

ARMING THE SYSTEM

To rearm the **SRS** system, turn the ignition switch to the **OFF** position. Connect both battery cables starting with the positive cable first.

CLOCKSPRING CENTERING

1. Be sure to align the spiral cable correctly when installing the steering wheel. Make sure that the spiral cable is in the neutral position.

➡**The neutral position is detected by turning to the left 2.6 revolutions from** the right end position and ending with the knob at the top. The spiral cable may snap due to steering operation if the cable is installed incorrectly. Also, with the steering linkage disconnected the cable may snap by turning the steering wheel beyond the limited number of turns (2.6 from the neutral position to both the left and right).

DRIVETRAIN

AUTOMATIC TRANSMISSION ASSEMBLY

REMOVAL & INSTALLATION

2WD

See Figures 6 through 10.

➡**Before removing the transmission remove the crankshaft position sensor (POS) from the transmission assembly.**

1. Before servicing the vehicle, refer to the Precautions Section.

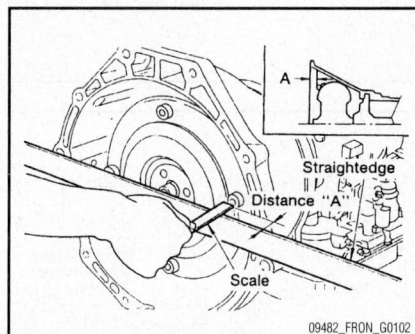

Fig. 6 Torque converter installation measurement "A"

2. Disconnect the negative battery cable.

3. Raise and support the vehicle safely. Drain the transmission fluid.

4. Remove the transmission fluid indicator. Remove the left fender protector.

5. Remove the crankshaft position sensor (POS) from the transmission. Do not disassemble it. Do not place it in an area affected by magnetism.

6. Remove the engine under covers. Remove the front crossmember. Remove the starter. Remove the rear driveshaft.

7. If equipped with a 4.0L engine, remove the left and right exhaust tubes.

8. Remove the selector control cable and bracket from the transmission. Disconnect and plug the cooler lines.

9. Remove the dust cover from the torque converter housing. Remove the torque converter to flex plate retaining bolts. There are four of them.

➡**Always rotate the crankshaft in the clockwise direction as viewed from the front of the engine.**

10. Support the transmission using a suitable jack.

11. Remove the insulator retaining nuts.

Remove the crossmember retaining bolts. Remove the crossmember.

➡**Be sure that the transmission is properly supported.**

12. Tilt the transmission slightly to gain clearance between the body and the transmission, and then remove the air breather hose and the breather tube.

13. Disconnect the transmission harness connector. Remove the wiring harness from the retainers. Remove the fluid indicator pipe.

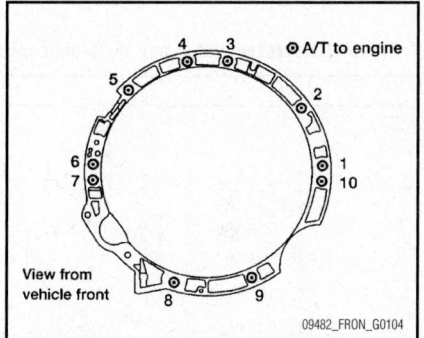

Fig. 8 4.0L engine transmission bolt tightening sequence

Bolt No.	1	2	3
Quantity	4	1	6
Bolt length " ℓ " mm (in)	60 (2.36)		65 (2.56)
Tightening torque N·m (kg-m, ft-lb)	35 (3.6, 26)		75 (7.7, 55)

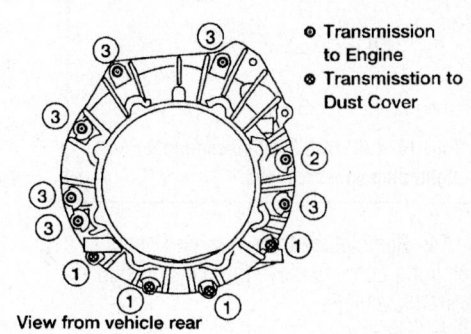

Fig. 7 2.5L engine transmission bolt tightening sequence

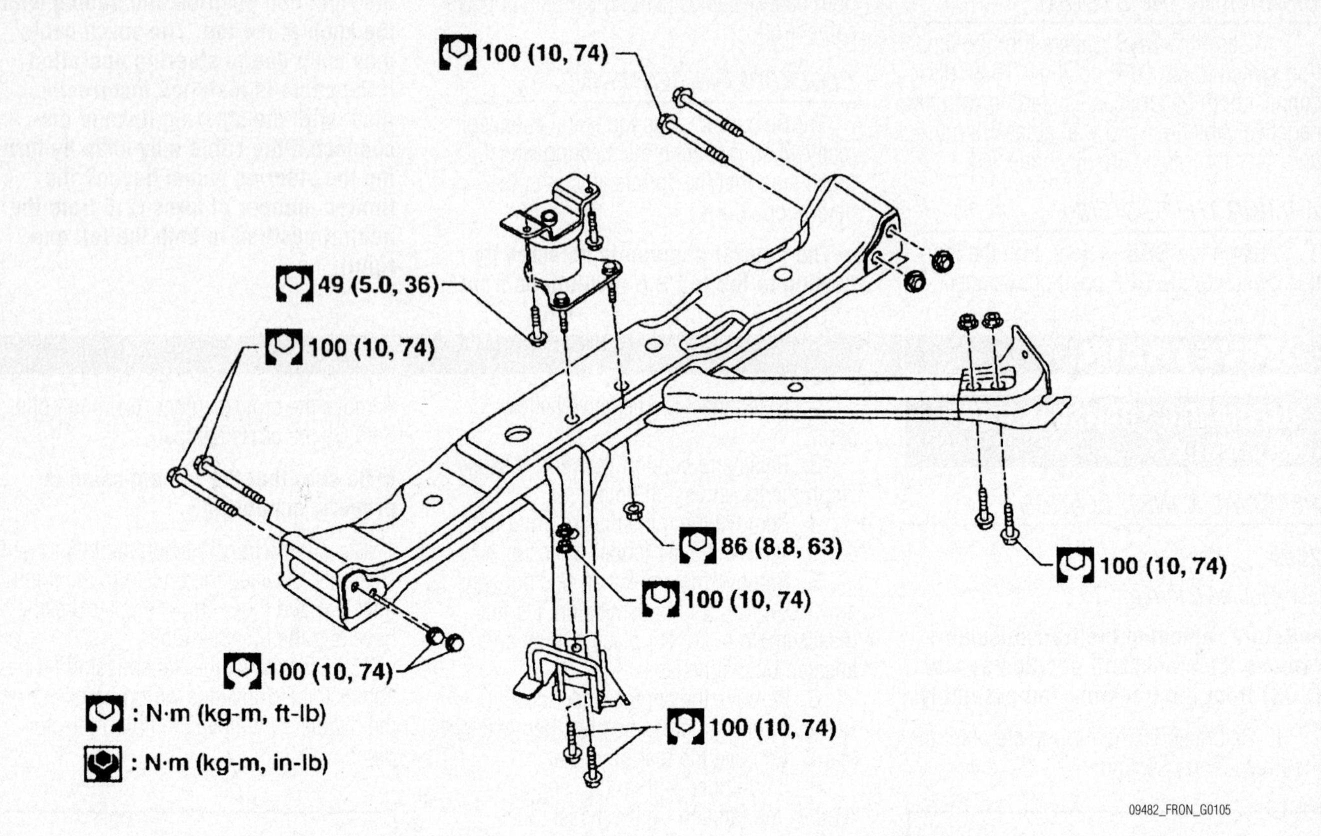

100 (10, 74)
49 (5.0, 36)
100 (10, 74)
86 (8.8, 63)
100 (10, 74)
100 (10, 74)
100 (10, 74)
100 (10, 74)
: N·m (kg-m, ft-lb)
: N·m (kg-m, in-lb)
09482_FRON_G0105

Fig. 9 2.5L crossmember bolt tightening specifications

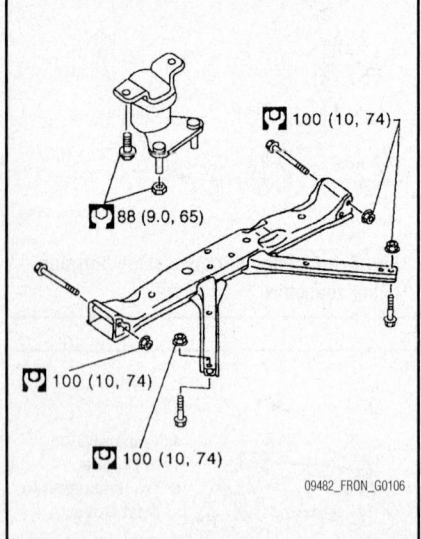

100 (10, 74)
88 (9.0, 65)
100 (10, 74)
100 (10, 74)
09482_FRON_G0106

Fig. 10 4.0L with 2WD crossmember bolt tightening specifications

14. Remove the transmission to engine retaining bolts. Remove the transmission from the vehicle.

To install:

15. Installation is the reverse of the removal procedure.

16. After installing the torque converter to the transmission, check dimension "A" to be sure it is within specification. Specification is 0.98 inch or more.

17. Torque the transmission to engine retaining bolts to specification and in the proper sequence. Torque bolts to 55 ft. lbs if equipped with 4.0L engine. If equipped with 2.5L engine, see illustration.

4WD

See Figures 11 through 13.

➡Before removing the transmission remove the crankshaft position sensor (POS) from the transmission assembly.

1. Before servicing the vehicle, refer to the Precautions Section.
2. Disconnect the negative battery cable.
3. Raise and support the vehicle safely. Drain the transmission fluid.
4. Remove the transmission fluid indicator. Remove the left fender protector.
5. Remove the crankshaft position sensor (POS) from the transmission. Do not disassemble it. Do not place it in an area affected by magnetism.
6. On Frontier, remove the air dam.
7. Remove the engine under covers. Remove the front crossmember. Remove the

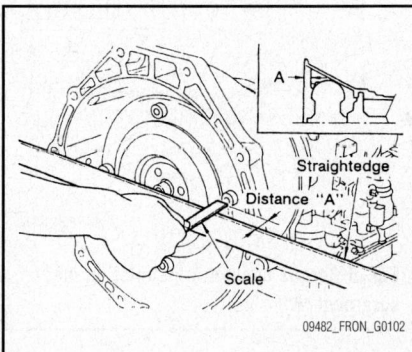

A
Straightedge
Distance "A"
Scale
09482_FRON_G0102

Fig. 11 Torque converter installation measurement "A"

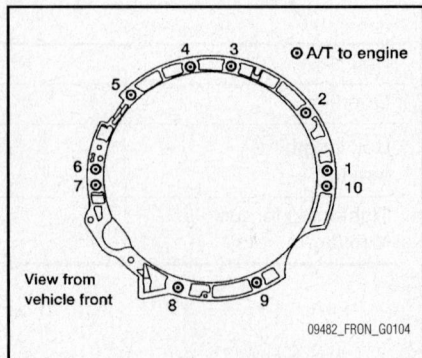

4
3
A/T to engine
5
2
6
7
1
10
View from
vehicle front
8
9
09482_FRON_G0104

Fig. 12 4.0L engine transmission bolt tightening sequence

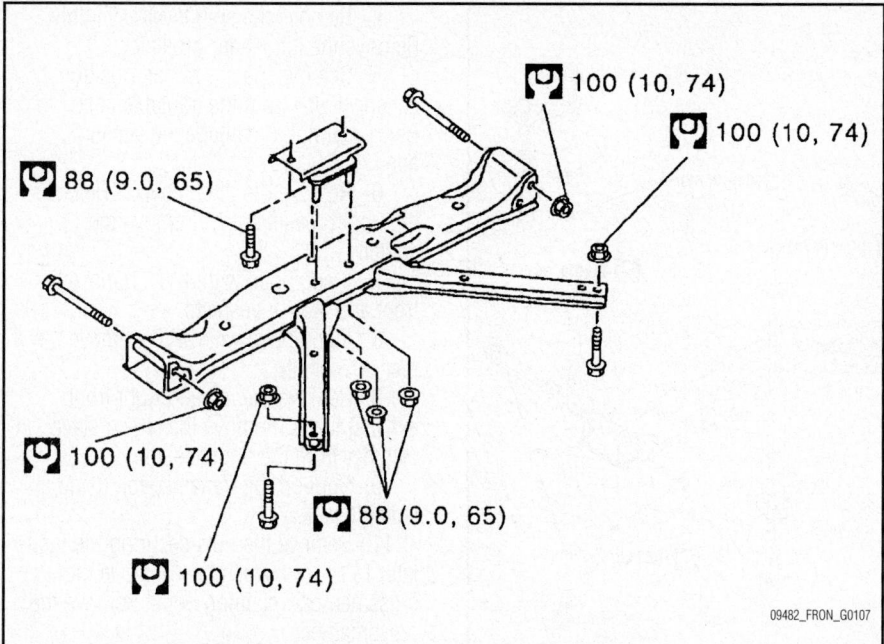

Fig. 13 4.0L with 4WD crossmember bolt tightening specifications

starter. Remove the front and rear driveshafts.

8. Remove the left and right exhaust tubes.

9. Remove the selector control cable and bracket from the transmission. Disconnect and plug the cooler lines.

10. Remove the dust cover from the torque converter housing. Remove the torque converter to flex plate retaining bolts. There are four of them.

➡**Always rotate the crankshaft in the clockwise direction as viewed from the front of the engine.**

11. Support the transmission using a suitable jack.

12. Remove the insulator retaining nuts.

Remove the crossmember retaining bolts. Remove the crossmember.

➡**Be sure that the transmission is properly supported.**

13. Tilt the transmission slightly to gain clearance between the body and the transmission, and then remove the air breather hose and the breather tube.

14. Disconnect the transmission harness connector. Disconnect the 4LO switch connector, wait detection switch connector, ATP switch connector and the transfer control device connector.

15. Remove the wiring harness from the retainers. Remove the fluid indicator pipe. Plug any openings.

16. Remove the transmission to engine

retaining bolts. Remove the transmission from the vehicle, with the transfer case attached.

17. As required, remove the transfer case from the transmission.

To install:

18. Installation is the reverse of the removal procedure.

19. After installing the torque converter to the transmission, check dimension "A" to be sure it is within specification. Specification is 0.98 inch or more.

20. Torque the transmission to engine retaining bolts to specification and in the proper sequence. Torque bolts to 55 ft. lbs.

MANUAL TRANSMISSION ASSEMBLY

REMOVAL & INSTALLATION

Five Speed (FS5R31A)

See Figures 14 and 15.

1. Before servicing the vehicle, refer to the Precautions Section.

2. Disconnect the negative battery cable.

3. Raise and support the vehicle safely. Drain the transmission fluid.

4. Remove the shift lever assembly. Remove the rear driveshaft. Remove the gusset.

5. Disconnect the heated oxygen sensor connector and remove the wire harness from the transmission.

6. Disconnect the backup light switch and the PNP switch connectors.

7. Remove the clutch slave cylinder from the transmission. Remove the starter.

8. Support the transmission using a suitable jack.

Bolt No.	1	2	3
Quantity	4	1	6
Bolt length " ℓ " mm (in)	60 (2.36)		65 (2.56)
Tightening torque N·m (kg-m, ft-lb)	34.3 (3.5, 25)		75 (7.7, 55)

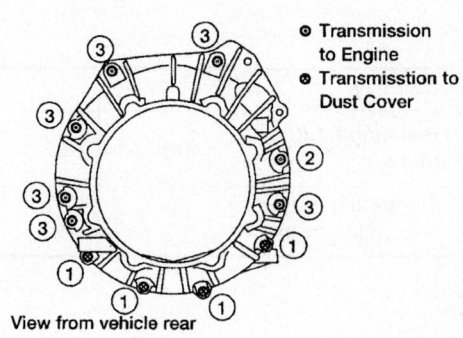

⊘ Transmission to Engine
⊗ Transmisstion to Dust Cover

View from vehicle rear

Fig. 14 Five speed transmission bolt tightening sequence

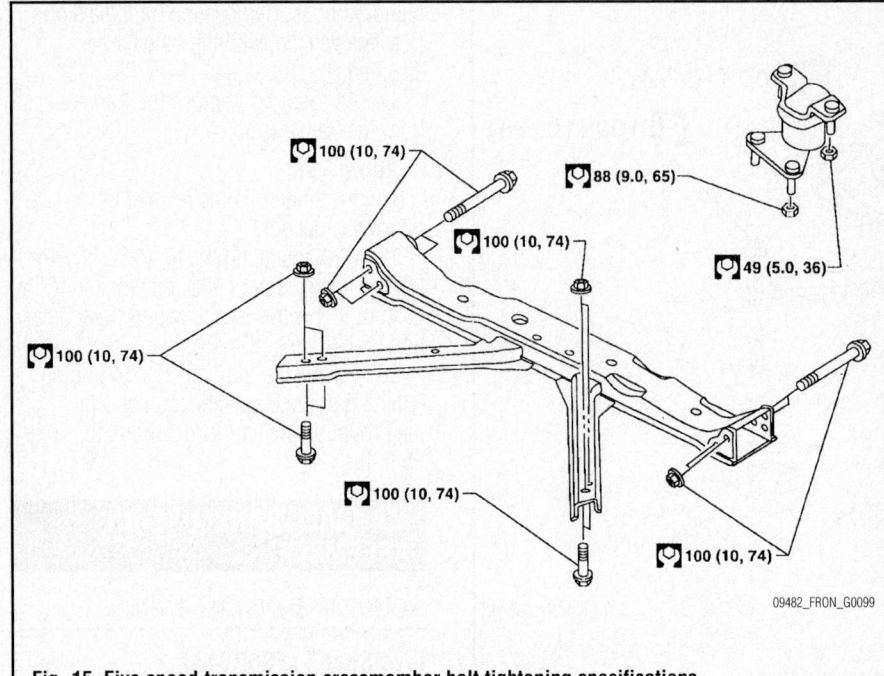

Fig. 15 Five speed transmission crossmember bolt tightening specifications

09482_FRON_G0099

9. Remove the transmission dust cover. Remove the transmission to engine retaining bolts.

10. Remove the nuts securing the insulator to the crossmember. Remove the crossmember retaining bolts. Remove the crossmember.

➡ **Be sure that the transmission is properly supported.**

11. Remove the air breather hose and the breather tube.

12. Separate the transmission from the engine and remove it from the vehicle.

To install:

13. Installation is the reverse of the removal procedure.

14. Torque the transmission to engine retaining bolts to specification and in the proper sequence.

15. Start the engine and check for leaks, correct as required.

16. Road test the vehicle and check for vibrations, correct as required.

Six Speed (FS6R31A)

See Figures 16 and 17.

1. Before servicing the vehicle, refer to the Precautions Section.

2. Disconnect the negative battery cable.

3. Raise and support the vehicle safely. Drain the transmission fluid.

4. Remove the shift lever assembly. Remove the left fender protector.

5. Remove the crankshaft position sensor (POS) from the transmission. Be careful not to damage the sensor edge.

6. Remove the undercovers. Remove the front crossmember. Remove the starter.

7. If equipped with 4WD, remove the front and rear driveshafts.

8. If equipped with 2WD, remove the rear driveshaft.

9. Remove the left and right front exhaust tubes. Remove the clutch slave cylinder from the transmission.

10. Support the transmission using a suitable jack.

11. Remove the nuts securing the insulator to the crossmember. Remove the crossmember retaining bolts. Remove the crossmember.

➡ **Be sure that the transmission is properly supported.**

12. Tilt the transmission slightly to gain clearance between the body and the transmission, and then remove the air breather hose and the breather tube.

13. Disconnect the backup light electrical connector and the PNP switch connector.

14. If equipped with 4WD, disconnect the 4LO switch connector, wait detection switch connector, ATP switch connector and the transfer control device connector.

15. Remove the wiring harness from the retainers.

16. Remove the transmission to engine retaining bolts. Separate the transmission from the engine and remove it from the vehicle.

Quantity	10
Bolt length "ℓ" mm (in)	65 (2.56)
Tightening torque N·m (kg-m, ft-lb)	75 (7.7, 55)

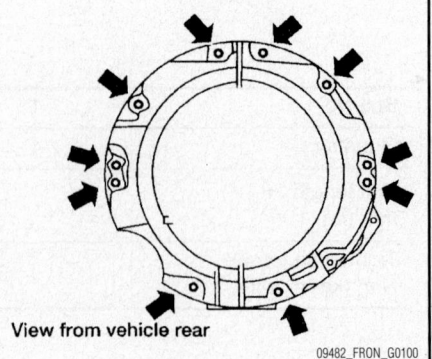

View from vehicle rear

09482_FRON_G0100

Fig. 16 Six speed transmission bolt tightening sequence

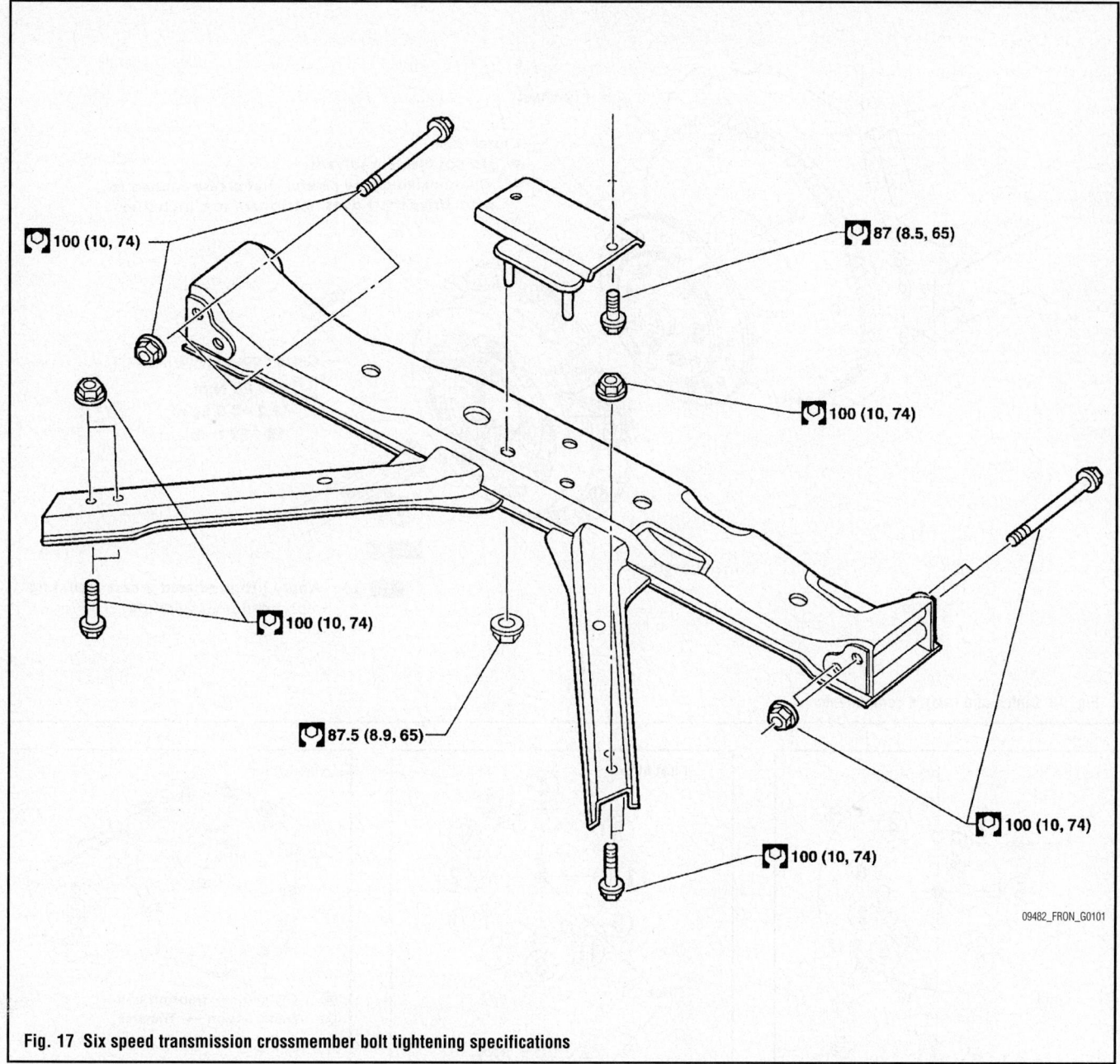

100 (10, 74)

87 (8.5, 65)

100 (10, 74)

100 (10, 74)

87.5 (8.9, 65)

100 (10, 74)

100 (10, 74)

09482_FRON_G0101

Fig. 17 Six speed transmission crossmember bolt tightening specifications

To install:

17. Installation is the reverse of the removal procedure.

18. Torque the transmission to engine retaining bolts to specification and in the proper sequence.

CLUTCH

REMOVAL & INSTALLATION

See Figures 18 through 20.

1. Before servicing the vehicle, refer to the Precautions Section.

2. Remove or disconnect the following:
   - Negative battery cable

- Transmission
- Pressure plate. Loosen the bolts evenly in ½ turn steps.
- Clutch disc

To install:
3. Install or connect the following:
   - Clutch disc and pressure plate.

➡On vehicles with 5 speed transmission tighten the pressure plate bolts in sequence to 11 ft. lbs. and then to 29 ft. lbs. On vehicles with 6 speed transmission tighten the pressure plate bolts in sequence to 11 ft. lbs. and then to 19 ft. lbs.

- Transmission
- Negative battery cable

BLEEDING

1. Before servicing the vehicle, refer to the Precautions Section.

➡Do not use a vacuum assist or any other type of power bleeder on this system. Use of a vacuum assist or power bleeder will not purge all the air from the system.

2. Fill the system with the proper grade and type fluid.

3. Have an assistant pump the clutch pedal slowly several times and hold it depressed.

4. Open the slave cylinder bleeder screw and allow air to escape.

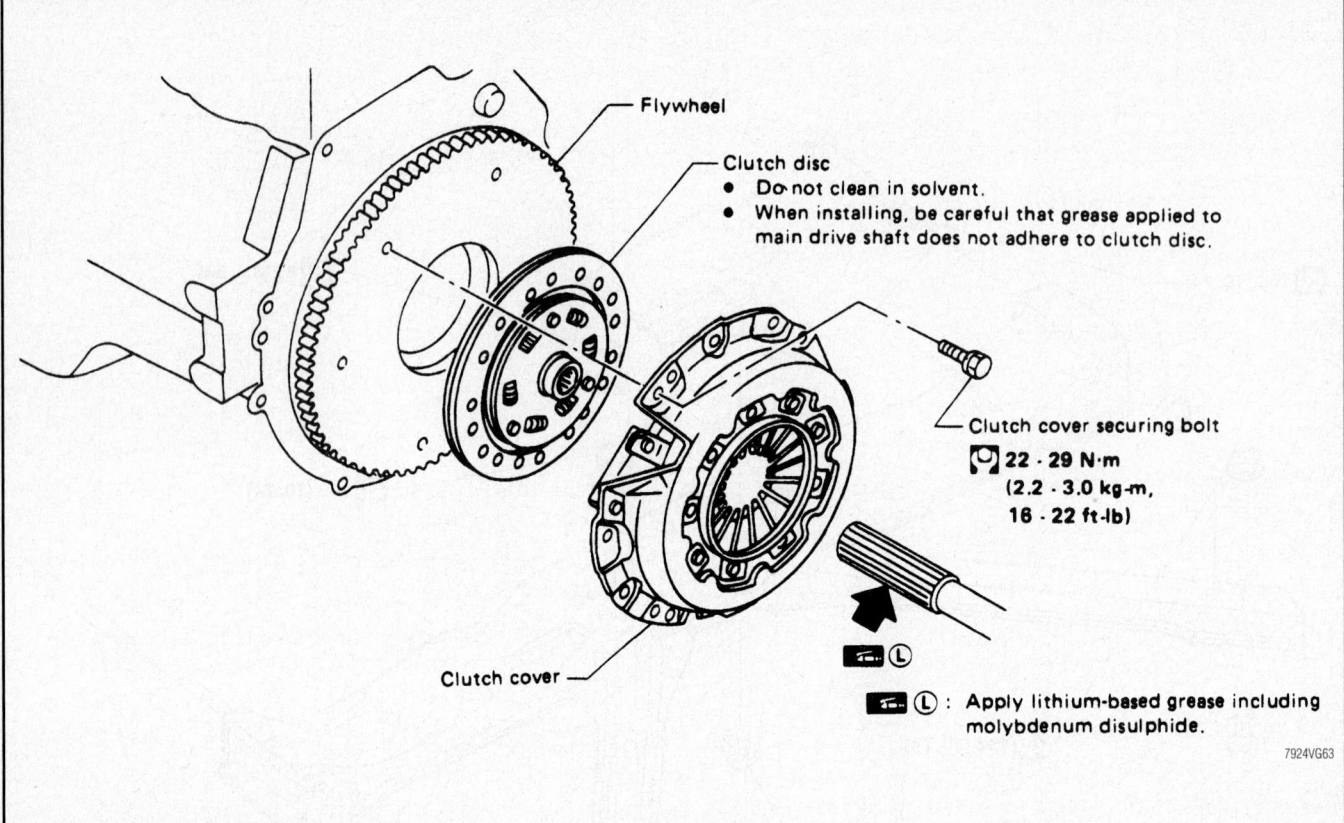

Flywheel

Clutch disc
- Do not clean in solvent.
- When installing, be careful that grease applied to main drive shaft does not adhere to clutch disc.

Clutch cover securing bolt

22 · 29 N·m
(2.2 · 3.0 kg-m,
16 · 22 ft-lb)

Clutch cover

: Apply lithium-based grease including molybdenum disulphide.

7924VG63

Fig. 18 Clutch and related components

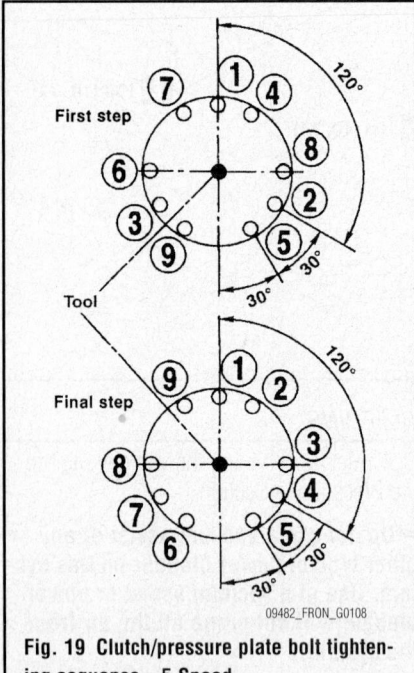

Fig. 19 Clutch/pressure plate bolt tightening sequence—5 Speed

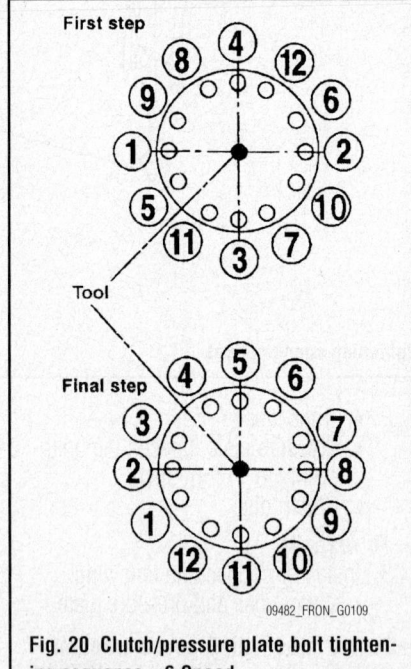

Fig. 20 Clutch/pressure plate bolt tightening sequence—6 Speed

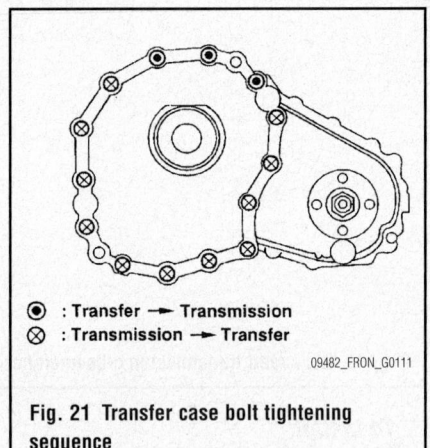

⊙ : Transfer → Transmission
⊗ : Transmission → Transfer

09482_FRON_G0111

Fig. 21 Transfer case bolt tightening sequence

5. Close the bleeder screw before releasing the clutch pedal.

6. Repeat until all air is purged from the clutch hydraulic system.

7. Refill the reservoir to the full mark.

TRANSFER CASE ASSEMBLY

REMOVAL & INSTALLATION

TX15B

See Figure 21.

1. Before servicing the vehicle, refer to the Precautions Section.

2. Disconnect the negative battery cable.

3. Switch the 4WD switch to 2WD. Set the transfer case to 2WD.

4. Raise and support the vehicle safely.

5. Remove the undercovers. Drain the transfer case fluid.

6. Remove the center exhaust tube and main muffler.

7. Remove the front and rear driveshafts. Install plug in rear oil seal.

8. Remove the transmission-to-cross-member bolts. Properly support the transmission and transfer case assembly, using a suitable jack.

9. Remove the transmission crossmember.

➡ **Support the transmission and transfer case using two suitable jacks while removing the transmission crossmember.**

10. Disconnect the ATP electrical connector, the 4LO switch connector, the wait detection switch and the transfer control device.

11. Disconnect each air breather hose from the transfer control device and the breather tube.

12. Remove the transfer case to transmission retaining bolts.

➡ **Support the transmission and transfer case, using a suitable jack.**

13. Remove the transfer case from the vehicle.

To install:

14. Installation is the reverse of the removal procedure.

15. Tighten the transfer case to transmission retaining bolts to specification and in the proper sequence. Specification is 27 ft. lbs.

16. Start the engine and check for leaks, correct as required.

FRONT HALFSHAFT

REMOVAL & INSTALLATION

1. Before servicing the vehicle, refer to the Precautions Section.

2. Raise and support the vehicle safely. Remove the tire and wheel assembly.

3. Remove the rear engine cover.

4. Remove the wheel sensor harness from the mount on the knuckle. Disconnect the harness connector.

➡ **Do not pull on the wheel sensor harness.**

5. Remove the wheel hub and bearing assembly.

➡ **It is not necessary to remove the wheel speed sensor from the wheel hub when the wheel hub is not being replaced. Carefully feed the sensor harness through the hole in the splash shield.**

6. Separate the upper link ball joint stud from the steering knuckle using

tool ST29020001 (J-24319-01) or equivalent.

7. Remove the halfshaft assembly from the vehicle by prying the halfshaft from the front final drive using the proper tool.

To install:

8. Installation is the reverse of the removal procedure.

9. Be sure to use a new differential side oil seal.

REAR AXLE SHAFT, BEARING & SEAL

REMOVAL & INSTALLATION

See Figures 22 through 24.

1. Before servicing the vehicle, refer to the Precautions Section.

2. Remove or disconnect the following:
- Rear wheel and tire assembly
- Wheel speed sensor
- Brake rotor
- Brake caliper assembly
- Parking brake cable
- Brake fluid line
- Bearing cage and backing plate bolts
- Axle shaft assembly
- Axle seal
- Wheel speed sensor rotor
- Snap ring and shim washer
- Bearing ring retainer
- back plate and torque member
- Axle bearing studs
- Wheel bearing
- Grease catcher

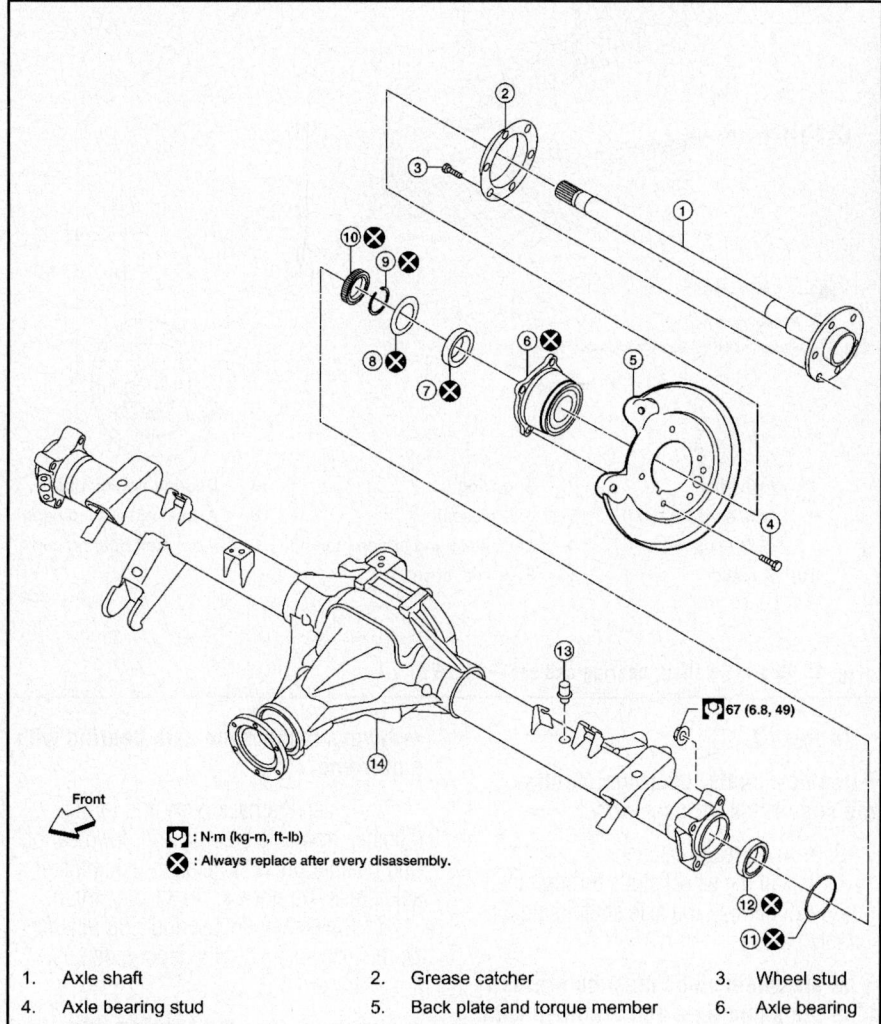

Front

⬛ : N·m (kg-m, ft-lb)
❌ : Always replace after every disassembly.

1.	Axle shaft	2.	Grease catcher	3.	Wheel stud
4.	Axle bearing stud	5.	Back plate and torque member	6.	Axle bearing
7.	Bearing ring retainer	8.	Shim washer	9.	Snap ring
10.	ABS sensor rotor	11.	O-ring	12.	Axle oil seal
13.	Breather	14.	Rear final drive		

22140_FRON_G0001

Fig. 22 Rear axle shaft, bearing and seal—C200

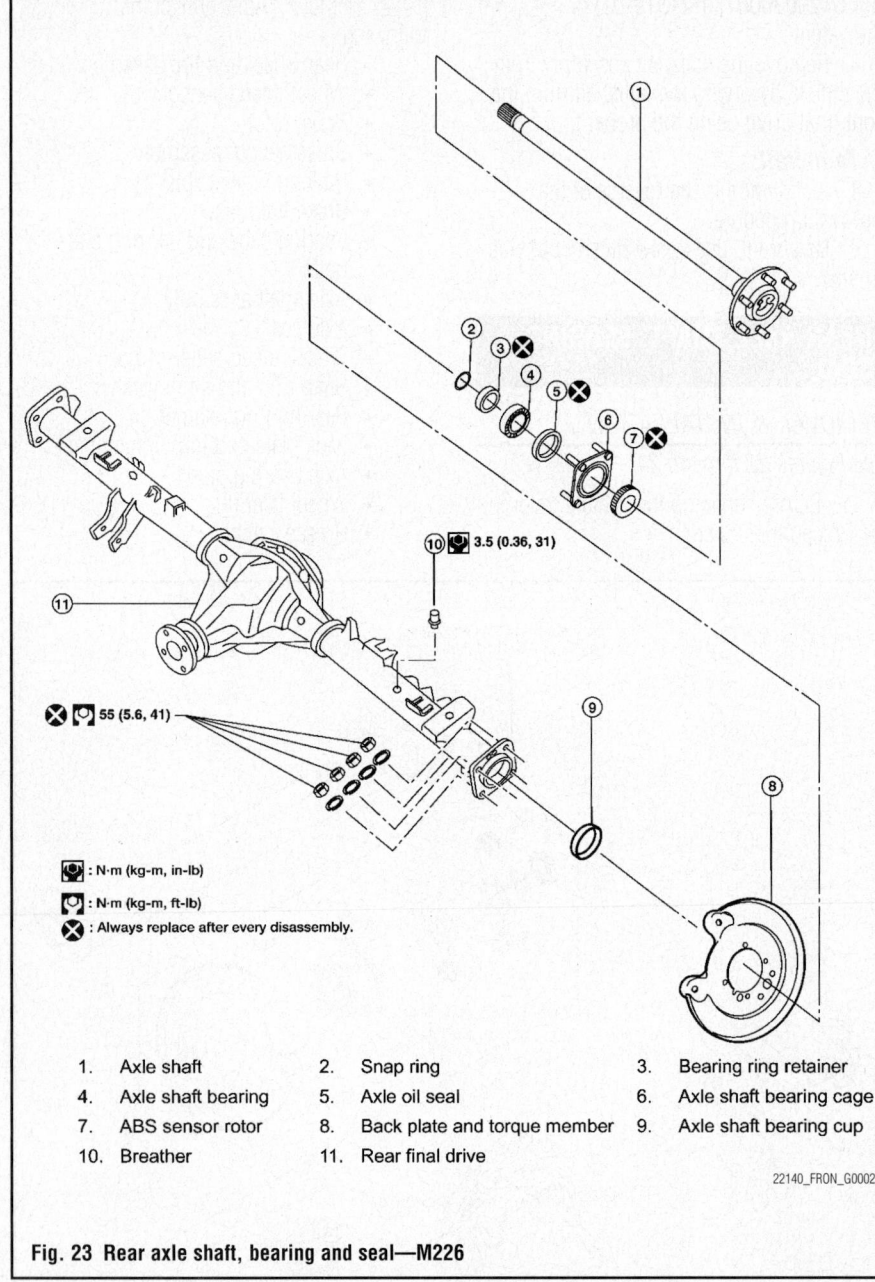

1. Axle shaft
2. Snap ring
3. Bearing ring retainer
4. Axle shaft bearing
5. Axle oil seal
6. Axle shaft bearing cage
7. ABS sensor rotor
8. Back plate and torque member
9. Axle shaft bearing cup
10. Breather
11. Rear final drive

22140_FRON_G0002

Fig. 23 Rear axle shaft, bearing and seal—M226

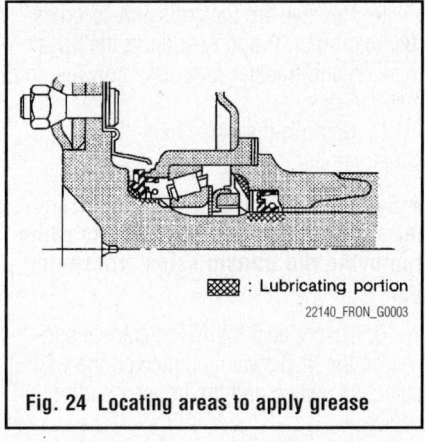

: Lubricating portion

22140_FRON_G0003

Fig. 24 Locating areas to apply grease

To install:

➡**Use new seals, bearings, circlips and snap rings for assembly.**

3. Install grease catcher

4. Install the wheel studs through the grease catcher into the axle shaft using a suitable press.

➡**All six wheel studs must be pressed on at the same time and are flush with the grease catcher when installed.**

5. Position the axle bearing on the back plate and torque member.

6. Install the axle bearing studs using a suitable press to attach the axle bearing to the back plate and torque member.

➡**Always replace the axle bearing with a new one.**

7. Install the back plate and torque member, new axle bearing and new bearing ring retainer on the axle shaft using a suitable press. Do not exceed 11 tons force.

8. Press the new bearing ring retainer on the axle shaft with the taper side positioned toward press.

➡**Always replace the bearing ring retainer with a new one.**

9. Select the correct size shim washer. Select the size of shim washer so that the installed snap ring to shim washer clearance is 0.008 inch or less.

10. Install a new snap ring on the axle shaft.

11. Do not over spread the snap ring when installing, measure the outer diameter of the snap ring after installation and replace if the snap ring outer diameter exceeds 1.87 inch maximum.

12. Check the snap ring to shim washer clearance. Repeat previous steps as necessary.

13. Perform break-in rotation of the wheel bearing.
   • Rotate the wheel bearing in the forward direction for a minimum of 10 revolutions at 50–70 RPM.
   • Rotate the wheel bearing in the reverse direction for a minimum of 10 revolutions at 50–70 RPM.

14. Measure the rotational torque of the wheel bearing. Rotational torque should be 16 in. lb. at 8–12 RPM.

15. Inspect that the wheel bearing is free from axial play relative to the axle shaft.

16. Install a new ABS sensor rotor on the axle shaft with notch side away from press using a suitable press.

➡**Always replace the ABS sensor rotor with a new one.**

17. Install new axle seal in housing.

18. Apply multi-purpose grease to the recess of axle case end as shown in illustration.

19. Insert Tool J-34296 into the new axle oil seal as a guide. Ensure tool ends do not overlap.

20. Insert the axle shaft assembly. Tighten the axle shaft nuts evenly in a criss-cross pattern to specification. Remove the Tool when the axle shaft assembly is approximately 90 percent inserted to protect the new axle oil seal.

21. Install parking brake assembly, rear caliper assembly and ABS wheel sensor.

ENGINE COOLING

THERMOSTAT

REMOVAL & INSTALLATION

2.5L Engine

See Figures 25 and 26.

➡**Never remove the radiator cap when the engine is hot. Serious burns could occur from high-pressure engine coolant escaping from the radiator.**

1. Be sure the engine is cold.
2. Disconnect the negative battery cable.

3. Drain the coolant. Properly disposed of used coolant.
4. Disconnect the lower radiator hose at the water inlet side (engine side).
5. Remove the water inlet retaining bolts.
6. Remove the water inlet and the thermostat.

To install:

7. Installation is the reverse of the removal procedure.
8. Be sure to apply a continuous bead of the proper grade and type RTV sealant to the housing.

9. Install the thermostat with the rubber ring groove positioned to fit the thermostat flange (the whole circumference).
10. Install the thermostat with the jiggle valve facing upward.

➡**The position may deviate within a range of 20 degrees.**

11. Be sure to refill the cooling using the proper grade and type engine coolant.
12. Start the engine and check for leaks.
13. Start the engine and allow it to reach operation temperature. Recheck the coolant level, fill as required.

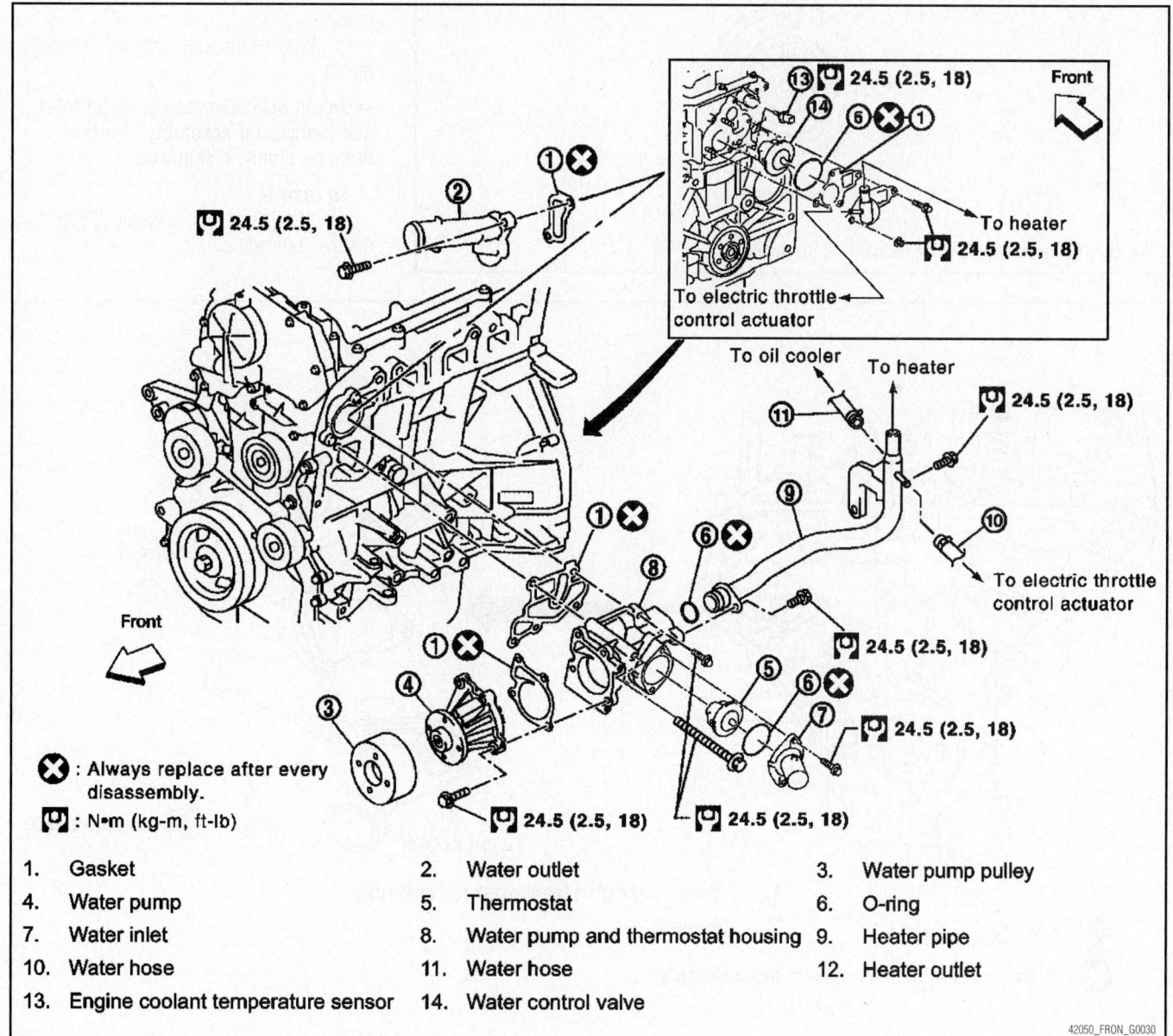

⊗ : Always replace after every disassembly.

⊡ : N•m (kg-m, ft-lb)

1.	Gasket	2.	Water outlet	3.	Water pump pulley
4.	Water pump	5.	Thermostat	6.	O-ring
7.	Water inlet	8.	Water pump and thermostat housing	9.	Heater pipe
10.	Water hose	11.	Water hose	12.	Heater outlet
13.	Engine coolant temperature sensor	14.	Water control valve		

42050_FRON_G0030

Fig. 25 Thermostat assembly and related components—2.5L engine

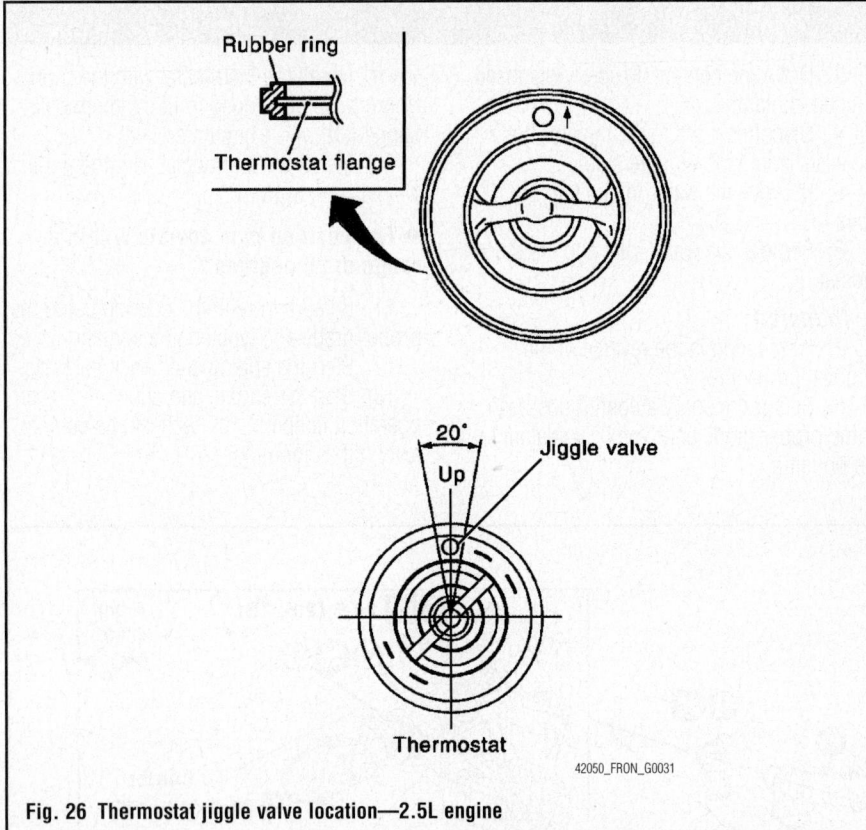

Fig. 26 Thermostat jiggle valve location—2.5L engine

42050_FRON_G0031

4.0L Engine

See Figures 27 and 28.

➡️**Never remove the radiator cap when the engine is hot. Serious burns could occur from high-pressure engine coolant escaping from the radiator.**

1. Be sure the engine is cold.
2. Disconnect the negative battery cable.
3. Drain the coolant. Properly disposed of used coolant.
4. Remove the air duct and air cleaner case.
5. Disconnect radiator hose (lower) and oil cooler hose from water inlet and thermostat assembly.
6. Remove the water inlet retaining bolts.
7. Remove the water inlet and the thermostat.

➡️**Do not disassemble the water inlet and thermostat assembly. Replace them as a unit, if required.**

To install:

8. Installation is the reverse of the removal procedure.

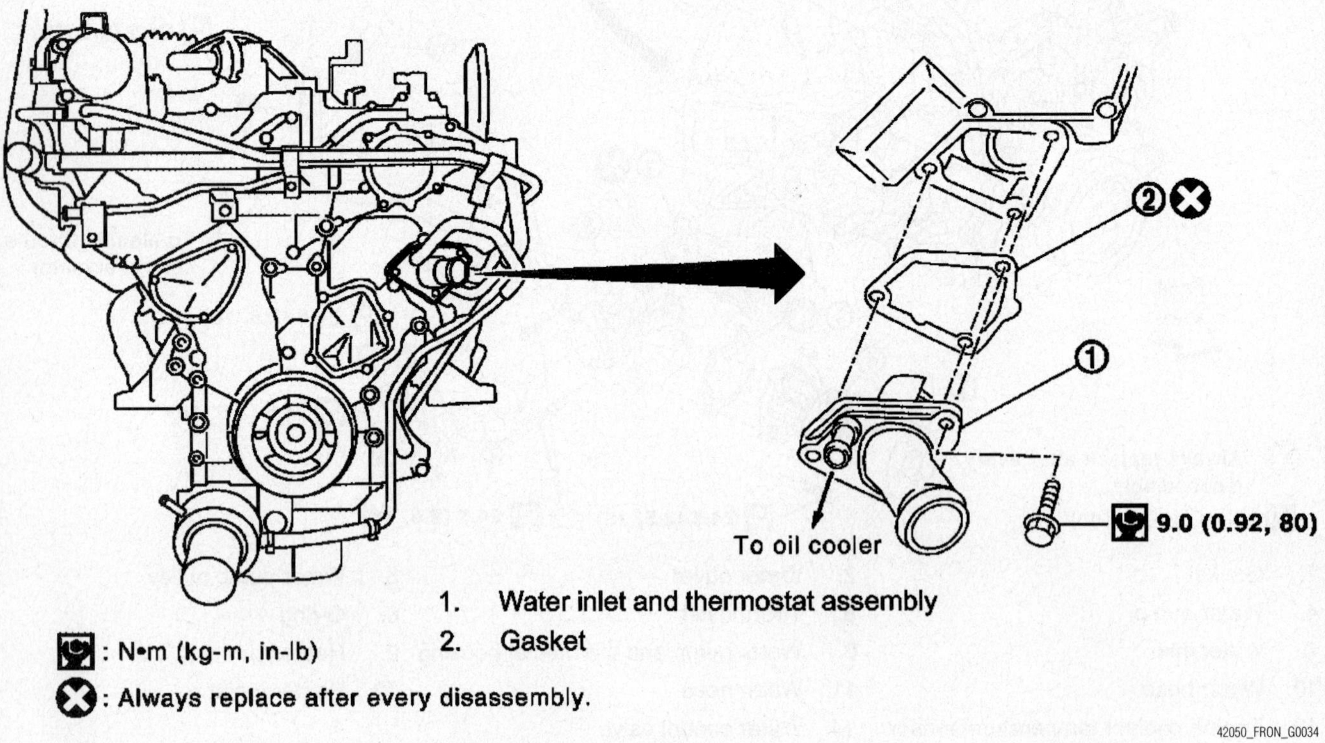

To oil cooler

9.0 (0.92, 80)

1. Water inlet and thermostat assembly
2. Gasket

🔧 : N•m (kg-m, in-lb)

❌ : Always replace after every disassembly.

42050_FRON_G0034

Fig. 27 Thermostat assembly and related components—4.0L engine

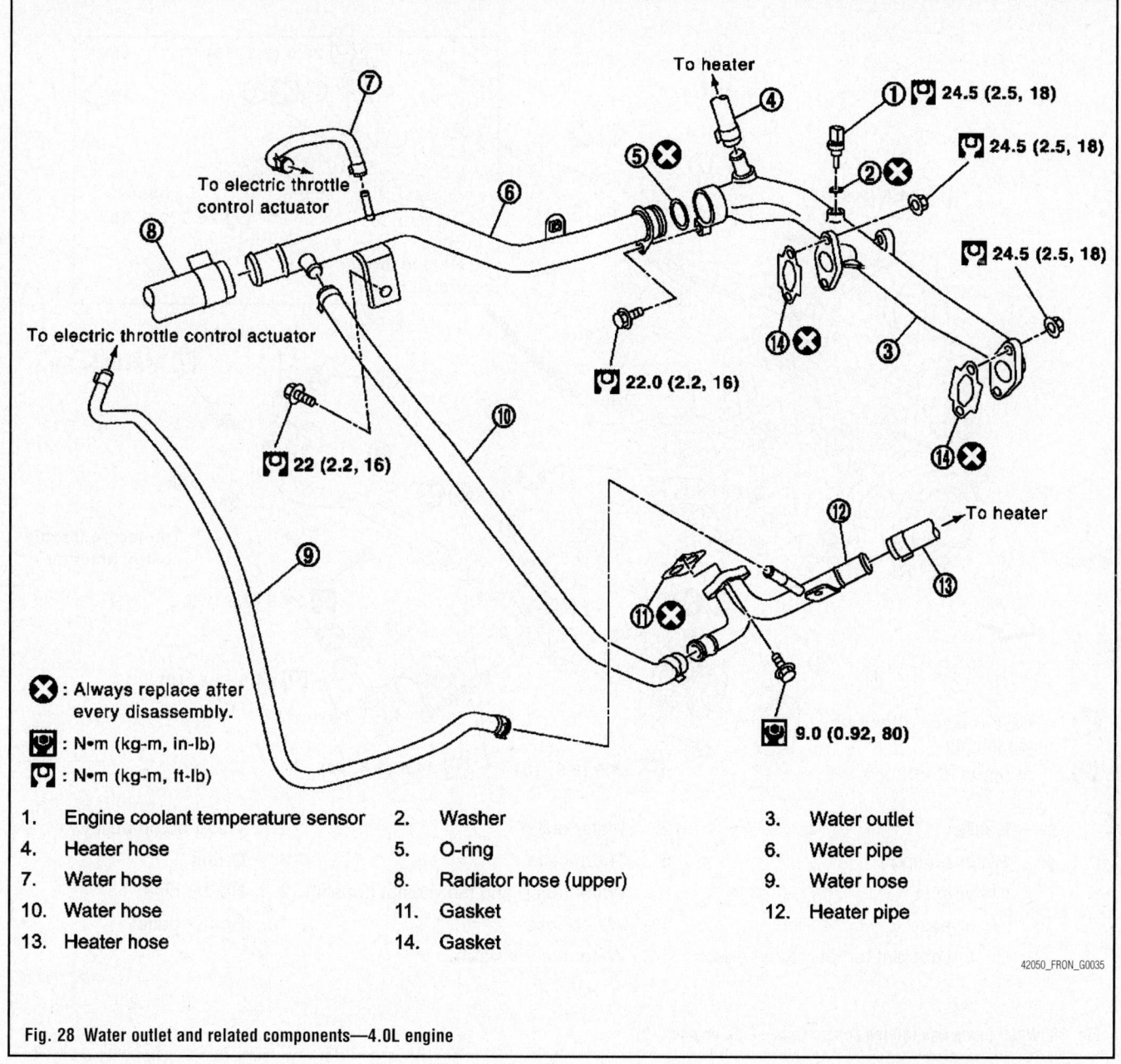

To heater

To electric throttle
control actuator

To electric throttle control actuator

① ⟦⟧ 24.5 (2.5, 18)

⟦⟧ 24.5 (2.5, 18)

⟦⟧ 24.5 (2.5, 18)

⟦⟧ 22.0 (2.2, 16)

⟦⟧ 22 (2.2, 16)

To heater

⟦⟧ 9.0 (0.92, 80)

⊗ : Always replace after every disassembly.

⟦⟧ : N•m (kg-m, in-lb)

⟦⟧ : N•m (kg-m, ft-lb)

1. Engine coolant temperature sensor	2. Washer	3. Water outlet
4. Heater hose	5. O-ring	6. Water pipe
7. Water hose	8. Radiator hose (upper)	9. Water hose
10. Water hose	11. Gasket	12. Heater pipe
13. Heater hose	14. Gasket	

42050_FRON_G0035

Fig. 28 Water outlet and related components—4.0L engine

9. Be sure to refill the cooling using the proper grade and type engine coolant.

10. Start the engine and check for leaks.

11. Start the engine and allow it to reach operation temperature. Recheck the coolant level, fill as required.

WATER PUMP

REMOVAL & INSTALLATION

2.5L Engine

See Figure 29.

1. Before servicing the vehicle, refer to the Precautions Section.

2. Disconnect the negative battery cable. Drain the cooling system.

3. Remove the air duct. Remove the drive belt.

4. Remove the upper and lower radiator hoses.

5. Remove the cooling fan and the water pump pulley.

6. Remove the water pump retaining bolts. Remove the water pump from the engine.

To install:

7. Installation is the reverse of the removal procedure.

8. Be sure to use new gaskets and O-rings, as required.

➡**When inserting heater pipe end into water pump and thermostat housing,**

apply a neutral detergent to O ring. **Then insert it immediately.**

9. Be sure to fill the cooling system with the proper grade and type engine coolant.

10. Start the engine and check for leaks.

4.0L Engine

See Figures 30 and 31.

1. Before servicing the vehicle, refer to the Precautions Section.

2. Disconnect the negative battery cable. Drain the cooling system.

3. Remove the air dam and undercover.

4. Remove air duct and resonator.

5. Remove the drive belts.

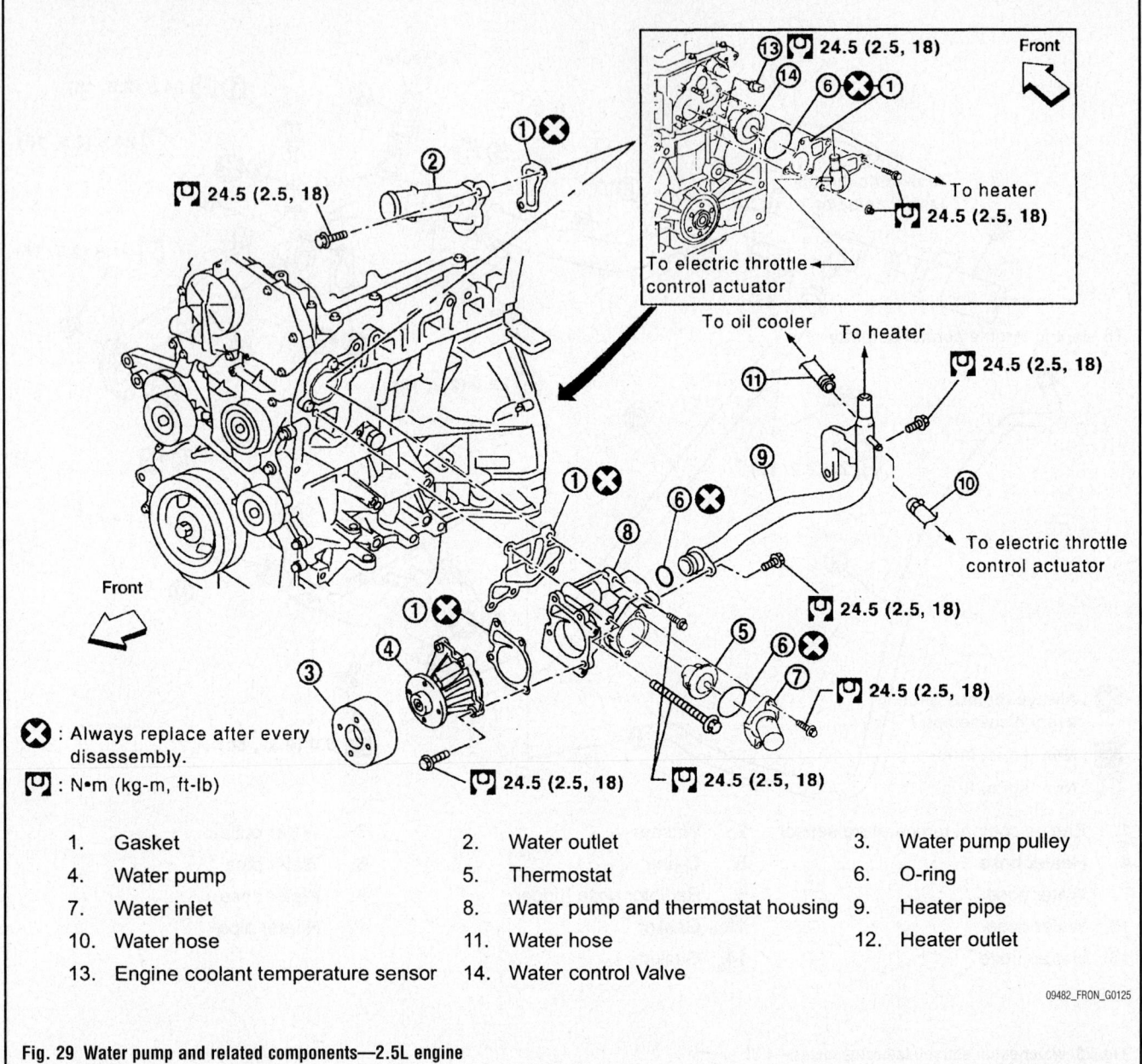

24.5 (2.5, 18)

To heater
24.5 (2.5, 18)

To electric throttle
control actuator

Front

To oil cooler   To heater
24.5 (2.5, 18)

To electric throttle
control actuator

24.5 (2.5, 18)

24.5 (2.5, 18)

24.5 (2.5, 18)

❌ : Always replace after every
disassembly.
[🔧] : N•m (kg-m, ft-lb)

1. Gasket	2. Water outlet	3. Water pump pulley
4. Water pump	5. Thermostat	6. O-ring
7. Water inlet	8. Water pump and thermostat housing	9. Heater pipe
10. Water hose	11. Water hose	12. Heater outlet
13. Engine coolant temperature sensor	14. Water control Valve	

09482_FRON_G0125

Fig. 29 Water pump and related components—2.5L engine

6. Remove the radiator upper hose. Remove the cooling fan.

7. Remove the chain tensioner cover and water pump cover from the front timing case, using tool KV10111100 (J-37228) or equivalent.

8. To remove the timing chain tensioner (primary), loosen the clip of the timing chain tensioner (primary) and release the plunger stopper. Insert the plunger into the tensioner body by pressing the slack guide. Keep the slack guide pressed and hold the plunger in by pushing the stopper pin through the tensioner body hole and plunger groove. Turn the crankshaft pulley clockwise so that the timing chain on the timing chain tensioner (primary) side is

loose. Remove the bolts and remove the timing chain tensioner (primary).

➡**Be careful not to drop the bolts inside the timing chain case.**

9. Remove the three water pump retaining bolts. Secure a gap between the water pump gear and the timing chain, by turning the crankshaft pulley counterclockwise until timing chain looseness on the water pump sprocket is at its maximum point.

10. Screw M8 bolts approximately 1.97 inch in length into the water pumps upper and lower bolt holes until they reach the timing chain case.

11. Alternately tighten each bolt for a half turn and pull out the water pump.

➡**Pull the pump straight out while preventing the vane from contacting the socket in the installation area. Remove the pump without causing the sprocket to contact the timing chain.**

12. Remove the M8 bolts. Remove and discard the O-rings.

To install:

13. Installation is the reverse of the removal procedure.

14. Be sure to use new gaskets and O-rings, as required. Apply engine oil to New O-rings before installation. Locate O-ring with White paint mark in forward groove.

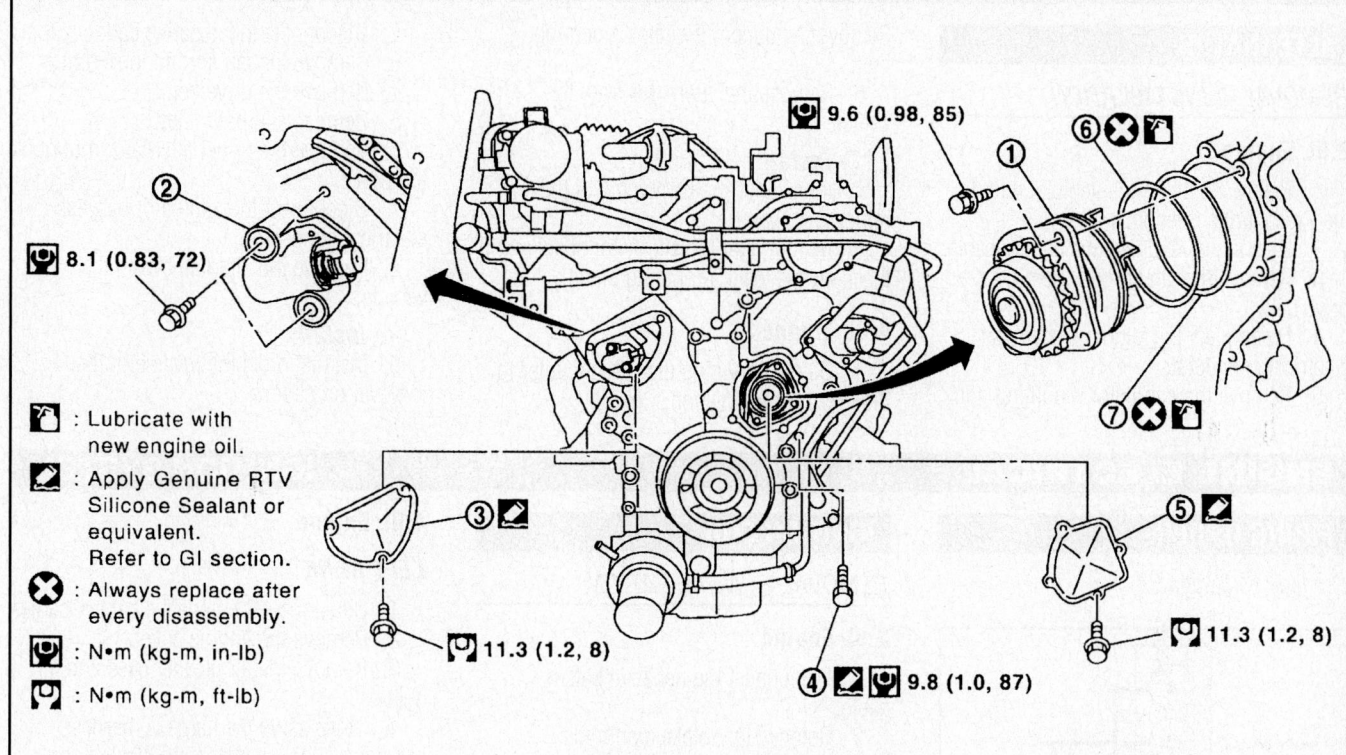

: Lubricate with new engine oil.

: Apply Genuine RTV Silicone Sealant or equivalent. Refer to GI section.

: Always replace after every disassembly.

: N•m (kg-m, in-lb)

: N•m (kg-m, ft-lb)

1. Water pump
2. Timing chain tensioner (primary)
3. Chain tensioner cover
4. Water drain plug (front)
5. Water pump cover
6. O-ring
7. O-ring

8.1 (0.83, 72)

9.6 (0.98, 85)

11.3 (1.2, 8)

9.8 (1.0, 87)

11.3 (1.2, 8)

09482_FRON_G0016

Fig. 30 Water pump and related components—4.0L engine

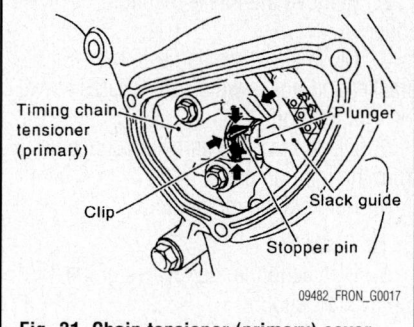

Timing chain tensioner (primary)

Plunger

Clip

Slack guide

Stopper pin

09482_FRON_G0017

Fig. 31 Chain tensioner (primary) cover removal—4.0L engine

15. When installing the water pump make sure that the timing chain and water pump sprocket are engaged. Tighten the bolts alternately and evenly to specification.

16. Before installing the chain tensioner cover and the water pump cover be sure to apply a continuous bead of sealant to the mating surfaces of the covers.

➡**Do not allow the sealant to set for more than five minutes before installing the covers.**

17. Be sure to fill the cooling system with the proper grade and type engine coolant.

18. Start the engine and check for leaks.

19. Let the engine idle for about three minutes than rev it up to 3,000 rpm's under a no load condition to purge air from the high pressure chamber of the chain tensioner. The engine may produce a rattling noise. This indicates that air still remains in the chamber and is not a matter of concern.

ENGINE ELECTRICAL

ALTERNATOR

REMOVAL & INSTALLATION

2.5L Engine

1. Before servicing the vehicle, refer to the Precautions Section.
2. Disconnect the negative battery cable.
3. Remove the fan shroud. Remove the drive belt.
4. Disconnect the alternator harness electrical connectors.
5. Remove the alternator mounting nut.

Remove the upper alternator mounting bolt.
6. Remove the alternator from the vehicle.

To install:
7. Installation is the reverse of the removal procedure.
8. Be sure that the alternator spacer is in place on the lower mounting stud.

4.0L Engine

1. Before servicing the vehicle, refer to the Precautions Section.

CHARGING SYSTEM

2. Disconnect the negative battery cable.
3. Remove the fan shroud, on Frontier.
4. Remove the drive belt.
5. Remove alternator stay.
6. Remove the upper alternator mounting bolt.
7. Disconnect the alternator harness electrical connectors.
8. Remove the alternator from the vehicle.

To install:
9. Installation is the reverse of the removal procedure.

ENGINE ELECTRICAL

FIRING ORDER

See Figures 32 and 33.

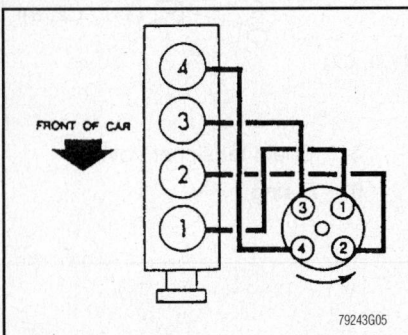

FRONT OF CAR

79243G05

Fig. 32 2.4L Engine
Firing order: 1–3–4–2
Distributorless ignition system

IGNITION COIL

REMOVAL & INSTALLATION

2.5L Engine

1. Disconnect the negative battery cable.
2. Remove the intake manifold.

➡ **If just removing the number one spark plug, it is not necessary to remove the intake manifold.**

3. Disconnect the harness connector from the ignition coil.
4. Remove the ignition coil.

To install:
5. Installation is the reverse of the removal procedure.

IGNITION SYSTEM

4.0L Engine

LEFT BANK

1. Disconnect the negative battery cable.
2. Remove the engine cover.
3. Remove the air cleaner case and air duct.
4. Move aside the harness, harness bracket and hoses which are located above the ignition coil.
5. Disconnect the harness connector from the ignition coil.
6. Remove the ignition coil.

To install:
7. Installation is the reverse of the removal procedure.

RIGHT BANK

1. Disconnect the negative battery cable.
2. Remove the intake manifold collector.
3. Move aside the harness, harness bracket and hoses which are located above the ignition coil.
4. Disconnect the harness connector from the ignition coil.
5. Remove the ignition coil.

To install:
6. Installation is the reverse of the removal procedure.

IGNITION TIMING

ADJUSTMENT

Ignition timing is controlled by the ECM. No adjustment is possible.

SPARK PLUGS

REMOVAL & INSTALLATION

See Figure 34.

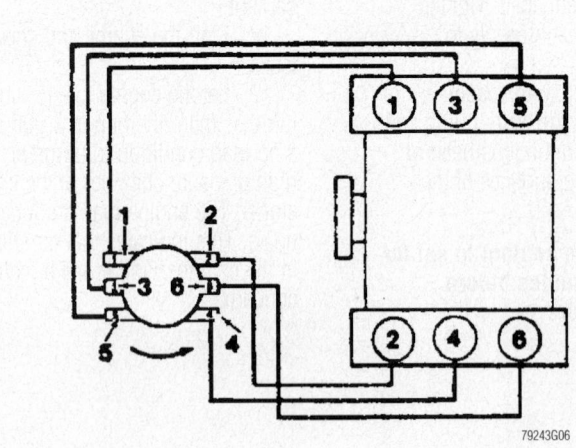

79243G06

Fig. 33 3.3L Engine
Firing order: 1–2–3–4–5–6
Distributorless ignition system

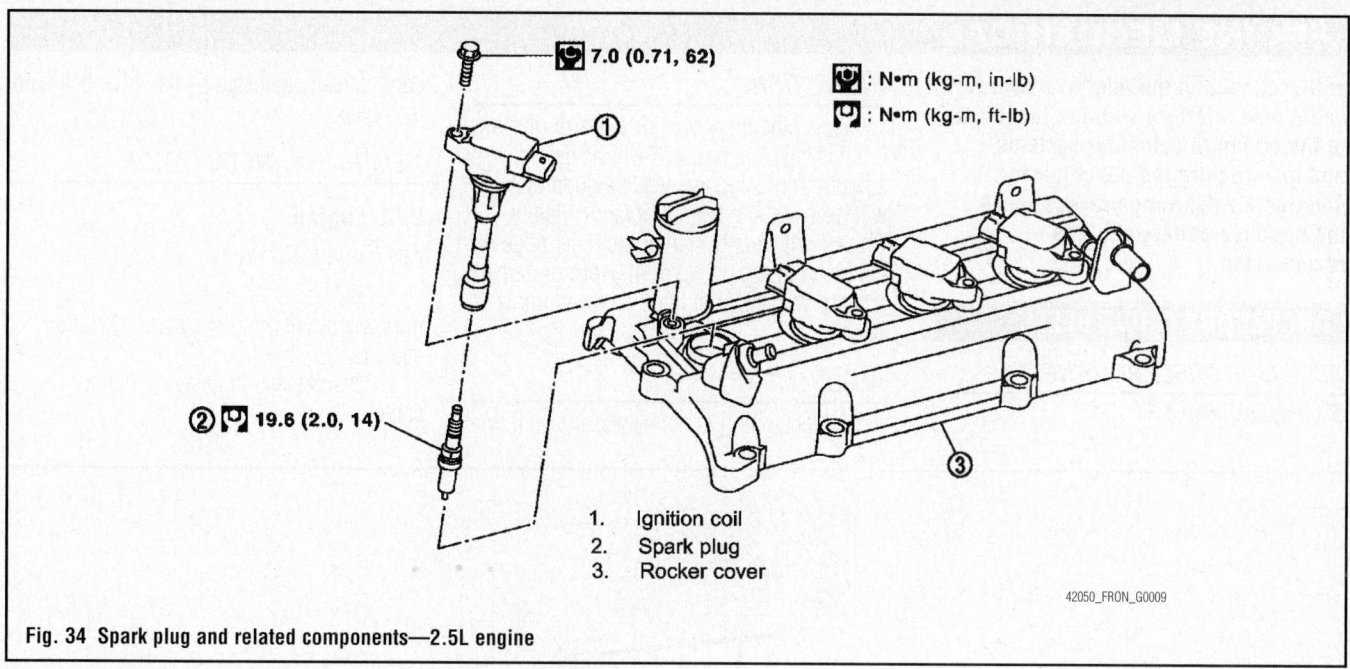

Fig. 34 Spark plug and related components—2.5L engine

1. Ignition coil
2. Spark plug
3. Rocker cover

2.5L Engine

1. Disconnect the negative battery cable.
2. Remove the intake manifold.

➡ **If just removing the number one spark plug, it is not necessary to remove the intake manifold.**

3. Disconnect the harness connector from the ignition coil.
4. Remove the ignition coil.
5. Remove the spark plug using a spark plug socket and wrench.

To install:

6. Installation is the reverse of the removal procedure.

4.0L Engine

See Figure 35.

1. Disconnect the negative battery cable.
2. Remove the intake manifold collector.
3. Remove the ignition coil.
4. Remove the spark plug using a spark plug socket and wrench.

To install:

5. Installation is the reverse of the removal procedure.

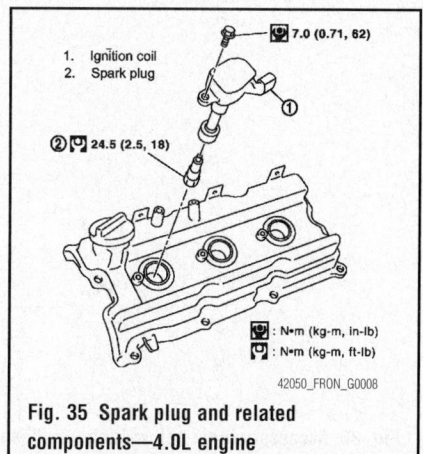

1. Ignition coil
2. Spark plug

Fig. 35 Spark plug and related components—4.0L engine

ENGINE ELECTRICAL

STARTER

REMOVAL & INSTALLATION

1. Before servicing the vehicle, refer to the Precautions Section.
2. Remove or disconnect the following:
   • Negative battery cable
   • Engine under cover
   • On vehicles with 2.5L engine, remove the air cleaner cover and the air cleaner to intake manifold collector duct.
   • On vehicles with 4.0L engine, remove the exhaust manifold cover to gain access to the starter retaining bolts
   • Starter harness connectors
   • Starter bolts
   • Starter motor

STARTING SYSTEM

To install:

3. Install or connect the following:
   • Starter motor
   • air cleaner cover and the air cleaner to intake manifold collector duct, if equipped
   • Exhaust manifold cover, if equipped
   • Starter harness connectors
   • Engine under cover
   • Negative battery cable

ENGINE MECHANICAL

➡**Disconnecting the negative battery cable may interfere with the functions of the on board computer systems and may require the computer to undergo a relearning process, once the negative battery cable is reconnected.**

ACCESSORY DRIVE BELTS

ACCESSORY BELT ROUTING

See Figures 36 and 37.

INSPECTION

Inspect the drive belt for signs of glazing or cracking. A glazed belt will be perfectly smooth from slippage, while a good belt will have a slight texture of fabric visible. Cracks will usually start at the inner edge of the belt and run outward. All worn or damaged drive belts should be replaced immediately.

ADJUSTMENT

Belt tensioning is not necessary, as it is automatically adjusted by the drive belt auto tensioner.

REMOVAL & INSTALLATION

2.5L Engine

See Figures 38 and 39.

1. Before servicing the vehicle, refer to the precautions in the beginning of this section.
2. Disconnect the negative battery cable.

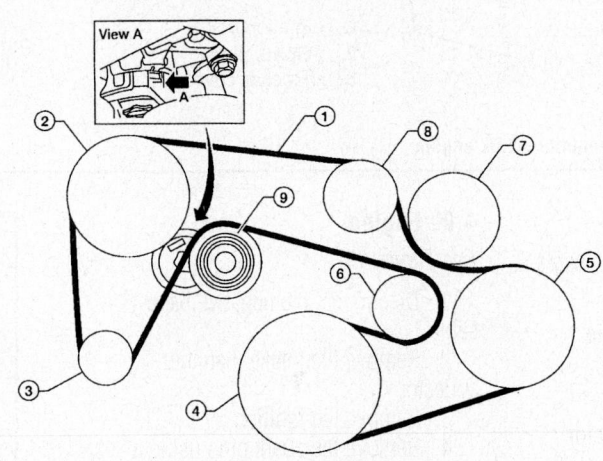

1. Drive belt	2. Power steering oil pump pulley	3. Generator pulley
4. Crankshaft pulley	5. A/C compressor (if equipped) or idler pulley	6. Idler pulley
7. Water pump	8. Idler pulley	9. Drive belt auto-tensioner

09482_FRON_G0005

Fig. 36 Accessory drive belt routing—2.5L engine

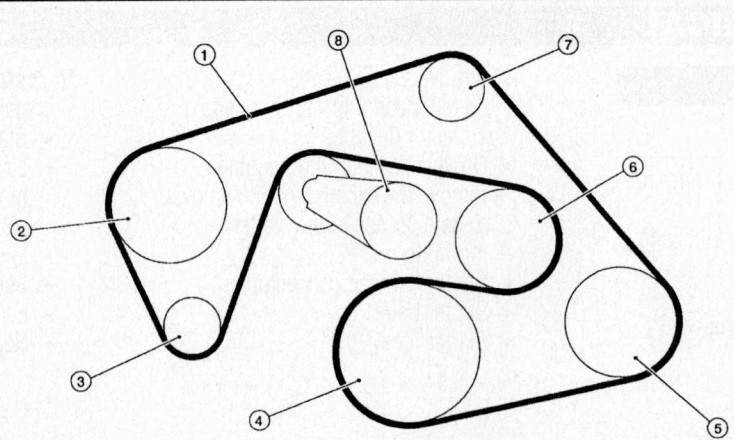

1. Drive belt	2. Power steering pump pulley	3. Generator pulley
4. Crankshaft pulley	5. A/C compressor	6. Cooling fan pulley
7. Idler pulley	8. Drive belt auto-tensioner	

09482_FRON_G0006

Fig. 37 Accessory drive belt routing—4.0L engine

3. Install tool J-46535, or equivalent on the auto tensioner pulley bolt. Move it in the direction shown in the illustration.

✳✳ CAUTION

Do not place your hand in a location where pinching may occur if the holding tool accidentally comes off.

➡**Do not loosen the auto tensioner pulley bolt. (Do not turn it counterclockwise). If turned counterclockwise, the complete auto tensioner must be replaced as a unit, including the pulley.**

4. Remove the drive belt.

To install:
5. Installation is the reverse of the removal procedure.

4.0L Engine

See Figures 40 and 41.

1. Before servicing the vehicle, refer to the precautions in the beginning of this section.
2. Disconnect the negative battery cable.
3. Remove the air duct and resonator assembly (inlet).
4. Rotate the drive belt auto tensioner in the direction shown in the illustration.

✳✳ CAUTION

Do not place your hand in a location where pinching may occur if the holding tool accidentally comes off.

5. Remove the drive belt.

To install:
6. Installation is the reverse of the removal procedure.

CAMSHAFT AND VALVE LIFTERS

REMOVAL & INSTALLATION

2.5L Engine

See Figures 42 through 52.

➡**The procedure below describes removal and installation of the camshaft without removing the front cover. If the front cover is removed refer to timing chain removal and installation.**

1. Before servicing the vehicle, refer to the Precautions Section.
2. Properly relieve the fuel system pressure.
3. Disconnect the negative battery cable. Drain the cooling system.

4. Remove the intake manifold.
5. Disconnect the PCV hose from the rocker cover. Remove the ignition coil.
6. Remove the PCV valve and O-ring from the rocker cover, if necessary.
7. Remove the oil filler cap from the rocker cover, if necessary.
8. Remove the rocker cover retaining bolts. Be sure to remove the bolts by reversing the order of the tightening torque sequence.
9. Remove the rocker cover. Discard the gasket.
10. Remove the drive belt. Disconnect and remove the camshaft position sensor (PHASE).
11. Disconnect the IVT control solenoid electrical connector.
12. Disconnect the ground electrical connectors from the front cover.
13. Remove the IVT control solenoid retaining bolts. Be sure to remove the bolts by reversing the order of the tightening torque sequence.
14. Remove the cover by cutting the sealant using tool KV10111100 (J-37228) or equivalent.
15. Position the number one cylinder on its compression stroke by rotating the crankshaft pulley clockwise. Align the

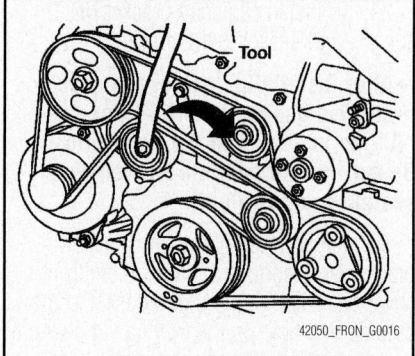

Fig. 38 Drive belt tension tool installation and removal direction—2.5L engine

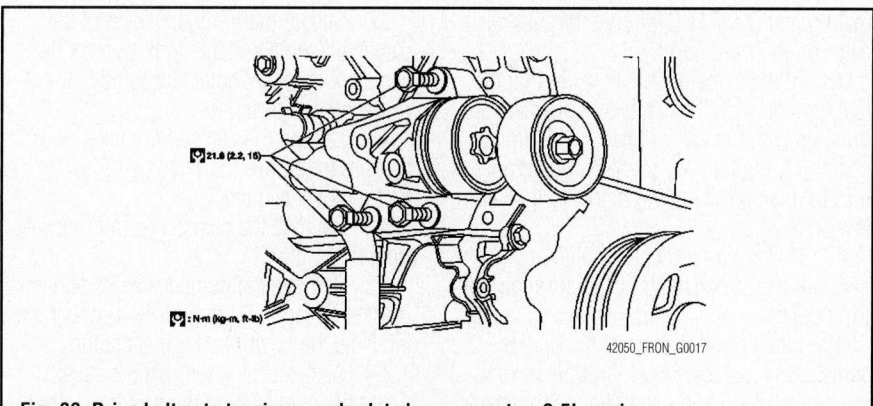

Fig. 39 Drive belt auto tensioner and related components—2.5L engine

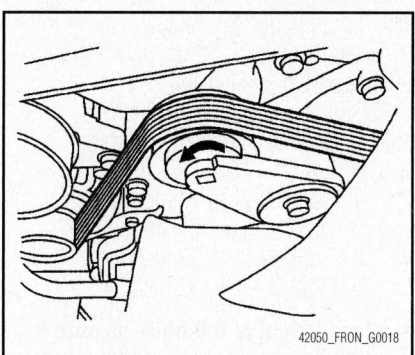

Fig. 40 Drive belt tension tool installation and removal direction—4.0L engine

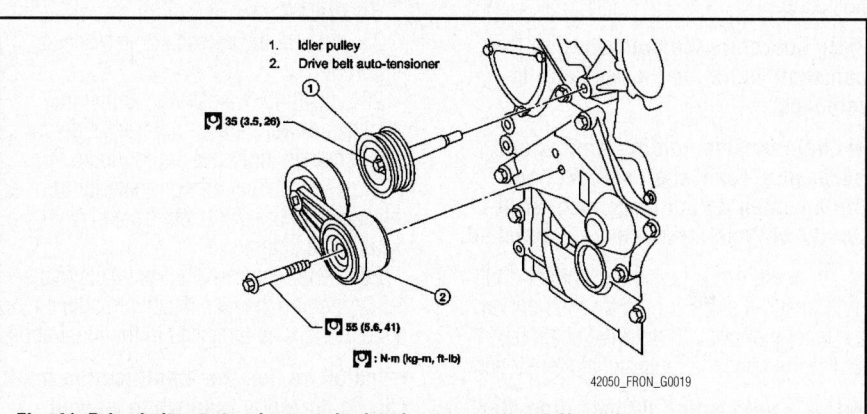

Fig. 41 Drive belt auto tensioner and related components—4.0L engine

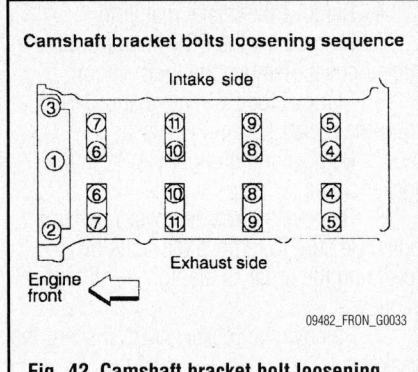

Fig. 42 Camshaft bracket bolt loosening sequence—2.5L engine

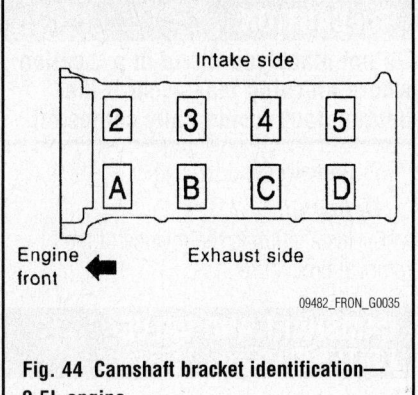

Fig. 44 Camshaft bracket identification—2.5L engine

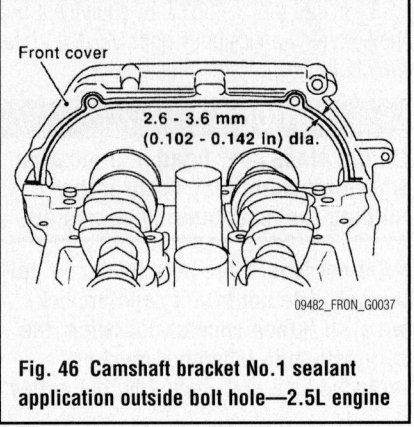

Fig. 46 Camshaft bracket No.1 sealant application outside bolt hole—2.5L engine

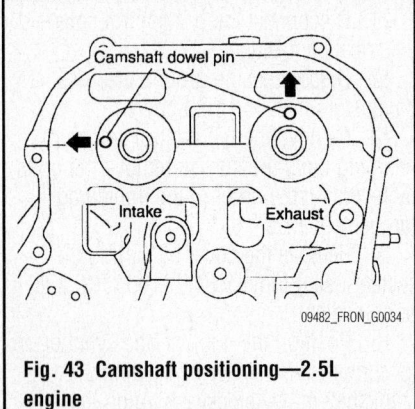

Fig. 43 Camshaft positioning—2.5L engine

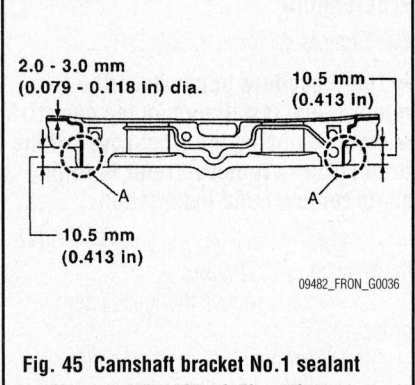

Fig. 45 Camshaft bracket No.1 sealant application point "A"—2.5L engine

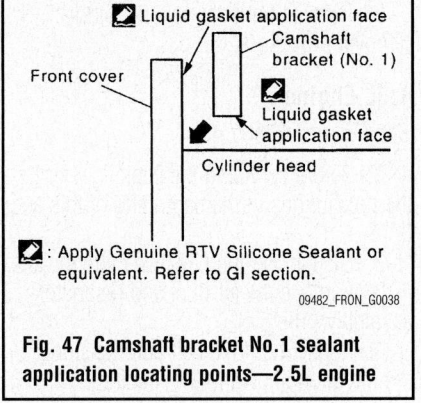

Fig. 47 Camshaft bracket No.1 sealant application locating points—2.5L engine

mating marks for TDC with the timing indicator on the front cover.

16. At the same time make sure that the mating marks on the camshaft sprockets are lined up with the yellow links in the timing chain. If not rotate the crankshaft one more turn to line up the mating marks to the yellow links.

17. Pull the timing chain guide out between the camshaft sprockets through the front cover.

18. Line up the mating marks on the camshaft sprockets with the yellow links in the timing chain and paint an indelible mating mark on the sprocket and timing chain link plate.

➡**Do not rotate the crankshaft or the camshaft while the timing chain is removed.**

➡**Chain tension holding work is not necessary. Crankshaft sprocket and timing chain do not disconnect structurally while the front cover is attached.**

19. Push in the tensioner plunger and hold. Insert a stopper pin into the hole on the tensioner body to hold the chain tensioner. Remove the timing chain tensioner.

➡**Use a wire with 0.02 inch diameter for a stopper pin.**

20. Secure the hexagonal part of the camshaft with a suitable tool. Loosen the camshaft sprocket bolts and remove the camshaft sprockets.

21. Loosen the camshaft bracket bolts. Be sure to remove the bolts by following the bolt removal sequence.

22. Remove the camshafts and brackets from the engine.

23. Remove the number one camshaft bracket by tapping lightly with a rubber mallet. Note the positions for installation.

24. As necessary, remove the valve lifters. Be sure to keep them in the proper order for installation.

To install:

25. Inspect the camshafts, replace as required.

26. Install the camshafts so that the camshaft dowel pins on the front side are positioned as indicated in the illustration.

27. Remove any foreign material from the camshaft bracket backside and from the cylinder head face.

28. Install the camshaft brackets (No.2–No.5) aligning the identification marks on the upper surface as indicated in the illustration.

➡**Install so that the identification mark can be correctly read when viewed from the exhaust side.**

29. To install camshaft bracket NO.1, apply liquid gasket to the bracket.

➡**After installation be sure to wipe excessive gasket material from part "A", as indicated in the illustration. Be sure to use genuine RTV silicone sealant or equivalent.**

30. Apply liquid gasket to camshaft bracket No.1 contact surface on the front cover backside. Apply liquid gasket to the outside bolt hole on the front cover. Be sure to use genuine RTV silicone sealant or equivalent.

31. Locate camshaft bracket No.1 near installation position and install it without disturbing the liquid gasket applied to the surfaces. Be sure to use genuine RTV silicone sealant or equivalent.

32. Tighten the camshaft bracket bolts in the proper sequence and to specification.
   a. Bolts 9–11 to 17 inch lbs.
   b. Bolts 1–8 to 17 inch lbs.
   c. Bolts 1–11 to 52 inch lbs.
   d. Bolts 1–11 to 92 inch lbs.

➡**After tightening the bolts be sure to wipe off any excessive liquid gasket. Be sure to use genuine RTV silicone sealant or equivalent.**

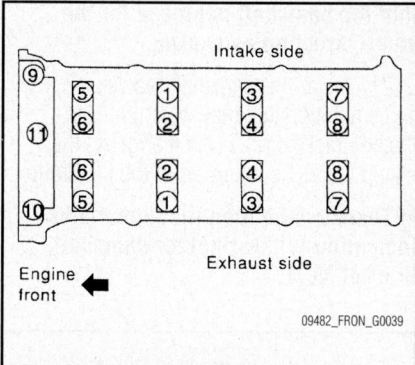

Fig. 48 Camshaft bracket bolt torque sequence—2.5L engine

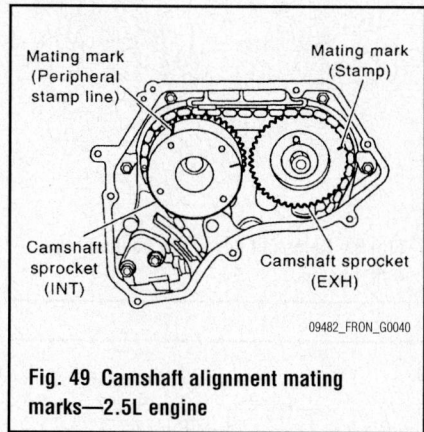

Fig. 49 Camshaft alignment mating marks—2.5L engine

33. Install the camshaft position sensor. Install the camshaft sprockets.

➡Install them by aligning the mating marks on each camshaft sprocket with the paint marks on the timing chain link plates, which were made during removal.

❋❋ CAUTION

Aligned mating marks could slip. Therefore, after matching them, hold the timing chain in place by hand. Before and after installing the chain tensioner, make sure again that the mating marks have not slipped.

34. Install the chain tensioner. After installation, pull the stopper pin off completely, and make sure that the chain tensioner plunger is released.

➡Before installation of the chain tensioner, it is possible to rematch the marks on the timing chain with new ones on each sprocket.

35. Install the chain guide. Install oil rings to the camshaft sprocket (INT) insertion points on backside of intake valve timing control cover. Install the O-ring to the front cover.
36. Apply a 0.083–0.122 inch diameter

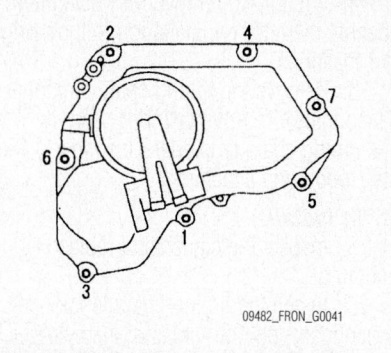

Fig. 50 Intake valve timing control valve cover bolt torque sequence—2.5L engine

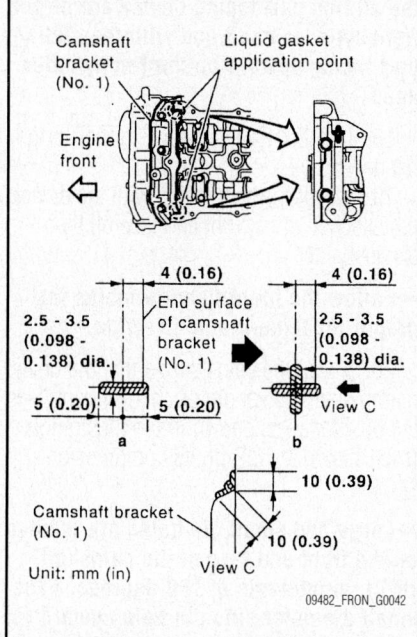

Fig. 51 Rocker cover sealant application locating points—2.5L engine

bead of liquid gasket to the intake valve timing control cover. Be sure to use genuine RTV silicone sealant or equivalent.
37. Install the cover. Tighten the bolts in the proper sequence. Connect the ground cables and install the harness clip.
38. Check and adjust the valve clearance, as required.

➡If hydraulic pressure inside the timing chain tensioner drops after removal/installation, slack in the guide may generate a pounding noise during and just after engine start. This is normal the noise will stop after hydraulic pressure rises.

39. Continue the installation in the reverse order of the removal procedure.
40. Apply liquid gasket, be sure to use genuine RTV silicone sealant or equivalent, to the positions shown in the illustration.

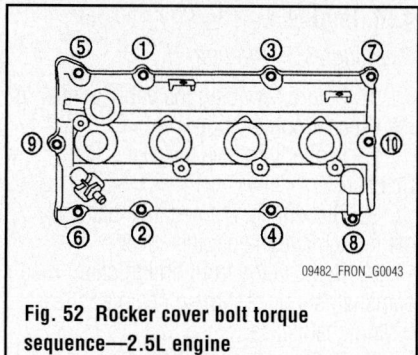

Fig. 52 Rocker cover bolt torque sequence—2.5L engine

Refer to figure "a" to apply liquid gasket to joint part of camshaft bracket No.1 and cylinder head. Refer to figure "b" to apply liquid gasket in 90 degrees to figure "b".
41. Install the rocker cover. Torque the retaining bolts to 18 inch lbs and then to 73 inch lbs, in the proper sequence.
42. Inspect the camshaft sprocket (INT) oil groove.

➡Perform this inspection only when DTC P0011 or DTC P0021 are detected in self diagnostic results of CONSULT-II.

43. Be sure the engine is cold. Check and adjust oil level, as required.
44. Properly release the fuel system pressure. Disconnect the ignition coil and injector harness connectors.

➡This is being done to prevent the engine from unintentionally being started while checking.

45. Remove the intake valve timing control solenoid valve.
46. Crank the engine, and then make sure that engine oil comes out from camshaft bracket (No.1) oil hole.

❋❋ WARNING

Be careful not to touch rotating parts, (drive belt, idler pulley, crankshaft pulley etc) as injury could result.

➡Oil may squirt from the intake valve timing control solenoid valve installation hole during engine cranking. Use a shop towel to prevent oil from squirting on engine components.

47. Clean the oil groove between the oil strainer and the intake timing control solenoid valve if engine oil does not come out from camshaft bracket (No.1) oil hole.
48. Remove the components between the intake valve timing control solenoid valve and the camshaft sprocket (INT). Check each oil groove for clogging.
49. After inspection install any removed components.

4.0L Engine

See Figures 53 through 61.

1. Before servicing the vehicle, refer to the Precautions Section.

2. Properly relieve the fuel system pressure.

3. Disconnect the negative battery cable. Remove the engine cover.

4. Remove the front timing chain case, camshaft sprocket, timing chain and rear timing chain case.

5. Remove the camshaft position sensor (PHASE) from the cylinder head back side.

➡**Handle carefully to avoid dropping and shocks. Do not disassemble. Do not place in a location where the sensor can be exposed to magnetism.**

6. Remove the intake manifold collector.

7. Separate the engine harness and remove their brackets from the rocker covers. Remove the harness bracket from the cylinder head, if necessary.

8. Remove the ignition coil. Remove the PCV hoses. Remove the oil filler cap, if necessary.

9. Loosen the rocker cover retaining bolts, in the reverse order of the tightening sequence.

10. Remove the rocker covers from the engine.

11. Remove the intake valve timing control solenoid valves. Discard the gaskets.

12. Mark the camshaft brackets and bolts for reinstallation. Remove the camshaft bracket bolts. Be sure to remove the bolts by reversing the order of the tightening torque sequence and in several steps.

13. Remove the camshafts.

14. If required, remove the valve lifters. Identify them for reinstallation in their original locations.

15. Remove the timing chain tensioner (secondary) from the cylinder head. Remove the timing chain tensioner (secondary) with its stopper pin attached.

To install:

16. Inspect the camshafts, replace as required.

17. Install the timing chain tensioners (secondary) on both sides of the cylinder head. Be sure to use new O-rings.

➡**Install the tensioner with its stopper pin attached. Install the tensioner with the sliding part facing downward on the right cylinder head and with the sliding part facing upward on the left cylinder head.**

18. Install the valve lifters, in their original bores.

19. Install the camshafts, with the dowel pin attached to its front end face on the exhaust side.

➡**Follow the identification marks for proper placement and direction.**

20. Install the camshaft so that the dowel pin hole and dowel pin on the front end face are positioned as shown in the illustration (No.1 piston at TDC on its compression stroke).

➡**Large and small pin holes are located on the front end face of the camshaft (INT), at intervals of 180 degrees. Face small diameter side pin hole upward (in cylinder head upper face direction).**

➡**Though the camshaft does not stop at the portion as shown, for placement of the cam nose, it is generally accepted**

that the camshaft is placed for the same direction as shown.

21. Install the camshaft brackets in the same position that they were removed. Install brackets No.2–No.4 aligning the stamp marks as indicated in the illustration.

➡**There are no identification marks indicating left or right for camshaft bracket No.1.**

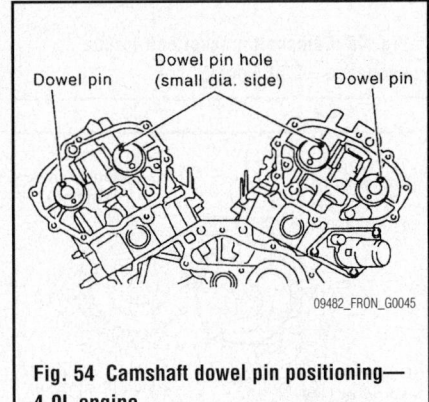

Fig. 54 Camshaft dowel pin positioning—4.0L engine

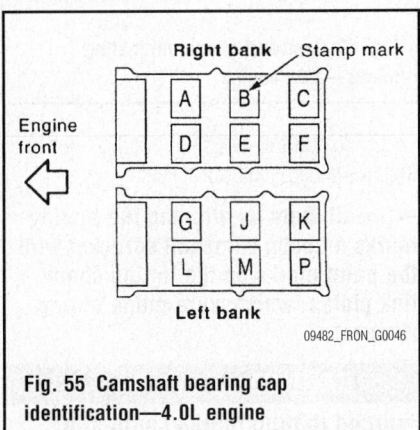

Fig. 55 Camshaft bearing cap identification—4.0L engine

Bank	INT/EXH	Dowel pin	Paint marks		Identification mark
			M1	M2	
RH	INT	No	Green	No	RE
	EXH	Yes	No	White	RE
LH	INT	No	Green	No	LH
	EXH	Yes	No	White	LH

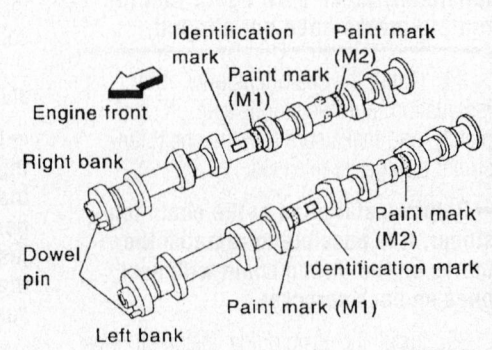

Fig. 53 Camshaft identification—4.0L engine

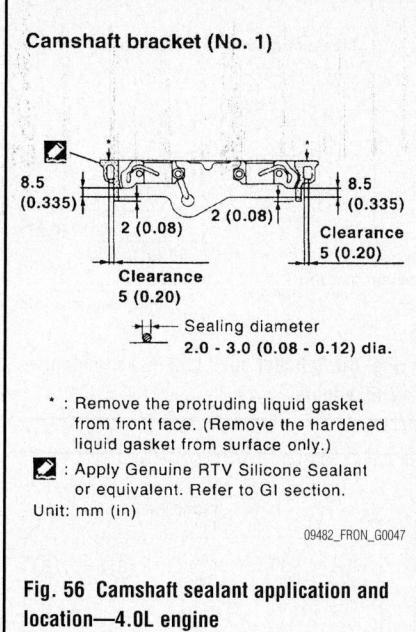

Fig. 56 Camshaft sealant application and location—4.0L engine

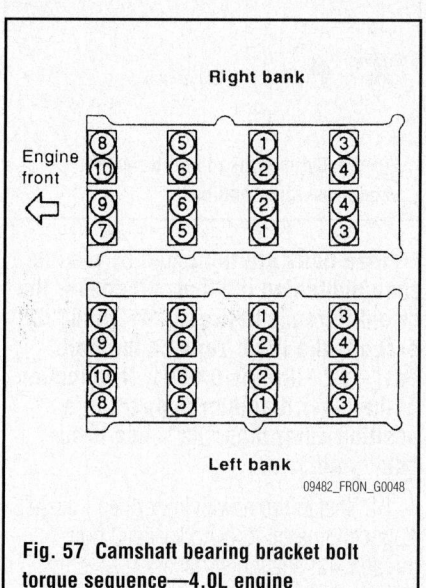

Fig. 57 Camshaft bearing bracket bolt torque sequence—4.0L engine

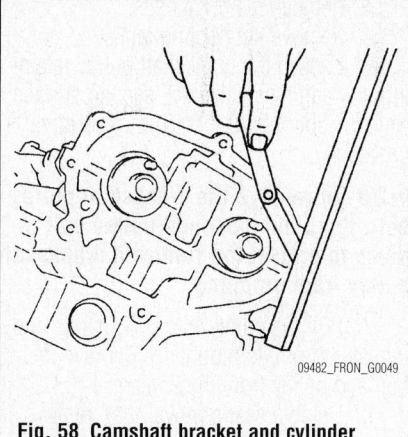

Fig. 58 Camshaft bracket and cylinder head measurement—4.0L engine

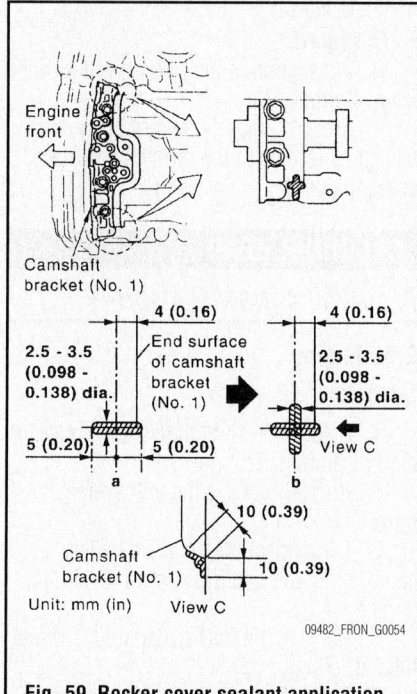

Fig. 59 Rocker cover sealant application locating points—4.0L engine

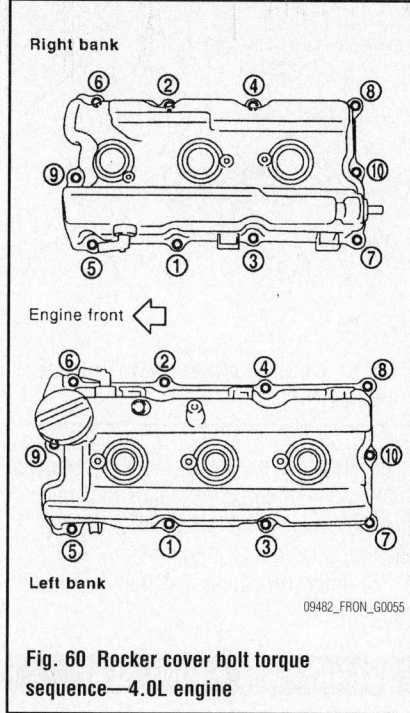

Fig. 60 Rocker cover bolt torque sequence—4.0L engine

22. Apply liquid gasket to the mating surfaces of camshaft bracket No.1 as shown in the illustration on both the left and right cylinder heads. Be sure to use genuine RTV sealant, or equivalent.

23. Tighten the camshaft bracket bolts in the proper sequence and to specification.
   a. Bolts 7–10 to 17 inch lbs.
   b. Bolts 1–6 to 17 inch lbs.
   c. All bolts to 52 inch lbs.
   d. All bolts to 92 inch lbs.

24. Measure the difference in levels between the front end faces of the camshaft bracket No.1 and the cylinder head. Specification should be -0.0055–0.0055 inch. If not within specification, reinstall camshaft bracket No.1.

➡**Measure two positions (both intake and exhaust side) for a single bank.**

25. Check and adjust valve clearance, as required.

26. Apply liquid gasket, be sure to use genuine RTV silicone sealant or equivalent, to the positions shown in the illustration. Refer to figure "a" to apply liquid gasket to joint part of camshaft bracket No.1 and cylinder head. Refer to figure "b" to apply liquid gasket to the figure "a" squarely.

27. Install the rocker cover. Torque the retaining bolts to 17 inch lbs and then to 74 inch lbs, in the proper sequence.

28. Continue the installation in the reverse order of the removal procedure.

29. Inspect the camshaft sprocket (INT) oil groove.

➡**Perform this inspection only when DTC P0011 or DTC P0021 are detected in self diagnostic results of CONSULT-II.**

30. Be sure the engine is cold. Check and adjust oil level, as required.

31. Properly release the fuel system pressure. Disconnect the ignition coil and injector harness connectors.

➡**This is being done to prevent the engine from unintentionally being started while checking.**

32. Remove the intake valve timing control solenoid valve.

33. Crank the engine, and then make sure that engine oil comes out from camshaft bracket (No.1) oil hole.

✱✱ WARNING

Be careful not to touch rotating parts, (drive belt, idler pulley, crankshaft pulley etc) as injury could result.

➡**Oil may squirt from the intake valve timing control solenoid valve installation hole during engine cranking. Use a shop towel to prevent oil from squirting on engine components.**

34. Clean the oil groove between the oil strainer and the intake timing control solenoid valve if engine oil does not come out from camshaft bracket (No.1) oil hole.

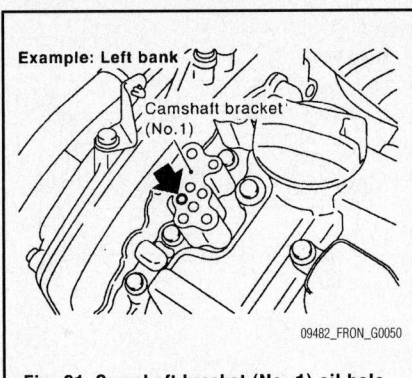

Fig. 61 Camshaft bracket (No. 1) oil hole location

35. Remove the components between the intake valve timing control solenoid valve and the camshaft sprocket (INT). Check each oil groove for clogging.

36. After inspection install any removed components.

CRANKSHAFT FRONT SEAL

REMOVAL & INSTALLATION

2.5L Engine

1. Before servicing the vehicle, refer to the Precautions Section.
2. Disconnect the negative battery cable.
3. Remove the engine undercover.
4. Remove the fan shroud. Remove the cooling fan.
5. Remove the drive belts.
6. Hold the crankshaft pulley with a suitable tool. Loosen the crankshaft pulley retaining bolt.
7. Pull the pulley out about 0.39 inch. Remove the crankshaft pulley bolt.
8. Attach a pulley puller in the M6 thread hole on the crankshaft pulley. Remove the crankshaft pulley.
9. Using a seal removal tool, remove the oil seal from its mounting.

➡ **Be careful not to damage the front cover and/or the crankshaft.**

To install:

10. Installation is the reverse order of the removal procedure.
11. Press fit the seal until it is flush with the front end surface of the front cover, using the proper tools.

4.0L Engine

1. Before servicing the vehicle, refer to the Precautions Section.
2. Disconnect the negative battery cable.
3. Remove the engine undercover.

4. Remove the drive belts.
5. Remove the cooling fan.
6. Loosen the crankshaft pulley retaining bolt and locate the bolt seating surface, which is about 0.39 inch from its original position.

➡ **Do not remove the crankshaft pulley bolt. Keep the loosened pulley bolt in place to protect the removed crankshaft pulley from dropping.**

7. Pull the pulley with both hands and remove it from its mounting. Remove the bolt and pulley from the engine.
8. Using a seal removal tool, remove the oil seal from its mounting.

➡ **Be careful not to damage the front cover and/or the crankshaft.**

To install:

9. Installation is the reverse order of the removal procedure.
10. Press fit until the height of the front oil seal is level with the mounting surface, using the proper tools.

CYLINDER HEAD

REMOVAL & INSTALLATION

2.5L Engine

See Figures 62 and 63.

1. Before servicing the vehicle, refer to the Precautions Section.
2. Properly relieve the fuel system pressure.
3. Disconnect the negative battery cable. Drain the cooling system. Drain the engine oil.
4. Remove the intake manifold and fuel tube assembly.
5. Remove the fuel injector and fuel tube assembly.
6. Remove the exhaust manifold and the three way catalyst.
7. Remove the water outlet. Remove the heater outlet.
8. Remove the front cover and the timing chain.
9. Remove the camshafts.
10. Remove the cylinder head retaining bolts. Be sure to remove the bolts by reversing the order of the tightening torque sequence.
11. Remove the cylinder head from the engine. Discard the gasket.

To install:

12. Installation is the reverse of the removal procedure.
13. Be sure to inspect the cylinder head bolts. Replace as required.

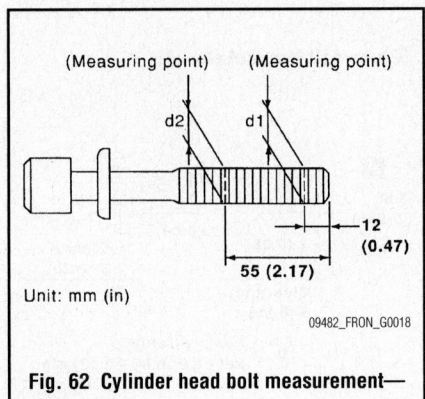

Fig. 62 Cylinder head bolt measurement—2.5L engine

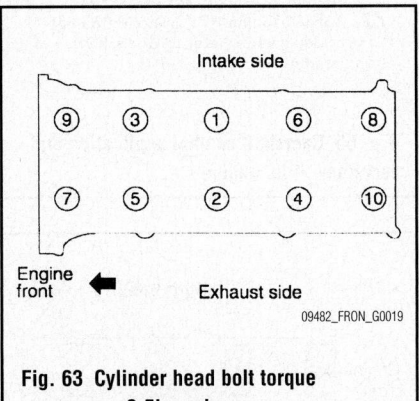

Fig. 63 Cylinder head bolt torque sequence—2.5L engine

➡ **Head bolts are tightened by plastic zone tightening method. Whenever the size difference between "d1" and "d2" exceeds the limit, replace the bolt. "d1"-"d2" limit is 0.0091. If reduction of the outer diameter appears in a position other than "d2", use it the "d2" point.**

14. Install the new cylinder head gasket. Apply engine oil to cylinder head bolt threads and seating surfaces. Torque the cylinder head bolts to specification and in the proper sequence.
15. Be sure to fill the cooling system with the proper grade and type engine coolant.
16. Be sure to fill the engine with the proper grade and type motor oil.
17. Start the engine and check for leaks.

4.0L Engine

See Figures 64 through 67.

1. Before servicing the vehicle, refer to the Precautions Section.
2. Properly relieve the fuel system pressure.
3. Disconnect the negative battery cable. Drain the cooling system.

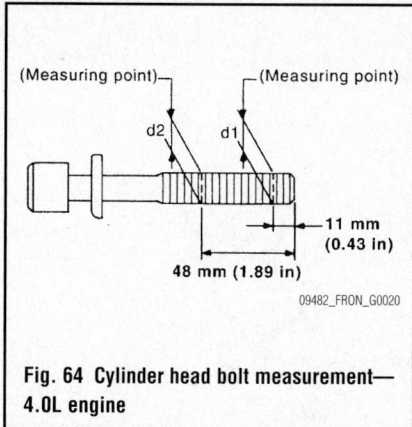

Fig. 64 Cylinder head bolt measurement—4.0L engine

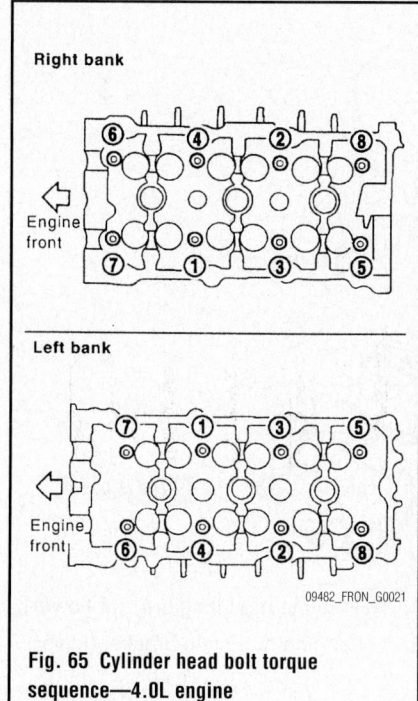

Fig. 65 Cylinder head bolt torque sequence—4.0L engine

4. Remove the camshaft.

5. Remove the intake manifold.

6. Remove the exhaust manifold.

7. Remove the water inlet and thermostat assembly.

8. Remove the water outlet, water pipe and heater pipe.

9. Remove the cylinder head retaining bolts. Be sure to remove the bolts by reversing the order of the tightening torque sequence.

10. Remove the cylinder head from the engine. Discard the gasket.

To install:

11. Installation is the reverse of the removal procedure.

12. Be sure to inspect the cylinder head bolts. Replace as required.

➡**Head bolts are tightened by plastic zone tightening method. Whenever the**

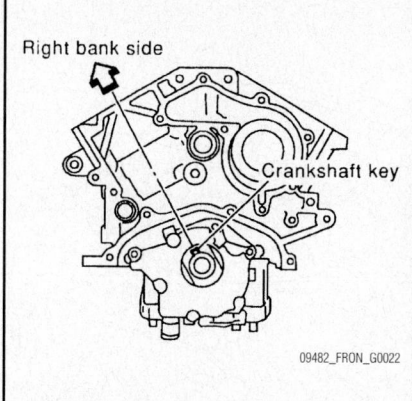

Fig. 66 Cylinder head and crankshaft key alignment—4.0L engine

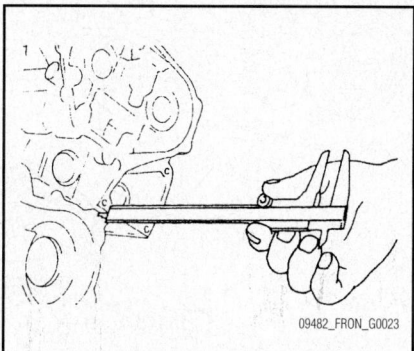

Fig. 67 Cylinder head to cylinder block installation measurement—4.0L engine

size difference between "d1" and "d2" exceeds the limit, replace the bolt. "d1"-"d2" limit is 0.0043. If reduction of the outer diameter appears in a position other than "d2", use it the "d2" point.

13. Install the new cylinder head gasket. Turn the crankshaft until the number one piston is at TDC.

➡**The crankshaft key should line up with the right bank center line, see illustration.**

14. Torque the cylinder head bolts to specification and in the proper sequence.

15. Measure the distance between the front end faces of the cylinder block and the cylinder head on both the left and right banks. If the measured value is not within specification reinstall the cylinder head. Specification is 0.555–0.587 inch.

16. Be sure to fill the cooling system with the proper grade and type engine coolant.

17. Start the engine and check for leaks.

ENGINE ASSEMBLY

REMOVAL & INSTALLATION

Frontier

2.5L ENGINE

See Figure 68.

➡**Be sure to disarm the SRS system, prior to working on the vehicle. Turn the ignition switch OFF, disconnect both battery cables and wait at least three minutes before starting any work.**

1. Before servicing the vehicle, refer to the Precautions Section.

2. Properly release the fuel system pressure. Disconnect the negative battery cable.

3. Drain the radiator. Drain the engine oil. Drain the automatic transmission fluid, if equipped.

4. Matchmark and remove the hood.

5. Remove the air duct and the air cleaner assembly.

6. Disconnect the vacuum hose between the vehicle and the engine and position it to the side.

7. Remove the coolant hoses. Remove the radiator.

8. Remove the drive belts. Remove the cooling fan.

9. Disconnect the engine electrical harness from the engine side and position it to the side.

10. Disconnect the engine harness ground wires.

11. Disconnect the power steering reservoir tank and position it to the side.

12. Remove the power steering pump from the engine and position it to the side.

13. Remove the air conditioning compressor retaining bolts. Position the air conditioning compressor to the side.

14. Disconnect the brake booster vacuum line. Disconnect the EVAP line.

15. Disconnect the fuel line hoses at the engine side connection.

16. Disconnect and plug the heater hoses at the cowl.

17. Remove the automatic transmission oil level indicator stick and tube assembly, if equipped.

18. Remove the three way catalyst.

19. Properly support the engine using an engine support tool.

20. Remove the transmission.

21. Connect a suitable engine lifting fixture and remove the engine from the vehicle.

➡**Before lifting the engine check to be**

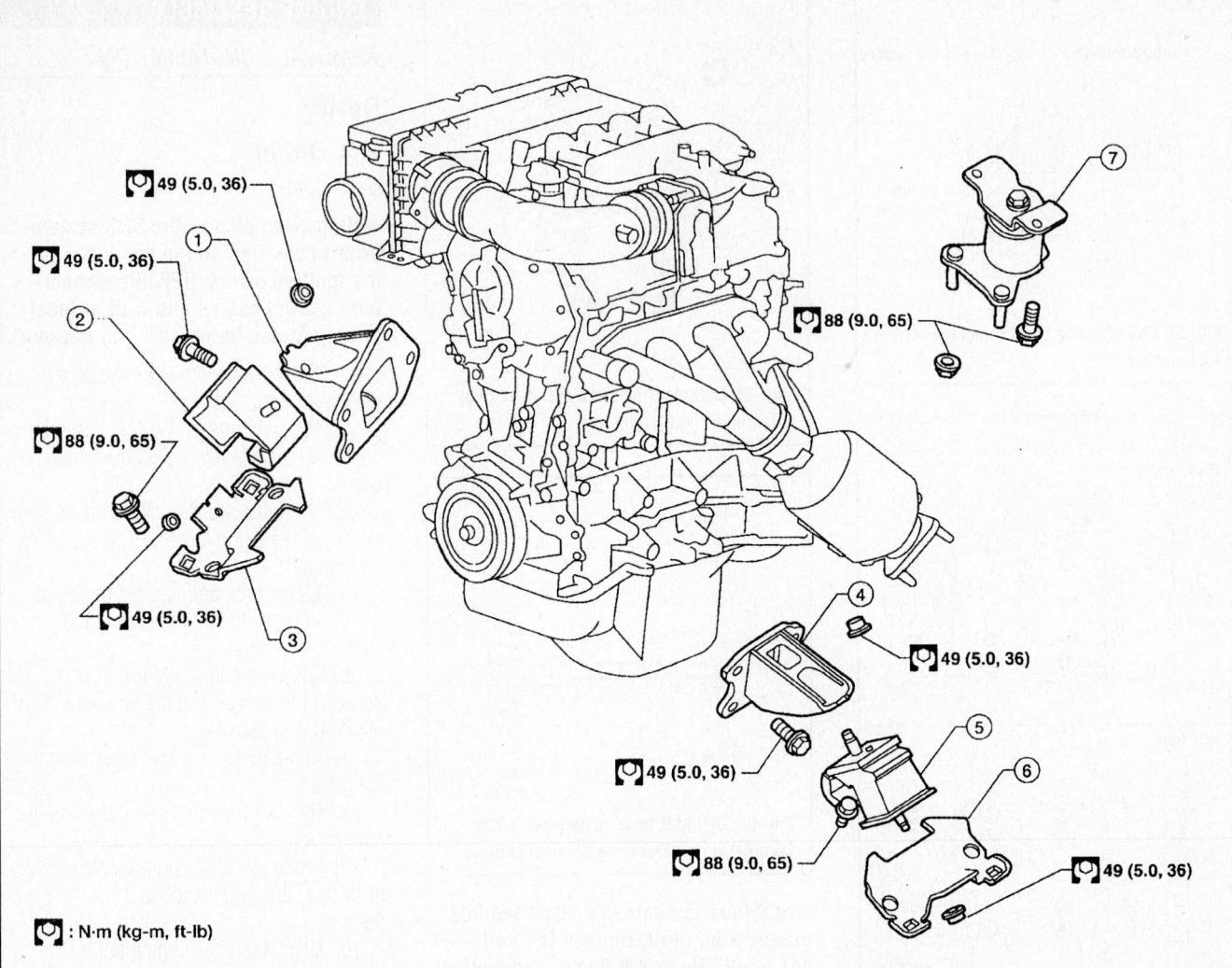

49 (5.0, 36)

49 (5.0, 36)

88 (9.0, 65)

49 (5.0, 36)

88 (9.0, 65)

49 (5.0, 36)

49 (5.0, 36)

88 (9.0, 65)

49 (5.0, 36)

: N·m (kg-m, ft-lb)

1. RH engine mounting bracket (upper)
2. RH engine mounting insulator
3. RH engine mounting bracket (lower)
4. LH engine mounting bracket (upper)
5. LH engine mounting insulator
6. LH engine mounting bracket (lower)
7. Rear engine mounting insulator

09482_FRON_G0007

Fig. 68 Engine mounts and related components—2.5L engine

sure that all necessary electrical connections, vacuum lines and grounding wires have been disconnected, so as not to interfere with the engine removal. Also, check that all mounting bolts have been removed before lifting the engine from the vehicle.

➡Be careful not to damage the drive plate. Avoid deforming and damaging the signal plate teeth, on the drive plate. If the drive plate is removed from the engine, position it with the signal plate surface facing other than downward. Keep magnetic materials away from the signal plate.

To install:

22. Installation is the reverse of the removal procedure.
23. Be sure to fill the engine with the proper grade and type engine oil and engine coolant.
24. Be sure to fill the automatic transmission with the proper grade and type transmission fluid, if equipped.
25. Start the engine and check for leaks, correct as required.

4.0L ENGINE

See Figure 69.

➡Be sure to disarm the SRS system, prior to working on the vehicle. Turn

the ignition switch OFF, disconnect both battery cables and wait at least three minutes before starting any work.

1. Before servicing the vehicle, refer to the Precautions Section.
2. Properly release the fuel system pressure. Disconnect the negative battery cable.
3. Drain the radiator. Drain the engine oil. Drain the automatic transmission fluid, if equipped.
4. Matchmark and remove the hood.
5. Remove the engine cover. Remove the air duct and the air cleaner assembly.
6. Disconnect the vacuum hose

between the vehicle and the engine and position it to the side.

7. Remove the coolant hoses. Remove the radiator.

8. Remove the drive belts. Remove the cooling fan.

9. Disconnect the engine electrical harness from the engine side and position it to the side.

10. Disconnect the engine harness ground wires.

11. Disconnect the power steering reservoir tank and position it to the side.

12. Remove the power steering pump from the engine and position it to the side.

13. Remove the air conditioning compressor retaining bolts. Position the air conditioning compressor to the side.

14. Disconnect the brake booster vacuum line. Disconnect the EVAP line.

15. Disconnect the fuel line hoses at the engine side connection.

16. Disconnect and plug the heater hoses at the cowl.

17. Remove the automatic transmission oil level indicator stick and tube assembly, if equipped.

18. If equipped with 4WD, remove the final drive assembly.

19. Remove the three way catalyst.

20. Properly support the engine using an engine support tool.

21. Remove the transmission.

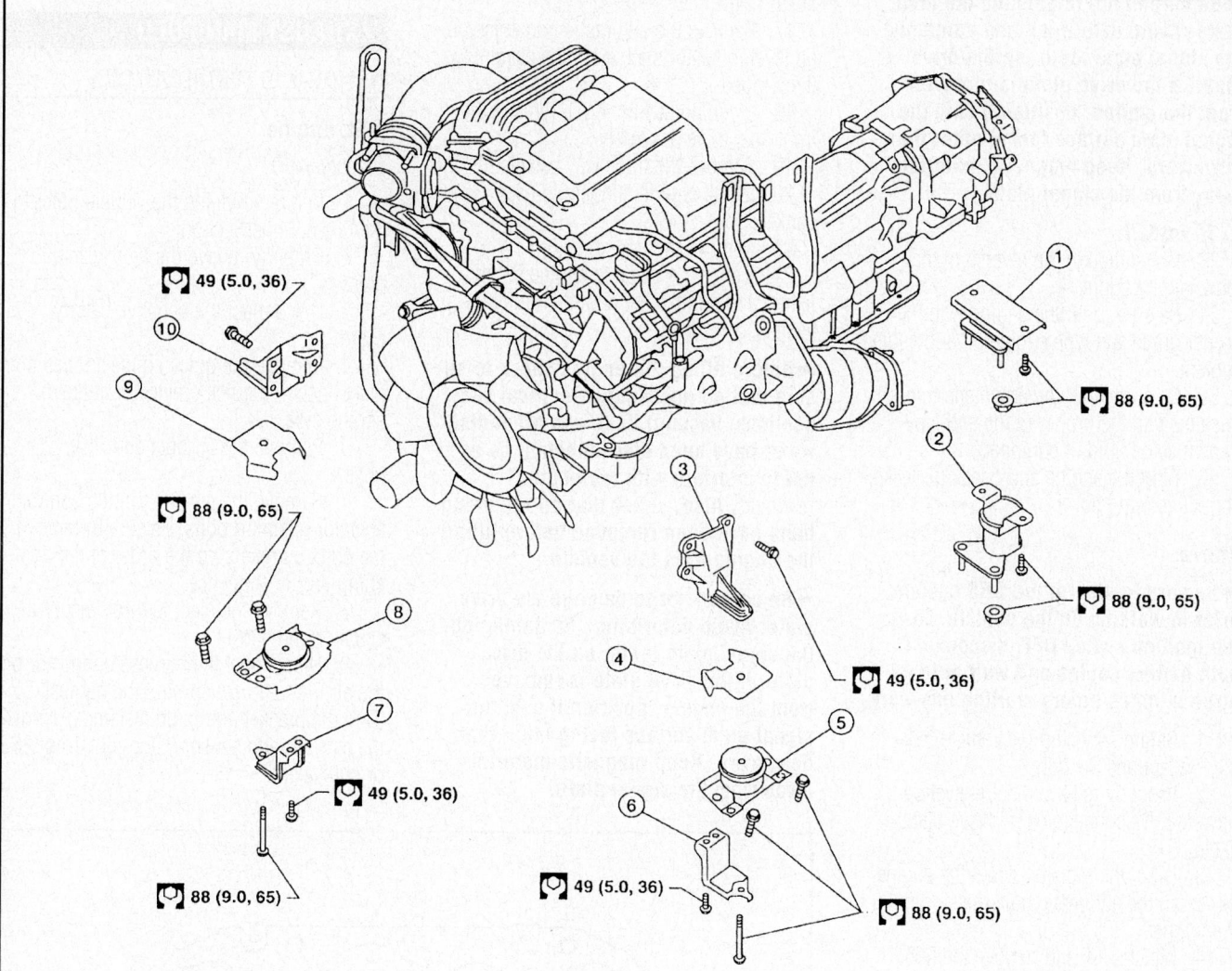

: N·m (kg-m, ft-lb)

1.	Rear engine mounting insulator 4x4	2.	Rear engine mounting insulator 4x2	3.	LH engine mounting bracket (upper)
4.	LH heat shield plate	5.	LH engine mounting insulator	6.	LH engine mounting bracket (lower)
7.	RH engine mounting bracket (lower)	8.	RH engine mounting insulator (upper)	9.	RH heat shield plate
10.	RH engine mounting bracket (upper)				

Fig. 69 Engine mounts and related components—4.0L engine

09482_FRON_G0008

22. Connect a suitable engine lifting fixture and remove the engine from the vehicle.

➡️Before lifting the engine check to be sure that all necessary electrical connections, vacuum lines and grounding wires have been disconnected, so as not to interfere with the engine removal. Also, check that all mounting bolts have been removed before lifting the engine from the vehicle.

➡️Be careful not to damage the drive plate. Avoid deforming and damaging the signal plate teeth, on the drive plate. If the drive plate is removed from the engine, position it with the signal plate surface facing other than downward. Keep magnetic materials away from the signal plate.

To install:

23. Installation is the reverse of the removal procedure.
24. Be sure to fill the engine with the proper grade and type engine oil and engine coolant.
25. Be sure to fill the automatic transmission with the proper grade and type transmission fluid, if equipped.
26. Start the engine and check for leaks, correct as required.

Xterra

➡️Be sure to disarm the SRS system, prior to working on the vehicle. Turn the ignition switch OFF, disconnect both battery cables and wait at least three minutes before starting any work.

1. Before servicing the vehicle, refer to the Precautions Section.
2. Properly release the fuel system pressure. Disconnect the negative battery cable.
3. Drain the radiator. Drain the engine oil. Drain the automatic transmission fluid, if equipped.
4. Matchmark and remove the hood.
5. Remove the engine cover. Remove the air duct and the air cleaner assembly.
6. Disconnect the vacuum hose between the vehicle and the engine and position it to the side.
7. Remove the coolant hoses. Remove the radiator.
8. Remove the drive belts. Remove the cooling fan.
9. Disconnect the engine electrical harness from the engine side and position it to the side.
10. Disconnect the engine harness ground wires.

11. Disconnect the power steering reservoir tank and position it to the side.
12. Remove the power steering pump from the engine and position it to the side.
13. Remove the air conditioning compressor retaining bolts. Position the air conditioning compressor to the side.
14. Disconnect the brake booster vacuum line. Disconnect the EVAP line.
15. Disconnect the fuel line hoses at the engine side connection.
16. Disconnect and plug the heater hoses at the cowl.
17. Remove the automatic transmission oil level indicator stick and tube assembly, if equipped.
18. If equipped with 4WD, remove the front final drive assembly.
19. Remove the three way catalyst.
20. Install engine slingers to each bank.
21. Remove the transmission.
22. Connect a suitable engine lifting fixture and remove the engine from the vehicle.

➡️Before lifting the engine check to be sure that all necessary electrical connections, vacuum lines and grounding wires have been disconnected, so as not to interfere with the engine removal. Also, check that all mounting bolts have been removed before lifting the engine from the vehicle.

➡️Be careful not to damage the drive plate. Avoid deforming and damaging the signal plate teeth, on the drive plate. If the drive plate is removed from the engine, position it with the signal plate surface facing other than downward. Keep magnetic materials away from the signal plate.

To install:

23. Installation is the reverse of the removal procedure.
24. Be sure to fill the engine with the proper grade and type engine oil and engine coolant.
25. Be sure to fill the automatic transmission with the proper grade and type transmission fluid, if equipped.
26. Bleed air from cooling system.
27. Start the engine and check for leaks, correct as required.

EXHAUST MANIFOLD

REMOVAL & INSTALLATION

2.5L Engine

See Figure 70.

1. Before servicing the vehicle, refer to the Precautions Section.
2. Properly relieve the fuel system pressure.
3. Disconnect the negative battery cable.
4. Remove the quick connector cap and disconnect the quick connector at the engine side.
5. Remove the air duct and PCV hose.
6. Remove the electric throttle control actuator retaining bolts. Be sure to remove the bolts by reversing the order of the tightening torque sequence.
7. Remove the electric throttle control actuator and gasket.
8. Disconnect the harness connector of the air fuel ratio sensor and the harness from the bracket and middle clamp. Remove the air fuel ratio sensor using tool J-44626, or equivalent.

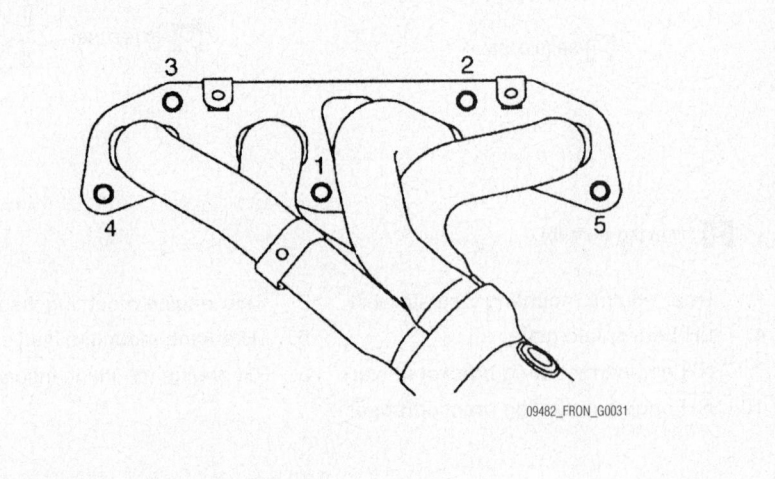

Fig. 70 Exhaust manifold bolt torque sequence—2.5L engine

09482_FRON_G0031

➡Be careful not to damage the air fuel ratio sensor. Discard the sensor if it has been dropped from a height of more than 19.7 inches on to a hard surface.

9. Remove the front exhaust tube. Remove the exhaust manifold cover.

10. Remove the bracket between the exhaust manifold three way catalyst assembly and the transmission assembly.

11. Remove the exhaust manifold retaining bolts. Be sure to remove the bolts by reversing the order of the tightening torque sequence.

12. Remove the exhaust manifold from the engine. Discard the gasket.

To install:

13. Installation is the reverse of the removal procedure.

14. Be sure to use new gaskets.

15. Be sure to tighten the exhaust manifold retaining bolts to specification and in the proper sequence.

➡Before installing a new air fuel sensor apply anti seize lubricant to the threads. Do not over torque the sensor, doing so may cause damage to the sensor resulting in the MIL light coming on.

➡See throttle valve closed position learning and idle air volume learning procedures, for relearning information.

4.0L Engine

LEFT SIDE

See Figure 71.

1. Before servicing the vehicle, refer to the Precautions Section.

2. Disconnect the negative battery cable.

3. Remove air duct, PCV hose (between air duct and rocker cover) and electric throttle control actuator.

4. Disconnect the harness connector and remove the heated oxygen sensor.

➡Be careful not to damage the air fuel ratio sensor. Discard the sensor if it has been dropped from a height of more than 19.7 inches on to a hard surface.

5. Remove the front exhaust tube.

6. Remove the exhaust manifold cover.

7. Remove bracket between exhaust manifold-three way catalyst assembly and transmission assembly.

8. Remove the exhaust manifold retaining bolts. Be sure to remove the bolts by reversing the order of the tightening torque sequence.

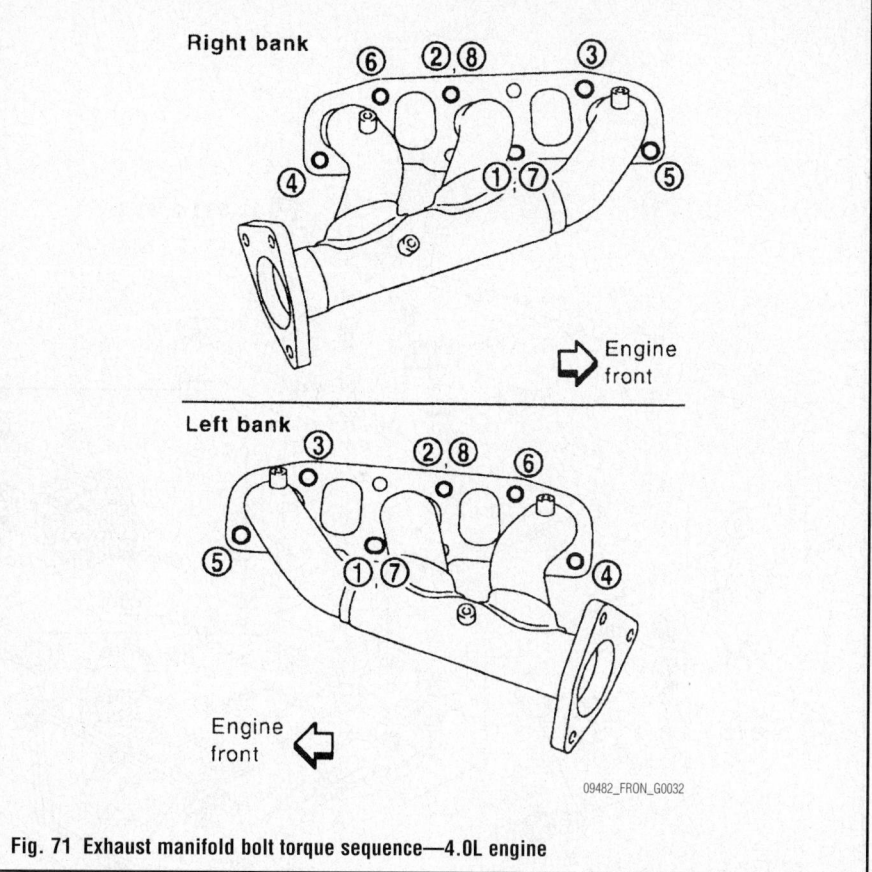

Fig. 71 Exhaust manifold bolt torque sequence—4.0L engine

09482_FRON_G0032

9. Remove the exhaust manifold from the engine. Discard the gasket.

To install:

10. Installation is the reverse of the removal procedure.

11. Be sure to use new gaskets.

12. Be sure to tighten the exhaust manifold retaining bolts to specification and in the proper sequence.

➡Before installing a new air fuel sensor and heated oxygen sensor, clean threads and apply anti seize lubricant to the threads. Do not over torque the sensor, doing so may cause damage to the sensor resulting in the MIL light coming on.

RIGHT SIDE

See Figure 71.

1. Before servicing the vehicle, refer to the Precautions Section.

2. Disconnect the negative battery cable.

3. Remove the engine from the vehicle. Position the assembly in a suitable holding fixture.

4. Remove the exhaust manifold retaining bolts. Be sure to remove the bolts by reversing the order of the tightening torque sequence.

➡Disregard the numerical order of No.7 and No.8 in the removal process.

5. Discard the gaskets.

To install:

6. Installation is the reverse of the removal procedure.

7. Be sure to use new gaskets.

8. Be sure to tighten the exhaust manifold retaining bolts to specification and in the proper sequence.

➡Before installing a new air fuel sensor and heated oxygen sensor apply anti seize lubricant to the threads. Do not over torque the sensor, doing so may cause damage to the sensor resulting in the MIL light coming on.

INTAKE MANIFOLD

REMOVAL & INSTALLATION

2.5L Engine

See Figures 72 through 74.

1. Before servicing the vehicle, refer to the Precautions Section.

2. Properly relieve the fuel system pressure.

3. Disconnect the negative battery cable. Drain the cooling system.

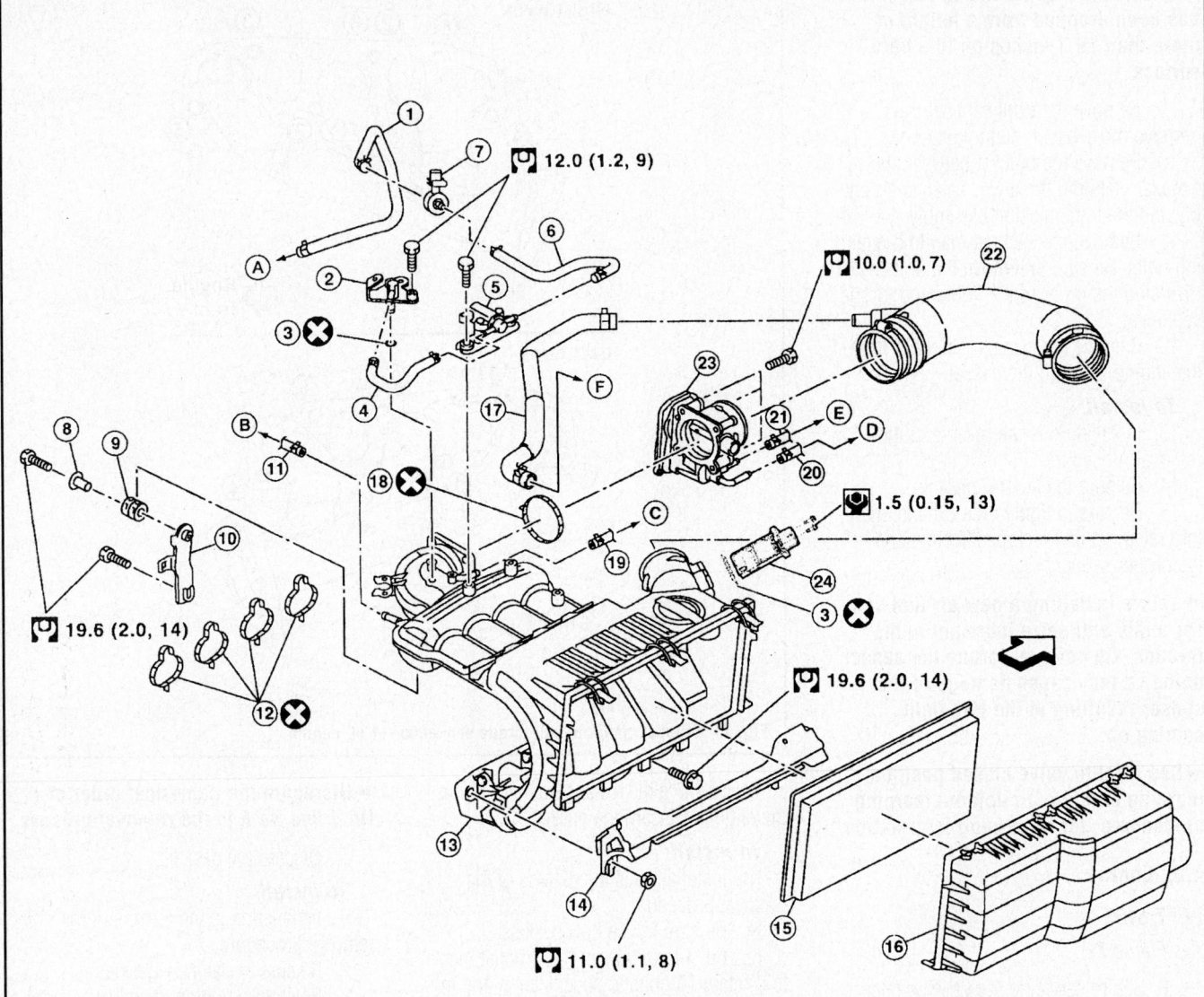

1. Vacuum hose
2. Vacuum hose adapter
3. O–ring
4. Vacuum hose
5. EVAP canister purge volume control solenoid valve
6. Vacuum hose
7. Service port
8. Collar
9. Grommet
10. Intake manifold support
11. Vacuum hose
12. Gasket
13. Intake manifold
14. Fuel tube protector
15. Air cleaner
16. Air cleaner case
17. PCV hose
18. Gasket
19. PCV hose
20. Water hose
21. Water hose
22. Air duct
23. Electric throttle control actuator
24. Mass air flow sensor
A. To vacuum pipe (EVAP canister)
B. To brake booster
C. To PCV valve
D. To heater outlet
E. To heater pipe
F. To rocker cover
→ Engine front

09482_FRON_G0027

Fig. 72 Intake manifold and related components—2.5L engine

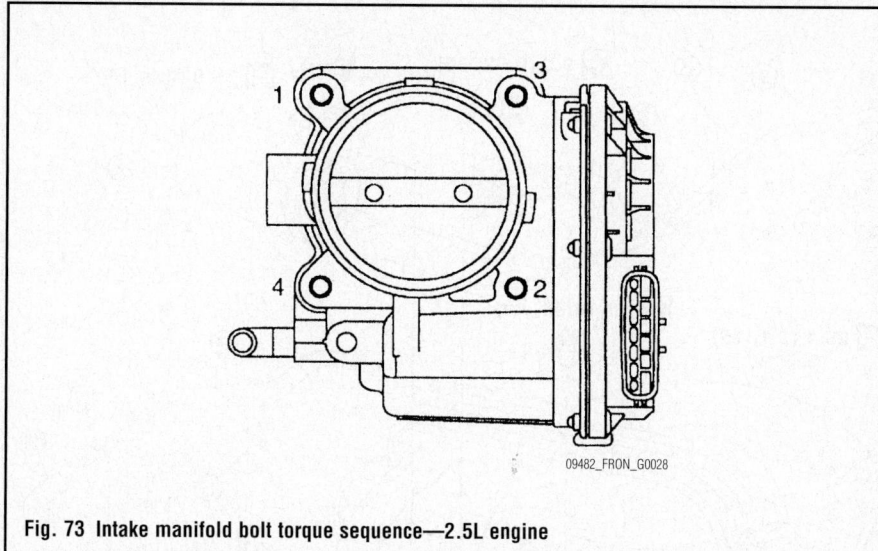

Fig. 73 Intake manifold bolt torque sequence—2.5L engine

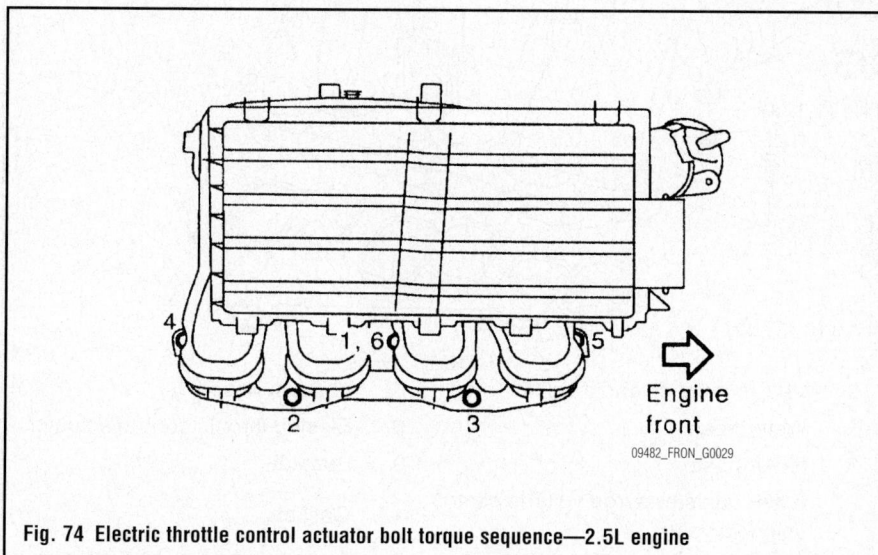

Fig. 74 Electric throttle control actuator bolt torque sequence—2.5L engine

4. Remove the air cleaner case, air cleaner and air duct.

5. Disconnect and plug the water hoses from the electric throttle control actuator.

6. Remove the mass air flow sensor from the intake manifold. Remove the quick connector cap and disconnect the quick connector at the engine side.

7. Remove the electric throttle control actuator harness connector and retaining bolts. Be sure to remove the bolts by reversing the order of the tightening torque sequence.

8. Remove the electric throttle control actuator and gasket.

9. Disconnect the harness, vacuum hoses and PCV hoses from the intake manifold and position them to the side.

10. Remove intake manifold support.

11. Remove the intake manifold retaining bolts. Be sure to remove the bolts by

reversing the order of the tightening torque sequence.

12. Remove the intake manifold, fuel tube protector and gasket from the engine.

13. As necessary, remove the EVAP canister purge volume solenoid valve and vacuum hose adapter from the intake manifold.

14. Disconnect the sub frame harness from the fuel injectors. Remove the fuel tube and fuel injector assembly from the intake manifold, if required.

To install:

15. Installation is the reverse of the removal procedure.

16. Be sure to use new gaskets.

17. Be sure to tighten the intake manifold retaining bolts to specification and in the proper sequence.

➡**Refer to the torque sequence illustration, No.6 means double tightening of bolt No.1. M8xM38mm (1.50 inches)**

are green in color (No1, No.6). M8xM35mm (1.38 inch) (No.2, No.3). Nut (No.4, No.5).

18. Be sure to tighten the electric throttle control actuator retaining bolts to specification and in the proper sequence.

➡**See throttle valve closed position learning and idle air volume learning procedures, for relearning information.**

19. Be sure to fill the cooling system with the proper grade and type engine coolant.

20. Start the engine and check for leaks.

4.0L Engine

See Figures 75 through 78.

➡**Upper intake manifold is also referred to as intake manifold collector.**

1. Before servicing the vehicle, refer to the Precautions Section.

2. Properly relieve the fuel system pressure.

3. Disconnect the negative battery cable. Drain the cooling system.

4. Remove the engine cover. Remove the air cleaner case (upper) with the mass air flow sensor and air duct assembly.

5. Disconnect the water hoses from the electric throttle control actuator. Disconnect the harness connector.

6. Remove the electric throttle control actuator retaining bolts. Be sure to remove the bolts by reversing the order of the tightening torque sequence.

7. Remove the electric throttle control actuator.

8. Remove the brake booster vacuum hose and the PCV hose. Remove the intake manifold collector support.

9. Disconnect the EVAP hoses and harness connector from the EVAP canister purge volume control solenoid valve. Remove the EVAP canister purge volume control solenoid valve.

10. Remove the VIAS control solenoid valve and vacuum tank.

11. Remove the intake manifold collector retaining bolts. Be sure to remove the bolts by reversing the order of the tightening torque sequence.

12. Remove the intake manifold collector from the engine.

13. Remove the fuel tube and fuel injector assembly.

14. Remove the intake manifold retaining bolts. Be sure to remove the bolts by reversing the order of the tightening torque sequence.

15. Remove the intake manifold from the engine.

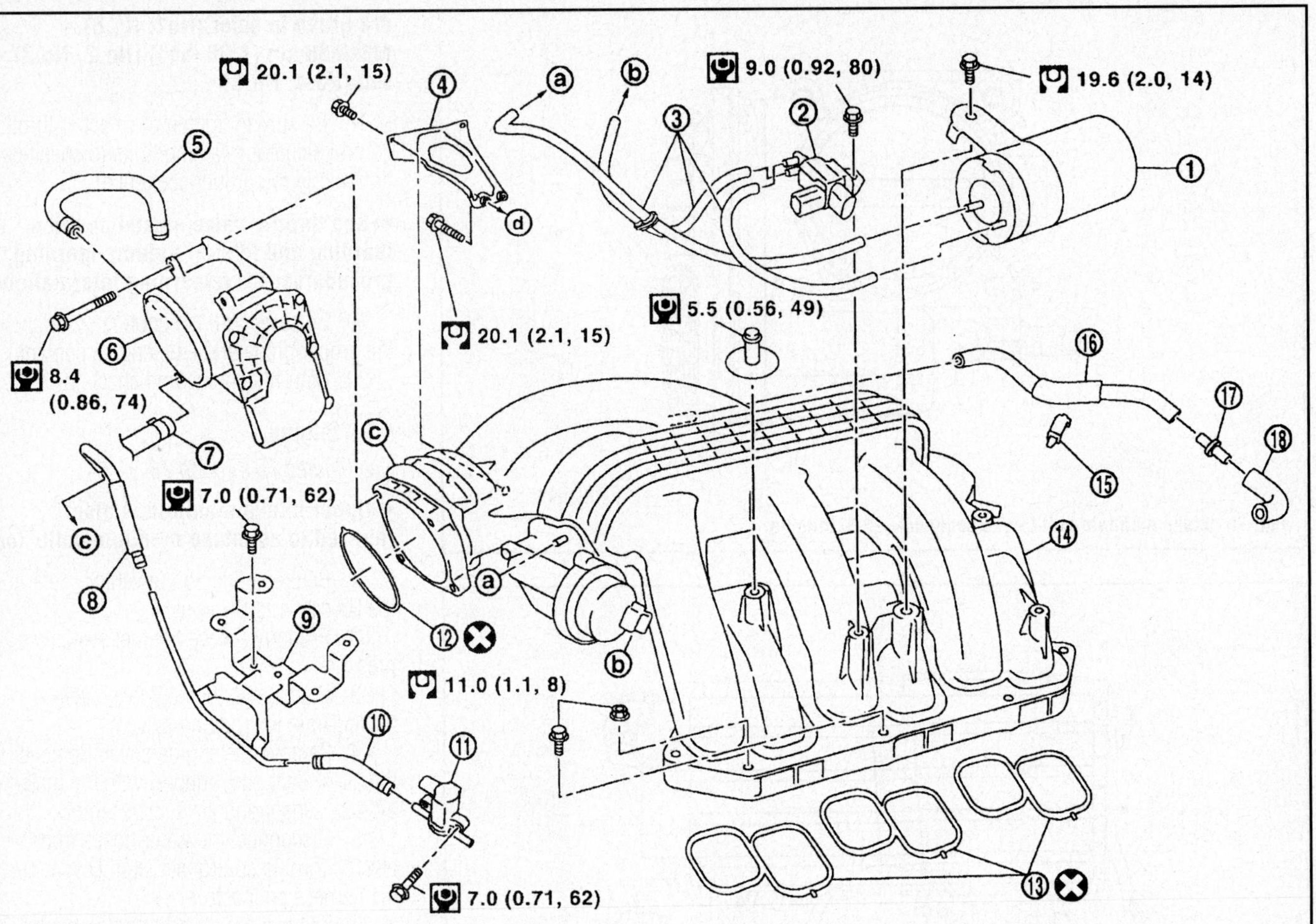

🔧 20.1 (2.1, 15)   🔧 9.0 (0.92, 80)   🔧 19.6 (2.0, 14)

🔧 20.1 (2.1, 15)

🔧 5.5 (0.56, 49)

🔧 8.4 (0.86, 74)

🔧 7.0 (0.71, 62)

🔧 11.0 (1.1, 8)

🔧 7.0 (0.71, 62)

1.	Vacuum tank	2.	VIAS control solenoid valve	3.	Vacuum hose
4.	Intake manifold collector support	5.	Water hose	6.	Electric throttle control actuator
7.	Water hose	8.	EVAP hose	9.	Bracket
10.	EVAP hose	11.	EVAP canister purge volume control solenoid valve	12.	Gasket
13.	Gasket	14.	Intake manifold collector	15.	Clip
16.	PCV hose	17.	Connector	18.	PCV hose
a.	To intake manifol collector	b.	To power valve	c.	To throttle body
d.	To cylinder head (RH bank)				

09482_FRON_G0024

Fig. 75 Intake manifold collector and related components—4.0L engine

To install:

16. Installation is the reverse of the removal procedure.

17. Be sure to use new gaskets.

18. Be sure to tighten the intake manifold retaining bolts to specification and in the proper sequence in two or more steps.

19. Be sure to tighten the intake manifold collector retaining bolts to specification and in the proper sequence.

20. Be sure to tighten the electric throttle control actuator retaining bolts to specification and in the proper sequence.

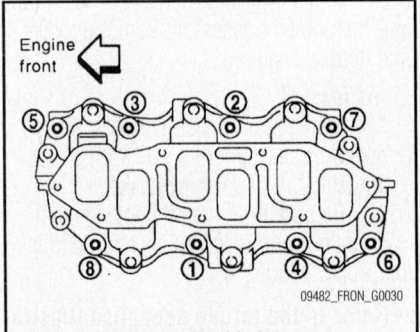

09482_FRON_G0030

Fig. 76 Intake manifold bolt torque sequence—4.0L engine

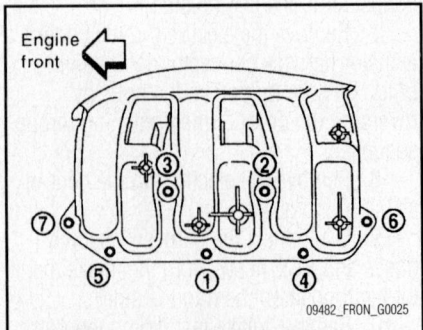

09482_FRON_G0025

Fig. 77 Intake manifold collector bolt torque sequence—4.0L engine

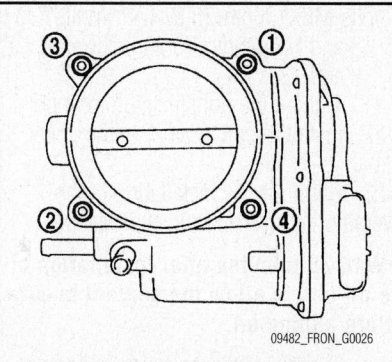

Fig. 78 Electric throttle control actuator bolt torque sequence—4.0L engine

➡ See throttle valve closed position learning and idle air volume learning procedures, for relearning information.

OIL PAN

REMOVAL & INSTALLATION

2.5L Engine

See Figure 79.

1. Before servicing the vehicle, refer to the Precautions Section.
2. Disconnect the negative battery cable.
3. Remove the engine under cover. Drain the engine oil.
4. If equipped with automatic transmission, remove the fluid cooler tube.
5. Loosen the oil pan retaining bolts, in the reverse order of the installation sequence.
6. Insert a seal cutter tool between the oil pan and the cylinder block, and slide it by tapping on the side of the tool with a hammer.
7. Remove the oil pan from the engine.

To install:

8. Be sure to clean all the oil gasket material from both the oil pan and the cylinder block surfaces, using the proper tools.

9. Apply a continuous bead of sealant 0.138–0.177 in. (3.5–4.5mm) to the oil pan mating surface.
10. Install the oil pan to the cylinder block. This must be done within 5 minutes after applying the liquid gasket.
11. Torque the bolts to specification and in the proper sequence.
12. Continue the installation in the reverse order of the removal procedure.

➡ Wait 30 minutes after installation of the oil pan to allow the sealant to cure before adding oil.

13. Fill the crankcase to the correct level.
14. Start the engine and check for leaks.

4.0L Engine

LOWER

See Figure 80.

1. Before servicing the vehicle, refer to the Precautions Section.
2. Disconnect the negative battery cable.
3. Remove the engine under cover. Drain the engine oil.
4. Loosen the oil pan retaining bolts, in the reverse order of the installation sequence.
5. Insert a seal cutter tool between the oil pan and the cylinder block, and slide it by tapping on the side of the tool with a hammer.
6. Remove the oil pan from the engine.

To install:

7. Be sure to clean all the oil gasket material from both the oil pan and the cylinder block surfaces, using the proper tools.
8. Apply a continuous bead of sealant 0.138–0.177 in. (3.5–4.5mm) to the oil pan mating surface. Be sure to use genuine RTV sealant, or equivalent.

9. Install the oil pan to the cylinder block. This must be done within 5 minutes after applying the liquid gasket.
10. Torque the bolts to specification and in the proper sequence.
11. Continue the installation in the reverse order of the removal procedure.

➡ Wait 30 minutes after installation of the oil pan to allow the sealant to cure before adding oil.

12. Fill the crankcase to the correct level.
13. Start the engine and check for leaks.

UPPER

See Figures 81 and 82.

1. Before servicing the vehicle, refer to the Precautions Section.
2. Disconnect the negative battery cable.
3. Remove the air duct. Remove the engine under cover.
4. Drain the engine oil. Drain the engine coolant.
5. Remove the final drive, if equipped with 4WD.
6. Disconnect the steering gear lower shaft joint bolt and steering gear nuts and bolts, position the assembly out of the way.
7. Remove the starter.
8. Disconnect the automatic transmission fluid cooler brackets, if equipped and position them out of the way.
9. Remove the oil filter, as necessary. Remove the oil cooler.
10. Remove the lower oil pan. Remove the oil strainer.
11. Remove the transmission joint bolts which pierce the oil pan.
12. Remove the rear cover plate.
13. Loosen the upper oil pan retaining bolts, in the reverse order of the installation sequence.

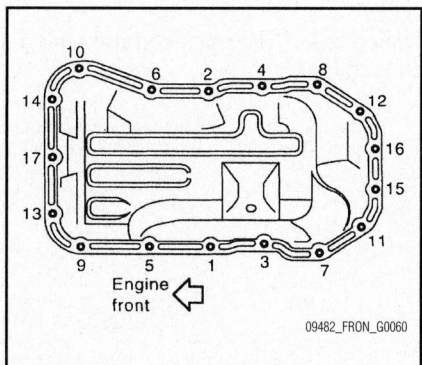

Fig. 79 Oil pan bolt torque sequence— 2.5L engine

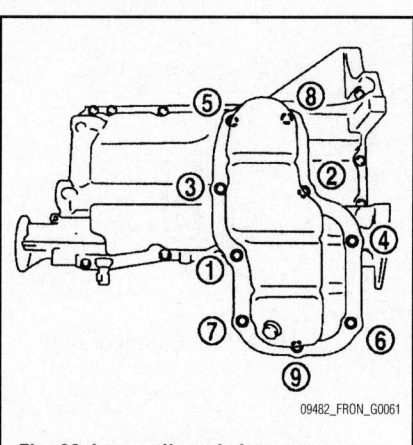

Fig. 80 Lower oil pan bolt torque sequence—4.0L engine

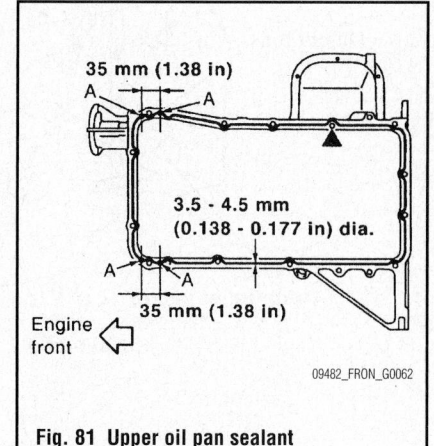

Fig. 81 Upper oil pan sealant application—4.0L engine

14. Insert a seal cutter tool between the oil pan and the cylinder block, and slide it by tapping on the side of the tool with a hammer.

15. Remove the oil pan from the engine. Remove the O-rings from the bottom lower cylinder block and oil pump.

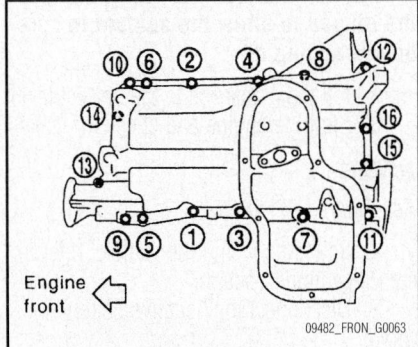

Fig. 82 Upper oil pan bolt torque sequence—4.0L engine

To install:

16. Be sure to clean all the oil gasket material from both the oil pan and the cylinder block surfaces, using the proper tools.

17. Install new O-rings on the bottom lower cylinder block and oil pump.

18. Apply a continuous bead of sealant 0.138–0.177 in. (3.5–4.5mm) to the lower cylinder block mating surfaces of the upper oil pan. Be sure to use genuine RTV sealant, or equivalent.

➡ **For bolt holes marked with a solid black triangle, apply liquid gasket outside the hole. Apply a bead of sealant (0.177–0.217 inch diameter) to area "A".**

19. Install the upper oil pan. This must be done within 5 minutes after applying the liquid gasket.

20. Torque the bolts to specification and in the proper sequence. There are two types of bolts M8X100mm (3.97 inch) bolts 7, 11, 12, 13 and M8X25mm (0.98 inch) except 7, 11, 12 and 13.

21. Tighten the transmission joint bolts.

22. Install the oil strainer to the upper oil pan.

23. Continue the installation in the reverse order of the removal procedure.

➡ **Wait 30 minutes after installation of the oil pan to allow the sealant to cure before adding oil.**

24. Fill the crankcase to the correct level.

25. Start the engine and check for leaks.

OIL PUMP

REMOVAL & INSTALLATION

4.0L Engine

See Figures 83 and 84.

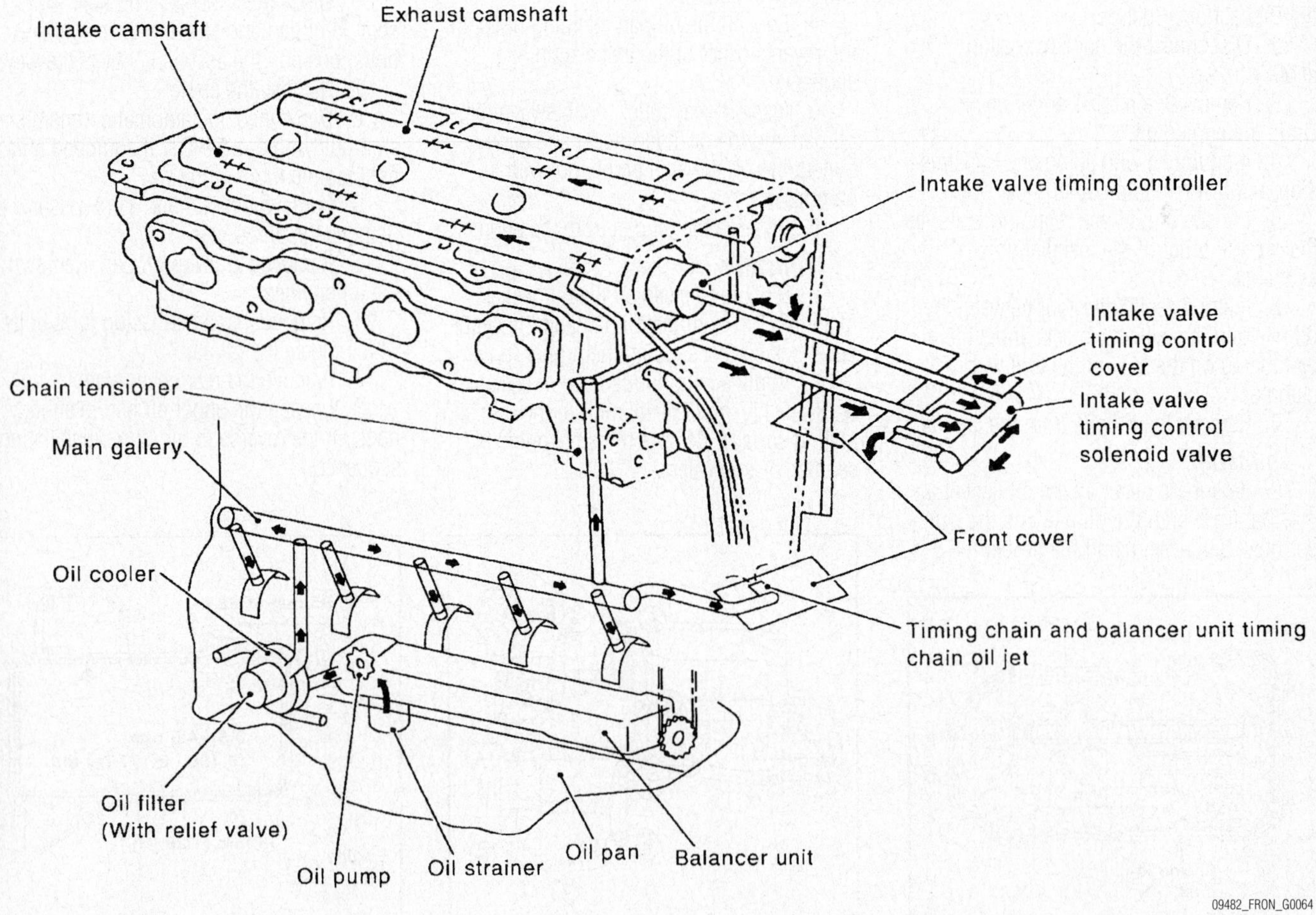

Intake camshaft

Exhaust camshaft

Intake valve timing controller

Intake valve timing control cover

Intake valve timing control solenoid valve

Front cover

Timing chain and balancer unit timing chain oil jet

Chain tensioner

Main gallery

Oil cooler

Oil filter (With relief valve)

Oil pump

Oil strainer

Oil pan

Balancer unit

09482_FRON_G0064

Fig. 83 Oil pump and related components—2.5L engine

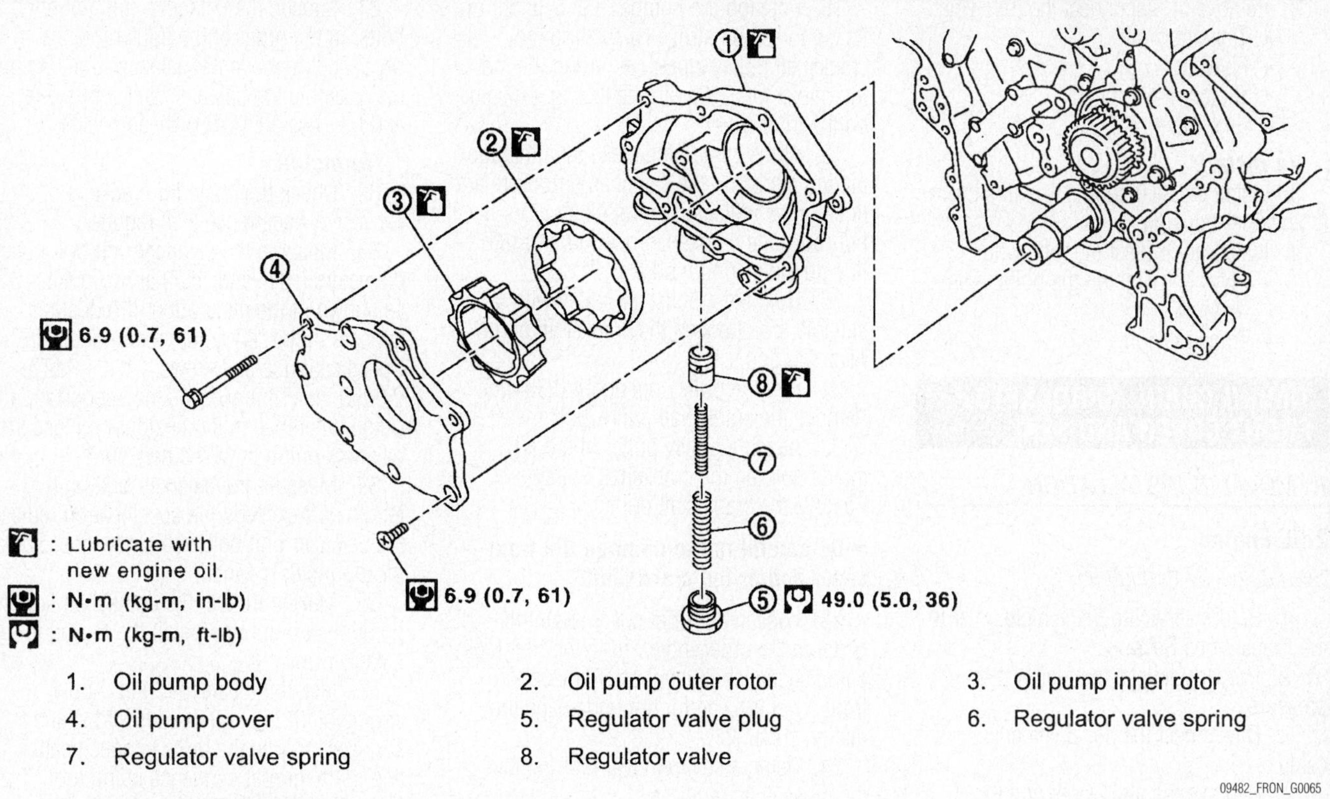

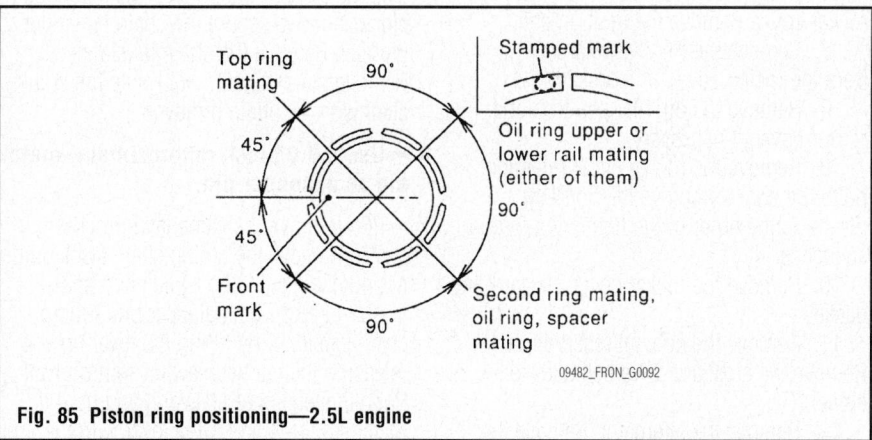

Fig. 84 Oil pump and related components—4.0L engine

: Lubricate with new engine oil.

: N·m (kg-m, in-lb)

: N·m (kg-m, ft-lb)

6.9 (0.7, 61)

6.9 (0.7, 61)

49.0 (5.0, 36)

1. Oil pump body
2. Oil pump outer rotor
3. Oil pump inner rotor
4. Oil pump cover
5. Regulator valve plug
6. Regulator valve spring
7. Regulator valve spring
8. Regulator valve

09482_FRON_G0065

1. Before servicing the vehicle, refer to the Precautions Section.
2. Disconnect the negative battery cable. Drain the engine oil. Drain the engine coolant.
3. Remove the lower oil pan.
4. Remove the upper oil pan.
5. Remove the front timing chain case and timing chain (primary).
6. Remove the oil pump from the engine

To install:

7. Installation is the reverse of the removal procedure.

➡**Wait 30 minutes after installation of the oil pan to allow the sealant to cure before adding oil.**

8. Fill the crankcase to the correct level.
9. Start the engine and check for leaks.

PISTON AND RING

POSITIONING

See Figures 85 and 86.

REAR MAIN SEAL

REMOVAL & INSTALLATION

1. Before servicing the vehicle, refer to the Precautions Section.

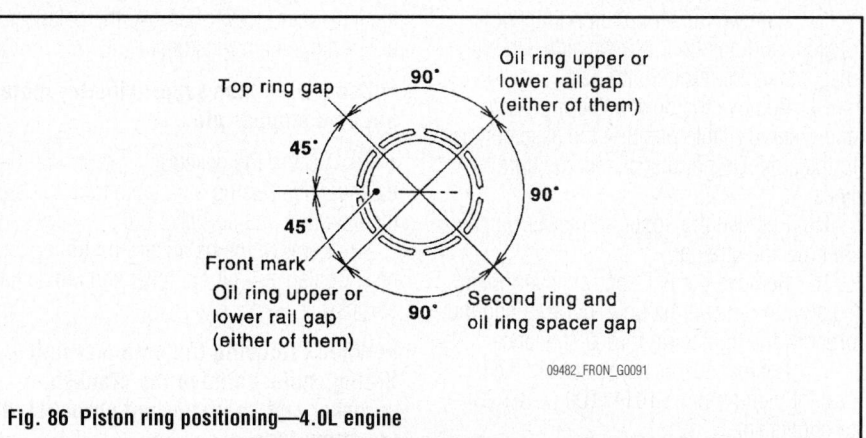

Fig. 85 Piston ring positioning—2.5L engine

Fig. 86 Piston ring positioning—4.0L engine

2. Remove or disconnect the following:
- Transmission
- Flywheel
- Clutch, if equipped
- Rear main seal

To install:

3. Install the seal so that it is flush with the retainer housing.

4. Install or connect the following:
- Flywheel. Tighten the bolts to specification.
- Transmission

TIMING CHAIN, SPROCKETS, FRONT COVER AND SEAL

REMOVAL & INSTALLATION

2.5L Engine

See Figures 87 through 95.

1. Before servicing the vehicle, refer to the Precautions Section.

2. Properly relieve the fuel system pressure.

3. Disconnect the negative battery cable.

4. Remove the air cleaner and the air duct assembly.

5. Remove the spark plugs.

6. Disconnect the PCV hose from the rocker cover. Remove the ignition coil.

7. Remove the PCV valve and O-ring from the rocker cover, if necessary.

8. Remove the oil filler cap from the rocker cover, if necessary.

9. Remove the rocker cover retaining bolts. Be sure to remove the bolts by reversing the order of the tightening torque sequence.

10. Remove the rocker cover. Discard the gasket.

11. Remove the coolant reservoir tank. Remove the auxiliary drive belt auto-tensioner.

12. Remove the alternator. Remove the strut tower brace.

13. Remove the air conditioning compressor and position it to the side. Do not disconnect the refrigerant lines.

14. Remove the power steering pump and reservoir tank; position the assembly to the side. Do not disconnect the fluid lines.

15. Remove the upper and lower oil pan. Remove the strainer.

16. Remove the IVT control cover bolts. Be sure to remove the bolts by reversing the order of the tightening torque sequence.

17. Remove the cover by cutting the sealant using tool KV10111100 (J-37228) or equivalent.

18. Position the number one cylinder on its compression stroke by rotating the crankshaft pulley clockwise. Align the mating marks for TDC with the timing indicator on the front cover.

19. At the same time make sure that the mating marks on the camshaft sprockets are lined up as indicated in the illustration. If not rotate the crankshaft one more turn to line up the mating marks.

20. Hold the crankshaft pulley with a suitable tool. Loosen the crankshaft pulley retaining bolt.

21. Pull the pulley out about 0.39 inch. Remove the crankshaft pulley bolt.

22. Attach a pulley puller in the M6 thread hole on the crankshaft pulley. Remove the crankshaft pulley.

➡ **Be careful not to damage the front cover and/or the crankshaft.**

23. Loosen the front cover retaining bolts, in the order indicated in the bolt loosening sequence illustration. Remove the front cover. Be careful not to damage the mating surfaces.

24. Using a seal removal tool, remove the oil seal, as required.

25. To remove the timing chain, push in on the chain tensioner plunger. Insert a stopper pin into the hole on the chain tensioner body to secure the chain tensioner plunger. Remove the chain tensioner. Remove the chain. Do not rotate the crankshaft with the chain removed.

➡ **Use a 0.02 inch (approximate) metal pin as a stopper pin.**

26. Remove the camshaft sprockets.

27. Remove the timing chain slack guide, timing chain tensioner guide and spacer.

28. Remove the balancer unit timing chain tensioner by lifting the lever up and releasing the ratchet claw for return proof. Push the tensioner sleeve in and hold it. Matching the hole on the lever with the one on the body, insert a stopper pin to secure the tensioner sleeve. Remove the balancer unit timing chain tensioner.

➡ **Use a 0.04 inch (approximate) metal pin as a stopper pin.**

29. Secure the hexagonal portion of the balancer shaft using a suitable tool. Loosen the balancer unit sprocket bolt.

30. Remove the balancer unit timing chain, balancer unit sprocket and crankshaft sprocket.

➡ **When removing the balancer unit timing chain, remove the crankshaft sprocket and balancer unit sprocket at the same time.**

31. Loosen the balancer unit mounting bolts, in the order of the tightening sequence. Remove the balancer unit. Do not disassemble the balancer unit. Bolts one and four use a E14 Torx®head socket.

➡ **To install:**

32. Check the chain for cracks and excessive wear, replace as required.

33. Measure the balancer unit bolt outer diameters ("d1" and "d2") at two positions, as shown in the illustration. If reduction appears in the "A" range, regard it as "d2". Specification is as follows: ("d1" - "d2"): 0.0059 inch. If it exceeds the specification (large difference in dimensions) replace the balancer unit bolt with a new one.

34. Measure the balancer bolt unit length. If it exceeds the specification replace the balancer unit bolt with a new one. Specification is 6.974 inch.

35. Make sure that the crankshaft key is pointing straight up. Install the O-ring to the balancer unit.

36. Install the balancer unit. Apply engine oil to the bolt threads and sealing surfaces. Torque the bolts to specification and in the proper sequence using tool KV10112100 (BT8653-A) or equivalent.
   - a. Step 1: bolts 1–4 to 35 ft. lbs.
   - b. Step 2: bolts 1–4 100 degrees clockwise
   - c. Step 3: bolts 1–4 loosen in the reverse order of the tightening sequence to zero
   - d. Step 4: bolts 1–4 to 35 ft. lbs.
   - e. Step 5: bolts 1–4 100 degrees clockwise
   - f. Step 6: bolts 5–6 to 22 ft. lbs.

➡ **Check the tightening angle using a tool or a protractor. Do not make a judgment by visual check alone.**

37. Install the crankshaft sprocket, balancer unit sprocket and balancer timing chain.

38. Make sure that the crankshaft sprocket is positioned with the mating marks on the cylinder block and crankshaft sprocket meeting at the top.

39. Install it by aligning the mating marks on each sprocket and balancer unit timing chain.

40. Secure the hexagonal portion of the balancer shaft using a suitable tool. Tighten the balancer unit sprocket bolt to specification.

➡ **Install the crankshaft sprocket, balancer unit sprocket and balancer unit timing chain at the same time.**

41. Install the balancer unit timing chain tensioner.

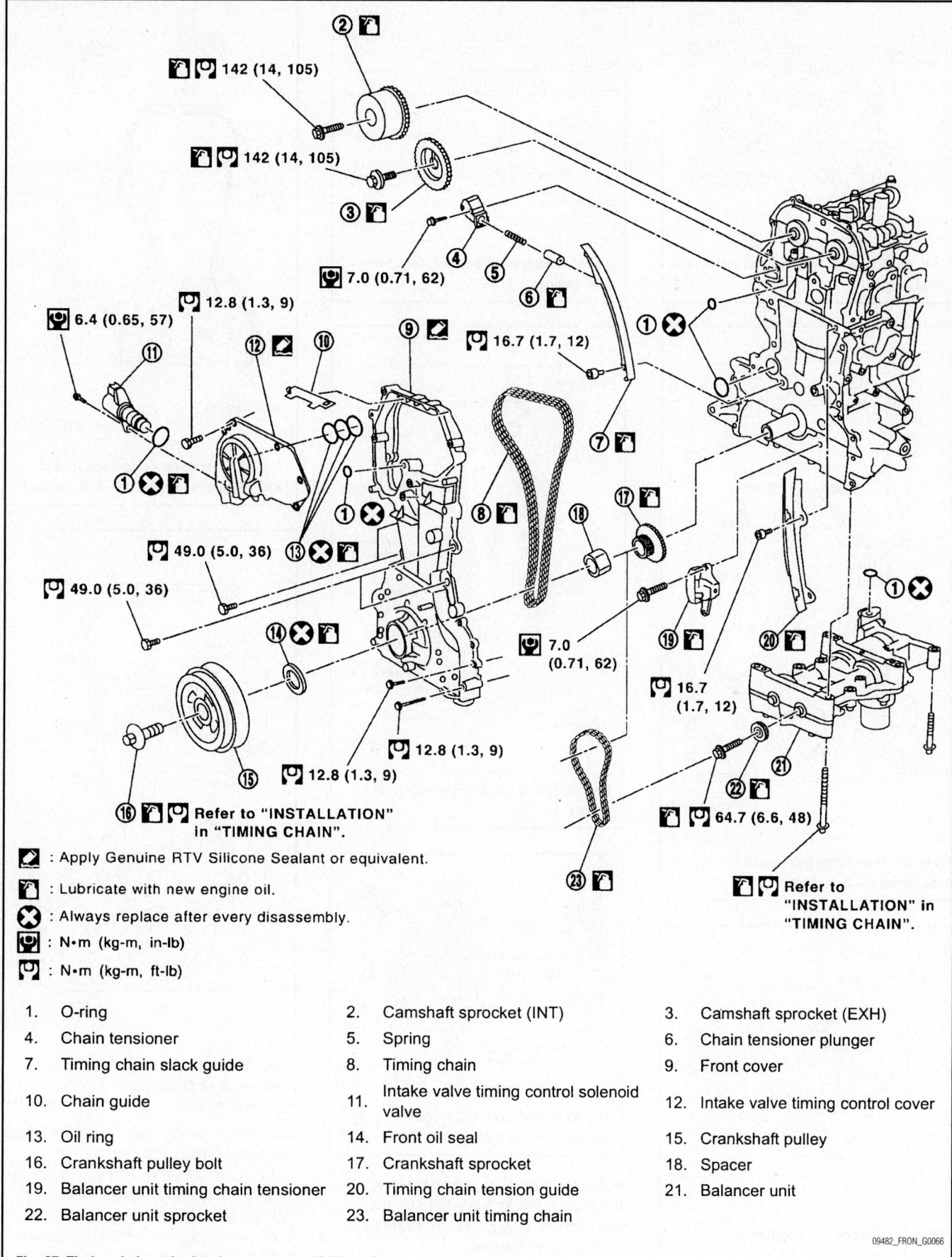

: Apply Genuine RTV Silicone Sealant or equivalent.

: Lubricate with new engine oil.

: Always replace after every disassembly.

: N•m (kg-m, in-lb)

: N•m (kg-m, ft-lb)

1. O-ring	2. Camshaft sprocket (INT)	3. Camshaft sprocket (EXH)
4. Chain tensioner	5. Spring	6. Chain tensioner plunger
7. Timing chain slack guide	8. Timing chain	9. Front cover
10. Chain guide	11. Intake valve timing control solenoid valve	12. Intake valve timing control cover
13. Oil ring	14. Front oil seal	15. Crankshaft pulley
16. Crankshaft pulley bolt	17. Crankshaft sprocket	18. Spacer
19. Balancer unit timing chain tensioner	20. Timing chain tension guide	21. Balancer unit
22. Balancer unit sprocket	23. Balancer unit timing chain	

09482_FRON_G0066

Fig. 87 Timing chain and related components—2.5L engine

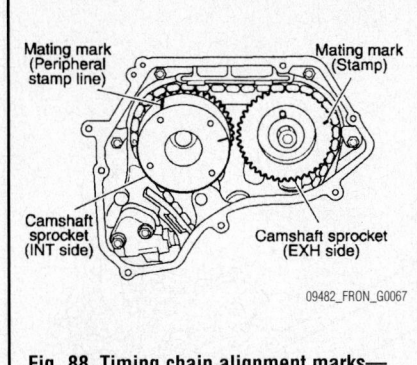

Fig. 88 Timing chain alignment marks—2.5L engine

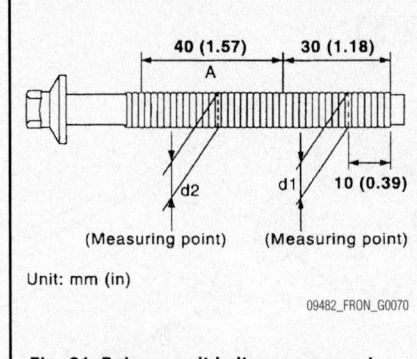

Fig. 91 Balance unit bolt measurement—2.5L engine

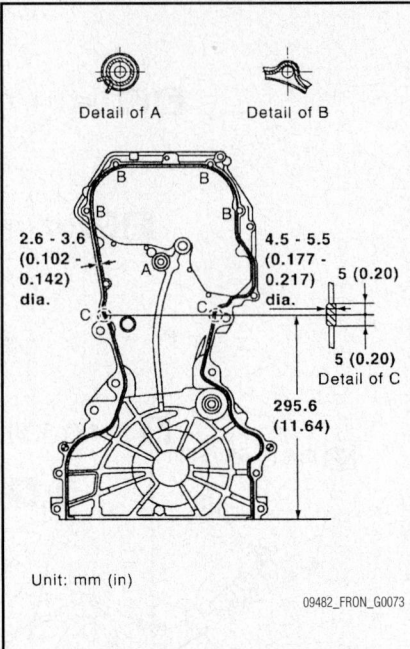

Fig. 94 Front cover sealant application with respect to positioning—2.5L engine

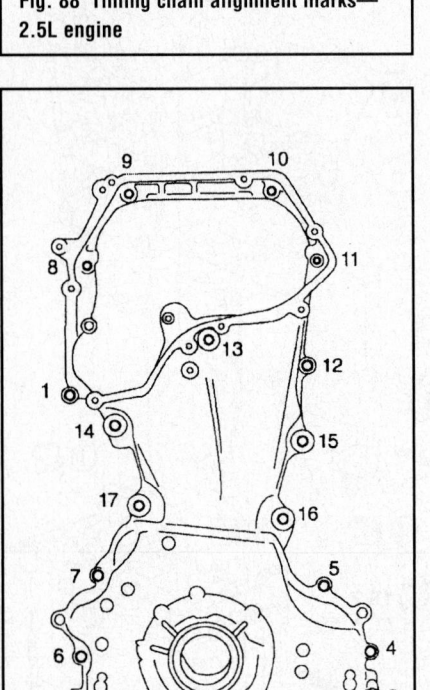

Fig. 89 Front cover bolt removal sequence—2.5L engine

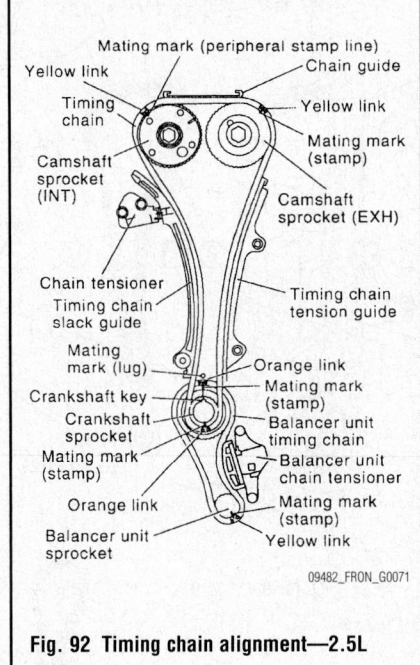

Fig. 92 Timing chain alignment—2.5L engine

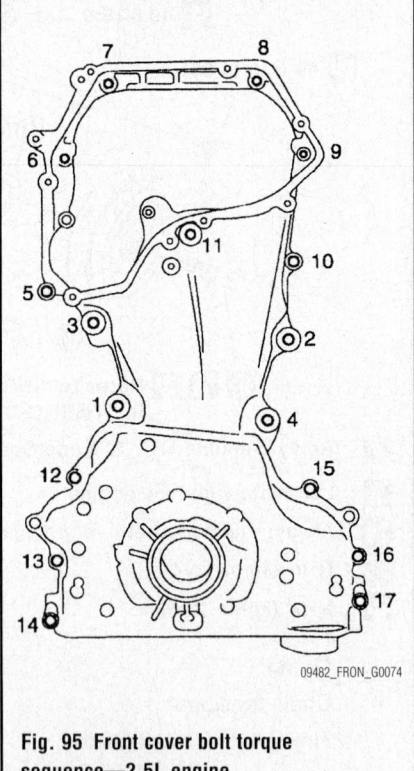

Fig. 95 Front cover bolt torque sequence—2.5L engine

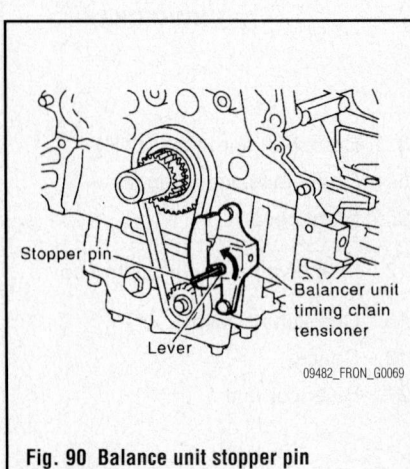

Fig. 90 Balance unit stopper pin installation—2.5L engine

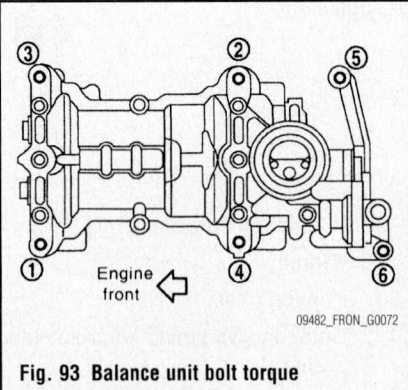

Fig. 93 Balance unit bolt torque sequence—2.5L engine

➡ **After installation, make sure that the mating marks have not slipped. Remove the stopper pin and release the tensioner sleeve.**

42. Align the mating marks on each

sprocket and timing chain. Install the timing chain and related parts.

43. Before and after installing the chain tensioner, check again to be sure that the mating marks have not slipped.

44. After installing the chain tensioner, remove the stopper pin. Make sure that the tensioner moves freely.

➡️**After the mating marks are aligned, keep them aligned by holding them with your hand. To avoid skipped teeth, do not rotate the crankshaft and camshaft until the cover is installed.**

➡️**Before installing the chain tensioner, it is possible to change the position of the mating mark on the timing chain for that on each sprocket for alignment.**

45. Install the front cover oil seal. Install O-rings to the cylinder head and the cylinder block,

46. Apply a continuous bead of liquid gasket to the front cover. Be sure to use genuine RTV sealant, or equivalent.

➡️**Sealant application instructions differ depending on position, refer to the illustration for positioning. Detail "A", cross over the start of the application and the end. Detail "B", apply liquid gasket outside of the bolt holes. For all bolt holes other than "B", apply to the inside. Detail "C", between here only, apply a bead of sealant 0.177–0.217 inch diameter.**

47. Make sure that the mating marks of the chain and each sprocket are still aligned.

48. Install the front cover. Torque the retaining bolts to specification and in the proper sequence. Bolt position 5, 10, 14 and 17: 45mm (1.77 inch). Except the above (except 1 to 4): 20mm (0.79 inch).

   a. M6 bolts: 9 ft. lbs.

   b. M10 bolts: 36 ft. lbs.

   c. After all bolts are tightened, retighten them to specification and in the proper sequence

➡️**Be sure to wipe off any excess liquid gasket leaking to the surface for installing the oil pan.**

49. Install the chain guide between the camshaft sprockets.

50. Install the oil rings to the camshaft sprocket (INT) insertion points on backside of the intake valve timing control cover. Install the O-ring to the front cover.

51. Apply a continuous bead of liquid gasket, 0.122 inch in diameter, to the front cover. Be sure to use genuine RTV sealant, or equivalent.

52. Install the cover. Tighten the bolts in the proper sequence to specification.

53. Install the intake valve timing control solenoid valve to the intake valve timing control cover, if removed.

54. Connect the ground cables, and install the harness clip.

55. Install the crankshaft pulley. Torque the retaining bolt to specification.

56. When installing the rocker cover, apply liquid gasket, be sure to use genuine RTV silicone sealant or equivalent, to the positions shown in the illustration. Refer to figure "a" to apply liquid gasket to joint part of camshaft bracket No.1 and cylinder head. Refer to figure "b" to apply liquid gasket in 90 degrees to figure "b".

57. Install the rocker cover. Torque the retaining bolts to 18 inch lbs and then to 73 inch lbs, in the proper sequence.

58. Continue the installation in the reverse order of the removal procedure.

➡️**If hydraulic pressure inside the timing chain tensioner drops after removal/installation, slack in the guide may generate a pounding noise during and just after engine start. This is normal the noise will stop after hydraulic pressure rises.**

4.0L Engine

See Figures 96 through 111.

➡️**The procedure below describes the removal and installation of the front timing case and timing chain related parts and rear timing chain case, when the upper oil pan needs to be removed or installed. When only the timing chain (primary) is being removed it is not necessary to remove the rocker covers.**

1. Before servicing the vehicle, refer to the Precautions Section.

2. Properly relieve the fuel system pressure.

3. Disconnect the negative battery cable.

4. Remove the engine cover. Drain the engine oil. Drain the engine coolant.

5. Remove the upper and lower oil pans.

6. Remove the radiator cooling fan assembly. Remove the drive belts.

7. Separate the engine wiring harnesses by removing their brackets from the front timing chain case.

8. Remove the power steering pump from the bracket with the fluid hoses attached. Position the assembly to the side. Do not disconnect the hoses. Remove the bracket.

9. Remove the alternator. Remove the water bypass hose, water hose clamp and idler pulley bracket from the front timing chain case.

10. Remove the left and right intake valve timing control covers. Loosen the

bolts in the reverse order of the tightening sequence. Use tool KV10111100 (J-37228) or equivalent to cut the liquid gasket seal.

➡️**The shaft is internally jointed with the camshaft sprocket (INT) center hole. When removing, keep it horizontal until it is completely disconnected.**

11. Remove the collared O-rings from the front timing chain case on both the left and right side.

12. Remove the intake manifold collector.

13. Separate the engine harness and remove their brackets from the rocker covers. Remove the harness bracket from the cylinder head, if necessary.

14. Remove the ignition coil. Remove the PCV hoses. Remove the oil filler cap, if necessary.

15. Loosen the rocker cover retaining bolts, in the reverse order of the tightening sequence.

16. Remove the rocker covers from the engine.

➡️**When only the timing chain (primary) is being removed it is not necessary to remove the rocker covers.**

17. Set the No.1 cylinder at TDC of its compression stroke by rotating the crankshaft pulley clockwise to align the timing mark (grooved line without color) with the timing indicator. Make sure that the intake and exhaust cam noses on No.1 cylinder (engine front side on right bank) are in alignment as shown in the illustration. If not, rotate the crankshaft in the clockwise direction 360 degrees.

➡️**When only the timing chain (primary) is removed, the rocker cover does not need to be removed. To be sure that the No.1 cylinder is set at TDC on the compression stroke, remove the front timing chain case cover first, then check the mating marks on the camshaft sprockets.**

18. Remove the starter. Position tool KV10117700 (J-44716) or equivalent.

19. Loosen the crankshaft pulley retaining bolt and locate the bolt seating surface, which is about 0.39 inch from its original position.

➡️**Do not remove the crankshaft pulley bolt. Keep the loosened pulley bolt in place to protect the removed crankshaft pulley from dropping.**

20. Pull the pulley with both hands and remove it from its mounting. Remove the bolt and pulley from the engine.

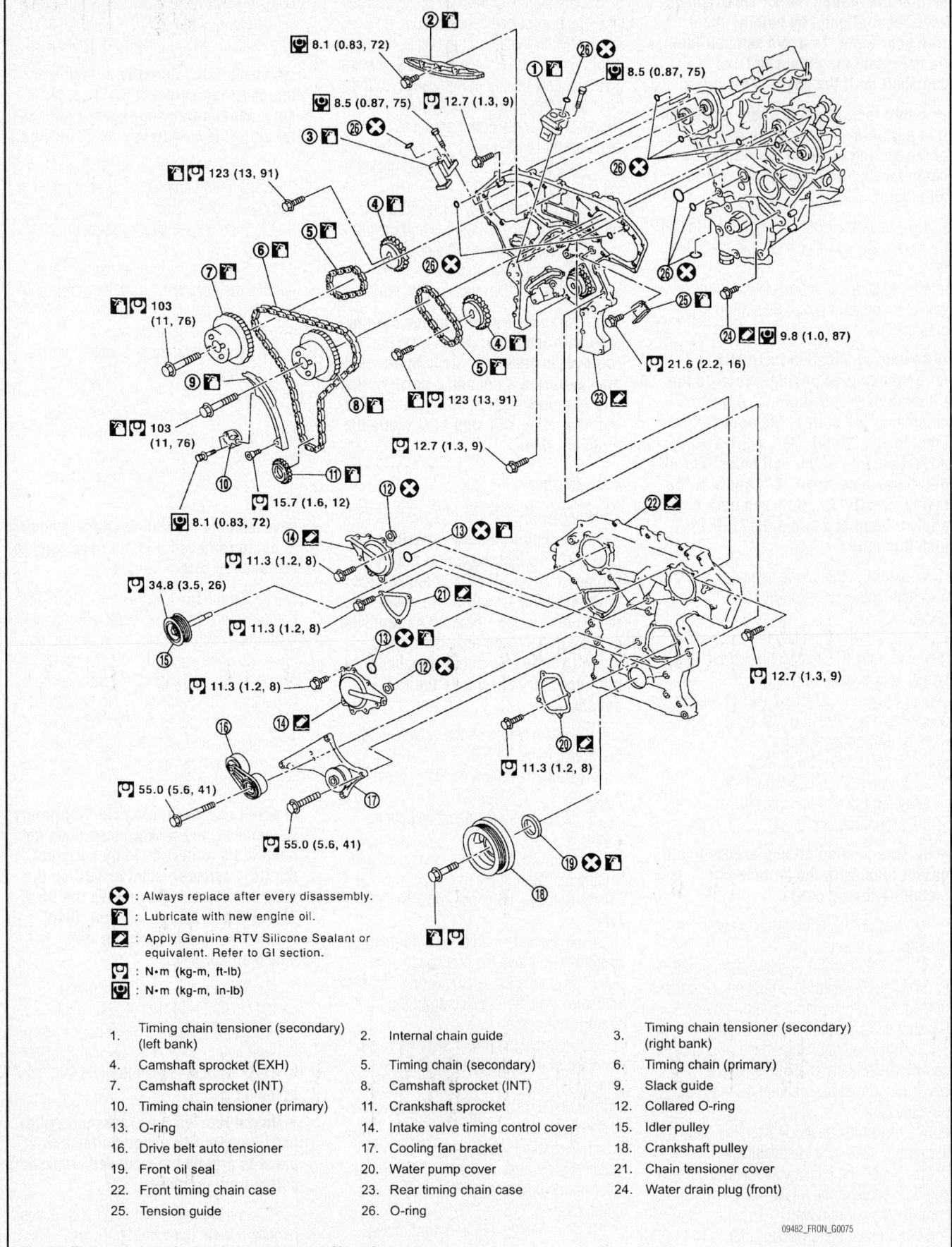

8.1 (0.83, 72)
8.5 (0.87, 75)
8.5 (0.87, 75)
12.7 (1.3, 9)
123 (13, 91)
103 (11, 76)
103 (11, 76)
8.1 (0.83, 72)
15.7 (1.6, 12)
9.8 (1.0, 87)
21.6 (2.2, 16)
123 (13, 91)
12.7 (1.3, 9)
11.3 (1.2, 8)
34.8 (3.5, 26)
11.3 (1.2, 8)
11.3 (1.2, 8)
12.7 (1.3, 9)
11.3 (1.2, 8)
55.0 (5.6, 41)
55.0 (5.6, 41)

⊗ : Always replace after every disassembly.
🛢 : Lubricate with new engine oil.
✍ : Apply Genuine RTV Silicone Sealant or equivalent. Refer to GI section.
: N•m (kg-m, ft-lb)
: N•m (kg-m, in-lb)

1.	Timing chain tensioner (secondary) (left bank)	2.	Internal chain guide	3.	Timing chain tensioner (secondary) (right bank)
4.	Camshaft sprocket (EXH)	5.	Timing chain (secondary)	6.	Timing chain (primary)
7.	Camshaft sprocket (INT)	8.	Camshaft sprocket (INT)	9.	Slack guide
10.	Timing chain tensioner (primary)	11.	Crankshaft sprocket	12.	Collared O-ring
13.	O-ring	14.	Intake valve timing control cover	15.	Idler pulley
16.	Drive belt auto tensioner	17.	Cooling fan bracket	18.	Crankshaft pulley
19.	Front oil seal	20.	Water pump cover	21.	Chain tensioner cover
22.	Front timing chain case	23.	Rear timing chain case	24.	Water drain plug (front)
25.	Tension guide	26.	O-ring		

09482_FRON_G0075

Fig. 96 Timing chain and related components—4.0L engine

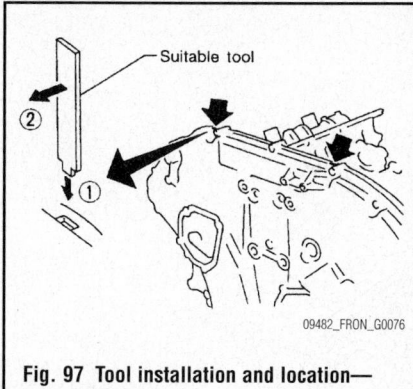

Fig. 97 Tool installation and location—4.0L engine

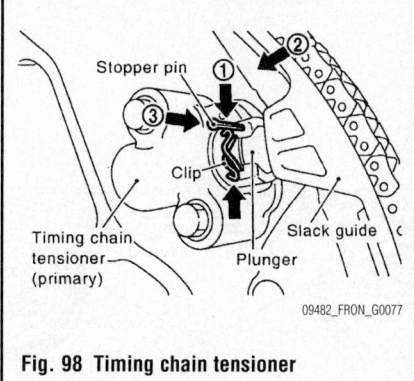

Fig. 98 Timing chain tensioner (primary)—4.0L engine

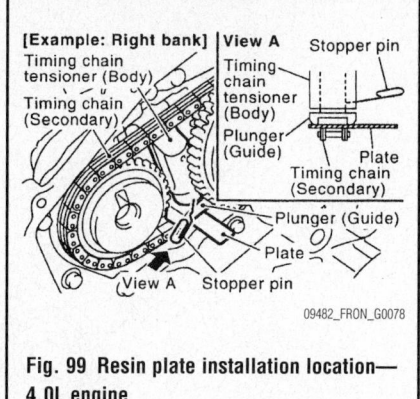

Fig. 99 Resin plate installation location—4.0L engine

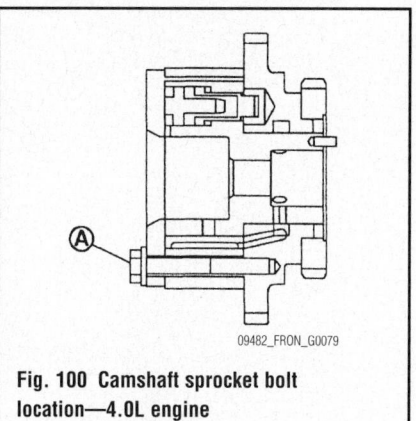

Fig. 100 Camshaft sprocket bolt location—4.0L engine

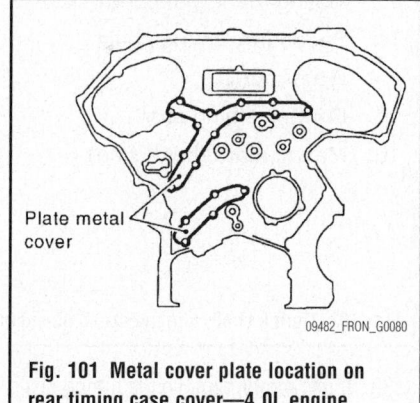

Fig. 101 Metal cover plate location on rear timing case cover—4.0L engine

21. Loosen and remove the two bolts of the upper oil pan.

22. Loosen the front timing chain cover retaining bolts in the reverse order of the tightening sequence.

23. Insert a suitable tool in the notch at the top of the front timing chain case and pry off the case by moving the tool as shown in the illustration. Use tool KV10111100 (J-37228) or equivalent to cut the liquid gasket seal.

➡**Do not use a screwdriver or something similar. After removal handle the front timing chain cover case carefully so it does not tilt, cant or warp under a load.**

24. Remove the O-rings from the rear timing chain case.

25. Remove the water pump cover and chain tensioner cover from the front timing chain case cover, as required.

26. Remove the oil seal from the front timing chain case cover, as required.

27. Remove the timing chain tensioner (primary) by loosening the clip of the timing chain tensioner (primary) and release the plunger stopper. Insert the plunger into the tensioner body by pressing the slack guide. Keep the slack guide pressed and hold the plunger in by pushing the stopper pin through the tensioner body hole and the plunger groove. Remove the bolts and remove the timing chain tensioner (primary).

28. Remove the internal chain guide, tension guide and slack guide.

➡**The tension guide can be removed after removing the timing chain (primary).**

29. Remove the timing chain (primary) and the crankshaft sprocket.

➡**After removing the timing chain (primary), do not turn the crankshaft and camshaft separately or the valves will strike the piston heads.**

30. To remove the timing chain (secondary) and camshaft sprockets, attach a suitable stopper pin to the right and left timing chain tensioner (secondary).

➡**Use a 0.02 inch (approximate) metal pin as a stopper pin.**

31. Remove the camshafts. Remove the valve lifters. Identify them for reinstallation in their original locations.

32. Remove the camshaft sprocket (INT and EXH) bolts. Secure the hexagonal portion of the camshaft using a wrench to loosen the bolts.

➡**Do not loosen the bolts with securing anything other than the camshaft hexagonal portion or with tensioning the timing chain.**

33. To remove the timing chain (secondary) together with the camshaft sprockets, turn the crankshaft slightly to secure slackness of the timing chain on the timing chain tensioner (secondary) side.

34. Insert a 0.020 inch thick metal or resin plate between the timing chain and timing chain plunger (guide). Remove the timing chain (secondary) together with the camshaft sprockets with the timing chain loose from the guide groove.

✳✳ CAUTION

Be careful of the plunger coming off when removing the timing chain (secondary). This is because the plunger of the timing chain tensioner (secondary) moves during operation, leading to coming off its fixed stopper pin.

➡**The camshaft sprocket (INT) is a one piece integrated design sprocket for the timing chain (primary) and for the timing chain (secondary). When handling the sprocket avoid shock to the sprocket. Do not disassemble or loosen bolt "A", as shown in the illustration.**

35. Remove the water pump.

36. Remove the rear timing chain case cover bolts, in the reverse order of the tightening sequence. Using the proper tool cut the liquid gasket sealant seal. Remove the cover.

➡**Do not remove the metal cover of the oil passage. After removal handle the case carefully so it does not tilt, or warp under a load.**

37. Remove the O-rings from the cylinder head and No.1 camshaft bracket. Remove the O-rings from the cylinder block.

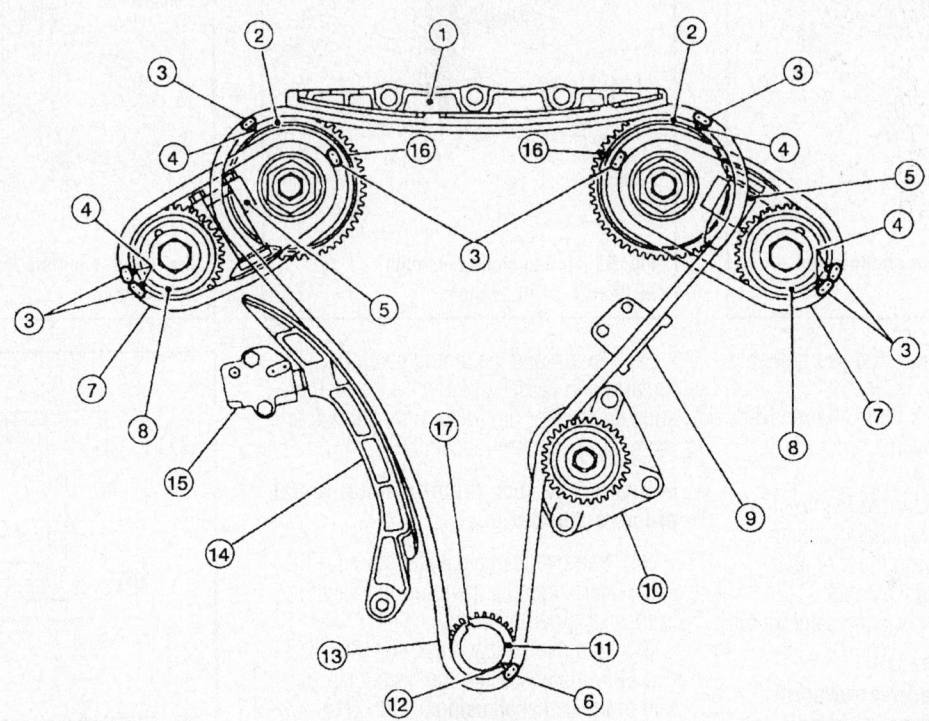

1. Internal chain guide
2. Camshaft sprocket (intake)
3. Mating mark (copper link)
4. Mating mark (punched)
5. Secondary timing chain tensioner
6. Mating mark (yellow link)
7. Secondary timing chain
8. Camshaft sprocket (exhaust)
9. Tensioner guide
10. Water pump
11. Crankshaft sprocket
12. Mating mark (notched)
13. Primary timing chain
14. Slack guide
15. Primary timing chain tensioner
16. Mating mark (back side)
17. Crankshaft key

09482_FRON_G0081

Fig. 102 Timing chain alignment—4.0L engine

38. If necessary, remove the timing chain tensioners (secondary) from the cylinder head by first removing the No.1 camshaft bracket. Remove the timing chain tensioners (secondary) with the stopper pin attached.

To install:

39. Check the chain for cracks and excessive wear, replace as required.

40. Be sure to remove all old gasket material from bolts and bolt holes.

41. If removed install the timing chain tensioners (secondary) to the cylinder head.

42. Install camshaft brackets No.1.

43. To install the rear timing chain case cover, first install new O-rings to the cylinder block, Install new O-rings to the cylinder head and camshaft bracket No.1.

44. Apply liquid gasket sealant to the rear timing chain case back side, as shown in the illustration. Be sure to use genuine RTV sealant, or equivalent.

➡**For "A" in the figure, completely wipe out excessive liquid gasket extended on a portion touching at engine coolant. Apply liquid gasket on the installation position of the water pump and cylinder head very completely.**

45. Align the rear timing case with dowel pins (right and left) on the cylinder block. Install the rear timing chain case. Make sure that the O-rings stay in place during installation to the cylinder block, cylinder head and camshaft bracket No.1.

46. Tighten the bolts to specification and in the proper sequence.

a. Bolt length: 0.79 inch. Bolt position: 1,2, 3, 6, 7, 8, 9 and 10.

b. Bolt length: 0.63 inch. Bolt position: except 1, 2, 3, 6, 7, 8, 9 and 10.

c. Torque bolts to 9 ft. lbs.

d. After all bolts are tightened, retighten them to specification and in the proper sequence

➡**Be sure to wipe off any excess liquid gasket leaking to the surface for installing the oil pan.**

47. After installing the rear timing case, check the surface height deference between the rear timing chain case and the lower cylinder block. Specification should be

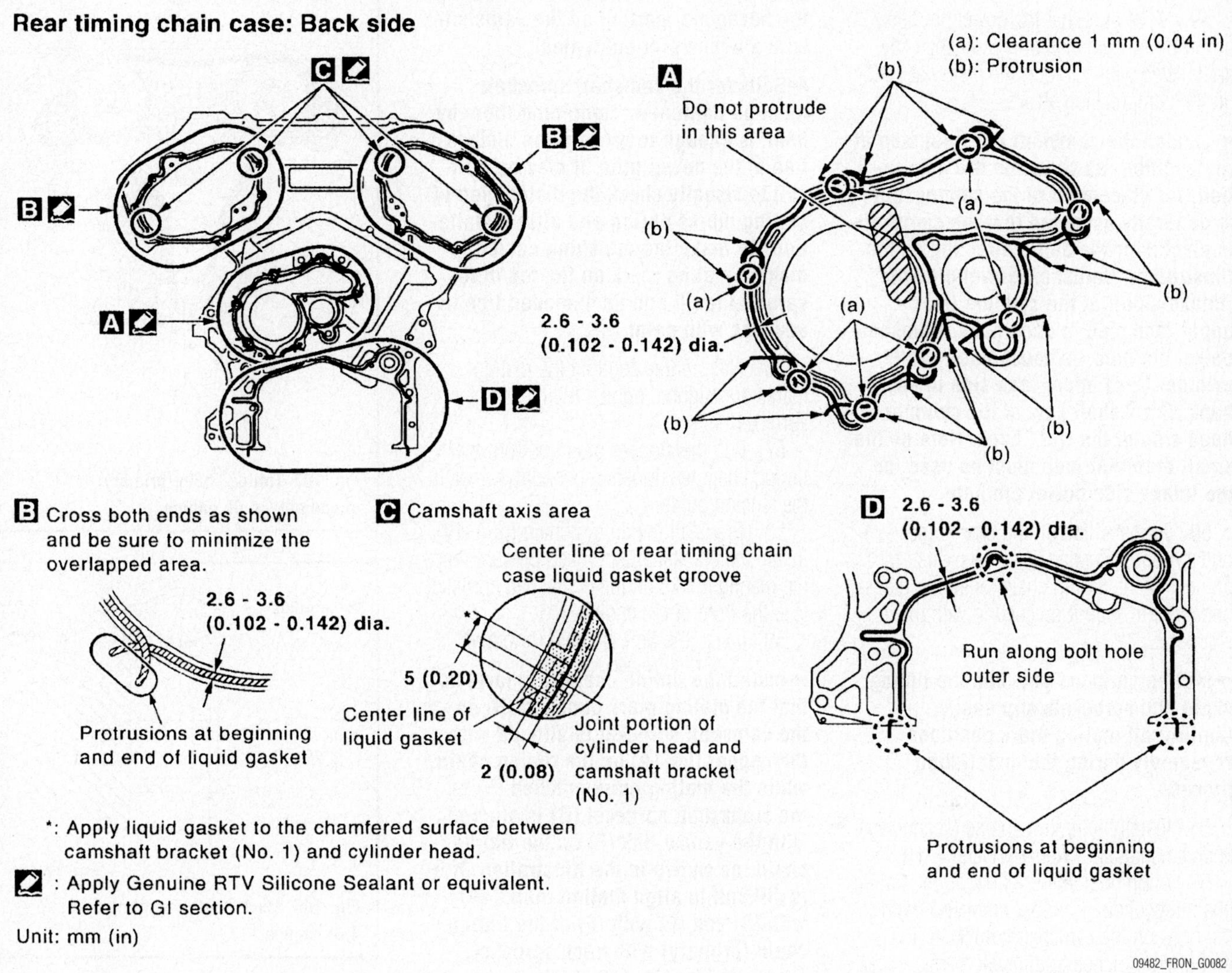

Rear timing chain case: Back side

C ✎

B ✎

B ✎

A ✎

D ✎

2.6 - 3.6
(0.102 - 0.142) dia.

(a): Clearance 1 mm (0.04 in)
(b): Protrusion

Do not protrude
in this area

(b)
(a)
(b)
(a)
(a)
(b)
(b)
(b)
(b)

B Cross both ends as shown
and be sure to minimize the
overlapped area.

2.6 - 3.6
(0.102 - 0.142) dia.

Protrusions at beginning
and end of liquid gasket

C Camshaft axis area

Center line of rear timing chain
case liquid gasket groove

5 (0.20)

Center line of
liquid gasket

2 (0.08)

Joint portion of
cylinder head and
camshaft bracket
(No. 1)

D 2.6 - 3.6
(0.102 - 0.142) dia.

Run along bolt hole
outer side

Protrusions at beginning
and end of liquid gasket

*: Apply liquid gasket to the chamfered surface between
camshaft bracket (No. 1) and cylinder head.

✎ : Apply Genuine RTV Silicone Sealant or equivalent.
Refer to GI section.

Unit: mm (in)

09482_FRON_G0082

Fig. 103 Rear timing chain cover sealant application—4.0L engine

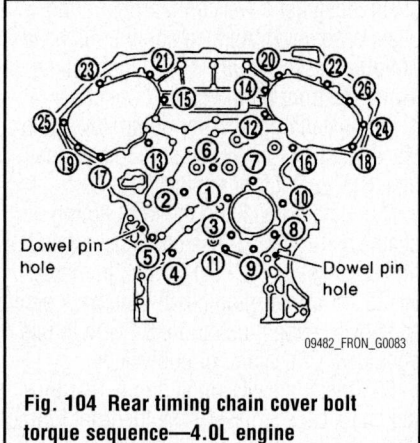

09482_FRON_G0083

**Fig. 104 Rear timing chain cover bolt
torque sequence—4.0L engine**

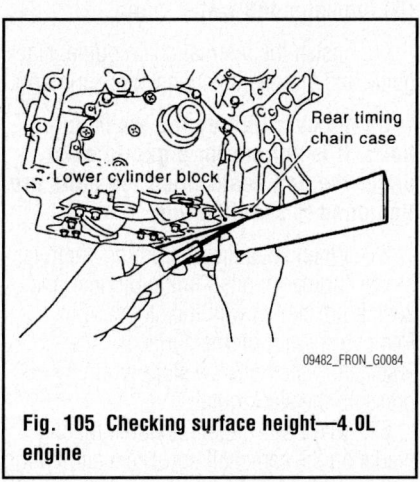

Rear timing
chain case

Lower cylinder block

09482_FRON_G0084

**Fig. 105 Checking surface height—4.0L
engine**

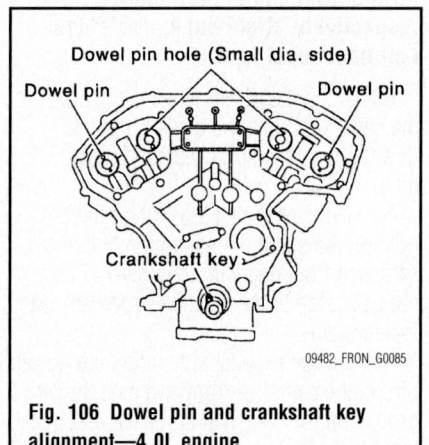

Dowel pin hole (Small dia. side)
Dowel pin
Dowel pin

Crankshaft key

09482_FRON_G0085

**Fig. 106 Dowel pin and crankshaft key
alignment—4.0L engine**

-0.0094–0.0055 inch. If not within specification, repeat the installation procedure.

48. Install the water pump, using new O-rings.

49. Make sure that the dowel pin hole, dowel pin of camshaft and crankshaft key are located with number one piston at TDC on the compression stroke.

➡Though the camshaft does not stop at the position, as shown in the illustration, for placement of the cam nose it is generally accepted that the camshaft is placed for the same direction as the illustration. Camshaft dowel pin hole (intake side): at the cylinder head upper face side in each bank. Camshaft dowel pin hole (exhaust side): at the cylinder head upper face side in each bank. Crankshaft key: at the cylinder head side of the right bank. Hole on the small diameter side must be used for the intake side dowel pin hole.

50. To install the timing chains (secondary) and camshaft sprockets, push the plunger of the timing chain tensioner (secondary) and keep it pressed in with the stopper pin.

➡Mating surfaces between the timing chain and sprockets slip easily. Confirm all mating mark positions repeatedly during the installation process.

51. Install the timing chains (secondary) and camshaft sprockets (INT and EXH).

52. Align the mating marks on the timing chain (secondary) cooper color link, with the ones on the camshaft sprockets (INT and EXH) punched and install them.

➡Mating marks for the camshaft sprocket (INT) are on the back side of the camshaft sprocket (secondary). There are two types of mating marks, circle and oval. They should be used for the right and the left banks, respectively. Right bank: circle type. Left bank: oval type.

53. Align the dowel pin and pin hole on the camshafts with the groove and the dowel pin on the sprockets, and install them.

54. On the exhaust side, align the pin hole on the small diameter side of the camshaft front end with the dowel pin on the back side of the camshaft sprocket, and install them.

55. On the exhaust side, align the dowel pin on the camshaft front end with the pin groove on the camshaft sprocket, and install them.

➡In case that the positions of each mating mark and each dowel pin will not fit on the mating marks, make a fine adjustment to the position holding the hexagonal portion on the camshaft with a wrench, or equivalent.

➡Bolts for the camshaft sprockets must be tightened. Tightening them by hand is enough to prevent the dislocation of the dowel pins. It may be difficult to visually check the dislocation of mating marks during and after installation. To make the matching easier, make a mating mark on the top of the sprocket teeth and its extended line in advance with paint.

56. After confirming that the mating marks are aligned, tighten the camshaft sprocket bolts.

57. Pull the stopper pins out from the timing chain tensioners (secondary). Install the tension guide.

58. To install the timing chain (primary), install the crankshaft sprocket. Be sure that the mating marks on the crankshaft sprocket face the front of the engine.

59. Install the timing chain (primary).

➡Install the timing chain (primary) so that the mating mark punched (B) on the camshaft sprocket is aligned with the copper link (A) on the timing chain, while the mating mark notched (E) on the crankshaft sprocket (D) is aligned with the yellow link (F) on the timing chain, as shown in the illustration. If it is difficult to align mating marks (A) with (B) and (E) with (F) of the timing chain (primary) with each sprocket, gradually turn the camshaft using a wrench on the hexagonal portion to align it with the timing marks. During alignment be careful to prevent dislocation of the mating marks alignments of the timing chains (secondary). Note (G) indicates the water pump.

60. Install the internal chain guide, slack guide and timing chain tensioner (primary).

➡Do not over tighten the slack guide bolts. It is normal for a gap to exist under the bolt seats when the bolts are tightened to specification.

61. When installing the timing chain tensioner (primary), push in the plunger and keep it pressed in with the stopper pin. Remove any dirt on the surfaces. After installation, pull out the stopper pin by pressing the slack guide.

62. Make sure, again, that the mating marks on the camshaft sprockets and timing

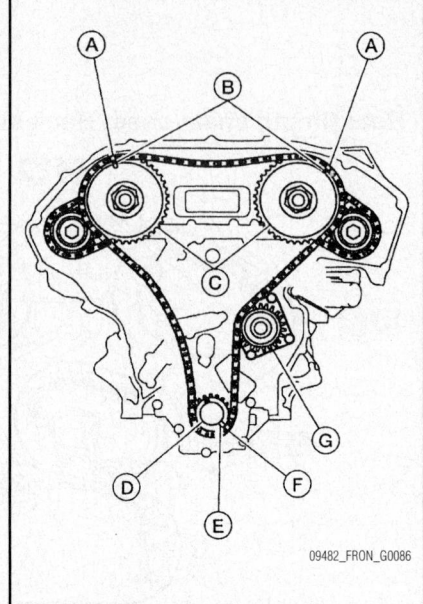

Fig. 107 Timing chain (primary) alignment—4.0L engine

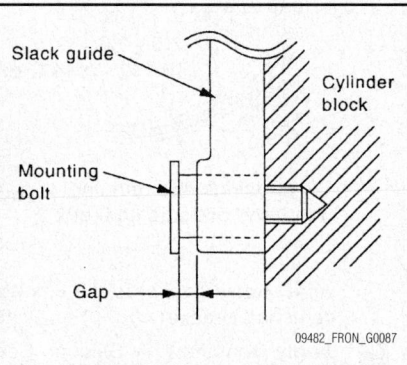

Fig. 108 Slack guide mounting bolt gap— 4.0L engine

chain have not slipped out of alignment. Install new O-rings on the rear timing chain case.

63. Install a new front seal in the front timing chain case cover.

64. Install the water pump cover and chain tensioner cover to the front timing chain case cover. Apply a continuous bead of liquid gasket (0.091–0.130 inch diameter) to the front timing chain case cover before installing the water pump cover and chain tensioner cover. Be sure to use genuine RTV sealant, or equivalent.

65. Before installing the front timing chain case cover apply a continuous bead of liquid gasket (0.102–0.142 inch in diameter) to the front timing chain case back side, as shown in the illustration. Be sure to use genuine RTV sealant, or equivalent.

66. Install new O-rings on the rear timing chain case. To assemble the front timing chain case cover, fit the lower end of the

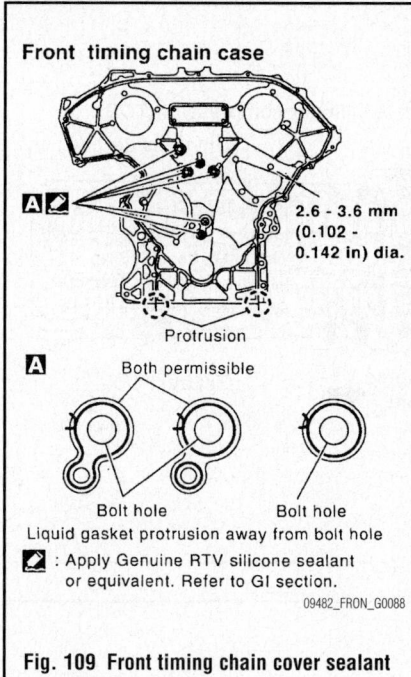

Fig. 109 Front timing chain cover sealant application—4.0L engine

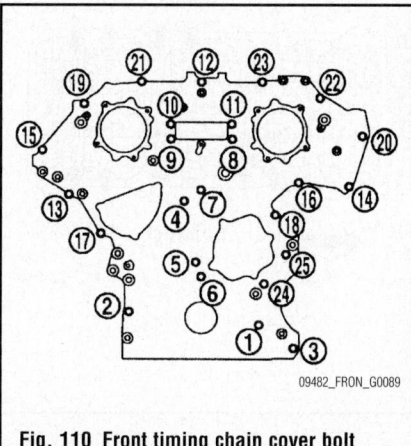

Fig. 110 Front timing chain cover bolt torque sequence—4.0L engine

front timing chain case tightly onto the top face of the oil pan (upper). From the fitting point, make entire front timing chain case contact rear timing chain case completely.

➡ **Since the front timing chain case cover is offset for difference of holt holes; tighten the bolts temporarily while holding the front timing chain case cover from the front and the top. Now insert a dowel pin while holding the front timing chain case cover from the front and the top.**

67. Once the cover is installed, torque the retaining bolts to specification and in the proper sequence. There are four different types of bolts.

a. Bolt diameter: 0.39 inch. Bolt position: 1–5. Torque to 41 ft. lbs.

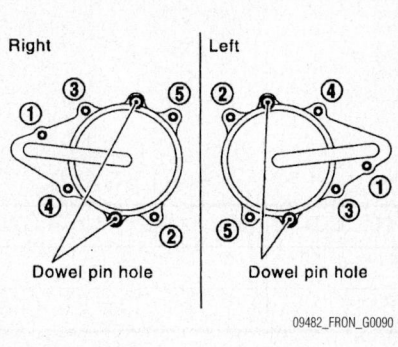

Fig. 111 Right and left intake valve timing control cover bolt torque sequence—4.0L engine

b. Bolt diameter: 0.24 inch. Bolt position: 6–25. Torque to 9 ft. lbs.

c. After all bolts are tightened, retighten them to specification and in the proper sequence

68. Install the two bolts in the oil pan (upper). Torque to 16 ft. lbs.

69. Install new seal rings in the shaft grooves of the right and left intake valve timing control covers.

70. Apply a continuous bead of liquid gasket (0.083–0.122 inch in diameter) to the covers. Be sure to use genuine RTV sealant, or equivalent.

71. Install new collared O-rings in the front timing chain case oil hole (left and right sides). Be careful not to move the seal ring from the installation groove, align the dowel pins on the front timing chain case with the holes to install the intake valve timing control covers.

72. Tighten the bolts in sequence and to specification.

73. Install the crankshaft pulley. Torque to specification

74. Install the upper and lower oil pans.

75. Install the intake manifold collector.

76. Before installing the rocker cover, apply liquid gasket, be sure to use genuine RTV silicone sealant or equivalent, to the positions shown in the illustration. Refer to figure "a" to apply liquid gasket to joint part of camshaft bracket No.1 and cylinder head. Refer to figure "b" to apply liquid gasket to the figure "a" squarely.

77. Install the rocker cover. Torque the retaining bolts to 17 inch lbs and then to 74 inch lbs, in the proper sequence.

78. Continue the installation in the reverse order of the removal procedure.

➡ **If hydraulic pressure inside the timing chain tensioner drops after removal/installation, slack in the guide**

may generate a pounding noise during and just after engine start. This is normal the noise will stop after hydraulic pressure rises.

VALVE LASH

ADJUSTMENT

2.5L Engine

See Figures 112 through 114.

1. Before servicing the vehicle, refer to the Precautions Section.

2. Disconnect the negative battery cable. Drain the cooling system.

3. Remove the intake manifold.

4. Disconnect the PCV hose from the rocker cover. Remove the ignition coil.

5. Remove the PCV valve and O-ring from the rocker cover, if necessary.

6. Remove the oil filler cap from the rocker cover, if necessary.

7. Remove the rocker cover retaining bolts. Be sure to remove the bolts by reversing the order of the tightening torque sequence.

8. Remove the rocker cover. Discard the gasket.

9. Remove the undercover. Remove the lower radiator shroud.

10. Set the No.1 cylinder at TDC of its compression stroke by rotating the crankshaft pulley clockwise to align the TDC mark to the timing indicator on the front cover. At the same time make sure that both the intake and exhaust cam noses of the NO.1 cylinder face outward, as indicated by the arrows in the illustration. If not, rotate the crankshaft in the clockwise direction 360 degrees.

11. Use a feeler gauge and measure the clearance between the valve lifter and the camshaft.

12. With the No.1 piston at TDC, refer to the illustration and measure the valve

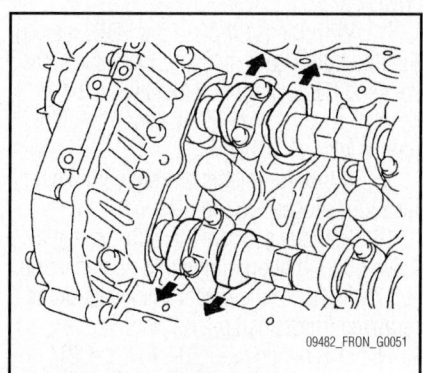

Fig. 112 No.1 cylinder at TDC (compression stroke)—2.5L engine

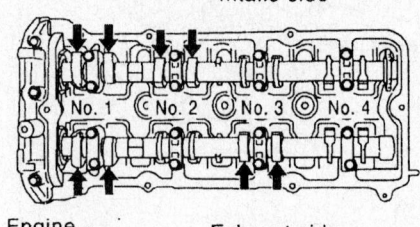

Measuring position		No. 1 CYL.	No. 2 CYL.	No. 3 CYL.	No. 4 CYL.
No. 1 cylinder at compression TDC	INT	×	×		
	EXH	×		×	

Fig. 113 Valve adjustment measurement No.1 cylinder at TDC (compression stroke)—2.5L engine

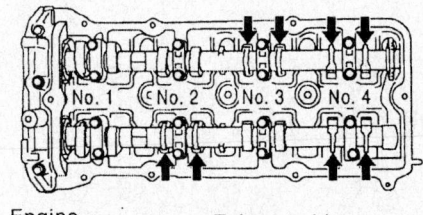

Measuring position		No. 1 CYL.	No. 2 CYL.	No. 3 CYL.	No. 4 CYL.
No. 4 cylinder at compression TDC	INT			×	×
	EXH		×		×

Fig. 114 Valve adjustment measurement No.4 cylinder at TDC (compression stroke)—2.5L engine

clearances at the locations marked with an "X". The "X" locations are indicated in the illustration with an arrow.

13. Rotate the crankshaft pulley clockwise 360 degrees and align the TDC mark to the timing indicator on the front cover.

14. With the No.4 piston at TDC, refer to the illustration and measure the valve clearances at the locations marked with an "X". The "X" locations are indicated in the illustration with an arrow.

15. If measurements are not within specification, proceed to the next step.

16. Remove the camshaft. Remove the valve lifters that are not within specification.

17. Measure the center thickness of the removed lifters, using a micrometer.

18. Use the equation (t=t1+(C1-C2) to calculate valve lifter thickness for replacement.

➡t= valve lifter thickness to be replaced. t1= removed valve lifter thickness. C1= measured valve clearance. C2= standard valve clearance.

19. Thickness of the new valve lifter can be identified by the stamp mark on the reverse side (inside the cylinder). The stamp mark "696" indicates 6.96 mm (0.2740 inch) thickness.

➡Available thickness of a valve lifter ranges from 6.96–7.46 mm (0.2740–0.2937 inch) in steps of 0.02 mm (0.0008 inch). There are 26 different sizes.

20. Install the selected valve lifters.
21. Install the camshaft.
22. Manually rotate the crankshaft pulley in the clockwise direction a few rotations.
23. Check the valve clearance and be sure it is within specification.

24. When installing the rocker cover, apply liquid gasket, be sure to use genuine RTV silicone sealant or equivalent, to the positions shown in the illustration. Refer to figure "a" to apply liquid gasket to joint part of camshaft bracket No.1 and cylinder head. Refer to figure "b" to apply liquid gasket in 90 degrees to figure "b".

25. Install the rocker cover. Torque the retaining bolts to 18 inch lbs and then to 73 inch lbs, in the proper sequence.

26. Continue the installation in the reverse of the removal procedure.

4.0L Engine

See Figures 115 through 118.

1. Before servicing the vehicle, refer to the Precautions Section.

2. Disconnect the negative battery cable. Remove the engine under cover.

3. Remove the intake manifold collector.

4. Separate the engine harness and remove their brackets from the rocker covers. Remove the harness bracket from the cylinder head, if necessary.

5. Remove the ignition coil. Remove the PCV hoses. Remove the oil filler cap, if necessary.

6. Loosen the rocker cover retaining bolts, in the reverse order of the tightening sequence.

7. Remove the rocker covers from the engine.

8. Set the No.1 cylinder at TDC of its compression stroke by rotating the crankshaft pulley clockwise to align the timing mark (grooved line without color) with the timing indicator. Make sure that the intake

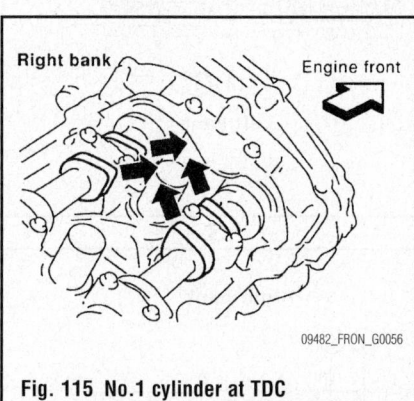

Fig. 115 No.1 cylinder at TDC (compression stroke)—4.0L engine

and exhaust cam noses on No.1 cylinder (engine front side on right bank) are in alignment as shown in the illustration. If not, rotate the crankshaft in the clockwise direction 360 degrees.

9. Use a feeler gauge and measure the clearance between the valve lifter and the camshaft.

10. With the No.1 piston at TDC, refer to the illustration and measure the valve clearances at the locations marked with an "X". The "X" locations are indicated in the illustration with an arrow.

11. Rotate the crankshaft pulley clockwise 240 degrees (when viewed from the engine front) to align No.3 cylinder at TDC on the compression stroke.

➡**The crankshaft pulley bolt flange has a stamped line every 60 degrees, which can be used as a guide to rotation angle.**

12. With the No.3 piston at TDC, refer to the illustration and measure the valve clearances at the locations marked with an "X". The "X" locations are indicated in the illustration with an arrow.

13. Rotate the crankshaft pulley clockwise 240 degrees (when viewed from the engine front) to align No.5 cylinder at TDC on the compression stroke.

➡**The crankshaft pulley bolt flange has a stamped line every 60 degrees,**

which can be used as a guide to rotation angle.

14. With the No.5 piston at TDC, refer to the illustration and measure the valve clearances at the locations marked with an "X". The "X" locations are indicated in the illustration with an arrow.

15. If measurements are not within specification, proceed to the next step.

16. Remove the camshaft. Remove the valve lifters that are not within specification.

17. Measure the center thickness of the removed lifters, using a micrometer.

18. Use the equation (t=t1+(C1-C2) to calculate valve lifter thickness for replacement.

➡**t= valve lifter thickness to be replaced. t1= removed valve lifter thickness. C1= measured valve clearance. C2= standard valve clearance.**

19. Intake valve lifter thickness of the new valve lifter can be identified by the stamp mark on the reverse side (inside the cylinder). The stamp mark "788U" indicates 7.88 mm (0.3102 inch) thickness.

➡**Available thickness of a valve lifter ranges from 7.88–8.40 mm (0.3102–0.3307 inch) in steps of 0.02 mm (0.0008 inch). There are 27 different sizes.**

20. Exhaust valve lifter thickness of the new valve lifter can be identified by the

Measuring position (right bank)		No. 1 CYL.	No. 3 CYL.	No. 5 CYL.
No. 1 cylinder at compression TDC	EXH		×	
	INT	×		
Measuring position (left bank)		No. 2 CYL.	No. 4 CYL.	No. 6 CYL.
No. 1 cylinder at compression TDC	INT			×
	EXH	×		

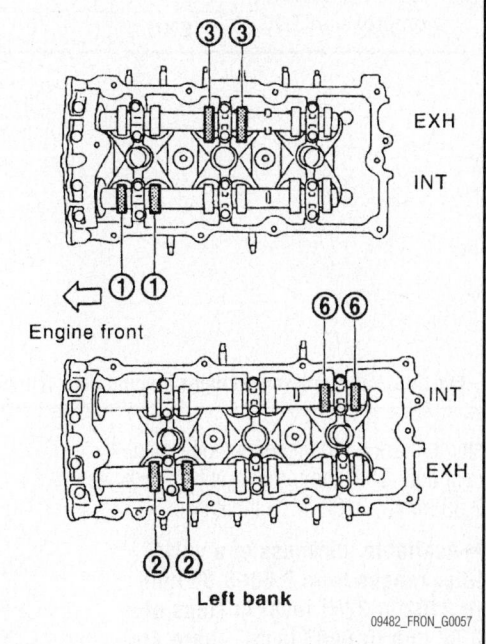

Fig. 116 Valve adjustment measurement No.1 cylinder at TDC (compression stroke)—4.0L engine

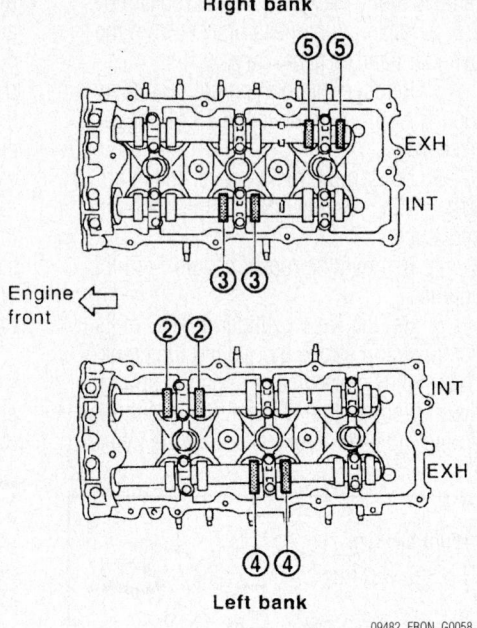

Measuring position (right bank)		No. 1 CYL.	No. 3 CYL.	No. 5 CYL.
No. 3 cylinder at compression TDC	EXH			×
	INT		×	
Measuring position (left bank)		No. 2 CYL.	No. 4 CYL.	No. 6 CYL.
No. 3 cylinder at compression TDC	INT	×		
	EXH		×	

09482_FRON_G0058

Fig. 117 Valve adjustment measurement No.3 cylinder at TDC (compression stroke)—4.0L engine

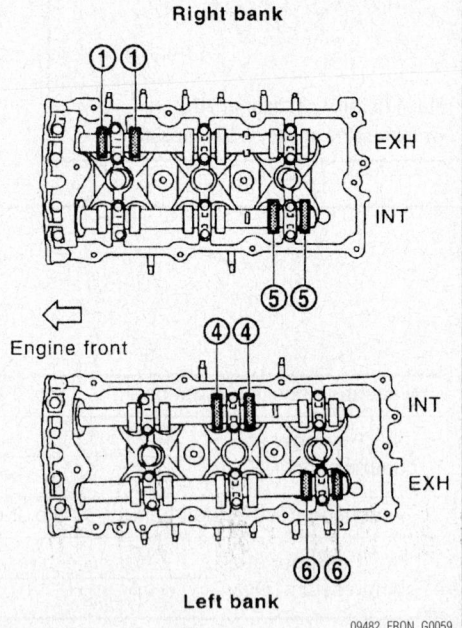

Measuring position (right bank)		No. 1 CYL.	No. 3 CYL.	No. 5 CYL.
No. 5 cylinder at compression TDC	EXH	×		
	INT			×
Measuring position (left bank)		No. 2 CYL.	No. 4 CYL.	No. 6 CYL.
No. 5 cylinder at compression TDC	INT		×	
	EXH			×

09482_FRON_G0059

Fig. 118 Valve adjustment measurement No.5 cylinder at TDC (compression stroke)—4.0L engine

stamp mark on the reverse side (inside the cylinder). The stamp mark "N788" indicates 7.88 mm (0.3102 inch) thickness.

➡**Available thickness of a valve lifter ranges from 7.88–8.36 mm (0.3102–0.3291 inch) in steps of 0.02 mm (0.0008 inch). There are 25 different sizes.**

21. Install the selected valve lifters.

22. Install the camshaft.

23. Manually rotate the crankshaft pulley in the clockwise direction a few rotations.

24. Check the valve clearance and be sure it is within specification.

25. When installing the rocker cover, apply liquid gasket, be sure to use genuine RTV silicone sealant or equivalent, to the positions shown in the illustration. Refer to

figure "a" to apply liquid gasket to joint part of camshaft bracket No.1 and cylinder head. Refer to figure "b" to apply liquid gasket to the figure "a" squarely.

26. Install the rocker cover. Torque the retaining bolts to 17 inch lbs and then to 74 inch lbs, in the proper sequence.

27. Continue the installation in the reverse of the removal procedure.

ENGINE PERFORMANCE & EMISSION CONTROL

COMPONENT LOCATIONS

See Figures 119 and 120.

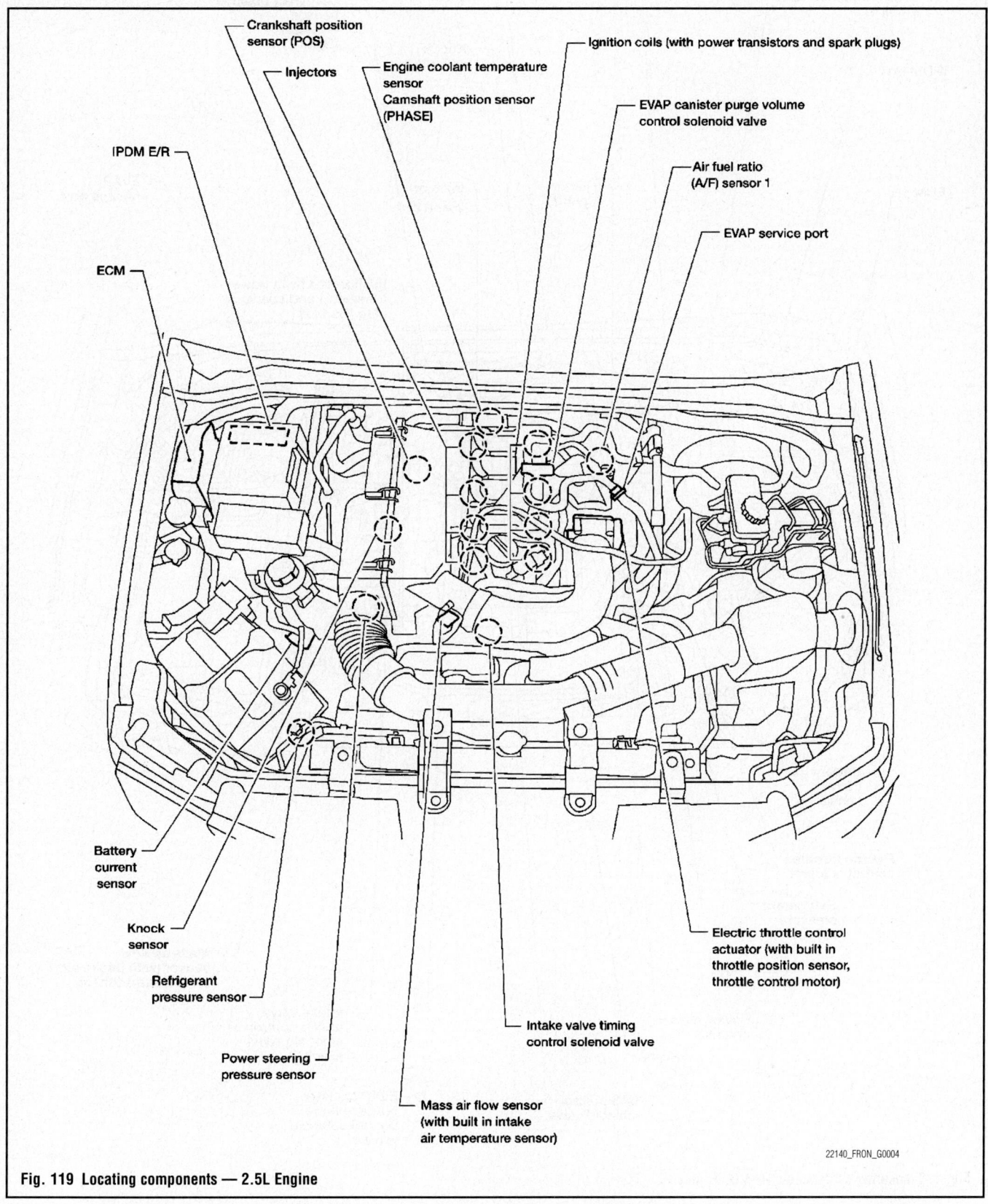

Crankshaft position sensor (POS)

Injectors

Engine coolant temperature sensor
Camshaft position sensor (PHASE)

Ignition coils (with power transistors and spark plugs)

EVAP canister purge volume control solenoid valve

Air fuel ratio (A/F) sensor 1

EVAP service port

IPDM E/R

ECM

Battery current sensor

Knock sensor

Refrigerant pressure sensor

Power steering pressure sensor

Mass air flow sensor (with built in intake air temperature sensor)

Intake valve timing control solenoid valve

Electric throttle control actuator (with built in throttle position sensor, throttle control motor)

22140_FRON_G0004

Fig. 119 Locating components — 2.5L Engine

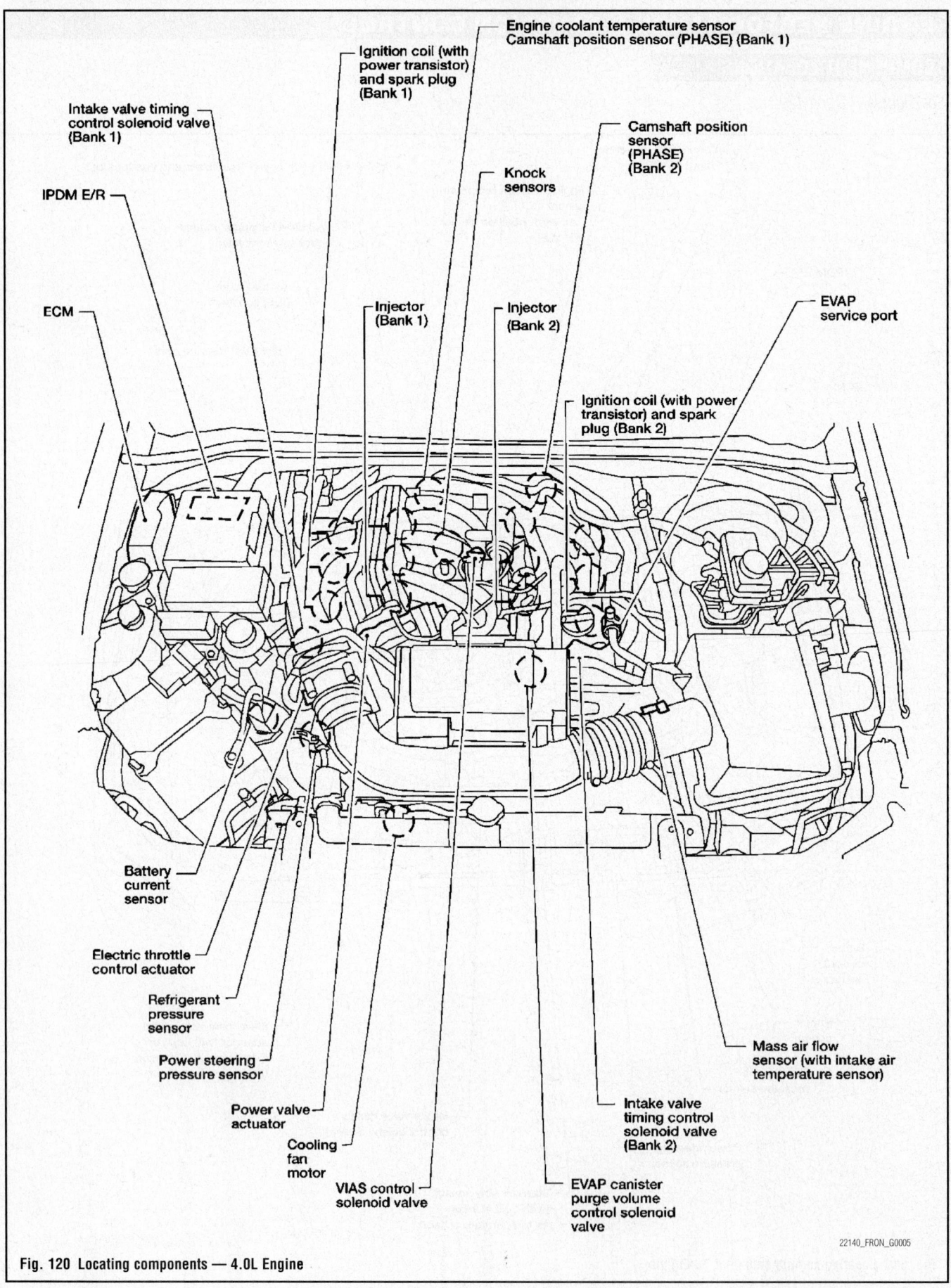

Intake valve timing control solenoid valve (Bank 1)

IPDM E/R

ECM

Engine coolant temperature sensor
Camshaft position sensor (PHASE) (Bank 1)

Ignition coil (with power transistor) and spark plug (Bank 1)

Knock sensors

Camshaft position sensor (PHASE) (Bank 2)

Injector (Bank 1)

Injector (Bank 2)

EVAP service port

Ignition coil (with power transistor) and spark plug (Bank 2)

Battery current sensor

Electric throttle control actuator

Refrigerant pressure sensor

Power steering pressure sensor

Power valve actuator

Cooling fan motor

VIAS control solenoid valve

Intake valve timing control solenoid valve (Bank 2)

EVAP canister purge volume control solenoid valve

Mass air flow sensor (with intake air temperature sensor)

22140_FRON_G0005

Fig. 120 Locating components — 4.0L Engine

CAMSHAFT POSITION (CMP) SENSOR

LOCATION

See Figures 121 and 122.

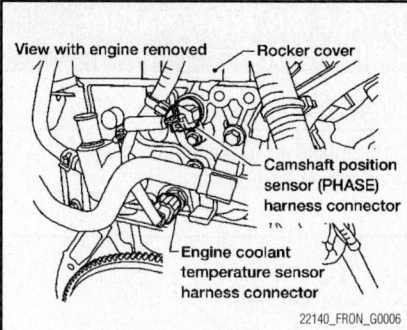

Fig. 121 Locating camshaft position sensor and engine coolant sensor — 2.5L Engine

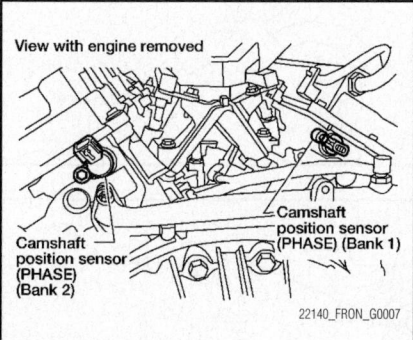

Fig. 122 Locating camshaft position sensor — 4.0L Engine

REMOVAL & INSTALLATION

1. Loosen the fixing bolt of the sensor.
2. Disconnect camshaft position sensor (PHASE) harness connector.
3. Remove the sensor.

To install:

4. Installation is the reverse of the removal procedure.

CRANKSHAFT POSITION (CKP) SENSOR

LOCATION

See Figures 123 and 124.

REMOVAL & INSTALLATION

1. Loosen the fixing bolt of the sensor.
2. Disconnect crankshaft position sensor (POS) harness connector.
3. Remove the sensor.

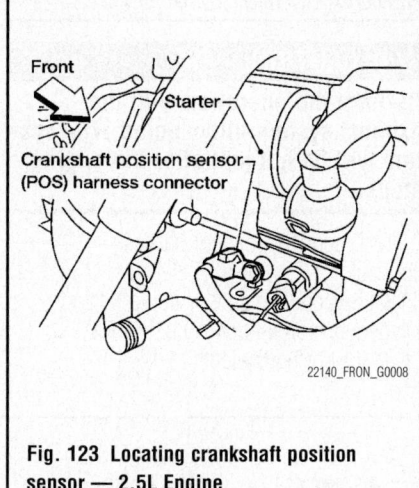

Fig. 123 Locating crankshaft position sensor — 2.5L Engine

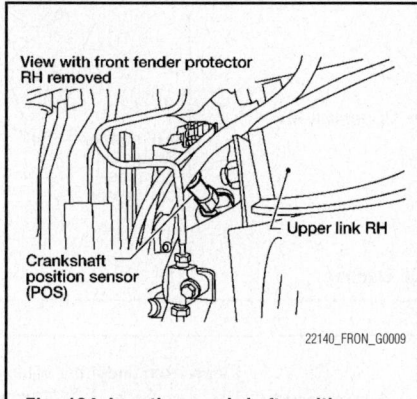

Fig. 124 Locating crankshaft position sensor — 4.0L Engine

To install:

4. Installation is the reverse of the removal procedure.

ELECTRONIC CONTROL MODULE (ECM)

LOCATION

See Figure 125.

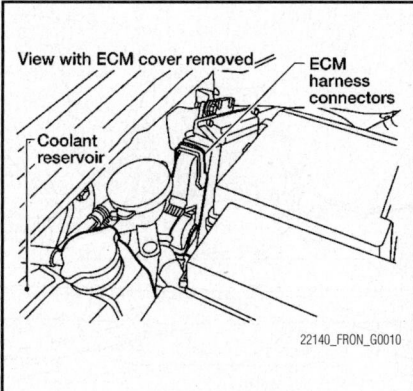

Fig. 125 Locating electronic control module

ENGINE COOLANT TEMPERATURE (ECT) SENSOR

LOCATION

See Figures 121 and 126.

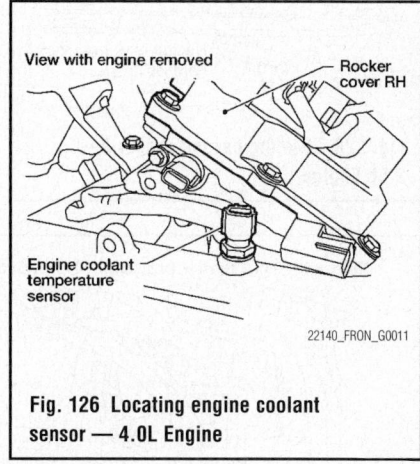

Fig. 126 Locating engine coolant sensor — 4.0L Engine

REMOVAL & INSTALLATION

✳✳ CAUTION

Do not remove sensor when engine or coolant is hot.

1. Loosen the fixing bolt of the sensor.
2. Disconnect engine coolant temperature sensor (ECT) harness connector.
3. Remove the sensor.

To install:

4. Installation is the reverse of the removal procedure.

HEATED OXYGEN (HO2S) SENSOR

LOCATION

See Figures 127 through 130.

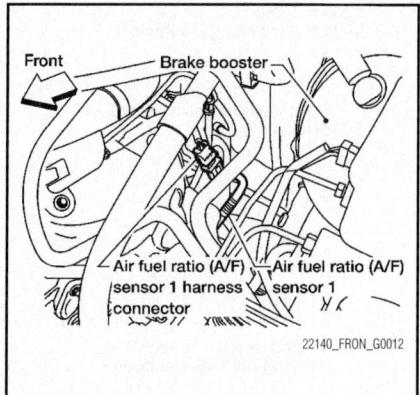

Fig. 127 Locating air fuel sensor — 2.5L Engine

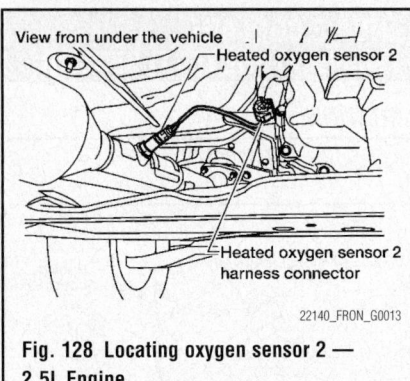

Fig. 128 Locating oxygen sensor 2 — 2.5L Engine

REMOVAL & INSTALLATION

❊❊ CAUTION

Perform the operation with the exhaust system fully cooled. The system will be hot just after the engine stops.

1. Disconnect sensor harness connector.

2. Remove the sensor using heated oxygen sensor wrench KV10114400 (J-38365) or equivalent.

To install:

3. Installation is the reverse of the removal procedure.

➡Clean exhaust system threads before installing sensor.

INTAKE AIR TEMPERATURE (IAT) SENSOR

LOCATION

The intake air temperature sensor is built into mass air flow sensor.

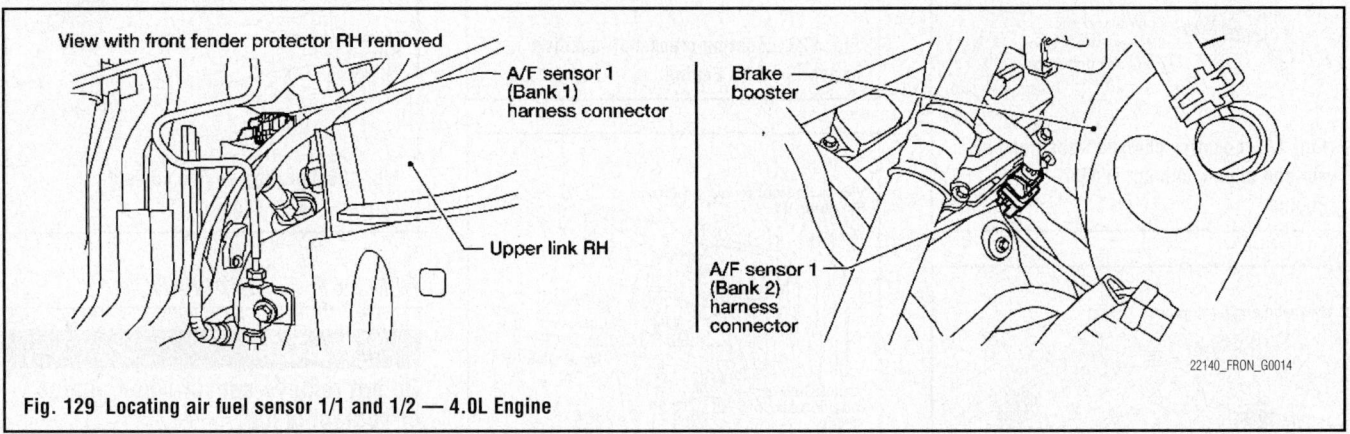

Fig. 129 Locating air fuel sensor 1/1 and 1/2 — 4.0L Engine

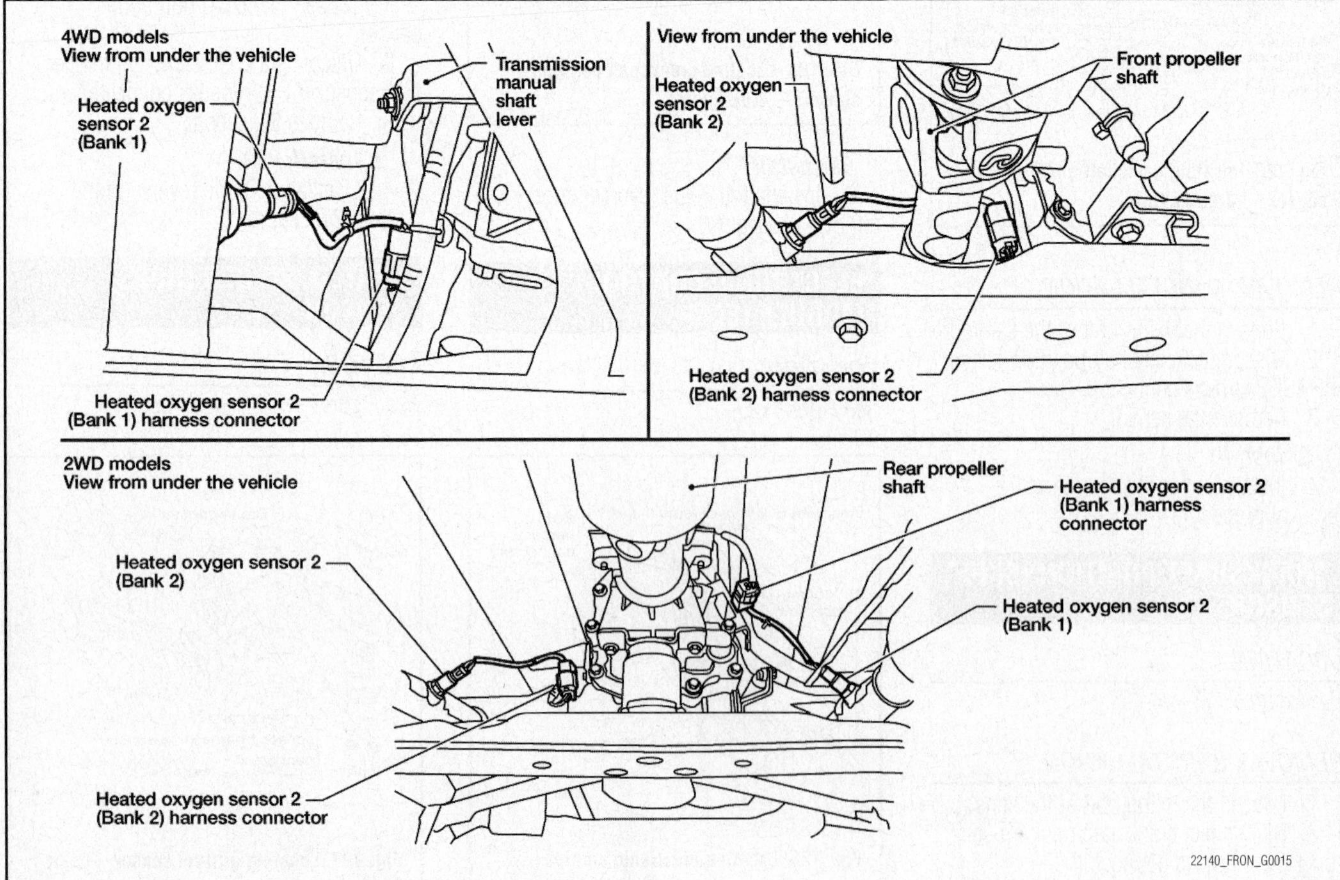

Fig. 130 Locating oxygen sensor 21/1 and 2/2 — 4.0L Engine

REMOVAL & INSTALLATION

See MAF sensor.

KNOCK SENSOR (KS)

LOCATION

See Figures 131 and 132.

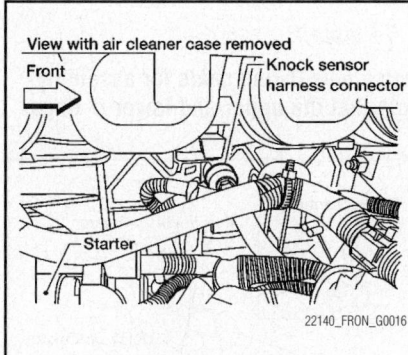

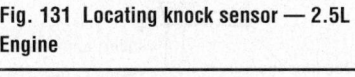

Fig. 131 Locating knock sensor — 2.5L Engine

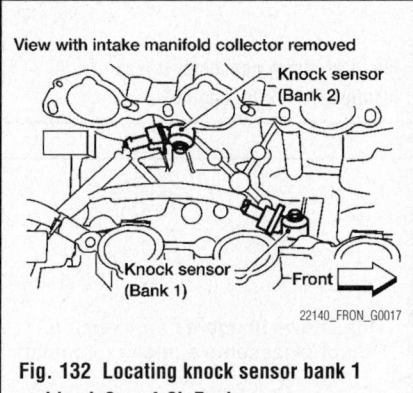

Fig. 132 Locating knock sensor bank 1 and bank 2 — 4.0L Engine

REMOVAL & INSTALLATION

1. On 4.0L engines, remove intake collector.
2. Remove sensor harness. Remove sensor.

To install:

3. Installation is the reverse of the removal procedure.

➡**Use care when installing sensor. Do not use any knock sensors that have been dropped or physically damaged. Use only new ones. Torque sensor properly.**

MASS AIR FLOW (MAF) SENSOR

LOCATION

See Figures 133 and 134.

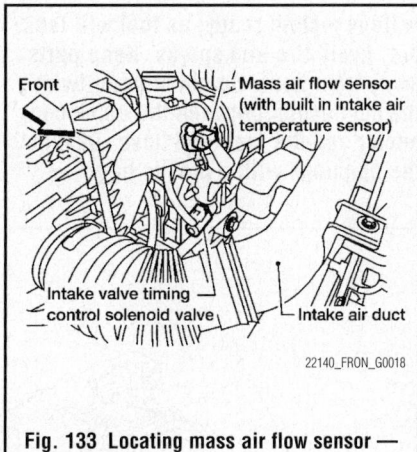

Fig. 133 Locating mass air flow sensor — 2.5L Engine

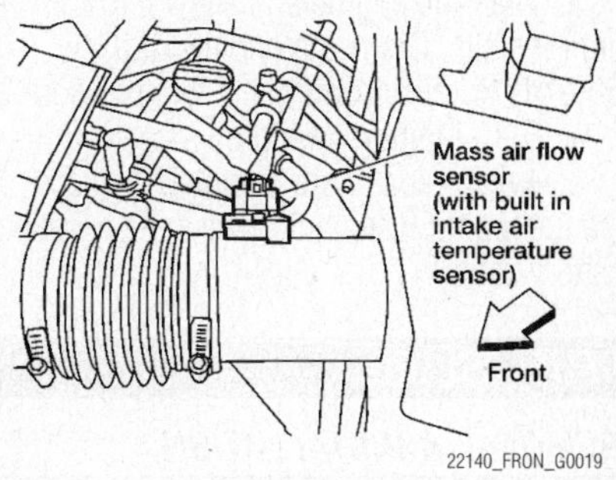

Fig. 134 Locating mass air flow sensor — 4.0L Engine

REMOVAL & INSTALLATION

See Air Filter.

THROTTLE POSITION SENSOR (TPS)

LOCATION

Electric throttle control actuator consists of throttle control motor, throttle position sensor, etc. The throttle position sensor responds to the throttle valve movement. The throttle position sensor has two sensors.

REMOVAL & INSTALLATION

The electric throttle control actuator must be replaced as a unit. See Intake Manifold in the Engine Mechanical Section.

FUEL   GASOLINE FUEL INJECTION SYSTEM

FUEL SYSTEM SERVICE PRECAUTIONS

Safety is the most important factor when performing not only fuel system maintenance but any type of maintenance. Failure to conduct maintenance and repairs in a safe manner may result in serious personal injury or death. Maintenance and testing of the vehicle's fuel system components can be accomplished safely and effectively by adhering to the following rules and guidelines.

• To avoid the possibility of fire and personal injury, always disconnect the negative battery cable unless the repair or test procedure requires that battery voltage be applied.

• Always relieve the fuel system pressure prior to disconnecting any fuel system component (injector, fuel rail, pressure regulator, etc.), fitting or fuel line connection. Exercise extreme caution whenever relieving fuel system pressure to avoid exposing skin, face and eyes to fuel spray. Please be advised that fuel under pressure may penetrate the skin or any part of the body that it contacts.

• Always place a shop towel or cloth around the fitting or connection prior to loosening to absorb any excess fuel due to spillage. Ensure that all fuel spillage (should it occur) is quickly removed from engine surfaces. Ensure that all fuel soaked cloths or towels are deposited into a suitable waste container.

• Always keep a dry chemical (Class B) fire extinguisher near the work area.

• Do not allow fuel spray or fuel vapors to come into contact with a spark or open flame.

• Always use a back-up wrench when loosening and tightening fuel line connection fittings. This will prevent unnecessary stress and torsion to fuel line piping.

• Always replace worn fuel fitting O-rings with new Do not substitute fuel hose or equivalent where fuel pipe is installed.

Before servicing the vehicle, make sure to also refer to the precautions in the beginning of this section as well.

RELIEVING FUEL SYSTEM PRESSURE

1. Before servicing the vehicle, refer to the Precautions Section.

2. Remove the fuel pump fuse from the panel.

3. Start the engine and allow it to run until it stalls. Crank the engine for a few seconds to relieve additional fuel pressure.

4. Turn ignition switch off.

5. When repairs are complete, replace the fuel pump fuse and connect the negative battery cable.

FUEL FILTER

REMOVAL & INSTALLATION

Fuel Filter is serviced with fuel pump and sending unit assembly. See Fuel Pump.

FUEL INJECTORS

REMOVAL & INSTALLATION

2.5L Engine

See Figures 135 and 136.

1. Before servicing the vehicle, refer to the Precautions Section.

2. Properly relieve the fuel system pressure.

3. Disconnect the negative battery cable. Remove the fuel filler cap.

4. Remove the quick connector cap (engine side). With the sleeve side of the quick connector release facing the quick connector, install the quick connector release on to the tube. Insert the quick connector release into the quick connector until the sleeve contacts and goes no further. Hold the quick connector release in that position.

➡**Disconnect the quick connector using tool J-45488, or equivalent, not by picking out the retainer tabs. Inserting the quick connector hard will not disconnect the quick connector. Hold the quick connector release where it contacts and goes no further.**

5. Draw and pull out the quick connector straight from the fuel tube. Grasp the quick connector holding "A" in the illustration. Do not pull with lateral force applied and the O-ring inside the quick connector could be damaged.

➡**Have a cloth ready, as fuel will leak out. Avoid fire and sparks. Keep parts away from heat. Do not bend or twist the connection between the quick connector and the fuel feed hose. Cover the openings with a plastic bag.**

6. Remove the intake manifold.

7. Disconnect the sub harness for the fuel injector.

8. Loosen the retaining bolts. Remove the fuel tube and fuel injector assembly.

9. To remove the fuel injectors from the fuel tube, open and remove the clip. Remove the injector by pulling it straight out.

To install:

➡**Use new O-ring seals for assembly. Note that the upper and lower O-rings**

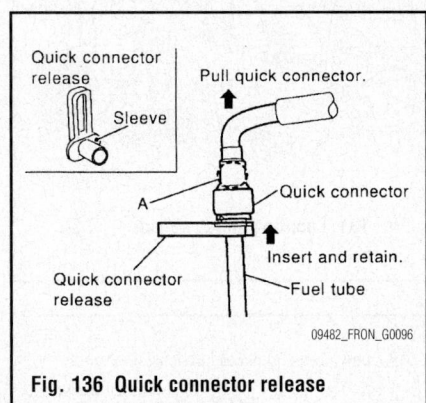

Fig. 136 Quick connector release location "A"—2.5L engine

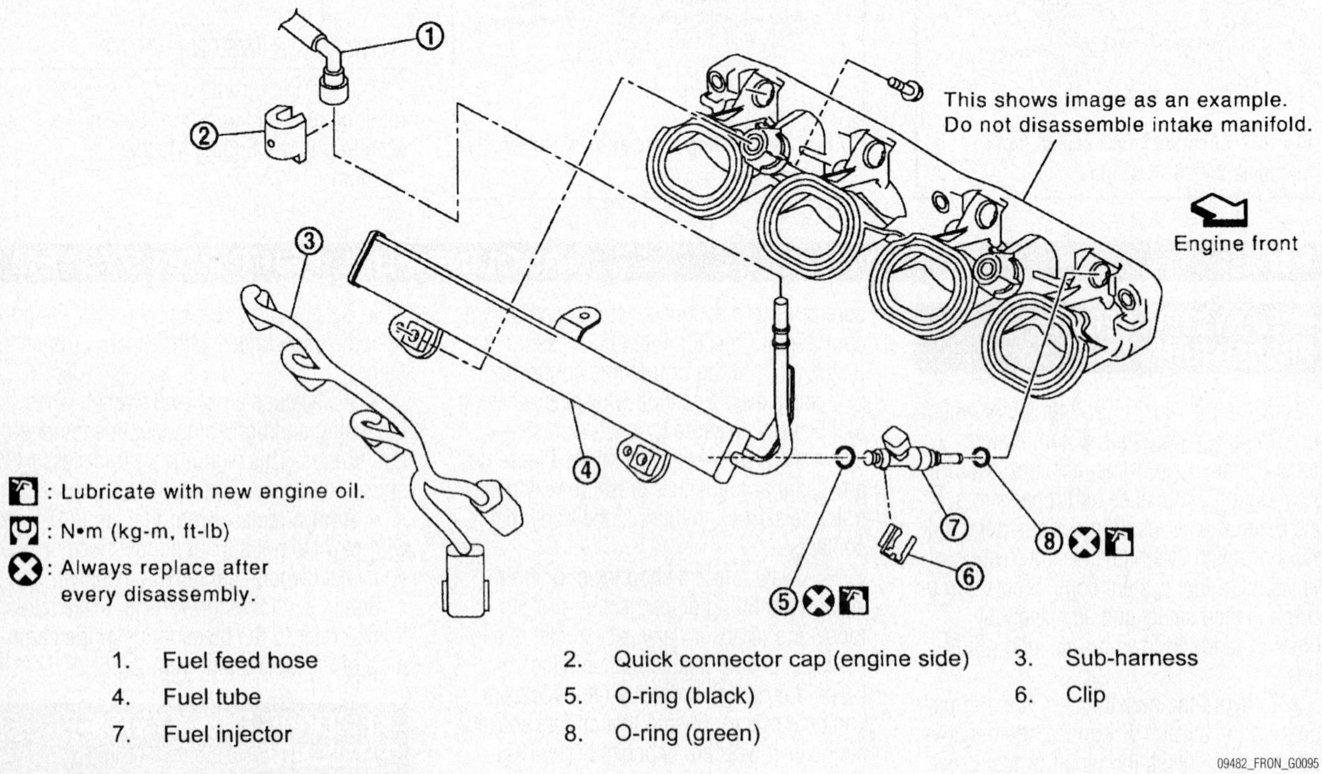

: Lubricate with new engine oil.

: N•m (kg-m, ft-lb)

❌ : Always replace after every disassembly.

1.	Fuel feed hose	2.	Quick connector cap (engine side)	3.	Sub-harness
4.	Fuel tube	5.	O-ring (black)	6.	Clip
7.	Fuel injector	8.	O-ring (green)		

09482_FRON_G0095

Fig. 135 Fuel injector tube and related components—2.5L engine

are different. Do not confuse them.
Fuel tube side: Black. Nozzle side:
Green.

10. Installation is the reverse of the removal procedure.

11. When installing the fuel feed tube be sure to torque the retaining bolts to 9 ft. lbs and then to 21 ft. lbs. in an alternating order.

12. Turn the ignition switch ON, but do not start the engine. Check the fuel lines and hose connections for leaks while applying fuel pressure to the system.

13. Start the engine and check for fuel leaks, correct as required.

4.0L Engine

See Figure 137.

1. Before servicing the vehicle, refer to the Precautions Section.

2. Properly relieve the fuel system pressure.

3. Disconnect the negative battery cable. Remove the fuel filler cap.

4. Remove the intake manifold collector.

5. Remove the quick connector cap (engine side). With the sleeve side of the quick connector release facing the quick

connector, install the quick connector release on to the tube. Insert the quick connector release into the quick connector until the sleeve contacts and goes no further. Hold the quick connector release in that position.

➡**Disconnect the quick connector using tool J-45488, or equivalent, not by picking out the retainer tabs. Inserting the quick connector hard will not disconnect the quick connector. Hold the quick connector release where it contacts and goes no further.**

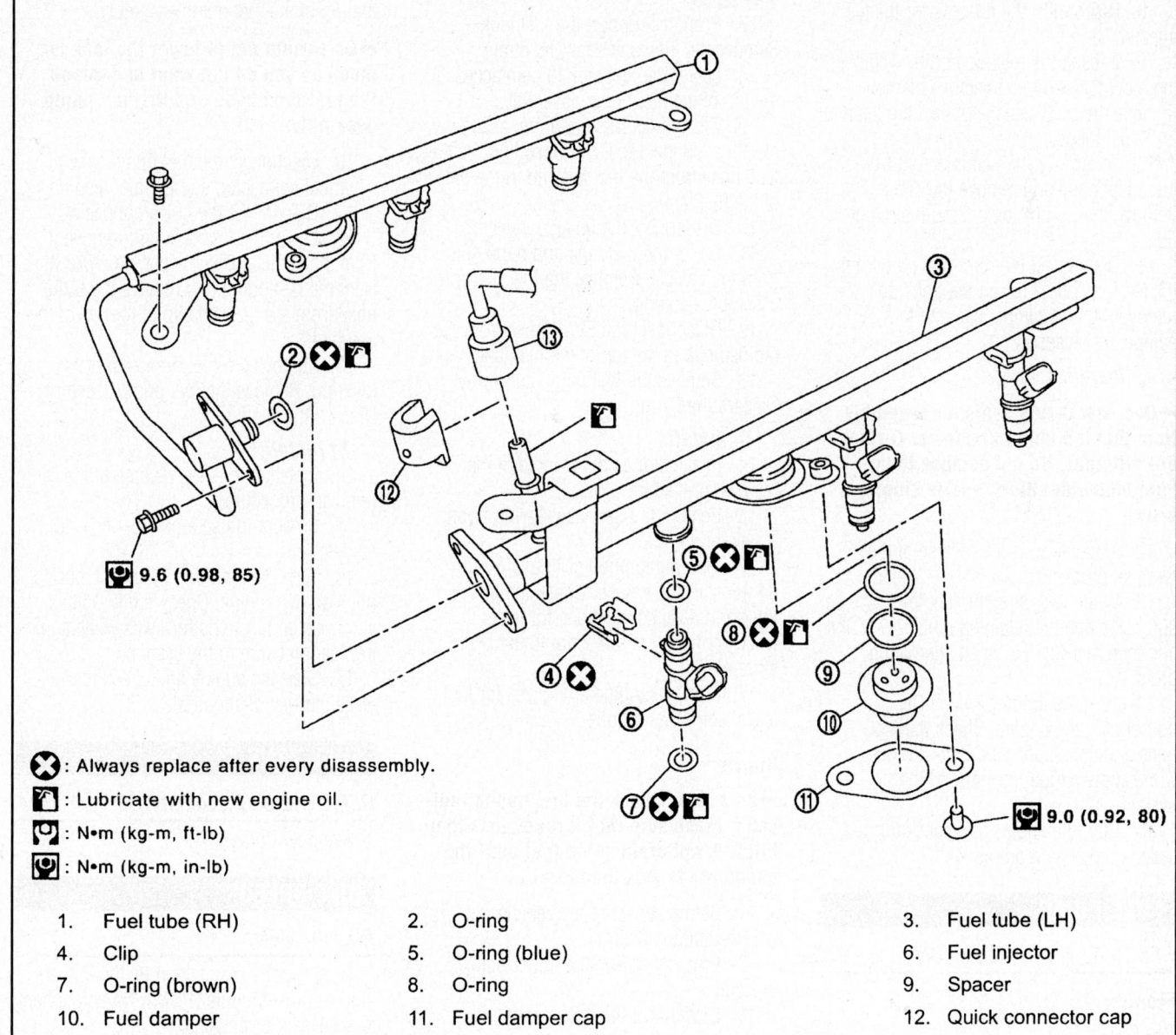

❌ : Always replace after every disassembly.

🛢 : Lubricate with new engine oil.

🔧 : N•m (kg-m, ft-lb)

🔧 : N•m (kg-m, in-lb)

1.	Fuel tube (RH)	2.	O-ring	3.	Fuel tube (LH)
4.	Clip	5.	O-ring (blue)	6.	Fuel injector
7.	O-ring (brown)	8.	O-ring	9.	Spacer
10.	Fuel damper	11.	Fuel damper cap	12.	Quick connector cap
13.	Fuel feed hose				

09482_FRON_G0097

Fig. 137 Fuel injector tube and related components—4.0L engine

6. Draw and pull out the quick connector straight from the fuel tube. Grasp the quick connector holding "A" in the illustration. Do not pull with lateral force applied and the O-ring inside the quick connector could be damaged.

➡**Have a cloth ready, as fuel will leak out. Avoid fire and sparks. Keep parts away from heat. Do not bend or twist the connection between the quick connector and the fuel feed hose. Cover the openings with a plastic bag.**

7. Remove the PCV hose between the rocker covers.

8. Disconnect the harness for the fuel injector.

9. Loosen the retaining bolts. Remove the fuel tube and fuel injector assembly. Remove the bolts which connect the left and right fuel tubes.

10. To remove the fuel injectors from the fuel tube, open and remove the clip. Remove the injector by pulling it straight out.

11. Disconnect the right fuel tube from the left fuel tube. Loosen the bolts, to remove the fuel damper cap and fuel damper, if necessary.

To install:

➡**Use new O-ring seals for assembly. Note that the upper and lower O-rings are different. Do not confuse them. Fuel tube side: Blue. Nozzle side: Brown.**

12. Installation is the reverse of the removal procedure.

13. When installing the fuel feed tube be sure to torque the retaining bolts to 7 ft. lbs and then to 16 ft. lbs. in an alternating order.

14. Turn the ignition switch ON, but do not start the engine. Check the fuel lines and hose connections for leaks while applying fuel pressure to the system.

15. Start the engine and check for fuel leaks, correct as required.

FUEL PUMP

REMOVAL & INSTALLATION

Frontier
See Figure 138.

➡**Be sure to check the fuel gauge indicator. Make sure that it reads less than FULL. If not drain some fuel until the gauge reads less than FULL.**

1. Before servicing the vehicle, refer to the Precautions Section.

2. Properly relieve the fuel system pressure.

3. Disconnect the negative battery cable.

4. Remove the fuel filler cap. Remove the left rear tire and wheel assembly.

5. Remove the fuel tank shield.

6. Remove the fuel filler pipe shield. Disconnect the fuel filler hose from the fuel filler pipe.

7. Properly support the fuel tank. Remove the fuel tank retaining straps.

8. Lower the fuel tank to gain access to the top of the fuel pump assembly.

9. Disconnect the fuel pump assembly electrical connector, EVAP hose and the fuel feed hose from the molded clip in the side of the fuel tank.

10. Disconnect the quick connector.

11. Lower the fuel tank and remove it from the vehicle. Remove the fuel pump assembly lockring.

12. Disconnect the EVAP hose from the molded clip in the top of the fuel tank.

13. Remove the fuel pump assembly. Discard the O-ring.

To install:

14. Installation is the reverse of the removal procedure.

15. Be sure to use a new O-ring upon installation.

16. Turn the ignition switch ON, but do not start the engine. Check the fuel lines and hose connections for leaks while applying fuel pressure to the system.

17. Start the engine and check for fuel leaks, correct as required.

Xterra

➡**Be sure to check the fuel gauge indicator. Make sure that it reads less than FULL. If not drain some fuel until the gauge reads less than FULL.**

1. Before servicing the vehicle, refer to the Precautions Section.

2. Properly relieve the fuel system pressure.

3. Disconnect the negative battery cable.

4. Remove the fuel filler cap. Remove the left rear tire and wheel assembly.

5. Disconnect the lower fuel filler hose from the fuel tank, the EVAP hose and the vent pipe quick connector.

➡**Disconnect the fuel feed hose from the molder clip in the side of the fuel tank.**

6. Remove the four tank shield retaining bolts. Remove the tank shield.

7. Remove the driveshaft.

8. Properly support the fuel tank. Remove the three fuel tank retaining strap bolts. Remove the fuel tank straps.

9. Lower the fuel tank to gain access to the top of the fuel pump assembly.

➡**Be careful not to lower the tank too much as you do not want to damage the fuel feed hose and the fuel pump assembly.**

10. Disconnect the fuel pump assembly electrical connector, and the fuel feed hose.

11. Disconnect the quick connector.

12. Lower the fuel tank and remove it from the vehicle. Disconnect the EVAP hose from the fuel pump and remove the EVAP hose from the molded clip in the top of the fuel tank.

13. Remove the fuel pump assembly lockring. Remove the fuel pump assembly. Discard the O-ring.

To install:

14. Installation is the reverse of the removal procedure.

15. Be sure to use a new O-ring upon installation.

16. Turn the ignition switch ON, but do not start the engine. Check the fuel lines and hose connections for leaks while applying fuel pressure to the system.

17. Start the engine and check for fuel leaks, correct as required.

FUEL TANK

REMOVAL & INSTALLATION

See Fuel Pump.

IDLE SPEED

ADJUSTMENT

Idle speed is maintained by the Powertrain Control Module (PCM). No adjustment is necessary or possible.

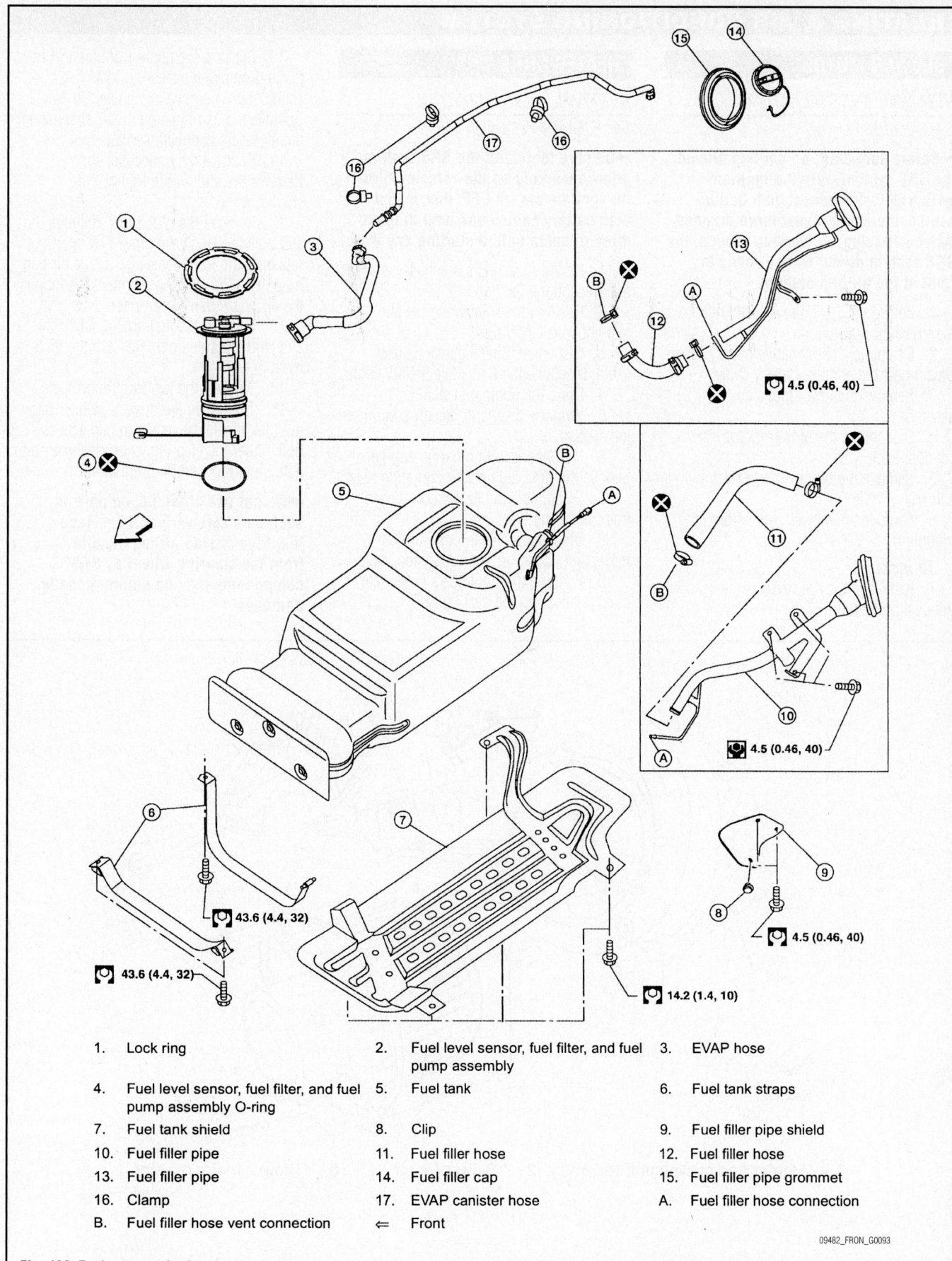

1.	Lock ring	
2.	Fuel level sensor, fuel filter, and fuel pump assembly	
3.	EVAP hose	
4.	Fuel level sensor, fuel filter, and fuel pump assembly O-ring	
5.	Fuel tank	
6.	Fuel tank straps	
7.	Fuel tank shield	
8.	Clip	
9.	Fuel filler pipe shield	
10.	Fuel filler pipe	
11.	Fuel filler hose	
12.	Fuel filler hose	
13.	Fuel filler pipe	
14.	Fuel filler cap	
15.	Fuel filler pipe grommet	
16.	Clamp	
17.	EVAP canister hose	
A.	Fuel filler hose connection	
B.	Fuel filler hose vent connection	
⇐	Front	

Fig. 138 Fuel pump and related components

09482_FRON_G0093

HEATING & AIR CONDITIONING SYSTEM

BLOWER MOTOR

REMOVAL & INSTALLATION

See Figure 139.

➡**Before servicing, or working around, the SRS system, turn the ignition switch OFF, disconnect both battery cables and wait at least three minutes. When servicing, or working around, the SRS system do not work directly in front of the air bag module.**

1. Before servicing the vehicle, refer to the Precautions Section.

2. Disconnect the negative battery cable. Disconnect the positive battery cable.

3. Remove the lower glove box assembly.

4. Disconnect the blower motor electrical connector.

5. Remove the blower motor retaining screws.

6. Remove the blower motor from its mounting.

To install:

7. Installation is the reverse of the removal procedure.

HEATER CORE

REMOVAL & INSTALLATION

See Figures 140 through 146.

➡**Be sure to disarm the SRS system, prior to working on the vehicle. Turn the ignition switch OFF, disconnect both battery cables and wait at least three minutes before starting any work.**

1. Before servicing the vehicle, refer to the Precautions Section.

2. Position the front wheels in the straight ahead direction.

3. Disconnect the negative battery cable. Disconnect the positive battery cable.

4. Drain the cooling system.

5. Properly discharge the air conditioning system.

6. If equipped with the 4.0L engine, remove the right side heater core pipe nuts.

7. Disconnect the heater core hoses from the heater core.

8. Disconnect the air conditioning refrigerant lines from the expansion valve.

9. Position the front seats in the rearmost position on the seat tracks.

10. Remove the upper front pillar trim panel. Remove the steering lock escutcheon. Remove the cluster lid "A". Remove the combination meter. Disconnect the electrical connections.

11. Remove the optical sensor. Remove the audio unit. Remove the cluster lid "D".

12. Remove the glove box. Remove the two bolts, through the glove box opening, retaining the front passenger's side air bag module to the steering member. Disconnect the air bag module connectors.

13. Remove the instrument stay right side and left side bolts. Remove the instrument panel.

14. Remove the two front floor ducts.

15. To remove the driver's side air bag module, locate the retaining clip access hole under the steering wheel. Insert a suitable blunt tool (4mm–6mm in size)

➡**Do not use sharp edged objects, such as a screwdriver, to release the driver's side airbag module from the steering wheel as SRS components may be unintentionally damaged.**

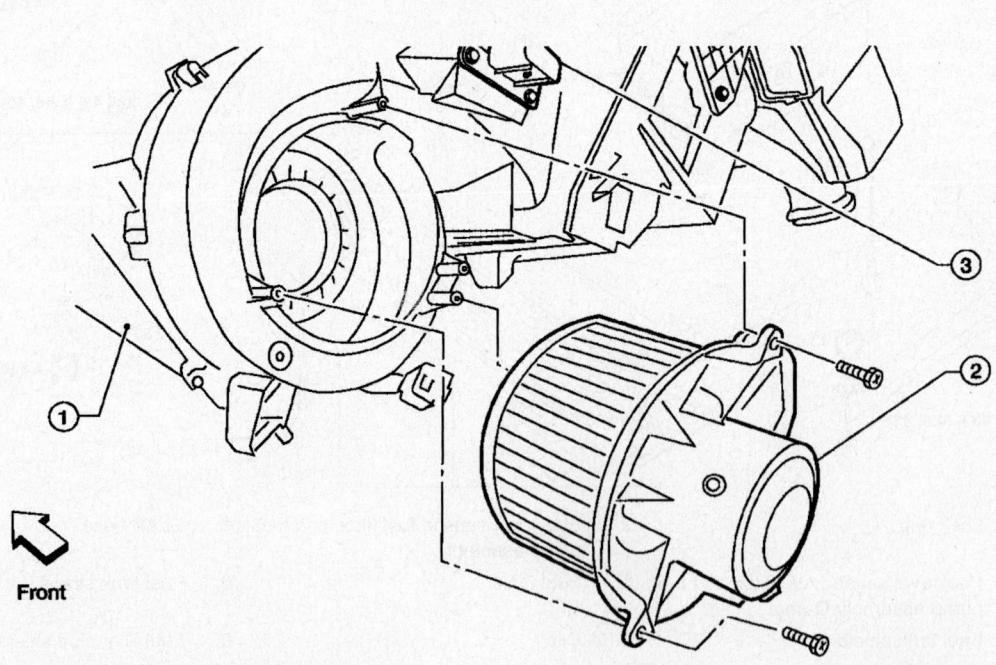

Front

1. Heater and cooling unit assembly 2. Blower motor 3. Blower motor resistor

42050_FRON_G0071

Fig. 139 Blower motor an related components

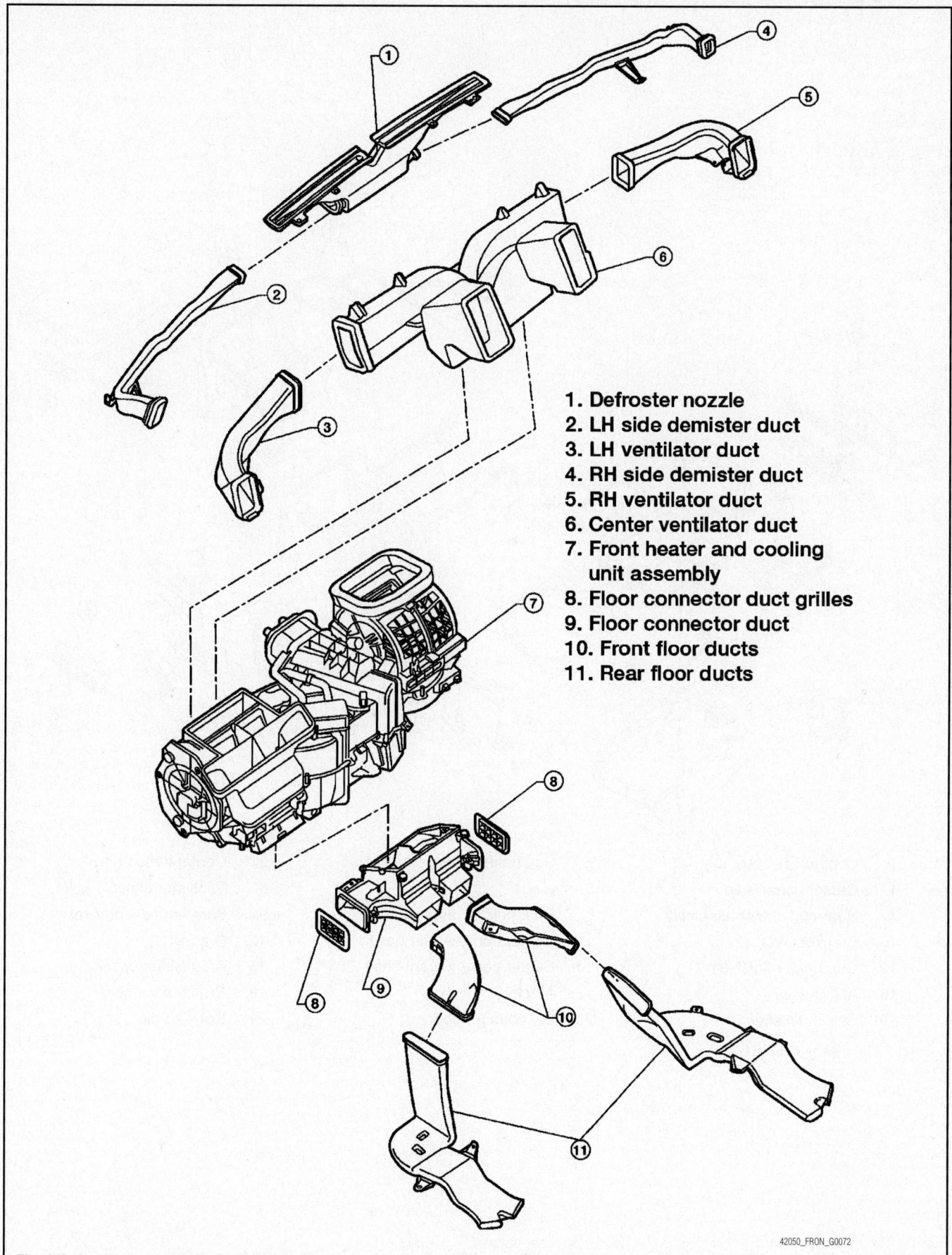

1. Defroster nozzle
2. LH side demister duct
3. LH ventilator duct
4. RH side demister duct
5. RH ventilator duct
6. Center ventilator duct
7. Front heater and cooling unit assembly
8. Floor connector duct grilles
9. Floor connector duct
10. Front floor ducts
11. Rear floor ducts

42050_FRON_G0072

Fig. 140 Duct work surrounding the heater/cooling unit

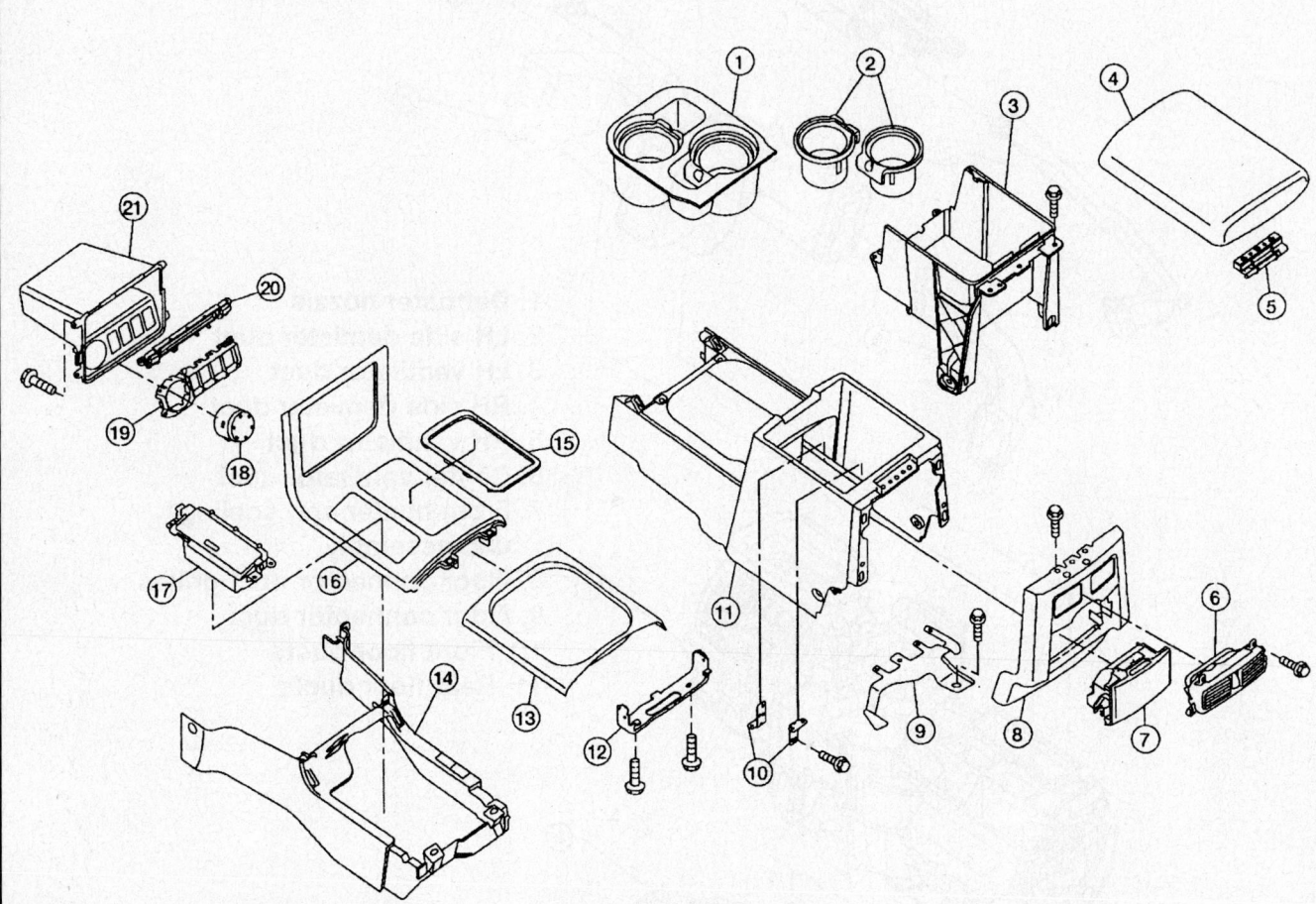

1.	Cup holder assembly	2.	Cup holder insert	3.	Center console bin
4.	Center console lid	5.	Hinge	6.	Ventilator console grille
7.	Rear cup holder assembly	8.	Rear finisher assembly	9.	Wire harness bracket
10.	Bracket DVD	11.	Center console rear base	12.	Bracket
13.	Cup holder finisher	14.	Center console front base	15.	A/T finisher bezel
16.	A/T finisher	17.	Ash tray	18.	Switch assembly
19.	Switch finisher	20.	CD changer door	21.	Console bin

09482_FRON_G0011

Fig. 141 Center console and related components

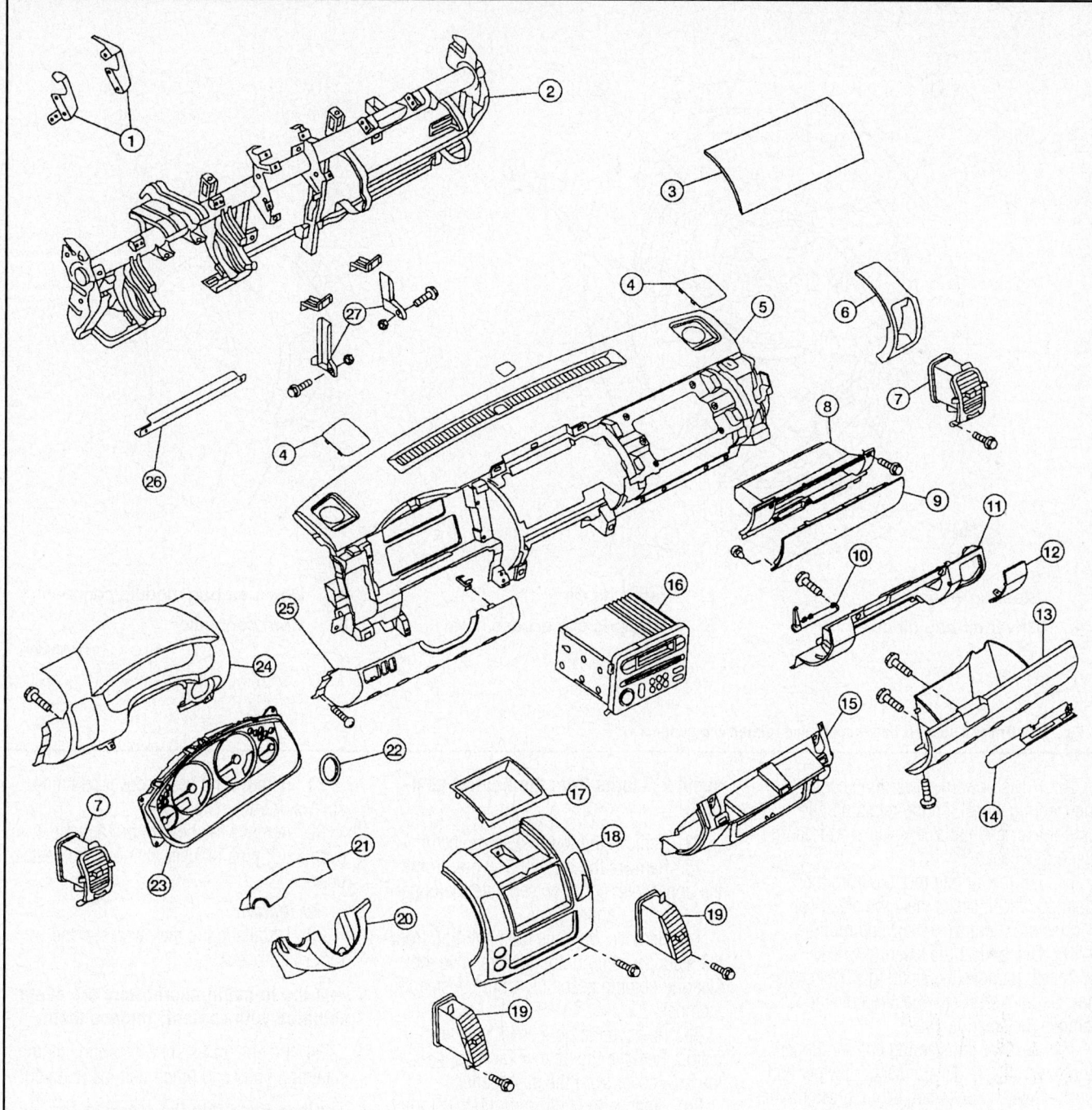

1. Display unit bracket RH/LH
2. Steering member assembly
3. Passenger air bag module cover
4. Speaker grille RH/LH
5. Instrument panel and pad assembly
6. Instrument side finisher
7. Side ventilator assembly RH/LH
8. Upper glove box bin
9. Upper glove box door
10. Lower glove box damper assembly
11. Lower instrument panel RH
12. Fuse block cover
13. Lower glove box assembly
14. Lower glove box latch assembly
15. Cluster lid D
16. Audio unit
17. Storage tray
18. Cluster lid C
19. Center ventilator assembly RH/LH
20. Steering column cover lower
21. Steering column cover upper
22. Steering lock escutcheon
23. Combination meter
24. Cluster lid A
25. Lower instrument panel LH
26. Knee protector brace
27. Instrument stay RH/LH

09482_FRON_G0009

Fig. 142 Instrument panel and related components

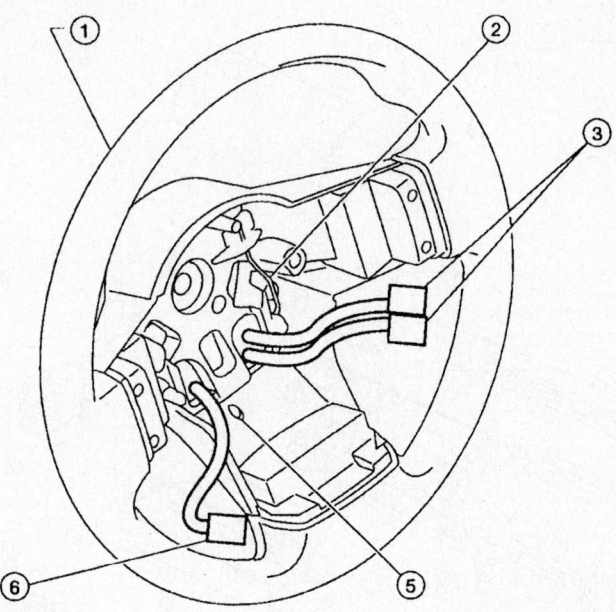
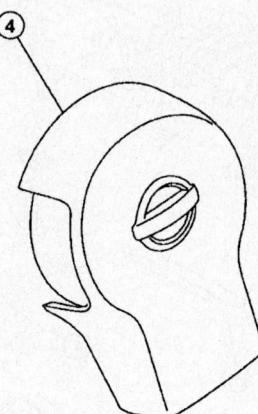

1. Steering wheel
4. Driver air bag module
2. Retaining clip
5. Retaining clip access hole
3. Driver air bag module connectors
6. Horn connector

09482_FRON_G0010

Fig. 143 Driver's side air bag module and related components

16. Press upward, toward the center of the steering wheel, on the retaining clip until the air bag module is released from the steering wheel.

17. Lift the air bag module from the steering wheel. Disconnect the electrical connectors. Remove the air bag module.

18. Disconnect the steering wheel switches. Remove the steering wheel center nut. Using a steering wheel removal tool, remove the steering wheel.

19. Remove the steering column upper and lower covers. Disconnect the wiper and washer switch connector. While pressing the tabs, pull the wiper and washer switch away from the spiral cable to remove it.

20. Disconnect the light and turn signal switch connector. While pressing the tabs, pull the light and turn signal switch toward the driver's door to remove it.

21. Remove the screws. While pressing the tab, pull the spiral cable away from the steering column assembly. Disconnect the electrical connectors.

➡**With the steering linkage disconnected, the spiral cable may snap by turning the steering wheel beyond the limited number of turns. The spiral cable can be turned counterclockwise**

about 2.5 turns from the neutral position.

22. Remove the lower knee protector.

23. Remove the locknut and bolt from the upper joint and then separate the upper joint from the upper shaft.

24. Remove the three nuts and bolt from the steering column and then remove the steering column assembly from the steering member.

25. Remove the hole cover seal and clamp. Remove the hole cover nuts, remove the hole cover from the dash panel.

26. Remove the bolt from the lower joint of the lower joint shaft and remove the lower joint shaft from the vehicle.

27. Disconnect the instrument panel wire harness at the right and left in-line connector brackets, and the fuse block (SMJ) electrical connectors.

28. Remove the covers and then remove the three steering member bolts from each side to disconnect the steering member from the vehicle body.

29. Remove the heater/evaporator case assembly with it attached to the steering member from the vehicle.

30. Separate the steering member from the heater/evaporator unit.

31. Remove the heater cover retaining screws. Remove the cover.

32. Remove the heater core and the evaporator pipe bracket. Remove the heater core.

To install:

33. Installation is the reverse of the removal procedure.

➡**If the in-cabin microfilters are contaminated with coolant, replace them.**

34. Be sure to use new steering column retaining bolts and pinch bolt, as required.

➡**When installing the steering column, finger tighten all of the lower bracket and joint bolts and then tighten them to specification. Do not apply undue stress to the steering column.**

35. With the wheels in the straight ahead position align the slit of the lower joint with the projection on the dust cover. Insert the joint until surface "A" contacts surface "B"

36. Be sure to align the spiral cable correctly when installing the steering wheel. Make sure that the cable is in the neutral position. The neutral position is detected by turning left 2.6 revolutions from the right

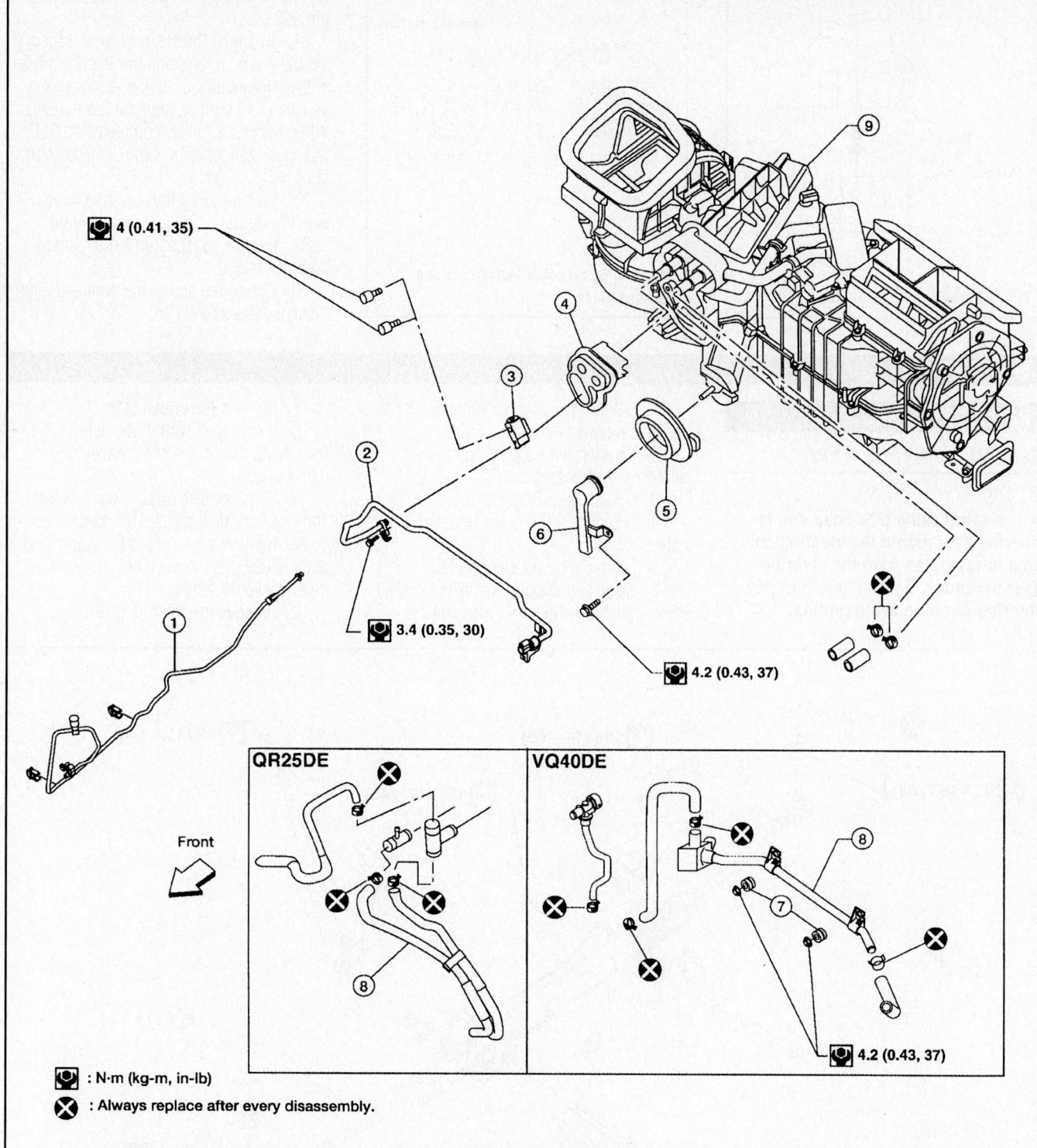

4 (0.41, 35)

3.4 (0.35, 30)

4.2 (0.43, 37)

Front

QR25DE

VQ40DE

4.2 (0.43, 37)

⊙ : N·m (kg-m, in-lb)

✗ : Always replace after every disassembly.

1. High-pressure A/C pipe	2. Low-pressure A/C pipe	3. Expansion valve
4. Heater core and evaporator pipes grommet	5. A/C drain hose grommet	6. A/C drain hose
7. Heater core pipe mounts	8. Heater core pipes	9. Heater and cooling unit assembly

09482_FRON_G0012

Fig. 144 Heater/evaporator core and related components

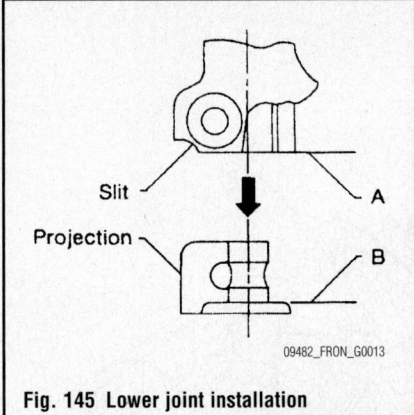

Fig. 145 Lower joint installation

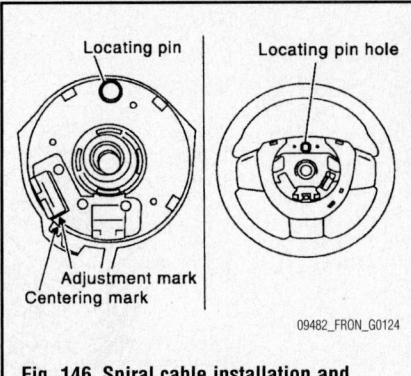

Fig. 146 Spiral cable installation and locating point

end position and ending with the locating pin at the top.

37. To adjust the steering angle sensor neutral position, position the steering wheel in the straight ahead position and rive the vehicle at 10 mph or more for ten minutes. When the procedure is complete, the SLP indicator lamp and the VDC OFF indicator lamp will turn off.

38. Be sure to fill the cooling system with the proper grade and type coolant.

39. Be sure to recharge the air conditioning system.

40. Check and adjust the front end alignment, as necessary.

STEERING

POWER STEERING GEAR

REMOVAL & INSTALLATION

See Figures 147 and 148.

➡ **The spiral cable may snap due to steering operation if the steering column is separated from the steering gear assembly. Be sure to secure the steering wheel to avoid turning.**

1. Before servicing the vehicle, refer to the Precautions Section.
2. Position the front wheels in the straight ahead position.
3. Disarm the SRS system.
4. Disconnect the negative battery cable.
5. Drain the power steering fluid.
6. Raise and support the vehicle safely. Remove the tire and wheel assemblies.

7. Remove the undercover.
8. If equipped with 4WD, remove the final drive, then support the halfshafts, using wire.
9. Remove the stabilizer bar brackets, and position the stabilizer bar aside.
10. Remove and discard the cotter pins at the steering outer sockets. Loosen the outer socket locknuts.
11. Remove the steering gear outer

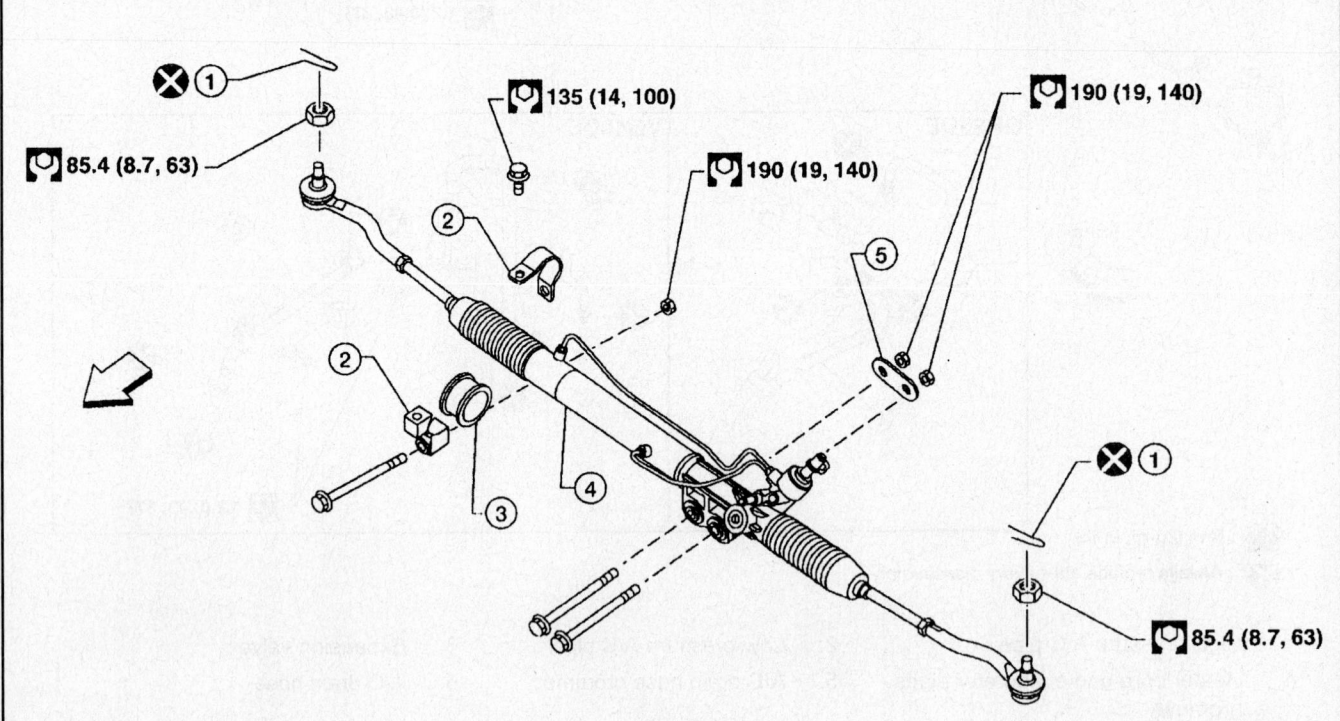

1.	Cotter pin	2.	Mounting bracket	3.	Mounting insulator
4.	Steering gear assembly	5.	Washer	⇐	Front

Fig. 147 Power steering gear and related components

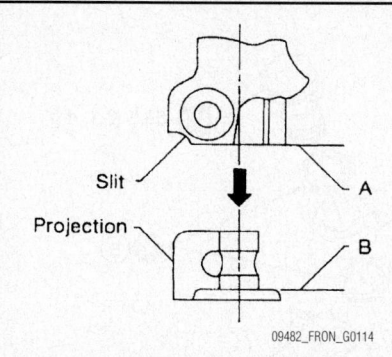

Fig. 148 Power steering gear lower joint installation alignment

sockets from the steering knuckles, using tool HT72520000 (J-25730-A) or equivalent.

12. Disconnect and plug the power steering fluid lines at the steering gear.

13. Remove the bolt from the lower joint of the lower joint assembly. Separate the lower joint from the steering gear assembly. Be careful not to damage the lower joint.

14. Remove the steering gear retaining nuts and bolts. Remove the steering gear from the vehicle.

To install:

15. With the steering wheel in the straight ahead position, align the slit of the lower joint with the projection on the dust cover. Insert the joint until both surfaces contact each other.

16. Continue the installation in the reverse order of the removal procedure.

17. Check and adjust the front alignment, as required.

18. Bleed the power steering system.

19. Fill the power steering pump with the proper grade and type fluid.

➡After removing/installing or replacing steering and suspension components which effect wheel alignment or after adjusting wheel alignment, or the steering angle sensor or the ABS actuator electrical unit be sure to adjust the neutral position of the steering angle sensor before running the vehicle.

20. Position the steering wheel in the straight ahead position.

21. When this procedure is complete the SLP indicator lamp and the VDC OFF indicator lamp will turn off.

POWER STEERING PUMP

REMOVAL & INSTALLATION

See Figures 149 and 150.

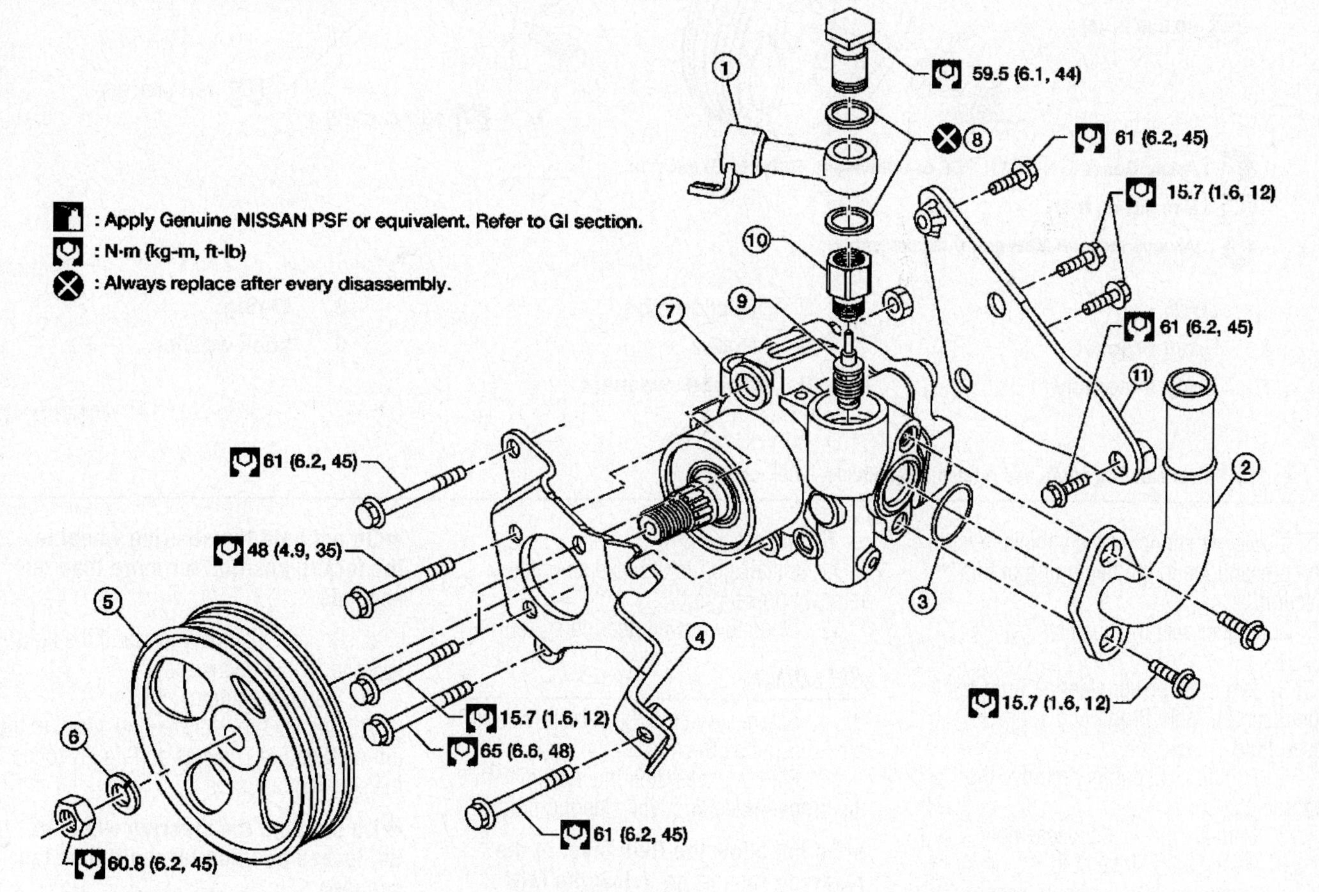

🔧 : Apply Genuine NISSAN PSF or equivalent. Refer to GI section.

🔧 : N·m (kg-m, ft-lb)

✖ : Always replace after every disassembly.

59.5 (6.1, 44)
61 (6.2, 45)
15.7 (1.6, 12)
61 (6.2, 45)
61 (6.2, 45)
48 (4.9, 35)
15.7 (1.6, 12)
65 (6.6, 48)
15.7 (1.6, 12)
61 (6.2, 45)
60.8 (6.2, 45)

1. Joint	2. Suction pipe	3. O-ring
4. Front bracket	5. Pulley	6. Lock washer
7. Body assembly	8. Copper washers	9. Flow control valve and spring
10. Connector	11. Rear bracket	

Fig. 149 Power steering pump and related components—2.5L engine

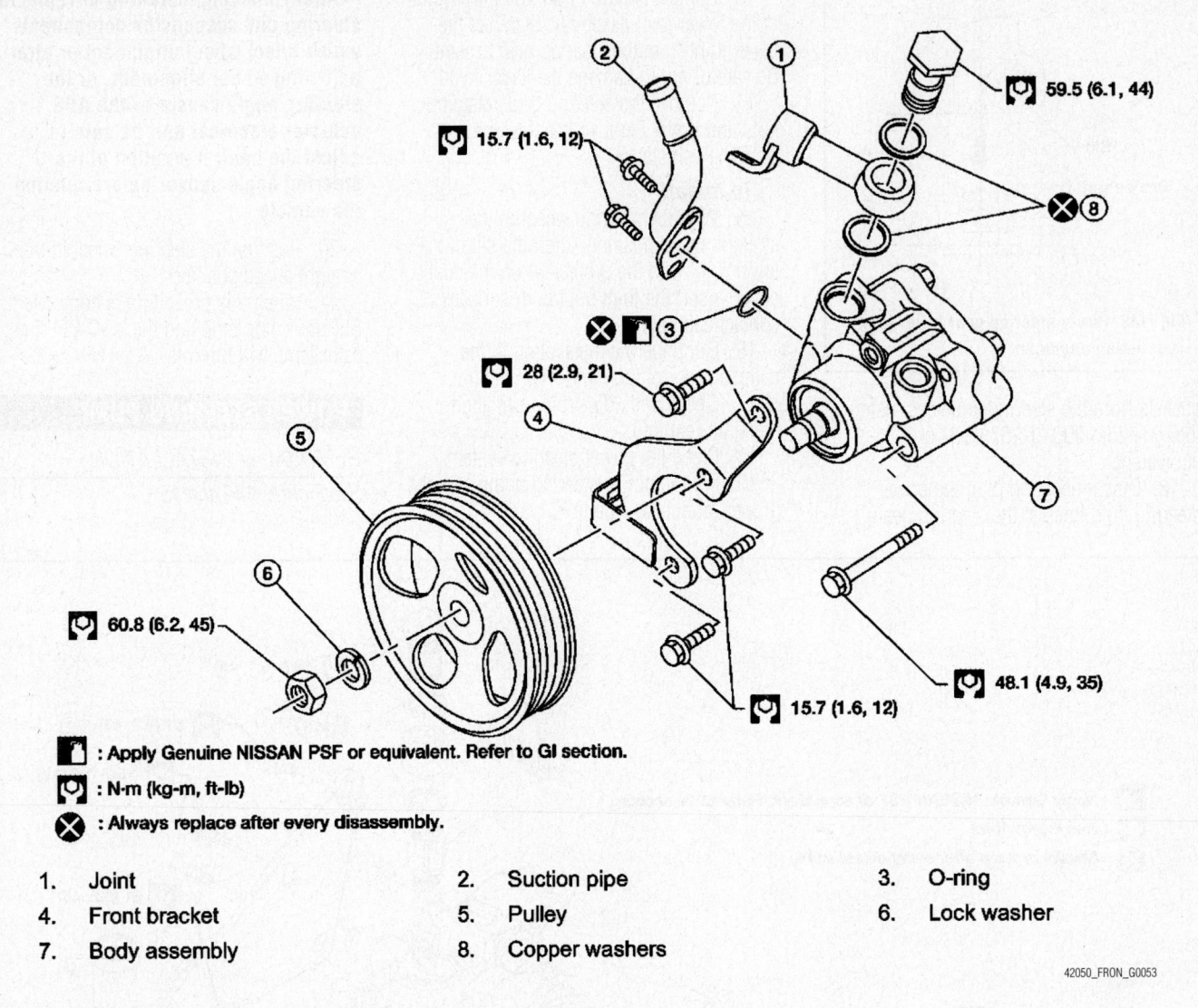

: Apply Genuine NISSAN PSF or equivalent. Refer to GI section.

: N·m (kg-m, ft-lb)

: Always replace after every disassembly.

1. Joint
2. Suction pipe
3. O-ring
4. Front bracket
5. Pulley
6. Lock washer
7. Body assembly
8. Copper washers

42050_FRON_G0053

Fig. 150 Power steering pump and related components—4.0L engine

1. Before servicing the vehicle, refer to the precautions in the beginning of this section.
2. Disconnect the negative battery cable.
3. Drain the power steering fluid from the reservoir tank. Properly dispose of used fluid.
4. On the 4.0L engine, remove the engine cover.
5. Remove the air duct assembly.
6. Remove the drive belt.
7. Disconnect the pressure sensor electrical connector.
8. Disconnect and plug the fluid lines.
9. Remove the pump retaining bolts.
10. Remove the pump from the vehicle.

To install:
11. Installation is the reverse of the removal procedure.
12. Bleed the power steering system.

BLEEDING

1. Before servicing the vehicle, refer to the Precautions Section.
2. Fill the power steering system with the proper grade and type steering fluid.

➡**Do not allow the fluid level in the reservoir tank to go below the MIN level line. Check and add fluid as needed.**

3. Raise and safely support the vehicle.
4. Quickly turn the steering wheel to the full right and left detents and lightly touch the steering stoppers.

➡**Do not hold the steering wheel in the locked position for more than ten seconds.**

5. Repeat this operation until the fluid level no longer decreases.
6. Start the engine.
7. Quickly turn the steering wheel to the full right and left detents and lightly touch the steering stoppers.

➡**Do not hold the steering wheel in the locked position for more than ten seconds.**

8. Check for air bubbles or cloudy fluid. If found, repeat the bleeding procedure.
9. Stop the engine and check the fluid level. Correct as required.

See Figure 151.

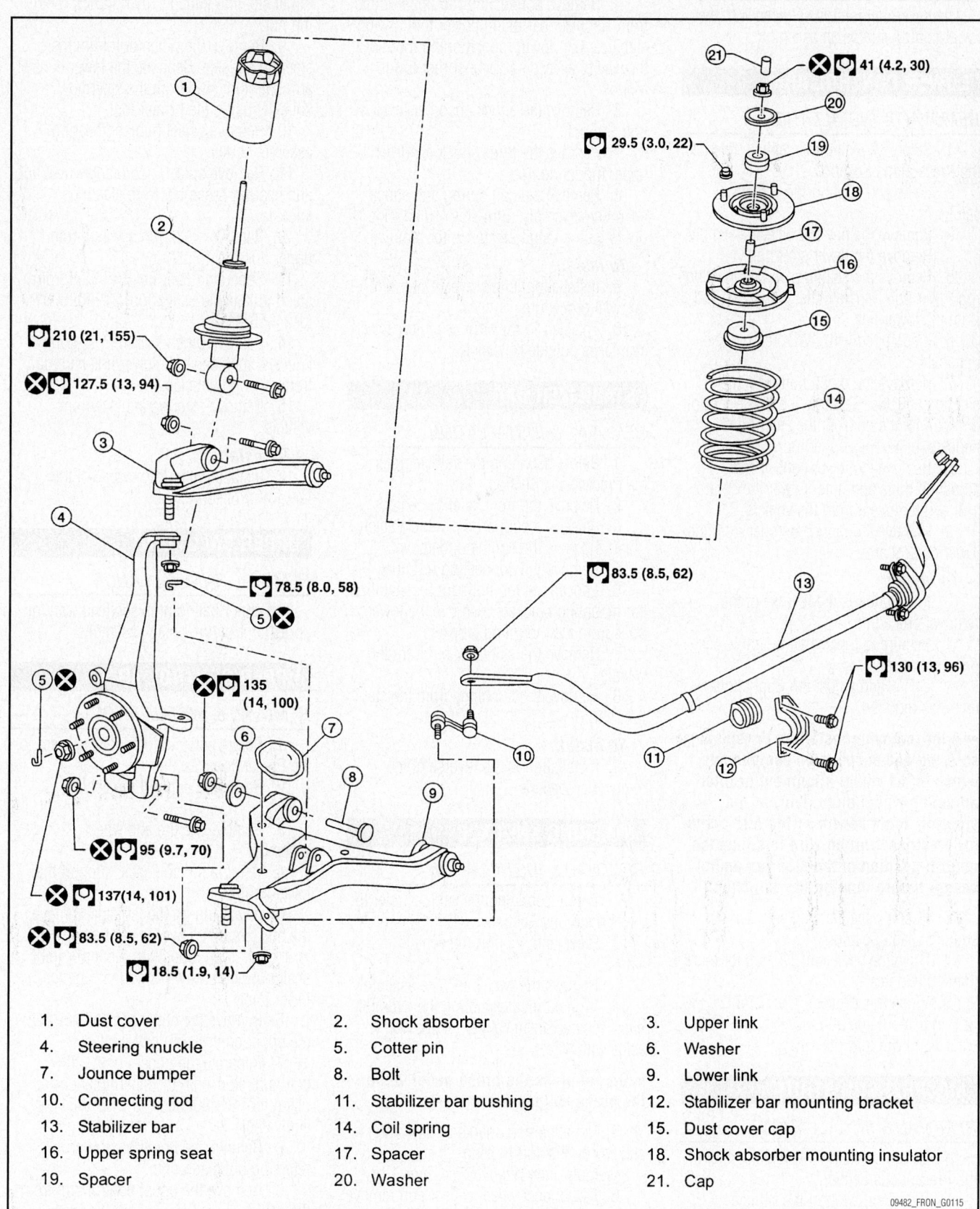

1.	Dust cover	2.	Shock absorber	3.	Upper link
4.	Steering knuckle	5.	Cotter pin	6.	Washer
7.	Jounce bumper	8.	Bolt	9.	Lower link
10.	Connecting rod	11.	Stabilizer bar bushing	12.	Stabilizer bar mounting bracket
13.	Stabilizer bar	14.	Coil spring	15.	Dust cover cap
16.	Upper spring seat	17.	Spacer	18.	Shock absorber mounting insulator
19.	Spacer	20.	Washer	21.	Cap

09482_FRON_G0115

Fig. 151 Front suspension and related components

LOWER BALL JOINT

REMOVAL & INSTALLATION

The lower ball joint is serviced with the lower control arm as an assembly.

LOWER CONTROL ARM

REMOVAL & INSTALLATION

1. Before servicing the vehicle, refer to the Precautions Section.
2. Raise and support the vehicle safely.
3. Remove the tire and wheel assembly.
4. Remove the lower strut bolt.
5. Remove the stabilizer bar connecting rod lower nut. Separate the connecting rod from the lower link.
6. If equipped with 4WD, remove the halfshaft.
7. Remove the pinch bolt from the steering knuckle. Separate the lower control arm ball joint stud from the steering knuckle, using the proper tool.
8. Remove the lower control arm adjusting bolts and nuts. Lower the control arm and remove it from the vehicle.
9. Remove the jounce bumper from the lower control arm.

To install:

10. Installation is the reverse of the removal procedure.
11. Be sure to replace all wearable components, as required.
12. Check and adjust the front alignment, as required.

➡ After removing/installing or replacing steering and suspension components which effect wheel alignment or after adjusting wheel alignment, or the steering angle sensor or the ABS actuator electrical unit be sure to adjust the neutral position of the steering angle sensor before running the vehicle.

13. Position the steering wheel in the straight ahead position.
14. Drive the vehicle at 10 mph for more than 10 minutes.
15. When this procedure is complete the SLP indicator lamp and the VDC OFF indicator lamp will turn off.

MACPHERSON STRUT

REMOVAL & INSTALLATION

1. Before servicing the vehicle, refer to the Precautions Section.
2. Raise and support the vehicle safely.

3. Remove the wheel and tire assembly.
4. Support the lower link using a suitable jack.
5. Remove connecting rod upper joints from stabilizer bar using power tool. Swing stabilizer bar down, repositioning it out of the way to access shock absorber lower mount.
6. Remove the shock absorber lower bolt and nut.
7. Remove the three shock absorber upper mounting nuts.
8. Remove the coil spring and shock absorber assembly. Turn steering knuckle out to gain enough clearance for removal.

To install:

9. Installation is the reverse of the removal procedure.
10. The step in the strut assembly lower seat faces outside of vehicle.

STABILIZER BAR

REMOVAL & INSTALLATION

1. Before servicing the vehicle, refer to the Precautions Section.
2. Remove the front valance center.
3. Raise and support the vehicle safely.
4. Remove the engine undercover.
5. Remove the connecting rod nuts.
6. Loosen the top bolts for the stabilizer bar mounting brackets. Remove the lower bolts from the mounting brackets.
7. Remove the stabilizer bar from the vehicle.
8. Remove the bushings from the stabilizer bar.

To install:

9. Installation is the reverse of the removal procedure.

STEERING KNUCKLE

REMOVAL & INSTALLATION

1. Before servicing the vehicle, refer to the Precautions Section.
2. Raise and support the vehicle safely.
3. Remove the wheel and tire assembly.
4. Without disassembling the hydraulic lines, remove brake caliper. Reposition it aside with wire.

➡ Do not press the brake pedal while the brake caliper is removed.

5. Put alignment marks on disc rotor and wheel hub and bearing assembly, then remove disc rotor.
6. Disconnect wheel sensor and remove bracket from steering knuckle.

7. On 4WD models, remove cotter pin, then remove lock nut from drive shaft.
8. Remove steering outer socket cotter pin at steering knuckle, then loosen mounting nut.
9. Remove the pinch bolt from the steering knuckle. Separate the lower control arm ball joint stud from the steering knuckle, using the proper tool.
10. Remove wheel hub and bearing assembly bolts.
11. Remove splash guard and wheel hub and bearing assembly from steering knuckle.
12. Remove cotter pin and nut from upper link ball joint.
13. Separate upper link ball joint from steering knuckle using Tool ST29020001 (J-24319-01).
14. Remove pinch bolt from steering knuckle, then separate lower link ball joint from steering knuckle.
15. Remove steering knuckle from vehicle.

To install:

16. Installation is the reverse of the removal procedure.

UPPER BALL JOINT

REMOVAL & INSTALLATION

The upper ball joint is serviced with the upper control arm as an assembly.

UPPER CONTROL ARM

REMOVAL & INSTALLATION

1. Before servicing the vehicle, refer to the Precautions Section.
2. Raise and support the vehicle safely.
3. Remove the tire and wheel assembly.
4. Using a suitable jack, support the lower control arm.
5. If working on the left side, remove the bolt from the lower joint of the lower joint shaft, then reposition the lower joint shaft out of the way. Do not damage the lower joint.
6. Remove the cotter pin and nut from the upper control arm ball joint.
7. Separate the upper control arm ball joint stud from the steering knuckle, using tool ST29020001 (J-24319-01) or equivalent.
8. Remove the upper control arm retaining bolts and nuts.
9. Remove the upper control arm from the vehicle.

To install:

10. Installation is the reverse of the removal procedure.

11. Be sure to replace all wearable components, as required.

12. Check and adjust the front alignment, as required.

➡**After removing/installing or replacing steering and suspension components which effect wheel alignment or after adjusting wheel alignment, or the steering angle sensor or the ABS actuator electrical unit be sure to adjust the neutral position of the steering angle sensor before running the vehicle.**

13. Position the steering wheel in the straight ahead position.

14. Drive the vehicle at 10 mph for more than 10 minutes.

15. When this procedure is complete the SLP indicator lamp and the VDC OFF indicator lamp will turn off.

WHEEL HUB AND BEARING

REMOVAL & INSTALLATION

See Figure 152.

1. Before servicing the vehicle, refer to the Precautions Section.

2. Raise and support the vehicle safely.

3. Remove the tire and wheel assembly.

4. Remove the caliper and position it to the side with wire. Do not disconnect the brake fluid line.

➡**Do not press the brake pedal while the brake caliper is removed.**

5. Matchmark the brake rotor and the wheel hub. Remove the brake rotor.

6. Remove the cotter pin. Remove the lock nut.

7. Remove the halfshaft from the wheel hub and bearing assembly.

8. Remove the wheel sensor from the hub and bearing assembly. Do not pull on the wheel sensor harness.

9. Remove the wheel hub and bearing assembly bolts.

10. Remove the splash guard. Remove the wheel hub and bearing assembly from the steering knuckle.

➡**Carefully remove the wheel sensor and harness through the hole in the splash guard.**

To install:

11. Inspect the wheel sensor O-ring, replace the wheel speed sensor assembly, as required.

12. Installation is the reverse of the removal procedure.

13. Be sure to use new bolts when installing the wheel hub and bearing assembly.

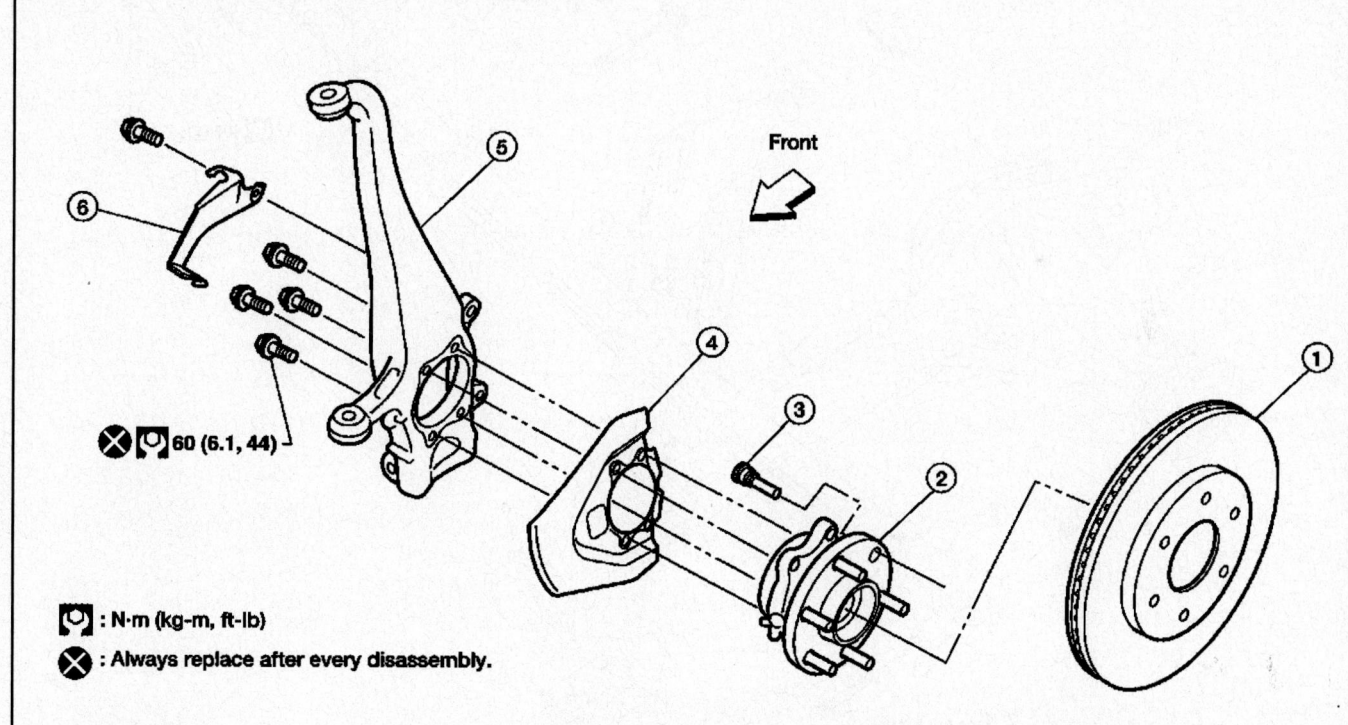

60 (6.1, 44)

⊡ : N·m (kg-m, ft-lb)

✕ : Always replace after every disassembly.

1. Disc rotor	2. Wheel hub and bearing assembly	3. Wheel stud
4. Splash guard	5. Steering knuckle	6. Wheel sensor bracket

42050_FRON_G0104

Fig. 152 Hub and bearing assembly

SUSPENSION   REAR SUSPENSION

See Figure 153.

STABILIZER BAR

REMOVAL & INSTALLATION

Xterra

1. Disconnect the stabilizer bar ends from the connecting rods using power tool.
2. Remove the stabilizer bar clamps using power tool, and remove the bushings.
3. Remove the stabilizer bar.

To install:

4. Installation is the reverse of the removal procedure.
5. Install the stabilizer bar clamp and bushing so they are positioned outside of the crimp ring on the stabilizer bar.

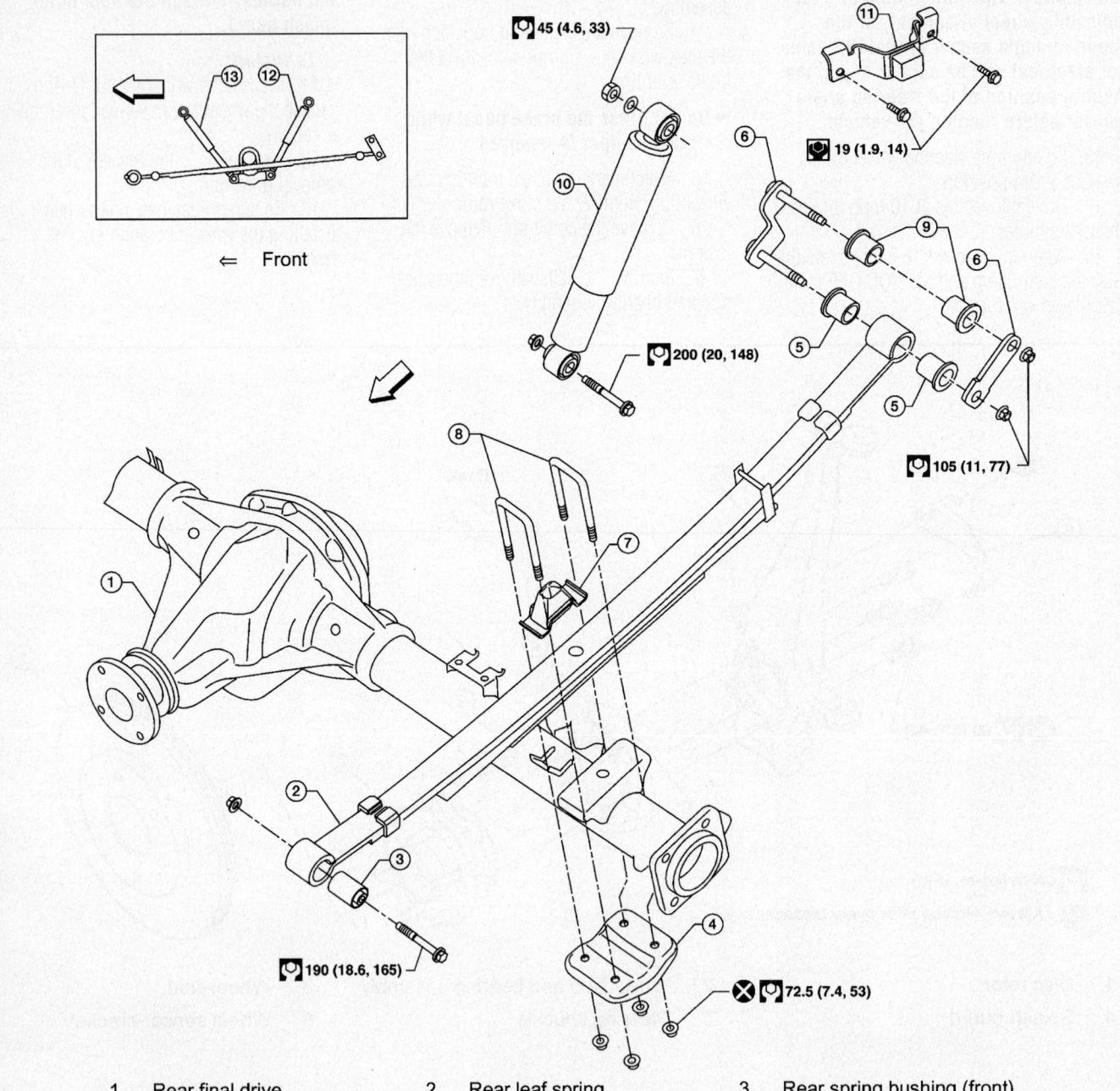

⇐ Front

1. Rear final drive
2. Rear leaf spring
3. Rear spring bushing (front)
4. Rear spring pad
5. Rear spring bushing (rear)
6. Rear spring shackle
7. Bumper
8. Rear spring clip U-bolts
9. Rear spring shackle bushing
10. Shock absorber
11. Bumper
12. Shock absorber (left side)
13. Shock absorber (right side)

22140_FRON_G0020

Fig. 153 Rear suspension and related components

NISSAN

Murano

SPECIFICATIONS AND MAINTENANCE CHARTS

ENGINE AND VEHICLE IDENTIFICATION

		Engine						Model Year	
Code	Liters (cc)	Cu. In.	Cyl.	Fuel Sys.	Engine	Eng. Mfg.		Code	Year
VQ35DE	3.5 (3498)	213.45	6	MFI	DOHC	Nissan		6	2006
								7	2007

MFI: Multi-port Fuel Injection

DOHC: Double Overhead Camshaft

22140_MURA_C0001

GENERAL ENGINE SPECIFICATIONS

Year	Model	Engine Displacement Liters	Engine ID	Net Horsepower @ rpm	Net Torque @ rpm (ft. lbs.)	Bore x Stroke (in.)	Com-pression Ratio	Oil Pressure @ rpm
2006	Murano	3.5	VQ35DE	245@5800	246@4400	3.76X3.20	10.3:1	43@2000
2007	Murano	3.5	VQ35DE	240@5800	244@4400	3.76X3.20	10.0:1	43@2000

22140_MURA_C0002

ENGINE TUNE-UP SPECIFICATIONS

Year	Engine Displacement Liters	Engine ID	Spark Plug Gap (in.)	Ignition Timing (deg.)	Fuel Pump (psi)	Idle Speed RPM	Valve Clearance (in.) In.	Valve Clearance (in.) Ex.
2006	3.5	VQ35DE	0.043	15 ①	51 ②	600-700	③	④
2007	3.5	VQ35DE	0.043	15 ①	51 ②	600-700	③	④

NOTE: The Vehicle Emission Control Information label often reflects specification changes made during production. The label figures must be used

①: Before top dead center

② At idle

③ 0.010-0.013 cold

　　0.012-0.016 hot

④ 0.011-0.015 cold

　　0.012-0.017 hot

22140_MURA_C0003

CAPACITIES

Year	Model	Engine Displacement Liters	Engine ID	Engine Oil with Filter (qts.)	Transmission (pts.)	Transfer Case (pts.)	Drive Axle Front (pts.)	Drive Axle Rear (pts.)	Fuel Tank (gal.)	Cooling System (qts.)
2006	Murano	3.5	VQ35DE	4 ①	21.5	0.63	—	1.5	21.6	9.75
2007	Murano	3.5	VQ35DE	4 ①	21.5	0.63	—	1.1	21.6	9.75

NOTE: All capacities are approximate. Add fluid gradually and check to be sure a proper fluid level is obtained.

① Dry engine (Overhaul) 5 quarts

22140_MURA_C0005

FLUID SPECIFICATIONS

Year	Model	Engine Displ. Liters (VIN)	Engine Oil	Man. Trans.	Auto. Trans.	Drive Axle Front	Drive Axle Rear	Trans. Case	Power Steering Fluid	Brake Master Cylinder	Cooling System
2006	Murano	VQ35DE	①	—	②	③	③	③	NISSAN PSF	DOT 3	④
			⑤	—	②	③	③	③	NISSAN PSF	DOT 3	④
2007	Murano	VQ35DE	①	—	②	③	③	③	NISSAN PSF	DOT 3	④
			⑤	—	②	③	③	③	NISSAN PSF	DOT 3	④

DOT: Department Of Transpotation

① API Certification Mark 1

  API grade SG/SH, Energy Conserving I & II or

  API grade SJ or SL, Energy Conserving*1

  API grade SG/SH, Energy Conserving I & II or

  API grade SJ or SL, Energy Conserving

  ILSAC grade GF-I, GF-II & GF-III

  SAE 5W-30

② Nissan CVT Fluid NS-2 2

③ GL-5 80W90

④ Nissan Long Life Antifreeze/Coolant

⑤ API Certification Mark

  API grade SJ or SL, Energy Conserving

  ILSAC grade GF-II & GF-III

  SAE 5W-30

22140_MURA_C0004

VALVE SPECIFICATIONS

Year	Engine Displacement Liters	Engine ID	Seat Angle (deg.)	Face Angle (deg.)	Spring Test Pressure (lbs. @ in.)	Spring Installed Height (in.)	Stem-to-Guide Clearance (in.) Intake	Stem-to-Guide Clearance (in.) Exhaust	Stem Diameter (in.) Intake	Stem Diameter (in.) Exhaust
2006	3.5	VQ35DE	45.15-45.45	45	42.3@1.467	1.457	0.0008-0.0021	0.0012-0.0025	0.2348-0.2354	0.2344-0.2350
2007	3.5	VQ35DE	45.15-45.45	45	45.4@1.457	1.457	0.0008-0.0021	0.0012-0.0025	0.2348-0.2354	0.2344-0.2350

22140_MURA_C0006

CAMSHAFT SPECIFICATIONS

All measurements are given in inches.

Year	Engine Displ. Liters	Engine VIN	Journal Dia.	Brg. Oil Clearance	Shaft End-play	Runout	Lobe Height Intake	Exhaust
2006	3.5	VQ35DE	①	②	0.0045-0.0074	③	1.7663-1.7738	1.7663-1.7738
2007	3.5	VQ35DE	①	②	0.0045-0.0074	③	1.7663-1.7738	1.7663-1.7738

① No.1: 1.0211-1.0218
No.2, No.3, No.4: 0.9230-0.9238

② No.1: 1.0018-1.0034
No.2, No.3, No.4: 0.0014-0.0030

③ Less then 0.001 (0.02 mm)

22140_MURA_C0007

CRANKSHAFT AND CONNECTING ROD SPECIFICATIONS

All measurements are given in inches.

Year	Engine Displacement Liters	Engine ID	Crankshaft Main Brg. Journal Dia.	Main Brg. Oil Clearance	Shaft End-play	Thrust on No.	Connecting Rod Journal Diameter	Oil Clearance	Side Clearance
2006	3.5	VQ35DE	①	0.0014-0.0018	0.0039-0.0098	4	②	0.0013-0.0023	0.0079-0.0138
2007	3.5	VQ35DE	①	0.0014-0.0018	0.0040 0.0098	4	②	0.0013-0.0023	0.0080-0.0138

① There are 24 different grades, ranging from A (2.3612) to 7 (2.3603)

② Grade 0: 0.0591-0.0592. Identification color: black

Grade 1: 0.0592-0.0593. Identification color: brown

Grade 2: 0.0593-0.0594. Identification color: green

22140_MURA_C0008

PISTON AND RING SPECIFICATIONS

All measurements are given in inches.

Year	Engine Displacement Liters	Engine ID	Piston Clearance	Ring Gap Top Comp.	Bottom Comp.	Oil Control	Ring Side Clearance Top Comp.	Bottom Comp.	Oil Control
2006	3.5	VQ35DE	0.0004-0.0012	0.0091-0.0130	0.0130-0.0189	0.0079-0.0197	0.0018-0.0031	0.0012-0.0028	0.0026-0.0053
2007	3.5	VQ35DE	0.0004-0.0012	0.0091-0.0130	0.0130-0.0189	0.0080-0.0200	0.0018-0.0031	0.0012-0.0028	0.0026-0.0053

22140_MURA_C0009

TORQUE SPECIFICATIONS

All readings in ft. lbs.

Year	Engine Displacement Liters	Engine ID	Cylinder Head Bolts	Main Bearing Bolts	Rod Bearing Bolts	Crankshaft Damper Bolts	Driveplate Bolts	Manifold Intake	Manifold Exhaust	Spark Plugs	Oil Pan Drain Plug
2006	3.5	VQ35DE	①	②	③	④	65	⑤	21-24	14-22	25
2007	3.5	VQ35DE	①	②	③	④	61-69	⑤	21-24	14-22	25

① Step 1: 72 ft. lbs.

Step 2: Loosen all bolts completely

Step 3: 29 ft. lbs.

Step 4: +90 degrees

Step 5: +90 degrees

② Step 1: 26 ft. lbs.

Step 2: +90 degrees

③ Step 1: 14 ft. lbs.

Step 2: +90 degrees

④ 33 ft. lbs. +90 degrees

⑤ Step 1: 5 ft. lbs

Step 2: 21 ft. lbs.

22140_MURA_C0010

WHEEL ALIGNMENT

Year	Model		Caster Range (+/-Deg.)	Caster Preferred Setting (Deg.)	Camber Range (+/-Deg.)	Camber Preferred Setting (Deg.)	Toe-in (in.)	Kingpin Inclination (Deg.)
2006	Murano	F	1.75/3.25	+2.50	-1.08/+.42	-0.33	0.02+/-0.04	14.33
		R	—	—	-1.27/-0.27	-0.77	0.126	—
2006	Murano AWD	F	1.83/3.33	+2.58	-1.08/+.42	-0.33	0.02+/-0.04	14.33
		R	—	—	-1.27/-0.27	-0.77	0.126	—
2007	Murano	F	1.75/3.25	+2.50	-2.571428571	-0.33	0.02+/-0.04	14.33
		R	—	—	-1.27/-0.27	-0.77	0.126	—
2007	Murano AWD	F	1.83/3.33	+2.58	-1.08/+.42	-0.33	0.02+/-0.04	14.33
		R	—	—	-1.27/-0.27	-0.77	0.126	—

Note: Specifications are taken with the following Fuel, radiator coolant and engine oil full. Spare tire, jack, hand tools and mats are in designated positions.

22140_MURA_C0011

TIRE, WHEEL AND BALL JOINT SPECIFICATIONS

Year	Model	OEM Tires Standard	OEM Tires Optional	Tire Pressures (psi) Front	Tire Pressures (psi) Rear	Wheel Size	Ball Joint Inspection	Lug Nut Torque (ft. lbs.)
2006	Murano	P235/65SR18	none	33	33	7.5JJ	①	80
2007	Murano	P235/65SR18	none	33	33	7.5JJ	①	90

OEM: Original Equipment Manufacturer

PSI: Pounds Per Square Inch

① 0 (0mm) inches axial end play

22140_MURA_C0012

BRAKE SPECIFICATIONS
All measurements in inches unless noted

Year	Model		Brake Disc			Minimum Lining Thickness	Brake Caliper	
			Original Thickness	Minimum Thickness	Maximum Runout		Bracket Bolts (ft. lbs.)	Mounting Bolts (ft. lbs.)
2006	Murano	F	1.102	1.024	0.0016	0.079	101-129	17-22
		R	0.630	0.551	0.0020	0.079	53-71	28-35
2007	Murano	F	1.102	1.024	0.0016	0.079	101-129	17-22
		R	0.630	0.551	0.0020	0.079	53-71	28-35

F: Front

R: Rear

22140_MURA_C0013

SCHEDULED MAINTENANCE INTERVALS
Nissan—Murano 2006-07

TO BE SERVICED	TYPE OF SERVICE	3.75	7.5	15	22.5	30	37.5	45	52.5	60
Engine oil & filter	R	✓	✓	✓	✓	✓	✓	✓	✓	✓
Brake lines & cables	S/I			✓		✓		✓		✓
Brake pads, discs	I			✓		✓		✓		✓
Driveshaft boots & propeller shaft	L/I			✓		✓		✓		✓
CVT, transfer case and differential fluid	I			✓		✓		✓		✓
Air cleaner filter	R					✓				✓
Drive belt(s) ①	S/I									
Engine coolant ②	R									✓
Spark plugs	R			Platinum plugs, every 105,000 miles						
Cabin air filter	R			✓		✓		✓		✓
Exhaust system	I				✓					✓
Fuel lines	S/I					✓				✓
Steering gear, linkage, axle & suspension parts	I			✓		✓		✓		✓
Tires (rotate)	S/I			every 5,000-6,000 miles						
Vapor lines	S/I					✓				✓

R: Replace   S/I: Service or Inspect   L: Lubricate

① First a 60,000, then every 15,000 miles

② After 60,000, replace every 30,000

FREQUENT OPERATION MAINTENANCE (SEVERE SERVICE)

If a vehicle is operated under any of the following conditions it is considered severe service:

- Extremely dusty areas.

- 50% or more of the vehicle operation is in 32°C (90°F) or higher temperatures, or constant temperatures below 0°C (32°F).

- Prolonged idling (vehicle operation in stop and go traffic).

- Frequent short running periods (engine does not warm to normal operating temperatures).

- Police, taxi, delivery usage or trailer towing usage.

Oil & oil filter: replace every 3750 miles.

Brake pads, discs, drums & linings: service or inspect every 7500 miles.

Driveshaft boots & propeller shaft: service or inspect every 7500 miles.

Exhaust system: service or inspect every 7500 miles.

Steering gear (box) & linkage, (steering damper-4x4), axle & suspension parts: service or inspect every 7500 miles.

Steering linkage ball joints & front suspension ball joints: service or inspect every 7500 miles.

22140_MURA_C0014

PRECAUTIONS

Before servicing any vehicle, please be sure to read all of the following precautions, which deal with personal safety, prevention of component damage, and important points to take into consideration when servicing a motor vehicle:

• Never open, service or drain the radiator or cooling system when the engine is hot; serious burns can occur from the steam and hot coolant.

• Observe all applicable safety precautions when working around fuel. Whenever servicing the fuel system, always work in a well-ventilated area. Do not allow fuel spray or vapors to come in contact with a spark, open flame, or excessive heat (a hot drop light, for example). Keep a dry chemical fire extinguisher near the work area. Always keep fuel in a container specifically designed for fuel storage; also, always properly seal fuel containers to avoid the possibility of fire or explosion. Refer to the additional fuel system precautions later in this section.

• Fuel injection systems often remain pressurized, even after the engine has been turned OFF. The fuel system pressure must be relieved before disconnecting any fuel lines. Failure to do so may result in fire and/or personal injury.

• Brake fluid often contains polyglycol ethers and polyglycols. Avoid contact with the eyes and wash your hands thoroughly after handling brake fluid. If you do get brake fluid in your eyes, flush your eyes with clean, running water for 15 minutes. If eye irritation persists, or if you have taken brake fluid internally, IMMEDIATELY seek medical assistance.

• The EPA warns that prolonged contact with used engine oil may cause a number of skin disorders, including cancer. You should make every effort to minimize your exposure to used engine oil. Protective gloves should be worn when changing oil. Wash your hands and any other exposed skin areas as soon as possible after exposure to used engine oil. Soap and water, or waterless hand cleaner should be used.

• All new vehicles are now equipped with an air bag system, often referred to as a Supplemental Restraint System (SRS) or Supplemental Inflatable Restraint (SIR) system. The system must be disabled before performing service on or around system components, steering column, instrument panel components, wiring and sensors. Failure to follow safety and disabling procedures could result in accidental air bag deployment, possible personal injury and unnecessary system repairs.

• Always wear safety goggles when working with, or around, the air bag system. When carrying a non-deployed air bag, be sure the bag and trim cover are pointed away from your body. When placing a non-deployed air bag on a work surface, always face the bag and trim cover upward, away from the surface. This will reduce the motion of the module if it is accidentally deployed. Refer to the additional air bag system precautions later in this section.

• Clean, high quality brake fluid from a sealed container is essential to the safe and proper operation of the brake system. You should always buy the correct type of brake fluid for your vehicle. If the brake fluid becomes contaminated, completely flush the system with new fluid. Never reuse any brake fluid. Any brake fluid that is removed from the system should be discarded. Also, do not allow any brake fluid to come in contact with a painted surface; it will damage the paint.

• Never operate the engine without the proper amount and type of engine oil; doing so WILL result in severe engine damage.

• Timing belt maintenance is extremely important. Many models utilize an interference-type, non-freewheeling engine. If the timing belt breaks, the valves in the cylinder head may strike the pistons, causing potentially serious (also time-consuming and expensive) engine damage. Refer to the maintenance interval charts for the recommended replacement interval for the timing belt, and to the timing belt section for belt replacement and inspection.

• Disconnecting the negative battery cable on some vehicles may interfere with the functions of the on-board computer system(s) and may require the computer to undergo a relearning process once the negative battery cable is reconnected.

• When servicing drum brakes, only disassemble and assemble one side at a time, leaving the remaining side intact for reference.

• Only an MVAC-trained, EPA-certified automotive technician should service the air conditioning system or its components.

BRAKES

GENERAL INFORMATION

The Anti-lock Brake System (ABS) module simultaneously manages the anti-lock braking, traction control and engine control systems to maintain vehicle control during deceleration and acceleration.

ANTI-LOCK BRAKE SYSTEM (ABS)

When the ignition switch is in the RUN position, the module carries out a preliminary electrical checks and, at approximately 12 MPH (20 km/h), the hydraulic pump motor is turned on for approximately one half-second. Any malfunction of the anti-lock brake system disables the traction control and stability assist (if equipped) and the anti-lock brake warning indicator illuminates. However, the power-assist braking system functions normally.

BRAKES

BLEEDING THE BRAKE SYSTEM

BLEEDING PROCEDURE

> ❋❋ **WARNING**
>
> Use of any other than the approved DOT 3 brake fluid will cause permanent damage to brake components and will render the brakes inoperative. Failure to follow these instructions may result in personal injury.

• Brake fluid contains polyglycol ethers and polyglycols. Avoid contact with eyes. Wash hands thoroughly after handling. If brake fluid contacts eyes, flush eyes with running water for 15 minutes. Get medical attention if irritation persists. If taken internally, drink water and induce vomiting. Get medical attention immediately. Failure to follow these instructions may result in personal injury.

> ❋❋ **WARNING**
>
> Do not allow the brake master cylinder reservoir to run dry during the bleeding operation. Keep the master cylinder reservoir filled with the specified brake fluid. Never reuse the brake fluid that has been drained from the hydraulic system.

> ❋❋ **WARNING**
>
> Brake fluid is harmful to painted and plastic surfaces. If brake fluid is spilled onto a painted or plastic surface, immediately wash it with water.

> ❋❋ **WARNING**
>
> When any part of the hydraulic system has been disconnected or a new component is installed, air may enter the system, causing spongy brake pedal action. This requires the bleeding of the hydraulic system after it has been correctly connected.

BRAKES

FRONT DISC BRAKES

> ❋❋ **CAUTION**
>
> Dust and dirt accumulating on brake parts during normal use may contain asbestos fibers from production or aftermarket brake linings. Breathing excessive concentrations of asbestos fibers can cause serious bodily harm. Exercise care when servicing brake parts. Do not sand or grind brake lining unless equipment used is designed to contain the dust residue. Do not clean brake parts with compressed air or by dry brushing. Cleaning should be done by dampening the brake components with a fine mist of water, then wiping the brake components clean with a dampened cloth. Dispose of cloth and all residue containing asbestos fibers in an impermeable container with the appropriate label. Follow practices prescribed by the Occupational Safety and Health Administration (OSHA) and the Environmental Protection Agency (EPA) for the handling, processing, and disposing of dust or debris that may contain asbestos fibers.

BRAKE CALIPER

REMOVAL & INSTALLATION

See Figure 1.

1. Before servicing the vehicle, refer to the Precautions Section.
2. Raise and support the vehicle safely, remove the tire and wheel assembly.
3. Drain brake fluid.

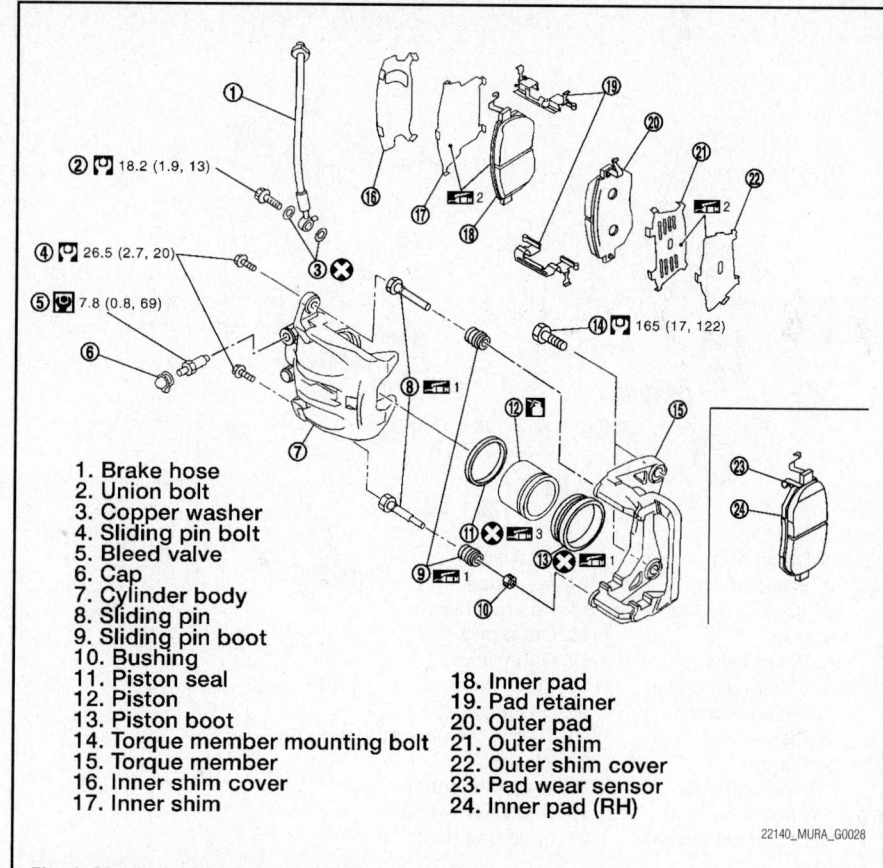

1. Brake hose
2. Union bolt
3. Copper washer
4. Sliding pin bolt
5. Bleed valve
6. Cap
7. Cylinder body
8. Sliding pin
9. Sliding pin boot
10. Bushing
11. Piston seal
12. Piston
13. Piston boot
14. Torque member mounting bolt
15. Torque member
16. Inner shim cover
17. Inner shim
18. Inner pad
19. Pad retainer
20. Outer pad
21. Outer shim
22. Outer shim cover
23. Pad wear sensor
24. Inner pad (RH)

22140_MURA_G0028

Fig. 1 Disc brake caliper and related components

4. Remove union bolts and torque member bolts, and remove brake caliper assembly.

To install:

5. Install caliper assembly to the vehicle, and tighten bolts to122 ft. lbs. (165 Nm).
6. Install brake hose to caliper assembly and tighten union bolts to 13 ft. lbs. (18.2 Nm).

➡Do not reuse the copper washer for union bolts. Attach brake hose to the brake hose mounting boss.

7. Refill the brake fluid to the proper level.
8. Install the tires to the vehicle.

DISC BRAKE PADS

REMOVAL & INSTALLATION

1. Before servicing the vehicle, refer to the Precautions Section.
2. Raise and support the vehicle safely. Remove the tire and wheel assembly.
3. Remove the sliding pin bolt (top).

4. Suspend cylinder body with a wire, and remove pads, pad retainers, shims from torque member.

To install:

5. Apply PBC (Poly Butyl Cuprysil) grease or silicon based grease to the rear of the pad and to both sides of the shim, and attach the inner shim and shim cover to the inner pad, and the outer shim and outer shim cover to the outer pad.

6. Attach the pad retainer and pad to the torque member.
7. Push the piston in so that the pad is attached and attach the cylinder body to the torque member.
8. Install the sliding pin bolt (one on top) and tighten to 20 ft. lbs. (26.6 Nm).
9. Check brake for drag.
10. Install the tires.

BRAKES

BRAKE CALIPER

REMOVAL & INSTALLATION
See Figure 2.

1. Before servicing the vehicle, refer to the Precautions Section.
2. Raise and support the vehicle safely. Remove the tire and wheel assembly.
3. Drain brake fluid.

REAR DISC BRAKES

4. Remove union bolts and torque member bolts, and remove brake caliper assembly.

To install:

5. Install disc rotor.
6. Install caliper assembly to the vehicle, and tighten bolts to 62 ft. lbs. (84.3 Nm).
7. Install brake hose to caliper assembly and tighten union bolts to 13 ft. lbs. (18.2 Nm).

➡**Do not reuse the copper washer for union bolts. Attach brake hose to the brake hose mounting boss.**

8. Refill new brake fluid.
9. Install the tires to the vehicle.

DISC BRAKE PADS

REMOVAL & INSTALLATION

1. Before servicing the vehicle, refer to the Precautions Section.
2. Raise and support the vehicle safely. Remove the tire and wheel assembly.
3. Remove sliding pin bolt (top).
4. Suspend cylinder body with a wire, and remove pads, pad retainers, shims from torque member.

To install:

5. Apply PBC (Poly Butyl Cuprysil) grease or silicon based grease to the rear of the pad and to both sides of the shim, and attach the inner shim and shim cover to the inner pad, and the outer shim and outer shim cover to the outer pad.
6. Attach the pad retainer and pad to the torque member.
7. Push the piston in so that the pad is attached and attach the cylinder body to the torque member.
8. Install the sliding pin bolt (one on top) and tighten to 32 ft. lbs. (43.1 Nm).
9. Check brake for drag.
10. Install the tires.

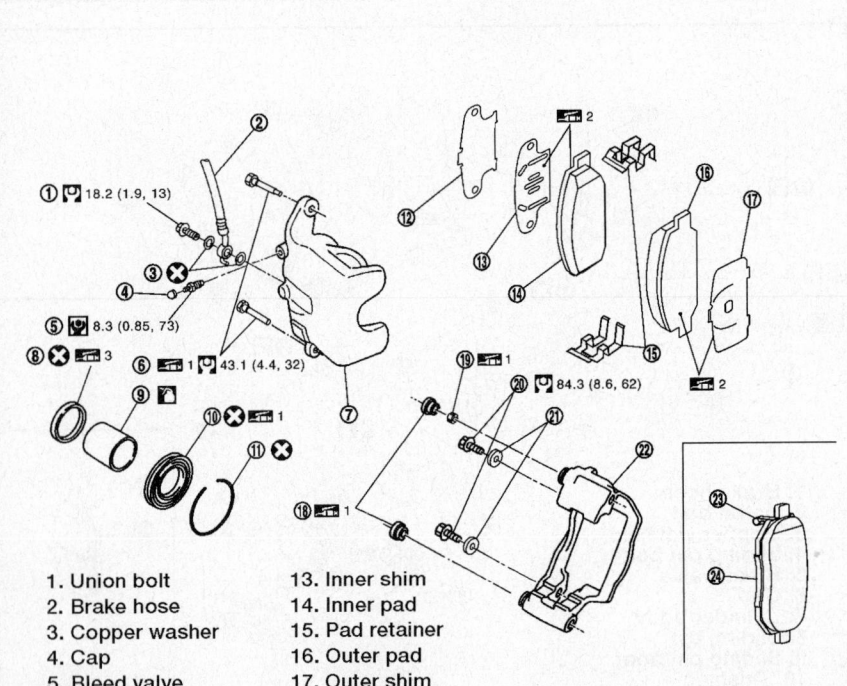

① 🔧 18.2 (1.9, 13)
③ ✕ ⬡
④
⑤ 🔧 8.3 (0.85, 73)
⑧ ✕ 🔩 3
⑥ 🔩 1 43.1 (4.4, 32)
⑨ 🔧
⑩ ✕ 🔩 1
⑪ ✕
⑱ 🔩 1
⑲ 🔩 1
⑳ 🔧 84.3 (8.6, 62)
㉑
🔩 2

1. Union bolt
2. Brake hose
3. Copper washer
4. Cap
5. Bleed valve
6. Sliding pin bolt
7. Cylinder body
8. Piston seal
9. Piston
10. Piston boot
11. Retaining ring
12. Inner shim cover
13. Inner shim
14. Inner pad
15. Pad retainer
16. Outer pad
17. Outer shim
18. Slide pin boot
19. Bushing
20. Torque member mounting bolt
21. Washer
22. Torque member
23. Pad wear sensor
24. Inner pad (RH)

22140_MURA_G0029

Fig. 2 Rear disc brake caliper and related components

BRAKES

PARKING BRAKE

PARKING BRAKE CABLES

ADJUSTMENT

See Figure 3.

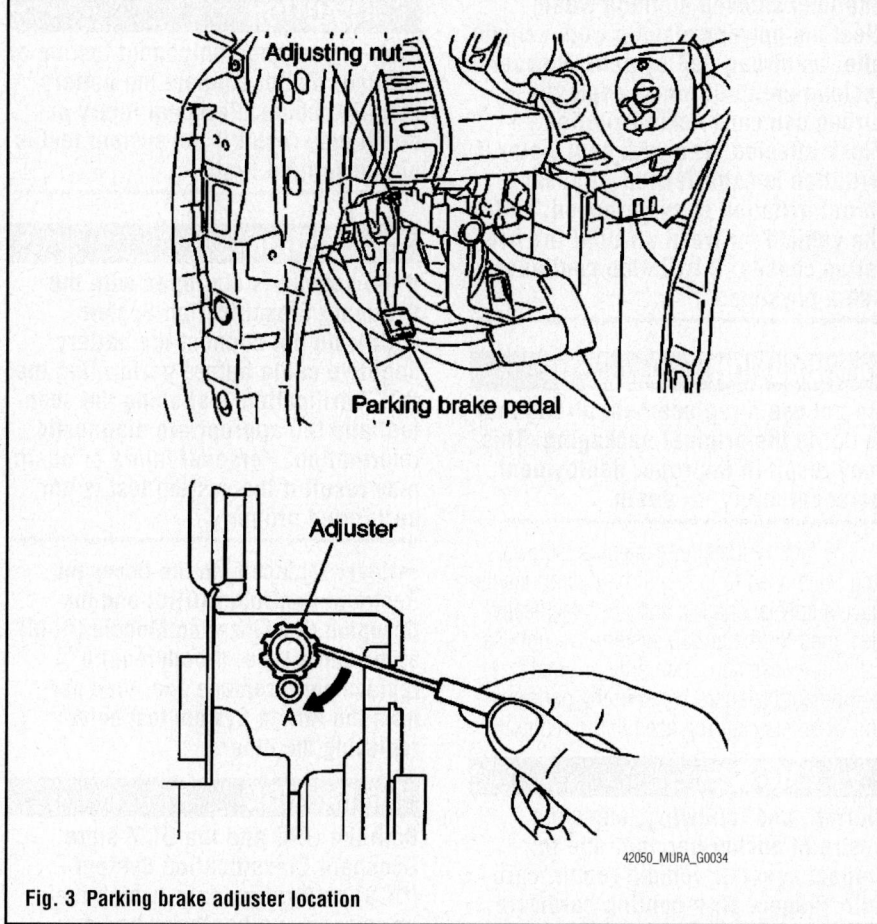

Fig. 3 Parking brake adjuster location

1. Before servicing the vehicle, refer to the Precautions Section.
2. Raise and safely support the vehicle.
3. Remove the rear wheels.
4. Insert a deep socket wrench to rotate the adjusting nut and loosen the cable. Then, return the pedal.
5. Using a couple of lug nuts, install them on the rotor to prevent it from tilting.
6. Remove the adjusting hole plug on the rotor. Using a flat bladed tool, turn the adjuster in direction "A" until the rotor is locked.
7. After locking the rotor turn the adjuster in the opposite direction (5–6 notches).

8. Rotate the rotor to make sure there is no drag. Install the adjusting plug cap.
9. Adjust the cable as follows. Operate the pedal 10 or more times with a force of 110 lbs.

10. Rotate the adjusting nut with a deep socket to adjust the pedal stroke.

➡**Do not reuse the adjusting nut after removing it.**

11. When the parking brake cable is operated with a force of 44 lbs, make sure the stroke is within the specified number of notches (3–4). Listen and count the ratcheting "click."
12. With the pedal completely returned, make sure there is no drag on the rear brake.
13. Correct as required.

PARKING BRAKE SHOES

ADJUSTMENT

1. Adjust the parking brake pedal stroke to 4–5 clicks fully depressed. Insert a deep socket wrench to rotate the adjusting nut and loosen the cable sufficiently. Then, return pedal.
2. Remove the wheels.
3. Using a lug nut, secure the rotor to hub to prevent it from tilting.
4. Remove the adjusting hole plug. Using a flat-bladed screwdriver, turn the adjuster clockwise until the rotor is locked. After locking, turn the adjuster in the opposite direction by 5 or 6 notches.
5. Rotate the rotor to make sure that there is no drag. Then install the adjusting hole plug.
6. After adjusting the clearance of rear shoes, with no drag on rear brake, adjust the cable as follows:
   a. Operate the pedal 10 or more times with a force of 490 N (110 lbs.).
   b. Depress the pedal until a deep socket can be inserted.
   Insert the deep socket, and rotate the adjusting nut to adjust the pedal stroke.

➡**Do not reuse the adjusting nut.**

   c. When the parking brake pedal is operated with a force of 200 N (45 lbs.), make sure the stroke is 4–5 notches. (Check it by listening and counting the ratchet clicks.)
   d. With the parking brake pedal completely returned, make sure there is no drag on the rear brake.
7. Perform the parking brake break-in operation as follows: Safely, drive forward at approximately 40 km/h (25 MPH) with the parking brake set with a force of approx. 200 N (45 lbs.) for about 30 seconds.
8. After the break-in operation, check the pedal stroke of parking brake. Readjust if necessary.

➡**To prevent lining from getting too hot, allow a cool off period of approximately 5 minutes after every break-in operation.**

GENERAL INFORMATION

✴ CAUTION

Some vehicles are equipped with an air bag system. The system must be disarmed before performing service on, or around, system components, the steering column, instrument panel components, wiring and sensors. Failure to follow the safety precautions and the disarming procedure could result in accidental air bag deployment, possible injury and unnecessary system repairs.

SERVICE PRECAUTIONS

✴ CAUTION

Disconnect and isolate the battery negative cable before beginning any airbag system component diagnosis, testing, removal, or installation procedures. Allow system capacitor to discharge for two minutes before beginning any component service. This will disable the airbag system. Failure to disable the airbag system may result in accidental airbag deployment, personal injury, or death.

✴ CAUTION

Do not place an intact undeployed airbag face down on a solid surface. The airbag will propel into the air if accidentally deployed and may result in personal injury or death.

✴ CAUTION

When carrying or handling an undeployed airbag, the trim side (face) of the airbag should be pointing towards the body to minimize possibility of injury if accidental deployment occurs. Failure to do this may result in personal injury or death.

✴ CAUTION

Replace airbag system components with OEM replacement parts. Substitute parts may appear interchangeable, but internal differences may result in inferior occupant protection.

Failure to do so may result in occupant personal injury or death.

✴ CAUTION

Wear safety glasses, rubber gloves, and long sleeved clothing when cleaning powder residue from vehicle after an airbag deployment. Powder residue emitted from a deployed airbag can cause skin irritation. Flush affected area with cool water if irritation is experienced. If nasal or throat irritation is experienced, exit the vehicle for fresh air until the irritation ceases. If irritation continues, see a physician.

✴ CAUTION

Do not use a replacement airbag that is not in the original packaging. This may result in improper deployment, personal injury, or death.

The factory installed fasteners, screws and bolts used to fasten airbag components have a special coating and are specifically designed for the airbag system. Do not use substitute fasteners. Use only original equipment fasteners listed in the parts catalog when fastener replacement is required.

✴ CAUTION

During, and following, any child restraint anchor service, due to impact event or vehicle repair, carefully inspect all mounting hardware, tether straps, and anchors for proper installation, operation, or damage. If a child restraint anchor is found damaged in any way, the anchor must be replaced. Failure to do this may result in personal injury or death.

✴ WARNING

Deployed and non-deployed airbags may or may not have live pyrotechnic material within the airbag inflator.

✴ WARNING

Do not dispose of driver/passenger/curtain airbags or seat belt tensioners unless you are sure of complete deployment. Refer to the Hazardous Substance Control System for proper disposal.

✴ WARNING

Dispose of deployed airbags and tensioners consistent with state, provincial, local, and federal regulations.

✴ CAUTION

After any airbag component testing or service, do not connect the battery negative cable. Personal injury or death may result if the system test is not performed first.

✴ CAUTION

If the vehicle is equipped with the Occupant Classification System (OCS), do not connect the battery negative cable before performing the OCS Verification Test using the scan tool and the appropriate diagnostic information. Personal injury or death may result if the system test is not performed properly.

➡ Never replace both the Occupant Restraint Controller (ORC) and the Occupant Classification Module (OCM) at the same time. If both require replacement, replace one, then perform the Airbag System test before replacing the other.

✴ CAUTION

Both the ORC and the OCM store Occupant Classification System (OCS) calibration data, which they transfer to one another when one of them is replaced. If both are replaced at the same time, an irreversible fault will be set in both modules and the OCS may malfunction and cause personal injury or death.

✴ CAUTION

If equipped with OCS, the Seat Weight Sensor is a sensitive, calibrated unit and must be handled carefully. Do not drop or handle roughly. If dropped or damaged, replace with another sensor. Failure to do so may result in occupant injury or death.

If equipped with OCS, the front passenger seat must be handled carefully as well. When removing the seat, be careful when setting on floor not to drop. If dropped, the sensor may be inoperative, could result in occupant injury, or possibly death.

✳✳ CAUTION

If equipped with OCS, when the passenger front seat is on the floor, no one should sit in the front passenger seat. This uneven force may damage the sensing ability of the seat weight

sensors. **If sat on and damaged, the sensor may be inoperative, could result in occupant injury, or possibly death.**

DISARMING THE SYSTEM

To disarm the **SRS** system turn the ignition switch to the **OFF** position. Then, disconnect both battery cables starting

with the negative cable first and wait at least 3 minutes after the cables are disconnected.

ARMING THE SYSTEM

To rearm the **SRS** system, turn the ignition switch to the **OFF** position. Connect both battery cables starting with the positive cable first.

DRIVETRAIN

AUTOMATIC TRANSAXLE ASSEMBLY

REMOVAL & INSTALLATION

See Figures 4 through 9.

The engine and transaxle are removed as an assembly from the vehicle.

1. Before servicing the vehicle, refer to the Precautions Section.

2. Disconnect the negative battery cable. Remove the engine under cover.

3. Remove the air guide. Remove the exhaust front tube.

4. Remove dust cover from converter housing part.

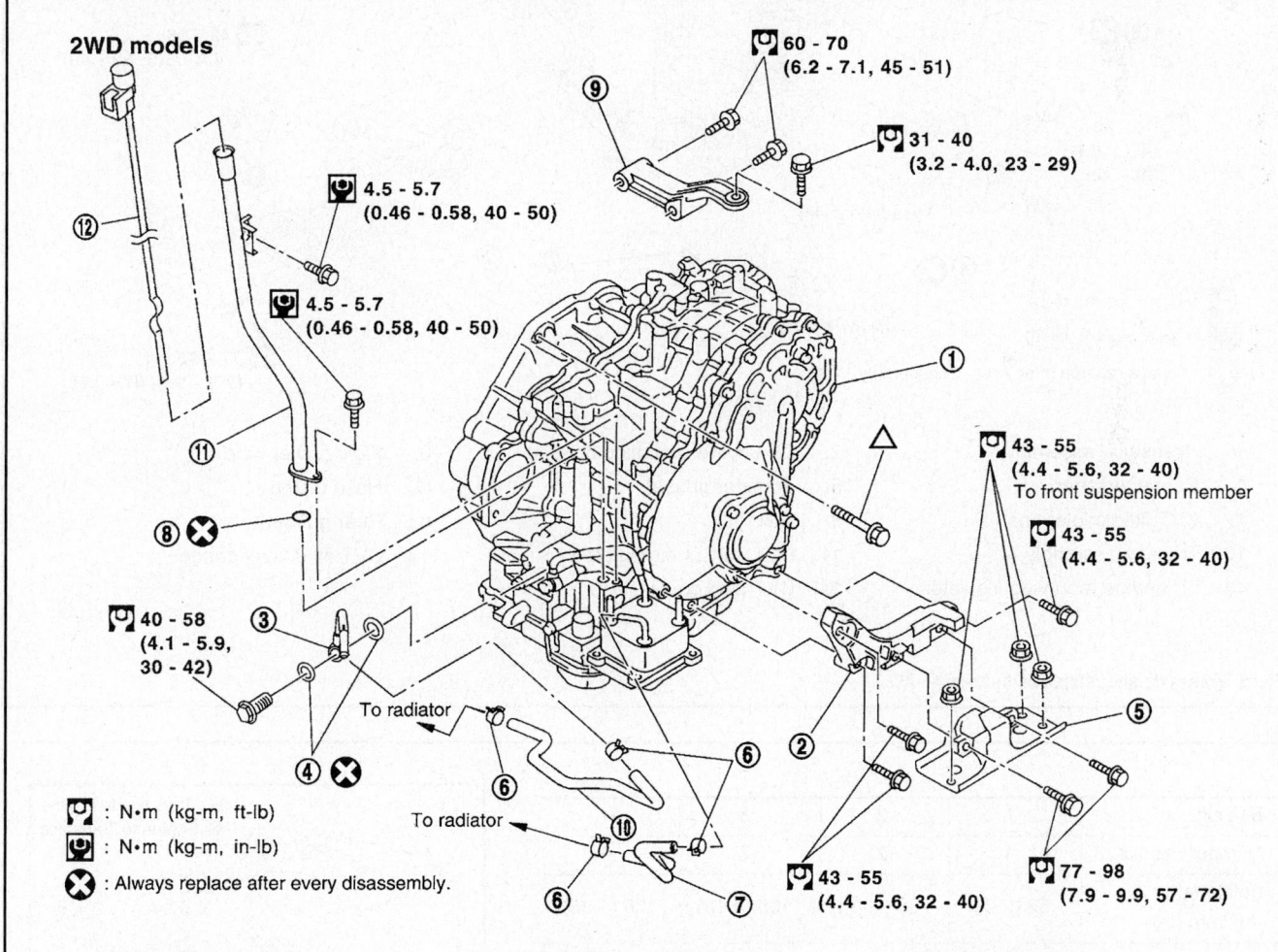

1. Transaxle assembly
2. LH engine mounting bracket
3. Fluid cooler tube
4. Copper washer
5. LH engine mounting insulator
6. Hose clamp
7. CVT fluid cooler hose
8. O-ring
9. Rear gusset
10. CVT fluid cooler hose
11. CVT fluid charging pipe
12. CVT fluid level gauge

42356-MURA-G53

Fig. 4 Transaxle and related components—2WD

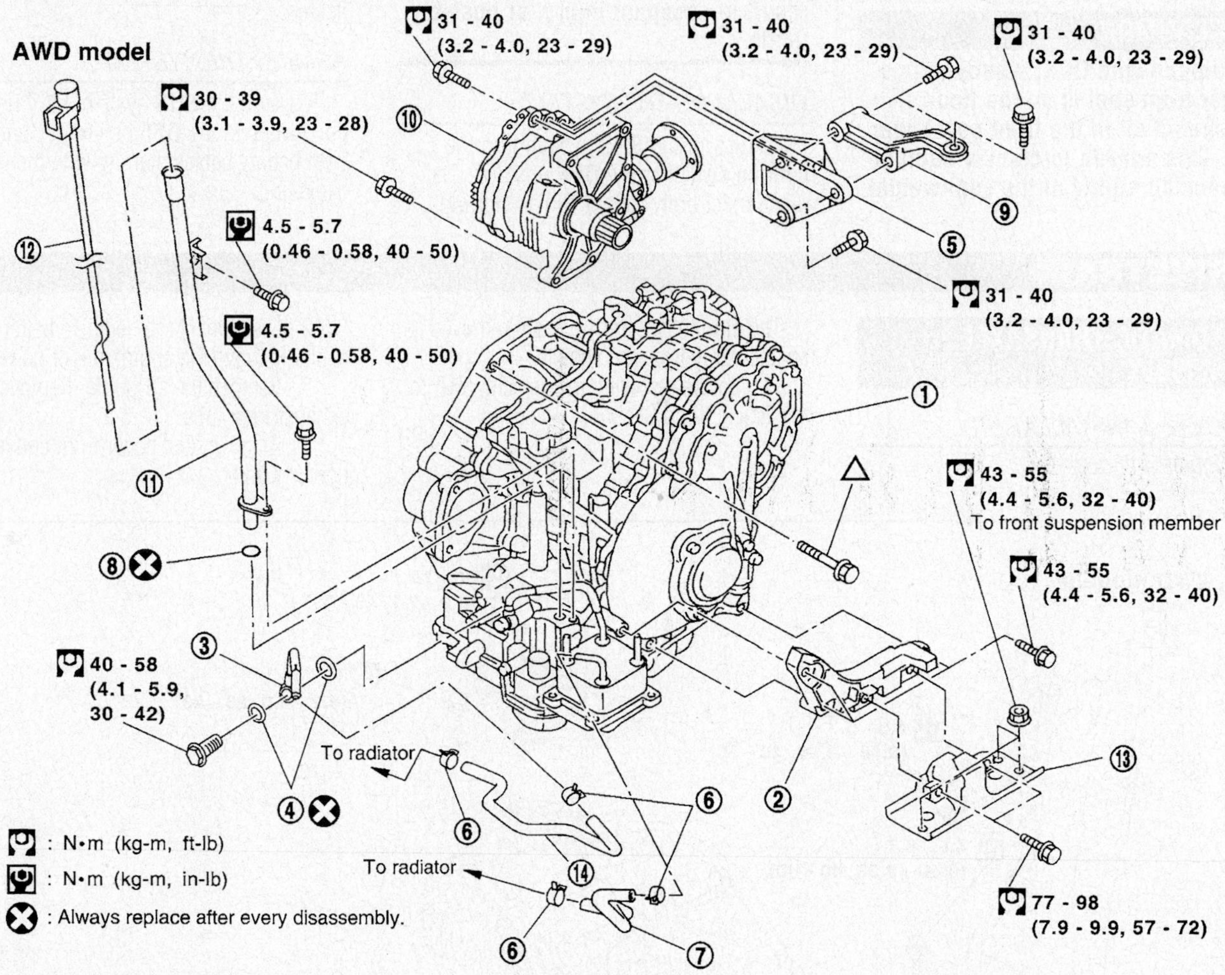

AWD model

⊙ 31 - 40
(3.2 - 4.0, 23 - 29)

⊙ 31 - 40
(3.2 - 4.0, 23 - 29)

⊙ 31 - 40
(3.2 - 4.0, 23 - 29)

⊙ 30 - 39
(3.1 - 3.9, 23 - 28)

⊙ 31 - 40
(3.2 - 4.0, 23 - 29)

⊙ 4.5 - 5.7
(0.46 - 0.58, 40 - 50)

⊙ 4.5 - 5.7
(0.46 - 0.58, 40 - 50)

⊙ 43 - 55
(4.4 - 5.6, 32 - 40)
To front suspension member

⊙ 43 - 55
(4.4 - 5.6, 32 - 40)

⊙ 40 - 58
(4.1 - 5.9, 30 - 42)

To radiator

To radiator

⊙ 77 - 98
(7.9 - 9.9, 57 - 72)

⊙ : N·m (kg-m, ft-lb)

⊙ : N·m (kg-m, in-lb)

⊗ : Always replace after every disassembly.

1. Transaxle assembly
2. LH engine mounting bracket
3. Fluid cooler tube
4. Copper washer
5. Transfer gusset
6. Hose clamp
7. CVT fluid cooler hose
8. O-ring
9. Rear gusset
10. Transfer assembly
11. CVT fluid charging pipe
12. CVT fluid level gauge
13. LH engine mounting insulator
14. CVT fluid cooler hose

42356-MURA-G54

Fig. 5 Transaxle and related components—AWD

Bolt No.	1	2	3	4
Number of bolts	1	2	2	4
Bolt length "ℓ" mm (in)	52 (2.05)	36 (1.42)	105 (4.13)	35 (1.38)
Tightening torque N-m (kg-m, ft-lb)	70 - 79 (7.1 - 8.0, 51 - 58)			42-52 (4.3 - 5.3, 31 - 38)

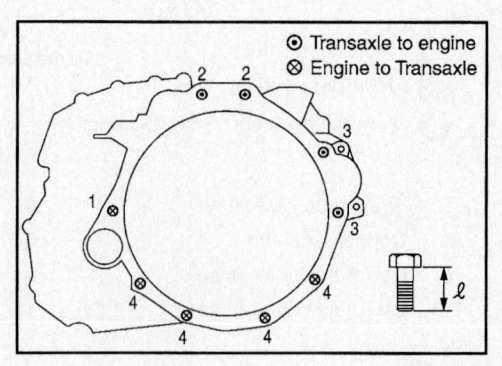

⊙ Transaxle to engine
⊗ Engine to Transaxle

42356-MURA-G55

Fig. 6 Transaxle bolt location and torque sequence

Fig. 7 Transaxle bolt removal location— 2WD

5. Turn crankshaft clockwise and remove the four tightening nuts for drive plate and torque converter.

6. Remove the four lower transaxle retaining bolts, 2WD vehicles. Remove the six lower transaxle retaining bolts, AWD vehicles. See illustration for location.

7. Remove the engine and transaxle from the vehicle, as an assembly.

8. Remove the halfshaft.

➥**Be sure to replace the new differential side oil seal every removal of halfshaft.**

9. Remove the transfer case gusset. (AWD vehicles)

10. Remove the transfer case assembly.

➥**Be sure to replace the new differential side oil seal (converter housing side only) whenever the transfer case is removed.**

11. Remove the filler pipe. Discard the O-ring.

12. Disconnect the harness connector and wire harness.

13. Remove the Crankshaft Position Sensor (POS) sensor, from engine assembly.

14. Remove the starter.

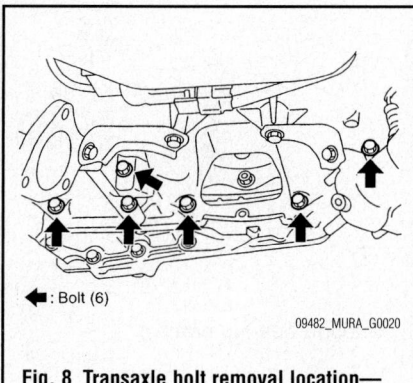

Fig. 8 Transaxle bolt removal location— AWD

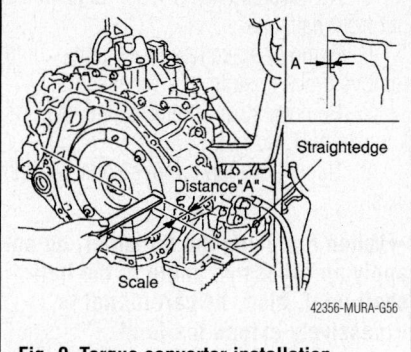

Fig. 9 Torque converter installation measurement

15. Remove CVT fluid cooler valve assembly. (With CVT fluid cooler tube assembly and heater hose).

16. Install the slinger to transaxle assembly.

17. Remove the rear gusset.

18. Remove the left engine mounting bracket and the left engine mounting insulator.

19. Remove the front suspension member from the engine/transaxle assembly.

20. Remove the transaxle assembly bolts.

21. Separate the transaxle from the engine with a hoist. Secure torque converter to prevent it from dropping.

➥**After installing a torque converter to a transaxle, be sure to check dimension A to ensure it is within the reference value limit. Specification should be 0.55 inch or more.**

To install:

22. Installation is the reverse of the removal. Note the following:
- Screw and set the locator into the stud bolts for the torque converter locate
- Rotate the torque converter to allow the locator to go down

- Rotate the drive plate so that the hole of the drive plate locator faces down
- Installing transaxle assembly from engine assembly with a hoist
- When installing fluid cooler tube to transaxle assembly, transaxle assembly the part with the tube aligned with the rib
- When installing the CVT fluid cooler valve assembly to the engine, torque the bolts to: 28–32 Nm (21–23 ft. lbs.)
- Align the positions of tightening nuts for drive plate with those of the torque converter, and temporarily tighten the nuts. Then, tighten the nuts to the specified torque.
- Install POS sensor
- After completing installation, check for fluid leakage, fluid level, and the positions of CVT
- When replacing the CVT assembly, erase EEP ROM in TCM

FRONT HALFSHAFT

REMOVAL & INSTALLATION

Left Side

See Figure 10.

1. Before servicing the vehicle, refer to the Precautions Section.

2. Raise and support the vehicle safely.

3. Remove the tire and wheel assembly.

4. Remove wheel sensor from steering knuckle.

5. Remove cotter pin. Then remove lock nut from halfshaft.

6. Remove brake hose lock plate. Then remove brake hose from strut assembly.

7. Remove strut assembly and steering knuckle bolt and nut.

8. Using a puller, remove halfshaft from steering knuckle.

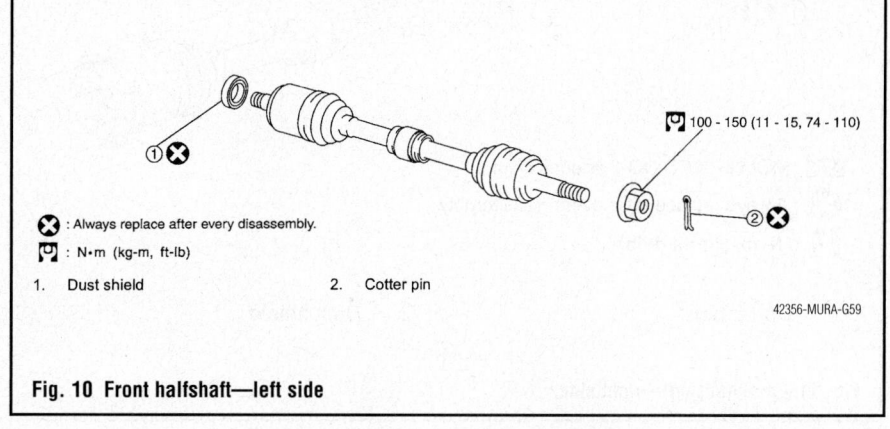

✕ : Always replace after every disassembly.

🔧 : N•m (kg-m, ft-lb)

1. Dust shield   2. Cotter pin

Fig. 10 Front halfshaft—left side

➡**When removing the halfshaft, do not apply an excessive angle to the half-shaft joint. Also, be careful not to excessively extend the joint.**

9. Remove halfshaft from the transaxle, using special tool KV40107500, or equivalent.

To install:

10. Installation is the reverse of the removal procedure.

11. Be sure to replace the differential side oil seal.

12. Be sure to replace all non reusable components with new ones.

➡**In order to prevent damage to the transaxle side oil seal, first install an oil seal protector tool onto the oil seal before inserting the halfshaft. Slide the halfshaft into the slide joint and tap with a hammer, to install securely.**

13. Be sure that the circlip is secured properly.

Right Side

2WD Models

See Figure 11.

1. Before servicing the vehicle, refer to the Precautions Section.

2. Raise and support the vehicle safely.

3. Remove the tire and wheel assembly.

4. Remove wheel sensor from steering knuckle.

5. Remove cotter pin. Then remove lock nut from halfshaft.

6. Remove brake hose lock plate. Then remove brake hose from strut assembly.

7. Remove strut assembly and steering knuckle bolt and nut.

8. Using a puller, remove halfshaft from axle.

➡**When removing the halfshaft, do not apply an excessive angle to the half-shaft joint. Also, be careful not to excessively extend the joint.**

9. Remove support bearing bolts, and pull halfshaft from transaxle. Pry off half-shaft from transaxle.

To install:

10. Installation is the reverse of the removal procedure.

11. Be sure to replace the differential side oil seal.

12. Be sure to replace all non reusable components with new ones.

AWD Models

See Figure 11.

1. Before servicing the vehicle, refer to the Precautions Section.

2. Raise and support the vehicle safely.

3. Remove the tire and wheel assembly.

4. Remove the wheel sensor from steering knuckle.

5. Remove the brake hose lock plate. Then remove brake hose from strut assembly.

6. Remove the cotter pin. Then remove the lock nut from halfshaft.

7. Remove the strut assembly and steering knuckle bolt and nut.

8. Using a puller, remove the halfshaft from the axle.

➡**When removing the halfshaft, do not apply an excessive angle to the half-shaft joint. Also, be careful not to excessively extend the joint.**

9. Remove the halfshaft from the transaxle, using special tool KV40107500, or equivalent.

To install:

10. Installation is the reverse of the removal procedure.

11. Be sure to replace the differential side oil seal.

12. Be sure to replace all non reusable components with new ones.

REAR AXLE SHAFT, BEARING & SEAL

REMOVAL & INSTALLATION

See Figure 12.

1. Before servicing the vehicle, refer to the Precautions Section.

2. Raise and support the vehicle safely.

3. Remove the tire and wheel assembly.

4. Remove the wheel sensor from axle. Do not pull on the wheel sensor harness.

5. Remove the cotter pin. Then remove lock nut from halfshaft.

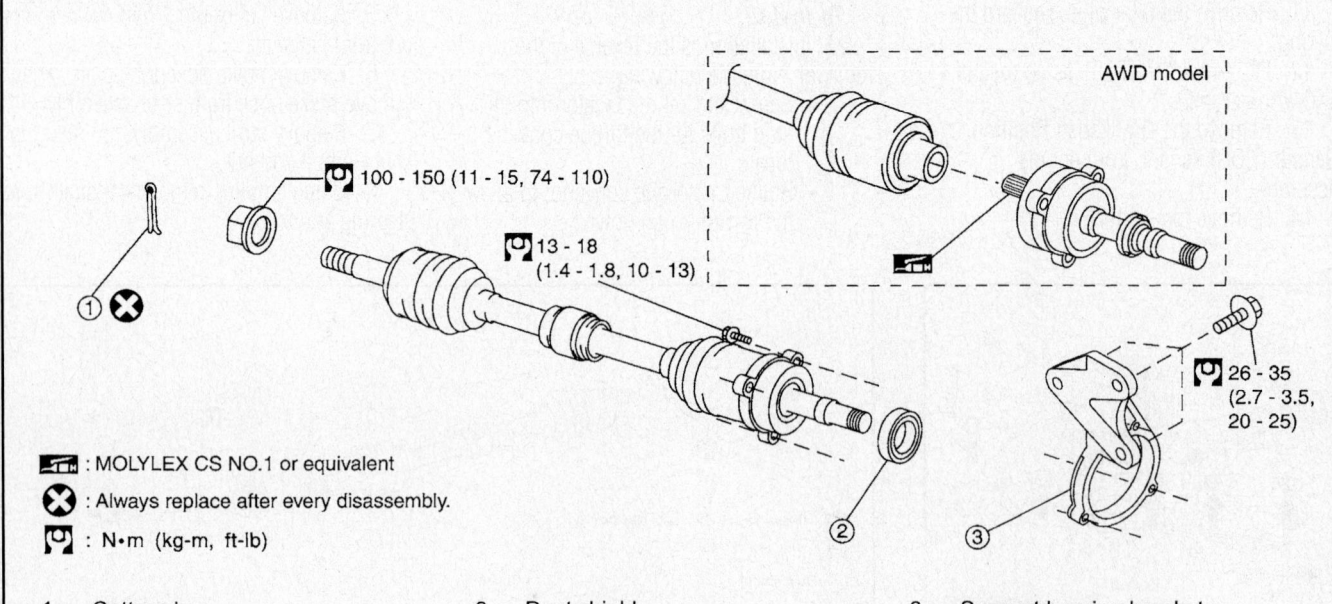

100 - 150 (11 - 15, 74 - 110)

13 - 18 (1.4 - 1.8, 10 - 13)

AWD model

26 - 35 (2.7 - 3.5, 20 - 25)

: MOLYLEX CS NO.1 or equivalent

: Always replace after every disassembly.

: N•m (kg-m, ft-lb)

1. Cotter pin   2. Dust shield   3. Support bearing bracket

42356-MURA-G60

Fig. 11 Front halfshaft—right side

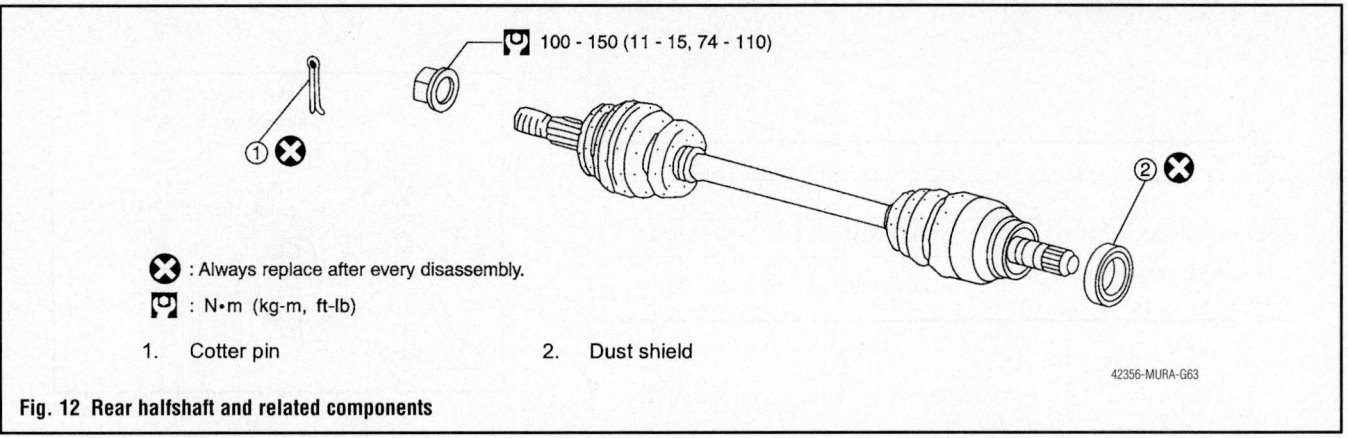

100 - 150 (11 - 15, 74 - 110)

⊗ : Always replace after every disassembly.

⊡ : N•m (kg-m, ft-lb)

1.   Cotter pin   2.   Dust shield

42356-MURA-G63

Fig. 12 Rear halfshaft and related components

6. Remove the parking cable and parking brake shoe from back plate.

7. Remove the wheel hub and bearing assembly bolts, then remove wheel hub and bearing assembly from axle.

8. Use a wheel wrench or other tool to remove halfshaft from final drive.

To install:

9. Installation is the reverse of the removal procedure.

10. Be sure to replace all non reusable components with new ones.

➡**In order to prevent damage to the final drive oil seal, first install an oil seal protector tool onto the oil seal before inserting the halfshaft. Slide the halfshaft into the slide joint and tap with a hammer, to install securely.**

11. Be sure that the circlip is secured properly.

REAR PINION SEAL

REMOVAL & INSTALLATION

1. Before servicing the vehicle, refer to the Precautions Section.

2. Raise and support the vehicle safely.

3. Remove the propeller shaft.

4. Put a mark on the end of the drive pinion corresponding to the position mark on the final drive companion flange.

5. Using the drive pinion flange wrench, Remove companion flange nut.

6. Using the puller, remove the companion flange.

7. Using the side bearing outer race puller, remove front oil seal.

To install:

8. Apply multi-purpose grease to sealing lips of oil seal. Press front oil seal into carrier with tool.

9. Align the matching mark of drive pinion with the matching mark of companion flange, then install the companion flange.

10. Apply oil or grease on the screw part of drive pinion and the seating surface of companion flange nut.

11. Install the companion flange nut with tool.

➡**Never reuse the companion flange nut, always use a new one.**

12. Install the propeller shaft.

TRANSFER CASE ASSEMBLY

REMOVAL & INSTALLATION

See Figures 13 and 14.

The engine and transaxle assembly must first be removed from the vehicle before the transfer case can be removed.

1. Before servicing the vehicle, refer to the Precautions Section.

2. Remove the engine and transaxle assembly, from the vehicle.

3. Remove the gusset mounting bolts, and then remove gusset from engine and transaxle.

4. Remove the transfer case mounting bolts and separate transfer case from transaxle.

➡**After removing the transfer case from transaxle, be sure to replace differential side oil seal of the transaxle side with new one.**

To install:

5. Installation is the reverse of the removal procedure.

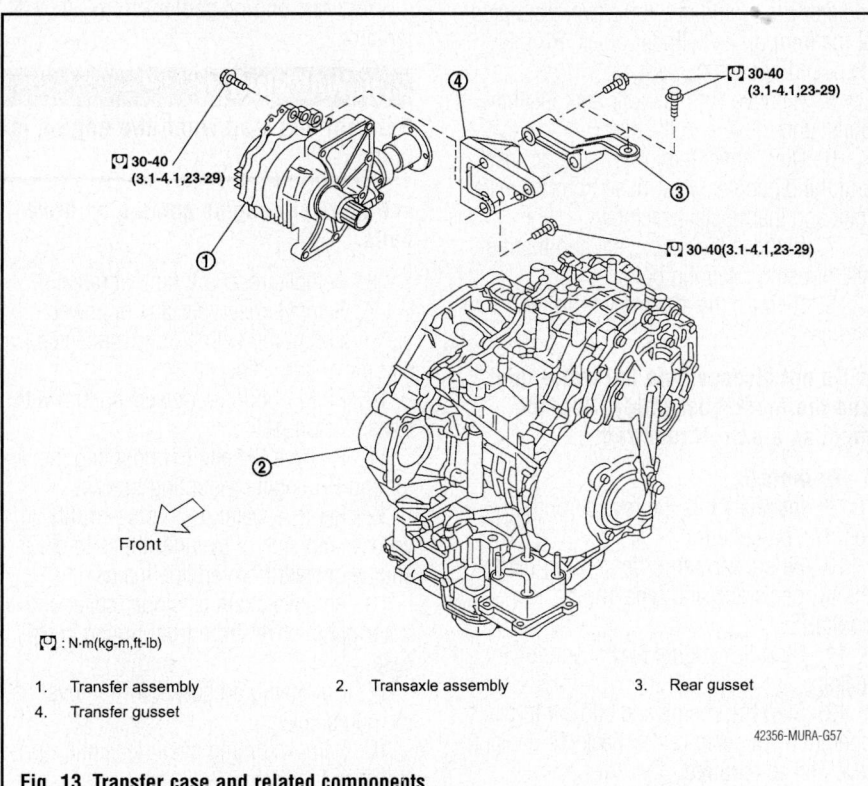

⊡ 30-40
(3.1-4.1,23-29)

⊡ 30-40
(3.1-4.1,23-29)

⊡ 30-40(3.1-4.1,23-29)

Front

⊡ : N•m(kg-m,ft-lb)

1.   Transfer assembly   2.   Transaxle assembly   3.   Rear gusset
4.   Transfer gusset

42356-MURA-G57

Fig. 13 Transfer case and related components

Bolt No.	1	2
Quantity	4	2
Nominal length mm (in)	65 (2.56)	40 (1.57)
Tightening torque [N·m (kg·m, ft.-lb.)]	29.4 - 39.2 (3.0 - 3.9, 22 - 28)	

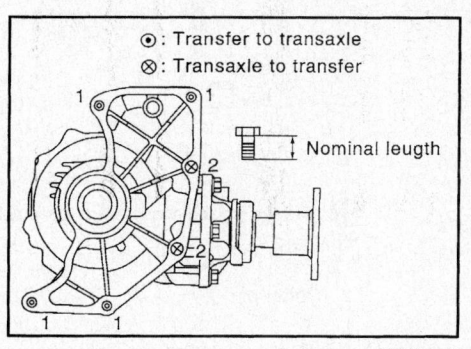

⊙ : Transfer to transaxle
⊗ : Transaxle to transfer

Nominal leugth

42356-MURA-G58

Fig. 14 Transfer case bolt location and torque sequence

ENGINE COOLING

THERMOSTAT

REMOVAL & INSTALLATION

➡ **Never remove the radiator cap when the engine is hot. Serious burns could occur from high-pressure engine coolant escaping from the radiator.**

1. Be sure the engine is cold.
2. Disconnect the negative battery cable.
3. Remove the front engine cover.
4. Drain the engine coolant using the radiator drain plug and the water drain plug at the front of the cylinder block. Properly disposed of used coolant.
5. Remove the reservoir tank retaining bolts, and move it to the side.
6. Disconnect the lower radiator hose and the oil cooler water hose from the water inlet and thermostat assembly.
7. Remove the water inlet and thermostat housing retaining bolts.
8. Remove the assembly from the engine.

➡ **Do not disassemble the water inlet and thermostat assembly. Replace them as a unit, if required.**

To install:
9. Installation is the reverse of the removal procedure.
10. Be sure to refill the cooling using the proper grade and type engine coolant.
11. Start the engine and check for leaks.
12. Start the engine and allow it to reach operation temperature. Recheck the coolant level, fill as required.

WATER PUMP

REMOVAL & INSTALLATION

See Figures 15 through 25.

1. Before servicing the vehicle, refer to the service precautions.
2. Disconnect the battery cable from the negative terminal.
3. Remove engine cover .
4. Remove air duct (inlet) and air cleaner case assembly.
5. Remove front engine undercover.
6. Remove drive belts.
7. Drain engine coolant from radiator.

❋❋ WARNING
Perform this step when the engine is cold.

➡ **Never spill engine coolant on drive belts.**

8. Remove reservoir tank of radiator.
9. Remove reservoir tank of power steering oil pump with piping connected, and move it to aside.
10. Support oil pan (lower) bottom with transmission jack.
11. Remove RH engine mounting insulator and RH engine mounting bracket.
12. Remove water drain plug (front) on water pump side of cylinder block to drain engine coolant from engine inside.
13. Remove chain tensioner cover and water pump cover from front timing chain case.
14. Cut the liquid gasket for removal on the pump cover.
15. Remove timing chain tensioner (primary) as follows:

a. Remove lower mounting bolt.

❋❋ WARNING
Be careful not to drop mounting bolt inside timing chain case.

b. Loosen upper mounting bolt slowly, and then turn chain tensioner (primary) on the mounting bolt so that plunger is fully expanded.

➡ **Even if plunger is fully expanded, it is not dropped from the body of timing chain tensioner (primary).**

c. Turn crankshaft pulley clockwise so that timing chain on the timing chain tensioner (primary) side is loose.
d. Remove upper mounting bolt, and then remove timing chain tensioner (primary).

❋❋ WARNING
Be careful not to drop mounting bolt inside timing chain case.

16. Remove water pump as follows:
a. Remove three water pump mounting bolts. Secure a gap between water pump gear and timing chain, by turning crankshaft pulley counterclockwise until timing chain looseness on water pump sprocket becomes maximum.
b. Screw M8 bolts: pitch: 0.0492 in (1.25 mm) length: approx. 1.97 in (50 mm); into water pumps upper and lower mounting bolt holes until they reach timing chain case. Then, alternately tighten each bolt for a half turn, and pull out water pump.
c. Pull straight out while preventing vane from contacting socket in installation area.

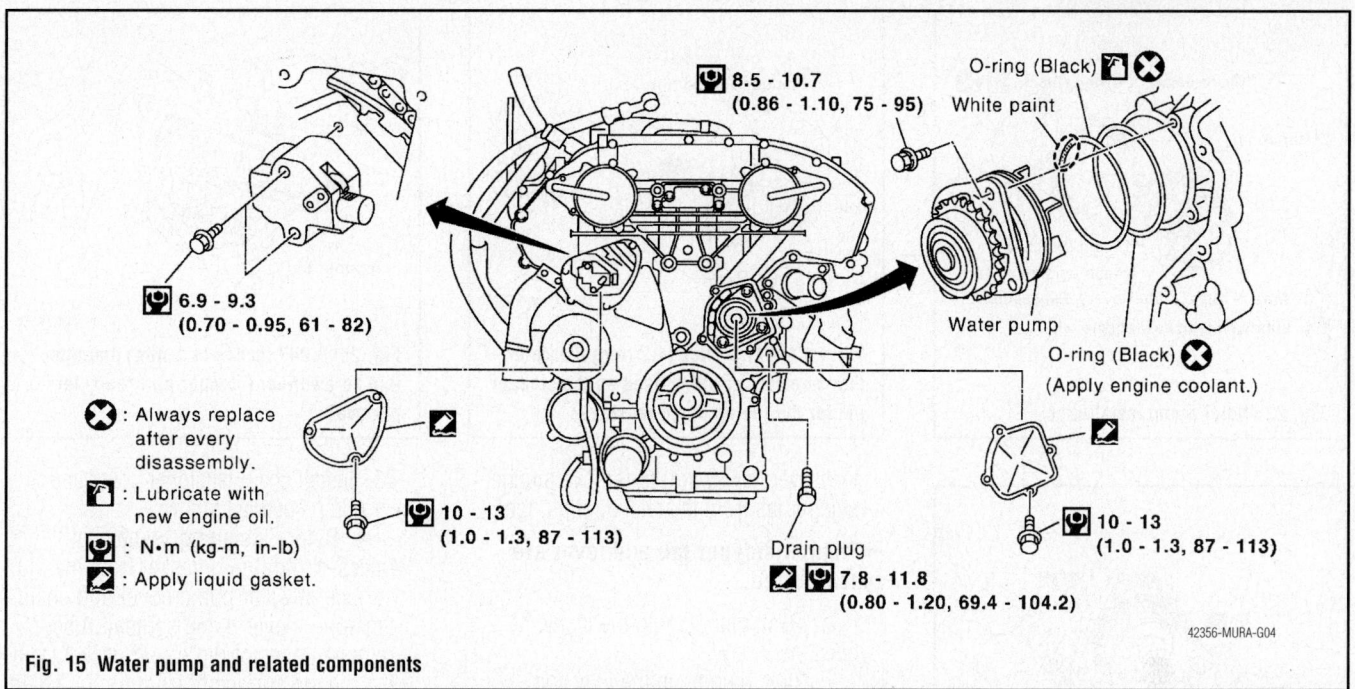

Fig. 15 Water pump and related components

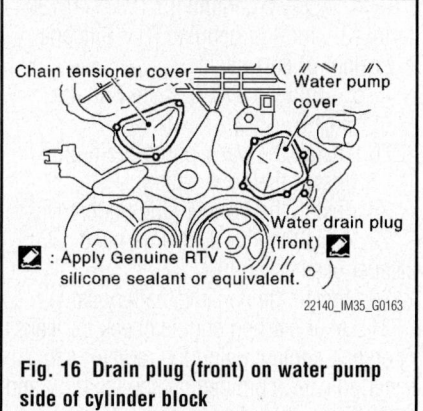

Fig. 16 Drain plug (front) on water pump side of cylinder block

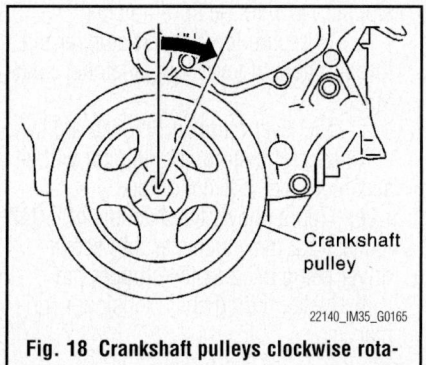

Fig. 18 Crankshaft pulleys clockwise rotation to loosen the timing chain—(primary) side

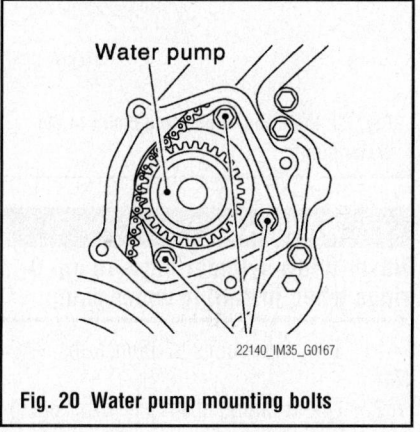

Fig. 20 Water pump mounting bolts

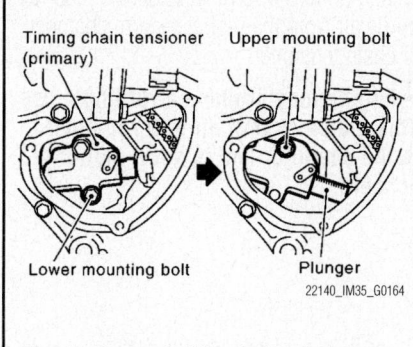

Fig. 17 Removal of the timing chain tensioner (primary)

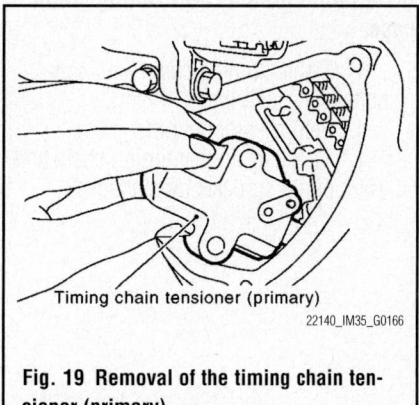

Fig. 19 Removal of the timing chain tensioner (primary)

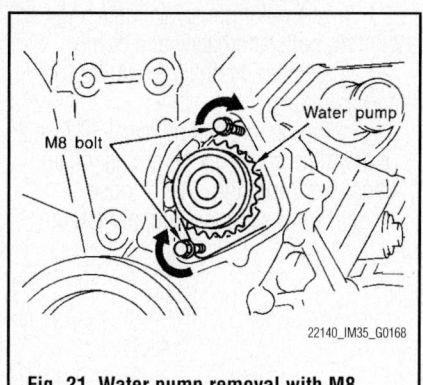

Fig. 21 Water pump removal with M8 bolts

  d. Remove water pump without causing sprocket to contact timing chain.

  e. Remove M8 bolts and O-rings from water pump.

✳✳ WARNING

Never disassemble water pump.

To install:
  17. Install new O-rings to water pump.

  18. Apply engine oil and engine coolant to O-rings.

  19. Locate O-ring with white paint mark to engine front side.

  20. Install the water pump.

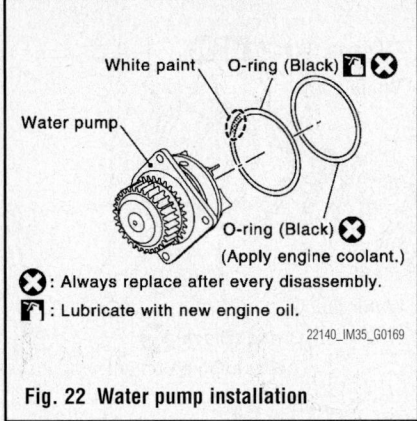

Fig. 22 Water pump installation

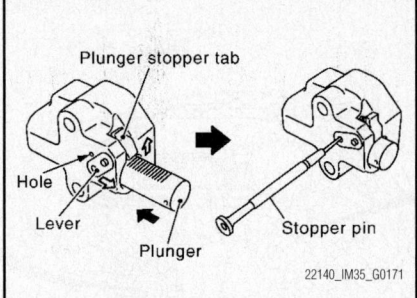

Fig. 24 0.047 inches (1.2 mm) diameter thin screwdriver being used as the stopper pin for the timing chain tensioner

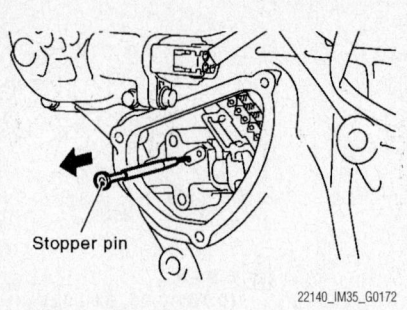

Fig. 25 0.047 inches (1.2 mm) diameter thin screwdriver (stopper pin) ready for removal

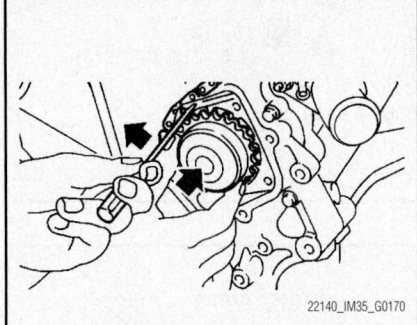

Fig. 23 Installing the timing chain to the water pump

※※ WARNING

Never allow cylinder block to nip O-rings when installing water pump.

21. Torque the bolts to 85 inch lbs. (9.6 Nm)

22. Check timing chain and water pump sprocket are engaged.

23. Insert water pump by tightening mounting bolts alternately and evenly.

24. Install timing chain tensioner (primary) as follows:

a. Turn crankshaft pulley clockwise so that timing chain on the timing. chain tensioner (primary) side is loose.

b. Pull the plunger stopper tab up (or turn lever downward) to remove plunger stopper tab from the ratchet of plunger.

➡**Plunger stopper tab and lever are synchronized.**

c. Push plunger into the inside of tensioner body.

d. Hold plunger in the fully compressed position by engaging plunger stopper tab with the tip of ratchet.

e. To secure lever, insert stopper pin through hole of lever into tensioner body hole.

f. The lever parts and the tab are synchronized. Therefore, the plunger will be secured under this condition.

g. Figure shows the example of 0.047 inches (1.2 mm) diameter thin screwdriver being used as the stopper pin.

h. Install timing chain tensioner (primary).

➡**Remove dust and foreign material completely from backside of timing chain tensioner (primary) and from installation area of rear timing chain case**

i. Tighten timing chain tensioner bolts to 72 inch lbs. (8.1 Nm).

j. Remove stopper pin.

k. Check again that timing chain and water pump sprocket are engaged.

25. Install chain tensioner cover and water pump cover as follows:

a. Before installing, remove all traces of old liquid gasket from mating surface of water pump cover and chain tensioner cover using scraper. Also, remove traces of old liquid gasket from the mating surface of front timing chain case.

b. Apply a continuous 0.091–0.130 in. (2–3mm) of genuine RTV Silicone sealant or equivalent.

c. Tighten cover bolts to 8 ft. lbs. (11.3 Nm).

26. Install the water pump drain plug.

27. Install the drive belts.

28. Install the front engine undercover.

29. Install the air duct (inlet) and air cleaner case assembly.

30. Install the negative battery cable.

31. After starting engine check for leaks of engine coolant using the radiator cap tester adapter (commercial service tool) and the radiator cap tester (commercial service tool), let idle for three minutes, then rev engine up to 3,000 RPM under no load to purge air from the high-pressure chamber of chain tensioner.

➡**Engine may produce a rattling noise. This indicates that air still remains in the chamber and is not a matter of concern.**

ENGINE ELECTRICAL   **CHARGING SYSTEM**

ALTERNATOR

REMOVAL & INSTALLATION

See Figure 26.

1. Before servicing the vehicle, refer to the Precautions Section.
2. Drain the cooling system.
3. Remove or disconnect the following:
   - Negative battery cable
   - Alternator harness connectors
   - Engine right side under cover
   - Radiator
   - Remove alternator and air conditioner compressor belt.
   - Idler pulley
   - Alternator

To install:

4. Install or connect the following:
   - Alternator
   - Idler pulley
   - Alternator belt. Tighten the through-bolts to 18–23 ft. lbs. (25–31 Nm).
   - Engine under cover
   - Radiator
   - Alternator harness connectors
   - Negative battery cable
5. Refill the cooling system, using the proper grade and type coolant.

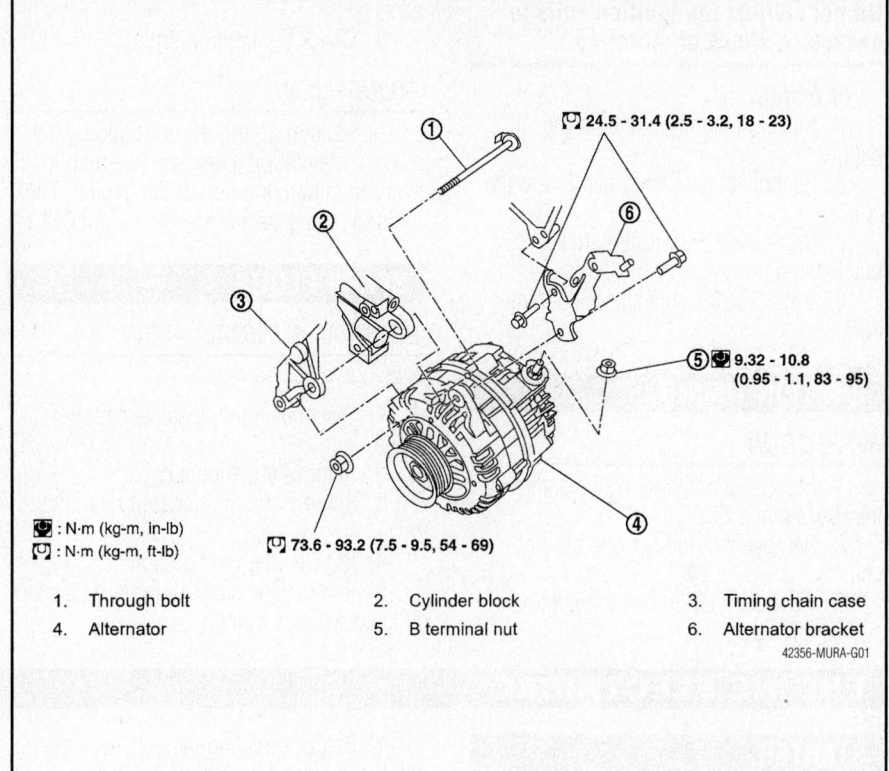

⚙ : N·m (kg-m, in-lb)
🔧 : N·m (kg-m, ft-lb)

🔧 73.6 - 93.2 (7.5 - 9.5, 54 - 69)

🔧 24.5 - 31.4 (2.5 - 3.2, 18 - 23)

⚙ 9.32 - 10.8 (0.95 - 1.1, 83 - 95)

1. Through bolt	2. Cylinder block	3. Timing chain case
4. Alternator	5. B terminal nut	6. Alternator bracket

42356-MURA-G01

Fig. 26 Alternator and related components

ENGINE ELECTRICAL   **DISTRIBUTORLESS IGNITION SYSTEM**

FIRING ORDER

See Figure 27.

79233G02

Fig. 27 3.5L Engines

IGNITION COIL

REMOVAL & INSTALLATION

See Figure 28.

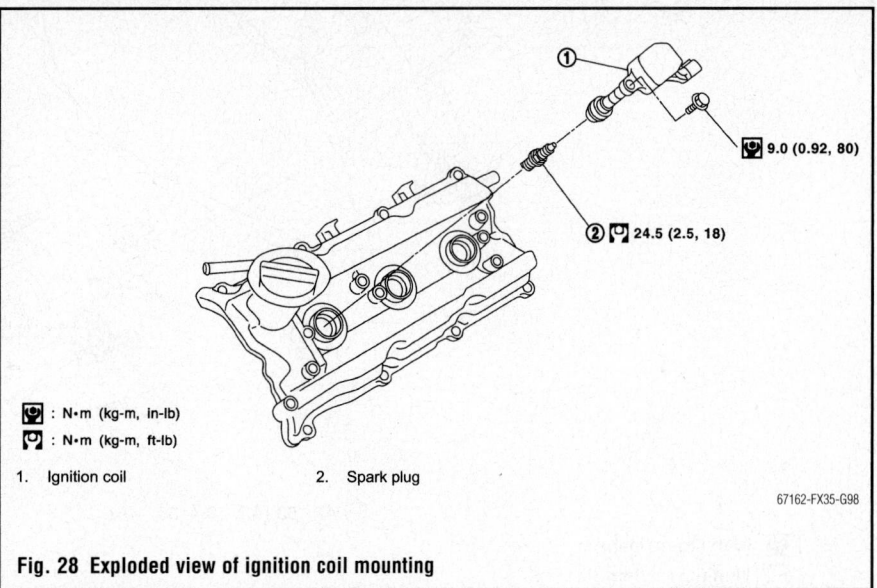

⚙ : N·m (kg-m, in-lb)
🔧 : N·m (kg-m, ft-lb)

⚙ 9.0 (0.92, 80)

🔧 24.5 (2.5, 18)

1. Ignition coil	2. Spark plug

67162-FX35-G98

Fig. 28 Exploded view of ignition coil mounting

1. Before servicing the vehicle, refer to the Precautions Section.
2. Remove the engine cover.
3. Remove the air duct (for ignition coil of left bank side).
4. Move aside the wiring harness, wiring harness bracket, and hoses located above ignition coil.
5. Disconnect the wiring harness connector from the ignition coil.

6. Remove the ignition coil.

☀☀ CAUTION

Do not subject the ignition coils to excessive shock or vibration.

To install:

7. Install the ignition coil on the engine.

8. Reconnect the wiring harness to the coil.

9. Reposition the wiring harness, bracket and hoses.

10. Install the air duct and the engine cover.

IGNITION TIMING

INSPECTION

1. Before servicing the vehicle, refer to the Precautions Section.

2. Remove the number one ignition coil.

3. Connect the number one ignition

coil and spark plug with a suitable high-tension wire.

4. Attach the timing light clamp to the wire.

5. Check the ignition timing.

ADJUSTMENT

The ignition timing is controlled by the ECM to maintain the best air-fuel ratio for every running condition of the engine. The ignition timing data is stored in the ECM.

SPARK PLUGS

REMOVAL & INSTALLATION

See Figure 29.

1. Disconnect the negative battery cable.

2. Remove the engine cover.

3. Remove the ignition coil retaining bolt.

4. Remove the ignition coil.

5. Remove the spark plug using a spark plug socket and wrench.

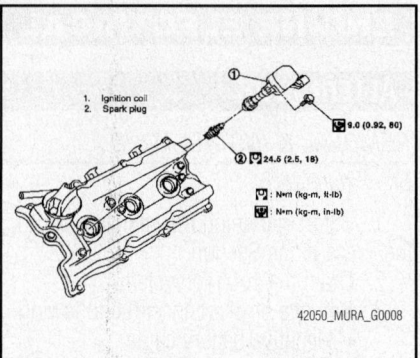

Fig. 29 Spark plug and related components

To install:

6. Be sure the spark plug gap is to specification (0.043 in).

7. Carefully install the spark plug and torque to specification, 18 ft. lbs.

8. Install the ignition coil, torque the retaining bolt to 80 inch lbs.

9. Install the engine cover.

10. Connect the negative battery cable.

ENGINE ELECTRICAL

STARTER

REMOVAL & INSTALLATION

See Figure 30.

1. Before servicing the vehicle, refer to the Precautions Section.

2. Remove or disconnect the following:
- Battery
- Air intake duct

STARTING SYSTEM

- Battery bracket
- Transaxle dipstick tube on AWD vehicles. Be sure to drain the transaxle fluid.
- S connector

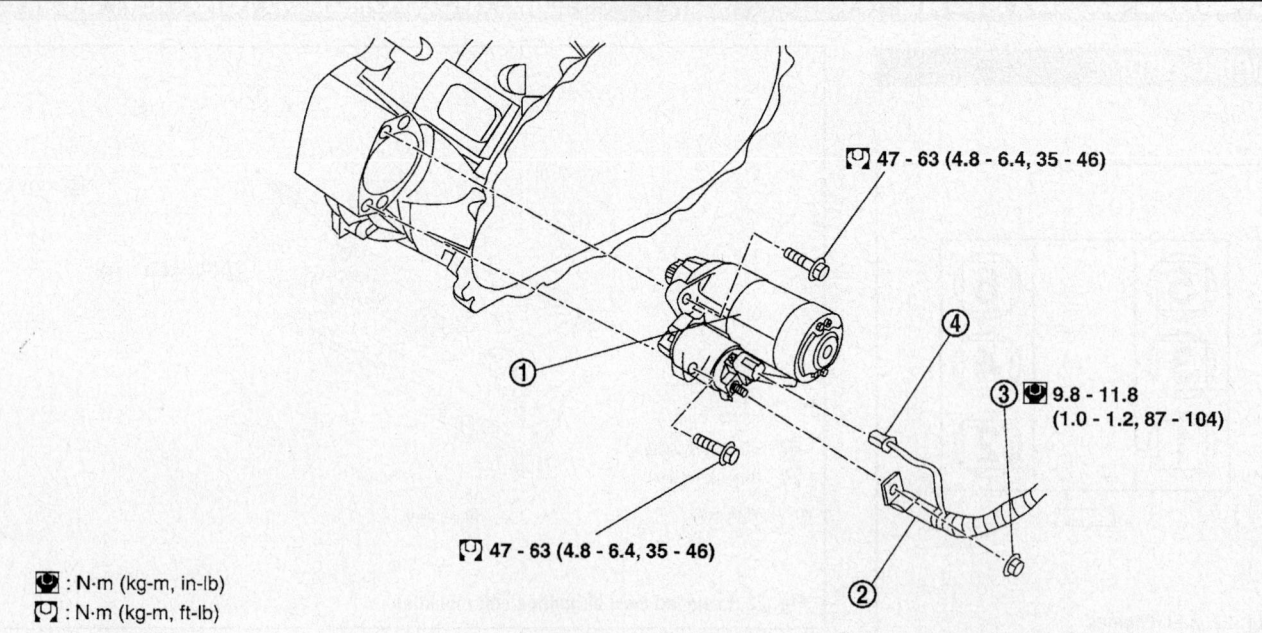

: N·m (kg-m, in-lb)
: N·m (kg-m, ft-lb)

1. Starter motor
2. B terminal harness
3. B terminal nut
4. S connector

Fig. 30 Starter and related components

- B terminal nut
- Starter motor mounting bolts
- Starter motor to the direction of upper side the vehicle

To install:

3. Installation is the reverse of the removal procedure. Observe the following toques:

- B terminal nut: 9.8–11.8 Nm (87–104 inch lbs.)
- Starter motor mounting bolt: 47–63 Nm (35–46 ft. lbs.)

- Battery bracket mounting bolt: 14–20 Nm (10–15 ft. lbs.)

SOLENOID OR RELAY REPLACEMENT

The solenoid is not serviceable, the starter must be serviced as an assembly.

ENGINE MECHANICAL

ACCESSORY DRIVE BELTS

ACCESSORY BELT ROUTING

See Figure 31.

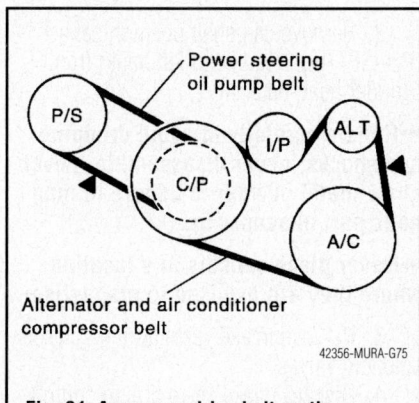

Fig. 31 Accessory drive belt routing

Refer to the accompanying illustration for drive belt routing.

INSPECTION

See Figure 32.

✳✳ WARNING

Under no circumstances should the accessory drive belt, tensioner or pulleys be lubricated as potential damage to the belt material and tensioner damping mechanism will occur. Do not apply any fluids or belt dressing to the accessory drive belt or pulleys.

The water pump drive belt is on back of engine. It is driven off the rear cam pulley and does not have any adjustments.

Visually inspect the belt for obvious signs of mechanical damage:

- Drive belt cracking/chunking/wear
- Belt/pulley contamination
- Incorrectly routed belt
- Pulley misalignment or excessive pulley runout
- Loose or mislocated hardware
- Incorrectly routed power steering tubes (rubbing)

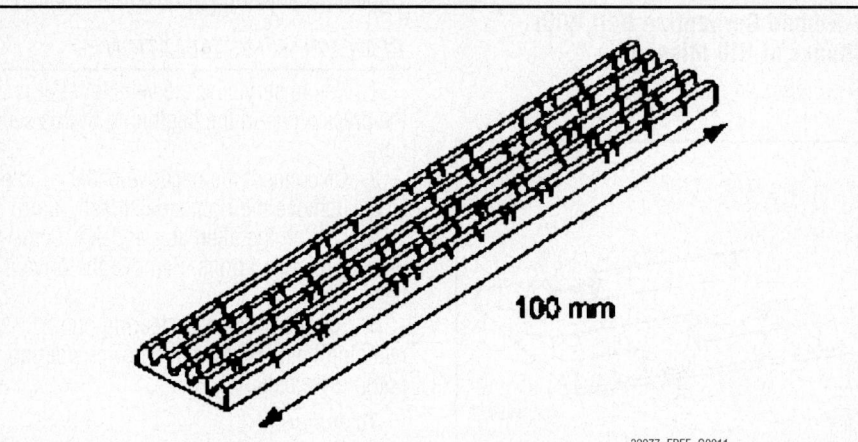

Fig. 32 Up to 15 cracks in a rib over a distance of 4 inches (100mm) can be considered acceptable. If cracks exceed this standard, install a new belt

➡️**Up to 15 cracks in a rib over a distance of 4 inches (100mm) can be considered acceptable. If damage exceeds the acceptable limit or any chunks are found to be missing from the ribs, a new belt must be installed.**

1. Check the belt for cracks. Up to 15 cracks in a rib over a distance of 4 inches (100mm) can be considered acceptable. If cracks exceed this standard, install a new belt.

V-Ribbed Serpentine Belt With Piling

See Figure 33.

➡️**Piling is an excessive buildup in the V-grooves of the belt.**

The condition of the V-ribbed drive belt should be compared against the illustration and appropriate action taken.

1. Small scattered deposits of rubber material. This is not a concern, therefore, installation of a new belt is not required.

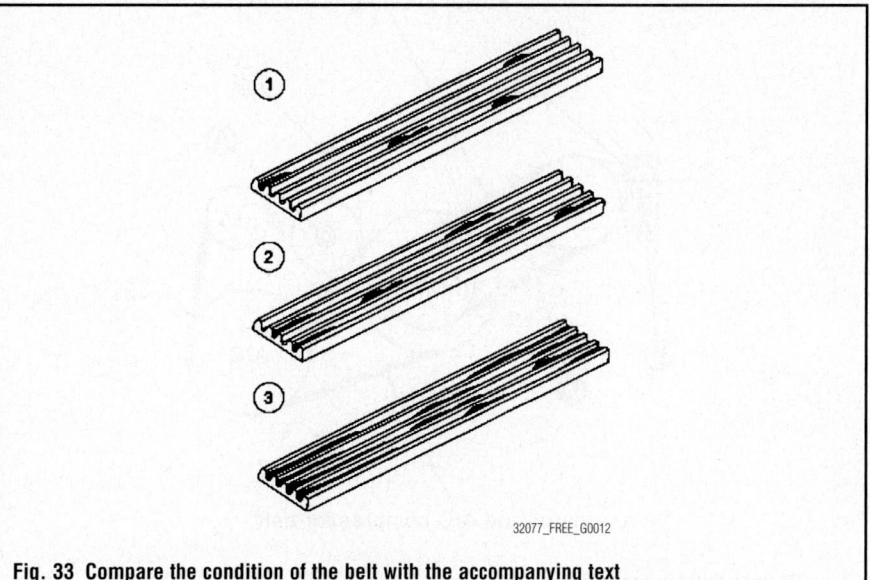

Fig. 33 Compare the condition of the belt with the accompanying text

2. Longer deposit areas building up to 50 percent of the rib height. This is not considered a concern but it can result in excessive noise. If noise is apparent, install a new belt.

3. Heavy deposits building up along the grooves resulting in a possible noise and belt stability concern. If heavy deposits are apparent, install a new belt.

V-Ribbed Serpentine Belt With Chunks of Rib Missing

See Figure 34.

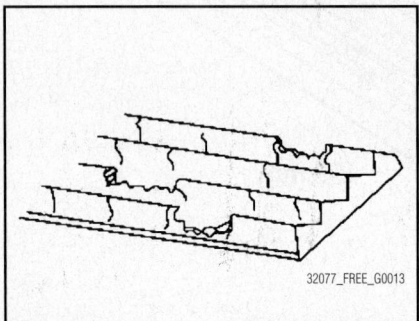

32077_FREE_G0013

Fig. 34 Replace the belt if missing chunks are found during inspection

There should be no chunks missing from the belt ribs. If the belt shows any evidence of this, install a new accessory drive belt.

ADJUSTMENT

Alternator and A/C compressor Belt

See Figure 35.

1. Before servicing the vehicle, refer to the precautions in the beginning of this section.

2. Disconnect the negative battery cable.

3. Remove the right side splash guard.

4. Loosen the idler pulley locknut (A) and adjust the tension by turning the adjusting nut (B).

5. Tighten the locknut (A) to 26 ft. lbs. Tighten the adjusting nut (B) to 48 inch lbs.

REMOVAL & INSTALLATION

1. Before servicing the vehicle, refer to the precautions in the beginning of this section.

2. Disconnect the negative battery cable.

3. Remove the right side splash guard.

4. Loosen the alternator and A/C compressor retaining bolts. Remove the drive belt.

5. Loosen the power steering pump retaining bolts. Remove the power steering pump drive belt.

To install:

6. Installation is the reverse of the removal procedure.

7. Check to ensure that the belt tension is within specification.

CAMSHAFT AND VALVE LIFTERS

REMOVAL & INSTALLATION

See Figures 36 through 49.

1. Remove front timing chain case, camshaft sprocket, timing chain and rear timing chain case.

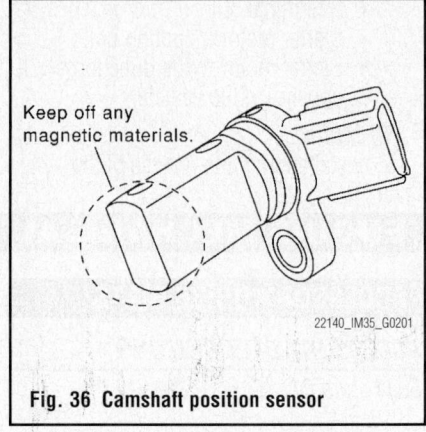

Keep off any magnetic materials.

22140_IM35_G0201

Fig. 36 Camshaft position sensor

2. Remove camshaft position sensor (PHASE) (right bank and left bank) from cylinder head back side.

➡**Handle carefully to avoid dropping and shocks, never disassemble, never allow metal powder to adhere to magnetic part at sensor tip.**

➡**Never place sensors in a location where they are exposed to magnetism.**

3. Remove intake valve timing control solenoid valves.

4. Discard intake valve timing control solenoid valve gaskets and use new gaskets for installation.

5. Remove camshaft brackets.

6. Mark camshafts, camshaft brackets and bolts so they are placed in the same position and direction for installation.

7. Equally loosen camshaft bracket bolts in several steps in reverse order as shown in the figure.

8. Remove camshaft.

9. Remove valve lifter.

10. Identify installation positions, and store them without mixing them up.

11. Remove timing chain tensioner (secondary) from cylinder head.

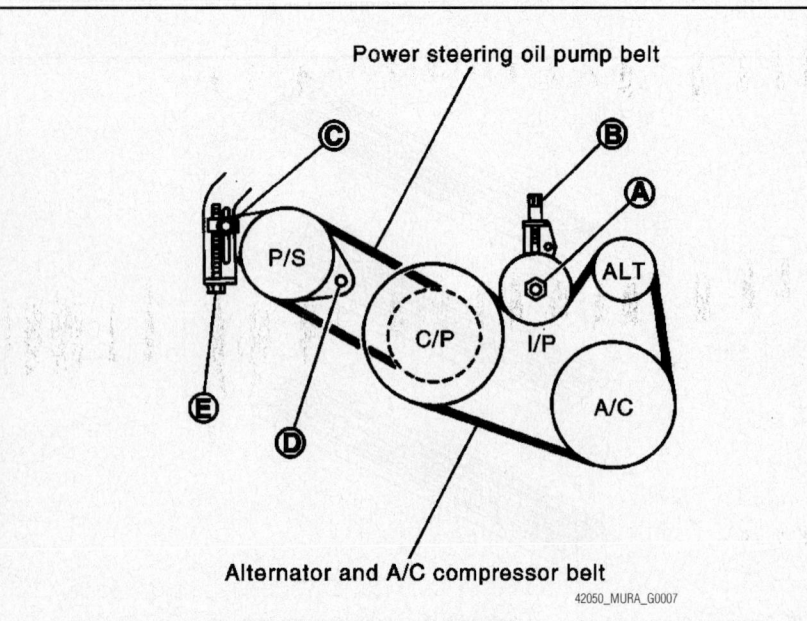

Power steering oil pump belt

Alternator and A/C compressor belt

42050_MURA_G0007

Fig. 35 Drive belt tension locating points

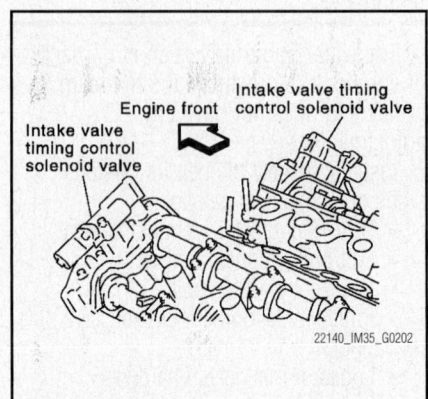

Engine front

Intake valve timing control solenoid valve

Intake valve timing control solenoid valve

Intake valve timing control solenoid valve

22140_IM35_G0202

Fig. 37 Intake valve timing control solenoid valves

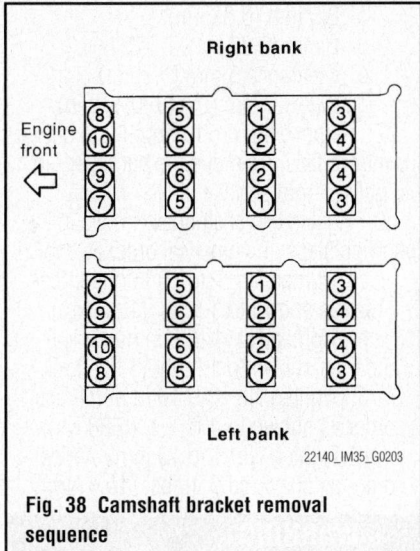

Fig. 38 Camshaft bracket removal sequence

12. Remove timing chain tensioner (secondary) with its stopper pin attached.

➡**Stopper pin should be attached when timing chain (secondary) is removed.**

To install:

13. Install timing chain tensioners (secondary) on both sides of cylinder head.

14. Install timing chain tensioner with its stopper pin attached.

15. Install timing chain tensioner with sliding part facing downward on right-side cylinder head, and with sliding part facing upward on left-side cylinder head.

16. Install new O-ring as shown in the figure.

17. Install valve lifter in the original position.

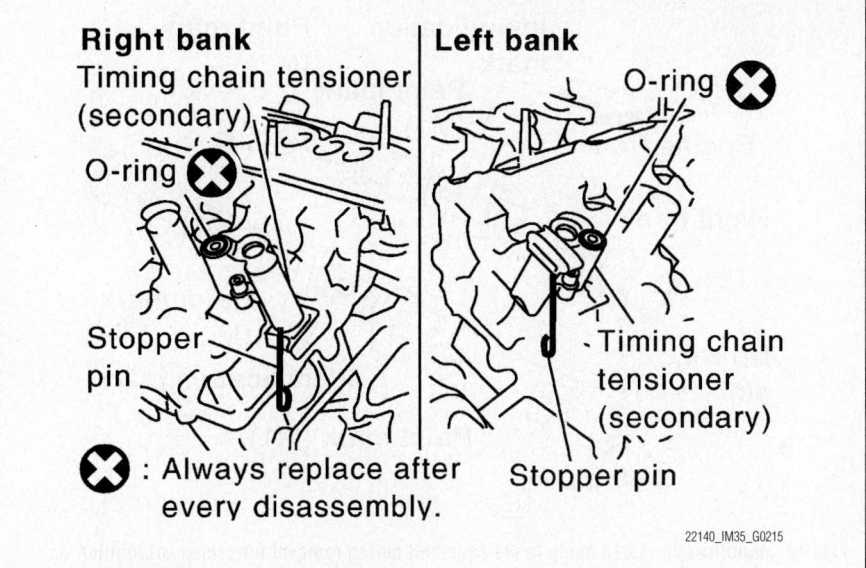

Fig. 40 Chain tensioners (secondary) on both sides of cylinder head

18. Install camshafts-Install camshaft with dowel pin attached to its front end face on the exhaust side.

19. Follow your identification marks made during removal, or follow the identification marks that are present on new camshafts for proper placement and direction.

20. Install camshaft so that dowel pin hole and dowel pin on front end face are positioned as shown in the figure. (No. 1 cylinder TDC on its compression stroke).

➡**Large and small pin holes are located on front end face of camshaft (INT), at intervals of 180 degrees. Face**

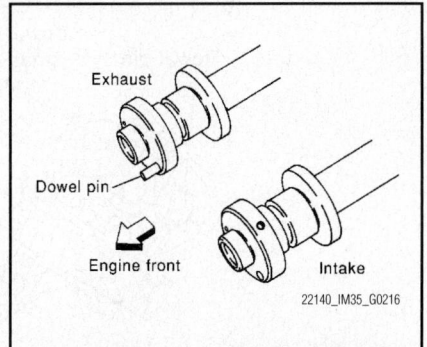

Fig. 41 Camshaft with dowel pin attached to its front end face on the exhaust side

small diameter side pin hole upward (in cylinder head upper face direction).

21. Though camshaft does not stop at the portion as shown in the figure, for the placement of cam nose, it is generally accepted camshaft is placed for the same direction of the figure.

22. Install camshaft brackets.

23. Remove foreign material completely from camshaft bracket backside and from cylinder head installation face.

24. Install camshaft bracket in original position and direction as shown in the figure.

25. Install camshaft brackets (No. 2 to 4) aligning the stamp marks as shown in the figure.

➡**There are no identification marks indicating left and right for camshaft bracket (No. 1).**

26. Apply liquid gasket to mating surface of camshaft bracket (No. 1) as shown on both right bank and left bank as follows.

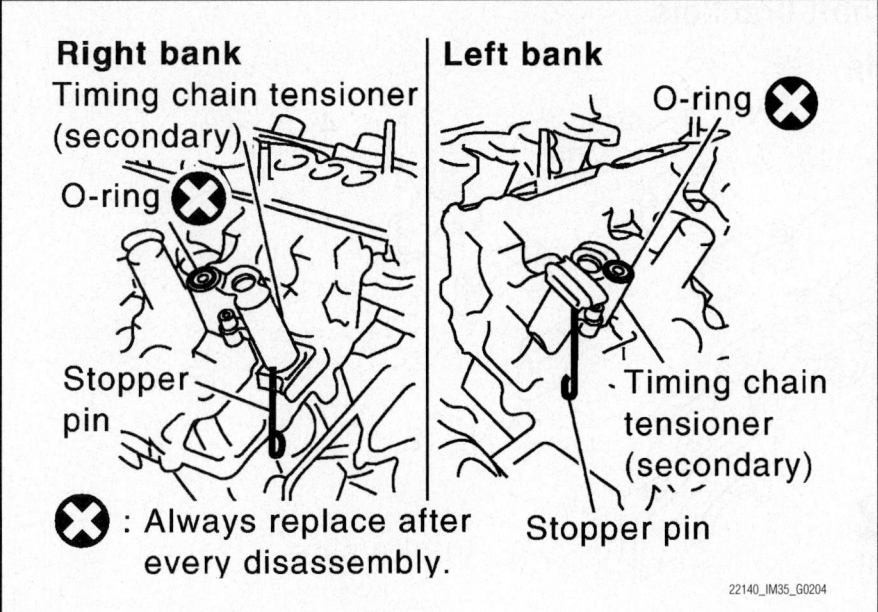

Fig. 39 Timing chain tensioner (secondary)

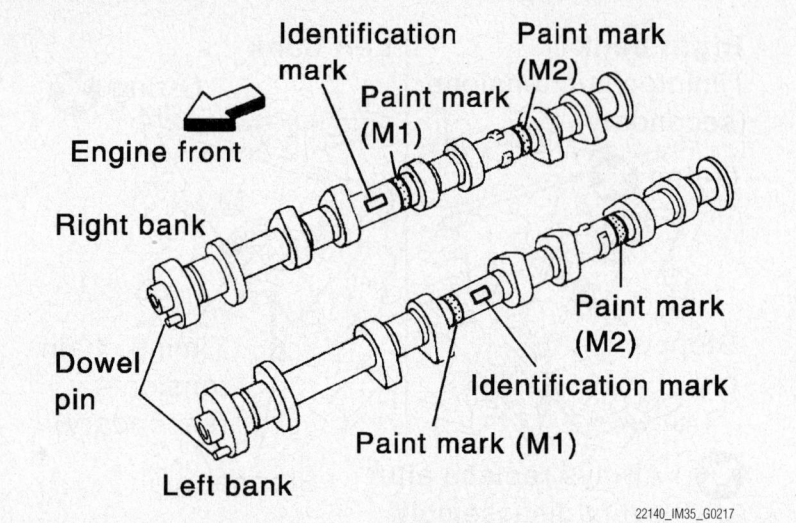

Fig. 42 Identification marks made to the camshaft during removal for proper orientation on assembly

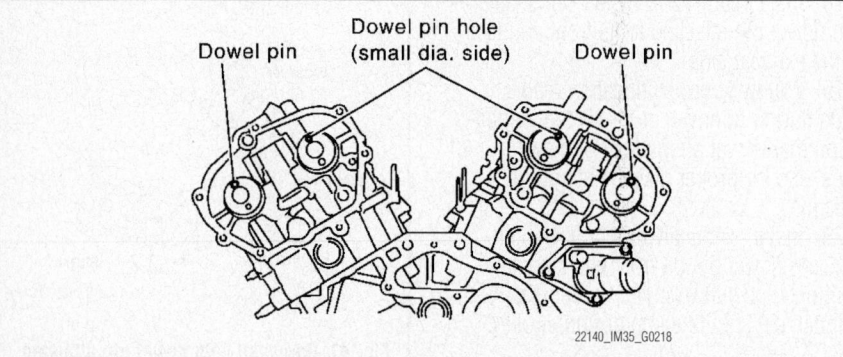

Fig. 43 Dowel pin hole and dowel pin on front end face positioning for TDC on its compression stroke

a. 8.5 mm (0.335 in).
b. 2 mm (0.08 in).
c. Clearance 5 mm (0.20 in).
d. 2.0–3.0 mm (0.079–0.118 in).

27. Remove the protruding liquid gasket from front face. (Remove the hardened liquid gasket from surface only).

28. Tighten camshaft bracket bolts in the following steps, in numerical order as shown.

a. Tighten No. 7 to 10 in numerical order as shown to 1 ft. lb. (1.96 Nm).

b. Tighten No. 1 to 6 in numerical order as shown to 1 ft. lb. (1.96 Nm).

c. Tighten No. 1 to 10 in numerical order as shown to 4 ft. lbs. (5.88 Nm).

d. Tighten No. 1 to 10 in numerical order as shown to 8 ft. lbs. (10.4 Nm).

❋❋ WARNING

After tightening mounting bolts of camshaft brackets (No. 1), be sure to wipe off excessive liquid gasket from the parts list below.

- Mating surface of rocker cover
- Mating surface of rear timing chain case

29. Measure difference in levels between front end faces of camshaft bracket (No. 1) and cylinder head.

30. Measurement standard 0.0055–0.0055 inches (0.14– 0.14 mm)

31. Measure two positions (both intake and exhaust side) for a single bank.

32. If the measured value is out of the standard, re-install camshaft bracket (No. 1).

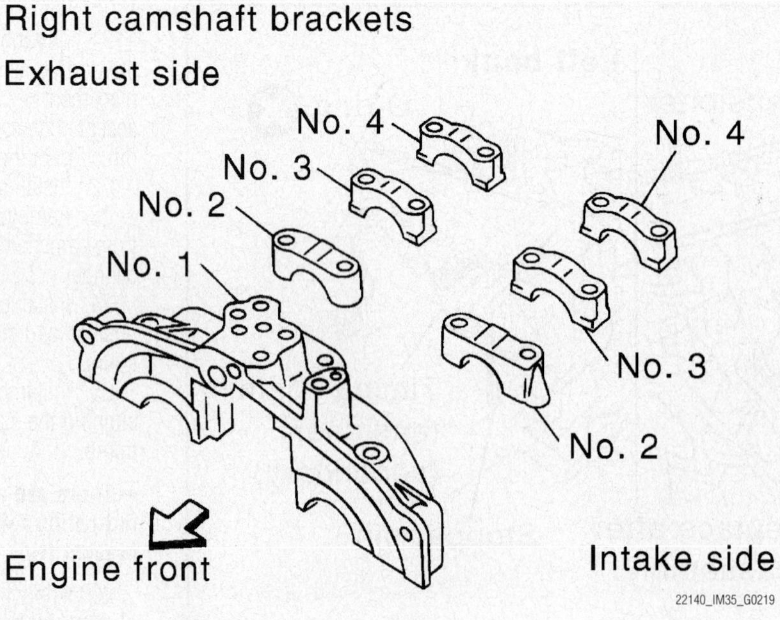

Fig. 44 Right camshaft bracket positions and direction

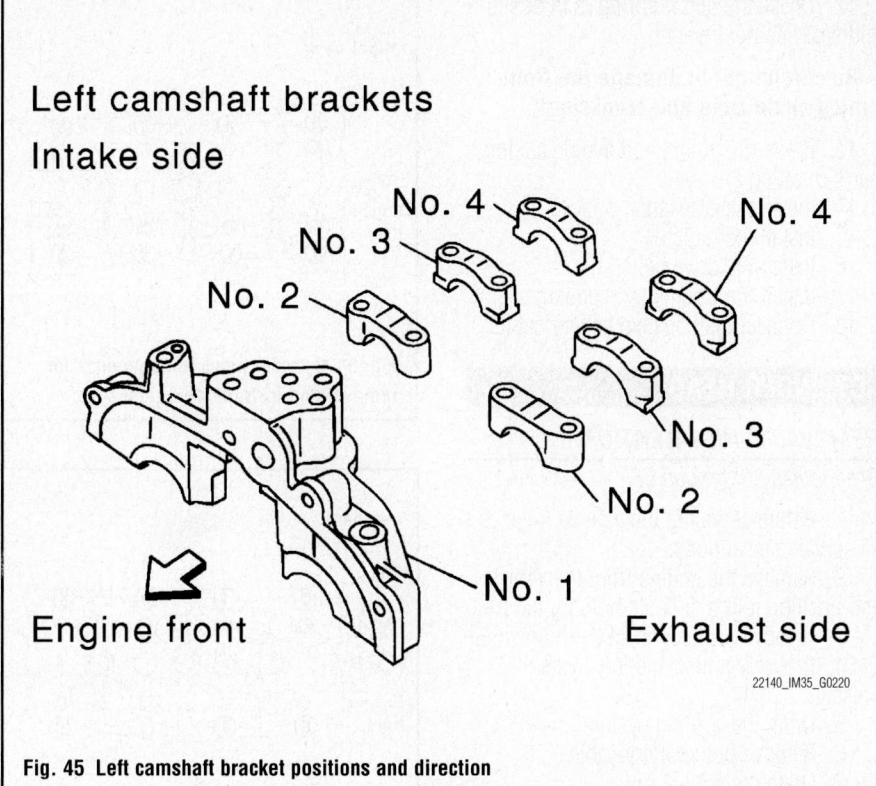

Fig. 45 Left camshaft bracket positions and direction

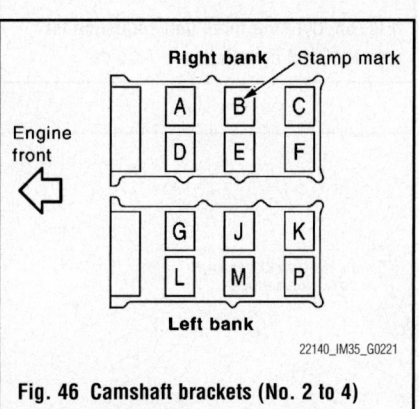

Fig. 46 Camshaft brackets (No. 2 to 4) aligning sequence

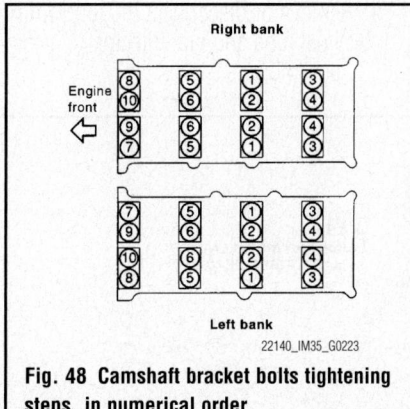

Fig. 48 Camshaft bracket bolts tightening steps, in numerical order

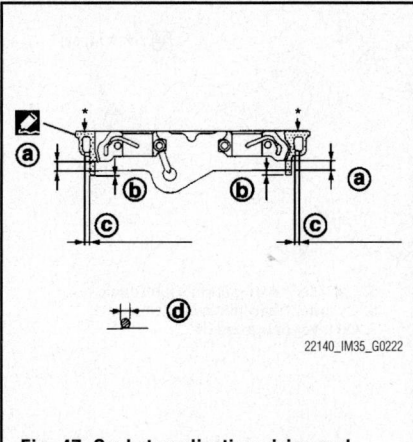

Fig. 47 Gasket application sizing and locations

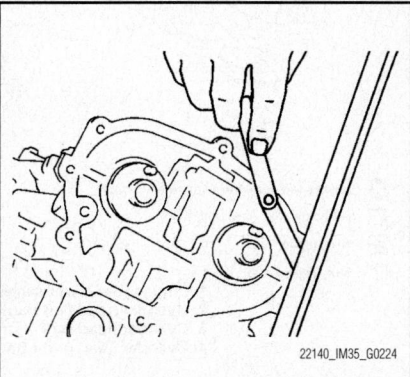

Fig. 49 Difference in levels between the front end faces of the camshaft bracket (No. 1) and the cylinder head

33. Remove valve lifters at the locations that are out of the standard.

34. Measure the center thickness of the removed valve lifters with a micrometer.

35. Inspect and adjust the valve clearance.

CAMSHAFT BEARING REPLACEMENT

There are no separate camshaft bearings, the camshaft caps are the bearing surfaces.

CRANKSHAFT DAMPER

REMOVAL & INSTALLATION

See Figure 50.

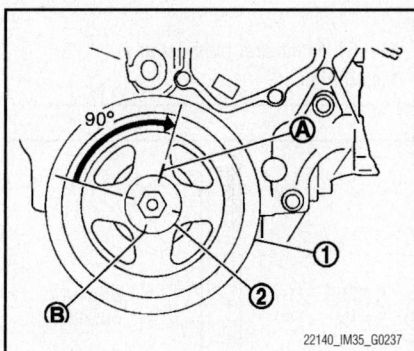

Fig. 50 Crankshaft pulley bolt tightening sequence

1. Before servicing the vehicle, refer to all service precautions.

2. Remove the negative battery cable.

3. Remove the front engine undercover (power tool).

4. Remove the drive belts.

5. Loosen crankshaft pulley bolt and rotate bolt seating surface at 10 mm (0.39 in) from its original position.

6. Place suitable puller tab on holes of crankshaft pulley, and pull crankshaft pulley through.

✳✳ WARNING

Never put puller tab on crankshaft pulley internal damper, as this will damage internal damper.

7. Installation is the reverse of installation.

8. Tighten the crankshaft pulley to 33 ft. lbs. (44.1 Nm).

9. Place a paint mark (A) on crankshaft pulley (1) aligning with the angle mark (B) on crankshaft pulley bolt (2). Tighten the bolt 90 degrees (angle tightening).

10. Rotate crankshaft pulley in normal direction (clockwise when viewed from front) to confirm it turns smoothly.

CRANKSHAFT FRONT SEAL

REMOVAL & INSTALLATION

See Figures 51 and 52.

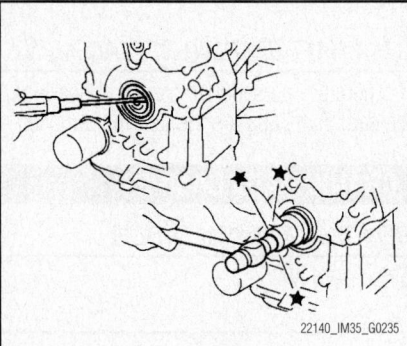

Fig. 51 Front seal removal and installation

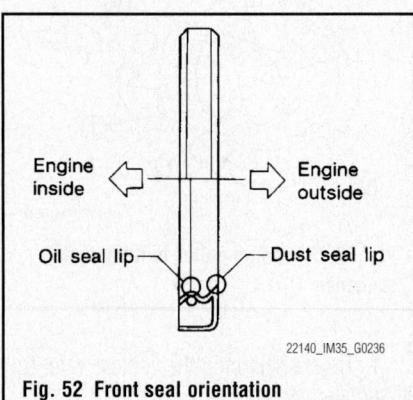

Fig. 52 Front seal orientation

1. Before servicing the vehicle, refer to all service precautions.
2. Remove the negative battery cable.
3. Remove the front engine undercover (power tool).
4. Remove the radiator 4.5L Engines.
5. Remove the drive belts.
6. Remove the crankshaft pulley.
7. Remove front oil seal using a suitable tool.

➡**Be careful not to damage front timing chain case and crankshaft.**

To install:

8. Apply new engine oil to both oil seal lip and dust seal lip of new front oil seal.
9. Install front oil seal.
10. Install front oil seal so that each seal lip is oriented as shown in the figure.
11. Using a suitable drift, press-fit until the height of front oil seal is level with the mounting surface.

➡**Suitable drift: outer diameter 60 mm (2.36 in), inner diameter 50 mm (1.97 in).**

12. Check the garter spring is in position and seal lips not inverted.

➡**Be careful not to damage the front timing chain case and crankshaft.**

13. Press-fit straight and avoid causing burrs or tilting oil seal.
14. Install the crankshaft pulley.
15. Install the radiator.
16. Install the drive belts.
17. Install the front engine undercover.
18. Connect the negative battery cable.

CYLINDER HEAD

REMOVAL & INSTALLATION

See Figures 53 through 58.

1. Before servicing the vehicle, refer to all service precautions.
2. Remove the engine from the vehicle and position it in a suitable holding fixture.
3. Remove or disconnect the following:
4. Properly relieve the fuel system pressure.
5. Drain the cooling system.
6. Remove both battery cables.
7. Drain the engine oil.
8. Remove the camshafts.
9. Remove or disconnect the following:
   • Fuel tube and fuel injector assembly

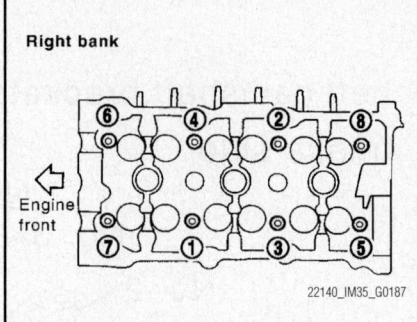

Fig. 54 Cylinder head bolt sequence for removal and installation—right side

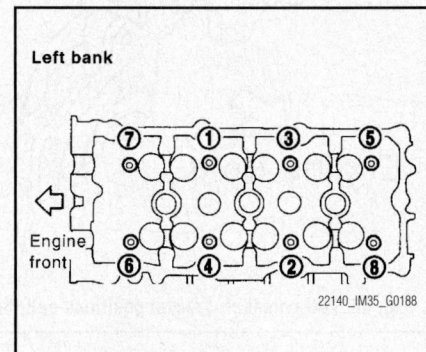

Fig. 55 Cylinder head bolt sequence for removal and installation—left side

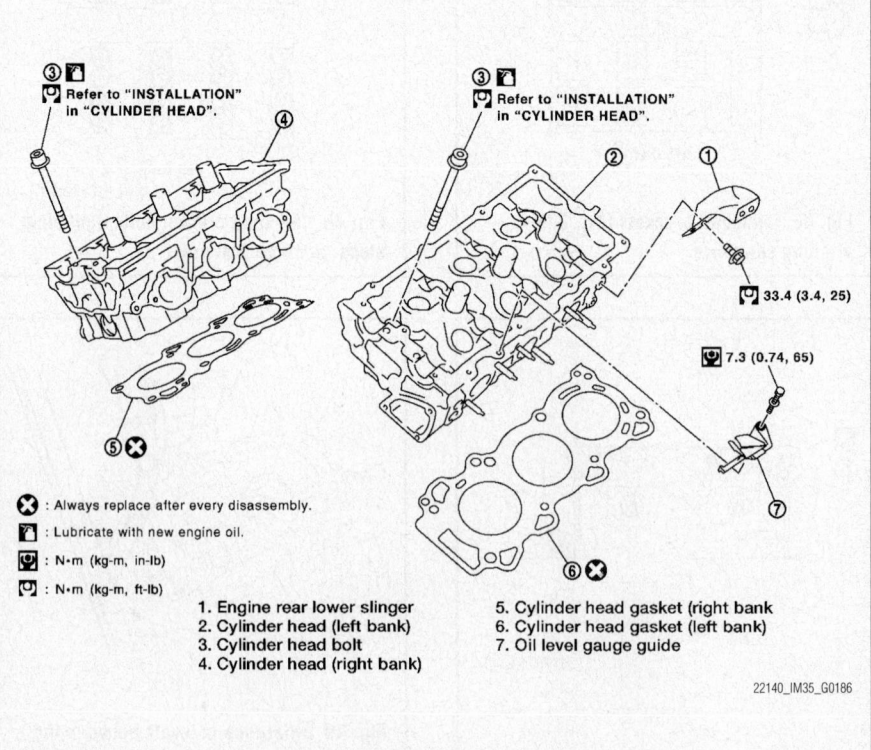

○ : Always replace after every disassembly.
▢ : Lubricate with new engine oil.
▢ : N·m (kg-m, in-lb)
▢ : N·m (kg-m, ft-lb)

1. Engine rear lower slinger
2. Cylinder head (left bank)
3. Cylinder head bolt
4. Cylinder head (right bank)
5. Cylinder head gasket (right bank)
6. Cylinder head gasket (left bank)
7. Oil level gauge guide

Fig. 53 Cylinder head assemblies—exploded view

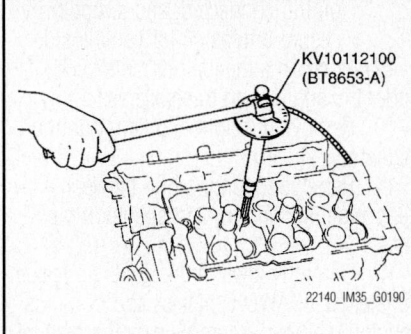

Fig. 56 Torque wrench (commercial service tool: J24239-01)

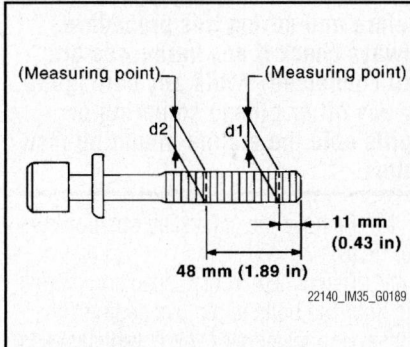

Fig. 57 Cylinder head bolts plastic zone tightening size specifications

- Intake manifold
- Exhaust manifolds
- Water inlet and thermostat assembly
- Water outlet, water pipe and heater pipe
- Cylinder head bolts in order shown
- Cylinder head gaskets

To install:

10. Install new cylinder head gaskets.
11. Turn crankshaft until No. 1 piston is set at TDC.
    - Crankshaft key should line up with the right bank cylinder center line as shown in the figure.
12. Install cylinder head, follow the steps below to tighten the cylinder head bolts in numerical order as shown in the figure.
13. Use (commercial service tool: J24239-01) shown in figure below.

➡**If cylinder head bolts are re-used, check their outer diameters before installation. Refer to "Cylinder Head Bolts Outer Diameter" in the figure below.**

14. Apply new engine oil to threads and seat surfaces of cylinder head bolts.
15. Tighten all cylinder head bolts to 72 ft. lbs. (98.1 Nm).

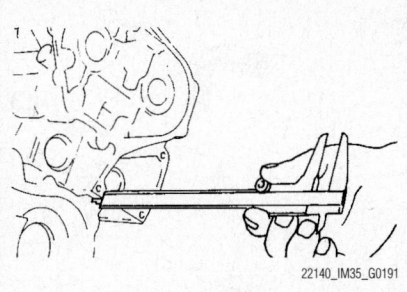

Fig. 58 Measurement of distance between front end faces of cylinder block and cylinder head

16. Completely loosen all cylinder head bolts in reverse order of tightening sequence.
17. Tighten all cylinder head bolts to 29 ft. lbs. (39.2 Nm) following tightening sequence.
18. Turn all cylinder head bolts 90 degrees clockwise (angle tightening).

➡**Check the tightening angle by using the angle wrench.**

19. Turn all cylinder head bolts 90 degrees clockwise again (angle tightening) following tightening sequence.
20. After installing cylinder head, measure distance between front end faces of cylinder block and cylinder head (right bank and left bank).
21. Distance must be between 0.555–0.587 inches (14.1–14.9 mm)
22. If measured value is out of the standard, re-install cylinder head.
23. Install or connect the following:
    - Water outlet, water pipe and heater pipe
    - Water inlet and thermostat assembly
    - Exhaust manifolds
    - Intake manifold
    - Fuel tube and fuel injector assembly
24. Install the camshafts.
25. Connect the battery cable.
26. Fill the cooling system with the proper coolant.
27. Fill the crankcase with oil.

➡**Before starting engine, check oil/fluid levels including engine coolant and engine oil. If less than required quantity, fill to the specified level**

28. Start engine. With engine speed increased, check again for fuel leakage at connection points.
29. Run engine to check for unusual noise and vibration.

30. Warm up engine thoroughly to check there is no leakage of fuel, exhaust gases, or any oil/fluids including engine oil and engine coolant.
31. Bleed air from lines and hoses of applicable lines, such as in cooling system.
32. After cooling down engine, again check oil/fluid levels including engine oil and engine coolant. Refill to the specified level, if necessary.

ENGINE ASSEMBLY

REMOVAL & INSTALLATION

See Figures 59 and 60.

1. Before servicing the vehicle, refer to the Precautions Section.
2. Properly release the fuel pressure.
3. Evacuate the A/C system.
4. Drain the engine oil.
5. Drain the cooling system.
6. Disconnect the negative battery cable.

✳✳ WARNING

Perform this step when engine is cold.

➡**Never spill engine coolant on drive belts**

7. Drain the transaxle fluid.
8. Matchmark and remove the hood.
9. Remove the engine cover, and the splash guards.
10. Drain engine coolant.
11. Remove or disconnect the following:
    - Battery and tray
    - Air inlet duct
    - Air duct and air cleaner case (upper) assembly with mass air flow sensor
    - Power brake booster vacuum hose
    - Drive belts
    - Radiator assembly, coolant reservoir, and system hoses
    - Front windshield wiper arm
    - Engine room harness from the ECM side
    - Heater hoses
    - Wheel and tires
    - A/C compressor with piping connected, and temporarily secure it aside
    - Fuel hose quick connector at vehicle piping side
    - Transaxle shift control cable.
    - Starter motor
    - Front exhaust tube
    - Reservoir tank for the power steering from engine compartment bracket and position it aside.

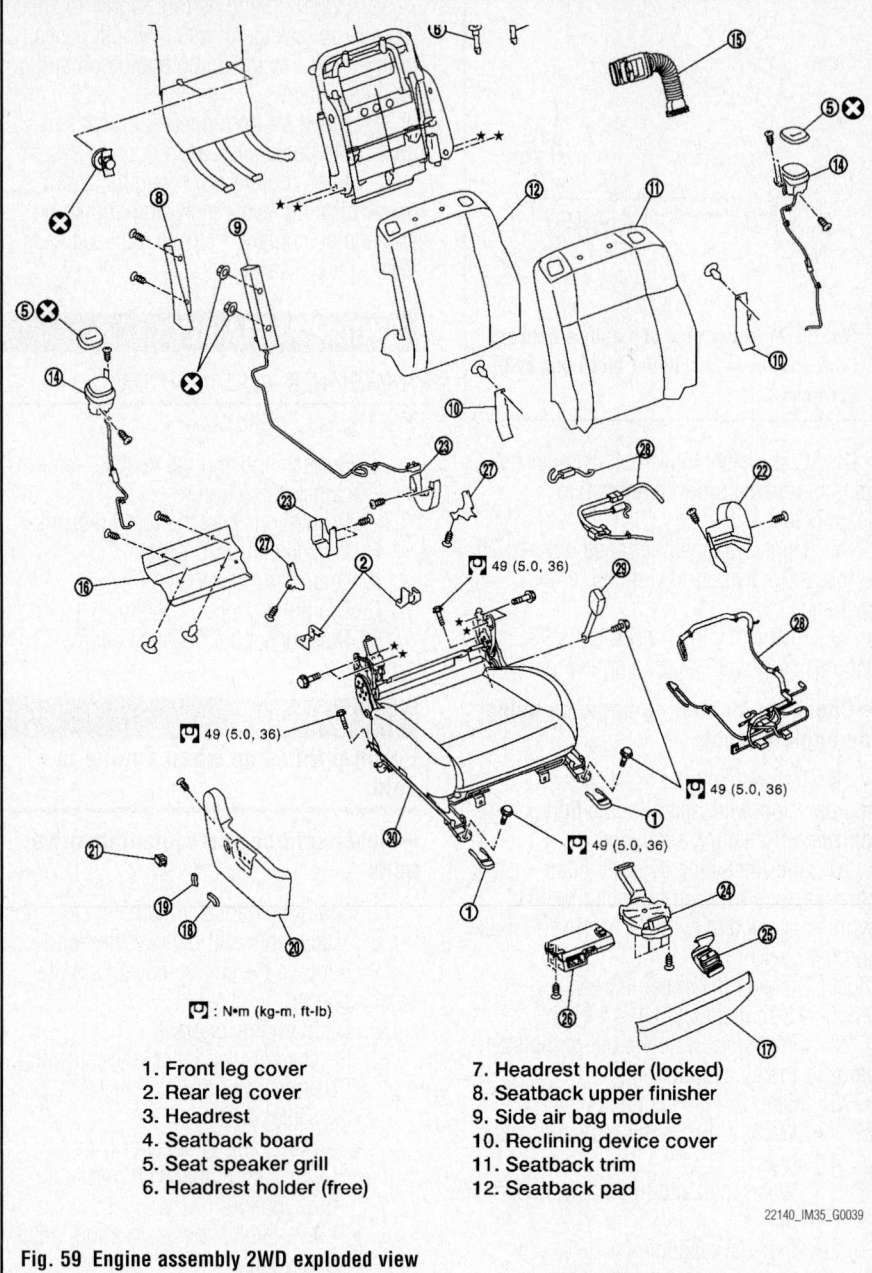

49 (5.0, 36)

49 (5.0, 36)

49 (5.0, 36)

49 (5.0, 36)

: N•m (kg-m, ft-lb)

1. Front leg cover
2. Rear leg cover
3. Headrest
4. Seatback board
5. Seat speaker grill
6. Headrest holder (free)
7. Headrest holder (locked)
8. Seatback upper finisher
9. Side air bag module
10. Reclining device cover
11. Seatback trim
12. Seatback pad

22140_IM35_G0039

Fig. 59 Engine assembly 2WD exploded view

1. Rear Display Unit
A. Plastic nuts
B. Plastic nuts
*Arrow shows the front of the vehicle

22140_IM35_G0040

Fig. 60 Engine assembly AWD exploded view

piping connected, and temporarily secure it to aside for vehicle side

12. Position a manual lift table caddy under the engine and transaxle assembly.

13. Remove the right engine mounting insulator

14. Remove mounting bolt between transverse link and front suspension member

15. Carefully lower the engine, transaxle, transfer case (AWD vehicles) and front suspension member assembly with the manual lift table caddy, avoiding interference with the vehicle body.

⁕⁕ WARNING

Before and during this procedure, always check if any harnesses are left connected. Avoid any damage to, or any oil or grease smearing or spills onto the engine mounting insulators.

16. Remove the crankshaft position sensor (POS).

17. Disconnect front suspension mounting nuts and bolts to remove engine, transaxle, transfer case (AWD vehicles) and front suspension member assembly as a unit.

18. Separate the engine, transaxle and transfer case (AWD vehicles) assembly and front suspension member.

To install:

19. Carefully raise the jack, or lower the lift to install the engine, the transmission assembly and front suspension member.

20. Install the front suspension member mounting bolts and nuts.

21. Tighten the front suspension member mounting bolts and nuts to 36 ft. lbs. (49 Nm).

22. Install the rear engine mounting member bolts.

23. Tighten the rear engine mounting member bolts and nuts to 36 ft. lbs. (49 Nm).

24. Remove the lift table caddy (commercial service tool) or equivalently rigid tool.

25. Install the transverse links mounting bolts at knuckle side.

26. Tighten the transverse links mounting bolts at knuckle side to 100 ft. lbs. (136 Nm).

27. Install the steering outer sockets from steering knuckle.

28. Tighten the steering outer sockets from steering knuckle to 25 ft. lbs. (34.4 Nm).

29. Install the lower ends of left and right strut from transverse link.

- Power steering gear from steering lower joint
- Steering outer socket from steering knuckle
- Stabilizer connecting rod
- Propeller shaft (AWD vehicles)

➡ **Mach mark the shaft and flange for reassembly.**

- Left and right front halfshafts (AWD)
- Power steering piping from power steering oil cooler
- Remove strut assembly and steering knuckle fixing nuts and bolts
- Front brake cylinder body with

30. Tighten the lower ends of left and right strut from transverse link to 79 ft. lbs. (107 Nm).

31. Install the front stabilizer at transverse link side.

32. Tighten the front stabilizer at transverse link side to 66 ft. lbs. (9 Nm).

33. Install the transmission joint bolts which pierce at oil pan (upper) lower rear side.

34. Tighten the transmission joint bolts which pierce at the oil pan (upper) lower rear side to 38 ft. lbs. (51 Nm).

35. Install the bolts attaching the drive plate to the torque converter.

36. Tighten the bolts attaching the drive plate to the torque converter to 38 ft. lbs. (51 Nm).

37. Install the rear plate cover from oil pan (upper).

38. Tighten the rear plate cover from oil pan (upper) to 41 ft. lbs. (55 Nm).

39. Install the rear propeller shaft using the matchmarks.

40. Tighten the rear propeller shaft to 55 ft. lbs. (7 Nm).

41. Install the steering lower shaft.

42. Tighten the steering lower shaft to 20 ft. lbs. (26.5 Nm).

43. Install the steering lower joint at power steering gear assembly side.

44. Tighten the steering lower joint at power steering gear assembly side to 20 ft. lbs. (26.5 Nm).

45. Install the three way catalyst and exhaust front tube.

46. Tighten the three way catalyst and exhaust front tube to 43 ft. lbs. (57.9 Nm).

47. Install or connect the following:
- Engine room harnesses to engine room side
- Engine room harness connectors at unit sides TCM, ECM and other
- Passenger-side kicking plate, dash side finisher, and glove box
- Reservoir tank of power steering oil pump and piping
- Fuel feed hose (with damper) and EVAP hose
- Grounding cable
- Battery positive cable at vehicle side
- Brake booster vacuum hose
- A/C piping to A/C compressor
- Wire bonding (between vehicle to left bank cylinder head)
- Heater hose from vehicle-side
- Radiator hoses (upper and lower)
- Air duct and air cleaner case assembly

- Cowl top cover (RH)
- Front and rear engine undercover
- Front road wheel and tires
- Engine cover
- Engine room cover (RH and LH)
- Hood
- Negative battery cable

48. Fill and bleed the cooling system.

49. Fill the engine with clean oil.

50. Fill the transaxle to the proper level.

51. Start the vehicle, check for leaks and repair if necessary.

EXHAUST MANIFOLD

REMOVAL & INSTALLATION

See Figures 61 and 62.

1. Before servicing the vehicle, refer to the Precautions Section.

2. Disconnect the negative battery cable. Drain the cooling system.

3. Remove the engine cover. Remove the undercover.

4. Remove the radiator cover grilles, air duct (inlet) and air cleaner cases (upper) with air flow sensor and air duct assembly.

5. Remove the front wiper arm. Remove the upper and lower extension cowl top.

6. Remove the intake manifold upper and lower collectors.

7. Remove the exhaust front tube mounting bracket. Remove the exhaust front tube. Remove the heat insulator.

8. Disconnect the harness connector and remove the air fuel ratio sensor on both sides. Be sure to use the proper removal tool. Matchmark the sensors to aid in reinstallation.

✽✽ WARNING

Be careful not to damage the air flow ratio sensor. Discard any heated oxygen sensor which has been dropped from a height of more than 0.5 m (19.7 inches) onto a hard surface such as a concrete floor; replace with a new sensor.

9. Disconnect the harness connector and remove the heated oxygen sensor on both sides. Be sure to use the proper removal tool. Matchmark the sensors to aid in reinstallation.

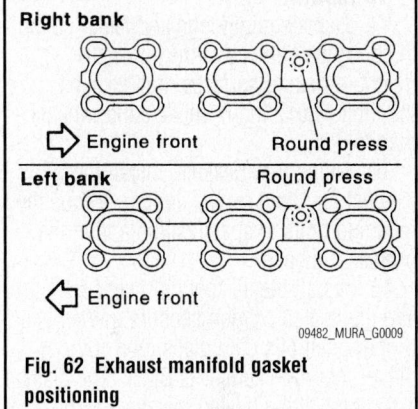

Fig. 62 Exhaust manifold gasket positioning

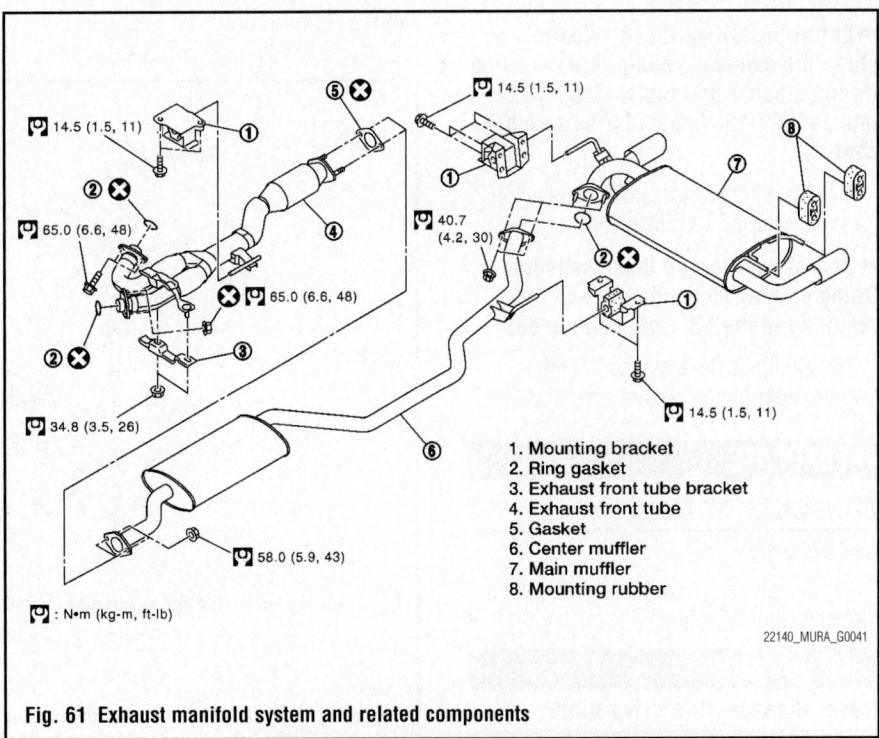

1. Mounting bracket
2. Ring gasket
3. Exhaust front tube bracket
4. Exhaust front tube
5. Gasket
6. Center muffler
7. Main muffler
8. Mounting rubber

Fig. 61 Exhaust manifold system and related components

⁂ WARNING

Be careful not to damage heated oxygen sensor. Discard any heated oxygen sensor which has been dropped from a height of more than 0.5 m (19.7 inches) onto a hard surface such as a concrete floor; replace with a new sensor.

10. Remove the exhaust manifold covers and the three way catalyst covers.

11. Remove bolts in the reverse order of illustration to remove three way catalyst supports.

12. Remove the left and right three way catalysts by loosening the bolts first and then removing the nuts.

13. Loosen the nuts in the reverse order of the torque sequence and remove the exhaust manifolds from the engine.

To install:

14. Use a straightedge and feeler gauge to check the flatness of the exhaust manifold mating surfaces. If it exceeds the limit (0.012 inch), replace the exhaust manifold.

15. Using new gaskets, install the exhaust manifold on the engine. Torque the retaining bolts to specification and in the proper sequence.

16. Install the air fuel ratio sensors and the heated oxygen sensors in their proper positions. The glass tube color of the air fuel ratio sensor is Black. The glass tube color of the heated oxygen sensor is White.

➡**Before installing these sensors, clean the exhaust system threads using thread cleaner and tool J43897-18 or tool J43897-12. Apply anti-seize lubricant.**

17. Install the sensors. See illustration for proper torque specifications.

➡**Do not over torque these sensors. Doing so may cause damage to them, resulting in the ML light coming on.**

18. Continue the installation in the reverse order of the removal procedure.

FLEXPLATE

REMOVAL & INSTALLATION

See Figure 63.

1. Before servicing the vehicle, refer to the Precautions Section.

⁂ WARNING

Never disassemble drive plate.

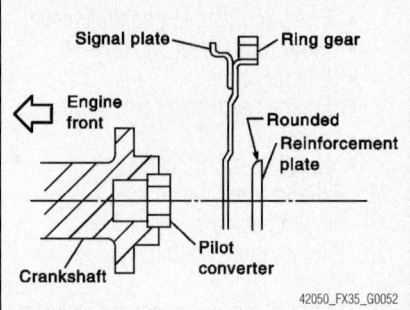

Fig. 63 Flexplate alignment and related components

⁂ WARNING

Never place drive plate with signal plate facing down.

➡**When handling signal plate, take care not to damage or scratch it.**

➡**Handle signal plate in a manner that prevents it from becoming magnetized.**

2. Remove the transmission.

3. Remove the flexplate retaining bolts.

4. Remove the flexplate from the engine.

To install:

5. When installing the flexplate be sure to correctly align the crankshaft side guide pin and drive side guide pin hole.

➡**If not correctly aligned the engine will run rough and turn on the MIL light.**

6. Install the flexplate and reinforcement plate, as shown in the illustration.

7. Hold the ring gear with the ring gear stopper, using tool SST:KV10117700 or equivalent.

8. Tighten the mounting bolts to specification and in a crosswise sequence in several passes.

➡**Be sure that the dowel pin is installed at the rear end of the crankshaft.**

9. Continue the installation in the reverse order of the removal procedure.

10. Tighten mounting bolts diagonally order.

11. Tighten the bolts to 65 ft.lbs. (88.2 Nm).

INTAKE MANIFOLD

REMOVAL & INSTALLATION

See Figures 64 and 65.

1. Before servicing the vehicle, refer to the Precautions Section.

2. Disconnect the negative battery cable.

3. Properly relieve the fuel system pressure.

4. Remove the upper and lower intake manifold collectors.

5. Remove the fuel tube and fuel injector assembly.

6. Loosen bolts and nuts in reverse order of illustration to remove intake manifold assembly from the engine.

To install:

7. Using straightedge and feeler gauge,

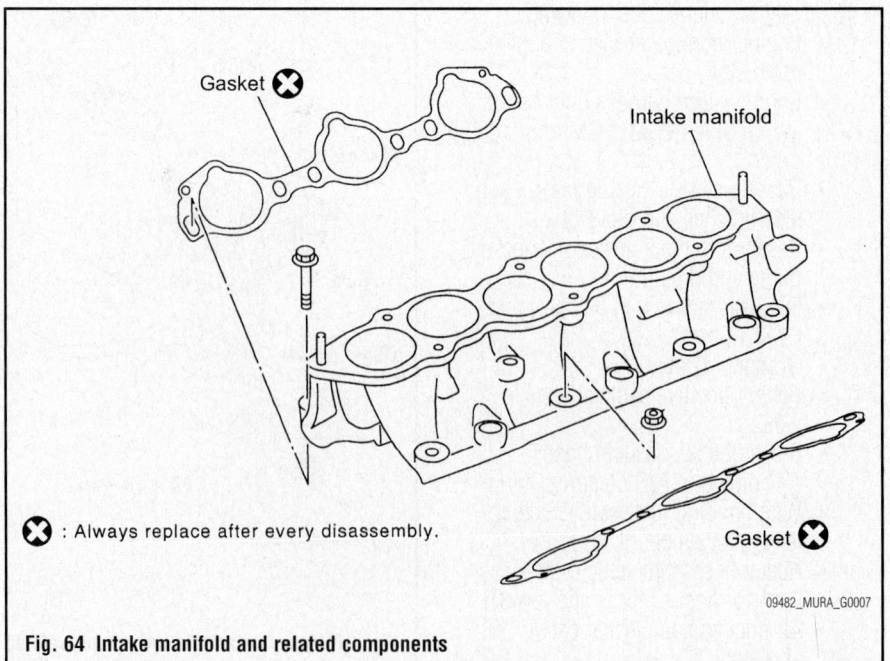

❌ : Always replace after every disassembly.

Fig. 64 Intake manifold and related components

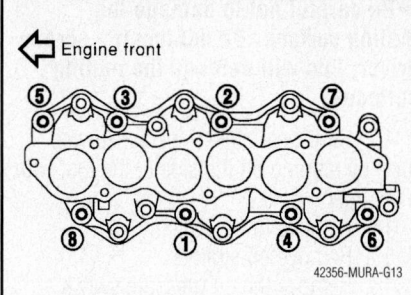

Fig. 65 Intake manifold bolt torque sequence

inspect the surface distortion of each surface on the intake manifold. If it exceeds the limit (0.004 inch), replace the intake manifold.

8. Installation is the reverse of the removal procedure.

9. If the intake manifold stud bolts were removed, install them and torque to 8 inch lbs.

10. Torque the intake manifold retaining bolts to specification and in the proper sequence.

11. Be sure to fill the engine with the proper grade and type coolant.

12. Start the engine and check for leaks, correct as required.

OIL PAN

REMOVAL & INSTALLATION

2WD Models

See Figures 66 through 71.

1. Before servicing the vehicle, refer to the Precautions Section.

➡**When removing the upper oil pan**

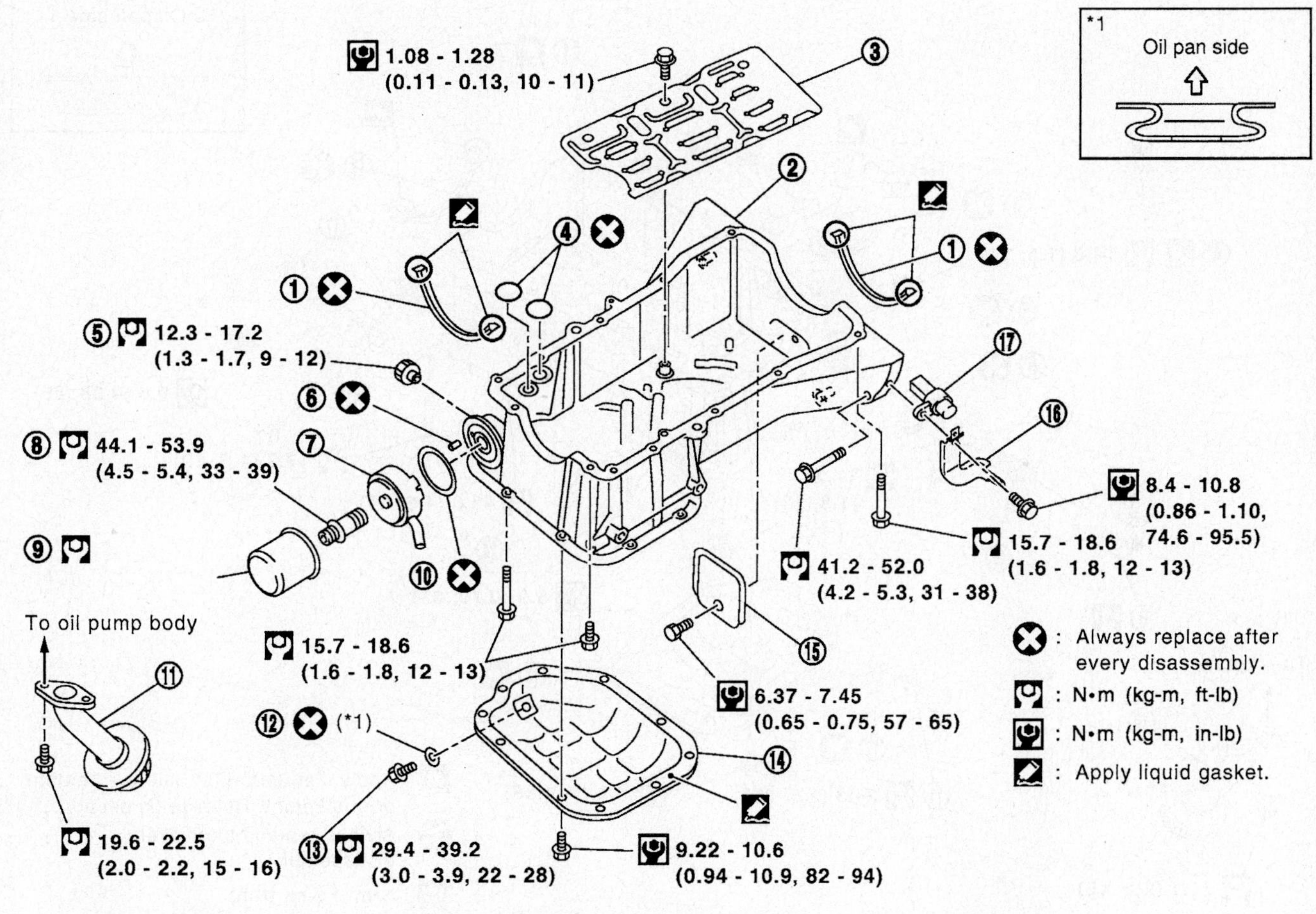

*1 Oil pan side ⬆

1.08 - 1.28 (0.11 - 0.13, 10 - 11)

5 12.3 - 17.2 (1.3 - 1.7, 9 - 12)

8 44.1 - 53.9 (4.5 - 5.4, 33 - 39)

9

To oil pump body

15.7 - 18.6 (1.6 - 1.8, 12 - 13)

12 ❌ (*1)

19.6 - 22.5 (2.0 - 2.2, 15 - 16)

13 29.4 - 39.2 (3.0 - 3.9, 22 - 28)

41.2 - 52.0 (4.2 - 5.3, 31 - 38)

6.37 - 7.45 (0.65 - 0.75, 57 - 65)

9.22 - 10.6 (0.94 - 10.9, 82 - 94)

8.4 - 10.8 (0.86 - 1.10, 74.6 - 95.5)

15.7 - 18.6 (1.6 - 1.8, 12 - 13)

❌ : Always replace after every disassembly.

🔧 : N•m (kg-m, ft-lb)

🔧 : N•m (kg-m, in-lb)

✏ : Apply liquid gasket.

1. Gasket
2. Upper oil pan
3. Baffle plate
4. O-ring
5. Oil pressure switch
6. Relief valve
7. Oil cooler
8. Oil cooler connector
9. Oil filter
10. Gasket
11. Oil strainer
12. Gasket
13. Drain plug
14. Lower oil pan
15. Rear cover plate
16. Heated oxygen sensor (bank 2) harness clamp (2WD models)
17. Crankshaft position sensor (POS)

42356-mura-G33

Fig. 66 Oil pan and related components

from the engine, first remove the crankshaft position sensor (POS). Be careful not to damage sensor edges or signal plate teeth.

2. Remove engine cover.
3. Remove right splash guard.
4. Remove the front right road wheel and tire.
5. Drain engine oil.
6. Drain engine coolant.
7. Remove oil filter.
8. Remove oil cooler and water pipes.

9. Remove all drive belts.
10. Remove A/C compressor with piping connected, and temporarily secure it aside.
11. Remove exhaust front tube.
12. Remove the heated oxygen sensor 2 (bank 2) and remove the three way catalyst (manifold) (bank 2) from the exhaust manifold.
13. Loosen lower oil pan bolts in reverse order of the installation sequence.
14. Insert a seal cutter (special service tool) between the lower oil pan and the upper oil pan.

→Be careful not to damage the mating surface. Do not insert a screwdriver; this will damage the mating surfaces.

15. Slide seal cutter (special service tool) by tapping on the side of the tool with a hammer.
16. Remove lower oil pan.
17. Remove oil strainer.
18. Remove the oil pressure switch.
19. If not already removed, remove crankshaft position sensor (POS).

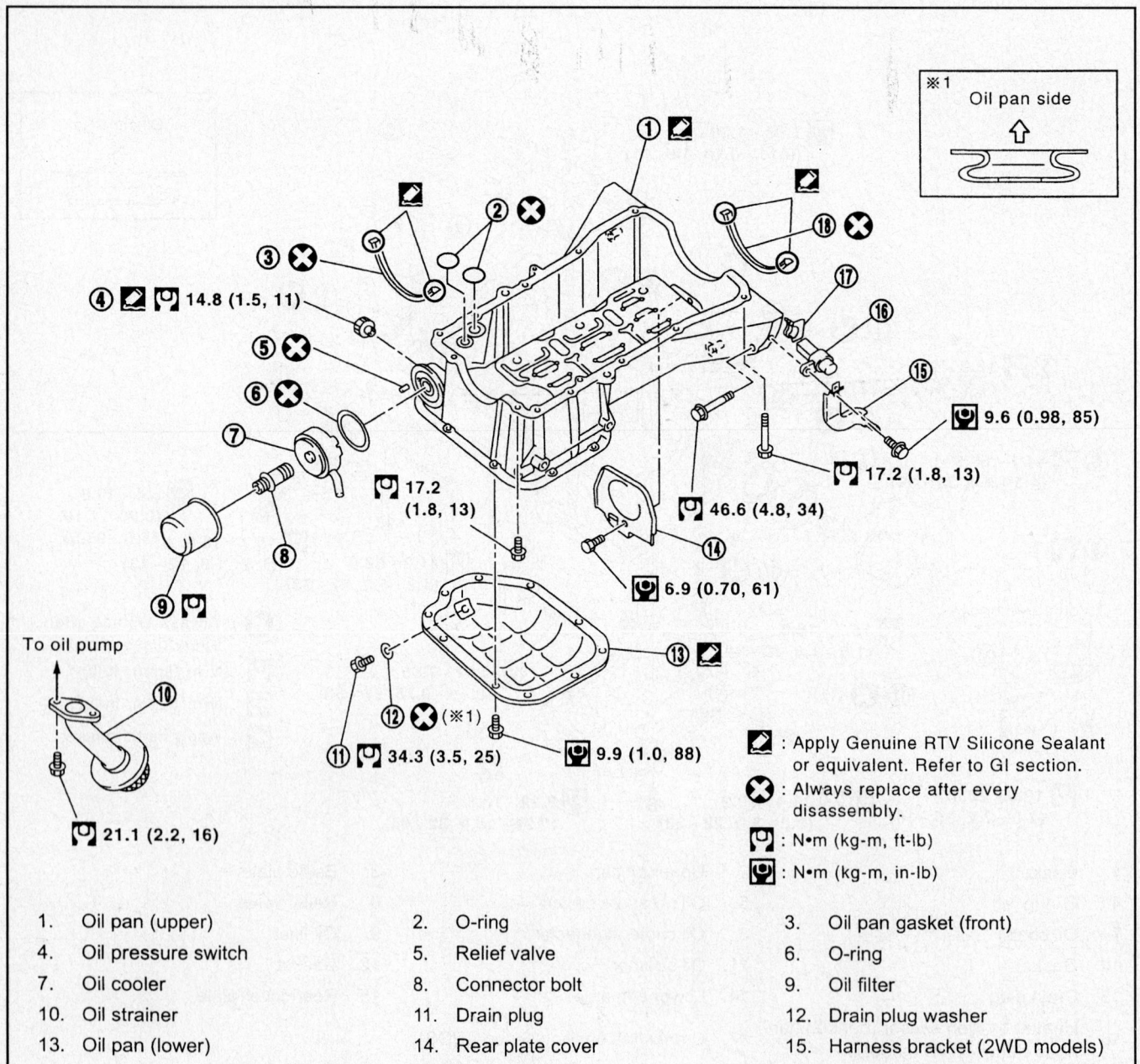

14.8 (1.5, 11)
17.2 (1.8, 13)
9.6 (0.98, 85)
17.2 (1.8, 13)
46.6 (4.8, 34)
6.9 (0.70, 61)
34.3 (3.5, 25)
9.9 (1.0, 88)
21.1 (2.2, 16)

To oil pump

※1 Oil pan side

: Apply Genuine RTV Silicone Sealant or equivalent. Refer to GI section.
: Always replace after every disassembly.
: N•m (kg-m, ft-lb)
: N•m (kg-m, in-lb)

1. Oil pan (upper)	2. O-ring	3. Oil pan gasket (front)
4. Oil pressure switch	5. Relief valve	6. O-ring
7. Oil cooler	8. Connector bolt	9. Oil filter
10. Oil strainer	11. Drain plug	12. Drain plug washer
13. Oil pan (lower)	14. Rear plate cover	15. Harness bracket (2WD models)
16. Crankshaft position sensor (POS)	17. Seal rubber	18. Oil pan gasket (rear)

09482_MURA_G0015

Fig. 67 Oil pan and related components

➡**Handle carefully to avoid dropping and shocks. Do not disassemble. Do not allow metal powder to adhere to magnetic part at sensor tip. Do not place sensors in a location where they are exposed to magnetism.**

20. Remove the four engine-to-transaxle bolts.

21. Remove upper oil pan. Loosen bolts in reverse order of the installation sequence.

22. Insert an appropriate size tool into the notch of the upper oil pan shown (1). Pry off the upper oil pan by moving the tool up and down shown (2).

23. Remove O-rings from the bottom of the cylinder block and oil pump body.

24. Remove oil pan gasket.

To install:

Note the following:
- Use a scraper to remove old liquid gasket from mating surfaces
- Also remove the old liquid gasket from mating surface of the cylinder block
- Remove the old liquid gasket from the bolt holes and threads

➡**Do not scratch or damage the mating surfaces when cleaning off the old liquid gasket.**

- Apply Genuine RTV Silicone Sealant or equivalent, to the front timing chain case gasket and the rear oil seal retainer gasket shown
- To install, align protrusion of oil pan gasket with notches of front timing chain case and rear oil seal retainer
- Install oil pan gasket with smaller arc to front timing chain case side
- Install new O-rings on the cylinder block and oil pump side
- Apply a continuous bead of sealant to the cylinder block mating surface of the upper oil pan to a limited

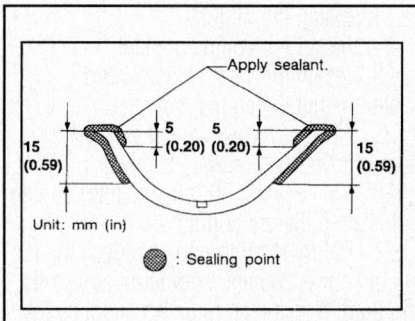

Fig. 68 RTV sealer application at the timing case

portion shown. Use RTV silicone sealant or equivalent
- For bolt holes with marks (5 locations), apply liquid gasket outside the holes
- Apply a bead of 4.5 to 5.5 mm (0.177 to 0.217 in) diameter to area "A"
- Attaching within 5 minutes after coating
- Install the upper oil pan. Torque the bolts to specification and in the proper sequence. There are two types of mounting bolts.

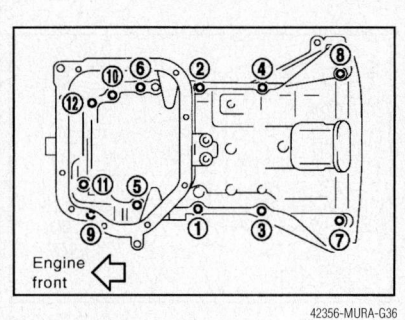

Fig. 69 Upper oil pan bolt torque sequence

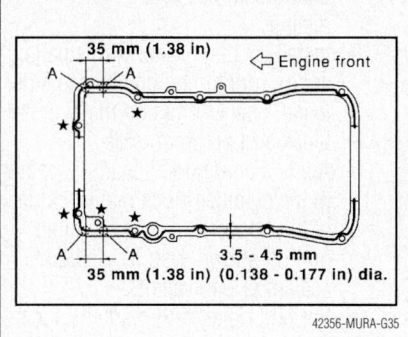

Fig. 70 RTV sealer application on the pan

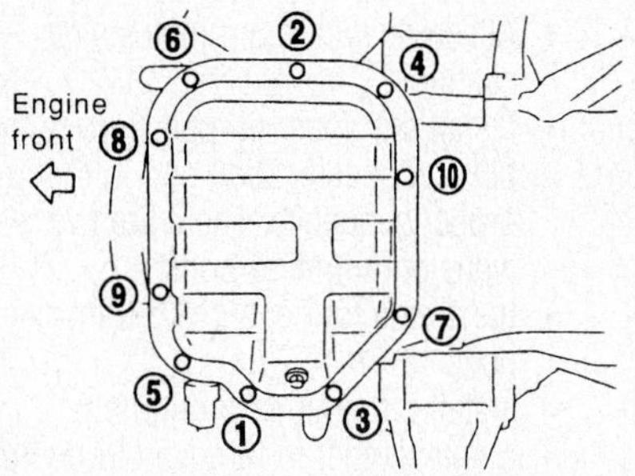

- Install the four engine-to-transaxle bolts
- Install oil strainer to oil pump
- Use a scraper to remove old liquid gasket from mating surfaces. Also, remove old liquid gasket from mating surface of upper oil pan.
- Apply a continuous bead of sealant to the lower oil pan. Use RTV silicone sealant. Be sure the sealant is 4.5–5.5 mm (0.177–0.217 inch) wide. Attach within 5 minutes after coating.
- Install lower oil pan. Torque the bolts to specification and in the proper sequence.
- Install oil pan drain plug.

25. Install the heated oxygen sensor 2 (bank 2) and the three way catalyst.

26. Install the A/C compressor with piping.

27. Install the drive belts.

28. Install the oil cooler and water pipes.

29. Install the oil filter.

30. Install the engine coolant.

31. Install the right splash guard.

32. Install the engine cover.

33. Heated oxygen sensor 2 (bank 2) and remove the three way catalyst

34. Wait at least 30 minutes after oil pan is installed, before adding engine oil.

35. Before starting engine, check the levels of engine coolant, engine oil and working fluid. If less than required quantity, fill to the specified level.

36. Use procedure below to check for fuel leakage.

37. Turn ignition switch ON (with engine stopped). With fuel pressure applied to fuel piping, check for fuel leakage at connection points.

38. Start engine. With engine speed increased, check again for fuel leakage at connection points.

39. Run engine to check for unusual noise and vibration.

40. Warm up engine thoroughly to make sure there is no leakage of engine coolant, engine oil and working fluid, fuel and exhaust gas.

41. Bleed air from passages in pipes and tubes of applicable lines, such as in cooling system.

42. After cooling down engine, again check amounts of engine coolant, engine oil and working fluid. Refill to specified level, if necessary.

AWD Models

See Figures 72 through 75.

1. Before servicing the vehicle, refer to the Precautions Section.

Fig. 71 Lower oil pan bolt torque sequence

➥When removing the upper oil pan from the engine, first remove the crankshaft position sensor (POS). Be careful not to damage sensor edges or signal plate teeth.

2. Remove engine assembly from vehicle, and separate front suspension member, transaxle and transfer case assembly from engine.

3. Remove the engine and mount it onto the engine stand.

4. Drain engine oil.

5. Remove oil filter.

6. Remove oil cooler and water pipes.

7. Remove the heated oxygen sensor 2 (bank 2) and remove the three way catalyst (manifold) (bank 2) from the exhaust manifold.

8. Loosen the lower oil pan bolts in reverse order of the installation sequence.

9. Insert a seal cutter (special service tool) between the lower oil pan and the upper oil pan.

➥Be careful not to damage the mating surface. Do not insert a screwdriver, this will damage the mating surfaces.

10. Slide seal cutter (special service tool) by tapping on the side of the tool with a hammer. Remove lower oil pan.

11. Remove oil strainer.

12. Remove the oil pressure switch.

13. Remove upper oil pan. Loosen bolts in reverse order of the installation sequence.

14. Insert an appropriate size tool into the notch of the upper oil pan shown (1). Pry off the upper oil pan by moving the tool up and down shown (2).

15. Remove O-rings from the bottom of the cylinder block and oil pump body.

16. Remove oil pan gasket.

To install:
Installation is the reverse of removal. Note the following:
- Use a scraper to remove old liquid gasket from mating surfaces
- Also remove the old liquid gasket from mating surface of the cylinder block
- Remove the old liquid gasket from the bolt holes and threads

➥Do not scratch or damage the mating surfaces when cleaning off the old liquid gasket.

- Apply Genuine RTV Silicone Sealant or equivalent, to the front timing chain case gasket and the rear oil seal retainer gasket shown
- To install, align protrusion of oil pan gasket with notches of front

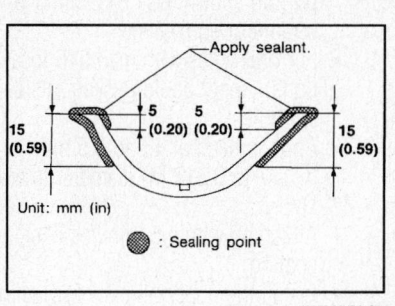

Fig. 72 RTV sealer application at the timing case

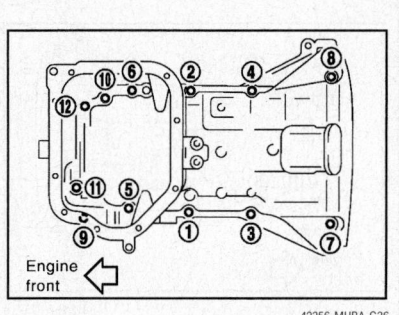

Fig. 73 Upper oil pan bolt torque sequence

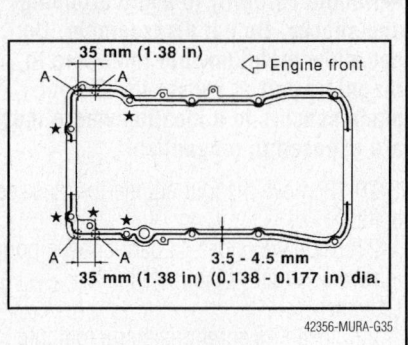

Fig. 74 RTV sealer application on the pan

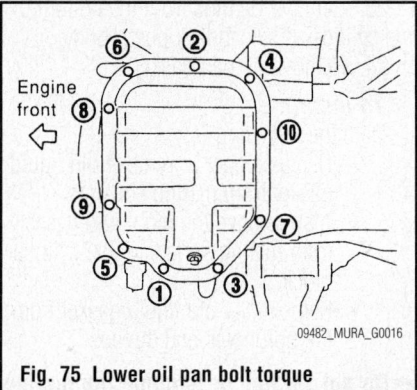

Fig. 75 Lower oil pan bolt torque sequence

timing chain case and rear oil seal retainer
- Install oil pan gasket with smaller arc to front timing chain case side
- Install new O-rings on the cylinder block and oil pump side
- Apply a continuous bead of sealant to the cylinder block mating surface of the upper oil pan to a limited portion shown. Use RTV silicone sealant or equivalent.
- For bolt holes with marks (5 locations), apply liquid gasket outside the holes
- Apply a bead of 4.5 to 5.5 mm (0.177 to 0.217 in) diameter to area "A"
- Attaching within 5 minutes after coating
- Install the upper oil pan. Torque the bolts to specification and in the proper sequence. There are two types of mounting bolts
- Install the four engine-to-transaxle bolts
- Install oil strainer to oil pump
- Use a scraper to remove old liquid gasket from mating surfaces. Also, remove old liquid gasket from mating surface of upper oil pan

- Apply a continuous bead of sealant to the lower oil pan. Use RTV silicone sealant. Be sure the sealant is 4.5–5.5 mm (0.177–0.217 inch) wide. Attach within 5 minutes after coating
- Install lower oil pan. Torque the bolts to specification and in the proper sequence
- Install oil pan drain plug

17. Install the heated oxygen sensor 2 (bank 2) and the three way catalyst.

18. Install the A/C compressor with piping.

19. Install the drive belts.

20. Install the oil cooler and water pipes.

21. Install the oil filter.

22. Install the engine coolant.

23. Install the right splash guard.

24. Install the engine cover.

25. Heated oxygen sensor 2 (bank 2) and remove the three way catalyst.

26. Wait at least 30 minutes after oil pan is installed, before adding oil.

27. Before starting engine, check the levels of engine coolant, engine oil and working fluid. If less than required quantity, fill to the specified level.

28. Use procedure below to check for fuel leakage.

29. Turn ignition switch ON (with engine stopped). With fuel pressure applied to fuel piping, check for fuel leakage at connection points.

30. Start engine. With engine speed increased, check again for fuel leakage at connection points.

31. Run engine to check for unusual noise and vibration.

32. Warm up engine thoroughly to make sure there is no leakage of engine coolant, engine oil and working fluid, fuel and exhaust gas.

33. Bleed air from passages in pipes and tubes of applicable lines, such as in cooling system.

34. After cooling down engine, again check amounts of engine coolant, engine oil and working fluid. Refill to specified level, if necessary.

OIL PUMP

REMOVAL & INSTALLATION

See Figure 76.

1. Before servicing the vehicle, refer to the Precautions Section.

2. Remove the upper and lower oil pans.

3. Remove the oil strainer.

4. Remove front timing chain case and timing chain (primary).

5. Remove oil pump assembly.

To install:

6. Installation is the reverse of the removal procedure.

7. Be sure to use new gaskets.

8. When installing, align crankshaft flat faces with inner rotor flat faces.

MAIN BEARING TORQUE SEQUENCE

See Figures 77 and 78.

1. Install main bearing cap bolts in numerical order as shown in the figure as follows:

   a. Inspect the outer diameter of main bearing cap bolt.

   b. Apply new engine oil to threads and seat surfaces of main bearing cap bolts.

   c. Tighten main bearing cap bolts in several different steps.

   d. Tighten main bearing cap bolts to 26 ft. lbs. (35.3 Nm).

   e. Turn all main bearing cap bolts 90 degrees clockwise (angle tightening).

2. Check tightening angle. Never make judgment by visual inspection.

➡**After installing main bearing cap bolts, check that crankshaft can be rotated smoothly by hand.**

➡**Check the crankshaft end play.**

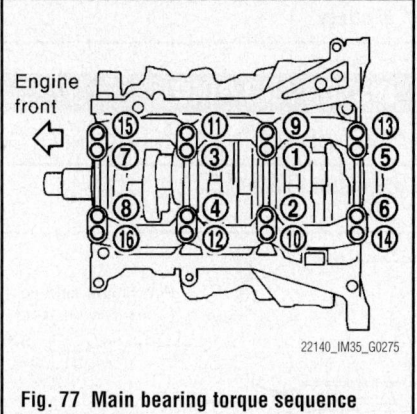

Fig. 77 Main bearing torque sequence

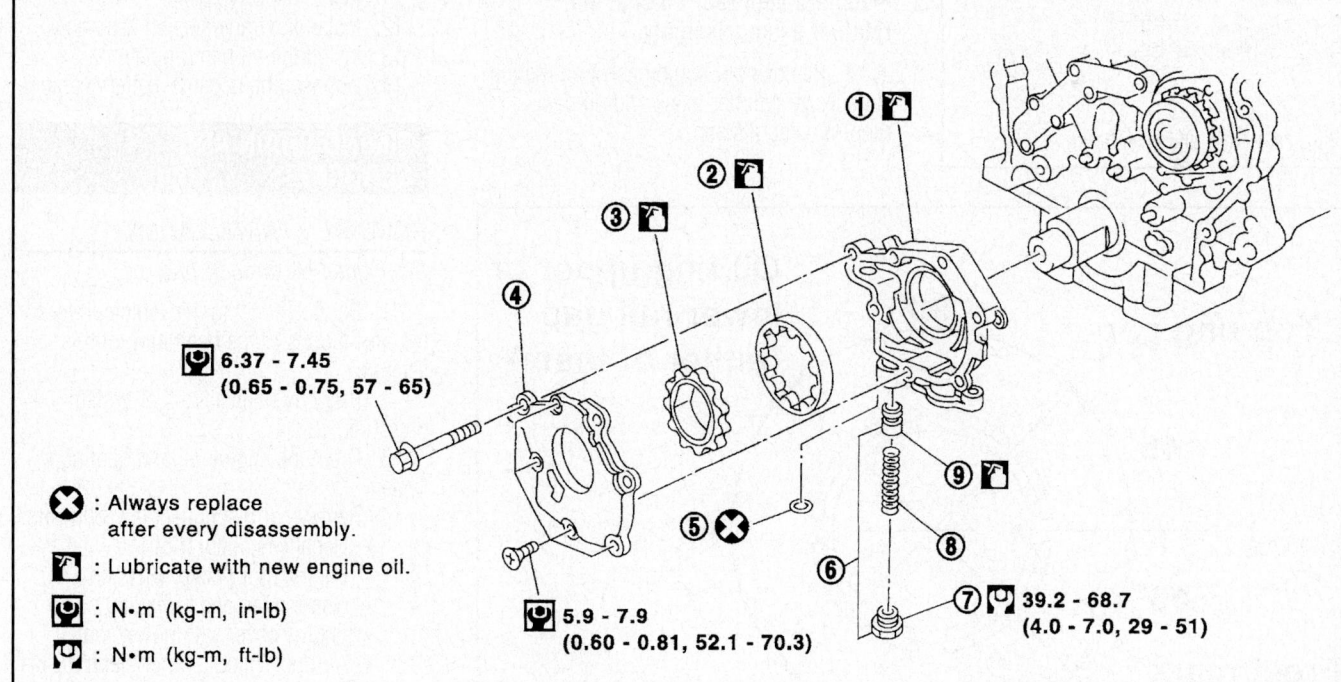

6.37 - 7.45
(0.65 - 0.75, 57 - 65)

❌ : Always replace after every disassembly.

🔧 : Lubricate with new engine oil.

🔧 : N•m (kg-m, in-lb)

🔧 : N•m (kg-m, ft-lb)

5.9 - 7.9
(0.60 - 0.81, 52.1 - 70.3)

39.2 - 68.7
(4.0 - 7.0, 29 - 51)

1. Oil pump body	2. Outer rotor	3. Inner rotor
4. Oil pump cover	5. O-ring	6. Regulator valve set
7. Regulator valve plug	8. Spring	9. Regulator valve

42356-MURA-G37

Fig. 76 Oil pump and related components

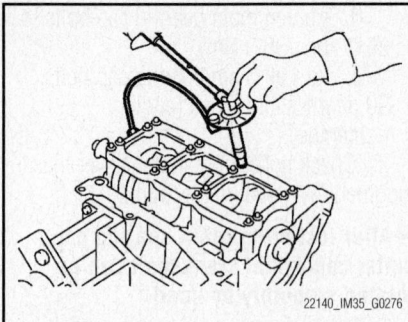

Fig. 78 Angle wrench Special Service Tool (SST) to check tightening angle of main bearings

PISTON AND RING

POSITIONING

See Figures 79 and 80.

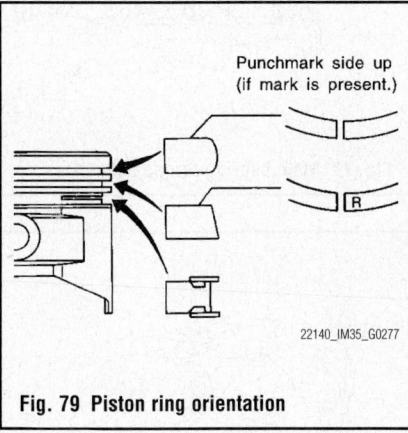

Fig. 79 Piston ring orientation

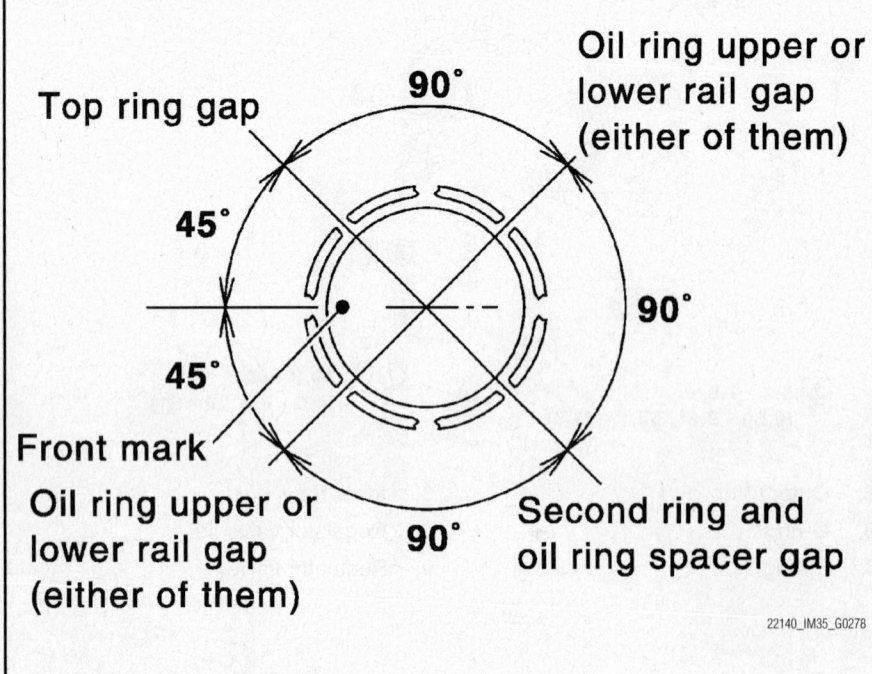

Fig. 80 Piston ring positioning

1. If there is stamped mark on ring, mount it with marked side up.

2. If there is no stamp on ring, no specific orientation is required for installation.

3. Stamped mark:

4. Top ring : -

5. Second ring : R

6. Position each ring with the gap as shown in the figure referring to the piston front mark.

REAR MAIN SEAL

REMOVAL & INSTALLATION

See Figure 81.

1. Before servicing the vehicle, refer to the service precaution.

2. Disconnect the negative battery cable.

3. Remove oil pan (upper).

4. Remove transmission assembly.

5. Remove drive plate.

6. Cut away liquid gasket and remove rear oil seal retainer.

✳✳ WARNING

Be careful not to damage mounting surface.

➡**Regard both rear oil seal and retainer as an assembly.**

7. Remove old liquid gasket on mating surfaces of cylinder block and oil pan (upper) using a scraper.

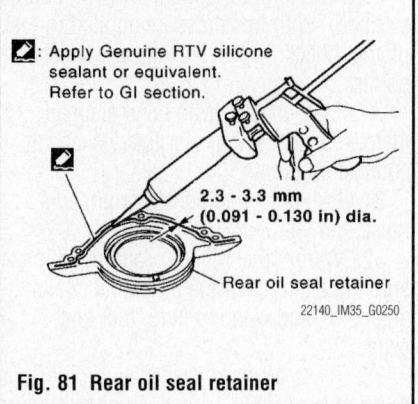

Fig. 81 Rear oil seal retainer

To install:

8. Apply new engine oil to both oil seal lip and dust seal lip of new rear oil seal retainer.

9. Apply a continuous bead of liquid gasket with the tube presser.

➡ **Assembly should be done within 5 minutes after coating.**

10. Install rear oil seal retainer to cylinder block.

➡**Check the garter spring is in position and seal lips not inverted.**

11. Install the drive plate.

12. Install the transmission assembly.

13. Install the oil pan (upper).

14. Connect the negative battery cable.

TIMING CHAIN AND SPROCKETS

REMOVAL & INSTALLATION

See Figures 82 through 118.

1. Before servicing the vehicle, refer to the precautions in the beginning of this section.

2. Properly relieve the fuel system pressure.

3. Drain the engine oil and cooling system.

4. Remove or disconnect the following:
   • Negative battery cable
   • Engine room cover (RH and LH).
   • Cooling fan and radiator
   • Engine cover with power tool
   • Air duct (inlet) and air cleaner case assembly
   • Front and rear engine undercover with power tool
   • Radiator hose (upper and lower) and A/T fluid cooler hose
   • Engine harnesses removing their brackets from front timing chain case
   • Drive belts

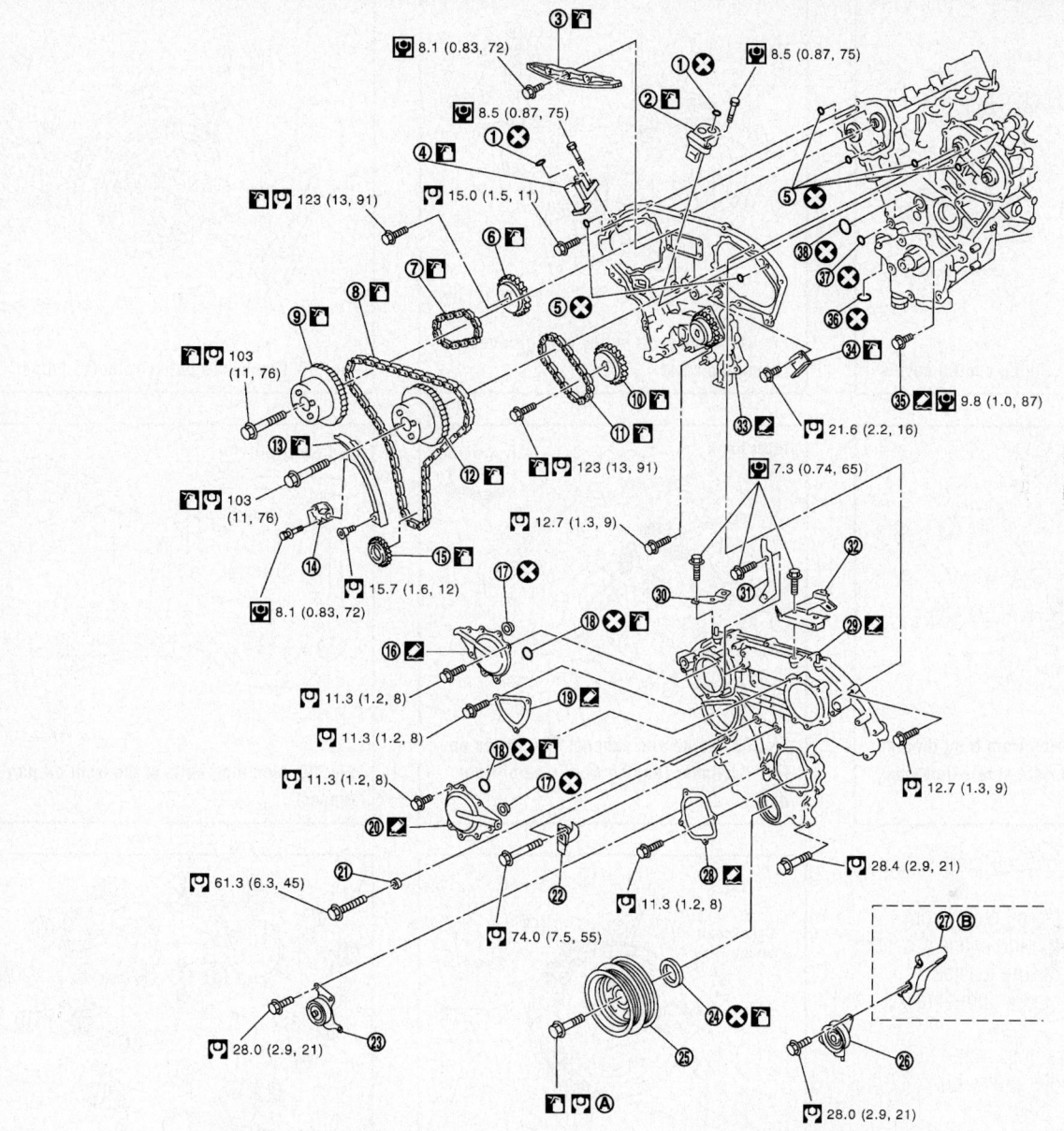

1. O-ring
2. Timing chain tensioner (secondary)
3. Internal chain guide
4. Timing chain tensioner (secondary)
5. O-ring
6. Camshaft sprocket (EXH)
7. Timing chain (secondary)
8. Timing chain (primary)
9. Camshaft sprocket (INT)
10. Camshaft sprocket (EXH)
11. Timing chain (secondary)
12. Camshaft sprocket (INT)
13. Slack guide
14. Timing chain tensioner (primary)
15. Crankshaft sprocket
16. Intake valve timing control cover
17. Collared O-ring
18. Seal ring

22. Water hose clamp
23. Idler pulley
24. Front oil seal
25. Crankshaft pulley
26. Idler pulley
27. A/C compressor bracket
28. Water pump cover
29. Front timing chain case
30. Bracket
31. Bracket
32. Bracket
33. Rear timing chain case
34. Tension guide
35. Water drain plug (front side)
36. O-ring
37. O-ring
38. O-ring

22140_IM35_G0296

Fig. 82 Exploded view of the timing chain and sprockets

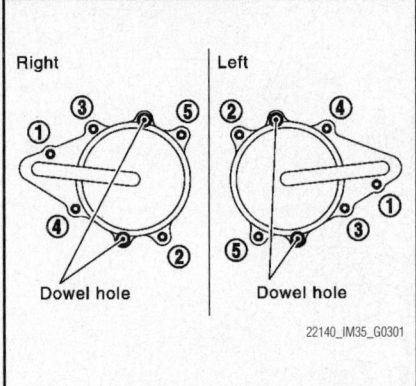

Fig. 83 Intake valve timing control covers

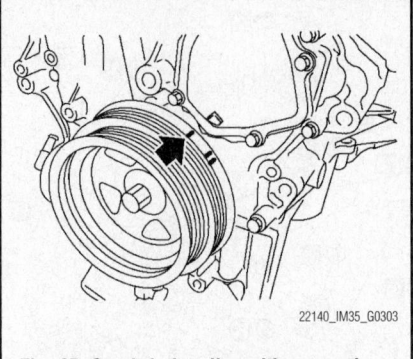

Fig. 85 Crankshaft pulley with grooved timing indicator

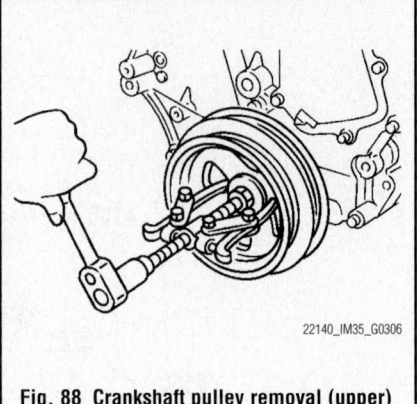

Fig. 88 Crankshaft pulley removal (upper)

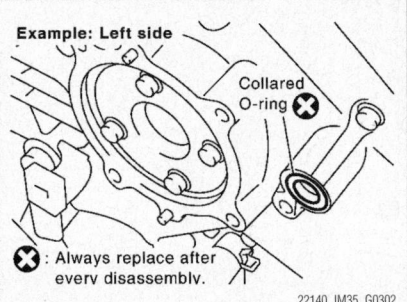

Fig. 84 Collared O-ring from front timing chain case (left and right side)—left side shown

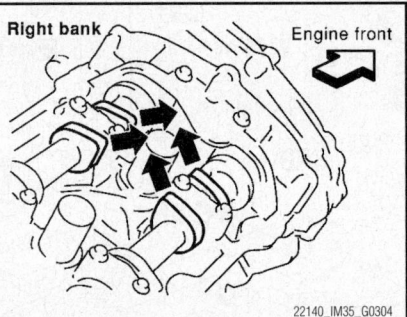

Fig. 86 Intake and exhaust cam noses on No. 1 cylinder (engine front side of right bank)

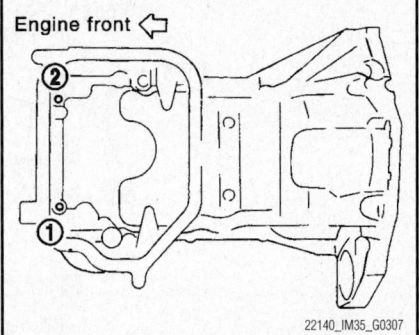

Fig. 89 Mounting bolts of the front oil pan (upper)

- Intake manifold collectors (upper and lower)
- Power steering oil pump from bracket with piping connected, and temporarily secure it aside
- Power steering oil pump bracket
- Alternator
- Water bypass hose, water hose clamp and idler pulley bracket from front timing chain case
- Intake valve timing control covers

5. Loosen mounting bolts in reverse order as shown in the figure.

6. Shaft is internally jointed with camshaft sprocket (INT) center hole. When removing, keep it horizontal until it is completely disconnected.

➡**Use the seal cutter SST: KV10111100 (J37228) to cut liquid gasket for removal.**

7. Remove or disconnect the following:
- Collared O-ring from front timing chain case (left and right side).
- Rocker covers (right bank and left bank)

8. Obtain No. 1 cylinder at TDC of its compression stroke as follows:
a. Rotate crankshaft pulley clockwise to align timing mark (grooved line without color) with timing indicator.

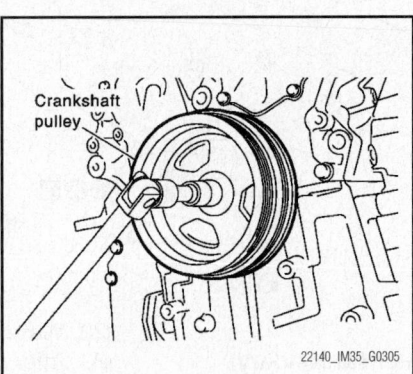

Fig. 87 Crankshaft pulley

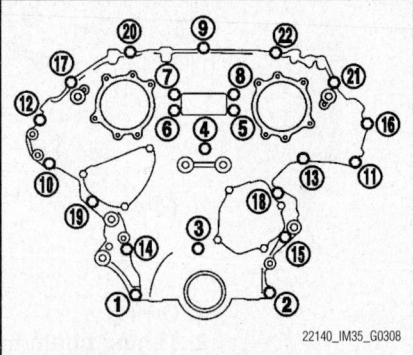

Fig. 90 Front timing chain cover bolt locations key

b. Check that intake and exhaust cam noses on No. 1 cylinder (engine front side of right bank) are located as shown in the figure. If not, turn crankshaft one revolution (360 degrees) and align as shown in the figure.

9. Remove crankshaft pulley as follows:
a. Remove rear cover plate (2WD models) or starter motor (AWD models) and set ring gear stopper.
b. Loosen crankshaft pulley bolt and locate bolt seating surface as 10 mm (0.39 in) from its original position.

➡**Never remove crankshaft pulley bolt**

as it will be used as a supporting point for suitable puller.

c. Place suitable puller tab on holes of crankshaft pulley, and pull crankshaft pulley through.

✳✳ WARNING
Never put suitable puller tab on crankshaft pulley, as this will damage internal damper.

10. Remove or disconnect the following:
- Oil pan (lower)
- Two mounting bolts in front of oil

pan (upper) reverse order shown in the figure
- Mounting bolts of the front timing cover with power tool in reverse order as shown in the figure.

11. Insert suitable tool into the notch at the top of front timing chain case as shown (1).

12. Pry off case by moving a tool as shown (2).

➡**Never use a screwdrivers or something similar.**

➡**After removal, handle front timing chain case carefully so it does not tilt, cant, or warp under a load.**

➡**Cut gasket to ease removal.**

13. Remove or disconnect the following:
- O-rings from rear timing chain case

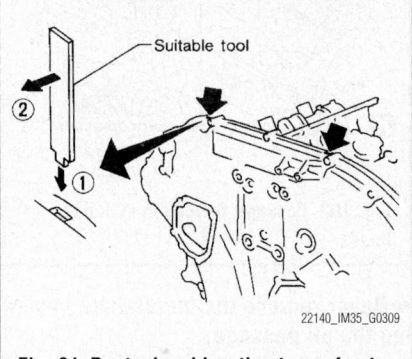

Fig. 91 Pry tool and location to pry front cover from motor

- Oil pan gasket (front)
- Water pump cover
- Chain tensioner cover from front timing chain case
- Front oil seal from front timing chain case
- Timing chain tensioner (primary)
- Lower mounting bolt.

14. Loosen upper mounting bolt slowly, and then turn timing chain tensioner (primary) on the mounting bolt so that the plunger is fully expanded.

➡**Even if plunger is fully expanded, it will not dropped from the body of timing chain tensioner (primary).**

15. Remove upper mounting bolt, and then remove timing chain tensioner (primary).

16. Remove internal chain guide, tension guide and slack guide.

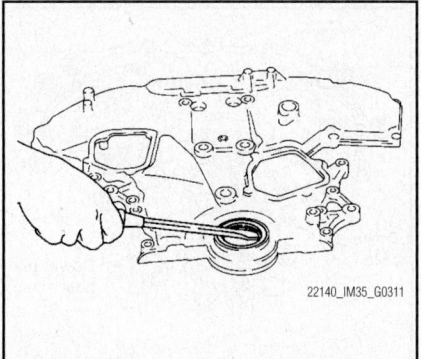

Fig. 93 Front oil seal front timing chain case

➡**Tension guide can be removed after removing timing chain (primary).**

17. Remove timing chain (primary) and crankshaft sprocket.

✳✳ WARNING

After removing timing chain (primary), never turn crankshaft and camshaft separately, or valves will strike the piston heads.

a. Remove timing chain (secondary) and camshaft sprockets as follows:

b. Attach suitable stopper pin to the right and left timing chain tensioners (secondary).

➡**Use approximately 0.5 mm (0.020 in) dia. hard metal pin as a stopper pin.**

c. Remove camshaft sprocket (INT and EXH) mounting bolts.

d. Secure the hexagonal portion of camshaft using a wrench to loosen mounting bolts.

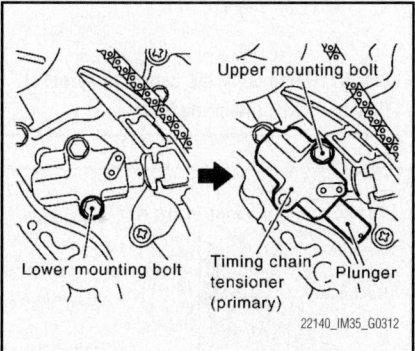

Fig. 94 Primary timing chain tensioner and bolt locations

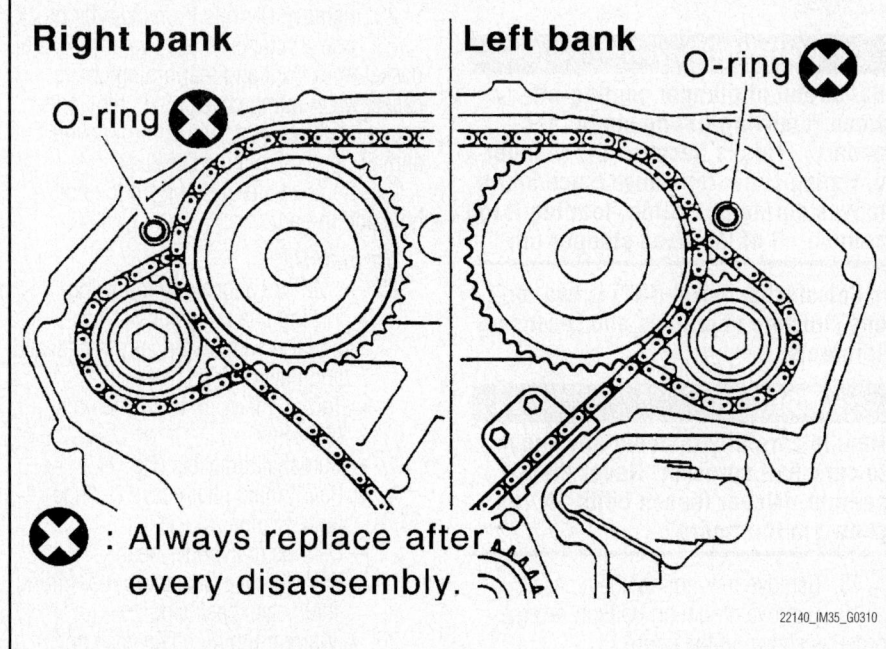

Right bank   **Left bank**

O-ring ✖   O-ring ✖

✖ : Always replace after every disassembly.

Fig. 92 Oil seal rear timing chain case

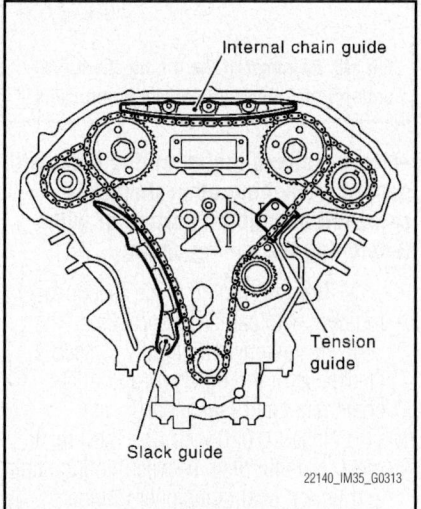

Fig. 95 Internal chain guide, tension guide and slack guide

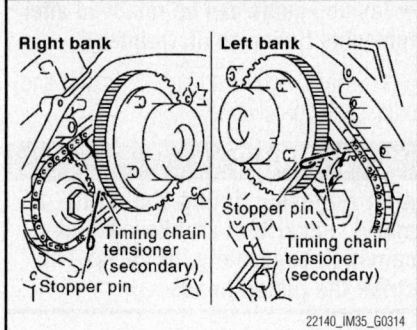

Fig. 96 Removal of timing chain tensioner (secondary)

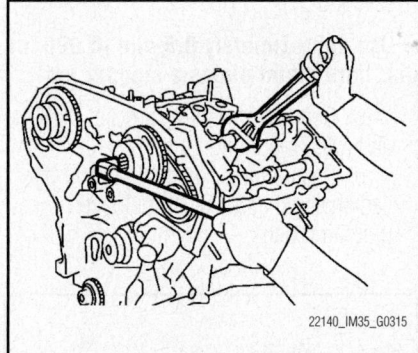

Fig. 97 Removal of the camshaft sprocket (INT and EXH) mounting bolts

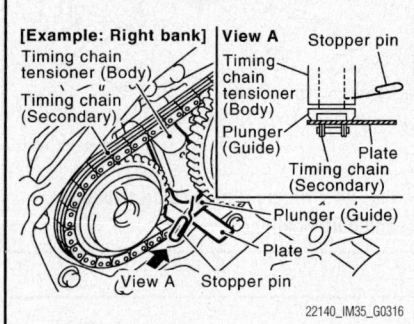

Fig. 98 Removal of the timing chain (secondary) together with camshaft sprockets

➡Never loosen the mounting bolts with securing anything other than the camshaft hexagonal portion or with tensioning the timing chain.

   e. Remove timing chain (secondary) together with camshaft sprockets.

   f. Turn camshaft slightly to secure slackness of timing chain on timing chain tensioner (secondary) side.

   g. Insert 0.020 inch (0.5 mm) thick metal or resin plate between timing chain and timing chain tensioner plunger (guide). Remove timing chain (secondary) together with camshaft sprockets

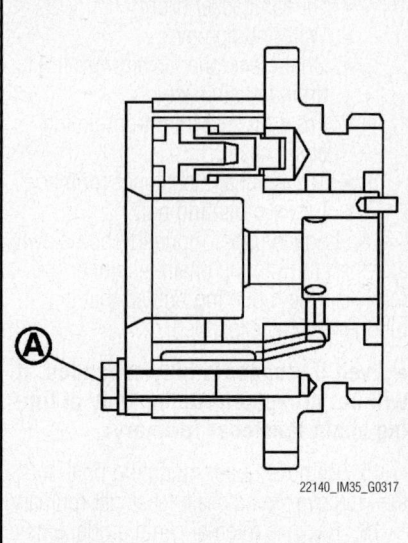

Fig. 99 Camshaft sprocket (intake)

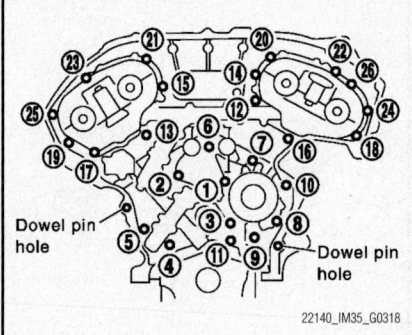

Fig. 100 Rear timing chain case mounting bolts locations

with timing chain loose from the guide groove.

✷✷ WARNING

Be careful of plunger coming-off when removing timing chain (secondary). This is because the plunger of timing chain tensioner (secondary) moves during operation, leading it to coming-off of the fixed stopper pin.

➡Camshaft sprocket (INT) is two-for-one structure of primary and secondary sprockets.

✷✷ WARNING

Handle carefully to avoid any shock to camshaft sprocket, Never disassemble. (Never loosen bolts "A" as shown in the figure).

  18. Remove rear timing chain case.
  19. Remove mounting bolts in reverse order as shown in the figure.
  20. Cut gasket to remove the case.

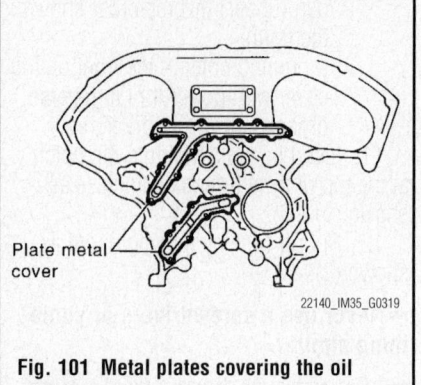

Fig. 101 Metal plates covering the oil passages

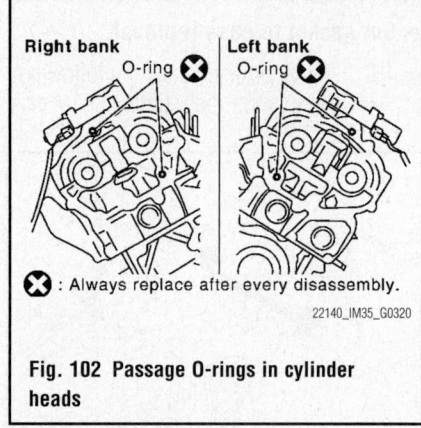

Fig. 102 Passage O-rings in cylinder heads

➡Never remove the metal plate covering the oil passage.

➡After removal, handle rear timing chain case carefully so it does not tilt, cant, or warp under a load.

  21. Remove O-rings from cylinder head.
  22. Remove O-rings from cylinder block.
  23. Use a scraper to remove all traces of gasket from front and rear timing chain cases, water pump cover, chain tensioner cover, intake valve timing control covers. and all opposite mating surfaces.
  24. Remove old gasket from the bolt hole and thread.

To install:
  25. Install or connect the following:
- Timing chain tensioners (secondary) with a stopper pin attached and new O-rings.
- Torque the bolts to 61 inch lbs. (6.9 Nm)
- Camshaft brackets (No. 1).
- Rear timing chain case O-rings onto cylinder block
- O-rings to cylinder head
- RTV Silicone Sealant to rear timing chain case back side

  26. Align rear timing chain case and water pump assembly with dowel pins (right

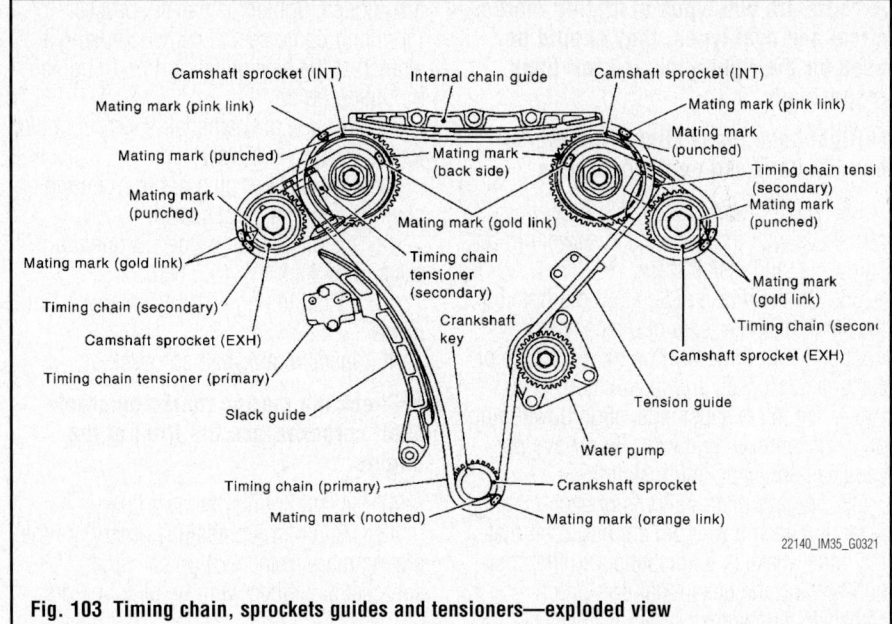

Fig. 103 Timing chain, sprockets guides and tensioners—exploded view

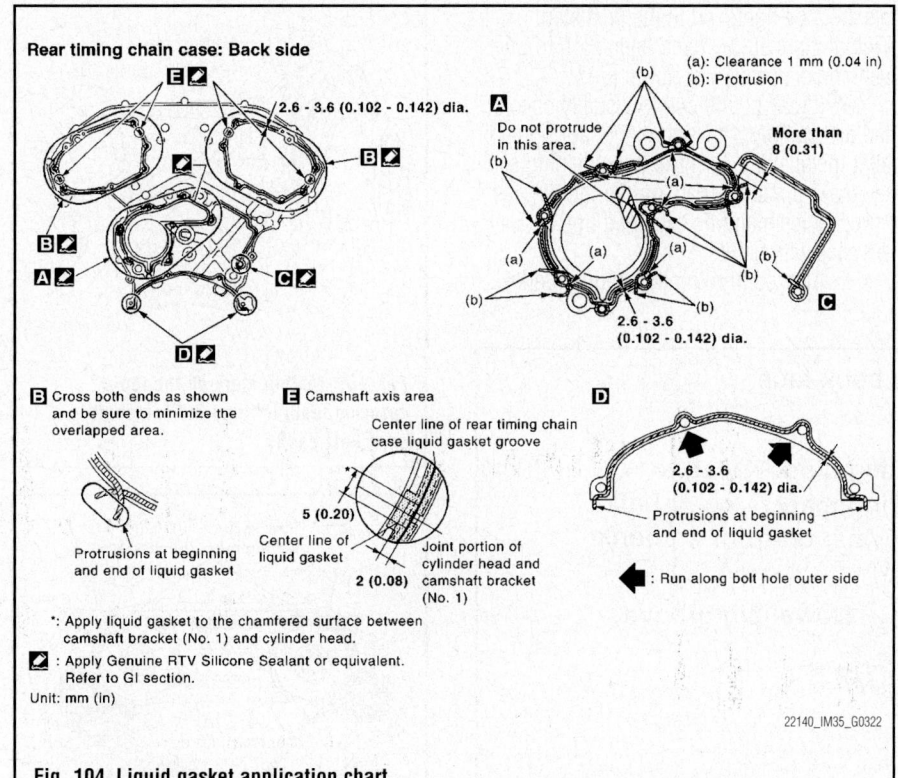

Fig. 104 Liquid gasket application chart

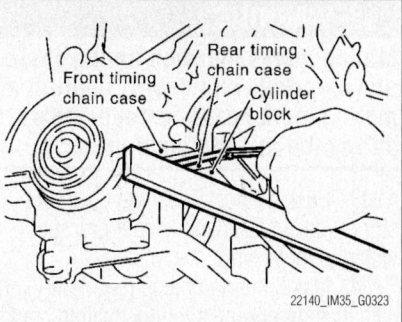

Fig. 105 Checking surface height difference between rear timing chain case and cylinder block

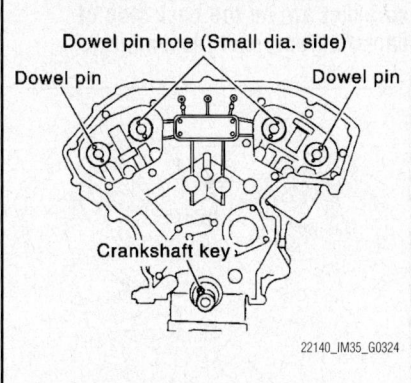

Fig. 106 Dowel pin hole, dowel pin and crankshaft key alignment for No. 1 cylinder at compression TDC

case, check the surface height difference between the following parts on the oil pan (upper) mounting surface.

30. Rear timing chain case and cylinder block:–0.0094–0.0055 inch (–0.24–0.14 mm).

➡ **If the measurement is not within the standard, repeat the installation procedure.**

31. Install water pump with new O-rings.

32. Check that the dowel pin hole, dowel pin and crankshaft key are located as shown in the figure. (No. 1 cylinder at compression TDC)

➡ **Though the camshaft does not stop at the position as shown in the figure, to show the placement of cam nose, it is generally accepted that the camshaft is placed in the same direction for the figure.**

✳✳ WARNING

Hole on small diameter side must be used for the intake side dowel pin hole, Never misidentify (ignore big diameter. side).

bank and left bank) on cylinder block and install rear timing chain case.

➡ **Check O-rings stay in place during installation to cylinder block and cylinder head.**

27. Tighten mounting bolts in numerical order as shown in the figure.

28. There are two types of mounting bolts. Refer to the following for locating bolts.

a. Bolt length: 0.79 inches (20 mm), Bolt position: 1, 2, 3, 6, 7, 8, 9, and 10

b. Bolt length: 0.63 inches (16 mm) Bolt position: Except the above 12, 13, 14, 15, 16, 17, 18, 19, 20, 21, 22, 23, 24, 25, 26:

c. Tighten 20mm bolts to 11 ft. lbs. (15 Nm).

d. Tighten 16mm bolts to 9 ft. lbs. (12.7 Nm).

29. After installing rear timing chain

※※ WARNING

Mating marks between timing chain and sprockets slip easily. Confirm all mating mark positions repeatedly during the installation process.

33. Push plunger of timing chain tensioner (secondary) and keep it pressed in with a stopper pin.

34. Install timing chains (secondary) and camshaft sprockets, align the mating marks on timing chain (secondary) (gold link) with the ones on the intake and exhaust camshaft sprockets (punched), and install them.

➡**Mating marks for intake camshaft sprocket are on the back side of camshaft sprocket (secondary).**

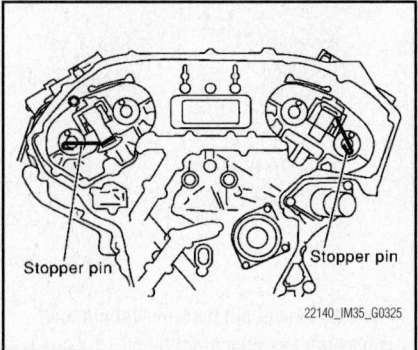

Fig. 107 Locking the plunger of the timing chain tensioner (secondary)

➡**There are two types of mating marks, circle and oval types, they should be used for the right bank and left bank, respectively.**

➡**Right bank, uses circle type marks and left bank use oval type marks.**

35. Align dowel pin and pin hole on camshafts with the groove and dowel pin on sprockets, and install them.

36. On the intake side, align pin hole on the small diameter side of the camshaft front end with dowel pin on the back side of camshaft sprocket, and install them.

37. On the exhaust side, align dowel pin on camshaft front end with pin groove on camshaft sprocket, and install them.

38. In case that positions of each mating mark and each dowel pin are not fit on mating parts, make fine adjustment to the position holding the hexagonal portion on camshaft with wrench or equivalent.

39. Mounting bolts for camshaft sprockets must be tightened in the next step. Tightening them by hand is enough to prevent the dislocation of dowel pins.

40. It may be difficult to visually check the dislocation of mating marks during and after installation. To make the matching easier, make a mating mark on the top of sprocket teeth and its extended line in advance with paint

41. After confirming the mating marks

are aligned, tighten camshaft sprocket mounting bolts, secure camshaft using a wrench at the hexagonal portion to tighten mounting bolts.

42. Tighten the camshaft sprockets bolts to 112 ft. lbs. (152 Nm).

43. Pull stopper pins out from timing chain tensioners (secondary).

44. Install tension guide. Torque the bolts to 61 inch lbs. (6.9 Nm).

45. Install timing chain (primary) as follows:

46. Install crankshaft sprocket.

➡**Check the mating marks on crankshaft sprocket face the front of the engine.**

47. Install timing chain (primary).

48. Install timing chain (primary) so the mating mark (punched) on camshaft sprocket is aligned with the pink link on

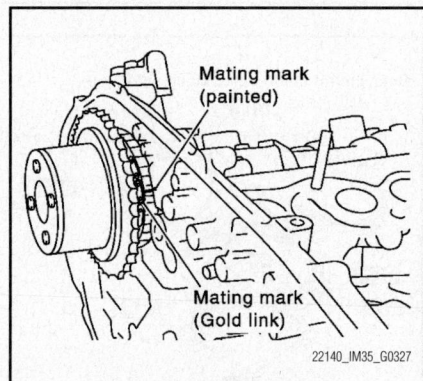

Fig. 109 Mating mark on the top of sprocket teeth to align the camshaft sprockets easily

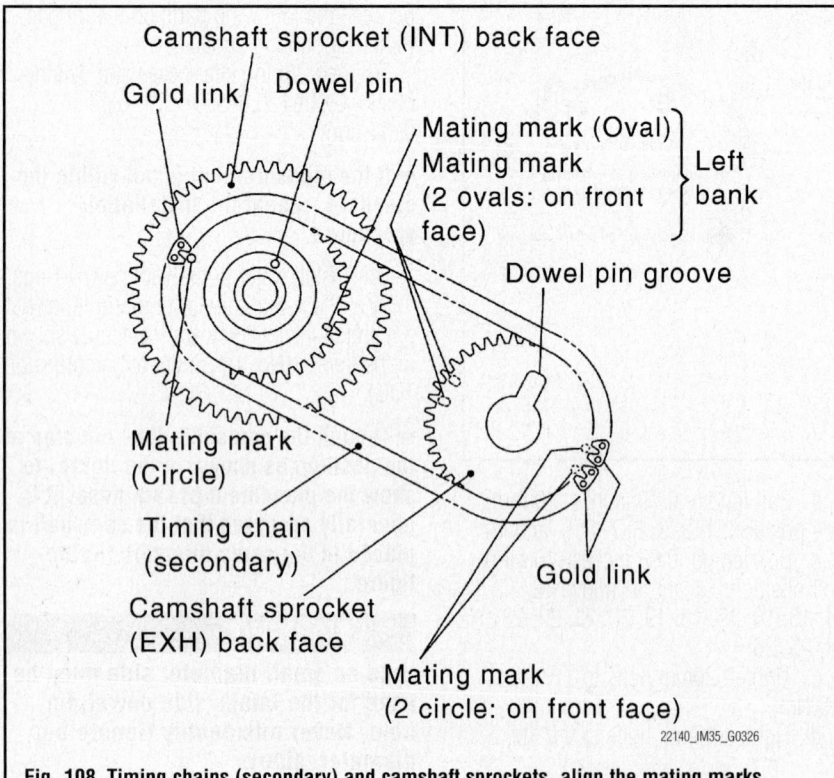

Fig. 108 Timing chains (secondary) and camshaft sprockets, align the mating marks

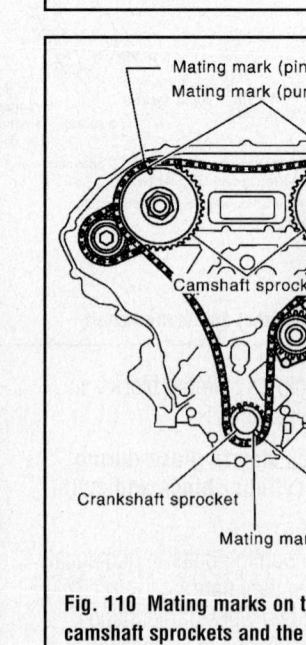

Fig. 110 Mating marks on the top of the camshaft sprockets and the crankshaft sprocket

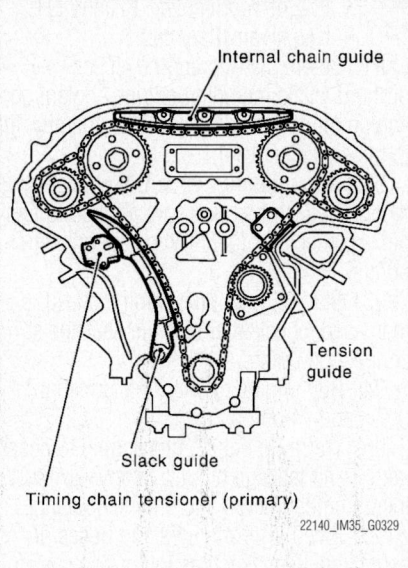

Fig. 111 Internal chain guide, slack guide and timing chain tensioner (primary)

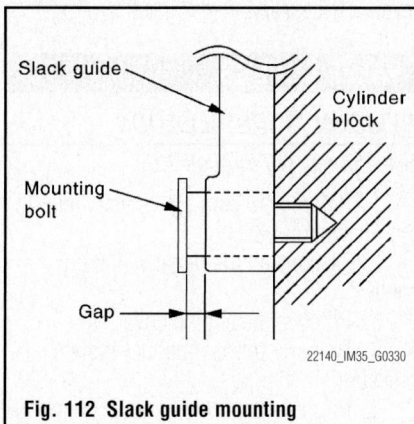

Fig. 112 Slack guide mounting

timing chain, while the mating mark (notched) on crankshaft sprocket is aligned with the orange one on timing chain, as shown in the figure.

49. When it is difficult to align mating marks of timing chain (primary) with each sprocket, gradually turn camshaft using wrench on the hexagonal portion to align it with the mating marks.

➡ **During alignment, be careful to prevent dislocation of mating mark and alignments of timing chains (secondary).**

50. Install internal chain guide, slack guide and timing chain tensioner (primary).
51. Torque the bolts to 12 ft. lbs. (16 Nm).

✳✳ WARNING

Never over tighten slack guide mounting bolts. It is normal for a gap to exist under the bolt seats when

mounting bolts are tightened to the specification.

52. Install the timing chain tensioner (primary).
53. Torque the bolts to 61 inch lbs. (6.9 Nm).
54. Pull plunger stopper tab up (or turn lever downward) so as to remove plunger stopper tab from the ratchet of plunger.

➡ **Plunger stopper tab and lever are synchronized.**

55. Push plunger into the inside of tensioner body.
56. Hold plunger in the fully compressed position by engaging plunger stopper tab with the tip of ratchet.
57. To secure lever, insert stopper pin through hole of lever into tensioner body hole, the lever parts and the tab are synchronized, therefore, the plunger will be secured under this condition.
58. Install timing chain tensioner (primary).

➡ **Remove any dirt and foreign materials completely from the back and the mounting surfaces of timing chain tensioner (primary).**

59. Torque the bolts to 61 inch lbs. (6.9 Nm) and remove the pins.
60. Check again that the mating marks on sprockets and timing chain have not slipped out of alignment.
61. Install new O-rings on rear timing chain case.
62. Install new front oil seal on front timing chain case.
63. Apply new engine oil to both oil seal lip and dust seal lip.
64. Using a suitable drift outer diameter: 60 mm (2.36 in), press fit oil seal until it becomes flush with front timing chain case end face.
65. Check the garter spring is in position and seal lip is not inverted.
66. Install water pump cover and chain tensioner cover to front timing chain case.
67. Apply a continuous bead of RTV Silicone) to the front timing chain case as shown in the figure.
68. Install front timing chain case as follows:
69. Apply a continuous bead of RTV Silicone) to the front timing chain case backside as shown in the figure.
70. Install the front timing chain case as to fit its dowel pin holes together with the dowel pin on the rear timing chain case.
71. Tighten mounting bolts to the specified torque in numerical order:

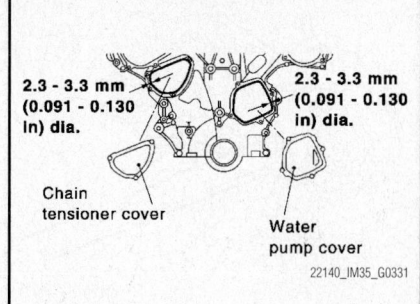

Fig. 113 Water pump cover and chain tensioner cover to front timing chain case sealant guide

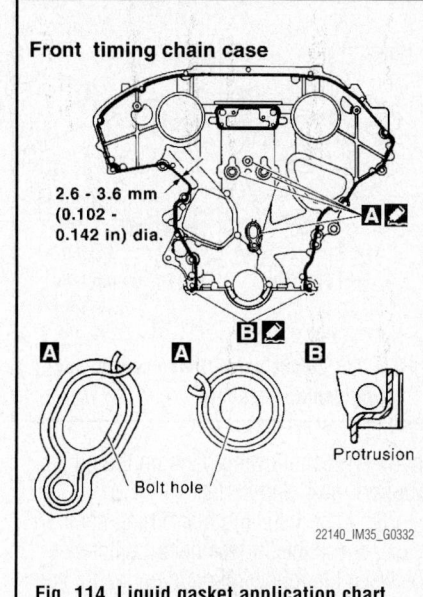

Fig. 114 Liquid gasket application chart for the front timing chain case back side

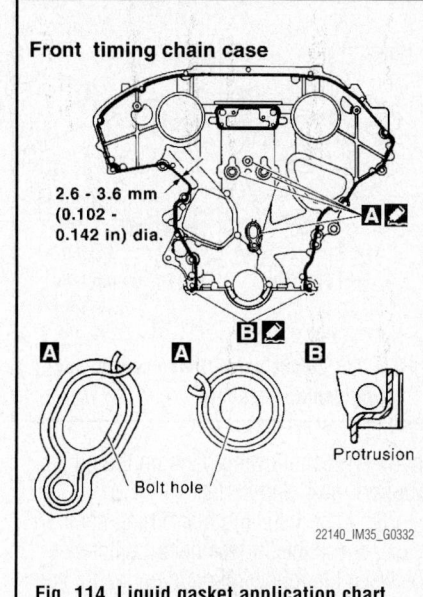

Fig. 115 Front timing chain case bolts location key

a. 21 ft. lbs. (28.4 Nm). For the M8 bolts 1and 2.
b. 9 ft. lbs. (27.7 Nm). For the M6 bolts except the above.

72. After installing front timing chain case, check the surface height difference

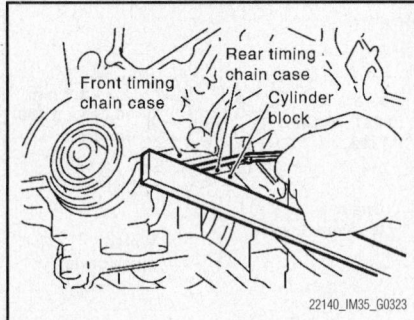

Fig. 116 Surface height difference between rear timing chain case and cylinder block

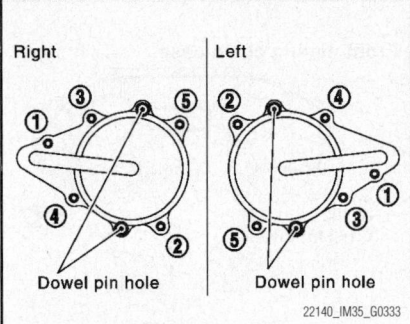

Fig. 117 Intake valve timing control covers tightening sequence

between the following parts on the oil pan (upper) mounting surface.

73. After installing rear timing chain case, check the surface height difference between the following parts on the oil pan (upper) mounting surface.

74. Rear timing chain case and cylinder block:–0.0055–0.0055 inch (–0.14–0.14 mm).

➡**If the measurement is not within the standard, repeat the installation procedure.**

75. Install right and left intake valve timing control covers as follows:

76. Install new seal rings in shaft grooves.

77. Apply a continuous bead of RTV Silicone Sealant to the intake valve timing control covers 0.083–0.122 inch (2.1–3.1 mm) diameter.

78. Install new collared O-rings in front timing chain case oil hole (left and right sides).

79. Being careful not to move seal ring from the installation groove, align dowel pins on front timing chain case with holes to install intake valve timing control covers.

80. Tighten mounting bolts in numerical order as shown in the figure

81. Install oil pans (upper and lower).

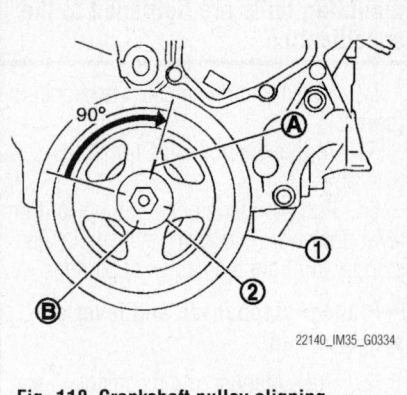

Fig. 118 Crankshaft pulley aligning

82. Install rocker covers (right bank and left bank).

83. Install crankshaft pulley as follows:

84. Hold the crankshaft using the ring gear stopper SST: KV10117700 (J44716).

85. Install crankshaft pulley, taking care not to damage front oil seal.

86. When press-fitting crankshaft pulley with plastic hammer, tap on its center portion (not circumference).

87. Tighten crankshaft pulley bolt to 33 ft. lbs. (44.1 Nm).

88. Place a paint mark (A) on crankshaft pulley (1) aligning with the angle mark (B) on crankshaft pulley bolt (2) Tighten the bolt 90 degrees (angle tightening).

89. Rotate crankshaft pulley in normal direction (clockwise when viewed from front) to confirm it turns smoothly.

90. Install or connect the following:

- Rocker covers (right bank and left bank)
- Rear cover plate (2WD models) or starter motor (AWD models)
- Water bypass hose, water hose clamp and idler pulley bracket from front timing chain case
- Alternator
- Power steering oil pump bracket
- Power steering oil pump to bracket with piping
- Intake manifold collectors (upper and lower)
- Drive belts
- Front and rear engine undercover with power tool
- Radiator hose (upper and lower) and A/T fluid cooler hose
- Engine harnesses installing their brackets to the front timing chain case
- Air duct (inlet) and air cleaner case assembly
- Engine cover
- Cooling fan and radiator

- Engine room cover (RH and LH).
- Negative battery cable

91. Before starting engine, check oil/fluid levels including engine coolant and engine oil. If less than required. quantity, fill to the specified level.

92. Turn ignition switch **ON** (with engine stopped). With fuel pressure applied to fuel piping, check for fuel leakage at connection points.

93. Start engine. With engine speed increased, check again for fuel leakage at connection points.

94. Run engine to check for unusual noise and vibration.

95. Warm up engine thoroughly to check there is no leakage of fuel, or any oil/fluids including engine oil and engine coolant.

96. Bleed air from lines and hoses of applicable lines, such as in cooling system.

97. After cooling down engine, again check oil/fluid levels including engine oil and engine coolant. Refill to the specified level, if necessary.

VALVE COVERS

REMOVAL & INSTALLATION

See Figures 119 through 122.

1. Before servicing the vehicle, refer to the Precautions Section.

2. Disconnect the negative battery cable.

3. Remove the engine cover.

4. Properly release the fuel system pressure.

5. Properly drain the engine coolant. Be sure the engine is cold before performing this operation.

6. Remove the front wiper arm and extension cowl top panel (lower and upper).

7. Remove the upper and lower intake manifold collectors.

8. Separate the engine harness by removing the brackets from the valve covers.

9. Remove the ignition coil.

10. Remove the PCV hoses from the valve cover. Remove the PCV valve and O-ring from the valve cover, right bank if necessary.

11. Remove the oil filler cap and oil catcher from the valve cover, left bank if necessary.

12. Loosen the valve cover retaining bolts. Reverse the order shown in the illustration.

13. Remove the valve cover from the engine.

To install:

14. Use a scraper and remove all old

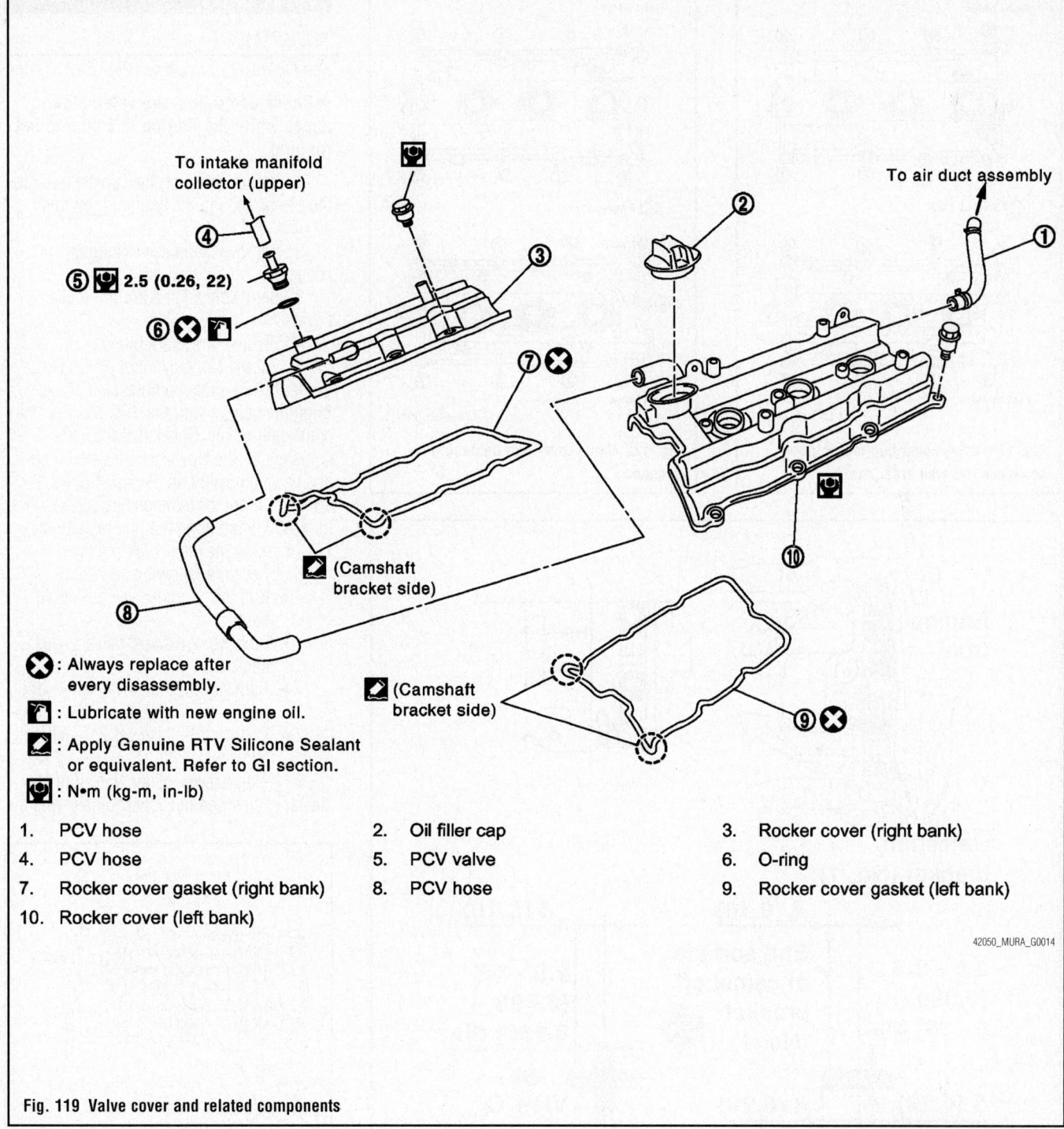

To intake manifold collector (upper)

To air duct assembly

④

⑤ 2.5 (0.26, 22)

⑥ ✕

② Oil filler cap

③

①

⑦ ✕

(Camshaft bracket side)

⑧

⑩

(Camshaft bracket side)

⑨ ✕

✕ : Always replace after every disassembly.

: Lubricate with new engine oil.

: Apply Genuine RTV Silicone Sealant or equivalent. Refer to GI section.

: N•m (kg-m, in-lb)

1. PCV hose
2. Oil filler cap
3. Rocker cover (right bank)
4. PCV hose
5. PCV valve
6. O-ring
7. Rocker cover gasket (right bank)
8. PCV hose
9. Rocker cover gasket (left bank)
10. Rocker cover (left bank)

42050_MURA_G0014

Fig. 119 Valve cover and related components

gasket material from the cylinder head and camshaft bracket (No. 1).

➡**Be sure not to scratch the mating surfaces with the scraper when removing the gasket material.**

15. Apply genuine RTV silicone sealant or equivalent to the joints of the valve cover, cylinder head and camshaft bracket (NO. 1) as shown in the illustration.

16. Refer to figure "A" in the illustration

to apply liquid gasket to the joint part of the camshaft (No. 1) and cylinder head.

17. Refer to figure "B" in the illustration to apply liquid gasket to the figure "A" squarely.

18. Install a new valve cover gasket to the valve cover.

19. Install the valve cover. Torque the retaining bolts to 17 inch lbs. and then to 74 inch lbs., in the proper sequence as shown in the illustration.

➡**Be sure that the gasket has not dropped from the grove prior to installation.**

20. Continue the installation in the reverse order of the removal procedure.

21. When installing the PCV hose, insert it 0.98–1.18 inch from the connector end.

22. When installing the PCV hose between the right and left valve covers be sure the identification paint mark is facing upward (right cover side).

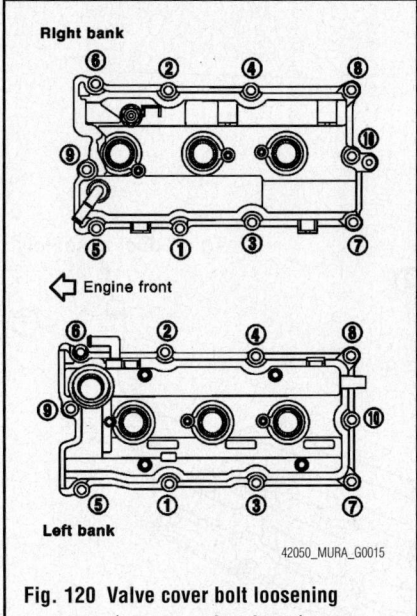

Fig. 120 Valve cover bolt loosening sequence (reverse order shown)

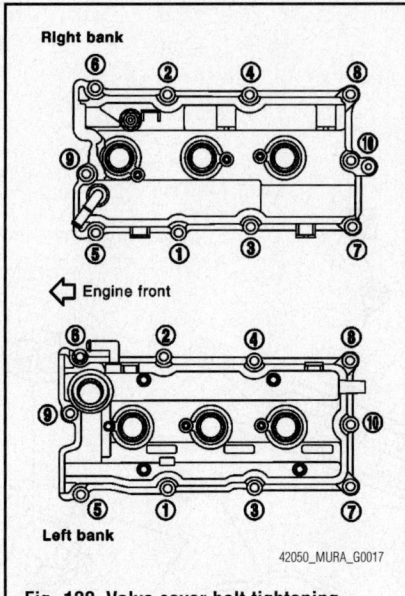

Fig. 122 Valve cover bolt tightening sequence

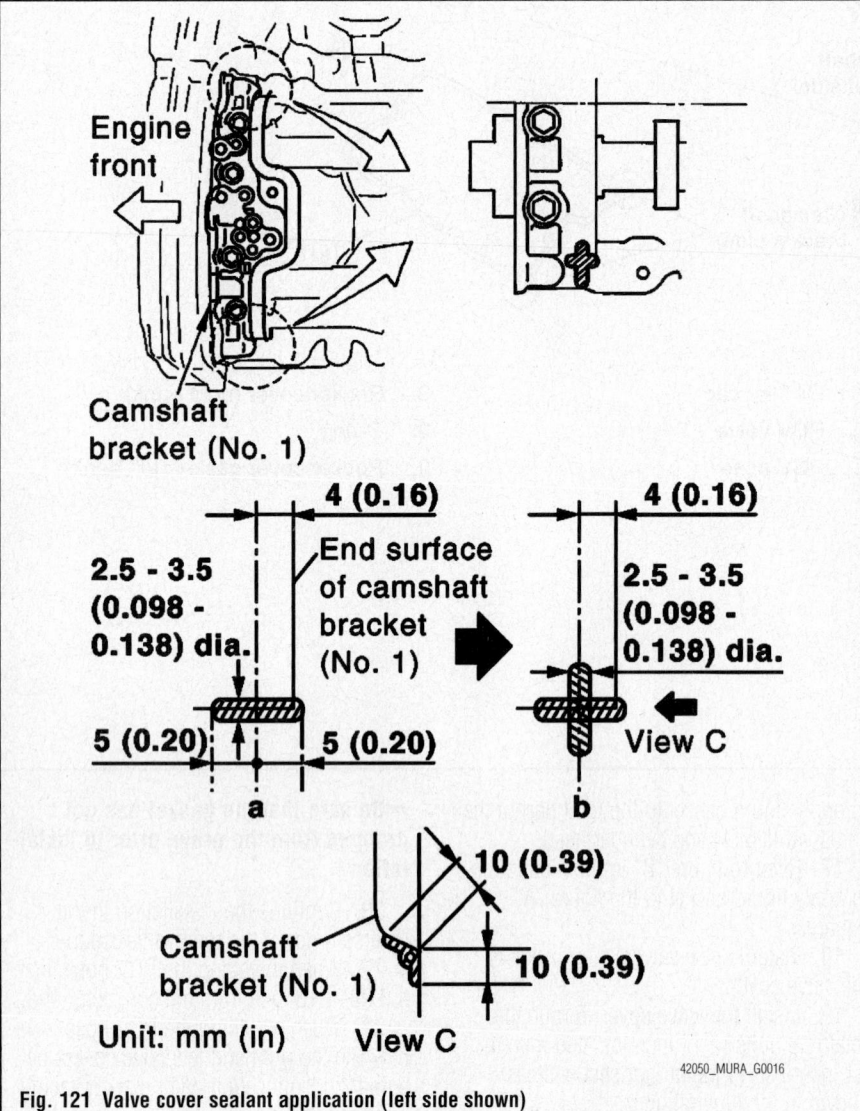

Fig. 121 Valve cover sealant application (left side shown)

VALVE LASH

ADJUSTMENT

See Figures 123 through 125.

➡ **Check and adjust the valve clearances while the engine is cold and not running.**

1. Before servicing the vehicle, refer to the precautions in the beginning of this section.

2. Remove the intake manifold collector.

3. Remove the left and right rocker covers.

4. Remove the spark plugs.

5. Set the No. 1 cylinder at Top Dead Center (TDC) on its compression stroke. Align the pointer with the TDC mark on the crankshaft pulley. Check that the valve adjusters on the No. 1 cylinder are loose and valve adjusters on the No. 4 cylinder are tight. If not, turn the crankshaft 1 revolution (360°) and align the pointer with the TDC mark on the crankshaft pulley.

6. Check the following valves:
   - Both No. 1 intake valves—right bank
   - Both No. 2 exhaust valves—left bank
   - Both No. 3 exhaust valves—right bank
   - Both No. 6 intake valves—left bank

7. Using a feeler gauge, measure the clearance between the valve adjuster and the

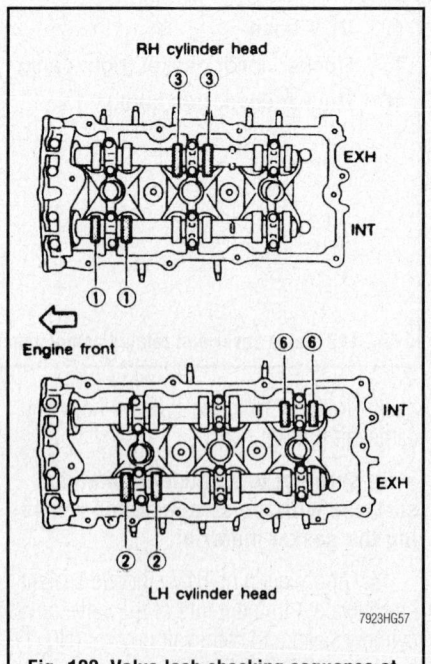

Fig. 123 Valve lash checking sequence at TDC of cylinder No. 1—3.5L Engines

camshaft. Record any valve clearance measurements that are out of specification. Intake valve clearance (cold) is 0.010–0.013 in. (0.26–0.34mm) and exhaust valve clearance (cold) is 0.011–0.015 in. (0.29–0.37mm).

8. Turn the crankshaft 240° and set the No. 3 cylinder to TDC of its compression stroke.

9. Check the following valves:
- Both No. 2 intake valves—left bank
- Both No. 3 intake valves—right bank
- Both No. 4 exhaust valves—left bank
- Both No. 5 exhaust valves—right bank

10. Using a feeler gauge, measure the clearance between the valve adjuster and the camshaft. Record any valve clearance measurements that are out of specification. Intake valve clearance (cold) is 0.010–0.013 in. (0.26–0.34mm) and exhaust valve clearance (cold) is 0.011–0.015 in. (0.29–0.37mm).

11. Turn the crankshaft 240° and set the No. 5 cylinder to TDC of its compression stroke.

12. Check the following valves:
- Both No. 1 exhaust valves—right bank
- Both No. 4 intake valves—left bank
- Both No. 5 intake valves—right bank
- Both No. 6 exhaust valves—left bank

13. Using a feeler gauge, measure the clearance between the valve adjuster and the camshaft. Record any valve clearance measurements that are out of specification. Intake valve clearance (cold) is 0.010–0.013 in. (0.26–0.34mm) and exhaust valve clearance (cold) is 0.011–0.015 inches (0.29–0.37mm).

14. If all the valve clearances are within specification, install the cylinder head cover,

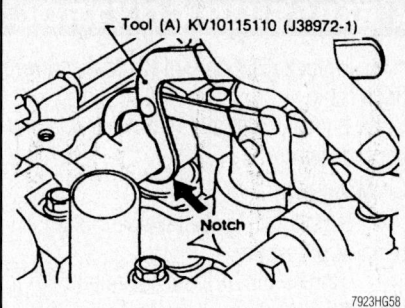

Fig. 124 Install the depressor tool around the camshaft being careful not to damage the surfaces—3.5L Engines shown

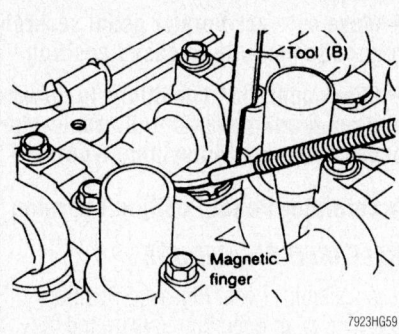

Fig. 125 Use a magnet to remove the shim from the adjuster. Sometimes a shot of compressed air can help lift the shim up—3.5L Engines shown

spark plugs, and the intake manifold collector.

15. If an adjustment is necessary, adjust the valve clearance while engine is cold by removing the adjusting shim. The adjusting shim can be removed by using the following procedures:

a. Turn the crankshaft so the camshaft lobe of the valve to be adjusted is pointed straight up.

b. Turn the adjuster so the notch is pointed towards the center of the cylinder head; this will facilitate the shim removal process.

c. Using a depressor tool No. KV10115110 push down on the adjuster and insert a keeper tool on the edge of the adjuster to keep the adjuster in the depressed position.

d. Remove the depressor tool and remove the shim with a magnet.

➡Compressed air can be blown into the hole of the adjuster to separate the adjusting shim from the adjuster.

16. Determine the replacement adjusting shim size by using the following procedures and formula:

a. Using a micrometer determine thickness of the removed shim.

b. Calculate the thickness of a new adjusting shim so valve clearance is within the specified values.
- t_1= thickness of the removed shim
- t= thickness of the new shim
- C_1= measured valve clearance
- C_2= Standard valve clearance:
- Calculate the Intake Shim as follows: $t = t_1 + (c_1 - C_2) - 0.012$ in. (0.30mm)
- Calculate the Exhaust Shim as follows: $N = R + M - 0.013$ in. (0.33mm)

17. Shims are available in 64 sizes from 0.0913–0.1161 in. (2.32–2.95mm) in steps of 0.004 in. (0.01mm). The thickness is stamped on the shim; this side is always installed facing down. Select new shims with thickness as close as possible to calculated valve and install it in the adjuster.

18. Install the new shim onto the adjuster.

19. Depress the adjuster and remove the keeper tool. Remove the depressor tool and recheck the valve clearance. Repeat this procedure for any other valves requiring adjustment.

20. When all valve adjustments are finished, install the cylinder head cover, spark plugs, and the intake manifold collector.

ENGINE PERFORMANCE & EMISSION CONTROL

ACCELERATOR PEDAL POSITION (APP) SENSOR

LOCATION

See Figures 126 and 127.

The accelerator pedal position sensor is installed on the upper end of the accelerator pedal assembly.

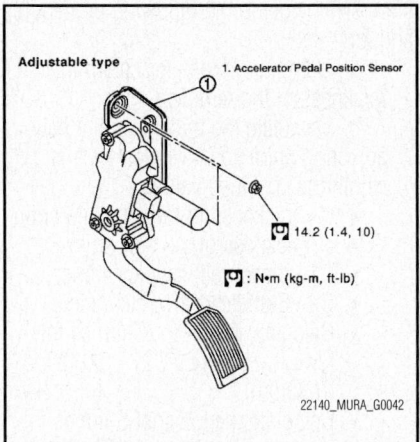

Fig. 126 Accelerator Pedal Position (APP) Sensor - adjustable type

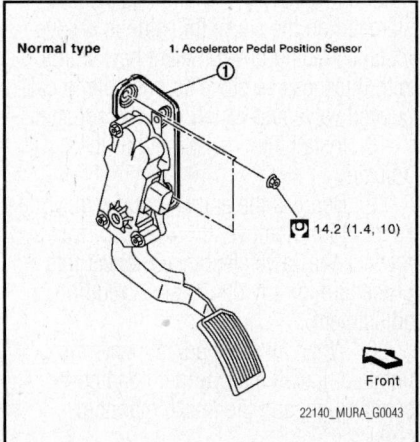

Fig. 127 Accelerator Pedal Position (APP) Sensor – normal type

REMOVAL & INSTALLATION

1. Move accelerator and brake pedals to the front most position (Adjustable type).
2. Turn ignition switch **OFF** and disconnect negative battery terminal.
3. Disconnect accelerator pedal position sensor harness connector.
4. Disconnect drive cable from accelerator pedal assembly (Adjustable type).

5. Unlock (1) then pull (2) to disconnect drive cable.
6. Loosen nuts, and remove accelerator pedal assembly.

To install:

7. Installation is the reverse of the removal procedure.
8. Tighten the APP sensor nuts to 10 ft. lbs (14.2 Nm).

➡ **Make sure accelerator pedal moves smoothly within the whole operation range when it is fully depressed and released.**

➡ **Make sure accelerator pedal securely returns to the fully released position.**

➡ **Check operation conditions in forward and rearward movement of accelerator pedal assembly (Adjustable type).**

Accelerator Pedal Position Learning

RELEASED PROCEDURE

Accelerator Pedal Released Position Learning is an operation to learn the fully released position of the accelerator pedal by monitoring the accelerator pedal position sensor output signal. It must be performed each time harness connector of accelerator pedal position sensor or ECM is disconnected.

1. Make sure that accelerator pedal is fully released.
2. Turn ignition switch ON and wait at least 2 seconds.
3. Turn ignition switch OFF and wait at least 10 seconds.
4. Turn ignition switch ON and wait at least 2 seconds.
5. Turn ignition switch OFF and wait at least 10 seconds.

CLOSED PROCEDURE

Throttle Valve Closed Position Learning is an operation to learn the fully closed position of the throttle valve by monitoring the throttle position sensor output signal. It must be performed each time harness connector of electric throttle control actuator or ECM is disconnected.
OPERATION PROCEDURE
1. Make sure that accelerator pedal is fully released.
2. Turn ignition switch ON.
3. Turn ignition switch OFF and wait at least 10 seconds.

➡ **Make sure that throttle valve moves during above 10 seconds by confirming the operating sound.**

CAMSHAFT POSITION (CMP) SENSOR

LOCATION

See Figures 128 and 129.

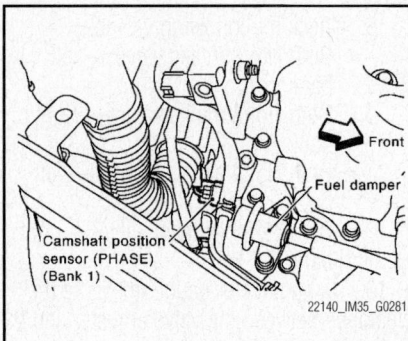

Fig. 128 The camshaft position sensor bank 1

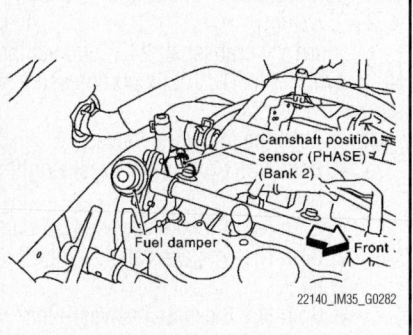

Fig. 129 The camshaft position sensor bank 1

REMOVAL & INSTALLATION

1. Loosen the fixing bolt of the sensor.
2. Disconnect Camshaft Position (CMP) sensor (PHASE) harness connector.
3. Remove the sensor.

To install:

4. Install the CMP sensor.
5. Install the CMP connector.
6. Tighten the bolt to 7.5 inch lbs. (10 Nm).

CRANKSHAFT POSITION (CKP) SENSOR

LOCATION

The Crankshaft Position (CKP) Sensor is located on the oil pan housing facing the gear teeth (cogs) of the signal plate.

REMOVAL & INSTALLATION

See Figure 130.

1. Loosen the fixing bolt of the sensor.
2. Disconnect CKP harness connector.
3. Remove the sensor.
4. Visually check the sensor for chipping.

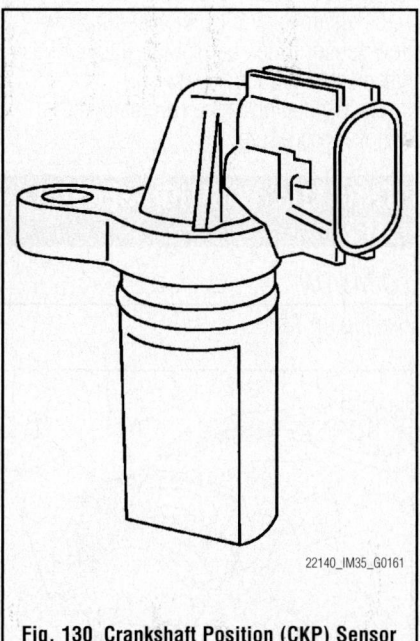

Fig. 130 Crankshaft Position (CKP) Sensor

ELECTRONIC CONTROL MODULE (ECM)

LOCATION

The ECM (1) is located behind the passenger side instrument lower panel. For this inspection, remove passenger side instrument lower panel.

REMOVAL & INSTALLATION

See Figure 131.

1. Remove ECM harness connector.

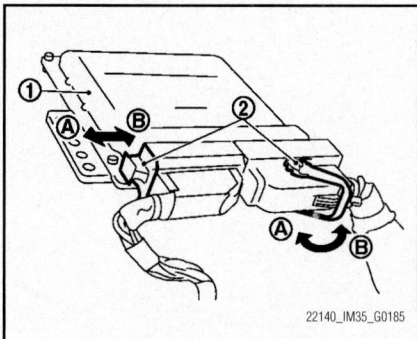

Fig. 131 ECM removal with harness and connector

2. When disconnecting ECM harness connector, loosen (A) it with levers (2) as far as they will go as shown in the figure.

ENGINE COOLANT TEMPERATURE (ECT) SENSOR

LOCATION

See Figure 132.

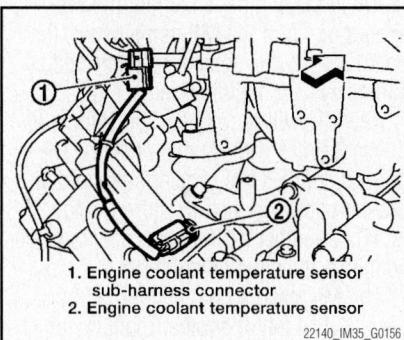

1. Engine coolant temperature sensor sub-harness connector
2. Engine coolant temperature sensor

Fig. 132 Engine Coolant Temperature (ECT) sensor location

REMOVAL & INSTALLATION

1. Before servicing the vehicle, refer to the Precautions Section.
2. Disconnect the negative battery cable.
3. Drain the cooling system.
4. Remove the sensor electrical connector.
5. Remove the sensor from its mounting.

To install:

6. Installation is the reverse of the removal procedure.

HEATED OXYGEN (HO2S) SENSOR

LOCATION

See Figure 133.

REMOVAL & INSTALLATION

1. Using the heated oxygen sensor wrench (SST), remove heated oxygen sensor 1 (right bank and left bank).

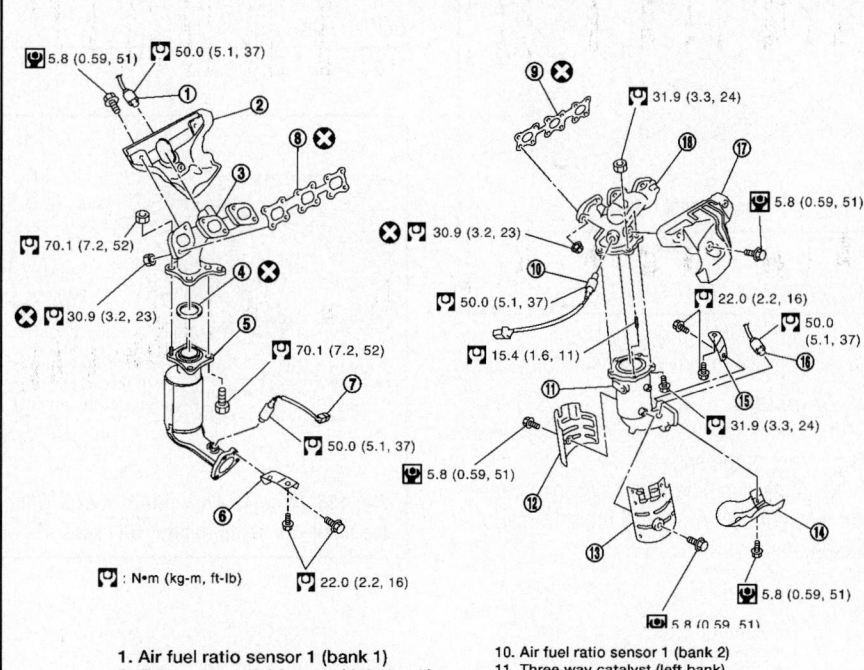

1. Air fuel ratio sensor 1 (bank 1)
2. Exhaust manifold cover (right bank)
3. Exhaust manifold (right bank)
4. Ring gasket
5. Three way catalyst (right bank)
6. Three way catalyst support (right bank)
7. Heated oxygen sensor 2 (bank 1)
8. Gasket
9. Gasket
10. Air fuel ratio sensor 1 (bank 2)
11. Three way catalyst (left bank)
12. Three way catalyst cover
13. Three way catalyst cover
14. Three way catalyst cover
15. Three way catalyst support (left bank)
16. Heated oxygen sensor 2 (bank 2)
17. Exhaust manifold cover (left bank)
18. Exhaust manifold (left bank)

Fig. 133 Exhaust system exploded view showing the HO2S sensor locations and locations of the Air Fuel (A/F) sensors

To install:

2. Installation is the reverse of the removal procedure. Tighten to 37 ft. lbs. (50 Nm).

INTAKE AIR TEMPERATURE (IAT) SENSOR

LOCATION

See Mass Air Flow (MAF) Sensor in this section.

REMOVAL & INSTALLATION

See Mass Air Flow Sensor removal and installation in this section.

KNOCK SENSOR (KS)

LOCATION

See Figure 134.

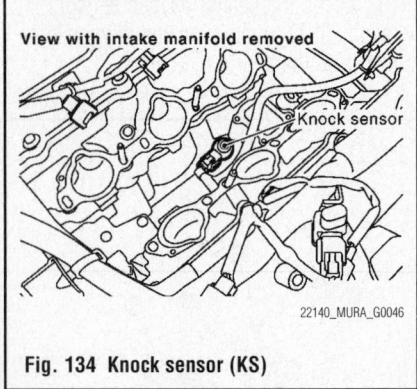

Fig. 134 Knock sensor (KS)

REMOVAL & INSTALLATION

1. Remove the intake manifold
2. Remove the knock sensor connector.

To install:

3. Install knock sensor so that connector faces front of the engine.
4. After installing knock sensor, connect harness connector, and lay it out to rear of the engine.

➡ **Make sure that knock sensor does not interfere with other parts.**

5. Tighten the knock sensor to 20 ft. lbs. (27 Nm).

MALFUNCTION INDICATOR LIGHT (MIL)

RESET PROCEDURES

When the engine is started, the MIL should go off. If the MIL remains on, the on board diagnostic, system has detected an engine system malfunction.

The MIL will go off after the vehicle is driven 3 times with no malfunction. The drive is counted only when the recorded driving pattern is met (as stored in the ECM). If another malfunction occurs while counting, the counter will reset.

The MIL can be commended off by the CONSULT-II. If another malfunction occurs while counting, the counter will reset.

MASS AIR FLOW (MAF) SENSOR

LOCATION

See Figure 135.

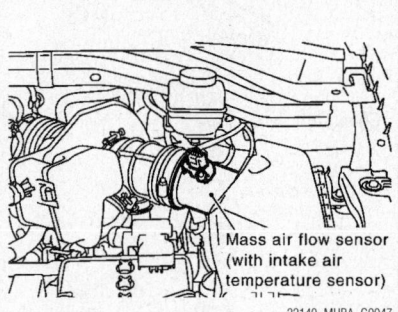

Fig. 135 Mass Air Flow (MAF) sensor with the Intake Air Temperature (IAT) sensor

REMOVAL & INSTALLATION

See Figure 135.

1. Remove engine room cover (RH and LH).
2. Remove air duct (inlet).
3. Disconnect mass air flow sensor harness connector.
4. Disconnect PCV hose.
5. Remove air cleaner case/mass air flow sensor assembly and air duct/air hose disconnecting their joints.
6. Installation is the reverse of the removal procedure.

THROTTLE POSITION SENSOR (TPS)

LOCATION

See Figure 136.

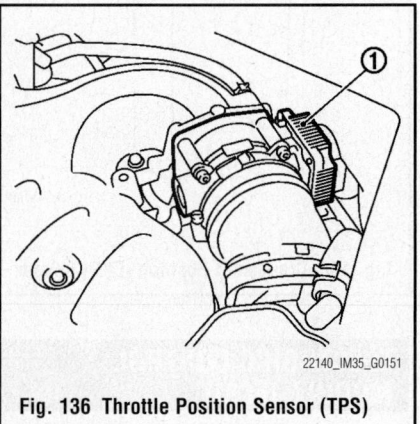

Fig. 136 Throttle Position Sensor (TPS)

REMOVAL & INSTALLATION

1. Disconnect electric throttle control actuator harness connector.
2. Remove rubber duct clamp
3. Remove mounting screws and remove the Throttle Position Sensor (TPS).
4. Installation is the reverse of the removal procedure.

FUEL SYSTEM
GASOLINE FUEL INJECTION SYSTEM

FUEL SYSTEM SERVICE PRECAUTIONS

Safety is the most important factor when performing not only fuel system maintenance but any type of maintenance. Failure to conduct maintenance and repairs in a safe manner may result in serious personal injury or death. Maintenance and testing of the vehicle's fuel system components can be accomplished safely and effectively by adhering to the following rules and guidelines.

• To avoid the possibility of fire and personal injury, always disconnect the negative battery cable unless the repair or test procedure requires that battery voltage be applied.

• Always relieve the fuel system pressure prior to disconnecting any fuel system component (injector, fuel rail, pressure regulator, etc.), fitting or fuel line connection. Exercise extreme caution whenever relieving fuel system pressure to avoid exposing skin, face and eyes to fuel spray. Please be advised that fuel under pressure may penetrate the skin or any part of the body that it contacts.

• Always place a shop towel or cloth around the fitting or connection prior to loosening to absorb any excess fuel due to spillage. Ensure that all fuel spillage (should it occur) is quickly removed from engine surfaces. Ensure that all fuel soaked cloths or towels are deposited into a suitable waste container.

• Always keep a dry chemical (Class B) fire extinguisher near the work area.

• Do not allow fuel spray or fuel vapors to come into contact with a spark or open flame.

• Always use a back-up wrench when loosening and tightening fuel line connection fittings. This will prevent unnecessary stress and torsion to fuel line piping.

• Always replace worn fuel fitting O-rings with new. Do not substitute fuel hose or equivalent where fuel pipe is installed.

Before servicing the vehicle, make sure to also refer to the precautions in the beginning of this section as well.

• Always use a back-up wrench when loosening and tightening fuel line connection fittings. This will prevent unnecessary stress and torsion to fuel line piping.

• Always replace worn fuel fitting O-rings with new. Do not substitute fuel hose or equivalent where fuel pipe is installed.

RELIEVING FUEL SYSTEM PRESSURE

• Always relieve the fuel system pressure prior to disconnecting any fuel system component (injector, fuel rail, pressure regulator, etc.), fitting or fuel line connection. Exercise extreme caution whenever relieving fuel system pressure to avoid exposing skin, face and eyes to fuel spray. Please be advised that fuel under pressure may penetrate the skin or any part of the body that it contacts.

• Always place a shop towel or cloth around the fitting or connection prior to loosening to absorb any excess fuel due to spillage. Ensure that all fuel spillage (should it occur) is quickly removed from engine surfaces. Ensure that all fuel soaked cloths or towels are deposited into a suitable waste container.

• Always keep a dry chemical (Class B) fire extinguisher near the work area.

• Do not allow fuel spray or fuel vapors to come into contact with a spark or open flame.

With CONSULT-III

1. Turn ignition switch **ON**.
2. Perform **"FUEL PRESSURE RELEASE"** in **"WORK SUPPORT"** mode with CONSULT-III.
3. Start engine.
4. After engine stalls, crank it two or three times to release all fuel pressure.
5. Turn ignition switch **OFF**.

Without CONSULT-III

See Figure 137.

1. Remove fuel pump fuse (1) located in IPDM E/R (2).
2. Start engine.

22140_IM35_G0280

Fig. 137 Fuel pump fuse (1) located in IPDM E/R (2)

3. After engine stalls, crank it two or three times to release all fuel pressure.
4. Turn ignition switch **OFF**.
5. Reinstall fuel pump fuse after servicing fuel system.

FUEL FILTER

REMOVAL & INSTALLATION

The filter is part of the pump and sender assembly. Service can only be done by removing the sender - pump assembly.

FUEL PUMP

REMOVAL & INSTALLATION
See Figure 138.

1. Before servicing the vehicle, refer to the Precautions Section.
2. Relieve the fuel system pressure.
3. Open the fuel filler lid.
4. Open the filler cap and release the pressure inside the fuel tank.
5. Remove or disconnect the following:

• Rear seat cushion trim and pad bolts, then lift up rear seat cushion.

• Inspection hole cover for main and sub fuel level sensor unit by turning clips counterclockwise by 90°.

• Harness connector and quick connectors for EVAP/Vent line hose and fuel feed tube. Disconnect EVAP/Vent line hose connector (push in tubs and pull out).

• Remove the retainer for main and sub fuel level sensor unit with fuel tank lock ring wrench (SST) by turning counterclockwise.

• Remove main fuel level sensor unit, fuel filter and fuel pump assembly, and sub fuel level sensor unit. Raise the main fuel level sensor unit, fuel filter and fuel pump assembly, and disconnect the fuel hose connector (push in tabs and pull out) and sub fuel level sensor unit harness connector. Raise and release the sub fuel level sensor unit to remove.

To install:
6. Installation is the reverse of removal. Note the following:

• Connect fuel hose connector (push in until it stops) and sub fuel level sensor unit harness connector.

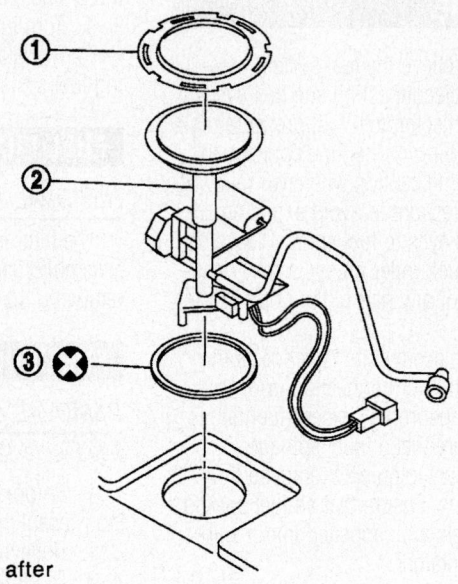

Left side

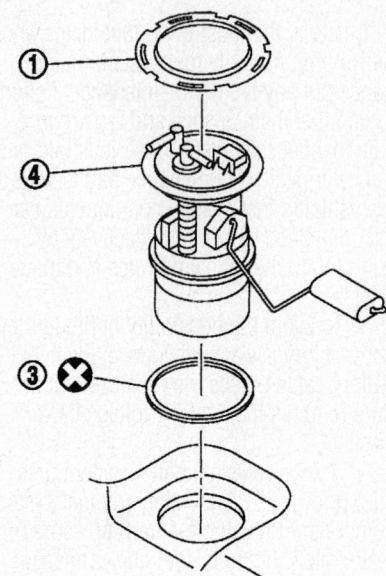

✖ : Always replace after every disassembly.

1. Retainer
2. Sub fuel level sensor unit
3. O-ring
4. Main fuel level sensor unit, fuel filter and fuel pump assembly

42356-MURA-G49

Fig. 138 Fuel pump and related components

- Align the direction mark on main and sub fuel level sensor unit with that on fuel tank as shown in the illustration.
- Install the inspection hole cover with front mark (arrow) facing front of the vehicle (Both for right and left). Lock the clips by turning clockwise.

7. Connect the quick connector as follows.

- Check the connection for damage or any foreign materials.
- Align the connector with the tube, then insert the connector straight into the tube until a click is heard.
- After connecting, make sure that the connection is secure by following method.
- Pull the tube and the connector to make sure they are securely connected. Visually confirm that the two retainer tabs are connected to the connector.

8. Turn ignition switch ON (with engine stopped), then check connections for leaks by applying fuel pressure to fuel piping.

9. Start the engine and let it idle and make sure there are no fuel leaks at the fuel system connections.

FUEL TANK

REMOVAL & INSTALLATION

See Figures 139.

1. Before servicing the vehicle, refer to the Precautions Section.
2. Relieve the fuel system pressure.
3. Open the fuel filler lid.
4. Open the filler cap and release the pressure inside the fuel tank.
5. Remove or disconnect the following:
6. Remove the fuel level sensor unit, fuel filter and fuel pump assembly.
7. Remove center muffler
8. Remove rear propeller shaft (AWD models).
9. Remove parking brake cables.
10. Remove rear final drive assembly (AWD models).

11. Using a transmission jack, support the bottom of the rear final drive assembly.

12. Remove mounting nuts on the rear suspension member, and lower the transmission jack carefully until the rear suspension member is removed from stud bolts on the vehicle (AWD models).

13. Remove fuel tank protector.

14. Disconnect fuel filler hose, EVAP/Vent line hose and EVAP (Recirculation) hose at the other end than fuel tank side.

15. Support the lower part of fuel tank with transmission jack.

➡**Support the position that fuel tank mounting bands do not engage.**

16. Remove fuel tank mounting bands.

17. Supporting with hands, descend transmission jack carefully, and remove fuel tank.

➡**Make sure that all connection points have been disconnected, confirm there is no interference with vehicle.**

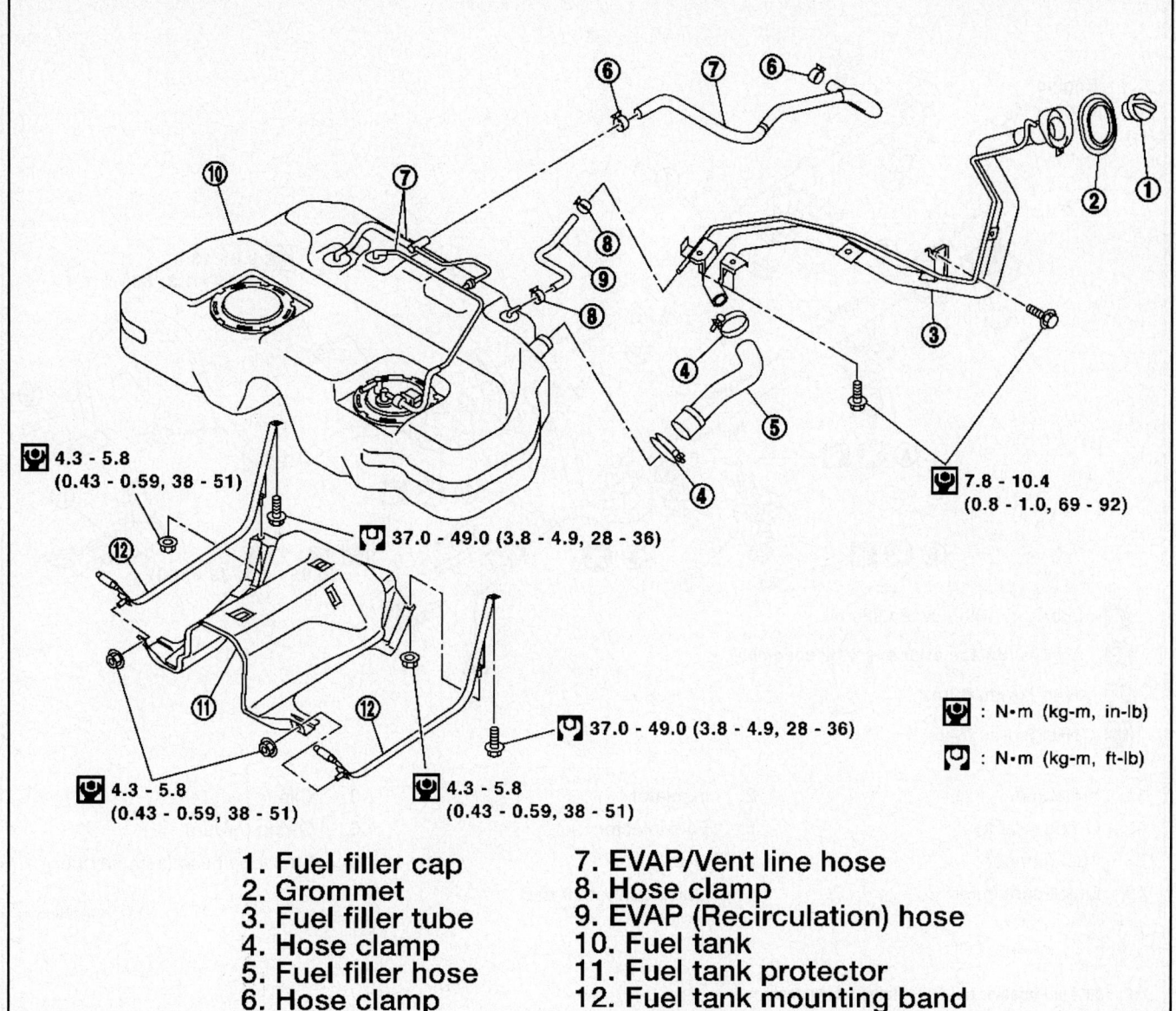

4.3 - 5.8
(0.43 - 0.59, 38 - 51)

37.0 - 49.0 (3.8 - 4.9, 28 - 36)

7.8 - 10.4
(0.8 - 1.0, 69 - 92)

37.0 - 49.0 (3.8 - 4.9, 28 - 36)

4.3 - 5.8
(0.43 - 0.59, 38 - 51)

4.3 - 5.8
(0.43 - 0.59, 38 - 51)

: N•m (kg-m, in-lb)

: N•m (kg-m, ft-lb)

1. Fuel filler cap
2. Grommet
3. Fuel filler tube
4. Hose clamp
5. Fuel filler hose
6. Hose clamp
7. EVAP/Vent line hose
8. Hose clamp
9. EVAP (Recirculation) hose
10. Fuel tank
11. Fuel tank protector
12. Fuel tank mounting band

22140_MURA_G0048

Fig. 139 Fuel tank and components exploded view

To install:

18. Install the fuel tank.

19. Install the fuel tank mounting bands.

20. Tighten the fuel tank mounting bands to 36 ft. lbs. (37 Nm).

21. Install the rear suspension member, rear final drive assembly (AWD models).

22. Install the parking brake cables.

23. Install the rear propeller shaft (AWD models).

24. Install the center muffler.

25. Install the fuel level sensor unit, fuel filter and fuel pump assembly.

26. Add ¼ tank of fuel for checks outlined below.

Inspection After Installation

1. Make sure there is no fuel leakage at connections in the following steps.

2. Turn ignition switch "ON" (with engine stopped), and check connections for leakage by applying fuel pressure to fuel piping.

3. Start engine and rev it up and make sure there are no fuel leaks at the fuel system tube and hose connections.

4. After removing/installing rear suspension assembly, make sure to adjust wheel alignment and then, adjust neutral position of steering angle sensor.

FUEL RAIL & INJECTORS

REMOVAL & INSTALLATION

See Figures 140 through 143.

1. Before servicing the vehicle, refer to the Precautions Section.

2. Remove the engine cover.

3. Properly release the fuel pressure.

4. Remove the radiator cover grille, air duct (inlet), air cleaner case, air duct assembly and mass air flow sensor.

5. Disconnect the electric throttle control actuator and engine coolant hoses. Disconnect vacuum hose, all fuel injector

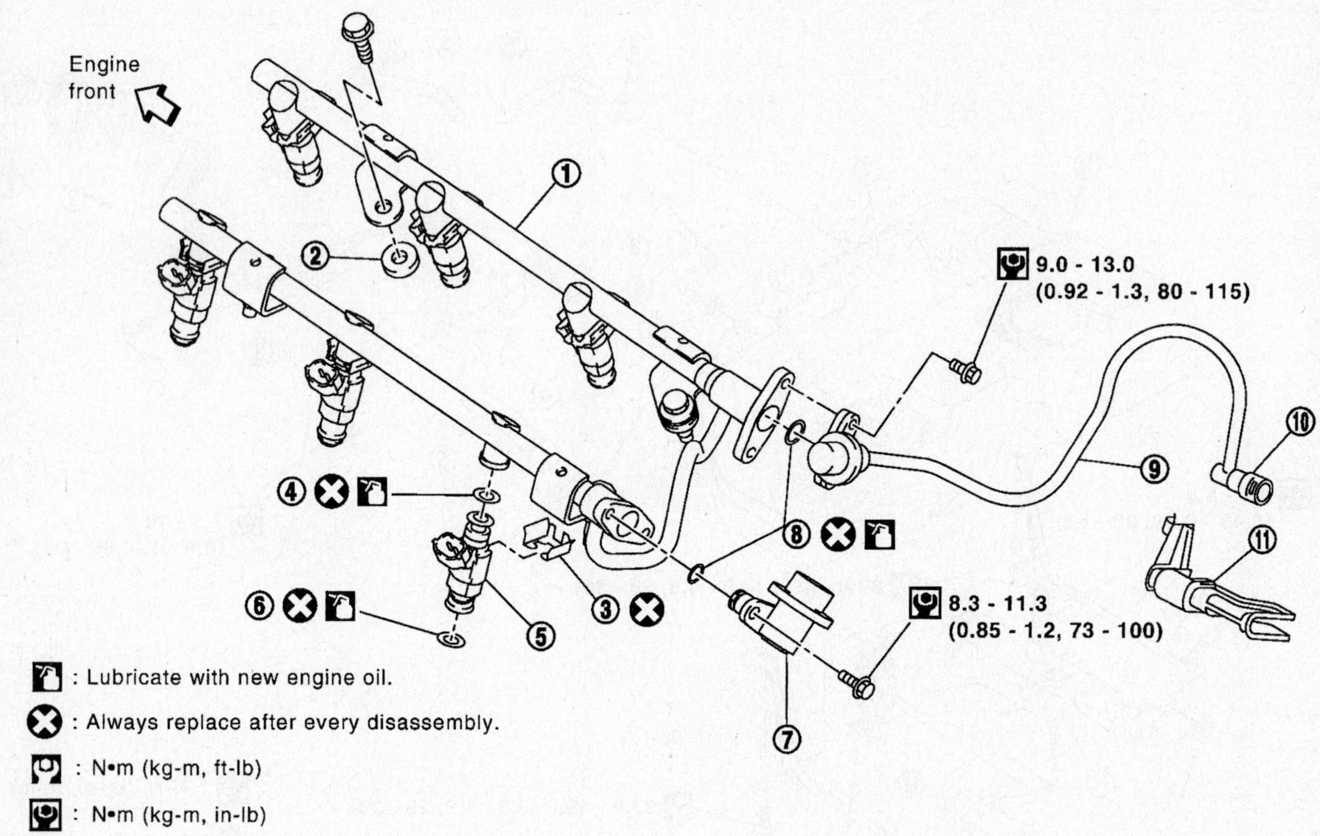

Engine
front

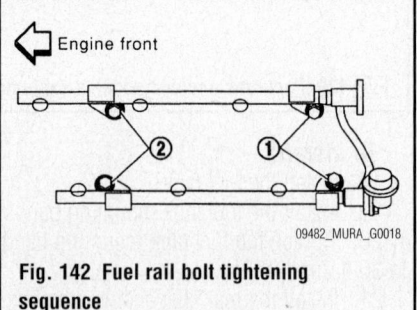

🔧 9.0 - 13.0
(0.92 - 1.3, 80 - 115)

④ ✖ 🛢
⑥ ✖ 🛢
⑤
③ ✖

⑩
⑨
⑪

⑧ ✖ 🛢

🔧 8.3 - 11.3
(0.85 - 1.2, 73 - 100)

⑦

🛢 : Lubricate with new engine oil.

✖ : Always replace after every disassembly.

🔧 : N•m (kg-m, ft-lb)

🔧 : N•m (kg-m, in-lb)

1. Fuel tube	2. Insulator	3. Clip
4. O-ring (black)	5. Fuel injector	6. O-ring (green)
7. Fuel damper	8. O-ring	9. Fuel feed hose (with damper)
10. Quick connector	11. Quick connector cap	

42356-MURA-G50

Fig. 140 Fuel injector rail and related components

electrical connectors, and PCV hose. Remove the vacuum tank from intake manifold collector (lower). Disconnect the power steering hose bracket.

6. Remove the intake manifold upper and lower collectors.

➡**The intake manifold collector (upper) should be moved aside with water hoses connected.**

7. Remove the fuel feed hose (with damper) from fuel tube.

8. Disconnect the fuel feed hose (with damper) quick connector at vehicle piping side. When separating fuel feed hose and centralized under-floor piping connection, disconnect quick connector with the following procedure.

a. Remove the quick connector cap from quick connector.

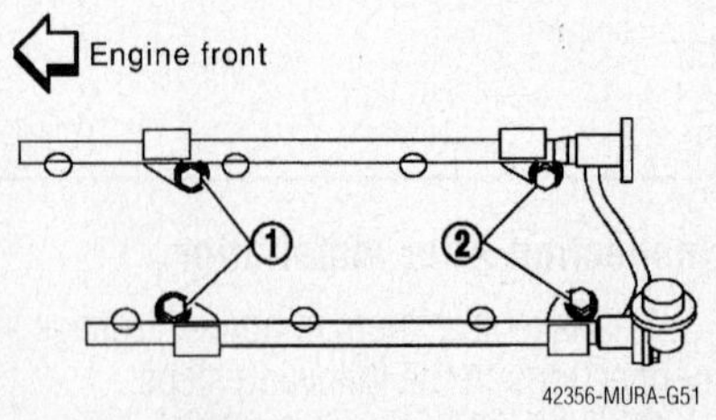

Engine front

① ②

42356-MURA-G51

Fig. 141 Fuel rail bolt tightening sequence

Engine front

② ①

09482_MURA_G0018

Fig. 142 Fuel rail bolt tightening sequence

b. Disconnect the quick connector from centralized under-floor piping.

9. Remove the harness connector from fuel injector.

10. Loosen the mounting bolts in the same sequence as the installation sequence. Remove fuel tube and fuel injector assembly.

11. Remove the fuel injector from fuel tube with following procedure.

a. Open and remove the clip.

b. Remove the fuel injector from the fuel tube by pulling straight.

12. Remove the fuel damper from fuel tube.

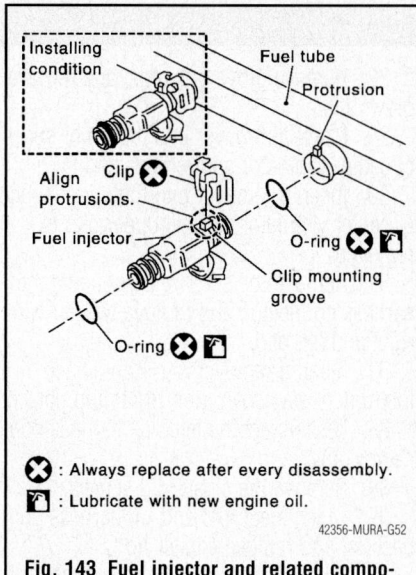

: Always replace after every disassembly.

: Lubricate with new engine oil.

42356-MURA-G52

Fig. 143 Fuel injector and related components

To install:

13. Install the fuel damper. Insert fuel damper straight into fuel tube. Tighten mounting bolts evenly in turn. After tightening the mounting bolts, make sure that there is no gap between flange and fuel tube. When handling O-rings, be careful of the following:

- Handle O-ring with bare hands. Never wear gloves.
- Lubricate O-ring with new engine oil.
- Do not clean O-ring with solvent.
- Make sure that O-ring and its mating part are free of foreign material.
- When installing the O-rings, be careful not to scratch it with tool or fingernails. Also, be careful not to twist or stretch O-ring. If O-ring was stretched while it was being attached, do not insert it quickly into fuel tube.

14. Install O-rings on the fuel injector. Upper and lower O-ring are different.

15. Install the fuel injector to fuel tube with the following procedure.

a. Insert the clip into clip mounting groove on fuel injector. Insert clip so that lug "A" of fuel injector matches notch "A" of the clip.

➡Do not reuse the clip. Replace it with a new one. Be careful to keep clip from interfering with O-ring. If interference occurs, replace O-ring.

b. Insert the fuel injector into fuel tube with clip attached. Insert it while matching it to the axial center. Insert the fuel injector so that lug "B" of fuel tube matches notch "B" of the clip. Make sure that fuel tube flange is securely fixed in flange groove on clip.

c. Make sure that installation is complete by checking that fuel injector does not rotate or come off.

16. Tighten the mounting bolts in two steps (first to 7 ft. lbs and then to 17 ft. lbs.) in the proper sequence.

17. Connect the fuel injector harness.

18. Install the intake manifold collector (upper and lower).

19. Connect the quick connector between fuel feed hose (with damper) and centralized under-floor piping connection with the following procedure:

a. Check the connection for damage and foreign materials.

b. Align the quick connector with the tube, then insert the connector straight into the tube until a click is heard.

c. After connecting the quick connector, use the following method to make sure it is full connected. Visually confirm that the two retainer tabs are connected to the connector. Pull the tube and the connector to make sure they are securely connected.

d. Install quick connector cap to quick connector connection. Install quick connector cap with arrow on surface facing in direction of quick connector.

➡If cap cannot be installed smoothly, quick connector may have not been installed correctly. Check connection again.

20. The remainder of installation is the reverse of removal.

21. Turn the ignition switch ON (with engine stopped). With fuel pressure applied to fuel piping, check for fuel leakage at connection points.

22. Start the engine. With engine speed increased, check again for fuel leakage at connection points.

IDLE SPEED

ADJUSTMENT

The idle speed is controlled by the ECM and no adjustment is possible.

THROTTLE BODY

REMOVAL & INSTALLATION

See Figure 144.

1. Remove the engine cover (1) with power tool

2. Disconnect the water hoses from intake manifold collector (upper), attach blind plug to prevent engine coolant leakage.

➡Perform this step when the engine is cold.

➡Never spill the engine coolant on drive belts.

3. Remove the air cleaner case and air duct.

4. Remove the electric throttle control actuator as follows:

a. Disconnect the harness connector.

b. Loosen the mounting bolts in reverse order as shown in the figure.

➡Never disassemble the throttle body.

To install:

5. Install the gasket with positioning no-protrusion surface upward or downward.

6. Tighten in numerical order as shown in the figure.

7. Tighten to 75 inch lbs. (8.5 Nm).

➡Perform the "Throttle Valve Closed Position Learning" when harness connector of electric throttle control actuator is disconnected

Throttle Valve Closed Position Learning

➡Throttle Valve Closed Position Learning is an operation to learn the fully closed position of the throttle valve by monitoring the throttle position sensor output signal. It must be performed each time harness connector of electric throttle control actuator or ECM is disconnected.

1. Make sure that the accelerator pedal is fully released.

2. Turn the ignition switch **ON**.

3. Turn the ignition switch **OFF** and wait at least 10 seconds.

4. Make sure that the throttle valve moves during above 10 seconds by confirming the operating sound.

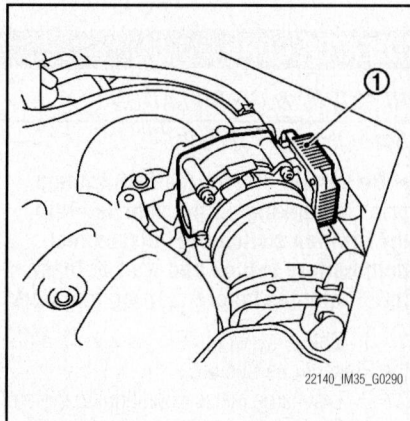

22140_IM35_G0290

Fig. 144 Throttle body control actuator (1) harness connector

HEATING & AIR CONDITIONING SYSTEM

BLOWER MOTOR

REMOVAL & INSTALLATION

See Figure 145.

➡ **Before servicing, or working around, the SRS system, turn the ignition switch OFF, disconnect both battery cables and wait at least three minutes. When servicing, or working around, the SRS system do not work directly in front of the air bag module.**

1. Before servicing the vehicle, refer to the Precautions Section.
2. Disconnect the negative battery cable.
3. Remove the lower instrument panel, passenger side.
4. Disconnect the blower motor electrical connector.
5. Remove the blower motor retaining screws.
6. Remove the blower motor from its mounting.

   To install:
7. Installation is the reverse of the removal procedure.

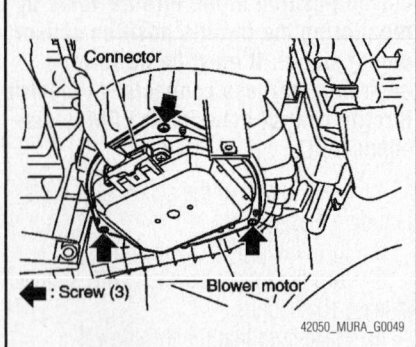

Fig. 145 Blower motor and related components

HEATER CORE

REMOVAL & INSTALLATION

See Figures 146 through 148.

➡ **Be sure to disarm the SRS system, prior to working on the vehicle. Turn the ignition switch OFF, disconnect both battery cables and wait at least three minutes before starting any work.**

1. Before servicing the vehicle, refer to the Precautions Section.
2. Discharge the air conditioning system.
3. Disconnect both battery cables.
4. Drain the coolant from cooling system.

5. Remove both right/left wiper arms.
6. Remove the cowl top seal rubber.
7. Remove the clips from cowl top cover (right) and remove cowl top cover (right).
8. Remove the clips from cowl top cover (left) and remove cowl top cover (left)
9. Remove the washer nozzles and hose from cowl top cover.
10. Remove the cowl top cover.
11. Disconnect evaporator-side one-touch joints.

   a. Install a disconnector tool (High-pressure side: 92530-89908, Low pressure side: 92530-89916) on A/C piping.
   b. Slide a disconnector toward vehicle front until it clicks.
   c. Slide the A/C piping toward vehicle front and disconnect it.

✳✳ WARNING

Seal connection opening of piping with a cap or vinyl tape to avoid exposure to atmosphere.

12. Disconnect the two heater hoses from heater core.
13. Remove the fuse lid.
14. Remove instrument driver lower panel screws.
15. Remove the data link connector.
16. Pull to disengage metal clip by removing panel in horizontal direction.
17. Disconnect in-vehicle sensor and each electrical parts.
18. Remove the bolts, and remove hood lock opener.
19. Remove the tilt lever knob screws.
20. Remove the knob by picking it up and pulling it out. Using a remover, ply and remove tilt lever mask.
21. Remove steering column cover screws.
22. Disengage the tab, and remove steering column cover. After removing combination meter screws, remove harness connector.
23. Using a remover, pry and remove side ventilator assembly (left/right).
24. Disconnect the aiming switch harness connector and VDC switch harness connector only left side.
25. Insert a remover into lower space of instrument side finisher (left/right) and remove by lifting.
26. Remove the screws and glove box striker, disconnect connectors, and remove instrument passenger lower panel assembly.
27. Detach the damper from glove box right side.

28. Remove glove box pins, and remove glove box.
29. Insert a remover into the lower space of center ventilator and remove by lifting.
30. Insert a remover into the upper space of center ventilator and the upper clip is removed.
31. Remove the screws. Disconnect the harness connector, and remove tweeter with right and left part.
32. Insert a remover into front space of instrument stay cover (left/right) and detach.
33. Disconnect the left side harness connector only.
34. Remove the cluster lid screws.
35. Disconnect A/C and AV harness connectors, and remove cluster lid C.
36. Remove the display unit screws.
37. Disconnect harness connector and remove display.
38. Using a remover, disengage the ignition key finisher metal clips
39. Disconnect the harness connector.
40. Using a remover, pry and remove the instrument mask.
41. Disconnect the harness connector.
42. Remove screw. Disengage the metal clip and remove instrument driver upper panel.
43. Remove the bolt and screws and remove the front passenger air bag module.
44. Disconnect the metal clips, then remove instrument passenger upper panel.
45. Disconnect the harness connector.
46. Pull to inside of vehicle disconnect metal clips and remove front pillar garnish.
47. Remove bolts and screws, and remove the instrument panel from passenger door opening portion.
48. Remove the tweeter and sensor harness clip are removed from the duct.
49. Remove the instrument panel assembly.
50. Remove the ECM with bracket attached.
51. Remove the nuts (2), then bolts (2) and screw (1), then remove blower unit.

➡ **Move the blower unit to the right and remove locating pin (1) and joint. Then remove blower unit downward.**

52. Disconnect intake door motor connector and blower fan motor connector.
53. Remove the harness clips (2) from blower unit.
54. Remove the blower unit.
55. Remove the clips from vehicle harness from steering member.
56. Remove the instrument stays (driver side and passenger side).

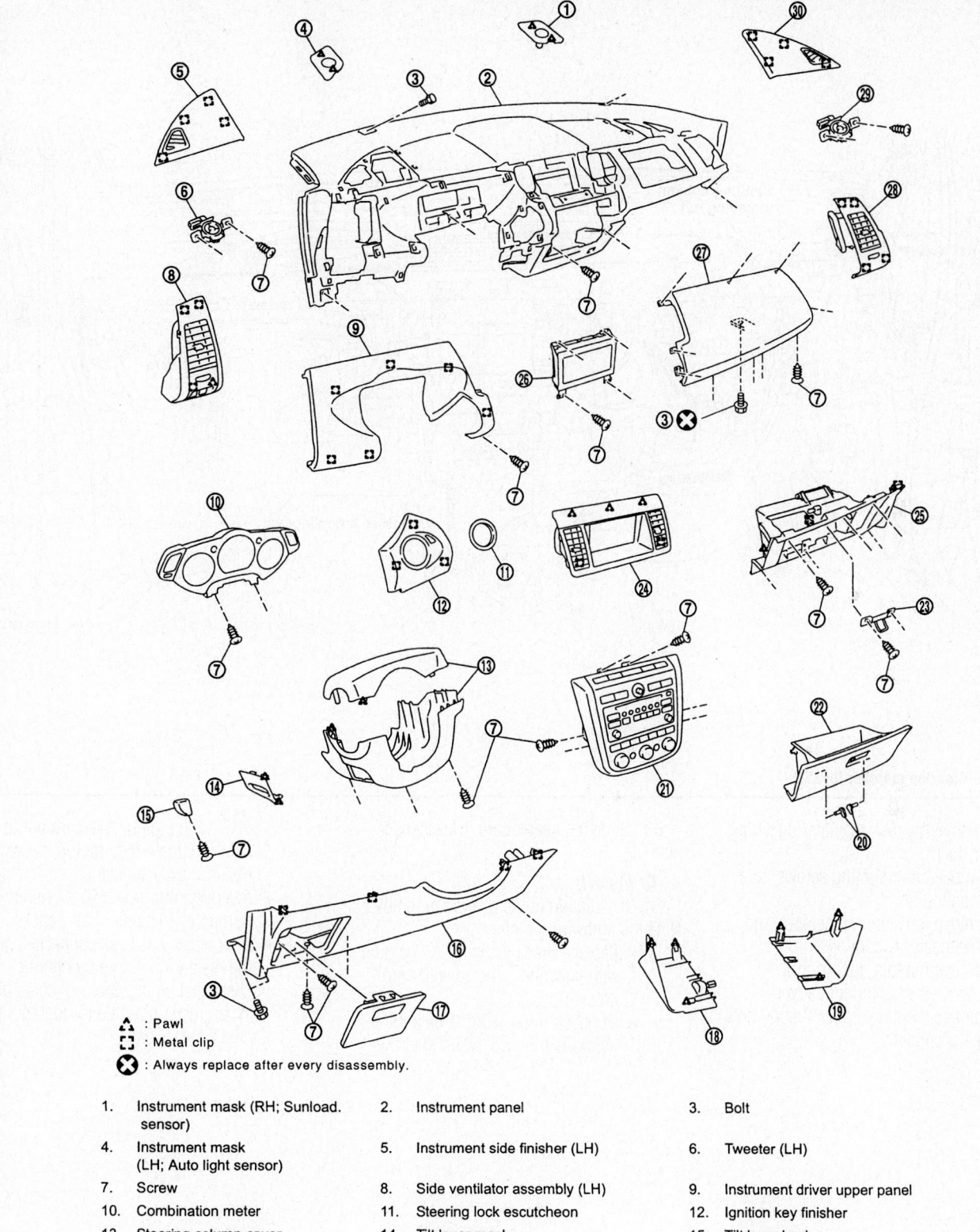

△ : Pawl

⟦ ⟧ : Metal clip

⊗ : Always replace after every disassembly.

1.	Instrument mask (RH; Sunload. sensor)	2.	Instrument panel	3.	Bolt
4.	Instrument mask (LH; Auto light sensor)	5.	Instrument side finisher (LH)	6.	Tweeter (LH)
7.	Screw	8.	Side ventilator assembly (LH)	9.	Instrument driver upper panel
10.	Combination meter	11.	Steering lock escutcheon	12.	Ignition key finisher
13.	Steering column cover	14.	Tilt lever mask	15.	Tilt lever knob
16.	Instrument driver lower panel	17.	Fuse lid	18.	Instrument stay cover (LH)
19.	Instrument stay cover (RH)	20.	Glove box pin	21.	Cluster lid C
22.	Glove box assembly	23.	Glove box striker	24.	Center ventilator
25.	Instrument passenger lower panel	26.	Display	27.	Instrument passenger upper panel
28.	Side ventilator assembly (RH)	29.	Tweeter (RH)	30.	Instrument side finisher (RH)

09482_MURA_G0004

Fig. 146 Instrument panel and related components

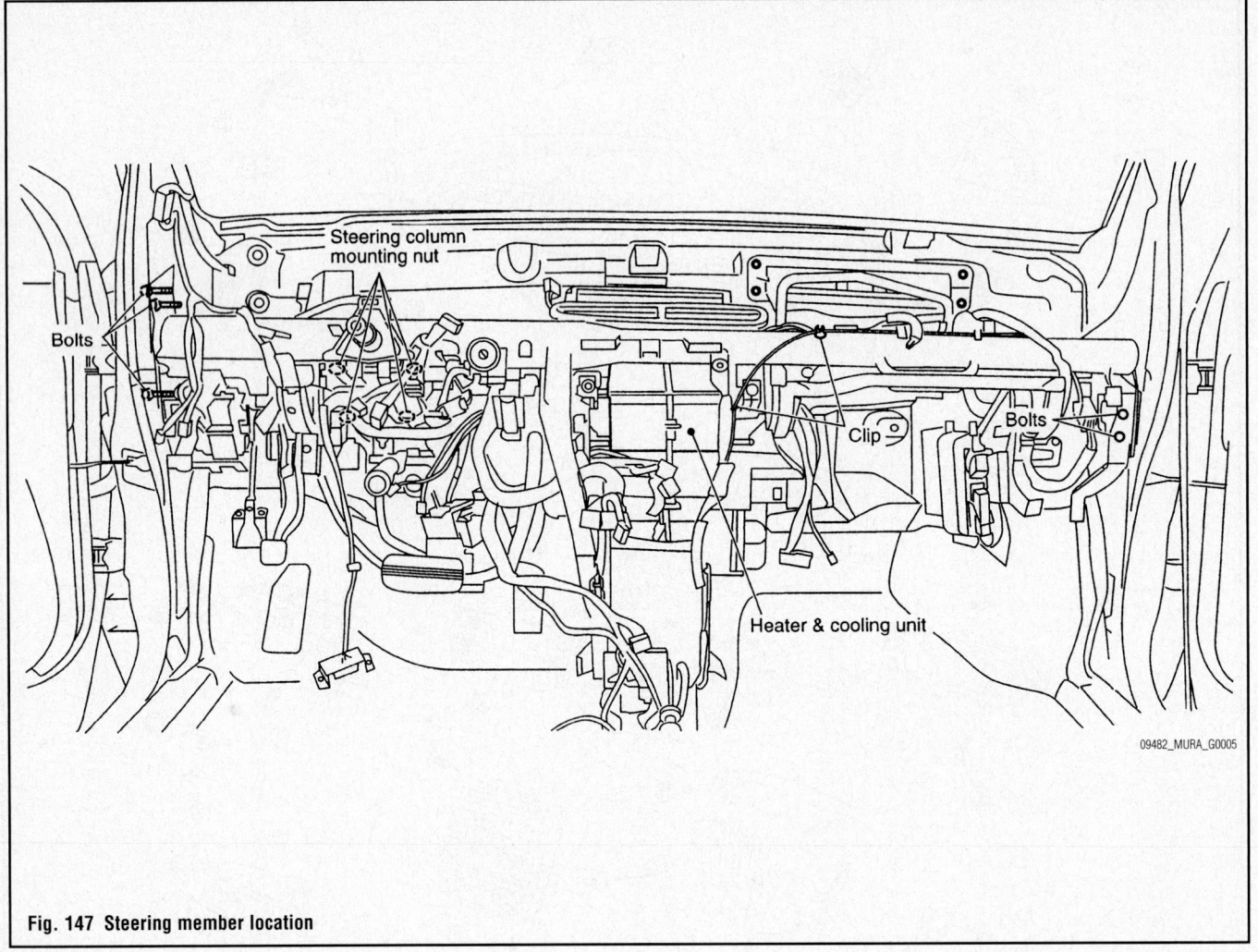

Steering column
mounting nut

Bolts

Clip

Bolts

Heater & cooling unit

09482_MURA_G0005

Fig. 147 Steering member location

57. Remove the rear ventilator duct1 and front floor duct.

58. Remove the mounting screws from heater & cooling unit.

59. Remove the steering member, and then remove heater & cooling unit.

60. Remove the foot duct (right).

61. Remove the heater core cover.

62. Remove the heater pipe support and heater pipe grommet.

63. Slide the heater core to passenger side.

To install:

64. Installation is the reverse of removal. Note the following points:

- Replace the O-rings for A/C piping with new ones, coated with compressor oil.
- Connection point for female-side piping is thin. So, when inserting male-side piping, take care not to deform female-side piping. Slowly insert in axial direction.
- Insert the one-touch joint connection point securely until it clicks.
- After piping has been connected, pull the male-side piping by hand to check that piping does not come off.
- When recharging the refrigerant, check for leaks.

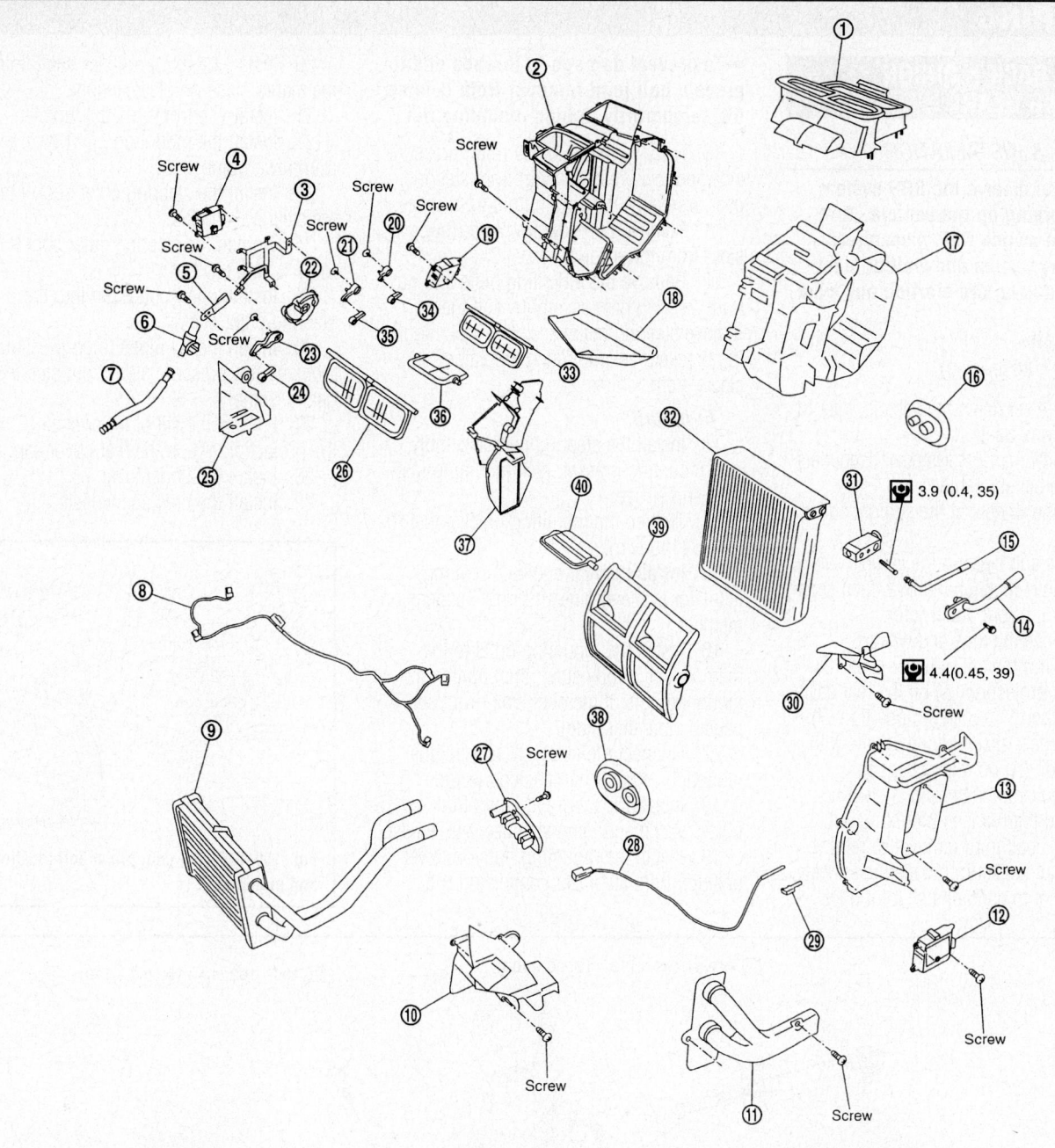

1. Adaptor duct	2. Heater & cooling case (left)	3. Mode door motor bracket
4. Mode door motor	5. Instrument lower cover bracket	6. Aspirator
7. Aspirator duct	8. Sub harness	9. Heater core
10. Foot duct (right)	11. Heater core cover	12. Air mix door motor (passenger side)
13. Evaporator cover	14. Low-pressure pipe 2	15. High-pressure pipe 2
16. Cooler pipe grommet	17. Heater & cooling case (right)	18. Insulator
19. Air mix door motor (driver side)	20. Defroster door lever	21. Max. cool door lever
22. Main link	23. Ventilator door lever	24. Ventilator door link
25. Foot duct (left)	26. Ventilator door	27. Heater pipe support
28. Intake sensor	29. Intake sensor bracket	30. Adaptor cover
31. Expansion valve	32. Evaporator	33. Defroster door
34. Defroster door link	35. Max. cool door link	36. Max. cool door (left)
37. Center case	38. Heater pipe grommet	39. Air mix door
40. Max. cool door (right)		

09482_MURA_G0006

Fig. 148 Heater/evaporator core and related components

STEERING

POWER RACK & PINION STEERING GEAR

REMOVAL & INSTALLATION

➡Be sure to disarm the SRS system, prior to working on the vehicle. Turn the ignition switch OFF, disconnect both battery cables and wait at least three minutes before starting any work.

2WD Models

See Figures 149 and 150.

1. Before servicing the vehicle, refer to the Precautions Section.
2. Disarm the SRS system. Disconnect the negative battery cable.
3. Set the wheels in the straight ahead position.
4. Raise and support the vehicle safely.
5. Remove locknut and bolt, then separate lower joint from upper joint.
6. Remove the tires and wheels.
7. Confirm the slit of lower shaft (C) fits with the projection (A) on the rear cover cap (1), furthermore marking position (B) on steering gear assembly nearly fits with the projection (A) on rear cover cap.
8. Remove the cotter pin at steering knuckle, and then loosen mounting nut.
9. Use a ball joint remover to remove steering outer socket from steering knuckle. Be careful not to damage ball joint boot.

➡To prevent damage to threads and to prevent ball joint remover from coming off, temporarily tighten mounting nut.

10. Remove the oil pipes (high pressure side and low pressure side) from steering gear assembly, then drain fluid from pipes.
11. Remove the mounting bolt (lower side) from lower joint.
12. Remove the mounting bolts and nut from steering gear assembly, and then remove steering gear assembly, rack mounting bracket, rack mounting insulator and sleeve from vehicle.

To install:

13. Install the steering gear assembly, rack mounting bracket, rack mounting insulator and sleeve.
14. Tighten the mounting bracket to 110 ft. lbs. (149 Nm).
15. Install the rear engine mounting insulator to the engine and front suspension member.
16. Install the mounting bolts to the member stay (body side), then tighten the mounting nuts of member stay (front suspension member side).
17. Connect the electrical rear engine mounting actuator harness connector.
18. Install the mounting bolts from stabilizer clamp and hang stabilizer on vehicle.
19. Install the mounting nuts on lower position from stabilizer connecting rod.

20. Install the rear propeller shaft lining the marks made on disassembly.
21. Install the front exhaust tube.
22. Install the mounting bolt (lower side) from lower joint.
23. Install the steering outer socket to the steering knuckle.
24. Tighten the steering outer socket nut to 37 ft. lbs. (50 Nm).
25. Install a new cotter pin into the steering outer socket.
26. Install the oil pipes (high pressure side and low pressure side) from steering gear assembly.
27. Line up the slit of lower shaft (C) with the projection (A) on the rear cover cap.
28. Install the undercover.
29. Install the tires and wheels.

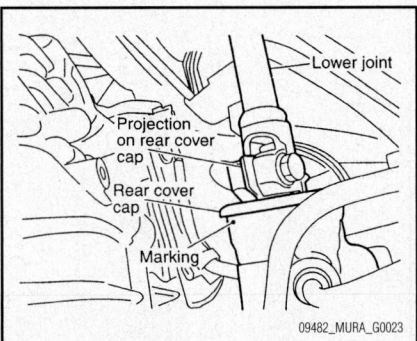

Fig. 150 Steering gear pinch bolt location and alignment

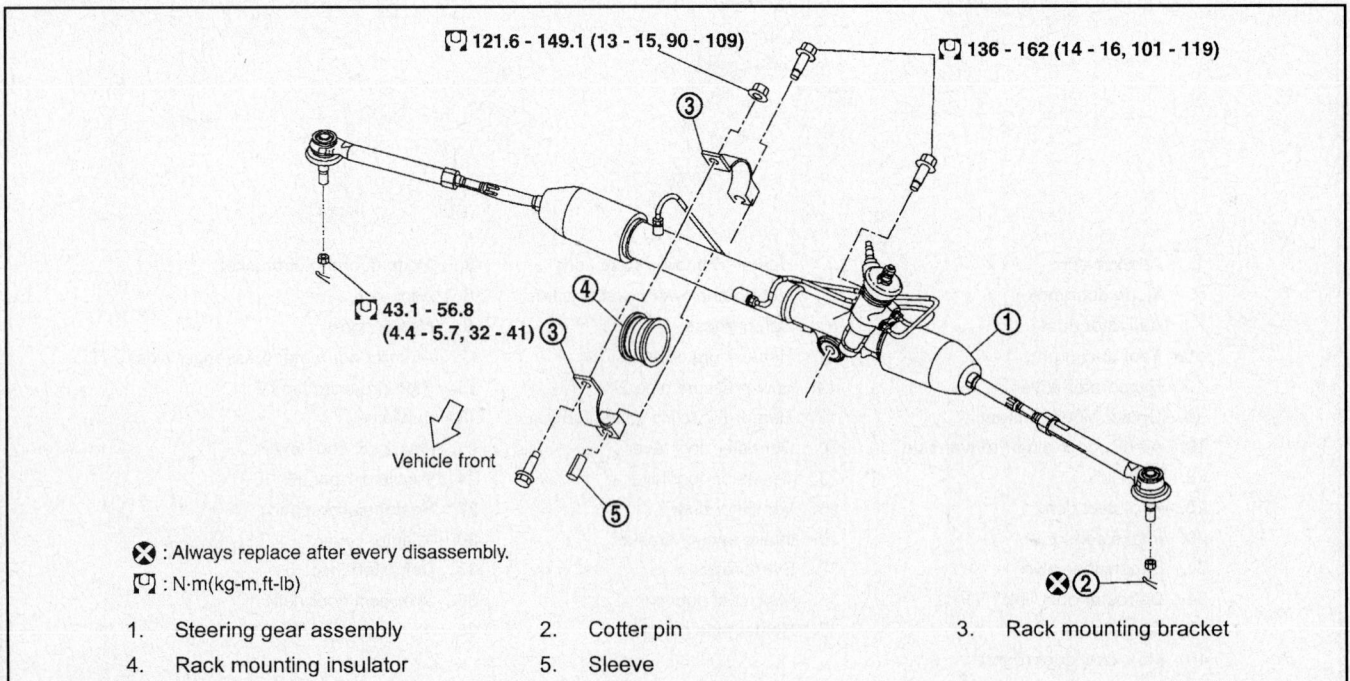

🗙 : Always replace after every disassembly.

🔧 : N·m(kg-m,ft-lb)

1. Steering gear assembly
2. Cotter pin
3. Rack mounting bracket
4. Rack mounting insulator
5. Sleeve

Fig. 149 Steering gear and related components

30. Be sure to replace all non reusable components with new ones.

31. When the steering wheel is set in the straight ahead direction, confirm slit of lower joint fits with the projection on rear cover cap, also the matchmarks on steering gear assembly nearly fits with the projection on rear cover cap.

32. After the installation, fill with P/S fluid and bleed air from piping.

33. Check if steering the wheel turns smoothly when it is turned several times fully to the end of the left and right.

34. Be sure to replace all non reusable components with new ones.

➡️**When the steering wheel is set in the straight ahead direction, confirm slit of lower joint fits with the projection on rear cover cap, also the matchmarks on steering gear assembly fit with the projection on rear cover cap.**

35. After installation, bleed air from piping.

36. Check if steering wheel turns smoothly when it is turned several times fully to the end of the left and right.

AWD Models

See Figures 151 and 152.

1. Before servicing the vehicle, refer to the Precautions Section.

2. Disarm the SRS system. Disconnect the negative battery cable.

3. Set the wheels in the straight ahead position.

4. Raise and support the vehicle safely.

5. Remove the locknut and bolt, then separate lower joint from upper joint.

6. Remove the tires and wheels.

7. Remove the undercover.

8. Confirm the slit of lower shaft (C) fits with the projection (A) on the rear cover cap (1), furthermore marking position (B) on steering gear assembly nearly fits with the projection (A) on rear cover cap.

9. Remove the oil pipes (high pressure side and low pressure side) from steering gear assembly, then drain fluid from pipes.

10. Remove the cotter pin at steering knuckle, then loosen mounting nut.

11. Use a ball joint remover to remove steering outer socket from steering knuckle. Be careful not to damage ball joint boot.

➡️**To prevent damage to threads and to prevent ball joint remover from coming off, and temporarily tighten mounting nut.**

12. Remove mounting bolt (lower side) from lower joint.

❄️ WARNING

Spiral cable may snap due to steering operation if steering column is separated from steering gear assembly. Therefore, fix steering wheel with a string to avoid turns.

13. Remove the front exhaust tube.

14. Mach mark the propeller shaft for proper alignment on reassembly.

15. Remove rear propeller shaft.

16. Remove the mounting nuts on lower position from stabilizer connecting rod.

17. Remove the mounting bolts from stabilizer clamp and hang stabilizer on vehicle.

18. Remove the electrical rear engine mounting actuator harness connector.

19. Disconnect the electrical rear engine mounting actuator harness connector.

20. Set jack under engine and front suspension member.

21. Remove the mounting bolts from rear engine mounting insulator.

22. Loosen the mounting nuts of front suspension member (front side).

23. Remove the mounting bolts from member stay (body side), then loosen

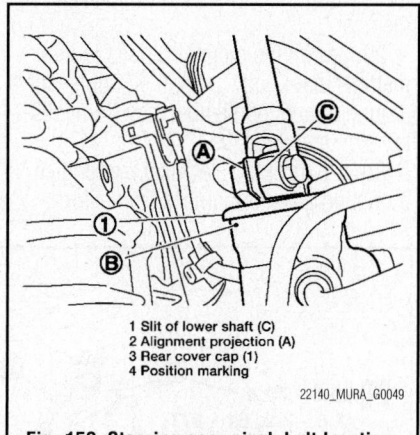

1 Slit of lower shaft (C)
2 Alignment projection (A)
3 Rear cover cap (1)
4 Position marking

22140_MURA_G0049

Fig. 152 Steering gear pinch bolt location and alignment

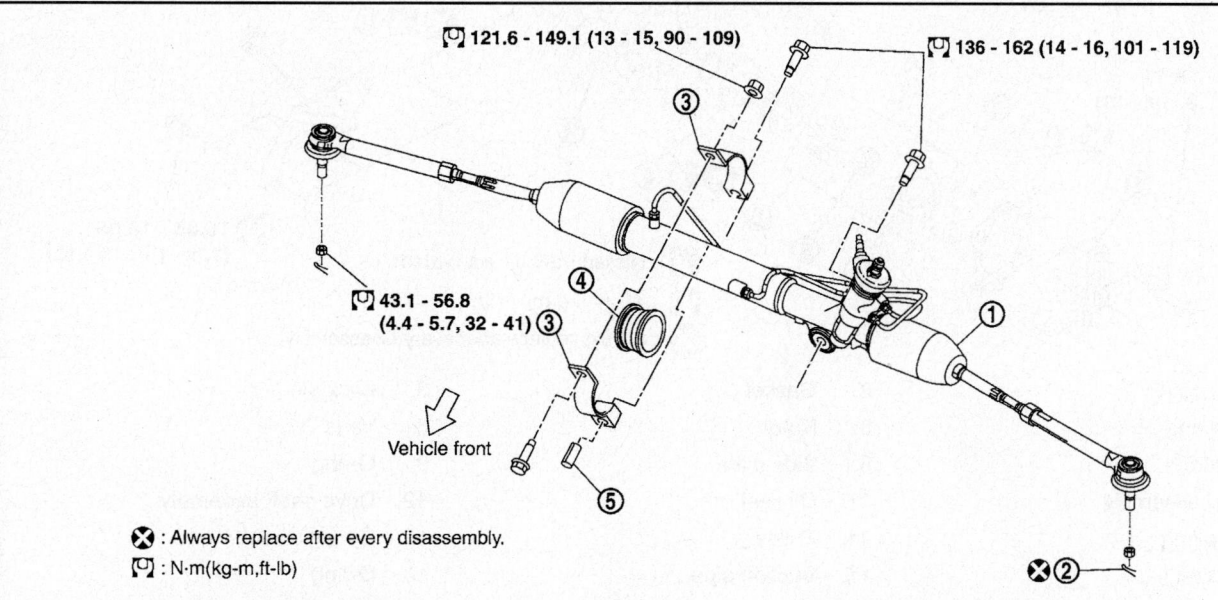

⏱️ 121.6 - 149.1 (13 - 15, 90 - 109)   ⏱️ 136 - 162 (14 - 16, 101 - 119)

⏱️ 43.1 - 56.8
(4.4 - 5.7, 32 - 41)

Vehicle front

❌ : Always replace after every disassembly.
⏱️ : N·m(kg-m,ft-lb)

1. Steering gear assembly
4. Rack mounting insulator
2. Cotter pin
5. Sleeve
3. Rack mounting bracket

42356-MURA-G68

Fig. 151 Steering gear and related components

mounting nuts of member stay (front suspension member side).

24. Move the jack down slowly (front suspension member side) to remove rear engine mounting insulator from engine and front suspension member.

25. Remove the mounting bolts and nut from steering gear assembly, and then remove steering gear assembly, rack mounting bracket, rack mounting insulator and sleeve from vehicle.

To install:

26. Install the steering gear assembly, rack mounting bracket, rack mounting insulator and sleeve.

27. Tighten the mounting bracket to 110 ft. lbs. (149 Nm).

28. Install the rear engine mounting insulator to the engine and front suspension member.

29. Install the mounting bolts to the member stay (body side), then tighten the mounting nuts of member stay (front suspension member side).

30. Connect the electrical rear engine mounting actuator harness connector.

31. Install the mounting bolts from stabilizer clamp and hang stabilizer on vehicle.

32. Install the mounting nuts on lower position from stabilizer connecting rod.

33. Install the rear propeller shaft lining the marks made on disassembly.

34. Install the front exhaust tube.

35. Install the mounting bolt (lower side) from lower joint.

36. Install the steering outer socket to the steering knuckle.

37. Tighten the steering outer socket nut to 37 ft. lbs. (50 Nm).

38. Install a new cotter pin into the steering outer socket.

39. Install the oil pipes (high pressure side and low pressure side) from steering gear assembly.

40. Line up the slit of lower shaft (C) with the projection (A) on the rear cover cap.

41. Install the undercover.

42. Install the tires and wheels.

43. Be sure to replace all non reusable components with new ones.

44. When steering wheel is set in the straight ahead direction, confirm slit of

lower joint fits with the projection on rear cover cap, also the matchmarks on steering gear assembly nearly fits with the projection on rear cover cap.

45. After installation, fill with P/S fluid and bleed air from piping.

46. Check if steering wheel turns smoothly when it is turned several times fully to the end of the left and right.

POWER STEERING PUMP

REMOVAL & INSTALLATION

See Figure 153.

1. Before servicing the vehicle, refer to the precautions in the beginning of this section.

2. Disconnect the negative battery cable.

3. Raise and support the vehicle safely. Remove the tire and wheel assemblies.

4. Remove the side splash guard.

5. Remove the heat insulator.

6. Loosen the adjusting screw and the pump retaining bolt. Remove the belt.

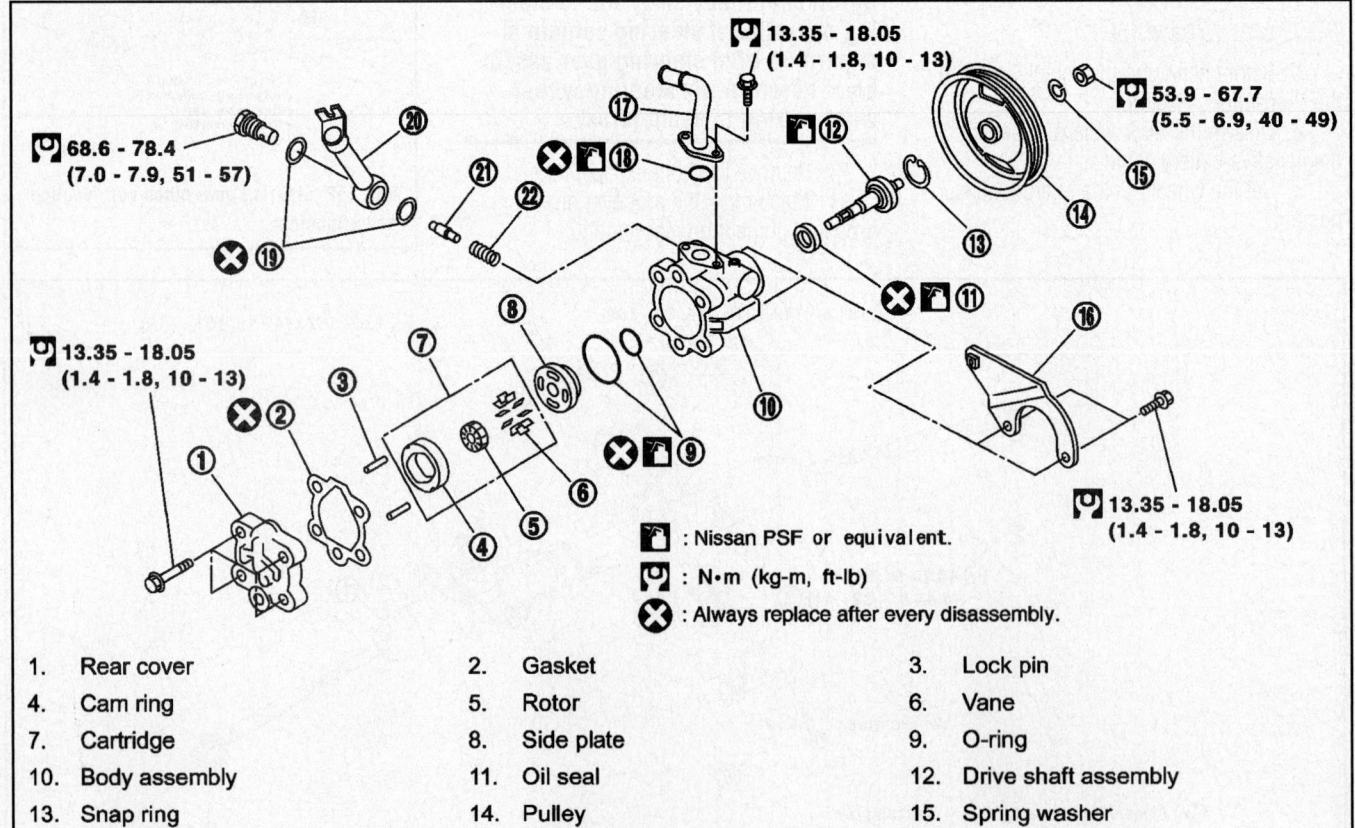

: Nissan PSF or equivalent.

: N·m (kg-m, ft-lb)

: Always replace after every disassembly.

1.	Rear cover	2.	Gasket	3.	Lock pin
4.	Cam ring	5.	Rotor	6.	Vane
7.	Cartridge	8.	Side plate	9.	O-ring
10.	Body assembly	11.	Oil seal	12.	Drive shaft assembly
13.	Snap ring	14.	Pulley	15.	Spring washer
16.	Bracket	17.	Suction pipe	18.	O-ring
19.	Washer	20.	Joint	21.	Flow control valve
22.	Spring				

Fig. 153 Power steering pump and related components

42050_MURA_G0033

7. Drain the power steering fluid.

8. Disconnect and plug the power steering hose lines.

9. Remove the power steering pump retaining bolts.

10. Remove the power steering pump bolt from the vehicle.

To install:

11. Installation is the reverse of the removal procedure.

12. Bleed the power steering system.

13. Adjust the belt tension, as required.

BLEEDING

1. Before servicing the vehicle, refer to the Precautions Section.

2. Fill the power steering system with the proper grade and type steering fluid.

➥**Do not allow the fluid level in the reservoir tank to go below the MIN level line. Check and add fluid as needed.**

3. Raise and safely support the vehicle.

4. Quickly turn the steering wheel to the full right and left detents and lightly touch the steering stoppers.

➥**Do not hold the steering wheel in the locked position for more than ten seconds.**

5. Repeat this operation until the fluid level no longer decreases.

6. Start the engine.

7. Quickly turn the steering wheel to the full right and left detents and lightly touch the steering stoppers.

➥**Do not hold the steering wheel in the locked position for more than ten seconds.**

8. Check for air bubbles or cloudy fluid. If found, repeat the bleeding procedure.

9. Stop the engine and check the fluid level. Correct as required.

STEERING LINKAGE

REMOVAL & INSTALLATION

See Figures 154 through 158.

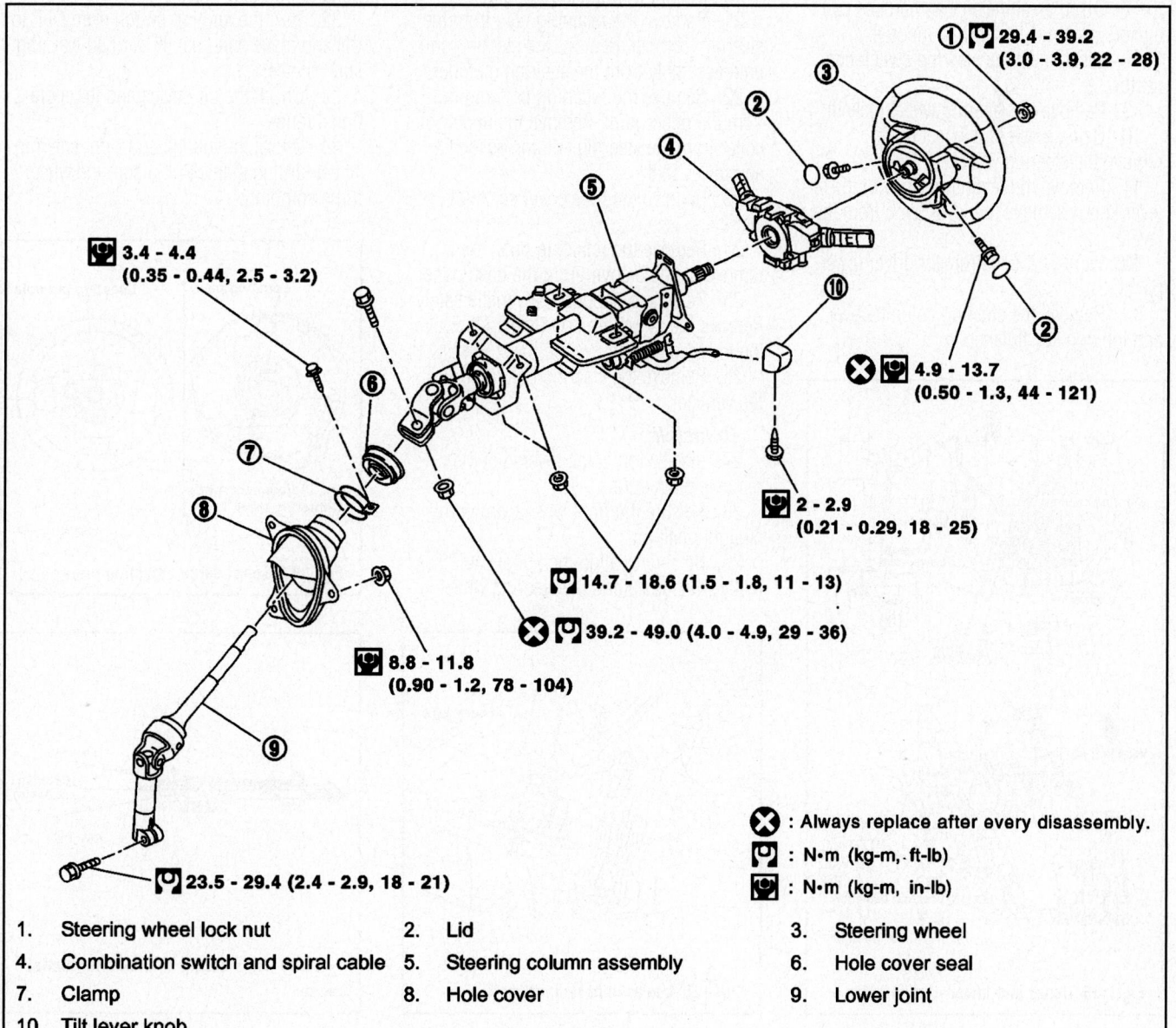

① 🔧 29.4 - 39.2
(3.0 - 3.9, 22 - 28)

🔧 3.4 - 4.4
(0.35 - 0.44, 2.5 - 3.2)

✖ 🔧 4.9 - 13.7
(0.50 - 1.3, 44 - 121)

🔧 2 - 2.9
(0.21 - 0.29, 18 - 25)

🔧 14.7 - 18.6 (1.5 - 1.8, 11 - 13)

✖ 🔧 39.2 - 49.0 (4.0 - 4.9, 29 - 36)

🔧 8.8 - 11.8
(0.90 - 1.2, 78 - 104)

🔧 23.5 - 29.4 (2.4 - 2.9, 18 - 21)

✖ : Always replace after every disassembly.

🔧 : N•m (kg-m, ft-lb)

🔧 : N•m (kg-m, in-lb)

1. Steering wheel lock nut	2. Lid	3. Steering wheel
4. Combination switch and spiral cable	5. Steering column assembly	6. Hole cover seal
7. Clamp	8. Hole cover	9. Lower joint
10. Tilt lever knob		

Fig. 154 Steering column and related components

42050_MURA_G0027

➡**Before servicing, or working around, the SRS system, turn the ignition switch OFF, disconnect both battery cables and wait at least three minutes. When servicing, or working around, the SRS system do not work directly in front of the air bag module.**

1. Before servicing the vehicle, refer to the Precautions Section.

2. Position the front wheels in the straight ahead position.

3. Disconnect the negative battery cable. Disconnect the positive battery cable.

4. Remove the side lids, located on the steering wheel.

5. Remove the right and left side special (Torx) bolts.

6. Lift up on the air bag module.

7. Disconnect the air bag harness connectors. Remove the air bag module.

8. Disconnect the steering switch connector.

9. Remove the steering wheel locknut.

10. Using a steering wheel puller, remove the steering wheel.

11. Remove the retaining screw of the tilt lever knob. Remove the lever knob from the lever.

12. Remove the instrument driver lower panel.

13. Remove the steering column cover and ignition key finisher.

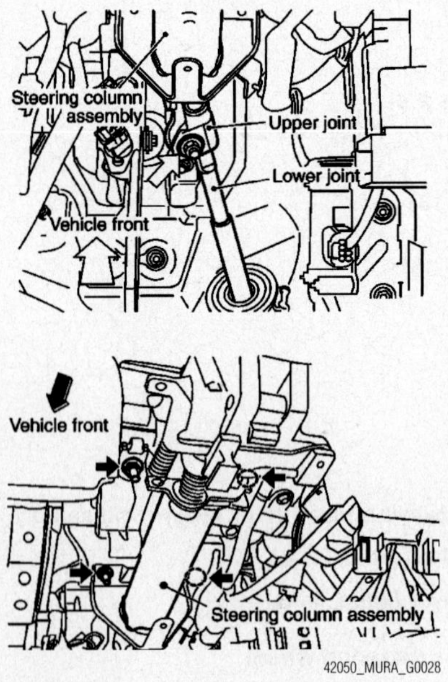

14. Remove the NATS antenna amp.

15. Remove the retaining screws holding the knee protector in place. Remove the knee protector.

16. Loosen the spiral cable retaining screws. Remove the spiral cable.

➡**Do not disassemble the spiral cable. Do not apply lubricant to the spiral cable.**

17. Disconnect the horn switch connector, and then the spiral cable connector.

18. Remove the combination switch.

19. Disconnect the harness connector from each switch on the steering column shaft and separate the vehicle side harness from it.

20. Remove the locknut, then separate the lower joint from the upper joint.

21. Remove the retaining nuts from the steering member. Remove the steering column assembly from the steering member.

22. Remove the retaining bolt and nut from the upper joint. Remove the upper joint collar from the steering column assembly and lower shaft.

23. Remove the hole cover seal and clamp.

24. Remove the retaining nuts, then remove the hole cover from the dash panel.

25. Raise and support the vehicle safely. Remove the mounting bolt (lower side) of the lower joint. Remove the lower joint.

26. Remove the steering column from the vehicle.

To install:

27. Installation is the reverse of the removal procedure.

28. Be sure the front wheels are in the straight position.

29. Be sure to align the spiral cable correctly when installing the steering wheel.

Make sure that the spiral cable is in the neutral position.

➡**The neutral position is detected by turning to the left 2.5 revolutions from the right end position and ending with the knob at the top. The spiral cable may snap due to steering operation if the cable is installed incorrectly. Also, with the steering linkage disconnected the cable may snap by turning the steering wheel beyond the limited number of turns (2.5 from the neutral position to both the left and right).**

30. Continue the installation in the reverse order of the removal procedure.

31. When reinstalling the air bag module, be sure to use new bolts. Tighten the bolts to 8 ft. lbs.

32. Turn the ignition switch from OFF to ON and make sure that the air bag warning lamp blinks.

33. Check the tilt device and its operational range.

34. Check the full left and right steering wheel turning diameter. Be sure the wheel turns smoothly.

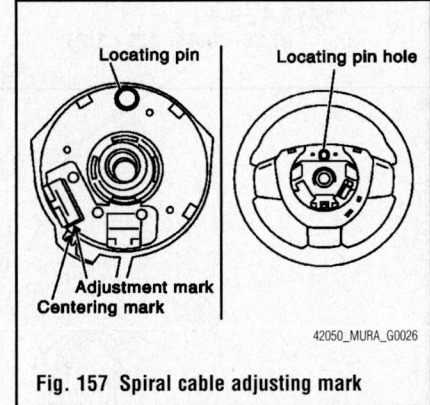

Fig. 157 Spiral cable adjusting mark

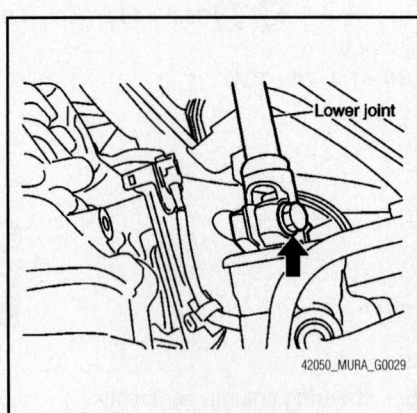

Fig. 155 Upper and lower joint location

Fig. 156 Lower joint removal bolt

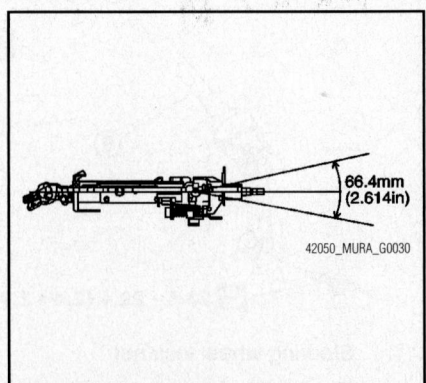

Fig. 158 Tilt steering column operational range

See Figure 159.

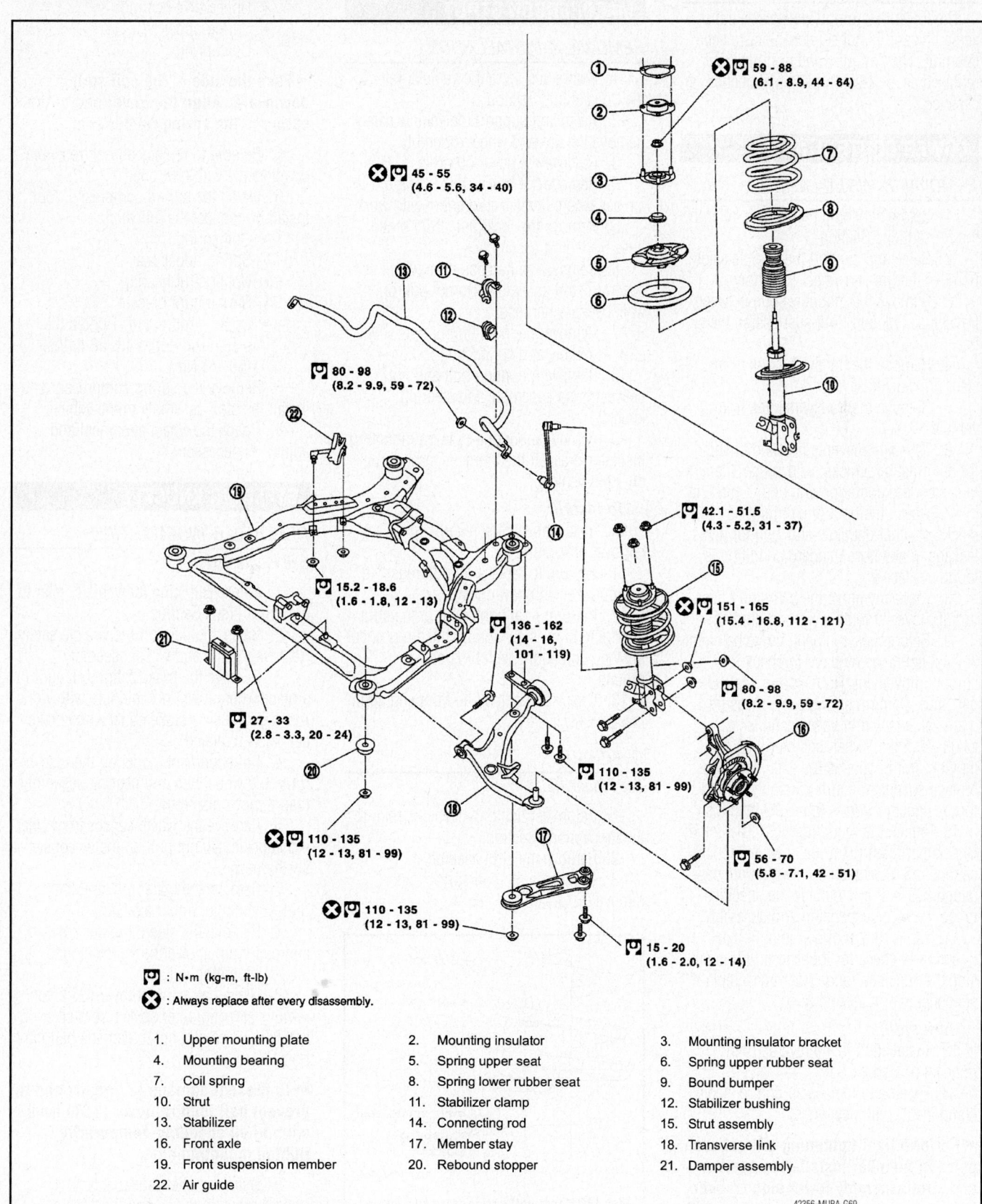

⊗ ⚙ 59 - 88
(6.1 - 8.9, 44 - 64)

⊗ ⚙ 45 - 55
(4.6 - 5.6, 34 - 40)

⚙ 80 - 98
(8.2 - 9.9, 59 - 72)

⚙ 15.2 - 18.6
(1.6 - 1.8, 12 - 13)

⚙ 42.1 - 51.5
(4.3 - 5.2, 31 - 37)

⊗ ⚙ 151 - 165
(15.4 - 16.8, 112 - 121)

⚙ 136 - 162
(14 - 16,
101 - 119)

⚙ 80 - 98
(8.2 - 9.9, 59 - 72)

⚙ 110 - 135
(12 - 13, 81 - 99)

⚙ 27 - 33
(2.8 - 3.3, 20 - 24)

⚙ 56 - 70
(5.8 - 7.1, 42 - 51)

⊗ ⚙ 110 - 135
(12 - 13, 81 - 99)

⊗ ⚙ 110 - 135
(12 - 13, 81 - 99)

⚙ 15 - 20
(1.6 - 2.0, 12 - 14)

⚙ : N•m (kg-m, ft-lb)

⊗ : Always replace after every disassembly.

1. Upper mounting plate	2. Mounting insulator	3. Mounting insulator bracket
4. Mounting bearing	5. Spring upper seat	6. Spring upper rubber seat
7. Coil spring	8. Spring lower rubber seat	9. Bound bumper
10. Strut	11. Stabilizer clamp	12. Stabilizer bushing
13. Stabilizer	14. Connecting rod	15. Strut assembly
16. Front axle	17. Member stay	18. Transverse link
19. Front suspension member	20. Rebound stopper	21. Damper assembly
22. Air guide		

42356-MURA-G69

Fig. 159 Front suspension and related components

LOWER BALL JOINT

REMOVAL & INSTALLATION

The lower ball joint is part of the transverse link and is not serviceable as a separate unit. The ball joint and arm must be replaced as an assembly if a malfunction is detected.

LOWER CONTROL ARM

REMOVAL & INSTALLATION

1. Before servicing the vehicle, refer to the Precautions Section.
2. Raise and support the vehicle safely. Remove the tire and wheel assembly.
3. Remove the mounting bolt between transverse link and front suspension member.
4. Remove the transverse link from steering knuckle.
5. Remove the transverse link from vehicle.
6. Check transverse link and bushing for deformation, cracks, or damage. If any non-standard condition is found, replace it.
7. Check the boot of ball joint for cracks or other damage, and also for grease leakage. If any non-standard condition is found, replace it.
8. Manually move the ball stud to confirm it moves smoothly with no binding.
9. Before measurement, move ball joint at least ten times by hand to check for smooth movement. Hook spring scale at ball stud. Confirm spring scale measurement value is within specifications (3.08–20.5 lbs.) when ball stud begins moving. If it is outside the specified range, replace suspension arm assembly. Swing torque specification is 0.5–0.30 inch lbs.
10. Attach the mounting nut to ball stud. Check that rotating torque is within specifications 0.5–0.30 inch lbs.) with a preload, gauge (SST). If it is outside the specified range, replace suspension arm assembly.
11. Move the tip of ball joint in axial direction to check for looseness. If it is outside the specified range (0.0 inch), replace suspension arm assembly.

To install:

12. Installation is the reverse of the removal procedure.
13. Be sure to replace all non reusable components with new ones.

➡ **Perform final tightening of front suspension member installation position and strut assembly lower side (rubber bushing) under unladen conditions with tires on level ground.**

14. Check the wheel alignment and adjust, as necessary.

MACPHERSON STRUT

REMOVAL & INSTALLATION

1. Before servicing the vehicle, refer to the Precautions Section.
2. Raise and support the vehicle safely. Remove the tire and wheel assembly.
3. Remove the cowl top grille.
4. Remove the brake caliper. Hang it in a place where it will not interfere with work.
5. Remove the lock plate from brake hose from strut assembly.
6. Remove the harness from wheel sensor from strut assembly. Do not pull on wheel sensor harness.
7. Remove the mounting nut between strut assembly and connecting rod.
8. Remove mounting bolt and nut between strut assembly and steering knuckle.
9. Remove mounting nuts on mounting insulator bracket, then remove strut assembly from vehicle.

To install:

10. Installation is the reverse of the removal procedure.
11. Be sure to replace all non reusable components with new ones.
12. Perform final tightening of the strut assembly lower side (rubber bushing) under unladen conditions with tires on level ground.
13. Check and adjust the front end alignment, as necessary.

OVERHAUL

See Figure 160.

1. Before servicing the vehicle, refer to the Precautions Section.
2. Remove the strut assembly.
3. Compress the coil spring and remove the piston rod nut.

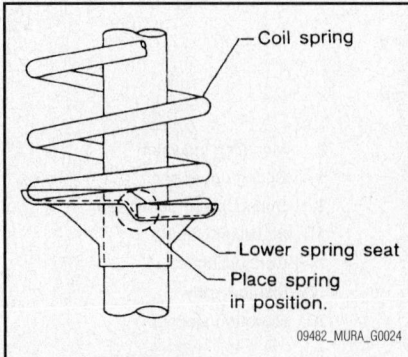

09482_MURA_G0024

Fig. 160 Front coil spring-to-lower spring seat alignment

4. Remove or disconnect the following:
- Upper strut mount
- Strut mount bracket
- Upper strut bearing
- Spring upper seat
- Coil spring

➡ **Face the side of the coil spring downward. Align the lower end of the spring to the spring rubber seat.**

5. Be sure to replace all non reusable components with new ones.
6. Install the spring compressor tool. Install or connect the following:
- Coil spring
- Spring upper seat
- Upper strut bearing
- Strut mount bracket
- Upper strut mount. Tighten the piston rod nut to 43–58 ft. lbs. (59–78 Nm).
7. Remove the spring compressor and install the strut assembly to the vehicle.
8. Check the wheel alignment and adjust, as necessary.

STEERING KNUCKLE

REMOVAL & INSTALLATION

See Figure 161.

1. Before servicing the vehicle, refer to the Precautions Section.
2. Raise and support the vehicle safely. Remove the tire and wheel assembly.
3. Remove the brake caliper. Hang it in a place where it will not interfere with work. Avoid depressing brake pedal while brake caliper is removed.
4. Put alignment marks on the disc rotor and wheel hub and bearing assembly, then remove disc rotor.
5. Remove the wheel sensor from steering knuckle. Do not pull on the wheel sensor harness.
6. Remove the cotter pin, and then remove lock nut from halfshaft.
7. Remove the steering outer socket and cotter pin at steering knuckle, then loosen mounting nut.
8. Use a ball joint remover (SST) to remove steering outer socket from steering knuckle. Be careful not to damage ball joint boot.

➡ **To prevent damage to threads and to prevent ball joint remover (SST) from coming off suddenly, temporarily tighten mounting nut.**

9. Using a puller (suitable tool), remove wheel hub and bearing assembly from halfshaft.

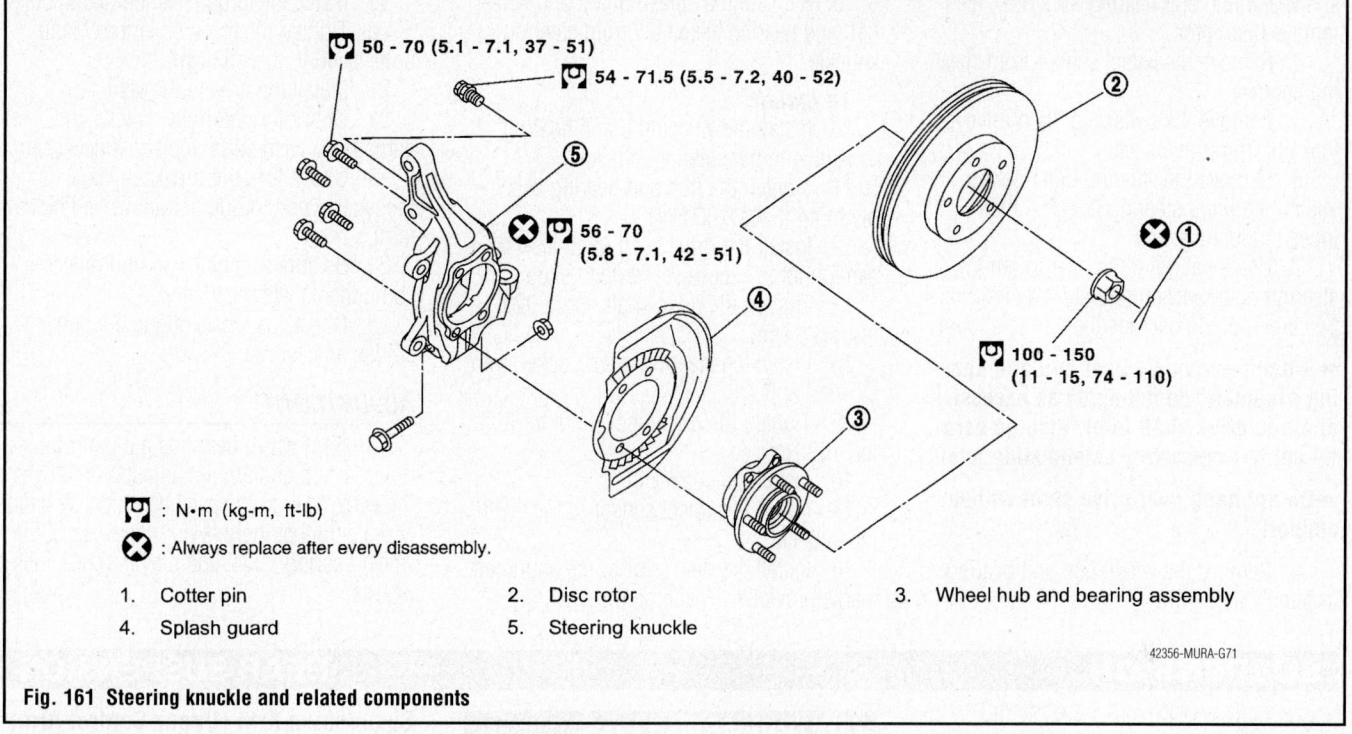

50 - 70 (5.1 - 7.1, 37 - 51)

54 - 71.5 (5.5 - 7.2, 40 - 52)

②

⑤

56 - 70
(5.8 - 7.1, 42 - 51)

④

✖ ①

100 - 150
(11 - 15, 74 - 110)

③

: N•m (kg-m, ft-lb)

✖ : Always replace after every disassembly.

1. Cotter pin
2. Disc rotor
3. Wheel hub and bearing assembly
4. Splash guard
5. Steering knuckle

42356-MURA-G71

Fig. 161 Steering knuckle and related components

10. Remove the wheel hub and bearing assembly bolt.

11. Remove the splash guard and wheel hub and bearing assembly from steering knuckle.

12. Remove the strut assembly and steering knuckle bolts and nuts.

13. Remove the transverse link and steering knuckle bolt and nut.

14. Remove the steering knuckle from vehicle.

To install:

15. Installation is the reverse of the removal procedure.

16. Check for deformity, cracks and damage on each parts, replace if necessary.

17. Check for boot breakage, axial looseness, and torque of transverse link ball joint.

18. Be sure to replace all non reusable components with new ones.

19. Check the wheel alignment and adjust, as necessary.

STABILIZER BAR

REMOVAL & INSTALLATION

2WD Models

See Figure 162.

1. Before servicing the vehicle, refer to the Precautions Section.

2. Raise and support the vehicle safely. Remove the tire and wheel assembly.

3. Remove the power steering gear assembly.

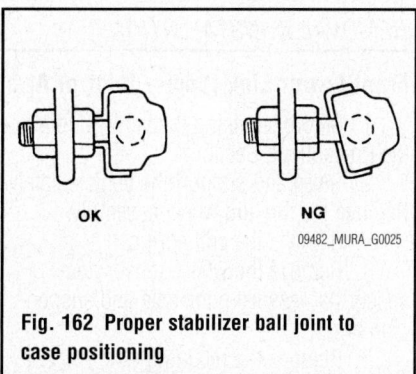

OK   NG

09482_MURA_G0025

Fig. 162 Proper stabilizer ball joint to case positioning

4. Remove the stabilizer connecting rod lower nut.

5. Separate the stabilizer bar and stabilizer connecting rod, using the proper tools.

6. Remove the stabilizer clamp mounting bolts. Remove the stabilizer bar from the vehicle.

To install:

7. Installation is the reverse of the removal procedure.

8. Stabilizer clamp tightening order, is as follows: Left side front, right side rear, right side front and left side rear.

➡**The stabilizer bar uses a pillow ball type connecting rod. Position the ball joint with the case on the pillow ball head parallel to the stabilizer bar.**

9. Check the wheel alignment and adjust, as necessary.

AWD Models

1. Before servicing the vehicle, refer to the Precautions Section.

2. Raise and support the vehicle safely. Remove the tire and wheel assembly.

3. Remove the power steering gear assembly.

4. Remove the stabilizer bar from the vehicle.

To install:

5. Installation is the reverse of the removal procedure.

6. Be sure to replace all non reusable components with new ones.

7. Stabilizer clamp tightening order, is as follows: Left side front, right side rear, right side front and left side rear.

➡**The stabilizer bar uses a pillow ball type connecting rod. Position the ball joint with the case on the pillow ball head parallel to the stabilizer bar.**

8. Check the wheel alignment and adjust, as necessary.

WHEEL HUB AND BEARING

REMOVAL & INSTALLATION

1. Remove tires from vehicle with power tool.

2. Remove torque member fixing bolts with power tool. Hang torque member in a place where it will not interfere with work.

3. Put alignment marks on disc rotor

and wheel hub and bearing assembly, then remove disc rotor.

4. Remove the wheel sensor from steering knuckle.

5. Remove the cotter pin, then remove lock nut from drive shaft.

6. Remove the steering outer socket and cotter pin at steering knuckle, then loosen mounting nut.

7. Use a ball joint remover to remove steering outer socket from steering knuckle. Be careful not to damage the ball joint boot.

➡**When removing wheel hub and bearing assembly, do not apply an excessive angle to drive shaft joint, also be careful not to excessively extend slide joint.**

➡**Do not hang over drive shaft without support.**

8. Remove the wheel hub and bearing assembly fixing bolt.

9. Remove the splash guard and wheel hub and bearing assembly from steering knuckle.

To install:

10. Install the wheel hub and bearing assembly to the steering knuckle.

11. Tighten the hub and bearing assembly to 44 ft. lbs. (60 Nm).

12. Install the drive shaft joint, also be careful not to excessively extend slide joint.

13. Tighten the drive shaft nut to 92 ft. lbs. (125 Nm).

14. Install the steering outer socket from steering knuckle.

15. Tighten the outer socket nut to 46 ft. lbs. (63 Nm).

16. Install the cotter pin.

17. Install the wheel sensor to the steering knuckle.

18. Install the disc rotor to the alignment marks previously made on disassembly.

19. Install the torque member fixing bolts.

20. Tighten the torque member fixing bolts to 46 ft. lbs. (63 Nm).

21. Install the tires to the vehicle.

22. Check for deformity, cracks and damage on each parts, replace if necessary.

23. Check for boot breakage, axial looseness, and torque of transverse link ball joint.

24. Be sure to replace all non reusable components with new ones.

25. Check the wheel alignment and adjust, as necessary.

ADJUSTMENT

The front wheel bearings are part of a unitized hub and are not adjustable. Move the wheel hub in the axial direction by hand. Make sure that there is no looseness of the wheel bearing. Axial end play is 0.002 inch or less.

SUSPENSION

See Figure 163.

COIL SPRING

REMOVAL & INSTALLATION

See Figure 164.

1. Before servicing the vehicle, refer to the Precautions Section.

2. Raise and support the vehicle safely. Remove the tire and wheel assembly.

3. Set jack under rear lower link.

4. Loosen bolt and nut between rear lower link and suspension member, and then remove bolt and nut between rear axle and rear lower link.

5. Slowly lower jack, then remove upper seat, coil spring and rubber seat from rear lower link.

6. Remove bolt and nut between rear suspension member and rear lower link.

To install:

7. Installation is the reverse of the removal procedure.

8. Be sure to replace all non reusable components with new ones.

➡**Perform final tightening of the rear suspension member and axle installation position (rubber bushing) under unladen conditions with tires on level ground.**

➡**Insert bracket tabs (3) and the inside protrusion on upper seat into each other beforehand as shown in the illustration. Match up rubber seat indentions and rear lower link grooves and attach.**

CONTROL ARMS/LINKS

REMOVAL & INSTALLATION

Front Lower Link (Lower Control Arm)

1. Before servicing the vehicle, refer to the Precautions Section.

2. Raise and support the vehicle safely. Remove the tire and wheel assembly.

3. Remove the coil spring.

4. Remove the wheel sensor and sensor harness from the axle and suspension arm.

5. Remove the retaining bolt in the lower side of the shock absorber.

6. Remove the stabilizer bushing and clamp from the suspension member.

7. Remove the retaining bolt and nut between the front lower link and the suspension member.

8. Remove the retaining bolt and nut between the front lower link and the axle.

9. Remove the front lower link from the vehicle.

To install:

10. Installation is the reverse of the removal procedure.

11. Be sure to replace all non reusable components with new ones.

➡**Perform final tightening of the rear suspension member and axle installation position under unladen conditions with tires on level ground.**

12. Check and adjust the rear alignment, as necessary.

REAR SUSPENSION

Suspension Arm (Upper Control Arm)

1. Before servicing the vehicle, refer to the Precautions Section.

2. Raise and support the vehicle safely. Remove the tire and wheel assembly.

3. Remove the coil spring.

4. Remove the wheel sensor and sensor harness from the axle and suspension arm.

5. Remove the stabilizer connecting rod mounting bracket from the suspension arm.

6. Position a jack under the front lower link.

7. Remove the fuel filler tube retaining bolt, left side only.

8. Remove the retaining nuts and bolts between the suspension arm and the rear suspension member.

9. Remove the cotter pin from the suspension arm ball joint. Loosen the nut.

10. Use a ball joint remover tool and remove the suspension arm from the axle. Be careful not to damage the ball joint boot.

11. Remove the suspension arm from the vehicle.

To install:

12. Installation is the reverse of the removal procedure.

13. Be sure to replace all non reusable components with new ones.

➡**Perform final tightening of the rear suspension member installation position (rubber bushing) under unladen conditions with tires on level ground.**

14. Check and adjust the rear alignment, as necessary.

Radius Arm

1. Before servicing the vehicle, refer to the Precautions Section.

2. Raise and support the vehicle safely. Remove the tire and wheel assembly.

3. Remove the coil spring.

4. Remove the wheel sensor and sensor harness from the axle and suspension arm.

5. Remove the retaining bolt in the lower side of the shock absorber.

6. Remove the retaining bolt and nut in the axle side of the front lower link.

7. Loosen the retaining bolt and nut of the front lower link in the side of the suspension member.

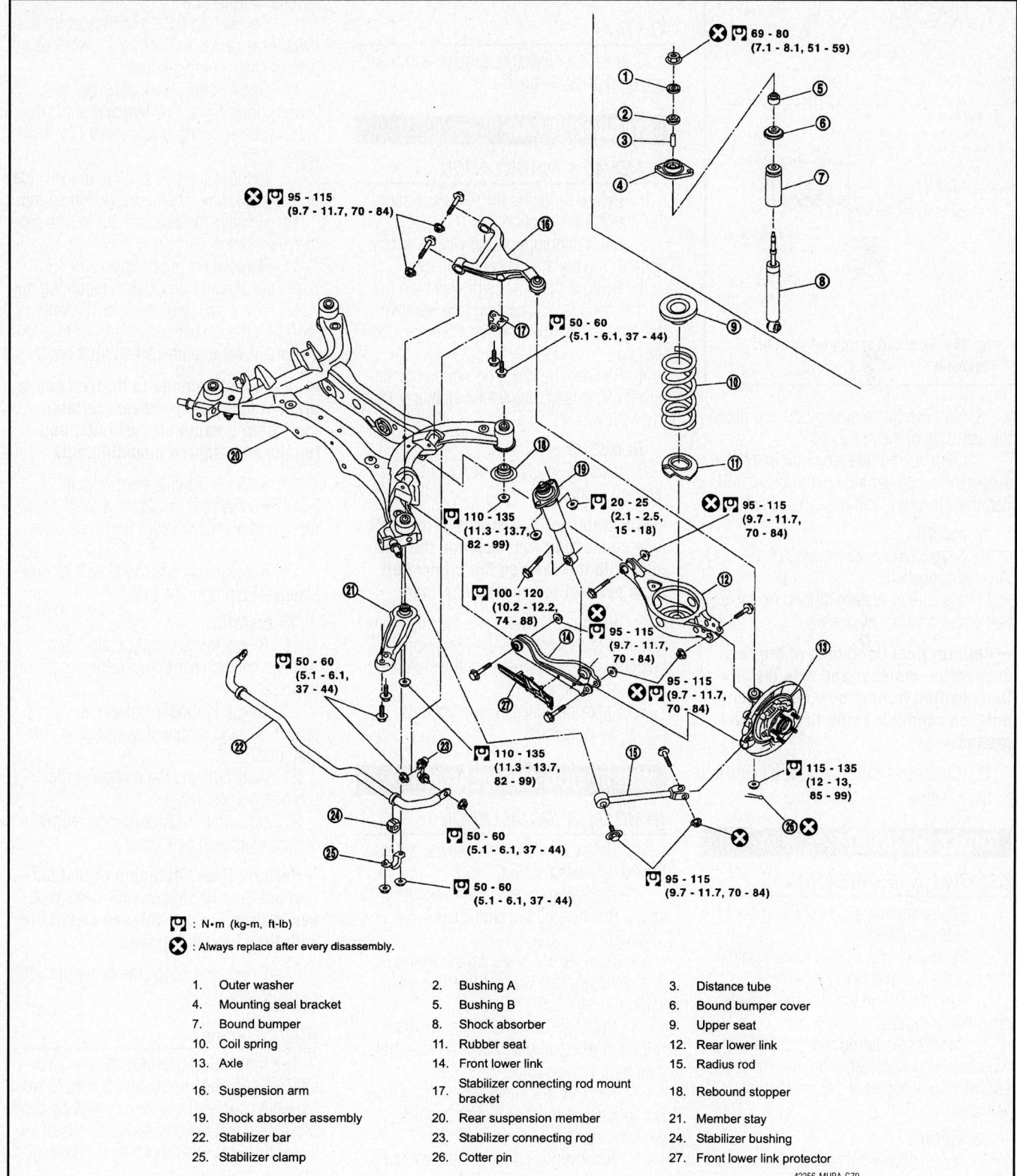

⊡ : N·m (kg-m, ft-lb)

❌ : Always replace after every disassembly.

1.	Outer washer	2.	Bushing A	3.	Distance tube
4.	Mounting seal bracket	5.	Bushing B	6.	Bound bumper cover
7.	Bound bumper	8.	Shock absorber	9.	Upper seat
10.	Coil spring	11.	Rubber seat	12.	Rear lower link
13.	Axle	14.	Front lower link	15.	Radius rod
16.	Suspension arm	17.	Stabilizer connecting rod mount bracket	18.	Rebound stopper
19.	Shock absorber assembly	20.	Rear suspension member	21.	Member stay
22.	Stabilizer bar	23.	Stabilizer connecting rod	24.	Stabilizer bushing
25.	Stabilizer clamp	26.	Cotter pin	27.	Front lower link protector

Fig. 163 Rear suspension and related components

42356-MURA-G70

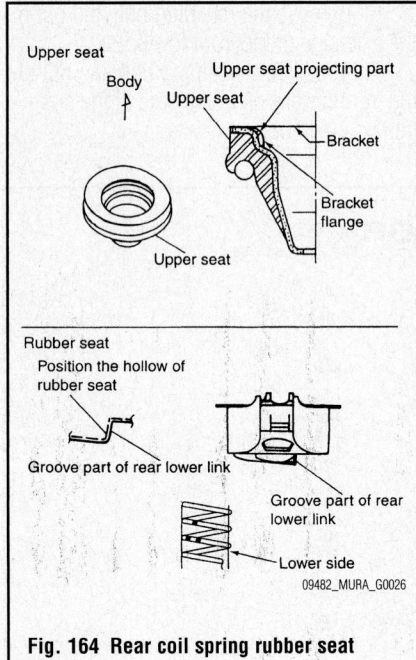

Fig. 164 Rear coil spring rubber seat alignment

8. Remove the retaining bolt and nut in the axle side of the radius rod.

9. Remove the retaining bolt in the rear suspension member side of the radius rod. Remove the radius rod from the vehicle.

To install:

10. Installation is the reverse of the removal procedure.

11. Be sure to replace all non reusable components with new ones.

➡**Perform final tightening of the rear suspension member and axle installation position (rubber bushing) under unladen conditions with tires on level ground.**

12. Check and adjust the rear alignment, as necessary.

SHOCK ABSORBER

REMOVAL & INSTALLATION

1. Before servicing the vehicle, refer to the Precautions Section.

2. Raise and support the vehicle safely. Remove the tire and wheel assembly.

3. Remove bolt in lower side of shock absorber assembly.

4. Remove mounting seal bracket nuts from shock absorber upper side and remove shock absorber assembly from vehicle.

To install:

5. Installation is the reverse of the removal procedure.

6. Be sure to replace all non reusable components with new ones.

➡**Perform final tightening of the shock absorber lower side (rubber bushing) under unladen conditions with tires on level ground.**

TESTING

Check for oil leakage, damage and breakage of installation positions.

STABILIZER BAR

REMOVAL & INSTALLATION

1. Before servicing the vehicle, refer to the Precautions Section.

2. Raise and support the vehicle safely. Remove the tire and wheel assembly.

3. Remove the lower side retaining nut on the stabilizer connecting rod. Remove the stabilizer connecting rod from the stabilizer bar.

4. Remove the retaining nut on the stabilizer clamp and remove the stabilizer from the vehicle.

To install:

5. Installation is the reverse of the removal procedure.

➡**The stabilizer bar uses a pillow ball type connecting rod. Position the ball joint with the case on the pillow ball head parallel to the stabilizer bar.**

6. When the bushing and the clamp are installed to the stabilizer bar, position the bushing and clamp inside of the side slip prevention clamp.

7. Check the wheel alignment and adjust, as necessary.

WHEEL HUB AND BEARING

REMOVAL & INSTALLATION

1. Before servicing the vehicle, refer to the Precautions Section.

2. Raise and support the vehicle safely. Remove the tire and wheel assembly.

3. Remove the brake caliper. Hang it in a place where it will not interfere with work.

4. Put alignment marks on the disc rotor and wheel hub and bearing assembly, then remove disc rotor.

5. Remove the wheel sensor from axle. Do not pull on the wheel speed sensor harness.

6. Remove the parking cable and parking brake shoe from back plate.

7. Remove the cotter pin. Then remove lock nut from halfshaft. (AWD vehicles)

8. Using a puller (suitable tool), remove wheel hub and bearing assembly from halfshaft. (AWD vehicles)

9. Remove the wheel hub and bearing assembly from axle.

10. Loosen the bolts and nuts of front lower link, radius rod and rear lower link in side of suspension member.

11. Remove the shock absorber bolt (lower), front lower link bolt and nut (axle-side) while supporting rear lower link with jack.

12. Remove the bolt and nut in axle side of rear lower link. Then remove coil spring.

13. Remove the bolt and nut in axle side of radius rod.

14. Remove the suspension arm and cotter pin at axle, then loosen mounting nut.

15. Use a ball joint remover (suitable tool) to remove suspension arm from axle. Be careful not to damage ball joint boot.

➡**To prevent damage to threads and to prevent ball joint remover (suitable tool) from coming off suddenly, and temporarily tighten mounting nut.**

16. Remove the axle from vehicle.

17. Remove the nuts from anchor block, then remove anchor block and back plate from axle.

18. Remove the axle cap (2WD) or dust shield (AWD) from the axle.

To install:

19. Check for deformity, cracks and damage on each parts, replace if necessary.

20. Check for boot breakage, axial looseness, and torque of suspension arm ball joint.

21. Installation is the reverse of the removal procedure.

22. Be sure to replace all non reusable components with new ones.

➡**Perform final tightening of installation position of suspension links (rubber bushing) under unladen conditions with tires on level ground.**

23. Check and adjust the rear alignment, as necessary.

ADJUSTMENT

The front wheel bearings are part of a unitized hub and are not adjustable. Move the wheel hub in the axial direction by hand. Make sure that there is no looseness of the wheel bearing. Axial end play is 0.002 inch or less.

SPECIFICATIONS AND MAINTENANCE CHARTS

ENGINE AND VEHICLE IDENTIFICATION

		Engine							Model Year	
Code ①	Liters (cc)	Cu. In.	Cyl.	Fuel Sys.	Engine	Eng. Mfg.		Code ②	Year
CK	4.0 (3954)	241	6	MFI	DOHC	Nissan		6	2006
PK	5.6 (5552)	339	8	MFI	DOHC	Nissan		7	2007
								8	2008

MFI: Multi-port Fuel Injection

DOHC: Double Overhead Camshafts

① Second and third digits of the Vehicle Identification Number (VIN)

② Tenth digit of the Vehicle Identification Number (VIN)

22140_PATH_C0001

GENERAL ENGINE SPECIFICATIONS

Year	Model	Engine Displacement Liters (cc)	Engine ID/VIN	Fuel System Type	Net Horsepower @ rpm	Net Torque @ rpm (ft. lbs.)	Bore x Stroke (in.)	Compression Ratio	Oil Pressure @ rpm
2006	Pathfinder	4.0 (3954)	VQ40DE	MFI	270@5600	291@4000	3.76X3.62	9.7:1	43@2000
2007	Pathfinder	4.0 (3954)	VQ40DE	MFI	266@5600	291@4000	3.76X3.62	9.7:1	43@2000
2008	Pathfinder	4.0 (3954)	VQ40DE	MFI	266@5600	288@4000	3.76X3.62	9.7:1	43@2000
2008	Pathfinder	5.6 (5552)	VQ56DE	MFI	310@5200	288@4000	3.86X3.62	9.8:1	43@2000

MFI: Multi-port Fuel Injection

22140_PATH_C0002

ENGINE TUNE-UP SPECIFICATIONS

Year	Engine Displacement Liters (cc)	Engine ID/VIN	Spark Plug Gap (in.)	Ignition Timing (deg.) MT	Ignition Timing (deg.) AT	Fuel Pump (psi)	Idle Speed (rpm) MT	Idle Speed (rpm) AT ②	Valve Clearance (in.) In.	Valve Clearance (in.) Ex.
2006	4.0 (3954)	VQ40DE	0.043	NA	15B	51	NA	575-675	HYD	HYD
2007	4.0 (3954)	VQ40DE	0.043	NA	15B	51	NA	575-675	HYD	HYD
2008	4.0 (3954)	VQ40DE	0.043	NA	15B	51	NA	575-675	HYD	HYD
2008	5.6 (5552)	VQ56DE	0.043	NA	15B	51	NA	575-675	HYD	HYD

NOTE: The Vehicle Emission Control Information label often reflects specification changes made during production. The label figures must be used if they differ from those in this chart.

B: Before top dead center

HYD: Hydraulic

① System pressure at idle with vacuum hose connected

   Should increase to 43 psi when disconnected

② Automatic transmission in Neutral or Park

22140_PATH_C0003

CAPACITIES

Year	Model	Engine Displacement Liters (cc)	Engine ID/VIN	Engine Oil with Filter (qts.)	Transmission (pts.) 5-Spd	Transmission (pts.) Auto.	Transfer Case (pts.)	Drive Axle Front (pts.)	Drive Axle Rear (pts.)	Fuel Tank (gal.)	Cooling System (qts.)
2006	Pathfinder	4.0 (3954)	VQ40DE	5.4	—	21.8	①	1.75	3.0	21.1	②
2007	Pathfinder	4.0 (3954)	VQ40DE	5.4	—	21.8	①	1.75	3.0	21.1	②
2008	Pathfinder	4.0 (3954)	VQ40DE	5.4	—	21.8	①	1.75	3.0	21.1	②
2008	Pathfinder	5.6 (5552)	VQ56DE	6.5	—	21.8	①	3.37	3.75	21.1	②

NOTE: All capacities are approximate. Add fluid gradually and check to be sure a proper fluid level is obtained.

① Part time: 4.25 pts; full time: 6.25 pts.

② Without rear A/C: 11 qts.; with rear A/C: 14 qts.

22140_PATH_C0004

FLUID SPECIFICATIONS

Year	Model	Engine Displ. Liters (VIN)	Engine Oil	Man. Trans.	Auto. Trans.	Drive Axle Front	Drive Axle Rear	Transfer Case	Power Steering Fluid	Brake Master Cylinder	Cooling System
2006	Pathfinder	4.0 (CK)	API Cert.	NA	①	②	②	③	④	⑤	⑥
2007	Pathfinder	4.0 (CK)	API Cert.	NA	①	②	②	③	④	⑤	⑥
2008	Pathfinder	4.0 (CK)	API Cert.	NA	①	②	②	③	④	⑤	⑥
2008	Pathfinder	5.6 (PK)	API Cert.	NA	①	②	②	③	④	⑤	⑥

① Nissan Matic J ATF

② API GL-5 75W-90 Gear Oil

③ Nissan Matic D ATF

④ Nissan Power Steering Fluid

⑤ Nissan Super Heavy Duty DOT 3

⑥ Nissan Long Life antifreeze or equivalent

22140_PATH_C0012

VALVE SPECIFICATIONS

Year	Engine Displacement Liters	Engine ID/VIN	Seat Angle (deg.)	Face Angle (deg.)	Spring Test Pressure (lbs. @ in.)	Spring Installed Height (in.)	Stem-to-Guide Clearance (in.) Intake	Stem-to-Guide Clearance (in.) Exhaust	Stem Diameter (in.) Intake	Stem Diameter (in.) Exhaust
2006	4.0	VQ40DE CK	45.15-45.45	45	37-42 @1.457	1.457	0.0008-0.0021	0.0012-0.0025	0.2348-0.2354	0.2344-0.2350
2007	4.0	VQ40DE CK	45.15-45.45	45	37-42 @1.457	1.457	0.0008-0.0021	0.0012-0.0025	0.2348-0.2354	0.2344-0.2350
2008	4.0	VQ40DE CK	45.15-45.45	45	37-42 @1.457	1.457	0.0008-0.0021	0.0012-0.0025	0.2348-0.2354	0.2344-0.2350
2008	5.6	VQ56DE PK	45.15-45.45	45	37-42 @1.457	1.457	0.0008-0.0021	0.0012-0.0025	0.2348-0.2354	0.2344-0.2350

22140_PATH_C0006

CAMSHAFT AND BEARING SPECIFICATIONS CHART

All measurements are given in inches.

Year	Engine Displ. Liters	Engine ID/VIN	Journal Diameter	Brg. Oil Clearance	Shaft End-play	Runout	Journal Bore	Lobe Height Intake	Lobe Height Exhaust
2006	4.0	VQ40DE	①	②	0.0045-0.0074	0.001	③	1.7900-1.7921	1.7746-1.7821
2007	4.0	VQ40DE	①	②	0.0045-0.0074	0.001	③	1.7900-1.7921	1.7746-1.7821
2008	4.0	VQ40DE	①	②	0.0045-0.0074	0.001	③	1.7900-1.7921	1.7746-1.7821
2008	5.6	VQ56DE	1.0217-1.0224	0.0012-0.0028	0.0045-0.0074	0.0008	1.0236-1.0244	1.7663-1.7738	1.7746-1.7821

① No. 1: 1.0211-1.0218

   Nos. 2-4: 0.9230-0.9238

② No. 1: 0.0018-0.0034

   Nos. 2-4: 0.0014-0.0030

③ No. 1: 1.0236-1.0244

   Nos. 2-4: 0.9252-0.9260

22140_PATH_C0014

CRANKSHAFT AND CONNECTING ROD SPECIFICATIONS

All measurements are given in inches.

Year	Engine Displacement Liters	Engine ID/VIN	Crankshaft Main Brg. Journal Dia.	Crankshaft Main Brg. Oil Clearance	Crankshaft Shaft End-play	Crankshaft Thrust on No.	Connecting Rod Journal Diameter	Connecting Rod Oil Clearance	Connecting Rod Side Clearance
2006	4.0	VQ40DE	①	0.0014-0.0018	0.0039-0.0098	4	NA	0.0013-0.0023	0.0079-0.0138
2007	4.0	VQ40DE	①	0.0014-0.0018	0.0039-0.0098	4	NA	0.0013-0.0023	0.0079-0.0138
2008	4.0	VQ40DE	①	0.0014-0.0018	0.0039-0.0098	4	NA	0.0013-0.0023	0.0079-0.0138
2008	5.6	VQ56DE	②	③	0.0039-0.0102	4	NA	0.0008-0.0015	0.0079-0.0157

NA - Not Available

① There are 24 different grades, ranging from A (2.7549) to 7 (2.7540)

② Journal No. 2, 3 and 4: There are 24 different grades, ranging from A (2.582) to 2 (2.5174)

③ No. 1 and 5: 0.00004-0.0004

   No. 2, 3 and 4: 0.0003-0.0007

22140_PATH_C0005

PISTON AND RING SPECIFICATIONS

All measurements are given in inches.

Year	Engine Displacement Liters	Engine ID/VIN	Piston Clearance	Ring Gap			Ring Side Clearance		
				Top Comp.	Bottom Comp.	Oil Control	Top Comp.	Bottom Comp.	Oil Control
2006	4.0	VQ40DE CK	0.0004-0.0012	0.0091-0.0130	0.0130-0.0189	0.0079-0.0197	0.0018-0.0031	0.0012-0.0028	0.0026-0.0053
2007	4.0	VQ40DE CK	0.0004-0.0012	0.0091-0.0130	0.0130-0.0189	0.0079-0.0197	0.0018-0.0031	0.0012-0.0028	0.0026-0.0053
2008	4.0	VQ40DE CK	0.0004-0.0012	0.0091-0.0130	0.0130-0.0189	0.0079-0.0197	0.0018-0.0031	0.0012-0.0028	0.0026-0.0053
2008	5.6	VQ56DE PK	0.0004-0.0012	0.0091-0.0130	0.098-0.0157	0.0079-0.0236	0.0014-0.0033	0.0012-0.0028	0.0006-0.0020

22140_PATH_C0007

TORQUE SPECIFICATIONS

All readings in ft. lbs.

Year	Engine Displacement Liters	Engine ID/VIN	Cylinder Head Bolts	Main Bearing Bolts	Rod Bearing Bolts	Crankshaft Damper Bolts	Flywheel Bolts	Manifold		Spark Plugs	Lug Nuts
								Intake	Exhaust		
2006	4.0	VQ40DE	①	②	③	④	65	⑤	22	18	98
2007	4.0	VQ40DE	①	②	③	④	65	⑤	22	18	98
2008	4.0	VQ40DE	①	②	③	⑥	65	⑤	22	18	98
2008	5.6	VQ56DE	⑦	⑧	⑨	⑩	65	⑤	25	18	98

① Step 1: 72 ft. lbs.

　Step 2: Loosen all bolts completely

　Step 3: 29 ft. lbs.

　Step 4: +90 degrees

　Step 5: +90 degrees

② Step 1: 26 ft. lbs.

　Step 2: +90 degrees

③ Step 1: 14 ft. lbs.

　Step 2: +90 degrees

④ 33 ft. lbs. +60 degrees

⑤ Step 1: 5 ft. lbs.

　Step 2: 20-23 ft. lbs.

⑥ 33 ft. lbs. +84 - 90 degrees

⑦ Step 1: 72 ft. lbs.

　Step 2: Loosen all bolts completely

　Step 3: 33 ft. lbs.

　Step 4: +60 degrees

　Step 5: +60 degrees

⑧ Step 1 Bolts 1-10 29 ft. lbs.

　Step 2: Bolts 11 - 20: 22 ft. lbs

　Step 3: Bolts 1 - 10: An additional 30 degrees

　Step 4: Bolts 11 - 20: an additional 30 degrees

⑨ Step 1: 11 ft. lbs.

　Step 2: +90 degrees

⑩ 69 ft. lbs. +90 degrees

22140_PATH_C0008

WHEEL ALIGNMENT

Year	Model		Caster Range (+/-Deg.)	Caster Preferred Setting (Deg.)	Camber Range (+/-Deg.)	Camber Preferred Setting (Deg.)	Toe-in (in.)
2006	Pathfinder	2WD	0.75	3.00	0.75	0.3	0.12+/-0.04
		4WD	0.75	2.75	0.75	0.5	0.12+/-0.04
2007	Pathfinder	2WD	0.75	3.00	0.75	0.25	0.12+/-0.04
		4WD	0.75	2.75	0.75	0.5	0.12+/-0.04
2008	Pathfinder	2WD	0.75	3.00	0.75	0.25	0.12+/-0.04
		4WD	0.75	2.75	0.75	0.5	0.12+/-0.04

22140_PATH_C0009

TIRE, WHEEL AND BALL JOINT SPECIFICATIONS

Year	Model	OEM Tires Standard	OEM Tires Optional	Tire Pressures (psi) Front	Tire Pressures (psi) Rear	Wheel Size	Ball Joint Inspection
2006	Pathfinder S	P245/75R16	none	35	35	16x7JJ	①
	Pathfinder SE	P265/70R16	P265/75R16	35	35	16x7JJ	①
	Pathfinder LE	P265/75R17	none	35	35	17x7.5JJ	①
2007	Pathfinder S	P245/75R16	none	35	35	16x7JJ	①
	Pathfinder SE	P265/70R16	P265/75R16	35	35	16x7JJ	①
	Pathfinder LE	P265/75R17	none	35	35	17x7.5JJ	①
2008	Pathfinder S	P245/75R16	none	35	35	16x7JJ	①
	Pathfinder SE (V6)	P265/65R17	none	35	35	17x7.5JJ	①
	Pathfinder SE (V8)	P265/60R18	none	35	35	18x8JJ	①
	Pathfinder OR	P265/75R16	none	35	35	16x7JJ	①
	Pathfinder LE (V6)	P265/65R17	none	35	35	17x7.5JJ	①
	Pathfinder LE (V8)	P265/60R18	none	35	35	18x8JJ	①

OEM: Original Equipment Manufacturer

PSI: Pounds Per Square Inch

① Axial play

  Upper: 0

  Lower: 0.008 in.

22140_PATH_C0010

BRAKE SPECIFICATIONS
All measurements in inches unless noted

Year	Model	Brake Disc Original Thickness	Brake Disc Minimum Thickness	Brake Disc Maximum Runout	Brake Drum Diameter Original Inside Diameter	Brake Drum Diameter Max. Wear Limit	Brake Drum Diameter Maximum Machine Diameter	Minimum Lining Thickness Front	Minimum Lining Thickness Rear	Brake Caliper Bracket Bolts (ft. lbs.)	Brake Caliper Mounting Bolts (ft. lbs.)
2006	Pathfinder	1.102	1.024	0.0006	NA	NA	NA	0.079	0.079	①	①
2007	Pathfinder	1.102	1.024	0.0006	NA	NA	NA	0.079	0.079	①	①
2008	Pathfinder	1.102	1.024	0.0006	NA	NA	NA	0.079	0.079	①	①

NA: Not Available

① Torque member mounting bolt: 127-134

  Main pin bolt: 24-31

22140_PATH_C0011

SCHEDULED MAINTENANCE INTERVALS
2006-08 NISSAN PATHFINDER

TO BE SERVICED	TYPE OF SERVICE	VEHICLE MILEAGE INTERVAL (x1000)											
		7.5	15	22.5	30	37.5	45	52.5	60	67.5	75	82.5	90
Engine oil & filter	R	✓	✓	✓	✓	✓	✓	✓	✓	✓	✓	✓	✓
	S/I		✓		✓		✓		✓		✓		✓
Brake lines & cables	I		✓		✓		✓		✓		✓		✓
Brake pads& rotors	L/I		✓		✓		✓		✓		✓		✓
Driveshaft boots & propeller	I				✓				✓				
Automatic & transfer case	I		✓		✓		✓		✓				
LSD gear oil	I		✓		✓		✓		✓		✓		✓
Front wheel bearing grease (4x4)	R				✓				✓				
Timing belt	R	Replace every 105,000 miles											
Air cleaner filter	R				✓				✓				✓
Engine coolant	R								✓				✓
Spark plugs	R	Replace every 105,000 miles											
Drive belt(s)	S/I								✓		✓		✓
Cabin air filter	R		✓		✓		✓		✓		✓		✓
Exhaust system	I				✓				✓				✓
Fuel lines	S/I				✓				✓				
Steering gear (box) & linkage, axle & suspension parts	I				✓				✓				✓
Vapor lines	S/I				✓				✓				✓

R: Replace   S/I: Service or Inspect   L: Lubricate

FREQUENT OPERATION MAINTENANCE (SEVERE SERVICE)

If a vehicle is operated under any of the following conditions it is considered severe service:

- Extremely dusty areas.

- 50% or more of the vehicle operation is in 32°C (90°F) or higher temperatures, or constant operation in temperatures below 0°C (32°F).

- Prolonged idling (vehicle operation in stop and go traffic).

- Frequent short running periods (engine does not warm to normal operating temperatures).

- Police, taxi, delivery usage or trailer towing usage.

Oil & oil filter: replace every 3750 miles.

Brake pads, discs, drums & linings: service or inspect every 7500 miles.

Driveshaft boots & propeller shaft: service or inspect every 7500 miles.

Exhaust system: service or inspect every 7500 miles.

Steering gear (box) & linkage, (steering damper-4x4), axle & suspension parts: service or inspect every 7500 miles.

Steering linkage ball joints & front suspension ball joints: service or inspect every 7500 miles.

22140_PATH_C0013

PRECAUTIONS

Before servicing any vehicle, please be sure to read all of the following precautions, which deal with personal safety, prevention of component damage, and important points to take into consideration when servicing a motor vehicle:

• Never open, service or drain the radiator or cooling system when the engine is hot; serious burns can occur from the steam and hot coolant.

• Observe all applicable safety precautions when working around fuel. Whenever servicing the fuel system, always work in a well-ventilated area. Do not allow fuel spray or vapors to come in contact with a spark, open flame, or excessive heat (a hot drop light, for example). Keep a dry chemical fire extinguisher near the work area. Always keep fuel in a container specifically designed for fuel storage; also, always properly seal fuel containers to avoid the possibility of fire or explosion. Refer to the additional fuel system precautions later in this section.

• Fuel injection systems often remain pressurized, even after the engine has been turned **OFF**. The fuel system pressure must be relieved before disconnecting any fuel lines. Failure to do so may result in fire and/or personal injury.

• Brake fluid often contains polyglycol ethers and polyglycols. Avoid contact with the eyes and wash your hands thoroughly after handling brake fluid. If you do get brake fluid in your eyes, flush your eyes with clean, running water for 15 minutes. If eye irritation persists, or if you have taken brake fluid internally, IMMEDIATELY seek medical assistance.

• The EPA warns that prolonged contact with used engine oil may cause a number of skin disorders, including cancer. You should make every effort to minimize your exposure to used engine oil. Protective gloves should be worn when changing oil. Wash your hands and any other exposed skin areas as soon as possible after exposure to used engine oil. Soap and water, or waterless hand cleaner should be used.

• All new vehicles are now equipped with an air bag system, often referred to as a Supplemental Restraint System (SRS) or Supplemental Inflatable Restraint (SIR) system. The system must be disabled before performing service on or around system components, steering column, instrument panel components, wiring and sensors. Failure to follow safety and disabling procedures could result in accidental air bag deployment, possible personal injury and unnecessary system repairs.

• Always wear safety goggles when working with, or around, the air bag system. When carrying a non-deployed air bag, be sure the bag and trim cover are pointed away from your body. When placing a non-deployed air bag on a work surface, always face the bag and trim cover upward, away from the surface. This will reduce the motion of the module if it is accidentally deployed. Refer to the additional air bag system precautions later in this section.

• Clean, high quality brake fluid from a sealed container is essential to the safe and proper operation of the brake system. You should always buy the correct type of brake fluid for your vehicle. If the brake fluid becomes contaminated, completely flush the system with new fluid. Never reuse any brake fluid. Any brake fluid that is removed from the system should be discarded. Also, do not allow any brake fluid to come in contact with a painted surface; it will damage the paint.

• Never operate the engine without the proper amount and type of engine oil; doing so WILL result in severe engine damage.

• Timing belt maintenance is extremely important. Many models utilize an interference-type, non-freewheeling engine. If the timing belt breaks, the valves in the cylinder head may strike the pistons, causing potentially serious (also time-consuming and expensive) engine damage. Refer to the maintenance interval charts for the recommended replacement interval for the timing belt, and to the timing belt section for belt replacement and inspection.

• Disconnecting the negative battery cable on some vehicles may interfere with the functions of the on-board computer system(s) and may require the computer to undergo a relearning process once the negative battery cable is reconnected.

• When servicing drum brakes, only disassemble and assemble one side at a time, leaving the remaining side intact for reference.

• Only an MVAC-trained, EPA-certified automotive technician should service the air conditioning system or its components.

BRAKES

GENERAL INFORMATION

PRECAUTIONS

• Certain components within the ABS system are not intended to be serviced or repaired individually.

• Do not use rubber hoses or other parts not specifically specified for and ABS system. When using repair kits, replace all parts included in the kit. Partial or incorrect repair may lead to functional problems and require the replacement of components.

• Lubricate rubber parts with clean, fresh brake fluid to ease assembly. Do not use shop air to clean parts; damage to rubber components may result.

• Use only DOT 3 brake fluid from an unopened container.

• If any hydraulic component or line is removed or replaced, it may be necessary to bleed the entire system.

• A clean repair area is essential. Always clean the reservoir and cap thoroughly before removing the cap. The slightest amount of dirt in the fluid may plug an orifice and impair the system function. Perform repairs after components have been thoroughly cleaned; use only denatured alcohol

ANTI-LOCK BRAKE SYSTEM (ABS)

to clean components. Do not allow ABS components to come into contact with any substance containing mineral oil; this includes used shop rags.

• The Anti-Lock control unit is a microprocessor similar to other computer units in the vehicle. Ensure that the ignition switch is **OFF** before removing or installing controller harnesses. Avoid static electricity discharge at or near the controller.

• If any arc welding is to be done on the vehicle, the control unit should be unplugged before welding operations begin.

BRAKES · BLEEDING THE BRAKE SYSTEM

BLEEDING PROCEDURE

BLEEDING PROCEDURE

❊❊ CAUTION

While bleeding, monitor the master cylinder brake fluid level.

1. Turn ignition switch OFF and disconnect battery negative cable.
2. Connect a vinyl tube to the rear right bleed valve.
3. Fully depress brake pedal 4 to 5 times.
4. With brake pedal depressed, loosen bleed valve to let the air out, and then tighten it immediately.

5. Repeat steps 3 and 4 until no more air comes out.
6. Perform steps 2 to 5 at each wheel, with master cylinder reservoir tank filled at least half way, bleed air from the front left, rear left, and front right bleed valve, in that order.

BRAKES · FRONT DISC BRAKES

❊❊ CAUTION

Dust and dirt accumulating on brake parts during normal use may contain asbestos fibers from production or aftermarket brake linings. Breathing excessive concentrations of asbestos fibers can cause serious bodily harm. Exercise care when servicing brake parts. Do not sand or grind brake lining unless equipment used is designed to contain the dust residue. Do not clean brake parts with compressed air or by dry brushing. Cleaning should be done by dampening the brake components with a fine mist of water, then wiping the brake components clean with a dampened cloth. Dispose of cloth and all residue containing asbestos fibers in an impermeable container with the appropriate label. Follow practices prescribed by the Occupational Safety and Health Administration (OSHA) and the Environmental Protection Agency (EPA) for the handling, processing, and disposing of dust or debris that may contain asbestos fibers.

BRAKE CALIPER

REMOVAL & INSTALLATION
See Figure 1.

1. Remove wheel and tire.
2. Drain brake fluid as necessary.

➡**Do not remove union bolt unless removing cylinder body from vehicle.**

3. Remove union bolt as necessary and torque member bolts, then remove cylinder body from the vehicle.
4. Position cylinder body aside using suitable wire, as necessary.

5. When servicing brake caliper, remove sliding pin bolts and caliper from torque member.
6. Remove torque member.
7. Remove disc rotor.

To install:

❊❊ CAUTION

Refill with new brake fluid. Do not reuse drained brake fluid.

8. Install disc rotor.
9. Install torque member and tighten to specification.
10. Install sliding pin bolts, if removed.
11. Install cylinder body, then tighten sliding pin bolts to 32 ft. lbs. (44 Nm).

❊❊ CAUTION

When attaching cylinder body to the vehicle, wipe any oil off knuckle spindle, washers and cylinder body attachment surfaces.

12. Install brake hose to cylinder body, if removed, then tighten union bolt to 24 ft. lbs. (32 Nm).

❊❊ CAUTION

Do not reuse copper washers for union bolt. Attach brake hose to cylinder body together with union bolt and washers.

13. Refill with new brake fluid as necessary and bleed air.
14. Install wheel and tire.

DISC BRAKE PADS

REMOVAL & INSTALLATION

1. Remove wheel and tire.

2. Remove master cylinder reservoir cap.
3. Remove lower sliding pin bolt.
4. Suspend cylinder body with a wire and remove pads, shim, shim covers, and retainers from torque member.

To install:
5. Apply Molykote AS880N grease between outer brake pad plate and shim, then attach shim and shim covers to brake pads.
6. Attach pad retainer to torque member, then install brake pad and shim assemblies.

❊❊ CAUTION

When attaching pad retainer, attach it firmly so that it is flush with torque member.

7. Push pistons into cylinder body.

➡**Using a disc brake piston tool (commercial service tool), etc., makes it easier to push in piston.**

❊❊ CAUTION

By pushing in piston, brake fluid returns to master cylinder reservoir tank. Watch the level of the surface of reservoir tank.

8. Remove wire then swing cylinder body down over brake pad assemblies.
9. Install lower sliding pin bolt and tighten to 32 ft. lbs. (44 Nm).
10. Check brake for drag.
11. Inspect fluid level, then install master cylinder reservoir cap.
12. Install wheel and tire.

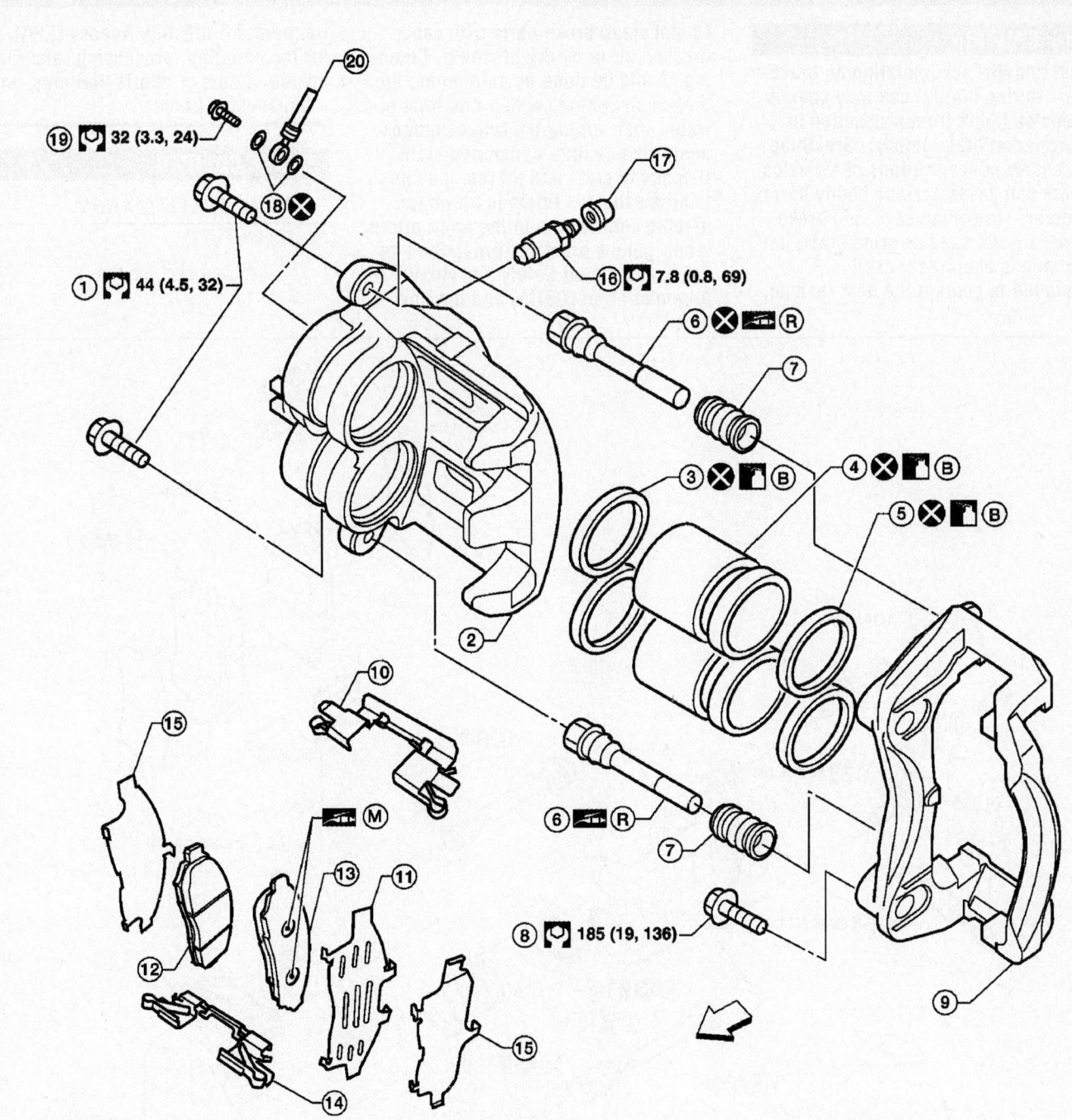

1. Sliding pin bolt
2. Cylinder body
3. Piston seal
4. Piston
5. Piston boot
6. Sliding pin
7. Sliding pin boot
8. Torque member bolt
9. Torque member
10. Upper pad retainer
11. Outer shim
12. Inner brake pad
13. Outer brake pad
14. Lower pad retainer
15. Shim cover
16. Bleed valve
17. Cap
18. Copper washers
19. Union bolt
20. Brake hose
⇐ Front

22140_PATH_G0008

Fig. 1 Exploded view of front brake caliper assembly

BRAKES

REAR DISC BRAKES

✳✳ CAUTION

Dust and dirt accumulating on brake parts during normal use may contain asbestos fibers from production or aftermarket brake linings. Breathing excessive concentrations of asbestos fibers can cause serious bodily harm. Exercise care when servicing brake parts. Do not sand or grind brake lining unless equipment used is designed to contain the dust residue.

Do not clean brake parts with compressed air or by dry brushing. Cleaning should be done by dampening the brake components with a fine mist of water, then wiping the brake components clean with a dampened cloth. Dispose of cloth and all residue containing asbestos fibers in an impermeable container with the appropriate label. Follow practices prescribed by the Occupational Safety and Health Administration (OSHA) and the Environmental Protection Agency (EPA) for the handling, processing, and disposing of dust or debris that may contain asbestos fibers.

BRAKE CALIPER

REMOVAL & INSTALLATION

See Figure 2.

1. Remove wheel and tire.
2. Drain brake fluid.

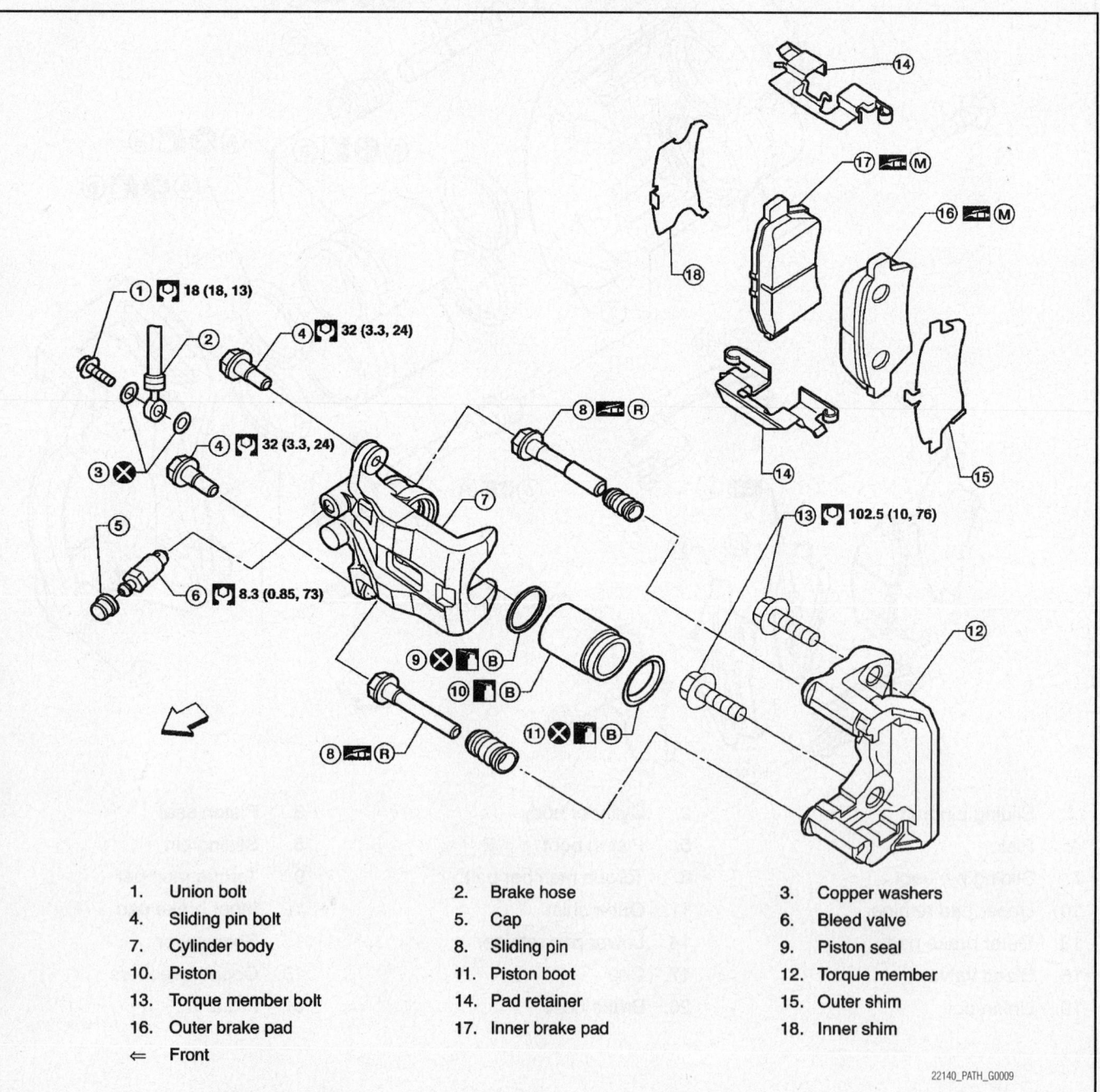

1.	Union bolt	2.	Brake hose	3.	Copper washers
4.	Sliding pin bolt	5.	Cap	6.	Bleed valve
7.	Cylinder body	8.	Sliding pin	9.	Piston seal
10.	Piston	11.	Piston boot	12.	Torque member
13.	Torque member bolt	14.	Pad retainer	15.	Outer shim
16.	Outer brake pad	17.	Inner brake pad	18.	Inner shim
⇐	Front				

22140_PATH_G0009

Fig. 2 Exploded view of rear brake caliper assembly—2006 Models

3. Remove union bolt and mounting bolts, and remove cylinder body.
4. Remove torque member.
5. Remove disc rotor.

To install:

❋❋ CAUTION

Refill with new brake fluid. Do not reuse drained brake fluid.

6. Install disc rotor.
7. Install torque member and tighten to 76 ft. lbs. (102.5 Nm).
8. Install cylinder body to the vehicle, and tighten mounting bolts to 24 ft. lbs. (32 Nm).

❋❋ CAUTION

Before installing cylinder body to the vehicle, wipe off mounting surface of cylinder body.

9. Install brake hose to cylinder body and tighten union bolt to 13 ft. lbs. (18 Nm) for 2006—2007 models; 19 ft. lbs. (26 Nm) for 2008 models.

❋❋ CAUTION

Do not reuse copper washer for union bolt. Securely attach brake hose to protrusion on cylinder body.

10. Refill new brake fluid and bleed air.
11. Install wheels and tires to the vehicle.

DISC BRAKE PADS

REMOVAL & INSTALLATION

1. Remove wheel and tire.
2. Remove master cylinder reservoir cap.
3. Remove lower sliding pin bolt.
4. Suspend cylinder body with a wire and remove pads, shim, shim covers, and retainers from torque member.

To install:

5. Apply Molykote AS880N grease between outer brake pad plate and shim, then attach shim and shim covers to brake pads.
6. Attach pad retainer to torque member, then install brake pad and shim assemblies.

❋❋ CAUTION

When attaching pad retainer, attach it firmly so that it is flush with torque member.

7. Push pistons into cylinder body.

➡ Using a disc brake piston tool (commercial service tool), etc., makes it easier to push in piston.

❋❋ CAUTION

By pushing in piston, brake fluid returns to master cylinder reservoir tank. Watch the level of the surface of reservoir tank.

8. Remove wire then swing cylinder body down over brake pad assemblies.
9. Install lower sliding pin bolt and tighten to 76 ft. lbs. (102.5 Nm).
10. Check brake for drag.
11. Inspect fluid level, then install master cylinder reservoir cap.
12. Install wheel and tire.

BRAKES   REAR DRUM BRAKES

❋❋ CAUTION

Dust and dirt accumulating on brake parts during normal use may contain asbestos fibers from production or aftermarket brake linings. Breathing excessive concentrations of asbestos fibers can cause serious bodily harm. Exercise care when servicing brake parts. Do not sand or grind brake lin- ing unless equipment used is designed to contain the dust residue. Do not clean brake parts with compressed air or by dry brushing. Cleaning should be done by dampening the brake components with a fine mist of water, then wiping the brake components clean with a dampened cloth. Dispose of cloth and all residue containing asbestos fibers in an impermeable container with the appropriate label. Follow practices prescribed by the Occupational Safety and Health Administration (OSHA) and the Environmental Protection Agency (EPA) for the handling, processing, and disposing of dust or debris that may contain asbestos fibers.

BRAKES   PARKING BRAKE

PARKING BRAKE SHOES

REMOVAL & INSTALLATION

1. Remove the wheel and tire assembly.
2. Remove the rear brake caliper, without disconnecting the hydraulic hose.
3. Reposition the rear brake caliper aside using suitable wire.

➡ Do not depress the brake pedal while the brake caliper is removed.

➡ Remove the disc rotor only with the parking brake pedal completely disengaged.

4. Remove the rear disc rotor.
5. Remove the rear drive shaft.
   a. Move the A/T select lever to the N position and release the parking brake.
   b. Put matching marks on the rear propeller shaft flange yoke and the rear final drive companion flange as shown.

❋❋ CAUTION

For matching marks, use paint. Never damage the rear propeller shaft flange yoke or the companion flange.

   c. Remove the bolts, then remove the propeller shaft from the rear final drive and A/T or transfer.
6. Disconnect wheel sensor at harness connector. Then remove wheel sensor wire from grommet mounts.
7. Remove wheel hub and bearing assembly.
   a. Remove the cotter pin, then remove the rear drive shaft nut.
   b. Discard the cotter pin, use a new one for installation.
   c. Remove the rear drive shaft.
   d. Remove the four rear wheel hub and bearing assembly bolts.
   e. Discard the four rear wheel hub and bearing assembly bolts, use new ones for installation.
   f. Remove the rear wheel hub and bearing assembly.

➡ Withdraw wheel sensor harness through back plate when removing wheel hub and bearing assembly.

8. Remove the return springs.
9. Remove the adjuster.
10. Remove the retainers, anti-rattle pins and shoes.

11. Disconnect the parking brake cable from the toggle lever.

12. Remove back plate.

To install:

13. Installation is in the reverse order of removal.

14. Apply brake grease to the specified points during assembly.

15. Install adjuster so that threaded part expands when rotating it in the direction shown by the arrow.

16. Shorten adjuster by rotating it in the opposite direction as shown by the arrow.

→**After replacing brake shoes or disc rotors, or if parking brake does not function well, perform break-in operation as follows.**

17. Adjust parking brake pedal stroke.

18. Perform parking brake burnishing operation by driving the vehicle forward under the following conditions:

✲✲ CAUTION

To prevent lining from getting too hot, allow a cool off period of approximately 5 minutes after every break-in operation. _ Do not perform excessive break-in operations, because it may cause uneven or early wear of lining.

19. After burnishing operation, check parking brake pedal stroke. Readjust if it is now longer than the specified stroke.

CHASSIS ELECTRICAL

GENERAL INFORMATION

✲✲ CAUTION

These vehicles are equipped with an air bag system. The system must be disarmed before performing service on, or around, system components, the steering column, instrument panel components, wiring and sensors. Failure to follow the safety precautions and the disarming procedure could result in accidental air bag deployment, possible injury and unnecessary system repairs.

SERVICE PRECAUTIONS

Disconnect and isolate the battery negative cable before beginning any airbag system component diagnosis, testing, removal, or installation procedures. Allow system capacitor to discharge for two minutes before beginning any component service. This will disable the airbag system. Failure to disable the airbag system may result in accidental airbag deployment, personal injury, or death.

Do not place an intact undeployed airbag face down on a solid surface. The airbag will propel into the air if accidentally deployed and may result in personal injury or death.

When carrying or handling an undeployed airbag, the trim side (face) of the airbag should be pointing towards the body to minimize possibility of injury if accidental deployment occurs. Failure to do this may result in personal injury or death.

Replace airbag system components with OEM replacement parts. Substitute parts may appear interchangeable, but internal differences may result in inferior occupant protection. Failure to do so may result in occupant personal injury or death.

Wear safety glasses, rubber gloves, and long sleeved clothing when cleaning pow-

AIR BAG (SUPPLEMENTAL RESTRAINT SYSTEM)

der residue from vehicle after an airbag deployment. Powder residue emitted from a deployed airbag can cause skin irritation. Flush affected area with cool water if irritation is experienced. If nasal or throat irritation is experienced, exit the vehicle for fresh air until the irritation ceases. If irritation continues, see a physician.

Do not use a replacement airbag that is not in the original packaging. This may result in improper deployment, personal injury, or death.

The factory installed fasteners, screws and bolts used to fasten airbag components have a special coating and are specifically designed for the airbag system. Do not use substitute fasteners. Use only original equipment fasteners listed in the parts catalog when fastener replacement is required.

During, and following, any child restraint anchor service, due to impact event or vehicle repair, carefully inspect all mounting hardware, tether straps, and anchors for proper installation, operation, or damage. If a child restraint anchor is found damaged in any way, the anchor must be replaced. Failure to do this may result in personal injury or death.

Deployed and non-deployed airbags may or may not have live pyrotechnic material within the airbag inflator.

Do not dispose of driver/passenger/curtain airbags or seat belt tensioners unless you are sure of complete deployment. Refer to the Hazardous Substance Control System for proper disposal.

Dispose of deployed airbags and tensioners consistent with state, provincial, local, and federal regulations.

After any airbag component testing or service, do not connect the battery negative cable. Personal injury or death may result if the system test is not performed first.

If the vehicle is equipped with the Occupant Classification System (OCS), do not connect the battery negative cable before performing the OCS Verification Test using the scan tool and the appropriate diagnostic information. Personal injury or death may result if the system test is not performed properly.

Never replace both the Occupant Restraint Controller (ORC) and the Occupant Classification Module (OCM) at the same time. If both require replacement, replace one, then perform the Airbag System test before replacing the other.

Both the ORC and the OCM store Occupant Classification System (OCS) calibration data, which they transfer to one another when one of them is replaced. If both are replaced at the same time, an irreversible fault will be set in both modules and the OCS may malfunction and cause personal injury or death.

If equipped with OCS, the Seat Weight Sensor is a sensitive, calibrated unit and must be handled carefully. Do not drop or handle roughly. If dropped or damaged, replace with another sensor. Failure to do so may result in occupant injury or death.

If equipped with OCS, the front passenger seat must be handled carefully as well. When removing the seat, be careful when setting on floor not to drop. If dropped, the sensor may be inoperative, could result in occupant injury, or possibly death.

If equipped with OCS, when the passenger front seat is on the floor, no one should sit in the front passenger seat. This uneven force may damage the sensing ability of the seat weight sensors. If sat on and damaged, the sensor may be inoperative, could result in occupant injury, or possibly death.

DISARMING THE SYSTEM

Before servicing the SRS, turn ignition switch OFF, disconnect both battery cables and wait at least 3 minutes.

ARMING THE SYSTEM

Connect the battery cables. Wait 2 minutes before performing any service.

DRIVETRAIN

AUTOMATIC TRANSMISSION ASSEMBLY

REMOVAL & INSTALLATION

2WD Models
See Figure 3.

> ✳✳ **CAUTION**
>
> **When removing the A/T assembly from engine, first remove the crankshaft position sensor (POS) from the A/T assembly.**

1. Disconnect the negative battery terminal.
2. Remove the A/T fluid indicator.
3. Remove the LH fender protector.
4. Remove the crankshaft position sensor (POS) from the A/T assembly.

> ✳✳ **CAUTION**
>
> **Do not subject it to impact by dropping or hitting it. Do not disassemble. Do not allow metal filings or foreign material to get on the sensor front edge magnetic area. Do not place in an area affected by magnetism.**

5. Remove the undercovers.
6. Partially drain the A/T fluid.
7. Remove the front crossmember.
8. Remove the starter motor.
9. Remove the rear propeller shaft.
10. Remove the left and right front exhaust tubes.
11. Remove the A/T selector control cable and bracket from the A/T.
12. Disconnect the A/T fluid cooler tubes from the A/T assembly.
13. Remove the dust cover from the converter housing.
14. Turn the crankshaft to access and remove the four bolts for the drive plate and torque converter.

> ✳✳ **CAUTION**
>
> **When turning the crankshaft, turn it clockwise as viewed from the front of the engine.**

15. Support the A/T assembly using a transmission jack.

> ✳✳ **CAUTION**
>
> **When setting the transmission jack, be careful not to allow it to collide against the drain plug.**

16. Remove the nuts securing the insulator to the crossmember.
17. Remove the crossmember.
18. Tilt the transmission slightly to gain clearance between the body and the transmission, then disconnect the air breather hose.
19. Disconnect the A/T assembly harness connector.
20. Remove the wiring harness from the retainers.
21. Remove the A/T fluid indicator pipe.
22. Plug any openings such as the A/T fluid indicator pipe hole.
23. Remove the A/T assembly to engine bolts.
24. Remove A/T assembly from the vehicle.

> ✳✳ **CAUTION**
>
> **Secure the torque converter to prevent it from dropping. Secure the A/T assembly the transmission jack.**

To install:
25. Installation is in the reverse order of the removal.
26. When installing transmission to the engine, tighten the bolts to the specified torque using sequence shown. Torque to 55 ft. lbs. (74 Nm).

> ✳✳ **CAUTION**
>
> **When replacing an engine or transmission you must make sure the dowels are installed correctly during reassembly. Improper alignment caused by missing dowels may cause vibration, oil leaks or breakage of drivetrain components.**

27. Align the positions of bolts for drive plate with those of the torque converter, and temporarily tighten the bolts.

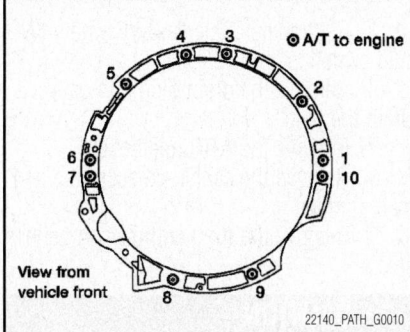

Fig. 3 Transmission-to-engine bolt tightening sequence

Then, tighten the bolts with the specified torque.

> ✳✳ **CAUTION**
>
> **When turning crankshaft, turn it clockwise as viewed from the front of the engine. After converter is installed to drive plate, rotate crankshaft several turns and check to be sure that transmission rotates freely without binding.**

28. Install crankshaft position sensor (POS).
29. After completing installation, check fluid leakage, fluid level, and the positions of A/T. R

4WD Models
See Figure 4.

> ✳✳ **CAUTION**
>
> **When removing the A/T assembly from engine, first remove the crankshaft position sensor (POS) from the A/T assembly.**

1. Disconnect the negative battery terminal.
2. Remove the A/T fluid indicator.
3. Remove the LH fender protector.
4. Remove the crankshaft position sensor (POS) from the A/T assembly.

> ✳✳ **CAUTION**
>
> **Do not subject it to impact by dropping or hitting it. Do not disassemble. Do not allow metal filings or foreign material to get on the sensor's front edge magnetic area. Do not place in an area affected by magnetism.**

5. Remove the undercovers.
6. Partially drain the A/T fluid.
7. Remove the front crossmember.
8. Remove the starter motor.
9. Remove the front and rear propeller shafts.
10. Remove the left and right front exhaust tubes.
11. Remove the A/T selector control cable and bracket from the A/T.
12. Disconnect the fluid cooler tubes from the A/T assembly.
13. Remove the dust cover from the converter housing.
14. Turn the crankshaft to access and remove the four bolts for the drive plate and torque converter.

✳✳ CAUTION

When turning the crankshaft, turn it clockwise as viewed from the front of the engine.

15. Support the A/T assembly using a transmission jack.

✳✳ CAUTION

When setting the transmission jack, be careful not to allow it to collide against the drain plug.

16. Remove the nuts securing the insulator to the crossmember.
17. Remove the crossmember.
18. Tilt the transmission slightly to gain clearance between the body and the transmission, then disconnect the air breather hose.
19. Disconnect the following:
   - A/T assembly harness connector
   - Neutral-4LO switch connector (ATX14B only)
   - 4LO switch connector (TX15B only)
   - Wait detection switch connector
   - ATP switch connector
   - Transfer motor connector (ATX14B only)
   - Control valve assembly connector (ATX14B only)
   - Transfer control device connector
20. Remove the wiring harness from the retainers.
21. Remove the A/T fluid indicator pipe.
22. Plug any openings such as the fluid charging pipe hole.
23. Remove the A/T assembly to engine bolts.
24. Remove A/T assembly with transfer from the vehicle.

✳✳ CAUTION

Secure the torque converter to prevent it from dropping. Secure the A/T assembly to the transmission jack.

25. Remove the transfer from the A/T assembly.

To install:
26. Installation is in the reverse order of the removal.
27. When installing transmission to the engine, tighten the bolts to the specified torque using sequence shown. Torque to 55 ft. lbs. (74 Nm).

✳✳ CAUTION

When replacing an engine or transmission you must make sure the

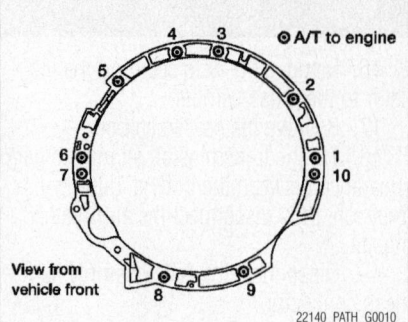

Fig. 4 Transmission to engine bolt tightening sequence

dowels are installed correctly during reassembly. Improper alignment caused by missing dowels may cause vibration, oil leaks or breakage of drivetrain components.

28. Align the positions of bolts for drive plate with those of the torque converter, and temporarily tighten the bolts. Then, tighten the bolts with the specified torque.

✳✳ CAUTION

When turning crankshaft, turn it clockwise as viewed from the front of the engine. After converter is installed to drive plate, rotate crankshaft several turns and check to be sure that transmission rotates freely without binding.

29. Install crankshaft position sensor (POS).
30. After completing installation, check fluid leakage, fluid level, and the positions of A/T.

TRANSFER CASE ASSEMBLY

REMOVAL & INSTALLATION

ATX14B

See Figure 5.

1. Set transfer state as 2WD when 4WD shift switch is at 2WD.
2. Remove the drain plug and gasket. Drain the fluid.
3. Remove the A/T undercover.
4. Remove the center exhaust tube and main muffler.
5. Remove the front and rear propeller shafts.

✳✳ CAUTION

Do not damage spline, sleeve yoke and rear oil seal when removing rear propeller shaft.

➡ Insert a plug into the rear oil seal after removing the rear propeller shaft.

6. Remove the A/T nuts from the A/T crossmember.
7. Position two suitable jacks under the A/T and transfer assembly.
8. Remove the crossmember.

✳✳ WARNING

Support A/T and transfer assembly using two suitable jacks while removing crossmember.

9. Disconnect the electrical connectors from the following:
   - ATP switch
   - Neutral 4LO switch
   - Wait detection switch
   - Transfer motor
   - Transfer control device
   - Transfer terminal cord assembly
10. Disconnect the air breather hoses from the following:
   - Actuator
   - Breather tube (transfer)
   - Transfer motor
11. Remove the transfer control device from the extension housing.
12. Remove the transfer to A/T and A/T to transfer bolts.

✳✳ WARNING

Support transfer assembly with suitable jack while removing it.

13. Remove the transfer assembly.

To install:
14. Installation is in the reverse order of removal.
15. Tighten the bolts to 27 ft. lbs. (36 Nm).
16. Fill the transfer with new fluid.
17. Check the transfer fluid.
18. Start the engine for one minute.

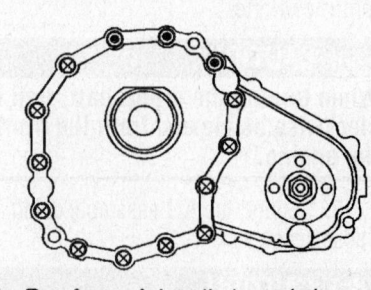

● : Transfer → Automatic transmission
⊗ : Automatic transmission → Transfer

Fig. 5 Transfer case bolt locations

Then stop the engine and recheck the transfer fluid.

TX15B

See Figure 6.

1. Switch 4WD shift switch to 2WD and set transfer assembly to 2WD.
2. Drain the transfer fluid.
3. Remove the A/T undercover.
4. Remove the center exhaust tube and main muffler.
5. Remove the front and rear propeller shafts.

✳✳ CAUTION

Do not damage spline, sleeve yoke and rear oil seal when removing rear propeller shaft.

➡**Insert a plug into the rear oil seal after removing the rear propeller shaft.**

6. Remove the A/T bolts.
7. Position two suitable jacks under the A/T and transfer assembly.
8. Remove the A/T crossmember.

✳✳ WARNING

Support A/T and transfer assembly using two suitable jacks while removing A/T crossmember.

9. Disconnect the electrical connectors from the following:
   - ATP switch
   - 4LO switch
   - Wait detection switch
   - Transfer control device
10. Disconnect each air breather hose from the following:
   - Transfer control device
   - Breather tube (transfer)
11. Remove the transfer to A/T and A/T to transfer bolts.

✳✳ WARNING

support transfer assembly with suitable jack while removing it.

12. Remove the transfer assembly.

✳✳ CAUTION

Do not damage rear oil seal (A/T).

To install:

13. Installation is in the reverse order of removal.
14. Tighten the bolts to 27 ft. lbs. (36 Nm).
15. Fill the transfer with new fluid.
16. Check the transfer fluid.

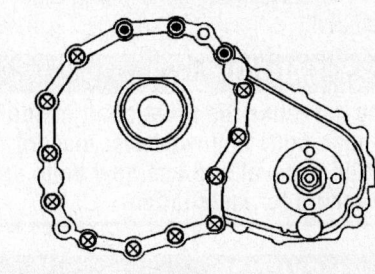

◉ : Transfer ➔ Automatic transmission
⊗ : Automatic transmission ➔ Transfer

22140_PATH_G0011

Fig. 6 Transfer case bolt locations

17. Start the engine for one minute. Then stop the engine and recheck the transfer fluid.
18. After the installation, check the 4WD shift indicator pattern. If NG, adjust the position between the transfer assembly and transfer control unit.

FRONT DRIVESHAFT

REMOVAL & INSTALLATION

1. Put matching marks on the front propeller shaft flange yoke and the front final drive companion flange.

✳✳ CAUTION

For matching marks, use paint. Never damage the flange yoke and companion flange of the front final drive.

2. Put matching marks on the front propeller shaft flange yoke and the transfer companion flange.
3. Remove the bolts and then remove the front propeller shaft from the front final drive and transfer.
4. Installation is in the reverse order of removal.
5. After installation, check for vibration by driving the vehicle.

✳✳ CAUTION

Do not reuse the bolts and nuts. Always install new ones.

FRONT HALFSHAFT

REMOVAL & INSTALLATION

1. Remove wheel and tire.
2. Remove rear engine under cover.
3. Remove wheel sensor harness from mount on knuckle, then disconnect wheel sensor harness connector.

✳✳ CAUTION

Do not pull on wheel sensor harness.

4. Remove wheel hub and bearing assembly

➡**It is not necessary to remove wheel sensor from wheel hub when wheel hub is not being replaced.**

5. Carefully feed wheel sensor harness through hole in splash shield.
6. Separate upper link ball joint stud from steering knuckle.
7. Support lower link with jack.
8. Remove drive shaft assembly.
   a. Pry drive shaft front final drive using suitable tool.
9. Installation is in the reverse order of removal.

REAR AXLE SHAFT, BEARING & SEAL

REMOVAL & INSTALLATION

1. Remove the wheel and tire assembly.
2. Remove the rear brake caliper, without disconnecting the hydraulic hose.
3. Reposition the rear brake caliper aside using suitable wire

➡**Do not depress the brake pedal while the brake caliper is removed.**

4. Remove the rear disc rotor.
5. Remove the cotter pin, then remove the rear drive shaft nut.

➡**Discard the cotter pin, use a new one for installation.**

6. Remove the rear drive shaft.
7. Remove the four rear wheel hub and bearing assembly bolts.

✳✳ CAUTION

Discard the four rear wheel hub and bearing assembly bolts, use new ones for installation.

8. Remove the rear wheel hub and bearing assembly.

To install:

9. Installation is in the reverse order of removal.

➡**Use a new cotter pin for installation.**

✳✳ CAUTION

Use new rear wheel hub and bearing assembly bolts for installation.

REAR HALFSHAFT

REMOVAL & INSTALLATION

1. Remove the wheel and tire assembly.

2. Remove the cotter pin and discard, then remove the lock nut from the drive shaft.

⊗ CAUTION

Do not reuse the cotter pin, discard after removal and use a new cotter pin for installation.

3. Remove the six rear drive shaft bolts from the rear final drive assembly flange.

✳ CAUTION

Do not reuse the rear drive shaft bolts, discard after removal and use new bolts for installation.

4. Separate the rear drive shaft from the rear wheel hub and bearing assembly by lightly tapping the end of the rear drive shaft with a suitable hammer and wood block. If it is difficult to separate, use a suitable puller.
5. Remove the rear drive shaft.

⊗ CAUTION

When removing the rear drive shaft, do not bend at an excessive angle to the rear drive shaft joint. Do not excessively extend the slide joint.

---

6. Installation is in the reverse order of removal.

✳✳ CAUTION

Do not reuse the drive shaft inside flange bolts and washers, discard after removal and use new bolts and washers for installation.

✳✳ CAUTION

Do not reuse the cotter pin, discard after removal and use a new cotter pin for installation.

REAR PINION SEAL

REMOVAL & INSTALLATION

1. Remove the rear propeller shaft.
2. Put a matching mark on the end of the drive pinion in line with the matching mark B on the companion flange.

✳✳ CAUTION

Use paint to make the matching mark on the drive pinion. Do not damage the companion flange or drive pinion.

➡ **The matching mark B on the final drive companion flange indicates the maximum vertical runout position.**

3. Remove the drive pinion lock nut.
4. Remove the companion flange using suitable tool.
5. Remove the front oil seal.

To install:
6. Install the front oil seal.

✳✳ CAUTION

Do not reuse oil seal. Do not incline oil seal when installing. Apply multi-purpose grease onto oil seal lips, and gear oil onto the circumference of oil seal.

7. Align the matching mark of the drive pinion with the matching mark B of the companion flange, then install the companion flange.
8. Install the drive pinion lock nut.

✳✳ CAUTION

Do not reuse drive pinion lock nut.

9. Install the rear propeller shaft.

ENGINE COOLING

THERMOSTAT

REMOVAL & INSTALLATION

1. Completely drain engine coolant.

⊗ CAUTION

Perform this step when engine is cold. Do not spill engine coolant on drive belts.

2. Remove air duct and air cleaner case.
3. Disconnect radiator hose (lower) and oil cooler hose from water inlet and thermostat assembly.
4. Remove water inlet and thermostat assembly.

✳✳ CAUTION

Do not disassemble water inlet and thermostat assembly. Replace them as a unit, if necessary.

5. Installation is in the reverse order of removal, paying attention to the following.

⊗ CAUTION

Be careful not to spill engine coolant over engine room. Use rag to absorb engine coolant.

---

6. Start and warm up engine. Visually check there are no leaks of engine coolant.

WATER PUMP

REMOVAL & INSTALLATION

V6 Engine
See Figure 7.

✳✳ CAUTION

When removing water pump assembly, be careful not to get engine coolant on timing chain and drive belts. Water pump cannot be disassembled and should be replaced as a unit. After installing water pump, connect hose and clamp securely, then check for leaks.

1. Remove air dam. (2008 models only)
2. Remove undercover.
3. Remove air duct and resonator. (2008 models only)
4. Remove drive belts.
5. Drain engine coolant.

✳✳ CAUTION

Perform this step when engine is

---

cold. Do not spill engine coolant on drive belts.

6. Remove radiator hoses (upper and lower) and cooling fan assembly.
7. Remove chain tensioner cover and water pump cover from front timing chain case.
8. Remove timing chain tensioner (primary) as follows:
   a. Loosen clip of timing chain tensioner (primary), and release plunger stopper. (1)
   b. Insert plunger into tensioner body by pressing slack guide. (2)
   c. Keep slack guide pressed and hold plunger in by pushing stopper pin through the tensioner body hole and plunger groove. (3)
   d. Turn crankshaft pulley clockwise so that timing chain on the timing chain tensioner (primary) side is loose.
   e. Remove bolts and remove timing chain tensioner (primary).

✳✳ CAUTION

Be careful not to drop bolts inside timing chain case.

9. Remove water pump as follows:

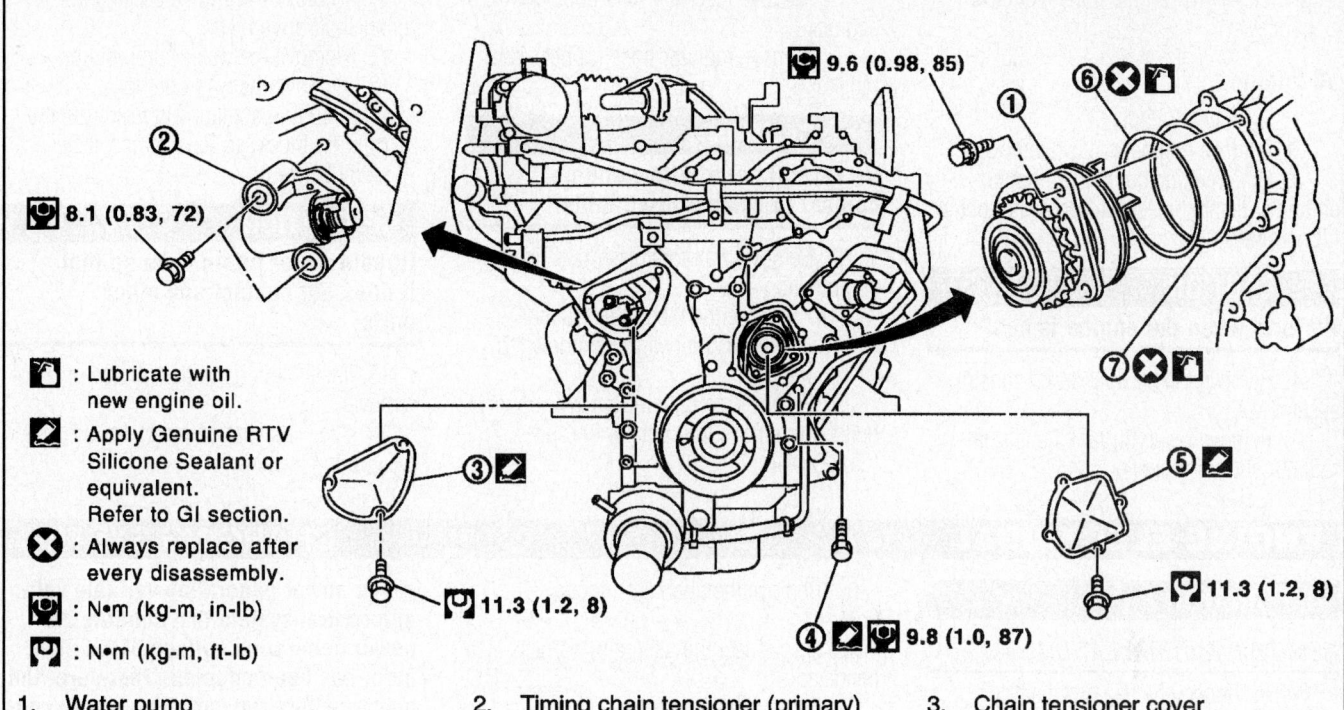

9.6 (0.98, 85)

8.1 (0.83, 72)

: Lubricate with new engine oil.

: Apply Genuine RTV Silicone Sealant or equivalent. Refer to GI section.

: Always replace after every disassembly.

: N•m (kg-m, in-lb)

: N•m (kg-m, ft-lb)

11.3 (1.2, 8)

9.8 (1.0, 87)

11.3 (1.2, 8)

1. Water pump
2. Timing chain tensioner (primary)
3. Chain tensioner cover
4. Water drain plug (front)
5. Water pump cover
6. O-ring
7. O-ring

22140_PATH_G0012

Fig. 7 Water pump component locations

a. Remove three water pump bolts. Secure a gap between water pump gear and timing chain, by turning crankshaft pulley counterclockwise until timing chain looseness on water pump sprocket becomes maximum.

b. Screw M8 bolts [pitch: 1.25 mm (0.049 in) length: approx. 50 mm (1.97 in)] into water pumps upper and lower bolt holes until they reach timing chain case. Then, alternately tighten each bolt for a half turn, and pull out water pump.

✸✸ CAUTION

Pull straight out while preventing vane from contacting socket in installation area. Remove water pump without causing sprocket to contact timing chain.

c. Remove M8 bolts and O-rings from water pump.

✸✸ CAUTION

Do not disassemble water pump.

➡Do not reuse O-rings.

To install:
10. Install new O-rings to water pump.

➡Apply engine oil to O-rings. Locate O-ring with white paint mark to engine front side.

11. Install water pump.

✸✸ CAUTION

Do not allow timing chain case to pinch O-rings when installing water pump.

a. Make sure that timing chain and water pump sprocket are engaged.

b. Insert water pump by tightening bolts alternately and evenly.

12. Install timing chain tensioner (primary) as follows:

a. Remove dust and foreign material completely from backside of timing chain tensioner (primary) and from installation area of rear timing chain case.

b. Turn crankshaft pulley clockwise so that timing chain on the timing chain tensioner (primary) side is loose.

c. Install timing chain tensioner (primary) with its stopper pin attached.

✸✸ CAUTION

Be careful not to drop bolts inside timing chain case.

d. Remove stopper pin.

e. Make sure again that timing chain and water pump sprocket are engaged.

13. Install chain tensioner cover and water pump cover as follows:

a. Before installing, remove all traces of old liquid gasket from mating surface of water pump cover and chain tensioner cover using scraper. Also remove traces of old liquid gasket from the mating surface of front timing chain case.

b. Apply a continuous bead of liquid gasket, to mating surface of chain tensioner and water pump cover.

✸✸ CAUTION

Attaching should be done within 5 minutes after coating.

c. Tighten bolts to specified torque.
14. Refill engine coolant system.
15. Installation of the remaining components is in the reverse order of removal after this step.

a. After starting engine, let idle for three minutes, then rev engine up to 3,000 rpm under no load to purge air from the high-pressure chamber of chain tensioner. Engine may produce a rattling noise. This indicates that air still remains in the chamber and is not a matter of concern.

V8 Engine

1. Remove air dam.
2. Remove engine front undercover.
3. Drain engine coolant so that no engine coolant comes out from water pump fitting hole.

> ✳ **CAUTION**
>
> **Perform when the engine is cold.**

4. Remove the air duct and resonator assembly.
5. Remove reservoir tank hose from radiator shroud (upper).

6. Remove reservoir tank hose from engine.
7. Remove radiator hose (upper) from radiator.

> ✳✳ **CAUTION**
>
> **Be careful not to allow engine coolant to contact drive belts.**

8. Remove the radiator shroud (lower) and position aside.
   a. Release the tabs, pull radiator shroud (lower) rearwards and down to remove.
9. Remove the radiator shroud (upper) bolts and remove the radiator shroud (upper) (A).

10. Remove the engine cooling fan (crankshaft driven type).
11. Remove the water pump pulley.
12. Remove the water pump.
   a. Engine coolant will leak from the cylinder block, so have a receptacle ready below.

> ✳✳ **CAUTION**
>
> **Handle water pump vane so that it does not contact any other parts.**

13. Installation is in the reverse order of removal.
14. After installation bleed the air from the cooling system.

ENGINE ELECTRICAL

ALTERNATOR

REMOVAL & INSTALLATION

1. Disconnect the negative battery terminal.
2. Remove the fan shroud.
3. Remove the drive belt.
4. Remove alternator stay.

5. Remove the alternator upper bolt.
6. Disconnect the alternator harness connectors.
7. Remove the alternator.
8. Installation is in the reverse order of removal.
9. Install the alternator and check tension of drive belt.

CHARGING SYSTEM

➡ **The power generation variable voltage control system that controls the power generation voltage of the alternator has been adopted. Therefore, the power generation variable voltage control system inspection should be performed after replacing the alternator in order to ensure that the system operates normally.**

ENGINE ELECTRICAL

FIRING ORDER

See Figures 8 and 9.

Firing order for V6 engine is
1–2–3–4–5–6
Firing order for V8 engine is
1–8–7–3–6–5–4–2

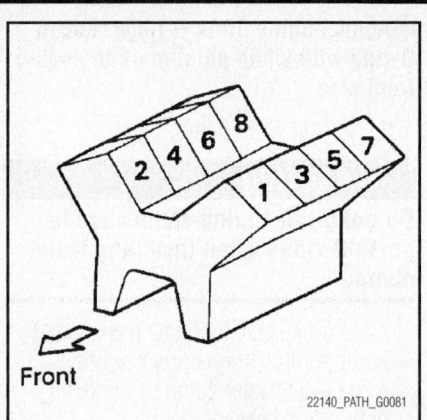

Fig. 9 Cylinder number arrangement—V8

IGNITION COIL

REMOVAL & INSTALLATION

V6 Engine

Left Bank

1. Remove engine cover.
2. Remove air cleaner case and air duct. (At the left bank side, remove ignition coil).
3. Move aside harness, harness bracket, and hoses located above ignition coil.

IGNITION SYSTEM

4. Disconnect harness connector from ignition coil.
5. Remove ignition coil.

> ✳✳ **CAUTION**
>
> **Do not shock it.**

6. Installation is in the reverse order of removal.

Right Bank

1. Remove intake manifold collector.
2. Move aside harness, harness bracket, and hoses located above ignition coil.
3. Disconnect harness connector from ignition coil.
4. Remove ignition coil.

> ✳✳ **CAUTION**
>
> **Do not shock it.**

5. Installation is in the reverse order of removal.

V8 Engine

1. Remove the engine room cover.
2. Remove the air duct and resonator assembly.

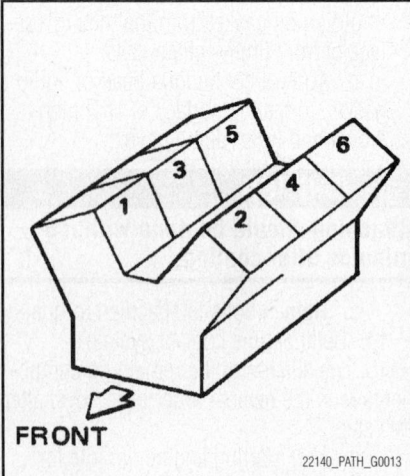

Fig. 8 Cylinder number arrangement—V6

3. Disconnect the harness connector from the ignition coil.

4. Remove the ignition coil.

✳✳ CAUTION

Do not shock ignition coil.

5. Installation is in the reverse order of removal.

IGNITION TIMING

ADJUSTMENT

Ignition timing is controlled by the ECM. No adjustment is necessary or possible.

SPARK PLUGS

REMOVAL & INSTALLATION

V6 Engine

1. Remove engine cover.

2. Remove air cleaner case and air duct. (At the left bank side, remove ignition coil).

3. Move aside harness, harness bracket, and hoses located above ignition coil.

4. Disconnect harness connector from ignition coil.

5. Remove ignition coil.

6. Remove the spark plug.

7. Installation is in the reverse order of removal.

V8 Engine

1. Remove the engine room cover.

2. Remove the air duct and resonator assembly.

3. Disconnect the harness connector from the ignition coil.

4. Remove the ignition coil.

✳✳ CAUTION

Do not shock ignition coil.

5. Remove spark plug.

6. Installation is in the reverse order of removal.

ENGINE ELECTRICAL

STARTER

REMOVAL & INSTALLATION

V6 Engine

See Figure 10.

1. Disconnect the negative battery terminal.

2. Remove engine undercover.

3. Remove exhaust manifold cover from exhaust manifold (right bank) to gain access to starter cover bolts.

4. Remove starter cover bolts and starter cover.

5. Disconnect terminal "1" connector and terminal "2" nut.

6. Remove the two starter bolts.

7. Remove the starter.

8. Installation is in the reverse order of removal.

V8 Engine

1. Remove the intake manifold.

2. Disconnect terminal "1" connector and terminal "2" nut.

3. Remove the two starter bolts.

4. Remove the starter.

5. Installation is in the reverse order of removal.

STARTING SYSTEM

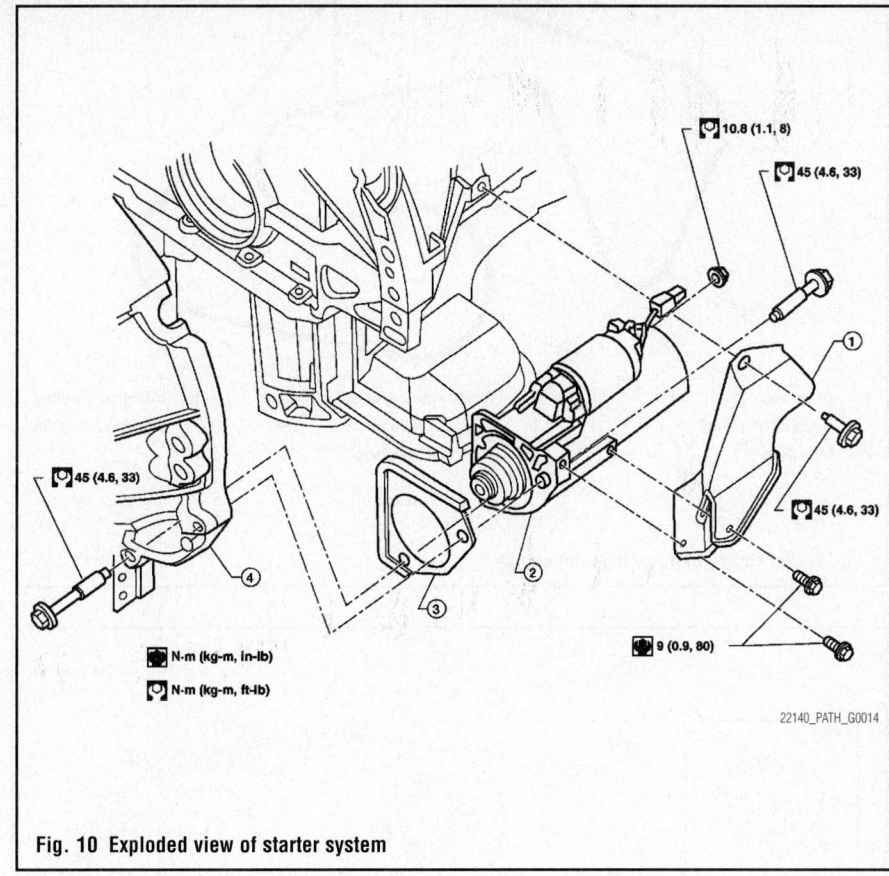

Fig. 10 Exploded view of starter system

ENGINE MECHANICAL

➡ Disconnecting the negative battery cable may interfere with the functions of the on board computer systems and may require the computer to undergo a relearning process, once the negative battery cable is reconnected.

ACCESSORY DRIVE BELTS

ACCESSORY BELT ROUTING

See Figures 11 and 12.

INSPECTION

Visually check entire belt for wear, damage or cracks.

ADJUSTMENT

Belt tensioning is not necessary, as it is automatically adjusted by auto tensioner.

REMOVAL & INSTALLATION

See Figure 13.

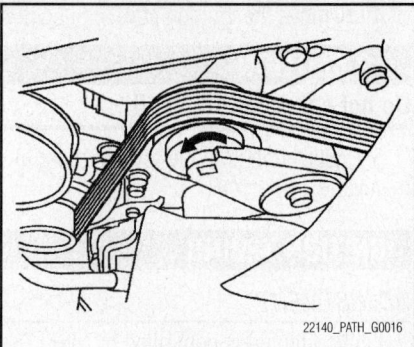

Fig. 13 Rotate the drive belt auto tensioner in the direction of arrow

1. Remove air duct and resonator assembly (inlet).
2. Rotate the drive belt auto tensioner in the direction of arrow (loosening direction of tensioner).

✳✳ CAUTION

Avoid placing hand in a location where pinching may occur if the tool accidentally comes off.

3. Remove the drive belt.
4. Installation is in the reverse order of removal.

✳✳ CAUTION

Make sure belt is securely installed around all pulleys.

CAMSHAFT AND VALVE LIFTERS

REMOVAL & INSTALLATION

V6 Engine

See Figures 14 through 19.

1. Remove front timing chain case, camshaft sprocket, timing chain and rear timing chain case.
2. Remove camshaft position sensor (PHASE) (right and left banks) from cylinder head back side.

✳✳ CAUTION

Handle carefully to avoid dropping and shocks. Do not disassemble. Do not allow metal powder to adhere to magnetic part at sensor tip. Do not place sensors in a location where they are exposed to magnetism.

3. Remove intake valve timing control solenoid valves.

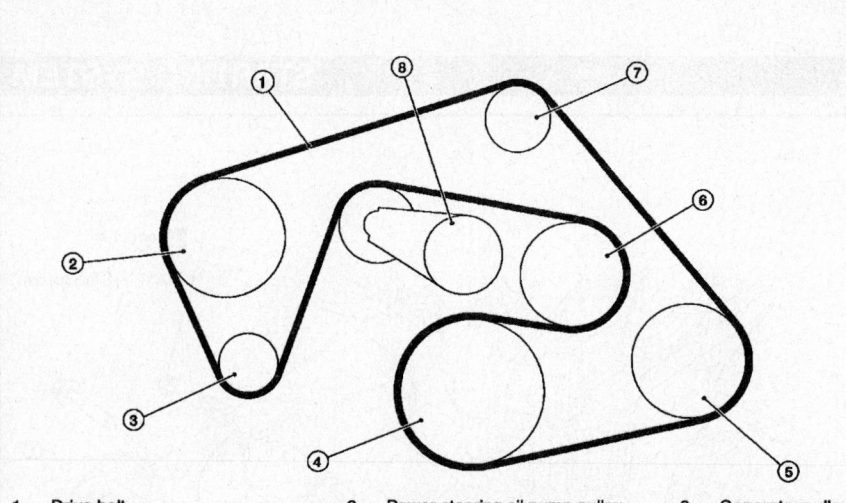

1.	Drive belt	2.	Power steering oil pump pulley	3.	Generator pulley
4.	Crankshaft pulley	5.	A/C compressor	6.	Cooling fan pulley
7.	Idler pulley	8.	Drive belt tensioner		

22140_PATH_G0015

Fig. 11 Accessory belt routing—V6 engine

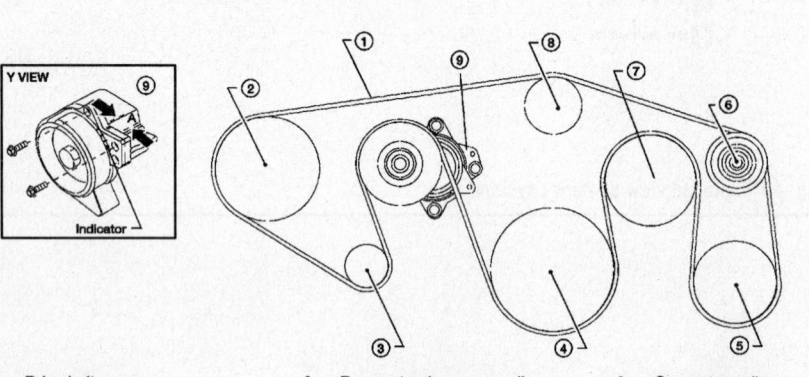

1.	Drive belt	2.	Power steering pump pulley	3.	Generator pulley
4.	Crankshaft pulley	5.	A/C compressor	6.	Idler pulley
7.	Cooling fan pulley	8.	Water pump pulley	9.	Drive belt auto tensioner
A.	Allowable working range				

22140_PATH_G0082

Fig. 12 Accessory belt routing—V8 engine

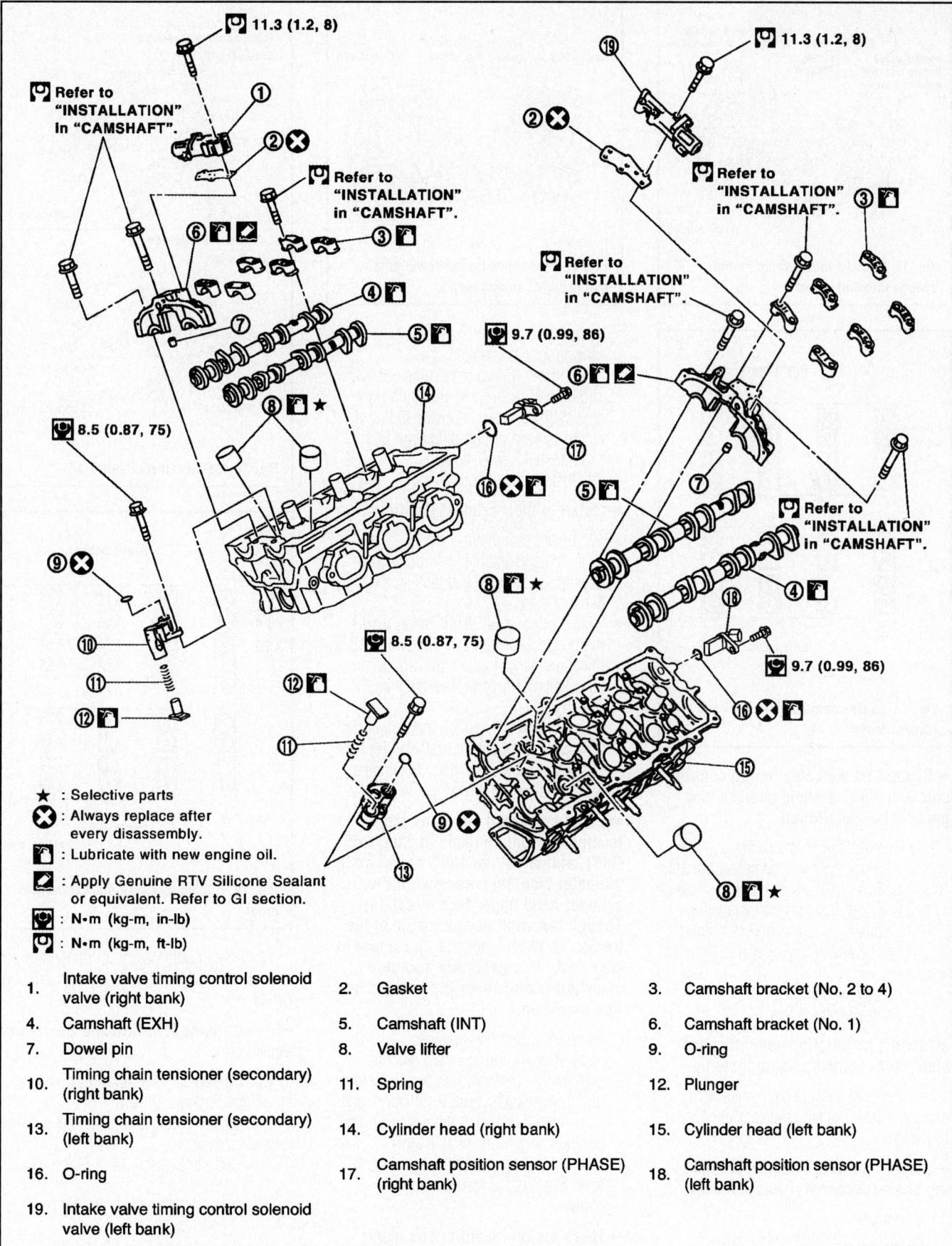

★ : Selective parts

❌ : Always replace after every disassembly.

🛢 : Lubricate with new engine oil.

✏ : Apply Genuine RTV Silicone Sealant or equivalent. Refer to GI section.

🔧 : N·m (kg-m, in-lb)

🔧 : N·m (kg-m, ft-lb)

1. Intake valve timing control solenoid valve (right bank)
2. Gasket
3. Camshaft bracket (No. 2 to 4)
4. Camshaft (EXH)
5. Camshaft (INT)
6. Camshaft bracket (No. 1)
7. Dowel pin
8. Valve lifter
9. O-ring
10. Timing chain tensioner (secondary) (right bank)
11. Spring
12. Plunger
13. Timing chain tensioner (secondary) (left bank)
14. Cylinder head (right bank)
15. Cylinder head (left bank)
16. O-ring
17. Camshaft position sensor (PHASE) (right bank)
18. Camshaft position sensor (PHASE) (left bank)
19. Intake valve timing control solenoid valve (left bank)

Fig. 14 Exploded view of camshaft components—V6 Engine

22140_PATH_G0017

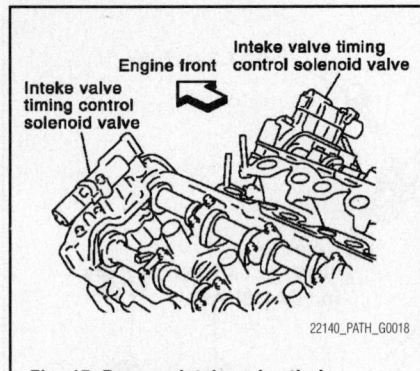

Fig. 15 Remove intake valve timing control solenoid valves

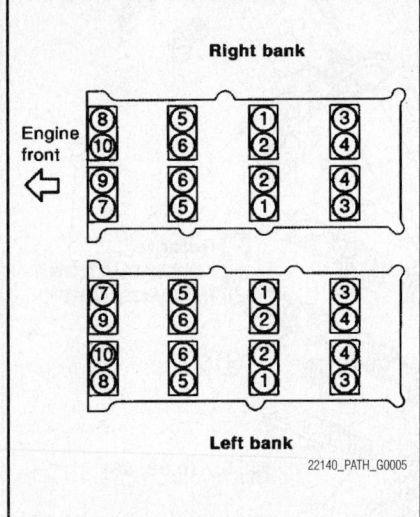

Fig. 16 Loosen camshaft bracket bolts in reverse order

➡**Discard intake valve timing control solenoid valve gaskets and use new gaskets for installation.**

4. Remove camshaft brackets.

   a. Mark camshafts, camshaft brackets and bolts so they are placed in the same position and direction for installation.

   b. Equally loosen camshaft bracket bolts in several steps in reverse order as shown.

5. Remove camshafts.

6. Remove valve lifters.

➡**Identify installation positions, and store them without mixing them up.**

7. Remove timing chain tensioner (secondary) from cylinder head with its stopper pin attached.

➡**Stopper pin was attached when timing chain (secondary) was removed.**

To install:

8. Install timing chain tensioners (secondary) on both sides of cylinder head.

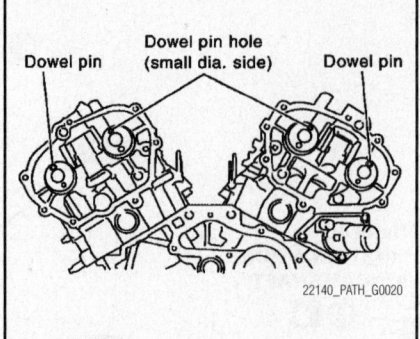

Fig. 17 Proper position of dowel pins during camshaft installation

   a. Install timing chain tensioner with its stopper pin attached.

   b. Install timing chain tensioner with sliding part facing downward on right-side cylinder head, and with sliding part facing upward on left-side cylinder head.

   c. Install new O-rings.

9. Install valve lifters.

➡**Install in their original positions.**

10. Install camshafts.

   a. Install camshaft with dowel pin attached to its front end face on the exhaust side.

   b. Follow your identification marks made during removal, or follow the identification marks that are present on new camshafts for proper placement and direction.

   c. Install camshaft so that dowel pin hole and dowel pin on front end face are positioned as shown. (No. 1 cylinder TDC on its compression stroke).

➡**Large and small pin holes are located on front end face of camshaft (INT), at intervals of 180°. Face small diameter side pin hole upward (in cylinder head upper face direction). Though camshaft does not stop at the portion as shown, for the placement of cam nose, it is generally accepted camshaft is placed for the same direction as shown.**

11. Install camshaft brackets.

   a. Remove foreign material completely from camshaft bracket backside and from cylinder head installation face.

   b. Install camshaft bracket in original position and direction as shown.

   c. Install camshaft brackets (No. 2 to 4) aligning the stamp marks as shown.

➡**There are no identification marks indicating left and right for camshaft bracket (No. 1).**

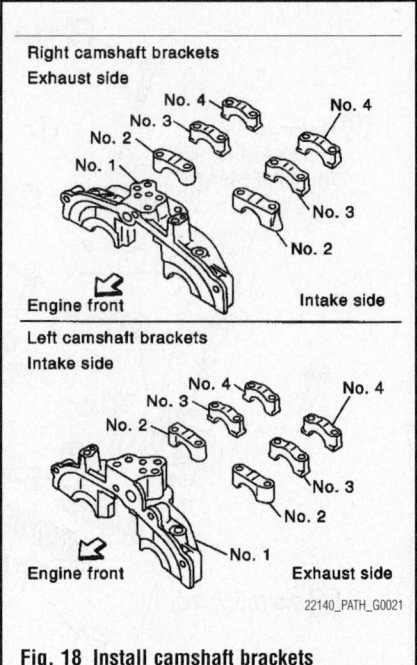

Fig. 18 Install camshaft brackets

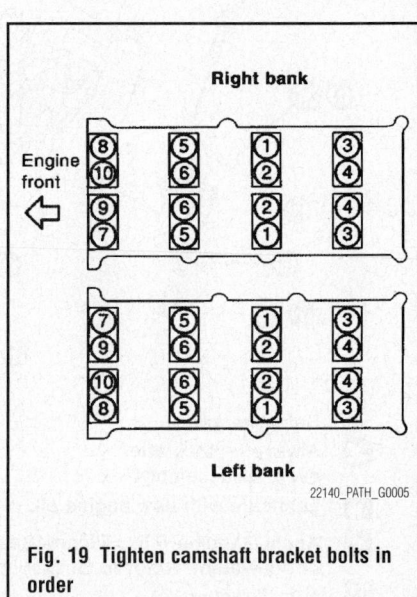

Fig. 19 Tighten camshaft bracket bolts in order

12. Apply liquid gasket to mating surface of camshaft bracket (No. 1) on right and left banks.

➡**Use Genuine RTV Silicone Sealant or equivalent.**

13. Tighten camshaft bracket bolts in numerical order as shown.

14. Tighten camshaft bracket bolts in the following steps:

   • Step 1: Bolts 7–10: 17 inch lbs, (1.96 Nm)
   • Step 2: Bolts 1–6: 17 inch lbs, (1.96 Nm)
   • Step 3: All bolts: 52 inch lbs. (5.88 Nm)

- Step 4: All bolts: 92 inch lbs. (10.4 Nm)

15. Measure the difference in levels between front end faces of camshaft bracket (No. 1) and cylinder head.

   a. Standard: –0.0055 to 0.0055 inches (–0.14 to 0.14 mm)

   b. Measure two positions (both intake and exhaust side) for a single bank.

   c. If the measured value is out of the standard, re-install camshaft bracket (No. 1).

16. Check and adjust the valve clearance.

17. Installation of the remaining components is in the reverse order of removal.

V8 Engine

See Figures 20 through 37.

➡**Do not remove the engine assembly to perform this procedure.**

1. Remove the RH bank and LH bank rocker covers.

2. Obtain compression TDC of No. 1 cylinder as follows:

   a. Turn the crankshaft pulley clockwise to align the TDC identification notch

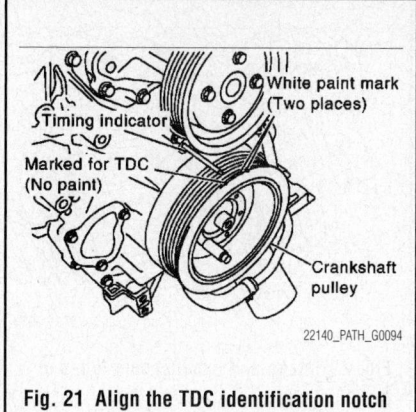

Fig. 21 Align the TDC identification notch

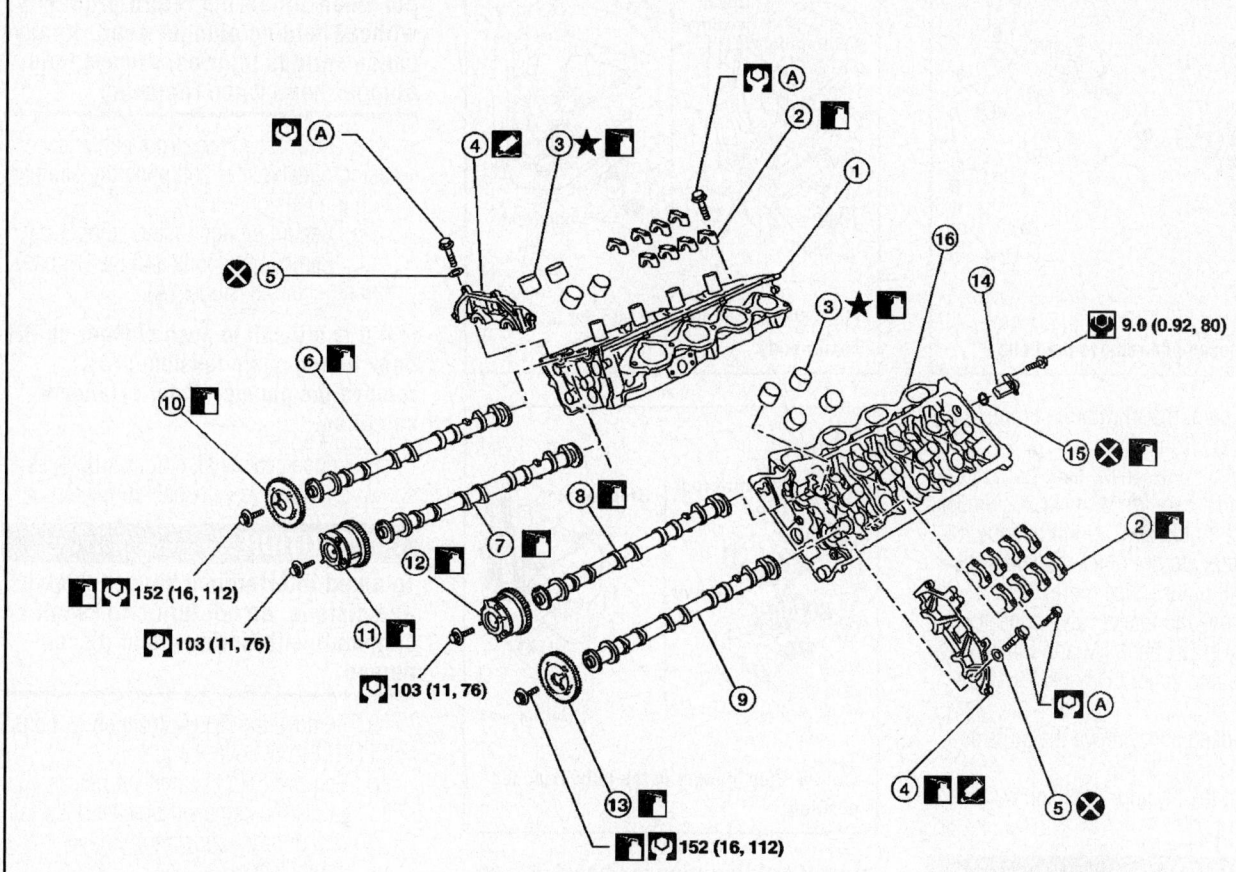

1. Cylinder head RH bank
2. Camshaft bracket (No. 2, 3, 4, 5)
3. Valve lifter
4. Camshaft bracket (No. 1)
5. Seal washer
6. Camshaft RH bank (EXH)
7. Camshaft RH bank (INT)
8. Camshaft LH bank (INT)
9. Camshaft LH bank (EXH)
10. Camshaft sprocket RH bank (EXH)
11. Camshaft sprocket RH bank (INT) (VTC)
12. Camshaft sprocket LH bank (INT) (VTC)
13. Camshaft sprocket LH bank (EXH)
14. Camshaft position sensor (PHASE)
15. O-ring
16. Cylinder head LH bank

Fig. 20 Exploded view of camshaft components—V8 Engine

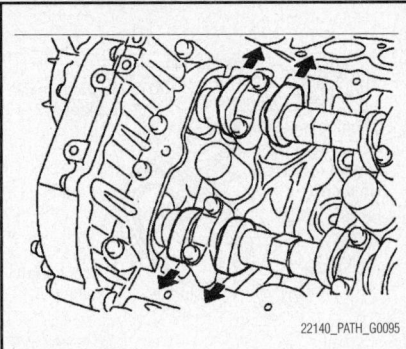

Fig. 22 Intake and exhaust cam lobes of No. 1 cylinder

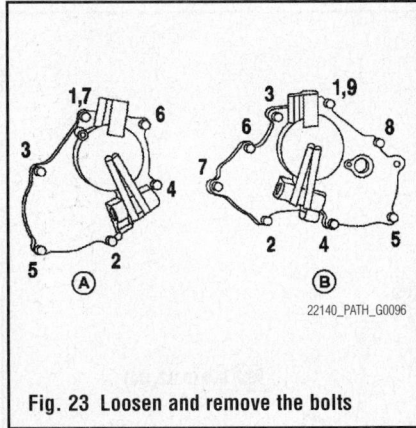

Fig. 23 Loosen and remove the bolts

(without paint mark) with the timing indicator on the front cover.

    b. At this time, make sure both intake and exhaust cam lobes of No. 1 cylinder (top front on LH bank) point outside.

    c. If they do not point outside, turn crankshaft pulley once more.

    3. Remove the intake valve control solenoid cover RH bank (A) and intake valve control solenoid cover LH bank (B) as follows:

    a. Loosen and remove the bolts as shown.

    b. Cut the liquid gasket and remove the covers.

✳✳ CAUTION
Do not damage mating surfaces.

    4. Paint alignment marks on the RH bank (A) timing chain links (C) and LH bank (B) timing chain links (D) and align with the camshaft sprocket alignment marks (E) and (F).

    5. Remove the LH bank timing chain tensioner using the following steps.

✳✳ WARNING
Plunger, spring, and spring seat pop out when squeezing return-proof clip

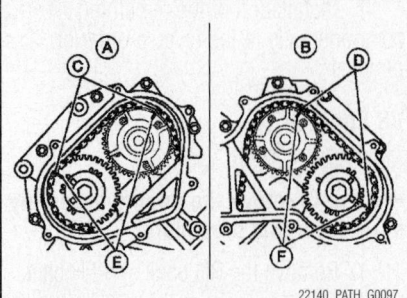

Fig. 24 Paint alignment marks on the RH bank (A) timing chain links (C) and LH bank (B) timing chain links

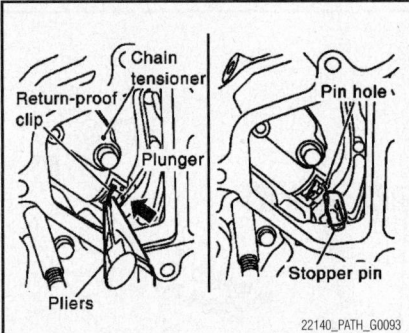

Fig. 25 Push the plunger into the tensioner body

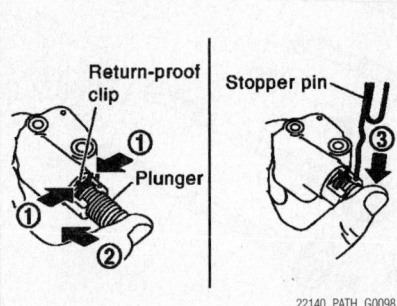

Fig. 26 Stop plunger in the fully extended position

without holding plunger head. It may cause serious injuries. Always hold plunger head when removing.

    a. Squeeze return-proof clip ends using suitable tool and push the plunger into the tensioner body.

    b. Secure plunger using stopper pin.

→**Stopper pin is made from hard wire approximately 1 mm (0.04 in) in diameter.**

    c. Remove the bolts and the timing chain tensioner.

→**Stop plunger in the fully extended position using return-proof clip (1) if stopper pin is removed. Push the plunger (2) into the tensioner body while squeezing the return-proof clip (1). Secure it using stopper pin (3).**

    6. Remove the RH bank timing chain tensioner cover from the front cover.

✳✳ CAUTION
Do not damage mating surfaces.

    7. Remove the RH bank timing chain tensioner using the following steps.

✳✳ WARNING
Plunger, spring, and spring seat pop out when squeezing return-proof clip without holding plunger head. It may cause serious injuries. Always hold plunger head when removing.

    a. Squeeze return-proof clip ends using suitable tool and push the plunger into the tensioner body.

    b. Secure plunger using stopper pin.

    c. Remove the bolts and the RH bank timing chain tensioner (A).

→**If it is difficult to push plunger on RH bank timing chain tensioner (A), remove the plunger under extended condition.**

    8. Loosen camshaft sprocket bolts as shown and remove camshaft sprockets.

✳✳ CAUTION
To avoid interference between valves and pistons, do not turn crankshaft or camshaft with timing chain disconnected.

    9. Remove the RH (A) front cover bolts and LH (B) front cover bolts.

    10. Remove RH (A) camshaft bracket bolts and LH (C) camshaft bracket bolts in the reverse of order shown to remove camshaft brackets.

    a. Remove No. 1 camshaft bracket.

→**The bottom and front surface of bracket will be stuck because of liquid gasket.**

    11. Remove the camshaft.

    12. Remove the valve lifters if necessary.

→**Correctly identify location where each part is removed from. Keep parts organized to avoid mixing them up.**

    To install:

    13. Install the valve lifters if removed.

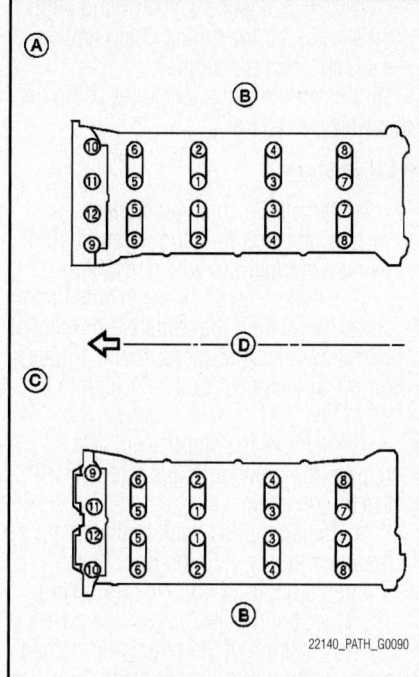

Fig. 27 Remove RH (A) camshaft bracket bolts and LH (C) camshaft bracket bolts in the reverse of order shown to remove camshaft brackets

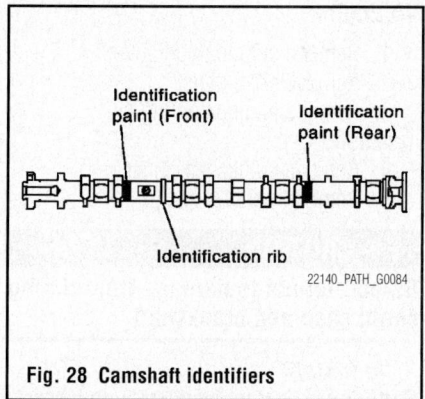

Fig. 28 Camshaft identifiers

→**Install removed parts in their original locations.**

14. Install the camshafts. Important details for identification of the RH and LH, and intake and exhaust.
- RH Bank: Intake: Front paint: Pink; Exhaust: Rear paint: Orange; Rib: Yes
- LH Bank: Intake: Front paint: Pink; Exhaust: Rear paint: Orange; Rib: No
- Install so that the RH bank (B) dowel pins (A) and LH bank (C) dowel pins (A) at the front of the camshaft face are in the direction shown.

15. Install the RH bank (B) and LH bank (D) camshaft brackets (A).

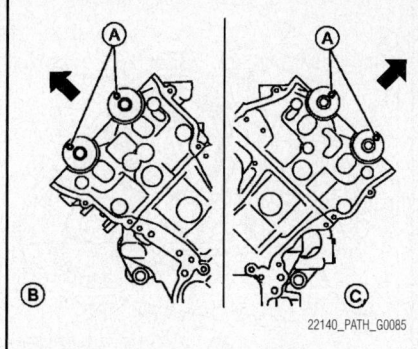

Fig. 29 Proper position for camshaft installation

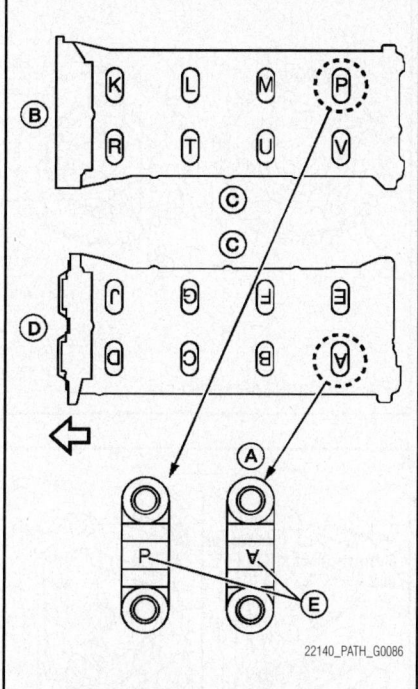

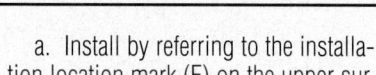

Fig. 30 Install the RH bank (B) and LH bank (D) camshaft brackets (A)

a. Install by referring to the installation location mark (E) on the upper surface.

b. Install so that the installation location mark (E) can be correctly read when viewed from the intake manifold side (C).

c. Install No. 1 camshaft bracket using the following procedure:
- C: 0.43 inches (11 mm)
- D: 0.079–0.118 inches (2.0–3.0 mm) diameter
- Apply liquid gasket to No. 1 camshaft bracket (A) and (B) as shown.

✳✳ CAUTION

After installation, be sure to wipe off any excessive liquid gasket outside

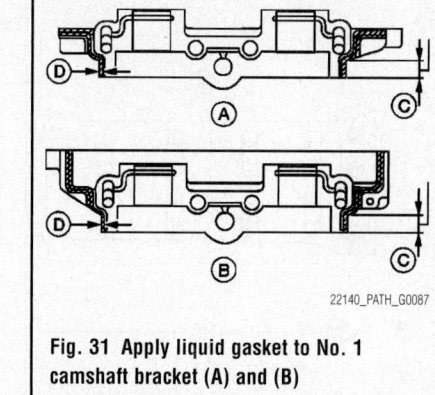

Fig. 31 Apply liquid gasket to No. 1 camshaft bracket (A) and (B)

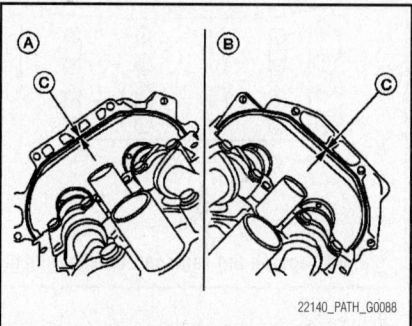

Fig. 32 Apply liquid gasket (C) to the back side of the LH (A) bank front cover and RH (B) bank front cover

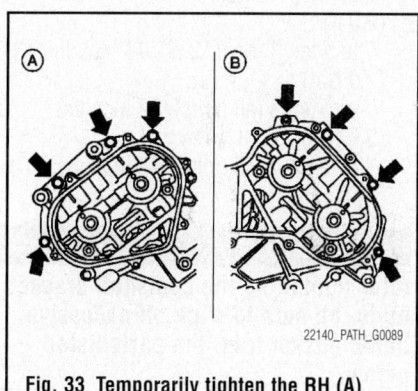

Fig. 33 Temporarily tighten the RH (A) and LH (B) front cover bolts

of application (C) and (D) both on RH and LH sides.

- Remove completely any excess of liquid gasket inside bracket.

d. Apply liquid gasket (C) to the back side of the LH (A) bank front cover and RH (B) bank front cover as shown.

e. C: 0.102–0.142 inches (2.6–3.6 mm) diameter

f. Position No. 1 camshaft bracket close to the mounting position, and then install it to prevent from touching liquid gasket applied to each surface.

g. Temporarily tighten the RH (A) and

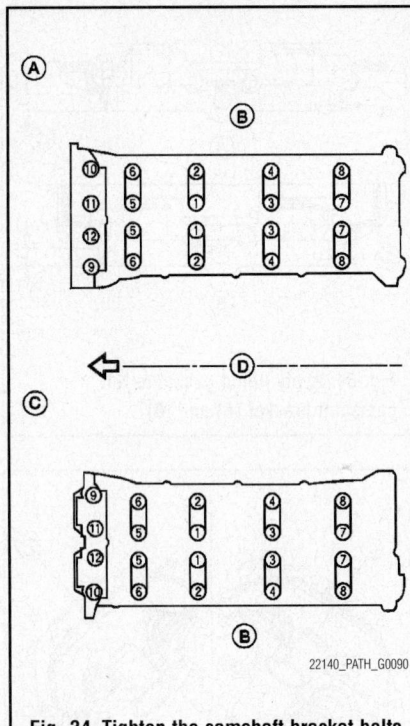

Fig. 34 Tighten the camshaft bracket bolts

LH (B) front cover bolts (4 for each bank) as shown.

16. Tighten the camshaft bracket bolts as follows:

    a. Step 1 (bolts 9–12): 17 inch lbs. (2.0 Nm)

    b. Step 2 (bolts 1–8): 17 inch lbs. (2.0 Nm)

    c. Step 3 (all bolts): 52 inch lbs. (5.9 Nm)

    d. Step 4 (all bolts): 92 inch lbs. (10.4 Nm)

✳✳ CAUTION

After tightening the camshaft bracket bolts, be sure to wipe off excessive liquid gasket from the parts listed below.

- Mating surface of rocker cover
- Mating surface of front cover
- A: RH bank
- B: Exhaust side
- C: LH bank
- D: Intake side

    e. Tighten the RH (A) and LH (B) front cover bolts (4 for each bank) to 8 ft. lbs. (11.0 Nm)

17. Install the camshaft sprockets using the following procedure:

➡ A: LH bank shown

    a. Install the camshaft sprockets aligning them with the matching marks painted on the timing chain (B) and the

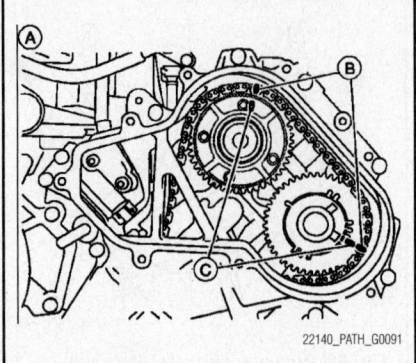

Fig. 35 Install the camshaft sprockets

Fig. 36 Camshaft sprockets

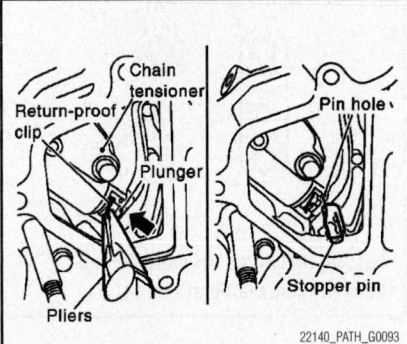

Fig. 37 Install the chain tensioner

camshaft sprockets (C) before removal. Align the camshaft sprocket key groove with the dowel pin on the camshaft front edge at the same time. Then temporarily tighten camshaft sprocket bolts.

    b. Install the intake VTC (A) and exhaust (B) side camshaft sprockets by selectively using the groove of the dowel pin according to the bank for the exhaust (B) side camshaft sprockets. (Common part used for both exhaust banks.)

    c. Lock the hexagonal part of the camshaft in the same way as for removal, and tighten the camshaft sprocket bolts.

    d. Check again that the timing alignment mark on the timing chain and on each sprocket are aligned.

18. Install the chain tensioner using the following procedure:

➡ LH is shown.

    a. Install the chain tensioner.

    b. Compress the plunger and hold it using a stopper pin when installing.

    c. Loosen the slack guide side timing chain by rotating the camshaft hexagonal part if mounting space is small. Torque for chain tensioner bolts: 61 inch lbs. (6.9 Nm).

    d. Remove the stopper pin and release the plunger to apply tension to the timing chain.

    e. Install the RH bank timing chain tensioner cover onto the front cover. Tighten bolts to 80 inch lbs. (9.0 Nm).

19. Check and adjust valve clearances.

20. Installation of the remaining components is in the reverse order of removal.

CRANKSHAFT FRONT SEAL

REMOVAL & INSTALLATION

V6 Engine

1. Remove engine undercover.
2. Remove drive belts.
3. Remove engine cooling fan assembly.
4. Remove crankshaft pulley.
5. Remove front oil seal.

✳✳ CAUTION

Be careful not to damage front timing chain case and crankshaft.

To install:

6. Apply new engine oil to both oil seal lip and dust seal lip of new front oil seal.

7. Install front oil seal.

    a. Install front oil seal so that each seal lip is oriented as shown.

    b. Press-fit until the height of front oil seal is level with the mounting surface using suitable tool.

    c. Suitable drift: outer diameter 2.36 inches (60 mm), inner diameter 1.97 inches (50 mm).

✳✳ CAUTION

Be careful not to damage front timing chain case and crankshaft. Press-fit straight and avoid causing burrs or tilting oil seal.

8. Installation is in the reverse order of removal after this step.

V8 Engine

See Figure 38.

1. Remove the air dam.
2. Remove the engine undercover.
3. Remove the air duct and resonator assembly and the air cleaner case (upper).
4. Remove the radiator assembly.
5. Remove the cooling fan (crankshaft driven type).
6. Remove the crankshaft pulley.
7. Remove the front oil seal using suitable tool.

❈❈ CAUTION

Do not damage front cover and oil pump drive spacer.

8. Apply new engine oil to both the oil seal lip and dust seal lip of the new front oil seal.
9. Install the front oil seal.
   a. Install the front oil seal so that each seal lip is oriented as shown.
   b. Press-fit until the front oil seal is level with the front cover using suitable tool.

❈❈ CAUTION

Do not damage front cover and crankshaft. Press—fit straight and avoid causing burrs or tilting oil seal.

10. Installation of the remaining components is in the reverse order of removal.

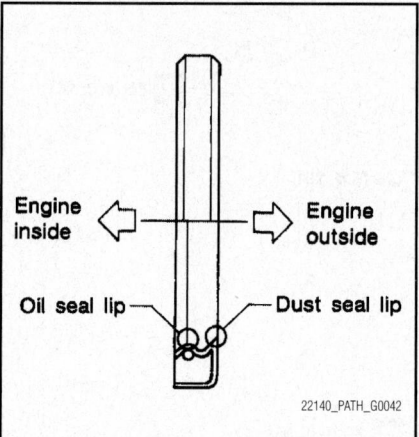

Fig. 38 Install the front oil seal so that each seal lip is oriented as shown.

CYLINDER HEAD

REMOVAL & INSTALLATION

V6 Engine

See Figure 39.

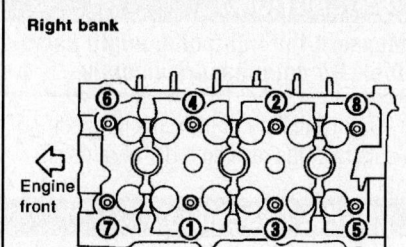

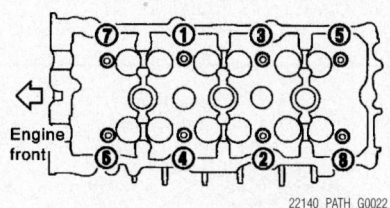

Fig. 39 Tighten cylinder head bolts in order as shown

1. Remove camshaft.
2. Remove intake manifold.
3. Remove exhaust manifold.
4. Remove water inlet and thermostat assembly.
5. Remove water outlet, water pipe and heater pipe.
6. Remove cylinder head bolts in reverse order of the tightening sequence to remove cylinder heads (right and left banks).
7. Remove cylinder head gaskets.

To install:

8. Install new cylinder head gasket.
9. Turn crankshaft until No. 1 piston is set at TDC.

➡ **Crankshaft key should line up with the right bank cylinder center line.**

10. Install cylinder head and tighten cylinder head bolts in numerical order as shown.

❈❈ CAUTION

If cylinder head bolts re-used, check their outer diameters before installation. Tighten cylinder head bolts using the following sequence:

- Step A: 72 ft. lbs. (98 Nm)
- Step B: Loosen bolts in the reverse order of tightening.
- Step C: 29 ft. lbs. (39.2 Nm)
- Step D: An additional 90° clockwise
- Step E: An additional 90° clockwise

11. After installing cylinder head, measure distance between front end faces of cylinder block and cylinder head (left and right banks).

➡ **If the measured value is out of the standard, re-install cylinder head.**

12. Installation of the remaining parts is in the reverse order of removal.

V8 Engine

See Figure 40.

1. Remove the engine assembly from the vehicle.
2. Remove the following components and related parts:
   - Drive belt auto tensioner drive belts and idler pulley
   - Thermostat housing and water piping
   - Fuel tube and fuel injector assembly
   - Starter
   - Rocker cover
   - Alternator and alternator bracket
   - Oil pan and oil strainer
3. Remove the crankshaft pulley, front cover, oil pump, and timing chain.
4. Remove the camshaft sprockets and camshafts.
5. Remove the cylinder head bolts in reverse order of the tightening sequence.

To install:

6. Install a new cylinder head gasket.
7. Install the cylinder head. Follow the steps below to tighten the bolts in the numerical order shown.

❈❈ CAUTION

If cylinder head bolts are re-used, check their diameters before installation.

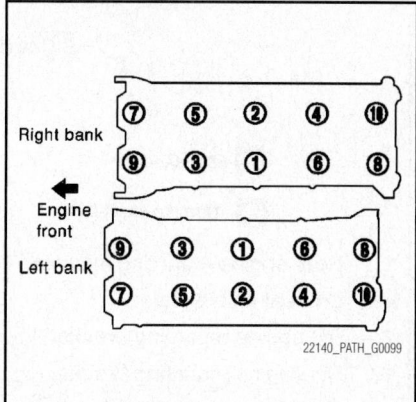

Fig. 40 Tighten the cylinder head bolts in the numerical order shown.

a. a. Apply engine oil to threads and seating surface of the bolts.

b. Tighten cylinder head bolts as follows:

- Step A: 72 ft. lbs. (98.1 Nm)
- Step B: Loosen all bolts in the reverse order of tightening
- Step C: 33 ft. lbs. (44.1 Nm)
- Step D: 60° clockwise
- Step E: An additional 60° clockwise

c. Measure the tightening angle using Tool Number KV10112100 (BT8653A).

✳✳ CAUTION

Measure the tightening angle using Tool. Do not measure visually.

8. Installation of the remaining components is in the reverse order of removal.

ENGINE ASSEMBLY

REMOVAL & INSTALLATION

V6 Engine

See Figures 41 through 43.

✳✳ WARNING

Situate vehicle on a flat and solid surface. Place chocks at front and back of rear wheels. For engines not equipped with engine slingers, attach proper slingers and bolts.

✳✳ CAUTION

Always be careful to work safely, avoid forceful or uninstructed operations. Do not start working until exhaust system and engine coolant

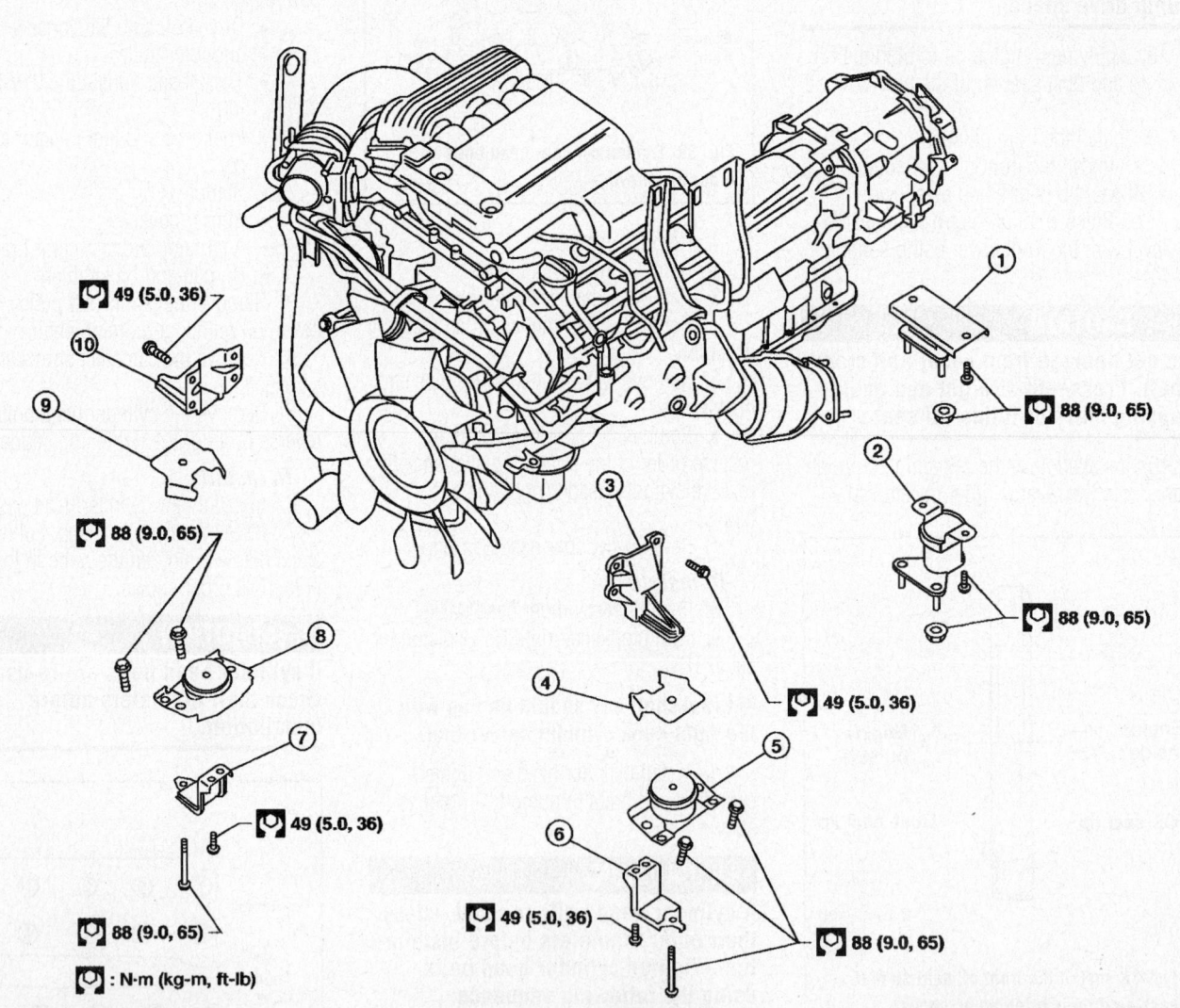

1. Rear engine mounting insulator 4x4
2. Rear engine mounting insulator 4x2
3. LH engine mounting bracket (upper)
4. LH heat shield plate
5. LH engine mounting insulator
6. LH engine mounting bracket (lower)
7. RH engine mounting bracket (lower)
8. RH engine mounting insulator (upper)
9. RH heat shield plate
10. RH engine mounting bracket (upper)

🔧 : N·m (kg-m, ft-lb)

Fig. 41 Exploded view of engine mounting components—V6 Engine

22140_PATH_G0023

are cooled sufficiently. If items or work required are not covered by the engine section, refer to the applicable sections. Always use the support point specified for lifting. Use either 2-point lift type or separate type lift. If board-on type is used for unavoidable reasons, support at the rear axle jacking point with transmission jack or similar tool before starting work, in preparation for the backward shift of center of gravity.

1. Drain engine coolant.

   a. 1. Turn ignition switch ON and set temperature control lever all the way to HOT position or the highest temperature position. Wait 10 seconds and turn ignition switch OFF.

   b. Remove the engine front undercover.

   c. Open the radiator drain plug at the bottom of the radiator, and remove the reservoir cap.

➡**This is the only step required when partially draining the cooling system (radiator only).**

   d. When draining all of the coolant in the system for engine removal or repair, it is necessary to drain the cylinder block.

   e. Remove the water drain plugs (A, B, C, D), and block heater (D) if equipped, to drain the cylinder block.

➡**For Canada, the water drain plug (D)**

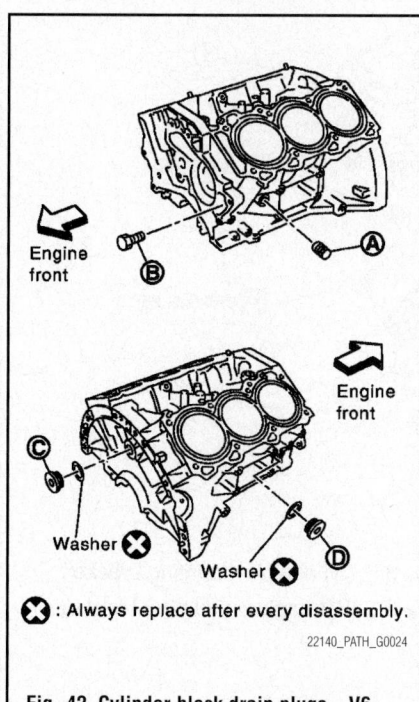

Fig. 42 Cylinder block drain plugs—V6

as shown, is not a water drain plug but a block heater.

2. Remove the reservoir tank to drain the engine coolant, then clean the reservoir tank before installing it.

3. Check the drained coolant for contaminants such as rust, corrosion or discoloration. If the coolant is contaminated, flush the engine cooling system.

4. Partially drain A/T fluid.

5. Release fuel pressure.

6. Remove the engine hood.

   a. Support the hood striker with suitable tool to prevent it from falling.

   b. Remove the hinge nuts from the hood to remove the hood assembly.

7. Remove engine room cover.

8. Remove the air duct and air cleaner case assembly.

9. Disconnect vacuum hose between vehicle and engine and set it aside.

10. Remove the radiator assembly and hoses.

11. Remove the drive belts.

12. Remove the engine cooling fan.

13. Disconnect the engine room harness from the engine side and set it aside for easier work.

14. Disconnect the engine harness grounds.

15. Disconnect the reservoir tank for power steering from engine and move it aside for easier work.

16. Disconnect power steering oil pump from engine. Move it from its location and secure with a rope for easier work.

17. Remove the A/C compressor bolts and set aside.

18. Disconnect brake booster vacuum line.

19. Disconnect EVAP line.

20. Disconnect the fuel hose at the engine side connection.

21. Disconnect the heater hoses at cowl, and install plugs to avoid leakage of engine coolant.

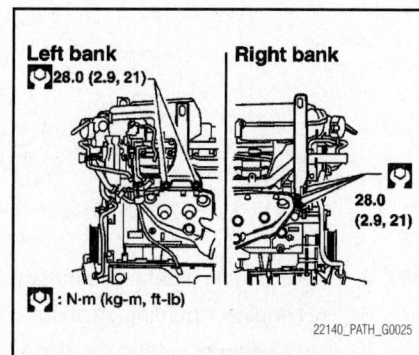

Fig. 43 Install engine slingers into left bank and right bank

22. Remove the A/T oil level indicator and indicator tube.

23. Remove front final drive assembly (4x4 only).

24. Remove three way catalyst.

25. Install engine slingers into left bank and right bank.

26. Remove transmission.

27. Lift with hoist and secure the engine in position.

28. Remove engine assembly from vehicle, avoiding interference with vehicle body.

✷✷ **CAUTION**

Before and during this lifting, always check if any harnesses are left connected.

29. Remove the parts that may restrict installation of engine to engine stand.

➡**The procedure is described assuming that you use an engine stand holding the surface, to which transmission is installed.**

30. Remove drive plate.

   a. Holding crankshaft pulley bolts, lock crankshaft to remove drive plate bolts.

   b. Loosen bolts diagonally.

✷✷ **CAUTION**

Be careful not to damage drive plate. Especially avoid deforming and damaging of signal plate teeth (circumference position). Place the drive plate with signal plate surface facing other than downward. Keep magnetic materials away from signal plate.

✷✷ **CAUTION**

Use an engine stand that has a load capacity [approximately 240kg (529 lb) or more] large enough for supporting the engine weight. If the load capacity of the stand is not adequate, remove the following parts beforehand to reduce the potential risk of overturning the stand:

- Remove fuel tube and fuel injector assembly.
- Remove intake manifold.
- Remove rocker cover.
- Other removable brackets.

✷✷ **CAUTION**

Before removing the hanging chains, make sure the engine stand is stable and there is no risk of overturning.

31. Remove alternator.
32. Remove engine mounting insulator bracket (upper).

To install:

33. Install engine mounting insulator bracket.
34. Install alternator.
35. Install any of the following if removed:
    - Fuel tube and fuel injector assembly.
    - Intake manifold.
    - Rocker cover.
    - Other removable brackets.
36. Install drive plate.
37. Install any removed parts that restricted installation of engine stand.
38. Install engine assembly.
39. Lower with hoist into vehicle.
40. Install transmission.
41. Remove engine slingers.
42. Install three way catalyst.
43. Install front final drive assembly (4x4).
44. Install the A/T oil level indicator and indicator tube.
45. Connect the heater hoses.
46. Connect the fuel hose.
47. Connect EVAP line.

48. Connect brake booster vacuum line.
49. Install the A/C compressor.
50. Connect power steering oil pump.
51. Connect the reservoir tank for power steering.
52. Connect the engine harness grounds.
53. Connect the engine room harness.
54. Install the engine cooling fan.
55. Install the drive belt.
56. Install the radiator assembly and hoses.
57. Connect the vacuum hose between vehicle and engine.
58. Install the air duct and air cleaner case assembly.
59. Install engine room cover.
60. Install the engine hood.
61. Fill A/T fluid.
62. Install the reservoir tank.
63. Install the water drain plugs and block heater, if equipped.
64. Close the radiator drain plug at the bottom of the radiator.
65. Install the engine front undercover.
66. Fill engine coolant.
67. Check the following after installation:
    - Before starting engine, check the levels of engine coolant, engine oil

and working fluid. If less than required quantity, fill to the specified level.
- Use procedure below to check for fuel leakage.
- Turn ignition switch ON (with engine stopped). With fuel pressure applied to fuel piping, check for fuel leakage at connection points.
- Start engine. With engine speed increased, check again for fuel leakage at connection points.
- Run engine to check for unusual noise and vibration.
- Warm up engine thoroughly to make sure there is no leakage of engine coolant, engine oil, working fluid, fuel and exhaust gas.
- Bleed air from passages in pipes and tubes of applicable lines, such as in cooling system.
- After cooling down engine, again check amounts of engine coolant, engine oil and working fluid. Refill to specified level, if necessary.

V8 Engine

See Figure 44.

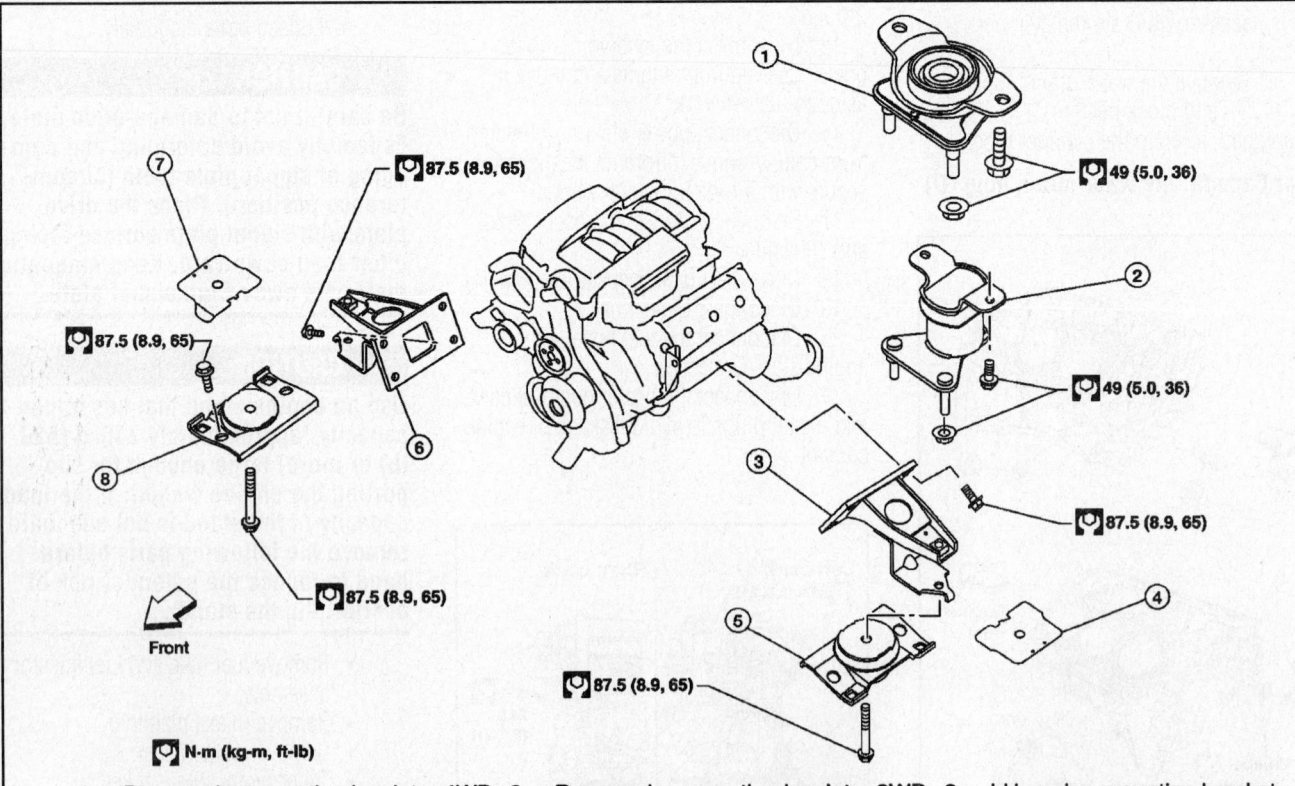

87.5 (8.9, 65)

49 (5.0, 36)

87.5 (8.9, 65)

49 (5.0, 36)

87.5 (8.9, 65)

87.5 (8.9, 65)

87.5 (8.9, 65)

Front

N·m (kg-m, ft-lb)

1. Rear engine mounting insulator 4WD
2. Rear engine mounting insulator 2WD
3. LH engine mounting bracket
4. LH heat shield plate
5. LH engine mounting insulator
6. RH engine mounting bracket
7. RH heat shield plate
8. RH engine mounting insulator

22140_PATH_G0100

Fig. 44 Exploded view of engine mounting components—V8

✳✳ WARNING

Situate vehicle on a flat and solid surface. Place chocks at front and back of rear wheels.

➡For engines not equipped with engine slingers, attach proper slingers and bolts.

✳✳ CAUTION

Heed the following cautions:

- Always be careful to work safely, avoid forceful or uninstructed operations.
- Do not start working until exhaust system and engine coolant are cooled sufficiently.
- If items or work required are not covered by the engine section, follow the applicable sections.
- Always use the support point specified for lifting.
- Use either 2-point lift type or separate type lift. If board-on type is used for unavoidable reasons, support at the rear axle jacking point with transmission jack or similar tool before starting work, in preparation for the backward shift of center of gravity.

1. Remove air dam.
2. Remove engine undercover.
3. Drain the engine coolant.
4. Partially drain the A/T fluid.
5. Drain the transfer fluid (4WD models).
6. Remove the engine hood.
   a. Support the hood striker with suitable tool to prevent it from falling.
   b. Remove the hinge nuts from the hood to remove the hood assembly.

✳✳ CAUTION

Operate with two workers, because of its heavy weight.

7. Release the fuel pressure.
8. Remove the engine room cover.
9. Remove the air duct and air cleaner case assembly.
10. Disconnect the vacuum hose between the vehicle and engine and set it aside.
11. Remove the radiator assembly and hoses.
12. Remove the drive belts.
13. Remove the fan blade.
14. Disconnect the engine room harness from the fuse box and set it aside.
15. Disconnect the ECM.
16. Disconnect the engine room harness from the engine side and set it aside.

17. Disconnect the engine harness grounds.
18. Disconnect the power steering reservoir tank from the engine and move it aside.
19. Disconnect the power steering oil pump from the engine.
20. Move it aside and secure it using suitable wire or rope.
21. Remove the A/C compressor bolts and set the compressor aside.
22. Disconnect the brake booster vacuum line.
23. Disconnect the EVAP line.
24. Disconnect the fuel hose at the engine side connection.
25. Disconnect the heater hoses at the cowl, and install plugs to avoid leakage of engine coolant.
26. Remove the A/T oil level indicator and indicator tube upper bolts.
27. Remove the front final drive assembly (4WD models).
28. Remove the exhaust manifolds.
29. Remove the automatic transmission.
30. Install the engine slingers into the left bank cylinder head (A) and right bank cylinder head (B).
31. Lift using a hoist and secure the engine in position.
32. Remove engine mounting insulator bolts.
33. Remove the engine assembly from the vehicle, avoid interference with the vehicle body.

✳✳ CAUTION

Before and during lifting, always check if any harnesses are left connected.

34. Remove the parts that may restrict installation of the engine to the engine stand.

➡This procedure is described assuming that you use an engine stand mounting to the surface to which the transmission mounts.

   a. Remove the drive plate.
   b. Holding the crankshaft pulley bolt, lock the crankshaft to remove the drive plate bolts.
   c. Loosen the bolts diagonally.

✳✳ CAUTION

Be careful not to damage the drive plate. Especially avoid deforming and damaging of the signal plate teeth (circumference position).

   d. Place the drive plate with the signal plate surface facing other than downward.

   e. Keep magnetic materials away from the signal plate.

✳✳ CAUTION

Use an engine stand that has a load capacity of approximately 529 lbs. (240 kg) or more, large enough for supporting the engine weight.

35. If the load capacity of the stand is not adequate, remove the following parts beforehand to reduce the potential risk of overturning the stand.
   a. Remove the fuel tube and fuel injector assembly.
   b. Remove the intake manifold.
   c. Remove the ignition coil.
   d. Remove the rocker cover.
   e. Other removable brackets.

✳✳ CAUTION

Before removing the hanging chains, make sure the engine stand is stable and there is no risk of overturning.

36. Remove the alternator.
37. Remove the engine mounting insulator and bracket.

To install:

38. Install the engine mounting insulator and bracket.
39. Install the alternator.
   a. Other removable brackets.
   b. Install the rocker cover.
   c. Install the ignition coil.
   d. Install the intake manifold.
   e. Install the fuel tube and fuel injector assembly.
   f. Install the drive plate.
40. Install the parts that may restrict installation of the engine to the engine stand.
41. Install the engine assembly from the vehicle, avoid interference with the vehicle body.
42. Install engine mounting insulator bolts.
43. Lift using a hoist and secure the engine in position.
44. Install the engine slingers into the left bank cylinder head (A) and right bank cylinder head (B).
45. Install the automatic transmission.
46. Install the exhaust manifolds.
47. Install the front final drive assembly (4WD models).
48. Install the A/T oil level indicator and indicator tube upper bolts.
49. Connect the heater hoses at the cowl, and install plugs to avoid leakage of engine coolant.
50. Connect the fuel hose at the engine side connection.

51. Connect the EVAP line.
52. Connect the brake booster vacuum line.
53. Install the A/C compressor bolts and set the compressor aside.
54. Connect the power steering oil pump from the engine.
55. Connect the power steering reservoir tank from the engine and move it aside.
56. Connect the engine harness grounds.
57. Connect the engine room harness from the engine side and set it aside.
58. Connect the ECM.
59. Connect the engine room harness from the fuse box and set it aside.
60. Install the fan blade.
61. Install the drive belts.
62. Install the radiator assembly and hoses.
63. Connect the vacuum hose between the vehicle and engine and set it aside.
64. Install the air duct and air cleaner case assembly.
65. Install the engine room cover.
66. Install the engine hood.

67. Fill the transfer fluid (4WD models).
68. Fill the A/T fluid.
69. Fill the engine coolant.
70. Install engine undercover.
71. Install air dam.

EXHAUST MANIFOLD

REMOVAL & INSTALLATION

V6 Engine

Left Bank

See Figures 45 through 49.

1. Remove air cleaner case and air duct.
2. Remove engine undercover.
3. Disconnect harness connector and remove heated oxygen sensor 2 on both banks.

✳✳ CAUTION

Be careful not to damage heated oxygen sensor 2. Discard any heated oxygen sensor 2 which has been dropped from a height of more than 20 inches (0.5 m) onto a hard surface such as a concrete floor; replace with a new sensor.

4. Remove center exhaust tube, main muffler and left front exhaust tube.
5. Remove exhaust manifold cover (left bank).
6. Disconnect harness connector and remove air fuel ratio sensor 1 (left bank).

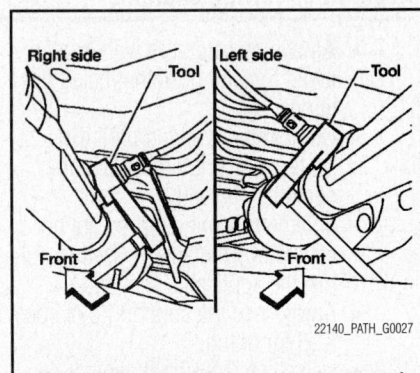

Fig. 46 Disconnect harness connector and remove heated oxygen sensor 2

---

1. **Exhaust manifold cover (right bank)**   2. **Exhaust manifold (right bank)**   3. **Gasket**
4. **Exhaust manifold cover (left bank)**   5. **Exhaust manifold (left bank)**

5.8 (0.59, 51)

30.5 (3.1, 22)

30.5 (3.1, 22)

5.8 (0.59, 51)

✗ : Always replace after every disassembly.
🔧 : N•m (kg-m, in-lb)
🔧 : N•m (kg-m, ft-lb)

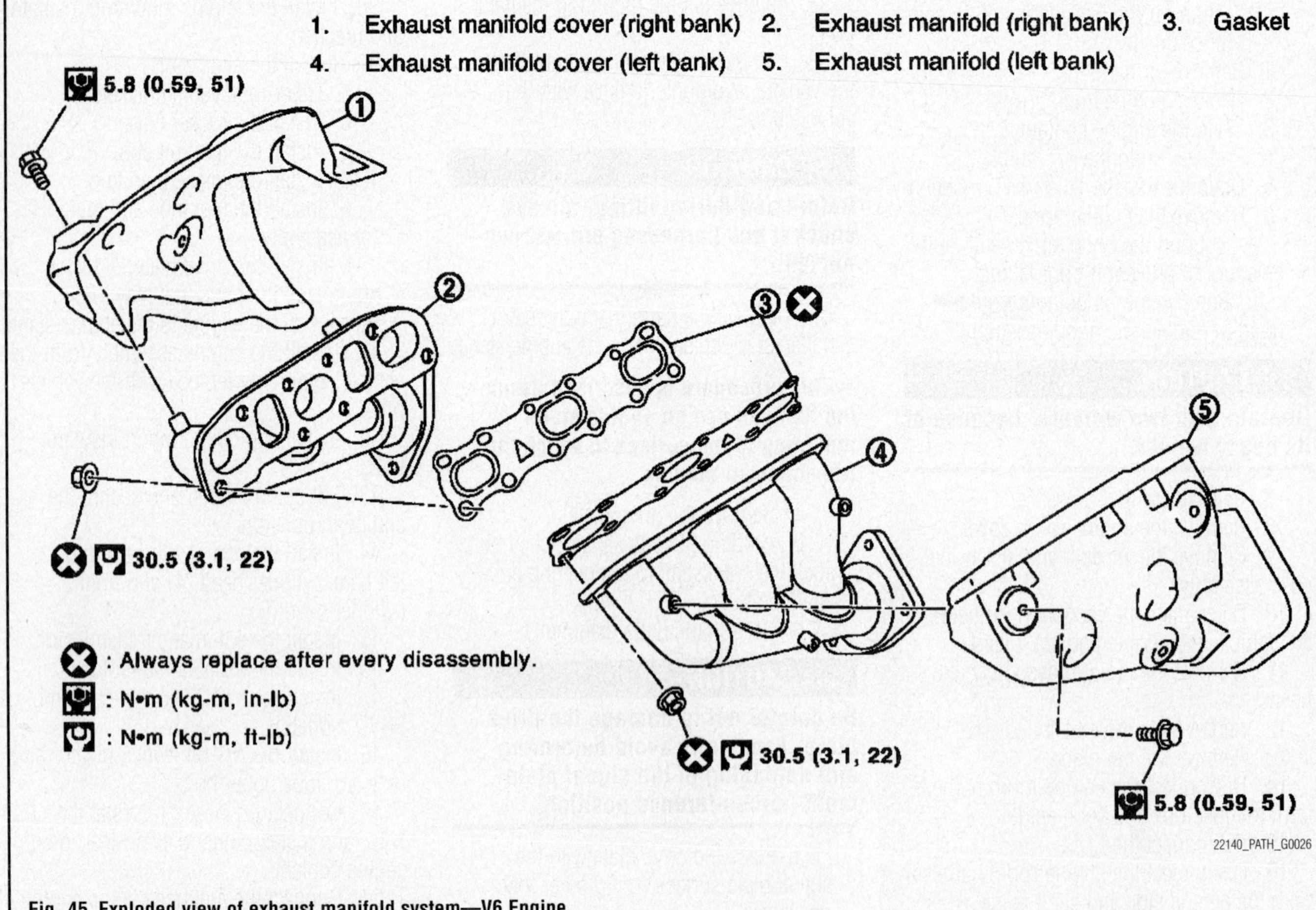

Fig. 45 Exploded view of exhaust manifold system—V6 Engine

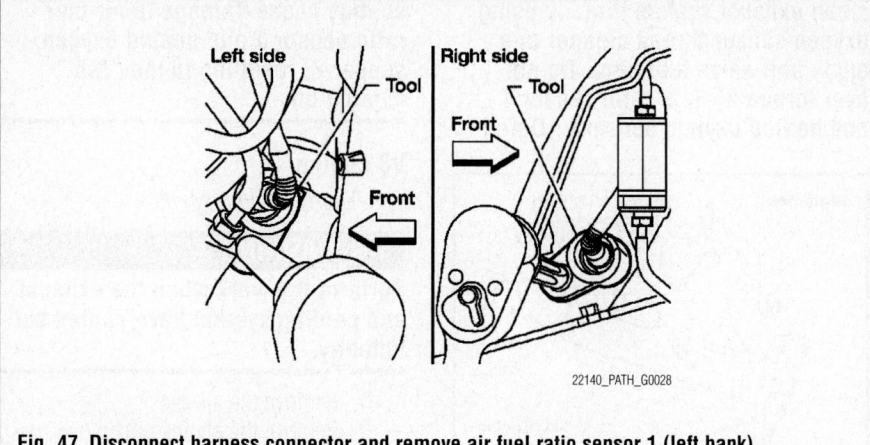

Fig. 47 Disconnect harness connector and remove air fuel ratio sensor 1 (left bank)

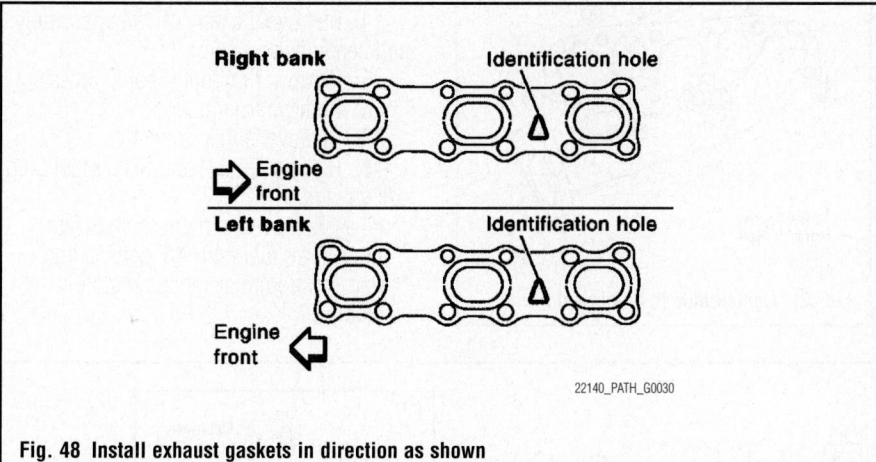

Fig. 48 Install exhaust gaskets in direction as shown

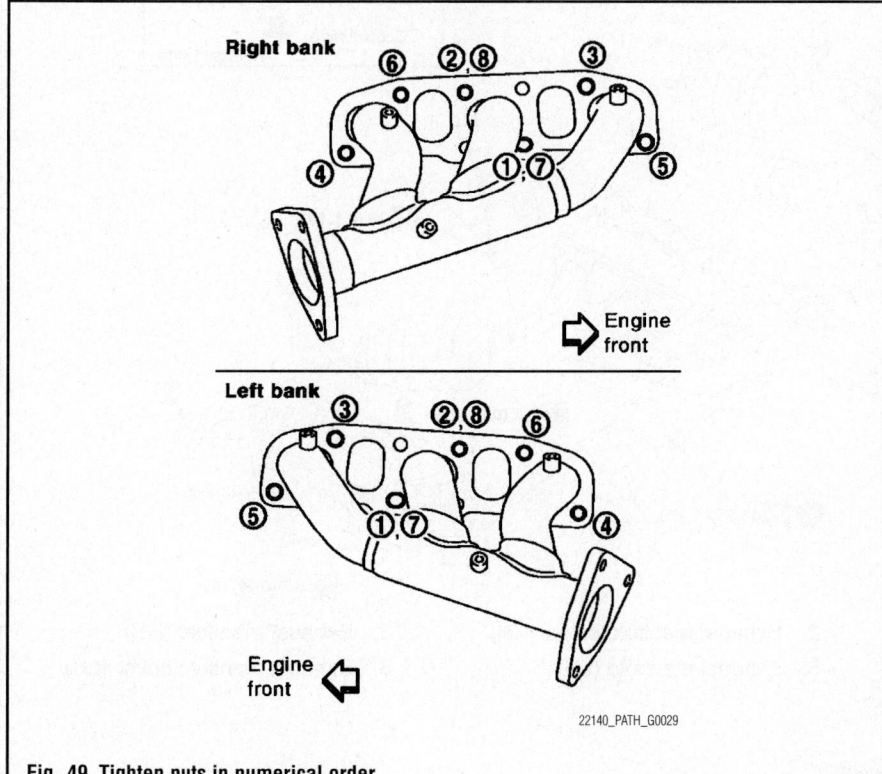

Fig. 49 Tighten nuts in numerical order

❊❊ CAUTION

Be careful not to damage air fuel ratio sensor 1. Discard any air fuel ratio sensor 1 which has been dropped from a height of more than 20 inches (0.5 m) onto a hard surface such as a concrete floor; replace with a new sensor.

   7. Remove three way catalyst (left bank).
   8. Loosen nuts in reverse order of the tightening sequence.

➡**Disregard the numerical order No. 7 and 8 in removal.**

   9. Remove gaskets.

   To install:
   10. Installation is in the reverse order of removal. Note the following:
- Install exhaust gaskets in direction as shown.
- If stud bolts were removed, install them and tighten to 11 ft. lbs. (14.7 Nm).
- Install exhaust manifold and tighten nuts in numerical order as shown.

➡**Tighten nuts No. 1 and 2 in two steps. The numerical order No. 7 and 8 shown second step.**

❊❊ CAUTION

Before installing a new air fuel ratio sensor 1 and heated oxygen sensor 2, clean exhaust system threads using oxygen sensor thread cleaner and apply anti-seize lubricant. Do not over torque air fuel ratio sensor 1 and heated oxygen sensor 2. Doing so may cause damage to air fuel ratio sensor 1 and heated oxygen sensor 2, resulting in the "MIL" coming on.

Right Bank

See Figures 50 and 51.

   1. Remove engine assembly.
   2. Loosen nuts in reverse order of the tightening sequence.

➡**Disregard the numerical order No. 7 and 8 in removal.**

   3. Remove gaskets.

   To install:
   4. Installation is in the reverse order of removal. Note the following:
- Install exhaust gaskets in direction as shown.

- If stud bolts were removed, install them and tighten to 11 ft. lbs. (14.7 Nm).
- Install exhaust manifold and tighten nuts in numerical order as shown.

➡Tighten nuts No. 1 and 2 in two steps. The numerical order No. 7 and 8 shown second step.

✳✳ CAUTION

Before installing a new air fuel ratio sensor 1 and heated oxygen sensor 2,

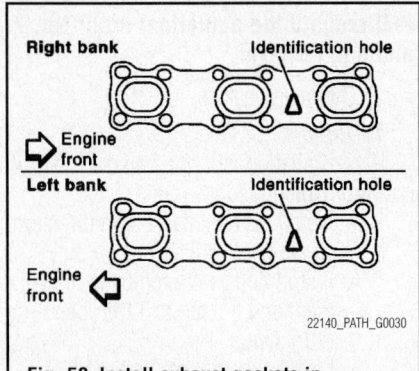

Fig. 50 Install exhaust gaskets in direction as shown

clean exhaust system threads using oxygen sensor thread cleaner and apply anti-seize lubricant. Do not over torque air fuel ratio sensor 1 and heated oxygen sensor 2. Doing

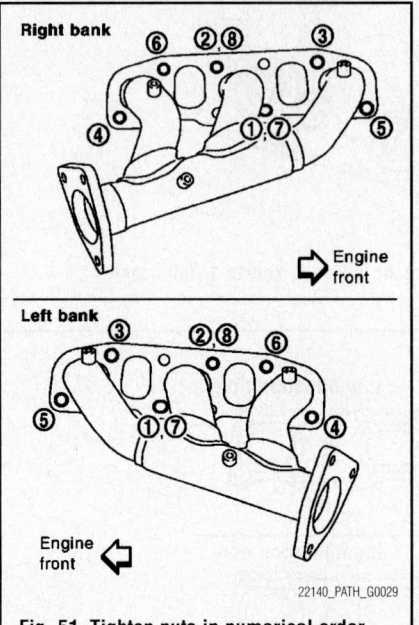

Fig. 51 Tighten nuts in numerical order

so may cause damage to air fuel ratio sensor 1 and heated oxygen sensor 2, resulting in the "MIL" coming on.

V8 Engine

See Figures 52 through 55.

✳✳ WARNING

Perform the work when the exhaust and cooling system have cooled sufficiently.

1. Remove the air dam.
2. Remove the engine undercover.
3. Remove front final drive assembly (4WD models).
4. Remove the main muffler assembly and center exhaust tube.
5. Remove the front exhaust tubes.
6. Remove front tires.
7. Remove fender protectors.
8. Remove the LH and RH air fuel ratio A/F sensors.

   a. Remove the harness connector of each air fuel ratio A/F sensor, and harness from bracket and middle clamp.

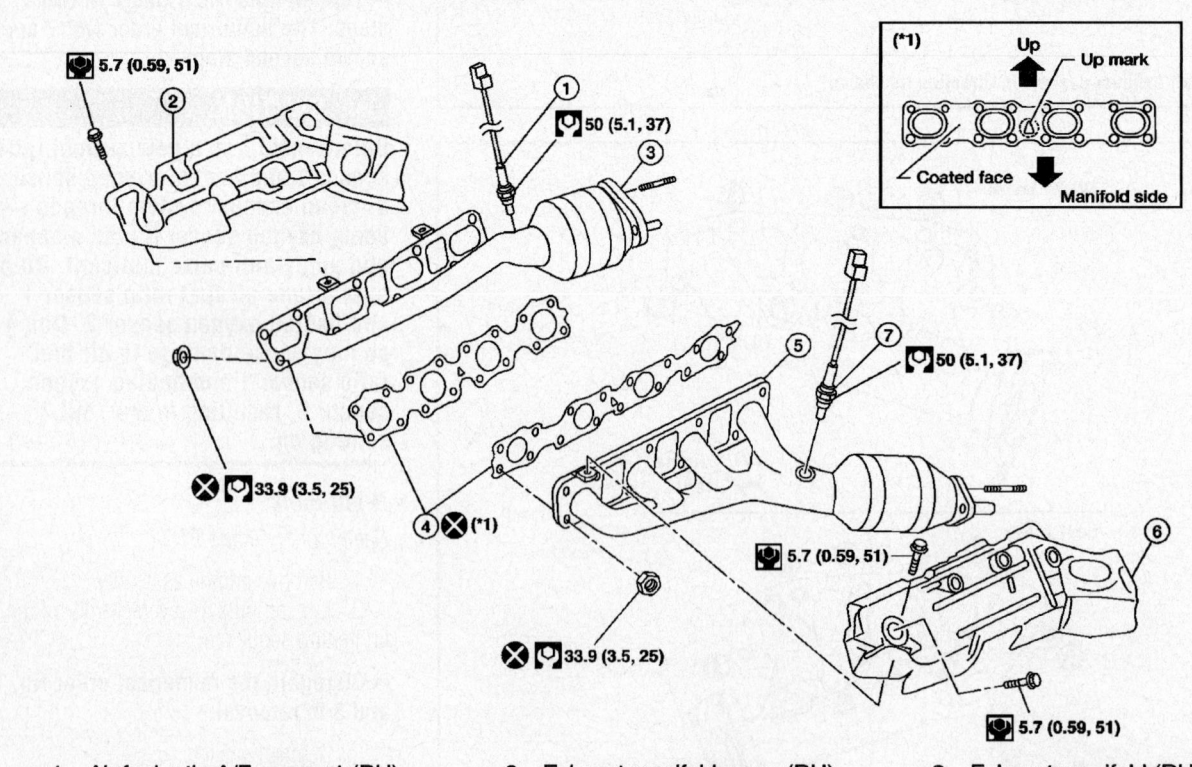

1. Air fuel ratio A/F sensor 1 (RH)
2. Exhaust manifold cover (RH)
3. Exhaust manifold (RH)
4. Gaskets
5. Exhaust manifold (LH)
6. Exhaust manifold cover (LH)
7. Air fuel ratio A/F sensor 1 (LH)

Fig. 52 Exploded view of exhaust manifold components

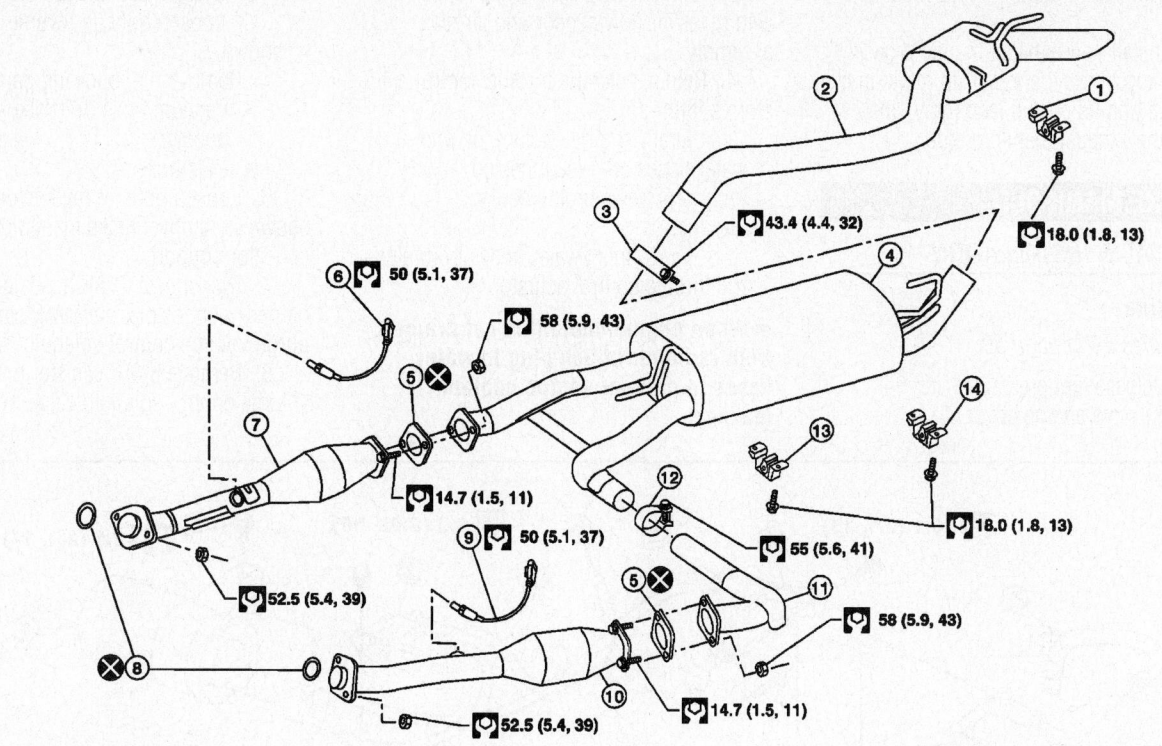

1.	Tailpipe hanger bracket	2.	Tailpipe	3. Clamp
4.	Main muffler	5.	Gasket	6. Heated oxygen sensor 2 (bank 2)
7.	Right front exhaust tube	8.	Ring gasket	9. Heated oxygen sensor 2 (bank 1)
10.	Left front exhaust tube	11.	Center exhaust tube	12. Clamp
13.	Muffler hanger bracket front	14.	Muffler hanger bracket rear	

22140_PATH_G0101

Fig. 53 Exploded view of exhaust system

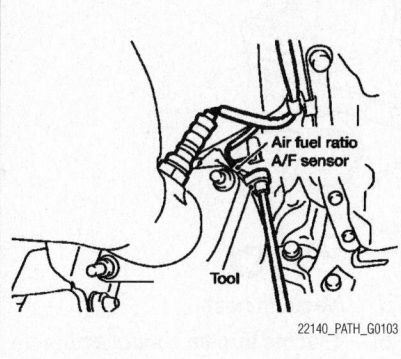

22140_PATH_G0103

Fig. 54 Remove the air fuel ratio A/F sensors from both LH and RH exhaust manifolds.

✳✳ CAUTION

Do not damage the air fuel ratio A/F sensors. Discard any air fuel ratio A/F sensor which has been dropped from a height of more than 19.7 inches (0.5m) onto a hard surface

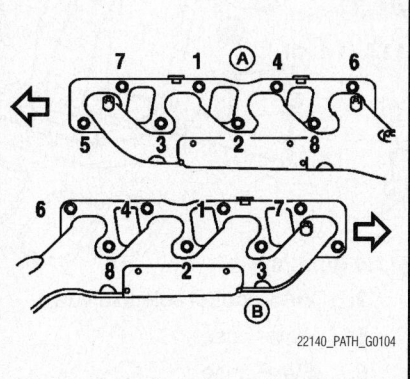

22140_PATH_G0104

Fig. 55 Exhaust manifold tightening sequence

such as a concrete floor. Replace it with a new one.

9. Support the engine using a suitable tool.

10. Remove the exhaust manifold (LH) (A) following the steps below.

a. Remove the engine mounting insulator.

b. Remove the exhaust manifold cover.

c. Remove the engine mounting bracket.

d. Loosen the nuts LH side in reverse order of the tightening sequence.

e. Remove the exhaust manifold (LH).

11. Remove the exhaust manifold (RH) (B) following the steps below.

a. Remove the engine mounting insulator.

b. Remove the exhaust manifold cover.

c. Remove the engine mounting bracket.

d. Remove the oil level gauge guide.

e. Loosen the nuts RH side in reverse order of the tightening sequence.

f. Remove the exhaust manifold (RH).

To install:

12. Installation is in the reverse order of removal.

13. Tighten the bolts in the sequence shown.

14. Install new exhaust manifold gasket with the top of the triangular up mark on it facing up and its coated face (gray side) toward the exhaust manifold side.

INTAKE MANIFOLD

REMOVAL & INSTALLATION

V6 Engine

See Figures 56 through 59.

1. Release fuel pressure.
2. Remove engine cover.

3. Remove air cleaner case (upper) with mass air flow sensor and air duct assembly.

4. Remove electric throttle control actuator as follows:

a. Drain engine coolant, or when water hoses are disconnected, attach plug to prevent engine coolant leakage.

b. Disconnect water hoses from electric throttle control actuator.

➡ **When engine coolant is not drained from radiator, attach plug to water hoses to prevent engine coolant leakage.**

c. Disconnect harness connector.

d. Loosen bolts in reverse order as shown.

5. Remove the following parts:
• Vacuum hose (to brake booster)
• PCV hose

6. Loosen bolts in reverse order as shown to remove intake manifold collector support.

7. Disconnect EVAP hoses and harness connector from EVAP canister purge volume control solenoid valve.

8. Remove EVAP canister purge volume control solenoid valve.

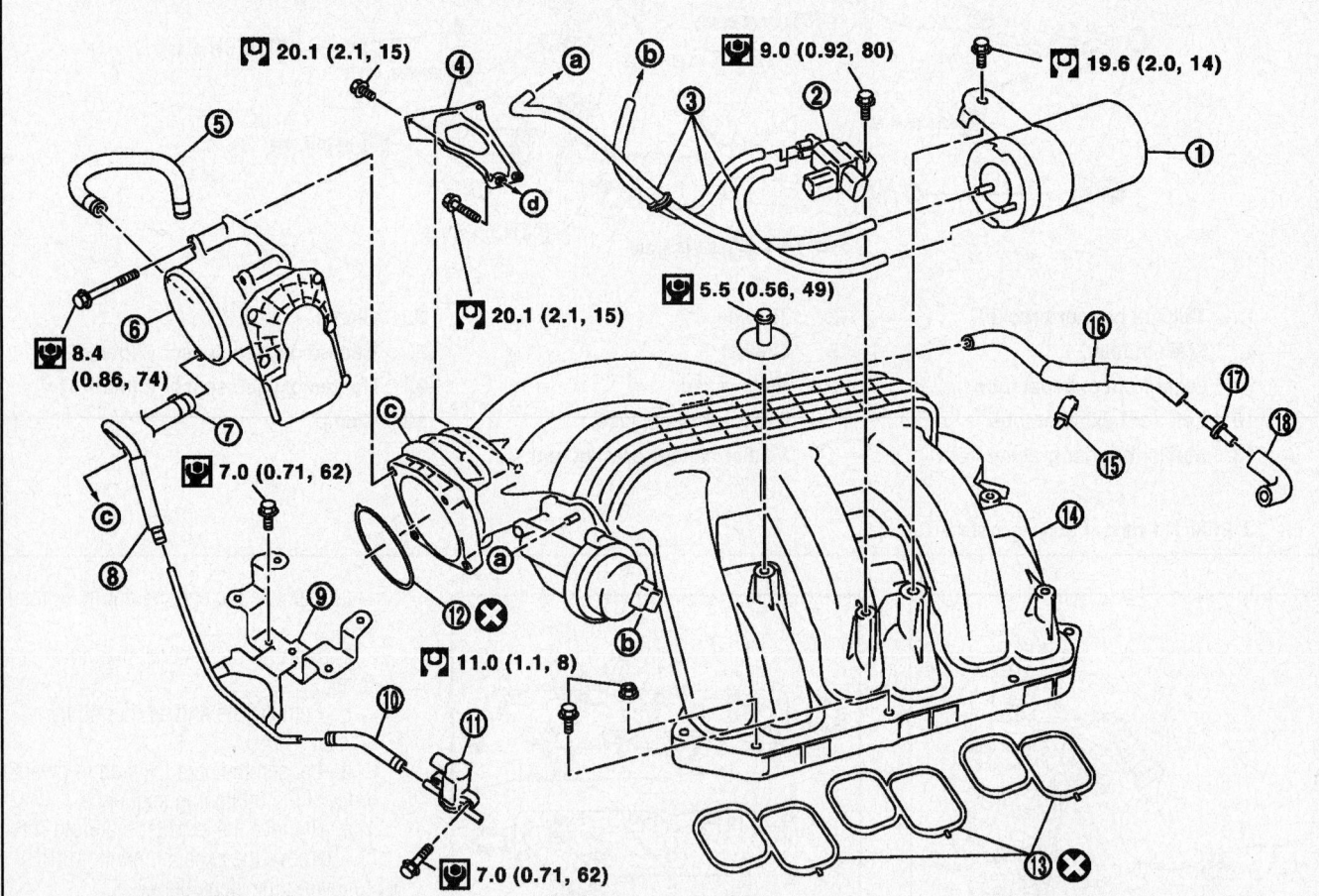

1.	Vacuum tank	2.	VIAS control solenoid valve	3.	Vacuum hose
4.	Intake manifold collector support	5.	Water hose	6.	Electric throttle control actuator
7.	Water hose	8.	EVAP hose	9.	Bracket
10.	EVAP hose	11.	EVAP canister purge volume control solenoid valve	12.	Gasket
13.	Gasket	14.	Intake manifold collector	15.	Clip
16.	PCV hose	17.	Connector	18.	PCV hose
a.	To intake manifol collector	b.	To power valve	c.	To throttle body
d.	To cylinder head (RH bank)				

22140_PATH_G0031

Fig. 56 Exploded view of intake manifold—V6 Engine

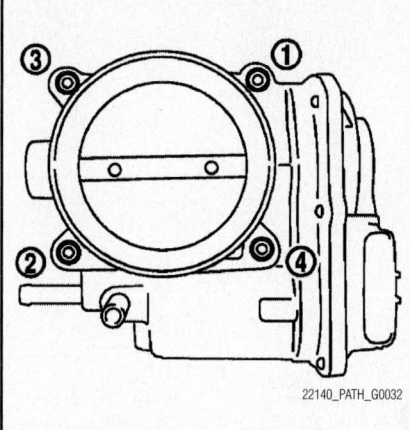

Fig. 57 Loosen bolts in reverse order as shown

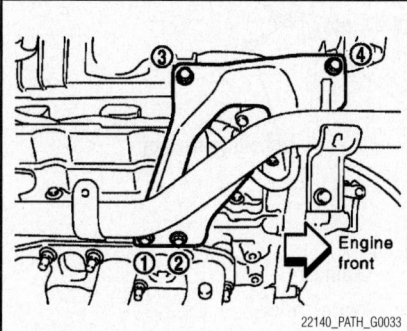

Fig. 58 Loosen bolts in reverse order as shown to remove intake manifold collector support

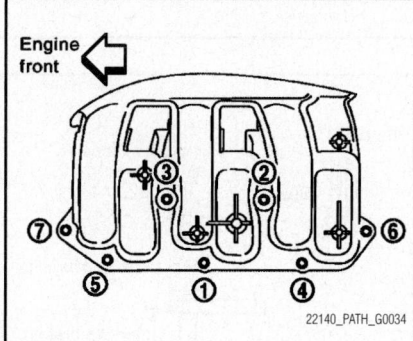

Fig. 59 Tighten nuts and bolts in order as shown

9. Remove VIAS control solenoid valve and vacuum tank.

10. Loosen nuts and bolts in reverse order as shown, and remove intake manifold collector.

11. Remove fuel tube and fuel injector assembly.

12. Loosen nuts and bolts in reverse order of the tightening sequence.

13. Remove gaskets.

To install:

14. Installation is in the reverse order of removal.

15. Note the following:
- If stud bolts were removed from cylinder head, install them and tighten to 8 ft. lbs. (11 Nm).
- Tighten all nuts and bolts to the specified torque in two or more steps in numerical order as shown.
- 1st step: 5 ft. lbs. (7.4 Nm)
- 2nd step and after: 21 ft. lbs. (29.0 Nm)

V8 Engine

See Figures 60 through 64.

1. Remove the air dam.
2. Remove the engine undercover.
3. Partially drain the engine coolant.

☀ WARNING

To avoid the danger of being scalded, never drain the engine coolant when the engine is hot.

4. Remove the engine room cover.
5. Release the fuel pressure.
6. Remove the air duct and resonator assembly.
7. Disconnect the fuel tube quick connector on the engine side.
   a. Remove quick connector cap (engine side only).
   b. With the sleeve side of Tool number 16641 6N210 (J45488) facing quick connector, install Tool onto fuel tube.
   c. Insert Tool into quick connector until sleeve contacts and goes no further. Hold the Tool in that position.

☀ CAUTION

Inserting the Tool hard will not disconnect quick connector. Hold Tool where it contacts and goes no further.

   d. Draw and pull out quick connector straight from fuel tube.

☀ CAUTION

Heed the following cautions:

- Pull quick connector holding "A" position in illustration.
- Do not pull with lateral force applied. O-ring inside quick connector may be damaged.
- Prepare container and cloth beforehand as fuel will leak out.
- Avoid fire and sparks.

- Be sure to cover openings of disconnected pipes with plug or plastic bag to avoid fuel leakage and entry of foreign materials.

8. Remove or disconnect harnesses, brackets, vacuum hose, vacuum gallery and PCV hose and tube from intake manifold.

9. Remove electric throttle control actuator by loosening bolts diagonally.

☀ CAUTION

Handle carefully to avoid any damage to the electric throttle control actuator. Do not disassemble.

10. Remove the fuel injectors and fuel tube assembly.

11. Loosen the bolts in reverse order of the tightening sequence.

12. Remove the intake manifold.

☀ CAUTION

Cover engine openings to avoid entry of foreign materials. Clean all gasket mating surfaces, do not reuse gaskets.

To install:

13. Installation is in the reverse order of removal.

14. Tighten the intake manifold bolts in numerical order as shown.

15. Install the EVAP canister purge control solenoid valve connector with it facing front of engine.

16. Tighten the electronic throttle control actuator bolts of the electric throttle control actuator equally and diagonally in several steps.

17. Install the water hose so that its overlap width for connection is between 1.06 inches (27 mm) and 1.26 inches (32 mm) (target: 1.06 inches or 27 mm).

18. Install quick connector as follows (the steps are the same for quick connectors on both engine side and vehicle side except for the quick connector cap).

19. Make sure no foreign substances are deposited in and around tube and quick connector, and they are not damaged.

20. Thinly apply new engine oil around the fuel tube from tip end to the spool end.

21. Align center to insert quick connector straight into fuel tube.
   a. Insert until the paint mark for engagement identification (white) goes completely inside quick connector so that you cannot see it from the straight side of the connected part. Use a mirror to check this where it is not possible to view directly from the straight side, such as quick connector on vehicle side.

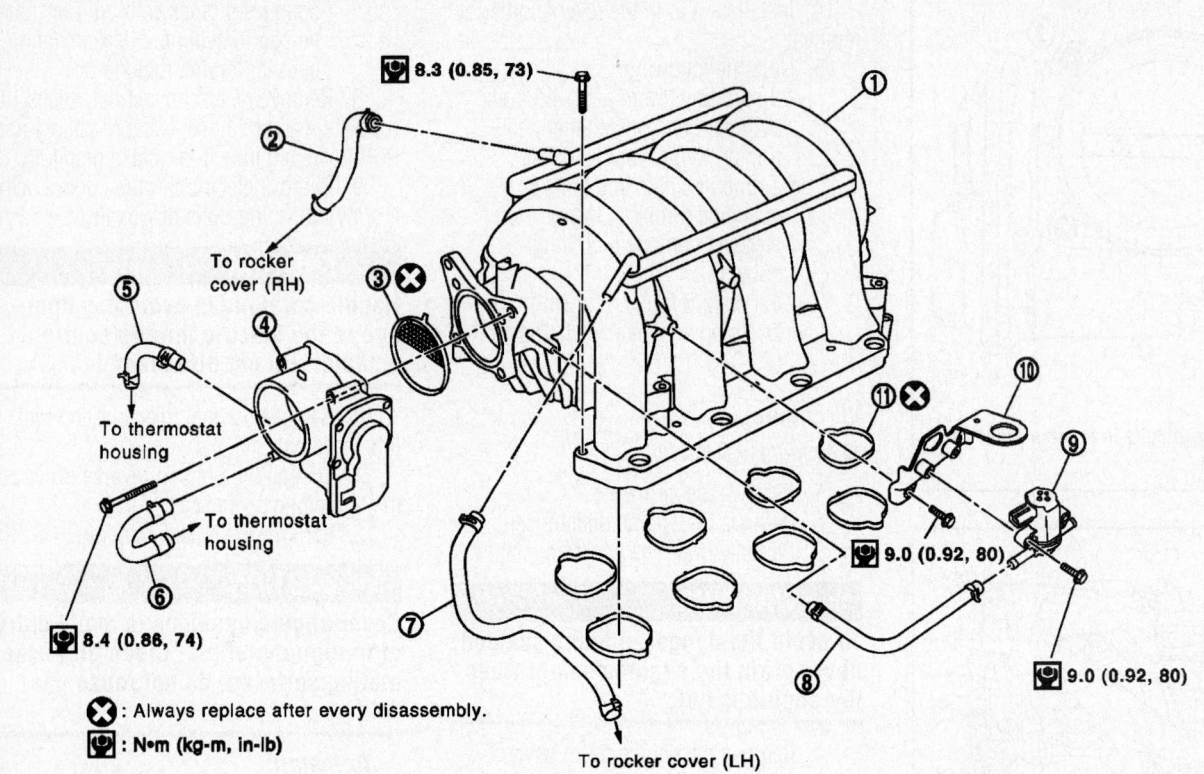

⊞ 8.3 (0.85, 73)

② To rocker cover (RH)

③ ✕

④

⑤

To thermostat housing

→ To thermostat housing

⑥

⊞ 8.4 (0.86, 74)

①

⑪ ✕

⑩

⑨

⊞ 9.0 (0.92, 80)

⊞ 9.0 (0.92, 80)

⑧

⑦

To rocker cover (LH)

✕ : Always replace after every disassembly.

⊞ : N•m (kg-m, in-lb)

1.	Intake manifold	2.	PCV hose	3.	Gasket
4.	Electric throttle control actuator	5.	Water hose	6.	Water hose
7.	PCV hose	8.	EVAP hose	9.	EVAP canister purge control solenoid valve
10.	Bracket	11.	Gasket		

22140_PATH_G0105

Fig. 60 Exploded view of intake manifold—V8 Engine

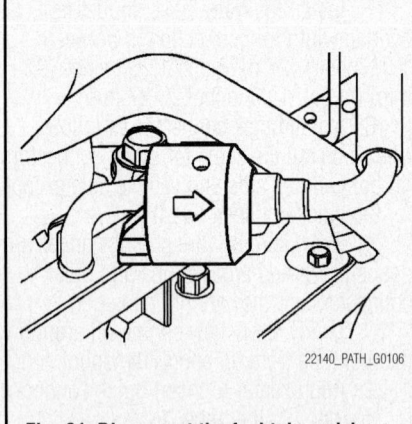

22140_PATH_G0106

Fig. 61 Disconnect the fuel tube quick connector on the engine side.

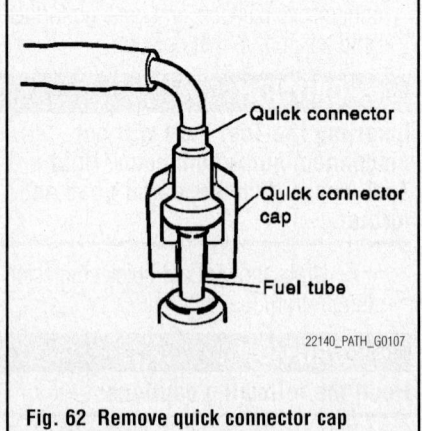

Quick connector

Quick connector cap

Fuel tube

22140_PATH_G0107

Fig. 62 Remove quick connector cap (engine side only)

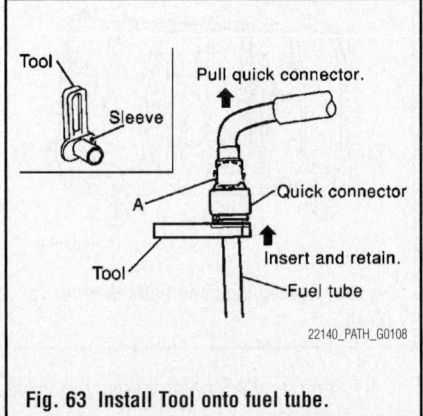

Tool

Sleeve

Pull quick connector.

A

Quick connector

Insert and retain.

Tool

Fuel tube

22140_PATH_G0108

Fig. 63 Install Tool onto fuel tube.

b. Insert fuel tube into quick connector until top spool is completely inside quick connector, and 2nd level spool exposes right below quick connector on engine side.

❊❊ CAUTION

Hold "A" position in illustration when inserting fuel tube into quick connector. Carefully align center to avoid inclined insertion to prevent

damage to O-ring inside quick connector. Insert until you hear a "click" sound and actually feel the engagement. To avoid misidentification of engagement with a similar sound, be sure to perform the next step.

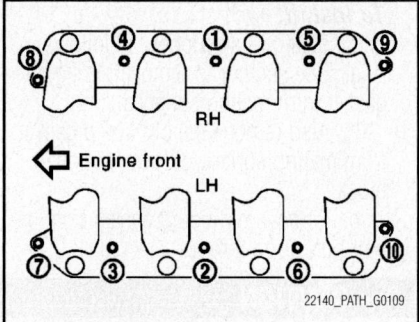

Fig. 64 Intake manifold tightening sequence

22. Pull quick connector by hand holding "A" position. Make sure it is completely engaged (connected) so that it does not come out from fuel tube.

➡**Recommended pulling force is 50 N (5.1 kg, 11.2 lb).**

23. Install the quick connector cap on the quick connector joint (on engine side only).

24. Install the fuel hose and tube to hose clamps.

25. Refill the engine coolant.

OIL PAN

REMOVAL & INSTALLATION

V6 Engine

Lower Oil Pan

See Figures 65 through 69.

✳✳ WARNING

To avoid the danger of being scalded, do not drain engine oil when engine is hot.

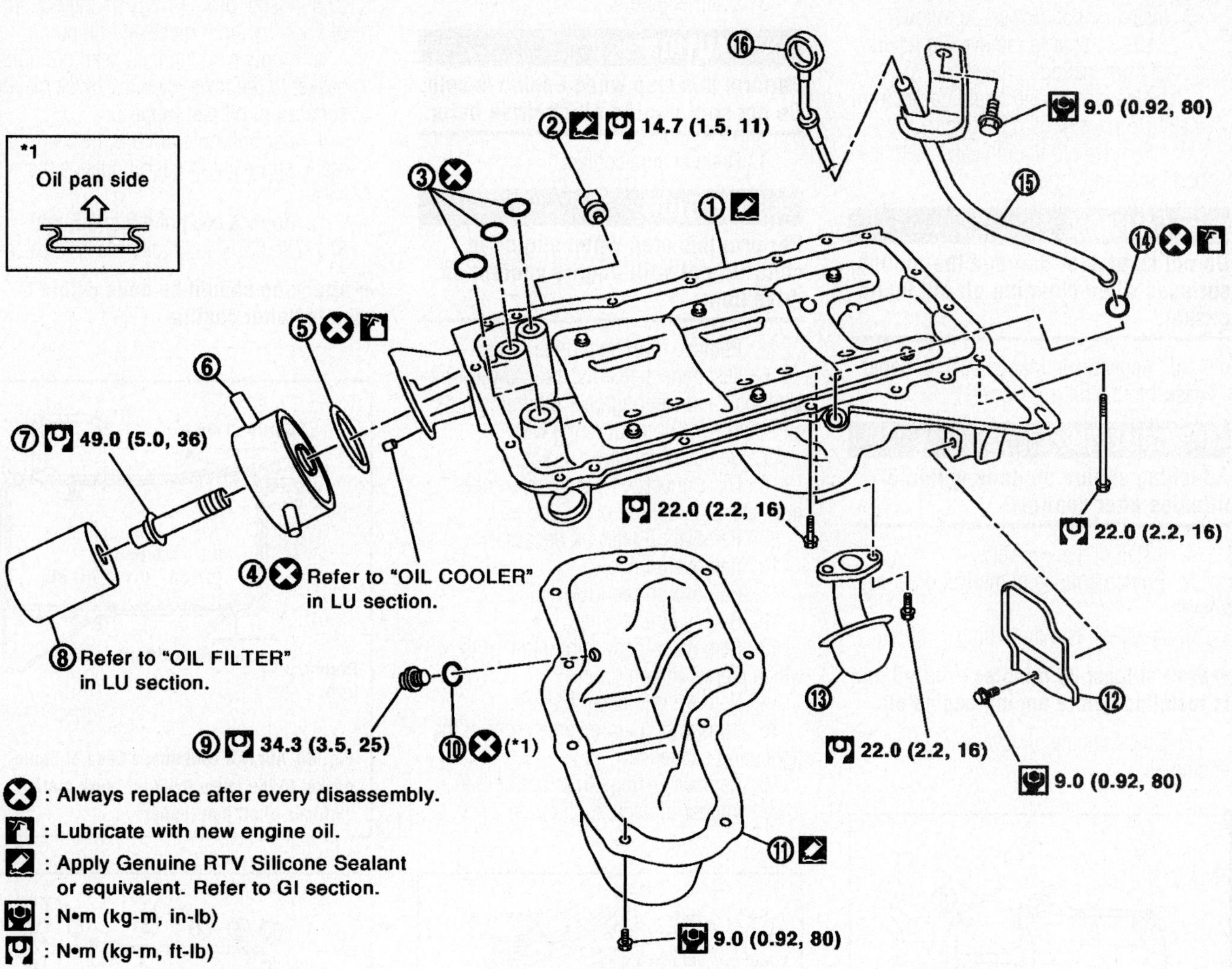

*1
Oil pan side

② ✏ 🔧 14.7 (1.5, 11)

🔧 9.0 (0.92, 80)

③ ✖

① ✏

⑤ ✖ 🛢

⑥

⑦ 🔧 49.0 (5.0, 36)

🔧 22.0 (2.2, 16)

🔧 22.0 (2.2, 16)

④ ✖ Refer to "OIL COOLER" in LU section.

⑧ Refer to "OIL FILTER" in LU section.

⑬

🔧 22.0 (2.2, 16)

⑫

🔧 9.0 (0.92, 80)

⑨ 🔧 34.3 (3.5, 25)

⑩ ✖ (*1)

⑪ ✏

✖ : Always replace after every disassembly.

🛢 : Lubricate with new engine oil.

✏ : Apply Genuine RTV Silicone Sealant or equivalent. Refer to GI section.

🔧 : N•m (kg-m, in-lb)

🔧 : N•m (kg-m, ft-lb)

🔧 9.0 (0.92, 80)

1.	Oil pan (upper)	2.	Oil pressure sensor
4.	Relief valve	5.	O-ring
7.	Connector bolt	8.	Oil filter
10.	Drain plug washer	11.	Oil pan (lower)
13.	Oil strainer	14	O-ring
16.	Oil level gauge		

3.	O-ring
6.	Oil cooler
9.	Drain plug
12.	Rear cover plate
15.	Oil level gauge guide

Fig. 65 Exploded view of oil pan assembly—V6

1. Drain engine oil.
2. Remove oil pan (lower) as follows:
   a. Loosen bolts in reverse order of the tightening sequence.
   b. Remove oil pan (lower).

✳ CAUTION

Be careful not to damage the mating surfaces. Do not insert screwdriver, this will damage the mating surfaces.

➡ **Slide seal cutter (1) by tapping on the side (2) of the tool with hammer.**

To install:

3. Install oil pan (lower) as follows:
   a. Use scraper to remove old liquid gasket from mating surfaces.
   b. Also remove old liquid gasket from mating surface of oil pan (upper).
   c. Remove old liquid gasket from the bolt holes and threads.

✳ CAUTION

Do not scratch or damage the mating surfaces when cleaning off old liquid gasket.

   d. Apply a continuous bead of liquid gasket to the oil pan (lower).

✳✳ CAUTION

Attaching should be done within 5 minutes after coating.

4. Install oil pan (lower).
5. Tighten bolts in numerical order as shown.
6. Install oil pan drain plug.

➡ **Wait at least 30 minutes after oil pan is installed before adding engine oil.**

7. Check engine oil level and adjust engine oil.

8. Start engine, and check there is no leak of engine oil.
9. Stop engine and wait for 10 minutes.
10. Check engine oil level again.

Upper Oil Pan

See Figures 67 through 69.

✳✳ WARNING

To avoid the danger of being scalded, do not drain engine oil when engine is hot.

1. Remove engine cover.
2. Remove air duct.
3. Drain engine oil.

✳✳ CAUTION

Perform this step when engine is cold. Do not spill engine oil on drive belts.

4. Drain engine coolant.

✳✳ CAUTION

Perform this step when engine is cold. Do not spill engine coolant on drive belts.

5. Remove front final drive (4X4).
6. Disconnect steering gear lower joint shaft bolt and steering gear mounting nuts and bolts, position out of the way.
7. Remove the starter motor.
8. Disconnect A/T fluid cooler tube brackets and position out of the way.
9. Remove oil filter, as necessary.
10. Remove oil cooler.
11. Remove oil pan (lower).
12. Remove oil strainer.
13. Remove transmission joint bolts which pierce oil pan (upper).
14. Remove rear cover plate.
15. Loosen bolts in reverse order of the tightening sequence.
16. Remove O-rings from bottom of lower cylinder block and oil pump.

To install:

17. Install oil pan (upper) as follows:
   a. Use scraper to remove old liquid gasket from mating surfaces.
   b. Also remove the old liquid gasket from mating surface of lower cylinder block.
   c. Remove old liquid gasket from the bolt holes and threads.

✳✳ CAUTION

Do not scratch or damage the mating surfaces when cleaning off old liquid gasket.

   d. Install new O-rings on the bottom of lower cylinder block and oil pump.
   e. Apply a continuous bead of liquid gasket to the lower cylinder block mating surfaces of oil pan (upper).
   f. For bolt holes with arrowhead mark, apply liquid gasket outside the hole.
   g. Apply a bead of 4.5 to 5.5 mm (0.177 to 0.217 in) in diameter to area "A".

➡ **Attaching should be done within 5 minutes after coating.**

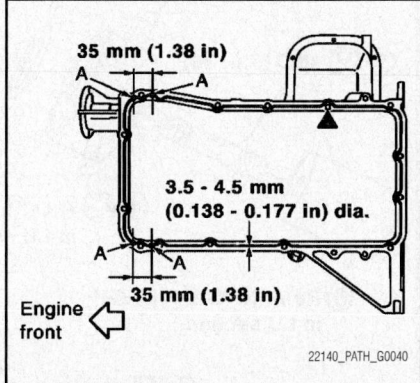

Fig. 68 Apply a continuous bead of liquid gasket to the lower cylinder block mating surfaces of oil pan (upper)

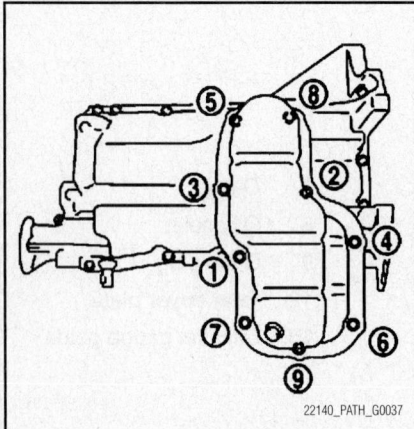

Fig. 66 Tighten bolts in numerical order

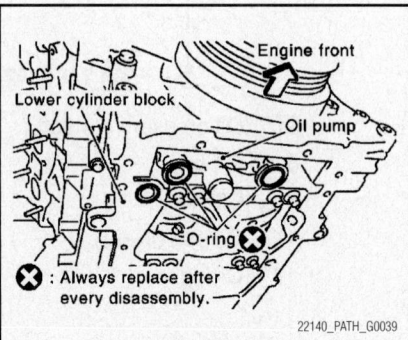

Fig. 67 Remove O-rings from bottom of lower cylinder block and oil pump

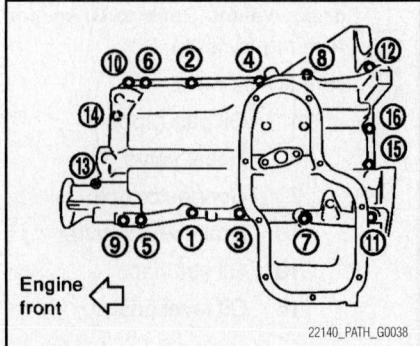

Fig. 69 Install and tighten bolts in numerical order

※ CAUTION

Install avoiding misalignment of both oil pan gaskets and O-rings.

18. Tighten bolts in numerical order as shown.

➡ **There are two types of bolts. Refer to the following for locating bolts.**

- M8x4 inches (100 mm): holes 7, 11, 12, 13

- M8x1 inch (25 mm): All other holes
19. Tighten transmission joint bolts.
20. Install oil strainer to oil pan (upper).
21. Install oil pan (upper).
22. Check engine oil level and adjust engine oil.
23. Start engine, and check there is no leak of engine oil.
24. Stop engine and wait for 10 minutes.
25. Check engine oil level again.

V8 Engine

See Figures 70 through 77.

※ WARNING

To avoid the danger of being scalded, never drain the engine oil when the engine is hot.

1. Remove the engine.
2. Remove the oil pan (lower) using the following steps.

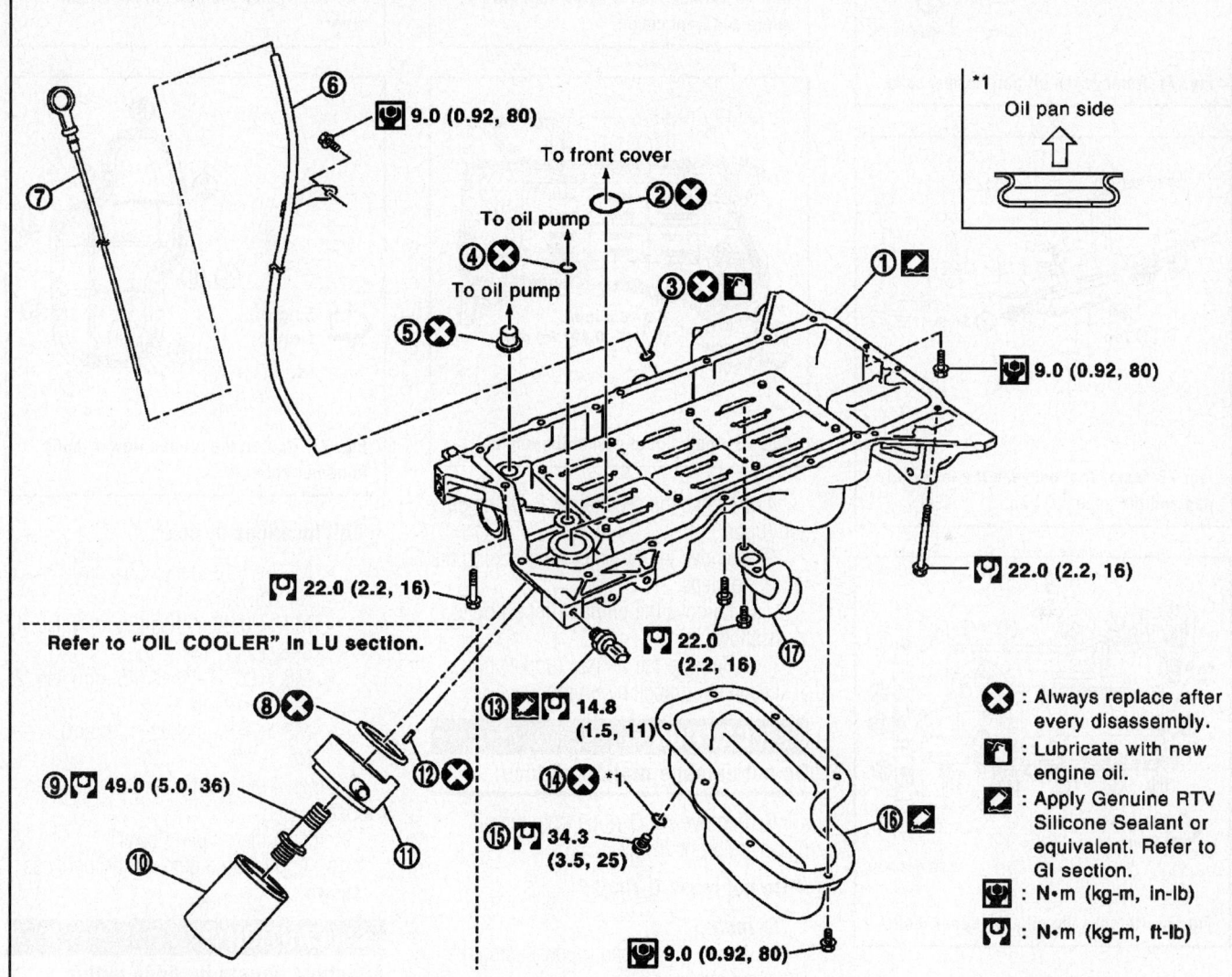

1. Oil pan (upper)
2. O-ring
3. O-ring
4. O-ring
5. O-ring (with collar)
6. Oil level gauge guide
7. Oil level gauge
8. O-ring
9. Connector bolt
10. Oil filter
11. Oil cooler
12. Relief valve
13. Oil pressure switch
14. Gasket
15. Drain plug
16. Oil pan (lower)
17. Oil strainer

Fig. 70 Exploded view of oil pan components

22140_PATH_G0110

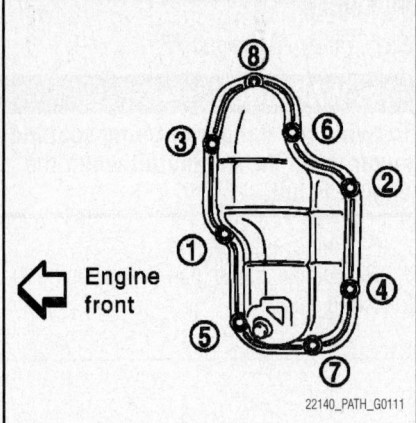

Fig. 71 Remove the oil pan (lower) bolts.

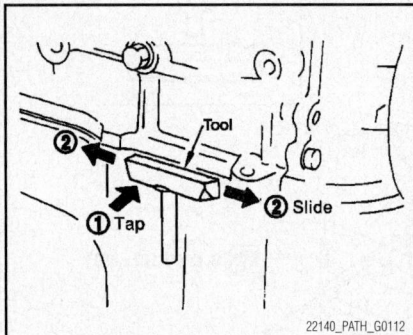

Fig. 72 Insert Tool between the lower oil pan and the upper oil pan.

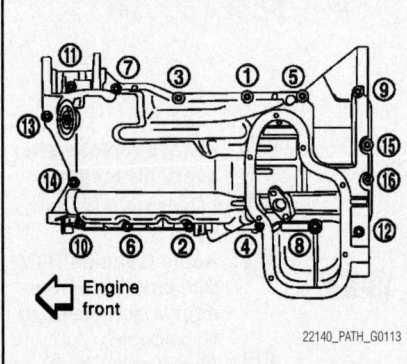

Fig. 73 Remove the oil pan (upper) bolts

a. Remove the oil pan (lower) bolts.

b. Insert Tool between the lower oil pan and the upper oil pan.

✴✴ CAUTION

Be careful not to damage the mating surface.

c. Tap seal cutter to insert it (1) and then slide it by tapping on the side (2) of the tool as shown.

3. Remove the oil cooler assembly.

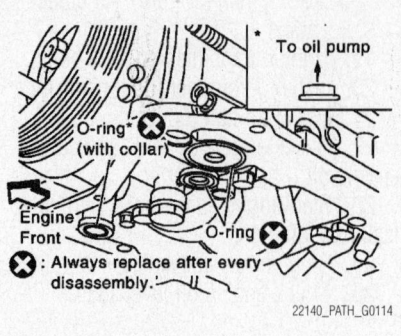

Fig. 74 Remove the O-rings from the oil pump and front cover

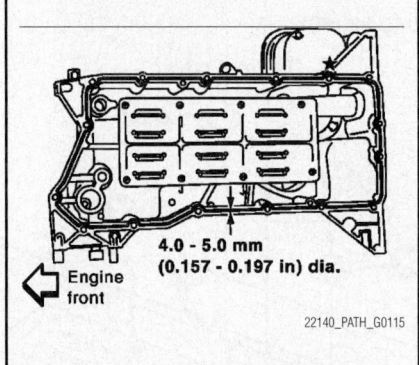

Fig. 75 Apply liquid gasket thoroughly

4. Remove the oil strainer from the oil pan (upper).

5. Remove the oil pan (upper) using the following steps.

a. Remove the oil pan (upper) bolts as shown.

b. Remove the oil pan (upper) from the cylinder block by prying.

✴✴ CAUTION

Do not damage mating surface.

6. Remove the O-rings from the oil pump and front cover.

➡**Do not reuse O-rings.**

To install:

7. Install the oil pan (upper) using the following steps.

a. Apply liquid gasket thoroughly as shown.

✴✴ CAUTION

Apply liquid gasket to outside of bolt hole for the hole shown by star (*).

b. Install new O-rings to the oil pump and front cover side.

c. Tighten the bolts in numerical order as shown.

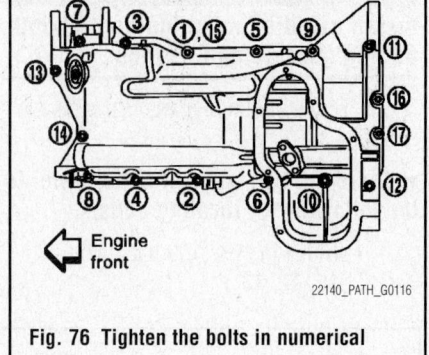

Fig. 76 Tighten the bolts in numerical order

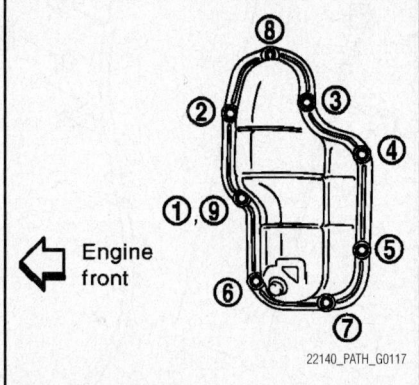

Fig. 77 Tighten the oil pan (lower) bolts in numerical order

➡**Bolt locations by size:**

- M6 × 1.18 inches (30 mm): No. 15, 16
- M8 × 0.98 inches (25 mm): No. 1, 3, 5, 7, 11, 13
- M8 × 1.77 inches (45 mm): No. 2, 4, 6, 8, 10, 14
- M8 × 4.84 inches (123 mm): No. 9, 12

8. Install the oil strainer to the oil pan (upper).

9. Install the oil pan (lower).

a. Apply liquid gasket thoroughly as shown.

✴✴ CAUTION

Attaching should be done within 5 minutes after coating.

b. Tighten the oil pan (lower) bolts in numerical order as shown.

10. Install the oil pan drain plug.

11. Install engine assembly.

➡**Do not fill the engine with oil for at least 30 minutes after oil pan is installed.**

12. Check engine oil level and add engine oil if necessary.

13. Start the engine, and check for leaks of engine oil.
14. Stop engine and wait for 10 minutes.
15. Check engine oil level again.

OIL PUMP

REMOVAL & INSTALLATION

V6 Engine

See Figure 78.

1. Remove oil pans (lower and upper).
2. Remove front timing chain case and timing chain (primary).
3. Remove oil pump assembly.

To install:

4. Installation is in the reverse order of removal, paying attention to the following.
5. When installing, align crankshaft flat faces with inner rotor flat faces.
6. Check the engine oil level.

7. Start engine, and check there are no leaks of engine oil.
8. Stop engine and wait for 10 minutes.
9. Check the engine oil level and add engine oil

V8 Engine

See Figure 79.

1. Remove front cover.
2. Remove the oil pump drive spacer.
3. Remove the oil pump assembly.

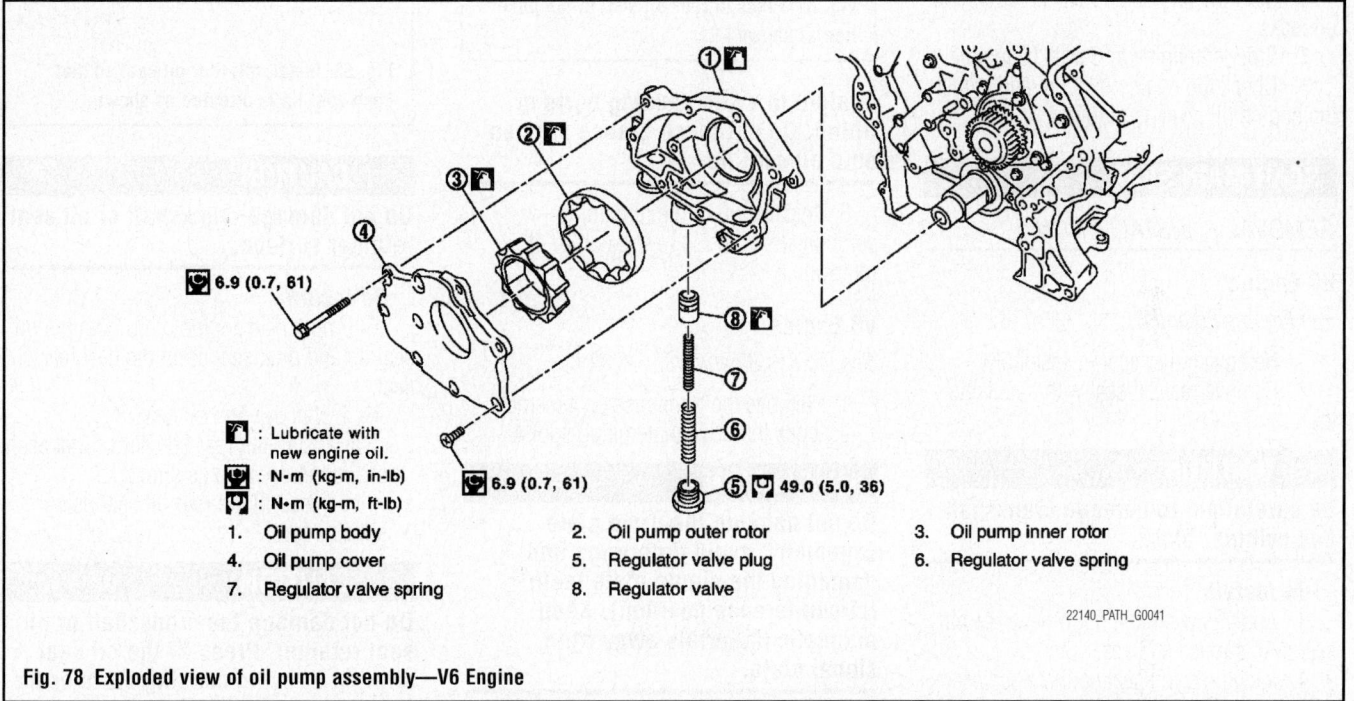

■ : Lubricate with new engine oil.
◉ : N•m (kg-m, in-lb)
◎ : N•m (kg-m, ft-lb)

1. Oil pump body
2. Oil pump outer rotor
3. Oil pump inner rotor
4. Oil pump cover
5. Regulator valve plug
6. Regulator valve spring
7. Regulator valve spring
8. Regulator valve

6.9 (0.7, 61)
6.9 (0.7, 61)
49.0 (5.0, 36)

22140_PATH_G0041

Fig. 78 Exploded view of oil pump assembly—V6 Engine

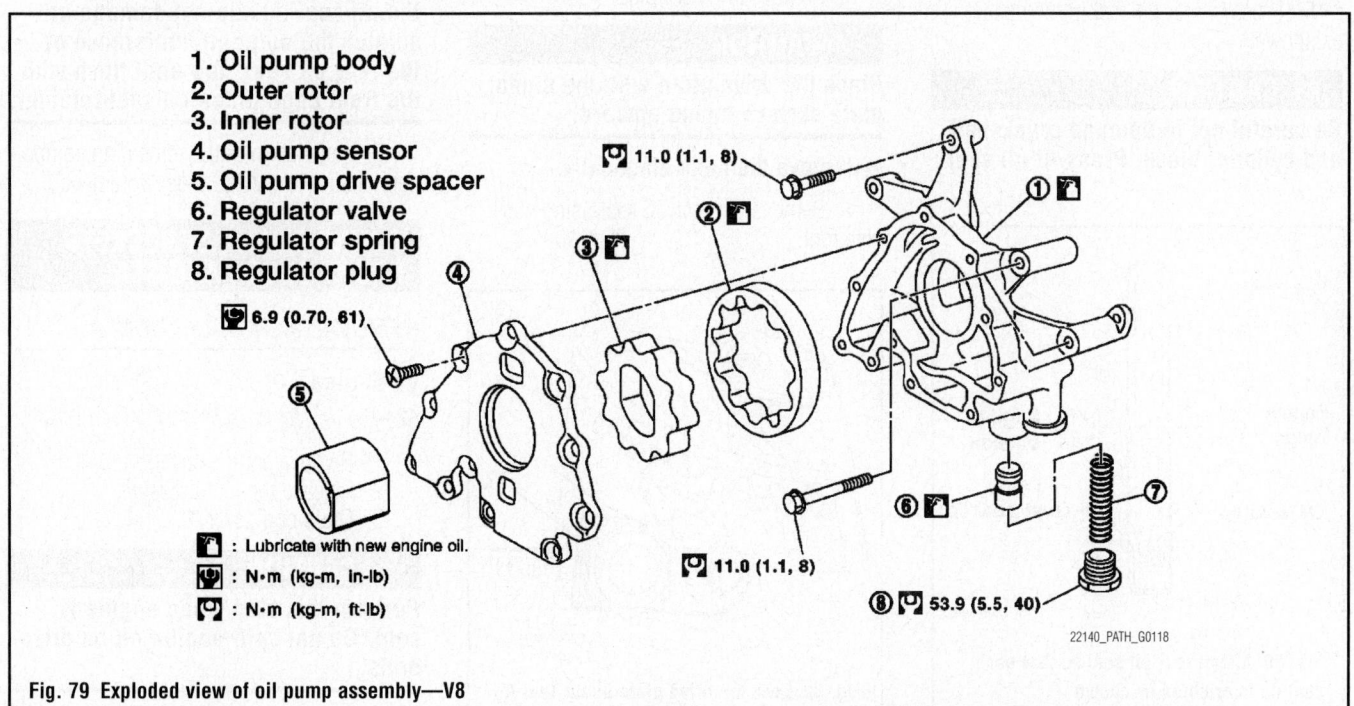

1. Oil pump body
2. Outer rotor
3. Inner rotor
4. Oil pump sensor
5. Oil pump drive spacer
6. Regulator valve
7. Regulator spring
8. Regulator plug

11.0 (1.1, 8)
6.9 (0.70, 61)
11.0 (1.1, 8)
53.9 (5.5, 40)

■ : Lubricate with new engine oil.
◉ : N•m (kg-m, in-lb)
◎ : N•m (kg-m, ft-lb)

22140_PATH_G0118

Fig. 79 Exploded view of oil pump assembly—V8

To install:

4. Installation is in the reverse order of removal, paying attention of the following:
- When inserting the oil pump drive spacer, align the crankshaft key and the flat face of the inner rotor.
- If they are not aligned, rotate the oil pump inner rotor by hand.
- Make sure that each part is aligned and tap lightly until it reaches the end.

5. Check the engine oil level.
6. Start the engine and check for engine oil leaks.
7. Stop the engine and wait 10 minutes.
8. Check the engine oil level and adjust the engine oil level as required.

REAR MAIN SEAL

REMOVAL & INSTALLATION

V6 Engine

See Figures 80 and 81.

1. Remove transmission assembly.
2. Remove rear oil seal with a suitable tool.

✳✳ CAUTION

Be careful not to damage crankshaft and cylinder block.

To install:

3. Apply new engine oil to new rear oil seal joint surface and seal lip.
4. Install rear oil seal so that each seal lip is oriented as shown.
5. Press in rear oil seal to the position as shown.

✳ CAUTION

Be careful not to damage crankshaft and cylinder block. Press-fit oil seal

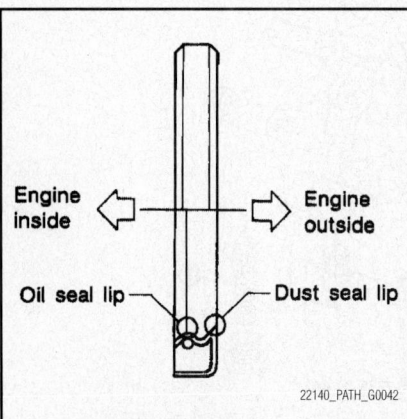

Fig. 80 Install rear oil seal so that each seal lip is oriented as shown

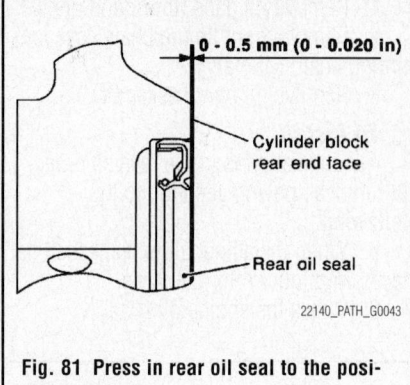

Fig. 81 Press in rear oil seal to the position as shown

straight to avoid causing burrs or tilting. Do not touch grease applied onto oil seal lip.

6. Installation of the remaining components is in the reverse order of removal.

V8 Engine

See Figures 82 and 83.

1. Remove the transmission assembly.
2. Lock the drive plate using Tool A.

✳✳ CAUTION

Do not damage the drive plate. Especially, avoid deforming and damaging the signal plate teeth (circumference position). Keep magnetic materials away from signal plate.

3. Remove the drive plate.

✳✳ CAUTION

Place the drive plate with the signal plate surface facing upward.

➡Remove the bolts diagonally.

4. Remove the rear oil seal using suitable tool.

Fig. 82 Lock the drive plate using Tool A

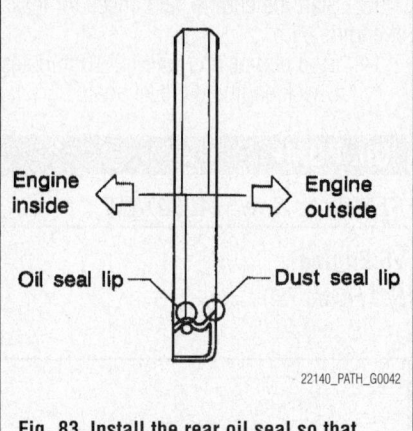

Fig. 83 Install the rear oil seal so that each seal lip is oriented as shown.

✳✳ CAUTION

Do not damage crankshaft or oil seal retainer surface.

To install:

5. Apply new engine oil to both the oil seal lip and dust seal lip of the new rear oil seal.
6. Install the rear oil seal.
   a. Install the rear oil seal so that each seal lip is oriented as shown.
   b. Press-fit the rear oil seal using suitable tool.

✳✳ CAUTION

Do not damage the crankshaft or oil seal retainer. Press-fit the oil seal straight to avoid causing burrs or tilting. Do not touch grease applied onto the oil seal lip. Do not damage or scratch the outer circumference of the rear oil seal. Tap until flush with the front edge of the oil seal retainer.

7. Installation of the remaining components is in the reverse order of removal.

TIMING CHAIN, SPROCKETS, FRONT COVER AND SEAL

REMOVAL & INSTALLATION

V6 Engine

See Figures 84 through 111.

1. Remove engine cover.
2. Release the fuel pressure.
3. Drain engine oil.

✳✳ CAUTION

Perform this step when engine is cold. Do not spill engine oil on drive belts.

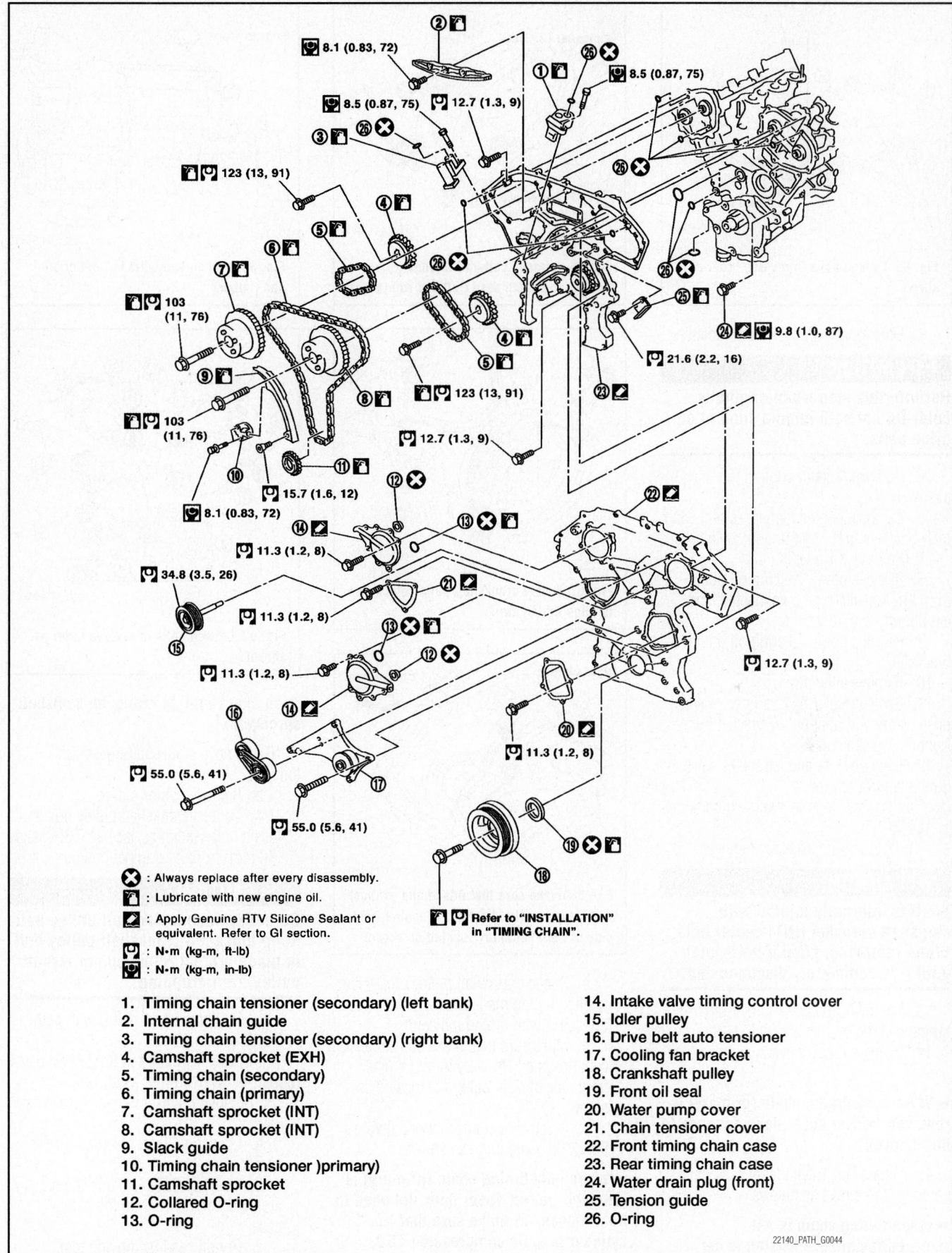

- ⊗ : Always replace after every disassembly.
- ⊡ : Lubricate with new engine oil.
- ⊿ : Apply Genuine RTV Silicone Sealant or equivalent. Refer to GI section.
- 🔧 : N·m (kg-m, ft-lb)
- 🔧 : N·m (kg-m, in-lb)

⊡ 🔧 Refer to "INSTALLATION" in "TIMING CHAIN".

1. Timing chain tensioner (secondary) (left bank)
2. Internal chain guide
3. Timing chain tensioner (secondary) (right bank)
4. Camshaft sprocket (EXH)
5. Timing chain (secondary)
6. Timing chain (primary)
7. Camshaft sprocket (INT)
8. Camshaft sprocket (INT)
9. Slack guide
10. Timing chain tensioner)primary)
11. Camshaft sprocket
12. Collared O-ring
13. O-ring
14. Intake valve timing control cover
15. Idler pulley
16. Drive belt auto tensioner
17. Cooling fan bracket
18. Crankshaft pulley
19. Front oil seal
20. Water pump cover
21. Chain tensioner cover
22. Front timing chain case
23. Rear timing chain case
24. Water drain plug (front)
25. Tension guide
26. O-ring

Fig. 84 Exploded view of timing chain assembly

22140_PATH_G0044

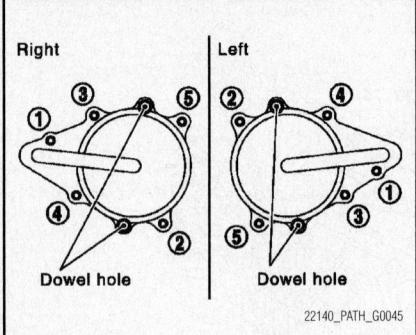

Fig. 85 Loosen bolts in reverse order as shown

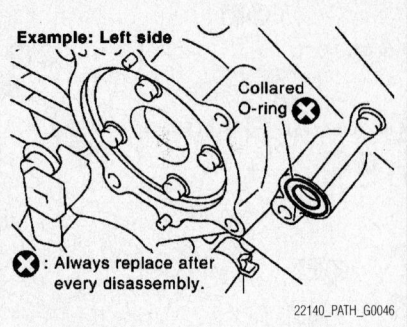

Fig. 86 Remove collared O-rings from front timing chain case (left and right side)

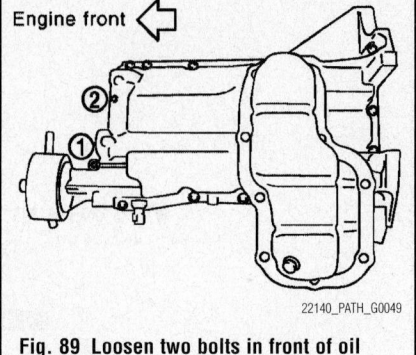

Fig. 89 Loosen two bolts in front of oil pan (upper)

4. Drain engine coolant from radiator.

❋❋ CAUTION

Perform this step when engine is cold. Do not spill engine coolant on drive belts.

5. Remove radiator cooling fan assembly.
6. Separate engine harnesses removing their brackets from front timing chain case.
7. Remove drive belts.
8. Remove power steering oil pump from bracket with piping connected, and temporarily secure it aside.
9. Remove power steering oil pump bracket.
10. Remove alternator.
11. Remove water bypass hose, water hose clamp and idler pulley bracket from front timing chain case.
12. Remove right and left intake valve timing control covers.
   a. Loosen bolts in reverse order as shown.
   b. Cut liquid gasket for removal.

❋❋ CAUTION

Shaft is internally jointed with camshaft sprocket (INT) center hole. When removing, keep it horizontal until it is completely disconnected.

13. Remove collared O-rings from front timing chain case (left and right side).
14. Remove rocker covers (right and left banks).

➡ **When only timing chain (primary) is removed, rocker cover does not need to be removed.**

15. Obtain No. 1 cylinder at TDC of its compression stroke as follows:

➡ **When timing chain is not removed/installed, this step is not required.**

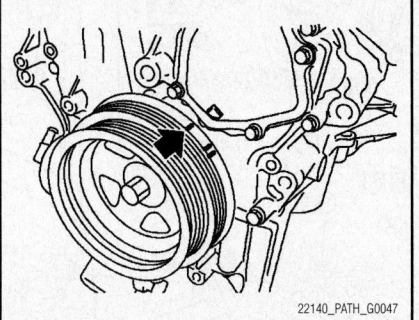

Fig. 87 Rotate crankshaft pulley clockwise to align timing mark

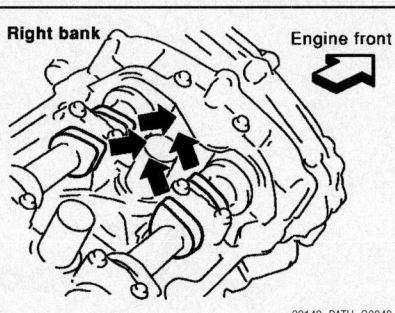

Fig. 88 Make sure that intake and exhaust cam noses on No. 1 cylinder (engine front side of right bank) are located as shown

   a. Rotate crankshaft pulley clockwise to align timing mark (grooved line without color) with timing indicator.
   b. Make sure that intake and exhaust cam noses on No. 1 cylinder (engine front side of right bank) are located as shown.
   c. If not, turn crankshaft one revolution (360°) and align as shown.

➡ **When only timing chain (primary) is removed, rocker cover does not need to be removed. To make sure that No. 1 cylinder is at its compression TDC, remove front timing chain case first.**

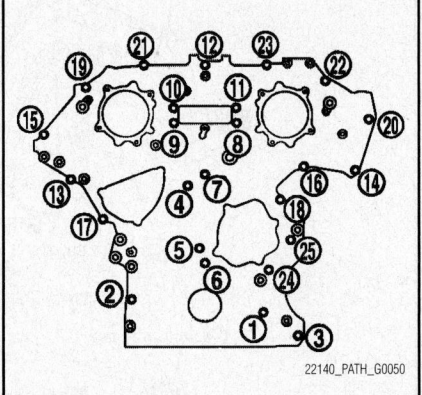

Fig. 90 Loosen bolts in reverse order as shown

Then check mating marks on camshaft sprockets.

16. Remove crankshaft pulley as follows:
   a. Remove starter motor.
   b. Loosen crankshaft pulley bolt and locate bolt seating surface as 0.39 inches (10 mm) from its original position.

❋❋ CAUTION

Do not remove crankshaft pulley bolt. Keep loosened crankshaft pulley bolt in place protect removed crankshaft pulley from dropping.

   c. Pull crankshaft pulley with both hands to remove it.
17. Loosen two bolts in front of oil pan (upper) in reverse order as shown.
18. Remove front timing chain case as follows:
   a. Loosen bolts in reverse order as shown.
   b. Insert suitable tool into the notch at the top of the front timing chain case.
   c. Pry off case by moving tool.
   d. Cut liquid gasket for removal.

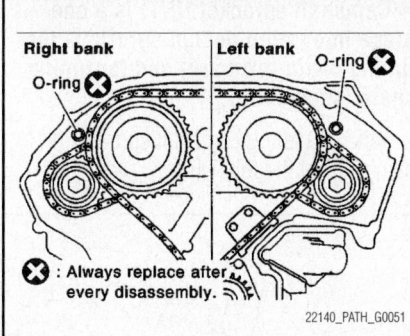

Fig. 91 Remove O-rings from rear timing chain case

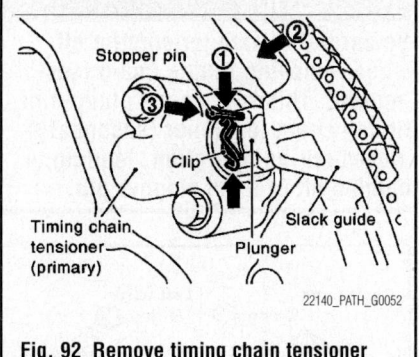

Fig. 92 Remove timing chain tensioner (primary)

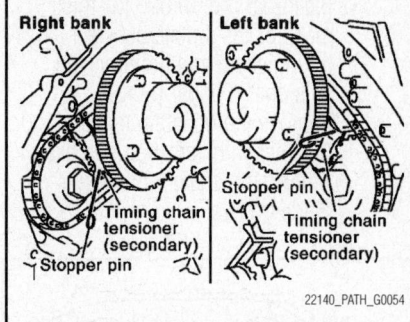

Fig. 94 Attach suitable stopper pin to the right and left timing chain tensioners (secondary)

> ✳✳ **CAUTION**
>
> **Do not use screwdriver or something similar. After removal, handle front timing chain case carefully so it does not tilt, cant, or warp under a load.**

19. Remove O-rings from rear timing chain case.
20. Remove water pump cover and chain tensioner cover from front timing chain case, if necessary.
21. Remove front oil seal from front timing chain case using suitable tool.

> ✳✳ **CAUTION**
>
> **Be careful not to damage front timing chain case.**

22. Use a scraper to remove all traces of old liquid gasket from front and rear timing chain cases and oil pan (upper), and liquid gasket mating surfaces.

> ✳✳ **CAUTION**
>
> **Be careful not to allow gasket fragments to enter oil pan.**

23. Remove old liquid gasket from bolt holes and threads.
24. Use a scraper to remove all traces of old liquid gasket from water pump cover, chain tensioner cover and intake valve timing control covers.
25. Remove timing chain tensioner (primary) as follows:
    a. Loosen clip of timing chain tensioner (primary), and release plunger stopper (1).
    b. Insert plunger into tensioner body by pressing slack guide (2).
    c. Keep slack guide pressed and hold plunger in by pushing stopper pin through the tensioner body hole and plunger groove (3).
    d. Remove bolts and remove timing chain tensioner (primary).

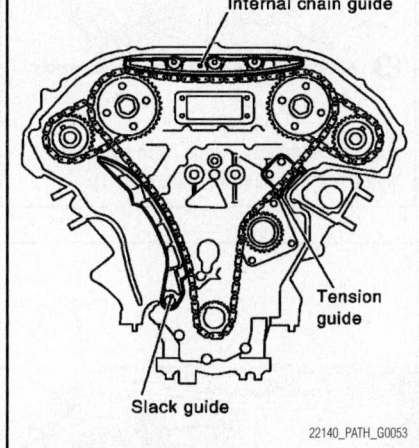

Fig. 93 Remove internal chain guide, tension guide and slack guide

26. Remove internal chain guide, tension guide and slack guide.

➡ **Tension guide can be removed after removing timing chain (primary).**

27. Remove timing chain (primary) and crankshaft sprocket.

> ✳✳ **CAUTION**
>
> **After removing timing chain (primary), do not turn crankshaft and camshaft separately, or valves will strike the piston heads.**

28. Remove timing chain (secondary) and camshaft sprockets as follows:
    a. Attach suitable stopper pin to the right and left timing chain tensioners (secondary).

➡ **Use approximately 0.02 inches (0.5 mm) diameter hard metal pin as a stopper pin.**

    b. Remove camshaft sprocket (INT and EXH) bolts.

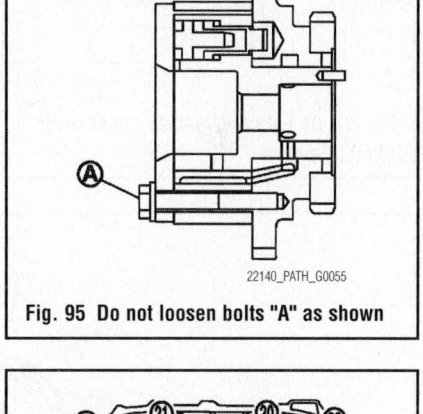

Fig. 95 Do not loosen bolts "A" as shown

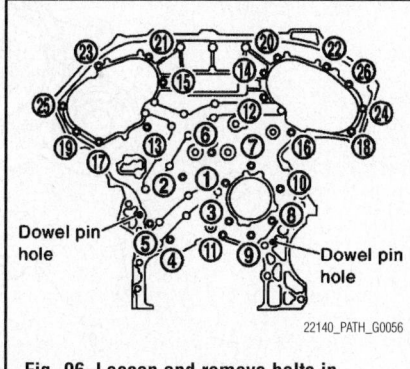

Fig. 96 Loosen and remove bolts in reverse order as shown

➡ **Secure the hexagonal portion of camshaft using wrench to loosen bolts.**

> ✳✳ **CAUTION**
>
> **Do not loosen bolts with securing anything other than the camshaft hexagonal portion or with tensioning the timing chain.**

    c. Remove timing chain (secondary) together with camshaft sprockets.
    • Turn camshaft slightly to secure slackness of timing chain on timing chain tensioner (secondary) side.

- Insert 0.5 mm (0.020 in)-thick metal or resin plate between timing chain and timing chain tensioner plunger (guide). Remove timing chain (secondary) together with camshaft sprockets with timing chain loose from guide groove.

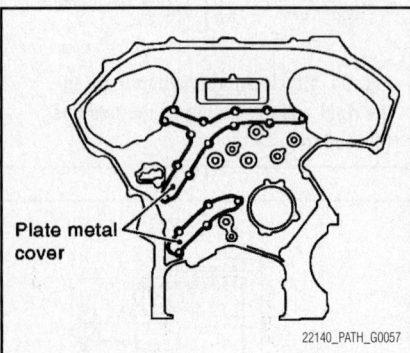

Fig. 97 Do not remove plate metal cover of oil passage

✳✳ CAUTION

Be careful of plunger coming off when removing timing chain (secondary). This is because plunger of timing chain tensioner (secondary) moves during operation, leading to coming off of fixed stopper pin.

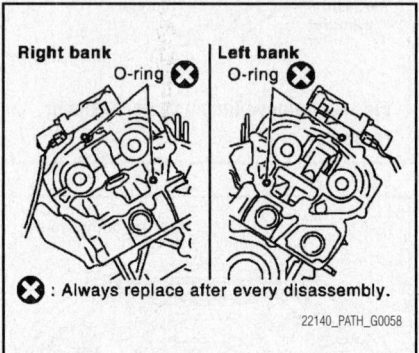

Fig. 98 Remove O-rings from cylinder head and camshaft bracket (No. 1)

➡Camshaft sprocket (INT) is a one piece integrated design sprockets for timing chain (primary) and for timing chain (secondary).

When handling camshaft sprocket (INT), be careful of the following:

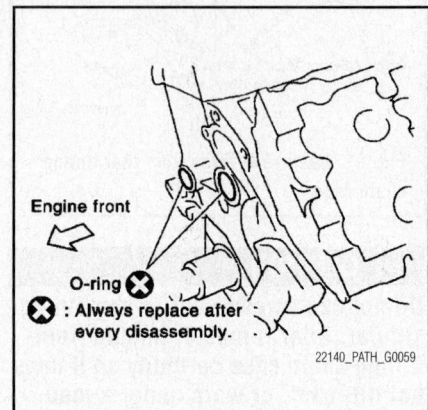

Fig. 99 Remove O-rings from cylinder block

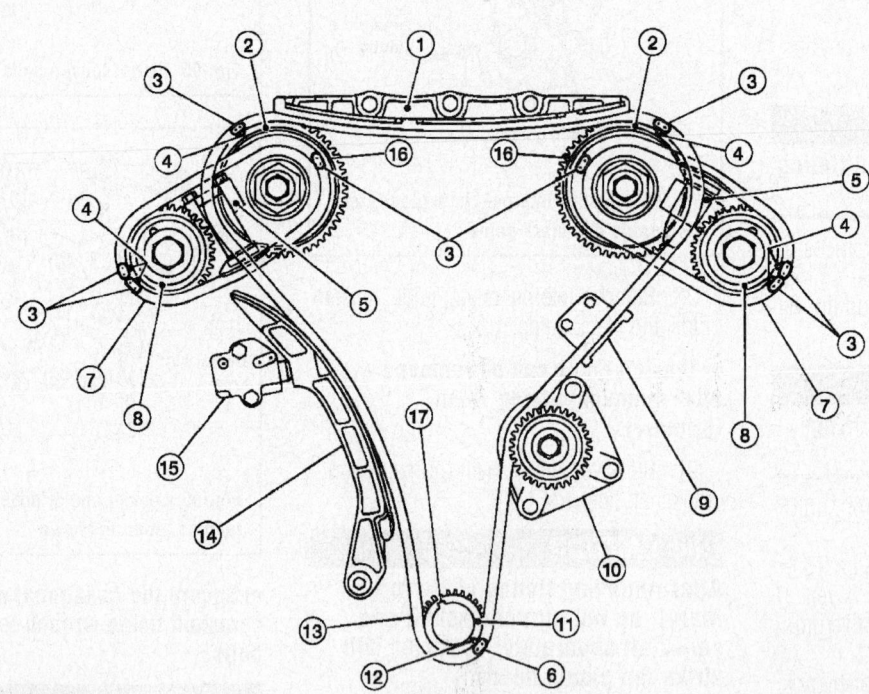

1. Internal chain guide	2. Camshaft sprocket (intake)	3. Mating mark (copper link)
4. Mating mark (punched)	5. Secondary timing chain tensioner	6. Mating mark (yellow link)
7. Secondary timing chain	8. Camshaft sprocket (exhaust)	9. Tensioner guide
10. Water pump	11. Crankshaft sprocket	12. Mating mark (notched)
13. Primary timing chain	14. Slack guide	15. Primary timing chain tensioner
16. Mating mark (back side)	17. Crankshaft key	

Fig. 100 Timing chain assembly with mating marks

※※ **CAUTION**

Handle carefully to avoid any shock to camshaft sprocket. Do not disassemble. (Do not loosen bolts "A" as shown).

29. Remove water pump.
30. Remove rear timing chain case as follows:
   a. Loosen and remove bolts in reverse order as shown.
   b. Cut liquid gasket using Tool and remove rear timing chain case.

※※ **CAUTION**

Do not remove plate metal cover of oil passage.

※※ **CAUTION**

After removal, handle rear timing chain case carefully so it does not tilt, cant, or warp under a load.

31. Remove O-rings from cylinder head and camshaft bracket (No. 1).
32. Remove O-rings from cylinder block.
33. Remove timing chain tensioners (secondary) from cylinder head if necessary.
   a. Remove camshaft brackets (No. 1).
   b. Remove timing chain tensioners (secondary) with stopper pin attached.
34. Use scraper to remove all traces of old liquid gasket from front and rear timing chain cases, and opposite mating surfaces. Remove old liquid gasket from bolt hole and thread.
35. Use scraper to remove all traces of liquid gasket from water pump cover, chain tensioner cover and intake valve timing control covers.
36. Check for cracks and any excessive wear at link plates and roller links of timing chain.
37. Replace timing chain as necessary.

   To install:

➡ **The figure below shows the relationship between the mating mark on each timing chain and that on the corresponding sprocket, with the components installed.**

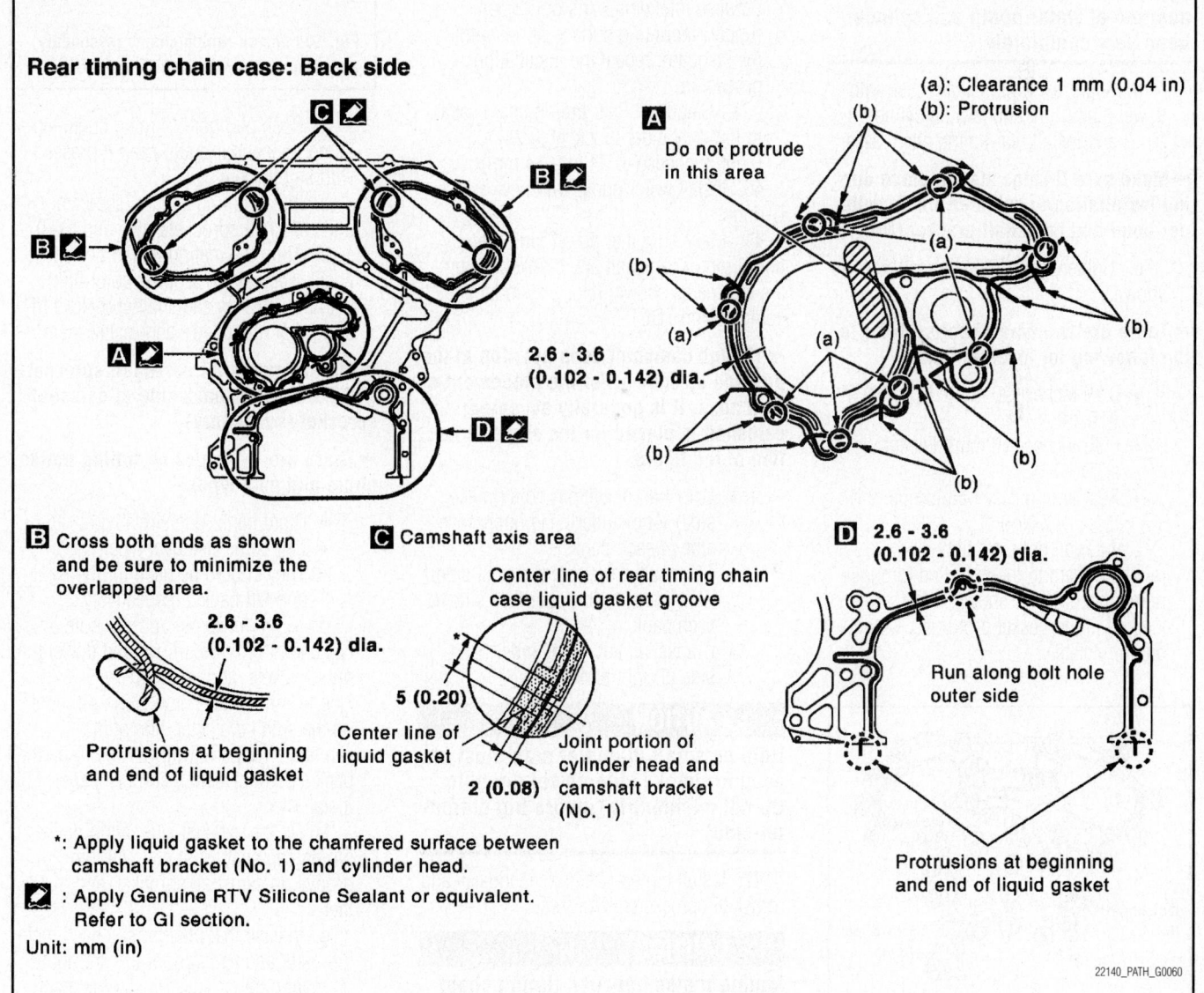

Rear timing chain case: Back side

(a): Clearance 1 mm (0.04 in)
(b): Protrusion

A Do not protrude in this area

2.6 - 3.6 (0.102 - 0.142) dia.

B Cross both ends as shown and be sure to minimize the overlapped area.

2.6 - 3.6 (0.102 - 0.142) dia.

Protrusions at beginning and end of liquid gasket

C Camshaft axis area

Center line of rear timing chain case liquid gasket groove

5 (0.20)

Center line of liquid gasket

2 (0.08)

Joint portion of cylinder head and camshaft bracket (No. 1)

D 2.6 - 3.6 (0.102 - 0.142) dia.

Run along bolt hole outer side

Protrusions at beginning and end of liquid gasket

*: Apply liquid gasket to the chamfered surface between camshaft bracket (No. 1) and cylinder head.

✎ : Apply Genuine RTV Silicone Sealant or equivalent. Refer to GI section.

Unit: mm (in)

22140_PATH_G0060

Fig. 101 Apply liquid gasket to rear timing chain case back side as shown

38. Install timing chain tensioners (secondary) to cylinder head if removed.

   a. Install timing chain tensioners (secondary) with stopper pin attached and new O-ring.

   b. Install camshaft brackets (No. 1).

39. Install rear timing chain case as follows:

   a. Install new O-rings onto cylinder block.

   b. Install new O-rings to cylinder head and camshaft bracket (No. 1).

   c. Apply liquid gasket to rear timing chain case back side as shown.

✳✳ CAUTION

For "A" in the figure, completely wipe out liquid gasket extended on a portion touching at engine coolant. Apply liquid gasket on installation position of water pump and cylinder head very completely

   d. Align rear timing chain case with dowel pins (right and left) on cylinder block and install rear timing chain case.

➡**Make sure O-rings stay in place during installation to cylinder block, cylinder head and camshaft bracket (No. 1).**

   e. Tighten bolts in numerical order as shown.

➡**There are two type of bolts. Refer to the following for locating bolts.**

- 0.79 inches (20 mm): 1, 2, 3, 6, 7, 8, 9, 10
- 0.63 inches (16 mm): Except the above
- Rear timing case bolt torque: 9 ft lbs. (12.7 Nm)

   f. After all bolts are tightened, retighten them to the specified torque in numerical order as shown.

   g. If liquid gasket protrudes, wipe it off immediately.

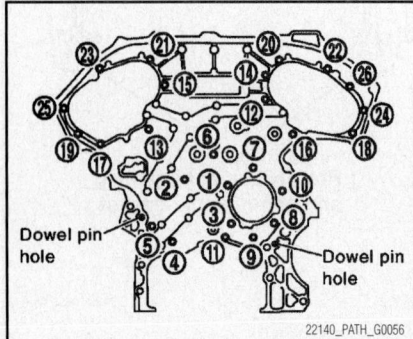

Fig. 102 Tighten bolts in numerical order as shown

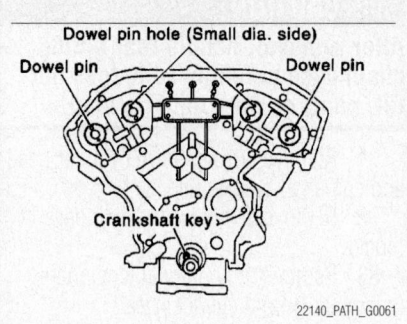

Fig. 103 Make sure that dowel pin hole, dowel pin of camshaft and crankshaft key are located as shown. (No. 1 cylinder at compression TDC)

   h. After installing rear timing chain case, check the surface height difference between following parts on oil pan (upper) mounting surface. If not within the standard, repeat the installation procedure.

   i. Standard: Rear timing chain case to lower cylinder block: -0.0094 to 0.0055 inches (-0.24 to 0.14 mm)

40. Install water pump with new O-rings.

41. Make sure that dowel pin hole, dowel pin of camshaft and crankshaft key are located as shown. (No. 1 cylinder at compression TDC)

➡**Though camshaft does not stop at the position as shown, for the placement of cam nose, it is generally accepted camshaft is placed for the same direction of the figure.**

- Camshaft dowel pin hole (intake side): At cylinder head upper face side in each bank.
- Camshaft dowel pin (exhaust side): At cylinder head upper face side in each bank.
- Crankshaft key: At cylinder head side of right bank.

✳✳ CAUTION

Hole on small diameter side must be used for intake side dowel pin hole. Do not misidentify (ignore big diameter side).

42. Install timing chains (secondary) and camshaft sprockets as follows:

✳✳ CAUTION

Mating marks between timing chain and sprockets slip easily. Confirm all mating mark positions repeatedly during the installation process.

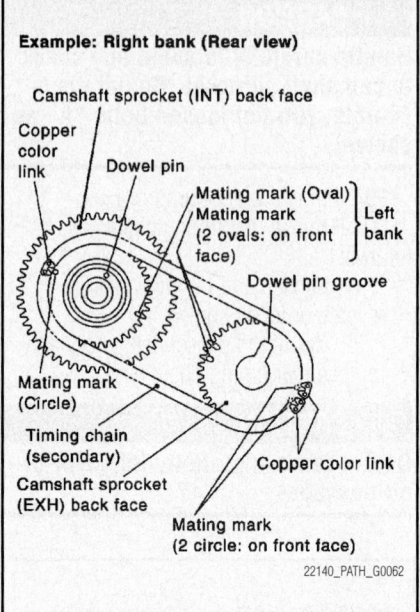

Fig. 104 Install timing chains (secondary) and camshaft sprockets (INT and EXH)

   a. Push plunger of timing chain tensioner (secondary) and keep it pressed in with stopper pin.

   b. Install timing chains (secondary) and camshaft sprockets (INT and EXH).

   c. Align the mating marks on timing chain (secondary) (copper color link) with the ones on camshaft sprockets (INT and EXH) (punched), and install them.

➡**Mating marks for camshaft sprocket (INT) are on the back side of camshaft sprocket (secondary).**

➡**There are two types of mating marks, circle and oval types.**

- Right bank: Use circle type.
- Left bank: Use oval type.
- They should be used for the right and left banks, respectively.

   d. Align dowel pin and pin hole on camshafts with the groove and dowel pin on sprockets, and install them.

   e. On the intake side, align pin hole on the small diameter side of the camshaft front end with dowel pin on the back side of camshaft sprocket, and install them.

   f. On the exhaust side, align dowel pin on camshaft front end with pin groove on camshaft sprocket, and install them.

   g. In case that positions of each mating mark and each dowel pin are not fit on mating parts, make fine adjustment to the position holding the hexagonal portion on camshaft with wrench or equivalent.

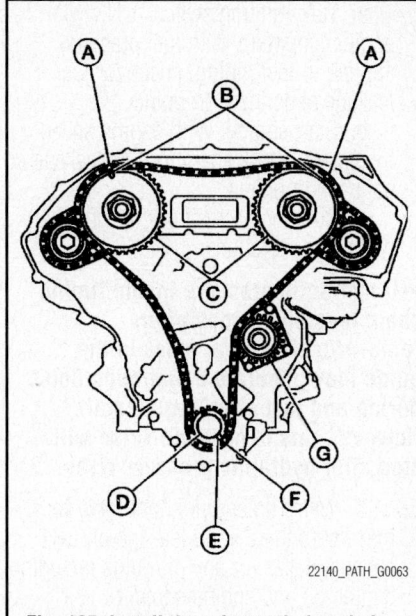

Fig. 105 Install the primary timing chain

h. Bolts for camshaft sprockets must be tightened in the next step. Tightening them by hand is enough to prevent the dislocation of dowel pins.

i. It may be difficult to visually check the dislocation of mating marks during and after installation. To make the matching easier, make a mating mark on the top of sprocket teeth and its extended line in advance with paint.

j. After confirming the mating marks are aligned, tighten camshaft sprocket bolts.

➡**Secure camshaft using wrench at the hexagonal portion to tighten bolts.**

k. Pull stopper pins out from timing chain tensioners (secondary).

43. Install tension guide.

44. Install timing chain (primary) as follows:

a. Install crankshaft sprocket.

➡**Make sure the mating marks on crankshaft sprocket face the front of engine.**

b. Install the primary timing chain.
• Water pump (G).
• Install primary timing chain so the mating mark punched (B) on camshaft sprocket is aligned with the copper link (A) on the timing chain, while the mating mark notched (E) on the crankshaft sprocket (D) is aligned with the yellow link (F) on the timing chain, as shown.
• When it is difficult to align mating marks (A) with (B) and (E) with (F)

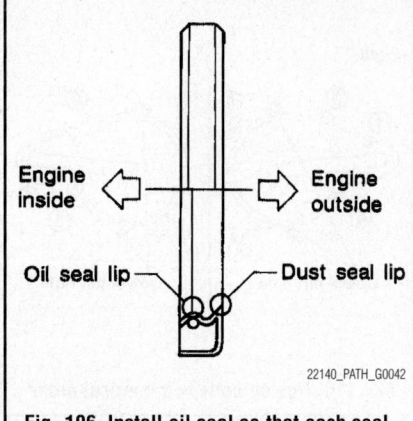

Fig. 106 Install oil seal so that each seal lip is oriented as shown

of the primary timing chain with each sprocket, gradually turn the camshaft using a wrench on the hexagonal portion to align it with the mating marks.
• During alignment, be careful to prevent dislocation of mating mark alignments of the secondary timing chains.

45. Install internal chain guide, slack guide and timing chain tensioner (primary).

⊛⊛ **CAUTION**

Do not overtighten slack guide bolts. It is normal for a gap to exist under the bolt seats when bolts are tightened to specification.

• When installing timing chain tensioner (primary), push in plunger and keep it pressed in with stopper pin.
• Remove any dirt and foreign materials completely from the back and the mounting surfaces of timing chain tensioner (primary).
• After installation, pull out stopper pin by pressing slack guide.

46. Make sure again that the mating marks on camshaft sprockets and timing chain have not slipped out of alignment.

47. Install new O-rings on rear timing chain case.

48. Install new front oil seal on front timing chain case.

a. Apply new engine oil to both oil seal lip and dust seal lip.

b. Install it so that each seal lip is oriented as shown.

c. Press-fit oil seal until it becomes flush with front timing chain case end face using suitable drift with outer diameter: 2.36 inches (60 mm).

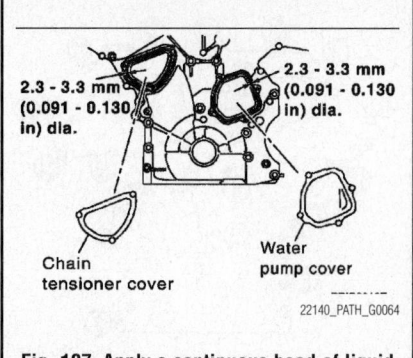

Fig. 107 Apply a continuous bead of liquid gasket to front timing chain

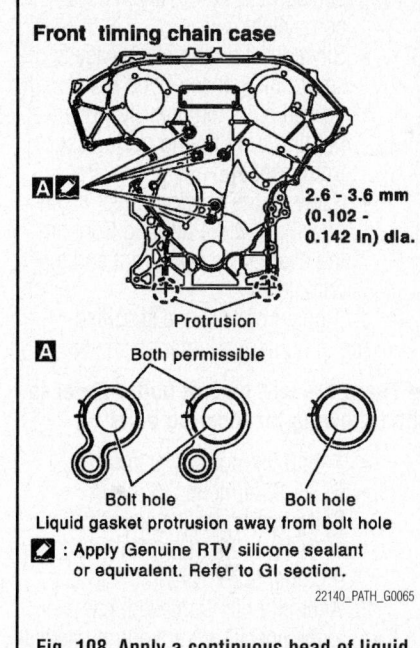

Fig. 108 Apply a continuous bead of liquid gasket to front timing chain case back side

d. Make sure the garter spring is in position and seal lip is not inverted.

49. Install water pump cover and chain tensioner cover to front timing chain case.

a. Apply a continuous bead of liquid gasket to front timing chain case as shown.

50. Install front timing chain case as follows:

a. Apply a continuous bead of liquid gasket to front timing chain case back side as shown.

b. Install new O-rings on rear timing chain case.

c. Assemble front timing chain case as follows:
• Fit lower end of front timing chain case tightly onto top face of oil pan (upper). From the fitting point, make entire front timing chain case

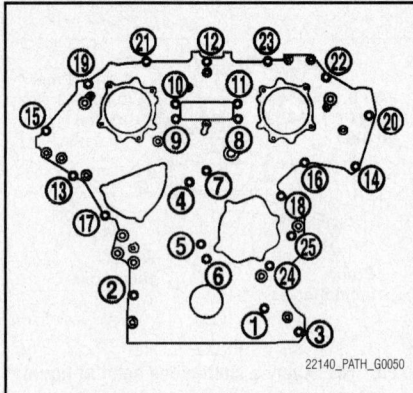

Fig. 109 Tighten bolts in numerical order

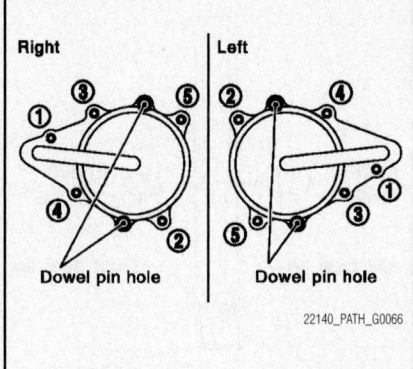

Fig. 110 Tighten bolts in numerical order

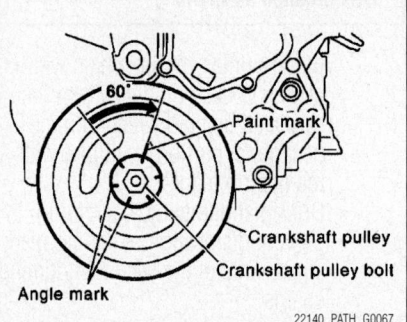

Fig. 111 Put a paint mark on crankshaft pulley aligning with angle mark on crankshaft pulley bolt

contact rear timing chain case completely.

- Since front timing chain case is offset for difference of bolt holes, tighten bolts temporarily while holding front timing chain case from front and top.
- Same as the previous step, insert dowel pin while holding front timing chain case from front and top completely.

d. Tighten bolts to the specified torque in numerical order as shown.

➡There are two type of bolts. Refer to the following for locating bolts.

- 1–5: 0.39 inches (10 mm)
- 6–25: 0.24 inches (6 mm)
- Bolt position torque specification: 1–5: 41 ft. lbs. (55.0 Nm); 6–25: 9 ft. lbs. (12.7 Nm)

e. After all bolts tightened, retighten them to the specified torque in numerical order as shown.

51. Install two bolts in front of oil pan (upper).

52. Install right and left intake valve timing control covers as follows:

a. Install new seal rings in shaft grooves.

b. Apply a continuous bead of liquid gasket to intake valve timing control covers.

c. Install new collared O-rings in front timing chain case oil hole (left and right sides).

d. Being careful not to move seal ring from the installation groove, align dowel pins on front timing chain case with the holes to install intake valve timing control covers.

e. Tighten bolts in numerical order as shown.

53. Install crankshaft pulley as follows:

a. Install crankshaft pulley, taking care not to damage front oil seal.

➡When press-fitting crankshaft pulley with plastic hammer, tap on its center portion (not circumference).

b. Tighten crankshaft pulley bolt. Crankshaft bolt torque: 33ft. lbs. (44.1 Nm)

c. Put a paint mark on crankshaft pulley aligning with angle mark on crankshaft pulley bolt. Then, further retighten bolt by 60°?(equivalent to one graduation).

54. Rotate crankshaft pulley in normal direction (clockwise when viewed from front) to confirm it turns smoothly.

55. Install oil pans (upper and lower).

56. Install rocker covers (right and left banks).

57. Installation of the remaining components is in the reverse order of removal after this step.

58. The following are procedures for checking fluid leaks, lubricant leaks and exhaust gases leaks.

a. Before starting engine, check oil/fluid levels including engine coolant and engine oil. If less than required quantity, fill to the specified level.

59. Use procedure below to check for fuel leakage.

a. Turn ignition switch "ON" (with engine stopped). With fuel pressure applied to fuel piping, check for fuel leakage at connection points.

b. Start engine. With engine speed increased, check again for fuel leakage at connection points.

c. Run engine to check for unusual noise and vibration.

➡If hydraulic pressure inside timing chain tensioner drops after removal/installation, slack in the guide may generate a pounding noise during and just after engine start. However, this is normal. Noise will stop after hydraulic pressure rises.

d. Warm up engine thoroughly to make sure there is no leakage of fuel, exhaust gases, or any oil/fluids including engine oil and engine coolant.

e. Bleed air from lines and hoses of applicable lines, such as in cooling system.

f. After cooling down engine, again check oil/fluid levels including engine oil and engine coolant. Refill to the specified level, if necessary.

V8 Engine

See Figures 112 through 129.

➡To remove timing chain and associated parts, start with those on the LH bank. The procedure for removing parts on the RH bank is omitted because it is the same as that for removal on the LH bank.

To install timing chain and associated parts, start with those on the RH bank. The procedure for installing parts on the LH bank is omitted because it is the same as that for installation on the RH bank.

1. Remove the engine assembly from the vehicle.

2. Remove the following components and related parts:

- Drive belt auto tensioner and idler pulley.
- Thermostat housing and water hose.
- Power steering oil pump bracket.
- Oil pan (lower), (upper) and oil strainer.
- Ignition coil.
- Rocker cover.

3. Remove the intake valve control solenoid valve cover (RH) (A) and intake valve control solenoid valve cover (LH) (B) as follows:

a. Loosen and remove the bolts as shown.

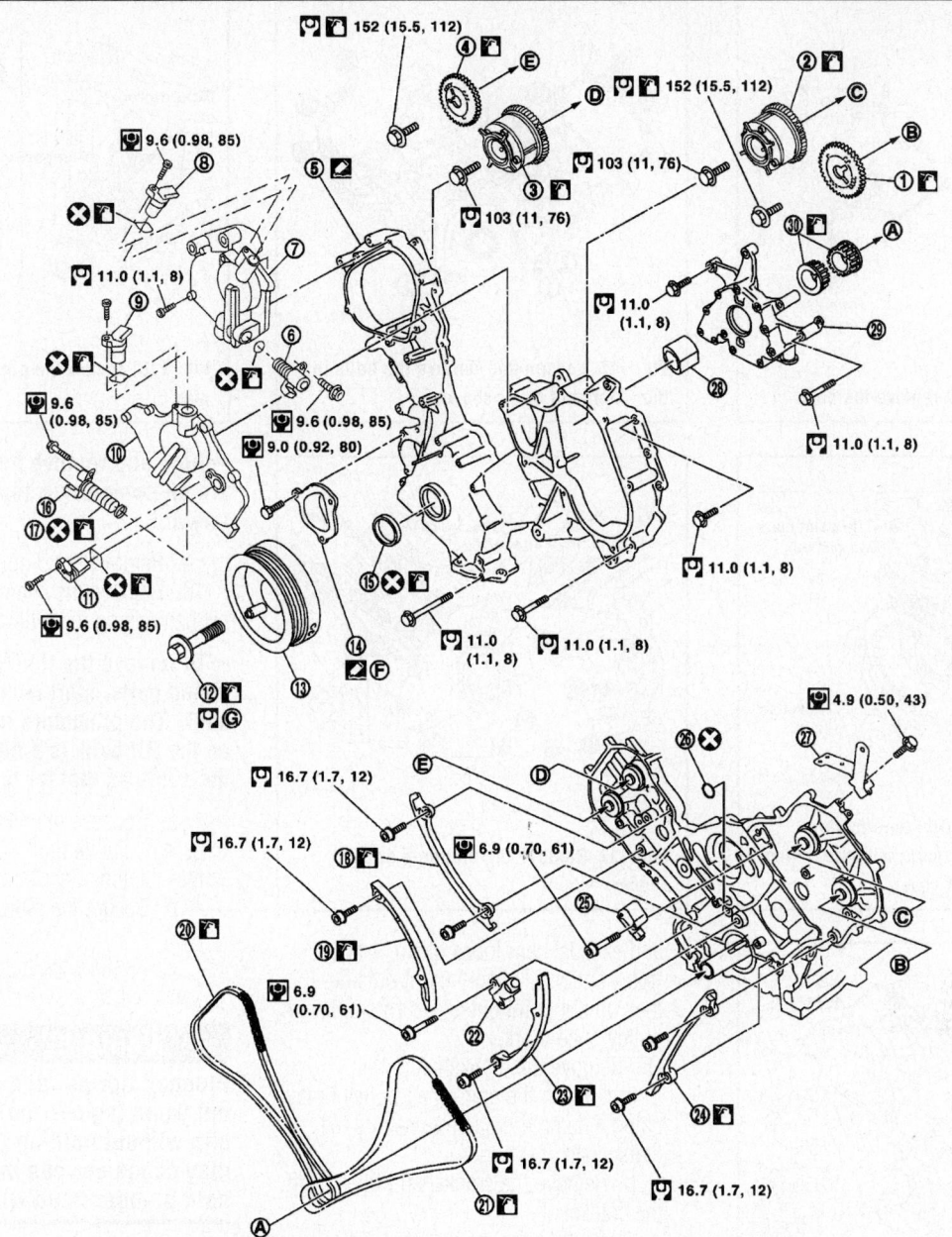

1. Camshaft sprocket LH bank (EXH)
2. Camshaft sprocket LH bank (INT) (VTC)
3. Camshaft sprocket RH bank (INT) (VTC)
4. Camshaft sprocket RH bank (EXH)
5. Front cover
6. Intake valve control solenoid valve (LH)
7. Intake valve control solenoid valve cover (LH)
8. Intake valve timing control position sensor (LH)
9. Intake valve timing control position sensor (RH)
10. Intake valve control solenoid valve cover (RH)
11. Camshaft position sensor (PHASE)
12. Crankshaft pulley bolt
13. Crankshaft pulley
14. Chain tensioner cover
15. Front oil seal
16. Intake valve control solenoid valve (RH)
17. O-ring
18. Timing chain tension guide RH bank

19. Timing chain slack guide (RH)
20. Timing chain LH bank
21. Timing chain (RH)
22. Chain tensioner (RH)
23. Timing chain slack guide LH bank
24. Timing chain tension guide LH bank
25. Chain tensioner (LH)
26. O-ring
27. Bracket
28. Oil pump drive spacer
29. Oil pump assembly
30. Crankshaft sprocket
A. To crankshaft
B. To camshaft LH bank (EXH)
C. To camshaft LH bank (INT) (VTC)
D. To camshaft RH bank (INT) (VTC)
E. To camshaft RH bank (EXH)
F. Apply sealant to mating side

22140_PATH_G0120

Fig. 112 Exploded view of timing chain, cover and sprocket components

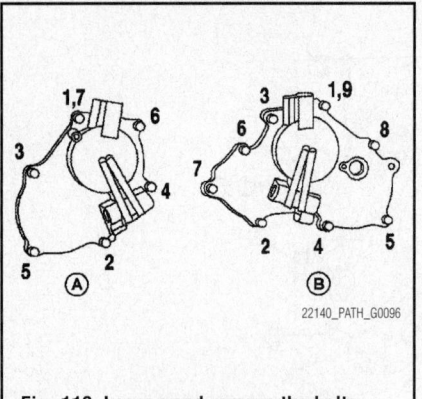

Fig. 113 Loosen and remove the bolts

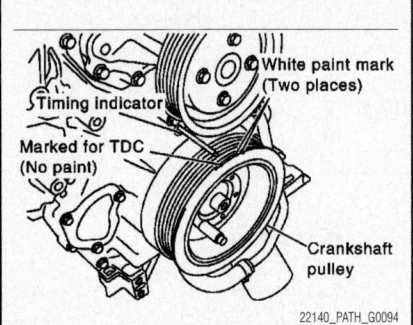

Fig. 114 Align the TDC identification notch (without paint mark) with the timing indicator on the front cover

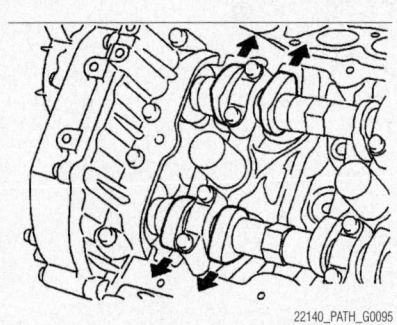

Fig. 115 Make sure both intake and exhaust cam lobes of No. 1 cylinder (top front on LH bank) point outside

b. Cut the liquid gasket and remove the covers.

** CAUTION**
Do not damage mating surfaces.

4. Obtain compression TDC of No. 1 cylinder as follows:
a. Turn the crankshaft pulley clockwise to align the TDC identification notch (without paint mark) with the timing indicator on the front cover.
b. At this time, make sure both intake

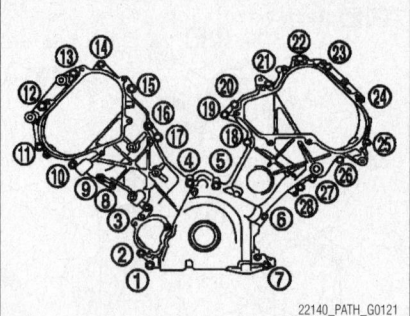

Fig. 116 Loosen and remove the bolts in the reverse of order shown

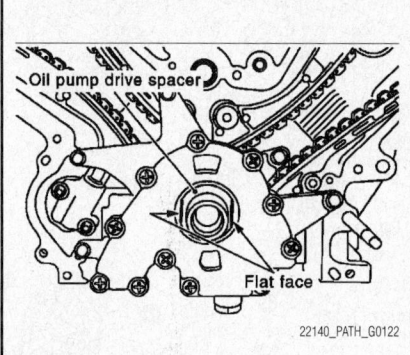

Fig. 117 Remove the oil pump drive spacer

and exhaust cam lobes of No. 1 cylinder (top front on LH bank) point outside. If they do not point outside, turn crankshaft pulley once more.
5. Remove the crankshaft pulley.
a. Loosen the crankshaft pulley bolts using a hammer handle to secure the crankshaft.
b. Remove the crankshaft pulley from the crankshaft.
c. Remove the crankshaft pulley.

➡**The dimension between the centers of the two bolt holes is 61 mm (2.40 in).**

6. Remove the front cover.
a. Loosen and remove the bolts in the reverse of order shown.
b. Cut the liquid gasket and remove the covers using Tool.

** CAUTION**
Do not damage mating surfaces.

7. Remove the front oil seal using suitable tool.

** CAUTION**
Do not damage front cover.

8. Remove the oil pump drive spacer.

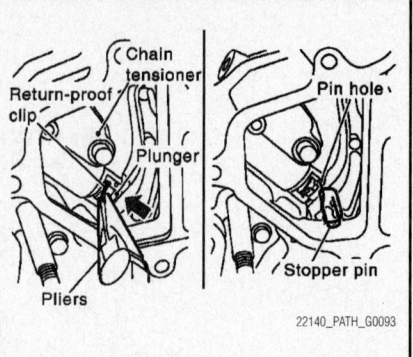

Fig. 118 Secure the plunger using stopper pin

➡**Hold and remove the flat space of the oil pump drive spacer by pulling it forward.**

9. Remove the oil pump.
10. Remove the chain tensioner on the LH bank using the following steps.

➡**To remove the timing chain and associated parts, start with those on the LH bank. The procedure for removing parts on the RH bank is omitted because it is the same as that for the LH bank.**

a. Squeeze the return-proof clip ends using suitable tool and push the plunger into the tensioner body.
b. Secure the plunger using stopper pin.
c. Remove the bolts and chain tensioner.

** WARNING**
Plunger, spring, and spring seat pop out when (squeezing) return-proof clip without holding plunger head. It may cause serious injuries. Always hold plunger head when removing.

➡**Stop the plunger in the fully extended position by using the return-proof clip (1) if the stopper pin is removed. Push the plunger (2) into the tensioner body while squeezing the return-proof clip (1). Secure it using stopper pin (3).**

11. Remove the timing chain tension guide and timing chain slack guide.
12. Remove the timing chain and crankshaft sprocket.
13. Loosen the camshaft sprocket bolts as shown and remove the camshaft sprocket.

** CAUTION**
To avoid interference between valves and pistons, do not turn crankshaft or camshaft when timing chain is disconnected.

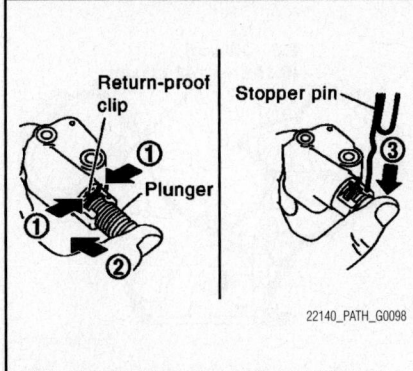

Fig. 119 Stop the plunger in the fully extended position

14. Repeat the same procedure to remove the RH timing chain and associated parts.

To install:

➡The above figure shows the relationship between the mating mark on each timing chain and that of the corresponding sprocket, with the components installed.

To install the timing chain and associated parts, start with those on the RH bank. The procedure for installing parts on the LH bank is omitted because it is the same as that for installation on the RH bank.

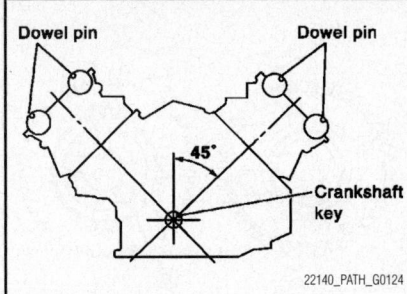

Fig. 121 Make sure the crankshaft key and RH bank camshaft dowel pin and LH bank camshaft dowel pin are facing in the direction

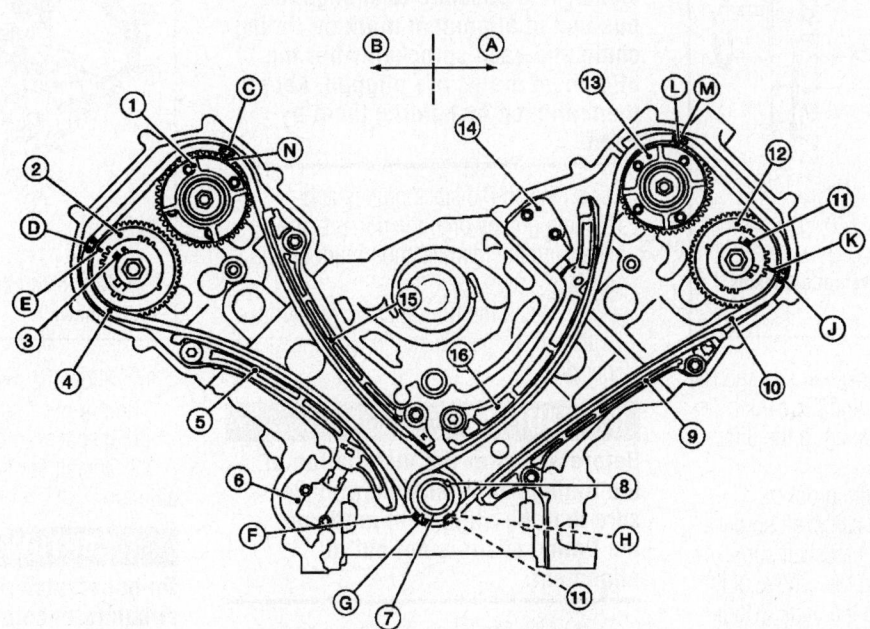

1. RH bank Camshaft sprocket (INT) (VTC)	2. RH bank Camshaft sprocket (EXH)	3. RH bank camshaft dowel pin
4. Timing chain	5. RH bank Timing chain slack guide	6. Primary timing chain tensioner
7. Crankshaft sprocket	8. Crankshaft key	9. LH Timing chain tension guide
10. Timing chain	11. LH Camshaft dowel pin	12. LH bank Camshaft sprocket (EXH)
13. LH bank Camshaft sprocket (INT) (VTC)	14. Secondary timing chain tensioner	15. RH bank timing chain tension guide
16. LH timing chain slack guide	A. LH bank	B. RH bank
C. Alignment mark (Link color: copper)	D. Alignment mark (Link color: copper)	E. Alignment mark (Identification mark)
F. Alignment mark for LH bank (Notch)	G. Alignment mark for LH bank (Link color: Yellow)	H. Alignment mark for RH bank (Link color: Yellow)
J. Alignment mark (Link color: copper)	K. Alignment mark (Identification mark)	L. Alignment mark (Identification mark)
M. Alignment mark (Link color: copper)	N. Alignment mark (Identification mark)	

Fig. 120 Relationship between the mating mark on each timing chain and that of the corresponding sprocket

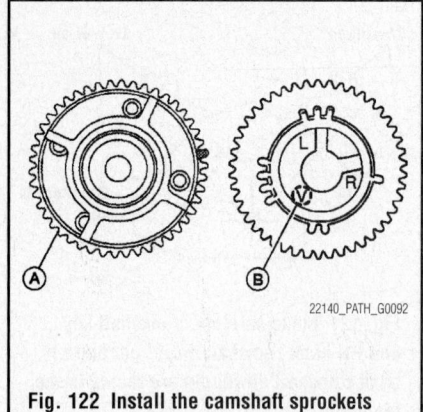

Fig. 122 Install the camshaft sprockets

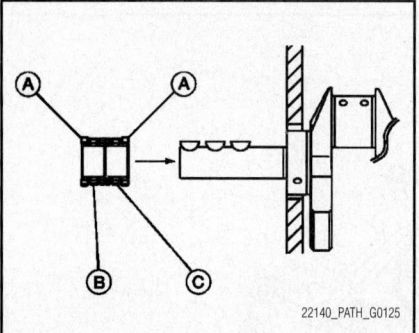

Fig. 123 Install the crankshaft sprockets for both banks

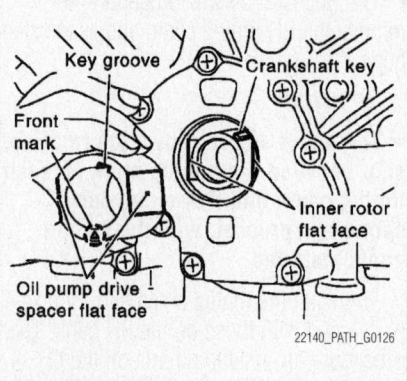

Fig. 124 Install the oil pump drive spacer

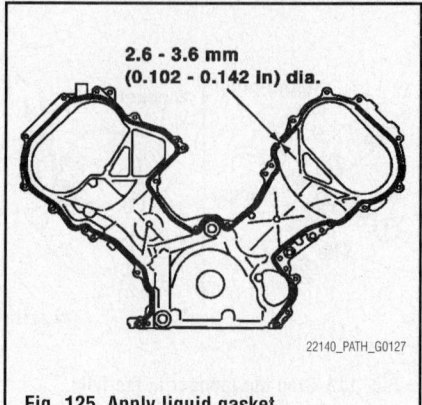

Fig. 125 Apply liquid gasket

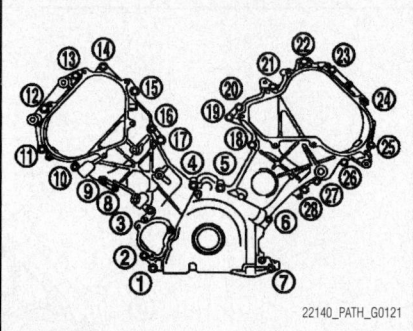

Fig. 126 Install the bolts in the numerical order shown

15. Make sure the crankshaft key and RH bank camshaft dowel pin and LH bank camshaft dowel pin are facing in the direction shown.

16. Install the camshaft sprockets.

a. Install the intake camshaft sprocket (VTC) (A) and exhaust camshaft sprockets (B) by selectively using the groove of the dowel pin according to the bank. (Common part used for both exhaust banks.)

b. Lock the hexagonal part of the camshaft in the same way as for removal, and tighten the bolts.

- A: Intake
- B = V: Exhaust

17. Install the crankshaft sprockets for both banks.

a. Install LH bank crankshaft sprocket (B) and RH bank crankshaft sprocket (C) so that their flange side (A) (the larger diameter side without teeth) faces in the direction shown.

➡The same parts are used but facing directions are different.

18. Install the timing chains and associated parts.

a. Align the alignment mark on each sprocket and the timing chain for installation.

✳✳ CAUTION

Before installing timing chain tensioner, it is possible to change the position of alignment mark on timing chain and each sprocket. After the alignment marks are aligned, keep them aligned by holding them by hand.

b. Install the slack guides and tension guides onto the correct side by checking the identification mark on the surface.

c. Install the timing chain tensioner with the plunger locked in with the stopper pin.

✳✳ CAUTION

Before and after the installation of the timing chain tensioner, make sure that the alignment mark on the timing chain is not out of alignment.

d. After installing the timing chain tensioner, remove the stopper pin to release the tensioner. Make sure the tensioner is released.

e. To avoid chain-link skipping of the timing chain, do not move crankshaft or camshafts until the front cover is installed.

19. In the same way as for the RH bank, install the timing chain and associated parts on the LH bank.

20. Install the oil pump.

21. Install the oil pump drive spacer as follows:

a. Install so that the front mark on the front edge of the oil pump drive spacer faces the front of the engine.

b. Insert the oil pump drive spacer according to the directions of the crankshaft key and the two flat surfaces of the oil pump inner rotor.

c. If the positional relationship does

not allow the insertion, rotate the oil pump inner rotor to allow the oil pump drive spacer to be inserted.

22. Install the front oil seal using suitable tool.

✳✳ CAUTION

Do not scratch or make burrs on the circumference of the oil seal.

23. Install the chain tensioner cover.

24. Install the front cover as follows:

a. Install a new O-ring on the cylinder block.

b. Apply liquid gasket as shown.

c. Check again that the timing alignment marks on the timing chain and on each sprocket are aligned. Then install the front cover.

d. Install the bolts in the numerical order shown.

e. After tightening, re-tighten to the specified torque.

✳✳ CAUTION

Be sure to wipe off any excessive liquid gasket leaking onto surface mating with oil pan.

25. Install the Intake valve control solenoid valve cover (RH) (A) and Intake valve

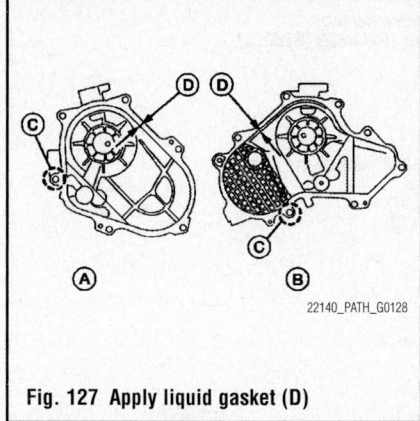

Fig. 127 Apply liquid gasket (D)

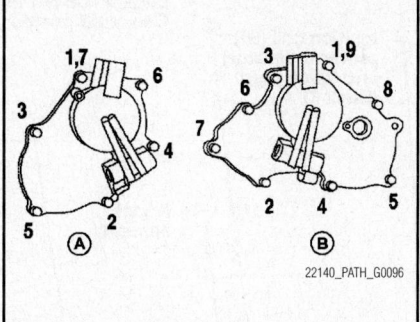

Fig. 128 Install the bolts in the numerical order shown

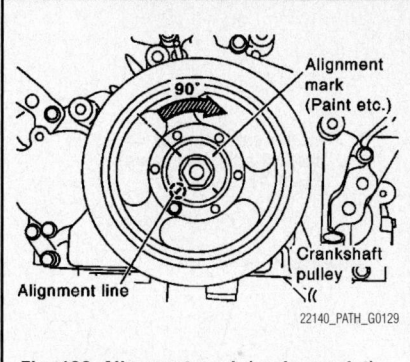

Fig. 129 Alignment mark (such as paint) on the crankshaft pulley

control solenoid valve cover (LH) (B) as follows:

    a. Cross mark (C) that cannot be seen after assembly.

    b. Apply liquid gasket (D) as shown.

✳✳ CAUTION

The start and end of the application of the liquid gasket should be crossed at a position that cannot be seen after attaching the Intake valve control solenoid valve cover.

    c. Install the bolts in the numerical order shown.

26. Install the crankshaft pulley.

    a. Install the key of the crankshaft.

    b. Insert the pulley by lightly tapping it.

✳✳ CAUTION

Do not tap pulley on the side surface where the belt is installed (outer circumference).

27. Tighten the crankshaft pulley bolt.

    a. Lock the crankshaft using suitable tool, then tighten the bolt.

    b. Perform the following steps for angular tightening:

    c. Apply engine oil onto the threaded parts of the bolt and seating area.

    d. Select the one most visible notch of the four on the bolt flange. Corresponding to the selected notch, put a alignment mark (such as paint) on the crankshaft pulley.

28. Rotate the crankshaft pulley in normal direction (clockwise when viewed from engine front) to check for parts interference.

29. Installation of the remaining components is in the reverse of order of removal.

ENGINE PERFORMANCE & EMISSION CONTROL

COMPONENT LOCATIONS

V6 Engine

See Figure 130.

V8 Engine

See Figure 131.

CAMSHAFT POSITION (CMP) SENSOR

LOCATION

Camshaft position sensors are located on the right and left bank cylinder heads at the back side.

REMOVAL & INSTALLATION

1. Loosen the fixing bolt of the sensor.
2. Disconnect the electrical connector.
3. Remove the bolt securing the sensor.
4. Installation is the reverse of removal.

CRANKSHAFT POSITION (CKP) SENSOR

LOCATION

The crankshaft position sensor (POS) is located on the A/T assembly facing the gear teeth (cogs) of the signal plate.

REMOVAL & INSTALLATION

1. Loosen the fixing bolt of the sensor.
2. Disconnect crankshaft position sensor (POS) harness connector.
3. Remove the sensor.
4. Visually check the sensor for chipping.

ELECTRONIC CONTROL MODULE (ECM)

LOCATION

The Electronic Control Module (ECM) is located in the engine room passenger side behind reservoir tank.

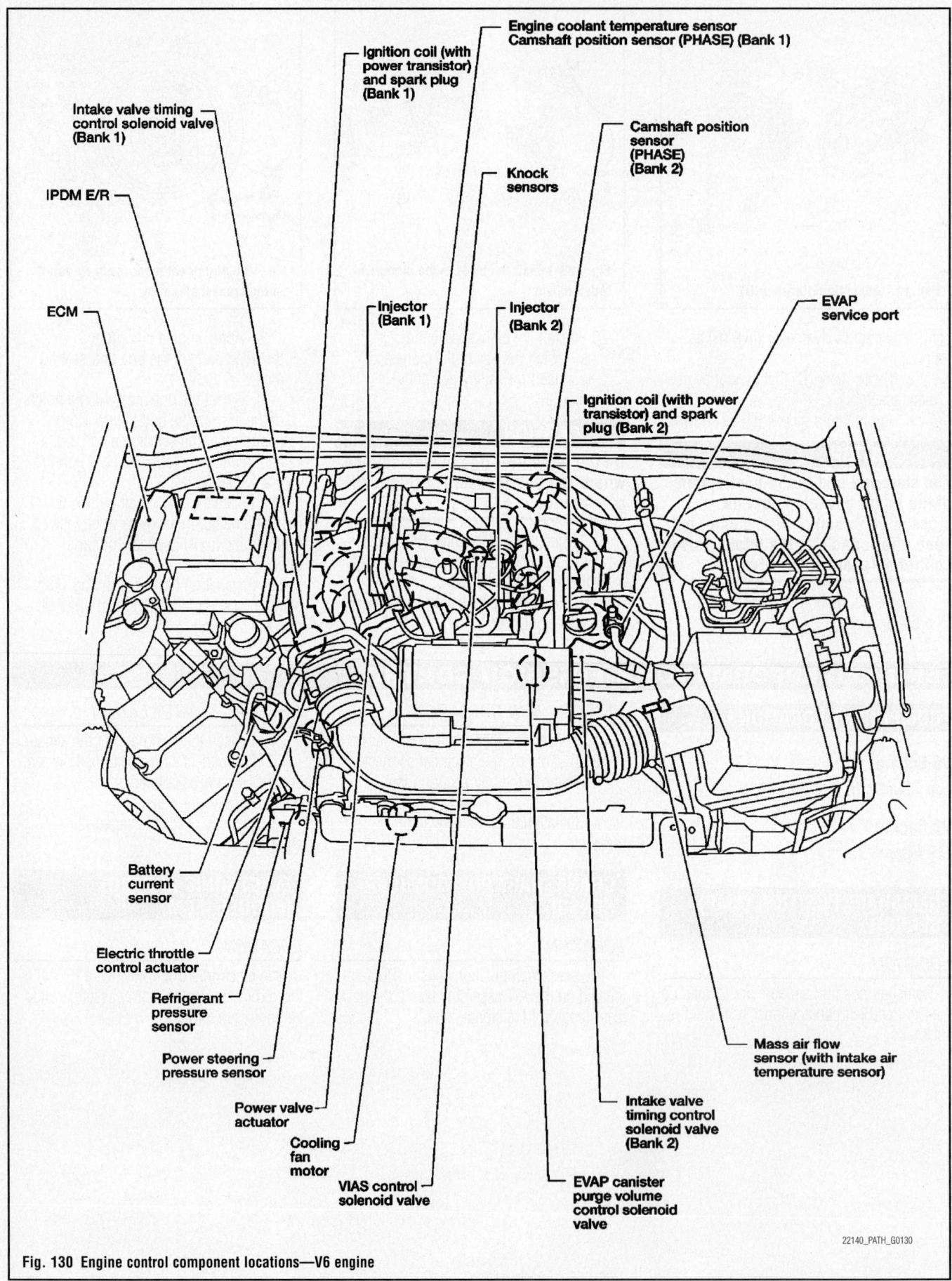

Fig. 130 Engine control component locations—V6 engine

22140_PATH_G0130

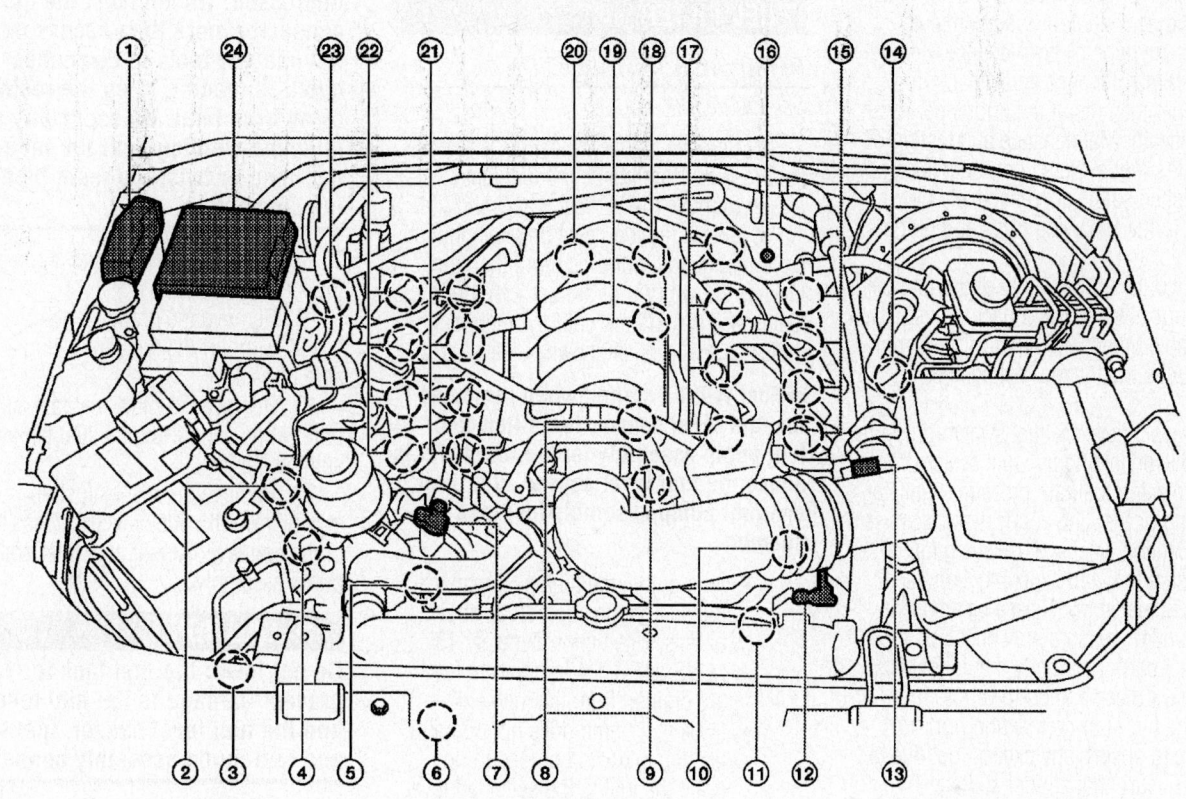

1. ECM
2. Battery current sensor
3. Refrigerant pressure sensor
4. Power steering pressure sensor
5. Intake valve timing control solenoid valve (bank 2)
6. Cooling fan motor
7. Intake valve timing control position sensor (bank 2)
8. Engine coolant temperature sensor
9. Electric throttle control actuator
10. Intake valve timing control position sensor (bank 1)
11. Intake valve timing control solenoid valve (bank 1)
12. Camshaft position sensor (PHASE)
13. Mass air flow sensor (with intake air temperature sensor)
14. A/F sensor 1 (bank 1)
15. Ignition coil (with power transistor) and spark plug (bank 1)
16. EVAP service port
17. Fuel injector (bank 1)
18. Knock sensor (bank 1)
19. EVAP canister purge volume control solenoid valve
20. Knock sensor (bank 2)
21. Fuel injector (bank 2)
22. Ignition coil (with power transistor) and spark plug (bank 2)
23. A/F sensor 1 (bank 2)
24. IPDM E/R

22140_PATH_G0131

Fig. 131 Engine control component locations—V8 engine

FUEL SYSTEM SERVICE PRECAUTIONS

Safety is the most important factor when performing not only fuel system maintenance but any type of maintenance. Failure to conduct maintenance and repairs in a safe manner may result in serious personal injury or death. Maintenance and testing of the vehicle's fuel system components can be accomplished safely and effectively by adhering to the following rules and guidelines.

• To avoid the possibility of fire and personal injury, always disconnect the negative battery cable unless the repair or test procedure requires that battery voltage be applied.

• Always relieve the fuel system pressure prior to disconnecting any fuel system component (injector, fuel rail, pressure regulator, etc.), fitting or fuel line connection. Exercise extreme caution whenever relieving fuel system pressure to avoid exposing skin, face and eyes to fuel spray. Please be advised that fuel under pressure may penetrate the skin or any part of the body that it contacts.

• Always place a shop towel or cloth around the fitting or connection prior to loosening to absorb any excess fuel due to spillage. Ensure that all fuel spillage (should it occur) is quickly removed from engine surfaces. Ensure that all fuel soaked cloths or towels are deposited into a suitable waste container.

• Always keep a dry chemical (Class B) fire extinguisher near the work area.

• Do not allow fuel spray or fuel vapors to come into contact with a spark or open flame.

• Always use a back-up wrench when loosening and tightening fuel line connection fittings. This will prevent unnecessary stress and torsion to fuel line piping.

• Always replace worn fuel fitting O-rings with new Do not substitute fuel hose or equivalent where fuel pipe is installed.

Before servicing the vehicle, make sure to also refer to the precautions in the beginning of this section as well.

RELIEVING FUEL SYSTEM PRESSURE

1. Remove fuel pump fuse located in IPDM E/R.
2. Start engine.
3. After engine stalls, crank it two or three times to release all fuel pressure.

4. Turn ignition switch OFF.
5. Reinstall fuel pump fuse after servicing fuel system.

FUEL FILTER

REMOVAL & INSTALLATION

See Figure 132.

1. Remove the fuel filler cap to release the pressure from inside the fuel tank.
2. Remove the LH rear wheel and tire.
3. Check the fuel level on level gauge. If the fuel gauge indicates more than the level as shown (full or almost full), drain the fuel from the fuel tank until the fuel gauge indicates the level as shown, or less.

➡**Fuel will be spilled when removing the fuel level sensor, fuel filter, and fuel pump assembly for the fuel level is above the fuel level sensor, fuel filter, and fuel pump assembly fuel tank opening.**

• As a guide, the fuel level reaches the fuel gauge position as shown, or less, when approximately 15 (4 US gal, 3 1/4 Imp gal) of fuel are drained from the fuel tank.
• If the fuel pump does not operate, use the following procedure to drain the fuel to the specified level.
  a. Insert a suitable hose of less than 15 mm (0.59 in) diameter into the fuel filler pipe through the fuel filler opening to drain the fuel from fuel filler pipe.
  b. Remove the fuel filler pipe shield.
  c. Disconnect the fuel filler hose from the fuel filler pipe.
  d. Insert a suitable hose into the fuel tank through the fuel filler hose to drain the fuel from the fuel tank.

4. Release the fuel pressure from the fuel lines.
5. Disconnect the battery negative terminal.
6. Disconnect the lower fuel filler hose from the fuel tank, the EVAP hose, and the vent pipe quick connector.

• Disconnect the fuel feed hose from the molded clip in the side of the fuel tank.
• Disconnect the quick connector as follows:
• Hold the sides of the connector, push in the tabs and pull out the tube.
• If the connector and the tube are stuck together, push and pull several times until they start to move. Then disconnect them by pulling.

❊❊ CAUTION

The quick connector can be disconnected when the tabs are completely depressed. Do not twist the quick connector more than necessary. Do not use any tools to disconnect the quick connector. Keep the resin tube away from heat. Be especially careful when welding near the tube. Do not bend or twist the resin tube during connection.

7. Remove the four bolts and remove the fuel tank shield.
8. Remove the propeller shaft.
9. Support the fuel tank using a suitable lift jack.
10. Remove the three fuel tank strap bolts while supporting the fuel tank with a suitable lift jack.
11. Remove the fuel tank straps and slowly lower the fuel tank to access the top of the fuel level sensor, fuel filter and fuel pump assembly.

❊❊ CAUTION

Do not lower the fuel tank too far to prevent damage to the fuel feed hose and the fuel level sensor, fuel filter and fuel pump assembly connector

12. Disconnect the fuel level sensor, fuel filter, and fuel pump assembly electrical connector, and the fuel feed hose.
  a. Disconnect the quick connector as follows:
  • Hold the sides of the connector, push in the tabs and pull out the tube.
  • If the connector and the tube are stuck together, push and pull several times until they start to move. Then disconnect them by pulling.

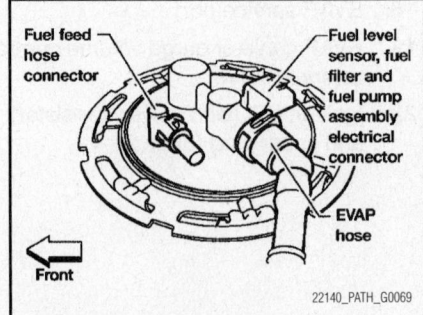

22140_PATH_G0069

Fig. 132 Disconnect the fuel level sensor, fuel filter, and fuel pump assembly electrical connector, and the fuel feed hose

⁂ CAUTION

The quick connector can be discon-nected when the tabs are completely depressed. Do not twist the quick connector more than necessary. Do not use any tools to disconnect the quick connector. Do not bend or twist the resin tube during connection.

13. Lower the fuel tank using a suitable lift jack and remove the fuel tank.

14. Disconnect the EVAP hose from the fuel pump and remove the EVAP hose from the molded clip in the top of the fuel tank.

15. Remove the lock ring.

16. Remove the fuel level sensor, fuel fil-ter, and fuel pump assembly.

⁂ CAUTION

Do not bend the float arm during removal. Avoid impacts such as drop-ping when handling the components.

To install:

17. Installation is in the reverse order of removal.

18. Connect the quick connector as follows:

- Check the connection for any dam-age or foreign materials.
- Align the connector with the pipe, then insert the connector straight into the pipe until a click is heard.
- Pull the tube and the connector to make sure they are securely con-nected.
- Visually inspect the connector to make sure the two retainer tabs are securely connected.

⁂ CAUTION

Do not bend the float arm during installation. Avoid impacts such as dropping when handling the compo-nents.

19. Turn the ignition switch ON but do not start engine, then check the fuel pipe and hose connections for leaks while apply-ing fuel pressure.

20. Start the engine and rev it above idle, then check that there are no fuel leaks at any of the fuel pipe and hose connec-tions.

FUEL INJECTORS

REMOVAL & INSTALLATION

V6 Engine

See Figures 133 through 135.

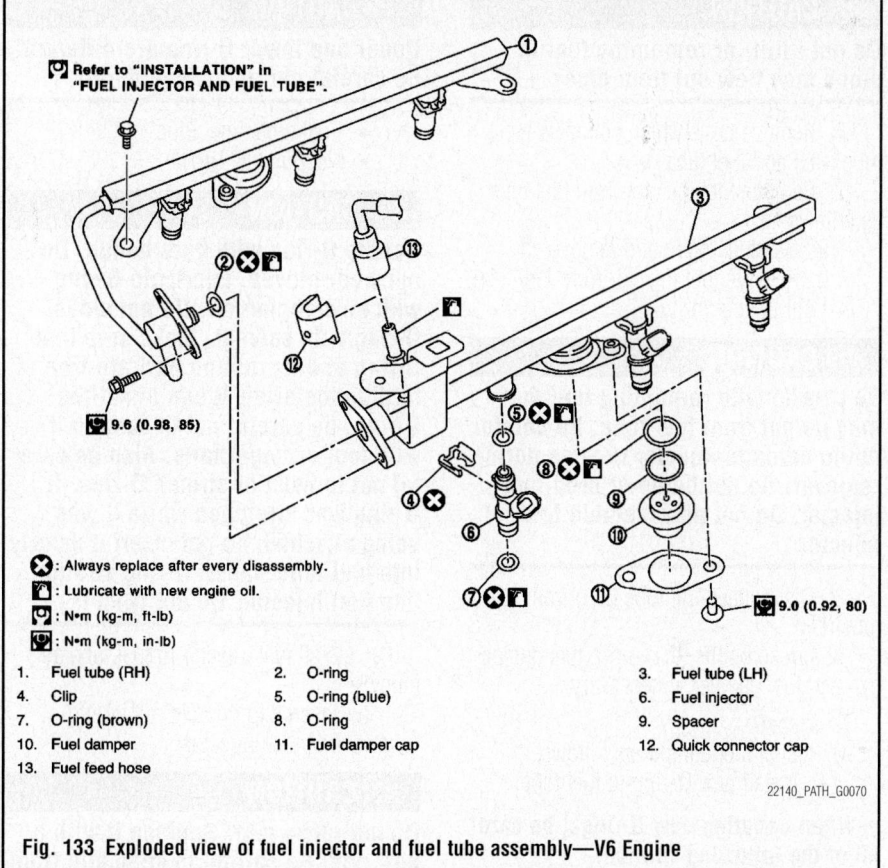

⊗ : Always replace after every disassembly.
■ : Lubricate with new engine oil.
⏧ : N•m (kg-m, ft-lb)
⏧ : N•m (kg-m, in-lb)

1. Fuel tube (RH)
2. O-ring
3. Fuel tube (LH)
4. Clip
5. O-ring (blue)
6. Fuel injector
7. O-ring (brown)
8. O-ring
9. Spacer
10. Fuel damper
11. Fuel damper cap
12. Quick connector cap
13. Fuel feed hose

9.6 (0.98, 85)   9.0 (0.92, 80)

22140_PATH_G0070

Fig. 133 Exploded view of fuel injector and fuel tube assembly—V6 Engine

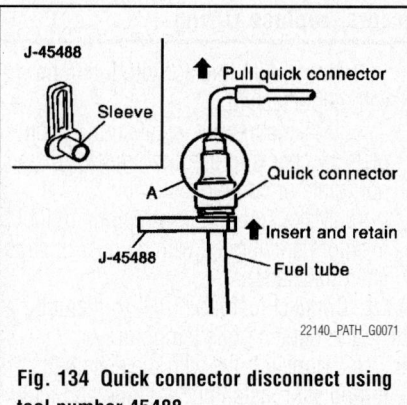

J-45488
Sleeve
Pull quick connector
Quick connector
A
Insert and retain
J-45488
Fuel tube

22140_PATH_G0071

Fig. 134 Quick connector disconnect using tool number 45488

1. Remove intake manifold collector.

⁂ CAUTION

Perform this step when engine is cold.

2. Disconnect the fuel quick connector on the engine side.

   a. Using Tool number -45488, per-form the following steps to disconnect the quick connector.

   b. Remove quick connector cap.

   c. With the sleeve side of Tool facing quick connector, install Tool onto fuel tube.

   d. Insert Tool into quick connector until sleeve contacts and goes no further. Hold the Tool on that position.

⁂ CAUTION

Inserting the Tool hard will not dis-connect quick connector. Hold Tool where it contacts and goes no further.

   e. Pull the quick connector straight out from the fuel tube.

⁂ CAUTION

Pull quick connector holding it at the A position. Do not pull with lateral force applied. O-ring inside quick connector may be damaged. Prepare container and cloth beforehand as fuel will leak out. Avoid fire and sparks. Be sure to cover openings of disconnected pipes with plug or plas-tic bag to avoid fuel leakage and entry of foreign materials.

3. Remove PCV hose between rocker covers (right and left banks).

4. Disconnect harness connector from fuel injector.

5. Loosen bolts in reverse order of the tightening sequence, and remove fuel tube and fuel injector assembly.

✳✳ CAUTION

Do not tilt it, or remaining fuel in pipes may flow out from pipes.

6. Remove bolts which connects fuel tube (RH) and fuel tube (LH).

7. Remove fuel injector from fuel tube as follows:

　a. Carefully open and remove clip.

　b. Remove fuel injector from fuel tube by pulling straight.

✳✳ CAUTION

Be careful with remaining fuel that may go out from fuel tube. Be careful not to damage injector nozzles during removal. Do not bump or drop fuel injector. Do not disassemble fuel injector.

8. Disconnect fuel tube (RH) from fuel tube (LH).

9. Loosen bolts, to remove fuel damper cap and fuel damper, if necessary.

To install:

10. Install fuel damper as follows:

　a. Install new O-ring to fuel tube.

➡When handling new O-rings, be careful of the following caution:

✳ CAUTION

Handle O-ring with bare hands. Do not wear gloves. Lubricate O-ring with new engine oil. Do not clean O-ring with solvent. Make sure that O-ring and its mating part are free of foreign material. When installing O-ring, be careful not to scratch it with tool or fingernails. Also be careful not to twist or stretch O-ring. If O-ring was stretched while it was being attached, do not insert it quickly into fuel tube. Insert new O-ring straight into fuel tube. Do not twist it.

　b. Install spacer to fuel damper.

　c. Insert fuel damper straight into fuel tube.

✳ CAUTION

Insert straight, making sure that the axis is lined up. Do not pressure-fit with excessive force.

　d. Tighten bolts evenly in turn.

➡After tightening bolts, make sure that there is no gap between fuel damper cap and fuel tube.

11. Install new O-rings to fuel injector, paying attention to the following.

✳✳ CAUTION

Upper and lower O-ring are different. Be careful not to confuse them.

- Fuel tube side: Blue
- Nozzle side: Brown

✳✳ CAUTION

Handle O-ring with bare hands. Do not wear gloves. Lubricate O-ring with new engine oil. Do not clean O-ring with solvent. Make sure that O-ring and its mating part are free of foreign material. When installing O-ring, be careful not to scratch it with tool or fingernails. Also be careful not to twist or stretch O-ring. If O-ring was stretched while it was being attached, do not insert it quickly into fuel tube. Insert O-ring straight into fuel injector. Do not twist it.

12. Install fuel injector to fuel tube as follows:

　a. Insert clip into clip mounting groove on fuel injector.

✳✳ CAUTION

Do not reuse clip. Replace it with a new one. Be careful to keep clip from interfering with O-ring. If interference occurs, replace O-ring.

　b. Insert fuel injector into fuel tube with clip attached.

　c. Make sure that installation is complete by checking that fuel injector does not rotate or come off.

　d. Make sure that protrusions of fuel injectors are aligned with cutouts of clips after installation.

13. Connect fuel tube (RH) to fuel tube (LH), and tighten bolts temporarily.

　a. Tighten bolts with the specified torque after installing fuel tube and fuel injector assembly.

✳✳ CAUTION

Handle O-ring with bare hands. Do not wear gloves. Lubricate O-ring with new engine oil. Do not clean O-ring with solvent. Make sure that O-ring and its mating part are free of foreign material. When installing O-ring, be careful not to scratch it with tool or fingernails. Also be careful not to twist or stretch O-ring. If O-ring was stretched while it was being attached, do not insert it quickly into fuel tube. Insert new O-ring straight into fuel tube. Do not twist it.

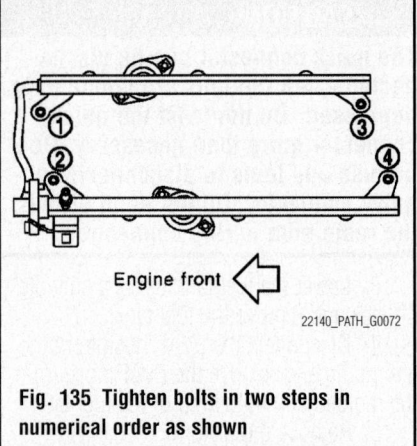

Engine front

22140_PATH_G0072

Fig. 135 Tighten bolts in two steps in numerical order as shown

14. Install fuel tube and fuel injector assembly to intake manifold.

✳✳ CAUTION

Be careful not to let tip of injector nozzle come in contact with other parts.

15. Tighten bolts in two steps in numerical order as shown.

　a. Fuel injector tube assembly bolts: 1st step: 7 ft. lbs. (10.1 Nm); 2nd step: 16 ft. lbs. (22.0 Nm)

16. Tighten bolts which connects fuel tube (RH) and fuel tube (LH) with the specified torque.

17. Connect fuel injector harness connector.

18. Install intake manifold collector.

19. Installation of the remaining components is in the reverse order of removal.

20. Turn ignition switch "ON" (with engine stopped). With fuel pressure applied to fuel piping, check for fuel leakage at connection points.

➡Use mirrors for checking at points out of clear sight.

21. Start engine. With engine speed increased, check again for fuel leakage at connection points.

V8 Engine

See Figures 136 through 140.

1. Disconnect the negative battery terminal.

2. Remove the engine room cover.

3. Release the fuel pressure.

4. Remove the air duct and resonator assembly.

5. Disconnect the fuel injector harness connectors.

6. Disconnect the fuel hose assembly from the fuel tubes (RH and LH).

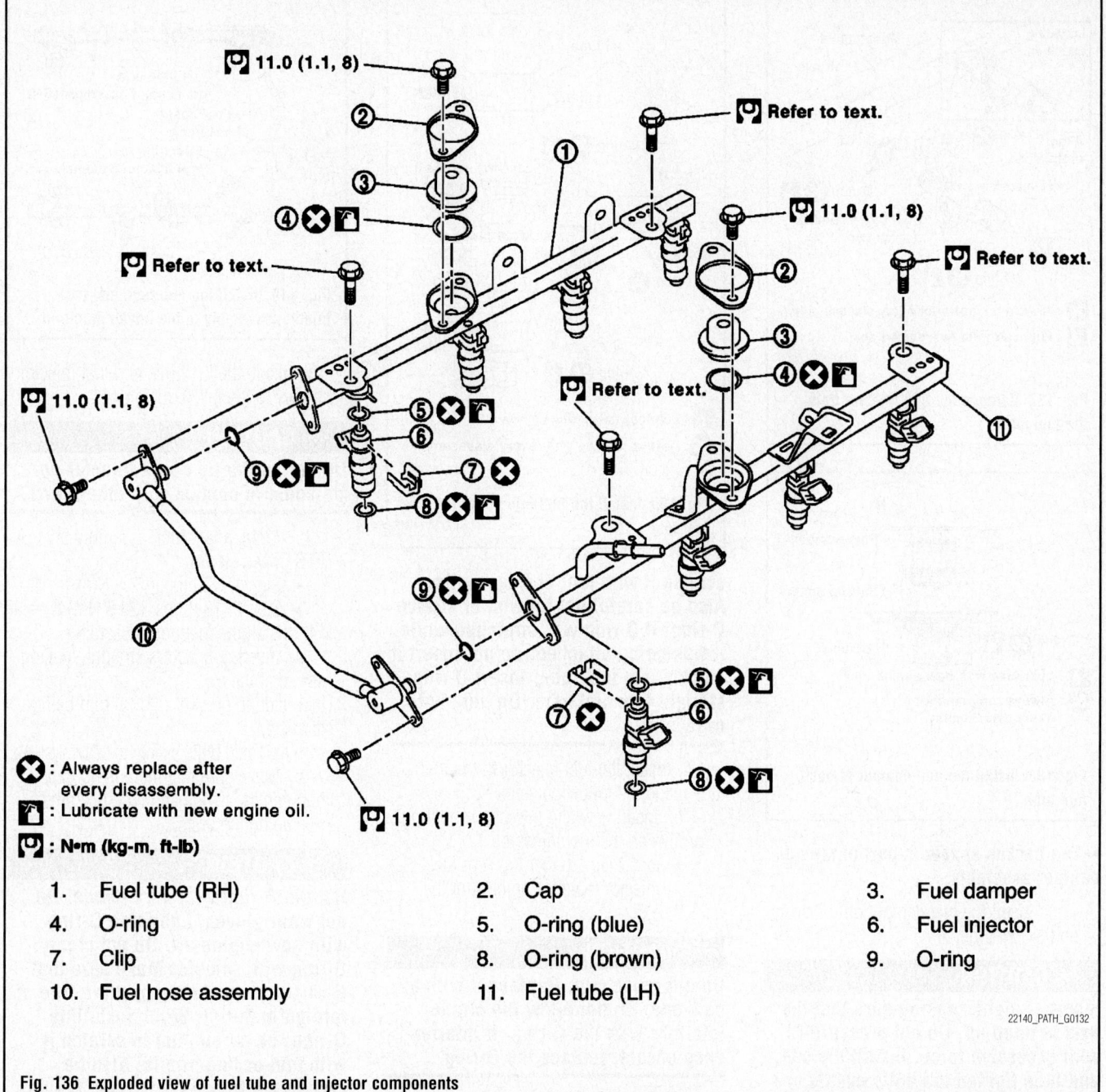

Fig. 136 Exploded view of fuel tube and injector components

⊗ : Always replace after every disassembly.

🛢 : Lubricate with new engine oil.

[🔧] : N•m (kg-m, ft-lb)

1. Fuel tube (RH)
2. Cap
3. Fuel damper
4. O-ring
5. O-ring (blue)
6. Fuel injector
7. Clip
8. O-ring (brown)
9. O-ring
10. Fuel hose assembly
11. Fuel tube (LH)

22140_PATH_G0132

✳✳ CAUTION

While hoses are disconnected, plug them to prevent fuel from draining. Do not separate the fuel connector and fuel hose.

7. Remove the fuel injectors with the fuel tube assembly.

8. Remove the fuel injector from the fuel tube using the following steps.

a. Spread open and remove the clip.

b. Remove the fuel injector from the fuel tube by pulling straight out.

✳✳ CAUTION

Be careful with remaining fuel that may leak out from fuel tube. Do not damage injector nozzles during removal. Do not bump or drop fuel injectors. Do not disassemble fuel injectors.

9. Remove the fuel damper from each fuel tube.

10. Install the fuel damper to each fuel tube using the following steps.

a. Apply engine oil to the new O-ring and set it into the cup of the fuel tube.

✳✳ CAUTION

Handle O-ring with bare hands. Never wear gloves. Lubricate new O-ring with new engine oil. Do not clean O-ring with solvent. Make sure that O-ring and its mating part are free of foreign material. When installing O-ring, do not scratch it with tool or fingernails. Do not twist or stretch the O-ring.

b. Make sure that the backup spacer is in the O-ring connecting surface of the fuel damper.

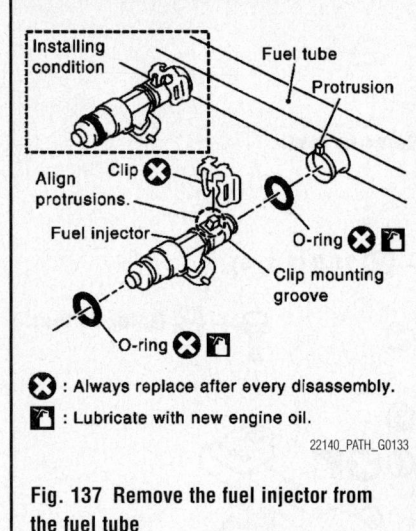

⊗ : Always replace after every disassembly.

▨ : Lubricate with new engine oil.

22140_PATH_G0133

Fig. 137 Remove the fuel injector from the fuel tube

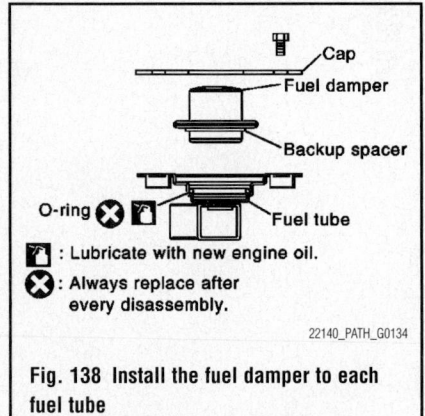

▨ : Lubricate with new engine oil.

⊗ : Always replace after every disassembly.

22140_PATH_G0134

Fig. 138 Install the fuel damper to each fuel tube

→The backup spacer is part of the fuel damper assembly.

c. Insert the fuel damper until it seats on the fuel tube.

✳✳ CAUTION
Insert straight, making sure that the axis is lined up. Do not pressure-fit with excessive force. Install the cap, and then tighten the bolts evenly. After tightening the bolts, make sure that there is no gap between the cap and fuel tube.

11. Install new O-rings to the fuel injector paying attention to the items below.

✳✳ CAUTION
Upper and lower O-rings are different colors. Handle O-ring with bare hands. Never wear gloves. Lubricate new O-ring with new engine oil. Do not clean O-ring with solvent. Make sure that O-ring and its mating part are free of foreign material. When installing O-ring, be careful not to

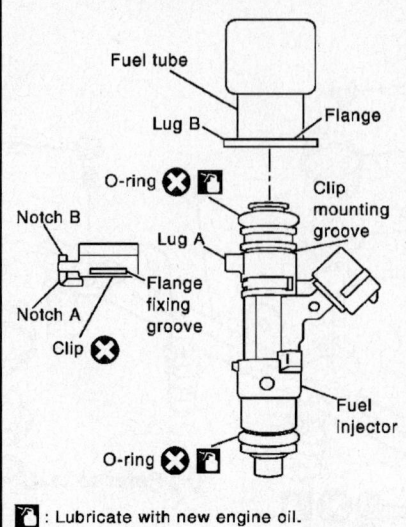

▨ : Lubricate with new engine oil.

⊗ : Always replace after every disassembly.

22140_PATH_G0135

Fig. 139 Install the fuel injector to the fuel tube

scratch it with tool or fingernails. Also be careful not to twist or stretch O-ring. If O-ring was stretched while it was being attached, do not insert it quickly into fuel tube. Insert O-ring straight into fuel tube. Do not angle or twist it.

12. Install the fuel injector to the fuel tube using the following steps.

a. Insert new clip into clip mounting groove on the fuel injector.

- Insert clip so that lug A of fuel injector matches notch A of the clip.

✳✳ CAUTION
Do not reuse clip. Replace it with a new one. Do not allow the clip to interfere with the O-ring. If interference occurs, replace the O-ring.

b. Insert the fuel injector into the fuel tube with the clip attached.

- Insert it while matching it to the axial center.
- Insert fuel injector so that lug B of fuel tube matches notch B of the clip.
- Make sure that the fuel tube flange is securely seated in the flange fixing groove on the clip.

c. Make sure that installation is complete by checking that the fuel injector does not rotate or come off.

- Make sure that the protrusions of the fuel injectors are aligned with the cutouts of the clips after installation.

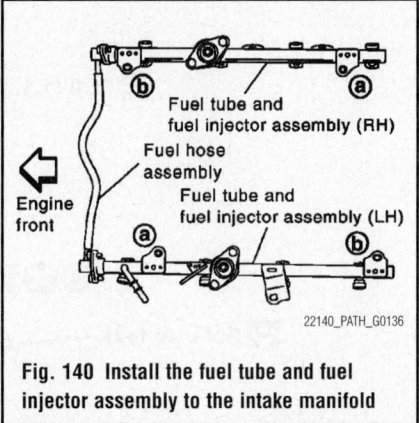

22140_PATH_G0136

Fig. 140 Install the fuel tube and fuel injector assembly to the intake manifold

13. Install the fuel tube and fuel injector assembly to the intake manifold.

✳✳ CAUTION
Do not let the tip of the injector nozzle come in contact with other parts.

a. Tighten fuel tube assembly bolts A to B in two steps.
- Step 1: 9 ft. lbs. (12.8 Nm)
- Step 2: 18 ft. lbs. (24.5 Nm)

14. Install the fuel hose assembly.

a. Insert connectors straight, making sure that the axis is lined up with fuel tube side to prevent O-ring from being damaged.

b. Tighten bolts evenly in several steps.

c. Make sure that there is no gap between the flange and fuel tube after tightening the bolts.

✳✳ CAUTION
Handle O-ring with bare hands. Do not wear gloves. Lubricate O-ring with new engine oil. Do not clean O-ring with solvent. Make sure that O-ring and its mating part are free of foreign material. When installing O-ring, be careful not to scratch it with tool or fingernails. Also be careful not to twist or stretch O-ring. If O-ring was stretched while it was being attached, do not insert it quickly into fuel tube. Insert new O-ring straight into fuel tube. Do not twist it.

15. Installation of the remaining components is in the reverse order of removal.

16. After installing the fuel tubes, make sure there are no fuel leaks at the connections using the following steps.

17. Apply fuel pressure to the fuel lines by turning ignition switch ON (with engine stopped). Then check for fuel leaks at the connections.

➡**Use mirrors for checking on hidden points.**

18. Start the engine and rev it up and check for fuel leaks at the connections.

✳✳ CAUTION

Do not touch the engine immediately after stopping, as engine becomes extremely hot.

FUEL PUMP

REMOVAL & INSTALLATION
See Figure 141.

1. Remove the fuel filler cap to release the pressure from inside the fuel tank.
2. Remove the LH rear wheel and tire.
3. Check the fuel level on level gauge. If the fuel gauge indicates more than the level as shown (full or almost full), drain the fuel from the fuel tank until the fuel gauge indicates the level as shown, or less.

➡**Fuel will be spilled when removing the fuel level sensor, fuel filter, and fuel pump assembly for the fuel level is above the fuel level sensor, fuel filter, and fuel pump assembly fuel tank opening.**

- As a guide, the fuel level reaches the fuel gauge position as shown, or less, when approximately 15 (4 US gal, 3 1/4 Imp gal) of fuel are drained from the fuel tank.
- If the fuel pump does not operate, use the following procedure to drain the fuel to the specified level.
a. Insert a suitable hose of less than 15 mm (0.59 in) diameter into the fuel filler pipe through the fuel filler opening to drain the fuel from fuel filler pipe.
b. Remove the fuel filler pipe shield.
c. Disconnect the fuel filler hose from the fuel filler pipe.
d. Insert a suitable hose into the fuel tank through the fuel filler hose to drain the fuel from the fuel tank.
4. Release the fuel pressure from the fuel lines.
5. Disconnect the battery negative terminal.
6. Disconnect the lower fuel filler hose from the fuel tank, the EVAP hose, and the vent pipe quick connector.
- Disconnect the fuel feed hose from the molded clip in the side of the fuel tank.
- Disconnect the quick connector as follows:

- Hold the sides of the connector, push in the tabs and pull out the tube.
- If the connector and the tube are stuck together, push and pull several times until they start to move. Then disconnect them by pulling.

✳✳ CAUTION

The quick connector can be disconnected when the tabs are completely depressed. Do not twist the quick connector more than necessary. Do not use any tools to disconnect the quick connector. Keep the resin tube away from heat. Be especially careful when welding near the tube. Do not bend or twist the resin tube during connection.

7. Remove the four bolts and remove the fuel tank shield.
8. Remove the propeller shaft.
9. Support the fuel tank using a suitable lift jack.
10. Remove the three fuel tank strap bolts while supporting the fuel tank with a suitable lift jack.
11. Remove the fuel tank straps and slowly lower the fuel tank to access the top of the fuel level sensor, fuel filter and fuel pump assembly.

✳✳ CAUTION

Do not lower the fuel tank too far to prevent damage to the fuel feed hose and the fuel level sensor, fuel filter and fuel pump assembly connector

12. Disconnect the fuel level sensor, fuel filter, and fuel pump assembly electrical connector, and the fuel feed hose.
a. Disconnect the quick connector as follows:
- Hold the sides of the connector, push in the tabs and pull out the tube.
- If the connector and the tube are

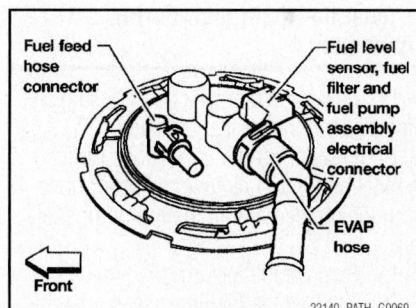

Fuel feed hose connector — Fuel level sensor, fuel filter and fuel pump assembly electrical connector — EVAP hose

Front

22140_PATH_G0069

Fig. 141 Disconnect the fuel level sensor, fuel filter, and fuel pump assembly electrical connector, and the fuel feed hose

stuck together, push and pull several times until they start to move. Then disconnect them by pulling.

✳✳ CAUTION

The quick connector can be disconnected when the tabs are completely depressed. Do not twist the quick connector more than necessary. Do not use any tools to disconnect the quick connector. Do not bend or twist the resin tube during connection.

13. Lower the fuel tank using a suitable lift jack and remove the fuel tank.
14. Disconnect the EVAP hose from the fuel pump and remove the EVAP hose from the molded clip in the top of the fuel tank.
15. Remove the lock ring.
16. Remove the fuel level sensor, fuel filter, and fuel pump assembly.

✳✳ CAUTION

Do not bend the float arm during removal. Avoid impacts such as dropping when handling the components.

To install:
17. Installation is in the reverse order of removal.
18. Connect the quick connector as follows:
- Check the connection for any damage or foreign materials.
- Align the connector with the pipe, then insert the connector straight into the pipe until a click is heard.
- Pull the tube and the connector to make sure they are securely connected.
- Visually inspect the connector to make sure the two retainer tabs are securely connected.

✳✳ CAUTION

Do not bend the float arm during installation. Avoid impacts such as dropping when handling the components.

19. Turn the ignition switch ON but do not start engine, then check the fuel pipe and hose connections for leaks while applying fuel pressure.
20. Start the engine and rev it above idle, then check that there are no fuel leaks at any of the fuel pipe and hose connections.

FUEL TANK

REMOVAL & INSTALLATION
See Figure 142.

1. Remove the fuel filler cap to release the pressure from inside the fuel tank.

2. Remove the LH rear wheel and tire.

3. Check the fuel level on level gauge. If the fuel gauge indicates more than the level as shown (full or almost full), drain the fuel from the fuel tank until the fuel gauge indicates the level as shown, or less.

➡**Fuel will be spilled when removing the fuel level sensor, fuel filter, and fuel pump assembly for the fuel level is above the fuel level sensor, fuel filter, and fuel pump assembly fuel tank opening.**

- As a guide, the fuel level reaches the fuel gauge position as shown, or less, when approximately 15 (4 US gal, 3 1/4 Imp gal) of fuel are drained from the fuel tank.
- If the fuel pump does not operate, use the following procedure to drain the fuel to the specified level.
  a. Insert a suitable hose of less than 15 mm (0.59 in) diameter into the fuel filler pipe through the fuel filler opening to drain the fuel from fuel filler pipe.
  b. Remove the fuel filler pipe shield.
  c. Disconnect the fuel filler hose from the fuel filler pipe.
  d. Insert a suitable hose into the fuel tank through the fuel filler hose to drain the fuel from the fuel tank.

4. Release the fuel pressure from the fuel lines.

5. Disconnect the battery negative terminal.

6. Disconnect the lower fuel filler hose from the fuel tank, the EVAP hose, and the vent pipe quick connector.

- Disconnect the fuel feed hose from the molded clip in the side of the fuel tank.
- Disconnect the quick connector as follows:
- Hold the sides of the connector, push in the tabs and pull out the tube.
- If the connector and the tube are stuck together, push and pull several times until they start to move. Then disconnect them by pulling.

✳✳ CAUTION

The quick connector can be disconnected when the tabs are completely depressed. Do not twist the quick connector more than necessary.

Do not use any tools to disconnect the quick connector. Keep the resin tube away from heat. Be especially careful when welding near the tube. Do not bend or twist the resin tube during connection.

7. Remove the four bolts and remove the fuel tank shield.

8. Remove the propeller shaft.

9. Support the fuel tank using a suitable lift jack.

10. Remove the three fuel tank strap bolts while supporting the fuel tank with a suitable lift jack.

11. Remove the fuel tank straps and slowly lower the fuel tank to access the top of the fuel level sensor, fuel filter and fuel pump assembly.

✳✳ CAUTION

Do not lower the fuel tank too far to prevent damage to the fuel feed hose and the fuel level sensor, fuel filter and fuel pump assembly connector

12. Disconnect the fuel level sensor, fuel filter, and fuel pump assembly electrical connector, and the fuel feed hose.

  a. Disconnect the quick connector as follows:
- Hold the sides of the connector, push in the tabs and pull out the tube.
- If the connector and the tube are stuck together, push and pull several times until they start to move. Then disconnect them by pulling.

✳✳ CAUTION

The quick connector can be disconnected when the tabs are completely depressed. Do not twist the quick connector more than necessary. Do not use any tools to disconnect the quick connector. Do not bend or twist the resin tube during connection.

13. Lower the fuel tank using a suitable lift jack and remove the fuel tank.

14. Disconnect the EVAP hose from the fuel pump and remove the EVAP hose from the molded clip in the top of the fuel tank.

15. Remove the lock ring.

16. Remove the fuel level sensor, fuel filter, and fuel pump assembly.

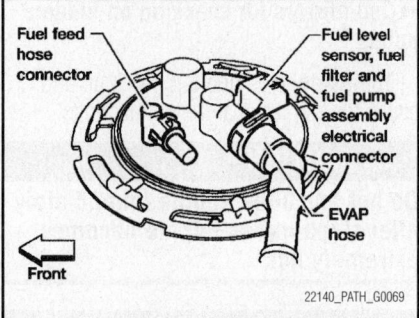

Fig. 142 Disconnect the fuel level sensor, fuel filter, and fuel pump assembly electrical connector, and the fuel feed hose

✳✳ CAUTION

Do not bend the float arm during removal. Avoid impacts such as dropping when handling the components.

To install:

17. Installation is in the reverse order of removal.

18. Connect the quick connector as follows:

- Check the connection for any damage or foreign materials.
- Align the connector with the pipe, then insert the connector straight into the pipe until a click is heard.
- Pull the tube and the connector to make sure they are securely connected.
- Visually inspect the connector to make sure the two retainer tabs are securely connected.

✳✳ CAUTION

Do not bend the float arm during installation. Avoid impacts such as dropping when handling the components.

19. Turn the ignition switch ON but do not start engine, then check the fuel pipe and hose connections for leaks while applying fuel pressure.

20. Start the engine and rev it above idle, then check that there are no fuel leaks at any of the fuel pipe and hose connections.

IDLE SPEED

ADJUSTMENT

Idle speed is controlled by the ECM. No Adjustment is necessary or possible.

HEATING & AIR CONDITIONING SYSTEM

BLOWER MOTOR

REMOVAL & INSTALLATION

1. Remove the lower glove box assembly.
2. Disconnect the front blower motor electrical connector.
3. Remove the three screws and remove the front blower motor.
4. Installation is in the reverse order of removal.

HEATER CORE

REMOVAL & INSTALLATION

1. Remove the front heater and cooling unit assembly.
   a. Discharge the refrigerant from the A/C system.
   b. Drain the coolant from the engine cooling system.
   c. Remove the front heater core pipes RH nut.
   d. Disconnect the front heater core hoses from the front heater core.
   e. Disconnect the high- and low-pressure A/C pipes from the front expansion valve.
   f. Move the two front seats to the rearmost position on the seat track.
   g. Remove the instrument panel and console panel.
   h. Remove the two front floor ducts.
   i. Remove the steering column.
   j. Disconnect the instrument panel wire harness at the RH and LH in-line connector brackets, and the fuse block (SMJ) electrical connectors.
   k. Remove the covers then remove the three steering member bolts from each side to disconnect the steering member from the vehicle body.
   l. Remove the front heater and cooling unit assembly with it attached to the steering member, from the vehicle.

✳✳ CAUTION

Use care not to damage the seats and interior trim panels when removing the front heater and cooling unit assembly with it attached to the steering member. Use suitable plugs on the heater core pipes to prevent coolant leakage.

   m. Remove the front heater and cooling unit assembly from the steering member.
2. Remove the three screws and remove the front heater core cover.
3. Remove the front heater core and evaporator pipe bracket.
4. Remove the front heater core.

➡ **If the in-cabin micro filters are contaminated from coolant leaking from the front heater core, replace the in-cabin micro filters with new ones before installing the new front heater core.**

5. Installation is in the reverse order of removal.

STEERING

POWER STEERING GEAR

REMOVAL & INSTALLATION
See Figures 143 through 146.

✳✳ CAUTION

Spiral cable may snap due to steering operation if the steering column is separated from the steering gear assembly. Therefore secure the steering wheel to avoid turning.

1. Set front wheels in the straight-ahead position.
2. Remove the front tires from the vehicle.
3. Remove the undercover.
4. On 4WD models, remove the front final drive, then support the drive shafts, using suitable wire.
5. Remove the stabilizer bar brackets and reposition the stabilizer bar.
6. Remove the cotter pins at the steering outer sockets.

✳✳ CAUTION

Do not reuse the cotter pins.

7. Loosen the outer socket nuts.
8. Remove the steering outer sockets from the steering knuckles, then remove the nuts.

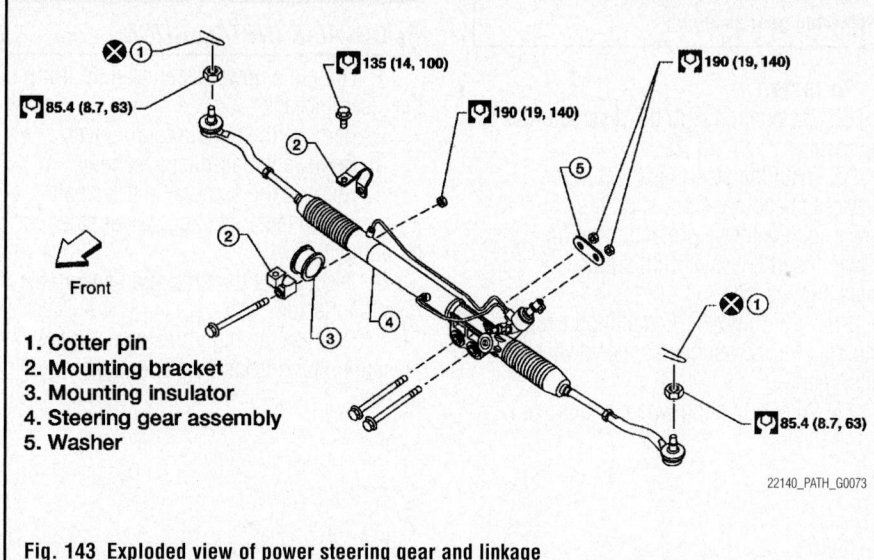

1. Cotter pin
2. Mounting bracket
3. Mounting insulator
4. Steering gear assembly
5. Washer

22140_PATH_G0073

Fig. 143 Exploded view of power steering gear and linkage

✳✳ CAUTION

Do not damage the outer socket boots. Do not damage the outer socket threads. Thread the ball joint nut onto the end of the outer socket during removal.

9. Remove the high-pressure and low-pressure piping from the steering gear assembly, then drain the fluid from the piping.

10. Remove the bolt from the lower joint of the lower joint shaft, then separate the lower joint from the steering gear assembly.

✳✳ CAUTION

Do not damage the lower joint.

11. Remove the nuts and bolts of the steering gear assembly, then remove the steering gear assembly from the vehicle.

Fig. 144 Remove the bolt from the lower joint of the lower joint shaft

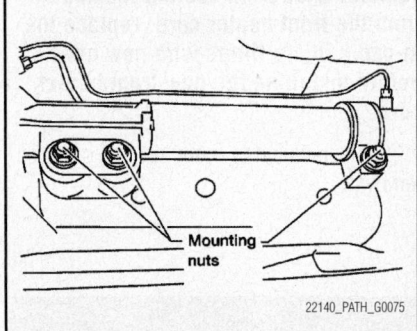

Fig. 145 Remove the nuts and bolts of the steering gear assembly

To install:

12. Installation is in the reverse order of removal.

13. With the steering wheel in the straight ahead position, align the slit of the lower joint with the projection on the dust cover. Insert the joint until surface "A" contacts surface "B".

14. After removing/installing or replacing steering components, check wheel alignment.

15. After adjusting wheel alignment,

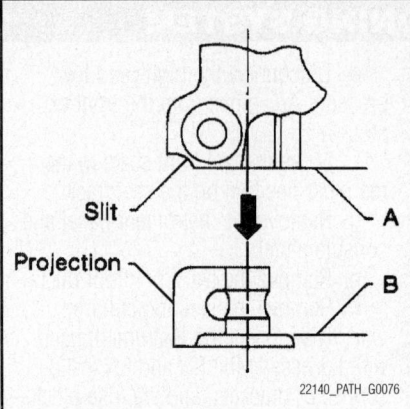

Fig. 146 Insert the joint until surface "A" contacts surface "B"

adjust neutral position of the steering angle sensor.

16. Bleed the air from the steering hydraulic system.

17. Check that the steering wheel turns smoothly to the left and right locks.

18. Check that the number of turns are the same from the straight-forward position to the left and right locks.

19. Check that the steering wheel is in the neutral position when driving straight ahead.

POWER STEERING PUMP

REMOVAL & INSTALLATION

1. Drain the power steering fluid from the reservoir tank.

2. Remove the engine room cover.

3. Remove the air duct assembly.

4. Remove the serpentine drive belt from the auto tensioner and power steering oil pump.

5. Disconnect the pressure sensor electrical connector.

6. Remove the high pressure and low pressure piping from the power steering oil pump.

7. Remove the power steering oil pump bolts, then remove the power steering pump.

8. Installation is in the reverse order of removal.

9. After installation, bleed the air from the hydraulic circuit thoroughly.

BLEEDING

➡**When the vehicle is stationary or while the steering wheel is being turned slowly, some noise may be heard from the oil pump or gear. This noise is normal and does not affect any system.**

1. Check for fluid leakage.

2. Start the engine and turn the steering wheel fully to the right and left several times.

✳✳ CAUTION

Do not allow steering fluid reservoir tank to go below the MIN level line. Check tank frequently and add fluid as needed.

3. Run the engine at idle speed. Hold the steering wheel at each "locked" position for three seconds.

✳✳ CAUTION

Do not hold steering wheel in the locked position for more than 10 seconds. (There is the possibility that oil pump may be damaged.)

4. Repeat step 3 several times at about three second intervals.

5. Check for air bubbles, cloudy fluid and fluid leakage.

6. If air bubbles or cloudiness exists, perform steps 3 and 4 again until air bubbles and cloudiness do not exist.

7. Stop the engine and check fluid level.

SUSPENSION

LOWER BALL JOINT

REMOVAL & INSTALLATION

➡ **The ball joints are part of the upper and lower links.**

LOWER CONTROL LINK

REMOVAL AND & INSTALLATION

See Figure 147.

1. Remove the wheel and tire.
2. Remove lower shock absorber bolt and nut.

3. Remove stabilizer bar connecting rod lower nut, then separate connecting rod from lower link.
4. For 4WD models, remove the drive shaft.
5. Remove pinch bolt from steering knuckle, then separate lower link ball joint from steering knuckle.
6. Remove lower link adjusting bolts and nuts, then the lower link.

➡ **Some vehicles may be equipped with straight (non-adjustable) lower link bolts and washers. In order to adjust**

camber and caster on these vehicles, **first replace the lower link bolts and washers with adjustable (cam) bolts and washers.**

7. Remove the jounce bumper from the lower link.
8. Installation is in the reverse order of removal.
9. Tighten all nuts and bolts to specification.
10. After installation, check that the front wheel alignment is within specification.

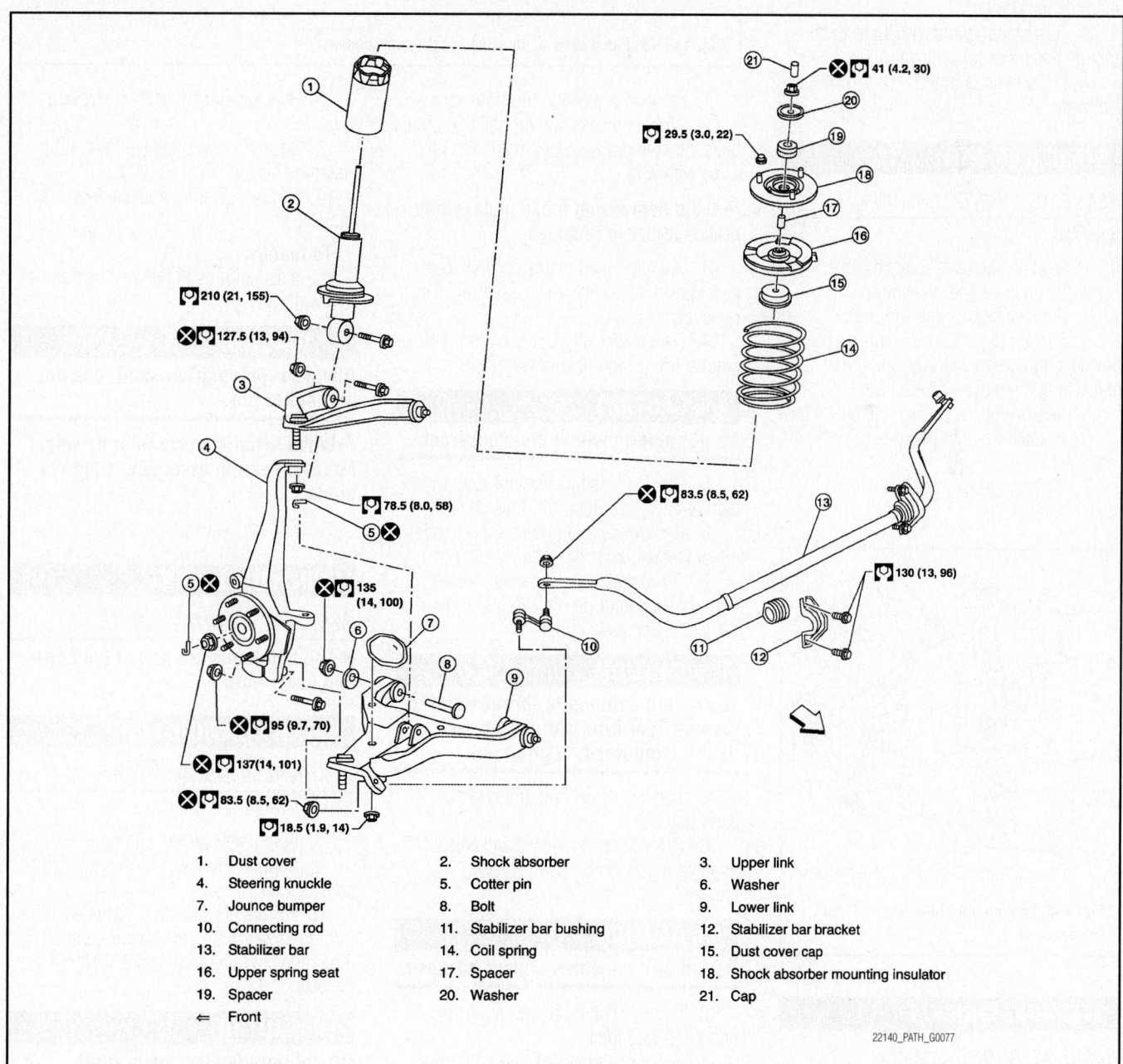

1.	Dust cover	2.	Shock absorber	3.	Upper link
4.	Steering knuckle	5.	Cotter pin	6.	Washer
7.	Jounce bumper	8.	Bolt	9.	Lower link
10.	Connecting rod	11.	Stabilizer bar bushing	12.	Stabilizer bar bracket
13.	Stabilizer bar	14.	Coil spring	15.	Dust cover cap
16.	Upper spring seat	17.	Spacer	18.	Shock absorber mounting insulator
19.	Spacer	20.	Washer	21.	Cap
⇐	Front				

22140_PATH_G0077

Fig. 147 Exploded view of front suspension

MACPHERSON STRUT

REMOVAL & INSTALLATION

1. Remove the wheel and tire.
2. Support the lower link using a suitable jack.
3. Remove connecting rod upper joints from stabilizer bar.
4. Swing stabilizer bar down, repositioning it out of the way to access shock absorber lower mount.
5. Remove the shock absorber lower bolt and nut.
6. Remove the three shock absorber upper mounting nuts.
7. Remove the coil spring and shock absorber assembly.
8. Turn steering knuckle out to gain enough clearance for removal.
9. Installation is in the reverse order of removal.

STABILIZER BAR

REMOVAL & INSTALLATION

See Figure 148.

1. Remove the front valance center.
2. Remove engine undercover.
3. Remove connecting rod nuts.
4. Loosen top bolts for stabilizer bar brackets, then remove lower bolts from brackets and remove stabilizer bar.
5. Remove bushings from stabilizer bar.
6. Installation is in the reverse order of removal.
7. Tighten all nuts and bolts to specification.

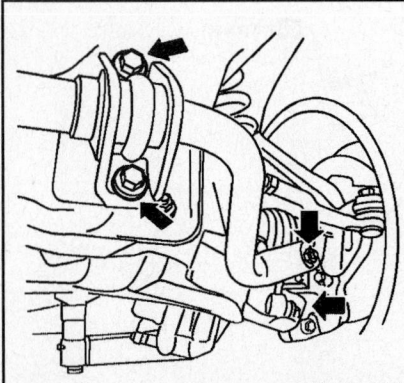

Fig. 148 Loosen top bolts for stabilizer bar brackets

STEERING KNUCKLE

REMOVAL & INSTALLATION

See Figure 149.

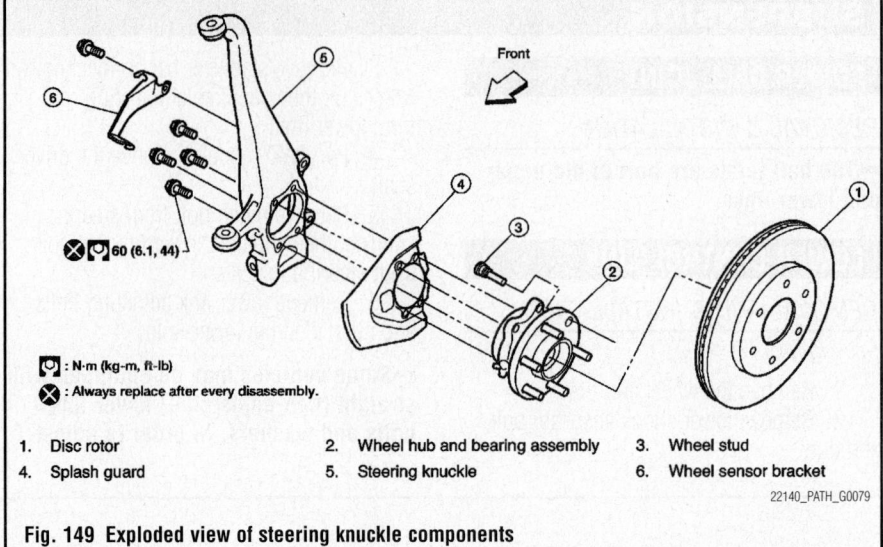

⊗ 🔧 60 (6.1, 44)

🔧 : N·m (kg-m, ft-lb)

⊗ : Always replace after every disassembly.

1. Disc rotor	2. Wheel hub and bearing assembly	3. Wheel stud
4. Splash guard	5. Steering knuckle	6. Wheel sensor bracket

22140_PATH_G0079

Fig. 149 Exploded view of steering knuckle components

1. Remove wheel and tire from vehicle.
2. Without disassembling the hydraulic lines, remove brake caliper. Reposition it aside with wire.

➡ **Avoid depressing brake pedal while brake caliper is removed.**

3. Put alignment marks on disc rotor and wheel hub and bearing assembly, then remove disc rotor.
4. Disconnect wheel sensor and remove bracket from steering knuckle.

✳✳ CAUTION

Do not pull on wheel sensor harness.

5. On 4WD models, remove cotter pin, then remove lock nut from drive shaft.
6. Remove steering outer socket cotter pin at steering knuckle, then loosen nut.
7. Disconnect steering outer socket from steering knuckle. Be careful not to damage outer socket boot.

✳✳ CAUTION

To prevent damage to threads and to prevent Tool from coming off suddenly, temporarily tighten nut.

8. Remove wheel hub and bearing assembly bolts.
9. Remove splash guard and wheel hub and bearing assembly from steering knuckle.

✳✳ CAUTION

Do not pull on wheel sensor harness.

10. Remove cotter pin and nut from upper link ball joint.
11. Separate upper link ball joint from steering knuckle.

12. Remove pinch bolt from steering knuckle.
13. Separate lower link ball joint from steering knuckle.
14. Remove steering knuckle from vehicle.

To install:

15. Installation is in the reverse order of removal.

✳✳ CAUTION

Always replace drive shaft lock nut and cotter pin.

➡ **When installing disc rotor on wheel hub and bearing assembly, align the marks.**

16. Perform wheel alignment.

UPPER BALL JOINT

REMOVAL & INSTALLATION

➡ **The ball joints are part of the upper and lower links.**

UPPER CONTROL LINK

REMOVAL & INSTALLATION

See Figure 147.

1. Remove the wheel and tire.
2. Support the lower link using a suitable jack.
3. For the LH side only, remove the bolt from the lower joint of the lower joint shaft, then reposition the lower joint shaft out of the way.

✳✳ CAUTION

Do not damage the lower joint.

4. Remove cotter pin and nut from upper link ball joint.

5. Separate upper link ball joint stud from steering knuckle.

6. Remove upper link bolts and nuts.

To install:

7. Installation is in the reverse order of removal.

8. Tighten all nuts and bolts to specification.

✽✽ CAUTION

Always replace drive shaft lock nut and cotter pin.

9. After installation, check that the front wheel alignment is within specification.

WHEEL BEARINGS

REMOVAL & INSTALLATION

1. Remove wheel and tire.

2. Without disassembling the hydraulic lines, remove caliper torque member bolts.

3. Reposition brake caliper aside with wire.

✽✽ CAUTION

Do not press brake pedal while brake caliper is removed.

4. Put alignment mark on disc rotor and wheel hub and bearing assembly, then remove disc rotor.

5. Remove cotter pin, then remove lock nut from drive shaft.

6. Remove drive shaft from wheel hub and bearing assembly.

7. Remove wheel sensor from wheel hub and bearing assembly.

a. Inspect the wheel sensor O-ring, replace the wheel sensor assembly if damaged.

b. Clean the wheel sensor hole and mounting surface with a suitable brake cleaner and clean lint-free shop rag. Be careful that dirt and debris do not enter the axle bearing area.

c. Apply a coat of suitable grease to the wheel sensor O-ring and mounting hole.

✽✽ CAUTION

Do not pull on the wheel sensor harness.

8. Remove wheel hub and bearing assembly bolts.

9. Remove splash guard and wheel hub and bearing assembly from steering knuckle.

10. Carefully remove wheel sensor and harness through hole in splash guard.

To install:

11. Installation is in the reverse order of removal.

12. Use new bolts when installing the wheel hub and bearing assembly.

13. When installing disc rotor on wheel hub and bearing assembly, position the disc rotor according to alignment mark.

SUSPENSION

COIL SPRING

REMOVAL & INSTALLATION
See Figure 150.

1. Remove the wheel and tire assembly.

2. If removing the LH rear lower link and coil spring, remove the spare wheel and tire assembly.

3. Set a suitable jack to relieve the coil spring tension and support the rear lower link.

4. Loosen the rear lower link adjusting bolt and nut connected to the rear suspension member without removing the adjusting bolt and nut.

5. Remove the rear lower link pinch bolt and nut from the knuckle.

6. Slowly lower the rear lower link using the suitable jack to release the coil spring tension.

7. Remove the upper rubber seat, coil spring and lower rubber seat from the rear lower link.

To install:

8. Installation is in the reverse order of removal.

a. When installing the upper and lower rubber seats for the rear coil springs, the arrow embossed on the rubber seats must point out toward the wheel and tire assembly.

b. Perform the final tightening of the rear lower link nuts and bolts

(with rubber bushings) under no—load conditions with tires on level ground.

c. Tighten the nuts and bolts to specification.

d. Check the wheel alignment.

FRONT LOWER CONTROL LINK

REMOVAL & INSTALLATION
See Figure 150.

1. Remove the wheel and tire assembly.

2. Remove the stabilizer bar.

3. Set a suitable jack under the rear lower link to relieve the coil spring tension.

4. Remove the shock absorber lower end bolt.

5. Remove the adjusting bolt and nut, and the bolt and nut, from the front lower link and rear suspension member.

6. Remove the front lower link pinch bolt and nut on the knuckle side.

7. Disconnect the front lower link from the knuckle using a soft hammer.

✽✽ CAUTION

Do not damage the ball joint with the soft hammer.

8. Remove the front lower link.

9. Installation is in the reverse order of removal.

a. Tighten the nuts and bolts to specification.

b. Perform the final tightening of the

REAR SUSPENSION

front lower link nuts and bolts (with rubber bushings) under no—load conditions with tires on level ground.

c. Check the wheel alignment.

REAR LOWER CONTROL LINK

REMOVAL & INSTALLATION
See Figure 150.

1. Remove the wheel and tire assembly.

2. If removing the LH rear lower link and coil spring, remove the spare wheel and tire assembly.

3. Set a suitable jack to relieve the coil spring tension and support the rear lower link.

4. Loosen the rear lower link adjusting bolt and nut connected to the rear suspension member without removing the adjusting bolt and nut.

5. Remove the rear lower link pinch bolt and nut from the knuckle.

6. Slowly lower the rear lower link using the suitable jack to release the coil spring tension.

7. Remove the upper rubber seat, coil spring and lower rubber seat from the rear lower link.

8. Remove the rear lower link adjusting bolt and nut from the rear suspension member.

9. Remove the rear lower link.

To install:

10. Installation is in the reverse order of removal.

a. When installing the upper and lower rubber seats for the rear coil springs, the arrow embossed on the rubber seats must point out toward the wheel and tire assembly.

b. Perform the final tightening of the rear lower link nuts and bolts (with rubber bushings) under no—load conditions with tires on level ground.

c. Tighten the nuts and bolts to specification.

d. Check the wheel alignment.

STABILIZER BAR

REMOVAL & INSTALLATION

See Figure 150.

1. Disconnect the stabilizer bar ends from the connecting rods.

2. Remove the stabilizer bar clamps, and remove the stabilizer bar bushings.

3. Remove the stabilizer bar.

4. Installation is in the reverse order of removal.

5. Install the stabilizer bar bushings and clamps so they are positioned outside of the sideslip prevention clamp on the stabilizer bar.

WHEEL BEARINGS

REMOVAL & INSTALLATION

1. Remove the wheel and tire assembly.

2. Remove the rear brake caliper, without disconnecting the hydraulic hose.

3. Reposition the rear brake caliper aside using suitable wire.

✳✳ CAUTION

Do not depress the brake pedal while the brake caliper is removed.

4. Remove the rear disc rotor.

5. Remove the cotter pin, then remove the rear drive shaft nut.

➡**Discard the cotter pin, use a new one for installation.**

6. Remove the rear drive shaft.

7. Remove the four rear wheel hub and bearing assembly bolts.

➡**Discard the four rear wheel hub and bearing assembly bolts, use new ones for installation.**

8. Remove the rear wheel hub and bearing assembly.

9. Installation is in the reverse order of removal.

a. Use a new cotter pin for installation.

b. Use new rear wheel hub and bearing assembly bolts for installation.

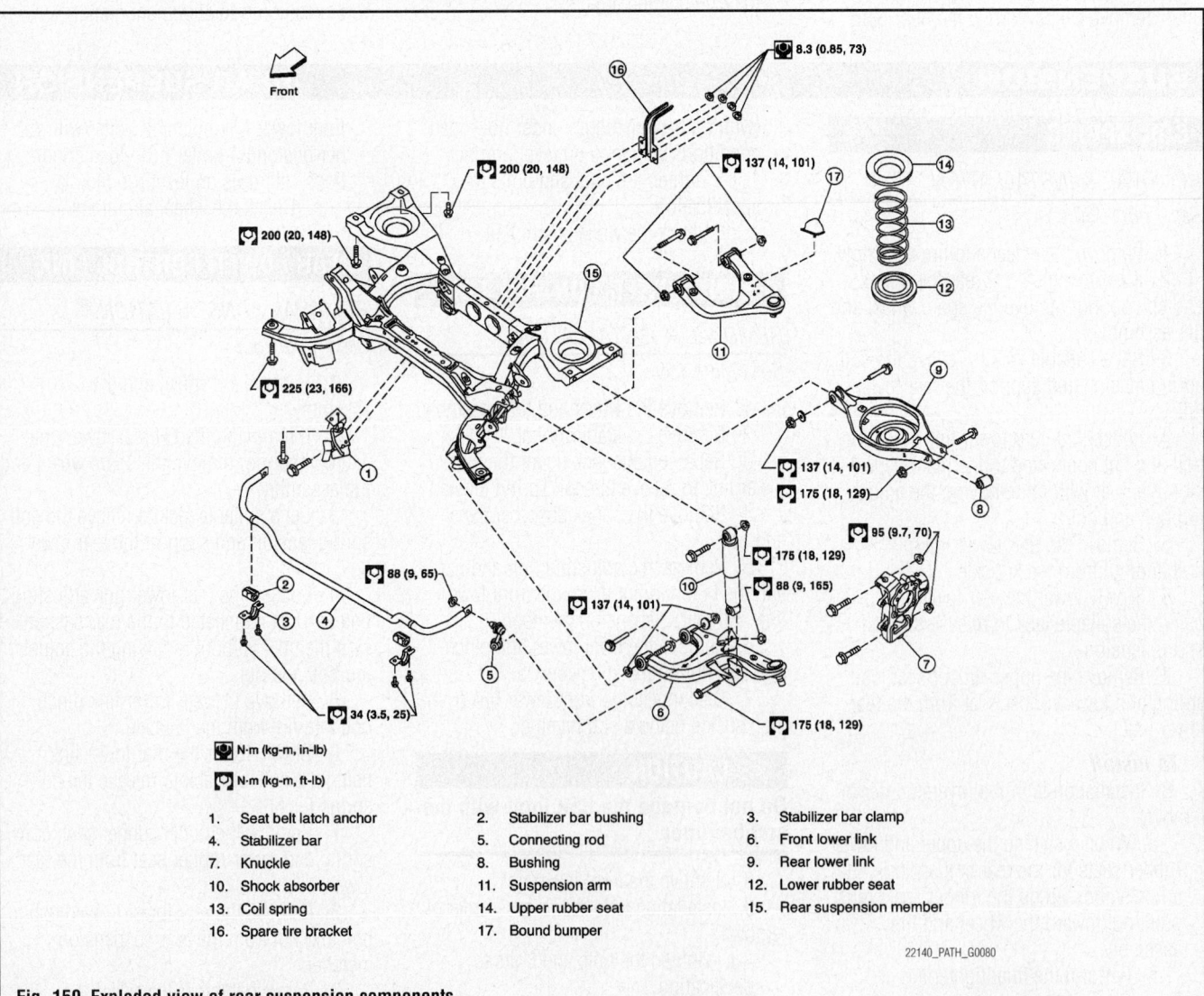

| N·m (kg-m, in-lb) |
| N·m (kg-m, ft-lb) |

1.	Seat belt latch anchor	2.	Stabilizer bar bushing	3.	Stabilizer bar clamp
4.	Stabilizer bar	5.	Connecting rod	6.	Front lower link
7.	Knuckle	8.	Bushing	9.	Rear lower link
10.	Shock absorber	11.	Suspension arm	12.	Lower rubber seat
13.	Coil spring	14.	Upper rubber seat	15.	Rear suspension member
16.	Spare tire bracket	17.	Bound bumper		

22140_PATH_G0080

Fig. 150 Exploded view of rear suspension components

SPECIFICATIONS AND MAINTENANCE CHARTS

ENGINE AND VEHICLE IDENTIFICATION

Engine							Model Year	
Code ①	Liters (cc)	Cu. In.	Cyl.	Fuel Sys.	Engine Type	Eng. Mfg.	Code ②	Year
QR25DE	2.5 (2595)	153.8	4	MPI	DOHC	Nissan	8	2008

SFI: Sequential Fuel Injection

DOHC: Double Overhead Camshaft

① Stamped on the left side of the engine block

② 10th digit of the Vehicle Identification Number (VIN)

22140_ROGU_C0001

GENERAL ENGINE SPECIFICATIONS

Year	Model	Engine Displacement Liters	Engine Series ID	Net Horsepower @ rpm	Net Torque @ rpm (ft. lbs.)	Bore x Stroke (in.)	Compression Ratio	Oil Pressure @ rpm
2008	Rogue	2.5	QR25DE	170@6000	175@4400	3.50x3.94	9.5:1	60@3000

22140_ROGU_C0002

ENGINE TUNE-UP SPECIFICATIONS

Year	Engine Displacement Liters	Engine ID	Spark Plug Gap (in.)	Ignition Timing (deg.)	Fuel Pump (psi)	Idle Speed (rpm)	Valve Clearance Intake	Valve Clearance Exhaust
2008	2.5	QR25DE	0.043	15B ①	51	650-750	0.0012-0.0016	0.0012-0.0017

NOTE: The Vehicle Emission Control Information label often reflects specification changes made during production.

The label figures must be used if they differ from those in this chart.

B: Before top dead center

① Plus or minus 5 degrees

22140_ROGU_C0003

CAPACITIES

Year	Model	Engine Displacement Liters	Engine ID	Engine Oil with Filter (qts.)	Auto Trans. (qts.)	Transfer Case (pts)	Rear Drive Axle (pts.) ①	Fuel Tank (gal.)	Cooling System (qts.)
2008	Rogue	2.5	QR25DE	4.7	②	0.8	0.6	15.8	8.5 ③

① Synthetic GL-5 (75W-90) or equivalent

② 2WD (9qts) and AWD (10 qts)

③ The use of genuine Nissan engine coolant is recommended or similar ethylene glycol based non-silicate, non-amine, non- nitrite, and non- borat coolant

22140_ROGU_C0005

FLUID SPECIFICATIONS

Year	Model	Engine Displacement Liters	Engine ID/VIN	Engine Oil	Auto. Trans. ①	Drive Axle ②	Power Steering Fluid	Brake Master Cylinder	Engine Coolant ③
2008	Rogue	2.5	QR25DE	5W-30	Nissan	80W-90	NA	DOT 3	Nissan

NA: Not Available

DOT: Department Of Transpotation

① Using trasmission fluid other than genuine Nissan CVT fluid NS-2 will damage the CVT

② GL-5 (80W-90) or equivalent

③ The use of genuine Nissan engine coolant is recommended

22140_ROGU_C0004

VALVE SPECIFICATIONS

Year	Engine Displacement Liters	Engine ID	Seat Angle (deg.)	Face Angle (deg.)	Inner Spring free length (in.)	Spring Installed Height (in.)	Stem-to-Guide Clearance (in.) Intake	Stem-to-Guide Clearance (in.) Exhaust	Stem Diameter (in.) Intake	Stem Diameter (in.) Exhaust
2008	2.5	QR25DE	45	40.5	①	1.390	0.0008-0.0021	0.0012-0.0025	0.2348-0.2354	0.2344 0.2350

① Intake (1.7213-1.7291) and Exhuast (1.7831-1.7909)

22140_ROGU_C0006

CAMSHAFT AND BEARING SPECIFICATIONS CHART
All measurements are given in inches.

Year	Engine Displ. Liters	Engine ID/VIN	Journal Dia.	Brg. Oil Clearance	Shaft End-play	Runout	Journal Bore	Lobe Height Intake	Lobe Height Exhaust
2008	2.5	QR25DE	①	0.0018-0.0034	0.0045-0.0074	0.0008	NA	1.7644-1.7718	1.7313-1.7388

NA: Not Available

① Journal 1: 1.0998-1.1006
All Others: 0.9252-0.9260

22140_ROGU_C0008

CRANKSHAFT AND CONNECTING ROD SPECIFICATIONS
All measurements are given in inches.

Year	Engine Displacement Liters	Engine ID	Main Brg. Journal Dia.	Main Brg. Oil Clearance	Shaft End-play	Thrust on No.	Journal Diameter	Oil Clearance	Side Clearance
2008	2.5	QR25DE	2.1636 2.1645	①	0.0039-0.0102	3	2.0863-2.0866	0.0011-0.0018	0.0079-0.0138

① Nos. 1, 3 and 5: 0.0005-0.0009 in.
Nos. 2 and 4: 0.0007-0.0011

22140_ROGU_C0007

PISTON AND RING SPECIFICATIONS
All measurements are given in inches.

Year	Engine Displ. Liters	Engine ID	Piston Clearance	Ring Gap Top Comp.	Ring Gap Bottom Comp.	Ring Gap Oil Control	Ring Side Clearance Top Comp.	Ring Side Clearance Bottom Comp.	Ring Side Clearance Oil Control
2008	2.5	QR25DE	0.0004-0.0012	0.0083-0.0122	0.0126-0.0185	0.0079-0.0236	0.0018-0.0031	0.0012-0.0028	0.0026-0.0053

22140_ROGU_C0009

TORQUE SPECIFICATIONS
All readings in ft. lbs.

Year	Engine Displacement Liters	Engine ID	Cylinder Head Bolts	Main Bearing Bolts	Rod Bearing Bolts	Crankshaft Damper Bolts	Flywheel Bolts	Manifold Intake	Manifold Exhaust	Spark Plugs	Oil Pan Drain Plug
2008	2.5	QR25DE	①	②	③	④	76-83	13-15	29-32	18	25

① Step 1: 72 ft. lbs.
  Step 2: Loosen completely, then retorque to 26-32 ft. lbs.
  Step 3: Turn each bolt, in sequence, an additional 75-80 degrees
  Step 4: Turn each bolt, in sequence, an additional 75-80 degrees

② Bolt Nos. 1-10:
  Step 1: 27-31 ft. lbs.
  Step 2: Torque an additional 60-65 degrees
  Bolt Nos. 11-14: Torque last, to 17-20 ft. lbs.

③ Step 1: 14-15 ft. lbs.
  Step 2: 85-95 degrees

④ Step 1: 29-36 ft. lbs.
  Step 2: 60-66 degrees

22140_ROGU_C0010

WHEEL ALIGNMENT

Year	Model		Caster Range (+/-Deg.)	Caster Preferred Setting (Deg.)	Camber Range (+/-Deg.)	Camber Preferred Setting (Deg.)	Toe-in (in.)
2008	Rogue	F	0.75	+5.67	0.75	①	0.04+/-0.08
		R	NA	NA	0.75	-0.92	0.12+/-0.08

NA: Not Applicable

F: Front

R: Rear

① Left side front camber -0.15
  Right side front front camber -0.30

22140_ROGU_C0011

TIRE, WHEEL AND BALL JOINT SPECIFICATIONS

| Year | Model | OEM Tires | | Tire Pressures (psi) | | Wheel Size | Ball Joint Inspection | Lug Nut Torque (ft. lbs.) |
		Standard	Optional	Front	Rear			
2008	Rogue	P215/70R16	225/60R17	33	33	6.5-JJ and 7-JJ	①	80

OEM: Original Equipment Manufacturer

PSI: Pounds Per Square Inch

STD: Standard

OPT: Optional

① Replace if any measurable movement is found.

22140_ROGU_C0012

BRAKE SPECIFICATIONS

All measurements in inches unless noted

| Year | Model | | Brake Disc | | | Minimum Lining Thickness | Brake Caliper | |
			Original Thickness	Minimum Thickness	Maximum Runout		Bracket Bolts (ft. lbs.)	Mounting Bolts (ft. lbs.)
2008	Rogue	F	1.024	0.945	0.0020	0.079	122	25
		R	0.630	0.551	0.0020	0.059	62	32

F: Front

R: Rear

22140_ROGU_C0013

SCHEDULED MAINTENANCE INTERVALS
Nissan—Rogue

TO BE SERVICED	TYPE OF SERVICE	VEHICLE MILEAGE INTERVAL (x1000)												
		7.5	15	22.5	30	37.5	45	52.5	60	67.5	75	82.5	90	97.5
Engine oil & filter	R	✓	✓	✓	✓	✓	✓	✓	✓	✓	✓	✓	✓	✓
Brake lines & cables	S/I		✓		✓		✓		✓		✓		✓	
Brake pads, discs, drums & linings	S/I		✓		✓		✓		✓		✓		✓	
Cabin air filter					✓				✓				✓	
Driveshaft boots	S/I		✓		✓		✓		✓		✓		✓	
Exhaust system	S/I				✓				✓				✓	
Transfer oil and differential gear oil	S/I		✓		✓		✓		✓		✓		✓	
Air cleaner filter	R				✓				✓				✓	
Spark plugs (except platinum)	R				✓				✓				✓	
Spark plugs (platinum tip)	R								✓					
Idle RPM	S/I				✓				✓					
Steering gear & linkage, axle & suspension parts	S/I				✓				✓				✓	
Engine coolant	R								✓					
Timing belt	R								✓					
Drive belts	S/I								✓					
Propeller shaft (AWD models)														
Fuel lines	S/I								✓					
Vapor lines	S/I								✓					

R: Replace   S/I: Service or Inspect

FREQUENT OPERATION MAINTENANCE (SEVERE SERVICE)

If a vehicle is operated under any of the following conditions it is considered severe service:

- Extremely dusty areas.
- 50% or more of the vehicle operation is in 32°C (90°F) or higher temperatures, or constant operation in temperatures below 0°C (32°F).
- Prolonged idling (vehicle operation in stop and go traffic).
- Frequent short running periods (engine does not warm to normal operating temperatures).
- Police, taxi, delivery usage or trailer towing usage.

Oil & oil filter: change every 3750 miles.

Brake pads & discs: service or inspect every 7500 miles.

Driveshaft boots: service or inspect every 7500 miles.

Exhaust system: service or inspect every 7500 miles.

Steering gear & linkage, axle & suspension parts: service or inspect every 7500 miles.

Steering linkage ball joints & front suspension ball joints: service or inspect every 7500 miles.

Air cleaner filter: service or inspect every 15,000 miles.

22140_ROGU_C0014

PRECAUTIONS

Before servicing any vehicle, please be sure to read all of the following precautions, which deal with personal safety, prevention of component damage, and important points to take into consideration when servicing a motor vehicle:

• Never open, service or drain the radiator or cooling system when the engine is hot; serious burns can occur from the steam and hot coolant.

• Observe all applicable safety precautions when working around fuel. Whenever servicing the fuel system, always work in a well-ventilated area. Do not allow fuel spray or vapors to come in contact with a spark, open flame, or excessive heat (a hot drop light, for example). Keep a dry chemical fire extinguisher near the work area. Always keep fuel in a container specifically designed for fuel storage; also, always properly seal fuel containers to avoid the possibility of fire or explosion. Refer to the additional fuel system precautions later in this section.

• Fuel injection systems often remain pressurized, even after the engine has been turned **OFF**. The fuel system pressure must be relieved before disconnecting any fuel lines. Failure to do so may result in fire and/or personal injury.

• Brake fluid often contains polyglycol ethers and polyglycols. Avoid contact with the eyes and wash your hands thoroughly after handling brake fluid. If you do get brake fluid in your eyes, flush your eyes with clean, running water for 15 minutes. If eye irritation persists, or if you have taken brake fluid internally, IMMEDIATELY seek medical assistance.

• The EPA warns that prolonged contact with used engine oil may cause a number of skin disorders, including cancer. You should make every effort to minimize your exposure to used engine oil. Protective gloves should be worn when changing oil. Wash your hands and any other exposed skin areas as soon as possible after exposure to used engine oil. Soap and water, or waterless hand cleaner should be used.

• All new vehicles are now equipped with an air bag system, often referred to as a Supplemental Restraint System (SRS) or Supplemental Inflatable Restraint (SIR) system. The system must be disabled before performing service on or around system components, steering column, instrument panel components, wiring and sensors. Failure to follow safety and disabling procedures could result in accidental air bag deployment, possible personal injury and unnecessary system repairs.

• Always wear safety goggles when working with, or around, the air bag system. When carrying a non-deployed air bag, be sure the bag and trim cover are pointed away from your body. When placing a non-deployed air bag on a work surface, always face the bag and trim cover upward, away from the surface. This will reduce the motion of the module if it is accidentally deployed. Refer to the additional air bag system precautions later in this section.

• Clean, high quality brake fluid from a sealed container is essential to the safe and proper operation of the brake system. You should always buy the correct type of brake fluid for your vehicle. If the brake fluid becomes contaminated, completely flush the system with new fluid. Never reuse any brake fluid. Any brake fluid that is removed from the system should be discarded. Also, do not allow any brake fluid to come in contact with a painted surface; it will damage the paint.

• Never operate the engine without the proper amount and type of engine oil; doing so WILL result in severe engine damage.

• Disconnecting the negative battery cable on some vehicles may interfere with the functions of the on-board computer system(s) and may require the computer to undergo a relearning process once the negative battery cable is reconnected.

• When servicing Parking brake shoes, only disassemble and assemble one side at a time, leaving the remaining side intact for reference.

• Only an MVAC-trained, EPA-certified automotive technician should service the air conditioning system or its components.

BRAKES

ANTI-LOCK BRAKE SYSTEM (ABS)

GENERAL INFORMATION

PRECAUTIONS

• Certain components within the ABS system are not intended to be serviced or repaired individually.

• Do not use rubber hoses or other parts not specifically specified for and ABS system. When using repair kits, replace all parts included in the kit. Partial or incorrect repair may lead to functional problems and require the replacement of components.

• Lubricate rubber parts with clean, fresh brake fluid to ease assembly. Do not use shop air to clean parts; damage to rubber components may result.

• Use only DOT 3 brake fluid from an unopened container.

• If any hydraulic component or line is removed or replaced, it may be necessary to bleed the entire system.

• A clean repair area is essential. Always clean the reservoir and cap thoroughly before removing the cap. The slightest amount of dirt in the fluid may plug an orifice and impair the system function. Perform repairs after components have been thoroughly cleaned; use only denatured alcohol to clean components. Do not allow ABS components to come into contact with any substance containing mineral oil; this includes used shop rags.

• The Anti-Lock control unit is a microprocessor similar to other computer units in the vehicle. Ensure that the ignition switch is **OFF** before removing or installing controller harnesses. Avoid static electricity discharge at or near the controller.

• If any arc welding is to be done on the vehicle, the control unit should be unplugged before welding operations begin.

BRAKES | **BLEEDING THE BRAKE SYSTEM**

BLEEDING PROCEDURE

BLEEDING PROCEDURE

See Figure 1.

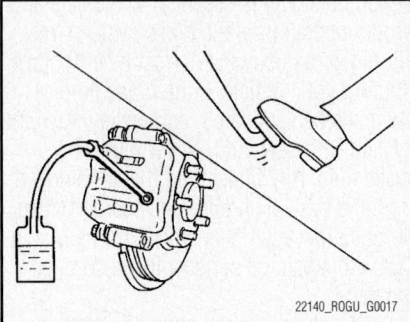

Fig. 1 Loosen the bleeder valve, slowly depress the brake pedal to the full stroke, and then release the pedal.

1. Check that there is no foreign material in the reservoir tank, and refill with new brake fluid.
2. Loosen the bleeder valve, slowly depress the brake pedal to the full stroke and shut off bleeder and then release the brake pedal.
3. Repeat this operation at intervals of 2 or 3 seconds until all brake fluid is discharged.
4. Then close the bleeder valve with the brake pedal depressed.
5. Repeat the same work on each wheel.
6. Check brake fluid and add if needed.

MASTER CYLINDER BLEEDING

See Figure 2.

1. Loosen the master cylinder lines, slowly depress the brake pedal to the full stroke, tighten the lines and then release the brake pedal.

2. Repeat this operation at intervals of 2 or 3 seconds until all brake fluid is discharged.
3. Then tighten the lines with the brake pedal depressed. When system is bleed tighten the lines to 13 ft. lbs. (18.2 Nm).
4. Check brake fluid and add if needed.

BRAKE LINE BLEEDING

1. Loosen the brake line in question , slowly depress the brake pedal to the full stroke, tighten the line and then release the brake pedal.
2. Repeat this operation at intervals of 2 or 3 seconds until all brake fluid is discharged.
3. Then close the bleeder valve with the brake pedal depressed.
4. Repeat the same work on each wheel.
5. Check brake fluid and add if needed.

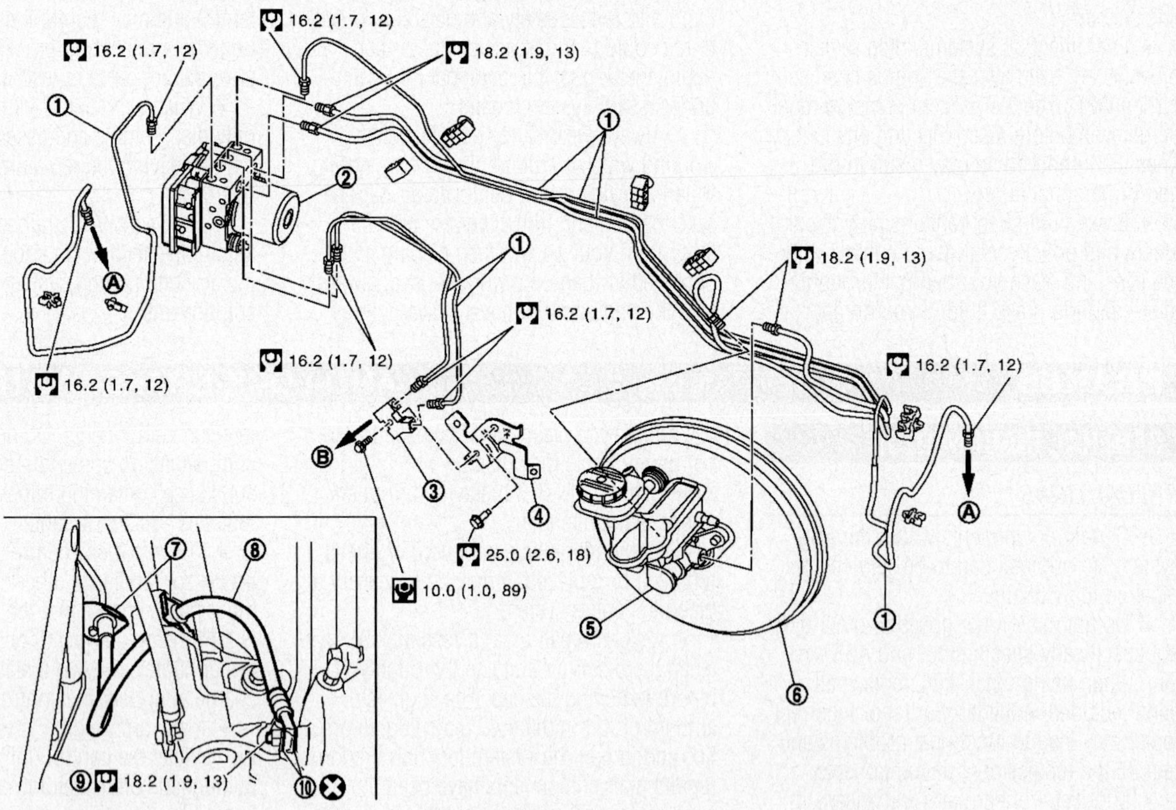

1. Brake tube
2. ABS actuator and electric unit (control unit)
3. Connector
4. Connector bracket
5. Master cylinder assembly
6. Brake booster
7. Lock plate
8. Brake hose
9. Union bolt
10. Copper washer
A. To front brake hose
B. To rear brake tube

Fig. 2 Master cylinder and related components view—Rogue

⁕⁕ CAUTION

Dust and dirt accumulating on brake parts during normal use may contain asbestos fibers from production or aftermarket brake linings. Breathing excessive concentrations of asbestos fibers can cause serious bodily harm. Exercise care when servicing brake parts. Do not sand or grind brake lining unless equipment used is designed to contain the dust residue. Do not clean brake parts with compressed air or by dry brushing. Cleaning should be done by dampening the brake components with a fine mist of water, then wiping the brake components clean with a dampened cloth. Dispose of cloth and all residue containing asbestos fibers in an impermeable container with the appropriate label. Follow practices prescribed by the Occupational Safety and Health Administration (OSHA) and the Environmental Protection Agency (EPA) for the handling, processing, and disposing of dust or debris that may contain asbestos fibers.

BRAKE CALIPER

REMOVAL & INSTALLATION

See Figure 3.

1. Before servicing the vehicle, refer to the Precautions Section.
2. Remove tires with power tool.
3. Fix the disc rotor using wheel nuts.
4. Drain the brake fluid.
5. Remove union bolt and copper washers, and then disconnect brake hose from caliper assembly.
6. Remove the torque member mounting bolts, and remove brake caliper assembly.
7. Remove the disc brake rotor. Put matching marks on the wheel hub and bearing assembly and the disc rotor before removing the disc rotor.

To install:

8. Install the disc brake rotor to the original position.
9. Install the brake caliper assembly to the vehicle and tighten the torque member mounting bolts to 122 ft. lbs. (165 Nm).
10. Install the brake hose and copper washers to the brake caliper assembly, and tighten union bolt to 13 ft. lbs. (18.2 Nm).
11. Refill with new brake fluid and perform the air bleeding.

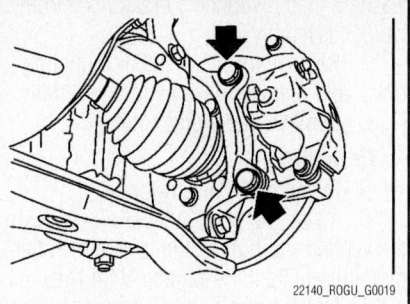

Fig. 3 Remove the torque member mounting bolts.

12. Check that no drag is present for the front disc brake.
13. Install the tires.

DISC BRAKE PADS

REMOVAL & INSTALLATION

See Figure 4.

1. Before servicing the vehicle, refer to the Precautions Section.
2. Remove tires with power tool.
3. Remove lower sliding pin bolt.
4. Suspend the cylinder body with suitable wire so that the brake hose will not stretch.

5. Then remove the brake pad from the torque member.

To install:

6. Install the pad retainer to the torque member if the pad retainers have been removed.
7. Apply copper based brake grease and install them to the brake pad.
8. Install the cylinder body and brake pads to the torque member.
9. Use a disc brake piston tool to easily press the piston back into the caliper.
10. Install the lower sliding pin bolt and tighten it to the specified torque.
11. Depress the brake pedal several times to check that no drag feel is present for the front disc brake.
12. Brake burnishing procedure is as follows:
     a. Drive the vehicle on a straight, flat road.
     b. Depress the brake pedal with the power to stop vehicle within 3 to 5 seconds until the vehicle stops.
     c. Drive without depressing the brakes for a few minutes to cool down the brake system.
     d. Repeat steps A—C until pad and disc rotor are securely fitted.

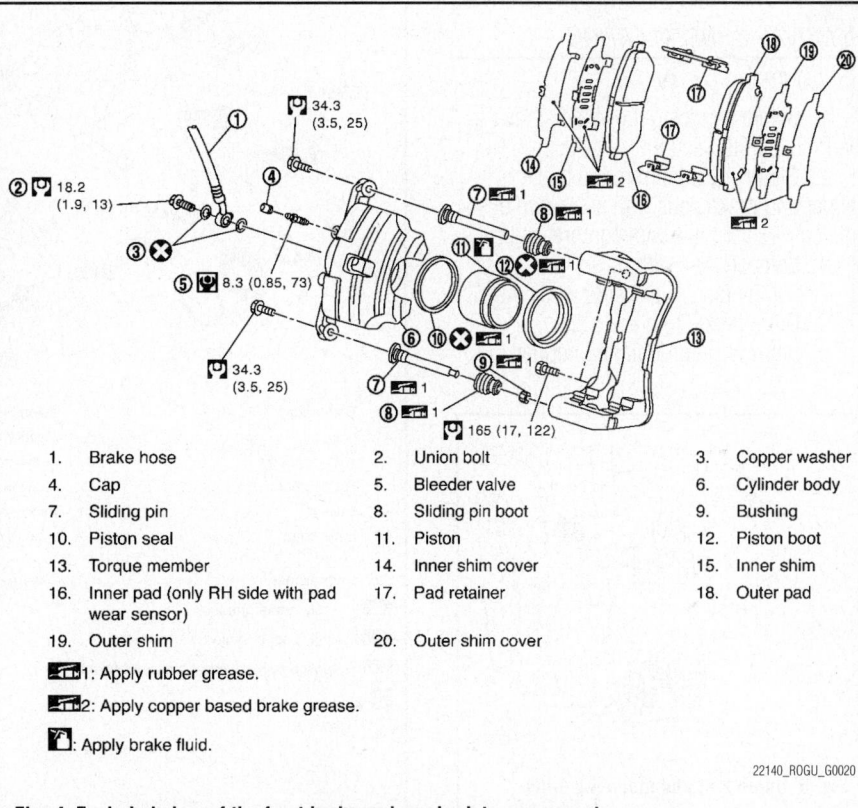

1.	Brake hose	2.	Union bolt	3.	Copper washer
4.	Cap	5.	Bleeder valve	6.	Cylinder body
7.	Sliding pin	8.	Sliding pin boot	9.	Bushing
10.	Piston seal	11.	Piston	12.	Piston boot
13.	Torque member	14.	Inner shim cover	15.	Inner shim
16.	Inner pad (only RH side with pad wear sensor)	17.	Pad retainer	18.	Outer pad
19.	Outer shim	20.	Outer shim cover		

⊏▥◻1: Apply rubber grease.

⊏▥◻2: Apply copper based brake grease.

◻: Apply brake fluid.

Fig. 4 Exploded view of the front brake pads and relate components

BRAKES

REAR DISC BRAKES

✳✳ CAUTION

Dust and dirt accumulating on brake parts during normal use may contain asbestos fibers from production or aftermarket brake linings. Breathing excessive concentrations of asbestos fibers can cause serious bodily harm. Exercise care when servicing brake parts. Do not sand or grind brake lining unless equipment used is designed to contain the dust residue. Do not clean brake parts with compressed air or by dry brushing. Cleaning should be done by dampening the brake components with a fine mist of water, then wiping the brake components clean with a dampened cloth. Dispose of cloth and all residue containing asbestos fibers in an impermeable container with the appropriate label. Follow practices prescribed by the Occupational Safety and Health Administration (OSHA) and the Environmental Protection Agency (EPA) for the handling, processing, and disposing of dust or debris that may contain asbestos fibers.

BRAKE CALIPER

REMOVAL & INSTALLATION

See Figure 5.

1. Before servicing the vehicle, refer to the Precautions Section.
2. Clean any dust from the brake caliper and brake pads with a vacuum dust collector. Never blow with compressed air.
3. Remove tires with power tool.
4. Fix the disc rotor using wheel nuts.
5. Drain brake fluid.
6. Remove union bolt and copper

washers then disconnect brake hose from caliper assembly.

7. Remove torque member mounting bolts, and remove brake caliper assembly.
8. Remove brake disc rotor.

To install:

9. Install brake disc rotor.
10. Install the brake caliper assembly to the vehicle and tighten the torque member mounting bolts to 64 ft. lbs. (84.4 Nm).
11. Install brake hose and copper washers to brake caliper assembly, and tighten union bolt to 13 ft. lbs. (18.2 Nm).
12. Refill with new brake fluid and perform the air bleeding.
13. Check that no drag is present for the rear disc brake.
14. Install the tires.

DISC BRAKE PADS

REMOVAL AND INSTALLATION

See Figure 6.

1. Before servicing the vehicle, refer to the Precautions Section.
2. Remove tires with power tool.
3. Remove lower sliding pin bolt.
4. Suspend the cylinder body with suitable wire so that the brake hose will not stretch.
5. Then remove the brake pad from the torque member.

To install:

6. Install the pad retainer to the torque member if the pad retainers have been removed.
7. Apply copper based brake grease and install them to the brake pad.
8. Install the cylinder body and brake pads to the torque member.
9. Use a disc brake piston tool to easily press the piston back into the caliper.
10. Install the lower sliding pin bolt and tighten it to 32 ft. lbs. (43.2 Nm).
11. Depress the brake pedal several

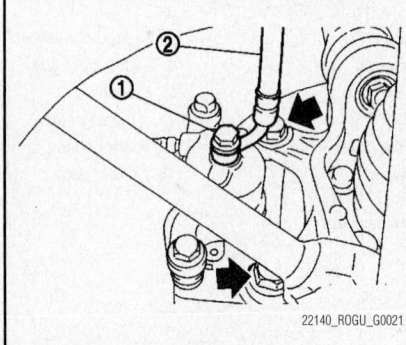

Fig. 5 Union bolt and mounting bolts shown

22140_ROGU_G0021

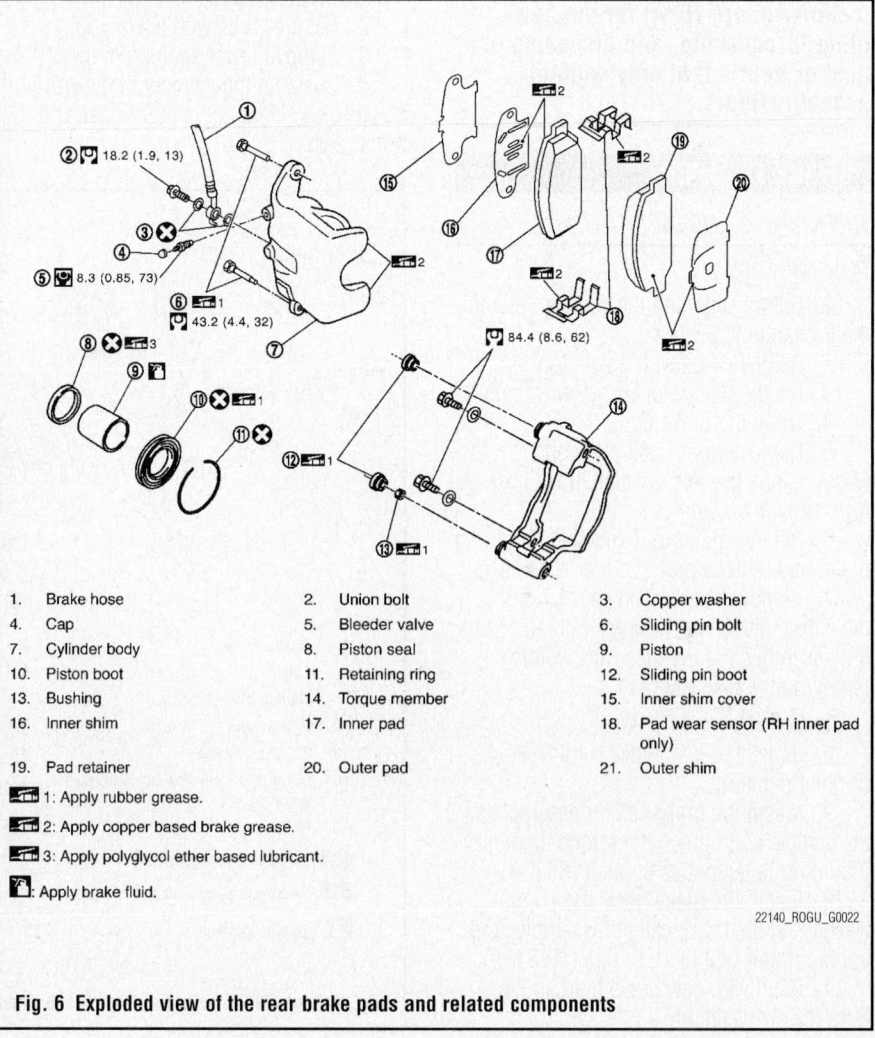

1.	Brake hose	2.	Union bolt	3.	Copper washer
4.	Cap	5.	Bleeder valve	6.	Sliding pin bolt
7.	Cylinder body	8.	Piston seal	9.	Piston
10.	Piston boot	11.	Retaining ring	12.	Sliding pin boot
13.	Bushing	14.	Torque member	15.	Inner shim cover
16.	Inner shim	17.	Inner pad	18.	Pad wear sensor (RH inner pad only)
19.	Pad retainer	20.	Outer pad	21.	Outer shim

1: Apply rubber grease.
2: Apply copper based brake grease.
3: Apply polyglycol ether based lubricant.
: Apply brake fluid.

22140_ROGU_G0022

Fig. 6 Exploded view of the rear brake pads and related components

times to check that no drag feel is present for the front disc brake.

12. Install the tires.

13. Brake burnishing procedure is as follows:

a. Drive the vehicle on a straight, flat road.

b. Depress the brake pedal with the power to stop vehicle within 3 to 5 seconds until the vehicle stops.

c. Drive without depressing the brakes for a few minutes to cool down the brake system.

d. Repeat steps A—C until pad and disc rotor are securely fitted.

| **BRAKES** | **PARKING BRAKE** |

PARKING BRAKE SHOES

REMOVAL & INSTALLATION

See Figures 7 and 8.

1. Before servicing the vehicle, refer to the Precautions Section.

2. Remove rear tires with power tool.

3. Remove disc rotor with parking brake completely in the released position.

➡Remove rear tires with power tool.

4. Put matching marks on the disc rotor and the wheel hub and bearing assembly when reusing the disc rotor.

5. If disc rotor cannot be removed, remove as follows:

a. Fix the disc rotor with wheel nuts and remove the adjusting hole plug.

b. Using suitable tool, rotate adjuster (1) in direction (B) to retract and loosen parking brake shoe.

6. Remove anti-rattle pins, springs, and return springs.

7. Remove brake strut, adjuster, parking brake shoes and toggle lever.

To install:

8. Install in the reverse order of removal.

9. Apply PBC (Poly Butyl Cuprysil)

grease or silicone-based grease to the back plate and brake shoe.

10. Assemble adjusters so that threaded part is expanded when rotated.

11. Shorten adjuster by rotating it.

12. When disassembling, apply PBC (Poly Butyl Cuprysil) grease or silicone-based grease to threads.

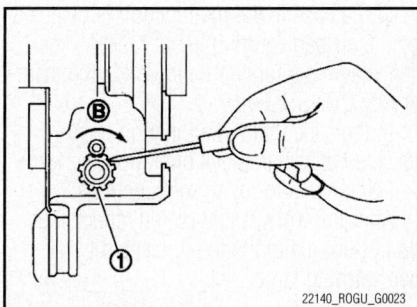

22140_ROGU_G0023

Fig. 7 Using suitable tool, rotate adjuster (1) in direction (B) to retract and loosen parking brake shoe

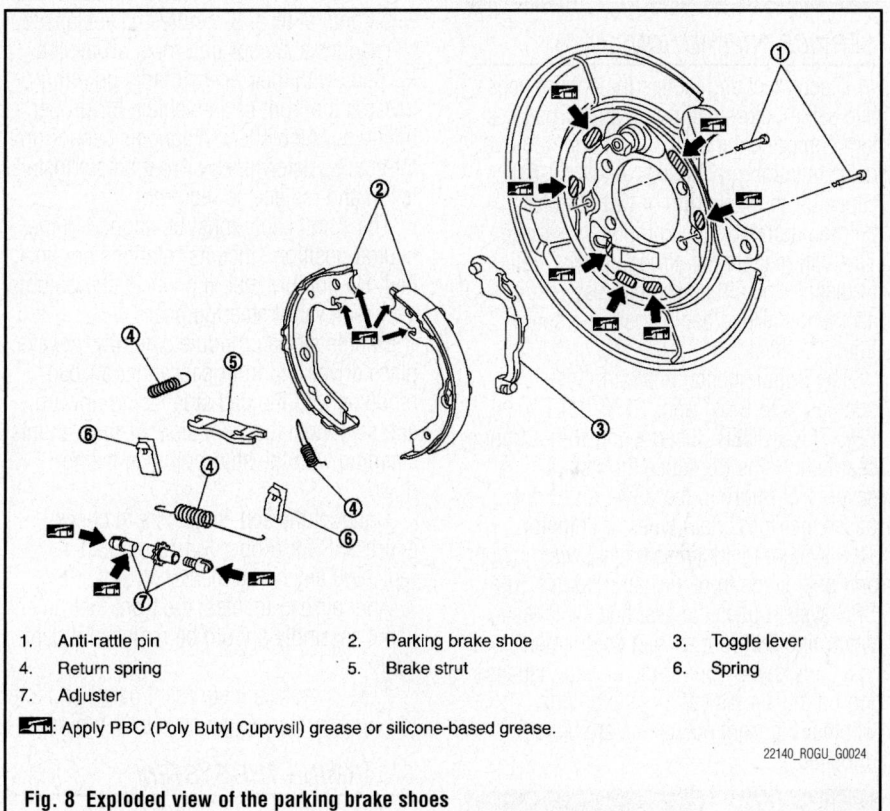

1.	Anti-rattle pin	2.	Parking brake shoe	3.	Toggle lever
4.	Return spring	5.	Brake strut	6.	Spring
7.	Adjuster				

▰: Apply PBC (Poly Butyl Cuprysil) grease or silicone-based grease.

22140_ROGU_G0024

Fig. 8 Exploded view of the parking brake shoes

CHASSIS ELECTRICAL

AIR BAG (SUPPLEMENTAL RESTRAINT SYSTEM)

GENERAL INFORMATION

✳✳ CAUTION

These vehicles are equipped with an air bag system. The system must be disarmed before performing service on, or around, system components, the steering column, instrument panel components, wiring and sensors. Failure to follow the safety precautions and the disarming procedure could result in accidental air bag deployment, possible injury and unnecessary system repairs.

SERVICE PRECAUTIONS

Disconnect and isolate the battery negative cable before beginning any airbag system component diagnosis, testing, removal, or installation procedures. Allow system capacitor to discharge for two minutes before beginning any component service. This will disable the airbag system. Failure to disable the airbag system may result in accidental airbag deployment, personal injury, or death.

The Supplemental Restraint System such as "AIR BAG" and "SEAT BELT PRE-TENSIONER", used along with a front seat belt, helps to reduce the risk or severity of injury to the driver and front passenger for certain types of collision. This system includes seat belt switch inputs and dual stage front air bag modules. The SRS system uses the seat belt switches to determine the front air bag deployment, and may only deploy one front air bag, depending on the severity of a collision and whether the front occupants are belted or unbelted.

✳✳ CAUTION

To avoid rendering the SRS inoperative, which could increase the risk of personal injury or death in the event of a collision which would result in air bag inflation, all maintenance must be performed by an authorized NISSAN/INFINITI dealer.

✳✳ CAUTION

Improper maintenance, including incorrect removal and installation of the SRS, can lead to personal injury caused by unintentional activation of the system.

✳✳ CAUTION

Do not use electrical test equipment on any circuit related to the SRS unless instructed to in this Service Manual. SRS wiring harnesses can be identified by yellow and/or orange harnesses or harness connectors.

Before servicing the SRS, turn ignition switch OFF, disconnect both battery cables and wait at least 3 minutes. For approximately 3 minutes after the cables are removed, it is still possible for the air bag and seat belt pretensioner to deploy. Therefore, do not work on any SRS connectors or wires until at least 3 minutes have elapsed.

Diagnosis sensor unit must always be installed with their arrow marks pointing towards the front of the vehicle for proper operation. Also check diagnosis sensor unit for cracks, deformities or rust before installation and replace as required.

The spiral cable must be aligned in the neutral position since its rotations are limited. Do not turn steering wheel and column after removal of steering gear.

Handle air bag module carefully. Always place driver and front passenger air bag modules with the pad side facing upward and seat mounted front side air bag module standing with the stud bolt side facing down.

Conduct the self-diagnosis to check entire SRS for proper functioning after replacing any components.

After air bag inflates, the front instrument panel assembly should be replaced if damaged.

Always replace instrument panel pad following front passenger air bag deployment.

DISARMING THE SYSTEM

Before servicing the SRS, turn ignition switch OFF, disconnect both battery cables and wait at least 3 minutes.

✳✳ CAUTION

For approximately 3 minutes after the cables are removed, it is still possible for the air bag and seat belt pretensioner to deploy. Therefore, do not work on any SRS connectors or wires until at least 3 minutes have elapsed.

ARMING THE SYSTEM

Reconnect the battery cables to rearm the SRS system.

CLOCKSPRING CENTERING

See Figure 9.

1. Before servicing the vehicle, refer to the Precautions Section.
2. The spiral cable may snap during steering operation if the cable is installed in an improper position.
3. The neutral position is set as follows:
   - Turn carefully the spiral cable clockwise to the end position.
   - Then turn it counterclockwise (about 2 and a half turns) and stop turning at the point (B) on which the stopper insertion holes are in the same position.
   - The service part is installed in the neutral position by the stopper and can be set without adjusting after the stopper is removed.

✳✳ WARNING

Do not over turn the spiral cable or go beyond number of turns required. (These will cause the cable to snap.)

- Adjust the spiral cable locating pin (A) to the steering wheel locating pin hole (C).

In the case that a malfunction is detected by the air bag warning lamp, reset by the self-diagnosis function and delete the memory by CONSULT-III.

In the case that a malfunction is still detected after the above operation, perform self-diagnosis to repair malfunctions.

After the work is completed, check that no system malfunction is detected by air bag warning lamp.

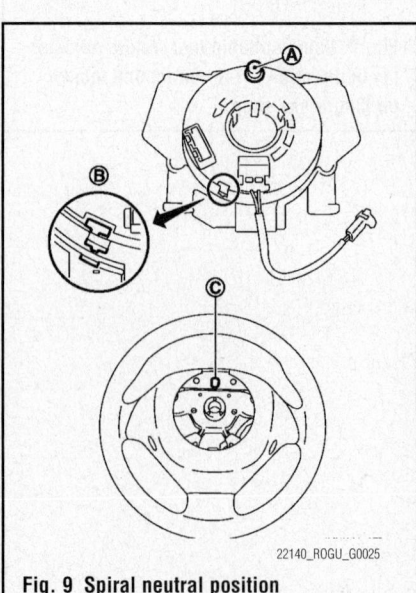

22140_ROGU_G0025

Fig. 9 Spiral neutral position

DRIVETRAIN

AUTOMATIC TRANSAXLE ASSEMBLY

REMOVAL & INSTALLATION

2WD Models

See Figures 10 through 12.

1. Before servicing the vehicle, refer to the Precautions Section.
2. Remove air duct (inlet).
3. Remove air cleaner case.
4. Remove engine under cover with power tool.
5. Drain the engine coolant.
6. Remove the CVT fluid level gauge.
7. Remove the CVT fluid charging pipe from transaxle assembly.
8. Remove O-ring from CVT fluid charging pipe.
9. Disconnect the fluid cooler hose from transaxle assembly (with fluid cooler only).
10. Disconnect the following harness

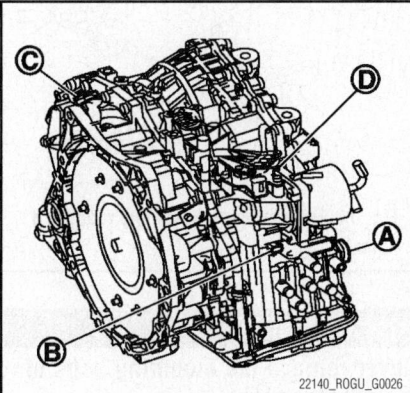

Fig. 10 Harness connectors and wiring harnesses (A)—(D)

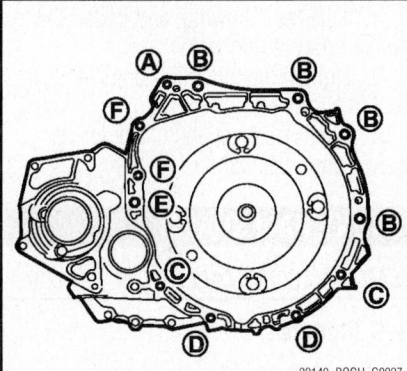

Fig. 11 Attach the fixing bolts in accordance with the graphics

connectors and wiring harnesses as follows:
- Remove the CVT unit connector (A).
- Primary speed sensor connector (B).
- Secondary speed sensor connector (C).
- PNP switch connector (D).

11. Remove the harness and clip from transaxle assembly.
12. Remove CVT water hoses.
13. Remove control cable from bracket.
14. Remove control cable bracket.
15. Remove the starter motor.
16. Remove rear plate cover.
17. Turn crankshaft, and remove the four nuts for the drive plate and torque converter.

✳✳ WARNING

When turning the crankshaft, turn it clockwise as viewed from the front of the engine.

18. Remove the exhaust front tube.
19. Remove the front drive shafts.
20. Remove front suspension member from vehicle.
21. Support the transaxle assembly with a transmission jack. When setting the transmission jack, be careful not to collide against drain plug.
22. Support engine assembly with a transmission jack. When setting the transmission jack, be careful not to collide against drain plug.
23. Remove engine mounting insulator (LH).
24. Remove bolts fixing transaxle assembly to engine assembly.
25. Remove the transaxle assembly from vehicle.

✳✳ WARNING

Secure the torque converter to prevent it from dropping. Secure

transaxle assembly to a transmission jack.

To install:

26. Note the following, and install in the reverse order of removal.
27. When installing transaxle assembly to the engine assembly, attach the fixing bolts in accordance with the graphic.
28. When not using drive plate location guide, rotate torque converter so that the stud bolt for mounting the drive plate location guide of torque converter aligns with the mounting position of starter motor.

AWD Models

See Figure 13.

1. Before servicing the vehicle, refer to the Precautions Section.
2. Remove the battery and battery bracket.
3. Remove air breather hose.
4. Remove air duct (inlet).
5. Remove air cleaner case.
6. Remove engine under cover with power tool.
7. Drain engine coolant.
8. Remove the CVT fluid level gauge.
9. Remove the CVT fluid charging pipe from transaxle assembly.
10. Remove O-ring from CVT fluid charging pipe.
11. Disconnect the fluid cooler hose from transaxle assembly (with fluid cooler only).
12. Disconnect the following harness connectors and wiring harnesses as follows:
- Disconnect the CVT unit connector (A).
- Primary speed sensor connector (B).
- Secondary speed sensor connector (C).
- PNP switch connector (D).

Insertion direction	Transaxle assembly to engine assembly		Engine assembly to transaxle assembly			
Bolt position	A	B	C	D	E	F
Number of bolts	1	4	2	2	1	2
Bolt length mm (in)	45 (1.77)		45 (1.77)	35 (1.38)	45 (1.77)	60 (2.36)
Tightening torque N·m (kg-m, ft-lb)	35.3 (3.6, 26)	74.5 (7.6, 55)	42.6 (4.3, 31)		74.5 (7.6, 55)	50 (5.1, 37)

Fig. 12 Attach the fixing bolts in accordance with the chart

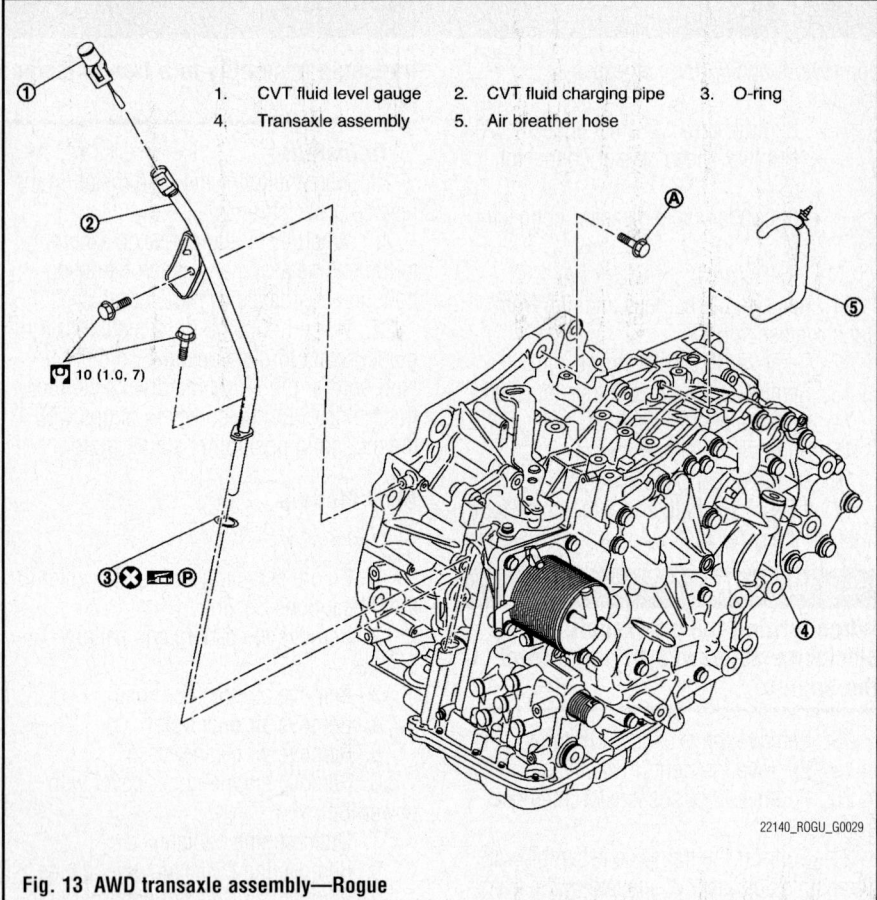

1. CVT fluid level gauge
2. CVT fluid charging pipe
3. O-ring
4. Transaxle assembly
5. Air breather hose

10 (1.0, 7)

22140_ROGU_G0029

Fig. 13 AWD transaxle assembly—Rogue

13. Remove harness and clip from the transaxle assembly.

14. Remove CVT water hoses. (if equipped)

15. Remove control cable from bracket.

16. Remove control cable bracket.

17. Remove the starter motor.

18. Remove rear plate cover.

19. Turn crankshaft, and remove the four tightening nuts for the drive plate and torque converter.

➡ **When turning crankshaft, turn it clockwise as viewed from the front of the engine.**

20. Remove exhaust front tube.

21. Separate the propeller shaft.

22. Remove front drive shafts.

23. Remove front suspension member from vehicle.

24. Remove transfer assembly from transaxle assembly with power tool.

25. Support the transaxle assembly with a transmission jack.

❊❊ **WARNING**

When setting the transmission jack, be careful not to collide against drain plug.

26. Support engine assembly with a transmission jack.

27. Remove engine mounting insulator (LH).

28. Remove bolts fixing transaxle assembly to engine assembly.

29. Remove the transaxle assembly from the vehicle.

❊❊ **WARNING**

Secure the torque converter to prevent it from dropping. Secure transaxle assembly to a transmission jack.

To install:

30. Note the following, and install in the reverse order of removal.

31. Check the fitting of dowel pin when installing transaxle assembly to engine assembly.

32. Align the position of tightening nuts for the drive plate with those of the torque converter, and temporarily tighten the nuts. Then, tighten the nuts to 38 ft. lbs. (58 Nm).

33. When tightening the nuts for the torque converter after fixing the crankshaft pulley bolts, confirm the tightening torque of the crankshaft pulley mounting bolts.

34. Rotate crankshaft several turns and check that transaxle rotates freely without binding after converter is installed to drive plate.

35. Never reuse O-ring.

36. Apply grease to O-ring.

TRANSFER CASE ASSEMBLY

REMOVAL & INSTALLATION

See Figure 14.

1. Before servicing the vehicle, refer to the Precautions Section.

2. Remove the exhaust front tube.

3. Remove the exhaust center tube.

4. Separate the rear propeller shaft.

5. Remove right side drive shaft and support bearing bracket.

6. Remove the mounting bolts of transaxle assembly and transfer assembly.

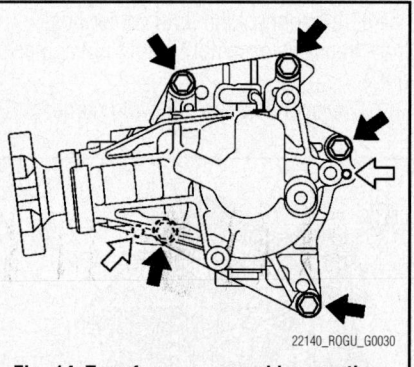

22140_ROGU_G0030

Fig. 14 Transfer case assembly mounting bolts

❊❊ **WARNING**

Never remove the mounting bolts of the adapter case.

7. Remove transfer assembly from the vehicle.

To install:

8. Note the following, and install in the reverse order of removal.

9. Tighten the mounting bolts to 32 ft. lbs. (44 Nm).

10. Check oil level and check for oil leakage after installation.

FRONT HALFSHAFT

REMOVAL & INSTALLATION

2WD Models

Left Side Axle

See Figure 15.

1. Before servicing the vehicle, refer to the Precautions Section.

2. Remove tires with power tool.

3. Remove wheel sensor from steering knuckle.

✳✳ WARNING

Never pull on the wheel sensor harness.

4. Remove lock plate from strut assembly.

5. Remove torque member mounting bolts with power tool. Hang torque member not to interfere with work.

6. Remove the brake disc rotor. Put matching marks on the wheel hub and bearing assembly and the disc rotor before removing the disc rotor.

➡**Never drop brake disc rotor.**

7. Remove cotter pin, and then loosen hub lock nut.

8. Patch hub lock nut with a piece of wood. Hammer the wood to disengage wheel hub and bearing assembly from drive shaft.

✳✳ WARNING

Never place drive shaft joint at an extreme angle. Also be careful not to overextend slide joint. Never allow drive shaft to hang down without support for housing (or joint sub-assembly), shaft and the other parts.

9. Use suitable puller if wheel hub and drive shaft cannot be separated even after performing the above procedure.

10. Remove the hub lock nut.

11. Remove transverse link from steering knuckle.

12. Loosen the nut of steering outer socket.

13. Remove steering outer socket (1) from steering knuckle (2) using the ball joint remover so as to not damage ball joint boot (3).

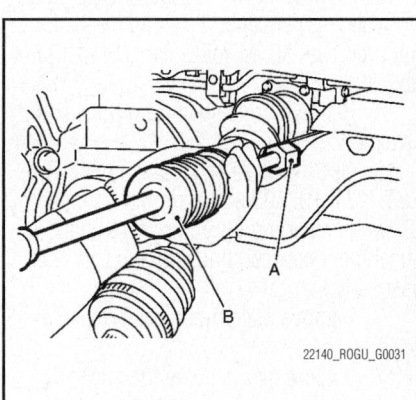

22140_ROGU_G0031

Fig. 15 Axle removal shown

14. Remove drive shaft from transaxle assembly.

15. Use the drive shaft attachment (A) SST: KV40107500 and a sliding hammer (B) while inserting tip of the drive shaft attachment between housing and transaxle assembly.

➡**Never place drive shaft joint at an extreme angle when removing drive shaft. Also be careful not to overextend slide joint.**

To install:

16. Note the following, and install in the reverse order of removal.

17. Place the axle protector (A) SST: KV38107900 onto transaxle assembly to prevent damage to the oil seal while inserting drive shaft. Slide drive shaft sliding joint and tap with a hammer to install securely.

✳✳ WARNING

Make sure that circular clip is completely engaged.

Right Side Axle

See Figure 16.

1. Before servicing the vehicle, refer to the Precautions Section.

2. Remove tires with power tool.

3. Remove wheel sensor from steering knuckle.

✳✳ WARNING

Never pull on the wheel speed sensor harness.

4. Remove the lock plate from strut assembly.

5. Remove torque member mounting bolts with power tool. Hang torque member not to interfere with work.

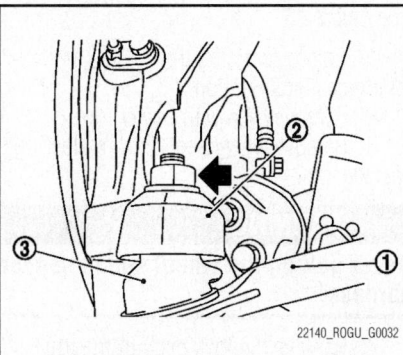

22140_ROGU_G0032

Fig. 16 Remove steering outer socket (1) from steering knuckle (2) using the ball joint remover so as not to damage ball joint boot (3)

6. Remove the brake disc rotor. Put matching marks on the wheel hub and bearing assembly and the disc rotor before removing the disc rotor.

➡**Never drop brake disc rotor.**

7. Remove cotter pin, and then loosen hub lock nut.

8. Patch hub lock nut with a piece of wood. Hammer the wood to disengage wheel hub and bearing assembly from drive shaft.

✳✳ WARNING

Never place drive shaft joint at an extreme angle. Also be careful not to overextend slide joint. Never allow drive shaft to hang down without support for housing (or joint sub-assembly), shaft and the other parts.

9. Use suitable puller if wheel hub and drive shaft cannot be separated even after performing the above procedure.

10. Remove the hub lock nut.

11. Remove transverse link from steering knuckle.

12. Loosen the nut of steering outer socket.

13. Remove steering outer socket (1) from steering knuckle (2) using the ball joint remover so as not to damage ball joint boot (3).

14. Remove drive shaft from transaxle assembly.

15. Use the drive shaft attachment (A) SST: KV40107500 and a sliding hammer (B) while inserting tip of the drive shaft attachment between housing and transaxle assembly.

16. Remove plate mounting bolts and plate.

17. If necessary, remove the support bearing bracket mounting bolts and the support bearing bracket.

18. Remove drive shaft from transaxle assembly.

19. Use the drive shaft attachment SST: KV40107500 and a sliding hammer while inserting tip of the drive shaft attachment between housing and transaxle assembly.

✳✳ WARNING

Never place drive shaft joint at an extreme angle when removing drive shaft. Also be careful not to overextend slide joint.

To install:

20. Note the following, and install in the reverse order of removal.

21. Place the axle protector onto transaxle assembly to prevent damage to the

oil seal while inserting drive shaft. Slide drive shaft sliding joint and tap with a hammer to install securely.

22. Tighten the axle nut to 92 ft. lbs. (125 Nm). Install the cotter pin.

❋❋ WARNING

Make sure that circular clip is completely engaged.

AWD Models

Left Side Axle

See Figure 17.

1. Before servicing the vehicle, refer to the Precautions Section.
2. Remove tires with power tool.
3. Remove wheel sensor from steering knuckle.
4. Never pull on wheel sensor harness.
5. Remove lock plate from strut assembly.
6. Remove torque member mounting bolts with power tool. Hang torque member not to interfere with work.
7. Remove disc brake rotor. Put matching marks on the wheel hub and bearing assembly and the disc rotor before removing the disc rotor. Never drop the disc rotor.
8. Remove cotter pin, and then loosen hub lock nut with power tool.
9. Patch hub lock nut with a piece of wood. Hammer the wood to disengage wheel hub and bearing assembly from drive shaft.

❋❋ WARNING

Never place drive shaft joint at an extreme angle. Also be careful not to overextend slide joint. Never allow drive shaft to hang down without support for housing (or joint subassembly), shaft and the other parts.

10. Use suitable puller if wheel hub and drive shaft cannot be separated even after performing the above procedure.
11. Remove the hub lock nut.
12. Remove transverse link from steering knuckle.
13. Loosen the nut of steering outer socket.
14. Remove steering outer socket from steering knuckle using the ball joint remover so as not to damage ball joint boot.
15. Remove plate mounting bolts and plate.
16. If necessary, remove the support bearing bracket mounting bolts and the support bearing bracket.

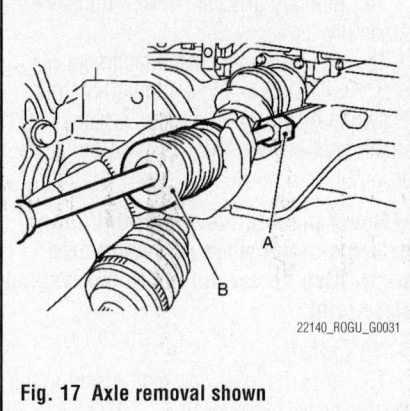

Fig. 17 Axle removal shown

17. Remove drive shaft from transaxle assembly.
18. Use a drive shaft attachment and a sliding hammer while inserting tip of the drive shaft attachment between housing and transaxle assembly.

To install:

19. Note the following, and install in the reverse order of removal.

➡**Always replace the differential side oil seal with new one when installing drive shaft.**

20. Place the seal protector onto transaxle assembly to prevent damage to the oil seal while inserting drive shaft. Slide drive shaft sliding joint and tap with a hammer to install securely.

❋❋ WARNING

Make sure that circular clip is completely engaged.

21. Slide drive shaft sliding joint and tap with a hammer to install securely.
22. Tighten the axle nut to 92 ft. lbs. (125 Nm). Install the cotter pin.

Right Side Axle

See Figure 18.

1. Before servicing the vehicle, refer to the Precautions Section.
2. Remove tires with power tool.
3. Remove wheel sensor from steering knuckle.

❋❋ WARNING

Never pull on the wheel speed sensor harness.

4. Remove the lock plate from strut assembly.
5. Remove torque member mounting bolts with power tool. Hang torque member not to interfere with work.
6. Remove the brake disc rotor. Put

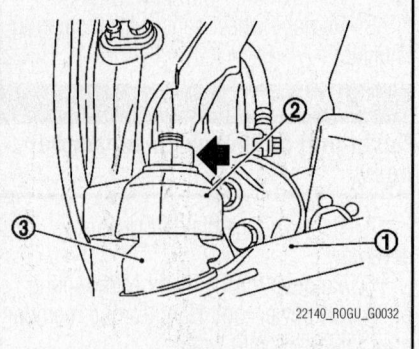

Fig. 18 Remove steering outer socket (1) from steering knuckle (2) using the ball joint remover so as not to damage ball joint boot (3)

matching marks on the wheel hub and bearing assembly and the disc rotor before removing the disc rotor.

➡**Never drop brake disc rotor.**

7. Remove cotter pin, and then loosen hub lock nut.
8. Patch hub lock nut with a piece of wood. Hammer the wood to disengage wheel hub and bearing assembly from drive shaft.

❋❋ WARNING

Never place drive shaft joint at an extreme angle. Also be careful not to overextend slide joint. Never allow drive shaft to hang down without support for housing (or joint subassembly), shaft and the other parts.

9. Use suitable puller if wheel hub and drive shaft cannot be separated even after performing the above procedure.
10. Remove the hub lock nut.
11. Remove transverse link from steering knuckle.
12. Loosen the nut of steering outer socket.
13. Remove steering outer socket (1) from steering knuckle (2) using the ball joint remover so as not to damage ball joint boot (3).
14. Remove drive shaft from transaxle assembly.
15. Use the drive shaft attachment (A) SST: KV40107500 and a sliding hammer (B) while inserting tip of the drive shaft attachment between housing and transaxle assembly.
16. Remove plate mounting bolts and plate.
17. If necessary, remove the support bearing bracket mounting bolts and the support bearing bracket.

18. Remove drive shaft from transaxle assembly.

19. Use the drive shaft attachment SST: KV40107500 and a sliding hammer while inserting tip of the drive shaft attachment between housing and transaxle assembly.

※ WARNING

Never place drive shaft joint at an extreme angle when removing drive shaft. Also be careful not to overextend slide joint.

To install:

20. Note the following, and install in the reverse order of removal.

21. Place the axle protector (A) SST: KV38107900 onto transaxle assembly to prevent damage to the oil seal while inserting drive shaft. Slide drive shaft sliding joint and tap with a hammer to install securely.

22. Tighten the axle nut to 92 ft. lbs. (125 Nm). Install the cotter pin.

※ WARNING

Make sure that circular clip is completely engaged.

CV-JOINTS OVERHAUL

See Figures 19 through 24.

1. Before servicing the vehicle, refer to the Precautions Section.

2. Remove the left axle assembly.

3. Remove snap ring (1), and then remove spider assembly from shaft.

4. Remove boot from shaft.

5. Remove the circular clip from housing (left side).

6. Remove dust shield from housing.

7. Clean the old grease on housing with paper waste.

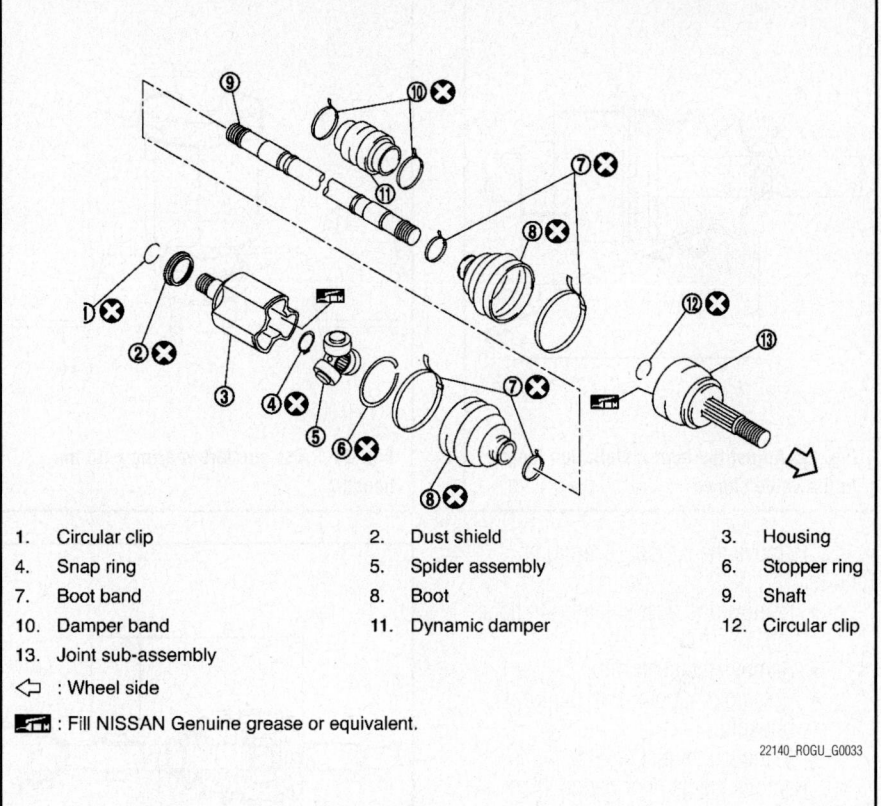

1.	Circular clip	2.	Dust shield	3.	Housing
4.	Snap ring	5.	Spider assembly	6.	Stopper ring
7.	Boot band	8.	Boot	9.	Shaft
10.	Damper band	11.	Dynamic damper	12.	Circular clip
13.	Joint sub-assembly				

◁ : Wheel side

: Fill NISSAN Genuine grease or equivalent.

22140_ROGU_G0033

Fig. 20 Exploded view of the left axle—Rogue

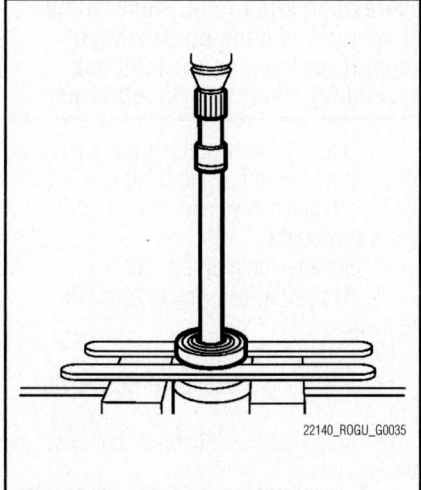

22140_ROGU_G0035

Fig. 19 Press out support bearing from housing

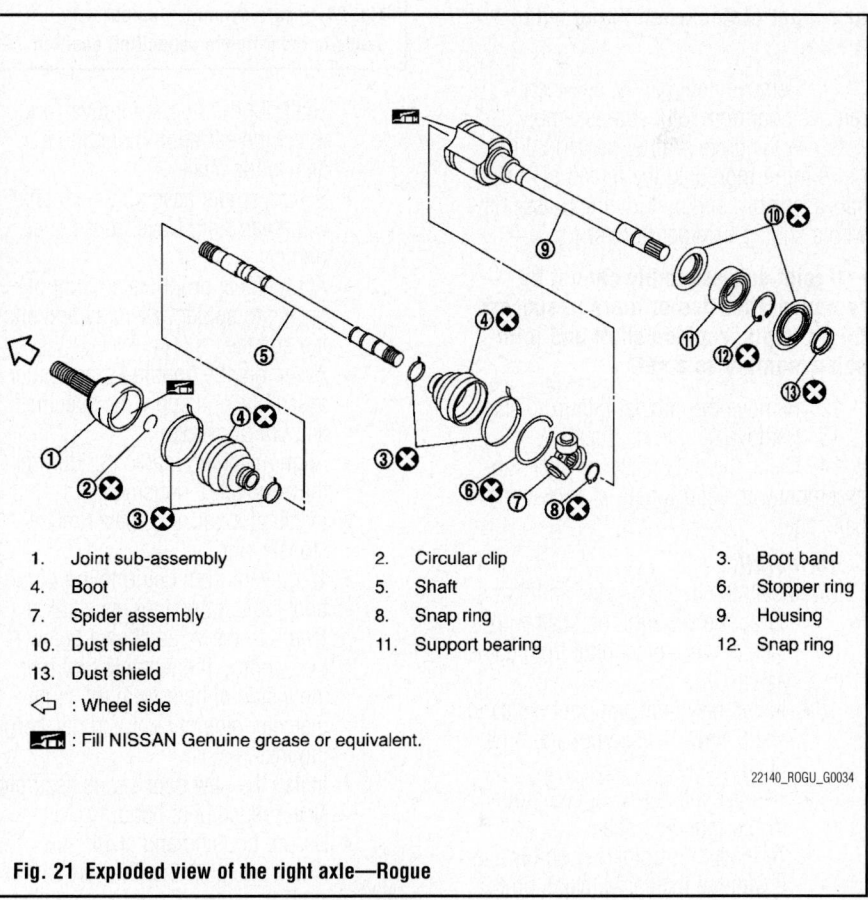

1.	Joint sub-assembly	2.	Circular clip	3.	Boot band
4.	Boot	5.	Shaft	6.	Stopper ring
7.	Spider assembly	8.	Snap ring	9.	Housing
10.	Dust shield	11.	Support bearing	12.	Snap ring
13.	Dust shield				

◁ : Wheel side

: Fill NISSAN Genuine grease or equivalent.

22140_ROGU_G0034

Fig. 21 Exploded view of the right axle—Rogue

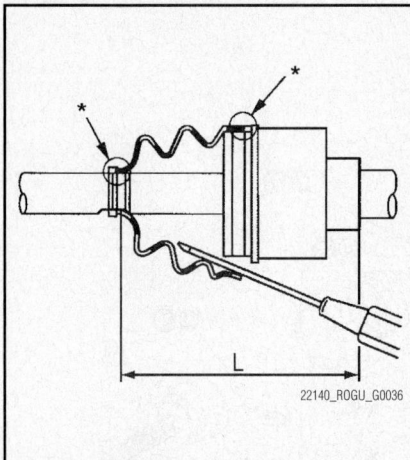

Fig. 22 Adjust the boot installation length to the value shown

8. Remove the support bearing as follows:
- Remove the dust shield from housing.
- Remove the snap ring.
- Press out support bearing from housing.
- Remove the dust shield.

9. Remove the damper bands, then remove dynamic damper from shaft.

10. Fix shaft with a vise.

➡**Protect the shaft using aluminum or copper plates when fixing with a vise.**

11. Remove boot bands, and then remove boot from joint sub-assembly.

Screw the drive shaft puller (A) 30 mm (1.18 in) or more into the thread of joint sub-assembly, and pull joint sub-assembly with a sliding hammer from shaft.

➡**If joint sub-assembly cannot be removed after five or more unsuccessful attempts, replace shaft and joint sub assembly as a set.**

12. Remove the circular clip from shaft.
13. Remove boot from shaft.
14. Clean the old grease on joint sub-assembly with paper waste while rotating ball cage.

To install:
15. Install the transaxle side as follows:
- Wrap the serration on shaft with tape (A) to protect boot from damage.
- Install new boot and boot bands to shaft. Never reuse boot and boot band.
- Remove the tape wrapped around the serration on shaft.
- To install the spider assembly align it with the matching marks on the

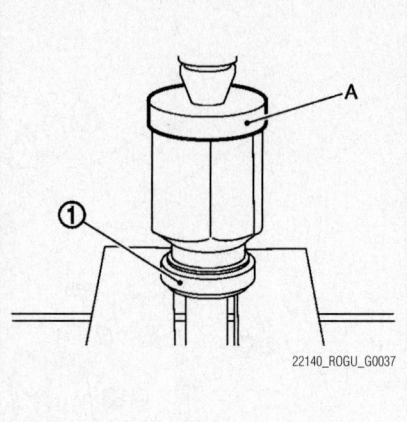

Fig. 23 Press support bearing onto the housing

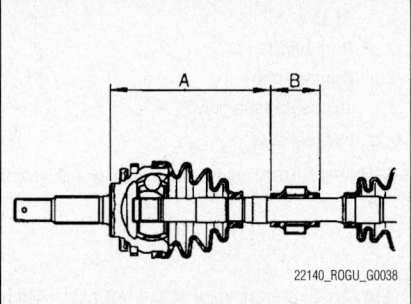

Fig. 24 Secure dynamic damper with bands in the following specified position

shaft put during the removal, and direct the serration mounting surface to the shaft.
- Secure spider assembly onto shaft with snap ring. Never reuse snap ring.
- Apply the appropriate amount of grease to spider assembly and sliding surface.
- Assemble the housing onto spider assembly, and apply the specified amount of grease.
- Align matching marks put during the removal of housing.
- Install stopper ring. Never reuse stopper ring.
- To prevent from deformation of the boot, adjust the boot installation length to the value shown below (L) by inserting the suitable tool into the inside of boot from the large diameter side of boot and discharging inside air.
- Install the new boot bands securely. Never reuse boot band.
- Secure housing and shaft, and then make sure that they are in the

correct position when rotating boot. Install them with new boot band when the mounting positions become incorrect.
- Install the dust shield (left side). Never reuse the dust shield.
- Install the circular clip to housing (left side). Never reuse circular clip.
- Support Bearing
- Install dust shield on housing. Never reuse dust shield.

16. Install the support bearing as follows:
- Press support bearing (1) onto housing using the suitable tool (A).
- Install the snap ring.
- Install the dust shields

17. Install the dynamic damper as follows:
- Secure dynamic damper with bands in the following specified position when installing.

REAR HALFSHAFT

REMOVAL & INSTALLATION

See Figure 25.

1. Before servicing the vehicle, refer to the Precautions Section.
2. Remove tires with power tool.
3. Remove torque member mounting bolts with power tool. Hang torque member not to interfere with work.
4. Remove disc rotor.
5. Remove cotter pin, then loosen hub lock nut with power tool.
6. Patch hub lock nut with a piece of wood. Hammer the wood to disengage the wheel hub and bearing assembly from drive shaft.

❊❊ WARNING

Never place drive shaft joint at an extreme angle. Also be careful not to overextend slide joint. Never allow drive shaft to hang down without support for housing (or joint sub-assembly), shaft and the other parts.

7. Use a suitable puller if the wheel hub and bearing assembly and drive shaft cannot be separated even after performing the above procedure.
8. Remove the hub lock nut.
9. Remove wheel sensor from axle housing.
10. Remove the stabilizer link.
11. Set suitable jack under suspension arm.
12. Remove shock absorber from suspension arm.
13. Remove the upper link from suspension arm with power tool.

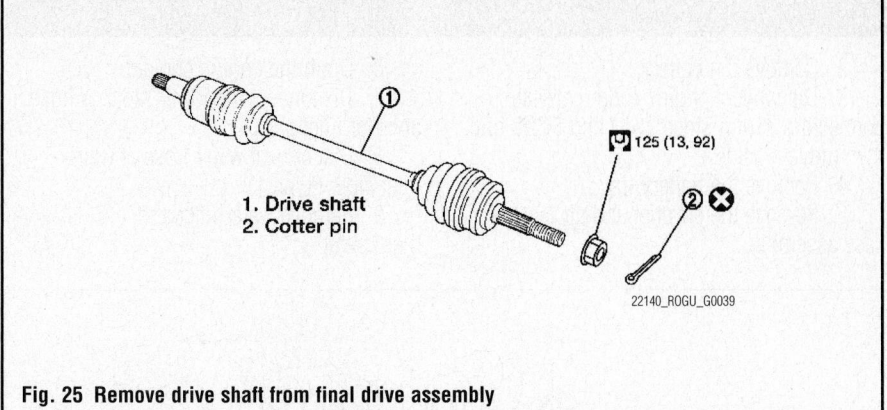

1. Drive shaft
2. Cotter pin

125 (13, 92)

22140_ROGU_G0039

Fig. 25 Remove drive shaft from final drive assembly

14. Remove lower link from suspension arm with power tool.

15. Remove drive shaft from final drive assembly.

16. Note the following, and install in the reverse order of removal.

17. Align the matching marks made during removal when reusing the disc rotor.

18. Tighten the axle nut to 92 ft. lbs. (125 Nm).

19. Perform final tightening of bolts and nuts at suspension arm (rubber bushing), under unladen conditions with tires on level ground.

REAR PINION SEAL

REMOVAL & INSTALLATION

See Figures 26 through 28.

1. Before servicing the vehicle, refer to the Precautions Section.

2. Remove rear propeller shaft.

3. Put matching mark on the thread edge of electric controlled coupling. The matching mark should be in line with the matching mark on companion flange.

4. Remove the companion flange lock nut, using a flange wrench (commercial service tool).

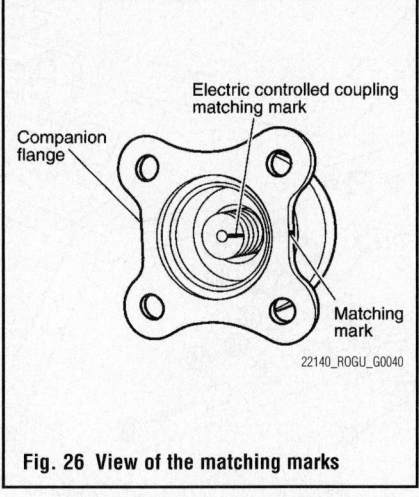

Electric controlled coupling matching mark

Companion flange

Matching mark

22140_ROGU_G0040

Fig. 26 View of the matching marks

5. Then remove the companion flange.

6. Remove the front oil seal from coupling cover, using a flat-bladed screwdriver.

To install:

7. Apply multi-purpose grease onto oil seal lips, and gear oil onto the circumference of oil seal.

8. Install front oil seal until it becomes flush with the coupling cover.

9. Align the matching mark of electric controlled coupling with the matching mark

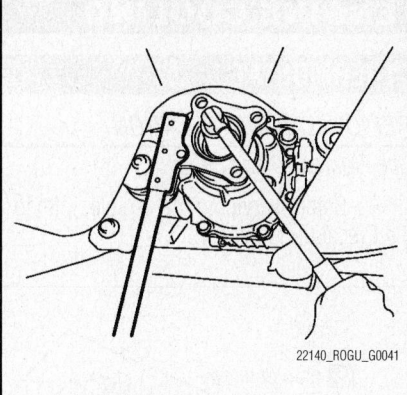

22140_ROGU_G0041

Fig. 27 Remove the companion flange lock nut

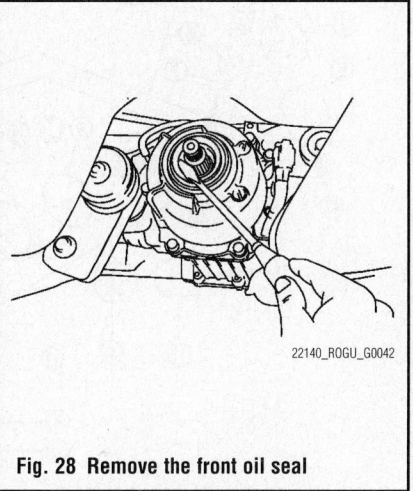

22140_ROGU_G0042

Fig. 28 Remove the front oil seal

of companion flange, then install the companion flange.

10. Install companion flange lock nut with a flange wrench (commercial service tool), tighten to 103 ft lbs. (140.3 Nm). Never reuse companion flange lock nut.

11. Install rear propeller shaft.

12. When oil leaks while removing, check oil level after the installation.

ENGINE COOLING

THERMOSTAT

REMOVAL & INSTALLATION

See Figures 29 and 30.

1. Before servicing the vehicle, refer to the Precautions Section.

2. Remove the battery.
3. Disconnect engine room harness connectors at unit sides TCM and ECM, and then move it aside.
4. Remove the battery tray.
5. Remove the air duct and air cleaner case assembly.

6. Drain the engine coolant.
7. Disconnect the lower radiator hose at water inlet side.
8. Disconnect water hose at water inlet side. (Type 1)
9. Remove water inlet and thermostat.

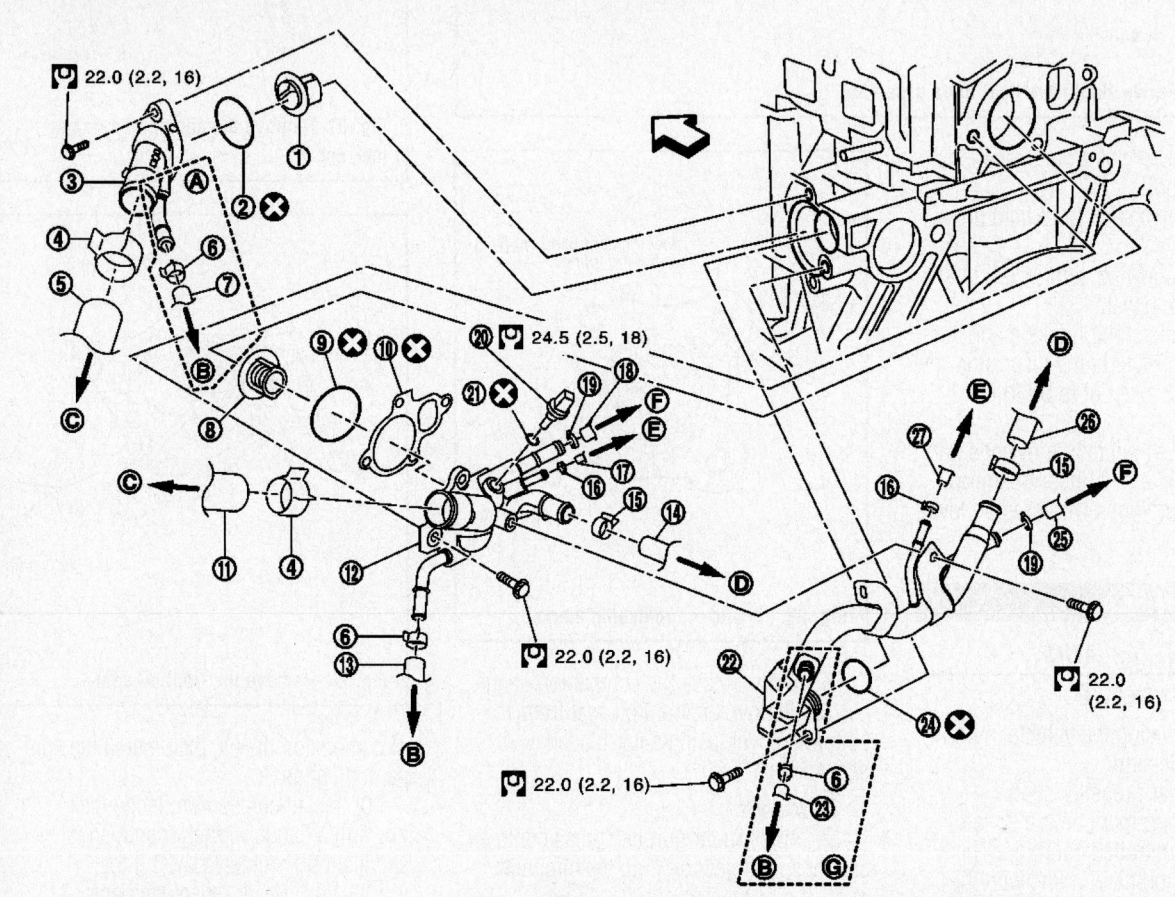

1. Thermostat	2. O-ring	3. Water inlet
4. Clamp	5. Radiator hose (lower)	6. Clamp
7. Water hose	8. Water control valve	9. O-ring
10. Gasket	11. Radiator hose (upper)	12. Water control valve housing (water outlet)
13. Water hose	14. Heater hose	15. Clamp
16. Clamp	17. Water hose	18. Water hose
19. Clamp	20. Engine coolant temperature sensor	21. Washer
22. Heater pipe	23. Water hose	24. O-ring
25. Water hose	26. Heater hose	27. Water hose
A. Type 1	B. To CVT fluid cooler	C. To radiator
D. To heater	E. To electric throttle control actuator	F. To oil cooler
G. Type 2		

⇦ : Engine front

22140_ROGU_G0043

Fig. 29 Exploded view of the thermostat, water control valve and related components—Rogue

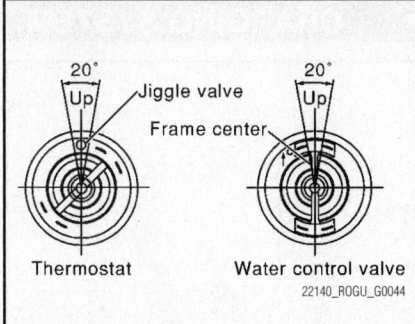

Fig. 30 Thermostat and water control valve positioning—Rogue

10. Remove the water control valve with the following procedure:
- Disconnect radiator hose (upper) at water control valve housing (water outlet) side.
- Disconnect harness connector from engine coolant temperature sensor.
- Remove the CVT fluid level gauge and CVT fluid charging pipe.
- Disconnect water hoses.
- Disconnect the air fuel ratio sensor 1 and heated oxygen sensor 2 harness connectors, and remove harness clips from heater pipe.
- Remove the heater pipe and heater hose.
- After removing the water control valve housing (water outlet), remove water control valve.

To install:
11. Note the following, and install in the reverse order of removal.
12. Refer to the exploded view for tightening specifications.
13. Install thermostat and water control valve with making rubber ring groove fit to thermostat flange and water control valve flange.

➡Same procedure is applied for installation of thermostat.

14. Install the thermostat with jiggle valve facing upwards. (The position deviation may be within the range of 20 degrees as shown in the figure.)
15. Install water control valve with the arrow facing up and the frame center part facing upwards. (The position deviation may be within the range of 20 degrees as shown in the figure.)

WATER PUMP

REMOVAL & INSTALLATION
See Figure 31.

1. Before servicing the vehicle, refer to the Precautions Section.
2. Drain the engine coolant
3. Remove the following parts:
- Drive belt
- Drive belt auto-tensioner
- Alternator
- Remove water pump.
4. Engine coolant will leak from the cylinder block, so have a receptacle ready below.
5. Remove water pump housing with the following procedure:
- Remove exhaust manifold cover.
- Remove oil level gauge and oil level gauge guide.
- Remove the mounting bolts for water pipe.
- Remove water pump housing.
- Remove exhaust manifold and three way catalyst assembly.
- Remove the water pipe.

To install:
6. Note the following, and install in the reverse order of removal.
7. When inserting water pipe end into cylinder block, apply a neutral detergent to O-ring.
8. Then insert it immediately.
9. Tighten water pump mounting bolts to 18 ft. lbs. (24.5 Nm).
10. Refer to the exploded water pump view for additional tightening specification.

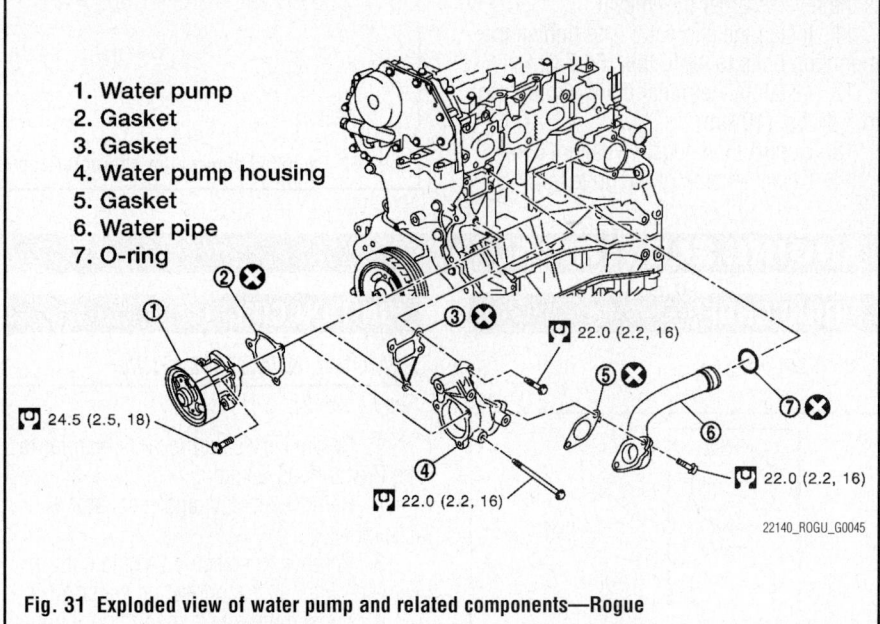

1. Water pump
2. Gasket
3. Gasket
4. Water pump housing
5. Gasket
6. Water pipe
7. O-ring

24.5 (2.5, 18)
22.0 (2.2, 16)
22.0 (2.2, 16)
22.0 (2.2, 16)

Fig. 31 Exploded view of water pump and related components—Rogue

ENGINE ELECTRICAL

CHARGING SYSTEM

ALTERNATOR

REMOVAL & INSTALLATION

See Figure 32.

1. Before servicing the vehicle, refer to the Precautions Section.
2. Disconnect the battery cable from the negative terminal.
3. Remove drive belt.
4. Disconnect the alternator connector (A).
5. Remove the "B" terminal nut (B) and "B" terminal harness.
6. Remove harness bracket (C).
7. Remove the upper alternator mounting bolt (D), using power tools.
8. Remove lower alternator mounting bolt (E), using power tools.
9. Remove the alternator upward from the vehicle.

To install:

10. Note the following, and installation is the reverse order of removal.
11. Install the alternator, and tighten the mounting bolts to 48 ft. lbs. (64.7 Nm).
12. Install the terminal nut B and tighten to 7 ft. lbs. (10 Nm).
13. Connect the negative battery cable.
14. Check tension of the drive belt.

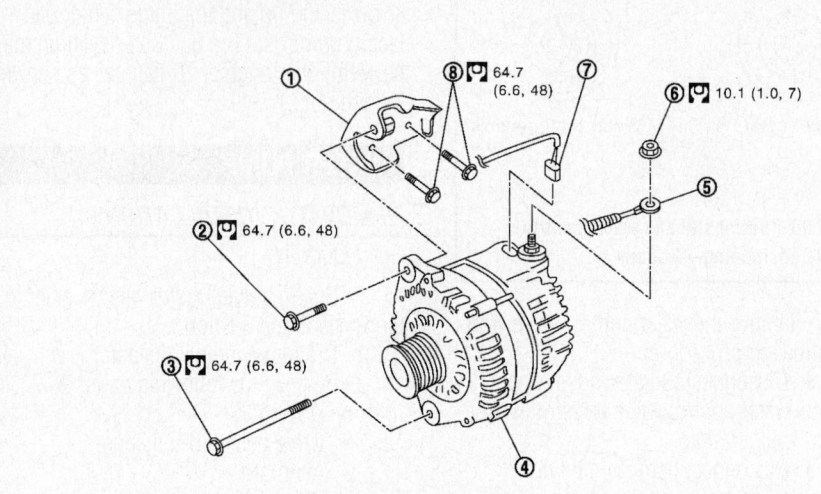

1.	Alternator bracket	2.	Upper alternator mounting bolt	3.	Lower alternator mounting bolt
4.	Alternator	5.	B terminal harness	6.	B terminal nut
7.	Alternator connector	8.	Alternator bracket mounting bolts		

22140_ROGU_G0046

Fig. 32 Exploded view of the alternator and related mounting components—Rogue

ENGINE ELECTRICAL

IGNITION SYSTEM

FIRING ORDER

See Figure 33.

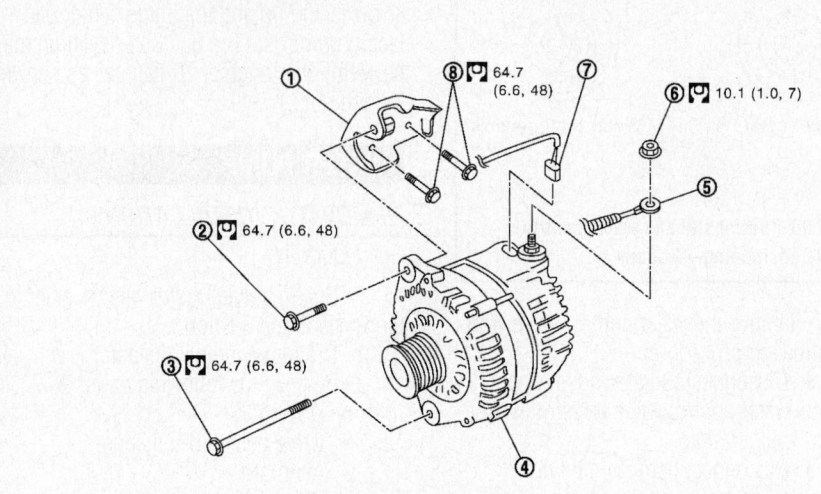

Front of Engine

22140_ROGU_G0047

Fig. 33 2.5L (QR25DE) Engine Firing order: 1-3-4-2
Distributor less ignition system (one coil on each cylinder)

IGNITION COIL

REMOVAL & INSTALLATION

See Figure 34.

1. Before servicing the vehicle, refer to the Precautions Section.
2. Remove air duct and resonator assembly.
3. Remove the electric throttle control actuator without disconnecting water hose.
4. Loosen the intake manifold mounting bolts and nuts.
5. Remove the intake manifold.
6. Disconnect harness connector from ignition coil.
7. Support the bottom surface of engine using a transmission jack.
8. Remove ground cable and harness from engine mounting bracket (RH).
9. Remove the ignition coil.

✳✳ WARNING

Never drop or shock the ignition coil. Never disassemble the ignition coil.

10. Disconnect PCV hose from rocker cover.
11. Remove engine mounting bracket (RH).
12. Remove PCV valve and O-ring from rocker cover, if necessary.
13. Remove oil filler cap from rocker cover if needed.

To install:

14. Note the following, and installation is the reverse order of removal.
15. Tighten the ignition coil mounting bolt to 62 ft. lbs. 7.0 Nm).

IGNITION TIMING

ADJUSTMENT

See Figures 35 and 36.

The ignition timing is not adjustable. If not within specifications, further diagnostic inspection is required. The following procedure is for viewing the ignition timing setting.

Visually check the air cleaner, intake hoses, ducts, Exhaust Gas Recirculation

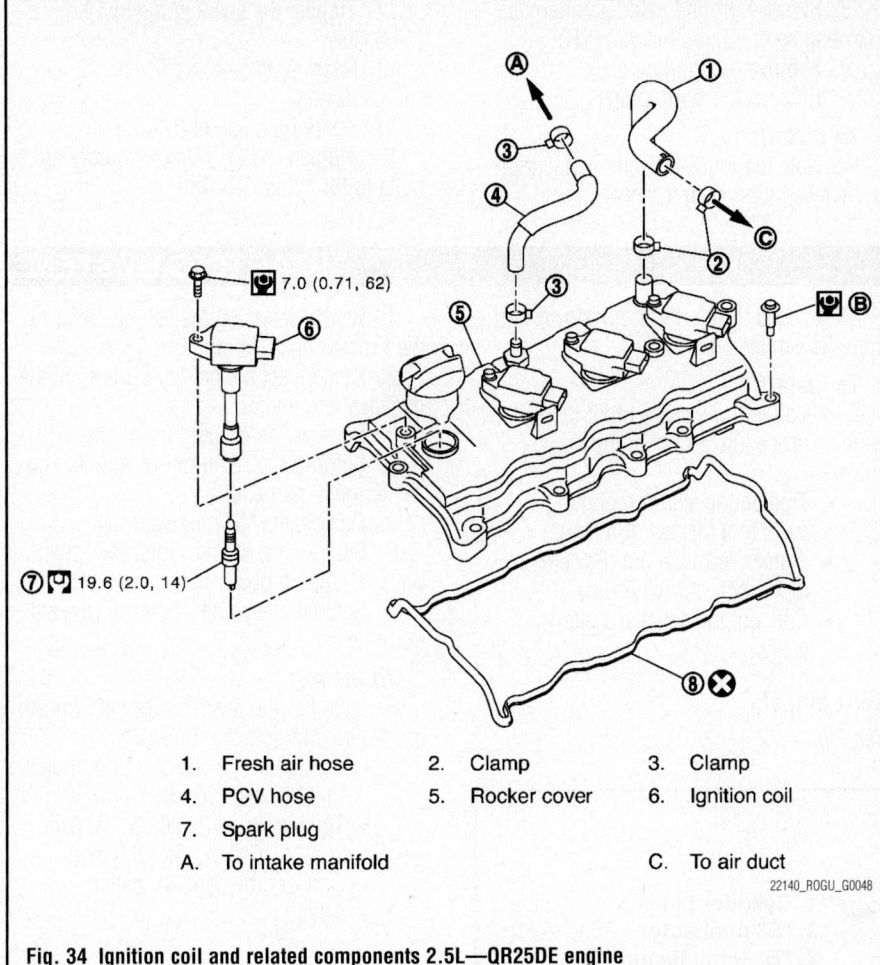

1. Fresh air hose
2. Clamp
3. Clamp
4. PCV hose
5. Rocker cover
6. Ignition coil
7. Spark plug
A. To intake manifold
C. To air duct

22140_ROGU_G0048

Fig. 34 Ignition coil and related components 2.5L—QR25DE engine

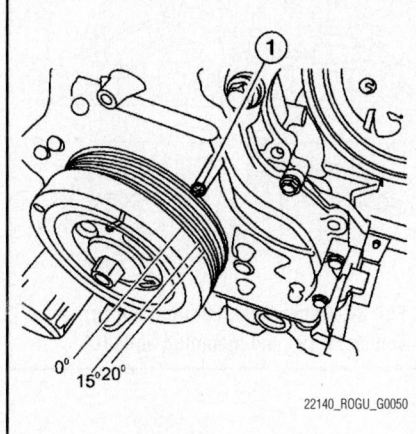

22140_ROGU_G0050

Fig. 35 Locate the timing marks on the crankshaft pulley

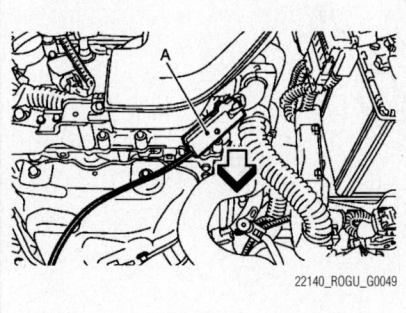

22140_ROGU_G0049

Fig. 36 Attach a timing light to the engine to number 1 cylinder ignition coil wire

(EGR) valve operation and electrical connections prior to the adjustment of the ignition timing. Correct or repair any problem as required. Be sure to inspect the throttle valve and Throttle Position (TP) sensor for proper operation.

1. Before servicing the vehicle, refer to the Precautions Section.

2. Locate the timing marks on the crankshaft pulley and the front of the engine.

3. Clean the timing marks.

4. The ignition timing specifications are as follows:
   • 10–20 degrees Before Top Dead Center (BTDC)

5. Using chalk or white paint, color the mark on the crankshaft pulley and the mark on the scale, which will indicate the correct timing when aligned with the notch on the crankshaft pulley.

6. Attach a tachometer to the engine.

7. Attach a timing light to the engine to number 1 cylinder ignition coil wire.

8. Turn **OFF** all the electrical equipment and accessories.

9. Check to be sure all of the wires clear the fan, then, start the engine and allow it to reach normal operating temperatures.

10. Block the front wheels and set the parking brake. Shift the transmission into **NEUTRAL**

11. Perform the following procedures:
   a. Race the engine at 2000 rpm for about 2 minutes under a no-load condition; be sure all of the accessories are turned **OFF**.
   b. Perform on board engine diagnostics and repair any fault code.
   c. Race the engine at 2000 rpm for about 2 minutes under a no-load condition.
   d. Turn the engine **OFF** and disconnect the TP sensor.
   e. Start and race the engine 2–3 times under no-load, then run the engine at idle speed.

12. Aim the timing light at the timing marks. If the marks on the pulley and the engine are aligned when the light flashes, the timing is correct. Turn the engine **OFF** and remove the tachometer and the timing light. If the marks are not in alignment, proceed with the following steps.

13. Turn the engine **OFF**.

14. Check the Camshaft Position (CMP) sensor (PHASE), Crankshaft Position (CKP) sensor (REF) and CKP sensor (POS). Replace if necessary.

15. Check that all the timing chain and gears are correctly aligned.

16. If the ignition timing is still not correct, substitute a known good Electronic Control Module (ECM).

➡The ECM may be the cause of the problem but this is rarely the case.

17. Turn the engine **OFF** and remove the tachometer and the timing light.

SPARK PLUGS

REMOVAL & INSTALLATION

1. Before servicing the vehicle, refer to the Precautions Section.

2. Remove air duct and resonator assembly.

3. Remove the electric throttle control actuator without disconnecting water hose.

4. Loosen the intake manifold mounting bolts and nuts.

5. Remove the intake manifold.

6. Disconnect harness connector from ignition coil.

7. Support the bottom surface of engine using a transmission jack.

8. Remove ground cable and harness from engine mounting bracket (RH).

9. Remove the ignition coils.

10. Remove the spark plugs.

To install:

11. Note the following, and installation is the reverse order of removal.

12. Tighten the spark plugs to 14 ft. lbs. (19.6 Nm).

13. Spark plug type: (NGK) DILKAR6A-11

14. Spark plug gap is 0.043 inch.

15. Tighten the ignition coil mounting bolts to 62 ft. lbs. 7.0 Nm).

ENGINE ELECTRICAL

STARTER

REMOVAL & INSTALLATION

2WD Models

See Figure 37.

1. Before servicing the vehicle, refer to the Precautions Section.

2. Disconnect the battery cable from the negative terminal.

3. Remove the terminal nut and terminal harness.

4. Disconnect "S" connector.

5. Remove the starter motor mounting bolts, using power tools.

6. Remove the starter motor downward from the vehicle.

To install:

7. Install in the reverse order of removal procedure and note the following:

- Tighten the starter mounting bolts to 37 ft. lbs. (50 Nm).
- Tighten terminal nut (B) carefully to 76 inch. lbs. (8.6 Nm).
- Connect the negative battery cable.

AWD Models

See Figure 38.

STARTING SYSTEM

1. Before servicing the vehicle, refer to the Precautions Section.

2. Disconnect the battery cable from the negative terminal.

3. Remove the exhaust front tube.

4. Remove the "B" terminal nut (A) and "B" terminal harness.

5. Disconnect "S" connector (B).

6. Remove the starter motor mounting bolts (C), using power tools.

7. Remove the starter motor downward from the vehicle.

To install:

8. Install in the reverse order of removal procedure and note the following:

- Tighten the starter mounting bolts to 37 ft. lbs. (50 Nm).
- Tighten the terminal nut (B) carefully to 76 inch. lbs. (8.6 Nm).
- Connect the negative battery cable.

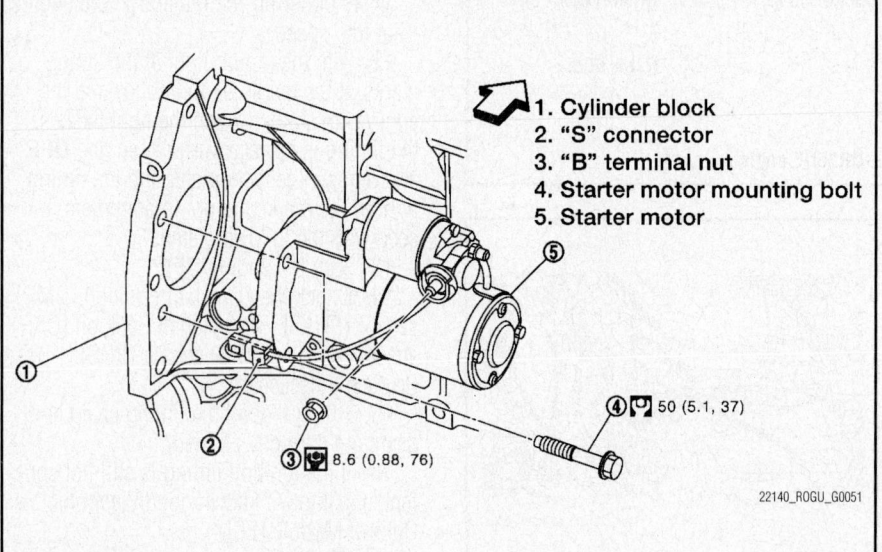

1. Cylinder block
2. "S" connector
3. "B" terminal nut
4. Starter motor mounting bolt
5. Starter motor

50 (5.1, 37)

8.6 (0.88, 76)

22140_ROGU_G0051

Fig. 37 View of starter motor mounting

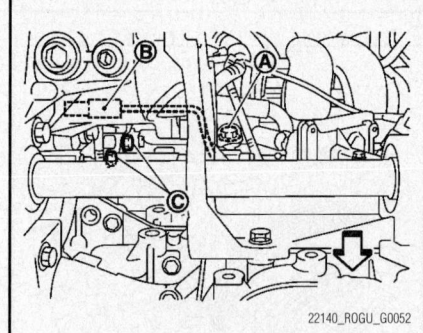

22140_ROGU_G0052

Fig. 38 Remove the terminal nut (A), connector (B) and mounting bolts (C)

ENGINE MECHANICAL

➡Disconnecting the negative battery cable may interfere with the functions of the on board computer systems and may require the computer to undergo a relearning process, once the negative battery cable is reconnected.

ACCESSORY DRIVE BELT

ACCESSORY BELT ROUTING

See Figure 39.

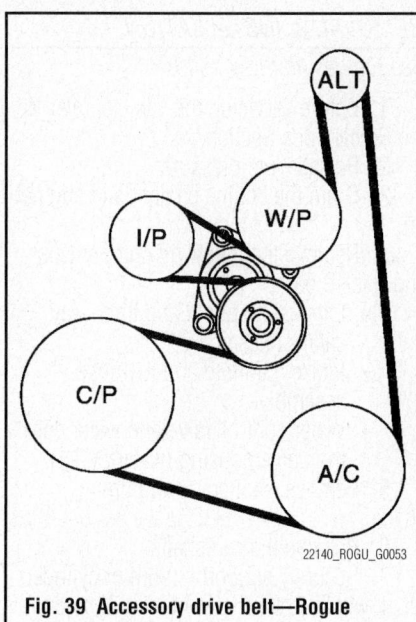

Fig. 39 Accessory drive belt—Rogue

INSPECTION

Inspect the drive belt for signs of glazing or cracking. A glazed belt will be perfectly smooth from slippage, while a good belt will have a slight texture of fabric visible. Cracks will usually start at the inner edge of the belt and run outward. All worn or damaged drive belts should be replaced immediately.

ADJUSTMENT

Belt tension is not manually adjustable, it is automatically adjusted by the drive belt auto-tensioner.

REMOVAL & INSTALLATION

See Figure 40.

1. Before servicing the vehicle, refer to the Precautions Section.
2. Remove front wheel and tire (RH).
3. Remove front fender protector (RH).
4. Hold the hexagonal part in center of drive belt auto-tensioner pulley with a box wrench securely. Then move the wrench

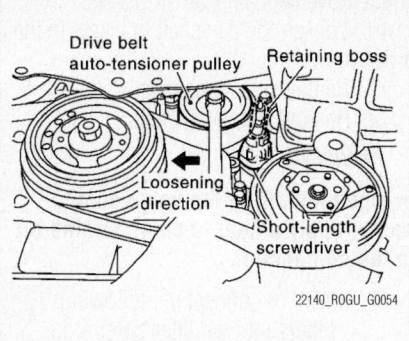

Fig. 40 Drive belt removal direction shown

handle in the direction of arrow (loosening direction of tensioner).

❊❊ WARNING

Avoid placing hand in a location where pinching may occur if the holding tool accidentally comes off. Never loosen the hexagonal part in center of drive belt auto-tensioner pulley (Never turn it counterclockwise). If turned counterclockwise, the complete drive belt auto-tensioner must be replaced as a unit, including the pulley.

5. Insert a rod approximately 6 mm (0.24 in) in diameter such as short-length screwdriver into the hole of the retaining boss to fix drive belt auto-tensioner pulley.
6. Loosen drive belt from water pump pulley in sequence, and remove it.

To install:
7. Hold the hexagonal part in center of drive belt auto-tensioner pulley with a box wrench securely. Then move the wrench handle in the direction of arrow (loosening direction of tensioner).
8. Insert a rod approximately 6 mm (0.24 in) in diameter such as short-length screwdriver into the hole of retaining boss to fix drive belt auto-tensioner pulley.
9. Hook drive belt onto all pulleys except for water pump, and then onto water pump pulley finally.
10. Confirm drive belt is completely set to pulleys.
11. Release the drive belt auto-tensioner, and apply tension to drive belt.
12. Turn the crankshaft pulley clockwise several times to equalize tension between each pulley.
13. Confirm tension of drive belt at

indicator (notch on fixed side) is within the possible use range.

CAMSHAFT AND VALVE LIFTERS

REMOVAL & INSTALLATION

See Figures 41 through 45.

1. Before servicing the vehicle, refer to the Precautions Section.
2. Relieve the fuel system pressure.
3. Drain the coolant from the engine and radiator.
4. Remove or disconnect the following:
   • Negative battery cable
   • Engine undercover
   • Right front wheel
   • Valve cover
   • Drive belt
   • Coolant reservoir
   • Variable timing control solenoid connector
   • Camshaft position sensor
   • Intake valve timing control cover in reverse of the tightening sequence

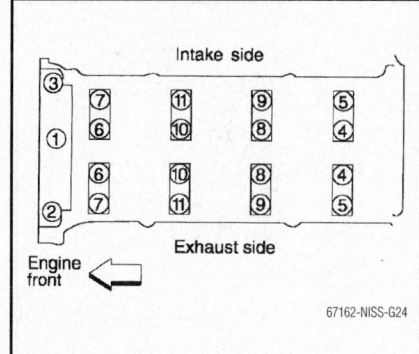

Fig. 41 Camshaft bearing cap removal sequence—2.5L engines

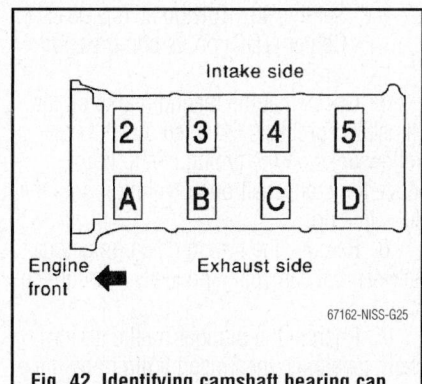

Fig. 42 Identifying camshaft bearing cap installation marks—2.5L engines

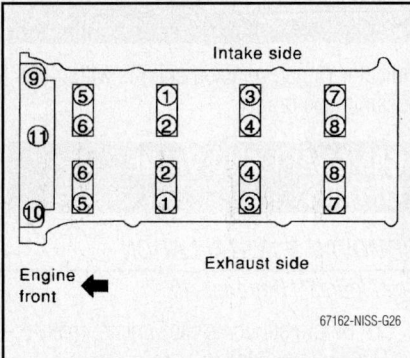

Fig. 43 Camshaft bearing cap tightening sequence—2.5L engines

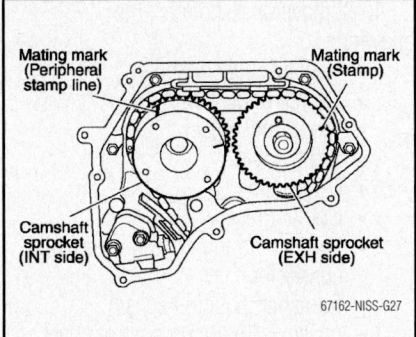

Fig. 44 Camshaft sprocket alignment marks—2.5L engines

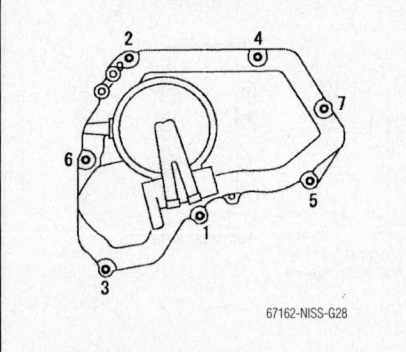

Fig. 45 Intake valve timing cover tightening sequence—2.5L engines

- Set the No. 1 piston at Top Dead Center (TDC) on its compression stroke
5. Check that the mating marks on the camshaft sprockets are lined up with the Yellow links on the timing chain. If not, rotate the crankshaft one revolution until the links line up.
6. Remove the timing chain guide out between the camshaft sprockets through the front cover.
7. Push in the plunger on the timing chain tensioner, then insert a pin to hold the tensioner retracted.
8. Remove the timing chain tensioner.

9. While holding the flat of the camshaft with an open end wrench, loosen and remove the camshaft sprockets.
10. Loosen the camshaft brackets in the order shown.
11. Remove the camshafts.
12. Remove the valve lifters.

To install:

➡**When installing the valve components, apply a coat of clean engine oil to the component.**

13. Install or connect the following:
   - Lifters into the lifter bores from which they were removed
   - Valve shims to the lifters from which they came
14. Install the camshafts so the dowel pin on the intake camshaft is placed at the three o'clock position and the exhaust camshaft pin is placed at the twelve o'clock position.
15. Install the camshaft caps with the identifying marks placed in the correct position as shown.
16. Place a bead of sealant of the No. 1 bearing cap bottom edge.
17. Place a bead of sealant on the back side of the front cover where the No. 1 bearing cap lines up.
18. Apply sealant to the bolt hole on the front cover.
19. Install the No. 1 bearing cap so the sealant joints line up.
20. Tighten the camshaft bearing caps bolts in the proper sequence as follows:
   a. Step 1 bolts 9–11: 1 ft. lbs. (2 Nm).
   b. Step 2 bolts 1–8: 1 ft. lbs. (2 Nm).
   c. Step 3 bolts 1–11: 4 ft. lbs. (6 Nm).
   d. Step 4 bolts 1–11: 10 ft. lbs. (8 Nm).
21. Install the camshaft sprockets so the mating marks are aligned as shown.
22. Install or connect the following:
   - Timing chain tensioner and remove the pin
   - Intake valve timing control cover and tighten the bolts in the sequence shown.
   - Check and adjust the valve clearance as necessary.
   - Timing chain guide between the camshaft sprockets through the front cover
   - Camshaft position sensor
   - Variable timing control solenoid connector
   - Coolant reservoir
   - Drive belt
   - Valve cover
   - Right front wheel

- Negative battery cable
- Engine undercover
23. Fill the cooling system.
24. Start the vehicle, check for leaks and repair if necessary.

CRANKSHAFT FRONT SEAL

REMOVAL & INSTALLATION

For the front crankshaft seal, refer to the timing chain, sprockets, front cover and seal.

CYLINDER HEAD

REMOVAL & INSTALLATION

See Figures 46 through 49.

1. Before servicing the vehicle, refer to the Precautions Section.
2. Release fuel pressure.
3. Drain the engine coolant and engine oil.
4. Remove the following components and related parts:
   - Exhaust manifold and three way catalyst assembly
   - Intake manifold and fuel tube assembly
   - Water control valve and water control valve housing (water outlet)
5. Remove front cover and timing chain.
6. Remove the camshafts.
7. Securely support bottom of cylinder block with a jack or equivalent tool, and release the hoist that was supporting it.
8. Remove cylinder head loosening bolts in reverse order as shown in the figure.(Refer to tightening sequence)
9. Using TORX® socket (size E20), loosen cylinder head bolts.
10. Remove the cylinder head gasket.

To install:

11. Carefully clean the engine block and cylinder head, check to see if cylinder head surface is warped.

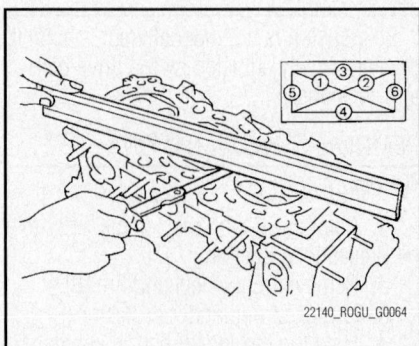

Fig. 46 Cylinder head distortion measurement locations

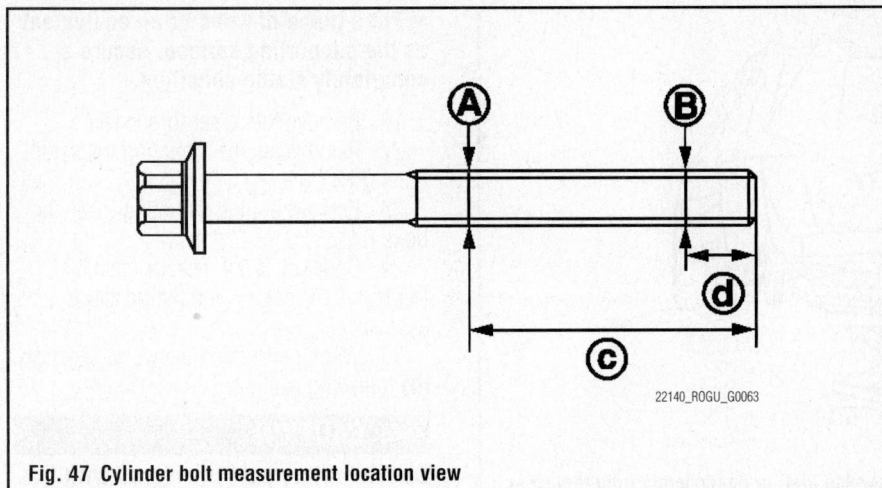

22140_ROGU_G0063

Fig. 47 Cylinder bolt measurement location view

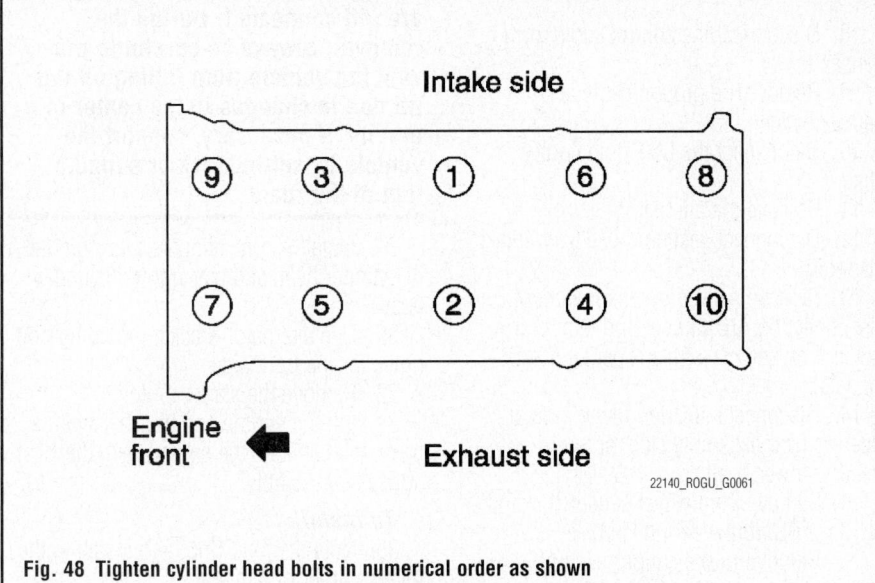

Intake side

Engine front ← Exhaust side

22140_ROGU_G0061

Fig. 48 Tighten cylinder head bolts in numerical order as shown

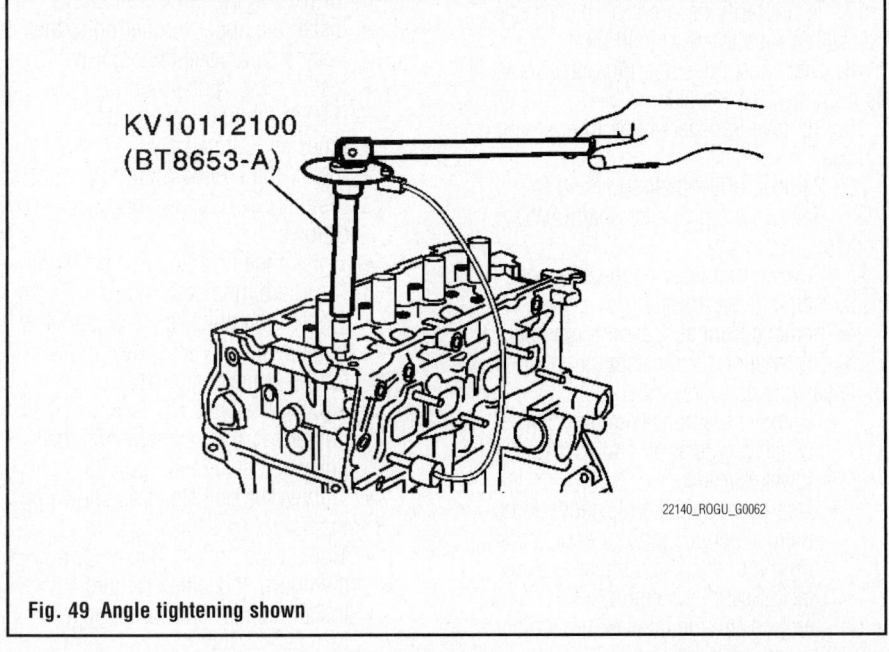

KV10112100
(BT8653-A)

22140_ROGU_G0062

Fig. 49 Angle tightening shown

12. At each of several locations on bottom surface of cylinder head, measure the distortion in six directions.

13. Cylinder head bolts are tightened by plastic zone tightening method. Whenever the size difference between (A) and (B) exceeds the limit, replace them with new one.

14. Limits are as follows:
- Limit ("B"–"A") : 0.23 mm (0.0091 in)

15. c : 55.0 mm (2.165 in)

16. d : 12.0 mm (0.472 in)

17. Install the cylinder head gasket.

18. Tighten cylinder head bolts in numerical order as shown in figure with the following procedure, and install cylinder head.

✳✳ WARNING

If cylinder head bolts are reused, check their outer diameters before installation.

19. Apply new engine oil to threads and seating surface of mounting bolts.

20. Tighten all bolts.

21. Turn all bolts 60 degrees clockwise (angle tightening).

22. Completely loosen.

23. In this step, loosen bolts in reverse order of that indicated in the figure.

24. Tighten all bolts as follows:
- Turn all bolts 75 degrees clockwise (angle tightening).
- Turn all bolts 75 degrees clockwise again (angle tightening).

✳✳ WARNING

Check and confirm the tightening angle by using an angle wrench (SST) or protractor. Avoid judgment by visual inspection without the tool.

25. Installation is in the reverse order of removal procedure after this step.

ENGINE ASSEMBLY

REMOVAL & INSTALLATION
See Figures 50 through 52.

1. Before servicing the vehicle, refer to the Precautions Section.

2. Release fuel pressure.

3. Drain engine coolant from radiator.

4. Remove the following parts:
- Air duct and the air cleaner case assembly
- Battery and battery tray
- Engine undercover

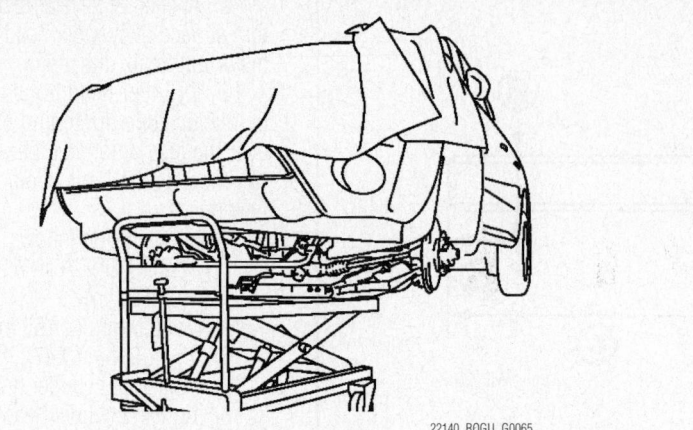

Fig. 50 Use a manual lift table caddy (commercial service tool) or equivalently rigid tool such as a transmission jack.

- Radiator hose (upper and lower) and cooling fan assembly
- Front road wheels and tires
- Front fender protector (RH and LH)
- Exhaust front tube

5. Disconnect all connections of engine harness around the engine mounting insulator (LH), and then temporarily secure the engine harness into the engine side.

6. Disconnect fuel feed hose at engine side.

7. Disconnect the heater hoses.

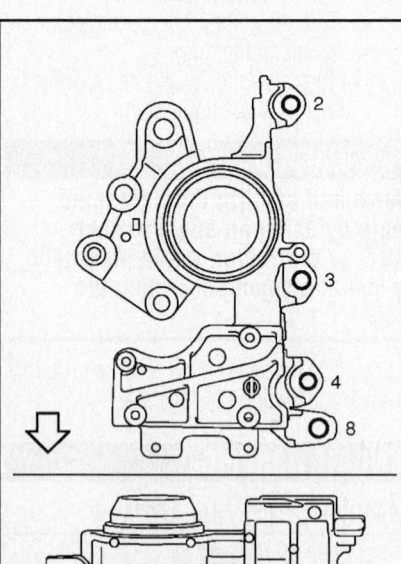

Fig. 51 Tighten the bolts as stated in the document

8. Disconnect the control cable from transaxle.

9. Remove the ground cable at transaxle side.

10. Disconnect the CVT fluid cooler hose.

11. Remove the alternator.

12. Disconnect vacuum hose from intake manifold.

13. Remove A/C compressor without disconnecting A/C piping, and temporarily fasten it on vehicle with a rope (with A/C models).

14. Disconnect steering lower joint at steering gear assembly side, and release steering lower shaft.

15. Remove front wheel sensor (LH and RH) for ABS from steering knuckle

16. Remove brake caliper assembly with piping connected from steering knuckle.

17. Temporarily secure it on the vehicle side with a rope to avoid load on it.

18. Disconnect the steering outer sockets from steering knuckle.

19. Remove transverse link from steering knuckle.

20. Remove drive shafts (LH and RH).

21. Remove the propeller shaft (AWD models).

22. Remove stabilizer connecting rod.

23. Remove rear torque rod.

24. Remove front suspension member.

25. Preparation for the separation work of transaxle is as follows:
- Remove the transaxle joint bolts which pierce at oil pan (upper) lower rear side.
- Use a manual lift table caddy (commercial service tool) or equivalently rigid tool such as a transmission jack. Securely support bottom of the engine and the transaxle assembly.

➡Put a piece of wood or an equivalent as the supporting surface, secure a completely stable condition.

26. Remove the upper torque rod.

27. Remove engine mounting insulator bolts (RH).

28. Remove engine mounting insulator bolts (LH).

29. Carefully lower jack, or raise lift to remove the engine and the transaxle assembly.

30. When performing the work, observe the following caution.

✳✳ CAUTION

Check that no part interferes with the vehicle side. Before and during this lifting, always check if any harnesses are left connected. During the removal, always be careful to prevent the vehicle from falling off the lift due to changes in the center of gravity. If necessary, support the vehicle by setting jack or suitable tool at the rear.

31. Install engine slingers into front left of cylinder head and rear right of cylinder head.

32. Use alternator bracket mounting bolt holes for the front side.

33. Remove the starter motor.

34. Lift the engine and transaxle with a hoist and separate the engine from the transaxle assembly.

To install:

35. Note the following, and install in the reverse order of removal:
- Install the engine mounting insulator (RH) to the body temporarily.
- Install the upper torque rod to the body side bracket temporarily.
- Install the engine mounting stay (LH) to the body and tighten. (specified torque)
- Install the engine mounting insulator (LH) to the body as follows:
- Tighten the bolt No. 1 as shown in the figure. (temporarily)
- Tighten the bolts No. 2, 3, and 4 in numerical order as shown in the figure. (specified torque)
- Tighten the bolts No. 5, 6, and 7 in numerical order as shown in the figure. (specified torque)
- Tighten the bolt No. 1 as shown in the figure. (specified torque)
- Tighten the nut No. 8 as shown in the figure. (specified torque)
- Install the rear bracket to the

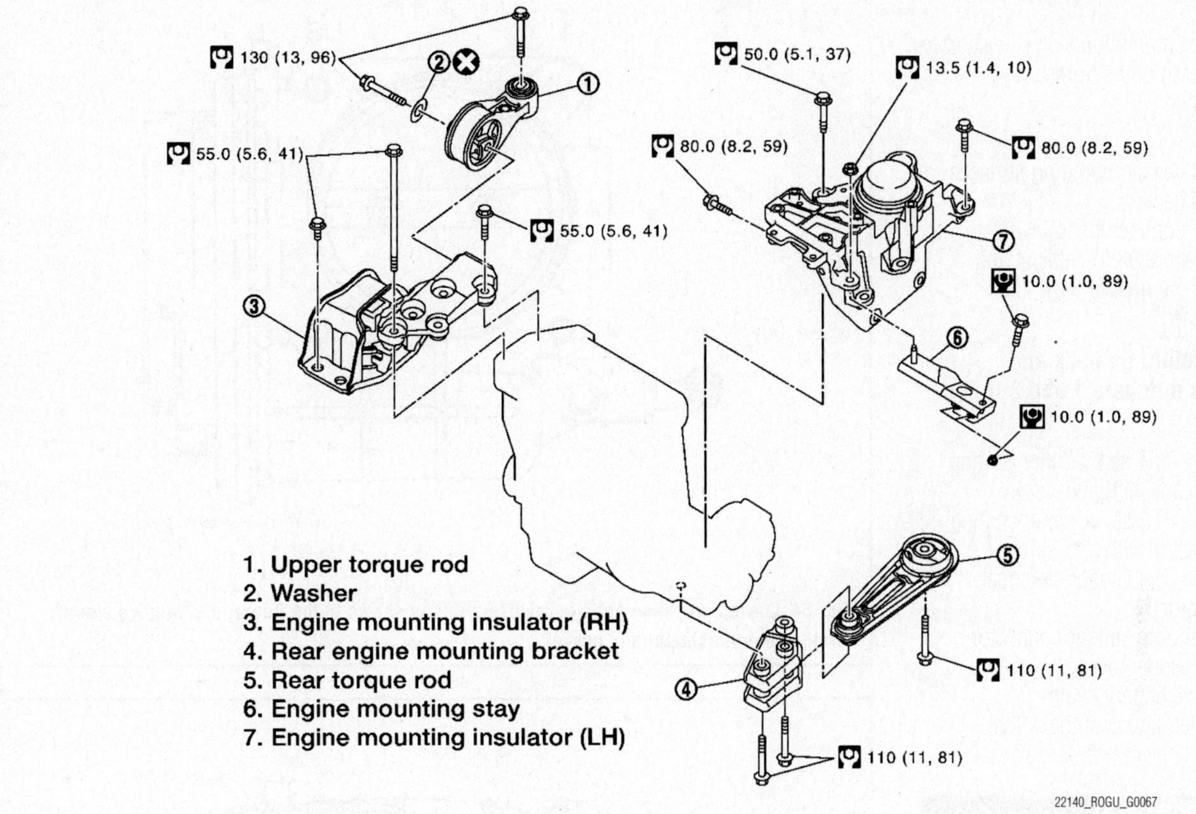

130 (13, 96)

②✗

① ①

50.0 (5.1, 37)

13.5 (1.4, 10)

55.0 (5.6, 41)

80.0 (8.2, 59)

80.0 (8.2, 59)

55.0 (5.6, 41)

⑦

③

10.0 (1.0, 89)

⑥

10.0 (1.0, 89)

⑤

1. Upper torque rod
2. Washer
3. Engine mounting insulator (RH)
4. Rear engine mounting bracket
5. Rear torque rod
6. Engine mounting stay
7. Engine mounting insulator (LH)

④

110 (11, 81)

110 (11, 81)

22140_ROGU_G0067

Fig. 52 Exploded view of the engine mounts and tightening specifications

transaxle and tighten. (specified torque)

36. Before starting engine, check oil/fluid levels including engine coolant and engine oil. If less than required quantity, fill to the specified level.

37. Use procedure below to check for fuel leakage:

a. Turn ignition switch "ON" (with engine stopped). With fuel pressure applied to fuel piping, check for fuel leakage at connection points.

b. Start the engine. With engine speed increased, check again for fuel leakage at connection points.

c. Run engine to check for unusual noise and vibration.

d. Warm up the engine thoroughly to check there is no leakage of fuel, exhaust gases, or any oil/fluids including engine oil and engine coolant.

e. Bleed air from lines and hoses of applicable lines, such as in cooling system.

f. After cooling down engine, again check oil/fluid levels including engine oil and engine coolant. Refill to the specified level, if necessary.

EXHAUST MANIFOLD

REMOVAL & INSTALLATION

1. Before servicing the vehicle, refer to the Precautions Section.

2. Remove or disconnect the following:

- Negative battery cable
- Engine undercover
- Air/fuel ratio sensor connector
- Oxygen (O_2) sensor electrical connector
- Air/fuel ratio sensor

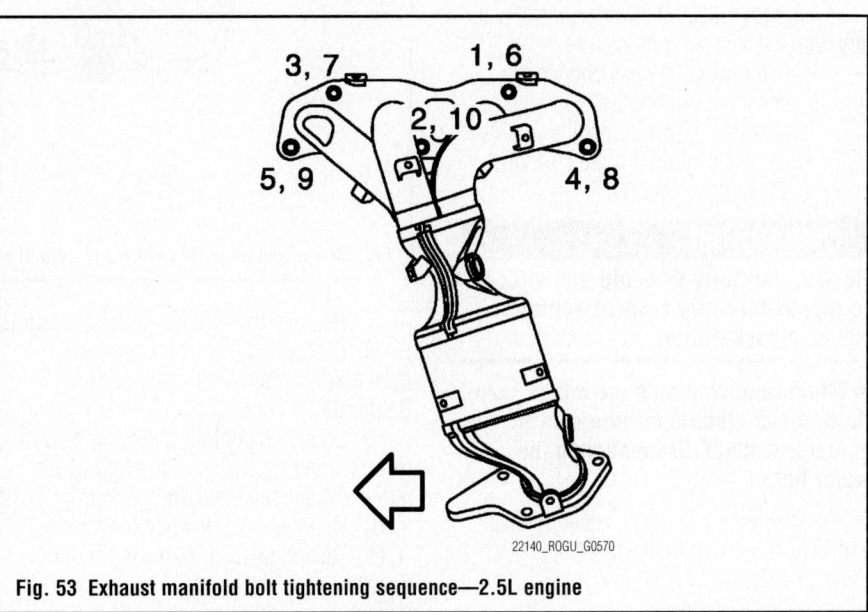

3, 7   1, 6

2, 10

5, 9   4, 8

22140_ROGU_G0570

Fig. 53 Exhaust manifold bolt tightening sequence—2.5L engine

- Oxygen (O₂) sensor
- Upper and lower exhaust manifold covers
- Exhaust manifold/catalyst assembly in reverse of the tightening sequence

To install:

3. Clean all gasket mounting surfaces and install new gaskets.

4. Install or connect the following:
- Exhaust manifold/catalyst and torque the nuts to 31 ft. lbs. (42 Nm)

➡ **After tightening, go back and retighten nuts numbers 1 and 3 to the specification.**

- Air/fuel ratio sensor after coating the threads with anti-seize
- Oxygen (O₂) sensor after coating the threads with anti-seize
- Oxygen (O₂) sensor electrical connector
- Air/fuel ratio sensor connector
- Engine undercover
- Negative battery cable

5. Start the engine and check for exhaust leaks.

NTAKE MANIFOLD

REMOVAL & INSTALLATION

See Figures 54 through 56.

1. Before servicing the vehicle, refer to the Precautions Section.
2. Release the fuel pressure.
3. Remove the cowl top cover.
4. Remove air cleaner case and mass air flow sensor assembly and air duct and resonator assembly.
5. Remove electric throttle control actuator with the following procedure:
- Disconnect harness connector.
- Loosen mounting bolts in reverse order as shown in the figure, and remove electric throttle control actuator and gasket.

❄ WARNING

Handle carefully to avoid any shock to electric throttle control actuator. Never disassemble.

➡ **When removing only the intake manifold, move electric throttle control actuator without disconnecting the water hose.**

6. Disconnect harness, vacuum hose and PCV hose from intake manifold, and move them aside.

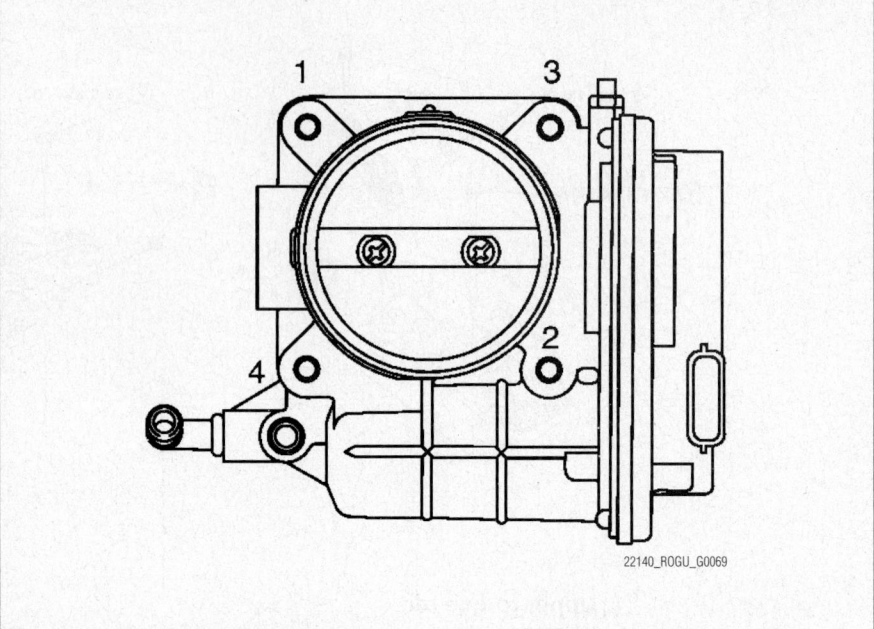

Fig. 54 Loosen mounting bolts in reverse order as shown in the figure, and remove electric throttle control actuator and gasket.

22140_ROGU_G0069

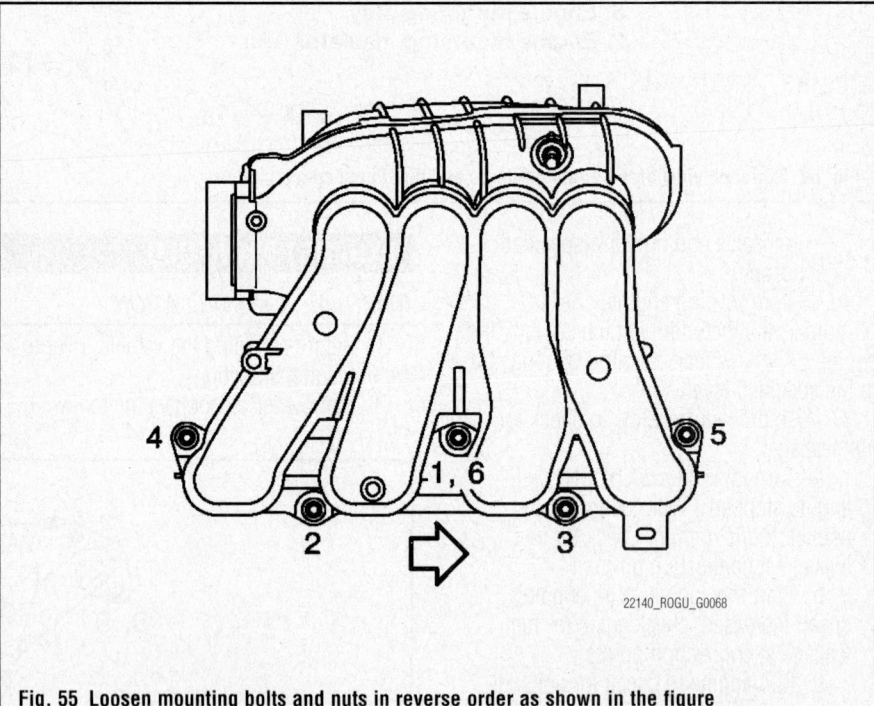

22140_ROGU_G0068

Fig. 55 Loosen mounting bolts and nuts in reverse order as shown in the figure

7. Remove the intake manifold support.
8. Disconnect harness connector from tumble control valve motor. (For California)
9. Loosen mounting bolts and nuts in reverse order as shown in the figure, and remove intake manifold and gasket.
10. Disregard No. 6 when loosening.
11. Disconnect sub-harness from fuel injector.
12. Remove the fuel tube and fuel injector assembly from intake manifold adaptor.
13. Remove the EVAP canister purge volume control solenoid valve from intake manifold, if necessary.

To install:

14. Note the following, and install in the reverse order of removal.
15. If stud bolts were removed, install them.

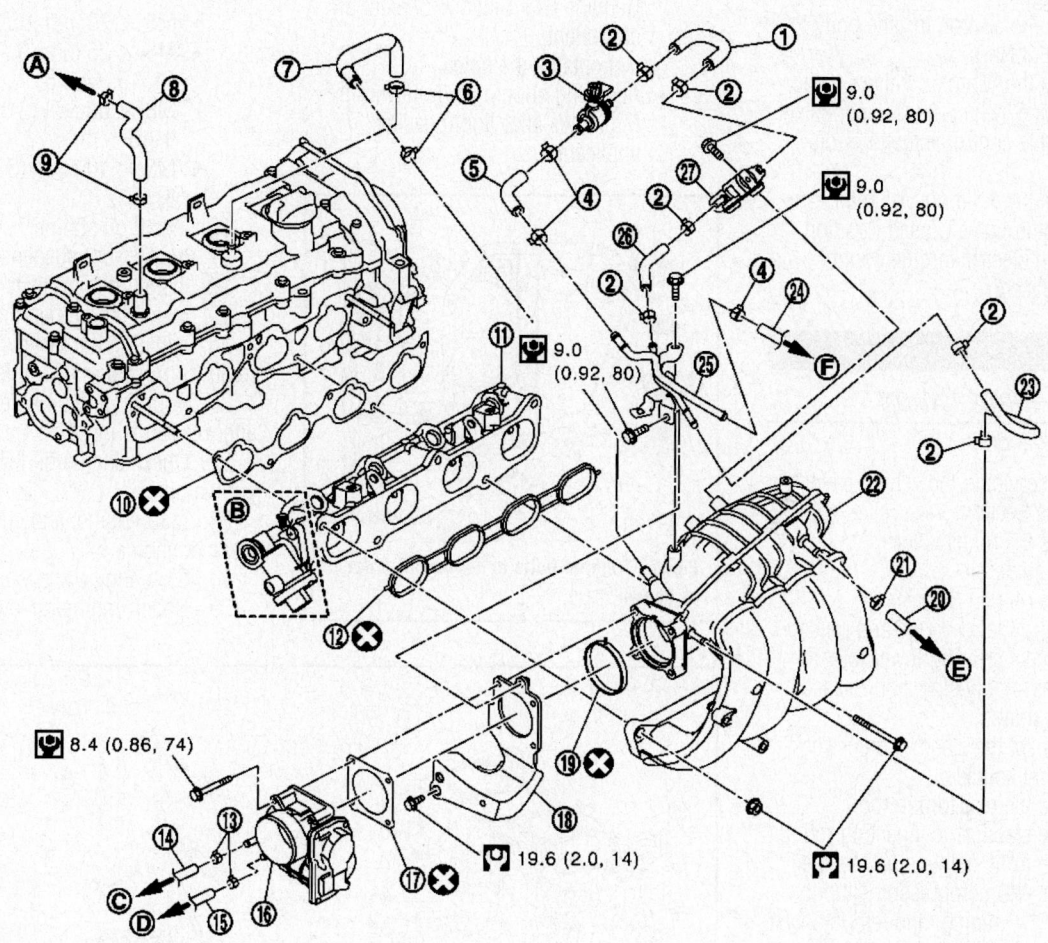

Fig. 56 Exploded view of intake manifold—Rogue

1. EVAP hose	2. Clamp	3. EVAP service port
4. Clamp	5. EVAP hose	6. Clamp
7. PCV hose	8. Fresh air hose	9. Clamp
10. Gasket	11. Intake manifold adapter	12. Gasket
13. Clamp	14. Water hose	15. Water hose
16. Electric throttle control actuator	17. Gasket	18. Intake manifold support
19. Gasket	20. Vacuum hose	21. Clamp
22. Intake manifold	23. EVAP hose	24. EVAP hose
25. EVAP tube	26. EVAP hose	27. EVAP canister purge volume control solenoid valve
A. To air duct	B. For California	C. To heater pipe
D. To water outlet	E. To brake booster	F. To vacuum pipe (canister)

22140_ROGU_G0070

16. No. 6 means double tightening of bolt No. 1.

17. Tighten the mounting bolts equally to 14 ft. lbs. (19.6 Nm), and diagonally in several steps and in numerical order as shown in the figure.

18. Tighten the electric throttle body to 74 inch. lbs. (8.4 Nm).

19. Perform the "Throttle Valve Closed Position Learning" when harness connector of electric throttle control actuator is disconnected.

20. Perform the "Idle Air Volume Learning" and "Throttle Valve Closed Position Learning" when electric throttle control actuator is replaced.

OIL PAN

REMOVAL & INSTALLATION

See Figures 57 and 58.

1. Before servicing the vehicle, refer to the Precautions Section.
2. Remove the undercover.
3. Drain engine oil.
4. Remove oil pan (lower).
5. Remove oil level gauge and guide.
6. Disconnect steering lower joint at steering gear assembly side, and release steering lower shaft.
7. Disconnect the steering outer sockets from steering knuckle.
8. Remove the rear torque rod.
9. Remove stabilizer connecting rod.
10. Remove front suspension member.
11. Remove A/C compressor without disconnecting A/C piping, and temporarily fasten it on vehicle with a rope.
12. Remove oil pan (upper) with the following procedure:

- Loosen bolts in reverse order as shown in the figure.
- Insert seal cutter (SST) between oil pan (upper) and lower cylinder block, and slide it by tapping on the side of the tool with a hammer.

➡**Be careful not to damage the mating surface.**

- Remove O-rings at front cover side.

To install:

13. Install oil pan (upper) with the following procedure:

- Use a scraper to remove old liquid gasket from mating surfaces.
- Also remove the old liquid gasket from mating surface of cylinder block.
- Remove old liquid gasket from the bolt holes and threads.

➡**Never scratch or damage the mating surfaces when cleaning off old liquid gasket.**

- Apply a continuous bead of liquid gasket with a tube presser. Use Genuine RTV Silicone Sealant or equivalent.
gasket outside the holes.
- Attaching should be done within 5 minutes after liquid gasket application.

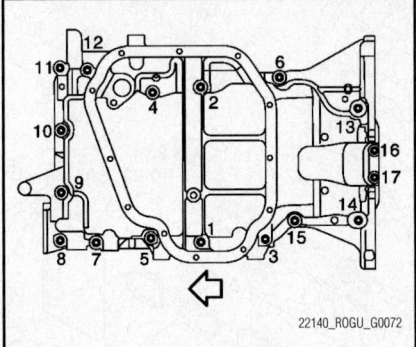

Fig. 57 Tighten bolts in numerical order as shown

- Install new O-rings at front cover side.

14. Tighten bolts in numerical order as shown in the figure.

15. Refer to the following for locating bolts:

- M6 × 20 mm (0.79 in) : No. 16, 17
- M8 × 25 mm (0.98 in) : No. 4, 6, 11, 13, 14, 15
- M8 × 60 mm (2.36 in) : No. 7, 8, 9, 10
- M8 × 100 mm (3.94 in) : No. 1, 2, 3, 5, 12

16. Install oil strainer.
17. Install front suspension member.
18. Install oil pan (lower).
19. Install the oil pan drain plug.
20. Install in the reverse order of removal procedure after this step.
21. Pour engine oil at least 30 minutes after oil pan is installed.
22. Check engine oil level and adjust engine oil.
23. Start engine, and check there is no leaks of engine oil.
24. Stop engine and wait for 10 minutes.
25. Check engine oil level again.

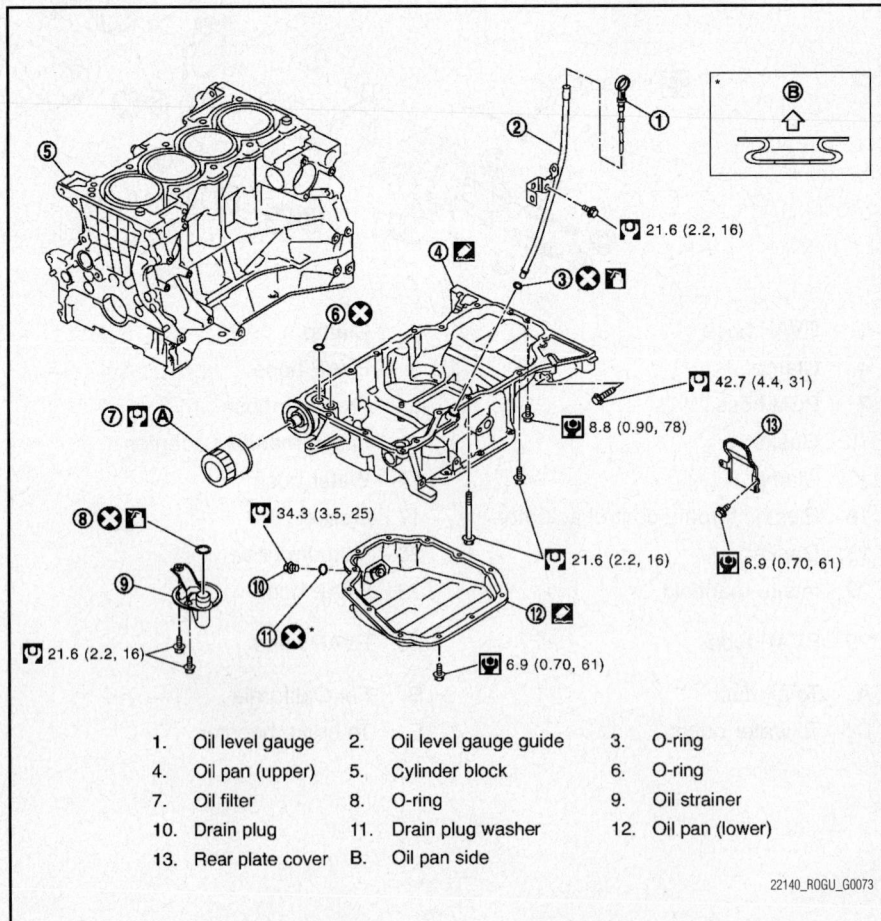

1. Oil level gauge
2. Oil level gauge guide
3. O-ring
4. Oil pan (upper)
5. Cylinder block
6. O-ring
7. Oil filter
8. O-ring
9. Oil strainer
10. Drain plug
11. Drain plug washer
12. Oil pan (lower)
13. Rear plate cover
B. Oil pan side

Fig. 58 Exploded view of upper and lower oil pan and mounting bolt tightening specifications

OIL PUMP

REMOVAL & INSTALLATION

See Figure 59.

1. Before servicing the vehicle, refer to the Precautions Section.
2. Drain the engine oil
3. Remove or disconnect the following:
   - Negative battery cable
   - Engine front cover (Refer to timing chain, front cover and seal)
   - Remove oil pump cover
   - Oil pump is built into front cover

To install:

4. Install or connect the following:
   - Install the oil pump cover
   - Oil pump is built into front cover
   - Apply a continuous bead of liquid gasket with a tube presser to front cover
   - Front cover. Tighten the M10 the bolts to 36 ft. lbs. and the M6 bolts to 12 ft. lbs. (12.8 Nm).
   - Negative battery cable
5. Fill the engine with clean oil.
6. Start the vehicle, check for leaks and repair if necessary

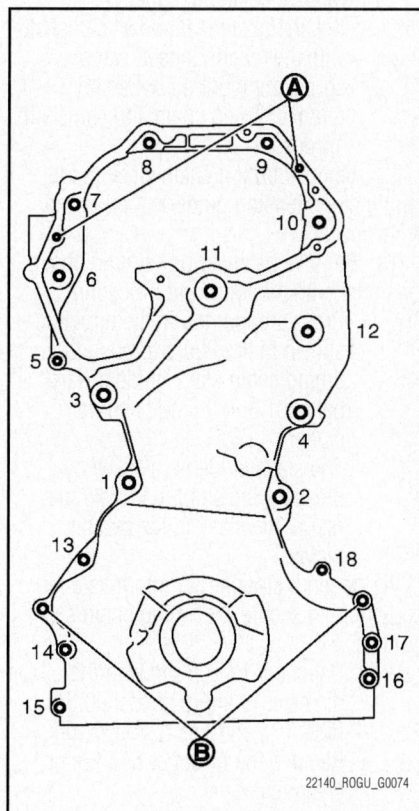

Fig. 59 Front cover tightening sequence with dowel alignment shown (A) and (B) — 2.5L engine

PISTON AND RING

POSITIONING

See Figures 60 through 62.

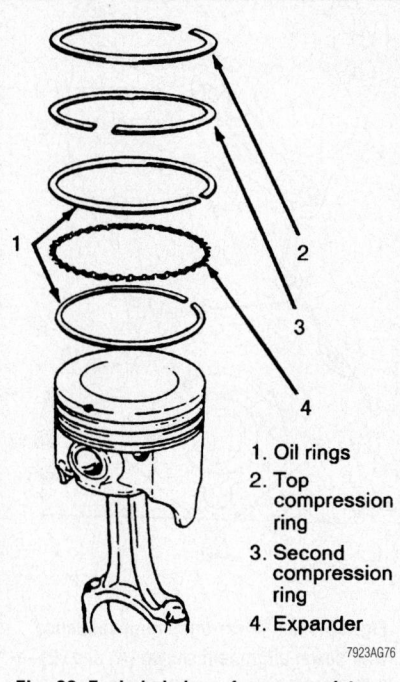

1. Oil rings
2. Top compression ring
3. Second compression ring
4. Expander

Fig. 60 Exploded view of common piston ring mounting

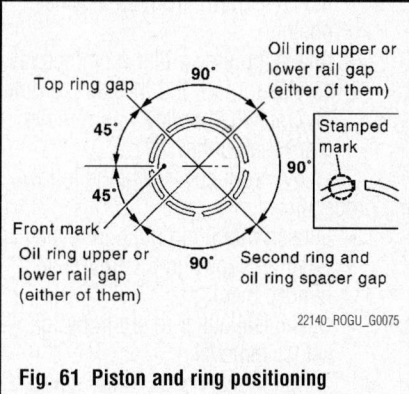

Fig. 61 Piston and ring positioning

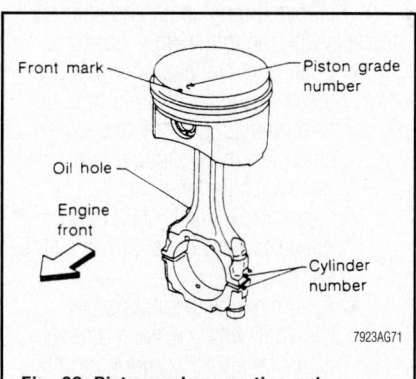

Fig. 62 Piston and connecting rod assembly positioning

REAR MAIN SEAL

REMOVAL & INSTALLATION

See Figure 63.

1. Before servicing the vehicle, refer to the Precautions Section.
2. Remove the transaxle assembly.
3. Remove the drive plate.
4. Remove rear oil seal with a suitable tool.

To install:

5. Apply new engine oil to new rear oil seal joint surface and seal lip.
6. Install the rear oil seal.
7. Press-fit rear oil seal with a suitable drift [outer diameter 4.02 inch. (102 mm), inner diameter 3.39 inch. (86 mm).

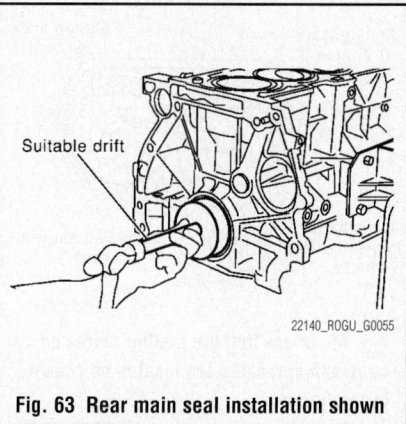

Suitable drift

Fig. 63 Rear main seal installation shown

TIMING CHAIN, SPROCKETS, FRONT COVER AND SEAL

REMOVAL & INSTALLATION

See Figures 64 through 71.

1. Before servicing the vehicle, refer to the Precautions Section.
2. Disconnect the negative battery cable.
3. Remove the following parts:
   - PCV hose:
   - Intake manifold
   - Ignition coil
   - Drive belt
   - Drive belt auto-tensioner
   - Remove engine mounting bracket (RH)
   - Remove rocker cover
   - Remove oil pan (lower)
   - Remove the oil pan (upper), and oil strainer
   - Remove intake valve timing control cover
   - Loosen bolts in reverse order as shown in the figure

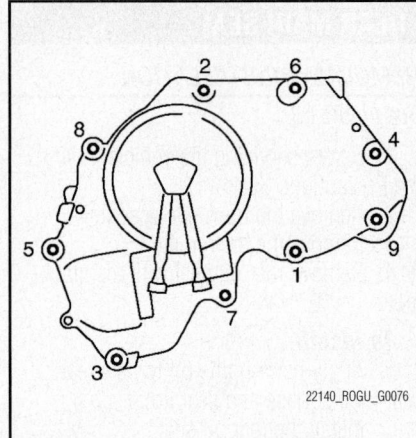

Fig. 64 Remove intake valve timing control cover in reverse order as shown

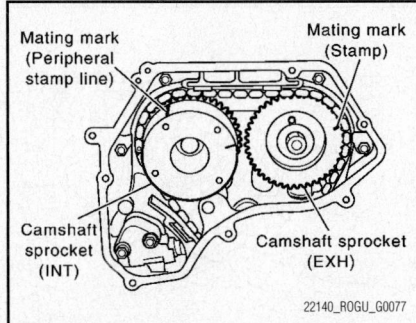

Fig. 65 Check that the mating marks on camshaft sprockets are located as shown in the figure.

- Use a seal cutter tool to cut liquid gasket for removal

※※ WARNING

Be careful not to damage mounting surface.

4. Pull chain guide between camshaft sprockets out through front cover.

5. Set No. 1 cylinder at TDC on its compression stroke with the following procedure:
- Rotate crankshaft pulley clockwise and align TDC mark to timing indicator on front cover.
- At the same time, check that the mating marks on camshaft sprockets are located as shown in the figure.
- If not, rotate crankshaft pulley one more turn to align mating marks to the positions in the figure.

6. Remove crankshaft pulley with the following procedure:
- Fix crankshaft pulley with a pulley holder (commercial service tool), loosen crankshaft pulley bolt, and

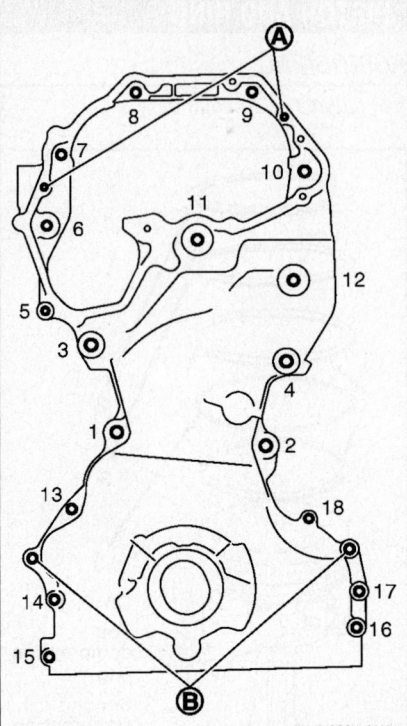

Fig. 66 Front cover tightening sequence with dowel alignment shown (A) and (B) — 2.5L engine

locate bolt seating surface at 0.39 inch. (10 mm), from its original position.
- Attach a pulley puller (commercial service tool) in the M 6 thread hole on crankshaft pulley, and remove crankshaft pulley.

7. Remove front cover with the following procedure:
- Loosen mounting bolts in reverse order as shown in the figure, and remove them.
- Use a seal cutter to cut liquid gasket for removal.

8. If front oil seal needs to be replaced, lift it with a suitable tool, and remove it.

9. Remove timing chain and camshaft sprockets with the following procedure:
- Push in chain tensioner plunger. Insert a stopper pin into hole on chain tensioner body to secure chain tensioner plunger and remove chain tensioner.
- Use approximately 0.02 in 0.5 mm diameter. hard metal pin as a stopper pin.
- Secure the hexagonal part of camshaft with a wrench. Loosen camshaft sprocket mounting bolts and remove timing chain and camshaft sprockets.

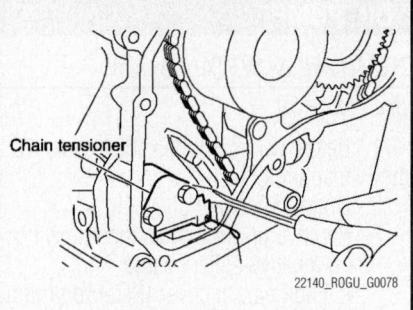

Fig. 67 Press stopper tab (A) in the direction shown in the figure to push remove chain tensioner

※※ WARNING

Never rotate the crankshaft or camshaft while timing chain is removed. It causes interference between valve and piston.

10. Remove timing chain slack guide, timing chain tension guide and oil pump drive spacer.

11. Remove balancer unit timing chain tensioner with the following procedure:
- Press stopper tab (A) in the direction shown in the figure to push remove chain tensioner.
- Secure hexagonal part of camshaft with a wrench. Loosen camshaft sprocket mounting bolts and remove timing chain and camshaft sprockets.

12. Remove timing chain slack guide, timing chain tension guide and oil pump drive spacer.

13. Remove balancer unit timing chain tensioner with the following procedure:
- Press stopper tab in the direction shown in the figure to push the timing chain slack guide toward balancer unit timing chain tensioner.
- The slack guide is released by pressing the stopper tab. As the result, the slack guide can be moved.

14. Insert a stopper pin into tensioner body hole to secure the timing chain slack guide.
- Use a hard metal pin with the diameter of approximately 0.047 inch. (1.2 mm) as a stopper pin.
- Remove the balancer unit timing chain tensioner.

15. When the holes on lever and tensioner body cannot be aligned, align these holes by slightly moving the slack guide.

16. Remove the balancer unit timing chain and crankshaft sprocket.

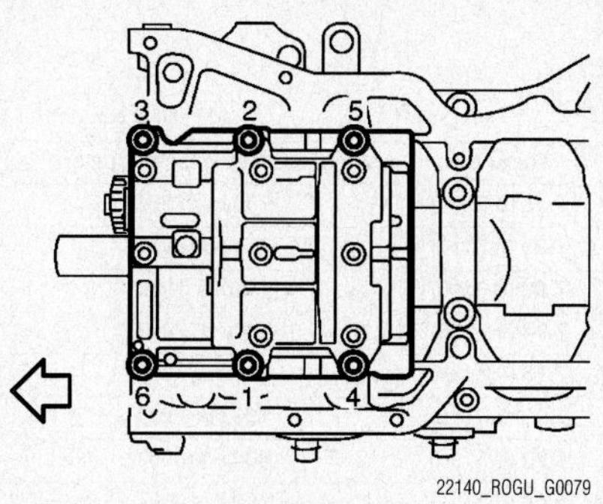

Fig. 68 Loosen mounting bolts in reverse order as shown in the figure

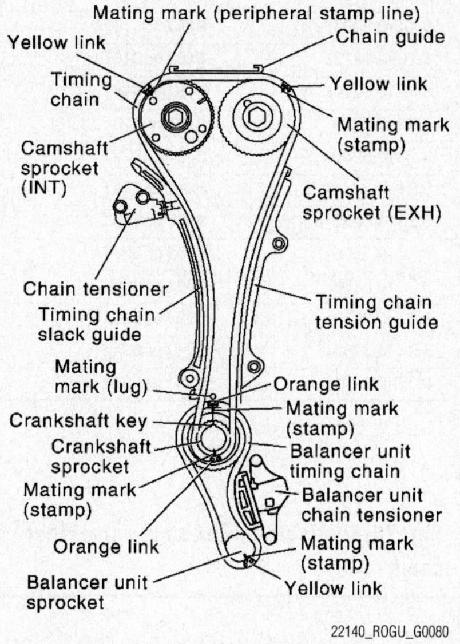

Fig. 69 The figure shows the relationship between the mating mark on each timing chain and that on the corresponding sprocket

17. Loosen mounting bolts in reverse order as shown in the figure, and remove balancer unit.

⁂ WARNING

Never disassemble balancer unit.

18. The figure shows the relationship between the mating mark on each timing chain and that on the corresponding sprocket, with the components installed.

19. Check that crankshaft key points straight up.

20. Tighten mounting bolts in numerical order as shown in the removal figure with the following procedure, and install balancer unit.

a. Apply new engine oil to threads and seat surfaces of mounting bolts.

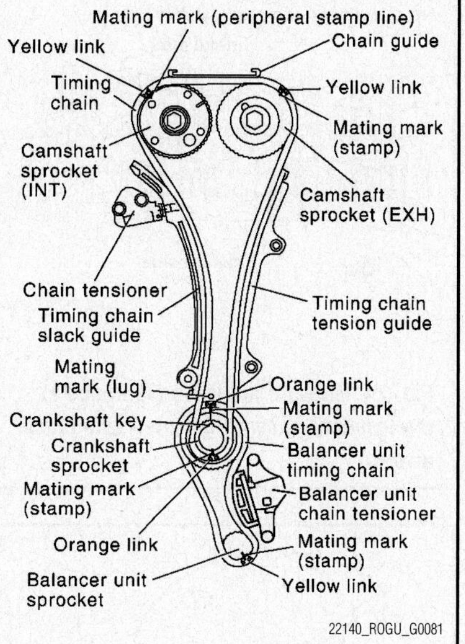

Fig. 70 Balancer chain alignment marks shown

b. Tighten No. 1– 5 bolts to 32 ft. lbs. (42 Nm).

c. Tighten the No. 6 bolt to 27 ft. lbs. (3.7 Nm).

d. Turn No. 1 to 5 bolts 120 degrees clockwise (angle tightening).

➡Use the angle wrench [SST: KV10112100 (BT8653-A)] (A) to check tightening angle. Never make judgment by visual inspection.

e. Turn No. 6 bolt 90 degrees clockwise (angle tightening).

Completely loosen all bolts.

21. In this step, loosen bolts in reverse order as shown in the figure.

a. Repeat the step b to e.

22. Install crankshaft sprocket and balancer unit timing chain.

23. Check that crankshaft sprocket is positioned with mating marks on cylinder block and crankshaft sprocket meeting at the top.

24. Install it by aligning mating marks on each sprocket and balancer unit timing chain.

25. Install the balancer unit timing chain tensioner.

26. Be careful not to let mating marks of each sprocket and timing chain slip.

27. After installation, check the mating marks have not slipped, then remove stopper pin and release tensioner sleeve.

28. Install the timing chain and related parts.

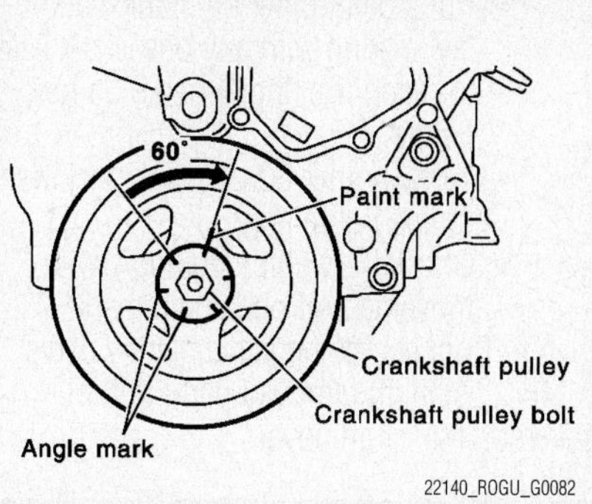

Fig. 71 Crankshaft pulley angle marks

29. Install by aligning mating marks on each sprocket and timing chain.

30. Before and after installing chain tensioner, check again to check that mating marks have not slipped.

31. After installing chain tensioner, remove stopper pin, and check that tensioner moves freely.

32. Apply a continuous bead of liquid gasket with a tube presser to front cover

33. Install the front cover.

34. Refer to cover removal and tighten the M10 the bolts to 36 ft. lbs. and the M6 bolts to 12 ft. lbs. (12.8 Nm).

35. Install chain guide between camshaft sprockets.

36. Install intake valve timing control cover with the following procedure:

37. Install intake valve timing control solenoid valves to intake valve timing control cover if removed.

38. Install new oil rings to the camshaft sprocket (INT) insertion points on backside of intake valve timing control cover.

39. Install new O-ring to front cover.

40. Apply a continuous bead of liquid gasket with a tube presser to intake valve timing control cover. Use Genuine RTV Silicone Sealant or equivalent.

➡Attaching should be done within 5 minutes after liquid gasket application.

41. Tighten mounting bolts in numerical order as shown in the figure.

42. Insert the crankshaft pulley by aligning with crankshaft key.

43. When inserting crankshaft pulley with a plastic hammer, tap on its center portion (not circumference).

44. Perform angle tightening with the following procedure:

- Apply new engine oil to thread and seat surfaces of crankshaft pulley bolt.
- Tighten crankshaft pulley bolt.

- Put a paint mark on crankshaft pulley, mating with any one of six easy to recognize angle marks on bolt flange.
- Turn another 60 degrees clockwise (angle tightening).
- Check the tightening angle with movement of one angle mark.
45. Connect the negative battery cable.
46. Install all removed parts in the reverse order of removal.

VALVE LASH

ADJUSTMENT

See Figures 72 through 76.

Inspection: Perform inspection as follows after removal, installation or replacement of camshaft or valve-related parts, or if there is unusual engine conditions regarding valve clearance.
1. Start the engine and warm it up.
2. Stop the engine.
3. Remove rocker cover.
4. Remove splash guard on RH fender protector.
5. Measure the valve clearance with the following procedure:
- Set the No. 1 cylinder at TDC of its compression stroke.

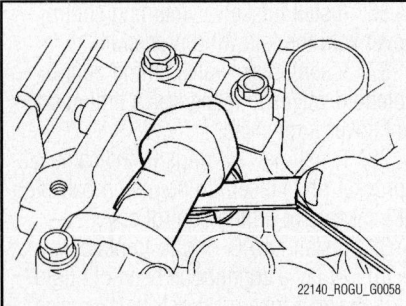

Fig. 72 Use a feeler gauge, measure the clearance between valve lifter and camshaft

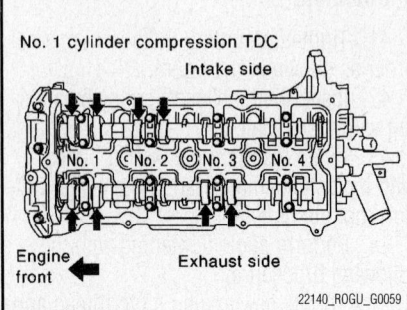

No. 1 cylinder compression TDC
Intake side
No. 1 No. 2 No. 3 No. 4
Engine front
Exhaust side
22140_ROGU_G0059

Fig. 73 Measure the valve clearances at locations marked (X)

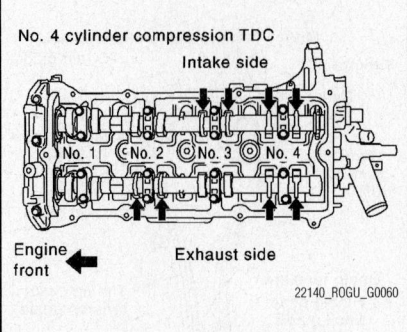

No. 4 cylinder compression TDC
Intake side
No. 1 No. 2 No. 3 No. 4
Engine front
Exhaust side
22140_ROGU_G0060

Fig. 74 Measure the valve clearance at the locations shown (indicated with black arrows)

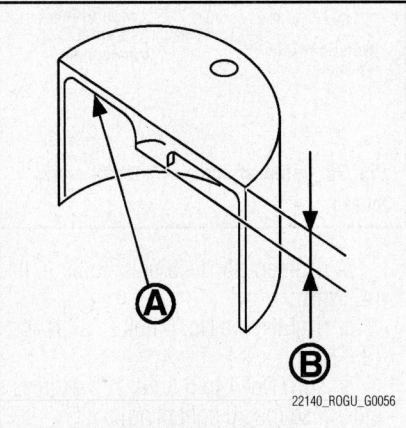

22140_ROGU_G0056

Fig. 75 Valve lifter identification, stamp mark (A), thickness (B)

	Unit: mm (in)
Thickness	Identification (stamped) mark
7.88 (0.3102)	788J or 788T
7.90 (0.3110)	790J or 790T
7.92 (0.3118)	792J or 792T
7.94 (0.3126)	794J or 794T
7.96 (0.3134)	796J or 796T
7.98 (0.3142)	798J or 798T
8.00 (0.3150)	800J or 800T
8.02 (0.3157)	802J or 802T
8.04 (0.3165)	804J or 804T
8.06 (0.3173)	806J or 806T
8.08 (0.3181)	808J or 808T
8.10 (0.3189)	810J or 810T
8.12 (0.3197)	812J or 812T
8.14 (0.3205)	814J or 814T
8.16 (0.3213)	816J or 816T
8.18 (0.3220)	818J or 818T
8.20 (0.3228)	820J or 820T
8.22 (0.3236)	822J or 822T
8.24 (0.3244)	824J or 824T
8.26 (0.3252)	826J or 826T
8.28 (0.3260)	828J or 828T
8.30 (0.3268)	830J or 830T
8.32 (0.3276)	832J or 822T
8.34 (0.3283)	834J or 834T
8.36 (0.3291)	836J or 836T

22140_ROGU_G0057

Fig. 76 Available thickness of valve lifter chart

- Rotate crankshaft pulley clockwise and align TDC mark to timing indicator on front cover.
- At the same time, check that both intake and exhaust cam noses of No. 1 cylinder face outside as shown in the figure.
- If they do not face outside, rotate crankshaft pulley once more (360 degrees).
- Use a feeler gauge, measure the clearance between valve lifter and camshaft.
- By referring to the figure, measure the valve clearances at locations marked (X) as shown in the table below (locations indicated with black arrow in the figure) with a feeler gauge.
- No. 1 cylinder compression TDC Rotate crankshaft pulley one revolution (360 degrees) and align TDC mark to timing indicator on front cover.
- By referring to the figure, measure the valve clearance at locations

marked (X) as shown in the table below (locations indicated with black arrow in the figure) with a feeler gauge.
- No. 4 cylinder compression TDC
6. If out of standard, perform adjustment.

Adjustment
7. Perform adjustments depending on selected head thickness of valve lifter.
8. Remove the camshaft.
9. Remove valve lifters at the locations that are out of the standard.
10. Measure the center thickness of the removed valve lifters with a micrometer (A).
11. Use the equation below to calculate valve lifter thickness for replacement.
- Valve lifter thickness calculation:
  $t = t1 + (C1 - C2)$
- t = Valve lifter thickness to be replaced
- t1 = Removed valve lifter thickness
- C1 = Measured valve clearance
- C2 = Standard valve clearance:
- Intake : 0.28 mm (0.011 in)
- Exhaust : 0.30 mm (0.012 in)

12. Thickness of new valve lifter (B) can be identified by stamp mark (A) on the reverse side (inside the cylinder).

13. Stamp mark "788" indicates 0.3102 inch. (7.88 mm) in thickness.

➡ **Available thickness of valve lifter: 26 sizes range 0.3102 to 0.3299 inch.**

(7.88 to 8.38 mm) (7.88 to 8.38 mm) in steps of 0.02 mm (0.0008 in) (when manufactured at factory).

14. Install the selected valve lifter.
15. Install the camshaft.
16. Manually rotate the crankshaft pulley a few rotations.

17. Check that the valve clearances for cold engine are within specifications by referring to the specified values.

18. Install all removed parts in the reverse order of removal.

19. Warm up the engine, and check for unusual noise and vibration.

ENGINE PERFORMANCE & EMISSION CONTROL

COMPONENT LOCATIONS

See Figures 77 and 78.

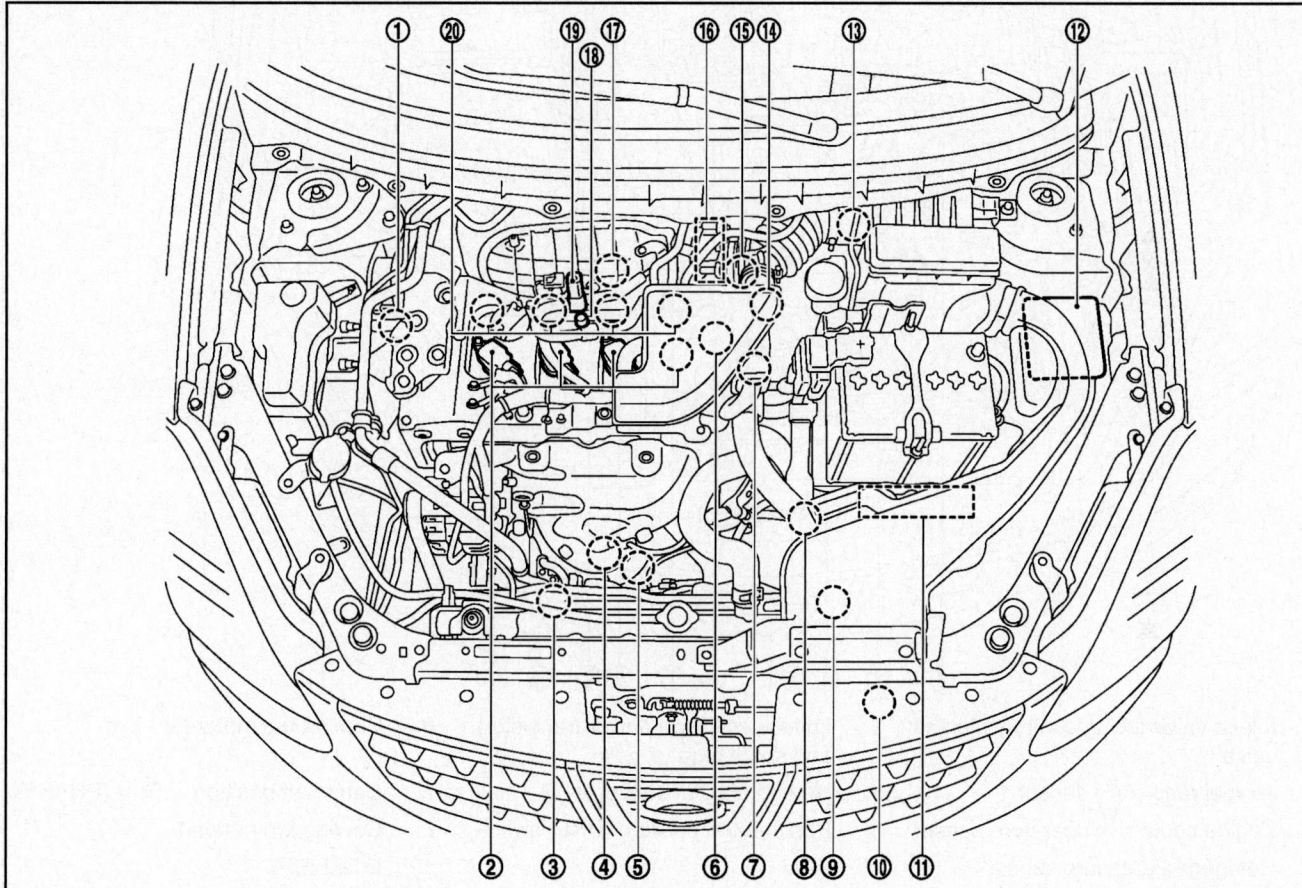

1. Intake valve timing control solenoid valve	2. Ignition coil (with power transistor) and spark plug	3. Cooling fan motor-2
4. Air fuel ratio (A/F) sensor 1	5. Heated oxygen sensor 2	6. Camshaft position sensor (PHASE)
7. Engine coolant temperature sensor	8. Park/neutral position (PNP) switch	9. Cooling fan motor-1
10. Refrigerant pressure sensor	11. ECM	12. IPDM E/R
13. Mass air flow sensor (with intake air temperature sensor)	14. Tumble control valve actuator	15. Crankshaft position sensor (POS)
16. Electric throttle control actuator (with built in throttle position sensor and throttle control motor)	17. Knock sensor	18. EVAP service port
19. EVAP canister purge volume control solenoid valve	20. Fuel injector	

22140_ROGU_G0083

Fig. 77 Engine component Locations—Rogue—2.5L engine

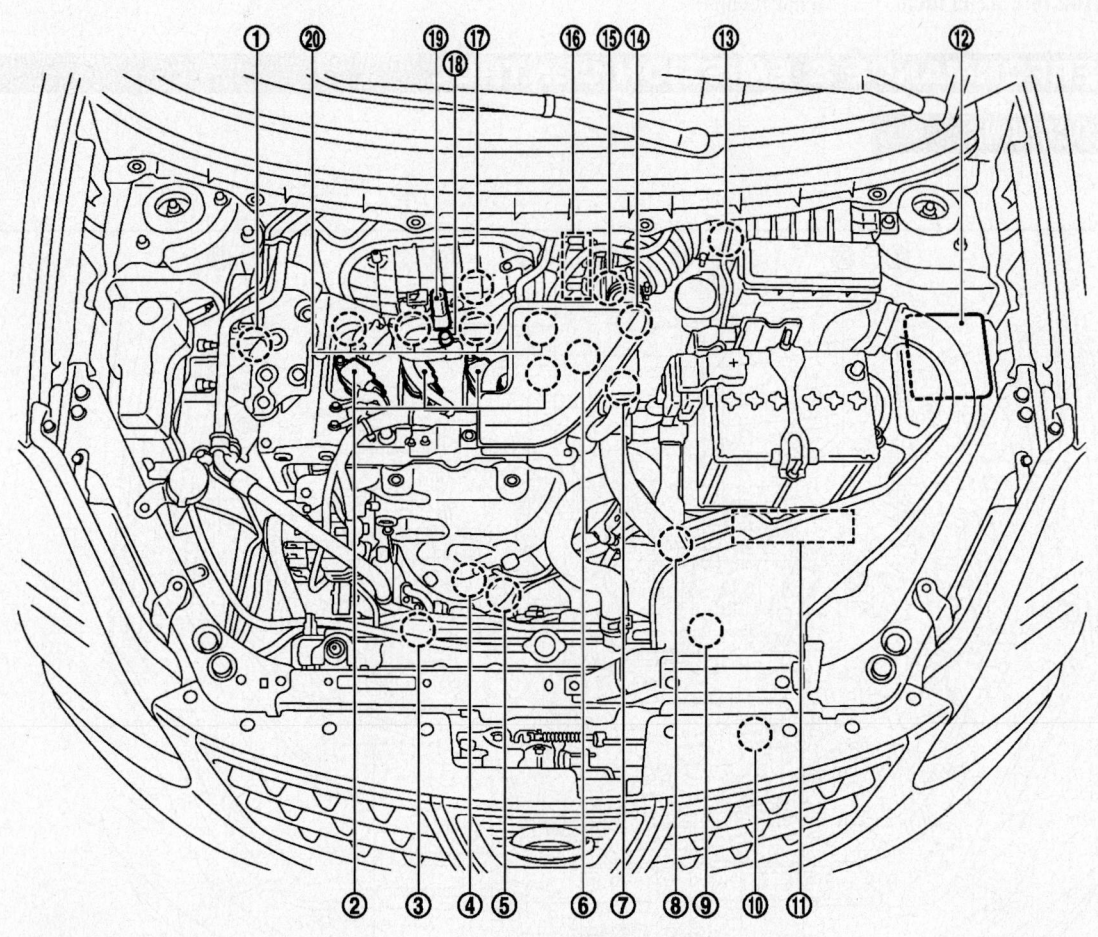

1. Intake valve timing control solenoid valve
2. Ignition coil (with power transistor) and spark plug
3. Cooling fan motor-2
4. Air fuel ratio (A/F) sensor 1
5. Heated oxygen sensor 2
6. Camshaft position sensor (PHASE)
7. Engine coolant temperature sensor
8. Park/neutral position (PNP) switch
9. Cooling fan motor-1
10. Refrigerant pressure sensor
11. ECM
12. IPDM E/R
13. Mass air flow sensor (with intake air temperature sensor)
14. Tumble control valve actuator
15. Crankshaft position sensor (POS)
16. Electric throttle control actuator (with built in throttle position sensor and throttle control motor)
17. Knock sensor
18. EVAP service port
19. EVAP canister purge volume control solenoid valve
20. Fuel injector

22140_ROGU_G0084

Fig. 78 Engine component Locations (California Models)—Rogue—2.5L engine

CAMSHAFT POSITION (CMP) SENSOR

LOCATION

The camshaft position sensor is located underneath the air cleaner duct resonator, and is mounted in the valve cover.

REMOVAL & INSTALLATION

1. Remove the air cleaner duct and resonator.
2. Disconnect the crankshaft position sensor harness connector.
3. Remove the camshaft position sensor mounting bolt.
4. Remove the camshaft sensor.

To install:

5. Install camshaft sensor, add oil to O-ring.
6. Install the mounting bolt and tighten to 62 inch. lbs. (7 Nm).
7. Reconnect the crankshaft position sensor harness connector.
8. Install the air cleaner duct and resonator.

CRANKSHAFT POSITION (CKP) SENSOR

LOCATION

The crankshaft position sensor is located on the oil pan facing the gear teeth (cogs) of the signal plate. It detects the fluctuation of the engine revolution.

REMOVAL & INSTALLATION

1. Disconnect the crankshaft position sensor harness connector.
2. Remove the camshaft position sensor mounting bolt.
3. Remove the camshaft sensor.

To install:

4. Install camshaft sensor, add oil to O-ring.
5. Install the mounting bolt and tighten to 62 inch. lbs. (7 Nm).
6. Reconnect the crankshaft position sensor harness connector

ELECTRONIC CONTROL MODULE (ECM)

LOCATION

The ECM is located in front of the battery and underneath the fresh air intake duct.

REMOVAL & INSTALLATION

1. Disconnect the negative battery cable.

2. Remove the air intake duct.
3. Carefully remove the ECM harness connectors.
4. Remove the ECM mounting bolts and the ECM

To install:

5. Install the ECM and mounting bolts, tighten to 62 inch. lbs. (7 Nm).
6. Carefully install the ECM harness connectors.
7. Install the air intake duct.
8. Connect the negative battery cable.

ENGINE COOLANT TEMPERATURE (ECT) SENSOR

LOCATION

The ECT is mounted in the water control valve housing (water outlet) on the right side of the cylinder head.

REMOVAL & INSTALLATION

1. Remove the battery.
2. Remove the battery tray
3. Remove air duct and air cleaner case assembly.
4. Disconnect the ECT harness connector.
5. Remove the ECT sensor

To install:

6. Install the ECT and tighten to 18 ft. lbs. (24.5 Nm)
7. Reconnect the ECT harness connector.
8. Install air duct and air cleaner case assembly.
9. Install the battery tray.
10. Install the battery.

HEATED OXYGEN (HO2S) SENSOR

LOCATION

The heated O2 sensor is mounted behind the exhaust manifold and three way catalyst assembly.

California models have an additional heated O2 sensor behind the number 2 sensor.

REMOVAL & INSTALLATION

1. Raise the vehicle.
2. Remove the front engine under cover.
3. Remove the heated O2 sensor harness connector.
4. Remove the heated O2 sensor.

To install:

5. Install the heated O2 sensor and tighten to 37 ft. lbs. (50 Nm).

6. Install the heated O2 sensor harness connector.
7. Install the front engine under cover.
8. Lower the vehicle.

INTAKE AIR TEMPERATURE (IAT) SENSOR

LOCATION

The Mass air flow sensor (with intake air temperature sensor) is mounted on top of the air filter housing.

REMOVAL & INSTALLATION

1. Remove the mass air flow sensor (with intake air temperature sensor) harness connector.
2. Remove the retaining screws and remove the sensor.

To install:

3. Install the sensor and tighten the retaining screws.
4. Install the mass air flow sensor (with intake air temperature sensor) harness connector.

KNOCK SENSOR (KS)

LOCATION

The knock sensor is mounted in the engine block on the intake side.

REMOVAL & INSTALLATION

1. Raise the vehicle and remove the engine under cover.
2. Remove the knock sensor harness connector.
3. Remove the mounting bolt and knock sensor.

To install:

4. Install the knock sensor and mounting bolt. Tighten to 16 ft. lbs. (21.1 Nm).
5. Install the knock sensor harness connector.
6. Install the engine under cover.
7. Lower the vehicle

MASS AIR FLOW (MAF) SENSOR

LOCATION

The Mass air flow sensor is mounted on top of the air filter housing.

REMOVAL & INSTALLATION

1. Remove the mass air flow sensor harness connector.
2. Remove the retaining screws and remove the sensor.

To install:

3. Install the sensor and tighten the retaining screws.

4. Install the mass air flow sensor harness connector.

THROTTLE POSITION SENSOR (TPS)

LOCATION

The throttle position sensor is integral to the electric throttle control actuator. The throttle control actuator is mounted at the front of the intake manifold.

REMOVAL & INSTALLATION

See Figure 79.

1. Remove the air intake duct.
2. Disconnect harness connector.
3. Disconnect water hose.
4. Loosen the throttle body assembly mounting bolts in reverse order as shown in the figure.

To install:

5. Install the throttle body assembly with a new gasket.
6. Tighten the mounting bolts in sequence to 74 inch. lbs. (8.4 Nm)
7. Reconnect the water hose.
8. Reconnect the harness connector.
9. Reconnect the air intake duct.

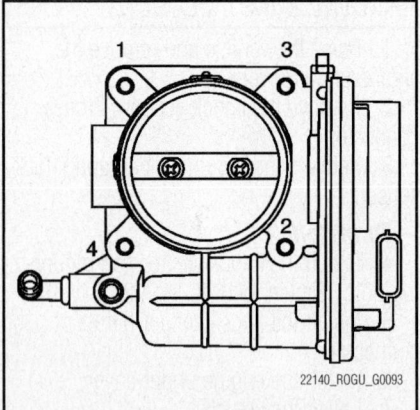

Fig. 79 Loosen mounting bolts in reverse order as shown

VEHICLE SPEED SENSOR (VSS)

LOCATION

See Figures 80 and 81.

The primary speed sensor is located on the left side of the transaxle. The secondary speed sensor is located at the top right of the transaxle.

REMOVAL & INSTALLATION

1. Disconnect the battery cable from negative terminal.
2. Disconnect the primary speed sensor connector.
3. Remove the primary speed sensor.
4. Remove O-ring from primary speed sensor.

To install:

5. Install the primary speed sensor and tighten the mounting bolt to 52 inch. lbs. (5.9 Nm).
6. Reconnect the primary speed sensor connector.

7. Connect the negative battery cable.

➡**Never reuse O-ring. Apply CVT fluid to O-ring.**

8. After completing installation, check for CVT fluid leakage and check CVT fluid level.

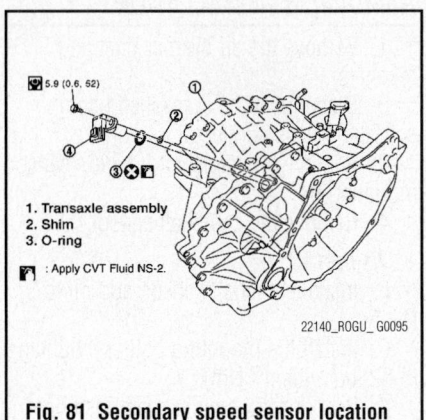

1. Transaxle assembly
2. Shim
3. O-ring

: Apply CVT Fluid NS-2

Fig. 81 Secondary speed sensor location

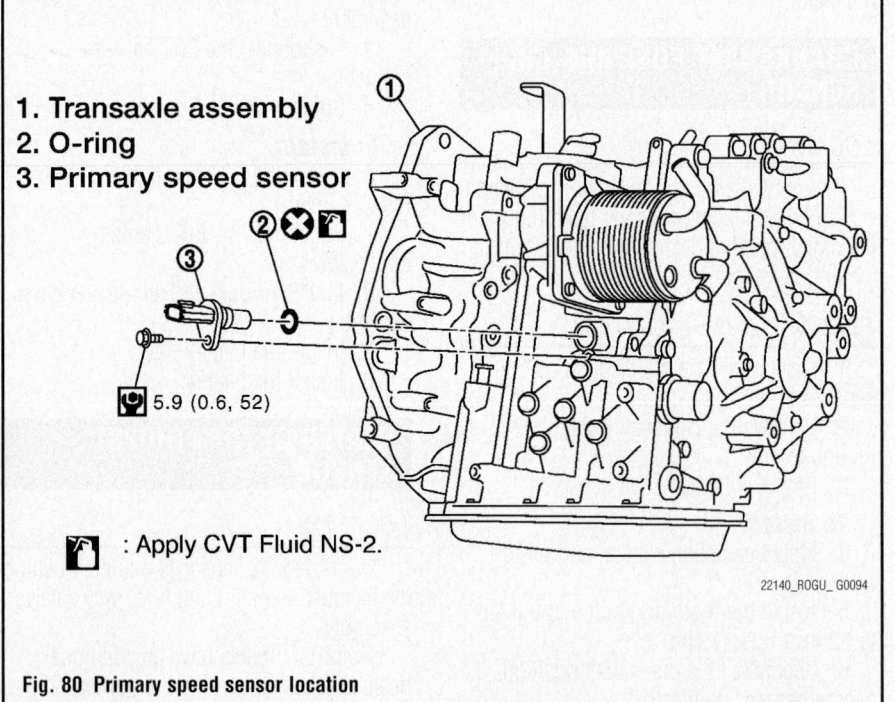

1. **Transaxle assembly**
2. **O-ring**
3. **Primary speed sensor**

5.9 (0.6, 52)

: Apply CVT Fluid NS-2.

Fig. 80 Primary speed sensor location

FUEL SYSTEM SERVICE PRECAUTIONS

1. When replacing fuel line parts, be sure to observe the following:
- Put a "CAUTION: FLAMMABLE" sign in the workshop.
- Be sure to work in a well ventilated area and furnish workshop with a CO2 fire extinguisher.
- Never smoke while servicing fuel system. Keep open flames and sparks away from the work area.
- Put drained fuel in an explosion-proof container and put the lid on securely. Keep the container in safe area.
- Release fuel pressure from the fuel lines.
- Disconnect the battery cable from the negative terminal.
- Always replace O-ring and clamps with new ones.
- Never bend or twist tubes when they are being installed.
- Never tighten hose clamps excessively to avoid damaging hoses.

2. After installing tubes, check that there is no fuel leakage at connections in the following steps.

3. Apply fuel pressure to fuel lines with turning ignition switch "ON" (with engine stopped). Then check for fuel leakage at connections.

4. Start engine and rev it up and check for fuel leakage at connections.

5. Use only a genuine NISSAN fuel filler cap as a replacement. If an incorrect fuel filler cap is used, the "MIL" may come on.

RELIEVING FUEL SYSTEM PRESSURE

1. Remove fuel pump fuse located in IPDM E/R. or unplug the fuel pump connector.

2. Start the engine.

3. After engine stalls, crank it two or three times to release all fuel pressure.

4. Turn ignition switch OFF.

5. Reinstall fuel pump fuse after servicing fuel system.

FUEL FILTER

REMOVAL & INSTALLATION

The fuel filter is integral to the fuel pump assembly.

FUEL INJECTORS

REMOVAL & INSTALLATION

See Figures 82 through 84.

1. Before servicing the vehicle, refer to the Precautions Section.

2. Remove fuel pump fuse located in IPDM E/R.

3. Start the engine.

4. After engine stalls, crank it two or three times to release all fuel pressure.

5. Turn ignition switch OFF.

6. Disconnect the negative battery cable.

7. Reinstall fuel pump fuse after servicing fuel system.

✳✳ WARNING

Disconnect quick connector by using quick connector release, not by picking out retainer tabs.

8. Disconnect quick connector (A) with the following procedure:
- Remove the quick connector cap.
- With the sleeve side of quick connector release facing quick connector, install quick connector release onto fuel tube.
- Insert quick connector release into quick connector until sleeve contacts and goes no further. Hold quick connector release on that position.

➡ **Inserting quick connector release hard will not disconnect quick connector.**

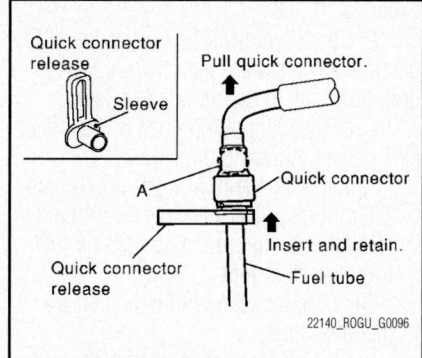

Fig. 82 Pull the quick connector holding (A)

Hold quick connector release where it contacts and goes no further.

- Draw and pull out quick connector straight from fuel tube.
- Pull the quick connector holding (A) position in the figure.

✳✳ WARNING

Never pull with lateral force applied. O-ring inside quick connector may be damaged.

- Prepare container and cloth beforehand as fuel will leak out.
Keep clean the connecting portion and to avoid damage and foreign materials, cover them completely with plastic bags or something similar.

9. Remove the intake manifold.

10. Disconnect sub-harness for fuel injector.

11. Remove the fuel tube and fuel injector assembly.

12. Loosen the mounting bolts (1) and (2) in reverse order as shown in the figure.

13. Remove fuel injector from fuel tube with the following procedure:
- Open and remove clip (1).

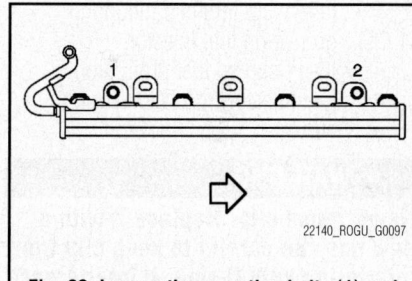

Fig. 83 Loosen the mounting bolts (1) and (2) in reverse order

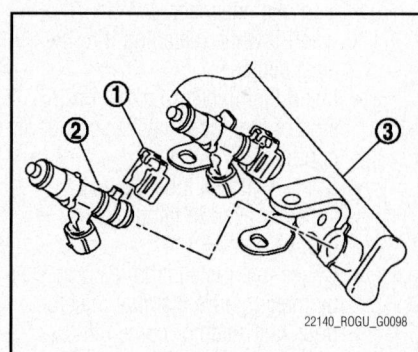

Fig. 84 Open and remove clip (1) remove fuel injector (2) from fuel tube (3)

- Remove fuel injector (2) from fuel tube (3) by pulling straight.

To install:

14. Note the following, and install O-rings to fuel injector as follows:
- Upper and lower O-rings are different. Be careful not to confuse them.
  a. Except for California:
  b. Fuel tube side : Blue
  c. Nozzle side : Brown
  d. For California:
  e. Fuel tube side : Black
  f. Nozzle side : Green
- Handle O-ring with bare hands. Never wear gloves.
- Lubricate O-ring with new engine oil.
- Never clean O-ring with solvent.
- Check that O-ring and its mating part are free of foreign material.
- When installing O-ring, be careful not to scratch it with tool or fingernails. Also be careful not to twist or stretch O-ring.
- If O-ring was stretched while it was being attached, never insert it quickly into fuel tube.
- Insert O-ring straight into fuel tube. Never uncenter or twist it.

15. Install fuel injector to fuel tube with the following procedure:
- Insert clip into clip mounting groove on fuel injector.
- Insert clip so that protrusion of fuel injector matches cutout of clip.

❈❈ WARNING

Never reuse clip. Replace it with a new one. Be careful to keep clip from interfering with O-ring. If interference occurs, replace O-ring.

- Insert fuel injector into fuel tube with clip attached.
- Insert it while matching it to the axial center.
- Insert fuel injector so that protrusion of fuel tube matches cutout of clip.
- Check that fuel tube flange is securely fixed in flange fixing groove on clip.
- Check that installation is complete by making sure that fuel injector does not rotate or come off.

16. Install fuel tube and fuel injector assembly with the following procedure:
- Insert the tip of each fuel injector into intake manifold adapter.
- Tighten mounting bolts in numerical order

17. Connect sub-harness for fuel injector.
18. Install the intake manifold.
19. Note the following, and connect quick connector to install fuel feed hose.
   a. Check the connection for foreign material and damage.
   b. Align center to insert quick connector straightly into fuel tube.
   Insert fuel tube into quick connector until the top spool on fuel tube is inserted completely and the second level spool is positioned slightly below quick connector bottom end.
20. Hold (A) position in the figure when inserting fuel tube into quick connector.
21. Carefully align center to avoid inclined insertion to prevent damage to O-ring inside quick connector.
22. Insert until you hear a (click sound) and actually feel the engagement.
23. To avoid misidentification of engagement with a similar sound, be sure to perform the next step.
24. Before clamping fuel feed hose with hose clamps, pull quick connector hard by hand holding (A) position.
25. Check it is completely engaged (connected) so that it does not come out from fuel feed tube.
26. Install the quick connector cap to quick connector connection.
27. Install so that the arrow mark on the side faces up.
28. Install fuel feed hose to hose clamp.
29. Install in the reverse order of removal procedure after this step.

FUEL PUMP

REMOVAL & INSTALLATION

See Figures 85 through 88.

1. Before servicing the vehicle, refer to the Precautions Section.
2. Disconnect the negative battery cable.
3. Check fuel level on fuel gauge. If fuel gauge indicates more than ¾ of a tank drain fuel tank to around a half a tank.
4. In case fuel pump does not operate, perform the following procedure:
   a. Insert hose of less than 0.79 inch. (20 mm) diameter into fuel filler tube through fuel filler opening to draw fuel from fuel filler tube.
   b. Disconnect fuel filler hose from fuel filler tube.
   c. Insert hose into fuel tank through fuel filler hose to draw fuel from fuel tank.
5. Release the fuel pressure from the fuel lines.

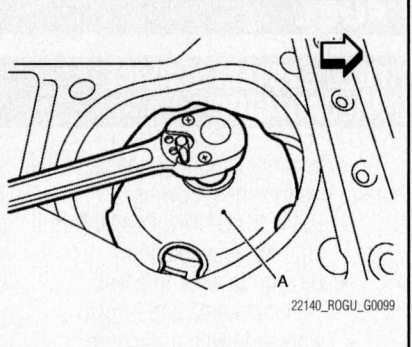

Fig. 85 Remove the lock ring with a lock ring wrench.

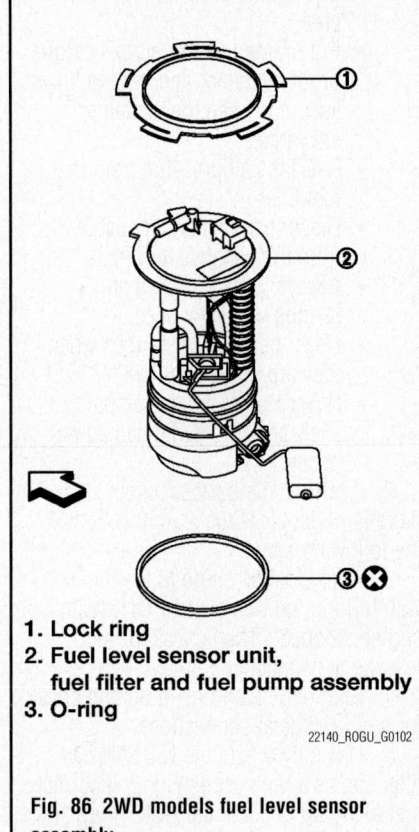

1. Lock ring
2. Fuel level sensor unit, fuel filter and fuel pump assembly
3. O-ring

Fig. 86 2WD models fuel level sensor assembly

6. Open the fuel filler lid.
7. Open fuel filler cap and release the pressure inside the fuel tank.
8. Remove rear seat cushion.
9. Remove inspection hole cover.
10. Using a screwdriver, remove it by turning clips clockwise by 90 degrees.
11. Disconnect harness connector and fuel line quick connector.
12. Using lock ring wrench [SST: KV991J0090 (J-46214)] (A), remove lock ring.

➡ **For reference when installing, put a matching mark on lock ring, fuel pump assembly and fuel tank. Raise fuel**

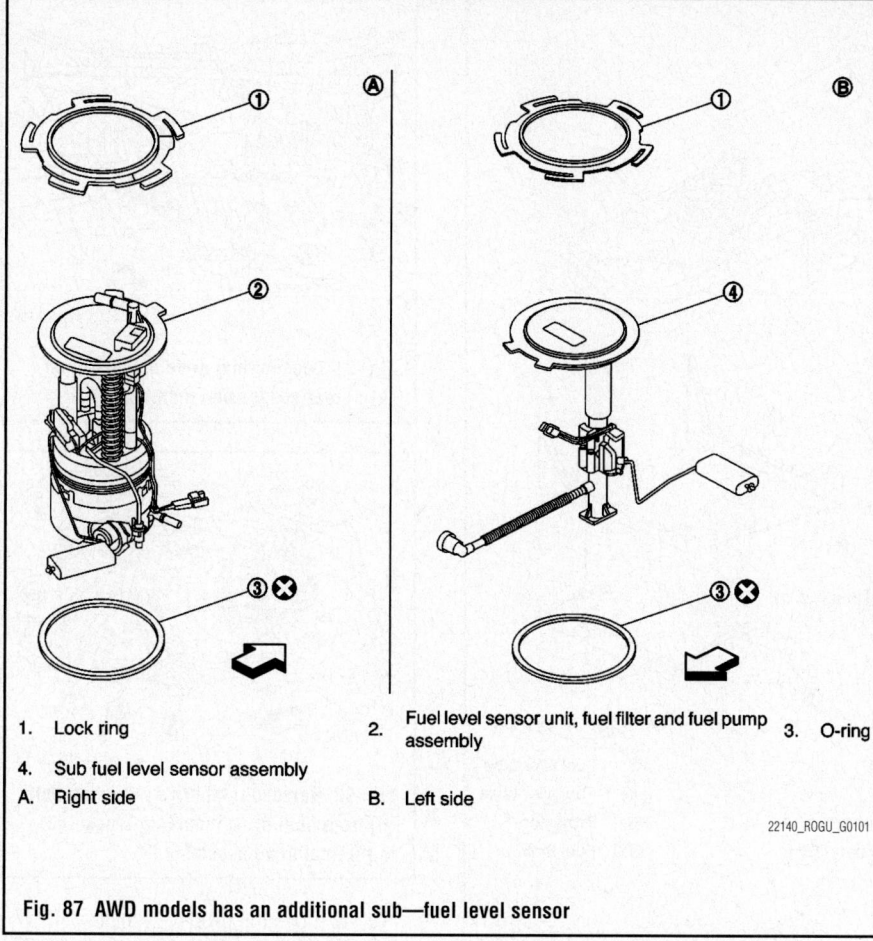

1. Lock ring
2. Fuel level sensor unit, fuel filter and fuel pump assembly
3. O-ring
4. Sub fuel level sensor assembly
A. Right side
B. Left side

22140_ROGU_G0101

Fig. 87 AWD models has an additional sub—fuel level sensor

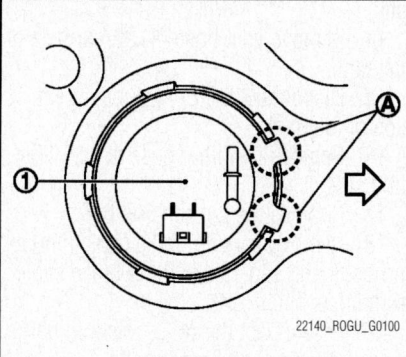

22140_ROGU_G0100

Fig. 88 Align (A) with vehicle front as shown in the figure. Install fuel level sensor unit (1) to fuel tank

level sensor unit, fuel filter and fuel pump assembly, and disconnect fuel tube and harness connector.

✳✳ WARNING

Never bend float arm during removal. Never pollute the inside by residue fuel. Draw out avoiding inclination by supporting with a cloth. Never cause impacts such by dropping when handling components.

To install:

13. Install O-ring to fuel tank without any twist.

14. Align (A) with vehicle front as shown in the figure. Install fuel level sensor unit (1) to fuel tank.

15. Connect quick connector of fuel feed tube in the following procedures:

a. Check the connection for damage or any foreign materials.

b. Align the connector with the tube, then insert the connector straight into the tube until a "click" sound is heard.

16. After connecting, check that the connection is secured with following procedures:

a. Visually confirm that the two tabs are connected to the connector.

b. Pull the tube and the connector to check that they are securely connected.

17. Connect the negative battery cable.

18. Before installing inspection hole cover, check that the connecting part has no fuel leakage.

19. Install inspection hole covers with the front mark (arrow) facing front of vehicle.

20. Lock clips by turning counterclockwise.

21. Install the rear seat cushion.
22. Install the fuel cap.

FUEL TANK

REMOVAL & INSTALLATION

2WD Models

See Figures 89 and 90.

1. Before servicing the vehicle, refer to the Precautions Section.

2. Disconnect the negative battery cable.

3. Check the fuel level on the fuel gauge. If fuel gauge indicates more than ¾ of a tank drain fuel tank to around a half a tank.

4. In case fuel pump does not operate, perform the following procedure:

a. Insert hose of less than 0.79 inch. (20 mm) diameter into fuel filler tube through fuel filler opening to draw the fuel from fuel filler tube.

b. Disconnect fuel filler hose from fuel filler tube.

c. Insert hose into fuel tank through fuel filler hose to draw fuel from fuel tank.

5. Release the fuel pressure from the fuel lines.

6. Open the fuel filler lid.

7. Open fuel filler cap and release the pressure inside the fuel tank.

8. Remove rear seat cushion.

9. Remove inspection hole cover.

10. Disconnect fuel lines and harness connectors.

11. Remove the muffler assembly.

12. Remove the protector from the fuel tank.

13. Remove the vent hose at rear side of fuel tank.

14. Disconnect the EVAP tube at rear side of fuel tank.

15. Remove fuel filler hose at fuel filler tube side.

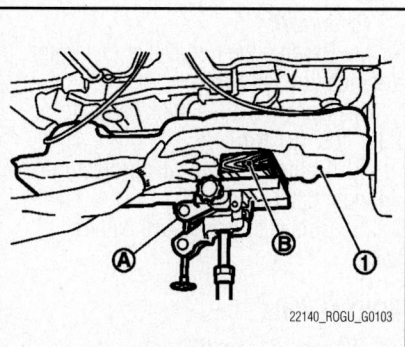

22140_ROGU_G0103

Fig. 89 Fuel tank (1) supported with transmission jack (A) and block of wood (B)

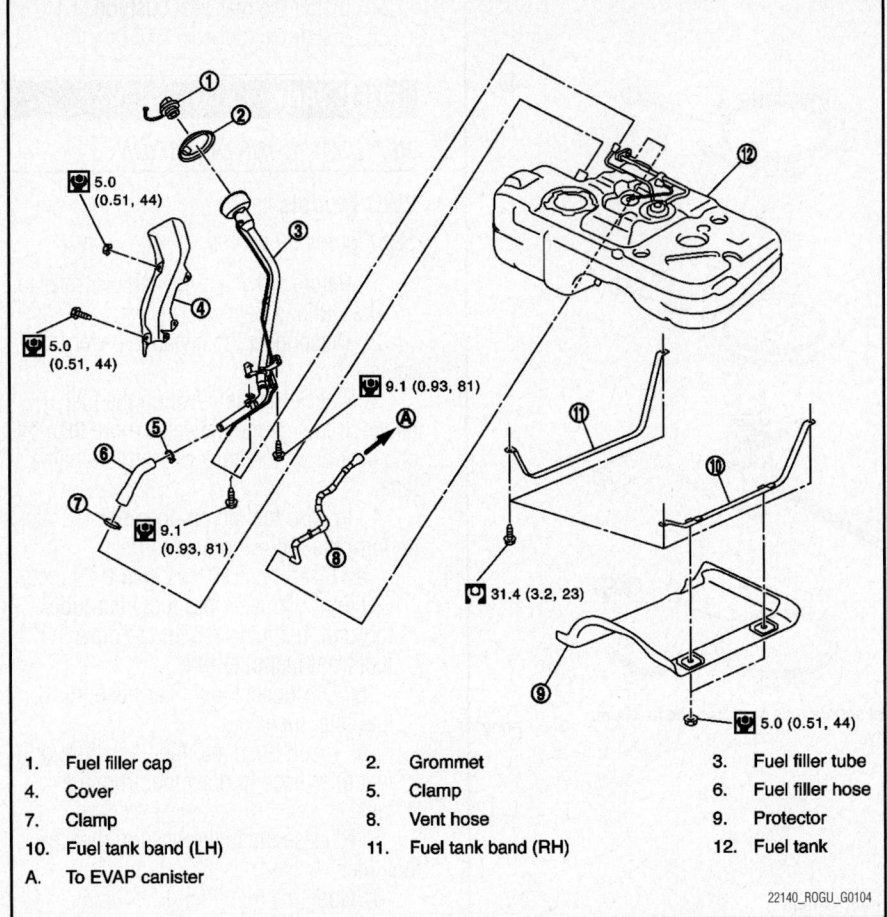

1. Fuel filler cap
2. Grommet
3. Fuel filler tube
4. Cover
5. Clamp
6. Fuel filler hose
7. Clamp
8. Vent hose
9. Protector
10. Fuel tank band (LH)
11. Fuel tank band (RH)
12. Fuel tank
A. To EVAP canister

22140_ROGU_G0104

Fig. 90 Exploded view of the 2WD fuel tank—Rogue

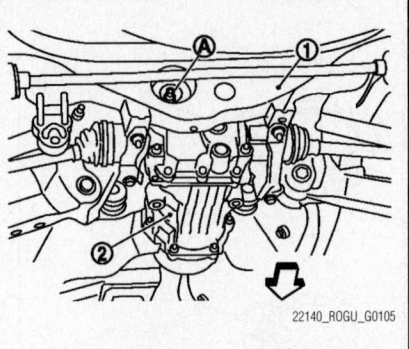

22140_ROGU_G0105

Fig. 91 Loosen final drive mounting nut (A) at rear suspension member (1)

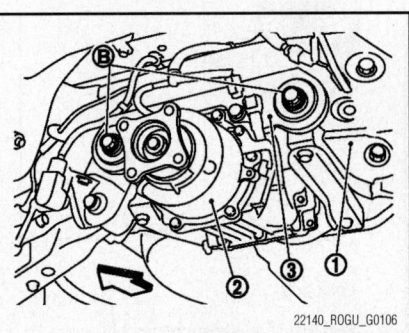

22140_ROGU_G0106

Fig. 92 Remove final drive mounting bolts (B) from final drive mounting bracket (3) to tilt final drive assembly (2)

16. Remove the suspension bar.

17. Remove parking brake cable mounting bolts and separate the parking brake cable from suspension arm.

18. Support center of fuel tank (1) with transmission jack (A).

19. Securely support the fuel tank with a piece of wood (B).

20. Remove fuel tank band (RH and LH).

21. Lower the transmission jack carefully to remove fuel tank while holding it by hand.

To install:

22. Reverse the removal at this point and note the following.

23. Tighten the fuel tank bands to 23 ft. lbs. (31 Nm).

24. Tighten the fuel tank protector to 44 inch. lbs. (5 Nm).

25. Connect the negative battery cable.

AWD Models

See Figures 91 and 92.

1. Before servicing the vehicle, refer to the Precautions Section.

2. Disconnect the negative battery cable.

3. Check the fuel level on the fuel gauge. If fuel gauge indicates more than ¾ of a tank drain fuel tank to around a half a tank.

4. In case fuel pump does not operate, perform the following procedure:

   a. Insert hose of less than 0.79 inch. (20 mm) diameter into fuel filler tube through fuel filler opening to draw the fuel from fuel filler tube.

   b. Disconnect fuel filler hose from fuel filler tube.

   c. Insert hose into fuel tank through fuel filler hose to draw fuel from fuel tank.

5. Release the fuel pressure from the fuel lines.

6. Open the fuel filler lid.

7. Open fuel filler cap and release the pressure inside the fuel tank.

8. Remove rear seat cushion.

9. Remove inspection hole cover.

10. Disconnect fuel lines and harness connectors.

11. Remove the muffler assembly.

12. Remove the propeller shaft.

13. Remove the protector from the fuel tank.

14. Remove vent hose (1) at rear side of fuel tank.

15. Disconnect the EVAP tube at rear side of fuel tank.

16. Remove fuel filler hose at fuel filler tube side.

17. Remove the suspension bar.

18. Remove parking brake cable mounting bolts and separate parking brake cable from suspension arm.

19. Disconnect the AWD solenoid harness connector and harness clip.

20. Loosen final drive mounting nut (A) at rear suspension member (1).

➡**Never remove final drive mounting nut.**

21. Remove final drive mounting bolts (B) from final drive mounting bracket (3) to tilt final drive assembly (2).

➡**Final drive assembly does not have to be removed from the vehicle.**

22. Support center of fuel tank with transmission jack.

23. Securely support the fuel tank with a piece of wood.

24. Remove fuel tank band (RH and LH).

25. Lower the transmission jack carefully to remove fuel tank while holding it by hand.

To install:

26. Reverse the removal procedure at this point and note the following;

27. Tighten the fuel tank bands to 23 ft. lbs. (31 Nm).

28. Tighten the fuel tank protector to 44 inch. lbs. (5 Nm).

29. Connect the negative battery cable.

IDLE SPEED

ADJUSTMENT

Idle speed is maintained by the Electronic Control Module (ECM). No adjustment is necessary or possible.

THROTTLE BODY

REMOVAL & INSTALLATION

See Figure 93.

1. Before servicing the vehicle, refer to the Precautions Section.

2. Remove the air intake duct.

3. Disconnect harness connector.

4. Disconnect water hose.

5. Loosen the throttle body assembly mounting bolts in reverse order as shown in the figure.

To install:

6. Install the throttle body assembly with a new gasket.

7. Tighten the mounting bolts in sequence to 74 inch. lbs. (8.4 Nm)

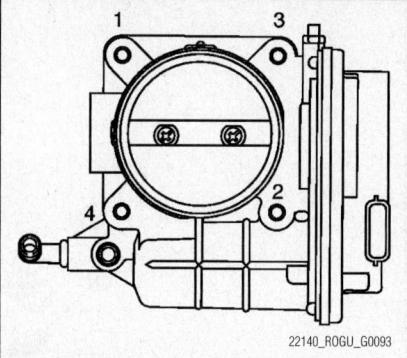

22140_ROGU_G0093

Fig. 93 Loosen mounting bolts in reverse order as shown

8. Reconnect the water hose.

9. Reconnect the harness connector.

10. Reconnect the air intake duct.

HEATING & AIR CONDITIONING SYSTEM

BLOWER MOTOR

REMOVAL & INSTALLATION

See Figures 94 through 96.

1. Before servicing the vehicle, refer to the Precautions Section.

2. Use a refrigerant collecting equipment (for HFC-134a) to discharge the refrigerant.

3. Drain engine coolant from cooling system.

4. Remove cowl top cover.

5. Remove the mounting nut, and lower dash insulator a position without the hindrance for work (as shown in the figure).

6. Remove mounting bolt, and then disconnect low-pressure flexible hose and high-pressure pipe from expansion valve.

➡**Cap or wrap the joint of the A/C piping and expansion valve with suitable material such as vinyl tape to avoid the entry of air.**

7. Remove clamps, and then remove the heater hoses from heater core.

8. Remove instrument panel as follows:

- Remove cluster lid.
- Using remover tool, release cluster lid fixing pawls and clips, from lower to upper, from Instrument panel.
- Release harness connectors.
- Remove screws of center console front side.
- Remove instrument lower covers (RH / LH).
- Pull from the rear of instrument lower cover to release rear pawls, use remover tool to release the upper clips.
- Pull backward to release instrument lower cover from instrument panel.
- Remove center console front fixing screws and clips.
- Remove center console rear fixing screws. Then move forward the front seats if necessary.
- Lift up the center console, and then disconnect harness connectors.
- Remove center console assembly.

9. Remove center speaker grille. (with BOSE audio)

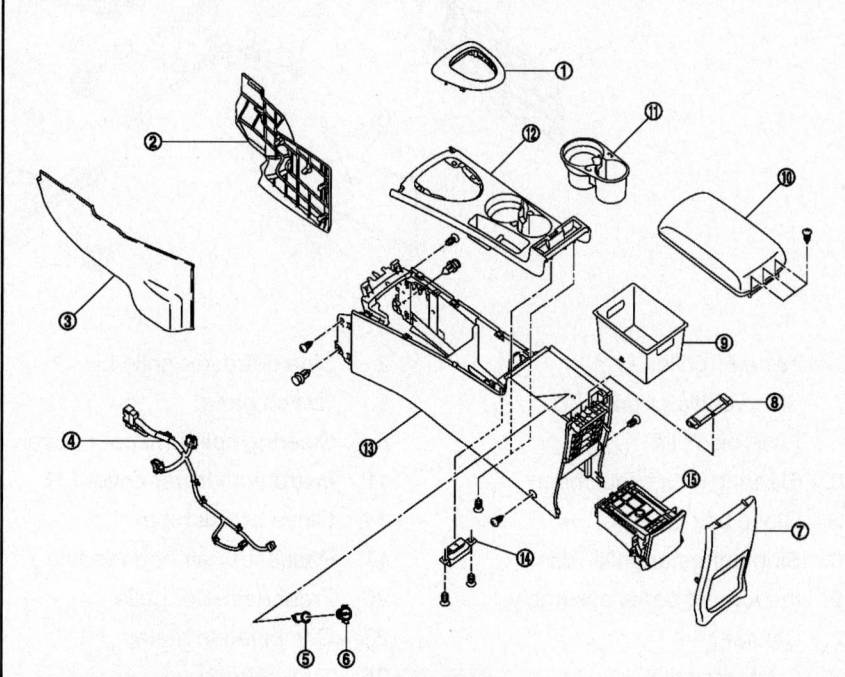

1.	Console finisher assembly	2.	Instrument lower cover RH	3.	Instrument lower cover LH
4.	Console harness assembly	5.	Power socket	6.	Socket knob
7.	Console rear finisher assembly	8.	Console mask	9.	Console pocket
10.	Console lid assembly	11.	Center cup holder assembly	12.	Console upper finisher
13.	Console body assembly	14.	Inside key antenna	15.	Rear cup holder assembly

22140_ROGU_G0109

Fig. 94 Explode view of center console

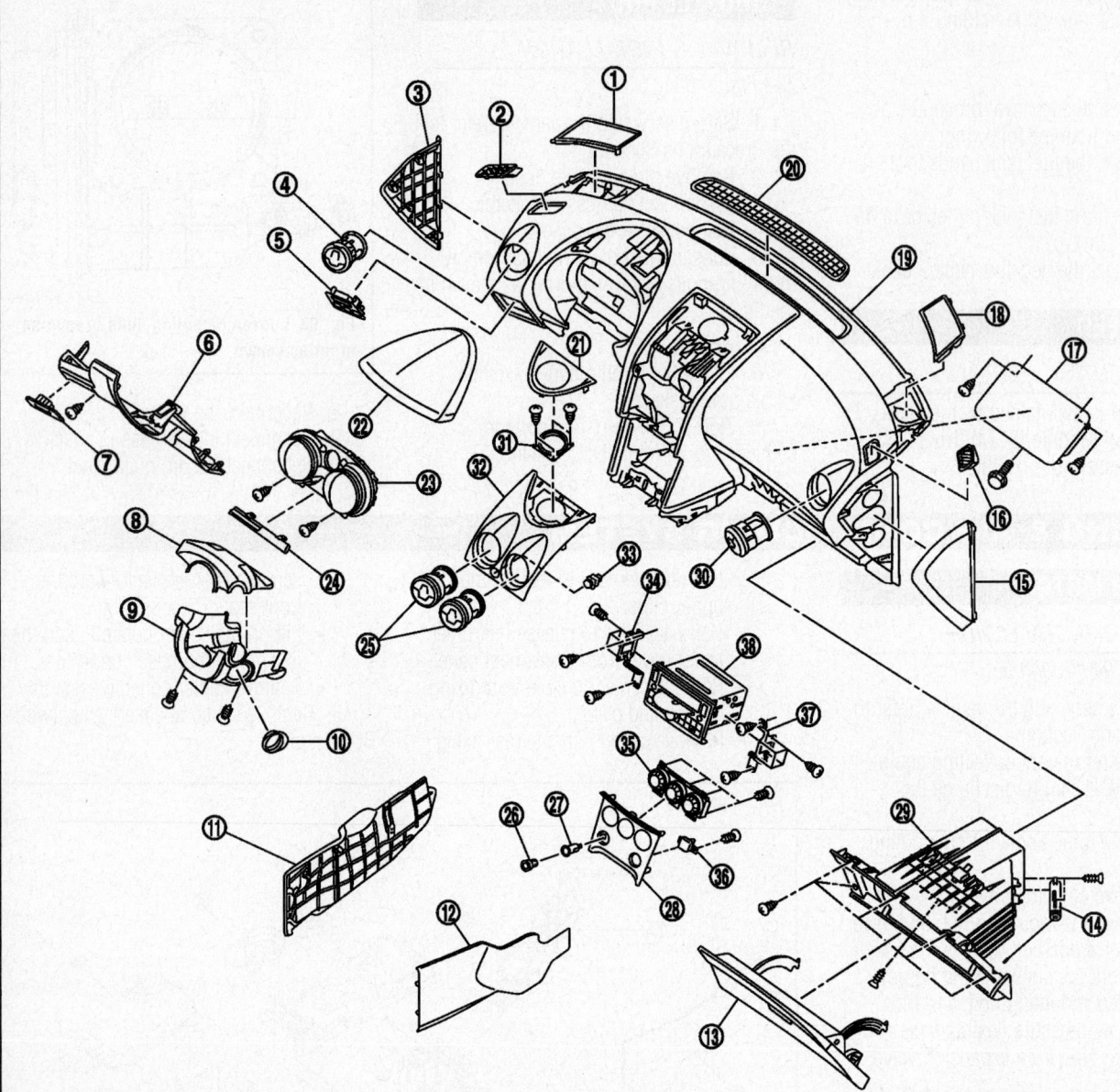

1.	Tweeter grille LH	2.	Side defroster grille LH	3.	Instrument side finisher LH
4.	Side ventilator grille LH	5.	Switch panel	6.	Instrument driver lower cover
7.	Fuse block lid	8.	Steering column upper cover	9.	Steering column lower cover
10.	Steering lock escutcheon	11.	Instrument lower cover LH	12.	Instrument lower cover RH
13.	Glove box lid	14.	Glove box dumper	15.	Instrument side finisher RH
16.	Side defroster grille RH	17.	Passenger air bag module	18.	Tweeter grille RH
19.	Instrument panel assembly	20.	Front defroster grille	21.	Center speaker grille
22.	Cluster lid A	23.	Combination meter	24.	Steering column finisher
25.	Center ventilator grille	26.	Socket knob	27.	Power socket
28.	Cluster lid D	29.	Glove box cover assembly	30.	Side ventilator grille RH
31.	Center speaker	32.	Cluster lid C	33.	Hazard switch

22140_ROGU_G0108

Fig. 95 Exploded view of instrument panel

10. Remove center speaker grille fixing pawls with remover tool.

11. Pull up the center speaker grille from the cluster lid.

12. Remove center speaker. (with BOSE audio)

13. Remove cluster lid.

14. Remove cluster lid fixing pawls and metal clips with remover tool.

15. Pull back cluster lid from lower to upper part.

16. Release the harness connector.

17. Remove the audio unit.

18. Remove the audio unit mounting screws.

19. Pull back the audio unit.

20. Disconnect the antenna feeder and harness connectors.

21. Remove instrument side finisher LH.

22. Insert a remover tool into lower space, and disengage instrument side finisher LH fixing pawls.

23. Pull back instrument side finisher.

24. Remove front body side welt LH.

25. Remove front pillar garnish LH.

26. Remove the tweeter grille LH.

27. Disengage the tweeter grille fixing pawls with remover tool and pull up.

28. Remove instrument driver lower panel.

29. Remove instrument driver lower panel mounting screw.

30. Pull back instrument driver lower panel (1).

31. Release date link connector (pawl) then remove it from instrument driver lower panel.

32. Release hood opener cable.

33. Disconnect harness clamp.

34. Remove knee protector mounting bolts with power tool, and then remove knee protector.

35. Remove the steering wheel.

36. Remove steering column covers.

37. Release the steering column handle.

38. Remove steering lock bezel

39. Remove the steering column lower cover fixing screws.

40. Pull up steering column upper cover, and then remove steering column upper cover.

41. Pull down steering column lower cover, and then remove steering column lower cover.

42. Release steering column finisher fixing pawls, and then remove steering column finisher.

43. Remove the combination switch.

44. Remove cluster lid.

45. Pull back cluster lid, and disengage metal clips.

46. Remove the cluster lid.

47. Remove the combination meter.

48. Remove the combination meter fixing screws.

49. Pull back combination meter.

50. Disconnect harness connector.

51. Remove the switch panel.

52. Remove switch panel fixing pawls with remover tool.

53. Pull back switch panel.

54. Release harness connectors.

55. Remove instrument side finisher RH.

56. Insert a remover tool into lower space, and disengage instrument side finisher RH fixing pawls.

57. Pull back instrument side finisher

58. Remove body side welt RH.

59. Remove front pillar garnish RH.

60. Remove the tweeter grille RH.

61. Disengage the tweeter grille (1) fixing pawls with remover tool

62. Pull up the tweeter grille RH.

63. Remove glove box assembly.

64. Open the glove box lid.

65. Remove the fixing screws.

66. Pull glove box assembly.

67. Disconnect glove box lamp harness connector.

68. Disconnect passenger air bag module connector.

69. Remove passenger air bag module fixing bolt.

70. Remove instrument panel assembly mounting screws and bolts.

71. Release floor harness clamps from instrument panel.

72. Release glove box lamp harness clamp from instrument panel.

73. Remove instrument panel assembly.

➡**When removing instrument panel, 2 workers are required so as to prevent it from dropping.**

74. Remove the following parts after removing instrument panel & pad:
- Passenger air bag module
- Center ventilator grille
- Center ventilator duct
- Side ventilator grilles
- Side ventilator ducts
- Side defroster grilles
- Side defroster ducts

75. Remove center ventilator duct.

76. Remove rear foot duct.

77. Remove rear foot duct.

78. Remove mounting screws and then remove BCM with bracket attached.

79. Remove ground bolts from steering member.

80. Remove the mounting screws from unit assembly.

81. Remove the steering column mounting nuts

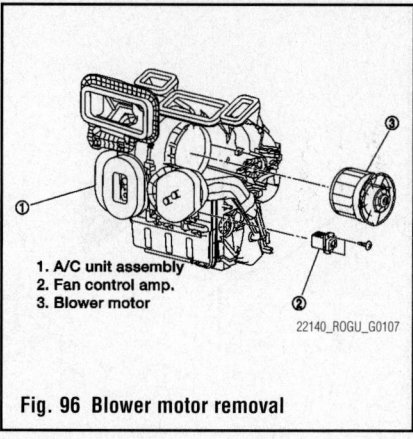

1. A/C unit assembly
2. Fan control amp.
3. Blower motor

22140_ROGU_G0107

Fig. 96 Blower motor removal

82. Remove the steering member mounting bolts and then remove steering member.

83. Remove the A/C unit assembly.

84. Disconnect the blower motor connector.

85. Press flange holding hook.

86. Then turn the blower motor counterclockwise.

87. Pull outside and remove blower motor.

88. Install in the reverse order of removal procedure and note the following:

89. Replace the O-rings with new ones. Then apply compressor oil to them when installing.

90. Install antifreeze and bleed the cooling system.

91. Vacuum and recharge the A/C refrigerant

92. Check the A/C system for leaks.

HEATER CORE

REMOVAL & INSTALLATION

See Figures 97 and 98.

1. Before servicing the vehicle, refer to the Precautions Section.

2. Use a refrigerant collecting equipment (for HFC-134a) to discharge the refrigerant.

3. Drain engine coolant from cooling system.

4. Remove cowl top cover.

5. Remove the mounting nut, and lower dash insulator a position without the hindrance for work (as shown in the figure).

6. Remove mounting bolt, and then disconnect low-pressure flexible hose and high-pressure pipe from expansion valve.

➡**Cap or wrap the joint of the A/C piping and expansion valve with suitable material such as vinyl tape to avoid the entry of air.**

7. Remove clamps, and then remove the heater hoses from heater core.

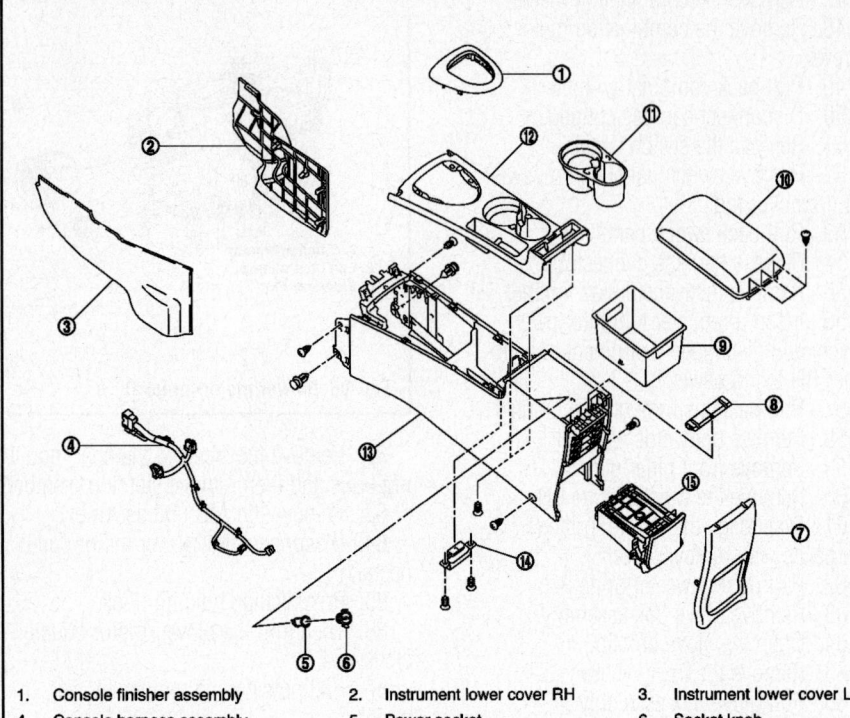

1. Console finisher assembly
2. Instrument lower cover RH
3. Instrument lower cover LH
4. Console harness assembly
5. Power socket
6. Socket knob
7. Console rear finisher assembly
8. Console mask
9. Console pocket
10. Console lid assembly
11. Center cup holder assembly
12. Console upper finisher
13. Console body assembly
14. Inside key antenna
15. Rear cup holder assembly

22140_ROGU_G0109

Fig. 97 Explode view of center console

8. Remove instrument panel as follows:
- Remove cluster lid.
- Using remover tool, release cluster lid fixing pawls and clips, from lower to upper, from Instrument panel.
- Release harness connectors.
- Remove screws of center console front side.
- Remove instrument lower covers (RH / LH).
- Pull from the rear of instrument lower cover to release rear pawls. Use remover tool to release the upper clips.
- Pull backward to release instrument lower cover from instrument panel.
- Remove center console front fixing screws and clips.
- Remove center console rear fixing screws. Then move forward the front seats if necessary.
- Lift up the center console, and then disconnect harness connectors.
- Remove center console assembly.

9. Remove center speaker grille. (with BOSE audio)

10. Remove center speaker grille fixing pawls with remover tool.

11. Pull up the center speaker grille from the cluster lid.

12. Remove center speaker. (with BOSE audio)

13. Remove cluster lid.

14. Remove cluster lid fixing pawls and metal clips with remover tool.

15. Pull back cluster lid from lower to upper part.

16. Release the harness connector.

17. Remove the audio unit.

18. Remove the audio unit mounting screws.

19. Pull back the audio unit.

20. Disconnect the antenna feeder and harness connectors.

21. Remove instrument side finisher LH.

22. Insert a remover tool into lower space, and disengage instrument side finisher LH fixing pawls.

23. Pull back instrument side finisher.

24. Remove front body side welt LH.

25. Remove front pillar garnish LH.

26. Remove the tweeter grille LH.

27. Disengage the tweeter grille fixing pawls with remover tool and pull up.

28. Remove instrument driver lower panel.

29. Remove instrument driver lower panel mounting screw.

30. Pull back instrument driver lower panel (1).

31. Release date link connector (pawl) then remove it from instrument driver lower panel.

32. Release hood opener cable.

33. Disconnect harness clamp.

34. Remove knee protector mounting bolts with power tool, and then remove knee protector.

35. Remove the steering wheel.

36. Remove steering column covers.

37. Release the steering column handle.

38. Remove steering lock bezel

39. Remove the steering column lower cover fixing screws.

40. Pull up steering column upper cover, and then remove steering column upper cover.

41. Pull down steering column lower cover, and then remove steering column lower cover.

42. Release steering column finisher fixing pawls, and then remove steering column finisher.

43. Remove the combination switch.

44. Remove cluster lid.

45. Pull back cluster lid, and disengage metal clips.

46. Remove the cluster lid.

47. Remove the combination meter.

48. Remove the combination meter fixing screws.

49. Pull back combination meter.

50. Disconnect harness connector.

51. Remove the switch panel.

52. Remove switch panel fixing pawls with remover tool.

53. Pull back switch panel.

54. Release harness connectors.

55. Remove instrument side finisher RH.

56. Insert a remover tool into lower space, and disengage instrument side finisher RH fixing pawls.

57. Pull back instrument side finisher

58. Remove body side welt RH.

59. Remove front pillar garnish RH.

60. Remove the tweeter grille RH.

61. Disengage the tweeter grille (1) fixing pawls with remover tool

62. Pull up the tweeter grille RH.

63. Remove glove box assembly.

64. Open the glove box lid.

65. Remove the fixing screws.

66. Pull glove box assembly.

67. Disconnect glove box lamp harness connector.

68. Disconnect the passenger air bag module connector.

69. Remove the passenger air bag module fixing bolt.

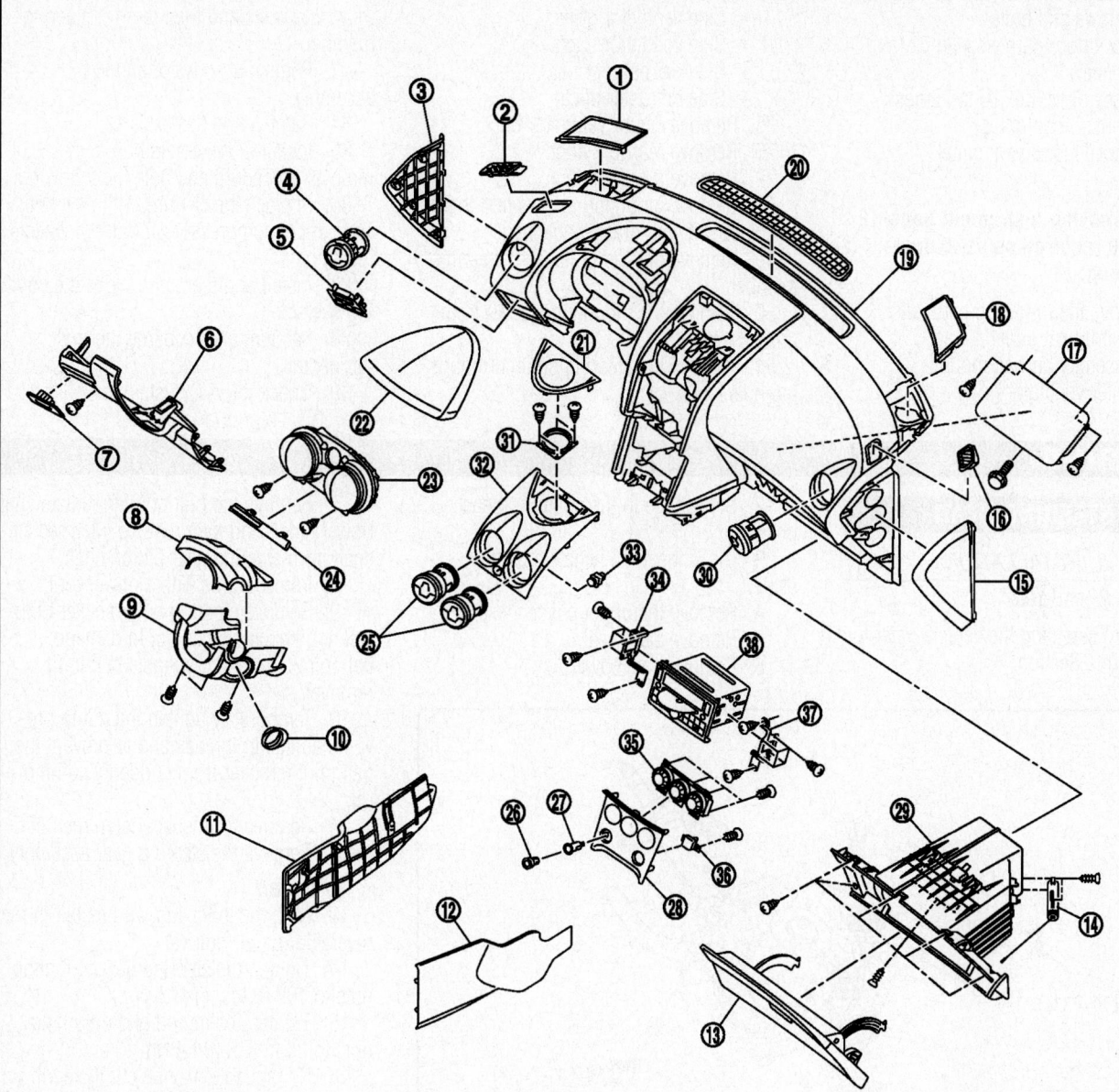

Fig. 98 Exploded view of instrument panel

1. Tweeter grille LH	2. Side defroster grille LH	3. Instrument side finisher LH
4. Side ventilator grille LH	5. Switch panel	6. Instrument driver lower cover
7. Fuse block lid	8. Steering column upper cover	9. Steering column lower cover
10. Steering lock escutcheon	11. Instrument lower cover LH	12. Instrument lower cover RH
13. Glove box lid	14. Glove box dumper	15. Instrument side finisher RH
16. Side defroster grille RH	17. Passenger air bag module	18. Tweeter grille RH
19. Instrument panel assembly	20. Front defroster grille	21. Center speaker grille
22. Cluster lid A	23. Combination meter	24. Steering column finisher
25. Center ventilator grille	26. Socket knob	27. Power socket
28. Cluster lid D	29. Glove box cover assembly	30. Side ventilator grille RH
31. Center speaker	32. Cluster lid C	33. Hazard switch

22140_ROGU_G0108

70. Remove instrument panel assembly mounting screws and bolts.

71. Release floor harness clamps from instrument panel.

72. Release glove box lamp harness clamp from instrument panel.

73. Remove instrument panel assembly.

➡ **When removing instrument panel, 2 workers are required so as to prevent it from dropping.**

74. Remove the following parts after removing instrument panel & pad:
- Passenger air bag module
- Center ventilator grille
- Center ventilator duct
- Side ventilator grilles
- Side ventilator ducts
- Side defroster grilles
- Side defroster ducts

75. Remove center ventilator duct.

76. Remove rear foot duct.

77. Remove rear foot duct.

78. Remove mounting screws and then remove BCM with bracket attached.

79. Remove ground bolts from steering member.

80. Remove the mounting screws from unit assembly.

81. Remove the steering column mounting nuts

82. Remove the steering member mounting bolts and then remove steering member.

83. Remove the Heater and A/C assembly.

84. Remove the heater core.

85. Install in the reverse order of removal procedure and note the following:

86. Replace the O-rings with new ones. Then apply compressor oil to them when installing.

87. Install antifreeze and bleed the cooling system.

88. Vacuum and recharge the A/C refrigerant

89. Check the A/C system for leaks.

STEERING

POWER STEERING GEAR

REMOVAL & INSTALLATION

See Figures 99 and 100.

1. Before servicing the vehicle, refer to the Precautions Section.

2. Set vehicle to the straight-ahead position.

3. Disconnect the negative battery cable.

4. Remove the upper cover.

5. Remove dash seal.

6. Remove hole cover.

7. Remove the bolt of intermediate shaft (lower side), and then remove intermediate shaft from steering gear pinion shaft.

8. Remove tires with a power tool.

9. Remove steering outer socket from steering knuckle so as not to damage ball joint boot using suitable ball joint remover.

10. Temporarily tighten the nut to prevent damage to threads and to prevent the ball joint remover from suddenly coming off.

11. Remove front suspension member.

12. Remove the steering gear assembly.

To install:

13. Note the following, and install in the reverse order of removal.

14. Tighten the steering gear mounting nuts to 109 ft. lbs. (147.5 Nm)

15. Tighten the tie rod end mounting nuts to 25 ft. lbs. (34.4 Nm).

16. Spiral cable may be cut if steering wheel turns while separating steering column assembly and steering gear assembly. Be sure to secure steering wheel using string to avoid turning.

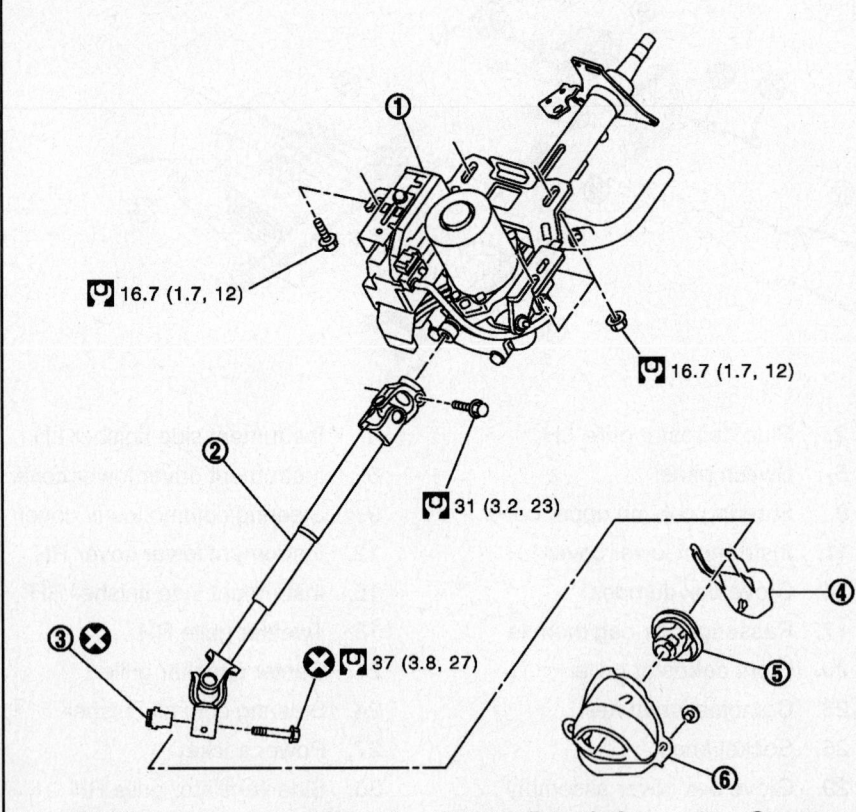

| 1. | Steering column assembly | 2. | Intermediate shaft | 3. | Cam nut |
| 4. | Upper cover | 5. | Dash seal | 6. | Hole cover |

22140_ROGU_G0111

Fig. 99 Remove the bolt of intermediate shaft (lower side)

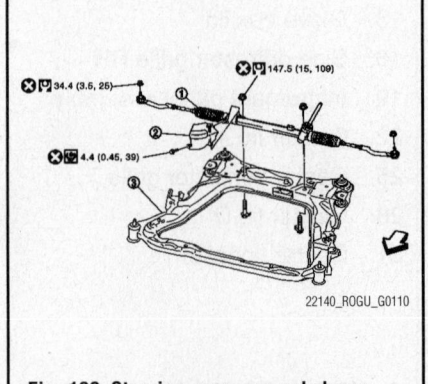
22140_ROGU_G0110

Fig. 100 Steering gear removal shown

17. Check each part of dash seal for damage or other malfunctions. Replace if there are.

18. Perform final tightening of nuts and bolts on each part under unladen conditions with tires on level ground when removing steering gear assembly.

19. Check wheel alignment.

20. Adjust the neutral position of the steering angle sensor as follows:

21. Stop the vehicle with front wheels in straight-ahead position.

22. On the CONSULT-III screen, touch "WORK SUPPORT" and "ST ANG SEN ADJUSTMENT" in order.

23. Touch "START".

➡**Do not touch steering wheel while adjusting steering angle sensor.**

24. After approximately 10 seconds, touch "END".

25. After approximately 60 seconds, it ends automatically.

26. Turn ignition switch OFF, then turn it ON again.

27. Run vehicle with front wheels in straight-ahead position, then stop.

28. Select "DATA MONITOR". Then make sure "STR ANGLE SIG" is within 0±2.5°.

SUSPENSION
FRONT SUSPENSION

COIL SPRING

REMOVAL & INSTALLATION

Refer to strut Removal & Installation.

LOWER BALL JOINT

REMOVAL & INSTALLATION

The Lower ball joint is integral to the lower control arm.

LOWER CONTROL ARM

REMOVAL & INSTALLATION

1. Before servicing the vehicle, refer to the Precautions Section.

2. Before servicing the vehicle, refer to the Precautions Section.

3. Remove or disconnect the following:
   - Front wheels
   - Disc brake caliper from the steering knuckle
   - Cotter pin and loosen the wheel bearing locknut
   - Cotter pin and the castle nut from the tie rod ball joint. Separate the tie rod with a suitable puller.
   - 2 bolts that secure the lower portion of the strut to the steering knuckle

4. Using a plastic or rubber mallet and tap on the loosened wheel bearing locknut to loosen the halfshaft in the knuckle.

5. Remove the locknut and remove the halfshaft from the steering knuckle. Be sure to cover the CV-joints with a shop rag.

6. Remove the nut that secures the stabilizer link to the lower control arm.

7. Remove the stabilizer from the lower control arm.

8. Note the positioning of the washers and spacers for reassembly.

9. Cotter pin and castle nut from the lower ball joint.

10. The lower ball joint from the knuckle.

11. Remove the knuckle from the vehicle.

12. Mounting nuts/bolts that secure the lower control arm to the frame

13. Control arm from the vehicle.

To install:

14. Install the lower control arm assembly and torque mounting bolts/nuts as follows:
   - Through bolt and nut: 104 ft. lbs. (141.5 Nm).
   - 2 saddle bracket mounting bolts: 126ft. lbs. (171 Nm).

15. Install or connect the following:
   - Install the steering knuckle to the lower ball joint and torque the castle nut to 54 ft. lbs. (74 Nm).
   - Install a new cotter pin.
   - Stabilizer link to the lower control arm and torque the nut to 62ft. lbs. (83.5 Nm).
   - Halfshaft through the wheel bearing
   - Wheel bearing locknut. Do not torque the locknut at this time.
   - Steering knuckle to the strut and torque the bolts to 104 ft. lbs. (141.5 Nm).
   - Tighten the tie rod end mounting nuts to 25 ft. lbs. (34.4 Nm).
   - Install a new cotter pin.
   - Disc brake caliper to the steering knuckle

16. Tighten the halfshaft mounting nut (hub nut) and torque the nut to 92 ft. lbs. (125Nm). It may be necessary to have an assistant hold the brake pedal while tightening the locknut. Install the adjusting cap and a new cotter pin.

17. Install the front wheels, lower the vehicle and perform a front end alignment.

MACPHERSON STRUT AND COIL SPRING

REMOVAL & INSTALLATION
See Figure 101.

1. Before servicing the vehicle, refer to the Precautions Section.

2. Remove the front tires.

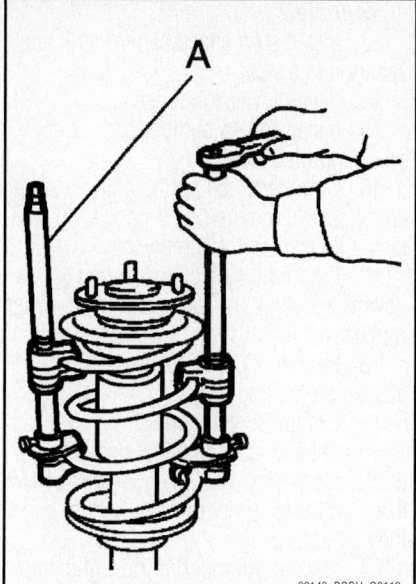

Fig. 101 Compress coil spring between strut mounting bearing and lower rubber seat on strut assembly

22140_ROGU_G0112

3. Remove the lock plat.

4. Remove cap and mounting nut on the upper side of stabilizer connecting rod, and then remove stabilizer connecting rod from strut assembly with power tool.

5. Separate the steering knuckle from strut assembly.

6. Remove mounting bolts of strut mounting insulator, and then remove strut assembly with power tool.

7. Install strut attachment to strut assembly and secure it in a vise.

8. Using a spring compressor (A) (commercial service tool), compress coil spring between strut mounting bearing and lower rubber seat (on strut assembly) until coil spring with a spring compressor is free.

✳✳ WARNING

Be sure a spring compressor is securely attached to coil spring. Compress coil spring.

9. Make sure coil spring with a spring compressor between strut mounting bearing and lower rubber seat (strut assembly) is free.

10. And then remove piston rod lock nut while securing the piston rod tip so that piston rod does not turn.

11. Remove strut mounting insulator and strut mounting bearing, and bound bumper from strut.

12. After remove coil spring with a spring compressor, and then gradually release a spring compressor.

To install:

13. Install strut attachment to strut and secure it in a vise.

14. Install lower rubber seat.

15. Install bound bumper onto strut mounting insulator.

16. Compress coil spring using a spring compressor (commercial service tool), and install it onto strut assembly.

17. Face tube side of coil spring (1) downward. Align the lower end (A) to lower rubber seat (2).

18. Be sure a compressor is securely attached to coil spring.

19. Compress coil spring.

20. Set coil spring so that its paint marks are aligned with the positions of 1.75 turns and 2.75 turns from the bottom end of the coil spring.

21. Install strut mounting bearing and strut mounting insulator with bound bumper to strut.

22. Installation position of strut mounting insulator is shown in the figure.

23. Secure piston rod tip so that piston rod does not turn, then tighten piston rod lock nut with specified torque.

24. Loosen while making sure coil spring attachment position does not move.

25. Remove the strut attachment from strut assembly.

26. Reverse the removal procedure and note the following:

- Tighten the lower strut knuckle bolt to 104 ft. lbs. (141 Nm).
- Tighten the upper strut mount nuts to 14 ft. lbs. (19 Nm).
- Tighten the strut rod mounting nut to 58ft. lbs. (79 Nm).

27. Install the front wheels, lower the vehicle and perform a front end alignment.

STABILIZER BAR

REMOVAL & INSTALLATION

See Figures 102 and 103.

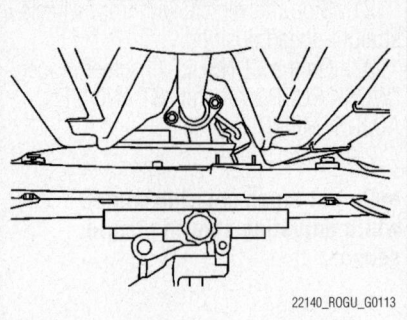

Fig. 102 Set suitable jack under front suspension member

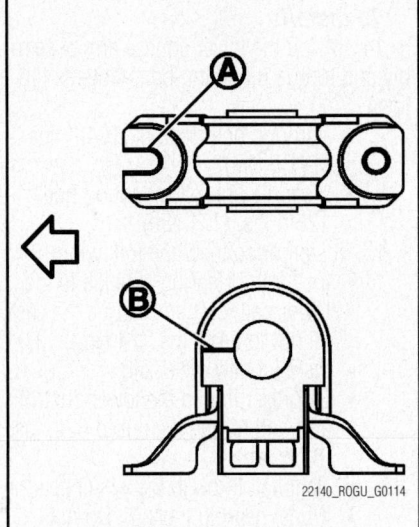

Fig. 103 Stabilizer bushing installation view

1. Before servicing the vehicle, refer to the Precautions Section.

2. Remove the front tires.

3. Remove engine under cover from vehicle.

4. Remove the steering outer socket from steering knuckle.

5. Remove stabilizer connecting rod.

6. Remove rear torque rod.

7. Separate the intermediate shaft from steering gear.

8. Set suitable jack under front suspension member.

9. Remove front suspension member stay from vehicle.

10. Gradually lower jack front suspension member in order to remove stabilizer mounting bolts.

11. Remove mounting bolts of stabilizer clamp, and then remove stabilizer clamp and stabilizer bushing from front suspension member.

12. Remove stabilizer bar.

To install:

13. Note the following, and install in the reverse order of removal.

14. Tighten the steering column shaft mounting bolt to 25 ft. lbs. (35 Nm)

15. Tighten the stabilizer bracket mounting bolts to 25 ft. lbs. (34 Nm).

16. Tighten the stabilizer link mounting nuts to 62 ft. lbs. (83.5 Nm).

17. Install stabilizer clamp that notch (A) becomes vehicle front side

18. Install stabilizer bushing that slit (B) becomes vehicle front side.

19. Install the front wheels, lower the vehicle and perform a front end alignment.

STEERING KNUCKLE

REMOVAL & INSTALLATION

See Figure 104.

1. Before servicing the vehicle, refer to the Precautions Section.

2. Remove tires with power tool.

3. Remove wheel sensor from steering knuckle.

❋❋ WARNING

Never pull on the wheel sensor harness.

4. Remove lock plate from strut assembly.

5. Remove torque member mounting bolts with power tool. Hang torque member not to interfere with work.

6. Remove disc rotor. Put matching marks on the wheel hub and bearing assembly and the disc rotor before removing the disc rotor.

7. Remove cotter pin, and then loosen hub lock nut with power tool.

disengage wheel hub and bearing assembly from drive shaft.

8. Remove the hub lock nut.

9. Use suitable puller, if wheel hub and bearing assembly and driveshaft cannot be separated even after performing the above procedure.

10. Remove wheel hub and bearing assembly.

11. Remove wheel hub and bearing assembly, and then remove splash guard.

12. Remove transverse link from steering knuckle.

13. Remove the steering knuckle from strut assembly.

14. Loosen the nut of steering outer socket.

15. Remove steering outer socket (1) from steering knuckle (2) using the ball

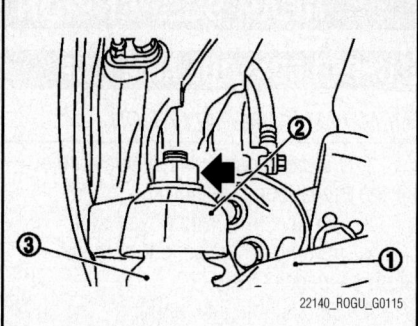

Fig. 104 Remove steering outer socket (1) from steering knuckle (2) using the ball joint remover so as not to damage ball joint boot (3)

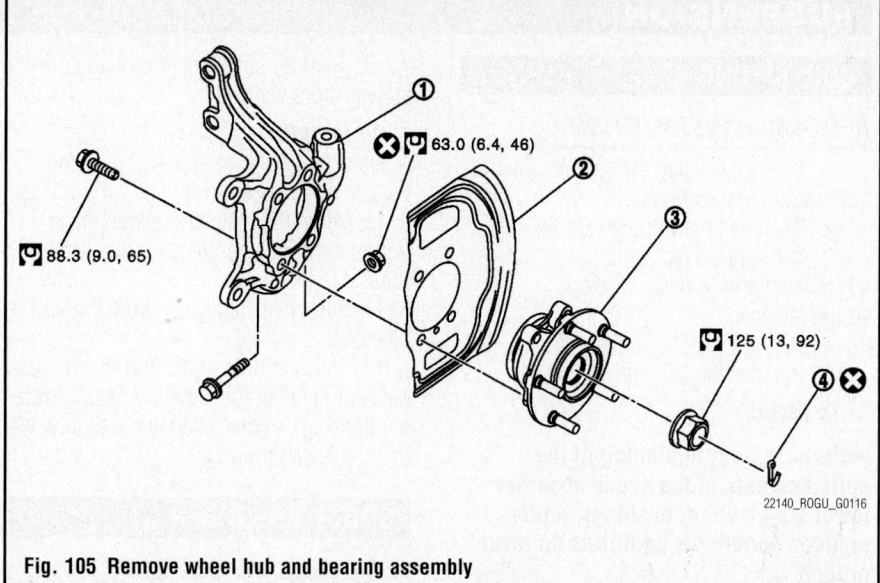

Fig. 105 Remove wheel hub and bearing assembly

joint remover so as not to damage ball joint boot (3).

16. Temporarily tighten the nut to prevent damage to threads and to prevent the ball joint remover from suddenly coming off.

17. Remove the steering knuckle from vehicle.

To install:

18. Note the following, and install in the reverse order of the removal.

19. Align the matching marks made during removal when reusing the disc rotor.

20. Install removed wheel hub and bearing assembly and steering knuckle and perform the final tightening of each part under unladen conditions on the level surface.

21. Tighten the stabilizer link mounting nuts to 62 ft. lbs. (83.5 Nm).

22. Tighten the lower strut knuckle bolt to 104 ft. lbs. (141 Nm).

23. Tighten the tie rod end mounting nuts to 25 ft. lbs. (34.4 Nm).

24. Tighten the axle nut to 92 ft. lbs. (125 Nm).

25. Install the front wheels, lower the vehicle and perform a front end alignment.

WHEEL HUB AND BEARING

REMOVAL & INSTALLATION

See Figure 105.

1. Before servicing the vehicle, refer to the Precautions Section.

2. Remove tires with power tool.

3. Remove wheel sensor from steering knuckle.

4. Remove lock plate from strut assembly.

5. Remove torque member mounting bolts with power tool. Hang torque member not to interfere with work.

6. Remove the disc rotor. Put matching marks on the wheel hub and bearing assembly and the disc rotor before removing the disc rotor.

7. Remove cotter pin, and then loosen hub lock nut with power tool.

8. Patch hub lock nut with a piece of wood. Hammer the wood to disengage wheel hub and bearing assembly from drive shaft.

9. Remove the hub lock nut.

✳✳ WARNING

Never place drive shaft joint at an extreme angle. Also be careful not to overextend slide joint. Never allow drive shaft to hang down without support for housing (or joint sub-assembly), shaft and the other parts.

10. Use suitable puller, if wheel hub and bearing assembly and drive shaft cannot be separated even after performing the above procedure.

11. Remove wheel hub and bearing assembly.

To install:

12. Install the hub and bearing assembly. Tighten the bolts to 65 ft. lbs. (83 Nm).

13. Install the hub lock nut and tighten to 92 ft. lbs. (125 Nm).

14. Install a new cotter pin.

15. Reverse the removal procedure at this point.

COIL SPRING

REMOVAL & INSTALLATION

1. Before servicing the vehicle, refer to the Precautions Section.
2. Remove tires with power tool.
3. Set suitable jack under suspension arm to relieve the coil spring tension.
4. Remove the shock absorber.
5. Remove the coil spring.

To install:

➡ **Perform final tightening of the bolts and nuts at the shock absorber lower side (rubber bushing), under unladen conditions with tires on level ground.**

6. Install the coil spring.
7. Raise the suspension arm to line up shock mounting holes.
8. Install the shock absorber.
9. Tighten the top mounting nut and bolt to 89 ft. lbs. (120 Nm).
10. Tighten the bottom mounting nut and bolt to 89 ft. lbs. (120 Nm).

LOWER CONTROL ARM

REMOVAL & INSTALLATION

See Figures 106.

1. Before servicing the vehicle, refer to the Precautions Section.
2. Remove torque member mounting bolts. Hang torque member where it does not interfere with work

➡ **Put matching marks on the wheel hub and bearing assembly and the disc rotor before removing the disc rotor.**

3. Remove wheel sensor and sensor harness from axle housing and lower link.

�֎ WARNING

Never pull on the wheel speed sensor harness.

4. Remove the parking brake cable mounting bolt.
5. Separate the brake tube from the brake hose, and remove lock plate.
6. Remove the stabilizer link.
7. Set suitable jack under suspension arm to relieve the coil spring tension.
8. Remove coil spring from suspension arm.

9. Remove suspension arm and arm stopper from vehicle.

To install:
Note the following and, install in the reverse order of removal.

10. Align the matching marks made during removal when reusing the disc rotor.
11. After installation, perform the air bleeding.
12. Perform final tightening of rear suspension member installation position (rubber bussing), under unladen conditions with tires on level ground.

SHOCK ABSORBER

REMOVAL & INSTALLATION

1. Before servicing the vehicle, refer to the Precautions Section.
2. Remove tires with power tool.
3. Set suitable jack under suspension arm to relieve the coil spring tension.
4. Remove the shock absorber.

To install:

➡ **Perform final tightening of the bolts and nuts at the shock absorber lower side (rubber bushing), under unladen conditions with tires on level ground.**

5. Raise the suspension arm to line up shock mounting holes.
6. Install the shock absorber.
7. Tighten the top mounting nut and bolt to 89 ft. lbs. (120 Nm).
8. Tighten the bottom mounting nut and bolt to 89 ft. lbs. (120 Nm).

STABILIZER BAR

REMOVAL & INSTALLATION

1. Before servicing the vehicle, refer to the Precautions Section.
2. Remove the stabilizer links.
3. Remove the main muffler.
4. Remove the mounting nuts on stabilizer clamp and stabilizer bar from suspension member.

To install:
5. Install the mounting nuts on stabilizer clamp and stabilizer bar to suspension member.
6. Tighten the stabilizer clamp nuts to 26 ft. lbs. (36 Nm).
7. Install the main muffler.
8. Tighten the stabilizer link mounting bolts to 81 ft. lbs. (110 Nm).

UPPER CONTROL ARM LINK

REMOVAL & INSTALLATION

1. Before servicing the vehicle, refer to the Precautions Section.
2. Remove tires with power tool.
3. Remove wheel sensor harness from suspension arm.
4. Set suitable jack under suspension arm to relieve the coil spring tension.
5. Remove the upper link from suspension arm with power tool.
6. Remove the upper link from suspension member with power tool.

To install:
7. Note the following, and install in the reverse order.

➡ **Perform final tightening of rear suspension member unladen conditions with tires on level ground.**

8. Install the upper control arm link.
9. Tighten the outer mounting bolt to 111 ft. lbs. (150 Nm).
10. Tighten the outer mounting bolt to 96 ft. lbs. (130 Nm).
11. Remove the jack.

WHEEL HUB AND BEARING

REMOVAL & INSTALLATION

See Figures 107.

1. Before servicing the vehicle, refer to the Precautions Section.
2. Remove tires with power tool.
3. Remove wheel sensor from axle housing.

✳✳ WARNING

Never pull on the wheel sensor harness.

4. Remove torque member mounting bolts. Hang torque member not to interfere with work.
5. Remove disc rotor. Put matching marks on the wheel hub and bearing assembly and the disc rotor before removing the disc rotor.
6. Remove cotter pin, and then loosen hub lock nut with power tool (AWD).
7. Patch hub lock nut with a piece of wood. Hammer the wood to disengage wheel hub and bearing assembly from drive shaft.
8. Remove the hub lock nut. (AWD)

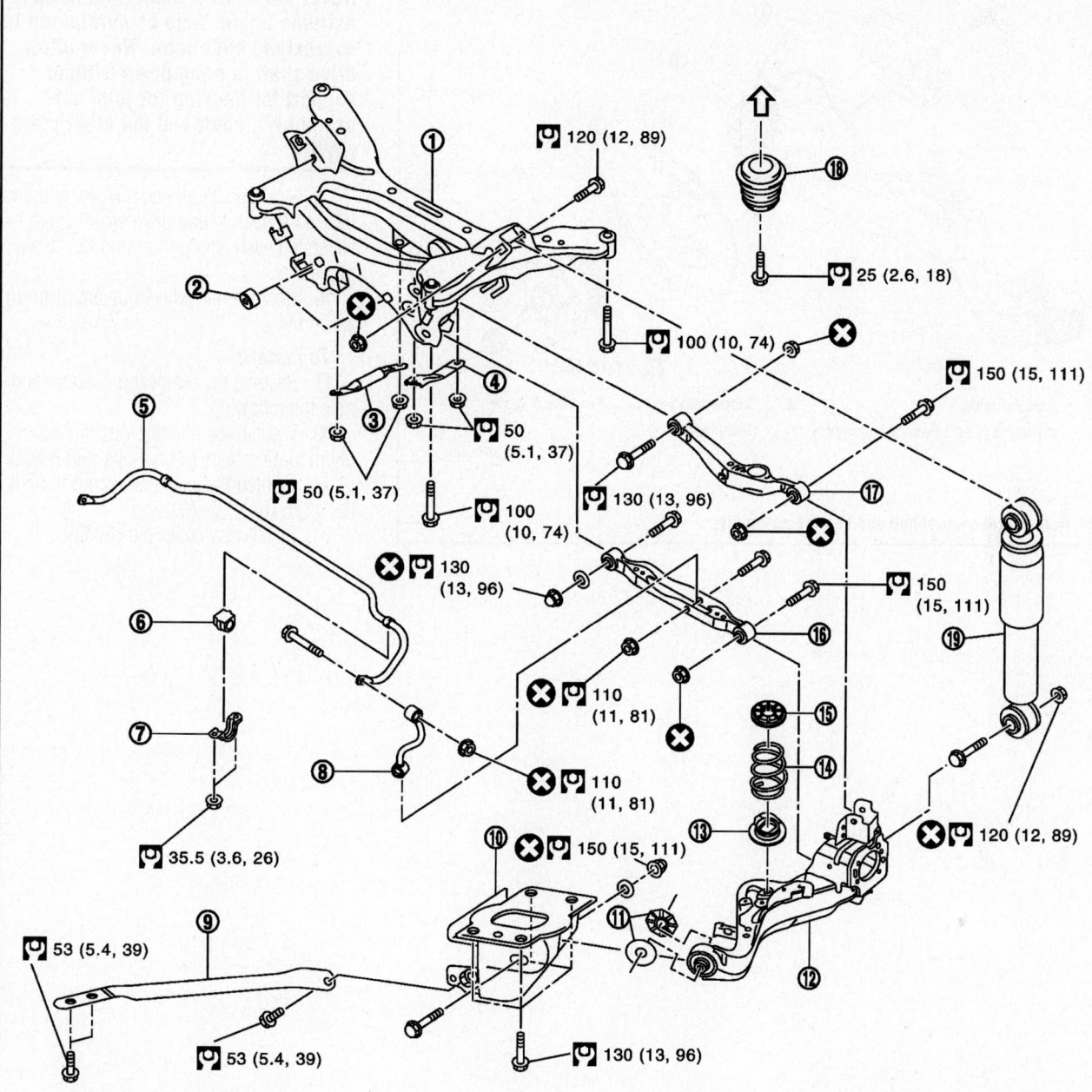

120 (12, 89)

18

25 (2.6, 18)

100 (10, 74)

150 (15, 111)

130 (13, 96)

17

50 (5.1, 37)

50 (5.1, 37)

100 (10, 74)

150 (15, 111)

130 (13, 96)

16

110 (11, 81)

150 (15, 111)

110 (11, 81)

19

120 (12, 89)

35.5 (3.6, 26)

15

14

13

11

12

53 (5.4, 39)

53 (5.4, 39)

130 (13, 96)

1. Rear suspension member	2. Suspension member protector (2WD)	3. Suspension member stay (right side)
4. Suspension member stay (left side)	5. Stabilizer bar	6. Stabilizer bushing
7. Stabilizer clamp	8. Stabilizer link	9. Suspension bar
10. Suspension arm bracket	11. Arm stopper	12. Suspension arm
13. Low rubber seat	14. Coil spring	15. Upper rubber seat
16. Lower link	17. Upper link	18. Bound bumper
19. Shock absorber		

22140_ROGU_G0117

Fig. 106 Exploded view of the rear suspension—Rogue

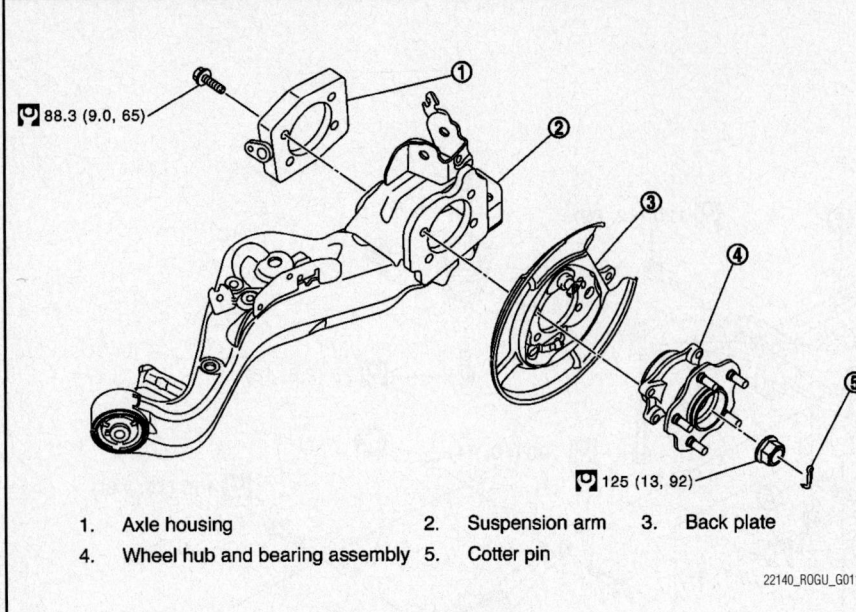

88.3 (9.0, 65)

125 (13, 92)

1. Axle housing
2. Suspension arm
3. Back plate
4. Wheel hub and bearing assembly
5. Cotter pin

22140_ROGU_G0118

Fig. 107 Remove rear wheel hub and bearing assembly

✳✳ WARNING

Never place drive shaft joint at an extreme angle. Also be careful not to overextend slide joint. Never allow drive shaft to hang down without support for housing (or joint sub-assembly), shaft and the other parts. (AWD)

9. Use suitable puller, if wheel hub and bearing assembly and drive shaft cannot be separated even after performing the above procedure.
10. Remove rear wheel hub and bearing assembly.

To install:
11. Reverse the removal procedure and note the following:
12. Tighten the hub and bearing assembly mounting bolts to 65 ft. lbs. (88.3 Nm).
13. Tighten the axle locking nut to 92 ft lbs. (125 Nm)—(AWD).
14. Install new cotter pin (AWD).

NISSAN

Titan

24

SPECIFICATIONS AND MAINTENANCE CHARTS

ENGINE AND VEHICLE IDENTIFICATION

Engine								Model Year	
Code ①	Liters (cc)	Cu. In.	Cyl.	Fuel Sys.	Engine	Eng. Mfg.		Code ②	Year
VK56DE	5.6 (5552)	338.8	8	MFI	DOHC	Nissan		7	2007
								8	2008

MFI: Multi-port Fuel Injection

DOHC: Double Overhead Camshafts

② 10th digit of the Vehicle Identification Number (VIN)

22140_TITA_C0001

GENERAL ENGINE SPECIFICATIONS

Year	Model	Engine Displacement Liters	Engine ID	Net Horsepower @ rpm	Net Torque @ rpm (ft. lbs.)	Bore x Stroke (in.)	Com- pression Ratio	Oil Pressure @ rpm
2007	Titan	5.6	VK56DE	305@4900	379@3600	3.86X3.62	9.8:1	43@2000
2008	Titan	5.6	VK56DE	305@4900	379@3600	3.86X3.62	9.8:1	43@2000

22140_TITA_C0002

ENGINE TUNE-UP SPECIFICATIONS

Year	Engine Displacement Liters	Engine ID	Spark Plug Gap (in.)	Ignition Timing	Fuel Pump (psi) ①	Idle Speed ②	Valve Clearance (in.) In.	Valve Clearance (in.) Ex.
2007	5.6	VK56DE	0.043	15B	51	650	0.010-0.013	0.011-0.015
2008	5.6	VK56DE	0.043	15B	51	650	0.010-0.013	0.011-0.015

NOTE: The Vehicle Emission Control Information label often reflects specification changes made during production. The label figures must be used if they differ from those in this chart.

B: Before top dead center

① System pressure at idle

② Automatic transmission in Neutral

22140_TITA_C0003

CAPACITIES

Year	Model	Engine Displacement Liters	Engine ID	Engine Oil with Filter (qts.)	Transmission (pts.)	Transfer Case (pts.)	Drive Axle		Fuel Tank (gal.) ①	Cooling System (qts.)
							Front (pts.)	Rear (pts.)		
2007	Titan	5.6	VK56DE	6.5	22.5	6.25	3.375	4.25	28.0	13
2008	Titan	5.6	VK56DE	6.5	22.5	6.25	3.375	4.25	28.0	13

NOTE: All capacities are approximate. Add fluid gradually and check to be sure a proper fluid level is obtained.

① Long wheel base (37 Gal.)

22140_TITA_C0005

FLUID SPECIFICATIONS

Year	Model	Engine Displ. Liters	Engine ID	Engine Oil	Auto Trans.	Transfer Case	Drive Axle		Power steering Fluid	Engine Coolant	Brake Fluid
							Front	Rear			
2007	Titan	5.6	VK56DE	5W30	①	②	80W90	75W90	Nissan	Nissan	DOT 3
2008	Titan	5.6	VK56DE	5W30	①	②	80W90	75W90	Nissan	Nissan	DOT 3

DOT: Department Of Transpotation

① Nissan Matic J ATF-2

① Nissan Matic D ATF

22140_TITA_C0004

VALVE SPECIFICATIONS

Year	Engine Displacement Liters	Engine ID	Seat Angle (deg.)	Face Angle (deg.)	Spring Test Pressure (lbs. @ in.)	Spring Installed Height (in.)	Stem-to-Guide Clearance (in.)		Stem Diameter (in.)	
							Intake	Exhaust	Intake	Exhaust
2007	5.6	VK56DE	45.15-45.45	45	37.0@1.457	1.9913	0.0008-0.0021	0.0012-0.0025	0.2348-0.2354	0.2344-0.2350
2008	5.6	VK56DE	45.15-45.45	45	37.0@1.457	1.9913	0.0008-0.0021	0.0012-0.0025	0.2348-0.2354	0.2344-0.2350

22140_TITA_C0006

CAMSHAFT AND BEARING SPECIFICATIONS CHART

All measurements are given in inches.

Year	Engine Displ. Liters	Engine ID/VIN	Journal Dia.	Brg. Oil Clearance	Shaft End-play	Runout	Journal Bore	Lobe Height Intake	Lobe Height Exhaust
2007	5.6	VK56DE	1.0218-1.0224	0.0012-0.0027	0.0045-0.0074	0.0008	1.0236-1.0244	1.7506-1.7581	1.7506-1.7581
2008	5.6	VK56DE	1.0218-1.0224	0.0012-0.0027	0.0045-0.0074	0.0008	1.0236-1.0244	1.7506-1.7581	1.7506-1.7581

22140_TITA_C0007

CRANKSHAFT AND CONNECTING ROD SPECIFICATIONS

All measurements are given in inches.

Year	Engine Displ. Liters	Engine ID	Crankshaft Main Brg. Journal Dia.	Crankshaft Main Brg. Oil Clearance	Crankshaft Shaft End-play	Crankshaft Thrust on No.	Connecting Rod Journal Diameter	Connecting Rod Oil Clearance	Connecting Rod Side Clearance
2007	5.6	VK56DE	①	②	0.0118	3	③	0.0008-0.0015	0.0079-0.0157
2008	5.6	VK56DE	①	②	0.0118	3	③	0.0008-0.0015	0.0079-0.0157

① There are 24 different grades, ranging from G (2.5183) to 9 (2.5173)

② No. 1 and 5: 0.00004-0.0004

   No. 2, 3 and 4: 0.0003-0.0007

③ Grade 0: 2.2441-2.2441
Grade 1: 2.2441-2.2442
Grade 2: 2.2442-2.2442
Grade 3: 2.2442-2.2443
Grade 4: 2.2443-2.2443
Grade 5: 2.2443-2.2443
Grade 6: 2.2443-2.2444
Grade 7: 2.2444-2.2444
Grade 8: 2.2444-2.2444
Grade 9: 2.2444-2.2445
Grade A: 2.2445-2.2445
Grade B: 2.2445-2.22446
Grade C: 2.2446-2.2446

22140_TITA_C0008

PISTON AND RING SPECIFICATIONS

All measurements are given in inches.

Year	Engine Displacement Liters	Engine ID	Piston Clearance	Ring Gap Top Comp.	Ring Gap Bottom Comp.	Ring Gap Oil Control	Ring Side Clearance Top Comp.	Ring Side Clearance Bottom Comp.	Ring Side Clearance Oil Control
2007	5.6	VK56DE	0.0004-0.0012	0.0091-0.0130	0.0091-0.0130	0.0079-0.0236	0.0014-0.0033	0.0012-0.0028	0.0006-0.0073
2008	5.6	VK56DE	0.0004-0.0012	0.0091-0.0130	0.0091-0.0130	0.0079-0.0236	0.0014-0.0033	0.0012-0.0028	0.0006-0.0073

22140_TITA_C0009

TORQUE SPECIFICATIONS
All readings in ft. lbs.

Year	Engine Displacement Liters	Engine ID	Cylinder Head Bolts	Main Bearing Bolts	Rod Bearing Bolts	Crankshaft Damper Bolts	Flywheel Bolts	Manifold		Spark Plugs	Oil Pan Drain Plug
								Intake	Exhaust		
2007	5.6	VK56DE	①	②	③	④	65	73	21	18	25
2008	5.6	VK56DE	①	②	③	④	65	73	21	18	25

① Step 1: 72 ft. lbs

   Step 2: Loosen all bolts completely

   Step 3: 33 ft. lbs.

   Step 4: +60 degrees

   Step 5: +60 degrees

② Step 1: Main Bolts to 29 ft. lbs.

   Step 2: Sub-bolts to 22 ft. lbs.

   Step 3: Main Bolts +40 degrees

   Step 4: Sub-Bolts +30 degrees

   Step 5: Side Bolts to 36 ft. lbs.

③ Step 1: 11 ft. lbs.

   Step 2: +90 degrees

④ Step 1: 65 ft. lbs.

   Step 2: +90 degrees

22140_TITA_C0010

WHEEL ALIGNMENT

Year	Model	Caster		Camber		Toe-in (in.)
		Range (+/-Deg.)	Preferred Setting (Deg.)	Range (+/-Deg.)	Preferred Setting (Deg.)	
2007	Titan ①	0.75	②	0.75	③	0.11+/-0.05
2008	Titan ①	0.75	②	0.75	③	0.11+/-0.05

① Assumes P245/70R17 tire

② 4x2: +3.27

   4x4: +2.37

③ 4x2: -0.12

   4x4: +0.43

22140_TITA_C0011

TIRE, WHEEL AND BALL JOINT SPECIFICATIONS

Year	Model	OEM Tires		Tire Pressures (psi)		Wheel Size	Ball Joint Inspection	Lug Nut Torque (ft. lbs.)
		Standard	Optional	Front	Rear			
2007	Titan	P245/70R17	①	35	35	17	②	98
2008	Titan	P265/70R18	③	35	35	17	②	98

OEM: Original Equipment Manufacturer

PSI: Pounds Per Square Inch

① P285/70R17

   P265/70R18

② Axial play

   Upper: 0

③ P275/70R18

   P275/60R20

22140_TITA_C0012

BRAKE SPECIFICATIONS

All measurements in inches unless noted

Year	Model		Brake Disc			Minimum Lining Thickness	Brake Caliper	
			Original Thickness	Minimum Thickness	Maximum Runout		Bracket Bolts (ft. lbs.)	Mounting Bolts (ft. lbs.)
2007	Titan	F	1.024	0.965	0.0016	0.039	155	32
		R	0.551	0.472	0.002	0.039	NA	24
2008	Titan	F	1.024	0.965	0.0016	0.039	155	32
		R	0.551	0.472	0.002	0.039	NA	24

NA: Not applicable

22140_TITA_C0013

SCHEDULED MAINTENANCE INTERVALS
Nissan - Titan

TO BE SERVICED	TYPE OF SERVICE	7.5	15	22.5	30	37.5	45	52.5	60
Engine oil & filter	R	✓	✓	✓	✓	✓	✓	✓	✓
Brake lines & cables	S/I		✓		✓		✓		✓
Brake pads and rotors	I		✓		✓		✓		✓
Driveshaft boots & propeller shaft (4x4)	L/I		✓		✓		✓		✓
Transmission, transfer & differential gear oil	I		✓		✓		✓		✓
Air cleaner filter	R				✓				✓
Engine coolant ①	R								✓
Spark plugs (Platinum)	R		Replace every 105,000 miles						
Drive belt(s) ②	S/I								✓
Cabin air filter	R		✓		✓		✓		✓
Exhaust system	I				✓				✓
Fuel lines	S/I				✓				✓
Fuel filter ③									
Steering gear (box) & linkage, axle & suspension parts	I				✓				✓
Vapor lines	S/I				✓				✓

R: Replace   S/I: Service or Inspect   L: Lubricate

① Coolant: After 60,000 miles, inspect every 30,000 miles.

② Drive Belts: After 60,000 miles, inspect every 15,000 miles. Replace belts if damaged.

③ Fuel Filter: Maintenance free item.

FREQUENT OPERATION MAINTENANCE (SEVERE SERVICE)

If a vehicle is operated under any of the following conditions it is considered severe service:

- Extremely dusty areas.

- Rough, muddy, or salt spread roads.

- 50% or more of the vehicle constant operation is in 32°C (90°F) or higher temperatures, or temperatures below 0°C (32°F).

- Prolonged idling (vehicle operation in stop and go traffic).

- Frequent short running periods (engine does not warm to normal operating temperatures).

- Police, taxi, delivery usage or trailer towing usage.

Oil & oil filter: replace every 3750 miles.

Brake pads, discs, drums & linings: service or inspect every 7500 miles.

Driveshaft boots & propeller shaft: service or inspect every 7500 miles.

Exhaust system: service or inspect every 7500 miles.

Steering gear (box) & linkage, (steering damper-4x4), axle & suspension parts: service or inspect every 7500 miles.

Steering linkage ball joints & front suspension ball joints: service or inspect every 7500 miles.

22140_TITA_C0014

PRECAUTIONS

Before servicing any vehicle, please be sure to read all of the following precautions, which deal with personal safety, prevention of component damage, and important points to take into consideration when servicing a motor vehicle:

• Never open, service or drain the radiator or cooling system when the engine is hot; serious burns can occur from the steam and hot coolant.

• Observe all applicable safety precautions when working around fuel. Whenever servicing the fuel system, always work in a well-ventilated area. Do not allow fuel spray or vapors to come in contact with a spark, open flame, or excessive heat (a hot drop light, for example). Keep a dry chemical fire extinguisher near the work area. Always keep fuel in a container specifically designed for fuel storage; also, always properly seal fuel containers to avoid the possibility of fire or explosion. Refer to the additional fuel system precautions later in this section.

• Fuel injection systems often remain pressurized, even after the engine has been turned **OFF**. The fuel system pressure must be relieved before disconnecting any fuel lines. Failure to do so may result in fire and/or personal injury.

• Brake fluid often contains polyglycol ethers and polyglycols. Avoid contact with the eyes and wash your hands thoroughly after handling brake fluid. If you do get brake fluid in your eyes, flush your eyes with clean, running water for 15 minutes. If eye irritation persists, or if you have taken brake fluid internally, IMMEDIATELY seek medical assistance.

• The EPA warns that prolonged contact with used engine oil may cause a number of skin disorders, including cancer. You should make every effort to minimize your exposure to used engine oil. Protective gloves should be worn when changing oil. Wash your hands and any other exposed skin areas as soon as possible after exposure to used engine oil. Soap and water, or waterless hand cleaner should be used.

• All new vehicles are now equipped with an air bag system, often referred to as a Supplemental Restraint System (SRS) or Supplemental Inflatable Restraint (SIR) system. The system must be disabled before performing service on or around system components, steering column, instrument panel components, wiring and sensors. Failure to follow safety and disabling procedures could result in accidental air bag deployment, possible personal injury and unnecessary system repairs.

• Always wear safety goggles when working with, or around, the air bag system. When carrying a non-deployed air bag, be sure the bag and trim cover are pointed away from your body. When placing a non-deployed air bag on a work surface, always face the bag and trim cover upward, away from the surface. This will reduce the motion of the module if it is accidentally deployed. Refer to the additional air bag system precautions later in this section.

• Clean, high quality brake fluid from a sealed container is essential to the safe and proper operation of the brake system. You should always buy the correct type of brake fluid for your vehicle. If the brake fluid becomes contaminated, completely flush the system with new fluid. Never reuse any brake fluid. Any brake fluid that is removed from the system should be discarded. Also, do not allow any brake fluid to come in contact with a painted surface; it will damage the paint.

• Never operate the engine without the proper amount and type of engine oil; doing so WILL result in severe engine damage.

• Timing belt maintenance is extremely important. Many models utilize an interference-type, non-freewheeling engine. If the timing belt breaks, the valves in the cylinder head may strike the pistons, causing potentially serious (also time-consuming and expensive) engine damage. Refer to the maintenance interval charts for the recommended replacement interval for the timing belt, and to the timing belt section for belt replacement and inspection.

• Disconnecting the negative battery cable on some vehicles may interfere with the functions of the on-board computer system(s) and may require the computer to undergo a relearning process once the negative battery cable is reconnected.

• When servicing drum brakes, only disassemble and assemble one side at a time, leaving the remaining side intact for reference.

• Only an MVAC-trained, EPA-certified automotive technician should service the air conditioning system or its components.

BRAKES

GENERAL INFORMATION

PRECAUTIONS

• Certain components within the ABS system are not intended to be serviced or repaired individually.

• Do not use rubber hoses or other parts not specifically specified for and ABS system. When using repair kits, replace all parts included in the kit. Partial or incorrect repair may lead to functional problems and require the replacement of components.

• Lubricate rubber parts with clean, fresh brake fluid to ease assembly. Do not use shop air to clean parts; damage to rubber components may result.

• Use only DOT 3 brake fluid from an unopened container.

• If any hydraulic component or line is removed or replaced, it may be necessary to bleed the entire system.

• A clean repair area is essential. Always clean the reservoir and cap thoroughly before removing the cap. The slightest amount of dirt in the fluid may plug an orifice and impair the system function. Perform repairs after components have been thoroughly cleaned; use only denatured alcohol

ANTI-LOCK BRAKE SYSTEM (ABS)

to clean components. Do not allow ABS components to come into contact with any substance containing mineral oil; this includes used shop rags.

• The Anti-Lock control unit is a microprocessor similar to other computer units in the vehicle. Ensure that the ignition switch is **OFF** before removing or installing controller harnesses. Avoid static electricity discharge at or near the controller.

• If any arc welding is to be done on the vehicle, the control unit should be unplugged before welding operations begin.

BRAKES / BLEEDING THE BRAKE SYSTEM

BLEEDING PROCEDURE

BLEEDING PROCEDURE

> ❋ **CAUTION**
>
> Carefully monitor the brake fluid level at the sub tank during bleeding operation.

> ❋ **CAUTION**
>
> Fill the sub tank with new brake fluid. Make sure it is full at all times while bleeding the air out of system.

➤ Place a container under the sub tank to avoid spilling brake fluid.

➤ Do not loosen the line fittings at the ABS actuator during air bleeding.

1. Turn ignition switch OFF and disconnect ABS actuator and control unit connector or negative battery terminal.
2. Connect a transparent vinyl tube and container to air bleeder valve.
3. Fully depress brake pedal several times.
4. With brake pedal depressed, open air bleeder valve to release air.
5. Close the air bleeder valve.
6. Release brake pedal slowly.
7. Tighten air bleeder valve to: 71 inch lbs. (8 Nm).
8. Repeat steps 2 through 7 until no more air bubbles come out of air bleeder valve.
9. Bleed the brake hydraulic system air bleeder valves in the following order:
   - Right rear brake
   - Left front brake
   - Left rear brake
   - Right front brake

BRAKES / FRONT DISC BRAKES

> ❋ **CAUTION**
>
> Dust and dirt accumulating on brake parts during normal use may contain asbestos fibers from production or aftermarket brake linings. Breathing excessive concentrations of asbestos fibers can cause serious bodily harm. Exercise care when servicing brake parts. Do not sand or grind brake lining unless equipment used is designed to contain the dust residue. Do not clean brake parts with compressed air or by dry brushing. Cleaning should be done by dampening the brake components with a fine mist of water, then wiping the brake components clean with a dampened cloth. Dispose of cloth and all residue containing asbestos fibers in an impermeable container with the appropriate label. Follow practices prescribed by the Occupational Safety and Health Administration (OSHA) and the Environmental Protection Agency (EPA) for the handling, processing, and disposing of dust or debris that may contain asbestos fibers.

BRAKE CALIPER

REMOVAL & INSTALLATION

See Figure 1.

1. Before servicing the vehicle, refer to the Precautions Section.
2. Drain brake fluid as necessary.
3. Remove or disconnect the following:
   - Wheel
   - Union bolt

- Caliper-to-torque member slide pins, or remove the caliper and torque member as an assembly.
- Brake caliper

To install:

4. Install or connect the following:
   - Brake caliper, tighten torque member bolts to 155 ft. lbs. (210 Nm);
   - the caliper slide pins to 32 ft. lbs. (44 Nm)
   - Union bolt and tighten to 13 ft. lbs. (18 Nm)
5. Fill the master cylinder and bleed the brake system.
6. Install the wheels.

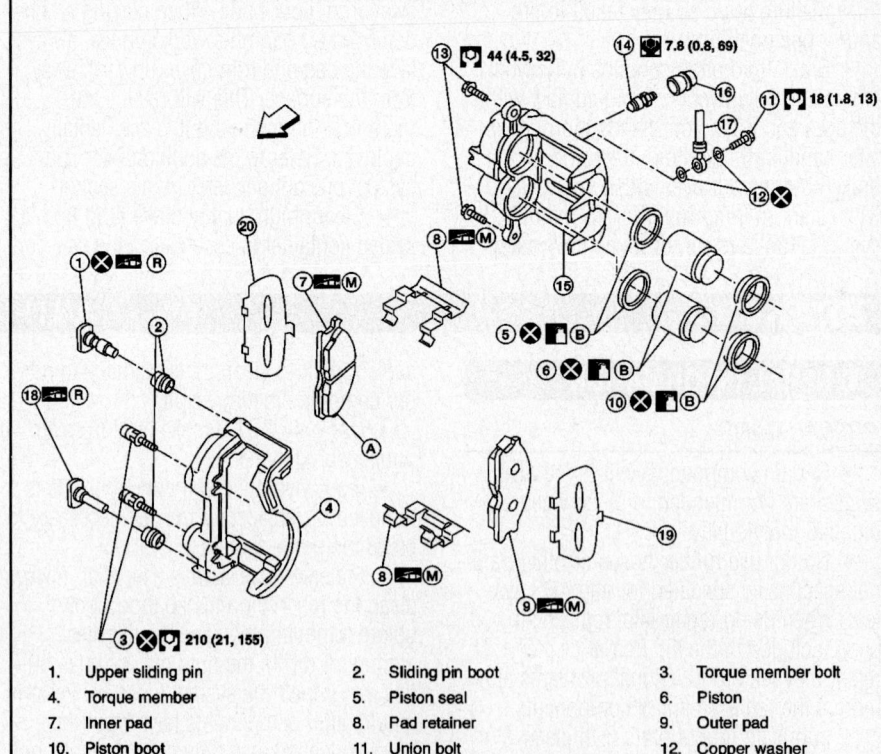

1.	Upper sliding pin	2.	Sliding pin boot	3.	Torque member bolt
4.	Torque member	5.	Piston seal	6.	Piston
7.	Inner pad	8.	Pad retainer	9.	Outer pad
10.	Piston boot	11.	Union bolt	12.	Copper washer
13.	Sliding pin bolt	14.	Bleed valve	15.	Cylinder body
16.	Cap	17.	Brake hose	18.	Lower sliding pin
19.	Outer shim	20.	Inner shim	A.	Wear indicator

22140_TITA_G0001

Fig. 1 Exploded view of the front brake caliper—Titan

DISC BRAKE PADS

REMOVAL & INSTALLATION
See Figure 2.

1. Before servicing the vehicle, refer to the Precautions Section.
2. Remove the front wheels.
3. Remove lower sliding pin bolt.
4. Suspend brake caliper with a remove and remove brake pad and shim from torque member.

To install:

✳✳ WARNING

By pushing in the pistons, brake fluid returns to master cylinder reservoir

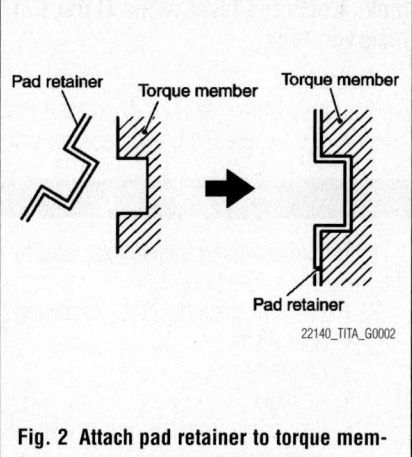

Fig. 2 Attach pad retainer to torque member

tank. Watch the level of the surface of reservoir tank.

5. Push pistons in so that the pad is firmly installed, using a suitable tool.
6. Mount the brake caliper to torque member.
7. Attach pad retainer to torque member.
8. Lubricate lower sliding pin bolt with a thin layer of silicone grease and install. Torque to 32 ft. lbs. (44 Nm).
9. Install the tires from vehicle with power tool.
10. Road test the vehicle.

BRAKES

✳✳ CAUTION

Dust and dirt accumulating on brake parts during normal use may contain asbestos fibers from production or aftermarket brake linings. Breathing excessive concentrations of asbestos fibers can cause serious bodily harm. Exercise care when servicing brake parts. Do not sand or grind brake lining unless equipment used is designed to contain the dust residue. Do not clean brake parts with compressed air or by dry brushing. Cleaning should be done by dampening the brake components with a fine mist of water, then wiping the brake components clean with a dampened cloth. Dispose of cloth and all residue containing asbestos fibers in an impermeable container with the appropriate label. Follow practices prescribed by the Occupational Safety and Health Administration (OSHA) and the Environmental Protection Agency (EPA) for the handling, processing, and disposing of dust or debris that may contain asbestos fibers.

BRAKE CALIPER

REMOVAL & INSTALLATION
See Figure 3.

1. Before servicing the vehicle, refer to the Precautions Section.
2. Remove tires from vehicle with power tool.
3. Remove the Brake hose mounting bolt and brake hose.

4. Remove the brake caliper assembly.

To install:
5. Install the brake caliper assembly and tighten mounting bolts to 24 ft. lbs. (32 Nm).
6. Install brake hose and tighten to 13 ft. lbs. (18 Nm).
7. Bleed the air from the brake caliper.
8. Install the tires from vehicle with power tool.
9. Road test the vehicle.

REAR DISC BRAKES

DISC BRAKE PADS

REMOVAL & INSTALLATION

1. Before servicing the vehicle, refer to the Precautions Section.
2. Remove tires from vehicle with power tool.
3. Remove the mounting bolt from the top mount.
4. Swing cylinder body open, and remove pads.

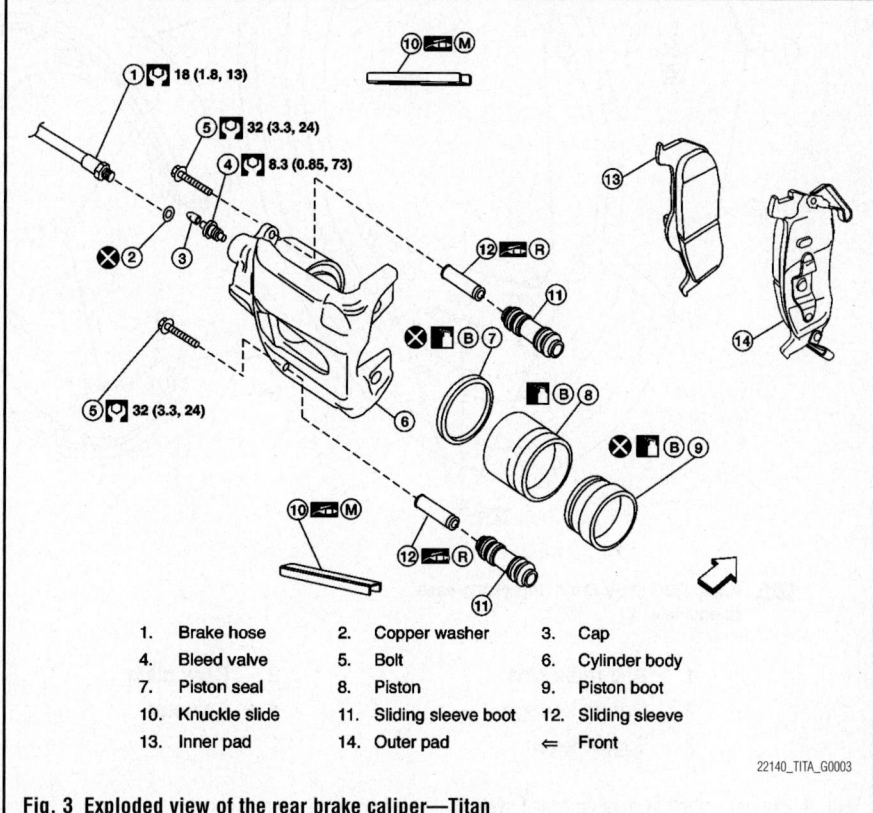

1.	Brake hose	2.	Copper washer	3.	Cap
4.	Bleed valve	5.	Bolt	6.	Cylinder body
7.	Piston seal	8.	Piston	9.	Piston boot
10.	Knuckle slide	11.	Sliding sleeve boot	12.	Sliding sleeve
13.	Inner pad	14.	Outer pad	⇐	Front

Fig. 3 Exploded view of the rear brake caliper—Titan

5. Push the piston in so that pad is firmly attached and mount cylinder body to torque member.

➡By pushing in the piston, brake fluid returns to master cylinder reservoir

tank. **Watch the level of the surface of reservoir tank.**

To install:

6. Apply Molykote (M-77) grease to the knuckle slide where the brake pads contact.

7. Install pads to cylinder body.
8. Install top mounting bolt and tighten to 24 ft. lbs. (32 Nm).
9. Check brake for drag.
10. Install tires to the vehicle.

BRAKES

PARKING BRAKE SHOES

See Figure 4.

REMOVAL & INSTALLATION

✳✳ WARNING

Clean the brakes with a vacuum dust collector to minimize the hazard of airborne particles or other materials.

➡**Remove the disc rotor only with the parking brake pedal completely in the released position.**

1. Before servicing the vehicle, refer to the Precautions Section.
2. Remove or disconnect the following:
   - Rear disc rotor
   - Return springs
   - Adjuster
   - Rear cable from the toggle lever, if necessary
   - Retainers, anti-rattle pins and shoes
3. Check shoe sliding surface on back plate for excessive wear and damage.
4. Check anti-rattle pins for excessive wear and corrosion.

PARKING BRAKE

5. Check the return springs for sagging.
6. Check the adjuster for rough operation.

To install:

7. To install, reverse removal procedure.
8. During installation, pay attention to the following:
   - Apply brake grease to the specified points during assembly
   - There is a difference between the adjusters orientation from left and right. Assemble the adjuster so the threaded part expands when rotating it

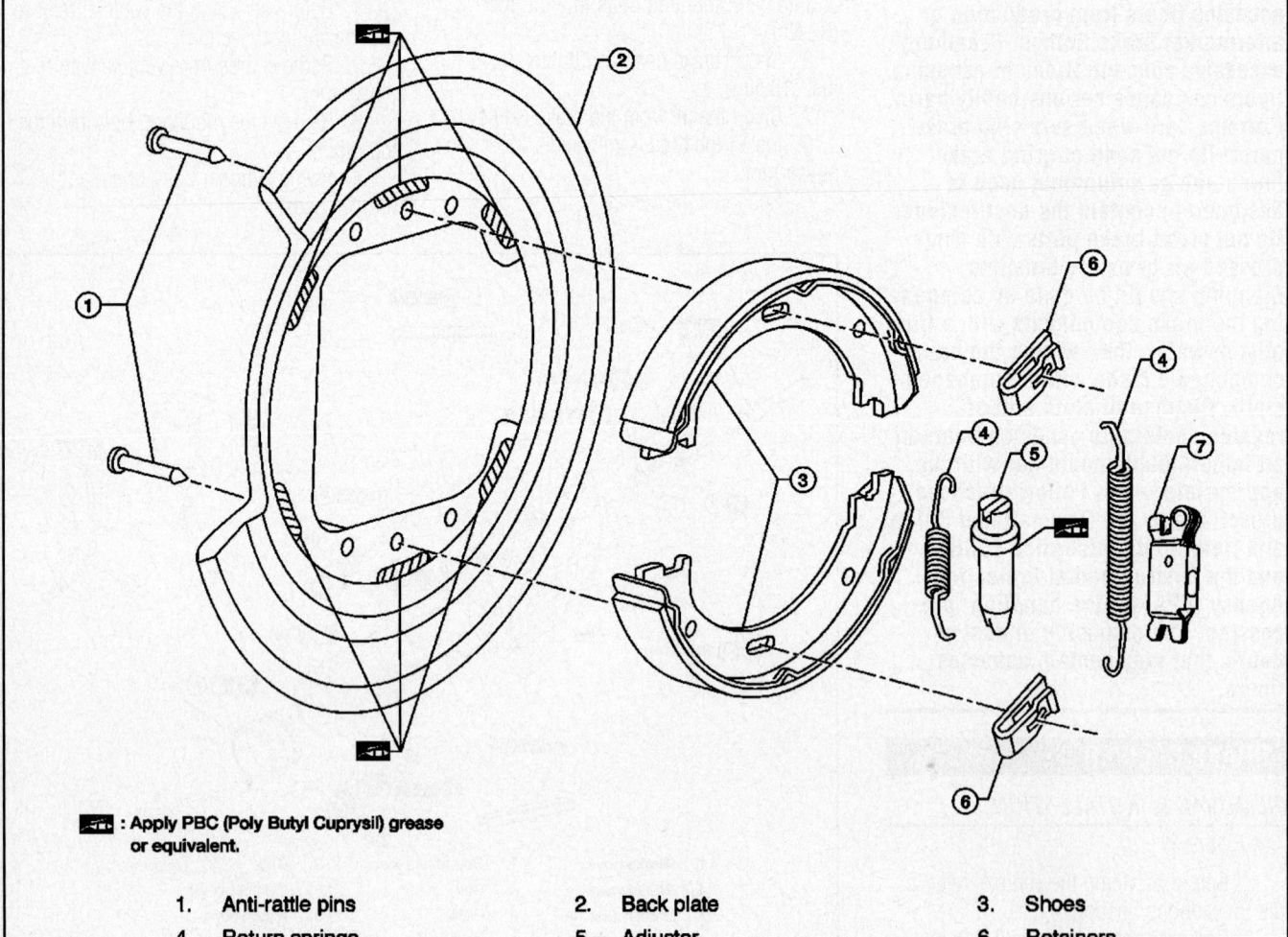

⬛ : Apply PBC (Poly Butyl Cuprysil) grease or equivalent.

1.	Anti-rattle pins	2.	Back plate	3.	Shoes
4.	Return springs	5.	Adjuster	6.	Retainers
7.	Toggle lever				

42050_TITA_G0029

Fig. 4 Exploded view of parking brake system

CHASSIS ELECTRICAL — AIR BAG (SUPPLEMENTAL RESTRAINT SYSTEM)

GENERAL INFORMATION

✳✳ CAUTION

These vehicles are equipped with an air bag system. The system must be disarmed before performing service on, or around, system components, the steering column, instrument panel components, wiring and sensors. Failure to follow the safety precautions and the disarming procedure could result in accidental air bag deployment, possible injury and unnecessary system repairs.

SERVICE PRECAUTIONS

Disconnect and isolate the battery negative cable before beginning any airbag system component diagnosis, testing, removal, or installation procedures. Allow system capacitor to discharge for two minutes before beginning any component service. This will disable the airbag system. Failure to disable the airbag system may result in accidental airbag deployment, personal injury, or death.

Do not place an intact undeployed airbag face down on a solid surface. The airbag will propel into the air if accidentally deployed and may result in personal injury or death.

When carrying or handling an undeployed airbag, the trim side (face) of the airbag should be pointing towards the body to minimize possibility of injury if accidental deployment occurs. Failure to do this may result in personal injury or death.

Replace the airbag system components with OEM replacement parts. Substitute parts may appear interchangeable, but internal differences may result in inferior occupant protection. Failure to do so may result in occupant personal injury or death.

Wear safety glasses, rubber gloves, and long sleeved clothing when cleaning powder residue from vehicle after an airbag deployment. Powder residue emitted from a deployed airbag can cause skin irritation. Flush affected area with cool water if irritation is experienced. If nasal or throat irritation is experienced, exit the vehicle for fresh air until the irritation ceases. If irritation continues, see a physician.

Do not use a replacement airbag that is not in the original packaging. This may result in improper deployment, personal injury, or death.

The factory installed fasteners, screws and bolts used to fasten airbag components have a special coating and are specifically designed for the airbag system. Do not use substitute fasteners. Use only original equipment fasteners listed in the parts catalog when fastener replacement is required.

During, and following, any child restraint anchor service, due to impact event or vehicle repair, carefully inspect all mounting hardware, tether straps, and anchors for proper installation, operation, or damage. If a child restraint anchor is found damaged in any way, the anchor must be replaced. Failure to do this may result in personal injury or death.

Deployed and non-deployed airbags may or may not have live pyrotechnic material within the airbag inflator.

Do not dispose of driver/passenger/curtain airbags or seat belt tensioners unless you are sure of complete deployment. Refer to the Hazardous Substance Control System for proper disposal.

Dispose of deployed airbags and tensioners consistent with state, provincial, local, and federal regulations.

After any airbag component testing or service, do not connect the battery negative cable. Personal injury or death may result if the system test is not performed first.

If the vehicle is equipped with the Occupant Classification System (OCS), do not connect the battery negative cable before performing the OCS Verification Test using the scan tool and the appropriate diagnostic information. Personal injury or death may result if the system test is not performed properly.

Never replace both the Occupant Restraint Controller (ORC) and the Occupant Classification Module (OCM) at the same time. If both require replacement, replace one, then perform the Airbag System test before replacing the other.

Both the ORC and the OCM store Occupant Classification System (OCS) calibration data, which they transfer to one another when one of them is replaced. If both are replaced at the same time, an irreversible fault will be set in both modules and the OCS may malfunction and cause personal injury or death.

If equipped with OCS, the Seat Weight Sensor is a sensitive, calibrated unit and must be handled carefully. Do not drop or handle roughly. If dropped or damaged, replace with another sensor. Failure to do so may result in occupant injury or death.

If equipped with OCS, the front passenger seat must be handled carefully as well. When removing the seat, be careful when setting on floor not to drop. If dropped, the sensor may be inoperative, could result in occupant injury, or possibly death.

If equipped with OCS, when the passenger front seat is on the floor, no one should sit in the front passenger seat. This uneven force may damage the sensing ability of the seat weight sensors. If sat on and damaged, the sensor may be inoperative, could result in occupant injury, or possibly death.

DISARMING THE SYSTEM

1. Before servicing the vehicle, refer to the Precautions Section.
2. Disconnect both battery cables.
3. Wait at least 3 minutes before working on the vehicle.

➡**The air bag system is designed to retain enough power to deploy the air bag for a short time after the battery has been disconnected.**

ARMING THE SYSTEM

1. Before servicing the vehicle, refer to the Precautions Section.
2. After repairs are complete, connect the negative battery cable.
3. Turn the ignition switch to the **ON** position and check the air bag warning light blinks for proper operation.

CLOCKSPRING CENTERING

See Figure 5.

1. Before servicing the vehicle, refer to the Precautions Section.
2. Align spiral cable correctly when installing steering wheel. Make sure that the spiral cable is in the neutral position.
3. The neutral position is detected by turning left 2.5 revolutions from the right end position and ending with the knob at the top.

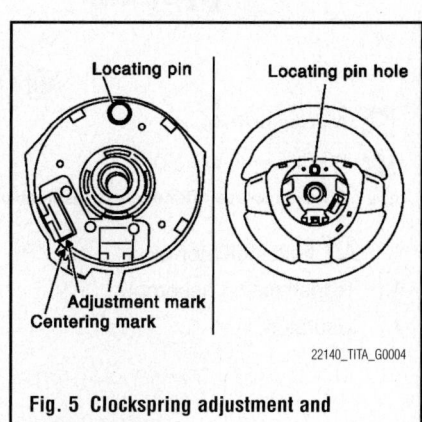

22140_TITA_G0004

Fig. 5 Clockspring adjustment and centering marks

DRIVETRAIN

AUTOMATIC TRANSMISSION ASSEMBLY

REMOVAL & INSTALLATION

2WD Models

See Figure 6.

1. Before servicing the vehicle, refer to the Precautions Section.
2. Remove or disconnect the following:
   - Negative battery cable
   - Engine cover
   - Transmission fluid indicator gauge
   - Engine splash guard
   - Exhaust front pipe
   - Center muffler
   - Rear drive shaft
   - Transmission control cable
   - Crankshaft position sensor
   - Fluid cooler tube
   - Dust cover from converter housing
3. Turning crankshaft clockwise, remove the four tightening bolts for drive plate and torque converter
4. Support the transmission with a suitable jack.
5. Remove or disconnect the following:
   - Transmission cross member
   - Air breather hose
   - Transmission assembly connector
   - Fluid indicator tube from transmission assembly
   - Transmission assembly to engine bolts
   - Transmission assembly from vehicle

To install:
6. Install or connect the following:
   - Transmission assembly into vehicle
   - Transmission assembly to engine bolts tightening to 83 ft. lbs. (113 Nm)

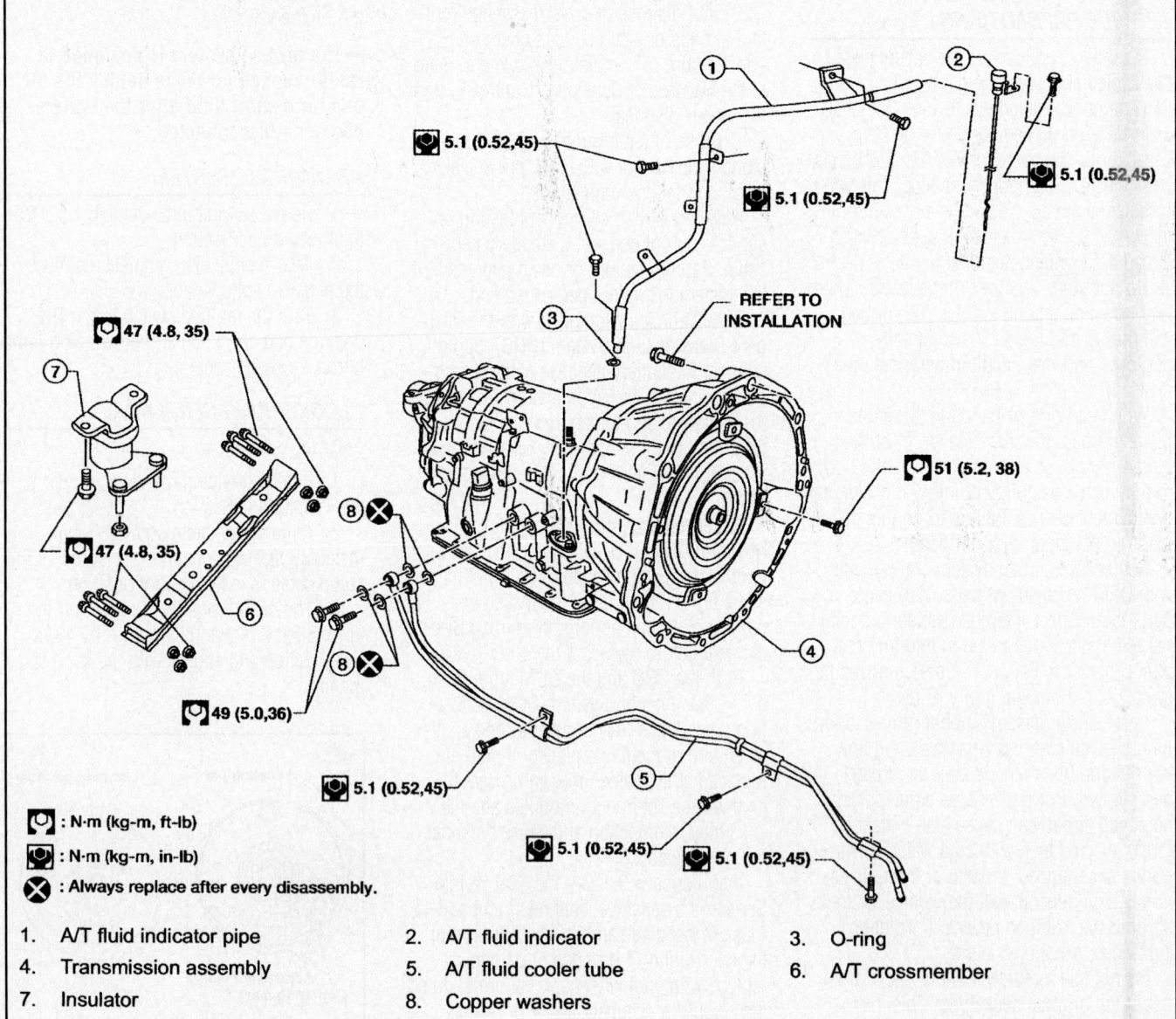

5.1 (0.52,45)

5.1 (0.52,45)

5.1 (0.52,45)

47 (4.8, 35)

47 (4.8, 35)

51 (5.2, 38)

49 (5.0,36)

REFER TO INSTALLATION

5.1 (0.52,45)

5.1 (0.52,45)

5.1 (0.52,45)

: N·m (kg-m, ft-lb)

: N·m (kg-m, in-lb)

: Always replace after every disassembly.

1. A/T fluid indicator pipe	2. A/T fluid indicator	3. O-ring
4. Transmission assembly	5. A/T fluid cooler tube	6. A/T crossmember
7. Insulator	8. Copper washers	

67170-ARMA-G39

Fig. 6 Transmission and related parts—2-wheel drive

- Fluid indicator tube to transmission assembly
- Transmission assembly connector
- Air breather hose
- Transmission cross member

7. Turning crankshaft clockwise, install the torque converter to drive plate.

➡**After torque converter is installed, rotate the crankshaft to ensure transmission rotates freely.**

8. Install or connect the following:
- Dust cover for converter housing
- Fluid cooler tube
- Crankshaft position sensor
- Transmission control cable

- Rear drive shaft
- Center muffler
- Exhaust front pipe
- Engine splash guard
- Transmission fluid indicator gauge
- Engine cover
- Negative battery cable

9. Start engine and check for leaks.

4WD Models

See Figure 7.

1. Before servicing the vehicle, refer to the Precautions Section.
2. Remove or disconnect the following:
- Disconnect the negative battery cable
- Engine cover

- Transmission fluid indicator gauge
- Engine splash guard
- Exhaust front pipe
- Center muffler
- Drive shaft
- Transmission control cable
- Crankshaft position sensor
- Fluid cooler tube
- Dust housing for torque converter

3. Turning the crankshaft clockwise, remove the four tightening bolts for drive plate and torque converter.

4. Support the transmission assembly with a suitable jack.

5. Remove the transmission cross member.

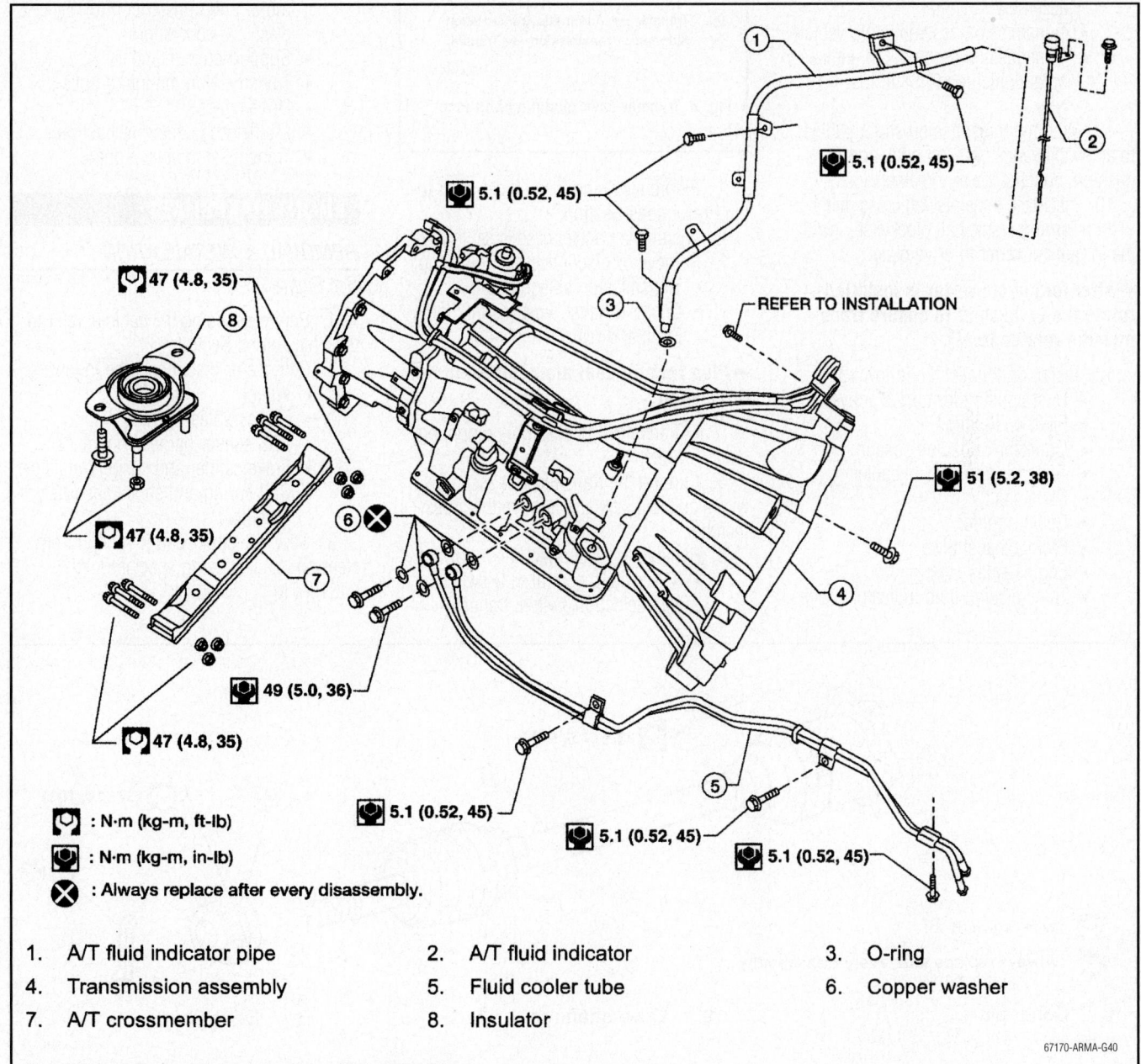

REFER TO INSTALLATION

: N·m (kg-m, ft-lb)

: N·m (kg-m, in-lb)

⊗ : Always replace after every disassembly.

1.	A/T fluid indicator pipe	2.	A/T fluid indicator	3.	O-ring
4.	Transmission assembly	5.	Fluid cooler tube	6.	Copper washer
7.	A/T crossmember	8.	Insulator		

Fig. 7 Transmission and related parts—with 4-wheel drive

67170-ARMA-G40

6. Tilt the transmission slightly to keep clearance between the body and the transmission assembly, then disconnect the air breather hose.

7. Remove or disconnect the following:
- Transmission assembly connector and transfer case connector
- Fluid indicator pipe
- Transmission assembly to engine bolts
- Transmission assembly, with transfer case attached, from vehicle
- Transmission assembly from transfer case

To install:

8. Install or connect the following:
- Transfer case to transmission assembly
- Transmission assembly into vehicle
- Transmission assembly to engine bolts tightening to 83 ft. lbs. (113 Nm)

9. With the transmission slightly tilted to allow clearance between body and transmission, connect the air breather hose.

10. Install the transmission cross member.

11. Turning crankshaft clockwise, install the torque converter to drive plate.

➡After torque converter is installed, rotate the crankshaft to ensure transmission rotates freely.

12. Install or connect the following:
- Dust housing for torque converter
- Fluid cooler tube
- Crankshaft position sensor
- Transmission control cable
- Drive shaft
- Center muffler
- Front exhaust pipe
- Engine splash guard
- Transmission fluid indicator gauge

- Engine cover
- Negative battery cable

13. Start engine and check for leaks.

TRANSFER CASE ASSEMBLY

REMOVAL & INSTALLATION

See Figure 8.

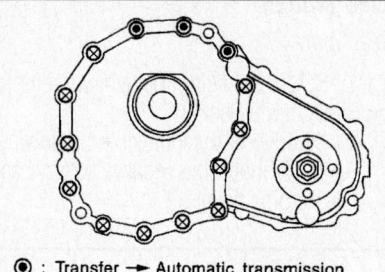

◉ : Transfer → Automatic transmission
⊗ : Automatic transmission → Transfer

67170-ARMA-G41

Fig. 8 Transfer case mounting bolt locations

1. Before servicing the vehicle, refer to the Precautions Section.

2. Ensure the transfer case is set to 2WD.

3. Remove or disconnect the following:
- Transmission splash guard
- Center exhaust pipe and muffler
- Front and rear drive shafts

➡Plug rear oil seal after removing rear drive shaft.

- Transmission assembly mounting bolts

4. Support the transmission assembly with a suitable jack and remove the crossmember.

5. Remove or disconnect the following:
- ATP switch, neutral 4LO switch, wait detection switch, transfer

motor and transfer control device electrical connectors
- Breather hoses
- Shift actuator from the extension housing
- Transfer case to transmission assembly bolts
- Transfer case assembly

To install:

6. Install or connect the following:
- Transfer case to transmission assembly bolts tightening to 26 ft. lbs. (36 Nm)
- Shift actuator
- Breather hoses
- ATP switch, neutral 4LO switch, wait detection switch, transfer motor and transfer control device electrical connectors
- Support crossmember
- Transmission mounting bolts
- Drive shafts
- Muffler and center exhaust pipe
- Transmission splash guard

FRONT HALFSHAFT

REMOVAL & INSTALLATION

See Figure 9.

1. Before servicing the vehicle, refer to the Precautions Section.

2. Remove or disconnect the following:
- Wheel
- Engine splash guard
- ABS sensor harness on knuckle
- Brake caliper and suspend it aside
- Coil spring and shock absorber assembly

3. Separate upper ball joint stud from steering knuckle using special tool J-24319-01.

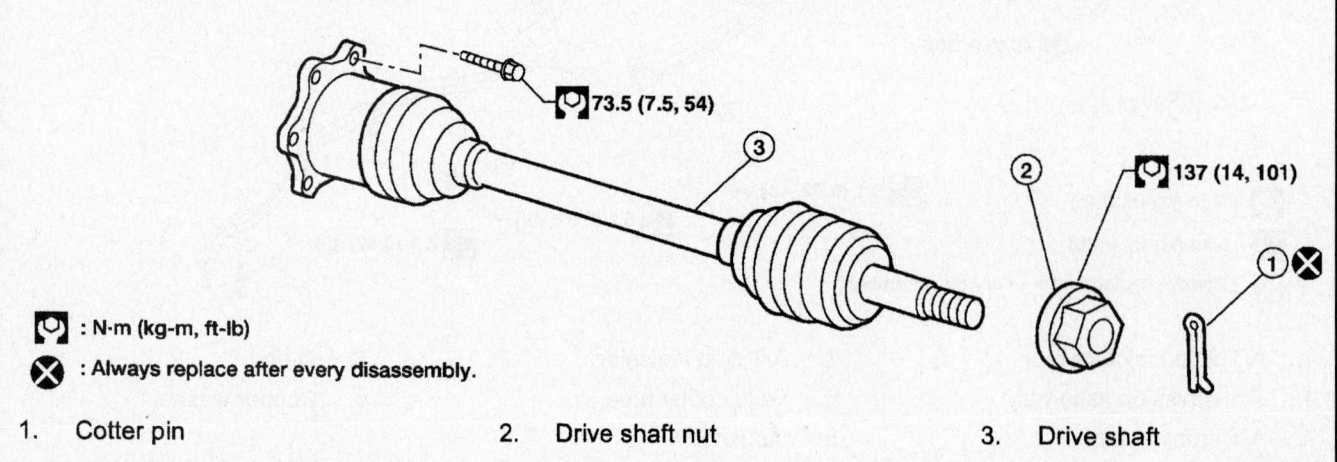

🔧 73.5 (7.5, 54)

🔧 137 (14, 101)

🔧 : N·m (kg-m, ft-lb)

⊗ : Always replace after every disassembly.

1. Cotter pin 2. Drive shaft nut 3. Drive shaft

67170-ARMA-G42

Fig. 9 Front halfshaft

4. Remove or disconnect the following:
- Cotter pin and halfshaft nut
- Halfshaft from front differential
- Halfshaft from hub and bearing assembly

To install:

5. Install or connect the following:
- Halfshaft into hub
- Halt shaft into front differential
- Halfshaft nut and tighten to 101 ft. lbs. and replace cotter pin
- Upper ball joint to steering knuckle
- Coil spring and shock absorber assembly
- Brake caliper
- ABS sensor
- Engine splash guard
- Wheel

CV-JOINTS OVERHAUL

Inner CV-Joint

See Figure 10.

1. Before servicing the vehicle, refer to the Precautions Section.
2. Remove the halfshaft from the vehicle.
3. Mount the halfshaft in a vise.

4. Remove the dust boot bands.
5. Remove the stopper ring with a flat-bladed screwdriver or suitable tool.
6. Remove the snap ring.
7. Disassemble the cage, ball and inner race assembly and dust boot for cleaning and inspection.

To install:

➡**Discard old dust boot, dust boot bands and snap ring and use new ones for assembly.**

8. Wrap the serrated part of the half-shaft with tape.
9. Install new dust boot and band onto halfshaft.
10. Remove tape from serrated part of halfshaft.
11. Install the cage, ball and inner race assembly.
12. Install the new snap ring.
13. Insert 4.50–5.3 oz of genuine NISSAN grease or equivalent onto the housing and also onto halfshaft.
14. Install the stopper ring onto the housing.
15. Install the dust boot into the grooves on joint sub-assembly.
16. Secure the big and small ends of the dust boot using new boot bands.

Outer CV-Joint

See Figure 11.

1. Before servicing the vehicle, refer to the Precautions Section.
2. Remove the halfshaft from the vehicle.
3. Mount halfshaft in a vise.
4. Remove the dust boot bands and dust boot from joint sub-assembly.
5. Insert a suitable puller into the threaded part of the halfshaft. Pull the joint sub-assembly off of the halfshaft as shown in figure.
6. Remove dust boot and circlip from halfshaft for cleaning and inspection.

To install:

➡**Discard old dust boot, dust boot bands and circlip and use new ones for assembly.**

7. Insert genuine NISSAN grease or equivalent into the joint sub-assembly until grease oozes from the ball groove and serration hole.
8. Wrap the serrated part of the half-shaft with tape.
9. Install new dust boot and band onto halfshaft.
10. Remove tape from serrated part of the halfshaft.

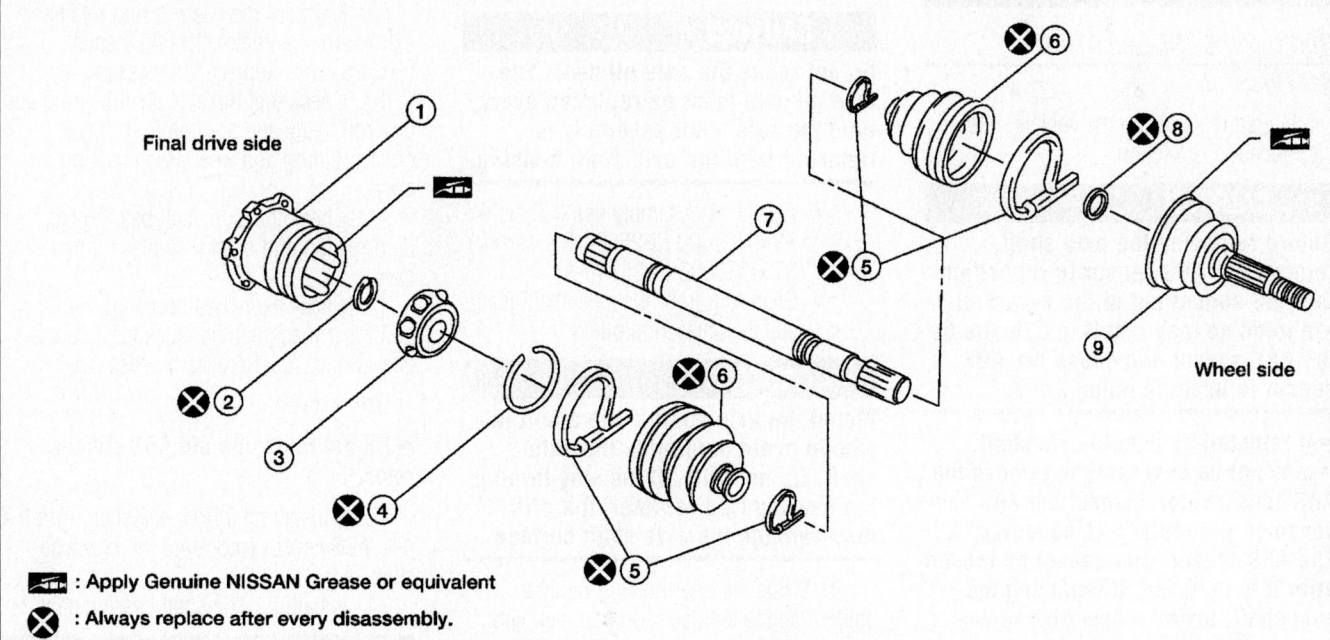

Final drive side

Wheel side

◤ : Apply Genuine NISSAN Grease or equivalent

✖ : Always replace after every disassembly.

1.	Housing	2.	Snap ring	3.	Ball cage, steel ball, iiner race assembly
4.	Stopper ring	5.	Boot band	6.	Boot
7.	Shaft	8.	Circlip	9.	Joint sub-assembly

67170-ARMA-G43

Fig. 10 Front halfshaft—exploded view

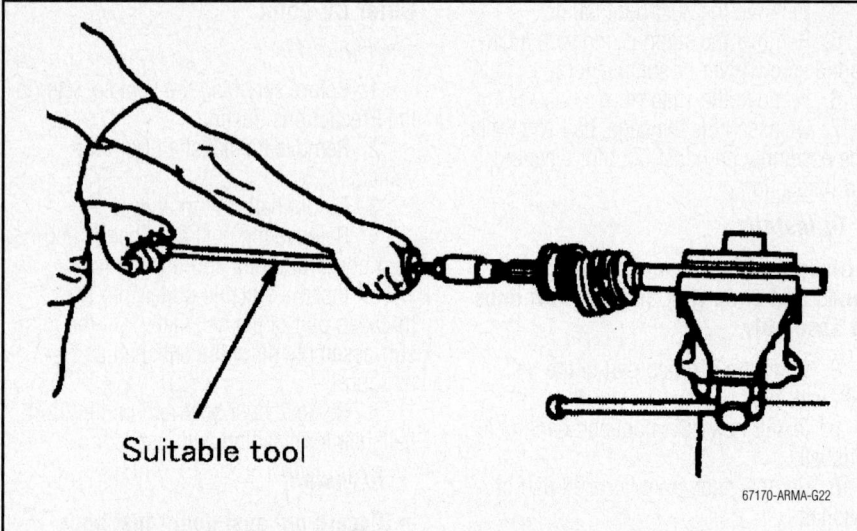

Suitable tool

67170-ARMA-G22

Fig. 11 Using a suitable puller to remove joint sub-assembly.

11. Press-fit the new circlip to the half-shaft.

12. Insert 5.1–5.8 oz of genuine NISSAN grease or equivalent into the joint sub-assembly and large end of boot.

13. Install the dust boot into the grooves on the joint sub-assembly.

14. Secure the big and small ends of the dust boot using new boot bands.

REAR AXLE SHAFT, BEARING & SEAL

REMOVAL & INSTALLATION

See Figure 12.

1. Before servicing the vehicle, refer to the Precautions Section.

✳✳ CAUTION

Before removing the axle shaft, remove the ABS sensor to reposition the ABS sensor out of the way. Failure to do so may result in damage to the ABS sensor and cause the ABS sensor to become inoperative.

➡ If reinstalling the old axle shaft, it may not be necessary to remove the ABS sensor rotor. Inspect the ABS sensor rotor and replace as necessary. The ABS sensor rotor cannot be reused after it is removed. If replacing the axle shaft, install a new ABS sensor rotor on the new axle shaft.

2. Before servicing the vehicle, refer to the Precautions Section.

3. Remove or disconnect the following:
- ABS sensor
- Rear brake rotor

- Parking brake assembly from the back plate
- Four axle shaft bearing cage nuts and lock washers

✳✳ CAUTION

The axle shaft bearing cup may stay in place in the axle shaft housing. Remove the cup carefully so as not to damage the inner surface of the axle shaft housing.

✳✳ CAUTION

Do not reuse the axle oil seal. The axle oil seal must be replaced every time the axle shaft assembly is removed from the axle shaft housing.

- Axle shaft assembly using Tools (A) KV40101000 (J-25604-01) and (B) ST36230000 (J-25840-A)
- Snap ring from the axle shaft using suitable snap ring pliers

✳✳ CAUTION

Mount the axle shaft using a soft jaw vise to avoid damaging the axle shaft. Do not drill all the way through the bearing ring retainer, the drill may damage the axle shaft surface.

4. Strike the bearing ring retainer using a suitable chisel and hammer, with the chisel positioned across the drilled hole. Break the bearing ring retainer to remove it.

✳✳ CAUTION

Do not heat or cut the axle shaft bearing or bearing ring retainer with

a torch during removal, doing so will damage the axle shaft.

5. Remove or disconnect the following:
- Axle shaft bearing cage studs using a suitable hammer or press

✳✳ CAUTION

Do not tighten the Tool against the axle shaft. Do not heat or cut the axle shaft bearing or bearing ring retainer with a torch during removal, doing so will damage the axle shaft.

- Axle shaft bearing off of the axle shaft using Tool ST30031000 and a suitable press

✳✳ CAUTION

Do not reuse the axle oil seal. The axle oil seal must be replaced every time the axle shaft assembly is removed from the axle shaft housing.

- Axle oil seal and discard
- wheel bearing cage

6. Clean and remove all nicks and burrs.

7. Check for straightness and distortion. Replace if necessary.

8. Inspect machined surfaces for evidence of overheating, damage and wear. Replace if necessary.

9. Measure the bearing ring retainer axle journal diameter: 1.5640 inches (39.726 mm). Replace if necessary.

10. Check that the axle shaft bearing and cup roll freely and are free from noise, cracks, pitting and wear. Replace if necessary.

11. Check the axle shaft bearing cage for deformation and cracks. Replace if necessary.

12. Check axel shaft housing exterior and inner machined surfaces for deformation and cracks. Replace if necessary.

To install:

➡ Do not reuse the old ABS sensor rotor.

13. If installing a new axle shaft, install a new ABS sensor rotor onto the new axle shaft.

14. Install the axle shaft bearing cage.

✳✳ CAUTION

Do not reuse the axle oil seal. The axle oil seal must be replaced every time the axle shaft assembly is removed from the axle shaft housing.

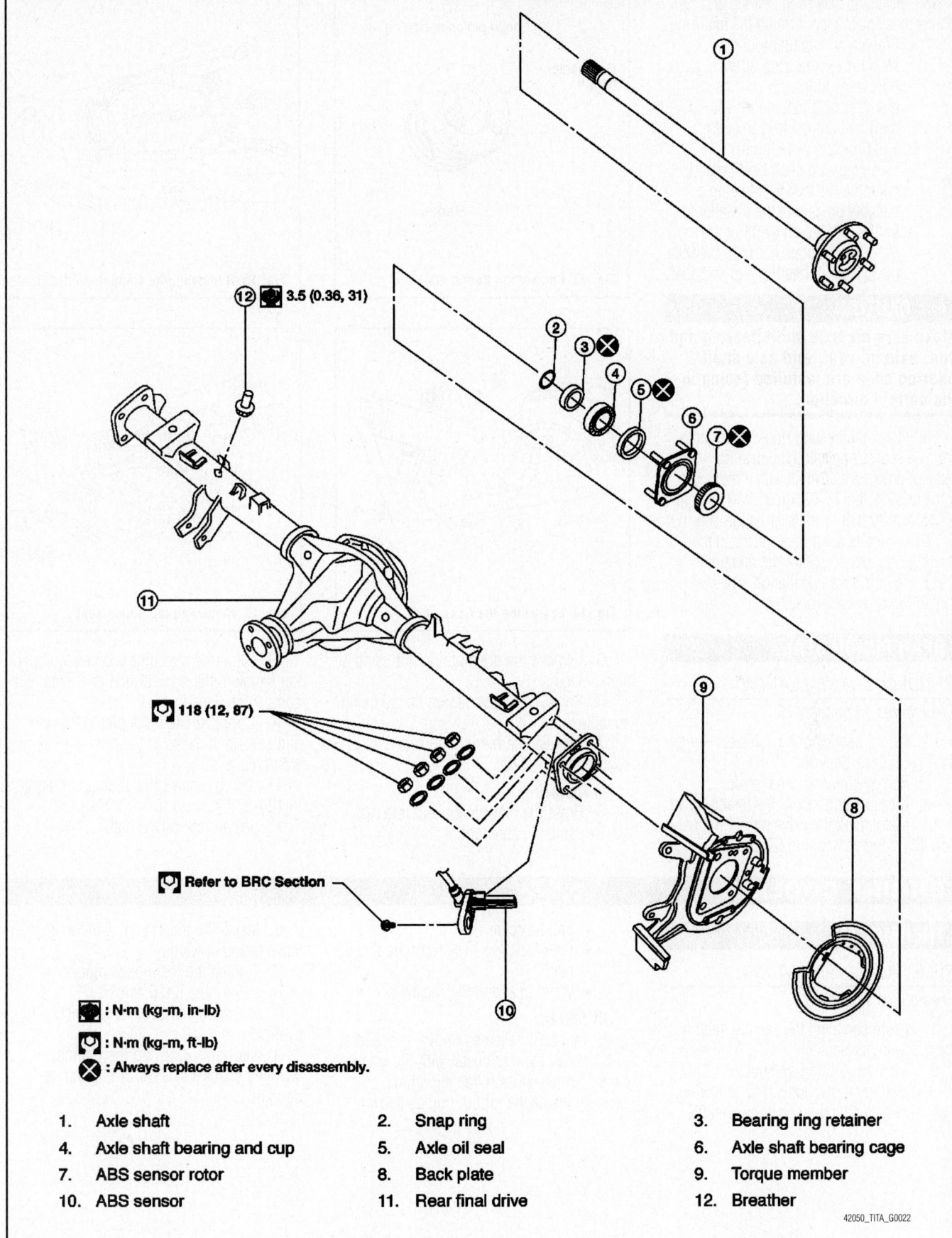

3.5 (0.36, 31)

118 (12, 87)

Refer to BRC Section

: N·m (kg-m, in-lb)

: N·m (kg-m, ft-lb)

: Always replace after every disassembly.

1.	Axle shaft	2.	Snap ring	3.	Bearing ring retainer
4.	Axle shaft bearing and cup	5.	Axle oil seal	6.	Axle shaft bearing cage
7.	ABS sensor rotor	8.	Back plate	9.	Torque member
10.	ABS sensor	11.	Rear final drive	12.	Breather

42050_TITA_G0022

Fig. 12 View of axle shaft, bearing, seal and components

15. Install the axle shaft bearing and cup on the axle shaft by performing the following:
- Prepare an installer tool from a steel tube measuring 30 inches (762 mm) long with an outside diameter of 2.125 inches (53.98 mm) and an inside diameter of 1.625 inches (41.28 mm).
- Press the axle shaft bearing and cup onto the axle shaft using a suitable press and the installer tool, until a .0015 inch (0.038 mm) feeler gauge does not fit in between the axle shaft bearing cup and seat.

✳✳ CAUTION

Make sure the axle shaft bearing and cup, axle oil seal, and axle shaft bearing cage are installed facing in the correct direction.

16. Install the bearing ring retainer onto the axle shaft by pressing the bearing ring retainer onto the axle shaft with a minimum force of 6992 lbs. (3172 kg, 31,100 N) until a .0015 inch (0.038 mm) feeler gauge does not fit between the bearing inner race and the bearing ring retainer in at least one point.

17. To complete installation, reverse remaining removal procedure.

REAR PINION SEAL

REMOVAL & INSTALLATION

See Figures 13 through 16.

1. Before servicing the vehicle, refer to the Precautions Section.
2. Remove the rear drive shaft.
3. Measure and record the total preload.
4. Matchmark the drive pinion to position 'B' on the companion flange.

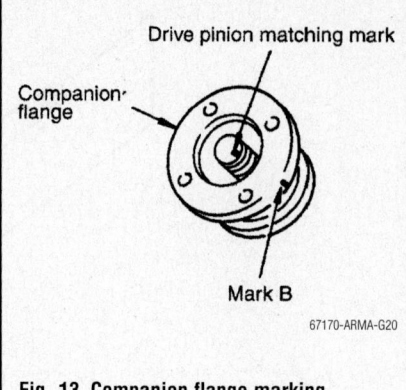

Fig. 13 Companion flange marking

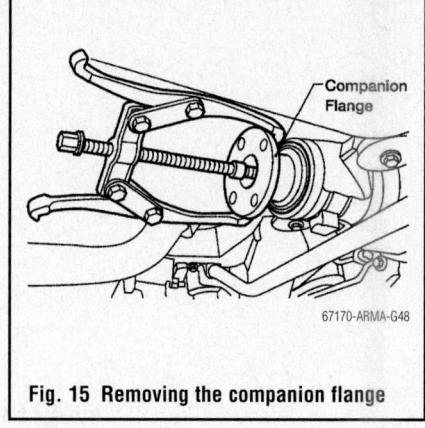

Fig. 15 Removing the companion flange

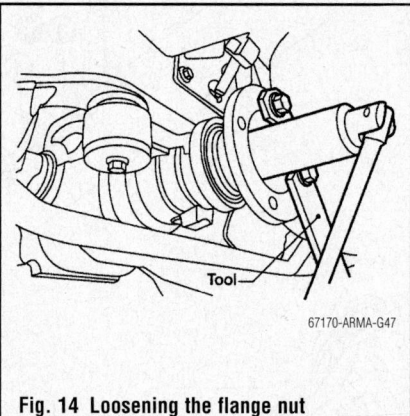

Fig. 14 Loosening the flange nut

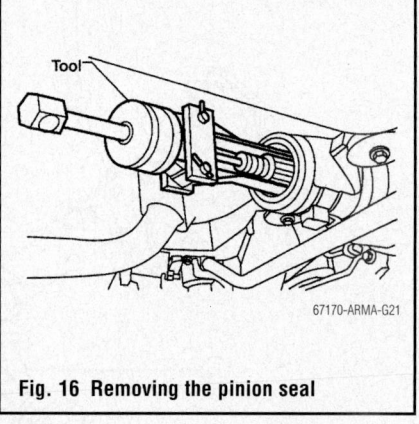

Fig. 16 Removing the pinion seal

5. Remove the drive pinion nut using suitable tool.
6. Remove the companion flange using suitable tool.
7. Remove the rear pinion seal using special tool J-34286.

To install:

8. Press the rear pinion seal into the carrier using suitable tool.

9. Align the matchmark on the companion flange to the drive pinion and install the companion flange.
10. Lubricate the drive pinion threads and seating surfaces of the drive pinion nut with grease.
11. Using a new drive pinion nut, tighten to 124-274 ft. lbs. (167-372 Nm).
12. Install rear drive shaft.

ENGINE COOLING

THERMOSTAT

REMOVAL & INSTALLATION

See Figures 17 and 18.

1. Before servicing the vehicle, refer to the Precautions Section.
2. Drain the cooling system.
3. Remove or disconnect the following:
- Air duct and resonator assembly
- Engine cover
- Water suction hose from the water inlet
- Water inlet and thermostat

To install:

4. To install, reverse removal procedure.
5. Install the thermostat with the whole circumference of each flange part fit securely inside the rubber ring as shown.

6. Install the thermostat with the jiggle valve facing upwards.
7. Tighten the thermostat mounting bolts to 15 ft. lbs. (20.6 Nm).
8. Install engine coolant and bleed the system.
9. Start and warm up the engine. Visually check for leaks of the engine coolant.

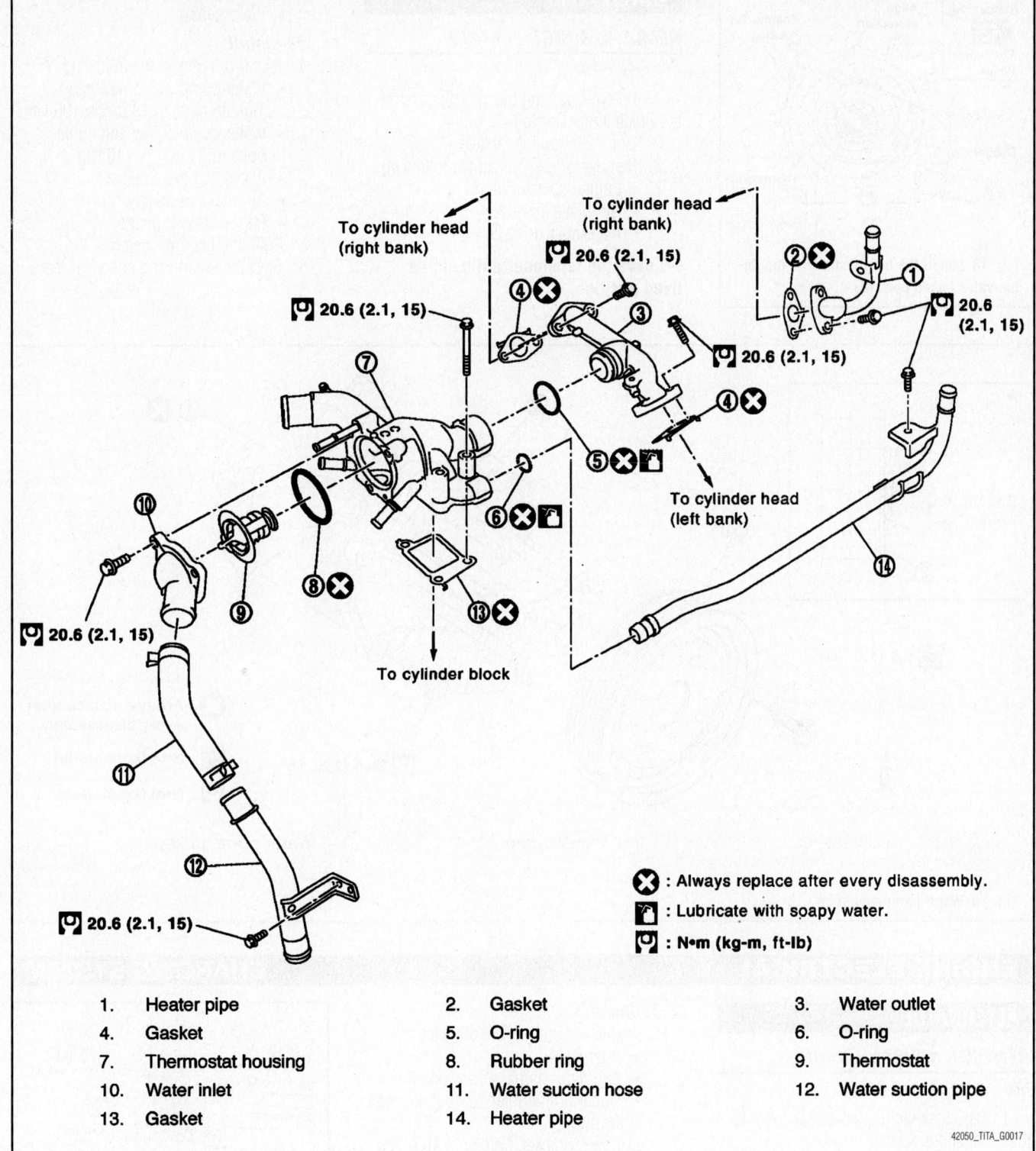

To cylinder head (right bank)

To cylinder head (right bank)

20.6 (2.1, 15)

20.6 (2.1, 15)

20.6 (2.1, 15)

20.6 (2.1, 15)

To cylinder head (left bank)

To cylinder block

20.6 (2.1, 15)

20.6 (2.1, 15)

: Always replace after every disassembly.

: Lubricate with soapy water.

: N•m (kg-m, ft-lb)

1.	Heater pipe	2.	Gasket	3.	Water outlet
4.	Gasket	5.	O-ring	6.	O-ring
7.	Thermostat housing	8.	Rubber ring	9.	Thermostat
10.	Water inlet	11.	Water suction hose	12.	Water suction pipe
13.	Gasket	14.	Heater pipe		

42050_TITA_G0017

Fig. 17 Exploded view of thermostat and components

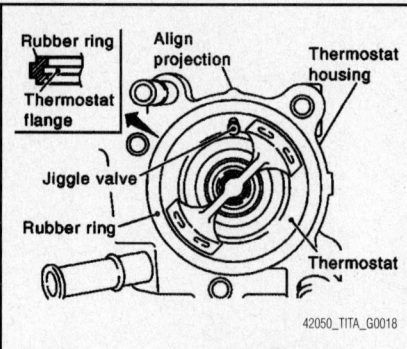

Fig. 18 Install the thermostat with the jiggle valve facing upwards.

WATER PUMP

REMOVAL & INSTALLATION

See Figure 19.

1. Before servicing the vehicle, refer to the Precautions Section.
2. Drain the cooling system.
3. Remove or disconnect the following:
   - Engine splash guard
   - Air intake assembly
   - Accessory drive belt

➡**Leave the tensioner pulley in its fixed position.**

- Water pump pulley
- Water pump

To install:

4. Install or connect the following:
   - Water pump with a new gasket. Tighten bolts to 18 ft. lbs. (25 Nm).
   - Water pump pulley and tighten bolts to 87 in. lbs. (10 Nm).
   - Accessory drive belt
   - Air intake assembly
   - Engine splash guard
5. Refill the cooling system.
6. Start the engine and check for leaks.

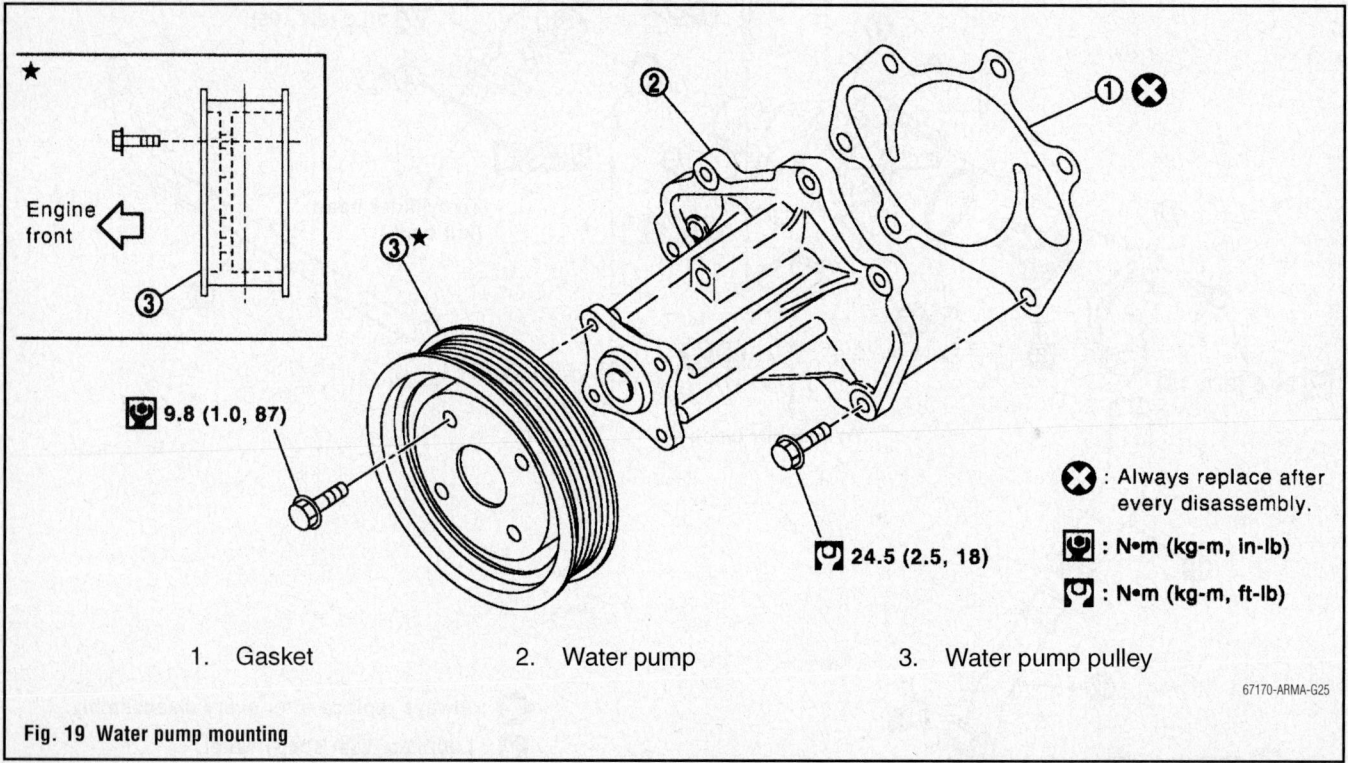

1. Gasket
2. Water pump
3. Water pump pulley

Fig. 19 Water pump mounting

ENGINE ELECTRICAL

ALTERNATOR

REMOVAL & INSTALLATION

See Figure 20.

1. Before servicing the vehicle, refer to the Precautions Section.
2. Remove or disconnect the following:
   - Negative battery cable
   - Fan shroud
   - Drive belt
   - Lower alternator bracket
   - Alternator upper bolt
   - Alternator harness connectors
   - Alternator

To install:

3. Install or connect the following:
   - Alternator
   - Alternator harness connectors
   - Upper bolt, tighten to 48 ft. lbs. (65 Nm)
   - Lower bracket, tighten to 16 ft. lbs (22 Nm)
   - Drive belt
   - Fan shroud
   - Negative battery cable

CHARGING SYSTEM

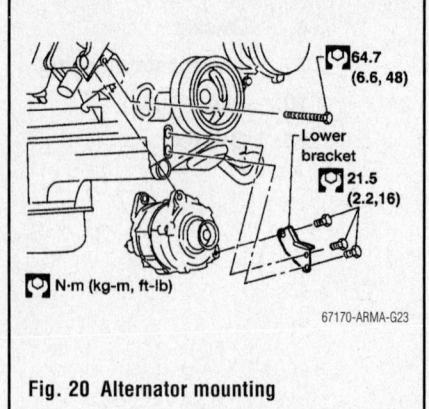

Fig. 20 Alternator mounting

ENGINE ELECTRICAL — IGNITION SYSTEM

FIRING ORDER

See Figure 21.

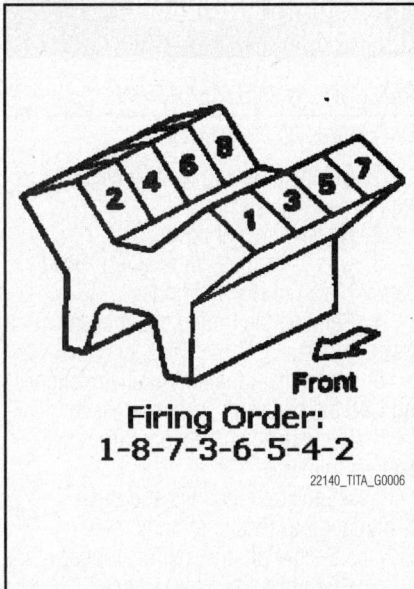

**Firing Order:
1-8-7-3-6-5-4-2**

22140_TITA_G0006

Fig. 21 5.6 (VK56DE) Engine firing order

IGNITION COIL

REMOVAL & INSTALLATION

See Figure 22.

1. Before servicing the vehicle, refer to the Precautions Section.

2. Remove the engine room cover using power tool.
3. Disconnect the harness connector from the ignition coil.
4. Remove the ignition coil.

To install:

5. Install and tighten the ignition coil to 85 inch. lbs. (10 Nm).
6. Install the engine room cover.

IGNITION TIMING

ADJUSTMENT

The ignition timing is controlled by the Engine Control Module (ECM). No adjustment is necessary or possible.

SPARK PLUGS

REMOVAL & INSTALLATION

See Figure 22.

1. Before servicing the vehicle, refer to the Precautions Section.
2. Remove the engine room cover using power tool.
3. Disconnect the harness connector from the ignition coil.
4. Remove the ignition coil.
5. Remove the spark plug.

To install:

6. Install the spark plug and tighten to 18 ft. lbs. (25 Nm).
7. Install and tighten the ignition coil to 85 inch. lbs. (10 Nm).
8. Install the engine room cover.

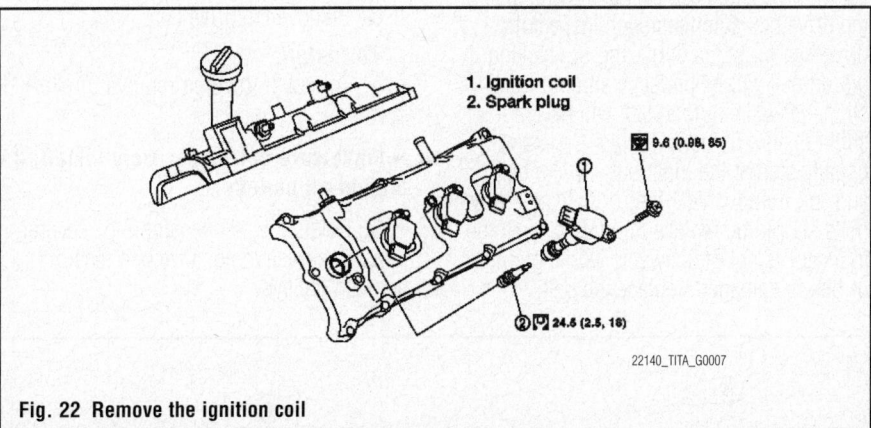

1. Ignition coil
2. Spark plug

9.6 (0.98, 85)

24.5 (2.5, 18)

22140_TITA_G0007

Fig. 22 Remove the ignition coil

ENGINE ELECTRICAL — STARTING SYSTEM

STARTER

REMOVAL & INSTALLATION

See Figure 23.

1. Before servicing the vehicle, refer to the Precautions Section.
2. Remove or disconnect the following:
   - Negative battery cable
   - Intake manifold
   - Starter harness connectors
   - Two starter bolts using power tools
   - Starter

To install:

3. To install, reverse removal procedure and note the following:
4. Tighten the starter mounting bolts to 34 ft. lbs. (47 Nm).
5. Connect the negative battery cable.

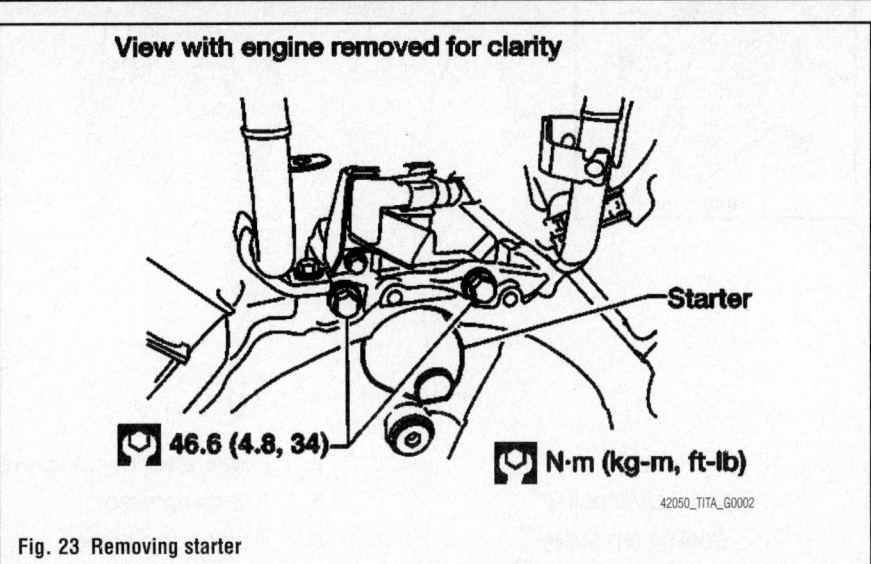

View with engine removed for clarity

Starter

46.6 (4.8, 34)

N·m (kg-m, ft-lb)

42050_TITA_G0002

Fig. 23 Removing starter

ENGINE MECHANICAL

➡ Disconnecting the negative battery cable may interfere with the functions of the on board computer systems and may require the computer to undergo a relearning process, once the negative battery cable is reconnected.

ACCESSORY DRIVE BELTS

ACCESSORY BELT ROUTING
See Figure 24.

INSPECTION

Remove the air duct and resonator assembly when inspecting drive belt. Make sure that indicator (single line notch) of each auto tensioner is within the allowable working range "A" (between three line notches) as shown in illustration. The indicator notch is located on the moving side of the drive belt auto tensioner. Inspect the drive belt for signs of glazing or cracking. A glazed belt will be perfectly smooth from slippage, while a good belt will have a slight texture of fabric visible. Cracks will usually start at the inner edge of the belt and run outward. All worn or damaged drive belts should be replaced immediately. If the indicator is out of allowable working range or belt is damaged, replace the belt.

ADJUSTMENT

There is no manual drive belt tension adjustment. The drive belt tension is automatically adjusted by the drive belt auto tensioner.

REMOVAL & INSTALLATION

> ✳✳ CAUTION
>
> **Avoid placing hand in a location where pinching may occur if the holding tool accidentally comes off.**

1. Remove the air duct and resonator assembly. Remove the air duct and resonator assembly.
2. Install Tool (J-46535) on drive belt auto tensioner pulley bolt, move in the direction of arrow (loosening direction of tensioner) as shown in illustration
3. Remove the drive belt.

To install:

4. To install, reverse removal procedure.

➡ **Make sure belt is securely installed around all pulleys.**

5. Rotate the crankshaft pulley several turns clockwise to equalize belt tension between pulleys.

6. Make sure belt tension is within the allowable working range, using the indicator notch on the drive belt auto tensioner.

CAMSHAFT AND VALVE LIFTERS

REMOVAL & INSTALLATION
See Figures 25 through 27.

1. Before servicing the vehicle, refer to the Precautions Section.
2. Remove rocker cover.
3. Obtain compression on Top Dead Center (TDC) of No. 1 cylinder.
4. Remove the timing chain case cover.
5. Matchmark the timing chain, aligning with the camshaft sprocket marks.
6. Remove chain tensioner from left bank as follows:
   a. Squeeze end clips and push plunger into tensioner body.
   b. Secure plunger using stopper pin.
   c. Remove chain tensioner.
7. Remove chain tensioner from right bank as follows:
   a. Remove chain tensioner cover using special tool J-37228.
   b. Squeeze end clips and push plunger into tensioner body.

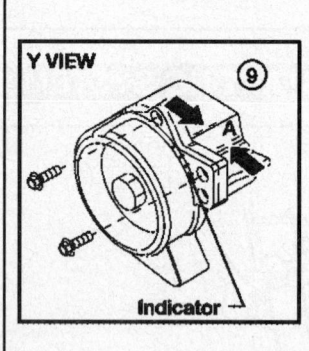

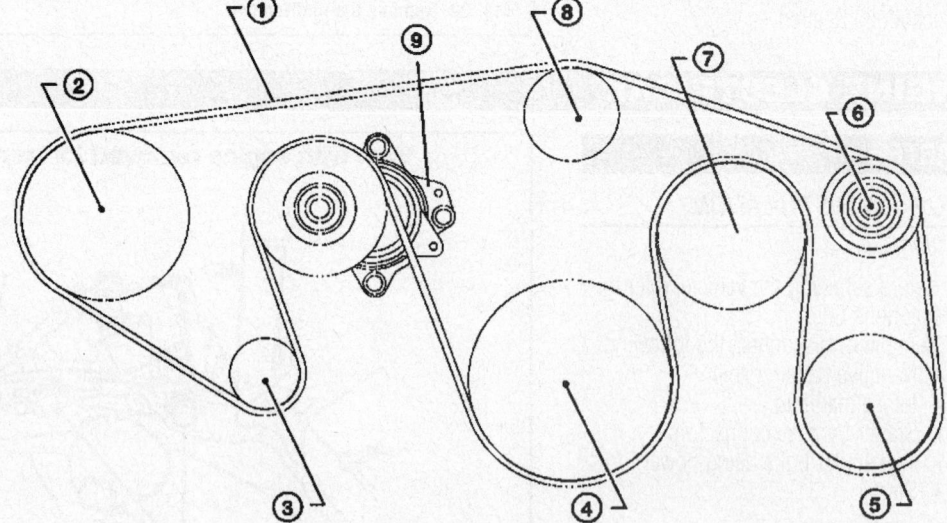

1.	Drive belt	2.	Power steering pump pulley	3.	Generator pulley
4.	Crankshaft pulley	5.	A/C compressor	6.	Idler pulley
7.	Cooling fan pulley	8.	Water pump pulley	9.	Drive belt auto tensioner

42050_TITA_G0004

Fig. 24 Accessory drive belt routing

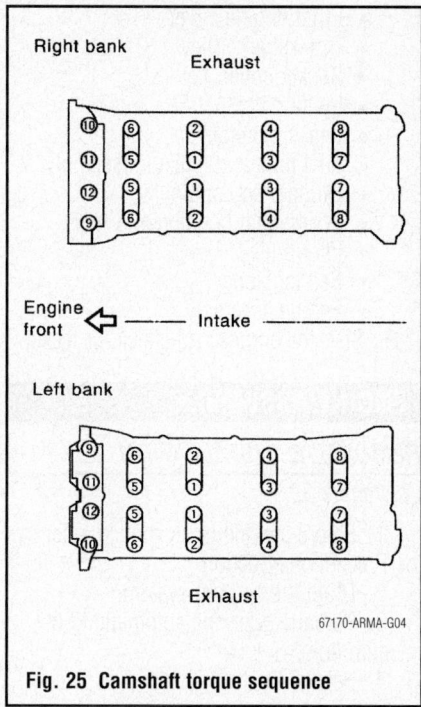

Fig. 25 Camshaft torque sequence

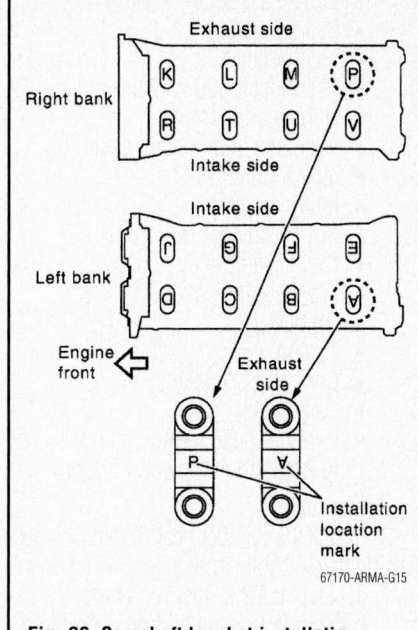

Fig. 26 Camshaft bracket installation markings

c. Secure plunger using stopper pin.

d. Remove chain tensioner.

8. With camshaft locked with a wrench, loosen bolts to remove camshaft sprocket.

9. Remove front cover bolts.

10. Remove camshaft brackets, removing bolts in reverse order shown in figure.

11. Remove the camshaft.

12. Remove valve lifters.

To install:

13. Install valve lifters.

14. Install camshaft, refer to table for correct placement.

15. Install camshaft brackets as follows:

a. Refer to location mark on upper surface of bracket.

b. Installation mark should be correctly read when viewed from intake side.

16. Install camshaft bracket #1 as follows:

a. Apply the liquid gasket to bracket and backside of front cover as shown in figure.

b. Carefully position and mount camshaft bracket #1.

c. Temporarily tighten the front cover bolts.

17. Tighten fixing bolts for camshaft brackets as follows:

a. Step 1: Bolts 9-12: 17 in. lbs. (1.9 Nm)

b. Step 2: Bolts 1-8: 17 in. lbs. (1.9 Nm)

c. Step 3: All bolts: 52 in. lbs. (5.9 Nm)

d. Step 4: All bolts: 92 in. lbs. (10 Nm)

18. Tighten front cover bolts to 8 ft. lbs. (11 Nm)

19. Install camshaft sprocket as follows:

a. Install the camshaft sprocket aligning matchmarks with timing chain. Align camshaft sprocket key groove with dowel pin on camshaft front edge.

b. Temporarily tighten bolts.

c. Lock the camshaft with a wrench and tighten the bolts.

20. Install chain tensioner as shown:

a. Install chain tensioner, compress plunger and hold with stopper pin.

b. Tighten chain tensioner bolts to 61 in. lbs. (7 Nm)

c. Remove stopper pin, release plunger and apply tension to timing chain.

d. Install chain tensioner front cover (Right-hand bank only) and tighten bolts to 80 in. lbs. (9 Nm).

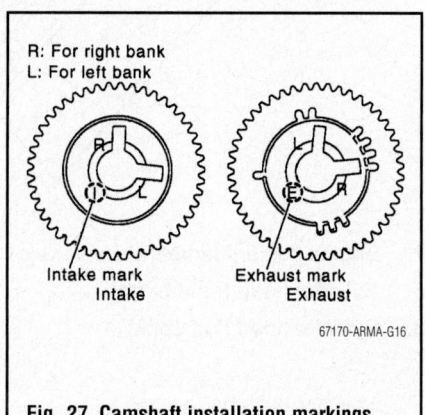

Fig. 27 Camshaft installation markings

21. Install the timing chain cover.

22. Install rocker cover.

CRANKSHAFT FRONT SEAL

REMOVAL & INSTALLATION

See Figures 28 and 29.

1. Before servicing the vehicle, refer to the Precautions Section.

2. Remove the engine.

3. Remove the crankshaft pulley.

a. Loosen the crankshaft pulley bolts using a hammer handle to secure the crankshaft.

b. Remove the crankshaft pulley from the crankshaft using tool.

c. Remove the crankshaft pulley using suitable tool. Set the bolts in the two bolt holes 0.04 inch.(M6 x 1.0 mm) on the front surface.

4. Remove the front oil seal using suitable tool.

To install:

5. Apply new engine oil to both the oil seal lip and dust seal lip of the new front oil seal.

6. Install the front oil seal.

7. Install the front oil seal so that each seal lip is oriented as shown.

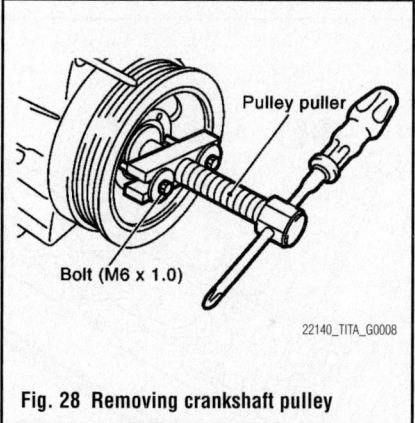

Fig. 28 Removing crankshaft pulley

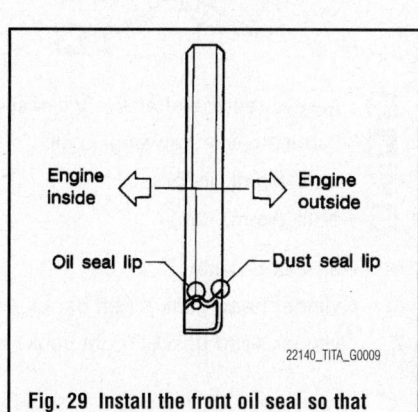

Fig. 29 Install the front oil seal so that each seal lip is oriented as shown

8. Press-fit until the height of the front oil seal is level with the mounting surface using suitable tool.

9. Installation of the remaining components is in the reverse order of removal.

CYLINDER HEAD

REMOVAL & INSTALLATION

See Figures 30 and 31.

1. Before servicing the vehicle, refer to the Precautions Section.
2. Remove or disconnect the following:

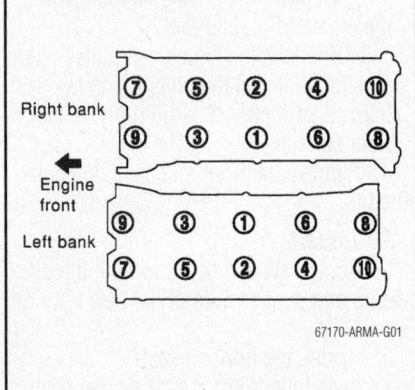

Fig. 30 Cylinder head torque sequence

- Engine assembly
- Belt tensioner
- Idler pulley
- Thermostat housing and hose
- Oil pan and strainer
- Fuel tube and injector assembly
- Intake manifold
- Ignition coil
- Rocker cover
- Crankshaft pulley
- Front engine cover
- Oil pump
- Timing chain
- Camshaft sprockets
- Camshafts
- Cylinder head, removing bolts in reverse order shown in figure

To install:

3. Install the cylinder head with a new gasket. Tighten the bolts in sequence as follows:
   a. Step 1: 72 ft. lbs. (98 Nm)
   b. Step 2: Loosen all bolts completely
   c. Step 3: 33 ft. lbs. (44 Nm)
   d. Step 4: Plus 60 degrees
   e. Step 5: Plus 60 degrees

4. Install or connect the following:
   - Camshaft
   - Camshaft sprockets
   - Timing chain
   - Oil pump

- Front engine cover
- Crankshaft pulley
- Rocker cover
- Ignition coil
- Intake manifold
- Fuel tube and injector assembly
- Oil pain and strainer
- Thermostat housing and hose
- Idler pulley
- Belt tensioner
- Engine assembly

5. Start the engine and check for leaks

ENGINE ASSEMBLY

REMOVAL & INSTALLATION

See Figure 32.

1. Before servicing the vehicle, refer to the Precautions Section.
2. Drain the cooling system.
3. Partially drain the automatic transmission fluid.
4. Relieve the fuel system pressure.
5. Remove or disconnect the following:
   - Hood
   - Cowl extension
   - Engine cover
   - Air intake assembly
   - Vacuum hose between vehicle and engine

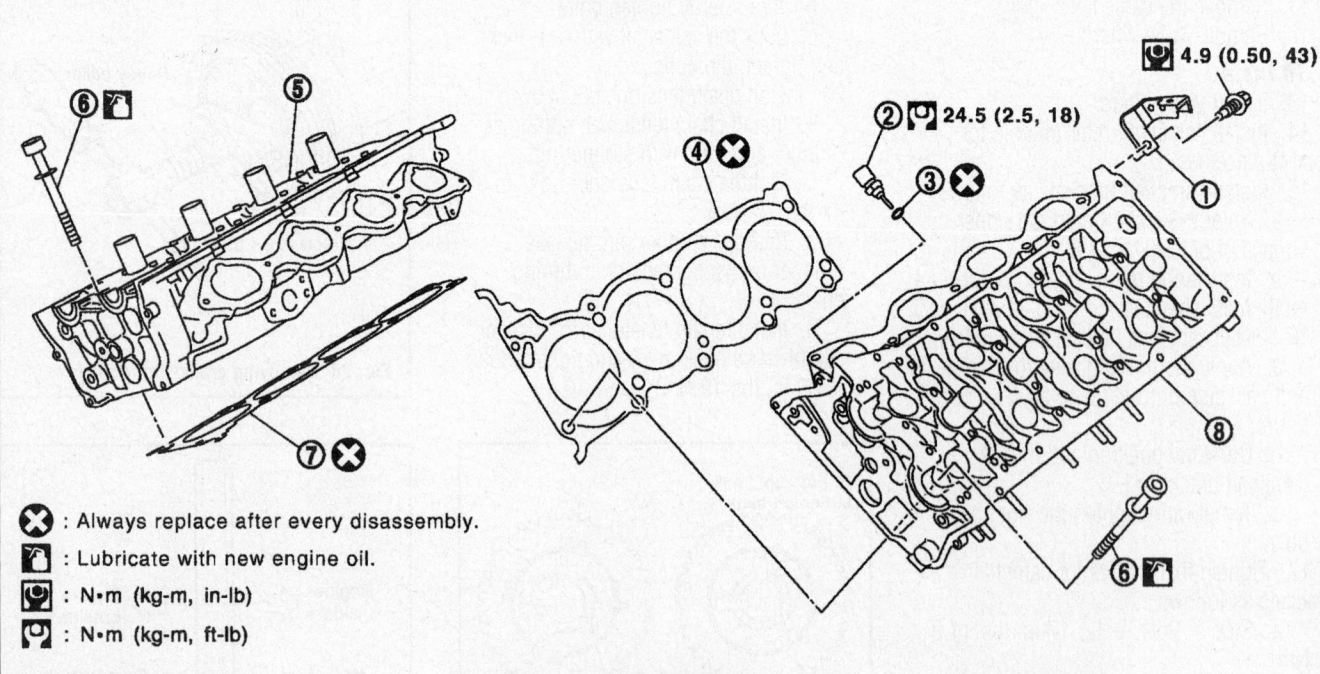

❌ : Always replace after every disassembly.

🛢 : Lubricate with new engine oil.

🔧 : N•m (kg-m, in-lb)

🔧 : N•m (kg-m, ft-lb)

1.	Harness bracket	2.	Engine coolant temperature sensor	3.	Washer
4.	Cylinder head gasket (left bank)	5.	Cylinder head (right bank)	6.	Cylinder head bolt
7.	Cylinder head gasket (right bank)	8.	Cylinder head (left bank)		

67170-ARMA-G28

Fig. 31 Cylinder heads and gaskets

- Radiator hoses
- Radiator
- Drive belts
- Engine fan
- Wiring harness
- ECM
- Power steering reservoir tank and oil pump
- A/C compressor
- Brake booster vacuum line
- EVAP line
- Fuel hose
- Heater hoses
- Transmission dipstick assembly
- Front final drive assembly (4WD models only)
- Exhaust manifolds

6. Install engine slings onto the left and right cylinder heads and tighten to 33 ft. lbs. (45 Nm).

7. Attach an engine hoist to slings and lift engine out of the vehicle

To install:

8. Lower engine into the vehicle and remove the engine slings.

9. Install or connect the following:
- Exhaust manifolds
- Front final drive assembly (4WD models only)
- Transmission dipstick assembly
- Heater hoses
- Fuel hose
- EVAP line
- Brake booster vacuum line
- A/C compressor
- Power steering oil pump and reservoir tank
- ECM
- Wiring harnesses
- Engine fan
- Drive belts
- Radiator
- Radiator hoses
- Vacuum hose between engine and vehicle

- Air intake assembly
- Engine cover
- Cowl extension
- Hood

10. Refill the automatic transmission fluid.
11. Refill the cooling system.
12. Start the engine and check for leaks.

EXHAUST MANIFOLD

REMOVAL & INSTALLATION

See Figures 33 and 34.

1. Before servicing the vehicle, refer to the Precautions Section.

2. Drain the cooling system.

3. Remove or disconnect the following:
- Air intake assembly
- Engine splash guard
- Radiator and radiator hoses
- Drive belts

4. Remove air fuel ratio sensors as follows:

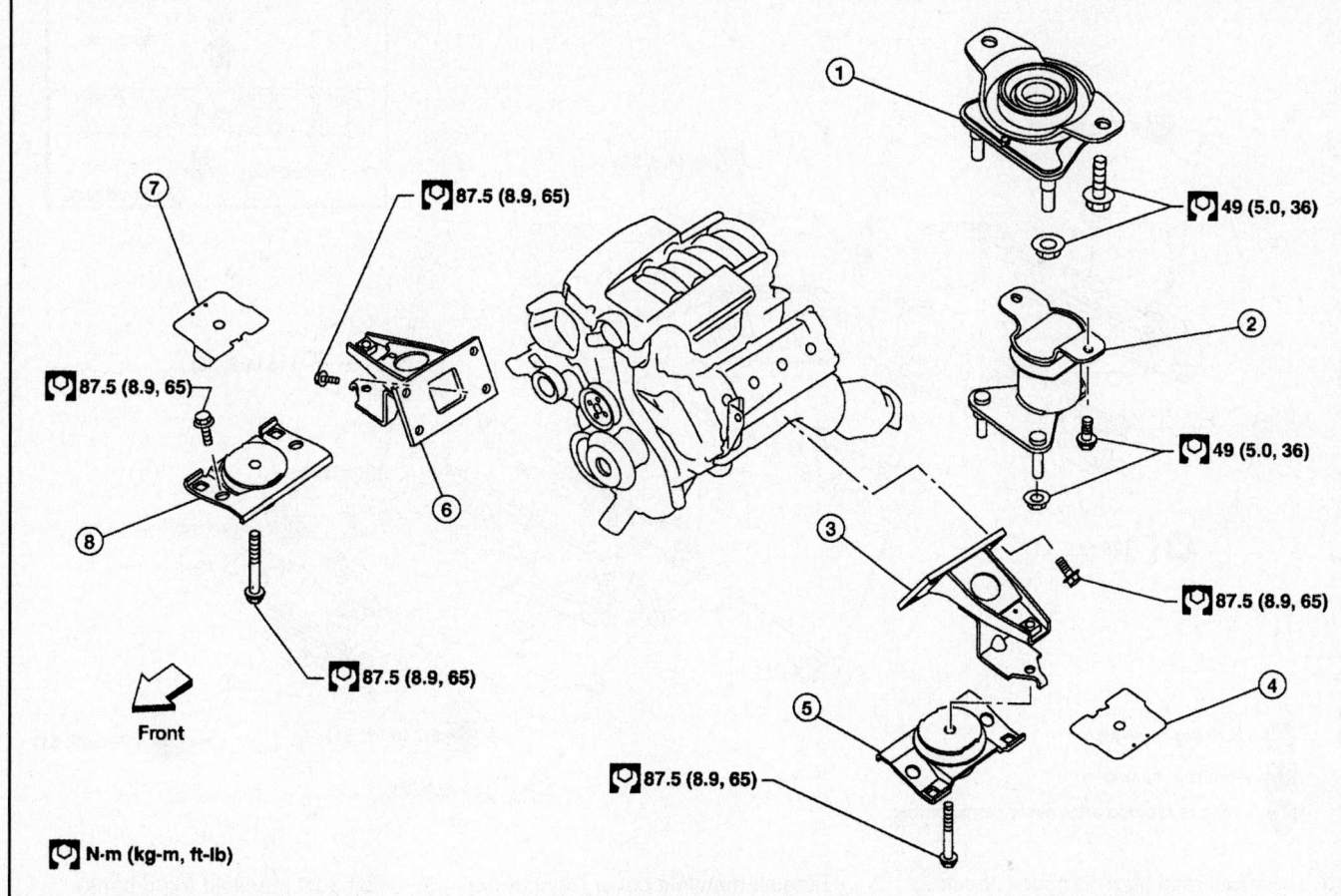

N·m (kg-m, ft-lb)

1. Rear engine mounting insulator 4x4
2. Rear engine mounting insulator 4x2
3. LH engine mounting bracket
4. LH Heat shield plate
5. LH engine mounting insulator
6. RH engine mounting bracket
7. RH Heat shield plate
8. RH engine mounting insulator

67170-ARMA-G24

Fig. 32 Engine mounts

Left bank
Engine front
Right bank
Engine front
67170-ARMA-G03

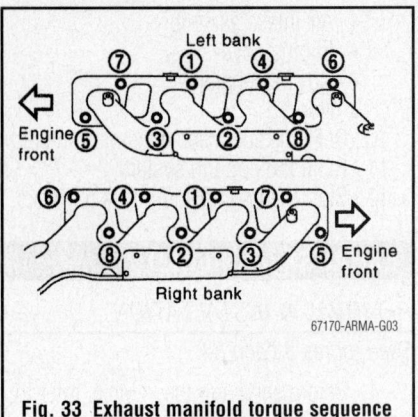

Fig. 33 Exhaust manifold torque sequence

a. Remove engine cover.
b. Remove wiring harness from each sensor
c. Remove sensors, using special tool J-38356

5. Remove front cross bar
6. Remove left exhaust manifold as follows:
a. Remove the exhaust front tube.
b. Remove the exhaust manifold cover.
c. Loosen the nuts in reverse order shown in figure.
d. Remove studs from position 2, 4, 6, and 8 and remove manifold.
7. Remove right exhaust manifold as follows:
a. Remove the exhaust front tube.
b. Remove the oil level gauge guide.
c. Remove the exhaust manifold cover.
d. Loosen the nuts in reverse order shown in figure.
e. Remove studs from position 2, 4, 6, and 8 and remove manifold.

To install:

8. Install or connect the following:
- The exhaust manifold gasket with triangle mark facing up and coated (gray) face toward exhaust manifold.
- Exhaust manifold, tightening the nuts as shown in figure
- Exhaust manifold cover
- Oil level gauge guide (right side only)
- Exhaust front tube
- Front cross bar
- Air fuel ratio sensors, with anti-seize lubricant
- Engine cover
- Drive belts
- Radiator and radiator hoses
- Engine splash guard
- Air intake assembly

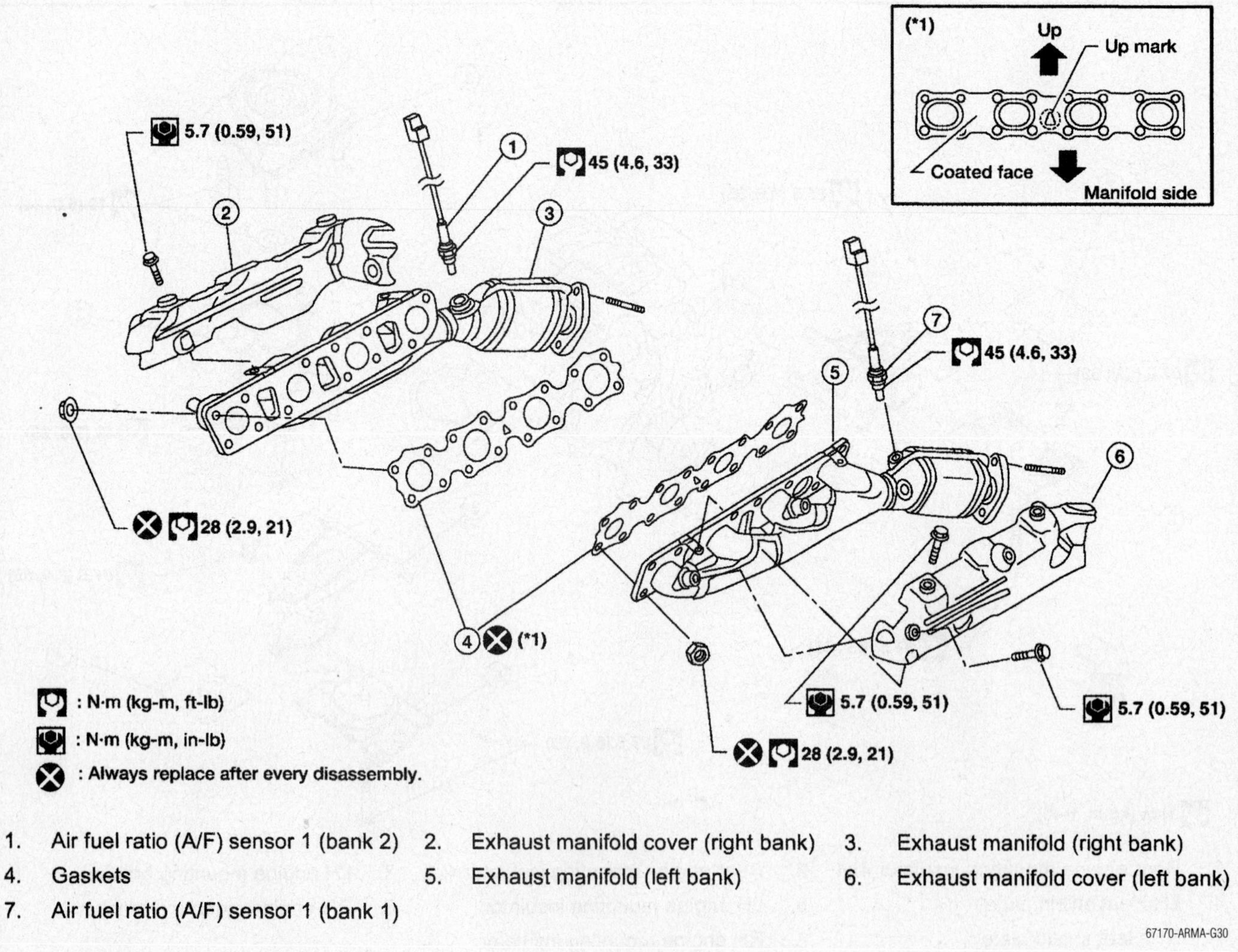

(*1)
Up
Up mark
Coated face
Manifold side
5.7 (0.59, 51)
45 (4.6, 33)
28 (2.9, 21)
(*1)
45 (4.6, 33)
5.7 (0.59, 51)
5.7 (0.59, 51)
28 (2.9, 21)
: N·m (kg-m, ft-lb)
: N·m (kg-m, in-lb)
: Always replace after every disassembly.
1. Air fuel ratio (A/F) sensor 1 (bank 2)
2. Exhaust manifold cover (right bank)
3. Exhaust manifold (right bank)
4. Gaskets
5. Exhaust manifold (left bank)
6. Exhaust manifold cover (left bank)
7. Air fuel ratio (A/F) sensor 1 (bank 1)
67170-ARMA-G30

Fig. 34 Exhaust manifolds and related parts

9. Refill the cooling system
10. Start engine and check for leaks.

INTAKE MANIFOLD

REMOVAL & INSTALLATION

See Figures 35 and 36.

1. Before servicing the vehicle, refer to the Precautions Section.
2. Drain the cooling system.
3. Relieve the fuel system pressure.
4. Remove or disconnect the following:

- Engine cover
- Air intake assembly
- Fuel tube quick connector using special tool J-45488

- Wiring harnesses and brackets from manifold
- Vacuum hoses
- PCV hose and tube
- Electric throttle control actuator, loosening bolts diagonally
- Fuel injectors
- Fuel tube assembly
- Intake manifold, removing bolts in reverse order shown in figure

To install:

5. Install the intake manifold with new gaskets. Tighten the bolts in order as shown.
6. Install or connect the following:

- Fuel tube assembly
- Fuel injectors
- Electronic throttle control actuator,

tightening the bolts in several steps
- PCV hose
- Vacuum hoses
- Wiring harnesses

7. Connect the fuel tube as follows:

a. Apply a thin layer of engine oil on the tube from tip end to spool end.

b. Insert tube into quick connector past the white identification mark

c. Insert tube into quick connector until top spool is completely inside the connector and 2nd level spool is exposed right below the connector.

d. Pull slightly on the quick connector to ensure it is fully engaged.

e. Install quick connector cap on quick connector joint.

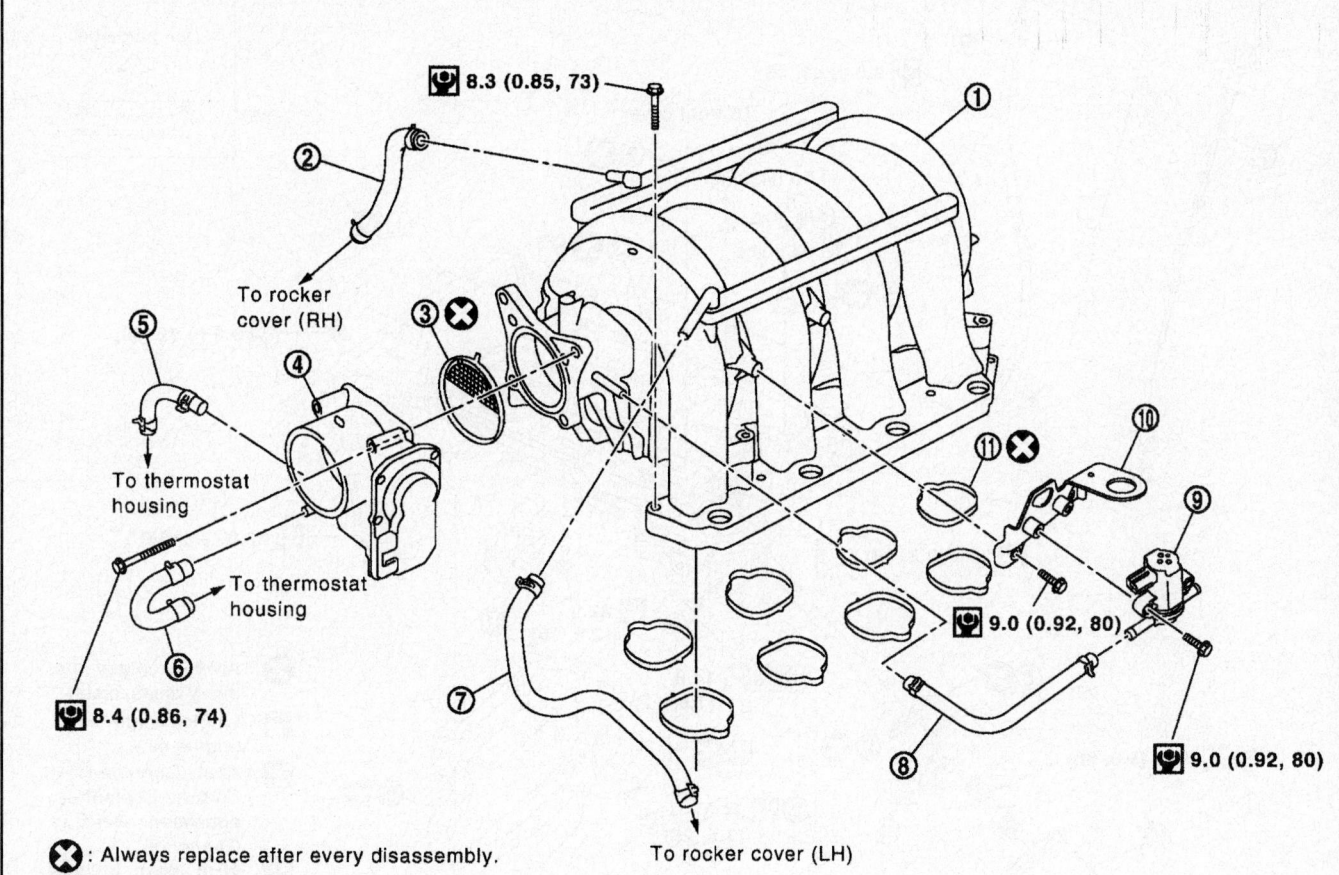

⊗ : Always replace after every disassembly.

🔧 : N•m (kg-m, in-lb)

1.	Intake manifold	2.	PCV hose	3.	Gasket
4.	Electric throttle control actuator	5.	Water hose	6.	Water hose
7.	PCV hose	8.	EVAP hose	9.	EVAP canister purge control solenoid valve
10.	Bracket	11.	Gasket		

67170-ARMA-G29

Fig. 35 Intake manifold and related parts

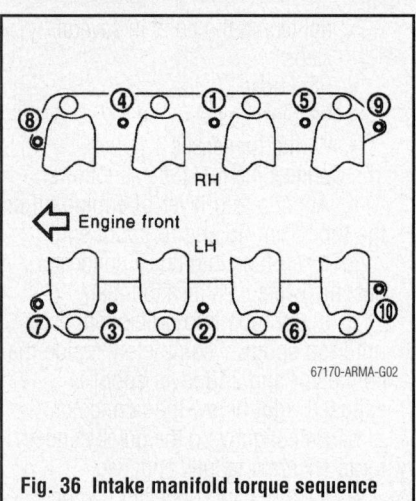

Fig. 36 Intake manifold torque sequence

8. Install or connect the following:
- Air intake assembly
- Engine cover
9. Refill the cooling system.
10. Start engine and check for leaks.

OIL PAN

REMOVAL & INSTALLATION

See Figures 37 through 39.

1. Before servicing the vehicle, refer to the Precautions Section.
2. Remove the engine assembly.
3. Remove the lower oil pan loosening bolts in reverse order shown in figure.
4. Remove oil strainer from upper oil pan.

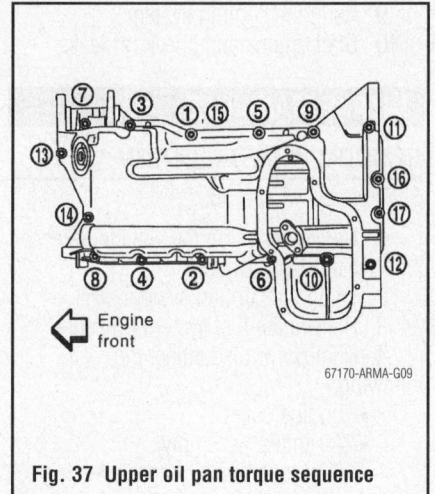

Fig. 37 Upper oil pan torque sequence

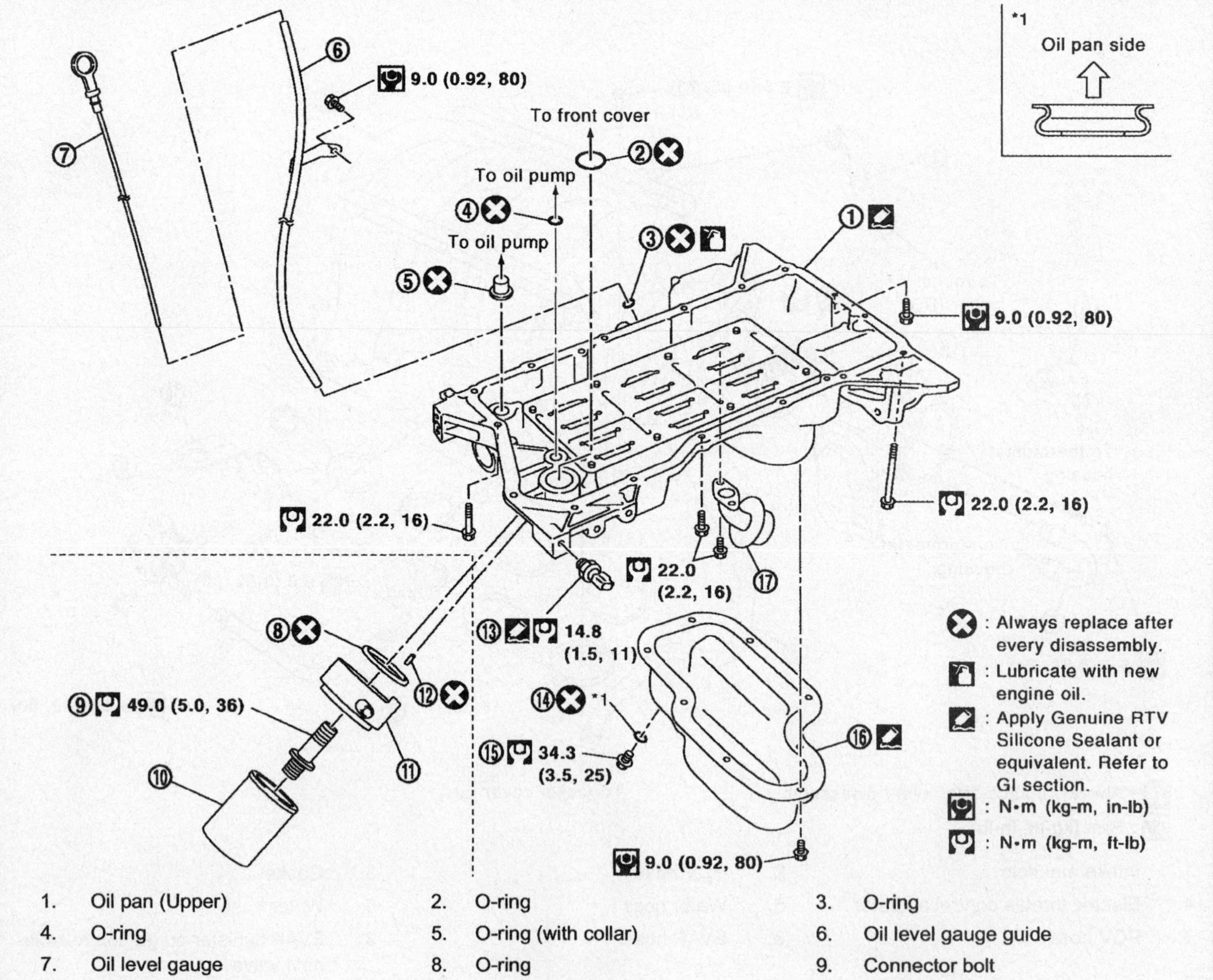

*1
Oil pan side

To front cover
To oil pump
To oil pump

9.0 (0.92, 80)

9.0 (0.92, 80)

22.0 (2.2, 16)

22.0 (2.2, 16)

22.0 (2.2, 16)

14.8 (1.5, 11)

49.0 (5.0, 36)

34.3 (3.5, 25)

9.0 (0.92, 80)

❌ : Always replace after every disassembly.

🛢 : Lubricate with new engine oil.

✏ : Apply Genuine RTV Silicone Sealant or equivalent. Refer to GI section.

⚙ : N•m (kg-m, in-lb)

⚙ : N•m (kg-m, ft-lb)

1.	Oil pan (Upper)	2.	O-ring	3.	O-ring
4.	O-ring	5.	O-ring (with collar)	6.	Oil level gauge guide
7.	Oil level gauge	8.	O-ring	9.	Connector bolt
10.	Oil filter	11.	Oil cooler	12.	Relief valve
13.	Oil pressure switch	14.	Gasket	15.	Drain plug
16.	Oil pan (Lower)	17.	Oil strainer		

Fig. 38 Oil pan and related parts

67170-ARMA-G31

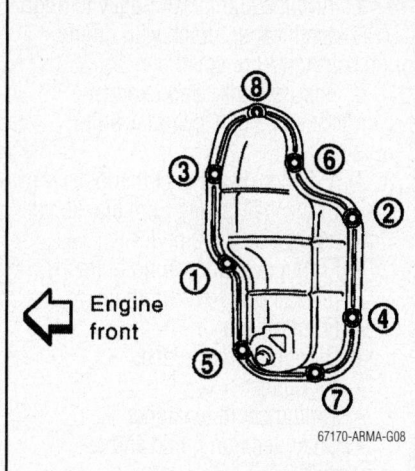

Fig. 39 Lower oil pan torque sequence

5. Gently pry and remove upper oil pan from engine block.

To install:

6. Apply liquid gasket to upper oil pan mating surfaces.

7. Install new O-rings to oil pump and front cover side.

8. Tighten upper oil pan bolts in following numerical order:
   a. No. 15, 16
   b. No. 1, 3, 5, 7, 11, 13
   c. No. 2, 4, 6, 8, 10, 14
   d. No. 9, 12
9. Install or connect the following:
   • Rear plate cover
   • Oil strainer to upper oil pan
   • Lower oil pan, tightening bolts in order shown in figure

OIL PUMP

REMOVAL & INSTALLATION

See Figure 40.

1. Before servicing the vehicle, refer to the Precautions Section.
2. Remove or disconnect the following:
   • Timing chain cover
   • Oil pump drive spacer
   • Oil pump

To install:

3. Install or connect the following:
   • Oil pump
   • Oil pump drive spacer
   • Timing chain cover

PISTON AND RING

POSITIONING

See Figures 41 and 42.

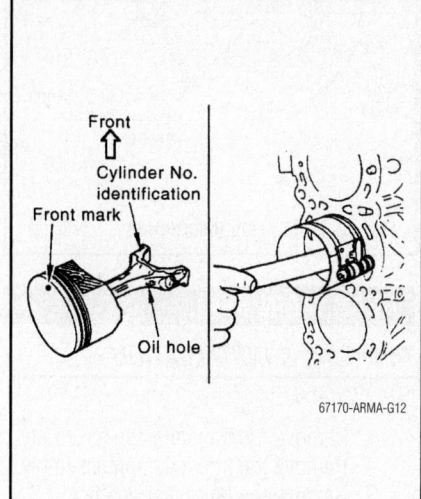

Fig. 41 Piston and rod positioning and identification

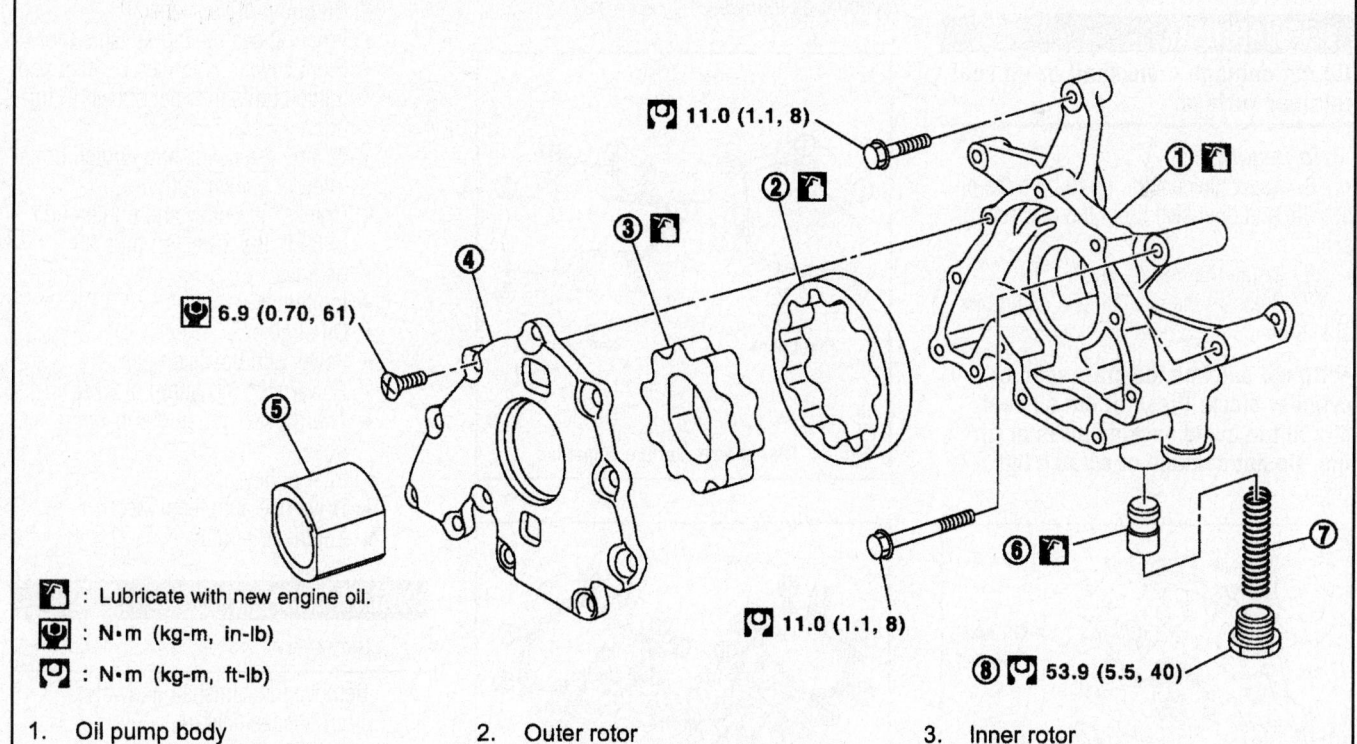

: Lubricate with new engine oil.

: N•m (kg-m, in-lb)

: N•m (kg-m, ft-lb)

1. Oil pump body
2. Outer rotor
3. Inner rotor
4. Oil pump cover
5. Oil pump drive spacer
6. Regulator valve
7. Regulator spring
8. Regulator plug

Fig. 40 Oil pump exploded view

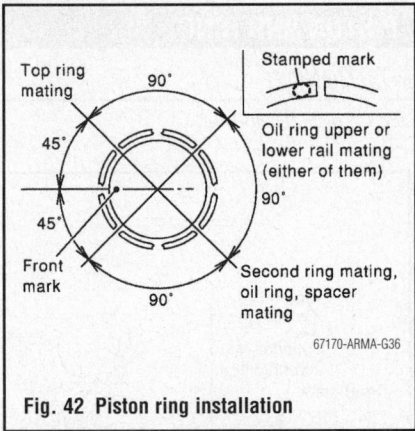

Fig. 42 Piston ring installation

REAR MAIN SEAL

REMOVAL & INSTALLATION

See Figure 43.

1. Remove the transmission assembly.
2. Remove the transmission assembly.
3. Remove the bolts diagonally.
4. Remove the drive plate.
5. Place the drive plate with the signal plate surface facing upward.
6. Remove the engine rear plate.
7. Remove the rear oil seal using suitable tool.

❋❋ WARNING

Do not damage crankshaft or oil seal retainer surface.

To install:

8. Apply new engine oil to both the oil seal lip and dust seal lip of the new rear oil seal.
9. Install the rear oil seal.
10. Press-fit the rear oil seal using suitable tool.

➡**Do not damage the crankshaft or cylinder block. Press-fit the oil seal straight to avoid causing burrs or tilting. Do not damage or scratch the**

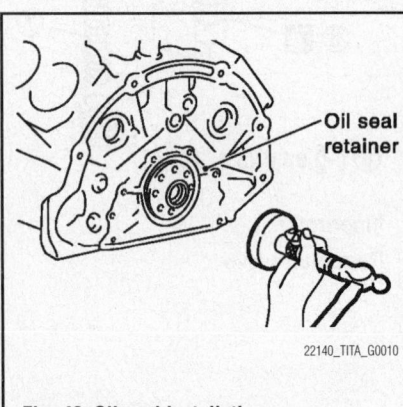

Fig. 43 Oil seal installation

outer circumference of the rear oil seal.

11. Tap until flattened with the front edge of the oil seal retainer.
12. Installation of the remaining components is in the reverse order of removal.

TIMING CHAIN, SPROCKETS, FRONT COVER AND SEAL

REMOVAL & INSTALLATION

See Figures 44 through 46.

1. Before servicing the vehicle, refer to the Precautions Section.
2. Remove or disconnect the following:
   - Engine assembly
   - Drive belt auto tensioner
   - Idler pulley
   - Thermostat housing and water hose
   - Power steering pump bracket
   - Oil pan (upper and lower)
   - Oil strainer
   - Ignition coil
   - Rocker cover
   - Timing chain case cover, loosening bolts in reverse order shown in figure
3. Obtain compression TDC of No. 1 cylinder as follows:

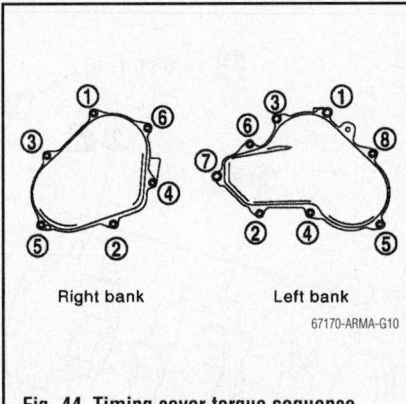

Fig. 44 Timing cover torque sequence

   a. Turn the crankshaft pulley to align TDC identification notch with timing indicator on front cover.
   b. Ensure intake and exhaust cam lobes of No. 1 cylinder point outside.
4. Remove or disconnect the following:
   - Crankshaft pulley from crankshaft using a suitable puller
   - Front cover, loosening bolts in reverse order shown in figure
   - Front oil seal
   - Oil pump drive spacer
   - Oil pump
   - Timing chain tensioner
   - Chain tension guide and slack guide
   - Timing chain
   - Camshaft sprocket

To install:

5. Ensure that the crankshaft key and dowel pin of each camshaft are facing the same direction.
6. Install or connect the following:
   - Camshaft sprockets
   - Timing chain
   - Chain tension guide and slack guide
   - Oil pump
   - Oil pump drive spacer
   - Front oil seal, using suitable tool
   - Front cover, using new O-rings and tighten bolts in order shown in figure
   - Chain case cover, and tighten bolts in order shown in figure
   - Crankshaft pulley and tighten bolt to 69 ft. lbs. (93 Nm) plus 90 degrees
   - Ignition coil
   - Oil strainer
   - Lower and upper oil pan
   - Power steering pump bracket
   - Thermostat housing and water hose
   - Idler pulley
   - Drive belt auto tensioner
   - Engine assembly

VALVE LASH

ADJUSTMENT

1. Remove the camshaft and valve lifter(s) out of specification.
2. Install the replacement valve lifter(s).
3. Install the camshaft.
4. Manually turn the crankshaft pulley several turns.
5. Recheck valve clearances with engine at operating temperature.

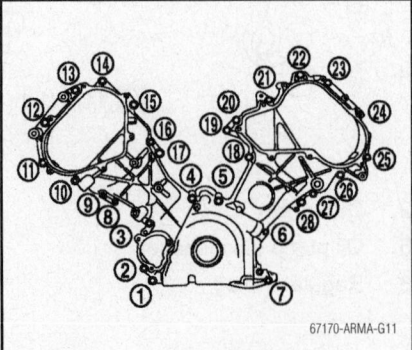

Fig. 45 Front cover torque sequence

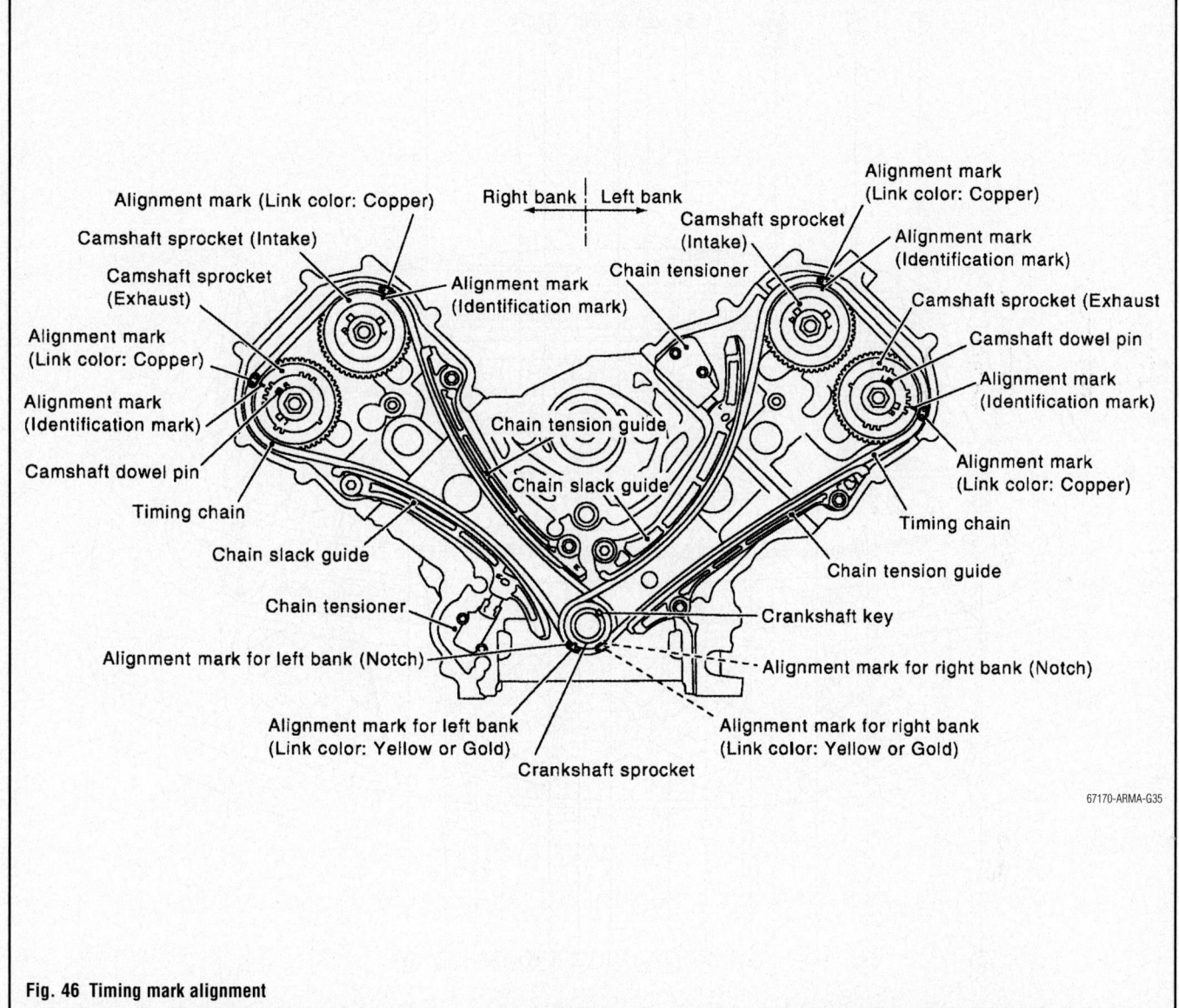

Right bank | Left bank

Alignment mark (Link color: Copper)
Camshaft sprocket (Intake)
Camshaft sprocket (Exhaust)
Alignment mark (Link color: Copper)
Alignment mark (Identification mark)
Camshaft dowel pin
Timing chain
Chain slack guide
Chain tensioner
Alignment mark for left bank (Notch)
Alignment mark for left bank (Link color: Yellow or Gold)
Crankshaft sprocket

Alignment mark (Identification mark)
Chain tensioner
Chain tension guide
Chain slack guide

Alignment mark (Link color: Copper)
Camshaft sprocket (Intake)
Alignment mark (Identification mark)
Camshaft sprocket (Exhaust)
Camshaft dowel pin
Alignment mark (Identification mark)
Alignment mark (Link color: Copper)
Timing chain
Chain tension guide
Crankshaft key
Alignment mark for right bank (Notch)
Alignment mark for right bank (Link color: Yellow or Gold)

67170-ARMA-G35

Fig. 46 Timing mark alignment

ENGINE PERFORMANCE & EMISSION CONTROL

COMPONENT LOCATIONS

See Figures 47 through 51.

AIR FUEL RATIO (AF) SENSOR

LOCATION

See Figure 52.

The A/F sensors are located in the left and right exhaust manifold assembly.

REMOVAL & INSTALLATION

1. Remove the air fuel ratio A/F sensors (bank 1, bank 2).

2. Follow steps below to remove each air fuel ratio A/F sensor.

3. Remove the harness connector of each air fuel ratio A/F sensor, and harness from bracket and middle clamp.

4. Remove the air fuel ratio A/F sensors from both left and right exhaust manifolds using Tool.

5. Reverse the removal procedure and note the following:
- Tighten the A/F sensor to 37 ft. lbs. 50 (Nm).
- Keep harness away from manifold assembly.

CAMSHAFT POSITION (CMP) SENSOR

LOCATION

The camshaft position sensor is located on the right front of the timing cover, facing the engine.

REMOVAL & INSTALLATION

1. Remove the engine cover.
2. Remove air intake duct.
3. Disconnect the camshaft position sensor.
4. Remove the bolt and the camshaft position sensor.

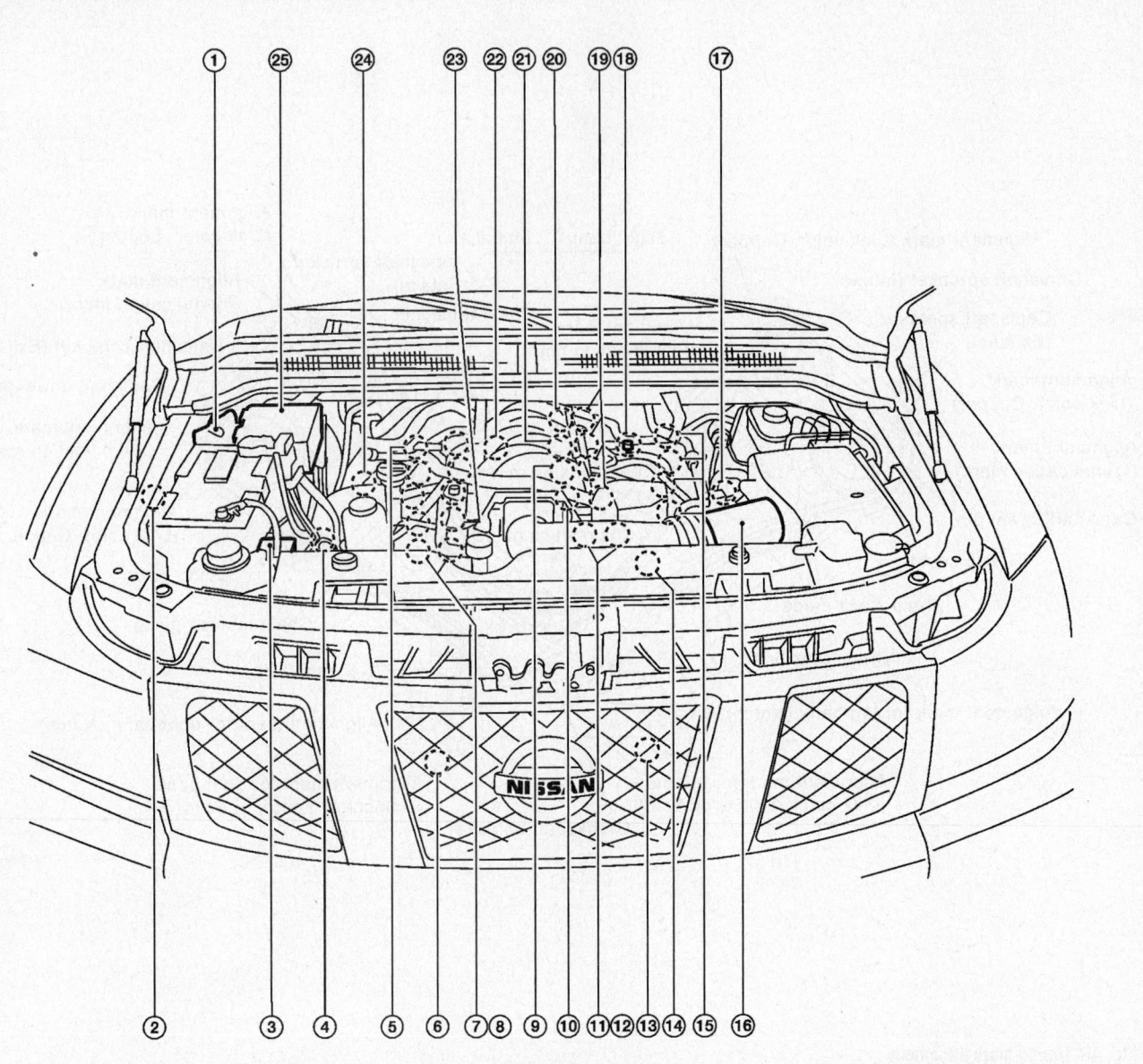

1. ECM
2. Dropping resistor (FFV models only)
3. Battery current sensor
4. Power steering pressure sensor
5. Ignition coil (with power transistor) and spark plug (bank 2)
6. Refrigerant pressure sensor
7. Intake valve timing control position sensor (bank 2)
8. Intake valve timing control solenoid valve (bank 2)
9. Engine coolant temperature sensor
10. Electric throttle control actuator
11. Intake valve timing control position sensor (bank 1)
12. Intake valve timing control solenoid valve (bank 1)
13. Cooling fan motor
14. Camshaft position sensor (PHASE)
15. Ignition coil (with power transistor) and spark plug (bank 1)
16. Mass air flow sensor (with intake air temperature sensor)
17. A/F sensor 1 (bank 1)
18. EVAP service port
19. Fuel injector (bank 1)
20. Knock sensor (bank 1)
21. EVAP canister purge volume control solenoid valve
22. Knock sensor (bank 2)
23. Fuel injector (bank 2)
24. A/F sensor 1 (bank 2)
25. IPDM E/R

22140_TITA_G0013

Fig. 47 Engine components parts location

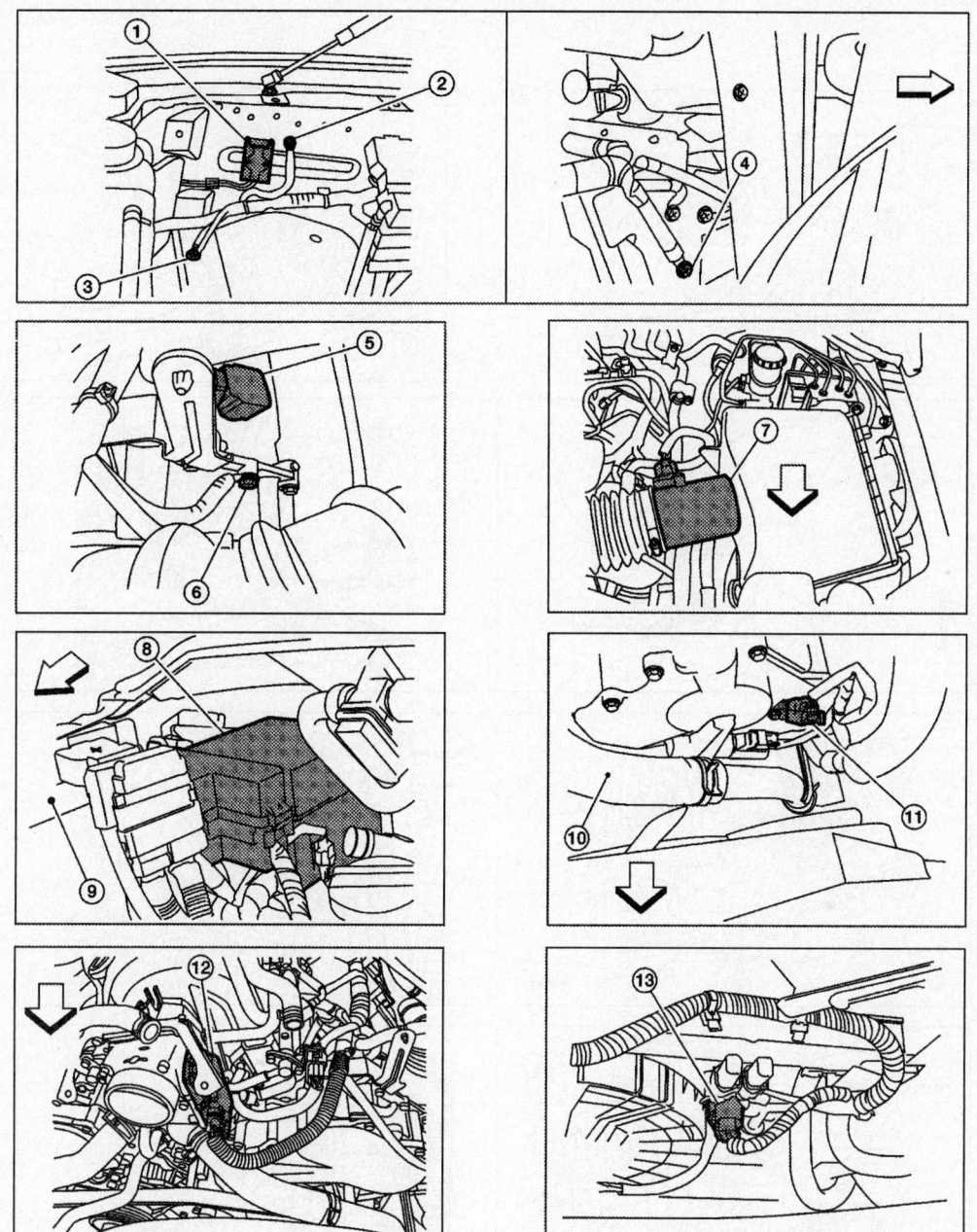

1. Dropping resistor (FFV models only) (view with battery removed)
2. Body ground (view with battery removed)
3. Body ground (view with battery removed)
4. Body ground
5. No.1 ignition coil
6. Engine ground
7. Mass air flow sensor (with intake air temperature sensor)
8. IPDM E/R
9. Battery
10. Radiator hose
11. Camshaft position sensor (PHASE)
12. Electric throttle control actuator (view with intake air duct removed)
13. Cooling fan motor harness connector

◁ : Vehicle front

22140_TITA_G0014

Fig. 48 Engine control systems location (1)

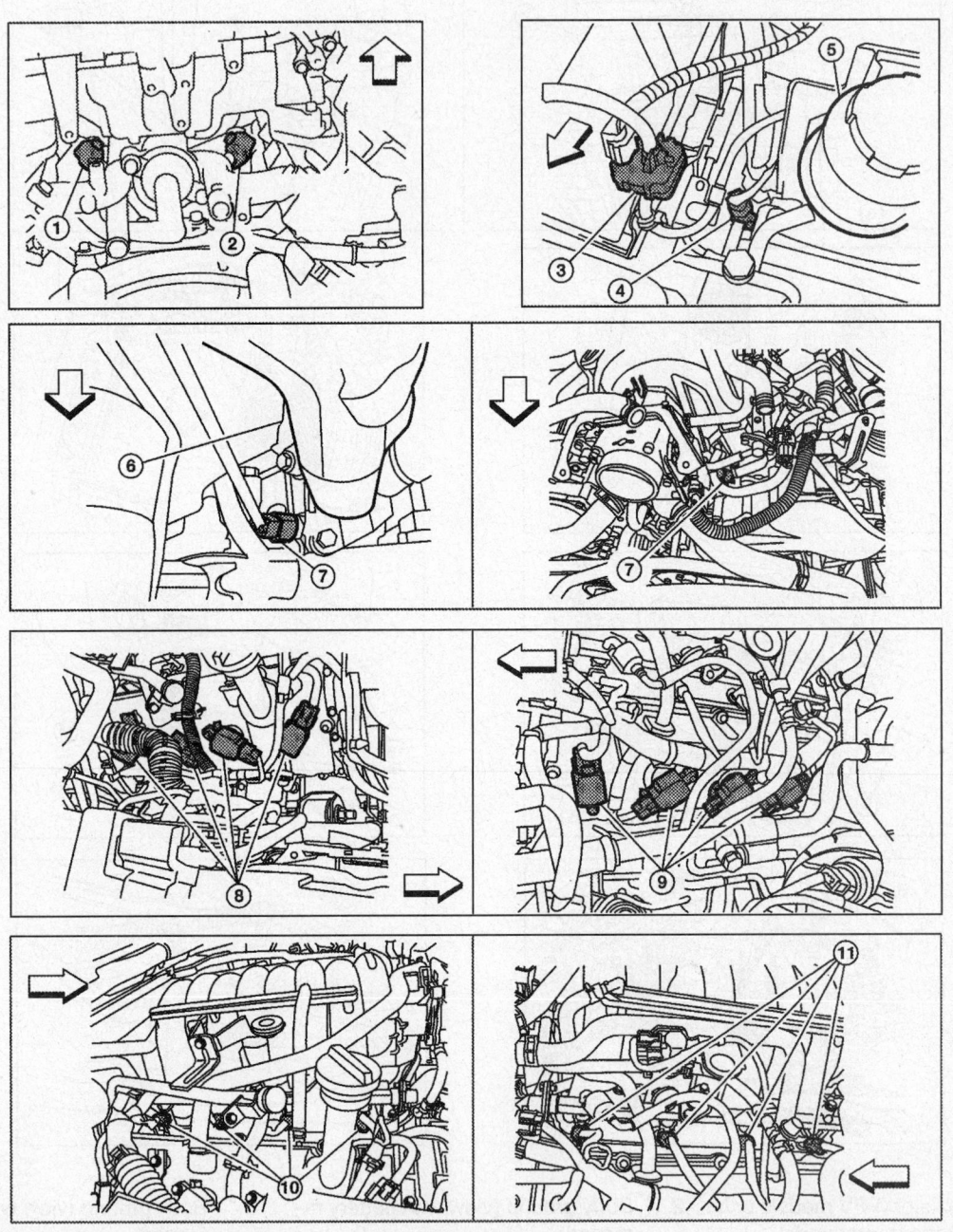

1. Knock sensor (bank 1) (view with engine removed)
2. Knock sensor (bank 2) (view with engine removed)
3. Battery current sensor
4. Power steering pressure sensor
5. Power steering fluid reservoir
6. Intake manifold
7. Engine coolant temperature sensor
8. Ignition coils (with power transistor)
9. Ignition coil (with power transistor)
10. Injector harness connectors (bank 2)
11. Injector harness connectors (bank 1)
⇦ : Vehicle front

22140_TITA_G0015

Fig. 49 Engine control systems location (2)

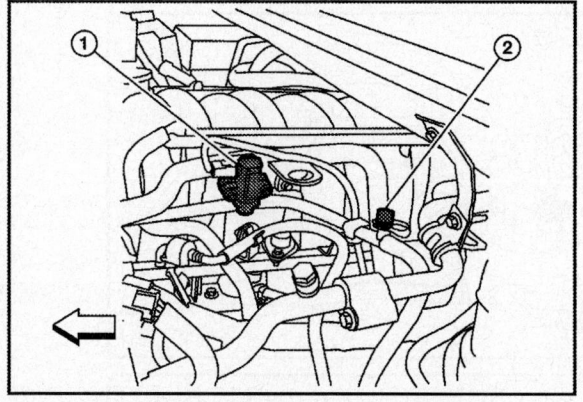
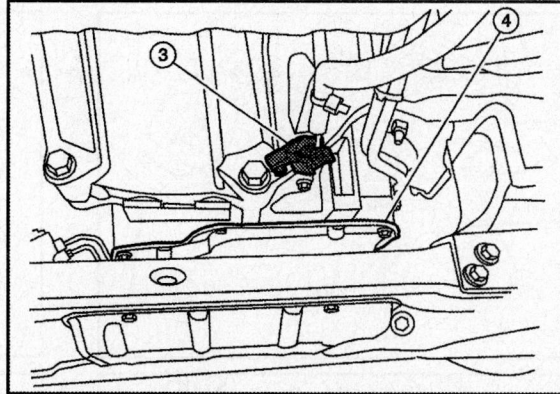

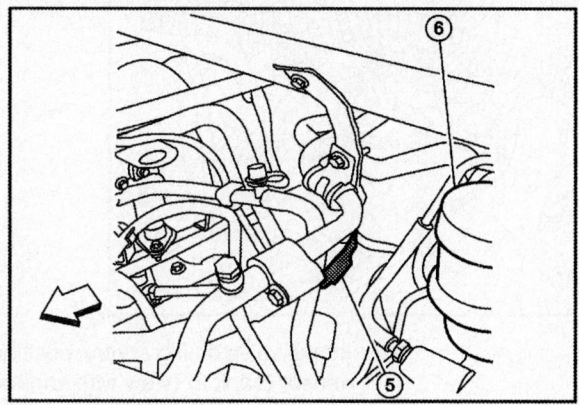
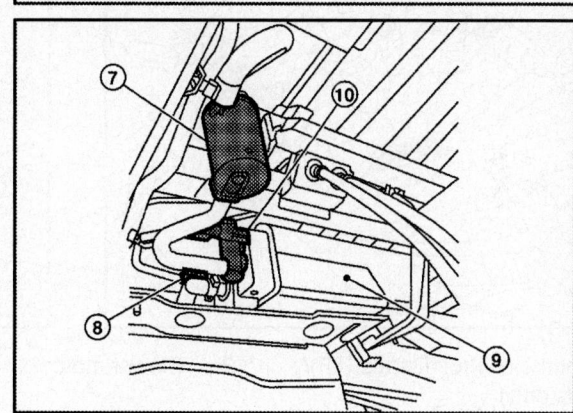

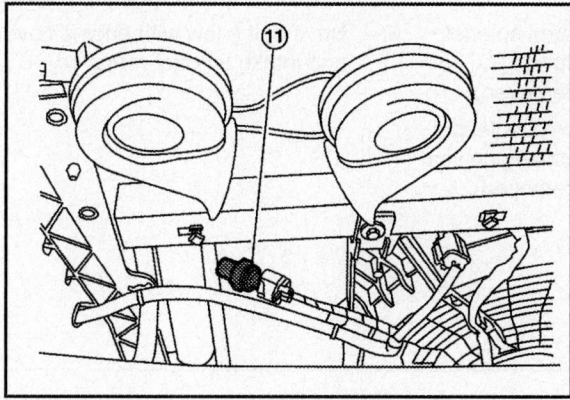

1. EVAP canister purge volume control solenoid valve (view with engine cover removed)
2. EVAP service port
3. Crankshaft position sensor (POS) (view from under the vehicle)

4. Engine oil pan
5. Condenser-1
6. Brake fluid reservoir

7. Drain filter (view from under vehicle)
8. EVAP control system pressure sensor
9. EVAP canister

10. EVAP canister vent control valve (view with fuel tank removed)
11. Refrigerant pressure sensor (view with front grille removed)

◁ : Vehicle front

22140_TITA_G0016

Fig. 50 Engine control systems location (3)

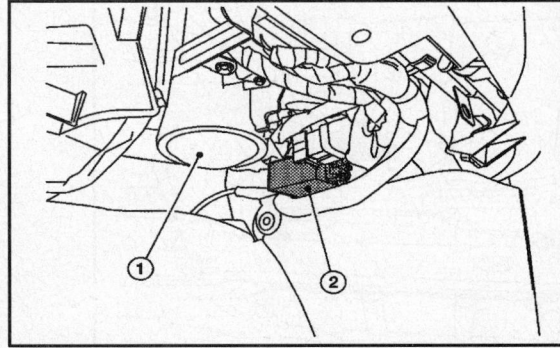
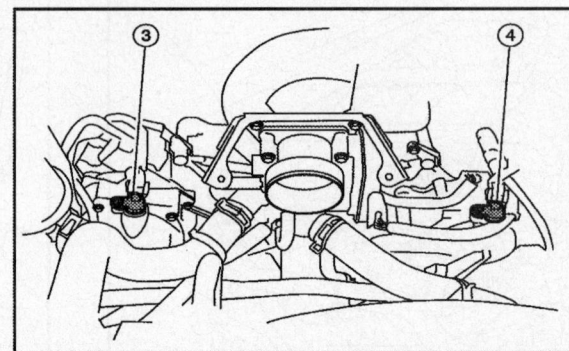
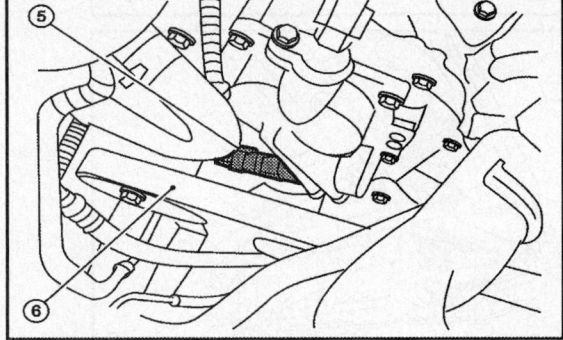
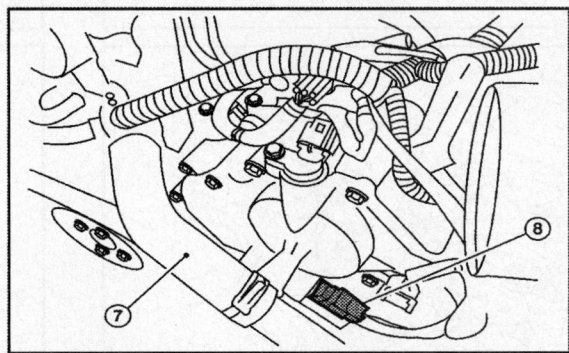

1. Fuel pump control module (FFV models only)
2. Blower motor
3. Intake valve timing control position sensor (bank 2) (view with engine cover and intake air duct removed)
4. Intake valve timing control position sensor (bank 1) (view with engine cover and intake air duct removed)
5. Intake valve timing control solenoid valve (bank 2) (view with engine cover and intake air duct removed)
6. Drive belt (view with engine cover and intake air duct removed)
7. Radiator hose (view with engine cover and intake air duct removed)
8. Intake valve timing control solenoid valve (bank 1) (view with engine cover and intake air duct removed)

22140_TITA_G0017

Fig. 51 Engine control systems location (2)

To install:
5. Install the camshaft sensor and tighten the bolt.
6. Reconnect the camshaft electrical sensor.
7. Install the air intake duct.
8. install the engine cover.

CRANKSHAFT POSITION (CKP) SENSOR

LOCATION

The crankshaft position sensor is mounted above the oil pan at the rear of the engine.

REMOVAL & INSTALLATION

1. Safely raise the vehicle on a lift.
2. Disconnect the crankshaft position sensor connector
3. Remove the mounting bolt and crankshaft position sensor.

To install:
4. Install the crankshaft position sensor and tighten the mounting bolt.
5. Reconnect the crankshaft position sensor connector.
6. Lower the vehicle.

ELECTRONIC CONTROL MODULE (ECM)

LOCATION

ECM is located in the engine room passenger side behind battery.

REMOVAL & INSTALLATION

1. Disconnect the battery cables.
2. Remove the battery.
3. Carefully remove the ECM harness connectors.
4. Remove the ECM mounting bolts and the ECM

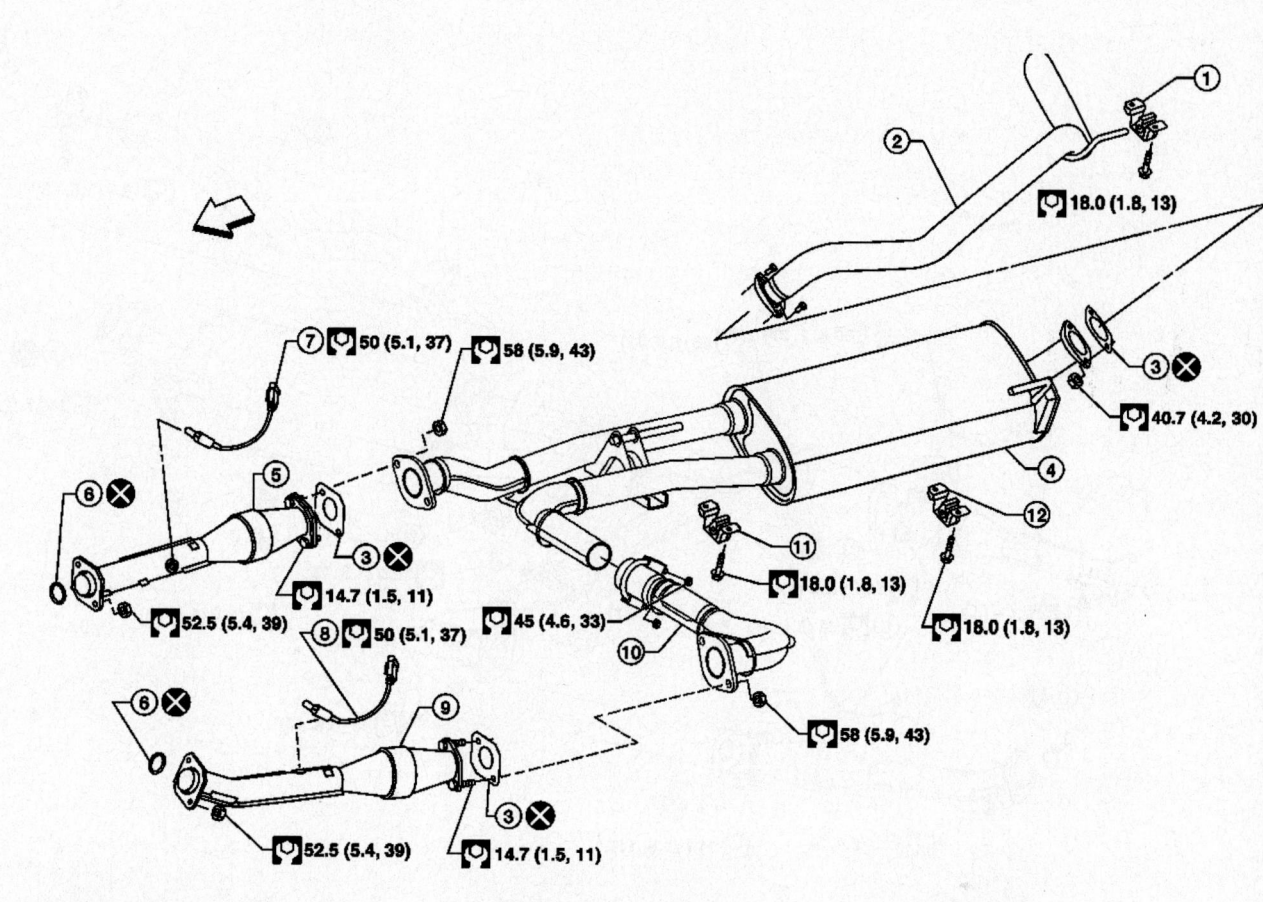

18.0 (1.8, 13)

50 (5.1, 37)   58 (5.9, 43)

40.7 (4.2, 30)

14.7 (1.5, 11)

52.5 (5.4, 39)   50 (5.1, 37)   45 (4.6, 33)   18.0 (1.8, 13)

18.0 (1.8, 13)

58 (5.9, 43)

52.5 (5.4, 39)   14.7 (1.5, 11)

1. Tailpipe hanger bracket	2. Tailpipe	3. Gasket
4. Main muffler	5. Right front exhaust tube	6. Ring gasket
7. Heated oxygen sensor 2 (bank 2)	8. Heated oxygen sensor 2 (bank 1)	9. Left front exhaust tube
10. Center exhaust tube	11. Muffler hanger bracket front	12. Muffler hanger bracket rear

22140_TITA_G0022

Fig. 52 Exhaust manifold and A/F sensor view

To install:

5. Install the ECM and mounting bolts, tighten to 62 inch. lbs. (7 Nm).
6. Carefully install the ECM harness connectors.
7. Install the battery.
8. Reconnect the battery cables.

ENGINE COOLANT TEMPERATURE (ECT) SENSOR

LOCATION

The Engine Coolant Temperature (ECT) Sensor is mounted in the front of the intake manifold. It is just to the right of the throttle body.

REMOVAL & INSTALLATION

1. Remove the engine cover.
2. Remove the intake air duct.
3. Disconnect the harness connector.
4. Remove the ECT sensor.

To install:

5. Install the ECT sensor and carefully tighten.
6. Reconnect the harness connector.
7. Install the intake air duct.

8. Install the engine cover.
9. Refill the engine coolant.

HEATED OXYGEN (HO2S) SENSOR

LOCATION

See Figure 53.

The bank (1) sensor (2) and bank (2) Sensor (2) heated oxygen sensors are located after the exhaust manifold converter assembly.

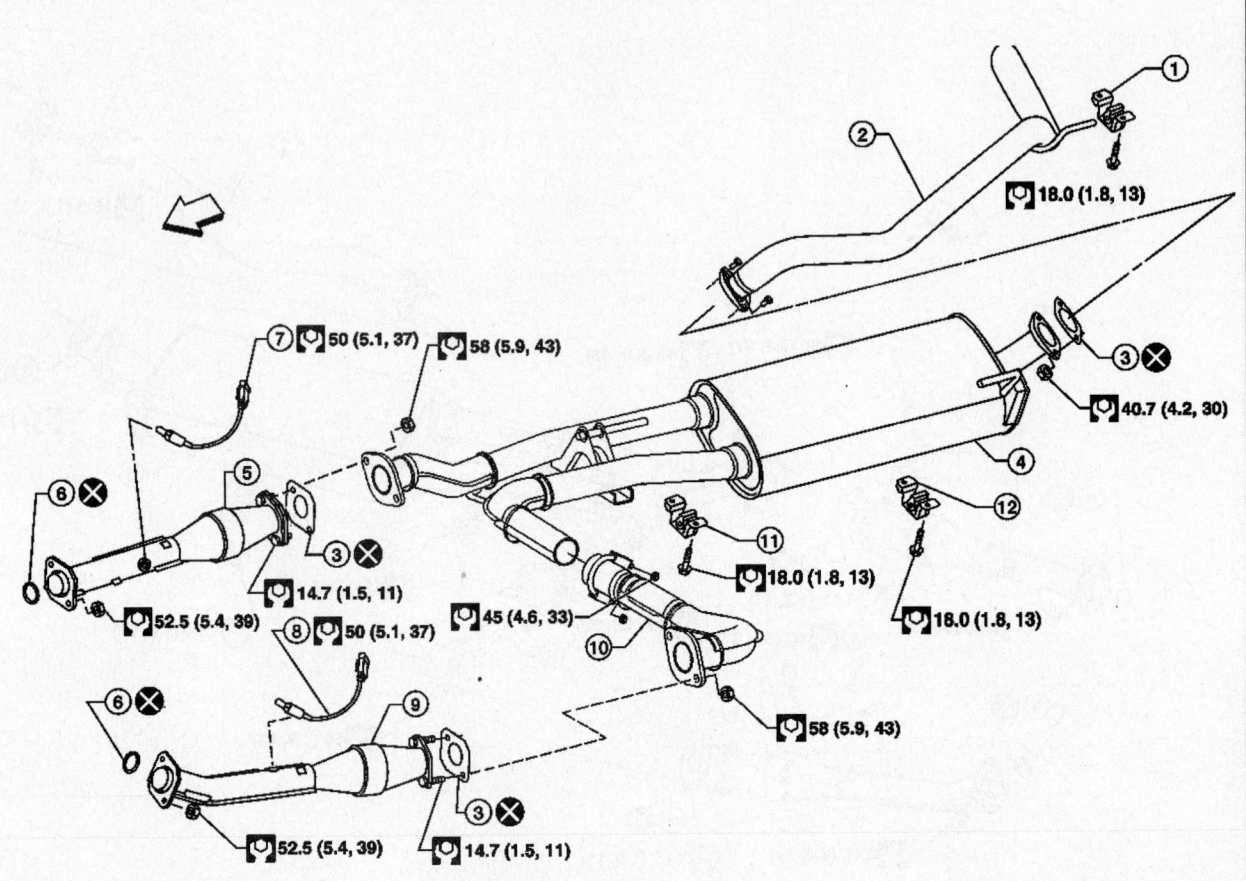

1. Tailpipe hanger bracket
2. Tailpipe
3. Gasket
4. Main muffler
5. Right front exhaust tube
6. Ring gasket
7. Heated oxygen sensor 2 (bank 2)
8. Heated oxygen sensor 2 (bank 1)
9. Left front exhaust tube
10. Center exhaust tube
11. Muffler hanger bracket front
12. Muffler hanger bracket rear

22140_TITA_G0022

Fig. 53 Exploded view of O2 sensors and rear exhaust system

REMOVAL & INSTALLATION

1. Unplug the heated O2 sensor harness.
2. Using an O2 wrench remove the O2 sensor.

➡**Lower the exhaust in needed.**

To install:

3. Install the O2 sensor and tighten to 37 ft. lbs. (50 Nm).
4. Install the harness connector.
5. Keep the harness connector and wiring away from exhaust system.

INTAKE AIR TEMPERATURE (IAT) SENSOR

LOCATION

The Intake Air Temperature (IAT) Sensor is integral to the mass air flow sensor. And is mounted on the air filter housing lid.

REMOVAL & INSTALLATION

1. Remove the engine room cover.
2. Remove the Intake Air Temperature (IAT)/(MAF Sensor harness.
3. Remove the mounting screws and the (IAT)/(MAF sensor.

To install:

4. Install the (IAT)/(MAF sensor.
5. Install the harness connector.
6. Install the engine room cover.

KNOCK SENSOR (KS)

LOCATION

The Knock Sensors (KS) are mounted under the intake manifold on the cylinder block.

REMOVAL & INSTALLATION

See Figures 54 and 55.

1. Partially drain the engine coolant.

☀☀ CAUTION

To avoid the danger of being scalded, never drain the engine coolant when the engine is hot.

2. Remove the engine room cover using power tool.
3. Release the fuel pressure.
4. Remove the air duct and resonator assembly.
5. Disconnect the fuel tube quick connector on the engine side.

Remove or disconnect harnesses, brackets, vacuum hose, vacuum gallery and PCV hose and tube from intake manifold.

6. Remove electric throttle control actuator by loosening bolts diagonally.

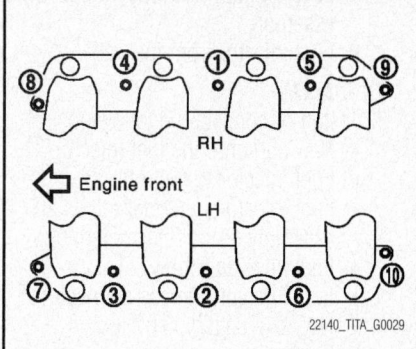

Fig. 54 Loosen the bolts in reverse order shown

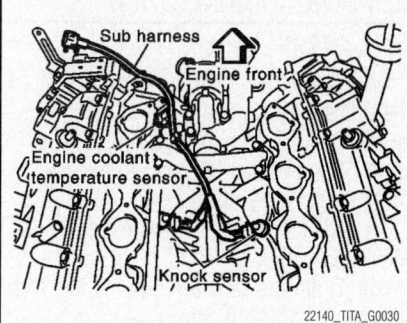

Fig. 55 Remove knock sensor and sub harness

➡**Handle carefully to avoid any damage to the electric throttle control actuator. Do not disassemble.**

7. Remove the fuel injectors and fuel tube assembly.
8. Loosen the bolts in reverse order shown.
9. Remove the intake manifold.
10. Cover engine openings to avoid entry of foreign materials.
11. Clean all gasket mating surfaces, do not reuse gaskets.
12. Remove knock sensor and sub harness.

To install:

13. Reverse the removal procedure and note the following:
    - Install the knock sensor and sub harness, tighten the mounting bolts to 16 ft. lbs. (21 Nm).
14. Tighten the intake manifold in sequence shown for removal. Tighten the bolts to 73 inch. lbs. (8.3 Nm).
15. Install engine coolant and check for leaks.

MASS AIR FLOW (MAF) SENSOR

LOCATION

The Mass Air Flow (MAF) Sensor is mounted on the air filter housing.

REMOVAL & INSTALLATION

1. Remove the engine room cover.
2. Remove the Intake Air Temperature (IAT)/(MAF Sensor harness.
3. Remove the mounting screws and the (IAT)/(MAF sensor.

To install:

4. Install the (IAT)/(MAF sensor.
5. Install the harness connector.
6. Install the engine room cover.

THROTTLE POSITION SENSOR (TPS)

LOCATION

The throttle position sensor is integral to the electric throttle control actuator. The throttle control actuator is mounted at the front of the intake manifold.

REMOVAL & INSTALLATION

See Figure 56.

1. Remove the air intake duct.
2. Disconnect harness connector.
3. Disconnect water hoses.
4. Loosen the throttle body assembly mounting bolts in reverse order as shown in the figure.

To install:

5. Install the throttle body assembly with a new gasket.
6. Tighten the mounting bolts in sequence to 74 inch. lbs. (8.4 Nm)
7. Reconnect the water hose.
8. Reconnect the harness connector.
9. Reconnect the air intake duct.

VEHICLE SPEED SENSOR (VSS)

LOCATION

The transmission speed sensor is located at the rear of the transmission case, under the tail shaft. For (4WD) models look under the transfer case.

REMOVAL & INSTALLATION

1. Disconnect the speed sensor harness.
2. Remove the mounting bolt and the speed sensor.

To install:

3. Apply a small amount of transmission fluid to the speed sensor O-ring.
4. Install the speed sensor and tighten the mounting bolt to 51 inch. lbs. (5.8 Nm).

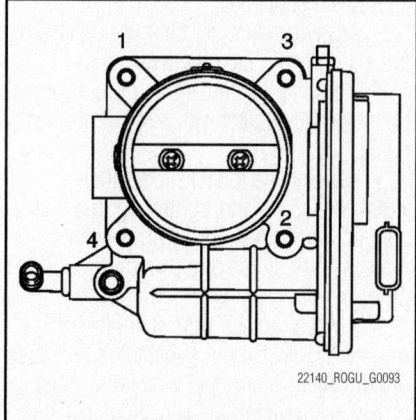

Fig. 56 Loosen mounting bolts in reverse order as shown

FUEL SYSTEM SERVICE PRECAUTIONS

Safety is the most important factor when performing not only fuel system maintenance but any type of maintenance. Failure to conduct maintenance and repairs in a safe manner may result in serious personal injury or death. Maintenance and testing of the vehicle's fuel system components can be accomplished safely and effectively by adhering to the following rules and guidelines.

• To avoid the possibility of fire and personal injury, always disconnect the negative battery cable unless the repair or test procedure requires that battery voltage be applied.

• Always relieve the fuel system pressure prior to disconnecting any fuel system component (injector, fuel rail, pressure regulator, etc.), fitting or fuel line connection. Exercise extreme caution whenever relieving fuel system pressure to avoid exposing skin, face and eyes to fuel spray. Please be advised that fuel under pressure may penetrate the skin or any part of the body that it contacts.

• Always place a shop towel or cloth around the fitting or connection prior to loosening to absorb any excess fuel due to spillage. Ensure that all fuel spillage (should it occur) is quickly removed from engine surfaces. Ensure that all fuel soaked cloths or towels are deposited into a suitable waste container.

• Always keep a dry chemical (Class B) fire extinguisher near the work area.

• Do not allow fuel spray or fuel vapors to come into contact with a spark or open flame.

• Always use a back-up wrench when loosening and tightening fuel line connection fittings. This will prevent unnecessary stress and torsion to fuel line piping.

• Always replace worn fuel fitting O-rings with new Do not substitute fuel hose or equivalent where fuel pipe is installed.

Before servicing the vehicle, make sure to also refer to the precautions in the beginning of this section as well.

1. When replacing fuel line parts, be sure to observe the following:
   • Put a "CAUTION: FLAMMABLE" sign in the workshop.
   • Be sure to work in a well ventilated area and furnish workshop with a CO2 fire extinguisher.

• Never smoke while servicing fuel system. Keep open flames and sparks away from the work area. fuel line parts, perform out the following procedures:
• Put drained fuel in an explosion-proof container and put the lid on securely. Keep the container in safe area.
• Release fuel pressure from the fuel lines.
• Disconnect the battery cable from the negative terminal.
• Always replace O-ring and clamps with new ones.
• Never bend or twist tubes when they are being installed.
• Never tighten hose clamps excessively to avoid damaging hoses.

2. After installing tubes, check that there is no fuel leakage at connections in the following steps.

3. Apply fuel pressure to fuel lines with turning ignition switch "ON" (with engine stopped). Then check for fuel leakage at connections.

4. Start engine and rev it up and check for fuel leakage at connections.

5. Use only a genuine NISSAN fuel filler cap as a replacement. If an incorrect fuel filler cap is used, the "MIL" may come on.

RELIEVING FUEL SYSTEM PRESSURE

WITH CONSULT-II

1. Turn ignition switch **ON**.
2. Perform "FUEL PRESSURE RELEASE" in "WORK SUPPORT" mode with CONSULT-II.
3. Start the engine.
4. After engine stalls, turn over the engine two or three times to release all fuel pressure.
5. Turn ignition switch **OFF**.

WITHOUT CONSULT-II

1. Remove fuel pump fuse located in IPDM E/R.
2. Start the engine.
3. After the engine stalls, turn over engine two or three times to release all fuel pressure.
4. Turn ignition switch **OFF**.
5. Reinstall fuel pump fuse after servicing fuel system.

FUEL FILTER

REMOVAL & INSTALLATION

The fuel filter is integral to the fuel pump assembly.

FUEL INJECTORS

REMOVAL & INSTALLATION

See Figure 57.

1. Before servicing the vehicle, refer to the Precautions Section.
2. Remove engine cover
3. Relieve fuel system pressure.
4. Remove or disconnect the following:
   • Negative battery cable
   • Fuel injector harness connectors
   • Fuel hose assembly from right and left fuel rails
   • Fuel injectors with fuel rail as an assembly
   • Fuel injector from fuel rail

To install:
5. Install or connect the following:
   • New clip onto the fuel injector
   • Fuel injector to fuel rail
   • Fuel injectors and fuel rail as an assembly to the intake manifold
   • Fuel hose assembly
   • Fuel injector harness connectors
   • Negative battery cable
   • Engine cover
6. Start engine and check for leaks.

FUEL PUMP

REMOVAL & INSTALLATION

See Figure 58.

1. Before servicing the vehicle, refer to the Precautions Section.
2. Relieve the fuel system pressure.
3. Remove fuel filler cap to release pressure from inside tank.
4. Disconnect fuel filler hose from fuel filler pipe.
5. Drain fuel tank through the fuel filler opening using a suitable hose.
6. Disconnect the following:
   • Fuel pump line protector
   • EVAP hose
   • Fuel level sensor
   • Fuel filter
   • Fuel pump wiring harness
   • Fuel supply hose
7. Using a suitable jack to support the fuel tank, remove the strap bolts and remove the fuel tank from the vehicle.

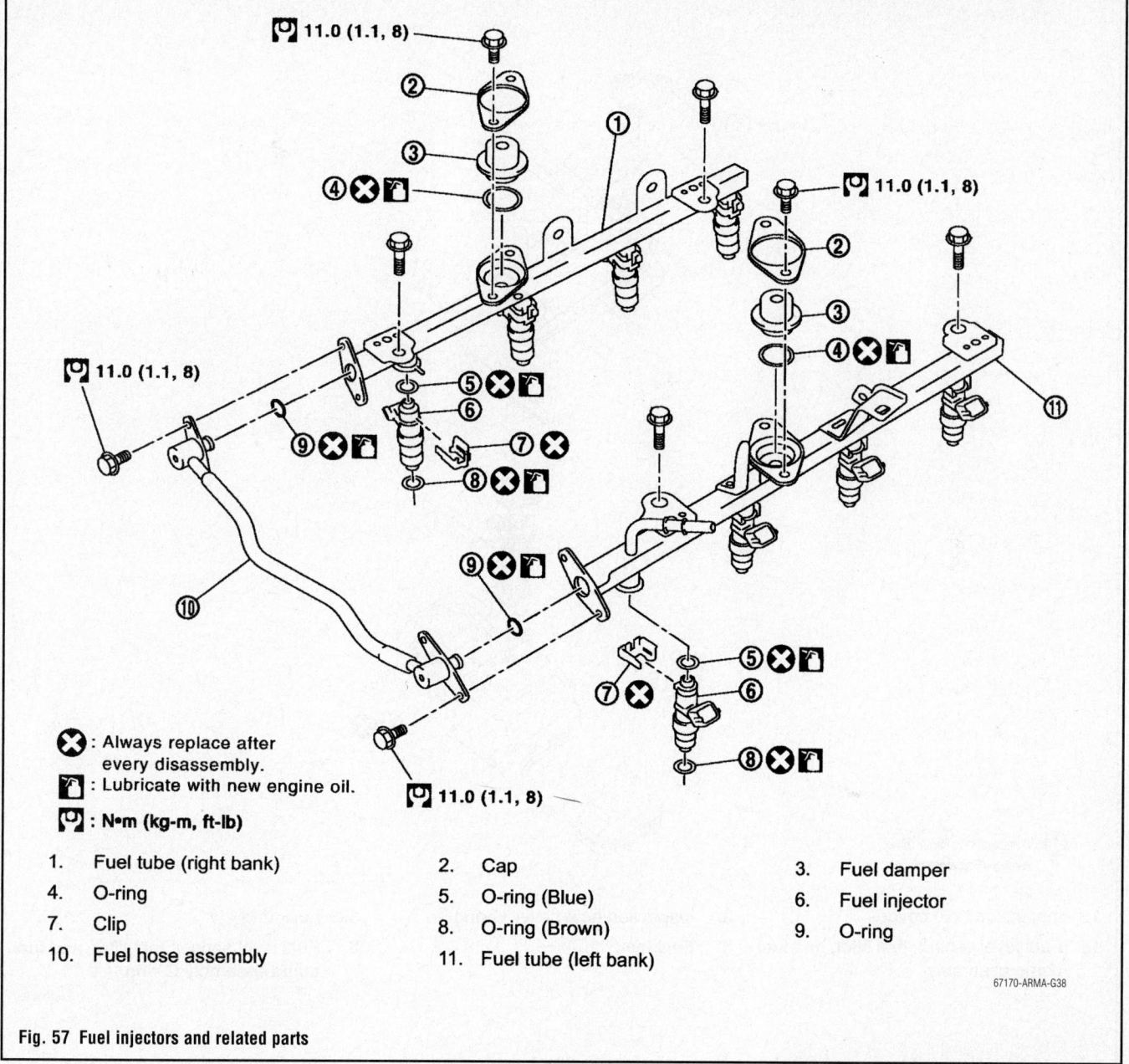

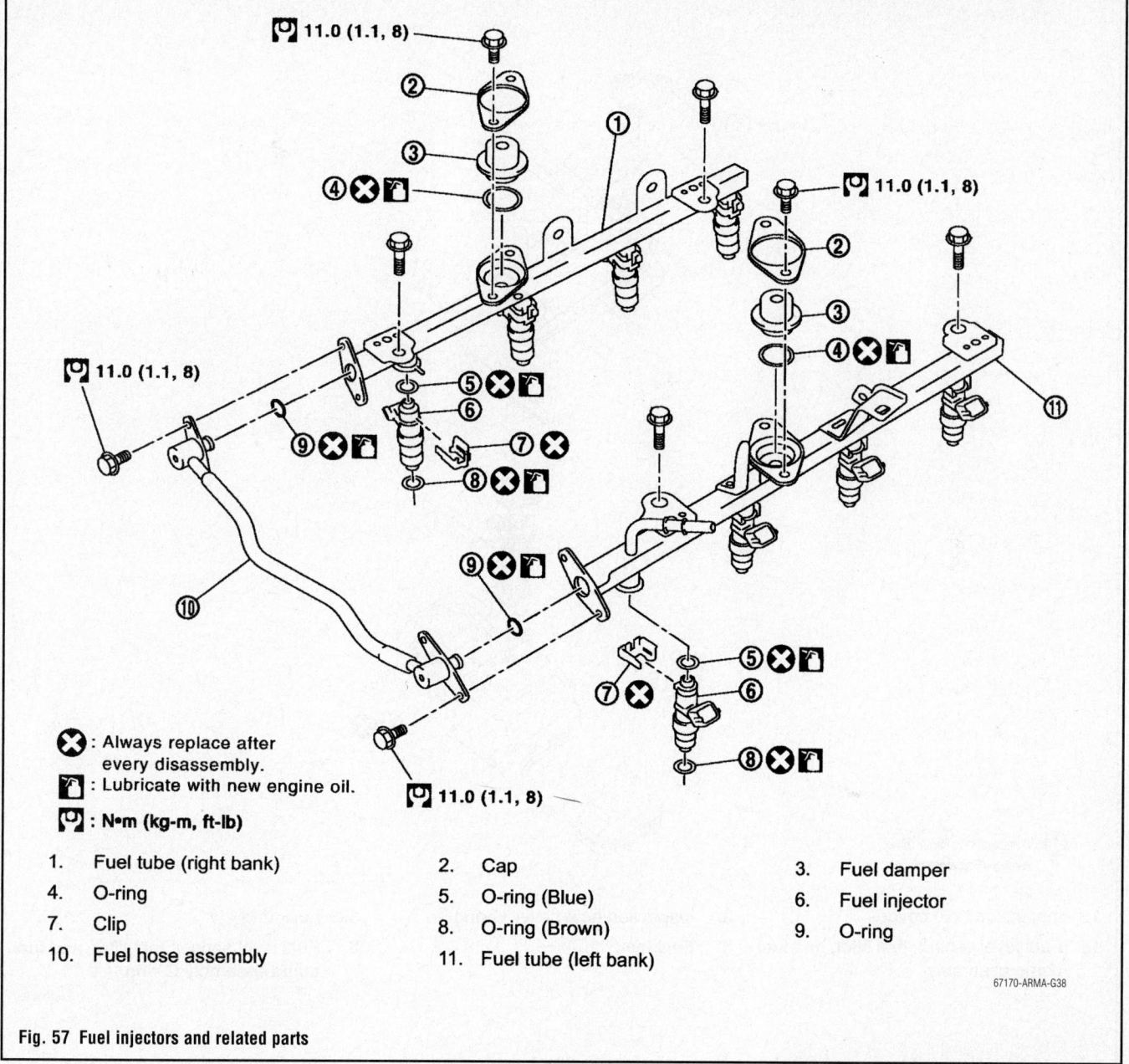: Always replace after every disassembly.

: Lubricate with new engine oil.

: N•m (kg-m, ft-lb)

1. Fuel tube (right bank)
2. Cap
3. Fuel damper
4. O-ring
5. O-ring (Blue)
6. Fuel injector
7. Clip
8. O-ring (Brown)
9. O-ring
10. Fuel hose assembly
11. Fuel tube (left bank)

67170-ARMA-G38

Fig. 57 Fuel injectors and related parts

8. Remove the lock ring using special tool J-46536.
9. Remove the following:
   • Fuel level sensor
   • Fuel filter
   • Fuel pump assembly

To install:

10. Install or connect the following:
   • Fuel pump assembly, using new O-ring
   • Fuel filter, using new filter
   • Fuel level sensor, using new sensor
   • Fuel pump assembly lock ring
   • Fuel tank
   • Fuel supply hose
   • Fuel pump wiring harness
   • EVAP hose

   • Fuel pump line protector
   • Fuel filler pipe
11. Start engine and check for leaks.

FUEL TANK

REMOVAL & INSTALLATION

See Figure 59.

✳✳ WARNING

When replacing fuel line parts, be sure to work in a well ventilated area and furnish workshop with a CO2 fire extinguisher. Do not smoke while servicing fuel system. Keep open flames and sparks away from the work area.

✳✳ CAUTION

Always replace O-rings and clamps with new ones. Do not kink or twist hoses when they are being installed. Do not tighten hose clamps excessively to avoid damaging hoses. Tighten high-pressure rubber hose clamp so that clamp end is 0.12 inches (3 mm) from hose end. Tightening torque specifications are the same for all rubber hose clamps. Ensure that screw does not contact adjacent parts.

1. Before servicing the vehicle, refer to the Precautions Section.

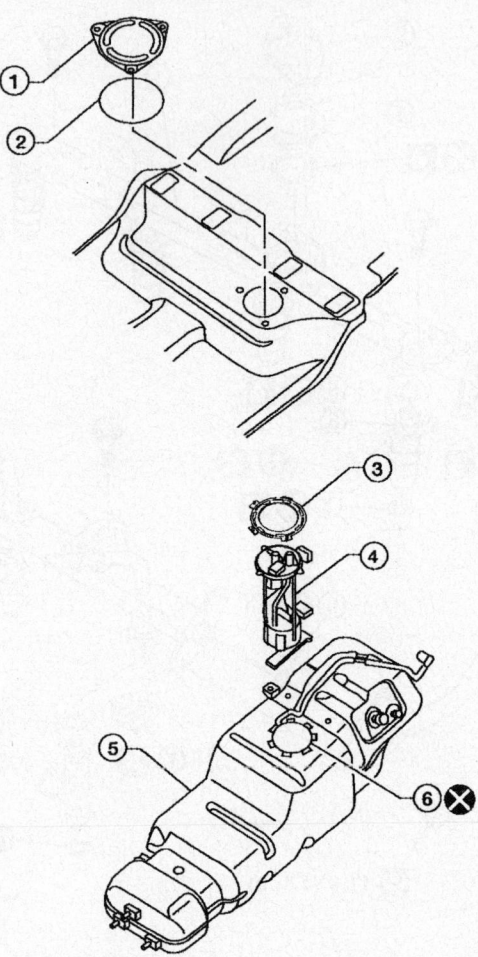

⊗ : Always replace after
every disassembly

1. Inspection hole cover
2. Inspection hole cover O-ring
3. Lock ring
4. Fuel level sensor, fuel filter, and fuel pump assembly
5. Fuel tank
6. Fuel level sensor, fuel filter, and fuel pump assembly O-ring

67170-ARMA-G37

Fig. 58 Fuel pump and related parts

2. Remove the fuel filler cap to release the pressure from inside the fuel tank.

3. Check the fuel level on level gauge. If the fuel gauge indicates more than the level as full or almost full, drain the fuel from the fuel tank until the fuel gauge indicates the level as less.

4. If the fuel pump does not operate, use the following procedure to drain the fuel:

- Insert a suitable hose of less than 15 mm (0.59 in) diameter into the fuel filler pipe through the fuel filler opening to drain the fuel from fuel filler pipe.
- Remove the LH rear wheel and tire.

- Remove the fuel filler pipe shield.
- Disconnect the fuel filler hose from the fuel filler pipe and disconnect the vent hose quick connector.
- Insert a suitable hose into the fuel tank through the fuel filler hose to drain the fuel from the fuel tank.

5. Release the fuel pressure from the fuel lines.

6. Disconnect the negative battery cable.

7. Remove the three nuts and remove fuel line pump protector.

8. Disconnect the EVAP hose at the EVAP canister.

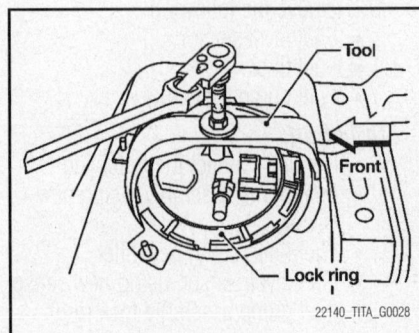

22140_TITA_G0028

Fig. 59 Remove the lock ring using Tool: J-46536

9. Disconnect the fuel level sensor, fuel filter, and fuel pump assembly electrical connector, and the fuel feed hose.

✳✳ CAUTION

Observe the following when disconnecting the quick-connectors:

- The tube can be removed when the tabs are completely depressed. Do not twist it more than necessary.
- Do not use any tools to remove the quick connector.
- Keep the resin tube away from heat. Be especially careful when welding near the tube.
- Prevent liquid acids, such as battery electrolyte, from getting on the resin tube.
- Do not bend or twist the tube during installation and removal.
- Only when the tube is replaced, remove the remaining retainer on the tube or fuel level sensor, fuel filter, and fuel pump assembly.
- When the tube or fuel level sensor, fuel filter, and fuel pump assembly is replaced, also replace the retainer with a new one (green colored retainer).
- To keep the connecting portion clean and to avoid damage and foreign materials, cover them completely with plastic bags or something similar.

10. Disconnect the quick-connectors by performing the following:
- Hold the sides of the connector, push in tabs and pull out the tube.
- If the connector and the tube are stuck together, push and pull several times until they start to move. Then disconnect them by pulling.

11. Remove the four bolts and remove the fuel tank shield using power tool.

12. Disconnect fuel filler hose at the fuel tank side.

13. Remove the fuel tank strap bolts while supporting the fuel tank with a suitable lift jack.

14. Lower the fuel tank using a suitable lift jack and remove it from the vehicle.

15. If necessary, remove the lock ring using Tool: J-46536.

16. If necessary, remove the fuel level sensor, fuel filter, and fuel pump assembly. Discard the fuel level sensor, fuel filter, and fuel pump assembly O-ring.

To install:

17. To install, reverse removal procedure and note the following:

   a. Check the connection for damage or any foreign materials.

   b. Align the connector with the tube, then insert the connector straight into the tube until a click is heard

   c. After the tube is connected, make sure the connection is secure by pulling on the tube and the connector to make sure they are securely connected.

   d. Tighten the fuel tank strap mounting bolts to 32 ft. lbs. (44 Nm).

18. Turn the ignition switch ON but do not start engine, then check the fuel pipe and hose connections for leaks while applying fuel pressure to the system.

19. Start the engine and rev it above idle speed, then check that there are no fuel leaks at any of the fuel pipe and hose connections.

IDLE SPEED

ADJUSTMENT

Idle speed is maintained by the Powertrain Control Module (PCM). No adjustment is necessary or possible.

THROTTLE BODY

REMOVAL & INSTALLATION

See Figure 60.

1. Remove the engine cover.
2. Remove the air intake duct.
3. Disconnect harness connector.
4. Disconnect water hoses.
5. Loosen the throttle body assembly mounting bolts in reverse order as shown in the figure.

To install:

6. Install the throttle body assembly with a new gasket.
7. Tighten the mounting bolts in sequence to 74 inch. lbs. (8.4 Nm)
8. Reconnect the water hose.
9. Reconnect the harness connector.
10. Reconnect the air intake duct.
11. Install the engine cover.

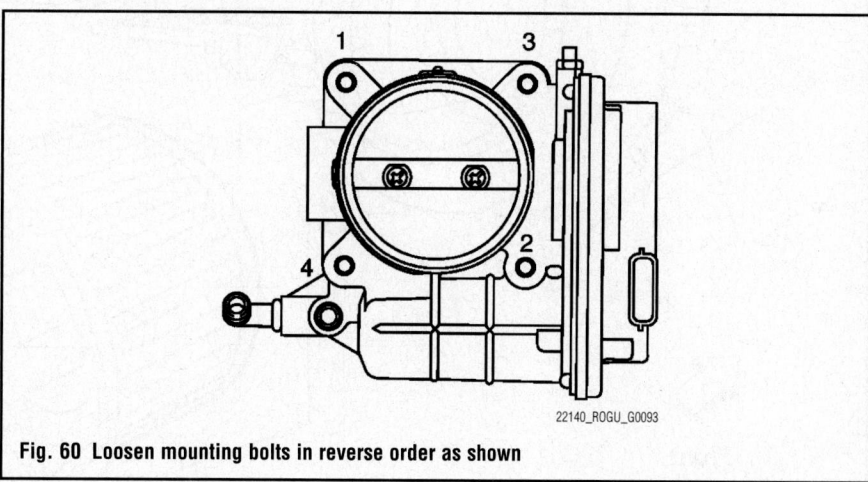

22140_ROGU_G0093

Fig. 60 Loosen mounting bolts in reverse order as shown

HEATING & AIR CONDITIONING SYSTEM

BLOWER MOTOR

REMOVAL & INSTALLATION

See Figure 61.

1. Remove the glove box assembly.
2. Disconnect the blower motor electrical connector.
3. Remove the three screws and remove the blower motor.

To install:

4. Install the three screws and Install the blower motor.

5. Reconnect the blower motor electrical connector.
6. Install the glove box assembly.

HEATER CORE

REMOVAL & INSTALLATION

See Figures 62 and 63.

1. Before servicing the vehicle, refer to the Precautions Section.
2. Discharge the A/C system.
3. Drain the cooling system.

4. Remove or disconnect the following:
- Right and left wiper arms
- Cowl top seal
- Left and right-hand cowl top covers
- Cowl top extension brackets
- Wiper motor and connecting rod linkage
- Windshield washer tube
- A/C low pressure pipe bracket from the cowl top extension
- Drain tube from each side of the cowl top extension

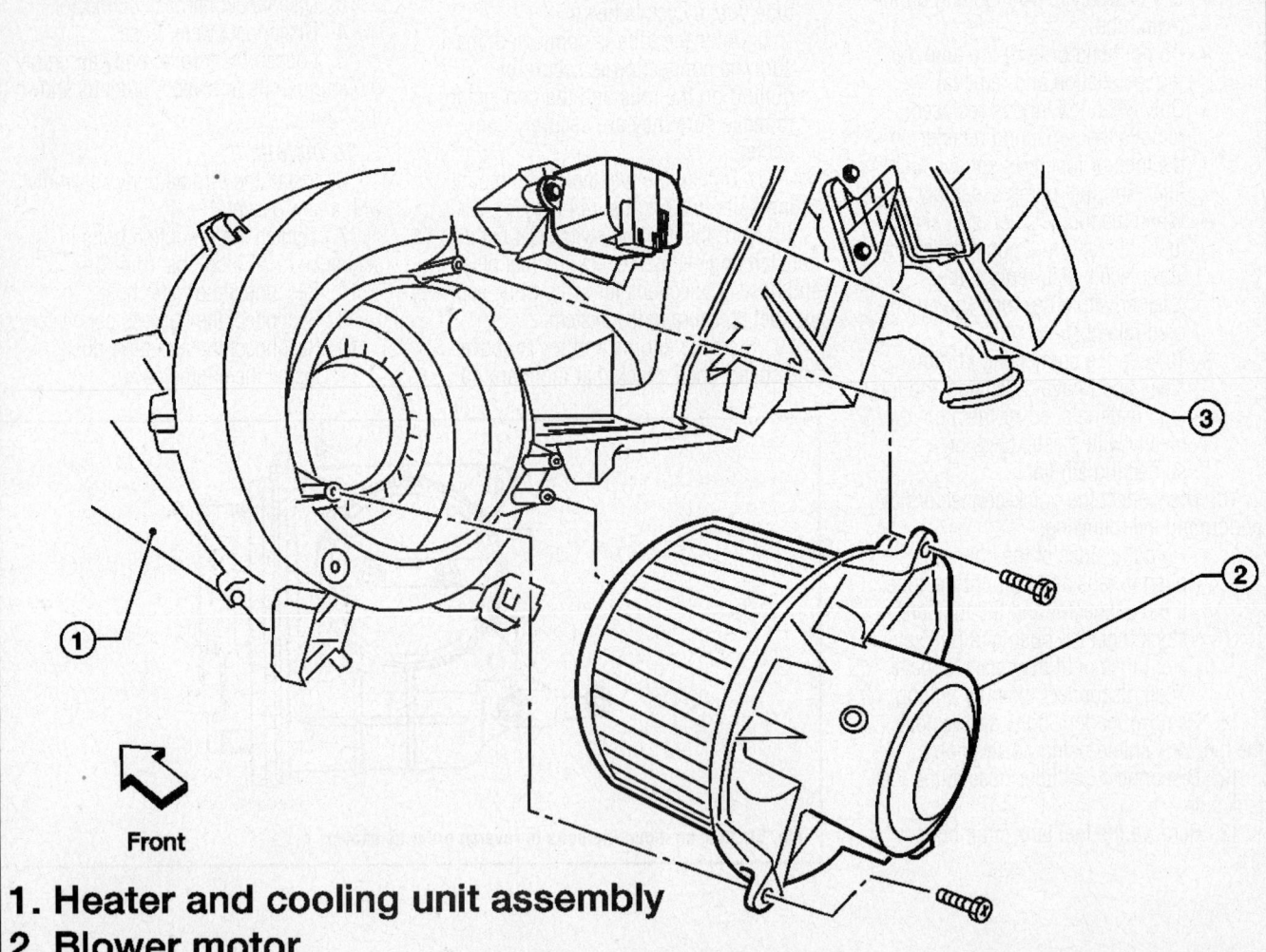

Front

1. Heater and cooling unit assembly
2. Blower motor
3. Variable blower control or front blower motor resistor if equipped

22140_TITA_G0033

Fig. 61 Blower motor unit assembly

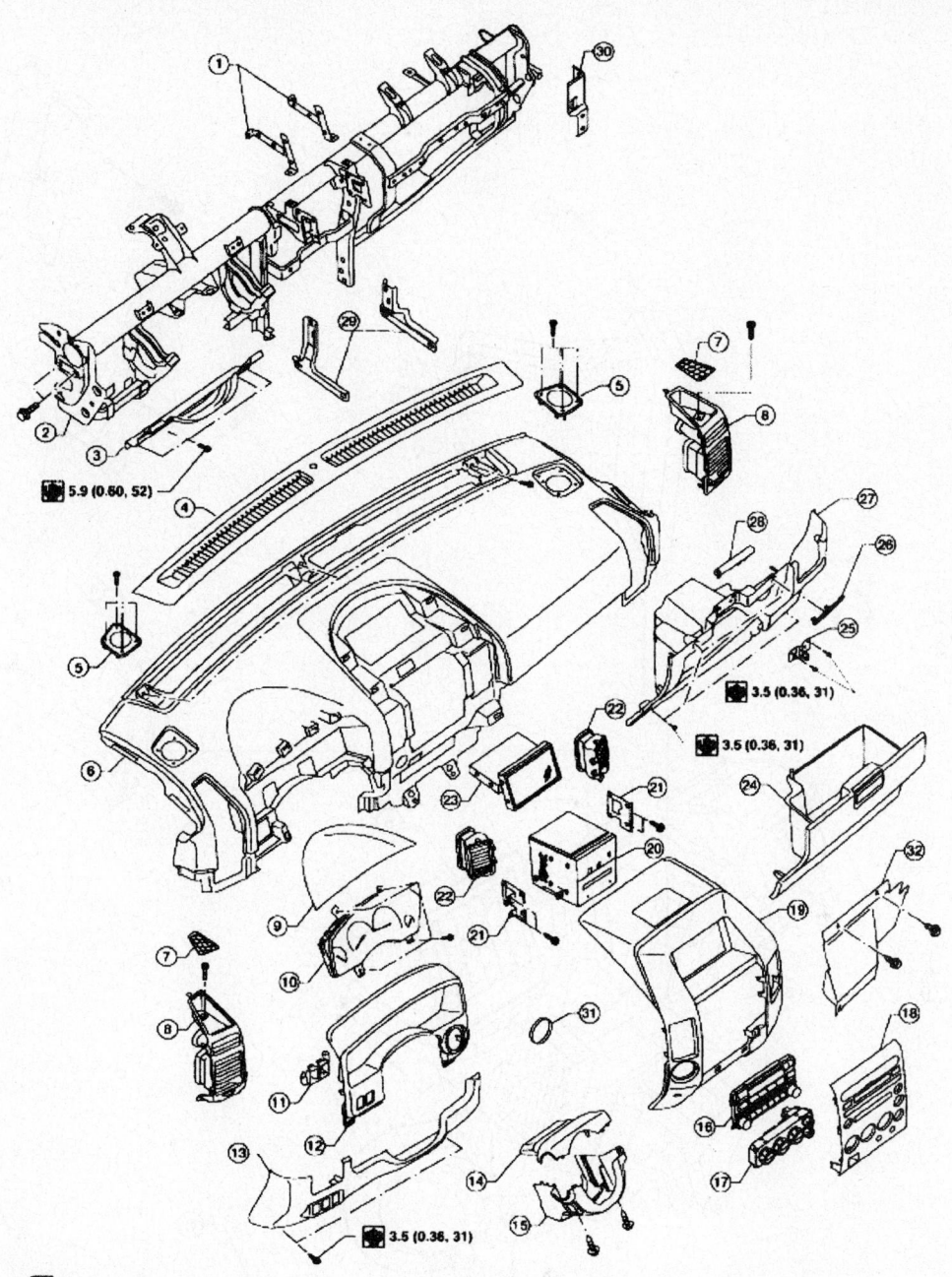

5.9 (0.60, 52)

3.5 (0.36, 31)

3.5 (0.36, 31)

3.5 (0.36, 31)

N·m (kg-m, in-lb)

1. Display unit bracket RH/LH	2. Steering member assembly	3. Lower knee protector
4. Defroster grille	5. Speaker grille RH/LH	6. Instrument panel and pad assembly
7. Deck pocket mat RH/LH	8. Side ventilator assembly RH/LH	9. Meter cover
10. Combination meter	11. Switch assembly	12. Cluster lid A
13. Lower instrument panel LH	14. Upper steering column cover	15. Lower steering column cover
16. Audio display switch assembly	17. Front air control	18. Cluster lid C
19. Cluster lid D	20. Audio unit	21. Radio Bracket RH/LH
22. Center ventilator assembly RH/LH	23. Display assembly	24. Glove box
25. Glove box lid striker	26. Fuse block cover	27. Lower instrument panel RH
28. Glove box damper	29. Instrument stay RH/LH	30. Instrument side bracket
31. Key cylinder escutcheon	32. Lower instrument panel RH	

09482_TITAN_G0001

Fig. 62 Exploded view of the Instrument Panel

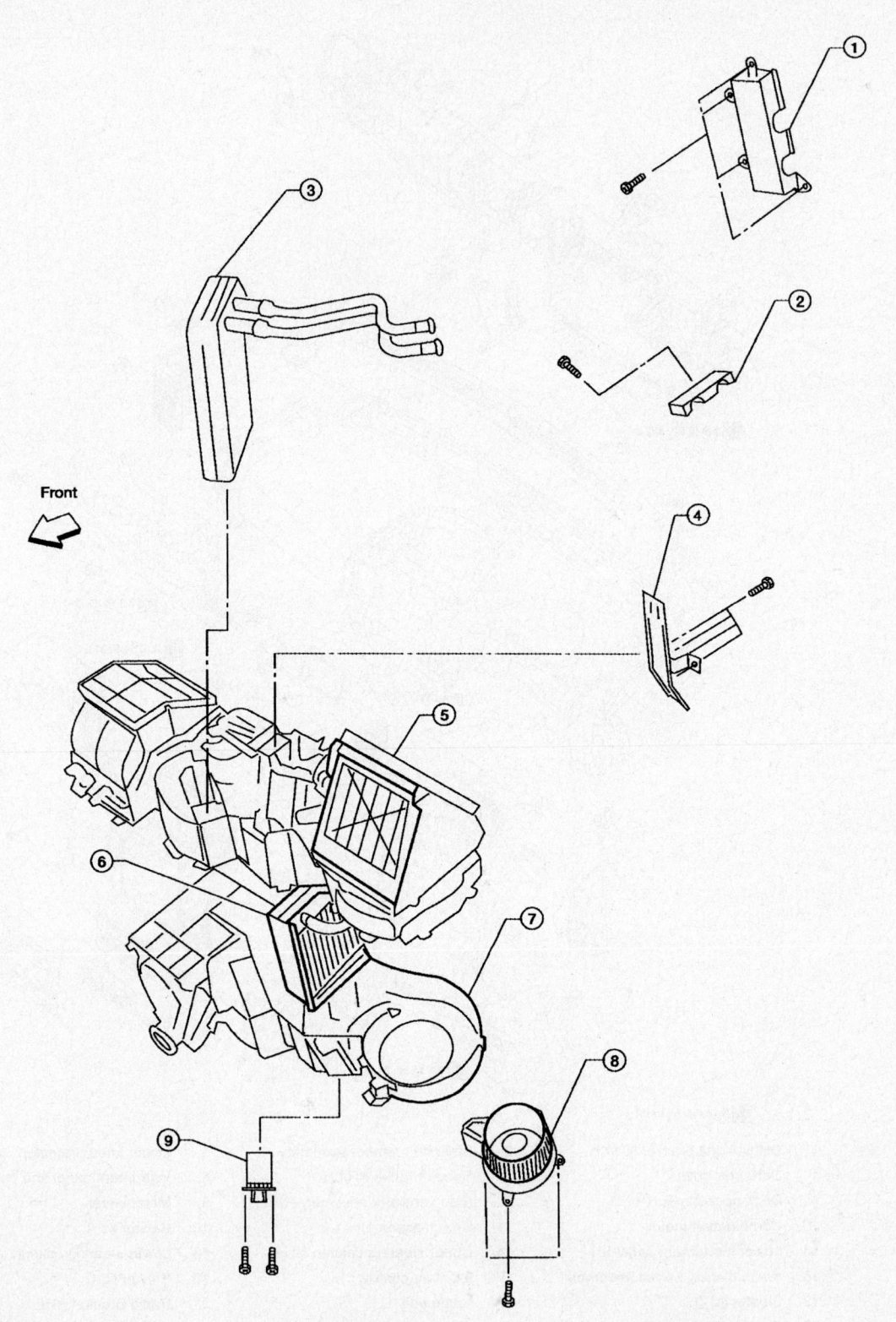

Front

1. Heater core cover
2. Heater core pipe bracket
3. Heater core
4. Upper bracket
5. Upper heater and cooling unit case
6. A/C evaporator
7. Lower heater and cooling unit case
8. Blower motor
9. Variable blower control

67170-ARMA-G26

Fig. 63 Front heater/AC assembly

- Cowl top extension
- All necessary exhaust system components
- Front heater hoses from the heater core
- High/Low pressure pipes
- Center console
- Steering column
- Gauge cluster
- Defroster grille
- Left-hand side ventilator assembly

- Right-hand assist grip and windshield trim
- Passenger air bag module
- Instrument and console panels
- Instrument panel wiring harness
- Steering member from each side of the vehicle body

5. Remove the heater assembly from the vehicle, attached to the steering gear assembly.

6. Remove the heating and cooling unit assembly from the steering assembly.

7. Remove the four bolts and remove the upper bracket.

8. Remove the four bolts and remove the heater core cover.

9. Remove the heater core pipe bracket.

10. Remove the heater core.

11. Installation is the reverse order of removal.

12. Refill the cooling system.

13. Recharge the A/C system.

STEERING

POWER STEERING GEAR

REMOVAL & INSTALLATION

See Figure 64.

1. Before servicing the vehicle, refer to the Precautions Section.

2. Ensure the wheels are in the straight-ahead position.

3. Remove or disconnect the following:
   - Wheels
   - Engine splash guard

4. On 4-wheel drive models only, remove front final drive and support the drive shafts.

5. Remove cotter pin at steering outer socket and loosen mounting nut.

6. Remove steering outer socket from steering knuckle using special tool J-25730-A.

7. On 2-wheel drive models only, remove stabilizer bar mounting bolts and secure the stabilizer bar.

8. Remove or disconnect the following:
   - Oil pipes from steering gear assembly
   - Lower joint mounting bolt from lower shaft
   - Mounting bolts and nuts from steering gear assembly
   - Steering gear assembly

To install:

9. Install or connect the following:
   - Steering gear assembly, tighten nuts to 140 ft. lbs. (190 Nm)
   - Lower joint mounting bolt
   - Oil pipes to steering gear assembly
   - Stabilizer bar, 2 wheel-drive models only
   - Steering outer socket to steering knuckle, tighten nut to 63 ft. lbs. (86 Nm)
   - Front final drive, 4-wheel drive models only
   - Engine splash guard
   - Wheels

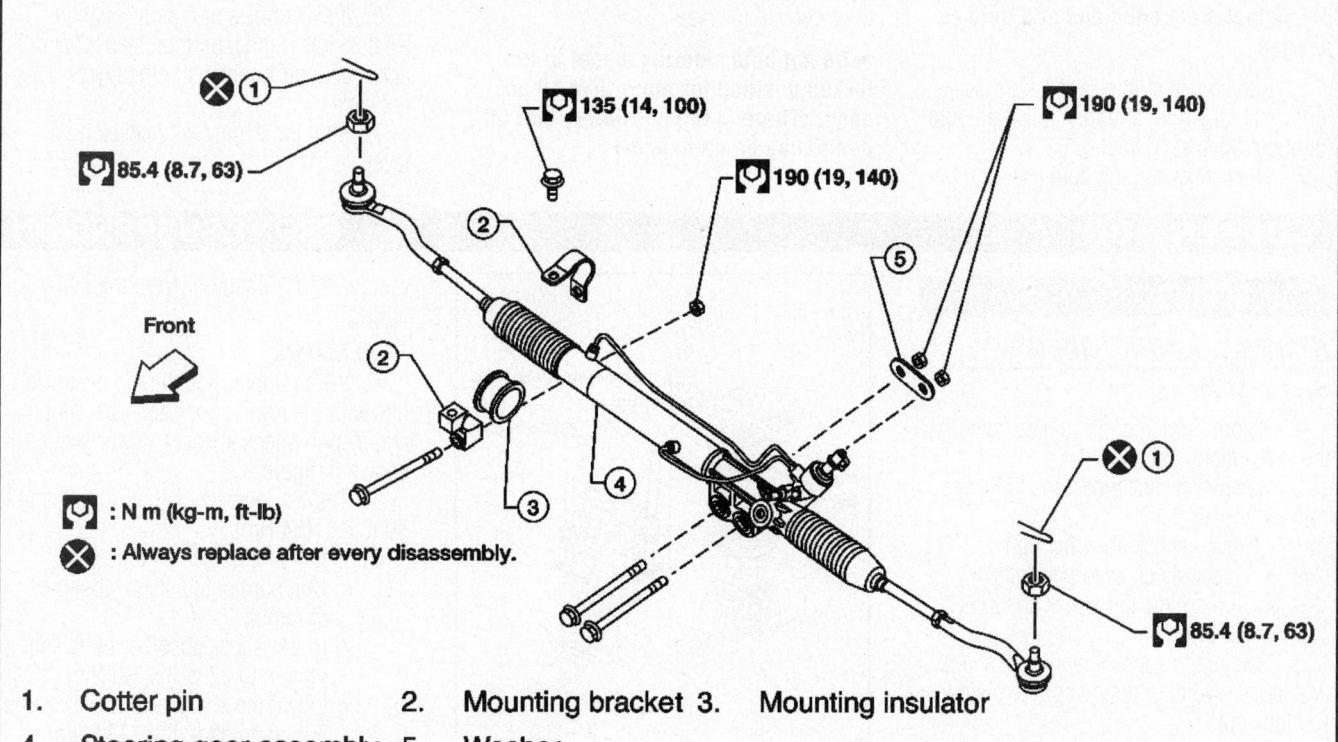

1. Cotter pin
2. Mounting bracket
3. Mounting insulator
4. Steering gear assembly
5. Washer

: N m (kg-m, ft-lb)

: Always replace after every disassembly.

22140_TITA_G0031

Fig. 64 Steering gear and related parts

10. Check the wheel alignment and adjust as necessary.

POWER STEERING PUMP

REMOVAL & INSTALLATION

See Figure 65.

1. Drain power steering fluid from reservoir tank.
2. Remove air duct assembly.
3. Remove power steering reservoir tank.
4. Remove serpentine drive belt from auto tensioner and power steering pump.
5. Disconnect pressure sensor electrical connector.
6. Remove the high pressure and low pressure piping from power steering oil pump.
7. Remove bolts, then remove power steering pump.
8. To install, reverse removal procedure.
9. Bleed power steering system of air.

BLEEDING

Recommended fluid is Genuine NISSAN PSF or equivalent.

1. Stop engine, and then turn steering wheel fully to right and left several times.

➡**Do not allow steering fluid reservoir tank to go below the MIN level line. Check tank frequently and add fluid as needed.**

2. Run engine at idle speed. Turn steering wheel fully right and then fully left, hold for about three seconds.
3. Then check for any fluid leaks.

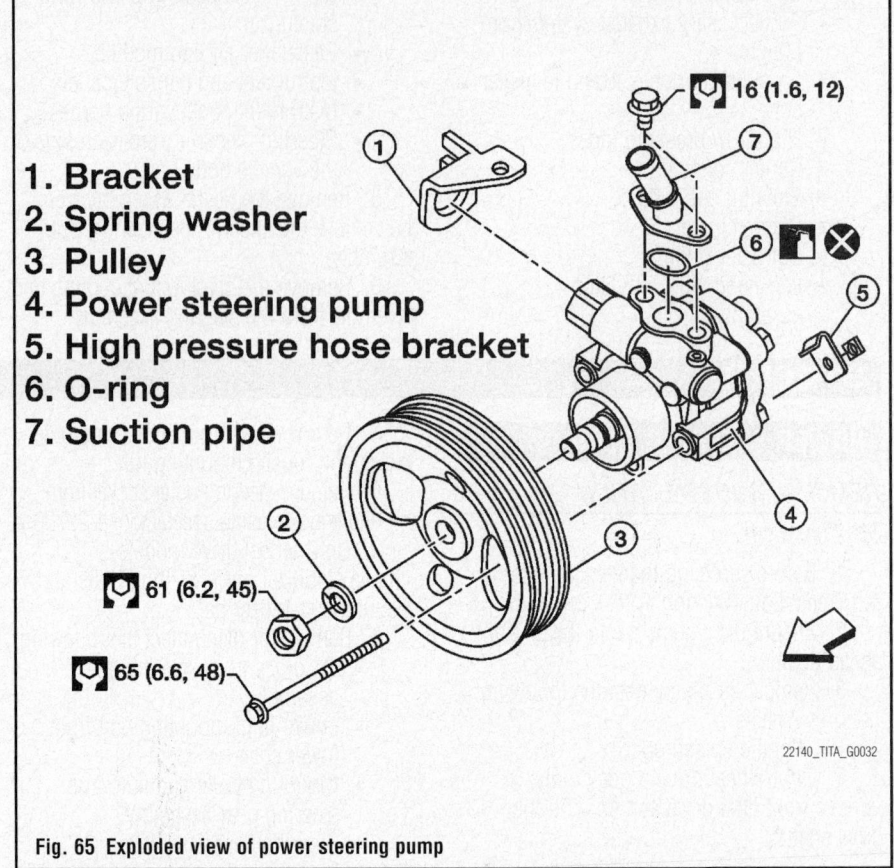

1. Bracket
2. Spring washer
3. Pulley
4. Power steering pump
5. High pressure hose bracket
6. O-ring
7. Suction pipe

16 (1.6, 12)

61 (6.2, 45)

65 (6.6, 48)

22140_TITA_G0032

Fig. 65 Exploded view of power steering pump

4. Repeat step 2 several times at about three second intervals.

➡**Do not hold steering wheel in the locked position for more than 10 seconds. (There is the possibility that oil pump may be damaged.)**

5. Check for air bubbles or cloudy fluid.
6. If air bubbles or cloudiness still exists, stop engine, perform steps 2 and 3 again until air bubbles or cloudiness does not exist.
7. Stop the engine, and check fluid level.

SUSPENSION

FRONT SUSPENSION

COIL SPRING

REMOVAL & INSTALLATION

See Figure 66.

1. Before servicing the vehicle, refer to the Precautions Section.
2. Remove or disconnect the following:
   • Wheel
   • Lower shock absorber bolt
   • Upper shock absorber bolts
   • Coil spring and shock absorber assembly
3. Secure the shock absorber in a vice and loosen (without removing) the piston rod lock nut.
4. Install a spring compressor and tighten until the shock absorber mounting insulator can be turned by hand.
5. Remove piston rod lock nut and

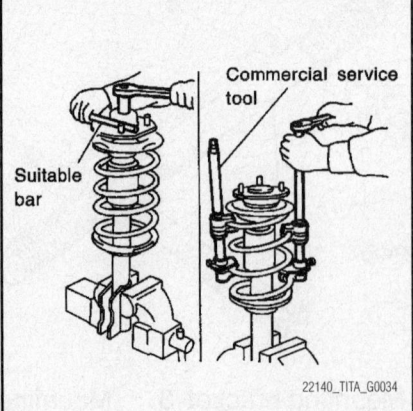

Suitable bar

Commercial service tool

22140_TITA_G0034

Fig. 66 Coil spring removal with compressing tool

remove shock absorber from the coil spring.

To install:

6. Install upper mounting insulator in line with the lower shock absorber mount and step in shock absorber lower seat as shown in figure.
7. Tighten the new piston rod lock nut to 40 ft. lbs. (54 Nm).
8. Install or connect the following:
   • Coil spring and shock absorber assembly
   • Upper shock absorber bolts and tighten to 22 ft. lbs (30 Nm)
   • Lower the shock absorber bolt and tighten to 99 ft. lbs. (134 Nm)
   • Wheel
9. Check wheel alignment and adjust as necessary.

LOWER BALL JOINT

REMOVAL & INSTALLATION

The lower ball joints are integral to the lower control arm. They may be referred to as a lower link.

LOWER CONTROL ARM

REMOVAL & INSTALLATION

See Figures 67 and 68.

1. Remove the wheel and tire using power tool.
2. Remove lower shock absorber bolt.
3. Remove stabilizer bar connecting rod lower nut using power tool, then separate connecting rod from lower link.
4. Remove drive shaft, if equipped.
5. Remove pinch bolt from steering knuckle using power tool, then separate lower link ball joint from steering knuckle.
6. Remove lower link cam bolts (1) and nuts, then the lower link (2).

   To install:

7. Install all removed parts in the reverse order of removal procedure and note the following:

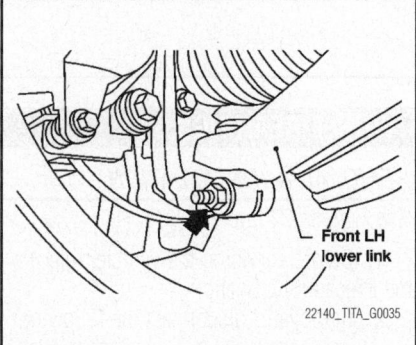

Front LH
lower link

22140_TITA_G0035

Fig. 67 Remove pinch bolt from steering knuckle

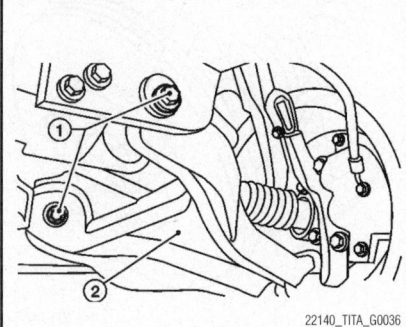

22140_TITA_G0036

Fig. 68 Remove lower link cam bolts

8. Tighten the cam nuts only in a relaxed position to 98 ft. lbs. (133 Nm). Cam bolts are for adjustment.
9. Tighten the steering knuckle pinch bolt to 70 ft. lbs. (95 (Nm).
10. After installation, check that the front wheel alignment is within specification.

STABILIZER BAR

REMOVAL & INSTALLATION

See Figure 69.

1. Before servicing the vehicle, refer to the Precautions Section.
2. Remove or disconnect the following:
   • Engine undercovers
   • Stabilizer bar mounting bracket bolts and connecting rod nuts using power tool
   • Bushings from stabilizer bar
3. Check stabilizer bar for twist and deformation. Replace if necessary. Check rubber bushing for cracks, wear and deterioration. Replace if necessary.

   To install:

4. To install, reverse removal procedure and note the following:
   a. Tighten stabilizer bar bracket mounting bolts to 94 ft. lbs. (128 Nm).
   b. Tighten the connecting rod nuts and tighten to 62 ft. lbs. (84 Nm).

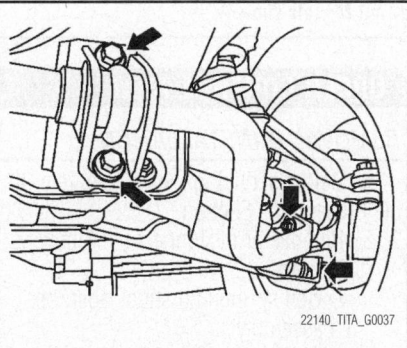

22140_TITA_G0037

Fig. 69 Stabilizer bar mounting bracket bolts and connecting rod nuts

STEERING KNUCKLE

REMOVAL & INSTALLATION

See Figure 70.

1. Before servicing the vehicle, refer to the Precautions Section.
2. Remove or disconnect the following:
   • Wheel
   • Engine splash guard

➡**Disconnect wheel sensor harness connector. Do not remove wheel sensor**

from wheel hub and bearing assembly for this procedure.

• Wheel and hub assembly (Remove cotter pin, then remove drive shaft nut 4WD models)

✷✷ CAUTION

Be careful not to damage ball joint boot. Temporarily tighten nut to prevent damage to threads and to prevent Tool from coming off.

• Steering outer socket from steering knuckle using Tool HT72520000 (J-25730-A)
• Coil spring and shock absorber assembly

➡**Support the lower link using a suitable jack.**

• Cotter pin and nut from upper link ball joint and discard the cotter pin
3. Separate the upper link ball joint from steering knuckle using Tool ST29020001 (J-24319-01).
4. Remove pinch bolt from steering knuckle using power tool, then separate lower link ball joint from steering knuckle.
5. Remove the steering knuckle from vehicle.
6. Check for deformity, cracks and damage on each part, replace if necessary.

   To install:

7. To install, reverse removal procedure and note the following:
   a. Tighten the steering knuckle pinch bolt to 70 ft. lbs. (95 (Nm).
   b. Tighten the upper control arm ball joint nut to 58 ft. lbs. (79 Nm).
   c. Use a new cotter pins for installation of lock nuts.
   d. For 4WD models tighten the axle nut 101 ft. lbs. (137 Nm).

STRUT & SPRING ASSEMBLY

REMOVAL & INSTALLATION

1. Before servicing the vehicle, refer to the Precautions Section.
2. Remove or disconnect the following:
   • Wheel
   • Lower shock absorber bolt
   • Upper shock absorber bolts
   • Coil spring and shock absorber assembly
3. Secure the shock absorber in a vice and loosen (without removing) the piston rod lock nut.
4. Install a spring compressor and tighten until the shock absorber mounting insulator can be turned by hand.

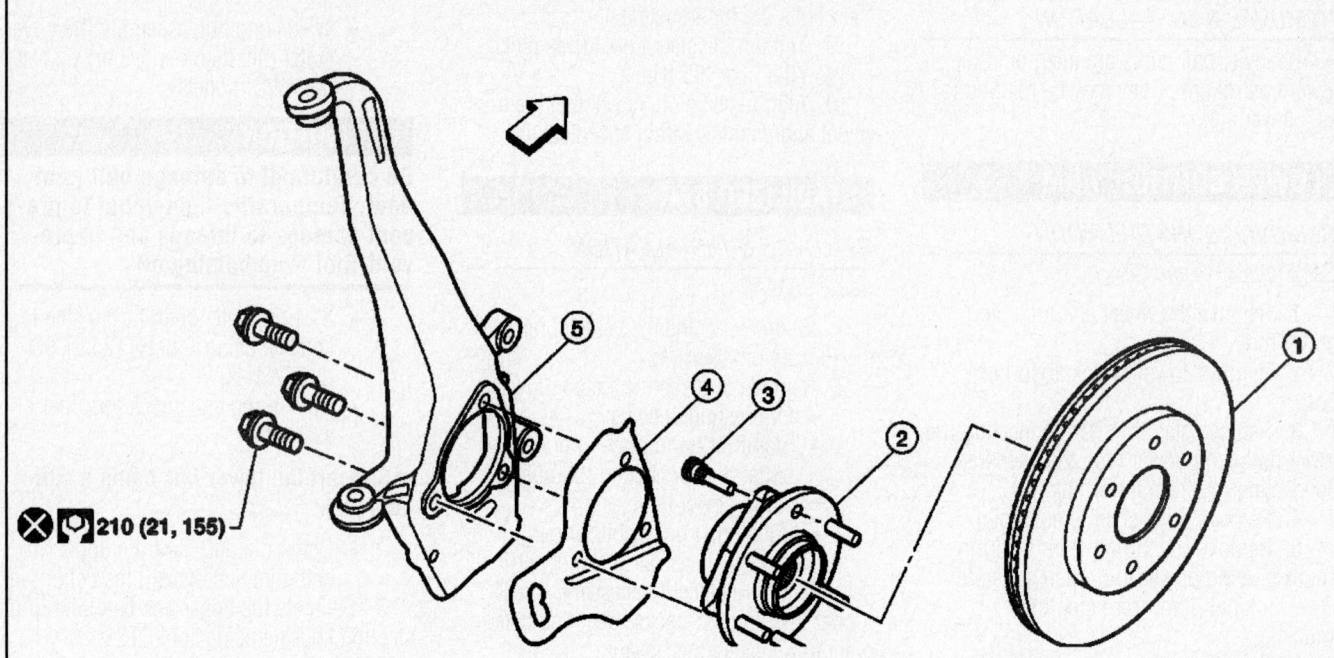

1. Disc rotor
4. Splash guard
2. Wheel hub and bearing assembly
5. Steering knuckle
3. Wheel stud
⇐ Front

⊗ ⟁ 210 (21, 155)

42050_TITA_G0025

Fig. 70 View of steering knuckle and components (2WD Models shown)

5. Remove piston rod lock nut and remove shock absorber from the coil spring.

To install:

6. Install upper mounting insulator in line with the lower shock absorber mount and step in shock absorber lower seat as shown in figure.

7. Tighten the new piston rod lock nut to 40 ft. lbs. (54 Nm).

8. Install or connect the following:
   • Coil spring and shock absorber assembly
   • Upper shock absorber bolts and tighten to 22 ft. lbs (30 Nm)
   • Lower the shock absorber bolt and tighten to 99 ft. lbs. (134 Nm)
   • Wheel

9. Check wheel alignment and adjust as necessary.

UPPER BALL JOINT

REMOVAL & INSTALLATION

The upper ball joints are integral to the upper control arm. They may be referred to as an upper link.

UPPER CONTROL ARM

REMOVAL & INSTALLATION

1. Before servicing the vehicle, refer to the Precautions Section.

2. Remove or disconnect the following:
   • Wheel
   • Coil spring and shock absorber assembly
   • Cotter pin and nut from upper ball joint

3. Separate upper ball joint stud from steering knuckle using special tool J-24319-01.

4. Remove the following:
   • Upper control arm mounting bolts
   • Upper arm

To install:

5. Install or connect the following:
   • Upper control arm and tighten bolts to 107 ft. lbs. (145 Nm)
   • Upper ball joint with new cotter pin and tighten nut to 58 ft. lbs. (79 Nm)
   • Coil spring and shock absorber assembly
   • Wheel

WHEEL BEARINGS

REMOVAL & INSTALLATION

See Figures 71 and 72.

1. Before servicing the vehicle, refer to the Precautions Section.

2. Remove or disconnect the following:

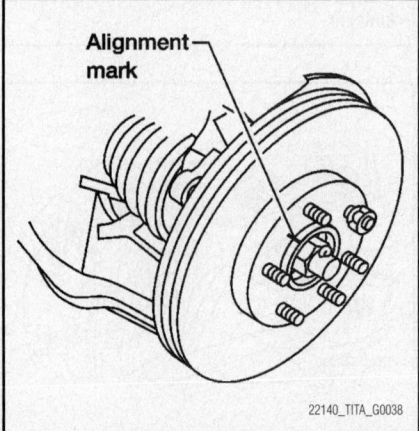

Alignment mark

22140_TITA_G0038

Fig. 71 Install a matchmark on the brake rotor and to the wheel hub (4WD) shown

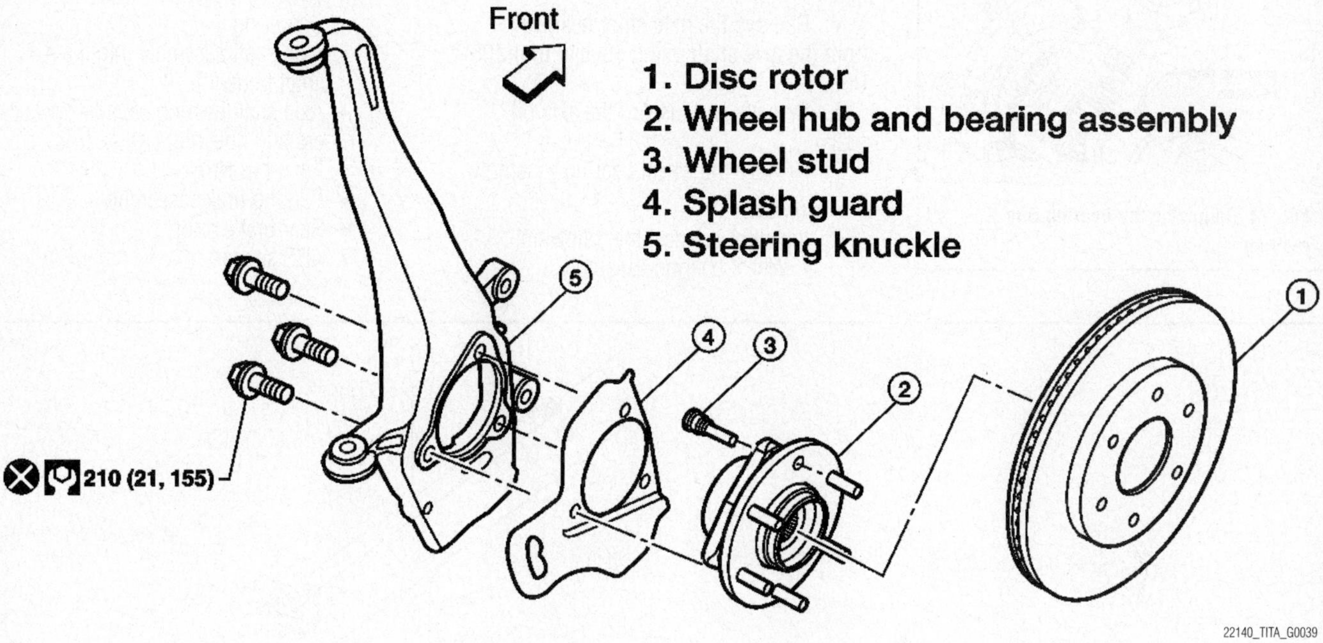

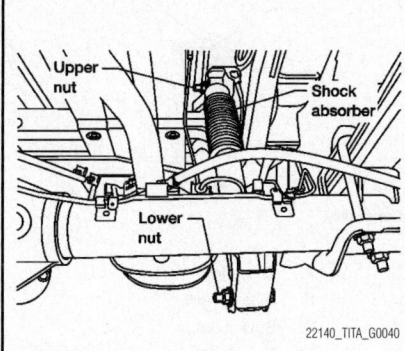

1. Disc rotor
2. Wheel hub and bearing assembly
3. Wheel stud
4. Splash guard
5. Steering knuckle

Front

⊗ 🔧 210 (21, 155)

Fig. 72 Exploded view of the hub and bearing assembly

- Wheel
- Engine splash guard
- Brake caliper without disconnecting the hydraulic lines, and reposition aside with wire

3. Install a matchmark on the brake rotor and to the wheel hub and remove the brake rotor.

4. Remove or disconnect the following:
- Cotter pin and lock nut from drive shaft (4WD)

- Drive shaft from wheel hub and bearing assembly
- ABS sensor
- Wheel hub and bearing assembly bolts
- Wheel hub and bearing assembly

To install:
5. Install or connect the following:
- Wheel hub and bearing assembly,

using new bolts and tighten to 155 ft. lbs. (210 Nm)
- ABS sensor
- Drive shaft to wheel hub and bearing assembly
- Cotter pin and lock nut and tighten to 101 ft. lbs. (137 Nm)
- Brake rotor
- Brake caliper
- Engine splash guard
- Wheel

SUSPENSION

SHOCK ABSORBER

REMOVAL & INSTALLATION
See Figure 73.

1. Before servicing the vehicle, refer to the Precautions Section.
2. Support the rear differential with a suitable jack.
3. Remove the upper and lower shock absorber mounting bolts.
4. Remove the shock absorber.

To install:
5. Install the shock absorber and tighten the upper and lower mounting bolts to 111 ft. lbs. (150 Nm).

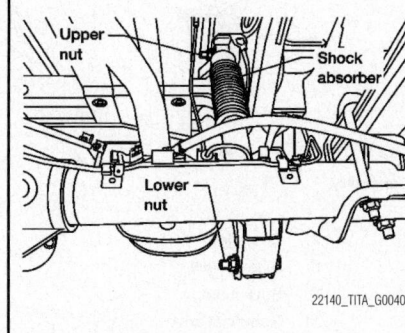

Fig. 73 Remove the upper and lower shock absorber mounting bolts

REAR SUSPENSION

WHEEL BEARINGS

REMOVAL & INSTALLATION
See Figures 74 and 75.

1. Before servicing the vehicle, refer to the Precautions Section.
2. Remove or disconnect the following:
- ABS sensor
- Rear brake rotor
- Parking brake assembly
- Four axle shaft bearing cage nuts and lock washers
- Axle shaft assembly using special tool J-25604-01 and J-25840-A
- Snap ring from axle shaft

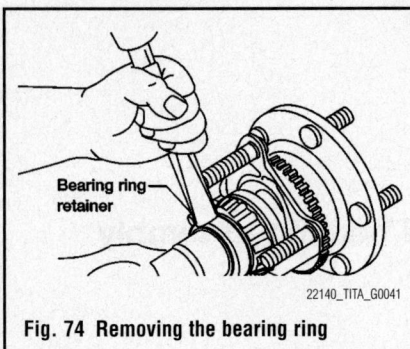

Fig. 74 Removing the bearing ring retainer

Bearing ring retainer

22140_TITA_G0041

3. Remove the bearing ring retainer by drilling 3/4 of thickness of the ring and using a hammer and chisel to break the ring free.

4. Remove the axle shaft bearing from the axle shaft using special tool 205-D002.

5. Remove and discard the axle oil seal.

6. Remove the wheel bearing assembly.

To install:

7. Install or connect the following:
- Wheel bearing assembly
- New axle oil seal
- Axle shaft bearing on axle shaft
- Bearing ring retainer onto the axle shaft
- Snap ring
- Axle shaft assembly into the axle shaft housing
- Axle shaft bearing cage lock washers and nuts and tighten to 87 ft. lbs. (118 Nm)
- Parking brake assembly
- Rear brake rotor
- ABS sensor

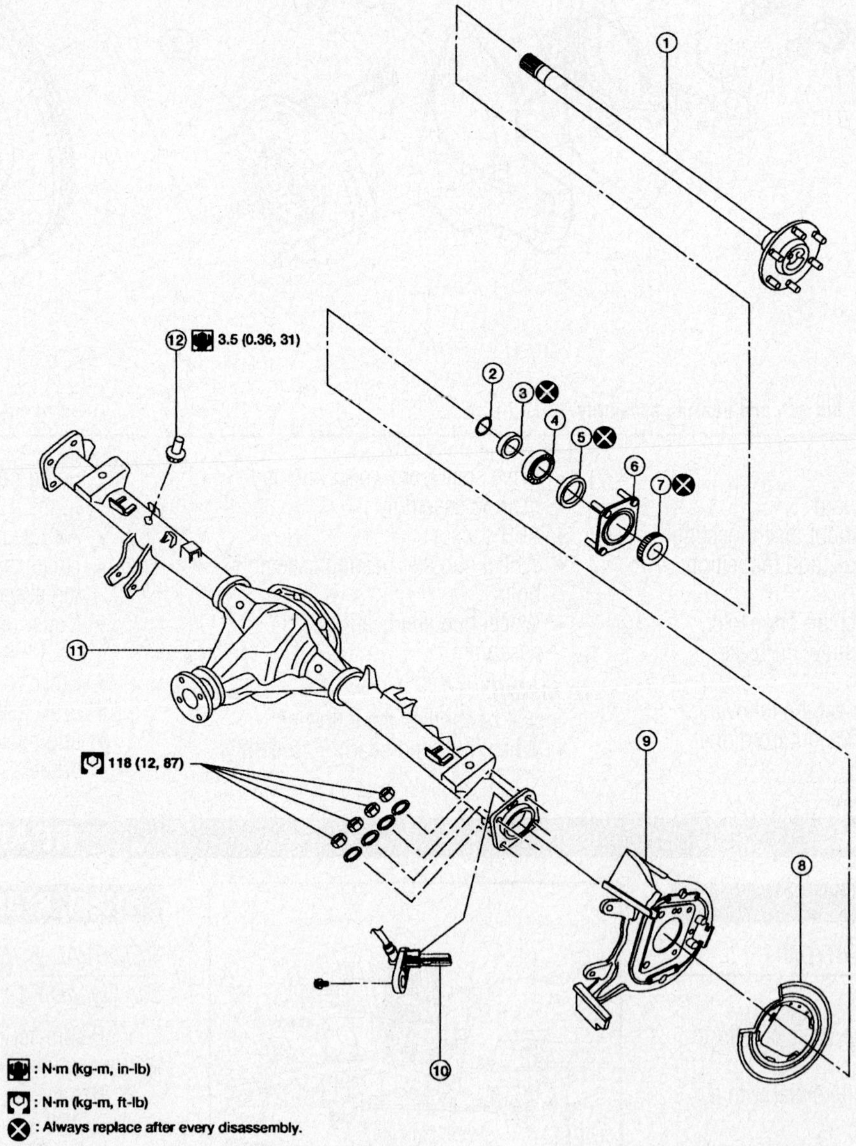

12 3.5 (0.36, 31)

118 (12, 87)

: N·m (kg-m, in-lb)

: N·m (kg-m, ft-lb)

: Always replace after every disassembly.

1. Axle shaft	2. Snap ring	3. Bearing ring retainer
4. Axle shaft bearing and cup	5. Axle oil seal	6. Axle shaft bearing cage
7. ABS sensor rotor	8. Back plate	9. Torque member
10. ABS sensor	11. Rear final drive	12. Breather

67170-TITA-G95

Fig. 75 Rear axle shaft and related parts

SPECIFICATIONS AND MAINTENANCE CHARTS

ENGINE AND VEHICLE IDENTIFICATION

Engine							Model Year	
Code ①	Liters (cc)	Cu. In.	Cyl.	Fuel Sys.	Engine Type	Eng. Mfg.	Code ②	Year
MR18DE	1.8 (1797)	109.65	4	SFI D4S	DOHC	Nissan	7	2007
MR18DE	1.8 (1797)	109.65	4	SFI D4S	DOHC	Nissan	8	2008

SFI: Sequential Fuel Injection

DOHC: Double Overhead Camshaft

NA: Information not available

① Stamped on the left side of the engine block

② 10th digit of the Vehicle Identification Number (VIN)

22140_VERS_C0001

GENERAL ENGINE SPECIFICATIONS

Year	Model	Engine Displaceme Liters	Engine Series ID	Net Horsepower @ rpm	Net Torque @ rpm (ft. lbs.)	Bore x Stroke (in.)	Com- pression Ratio	Oil Pressure @ rpm
2007	Versa	1.8	MR18DE	122@5200	127@4800	3.31x3.19	9.9:1	29 plus@2000
2008	Versa	1.8	MR18DE	122@5200	127@4800	3:31x3.19	9.9:1	29 plus@2000

22140_VERS_C0002

ENGINE TUNE-UP SPECIFICATIONS

Year	Engine Displacement Liters	Engine ID	Spark Plug Gap (in.)	Ignition Timing (deg.)*	Fuel Pump (psi)	Idle Speed (rpm)	Valve Clearance	
							Intake	Exhaust
2007	1.8	MR18DE	0.039-0.043	8-18B	N/A	650-750	0.010-0.013	0.011-0.015
2008	1.8	MR18DE	0.039-0.043	8-18B	N/A	650-750	0.010-0.013	0.011-0.015

NOTE: The Vehicle Emission Control Information label often reflects specification changes made during production.

The label figures must be used if they differ from those in this chart.

22140_VERS_C0003

CAPACITIES

| Year | Model | Engine Displacement Liters | Engine ID | Engine Oil with Filter (qts.) | Transmission (pts.) | | | Drive Axle | | Fuel Tank (gal.) | Cooling System (qts.) |
					6-Spd	Auto.	CVT	Front (pts.)	Rear (pts.)		
2007	Versa	1.8	MR18DE	4.1	4.25	16.75	17.5	—	—	13.7	7.25
2008	Versa	1.8	MR18DE	4.1	4.25	16.75	17.5	—	—	13.7	7.25

22140_VERS_C0004

VALVE SPECIFICATIONS

| Year | Engine Displacement Liters | Engine ID | Seat Angle (deg.) | Face Angle (deg.) | Spring Test Pressure (lbs. @ in.) | Spring Free-Length (in.) | Stem-to-Guide Clearance (in.) | | Stem Diameter (in.) | |
							Intake	Exhaust	Intake	Exhaust
2007	1.8	MR18DE	45	45	①	1.7677-1.7755	0.0008-0.0021	0.0012-0.0025	1.3310-1.3430	1.0870-1.0980
2008	1.8	MR18DE	45	45	①	1.7677-1.7755	0.0008-0.0021	0.0012-0.0025	1.3310-1.3430	1.0870-1.0980

① Intake 34-39 lbs. @1.390 inches. Exhaust 31-35 lbs. @1.390 inches.

22140_VERS_C0005

CRANKSHAFT AND CONNECTING ROD SPECIFICATIONS

All measurements are given in inches.

| Year | Engine Displacement Liters | Engine ID | Crankshaft | | | | Connecting Rod | | |
			Main Brg. Journal Dia.	Main Brg. Oil Clearance	Shaft End-play	Thrust on No.	Journal Diameter	Oil Clearance	Side Clearance
2007	1.8	MR18DE	2.0456-2.0464*	①	0.0039-0.0102	3	1.7304-1.7311*	0.0015-0.0019	0.0079-0.0138
2008	1.8	MR18DE	2.0456-2.0464*	①	0.0039-0.0102	3	1.7304-1.7311*	0.0015-0.0019	0.0079-0.0138

* Based upon grade

① Journal No. 1, 4 and 5: 0.0009 - 0.0013 inch
  Remaining journals: 0.0005 - 0.0009 inch

22140_VERS_C0006

PISTON AND RING SPECIFICATIONS
All measurements are given in inches.

| Year | Engine Displ. Liters | Engine ID | Piston Clearance | Ring Gap | | | Ring Side Clearance | | |
				Top Comp.	Bottom Comp.	Oil Control	Top Comp.	Bottom Comp.	Oil Control
2007	1.8	MR18DE	0.0008-0.0016	0.008-0.012	0.020-0.026	0.006-0.018	0.002-0.003	0.001-0.003	0.001-0.007
2008	1.8	MR18DE	0.0008-0.0016	0.008-0.012	0.020-0.026	0.006-0.018	0.002-0.003	0.001-0.003	0.001-0.007

22140_VERS_C0007

TORQUE SPECIFICATIONS
All readings in ft. lbs.

| Year | Engine Displacement Liters | Engine ID | Cylinder Head Bolts | Main Bearing Bolts | Rod Bearing Bolts | Crankshaft Damper Bolts | Flywheel Bolts | Manifold | | Spark Plugs | Oil Pan Drain Plug |
								Intake	Exhaust		
2007	1.8	MR18DE	①	②	③	④	80	20	25	13	⑤
2008	1.8	MR18DE	①	②	③	④	80	20	25	13	⑤

① Step 1: 20 ft. lbs.
   Step 2: Tighten an additional 100 degrees
   Step 3: Loosen to 0 ft. lbs.
   Step 4: 30 ft. lbs.
   Step 5: Tighten an additional 100 degrees
   Step 6: Tighten an additional 100 degrees

② Step 1: 25 ft. lbs.
   Step 2: Plus 60 degrees

③ Step 1: 20 ft. lbs.
   Step 2: Loosen to 0 ft. lbs.
   Step 3: Retighten to 14 ft. lbs.

④ Step 1: 51 ft. lbs.
   Step 2: Loosen to 0 ft. lbs.
   Step 3: Retighten to 22 ft. lbs.
   Step 4: Plus 60 degrees

⑤ Upper: 19 ft. lbs.
   Lower: 7 ft. lbs.

22140_VERS_C0008

WHEEL ALIGNMENT

| Year | Model | | Caster | | Camber | | Toe-in (in.) | Steering Axis Inclination (Deg.) |
			Range (+/-Deg.)	Preferred Setting (Deg.)	Range (+/-Deg.)	Preferred Setting (Deg.)		
2007	Versa	RF	0.75	+4.83	0.75	-0.33	0.04+/-0.04	—
		LF	0.75	+4.67	0.75	-0.17	0.04+/-0.04	—
		Rear	—	—	0.75	-1.52	0.2+/-0.15	—
2008	Versa	RF	0.75	+4.83	0.75	-0.33	0.04+/-0.04	—
		LF	0.75	+4.67	0.75	-0.17	0.04+/-0.04	—
		Rear	—	—	0.75	-1.52	0.2+/-0.15	—

22140_VERS_C0009

TIRE, WHEEL AND BALL JOINT SPECIFICATIONS

Year	Model	OEM Tires		Tire Pressures (psi)		Wheel Size	Ball Joint Inspection	Lug Nut Torque (ft. lbs.)
		Standard	Optional	Front	Rear			
2007	Versa	P185/65R15	N/A	33	33	5.5	①	83
2008	Versa	P185/65R15	N/A	33	33	5.5	①	83

OEM: Original Equipment Manufacturer

PSI: Pounds Per Square Inch

STD: Standard

OPT: Optional

① Replace if any measurable movement is found.

22140_VERS_C0010

BRAKE SPECIFICATIONS
All measurements in inches unless noted

Year	Model		Brake Disc			Minimum Lining Thickness	Brake Caliper	
			Original Thickness	Minimum Thickness	Maximum Runout		Bracket Bolts (ft. lbs.)	Mounting Bolts (ft. lbs.)
2007	Versa	F	0.374	0.079	0.0016	0.866	62	20
		R	—	—	—	—	—	—
2008	Versa	F	0.374	0.079	0.0016	0.866	62	20
		R	—	—	—	—	—	—

F: Front

R: Rear

22140_VERS_C0011

SCHEDULED MAINTENANCE INTERVALS
NISSAN—VERSA

TO BE SERVICED	TYPE OF SERVICE	7.5	15	22.5	30	37.5	45	52.5	60
Engine oil and filter ①	R	✓	✓	✓	✓	✓	✓	✓	✓
Brake lines & cables	S/I		✓		✓		✓		✓
Brake pads and rotors	I	✓	✓	✓	✓	✓	✓	✓	✓
Driveshaft boots	L/I	✓	✓	✓	✓	✓	✓	✓	✓
Automatic or manual transaxle	R				✓				✓
CVT transaxle	R								✓
Air cleaner filter	R				✓				✓
Engine coolant ②	R								✓
Spark plugs (Iridium)	R	Replace every 105,000 miles							
Drive belt(s) ③	S/I								✓
Cabin air filter	R		✓		✓		✓		✓
Exhaust system	I	✓	✓	✓	✓	✓	✓	✓	✓
Fuel lines	S/I								✓
Fuel filter ④									
Steering gear (box) & linkage, axle & suspension parts	I	✓	✓	✓	✓	✓	✓	✓	✓
Vapor lines	S/I								✓

R: Replace   S/I: Service or Inspect   L: Lubricate

① Engine oil replaced every 3750 miles

② Coolant: After 60,000 miles, inspect every 30,000 miles.

③ Drive Belts: After 60,000 miles, inspect every 15,000 miles. Replace belts if damaged.

④ Fuel Filter: Maintenance free item.

FREQUENT OPERATION MAINTENANCE (SEVERE SERVICE)

 If a vehicle is operated under any of the following conditions it is considered severe service:

- Extremely dusty areas.

- Rough, muddy, or salt spread roads.

- 50% or more of the vehicle constant operation is in 32°C (90°F) or higher temperatures, or temperatures below 0°C (32°F).

- Prolonged idling (vehicle operation in stop and go traffic).

- Frequent short running periods (engine does not warm to normal operating temperatures).

- Police, taxi, delivery usage or trailer towing usage.

Brake pads, discs, drums & linings: service or inspect every 7500 miles.

Driveshaft boots & propeller shaft: service or inspect every 7500 miles.

Exhaust system: service or inspect every 7500 miles.

22140_VERS_C0012

PRECAUTIONS

Before servicing any vehicle, please be sure to read all of the following precautions, which deal with personal safety, prevention of component damage, and important points to take into consideration when servicing a motor vehicle:

• Never open, service or drain the radiator or cooling system when the engine is hot; serious burns can occur from the steam and hot coolant.

• Observe all applicable safety precautions when working around fuel. Whenever servicing the fuel system, always work in a well-ventilated area. Do not allow fuel spray or vapors to come in contact with a spark, open flame, or excessive heat (a hot drop light, for example). Keep a dry chemical fire extinguisher near the work area. Always keep fuel in a container specifically designed for fuel storage; also, always properly seal fuel containers to avoid the possibility of fire or explosion. Refer to the additional fuel system precautions later in this section.

• Fuel injection systems often remain pressurized, even after the engine has been turned **OFF**. The fuel system pressure must be relieved before disconnecting any fuel lines. Failure to do so may result in fire and/or personal injury.

• Brake fluid often contains polyglycol ethers and polyglycols. Avoid contact with the eyes and wash your hands thoroughly after handling brake fluid. If you do get brake fluid in your eyes, flush your eyes with clean, running water for 15 minutes. If eye irritation persists, or if you have taken

brake fluid internally, IMMEDIATELY seek medical assistance.

• The EPA warns that prolonged contact with used engine oil may cause a number of skin disorders, including cancer. You should make every effort to minimize your exposure to used engine oil. Protective gloves should be worn when changing oil. Wash your hands and any other exposed skin areas as soon as possible after exposure to used engine oil. Soap and water, or waterless hand cleaner should be used.

• All new vehicles are now equipped with an air bag system, often referred to as a Supplemental Restraint System (SRS) or Supplemental Inflatable Restraint (SIR) system. The system must be disabled before performing service on or around system components, steering column, instrument panel components, wiring and sensors. Failure to follow safety and disabling procedures could result in accidental air bag deployment, possible personal injury and unnecessary system repairs.

• Always wear safety goggles when working with, or around, the air bag system. When carrying a non-deployed air bag, be sure the bag and trim cover are pointed away from your body. When placing a non-deployed air bag on a work surface, always face the bag and trim cover upward, away from the surface. This will reduce the motion of the module if it is accidentally deployed. Refer to the additional air bag system precautions later in this section.

• Clean, high quality brake fluid from a

sealed container is essential to the safe and proper operation of the brake system. You should always buy the correct type of brake fluid for your vehicle. If the brake fluid becomes contaminated, completely flush the system with new fluid. Never reuse any brake fluid. Any brake fluid that is removed from the system should be discarded. Also, do not allow any brake fluid to come in contact with a painted surface; it will damage the paint.

• Never operate the engine without the proper amount and type of engine oil; doing so WILL result in severe engine damage.

• Timing belt maintenance is extremely important. Many models utilize an interference-type, non-freewheeling engine. If the timing belt breaks, the valves in the cylinder head may strike the pistons, causing potentially serious (also time-consuming and expensive) engine damage. Refer to the maintenance interval charts for the recommended replacement interval for the timing belt, and to the timing belt section for belt replacement and inspection.

• Disconnecting the negative battery cable on some vehicles may interfere with the functions of the on-board computer system(s) and may require the computer to undergo a relearning process once the negative battery cable is reconnected.

• When servicing drum brakes, only disassemble and assemble one side at a time, leaving the remaining side intact for reference.

• Only an MVAC-trained, EPA-certified automotive technician should service the air conditioning system or its components.

BRAKES

GENERAL INFORMATION

PRECAUTIONS

• Certain components within the ABS system are not intended to be serviced or repaired individually

• Do not use rubber hoses or other parts not specifically specified for and ABS system. When using repair kits, replace all parts included in the kit. Partial or incorrect repair may lead to functional problems and require the replacement of components

• Lubricate rubber parts with clean, fresh brake fluid to ease assembly. Do not use shop air to clean parts; damage to rubber components may result

• Use a flare nut wrench when removing a brake tube and use a flare nut torque wrench when installing a brake tube

• When installing brake tubes and hoses, be sure to check torque

• Use only recommended DOT 3 brake fluid from an unopened container. Never reuse drained brake fluid

• If any hydraulic component or line is removed or replaced, it may be necessary to bleed the entire system

• A clean repair area is essential. Always clean the reservoir and cap thoroughly before removing the cap. The slightest amount of dirt in the fluid may plug an orifice and impair the system function. Perform repairs after components have been thoroughly cleaned; use only denatured alcohol to clean components. Do not allow components to come into contact with any substance containing mineral oil such as gasoline or kerosene; this includes used shop rags. They will ruin rubber parts of the hydraulic system

ANTI-LOCK BRAKE SYSTEM (ABS)

• The Anti-Lock control unit is a microprocessor similar to other computer units in the vehicle. Ensure that the ignition switch is **OFF** and disconnect the connectors of ABS actuator and electric unit (control unit) or the battery cable from the negative terminal before working. Avoid static electricity discharge at or near the controller

• Burnish the new braking surfaces after refinishing or replacing drums or rotors, after replacing pads or linings, or if a soft pedal occurs at very low mileage

• Be careful not to splash brake fluid on painted surface of body. If brake fluid is splashed on painted surfaces of body immediately wipe it off with cloth and then wash it away with water

• If any arc welding is to be done on the vehicle, the control unit should be unplugged before welding operations begin

BRAKES | BLEEDING THE BRAKE SYSTEM

BLEEDING PROCEDURE

BLEEDING PROCEDURE

1. Before servicing the vehicle, refer to the Precautions Section.
2. Connect a vinyl tube to the rear right bleed valve.
3. Fully depress brake pedal 4 to 5 times.
4. With brake pedal depressed, loosen bleed valve to let the air out, and then tighten it immediately.
5. Repeat the above 2 steps until no more air comes out.
6. Tighten bleed valve to 69 inch lbs. (7.8 Nm) for front disc brakes and 70 inch lbs. (7.9 Nm) for rear drum brakes.
7. Following the above 5 steps, with master cylinder reservoir tank filled at least half way, bleed air from the rear right, front left, rear left, and front right brake, in that order.

➡ **While bleeding, pay attention to master cylinder fluid level.**

➡ **Before working, disconnect connectors of ABS actuator and electric unit (control unit) or the battery cable from the negative terminal.**

BRAKES | FRONT DISC BRAKES

✳ CAUTION

Dust and dirt accumulating on brake parts during normal use may contain asbestos fibers from production or aftermarket brake linings. Breathing excessive concentrations of asbestos fibers can cause serious bodily harm. Exercise care when servicing brake parts. Do not sand or grind brake lining unless equipment used is designed to contain the dust residue. Do not clean brake parts with compressed air or by dry brushing. Cleaning should be done by dampening the brake components with a fine mist of water, then wiping the brake components clean with a dampened cloth. Dispose of cloth and all residue containing asbestos fibers in an impermeable container with the appropriate label. Follow practices prescribed by the Occupational Safety and Health Administration (OSHA) and the Environmental Protection Agency (EPA) for the handling, processing, and disposing of dust or debris that may contain asbestos fibers.

BRAKE CALIPER

REMOVAL & INSTALLATION

1. Before servicing the vehicle, refer to the Precautions Section.
2. Remove tires.
3. Secure disc rotor using wheel nuts. Put matching marks on wheel hub assembly and disc rotor, if it is necessary to remove disc rotor.
4. Drain brake fluid.
5. Remove union bolt, and then remove brake hose from caliper assembly.
6. Remove torque member mounting bolts from torque member, and remove caliper assembly from vehicle.

To install:

➡ **Before installing torque member to vehicle, wipe oil and grease on mounting surface of steering knuckle and torque member.**

7. Install caliper assembly to vehicle, and tighten mounting bolts to 62 ft. lbs (84 Nm).
8. Install brake hose to caliper assembly.
9. Refill with new brake fluid and bleed air.
10. Check front disc brake for drag.
11. Install tires to the vehicle.

DISC BRAKE PADS

REMOVAL & INSTALLATION

1. Before servicing the vehicle, refer to the Precautions Section.
2. Remove tires.
3. Remove sliding pin bolt (lower side).
4. Hang cylinder body with a wire, and remove pads, shims and pad retainers from torque member.

➡ **When removing pad retainer from torque member, lift pad retainer, so as not to deform it.**

To install:

5. Apply AS-880N grease or equivalent to the shims. Install shims to pads.

➡ **Securely install shims according to mounting direction of pads.**

6. Apply M7439 grease or equivalent to pad contact surface on pad retainers. Install pad retainers and pads to the torque member.

➡ **When installing pad retainer, attach it firmly so that it is not lifted up from torque member.**

7. Install cylinder body to torque member.

➡ **Use a disc brake piston tool (commercial service tool) to easily press to piston in.**

➡ **Check the brake fluid level in the reservoir tank for fluid level because brake fluid returns to master cylinder reservoir tank when pressing piston in.**

8. Install lower sliding pin bolt (lower side), and tighten it to 20 ft. lbs. (27 Nm).
9. Check brake for drag.
10. Install tires to the vehicle.

BRAKES

REAR DRUM BRAKES

BRAKE DRUM

REMOVAL & INSTALLATION

See Figure 1.

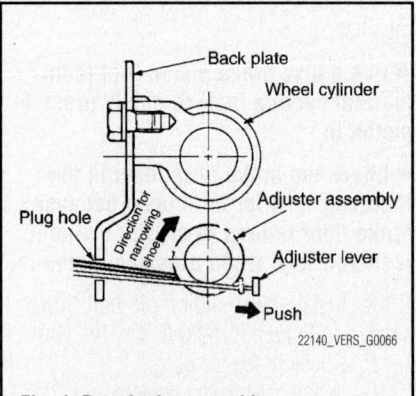

- Back plate
- Wheel cylinder
- Plug hole
- Direction for narrowing shoe
- Adjuster assembly
- Adjuster lever
- Push

22140_VERS_G0066

Fig. 1 Rear brake assembly

1. Before servicing the vehicle, refer to the Precautions Section.
2. Remove tire from the vehicle.
3. With the parking brake lever released, remove the brake drum. If it is difficult to remove brake drum, remove as follows:

   a. Press up adjuster lever with a wire or equivalent from plug hole (plug hole at the side of wheel cylinder) on the back plate as shown in the figure. Turn frame of adjuster assembly with a flat bladed screw driver in the direction that narrows frame to narrow enlarged brake shoe.

To install:

4. Installation is the reverse of removal.

   a. Inspect and adjust brakes.

BRAKE SHOES

REMOVAL & INSTALLATION

See Figure 1.

1. Before servicing the vehicle, refer to the Precautions Section.
2. Remove brake drum.
3. While pushing and rotating the retainer, pull out shoe hold pin, and remove shoe assembly. Do not damage the wheel cylinder boot.
4. Remove the parking brake rear cable from the operating lever. Do not bend the parking brake cable.
5. Disassemble the shoe assembly (shoe, springs, adjuster, adjuster lever).
6. Remove retainer ring with a tool to separate operating lever from brake shoe.

To install:

7. Install operating lever to brake shoe, if necessary.
8. Install retainer ring to operating lever, and crimp them until their contact points are met.
9. Apply NISSAN brake grease (KRF0000005) to brake shoes sliding surfaces (the shaded areas) and other parts on the back plate.
10. Apply NISSAN brake grease (KRF0000005) to the screw and confirm the differ-

ence between the right and left wheel for assembling when disassembled.

11. Assemble the shoe, adjuster, adjuster lever and springs to the shoe assembly.
12. Connect the parking brake rear cable to the operating lever.
13. Install the shoe assembly. After assembly, be sure that each part is installed properly. Do not damage the wheel cylinder piston boot.
14. Install the brake drum.

ADJUSTMENT

1. Before servicing the vehicle, refer to the Precautions Section.
2. Remove console mask cover.
3. Engage parking brake lever, then lift up the end of the trim on the lever to access the adjusting nut.
4. Insert a deep socket wrench onto adjusting nut. Rotate adjusting nut to fully loosen cable, and then release parking brake lever.
5. Depress the foot brake about 10 times and adjust the rear shoe clearance. Be sure to securely depress the foot brake.
6. Rotate brake drum to make sure that there is no drag.
7. Adjust parking brake cable with the following procedure.

   a. When replace parking brake cable, operate parking brake lever with a force of 110 ft. lbs. (490 Nm) about 10 times.

   b. Engage parking brake lever, then lift up the end of the trim on the lever to access the adjusting nut.

   c. Rotate adjusting nut to adjust parking brake lever stroke using a deep socket wrench.

   d. Operate parking brake lever with a force of 44 ft. lbs. (196 Nm), make sure the parking brake lever stroke is within the specified number of notches. (Check it by listening and counting ratchet clicks.)

   e. Make sure that there is no drag on rear brake with parking brake lever completely released.

8. Install console mask.

CHASSIS ELECTRICAL | AIR BAG (SUPPLEMENTAL RESTRAINT SYSTEM)

GENERAL INFORMATION

✳✳ CAUTION

These vehicles are equipped with an air bag system. The system must be disarmed before performing service on, or around, system components, the steering column, instrument panel components, wiring and sensors. Failure to follow the safety precautions and the disarming procedure could result in accidental air bag deployment, possible injury and unnecessary system repairs.

SERVICE PRECAUTIONS

Turn ignition switch **OFF**, disconnect both battery cables and wait at least 3 minutes before beginning any airbag system component diagnosis, testing, removal, or installation procedures. For approximately 3 minutes after the cables are removed, it is still possible for the air bag and seat belt pre-tensioner to deploy. Therefore, do not work on any SRS connectors or wires until at least 3 minutes have passed. This will disable the airbag system. Failure to disable the airbag system may result in accidental airbag deployment, personal injury, or death.

Do not use electrical test equipment on any circuit related to the SRS unless instructed to. SRS wiring harnesses can be identified by yellow and/or orange harnesses or harness connectors.

The air bag diagnosis sensor unit must always be installed with the arrow mark pointing toward the front of the vehicle for proper operation. Also check air bag diagnosis sensor unit for cracks, deformities or rust before installation and replace as required.

The spiral cable must be aligned with the neutral position since its rotations are limited. Do not attempt to turn steering wheel or column after removal of steering gear.

Always place the driver and front passenger air bag module face down on a solid surface, and the seat mounted front side air bag module standing with the stud bolt side facing down. The airbag will propel into the air if accidentally deployed and may result in personal injury or death.

When carrying or handling an undeployed airbag, the trim side (face) of the airbag should be pointing towards the body to minimize possibility of injury if accidental deployment occurs. Failure to do this may result in personal injury or death.

Replace airbag system components with OEM replacement parts. Substitute parts may appear interchangeable, but internal differences may result in inferior occupant protection. Failure to do so may result in occupant personal injury or death.

Wear safety glasses, rubber gloves, and long sleeved clothing when cleaning powder residue from vehicle after an airbag deployment. Powder residue emitted from a deployed airbag can cause skin irritation. Flush affected area with cool water if irritation is experienced. If nasal or throat irritation is experienced, exit the vehicle for fresh air until the irritation ceases. If irritation continues, see a physician.

Do not use a replacement airbag that is not in the original packaging. This may result in improper deployment, personal injury, or death.

The factory installed fasteners, screws and bolts used to fasten airbag components have a special coating and are specifically designed for the airbag system. Do not use substitute fasteners. Use only original equipment fasteners listed in the parts catalog when fastener replacement is required.

During, and following, any child restraint anchor service, due to impact event or vehicle repair, carefully inspect all mounting hardware, tether straps, and anchors for proper installation, operation, or damage. If a child restraint anchor is found damaged in any way, the anchor must be replaced. Failure to do this may result in personal injury or death.

Conduct self-diagnosis to check entire SRS for proper function after replacing any components.

After air bag inflates, the front instrument panel assembly should be replaced if damaged.

Do not reuse center pillar upper garnish if removed.

Deployed and non-deployed airbags may or may not have live pyrotechnic material within the airbag inflator.

Do not dispose of driver/passenger/side/curtain airbags or seat belt tensioners unless you are sure of complete deployment. Refer to the Hazardous Substance Control System for proper disposal.

Dispose of deployed airbags and tensioners consistent with state, provincial, local, and federal regulations.

After any airbag component testing or service, do not connect the battery negative cable. Personal injury or death may result if the system test is not performed first.

If the vehicle is equipped with the Occupant Classification System (OCS), do not connect the battery negative cable before performing the OCS Verification Test using the scan tool and the appropriate diagnostic information. Personal injury or death may result if the system test is not performed properly.

Replace control unit and passenger front seat cushion as an assembly.

Never replace both the Occupant Restraint Controller (ORC) and the Occupant Classification Module (OCM) at the same time. If both require replacement, replace one, then perform the Airbag System test before replacing the other.

Both the ORC and the OCM store Occupant Classification System (OCS) calibration data, which they transfer to one another when one of them is replaced. If both are replaced at the same time, an irreversible fault will be set in both modules and the OCS may malfunction and cause personal injury or death.

If equipped with OCS, the Seat Weight Sensor is a sensitive, calibrated unit and must be handled carefully. Do not drop or handle roughly. If dropped or damaged, replace with another sensor. Failure to do so may result in occupant injury or death.

If equipped with OCS, the front passenger seat must be handled carefully as well. When removing the seat, be careful when setting on floor not to drop. If dropped, the sensor may be inoperative, could result in occupant injury, or possibly death.

If equipped with OCS, when the passenger front seat is on the floor, no one should sit in the front passenger seat. This uneven force may damage the sensing ability of the seat weight sensors. If sat on and damaged, the sensor may be inoperative, could result in occupant injury, or possibly death.

DISARMING THE SYSTEM

1. Before servicing the vehicle, refer to the Precautions Section.

Disconnect the negative and positive battery cables, then wait at least 3 minutes. For approximately 3 minutes after the cables are removed, it is still possible for the air bag and seat belt pre-tensioner to deploy.

ARMING THE SYSTEM

1. Before servicing the vehicle, refer to the Precautions Section.

Connect the negative and positive battery cables.

CLOCKSPRING CENTERING

See Figure 2.

1. Before servicing the vehicle, refer to the Precautions Section.

2. Make sure spiral cable alignment and centering marks (B) are matched and in the neutral position. The neutral position is detected by turning left 2.6 revolutions from the right end position, ending with the locating pin (A) at the top. Place steering wheel in straight ahead position, then install it with the locating pin hole (C) directly over spiral cable locating pin.

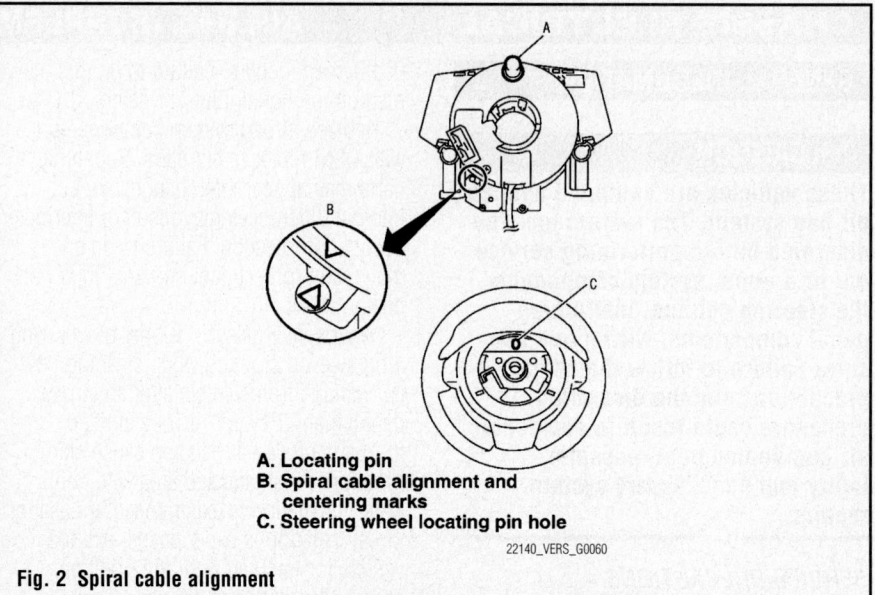

A. Locating pin
B. Spiral cable alignment and centering marks
C. Steering wheel locating pin hole

22140_VERS_G0060

Fig. 2 Spiral cable alignment

DRIVETRAIN

AUTOMATIC TRANSAXLE ASSEMBLY

REMOVAL & INSTALLATION

See Figures 3 through 6.

1. Before servicing the vehicle, refer to the Precautions Section.

2. Remove engine and transaxle assembly.

3. Disconnect the following connectors and remove the wire harness:

a. Turbine revolution sensor (power train revolution sensor) harness connector.

b. Terminal cord assembly harness connector.

c. PNP switch connector.

d. Revolution sensor harness connector.

4. Remove the four drive plate to torque converter bolts.

➡Rotate the crankshaft clockwise as viewed from front of engine for access to drive plate to torque converter bolts.

5. Put matchmarks on the drive plate and torque converter. For matching marks, use paint. Never damage the drive plate or torque converter.

6. Remove the transaxle to engine and engine to transaxle bolts.

7. Separate the transaxle from the engine. Secure torque converter to prevent it from dropping.

To install:

8. Installation is the reverse of removal, noting the following:

a. When replacing an engine or transmission you must make sure any dowels are installed correctly during re-assembly. Improper alignment caused by missing dowels may cause vibration,

oil leaks or breakage of drivetrain components.

➡Do not reuse O-rings and copper washers.

b. When turning crankshaft, turn it clockwise as viewed from the front of the engine.

c. When tightening the bolts for the torque converter while securing the crankshaft pulley bolt, be sure to confirm the tightening torque of the crankshaft pulley bolt.

d. After converter is installed to drive plate, rotate crankshaft several turns to check that transaxle rotates freely without binding.

e. When installing the torque converter to the transaxle measure distance A, it should measure 14.4 mm (0.567 in) or more.

f. Check the fitting of the dowel pins

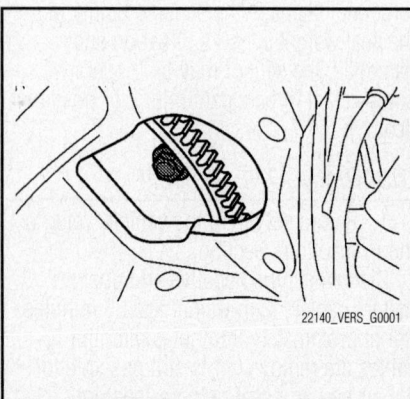

22140_VERS_G0001

Fig. 3 Drive plate torque converter bolts

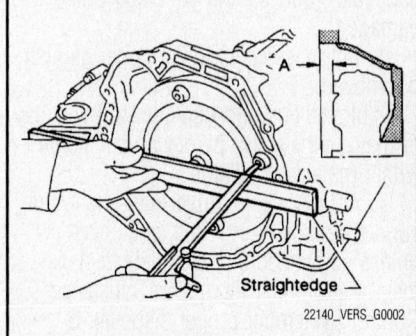

Straightedge

22140_VERS_G0002

Fig. 4 Torque converter to transaxle distance A

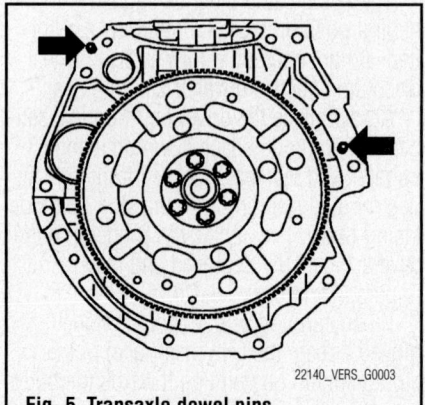

22140_VERS_G0003

Fig. 5 Transaxle dowel pins

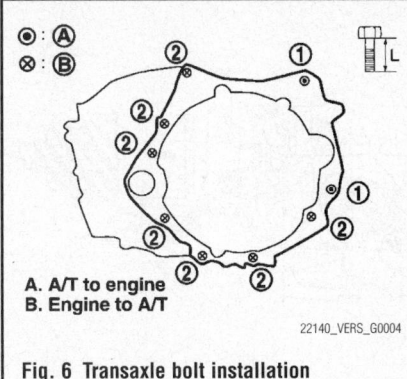

A. A/T to engine
B. Engine to A/T

22140_VERS_G0004

Fig. 6 Transaxle bolt installation sequence

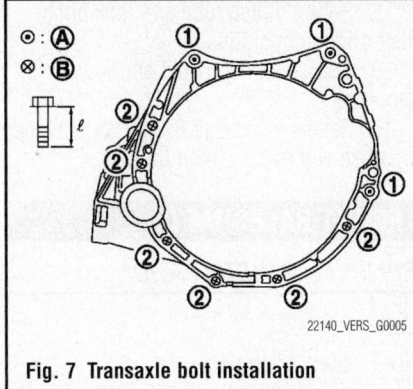

22140_VERS_G0005

Fig. 7 Transaxle bolt installation sequence

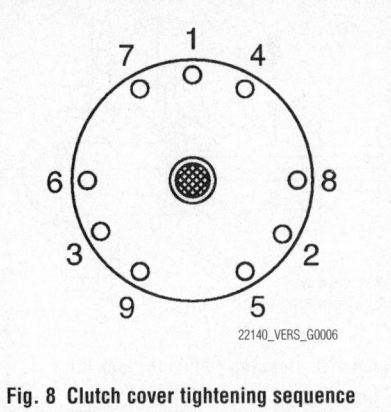

22140_VERS_G0006

Fig. 8 Clutch cover tightening sequence

when installing the transaxle assembly and the engine assembly.

g. When installing the transaxle to the engine, align the matching mark on the drive plate with the matching mark on the torque converter.

h. When securing the transaxle to the engine, attach the bolts and tighten to 46 ft. lbs. (62 Nm).

i. Align the positions for the bolts on drive plate with those of the torque converter, and temporarily tighten bolts. Then, tighten the bolts to 38 ft. lbs. (51 Nm).

j. After completing installation, check for A/T fluid leakage, A/T fluid level, and A/T positions.

MANUAL TRANSAXLE ASSEMBLY

REMOVAL & INSTALLATION
See Figure 7.

1. Before servicing the vehicle, refer to the Precautions Section.
2. Drain gear oil.
3. Drain clutch fluid and remove clutch tube from the concentric slave cylinder. Do not depress clutch pedal during removal procedure.
4. Remove engine and transaxle assembly.
5. Remove starter motor.
6. Remove transaxle assembly to engine bolts.
7. Separate transaxle assembly from engine.

To install:
8. If transaxle is removed from the vehicle, always replace concentric slave cylinder and install a new one to the clutch housing and tightening the bolts to 15 ft. lbs. (21 Nm).

➡**Do not reuse CSC.**

➡**Do not insert and operate CSC because piston and stopper of CSC components may fall off.**

9. The remainder of installation is the reverse of removal, noting the following:

a. When securing the transaxle to the engine, attach the bolts and tighten to 46 ft. lbs. (62 Nm).

➡**When installing transaxle assembly, be careful not to bring transaxle input shaft into contact with clutch cover.**

b. Bleed the air from the clutch hydraulic system.
c. After installation, check oil level, and check for leaks and loose mechanisms.

CLUTCH

REMOVAL & INSTALLATION
See Figure 8.

1. Before servicing the vehicle, refer to the Precautions Section.

➡**If transaxle assembly is removed from the vehicle, always replace Concentric slave cylinder (CSC). Return CSC insert to original position to remove transaxle assembly. Dust on clutch disc sliding parts may damage CSC seal and may cause clutch fluid leakage.**

➡**DO NOT apply any grease to the clutch disc facing, pressure plate surface and flywheel surface.**

2. Remove transaxle assembly from the vehicle.
3. Loosen clutch cover bolts evenly. Then remove clutch cover and clutch disc.

To install:
4. Clean clutch disc and input shaft splines to remove grease and dust caused by abrasion.

5. Apply recommended grease to clutch disc and input shaft splines.

➡**Be sure to apply grease to the points specified. Otherwise, noise, poor disengagement, or damage to the clutch may result. Excessive grease may cause slip or shudder. If it adheres to CSC seal, it will cause clutch fluid leakage. Wipe off excess grease.**

6. Install clutch disc using tool No. KV30101000.
7. Install clutch cover. Pre-tighten clutch cover bolts.
8. Tighten clutch cover bolts evenly in two steps to 19 ft. lbs. (26 Nm) in the order shown.
9. Install transaxle assembly.

BLEEDING
See Figures 9 through 11.

1. Before servicing the vehicle, refer to the Precautions Section.

➡**Do not use a vacuum assist or any other type of power bleeder on this system. Use of a vacuum assist or power bleeder will not purge all the air from the system.**

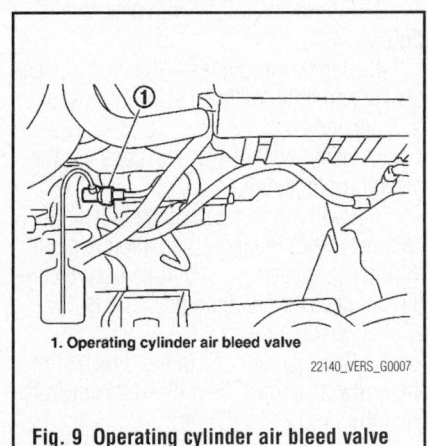

1. Operating cylinder air bleed valve

22140_VERS_G0007

Fig. 9 Operating cylinder air bleed valve

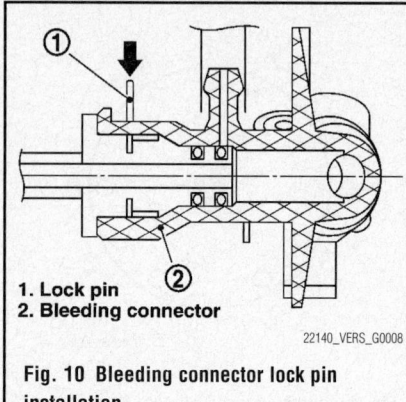

1. Lock pin
2. Bleeding connector

22140_VERS_G0008

Fig. 10 Bleeding connector lock pin installation

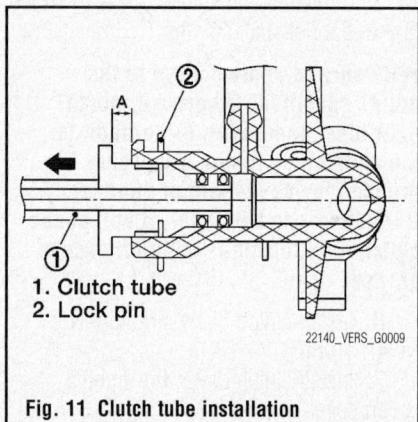

1. Clutch tube
2. Lock pin

22140_VERS_G0009

Fig. 11 Clutch tube installation

➡ **Carefully monitor fluid level in reservoir tank during bleeding operation.**

➡ **Keep painted surface of body and other parts free of clutch fluid. If it spills, wipe up immediately and wash the affected area with water.**

➡ **First bleed the air from the operating cylinder air bleed valve and then from the bleed connector air bleed valve.**

2. Connect a transparent vinyl tube and container to the air bleeder valve on the clutch operating cylinder.

3. Fill reservoir tank with new clutch fluid.

4. Depress and release the clutch pedal slowly and fully 15 times at an interval of 2 to 3 seconds.

5. With pedal depressed, continue the bleeding procedure.

6. Push the lock pin of the bleeding connector into the locked position. Hold it to prevent releasing clutch tube from bleeding connector when fluid pressure is applied in the tube.

7. Slide clutch tube in the direction of the arrow as shown, then bleed the air from the tube. A: 0.20 in (5 mm).

8. Return clutch tube and lock pin to their original positions.

9. Release clutch pedal and wait for 5 seconds.

10. Repeat steps 2 to 8 until no bubbles are observed in the clutch fluid.

FRONT HALFSHAFT

REMOVAL & INSTALLATION

Left

See Figures 12 and 13.

1. Before servicing the vehicle, refer to the Precautions Section.

2. Remove wheel and tire.

3. Remove wheel sensor from steering knuckle. Do not pull on wheel sensor harness.

4. Remove transverse link ball joint nut and bolt. Then, remove transverse link from steering knuckle.

5. Remove cotter pin, then loosen hub lock nut using power tool.

➡ **Temporarily leave the hub lock nut installed to prevent damage to threads.**

6. Separate the drive shaft from the wheel hub and bearing assembly by lightly tapping the end of the drive shaft using a hammer or suitable tool and wood block, and then remove hub lock nut.

7. Use a suitable puller if wheel hub and bearing assembly and drive shaft cannot be separated after performing the above procedure.

8. Remove the drive shaft from the wheel hub and bearing assembly.

➡ **Do not apply an excessive angle to drive shaft joint when removing from the wheel hub and bearing assembly,**

➡ **Do not excessively extend slide joint.**

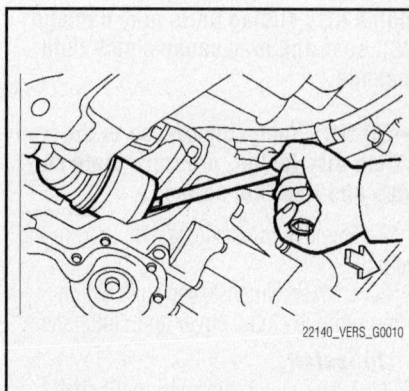

22140_VERS_G0010

Fig. 12 Drive shaft removal

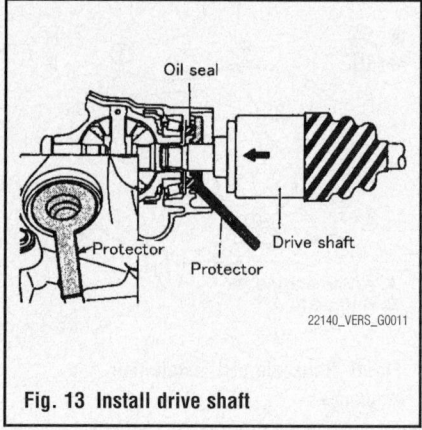

Fig. 13 Install drive shaft

➡ **Do not allow drive shaft to hang down. Support the entire drive shaft.**

9. Pry off drive shaft from transaxle assembly side as shown. Make sure that circlip is attached on the edge.

➡ **Do not apply an excessive angle to drive shaft joint when removing from the transaxle.**

➡ **Do not excessively extend slide joint.**

➡ **Do not allow drive shaft to hang down. Support the entire drive shaft.**

To install:

10. Installation is in the reverse order of removal, noting the following:

a. Tighten the drive shaft nut to 83 ft. lbs. (113 Nm).

➡ **Do not reuse non-reusable parts.**

b. In order to prevent damage to differential side oil seal, place tool No. KV38105500 (J-33904) onto oil seal before inserting drive shaft as shown. Slide drive shaft into slide joint and tap with a hammer to install securely.

c. Install new circlip on drive shaft in the circlip groove on transaxle side. Make sure the new circlip on the drive shaft is securely fastened.

d. After its insertion, try to pull the flange out of the slide joint by hand. If it pulls out, the circlip is not properly meshed with the transaxle side gear.

Right

See Figure 14.

1. Before servicing the vehicle, refer to the Precautions Section.

2. Remove wheel and tire.

3. Remove wheel sensor from steering knuckle. Do not pull on wheel sensor harness.

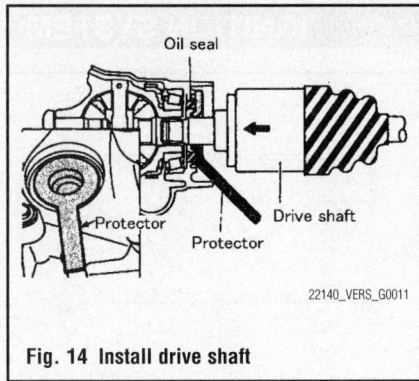

Fig. 14 Install drive shaft

4. Remove transverse link ball joint nut and bolt. Then, remove transverse link from steering knuckle.

5. Remove cotter pin, then loosen hub lock nut using power tool.

6. Temporarily leave the hub lock nut installed to prevent damage to threads.

7. Separate the drive shaft from the wheel hub and bearing assembly by lightly tapping the end of the drive shaft using a hammer or suitable tool and wood block, and then remove hub lock nut.

➡**Use a suitable puller if wheel hub and bearing assembly and drive shaft cannot be separated after performing the above procedure.**

8. Remove the drive shaft from the wheel hub and bearing assembly.

➡**Do not apply an excessive angle to drive shaft joint when removing from the wheel hub and bearing assembly,**

➡**Do not excessively extend slide joint.**

➡**Do not allow drive shaft to hang down. Support the entire drive shaft.**

9. Remove the plate bolts and plate.

10. Remove the drive shaft from the transaxle assembly.

To install:

11. Installation is in the reverse order of removal, noting the following:

a. Tighten the drive shaft nut to 83 ft. lbs. (113 Nm).

b. Tighten the plate bolts to 11 ft. lbs. (15 Nm).

➡**Do not reuse non-reusable parts.**

c. In order to prevent damage to differential side oil seal, place tool No. KV38106700 (J-34296) onto oil seal before inserting drive shaft as shown. Slide drive shaft into slide joint and tap with a hammer to install securely.

ENGINE COOLING

THERMOSTAT

REMOVAL & INSTALLATION

1. Before servicing the vehicle, refer to the Precautions Section.

2. Drain engine coolant.

3. Remove water inlet.

4. Remove thermostat.

To install:

5. Installation is in the reverse order of removal, noting the following:

➡**Replace the rubber ring with a new one.**

a. Install thermostat while making rubber ring groove fit to thermostat flange around the whole circumference.

b. Install thermostat into the thermostat housing with jiggle valve facing upwards.

WATER PUMP

REMOVAL & INSTALLATION

1. Before servicing the vehicle, refer to the Precautions Section.

2. Drain engine coolant.

3. Remove drive belt auto-tensioner.

4. Remove water pump.

➡**Handle water pump vane so that it does not contact any other parts.**

To install:

5. Installation is in the reverse order of removal.

6. Check for coolant leaks.

ENGINE ELECTRICAL

ALTERNATOR

REMOVAL & INSTALLATION

1. Before servicing the vehicle, refer to the Precautions Section.

2. Disconnect the battery cable from the negative terminal.

3. Remove drive belt.

4. Disconnect alternator connector.

5. Remove "B" terminal nut.

6. Remove alternator bolts.

7. Remove alternator assembly from the vehicle.

CHARGING SYSTEM

To install:

8. Installation is in the reverse order of removal.

➡**Be sure to tighten "B" terminal nut carefully.**

ENGINE ELECTRICAL

FIRING ORDER

The firing order is 1–3–4–2.

IGNITION COIL

REMOVAL & INSTALLATION

See Figures 15 and 16.

1. Before servicing the vehicle, refer to the Precautions Section.
2. Remove intake manifold.
3. Remove ignition coil.

➡**Handle it carefully and avoid impacts.**

10. Ignition coil with power transistor and spark plug
11. Fuel injector

22140_VERS_G0078

Fig. 15 Ignition coil location

➡**Never disassemble.**

4. Remove spark plug using suitable tool.

➡**Never drop or shock it.**

5. Remove rocker cover. Loosen bolts in reverse order of the tightening sequence.

To install:

6. Install rocker cover gasket to rocker cover.
7. Install rocker cover.
8. Tighten bolts in two steps separately in numerical order as shown. Step 1:

IGNITION SYSTEM

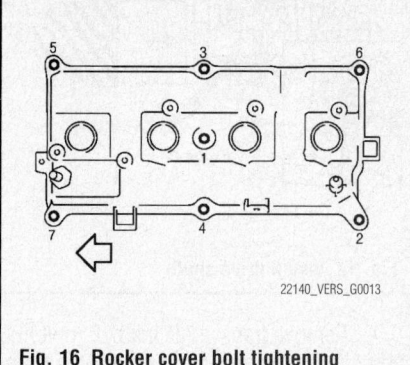

22140_VERS_G0013

Fig. 16 Rocker cover bolt tightening sequence

17 inch lbs. (1.96 Nm), Step 2: 73 inch lbs. (8.33 Nm)

9. Install spark plug using suitable tool. Spark plug part No. FXE20HR11.

➡**Never drop or shock it.**

10. Install ignition coil.
11. Install intake manifold.

IGNITION TIMING

ADJUSTMENT

Ignition timing adjustment is not possible.

SPARK PLUGS

REMOVAL & INSTALLATION

1. Before servicing the vehicle, refer to the Precautions Section.
2. Remove ignition coil.
3. Remove spark plug using suitable tool.

To install:

4. Installation is the reverse of removal.

ENGINE ELECTRICAL

STARTING SYSTEM

STARTER

REMOVAL & INSTALLATION

See Figure 17.

1. Before servicing the vehicle, refer to the Precautions Section.
2. Disconnect the battery negative terminal.
3. Remove air duct (inlet).
4. Remove reservoir tank.
5. Remove "S" terminal nut.
6. Remove "B" terminal nut.
7. Remove starter motor bolts.
8. Remove starter motor.

To install:

9. Installation is in the reverse order of removal.

➡ **Be sure to tighten "B" terminal nut carefully.**

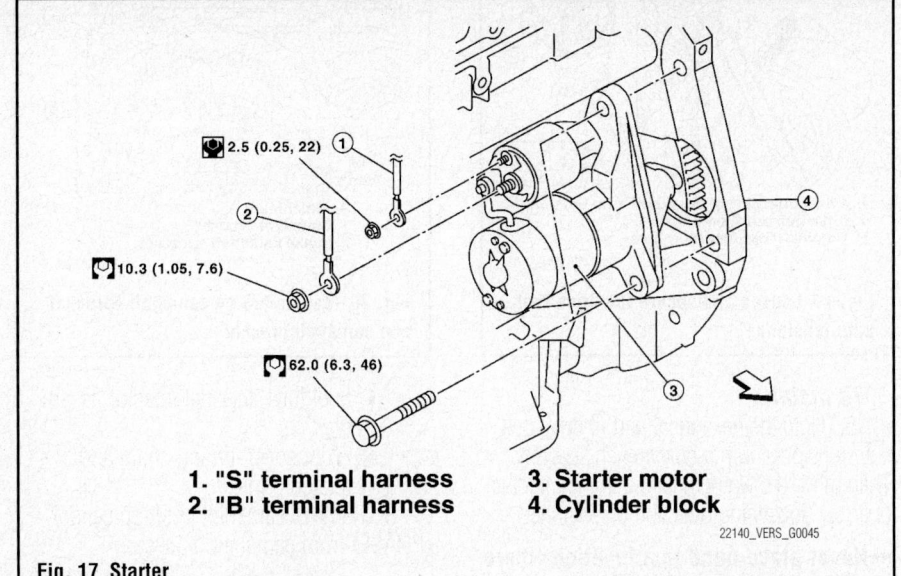

| 1. "S" terminal harness | 3. Starter motor |
| 2. "B" terminal harness | 4. Cylinder block |

22140_VERS_G0045

Fig. 17 Starter

ENGINE MECHANICAL

➡ **Disconnecting the negative battery cable may interfere with the functions of the on board computer systems and may require the computer to undergo a relearning process, once the negative battery cable is reconnected.**

ACCESSORY DRIVE BELTS

ACCESSORY BELT ROUTING

See Figure 18.

INSPECTION

1. Before servicing the vehicle, refer to the Precautions Section.

✳ CAUTION

Be sure to perform this step when the engine is stopped.

2. Make sure that the indicator (notch on fixed side) of drive belt auto-tensioner is within the possible use range.

➡ **Check the drive belt auto-tensioner indication when the engine is cold.**

➡ **When new drive belt is installed, the indicator (notch on fixed side) should be within the range.**

3. Visually check entire drive belt for wear, damage or cracks.
4. If the indicator (notch on fixed side) is out of the possible use range or belt is damaged, replace drive belt.

ADJUSTMENT

Belt tension is not necessary, as it is automatically adjusted by drive belt auto-tensioner.

REMOVAL & INSTALLATION

See Figure 19.

1. Before servicing the vehicle, refer to the Precautions Section.
2. Hold the hexagonal part of drive belt auto-tensioner with a wrench securely. Then move the wrench handle in the direction of arrow (loosening direction of tensioner).

➡ **Never place hand in a location where pinching may occur if the holding tool accidentally comes off.**

3. Insert a rod such as short-length screwdriver approximately 6 mm (0.24 in) in diameter into the hole of the retaining boss to hold the drive belt auto-tensioner.
4. Remove drive belt.

1. **Alternator**
2. **Drive belt auto–tensioner**
3. **Crankshaft pulley**
4. **A/C compressor (models with A/C)
   /Idler pulley (models without A/C)**
5. **Water pump**
6. **Drive belt**

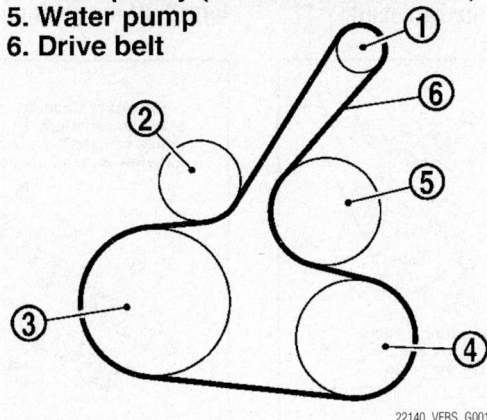

22140_VERS_G0015

Fig. 18 Accessory belt routing

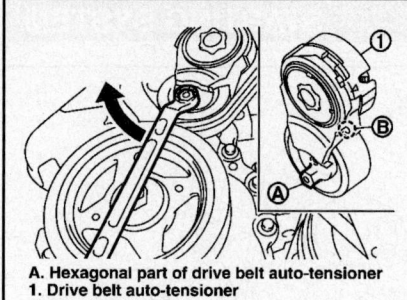

A. Hexagonal part of drive belt auto-tensioner
1. Drive belt auto-tensioner
B. Retaining rod insertion hole

22140_VERS_G0014

Fig. 19 Loosen and secure the drive belt auto-tensioner

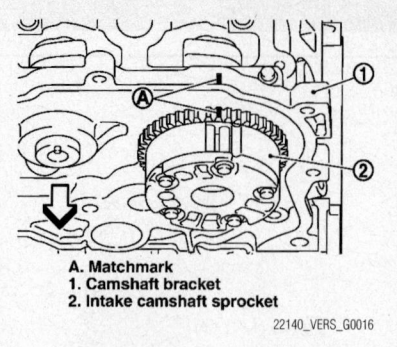

A. Matchmark
1. Camshaft bracket
2. Intake camshaft sprocket

22140_VERS_G0016

Fig. 20 Matchmark on camshaft sprocket and camshaft bracket

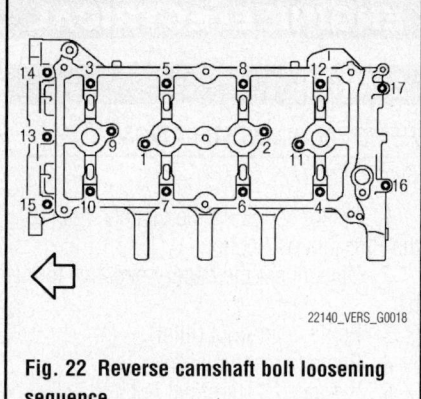

22140_VERS_G0018

Fig. 22 Reverse camshaft bolt loosening sequence

To install:

5. Hold the hexagonal part of drive belt auto-tensioner with a box wrench securely. Then move the wrench handle in the direction of arrow (loosening direction of tensioner).

➥**Never place hand in a location where pinching may occur if the holding tool accidentally comes off.**

6. Insert a rod such as short-length screwdriver approximately 6 mm (0.24 in) in diameter into the hole of retaining boss to fix drive belt auto-tensioner.

7. Install drive belt.

➥**Confirm drive belt is completely set to pulleys.**

➥**Check for engine oil, working fluid and engine coolant are not adhered to drive belt and each pulley groove.**

8. Release drive belt auto-tensioner, and apply tension to drive belt.

9. Turn crankshaft pulley clockwise several times to equalize tension between each pulley.

10. Confirm tension of drive belt at indicator (notch on fixed side) is within the possible use range.

CAMSHAFT AND VALVE LIFTERS

REMOVAL & INSTALLATION

See Figures 20 through 30.

1. Before servicing the vehicle, refer to the Precautions Section.

2. Release the fuel pressure.

3. Disconnect negative battery cable.

4. Remove the right front wheel.

5. Remove the right front fender protector.

6. Drain engine coolant.

7. Remove the following parts:
  • Intake manifold
  • Rocker cover

• Fuel tube and fuel injector assembly
• Front cover, timing chain and related parts

8. Remove camshaft position sensor (PHASE) from camshaft bracket.

➥**Handle carefully to avoid dropping and shocks.**

➥**Never disassemble.**

➥**Never allow metal powder to adhere to magnetic part at sensor tip.**

➥**Never place sensor in a location where it is exposed to magnetism.**

9. Put the matchmark on the intake camshaft sprocket and the camshaft bracket as shown.

10. Remove camshaft intake and exhaust sprockets.

11. Secure hexagonal part of camshaft with a wrench. Loosen camshaft sprocket bolts and remove the camshaft sprocket.

➥**Never rotate crankshaft or camshaft while timing chain is removed. It causes interference between valve and piston.**

➥**Never loosen the bolts with securing anything other than the camshaft**

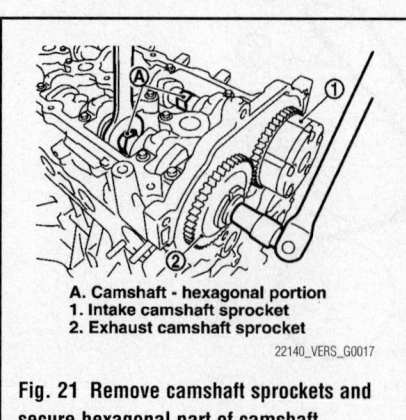

A. Camshaft - hexagonal portion
1. Intake camshaft sprocket
2. Exhaust camshaft sprocket

22140_VERS_G0017

Fig. 21 Remove camshaft sprockets and secure hexagonal part of camshaft

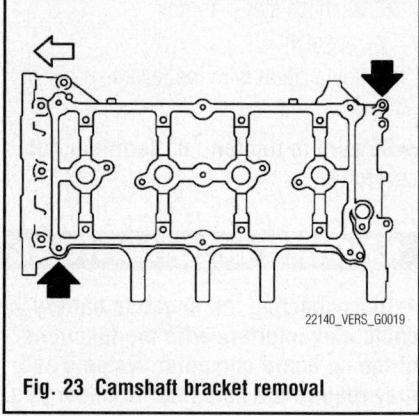

22140_VERS_G0019

Fig. 23 Camshaft bracket removal

hexagonal part or with tensioning the timing chain.

12. Loosen bolts in reverse order as shown.

13. Cut liquid gasket by prying the position shown, and then remove the camshaft bracket.

➥**Be careful not to damage the mating surface.**

➥**A more adhesive liquid gasket is applied compared to previous types when shipped, so it should not be forced off the position not specified.**

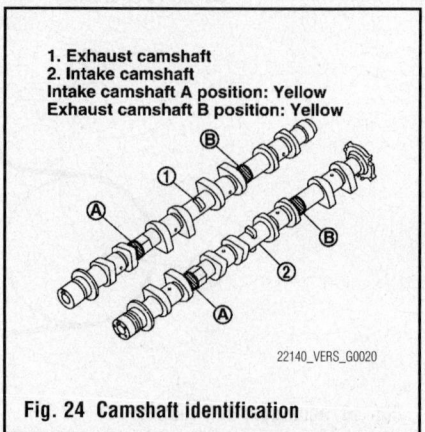

1. Exhaust camshaft
2. Intake camshaft
Intake camshaft A position: Yellow
Exhaust camshaft B position: Yellow

22140_VERS_G0020

Fig. 24 Camshaft identification

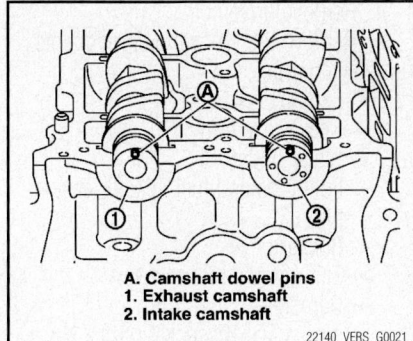

A. Camshaft dowel pins
1. Exhaust camshaft
2. Intake camshaft

22140_VERS_G0021

Fig. 25 Camshaft dowel pin installation positioning

14. Remove camshafts.
15. Remove valve lifters.

➡**Identify installed positions, and store them without mixing them up.**

To install:

16. Install valve lifters in the original positions.
17. Install camshafts.
18. Clean camshaft journal to remove any foreign material.

➡**Distinguish between the intake and the exhaust by looking at the different shapes of the front and rear ends of the camshaft or using the identification colors.**

19. Install camshafts so that camshaft dowel pins on the front side are positioned as shown.

➡**Though camshaft does not stop at the positions as shown, for the placement of cam nose, it is generally accepted camshaft is placed for the same direction as shown.**

20. Remove foreign material completely from camshaft bracket backside and from cylinder head installation face.
21. Apply liquid gasket to camshaft bracket as shown.

➡**Use Genuine Silicone RTV Sealant (Tool No. WS39930000) or equivalent.**

22. Install camshaft bracket bolts numerically as shown, noting the following:
   a. Note the 2 types of M6 bolts: bolts No. 13, No. 14, and No. 15 in the figure have a thread length of 57.5 mm (2.264 inch), and the remaining bolts have a thread length of 35 mm (1.378 inch).
   b. Tighten all the bolts in 3 steps:
   • Step 1: 17 inch lbs. (1.96 Nm)
   • Step 2: 52 inch lbs. (5.88 Nm)
   • Step 3: 84 inch lbs. (9.5 Nm)
23. Install the intake camshaft sprocket to the intake camshaft.

➡**When the installing the intake camshaft sprocket, refer to the match-**

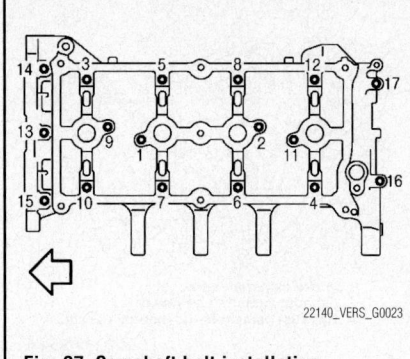

22140_VERS_G0023

Fig. 27 Camshaft bolt installation sequence

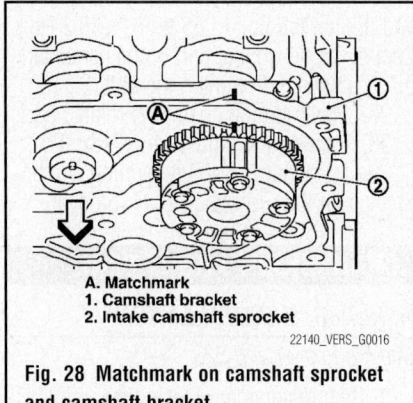

A. Matchmark
1. Camshaft bracket
2. Intake camshaft sprocket

22140_VERS_G0016

Fig. 28 Matchmark on camshaft sprocket and camshaft bracket

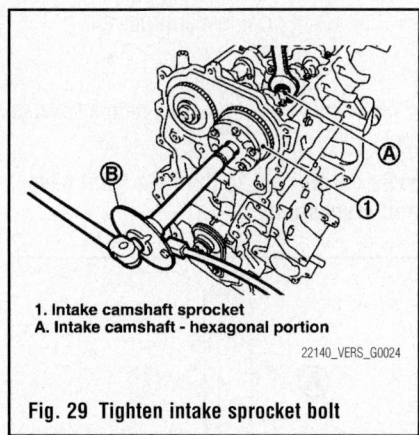

1. Intake camshaft sprocket
A. Intake camshaft - hexagonal portion

22140_VERS_G0024

Fig. 29 Tighten intake sprocket bolt

mark. **Securely align the knock pin and the pin hole, and then install.**

24. Tighten the intake camshaft sprocket bolt to 26 ft. lbs. (35 Nm). Secure the hexagonal part of the intake camshaft using a wrench to tighten the bolt.
25. Turn 67 degrees clockwise (angle tightening) using Tool No. KV10112100 (BT-8653-A).

❊❊ CAUTION

Never judge by visual inspection without an angle wrench.

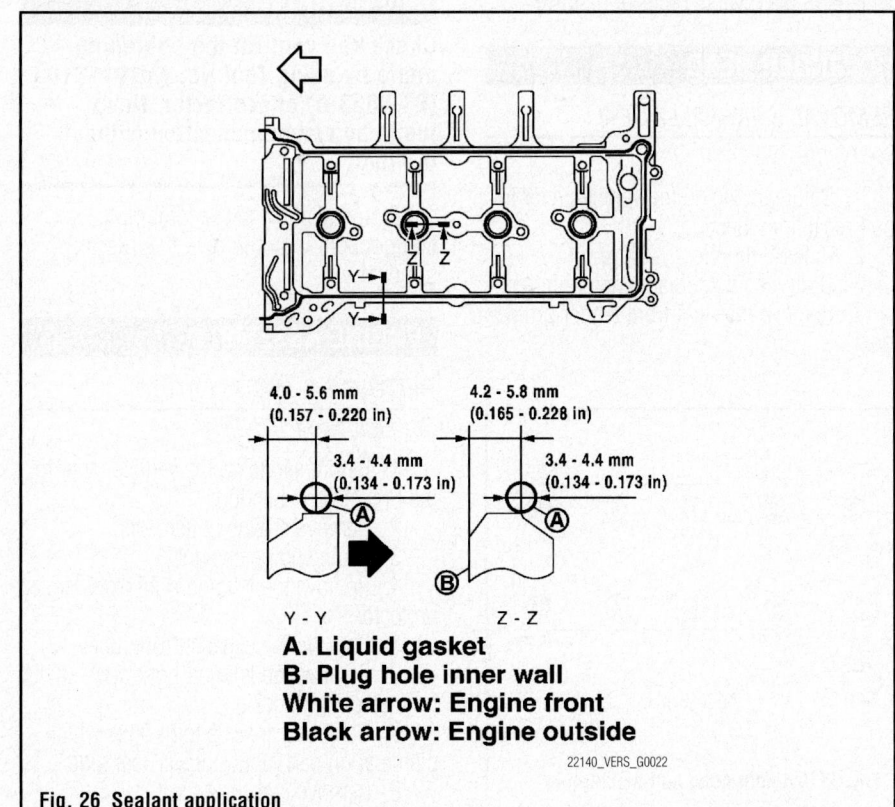

4.0 - 5.6 mm
(0.157 - 0.220 in)

4.2 - 5.8 mm
(0.165 - 0.228 in)

3.4 - 4.4 mm
(0.134 - 0.173 in)

3.4 - 4.4 mm
(0.134 - 0.173 in)

Y - Y   Z - Z

A. Liquid gasket
B. Plug hole inner wall
White arrow: Engine front
Black arrow: Engine outside

22140_VERS_G0022

Fig. 26 Sealant application

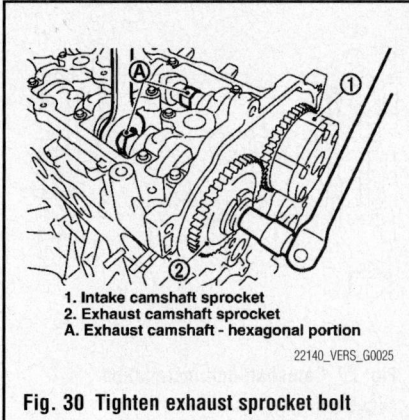

1. Intake camshaft sprocket
2. Exhaust camshaft sprocket
A. Exhaust camshaft - hexagonal portion

22140_VERS_G0025

Fig. 30 Tighten exhaust sprocket bolt

26. Install the exhaust camshaft sprocket and tighten the bolt to 65 ft. lbs. (88.2 Nm). Secure the hexagonal part of the camshaft using a wrench to tighten the bolt.

27. Install timing chain and related parts.

28. Inspect and adjust valve clearance.

29. Installation of the remaining components is in the reverse order of removal.

CRANKSHAFT FRONT SEAL

REMOVAL & INSTALLATION

See Figures 31 and 32.

1. Before servicing the vehicle, refer to the Precautions Section.

2. Remove the following parts:
   • Right front fender protector.
   • Drive belt.
   • Crankshaft pulley.

3. Remove front oil seal using a suitable tool.

➡**Be careful not to damage front cover and crankshaft.**

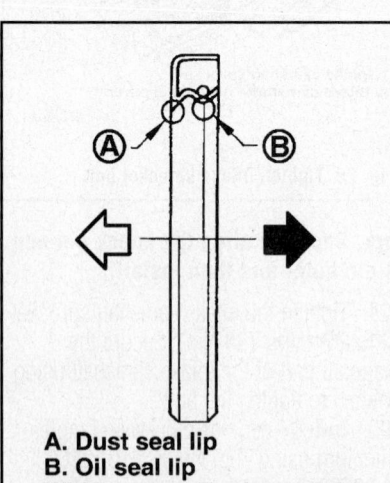

A. Dust seal lip
B. Oil seal lip
White arrow: Engine outside
Black arrow: Engine inside

22140_VERS_G0026

Fig. 31 Front oil seal lip installation positioning

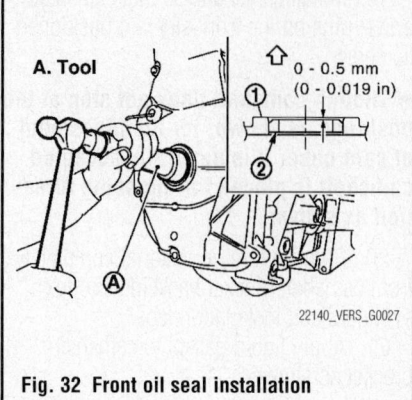

A. Tool

⇧ 0 - 0.5 mm
(0 - 0.019 in)

22140_VERS_G0027

Fig. 32 Front oil seal installation

To install:

4. Apply new engine oil to new front oil seal joint surface and seal lip.

5. Install front oil seal so that each seal lip is oriented as shown in the figure.

6. Install the front oil seal using a suitable tool with an outer diameter of 57 mm (2.24 in) and an inner diameter of 45 mm (1.77 in) to: within 0.3 mm (0.012 inch) toward engine front, within 0.5 mm (0.020 inch) toward engine rear.

➡**Press-fit oil seal straight to avoid causing burrs or tilting.**

➡**Do not touch grease applied on oil seal lip.**

7. Installation of the remaining components is in the reverse order of removal.

CYLINDER HEAD

REMOVAL & INSTALLATION

See Figure 33.

1. Before servicing the vehicle, refer to the Precautions Section.

2. Release the fuel pressure.

3. Drain engine coolant and engine oil.

4. Remove the right front fender protector.

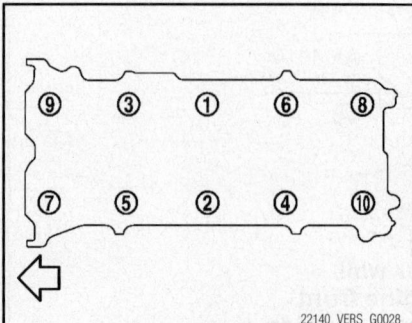

22140_VERS_G0028

Fig. 33 Cylinder head bolt installation sequence

5. Remove drive belt.

6. Remove the following components and related parts:
   • Exhaust manifold
   • Intake manifold
   • Water outlet
   • Fuel tube and fuel injector assembly
   • Rocker cover
   • Timing chain front cover
   • Camshaft

7. Using a TORX® socket (size E18), remove cylinder head, loosening the bolts in the reverse order of the tightening sequence.

8. Remove the cylinder head gasket.

To install:

9. Install the cylinder head gasket.

10. Apply new engine oil to threads and seating surface of bolts. If cylinder head bolts re-used, check their outer diameters before installation.

11. Install cylinder head, follow the steps below to tighten cylinder head bolts in numerical order as shown:

Step A: 30 ft. lbs. (40 Nm)
Step B: 100° clockwise
Step C: Loosen to 0 Nm in the reverse order of tightening.
Step D: 30 ft. lbs. (40 Nm)
Step E: 100° clockwise
Step F: 100° clockwise

✳✳ WARNING

Check and confirm the tightening angle by using Tool No. KV10112100 (BT-8653-A) or protractor. Never judge by visual inspection without the tool.

12. Installation of the remaining components is in the reverse order of removal.

ENGINE ASSEMBLY

REMOVAL & INSTALLATION

See Figures 34 and 35.

1. Before servicing the vehicle, refer to the Precautions Section.

2. Remove engine undercover.

3. Drain engine coolant.

4. Remove the left and right front fender protectors.

5. Remove the exhaust front tube.

6. Remove the left and right drive shafts from steering knuckle.

7. Remove transaxle joint bolts which pierce at oil pan (upper) lower rear side.

8. Remove rear torque rod.

9. Remove hood assembly.

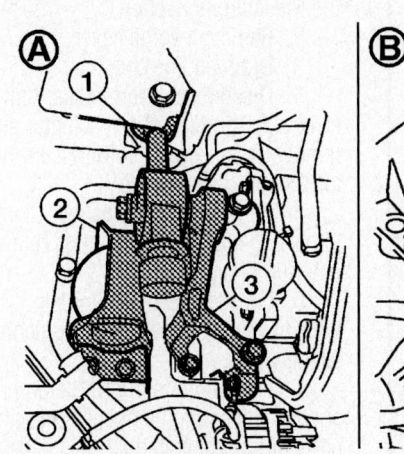 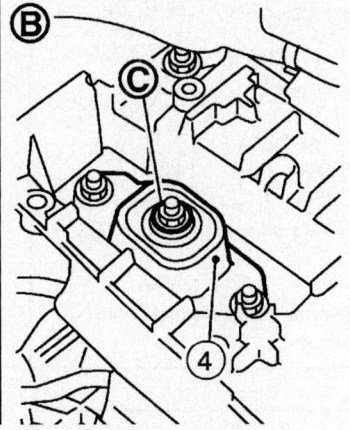

1. Right-hand torque rod
2. Right-hand engine insulator
3. Right-hand engine bracket
4. Left-hand engine insulator

A. Engine front side
B. Transaxle side
C. Engine through bolt-securing

22140_VERS_G0046

Fig. 34 Remove engine mounts

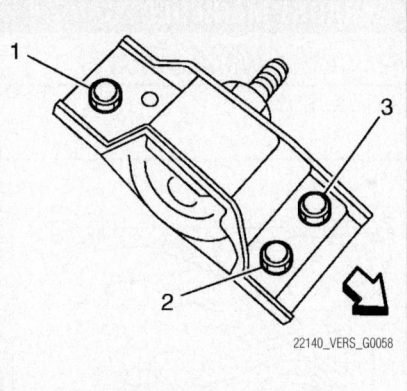

22140_VERS_G0058

Fig. 35 Right engine mounting insulator bolt tightening sequence

10. Remove cowl top cover and cowl top extension assembly.

11. Release fuel pressure.

12. Remove battery and battery tray.

13. Remove drive belt.

14. Remove air duct and air cleaner case assembly.

15. Remove cooling fan assembly.

16. Remove the upper and lower radiator hoses.

17. Disconnect A/T, CVT fluid cooler hoses.

18. Disconnect all connections of engine harness around the left engine mounting insulator, and then temporarily secure the engine harness into the engine side.

➡ **Protect connectors using a resin bag to protect against foreign materials during the operation.**

19. Disconnect fuel feed hose at engine side.

20. Disconnect heater hoses, and install plugs them to prevent engine coolant from draining.

21. Disconnect control cable from transaxle.

22. Remove ground cable at transaxle side.

23. Remove ground cable between front cover and vehicle.

24. Remove alternator.

25. Remove A/C compressor with piping connected from the engine. Temporarily secure it on the vehicle side with a rope to avoid putting load on it.

26. Remove the intake manifold to prevent the hanging chain from interfering.

27. Install engine slinger to cylinder head front left side and rear right side and support the engine position with a hoist.

28. Support engine and transaxle assembly with a hoist and secure the engine in appropriate position.

29. Use a manual lift table caddy or equivalently rigid tool such as a transmission jack. Securely support bottom of the engine and the transaxle, and simultaneously adjust hoist tension.

➡ **Put a piece of wood or something similar as the supporting surface, and secure a completely stable condition.**

30. Remove right hand torque rod, right engine insulator and right engine bracket.

31. Remove engine through bolt-securing nut.

32. Remove the engine and the transaxle assembly from the vehicle downward by carefully operating supporting tools.

➡ **During the operation, make sure that no part interferes with the vehicle side.**

➡ **Before and during this lifting, always check if any harnesses are left connected.**

➡ **During the removal operation, always be careful to prevent the**

vehicle from falling off the lift due to changes in the center of gravity.

➡ **If necessary, support the vehicle by setting jack or suitable tool at the rear.**

➡ **During operation, securely support the engine by placing a piece of wood under the engine oil pan and transaxle oil pan. Securely support the engine slingers with a hoist.**

33. When the engine hoisting is not performed simultaneously, install engine slinger to cylinder head front left side and rear right side.

To install:

➡ **Do not allow engine oil to get on engine mounting insulator. Be careful not to damage engine mounting insulator.**

34. Installation is the reverse of removal, noting the following torque specs:
- Right torque rod bolts: 103 ft. lbs. (140 Nm)
- Right engine mounting bracket bolt: 33 ft. lbs. (45 Nm)
- Rear torque rod bracket bolts: 59 ft. lbs. (80 Nm)
- Rear torque rod bolt: 59 ft. lbs. (80 Nm)
- Engine through bolt: 59 ft. lbs. (80 Nm)
- Left engine mounting bracket bolts: 36 ft. lbs. (48.5 Nm)
- Left engine mounting insulator nuts: 44 ft. lbs. (60 Nm), and 77 ft. lbs. (105 Nm)

a. Tighten the right engine mounting insulator bolts to 48 ft. lbs. (65 Nm), and 77 ft. lbs. (105 Nm) in the order shown.

b. Make sure that each mounting insulator is seated properly, and tighten nuts and bolts.

EXHAUST MANIFOLD

REMOVAL & INSTALLATION

See Figure 36.

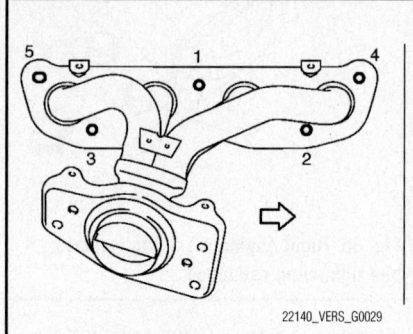

Fig. 36 Exhaust manifold bolt installation sequence

1. Before servicing the vehicle, refer to the Precautions Section.
2. Remove exhaust front tube.
3. Remove exhaust manifold cover.
4. Remove the A/F sensor 1, using Tool No. KV991J0050 (J-44626). Handle it carefully and avoid impacts.
5. Remove exhaust manifold side bolt of exhaust manifold stay.
6. Loosen nuts in reverse order of the tightening sequence and remove exhaust manifold.

➡ **Cover engine openings to avoid entry of foreign materials.**

To install:

7. Install exhaust manifold gasket.
8. Tighten exhaust manifold nuts to specification in two stages in the numerical order as shown.
9. Install exhaust manifold stay.
10. Install the A/F ratio sensor 1, using Tool No. KV991J0050 (J-44626).

➡ **Before installing a new A/F ratio sensor, clean the exhaust tube threads using suitable tool and approved anti-seize lubricant.**

➡ **Do not over-tighten the A/F ratio sensor. Doing so may damage the A/F ratio sensor, resulting in the MIL coming on.**

11. Installation of the remaining parts is in the reverse order of removal.

INTAKE MANIFOLD

REMOVAL & INSTALLATION

See Figures 37 through 39.

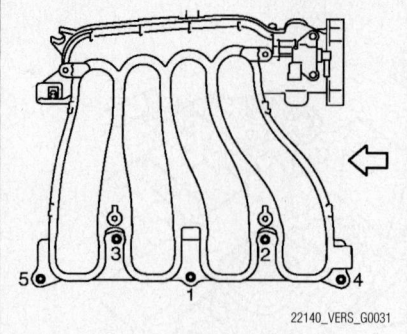

Fig. 37 Reverse intake manifold bolt removal sequence

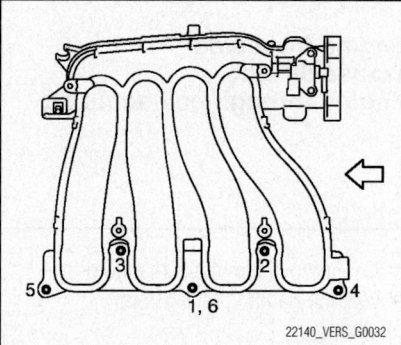

Fig. 38 Intake manifold bolt tightening sequence

1. Before servicing the vehicle, refer to the Precautions Section.
2. Remove engine cover.
3. Drain engine coolant.
4. Disconnect water hoses from the electronic throttle control actuator and remove the electronic throttle control actuator. Do not disassemble.
5. Remove oil level gauge. Cover the oil level gauge guide openings to avoid entry of foreign materials.
6. Loosen and remove intake manifold bolts in reverse order of the tightening sequence.
7. Loosen bolts in reverse order as shown. Cover engine openings to avoid entry of foreign materials.
8. Remove intake manifold.

To install:

9. Install intake manifold. Be sure the intake manifold gasket is seated correctly in groove of intake manifold.
10. Tighten bolts in numerical order as shown.
11. Tighten intake manifold bolt (A), and then tighten bolt (B).
12. Install electronic throttle control actuator.
13. Install the water hoses to electronic throttle control actuator as shown.

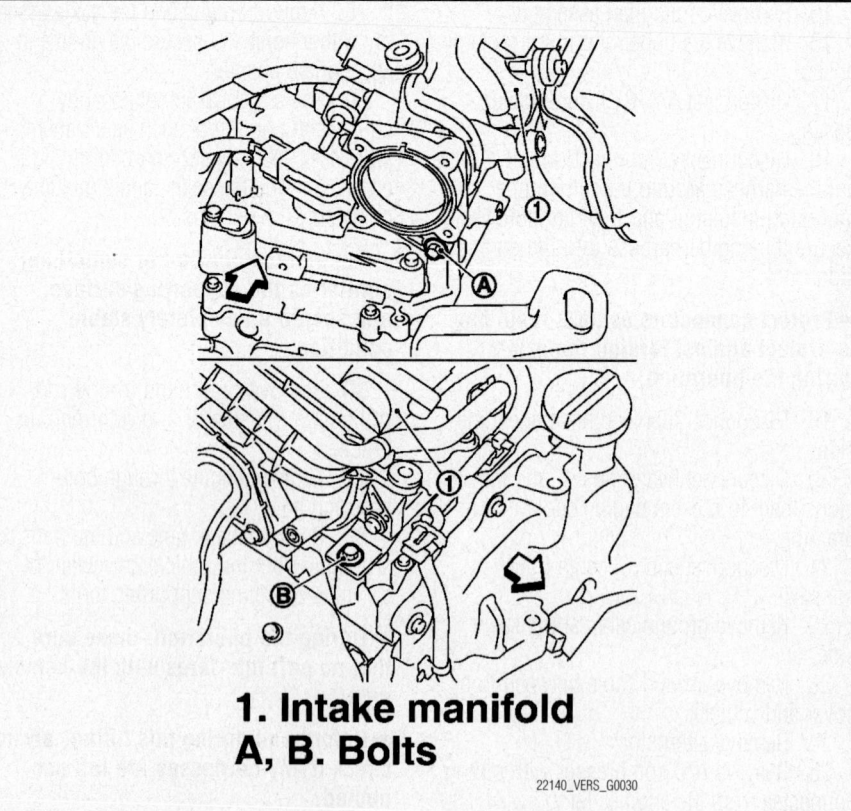

1. Intake manifold
A, B. Bolts

Fig. 39 Intake manifold bolt installation

14. Installation of the remaining components is in the reverse order of removal.

OIL PAN

REMOVAL & INSTALLATION

See Figures 40 through 46.

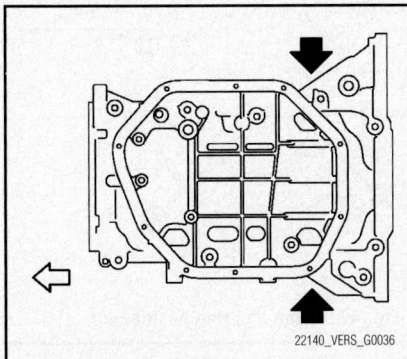

Fig. 40 Pry between the upper oil pan and cylinder block

1. Before servicing the vehicle, refer to the Precautions Section.
2. Drain engine oil.
3. Remove engine and transaxle assembly.

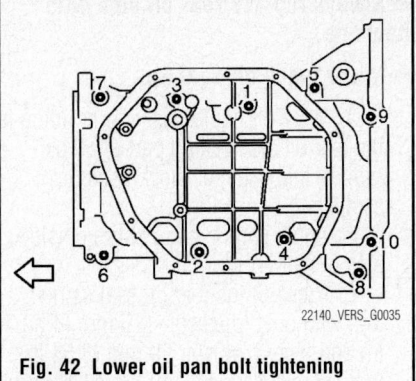

Fig. 42 Lower oil pan bolt tightening sequence

4. Remove oil filter using Tool No. KV10115801.

➡**When removing, prepare a shop cloth to absorb any engine oil leakage or spillage.**

5. Remove lower oil pan bolts in reverse of the tightening sequence.
6. After removing the bolts and nuts, separate the mating surface and remove the sealant using Tool No. KV10111100 (J-37228). Be careful not to damage the mating surfaces.
7. Remove the following parts:

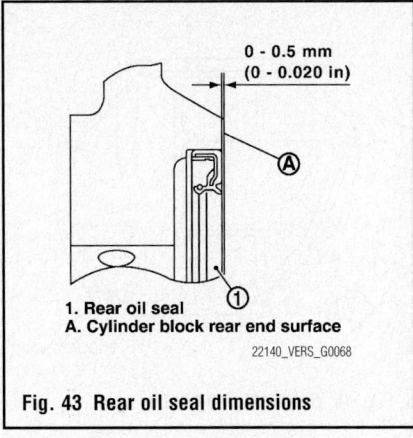

1. Rear oil seal
A. Cylinder block rear end surface

Fig. 43 Rear oil seal dimensions

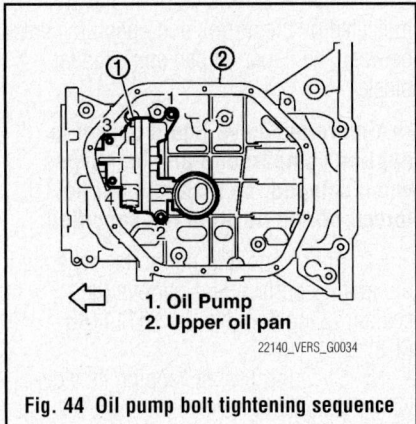

1. Oil Pump
2. Upper oil pan

Fig. 44 Oil pump bolt tightening sequence

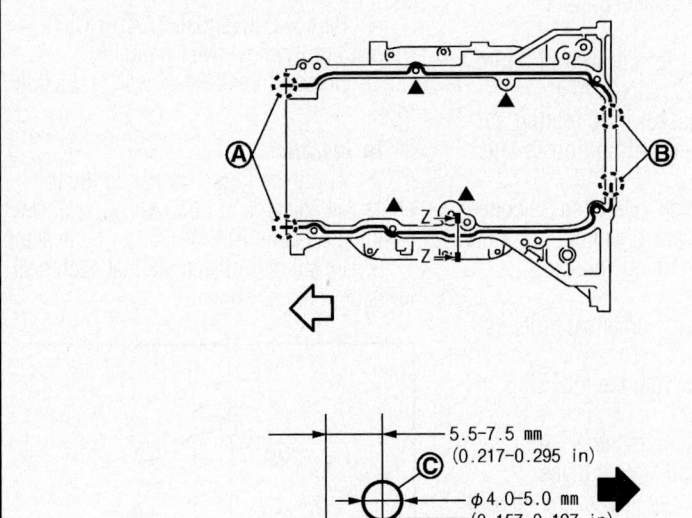

1: **Upper oil pan**
A: **2 mm protruded to outside**
B: **2 mm protruded to rear oil seal mounting side**
White arrow: **Engine front**
Black arrow: **Engine outside**
Black triangle positions: **Apply liquid gasket to outside of bolt hole**

Fig. 41 Sealant application

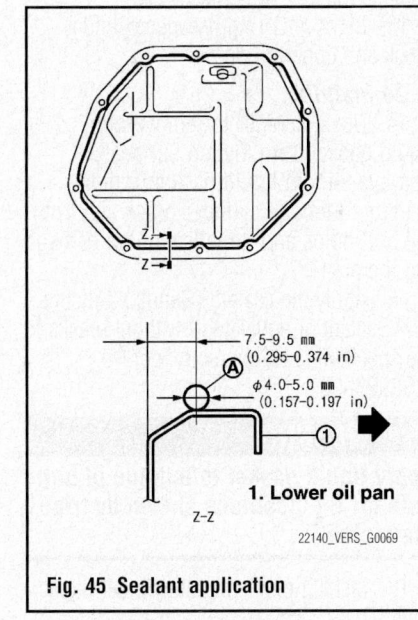

1. Lower oil pan

Fig. 45 Sealant application

- Flywheel (M/T models) or drive plate (A/T or CVT models)
- Front cover, timing chain, oil pump drive chain

8. Remove oil pump. Loosen bolts in reverse order as shown.
9. Remove oil pan (lower) bolts in reverse order of the tightening sequence.

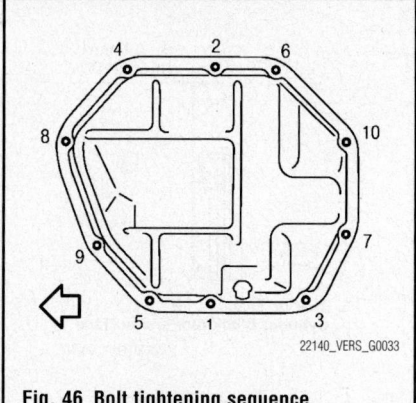

Fig. 46 Bolt tightening sequence

10. Insert a screwdriver into the area indicated by the arrows and open up a crack between the upper oil pan and cylinder block.

➡ **A more adhesive liquid gasket is applied compared to previous types when shipped, so it should not be forced off the position not specified.**

11. After removing the bolts, separate the mating surface and remove the sealant using Tool No. KV10111100 (J-37228).

12. Slide the Tool by tapping its side with a hammer to remove the lower oil pan from the upper oil pan. Be careful not to damage the mating surfaces.

13. Remove O-ring between cylinder block and upper oil pan.

To install:

14. Use a scraper to remove old liquid gasket from mating surfaces. Remove the old liquid gasket from the mating surface of cylinder block and from the bolt holes and threads without damaging the area.

15. Apply the sealant (Genuine Silicone RTV Sealant or equivalent) without breaks to the specified location using Tool No. WS39930000.

☀ WARNING

Apply liquid gasket to outside of bolt hole for the positions shown by triangle marks.

16. Install new O-ring at cylinder block side.

17. Tighten bolts in numerical order as shown.

18. Install rear oil seal with the following procedure:

➡ **The installation of rear oil seal should be completed within 5 minutes after installing oil pan (upper).**

➡ **Always replace rear oil seal with new one.**

➡ **Never touch oil seal lip.**

a. Wipe off liquid gasket protruding to the rear oil seal mating part of oil pan (upper) and cylinder block using a scraper.

b. Apply engine oil to entire outside area of rear oil seal.

c. Press-fit the rear oil seal using a drift with outer diameter 115 mm (4.53 in) and inner diameter 90 mm (3.54 in) (commercial service tool). Press-fit to the specified dimensions as shown.

➡ **The standard surface of the dimension is the rear end surface of cylinder block.**

➡ **Never touch the grease applied to the oil seal lip.**

➡ **Be careful not to damage the rear oil seal mounting part of oil pan (upper) and cylinder block or the crankshaft.**

➡ **Press-fit straight, making sure that rear oil seal does not curl or tilt.**

19. Install oil pump:
a. Tighten bolts in numerical order as shown.

20. Install oil pump sprocket, oil pump drive chain and other related parts if removed.

21. Use a scraper to remove old liquid gasket from mating surfaces. Also remove old liquid gasket from mating surface of oil pan (upper) and bolt holes and threads.

22. Apply the sealant (Genuine Silicone RTV Sealant or equivalent) without breaks to the specified location using Tool No. WS39930000.

23. Tighten bolts in numerical order as shown.

24. Install oil filter with the following procedure:
a. Remove foreign materials adhering to the oil filter installation surface.

b. Apply new engine oil to the oil seal contact surface of new oil filter.

c. Screw oil filter manually until it touches the installation surface, and then tighten it by ⅔ turn. Or tighten to 13 ft. lbs. (18 Nm).

25. The remainder of installation is the reverse of removal.

OIL PUMP

REMOVAL & INSTALLATION

See Timing Chain.

PISTON AND RING

POSITIONING

See Figure 47.

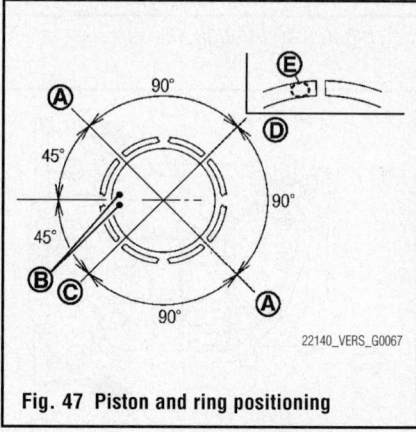

Fig. 47 Piston and ring positioning

REAR MAIN SEAL

REMOVAL & INSTALLATION

See Figures 48 and 49.

1. Before servicing the vehicle, refer to the Precautions Section.

2. Remove transaxle assembly.

3. Remove clutch cover and clutch disk (M/T models).

4. Remove drive plate (A/T or CVT models) or flywheel (M/T models).

5. Remove rear oil seal with a suitable tool.

To install:

6. Apply the liquid gasket lightly to entire outside area of new rear oil seal. Use Genuine Silicone RTV Sealant or equivalent.

7. Install rear oil seal so that each seal lip is oriented as shown.

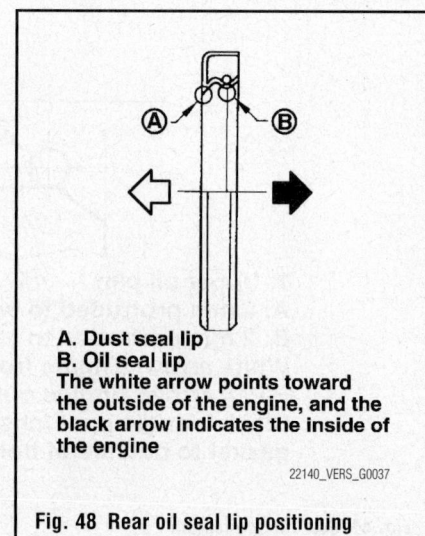

A. Dust seal lip
B. Oil seal lip
The white arrow points toward the outside of the engine, and the black arrow indicates the inside of the engine

Fig. 48 Rear oil seal lip positioning

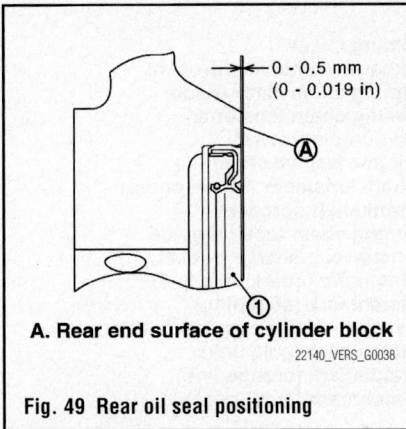

A. Rear end surface of cylinder block

22140_VERS_G0038

Fig. 49 Rear oil seal positioning

8. Install rear oil seal with a suitable tool with an outer diameter 115 mm (4.53 in) and inner diameter 90 mm (3.54 in).

➡**Be careful not to damage crankshaft and cylinder block.**

➡**Press-fit oil seal straight to avoid causing burrs or tilting.**

➡**Do not touch grease applied onto oil seal lip.**

9. Install rear oil seal to the position as shown.

➡**The standard surface of the dimension is the rear end surface of cylinder block.**

10. Installation of the remaining components is in the reverse order of removal.

TIMING CHAIN, SPROCKETS, FRONT COVER AND SEAL

REMOVAL & INSTALLATION

See Figures 50 through 60.

1. Before servicing the vehicle, refer to the Precautions Section.
2. Remove the right front wheel.
3. Remove the right front fender protector.

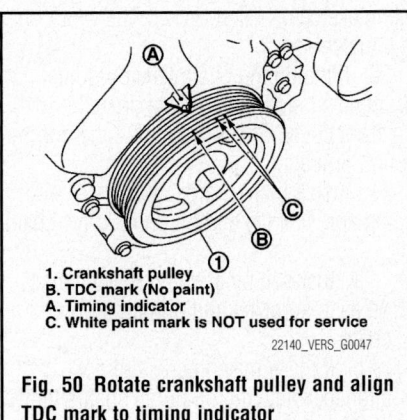

1. Crankshaft pulley
B. TDC mark (No paint)
A. Timing indicator
C. White paint mark is NOT used for service

22140_VERS_G0047

Fig. 50 Rotate crankshaft pulley and align TDC mark to timing indicator

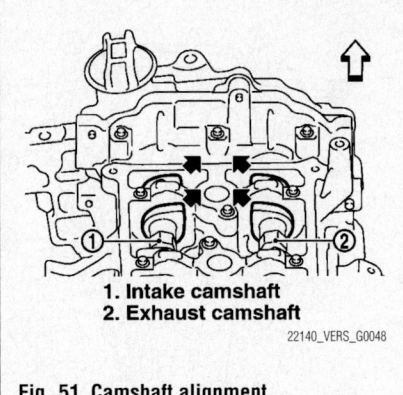

1. Intake camshaft
2. Exhaust camshaft

22140_VERS_G0048

Fig. 51 Camshaft alignment

4. Drain engine oil.
5. Remove the rocker cover.
6. Remove the drive belt.
7. Remove the water pump pulley.
8. Remove the ground cable.
9. Support the bottom surface of engine using a transmission jack, and then remove the right-hand engine bracket and insulator.
10. Set No. 1 cylinder at TDC on its compression stroke with the following procedure:

   a. Rotate crankshaft pulley (1) clockwise and align TDC mark (no paint) (B) to timing indicator (A) on front cover.

   b. At the same time, make sure that the cam noses of the No.1 cylinder are located as shown. If not, rotate crankshaft

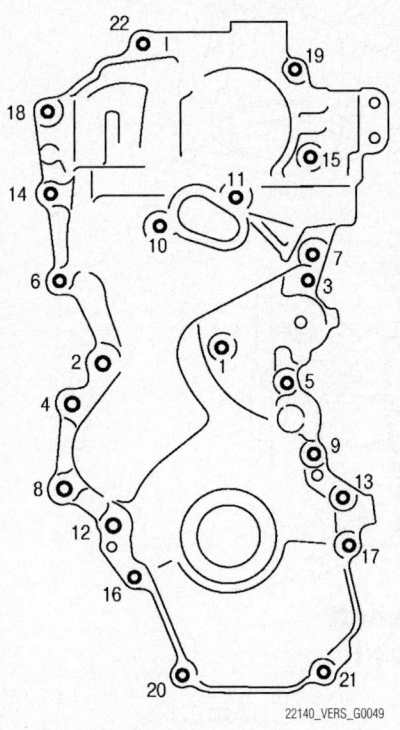

22140_VERS_G0049

Fig. 52 Reverse timing chain cover bolt removal sequence

pulley one revolution (360 degrees) and align as shown.

11. Hold crankshaft pulley using suitable tool, and loosen crankshaft pulley bolt. Locate bolt seating surface at 10 mm (0.39 in) from its original position.

➡**Never remove the crankshaft pulley bolt as it will be used as a supporting point for the pulley puller.**

12. Attach a pulley puller in the M6 thread hole on crankshaft pulley, and remove crankshaft pulley.

13. Remove the lower oil pan. When crankshaft sprocket, oil pump sprocket and other related parts are not removed, this step is unnecessary.

14. Remove the intake valve timing control solenoid valve.

15. Remove drive belt auto-tensioner.

16. Loosen bolts in reverse order as shown.

17. Cut liquid gasket by prying, and then remove the front cover.

➡**Be careful not to damage the mating surface.**

➡**A more adhesive liquid gasket is applied compared to previous types when shipped, so it should not be forced off the position not specified.**

18. Remove front oil seal from front cover. Lift up front oil seal using a suitable tool. Be careful not to damage front cover.

19. Push in timing chain tensioner plunger.

20. Insert a stopper pin into the body hole to retain the plunger in collapsed position. Use approximately 1.5 mm (0.059 in) diameter hard metal pin as a stopper pin.

21. Remove timing chain tensioner.

22. Remove timing chain slack guide (2), timing chain tension guide (3) and timing chain (1).

❊❊ WARNING

Never rotate each crankshaft and camshaft individually while timing chain is removed. It causes interference between valve and piston.

23. Fully lift up lever, and push the slack guide into the inside of chain tensioner (for oil pump). The slack guide is released by fully lifting the lever up. As the result, the slack guide can be moved.

24. Matching the hole on lever with the hole on tensioner body, insert a stopper pin to secure slack guide. Use approximately 1.0 mm (0.04 in) diameter. Hard metal pin as a stopper pin.

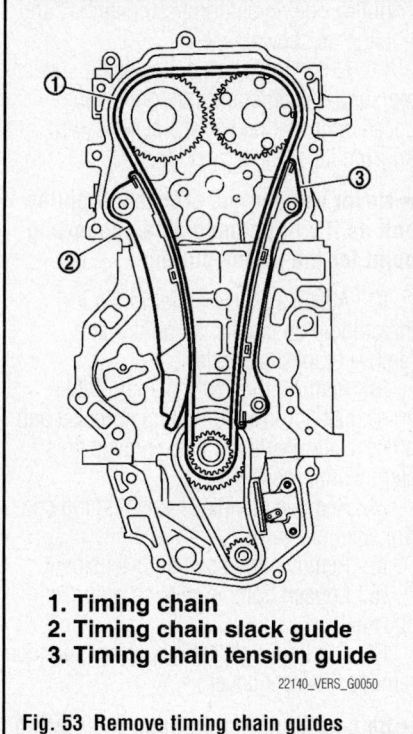

1. Timing chain
2. Timing chain slack guide
3. Timing chain tension guide

22140_VERS_G0050

Fig. 53 Remove timing chain guides

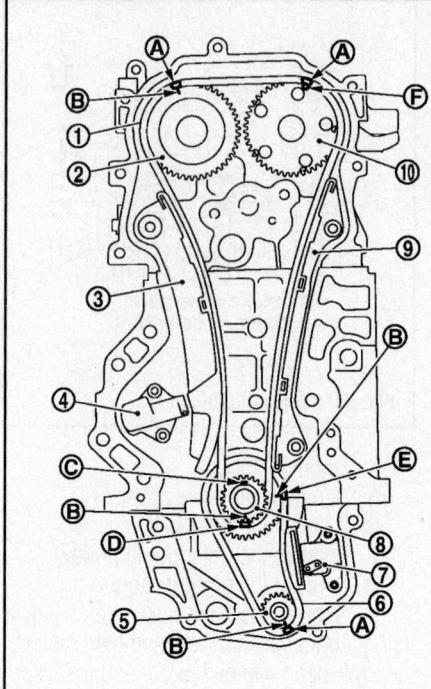

1. Timing chain
2. Exhaust camshaft sprocket
3. Timing chain slack guide
4. Timing chain tensioner
5. Oil pump sprocket
6. Oil pump drive chain
7. Chain tensioner (for oil pump)
8. Crankshaft sprocket
9. Timing chain tension guide
10. Intake camshaft sprocket
A. Matchmark (dark blue link)
B. Matchmark (stamping)
C. Crankshaft key position (straight up)
D. Matchmark (gold link)
E. Matchmark (orange link)
F. Matchmark (outer groove)

22140_VERS_G0052

Fig. 55 Timing chain positioning

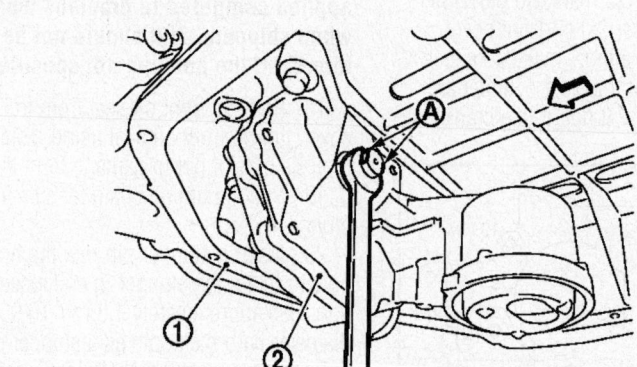

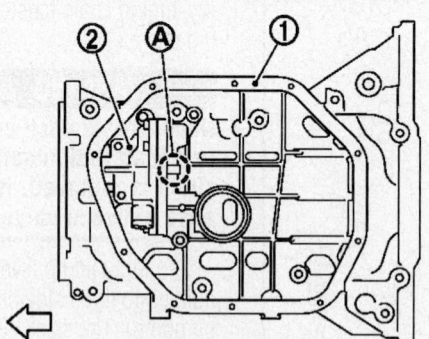

A. Oil pump shaft
1. Upper oil pan
2. Oil pump

22140_VERS_G0051

Fig. 54 Oil pump shaft and sprocket bolt removal

25. Remove chain tensioner (for oil pump). When the holes on lever and tensioner body cannot be aligned, align these holes by slightly moving the slack guide.

26. Hold the WAF part of oil pump shaft (A), and then loosen the oil pump sprocket bolt and remove them.

➡ **Secure the oil pump shaft with the WAF part. Never loosen the oil pump sprocket bolt by tightening the oil pump drive chain.**

27. Remove crankshaft sprocket, oil pump sprocket and oil pump drive chain as a set.

28. Remove timing chain tension guide (front cover side) from front cover if necessary.

To install:

29. Make sure that crankshaft key points straight up. There are two outer grooves in the intake camshaft sprocket. The wider one is a matchmark.

30. If the timing chain tension guide (front cover side) is removed, install it to the front cover. Check the joint condition by sound or feeling.

31. Install crankshaft sprocket (2), oil pump sprocket (3) and oil pump drive chain (1):

    a. Install it by aligning matchmarks on each sprocket and oil pump drive chain.

    b. If these matchmarks are not aligned, rotate the oil pump shaft slightly to correct the position.

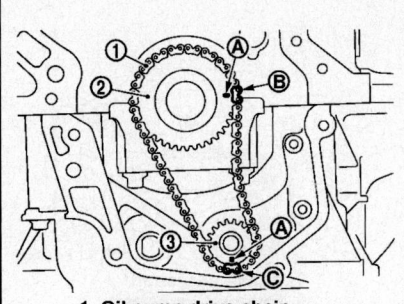

1. Oil pump drive chain
2. Crankshaft sprocket
3. Oil pump sprocket
A. Matchmark (stamping)
B. Matchmark (orange link)
C. Matchmark (dark blue link)

22140_VERS_G0053

Fig. 56 Install crankshaft sprocket, oil pump sprocket and oil pump drive chain

c. Check matchmark position of each sprocket after installing the oil pump drive chain.

32. Hold the WAF part of oil pump shaft (A), and then tighten the oil pump sprocket bolt.

33. Secure the oil pump shaft with the WAF part. Never loosen the oil pump sprocket bolt by tightening the oil pump drive chain.

34. Install chain tensioner (for oil pump):

a. Fix the plunger at the most compressed position using a stopper pin, and then install it.

b. Securely pull out the stopper pin after installing the chain tensioner (for oil pump).

c. Check matchmark position of oil pump drive chain and each sprocket again.

35. Align the matchmarks of each sprocket with the matchmarks of timing chain. There are two outer grooves in the intake camshaft sprocket. The wider one is a matchmark. If these matchmarks are not aligned, rotate the camshaft slightly by holding the hexagonal portion to correct the position.

➡ **Check matchmark position of each sprocket and timing chain again after installing the timing chain.**

36. Install the timing chain tension guide (3) and the timing chain slack guide (2).

37. Install timing chain tensioner:

a. Fix the plunger at the most compressed position using a stopper pin, and then install it.

b. Securely pull out the stopper pin after installing the timing chain tensioner.

38. Check matchmark position of timing chain and each sprocket again.

39. Apply new engine oil to new front oil seal joint surface.

40. Using a suitable tool install front oil seal so that each seal lip is oriented properly:

a. Press-fit front oil seal until it is flush with front end surface of front cover as shown below with a suitable tool. Within 0.3 mm (0.012 in) toward engine front, within 0.5 mm (0.020 in) toward engine rear.

➡ **Be careful not to damage front cover and crankshaft.**

➡ **Press-fit oil seal straight to avoid causing burrs or tilting.**

➡ **Never touch grease applied onto oil seal lip.**

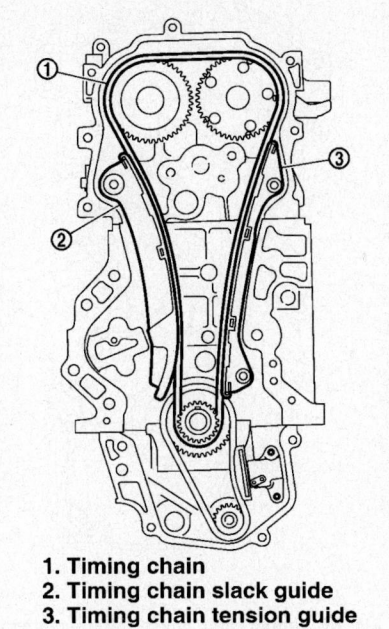

1. Timing chain
2. Timing chain slack guide
3. Timing chain tension guide

22140_VERS_G0050

Fig. 58 Install timing chain tension guides

41. Install new O-ring to cylinder block.

42. Apply the sealant without breaks to the specified location using Tool WS39930000. Tool. Use Genuine Silicone RTV Sealant or equivalent.

43. Make sure that matching marks of timing chain and each sprocket are still aligned.

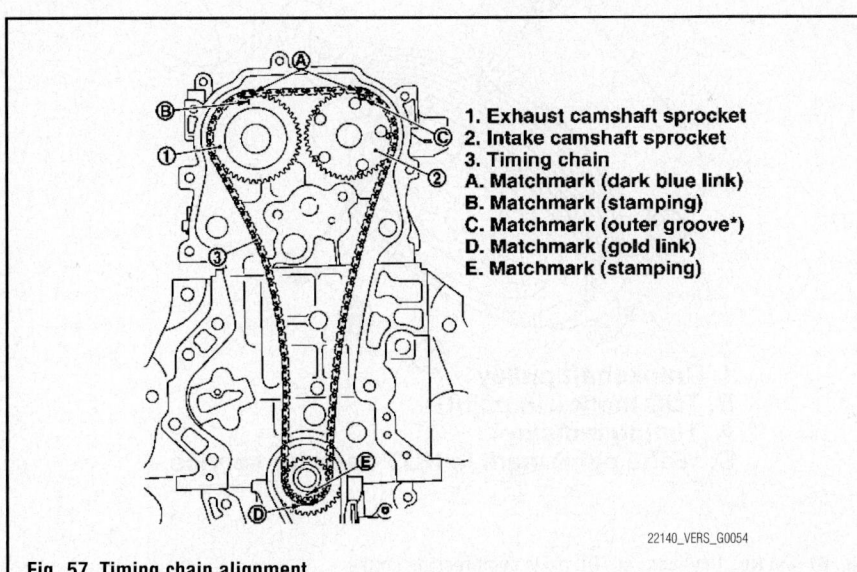

1. Exhaust camshaft sprocket
2. Intake camshaft sprocket
3. Timing chain
A. Matchmark (dark blue link)
B. Matchmark (stamping)
C. Matchmark (outer groove*)
D. Matchmark (gold link)
E. Matchmark (stamping)

22140_VERS_G0054

Fig. 57 Timing chain alignment

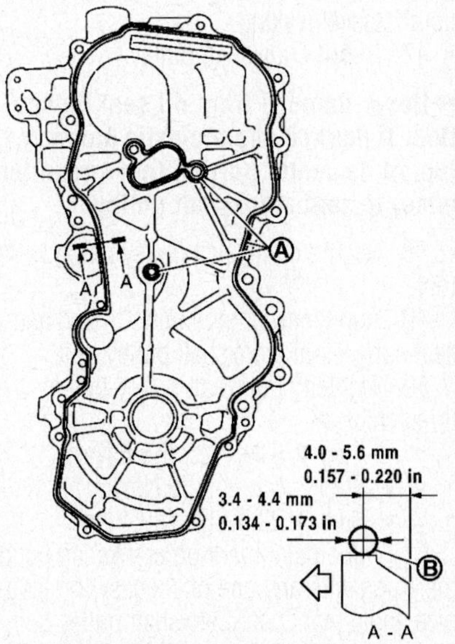

A. Liquid gasket application area

4.0 - 5.6 mm
0.157 - 0.220 in

3.4 - 4.4 mm
0.134 - 0.173 in

A - A

22140_VERS_G0055

Fig. 59 Sealant application

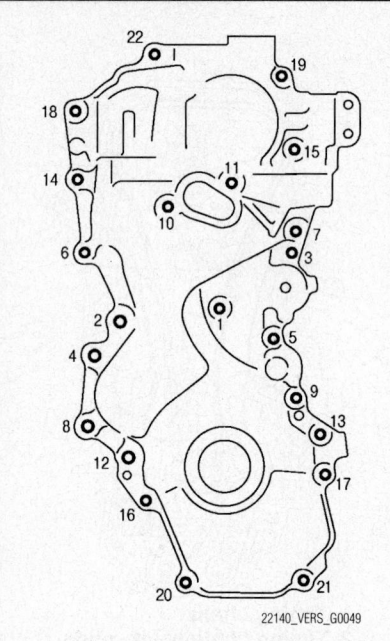

Fig. 60 Timing chain cover bolt tightening sequence

44. Install front cover, and tighten bolts in numerical order as shown.

➡**Attaching should be done within 5 minutes after liquid gasket application.**

45. Bolt installation positions:
- M6 bolts: No. 1
- M10 bolts: No. 6, 7, 10, 11, 14
- M12 bolts: No. 2, 4, 8, 12
- M8 bolts: Except the above

46. Tighten all bolts are in two stages to specified torque in numerical order as shown. Be sure to wipe off any excessive liquid gasket leaking.

47. Install crankshaft pulley.

➡**Never damage front oil seal lip section. If needed use a plastic hammer, tap on its center portion (not circumference) to seat crankshaft pulley.**

48. Secure crankshaft pulley using tool.

49. Apply new engine oil to thread and seat surfaces of crankshaft pulley bolt.

50. Tighten crankshaft pulley bolt in three steps:
- Step 1: 51 ft. lbs. (68.6 Nm)
- Step 2: 0 ft. lbs. (0 Nm)
- Step 3: 22 ft. lbs. (29.4 Nm)

51. Put a paint mark on crankshaft pulley matching with any one of six easy to recognize angle marks on crankshaft pulley bolt flange.

52. Turn another 60 degrees clockwise (angle tightening) using Tool No. KV10112100 (BT-8653-A).

53. Check the tightening angle with movement of one angle mark.

54. Make sure that crankshaft rotates clockwise smoothly.

55. Installation of the remaining components is in the reverse order of removal.

VALVE LASH

ADJUSTMENT

See Figures 61 through 64.

1. Before servicing the vehicle, refer to the Precautions Section.

2. Measure the valve clearance with the following procedure:

3. Set No. 1 cylinder at TDC of its compression stroke by rotating the crankshaft pulley (1) clockwise and align TDC mark (no paint) (B) to timing indicator (A) on front cover. At the same time, make sure that both intake and exhaust cam noses of No. 1 cylinder face inside. If they do not, rotate crankshaft pulley once more (360 degrees) and align.

4. Use a feeler gauge, measure the clearance between valve lifter and camshaft.

5. Intake valve clearance:
- Cold: 0.010–0.013 inches (0.26–0.34 mm)
- Hot: 0.012–0.016 inches (0.304–0.416 mm)

6. Exhaust valve clearance:
- Cold: 0.011–0.015 inches (0.29–0.37 mm)
- Hot: 0.012–0.017 inches (0.308–0.432 mm)

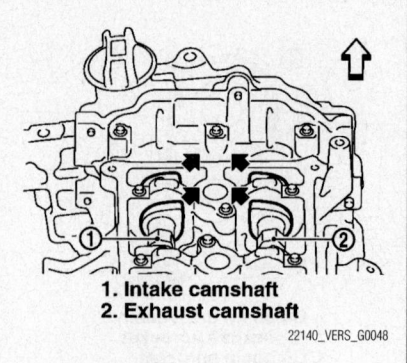

1. Intake camshaft
2. Exhaust camshaft

Fig. 62 Camshaft alignment, TDC

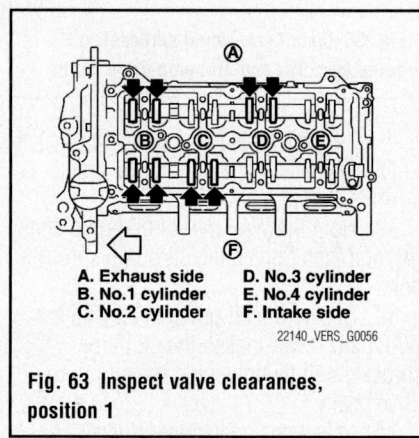

A. Exhaust side
B. No.1 cylinder
C. No.2 cylinder
D. No.3 cylinder
E. No.4 cylinder
F. Intake side

Fig. 63 Inspect valve clearances, position 1

7. Set No.4 cylinder at TDC of its compression stroke by rotating crankshaft pulley one revolution (360 degrees) and align TDC mark (no paint) to timing indicator on front cover.

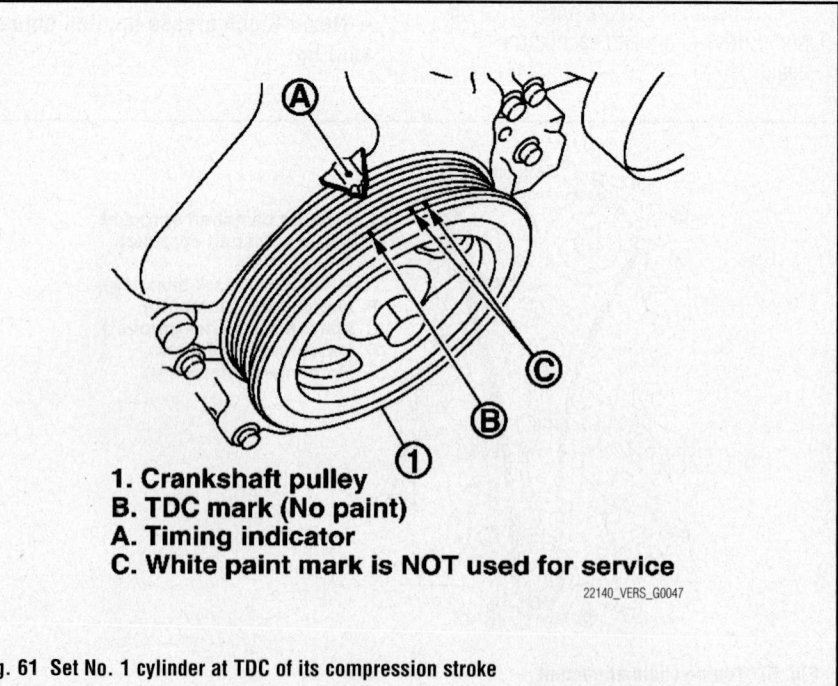

1. Crankshaft pulley
B. TDC mark (No paint)
A. Timing indicator
C. White paint mark is NOT used for service

Fig. 61 Set No. 1 cylinder at TDC of its compression stroke

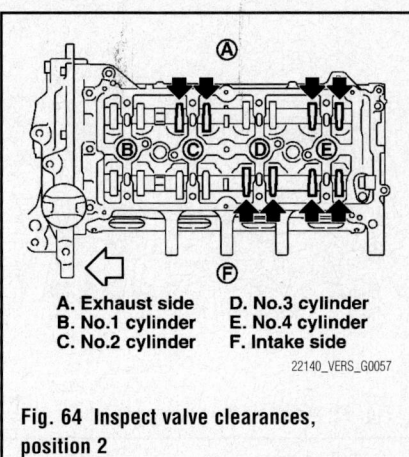

A. Exhaust side   D. No.3 cylinder
B. No.1 cylinder   E. No.4 cylinder
C. No.2 cylinder   F. Intake side

22140_VERS_G0057

Fig. 64 Inspect valve clearances, position 2

8. Use a feeler gauge, measure the clearance between valve lifter and camshaft.

9. If out of standard, perform adjustment. Perform adjustment depending on selected head thickness of valve lifter.

10. Remove camshaft.

11. Remove valve lifters at the locations that are out of the standard.

12. Measure the center thickness of the removed valve lifters with a micrometer.

13. Use the equation below to calculate valve lifter thickness for replacement:

a. Valve lifter thickness calculation: t=t1 + (C1 - C2), t=Valve lifter thickness to be replaced, t1=Removed valve lifter thickness, C1=Measured valve clearance, C2=Standard valve clearance: Intake : 0.30 mm (0.012 in), Exhaust : 0.33 mm (0.013 in).

b. Available thickness of valve lifter: 26 sizes range 3.00 to 3.50 mm (0.1181 to 0.1378 in) in steps of 0.02 mm (0.0008 in) (when manufactured at factory).

14. Install the selected valve lifter.

15. Install camshaft.

16. Install timing chain and related parts.

17. Manually rotate crankshaft pulley a few rotations.

18. Make sure that the valve clearances is within the standard.

ENGINE PERFORMANCE & EMISSION CONTROL

COMPONENT LOCATIONS

See Figures 65 through 74.

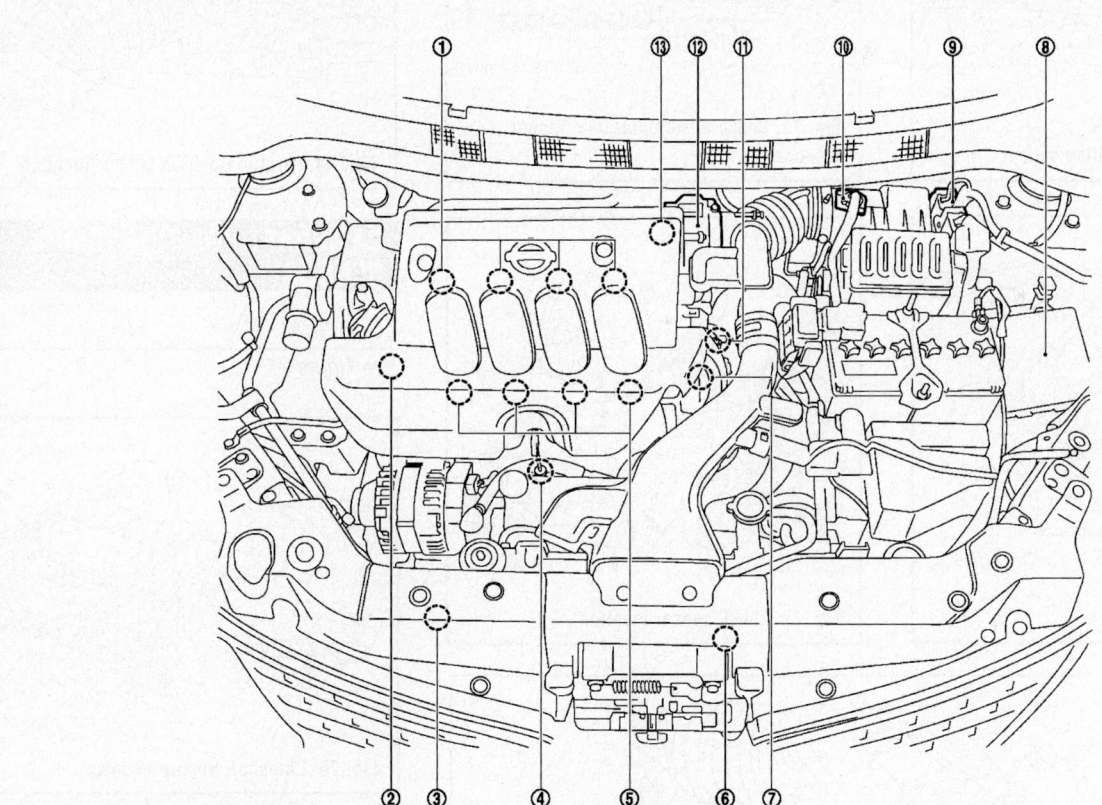

1. **Ignition coil (with power transistor) and spark plug**
2. **Intake valve timing control solenoid valve**
3. **Refrigerant pressure sensor**
4. **Knock sensor**
5. **Fuel injector**
6. **Cooling fan motor**
7. **Camshaft position sensor (PHASE)**
8. **IPDM E/R**
9. **ECM**
10. **Mass air flow sensor (with intake air temperature sensor)**
11. **Engine coolant temperature sensor**
12. **Electric throttle control actuator (with built-in throttle position sensor, throttle control motor)**
13. **EVAP canister purge volume control solenoid valve**

22140_VERS_G0061

Fig. 65 Engine control component locations

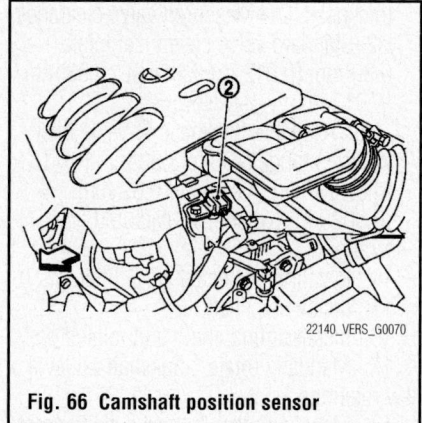

Fig. 66 Camshaft position sensor

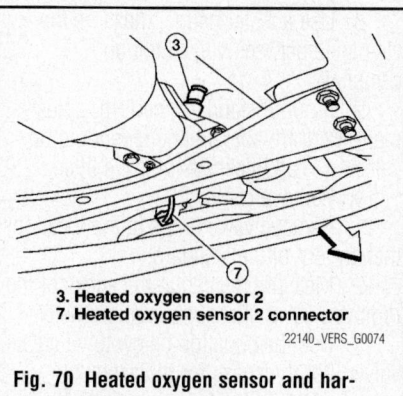

3. Heated oxygen sensor 2
7. Heated oxygen sensor 2 connector

Fig. 70 Heated oxygen sensor and harness connector locations

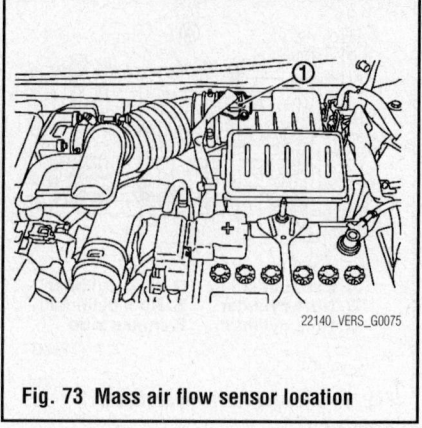

Fig. 73 Mass air flow sensor location

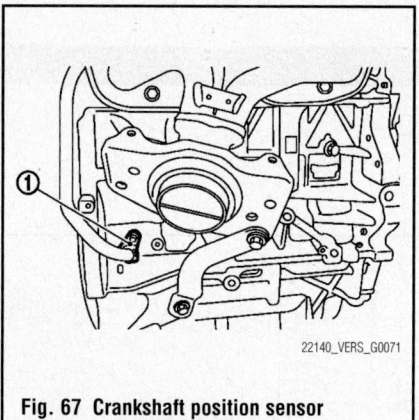

Fig. 67 Crankshaft position sensor

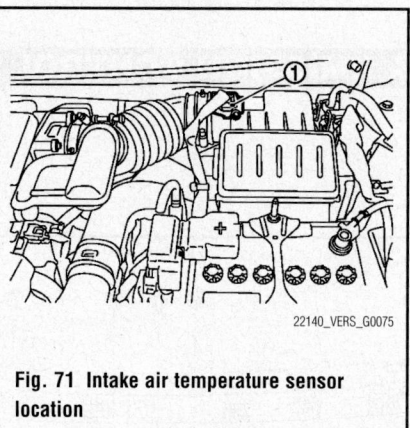

Fig. 71 Intake air temperature sensor location

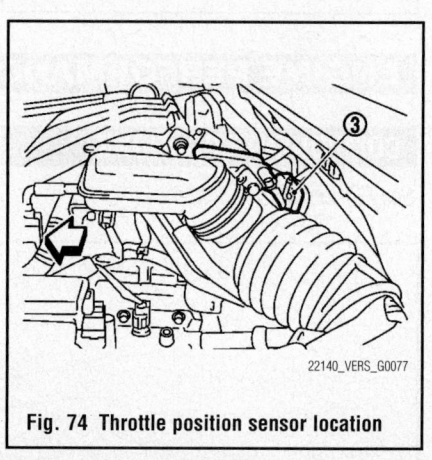

Fig. 74 Throttle position sensor location

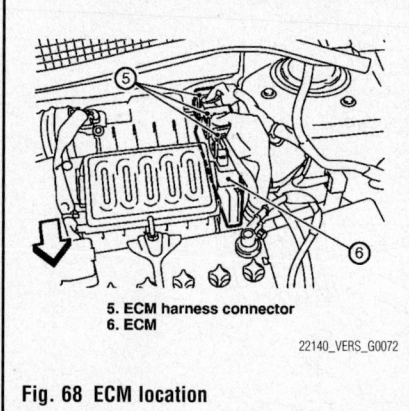

5. ECM harness connector
6. ECM

Fig. 68 ECM location

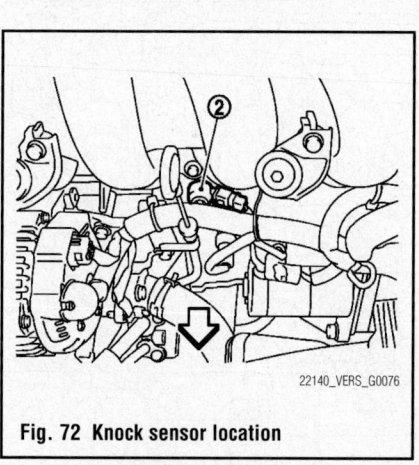

Fig. 72 Knock sensor location

CAMSHAFT POSITION (CMP) SENSOR

LOCATION
See Figure 75.

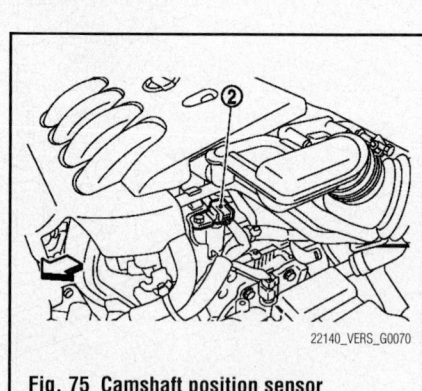

Fig. 75 Camshaft position sensor

REMOVAL & INSTALLATION
See Camshaft removal and installation procedure.

CRANKSHAFT POSITION (CKP) SENSOR

LOCATION
See Figure 76.

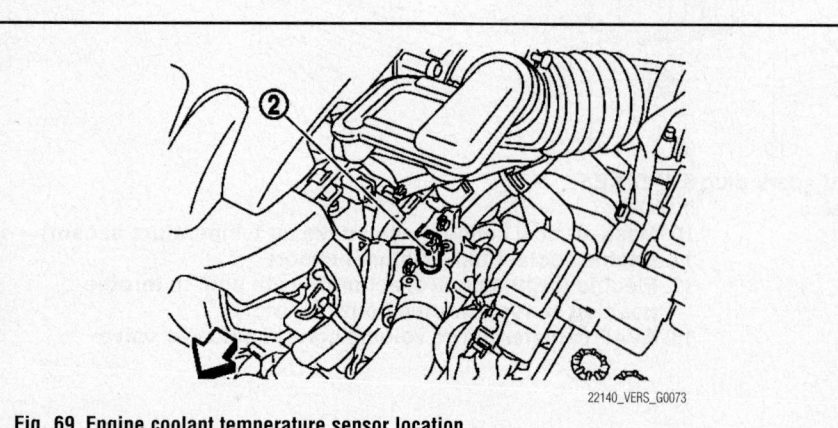

Fig. 69 Engine coolant temperature sensor location

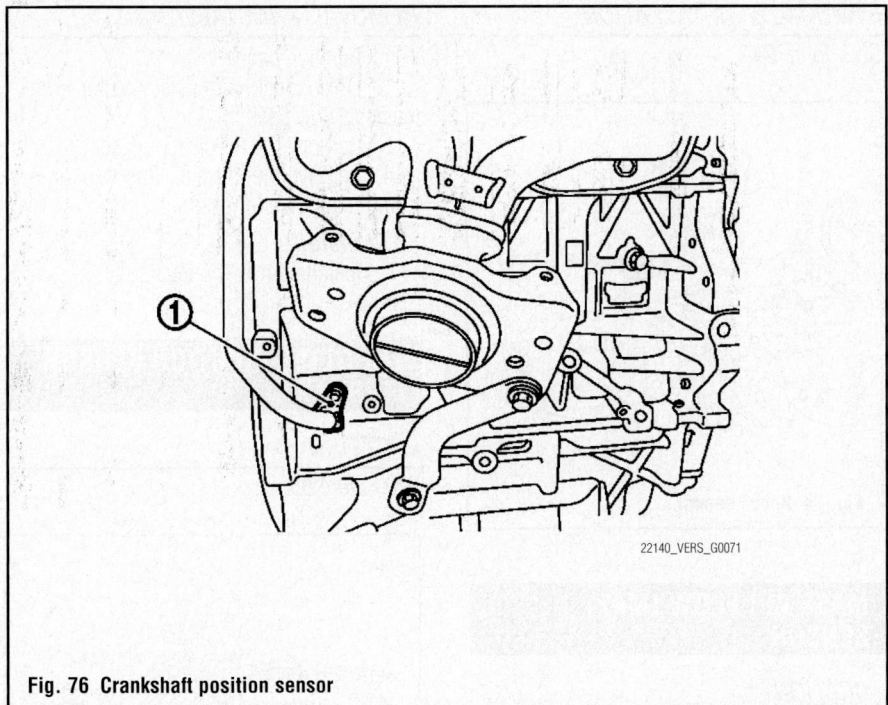

Fig. 76 Crankshaft position sensor

REMOVAL & INSTALLATION

See Figure 76.

1. The crankshaft position sensor (POS) is located on the cylinder block rear housing facing the gear teeth (cogs) of the signal plate at the end of the crankshaft.

ELECTRONIC CONTROL MODULE (ECM)

LOCATION

See Figure 77.

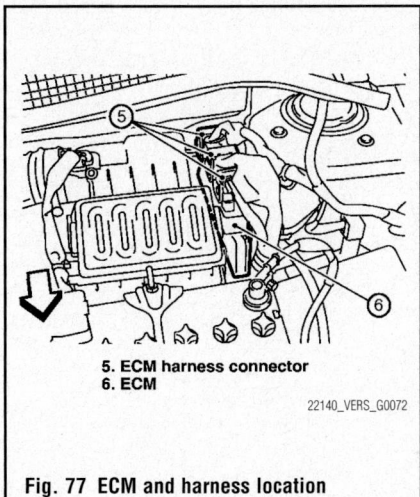

5. ECM harness connector
6. ECM

Fig. 77 ECM and harness location

REMOVAL & INSTALLATION

See Figure 78.

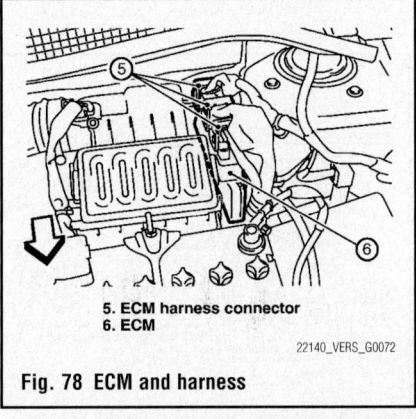

5. ECM harness connector
6. ECM

Fig. 78 ECM and harness

ENGINE COOLANT TEMPERATURE (ECT) SENSOR

LOCATION

See Figure 79.

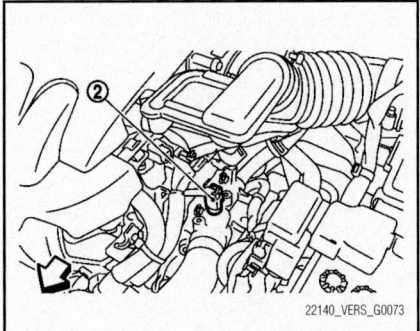

Fig. 79 Engine coolant temperature sensor location

REMOVAL & INSTALLATION

See Figure 79.

HEATED OXYGEN (HO2S) SENSOR

LOCATION

See Figure 80.

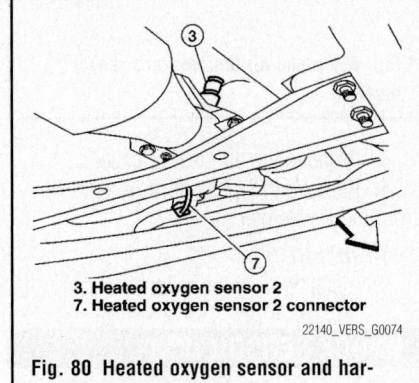

3. Heated oxygen sensor 2
7. Heated oxygen sensor 2 connector

Fig. 80 Heated oxygen sensor and harness connector locations

REMOVAL & INSTALLATION

See Figure 81.

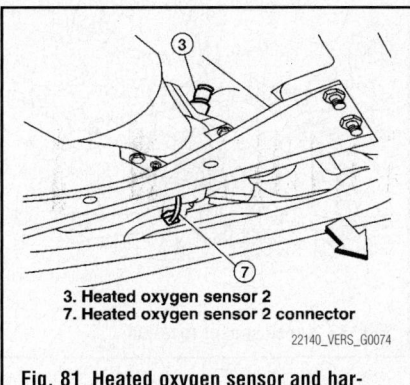

3. Heated oxygen sensor 2
7. Heated oxygen sensor 2 connector

Fig. 81 Heated oxygen sensor and harness connector

INTAKE AIR TEMPERATURE (IAT) SENSOR

LOCATION

See Figure 82.

The Intake Air Temperature Sensor is part of the Mass Air Flow Sensor.

REMOVAL & INSTALLATION

➡The intake air temperature sensor is located in the mass air flow sensor.

1. Remove battery.
2. Disconnect harness connector from mass air flow sensor.

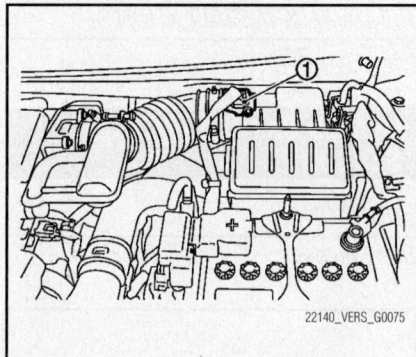

Fig. 82 Intake air temperature sensor location

3. Remove the air cleaner case.
4. Remove the mass air flow sensor from the air cleaner case.

To install:
5. Installation is the reverse of removal.

KNOCK SENSOR (KS)

LOCATION

See Figure 83.

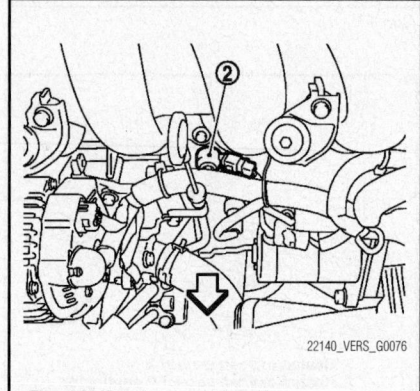

Fig. 83 Knock sensor location

REMOVAL & INSTALLATION

See Figure 84.

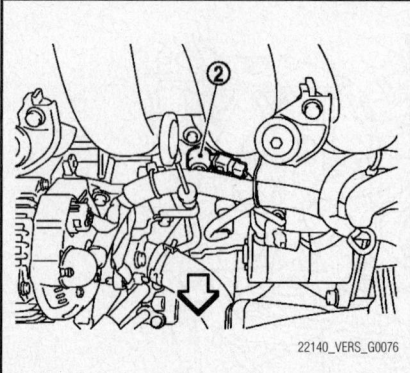

Fig. 84 Knock sensor

MASS AIR FLOW (MAF) SENSOR

LOCATION

See Figure 85.

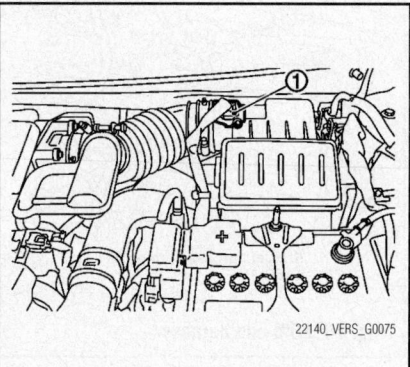

Fig. 85 Mass air flow sensor location

REMOVAL & INSTALLATION

1. Remove battery.
2. Disconnect harness connector from mass air flow sensor.
3. Remove the air cleaner case.
4. Remove the mass air flow sensor from the air cleaner case.

To install:
5. Installation is the reverse of removal.

THROTTLE POSITION SENSOR (TPS)

LOCATION

See Figure 86.

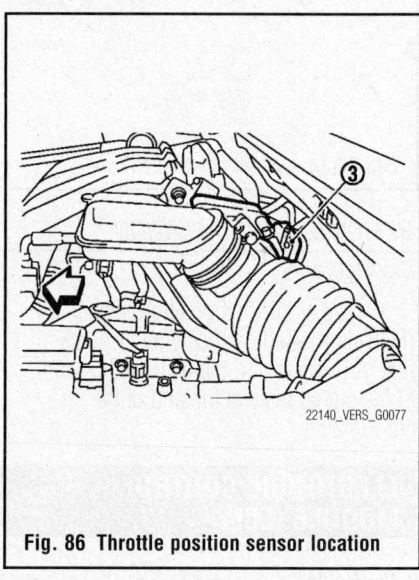

Fig. 86 Throttle position sensor location

REMOVAL & INSTALLATION

1. See Throttle body.

FUEL SYSTEM SERVICE PRECAUTIONS

Safety is the most important factor when performing not only fuel system maintenance but any type of maintenance. Failure to conduct maintenance and repairs in a safe manner may result in serious personal injury or death. Maintenance and testing of the vehicle's fuel system components can be accomplished safely and effectively by adhering to the following rules and guidelines.

• Put drained fuel in an explosion-proof container and put the lid on securely. Keep the container in safe area.

• Perform work on level surface.

• Use gasoline required by the regulations for octane number.

• To avoid the possibility of fire and personal injury, always disconnect the negative battery cable unless the repair or test procedure requires that battery voltage be applied.

• Always relieve the fuel system pressure prior to disconnecting any fuel system component (injector, fuel rail, pressure regulator, etc.), fitting or fuel line connection. Exercise extreme caution whenever relieving fuel system pressure to avoid exposing skin, face and eyes to fuel spray. Please be advised that fuel under pressure may penetrate the skin or any part of the body that it contacts.

• Always place a shop towel or cloth around the fitting or connection prior to loosening to absorb any excess fuel due to spillage. Ensure that all fuel spillage (should it occur) is quickly removed from engine surfaces. Ensure that all fuel soaked cloths or towels are deposited into a suitable waste container.

• Always keep a dry chemical (Class B) fire extinguisher near the work area.

• Do not allow fuel spray or fuel vapors to come into contact with a spark or open flame. Be sure to work in a well ventilated area.

• Always use a back-up wrench when loosening and tightening fuel line connection fittings. This will prevent unnecessary stress and torsion to fuel line piping.

• Do not tighten hose clamps excessively to avoid damaging hoses.

• Always replace worn fuel fitting O-rings with new. Do not substitute fuel hose or equivalent where fuel pipe is installed. Do not kink or twist tubes when they are being installed.

• After connecting fuel tube quick connectors, make sure quick connectors are secure. Ensure that connector and resin tube do not contact any adjacent parts.

• After installing tubes, make sure there is no fuel leakage at connections by applying fuel pressure to fuel lines by turning ignition switch ON (without starting the engine) and then checking for fuel leaks at the fuel tube connections.

• Use only a genuine NISSAN fuel filler cap as a replacement. If an incorrect fuel filler cap is used, the MIL may come on.

Before servicing the vehicle, make sure to also refer to the precautions in the beginning of this section as well.

RELIEVING FUEL SYSTEM PRESSURE

1. Before servicing the vehicle, refer to the Precautions Section.
2. Remove fuel pump fuse.
3. Start engine.
4. After engine stalls, crank it two or three times to release all fuel pressure.
5. Turn ignition switch OFF.

FUEL FILTER

REMOVAL & INSTALLATION

The fuel filter is part of the fuel pump assembly.

FUEL INJECTORS

REMOVAL & INSTALLATION

See Figures 87 through 89.

1. Before servicing the vehicle, refer to the Precautions Section.
2. Release the fuel pressure.
3. Remove quick connector cap from quick connector connection.
4. Disconnect fuel feed hose from hose clamp. There is no fuel return path.

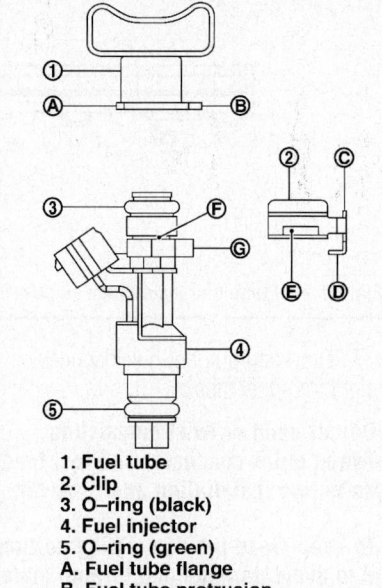

1. Fuel tube
2. Clip
3. O-ring (black)
4. Fuel injector
5. O-ring (green)
A. Fuel tube flange
B. Fuel tube protrusion
C. Clip cut-out
D. Clip cut-out
E. Clip flange fixing groove
F. Fuel injector clip groove
G. Fuel injector protrusion

22140_VERS_G0063

Fig. 88 Install fuel injector to fuel tube

5. With the sleeve side of quick connector release facing quick connector, install quick connector release onto fuel tube.

6. Insert quick connector release into quick connector until sleeve contacts and goes no further. Hold quick connector release on that position. Inserting quick connector release hard will not disconnect quick connector. Hold quick connector release where it contacts and goes no further.

10. **Ignition coil with power transistor and spark plug**
11. **Fuel injector**

22140_VERS_G0078

Fig. 87 Fuel injector location

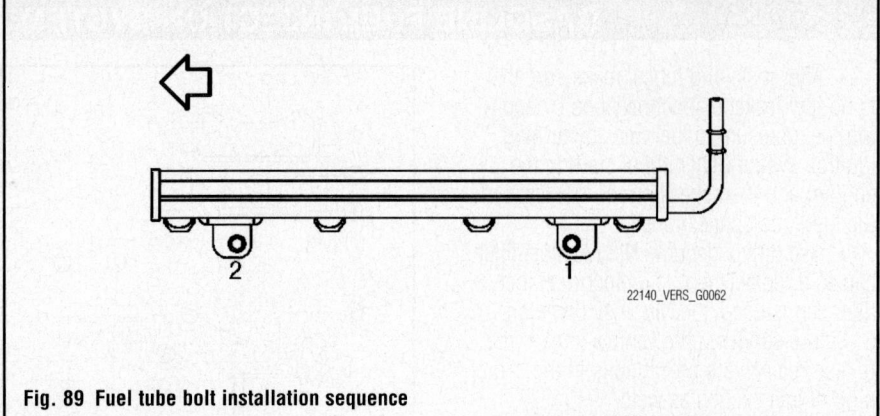

22140_VERS_G0062

Fig. 89 Fuel tube bolt installation sequence

7. Draw and pull out quick connector straight from fuel tube.

➡**Do not bend or twist connection between quick connector and fuel feed hose during installation and removal.**

➡**To keep clean the connecting portion and to avoid damage and foreign materials, cover them completely with plastic bags or something similar.**

8. Remove intake manifold.
9. Remove fuel tube. Loosen bolts in reverse order of the tightening sequence.
10. Remove the fuel tube and fuel injector assembly.

➡**When removing, be careful to avoid any interference with fuel injector.**

➡**Use a shop cloth to absorb any fuel leaks from fuel tube.**

11. Remove fuel injector from fuel tube with the following procedure:
    a. Open and remove clip.
    b. Remove fuel injector from fuel tube by pulling straight.

➡**Be careful with remaining fuel that may go out from fuel tube.**

➡**Be careful not to damage fuel injector nozzle during removal.**

➡**Never bump or drop fuel injector.**

➡**Never disassemble fuel injector.**

To install:
12. Install O-rings to fuel injector, noting the following:
    a. The upper and lower O-rings are different. Be careful not to confuse them.
    • Fuel tube side: Black
    • Nozzle side: Green
    b. Handle O-ring with bare hands. Never wear gloves.
    c. Lubricate O-ring with new engine oil.

    d. Never clean O-ring with solvent.
    e. Make sure that O-ring and its mating part are free of foreign material.
    f. When installing O-ring, be careful not to scratch it with tool or fingernails. Also be careful not to twist or stretch O-ring. If O-ring was stretched while it was being attached, never insert it quickly into fuel tube.
    g. Insert O-ring straight into fuel tube. Never twist it.
13. Install fuel injector to fuel tube with the following procedure:
    a. Insert clip into clip groove on fuel injector. Insert clip so that protrusion of fuel injector matches cutout of clip.

➡**Never reuse clip. Replace it with a new one.**

➡**Be careful to keep clip from interfering with O-ring. If interference occurs, replace O-ring.**

    b. Insert fuel injector into fuel tube with clip attached:
    • Insert it while matching it to the axial center
    • Insert fuel injector so that protrusion of fuel tube matches cut-out of clip
    • Make sure that fuel tube flange is securely fixed in flange fixing groove on clip
    c. Make sure that installation is complete by making sure that fuel injector does not rotate or come off.
14. Set fuel tube and fuel injector assembly at its position for installation on cylinder head. Be careful not to interfere with fuel injector nozzle.
15. Tighten bolts in numerical order as shown.
16. Installation of the remaining components is in the reverse order of removal.

FUEL PUMP

REMOVAL & INSTALLATION
See Figure 90.

1. Before servicing the vehicle, refer to the Precautions Section.
2. Check fuel level with the vehicle on a level surface. If the fuel gauge indicates more than the level as shown (7/8 full), drain fuel from the fuel tank until the fuel gauge indicates level as shown (7/8 full).
3. Open fuel door and unscrew the fuel filler cap to release the pressure inside the fuel tank.
4. Release the fuel pressure from the fuel lines.
5. Remove rear seat bottom.
6. Turn the four retainers 90° in a clockwise direction and remove the fuel pump inspection hole cover.
7. Disconnect electrical connector and fuel feed hose quick connector.
8. Disconnect the quick connector using the following procedure:
    a. Hold the sides of the connector, push in tabs and pull out the tube.
    b. If quick connector and tube on fuel pump assembly are stuck, push and pull several times until they move. Disconnect them by pulling.

➡**The tube can be removed when the tabs are completely depressed. Do not twist it more than necessary.**

➡**Do not use any tools to remove the quick connector.**

➡**Keep resin tube away from heat. Be especially careful when welding near the resin tube.**

➡**Prevent acid liquid such as battery electrolyte, from getting on resin tube.**

➡**Do not bend or twist resin tube during installation and removal.**

➡**To keep the connecting portion clean, free of foreign materials and to avoid damage, cover them completely with plastic bags or something similar.**

➡**Do not insert plug to prevent damage to O-ring in quick connector.**

9. Remove the lock ring using Tool KV991J0090 (J-46214).
10. Remove fuel level sensor unit, fuel filter and fuel pump assembly.

➡**Do not bend float arm during removal.**

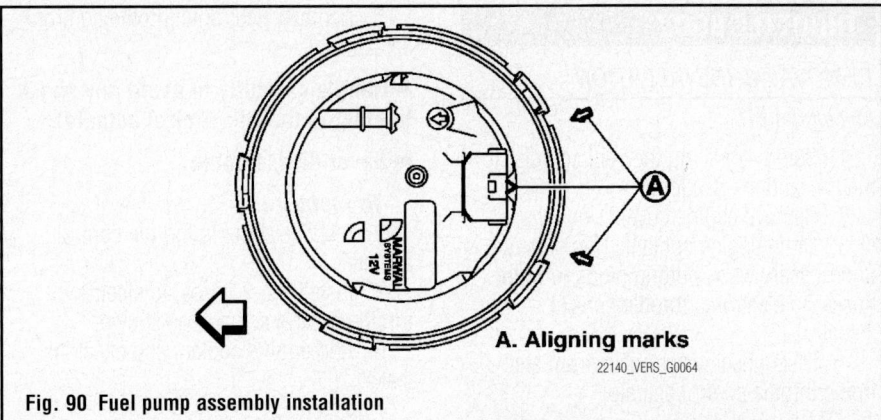

A. Aligning marks

22140_VERS_G0064

Fig. 90 Fuel pump assembly installation

➡Do not allow foreign materials to fall into fuel tank. Use a lint free cloth when handling components.

➡Avoid impacts such as dropping when handling components.

To install:

11. Installation is in the reverse order of removal, noting the following:

  a. Install O-ring to fuel tank without twisting.

  b. Install fuel level sensor unit with aligning mating marks on fuel tank and fuel level sensor unit.

  c. Turn the lock ring until the lock ring is fully rotated into the fuel tank lock tabs.

  d. Connect fuel feed tube quick connector by aligning the connector with the tube, then inserting the connector straight into the tube until it clicks.

➡Check the connection for damage or any foreign materials.

  e. After connecting, visually confirm that the two retainer tabs are secured to the connector and pull the tube and the connector to make sure they are securely connected.

  f. Connect electrical harness connector.

12. Before installing inspection hole cover, confirm that there are no fuel leaks, install inspection hole cover with the front mark (arrow) facing front of vehicle, and lock the clips by turning counterclockwise.

FUEL TANK

REMOVAL & INSTALLATION

1. Before servicing the vehicle, refer to the Precautions Section.

2. Drain fuel from fuel tank if necessary.

✳✳ CAUTION

Because fuel tank becomes unstable when installing/removing, fuel should be drained if the level exceeds specification.

➡Situate vehicle on a flat and solid surface.

3. Open fuel door and unscrew the fuel filler cap to release the pressure inside the fuel tank.

4. Release the fuel pressure from the fuel lines.

5. Remove rear seat bottom.

6. Turn the four retainers 90° in a clockwise direction and remove the fuel pump inspection hole cover.

7. Disconnect electrical connector and fuel feed hose quick connector.

8. Disconnect the quick connector using the following procedure:

  a. Hold the sides of the connector, push in tabs and pull out the tube.

  b. If quick connector and tube on fuel pump assembly are stuck, push and pull several times until they move. Disconnect them by pulling.

➡The tube can be removed when the tabs are completely depressed. Do not twist it more than necessary.

➡Do not use any tools to remove the quick connector.

➡Keep resin tube away from heat. Be especially careful when welding near the resin tube.

➡Prevent acid liquid such as battery electrolyte, from getting on resin tube.

➡Do not bend or twist resin tube during installation and removal.

➡To keep the connecting portion clean, free of foreign materials and to avoid damage, cover them completely

with plastic bags or something similar.

➡Do not insert plug to prevent damage to O-ring in quick connector.

9. Remove center exhaust tube.

10. Remove exhaust heat shields.

11. Disconnect parking brake cables from the lower surface of fuel tank and axle and position the parking brake cables out of the way.

12. Remove brake tube protector from rear axle.

13. Loosen fuel filler hose clamp and remove fuel filler hose from fuel tank.

➡Do not remove fuel filler hose from fuel filler tube. Do not allow contact with the suspension during installation. When removing fuel filler hose at the fuel filler tube, mark components for alignment.

14. Remove vent hose and EVAP hose at rear of fuel tank.

15. Disconnect vent hose and EVAP hose quick connectors using the following procedures:

  a. Pinch retaining tabs of vent hose quick connector and remove vent hose.

  b. Slide sleeve of EVAP hose quick connector and remove EVAP hose.

  c. If hoses are stuck, push and pull several times until they move freely, and disconnect.

➡The tube can be removed when the tabs are completely depressed. Do not twist it more than necessary.

➡Do not use any tools to remove the quick connector.

➡Keep resin tube away from heat. Be especially careful when welding near the resin tube.

➡Prevent acid liquid such as battery electrolyte, from getting on resin tube.

➡Do not bend or twist resin tube during installation and removal.

➡To keep the connecting portion clean, free of foreign materials and to avoid damage, cover them completely with plastic bags or something similar.

➡Do not insert plug to prevent damage to O-ring in quick connector.

16. Support center of fuel tank with transmission jack. Securely support the fuel tank with a suitable tool.

17. Remove fuel tank bands.

18. Lower transmission jack carefully to remove fuel tank while supporting it by

hand. Fuel tank may be in an unstable position because of the shape of fuel tank bottom. Be sure to support tank securely.

To install:

19. Installation is in the reverse order of removal, noting the following:

a. Check the EVAP canister connection for damage or any foreign materials, and align the connector with the tube, then insert the connector straight into the tube until it clicks. After connecting, make sure that the connection is secure by pulling the tube and the connector.

b. Install the fuel tank bands in the proper position by referring to the identification stamp mark "R" and "L" on the end.

c. Insert fuel filler hose to 35mm (1.38 inches). Be sure hose clamp is not placed on swelled area of fuel filler tube.

d. Check the EVAP hose connections for damage or foreign material, and align the matching quick connector with the center of EVAP hose, and insert quick connector straight until it clicks. Make sure connections are secure by pulling on quick connector and EVAP hose by hand.

➡ The tube can be removed when the tabs are completely depressed. Do not twist it more than necessary.

➡ Do not use any tools to remove the quick connector.

➡ Keep resin tube away from heat. Be especially careful when welding near the resin tube.

➡ Prevent acid liquid such as battery electrolyte, from getting on resin tube.

➡ Do not bend or twist resin tube during installation and removal.

➡ To keep the connecting portion clean, free of foreign materials and to avoid damage, cover them completely with plastic bags or something similar.

➡ Do not insert plug to prevent damage to O-ring in quick connector.

IDLE SPEED

ADJUSTMENT

Idle speed is maintained by the ECM. No adjustment is necessary or possible.

THROTTLE BODY

REMOVAL & INSTALLATION

See Figure 91.

1. Before servicing the vehicle, refer to the Precautions Section.
2. Remove engine cover.
3. Drain engine coolant. This step is unnecessary when putting plugs to water hoses (to electronic throttle control actuator).
4. Disconnect water hoses from electronic throttle control actuator.

5. Remove electronic throttle control actuator.

➡ Handle carefully to avoid any shock to electric throttle control actuator.

➡ Never disassemble.

To install:

6. Install electronic throttle control actuator.
7. Install water hoses, to electronic throttle control actuator as shown.
8. Add engine coolant and check for leaks.

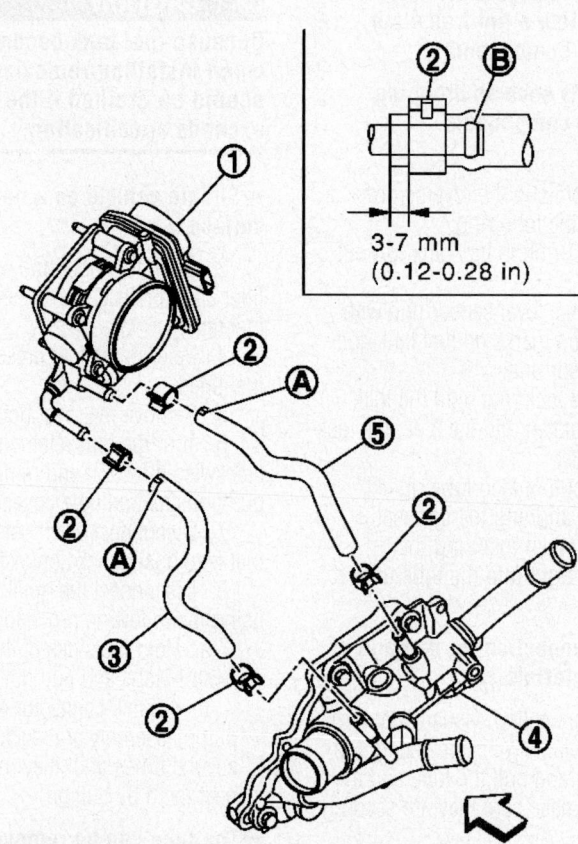

1. **Electric throttle control actuator**
2. **Clamp**
4. **Water outlet**
A. **Paint Mark**
B. **The clamp shall not interfere with the bulged section**

22140_VERS_G0065

Fig. 91 Install water hoses to electronic throttle control actuator

HEATING & AIR CONDITIONING SYSTEM

BLOWER MOTOR

REMOVAL & INSTALLATION

See Figure 92.

1. Before servicing the vehicle, refer to the Precautions Section.
2. Remove the instrument panel and pad.
3. Remove the right side ventilator duct.
4. Disconnect the blower motor connector.
5. Push the flange holding hook toward the blower motor, then rotate the blower motor clockwise and remove it from the A/C unit assembly.

➡**When blower fan and blower motor are assembled, the balance is adjusted, do not disassemble to replace the individual parts.**

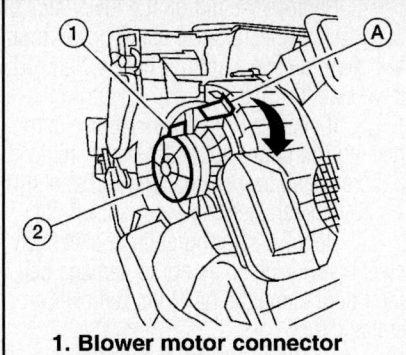

1. Blower motor connector
2. Blower motor
A. Flange holding hook

22140_VERS_G0043

Fig. 92 Blower motor

To install:

6. Installation is the reverse of removal, noting the following:

a. Rotate the blower motor until the blower motor flange holding hook locks securely in A/C unit assembly.

HEATER CORE

REMOVAL & INSTALLATION

1. Before servicing the vehicle, refer to the Precautions Section.
2. Remove the A/C unit assembly.
3. Remove the left foot duct.
4. Remove the heater pipe cover screw, and then remove the heater pipe cover.
5. Remove the heater pipe clip screw, and then remove the heater pipe clip.
6. Slide the heater core out of the A/C unit assembly.

To install:

7. Installation is the reverse of removal.

STEERING

POWER STEERING GEAR

REMOVAL & INSTALLATION

See Figure 93.

1. Before servicing the vehicle, refer to the Precautions Section.

❄❄ WARNING

Spiral cable may be cut if steering wheel turns while separating steering column assembly and steering gear assembly. Be sure to secure steering wheel using string to avoid turning.

2. Set vehicle to the straight-ahead position.
3. Remove bolt of intermediate shaft (lower side), and then remove intermediate shaft from steering gear pinion shaft.
4. Raise vehicle.
5. Remove tires from vehicle with a power tool.
6. Loosen steering outer socket mounting nut.
7. Remove steering outer socket from steering knuckle so as not to damage ball joint boot using the ball joint remover (suitable tool).

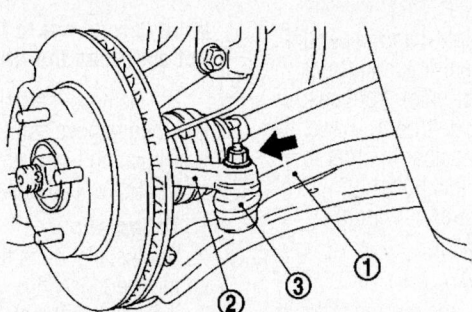

1. Steering outer socket
2. Steering knuckle
3. Ball joint boot

22140_VERS_G0044

Fig. 93 Power steering gear

➡**Temporarily tighten the nut to prevent damage to threads and to prevent the ball joint remover (suitable tool) from suddenly coming off.**

8. Remove front suspension member.
9. Remove mounting bolts and nuts of steering gear assembly.

To install:

10. Installation is in the reverse order of removal, noting the following:

a. Tighten the steering outer socket mounting nut 25 ft. lbs. (34.4 Nm).

b. Tighten the steering gear assembly mounting bolts and nuts 69 ft. lbs. (94 Nm).

c. Clean mounting surface on the body side of fire wall seal when installing steering gear assembly.

d. Check wheel alignment under unladen conditions with tires on level ground.

SUSPENSION   **FRONT SUSPENSION**

LOWER BALL JOINT

REMOVAL & INSTALLATION

See Steering Knuckle.

LOWER CONTROL ARM

REMOVAL & INSTALLATION

1. Before servicing the vehicle, refer to the Precautions Section.
2. Remove tires.
3. Remove transverse link ball joint nut and bolt. Then, remove transverse link from steering knuckle.
4. Remove transverse link nuts and bolts, then remove transverse link from front suspension member.

➡ **When removing the left-hand transverse link it may be necessary to lower the suspension member in order to remove bolts to avoid contact with the transaxle.**

    a. Set jack under front suspension member.
    b. Loosen the right-hand upper link bolts, left-hand upper link bolt (front suspension member side), front suspension member bolts (left/right). Lower the front suspension member in order to remove transverse link bolts.
5. Remove transverse link from vehicle.

To install:

6. Installation is in the reverse order of removal, noting the following:
    a. Install the transverse link ball joint nut and bolt to the steering knuckle and tighten to 41 ft. lbs. (55 Nm).
    b. Install the transverse link ball joint nut and bolt to the steering knuckle and tighten to front bolt to 88 ft. lbs. (119 Nm), and the rear nut and bolt to 102 ft. lbs. (138 (Nm).
    c. Perform final tightening of bolts and at the front suspension member installation position (rubber bushing) under unladen conditions with tires on level ground.

STABILIZER BAR

REMOVAL & INSTALLATION

1. Before servicing the vehicle, refer to the Precautions Section.
2. Separate intermediate shaft from steering gear pinion shaft.
3. Remove tires from vehicle using power tool.

4. Remove the nut on the lower side of stabilizer connecting rod using power tool, and then remove stabilizer connecting rod from stabilizer bar.
5. If necessary remove stabilize connecting rod upper nut using power tool. Separate stabilizer connecting rod and strut.
6. Loosen steering outer socket nut.
7. Remove steering outer socket from steering knuckle so as not to damage ball joint boot using the ball joint remover or suitable tool.

➡ **Temporarily tighten the nut to prevent damage to threads and to prevent the ball joint remover (suitable tool) from suddenly coming off.**

8. Remove rear torque rod.
9. Set jack under front suspension member.
10. Remove the bolts of member stay, and then remove member stay from vehicle.
11. Gradually lower front suspension member in order to remove stabilizer bolts.

➡ **Be careful not to lower it too far. (Do not over load the links.)**

12. Remove the bolts of stabilizer clamp, and then remove stabilizer clamp and stabilizer bushing from vehicle.
13. Remove stabilizer bar from vehicle.

To install:

14. Installation is the reverse of removal, noting the following:
    a. Install the stabilizer connecting rod and strut and tighten the nut to 27 ft. lbs. (37 Nm).
    b. Install the stabilizer clamp and tighten the bolts to 21 ft. lbs. (28 Nm).

STEERING KNUCKLE

REMOVAL & INSTALLATION

1. Before servicing the vehicle, refer to the Precautions Section.
2. Remove wheel and tire.
3. Without disassembling the hydraulic lines, remove the torque member bolts. Then reposition the torque member and brake caliper assembly aside with wire.

➡ **Do not depress brake pedal while brake caliper is removed.**

4. Put alignment marks on disc rotor and wheel hub and bearing assembly, then remove disc rotor.
5. Remove wheel sensor from steering knuckle. Do not pull on wheel sensor harness.

6. Loosen steering outer socket nut.
7. Remove steering outer socket from steering knuckle so as not to damage ball joint boot using ball joint remover or suitable tool.

➡ **Temporarily leave the outer socket nut installed to prevent damage to threads and to prevent the ball joint remover or suitable tool from suddenly coming off.**

8. Remove transverse link ball joint nut and bolt. Then, remove transverse link from steering knuckle.
9. Remove cotter pin, then loosen hub lock nut using power tool. Temporarily leave the hub lock nut installed to prevent damage to threads.
10. Separate the drive shaft from the wheel hub and bearing assembly by lightly tapping the end of the drive shaft using a hammer or suitable tool, and then remove hub lock nut.

➡ **Use a suitable puller if wheel hub and bearing assembly and drive shaft cannot be separated after performing the above procedure.**

11. Remove the drive shaft from the wheel hub and bearing assembly and support the drive shaft.

➡ **Do not apply an excessive angle to drive shaft joint when removing from the wheel hub and bearing assembly,**

➡ **Do not excessively extend slide joint.**

➡ **Do not allow drive shaft to hang down. Support the entire drive shaft.**

12. Remove wheel hub and bearing assembly bolts, and then remove splash guard and wheel hub and bearing assembly from steering knuckle.
13. Remove nuts and bolts, and then remove steering knuckle from strut assembly.

To install:

14. Installation is the reverse order of removal, noting the following:
    a. Tighten the transverse link ball joint nut and bolt to 41 ft. lbs. (55 Nm).
    b. Tighten the hub lock nut to 83 ft. lbs. (113 Nm).
    c. Perform the final tightening of each of parts under unladen conditions, which were removed when removing wheel hub and bearing assembly and steering knuckle. Check the wheel alignment.
    d. When installing disc rotor on wheel hub and bearing assembly, align the marks.

STRUT & SPRING ASSEMBLY

REMOVAL & INSTALLATION

1. Before servicing the vehicle, refer to the Precautions Section.
2. Remove cowl top panel.
3. Remove tires.
4. Remove harness of wheel sensor from strut assembly.

➡**Do not pull on wheel sensor harness.**

5. Remove brake hose lock plate.
6. Remove the nut on the upper side of stabilizer connecting rod using power tool, and then remove stabilizer connecting rod from strut assembly.
7. Remove nuts and bolts, and then remove steering knuckle from strut assembly.
8. Remove the strut mounting insulator bolts, then remove strut assembly from vehicle.

To install:

9. Installation is in the reverse order of removal, noting the following:
   a. Perform final tightening of bolts and nuts at the strut assembly lower side (rubber bushing) under unladen conditions with tires on level ground. Check wheel alignment.
   b. Check wheel sensor harness for proper connection.

WHEEL BEARINGS

REMOVAL & INSTALLATION

1. Before servicing the vehicle, refer to the Precautions Section.
2. Remove wheel and tire.
3. Without disassembling the hydraulic lines, remove the torque member bolts. Then reposition the torque member and brake caliper assembly aside with wire.

➡**Do not depress brake pedal while brake caliper is removed.**

4. Put alignment marks on disc rotor and wheel hub and bearing assembly, then remove disc rotor.
5. Remove wheel sensor from steering knuckle. Do not pull on wheel sensor harness.
6. Loosen steering outer socket nut.
7. Remove steering outer socket from steering knuckle so as not to damage ball joint boot using ball joint remover or suitable tool.

➡**Temporarily leave the outer socket nut installed to prevent damage to threads and to prevent the ball joint remover or suitable tool from suddenly coming off.**

8. Remove transverse link ball joint nut and bolt. Then, remove transverse link from steering knuckle.
9. Remove cotter pin, then loosen hub lock nut using power tool. Temporarily leave the hub lock nut installed to prevent damage to threads.
10. Separate the drive shaft from the

wheel hub and bearing assembly by lightly tapping the end of the drive shaft using a hammer or suitable tool, and then remove hub lock nut.

➡**Use a suitable puller if wheel hub and bearing assembly and drive shaft cannot be separated after performing the above procedure.**

11. Remove the drive shaft from the wheel hub and bearing assembly and support the drive shaft.

➡**Do not apply an excessive angle to drive shaft joint when removing from the wheel hub and bearing assembly,**

➡**Do not excessively extend slide joint.**

➡**Do not allow drive shaft to hang down. Support the entire drive shaft.**

12. Remove wheel hub and bearing assembly bolts, and then remove splash guard and wheel hub and bearing assembly from steering knuckle.
13. Remove nuts and bolts, and then remove steering knuckle from strut assembly.

To install:

14. Installation is the reverse of removal, noting the following:
   a. Perform the final tightening of each of parts under unladen conditions, which were removed when removing wheel hub and bearing assembly and steering knuckle. Check the wheel alignment.
   b. When installing disc rotor on wheel hub and bearing assembly, align the marks.

SUSPENSION

COIL SPRING

REMOVAL & INSTALLATION

1. Before servicing the vehicle, refer to the Precautions Section.
2. Remove rear tires.
3. Remove wheel sensor from wheel hub and bearing assembly.

➡**Do not pull on wheel sensor harness.**

4. Separate brake tube from wheel cylinder.
5. Set jack under rear suspension beam.
6. Remove shock absorber lower side bolt.

To install:

7. Installation is the reverse of removal, noting the following:
   a. Tighten the lower shock absorber side bolt to 91 ft. lbs. (124 Nm).

➡**When installing spring, be sure to securely install the spring end position aligned to flush of rear spring rubber seat (lower).**

SHOCK ABSORBER

REMOVAL & INSTALLATION

1. Before servicing the vehicle, refer to the Precautions Section.
2. Remove rear tires.
3. Remove wheel sensor from wheel hub and bearing assembly and rear suspension beam. Do not pull on wheel sensor harness.
4. Remove shock absorber mask from trunk side finisher using a flat-bladed screwdriver with its tip taped.
5. Set jack under rear suspension beam.

REAR SUSPENSION

6. Remove upper nut of the shock absorber, and then remove washer (upper), bushing (upper) from shock absorber.
7. Remove shock absorber lower side bolt.
8. Gradually lower the jack, and remove the bushing (lower), washer (lower), distance tube, bound bumper cover, bound bumper and shock absorber from vehicle.

To install:

9. Installation is the reverse of removal, noting the following:
   a. Tighten the lower shock absorber side bolt to 91 ft. lbs. (124 Nm).
   b. Tighten the upper shock absorber nut to 15 ft. lbs. (20 Nm).
   c. When installing body side bushing (upper), install the projection to the body side hole securely.

WHEEL BEARINGS

REMOVAL & INSTALLATION

1. Before servicing the vehicle, refer to the Precautions Section.
2. Remove tires.
3. Remove wheel sensor from wheel hub and bearing assembly. Do not pull on wheel sensor harness.
4. Remove the drum brake assembly.

5. Remove wheel hub and bearing assembly bolts, and then remove wheel hub and bearing assembly from vehicle.

To install:

6. Installation is the reverse of removal.

ADJUSTMENT

1. Before servicing the vehicle, refer to the Precautions Section.

2. The wheel hub assembly does not require maintenance. If any of the following symptoms are noted, replace the wheel hub assembly:

- Growling noise is emitted from the wheel hub bearing during operation.
- Wheel hub bearing drags or turns roughly.

INFINITI

Diagnostic Trouble Codes

DIAGNOSTIC TROUBLE CODES

OBD II VEHICLE APPLICATIONS

INFINITI

EX35
2008
- 3.5L V6 VQ35HR . . . . VIN A

FX35, FX45
2007–2008
- 3.5L V6 VQ35HR . . . . VIN A
- 4.5L V8 VK45DE . . . . VIN B

G35 & G35 Coupe
2007–2008
3.5L V6 VQ35HR . . . . VIN A
3.5L V6 VQ35DE . . . . VIN C

G37
2007–2008
3.7L V6 VQ37VHR . . . . VIN C

QX56
2007–2008
5.6L V8 VK56DE . . . . VIN A

OBD II Trouble Code List (P0xxx Codes)

DTC	Trouble Code Title, Conditions & Possible Causes
DTC: P0011 **2T ECM, MIL: Yes** **Years: 2007, 2008** **Models:** EX35, FX35, G35, G37, FX45, QX56 **Engines:** 3.5L, 3.7L, 4.5L, 5.6L **Transmissions:** All	**Intake Valve Timing Control Range/Performance (Bank 1)** Engine started and running at normal operating temperature at 1200-2000 RPM. Selector lever in D. Stop vehicle and idle. Then, accelerate to over 2000 RPM, while driving uphill. There is a gap between angle of target and phase-control angle degree. **Possible Causes:** • Crankshaft position sensor (POS) • Camshaft position sensor (PHASE) • Intake valve timing control solenoid valve • Accumulation of debris to the signal pick-up portion of the camshaft • Timing chain installation • Foreign matter caught in the oil groove for intake valve timing control
DTC: P0014 **2T ECM, MIL: Yes** **Years: 2007, 2008** **Models:** EX35, G35 **Engines:** 3.5L **Transmissions:** All	**Exhaust Valve Timing Control Range/Performance (Bank 1)** Engine running at normal operating temperature. Vehicle driven 60-70 MPH. Stop vehicle and idle. There is a gap between angle of target and phase-control angle degree. **Possible Causes:** • Crankshaft position sensor (POS) • Camshaft position sensor (PHASE) • Exhaust valve timing control solenoid valve • Exhaust valve control magnet retarder • Accumulation of debris to the signal pick-up portion of the camshaft • Timing chain installation • Exhaust valve timing control pulley
DTC: P0021 **2T ECM, MIL: Yes** **Years: 2007, 2008** **Models:** EX35, FX35, G35, G37, FX45, QX56 **Engines:** 3.5L, 3.7L, 4.5L, 5.6L **Transmissions:** All	**Intake Valve Timing Control Performance (Bank 2)** Engine started, engine running at idle speed, followed by a quick and steady acceleration in 1st gear to over 2000 RPM, and the ECM detected a problem in the operation of the IVT control during testing. **Possible Causes:** • Crankshaft position sensor (POS) • Camshaft position sensor (PHASE) • Intake valve timing control solenoid valve • Accumulation of debris to the signal pick-up portion of the camshaft • Timing chain installation • Foreign matter caught in the oil groove for intake valve timing control
DTC: P0024 **2T ECM, MIL: Yes** **Years: 2007, 2008** **Models:** EX35, G35 **Engines:** 3.5L **Transmissions:** All	**Exhaust Valve Timing Control Range/Performance (Bank 2)** Engine running at normal operating temperature. Vehicle driven 60-70 MPH. Stop vehicle and idle. There is a gap between angle of target and phase-control angle degree. **Possible Causes:** • Crankshaft position sensor (POS) • Camshaft position sensor (PHASE) • Exhaust valve timing control solenoid valve • Exhaust valve control magnet retarder • Accumulation of debris to the signal pick-up portion of the camshaft • Timing chain installation • Exhaust valve timing control pulley
DTC: P0031 **2T CCM, MIL: Yes** **Years: 2007, 2008** **Models:** EX35, FX35, G35, G37, FX45, QX56 **Engines:** 3.5L, 3.7L, 4.5L, 5.6L **Transmissions:** All	**A/F Sensor 1 Heater (Bank 1) Circuit Low Input** Engine started and idling. The current amperage in the A/F sensor 1 heater circuit is out of the normal range. (An excessively low voltage signal is sent to ECM through the A/F sensor 1 heater.) **Possible Causes:** • Harness or connectors (A/F sensor 1 heater circuit is open or shorted.) • A/F sensor 1 heater
DTC: P0032 **2T CCM, MIL: Yes** **Years: 2007, 2008** **Models:** EX35, FX35, G35, G37, FX45, QX56 **Engines:** 3.5L, 3.7L, 4.5L, 5.6L **Transmissions:** All	**A/F Sensor 1 Heater (Bank 1) Circuit High Input** Engine started and idling. The current amperage in the A/F sensor 1 heater circuit is out of the normal range. (An excessively high voltage signal is sent to ECM through the A/F sensor 1 heater.) **Possible Causes:** • Harness or connectors (A/F sensor 1 heater circuit is open or shorted.) • A/F sensor 1 heater

DTC	Trouble Code Title, Conditions & Possible Causes
DTC: P0037 **2T CCM, MIL:** Yes **Years:** 2007, 2008 **Models:** EX35, FX35, G35, G37, FX45, QX56 **Engines:** 3.5L, 3.7L, 4.5L, 5.6L **Transmissions:** All	**HO2S-12 (Bank 1 Sensor 2) Heater Circuit Low Input** Start engine and keep the engine speed between 3,500 and 4,000 RPM for at least 1 minute under no load. Let engine idle for 1 minute. The current amperage in the heated oxygen sensor 2 heater circuit is out of the normal range. (An excessively low voltage signal is sent to ECM through the heated oxygen sensor 2 heater.) **Possible Causes:** • Harness or connectors (heated oxygen sensor 2 heater circuit is open or shorted.) • Heated oxygen sensor 2 heater
DTC: P0038 **2T CCM, MIL:** Yes **Years:** 2007, 2008 **Models:** EX35, FX35, G35, G37, FX45, QX56 **Engines:** 3.5L, 3.7L, 4.5L, 5.6L **Transmissions:** All	**HO2S-12 (Bank 1 Sensor 2) Heater Circuit High Input** Start engine and keep the engine speed between 3,500 and 4,000 RPM for at least 1 minute under no load. Let engine idle for 1 minute. The current amperage in the heated oxygen sensor 2 heater circuit is out of the normal range. (An excessively high voltage signal is sent to ECM through the heated oxygen sensor 2 heater.) **Possible Causes:** • Harness or connectors (heated oxygen sensor 2 heater circuit is open or shorted.) • Heated oxygen sensor 2 heater
DTC: P0051 **2T CCM, MIL:** Yes **Years:** 2007, 2008 **Models:** EX35, FX35, G35, G37, FX45, QX56 **Engines:** 3.5L, 3.7L, 4.5L, 5.6L **Transmissions:** All	**A/F Sensor 1 Heater (Bank 2) Control Circuit Low** Engine started and idling. The current amperage in the A/F sensor 1 heater circuit is out of the normal range. (An excessively low voltage signal is sent to ECM through the A/F sensor 1 heater.) **Possible Causes:** • Harness or connectors (A/F sensor 1 heater circuit is open or shorted.) • A/F sensor 1 heater
DTC: P0052 **2T CCM, MIL:** Yes **Years:** 2007, 2008 **Models:** EX35, FX35, G35, G37, FX45, QX56 **Engines:** 3.5L, 3.7L, 4.5L, 5.6L **Transmissions:** All	**A/F Sensor 1 Heater (Bank 2) Circuit High** Engine started and idling. The current amperage in the A/F sensor 1 heater circuit is out of the normal range. (An excessively high voltage signal is sent to ECM through the A/F sensor 1 heater.) **Possible Causes:** • Harness or connectors (A/F sensor 1 heater circuit is open or shorted.) • A/F sensor 1 heater
DTC: P0057 **2T CCM, MIL:** Yes **Years:** 2007, 2008 **Models:** EX35, FX35, G35, G37, FX45, QX56 **Engines:** 3.5L, 3.7L, 4.5L, 5.6L **Transmissions:** All	**HO2S-22 (Bank 2 Sensor 2) Heater Circuit Low Input** Engine started, system voltage from 10.5-16.0v, and the ECM detected the HO2S-22 heater control circuit input (voltage) was excessively low (i.e., it was operating out of its normal range). **Possible Causes:** • HO2S heater control connector is damaged or loose (open) • HO2S heater power circuit is open (check the No. 8 fuse 15A) • HO2S heater assembly is damaged or has failed • ECM has failed
DTC: P0058 **2T ECM, MIL:** Yes **Years:** 2007, 2008 **Models:** EX35, FX35, G35, G37, FX45, QX56 **Engines:** 3.5L, 3.7L, 4.5L, 5.6L **Transmissions:** All	**HO2S-22 (Bank 2 Sensor 2) Heater Circuit High Input** Engine started, system voltage from 10.5-16.0v, and the ECM detected the HO2S-22 heater control circuit input (voltage) was excessively high (i.e., it was operating out of its normal range). **Possible Causes:** • HO2S heater control connector is damaged or loose (shorted) • HO2S heater control circuit is shorted to ground • HO2S heater assembly is damaged or has failed • ECM has failed
DTC: P0075 **2T ECM, MIL:** Yes **Years:** 2007, 2008 **Models:** EX35, FX35, G35, G37, FX45, QX56 **Engines:** 3.5L, 3.7L, 4.5L, 5.6L **Transmissions:** All	**IVT Control Solenoid Valve (Bank 1) Circuit Malfunction** An improper voltage is sent to the ECM through intake valve timing control solenoid valve. **Possible Causes:** • Harness or connectors (intake valve timing control solenoid valve circuit is open or shorted.) • Intake valve timing control solenoid valve
DTC: P0078 **2T ECM, MIL:** Yes **Years:** 2007, 2008 **Models:** EX35, G35 **Engines:** 3.5L **Transmissions:** All	**EVT Timing Control Magnet Retarder (Bank 1) Circuit Malfunction** An improper voltage is sent to the ECM through intake valve timing control solenoid valve. **Possible Causes:** • Harness or connectors (exhaust valve timing control magnet retarder circuit is open or shorted.) • Exhaust valve timing control magnet retarder

DTC	Trouble Code Title, Conditions & Possible Causes
DTC: P0081 **2T ECM, MIL: Yes** **Years: 2007, 2008** **Models:** EX35, FX35, G35, G37, FX45, QX56 **Engines:** 3.5L, 3.7L, 4.5L, 5.6L **Transmissions:** All	**IVT Control Solenoid Valve (Bank 1) Circuit Malfunction** An improper voltage is sent to the ECM through exhaust valve timing control magnet retarder. **Possible Causes:** • Harness or connectors (intake valve timing control solenoid valve circuit is open or shorted.) • Intake valve timing control solenoid valve
DTC: P0084 **2T ECM, MIL: Yes** **Years: 2007, 2008** **Models:** EX35, G35 **Engines:** 3.5L **Transmissions:** All	**EVT Timing Control Magnet Retarder (Bank 2) Circuit Malfunction** An improper voltage is sent to the ECM through intake valve timing control solenoid valve. **Possible Causes:** • Harness or connectors (exhaust valve timing control magnet retarder circuit is open or shorted.) • Exhaust valve timing control magnet retarder
DTC: P010A **1T CCM, MIL: Yes** **Years: 2007, 2008** **Models:** G37 **Engines:** 3.7L **Transmissions:** All	**Mass Air Pressure (MAP) Sensor Circuit** An excessively low voltage from the sensor is sent to ECM. Or, an excessively high voltage from the sensor is sent to ECM. **Possible Causes:** • Harness or connectors - sensor circuit is open or shorted. • Manifold absolute pressure (MAP) sensor
DTC: P010B **1T CCM, MIL: Yes** **Years: 2007, 2008** **Models:** EX35, G37 **Engines:** 3.5L, 3.7L **Transmissions:** All	**Mass Airflow Sensor (Bank 2) Signal Range/Performance** Engine started, engine running at light engine load, and the ECM detected the MAF sensor signal was excessively high; or with the engine running under high engine load, the ECM detected the MAF sensor signal was excessively low under these operating conditions. **Possible Causes:** • Harness or connectors (The sensor circuit is open or shorted.) • Mass air flow sensor • EVAP control system • Intake air temperature sensor • Intake air leaks
DTC: P010C **1T CCM, MIL: Yes** **Years: 2007, 2008** **Models:** EX35, G37 **Engines:** 3.5L, 3.7L **Transmissions:** All	**Mass Air Flow Sensor (Bank 2) Circuit Low Input** Key on or engine running; and the ECM detected an unexpected low voltage condition on the MAF sensor signal circuit. **Possible Causes:** • MAF sensor signal circuit is open or shorted • Intake air leaks • MAF sensor is damaged or has failed
DTC: P010D **1T CCM, MIL: Yes** **Years: 2007, 2008** **Models:** EX35, G37 **Engines:** 3.5L, 3.7L **Transmissions:** All	**Mass Air Flow Sensor (Bank 2) Circuit High Input** Key on or engine running; and the ECM detected an unexpected low voltage condition on the MAF sensor signal circuit. **Possible Causes:** • MAF sensor signal circuit is open or shorted • MAF sensor is damaged or has failed
DTC: P0101 **1T CCM, MIL: Yes** **Years: 2007, 2008** **Models:** EX35, FX35, G35, G37, FX45, QX56 **Engines:** 3.5L, 3.7L, 4.5L, 5.6L **Transmissions:** All	**Mass Airflow Sensor (Bank 1) Signal Range/Performance** Engine started, engine running at light engine load, and the ECM detected the MAF sensor signal was excessively high; or with the engine running under high engine load, the ECM detected the MAF sensor signal was excessively low under these operating conditions. **Possible Causes:** • Harness or connectors (The sensor circuit is open or shorted.) • Mass air flow sensor • EVAP control system • Intake air temperature sensor • Intake air leaks
DTC: P0102 **1T CCM, MIL: Yes** **Years: 2007, 2008** **Models:** EX35, FX35, G35, G37, FX45, QX56 **Engines:** 3.5L, 3.7L, 4.5L, 5.6L **Transmissions:** All	**Mass Air Flow Sensor (Bank 1) Circuit Low Input** Key on or engine running; and the ECM detected an unexpected low voltage condition on the MAF sensor signal circuit. **Possible Causes:** • MAF sensor signal circuit is open or shorted • Intake air leaks • MAF sensor is damaged or has failed

DTC	Trouble Code Title, Conditions & Possible Causes
DTC: P0103 **1T CCM, MIL: Yes** **Years:** 2007, 2008 **Models:** EX35, FX35, G35, G37, FX45, QX56 **Engines:** 3.5L, 3.7L, 4.5L, 5.6L **Transmissions:** All	**Mass Air Flow Sensor (Bank 1) Circuit High Input** Key on or engine running; and the ECM detected an unexpected low voltage condition on the MAF sensor signal circuit. **Possible Causes:** • MAF sensor signal circuit is open or shorted • MAF sensor is damaged or has failed
DTC: P0112 **2T CCM, MIL: Yes** **Years:** 2007, 2008 **Models:** EX35, FX35, G35, G37, FX45, QX56 **Engines:** 3.5L, 3.7L, 4.5L, 5.6L **Transmissions:** All	**Intake Air Temperature Sensor Circuit Low Input** Key on or engine running; and the ECM detected an unexpected low voltage condition on the IAT sensor signal circuit. **Possible Causes:** • IAT sensor connector is open or shorted • IAT sensor is damaged or has failed
DTC: P0113 **2T CCM, MIL: Yes** **Years:** 2007, 2008 **Models:** EX35, FX35, G35, G37, FX45, QX56 **Engines:** 3.5L, 3.7L, 4.5L, 5.6L **Transmissions:** All	**Intake Air Temperature Sensor Circuit High Input** Key on or engine running; and the ECM detected an unexpected high voltage condition on the IAT sensor signal circuit. **Possible Causes:** • IAT sensor connector is open or shorted • IAT sensor is damaged or has failed
DTC: P0117 **1T CCM, MIL: Yes** **Years:** 2007, 2008 **Models:** EX35, FX35, G35, G37, FX45, QX56 **Engines:** 3.5L, 3.7L, 4.5L, 5.6L **Transmissions:** All	**Engine Coolant Temperature Sensor Circuit Low Input** Key on or engine running; and the ECM detected an unexpected low voltage condition on the ECT sensor signal circuit. **Possible Causes:** • ECT sensor signal circuit is shorted to ground • ECT sensor is damaged or has failed
DTC: P0118 **1T CCM, MIL: Yes** **Years:** 2007, 2008 **Models:** EX35, FX35, G35, G37, FX45, QX56 **Engines:** 3.5L, 3.7L, 4.5L, 5.6L **Transmissions:** All	**Engine Coolant Temperature Sensor Circuit High Input** Key on or engine running; and the ECM detected an unexpected high voltage condition on the ECT sensor signal circuit. **Possible Causes:** • ECT sensor signal circuit is shorted to ground • ECT sensor is damaged or has failed
DTC: P0122 **1T CCM, MIL: Yes** **Years:** 2007, 2008 **Models:** EX35, FX35, G35, G37, FX45, QX56 **Engines:** 3.5L, 3.7L, 4.5L, 5.6L **Transmissions:** All	**Throttle Position Sensor 2 (Bank 1) Circuit Low Input** Engine started and idled. ECM detected an unexpected low voltage condition on the TP sensor signal. **Possible Causes:** • Harness or connectors (TP sensor 2 circuit is open or shorted.) • Electric throttle control actuator (TP sensor 2)
DTC: P0123 **1T CCM, MIL: Yes** **Years:** 2007, 2008 **Models:** EX35, FX35, G35, G37, FX45, QX56 **Engines:** 3.5L, 3.7L, 4.5L, 5.6L **Transmissions:** All	**Throttle Position Sensor 2 (Bank 1) Circuit High Input** Engine started and idled. ECM detected an unexpected high voltage condition on the TP sensor signal. **Possible Causes:** • Harness or connectors (TP sensor 2 circuit is open or shorted.) • Electric throttle control actuator (TP sensor 2)
DTC: P0125 **1T ECT, MIL: Yes** **Years:** 2007, 2008 **Models:** EX35, FX35, G35, G37, FX45, QX56 **Engines:** 3.5L, 3.7L, 4.5L, 5.6L **Transmissions:** All	**ECT Sensor Excessive Time To Enter Closed Loop** Voltage sent to ECM from the sensor is not practical, even when some time has passed after starting the engine. Engine coolant temperature is insufficient for closed loop fuel control. **Possible Causes:** • Harness or connectors (High resistance in the circuit) • Engine coolant temperature sensor • Thermostat

DTC	Trouble Code Title, Conditions & Possible Causes
DTC: P0127 **2T ECT, MIL: Yes** **Years: 2007, 2008** **Models:** EX35, FX35, G35, G37, FX45, QX56 **Engines:** 3.5L, 3.7L, 4.5L, 5.6L **Transmissions:** All	**Intake Air Temperature Sensor Range/Performance** Rationally incorrect voltage from the sensor is sent to ECM, compared with the voltage signal from engine coolant temperature sensor. **Possible Causes:** • Harness or connectors (sensor circuit is open or shorted) • Intake air temperature sensor
DTC: P0128 **2T ECT, MIL: Yes** **Years: 2007, 2008** **Models:** EX35, FX35, G35, G37, FX45, QX56 **Engines:** 3.5L, 3.7L, 4.5L, 5.6L **Transmissions:** All	**Thermostat Malfunction** The engine coolant temperature does not reach to specified temperature even though the engine has run long enough. **Possible Causes:** • Check the operation of the thermostat (it may be stuck open) • Leakage from sealing portion of thermostat • Engine coolant temperature sensor
DTC: P0130 **2T CCM, MIL: Yes** **Years: 2007, 2008** **Models:** EX35, FX35, G35, G37, FX45, QX56 **Engines:** 3.5L, 3.7L, 4.5L, 5.6L **Transmissions:** All	**A/F Sensor 1 (Bank 1) Circuit** The A/F signal computed by ECM from the A/F sensor 1 signal is constantly in the range other than approx. 2.2 V. Or, A/F signal computed by ECM from the A/F sensor 1 signal is constantly approx. 2.2 V. **Possible Causes:** • Harness or connectors (A/F sensor 1 circuit is open or shorted.) • A/F sensor 1
DTC: P0131 **2T CCM, MIL: Yes** **Years: 2007, 2008** **Models:** EX35, FX35, G35, G37, FX45, QX56 **Engines:** 3.5L, 3.7L, 4.5L, 5.6L **Transmissions:** All	**A/F Sensor 1 (Bank 1) Low Voltage** The A/F signal computed by ECM from the A/F sensor 1 signal is constantly approx. 0V. **Possible Causes:** • Harness or connectors (A/F sensor 1 circuit is open or shorted.) • A/F sensor 1
DTC: P0132 **2T CCM, MIL: Yes** **Years: 2007, 2008** **Models:** EX35, FX35, G35, G37, FX45, QX56 **Engines:** 3.5L, 3.7L, 4.5L, 5.6L **Transmissions:** All	**A/F Sensor 1 (Bank 1) High Voltage** The A/F signal computed by ECM from the A/F sensor 1 signal is constantly approx. 0.5V. **Possible Causes:** • Harness or connectors (A/F sensor 1 circuit is open or shorted.) • A/F sensor 1
DTC: P0133 **2T O2S, MIL: Yes** **Years: 2007, 2008** **Models:** EX35, FX35, G35, G37, FX45, QX56 **Engines:** 3.5L, 3.7L, 4.5L, 5.6L **Transmissions:** All	**A/F Sensor 1 (Bank 1) Slow Response** The response of the A/F signal computed by ECM from A/F sensor 1 signal takes more than the specified time. **Possible Causes:** • Harness or connectors (A/F sensor 1 circuit is open or shorted.) • A/F sensor 1 • A/F sensor 1 heater • Fuel pressure • Fuel injector • Intake air leaks • Exhaust gas leaks • PCV • Mass air flow sensor
DTC: P0137 **2T CCM, MIL: Yes** **Years: 2007, 2008** **Models:** EX35, FX35, G35, G37, FX45, QX56 **Engines:** 3.5L, 3.7L, 4.5L, 5.6L **Transmissions:** All	**HO2S-12 (Bank 1 Sensor 2) Circuit Low Voltage** The maximum voltage from the sensor is not to the specified voltage range. **Possible Causes:** • Harness or connectors (sensor circuit is open or shorted) • Heated oxygen sensor 2 • Fuel pressure • Fuel injector • Intake air leaks

DTC	Trouble Code Title, Conditions & Possible Causes
DTC: P0138 **2T CCM, MIL: Yes** **Years: 2007, 2008** **Models:** EX35, FX35, G35, G37, FX45, QX56 **Engines:** 3.5L, 3.7L, 4.5L, 5.6L **Transmissions:** All	**HO2S-12 (Bank 1 Sensor 2) Circuit High Voltage** An excessively high voltage from the sensor is sent to ECM. Or, The minimum voltage from the sensor is not reached to the specified voltage. **Possible Causes:** • Harness or connectors (sensor circuit is open or shorted) • Heated oxygen sensor 2 • Fuel pressure • Fuel injector
DTC: P0139 **2T O2S, MIL: Yes** **Years: 2007, 2008** **Models:** EX35, FX35, G35, G37, FX45, QX56 **Engines:** 3.5L, 3.7L, 4.5L, 5.6L **Transmissions:** All	**HO2S-12 (Bank 1 Sensor 2) Slow Response** It takes more time for the sensor to respond between rich and lean than the specified time. **Possible Causes:** • Harness or connectors (sensor circuit is open or shorted) • Heated oxygen sensor 2 • Fuel pressure • Fuel injector • Intake air leaks
DTC: P0150 **2T CCM, MIL: Yes** **Years: 2007, 2008** **Models:** EX35, FX35, G35, G37, FX45, QX56 **Engines:** 3.5L, 3.7L, 4.5L, 5.6L **Transmissions:** All	**A/F Sensor 1 (Bank 2) Circuit** The A/F signal computed by ECM from the A/F sensor 1 signal is constantly in the range other than approx. 2.2 V. Or, A/F signal computed by ECM from the A/F sensor 1 signal is constantly approx. 2.2 V. **Possible Causes:** • Harness or connectors (A/F sensor 1 circuit is open or shorted.) • A/F sensor 1
DTC: P0151 **2T CCM, MIL: Yes** **Years: 2007, 2008** **Models:** EX35, FX35, G35, G37, FX45, QX56 **Engines:** 3.5L, 3.7L, 4.5L, 5.6L **Transmissions:** All	**A/F Sensor 1 (Bank 2) Low Voltage** The A/F signal computed by ECM from the A/F sensor 1 signal is constantly approx. 0V. **Possible Causes:** • Harness or connectors (A/F sensor 1 circuit is open or shorted.) • A/F sensor 1
DTC: P0152 **2T CCM, MIL: Yes** **Years: 2007, 2008** **Models:** EX35, FX35, G35, G37, FX45, QX56 **Engines:** 3.5L, 3.7L, 4.5L, 5.6L **Transmissions:** All	**A/F Sensor 1 (Bank 2) High Voltage** The A/F signal computed by ECM from the A/F sensor 1 signal is constantly approx. 0.5V. **Possible Causes:** • Harness or connectors (A/F sensor 1 circuit is open or shorted.) • A/F sensor 1
DTC: P0153 **2T O2S, MIL: Yes** **Years: 2007, 2008** **Models:** EX35, FX35, G35, G37, FX45, QX56 **Engines:** 3.5L, 3.7L, 4.5L, 5.6L **Transmissions:** All	**A/F Sensor 1 (Bank 2) Slow Response** The response of the A/F signal computed by ECM from A/F sensor 1 signal takes more than the specified time. **Possible Causes:** • Harness or connectors (A/F sensor 1 circuit is open or shorted.) • A/F sensor 1 • A/F sensor 1 heater • Fuel pressure • Fuel injector • Intake air leaks • Exhaust gas leaks • PCV • Mass air flow sensor
DTC: P0157 **2T CCM, MIL: Yes** **Years: 2007, 2008** **Models:** EX35, FX35, G35, G37, FX45, QX56 **Engines:** 3.5L, 3.7L, 4.5L, 5.6L **Transmissions:** All	**HO2S-12 (Bank 2 Sensor 2) Circuit Low Voltage** The maximum voltage from the sensor is not to the specified voltage range. **Possible Causes:** • Harness or connectors (sensor circuit is open or shorted) • Heated oxygen sensor 2 • Fuel pressure • Fuel injector • Intake air leaks

DTC	Trouble Code Title, Conditions & Possible Causes
DTC: P0158 **2T CCM, MIL: Yes** **Years: 2007, 2008** **Models:** EX35, FX35, G35, G37, FX45, QX56 **Engines:** 3.5L, 3.7L, 4.5L, 5.6L **Transmissions:** All	**HO2S-12 (Bank 2 Sensor 2) Circuit High Voltage** An excessively high voltage from the sensor is sent to ECM. Or, The minimum voltage from the sensor is not reached to the specified voltage. **Possible Causes:** • Harness or connectors (sensor circuit is open or shorted) • Heated oxygen sensor 2 • Fuel pressure • Fuel injector
DTC: P0159 **2T O2S, MIL: Yes** **Years: 2007, 2008** **Models:** EX35, FX35, G35, G37, FX45, QX56 **Engines:** 3.5L, 3.7L, 4.5L, 5.6L **Transmissions:** All	**HO2S-12 (Bank 2 Sensor 2) Slow Response** It takes more time for the sensor to respond between rich and lean than the specified time. **Possible Causes:** • Harness or connectors (sensor circuit is open or shorted) • Heated oxygen sensor 2 • Fuel pressure • Fuel injector • Intake air leaks
DTC: P0171 **2T FUEL, MIL: Yes** **Years: 2007, 2008** **Models:** EX35, FX35, G35, G37, FX45, QX56 **Engines:** 3.5L, 3.7L, 4.5L, 5.6L **Transmissions:** All	**Fuel Trim Lean (Bank 1)** Fuel injection system does not operate properly. Or, the amount of mixture ratio compensation is too large (mixture ratio is too lean). **Possible Causes:** • Intake air leaks • A/F sensor 1 • Fuel injector • Exhaust gas leaks • Incorrect fuel pressure • Lack of fuel • Mass air flow sensor • Incorrect PCV hose connection
DTC: P0172 **2T FUEL, MIL: Yes** **Years: 2007, 2008** **Models:** EX35, FX35, G35, G37, FX45, QX56 **Engines:** 3.5L, 3.7L, 4.5L, 5.6L **Transmissions:** All	**Fuel Trim Rich (Bank 1)** Fuel injection system does not operate properly. • The amount of mixture ratio compensation is too large. (The mixture ratio is too rich.) **Possible Causes:** • A/F sensor 1 • Fuel injector • Exhaust gas leaks • Incorrect fuel pressure • Mass air flow sensor
DTC: P0174 **2T FUEL, MIL: Yes** **Years: 2007, 2008** **Models:** EX35, FX35, G35, G37, FX45, QX56 **Engines:** 3.5L, 3.7L, 4.5L, 5.6L **Transmissions:** All	**Fuel Trim Lean (Bank 2)** Fuel injection system does not operate properly. Or, the amount of mixture ratio compensation is too large (mixture ratio is too lean). **Possible Causes:** • Intake air leaks • A/F sensor 1 • Fuel injector • Exhaust gas leaks • Incorrect fuel pressure • Lack of fuel • Mass air flow sensor • Incorrect PCV hose connection
DTC: P0175 **2T FUEL, MIL: Yes** **Years: 2007, 2008** **Models:** EX35, FX35, G35, G37, FX45, QX56 **Engines:** 3.5L, 3.7L, 4.5L, 5.6L **Transmissions:** All	**Fuel Trim Rich (Bank 2)** Fuel injection system does not operate properly. • The amount of mixture ratio compensation is too large. (The mixture ratio is too rich.) **Possible Causes:** • A/F sensor 1 • Fuel injector • Exhaust gas leaks • Incorrect fuel pressure • Mass air flow sensor

DTC	Trouble Code Title, Conditions & Possible Causes
DTC: P0181 **2T CCM, MIL: Yes** **Years: 2007, 2008** **Models:** EX35, FX35, G35, G37, FX45, QX56 **Engines:** 3.5L, 3.7L, 4.5L, 5.6L **Transmissions:** All	**Fuel Tank Temperature Sensor Range/Performance** Rationally incorrect voltage from the sensor is sent to ECM, compared with the voltage signals from engine coolant temperature sensor and intake air temperature sensor. **Possible Causes:** • Fuel tank temperature sensor harness (circuit is open or shorted) • Fuel tank temperature sensor
DTC: P0182 **2T CCM, MIL: Yes** **Years: 2007, 2008** **Models:** EX35, FX35, G35, G37, FX45, QX56 **Engines:** 3.5L, 3.7L, 4.5L, 5.6L **Transmissions:** All	**Fuel Tank Temperature Sensor Circuit Low Input** An excessively low voltage from the sensor is sent to ECM. **Possible Causes:** • FTT harness or connectors (circuit is open or shorted • FTT sensor
DTC: P0183 **2T CCM, MIL: Yes** **Years: 2007, 2008** **Models:** EX35, FX35, G35, G37, FX45, QX56 **Engines:** 3.5L, 3.7L, 4.5L, 5.6L **Transmissions:** All	**Fuel Tank Temperature Sensor Circuit High Input** An excessively low voltage from the sensor is sent to ECM. **Possible Causes:** • FTT harness or connectors (circuit is open or shorted • FTT sensor
DTC: P0196 **2T CCM, MIL: Yes** **Years: 2007, 2008** **Models:** EX35, G37 **Engines:** 3.5L, 3.7L **Transmissions:** All	**Engine Oil Temperature Sensor Range/Performance** Rationally incorrect voltage from the sensor is sent to ECM, compared with the voltage signals from engine coolant temperature sensor and intake air temperature sensor. **Possible Causes:** • Harness or connectors (circuit is open or shorted • EOT sensor
DTC: P0197 **2T CCM, MIL: Yes** **Years: 2007, 2008** **Models:** EX35, G37 **Engines:** 3.5L, 3.7L **Transmissions:** All	**Engine Oil Temperature Sensor Low Voltage** Voltage signal from sensor to ECM is excessively low. **Possible Causes:** • Harness or connectors (circuit is open or shorted • EOT sensor
DTC: P0198 **2T CCM, MIL: Yes** **Years: 2007, 2008** **Models:** EX35, G37 **Engines:** 3.5L, 3.7L **Transmissions:** All	**Engine Oil Temperature Sensor High Voltage** Voltage signal from sensor to ECM is excessively high. **Possible Causes:** • Harness or connectors (circuit is open or shorted • EOT sensor
DTC: P0222 **1T CCM, MIL: Yes** **Years: 2007, 2008** **Models:** EX35, FX35, G35, G37, FX45 **Engines:** 3.5L, 3.7L, 4.5L **Transmissions:** All	**Throttle Position Sensor 1 (Bank 1) Signal Low Input** An excessively low input signal is sent from the sensor to the ECM. **Possible Causes:** • Harness or connectors (circuit is open or shorted • Electric throttle control actuator (TP sensor 1)
DTC: P0223 **1T CCM, MIL: Yes** **Years: 2007, 2008** **Models:** EX35, FX35, G35, G37, FX45 **Engines:** 3.5L, 3.7L, 4.5L **Transmissions:** All	**Throttle Position Sensor 1 (Bank 1) Signal High Input** An excessively high input signal is sent from the sensor to the ECM. **Possible Causes:** • Harness or connectors (circuit is open or shorted • Electric throttle control actuator (TP sensor 1)
DTC: P0227 **1T CCM, MIL: Yes** **Years: 2007, 2008** **Models**: EX35, G37 **Engines:** 3.5L, 3.7L **Transmissions:** All	**Throttle Position Sensor 2 (Bank 2) Circuit Low Input** Engine started and idled. ECM detected an unexpected low voltage condition on the TP sensor signal. **Possible Causes:** • Harness or connectors (TP sensor 2 circuit is open or shorted.) • Electric throttle control actuator (TP sensor 2)

DTC	Trouble Code Title, Conditions & Possible Causes
DTC: P0228 **1T CCM, MIL: Yes** **Years: 2007, 2008** **Models**: EX35, G37 **Engines**: 3.5L, 3.7L **Transmissions**: All	**Throttle Position Sensor 2 (Bank 2) Circuit High Input** Engine started and idled. ECM detected an unexpected high voltage condition on the TP sensor signal. **Possible Causes:** • Harness or connectors (TP sensor 2 circuit is open or shorted.) • Electric throttle control actuator (TP sensor 2)
DTC: P0300 **2T MISFIRE, MIL: Yes** **Years: 2007, 2008** **Models:** EX35, FX35, G35, G37, FX45, QX56 **Engines:** 3.5L, 3.7L, 4.5L, 5.6L **Transmissions:** All	**Multiple Misfire Detected** Start engine and warm it up to normal operating temperature. Turn ignition switch OFF and wait at least 10 seconds. Restart engine and let it idle for about 15 minutes. Check 1st trip DTC. **Possible Causes:** • Improper spark plug • Insufficient compression • Incorrect fuel pressure • The fuel injector circuit is open or shorted • Fuel injector • Intake air leak • The ignition signal circuit is open or shorted • Lack of fuel • Signal plate • A/F sensor 1 • Incorrect PCV hose connection
DTC: P0301 **2T MISFIRE, MIL: Yes** **Years: 2007, 2008** **Models:** EX35, FX35, G35, G37, FX45, QX56 **Engines:** 3.5L, 3.7L, 4.5L, 5.6L **Transmissions:** All	**Cylinder 1 Misfire Detected** Start engine and warm it up to normal operating temperature. Turn ignition switch OFF and wait at least 10 seconds. Restart engine and let it idle for about 15 minutes. Check 1st trip DTC. **Possible Causes:** • Improper spark plug • Insufficient compression • Incorrect fuel pressure • The fuel injector circuit is open or shorted • Fuel injector • Intake air leak • The ignition signal circuit is open or shorted • Lack of fuel • Signal plate • A/F sensor 1 • Incorrect PCV hose connection
DTC: P0302 **2T MISFIRE, MIL: Yes** **Years: 2007, 2008** **Models:** EX35, FX35, G35, G37, FX45, QX56 **Engines:** 3.5L, 3.7L, 4.5L, 5.6L **Transmissions:** All	**Cylinder 2 Misfire Detected** Start engine and warm it up to normal operating temperature. Turn ignition switch OFF and wait at least 10 seconds. Restart engine and let it idle for about 15 minutes. Check 1st trip DTC. **Possible Causes:** • Improper spark plug • Insufficient compression • Incorrect fuel pressure • The fuel injector circuit is open or shorted • Fuel injector • Intake air leak • The ignition signal circuit is open or shorted • Lack of fuel • Signal plate • A/F sensor 1 • Incorrect PCV hose connection

DTC	Trouble Code Title, Conditions & Possible Causes
DTC: P0303 **2T MISFIRE, MIL: Yes** **Years: 2007, 2008** **Models:** EX35, FX35, G35, G37, FX45, QX56 **Engines:** 3.5L, 3.7L, 4.5L, 5.6L **Transmissions:** All	**Cylinder 3 Misfire Detected** Start engine and warm it up to normal operating temperature. Turn ignition switch OFF and wait at least 10 seconds. Restart engine and let it idle for about 15 minutes. Check 1st trip DTC. **Possible Causes:** • Improper spark plug • Insufficient compression • Incorrect fuel pressure • The fuel injector circuit is open or shorted • Fuel injector • Intake air leak • The ignition signal circuit is open or shorted • Lack of fuel • Signal plate • A/F sensor 1 • Incorrect PCV hose connection
DTC: P0304 **2T MISFIRE, MIL: Yes** **Years: 2007, 2008** **Models:** EX35, FX35, G35, G37, FX45, QX56 **Engines:** 3.5L, 3.7L, 4.5L, 5.6L **Transmissions:** All	**Cylinder 4 Misfire Detected** Start engine and warm it up to normal operating temperature. Turn ignition switch OFF and wait at least 10 seconds. Restart engine and let it idle for about 15 minutes. Check 1st trip DTC. **Possible Causes:** • Improper spark plug • Insufficient compression • Incorrect fuel pressure • The fuel injector circuit is open or shorted • Fuel injector • Intake air leak • The ignition signal circuit is open or shorted • Lack of fuel • Signal plate • A/F sensor 1 • Incorrect PCV hose connection
DTC: P0305 **2T MISFIRE, MIL: Yes** **Years: 2007, 2008** **Models:** EX35, FX35, G35, G37, FX45, QX56 **Engines:** 3.5L, 3.7L, 4.5L, 5.6L **Transmissions:** All	**Cylinder 5 Misfire Detected** Start engine and warm it up to normal operating temperature. Turn ignition switch OFF and wait at least 10 seconds. Restart engine and let it idle for about 15 minutes. Check 1st trip DTC. **Possible Causes:** • Improper spark plug • Insufficient compression • Incorrect fuel pressure • The fuel injector circuit is open or shorted • Fuel injector • Intake air leak • The ignition signal circuit is open or shorted • Lack of fuel • Signal plate • A/F sensor 1 • Incorrect PCV hose connection
DTC: P0306 **2T MISFIRE, MIL: Yes** **Years: 2007, 2008** **Models:** EX35, FX35, G35, G37, FX45, QX56 **Engines:** 3.5L, 3.7L, 4.5L, 5.6L **Transmissions:** All	**Cylinder 6 Misfire Detected** Start engine and warm it up to normal operating temperature. Turn ignition switch OFF and wait at least 10 seconds. Restart engine and let it idle for about 15 minutes. Check 1st trip DTC. **Possible Causes:** • Improper spark plug • Insufficient compression • Incorrect fuel pressure • The fuel injector circuit is open or shorted • Fuel injector • Intake air leak • The ignition signal circuit is open or shorted • Lack of fuel • Signal plate • A/F sensor 1 • Incorrect PCV hose connection

DTC	Trouble Code Title, Conditions & Possible Causes
DTC: P0307 **2T MISFIRE, MIL: Yes** **Years: 2007, 2008** **Models:** FX45, QX56 **Engines:** 4.5L, 5.6L **Transmissions:** All	**Cylinder 7 Misfire Detected** Start engine and warm it up to normal operating temperature. Turn ignition switch OFF and wait at least 10 seconds. Restart engine and let it idle for about 15 minutes. Check 1st trip DTC. **Possible Causes:** • Improper spark plug • Insufficient compression • Incorrect fuel pressure • The fuel injector circuit is open or shorted • Fuel injector • Intake air leak • The ignition signal circuit is open or shorted • Lack of fuel • Signal plate • A/F sensor 1 • Incorrect PCV hose connection
DTC: P0308 **2T MISFIRE, MIL: Yes** **Years: 2007, 2008** **Models:** FX45, QX56 **Engines:** 4.5L, 5.6L **Transmissions:** All	**Cylinder 8 Misfire Detected** Start engine and warm it up to normal operating temperature. Turn ignition switch OFF and wait at least 10 seconds. Restart engine and let it idle for about 15 minutes. Check 1st trip DTC. **Possible Causes:** • Improper spark plug • Insufficient compression • Incorrect fuel pressure • The fuel injector circuit is open or shorted • Fuel injector • Intake air leak • The ignition signal circuit is open or shorted • Lack of fuel • Signal plate • A/F sensor 1 • Incorrect PCV hose connection
DTC: P0327 **2T CCM, MIL: No** **Years: 2007, 2008** **Models:** EX35, FX35, G35, G37, FX45, QX56 **Engines:** 3.5L, 3.7L, 4.5L, 5.6L **Transmissions:** All	**Knock Sensor Circuit Low Input (Bank 1)** An excessively low signal is sent from sensor to ECM. **Possible Causes:** • Harness or connectors (sensor circuit is open or shorted.) • Knock sensor)
DTC: P0328 **2T CCM, MIL: No** **Years: 2007, 2008** **Models:** EX35, FX35, G35, G37, FX45, QX56 **Engines:** 3.5L, 3.7L, 4.5L, 5.6L **Transmissions:** All	**Knock Sensor Circuit High Input (Bank 1)** An excessively high signal is sent from sensor to ECM. **Possible Causes:** • Harness or connectors (sensor circuit is open or shorted.) • Knock sensor)
DTC: P0332 **2T CCM, MIL: No** **Years: 2007, 2008** **Models:** EX35, FX35, G35, G37, FX45, QX56 **Engines:** 3.5L, 3.7L, 4.5L, 5.6L **Transmissions:** All	**Knock Sensor Circuit Low Input (Bank 2)** An excessively low signal is sent from sensor to ECM. **Possible Causes:** • Harness or connectors (sensor circuit is open or shorted.) • Knock sensor)
DTC: P0333 **2T CCM, MIL: No** **Years: 2007, 2008** **Models:** EX35, FX35, G35, G37, FX45, QX56 **Engines:** 3.5L, 3.7L, 4.5L, 5.6L **Transmissions:** All	**Knock Sensor Circuit High Input (Bank 2)** An excessively high signal is sent from sensor to ECM. **Possible Causes:** • Harness or connectors (sensor circuit is open or shorted.) • Knock sensor)

DTC	Trouble Code Title, Conditions & Possible Causes

DTC: P0335
2T CCM, MIL: Yes
Years: 2007, 2008
Models: EX35, FX35, G35, G37, FX45, QX56
Engines: 3.5L, 3.7L, 4.5L, 5.6L
Transmissions: All
	Crankshaft Position (CKP) Sensor (POS) Circuit Malfunction
	The crankshaft position sensor (POS) signal is not detected by the ECM during the first few seconds of engine cranking. The proper pulse signal from the crankshaft position sensor (POS) is not sent to ECM while the engine is running. The crankshaft position sensor (POS) signal is not in the normal pattern during engine running.
	Possible Causes:
	• Harness or connectors
	  • CKP sensor (POS) circuit is open or shorted.
	  • CMP sensor (PHASE) (bank 2) circuit is shorted.
	  • EVT control position sensor (bank 2) circuit is shorted.
	  • Battery current sensor circuit is shorted.
	  • APP sensor 2 circuit is shorted.
	  • EVAP control system pressure sensor circuit is shorted.
	  • Refrigerant pressure sensor circuit is shorted.
	• Crankshaft position sensor (POS)
	• Camshaft position sensor (PHASE) - (bank 2)
	• Exhaust valve timing control position sensor (bank 2)
	• Battery current sensor
	• Accelerator pedal position sensor
	• EVAP control system pressure sensor
	• Refrigerant pressure sensor
	• Signal plate

DTC: P0340 **2T CCM, MIL: Yes** **Years: 2007, 2008** **Models:** EX35, FX35, G35, G37, FX45, QX56 **Engines:** 3.5L, 3.7L, 4.5L, 5.6L **Transmissions:** All	**Camshaft Position (CMP) Sensor (Phase) Circuit Malfunction (Bank 1)** The cylinder No. signal is not sent to ECM for the first few seconds during engine cranking. The cylinder No. signal is not sent to ECM during engine running. The cylinder No. signal is not in the normal pattern during engine running. **Possible Causes:** • Harness or connectors - CMP sensor (PHASE) (bank 1) circuit is open or shorted. • Camshaft position sensor (PHASE) - (bank 1) • Camshaft (INT) • Starter motor • Starting system circuit • Dead (Weak) battery

DTC: P0345
2T CCM, MIL: Yes
Years: 2007, 2008
Models: EX35, FX35, G35, G37, FX45
Engines: 3.5L, 3.7L, 4.5L
Transmissions: All
	Camshaft Position (CMP) Sensor (Phase) Circuit Malfunction (Bank 2)
	The cylinder No. signal is not sent to ECM for the first few seconds during engine cranking.
	The cylinder No. signal is not sent to ECM during engine running.
	The cylinder No. signal is not in the normal pattern during engine running.
	Possible Causes:
	• Harness or connectors
	  • CMP sensor (PHASE) (bank 2) circuit is open or shorted.
	  • CKP sensor (POS) circuit is shorted.
	  • EVT control position sensor (bank 2) circuit is shorted.
	  • Battery current sensor circuit is shorted.
	  • (APP sensor 2 circuit is shorted.
	  • (EVAP control system pressure sensor circuit is shorted.
	  • Refrigerant pressure sensor circuit is shorted.
	• Camshaft position sensor (PHASE) - (bank 2)
	• Crankshaft position sensor (POS)
	• Exhaust valve timing control position sensor (bank 2)
	• Battery current sensor
	• Accelerator pedal position sensor
	• EVAP control system pressure sensor
	• Refrigerant pressure sensor
	• Camshaft (INT)
	• Starter motor
	• Starting system circuit
	• Dead (Weak) battery

DTC	Trouble Code Title, Conditions & Possible Causes
DTC: P0420 **1T CAT, MIL: Yes** **Years: 2007, 2008** **Models:** EX35, FX35, G35, G37, FX45, QX56 **Engines:** 3.5L, 3.7L, 4.5L, 5.6L **Transmissions:** All	**Catalyst Efficiency Below Normal (Bank 1)** Three-way catalyst (manifold) does not operate properly. Three-way catalyst (manifold) does not have enough oxygen storage capacity. **Possible Causes:** • Three way catalyst (manifold) • Exhaust tube • Intake air leaks • Fuel injector • Fuel injector leaks • Spark plug • Improper ignition timing
DTC: P0430 **1T CAT, MIL: Yes** **Years: 2007, 2008** **Models:** EX35, FX35, G35, G37, FX45, QX56 **Engines:** 3.5L, 3.7L, 4.5L, 5.6L **Transmissions:** All	**Catalyst Efficiency Below Normal (Bank 2)** Three-way catalyst (manifold) does not operate properly. Three-way catalyst (manifold) does not have enough oxygen storage capacity. **Possible Causes:** • Three way catalyst (manifold) • Exhaust tube • Intake air leaks • Fuel injector • Fuel injector leaks • Spark plug • Improper ignition timing
DTC: P0441 **2T EVAP, MIL: Yes** **Years: 2007, 2008** **Models:** EX35, FX35, G35, G37, FX45, QX56 **Engines:** 3.5L, 3.7L, 4.5L, 5.6L **Transmissions:** All	**EVAP System Incorrect Purge Flow Detected** EVAP control system does not operate properly, EVAP control system has a leak between intake manifold and EVAP control system pressure sensor. **Possible Causes:** • EVAP canister purge volume control solenoid valve stuck closed • EVAP control system pressure sensor and the circuit • Loose, disconnected or improper connection of rubber tube • Blocked rubber tube • Cracked EVAP canister • EVAP canister purge volume control solenoid valve circuit • Accelerator pedal position sensor • Blocked purge port • EVAP canister vent control valve
DTC: P0442 **2T EVAP, MIL: Yes** **Years: 2007, 2008** **Models:** EX35, FX35, G35, G37, FX45, QX56 **Engines:** 3.5L, 3.7L, 4.5L, 5.6L **Transmissions:** All	**EVAP System Small Leak Detected** EVAP control system has a leak, EVAP control system does not operate properly. **Possible Causes:** • Incorrect fuel tank vacuum relief valve • Incorrect fuel filler cap used • Fuel filler cap remains open or fails to close. • Foreign matter caught in fuel filler cap. • Leak is in line between intake manifold and EVAP canister purge volume control solenoid valve. • Foreign matter caught in EVAP canister vent control valve. • EVAP canister or fuel tank leaks • EVAP purge line (pipe and rubber tube) leaks • EVAP purge line rubber tube bent • Loose or disconnected rubber tube • EVAP canister vent control valve and the circuit • EVAP canister purge volume control solenoid valve and the circuit • Fuel tank temperature sensor • O-ring of EVAP canister vent control valve is missing or damaged • EVAP canister is saturated with water • EVAP control system pressure sensor • Fuel level sensor and the circuit • Refueling EVAP vapor cut valve • ORVR system leaks

DTC	Trouble Code Title, Conditions & Possible Causes
DTC: P0443 **2T CCM, MIL: Yes** **Years: 2007, 2008** **Models:** EX35, FX35, G35, G37, FX45, QX56 **Engines:** 3.5L, 3.7L, 4.5L, 5.6L **Transmissions:** All	**EVAP Canister Purge Volume Control Solenoid Circuit Malfunction** The canister purge flow is detected during the specified driving conditions, even when EVAP canister purge volume control solenoid valve is completely closed. **Possible Causes:** <ul><li>EVAP control system pressure sensor</li><li>EVAP canister purge volume control solenoid valve (valve is stuck open.)</li><li>EVAP canister vent control valve</li><li>EVAP canister</li><li>Hoses (Hoses are connected incorrectly or clogged.)</li></ul>
DTC: P0444 **2T CCM, MIL: Yes** **Years: 2007, 2008** **Models:** EX35, FX35, G35, G37, FX45, QX56 **Engines:** 3.5L, 3.7L, 4.5L, 5.6L **Transmissions:** All	**EVAP Canister Purge Volume Control Solenoid Circuit Low Input** Excessively low input signal is sent from solenoid to ECM. **Possible Causes:** <ul><li>Harness or connectors (solenoid valve circuit is open or shorted.)</li><li>EVAP canister purge volume control solenoid valve</li></ul>
DTC: P0445 **2T CCM, MIL: Yes** **Years: 2007, 2008** **Models:** EX35, FX35, G35, G37, FX45, QX56 **Engines:** 3.5L, 3.7L, 4.5L, 5.6L **Transmissions:** All	**EVAP Canister Purge Volume Control Solenoid Circuit High Input** Excessively high input signal is sent from solenoid to ECM. **Possible Causes:** <ul><li>Harness or connectors (solenoid valve circuit is open or shorted.)</li><li>EVAP canister purge volume control solenoid valve</li></ul>
DTC: P0447 **2T CCM, MIL: Yes** **Years: 2007, 2008** **Models:** EX35, FX35, G35, G37, FX45, QX56 **Engines:** 3.5L, 3.7L, 4.5L, 5.6L **Transmissions:** All	**EVAP Canister Vent Control Valve Circuit Malfunction** An improper voltage signal is sent to ECM through EVAP canister vent control valve. **Possible Causes:** <ul><li>Harness or connectors: the valve circuit is open or shorted</li><li>Canister vent control solenoid valve</li></ul>
DTC: P0448 **2T CCM, MIL: Yes** **Years: 2007, 2008** **Models:** EX35, FX35, G35, G37, FX45, QX56 **Engines:** 3.5L, 3.7L, 4.5L, 5.6L **Transmissions:** All	**EVAP Canister Vent Control Valve Circuit Malfunction** The EVAP canister vent control valve remains closed under certain driving conditions. **Possible Causes:** <ul><li>EVAP canister vent control valve</li><li>EVAP control system pressure sensor and the circuit</li><li>Blocked rubber tube to EVAP canister vent control valve</li><li>EVAP canister is saturated with water</li></ul>
DTC: P0451 **2T CCM, MIL: Yes** **Years: 2007, 2008** **Models:** EX35, FX35, G35, G37, FX45, QX56 **Engines:** 3.5L, 3.7L, 4.5L, 5.6L **Transmissions:** All	**EVAP System Pressure Sensor Range/Performance** ECM detects a sloshing signal from the EVAP control system pressure sensor. **Possible Causes:** <ul><li>Harness or connectors (EVAP control system pressure sensor circuit is shorted.)<ul><li>CKP sensor (POS) circuit is shorted.</li><li>CMP sensor (PHASE) (bank 2) circuit is shorted.</li><li>EVT control position sensor (bank 2) circuit is shorted.</li><li>Battery current sensor circuit is shorted.</li><li>APP sensor 2 circuit is shorted.</li><li>Refrigerant pressure sensor circuit is shorted.</li></ul></li><li>EVAP control system pressure sensor</li><li>Crankshaft position sensor (POS)</li><li>Camshaft position sensor (PHASE) - (bank 2)</li><li>Exhaust valve timing control position sensor (bank 2)</li><li>Battery current sensor</li><li>Accelerator pedal position sensor</li><li>Refrigerant pressure sensor</li></ul>

DTC	Trouble Code Title, Conditions & Possible Causes

DTC: P0452
2T CCM, MIL: Yes
Years: 2007, 2008
Models: EX35, FX35, G35, G37, FX45, QX56
Engines: 3.5L, 3.7L, 4.5L, 5.6L
Transmissions: All
	EVAP System Pressure Sensor Circuit Low Input
	An excessively low voltage signal is sent from the senor to the ECM.
	Possible Causes:
	• Harness or connectors (EVAP control system pressure sensor circuit is shorted.)
	   • CKP sensor (POS) circuit is shorted.
	   • CMP sensor (PHASE) (bank 2) circuit is shorted.
	   • EVT control position sensor (bank 2) circuit is shorted.
	   • Battery current sensor circuit is shorted.
	   • APP sensor 2 circuit is shorted.
	   • Refrigerant pressure sensor circuit is shorted.
	• EVAP control system pressure sensor
	• Crankshaft position sensor (POS)
	• Camshaft position sensor (PHASE) - (bank 2)
	• Exhaust valve timing control position sensor (bank 2)
	• Battery current sensor
	• Accelerator pedal position sensor
	• Refrigerant pressure sensor

DTC: P0453
2T CCM, MIL: Yes
Years: 2007, 2008
Models: EX35, FX35, G35, G37, FX45, QX56
Engines: 3.5L, 3.7L, 4.5L, 5.6L
Transmissions: All
	EVAP System Pressure Sensor Circuit High Input
	An excessively high voltage signal is sent from the senor to the ECM.
	Possible Causes:
	• Harness or connectors (EVAP control system pressure sensor circuit is shorted.)
	   • CKP sensor (POS) circuit is shorted.
	   • CMP sensor (PHASE) (bank 2) circuit is shorted.
	   • EVT control position sensor (bank 2) circuit is shorted.
	   • Battery current sensor circuit is shorted.
	   • APP sensor 2 circuit is shorted.
	   • Refrigerant pressure sensor circuit is shorted.
	• EVAP control system pressure sensor
	• Crankshaft position sensor (POS)
	• Camshaft position sensor (PHASE) - (bank 2)
	• Exhaust valve timing control position sensor (bank 2)
	• Battery current sensor
	• Accelerator pedal position sensor
	• Refrigerant pressure sensor
	• EVAP canister vent control valve
	• EVAP canister
	• Rubber hose from EVAP canister vent control valve to vehicle frame

DTC: P0455 **2T EVAP, MIL: Yes** **Years: 2007, 2008** **Models:** EX35, FX35, G35, G37, FX45, QX56 **Engines:** 3.5L, 3.7L, 4.5L, 5.6L **Transmissions:** All	**EVAP System Large Leak Detected** EVAP control system has a very large leak such as fuel filler cap fell off, EVAP control system does not operate properly. **Possible Causes:** • Fuel filler cap remains open or fails to close. • Incorrect fuel tank vacuum relief valve • Incorrect fuel filler cap used • Foreign matter caught in fuel filler cap. • Leak is in line between intake manifold and EVAP canister purge volume control solenoid valve. • Foreign matter caught in EVAP canister vent control valve. • EVAP canister or fuel tank leaks • EVAP purge line (pipe and rubber tube) leaks • EVAP purge line rubber tube bent. • Loose or disconnected rubber tube • EVAP canister vent control valve and the circuit • EVAP canister purge volume control solenoid valve and the circuit • Fuel tank temperature sensor • O-ring of EVAP canister vent control valve is missing or damaged. • EVAP control system pressure sensor • Refueling EVAP vapor cut valve • ORVR system leaks

DTC	Trouble Code Title, Conditions & Possible Causes
DTC: P0456 **2T EVAP, MIL: Yes** **Years: 2007, 2008** **Models:** EX35, FX35, G35, G37, FX45, QX56 **Engines:** 3.5L, 3.7L, 4.5L, 5.6L **Transmissions:** All	**EVAP System Very Small Leak Detected** EVAP system has a very small leak. EVAP system does not operate properly. **Possible Causes:** • Incorrect fuel tank vacuum relief valve • Incorrect fuel filler cap used • Fuel filler cap remains open or fails to close. • Foreign matter caught in fuel filler cap. • Leak is in line between intake manifold and EVAP canister purge volume control solenoid valve. • Foreign matter caught in EVAP canister vent control valve. • EVAP canister or fuel tank leaks • EVAP purge line (pipe and rubber tube) leaks • EVAP purge line rubber tube bent • Loose or disconnected rubber tube • EVAP canister vent control valve and the circuit • EVAP canister purge volume control solenoid valve and the circuit • Fuel tank temperature sensor • O-ring of EVAP canister vent control valve is missing or damaged • EVAP canister is saturated with water • EVAP control system pressure sensor • Refueling EVAP vapor cut valve • ORVR system leaks • Fuel level sensor and the circuit • Foreign matter caught in EVAP canister purge volume control solenoid valve
DTC: P0460 **2T CCM, MIL: Yes** **Years: 2007, 2008** **Models:** EX35, FX35, G35, G37, FX45, QX56 **Engines:** 3.5L, 3.7L, 4.5L, 5.6L **Transmissions:** All	**Fuel Level Sensor Circuit Noise** Even though the vehicle is parked, a signal being varied is sent from the fuel level sensor to ECM. **Possible Causes:** • Harness or connectors (CAN communication line is open or shorted) • Harness or connectors (sensor circuit is open or shorted) • Unified meter and A/C amp. • Fuel level sensor
DTC: P0461 **2T CCM, MIL: Yes** **Years: 2007, 2008** **Models:** EX35, FX35, G35, G37, FX45, QX56 **Engines:** 3.5L, 3.7L, 4.5L, 5.6L **Transmissions:** All	**Fuel Level Sensor Range/Performance** The output signal of the fuel level sensor does not change within the specified range even though the vehicle has been driven a long distance. **Possible Causes:** • Harness or connectors (CAN communication line is open or shorted) • Harness or connectors (sensor circuit is open or shorted) • Unified meter and A/C amp. • Fuel level sensor
DTC: P0462 **2T CCM, MIL: Yes** **Years: 2007, 2008** **Models:** EX35, FX35, G35, G37, FX45, QX56 **Engines:** 3.5L, 3.7L, 4.5L, 5.6L **Transmissions:** All	**Fuel Level Sensor Circuit Low Input** An excessively low input signal is sent by the sensor to the ECM. **Possible Causes:** • Harness or connectors (CAN communication line is open or shorted) • Harness or connectors (sensor circuit is open or shorted) • Unified meter and A/C amp. • Fuel level sensor
DTC: P0463 **2T CCM, MIL: Yes** **Years: 2007, 2008** **Models:** EX35, FX35, G35, G37, FX45, QX56 **Engines:** 3.5L, 3.7L, 4.5L, 5.6L **Transmissions:** All	**Fuel Level Sensor Circuit High Input** An excessively high input signal is sent by the sensor to the ECM. **Possible Causes:** • Harness or connectors (CAN communication line is open or shorted) • Harness or connectors (sensor circuit is open or shorted) • Unified meter and A/C amp. • Fuel level sensor
DTC: P0500 **2T CCM, MIL: Yes** **Years: 2007, 2008** **Models:** EX35, FX35, G35, G37, FX45, QX56 **Engines:** 3.5L, 3.7L, 4.5L, 5.6L **Transmissions:** All	**Vehicle Speed Sensor Circuit Malfunction** The vehicle speed signal sent to ECM is almost 0 km/h (0 MPH) even when vehicle is being driven. **Possible Causes:** • Harness or connectors (CAN communication line is open or shorted) • Harness or connectors (vehicle speed signal circuit is open or shorted) • Wheel sensor • Unified meter and A/C amp. • ABS actuator and electric unit (control unit)

DTC	Trouble Code Title, Conditions & Possible Causes
DTC: P0506 **2T CCM, MIL: Yes** **Years: 2007, 2008** **Models:** EX35, FX35, G35, G37, FX45, QX56 **Engines:** 3.5L, 3.7L, 4.5L, 5.6L **Transmissions:** All	**Idle Speed Control System RPM Lower Than Expected** Engine started, engine running at hot idle speed for 30 seconds, and the ECM detected the Idle Speed Control system engine speed was more than 200 RPM lower than the desired (control) idle speed. **Possible Causes:** • Electric throttle control actuator • Intake air leak • PCV system
DTC: P0507 **2T CCM, MIL: Yes** **Years: 2007, 2008** **Models:** EX35, FX35, G35, G37, FX45, QX56 **Engines:** 3.5L, 3.7L, 4.5L, 5.6L **Transmissions:** All	**Idle Speed Control System RPM Higher Than Expected** Engine started, engine running at hot idle speed for 30 seconds, and the ECM detected the Idle Speed Control system engine speed was more than 200 RPM higher than the desired (control) idle speed. **Possible Causes:** • Electric throttle control actuator • Intake air leak • PCV system
DTC: P0524 **2T CCM, MIL: Yes** **Years: 2007, 2008** **Models:** G37 **Engines:** 3.7L **Transmissions:** All	**Engine Oil Pressure Too Low** Engine oil pressure is low because there is a gap between angle of target and phase-control angle. **Possible Causes:** • Engine oil pressure or level too low • Crankshaft position sensor (POS) • Camshaft position sensor (PHASE) • Intake valve control solenoid valve • Accumulation of debris to the signal pick-up portion of the camshaft • Timing chain installation • Foreign matter caught in the oil groove for intake valve timing control
DTC: P0550 **2T CCM, MIL: Yes** **Years: 2007, 2008** **Models:** EX35, FX35, G35, G37, FX45, QX56 **Engines:** 3.5L, 3.7L, 4.5L, 5.6L **Transmissions:** All	**Power Steering Pressure Sensor Circuit Malfunction** An excessively low or high input signal is sent by the sensor to the ECM. **Possible Causes:** • Harness or connectors (sensor circuit is open or shorted) • Power steering pressure sensor

DTC: P0555
2T CCM, MIL: Yes
Years: 2007, 2008
Models: G37
Engines: 3.7L
Transmissions: All
	Brake Booster Pressure Sensor Circuit
	An excessively low voltage from the sensor is sent to ECM.
	Or, an excessively high voltage from the sensor is sent to ECM.
	Possible Causes:
	• Harness or connectors (sensor circuit is open or shorted.)
	  • CKP sensor (POS) circuit is shorted.
	  • APP sensor 2 circuit is shorted
	  • EVAP control system pressure sensor circuit is shorted.
	  • Refrigerant pressure sensor circuit is shorted.
	• Brake booster pressure sensor
	• Crankshaft position sensor (POS)
	• Accelerator pedal position sensor
	• EVAP control system pressure sensor
	• Refrigerant pressure sensor

DTC: P0603 **2T ECM, MIL: No** **Years: 2007, 2008** **Models:** EX35, FX35, G35, G37, FX45, QX56 **Engines:** 3.5L, 3.7L, 4.5L, 5.6L **Transmissions:** All	**ECM Power Supply** ECM back-up RAM system does not function properly. **Possible Causes:** • Harness or connectors [ECM power supply (back-up) circuit is open or shorted.] • ECM
DTC: P0605 **2T ECM, MIL: Yes** **Years: 2007, 2008** **Models:** EX35, FX35, G35, G37, FX45, QX56 **Engines:** 3.5L, 3.7L, 4.5L, 5.6L **Transmissions:** All	**ECM** ECM calculation function is malfunctioning. ECM EEP-ROM system is malfunctioning. ECM self shut-off function is malfunctioning. **Possible Causes:** • ECM

DTC	Trouble Code Title, Conditions & Possible Causes
DTC: P0607 **2T ECM, MIL: Yes** **Years: 2007, 2008** **Models:** EX35 **Engines:** 3.5L **Transmissions:** All	**ECM – CAN Communication Bus** When detecting error during the initial diagnosis of CAN controller of ECM. **Possible Causes:** • ECM

DTC: P0643
2T ECM, MIL: Yes
Years: 2007, 2008
Models: EX35, FX35, G35, G37, FX45, QX56
Engines: 3.5L, 3.7L, 4.5L, 5.6L
Transmissions: All
	Sensor Power Supply Circuit Short
	ECM detects a voltage of power source for sensor is excessively low or high.
	Possible Causes:
	  • Harness or connectors
	    • APP sensor 1 circuit is shorted.
	    • TP sensor circuit is shorted.
	    • CMP sensor (PHASE) (bank 1) circuit is shorted.
	    • EVT control position sensor (bank 1) circuit is shorted.
	    • PSP sensor circuit is shorted.
	  • Accelerator pedal position sensor
	  • Throttle position sensor
	  • Camshaft position sensor (PHASE) - (bank 1)
	  • Exhaust valve timing control position sensor (bank 1)
	  • Power steering pressure sensor

DTC: P0850
2T CCM, MIL: No
Years: 2007, 2008
Models: EX35, FX35, G35, G37, FX45, QX56
Engines: 3.5L, 3.7L, 4.5L, 5.6L
Transmissions: A/T
	PNP Switch
	The signal of the park/neutral position (PNP) signal is not changed in the process of engine starting and driving.
	Possible Causes:
	  • Harness or connectors - park/neutral position (PNP) signal circuit is open or shorted.
	  • TCM

OBD II Trouble Code List (P1xxx Codes)

DTC	Trouble Code Title, Conditions & Possible Causes

DTC: P100A
2T CCM, MIL: Yes
Years: 2007, 2008
Models: G37
Engines: 3.7L
Transmissions: All
	VVEL Response Malfunction (Bank 1)
	Actual event response to target is poor.
	Possible Causes:
	  • Harness or connectors (VVEL actuator motor circuit is open or shorted.)
	  • VVEL actuator motor
	  • VVEL actuator sub assembly
	  • VVEL ladder assembly
	  • VVEL control module

DTC: P100B
2T CCM, MIL: Yes
Years: 2007, 2008
Models: G37
Engines: 3.7L
Transmissions: All
	VVEL Response Malfunction (Bank 2)
	Actual event response to target is poor.
	Possible Causes:
	  • Harness or connectors (VVEL actuator motor circuit is open or shorted.)
	  • VVEL actuator motor
	  • VVEL actuator sub assembly
	  • VVEL ladder assembly
	  • VVEL control module

DTC: P1078
2T CCM, MIL: No
Years: 2007, 2008
Models: EX35, FX35, G35, FX45
Engines: 3.5L, 4.5L
Transmissions: A/T
	Exhaust Valve Timing Control Position Sensor (Bank 1) Circuit Malfunction
	An excessively high or low voltage from the sensor is sent to ECM.
	Possible Causes:
	  • Harness or connectors - EVT control position sensor (bank 1) circuit is open or shorted
	  • Exhaust valve timing control position sensor
	  • Crankshaft position sensor (POS)
	  • Camshaft position sensor (PHASE) (bank 1)
	  • Accumulation of debris to the signal pick-up portion of the camshaft

DTC	Trouble Code Title, Conditions & Possible Causes

DTC: P1084
2T CCM, MIL: No
Years: 2007, 2008
Models: EX35, FX35, G35, FX45
Engines: 3.5L, 4.5L
Transmissions: A/T
	Exhaust Valve Timing Control Position Sensor (Bank 2) Circuit Malfunction
	An excessively high or low voltage from the sensor is sent to ECM.
	Possible Causes:
	• Harness or connectors - EVT control position sensor (bank 2) circuit is open or shorted
	  • CKP sensor (POS) circuit is shorted.
	  • CMP sensor (PHASE) (bank 2) circuit is shorted.
	  • Battery current sensor circuit is shorted.
	  • APP sensor 2 circuit is shorted.
	  • EVAP control system pressure sensor circuit is shorted.
	  • Refrigerant pressure sensor circuit is shorted.
	• Exhaust valve timing control position sensor (bank 2)
	• Crankshaft position sensor (POS)
	• Camshaft position sensor (PHASE) (bank 2)
	• Battery current sensor
	• Accelerator pedal position sensor
	• EVAP control system pressure sensor
	• Refrigerant pressure sensor
	• Accumulation of debris to the signal pick-up portion of the camshaft

DTC: P1087 **2T CCM, MIL: Yes** **Years: 2007, 2008** **Models:** G37 **Engines:** 3.7L **Transmissions:** All	**VVEL Small Event Angle Malfunction (Bank 1)** Event angle of the VVEL control shaft is always small. **Possible Causes:** • Harness or connectors (VVEL actuator motor circuit is open or shorted.) • VVEL actuator motor • VVEL actuator sub assembly • VVEL ladder assembly • VVEL control module
DTC: P1088 **2T CCM, MIL: Yes** **Years: 2007, 2008** **Models:** G37 **Engines:** 3.7L **Transmissions:** All	**VVEL Small Event Angle Malfunction (Bank 2)** Event angle of the VVEL control shaft is always small. **Possible Causes:** • Harness or connectors (VVEL actuator motor circuit is open or shorted.) • VVEL actuator motor • VVEL actuator sub assembly • VVEL ladder assembly • VVEL control module
DTC: P1089 **2T CCM, MIL: Yes** **Years: 2007, 2008** **Models:** G37 **Engines:** 3.7L **Transmissions:** All	**VVEL Control Shaft Position Sensor Circuit (Bank 1)** An excessively low voltage from the sensor is sent to VVEL control module. Or, an excessively high voltage from the sensor is sent to VVEL control module. Or, a rationally incorrect voltage is sent to VVEL control module compared with the signals from VVEL control shaft position sensor 1 and VVEL control shaft position sensor 2. **Possible Causes:** • Harness or connectors (VVEL control shaft position sensor circuit is open or shorted.) • VVEL control shaft position sensor • VVEL control module
DTC: P1090 **2T CCM, MIL: Yes** **Years: 2007, 2008** **Models:** G37 **Engines:** 3.7L **Transmissions:** All	**VVEL System Performance (Bank 1)** Event angle difference between the actual and the target is detected. Or, abnormal current is sent to VVEL actuator motor. **Possible Causes:** • Harness or connectors (VVEL actuator motor circuit is open or shorted.) • VVEL actuator motor • VVEL actuator sub assembly • VVEL ladder assembly • VVEL control module

DTC	Trouble Code Title, Conditions & Possible Causes

DTC: P1091
2T CCM, MIL: Yes
Years: 2007, 2008
Models: G37
Engines: 3.7L
Transmissions: All
	VVEL Actuator Motor Relay Circuit
	VVEL control module detects the VVEL actuator motor relay is stuck OFF. Or, VVEL control module detects the VVEL actuator motor relay is stuck ON.
	Possible Causes:
	  • Harness or connectors (VVEL actuator motor relay circuit is open or shorted.)
	    • Abort circuit is open or shorted.
	  • VVEL actuator motor relay
	  • VVEL control module
	  • ECM

DTC: P1092
2T CCM, MIL: Yes
Years: 2007, 2008
Models: G37
Engines: 3.7L
Transmissions: All
	VVEL Control Shaft Position Sensor Circuit (Bank 2)
	An excessively low voltage from the sensor is sent to VVEL control module. Or, an excessively high voltage from the sensor is sent to VVEL control module. Or, a rationally incorrect voltage is sent to VVEL control module compared with the signals from VVEL control shaft position sensor 1 and VVEL control shaft position sensor 2.
	Possible Causes:
	  • Harness or connectors (VVEL control shaft position sensor circuit is open or shorted.)
	  • VVEL control shaft position sensor
	  • VVEL control module

DTC: P1093
2T CCM, MIL: Yes
Years: 2007, 2008
Models: G37
Engines: 3.7L
Transmissions: All
	VVEL System Performance (Bank 2)
	Event angle difference between the actual and the target is detected.
	Or, abnormal current is sent to VVEL actuator motor.
	Possible Causes:
	  • Harness or connectors (VVEL actuator motor circuit is open or shorted.)
	  • VVEL actuator motor
	  • VVEL actuator sub assembly
	  • VVEL ladder assembly
	  • VVEL control module

DTC: P1140
2T CCM, MIL: Yes
Years: 2007, 2008
Models: FX45, QX56
Engines: 4.5L, 5.6L
Transmissions: All
	IVT Control Position Sensor Circuit (Bank 1)
	An excessively high or low voltage from the sensor is sent to ECM.
	Possible Causes:
	  • Harness or connectors (intake valve timing control position sensor circuit is open or shorted)
	  • Intake valve timing control position sensor
	  • Crankshaft position sensor (POS)
	  • Camshaft position sensor (PHASE)
	  • Accumulation of debris to the signal pick-up portion of the camshaft sprocket

DTC: P1145
2T CCM, MIL: Yes
Years: 2007, 2008
Models: FX45, QX56
Engines: 4.5L, 5.6L
Transmissions: All
	IVT Control Position Sensor Circuit (Bank 2)
	An excessively high or low voltage from the sensor is sent to ECM.
	Possible Causes:
	  • Harness or connectors (intake valve timing control position sensor circuit is open or shorted)
	  • Intake valve timing control position sensor
	  • Crankshaft position sensor (POS)
	  • Camshaft position sensor (PHASE)
	  • Accumulation of debris to the signal pick-up portion of the camshaft sprocket

DTC: P1148
2T CCM, MIL: Yes
Years: 2007, 2008
Models: EX35, FX35, G35, G37, FX45, QX56
Engines: 3.5L, 3.7L, 4.5L, 5.6L
Transmissions: All
	Closed Loop Malfunction Detected (Bank 1)
	The closed loop control function for bank 1 does not operate even when vehicle is driving in the specified condition.
	Possible Causes:
	  • Harness or connectors - The A/F sensor 1 circuit is open or shorted.
	  • A/F sensor 1
	  • A/F sensor 1 heater

DTC: P1168
2T CCM, MIL: Yes
Years: 2007, 2008
Models: EX35, FX35, G35, G37, FX45, QX56
Engines: 3.5L, 3.7L, 4.5L, 5.6L
Transmissions: All
	Closed Loop Malfunction Detected (Bank 2)
	The closed loop control function for bank 1 does not operate even when vehicle is driving in the specified condition.
	Possible Causes:
	  • Harness or connectors - The A/F sensor 1 circuit is open or shorted.
	  • A/F sensor 1
	  • A/F sensor 1 heater

DTC	Trouble Code Title, Conditions & Possible Causes

DTC: P1211
2T CCM, MIL: No
Years: 2007, 2008
Models: EX35, FX35, G35, G37, FX45, QX56
Engines: 3.5L, 3.7L, 4.5L, 5.6L
Transmissions: All
	TCS Control Unit Malfunction
	ECM receives a malfunction information from "ABS actuator and electric unit (control unit)".
	Possible Causes:
	   • ABS actuator and electric unit (control unit)
	   • TCS related parts

DTC: P1212
2T CCM, MIL: No
Years: 2007, 2008
Models: EX35, FX35, G35, G37, FX45, QX56
Engines: 3.5L, 3.7L, 4.5L, 5.6L
Transmissions: All
	TCS Control Unit Malfunction
	ECM can not receive the information from "ABS actuator and electric unit (control unit)" continuously.
	Possible Causes:
	   • Harness or connectors - CAN communication line is open or shorted.
	   • ABS actuator and electric unit (control unit)
	   • Dead (weak) battery

DTC: P1217
1T CCM, MIL: Yes
Years: 2007, 2008
Models: EX35, FX35, G35, G37, FX45, QX56
Engines: 3.5L, 3.7L, 4.5L, 5.6L
Transmissions: All
	Engine Over-Temperature Condition
	Cooling fan does not operate properly (Overheat). Cooling fan system does not operate properly (Overheat). Engine coolant was not added to the system using the proper filling method. Engine coolant is not within the specified range.
	Possible Causes:
	   • Harness or connectors - The cooling fan circuit is open or shorted.
	   • IPDM E/R
	   • Cooling fan control module
	   • Cooling fan motor
	   • Radiator hose
	   • Radiator
	   • Radiator cap
	   • Water pump
	   • Thermostat

DTC: P1225 **2T CCM, MIL: No** **Years:** 2007, 2008 **Models:** EX35, FX35, G35, G37, FX45 **Engines:** 3.5L, 3.7L, 4.5L **Transmissions:** All	**Closed Throttle Position Learning Performance (Bank 1)** Closed throttle position learning value is excessively low. **Possible Causes:** • Electric throttle control actuator (TP sensor 1 and 2)
DTC: P1226 **2T CCM, MIL: No** **Years:** 2007, 2008 **Models:** EX35, FX35, G35, G37, FX45, QX56 **Engines:** 3.5L, 3.7L, 4.5L, 5.6L **Transmissions:** All	**Closed Throttle Position Learning Performance (Bank 1)** Closed throttle position learning value is not performing successfully or repeatedly. **Possible Causes:** • Electric throttle control actuator (TP sensor 1 and 2)

DTC: P1233
1T CCM, MIL: Yes
Years: 2007, 2008
Models: EX35, G37
Engines: 3.5L, 3.7L
Transmissions: All
	Electric Throttle Control Performance (Bank 2)
	Electric throttle control function does not operate properly.
	Possible Causes:
	   • Harness or connectors - Throttle control motor circuit is open or shorted
	   • Electric throttle control actuator

DTC: P1234 **2T CCM, MIL: No** **Years:** 2007, 2008 **Models:** EX35, G37 **Engines:** 3.5L, 3.7L **Transmissions:** All	**Closed Throttle Position Learning Performance (Bank 2)** Closed throttle position learning value is excessively low. **Possible Causes:** • Electric throttle control actuator (TP sensor 1 and 2)
DTC: P1235 **2T CCM, MIL: No** **Years:** 2007, 2008 **Models:** EX35, G37 **Engines:** 3.5L, 3.7L **Transmissions:** All	**Closed Throttle Position Learning Performance (Bank 2)** Closed throttle position learning value is not performing successfully or repeatedly. **Possible Causes:** • Electric throttle control actuator (TP sensor 1 and 2)

DTC	Trouble Code Title, Conditions & Possible Causes
DTC: P1236 **2T CCM, MIL: No** **Years: 2007, 2008** **Models:** EX35, G37 **Engines:** 3.5L, 3.7L **Transmissions:** All	**Throttle Control Motor Short (Bank 2)** ECM detects short in both circuits between ECM and throttle control motor. **Possible Causes:** • Harness or connectors (throttle control motor circuit is shorted) • Electric throttle control actuator (throttle control motor)
DTC: P1238 **2T CCM, MIL: No** **Years: 2007, 2008** **Models:** EX35, G37 **Engines:** 3.5L, 3.7L **Transmissions:** All	**Electric Throttle Control Actuator (Bank 2)** Electric throttle control actuator does not function properly due to the return spring malfunction. Or, Throttle valve opening angle in fail-safe mode is not in specified range. Or, ECM detect the throttle valve is stuck open. **Possible Causes:** • Electric throttle control actuator
DTC: P1239 **2T CCM, MIL: No** **Years: 2007, 2008** **Models:** EX35, G37 **Engines:** 3.5L, 3.7L **Transmissions:** All	**Throttle Position Sensor (Bank 2) Range/Performance** Rationally incorrect voltage is sent to ECM compared with the signals from TP sensor 1 and TP sensor 2. **Possible Causes:** • Harness or connector (TP sensor 1 and 2 circuit is open or shorted.) • Electric throttle control actuator (TP sensor 1 and 2)
DTC: P1290 **2T CCM, MIL: No** **Years: 2007, 2008** **Models:** EX35, G37 **Engines:** 3.5L, 3.7L **Transmissions:** All	**Throttle Control Motor Relay Circuit Open (Bank 2)** ECM detects a voltage of power source for throttle control motor is excessively low. **Possible Causes:** • Harness or connectors (throttle control motor relay circuit is open) • Throttle control motor relay
DTC: P1421 **2T CCM, MIL: No** **Years: 2007, 2008** **Models:** EX35, FX35, G35, G37, FX45, QX56 **Engines:** 3.5L, 3.7L, 4.5L, 5.6L **Transmissions:** All	**Cold Start Control** ECM does not control ignition timing and engine idle speed properly when engine is started with pre-warming up condition. **Possible Causes:** • Lack of intake air volume • Fuel injection system • ECM

DTC: P1550
2T CCM, MIL: No
Years: 2007, 2008
Models: EX35, G37, QX56
Engines: 3.5L, 3.7L, 5.6L
Transmissions: All
	Battery Current Sensor Range/Performance
	The output voltage of the battery current sensor remains within the specified range while engine is running.
	Possible Causes:
	• Harness or connectors (battery current sensor circuit is open or shorted.)
	  • CKP sensor (POS) circuit is shorted.
	  • CMP sensor (PHASE) (bank 2) circuit is shorted.
	  • EVT control position sensor (bank 2) circuit is shorted.
	  • APP sensor 2 circuit is shorted.
	  • EVAP control system pressure sensor circuit is shorted.
	  • Refrigerant pressure sensor circuit is shorted.
	• Battery current sensor
	• Crankshaft position sensor (POS)
	• Camshaft position sensor (PHASE) - (bank 2)
	• Exhaust valve timing control position sensor (bank 2)
	• Accelerator pedal position sensor
	• EVAP control system pressure sensor
	• Refrigerant pressure sensor

DTC	Trouble Code Title, Conditions & Possible Causes

DTC: P1551
2T CCM, MIL: No
Years: 2007, 2008
Models: EX35, G37, QX56
Engines: 3.5L, 3.7L, 5.6L
Transmissions: All
	Battery Current Sensor Low Input
	An excessively low voltage from the sensor is sent to ECM.
	Possible Causes:
	• Harness or connectors (battery current sensor circuit is open or shorted.)
	  • CKP sensor (POS) circuit is shorted.
	  • CMP sensor (PHASE) (bank 2) circuit is shorted.
	  • EVT control position sensor (bank 2) circuit is shorted.
	  • APP sensor 2 circuit is shorted.
	  • EVAP control system pressure sensor circuit is shorted.
	  • Refrigerant pressure sensor circuit is shorted.
	• Battery current sensor
	• Crankshaft position sensor (POS)
	• Camshaft position sensor (PHASE) - (bank 2)
	• Exhaust valve timing control position sensor (bank 2)
	• Accelerator pedal position sensor
	• EVAP control system pressure sensor
	• Refrigerant pressure sensor

DTC: P1552
2T CCM, MIL: No
Years: 2007, 2008
Models: EX35, G37, QX56
Engines: 3.5L, 3.7L, 5.6L
Transmissions: All
	Battery Current Sensor High Input
	An excessively high voltage from the sensor is sent to ECM.
	Possible Causes:
	• Harness or connectors (battery current sensor circuit is open or shorted.)
	  • CKP sensor (POS) circuit is shorted.
	  • CMP sensor (PHASE) (bank 2) circuit is shorted.
	  • EVT control position sensor (bank 2) circuit is shorted.
	  • APP sensor 2 circuit is shorted.
	  • EVAP control system pressure sensor circuit is shorted.
	  • Refrigerant pressure sensor circuit is shorted.
	• Battery current sensor
	• Crankshaft position sensor (POS)
	• Camshaft position sensor (PHASE) - (bank 2)
	• Exhaust valve timing control position sensor (bank 2)
	• Accelerator pedal position sensor
	• EVAP control system pressure sensor
	• Refrigerant pressure sensor

DTC: P1553
2T CCM, MIL: No
Years: 2007, 2008
Models: EX35, G37, QX56
Engines: 3.5L, 3.7L, 5.6L
Transmissions: All
	Battery Current Sensor Performance – High Output
	The signal voltage transmitted from the sensor to ECM is higher than the amount of the maximum power generation.
	Possible Causes:
	• Harness or connectors (battery current sensor circuit is open or shorted.)
	  • CKP sensor (POS) circuit is shorted.
	  • CMP sensor (PHASE) (bank 2) circuit is shorted.
	  • EVT control position sensor (bank 2) circuit is shorted.
	  • APP sensor 2 circuit is shorted.
	  • EVAP control system pressure sensor circuit is shorted.
	  • Refrigerant pressure sensor circuit is shorted.
	• Battery current sensor
	• Crankshaft position sensor (POS)
	• Camshaft position sensor (PHASE) - (bank 2)
	• Exhaust valve timing control position sensor (bank 2)
	• Accelerator pedal position sensor
	• EVAP control system pressure sensor
	• Refrigerant pressure sensor

DTC	Trouble Code Title, Conditions & Possible Causes
DTC: P1554 **2T CCM, MIL: No** **Years:** 2007, 2008 **Models:** EX35, G37, QX56 **Engines:** 3.5L, 3.7L, 5.6L **Transmissions:** All	**Battery Current Sensor Performance – Low Output** The signal voltage transmitted from the sensor to ECM is lower than the amount of the maximum power generation. **Possible Causes:** • Harness or connectors (battery current sensor circuit is open or shorted.) • CKP sensor (POS) circuit is shorted. • CMP sensor (PHASE) (bank 2) circuit is shorted. • EVT control position sensor (bank 2) circuit is shorted. • APP sensor 2 circuit is shorted. • EVAP control system pressure sensor circuit is shorted. • Refrigerant pressure sensor circuit is shorted. • Battery current sensor • Crankshaft position sensor (POS) • Camshaft position sensor (PHASE) - (bank 2) • Exhaust valve timing control position sensor (bank 2) • Accelerator pedal position sensor • EVAP control system pressure sensor • Refrigerant pressure sensor
DTC: P1564 **1T CCM, MIL: No** **Years:** 2007, 2008 **Models:** EX35, FX35, G35, G37, FX45, QX56 **Engines:** 3.5L, 3.7L, 4.5L, 5.6L **Transmissions:** All	**ASCD Steering Switch Circuit Malfunction** An excessively high voltage signal from the ASCD steering switch is sent to ECM. Or, ECM detects that input signal from the ASCD steering switch is out of the specified range. Or, ECM detects that the ASCD steering switch is stuck ON. **Possible Causes:** • Harness or connectors (switch circuit is open or shorted) • ASCD steering switch • ECM
DTC: P1564 **1T CCM, MIL: No** **Years:** 2007, 2008 **Models:** EX35, FX35, G35, G37, FX45, QX56 **Engines:** 3.5L, 3.7L, 4.5L, 5.6L **Transmissions:** All	**ICC Steering Switch Circuit Malfunction** An excessively high voltage signal from the ICC steering switch is sent to ECM. Or, ECM detects that input signal from the ICC steering switch is out of the specified range. Or, ECM detects that the ICC steering switch is stuck ON. **Possible Causes:** • Harness or connectors (switch circuit is open or shorted) • ICC steering switch • ECM
DTC: P1568 **1T CCM, MIL: No** **Years:** 2007, 2008 **Models:** EX35, FX35, G37, FX45, QX56 **Engines:** 3.5L, 3.7L, 4.5L, 5.6L **Transmissions:** All	**ICC Function** ECM detects a difference between signals from ICC sensor integrated unit is out of specified range. **Possible Causes:** • Harness or connectors (CAN communication line is open or shorted.) • ICC sensor integrated unit • ECM
DTC: P1572 **1T CCM, MIL: No** **Years:** 2007, 2008 **Models:** EX35, FX35, G37, FX45, QX56 **Engines:** 3.5L, 3.7L, 4.5L, 5.6L **Transmissions:** All	**ASCD Brake Switch Circuit Malfunction** When the vehicle speed is above 19 MPH, ON signals from the stop lamp switch and the ASCD brake switch are sent to the ECM at the same time. Or, ASCD brake switch signal is not sent to ECM for extremely long time while the vehicle is driving. **Possible Causes:** • Harness or connectors (stop lamp switch circuit is shorted.) • Harness or connectors (ASCD brake switch circuit is shorted.) • Stop lamp switch • ASCD brake switch • Incorrect stop lamp switch installation • Incorrect ASCD brake switch installation • ECM
DTC: P1572 **1T CCM, MIL: No** **Years:** 2007, 2008 **Models:** EX35, FX35, G37, FX45, QX56 **Engines:** 3.5L, 3.7L, 4.5L, 5.6L **Transmissions:** All	**ICC Brake Switch Circuit Malfunction** When the vehicle speed is above 19 MPH, ON signals from the stop lamp switch and the ICC brake switch are sent to the ECM at the same time. Or, ASCD brake switch signal is not sent to ECM for extremely long time while the vehicle is driving. **Possible Causes:** • Harness or connectors (stop lamp switch circuit is shorted.) • Harness or connectors (ICC brake switch circuit is shorted.) • Stop lamp switch • ICC brake switch • Incorrect stop lamp switch installation • Incorrect ICC brake switch installation • ECM

DTC	Trouble Code Title, Conditions & Possible Causes
DTC: P1574 **1T CCM, MIL: No** **Years: 2007, 2008** **Models:** EX35, FX35, G37, FX45, QX56 **Engines:** 3.5L, 3.7L, 4.5L, 5.6L **Transmissions:** All	**ASCD Vehicle Speed Sensor Signal Malfunction** ECM detects a difference between two vehicle speed signals is out of the specified range. **Possible Causes:** • Harness or connectors (CAN communication line is open or shorted.) • Unified meter and A/C amp. • ABS actuator and electric unit (control unit) • Wheel sensor • TCM • ECM
DTC: P1574 **1T CCM, MIL: No** **Years: 2007, 2008** **Models:** EX35, FX35, G37, FX45, QX56 **Engines:** 3.5L, 3.7L, 4.5L, 5.6L **Transmissions:** All	**ICC Vehicle Speed Sensor Signal Malfunction** ECM detects a difference between two vehicle speed signals is out of the specified range. **Possible Causes:** • Harness or connectors (CAN communication line is open or shorted.) • Unified meter and A/C amp. • ABS actuator and electric unit (control unit) • Wheel sensor • TCM • ECM
DTC: P1575 **1T CCM, MIL: No** **Years: 2007, 2008** **Models:** EX35 **Engines:** 3.5L **Transmissions:** All	**Input Speed Sensor (Turbine Revolution Sensor)** Turbine revolution sensor signal is different from the theoretical value calculated by ECM from revolution sensor signal and engine RPM signal. **Possible Causes:** • Harness or connectors (CAN communication line is open or shorted) • Harness or connectors (Turbine revolution sensor circuit is open or shorted) • TCM
DTC: P1606 **1T CCM, MIL: No** **Years: 2007, 2008** **Models:** EX35 **Engines:** 3.5L **Transmissions:** All	**VVEL Control Module** VVEL control module calculation function is malfunctioning. Or, VVEL EEP-ROM system is malfunctioning. **Possible Causes:** • VVEL control module
DTC: P1607 **1T CCM, MIL: No** **Years: 2007, 2008** **Models:** EX35 **Engines:** 3.5L **Transmissions:** All	**VVEL Control Module** VVEL control module internal circuit is malfunctioning. **Possible Causes:** • VVEL control module
DTC: P1608 **1T CCM, MIL: No** **Years: 2007, 2008** **Models:** EX35 **Engines:** 3.5L **Transmissions:** All	**VVEL Sensor Power Supply Circuit** VVEL control module detects a voltage of power source for sensor is excessively low or high. **Possible Causes:** • Harness or connectors (VVEL control shaft position sensor power supply circuit is open or shorted.) • VVEL control shaft position sensor • VVEL control module
DTC: P1715 **1T CCM, MIL: No** **Years: 2007, 2008** **Models:** EX35, FX35, G35, G37, FX45 **Engines:** 3.5L, 3.7L, 4.5L **Transmissions:** All	**MAF Sensor Circuit Range/Performance** A high voltage from the sensor is sent to ECM under light load driving condition. Or, a low voltage from the sensor is sent to ECM under heavy load driving condition. **Possible Causes:** • Harness or connectors (mass air flow sensor circuit is open or shorted.) • Intake air leaks • Mass air flow sensor • EVAP control system pressure sensor • Intake air temperature sensor
DTC: P1800 **1T CCM, MIL: No** **Years: 2007, 2008** **Models:** FX45 **Engines:** 4.5L **Transmissions:** All	**VIAS Control Solenoid Valve Circuit** An excessively low or high voltage signal is sent to ECM through the valve **Possible Causes:** • Harness or connectors (solenoid valve circuit is open or shorted.) • VIAS control solenoid valve

DTC	Trouble Code Title, Conditions & Possible Causes
DTC: P1805 **2T CCM, MIL: Yes** **Years:** 2007, 2008 **Models:** EX35, FX35, G35, G37, FX45, QX56 **Engines:** 3.5L, 3.7L, 4.5L, 5.6L **Transmissions:** All	**Brake Switch Circuit Malfunction** A brake switch signal is not sent to ECM for extremely long time while the vehicle is driving. **Possible Causes:** • Harness or connectors (sop lamp switch circuit is open or shorted.) • Stop lamp switch

OBD II Trouble Code List (P2xxx Codes)

DTC	Trouble Code Title, Conditions & Possible Causes
DTC: P2100 **2T CCM, MIL: No** **Years:** 2007, 2008 **Models:** EX35, FX35, G35, G37, FX45, QX56 **Engines:** 3.5L, 3.7L, 4.5L, 5.6L **Transmissions:** All	**Throttle Control Motor Relay Circuit Open (Bank 1)** ECM detects a voltage of power source for throttle control motor is excessively low. **Possible Causes:** • Harness or connectors (throttle control motor relay circuit is open) • Throttle control motor relay
DTC: P2101 **1T CCM, MIL: Yes** **Years:** 2007, 2008 **Models:** EX35, FX35, G35, G37, FX45, QX56 **Engines:** 3.5L, 3.7L, 4.5L, 5.6L **Transmissions:** All	**Electric Throttle Control Performance (Bank 1)** Electric throttle control function does not operate properly. **Possible Causes:** • Harness or connectors - Throttle control motor circuit is open or shorted • Electric throttle control actuator
DTC: P2103 **2T CCM, MIL: No** **Years:** 2007, 2008 **Models:** EX35, FX35, G35, G37, FX45, QX56 **Engines:** 3.5L, 3.7L, 4.5L, 5.6L **Transmissions:** All	**Throttle Control Motor Relay Short** ECM detects throttle control motor relay is stuck ON. **Possible Causes:** • Harness or connectors (throttle control motor relay circuit is shorted) • Throttle control motor relay
DTC: P2118 **2T CCM, MIL: No** **Years:** 2007, 2008 **Models:** EX35, FX35, G35, G37, FX45, QX56 **Engines:** 3.5L, 3.7L, 4.5L, 5.6L **Transmissions:** All	**Throttle Control Motor Short (Bank 1)** ECM detects short in both circuits between ECM and throttle control motor. **Possible Causes:** • Harness or connectors (throttle control motor circuit is shorted) • Electric throttle control actuator (throttle control motor)
DTC: P2119 **2T CCM, MIL: No** **Years:** 2007, 2008 **Models:** EX35, FX35, G35, G37, FX45, QX56 **Engines:** 3.5L, 3.7L, 4.5L, 5.6L **Transmissions:** All	**Electric Throttle Control Actuator (Bank 1)** Electric throttle control actuator does not function properly due to the return spring malfunction. Or, Throttle valve opening angle in fail-safe mode is not in specified range. Or, ECM detect the throttle valve is stuck open. **Possible Causes:** • Electric throttle control actuator
DTC: P2122 **1T CCM, MIL: Yes** **Years:** 2007, 2008 **Models:** EX35, FX35, G35, G37, FX45, QX56 **Engines:** 3.5L, 3.7L, 4.5L, 5.6L **Transmissions:** All	**Accelerator Pedal Position (APP) Sensor 1 Signal Low Input** An excessively low voltage from the APP sensor 1 is sent to ECM. **Possible Causes:** • Harness or connectors (APP sensor 1 circuit is open or shorted.) • Accelerator pedal position sensor (APP sensor 1)
DTC: P2123 **1T CCM, MIL: Yes** **Years:** 2007, 2008 **Models:** EX35, FX35, G35, G37, FX45, QX56 **Engines:** 3.5L, 3.7L, 4.5L, 5.6L **Transmissions:** All	**Accelerator Pedal Position (APP) Sensor 1 Signal High Input** An excessively high voltage from the APP sensor 1 is sent to ECM. **Possible Causes:** • Harness or connectors (APP sensor 1 circuit is open or shorted.) • Accelerator pedal position sensor (APP sensor 1)

DTC	Trouble Code Title, Conditions & Possible Causes
DTC: P2127 **1T CCM, MIL: Yes** **Years: 2007, 2008** **Models:** EX35, FX35, G35, G37, FX45, QX56 **Engines:** 3.5L, 3.7L, 4.5L, 5.6L **Transmissions:** All	**Accelerator Pedal Position (APP) Sensor 2 Signal Low Input** An excessively low voltage from the APP sensor 2 is sent to ECM. **Possible Causes:** • Harness or connectors (APP sensor 2 circuit is open or shorted.) • Accelerator pedal position sensor (APP sensor 2)
DTC: P2128 **1T CCM, MIL: Yes** **Years: 2007, 2008** **Models:** EX35, FX35, G35, G37, FX45, QX56 **Engines:** 3.5L, 3.7L, 4.5L, 5.6L **Transmissions:** All	**Accelerator Pedal Position (APP) Sensor 2 Signal High Input** An excessively high voltage from the APP sensor 2 is sent to ECM. **Possible Causes:** • Harness or connectors (APP sensor 2 circuit is open or shorted.) • Accelerator pedal position sensor (APP sensor 2)
DTC: P2132 **1T CCM, MIL: Yes** **Years: 2007, 2008** **Models:** EX35, G37 **Engines:** 3.5L, 3.7L **Transmissions:** All	**Throttle Position Sensor 1 (Bank 2) Signal Low Input** An excessively low input signal is sent from the sensor to the ECM. **Possible Causes:** • Harness or connectors (circuit is open or shorted • Electric throttle control actuator (TP sensor 1)
DTC: P2133 **1T CCM, MIL: Yes** **Years: 2007, 2008** **Models:** EX35, G37 **Engines:** 3.5L, 3.7L **Transmissions:** All	**Throttle Position Sensor 1 (Bank 2) Signal High Input** An excessively high input signal is sent from the sensor to the ECM. **Possible Causes:** • Harness or connectors (circuit is open or shorted • Electric throttle control actuator (TP sensor 1)
DTC: P2135 **2T CCM, MIL: No** **Years: 2007, 2008** **Models:** EX35, FX35, G35, G37, FX45, QX56 **Engines:** 3.5L, 3.7L, 4.5L, 5.6L **Transmissions:** All	**Throttle Position Sensor (Bank 1) Range/Performance** Rationally incorrect voltage is sent to ECM compared with the signals from TP sensor 1 and TP sensor 2. **Possible Causes:** • Harness or connector (TP sensor 1 and 2 circuit is open or shorted.) • Electric throttle control actuator (TP sensor 1 and 2)

DTC: P2138
1T CCM, MIL: Yes
Years: 2007, 2008
Models: EX35, FX35, G35, G37, FX45, QX56
Engines: 3.5L, 3.7L, 4.5L, 5.6L
Transmissions: All
	Accelerator Pedal Position (APP) Sensor Circuit Range/Performance
	Rationally incorrect voltage is sent to ECM compared with the signals from APP sensor 1 and APP sensor 2.
	Possible Causes:
	• Harness or connectors (APP sensor 2 circuit is open or shorted.)
	  • CKP sensor (POS) circuit is shorted.
	  • CMP sensor (PHASE) (bank 2) circuit is shorted.
	  • EVT control position sensor (bank 2) circuit is shorted.
	  • Battery current sensor circuit is shorted.
	  • EVAP control system pressure sensor circuit is shorted.
	  • Refrigerant pressure sensor circuit is shorted.
	• Accelerator pedal position sensor (APP sensor 2)
	• Crankshaft position sensor (POS)
	• Camshaft position sensor (PHASE) - (bank 2)
	• Exhaust valve timing control position sensor (bank 2)
	• Battery current sensor
	• EVAP control system pressure sensor
	• Refrigerant pressure sensor

DTC: P2A00 **1T CCM, MIL: Yes** **Years: 2007, 2008** **Models:** EX35, FX35, G35, G37, FX45, QX56 **Engines:** 3.5L, 3.7L, 4.5L, 5.6L **Transmissions:** All	**A/F Sensor 1 (Bank 1) Circuit Range/Performance** The output voltage computed by ECM from the A/F sensor 1 signal is shifted to the lean side for a specified period. Or, the A/F signal computed by ECM from the A/F sensor 1 signal is shifted to the rich side for a specified period. **Possible Causes:** • A/F sensor 1 • A/F sensor 1 heater • Fuel pressure • Fuel injector • Intake air leaks

DTC	Trouble Code Title, Conditions & Possible Causes
DTC: P2A03 **1T CCM, MIL: Yes** **Years: 2007, 2008** **Models:** EX35, FX35, G35, G37, FX45, QX56 **Engines:** 3.5L, 3.7L, 4.5L, 5.6L **Transmissions:** All	**A/F Sensor 1 (Bank 2) Circuit Range/Performance** The output voltage computed by ECM from the A/F sensor 1 signal is shifted to the lean side for a specified period. Or, the A/F signal computed by ECM from the A/F sensor 1 signal is shifted to the rich side for a specified period. **Possible Causes:** • A/F sensor 1 • A/F sensor 1 heater • Fuel pressure • Fuel injector • Intake air leaks

OBD II Trouble Code List (Uxxxx Codes)

DTC	Trouble Code Title, Conditions & Possible Causes
DTC: U1000 **1T CCM, MIL: Yes** **Years: 2007, 2008** **Models:** EX35, FX35, G35, FX45, QX56 **Engines:** 3.5L, 4.5L, 5.6L **Transmissions:** All	**Controller Area Network (CAN) Line Malfunction** When ECM is not transmitting or receiving CAN communication signal of OBD (emission related diagnosis) for 2 seconds or more. **Possible Causes:** • Harness or connectors (CAN communication line is open or shorted
DTC: U1001 **1T CCM, MIL: Yes** **Years: 2007, 2008** **Models:** EX35, FX35, G35, FX45, QX56 **Engines:** 3.5L, 4.5L, 5.6L **Transmissions:** All	**Controller Area Network (CAN) Line Malfunction** When ECM is not transmitting or receiving CAN communication signal other than OBD (emission related diagnosis) for 2 seconds or more. **Possible Causes:** • Harness or connectors (CAN communication line is open or shorted
DTC: U1003 **1T CCM, MIL: Yes** **Years: 2007, 2008** **Models:** G37 **Engines:** 3.7L **Transmissions:** All	**Controller Area Network (CAN) Communication Circuit Lost Communication** CAN communication signal other than OBD (emission related diagnosis) is not received between VVEL control module and ECM for 2 seconds or more. **Possible Causes:** • Harness or connectors (VVEL CAN communication line is open or shorted) • ECM • VVEL control module
DTC: U1010 **1T CCM, MIL: Yes** **Years: 2007, 2008** **Models:** EX35, FX35, G35, FX45, QX56 **Engines:** 3.5L, 4.5L, 5.6L **Transmissions:** All	**Controller Area Network (CAN) Bus** When detecting error during the initial diagnosis of CAN controller of ECM. **Possible Causes:** • ECM
DTC: U1011 **1T CCM, MIL: Yes** **Years: 2007, 2008** **Models:** G37 **Engines:** 3.7L **Transmissions:** All	**VVEL CAN Controller** When detecting error during the initial diagnosis of VVEL CAN controller of ECM. **Possible Causes:** • ECM
DTC: U1013 **1T CCM, MIL: Yes** **Years: 2007, 2008** **Models:** G37 **Engines:** 3.7L **Transmissions:** All	**Controller Area Network (CAN) Communication Circuit Lost Communication** CAN communication signal of OBD (emission related diagnosis) is not received between VVEL control module and ECM for 2 seconds or more. **Possible Causes:** • Harness or connectors (VVEL CAN communication line is open or shorted) • ECM • VVEL control module

DTC	Trouble Code Title, Conditions & Possible Causes

DTC: U1024
1T CCM, MIL: Yes
Years: 2007, 2008
Models: G37
Engines: 3.7L
Transmissions: All
	VVEL CAN Communication
	When VVEL control module cannot transmit/receive can communication signal from ECM.
	When detecting error during the initial diagnosis of CAN controller of VVEL control module.
	Possible Causes:
	    • Harness or connectors (VVEL CAN communication line is open or shorted)
	    • ECM
	    • VVEL control module

NISSAN

Diagnostic Trouble Codes

DIAGNOSTIC TROUBLE CODES

OBD II VEHICLE APPLICATIONS

NISSAN

350ZX
2007–2008
- 3.5L V6....................VIN 6

Altima
2007–2008
- 1.8L, 2.0L, 2.5L I4 ....VIN 1
- 3.5L V6....................VIN A

Armada
2007–2008
- 5.6L V8....................VIN 8

Frontier
2007–2008
- 1.8L, 2.0L, 2.5L I4 ....VIN 1
- 4.0L V8....................VIN 6

Maxima
2007–2008
- 3.5L V6....................VIN A

Murano
2007–2008
- 3.5L V6....................VIN A

Pathfinder
2007–2008
- 4.0L V8....................VIN A
- 5.6L V8....................VIN A

Quest
2008
- 3.5L V6....................VIN B

Rogue, Sentra
2008
- 1.8L, 2.0L, 2.5L I4 ....VIN 1

Sentra
2007–2008
- 2.0L I4 ....VIN A
- 1.8L, 2.0L, 2.5L I4 ....VIN C

Titan
2007–2008
- 5.6L V8....................VIN A

Versa
2007–2008
- 1.8L I4 ....VIN B

Xterra
2007–2008
- 4.0L V8....................VIN 6

OBD II Trouble Code List (P0xxx Codes)

DTC	Trouble Code Title, Conditions & Possible Causes
DTC: P0011 **1T PCM, MIL: Yes** **Years: 2007, 2008** **Models:** 350Z, Altima, Armada, Frontier, Maxima, Murano, Pathfinder, Quest, Rogue, Sentra, Titan, Versa, Xterra **Engines:** All **Transmissions:** All	**Intake Valve Timing Control Performance** Engine running at idle, no load. Gear in P or N. PCM determined there is a gap between angle of target and phase-control angle degree. **Possible Causes:** • Crankshaft position sensor (POS) • Camshaft position sensor (PHASE) • Intake valve timing control solenoid valve • Accumulation of debris to the signal pick-up portion of the camshaft • Timing chain installation • Foreign matter caught in the oil groove for intake valve timing control
DTC: P0014 **1T PCM, MIL: Yes** **Years: 2007, 2008** **Models:** 350Z **Engines:** 3.5L **Transmissions:** All	**Intake Valve Timing Control Performance** Engine running at idle, no load. PCM determined there is a gap between angle of target and phase-control angle degree. **Possible Causes:** • Crankshaft position sensor (POS) • Camshaft position sensor (PHASE) • Exhaust valve timing control position sensor • Exhaust valve timing control magnet retarder • Accumulation of debris to the signal pick-up portion of the camshaft • Timing chain installation • Exhaust valve timing control pulley assembly
DTC: P0021 **1T PCM, MIL: Yes** **Years: 2007, 2008** **Models:** 350Z, Altima, Armada, Frontier, Maxima, Murano, Pathfinder, Quest, Titan, Xterra **Engines:** 3.5L, 4.0L, 5.6L **Transmissions:** All	**Intake Valve Timing Control Performance** Engine running at idle, no load. PCM determined there is a gap between angle of target and phase-control angle degree. **Possible Causes:** • Crankshaft position sensor (POS) • Camshaft position sensor (PHASE) • Intake valve timing control solenoid valve • Accumulation of debris to the signal pick-up portion of the camshaft • Timing chain installation • Foreign matter caught in the oil groove for intake valve timing control
DTC: P0024 **1T PCM, MIL: Yes** **Years: 2007, 2008** **Models:** 350Z **Engines:** 3.5L **Transmissions:** All	**Intake Valve Timing Control Performance** Engine running at idle, no load. PCM determined there is a gap between angle of target and phase-control angle degree. **Possible Causes:** • Crankshaft position sensor (POS) • Camshaft position sensor (PHASE) • Exhaust valve timing control position sensor • Exhaust valve timing control magnet retarder • Accumulation of debris to the signal pick-up portion of the camshaft • Timing chain installation • Exhaust valve timing control pulley assembly
DTC: P0031 **1T PCM, MIL: Yes** **Years: 2007, 2008** **Models:** 350Z, Altima, Armada, Frontier, Maxima, Murano, Pathfinder, Quest, Rogue, Sentra, Titan, Versa, Xterra **Engines:** All **Transmissions:** All	**A/F Sensor 1 Heater Control Circuit Low Input** Engine at idle. The current amperage in the air fuel ratio (A/F) sensor 1 heater circuit is out of the normal range. (An excessively low voltage signal is sent to ECM through the A/F sensor 1 heater.) **Possible Causes:** • Harness or connectors (The A/F sensor 1 heater circuit is open or shorted.) • Air fuel ratio (A/F) sensor 1 heater
DTC: P0032 **1T PCM, MIL: Yes** **Years: 2007, 2008** **Models:** 350Z, Altima, Armada, Frontier, Maxima, Murano, Pathfinder, Quest, Rogue, Sentra, Titan, Versa, Xterra **Engines:** All **Transmissions:** All	**A/F Sensor 1 Heater Control Circuit High Input** Engine at idle. The current amperage in the air fuel ratio (A/F) sensor 1 heater circuit is out of the normal range. (An excessively high voltage signal is sent to ECM through the A/F sensor 1 heater.) **Possible Causes:** • Harness or connectors (The A/F sensor 1 heater circuit is shorted.) • Air fuel ratio (A/F) sensor 1 heater

DTC	Trouble Code Title, Conditions & Possible Causes
DTC: P0037 **1T PCM, MIL: Yes** **Years: 2007, 2008** **Models:** 350Z, Altima, Armada, Frontier, Maxima, Murano, Pathfinder, Quest, Rogue, Sentra, Titan, Versa, Xterra **Engines:** All **Transmissions:** All	**HO2S Heater Bank 1 Control Circuit Low** Engine speed: Below 3,600 RPM after the following conditions are met. Engine at normal operating temperature. Keeping the engine speed between 3,500 and 4,000 RPM for 1 minute and at idle for 1 minute under no load. Heater should be ON. An excessively low voltage signal is sent to ECM through the HO2S2 heater. **Possible Causes:** • Harness or connectors (The HO2S2 heater circuit is open or shorted.) • HO2S2 heater
DTC: P0038 **1T PCM, MIL: Yes** **Years: 2007, 2008** **Models:** 350Z, Altima, Armada, Frontier, Maxima, Murano, Pathfinder, Quest, Rogue, Sentra, Titan, Versa, Xterra **Engines:** All **Transmissions:** All	**HO2S Heater Bank 1 Control Circuit High** Engine speed: Below 3,600 RPM after the following conditions are met. Engine at normal operating temperature. Keeping the engine speed between 3,500 and 4,000 RPM for 1 minute and at idle for 1 minute under no load. Heater should be ON. An excessively high voltage signal is sent to ECM through the HO2S2 heater. **Possible Causes:** • Harness or connectors (The HO2S2 heater circuit is shorted.) • HO2S2 heater
DTC: P0043 **1T PCM, MIL: Yes** **Years: 2007, 2008** **Models:** Altima **Engines:** 2.5L (CA) **Transmissions:** All	**HO2S-3 Heater Control Circuit Low** Engine speed: Below 3,600 RPM after the following conditions are met. Engine at normal operating temperature. Keeping the engine speed between 3,500 and 4,000 RPM for 1 minute and at idle for 1 minute under no load. Heater should be ON. An excessively low voltage signal is sent to ECM through the HO2S2 heater. **Possible Causes:** • Harness or connectors (heated oxygen sensor 3 heater circuit is open or shorted.) • Heated oxygen sensor 3 heater
DTC: P0044 **1T PCM, MIL: Yes** **Years: 2007, 2008** **Models:** Altima **Engines:** 2.5L (CA) **Transmissions:** All	**HO2S-3 Heater Control Circuit High** Engine speed: Below 3,600 RPM after the following conditions are met. Engine at normal operating temperature. Keeping the engine speed between 3,500 and 4,000 RPM for 1 minute and at idle for 1 minute under no load. Heater should be ON. An excessively high voltage signal is sent to ECM through the HO2S2 heater. **Possible Causes:** • Harness or connectors (heated oxygen sensor 3 heater circuit is shorted.) • Heated oxygen sensor 3 heater
DTC: P0051 **1T PCM, MIL: Yes** **Years: 2007, 2008** **Models:** 350Z, Altima, Armada, Frontier, Maxima, Murano, Pathfinder, Quest, Titan, Xterra **Engines:** 3.5L, 4.0L, 5.6L **Transmissions:** All	**A/F Sensor 1 Heater Control Circuit Low Input** Engine at idle. The current amperage in the air fuel ratio (A/F) sensor 1 heater circuit is out of the normal range. (An excessively low voltage signal is sent to ECM through the A/F sensor 1 heater.) **Possible Causes:** • Harness or connectors (The A/F sensor 1 heater circuit is open or shorted.) • Air fuel ratio (A/F) sensor 1 heater
DTC: P0052 **1T PCM, MIL: Yes** **Years: 2007, 2008** **Models:** 350Z, Altima, Armada, Frontier, Maxima, Murano, Pathfinder, Quest, Titan, Xterra **Engines:** 3.5L, 4.0L, 5.6L **Transmissions:** All	**A/F Sensor 1 Heater Control Circuit High Input** Engine at idle. The current amperage in the air fuel ratio (A/F) sensor 1 heater circuit is out of the normal range. (An excessively high voltage signal is sent to ECM through the A/F sensor 1 heater.) **Possible Causes:** • Harness or connectors (The A/F sensor 1 heater circuit is shorted.) • Air fuel ratio (A/F) sensor 1 heater
DTC: P0057 **1T PCM, MIL: Yes** **Years: 2007, 2008** **Models:** 350Z, Altima, Armada, Frontier, Maxima, Murano, Pathfinder, Quest, Titan, Xterra **Engines:** 3.5L, 4.0L, 5.6L **Transmissions:** All	**HO2S Heater Bank 2 Control Circuit Low** Engine speed: Below 3,600 RPM after the following conditions are met. Engine at normal operating temperature. Keeping the engine speed between 3,500 and 4,000 RPM for 1 minute and at idle for 1 minute under no load. Heater should be ON. **Possible Causes:** • Harness or connectors (The HO2S2 heater circuit is open or shorted.) • HO2S2 heater

DTC	Trouble Code Title, Conditions & Possible Causes
DTC: P0058 **1T PCM, MIL: Yes** **Years: 2007, 2008** **Models:** 350Z, Altima, Armada, Frontier, Maxima, Murano, Pathfinder, Quest, Titan, Xterra **Engines:** 3.5L, 4.0L, 5.6L **Transmissions:** All	**HO2S Heater Bank 2 Control Circuit High** Engine speed: Below 3,600 RPM after the following conditions are met. Engine at normal operating temperature. Keeping the engine speed between 3,500 and 4,000 RPM for 1 minute and at idle for 1 minute under no load. Heater should be ON. An excessively high voltage signal is sent to ECM through the HO2S2 heater. **Possible Causes:** • Harness or connectors (The HO2S2 heater circuit is shorted.) • HO2S2 heater
DTC: P0075 **1T PCM, MIL: Yes** **Years: 2007, 2008** **Models:** 350Z, Altima, Armada, Frontier, Maxima, Murano, Pathfinder, Quest, Titan, Xterra **Engines:** 3.5L, 4.0L, 5.6L **Transmissions:** All	**IVT Control Solenoid Bank 1 Improper Voltage Signal** Shift lever: P or N (A/T), Neutral (M/T). Air conditioner switch: OFF. No load. At idle, should be 0 to 2 %; at 2000 RPM, should be 0 to 50 %. An improper voltage is sent to the ECM through intake valve timing control solenoid valve. **Possible Causes:** • Harness or connectors (The intake valve timing control solenoid valve circuit is open or shorted.) • Intake valve timing control solenoid valve
DTC: P0075 **1T PCM, MIL: Yes** **Years: 2007, 2008** **Models:** Altima, Frontier, Rogue, Sentra, Versa **Engines:** 1.8L, 2.0L, 2.5L **Transmissions:** All	**IVT Control Solenoid Improper Voltage Signal** Shift lever: P or N (A/T), Neutral (M/T). Air conditioner switch: OFF. No load. At idle, should be 0 to 2 %; at 2000 RPM, should be 0 to 50 %. An improper voltage is sent to the ECM through intake valve timing control solenoid valve. **Possible Causes:** • Harness or connectors (The intake valve timing control solenoid valve circuit is open or shorted.) • Intake valve timing control solenoid valve
DTC: P0078 **1T PCM, MIL: Yes** **Years: 2007, 2008** **Models:** 350Z **Engines:** 3.5L **Transmissions:** All	**IVT Control Solenoid Bank 1 Improper Voltage Signal** Shift lever: P or N (A/T), Neutral (M/T). Air conditioner switch: OFF. No load. At idle, should be 0 to 2 %; at 2000 RPM, should be 0 to 50 %. An improper voltage is sent to the ECM through exhaust valve timing control magnet retarder. **Possible Causes:** • Harness or connectors (The intake valve timing control magnet retarder circuit is open or shorted.) • Exhaust valve timing control magnet retarder
DTC: P0081 **1T PCM, MIL: Yes** **Years: 2007, 2008** **Models:** 350Z, Altima, Armada, Frontier, Maxima, Murano, Pathfinder, Quest, Titan, Xterra **Engines:** 3.5L, 4.0L, 5.6L **Transmissions:** All	**IVT Control Solenoid Bank 2 Improper Voltage Signal** Shift lever: P or N (A/T), Neutral (M/T). Air conditioner switch: OFF. No load. At idle, should be 0 to 2 %; at 2000 RPM, should be 0 to 50 %. An improper voltage is sent to the ECM through intake valve timing control solenoid valve. **Possible Causes:** • Harness or connectors (The intake valve timing control solenoid valve circuit is open or shorted.) • Intake valve timing control solenoid valve
DTC: P0085 **1T PCM, MIL: Yes** **Years: 2007, 2008** **Models:** 350Z **Engines:** 3.5L **Transmissions:** All	**IVT Control Solenoid Bank 2 Improper Voltage Signal** Shift lever: P or N (A/T), Neutral (M/T). Air conditioner switch: OFF. No load. At idle, should be 0 to 2 %; at 2000 RPM, should be 0 to 50 %. An improper voltage is sent to the ECM through exhaust valve timing control magnet retarder. **Possible Causes:** • Harness or connectors (The intake valve timing control magnet retarder circuit is open or shorted.) • Exhaust valve timing control magnet retarder
DTC: P010B **1T PCM, MIL: Yes** **Years: 2007, 2008** **Models:** 350Z **Engines:** 3.5L **Transmissions:** All	**Mass Airflow Sensor Bank 2 Improper Voltage Signal** A low voltage from the sensor is sent to ECM under heavy load driving condition. Or, a high voltage from the sensor is sent to ECM under light load driving condition. **Possible Causes:** • Harness or connectors (The sensor circuit is open or shorted.) • Mass air flow sensor • Intake air leaks • EVAP control system pressure sensor • Intake air temperature sensor
DTC: P010C **1T PCM, MIL: Yes** **Years: 2007, 2008** **Models:** 350Z **Engines:** 3.5L **Transmissions:** All	**Mass Airflow Sensor Bank 2 Low Voltage Signal** An excessively low voltage from the sensor is sent to ECM. **Possible Causes:** • Harness or connectors (The sensor circuit is open or shorted.) • Intake air leaks • Mass airflow sensor

DTC	Trouble Code Title, Conditions & Possible Causes
DTC: P010D **1T PCM, MIL: Yes** **Years: 2007, 2008** **Models:** 350Z **Engines:** 3.5L **Transmissions:** All	**Mass Airflow Sensor Bank 2 High Voltage Signal** An excessively high voltage from the sensor is sent to ECM. **Possible Causes:** • Harness or connectors (The sensor circuit is open or shorted.) • Mass airlow sensor
DTC: P0101 **1T PCM, MIL: Yes** **Years: 2007, 2008** **Models:** 350Z, Altima, Armada, Frontier, Maxima, Murano, Pathfinder, Quest, Titan, Xterra **Engines:** 3.5L, 4.0L, 5.6L **Transmissions:** All	**Mass Airflow Sensor Bank 1 Improper Voltage Signal** A low voltage from the sensor is sent to ECM under heavy load driving condition. Or, a high voltage from the sensor is sent to ECM under light load driving condition. **Possible Causes:** • Harness or connectors (The sensor circuit is open or shorted.) • Mass air flow sensor • Intake air leaks • EVAP control system pressure sensor • Intake air temperature sensor
DTC: P0101 **1T PCM, MIL: Yes** **Years: 2007, 2008** **Models:** Altima, Frontier, Rogue, Sentra, Versa **Engines:** 1.8L, 2.0L, 2.5L **Transmissions:** All	**Mass Airflow Sensor Improper Voltage Signal** A low voltage from the sensor is sent to ECM under heavy load driving condition. Or, a high voltage from the sensor is sent to ECM under light load driving condition. **Possible Causes:** • Harness or connectors (The sensor circuit is open or shorted.) • Mass air flow sensor • Intake air leaks • EVAP control system pressure sensor • Intake air temperature sensor
DTC: P0102 **1T PCM, MIL: Yes** **Years: 2007, 2008** **Models:** 350Z, Altima, Armada, Frontier, Maxima, Murano, Pathfinder, Quest, Titan, Xterra **Engines:** 3.5L, 4.0L, 5.6L **Transmissions:** All	**Mass Airflow Sensor Bank 1 Low Voltage Signal** An excessively low voltage from the sensor is sent to ECM. **Possible Causes:** • Harness or connectors (The sensor circuit is open or shorted.) • Intake air leaks • Mass airlow sensor
DTC: P0102 **1T PCM, MIL: Yes** **Years: 2007, 2008** **Models:** Altima, Frontier, Rogue, Sentra, Versa **Engines:** 1.8L, 2.0L, 2.5L **Transmissions:** All	**Mass Airflow Sensor Low Voltage Signal** An excessively low voltage from the sensor is sent to ECM. **Possible Causes:** • Harness or connectors (The sensor circuit is open or shorted.) • Intake air leaks • Mass airlow sensor
DTC: P0103 **1T PCM, MIL: Yes** **Years: 2007, 2008** **Models:** 350Z, Altima, Armada, Frontier, Maxima, Murano, Pathfinder, Quest, Titan, Xterra **Engines:** 3.5L, 4.0L, 5.6L **Transmissions:** All	**Mass Airflow Sensor Bank 1 High Voltage Signal** An excessively high voltage from the sensor is sent to ECM. **Possible Causes:** • Harness or connectors (The sensor circuit is open or shorted.) • Mass airlow sensor
DTC: P0103 **1T PCM, MIL: Yes** **Years: 2007, 2008** **Models:** Altima, Frontier, Rogue, Sentra, Versa **Engines:** 1.8L, 2.0L, 2.5L **Transmissions:** All	**Mass Airflow Sensor High Voltage Signal** An excessively high voltage from the sensor is sent to ECM. **Possible Causes:** • Harness or connectors (The sensor circuit is open or shorted.) • Mass airlow sensor

DTC	Trouble Code Title, Conditions & Possible Causes
DTC: P0112 **1T PCM, MIL: Yes** **Years: 2007, 2008** **Models:** 350Z, Altima, Armada, Frontier, Maxima, Murano, Pathfinder, Quest, Rogue, Sentra, Titan, Versa, Xterra **Engines:** All **Transmissions:** All	**Intake Air Temperature Sensor Low Voltage** An excessively low voltage from the sensor is sent to ECM. **Possible Causes:** • Harness or connectors (The sensor circuit is open or shorted.) • Intake air temperature sensor
DTC: P0113 **1T PCM, MIL: Yes** **Years: 2007, 2008** **Models:** 350Z, Altima, Armada, Frontier, Maxima, Murano, Pathfinder, Quest, Rogue, Sentra, Titan, Versa, Xterra **Engines:** All **Transmissions:** All	**Intake Air Temperature Sensor High Voltage** An excessively high voltage from the sensor is sent to ECM. **Possible Causes:** • Harness or connectors (The sensor circuit is open or shorted.) • Intake air temperature sensor
DTC: P0117 **1T PCM, MIL: Yes** **Years: 2007, 2008** **Models:** 350Z, Altima, Armada, Frontier, Maxima, Murano, Pathfinder, Quest, Rogue, Sentra, Titan, Versa, Xterra **Engines:** All **Transmissions:** All	**Engine Coolant Temperature Sensor Low Voltage** An excessively low voltage from the sensor is sent to ECM. **Possible Causes:** • Harness or connectors (The sensor circuit is open or shorted.) • Engine coolant temperature sensor
DTC: P0118 **1T PCM, MIL: Yes** **Years: 2007, 2008** **Models:** 350Z, Altima, Armada, Frontier, Maxima, Murano, Pathfinder, Quest, Rogue, Sentra, Titan, Versa, Xterra **Engines:** All **Transmissions:** All	**Engine Coolant Temperature Sensor High Voltage** An excessively high voltage from the sensor is sent to ECM. **Possible Causes:** • Harness or connectors (The sensor circuit is open or shorted.) • Engine coolant temperature sensor
DTC: P0122 **1T PCM, MIL: Yes** **Years: 2007, 2008** **Models:** 350Z, Altima, Armada, Frontier, Maxima, Murano, Pathfinder, Quest, Titan, Xterra **Engines:** 3.5L, 4.0L, 5.6L **Transmissions:** All	**Bank 1 Throttle Position Sensor 2 Low Voltage** An excessively low voltage from the sensor is sent to ECM. **Possible Causes:** • Harness or connectors (The TP sensor 2 circuit is open or shorted.) • Electronic throttle control actuator (TP Sensor 2)
DTC: P0122 **1T PCM, MIL: Yes** **Years: 2007, 2008** **Models:** Altima, Frontier, Rogue, Sentra, Versa **Engines:** 1.8L, 2.0L, 2.5L **Transmissions:** All	**Throttle Position Sensor 2 Low Voltage** An excessively low voltage from the sensor is sent to ECM. **Possible Causes:** • Harness or connectors (The TP sensor 2 circuit is open or shorted.) • Electronic throttle control actuator (TP Sensor 2)
DTC: P0123 **1T PCM, MIL: Yes** **Years: 2007, 2008** **Models:** 350Z, Altima, Armada, Frontier, Maxima, Murano, Pathfinder, Quest, Titan, Xterra **Engines:** 3.5L, 4.0L, 5.6L **Transmissions:** All	**Bank 1 Throttle Position Sensor 2 High Voltage** An excessively high voltage from the sensor is sent to ECM. **Possible Causes:** • Harness or connectors (The TP sensor 2 circuit is open or shorted.) • Electronic throttle control actuator (TP Sensor 2)

DTC	Trouble Code Title, Conditions & Possible Causes
DTC: P0123 **1T PCM, MIL: Yes** **Years: 2007, 2008** **Models:** Altima, Frontier, Rogue, Sentra, Versa **Engines:** 1.8L, 2.0L, 2.5L **Transmissions:** All	**Throttle Position Sensor 2 High Voltage** An excessively high voltage from the sensor is sent to ECM. **Possible Causes:** • Harness or connectors (The TP sensor 2 circuit is open or shorted.) • Electronic throttle control actuator (TP Sensor 2)
DTC: P0125 **2T ECT, MIL: Yes** **Years: 2007, 2008** **Models:** 350Z, Altima, Armada, Frontier, Maxima, Murano, Pathfinder, Quest, Rogue, Sentra, Titan, Versa, Xterra **Engines:** All **Transmissions:** All	**ECT Sensor Excessive Time To Enter Closed Loop** Voltage sent to ECM from the sensor is not practical, even when some time has passed after starting the engine. Engine coolant temperature is insufficient for closed loop fuel control. **Possible Causes:** • Harness or connectors (High resistance in the circuit) • Engine coolant temperature sensor • Thermostat
DTC: P0127 **2T ECT, MIL: Yes** **Years: 2007, 2008** **Models:** 350Z, Altima, Armada, Frontier, Maxima, Murano, Pathfinder, Quest, Rogue, Sentra, Titan, Versa, Xterra **Engines:** All **Transmissions:** All	**IAT Sensor Air Temperature Too High** Rationally incorrect voltage from the sensor is sent to ECM, compared with the voltage signal from engine coolant temperature sensor. **Possible Causes:** • Harness or connectors (High resistance in the circuit) • Intake air temperature sensor
DTC: P0128 **2T ECT, MIL: Yes** **Years: 2007, 2008** **Models:** 350Z, Altima, Armada, Frontier, Maxima, Murano, Pathfinder, Quest, Rogue, Sentra, Titan, Versa, Xterra **Engines:** All **Transmissions:** All	**Improper Thermostat Function** The engine coolant temperature does not reach to specified temperature even though the engine has run long enough. **Possible Causes:** • Leakage from sealing portion of thermostat • Engine coolant temperature sensor
DTC: P0130 **1T CCM, MIL: Yes** **Years: 2007, 2008** **Models:** 350Z, Altima, Armada, Frontier, Maxima, Murano, Pathfinder, Quest, Titan, Xterra **Engines:** 3.5L, 4.0L, 5.6L **Transmissions:** All	**A/F Sensor 1 (Bank 1) Circuit Malfunction** The A/F signal computed by ECM from the A/F sensor 1 signal is constantly in the range other than approx. 2.2V. **Possible Causes:** • A/F Sensor 1 circuit is open or shorted to ground • A/F Sensor 1
DTC: P0130 **1T CCM, MIL: Yes** **Years: 2007, 2008** **Models:** Altima, Frontier, Rogue, Sentra, Versa **Engines:** 1.8L, 2.0L, 2.5L **Transmissions:** All	**A/F Sensor 1 Closed Loop Fuel Control** The A/F signal computed by ECM from the A/F sensor 1 signal is constantly in the range other than approx. 2.2V. Or, the A/F signal computed by ECM from the A/F sensor 1 signal is constantly approx. 2.2V **Possible Causes:** • A/F sensor 1 is open or shorted to ground • A/F sensor 1
DTC: P0131 **1T CCM, MIL: Yes** **Years: 2007, 2008** **Models:** 350Z, Altima, Armada, Frontier, Maxima, Murano, Pathfinder, Quest, Titan, Xterra **Engines:** 3.5L, 4.0L, 5.6L **Transmissions:** All	**A/F Sensor 1 (Bank 1) Circuit Low Voltage** The A/F signal computed by ECM from the A/F sensor 1 signal is constantly approx. 0 V. **Possible Causes:** • A/F Sensor 1 circuit is open or shorted to ground • A/F Sensor 1

DTC	Trouble Code Title, Conditions & Possible Causes
DTC: P0131 **1T CCM, MIL: Yes** **Years: 2007, 2008** **Models:** Altima, Frontier, Rogue, Sentra, Versa **Engines:** 1.8L, 2.0L, 2.5L **Transmissions:** All	**A/F Sensor 1Circuit Low Voltage** The A/F signal computed by ECM from the A/F sensor 1 signal is constantly approx. 0 V. **Possible Causes:** • A/F Sensor 1 circuit is open or shorted to ground • A/F Sensor 1
DTC: P0132 **1T CCM, MIL: Yes** **Years: 2007, 2008** **Models:** 350Z, Altima, Armada, Frontier, Maxima, Murano, Pathfinder, Quest, Titan, Xterra **Engines:** 3.5L, 4.0L, 5.6L **Transmissions:** All	**A/F Sensor 1 (Bank 1) Circuit High Voltage** The A/F signal computed by ECM from the A/F sensor 1 signal is constantly approx. 5 V. **Possible Causes:** • A/F Sensor 1 circuit is open or shorted to ground • A/F Sensor 1
DTC: P0132 **1T CCM, MIL: Yes** **Years: 2007, 2008** **Models:** Altima, Frontier, Rogue, Sentra, Versa **Engines:** 1.8L, 2.0L, 2.5L **Transmissions:** All	**A/F Sensor 1 Circuit High Voltage** The A/F signal computed by ECM from the A/F sensor 1 signal is constantly approx. 5 V. **Possible Causes:** • A/F Sensor 1 circuit is open or shorted to ground • A/F Sensor 1
DTC: P0133 **1T CCM, MIL: Yes** **Years: 2007, 2008** **Models:** 350Z, Altima, Armada, Frontier, Maxima, Murano, Pathfinder, Quest, Titan, Xterra **Engines:** 3.5L, 4.0L, 5.6L **Transmissions:** All	**A/F Sensor 1 (Bank 1) Slow Response** The response of the A/F signal computed by ECM from A/F sensor 1 signal takes more than the specified time. **Possible Causes:** • Harness or connectors (The A/F sensor 1 circuit is open or shorted.) • Air fuel ratio (A/F) sensor 1 • Air fuel ratio (A/F) sensor 1 heater • Fuel pressure • Fuel injector • Intake air leaks • Exhaust gas leaks • PCV valve • Mass air flow sensor
DTC: P0133 **1T CCM, MIL: Yes** **Years: 2007, 2008** **Models:** Altima, Frontier, Rogue, Sentra, Versa **Engines:** 1.8L, 2.0L, 2.5L **Transmissions:** All	**A/F Sensor 1 Slow Response** The response of the A/F signal computed by ECM from A/F sensor 1 signal takes more than the specified time. **Possible Causes:** • Harness or connectors (The A/F sensor 1 circuit is open or shorted.) • Air fuel ratio (A/F) sensor 1 • Air fuel ratio (A/F) sensor 1 heater • Fuel pressure • Fuel injector • Intake air leaks • Exhaust gas leaks • PCV valve • Mass air flow sensor
DTC: P0137 **2T CCM, MIL: Yes** **Years: 2007, 2008** **Models:** 350Z, Altima, Armada, Frontier, Maxima, Murano, Pathfinder, Quest, Titan, Xterra **Engines:** 3.5L, 4.0L, 5.6L **Transmissions:** All	**HO2S-21 (Bank 2 Sensor 1) Circuit Low Voltage** Maximum voltage from the sensor has not reached the specified voltage. **Possible Causes:** • Harness or connectors (The sensor circuit is open or shorted) • Heated oxygen sensor 2 • Fuel pressure • Fuel injector • Intake air leaks

DTC	Trouble Code Title, Conditions & Possible Causes
DTC: P0137 **2T CCM, MIL: Yes** **Years: 2007, 2008** **Models:** Altima, Frontier, Rogue, Sentra, Versa **Engines:** 1.8L, 2.0L, 2.5L **Transmissions:** All	**HO2S-2 (Sensor 2) Circuit Low Voltage** Maximum voltage from the sensor has not reached the specified voltage. **Possible Causes:** • Harness or connectors (The sensor circuit is open or shorted) • Heated oxygen sensor 2 • Fuel pressure • Fuel injector • Intake air leaks
DTC: P0138 **2T CCM, MIL: Yes** **Years: 2007, 2008** **Models:** 350Z, Altima, Armada, Frontier, Maxima, Murano, Pathfinder, Quest, Titan, Xterra **Engines:** 3.5L, 4.0L, 5.6L **Transmissions:** All	**HO2S-21 (Bank 2 Sensor 1) Circuit High Voltage** An excessively high voltage from the sensor is sent to the ECM. **Possible Causes:** • Harness or connectors (The sensor circuit is open or shorted) • Heated oxygen sensor 2
DTC: P0138 **2T CCM, MIL: Yes** **Years: 2007, 2008** **Models:** Altima, Frontier, Rogue, Sentra, Versa **Engines:** 1.8L, 2.0L, 2.5L **Transmissions:** All	**HO2S-2 (Sensor 2) Circuit High Voltage** An excessively high voltage from the sensor is sent to the ECM. Or, the minimum voltage from the sensor is not reached to the specified voltage. **Possible Causes:** • Harness or connectors (The sensor circuit is open or shorted) • Heated oxygen sensor 2 • Fuel pressure • Fuel injector
DTC: P0139 **2T CCM, MIL: Yes** **Years: 2007, 2008** **Models:** 350Z, Altima, Armada, Frontier, Maxima, Murano, Pathfinder, Quest, Titan, Xterra **Engines:** 3.5L, 4.0L, 5.6L **Transmissions:** All	**HO2S-21 (Bank 2 Sensor 1) Slow Response** Engine at normal operating temperature. It takes more time for the sensor to respond between rich and lean than the specified time. **Possible Causes:** • Harness or connectors (The sensor circuit is open or shorted) • Heated oxygen sensor 2 • Fuel pressure • Fuel injector • Intake air leaks
DTC: P0139 **2T O2S, MIL: Yes** **Years: 2007, 2008** **Models:** Altima, Frontier, Rogue, Sentra, Versa **Engines:** 1.8L, 2.0L, 2.5L **Transmissions:** All	**HO2S-2 (Sensor 2) Slow Response** It takes more time for the sensor to respond between rich and lean than the specified time. **Possible Causes:** • Harness or connectors (The sensor circuit is open or shorted) • Heated oxygen sensor 2 • Fuel pressure • Fuel injector • Intake air leaks
DTC: P0143 **2T CCM, MIL: Yes** **Years: 2007, 2008** **Models:** Altima **Engines:** 2.5L (CA) **Transmissions:** All	**HO2S-3 (Sensor 3) Circuit High Voltage** The voltage from the sensor is beyond the specified voltage. **Possible Causes:** • Harness or connectors (sensor circuit is open or shorted) • Heated oxygen sensor 3 • Fuel pressure • Fuel injector
DTC: P0144 **2T CCM, MIL: Yes** **Years: 2007, 2008** **Models:** Altima **Engines:** 2.5L (CA) **Transmissions:** All	**HO2S-3 (Sensor 3) Circuit Low Voltage** The minimum voltage from the sensor is not reached to the specified voltage. **Possible Causes:** • Harness or connectors (sensor circuit is open or shorted) • Heated oxygen sensor 3 • Fuel pressure • Fuel injector

DTC	Trouble Code Title, Conditions & Possible Causes
DTC: P0145 **2T CCM, MIL: Yes** **Years: 2007, 2008** **Models:** Altima **Engines:** 2.5L (CA) **Transmissions:** All	**HO2S-3 (Sensor 3) Circuit Slow Response** It takes more time for the sensor to respond between rich and lean than the specified time. **Possible Causes:** • Harness or connectors (sensor circuit is open or shorted) • Heated oxygen sensor 3 • Fuel pressure • Fuel injector • Intake air leaks
DTC: P0146 **2T CCM, MIL: Yes** **Years: 2007, 2008** **Models:** Altima **Engines:** 2.5L (CA) **Transmissions:** All	**HO2S-3 (Sensor 3) Circuit No Activity Detected** Excessively high voltage is sent from the sensor to the ECM. **Possible Causes:** • Harness or connectors (sensor circuit is open or shorted) • Heated oxygen sensor 3
DTC: P0150 **1T CCM, MIL: Yes** **Years: 2007, 2008** **Models:** 350Z, Altima, Armada, Frontier, Maxima, Murano, Pathfinder, Quest, Titan, Xterra **Engines:** 3.5L, 4.0L, 5.6L **Transmissions:** All	**A/F Sensor 1 (Bank 2) Circuit Malfunction** The A/F signal computed by ECM from the A/F sensor 1 signal is constantly approx. 2.2V. **Possible Causes:** • A/F Sensor 1 circuit is open or shorted to ground • A/F Sensor 1
DTC: P0151 **1T CCM, MIL: Yes** **Years: 2007, 2008** **Models:** 350Z, Altima, Armada, Frontier, Maxima, Murano, Pathfinder, Quest, Titan, Xterra **Engines:** 3.5L, 4.0L, 5.6L **Transmissions:** All	**A/F Sensor 1 (Bank 2) Circuit Low Voltage** The A/F signal computed by ECM from the A/F sensor 1 signal is constantly approx. 0 V. **Possible Causes:** • A/F Sensor 1 circuit is open or shorted to ground • A/F Sensor 1
DTC: P0152 **1T CCM, MIL: Yes** **Years: 2007, 2008** **Models:** 350Z, Altima, Armada, Frontier, Maxima, Murano, Pathfinder, Quest, Titan, Xterra **Engines:** 3.5L, 4.0L, 5.6L **Transmissions:** All	**A/F Sensor 1 (Bank 2) Circuit High Voltage** The A/F signal computed by ECM from the A/F sensor 1 signal is constantly approx. 5 V. **Possible Causes:** • A/F Sensor 1 circuit is open or shorted to ground • A/F Sensor 1
DTC: P0153 **1T CCM, MIL: Yes** **Years: 2007, 2008** **Models:** 350Z, Altima, Armada, Frontier, Maxima, Murano, Pathfinder, Quest, Titan, Xterra **Engines:** 3.5L, 4.0L, 5.6L **Transmissions:** All	**A/F Sensor 1 (Bank 2) Slow Response** The response of the A/F signal computed by ECM from A/F sensor 1 signal takes more than the specified time. **Possible Causes:** • Harness or connectors (The A/F sensor 1 circuit is open or shorted.) • Air fuel ratio (A/F) sensor 1 • Air fuel ratio (A/F) sensor 1 heater • Fuel pressure • Fuel injector • Intake air leaks • Exhaust gas leaks • PCV valve • Mass air flow sensor
DTC: P0157 **2T CCM, MIL: Yes** **Years: 2007, 2008** **Models:** 350Z, Altima, Armada, Frontier, Maxima, Murano, Pathfinder, Quest, Titan, Xterra **Engines:** 3.5L, 4.0L, 5.6L **Transmissions:** All	**HO2S-22 (Bank 2 Sensor 2) Circuit Low Voltage** Maximum voltage from the sensor has not reached the specified voltage. **Possible Causes:** • Harness or connectors (The sensor circuit is open or shorted) • Heated oxygen sensor 2 • Fuel pressure • Fuel injector • Intake air leaks

DTC	Trouble Code Title, Conditions & Possible Causes
DTC: P0158 **2T CCM, MIL: Yes** **Years: 2007, 2008** **Models:** 350Z, Altima, Armada, Frontier, Maxima, Murano, Pathfinder, Quest, Titan, Xterra **Engines:** 3.5L, 4.0L, 5.6L **Transmissions:** All	**HO2S-22 (Bank 2 Sensor 2) Circuit High Voltage** An excessively high voltage from the sensor is sent to the ECM. **Possible Causes:** • Harness or connectors (The sensor circuit is open or shorted) • Heated oxygen sensor 2
DTC: P0159 **2T CCM, MIL: Yes** **Years: 2007, 2008** **Models:** 350Z, Altima, Armada, Frontier, Maxima, Murano, Pathfinder, Quest, Titan, Xterra **Engines:** 3.5L, 4.0L, 5.6L **Transmissions:** All	**HO2S-22 (Bank 2 Sensor 2) Slow Response** Engine at normal operating temperature. It takes more time for the sensor to respond between rich and lean than the specified time. **Possible Causes:** • Harness or connectors (The sensor circuit is open or shorted) • Heated oxygen sensor 2 • Fuel pressure • Fuel injector • Intake air leaks
DTC: P0171 **2T CCM, MIL: Yes** **Years: 2007, 2008** **Models:** 350Z, Altima, Armada, Frontier, Maxima, Murano, Pathfinder, Quest, Titan, Xterra **Engines:** 3.5L, 4.0L, 5.6L **Transmissions:** All	**Fuel Injection System (Bank 1) Too Lean** Fuel injection system does not operate properly. The amount of mixture ratio compensation is too large. (The mixture ratio is too lean.) **Possible Causes:** • Intake air leaks • Air fuel ratio (A/F) sensor 1 • Fuel injector • Exhaust gas leaks • Incorrect fuel pressure • Lack of fuel • Mass air flow sensor • Incorrect PCV hose connection
DTC: P0171 **2T CCM, MIL: Yes** **Years: 2007, 2008** **Models:** Altima, Frontier, Rogue, Sentra, Versa **Engines:** 1.8L, 2.0L, 2.5L **Transmissions:** All	**Fuel Injection System Too Lean** Fuel injection system does not operate properly. The amount of mixture ratio compensation is too large. (The mixture ratio is too lean.) **Possible Causes:** • Intake air leaks • Air fuel ratio (A/F) sensor 1 • Fuel injector • Exhaust gas leaks • Incorrect fuel pressure • Lack of fuel • Mass air flow sensor • Incorrect PCV hose connection
DTC: P0172 **2T CCM, MIL: Yes** **Years: 2007, 2008** **Models:** 350Z, Altima, Armada, Frontier, Maxima, Murano, Pathfinder, Quest, Titan, Xterra **Engines:** 3.5L, 4.0L, 5.6L **Transmissions:** All	**Fuel Injection System (Bank 1) Too Rich** Fuel injection system does not operate properly. The amount of mixture ratio compensation is too large. (The mixture ratio is too rich.) **Possible Causes:** • Intake air leaks • Air fuel ratio (A/F) sensor 1 • Fuel injector • Exhaust gas leaks • Incorrect fuel pressure • Lack of fuel • Mass air flow sensor • Incorrect PCV hose connection
DTC: P0172 **2T CCM, MIL: Yes** **Years: 2007, 2008** **Models:** Altima, Frontier, Rogue, Sentra, Versa **Engines:** 1.8L, 2.0L, 2.5L **Transmissions:** All	**Fuel Injection System Too Rich** Fuel injection system does not operate properly. The amount of mixture ratio compensation is too large. (The mixture ratio is too rich.) **Possible Causes:** • Intake air leaks • Air fuel ratio (A/F) sensor 1 • Fuel injector • Exhaust gas leaks • Incorrect fuel pressure • Lack of fuel • Mass air flow sensor • Incorrect PCV hose connection

DTC	Trouble Code Title, Conditions & Possible Causes
DTC: P0174 **2T CCM, MIL: Yes** **Years: 2007, 2008** **Models:** 350Z, Altima, Armada, Frontier, Maxima, Murano, Pathfinder, Quest, Titan, Xterra **Engines:** 3.5L, 4.0L, 5.6L **Transmissions:** All	**Fuel Injection System (Bank 2) Too Lean** Fuel injection system does not operate properly. The amount of mixture ratio compensation is too large. (The mixture ratio is too lean.) **Possible Causes:** • Intake air leaks • Air fuel ratio (A/F) sensor 1 • Fuel injector • Exhaust gas leaks • Incorrect fuel pressure • Lack of fuel • Mass air flow sensor • Incorrect PCV hose connection
DTC: P0175 **2T CCM, MIL: Yes** **Years: 2007, 2008** **Models:** 350Z, Altima, Armada, Frontier, Maxima, Murano, Pathfinder, Quest, Titan, Xterra **Engines:** 3.5L, 4.0L, 5.6L **Transmissions:** All	**Fuel Injection System (Bank 2) Too Rich** Fuel injection system does not operate properly. The amount of mixture ratio compensation is too large. (The mixture ratio is too rich.) **Possible Causes:** • Intake air leaks • Air fuel ratio (A/F) sensor 1 • Fuel injector • Exhaust gas leaks • Incorrect fuel pressure • Lack of fuel • Mass air flow sensor • Incorrect PCV hose connection
DTC: P0181 **2T CCM, MIL: Yes** **Years: 2007, 2008** **Models:** 350Z, Altima, Armada, Frontier, Maxima, Murano, Pathfinder, Quest, Rogue, Sentra, Titan, Versa, Xterra **Engines:** All **Transmissions:** All	**Fuel Tank Temperature (FTT) Sensor/Circuit Performance** Rationally incorrect voltage from the sensor is sent to ECM, compared with the voltage signals from engine coolant temperature sensor and intake air temperature sensor. **Possible Causes:** • Harness or connectors (The sensor circuit is open or shorted) • FTT sensor
DTC: P0182 **2T CCM, MIL: Yes** **Years: 2007, 2008** **Models:** 350Z, Altima, Armada, Frontier, Maxima, Murano, Pathfinder, Quest, Rogue, Sentra, Titan, Versa, Xterra **Engines:** All **Transmissions:** All	**Fuel Tank Temperature (FTT) Sensor Circuit Low Voltage** An excessively low voltage from the sensor is sent to ECM. **Possible Causes:** • Harness or connectors (The sensor circuit is open or shorted) • FTT sensor
DTC: P0183 **2T CCM, MIL: Yes** **Years: 2007, 2008** **Models:** 350Z, Altima, Armada, Frontier, Maxima, Murano, Pathfinder, Quest, Rogue, Sentra, Titan, Versa, Xterra **Engines:** All **Transmissions:** All	**Fuel Tank Temperature (FTT) Sensor Circuit High Voltage** An excessively high voltage from the sensor is sent to ECM. **Possible Causes:** • Harness or connectors (The sensor circuit is open or shorted) • FTT sensor
DTC: P0196 **2T CCM, MIL: Yes** **Years: 2007, 2008** **Models:** 350Z **Engines:** 3.5L **Transmissions:** All	**Engine Oil Temperature (EOT) Sensor Circuit Range/Performance** Rationally incorrect voltage from the sensor is sent to ECM, compared with the voltage signals from engine coolant temperature sensor and intake air temperature sensor. **Possible Causes:** • Harness or connectors (the sensor circuit is open or shorted) • EOT sensor

DTC	Trouble Code Title, Conditions & Possible Causes

DTC: P0197
2T CCM, MIL: Yes
Years: 2007, 2008
Models: 350Z
Engines: 3.5L
Transmissions: All
	Engine Oil Temperature (EOT) Sensor Circuit Low Voltage
	Excessively low voltage is sent from the sensor to ECM.
	Possible Causes:
	   • Harness or connectors (the sensor circuit is open or shorted)
	   • EOT sensor

DTC: P0198
2T CCM, MIL: Yes
Years: 2007, 2008
Models: 350Z
Engines: 3.5L
Transmissions: All
	Engine Oil Temperature (EOT) Sensor Circuit High Voltage
	Excessively high voltage is sent from the sensor to ECM.
	Possible Causes:
	   • Harness or connectors (the sensor circuit is open or shorted)
	   • EOT sensor

DTC: P0222
2T CCM, MIL: Yes
Years: 2007, 2008
Models: 350Z, Altima, Armada, Frontier, Maxima, Murano, Pathfinder, Quest, Titan, Xterra
Engines: 3.5L, 4.0L, 5.6L
Transmissions: All
	Throttle Position (TP) Sensor 1 (Bank 1) Circuit Low Voltage
	Excessively low voltage is sent from the sensor to ECM.
	Possible Causes:
	   • Harness or connectors (the sensor circuit is open or shorted)
	   • EOT sensor

DTC: P0222
2T CCM, MIL: Yes
Years: 2007, 2008
Models: Altima, Frontier, Rogue, Sentra, Versa
Engines: 1.8L, 2.0L, 2.5L
Transmissions: All
	Throttle Position (TP) Sensor 1 Circuit Low Voltage
	Excessively low voltage is sent from the sensor to ECM.
	Possible Causes:
	   • Harness or connectors (the sensor circuit is open or shorted)
	   • EOT sensor (TP sensor 1)

DTC: P0223
2T CCM, MIL: Yes
Years: 2007, 2008
Models: 350Z, Altima, Armada, Frontier, Maxima, Murano, Pathfinder, Quest, Titan, Xterra
Engines: 3.5L, 4.0L, 5.6L
Transmissions: All
	Throttle Position (TP) Sensor 1 (Bank 1) Circuit High Voltage
	Excessively high voltage is sent from the sensor to ECM.
	Possible Causes:
	   • Harness or connectors (the sensor circuit is open or shorted)
	   • EOT sensor (TP sensor 1)

DTC: P0223
2T CCM, MIL: Yes
Years: 2007, 2008
Models: Altima, Frontier, Rogue, Sentra, Versa
Engines: 1.8L, 2.0L, 2.5L
Transmissions: All
	Throttle Position (TP) Sensor 1 Circuit High Voltage
	Excessively high voltage is sent from the sensor to ECM.
	Possible Causes:
	   • Harness or connectors (the sensor circuit is open or shorted)
	   • EOT sensor (TP sensor 1)

DTC: P0300
2T MISFIRE, MIL: Yes
Years: 2007, 2008
Models: All
Engines: All
Transmissions: All
	Multiple Misfire Detected
	One Trip Detection Logic (Three Way Catalyst Damage): On the first trip that a misfire condition occurs that can damage the three way catalyst (TWC) due to overheating, the MIL will blink. When a misfire condition occurs, the ECM monitors the CKP sensor (POS) signal every 200 engine revolutions for a change. When the misfire condition decreases to a level that will not damage the TWC, the MIL will turn off. If another misfire condition occurs that can damage the TWC on a second trip, the MIL will blink. When the misfire condition decreases to a level that will not damage the TWC, the MIL will remain on. If another misfire condition occurs that can damage the TWC, the MIL will begin to blink again.
	Two Trip Detection Logic (Exhaust quality deterioration): For misfire conditions that will not damage the TWC (but will affect vehicle emissions), the MIL will only light when the misfire is detected on a second trip. During this condition, the ECM monitors the CKP sensor signal every 1,000 engine revolutions. A misfire malfunction can be detected on any one cylinder or on multiple cylinders.
	Possible Causes:
	   • Air leak in the intake manifold, or in the EGR or PCM system
	   • Base engine problem affecting two or more cylinders
	   • CMP or CKP sensor signals erratic or out of phase
	   • EGR valve stuck open, or EVAP purge system has failed
	   • Fuel delivery component fault that affects more than 1 cylinder (i.e., contaminated, dirty or sticking fuel injectors)
	   • Ignition system problem affecting two or more cylinders
	   • Vehicle driven while quite low or until it ran out of fuel

DTC	Trouble Code Title, Conditions & Possible Causes
DTC: P0301 **2T MISFIRE, MIL:** Yes **Years:** 2007, 2008 **Models:** All **Engines:** All **Transmissions:** All	**Cylinder 1 Misfire Detected** One Trip Detection Logic (Three Way Catalyst Damage): On the first trip that a misfire condition occurs that can damage the three way catalyst (TWC) due to overheating, the MIL will blink. When a misfire condition occurs, the ECM monitors the CKP sensor (POS) signal every 200 engine revolutions for a change. When the misfire condition decreases to a level that will not damage the TWC, the MIL will turn off. If another misfire condition occurs that can damage the TWC on a second trip, the MIL will blink. When the misfire condition decreases to a level that will not damage the TWC, the MIL will remain on. If another misfire condition occurs that can damage the TWC, the MIL will begin to blink again. Two Trip Detection Logic (Exhaust quality deterioration): For misfire conditions that will not damage the TWC (but will affect vehicle emissions), the MIL will only light when the misfire is detected on a second trip. During this condition, the ECM monitors the CKP sensor signal every 1,000 engine revolutions. A misfire malfunction can be detected on any one cylinder or on multiple cylinders. **Possible Causes:** • Air leak in the intake manifold, or in the EGR or PCM system • Base engine mechanical fault that affects only one cylinder • Fuel delivery component fault that affects only one cylinder (i.e., a contaminated, dirty or sticking fuel injector) • Ignition system problem (coil or plug) that affects one cylinder • Vehicle driven while quite low or until it ran out of fuel
DTC: P0302 **2T MISFIRE, MIL:** Yes **Years:** 2007, 2008 **Models:** All **Engines:** All **Transmissions:** All	**Cylinder 2 Misfire Detected** One Trip Detection Logic (Three Way Catalyst Damage): On the first trip that a misfire condition occurs that can damage the three way catalyst (TWC) due to overheating, the MIL will blink. When a misfire condition occurs, the ECM monitors the CKP sensor (POS) signal every 200 engine revolutions for a change. When the misfire condition decreases to a level that will not damage the TWC, the MIL will turn off. If another misfire condition occurs that can damage the TWC on a second trip, the MIL will blink. When the misfire condition decreases to a level that will not damage the TWC, the MIL will remain on. If another misfire condition occurs that can damage the TWC, the MIL will begin to blink again. Two Trip Detection Logic (Exhaust quality deterioration): For misfire conditions that will not damage the TWC (but will affect vehicle emissions), the MIL will only light when the misfire is detected on a second trip. During this condition, the ECM monitors the CKP sensor signal every 1,000 engine revolutions. A misfire malfunction can be detected on any one cylinder or on multiple cylinders. **Possible Causes:** • Air leak in the intake manifold, or in the EGR or PCM system • Base engine mechanical fault that affects only one cylinder • Fuel delivery component fault that affects only one cylinder (i.e., a contaminated, dirty or sticking fuel injector) • Ignition system problem (coil or plug) that affects one cylinder • Vehicle driven while quite low or until it ran out of fuel
DTC: P0303 **2T MISFIRE, MIL:** Yes **Years:** 2007, 2008 **Models:** All **Engines:** All **Transmissions:** All	**Cylinder 3 Misfire Detected** One Trip Detection Logic (Three Way Catalyst Damage): On the first trip that a misfire condition occurs that can damage the three way catalyst (TWC) due to overheating, the MIL will blink. When a misfire condition occurs, the ECM monitors the CKP sensor (POS) signal every 200 engine revolutions for a change. When the misfire condition decreases to a level that will not damage the TWC, the MIL will turn off. If another misfire condition occurs that can damage the TWC on a second trip, the MIL will blink. When the misfire condition decreases to a level that will not damage the TWC, the MIL will remain on. If another misfire condition occurs that can damage the TWC, the MIL will begin to blink again. Two Trip Detection Logic (Exhaust quality deterioration): For misfire conditions that will not damage the TWC (but will affect vehicle emissions), the MIL will only light when the misfire is detected on a second trip. During this condition, the ECM monitors the CKP sensor signal every 1,000 engine revolutions. A misfire malfunction can be detected on any one cylinder or on multiple cylinders. **Possible Causes:** • Air leak in the intake manifold, or in the EGR or PCM system • Base engine mechanical fault that affects only one cylinder • Fuel delivery component fault that affects only one cylinder (i.e., a contaminated, dirty or sticking fuel injector) • Ignition system problem (coil or plug) that affects one cylinder • Vehicle driven while quite low or until it ran out of fuel

DTC	Trouble Code Title, Conditions & Possible Causes
DTC: P0304 **2T MISFIRE, MIL: Yes** **Years: 2007, 2008** **Models:** All **Engines:** All **Transmissions:** All	**Cylinder 4 Misfire Detected** One Trip Detection Logic (Three Way Catalyst Damage): On the first trip that a misfire condition occurs that can damage the three way catalyst (TWC) due to overheating, the MIL will blink. When a misfire condition occurs, the ECM monitors the CKP sensor (POS) signal every 200 engine revolutions for a change. When the misfire condition decreases to a level that will not damage the TWC, the MIL will turn off. If another misfire condition occurs that can damage the TWC on a second trip, the MIL will blink. When the misfire condition decreases to a level that will not damage the TWC, the MIL will remain on. If another misfire condition occurs that can damage the TWC, the MIL will begin to blink again. Two Trip Detection Logic (Exhaust quality deterioration): For misfire conditions that will not damage the TWC (but will affect vehicle emissions), the MIL will only light when the misfire is detected on a second trip. During this condition, the ECM monitors the CKP sensor signal every 1,000 engine revolutions. A misfire malfunction can be detected on any one cylinder or on multiple cylinders. **Possible Causes:** • Air leak in the intake manifold, or in the EGR or PCM system • Base engine mechanical fault that affects only one cylinder • Fuel delivery component fault that affects only one cylinder (i.e., a contaminated, dirty or sticking fuel injector) • Ignition system problem (coil or plug) that affects one cylinder • Vehicle driven while quite low or until it ran out of fuel
DTC: P0305 **2T MISFIRE, MIL: Yes** **Years: 2007, 2008** **Models:** 350Z, Altima, Armada, Frontier, Maxima, Murano, Pathfinder, Quest, Titan, Xterra **Engines:** 3.5L, 4.0L, 5.6L **Transmissions:** All	**Cylinder 5 Misfire Detected** One Trip Detection Logic (Three Way Catalyst Damage): On the first trip that a misfire condition occurs that can damage the three way catalyst (TWC) due to overheating, the MIL will blink. When a misfire condition occurs, the ECM monitors the CKP sensor (POS) signal every 200 engine revolutions for a change. When the misfire condition decreases to a level that will not damage the TWC, the MIL will turn off. If another misfire condition occurs that can damage the TWC on a second trip, the MIL will blink. When the misfire condition decreases to a level that will not damage the TWC, the MIL will remain on. If another misfire condition occurs that can damage the TWC, the MIL will begin to blink again. Two Trip Detection Logic (Exhaust quality deterioration): For misfire conditions that will not damage the TWC (but will affect vehicle emissions), the MIL will only light when the misfire is detected on a second trip. During this condition, the ECM monitors the CKP sensor signal every 1,000 engine revolutions. A misfire malfunction can be detected on any one cylinder or on multiple cylinders. **Possible Causes:** • Air leak in the intake manifold, or in the EGR or PCM system • Base engine mechanical fault that affects only one cylinder • Fuel delivery component fault that affects only one cylinder (i.e., a contaminated, dirty or sticking fuel injector) • Ignition system problem (coil or plug) that affects one cylinder • Vehicle driven while quite low or until it ran out of fuel
DTC: P0306 **2T MISFIRE, MIL: Yes** **Years: 2007, 2008** **Models:** All **Engines:** All **Transmissions:** All	**Cylinder 6 Misfire Detected** One Trip Detection Logic (Three Way Catalyst Damage): On the first trip that a misfire condition occurs that can damage the three way catalyst (TWC) due to overheating, the MIL will blink. When a misfire condition occurs, the ECM monitors the CKP sensor (POS) signal every 200 engine revolutions for a change. When the misfire condition decreases to a level that will not damage the TWC, the MIL will turn off. If another misfire condition occurs that can damage the TWC on a second trip, the MIL will blink. When the misfire condition decreases to a level that will not damage the TWC, the MIL will remain on. If another misfire condition occurs that can damage the TWC, the MIL will begin to blink again. Two Trip Detection Logic (Exhaust quality deterioration): For misfire conditions that will not damage the TWC (but will affect vehicle emissions), the MIL will only light when the misfire is detected on a second trip. During this condition, the ECM monitors the CKP sensor signal every 1,000 engine revolutions. A misfire malfunction can be detected on any one cylinder or on multiple cylinders. **Possible Causes:** • Air leak in the intake manifold, or in the EGR or PCM system • Base engine mechanical fault that affects only one cylinder • Fuel delivery component fault that affects only one cylinder (i.e., a contaminated, dirty or sticking fuel injector) • Ignition system problem (coil or plug) that affects one cylinder • Vehicle driven while quite low or until it ran out of fuel

DTC	Trouble Code Title, Conditions & Possible Causes
DTC: P0307 **2T MISFIRE, MIL: Yes** **Years: 2007, 2008** **Models**: Armada, Pathfinder **Engines**: 5.6L **Transmissions**: All	**Cylinder 7 Misfire Detected** One Trip Detection Logic (Three Way Catalyst Damage): On the first trip that a misfire condition occurs that can damage the three way catalyst (TWC) due to overheating, the MIL will blink. When a misfire condition occurs, the ECM monitors the CKP sensor (POS) signal every 200 engine revolutions for a change. When the misfire condition decreases to a level that will not damage the TWC, the MIL will turn off. If another misfire condition occurs that can damage the TWC on a second trip, the MIL will blink. When the misfire condition decreases to a level that will not damage the TWC, the MIL will remain on. If another misfire condition occurs that can damage the TWC, the MIL will begin to blink again. Two Trip Detection Logic (Exhaust quality deterioration): For misfire conditions that will not damage the TWC (but will affect vehicle emissions), the MIL will only light when the misfire is detected on a second trip. During this condition, the ECM monitors the CKP sensor signal every 1,000 engine revolutions. A misfire malfunction can be detected on any one cylinder or on multiple cylinders. **Possible Causes:** • Air leak in the intake manifold, or in the EGR or PCM system • Base engine mechanical fault that affects only one cylinder • Fuel delivery component fault that affects only one cylinder (i.e., a contaminated, dirty or sticking fuel injector) • Ignition system problem (coil or plug) that affects one cylinder • Vehicle driven while quite low or until it ran out of fuel
DTC: P0308 **2T MISFIRE, MIL: Yes** **Years: 2007, 2008** **Models**: Armada, Pathfinder **Engines**: 5.6L **Transmissions**: All	**Cylinder 8 Misfire Detected** One Trip Detection Logic (Three Way Catalyst Damage): On the first trip that a misfire condition occurs that can damage the three way catalyst (TWC) due to overheating, the MIL will blink. When a misfire condition occurs, the ECM monitors the CKP sensor (POS) signal every 200 engine revolutions for a change. When the misfire condition decreases to a level that will not damage the TWC, the MIL will turn off. If another misfire condition occurs that can damage the TWC on a second trip, the MIL will blink. When the misfire condition decreases to a level that will not damage the TWC, the MIL will remain on. If another misfire condition occurs that can damage the TWC, the MIL will begin to blink again. Two Trip Detection Logic (Exhaust quality deterioration): For misfire conditions that will not damage the TWC (but will affect vehicle emissions), the MIL will only light when the misfire is detected on a second trip. During this condition, the ECM monitors the CKP sensor signal every 1,000 engine revolutions. A misfire malfunction can be detected on any one cylinder or on multiple cylinders. **Possible Causes:** • Air leak in the intake manifold, or in the EGR or PCM system • Base engine mechanical fault that affects only one cylinder • Fuel delivery component fault that affects only one cylinder (i.e., a contaminated, dirty or sticking fuel injector) • Ignition system problem (coil or plug) that affects one cylinder • Vehicle driven while quite low or until it ran out of fuel
DTC: P0327 **1T PCM, MIL: Yes** **Years: 2007, 2008** **Models:** 350Z, Altima, Armada, Frontier, Maxima, Murano, Pathfinder, Quest, Titan, Xterra **Engines:** 3.5L, 4.0L, 5.6L **Transmissions:** All	**Knock Sensor (KS) Sensor (Bank 1) Circuit Low Input** An excessively low voltage from the sensor is sent to ECM. **Possible Causes:** • Harness or connectors (the sensor circuit is open or shorted.) • Knock sensor
DTC: P0327 **1T PCM, MIL: Yes** **Years: 2007, 2008** **Models:** Altima, Frontier, Rogue, Sentra, Versa **Engines:** 1.8L, 2.0L, 2.5L **Transmissions:** All	**Knock Sensor (KS) Sensor Circuit Low Input** An excessively low voltage from the sensor is sent to ECM. **Possible Causes:** • Harness or connectors (the sensor circuit is open or shorted.) • Knock sensor
DTC: P0328 **1T PCM, MIL: Yes** **Years: 2007, 2008** **Models:** 350Z, Altima, Armada, Frontier, Maxima, Murano, Pathfinder, Quest, Titan, Xterra **Engines:** 3.5L, 4.0L, 5.6L **Transmissions:** All	**Knock Sensor (KS) Sensor (Bank 1) Circuit High Input** An excessively high voltage from the sensor is sent to ECM. **Possible Causes:** • Harness or connectors (the sensor circuit is open or shorted.) • Knock sensor

DTC	Trouble Code Title, Conditions & Possible Causes
DTC: P0328 **1T PCM, MIL: Yes** **Years: 2007, 2008** **Models:** Altima, Frontier, Rogue, Sentra, Versa **Engines:** 1.8L, 2.0L, 2.5L **Transmissions:** All	**Knock Sensor (KS) Sensor Circuit High Input** An excessively high voltage from the sensor is sent to ECM. **Possible Causes:** • Harness or connectors (the sensor circuit is open or shorted.) • Knock sensor
DTC: P0332 **1T PCM, MIL: Yes** **Years: 2007, 2008** **Models:** 350Z, Altima, Armada, Frontier, Maxima, Murano, Pathfinder, Quest, Titan, Xterra **Engines:** 3.5L, 4.0L, 5.6L **Transmissions:** All	**Knock Sensor (KS) Sensor (Bank 2) Circuit Low Input** An excessively low voltage from the sensor is sent to ECM. **Possible Causes:** • Harness or connectors (the sensor circuit is open or shorted.) • Knock sensor
DTC: P0333 **1T PCM, MIL: Yes** **Years: 2007, 2008** **Models:** 350Z, Altima, Armada, Frontier, Maxima, Murano, Pathfinder, Quest, Titan, Xterra **Engines:** 3.5L, 4.0L, 5.6L **Transmissions:** All	**Knock Sensor (KS) Sensor (Bank 2) Circuit High Input** An excessively high voltage from the sensor is sent to ECM. **Possible Causes:** • Harness or connectors (the sensor circuit is open or shorted.) • Knock sensor
DTC: P0335 **2T CCM, MIL: Yes** **Years: 2007, 2008** **Models:** 350Z, Altima, Armada, Frontier, Maxima, Murano, Pathfinder, Quest, Rogue, Sentra, Titan, Versa, Xterra **Engines:** All **Transmissions:** All	**Crankshaft Position Sensor Circuit Malfunction** Engine cranking for over 2 seconds, and the PCM did not detect a proper CKP sensor (1°) signal, or with Engine running, it did not detect a normal pattern of CKP sensor signals during the CCM test. **Possible Causes:** • CKP sensor signal circuit is open or shorted to ground • CKP sensor signal is shorted to VREF or system power • CKP sensor is damaged or has failed • PCM has failed
DTC: P0340 **2T CCM, MIL: Yes** **Years: 2007, 2008** **Models:** 350Z, Altima, Armada, Frontier, Maxima, Murano, Pathfinder, Quest, Titan, Xterra **Engines:** 3.5L, 4.0L, 5.6L **Transmissions:** All	**Camshaft Position Sensor (Bank 1) Circuit Malfunction** Engine cranking for over 2 seconds, and the PCM did not detect any CMP sensor signals, or with Engine running, the PCM did not detect a normal pattern of CMP signals during the CCM test. **Possible Causes:** • CMP sensor signal circuit is open or shorted to ground • CMP sensor signal is shorted to VREF or system power • CMP sensor is damaged or has failed • PCM has failed
DTC: P0340 **2T CCM, MIL: Yes** **Years: 2007, 2008** **Models:** Altima, Frontier, Rogue, Sentra, Versa **Engines:** 1.8L, 2.0L, 2.5L **Transmissions:** All	**Camshaft Position Sensor Circuit Malfunction** Engine cranking for over 2 seconds, and the PCM did not detect any CMP sensor signals, or with Engine running, the PCM did not detect a normal pattern of CMP signals during the CCM test. **Possible Causes:** • CMP sensor signal circuit is open or shorted to ground • CMP sensor signal is shorted to VREF or system power • CMP sensor is damaged or has failed • PCM has failed

DTC	Trouble Code Title, Conditions & Possible Causes
DTC: P0345 **2T CCM, MIL: Yes** **Years: 2007, 2008** **Models:** 350Z, Altima, Armada, Frontier, Maxima, Murano, Pathfinder, Quest, Titan, Xterra **Engines:** 3.5L, 4.0L, 5.6L **Transmissions:** All	**Camshaft Position Sensor (Bank 2) Circuit Malfunction** The cylinder No. signal is not sent to ECM for the first few seconds during engine cranking. The cylinder No. signal is not sent to ECM during engine running. The cylinder No. signal is not in the normal pattern during engine running. **Possible Causes:** • Harness or connectors • CMP sensor (PHASE) (bank 2) circuit is open or shorted. • CKP sensor (POS) circuit is shorted. • EVT control position sensor (bank 2) circuit is shorted. • APP sensor 2 circuit is shorted. • EVAP control system pressure sensor circuit is shorted. • Refrigerant pressure sensor circuit is shorted. • Camshaft position sensor (PHASE) (bank 2) • Crankshaft position sensor (POS) • Exhaust valve timing control position sensor (bank 2) • Accelerator pedal position sensor • EVAP control system pressure sensor • Refrigerant pressure sensor • Camshaft (INT) • Starter motor • Starting system circuit • Dead (Weak) battery
DTC: P0420 **1T CAT, MIL: Yes** **Years: 2007, 2008** **Models:** 350Z, Altima, Armada, Frontier, Maxima, Murano, Pathfinder, Quest, Titan, Xterra **Engines:** 3.5L, 4.0L, 5.6L **Transmissions:** All	**Catalyst Efficiency Below Normal (Bank 1)** Three way catalyst (manifold) does not operate properly. Three way catalyst (manifold) does not have enough oxygen storage capacity. **Possible Causes:** • Air leaks at the exhaust manifold or in the exhaust pipes • Catalytic converter is damaged, contaminated or has failed • Front HO2S or rear HO2S is contaminated with fuel or moisture • Front HO2S and/or the rear HO2S is loose in the mounting hole • Front HO2S older (aged) than the rear HO2S (HO2S is lazy)
DTC: P0420 **1T CAT, MIL: Yes** **Years: 2007, 2008** **Models:** Altima, Frontier, Rogue, Sentra, Versa **Engines:** 1.8L, 2.0L, 2.5L **Transmissions:** All	**Catalyst Efficiency Below Normal** Three way catalyst (manifold) does not operate properly. Three way catalyst (manifold) does not have enough oxygen storage capacity. **Possible Causes:** • Three way catalyst (manifold) • Exhaust tube • Intake air leaks • Fuel injector • Fuel injector leaks • Spark plug • Improper ignition timing
DTC: P0430 **1T CAT, MIL: Yes** **Years: 2007, 2008** **Models:** 350Z, Altima, Armada, Frontier, Maxima, Murano, Pathfinder, Quest, Titan, Xterra **Engines:** 3.5L, 4.0L, 5.6L **Transmissions:** All	**Catalyst Efficiency Below Normal (Bank 2)** Three way catalyst (manifold) does not operate properly. Three way catalyst (manifold) does not have enough oxygen storage capacity. **Possible Causes:** • Air leaks at the exhaust manifold or in the exhaust pipes • Catalytic converter is damaged, contaminated or has failed • Front HO2S or rear HO2S is contaminated with fuel or moisture • Front HO2S and/or the rear HO2S is loose in the mounting hole • Front HO2S older (aged) than the rear HO2S (HO2S is lazy)
DTC: P0441 **2T EVAP, MIL: Yes** **Years: 2007, 2008** **Models:** 350Z, Altima, Armada, Frontier, Maxima, Murano, Pathfinder, Quest, Rogue, Sentra, Titan, Versa, Xterra **Engines:** All **Transmissions:** All	**EVAP Control System Incorrect Purge Flow** EVAP control system does not operate properly, EVAP control system has a leak between intake manifold and EVAP control system pressure sensor. **Possible Causes:** • EVAP canister purge volume control solenoid valve stuck closed • EVAP control system pressure sensor and the circuit • Loose, disconnected or improper connection of rubber tube • Blocked rubber tube • Cracked EVAP canister • EVAP canister purge volume control solenoid valve circuit • Accelerator pedal position sensor • Blocked purge port • EVAP canister vent control valve

DTC	Trouble Code Title, Conditions & Possible Causes
DTC: P0442 **2T EVAP, MIL: Yes** **Years: 2007, 2008** **Models:** 350Z, Altima, Armada, Frontier, Maxima, Murano, Pathfinder, Quest, Rogue, Sentra, Titan, Versa, Xterra **Engines:** All **Transmissions:** All	**EVAP Control System Small Leak Detected** EVAP control system does not operate properly, EVAP control system has a leak between intake manifold and EVAP control system pressure sensor. **Possible Causes:** • Incorrect fuel tank vacuum relief valve • Incorrect fuel filler cap used • Fuel filler cap remains open or fails to close. • Foreign matter caught in fuel filler cap. • Leak is in line between intake manifold and EVAP canister purge volume control solenoid valve. • Foreign matter caught in EVAP canister vent control valve. • EVAP canister or fuel tank leaks • EVAP purge line (pipe and rubber tube) leaks • EVAP purge line rubber tube bent • Loose or disconnected rubber tube • EVAP canister vent control valve and the circuit • EVAP canister purge volume control solenoid valve and the circuit • Fuel tank temperature sensor • O-ring of EVAP canister vent control valve is missing or damaged • EVAP canister is saturated with water • EVAP control system pressure sensor • Fuel level sensor and the circuit • Refueling EVAP vapor cut valve • ORVR system leaks
DTC: P0443 **2T CCM, MIL: Yes** **Years: 2007, 2008** **Models:** 350Z, Altima, Armada, Frontier, Maxima, Murano, Pathfinder, Quest, Rogue, Sentra, Titan, Versa, Xterra **Engines:** All **Transmissions:** All	**EVAP Canister Purge Solenoid Circuit Malfunction** Engine started, Engine running at cruise speed under light engine load, system voltage from 11-16v, and the PCM detected an unexpected voltage condition on the Purge solenoid circuit, or it detected an invalid EVAP signal present when the Purge solenoid was commanded "on" and "off" during the CCM test. **Possible Causes:** • Purge solenoid control circuit open, shorted to ground or power • Purge solenoid is shorted to system power (B+) • Purge solenoid is damaged or has failed • PCM has failed
DTC: P0444 **2T CCM, MIL: Yes** **Years: 2007, 2008** **Models:** 350Z, Altima, Armada, Frontier, Maxima, Murano, Pathfinder, Quest, Rogue, Sentra, Titan, Versa, Xterra **Engines:** All **Transmissions:** All	**EVAP Canister Purge Solenoid Circuit Open** Engine: After warming up, shift lever to P or N (A/T) or Neutral (M/T) position. Air conditioner switch: OFF. No load. **Possible Causes:** • Harness or connectors (the solenoid valve circuit is open) • EVAP canister purge volume control solenoid valve
DTC: P0445 **2T CCM, MIL: Yes** **Years: 2007, 2008** **Models:** 350Z, Altima, Armada, Frontier, Maxima, Murano, Pathfinder, Quest, Rogue, Sentra, Titan, Versa, Xterra **Engines:** All **Transmissions:** All	**EVAP Canister Purge Solenoid Circuit Shorted** Engine: After warming up, shift lever to P or N (A/T) or Neutral (M/T) position. Air conditioner switch: OFF. No load. **Possible Causes:** • Harness or connectors (the solenoid valve circuit is shorted) • EVAP canister purge volume control solenoid valve
DTC: P0447 **2T CCM, MIL: Yes** **Years: 2007, 2008** **Models:** 350Z, Altima, Armada, Frontier, Maxima, Murano, Pathfinder, Quest, Rogue, Sentra, Titan, Versa, Xterra **Engines:** All **Transmissions:** All	**EVAP Canister Vent Control Valve Circuit Open** Ignition is ON. There is an improper voltage signal to the ECM through the vent control valve. **Possible Causes:** • Harness or connectors (the solenoid valve circuit is open or shorted) • EVAP canister vent control solenoid valve

DTC	Trouble Code Title, Conditions & Possible Causes
DTC: P0448 **2T CCM, MIL: Yes** **Years: 2007, 2008** **Models:** 350Z, Altima, Armada, Frontier, Maxima, Murano, Pathfinder, Quest, Rogue, Sentra, Titan, Versa, Xterra **Engines:** All **Transmissions:** All	**EVAP Canister Vent Control Valve Circuit Closed** EVAP canister vent control valve remains closed under specified driving conditions. **Possible Causes:** • EVAP canister vent control valve • EVAP control system pressure sensor and the circuit • Blocked rubber tube to EVAP canister vent control valve • EVAP canister is saturated with water

DTC: P0451
2T CCM, MIL: Yes
Years: 2007, 2008
Models: 350Z, Altima, Armada, Frontier, Maxima, Murano, Pathfinder, Quest, Rogue, Sentra, Titan, Versa, Xterra
Engines: All
Transmissions: All
	EVAP Control System Pressure Sensor Performance
	Ignition is ON. ECM detects a sloshing signal from the EVAP control system pressure sensor.
	Possible Causes:
	• Harness or connectors
	  • EVAP control system pressure sensor circuit is shorted.
	  • CKP sensor (POS) circuit is shorted.
	  • CMP sensor (PHASE) (bank 2) circuit is shorted.
	  • EVT control position sensor (bank 2) circuit is shorted.
	  • APP sensor 2 circuit is shorted.
	  • Refrigerant pressure sensor circuit is shorted.
	• EVAP control system pressure sensor
	• Crankshaft position sensor (POS)
	• Camshaft position sensor (PHASE) (bank 2)
	• Exhaust valve timing control position sensor (bank 2)
	• Accelerator pedal position sensor
	• Refrigerant pressure sensor

DTC: P0452
2T CCM, MIL: Yes
Years: 2007, 2008
Models: 350Z, Altima, Armada, Frontier, Maxima, Murano, Pathfinder, Quest, Rogue, Sentra, Titan, Versa, Xterra
Engines: All
Transmissions: All
	EVAP Control System Pressure Sensor Low Input
	Ignition is ON. An excessively low voltage is sent to the ECM from the sensor.
	Possible Causes:
	• Harness or connectors
	  • EVAP control system pressure sensor circuit is shorted.
	  • CKP sensor (POS) circuit is shorted.
	  • CMP sensor (PHASE) (bank 2) circuit is shorted.
	  • EVT control position sensor (bank 2) circuit is shorted.
	  • APP sensor 2 circuit is shorted.
	  • Refrigerant pressure sensor circuit is shorted.
	• EVAP control system pressure sensor
	• Crankshaft position sensor (POS)
	• Camshaft position sensor (PHASE) (bank 2)
	• Exhaust valve timing control position sensor (bank 2)
	• Accelerator pedal position sensor
	• Refrigerant pressure sensor

DTC: P0453
2T CCM, MIL: Yes
Years: 2007, 2008
Models: 350Z, Altima, Armada, Frontier, Maxima, Murano, Pathfinder, Quest, Rogue, Sentra, Titan, Versa, Xterra
Engines: All
Transmissions: All
	EVAP Control System Pressure Sensor High Input
	Ignition is ON. An excessively high voltage is sent to the ECM from the sensor.
	Possible Causes:
	• Harness or connectors
	  • EVAP control system pressure sensor circuit is shorted.
	  • CKP sensor (POS) circuit is shorted.
	  • CMP sensor (PHASE) (bank 2) circuit is shorted.
	  • EVT control position sensor (bank 2) circuit is shorted.
	  • APP sensor 2 circuit is shorted.
	  • Refrigerant pressure sensor circuit is shorted.
	• EVAP control system pressure sensor
	• Crankshaft position sensor (POS)
	• Camshaft position sensor (PHASE) (bank 2)
	• Exhaust valve timing control position sensor (bank 2)
	• Accelerator pedal position sensor
	• Refrigerant pressure sensor
	• EVAP canister vent control valve
	• EVAP canister
	• Rubber hose from EVAP canister vent control valve to vehicle frame

DTC	Trouble Code Title, Conditions & Possible Causes
DTC: P0455 **2T EVAP, MIL: Yes** **Years: 2007, 2008** **Models:** 350Z, Altima, Armada, Frontier, Maxima, Murano, Pathfinder, Quest, Rogue, Sentra, Titan, Versa, Xterra **Engines:** All **Transmissions:** All	**EVAP System Gross Leak Detected** EVAP control system gross leak detected. EVAP control system has a very large leak such as fuel filler cap fell off, EVAP control system does not operate properly. **Possible Causes:** • Fuel filler cap remains open or fails to close. • Incorrect fuel tank vacuum relief valve • Incorrect fuel filler cap used • Foreign matter caught in fuel filler cap. • Leak is in line between intake manifold and EVAP canister purge volume control solenoid valve. • Foreign matter caught in EVAP canister vent control valve. • EVAP canister or fuel tank leaks • EVAP purge line (pipe and rubber tube) leaks • EVAP purge line rubber tube bent. • Loose or disconnected rubber tube • EVAP canister vent control valve and the circuit • EVAP canister purge volume control solenoid valve and the circuit • Fuel tank temperature sensor • O-ring of EVAP canister vent control valve is missing or damaged. • EVAP control system pressure sensor • Refueling EVAP vapor cut valve • ORVR system leaks
DTC: P0456 **2T EVAP, MIL: Yes** **Years: 2007, 2008** **Models:** 350Z, Altima, Armada, Frontier, Maxima, Murano, Pathfinder, Quest, Rogue, Sentra, Titan, Versa, Xterra **Engines:** All **Transmissions:** All	**EVAP System Small Leak Detected** EVAP control system very small leak detected. EVAP control system has a very small leak. EVAP control system does not operate properly. **Possible Causes:** • Incorrect fuel tank vacuum relief valve • Incorrect fuel filler cap used • Fuel filler cap remains open or fails to close. • Foreign matter caught in fuel filler cap. • Leak is in line between intake manifold and EVAP canister purge volume control solenoid valve. • Foreign matter caught in EVAP canister vent control valve. • EVAP canister or fuel tank leaks • EVAP purge line (pipe and rubber tube) leaks • EVAP purge line rubber tube bent • Loose or disconnected rubber tube • EVAP canister vent control valve and the circuit • EVAP canister purge volume control solenoid valve and the circuit • Fuel tank temperature sensor • O-ring of EVAP canister vent control valve is missing or damaged • EVAP canister is saturated with water • EVAP control system pressure sensor • Refueling EVAP vapor cut valve • ORVR system leaks • Fuel level sensor and the circuit • Foreign matter caught in EVAP canister purge volume control solenoid valve
DTC: P0460 **2T CCM, MIL: Yes** **Years: 2007, 2008** **Models:** 350Z, Altima, Armada, Frontier, Maxima, Murano, Pathfinder, Quest, Titan, Xterra **Engines:** 3.5L, 4.0L, 5.6L **Transmissions:** All	**Fuel Level Sensor Circuit Noise** Even though the vehicle is parked, a signal being varied is sent from the fuel level sensor to ECM. **Possible Causes:** • Harness or connectors (the CAN communication line is open or shorted) • Harness or connectors (the sensor circuit is open or shorted) • Unified meter and A/C amp. • Fuel level sensor
DTC: P0460 **2T CCM, MIL: Yes** **Years: 2007, 2008** **Models:** Altima, Frontier, Rogue, Sentra, Versa **Engines:** 1.8L, 2.0L, 2.5L **Transmissions:** All	**Fuel Level Sensor Circuit Noise** Even though the vehicle is parked, a signal being varied is sent from the fuel level sensor to ECM. **Possible Causes:** • Harness or connectors (CAN communication line is open or shorted) • Harness or connectors (sensor circuit is open or shorted) • Combination meter • Fuel level sensor

DTC	Trouble Code Title, Conditions & Possible Causes
DTC: P0461 **2T CCM, MIL: Yes** **Years: 2007, 2008** **Models:** 350Z, Altima, Armada, Frontier, Maxima, Murano, Pathfinder, Quest, Rogue, Sentra, Titan, Versa, Xterra **Engines:** All **Transmissions:** All	**Fuel Level Sensor Range/Performance** Engine running, then after the vehicle traveled a distance of more than 30 miles, the PCM did not detect any change in the Fuel Level sensor signal (i.e., no change after driving for several miles). **Possible Causes:** • Harness or connectors (the CAN communication line is open or shorted) • Harness or connectors (the sensor circuit is open or shorted) • Combination meter • Unified meter and A/C amp. • Fuel level sensor
DTC: P0462 **2T CCM, MIL: Yes** **Years: 2007, 2008** **Models:** 350Z, Altima, Armada, Frontier, Maxima, Murano, Pathfinder, Quest, Rogue, Sentra, Titan, Versa, Xterra **Engines:** All **Transmissions:** All	**Fuel Level Sensor Circuit Low Input** An excessively low voltage signal is sent to the ECM from the sensor. **Possible Causes:** • Harness or connectors (the CAN communication line is open or shorted) • Harness or connectors (the sensor circuit is open or shorted) • Combination meter • Unified meter and A/C amp. • Fuel level sensor
DTC: P0463 **2T CCM, MIL: Yes** **Years: 2007, 2008** **Models:** 350Z, Altima, Armada, Frontier, Maxima, Murano, Pathfinder, Quest, Rogue, Sentra, Titan, Versa, Xterra **Engines:** All **Transmissions:** All	**Fuel Level Sensor Circuit High Input** An excessively high voltage signal is sent to the ECM from the sensor. **Possible Causes:** • Harness or connectors (the CAN communication line is open or shorted) • Harness or connectors (the sensor circuit is open or shorted) • Combination meter • Unified meter and A/C amp. • Fuel level sensor
DTC: P0500 **2T CCM, MIL: Yes** **Years: 2007, 2008** **Models:** 350Z, Altima, Armada, Frontier, Maxima, Murano, Pathfinder, Quest, Rogue, Sentra, Titan, Versa, Xterra **Engines:** All **Transmissions:** All	**Vehicle Speed Sensor Circuit Malfunction** The almost 0 MPH signal from vehicle speed sensor is sent to ECM even when vehicle is being driven. **Possible Causes:** • Harness or connectors (the CAN communication line is open or shorted) • Harness or connectors (the vehicle speed signal circuit is open or shorted) • Wheel sensor • Combination meter • Unified meter and A/C amp. • VDC/TCS/ABS control unit (with VDC models) • ABS actuator and electric unit (control unit) - (without VDC models)
DTC: P0506 **2T CCM, MIL: Yes** **Years: 2007, 2008** **Models:** 350Z, Altima, Armada, Frontier, Maxima, Murano, Pathfinder, Quest, Rogue, Sentra, Titan, Versa, Xterra **Engines:** All **Transmissions:** All	**Idle Speed Control (ISC) System Low RPM** Idle speed is less than 100 RPM below the target idle RPM. **Possible Causes:** • Electric throttle control actuator • Intake air leaks
DTC: P0507 **2T CCM, MIL: Yes** **Years: 2007, 2008** **Models:** 350Z, Altima, Armada, Frontier, Maxima, Murano, Pathfinder, Quest, Rogue, Sentra, Titan, Versa, Xterra **Engines:** All **Transmissions:** All	**Idle Speed Control (ISC) System High RPM** Idle speed is more than 200 RPM above the target idle RPM. **Possible Causes:** • Electric throttle control actuator • Intake air leaks • PCV system

DTC	Trouble Code Title, Conditions & Possible Causes
DTC: P0550 **2T CCM, MIL: Yes** **Years: 2007, 2008** **Models:** 350Z, Altima, Armada, Frontier, Maxima, Murano, Pathfinder, Quest, Rogue, Sentra, Titan, Versa, Xterra **Engines:** All (Exc. 2.0L) **Transmissions:** All	**Power Steering Pressure (PSP) Sensor Circuit Performance** An excessively low or high voltage from the sensor is sent to ECM. **Possible Causes:** • Harness or connectors (the sensor circuit is open or shorted) • Power steering pressure sensor
DTC: P0603 **2T CCM, MIL: Yes** **Years: 2007, 2008** **Models:** 350Z, Altima, Armada, Frontier, Maxima, Murano, Pathfinder, Quest, Rogue, Sentra, Titan **Engines:** 2.5L, 3.5L, 4.0L, 5.6L **Transmissions:** All	**ECM Power Supply Circuit** ECM back-up RAM system does not function properly. **Possible Causes:** • Harness or connectors (the ECM power supply circuit is open or shorted) • Power steering pressure sensor
DTC; P0605 **2T PCM, MIL: Yes** **Years: 2007, 2008** **Models:** 350Z, Altima, Armada, Frontier, Maxima, Murano, Pathfinder, Quest, Rogue, Sentra, Titan, Versa, Xterra **Engines:** All **Transmissions:** All	**ECM Internal Error** ECM calculation function is malfunctioning. ECM EEP-ROM system is malfunctioning. ECM self-shut off function is malfunctioning. **Possible Causes:** • ECM
DTC; P0607 **2T PCM, MIL: Yes** **Years: 2007, 2008** **Models:** Altima, Rogue, Sentra **Engines:** 2.5L **Transmissions:** All	**ECM CAN Communication Bus** When detecting error during the initial diagnosis of CAN controller of ECM. **Possible Causes:** • ECM

DTC; P0643
2T PCM, MIL: Yes
Years: 2007, 2008
Models: 350Z, Altima, Armada, Frontier, Maxima, Murano, Pathfinder, Quest, Rogue, Sentra, Titan, Versa, Xterra
Engines: All
Transmissions: All
	Sensor Power Supply Circuit Short
	ECM detects a voltage of power source for sensor is excessively low or high.
	Possible Causes:
	• Harness or connectors
	  • APP sensor 1 circuit is shorted.
	  • TP sensor circuit is shorted.
	  • CMP sensor (PHASE) (bank 1) circuit is shorted.
	  • EVT control position sensor (bank 1) circuit is shorted.
	  • PSP sensor circuit is shorted.
	• Accelerator pedal position sensor
	• Throttle position sensor
	• Camshaft position sensor (PHASE) (bank 1)
	• Exhaust valve timing control position sensor (bank 1)
	• Power steering pressure sensor

DTC; P0850 **2T PCM, MIL: Yes** **Years: 2007, 2008** **Models:** 350Z, Altima, Armada, Frontier, Maxima, Murano, Pathfinder, Quest, Rogue, Sentra, Titan, Versa, Xterra **Engines:** All **Transmissions:** All	**Park/Neutral Position (PNP) Switch** The signal of the park/neutral position (PNP) switch is not changed in the process of engine starting and driving. **Possible Causes:** • Harness or connectors: the park/neutral position (PNP) switch circuit is open or shorted. • Park/neutral position (PNP) switch • TCM (A/T models)

OBD II Trouble Code List (P1xxx Codes)

DTC	Trouble Code Title, Conditions & Possible Causes
DTC; P1078 **2T PCM, MIL: Yes** **Years: 2007, 2008** **Models:** 350Z **Engines:** 3.5L **Transmissions:** All	**Exhaust Valve Timing (EVT) Bank 1 Control Position Sensor** Engine at normal operating temperature and running at about 2500 RPM. Transmission in P or N. Accessories OFF (no load). An excessively high or low input signal to ECM is sent by the sensor. **Possible Causes:** • Harness or connectors: Exhaust valve timing control position sensor (bank 1) circuit is open or shorted • Exhaust valve timing control position sensor (bank 1) • Crankshaft position sensor (POS) • Camshaft position sensor (PHASE) (bank 1) • Accumulation of debris to the signal pick-up portion of the camshaft

DTC; P1084
2T PCM, MIL: Yes
Years: 2007, 2008
Models: 350Z
Engines: 3.5L
Transmissions: All
	Exhaust Valve Timing (EVT) Bank 2 Control Position Sensor
	Engine at normal operating temperature and running at about 2500 RPM. Transmission in P or N. Accessories OFF (no load). An excessively high or low input signal to ECM is sent by the sensor.
	Possible Causes:
	• Harness or connectors:
	  • EVT control position sensor (bank 2) circuit is shorted.
	  • CKP sensor (POS) circuit is shorted.
	  • CMP sensor (PHASE) (bank 2) circuit is open or shorted.
	  • APP sensor 2 circuit is shorted.
	  • EVAP control system pressure sensor circuit is shorted.
	  • Refrigerant pressure sensor circuit is shorted.
	• Exhaust valve timing control position sensor (bank 2)
	• Crankshaft position sensor (POS)
	• Camshaft position sensor (PHASE) (bank 2)
	• Accelerator pedal position sensor
	• EVAP control system pressure sensor
	• Refrigerant pressure sensor
	• Accumulation of debris to the signal pick-up portion of the camshaft

DTC: P1140 **2T CCM, MIL: Yes** **Years: 2007, 2008** **Models:** Armada, Pathfinder, Titan **Engines:** 5.6L **Transmissions:** All	**Intake Valve Timing Control Position Sensor Circuit (Bank 1)** An excessively high or low voltage from the sensor is sent to ECM. **Possible Causes:** • Harness or connectors (Intake valve timing control position sensor circuit is open or shorted) • Intake valve timing control position sensor • Crankshaft position sensor (POS) • Camshaft position sensor (PHASE) • Accumulation of debris to the signal pick-up portion of the camshaft sprocket
DTC: P1145 **2T CCM, MIL: Yes** **Years: 2007, 2008** **Models:** Armada, Pathfinder, Titan **Engines:** 5.6L **Transmissions:** All	**Intake Valve Timing Control Position Sensor Circuit (Bank 2)** An excessively high or low voltage from the sensor is sent to ECM. **Possible Causes:** • Harness or connectors (Intake valve timing control position sensor circuit is open or shorted) • Intake valve timing control position sensor • Crankshaft position sensor (POS) • Camshaft position sensor (PHASE) • Accumulation of debris to the signal pick-up portion of the camshaft sprocket
DTC: P1148 **2T CCM, MIL: Yes** **Years: 2007, 2008** **Models:** 350Z, Altima, Armada, Frontier, Maxima, Murano, Pathfinder, Quest, Titan, Xterra **Engines:** 3.5L, 4.0L, 5.6L **Transmissions:** All	**Closed Loop Malfunction Detected (Bank 1)** Engine running in closed loop for over 2 minutes, and the PCM detected that the engine was not operating in closed loop mode. **Possible Causes:** • A/FS or HO2S signal circuit is open or shorted to ground • A/FS or HO2S heater is damaged or has failed • A/FS or HO2S is damaged, contaminated or has failed • PCM has failed
DTC: P1148 **2T CCM, MIL: Yes** **Years: 2007, 2008** **Models:** Altima, Frontier, Rogue, Sentra, Versa **Engines:** 1.8L, 2.0L, 2.5L **Transmissions:** All	**Closed Loop Malfunction Detected** The closed loop control function for bank 1 does not operate even when vehicle is driving in the specified condition. The closed loop control function for bank 2 does not operate even when vehicle is driving in the specified condition. **Possible Causes:** • Harness or connectors (A/F sensor 1 circuit is open or shorted.) • A/F sensor 1 • A/F sensor 1 heater

DTC	Trouble Code Title, Conditions & Possible Causes
DTC: P1168 **2T CCM, MIL: Yes** **Years: 2007, 2008** **Models:** 350Z, Altima, Armada, Frontier, Maxima, Murano, Pathfinder, Quest, Titan, Xterra **Engines:** 3.5L, 4.0L, 5.6L **Transmissions:** All	**Closed Loop Malfunction Detected (Bank 2)** Engine started, engine runtime from 2-5 minutes, and the PCM detected the engine was not operating in closed loop during the test. **Possible Causes:** • A/FS or HO2S signal circuit is open or shorted to ground • A/FS or HO2S heater is damaged or has failed • A/FS or HO2S is damaged, contaminated or has failed • PCM has failed
DTC: P1211 **2T CCM, MIL: Yes** **Years: 2007, 2008** **Models:** 350Z, Altima, Armada, Frontier, Maxima, Murano, Pathfinder, Quest, Titan, Xterra **Engines:** 3.5L, 4.0L, 5.6L **Transmissions:** All	**TCS Control Unit Malfunction Detected** ECM receives a malfunction information from VDC/TCS/ABS control unit (with VDC models) or "ABS actuator and electric unit (control unit)" (without VDC models). **Possible Causes:** • VDC/TCS/ABS control unit (with VDC models) • ABS actuator and electric unit (control unit) (without VDC models) • TCS related parts
DTC: P1212 **2T CCM, MIL: Yes** **Years: 2007, 2008** **Models:** 350Z, Altima, Armada, Frontier, Maxima, Murano, Pathfinder, Quest, Titan, Xterra **Engines:** 3.5L, 4.0L, 5.6L **Transmissions:** All	**TCS Communication Line Not Receiving Data** ECM cannot receive information from VDC/TCS/ABS control unit (with VDC models) or "ABS actuator and electric unit (control unit)" (without VDC models) continuously. **Possible Causes:** • Harness or connectors (the CAN communication line is open or shorted.) • VDC/TCS/ABS control unit (with VDC models) • ABS actuator and electric unit (control unit) (without VDC models) • Dead or weak battery
DTC: P1217 **2T CCM, MIL: No** **Years: 2007, 2008** **Models:** 350Z, Altima, Armada, Frontier, Maxima, Murano, Pathfinder, Quest, Titan, Xterra **Engines:** 3.5L, 4.0L, 5.6L **Transmissions:** All	**Engine Over-Temperature Condition** Engine started, Engine running in closed loop for 3-5 minutes, and the PCM detected an engine overheated (engine over temperature) condition for too long a period during the CCM Rationality test. **Possible Causes:** • Engine coolant low or an incorrect coolant mixture exists • Engine cooling fan circuit(s) open or shorted to ground • Engine cooling fan is damaged or has failed • Check cooling system components (radiator hose, cap, etc.) • Check the thermostat operation (it may be stuck partly closed)
DTC: P1217 **2T CCM, MIL: No** **Years: 2007, 2008** **Models:** Altima, Frontier **Engines:** 1.8L, 2.0L, 2.5L **Transmissions:** All	**Engine Over-Temperature Condition** Cooling fan does not operate properly (Overheat). Cooling fan system does not operate properly (Overheat). Engine coolant was not added to the system using the proper filling method. Engine coolant is not within the specified range. **Possible Causes:** • Harness or connectors (cooling fan circuit is open or shorted.) • IPDM E/R (Cooling fan relay-1) • Cooling fan relays-2 and -3 • Cooling fan motor • Radiator hose • Radiator • Radiator cap • Reservoir tank • Water pump • Thermostat • Water control valve
DTC: P1225 **2T CCM, MIL: Yes** **Years: 2007, 2008** **Models:** 350Z, Altima, Armada, Frontier, Maxima, Murano, Pathfinder, Quest, Titan, Xterra **Engines:** 3.5L, 4.0L, 5.6L **Transmissions:** All	**TP Sensor (Bank 1) Closed Position Too Low Input** Closed throttle position value is extremely low. **Possible Causes:** • Electric throttle control actuator (TP sensor 1 and 2)

DTC	Trouble Code Title, Conditions & Possible Causes
DTC: P1225 **2T CCM, MIL: Yes** **Years: 2007, 2008** **Models:** Altima, Frontier **Engines:** 1.8L, 2.0L, 2.5L **Transmissions:** All	**TP Sensor Closed Position Too Low Input** Closed throttle position value is extremely low. **Possible Causes:** • Electric throttle control actuator (TP sensor 1 and 2)
DTC: P1226 **2T CCM, MIL: Yes** **Years: 2007, 2008** **Models:** 350Z, Altima, Armada, Frontier, Maxima, Murano, Pathfinder, Quest, Titan, Xterra **Engines:** 3.5L, 4.0L, 5.6L **Transmissions:** All	**TP Sensor (Bank 1) Learning Performance** Closed throttle position learning is not performing successfully or repeatedly **Possible Causes:** • Electric throttle control actuator (TP sensor 1 and 2)
DTC: P1226 **2T CCM, MIL: Yes** **Years: 2007, 2008** **Models:** Altima, Frontier **Engines:** 1.8L, 2.0L, 2.5L **Transmissions:** All	**TP Sensor Learning Performance** Closed throttle position learning is not performing successfully or repeatedly. **Possible Causes:** • Electric throttle control actuator (TP sensor 1 and 2)

DTC: P1233
2T CCM, MIL: Yes
Years: 2007, 2008
Models: 350Z
Engines: 3.5L
Transmissions: All
	Electric Throttle Control Performance (Bank 1)
	Throttle control function does not perform properly.
	Possible Causes:
	   • Electric throttle control motor circuit open or shorted
	   • Electric throttle control actuator

DTC: P1234 **2T CCM, MIL: Yes** **Years: 2007, 2008** **Models:** 350Z **Engines:** 3.5L **Transmissions:** All	**TP Sensor (Bank 2) Closed Position Too Low Input** Closed throttle position value is extremely low. **Possible Causes:** • Electric throttle control actuator (TP sensor 1 and 2)
DTC: P1235 **2T CCM, MIL: Yes** **Years: 2007, 2008** **Models:** 350Z **Engines:** 3.5L **Transmissions:** All	**TP Sensor (Bank 2) Closed Position Learning Performance** Closed throttle position learning is not performing successfully or repeatedly **Possible Causes:** • Electric throttle control actuator (TP sensor 1 and 2)

DTC: P1236
2T CCM, MIL: Yes
Years: 2007, 2008
Models: 350Z
Engines: 3.5L
Transmissions: All
	Throttle Control Motor Circuit Short (Bank 2)
	ECM detects a short in throttle control motor circuit.
	Possible Causes:
	   • Electric throttle control motor circuit shorted
	   • Electric throttle control actuator or motor

DTC: P1238 **2T CCM, MIL: Yes** **Years: 2007, 2008** **Models:** 350Z **Engines:** 3.5L **Transmissions:** All	**Electric Throttle Control Actuator Function (Bank 2)** Electric throttle control actuator does not function properly due to the return spring malfunction. Throttle valve opening angle in fail-safe mode is not in specified range. ECM detects the throttle valve is stuck open. **Possible Causes:** • Electric throttle control actuator

DTC: P1239
2T CCM, MIL: Yes
Years: 2007, 2008
Models: 350Z
Engines: 3.5L
Transmissions: All
	TP Sensor Circuit Range/Performance (Bank 2)
	Engine stopped. Shift lever: D (A/T) or 1st (M/T) position. Rationally incorrect voltage is sent to ECM compared with the signals from TP sensor 1 and TP sensor 2.
	Possible Causes:
	   • Harness or connector (TP sensor 1 and 2 circuit is open or shorted.)
	   • Electric throttle control actuator (TP sensor 1 and 2)

DTC	Trouble Code Title, Conditions & Possible Causes
DTC: P1290 **2T CCM, MIL: Yes** **Years: 2007, 2008** **Models:** 350Z **Engines:** 3.5L **Transmissions:** All	**Thottle Control Motor Relay Circuit Open (Bank 2)** Ignition switch is ON. ECM detects a voltage of power source for throttle control motor is excessively low. **Possible Causes:** • Harness or connector (throttle control motor relay circuit is open.) • Throttle control motor relay
DTC: P1402 **2T CCM, MIL: Yes** **Years: 2007, 2008** **Models:** Quest **Engines:** 3.5L **Transmissions:** All	**EGR Function** Engine at normal operating temperature. A/C OFF. Gear in P or N. No electrical load. EGR flow is detected under the condition that does not call for EGR. **Possible Causes:** • Harness or connectors (EGR volume control valve circuit is open or shorted.) • EGR volume control valve leaking or stuck open • EGR temperature sensor
DTC: P1421 **2T CCM, MIL: Yes** **Years: 2007, 2008** **Models:** 350Z, Altima, Armada, Frontier, Maxima, Murano, Pathfinder, Quest, Rogue, Sentra, Titan, Versa, Xterra **Engines:** All **Transmissions:** All	**Cold Start Control** ECM does not control ignition timing and engine idle speed properly when engine is started with pre-warming up condition. **P ssible Causes:** • Lack of intake air volume • Fuel injection system • ECM
DTC: P1550 **2T CCM, MIL: Yes** **Years: 2007, 2008** **Models:** 350Z, Altima, Armada, Frontier, Maxima, Murano, Pathfinder, Quest, Titan, Xterra **Engines:** 3.5L, 4.0L, 5.6L **Transmissions:** All	**Battery Current Sensor Circuit Range/Performance** The output voltage of the battery current sensor remains within the specified range while engine is running. **Possible Causes:** • Harness or connectors (sensor circuit is open or shorted) • Battery current sensor
DTC: P1551 **2T CCM, MIL: Yes** **Years: 2007, 2008** **Models:** 350Z, Altima, Armada, Frontier, Maxima, Murano, Pathfinder, Quest, Titan, Xterra **Engines:** 3.5L, 4.0L, 5.6L **Transmissions:** All	**Battery Current Sensor Circuit Low Voltage** Excessivley low voltage from the sensor is sent to the ECM. **Possible Causes:** • Harness or connectors (sensor circuit is open or shorted) • Battery current sensor
DTC: P1552 **2T CCM, MIL: Yes** **Years: 2007, 2008** **Models:** Altima, Armada, Frontier, Maxima **Engines:** 3.5L, 4.0L, 5.6L **Transmissions:** All	**Battery Current Sensor Circuit High Voltage** Excessivley high voltage from the sensor is sent to the ECM. **Possible Causes:** • Harness or connectors (sensor circuit is open or shorted) • Battery current sensor
DTC: P1553 **2T CCM, MIL: Yes** **Years: 2007, 2008** **Models:** 350Z, Altima, Armada, Frontier, Maxima, Murano, Pathfinder, Quest, Titan, Xterra **Engines:** 3.5L, 4.0L, 5.6L **Transmissions:** All	**Battery Current Sensor Performance** The signal voltage transmitted from the sensor to ECM is higher than the amount of the maximum power generation. **Possible Causes:** • Harness or connectors (sensor circuit is open or shorted) • Battery current sensor

DTC	Trouble Code Title, Conditions & Possible Causes
DTC: P1554 **2T CCM, MIL: Yes** **Years: 2007, 2008** **Models:** 350Z, Altima, Armada, Frontier, Maxima, Murano, Pathfinder, Quest, Titan, Xterra **Engines:** 3.5L, 4.0L, 5.6L **Transmissions:** All	**Battery Current Sensor Performance** The output voltage of the battery current sensor is lower than the specified value while the battery voltage is high enough. **Possible Causes:** • Harness or connectors (sensor circuit is open or shorted) • Battery current sensor
DTC: P1564 **2T CCM, MIL: Yes** **Years: 2007, 2008** **Models:** 350Z, Altima, Armada, Frontier, Maxima, Murano, Pathfinder, Quest, Rogue, Sentra, Titan, Versa, Xterra **Engines:** All **Transmissions:** All	**ASCD Steering Switch Malfunction** Ignition switch is ON. An excessively high voltage signal from the ASCD steering switch is sent to ECM. ECM detects that input signal from the ASCD steering switch is out of the specified range. ECM detects that the ASCD steering switch is stuck ON. **Possible Causes:** • Harness or connectors: switch circuit is open or shorted • ASCD steering switch • ECM
DTC: P1572 **2T CCM, MIL: Yes** **Years: 2007, 2008** **Models:** 350Z, Altima, Armada, Frontier, Maxima, Murano, Pathfinder, Quest, Rogue, Sentra, Titan, Versa, Xterra **Engines:** All **Transmissions:** All	**ASCD Brake Switch Malfunction** Ignition switch is ON. When the vehicle speed is above 19 MPH, ON signals from the stop lamp switch and the ASCD brake switch are sent to ECM at the same time. Or, ASCD brake switch signal is not sent to ECM for extremely long time while the vehicle is driving. **Possible Causes:** • Harness or connectors (stop lamp switch circuit is shorted.) • Harness or connectors (ASCD brake switch circuit is shorted.) • Harness or connectors (ASCD clutch switch circuit is shorted.) (M/T models) • Stop lamp switch • ASCD brake switch • ASCD clutch switch (M/T models) • Incorrect stop lamp switch installation • Incorrect ASCD brake switch installation • Incorrect ASCD clutch switch installation (M/T models) • ECM
DTC: P1574 **2T CCM, MIL: Yes** **Years: 2007, 2008** **Models:** 350Z, Altima, Armada, Frontier, Maxima, Murano, Pathfinder, Quest, Rogue, Sentra, Titan, Versa, Xterra **Engines:** All **Transmissions:** All	**ASCD Vehicle Speed Sensor** ECM detects a difference between the two vehicle speed signals is out of range. **Possible Causes:** • Harness or connectors (CAN communication line is open or shorted.) • Unified meter and A/C amp. • VDC/TCS/ABS control unit (with VDC models) • ABS actuator and electric unit (control unit) (without VDC models) • Wheel sensor • TCM (A/T models) • ECM
DTC: P1715 **2T CCM, MIL: Yes** **Years: 2007, 2008** **Models:** 350Z, Versa **Engines:** 1.8L, 3.5L **Transmissions:** All	**Input Speed Sensor (Turbine Revolution Sensor) Circuit Malfunction** Turbine revolution sensor signal is different from the theoretical value calculated by ECM from revolution sensor signal and engine RPM signal. **Possible Causes:** • Harness or connectors (CAN communication line is open or shorted) • Harness or connectors (Turbine revolution sensor circuit is open or shorted) • TCM
DTC: P1715 **2T CCM, MIL: Yes** **Years: 2007, 2008** **Models:** Altima, Armada, Frontier, Maxima, Murano, Pathfinder, Quest, Rogue, Sentra, Titan, Versa **Engines:** All **Transmissions:** All	**Input Speed Sensor (Primary Speed Sensor) (TCM Output)** Primary speed sensor signal is different from the theoretical value calculated by ECM from secondary speed sensor signal and engine RPM signal. **Possible Causes:** • Harness or connectors (CAN communication line is open or shorted) • Harness or connectors (Primary speed sensor circuit is open or shorted) • TCM

DTC	Trouble Code Title, Conditions & Possible Causes
DTC: P1720 **2T CCM, MIL: Yes** **Years: 2007, 2008** **Models:** Altima **Engines:** 3.5L **Transmissions:** All	**Vehicle Speed Sensor (TCM Output)** A difference between two vehicle speed signals is out of the specified range. **Possible Causes:** • Harness or connectors (Secondary speed sensor circuit is open or shorted.) • Harness or connectors (Wheel sensor circuit is open or shorted.) • TCM • Secondary speed sensor • ABS actuator and electric unit (control unit) • Wheel sensor • Combination meter
DTC: P1800 **2T CCM, MIL: Yes** **Years: 2007, 2008** **Models:** Altima, Armada, Frontier, Maxima, Pathfinder, Quest, Titan **Engines:** 3.5L, 4.0L, 5.6L **Transmissions:** All	**VIAS Control Solenoid Valve 1** An excessively low or high voltage signal is sent to ECM through the valve 1. **Possible Causes:** • Harness or connectors (solenoid valve 1 circuit is open or shorted.) • VIAS control solenoid valve 1
DTC: P1801 **2T CCM, MIL: Yes** **Years: 2007, 2008** **Models:** Altima, Armada, Frontier, Maxima, Pathfinder **Engines:** 3.5L, 4.0L, 5.6L **Transmissions:** All	**VIAS Control Solenoid Valve 2** An excessively low or high voltage signal is sent to ECM through the valve 2. **Possible Causes:** • Harness or connectors (solenoid valve 1 circuit is open or shorted.) • VIAS control solenoid valve 2
DTC: P1805 **2T CCM, MIL: Yes** **Years: 2007, 2008** **Models:** 350Z, Altima, Armada, Frontier, Maxima, Murano, Pathfinder, Quest, Rogue, Sentra, Titan, Versa, Xterra **Engines:** All **Transmissions:** All	**Brake Switch Loss of Signal** Ignition is ON. A brake switch signal is not sent to ECM for extremely long time while the vehicle is driving. **Possible Causes:** • Harness or connectors (stop lamp switch circuit is open or shorted.) • Stop lamp switch

OBD II Trouble Code List (P2xxx Codes)

DTC	Trouble Code Title, Conditions & Possible Causes

DTC: P2004
2T CCM, MIL: Yes
Years: 2007, 2008
Models: 350Z, Altima
Engines: 3.5L
Transmissions: All
	Tumble Control Valve Stuck
	The target angle of tumble control valve controlled by ECM and the input signal from tumble control valve position sensor is not in the normal range.
	Possible Causes:
	• Harness or connectors (Tumble control valve motor circuit is open or shorted.)
	  • Tumble control valve position sensor circuit is open or shorted.
	  • Accelerator pedal position sensor 2 circuit is shorted.
	  • Crankshaft position sensor (POS) circuit is shorted.
	  • EVAP control system pressure sensor circuit is shorted.
	  • Refrigerant pressure sensor circuit is shorted.
	• Tumble control valve actuator (Tumble control valve motor; Tumble control valve position sensor
	• Accelerator pedal position sensor (APP sensor 2)
	• Crankshaft position sensor (POS)
	• EVAP control system pressure sensor
	• Refrigerant pressure sensor

DTC	Trouble Code Title, Conditions & Possible Causes

DTC: P2014
2T CCM, MIL: Yes
Years: 2007, 2008
Models: 350Z, Altima
Engines: 3.5L
Transmissions: All
	Tumble Control Valve Position Sensor Circuit
	An excessively low or high voltage from the sensor is sent to ECM.
	Possible Causes:
	• Harness or connectors (Tumble control valve motor circuit is open or shorted.)
	  • Tumble control valve position sensor circuit is open or shorted.
	  • Accelerator pedal position sensor 2 circuit is shorted.
	  • Crankshaft position sensor (POS) circuit is shorted.
	  • EVAP control system pressure sensor circuit is shorted.
	  • Refrigerant pressure sensor circuit is shorted.
	• Tumble control valve actuator (Tumble control valve motor; Tumble control valve position sensor)
	• Accelerator pedal position sensor (APP sensor 2)
	• Crankshaft position sensor (POS)
	• EVAP control system pressure sensor
	• Refrigerant pressure sensor

DTC: P2100 **2T CCM, MIL: Yes** **Years: 2007, 2008** **Models:** 350Z, Altima, Armada, Frontier, Maxima, Murano, Pathfinder, Quest, Titan, Xterra **Engines:** 3.5L, 4.0L, 5.6L **Transmissions:** All	**Thottle Control Motor Relay Circuit Open (Bank 1)** Ignition switch is ON. ECM detects a voltage of power source for throttle control motor is excessively low. **Possible Causes:** • Harness or connector (throttle control motor relay circuit is open.) • Throttle control motor relay
DTC: P2100 **2T CCM, MIL: Yes** **Years: 2007, 2008** **Models:** Altima, Frontier, Rogue, Sentra, Versa **Engines:** 1.8L, 2.0L, 2.5L **Transmissions:** All	**Thottle Control Motor Relay Circuit Open** Ignition switch is ON. ECM detects a voltage of power source for throttle control motor is excessively low. **Possible Causes:** • Harness or connector (throttle control motor relay circuit is open.) • Throttle control motor relay
DTC: P2101 **2T CCM, MIL: Yes** **Years: 2007, 2008** **Models:** 350Z, Altima, Armada, Frontier, Maxima, Murano, Pathfinder, Quest, Titan, Xterra **Engines:** 3.5L, 4.0L, 5.6L **Transmissions:** All	**Electric Throttle Control Performance (Bank 2)** Electric throttle control function does not perform properly. **Possible Causes:** • Electric throttle control motor circuit open or shorted • Electric throttle control actuator
DTC: P2101 **2T CCM, MIL: Yes** **Years: 2007, 2008** **Models:** Altima, Frontier, Rogue, Sentra, Versa **Engines:** 1.8L, 2.0L, 2.5L **Transmissions:** All	**Electric Throttle Control Performance** Electric throttle control function does not perform properly. **Possible Causes:** • Electric throttle control motor circuit open or shorted • Electric throttle control actuator
DTC: P2103 **2T CCM, MIL: Yes** **Years: 2007, 2008** **Models:** 350Z, Altima, Armada, Frontier, Maxima, Murano, Pathfinder, Quest, Rogue, Sentra, Titan, Versa, Xterra **Engines:** All **Transmissions:** All	**Thottle Control Motor Relay Circuit Short** Ignition switch is ON. ECM detects throttle control motor relay is stuck ON. **Possible Causes:** • Harness or connector (throttle control motor relay circuit is shorted.) • Throttle control motor relay
DTC: P2118 **2T CCM, MIL: Yes** **Years: 2007, 2008** **Models:** 350Z, Altima, Armada, Frontier, Maxima, Murano, Pathfinder, Quest, Titan, Xterra **Engines:** 3.5L, 4.0L, 5.6L **Transmissions:** All	**Throttle Control Motor Circuit Short (Bank 1)** ECM detects a short in throttle control motor circuit. **Possible Causes:** • Electric throttle control motor circuit shorted • Electric throttle control actuator or motor

DTC	Trouble Code Title, Conditions & Possible Causes
DTC: P2118 **2T CCM, MIL: Yes** **Years: 2007, 2008** **Models:** Altima, Frontier, Rogue, Sentra, Versa **Engines:** 1.8L, 2.0L, 2.5L **Transmissions:** All	**Throttle Control Motor Circuit Short** ECM detects a short in throttle control motor circuit. **Possible Causes:** • Electric throttle control motor circuit shorted • Electric throttle control actuator or motor
DTC: P2119 **2T CCM, MIL: Yes** **Years: 2007, 2008** **Models:** 350Z, Altima, Armada, Frontier, Maxima, Murano, Pathfinder, Quest, Titan, Xterra **Engines:** 3.5L, 4.0L, 5.6L **Transmissions:** All	**Electric Throttle Control Actuator Function (Bank 1)** Electric throttle control actuator does not function properly due to the return spring malfunction. Throttle valve opening angle in fail-safe mode is not in specified range. ECM detects the throttle valve is stuck open. **Possible Causes:** • Electric throttle control actuator
DTC: P2119 **2T CCM, MIL: Yes** **Years: 2007, 2008** **Models:** Altima, Frontier, Rogue, Sentra, Versa **Engines:** 1.8L, 2.0L, 2.5L **Transmissions:** All	**Electric Throttle Control Actuator Function** Electric throttle control actuator does not function properly due to the return spring malfunction. Throttle valve opening angle in fail-safe mode is not in specified range. ECM detects the throttle valve is stuck open. **Possible Causes:** • Electric throttle control actuator
DTC: P2122 **2T CCM, MIL: Yes** **Years: 2007, 2008** **Models:** 350Z, Altima, Armada, Frontier, Maxima, Murano, Pathfinder, Quest, Rogue, Sentra, Titan, Versa, Xterra **Engines:** All **Transmissions:** All	**APP Sensor 1 Low Input** Ignition is ON, with engine OFF. An excessively low voltage from the APP sensor 1 is sent to ECM. **Possible Causes:** • Harness or connectors (APP sensor 1 circuit is open or shorted.) • Accelerator pedal position sensor (APP sensor 1)
DTC: P2123 **2T CCM, MIL: Yes** **Years: 2007, 2008** **Models:** 350Z, Altima, Armada, Frontier, Maxima, Murano, Pathfinder, Quest, Rogue, Sentra, Titan, Versa, Xterra **Engines:** All **Transmissions:** All	**APP Sensor 1 High Input** Ignition is ON, with engine OFF. An excessively high voltage from the APP sensor 1 is sent to ECM. **Possible Causes:** • Harness or connectors (APP sensor 1 circuit is open or shorted.) • Accelerator pedal position sensor (APP sensor 1)

DTC: P2127
2T CCM, MIL: Yes
Years: 2007, 2008
Models: 350Z, Altima, Armada, Frontier, Maxima, Murano, Pathfinder, Quest, Rogue, Sentra, Titan, Versa, Xterra
Engines: All
Transmissions: All
	APP Sensor 2 Circuit Low Input
	Ignition is ON, with engine OFF. An excessively low voltage from the APP sensor 2 is sent to ECM.
	Possible Causes:
	• Harness or connectors (APP sensor 2 dircuit is open or shorted.)
	  • CKP sensor (POS) circuit is open or shorted.
	  • CMP sensor (PHASE) (bank 2) circuit is shorted.
	  • EVT control position sensor (bank 2) circuit is shorted.
	  • EVAP control system pressure sensor circuit is shorted.
	  • Refrigerant pressure sensor circuit is shorted.
	• Accelerator pedal position sensor
	• Crankshaft position sensor (POS)
	• Camshaft position sensor (PHASE) (bank 2)
	• Exhaust valve timing control position sensor (bank 2)
	• EVAP control system pressure sensor
	• Refrigerant pressure sensor

DTC	Trouble Code Title, Conditions & Possible Causes

DTC: P2128
2T CCM, MIL: Yes
Years: 2007, 2008
Models: 350Z, Altima, Armada, Frontier, Maxima, Murano, Pathfinder, Quest, Rogue, Sentra, Titan, Versa, Xterra
Engines: All
Transmissions: All
	APP Sensor 2 Circuit High Input
	Ignition is ON, with engine OFF. An excessively high voltage from the APP sensor 2 is sent to ECM.
	Possible Causes:
	    • Harness or connectors (APP sensor 2 dircuit is open or shorted.)
	        • CKP sensor (POS) circuit is open or shorted.
	        • CMP sensor (PHASE) (bank 2) circuit is shorted.
	        • EVT control position sensor (bank 2) circuit is shorted.
	        • EVAP control system pressure sensor circuit is shorted.
	        • Refrigerant pressure sensor circuit is shorted.
	    • Accelerator pedal position sensor
	    • Crankshaft position sensor (POS)
	    • Camshaft position sensor (PHASE) (bank 2)
	    • Exhaust valve timing control position sensor (bank 2)
	    • EVAP control system pressure sensor
	    • Refrigerant pressure sensor

DTC: P2132
2T CCM, MIL: Yes
Years: 2007, 2008
Models: 350Z
Engines: 3.5L
Transmissions: All
	Throttle Position (TP) Sensor 1 (Bank 2) Circuit Low Voltage
	Excessively low voltage is sent from the sensor to ECM.
	Possible Causes:
	    • Harness or connectors (the sensor circuit is open or shorted)
	    • EOT sensor

DTC: P2133
2T CCM, MIL: Yes
Years: 2007, 2008
Models: 350Z
Engines: 3.5L
Transmissions: All
	Throttle Position (TP) Sensor 1 (Bank 2) Circuit High Voltage
	Excessively high voltage is sent from the sensor to ECM.
	Possible Causes:
	    • Harness or connectors (the sensor circuit is open or shorted)
	    • EOT sensor

DTC: P2135
2T CCM, MIL: Yes
Years: 2007, 2008
Models: 350Z, Altima, Armada, Frontier, Maxima, Murano, Pathfinder, Quest, Titan, Xterra
Engines: 3.5L, 4.0L, 5.6L
Transmissions: All
	TP Sensor Circuit Range/Performance (Bank 1)
	Engine stopped. Shift lever: D (A/T) or 1st (M/T) position. Rationally incorrect voltage is sent to ECM compared with the signals from TP sensor 1 and TP sensor 2.
	Possible Causes:
	    • Harness or connector (TP sensor 1 and 2 circuit is open or shorted.)
	    • Electric throttle control actuator (TP sensor 1 and 2)

DTC: P2135
2T CCM, MIL: Yes
Years: 2007, 2008
Models: Altima, Frontier, Rogue, Sentra, Versa
Engines: 1.8L, 2.0L, 2.5L
Transmissions: All
	TP Sensor Circuit Range/Performance
	Engine stopped. Shift lever: D (A/T) or 1st (M/T) position. Rationally incorrect voltage is sent to ECM compared with the signals from TP sensor 1 and TP sensor 2.
	Possible Causes:
	    • Harness or connector (TP sensor 1 and 2 circuit is open or shorted.)
	    • Electric throttle control actuator (TP sensor 1 and 2)

DTC: P2138
2T CCM, MIL: Yes
Years: 2007, 2008
Models: 350Z, Altima, Armada, Frontier, Maxima, Murano, Pathfinder, Quest, Rogue, Sentra, Titan, Versa, Xterra
Engines: All
Transmissions: All
	APP Sensor Circuit Range/Performance
	Ignition ON. Engine stopped. Rationally incorrect voltage is sent to ECM compared with the signals from APP sensor 1 and APP sensor 2.
	Possible Causes:
	    • Harness or connectors (APP sensor 1 and 2 dircuit is open or shorted.)
	        • APP sensor 2 dircuit is open or shorted
	        • CKP sensor (POS) circuit is open or shorted.
	        • CMP sensor (PHASE) (bank 2) circuit is shorted.
	        • EVT control position sensor (bank 2) circuit is shorted.
	        • EVAP control system pressure sensor circuit is shorted.
	        • Refrigerant pressure sensor circuit is shorted.
	    • Accelerator pedal position sensor 1 and/or 2
	    • Crankshaft position sensor (POS)
	    • Camshaft position sensor (PHASE) (bank 2)
	    • Exhaust valve timing control position sensor (bank 2)
	    • EVAP control system pressure sensor
	    • Refrigerant pressure sensor

DTC	Trouble Code Title, Conditions & Possible Causes
DTC: P2423 **2T CCM, MIL: Yes** **Years:** 2007, 2008 **Models:** Altima **Engines:** 2.5 L (CA) **Transmissions:** All	**HC Adsorption Catalyst Efficiency Below Threshhold** HC adsorption catalyst (under floor) does not operate properly. Or, HC adsorption catalyst (under floor) does not have enough oxygen storage capacity. **Possible Causes:** • HC adsorption catalyst (under floor) • Exhaust tube • Intake air leaks • Fuel injector • Fuel injector leaks • Spark plug • Improper ignition timing
DTC: P2A00 **2T CCM, MIL: Yes** **Years:** 2007, 2008 **Models:** 350Z, Altima, Armada, Frontier, Maxima, Murano, Pathfinder, Quest, Titan, Xterra **Engines:** 3.5L, 4.0L, 5.6L **Transmissions:** All	**A/F Sensor 1 (Bank 1) Circuit Range/Performance** Engine at normal operating temperature. Engine running at 2000 RPM. The output voltage computed by ECM from the A/F sensor 1 signal is shifted to the lean side for a specified period. The output voltage computed by ECM from the A/F sensor 1 signal is shifted to the rich side for a specified period. **Possible Causes:** • Air fuel ratio (A/F) sensor 1 • Air fuel ratio (A/F) sensor 1 heater • Fuel pressure • Fuel injector • Intake air leaks
DTC: P2A00 **2T CCM, MIL: Yes** **Years:** 2007, 2008 **Models:** Altima, Frontier, Rogue, Sentra, Versa **Engines:** 1.8L, 2.0L, 2.5L **Transmissions:** All	**A/F Sensor 1 Circuit Range/Performance** Engine at normal operating temperature. Engine running at 2000 RPM. The output voltage computed by ECM from the A/F sensor 1 signal is shifted to the lean side for a specified period. The output voltage computed by ECM from the A/F sensor 1 signal is shifted to the rich side for a specified period. **Possible Causes:** • Air fuel ratio (A/F) sensor 1 • Air fuel ratio (A/F) sensor 1 heater • Fuel pressure • Fuel injector • Intake air leaks
DTC: P2A03 **2T CCM, MIL: Yes** **Years:** 2007, 2008 **Models:** 350Z, Altima, Armada, Frontier, Maxima, Murano, Pathfinder, Quest, Titan, Xterra **Engines:** 3.5L, 4.0L, 5.6L **Transmissions:** All	**A/F Sensor 1 (Bank 2) Circuit Range/Performance** Engine at normal operating temperature. Engine running at 2000 RPM. The output voltage computed by ECM from the A/F sensor 1 signal is shifted to the lean side for a specified period. The output voltage computed by ECM from the A/F sensor 1 signal is shifted to the rich side for a specified period. **Possible Causes:** • Air fuel ratio (A/F) sensor 1 • Air fuel ratio (A/F) sensor 1 heater • Fuel pressure • Fuel injector • Intake air leaks

OBD II Trouble Code List (U1xx Codes)

DTC	Trouble Code Title, Conditions & Possible Causes
DTC: U1000 **2T CCM, MIL: Yes** **Years:** 2007, 2008 **Models:** 350Z, Altima, Armada, Frontier, Maxima, Murano, Pathfinder, Quest, Rogue, Sentra, Titan, Versa, Xterra **Engines:** All **Transmissions:** All	**CAN Communication Circuit** When ECM is not transmitting or receiving CAN communication signal of OBD (emission-related diagnosis) for 2 seconds or more. **Possible Causes:** • Harness or connectors (CAN communication line is open or shorted)

DTC	Trouble Code Title, Conditions & Possible Causes
DTC: U1001 **2T CCM, MIL: Yes** **Years: 2007, 2008** **Models:** 350Z, Altima, Armada, Frontier, Maxima, Murano, Pathfinder, Quest, Rogue, Sentra, Titan, Versa, Xterra **Engines:** All **Transmissions:** All	**CAN Communication Circuit** When ECM is not transmitting or receiving CAN communication signal of other than OBD (emission-related diagnosis) for 2 seconds or more. **Possible Causes:** • Harness or connectors (CAN communication line is open or shorted)
DTC: U1010 **2T CCM, MIL: Yes** **Years: 2007, 2008** **Models:** 350Z, Altima, Armada, Frontier, Maxima, Murano, Pathfinder, Quest, Rogue, Sentra, Titan, Versa, Xterra **Engines:** All **Transmissions:** All	**Control Unit (CAN)** When detecting error during the initial diagnosis of CAN controller of ECM. **Possible Causes:** • ECM

Commonly Used Abbreviations

2
2WD   Two Wheel Drive

4
4WD   Four Wheel Drive

A
A/C   Air Conditioning
ABDC   After Bottom Dead Center
ABS   Anti-lock Brakes
AC   Alternating Current
ACL   Air cleaner
ACT   Air Charge Temperature
AIR   Secondary Air Injection
ALCL   Assembly Line Communications Link
ALDL   Assembly Line Diagnostic Link
AT   Automatic Transaxle/Transmission
ATDC   After Top Dead Center
ATF   Automatic Transmission Fluid
ATS   Air Temperature Sensor
AWD   All Wheel Drive

B
BAP   Barometric Absolute Pressure
BARO   Barometric Pressure
BBDC   Before Bottom Dead Center
BCM   Body Control Module
BDC   Bottom Dead Center
BPT   Backpressure Transducer
BTDC   Before Top Dead Center
BVSV   Bimetallic Vacuum Switching Valve

C
CAC   Charge Air Cooler
CARB   California Air Resources Board
CAT   Catalytic Converter
CCC   Computer Command Control
CCCC   Computer Controlled Catalytic Converter
CCCI   Computer Controlled Coil Ignition
CCD   Computer Controlled Dwell
CDI   Capacitor Discharge Ignition
CEC   Computerized Engine Control
CFI   Continuous Fuel Injection
CIS   Continuous Injection System
CIS-E   Continuous Injection System - Electronic
CKP   Crankshaft Position
CL   Closed Loop
CMP   Camshaft Position
CPP   Clutch Pedal Position
CTOX   Continuous Trap Oxidizer System
CTP   Closed Throttle Position
CVC   Constant Vacuum Control
CYL   Cylinder

D
DBC   Dual Bed Catalyst
DC   Direct Current
DFI   Direct Fuel Injection
DIS   Distributorless Ignition System
DLC   Data Link Connector
DMM   Digital Multimeter
DOHC   Double Overhead Camshaft
DRB   Diagnostic Readout Box
DTC   Diagnostic Trouble Code
DTM   Diagnostic Test Mode
DVOM   Digital Volt/Ohmmeter

E
EBCM   Electronic Brake Control Module
ECM   Engine Control Module
ECT   Engine Coolant Temperature
ECU   Engine Control Unit or Electronic Control Unit
EDIS   Electronic Distributorless Ignition System
EEC   Electronic Engine Control
EEPROM   Electrically Erasable Programmable Read Only Memory
EFE   Early Fuel Evaporation •
EGR   Exhaust Gas Recirculation
EGRT   Exhaust Gas Recirculation Temperature
EGRVC   EGR Valve Control
EPROM   Erasable Programmable Read Only Memory
EVAP   Evaporative Emissions
EVP   EGR Valve Position

F
FBC   Feedback Carburetor
FEEPROM   Flash Electrically Erasable Programmable Read Only Memory
FF   Flexible Fuel
FI   Fuel Injection
FT   Fuel Trim
FWD   Front Wheel Drive

G
GND   Ground

H
HAC   High Altitude Compensation
HEGO   Heated Exhaust Gas Oxygen sensor
HEI   High Energy Ignition
HO2 Sensor   Heated Oxygen Sensor

I
IAC   Idle Air Control
IAT   Intake Air Temperature
ICM   Ignition Control Module
IFI   Indirect Fuel Injection
IFS   Inertia Fuel Shutoff
ISC   Idle Speed Control
IVSV   Idle Vacuum Switching Valve

Commonly Used Abbreviations

K

KOEO	Key On, Engine Off
KOER	Key ON, Engine Running
KS	Knock Sensor

M

MAF	Mass Air Flow
MAP	Manifold Absolute Pressure
MAT	Manifold Air Temperature
MC	Mixture Control
MDP	Manifold Differential Pressure
MFI	Multiport Fuel Injection
MIL	Malfunction Indicator Lamp or Maintenance
MST	Manifold Surface Temperature
MVZ	Manifold Vacuum Zone

N

NVRAM	Nonvolatile Random Access Memory

O

O2 Sensor	Oxygen Sensor
OBD	On-Board Diagnostic
OC	Oxidation Catalyst
OHC	Overhead Camshaft
OL	Open Loop

P

P/S	Power Steering
PAIR	Pulsed Secondary Air Injection
PCM	Powertrain Control Module
PCS	Purge Control Solenoid
PCV	Positive Crankcase Ventilation
PIP	Profile Ignition Pick-up
PNP	Park/Neutral Position
PROM	Programmable Read Only Memory
PSP	Power Steering Pressure
PTO	Power Take-Off
PTOX	Periodic Trap Oxidizer System

R

RABS	Rear Anti-lock Brake System
RAM	Random Access Memory
ROM	Read Only Memory
RPM	Revolutions Per Minute
RWAL	Rear Wheel Anti-lock Brakes
RWD	Rear Wheel Drive

S

SBC	Single Bed Converter
SBEC	Single Board Engine Controller
SC	Supercharger
SCB	Supercharger Bypass
SFI	Sequential Multiport Fuel Injection
SIR	Supplemental Inflatable Restraint
SOHC	Single Overhead Camshaft
SPL	Smoke Puff Limiter
SPOUT	Spark Output
SRI	Service Reminder Indicator
SRS	Supplemental Restraint System
SRT	System Readiness Test
SSI	Solid State Ignition
ST	Scan Tool
STO	Self-Test Output

T

TAC	Thermostatic Air Cleaner
TBI	Throttle Body Fuel Injection
TC	Turbocharger
TCC	Torque Converter Clutch
TCM	Transmission Control Module
TDC	Top Dead Center
TFI	Thick Film Ignition
TP	Throttle Position
TR Sensor	Transaxle/Transmission Range Sensor
TVV	Thermal Vacuum Valve
TWC	Three-way Catalytic Converter

V

VAF	Volume Air Flow, or Vane Air Flow
VAPS	Variable Assist Power Steering
VRV	Vacuum Regulator Valve
VSS	Vehicle Speed Sensor
VSV	Vacuum Switching Valve

W

WOT	Wide Open Throttle
WU-TWC	Warm Up Three-way Catalytic Converter

ENGLISH TO METRIC CONVERSION: TORQUE

To convert foot-pounds (ft. lbs.) to Newton-meters (Nm), multiply the number of ft. lbs. by 1.36
To convert Newton-meters (Nm) to foot-pounds (ft. lbs.), multiply the number of Nm by 0.7376

ft. lbs.	Nm	ft. lbs.	Nm	ft. lbs.	Nm	ft. lbs.	Nm
0.1	0.1	34	46.2	76	103.4	118	160.5
0.2	0.3	35	47.6	77	104.7	119	161.8
0.3	0.4	36	49.0	78	106.1	120	163.2
0.4	0.5	37	50.3	79	107.4	121	164.6
0.5	0.7	38	51.7	80	108.8	122	165.9
0.6	0.8	39	53.0	81	110.2	123	167.3
0.7	1.0	40	54.4	82	111.5	124	168.6
0.8	1.1	41	55.8	83	112.9	125	170.0
0.9	1.2	42	57.1	84	114.2	126	171.4
1	1.4	43	58.5	85	115.6	127	172.7
2	2.7	44	59.8	86	117.0	128	174.1
3	4.1	45	61.2	87	118.3	129	175.4
4	5.4	46	62.6	88	119.7	130	176.8
5	6.8	47	63.9	89	121.0	131	178.2
6	8.2	48	65.3	90	122.4	132	179.5
7	9.5	49	66.6	91	123.8	133	180.9
8	10.9	50	68.0	92	125.1	134	182.2
9	12.2	51	69.4	93	126.5	135	183.6
10	13.6	52	70.7	94	127.8	136	185.0
11	15.0	53	72.1	95	129.2	137	186.3
12	16.3	54	73.4	96	130.6	138	187.7
13	17.7	55	74.8	97	131.9	139	189.0
14	19.0	56	76.2	98	133.3	140	190.4
15	20.4	57	77.5	99	134.6	141	191.8
16	21.8	58	78.9	100	136.0	142	193.1
17	23.1	59	80.2	101	137.4	143	194.5
18	24.5	60	81.6	102	138.7	144	195.8
19	25.8	61	83.0	103	140.1	145	197.2
20	27.2	62	84.3	104	141.4	146	198.6
21	28.6	63	85.7	105	142.8	147	199.9
22	29.9	64	87.0	106	144.2	148	201.3
23	31.3	65	88.4	107	145.5	149	202.6
24	32.6	66	89.8	108	146.9	150	204.0
25	34.0	67	91.1	109	148.2	151	205.4
26	35.4	68	92.5	110	149.6	152	206.7
27	36.7	69	93.8	111	151.0	153	208.1
28	38.1	70	95.2	112	152.3	154	209.4
29	39.4	71	96.6	113	153.7	155	210.8
30	40.8	72	97.9	114	155.0	156	212.2
31	42.2	73	99.3	115	156.4	157	213.5
32	43.5	74	100.6	116	157.8	158	214.9
33	44.9	75	102.0	117	159.1	159	216.2

METRIC TO ENGLISH CONVERSION: TORQUE

To convert foot-pounds (ft. lbs.) to Newton-meters (Nm), multiply the number of ft. lbs. by 1.36
To convert Newton-meters (Nm) to foot-pounds (ft. lbs.), multiply the number of Nm by 0.7376

Nm	ft. lbs.	Nm	ft. lbs.	Nm	ft. lbs.	Nm	ft. lbs.	Nm	ft. lbs.
0.1	0.1	34	25.0	76	55.9	118	86.8	160	117.6
0.2	0.1	35	25.7	77	56.6	119	87.5	161	118.4
0.3	0.2	36	26.5	78	57.4	120	88.2	162	119.1
0.4	0.3	37	27.2	79	58.1	121	89.0	163	119.9
0.5	0.4	38	27.9	80	58.8	122	89.7	164	120.6
0.6	0.4	39	28.7	81	59.6	123	90.4	165	121.3
0.7	0.5	40	29.4	82	60.3	124	91.2	166	122.1
0.8	0.6	41	30.1	83	61.0	125	91.9	167	122.8
0.9	0.7	42	30.9	84	61.8	126	92.6	168	123.5
1	0.7	43	31.6	85	62.5	127	93.4	169	124.3
2	1.5	44	32.4	86	63.2	128	94.1	170	125.0
3	2.2	45	33.1	87	64.0	129	94.9	171	125.7
4	2.9	46	33.8	88	64.7	130	95.6	172	126.5
5	3.7	47	34.6	89	65.4	131	96.3	173	127.2
6	4.4	48	35.3	90	66.2	132	97.1	174	127.9
7	5.1	49	36.0	91	66.9	133	97.8	175	128.7
8	5.9	50	36.8	92	67.6	134	98.5	176	129.4
9	6.6	51	37.5	93	68.4	135	99.3	177	130.1
10	7.4	52	38.2	94	69.1	136	100.0	178	130.9
11	8.1	53	39.0	95	69.9	137	100.7	179	131.6
12	8.8	54	39.7	96	70.6	138	101.5	180	132.4
13	9.6	55	40.4	97	71.3	139	102.2	181	133.1
14	10.3	56	41.2	98	72.1	140	102.9	182	133.8
15	11.0	57	41.9	99	72.8	141	103.7	183	134.6
16	11.8	58	42.6	100	73.5	142	104.4	184	135.3
17	12.5	59	43.4	101	74.3	143	105.1	185	136.0
18	13.2	60	44.1	102	75.0	144	105.9	186	136.8
19	14.0	61	44.9	103	75.7	145	106.6	187	137.5
20	14.7	62	45.6	104	76.5	146	107.4	188	138.2
21	15.4	63	46.3	105	77.2	147	108.1	189	139.0
22	16.2	64	47.1	106	77.9	148	108.8	190	139.7
23	16.9	65	47.8	107	78.7	149	109.6	191	140.4
24	17.6	66	48.5	108	79.4	150	110.3	192	141.2
25	18.4	67	49.3	109	80.1	151	111.0	193	141.9
26	19.1	68	50.0	110	80.9	152	111.8	194	142.6
27	19.9	69	50.7	111	81.6	153	112.5	195	143.4
28	20.6	70	51.5	112	82.4	154	113.2	196	144.1
29	21.3	71	52.2	113	83.1	155	114.0	197	144.9
30	22.1	72	52.9	114	83.8	156	114.7	198	145.6
31	22.8	73	53.7	115	84.6	157	115.4	199	146.3
32	23.5	74	54.4	116	85.3	158	116.2	200	147.1
33	24.3	75	55.1	117	86.0	159	116.9	201	147.8

ENGLISH/METRIC CONVERSION: TEMPERATURE

To convert Fahrenheit (F°) to Celsius (C°), take F° temperature and subtract 32, multiply the result by 5 and divide the result by 9
To convert Celsius (C°) to Fahrenheit (F°), take C° temperature and multiply it by 9, divide the result by 5 and add 32

F°	C°	F°	C°	C°	F°	C°	F°
-40	-40.0	150	65.6	-38	-36.4	46	114.8
-35	-37.2	155	68.3	-36	-32.8	48	118.4
-30	-34.4	160	71.1	-34	-29.2	50	122
-25	-31.7	165	73.9	-32	-25.6	52	125.6
-20	-28.9	170	76.7	-30	-22	54	129.2
-15	-26.1	175	79.4	-28	-18.4	56	132.8
-10	-23.3	180	82.2	-26	-14.8	58	136.4
-5	-20.6	185	85.0	-24	-11.2	60	140
0	-17.8	190	87.8	-22	-7.6	62	143.6
1	-17.2	195	90.6	-20	-4	64	147.2
2	-16.7	200	93.3	-18	-0.4	66	150.8
3	-16.1	205	96.1	-16	3.2	68	154.4
4	-15.6	210	98.9	-14	6.8	70	158
5	-15.0	212	100.0	-12	10.4	72	161.6
10	-12.2	215	101.7	-10	14	74	165.2
15	-9.4	220	104.4	-8	17.6	76	168.8
20	-6.7	225	107.2	-6	21.2	78	172.4
25	-3.9	230	110.0	-4	24.8	80	176
30	-1.1	235	112.8	-2	28.4	82	179.6
35	1.7	240	115.6	0	32	84	183.2
40	4.4	245	118.3	2	35.6	86	186.8
45	7.2	250	121.1	4	39.2	88	190.4
50	10.0	255	123.9	6	42.8	90	194
55	12.8	260	126.7	8	46.4	92	197.6
60	15.6	265	129.4	10	50	94	201.2
65	18.3	270	132.2	12	53.6	96	204.8
70	21.1	275	135.0	14	57.2	98	208.4
75	23.9	280	137.8	16	60.8	100	212
80	26.7	285	140.6	18	64.4	102	215.6
85	29.4	290	143.3	20	68	104	219.2
90	32.2	295	146.1	22	71.6	106	222.8
95	35.0	300	148.9	24	75.2	108	226.4
100	37.8	305	151.7	26	78.8	110	230
105	40.6	310	154.4	28	82.4	112	233.6
110	43.3	315	157.2	30	86	114	237.2
115	46.1	320	160.0	32	89.6	116	240.8
120	48.9	325	162.8	34	93.2	118	244.4
125	51.7	330	165.6	36	96.8	120	248
130	54.4	335	168.3	38	100.4	122	251.6
135	57.2	340	171.1	40	104	124	255.2
140	60.0	345	173.9	42	107.6	126	258.8
145	62.8	350	176.7	44	111.2	128	262.4

LENGTH CONVERSION

To convert inches (in.) to millimeters (mm), multiply the number of inches by 25.4

To convert millimeters (mm) to inches (in.), multiply the number of millimeters by 0.04

Inches	Millimeters	Inches	Millimeters	Inches	Millimeters	Inches	Millimeters
0.0001	0.00254	0.005	0.1270	0.09	2.286	4	101.6
0.0002	0.00508	0.006	0.1524	0.1	2.54	5	127.0
0.0003	0.00762	0.007	0.1778	0.2	5.08	6	152.4
0.0004	0.01016	0.008	0.2032	0.3	7.62	7	177.8
0.0005	0.01270	0.009	0.2286	0.4	10.16	8	203.2
0.0006	0.01524	0.01	0.254	0.5	12.70	9	228.6
0.0007	0.01778	0.02	0.508	0.6	15.24	10	254.0
0.0008	0.02032	0.03	0.762	0.7	17.78	11	279.4
0.0009	0.02286	0.04	1.016	0.8	20.32	12	304.8
0.001	0.0254	0.05	1.270	0.9	22.86	13	330.2
0.002	0.0508	0.06	1.524	1	25.4	14	355.6
0.003	0.0762	0.07	1.778	2	50.8	15	381.0
0.004	0.1016	0.08	2.032	3	76.2	16	406.4

ENGLISH/METRIC CONVERSION: LENGTH

To convert inches (in.) to millimeters (mm), multiply the number of inches by 25.4
To convert millimeters (mm) to inches (in.), multiply the number of millimeters by 0.04

| Inches | | Millimeters | Inches | | Millimeters | Inches | | Millimeters |
Fraction	Decimal	Decimal	Fraction	Decimal	Decimal	Fraction	Decimal	Decimal
1/64	0.016	0.397	11/32	0.344	8.731	11/16	0.688	17.463
1/32	0.031	0.794	23/64	0.359	9.128	45/64	0.703	17.859
3/64	0.047	1.191	3/8	0.375	9.525	23/32	0.719	18.256
1/16	0.063	1.588	25/64	0.391	9.922	47/64	0.734	18.653
5/64	0.078	1.984	13/32	0.406	10.319	3/4	0.750	19.050
3/32	0.094	2.381	27/64	0.422	10.716	49/64	0.766	19.447
7/64	0.109	2.778	7/16	0.438	11.113	25/32	0.781	19.844
1/8	0.125	3.175	29/64	0.453	11.509	51/64	0.797	20.241
9/64	0.141	3.572	15/32	0.469	11.906	13/16	0.813	20.638
5/32	0.156	3.969	31/64	0.484	12.303	53/64	0.828	21.034
11/64	0.172	4.366	1/2	0.500	12.700	27/32	0.844	21.431
3/16	0.188	4.763	33/64	0.516	13.097	55/64	0.859	21.828
13/64	0.203	5.159	17/32	0.531	13.494	7/8	0.875	22.225
7/32	0.219	5.556	35/64	0.547	13.891	57/64	0.891	22.622
15/64	0.234	5.953	9/16	0.563	14.288	29/32	0.906	23.019
1/4	0.250	6.350	37/64	0.578	14.684	59/64	0.922	23.416
17/64	0.266	6.747	19/32	0.594	15.081	15/16	0.938	23.813
9/32	0.281	7.144	39/64	0.609	15.478	61/64	0.953	24.209
19/64	0.297	7.541	5/8	0.625	15.875	31/32	0.969	24.606
5/16	0.313	7.938	41/64	0.641	16.272	63/64	0.984	25.003
21/64	0.328	8.334	21/32	0.656	16.669	1/1	1.000	25.400
			43/64	0.672	17.066			

NOTES